P9-DGQ-673

8/89

For Reference

1989

DEC

Not to be taken from this room

CHILTON'S AUTO REPAIR MANUAL 1980-1987

Editorial Director	Alan F. Turner
Executive Director	Kerry A. Freeman, S.A.E.
Senior Editor	Richard J. Rivele, S.A.E.
Project Coordinator	Tony Molla, S.A.E.
Editorial Staff	John M. Baxter, S.A.E.
	Martin J. Gunther
	Michael A. Newsome
	W. Calvin Settle Jr., S.A.E.
	Richard T. Smith
Production Manager	John J. Cantwell
Managing Editor & Design	Dean F. Morgantini, S.A.E.
Art & Production Coordinator	Robin S. Miller
Supervisor Mechanical Paste-up	Margaret A. Stoner
Mechanical Artists	Cynthia Fiore, William Gaskins

OFFICERS

President	Lawrence A. Fornasieri
Vice President & General Manager	John P. Kushnerick

CHILTON BOOK COMPANY
Chilton Way Radnor, Pa. 19089

Manufactured in USA
© 1986 Chilton Book Company
ISBN 0–8019–7670–7
ISSN No. 0069–3634
Library of Congress Card Catalog No.76–648878

6 7 8 9 0   8 9 0

CAR MODELS

Car Sections

Unit Repair Sections

INDEX

		AMC	Chrysler FWD	Chrysler RWD	Ford FWD	Ford RWD	Ford Taurus/Sable	Buick RWD	Cadillac RWD	Corvette	Chevrolet RWD	Oldsmobile RWD
		1	2	3	4	5	6	7	8	9	10	11
1	Alternator R & I	C6	C53	C104	C144	C198	C268	C320	C350	C377	C410	C445
2	Alternator Overhaul	U41	U41	U41	U41	U41	U41	U41	U41	U41	U41	U41
3	Auto Transmission R & I	C29	C84	C123	C172	C233	C291	C333	C352	C388	C423	C460
4	Service	U275	U275	U275	U275	U275	U275	U275	U275	U275	U275	U275
5	Axle Shaft	C32	C84	C125	C173	C235	C292	C334	C363	C389	C424	C461
6	Ball Joints	C36	C87	C128	C174	C238	C296	C335	C364	C390	C426	C461
7	Brakes	C38	C90	C131	C176	C249	C299	C338	C366	C396	C428	C464
8	Camshaft	C23	C77	C117	C167	C225	C285	C330	C360	C420	C420	C457
9	Carburetor R & I	C9	C57	C107	C147	C202	—	C322	C354	C381	C413	C447
10	Specifications & Overhaul	U69	U69	U69	U69	U69	U69	U69	U69	U69	U69	U69
11	Clutch	C26	C80	C121	C170	C231	C288	C332	—	C384	C421	C459
12	Combination Switch	—	C91	C132	C178	—	C302	—	—	—	—	—
13	CV-Joint	—	C84	—	C173	—	C292	—	—	—	—	—
14	Cylinder Head	C20	C69	C115	C161	C218	C282	C327	C358	C418	C418	C454
15	Diesel Injection	—	—	—	C152	C206	—	C321	C352	—	—	C448
16	Diesel Service	—	—	—	U13	U13	—	U13	U13	—	—	U13
17	Differential	C31	—	C126	—	C236	—	C334	—	C393	C424	—
18	Distributor R & I	C6	C53	C104	C145	C199	C270	C320	C351	C378	C411	C446
19	Drive Axles	C32	C84	C125	C173	C235	C292	C333	C363	C388	C423	C461
20	Driveshaft	C33	—	C125	—	C235	—	C333	C363	C389	C423	C461
21	Electronic Ignition	C6	C53	C104	C145	C199	C270	C320	C351	C378	C411	C446
22	Distributor R & I	C6	C53	C104	C145	C199	C270	C320	C351	C378	C410	C446
23	Timing Adjustment	C7	C55	C106	C145	C200	C270	C320	C351	C378	C411	C446
24	Emission Controls	C13	C64	C111	C154	C211	C277	C324	C355	C382	C415	C452
25	Engine Cooling	C12	C63	C110	C153	C209	C276	C323	C354	C381	C414	C451
26	Engine R & I	C14	C64	C111	C155	C211	C277	C324	C355	C382	C416	C452
27	Engine Rebuilding	U189	U189	U189	U189	U189	U189	U189	U189	U189	U189	U189
28	Exhaust Manifold	C18	C65	C113	C159	C215	C279	C325	C356	C416	C416	C453
29	Firing Order	C3	C50	C101	C142	C190	C266	C314	C347	C375	C406	C440
30	Front Cover	C21	C71	C116	C165	C222	C283	C329	C359	C419	C419	C455
31	Front End Alignment	U379	U379	U379	U379	U379	U379	U379	U379	U379	U379	U379
32	Specifications	C5	C52	C103	C144	C195	C268	C319	C350	C377	C410	C445
33	Front Suspension	C33	C86	C127	C173	C237	C296	C335	C363	C389	C424	C461
34	Fuel Filter	C9	C56	C107	C147	C202	C272	C321	C352	C380	C412	C447
35	Fuel Injection	—	C60	C110	C151	C203	C271	C321	C352	C381	C414	C414
36	Fuel Pump	C9	C56	C106	C147	C201	C271	C321	C352	C379	C412	C447
37	Fuse Box	C44	C96	C138	C181	C264	C308	C343	C372	C399	C434	C467
38	Heater	C43	C94	C135	C180	C254	C305	C341	C370	C400	C432	C468
39	Idle Speed and Mixture	C10	C57	C107	C148	C203	C273	C322	C353	C381	C413	C447
40	Ignition Lock and Switch	C40	C92	C133	C178	C251	C302	C340	C369	C431	C431	C466
41	Ignition Timing	C7	C55	C106	C145	C200	C270	C320	C351	C378	C411	C446
42	Instrument Cluster	C41	C96	C138	C181	C260	C307	C344	C371	C399	C432	C467
43	Intake Manifold	C16	C65	C112	C157	C213	C279	C325	C356	C383	C417	C452
44	Lower Control Arm	C35	C87	C129	C174	C237	C297	C335	C363	C391	C426	C463
45	MacPherson Strut R & I	—	C86	—	C174	C240	C296	—	—	—	—	—
46	MacPherson Strut Overhaul	—	U311	—	U311	U311	U311	—	—	—	—	—
47	Maintenance	U2	U2	U2	U2	U2	U2	U2	U2	U2	U2	U2
48	Manual Steering Gear	C39	C92	—	C179	C251	—	—	—	—	C430	C464

Pontiac RWD	GM "A" & "X" Body	GM "C" Body	GM "E" & "K" Body	GM "F" Body	GM "H" Body RWD	GM "H" Body FWD	GM "J" Body	GM "N" Body	GM "P" Body	GM "T" Body	GM Spectrum	GM Sprint	GM Nova	
12	13	14	15	16	17	18	19	20	21	22	23	24	25	
C477	C509	C555	C587	C629	C673	C696	C716	C756	C784	C814	C844	C868	C892	1
U41	U41	U41	U41	U41	U41	U41	U41	U41	U41	U41	U41	U41	U41	2
C489	C537	C565	C597	C652	C681	C702	C737	C770	C796	C831	C854	C878	C902	3
U275	U275	U275	U275	U275	U275	U275	U275	U275	U275	U275	U275	U275	U275	4
C489	C538	C567	C599	C654	C681	C702	C738	C770	C798	C831	C854	C879	C903	5
C490	C539	C567	C605	C657	C682	C703	C742	C772	C798	C832	C856	C880	C904	6
C494	C542	C571	C607	C661	C684	C706	C745	C774	C803	C834	C857	C883	C905	7
C487	C530	C561	C595	C646	C677	C701	C731	C766	C794	C827	C851	C876	C899	8
C479	C513	C559	C589	C632	C675	—	C720	—	—	C816	C846	C870	C893	9
U69	U69	U69	U69	U69	U69	U69	U69	—	U69	U69	U69	U69	U69	10
C488	C534	—	—	C649	C679	—	C735	C768	C796	C829	C852	C877	C901	11
—	—	—	—	—	—	—	—	—	—	—	C861	C884	C907	12
—	C538	C567	C599	—	—	C702	C738	C771	C798	—	C854	C879	C903	13
C485	C527	C561	C591	C643	C677	C700	C729	C764	C791	C824	C850	C875	C897	14
C482	C514	—	C588	—	—	—	—	—	—	C817	—	—	—	15
U13	U13	—	U13	—	—	—	—	—	—	U13	—	—	—	16
—	—	—	C602	C653	C681	—	—	—	—	C831	—	—	—	17
C477	C510	C556	C588	C629	C673	C697	C717	C757	C785	C814	C845	C868	C892	18
C489	C538	C567	C599	C653	C681	C702	C738	C770	C798	C831	C854	C879	C903	19
C489	C538	C567	C599	C653	C681	—	—	—	—	C831	—	—	—	20
C477	C510	C556	C588	C629	C673	C697	C717	C757	C785	C814	C845	C868	C892	21
C477	C510	C556	C588	C629	C673	C697	C717	C757	C785	C814	C845	C868	C892	22
C477	C510	C557	C588	C630	C674	C697	C718	C758	C785	C815	C845	C869	C892	23
C483	C518	C560	C591	C636	C677	C699	C721	C761	C787	C820	C848	C873	C895	24
C482	C516	C559	C589	C635	C677	C698	C720	C760	C786	C819	C847	C872	C894	25
C483	C518	C561	C591	C636	C677	C699	C722	C762	C787	C821	C848	C873	C895	26
U189	U189	U189	U189	U189	U189	U189	U189	U189	U189	U189	U189	U189	U189	27
C484	C523	C563	C593	C637	C677	C699	C727	C763	C789	C823	C848	C873	C896	28
C473	C505	C552	C581	C620	C671	C694	C713	C754	C782	C811	C842	C867	C890	29
C485	C528	C561	C594	C644	C677	C700	C730	C765	C792	C825	C850	C877	C900	30
U379	U379	U379	U379	U379	U379	U379	U379	U379	U379	U379	U379	U379	U379	31
C476	C508	C555	C587	C628	C673	C696	C715	C756	C784	C813	C843	C868	C892	32
C490	C539	C567	C602	C655	C681	C703	C740	C771	C798	C832	C856	C880	C904	33
C478	C512	C559	C589	C631	C675	C698	C720	C760	C786	C816	C846	C870	C893	34
C482	C513	C558	C589	C634	—	C698	C720	C759	C786	—	—	—	—	35
C478	C512	C558	C588	C630	C674	C698	C718	C759	C786	C816	C846	C870	C893	36
C500	C550	C575	C611	C668	C692	C710	C752	C779	C807	C840	C864	C887	C908	37
C499	C546	C576	C612	C668	C689	C709	C749	C778	C806	C839	C864	C886	C908	38
C479	C513	C559	C588	C632	C676	C698	C720	—	C786	C816	C847	C870	C894	39
C497	C544	C574	C607	C664	C687	C707	C747	C776	C805	C837	C861	C884	C907	40
C477	C510	C557	C587	C630	C674	C697	C718	C758	C785	C815	C845	C869	C892	41
C498	C549	C575	C608	C665	C688	C709	C751	C779	C806	C837	C862	C887	C907	42
C484	C522	C561	C594	C638	C677	C699	C726	C763	C788	C822	C848	C873	C896	43
C492	C540	C568	C604	C658	C683	C703	C743	C773	C800	C832	C856	C880	C905	44
—	C539	C567	C602	—	—	C703	C740	C771	C801	—	C856	C880	C904	45
—	U311	U311	U311	U311	U311	U311	U311	U311	U311	U311	U311	U311	U311	46
U2	U2	U2	U2	U2	U2	U2	U2	U2	U2	U2	U2	U2	U2	47
C495	C545	—	—	C662	C685	C707	C747	C775	C804	C835	C859	C884	C906	48

INDEX

Pontiac RWD	GM "A" & "X" Body	GM "C" Body	GM "E" & "K" Body	GM "F" Body	GM "H" Body RWD	GM "H" Body FWD	GM "J" Body	GM "N" Body	GM "P" Body	GM "T" Body	GM Spectrum	GM Sprint	GM Nova	
12	13	14	15	16	17	18	19	20	21	22	23	24	25	
—	C536	—	—	—	—	—	C736	C769	C796	—	C853	C877	C902	1
—	U209	—	—	—	—	—	U209	U209	U209	—	—	U209	U209	2
C488	—	—	—	C651	C680	—	—	—	—	C830	—	—	—	3
U209	—	—	—	U209	U209	—	—	—	—	U209	U209	—	—	4
C494	C542	C571	C607	C661	C684	C706	C745	C774	C803	C834	C857	C883	C905	5
U325	U325	U325	U325	U325	U325	U325	U325	U325	U325 .	U325	U325	U325	U325	6
U275	U275	U275	U275	U275	U275	U275	C735	U275	U275	U275	—	—	—	7
C487	C531	C564	C595	C647	C677	C701	C733	C766	C794	C828	C851	C876	C900	8
U275	U275	U275	U275	U275	U275	U275	U275	U275	U275	U275	U275	U275	U275	9
C487	C534	C561	C596	C647	C677	C701	C734	C767	C795	C828	C852	C877	C900	10
C495	C542	C571	C607	C662	C685	C706	C746	C775	C803	C835	C858	C883	C906	11
C487	C531	C561	C591	C646	C677	C701	C733	C766	C795	C828	C851	C876	C900	12
C495	C545	C572	C607	C662	C685	C707	C747	C776	C804	C835	C860	—	C906	13
C496	C545	C572	C607	C663	C688	C708	C747	C776	—	C835	C860	—	C907	14
C482	C516	C559	C590	C635	C677	C698	C720	C760	C786	C819	C847	C872	C894	15
C499	C547	C575	C611	C667	C689	C710	C749	C779	C807	C839	C864	C886	C908	16
C489	C539	C567	C606	C654	C681	C702	C739	C773	C798	C831	C855	C880	C903	17
C487	C533	C561	C597	C648	C677	C701	C734	C767	C795	C828	C852	C877	C900	18
C492	C540	C569	C606	C658	C683	C704	C744	C773	C801	C833	C857	C882	C905	19
C477	C509	C556	C587	C629	C673	C697	C717	C757	C784	C814	C844	C868	C892	20
C485	C525	C561	C591	C642	C677	C700	C728	C763	C789	C823	C848	C874	C896	21
C470	C503	C551	C578	C614	C670	C693	C712	C753	C781	C810	C841	C865	C889	22
C490	C540	C567	C606	C656	C682	C703	C742	C773	C798	C832	C857	C882	C904	23
C470	C503	C551	C578	C614	C669	C693	C712	C753	C781	C810	C841	C866	C889	24
C489	C539	C567	C602	C655	C682	C703	C740	C771	C800	C832	C856	C880	C904	25
C492	C541	C569	C606	C659	C683	C704	C744	C773	C801	C833	C857	C883	C905	26
C477	C509	C556	C688	C629	C673	C697	C717	C757	C784	C814	C844	C868	C892	27
C495	C543	C571	C607	C662	C685	C706	C746	C775	C804	C835	C858	C884	C906	28
C483	C518	C560	C589	C636	C677	C699	C721	C761	C787	C820	C847	C872	C895	29
—	—	—	—	—	—	—	C731	—	—	C825	C850	C876	C899	30
C485	C528	C561	C594	C644	C677	C700	C730	C765	C792	C825	C850	C876	C898	31
C486	C529	C561	C595	C645	C677	C701	C730	C765	C793	C825	C850	C876	C899	32
U389	U389	U389	U389	U389	U389	U389	U389	U389	U389	U389	U389	U389	U389	33
—	—	—	C593	C642	—	—	C728	—	—	—	—	—	—	34
C496	C543	C573	C607	C663	C686	C707	C746	C776	C804	C836	C861	C884	C907	35
C489	—	C567	—	C653	C681	—	—	—	—	C831	—	—	—	36
C490	—	—	C604	C658	C681	—	—	—	C800	C833	—	—	—	37
C485	C526	C561	C591	C643	C677	C700	C728	C763	C791	C823	C850	C875	C897	38
C470	C503	C551	C578	C614	C669	C693	C712	C753	C781	C810	C841	C865	C889	39
C482	C516	C559	C590	C635	C677	C698	C721	C760	C786	C819	C847	C872	C895	40
U379	U379	U379	U379	U379	U379	U379	U379	U379	U379	U379	U379	U379	U379	41
U379	U379	U379	U379	U379	U379	U379	U379	U379	U379	U379	U379	U379	U379	42
C476	C508	C555	C587	C628	C673	C696	C715	C756	C784	C813	C843	C868	C892	43
C490	C540	C567	C603	C657	C683	C702	C739	C771	C801	C833	C855	C881	C904	44
C489	C541	C567	C606	C654	C681	C702	C745	C773	—	C834	C855	C880	C903	45
C494	C542	U325	U325	C661	U325	U325	C745	C775	U325	C834	C858	U325	U325	46
C499	C547	C575	C610	C667	C688	C709	C750	C778	C806	C838	C862	C886	C908	47

HOW TO USE THIS MANUAL

This manual is arranged in two sections:

Car Section

Car Sections are grouped by manufacturer and arranged in alphabetical order. The text and illustrations that comprise the service procedures in each Car Section are arranged in the following order of systems and components: Charging, Starting, Ignition, Fuel, and Cooling Systems; Emission Controls, Engine, Clutch, Manual Transmission, Automatic Transmission, Driveshaft and U-Joints, Rear Axle, Jacking and Hoisting, Front Suspension, Rear Suspension, Brakes, Steering, Instrument Panel, Windshield Wipers, Radio, and Heater. Specifications are always located at the front of each section. All illustrations are located as close as possible to the pertinent text. Procedures are for all models in the particular section unless specifically noted otherwise.

Unit Repair Section

The Unit Repair Section contains troubleshooting and overhaul procedures for the major components and systems of your car, and is intended to be used in conjunction with the Car Sections. For example, if your car's engine is misfiring and you do not know the cause, use the "Troubleshooting" portion of the Unit Repair Section to find the cause and its remedy. If the cause should prove to be defective piston rings which are allowing oil to foul the spark plugs, the remedy is to overhaul the engine. Turn to the proper Car Section to find the procedure for removing the engine from the car. After you have removed the engine, turn to "Engine Rebuilding" in the Unit Repair Section and follow the steps listed there to overhaul the engine.

Most Unit Repair Sections are arranged by brands, manufacturers, or types of components rather than models of cars, and all overhaul procedures begin with the component removed from the car. The reason for this division of material is economic. The steps involved in overhauling an engine, for example, are virtually the same for all engines, but the operation of removing the engine from the car varies greatly from model to model. By combining where possible and separating where necessary, we are able to publish the maximum amount of information.

Locating Information

The Table of Contents, at the front of the book, lists the beginning of each Car and Unit Repair Section in the manual. The Car Sections are grouped by manufacturer. There is also an alphabetical listing of names of cars, so that you can find the right section quickly, even if you are not sure of the manufacturer. Car Section pages are prefixed with the letter "C." Unit Repair Section pages are prefixed with the letter "U."

To find the page number for a particular Car Section, you need only look in the Table of Contents. Once you have found the proper section, you may wish to find where specific procedures are located in that section. Turn to the Index on the following page and read across the top of the grid until you reach the number that corresponds to the car section. When the proper column has been found, read down the side column to the procedure in question. The intersection of the two columns will provide the page number for the procedure.

Safety Notice

Proper service and repair procedures are vital to the safe, reliable operation of all motor vehicles, as well as the personal safety of those performing repairs. This manual outlines procedures for servicing and repairing vehicles using safe effective methods. The procedures contain many NOTES, CAUTIONS and WARNINGS which should be followed along with standard safety procedures to eliminate the possibility of personal injury or improper service which could damage the vehicle or compromise its safety.

It is important to note that repair procedures and techniques, tools and parts for servicing motor vehicles, as well as the skill and experience of the individual performing the work may vary widely. It is not possible to anticipate all of the conceivable ways or conditions under which vehicles may be serviced, or to provide cautions as to all of the possible hazards that may result. Standard and accepted safety precautions and equipment should be used when handling toxic or flammable fluids, and safety goggles or other protection should be used during cutting, grinding, chiseling, prying, or any other process that can cause material removal or projectiles.

Some procedures require the use of tools specially designed for a specific purpose. Before substituting another tool or procedure, you must be completely satisfied that neither your personal safety nor the performance of the vehicle will be endangered.

Part Numbers

Part numbers listed in this book are not recommendations by Chilton for any product by brand name. They are references that can be used with interchange manuals and aftermarket supplier catalogs to locate each brand supplier's discrete part number.

American Motors
AMX, Concord, Eagle, SX-4, Pacer, Spirit

YEAR IDENTIFICATION

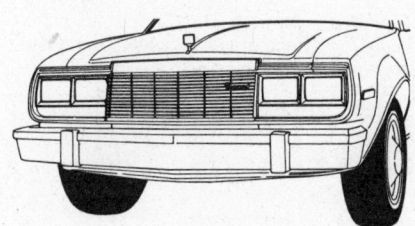

1980 Concord

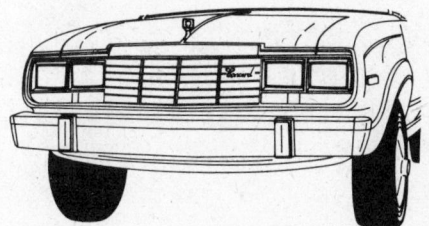

1981–83 Concord

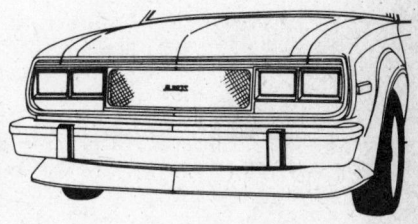

1980 AMX

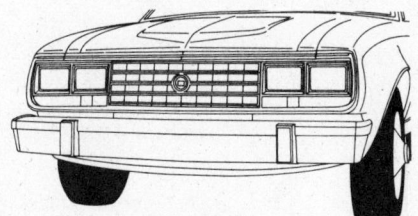

1980 Spirit

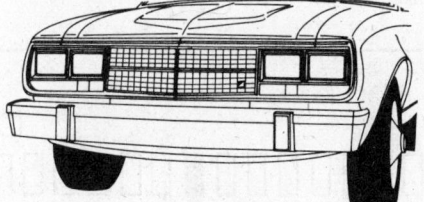

1981–83 Spirit

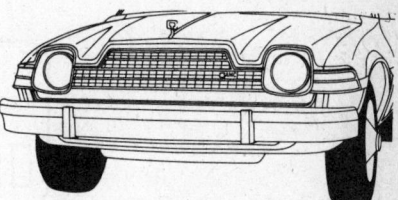

1980 Pacer

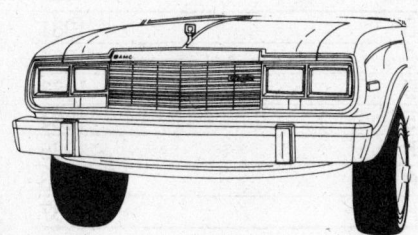

1980 Eagle

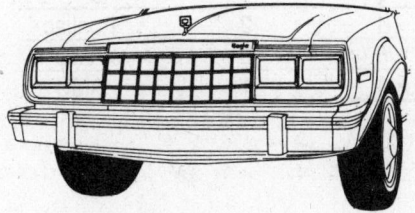

1981-87 Eagle

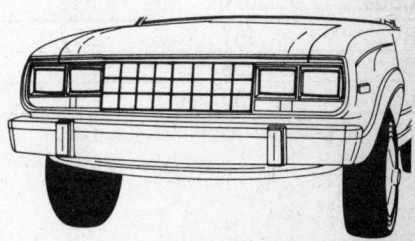

1981-87 Eagle SX-4

VEHICLE IDENTIFICATION NUMBER (VIN)

It is important for servicing and ordering parts to be certain of the vehicle and engine identification. The VIN (vehicle identification number) is a 13 or 17 digit number visible through the windshield on the driver's side of the dash and contains the vehicle and engine identification codes. It can be interpreted as follows:

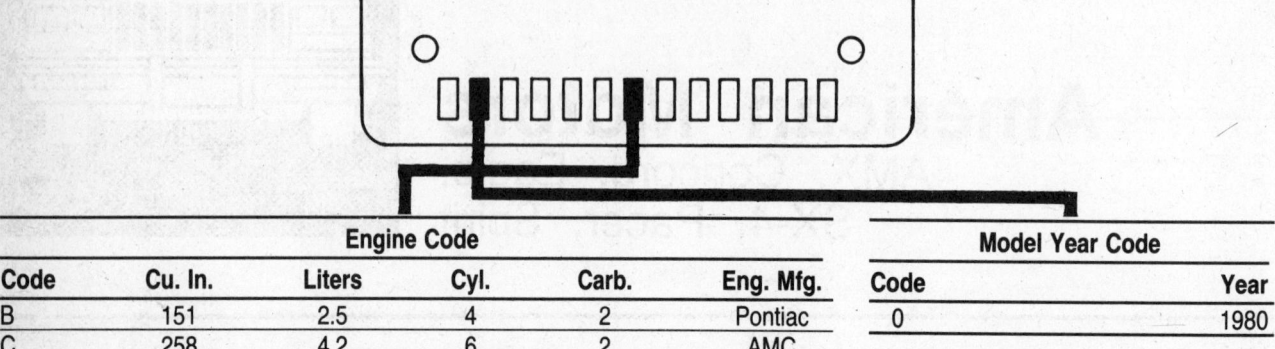

Engine Code						Model Year Code	
Code	Cu. In.	Liters	Cyl.	Carb.	Eng. Mfg.	Code	Year
B	151	2.5	4	2	Pontiac	0	1980
C	258	4.2	6	2	AMC		

The thirteen digit Vehicle Identification Number can be used to determine engine application and model year. The second digit indicates model year, and the seventh digit indicates engine code.

VEHICLE IDENTIFICATION NUMBER (VIN)

It is important for servicing and ordering parts to be certain of the vehicle and engine identification. The VIN (vehicle identification number) is a 13 or 17 digit number visible through the windshield on the driver's side of the dash and contains the vehicle and engine identification codes. It can be interpreted as follows:

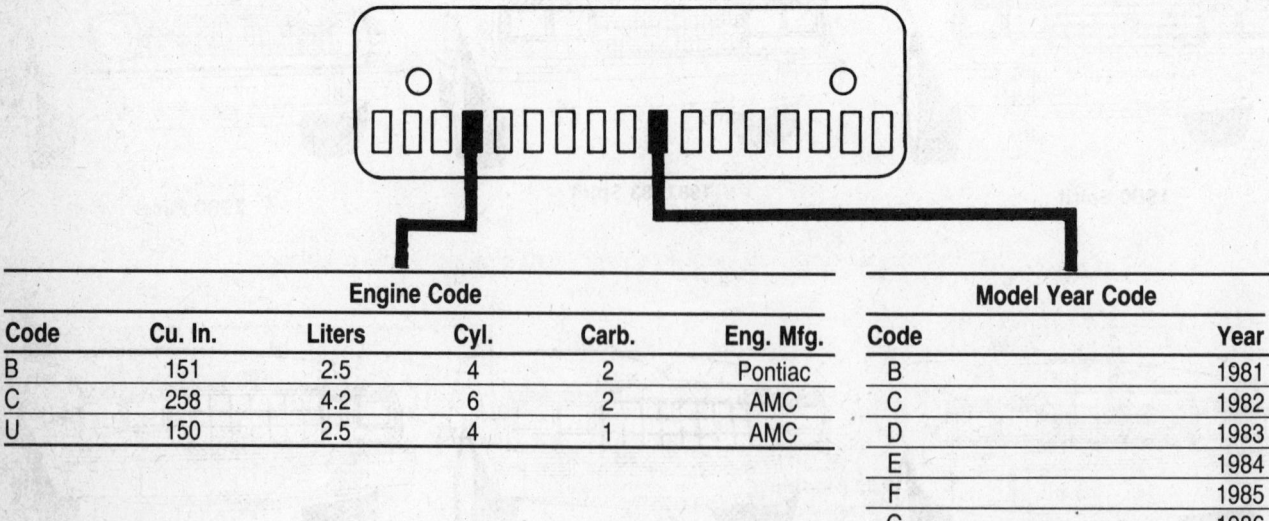

Engine Code						Model Year Code	
Code	Cu. In.	Liters	Cyl.	Carb.	Eng. Mfg.	Code	Year
B	151	2.5	4	2	Pontiac	B	1981
C	258	4.2	6	2	AMC	C	1982
U	150	2.5	4	1	AMC	D	1983
						E	1984
						F	1985
						G	1986
						H	1987

The seventeen digit Vehicle Identification Number can be used to determine engine identification and model year. The tenth digit indicates model year, and the fourth digit indicates engine code.

GENERAL ENGINE SPECIFICATIONS

Year	Eng. V.I.N. Code	Engine No. Cyl. Displacement cu. In.	Eng. Mfg.	Carburetor Type	Horsepower @ rpm ■	Torque @ rpm (ft lbs) ■	Bore × Stroke (in.)	Compression Ratio	Oil Pressure @ 2000 rpm
'80–'82	B	4-151	Pontiac	2 bbl	99 @ 4000	134 @ 2400	4.000 × 3.000	8.2:1 ①	36–41
	C	6-258	AMC	2 bbl	110 @ 3200	210 @ 1800	3.750 × 3.900	8.3:1	46
'83	B	4-151	Pontiac	2 bbl	99 @ 4000	134 @ 2400	4.000 × 3.000	8.3:1	36–41
	C	6-258	AMC	2 bbl	110 @ 3200	210 @ 1800	3.750 × 3.900	8.3:1	46
'84	U	4-150	AMC	1 bbl	83 @ 4200	116 @ 2600	3.876 × 3.188	9.2:1	40
'84–'87	C	6-258	AMC	2 bbl	110 @ 3200	210 @ 1800	3.750 × 3.900	9.2:1	46

■ Horsepower and torque are SAE net figures. They are measured at the rear of the transmission with all accessories installed and operating. Since the figures vary when a given engine is installed in different models, some are representative rather than exact.
① 1981–82 8.3:1

TUNE-UP SPECIFICATIONS

(When analyzing compression test results, look for uniformity among cylinders rather than specific pressures.)

Year	Engine No. Cyl. Displacement (cu. in.)	Eng. VIN Code	Carb	Eng. Mfg.	Spark Plugs Orig. Type	Gap (in.)	Distributor	Ignition Timing (deg) ▲ Man. Trans. ●	Ignition Timing (deg) ▲ Auto. Trans.	Intake Valve Opens ■ (deg.)	Fuel Pump Pressure (psi)	Idle Speed (rpm) ▲ Man. Trans.	Idle Speed (rpm) ▲ Auto. Trans.*
'80	4-151	B	2 bbl	Pontiac	R44TSX	.060	Electronic	10B(12B)	12B(10B)	33	6½–8	900	700
	6-258	C	2 bbl	AMC	N14LY③	.035	Electronic	6B	10B②	14½	4–5	700	600
'81–'82	4-151	B	2 bbl	Pontiac	R44TSX	.060	Electronic	10B④	12B②	25	6½–8	900	700
	6-258	C	2 bbl	AMC	RFN-14LY	.033	Electronic	⑤	⑥	9	5–6½	700	600
'83	4-151	B	2 bbl	Pontiac	R44TSX	.060	Electronic	10B④	12B②	25	6½–8	900	700
	6-258	C	2 bbl	AMC	RFN14LY	.033	electronic	⑤	⑥	9	5–6½	700	600
'84	4-150	U	1 bbl	AMC	RFN14LY	.035	Electronic	12B	12B	12	6½–8	750	750
	6-258	C	2 bbl	AMC	RFN14LY	.033	Electronic	⑤	⑥	9	5–6½	700	600
'85–'86	6-258	C	2 bbl	AMC	RFN14LY	.035	Electronic	9B①	9B①	9	5–6½	⑦	⑧
'87	All				See Underhood Specifications Sticker								

▲ If underhood emissions decal differs, always use the specifications on the decal.
* With transmission in Drive
● Figure in parentheses indicates California engine
■ All figures Before Top Dead Center
B Before Top Dead Center
TDC Top Dead Center (zero degrees)

① High Altitude—16B
② 8B—Eagle for Calif. and all Pacer
③ N13L—Eagle
④ Eagle except for Calif.—11B
⑤ Concord, Spirit—6B
   Eagle except for Calif.—8B
   Eagle Calif.—4B High Alt.—15B

⑥ Concord, Spirit—6B
   Eagle Except Calif.—8B
   Eagle Calif.—6B High Alt.—15B
⑦ 49 States—600
   California—680
   High Altitude—700
⑧ 49 States—680
   California—600
   High Altitude—650

FIRING ORDER

NOTE: To avoid confusion, always replace spark plug wires one at a time.

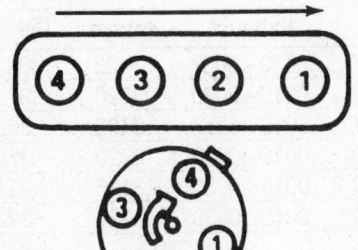

150 4 cylinder
Firing order: 4-3-2-1

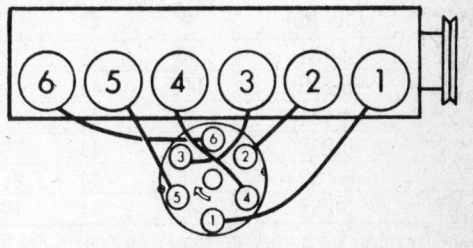

258 6 cylinder
Firing order: 1-5-3-6-2-4

FIRING ORDERS

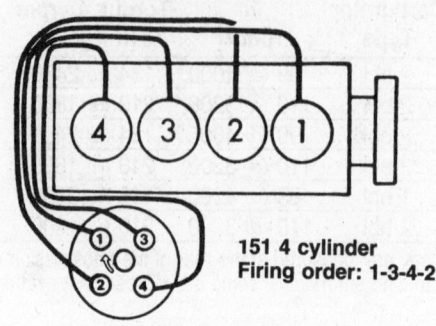

151 4 cylinder
Firing order: 1-3-4-2

CAPACITIES

Year	Engine No. Cyl. Displacement (cu. in.)	Model	Engine Crankcase Add 1 qt. for New Filter	Transmission (pts)			Drive Axle (pts.)	Fuel Tank (gal.)	Cooling System (qts.)	
				4-spd	5-spd	Auto.			w/heater	w/AC
'80	4-151	Spirit, Concord	3.0	3.5	—	17	3	22	6.5	6.5
	6-258	All Models	4.0	3.3②	—	17	3③	22	11①	14①
'81	4-151	Spirit, Concord	3.0	3.5	—	14.2	3	22	6.5	6.5
	4-151	Eagle	3.0	3.5	—	14.2	3③	22	6.5	6.5
	6-258	Spirit, Concord	4.0	3.5	—	17	3	22	11	14
	6-258	Eagle	4.0	3.5	—	17	3③	22	14	14
'82–'83	4-151	Spirit, Concord	3.0	4.0	4.5	14.2	3	22	6.5	6.5
	4-151	Eagle	3.0	3.5	4.0	14.2	3③	22	6.5	6.5
	6-258	Spirit, Concord	4.0	4.0	4.5	17	3	22	11	14
	6-258	Eagle	4.0	3.5	4.0	17	3③	22	14	14
'84–'87	4-150	Eagle	4.0	3.5	4.0	15.8	3③	22	10	10
	6-258	Eagle	4.0	3.5	4.0	17	3③	22	14	14

① 14.5 qts. in Pacer
② Eagle transfercase: 3.0 pts. until March 1980; 4.0 pts. thereafter
③ Front axle: 2.5 pts.

VALVE SPECIFICATIONS

Year	Engine No. Cyl. Displacement (cu. in.)	Seat Angle (deg.) ■	Face Angle (deg.) ●	Outer Spring Test Pressure (lbs. @ in.)	Spring Installed Height (in.)	Stem-to-Guide Clearance (in.)		Stem Diameter (in.)	
						Intake	Exhaust	Intake	Exhaust
'80–'82	4-151	46	45	176 @ 1.254	①	.0010–.0027	.0010–.0027	.3423	.3423
	6-258	44.5	44	195 @ 1.411	1¹³⁄₁₆	.0010–.0030	.0010–.0030	.3720	.3720
'83	4-151	46	45	176 @ 1.254	1	.0010–.0027	.0010–.0027	.3423	.3423
	6-258	44.5	44	195 @ 1.411	1¹³⁄₁₆	.0010–.0030	.0010–.0030	.3720	.3720
'84	4-150	45	44	212 @ 1.203	1¹¹⁄₃₂	.0010–.0030	.0010–.0030	.3115	.3115
'84–'87	6-258	44.5	44	195 @ 1.411	1¹³⁄₁₆	.0010–.0030	.0010–.0030	.3720	.3720

● Exhaust valve face angles are shown. All intake valve face angles are 29°, except 151 cu. in. engines.
■ Exhaust valve seat angles are shown. All intake valve seat angles are 30°, except 151 cu. in. engines.
① Intake: 2.057 in.; exhaust: 1.730 in.

CRANKSHAFT AND CONNECTING ROD SPECIFICATIONS

(All measurements are given in inches.)

| Year | Engine | Crankshaft | | | | Connecting Rod | | |
		Main Brg. Journal Dia.	Main Brg. Oil Clearance	Shaft End-Play	Thrust on No.	Journal Diameter	Oil Clearance	Side Clearance
'80–'83	4-151	2.2988	.0005–.0022	.0035–.0085	5	2.000	.0005–.0026	.017
'80–'87	6-258	2.4996–2.5001	.0010–.0030	.0015–.0065	3	2.0934–2.0955	.0010–.0030	.005–.014
'84	4-150	2.4996–2.5001	.0010–.0025	.0015–.0060	2	2.0934–2.0955	.0010–.0030	.010–.019

CAMSHAFT SPECIFICATIONS

(All measurements are given in inches.)

| Engine | Journal Diameter | | | | Bearing Clearance | Lobe Lift | | Camshaft End Play |
	1	2	3	4		Intake	Exhaust	
4-151	1.8690	1.8690	1.8690	—	.0007–.0027	.398	.398	.0015–.0050
6-258	2.0290–2.0300	2.0190–2.0200	2.0090–2.0100	1.9990–2.0000	.001–.003	.254①	.254①	0
4-150	2.0290–2.0300	2.0190–2.0200	2.0009–2.0100	1.9990–2.0000	.001–.003	.265	.265	0

① 258 2-bbl: .248

PISTON AND RING SPECIFICATIONS

(All measurements are given in inches.)

| Year | Engine Type/ Disp. | Piston-to-Bore Clearance | Ring Gap | | | Ring Side Clearance | | |
			Top Compression	Bottom Compression	Oil Control	Top Compression	Bottom Compression	Oil Control
'80–'83	4-151	.0025–.0033	.010–.022	.010–.028	.015–.055	.0030	.0030	.0000
'80	6-258	.0009–.0017	.010–.020	.010–.020	.010–.025	.0015–.0030	.0015–.0030	.0011–.0080
'81–'87	6-258	.0009–.0017	.010–.020	.010–.020	.010–.025	.0017–.0032	.0017–.0032	.0010–.0080
'84	4-150	.0009–.0017	.010–.020	.010–.020	.010–.025	.0017–.0032	.0017–.0032	.0010–.0080

WHEEL ALIGNMENT SPECIFICATIONS

| Year | Model | Caster | | Camber | | Toe-in (in.) | Steering Axis Inclin. (deg.) | Wheel Pivot Ratio | |
		Range (deg.)	Pref. Setting (deg.)	Range (deg.)	Pref. Setting (deg.)			Inner Wheel (deg.)	Outer Wheel (deg.)
'80	AMX	0–2½P	1P	0–¾P	¼P	1/16 to 3/16	7¾	2	22
	Spirit, Concord	0–2½P	1P	0–¾P	¼P	1/16 to 3/16	7¾	25	22
	Pacer	1P–3½P	2P	0–¾P	¼P	1/16 to 3/16	7¾	25	22
'80	Eagle	3P–5P	4P	3/8N–3/8P	0	1/16 to 3/16③	11½	—	—
'81–'83	Spirit, Concord	0–2½P	1P	④	②	1/16 to 3/16	7¾		
'81–'87	Eagle	2P–3P	2½P	1/8N–5/8P	3/8P	1/16 to 3/16③⑤	11½	—	—

N Negative
P Positive
— Specifications not available at time of publication
② Left: 3/8P; Right: 1/8P
③ Toe-out
④ Left: 3/4P to 1/8P; Right: 1/2P to ①P
⑤ 1982 Eagles 1/16 toe-out, 1/16 toe in

TORQUE SPECIFICATIONS
(All specifications in ft. lbs.)

Year	Engine	Cylinder Head Bolts	Connecting Rod Bearing Bolts	Main Bearing Bolts	Crank Pulley Bolt	Flywheel Bolts	Manifold	
							Intake	Exhaust
'80	6-258	95–115	30–35	75–85	70–90	95–120	18–28	18–28
'81–'87	6-258	80–90	30–35	75–85	70–90	95–115	18–28	18–28
'80–'83	4-151	80–103	27–33	62–68	157–163	65–71	34–40	36–42
'84	4-150	80–90	30–35	75–85	75–85	50–65	20–25	20–25

ENGINE ELECTRICAL

Delco-Remy or Bosch alternators are used on AMC models, depending on the engine. Most 1980 models, except Eagles with heated rear windows and fog lights, use Delco-Remy alternators; the Eagles use the Bosch unit. Most 1981 and later models use Delco-Remy alternators.

For information on alternator and regulator repair and troubleshooting, please refer to "Charging and Starting" in the Unit Repair Section. Voltage regulators that are built into the alternator require alternator disassembly for replacement.

External Regulator

REMOVAL & INSTALLATION

Disconnect the negative battery cable. Disconnect plug to the regulator. Remove the metal screws which hold the regulator in place, and lift off the regulator. Install in reverse order after cleaning the mounting surface around the screw holes.

Alternator

REMOVAL & INSTALLATION

1. Disconnect battery cables, negative cable first.
2. Disconnect and label the alternator wires or plug, then loosen adjusting bolt.
3. Remove the V-belt, mounting bolts and alternator.
4. To install, reverse the removal procedure.
5. There are several methods used for tightening the belt. Some alternator brackets have a hole through which you can insert a bar to pry out on the front alternator housing, others have a hole into which you can insert a ½ in. square socket drive to pull out on the alternator, and still others have a square boss around the adjusting bolt which takes a 1 in. open end wrench. If there are none of these systems, use a wood bar or broom handle to pry against the front alternator housing. The longest run of belt should deflect about ½ in. under moderate thumb pressure.

Starter

American Motors engines, except for the 4-150 and 4-151, are equipped with an integral positive engagement drive starter and a separate starter relay. Cars equipped with the 4-151 engine use Delco starters which do not have a relay mounted on top of the starter. Cars equipped with the 4-150 use a Bosch starter which have a relay similar to the Delco unit.

For additional information on starters, please refer to "Charging and Starting" in the Unit Repair section.

REMOVAL & INSTALLATION

Disconnect the negative battery cable first, then the battery and solenoid lead at the starter (if used). Raise the car and support it safely. Working underneath the car, remove the bolts which hold the starter to the bellhousing (and starter-to-engine brace on the 151), then remove the starter. Be sure to watch for and catch any shims that may be used behind the mounting bolts. Before installing the starter, make sure the mounting surfaces are free from burrs and dirt. Install the starter onto the housing together with any shims, and tighten the bolts to 18 ft. lbs. on six cylinder engines; Tighten the bolts to 17 ft. lbs. on the 151 cu. in. engine and 33 ft. lbs. on the 150 cu. in. engine. Clean the battery and solenoid terminal(s), install the cable(s).

Distributor

American Motors 150-4 and 258-6 engines use a Solid State Ignition (SSI) system. The 151-4 engines use the Delco-Remy High Energy Ignition (HEI) system.

REMOVAL

1. Bring the engine's No. 1 piston to TDC on the compression stroke. Remove the distributor cap, mark the position of the rotor relative to the distributor body and mark the body relative to the block. Remove the carburetor air cleaner if necessary, then remove the distributor primary wire and the distributor vacuum lines. Tag any disconnected wires or hoses for installation.
2. Remove the hold-down bolt and lift the distributor up out of the block.

NOTE: Don't turn the engine while the distributor is removed. This would complicate reinstalling the distributor.

3. The procedure for installation varies depending on whether or not the engine was turned while the distributor was out.

INSTALLATION

Engine Not Turned (All Except 4-150)

1. Turn the rotor about one-eighth turn past the mark indicating its original position. This allows the helical gears to align properly.
2. Align the distributor locating marks on the distributor body and the block, and drop the unit into place. The rotor should turn back to the mark as the unit seats and the gear meshes. Wiggle the rotor slightly to start the gear in mesh as necessary.

3. Tighten the hold-down bolt and reconnect the primary wire and the vacuum line. Install the cap.

4. Check the ignition timing. Theoretically, the timing shouldn't have changed if the distributor was reinstalled exactly in its original position, but it is always best to make sure.

4-150 Only

1. Rotate the engine until the No. 1 piston is at TDC compression.

2. Using a flat bladed screwdriver, in the distributor hole, rotate the oil pump gear so that the slot in the oil pump shaft is slightly past the 3 o'clock position, relative to the length of the engine block.

3. With the distributor cap removed, install the distributor with the rotor at the 5 o'clock position, relative to the oil pump gear shaft slot. When the distributor is comnpletely in place, the rotor should be at the 6 o'clock position. If not, remove the distributor and perform the entire procedure again.

4. Tighten the lockbolt.

Engine Turned (All Except 4-150)

This procedure is necessary to install a new distributor or if the engine has been turned while the distributor was out.

1. Place the No. 1 cylinder in firing position by turning the engine with a finger held over the No. 1 spark plug hole. No. 1 spark plug is the front one on a 4 or 6 cylinder. When compression is felt, turn the engine to align the TDC mark on the timing pointer with the notch on the crankshaft pulley.

2. Align the metal end of the rotor with the No. 1 spark plug wire in the distributor cap.

3. Turn the distributor body counterclockwise about one-eighth turn and set it into place in the engine. Wiggle the rotor slightly to start the gear in mesh as necessary.

4. The distributor should not be positioned so that the rotor is still pointing to the No. 1 wire location on the cap.

5. Tighten the hold-down bolt and reconnect the primary wire and the vacuum line. Replace the cap.

6. Check and adjust the ignition timing.

IGNITION TIMING ADJUSTMENT

A scale located on the timing chain cover and a notch milled into the vibration damper are used as references to set ignition timing. A magnetic

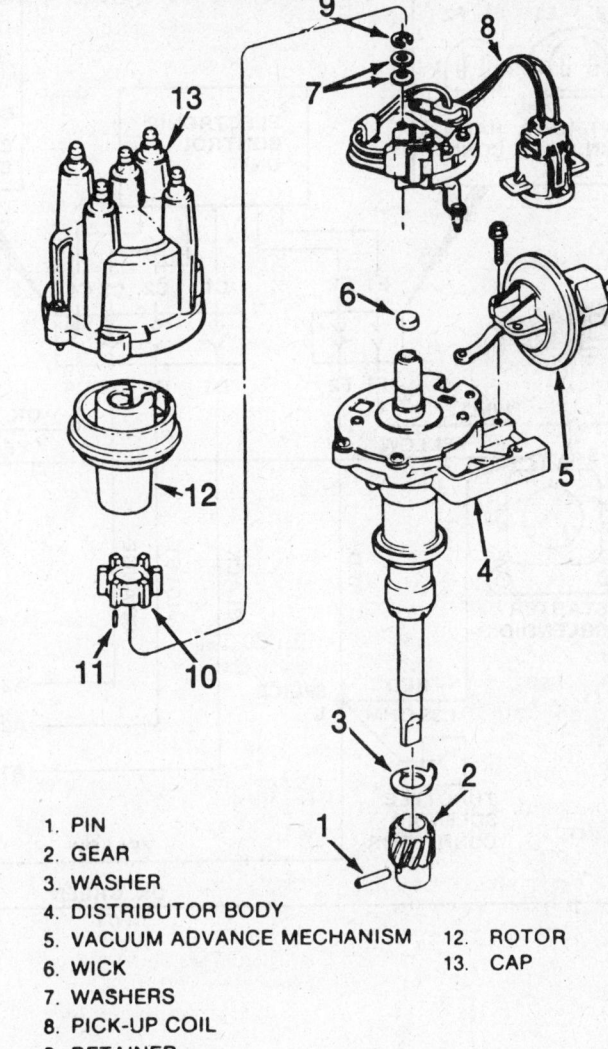

1. PIN
2. GEAR
3. WASHER
4. DISTRIBUTOR BODY
5. VACUUM ADVANCE MECHANISM
6. WICK
7. WASHERS
8. PICK-UP COIL
9. RETAINER
10. TRIGGER WHEEL
11. PIN
12. ROTOR
13. CAP

150 4 cylinder electronic distributor

timing probe socket is provided integral with the timing degree scale for use with a special magnetic timing probe. The probe socket is located at 9.5°ATDC and the equipment used is calibrated to compensate for the location. Do not use the timing probe socket as a reference point to check the ignition timing when using a conventional timing light. Timing instructions are given on the underhood emission control sticker. Check the sticker instructions before adjusting the ignition timing and follow those directions if they differ from the procedures given below.

NOTE: Connect a tachometer to the SSI ignition system in the conventional way; to the negative (distributor) side of the coil and to a ground. HEI distributor caps have

a "Tach" terminal. **Some tachometers may not work with an SSI or HEI ignition system and there is a possibility that some could be damaged. Check with the manufacturer of the tachometer to make sure it can be used.**

151-4 Engine

1. Disconnect the vacuum hose, at the distributor vacuum unit. Plug the vacuum line to prevent leakage.

2. Connect a timing light and a tachometer in accordance with the manufacturer's instructions. If the timing light has an advance control, be sure it is off.

3. Start the engine. Adjust the carburetor curb idle screw so the engine idles at the specified curb idle speed at operating temperature. If there is a

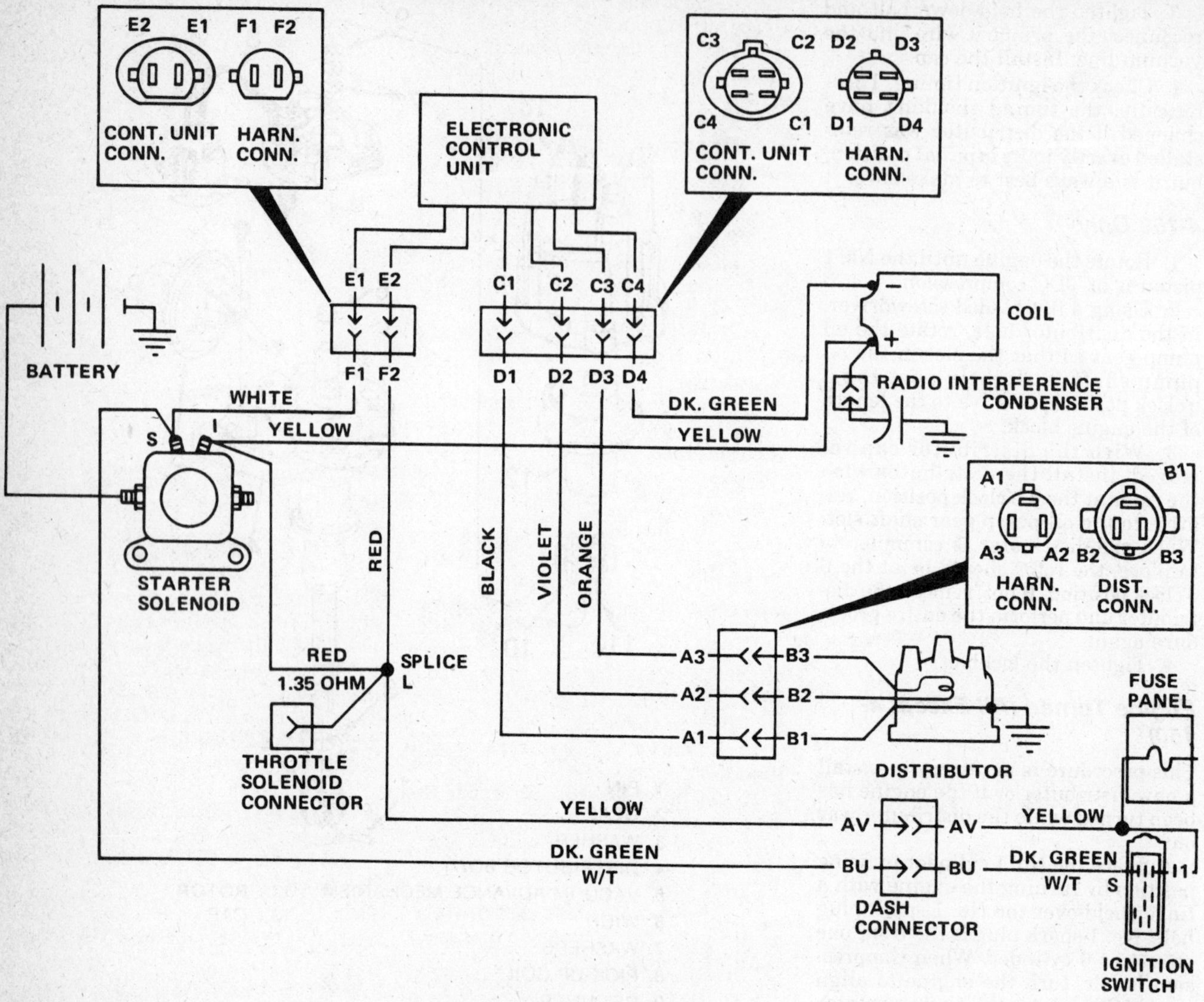

E2 E1 F1 F2

CONT. UNIT | HARN.
CONN. | CONN.

ELECTRONIC
CONTROL
UNIT

C3 C2 D2 D3

C4 C1 D1 D4

CONT. UNIT | HARN.
CONN. | CONN.

BATTERY

E1 E2 C1 C2 C3 C4

F1 F2 D1 D2 D3 D4

COIL

WHITE

YELLOW

RADIO INTERFERENCE
CONDENSER

DK. GREEN

YELLOW

STARTER
SOLENOID

RED

BLACK

VIOLET

ORANGE

A1   B1

A3 A2 B2 B3

HARN. | DIST.
CONN. | CONN.

RED
1.35 OHM

SPLICE
L

A3 — B3

A2 — B2

A1 — B1

THROTTLE
SOLENOID
CONNECTOR

DISTRIBUTOR

FUSE
PANEL

YELLOW

AV — AV

YELLOW

DK. GREEN
W/T

BU — BU

DK. GREEN
W/T   S   I1

DASH
CONNECTOR

IGNITION
SWITCH

Schematic of typical SSI ignition system

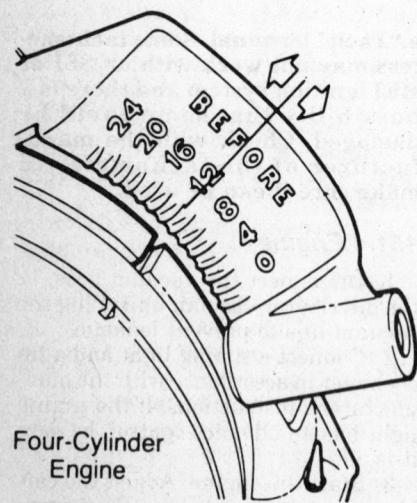

Four-Cylinder
Engine

150-4 timing degree scale and notch locations

throttle stop solenoid, disconnect the electrical connector. Aim the timing light at the pointer marks.

4. Adjust the timing by loosening the distributor clamp nut and rotating the distributor. Set the timing to the specifications given in the Tune-Up chart or on the underhood emission sticker.

NOTE: On some models, a white paint mark is applied to the scale for the specified, initial timing setting. Do not mistake this mark for TDC.

5. Check the timing again after tightening the distributor clamp.

6. Connect the vacuum hose(s) and set the idle speed to specifications.

150-4 and 258-6 Engines

1. Start the engine and allow it to

reach normal operating temperature, then switch the ignition off. Make sure the automatic transmission is in Park or the manual transmission is in Neutral.

2. On 150-4 engines, disconnect the three wire connector to the vacuum input switch. On 258-6 engines, disconnect the two wire connector (yellow and black wires) at the electronic ignition module. Install a jumper wire between the two wires at the module connector terminal.

3. Disconnect and plug the distributor vacuum advance hose.

4. Attach a timing light and tachometer to the engine according to the manufacturers instructions.

5. Start the engine and increase the engine speed to 1600 rpm while observing the timing marks with a timing light. If the timing light being used incorporates an advance control

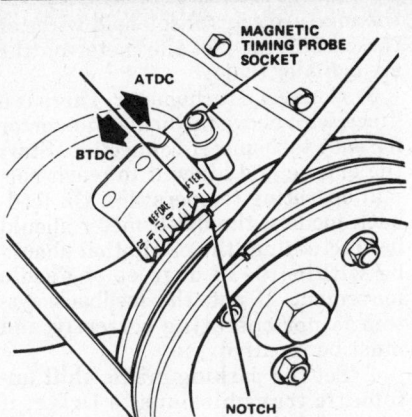

151-4 timing degree scale and notch locations

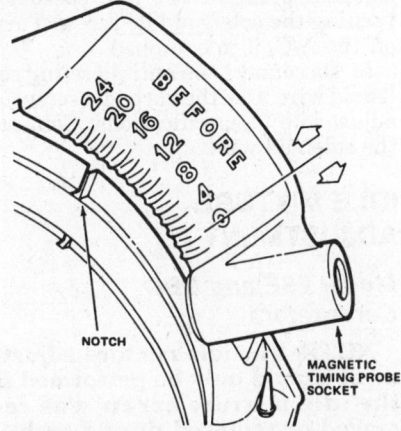

258-6 timing degree scale and notch locations

feature, make sure the control is in the OFF position.

6. Adjust the ignition timing to the specifications given in the Tune-Up chart or on the underhood emission sticker by loosening the distributor hold-down bolt and turning the distributor body. Exercise caution when working around a running engine; stay clear of all moving components, belts, etc. and do not wear loose clothing.

7. Once the timing has been set, reconnect the vacuum line and all electrical connectors.

FUEL SYSTEM

Fuel Pump

REMOVAL & INSTALLATION

All Engines

1. Disconnect the inlet and outlet fuel lines, and any vacuum lines.

2. Remove the two fuel pump body attaching nuts and lockwashers.

3. Pull the pump and gasket free of the engine. Make sure the mating surfaces of the fuel pump and the engine are clean. Scrape off any old gasket material and sealer before installation.

4. Cement a new gasket to the mounting flange of the fuel pump.

5. Position the fuel pump on the engine block so that the lever of the fuel pump rests on the fuel pump cam of the camshaft.

6. Secure the fuel pump to the block with the two capscrews and lockwashers.

7. Connect the intake and outlet fuel lines to the fuel pump, and any vacuum lines.

Fuel Filter

REMOVAL & INSTALLATION

American Motors uses an in-line fuel filter in the line from the carburetor to the fuel pump on all engines except the 151-4. All models (except the 151-4) also have a vapor return line from the filter to the tank.

1. Remove the air cleaner as necessary.

2. Put an absorbent rag under the filter to catch spillage.

3. Remove the hose clamps.

4. Remove the filter and short attaching hoses.

5. Assemble the new filter and hoses. If the filter has a return line, position the line at the top.

NOTE: The original equipment type hose clamps can't be reused with much success. It is much better to replace them with screw type clamps. If there is an arrow on the new filter, it must point toward the carburetor.

6. Fit the filter in place, tighten the clamps, start the engine and check for leaks. Discard the gas-soaked rag safely.

151-4 Only

The 151 cu. in. engine uses a replaceable filter located within the carburetor body at the fuel inlet fitting.

1. Place an absorbent cloth under the inlet fitting at the carburetor.

2. Using a suitable line wrench on the fuel line fitting and a backup wrench on the fuel inlet nut, loosen the fuel line and pull it out of the inlet nut. Unscrew the fuel inlet nut.

3. The filter will be pushed out part way by a spring located behind the filter. Remove the filter.

4. Install the new filter with the hole in the filter facing the inlet fit-

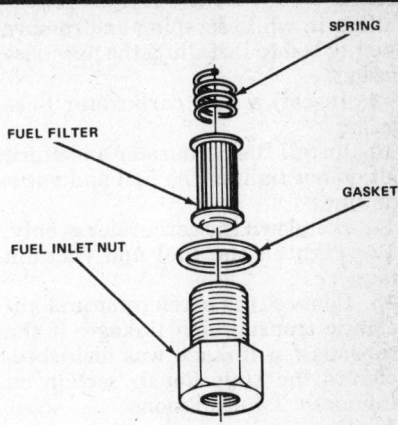

151-4 fuel filter assembly

ting. Install a new gasket on the fitting. Tighten to 18 ft. lbs.

Carburetor

Depending on the engine, four types of carburetors are used on AMC models. The 151-4 uses a Rochester 2SE on Federal models, or a Rochester E2SE on California models. The 150-4 uses a Carter YFA feedback carburetor and the 258-6 uses a Carter BBD feedback carburetor.

For further information on feedback carburetors, please refer to *Chilton's Guide To Fuel Injection And Feedback Carburetors.*

REMOVAL & INSTALLATION

NOTE: The following procedure is meant to be general in nature. Tag all hoses and electrical connections before removal. Take precautions to avoid the risk of fire as some fuel will remain in the carburetor float bowl. For all overhaul and adjustment procedures, see the Carburetor Unit Repair Section.

1. Remove the air cleaner.

2. Disconnect the fuel and vacuum lines, along with any electrical connectors.

3. Disconnect the choke rod.

4. Disconnect the accelerator linkage.

5. Disconnect the automatic transmission linkage.

6. Remove the mounting bolts and lift off the carburetor.

7. Remove the base gasket.

8. Before installation, make sure that the carburetor and manifold sealing surfaces are clean. Scrape off any old gasket material or sealer. Place a clean rag in the intake manifold to prevent gasket material from

falling in while scraping and remove the rag before installing the new base gasket.

9. Install a new carburetor base gasket.

10. Install the carburetor and start, but do not tighten the fuel and vacuum lines.

11. Bolt down the carburetor evenly.

12. Tighten the fuel and vacuum lines.

13. Connect the accelerator and automatic transmission linkage. If the transmission linkage was disturbed, refer to the Unit Repair section on Automatic Transmissions.

14. Connect the choke rod.

15. Install the air cleaner. Adjust the idle speed.

IDLE SPEED ADJUSTMENT

The engine and related systems must be operating normally before performing idle speed adjustments. The mixture adjustment should not have to be performed unless the mixture screw setting was altered during carburetor overhaul. Since automatic transmission vehicles are adjusted in Drive, set the parking brake firmly and do not accelerate the engine. Perform all procedures with the air cleaner installed or with the air cleaner removed and all vacuum hoses plugged. Make sure the ignition timing is correct before setting the idle speed.

Model YFA and BBD Carburetors

1. Connect a tachometer according to the manufacturer instructions. Start the engine and allow it to reach normal operating temperature. The air cleaner should be installed, automatic transmissions in Drive and manual transmissions in Neutral.

2. Remove the vacuum hose to the SOLE-VAC vacuum actuator unit and plug the hose. Disconnect the holding solenoid wire connector.

3. Adjust the curb idle speed by using the vacuum actuator adjustment screw on the throttle lever. This adjustment is made with all accessories off. Refer to the Tune-Up Specifications Chart or the underhood emission control sticker for the correct idle speed.

4. After the vacuum actuator adjustment is complete, leave the vacuum hose plugged and disconnected. Apply a direct vacuum source to the vacuum actuator to fully extend the throttle positioner.

5. Adjust the idle using the ¼ in. hex-head adjustment screw on the end of the SOLE-VAC unit. Adjust to specifications.

NOTE: The engine speed will vary 10–30 rpm during this mode due to being in the closed loop fuel control mode.

6. If equipped, turn the A/C on. Attach a jumper wire from the positive (+) battery cable to the holding solenoid terminal to energize it and hold the throttle open manually to allow the throttle positioner to fully extend.

7. If the holding solenoid is not within specifications, adjust the SOLE-VAC unit to obtain the correct idle rpm.

8. Disconnect the jumper wire, vacuum source and reconnect the hose to the vacuum actuator.

Model 2SE Carburetor

1. Start the engine and allow it to reach normal operating temperature. Set the parking brake firmly. Connect a tachometer to the ignition coil or the pigtail wire connector above the heater blower.

2. Disconnect the vacuum hose from the vacuum advance and plug the hose. Check and adjust the ignition timing if necessary and reconnect the vacuum advance.

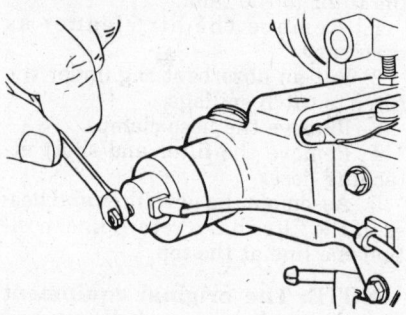

150-4 cylinder curb-idle speed adjustment

3. Disconnect the decelerator valve hose and canister purge hose. Plug the hoses and remove the air cleaner assembly.

4. If equipped with A/C, turn it on and open the throttle momentarily to make sure the solenoid arm is fully extended. Adjust the solenoid idle speed screw to obtain the rpm given in the Tune-Up Specifications Chart or on the underhood emission control sticker.

5. If not equipped with A/C, adjust the engine idle speed with the solenoid idle speed screw. Disconnect the solenoid and adjust the curb idle.

6. Install the air cleaner and reconnect all hoses.

Model E2SE Carburetor

1. Disconnect and plug the purge hose at the charcoal canister.

2. Connect a dwellmeter to the single light blue wire which is taped to

the mixture control solenoid wires at the carburetor. Set the meter on the six cylinder scale.

3. Connect a tachometer. There is a green wire above the heater fan motor for easy tachometer connection. Start the engine and allow it to reach normal operating temperature. On feedback models, the dwell meter should be fluctuating; the oscillation should be within 10–15 degrees of needle movement. If not, the feedback system is not operating correctly and must be repaired.

4. Set the parking brake, shift automatic transmissions to Drive; if equipped with air conditioning, turn it on. Open the throttle to extend the solenoid plunger. Set idle speed by turning the solenoid idle screw. Turn off the A/C, if so equipped.

5. Disconnect the anti-dieseling solenoid wire. Use the curb idle screw to adjust idle to specifications. Connect the solenoid wire.

IDLE MIXTURE ADJUSTMENT

Model 2SE and BBD Carburetors

NOTE: The idle mixture adjustment should only be performed if the idle mixture screw was removed or replaced during carburetor overhaul. Otherwise, the mixture is preset at the factory and requires no periodic adjustment. The carburetor may have to be removed to knock out the mixture screw tamper-proof plugs before any adjustment is attempted.

1. Connect a tachometer according to the manufacturer instructions, then start the engine and allow it to reach normal operating temperature.

2. Chock the wheels and set the parking brake firmly. Place automatic transmission in Drive or manual transmission in Neutral.

3. Adjust the idle speed as described above.

4. Using a suitable mixture adjusting tool, turn the mixture screw clockwise (lean) until a loss of engine rpm is noted.

5. Turn the mixture screw counterclockwise (rich) until the highest engine rpm is obtained. Do not turn the screw any further than the point at which the highest engine rpm is noted. This is referred to as the best lean idle. Engine speed will increase above curb idle speed an amount that approximately corresponds to the lead drop specifications.

6. As the final adjustment, turn the idle mixture screw clockwise in small increments until the specified speed

drop is noted. Refer to the underhood emission control sticker for idle drop specifications.

NOTE: If the final engine rpm differs more than 30 rpm plus or minus the original set curb idle speed, adjust the curb idle speed again as required. Replace the idle mixture plugs when adjustment is complete.

Model YFA Carburetor

The mixture control (MC) solenoid dwell is used as a reference for the adjustment of the mixture and is indicated on a dwell meter set on the six cylinder scale. With the engine at idle, it is normal for the MC solenoid to increase or decrease dwell between 10–15 degrees. The mixture solenoid dwell is an indication of the ON/OFF time ratio between the energized and de-energized time of the solenoid.

1. Remove the carburetor.
2. To remove the mixture screw plug, center punch it then drill a $\frac{1}{8}$ in. hole into the plug. Be careful not to damage the mixture screw while drilling. Install a self-tapping screw into the plug hole, then pull the screw out with pliers to remove the plug. Install the carburetor on the engine.
3. Connect a tachometer to the ignition coil TACH wire connector.
4. Connect a dwell meter to the MC solenoid test terminals in the diagnostic connector at terminals D2-14 and D2-7. Set the dwell meter on the six cylinder scale.
5. Place the automatic transmission in Park or the manual transmission in Neutral with the parking brake firmly set.
6. Disconnect and plug the canister purge vacuum hose at the canister.
7. Start the engine and operate it at fast idle speed for at least three minutes to allow the feedback system to enter closed loop operation. Allow the engine speed to stabilize at the specified idle rpm. Adjust the idle speed if necessary.
8. Adjust the idle mixture screw to obtain an average dwell reading of 30 degrees, within a span of 25–35 degrees. If the dwell is too low, slowly turn the mixture screw counterclockwise (out); if the dwell is too high, turn the mixture screw clockwise (in).

NOTE: Allow time for the system to react and stabilize after each movement of the adjustment screw. The system is very sensitive to adjustment.

9. If the dwell cannot be obtained by adjustment, check the carburetor idle circuit for air leaks or restrictions. Plug the idle mixture screws

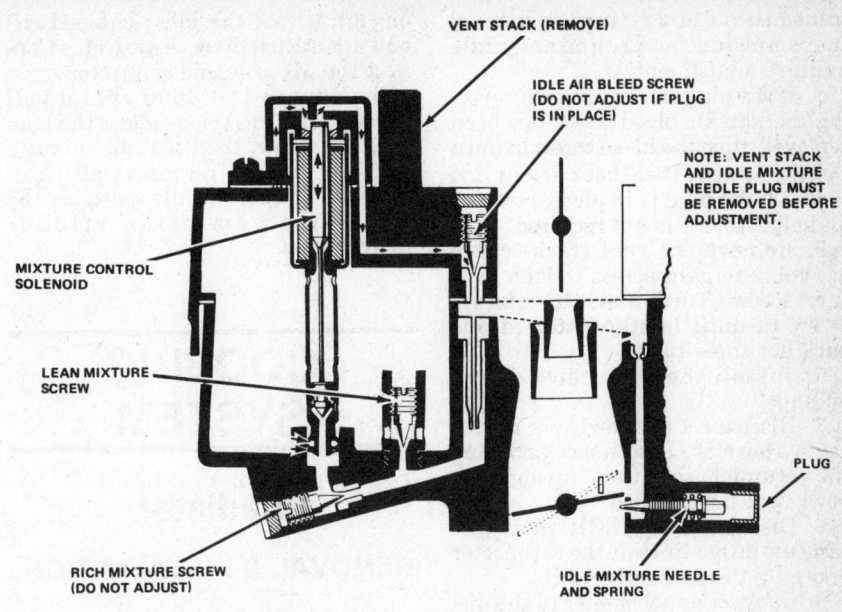

Cross section of typical F2SE carburetor showing adjustment points

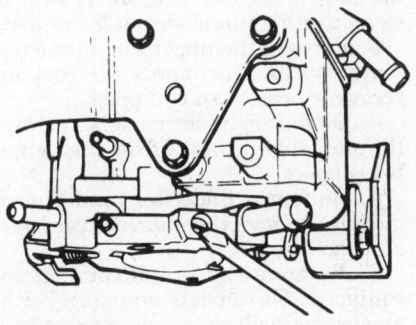

Idle mixture screw location on YFA carburetor

with RTV sealant when adjustments are complete.

Model E2SE Carburetor

Idle mixture is calibrated at the factory and should normally not need adjustment. Because the computer-controlled carburetor system is very complex, the procedure must be followed carefully. The mixture control (MC) solenoid dwell is used as a reference for the adjustment of the mixture and is indicated on a dwell meter set on the six cylinder scale. With the engine at idle, it is normal for the MC solenoid to increase or decrease dwell between 10–15 degrees. The mixture solenoid dwell is an indication of the ON/OFF time ratio between the energized and de-energized time of the solenoid.

1. Remove the carburetor and invert it in a suitable holding fixture.
2. Using a punch between the two locator points in the throttle body be-

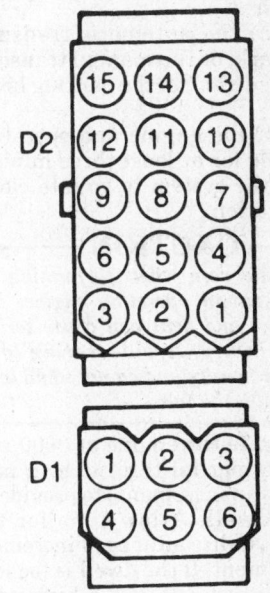

Diagnostic terminal locations

neath the idle mixture screw plug (manifold side), break out the throttle body to gain access to the mixture screw plug. Drive out the hardened steel plug.

NOTE: The plug will probably shatter when removal is attempted. It is not necessary to remove the plug completely. Remove the loose pieces to allow the use of adjusting tool J-28706 or a $\frac{3}{16}$ in. thin wall deep socket wrench.

3. Lightly seat the mixture screw then back out 3 turns with automatic

transmission, or $2\frac{1}{2}$ turns on manual transmission for preliminary idle mixture adjustment.

4. If the plug in the air horn covering the idle air bleed screw has been removed, turn the bleed screw in until lightly seated, then back it out $1\frac{1}{4}$ turns. If the plug is in place, removal and adjustment is not required.

5. Remove the vent stack-screen assembly to gain access to lean mixture screw. Turn the lean mixture screw in until lightly bottomed and back out three turns.

6. Install the carburetor on the engine.

7. Disconnect the bowl vent line at the carburetor. Disconnect and plug the vacuum hose at the T-fitting n the bowl vent line, if used.

8. Disconnect the EGR valve and canister purge lines at the carburetor and plug the carburetor ports.

9. Connect a tachometer to the distributor tach output and a dwell meter to the mixture control solenoid dwell test lead. There is a distributor tach wire above the heater fan motor (green wire).

10. Place the automatic transmission in Park, or the manual transmission in Neutral. Set the parking brake firmly.

11. Start the engine and operate it at fast idle for at least three minutes to allow the system to go into closed loop operation.

— CAUTION —

Use extreme care when performing adjustments on an operating engine. The fan, pulleys and belts can cause serious personal injury. Avoid wearing loose clothing or jewelry and do not stand in direct line with the fan.

12. Operate the engine at 3000 rpm and adjust partial throttle lean mixture screw in increments to provide 25 degrees dwell. Allow time for the dwell to stabilize after each incremental adjustment. If the dwell is too low, back the screw out; if too high, turn the screw in.

13. Check and adjust the idle speed if necessary.

14. Adjust the idle mixture screw to obtain an average dwell of 25 degrees. Again, if too low back the screw out; if too high turn the screw in. Allow time for the dwell to stabilize after each adjustment and remove the tool when checking the dwell reading. The system is very sensitive to adjustment. If unable to adjust to specifications, check the idle system for leaks, restrictions, etc.

15. Disconnect the mixture control solenoid connector and check for an engine speed change of at least 50 rpm. If the rpm does not change

enough, check the idle air bleed circuit for restrictions, leaks, etc. Connect the MC solenoid connector.

16. Repeat the 3000 rpm dwell check. If not correct, readjust the lean mixture screw, then the idle mixture screw. If correct, reconnect all vacuum lines and set the idle speed per the underhood emission sticker instructions.

COOLING SYSTEM

Radiator

REMOVAL & INSTALLATION

1. Raise the hood and remove the radiator cap. Be sure the engine is cold.

2. Drain the radiator. If the coolant appears to be clean, drain it into a clean container and save it for re-use.

3. Remove the upper and lower radiator hoses. Disconnect the coolant recovery hose, if so equipped.

4. On four cylinder models, remove the ambient air intake from the radiator support.

5. On four cylinder air-conditioned models, remove the charcoal canister and the bracket.

6. Remove the fan shroud, if so equipped. On models equipped with an electric cooling fan; disconnect the motor and sensor wiring harnesses, remove the motor, fan and support; or remove as an assembly with the radiator.

7. On automatic transmission models, disconnect and plug the fluid cooler lines. Remove battery on Pacers for access.

8. Remove the radiator attaching screws and bolts and lift out the radiator.

9. To install radiator, reverse sequence above.

Water Pump

REMOVAL & INSTALLATION

150-4 Engine

1. Disconnect all hoses at the pump.

2. Remove the drive belts.

3. Remove the fan shroud attaching screws.

4. Unbolt the fan and fan drive assembly, and remove along with the shroud. On some models it may be

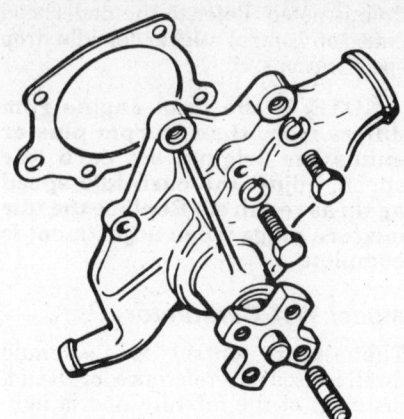

150 4 cylinder water pump and gasket

easier to turn the shroud $\frac{1}{2}$ turn, to remove it.

5. Unbolt and remove the pump.

— CAUTION —

Engines built for sale in California having a single, serpentine drive belt and viscous fan drive, use a reverse rotating pump and drive. These components are identified by the word REVERSE stamped on the drive cover and inner side of the fan, and REV cast into the water pump body. Never interchange standard rotating parts with these.

6. Installation is in the reverse order of removal. Always use a new gasket coated with sealer. Torque the water pump bolts to 13 ft. lbs., the fan bolts to 18 ft. lbs.

151-4 Engine

1. Drain the cooling system. Make sure the engine is cold.

2. Remove all drive belts.

3. Remove the fan and pump pulley.

4. Unbolt and remove the pump from the engine.

5. Clean the gasket surfaces, coat the new gasket with non-hardening type sealer and position the gasket on the block.

6. Coat the threaded areas of the bolts with waterproof sealer and install the pump. Torque the bolts to 25 ft. lbs.

7. Install the pulley and fan.

8. Install the drive belts. The belts should be adjusted so that a $\frac{1}{2}$ in. deflection is present when they are depressed mid-point along their longest straight run.

258-6 Engine

1. Drain the cooling system. Disconnect the negative cable from the battery.

2. Unfasten the radiator and the heater hoses at the pump.

3. Loosen the adjustment bolts

from the alternator and the power steering pump, if so equipped. Remove the V-belts.

4. Unfasten the fan ring securing bolts. Remove the fan and pump pulley assembly. Withdraw the fan ring (or shroud).

5. Remove the securing bolts from the water pump. Withdraw the pump along with its gasket.

6. Installation is in the reverse order of removal. Always clean all mating surfaces and use a new pump gasket. Bleed the radiator by running the engine and opening the heater control valve. Run the engine long enough to open the thermostat, then check the coolant level. The water pump mounting bolts should be tightened to 10–15 ft. lbs.

Thermostat

REMOVAL & INSTALLATION

The thermostat is located in the water outlet housing at the top or front of the cylinder head or in the hose, itself. Drain the coolant to a point below the thermostat. Disconnect the upper radiator hose and/or remove the bolts which hold the water outlet neck to the engine. Remove the thermostat.

When installing the thermostat, make certain the pellet or coil spring is facing the engine. Thermostats are marked on the outer flange with the proper installing direction. Replace the gasket between the thermostat and the housing cover. The bleed hole on the thermostats used on six cylinder engines must be installed up (at 12 o'clock), to prevent "burping" caused by trapped air. Refill the cooling system, and run the engine for a while with the heater on to bleed the system of air. Recheck the coolant level.

————— CAUTION —————
Tightening the housing bolts unevenly, or with the thermostat cocked in its recess, will cause the housing to crack.

EMISSION CONTROLS

NOTE: For a description and service procedures for the following emission control systems, please refer to the "Emission Control" section of the Unit Repair section. Not all vehicles have all systems.

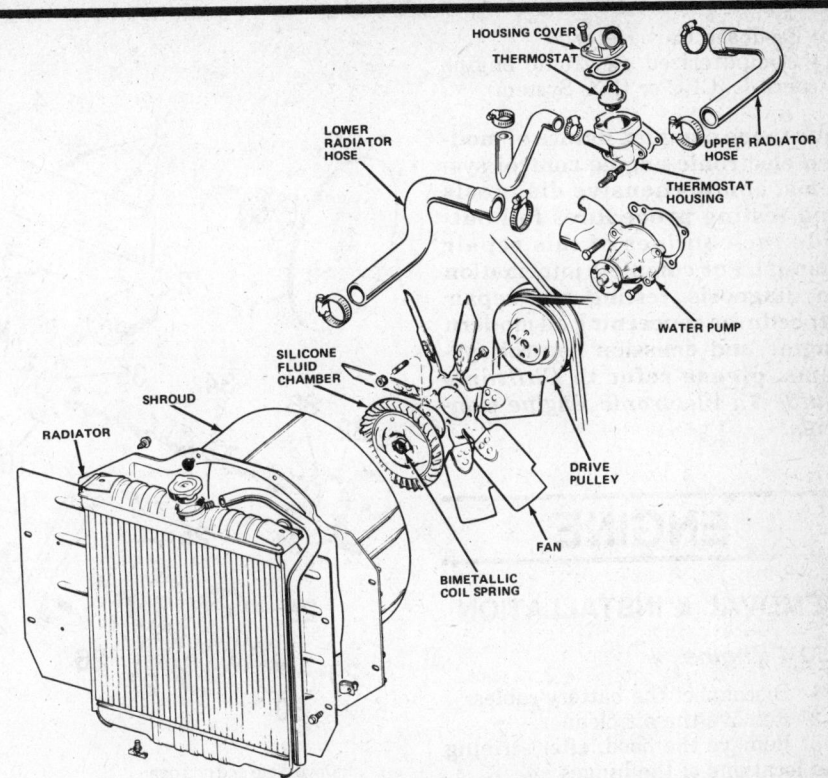

151-4 cooling system components

Vehicles built for operating at altitudes greater than 4,000 feet are equipped with special emission control devices. Some of these devices vary model to model. All models have as standard equipment the following emission control features:

• Air Guard system (air pump and components) or Pulsair system on 151-4
• Closed Positive Crankcase Ventilation (PCV) system
• Single diaphragm vacuum advance unit
• Fuel vapor control, canister storage
• Heated and thermostatically controlled air cleaner (TAC)
• Exhaust Gas Recirculation (EGR) valve
• Catalytic Converter
• Transmission Controlled Spark (TCS)

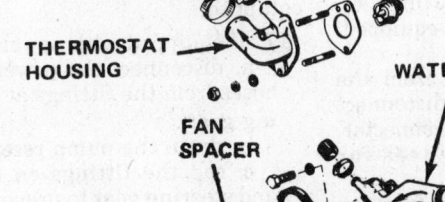

Cooling system components—6 cylinder engines

- Feedback Carburetor
- Computerized electronic engine controls (CEC or CCC System)

Due to the complex nature of modern electronic engine control systems, comprehensive diagnosis and testing procedures fall outside the confines of this repair manual. For complete information on diagnosis, testing and repair procedures concerning all modern engine and emission control systems, please refer to *Chilton's Guide To Electronic Engine Controls.*

ENGINE

REMOVAL & INSTALLATION

150-4 Engine

1. Disconnect the battery cables.
2. Remove the air cleaner.
3. Remove the hood, after scribing the locations of the hinges.

— CAUTION —

Allow the engine to cool. If the engine has been recently operated, use care to prevent scalding by hot coolant. The cooling system is pressurized.

4. Drain the radiator.
5. Remove the lower radiator hose.
6. Remove the upper radiator hose and coolant recovery hose.
7. Remove the fan shroud and disconnect the transmission fluid cooler tubing (automatic transmission).
8. Remove the radiator/condenser (if equipped with A/C).
9. Remove the fan assembly and install a $\frac{5}{16}$ x $\frac{1}{2}$ in. SAE capscrew through the fan pulley into the water pump flange, to maintain the pulley and water pump in alignment when the crankshaft is rotated.
10. Disconnect the heater hoses.
11. Disconnect the throttle linkages, cruise control cable (if so equipped) and throttle valve rod.
12. Disconnect the wires from the starter motor solenoid and disconnect CEC System wire harness connector. Tag all disconnected wires for reassembly.
13. Disconnect the fuel pipe from the fuel pump and plug the fuel line.
14. If equipped with air conditioning, remove the service valves and cap the compressor ports. Do not attempt to bleed the freon unless you are familiar with air conditioning systems. See the Caution above.
15. Disconnect the fuel return hose from the fuel filter.

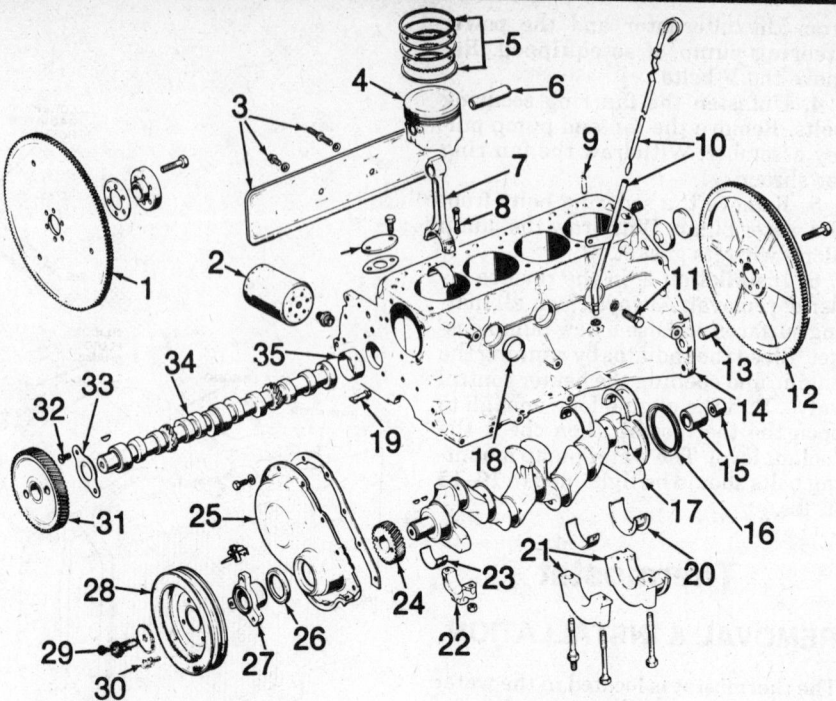

1. Drive plate and ring gear (automatic trans)
2. Oil filter
3. Push rod cover and bolts
4. Piston
5. Piston ring
6. Piston pin
7. Connecting rod
8. Connecting rod bolt
9. Dowel
10. Oil level indicator and tube
11. Block drain
12. Flywheel and ring gear (manual trans)
13. Dowel
14. Cylinder block
15. Pilot and/or converter bushing
16. Rear oil seal
17. Crankshaft
18. Block core plug
19. Timing gear oil nozzle
20. Main bearings
21. Main bearing caps
22. Connecting rod bearing cap
23. Connecting rod bearing
24. Crankshaft gear
25. Timing gear cover (front)
26. Timing gear cover oil seal
27. Crankshaft pulley hub
28. Crankshaft pulley
29. Crankshaft pulley hub bolt
30. Crankshaft pulley bolt
31. Crankshaft timing gear
32. Camshaft thrust plate screw
33. Camshaft thrust plate
34. Camshaft
35. Camshaft bearing
36. Oil pump driveshaft retainer plate, gasket and bolt

Exploded view of the 151 4 cylinder block assembly

16. Remove the power brake vacuum check valve from the booster, if so equipped.
17. If equipped with power steering:
   a. disconnect the power steering hoses from the fittings at the steering gear
   b. drain the pump reservoir
   c. cap the fittings on the hoses and steering gear to prevent foreign objects from entering the system.
18. Identify, tag and disconnect all necessary wire connectors and vacuum hoses.
19. Raise the vehicle and support it safely.
20. Remove the starter motor.
21. Disconnect the exhaust pipe from the manifold.

22. Remove the flywheel/converter housing access cover. On models with automatic transmission, mark the converter and drive place location and remove the converter-to-drive plate bolts.
23. Remove the upper flywheel/converter housing bolts and loosen the bottom bolts.
24. Remove the engine mount cushion-to-engine compartment bracket bolts.
25. Attach a lifting device to the engine.
26. Raise the engine off the front supports.
27. Place a support stand under the converter (or flywheel) housing.
28. Remove the remaining converter (or flywheel) housing bolts.

29. Lift the engine out of the engine compartment.

30. Installation is in the reverse order of removal.

151-4 Engine

The engine and transmission are removed as an assembly.

1. Disconnect the negative battery cable. Mark the hood hinge locations and remove the hood.

2. Drain the cooling system. Disconnect the hoses. If equipped with automatic transmission, disconnect and plug the coolant lines from the radiator. Remove the fan and shroud; remove the radiator. Disconnect the heater hose from the intake manifold.

3. If equipped with power steering, remove the pump and set it aside, without disconnecting any hoses. If equipped with air conditioning, unbolt the compressor and move it aside without disconnecting any hoses. Remove the evaporator-to-dryer line from the sill clips, but do not disconnect the line. Unbolt the condenser and move it aside, without disconnecting any hoses.

4. Disconnect and label the alternator harness, starter wires, vacuum hoses and electrical connections to the carburetor, carburetor linkage, and vacuum and electrical connections to the distributor. Disconnect the coolant and oil pressure sending unit wires.

5. Disconnect the oil dipstick tube from the exhaust manifold, and pull the tube from the block.

6. Raise and support the car. Remove the engine mount nuts from the crossmember. Remove the ground cable at the left mount bracket.

7. Support assembly with floorjack. Loosen the crossmember and lower it slightly. Remove the speedometer cable from the transmission. Remove the cooler tubes if equipped with automatic transmission. Remove the transmission linkage and backup light switch wiring. Remove the rear transmission mount.

8. Matchmark the driveshaft and remove.

9. Remove the crossmember.

10. Disconnect the exhaust pipe from the manifold.

11. Disconnect and plug the fuel line.

12. Check that all hoses and wires have been disconnected. Lower the car and support jack. Attach a chain to the engine rear bracket, and the air conditioner or alternator bracket, and raise and remove the engine/transmission assembly.

13. Installation is in the reverse order of removal.

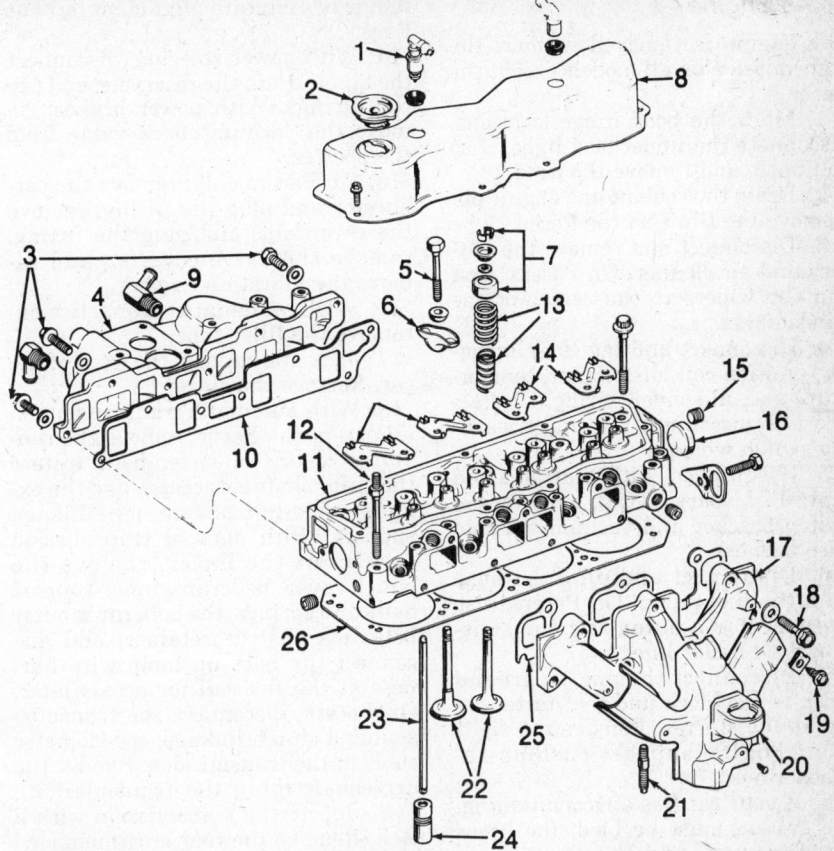

1. PCV valve	11. Cylinder head	20. Exhaust manifold heat shroud (heat shield)
2. Oil filler cap	12. Cylinder head stud bolt	21. Exhaust manifold to exhaust pipe stud
3. Intake manifold attaching bolts	13. Valve spring	
4. Intake manifold	14. Push rod guide	22. Valves
5. Rocker arm capscrew	15. Cylinder head plug	23. Push rod
6. Rocker arm	16. Cylinder head core plug	24. Tappet
7. Valve spring retainer assembly	17. Exhaust manifold	25. Exhaust manifold gasket
8. Cylinder head cover (rocker cover)	18. Exhaust manifold bolt	26. Cylinder head gasket
9. Coolant hose fitting	19. Oil level indicator tube attaching screw	
10. Intake manifold gasket		

Exploded view of the 151 4 cylinder head assembly

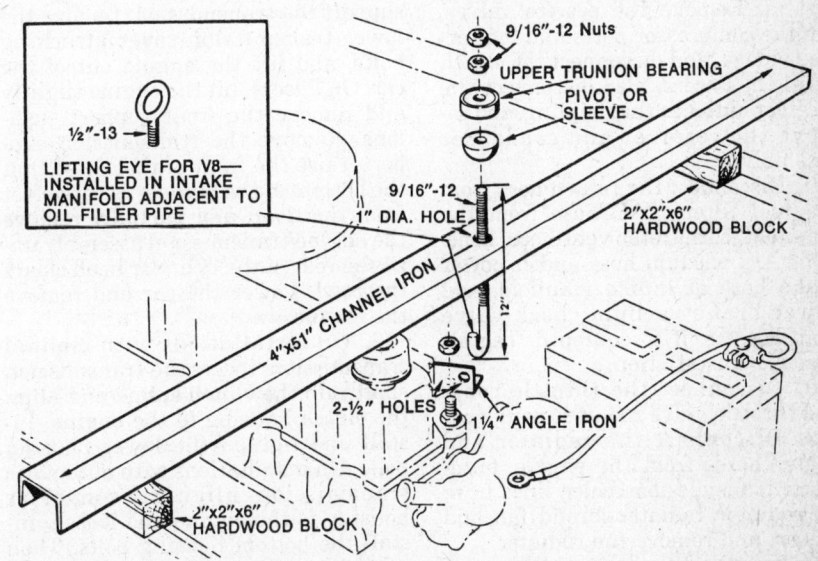

Lifting fixture can be fabricated as illustrated to facilitate oil pan and motor mount removal

258-6 Engine

The engine is removed without the transmission on all models except the Pacer.

1. Mark the hood hinge locations, disconnect the underhood light, if so equipped, and remove the hood.

2. Drain the coolant and engine oil. Remove the filter on the Pacer.

3. Disconnect and remove the battery and air cleaner. On Pacers, first run the wipers to the center of the windshield.

4. Disconnect and tag the alternator, ignition coil, distributor, temperature and oil sender wiring. On Pacers, also disconnect the brake warning switch wiring.

5. If equipped with transmission controlled spark (TCS), remove the switch bracket and vacuum solenoid wire harness.

6. Disconnect and plug the hose from the fuel pump. On Pacers, also disconnect the automatic transmission fluid cooler line.

7. Disconnect the engine ground strap at the block and the starter cable at the starter. Remove the right front engine support cushion-to-bracket bolt.

8. If your car has air conditioning, the system must be bled, the hoses disconnected, and the compressor removed.

--- CAUTION ---

Do not perform this operation if you are unfamiliar with A/C systems. Have the system bled by a qualified mechanic. Compressed refrigerant will freeze any surface it contacts, including your eyes. It also forms a poisonous gas in the presence of flame.

9. Bleed the refrigerant from the system. Remove the service valves, cap the compressor ports and the service valves, and disconnect the clutch wire. On Pacers, also disconnect the receiver outlet at the coupling, and remove the receiver and condenser assembly.

10. Disconnect the return hose from the fuel filter, TAC hose from the manifold, carburetor vent hose, heater or A/C vacuum hose and/or power brake hose at intake manifold, and power brake vacuum check valve from booster, if so equipped. Tag all hoses for installation.

10. Disconnect the throttle cable and throttle valve rod, if so equipped.

12. Disconnect the radiator and heater hoses from the engine, automatic transmission cooler lines from the radiator, radiator shroud, fan, and spacer, and remove the radiator.

13. Install a ⁵⁄₁₆ x ½ in. bolt through the fan pulley into the water pump

flange to maintain alignment (all but Pacer).

14. With power steering, disconnect the hoses, drain the reservoir, and cap the fittings. With power brakes, remove the vacuum check valve from the booster.

15. On Pacers only, remove the carburetor and plug the fitting, remove the carburetor and plug the fitting, remove the valve cover(s), and remove the vibration damper.

16. With automatic transmission, remove the filler tube.

17. Jack and support the front of the car. Remove the starter.

18. With automatic transmission on all except the Pacer, remove the converter cover, converter bolts (rotate the crankshaft for access), and the exhaust pipe-to-transmission linkage support. With manual transmission on all but the Pacer, remove the clutch cover, bellcrank inner support bolts and springs, the bellcrank, outer bellcrank-to-strut retainer, and disconnect the back-up lamp wire harness at the firewall for access later. On Pacers, disconnect the transmission and clutch linkage, speedometer cable at the transmission, remove the driveshaft (plug the transmission), and support the transmission with a jack. Remove the rear crossmember.

19. Attach the lifting device and support the engine. Remove the engine mount bolts.

20. Disconnect the exhaust pipe from the manifold.

21. On all but the Pacer, remove the upper converter or clutch housing bolts and loosen the lower bolts. Raise the car and move the jackstands to the jack pad area. Remove the A/C idler pulley and bracket, if so equipped. Lift the engine off the front supports, support the transmission, remove the lower transmission cover attaching bolts, and lift the engine out of the car. On Pacers, lift the engine slightly and remove the front support cushions. Remove the transmission support, raise the front of the car so that the bottom of the bumper is three feet from the floor, and partially remove the engine/transmission assembly until the rear of the cylinder head clears the cowl. Lower the car and remove the engine.

22. On installations with manual transmission, insert the transmission shaft into the clutch spline and align the clutch housing to the engine. Install and tighten the lower housing bolts. On installations with automatic transmission, align the converter housing to the engine and loosely install the bottom housing bolts. Then install the next higher bolts and tighten all four bolts. With both

transmissions, next remove the transmission support, lower the engine onto the mounts, and install the mounting bolts. The remainder of the installation is in the reverse order of removal. On Pacers, raise the car with a jack as in Step 21. Lower the engine/transmission assembly into the compartment. Raise the transmission into position with a jack and install the rear crossmember. Install the front engine support cushions. The remainder of the installation is in the reverse order of removal.

Intake Manifold

REMOVAL & INSTALLATION

150-4 Engine

NOTE: It is necessary to remove the carburetor from the intake manifold before the manifold is removed. After removing the carburetor from the intake manifold, it may be set to one side with vacuum hoses still attached.

1. If the engine has been recently operated, use care to prevent scalding by hot coolant. The system is pressurized. Remove the radiator cap and draincock to drain the coolant. Do not waste reusable coolant. If the coolant is acceptable for reuse, drain into a clean container.

2. Remove the air cleaner. Disconnect the fuel pipe, carburetor air horn vent hose, idle speed control vacuum hose, and wire connector.

3. Disconnect the coolant hoses from the intake manifold.

4. Disconnect the throttle cable from the bellcrank.

5. Disconnect the PCV valve vacuum hose from the intake manifold.

6. Remove the vacuum advance CTO valve vacuum hoses.

7. Disconnect the feedback system coolant temperature sender wire connector (located on the intake manifold).

8. Disconnect the vacuum hose from the EGR valve.

9. Disconnect the intake manifold electric heater wire connector.

10. Remove the power steering mounting bracket, if so equipped.

11. Detach the power steering pump and set aside, if so equipped.

12. Do not remove the hoses.

13. Disconnect the throttle valve linkage, if equipped with automatic transmission.

14. Disconnect the EGR valve tube from the intake manifold.

15. Remove the intake manifold attaching screws, nuts and clamps. Remove the intake manifold. Discard the gasket.

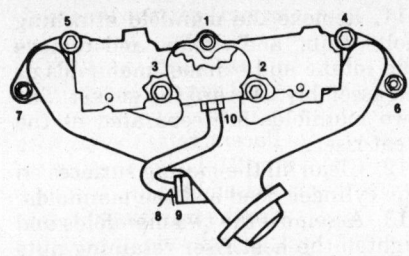

150-4 cylinder intake manifold bolt tightening sequence. Tightening torque is 23 ft. lbs.

16. Clean the mating surfaces of the manifold and cylinder head.

NOTE: If the manifold is being replaced, ensure all fittings, etc. are transferred to the replacement manifold.

17. Installation is in the reverse order of removal. Torque manifold bolts to 23 ft. lbs.

151-4 Engine

1. Remove the air cleaner. Allow the engine to cool, then drain the cooling system. Disconnect the heater hose from the intake manifold.

2. Disconnect and label the fuel line, all vacuum lines and electrical connectors from the carburetor, insulator and the intake manifold. Plug the fuel line.

3. Disconnect the throttle linkage.

4. Remove the carburetor and insulator.

5. Remove the alternator rear support bracket from the manifold.

6. Remove the A/C compressor, if so equipped.

——————— **CAUTION** ———————

If you are not familiar with air conditioning systems, have the freon bled by a qualified technician. Compressed refrigerant will freeze any surface it touches, including skin and eyeballs. It also forms a poisonous gas in the presence of an open flame.

7. Remove the intake manifold bolts and remove the manifold.

8. To install, place a new gasket against the cylinder head, then install the manifold in place by starting all bolts finger-tight.

9. Torque the intake manifold bolts to 25 ft. lbs. in two stages, using the torque sequence shown. The rest of the installation is in the reverse order of removal.

258-6 Engine

The intake manifold is mounted on the left-hand side of the engine and bolted to the cylinder head. A gasket is used between the intake manifold

1   CYLINDER HEAD
2   ROCKER ARMS
3   BRIDGE
4   PIVOT
5   PUSHROD
6   INTAKE MANIFOLD
7   EXHAUST MANIFOLD
8   CYL. HEAD BOLTS
9   CYLINDER HEAD
10 CYL. HEAD GASKET
11 INTAKE   MANIF.
    GASKET

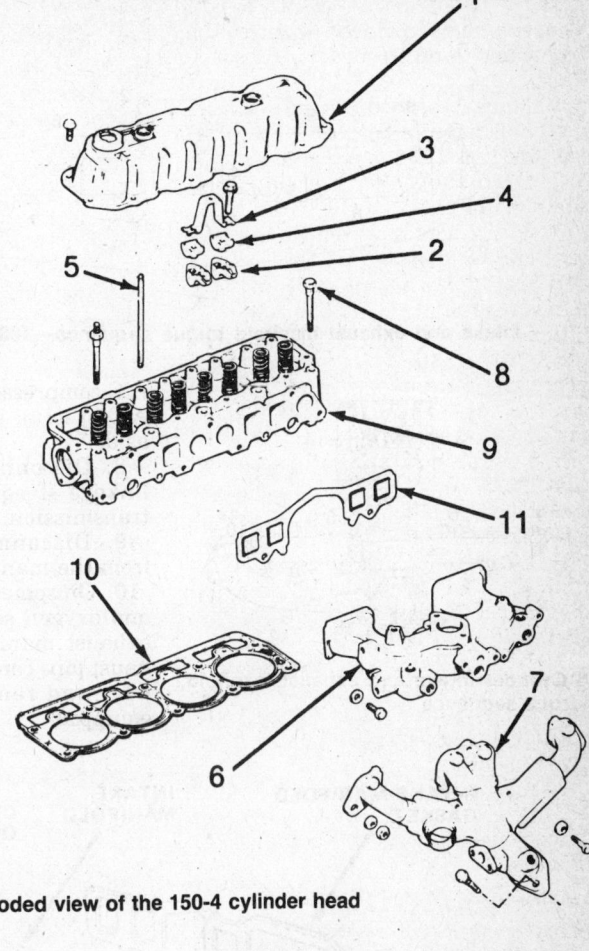

Exploded view of the 150-4 cylinder head

and the head; none is required for the exhaust manifold.

1. Remove the air cleaner. Disconnect the fuel line, vent hose, and solenoid wire, if so equipped.

2. Disconnect the accelerator cable from the accelerator bellcrank.

3. Disconnect the PCV vacuum hose from the intake manifold and the TCS solenoid and bracket, if so equipped.

4. Remove the spark CTO switch and EGR valve (or exhaust back pressure sensor) vacuum lines from each of these components.

5. Disconnect the hoses from the air pump and the injection manifold check valve. Disconnect the vacuum line from the diverter valve and remove the diverter valve with hoses, if so equipped.

6. Remove the air pump and power steering bracket (if so equipped) and remove the air pump. Move the power steering pump aside, out of the way, without disconnecting the hoses.

7. Remove the air conditioning drive belt idler assembly from the cylinder head, if so equipped. On some models it is necessary to remove the

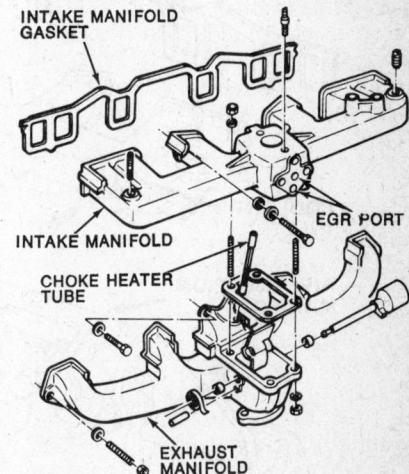

Typical 6 cylinder intake and exhaust manifold

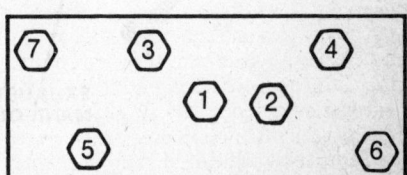

151-4 cylinder intake manifold bolt torque sequence

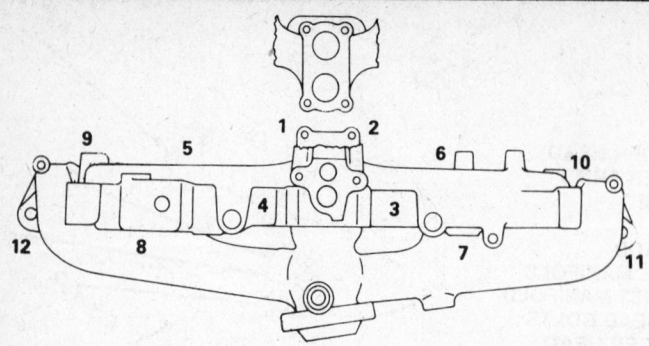

Intake and exhaust manifold torque sequence—1981 and later 6 cylinder

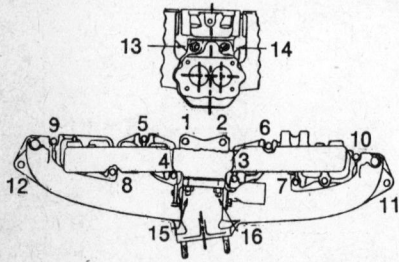

6 Cylinder intake and exhaust manifold torque sequence

A/C compressor. Do not discharge the A/C system; just lay the compressor aside.

8. Disconnect the throttle valve linkage if equipped with automatic transmission.

9. Disconnect the exhaust pipe from the manifold.

10. On some 1981 and later models, and oxygen sensor is screwed in the exhaust manifold just above the exhaust pipe connection. Disconnect the wire and remove the sensor, if so equipped.

11. Remove the manifold attaching bolts, nuts, and clamps and remove the intake and exhaust manifolds as an assembly. Discard the gasket. The two manifolds are separated at the heat riser.

12. Clean all the mating surfaces on the cylinder head and the manifolds.

13. Assemble the two manifolds and tighten the heat riser retaining nuts to 5 ft. lbs.

14. Position the manifold on the engine, together with a new intake manifold gasket, and tighten the manifold attaching bolts and nuts in the proper sequence to the specified torque.

15. Install the remaining components in the reverse order of removal. Adjust the automatic transmission throttle linkage, if so equipped. Adjust the drive belt tension.

Exhaust Manifold

REMOVAL & INSTALLATION

150-4 Engine

1. Remove the intake manifold.
2. Disconnect the EGR valve tube.
3. Disconnect the exhaust pipe from the exhaust manifold.
4. Disconnect the oxygen sensor wire connector.
5. Remove the sensor from the manifold if a replacement manifold is to be installed.
6. Remove the nuts from the end studs. Remove the exhaust manifold.
7. Installation is in the reverse order of removal. Torque manifold nuts to 23 ft. lbs., oxygen sensor to 35 ft. lbs.

151-4 Engine

1. Remove the air cleaner and the hot air tube.
2. Remove the Pulsair system from the exhaust manifold.
3. Disconnect the exhaust pipe from the manifold at the flange. Spray the bolts first with penetrating sealer, if necessary.
4. Remove the engine oil dipstick bracket bolt.
5. Remove the exhaust manifold bolts and remove the manifold from the head.
6. To install, place a new gasket

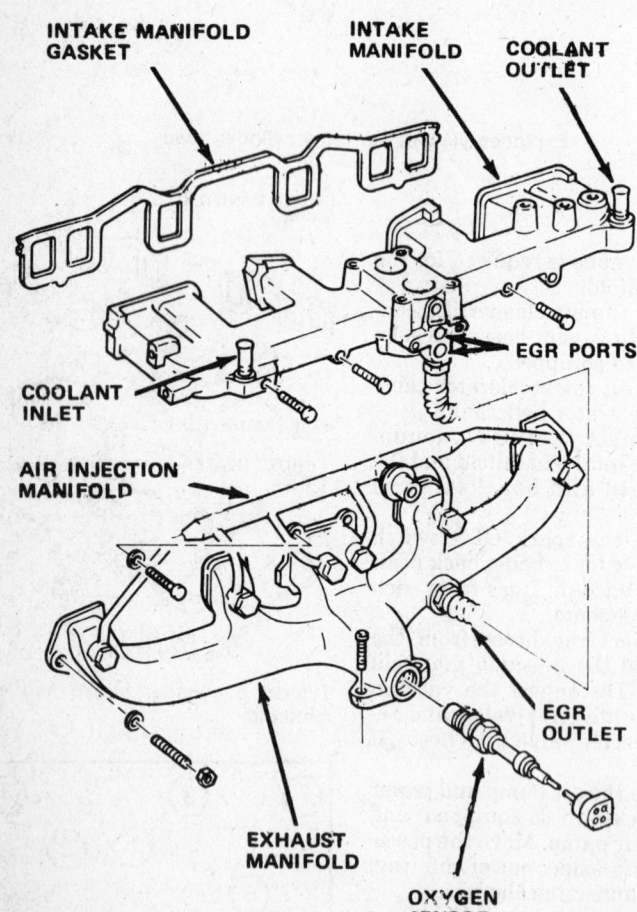

Intake and exhaust manifold assembly 1981 and later 6 cylinder

INTAKE MANIFOLD GASKET
INTAKE MANIFOLD
COOLANT OUTLET
EGR PORTS
COOLANT INLET
AIR INJECTION MANIFOLD
EGR OUTLET
EXHAUST MANIFOLD
OXYGEN SENSOR

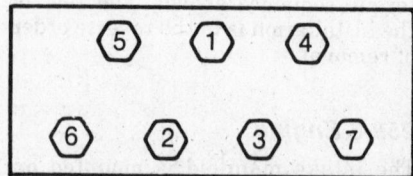

151-4 cylinder exhaust manifold bolt torque sequence

against the cylinder head, then install the exhaust manifold over it. Start all the bolts into the head finger-tight.

7. Torque the exhaust manifold bolts to 37 ft. lbs. in two stages, using the torque sequence illustrated.

8. Complete the installation in the reverse order of removal.

258-6 Engine

The exhaust manifold is removed along with the intake manifold; see previous instructions.

Valve System

ADJUSTMENT

All engines use hydraulic lifters, eliminating periodic valve adjustments. The cylinder heads have integral valve guides. Oversized valves are available in 0.003 and 0.005 in. sizes. To fit these, the valve guide bores must be enlarged with a reamer. As an alternate procedure, some automotive machine shops fit replacement valve guides which accept the standard size valves.

Rocker Assembly

REMOVAL & INSTALLATION

150-4 Engine

1. Remove the valve cover.
2. Remove the capscrews at each bridge and pivot assembly.
3. Alternately loosen the capscrews one turn at a time to avoid damaging the bridge.
4. Remove the bridges, pivots and corresponding pair of rocker arms.
5. Installation is in the reverse order of removal. Tighten capscrews to 19 ft. lbs.

151-4 Engine

1. Remove the valve cover.
2. Remove the rocker arm nut and rocker arm ball.
3. Lift the rocker arm off the stud. Always keep the rocker arm assemblies together and assemble them on the same stud.
4. Remove the pushrod from its bore. Make sure the rods are returned to their original bores, with the same end in the block.
5. Reverse the removal procedure to install. Lubricate all parts before installation. Tighten the rocker arm ball retaining nut to 20 ft. lbs.

258-6 Engine

The valve guides are integral with the head on all engines. The valve stem

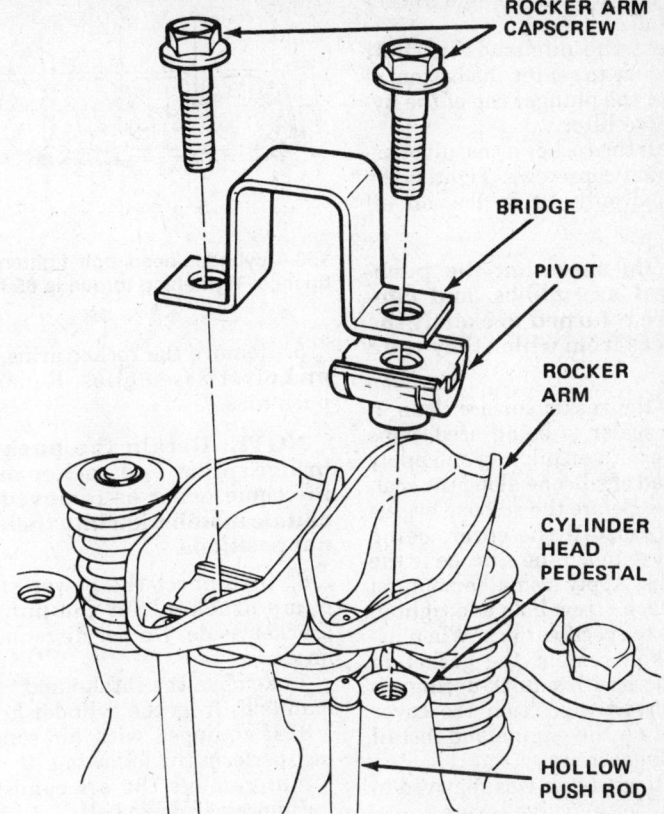

258 6 cylinder rocker arm assembly

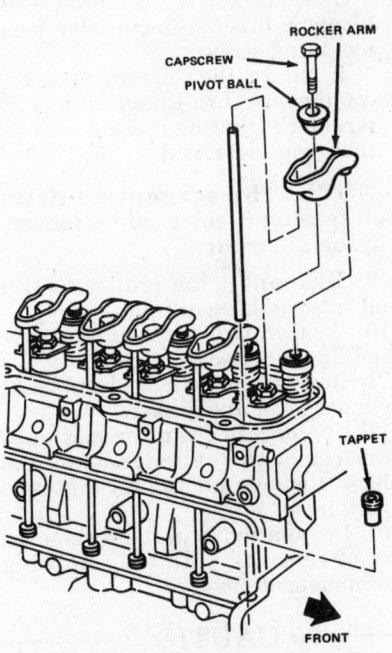

151 4 cylinder rocker arm assembly

oil deflectors should be replaced whenever valve service is performed.

American Motors engines do not have replaceable valve guides. If stem-to-guide clearance is excessive, guides must be reamed to the proper oversize. Three oversize valves are available with stems 0.003, 0.015 and 0.030 in. larger than standard diameter.

The intake and exhaust rocker arms for each cylinder pivot on a bridged pivot assembly bolted to the cylinder head. The pushrods are hollow to supply lubrication to the rocker arms. The pushrods act as guides to keep the rocker arms in alignment, so it is not abnormal for pushrods to rub slightly on the cylinder head.

NOTE: Be careful when ordering new valve train components not to get parts for the wrong year.

1. Remove any accessories which are in the way and remove the valve cover, complete with gasket.
2. Unscrew the rocker arm capscrews evenly to avoid breaking the bridge.
3. Remove the pivot assemblies, rocker arms, and pushrods.

NOTE: Be sure to keep all parts in the same order in which they were removed.

4. Clean all parts in solvent. Blow all oil passages in the rocker arms and pushrods dry with compressed air. Replace any deeply pitted rocker arms and scuffed or worn pushrods. If the pushrod is worn from lack of oil, re-

place it, its valve lifter and rocker arm, as well.

5. Insert the pushrods in their bores. Be sure to center the bottom of each rod in the plunger cap of the hydraulic valve lifter.

6. Install the rocker arms, pivot assemblies and capscrews. Tighten the capscrews evenly 19 ft. lbs. on all engines.

NOTE: Be sure that the pushrods, pivot assemblies, and capscrews are returned to exactly the same places from which they were removed.

7. Wipe the gasket surface clean. If a silicone sealer is being used, wipe the surface with an oily rag and apply a $\frac{1}{8}$ in. bead of silicone along the sealing surface. Before the silicone begins to harden, install the cover, being careful not to touch the silicone to the rocker arms. Apply a small amount of sealer to each screw hole and tighten the screws to specifications. When using a gasket, cement the gasket in several places with a quick-drying adhesive. Correctly position the cover and gasket on the engine and install the attaching screws.

8. Install whatever was removed to gain access to the valve covers.

Cylinder Head

REMOVAL & INSTALLATION

--- **CAUTION** ---

To prevent cylinder head warping don't loosen the head bolts until the engine is thoroughly cool. Do not remove block drain plugs or loosen radiator draincock with the system hot and under pressure, as serious burns from coolant can occur.

If the head sticks, operate the starter to loosen it by compression or rap it upward with a soft hammer. Do not force anything between the head and the block. Cylinder head bolts should be retorqued after the first 500 miles or so, unless a special AMC gasket is used. The special gasket doesn't require retorquing. Make sure to blow any coolant out of the cylinder head bolt holes before reassembly to prevent inaccurate torque readings.

150-4 Engine

--- **CAUTION** ---

Do not perform this procedure on a hot engine.

1. Disconnect the battery cables.
2. Drain the coolant and disconnect the hoses at the thermostat housing.
3. Remove the air cleaner.
4. Remove the valve cover.

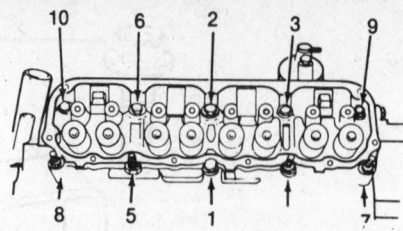

150-4 cylinder head bolt tightening sequence. Tightening torque is 85 ft. lbs.

5. Remove the rocker arms, bridge and pivot assemblies. Remove the push rods.

NOTE: Retain the push rods, bridge, pivot and rocker arms in the same order as removed to facilitate installation into their original positions.

6. Disconnect the power steering pump bracket. Set the pump and bracket aside. Do not disconnect the hoses.

7. Remove the intake and exhaust manifolds from the cylinder head.

8. If equipped with air conditioning, perform the following:
   a. remove the air conditioner compressor drive belt
   b. loosen the alternator drive belt
   c. remove the A/C compressor/alternator bracket-to-cylinder head mounting screw.
   d. remove the bolts from the A/C compressor (if so equipped) and alternator mounting bracket, and set the compressor aside.

NOTE: The serpentine drive belt tension is released by loosening the alternator.

9. Disconnect the ignition wires and remove the spark plugs.

10. Disconnect the temperature sending unit wire connector.

11. Remove the cylinder head bolts, cylinder head and gasket.

12. Thoroughly clean the machined surfaces of the cylinder head and block. Remove all gasket material and cement.

13. Installation is in the reverse order of removal, with the following recommendations.

--- **CAUTION** ---

Do not apply sealing compound to the cylinder head and block machined surfaces. Do not allow the sealing compound to enter the cylinder bores.

14. Apply an even coat of Perfect Seal sealing compound, or equivalent, to both sides of the replacement cylinder head gasket and position the gas-

ket on the cylinder block with word TOP facing upward.

15. Torque the head bolts to 85 ft. lbs. in the sequence illustrated.

151-4 Engine

1. Disconnect the negative battery cable. Drain the cooling system.

2. Disconnect the accelerator cable at the bellcrank, and the manifold vacuum and fuel lines at the carburetor.

3. Remove the intake and exhaust manifolds.

4. Remove the alternator and power steering pump. Unbolt the A/C compressor, if so equipped and move it aside without disconnecting any lines.

5. Disconnect all electrical connectors at the head.

6. Disconnect the radiator and heater hoses, and the battery ground strap.

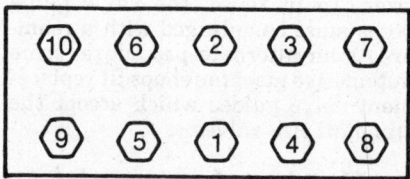

151-4 cylinder head bolt torque sequence

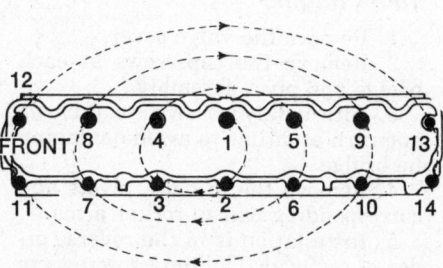

Cylinder head torque sequence for all 6 cylinder engines

7. Remove the spark plugs.

8. Remove the rocker arm cover, rocker arms, and push rods. Keep all parts in order.

9. Unbolt and remove the cylinder head.

10. Clean the gasket surfaces thoroughly.

11. Install a new gasket over the dowels, and then position the cylinder head.

12. Coat the head bolt threads with sealer and install finger-tight.

13. Tighten the bolts in sequence, in three equal steps to the specified torque.

14. Install all parts in the reverse order of removal.

258-6 Engine

NOTE: On Pacers, run the wipers to the center of the windshield to ease valve cover removal.

1. Drain the cooling system. Disconnect throttle linkage, fuel lines, water hoses, spark plug wires and vacuum line. Remove the air cleaner, PCV hose, and the temperature sender.

2. Remove the valve cover and its gasket. Remove the rocker arm assembly and the pushrods. With bridged pivots, loosen each bolt alternately, one turn at a time, to avoid damage. Keep the pushrods in order. If equipped with power steering, remove the power steering pump bracket and Air Guard pump, and set them aside. Don't disconnect the hoses.

3. Remove the intake and exhaust manifold assemblies from the head.

4. Disconnect the spark plug wires, and remove the plugs.

5. Disconnect the battery ground cable, the coil, and the coil bracket from the head. Disconnect the temperature sending unit wire.

6. If the vehicle is equipped with air conditioning, remove the drive belt idler pulley bracket from the cylinder head. Loosen the alternator drive belt and remove the bolts from the compressor mounting bracket. Set the compressor aside, all hoses attached.

7. Remove the bolts and remove the cylinder head from the block.

8. Clean the gasket surfaces of both the head and the block. Remove the carbon deposits from the top of each piston, and from the combustion chambers.

9. Check the head for straightness. If the head (or the block) is 0.008 in. out of true over its entire length, 0.001 in. in 1 in. or 0.002 in. in 6 in., the head requires resurfacing.

10. Use a new head gasket, and coat both of its sides with sealer. The word TOP on the gasket faces upward.

11. Tighten the head bolts in three stages, in the proper sequence and to the proper torque specification.

12. Complete the installation in the reverse order of removal. Refill the cooling system.

Vibration Damper

REMOVAL

All Engines

Remove the radiator core, all drive belts, and the fan. Remove the nut from the center of the pulley. The best way to do this is to affix a heavy wrench and rap it with a substantial

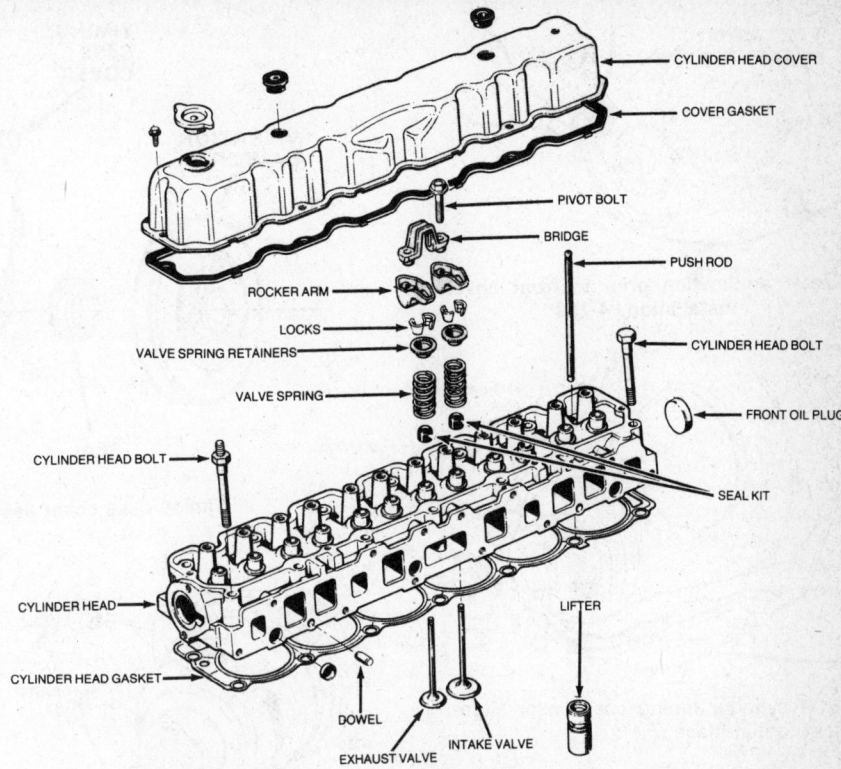

Exploded view of the 258 6 cylinder head assembly

Typical vibration damper removal

hammer. It may be necessary to lock up the engine at the flywheel to prevent crankshaft rotation. The nut must be unscrewed in the opposite direction of normal engine rotation. Using a puller, remove the pulley from the front of the crankshaft.

INSTALLATION

1. With key in crankshaft keyway, align damper hub keyway with crankshaft key and tap damper onto crankshaft.

2. Install damper retaining bolt and washer, and tighten bolt to specifications.

3. Install crankshaft pulley and retaining bolts, and tighten to specifications.

4. Install drive belt(s) and tighten to specified tension.

Timing Case Cover

REMOVAL & INSTALLATION

151-4 Engine

1. Remove the crankshaft hub.

2. Remove the oil pan-to-front cover screws.

3. Remove the front cover-to-block screws.

4. Pull the cover slightly forward, just enough to allow cutting of the oil pan front seal flush with the block on both sides.

5. Remove the front cover and attached portion of the pan seal.

6. Clean the gasket surfaces thoroughly.

7. Cut the tabs from the new oil pan front seal.

8. Install the seal on the front cover, pressing the tips into the holes provided.

9. Coat the new gasket with sealer and position it on the front cover.

10. Apply a 1/8 in. bead of silicone sealer to the joint formed at the oil pan and block.

11. Align the front cover seal with a centering tool and install the front cover. Tighten the screws to 7.5 ft. lbs. Install the crankshaft hub.

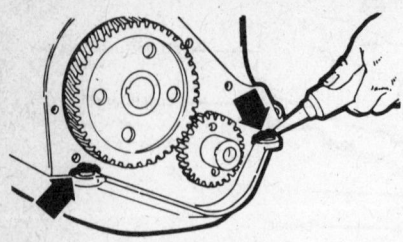

Sealer application prior to front cover installation, 4-151

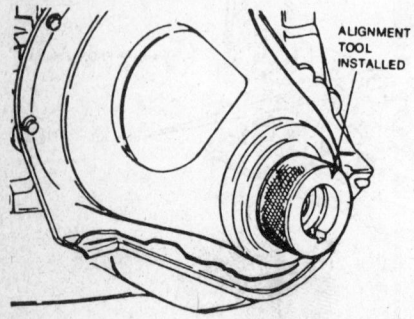

151-4 cylinder timing case cover alignment tool in place

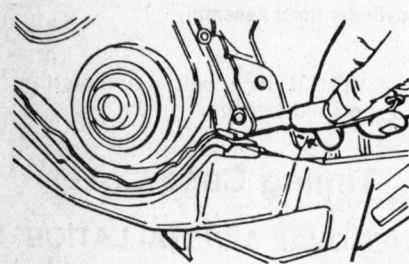

Cutting the pan gasket on the 4 cylinder engine

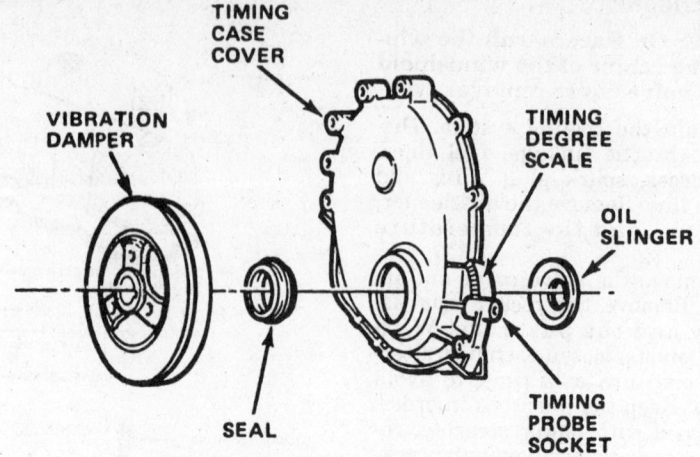

Timing case cover assembly—258 6 cylinder

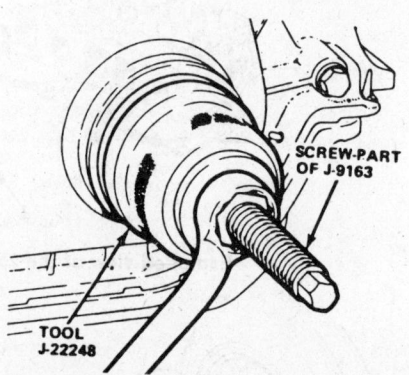

6 cylinder engine timing case cover seal installation

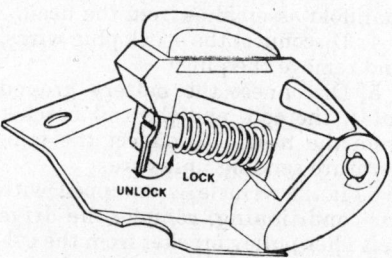

150-4 cylinder timing chain tensioner

150-4 and 258-6 Engines

1. Remove all V-belts, fan and pulley.
2. Remove vibration damper.
3. Remove oil pan-to-cover bolts and cover-to-block bolts.
4. Raise cover and pull oil pan front seal up far enough to extract the tabs from the holes in cover.

——————— CAUTION ———————

If this isn't done, the oil pan will have to be removed to get the seals into place.

5. Remove cover gasket from block; cut off seal tab flush with front face of block.
6. Clean all mating surfaces and remove oil seal.
7. Install a new front oil seal.
8. Install a new neoprene seal in front oil pan, cutting off protruding tabs to match original. Use sealer on the end tabs and the gasket surfaces.
9. Position cover on block and in-stall bolts. Align the front cover with a centering tool. Tighten cover bolts to 4–6 ft. lbs.; four lower bolts to 10–12 ft. lbs. Remove the centering tool.
10. Install vibration damper, tightening the bolt to the specified torque.

NOTE: Front oil seal can be installed with cover in place only if proper tool or equivalent is available.

Timing Chain or Gears

REMOVAL & INSTALLATION

150-4 Engine

1. Remove the timing case cover as previously outlined.
2. Rotate the crankshaft until the zero timing mark on the crankshaft sprocket is closest to and on center line with the mark on the cam sprocket.
3. Remove the oil slinger from the crankshaft.
4. Remove the camshaft retaining bolt and remove the sprockets and chain as an assembly.
5. Installation is in the reverse or-der of removal, with the following recommendations:
   a. Turn the tensioner lever to the unlock (down) position.
   b. Pull the tensioner block toward the tensioner lever to compress the spring. Hold the block and turn the tensioner lever to the lock (up) position.

151-4 Engine

The 151 uses timing gears instead of a chain and sprockets or a belt. The cam timing gear is pressed onto the camshaft. The camshaft must be removed to remove the gear, which must be pressed off the camshaft. See the camshaft removal procedure for details. The replacement cam gear must be pressed onto the camshaft. To replace the gear, first place the gear spacer ring and thrust plate over the end of the camshaft, then install the woodruff key. Press the camshaft gear onto the cam until it bottoms against the gear spacer ring. End clearance of the thrust plate must be 0.0015–0.0050 in. If less than 0.0015 in., the spacer ring must be replaced. If more than 0.0050 in., the thrust plate must be replaced.

258-6 Engine

1. Remove the drive belt(s).

2. Remove the engine fan and hub assembly.

3. Remove the vibration damper pulley and remove the vibration damper.

4. Remove the timing case cover. Remove the seal from the timing case cover, because the seal should be replaced every time the cover is removed from the engine.

5. Remove the camshaft sprocket retaining bolt and washer.

6. Turn the crankshaft until the 0 timing mark on the crankshaft sprocket is closest to and on a centerline with the timing pointer of the camshaft sprocket.

7. Remove the crankshaft sprocket, camshaft sprocket and timing chain as an assembly. Disassemble the chain and sprockets.

8. To install; assemble the timing chain, crankshaft sprocket, and camshaft sprocket with the timing marks aligned.

9. Install the assembly to the crankshaft and camshaft.

10. Install the camshaft sprocket retaining bolt and washer and tighten the bolt to 50 ft. lbs.

11. To ensure the correct installation of the timing chain, locate the timing mark of the camshaft sprocket at about the 1 o'clock position. This should place the timing mark on the crankshaft sprocket where the sprocket teeth mesh with the chain. There must be 15 timing chain pins between the timing marks of both sprockets.

Camshaft

REMOVAL & INSTALLATION

150-4 Engine

1. Drain and remove the radiator.

2. If so equipped, remove air conditioning condenser and receiver assembly as a charged unit.

3. Remove fuel pump, distributor and ignition wires.

4. Remove cylinder head cover and gasket.

5. Remove rocker arms, bridged pivot assemblies and pushrods. Be sure to replace these parts in the same order as removed.

6. Remove cylinder head and gasket and lifters.

7. Remove timing case cover.

8. Remove timing chain and sprockets as one assembly, being careful to rotate the crankshaft until the timing mark on the crankshaft sprocket is lined up with the timing pointer on the camshaft sprocket.

9. Remove the front bumper or grille as required.

10. Carefully remove the camshaft from the engine.

11. Installation is in the reverse order of removal.

151-4 Engine

1. Drain the cooling system.

2. Remove the radiator.

3. Remove the fan and water pump pulley.

4. Remove the grille and bumper if necessary for clearance.

5. Remove the rocker cover, rocker arms, and pushrods.

6. Remove the distributor, spark plugs, and fuel pump.

7. Remove the pushrod cover and gasket. Remove the lifters.

8. Remove the crankshaft hub and timing gear cover.

9. Remove the two camshaft thrust plate screws by working through the holes in the gear.

10. Remove the camshaft and gear assembly by pulling it through the front of the block. Take care not to damage the bearings.

11. Install in the reverse order. Torque the thrust plate screws to 6 ft. lbs.

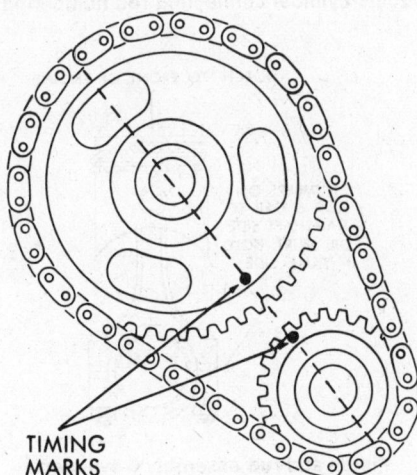

TIMING MARKS

Six-cylinder timing chain and sprockets

258-6 Engine

1. Drain the cooling system and remove the radiator. Remove the hood (Pacers only).

2. If the car is equipped with air conditioning, remove the condenser and the receiver unit as a charged assembly, only.

NOTE: Do not discharge the A/C system.

3. Remove the valve cover and gasket.

4. Remove the rocker arm assembly and the cylinder head. Remove the pushrods and tappets.

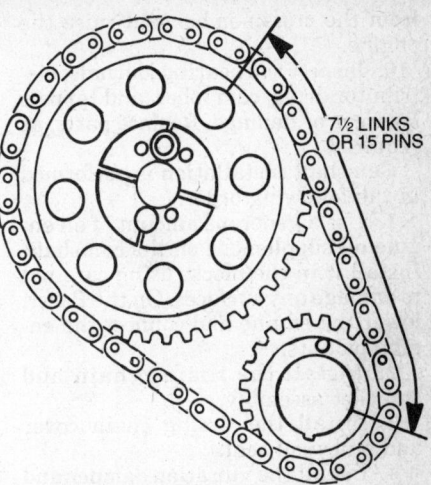

7½ LINKS OR 15 PINS

Correct timing chain installation—6 cylinder

NOTE: Pushrods and tappets should be kept in the proper order. They must be returned to their original places during assembly.

5. Remove the drivebelt(s), fan assembly, accessory pulley(s), vibration damper, and the timing chain cover.

6. Remove the fuel pump. Remove the distributor assembly, including spark plug wires.

7. Turn the crankshaft until the "O" timing mark on the crankshaft sprocket is nearest to, on a centerline with, and aligns with the timing pointer on the camshaft sprocket.

8. Remove the sprockets and the timing chain as an assembly.

9. Remove the front bumper and/or grille as necessary. Withdraw the camshaft through the opening. On the Pacer, unbolt the front engine mounts

A

150-4 cylinder timing mark (A) alignment

from the crossmember and raise the engine.

10. Inspect the bearing journals, distributor drive, cam lobes, and tappets for wear or damage. Replace parts, as required.

Camshaft installation is performed in the following order:

1. Use a generous amount of an engine oil supplement on the camshaft. Install it in the block, using care not to damage any surfaces. On the Pacer, lower the engine and connect the engine mounts.

2. Install the timing chain and sprocket assembly.

3. Install the timing chain cover and a new oil seal.

4. Install the vibration damper and the accessory drive pulley(s).

5. Install the engine fan assembly and the drive belt(s). Tighten the belts to the proper tension.

6. Install the fuel pump.

7. With the number one piston at TDC of its compression stroke, fit the distributor so that the rotor is aligned with the No. 1 terminal on the cap (distributor fully seated on the block). Install the cap and the spark plug wires.

8. Install the tappets, cylinder head, its gasket, valve train (pushrods in the same order as removed), valve cover and its gasket.

NOTE: All valve train components must be lubricated with engine oil supplement. The supplement must remain in the engine for at least the first 1000 miles. It does not require draining until the next regular oil change.

Pistons and Connecting Rods

IDENTIFICATION & POSITIONING

The piston and rod assemblies are installed from the top, and the dimple, notch, or dot, marked on the top of the piston, goes toward the front. On the 151-4 engine, the raised notch side of the rod (near the bearing end) must be 180° opposite the notch in the piston. On the six cylinder, the connecting rod numbers must go toward the camshaft.

Oil Pan

REMOVAL & INSTALLATION

150-4 Engine

1. Raise the vehicle and safely support it on jackstands. Disconnect the

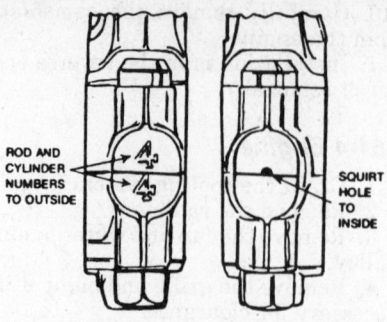

151-4 cylinder connecting rod and cylinder numbers, and squirt hole

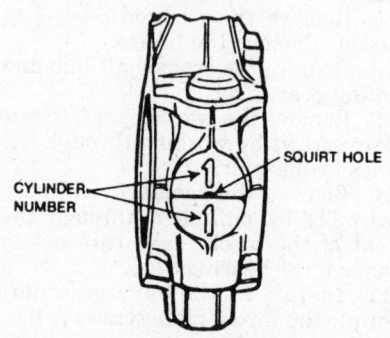

258-6 cylinder connecting rod numbering

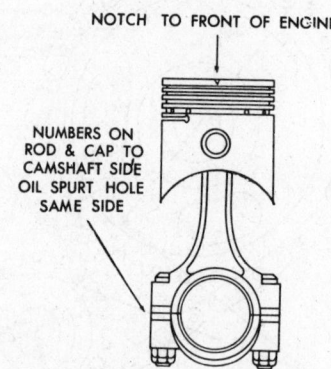

Piston and rod assembly 6 cylinder engine

negative battery cable. Drain the engine oil.

2. Remove the starter motor.

3. If clearance is insufficient, place a jack under the transmission bell housing. Disconnect the engine right support cushion bracket from the block and raise the engine to allow sufficient clearance for oil pan removal. Remove the flywheel access cover.

4. Remove the oil pan attaching bolts and remove the oil pan.

5. Remove the oil pan front and rear neoprene oil seals and the side gaskets. Thoroughly clean the gasket surfaces of the oil pan and the engine block. Remove all of the sludge and dirt from the oil pan sump.

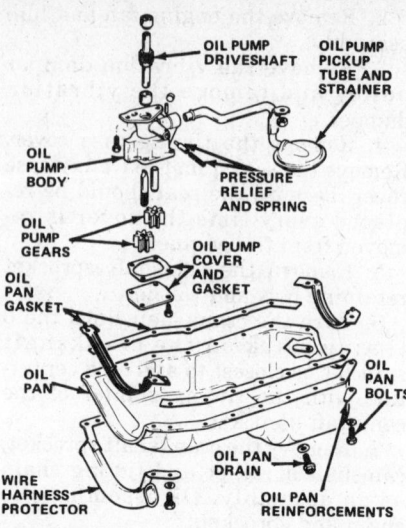

Oil pump and pan assembly—151 4 cylinder

6. Apply a generous amount of RTV silicone to the end tabs of a new oil pan front seal and install the seal onto the timing case cover.

7. Cement new oil pan side gaskets into position on the engine block and apply a generous amount of RTV silicone to the side gasket contacting surface of the seal end tabs.

8. Install the seal in the recess of the rear main bearing cap, making sure that it is fully sealed.

9. Coat the oil pan contacting surface of the front and rear oil pan seals with engine oil.

10. Install the oil pan and assemble the engine mount in the reverse order of removal.

151-4 Engine

1. Raise the car and support it with stands. Drain the oil.

2. Install an engine-lifting device and support the weight of the engine, while removing the engine bracket-to-mount cushion nuts. Loosen the strut and bracket screws.

3. Raise the engine approximately two inches and remove the crossmember-to-sill attaching parts.

4. Remove the steering gear idler bracket from the frame rail.

5. Pry the crossmember down and insert wooden blocks between the crossmember and the side sill on both sides.

6. Remove the oil pan, and clean the gasket from the mating surfaces of the block and oil pan.

7. To install the pan, thoroughly clean all the gasket mating surfaces.

8. Install a new rear oil pan gasket in the rear main bearing cap. Apply a small quantity of RTV silicone sealer into the depressions where the rear pan gasket engages the block.

9. Install a new front oil pan gasket onto the timing gear cover. Press the tips into the holes in the cover.

10. Install the side gaskets onto the block, not the oil pan. Retain them with a thin film of grease. Apply a $\frac{1}{4}$ in. long bead of RTV silicone sealer to the split lines of the front and side gaskets; the bead should be $\frac{1}{8}$ in. wide.

11. Install the oil pan onto the engine. The timing cover bolts should be installed last. They are installed at an angle; the holes will line up after the rest of the pan bolts have been snugged down. The bolts should be tightened to 6 ft. lbs. all around. The remainder of the installation is in the reverse order of removal.

258-6 Engine
EXCEPT PACER

1. Turn the steering wheel to full left lock. Support the engine with a hoist. Raise and support the car at the side sills. Disconnect the negative battery cable and engine ground strap cable.

2. Unbolt the steering idler arm at the side sill, and the engine cushions at the brackets.

3. Remove the sway bar, if so equipped. Remove the front cross member-to-side sill bolts and pull the crossmember down. Remove the right engine bracket. Loosen but do not remove the strut rods at the lower control arm.

4. Drain the engine oil.

5. Remove the starter.

6. Remove the oil pan bolts and pan. Remove the front and rear seals, and clean the gasket surfaces.

7. Install the new pan-to-timing cover front seal, and apply sealer to end tabs. Cement new pan side gaskets to the block, and apply sealer to the ends of the gasket.

8. Coat the inside surface of the new rear seal with soap, and apply sealer to end tabs. Install the seal in the rear main cap.

9. Coat front and rear seal contact surfaces with engine oil, and install the pan. The remainder of the installation is in the reverse order of removal.

PACER

1. Drain the engine oil.

2. Install an engine-lifting device and support the weight of the engine.

3. Disconnect the steering shaft flexible joint and hold it aside with a length of wire.

4. Raise and support the car.

5. Remove the front engine support through bolts.

6. Disconnect the front brake lines at the wheel cylinders.

7. Disconnect the upper ball joints from the spindles. Make sure the shock absorbers are attached securely.

8. Remove the upper control arm and move it aside.

9. Support the front crossmember with a jack.

10. Remove the nuts from the front crossmember rear mounts and swing the crossmember down and forward.

11. Follow Steps 5–9 of the preceding six cylinder procedure.

12. Install and assemble the remaining components in the reverse order of removal, tightening the $\frac{1}{4}$ in. oil pan screws to 7 ft. lbs., the $\frac{5}{16}$ in. oil pan screws to 11 ft. lbs., the crossmember attaching nuts to 50 ft. lbs., the upper control arm cross shaft bolt and nut to 60 ft. lbs., and the engine mount and steering shaft nuts to 25 ft. lbs. Fill the crankcase with oil and bleed the brakes.

Oil Pump
REMOVAL & INSTALLATION

——— CAUTION ———
Whenever the oil pump cover is removed or the pump diassembled, the pump must be primed by filling the spaces around the gears with petroleum jelly. Do not use grease.

150-4 Engine

1. Drain the oil and remove the oil pan.

2. Remove the oil pump retaining screws and separate the oil pump and gasket from the engine block.

3. Install in the reverse order of the above procedure.

151-4 Engine

1. Remove engine oil pan.

2. Remove the two bolts and one nut, and carefully lower the pump.

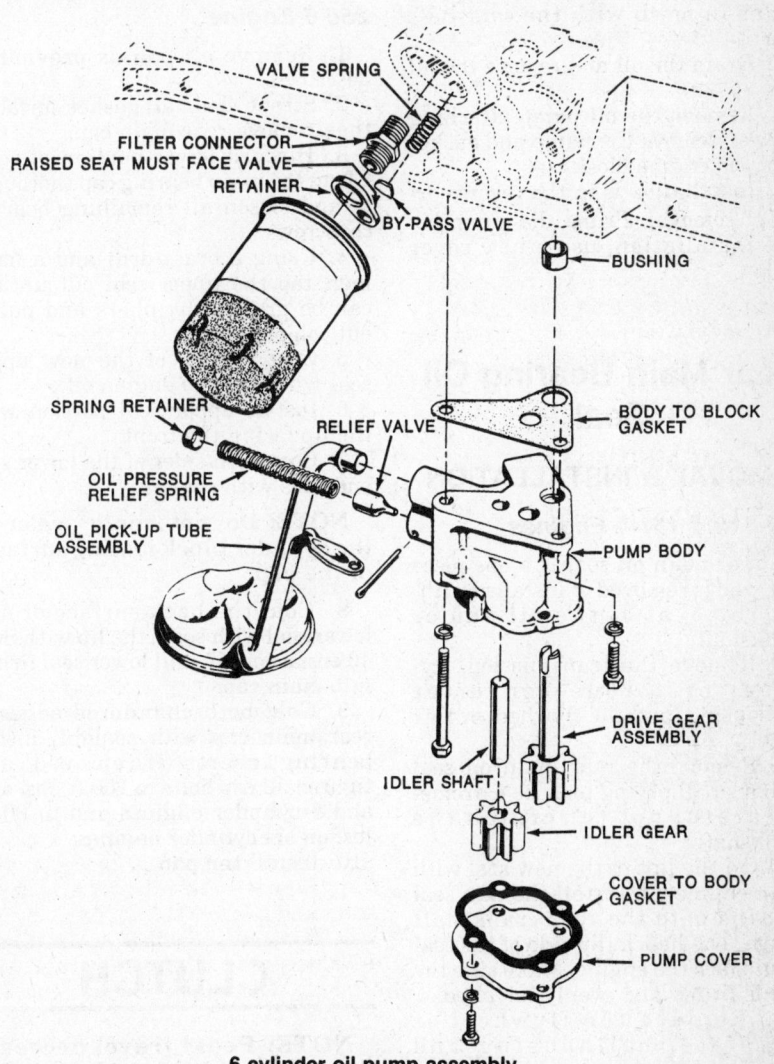

6 cylinder oil pump assembly

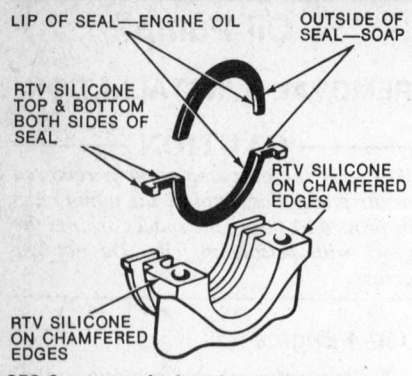

LIP OF SEAL—ENGINE OIL
OUTSIDE OF SEAL—SOAP
RTV SILICONE TOP & BOTTOM BOTH SIDES OF SEAL
RTV SILICONE ON CHAMFERED EDGES
RTV SILICONE ON CHAMFERED EDGES

258-6 rear main bearing installation details

3. Reinstall in reverse order. To ensure immediate oil pressure on start-up, the oil pump gear cavity can be packed with petroleum jelly.

258-6 Engine

The oil pump is driven by the distributor drive shaft. Oil pump replacement does not, however, affect distributor timing because the drive gear remains in mesh with the camshaft gear.

1. Drain the oil and remove the oil pan.
2. Remove the oil pump attaching screws. Remove the pump and gasket from the engine block.
3. Installation is in the reverse order of removal. Prime the pump before installation; use a new cover gasket.

Rear Main Bearing Oil Seal

REMOVAL & INSTALLATION

150-4 and 151-4 Engines

The rear main oil seal is a one piece unit, and is removed or installed without removal of the oil pan or crankshaft.

1. Remove the transmission, flywheel or torque converter bellhousing, and the flywheel or flex plate.
2. Remove the rear main oil seal with a small prying tool. Be extremely careful not to scratch the crankshaft.
3. Oil the lips of the new seal with clean engine oil. Install the new seal by hand onto the rear crankshaft flange. The helical lip side of the seal should face the engine. Make sure the seal is firmly and evenly installed.
4. Replace the flywheel or flexplate, bellhousing and transmission.

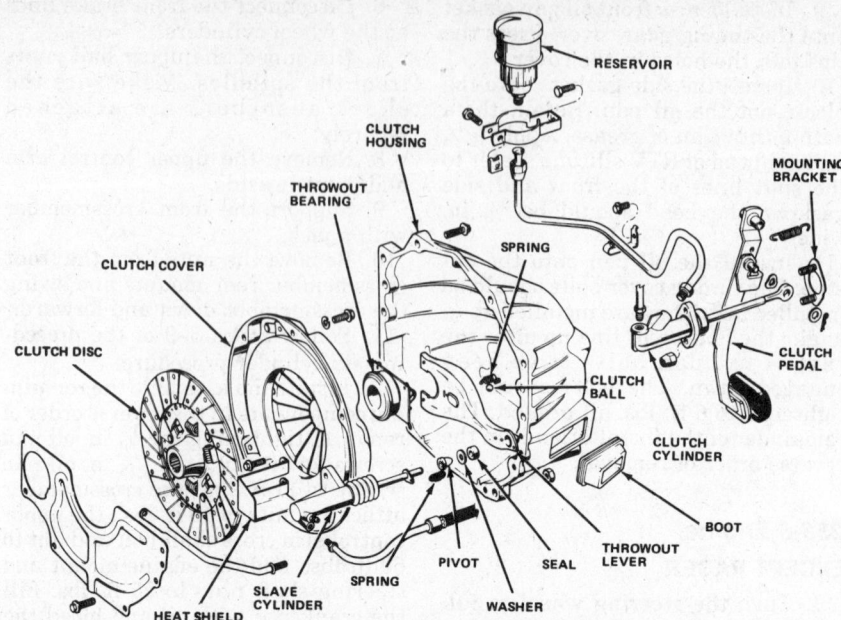

Exploded view of 151-4 cylinder clutch assembly

258-6 Engine

1. Remove oil pan as previously described.
2. Scrape clean all gasket surfaces, then remove rear main cap.
3. Discard lower portion of seal. Clean the main bearing cap thoroughly and loosen all remaining bearing capscrews.
4. Using a brass drift and a hammer, tap the upper seal out until it can be grasped by pliers and pulled out completely.
5. Coat the lip of the new upper seal with SAE 40 engine oil.
6. Install upper seal portion with the lip facing the front.
7. Coat both sides of the lower seal end tabs with sealant.

NOTE: Do not apply sealer to the cylinder block mating surfaces of the cap.

8. Coat the back surface of new lower seal with soap, the lip with SAE 40 engine oil. Install lower seal firmly into main cap.
9. Coat both chamfered edges of rear main cap with sealant, install bearing inserts (if removed) and tighten all cap bolts to 100 ft. lbs. on 4 and 8 cylinder engines and to 80 ft. lbs. on six cylinder engines.
10. Install the pan.

CLUTCH

NOTE: Pedal travel decrease due to normal wear of the lining

can be compensated for by adjusting the clutch pedal free-play, except for 1981 and later models, which have a hydraulically actuated clutch with no provision for adjustment.

PEDAL FREE-PLAY ADJUSTMENT

258-6 Engine

Adjust the free-play of the clutch pedal to $\frac{7}{8}$–$1\frac{1}{8}$ in. This is done by changing the length of the link between the throwout lever rod and the bellcrank assembly.

REMOVAL & INSTALLATION

150-4 Engine

1. Remove the transmission transfer case assembly as described in the Unit Repair section.
2. Mark the position of the clutch pressure plate in relation to the flywheel, for installation.
3. Loosen the pressure plate bolts evenly, a little at a time each! Failure to loosen the bolts evenly, in rotation, will cause warping of the pressure plate.
4. When all spring tension is relieved from the pressure plate, back out the bolts and remove the pressure plate and driven plate.
5. If the pilot bushing is equipped with a lubricating wick, remove it and soak it in clean engine oil.
6. Installation is in the reverse order of removal. Tighten the pressure

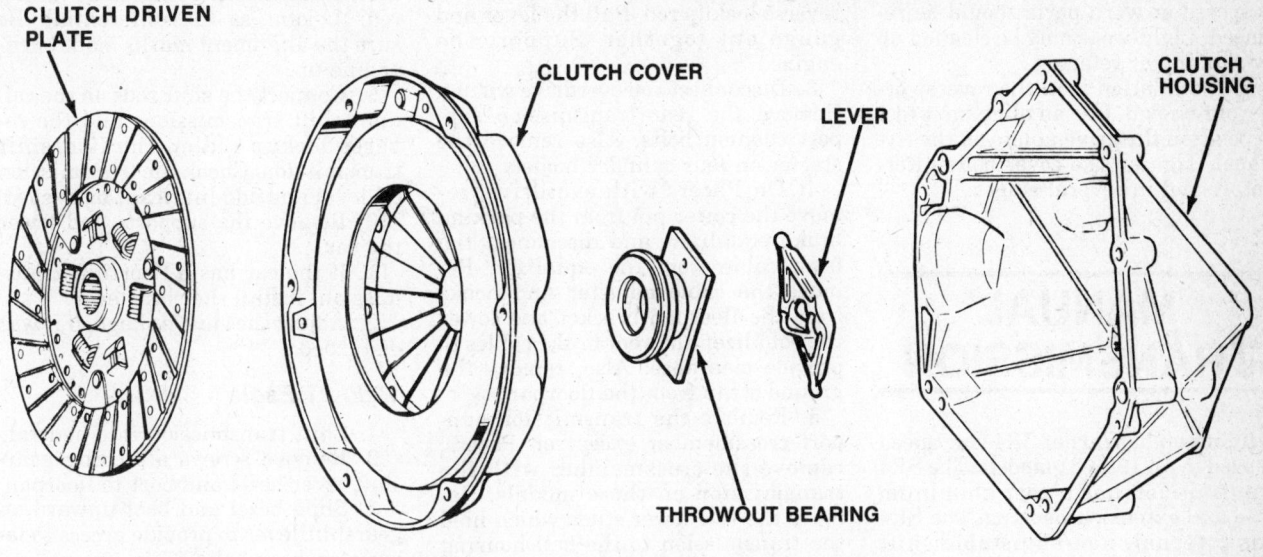

Exploded view of 150-4 cylinder clutch assembly

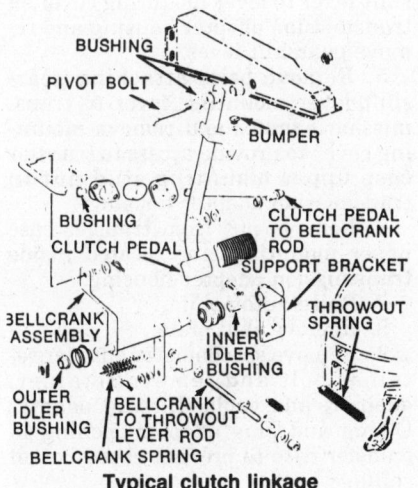

Typical clutch linkage

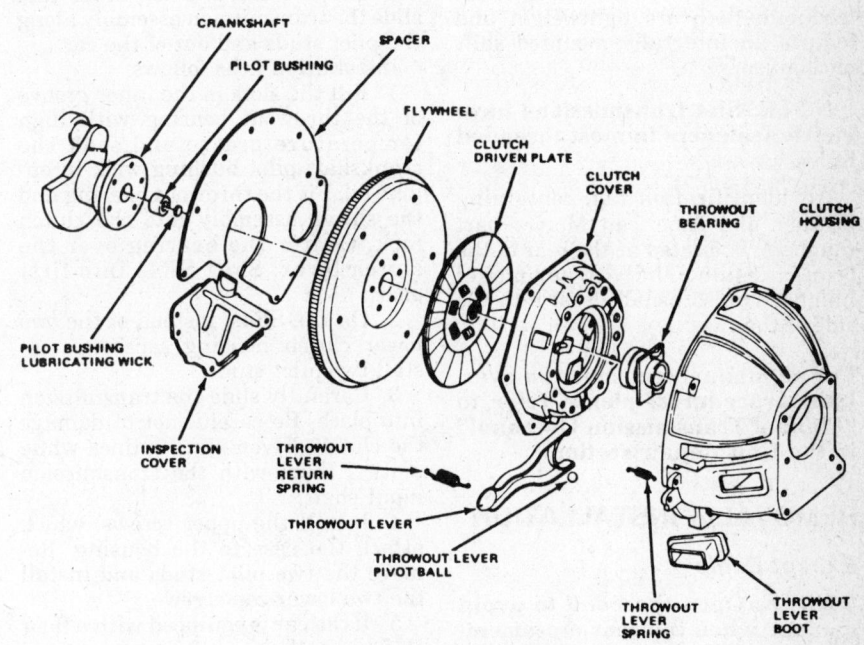

Typical 258-6 clutch assembly

plate bolts evenly, in rotation, in three or four different steps, to 23 ft. lbs.

151-4 Engine

1. Remove the starter, disconnect the slave cylinder spring at the throwout lever, and remove the transmission.

2. Remove the clutch housing-to-engine bolts. Remove the housing.

3. Remove the throwout bearing.

4. Matchmark the clutch cover and flywheel for installation. Loosen the clutch cover bolts alternately and evenly, to avoid distortion, and remove the clutch cover and disc.

5. Inspect the parts for signs of overheating (blue color), scoring, or abnormal wear. Overheated parts should be replaced. Deep scoring or wear may require replacement of the disc and cover, and refacing or replacement of the flywheel.

6. If the same cover is being used, place the disc and cover on the flywheel, aligning the marks made previously. Be sure the cover is engaged with the dowel pins. Install the cover bolts finger-tight.

7. Align the disc with an alignment tool.

8. Tighten the cover bolts alternately and evenly to 23 ft. lbs. Remove the alignment tool.

9. Install the throwout bearing, clutch housing, and transmission. The housing-to-engine bolts and transmission-to-housing bolts should be tightened to 54 ft.lbs.

258-6 Engine

1. Remove the transmission, starter motor and throwout bearing.

2. Disconnect the clutch linkage at the housing and remove the housing.

3. Mark the clutch cover and flywheel for reassembly.

4. Remove the clutch cover and the driven plate by loosening the bolts alternately and in several stages.

5. Remove the pilot bushing lubricating wick and soak the wick in engine oil.

6. Inspect the parts for signs of overheating (blue color), distortion,

scoring, or wear. Overheated or deeply scored or worn parts should be replaced. Light wear may be cleaned up by sanding or refacing.

7. Installation is in the reverse order of removal. Use an alignment tool to position the driven plate on the flywheel. Tighten the cover bolts alternately and in several stages.

MANUAL TRANSMISSION

A lightweight Warner SR4 four-speed is used in all 1980-81 models. The SR4 transmission has a cast aluminum case and extension housing. The SR4 has internal, non-adjustable shift linkage. In 1982 a new Warner (T4) four speed and an optional Warner (T5) five speed transmission were introduced. Both are lightweight and feature an integrally mounted shift mechanism.

NOTE: SR4 transmissions have metric fasteners in most threaded holes.

An identification tag, containing Warner and American Motors part numbers, is located at the rear of the transmission. The Warner model number is also usually cast into the side of the case.

For all manual transmission overhaul procedures, please refer to "Manual Transmission Overhaul" in the Unit Repair section.

REMOVAL & INSTALLATION

Except Eagle

NOTE: Open the hood to avoid damage when the rear crossmember is removed. If the overdrive and transmission are to be separated, first engage then disengage the overdrive with the clutch pedal depressed and the engine running.

1. Matchmark the driveshaft and rear axle yoke for correct installation. Split the gear universal joint and slide the driveshaft off the back of the transmission. Support the transmission with a jack.
2. Detach the column shift mechanism linkage to the transmission, and disconnect the clutch linkage and speedometer cable; disconnect the back-up light switch wiring and TCS switch wiring. On a floorshift, remove the shift lever. Remove the boot and

unbolt the lever. Detach the column reverse lockup rod. Pull the lever and gauge out together. Support the engine.

3. Disconnect the overdrive wiring. Remove the rear transmission support cushion bolts. Also remove the starter on four cylinder models.
4. On Pacers with overdrive, remove the cotter pin from the parking brake equalizer, and disconnect the front cable from the equalizer. Remove the cable adjuster and hooks from the floorpan bracket and lower the equalizer and rear brake cables to provide clearance. Also, remove the ground strap from the floorpan.
5. Remove the transmission support crossmember except on Pacers; remove the crossmember with the transmission on those models. Remove the two lower studs which hold the transmission to the bell housing and replace these two studs with two long pilot studs on SR4s.
6. Remove the two top studs and slide the transmission assembly along the pilot studs and out of the car.

Installation is as follows:

1. Fill the slots in the inner groove of the throwout bearing with high temperature grease and soak the crankshaft pilot bushing wick in engine oil. Fit the throwout bearing and the sleeve assembly into the clutch fork. Center the bearing over the clutch lever. Shift SR4s into first gear.
2. On the SR4s, instead of the two lower clutch housing capscrews install two pilot studs.
3. Carefully slide the transmission into place. Be careful not to damage the clutch driven plate splines while mating them with the transmission input shaft.
4. Install the upper screws, which attach the case to the housing. Remove the two pilot studs and install the two lower cap screws.
5. If the car is equipped with a floor shift, install the shift lever retainer and shift rods, if removed.
6. Attach the speedometer cable, connect the back-up light switch wires and the transmission controlled spark (TCS) wire, if so equipped. On the HR1, connect the clutch cable and adjust as necessary. Also install the inspection cover and the catalytic converter bracket bolts.
7. Raise the transmission. Attach the rear crossmember and support to the transmission. Fasten the crossmember to the side sills and finger-tighten the bolts. Install and tighten the crossmember-to-support bolts. Tighten the crossmember stud nuts. Install the parking brake cables and ground strap on Pacer.
8. Install the front U-joint yoke on

the transmission. Do the same for the rear U-joint at the differential. Be sure the alignment marks made earlier line up.

9. Connect the shift rods on the column shift transmissions and the reverse lockup rod on the floorshift transmission. Check the transmission oil level and add lubricant as needed.
10. Remove the supports and lower the car.
11. If the car has a floorshift transmission, install the shift lever.
12. Adjust the shift linkage, if it was disturbed.

1980-81 Eagle

1. Shift transmission into neutral.
2. Remove screws attaching gearshift lever bezel and boot to floorpan.
3. Slide bezel and boot upward on gearshift lever to provide access to lever attaching bolts.
4. Remove bolts attaching gearshift lever to lever mounting cover on transmission adapter housing and remove gearshift lever.
5. Remove bolts attaching gearshift lever mounting cover to transmission adapter and remove mounting cover to provide access to transfer case upper mounting stud nut in transmission adapter housing.
6. Remove nut from transfer case upper mounting stud located inside transmission adapter housing.
7. Raise automobile.
8. Remove skid plate.
9. Remove speedometer adapter retainer bolt and remove retainer, adapter, and cable. Discard adapter O-ring and plug adapter opening in transfer case to prevent excessive oil spillage.

NOTE: Mark the position of the speedometer adapter for assembly reference before removing it.

10. Mark propeller shafts and axle yokes for assembly alignment reference and disconnect propeller shafts at transfer case.
11. Disconnect backup lamp switch wire.
12. Place support stand under engine.
13. Support transmission and transfer case using transmission jack.
14. Remove rear crossmember.
15. Remove catalytic converter bracket from transfer case.
16. Remove bolts attaching transmission to clutch housing.
17. Remove transmission and transfer case as an assembly.
18. Remove nuts from transfer case mounting studs and remove transmission from transfer case.
19. Install transmission on trans-

fer case. Install and tighten all transfer case mounting stud nuts to 26 ft. lbs. torque.

20. Support transmission-transfer case assembly on transmission jack.

21. Align transmission clutch shaft with throwout bearing and clutch disc splines and seat transmission against clutch housing.

22. Install and tighten transmission-to-clutch housing attaching bolts to 55 ft. lbs. torque.

23. Connect propeller shafts to transfer case yokes. Tighten clamp strap bolts to 15 ft. lbs. torque.

24. Install rear crossmember. Tighten crossmember attaching bolts to 30 ft. lbs. torque.

25. Connect backup lamp switch wire.

26. Install replacement O-ring on speedometer adapter and install adapter and cable, and retainer. Tighten retainer bolt to 100 inch lbs. torque.

CAUTION

Do not attempt to reuse the original adapter O-ring. The ring is designed to swell in service to improve its sealing qualities and could be cut or torn during installation if reuse is attempted.

27. Attach catalytic converter bracket to transfer case. Tighten retaining nuts to 26 ft. lbs. torque.

28. Check and correct lubricant levels in transmission and transfer case, if necessary.

29. Install skid plate. Tighten skid plate attaching bolts to 30 ft. lbs. torque.

30. Remove stand used to support engine, and remove transmission jack, if not removed previously.

31. Lower the vehicle.

32. Clean mating surfaces of gearshift lever mounting cover and transmission adapter housing.

33. Apply RTV-type sealant to gearshift lever mounting cover and install cover on transmission adapter housing. Tighten cover bolts to 13 ft. lbs. torque.

34. Install gearshift lever on mounting cover. Be sure lever is engaged with shift rail before tightening lever attaching bolts. Tighten lever attaching bolts to 18 ft. lbs. torque.

35. Position gearshift lever boot and bezel in floorpan and install bezel attaching screws.

1982 and Later Eagle

1. Shift transmission into neutral.

2. Remove screws attaching gearshift lever bezel and boot to floorpan.

3. Slide bezel and boot upward on gearshift lever to provide access to lever attaching bolts.

4. Remove bolts attaching gearshift lever to lever mounting cover on transmission adapter housing and remove gearshift lever.

5. Remove bolts attaching gearshift lever mounting cover to transmission adapter and remove mounting cover to provide access to transfer case upper mounting stud nut in transmission adapter housing.

6. Remove nut from transfer case upper mounting stud located inside transmission adapter housing.

7. Raise the vehicle.

8. Remove the skid plate.

9. Remove speedometer adapter retainer bolt and remove retainer, adapter, and cable. Discard adapter O-ring and plug adapter opening in transfer case to prevent excessive oil spillage.

NOTE: Matchmark the position of the speedometer adapter for assembly alignment, before removing it.

10. Matchmark propeller shafts and axle yokes for assembly alignment, and disconnect propeller shafts at transfer case.

11. Disconnect backup lamp switch wire.

12. Place support stand under the engine.

13. Support transmission and transfer case using suitable transmission jack.

14. Remove rear crossmember.

15. Remove catalytic converter bracket from transfer case, and brace rod from racket.

16. Remove bolts attaching transmission to clutch housing.

17. Remove transmission and transfer case as an assembly.

18. Remove nuts from transfer case mounting studs and remove transmission from transfer case.

19. Install transmission on transfer case. Install and tighten all transfer case mounting stud nuts to 26 ft.lbs. torque.

20. Support transmission-transfer case assembly on suitable transmission jack.

21. Align transmission clutch shaft without throwout bearing and clutch disc splines and seat transmission against clutch housing.

22. Install and tighten transmission-to-clutch housing attaching bolts to 55 ft. lbs. torque.

23. Connect propeller shafts to transfer case yokes. Tighten clamp strap bolts to 15 ft. lbs. torque.

24. Install brace rod and rear crossmember. Tighten attaching bolts to 30 ft. lbs. torque.

25. Connect backup lamp switch wire.

26. Install replacement O-ring on speedometer adapter and install adapter and cable, and retainer. Tighten retainer bolt to 100 inch lbs. torque.

CAUTION

Do not attempt to reuse the original adapter O-ring. The ring is designed to swell in service to improve its sealing qualities and could be cut or torn during installation if reuse is attempted.

27. Attach catalytic converter bracket to transfer case. Tighten skid plate attaching bolts to 30 ft. lbs. torque.

28. Check and correct lubricant levels in transmission and transfer case, if necessary.

29. Install skid plate. Tighten skid plate attaching bolts to 30 ft. lbs. torque.

30. Remove stand used to support engine, and remove transmission jack, if not removed previously.

31. Lower the vehicle.

32. Clean mating surfaces of gearshift lever mounting cover and of transmission adapter housing.

33. Apply RTV-type sealant to gearshift lever mounting cover, install cover bolts and torque to 13 ft. lbs. torque.

34. Install gearshift lever on mounting cover. Before tightening lever attaching bolts make certain lever is engaged with shift rail. Tighten lever attaching bolts to 18 ft. lbs. torque.

35. Position gearshift lever boot and bezel in floorpan and install bezel attaching screws.

AUTOMATIC TRANSMISSION

AUTOMATIC TRANSMISSION IDENTIFICATION

Year	Engine	Transmission
'80–'87	4-150	Chrysler 904 ①
	4-151	Chrysler 904 ①
	6-258	Chrysler 904 ②
	6-258 (Eagle)	Chrysler 998

① Standard ratio
② Wide ratio

REMOVAL & INSTALLATION

4 Cylinder Engines

1. Open hood.

CAUTION

The hood must remain open to avoid damaging the hood and air cleaner when the rear crossmember is removed.

---

2. Disconnect fan shroud.
3. Remove bolt attaching transmission fill tube to engine.
4. Place gearshift lever in Neutral.
5. Raise and support car. Drain transmission fluid.
6. Mark propeller shaft and yoke for assembly alignment reference. On Eagle models, also remove skid plate.
7. Remove propeller shafts.
8. On Eagle models, disconnect exhaust system at exhaust manifold, loosen exhaust system hangers and move exhaust system as necessary to gain work space.
9. Remove starter motor. On Eagle models, also remove stiffening braces.
10. Remove speedometer adapter and cable assembly. Cover adapter bore after removal.
11. Disconnect gearshift and throttle linkage. On automobiles with column shift, remove bolt that mounts the linkage bellcrank bracket on the converter housing.
12. Remove cover at front of converter housing.
13. Mark converter drive plate and converter for assembly alignment reference.
14. To gain excess to and remove the bolts attaching the drive plate to the converter, you'll first have to rotate the crankshaft and drive plate. This can be done by using a ratchet handle and socket, or a box-end wrench, on the crankshaft's front pulley bolt.

NOTE: The crankshaft pulley bolt is a metric size bolt.

15. Support transmission using transmission jack. Retain transmission on jack using safety chain.

CAUTION

On Eagle models, both the transmission and transfer case must be properly supported on the transmission jack and retained with safety chain.

---

16. Disconnect oil cooler lines at transmission.
17. Remove bolt attaching rear support cushion to rear support cushion bracket (bracket is attached to transmission extension housing).
18. Remove rear crossmember attaching nuts and remove crossmember and support cushion as an assembly.
19. Place support stand under front of engine.
20. Remove bolts attaching catalytic converter support bracket to transmission, if so equipped.
21. Remove fill tube.

22. Remove bolts attaching transmission to engine.

NOTE: The transmission-to-engine block bolts are metric size bolts.

23. Move transmission (and transfer case, if so equipped) and converter back until clear of crankshaft.
24. Hold converter in position and lower transmission until transmission converter housing clears engine.

CAUTION

If the transmission was removed to correct a malfunction that generated sludge or heavy accumulations of metal or friction material particles, the oil cooler and cooler lines must be thoroughly flushed and the torque converter replaced. Do not attempt to flush the converter if it is contaminated.

---

25. If torque converter was removed, insert Pump Aligning Tool J-24033 into pump rotor and engage tool slots with pump rotor drive lugs.
26. Rotate aligning tool until hole in tool is vertical, then remove tool.
27. Rotate converter until pump drive slots in converter hub are vertical.
28. Carefully insert converter hub into oil pump. Be sure drive lugs of pump inner rotor are completely engaged with drive slots in converter hub.
29. Raise transmission and align converter with drive plate. Refer to alignment marks made during removal.

CAUTION

On Eagle models, both the transmission and transfer case must be supported on a transmission jack and retained with safety chain.

---

30. Move transmission forward and raise, lower, or tilt transmission to align transmission converter housing dowel holes with dowels in engine block.

NOTE: If the downward angle at the rear of the engine is not sufficient to permit transmission installation, raise the front of the engine to increase the downward angle.

31. Install two transmission-to-engine lower attaching bolts and tighten bolts evenly to pull transmission to engine.
32. Install drive plate-to-converter attaching bolts. Tighten bolts to 40 ft. lbs. torque.

NOTE: Coat threads of drive plate-to-converter attaching bolts with Loctite® 271 or equivalent.

33. Install remaining transmission-to-en-gine attaching bolts. Tighten bolts to 54 ft. lbs. torque.
34. Connect oil cooler lines.
35. Install propeller shaft using reference marks made during assembly.
36. Install rear crossmember and support cushion. Tighten attaching nuts to 30 ft. lbs. torque.
37. Install rear support cushion-to-support bracket bolt. Tighten bolt to 48 ft. lbs. torque.
38. Remove safety chain and transmission jack.
39. Install converter housing inspection cover.
40. Install starter motor. On Eagle models also install stiffening brace and skid plater.
41. Connect neutral start switch wires to switch terminal.
42. Connect gearshift and throttle linkage. On automobiles with column shift, position linkage bellcrank bracket on converter housing and install bracket attaching bolt.
43. Install speedometer cable and adapter assembly. Be sure adapter is correctly indexed.
44. Connect catalytic converter, if so equipped. On Eagle models, add correct quantity of transfer case lubricant.
45. Lower the vehicle.
46. Fill transmission to correct fluid level.
47. Check and adjust gearshift lever and throttle linkage if necessary.

NOTE: The gearshift lever adjusting trunnion is located at the lower end of the steering column.

48. Road test to check transmission operation.

6 Cylinder Engine

1. Disconnect fan shroud, if so equipped.
2. Disconnect transmission fill tube at upper bracket.
3. Open hood.

CAUTION

It is necessary that the hood be open to avoid damaging the hood and air cleaner when the rear crossmember is removed.

---

4. Raise and support car. Drain transmission fluid.
5. Remove inspection cover from converter housing.
6. On Spirit and Concord, remove screw attaching exhaust pipe clamp to exhaust pipe support bracket and slide clamp off bracket.
7. Remove transmission fill tube.
8. Remove starter. On Eagle models, also remove stiffening braces.
9. Mark propeller shaft(s) and

yoke(s) for assembly alignment reference.

10. Remove propeller shaft(s).

11. On Eagle models, disconnect exhaust pipe and move it aside for working clearance.

12. Remove speedometer adapter and cable assembly. Discard adapter and cable seals, they are not reuseable. Cover adapter bore after removal.

13. Disconnect gearshift and throttle linkage.

14. Disconnect wires at neutral start switch.

15. Mark converter drive plate and converter for assembly alignment reference.

16. To gain access to and remove the bolts attaching the drive plate to the converter, you'll first have to rotate the crankshaft and drive plate. This can be done by using a ratchet handle and socket, or a box-end wrench, on the crankshaft's front pulley bolt.

17. On Eagle models, remove skid plate and stiffening brace.

18. Support transmission (and transfer case on Eagle models) using a suitable transmission jack. Retain transmission on jack using safety chain.

19. Disconnect oil cooler lines at transmission.

20. Remove bolts attaching rear support cushion to transmission.

21. Remove rear crossmember.

22. Remove bolts attaching transmission to engine.

23. Move transmission and converter back to clear crankshaft.

24. Hold converter in position and lower assembly until converter housing clears engine.

─────── CAUTION ───────

If the transmission was removed to correct a malfunction that generated sludge or heavy accumulations of metal particles or friction material, the oil cooler and cooler lines must be flushed thoroughly and the torque converter replaced. Do not attempt to flush the converter if it is contaminated.

25. If torque converter was removed, insert Pump Aligning Tool J-24033 in pump rotor until rotor drive lugs engage slots in tool.

26. Rotate tool until drilled hole in tool is vertical and remove tool.

27. Rotate converter until pump drive slots in converter hub into pump. Be sure drive lugs of pump inner rotor are properly engaged in drive slots of converter hub.

28. Raise transmission (and transfer case on Eagle models) and align converter with drive plate. Refer to assembly alignment marks.

29. Move transmission forward.

30. Raise, lower, or tilt transmission to align converter housing pilot holes with dowels in engine.

31. Install two transmission lower attaching bolts and tighten bolts evenly to pull transmission to engine.

32. Install drive plate-to-converter attaching bolts.

33. Install remaining transmission attaching bolts and tighten all bolts to 28 ft. lbs. torque.

34. Connect oil cooler lines.

35. Install rear support cushion on transmission, if removed.

36. Install rear crossmember.

37. Remove safety chain and transmission jack.

38. Install inspection cover.

39. On Spirit and Concord, install exhaust pipe clamp on support bracket.

40. Install starter. On Eagle models, also install stiffening braces.

41. Connect wires to neutral start switch.

42. Connect gearshift and throttle linkage.

43. Install speedometer cable and adapter assembly. Be sure adapter is correctly indexed.

44. On Eagle models, connect exhaust pipes, attach stiffening brace and install skid plate.

45. Install propeller shaft(s). Refer to alignment marks made during removal. On Eagle models, add correct quantity of transfer case lubricant.

46. Lower automobile.

47. Fill transmission to correct level.

48. Check and adjust gearshift and throttle linkage, if necessary.

49. Road-test the vehicle to check transmission operation.

For further details on linkage and band adjustments, please refer to "Automatic Transmissions" in the Unit Repair section.

Transfer Case

For all transfer case overhaul procedures, please refer to "Transfer Cases" in the Unit Repair section.

REMOVAL & INSTALLATION

Manual Transmission

NOTE: Steps 1-6, below, pertain to models with the SR-4 transmission only.

1. Shift transmission into Neutral.

2. Remove screws attaching gearshift lever bezel to floorpan or console, if so equipped.

3. Slide bezel and boot upward on gearshift lever to provide access to lever attaching bolts.

4. Remove bolts attaching gearshift lever to lever mounting cover on transmission adapter housing and remove lever.

5. Remove bolts attaching gearshift lever mounting cover to transmission adapter housing and remove cover.

6. Remove nut from transfer case mounting stud located inside transmission adapter housing.

7. Raise and support the vehicle safely.

8. Remove skid plate and rear brace rod at transfer case.

9. Remove speedometer adapter retainer attaching bolt and remove retainer, adapter, and cable. Plug adapter opening in transfer case to prevent excessive oil spillage.

NOTE: Mark the position of the speedometer adapter for assembly reference before removing it.

10. Mark propeller shafts and axle yokes for assembly alignment reference and disconnect propeller shafts at transfer case.

11. On models so equipped, remove transfer case shift motor vacuum harness.

12. Support transfer case with a suitable transmission jack.

13. Remove nuts from transfer case mounting studs and remove transfer case.

14. Align transmission output and transfer case input shafts and install transfer case on transmission adapter housing.

15. Install and tighten transfer case mounting stud nuts to 33 ft. lbs. torque.

16. Remove jack used to support transfer case.

17. Align and connect propeller shafts to axle yokes. Tighten clamp strap bolts to 15 ft. lbs. torque.

18. Install replacement O-ring on speedometer adapter and install adapter and cable and retainer. Tighten retainer bolt to 100 inch lbs. torque.

─────── CAUTION ───────

Do not attempt to reuse the original adapter O-ring. The O-ring is designed to "swell" in service to improve its sealing qualities and it could be cut or torn during installation if reuse is attempted.

19. Install skid plate and rear brace rod. Torque retaining bolts to 30 ft. lbs. torque.

20. On models so equipped, install transfer case shift motor vacuum harness.

21. Check and correct lubricant levels in transmission and transfer case, if necessary.

22. Lower the vehicle.

23. Install nut on transfer case mounting stud located inside transmission adapter housing. Tighten nut to 33 ft. lbs. torque.

24. Install gearshift lever mounting cover on transmission adapter housing.

25. Install gearshift lever on mounting cover. Be sure lever is engaged with shift rail before tightening lever attaching bolts.

26. Position gearshift lever boot and bezel on floorpan or console, if so equipped, and install bezel attaching screws.

Automatic Transmission

1. Raise the vehicle and support it safely.

2. Support engine and transmission with support stand or transmission jack.

3. Disconnect catalytic converter support bracket at adapter housing.

4. Remove skid plate and rear brace rod at transfer case.

5. Remove speedometer cable and adapter from transfer case. Discard adapter O-ring, it is not reusable.

6. Matchmark propeller shafts and transfer case yokes for assembly reference.

7. Disconnect propeller shafts at yokes. Secure shafts to underside of vehicle.

8. Disconnect gearshift and throttle linkage at transmission.

9. Lower the rear crossmember.

10. Remove all transfer case-to-adapter housing stud nuts and remove transfer case.

11. Install transfer case on adapter housing. Be careful not to damage output shaft splines during installation.

12. Install transfer case-to-adapter housing stud nuts. Tighten nuts to 33 ft. lbs. torque.

13. Install rear crossmember and tighten the attaching nuts to 30 ft. lbs.

14. Install the rear brace rod.

15. Remove transmission jack or support stand.

16. Connect gearshift and throttle linkage to transmission.

17. Connect propeller shafts. Tighten clamp strap bolts to 15 ft. lbs. torque.

18. Install new O-ring on speedometer adapter and install adapter and cable in transfer case.

NOTE: Do not attempt to reuse the old adapter O-ring. O-ring is designed to swell in service to provide improved sealing qualities and could be cut or torn if reinstallation is attempted.

19. Install skid plate and stiffening brace, if so equipped. Tighten retaining bolts to 30 ft. lbs.

20. Connect catalytic converter support bracket to adapter housing.

21. Check transfer case lubricant level and transmission linkage adjustments to make certain they are correct.

22. Lower the automobile.

FRONT AXLE

Axle Shaft, Shaft Seal And Bearing

The procedure for replacing the axle shafts and seals on four-wheel drive models calls for removal of the axle first.

1. Raise and support the front of the car. Install protectors over the halfshaft boots. Remove the halfshaft-to-axle flange bolts, and tie the halfshafts out of the way.

2. Matchmark the driveshaft and the axle yoke. Remove the driveshaft.

3. Support the axle on stands. Remove the five axle-to-engine mounting bolts.

4. Lower the axle part way and remove the vent hose. Remove the axle.

5. Remove the differential cover and drain the oil. Remove the axle shaft C-clips.

6. Remove the axle shafts.

7. Carefully remove the shaft seal.

8. Two different bearings are used; the left side uses a ball bearing, and the right side uses a needle bearing. The ball bearing should be removed using a brass drift and a hammer; the needle bearing should be removed using a needle bearing removal tool.

NOTE: If the proper bearing removal tool is not available, remove the differential and remove the needle bearing using a $15/16$ inch socket and a three foot ratchet extension.

9. Install the bearings, using drivers of the appropriate type and size.

10. Oil the lips of the new seal and install into the housing using a driver of the correct size.

11. Install the axle shafts and C-clips.

12. Apply a bead of silicone seal to the differential cover and install the cover.

13. Fill the axle with 2.5 pints of 85W-90 GL-5 gear oil.

14. Move the axle into place under the car. Raise it sufficiently to connect the vent hose, then raise it fully into place and install the mounting bolts. Tighten to 50 ft. lbs.

15. Install the driveshaft, aligning the marks made during removal. Install the halfshaft-to-axle flange bolts, and tighten to 45 ft. lbs.

REAR AXLE

A letter code used to identify the axle ratio will be found on most differentials, stamped on the right axle tube housing boss, on the rear side, adjacent to the dowel hole. Some earlier cars have either a metal tag attached to one of the bolts of the differential housing cover or the code letter is stamped on the right differential housing cover flange. It may be necessary to remove the cover from the differential to locate the letter. The codes and the axle ratios are listed in dealer parts books.

Axle Shaft, Bearing and Seal

REMOVAL & INSTALLATION

1. The hub and drum are separate units and are removed after the wheel is removed. The hub and axle shaft are serrated together on the taper. An axle shaft key assures proper alignment during assembly.

2. With the wheel on the ground and the parking brake applied, remove and discard the axle shaft nut cotter pin and remove the nut. Raise the car and remove the wheel. Release the parking brakes and remove the drum.

3. Attach a puller to the rear hub and remove the hub. The use of a "knock-out" puller should be discouraged, since it may result in damage to the axle shaft or wheel bearings.

4. Disconnect the parking brake cable at the equalizer.

5. Disconnect the brake tube at the wheel cylinder and remove the brake support plate assembly, oil seal, and axle shims. Note that the axle shims are located on the left side only.

6. Using a screw type puller, remove the axle shaft and bearings from the axle housing.

--- **CAUTION** ---

On Twin-Grip axles, rotating the differential with one shaft removed will misalign the side gear splines, preventing installation of the replacement shaft.

---

7. Remove the axle shaft inner oil

seal and install new seals at assembly.

8. The bearing is a press fit and should be removed with an arbor press.

9. The axle shaft bearings have no provision for lubrication after assembly.

Before installing the bearings, they should be packed with a good quality wheel bearing lubricant.

10. Press the axle shaft bearings onto the axle shaft with the small diameter of the cone toward the outer (tapered) end of the shaft.

11. Soak the inner axle shaft seal in light lubricating oil. Coat the outer surface of the seal retainer with sealant.

12. Install the inner oil seal.

13. Install the axle shafts, indexing the splined end with the differential side gears.

14. Install the outer bearing cup.

15. Install the brake support plate. Sealant should be applied to the axle housing flange and to the brake support mounting plate.

16. Install the original shims, oil seal and brake support plate. Torque the nuts to 30–35 ft. lbs.

NOTE: The oil seal and retainer go between the axle housing flange and the brake support plate on 9 in. brakes. On 10 in. brakes, they go on the outside of the brake support plate.

17. To adjust the axle shaft endplay, strike the axle shafts with a lead mallet to seat the bearings. Install a dial indicator on the brake support plate and check the play while pushing and pulling the axle shaft. Endplay should be 0.004–0.008 in., with 0.006 in. desirable. Add shims to the left side only to decrease the play and remove shims to increase the play.

18. Slide the hub onto the axle shafts by aligning the serrations and the keyway on the hub with the axle shaft key.

19. Replace the hub and drum, install the wheel, lower the car onto the floor and tighten the axle shaft nut to 250 ft. lbs. If the cotter pin hole is not aligned with a castellation on the nut, tighten the nut to the next castellation.

NOTE: A new hub must be installed whenever a new axle shaft is installed. Install two thrust washers on the shaft. Tighten the new hub onto the shaft until the hub is 1.19 in. from the end of the shaft. Remove the nut and remove one thrust washer. Install the nut and torque to 250 ft. lbs. New hubs do not have serrations on the axle

shaft mating surface. The serrations are cut when the hub is installed onto the axle shaft.

20. Connect the parking brake cable at the equalizer.

21. Connect the brake tube at the wheel cylinder and bleed the brakes.

DRIVESHAFT AND U-JOINTS

A one-piece tubular driveshaft is used with a yoke at each end, to position the cross-and-roller tube universal joints.

NOTE: The driveshaft is a balanced unit; care must be used in handling. Do not bend or distort the tube or yokes, or vibration will result.

Driveshaft

REMOVAL & INSTALLATION

All Models Except Eagle

1. Matchmark and disassemble rear U-joint by removing nuts. Retention is by U-bolts or straps, depending on model.

2. Drop rear of driveshaft and slide front yoke out of transmission.

3. To install, reverse removal procedure, tightening U-joint nuts to 15 ft. lbs.

Eagle

Both driveshafts are secured at the transfer case end and the axle yoke end by straps. The straps are retained by Torx® head bolts.

1. Shift into Neutral. Raise and support the car.

2. Matchmark the driveshaft(s) at the transfer case and axle yoke for alignment reference.

3. Remove the retaining straps with a Torx® bit tool of the proper size. Remove the driveshaft(s).

4. To install, align the matchmarks made during removal to assure proper balance. Seat the universal joints in the yokes and install the straps, tightening to 17 ft. lbs.

UNIVERSAL JOINT OVERHAUL

For all U-Joint overhaul procedures, please refer to "U-Joint/CV-Joint Overhaul" in the Unit Repair section.

FRONT SUSPENSION

The front suspension on all models except Pacer is an independent, linked-type with the coil springs located between seats in the wheelwell panels and seats in the upper control arms. Rubber insulators between the springs and seats reduce noise transmission to the body. Direct acting, telescopic shock absorbers are located inside the coil springs and the control arms are attached to the body via rubber bushings.

The suspension system is a double ball joint design, both upper and lower control arms each having one joint. On all models, strut rods serve to support the lower control arms. Stabilizer bars are used on some models.

The Pacer front suspension is different from all other AMC cars. The coil spring is mounted between the two control arms; it is seated at the bottom on the lower control arm and at the top in the suspension/engine mount crossmember. The crossmember is isolated from the rest of the body structure by rubber mounting joints. The shock absorbers are mounted within the coil springs. The steering knuckle is attached to the upper and lower control arms by upper and lower ball joints. A front stabilizer bar is optional.

NOTE: The front end alignment must be checked after disassembly procedure.

Shock Absorber

REPLACEMENT

NOTE: When installing new shock absorbers, purge them of air by extending them in their normal position and compressing them while inverted. To this several times. It is normal for new shock absorbers to be more resistant to extension than to compression.

All Except Pacer

1. Remove the two lower shock absorber attaching nuts. Remove the washers and the grommets.

2. Remove the upper mounting bracket nuts and bolts.

3. Remove the bracket, complete with shock.

4. Remove the upper attaching nut and separate the shock from the mounting bracket.

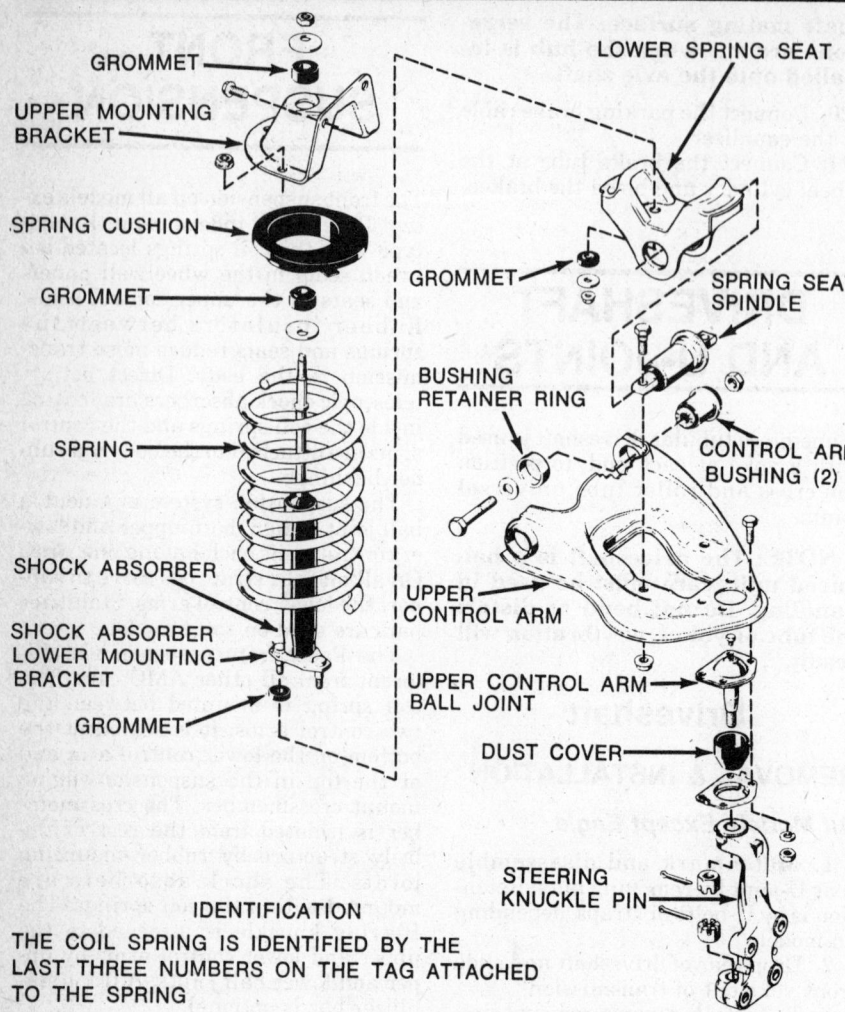

GROMMET

UPPER MOUNTING BRACKET

SPRING CUSHION

GROMMET

SPRING

SHOCK ABSORBER

SHOCK ABSORBER LOWER MOUNTING BRACKET

GROMMET

LOWER SPRING SEAT

GROMMET

SPRING SEAT SPINDLE

BUSHING RETAINER RING

CONTROL ARM BUSHING (2)

UPPER CONTROL ARM

UPPER CONTROL ARM BALL JOINT

DUST COVER

STEERING KNUCKLE PIN

IDENTIFICATION
THE COIL SPRING IS IDENTIFIED BY THE LAST THREE NUMBERS ON THE TAG ATTACHED TO THE SPRING

Upper control arm and shock absorber—all except Eagle and Pacer

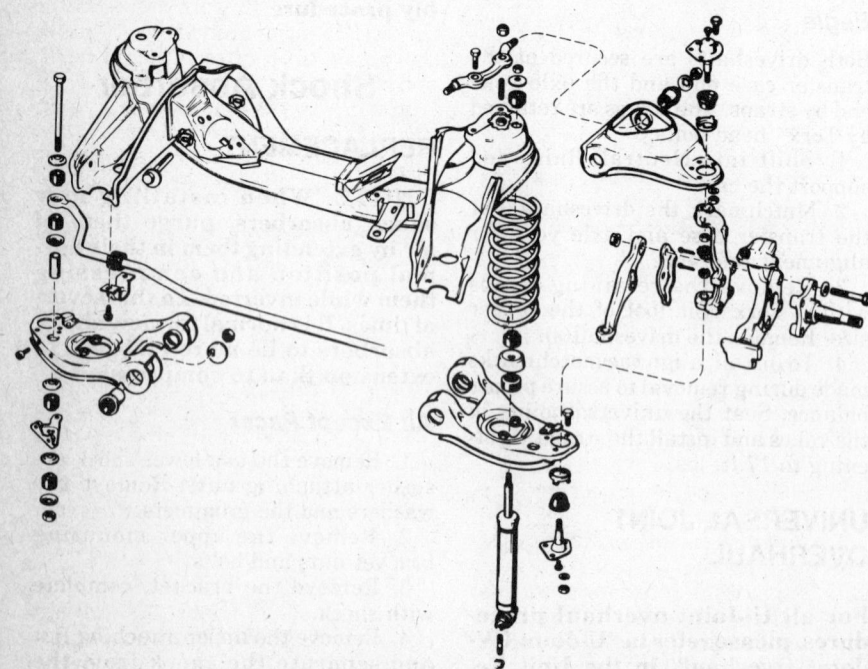

Exploded view of the Pacer front suspension

5. For adjustable shocks: To adjust the shock, compress the piston completely. Holding the upper part of the shock, turn the shock until the lower arrow is aligned with the desired setting. A click will be heard when the desired setting is reached.

6. To install, fit the grommets, washers, upper mounting bracket and nut on the shock in the reverse order of removal. Tighten the nut to 8 ft. lbs.

7. Fully extend the shock and install two grommets on the lower mounting stud.

8. Lower the shock through the hole in the wheelwell. Fit the lower attachment studs through the lower spring seat.

9. Install the grommets, washers, and nuts. Tighten the nuts to 15 ft. lbs.

10. Secure the upper mounting bracket with its attachment nuts and bolts. Tighten them to 20 ft. lbs.

Pacer

1. Remove the shock absorber upper locknut.

2. Raise the car and remove the nuts from the lower shock absorber mounting studs.

3. Remove the shock along with the lower grommet and jounce bumper retainer from the shock absorber piston rod.

4. Install the retainer on the new shock and the lower grommet on the piston rod.

5. Extend the piston to full length and insert the shock through the lower control arm.

6. Install the locknuts on the lower mounting studs and lower the car.

7. Install the grommet, retainer, and locknut on the piston rod, making sure the grommet seats properly in the hole in the crossmember.

Spring

REMOVAL & INSTALLATION

All Except Pacer

1. Remove the shock absorber.

2. Install a spring compressor through the upper spring seat opening and bolt it to the lower spring seat using the lower shock absorber mounting holes.

3. Remove the lower spring seat pivot retaining nuts, then tighten the compressor tool to compress the spring about 1 in.

4. Jack up the front of the car and support it on axle stands at the subframe, allowing the control arms to hang free.

5. Remove the front wheel and pull

the lower spring seat out and away from the car, then slowly release the spring tension and remove the coil spring and lower spring seat.

6. To install, place the spring compressor through the coil spring and tape the rubber spring cushion to the small-diameter end of the spring (upper).

7. Place the lower spring seat against the spring with the end of the coil against the formed shoulder in the seat. The shoulder and coil end face inward, toward the engine, when the spring is installed.

8. Place the spring up against the upper seat, then align the lower spring seat pivot so the retaining studs will enter the holes in the upper control arm.

9. Compress the coil spring and install the spring, then install the wheel and tire and lower the car to the floor to place weight on suspension.

10. Install and tighten lower spring seat spindle retaining nuts and tighten them to 35 ft. lbs. Remove the spring compressor and install the shock absorber.

Pacer

1. Disconnect the upper end of the shock absorber.

2. Raise the front end of the car and support it.

3. Disconnect the lower end of the shock absorber and remove it.

4. Disconnect the stabilizer bar at the lower control arm, if so equipped.

5. Remove the wheel, and caliper and rotor. Do not allow the brake hose to support the weight of the caliper; use a length of wire to suspend the caliper from the frame.

6. Remove the two bolts that attach the steering arm to the steering knuckle and move the steering arm aside.

7. Use a spring compressor to compress the coil spring.

—— CAUTION ——

Make sure the spring compressor is seated properly. A compressed spring suddenly released can cause serious injury.

8. Remove the cotter pin and nut from the lower ball joint stud and disengage the stud from the steering knuckle with a puller.

9. Move the steering knuckle, steering spindle, and support plate, or anchor plate assembly aside, to provide working clearance. Do not allow the brake hose to support the weight of these components. Use wire to hang the components from the upper control arm.

10. Move the lower control arm aside and remove the spring.

11. To install the front coil spring, position the upper end of the spring in the spring seat of the front crossmember. Align the cut-off end of the bottom coil with the formed shoulder in the spring seat. The top coil is flat and does not use an insulator. Use a floor jack or jackstand to support the spring until the spring compressor is installed. Install the spring compressor.

12. Assemble the remaining components of the front suspension in the reverse order of removal. Tighten the ball joint stud nut to 75 ft. lbs., the steering arm-to-knuckle attaching bolts to 55 ft. lbs., the shock absorber lower mounting nuts to 20 ft. lbs., and the stabilizer bar locknut to 8 ft. lbs.

Control Arm

REMOVAL & INSTALLATION

Upper Control Arm—Except Pacer

1. Remove the shock absorber and compress the coil spring approximately 2 in. using the procedure under "Front Spring Removal and Installation".

2. Jack up the front of the car and support the body on jackstands placed under subframes, allowing the control arms to hang freely.

3. Remove the wheel and the upper ball joint cotter pin and retaining nut.

4. Separate the ball joint stud from the steering knuckle using a ball joint removal tool.

5. Remove the inner pivot bolts, then remove the control arm.

6. To install, reverse the removal procedure. Do not tighten the pivot bolt nuts until the full weight of the car is on the wheels. The ball joint stud nut must be tightened to 75 ft. lbs., the lower spring seat pivot retaining nuts to 35 ft. lbs., and the control arm inner pivot bolts to 80 ft. lbs.

Upper Control Arm—Pacer

1. Raise and support the front of the vehicle.

2. Remove the wheel and tire.

3. Remove the cotter pin, locknut, and retaining nuts from the upper ball joint stud.

4. Loosen the stud from the steering knuckle with a ball joint removal tool.

5. Support the lower control arm with a floor jack.

6. Disengage the stud from the steering knuckle.

7. Remove the retaining nuts that attach the cross-shaft to the front crossmember and remove the upper control arm assembly.

8. Install the upper control arm in the reverse order of removal, tightening the cross-shaft retaining nuts to 80 ft. lbs., the upper ball joint stud nut to 75 ft. lbs., and if new bushings were installed, tighten the nuts to 60 ft. lbs. after the car is lowered to the floor.

Lower Control Arm—Except Eagle and Pacer

The inner end of the lower control arm is attached to a removable crossmember. The outer end is attached to the steering knuckle pin and ball joint assembly.

1. Jack up the car and support it on axle stands under the subframes.

2. Remove the caliper and rotor from the spindle, then disconnect the steering arm from the knuckle pin.

3. Remove the lower ball joint stud cotter pin and nut. Separate the ball joint from the knuckle pin using a ball joint removal tool.

4. Disconnect the sway bar from the control arm, then unbolt the strut rod. Remove the inner pivot bolt and the control arm.

5. To install, reverse the removal procedure; do not tighten the inner pivot bolt until the car weight is on the wheels. Tighten the ball joint retaining nut to 75 ft. lbs; strut rod bolts to 75 ft. lbs.; sway bar bolts to 8 ft. lbs.; steering arm bolts to 55 ft. lbs. and control arm inner pivot bolt to 110 ft. lbs.

Lower Control Arm—Eagle

1. Remove the wheel cover. Remove and discard the cotter pin. Remove the nut lock and the hub pin.

2. Raise and support the front of the car. Remove the wheel. Remove the brake caliper from the knuckle and suspend it from the body by a length of wire; do not allow it to hang by the hose. Remove the rotor.

3. Remove the lower ball joint cotter pin and retaining nut. Discard the cotter pin.

4. Separate the ball joint stud from the steering knuckle using a ball joint removal tool.

5. Remove the halfshaft flange bolts and remove the halfshaft.

6. Remove the strut rod-to-control arm bolts. Disconnect the stabilizer bar from the arm.

7. Remove the inner pivot bolt and remove the control arm.

8. To install, place the control arm into position and install the inner pivot bolt, but do not tighten the pivot bolt yet.

9. Install the ball joint stud into the steering knuckle. Install the nut and tighten to 75 ft. lbs. Continue to tighten until the holes align, and install a new cotter pin.

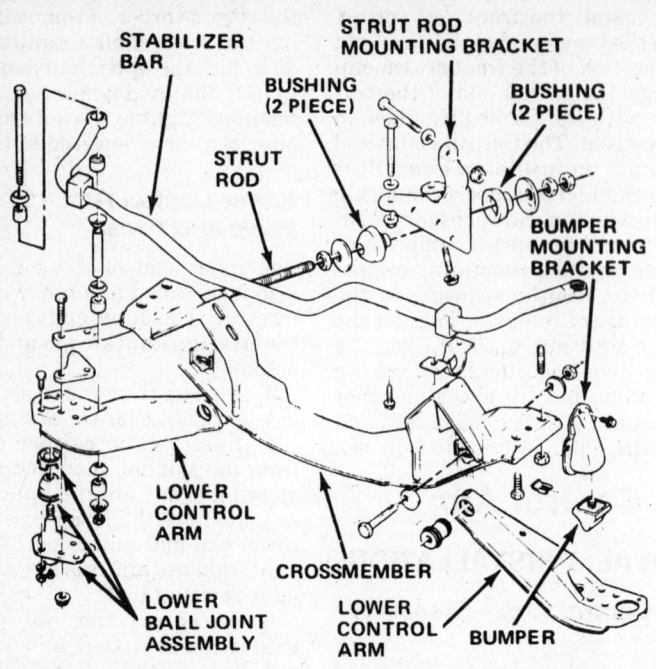

Lower control arm assembly—all except Eagle and Pacer

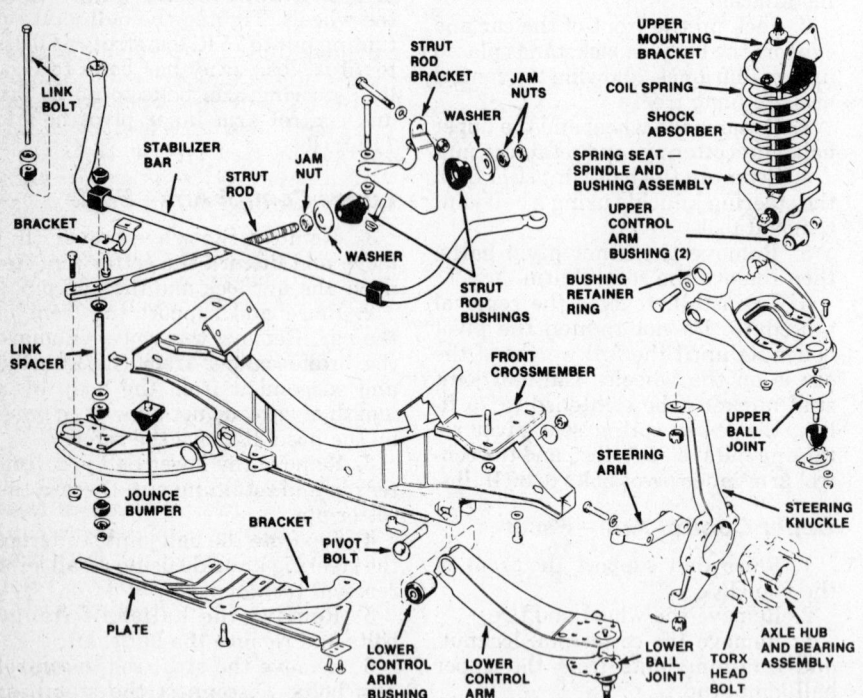

Exploded view of the Eagle front suspension

Lower Control Arm—Pacer

1. Disconnect the upper end of the shock absorber, raise the front end of the car and disconnect the lower end of the shock absorber, then remove the shock absorber.

2. Disconnect the stabilizer bar at the lower control arm, if so equipped.

3. Remove the wheel, brake drum, or caliper and rotor. Do not allow the brake hose to support the weight of the caliper. Use wire to support it from the frame.

4. Remove the two bolts attaching the steering arm to the steering knuckle and move the steering arm aside.

5. Install a spring compressor and compress the spring.

6. Remove the cotter pin and nut from the lower ball joint stud. Remove the ball joint from the steering knuckle using a ball joint removal tool.

7. Move the steering knuckle assembly out of the way. Support the assembly with wire from the upper control arm.

8. Remove the two pivot bolts that attach the lower arm to the front crossmember and remove the lower control arm.

9. Install the lower control arm in the reverse order of removal, tightening the ball joint stud nut to 75 ft. lbs., the steering arm attaching bolts to 55 ft. lbs., the shock absorber lower attaching nuts to 20 ft. lbs, the stabilizer bar locknut to 8 ft. lbs, and lastly, after the car has been lowered to the ground with the wheel and tire installed, tighten the lower control arm pivot bolts to 110 ft. lbs.

Ball Joints

INSPECTION

Except Pacer

NOTE: Before checking the upper ball joint, make certain the front wheel bearings are adjusted to specifications.

1. Jack up the front of the car and place jackstands under the frame side sills. The control arms must hang free if an accurate reading is to be obtained.

2. Check the lower ball joints by grasping the lower portion of the wheel and pulling it in and out.

3. If there is noticeable lateral freeplay, the lower ball joint is worn and must be replaced.

NOTE: On Eagles the lower ball joints and control arms must be replaced as assemblies.

10. Connect the stabilizer bar to the arm; tighten the bolts to 7 ft. lbs. Install the strut rod; tighten the bolts to 75 ft. lbs.

11. Install the halfshaft-to-axle flange bolts; tighten to 45 ft. lbs.

12. Place a jack under the lower control arm. Raise the jack carefully to compress the spring slightly. Tighten the control arm pivot bolt to 110 ft. lbs.

13. Install the rotor, caliper, and hub nut. Tighten the hub nut to 180 ft. lbs. Install the nut lock and a new cotter pin.

14. Install the wheel. Check and adjust the front end alignment as necessary.

4. To check the condition of the upper ball joint, place a dial indicator with its plunger against the tie scrub bead (just outside the whitewall).

5. Move the upper portion of the wheel and tire toward the car's center, while watching the dial indicator.

6. Move the wheel and tire back out while watching the indicator.

7. The upper ball joint should be replaced if its total movement is greater than 0.160 in.

NOTE: On 1980 Eagles the upper ball joints and control arms must be replaced as assemblies. On 1981 and later Eagles the upper ball joints are replaceable separately.

Pacer

1. Check that the front wheel bearings are adjusted properly.

2. Remove the lubrication plug from the lower ball joint. Insert a straight piece of stiff wire until it contacts the ball. Mark the wire even with the edge of the plug hole.

3. Measure from the end of the wire to the mark. If it exceeds $7/16$ in., the ball joint should be replaced.

4. Place a jack under the lower control arm and lift the wheel off the floor.

5. Push the top of the tire in and out. If there is any looseness, replace the upper ball joint.

6. Pry the upper control arm up and down. If there is any looseness, replace the upper ball joint.

REMOVAL & INSTALLATION

Lower Ball Joint

1. On all vehicles except Pacer, place a 2 x 4 x 5 in. block of wood on the side sill so that it supports the control arm.

2. Jack up the front end of the car and place jackstands underneath the frame side sills to support the body.

3. Remove the wheel, the caliper and rotor.

4. Disconnect the lower control arm strut rod, on models other than Pacer. Disconnect the stabilizer bar, if so equipped.

5. Separate the steering arm from the steering knuckle.

6. Remove the ball stud retaining nut, after removing its cotter pin.

7. Install a ball joint removal tool, then loosen the ball stud in the knuckle pin. Leave the tool in place on the stud.

8. Place a jackstand under the lower control arm.

9. Chisel the heads off the rivets which secure the ball joint to the con-

trol arm. Use a punch to remove the rivets.

10. Remove the tool from the ball stud.

11. Remove the ball stud from the knuckle pin and remove the joint from the control arm.

12. Position the new ball joint so that its securing holes align with the rivet holes in the control arm.

13. Loosely install the special $5/16$ in. bolts used to secure the ball joint.

CAUTION

Use only the hardened $5/16$ in. bolts supplied with the ball joint replacement kit; standard bolts are not strong enough.

---

14. Install the steering strut and stop on the lower control arm. Tighten their bolts to 75 ft. lbs.

15. Apply chassis grease to the steering stops and fit the knuckle pin and retaining nut on the ball stud; tighten the nut to 75 ft. lbs. Install a new cotter pin.

16. Complete the installation procedure in the reverse order of removal, then check front end alignment.

Upper Ball Joint

1. Perform Steps 1–3 of the "Lower Ball Joint Removal" procedure.

NOTE: On 1981 and later Eagle models temporarily reinstall two lugnuts to retain each brake rotor. This eliminates repositioning rotors and calipers prior to reassembly.

2. Perform Steps 6–9 of the "Lower Ball Joint Removal" procedure to the upper ball joint.

3. Separate the upper ball joint from the control arm.

4. Remove the ball joint puller from the knuckle pin.

5. Perform Steps 1–2 of the "Lower Ball Joint Installation" procedure.

6. Skip Step 3 and go on to Steps 4–5 of the "Lower Ball Joint Installation" procedure.

7. Complete the installation in the reverse order of removal and check front end alignment.

Wheel Bearings

Four wheel drive models have sealed, nonadjustable front hubs and bearings. There are darkened areas surrounding the bearing races in the hubs. These darkened areas are the result of a heat treatment process, they do not signify defects.

INSPECTION

Check to see that the inner cones of the bearings are free to "creep" on the

spindle. Polish and lubricate the spindle to allow "creeping" movement and to keep rust from forming.

ADJUSTMENT

1. With the tire and wheel removed and the car supported by a suitable and safe means, remove the dust cover from the spindle.

2. Remove the cotter pin and nut retainer.

3. Rotate the wheel while tightening the spindle nut to 20–25 ft. lbs.

4. Loosen the spindle nut $1/3$ of a turn.

5. Rotate the wheel while tightening the spindle nut to 6 inch lbs.

6. Fit the nut retainer over the spindle and align the slots in it with the cotter pin hole. Insert the cotter pin.

7. Install the dust cover.

REAR SUSPENSION

All models use a four or five-leaf semi-elliptic spring and live axle rear suspension. Shock absorbers are mounted at their lower ends to studs and are bayonet or stud type at their upper ends. Upper shock nuts are accessible by removing cover plates or by removing trunk floormat on some models, or by removing underbody brackets bolted to the trunk pan on others, such as on the Pacer and Concord.

Shock Absorber

REMOVAL & INSTALLATION

NOTE: When installing new shocks purge them of air by repeatedly extending them in their normal position and compressing them while inverted. It is normal for new shocks to be more resistant to extension than to compression.

1. Support the rear axle with jacks or a lift; this allows the weight of the car to compress the rear spring.

2. Remove the lower shock attachment.

3. Remove the access plate on the rear underbody panel, and remove the upper securing nut. It may be necessary to hold the top of the shock while unfastening the nut.

NOTE: Some models do not have an access plate. On these cars, remove the upper attach-

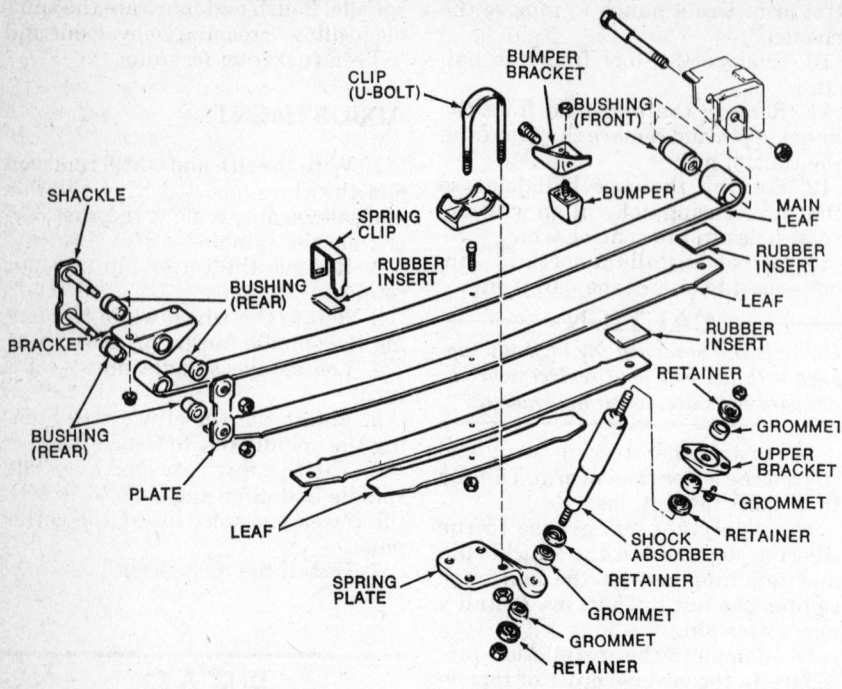

CLIP (U-BOLT)
BUMPER BRACKET
BUSHING (FRONT)
BUMPER
MAIN LEAF
SHACKLE
SPRING CLIP
RUBBER INSERT
RUBBER INSERT
BUSHING (REAR)
LEAF
BRACKET
RUBBER INSERT
BUSHING (REAR)
RETAINER
GROMMET
UPPER BRACKET
PLATE
GROMMET
LEAF
RETAINER
SHOCK ABSORBER
RETAINER
SPRING PLATE
GROMMET
GROMMET
RETAINER

Typical leaf spring rear suspension

ment plate complete as an assembly, from under the car.

4. Remove the shock from under the car.

5. Installation is in the reverse order of removal.

Spring

REMOVAL & INSTALLATION

1. Raise the car. Support the rear axle with jacks, or a lift, to take the load off the rear springs.

2. Disconnect the rear shock from the lower mounting stud.

3. Disconnect the axle U-bolts.

4. Remove the nut from the bolt which attaches the eye of the spring to the front mount. Remove the bolt.

5. On all models except Pacer, remove the nuts from the rear shackle. Remove the shackle.

6. On the Pacer, remove the nuts from the rear hanger bracket on the frame side sill and remove the spring. Remove the shackle nuts and the shackle after the spring is removed.

7. Installation is in the reverse order of removal.

BRAKES

All American Motors cars are equipped with tandem (dual reser-

voir) master cylinders. This allows one set of brakes to operate, should the other set fail. A switch in the system, connected to a warning light on the instrument panel, indicates a difference in pressure between the front and rear brake lines, thus signaling the failure of one brake system. Repair procedures for both the master cylinder and the switch are found in the Unit Repair Section.

All drum brakes have automatic brake adjusters. These compensate for lining wear by operating when the brakes are applied while the car is backing up. The automatic adjusting mechanism functions through the star wheel.

For additional brake system repair and service procedures, please refer to "Brakes" in the Unit Repair section.

Master Cylinder

REMOVAL & INSTALLATION

1. Disconnect the front and rear brake lines from the master cylinder. Both outlets must be plugged, to prevent fluid loss.

2. Remove the nuts which attached the master cylinder to the firewall or the power brake booster, if so equipped. On Pacers, remove the mounting bracket and the boot retainer plate.

3. On cars without power brakes,

disconnect the pedal push-rod at the brake pedal.

4. Remove the master cylinder from the car.

5. Installation is in the reverse order of removal. Remember to bleed the brake system once the master cylinder has been installed. See the Unit Repair section.

Power Brake Unit

REMOVAL & INSTALLATION

Disconnect the power brake clevis pin from the power unit operating rod at the linkage under the hood, or from the brake pedal inside the car, depending on which type is being serviced. Remove the vacuum hose from the check valve. Separate the master cylinder from the power unit. Do not disconnect the hydraulic lines from the master cylinder. Remove the power unit mounting bolts, and lift the unit from the car. Installation is in the reverse order of removal.

Parking Brake

ADJUSTMENT

1. To adjust the drum brakes, apply the brakes several times while backing up. Make one forward application for each reverse application to equalize the adjustment. Fully apply the parking brake about 10 times. Set the pedal on the first notch from the released position.

2. Block the front wheels and raise the rear wheels.

3. Tighten the cable at the equalizer so the wheels can just barely be turned forward. Be sure to hold the end of the cable screw to prevent the cable from turning.

4. Release the parking brake and check for rear brake drag. The wheels should rotate freely with the parking brake off.

STEERING

All models except the Pacer use Acklerman-type articulated linkage to interconnect the steering gear and front wheels. Pacers use rack and pinion steering with integral linkage to the front wheels.

Tie Rod End

REMOVAL & INSTALLATION

1. Raise and support the front of the car.

2. Remove the cotter pin and retaining nut from the tie rod end stud.

3. Mark the position of the tie rod end adjuster tube and inner tie rod, for reference.

4. Loosen the adjuster tube clamps.

5. Disconnect the tie rod end from the steering arm with a puller.

6. Remove the tie rod end from the adjuster tube. Count the number of turns so that the replacement part may be installed in the same position.

7. Install the replacement tie rod end in the adjuster tube, and insert the end stud in the steering arm. Tighten the nut to 35 ft. lbs, and install a new cotter pin. Do not loosen the nuts to align. Adjust the toe-in and tighten the clamps.

Manual Steering Gear

REMOVAL & INSTALLATION

Except Pacer

1. Remove flexible coupling bolts.
2. Remove Pitman arm, using puller J-5566-04 or equivalent.
3. Remove steering gear mounting screws and lower the steering gear.
4. Center steering gear with index mark up. Mark on shaft of flange must be aligned at assembly.
5. Insert the flexible coupling bolts into shaft flange. Tighten nuts to 20 ft. lbs. torque and pinch bolt to 30 ft. lbs.
6. Tighten gear mounting screws to 65 ft. lbs. and Pitman arm nut to 115 ft. lbs.

NOTE: After tightening Pitman shaft nut, stake thread at nut with a center punch to insure nut retention. Whenever the steering gear assembly is removed for replacement or overhaul, or the mounting bolts are loosened for any reason, the steering column MUST be realigned to the gear assembly. Slight misalignment of the steering column may cause increased steering effort and additional wear to the steering components.

Pacer

1. Unlock steering column.
2. Raise and support front of car.
3. Remove screws attaching reinforcement brace to front crossmember and left engine support bracket and remove brace.
4. Remove flexible coupling pinch bolt and disengage flexible coupling from steering gear pinion shaft.
5. Remove cotter pins and nuts from tie rod ends.
6. Disconnect tie-rod ends using tool J-26951.

7. Remove bolts attaching steering gear mounting clamp to right side of front crossmember.

NOTE: Before removing bolts, loosen them slightly to minimize clamp distortion.

8. Remove steering gear housing-to-crossmember nuts. Using a blunt punch remove bolts, washers, sleeves, and grommets.

9. Rotate bottom of gear housing toward front of car until pinion shaft is approximately parallel with skid plate. Slide gear assembly toward right side of car until housing and tube clear mounting plate and remove steering gear assembly.

10. Assemble grommets, sleeves, and washers and install on steering gear. Sleeves will hold grommets in place during assembly.

11. Position steering gear assembly on crossmember. Install tube and housing from right side of car. During installation, keep pinion shaft approximately parallel with mounting plate.

12. Install mounting clamp-to-crossmember attaching bolts. Hand-tighten bolts only.

13. Install steering gear housing-to-crossmember attaching bolts, washers, and nuts and tighten to 60 ft. lbs. torque.

14. Tighten mounting clamp-to-crossmember attaching bolts to 50 ft. lbs. torque.

15. Connect tie rod ends to steering arms. Tighten nuts to 50 ft. lbs. torque and install replacement cotter pins.

16. Align flat spline on pinion shaft with flat on flexible coupling and install coupling on shaft. Install pinchbolt and tighten to 30 ft. lbs. torque.

17. Install bolts attaching reinforcement brace to front crossmember and engine support bracket. Torque bolts to 30 ft. lbs.

18. Remove supports and lower car.

19. Check and correct toe-in adjustment, if necessary.

Power Steering Gear

REMOVAL & INSTALLATION

Except Pacer

1. Place wheels in straight-ahead position.
2. Position drain pan under steering gear.
3. Disconnect hoses at gear. Raise and secure hoses above pump fluid level to prevent excessive oil spillage and cap ends of hoses to keep out dirt.

4. Remove flexible coupling-to-intermediate shaft attaching nuts.

5. Raise the car and support it safely. On Eagle models, remove the skid plate, if so equipped; the left side crossmember-to-still support brace; and the stabilizer bar brackets from the frame.

6. Paint alignment marks on Pitman arm and Pitman shaft for assembly reference.

7. Remove Pitman arm using Puller Tool J-5566-04 or equivalent.

8. Remove steering gear mounting bolts and remove steering gear.

9. Center steering gear. Turn stub shaft (using flexible coupling) from stop-to-stop and count total number of turns; then turn back from either stop one-half total number of turns to center gear. At this point, flat on stub shaft should be facing upward.

10. Align flexible coupling and intermediate shaft flange.

11. Install gear mounting bolts in gear, install spacer on gear, and mount gear on frame side-sill. Tighten gear mounting bolts to 65 ft. lbs. torque.

12. Install and tighten flexible coupling nuts to 25 ft. lbs. torque.

13. Install Pitman arm, Index arm to shaft using alignment marks made during removal.

14. Install Pitman arm nut. Tighten nut to 115 ft. lbs. torque and stake nut to Pitman shaft in one place.

CAUTION

The Pitman arm nut must be staked to the shaft to retain it properly.

15. On Eagle models, install the stabilizer bar brackets, left side crossmember-to-sill support brace and skid plate, if so equipped. Lower the vehicle.

16. Align flexible coupling, if necessary.

17. Connect hoses to gear and tighten fittings to 25 ft. lbs. torque.

18. Fill pump reservoir with power steering fluid and bleed air from system.

Pacer

1. Unlock steering column.
2. Raise and support front of car.
3. Remove screws attaching reinforcement brace to front crossmember and left engine support bracket and remove brace.
4. Disconnect stabilizer bar at left-side lower control arm, if so equipped.
5. Remove bolts attaching stabilizer bar mounting clamps to frame rail brackets and move bar away from front crossmember.
6. Place support under stabilizer bar to prevent damaging the bolt at-

taching the bar to the right-side lower control arm.

7. Remove left-side frame rail clamp brackets.

8. Position drain pan under steering gear housing and disconnect power steering hoses at gear housing. Cap hoses and plug gear housing to keep dirt out.

9. Remove flexible coupling pinchbolt and disengage flexible coupling from steering gear pinion shaft.

10. Remove and discard cotter pins from tie rod and retaining nuts.

11. Disconnect tie rod ends using tool J-27951.

12. Remove bolts attaching steering gear mounting clamp to crossmember.

———— CAUTION ————
Before removing the bolts, loosen them slightly to minimize clamp distortion.

13. Remove steering gear housing-to-crossmember attaching bolt nuts.

14. Using a blunt punch remove bolts, washers, sleeves and grommets from steering gear housing.

15. Rotate bottom of gear housing toward front of car until pinion shaft is approximately parallel with skid plate. Slide gear assembly toward right side of car until housing and tube clear mounting plate and remove steering gear assembly.

16. Assemble and install grommets, sleeves, and washers on steering gear. Sleeves will hold grommets in place during assembly.

17. Position steering gear assembly on crossmember. Install gear from driver side of car.

NOTE: When installing the gear, keep the pinion shaft approximately parallel to the mounting plate.

18. Install mounting clamp bolts, and hand-tighten them only.

19. Install steering gear housing-to-crossmember attaching bolts, washers, and nuts and tighten to 60 ft. lbs. torque.

20. Tighten mounting clamp bolts to 48 ft. lbs. torque.

21. Connect tie rod ends to steering arms. Tighten tie rod end retaining nuts to 50 ft. lbs. torque and install replacement cotter pins.

22. Align flat spline of steering gear pinion shaft with index flat of flexible coupling and install coupling on pinion shaft. Install pinchbolt and tighten to 30 ft. lbs. torque.

23. Install bolts attaching reinforcement brace to front crossmember and engine support bracket. Tighten bolts to 30 ft. lbs. torque.

24. Connect power steering hoses to gear housing. Be sure that hoses do not touch brace or crossmember.

25. Install left-side frame rail clamp brackets.

26. Position stabilizer bar on frame rail brackets and install mounting clamp bolts. Hand tighten bolts only.

27. Install bolts, washers, and grommets attaching stabilizer bar to left-side lower control arm.

28. Tighten stabilizer bar mounting clamp bolts to 18 ft. lbs. torque.

29. Remove supports and lower car.

30. Fill power steering pump reservoir with power steering fluid.

31. Operate engine until fluid reaches normal operating temperature. Turn wheel right and left several times (do not hold wheel against steering stops). Stop engine and check fluid level. Add fluid as necessary.

Power Steering Pump

REMOVAL & INSTALLATION

1. Remove the fan belt.

2. Place a container under the pump to catch fluid. Remove the fuel vapor storage canister and six cylinder air cleaner, if necessary.

3. Disconnect the hoses and cap the outlets, so the power steering unit does not lose fluid. Remove the air pump belt.

4. On sixes with air conditioning, loosen the idler pulley adjusting bolt and idler pulley, air pump adjusting strap mounting bolt and remove the compressor drive belt from the idler pulley. Loosen the two nuts that attach the upper leg of the aluminum idler pulley mounting bracket to the cylinder head and remove the bolt that attaches the lower leg of the mounting bracket to the engine front cover.

5. On sixes, remove the nut from the air pump mounting stud, remove the power steering pump to engine front cover front adapter plate (do not unbolt the adapter plate from the pump), remove the long adjusting bolt that passes through the adapter plate, and remove the bolt hidden behind the flange in the rear adapter plate. Remove the pump, adapter plate and mounting bracket together.

On four cylinder models, remove the adjuster locknuts and washers which retain the pump and pivot bracket on the mounting bracket. All of the pump mounting bolts are metric, except for the $9/16$ in. adjuster locknuts. Move the pump and remove the belt. Remove the bolts which connect the front bracket to the rear bracket and engine block, and remove the pump complete with the pivot and front brackets.

6. After installation, fill the system with Dexron® AMC power steering fluid. Bleed the system of air by raising the front of the car, and by turning the wheels from side to side several times, without hitting the stops. Check the level frequently.

Steering Wheel

REMOVAL & INSTALLATION

1. Disconnect the battery and remove the horn button by one of the following methods:

    a. center button—lift upward

    b. trim cover—from the back side of the steering wheel remove the screws which hold the cover on. On "rim-blow" wheels, remove the center contact.

2. Remove the steering wheel center nut and washer. Before removing the wheel, note the position of the index marks on the wheel and the steering shaft. If none are present, paint an alignment mark on the shaft and wheel.

3. Remove the wheel with a puller. Installation is in the reverse order of removal. Tighten the steering wheel nut to 20 ft. lbs.

NOTE: Some shafts have metric threads. These can be identified by a groove in the shaft splines. Metric nuts are coded blue.

———— CAUTION ————
Do not hammer on the end of the steering shaft; hammering could shear the plastic retainers which maintain the rigidity of the energy-absorbing steering column.

Turn Signal Switch, Hazard Signal and Lock Cylinder

REMOVAL & INSTALLATION

1. Disconnect the ground cable from the battery. On cars with tilt steering wheels, place the column in the straight position. Remove the steering wheel.

2. Loosen the anti-theft cover attaching screws and remove the cover from the column. Do not hammer on the shaft. Do not remove the screws from the cover; they are attached to it with plastic retainers.

3. To remove the lockplate, a special compressing tool is required. This tool is an inverted U-shape with a hole for the shaft. The shaft nut is used to force it down. Depress the lockplate and pry the snap-ring from the groove in the steering shaft. Re-

move the tool, snap-ring, plate turn signal cam, upper bearing preload spring, and the thrust washer from the shaft.

4. Place the turn signal lever in the Right turn position and remove it.

5. Depress the hazard warning switch button and remove it by rotating it counterclockwise. Remove the package tray (if so equipped) and the lower trim panel.

6. Disconnect the wire harness connector block at its mounting bracket, which is located on the right side of the lower column. Remove the steering column mounting bracket attaching bolts. Remove the turn signal switch wiring harness protector from the bottom of the column.

NOTE: To ease the removal and replacement of the directional switch harness, tape the harness connector to the wire harness. This will prevent snagging while removing the wiring harness assembly through the steering column. Prepare the new turn signal switch harness in the same manner to ease installation.

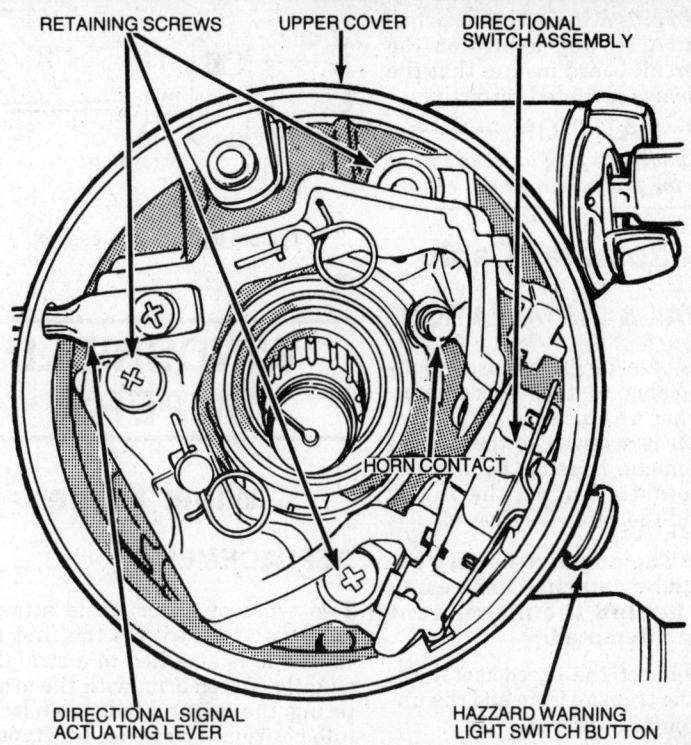

Turn signal switch

7. If the car (Concord and Spirit only) is equipped with a column-mounted automatic transmission selector, use a paper clip to depress the locktab that holds the shift quadrant light wire in the connector block.

8. Remove the switch attaching screws. Withdraw the switch and wire harness from the column.

9. Insert the key into the lock cylinder and turn the key to the ON position. Remove the warning buzzer switch and the contacts as an assembly using needlenose pliers. Take care not to let the contacts fall into the column.

10. Turn the key to the LOCK position and compress the lock cylinder retaining tab. Remove the lock cylinder. If the tab is not visible through the slot, knock the casting flash out of the slot.

To install:

1. Hold the lock cylinder sleeve and turn the lock cylinder counterclockwise until it contacts the stop.

2. Align the lock cylinder key with the keyway in the housing and slip the cylinder into the housing.

3. Lightly depress the cylinder against the sector, while turning it counterclockwise, until the cylinder and sector are engaged.

4. Depress the cylinder until the retaining tab engages, and the lock cylinder is secured.

5. In installing the turn signal switch, don't screw it in place until you are sure the actuating lever pivot

is properly seated and aligned in the top of the housing boss.

6. Install the turn signal lever and check the operation of the switch.

7. Install the thrust washer, spring and turn-signal cancelling cam on the steering shaft.

8. Align the lockplate and steering shaft splines, and position the lockplate so the turn signal camshaft protrudes from the "dogleg" opening in the lockplate.

9. Use snap-ring pliers to install the snap-ring on the end of the steering shaft.

10. Secure the anti-theft cover with its screws.

11. Install the button on the hazard warning switch. Install the steering wheel as detailed above.

Ignition Switch

REMOVAL & INSTALLATION

The ignition switch on all models is mounted on the lower steering column tube and is connected to the lock cylinder via a lock rod.

1. Place the key in OFF-LOCK position.

2. Remove switch mounting screws.

3. Disconnect the lock rod, remove harness connector and switch.

4. To install on the standard column, move the switch slide as far as it will go to the left (toward the wheel).

On the tilt-column, push the slide to the extreme right.

5. Position the lock rod into the hole on the switch slide.

6. Install the switch on the steering column. Be sure that the slide stays in its detent.

7. On the tilt-column, do not tighten the mounting screws. Instead, push the switch down the column, away from the steering wheel. This will remove any slack from the lock rod.

8. Tighten the switch mounting screws.

INSTRUMENT PANEL

To remove the various units from the instrument panel it is necessary to remove the bezels, overlays, housings, and crash pads. Numerous fasteners are hidden. Caution must be exercised not to damage or break the panel trim.

Current is supplied to the instruments and the instrument panel lights through a printed circuit which is attached to the rear of the instrument cluster. The disconnect plug is part of the panel wiring harness and connects to pins attached to the printed circuit. A keyway located on the printed circuit board insures that the plug is always mounted correctly.

— CAUTION —
Never pry under the plug to remove it, or damage to the printed circuit will result.

Speedometer Cable

REMOVAL & INSTALLATION

Two types of fasteners are used to attach the cable to the speedometer. One type has a knurled round captive nut, which is screwed to the rear of the speedometer housing. By depressing the plastic finger, the lug is raised, and the cable is released.

NOTE: The negative battery cable should be detached before any repairs behind the instrument panel are attempted.

1. Disconnect the speedometer cable from the transmission and the underbody routing brackets.
2. Disconnect the negative battery cable.
3. Remove the package tray, if so equipped.
4. If the speedometer cable connection at the rear of the speedometer can not be reached from under the dash, remove the instrument cluster bezel.
5. Remove the headlight switch overlay cover, if necessary for clearance.
6. Unscrew the speedometer cable, and remove the cable and the grommet from the dash panel.
7. Installation is in the reverse order of removal.

Headlight Switch

REMOVAL & INSTALLATION

Light switches are similar in all models. Some variation occurs in the shape and position of the nut mounting the switch on the dash.

1. Disconnect battery and remove the switch overlay cover attaching screws so the cover can be pulled forward.
2. With the switch in the ON position, press the release button on the switch and remove the knob and shaft.
3. Remove screws that attach switch or bracket to panel.
4. Reverse the above procedure for installation, positioning switch so that the shaft is lined up properly before tightening the bracket screws.

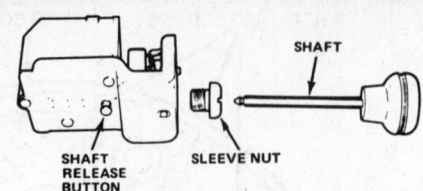

Typical light switch assembly

WINDSHIELD WIPERS

Wiper Blade

REPLACEMENT

Two types of wiper blade attaching methods are used. On the first type, the blade is attached to a straight or slightly curved arm, with the arm entering the wiper blade and locking into position. To release this type, depress the locking tab and remove the blade from the arm. The second type of blade is attached to a pin at a right angle to the arm. To release the pin, a tool is inserted into the wiper blade saddle to depress the spring clip and release the pin.

Wiper Motor

REMOVAL & INSTALLATION

Concord, Eagle and Spirit

1. Remove the wiper arms and blades.
2. Remove the screws holding the motor adapter plate on the dash panel.
3. Separate the wiper wiring harness connector at the motor.
4. Pull the motor and linkage out of the opening to expose the drive link-to-crank stud retaining clip. Raise up the lock tab of the clip with a screwdriver and slide the clip off the stud. Remove the wiper motor assembly.
5. Install the windshield wiper motor in the reverse order of removal.

Pacer

1. Remove the vacuum canister bracket and canister, if so equipped.
2. Disconnect the linkage drive arm from the motor output arm crankpin by removing the retaining clip.
3. On vehicles equipped with air conditioning:
   a. Remove the two nuts on the left side of the heater housing.

   b. Remove the one nut on the right side of the heater housing.
   c. Remove the screw from the heater housing support.
4. On vehicles not equipped with air conditioning:
   a. Remove the two nuts and one screw on the left side of the heater housing.
   b. Remove the one nut on the right side of the heater housing.
   c. Remove the screw from the heater housing support. Pull the heater housing forward.
5. Remove the wiper motor mounting plate attaching screws and remove the wiper motor assembly from the cowl.
6. Disconnect the two wire connectors from the wiper motor.
7. Remove the wiper motor attaching screws and remove the wiper motor.
8. Install the wiper motor following the reverse order of removal.

RADIO

The following precautions should be observed when working on a car radio:

1. Always observe the proper polarity of the power connections; i.e., positive (+) goes to the power source and negative (−) to ground (negative ground electrical system).
2. Never run the radio without a speaker; damage to the output transistors will result. If a replacement (or additional) speaker is used, be sure that it is of the correct impedance (ohms) for the radio. The proper impedance is stamped on the case of American Motors radios.
3. If a new antenna or antenna cable is used, adjust the antenna trimmer for the best reception of a weak AM station around 1400kc; the trimmer is located behind or above the tuning knob or in the radio case near the antenna lead. On tape player radios, it is in the cartridge slot.

REMOVAL & INSTALLATION

Concord, Spirit and Eagle

1. Disconnect the battery ground cable.
2. Pull off the radio knobs and remove shaft retaining nuts.
3. Remove the bezel retaining screws and remove the bezel. On models with A/C, remove the center housing of the instrument panel.
4. Disconnect the speaker, anten-

na, and power leads, and remove the radio.

5. Installation is the reverse of the above.

Pacer

1. Disconnect the negative battery cable.

2. Remove the radio knobs, attaching nuts, cluster bezel, and overlay cover.

3. Loosen the radio-to-instrument panel attaching the screw.

4. Lift the rear of the radio and pull forward slightly. Disconnect the electrical connections and the antenna and remove the radio.

5. Installation is the reverse of the above.

HEATER

NOTE: It is recommended, unless you are trained in air conditioning servicing procedures, that you do not disconnect any of the air conditioning refrigerant lines.

Heater Core

REMOVAL & INSTALLATION

Concord, Spirit, and Eagle

1. Disconnect the negative battery cable and drain 1 qt. of coolant.

2. Disconnect heater hoses and plug hoses and corer fittings.

3. Disconnect blower wires and remove motor and fan assembly.

4. Remove the housing attaching nut from the stud in the engine compartment.

5. Remove package shelf, if so equipped.

6. Disconnect wire at resistor, located below glove box.

7. Remove instrument panel center bezel, air outlet and duct, on A/C models.

8. Disconnect air and defroster cables from damper levers.

9. Remove right-side windshield pillar molding, the instrument panel upper sheet metal screws and the capscrew at the right door post.

10. Remove the right cowl trim panel and door sill plate.

11. Remove right kick panel and heater housing attaching screws.

12. Pull right side of instrument panel outward slightly and remove housing.

13. Remove core, defroster and blower housing.

14. Remove core from housing.

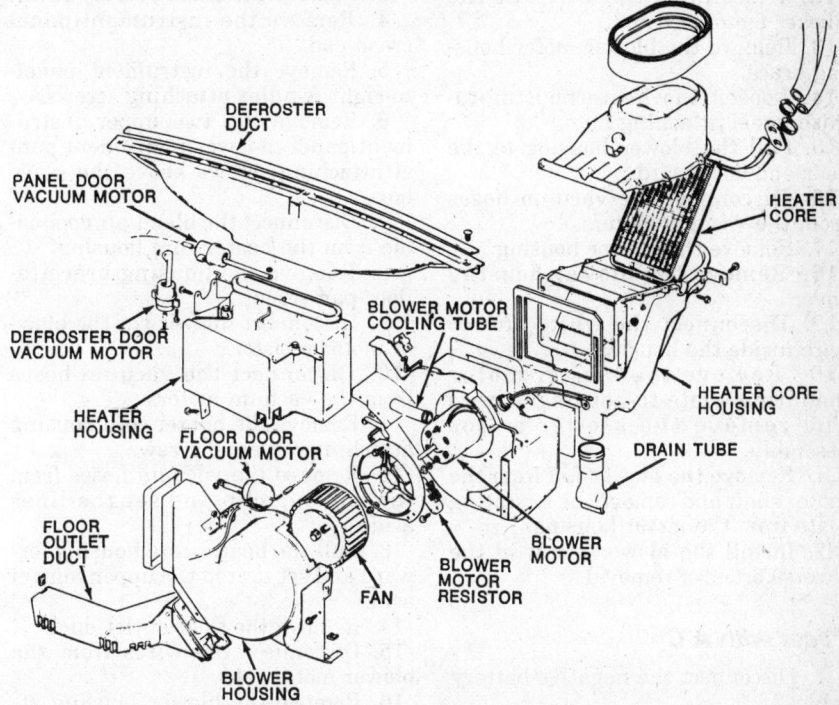

Pacer heater assembly—without air conditioning

15. Installation is in the reverse order of removal.

Pacer

1. Disconnect the negative battery cable. Drain about 2 quarts of coolant from the radiator.

2. Disconnect the heater hoses from the heater core tubes and install plugs in the heater hoses and core tubes.

3. Remove the vacuum hoses from the heater core housing cover clip and move the lines aside. With A/C, disconnect the outside air door vacuum hose from the vacuum motor.

4. Remove the heater core housing cover screws and disconnect the air blend door cable.

5. Disconnect the overcenter spring from the cover and remove the cover.

6. Remove the heater core-to-housing attaching screws and remove the heater core.

7. Install the heater core in the reverse order of removal.

Heater Blower

REMOVAL & INSTALLATION

Concord, Spirit and Eagle

1. Drain about 2 quarts of coolant from the radiator.

2. Disconnect the heater hoses from the heater core tubes and plug the core tubes.

3. Disconnect blower wires.

4. Remove retaining nut from cover and remove motor and fan assembly.

5. To install, reverse removal procedure.

Pacer without A/C

1. Disconnect the negative battery cable.

2. Remove the right side windshield finish molding.

3. Remove the instrument panel crash pad.

4. Remove the right scuff plate and cowl trim panel.

5. Remove the lower instrument panel-to-right A-pillar attaching screws.

6. Pull the instrument panel to the rear and replace the lower attaching screw in the right A-pillar. Allow the instrument panel to rest on the screw.

7. Remove the heater core housing attaching nuts and screw.

8. Remove the vacuum hoses from the heater core housing clip and set the lines aside.

9. Disconnect the blend-air door cable from the heater core housing.

10. Pull the heater core housing forward and set atop the upper control arm.

11. Remove the blower motor ground wire. Remove the blower motor housing attaching screw.

12. Disconnect the wires at the blower motor resistor.

13. Remove the blower motor housing brace.

14. Loosen the heaterhousing-to-dash panel attaching nuts.

15. Pull the blower housing to the rear and downward.

16. Disconnect the vacuum hoses from the vacuum motors.

17. Remove the blower housing.

18. Remove the blower housing cover.

19. Disconnect the white blower wire inside the housing.

20. Remove the blower motor mounting plate-to-housing screws and remove the blower motor assembly.

21. Remove the blower fan from the motor shaft and remove the mounting plate from the motor housing.

22. Install the blower motor in the reverse order of removal.

Pacer with A/C

1. Disconnect the negative battery cable.

2. Remove the right scuff plate and cowl trim panel.

3. Remove the radio overlay cover.

4. Remove the instrument panel crash pad.

5. Remove the instrument panel-to-right A-pillar attaching screws.

6. Remove the two upper instrument panel-to-lower instrument panel attaching screws above the glove box.

7. Disconnect the blend-air door cable from the heater core housing.

8. Remove the housing brace-to-floor-pan screw.

9. Disconnect the wire at the blower motor resistor.

10. Disconnect the vacuum hoses from the vacuum motors.

11. Remove the heater core housing attaching nuts and screw.

12. Remove the vacuum hoses from the housing clip and set the lines aside.

13. Pull the heater core housing forward and set it atop the upper control arm.

14. Remove the floor outlet duct.

15. Disconnect the wires from the blower motor relay.

16. Remove the blower housing attaching screw located in the engine compartment on the firewall.

17. Loosen the evaporator housing-to-firewall panel attaching nuts.

18. Remove the blower housing to firewall attaching screw.

19. Pull the blower housing to the rear and downward.

20. Pull the right side of the instrument panel to the rear and remove the blower housing from under the panel.

21. Remove the floor door vacuum motor attaching screws and motor to gain access to the blower housing cover attaching screws.

22. Remove the blower housing cover attaching screws and remove the cover.

23. Remove the blower motor mounting plate and remove the blower motor assembly.

24. Remove the blower fan from the motor shaft and the mounting plate from the body of the motor.

25. Install the motor in the reverse order of removal.

FUSE BOX LOCATION

- Pacer: On the left of the glove box
- All other models: Next to the parking brake mechanism

FUSIBLE LINKS

Location	Color	Circuit Protected	Years Used		
			1980–81	1982–83	1984–87
Starter relay battery terminal to main wire harness	Red	Complete wiring	X	X	X
Horn relay battery terminal to main wire harness	Pink	Horn circuit	X	X	X
Starter relay battery terminal to rear window defogger switch	Red	Rear window defogger, (1980–85 headlights)	X	X	X
Battery terminal of starter to harness	Pink	Rally Package Ammeter	X	X	
Block terminal to harness	Pink	Rally Package Ammeter			
Harness to dash conn.	Red	Headlight switch	X	X	
Battery terminal of starter solenoid to main wiring	Red	Ign./power option eng. comp. lamp, deck lid release			X
Battery terminal of start solenoid to main wiring	Red	Power Options			X
Alternator to battery terminal of starter solenoid	Red	Alternator output wire			X

Chrysler Corp.
Front Wheel Drive
Aries, Charger, Caravelle, Daytona, E Class, Horizon, Lancer, Laser, LeBaron GTS, New Yorker, Omni, Reliant, TC3, Turismo, Sundance, Shadow, O24, 400, 600,

YEAR IDENTIFICATION

1980 Omni

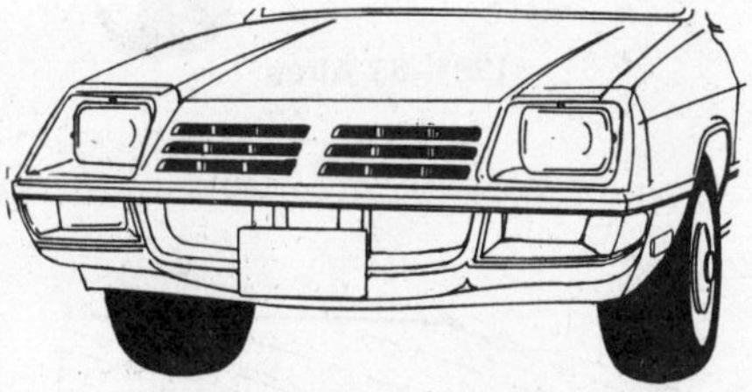

1980–83 Omni 024, Charger 2.2

1981 Omni

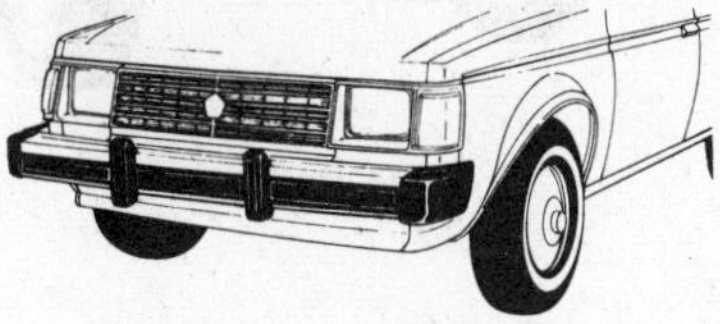

1982–83 Omni

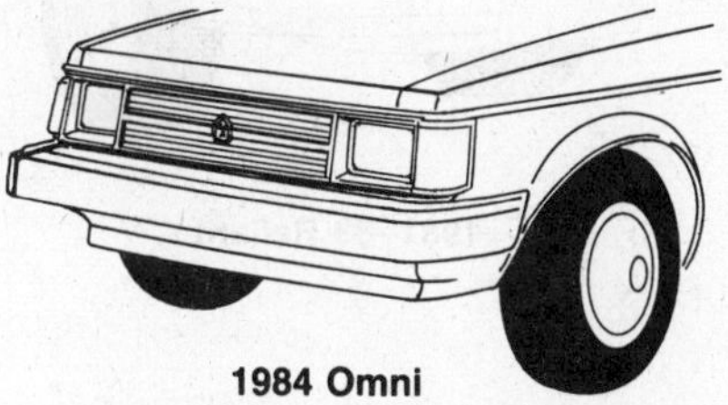

1984 Omni

1985–86 Omni GLH

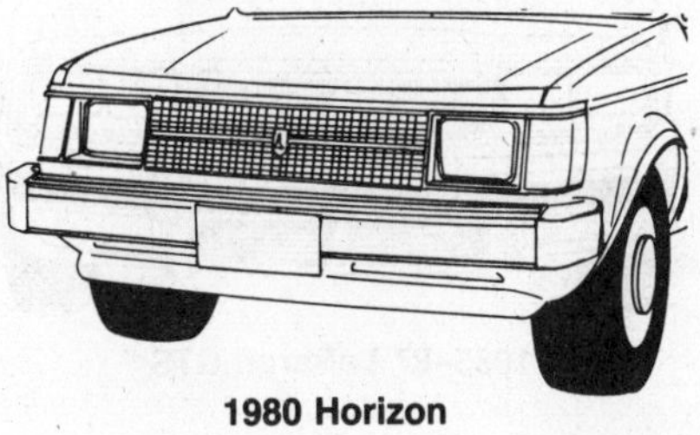

1980 Horizon

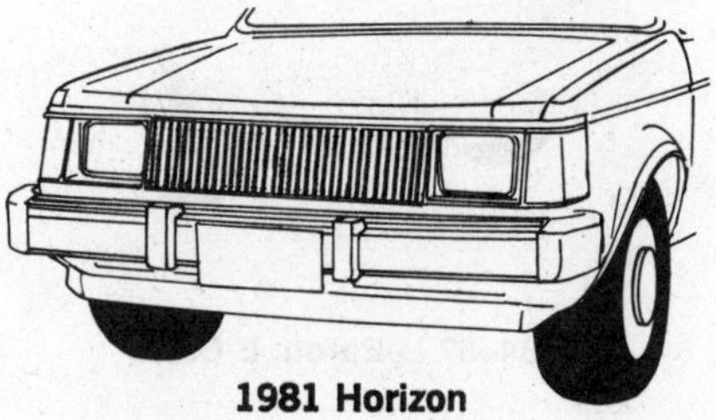

1981 Horizon

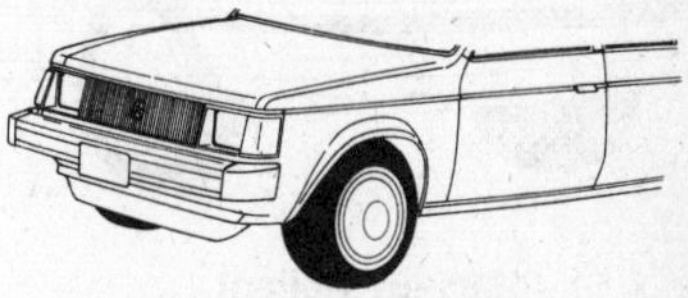

1982–83 Horizon

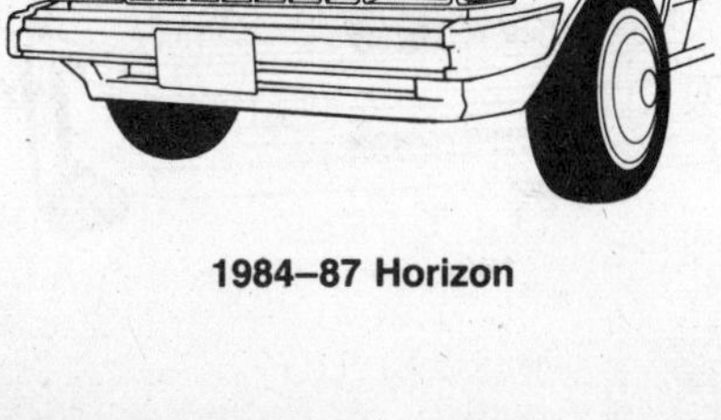

1984–87 Horizon

1980 Horizon TC3

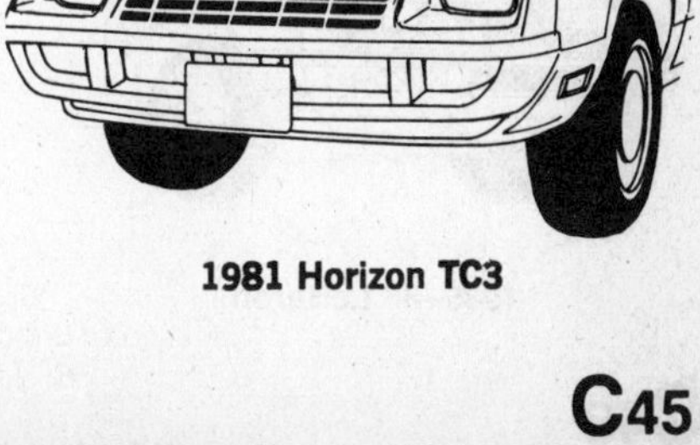

1981 Horizon TC3

YEAR IDENTIFICATION

1984 Turismo/Charger

1985–86 Turismo Duster

1985–87 Turismo

1985–87 Charger 2.2L

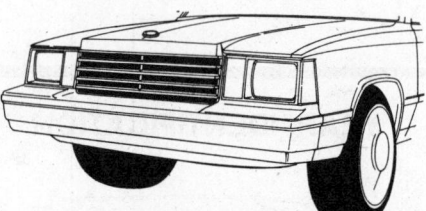

1981–83 Aires

1984 Aries

1985–87 Aries

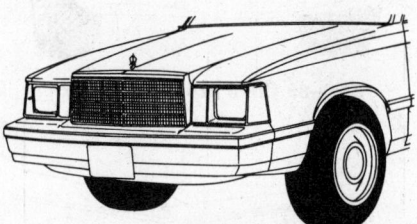

1981–83 Reliant

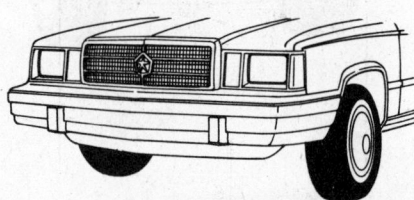

1984 Reliant

1985–87 Reliant

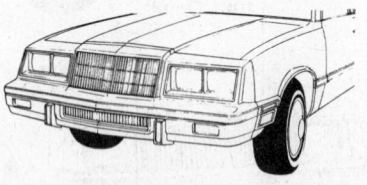

1984–87 LeBaron, E Class

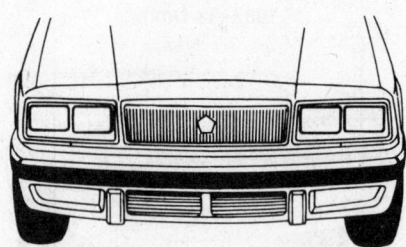

1985–87 LeBaron GTS

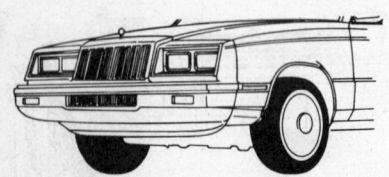

1985–86 LeBaron

1984–86 Shelby Charger

1984–86 Daytona

YEAR IDENTIFICATION

1984–86 Daytona Turbo

1984–86 Laser XE

1982–86 Dodge 400, 600

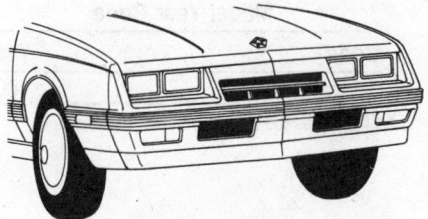

1985–87 Lancer

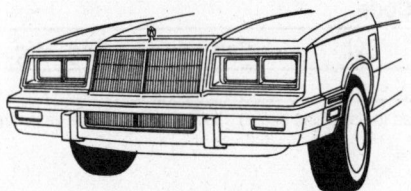

1983–86 New Yorker

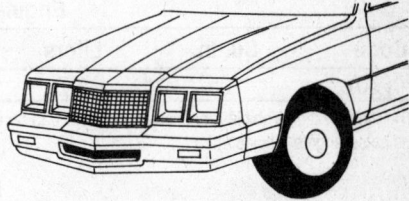

1985–86 Caravelle

1985–86 Lancer

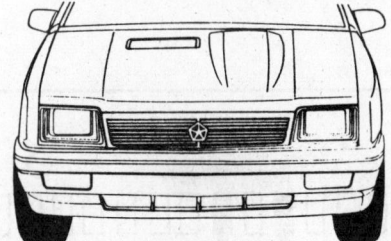

1987 Shadow

1987 Shadow D.C.

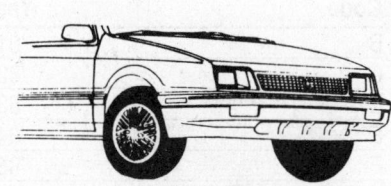

1987 Sundance

1987 Caravelle SE

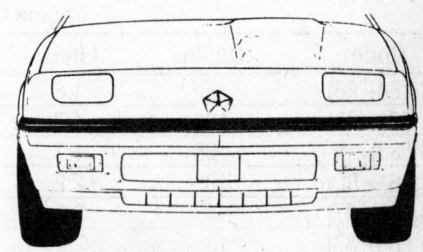

1987 Shelby Z

1987 Daytona Pacifica

1987 LeBaron

C47

VEHICLE IDENTIFICATION NUMBER (VIN)

It is important for servicing and ordering parts to be certain of the vehicle and engine identification. The VIN (vehicle identification number) is a 13 digit number visible through the windshield on the driver's side of the dash, and contains the vehicle and engine identification codes. It can be interpreted as follows:

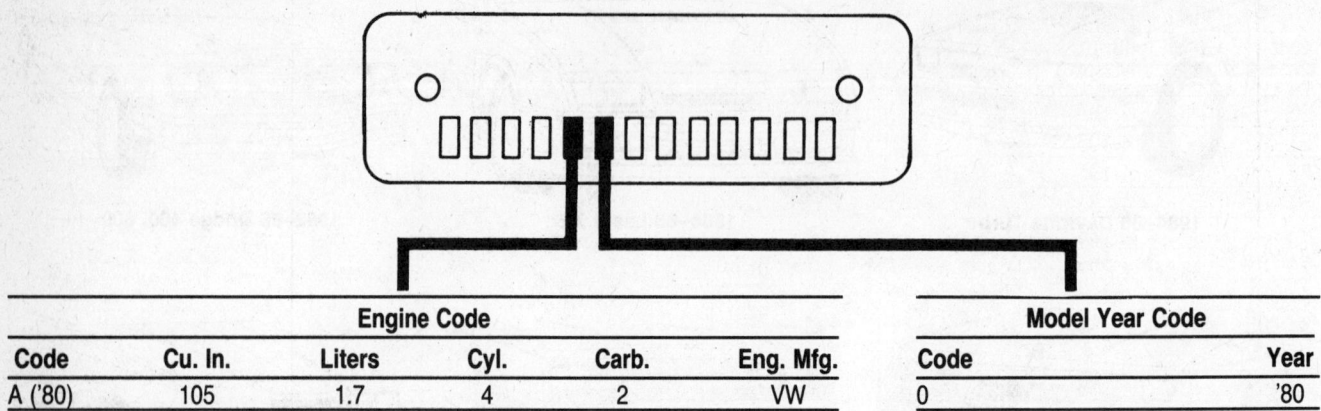

Engine Code						Model Year Code	
Code	Cu. In.	Liters	Cyl.	Carb.	Eng. Mfg.	Code	Year
A ('80)	105	1.7	4	2	VW	0	'80

The thirteen digit Vehicle Identification Number can be used to determine engine application and model year. The sixth digit indicates the model year, and the fifth digit indicates engine displacement.

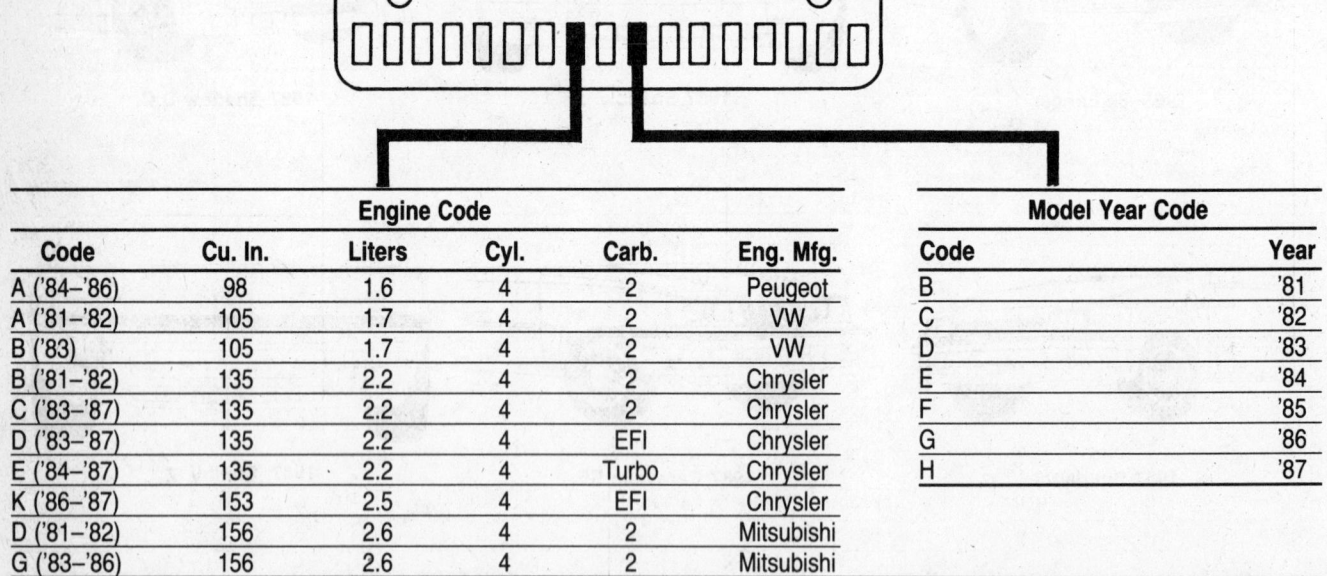

Engine Code						Model Year Code	
Code	Cu. In.	Liters	Cyl.	Carb.	Eng. Mfg.	Code	Year
A ('84–'86)	98	1.6	4	2	Peugeot	B	'81
A ('81–'82)	105	1.7	4	2	VW	C	'82
B ('83)	105	1.7	4	2	VW	D	'83
B ('81–'82)	135	2.2	4	2	Chrysler	E	'84
C ('83–'87)	135	2.2	4	2	Chrysler	F	'85
D ('83–'87)	135	2.2	4	EFI	Chrysler	G	'86
E ('84–'87)	135	2.2	4	Turbo	Chrysler	H	'87
K ('86–'87)	153	2.5	4	EFI	Chrysler		
D ('81–'82)	156	2.6	4	2	Mitsubishi		
G ('83–'86)	156	2.6	4	2	Mitsubishi		

The seventeen digit Vehicle Identification Number can be used to determine engine application and model year. The tenth indicates the model year, and the eighth digit identifies engine displacement.

GENERAL ENGINE SPECIFICATIONS

Year	Eng. VIN Code	Engine No. Cyl. Displ. cu. in.	Eng. Mfg.	Carb. Type	Horsepower @ rpm ■	Torque (ft. lbs.) @ rpm ■	Bore × Stroke (in.)	Compression Ratio	Oil Pressure (psi) @ 2000 rpm
'83–'86	A	4-98	Peugeot	2 bbl.	64 @ 4800	87 @ 2800	3.17 × 3.07	8.8:1	40–90 ①
'80	A	4-105	VW	2 bbl.	75 @ 5600	90 @ 3200	3.13 × 3.40	8.2:1	60–90
'81–'82	A	4-105	VW	2 bbl.	63 @ 4800	83 @ 3200	3.13 × 3.40	8.2:1	60–90
'83	B	4-105	VW	2 bbl.	63 @ 4800	83 @ 2400	3.13 × 3.40	8.2:1	60–90
'81–'82	B	4-135	Chrysler	2 bbl.	84 @ 4800	111 @ 2800	3.44 × 3.62	8.5:1	50
'83	C	4-135	Chrysler	2 bbl.	94 @ 5200	117 @ 3200	3.44 × 3.62	9.0:1	50
'84–'87	C	4-135	Chrysler	2 bbl.	96 @ 5200	119 @ 3200	3.44 × 3.62	9.5:1	50
'83–'87	D	4-135	Chrysler	EFI	99 @ 5600	121 @ 3200	3.44 × 3.62	9.5:1 ③	50
'84–'86	E	4-135	Chrysler	Turbo	146 @ 5200	168 @ 3600	3.44 × 3.62	8.5:1	50
'87	E	4-135	Chrysler	Turbo	146 @ 5200	170 @ 3600	3.44 × 3.62	8.1:1	50
'86–'87	K	4-153	Chrysler	EFI	100 @ 4800	133 @ 2800	3.44 × 4.09	9.0:1	80 ①
'81–'82	D	4-156	Mitsubishi	2 bbl.	92 @ 4500	131 @ 2500	3.59 × 3.86	8.2:1	58 ②
'83	G	4-156	Mitsubishi	2 bbl.	93 @ 5600	132 @ 2800	3.59 × 3.86	8.2:1	58 ②
'84–'86	G	4-156	Mitsubishi	2 bbl.	101 @ 5600	140 @ 2800	3.59 × 3.86	8.7:1	85 ②

■ Horsepower and torque are SAE net, with all accessories installed and operating. Figure may vary from model-to-model and is intended to be representative rather than exact.
① @ 3000 rpm
② @ 2500 rpm
③ 10:1—Shelby and Hi-Performance models

TUNE-UP SPECIFICATIONS

Year	VIN Code	Engine No. Cyl. Displ. (cu. in.)	Mfg.	Spark Plugs Orig. Type	Gap (in.)	Ignition Timing (deg.) Man. Trans.	Auto Trans.	Intake Valve Opens (deg.)■	Fuel Pump Pressure (psi)	Idle Speed (rpm) Man. Trans.	Auto Trans.	Valve Lash (in.) Intake	Exhaust
'84–86	A	4-98	Peugeot	RN-12Y	.035	⑥	⑥	16.5	4.5–6.0	850	850	.012C	.014C
'80	A	4-105	VW	RN-12Y	.035	15B	15B	14	4.4–5.8	900	900	.008–.012H	.016–.020H
'81–'82	A	4-105	VW	P65-PR4	.048 ②	12B ④	10B	14	4.5–6.0	900	900	.008–.012H	.016–.020H
'83	A	4-105	VW	65PR	.035	20B	12B	14	4.4–5.8	900	900	.008–.012H	.016–.020H
'81–'82	B	4-135	Chrysler	P65-PR4	.035	10B	10B	12	4.5–6.0	900	900	Hyd.	Hyd.
'84–'86	C	4-135	Chrysler	RN12YC	.035	⑥	⑥	10.5	4.5–6.0	800 ⑤	900	Hyd.	Hyd.
'84–'86	D	4-135	Chrysler	RN-12Y	.035	⑥	⑥	16	15	900	800	Hyd.	Hyd.
'84–'86	E	4-135	Chrysler	RN12YC	.035	⑥	⑥	10	55	900	800	Hyd.	Hyd.
'86	K	4-153	Chrysler	RN12YC	.035	⑥	⑥	12	15	800	700	Hyd.	Hyd.
'81–'82	D	4-156	Mitsubishi	P65-PR4	.040 ③	7B	7B	25	4.5–6.0	800 ①	800 ①	.006H	.010H
'83–'86	G	4-156	Mitsubishi	RN-12Y	.040 ③	—	⑥	25	4.5–6.0	—	800 ①	.006H	.010H
'87						See Underhood Specifications Sticker							

NOTE: The underhood specifications sticker often reflects tune-up specification changes made in production. Sticker figures must be used if they disagree with those in this chart. Part numbers in this chart are not recommendations by Chilton for any product by brand name.
■ Before top dead center
Hyd. Hydraulic
H Hot
C Cold
① 750 rpm-Canada
② .035-Canada
③ .030-Canada
④ 1982–20B
⑤ 900 rpm-Canada
⑥ Refer to emission control label on vehicle

FIRING ORDERS

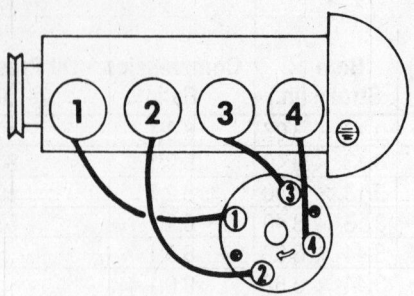

Chrysler Corp. 1.7L (1980–83)
Engine Firing Order: 1–3–4–2
Distributor Rotation: Clockwise

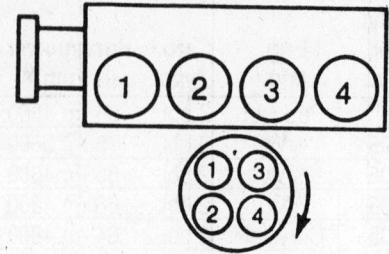

Chrysler Corp. 2.2L and 2.5L
Engine Firing Order: 1–3–4–2
Distributor Rotation: Clockwise

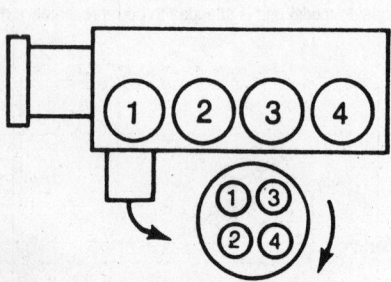

Chrysler Corp. (Mitsubishi) 2.6L
Engine Firing Order: 1-3-4-2
Distributor Rotation: Clockwise

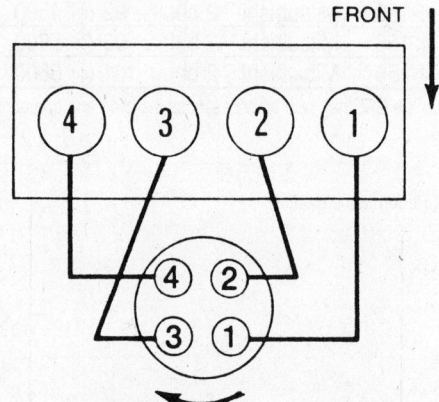

Chrysler Corp. 1.6L ('84 and Later)
Engine Firing Order: 1-3-4-2
Distributor Rotation: Clockwise

CAPACITIES

Year	Engine Type/ Displ. (cu. in.)	Crankcase Incl. Filter (qts.)	Transaxle Pints-to-Refill After Draining Manual	Transaxle Pints-to-Refill After Draining Automatic ①	Fuel Tank (gal.)	Cooling System (qts.) With Heater	Cooling System (qts.) With A/C
'84–'86	4-98	3.5	②	16	14⑥	6.8	6.8
'80	4-105	4	2.65	13.0	13	6.0	6.0
'81–'82	4-105	4	2.60	14.5	13	6.0	6.0
'83	4-105	4	②	16.75	13	6.0	6.0
'81–'82	4-135	4	2.60	15.0	13	7.5	7.5
'83–'87	4-135	4③	②	17.75	14⑥	9.0	9.0
'86–'87	4-153	4④	⑤	17.75	14⑥	9.0	9.0
'81–'82	4-156	5	3.75	15.0	13	8.5	8.5
'83–'86	4-156	5	—	17.75	14⑥	9.0	9.0

NOTE: If the starter motor is located on the radiator side of the engine the car is equipped with an A-412 manual transaxle; Use GL-4 Hypoid Lubricant. If the starter motor is located on the firewall side of the engine the car is equipped with an A-460 manual transaxle; Use only Dexron II lubricant.

① Includes torque converter. Approx. 6 pts without draining converter.
② 4 speed: 3.75–use Dexron® II lubricant
5 speed: 4.55–use Dexron® II lubricant
③ '84–'87 2.2 Turbo engines: 5 qts.
④ With or without filter
⑤ A460 (4-spd.)—4.0
A525 (5-spd.)—4.6
⑥ 13 gal—Charger, Horizon, Omni, Turismo

CAMSHAFT SPECIFICATIONS

All measurements in inches

Year	VIN Code	Engine Type/ Displ. (cu. in.)	Eng. Mfg.	Journal Diameter					Bearing Clearance	Lobe Lift		Camshaft End Play
				1	2	3	4	5		Intake	Exhaust	
'80–'82	A	4-105	VW	1.021– 1.023	1.021– 1.023	1.021– 1.023	1.021– 1.023	1.021– 1.023	0.0016– 0.003	0.406	0.406	0.006
'83	B	4-105	VW	1.021– 1.023	1.021– 1.023	1.021– 1.023	1.021– 1.023	1.021– 1.023	0.0016– 0.003	0.406	0.406	0.006
'81–'82	B	4-135	Chrysler	1.375– 1.376	1.375– 1.376	1.375– 1.376	1.375– 1.376	1.375– 1.376	0.010 max	0.430	0.430	0.006
'83–'87	C	4-135	Chrysler	1.375– 1.376	1.375– 1.376	1.375– 1.376	1.375– 1.376	1.375– 1.376	0.010 max	0.430	0.430	0.006
'83–'87	D	4-135	Chrysler	1.375– 1.376	1.375– 1.376	1.375– 1.376	1.375– 1.376	1.375– 1.376	0.010 max	0.430	0.430	0.006
'84–'87	E	4-135	Chrysler	1.375– 1.376	1.375– 1.376	1.375– 1.376	1.375– 1.376	1.375– 1.376	0.010 max	0.430	0.430	0.006
'86–'87	K	4-153	Chrysler	1.375– 1.376	1.375– 1.376	1.375– 1.376	1.375– 1.376	1.375– 1.376	0.010 max	0.430	0.430	0.006

NOTE: Specifications for the 1.6L and 2.6L engines are not available.

CRANKSHAFT AND CONNECTING ROD SPECIFICATIONS

All specifications in inches

Year	Engine Type/ Displ. (cu. in.)	Main Brg. Journal Dia.	Main Brg. Oil Clearance	Crankshaft End Play	Thrust on No.	Connecting Rod Journal Dia.	Rod Bearing Oil Clearance	Rod Bearing Side Clearance
'84–'86	4-98	2.046	.0009–.0031	.0035–.011	3	1.612	.001–.0025	.006–.009
'80–'82	4-105	2.124–2.128	.0008–.0030	.003–.007	3	1.809–1.813	.0011–.0034	.015
'83	4-105	2.1236–2.1244	.0008–.0030	.003–.007	3	1.809–1.813	.0011–.0034	.014
'81–'82	4-135	2.362–2.363	.0004–.0026	.002–.007	3	1.968–1.969	.0004–.0026	.005–.013
'84–'87	4-135	2.362–2.363	.0003–.0031①	.002–.007	3	1.968–1.969	.0008–.0034②	.005–.013
'86–'87	4-153	2.362–2.363	.0003–.0031①	.002–.007	3	1.968–1.969	.0008–.0034②	.005–.013
'82	4-156	2.3622	.0008–.0028	.002–.007	3	2.0866	.0008–.0028	.004–.010
'83–'86	4-156	2.3622	.0008–.0028	.002–.007	3	2.0866	.0008–.0028	.004–.010

① Turbo: .0004–.0023
② Turbo: .0008–.0031

VALVE SPECIFICATIONS

Year	Engine Type/ Displ. (cu. in.)	Seat Angle (deg.)	Face Angle (deg.)	Spring Test Pressure (lbs @ in.)	Spring Installed Height (in.)	Stem-to-Guide Clearance (in.)		Stem Diameter (in.)	
						Intake	Exhaust	Intake	Exhaust
'84–'86	4-98	45	45	—	1.65	.0005–.0018	.0013–.0026	.3140–.3146	.3132–.3138
'80	4-105	45	①	②	③	.020 max.	.027 max.	.3140	.3130
'81–'82	4-105	45	45	②	③	.0028	.0035	.314	.313
'81–'82	4-135	45	45.5	175 @ 1.22	1.65	.001–.003	.002–.004	.312–.313	.311–.312
'83–'86	4-135	45	45	150 @ 1.22④	1.65	.0009–.0026	.0030–.0047	.3124	.3103
'87	4-153	45	45	175 @ 1.22	1.65	.0009–.0026	.0030–.0047	.3124	.3103
'81–'82	4-156	43.75	45.25	34.1 @ 1.18	1.59	.001–.002	.002–.003	.315	.315
'86–'87	4-153	45	45	150 @ 1.22	1.65	.0009–.0026	.0030–.0047	.3124	.3103
'83–'86	4-156	45	45	61 @ 1.59	1.59	.0012–.0024	.0020–.0035	.315	.315

① Intake 45°33'
   Exhaust: 43°33'
② Outer: 101 @ .878
   Inner: 49 @ .720

③ Outer: 1.28
   Inner: 1.13
④ Hi-Performance: 175 @ 1.22

PISTON, RING AND PIN SPECIFICATIONS

All specifications in inches

Year	Engine Type Displ. (cu. in.)	Piston Clearance	Ring Gap: Top Compression	Ring Gap: Bottom Compression	Ring Gap: Oil Control	Ring Side Clearance: Top Compression	Ring Side Clearance: Bottom Compression	Ring Side Clearance: Oil Control	Pin Clearance In Piston
'84–'86	4-98	.0016–.0020	.012–.018	.012–.018	.010–.016	.0018–.0028	.0018–.0020	—	①
'80–'83	4-105	.0005–.0015	.012–.018	.012–.018	.016–.055	.0016–.0028	.0008–.0020	.008–.0020	.00004–.00035
'81–'85	4-135	.0005–.0015④	.011–.021②	.011–.021③	.015–.055	.0015–.0031	.0015–.0037	—	①
'86–'87	4-135	.0005–.0015④	.011–.021②	.011–.021③	.015–.055	.0015–.0031	.0015–.0037	.008 max	①
'86–'87	4-153	.0005–.0015	.011–.021	.011–.021	.015–.055	.0015–.0031	.0015–.0037	.008 max	①
'81–'86	4-156	.0008–.0016	.011–.018	.011–.018	.0078–.035	.0024–.0039	.0008–.0024	—	①

① Press fit
② Turbo: .010–.020
③ Turbo: .009–.018
④ Turbo: .0015–.0025

TORQUE SPECIFICATIONS

All readings in ft. lbs.

Year	Engine Type/ Displ. (cu. in.)	Cylinder Head Bolts	Connecting Rod Bearing Bolts	Main Bearing Bolts	Crankshaft Bolt	Flywheel-to-Crankshaft Bolts	Camshaft Cap Bolts
'84–'86	4-98	52	28	48	110	70	—
'80–'83	4-105	60①	35	47	58	60②	168④
'81–'85	4-135	③	40①	30①	50	65	165④
'86–'87	4-135	⑥	40①	30①	50	70	165④⑦
'86–'87	4-153	⑥	40①	30①	50	70	165④
'81–'86	4-156	69⑤	34	58	87	70	160④

① plus 1/4 turn more
② 50 with auto. trans.
③ For torque sequence - 30, 45, 45 + 1/4 turn more
④ Inch lbs.
⑤ Cold
⑥ For torque sequence: 45, 65, 65 + 1/4 turn more
⑦ 215 inch lbs.—1987

WHEEL ALIGNMENT SPECIFICATIONS

Caster is not adjustable

Year	Front Camber: Range (deg.)	Front Camber: Preferred	Rear Chamber: Range (deg.)	Rear Chamber: Preferred	Toe-Out (in.): Front	Toe-Out (in.): Rear
'80 Omni, Horizon	1/5N to 4/5P	1/3P	1 1/2N to 1/2N	1N	5/32 out to 1/8 in	5/32 out to 1/32 in
'81–'82 Omni, Horizon	1/4N to 3/4P	5/16P	1 1/2N to 3/4N	1/2N	7/32 out to 1/8 in	5/32 out to 11/32 in
'81–'82 Aries, Reliant	1/4N to 3/4P	5/16P	1N-0	1/2N	7/32 out to 1/8 in	3/16 out to 3/16 in
'82 LeBaron, Dodge 400	1/10N to 7/10N	3/10N	1N-0	1/2N	3/20 out to 1/10 in	3/16 out to 3/16 in
'83–'85 All Models	1/4N to 3/4P	5/16P	②	1/2N	7/32 out to 1/8 in	①
'86 All Models	1/4N to 3/4P	5/16P	③	1/2N	7/32 in to 1/8 out	④
'87 All Models	1/4N to 3/4P	5/16P	1 1/4N to 1/4P	1/2N	7/32 in to 1/8 out	5/16 out to 5/16 in

① Omni, Horizon, Turismo & Charger 5/32 out to 11/32 in
All others 3/16 out to 3/16 in.
② Omni, Horizon, Turismo & Charger 1-1/4 N to 1/4 N
Hi-Performance 1-1/8 N to 1/8 N
All others 1 N to O
③ Omni, Horizon, Turismo, Charger:1 1/4N to 1/4N
All others: 1 1/4N to 1/4N
④ Omni, Horizon, Turismo, Charger: 3/16 out to 13/32 in.
All others: 5/16 out to 5/16 in

ENGINE ELECTRICAL

Alternator

For further information on the charging system, please refer to "Charging And Starting" in the Unit Repair section.

SERVICE PRECAUTIONS

To prevent serious damage to the alternator and the rest of the charging system, the following precautions must be observed:

1. When installing a battery, make sure that the positive cable is connected to the positive terminal and the negative to the negative.

2. When jump-starting the vehicle with another battery, make sure that like terminals are connected. This also applies when using a battery charger.

3. Never operate the alternator with the battery disconnected or otherwise on an uncontrolled open circuit. Double-check to see that all connections are tight.

4. Do not short across or ground any alternator or regulator terminals.

5. Do not try to polarize the alternator.

6. Do not apply full battery voltage to the field connector.

7. Always disconnect the battery ground cable before disconnecting the alternator lead.

BELT TENSION ADJUSTMENT

For the belt tension adjustments, please refer to the alternator removal and installation procedures listed below.

REMOVAL & INSTALLATION

NOTE: Alternators are supplied by three manufacturers; Chrysler, Bosch and Mitsubishi.

1. Disconnect the negative battery cable.

2. Disconnect and tag the wires from the alternator.

3. Loosen the adjusting bolt, disconnect the drive belt(s), remove the supporting nuts/bolts and the alternator.

4. To install, check the clearance between the alternator housing and the engine mount (if more than 0.008 in. clearance exists, install shims)

and reverse the removal procedures. Adjust the belt tension to $\frac{1}{4}$–$\frac{3}{8}$ in. deflection between the longest span between two pulleys, under moderate thumb pressure.

NOTE: To perform the alternator adjustment on some models, it may be necessary to raise the vehicle and remove the splash shield.

Voltage Regulator

The internally or externally mounted electronic voltage regulator has no moving parts and requires no adjustment. If service is required, the regulator is replaced as a unit.

REMOVAL & INSTALLATION

NOTE: Several types of voltage regulators are used; external, internal and in-circuit (contained within the power/logic modules). The external and in-circuit types are used by Chrysler and Bosch; the internal type is used by Bosch and Mitsubishi.

External Type

1. Disconnect the negative battery cable.

2. Disconnect the wires from the regulator.

3. At the right side fender skirt, remove the sheet metal screws and the regulator.

4. To install, reverse the removal procedures.

In-Circuit Type

This type can ONLY be serviced by replacing the power/logic modules.

Internal Type
BOSCH

1. Disconnect the negative battery cable.

2. At the rear of the alternator, remove the two regulator securing screws.

3. Remove the regulator/brush holding assembly from the alternator.

4. To install, reverse the removal procedures.

MITSUBISHI

To remove the regulator, the alternator must be removed from the vehicle and disassembled. For disassembly procedures, please refer to "Charging and Starting" in the Unit Repair section.

Starter

Three makes of starters are used; Bosch, Mitsubishi and Nippondenso.

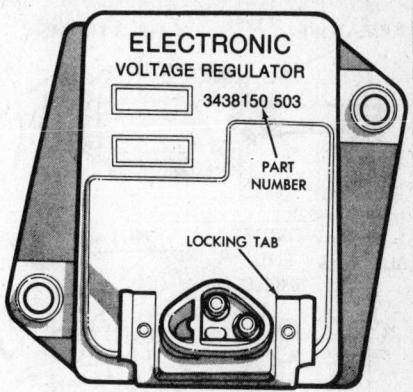

Voltage regulator

Removal and installation procedures are the same for all units. The solenoid is mounted on the starter motor. For further information, please refer to "Charging and Starting" in the Unit Repair section.

REMOVAL & INSTALLATION

1. Disconnect the negative battery cable.

2. If equipped, remove the heat shield clamp and the heat shield.

3. If equipped with a 2.2L or 2.5L engine, it may be necessary to loosen the air pump tube at the exhaust manifold and swivel the tube bracket away from the starter.

4. Disconnect the battery and solenoid wires from the starter.

5. Support the starter, then remove the bolts and the starter from the flywheel housing and the transaxle support bracket.

6. To install, reverse the removal procedures.

Distributor

The 1.6L, 1.7L, 2.2L and 2.5L engines are equipped with a computer controlled Hall Effect ignition system and use a three-pronged spark control computer connector at the distributor. The 2.6L engine electronic ignition system is slightly different from that used on the other models. To remove the spark plug cables from the distributor cap (1981 and later), remove the distributor cap, squeeze the inside retaining prongs together and pull out the cables.

REMOVAL & INSTALLATION

2.6L Engine

1. Disconnect the negative battery cable and the distributor pickup lead wire at the harness connector. Remove the spark plug cables from the distributor cap.

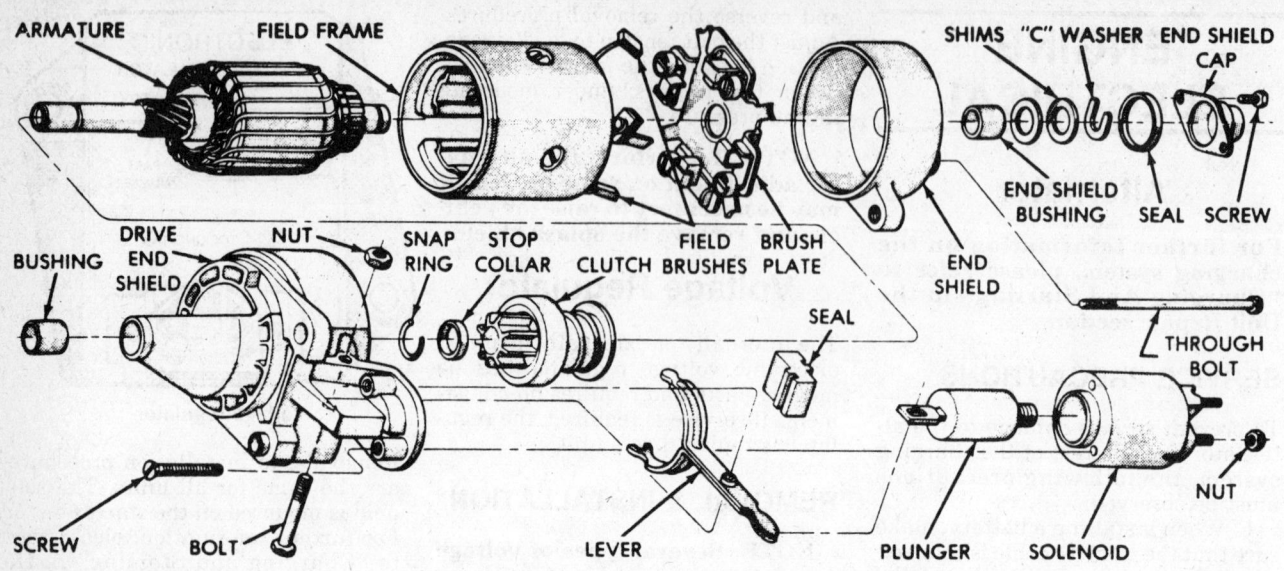

Exploded view of the Bosch starter

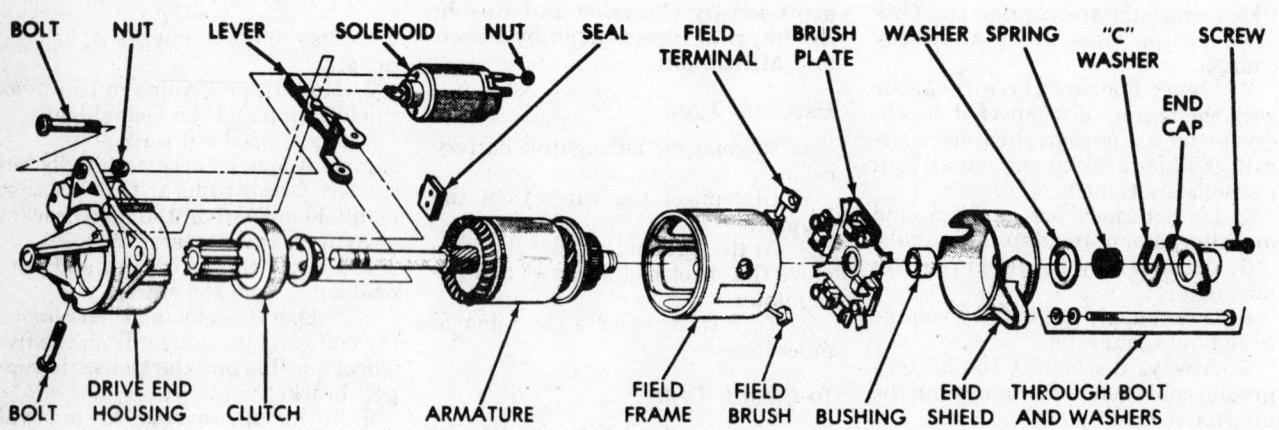

Exploded view of the Nippondenso starter

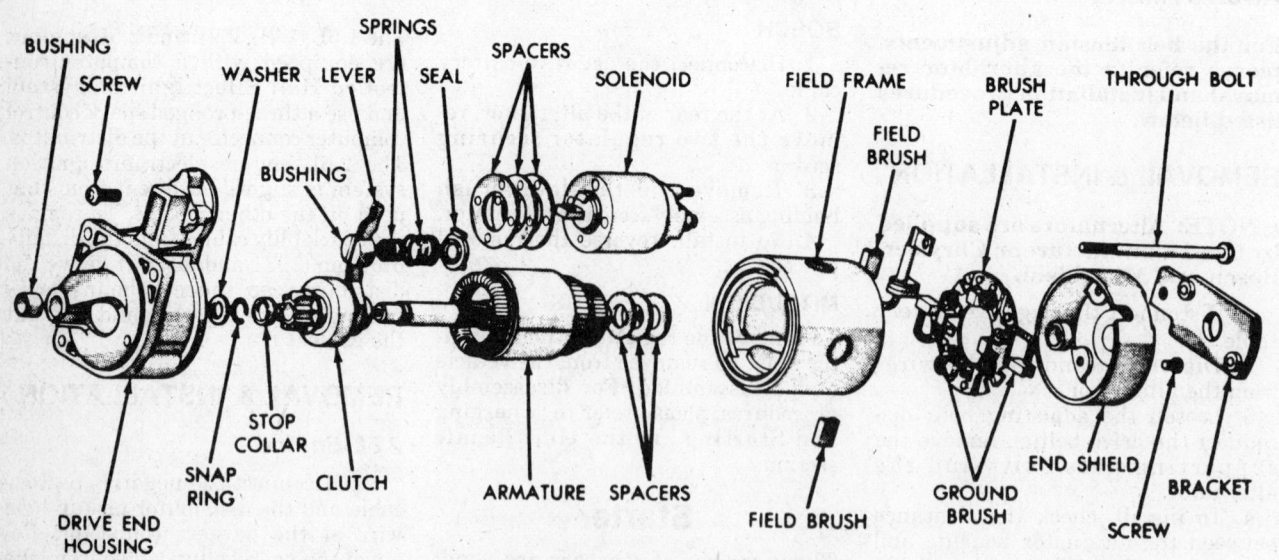

Exploded view of the Mitsubishi starter

2. Remove the vacuum hose from the the vacuum control unit.

3. Remove the hold-down nut and the distributor assembly.

4. To install the distributor:

a. Rotate the crankshaft until the No. 1 piston is at TDC on the compression stroke. Align the crankshaft pulley timing mark to the TDC mark on the timing plate.

b. Align the mating mark (line) with the mating punch mark on the distributor gear.

c. Install the distributor to the cylinder head. Align the distributor flange mating mark with the center of the distributor hold-down stud.

5. To complete the installation, reverse the removal procedures. Start the engine. Check and adjust the timing, if necessary.

All Other Engines

1. Disconnect the distributor pickup lead wire at the harness connector. If equipped, remove the the splash shield.

2. Remove the retaining screws and the distributor cap.

3. Rotate the engine crankshaft (in the direction of normal rotation) until No. 1 cylinder is at TDC on the compression stroke. Make a mark on the block and the distributor housing where the rotor points for installation reference.

4. Remove the hold-down bolt and the distributor.

5. Carefully lift the distributor from the engine. The shaft on some engines may rotate slightly as the distributor is removed.

6. If the engine has been cranked while the distributor was removed, perform the following procedures:

a. Rotate the crankshaft until the No. 1 piston is at TDC on the compression stroke. This will be indicated by the 0° mark on the flywheel or crank pulley aligning with the pointer on the clutch housing or engine front cover.

b. Position the rotor just ahead of the No. 1 terminal of the cap and lower the distributor into the engine.

c. With the distributor fully seated (make sure that the O-ring is seated), the rotor should be directly under the No. 1 terminal.

7. If the engine was not disturbed while the distributor was removed, lower the distributor into the engine, engaging the gears and making sure that the O-ring is properly seated in the block. The rotor should align with the marks made before removal.

8. Tighten the hold-down bolt and

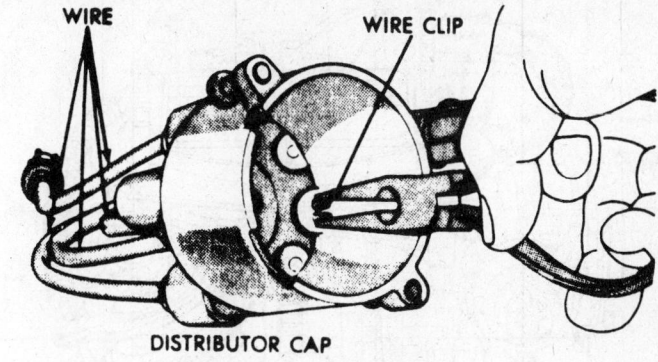

Removing the spark plug wires

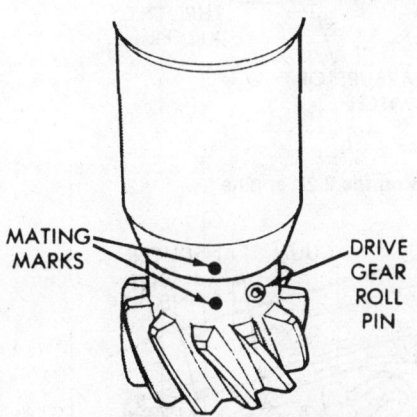

Mating the distributor timing marks

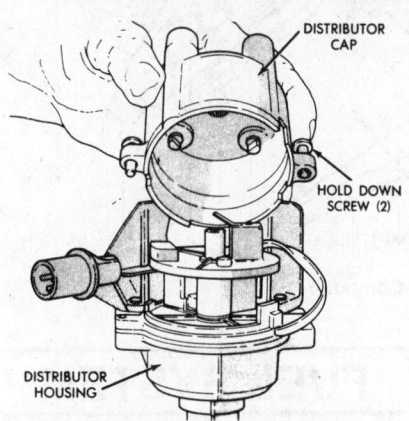

View of the distributor—1980–86

connect the wires (make sure they snap into place).

9. To complete the installation, reverse the removal procedures. Start the engine. Check and adjust the ignition timing, if necessary.

IGNITION TIMING ADJUSTMENT

The ignition is timed using the No. 1 cylinder at the left side of the engine, facing the vehicle. The exception is the 1.6L engine, where the No. 1 cylinder is on the right side (flywheel end).

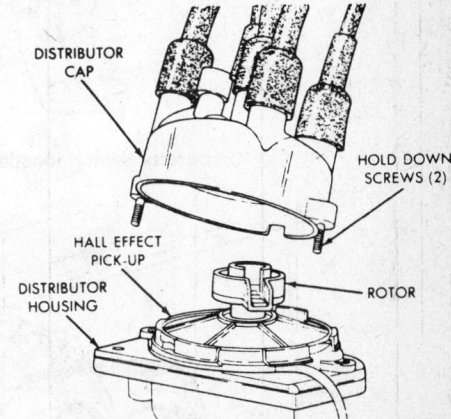

Exploded view of the distributor—1987 and later

1. Connect a timing light according to the manufacturer's instructions. Connect the red lead of a tachometer to the negative coil terminal and the black lead to ground; place the selector switch on the four cylinder position.

2. Operate the engine until the operating temperature is reached.

3. Momentarily open the throttle and release it. Make sure that there is no binding of the throttle linkage and the idle speed screw (on carburetor models) is against the stop.

NOTE: On EFI models, disconnect and reconnect the water temperature sensor connector on the thermostat housing. The loss of power lamp on the dash should turn on and stay on.

4. On carburetor models (equipped with a carburetor switch), connect a jumper wire between the carburetor switch and ground. At the Spark Control Computer (SCC), disconnect and plug the vacuum line.

5. Using the 1000 rpm scale, read the curb idle speed. If the rpm is higher than the specification, adjust the idle speed screw (on carburetor models) on top of the solenoid.

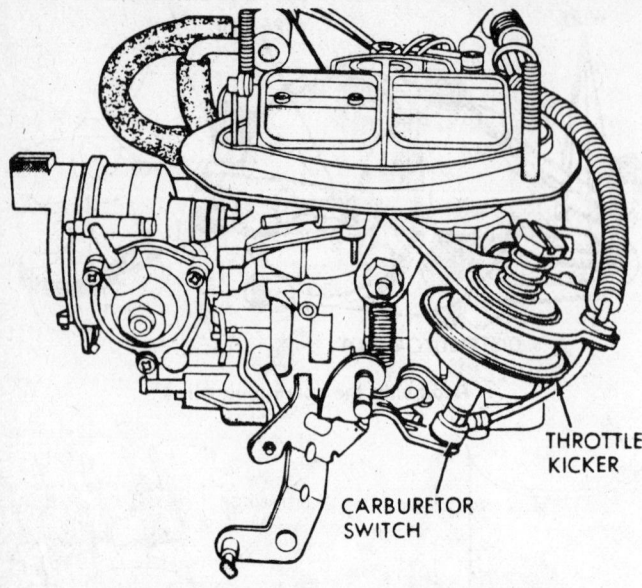

Carburetor switch location on the 2.2L engine

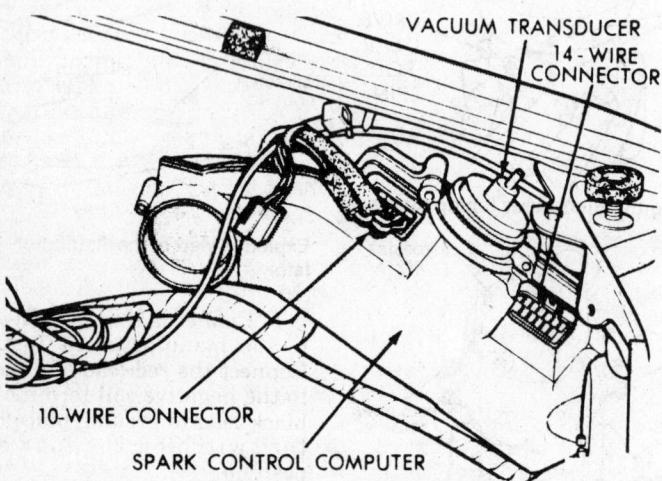

Spark Control Computer

NOTE: On 2.6L engines, coat both sides of the insulator and gasket with sealer.

Electric Fuel Pump

REMOVAL & INSTALLATION

NOTE: Refer to the "Fuel Pressure Release" procedures in this section and reduce the fuel pressure. The electric fuel pump is located in the fuel tank.

1. Raise and support the rear of the vehicle on jackstands. Release the pressure in the fuel system.
2. Disconnect the negative battery terminal.
3. Remove the fuel tank cap and the fuel tube-to-quarter panel screws.
4. Remove the draft tube cap from the sending unit, connect a siphon hose to the draft tube and siphon the fuel from the tank.
5. Disconnect the fuel pump wiring connector from the lock ring cap. Wrap a cloth around the fuel hose and remove the hose from the lock ring cap.
6. Disconnect and lower the fuel tank from the vehicle.
7. Using a non-metallic drift and a hammer, remove the lock ring by driving it counterclockwise.
8. Remove the fuel pump and O-ring seal from the tank. Check the in-tank filter and replace it, if necessary.
9. To install, use a new O-ring seal and reverse the removal procedures. Start the engine and check for leaks. If necessary, check the fuel pressure: EFI—14.5 psi or Turbo—55 psi.

Fuel Filter

Several types of fuel filters are used. One is located in the fuel tank of all the vehicles and is part of the fuel gauge unit assembly. Second is a sealed paper element, in-line unit; mechanical system—located on the engine (above the fuel pump) or electrical system—located in front of the fuel tank. Third is a paper element, in-line fuel filter/vapor separator, housed in a fuel reservoir. Normally, the in-tank filter does not require changing. If it does, the fuel tank must be removed.

REMOVAL & INSTALLATION

NOTE: If removing the filter on a EFI or an MFI system, release the pressure in the fuel system. See the procedure under Fuel Injection for details.

1. Place a rag around the fuel filter

6. Loosen the distributor hold-down bolt so the distributor can be rotated.

7. Aim the timing light or read the magnetic timing unit. For the 1.7L, 2.2L and 2.5L engines, remove the timing hole access cover on the clutch housing or read the magnetic timing unit. The timing marks, for the 1.6L and 2.6L engines, are on the crankshaft pulley and front cover. Carefully rotate the distributor until the timing marks are aligned.

8. Tighten the distributor hold down bolt and recheck the timing. Check and adjust the idle speed, if necessary.

9. Unplug and reconnect the vacuum hose to the Spark Control Computer.

10. To complete the installation, reverse the removal procedures.

FUEL SYSTEM

⸻ CAUTION ⸻
Whenever working on or around any part of the fuel system, take precautions to avoid the risk of fire.

Mechanical Fuel Pump

REMOVAL & INSTALLATION

1. Disconnect and plug the fuel lines at the fuel pump, located on the left side of the engine.
2. Remove the fuel pump-to-engine bolts.
3. Clean the gasket mounting surfaces with a suitable scraper.
4. To install, use new gaskets and reverse the removal procedures.

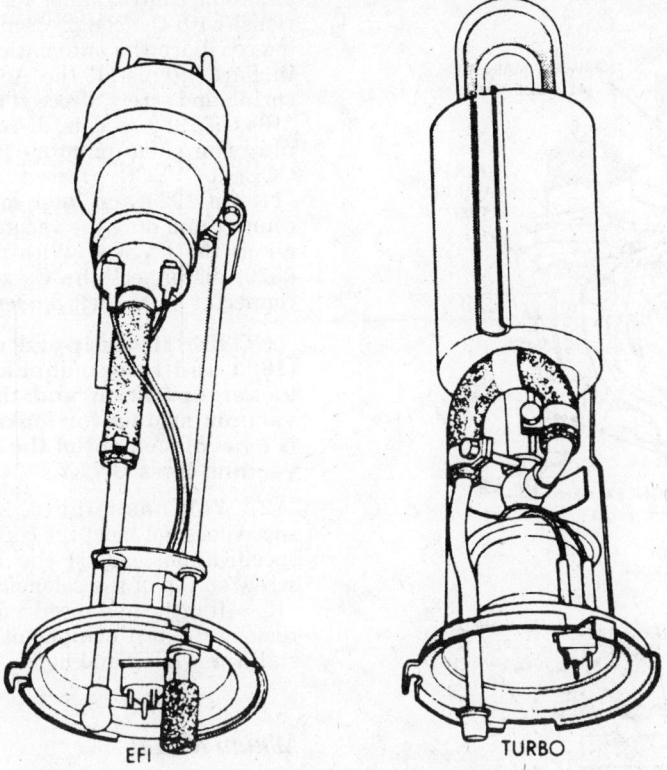

View of the EFI and the Turbo fuel pumps

hose(s)/tube(s) to catch the excess fuel.

2. Remove the clamps, the fuel hose(s)/tube(s) and the fuel filter (discard it).

3. To install a new filter, note the direction of fuel flow (usually indicated by an arrow) and reverse the removal procedures.

Carburetor

Due to the complex nature of modern feedback carburetors and fuel injection systems, comprehensive diagnosis and testing procedures fall outside the confines of this repair manual. For further information on diagnosis, testing and repair of feedback carburetors, please refer to *Chilton's Guide To Fuel Injection and Feedback Carburetors.*

REMOVAL & INSTALLATION

1. Disconnect the negative battery cable.

2. Remove the air cleaner and the fuel tank cap.

NOTE: On the 2.6L engine, remove the air intake housing and the carburetor protector. Drain the radiator and remove the coolant hoses from the carburetor.

3. At the carburetor, place a container under the the fuel inlet hose and disconnect the fuel line.

4. Disconnect the vacuum hoses and wiring connectors after marking their locations with tape for installation.

5. Disconnect the throttle linkage.

6. Remove the mounting bolts or nuts and the carburetor.

7. Clean the gasket mounting surfaces with a suitable scraper.

8. To install, use a new gasket and reverse the removal procedures.

IDLE SPEED ADJUSTMENT

5220/6520 Models

1. Set the parking brake and place the transaxle in Neutral.

2. Turn off the headlights, all accessories and the air conditioning.

3. Connect a tachometer to the engine following the manufacturer's instructions.

4. If equipped with a Fuel Control Computer (FCC), connect a jumper wire from the carburetor switch to ground.

5. Start the engine and allow it to reach normal operating temperature.

6. Refer to the emission control label in the engine compartment, then check and/or adjust the timing, if necessary.

7. Unplug the connector at the radiator fan and install a jumper wire so the fan will run continuously. Pull the PCV valve from the cylinder head cover and the $^3/_{16}$ in. dia. control hose at the canister and allow them to draw air.

8. On 1980 models, disconnect and plug the vacuum hoses to the EGR valve. Proceed to Step 11.

9. On 1981–82 models, disconnect and plug the vacuum hoses to the EGR valve and the $^3/_{16}$ in. dia. control hose at the canister. If equipped with A/C, turn on the A/C and open the throttle to energize the solenoid. If equipped with an automatic transaxle, place the transaxle in Drive. Remove the adjusting screw and spring from the top of the A/C solenoid. Insert a $^1/_8$ in. Allen wrench into the solenoid and adjust the idle speed (check

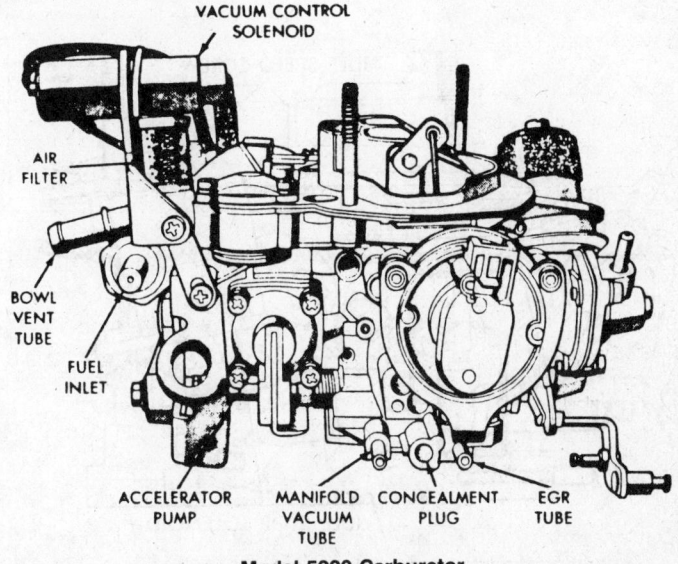

Model 5220 Carburetor

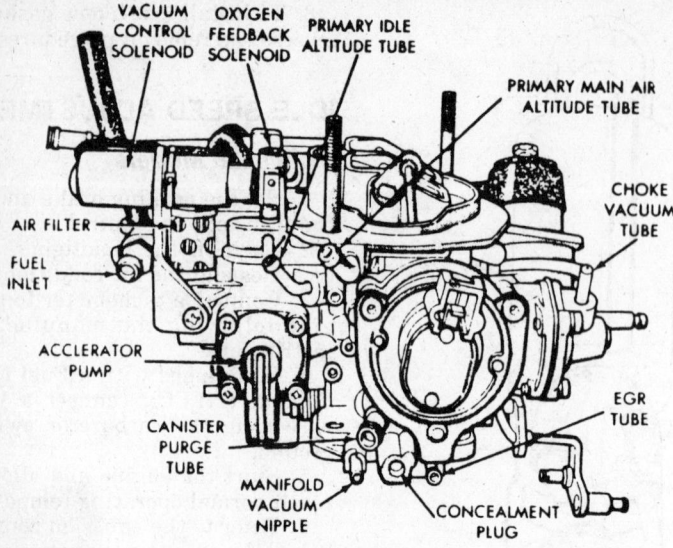

Model 6520 Carburetor

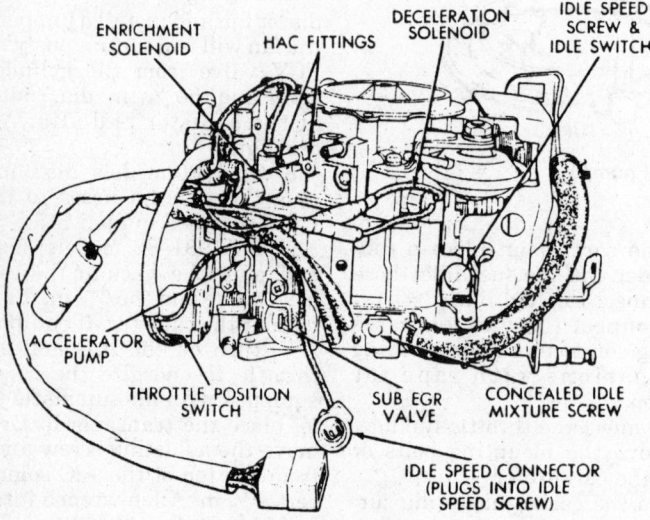

Mikuni Carburetor

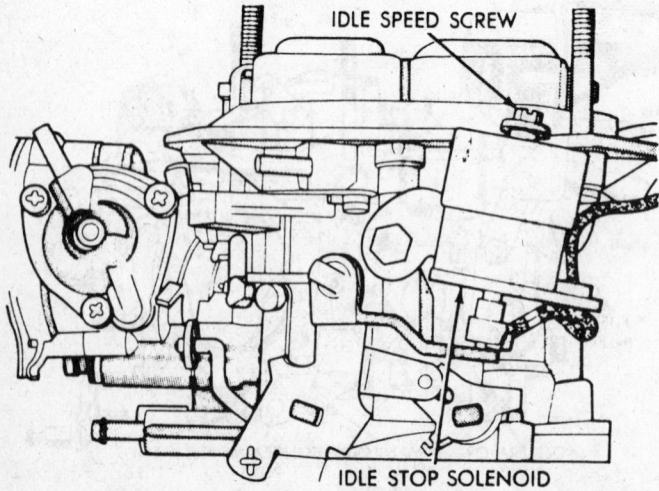

Adjusting the idle speed with the solenoid idle stop screw

emission control label for specifications) with the compressor clutch engaged. Place the automatic transaxle in Park. Reinstall the A/C solenoid spring and screw. Proceed to Step 11.

10. On 1983 models, disconnect and plug the vacuum connector at the CCEGR/CVSCC. Proceed to Step 11.

11. On 1984 and later models, disconnect and plug the vacuum connector at the CVSCC. With the Holley 6520, disconnect the O_2 sensor test connector on the left fender shield.

NOTE: If equipped with A/C (1983 and later), check the A/C kicker operation and the kicker vacuum system for leaks, service is by replacement of the kicker or vacuum lines ONLY.

12. Wait one minute, if the idle speed does not meet the curb idle rpm specifications, adjust the idle speed screw on top of the solenoid.

13. After the idle speed has been set, remove the test equipment and reinstall the wiring and hoses.

Mikuni Model

1. Set the parking brake and place the transaxle in Neutral.

2. Turn off all lights and accessories. Disconnect the cooling fan.

3. Connect a tachometer to the engine, following the manufacturer's instructions.

4. Start the engine, allow it to reach normal operating temperature and refer to the emission control label in the engine compartment.

5. Install a timing light; check the timing and adjust (if necessary). Remove the timing light.

6. Open the throttle, run the engine at 2500 rpm for 10 seconds. Return to idle, wait two minutes and read the idle rpm. If it is not the same as the curb idle specified on the emission control label, adjust the idle speed adjusting screw. The screw is accessible through the hole in the choke cover plate using a long, narrow screwdriver at a 45° angle inward.

NOTE: On 1984 and later models, remove the idle switch connector.

7. If equipped with A/C, set the temperature control lever to the coldest position and turn on the A/C. With the compressor running, set the engine speed to 900 rpm by turning the idle up adjusting screw.

8. Turn the engine off, disconnect the tachometer, reconnect the cooling fan and idle switch connector (if equipped).

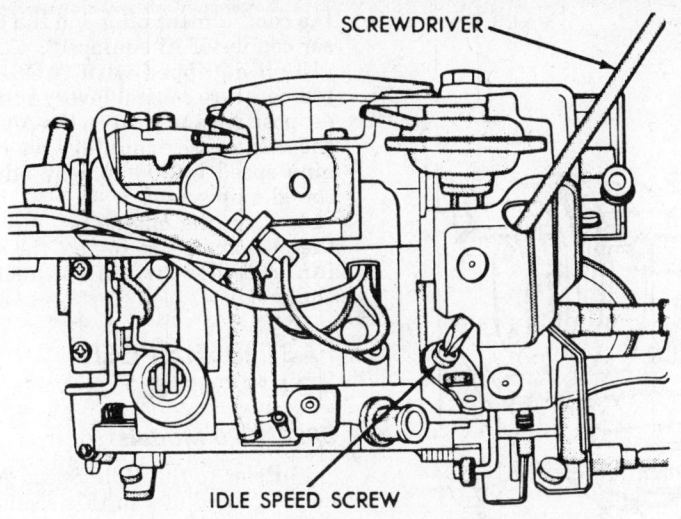

SCREWDRIVER

IDLE SPEED SCREW

Adjusting the idle speed screw

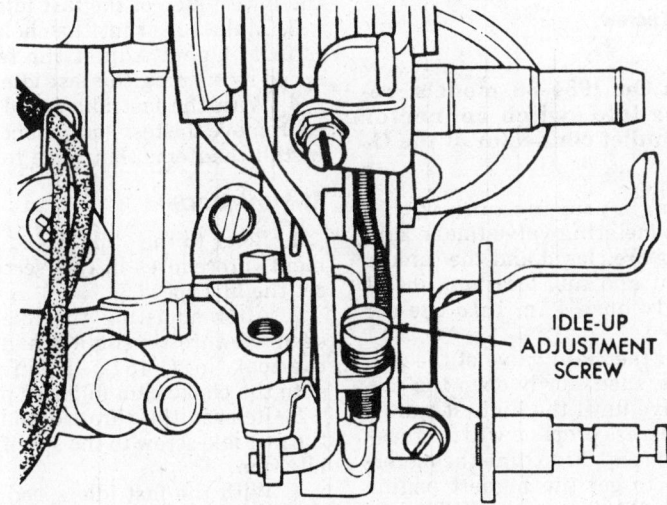

IDLE-UP
ADJUSTMENT
SCREW

Adjusting the idle-up screw

PROPANE ASSISTED IDLE SPEED AND AIR/FUEL MIXTURE ADJUSTMENT
5220/6520 Models

NOTE: On 1981 and later models, remove the air/fuel mixture concealment plug, by refering to "Carburetors" in the Unit Repair section.

1. Set the parking brake and place the transaxle in Neutral.
2. Turn off the headlights, all accessories and the A/C compressor.
3. Connect a tachometer to the engine following the manufacturer's instructions.
4. If equipped with a Fuel Control Computer (FCC), connect a jumper wire from the carburetor switch to ground.

5. Start the engine and allow it to reach normal operating temperature.
6. Refer to the emission control label in the engine compartment, then check and adjust the timing, if necessary.
7. Unplug the connector at the radiator fan and install a jumper so the fan will run continuously. Pull the PCV valve from the cylinder head cover and the $\frac{3}{16}$ in. dia. control hose at the canister, allow them to draw air.
8. Using a propane tank with a main and metering valve (make sure the valves are closed and the tank is an upright and safe location), install the supply hose to the carburetor.
9. On the 1980 models, disconnect and plug the vacuum hoses to the EGR valve. Disconnect the vacuum hose of the distributor at the 3-way

connector and replace it with the propane supply hose. Proceed to Step 12.
10. On 1981–82 models, disconnect and plug the vacuum hoses to the EGR valve and the $\frac{3}{16}$ in. dia. control hose at the canister. Disconnect the vacuum hose of the heated door sensor at the 3-way connector and replace it with the propane supply hose. If equipped with A/C, turn on the A/C and open the throttle to energize the solenoid. Remove the adjusting screw and spring from the top of the A/C solenoid. Insert a $\frac{1}{8}$ in. Allen wrench into the solenoid and adjust the idle speed (check emission control label for specifications) with the compressor clutch engaged. After adjustment, place the automatic transaxle in Park. Reinstall the A/C solenoid spring and screw. Proceed to Step 11.
11. On 1983 models, disconnect and plug the vacuum connector at the Coolant Controlled Exhaust Gas Recirculation/Coolant Vacuum Switch Cold Closed (CCEGR/CVSCC). Disconnect the vacuum hose of the heated door sensor at the 3-way connector and replace it with the propane supply hose. Proceed to Step 12.
12. On 1984 and later models, disconnect and plug the vacuum connector at the Coolant Vacuum Switch Cold Closed (CVSCC). If equipped with a Holley 6520, disconnect the O_2 sensor test connector on the left fender shield. Disconnect the vacuum hose of the heated door sensor at the 3-way connector and replace it with the propane supply hose.

NOTE: If equipped with A/C (1983–86), check the A/C kicker operation and the kicker vacuum system for leaks, service is by replacement of the kicker or vacuum lines ONLY.

13. Open the main valve of the propane tank, then slowly open the metering valve until the highest rpm is reached (excessive propane will decrease the engine rpm). Fine tune the metering valve to get the highest engine rpm. With the propane still flowing, adjust the idle speed screw (on top of the solenoid) to meet the propane rpm specified on the emission control label.
14. Again, fine tune the propane metering valve to get the highest rpm. If the engine speed changed, readjust the idle speed screw to meet the propane rpm specified on the emission control label.
15. Turn off the main valve of the propane tank, allow the engine to stabilize and adjust the air/fuel mixture screw (1981–86, use an Allen wrench) to achieve the specified idle rpm. Af-

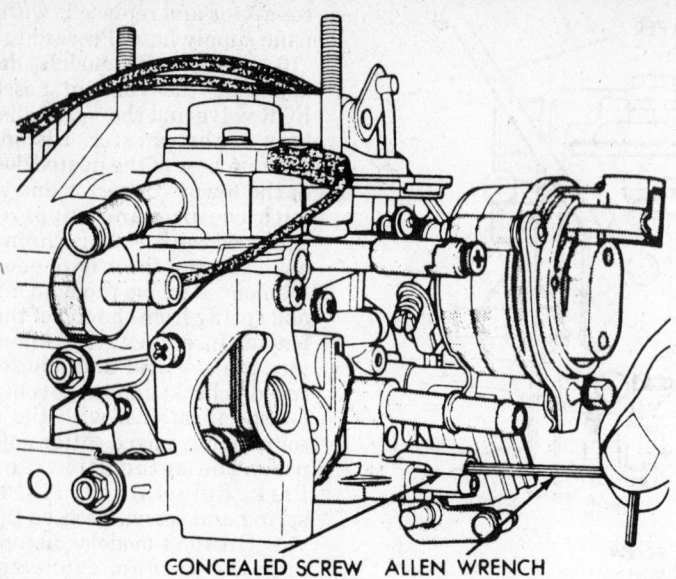

CONCEALED SCREW   ALLEN WRENCH

Adjusting the Air/Fuel mixture screw

ter adjustment, allow the engine to stabilize.

16. Turn On the main valve of the propane tank and fine tune the metering valve to achieve the highest engine rpm. If the highest rpm is 25 rpm or higher than the specified propane rpm, repeat the propane assisted adjustment steps; if the highest rpm is 25 rpm or less than the specified propane rpm, turn off the valves and remove the propane tank.

17. After the air/fuel mixture and idle speed have been set, remove the test equipment. Reinstall the concealment or limiter plugs, the wiring and hoses.

Mikuni Model

NOTE: On the 1981–86 models, remove the air/fuel mixture concealment plug, by refering to "Carburetors" in the Unit Repair section.

1. Set the parking brake and place the transaxle in Neutral.
2. Turn off all lights and accessories. Disconnect the cooling fan.
3. Connect a tachometer to the engine, following the manufacturer's instructions.
4. Start the engine, allow it to reach normal operating temperature and stop the engine.
5. Disconnect the negative battery cable, wait three seconds and reconnect it.
6. Start the engine, open the throttle, run the engine at 2500 rpm for 10 seconds, return to idle and wait two minutes. Refer to the emission control label in the engine compartment.

NOTE: On 1984–86 models, remove the idle switch connector and the bullet connector at the O_2 sensor.

7. Using a propane tank with a main and metering valve (make sure the valves are closed and the tank is an upright and safe location), insert the supply hose 4 in. into the air cleaner.

8. Open the main valve of the propane tank, then slowly open the metering valve until the highest rpm is reached (excess propane will decrease the engine rpm). Fine tune the metering valve to get the highest engine rpm. With the propane still flowing, adjust the idle speed screw to meet the propane rpm specified on the emission control label.

9. Again, fine tune the propane metering valve to get the highest rpm. If the engine speed changed, readjust the idle speed screw to meet the propane rpm specified on the emission control label.

10. Turn off the main valve of the propane tank, allow the engine to stabilize and adjust the air/fuel mixture screw to achieve the specified idle rpm. After adjustment, allow the engine to stabilize.

11. Turn on the main valve of the propane tank and fine tune the metering valve to achieve the highest engine rpm. If the highest rpm is 25 rpm or higher than the specified propane rpm, repeat the propane assisted adjustment steps; if the highest rpm is 25 rpm or less than the specified propane rpm, turn off the valves and remove the propane tank.

12. Reinstall the air cleaner duct,

the concealment plug and the O_2 sensor connector (if equipped).

13. If equipped with A/C, set the temperature control lever to the coldest position and turn on the A/C. With the compressor running, set the engine speed to 900 rpm by adjusting the idle up screw.

14. Turn the engine off, disconnect the tachometer, reconnect the cooling fan and idle switch connector (if equipped).

FAST IDLE SPEED ADJUSTMENT

5220/6520 Models

1. Refer to the "Idle Speed Adjustment" procedures in this section and set the idle speed.
2. After adjusting the idle speed, place the fast idle adjusting screw on the lowest step of the fast idle cam.
3. Make sure that the choke valve is fully Open. Adjust the fast idle speed by turning the fast idle screw.
4. With the fast idle speed adjusted, remove the test equipment and reinstall the items that were removed.

Mikuni Model

1. Refer to the "Idle Speed Adjustment" procedures in this section and set the idle speed.
2. After adjusting the idle speed, open the throttle slightly and install the tool No. C-4812 or equivalent, onto the choke cam follower pin.
3. Release the throttle and adjust the fast idle screw to the specified fast idle rpm.
4. With the fast idle speed adjusted, remove the adjusting tool and test equipment; then reinstall the items that were removed.

Fuel Injection

Two types of electronic (computer regulated) fuel injection systems are used; a single-point and a multi-point. Both systems utilize a digital computer (Logic Module) which responds to changing operating conditions by regulating the ignition timing, the air/fuel ratio, the emission control devices and the idle speed. In both systems, the sensors and switches are used to gather information for the Logic Module, which regulates the fuel injection.

The Single-Point Fuel Injection (SFI) system consists of a throttle body assembly (mounted on top of the intake manifold) which houses a fuel injector, a pressure regulator, a throttle position sensor and an automatic idle speed motor.

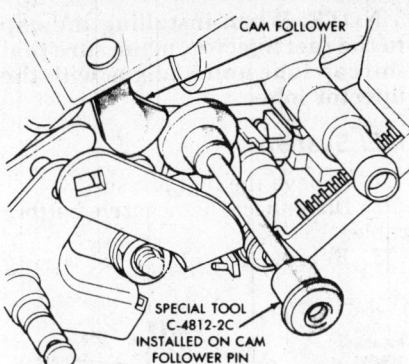

Installing tool C-4812-2C on the choke cam follower pin

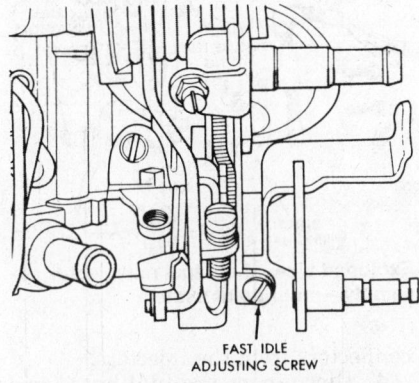

Adjusting the fast idle speed screw

The Multi-Point Fuel Injection (MFI) system utilizes a throttle body (mounted on top of the intake manifold) which houses a throttle position sensor and an automatic idle speed motor; a separate fuel injector is used for each cylinder.

Due to the complex nature of modern fuel injection systems, comprehensive testing and diagnosis procedures fall outside the confines of this repair manual. For complete information on diagnosis, testing and repair procedures concerning all modern fuel injection systems, please refer to *Chilton's Guide To Fuel Injection and Feedback Carburetors.*

FUEL PRESSURE RELEASE

The fuel system is under constant pressure which must be relieved before attempting any service procedures. Take precautions to avoid the risk of fire whenever working on or around any fuel system components.

1. To release the pressure, remove the fuel tank cap and the wiring connector from the fuel injector (SFI) or any fuel injector (MFI).

2. Connect a jumper wire from one fuel injector terminal to ground and a second jumper wire from the other injector terminal to the positive battery terminal for 10 seconds.

3. With the fuel pressure reduced, remove the jumper wires and continue with the fuel system service.

Throttle Body

REMOVAL & INSTALLTION

SFI System

1. Relieve the fuel pressure.
2. Disconnect the negative battery terminal.
3. Disconnect the fuel injector wiring connector and the 6-way connector from the throttle body.

4. Remove the ground wire from the 6-way connector and the air cleaner hose.
5. Remove the throttle cable, the speed control and the transmission kickdown cables, if equipped.
6. Remove the return spring and the vacuum hoses.
7. Loosen the fuel delivery and the return hose clamps, wrap a towel around each hose and twist the hose from the connection.
8. Remove the mounting screws and the throttle body from the vehicle.
9. To install, use a new gasket and reverse the removal procedures. Torque the throttle body-to-intake manifold screws to 17 ft. lbs. Start the engine and check for fuel leaks.

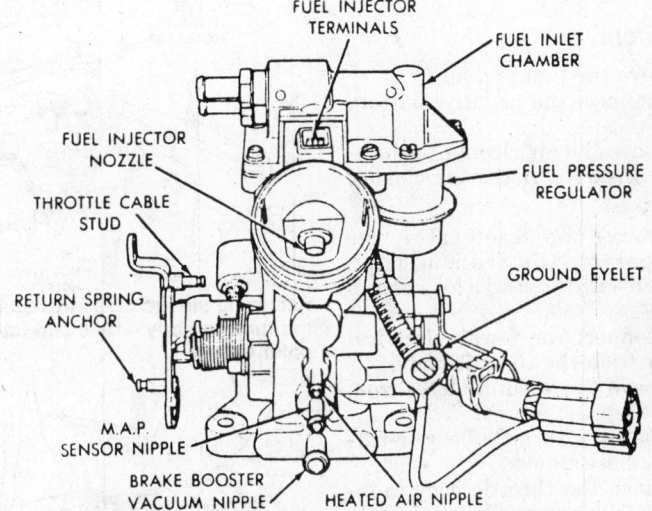

Single-Point Fuel Injection Throttle Body (1983–85)

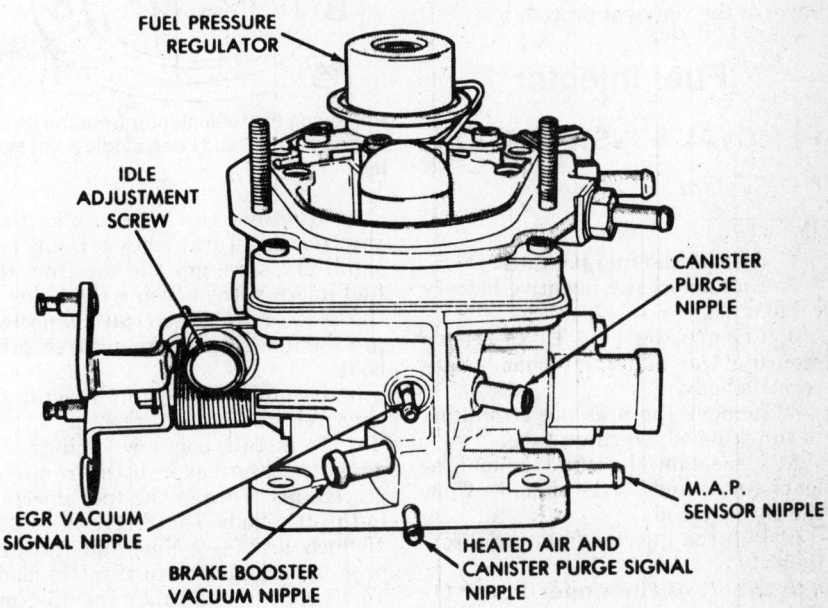

View of the single point throttle body—1986 and later

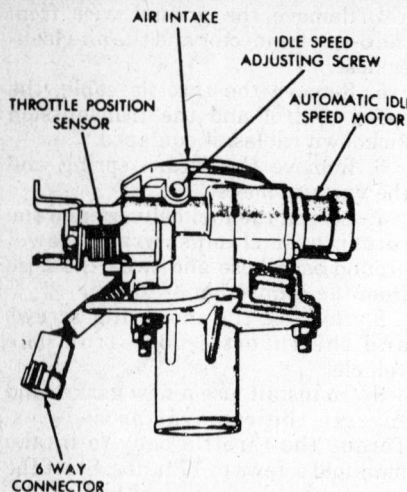

Multi-Point Fuel Injection Throttle Body

MFI System

1. Relieve the fuel pressure.
2. Disconnect the negative battery cable.
3. Remove the air cleaner-to-throttle body screws and the air cleaner adaptor.
4. Remove the accelerator, the speed control, the transmission kickdown cables and the return spring.
5. Disconnect the 6-way electrical connector from the throttle body.
6. Remove the vacuum hoses from the throttle body.
7. Loosen the throttle body-to-turbocharger hose clamp.
8. Remove the throttle body-to-intake manifold nuts and the throttle body from the vehicle.
9. To install, use a new gasket and reverse the removal procedures.

Fuel Injector

REMOVAL & INSTALLATION

SFI System

1983–85

1. Relieve the fuel pressure.
2. Disconnect the negative battery cable.
3. Remove the four Torx® screws securing the fuel inlet chamber-to-throttle body.
4. Remove the pressure regulator-to-throttle body vacuum tube.
5. Place a towel around the fuel inlet chamber and lift the chamber from the throttle body.
6. Pull the injector from the throttle body.
7. Remove the upper/lower O-rings, the snap ring, the seal and the washer from the fuel injector.

8. To install, use new O-rings, seal and washer, then reverse the removal procedures. Torque the inlet chamber screws to 35 inch lbs.

1986 and Later

1. Relieve the fuel pressure.
2. Disconnect the negative battery terminal and remove the air cleaner.
3. Remove the fuel pressure regulator-to-throttle body screws and pull the regulator from the throttle body. Remove and discard the O-ring and the gasket from the pressure regulator.

——— CAUTION ———

Be sure to place a towel around the inlet chamber to catch any fuel left in the system.

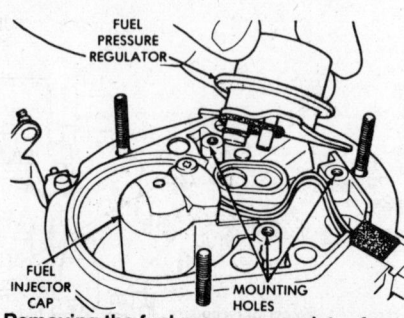

Removing the fuel pressure regulator from the throttle body—1986 and later, single point system

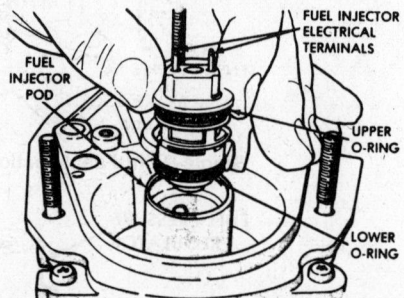

Removing the fuel injector from the throttle body—1986 and later, single point system.

4. Remove the injector cap-to-throttle body Torx® screws. Using two small pry bars, pry the cap from the fuel injector, then (using the holes in the side of the electrical connector) pry the fuel injector from the throttle body.
5. Remove the lower injector O-ring from the pod.
6. To install, use new O-rings and gaskets, then reverse the removal procedures. Torque the fuel injector-to-throttle body Torx® screws to 35–45 inch lbs. (4–5 Nm) and the fuel pressure regulator-to-throttle body Torx® screws to 40 inch lbs. (5 Nm). Start the engine and check for fuel leaks.

NOTE: When installing the cap to the fuel injector, make sure that the cap lobe notch aligns with the injector lobe.

MFI System

1. Relieve the fuel pressure.
2. Disconnect the negative battery cable.
3. Remove the electrical wiring

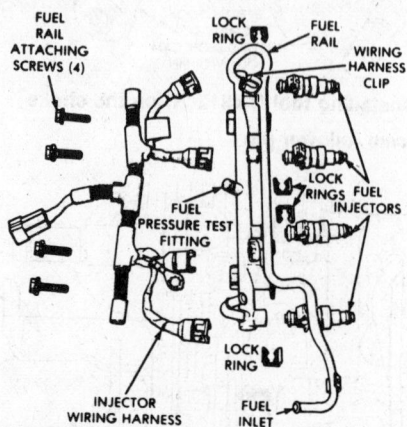

Exploded view of the fuel rail/injector assembly—multi point system

connectors from the injectors.
4. Disconnect the fuel hose from the fuel rail inlet and the pressure regulator vacuum hose from the intake manifold vacuum tree.
5. Remove the pressure regulator-to-intake manifold screws, the fuel rail-to-pressure regulator hose and the regulator.
6. Remove the fuel rail-to-valve cover bracket screw, the fuel injector heat shield clips, the fuel injector-to-intake manifold screws and pull the fuel rail/injector assembly from the intake manifold.
7. Remove the fuel injector-to-fuel rail clip and separate the injector from the fuel rail.
8. To install, use new gaskets and O-rings, then reverse the removal procedures. Start the engine and check for leaks.

IGNITION TIMING ADJUSTMENT

1. Using a timing light, connect it to the engine according to the manufacturers instructions.
2. Using a tachometer, connect it to the engine and place the selector on the proper cylinder position.
3. Start the engine and allow it to reach operating temperatures.
4. At the thermostat housing, disconnect and reconnect the water temperature sensor electrical connector; the loss of power light (on the dash)

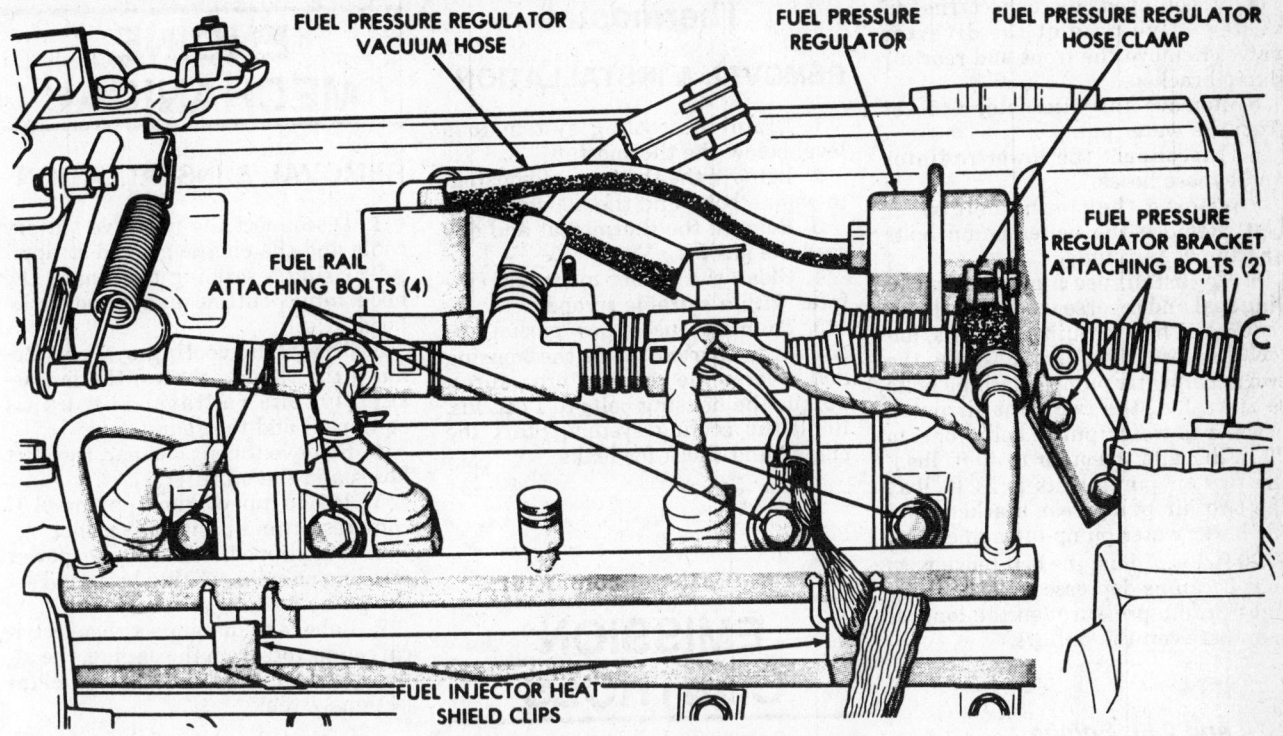

FUEL PRESSURE REGULATOR VACUUM HOSE

FUEL PRESSURE REGULATOR

FUEL PRESSURE REGULATOR HOSE CLAMP

FUEL RAIL ATTACHING BOLTS (4)

FUEL PRESSURE REGULATOR BRACKET ATTACHING BOLTS (2)

FUEL INJECTOR HEAT SHIELD CLIPS

View of the multi point fuel injection system

will come on and stay on. Check the engine speed.

5. Using the timing light, aim it at the timing hole in the bell housing or read the magnetic unit.

6. If timing adjustment is necessary, loosen the distributor hold down clamp and turn the distributor to adjust the timing.

7. Turn the engine off, then disconnect and reconnect the positive battery quick disconnect. Start the engine; the loss of power light should be off.

8. Turn the engine off, then turn the ignition switch ON and OFF several times to clear the fault codes.

ENGINE COOLING

The cooling system consists of a radiator, a fan shroud with A/C, overflow tank, water pump, thermostat, coolant temperature switch, electric fan and radiator fan switch (some engines). A radiator bypass system is used for faster warmup.

Radiator

REMOVAL & INSTALLATION

1. Move the temperature selector

to "FULL ON" and open the radiator drain cock.

2. When the coolant reserve tank is empty, remove the radiator cap.

3. On the 1.6L, 2.2L and 2.5L engines (1984 and later), remove the vacuum valve or plug on top of the thermostat housing.

4. Remove the clamps and hoses from the radiator.

5. If equipped with an automatic transaxle, disconnect and plug the fluid cooler lines.

6. Disconnect the fan motor and switch leads. Remove the fan and the fan support assembly.

7. Remove the upper shroud attaching screws, lift the shroud, separate it from the bottom shroud attaching clips and the radiator.

8. Remove the upper radiator attaching bolts.

9. On the 2.6L engine, disconnect the carburetor air duct and the engine block heater wire (if equipped).

10. On the 2.2L and 2.5L engine, remove the heater return hose from the radiator.

11. Lift the radiator from the engine compartment.

12. To install, reverse the removal procedures and add coolant until it meets the radiator cap seat. On the 1.6L, 2.2L and 2.5L engines (1984 and later), add coolant until it rises to the top of the hole in the thermostat housing, install the vacuum valve or plug and continue adding fluid until it meets the radiator cap seat. On all ve-

hicles, after the radiator cap is installed, fill the reserve tank to the "MAX" line and check the coolant level after 2–3 warm ups.

Water Pump

REMOVAL & INSTALLATION

1.6L Engine

1. Remove the radiator cap and drain the cooling system through the water pump drain plug.

2. Disconnect the pump-to-block coolant hose at the pump.

3. Loosen the water pump drive belt. Remove the retaining screws and the water pump pulley.

4. Remove the four pump-to-crankcase extension screws and the water pump assembly.

5. To install, use a new gasket and reverse the removal procedures. Torque the pump extension bolts to 9 ft. lbs. and the pump drain plug to 13 ft. lbs. Adjust the belt tension, so that it can be depressed $\frac{1}{4}$ in. (under light thumb pressure) on the longest span between the two pulleys.

1.7L Engine

1. Drain the cooling system.

2. Without discharging the system, move the A/C compressor from the engine brackets and set it aside.

3. Remove the alternator and the water pump pulley.

4. If equipped, disconnect the diverter valve hose at the diverter valve. Remove the front and rear air pump brackets.

5. Remove the alternator bracket from the water pump.

6. Disconnect the lower radiator and bypass hoses.

7. Loosen the timing belt cover bolt. Remove the water pump bolts and the water pump.

8. To install, use a new rubber O-ring seal and reverse the removal procedures. When torquing the bolts, follow this torque sequence: Torque the two upper water pump attaching bolts to 20 ft. lbs., the two front air pump bracket-to-water pump bolts (one to 20 ft. lbs. and the other to 40 ft. lbs.), the two air pump bolts to 24 ft. lbs., the two air pump rear brackets and the lower water pump-to-engine bolts to 20 ft. lbs. Adjust the belt tension, so that it can be depressed 1/4 in. (under light thumb pressure) on the longest span between the pulleys.

2.2L and 2.5L Engine

1. Drain the cooling system.

2. Remove the upper radiator hose.

3. Without discharging the A/C system, remove the compressor from the engine brackets and set it aside.

4. Remove the alternator and move it aside.

5. Disconnect the lower radiator and bypass hoses. Remove the water pump-to-engine screws and the water pump.

6. To install, reverse the removal procedures. Torque the top three retaining screws to 20 ft. lbs. and the lower screw to 50 ft. lbs. Adjust the belt tension, so that it can be depressed 1/4 in. (under light thumb pressure) on the longest span between the two pulleys.

2.6L Engine

1. Drain the cooling system.

2. Remove the radiator, bypass and heater hoses from the water pump.

3. Remove the drive pulley shield.

4. Remove the locking and pivot screws.

5. Remove the water pump drive belt and pump from the engine.

6. To install, use a new gasket and O-ring, then reverse the removal procedures. Torque the locking and pivot screws to 17 ft. lbs. Tighten the drive pulley shield to 9 ft. lbs. Adjust the belt tension, so that it can be depressed 1/4 in. (under light thumb pressure) on the longest span between the two pulleys.

Thermostat

REMOVAL & INSTALLATION

1. Drain the cooling system to a level below the thermostat.

2. Remove the thermostat housing-to-engine bolts and the housing.

3. Remove the thermostat and discard the gasket.

4. Clean the gasket mounting surfaces with a suitable scraper.

5. To install, use a new gasket, position the thermostat in the housing and reverse the removal procedures. Torque the housing bolts to 17 ft. lbs. Refill the cooling system. Start the engine and check for leaks.

EMISSION CONTROLS

Several systems are used and vary with each vehicle. Most do not require service, but those that do require the use of sophisticated testing equipment.

For further maintenance and service information on the emission control system, please refer to "Emission Controls" in the Unit Repair section.

Oxygen Sensor

REMOVAL & INSTALLATION

1. Remove the bullet connector from the tip of the sensor.

2. Using tool No. C-4589 or equivalent, remove the sensor from the exhaust manifold.

3. Clean the manifold threaded hole with a tap (18mm x 1.5 x 6E).

4. To install, coat the threads with an anti-seize compound and reverse the removal procedures.

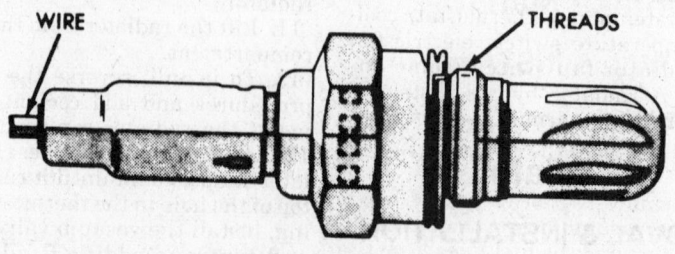

Oxygen sensor

ENGINE MECHANICAL

REMOVAL & INSTALLATION

1. Disconnect the negative battery cable and the engine ground straps.

2. Using a scribing tool, mark the hood hinge outline and remove the hood.

3. Drain the cooling system. Remove the radiator hoses, the fan assembly, the radiator shroud (if equipped) and the radiator.

4. Remove the air cleaner, the duct hoses and the oil filter.

5. If equipped, remove the A/C compressor and position it aside.

6. If equipped, remove the power steering pump mounting bolts and set the pump aside.

7. Label and disconnect the electrical connectors from the engine, the alternator and the carburetor or fuel injection system.

NOTE: If equipped with a fuel injection system, it will be necessary to reduce the pressure in the fuel system before disconnecting the fuel lines.

8. Disconnect the fuel line, the heater hoses and the accelerator linkage.

9. If equipped, disconnect the air pump lines and remove the pump.

10. Remove the alternator.

11. Disconnect the shift linkage(s), the clutch (if equipped) and speedometer cables.

12. Raise and support the front of the vehicle on jackstands.

13. If equipped with a manual transaxle, perform the following procedures:

   a. Disconnect the clutch cable.

   b. Remove the lower cover from the transaxle case.

   c. Remove the exhaust pipe-to-manifold bolts and separate the pipe from the manifold.

   d. Remove the starter and lay it aside.

e. Using a transmission holding tool, secure it to the transaxle.

f. If equipped, remove the anit-roll strut or damper (turbocharged) from the transaxle.

14. If equipped with an automatic transaxle, perform the following procedures:

a. Remove the lower cover from the transaxle case.

b. Remove the exhaust pipe-to-manifold bolts and separate the pipe from the manifold.

c. Remove the starter and lay it aside.

d. Mark the flex plate-to-torque converter, for installation purposes.

e. Remove the torque converter-to-flex plate bolts and separate the converter from the flex plate.

f. Using a "C" clamp, secure the bottom of the torque converter to the transaxle so that it will not come out.

g. Using a transmission holding tool, secure it to the transaxle.

15. Attach a lifting fixture and a vertical lift to the engine.

16. To lower the engine, separate the right side engine bracket from the yoke bracket. To raise the engine, remove the yoke/insulator long bolt.

NOTE: If the insulator-to-rail screws are to be removed, mark the position of the insulator on the side rail (to insure reinstallation).

17. Remove the transaxle-to-engine bolts, the front engine mount nut/bolt and the left insulator through bolt (from inside the wheelhouse) or the insulator bracket-to-transaxle bolts.

18. Lift the engine from the vehicle.

19. To install, reverse the removal procedures. Loosely install all of the mounting bolts and torque the engine-to-mount bolts to 40 ft. lbs. (54 Nm); the engine-to-transaxle bolts to 70 ft. lbs. (95 Nm) and the torque converter-to-flex plate bolts to 40 ft. lbs. (54 Nm). Refill the cooling system. Start the engine, allow it to reach normal operating temperature and check for leaks. Check the timing and adjust if necessary. Adjust the idle speed/mixture (if possible) and the transaxle linkage.

Intake Manifold

REMOVAL & INSTALLATION

1.6L Engine

1. Disconnect the negative battery cable.

2. Remove the air cleaner. Disconnect the vacuum hoses, the electrical connectors and the fuel lines from the carburetor.

3. Drain the cooling system. Disconnect the water box-to-heated intake manifold inlet hose, the intake manifold-to-heater hose and the intake manifold EGR valve tube.

4. Remove the eight manifold-to-head bolts and lift out the intake manifold. Remove the carburetor from the intake manifold.

5. Clean the gasket mounting surfaces with a suitable scraper.

6. To install, use new gaskets and reverse the removal procedures. Torque the intake manifold to 11 ft. lbs. Refill the cooling system. Start the engine and check for leaks.

NOTE: When torquing the intake manifold-to-cylinder head bolts, start in the center of the manifold and work outward.

2.2L and 2.5L Engines

NOTE: If equipped with a turbocharger, refer to "Turbocharger Removal & Installation" procedures in this section and remove the turbocharger. After the turbocharger has been removed, remove the intake manifold-to-cylinder head bolts and the intake manifold.

1. Disconnect the negative battery cable. Drain the cooling system.

2. Remove the air cleaner. Disconnect the vacuum lines, the electrical connectors and the fuel lines from the carburetor or the fuel injection system.

NOTE: If equipped with a fuel injection system, it will be necessary to reduce the pressure in the fuel system before disconnecting the fuel lines.

3. Remove the throttle linkage.

4. Loosen the power steering pump and remove the belt.

5. Remove the power brake vacuum hose from the intake manifold.

NOTE: If equipped with an AIR system, remove the coupling hose from the diverter valve-to-exhaust manifold air injection tube assembly.

6. Remove the water crossover hoses.

7. Raise and support the front of the vehicle on jackstands. Remove the exhaust pipe-to-exhaust manifold bolts and separate the pipe from the mainfold.

8. Remove the power steering pump assembly and set it aside.

9. Remove the intake manifold support bracket and the EGR tube.

NOTE: If equipped with an AIR system, remove the four injection

tube bolts and the air injection assembly.

10. Remove the intake manifold-to-cylinder head bolts. Lower the vehicle and remove the intake manifold.

11. Clean the gasket mating surfaces with a suitable scraper.

12. To install, use new gaskets and reverse the removal procedures. Torque the intake manifold-to-cylinder head bolts to 17 ft. lbs. (23 Nm). Refill the cooling system. Start the engine and check for leaks.

NOTE: When torquing the intake manifold-to-cylinder head bolts, start in the center of the manifold and work outward.

2.6L Engine

1. Disconnect the negative battery cable.

2. Drain the cooling system and disconnect the water pump-to-intake manifold hoses.

3. Disconnect the carburetor air horn and move aside.

4. Disconnect the vacuum hoses/throttle linkage from the carburetor and the intake manifold.

5. Disconnect the fuel inlet line at the fuel filter.

6. Remove the fuel filter and the fuel pump, move it aside.

7. Remove the intake manifold retaining nuts/washers and the manifold. Remove the carburetor from the intake manifold.

8. Clean the gasket mounting surfaces with a suitable scraper.

9. To install, use new gaskets and reverse the removal procedures. Torque the intake manifold retaining nuts to 12 ft. lbs. Refill the cooling system. Start the engine and check for leaks.

NOTE: When torquing the intake manifold-to-cylinder head bolts, start in the center of the manifold and work outward.

Exhaust Manifold

REMOVAL & INSTALLATION

1.6L Engine

1. Disconnect the negative battery cable.

2. Separate the carburetor air heater tube from the exhaust manifold heat stove.

3. Remove the O_2 sensor from the exhaust manifold. If equipped with an AIR system, disconnect the AIR pipe from the exhaust manifold.

4. Separate the EGR assembly from the exhaust manifold.

5. Raise and support the front of

the vehicle on jackstands. Remove the exhaust pipe-to-exhaust manifold nuts and separate the pipe from the manifold.

6. Remove the exhaust manifold-to-cylinder head nuts and the exhaust manifold. If equipped, remove the carburetor air heater from the exhaust manifold.

7. Clean the gasket mounting surfaces with a suitable scraper.

8. To install, use new gaskets and reverse the removal procedures. Torque the exhaust manifold to 15 ft. lbs.

NOTE: When torquing the exhaust manifold-to-cylinder head bolts, start in the center of the manifold and work outward.

2.2L and 2.5L Engines

NOTE: If equipped with a turbocharger, refer to "Turbocharger Removal & Installation" procedures in this section and remove the turbocharger. After the turbocharger has been removed, remove the exhaust manifold retaining nuts and the exhaust manifold.

1. Disconnect the negative battery cable. Drain the cooling system.

2. Remove the air cleaner. Disconnect the vacuum lines, the electrical connectors and the fuel lines from the carburetor.

3. Loosen the power steering pump and remove the belt.

NOTE: If equipped with an AIR system, remove the coupling hose from the diverter valve-to-exhaust manifold air injection tube assembly.

4. Remove the water crossover hoses.

5. Raise and support the front of the vehicle on jackstands. Remove the exhaust pipe from the exhaust manifold.

6. Remove the power steering pump assembly and set it aside.

7. Remove the intake manifold support bracket and the EGR tube.

NOTE: If equipped with an AIR system, remove the four injection tube bolts and the air injection assembly.

8. Lower the vehicle. Remove the exhaust manifold nuts and the exhaust manifold.

9. Clean the gasket mounting surfaces with a suitable scraper.

10. To install, use new gaskets and reverse the removal procedures. Torque the exhaust manifold to 17 ft. lbs. Refill the cooling system. Start the engine and check for leaks.

NOTE: When torquing the exhaust manifold-to-cylinder head bolts, start in the center of the manifold and work outward.

2.6L Engine

1. Disconnect the negative battery cable. Drain the cooling system.

2. Remove the air cleaner.

3. Remove the belt from the power steering pump.

4. Raise and support the front of the vehicle on jackstands. Remove the exhaust pipe from the manifold.

5. Disconnect the air injection tube assembly from the exhaust manifold and lower the vehicle.

6. Disconnect the air injection tube assembly from the air pump and move it aside.

7. Remove the power steering pump assembly and move it aside.

8. Remove the heat cowl from the exhaust manifold.

9. Remove the exhaust manifold-to-cylinder head nuts and the exhaust manifold assembly.

10. Remove the carburetor air heater from the manifold.

11. Separate the exhaust manifold from the catalytic converter by removing the retaining screws.

12. Clean the gasket mounting surfaces with a suitable scraper.

13. To install, use new gasket (coat the cylinder head side lightly with sealer) and reverse the removal procedures. Torque the exhaust manifold-to-catalytic converter mounting screws to 24 ft. lbs. and the exhaust manifold-to-cylinder head nuts to 12 ft. lbs. Refill the cooling system. Start the engine and check for leaks.

NOTE: When torquing the exhaust manifold-to-cylinder head bolts, start in the center of the manifold and work outward.

Combination Manifold

REMOVAL & INSTALLATION

1.7L Engine

1. Disconnect the negative battery cable.

2. Remove the air cleaner. Disconnect the vacuum lines, the electrical connectors and the fuel line from the carburetor.

3. Remove the throttle linkage.

4. Remove the power brake hose from the intake manifold.

5. Raise and support the front of the vehicle on jackstands. Remove the exhaust pipe from the exhaust manifold.

6. Remove the power steering pump and set it aside.

7. Remove the intake/exhaust manifold-to-cylinder head nuts/bolts.

8. Lower the vehicle. Remove the carburetor and the intake/exhaust manifold assembly.

9. Remove the carburetor, then separate the intake and exhaust manifolds.

10. Clean the gasket mounting surfaces with a suitable scraper.

11. To install, use new gaskets and reverse the removal procedures. Install a new gasket between the manifolds, assemble but DO NOT tighten the fasteners. Coat the manifold-to-cylinder head gaskets with sealer, assemble and torque the nuts/screws to 10–15 inch lbs. Torque the inboard manifold nuts to 12 ft. lbs. and the outboard nuts to 17 ft. lbs. Torque the manifolds-to-cylinder head nuts and screws to 17 ft. lbs.

NOTE: When torquing the exhaust manifold-to-cylinder head bolts, start in the center of the manifold and work outward.

Turbocharger

A turbocharger is used on the 2.2L engine (1984 and later). For service information, refer to "Turbocharging" in the Unit Repair section.

REMOVAL & INSTALLATION

1. Disconnect the negative battery cable and drain the cooling system.

2. Disconnect the exhaust pipe from the turbocharger and the O_2 sensor electrical connector.

3. Remove the turbocharger-to-engine support bracket.

4. Loosen the oil drain back hose clamps and slide the hose down on the engine nipple.

5. Disconnect the turbocharger coolant tube nut at the engine outlet (below the power steering pump bracket) and the support bracket.

6. Remove the air cleaner assembly, the throttle body adaptor, the hose, the air cleaner box and bracket.

7. Disconnect the accelerator linkage, the throttle body electrical connector and the vacuum hoses.

8. Loosen the throttle body-to-turbocharger inlet hose clamps.

9. Remove the three throttle body-to-intake manifold screws and the throttle body.

10. Loosen the turbocharger discharge hose end clamp ONLY (the center band retains the deswirler).

11. At the fuel rail, remove the hose retainer bracket screw, the four intake bracket screws from the intake manifold and the two bracket-to-heat shield bracket clips. Lift the fuel rail

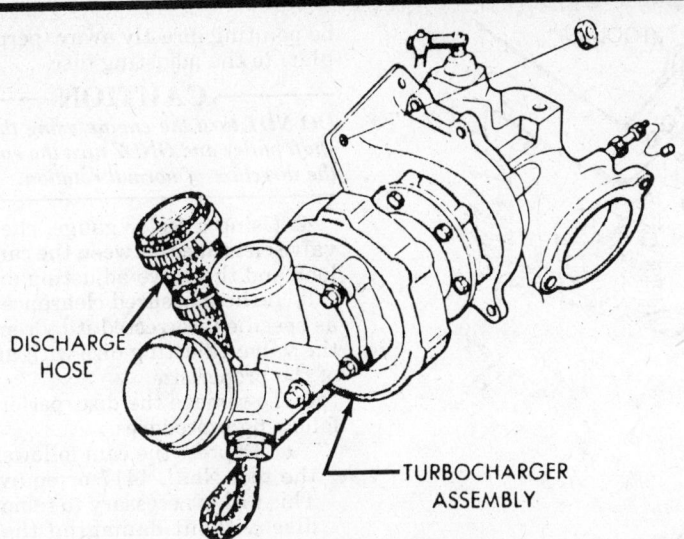

Turbocharger Assembly

DISCHARGE HOSE

TURBOCHARGER ASSEMBLY

(with the injectors, wiring harness and fuel lines intact) up and secure out of the way.

12. Disconnect the oil feed line from the turbocharger bearing housing.

13. Remove the three heat shield-to-intake manifold screws and the heat shield.

14. Disconnect the coolant return tube and hose assembly from the turbocharger and the water box. Remove the tube support bracket from the cylinder head and the assembly.

15. Remove the four turbocharger-to-exhaust manifold nuts. Lift the turbocharger off the studs, push down towards the passenger side, lift up and out of the engine compartment.

16. Clean the gasket mounting surfaces with a suitable scraper.

17. To install, use new gaskets and reverse the removal procedures.

TROUBLESHOOTING

For more information, please refer to "Turbocharging" in the Unit Repair section.

Rocker Arms/Shafts

REMOVAL & INSTALLATION

1.6L Engine

1. Refer to the "Cylinder Head Removal & Installation" procedures in this section, to remove the rocker arm assembly.

2. With the rocker arm assemblies removed from the engine, slide the end brackets, rocker arms and springs from the rocker arm shafts.

NOTE: Store the assembly parts in the correct order of removal, for the rocker arms are assembled in pairs.

3. Check all items for wear and/or distortion. Make sure the oil holes are clear.

4. To install, use new gaskets, lubricate the parts and reverse the removal procedures.

2.2L and 2.5L Engines

1. Remove the PCV module from the cylinder head cover by turning it counterclockwise.

2. Remove the cylinder head cover retaining bolts and the cover.

3. Rotate the crankshaft until the base circle of the camshaft journal is in contact with the rocker arm.

4. Using the valve spring compression tool No. 4682 or equivalent, depress the valve spring and slide out the rocker arm.

NOTE: Be careful not to dislodge the valve spring retainer locks, when compressing the valve springs.

5. To install, use new gaskets and reverse the removal procedures.

2.6L Engine

1. Remove the carburetor-to-cylinder head cover bracket.

2. Remove the cylinder head cover bolts and the cover.

3. Remove the camshaft bearing caps but DO NOT remove the bolts from the bearing caps.

4. Remove the camshaft bearing/rocker arm assembly from the cylinder head.

5. Remove the camshaft bearing bolts from the bearing caps. Slide the bearing caps, rocker arms and springs from the rocker arm shafts.

NOTE: Store the rocker arm assembly parts in the correct order of removal, for the rocker arms and bearing caps are directional. The left rocker arm shaft has two oil holes at the bottom; the right shaft has four oil holes at the bottom.

6. Check for wear and/or distortion, replace the damaged parts.

7. To install, lubricate the parts, use a new cover gasket and reverse the removal procedures. Starting with the center bearing cap and working toward both ends, torque the bearing caps to 7 ft. lbs. and retorque to 14 ft. lbs.

VALVE ADJUSTMENT

1.7L Engine

Valve adjustment is not required as a matter of routine maintenance. It is, however, necessary to check the valve clearance after any cylinder head repairs. Adjusting the clearance is a matter of substituting the discs located at the top of the cam follower. The discs are available in 0.05mm increments from 3.00–4.25mm. One disc is located in each follower. A special tool is required for disc removal and installation. Cold clearance should be 0.006–0.010 in. (0.15–0.25mm) intake

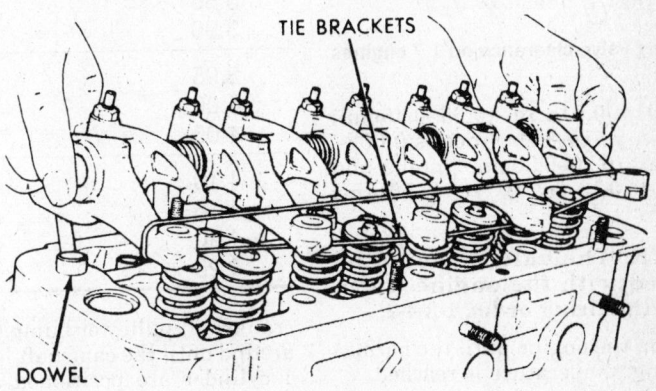

TIE BRACKETS

DOWEL

Removing the rocker arm assembly on the 1.6L engine

Removing the rocker arms on the 2.2L engine

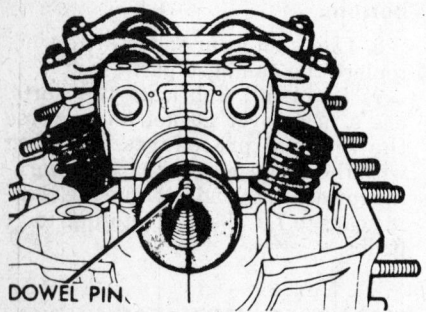
DOWEL PIN
Removing the rocker arm assembly on the 2.6L engine

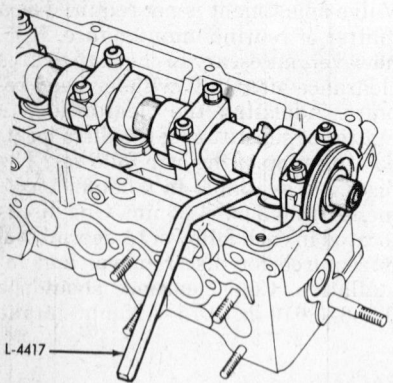
L-4417
Adjusting valve clearance on 1.7 engines

and 0.014–0.018 in. (0.35–0.45mm) exhaust; warm clearance is 0.008–0.012 in. (0.20–0.30mm) intake and 0.016–0.020 in. (0.40–0.50mm) exhaust.

NOTE: The valves should be checked with the engine warm and in the firing order 1-3-4-2.

1. Run the engine until the normal operating temperature is reached.
2. Remove the valve cover.
3. Use a socket wrench on the

VALVE ADJUSTING DISCS

Thickness (mm)	Part Number
3.00	5240946
3.05	5240945
3.10	5240944
3.15	5240943
3.20	5240942
3.25	5240941
3.30	5240573
3.35	5240574
3.40	5240575
3.45	5240576
3.50	5240577
3.55	5240578
3.60	5240579
3.65	5240580
3.70	5240581
3.75	5240582
3.80	5240583
3.85	5240584
3.90	5240585
3.95	5240586
4.00	5240587
4.05	5240588
4.10	5240589
4.15	5240590
4.20	5240591
4.25	5240592

crankshaft pulley or bump the engine around until the camshaft lobes of No. 1 cylinder are positioned as shown. Due to the design of the camshaft lobes, it is not necessary that the lobes

be pointing directly away (perpendicular) to the adjusting disc.

———— CAUTION ————
DO NOT turn the engine using the camshaft pulley and ONLY turn the engine in the direction of normal rotation.

4. Using a feeler gauge, check the valve clearance between the camshaft lobe and the valve adjusting disc.
5. If the measured clearance is not as specified, corrected it by replacing the valve adjusting disc with another of the proper size.
6. To remove the disc, perform the following procedures:
   a. Depress the cam follower with the tool No. L-4417 or equivalent. This tool is necessary to remove the disc without damaging the camshaft or cylinder head.
   b. Remove the valve adjusting disc with a magnet.
   c. Calculate the thickness of a new disc and install one of the proper size. Be sure the number indicating the thickness of the disc (mm) faces down when installed.
   d. Recheck the valve clearances.
7. Recheck or adjust the other valves in the same manner. Reinstall the valve cover.

1.6L Engine

NOTE: Valve clearance must be set, with the piston at the TDC of the compression stroke, with the engine cold.

1. Turn the crankshaft and watch the movement of the exhaust valves. When one is closing (moving upward), continue turning slowly until the inlet valve of the same cylinder, just starts to open. This is the "valve rocking" position. The piston in the opposite cylinder is then at TDC of the compression stroke and its valve clearance can be checked and adjusted. To check the valve clearances on No. 1 cylinder, position the valves of the (paired) cylinder No. 4 in rocking position, as follows:
   a. Observe the rockers on the (paired) cylinder No. 4. Turn the crankshaft until the exhaust valve rocker is moving upward (valve closing). Keep turning slowly until intake valve rocker just starts to

Valves 'Rocking' on Cylinder Number	Adjust Valves on Cylinder Number
4	1
2	3
1	4
3	2

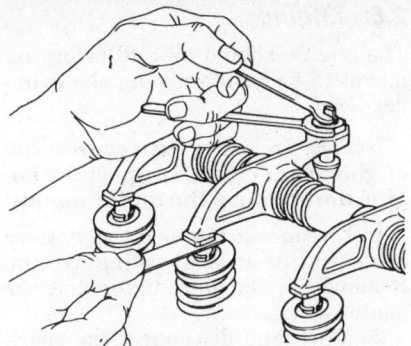

1.6 engine—adjusting valve clearance

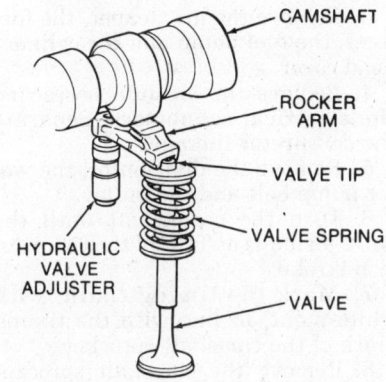

CAMSHAFT

ROCKER ARM

VALVE TIP

VALVE SPRING

HYDRAULIC VALVE ADJUSTER

VALVE

Hydraulic valve adjuster used on 2.2 engine

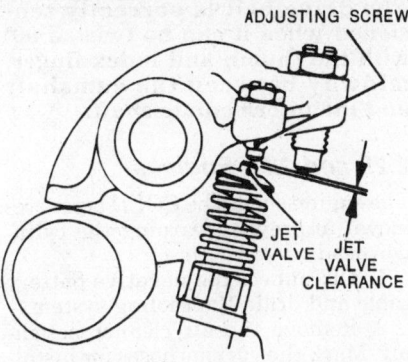

ADJUSTING SCREW

JET VALVE

JET VALVE CLEARANCE

Adjusting the jet valve on 2.6 engines

Adjusting valve lash on 2.6 engines

Exhaust Valve Closing	Adjust
No. 1 Cylinder	No. 4 Cylinder Valves
No. 2 Cylinder	No. 3 Cylinder Valves
No. 3 Cylinder	No. 2 Cylinder Valves
No. 4 Cylinder	No. 1 Cylinder Valves

move down (valve opening), then stop.

b. Check both valve clearances on No. 1 cylinder.

2. After checking both valve clearances, rotate the crankshaft ½ turn, the next cylinder in the firing order should have its valves "rocking" and the paired cylinder can be adjusted.

3. Adjust the valves as needed.

2.2L and 2.5L Engine

The 2.2L and 2.5L engines uses hydraulic lash adjusters. No periodic adjustment or checking is necessary.

2.6L Engine

The 2.6 engine has a jet valve located beside the intake valve of each cylinder.

NOTE: When adjusting the valve clearances, the jet valve must be adjusted before the intake valve.

1. Start the engine and allow it to reach normal operating temperature.

2. Stop the engine. Remove the air cleaner and its hoses. Remove any other cables, hoses, wires or etc., which are attached to the valve cover and remove the valve cover.

3. Disconnect the high tension coil-to-distributor wire at the coil.

4. Watch the rocker arms of the No. 1 cylinder and rotate the crankshaft until the exhaust valve is closing and the intake valve has just started to open. At this point, No. 4 cylinder will be at Top Dead Center (TDC) commencing its firing stroke.

5. Loosen the locknut of the No. 4 cylinder intake valve and back off the intake valve adjusting screw two or more turns.

6. Loosen the locknut on the jet valve adjusting screw.

7. Turn the jet valve adjusting screw counterclockwise, insert a 0.006 in. feeler gauge between the jet valve stem and the adjusting screw.

8. Tighten the adjusting screw until it touches the feeler gauge. Take care not to press on the valve while adjusting, for the jet valve spring is very weak.

NOTE: If the adjusting screw is tight, special care must be taken to avoid pressing down on the jet valve when adjusting the clearance or a false reading will result.

9. Tighten the locknut securely while holding the rocker arm adjusting screw with a screwdriver, to prevent it from turning.

10. Make sure that a 0.006 in. feeler gauge can be easily inserted between the jet valve and the rocker arm.

11. Adjust the No. 4 cylinder's intake valve to 0.006 in. and its exhaust valve to 0.010 in. Tighten the adjusting screw locknuts and recheck each clearance.

12. Perform Step 4 in conjunction with the chart to set up the remaining three cylinders for valve adjustments.

13. Replace the valve cover and all other components. Run the engine and check for oil leaks at the valve cover.

Cylinder Head

REMOVAL & INSTALLATION

1.6L Engine

The engine should be COLD before removal to prevent warping the cylinder head.

1. Remove the valve cover and any necessary vacuum lines. Drain the cooling system.

2. Release the cylinder head bolts evenly, beginning at the ends and working toward the center. The brackets supporting the rocker assembly are located on dowels and are retained by the head bolts. Only, the brackets No. 2 and 4 are pinned to the rocker arm shafts.

3. Tie the end brackets and remove the rocker assembly.

4. Remove the push rods and identify, so they can be installed in the same positions.

5. Remove the cylinder head.

6. Using a suitable scraper, clean the gasket mounting surfaces.

7. To install, use new gaskets and reverse the removal procedures. When installing the head, tighten the bolts progressively to 52 ft. lbs. Refill the cooling system. Run the engine to operating temperature and check for leaks. Allow it to cool to normal air temperature. Retorque the head bolts as needed.

NOTE: When installing the head gasket, the word "DESSUS" or "TOP" must be facing upward.

FRONT

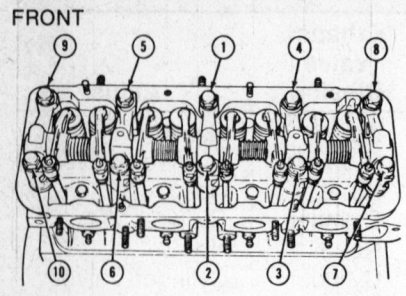

1.6 engine—head bolt tightening sequence

1.7L Engine

The engine should be COLD before removal to prevent warping the cylinder head.

1. Disconnect the negative battery cable. Drain the cooling system.

2. Remove the air cleaner assembly.

3. Disconnect all vacuum lines, hoses and wires from the head, manifold and carburetor. Disconnect the accelerator linkage.

4. Remove the distributor cap.

5. Disconnect the exhaust pipe from the exhaust manifold.

6. Remove the carburetor, the intake/exhaust manifolds and the upper portion of the front cover.

7. Turn the engine by hand until the timing gear marks are aligned.

8. Loosen the drive belt tensioner and slip the belt off the camshaft gear.

NOTE: The camshaft timing mark, located on the back of the camshaft sprocket, is properly positioned when it is aligned with the left corner of the camshaft cover, at the head.

9. If equipped with A/C, remove the compressor from the mounting bracket and move it aside. Remove the mounting brackets from the head.

10. Remove the valve cover, the gaskets/seals and the head bolts, by reversing the tightening sequence.

11. Lift off the head and discard the gasket. Clean the gasket mounting surfaces.

12. Clean the gasket mounting surfaces with a suitable scraper.

13. To install, use new gaskets/seals and reverse the removal procedures. When installing the head gasket, make sure the word "OBEN" (top) faces up. When positioning the head on the block, insert bolts No. 8 and 10 (see illustration) to align the head. Torque the cylinder head bolts in sequence to 30 ft. lbs., again to 60 ft. lbs. and add another ¼ turn. Refill the cooling system. Start the engine and check for leaks.

NOTE: Align the timing marks before installing the drive belt.

The drive belt is correctly tensioned when it can be twisted 90° with the thumb and index finger, midway between the camshaft and the intermediate shaft.

2.2L and 2.5L Engines

The engine should be COLD before removal to prevent warping the cylinder head.

1. Disconnect the negative battery cable and drain the cooling system.

2. Remove the air cleaner assembly. Mark the various hoses for installation identification.

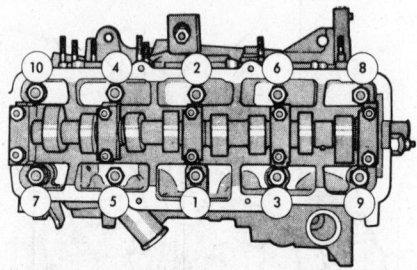

Cylinder head torquing sequence for 1.7L engine. 2.2L and 2.5L engines are similar

3. Disconnect the vacuum lines, hoses and wiring connectors from the manifold(s), carburetor or throttle body and cylinder head.

4. Disconnect the accelerator linkage, the converter and the exhaust pipe. Remove the intake and exhaust manifolds with the carburetor or throttle body.

5. Remove the upper part of the timing case (front cover).

6. Turn the engine by hand until the timing marks align (No. 1 piston at TDC).

7. Loosen the drive belt tensioner and remove the timing belt from the camshaft sprocket.

8. If equipped with A/C, remove the compressor mounting brackets and move the compressor aside.

9. Remove the valve cover, gasket/seals and the head bolts, by reversing the torquing sequence.

10. Lift off the cylinder head.

11. Clean the gasket mounting surfaces with a suitable scraper.

12. To install, use new gaskets/seals and reverse the removal procedures. Torque the cylinder head bolts, in the sequence to 30 ft. lbs., again to 45 ft. lbs. and another ¼ turn. Refill the cooling system. Start the engine and check for leaks.

NOTE: Align the timing marks before installing the drive belt. The drive belt is correctly tensioned when it can be twisted 90° with the thumb and index finger, midway between the cam and intermediate shafts.

2.6L Engine

The engine should be cold before removal to prevent warping the cylinder head.

NOTE: In order to remove the engine front cover, support the engine and remove the motor mount.

1. Disconnect the negative battery cable and drain the cooling system. Remove the upper radiator and the heater hoses.

2. Mark and disconnect the spark plug cables. Remove the carburetor-to-cylinder head cover bracket.

3. Remove the air cleaner, the fuel lines, the fuel pump and the cylinder head cover.

4. Remove the vacuum hoses and the electrical connectors. Separate the carburetor linkage.

5. Remove the distributor, the water pump belt and the pulley.

6. Turn the crankshaft until the No. 1 piston is at TDC of the compression stroke.

7. Mark the timing chain, with white paint, in line with the timing mark of the camshaft sprocket.

8. Remove the camshaft sprocket bolt, the sprocket, the distributor drive gear and the air feeder hoses (from under the vehicle).

9. Remove the power steering pump (set it aside), the ground wire and the dipstick tube.

10. Remove the exhaust manifold shield and separate the manifold from the converter.

11. Remove the cylinder head bolts, by reversing the torquing sequence. The head bolts should be loosened in 2–3 steps, to prevent head warpage.

12. Remove the cylinder head and gasket.

13. Clean the gasket mating surfaces with a suitable scraper.

14. To install, use new gaskets and seals, then reverse the removal procedures. Install the ten cylinder head bolts. Refer to the cylinder head bolt tightening sequence, torque the bolts to 35 ft. lbs. and again, to 69 ft. lbs. (COLD). Torque the two front (chain case cover) bolts to 11–15 ft. lbs. Align the timing marks and install the timing chain/camshaft sprocket. Torque the camshaft sprocket bolt to 40 ft. lbs. Very slowly turn the engine over two times to make sure the valve timing is correct. Fill the cooling system with coolant. Start the engine and check for leaks.

OVERHAUL

For all cylinder head overhaul procedures, please refer to "Engine Rebuilding" in the Unit Repair section.

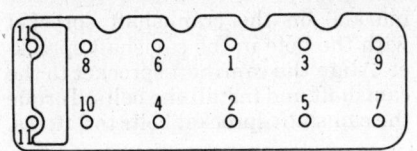

Cylinder head torque sequence—2.6 engines

Front Cover

REMOVAL & INSTALLATION

1.6L Engine

1. Remove the air pump and the alternator belts. Remove the air pump mounting bracket.
2. Raise the vehicle and remove the right inner splash shield.
3. Remove the crankshaft pulley bolt, washer and pulley.
4. Drain the cooling system through the water pump drain plug. Remove the water pump-to-front cover hose.
5. Slightly raise and carefully support the front cover end of the engine.
6. Remove the bolts securing the engine mount bracket-to-front cover and engine block.
7. Remove the two crankcase extension-to-front cover and front cover-to-engine block screws. Remove the front cover.

NOTE: Make sure that the two front cover-to-engine block screws, passing through the tubular locating dowels, DO NOT fall into the crankcase extension, when removing the front cover.

8. Clean the gasket mounting surfaces with a suitable scraper.
9. To install, use a new gasket and reverse the removal procedures. Torque the front cover-to-engine bolts to 9 ft. lbs. (7mm) and/or 15 ft. lbs. (8mm). and the crankshaft pulley bolt to 110 ft. lbs.

1.7L Engine

1. Disconnect the negative battery cable.
2. Remove the A/C compressor, the alternator, the power steering pump and the drive belts, then set the equipment aside.
3. Raise and support the front of the vehicle on jackstands, then remove the splash fender shield.
4. Remove the idler pulley assembly, the crankshaft pulley and the upper/lower front cover.
5. To install, reverse the removal procedures.

2.2L and 2.5L Engine

1. If equipped with A/C, remove the mounting bracket bolts and move the compressor aside. Remove the water pump and the power steering belts.
2. Remove the crankshaft pulley bolt and the pulley.
3. Remove the upper and lower timing belt covers.
4. To complete the installation, reverse the removal procedures. Torque the front cover-to-engine bolts to 40 inch lbs. (4 Nm) and the crankshaft pulley bolt to 21 ft. lbs. (28 Nm). Check the ignition timing.

2.6L Engine

1. Remove the negative battery cable.
2. Remove the alternator, the alternator bracket and the drive belt.
3. If equipped with A/C, remove the compressor bracket and move it aside.
4. Remove the power steering pump bracket and move the pump aside.
5. Remove the distributor retaining nut and move the distributor aside.
6. Raise and support the front of the vehicle on jackstands. Remove the right inner splash shield. Remove the crankshaft pulley bolt and the pulley.
7. Lower the vehicle and place a jack under the engine. Remove the right side engine mounting bolt and raise the engine slightly.
8. Remove the dip stick tube. Drain the engine oil. Remove the oil pan screws, the oil pan and the screen.
9. Remove the air cleaner, the spark plug wires and the cylinder head vacuum hose connections.
10. Remove the cylinder head cover screws, the cover and the two front cylinder head screws.
11. At the front of the engine, remove the timing indicator plate and the engine mounting plate screws.
12. Remove the front cover screws and the cover.
13. Clean the gasket mounting surfaces with a suitable scraper.
14. To install, use a new gaskets, sealant and reverse the removal procedures. Torque the front cover-to-engine bolts to 13 ft. lbs. (18 Nm) and the crankshaft pulley bolt to 87 ft. lbs. (118 Nm).

OIL SEAL REPLACEMENT

1.6L Engine

1. Remove the air pump and the alternator belts. Remove the air pump mounting bracket.
2. Raise and support the front of the vehicle on jackstands, then remove the right inner splash shield.
3. Remove the crankshaft pulley bolt, washer and pulley.

4. Install the seal remover tool No. C-748 or equivalent, over the crankshaft nose and turn, tightly, into the seal.
5. Tighten the thrust screw to remove the seal.

NOTE: If the front cover is removed from the engine, tap the side of the thrust screw to remove the seal.

6. Using the oil seal installation tool No. C-4761 or equivalent, drive the new seal into the front cover.
7. To complete the installation, reverse the removal procedures. Torque the crankshaft pulley bolt to 110 ft. lbs.

2.6L Engine

COVER INSTALLED

1. Refer to the "Front Cover, Removal and Installation" procedures in this section and remove the crankshaft pulley.
2. Using a small pry bar, pry the oil seal from the front cover; be careful not to damage the crankshaft sealing surface.
3. Using a new oil seal, lubricate the seal lips with engine oil, then drive it into the front cover with a seal driver tool until it seats in the cover.
4. To complete the installation, reverse the removal procedures. Torque the crankshaft pulley bolt to 87 ft. lbs. (118 Nm).

COVER REMOVED

1. Using a drift punch and a hammer, drive the oil seal from the front cover.
2. Using a new oil seal, lubricate the seal lips with engine oil, then drive it into the front cover with a seal driver tool until it seats in the cover.
3. To complete the installation, use a new gasket, sealant and reverse the removal procedures. Torque the front cover-to-engine bolts to 13 ft. lbs. (18 Nm) and the crankshaft pulley bolt to 87 ft. lbs. (118 Nm).

Timing Chain, Belt and Tensioner

ADJUSTMENT

1.6L Engine

No adjustment of the timing chain is necessary or possible; the chain is serviced by replacement only.

1. Refer to the "Front Cover Removal & Installation" procedures in this section and remove the front cover, then discard the gasket.
2. Hold a ruler even with the edge of a chain link.

3. Using a torque wrench and socket (installed on the top bolt), apply torque in the direction of the crankshaft rotation to take up the slack, to 30 ft. lbs. with the cylinder head installed or 15 ft. lbs. with the cylinder head removed. DO NOT allow the crankshaft to rotate during this procedures.

4. Clean the gasket mounting surfaces with a suitable scraper.

5. Apply the same torque (as specified above) in the reverse direction and note the amount of chain movement. If the chain movement exceeds $\frac{1}{8}$ in., install a new chain.

5. To install the front cover, reverse the removal procedures.

1.7L Engine

1. Refer to the "Front Cover, Removal and Installation" procedures in this section and remove the front cover.

2. Remove the spark plugs and rotate the crankshaft until the No. 1 piston is at the TDC position.

3. Place a belt tension tool No. L-4502 or equivalent, horizontally on the large hex of the timing belt ten-

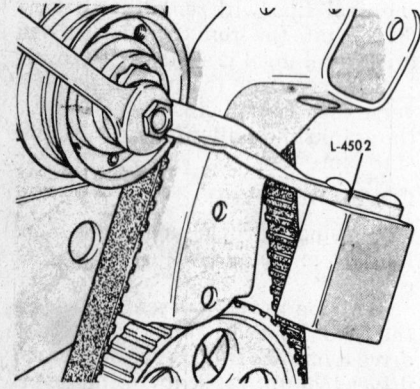

Adjusting drive belt tension—1.7 engine

sioner pulley and loosen the tensioner lock nut.

4. Reset the belt tension tool No. L-4502 or equivalent, index, if necessary, to have the axis within 15° of horizontal.

5. Turn the engine crankshaft clockwise from TDC, two revolutions, to TDC.

6. Tighten the locknut to 32 ft. lbs.

7. To complete the installation, reverse the removal procedures.

2.2L and 2.5L Engine

1. Refer to the "Front Cover, Removal and Installation" procedures in this section and remove the front covers.

2. While holding the large hex wrench on the tension pulley, loosen the pulley nut.

3. Adjust the tensioner by turning the large tensioner hex to the right.

NOTE: The tension is correct when the belt can be twisted 90° with the thumb and forefinger midway between the camshaft and the intermediate pulleys.

4. Tighten the tensioner locknut to 32 ft. lbs.

5. To complete the installation, reverse the removal procedures. Torque the front cover-to-engine bolts to 40 inch lbs. (4 Nm) and the crankshaft pulley bolt to 21 ft. lbs. (28 Nm).

2.6L Engine

No adjustment of the timing chain is necessary or possible; only inspection and replacement of the parts as service.

REMOVAL & INSTALLATION

1.6L Engine

1. Refer to the "Front Cover, Removal and Installation" procedures in this section and remove the front cover.

2. Rotate the camshaft sprocket so that one (of three) bolt heads is located at the top of a centerline drawn through the camshaft and crankshaft sprockets.

3. Check the timing chain for excessive wear (stretch) and replace it, if necessary.

4. Remove the timing gear bolts, timing gear and chain. Clean the gasket mounting surfaces.

5. To install the camshaft sprocket and the timing chain, align the

mark(s) on the crankshaft sprocket with the hole in the camshaft sprocket. Align the camshaft sprocket to the camshaft and install the bolts. Torque the camshaft sprocket bolts to 9 ft. lbs.

NOTE: If the crankshaft sprocket has two marks (side by side), align the camshaft sprocket between the two marks.

6. To complete the installation, apply sealer to the new front cover gasket and reverse the removal procedures. Torque the front cover-to-engine bolts to 9 ft. lbs. (7mm) and 15 ft. lbs. (8mm), then the crankshaft pulley bolt to 110 ft. lbs.

1.7L Engine

1. Refer to the "Front Cover, Removal and Installation" procedures in this section and remove the upper/lower front covers.

2. Lower the vehicle and place a jack under the engine.

3. Remove the right side engine mounting bolt and raise the engine slightly.

4. Loosen the timing belt tensioner and remove the timing belt.

NOTE: If reinstalling the timing belt, be sure to install it in the same direction of rotation.

5. To install, turn the crankshaft and intermediate sprockets until both markings on the sprockets are aligned.

6. Turn the camshaft sprocket until the mark on the sprocket is aligned with the cylinder head cover.

7. Install the timing belt and adjust the tension.

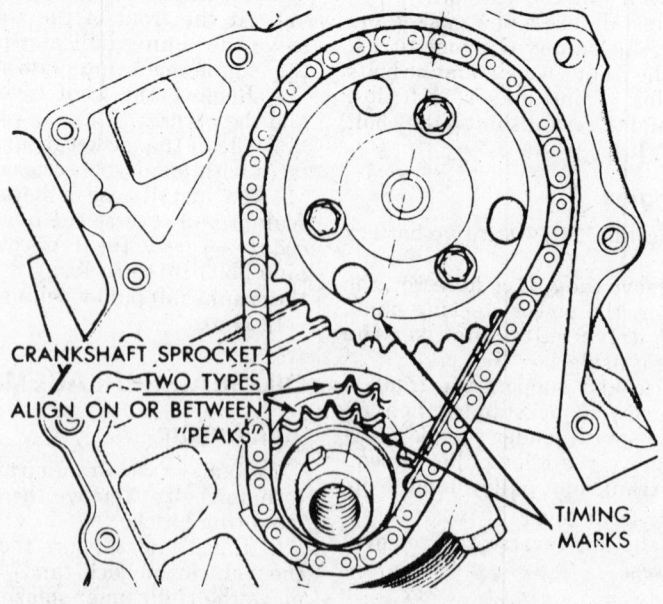

Aligning the timing marks for the 1.6L engine

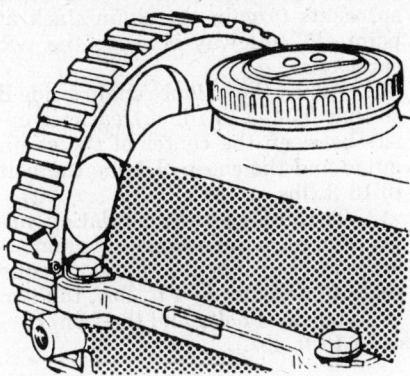

Camshaft gear positioning on 1.7 engines

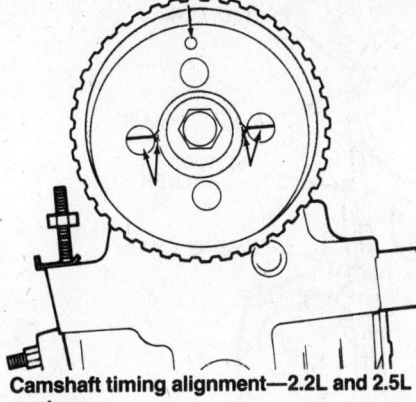

Camshaft timing alignment—2.2L and 2.5L engines

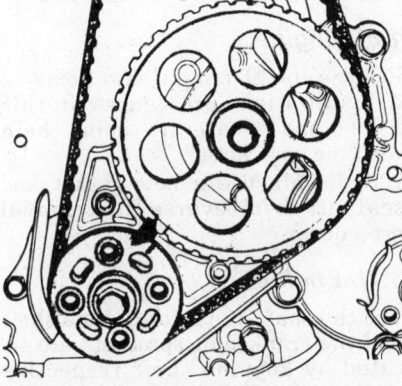

Aligning the timing marks of the crank shaft and the intermediate pulleys for the 1.7L engine

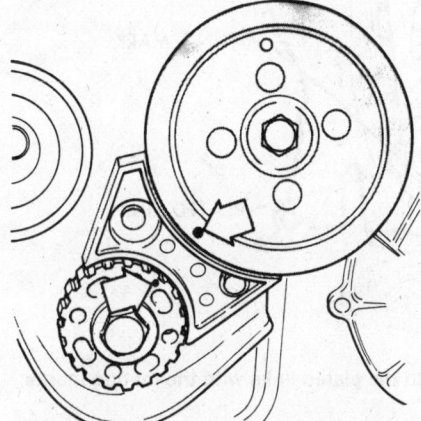

Crankshaft and intermediate shaft timing marks—2.2L and 2.5L engines

8. To complete the installation, reverse the removal procedures. Adjust the timing belt tension.

NOTE: If a whirring noise is heard from the timing belt with the engine running the belt is too tight.

2.2L and 2.5L Engine

1. Refer to the "Front Cover, Removal and Installation" procedures in this section and remove the front covers.

2. While holding the large hex wrench on the tension pulley, loosen the pulley nut.

3. Remove the timing belt from the tensioner.

4. Slide the belt off the three toothed sprockets.

5. Using the larger bolt on the crankshaft pulley, turn the engine until the No. 1 cylinder is at TDC of the compression stroke. At this point the valves for the No. 1 cylinder will be closed and the timing mark will be aligned with the pointer on the flywheel housing. Make sure the arrows on the camshaft sprocket are aligned with the camshaft cap/cylinder head line.

6. Verify that the dot mark on the crankshaft pulley aligns with the line mark on the intermediate shaft.

CAUTION
If the timing marks are not perfectly aligned, poor engine performance and engine damage will result.

7. Install the timing belt on the toothed sprockets.

8. Adjust the timing belt tension.

9. To complete the installation, reverse the removal procedures. Torque the tensioner locknut to 32 ft. lbs., the front cover-to-engine bolts to 40 inch lbs. (4 Nm) and the crankshaft pulley bolt to 21 ft. lbs. (28 Nm). Check the ignition timing.

2.6L Engine

The engine is equipped with two balance shafts which cancel the vertical vibrating force of the engine and the secondary vibrating forces, which include the sideways rocking of the engine due to the turning direction of the crankshaft and other rolling parts. The timing sprockets are linked by a duplex chain. The balance shaft chain assembly is mounted in the front of the timing chain assem-bly; the balance chain assembly must be removed to service the timing chain assembly.

1. Refer to the "Front Cover, Removal and Installation" procedures in this section and remove the front cover.

2. From the "Balance Shaft" chain system, remove the three chain guides: side (A), top (B) and bottom (C).

3. Remove the two sprocket screws, the crankshaft sprocket, the silent shaft sprockets and the drive chain.

4. Remove the camshaft sprocket and the distributor drive gear. Remove the camshaft sprocket holder and the right/left chain guides.

5. Depress the tensioner and remove the drive chain, the crankshaft and the camshaft sprockets.

6. Clean the gasket mounting surfaces. Inspect the parts for damage or wear and replace, if necessary.

Installation is performed in the following manner:

1. Position the dowel pin, of the camshaft sprocket end, at the top.

2. Install the camshaft sprocket holder, the right and left chain guides.

3. Turn the crankshaft until the No. 1 piston is at TDC of the compression stroke.

4. Install the tensioner spring and shoe to the oil pump body.

5. Install the timing chain on the camshaft and the crankshaft sprockets.

NOTE: Align the punch marks of the sprockets with the plated links of the timing chain.

6. With timing marks aligned, slide the camshaft sprocket onto the camshaft dowel pin and the crankshaft sprocket onto the crankshaft keyway.

7. At the camshaft sprocket, install the distributor gear dowel pin, the distributor gear, the washer and the camshaft bolt. Torque the camshaft sprocket bolt to 40 ft. lbs.

8. Install the balance shaft chain drive pulley on the crankshaft.

9. Install the balance chain to the oil pump and the balance shaft sprockets. Align the crankshaft sprocket to the balance chain.

NOTE: Align the punch marks of the sprockets with the plated links of the balance chain.

10. Torque screws of the oil pump and balance shaft sprockets to 25 ft. lbs. Loosely install the three balance chain guides.

11. Torque the bolts of the balance chain guides A and C to 13 ft. lbs. Shake the oil pump and balance shaft

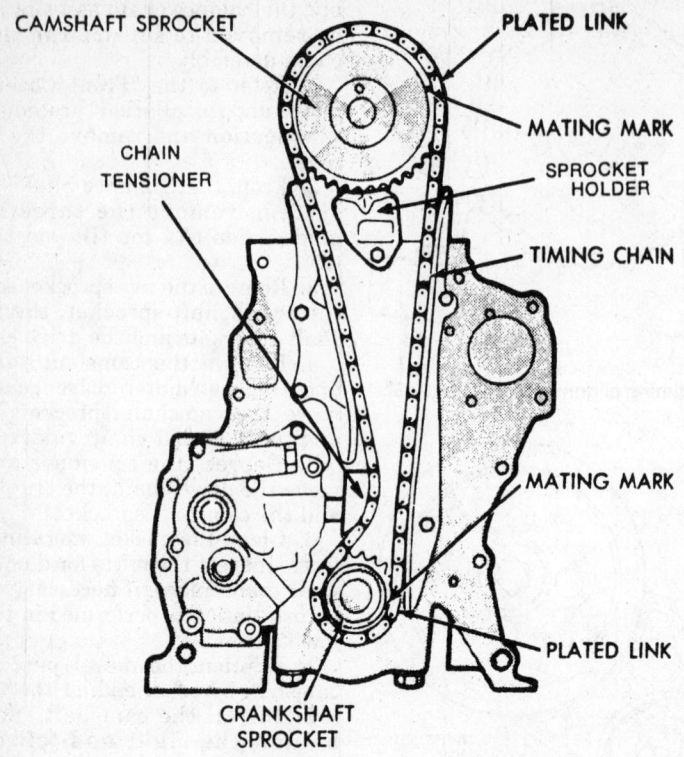

Timing chain installation on 2.6 engines. Align the plated links with the mating marks on the camshaft and crankshaft sprockets

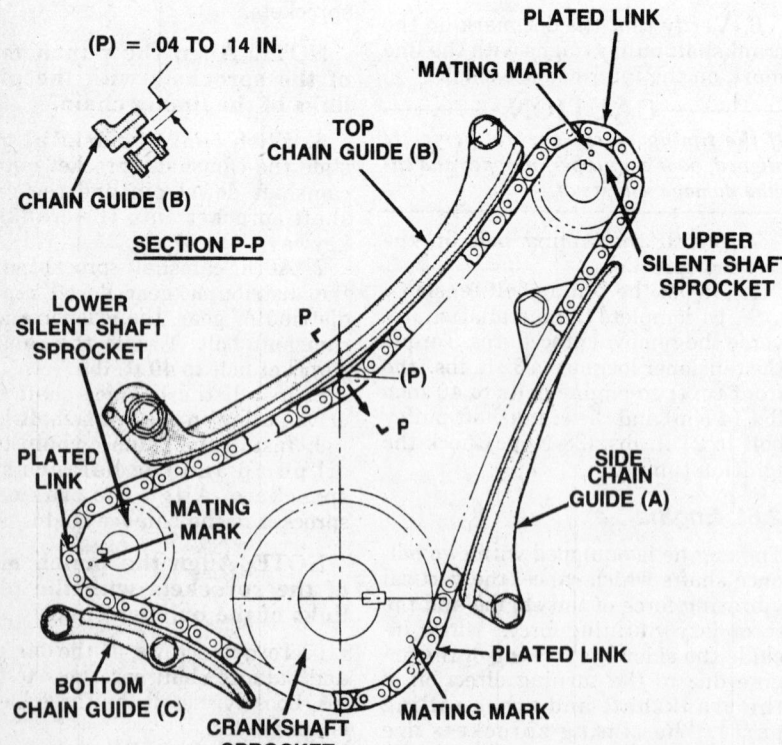

"Silent Shaft" balancing system on 2.6 engines

sprockets to gain the chain slack at point "P" (midway between the two sprockets).

12. Adjust the balance chain guide B so that 0.04–0.14 in. of clearance exists between the center of the chain guide and the chain. Torque the bolt to 13 ft. lbs.

13. To complete the installation, apply sealant to the new gaskets and reverse the removal procedures. Torque the oil pan bolts to 4 ft. lbs., the two cylinder head bolts and the front cover to 13 ft. lbs.

Timing Sprocket

REMOVAL & INSTALLATION

1.6L Engine

Refer to the "Timing Chain Removal & Installation" procedures, in this section and remove the timing chain with the sprocket(s).

2. To install, use a new gasket and sealant, then reverse the removal procedures.

1.7L Engine

The camshaft, the intermediate shaft and the crankshaft sprockets are located by keys on their respective shafts and each is retained by a bolt. Remove the center bolts and pull the sprockets from the shafts.

1. Refer to the "Front Cover and Belt Removal & Installation" procedures in this section and remove the timing belt.

2. Remove the crankshaft sprocket bolt. Connect the removal tool No. L-4524 or equivalent, to the crankshaft sprocket and pull the sprocket from the crankshaft.

3. Remove the crankshaft key. Using the oil seal removal tool No. L-4424 or equivalent, pry the oil seal from the crankshaft.

4. Using the seal installer tool No. L-4422 or equivalent, drive the new oil seal into the crankshaft housing.

5. To complete the installation, use new gaskets and reverse the removal procedures. Torque the crankshaft center bolt to 58 ft. lbs. Check the ignition timing.

2.2L and 2.5L Engines

1. Refer to the "Timing Belt, Removal and Installation" procedures in this section and remove the timing belt.

2. Remove the crankshaft sprocket bolt. Using the puller tool No. L-4685 or equivalent, and the button from tool No. L-4524 or equivalent, remove the crankshaft sprocket.

3. Using the tool No. C-4687 or equivalent, hold the camshaft and/or

intermediate sprocket, remove the center bolt(s) and the sprocket(s).

4. To install, use new gaskets and reverse the removal procedures. Torque the camshaft and intermediate sprocket bolts to 65 ft. lbs. and the crankshaft sprocket bolt to 50 ft. lbs. Adjust the timing belt tension.

2.6L Engines

Refer to the "Timing Chain Removal & Installation" procedures in this section and remove the timing chain and sprockets. To install, reverse the removal procedures.

OIL SEAL REPLACEMENT

1.7L Engine

The camshaft, the intermediate shaft and the crankshaft sprockets are located by keys on their respective shafts and each is retained by a bolt. Remove the center bolts and pull the sprockets from the shafts.

1. Refer to the "Timing Belt Removal & Installation" procedures in this section and remove the timing belt.

2. Remove the crankshaft sprocket bolt. Connect the removal tool No. L-4524 or equivalent, to the crankshaft sprocket and pull the sprocket from the crankshaft.

3. Remove the crankshaft key. Using the oil seal removal tool No. L-4424 or equivalent, pry the oil seal from the crankshaft.

4. Using the seal installer tool No. L-4422 or equivalent, drive the new crankshaft seal (lubricate the seal lips with oil) into the crankshaft housing until it seats.

5. To complete the installation, use new gaskets and reverse the removal procedures. Torque the crankshaft center bolt to 58 ft. lbs. Check the ignition timing.

2.2L and 2.5L Engines

1. Refer to the "Timing Sprocket, Removal & Installation" procedures in this section and remove the timing sprockets.

2. Using the tool No. C-4679 or equivalent, remove the crankshaft, the camshaft and/or intermediate seal(s).

3. To install the oil seals, polish the shafts with 400 grit emery paper, use the oil seal installation tool No. C-4680 or equivalent, and drive the seal into its seat.

NOTE: If the seal has a steel case, lightly coat the seal with Loctite Stud N' Bearing Mount® or its equivalent. If the seal case is rubber coated, apply soap and water to facilitate the installation.

4. To complete the installation, use new gaskets and reverse the removal procedures. Torque the camshaft and intermediate sprocket bolts to 65 ft. lbs. and the crankshaft sprocket bolt to 50 ft. lbs. Adjust the timing belt tension.

Balance Shafts

The engine is equipped with two "Balance Shafts" which cancel the vertical vibrating force of the engine and the secondary vibrating forces, which include the sideways rocking of the engine due to the turning direction of the crankshaft and other rolling parts. The timing sprockets are linked by a duplex chain.

REMOVAL & INSTALLATION

2.5L Engine—1986 and Later

The balance shaft chain assembly is installed in a carrier attached to the lower crankcase. The balance shaft chain assembly is located behind the timing belt assembly; the timing belt MUST BE removed before removing the balance chain assembly.

1. Refer to the "Timing Belt Removal & Installation" procedures in this section and remove the timing belt.

FASTENER TORQUE

Letter Code	Nm	Inch Lbs.	Ft. Lbs.
A	12	105	9
B	28	250	21
C	54	480	40
D①	41①	360①	30①
E	95	840	70
F	Plug—Loctite®		277
G	15	130	11

① Specified torque plus ¼ turn

2. Raise and support the front of the vehicle on jackstands. Remove the oil pan, the oil pickup, the crankshaft belt sprocket and the front crankshaft oil seal retainer.

3. Remove the balance shaft chain cover, the guide and the tensioner.

4. Remove the balance shaft sprocket-to-shaft bolt, the gear cover-to-balance shaft bolt and the crankshaft sprocket-to-crankshaft bolts, then the sprockets with the balance chain.

5. Remove the front gear cover-to-carrier housing stud, the gear cover and the balance shaft drive gears.

6. Remove the rear gear cover-to-carrier housing bolts, the rear cover and the balance from the rear of the carrier.

7. If necessary, remove the carrier housing-to-crankcase bolts and the housing.

8. To install the balance shaft/carrier assembly, perform the following procedures:

a. If the carrier housing is being installed, torque the carrier housing-to-crankcase bolts to 40 ft. lbs. (54 Nm).

b. Rotate the balance shafts until the keyways are facing upward (parallel to the vertical centerline of the engine).

c. Install the short hub gear on the sprocket driven shaft and the long hub gear on the gear driven shaft; make sure that the gear timing marks are aligned (facing each other).

d. Install the front gear cover and torque the front gear cover-to-carrier housing stud bolt to 8.5 ft. lbs. (12 Nm).

e. Install the balance chain sprocket and torque the sprocket-to-crankshaft bolts to 11 ft. lbs. (13 Nm).

f. Rotate the crankshaft to position the No. 1 cylinder on the TDC of the compression stroke; the timing marks on the chain sprocket should align with the parting line on the left side of the No. 1 main bearing cap.

g. Position the balance shaft sprocket into the balance chain so that the sprocket (yellow dot) timing mark mates with the chain (yellow) link.

h. Install the balance chain/sprocket assembly onto the crankshaft and the balance shaft. Torque the sprocket-to-shaft bolts to 21 ft. lbs. (28 Nm).

NOTE: If necessary to secure the crankshaft while tightening the bolts, place a block of wood between the crankcase and the crankshaft counterbalance.

i. Loosely install the chain tensioners and place a shim (0.039 in. x 2.75 in.) between the chain and the tensioner. Apply firm pressure (to reduce the chain slack) to the tensioner shoe. Torque the tensioner-to-front gear cover bolts to 8.5 ft. lbs. (12 Nm).

j. Install the chain cover and the rear cover to the carrier housing and torque the bolts to 8.5 ft. lbs. (12 Nm).

9. To complete the installation, use new gasket, sealant and reverse the removal procedures. Refill the crankcase. Adjust the timing belt tension.

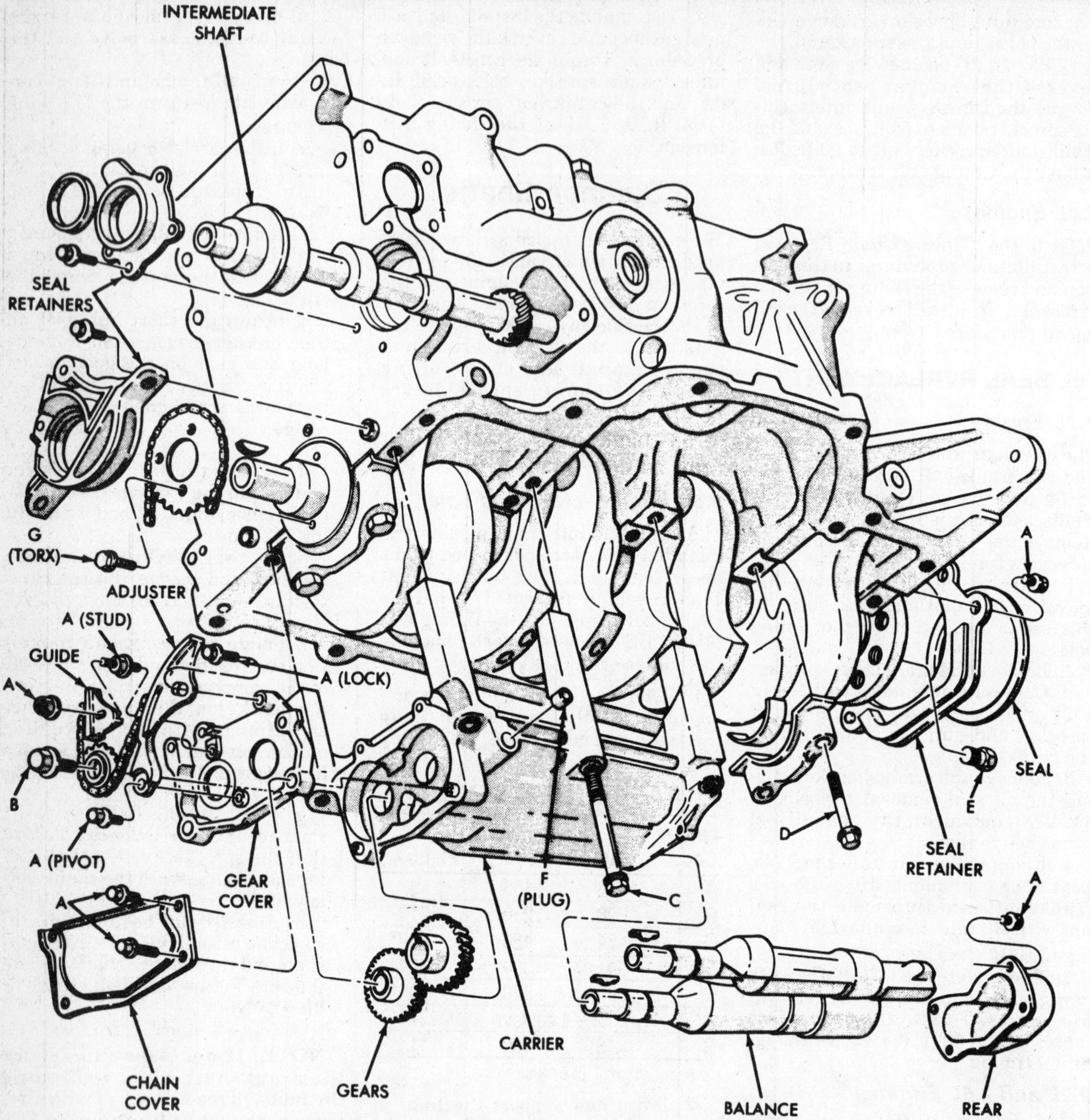

INTERMEDIATE SHAFT

SEAL RETAINERS

G (TORX)

ADJUSTER

A (STUD)

GUIDE

A

B

A (PIVOT)

A

CHAIN COVER

GEAR COVER

GEARS

A (LOCK)

CARRIER

F (PLUG)

C

D

E

SEAL

SEAL RETAINER

A

BALANCE SHAFTS

REAR COVER

Exploded view of the crankshaft, intermediate and balance shaft assemblies—2.5L engine

Check and/or adjust the engine timing.

2.6L Engine

The balance shaft chain assembly is mounted in the front of the timing chain assembly; the assembly must be removed to service the timing chain assembly.

RIGHT SIDE

1. Refer to the "Oil Pump Removal & Installation" procedures in this section and remove the oil pump.

2. Remove the oil pump adapter plate.

3. Pull the balance shaft through the front of the engine.

4. To install, reverse the removal procedures.

LEFT SIDE

1. Refer to the "Oil Pump Removal & Installation" procedures in this section. Remove the balance chain sprocket bolt, the sprocket, the chain and the spacer.

2. Remove the two balance shaft

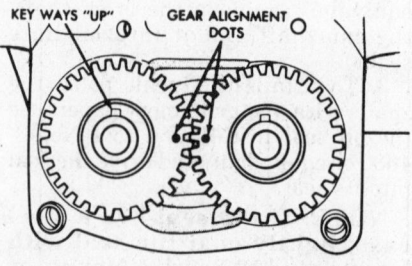

KEY WAYS "UP"   GEAR ALIGNMENT DOTS

Aligning the balance shaft gears—2.5L engine

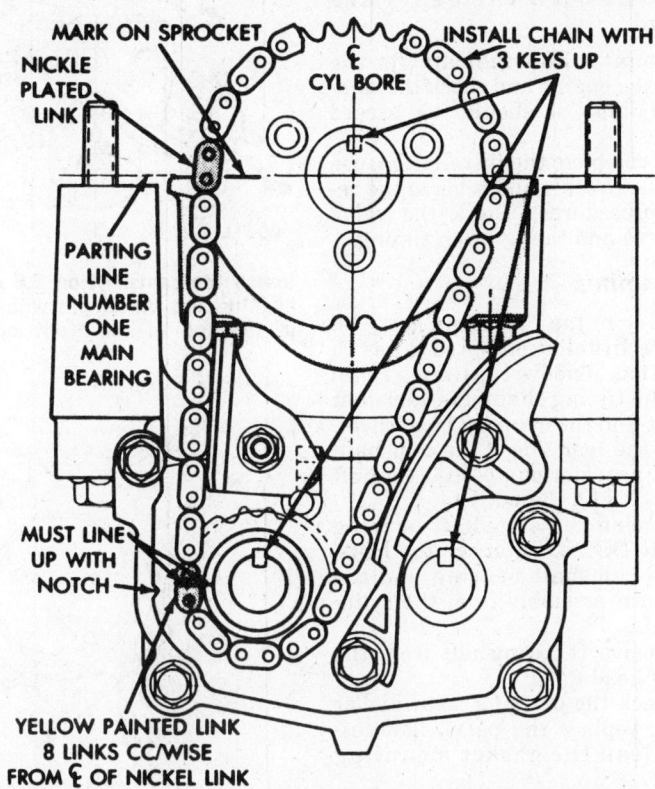

MARK ON SPROCKET

NICKLE PLATED LINK

INSTALL CHAIN WITH 3 KEYS UP

¢ CYL BORE

PARTING LINE NUMBER ONE MAIN BEARING

MUST LINE UP WITH NOTCH

YELLOW PAINTED LINK 8 LINKS CC/WISE FROM ¢ OF NICKEL LINK

Installing the balance shaft chain onto the crankshaft and balance shaft—2.5L engine

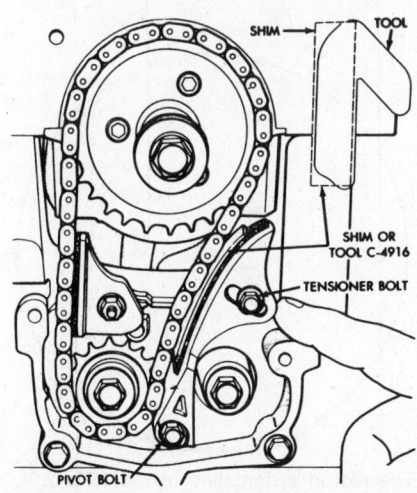

SHIM

TOOL

SHIM OR TOOL C-4916

TENSIONER BOLT

PIVOT BOLT

Adjusting the balance shaft chain tension—2.5L engine

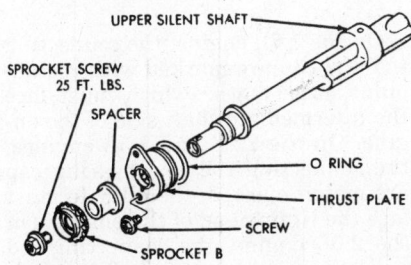

UPPER SILENT SHAFT

SPROCKET SCREW 25 FT. LBS.

SPACER

O RING

THRUST PLATE

SCREW

SPROCKET B

Upper "Silent Shaft" components on 2.6 engines

thrust plate bolts and pull the thrust plate from the engine.

3. Pull the balance shaft through the front of the engine.

4. To install, reverse the removal procedures. Torque the sprocket bolt to 25 ft. lbs.

Camshaft

REMOVAL & INSTALLATION

1.6L Engine

NOTE: Refer to the "Engine Removal & Installation" procedures in this section and remove the engine.

1. Remove the front cover, the timing chain and the camshaft sprocket.
2. Remove the cylinder head, the push rods and the valve tappets. Identify the tappets to ensure installation in their original positions.
3. Remove the fuel pump, the oil pump, the distributor and the drive housing. Mark the engine in relation to the drive slot.
4. Using a magnet, remove the distributor drive from halfshaft spindle.
5. Remove shaft drive gear circlip.

NOTE: Insert a shop towel in the drive gear cavity to insure that the circlip does not fall into the

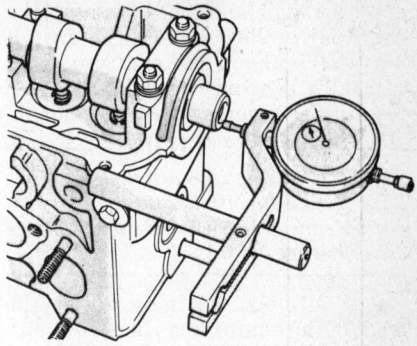

Checking camshaft end-play

crankcase during removal or installation.

6. Tap the halfshaft toward the pump side of crankcase until shaft drive gear is free from the spline and remove the gear/washer.
7. Remove the halfshaft from the pump side of the crankcase.
8. Remove the two camshaft thrust plate bolts and the thrust plate.
9. Carefully remove the camshaft from the front of the engine. Check for wear and/or damage, then replace the parts, as necessary.
10. To install, use new gaskets and reverse the removal procedures. Torque the thrust plate bolts to 11 ft. lbs. Install a dial indicator to the engine and check the camshaft end play (0.004–0.008 in.).

NOTE: When installing a new camshaft or tappets, add one pint of Chrysler Crankcase Conditioner, Part Number 3419130 or equivalent, to the engine oil. Retain oil mixture for a minimum of 500 miles. When replacing the camshaft, use a straight edge to check all tappet faces for wear. Replace tappets with negative crown or dishing.

1.7L Engine

1. Remove the timing belt covers and adjust the engine until the No. 1 cylinder is at the TDC of the compression stroke.
2. Remove the timing belt, the camshaft sprocket bolt and the sprocket.
3. Remove the air cleaner assembly and the cylinder head cover.
4. Remove the No. 1, 3 and 5 camshaft bearing caps.
5. Loosen caps No. 2 and 4, diagonally and remove the caps.
6. Carefully, lift out the camshaft. Check for wear and/or damage, replace the parts, if necessary.
7. Clean the gasket mating surfaces with a suitable scraper.
8. To install, lubricate the cam-

shaft journals and lobes, position the camshaft in the cylinder head, install a new oil seal and gaskets. Torque the camshaft bearing cap bolts to 14 ft. lbs., in the reverse order of removal.

—————— **CAUTION** ——————

All of the bearing caps are slightly offset. They should be installed so the numbers on the caps read right side up from the driver's seat.

9. Position a dial indicator to the front of the engine and check the camshaft end play, it should not exceed 0.006 in.

10. If necessary, replace the camshaft end plug in the cylinder head.

11. To complete the installation, reverse the removal procedures. Check the valve clearances and the ignition timing.

2.2L and 2.5L Engine

1. Remove the timing case covers and turn the crankshaft so that the No. 1 piston is at the TDC of the compression stroke.

2. Remove the timing belt, the camshaft sprocket bolt and the sprocket.

3. Remove the PCV module (by turning it counterclockwise) from the cylinder head cover. Remove the cylinder head cover screws and the cover.

4. Mark the rocker arms for installation identification and loosen the camshaft bearing bolts, several turns each.

5. Using a soft mallet, rap the rear of the camshaft, a few times, to break the bearing caps loose.

6. Remove the bolts and bearing caps, be careful that the camshaft does not cock. Cocking the camshaft could cause damage to the bearings.

7. Check the oil holes for blockages and the parts for wear and/or damage, replace the parts, if necessary. Clean the gasket mounting surfaces.

8. To install, lubricate the camshaft, place the bearing caps with No. 1 at the timing belt end and No. 5 at the transaxle end. The camshaft bearing caps are numbered and have arrows facing forward. Torque the camshaft bearing bolts to 18 ft. lbs.

NOTE: Apply RTV silicone gasket material to the No. 1 and 5 bearing caps. Install the bearing

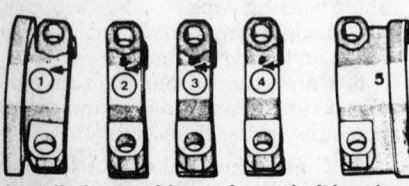

Installation positions of camshaft bearing caps

caps BEFORE the seals are installed.

9. Mount a dial indicator to the front of the engine and check the camshaft endplay, it should not exceed 0.006 in.

10. To complete the installation, use new gaskets/seals and reverse the removal procedures. Check the valve clearances and the ignition timing.

2.6L Engines

1. Refer to the "Timing Chain Removal & Installation" procedures in this section. Remove the timing chain cover, the timing chain, the camshaft sprocket and the cylinder head cover.

2. At the rear side of the camshaft, remove the water pump, the camshaft pulley bolt and the pulley.

3. Loosen the camshaft bearing bolts but DO NOT remove the bolts. Lift the camshaft bearing and the rocker arm assembly from the cylinder head.

4. Remove the camshaft from the cylinder head.

5. Check the parts for wear and/or damage, replace the parts, if necessary. Clean the gasket mounting surfaces.

6. To install, lubricate the parts, install new camshaft seals, the camshaft, the camshaft bearing and rocker arm assembly. Torque the camshaft bearing bolts to 7 ft. lbs. and then to 14 ft. lbs. in the following order: Bearing caps No. 3, 2, 4, front and rear.

7. To complete the installation, use new gaskets and reverse the removal procedures. Check the valve clearances and the ignition timing.

Pistons And Connecting Rods

For all piston and connecting rod overhaul procedures, please refer to "Engine Rebuilding" in the Unit Repair section.

INSTALLATION

ALL Engines

On all the engines, except the 1.6L engine, the piston crown is marked with an arrow or an indent, which must point toward the timing belt/chain end of the engine. On the 1.6L engine, the pistons are notched at the bottom of the piston skirt. Install the No. 1 and 3 pistons with the notch facing the flywheel and the No. 2 and 4 pistons with the notch facing the front cover. The connecting rod's crankshaft bearing oil hole must face the camshaft.

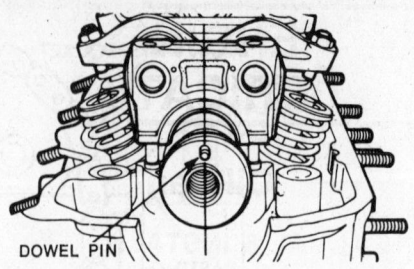

Install the camshaft on 2.6 engines by aligning the dowel pin with the notch in the top of the front bearing cap

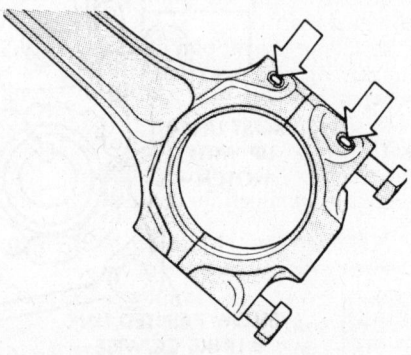

Matching the connecting rod with the cap on the 1.7L engine

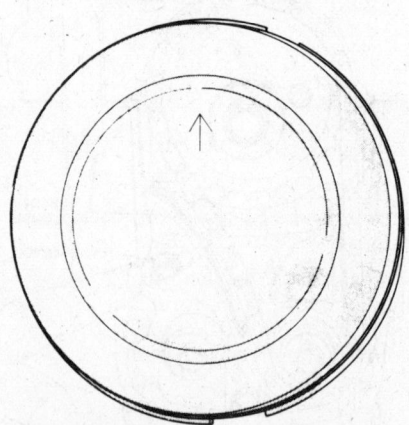

The piston is installed in the 1.7L, 2.2L and 2.5L engines with the arrow facing forward

On the 1.7L engine, the connecting rod and cap are marked with rectangular forge marks which must face the intermediate shaft side of the engine. On the 2.2L and 2.5L engines, the connecting rod and bearing cap are match marked, the oil hole must face the timing end of the engine. On the 2.6L engine, the connecting rod and bearing cap are match marked, which must face the right side of the engine.

ENGINE LUBRICATION

Oil Pan

REMOVAL & INSTALLATION

1. Place a catch pan under the engine, remove the drain plug and drain the crankcase.
2. Support the oil pan and remove the attaching bolts.
3. Lower the oil pan and discard the gaskets.
4. Clean the oil pan and the gasket mounting surfaces with a suitable scraper.
5. To install, use a new pan gasket with sealer and reverse the removal procedures. Refill the pan, start the engine and check for leaks.

NOTE: On the 1.6L engine, torque the oil pan bolts to 9 ft. lbs. (1984) or 7 ft. lbs. (1985–86). Torque the oil pan bolts to 6 ft. lbs. (1.7L engine), to 17 ft. lbs. (2.2L and 2.5L engines) or 5 ft. lbs. (2.6L engine).

Rear Main Bearing Oil Seal

REMOVAL & INSTALLATION

1.6L Engine

NOTE: The rear main seal removal is easier to accomplish if the engine is removed first.

1. Refer to the "Engine Removal & Installation" procedures in this section and remove the engine from the vehicle. Remove the flywheel or drive plate from the crankshaft.
2. Remove the rear oil seal housing-to-engine bolts and the housing.
3. Place the housing inner surface on two blocks of wood, allowing clearance for the seal removal.
4. Using the seal removal/installation tool No. C-4759 or equivalent, drive the seal from the housing.
5. To install, invert the seal housing and place it on a smooth surface. Using the seal removal/installation tool No. C-4759 or equivalent, drive the new seal into the housing until it seats.
6. To complete the installation, use a new gasket, lubricate the seal lips and reverse the removal procedures. Torque the seal mounting bolts to 9 ft. lbs.

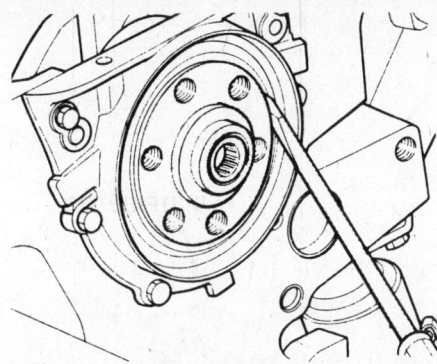

Removing the rear mail oil seal on the 1.7L, 2.2L and 2.5L engines

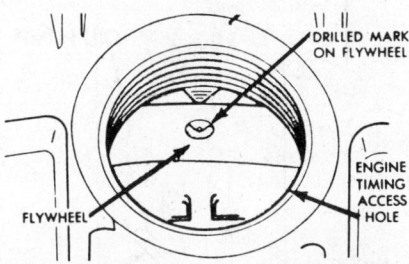

Drilled mark on 1.7 engine flywheels

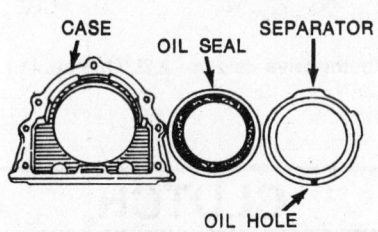

Rear main oil seal on 2.6 engines

1.7L, 2.2L and 2.5L Engines

The rear main seal is located in a housing on the rear of the block. To replace the seal the engine must be removed.

1. Refer to the "Engine Removal & Installation" procedures in this section and remove the engine from the vehicle. Remove the flywheel or the drive plate from the crankshaft.

CAUTION

Before removing the transaxle, align the dimple on the flywheel with the pointer on the flywheel housing. The transaxle will not mate with the engine during installation, unless this alignment is observed.

---

2. Using a small pry bar, carefully pry the old seal out of the oil seal housing.
3. Coat the new seal with Loctite Stud N' Bearing Mount (4057987) or equivalent and drive it into place, using the installation tool No. L-4425-1 or equivalent, (1.7L engine) or No. L-4681 or equivalent, (2.2L and 2.5L engines). Take care not to scratch the seal or the crankshaft.

4. To install, reverse the removal procedures.

2.6L Engine

The rear main oil seal is located in a housing on the rear of the block. To replace the seal, remove the transaxle and do the work from underneath the vehicle or remove the engine and do the work on the bench.

1. Remove the oil seal housing-to-engine bolts and housing from the engine.
2. Remove the separator from the oil seal housing.
3. Using a small pry bar, pry the old seal from the housing.
4. Clean the gasket mounting surfaces with a suitable scraper.
5. To install, use new gaskets/seals, lubricate the new oil seal and seat it in the housing. Install the separator with the oil hole facing down.
6. To complete the installation, reverse the removal procedures.

Oil Pump

The conventional lubrication system uses a gear type oil pump, with a pressure relief valve to prevent extreme pressure build up.

REMOVAL & INSTALLATION

1.6L Engine

The oil pump is located on the right side of the engine block.

1. Remove the oil filter.
2. Holding the oil pump cover and housing together, remove the seven mounting bolts and pull the assembly away from the engine.
3. Remove the gaskets. Clean the gasket mounting surfaces with a suitable scraper.
4. To install; prime the pump body with petroleum jelly. Use new gaskets and place sealer on the mounting bolt threads, then insert the two bolts to align the pump body with the housing. Align the pump with the engine (engage the driving shaft gear tongue with the slot in the halfshaft) and torque the mounting bolts to 9 ft. lbs. Start the engine and check for leaks.

1.7L, 2.2L and 2.5L Engines

1. Remove the oil pan.
2. On the 2.2L and 2.5L engines, remove the No. 3 bearing bolt (holding the pickup tube) and the tube.
3. Remove the oil pump mounting bolts. Pull the pump down and out of the engine.
4. To install, use new gaskets, prime the pump with petroleum jelly, align the pump with the distributor

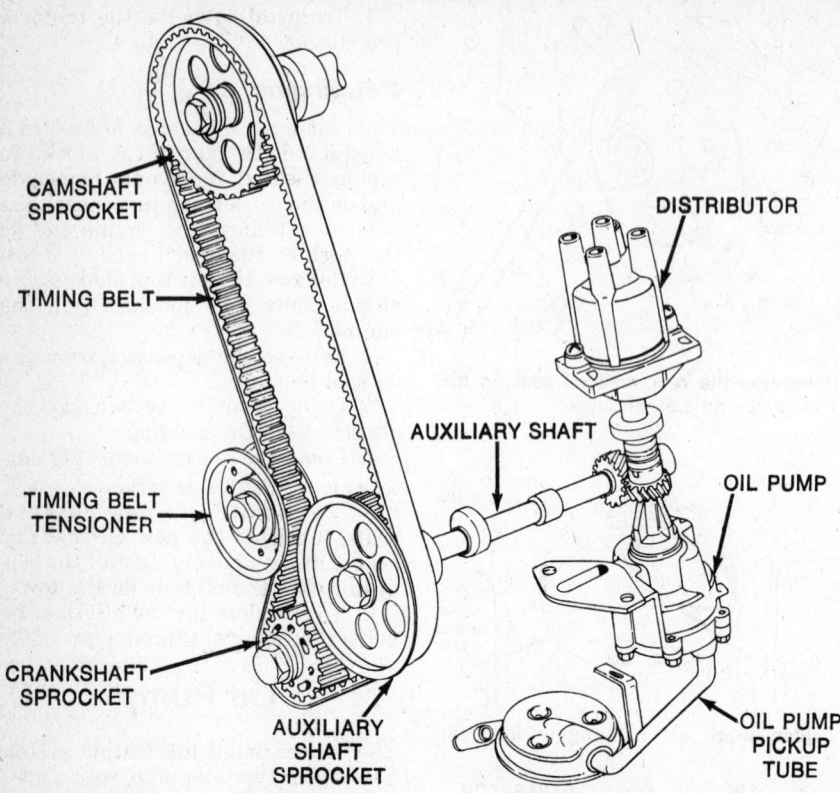

Timing belt, auxiliary shaft, oil pump and distributor drive details—2.2L (135 cu. in.) and 2.5L (153 cu. in.) engines

Labels: CAMSHAFT SPROCKET, TIMING BELT, TIMING BELT TENSIONER, CRANKSHAFT SPROCKET, AUXILIARY SHAFT SPROCKET, AUXILIARY SHAFT, DISTRIBUTOR, OIL PUMP, OIL PUMP PICKUP TUBE

REMOVAL & INSTALLATION

A-412 Transaxle

1. Refer to the "Manual Transaxle Removal & Installation" procedures in this section and remove the transaxle.

2. Diagonally loosen the flywheel-to-pressure plate bolts, 1–2 turns at a time to avoid warpage.

3. Remove the flywheel and clutch disc from the pressure plate.

4. Remove the retaining ring and the release plate.

5. Diagonally loosen the pressure plate-to-crankshaft bolts. Mark all of the parts for reassembly.

6. Remove the mounting bolts, the backing plate and pressure plate.

7. The flywheel and pressure plate surfaces should be cleaned thoroughly with a water dampened cloth.

8. To install, align the marks of the pressure plate, the backing plate and bolts. Coat the bolts with thread compound and torque them to 55 ft. lbs.

9. Install the release plate and the retaining ring.

10. Using the clutch disc installation tool No. L-4533 or equivalent, install the clutch disc and the flywheel on the pressure plate.

drive gear tongue and reverse the removal procedures. Torque the pump mounting bolts to 14 ft. lbs. (1.7L engine) and 17 ft. lbs. (2.2L and 2.5L engines).

2.6L Engine

1. Refer to the "Front Cover and Belt Removal & Installation" procedures in this section. Remove the timing belt cover, the silent chain guides, the silent chain/oil pump sprocket, the silent chain and the timing chain tensioner (if necessary).

2. Remove the silent shaft bolt (the bolt directly above the silent chain sprocket).

3. Remove the oil pump bolts and pull the pump housing straight forward. Remove the gaskets and the oil pump backing plate. Clean the gasket mounting surfaces.

4. To install the oil pump, use new gaskets/seals, align the mating marks of the oil pump gears, refill the pump with oil, install the pump and reverse the removal procedures. Torque the oil pump mounting bolts to 6 ft. lbs., the balance chain guide and sprocket bolts to 13 ft. lbs.

CLUTCH

CAUTION

When servicing the clutch assemblies, DO NOT create dust by sanding or cleaning the clutch parts (a water dampened cloth should be used). The clutch disc contains asbestos fibers which can create a health hazard by breathing during service operations. The clutch is a non-adjustable, dry disc unit; adjustment is provided at the adjustable sleeve in the pedal linkage.

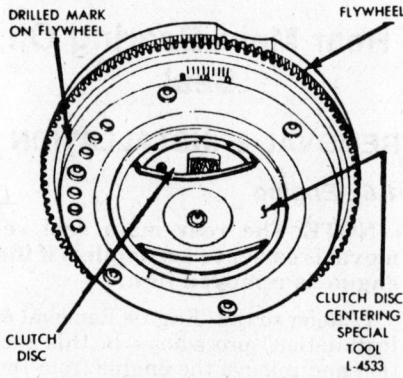

Using the special aligning tool

Labels: DRILLED MARK ON FLYWHEEL, FLYWHEEL, CLUTCH DISC, CLUTCH DISC CENTERING SPECIAL TOOL L-4533

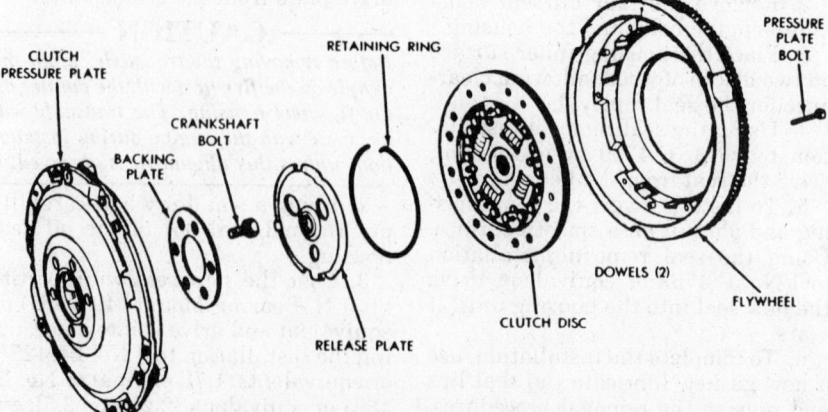

A-412 manual transaxle clutch—disassembled

Labels: CLUTCH PRESSURE PLATE, BACKING PLATE, CRANKSHAFT BOLT, RELEASE PLATE, RETAINING RING, CLUTCH DISC, DOWELS (2), FLYWHEEL, PRESSURE PLATE BOLT

CAUTION

Make certain the drilled mark on the flywheel is at the top, so that the two dowels on the flywheel align with the proper holes of the pressure plate.

11. To complete the installation, torque the six flywheel bolts to 15 ft. lbs. and reverse the removal procedures.

A-460, A-465 and A-525 Transaxles

1. Refer to the "Manual Transaxle Removal & Installation" procedures in this section and remove the transaxle.

2. Match mark the clutch/pressure plate cover and flywheel. Insert the clutch plate alignment tool No. C-4676 or equivalent, into the clutch disc hub.

3. Loosen the flywheel-to-pressure plate bolts diagonally, 1–2 turns at a time to avoid warpage.

4. Remove the pressure plate/clutch assembly from the flywheel.

5. To install, use clutch disc alignment tool No. C-4676 or equivalent, (to hold the clutch) and reverse the removal procedures. Torque the pressure plate/clutch assembly mounting bolts to the flywheel, a few turns at a time (diagonally), to 21 ft. lbs. After installation, adjust the clutch freeplay.

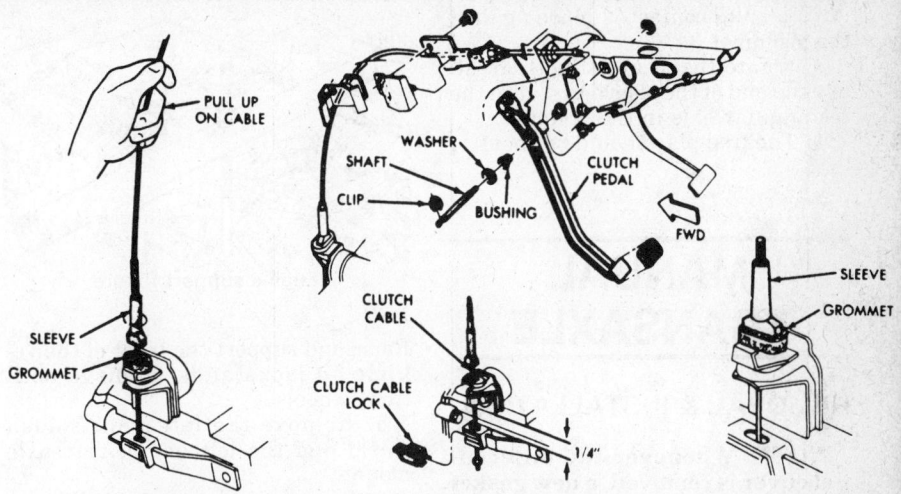

Adjusting the clutch pedal free-play—A-412 manual transaxle

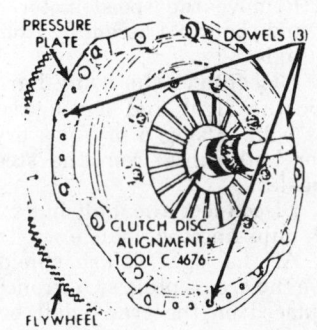

Clutch assembly and alignment tool

PEDAL HEIGHT/FREE-PLAY ADJUSTMENT

NOTE: The A-460, A-465 and A-525 transaxles are equipped with a self-adjusting clutch release mechanism. No adjustment is necessary.

A-412 Transaxle

1. Pull up on the clutch cable.
2. While holding the cable up, rotate the adjusting sleeve downward

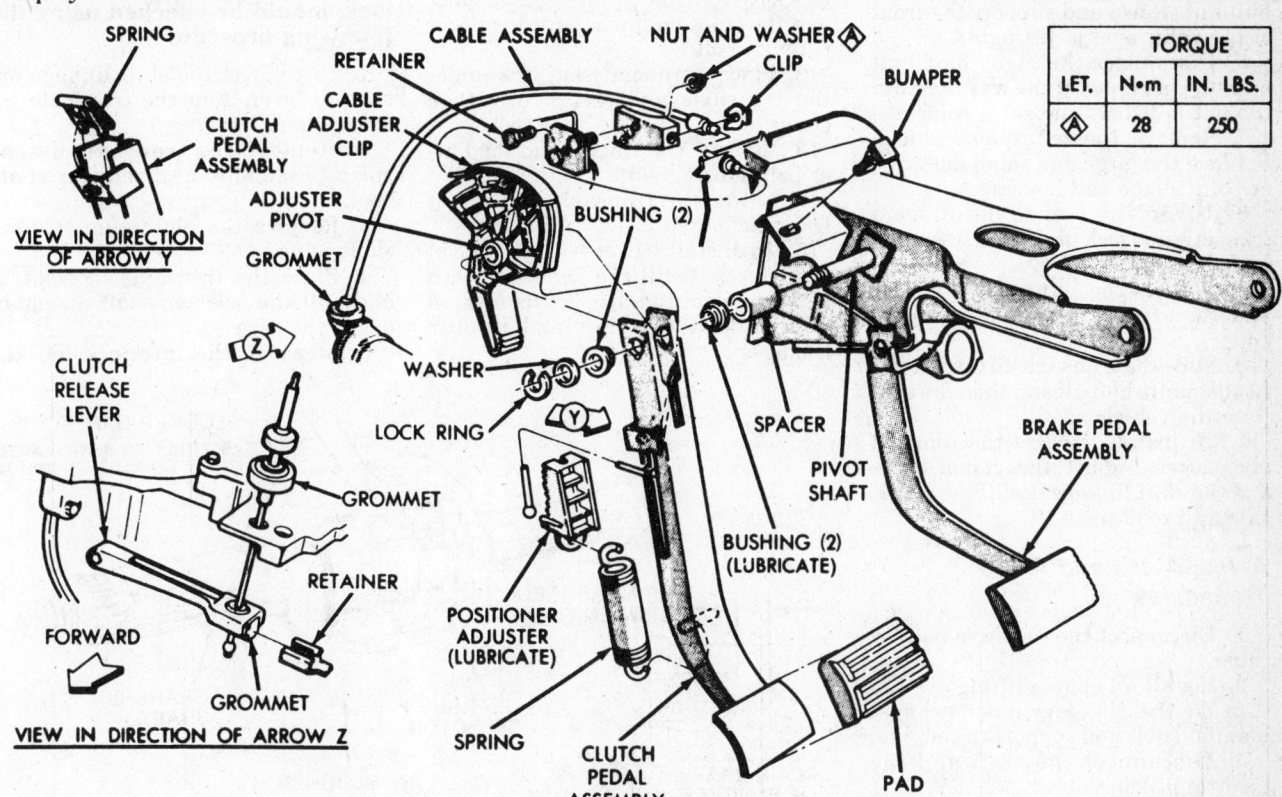

TORQUE		
LET.	N•m	IN. LBS.
Ⓐ	28	250

Manual transaxle self-adjusting clutch release mechanism—for all except A-412 model

until a snug contact is made against the grommet.

3. Rotate the sleeve slightly to allow the end of the sleeve to seat in the rectangular hole in the grommet.

4. The free-play should be about $\frac{1}{4}$ in.

MANUAL TRANSAXLE

REMOVAL & INSTALLATION

NOTE: Whenever the differential cover is removed, a new gasket should be formed using RTV sealant.

A-412 Transaxle

1. Disconnect the negative battery cable, the starter wires and the back-up light switch wire.

2. Remove the starter bolts and the starter.

3. Disconnect the shift linkage rods and the clutch cable.

4. Remove the speedometer cable retaining bolt and the cable from the transaxle.

5. Using a vertical hoist or a fabricated support fixture, support the engine from above. Loosen the left wheel hub nut. Raise and support the front of the vehicle on jackstands.

6. Disconnect the right halfshaft and support it out of the way. Remove the left halfshaft and set it aside.

7. Remove the left splash shield. Remove the large and small dust cover bolts at the bell housing.

8. Drain the transaxle. Place a transmission jack under the transaxle and chain in place.

9. Unbolt the left engine mount and remove the transaxle-to-engine bolts.

10. Slide the transaxle to the left until the mainshaft clears, then lower it from the vehicle.

11. To install, reverse the removal procedures. Adjust the clutch cable and the shift linkage. Refill the transaxle and road test.

A-460, A-465 and A-525 Transaxles

1. Disconnect the negative battery cable.

2. Install an engine lifting eye fixture on the No. 4 cylinder exhaust manifold bolt and support the engine.

3. Disconnect the shift and the throttle linkages.

4. Remove the hub castle lock nut and cotter pin from both front wheels.

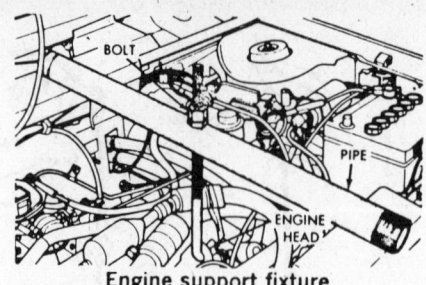
Engine support fixture

Raise and support the front of the vehicle on jackstands. Remove both front wheels.

5. Remove the left front splash shield and the left engine/transaxle mount.

6. Remove the two upper bell housing bolts.

7. Remove the speedometer cable bolt and the cable assembly from the transaxle.

8. Disconnect the sway bar. Remove the right and left lower ball joint bolts. Using a pry bar, pry the lower ball joints from the steering knuckles.

9. Disconnect the halfshafts from both hubs and the vehicle.

10. At the right splash shield, remove the access plug, so a wrench can be placed on the crankshaft bolt to turn the crankshaft.

11. Remove the engine mount bracket from the front crossmember and the front engine mount insulator through bolt.

12. Place a transmission jack under the transaxle and secure it with a chain.

13. Remove the starter and the lower bell housing bolts.

14. Lower the transaxle and pry it from the engine.

15. To install, reverse the removal procedures. Refill the transaxle with Dexron®II automatic transmission fluid. Adjust the shift and throttle linkages.

OVERHAUL

For all manual transaxle overhaul procedures, please refer to "Manual Transmission Overhaul" in the Unit Repair section.

SHIFT LINKAGE ADJUSTMENT

A-412 Transaxle

1. Place the transaxle in Neutral at the 3–4 position.

2. Loosen the shift tube clamp.

3. Place a $\frac{3}{4}$ in. spacer between the shift tube flange and the yoke at the shift base.

4. Tighten the shift tube clamp and remove the spacer.

NOTE: While torquing the shift tube clamp nut, no force should be exerted upward.

LOCKUP ADJUSTMENT

NOTE: It is possible for the A-412 transaxle to become locked in two gears at once. This will occur if the interlock blocker on the gearshift selector lever has spread apart. The result of operating with this condition may result in clutch or driveline failure. To correctly diagnose the problem, the interlock should be checked using the following procedures:

1. Disconnect the shift linkage operating lever from the transaxle selector shaft.

2. Remove the transaxle detent spring assembly and selector shaft boot.

3. Remove the aluminum selector shaft plug.

4. Place the transaxle in Neutral and pull the selector shaft assembly out of the case.

5. Measure the interlock blocker

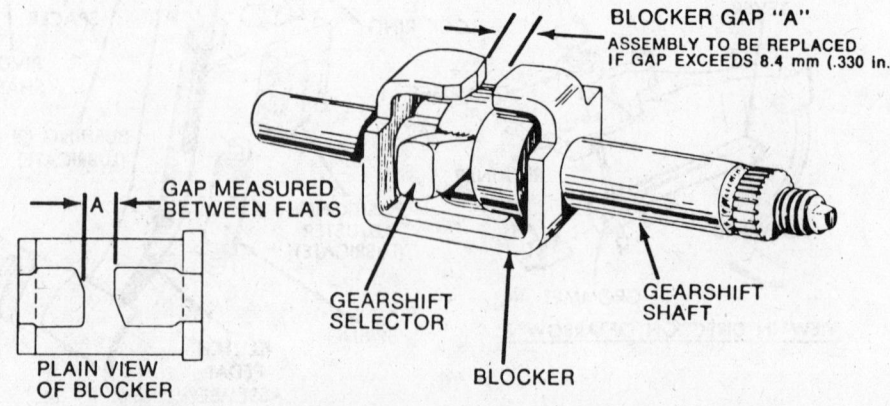

Checking the interlock blocker for failure on A-412 manual transaxles

gap "A", if the gap exceeds 0.330 in. replace the gearshift selector shaft assembly.

6. Apply a thick coating of chassis grease to the selector shaft shoulder at the threaded end and carefully insert the shaft through the selector shaft oil seal.

7. To install, reverse the removal procedures. Adjust the shift linkage.

A-460, A-465 and A-525 Transaxles

ROD OPERATED TYPE

1. Working over the left front fender, remove the lock pin from the transaxle selector shaft housing.

2. Reverse the lock pin (so the long end is down) and insert the lock pin into same threaded hole while pushing the selector shaft into the selector housing. A hole in the selector shaft will align with the lock pin, allowing the lock pin to be screwed into the housing. This operation locks the selector shaft in the 1–2 neutral position.

3. Raise and support the front of the vehicle on jackstands.

4. Loosen the clamp bolt that secures the gearshift tube to the gearshift connector.

5. Check to see that gearshift connector slides and turns freely in gearshift tube.

6. Position the shifter mechanism connector assembly so that the isolator is contacting the upstanding flange and the rib on the isolater is aligned in both directions with the hole in the blockout bracket. Hold the connector isolator in this position while tightening the clamp bolt on the gearshift tube to 14 ft. lbs. No significant force should be exerted on the linkage during this operation.

7. Lower the vehicle to the floor.

8. Remove the lock pin from the selector shaft housing and reinstall the lock pin (with the long end up) in the selector shaft housing. Tighten the lock pin to 9 ft. lbs.

9. Check the first/reverse shifting and the blockout into reverse.

CABLE OPERATED TYPE

1. Working over the left front fender, remove the lock pin from the transaxle selector shaft housing.

2. Reverse the lock pin (so the long end is down) and insert lock pin into same threaded hole while pushing the selector shaft into the selector housing. A hole in the selector shaft will align with the lock pin, allowing the lock pin to be screwed into the housing. This operation locks the selector shaft in the 1–2 neutral position.

3. Remove the gearshift knob, the

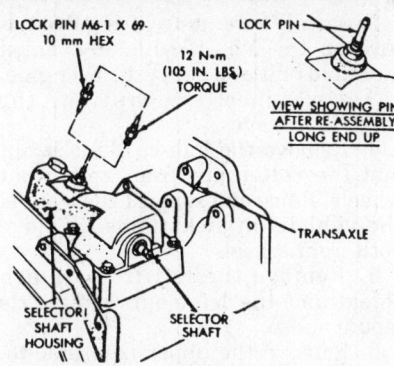

Transaxle pinned in the 1-2 neutral position to adjust gearshift linkage (rod or cable operated)

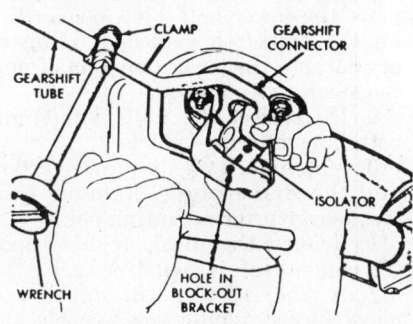

Adjusting gearshift linkage (rod operated)

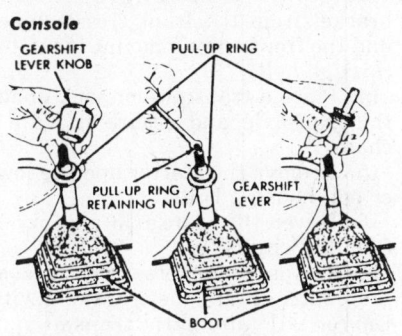

Gearshift knob, retaining nut and pull-up ring (Cable operated linkage)

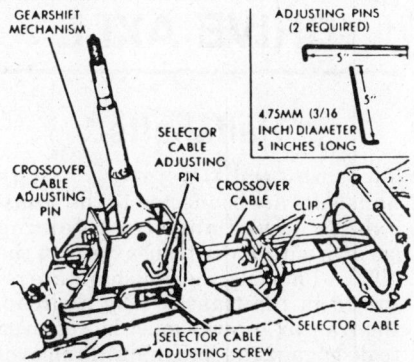

Fabricate (2) cable adjusting pins—all except Daytona, Laser, Shadow and Sundance

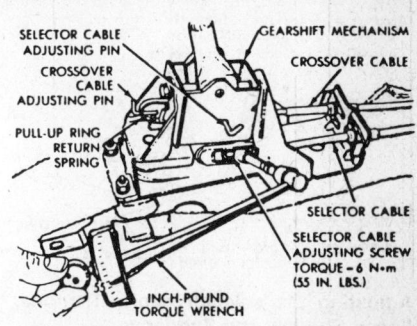

Adjusting the selector cable for all except Daytona, Laser, Shadow and Sundance

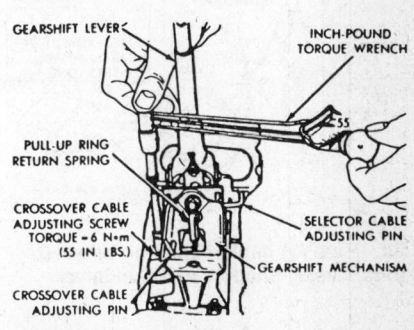

Adjusting the crossover cable for all except Daytona, Laser, Shadow and Sundance

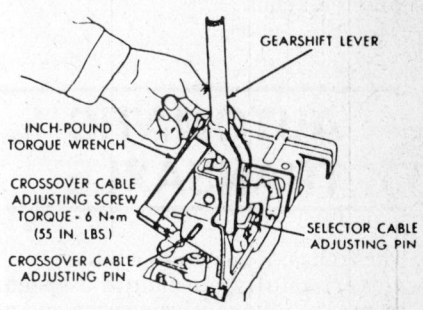

Adjusting the crossover cable—Daytona, Laser, Shadow and Sundance

retaining nut and the pull-up ring from the gearshift lever.

4. If necessary, remove the shift lever boot and console to expose the gearshift linkage.

5. Fabricate two cable adjusting pins: $\frac{3}{16}$ in. dia. x 5 in. long with a $\frac{1}{2}$ in. 90° bend at one end.

6. Place one pin in the hole provided at the right side and the other in the hole provided at the rear side of the shifting mechanism (make sure that the alignment holes match). Torque the selector (right side) and the crossover (left side) adjusting bolts, to 4–5 ft. lbs.

7. Remove the lock pin from the selector shaft housing and reinstall the lock pin (with the long end up) in the selector shaft housing. Torque the lock pin to 9 ft. lbs.

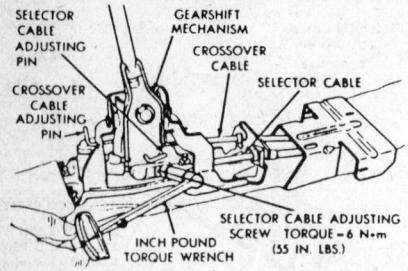

Adjusting the selector cable—Daytona, Laser, Shadow and Sundance

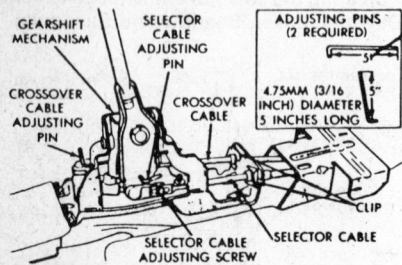

Fabricate (2) cable adjusting pins—Daytona, Laser, Shadow and Sundance

8. Check the first/reverse shifting and blockout into reverse.

9. Reinstall the console, the boot, the pull-up ring, the retaining nut and the knob.

AUTOMATIC TRANSAXLE

The transaxle combines a torque converter, a fully automatic 3-speed transaxle, a final drive gearing and a differential into a compact front wheel drive system. Officially, they are designated as the A-404, A-413, A-415 and A-470 Torqueflite Automatic Transaxles.

For the automatic transaxle linkage/band adjustments and the fluid/filter change procedures, please refer to "Automatic Transmissions" in the Unit Repair section.

REMOVAL & INSTALLATION

The transaxle can be removed with the engine installed in the vehicle but the transaxle and torque converter must be removed as an assembly. Otherwise the drive plate, pump bushing or oil seal could be damaged. The drive plate will not support a load—no weight should be allowed to bear on the drive plate.

1. Disconnect the negative battery cable.

2. Install an engine lifting eye fixture on the No. 4 cylinder exhaust manifold bolt and support the engine.

3. Disconnect the shift and the throttle linkages.

4. Remove the hub castle lock nut and the cotter pin from both front wheels. Raise and support the front of the vehicle on jackstands. Remove both front wheels.

5. Remove the left front splash shield and the left engine/transaxle mount.

6. Remove the upper transaxle-to-engine bolts.

7. Remove the speedometer cable bolt and the cable assembly from the transaxle.

8. Disconnect the sway bar. Remove the lower ball joint-to-steering knuckle bolts from both sides. Using a pry bar, pry the lower ball joints from the steering knuckles.

9. Disconnect the halfshafts from both hubs and the vehicle.

10. Matchmark the torque converter and the drive plate. Remove the torque converter mounting bolts.

11. Remove the oil cooler tubes and the neutral safety switch wire.

12. At the right splash shield, remove the access plug, so a wrench can be placed on the crankshaft bolt to turn the crankshaft.

13. Remove the engine mount bracket from the front crossmember and the front engine mount insulator through bolt.

14. Place a transmission jack under the transaxle and secure it with a chain.

15. Remove the starter and the lower bell housing bolts.

16. Lower the transaxle and pry it from the engine.

17. To install, reverse the removal procedures. Refill the transaxle with Dexron^k II automatic transmission fluid. Adjust the shift and throttle linkages.

DRIVE AXLE

Halfshafts

The inboard CV-joints of the halfshafts are connected to the transaxle by drive flanges or retaining springs within the transaxles. On the 1980–81 models, the halfshafts are retained in the transaxle by a circlip, the transaxle cover must be removed to disengage the halfshaft. If removal of the retained spring type halfshaft is required, the splines on the transaxle ends of both shafts can be easily

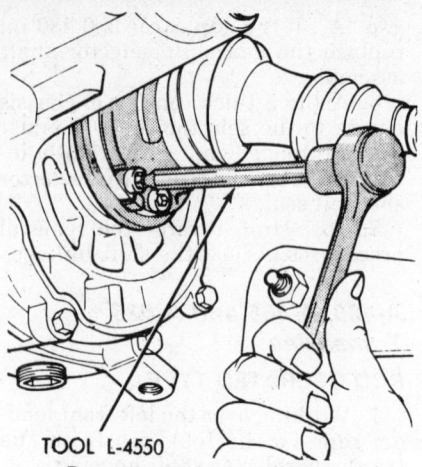

TOOL L-4550
Removing Allen head screws

pulled out without getting into the transaxle.

REMOVAL & INSTALLATION

1980–81 Automatic Transaxle

1. With the vehicle on the ground, remove the cotter pin, the castle hub nut lock, the spring washer, the hub nut and washer.

2. Drain the transaxle and remove the the cover plate.

3. Before removing the right side halfshaft, disconnect the speedometer cable and remove the cable assembly.

4. Rotate the halfshaft to expose the circlip tangs inside the transaxle. Using needle pliers, compress the circlip tangs and pry the halfshaft into the splined cavity.

5. Remove the ball joint stud-to-steering knuckle clamp bolt.

6. Using a medium pry bar, separate the ball joint stud from the steering knuckle, by prying against the knuckle leg and the control arm.

7. Separate the outer CV-joint splined shaft from the hub by holding the CV housing and moving the hub away. DO NOT pry on the slinger or outer CV-joint.

8. When removing the halfshaft from the transaxle, support the shaft at the CV-joints and pull on the inner shaft (DO NOT pull on the halfshaft).

9. To install, use a new gasket, sealant (on the transaxle cover) and reverse the removal procedures. Torque the transaxle cover bolts to 21 ft. lbs., the steering knuckle/ball joint bolt to 50 ft. lbs. and the hub nut to 180 ft. lbs. Refill the transaxle with Dexron^k II automatic transmission fluid.

All Other Transaxles

NOTE: Drain some of the oil from the transaxle to prevent oil

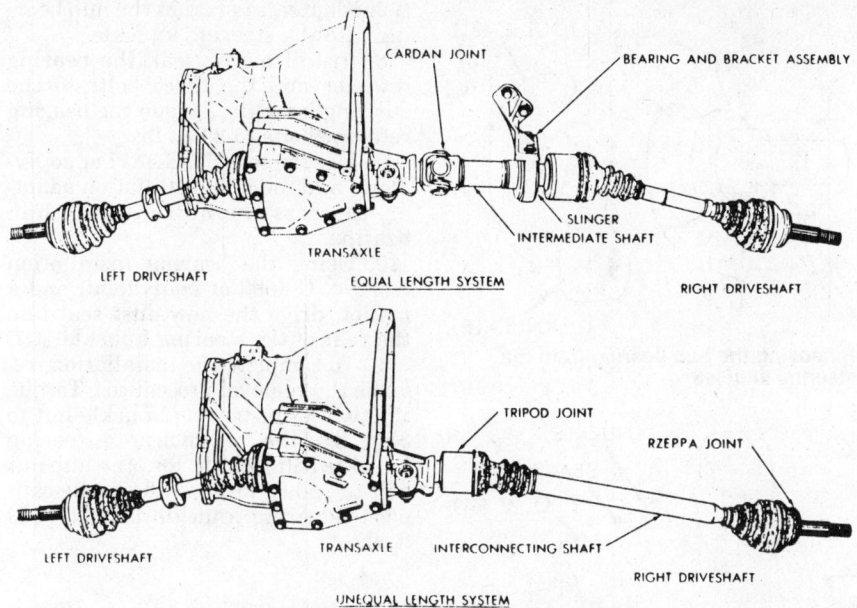

Front wheel drive axleshaft assemblies

spillage when the halfshafts are removed from the transaxle.

1. With the vehicle on the ground, remove the cotter pin, the castle hub nut lock, the spring washer, the hub nut and washer.
2. Before removing the right side halfshaft, disconnect the speedometer cable and remove the cable assembly.
3. Remove the ball joint stud-to-steering knuckle clamp bolt.
4. Using a medium pry bar, separate the ball joint stud from the steering knuckle, by prying against the knuckle leg and the control arm.
5. Separate the outer CV-joint splined shaft from the hub by holding the CV housing and moving the hub away. DO NOT pry on the slinger or the outer CV-joint.
6. When removing the halfshaft from the transaxle, support the shaft at the CV-joints and pull on the inner shaft (DO NOT pull on the halfshaft). If equipped with the flange type halfshaft, remove the six Allen head screws from the transaxle drive flange and separate the halfshaft from the transaxle.
7. To install, reverse the removal procedures. Torque the axle flange bolts (if equipped) to 37 ft. lbs., the steering knuckle/ball joint bolt to 70 ft. lbs. and the hub nut to 180 ft. lbs.

CV-JOINT OVERHAUL

For all CV-Joint overhaul procedures, please refer to "CV-Joint Overhaul" in the Unit Repair section.

Rear Axle Spindle

REMOVAL & INSTALLATION

Omni, Horizon, Charger, Turismo, TC3 and 024

1. Raise and support the rear of the vehicle on jackstands, then place a floor jack under the axle assembly.
2. Remove the wheel and tire assembly.
3. Remove the brake fittings and the brake line retaining clips.
4. Remove the parking brake cable adjusting nut. Slip the parking brake cable ball-ends from the connectors and through the brackets.
5. Remove the brake drum, the brake assembly/axle spindle retaining plate and the axle spindle. Use a piece of wire to hang the brake assembly retaining plate out of the way.
6. Remove the trailing arm and the shock absorber mounting bolts.
7. Remove the axle assembly from the vehicle.
8. To install, reverse the removal procedures.

All Other Models

1. Raise and support the rear of the vehicle on jackstands. Place a floor jack under the axle assembly.
2. Remove the wheel and tire assembly.
3. Remove the brake tube assembly from the brake hose at the trailing arm support bracket and the brake line lock clips.
4. Remove the parking brake cable at the adjustment and the cable housing at the floor pan bracket.

5. Remove the brake drum, the brake assembly/axle spindle retaining plate and the axle spindle. Use a piece of wire to hang the brake assembly retaining plate out of the way.

NOTE: If equipped with an automatic load leveling system, disconnect the link from the sensor track bar.

6. Remove the track bar end (support with a wire) and the shock absorber mounting bolts.
7. Lower the axle. Remove the spring and isolator assembly.
8. Support the pivot bushing end of the trailing arms and remove the pivot bushing hanger bracket-to-frame bolts.
9. Remove the axle assembly from the vehicle.
10. To install, reverse the removal procedures. Torque the backing plate and spindle assembly to 45 ft. lbs.

Front Axle Hub, Bearing And Seal

REMOVAL & INSTALLATION

1980–83

1. Refer to the "Steering Knuckle Removal & Installation" procedures in this section and remove the steering knuckle from the vehicle.
2. Using the hub removal tool No. C-4539 or equivalent, and the triangular adapter, attach them to the three rear threaded holes of the steering knuckle housing with the thrust button and the fabricated washer ($\frac{15}{16}$ in. dia. x $1\frac{1}{2}$ in. dia.), inside the hub bore.
3. Tighten the center bolt of the tool to press the hub from the steering knuckle. Remove the removal tools.
4. Using the thrust button from tool No. C-4539 or equivalent, and a wheel puller, remove the bearing outer race from the hub.
5. Remove the three bolts and the bearing retainer from the outside of the steering knuckle.
6. Carefully pry the bearing seal from the machined recess of the steering knuckle and clean the recess.
7. Using an arbor press, place the steering knuckle face down on two supports, place a $1\frac{5}{8}$ in. socket on the hub bearing and press the bearing from the steering knuckle. Discard the bearing.
8. To install, place the steering knuckle (face up) on two wooden blocks, under an arbor press. Place the new bearing in the steering knuckle and the installation tool No. L-4463 or equivalent, on the bearing.

Press the bearing into the steering knuckle.

9. Install the bearing retainer and the three bolts to the steering knuckle. Torque the bearing retainer bolts to 20 ft. lbs.

10. On an arbor press, place the hub (face down) on a wooden support, position the steering knuckle on the hub, a $1\frac{5}{8}$ in. socket on the hub bearing and press the assembly together.

11. Place a new seal on the rear of the steering knuckle. Using the bearing installation tool No. C-4698 or equivalent, and a mallet, drive the new dust seal into the steering knuckle. Lubricate the wear sleeve with multi-purpose grease and install on the new seal.

12. To complete the installation, reverse the removal procedures. Torque the tie rod-to-steering knuckle nut to 35 ft. lbs., the control arm-to-steering knuckle bolt to 50 ft. lbs., the hub nut bolt to 180 ft. lbs., the brake caliper-to-steering knuckle bolts to 160 ft. lbs. and strut-to-steering knuckle bolts to 45 ft. lbs., plus a $\frac{1}{4}$ turn.

1984 and Later

1. Remove the halfshaft from the steering knuckle assembly.

2. Attach the hub removal tool No. C-4811 or equivalent, and the triangular adapter, to the three rear threaded holes of the steering knuckle housing with the thrust button inside the hub bore.

3. Tighten the bolt, in the center of the tool, to press the hub from the steering knuckle. Remove the removal tools.

4. Remove the three bolts and the bearing retainer from the outside of the steering knuckle.

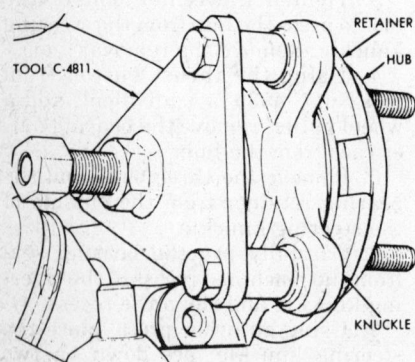

Removing the hub from the steering knuckle

5. Carefully pry the bearing seal from the machined recess of the steering knuckle and clean the recess.

6. Insert the tool No. C-4811 or equivalent, through the hub bearing and install bearing removal adapter

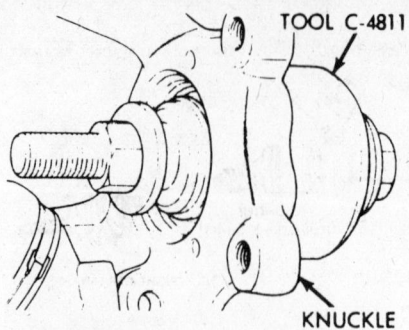

Removing the hub bearing from the steering knuckle

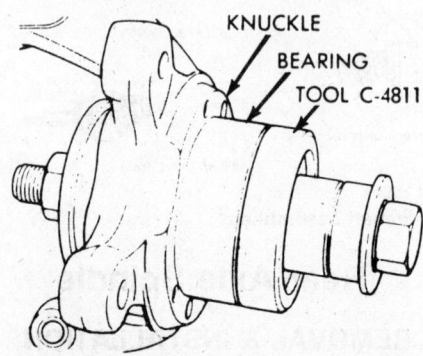

Installing the hub bearing to the steering knuckle

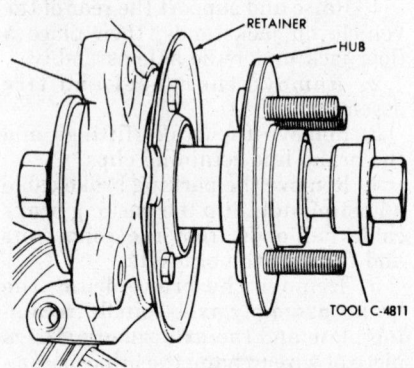

Installing the hub to the steering knuckle

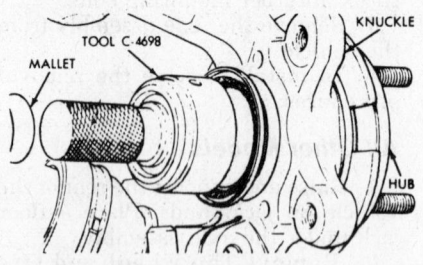

Installing the seal to the steering knuckle

to the outside of the steering knuckle, tighten the tool to press the hub bearing from the steering knuckle. Discard the bearing and the seal.

7. To install, use tool No. C-4811 or equivalent, and the bearing installa-

tion adapter, to press in the hub bearing into the steering knuckle.

8. Install a new seal, the bearing retainer and the three bolts to the steering knuckle. Torque the bearing retainer bolts to 20 ft. lbs.

9. Use the tool No. C-4811 or equivalent, and the hub installation adapter, to press the hub into the hub bearing.

10. Using the bearing installation tool No. C-4698 or equivalent, and a mallet, drive the new dust seal into the rear of the steering knuckle.

11. To complete the installation, reverse the removal procedures. Torque the tie rod-to-steering knuckle nut to 35 ft. lbs., the control arm-to-steering knuckle bolt to 70 ft. lbs., the hub nut bolt to 180 ft. lbs. and the brake caliper-to-steering knuckle bolts to 160 ft. lbs.

FRONT SUSPENSION

A MacPherson front suspension is used, with the vertical shock absorbers attached to the upper fender reinforcement and the steering knuckle. The lower control arms are attached, inboard, to a cross-member and outboard, to the steering knuckle through a ball joint, provided at the lower steering knuckle position.

MacPherson Strut

REMOVAL & INSTALLATION

1. Raise and support the front of the vehicle on jackstands. Remove the wheel/tire assembly.

2. If the original strut is to be assembled to the original knuckle, mark the cam adjusting bolt. Remove the cam adjusting bolt, the steering knuckle bolt and the brake hose bracket retaining screw.

3. Remove the strut-to-body mounting nuts and the strut.

5. To install, reverse the removal procedures. Position the steering knuckle in the strut and install the upper (cam) bolt. Index the cam bolt with the matchmarks. Torque the strut-to-body nuts to 20 ft. lbs., the wheel nuts to 80–95 ft. lbs., the brake hose bracket screw to 10 ft. lbs., the cam/steering knuckle bolts to 45 ft. lbs. (Omni and Horizon) or 75 ft. lbs. (all other models), plus a $\frac{1}{4}$ turn.

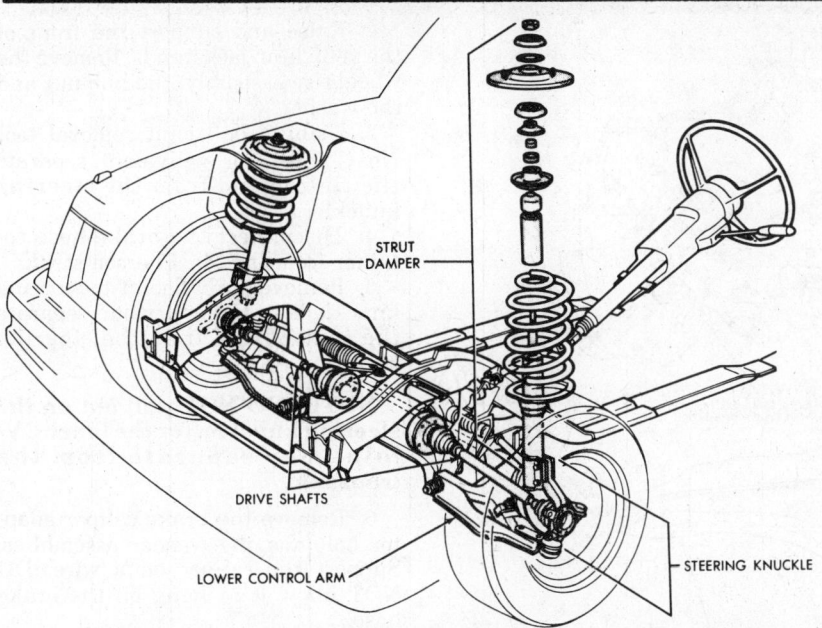

Front suspension—Omni and Horizon shown; other models similar

OVERHAUL

For the strut overhaul procedures, please refer to "Strut Overhaul" in the Unit Repair Section.

Ball Joint

INSPECTION

1980

1. Raise and support the front of the vehicle.
2. With the suspension fully extended (at full travel) clamp a dial indicator to the lower control arm with the plunger indexed against the steering knuckle leg.
3. Zero the dial indicator.
4. Use a stout bar to pry on the top of the ball joint housing-to-lower control arm bolt, with the bar tip under the steering knuckle leg.
5. Measure the axial travel of the steering knuckle leg in relation to the control arm by raising and lowering the steering knuckle.
6. If the travel is more than 0.050 in., the ball joint should be replaced.

1981 and Later

The lower ball joint is checked at the lube fitting. Shake the lube fitting, if it moves, the ball joint is worn and should be replaced.

REMOVAL & INSTALLATION

1980

NOTE: The ball joint housing is

If the grease nipple (arrow) in the ball joint wobbles or turns, the ball joint is worn and should be replaced.

bolted to the lower control arm with the joint stud retained in the steering knuckle by a clamp bolt.

1. Raise and support the front of the vehicle on jackstands.
2. Remove the steering knuckle-to-ball joint stud clamp bolt and separate the stud from the knuckle leg.
3. Remove the two bolts holding the ball joint housing-to-lower control arm.
4. Remove the ball joint housing.
5. To install, use a new ball joint and reverse the removal procedures. Torque the ball joint housing/control arm bolts to 60 ft. lbs. and the ball joint/steering knuckle bolt to 50 ft. lbs. Lower the vehicle.

1981 and Later

NOTE: On some models the ball joints are welded to the control arms and are not to be pressed out. To service the ball joint, remove and replace the ball joint and control arm assembly.

1. Raise and support the front of the vehicle on jackstands.
2. Remove the bolt and separate the ball joint from the steering knuckle. Remove the control arm pivot bolt and the control arm.
3. Pry the dust seal from the ball joint.
4. Using an arbor press, position a receiving cup tool No. C-4699-2 or equivalent, to support the lower control arm. Install a $1\frac{1}{16}$ in. deep socket over/against the stud joint upper housing and press the ball joint from the control arm.
5. To install, position a new ball joint housing into the control arm cavity.
6. Position the ball joint and the control arm in the press, with an installer tool No. C-4699-1 or equivalent, (under the ball joint) and the receiver cup tool No. C-4699-2 or equivalent, (over the ball joint stud). Align the assembly and press it together, until the housing ledge stops against the control arm cavity down flange.
7. Place a new seal over the ball joint stud, followed by a $1\frac{1}{2}$ in. socket, support the ball joint housing with installer tool No. C-4699-1 or equivalent, and press the seal onto the ball joint housing, until it seats against the control arm.
8. To complete the installation, reverse the removal procedures. Torque the control arm pivot bolt to 105 ft. lbs. and the ball joint-to-steering knuckle bolt to 70 ft. lbs.

Front Control Arms

REMOVAL & INSTALLATION

1. Raise and support the front of the vehicle on jackstands.
2. Remove the front inner pivot through-bolt, the stub strut nut, the retainer and the ball joint-to-steering knuckle clamp bolt.
3. Separate the ball joint stud from the steering knuckle by prying between the steering knuckle and the control arm.

—————— **CAUTION** ——————
Pulling the steering knuckle out from the vehicle, after releasing it from the ball joint, can separate the inner CV-joint.

4. Remove the sway bar-to-control arm nut and rotate the control arm

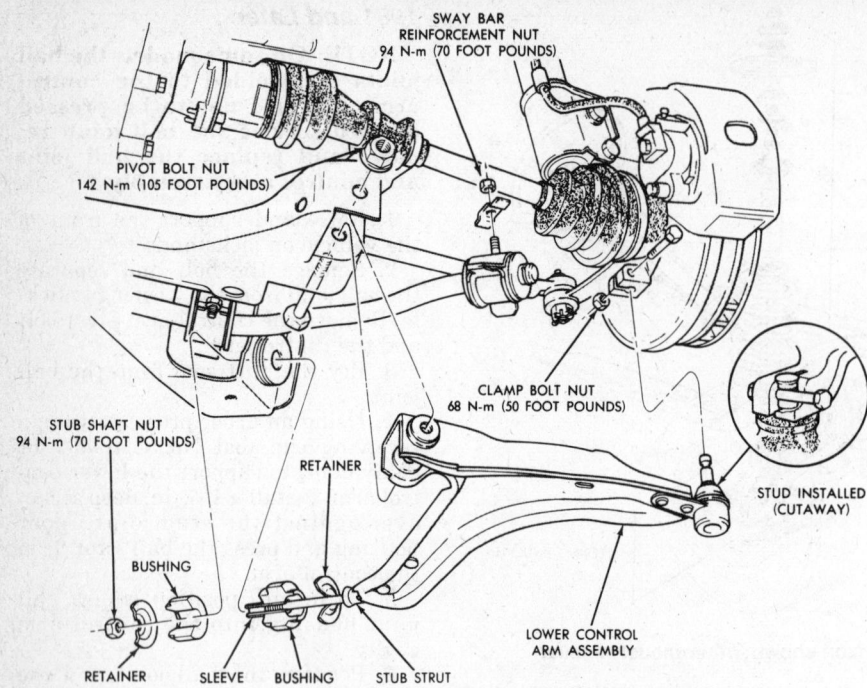

Typical lower control arm

over the sway bar. Remove the control arm from the stub strut bushing assembly.

NOTE: The substitution of fasteners with other than those of the grade originally used is not recommended.

5. To install, reverse the removal procedures. Torque ball joint stud-to-steering knuckle bolt to 70 ft. lbs. and the sway bar bracket nuts to 25 ft. lbs. Lower the vehicle. Torque the front pivot bolt to 105 ft. lbs. and the stub strut nut to 70 ft. lbs.

Front Wheel Bearing

ADJUSTMENT

1. Raise and support the front of the vehicle on jackstands.
2. Remove the grease cap, the cotter pin, the lock nut and the spring washer. Loosen the hub nut.
3. Apply pressure to the brakes.
4. Torque the hub nut to 180 ft. lbs.
5. To install, reverse the removal procedures. Lower the vehicle.

REMOVAL & INSTALLATION

Refer to the "Front Axle Hub, Bearing and Seal Removal & Installation" procedures in the "Drive Axle" division of this section and replace the front wheel bearing.

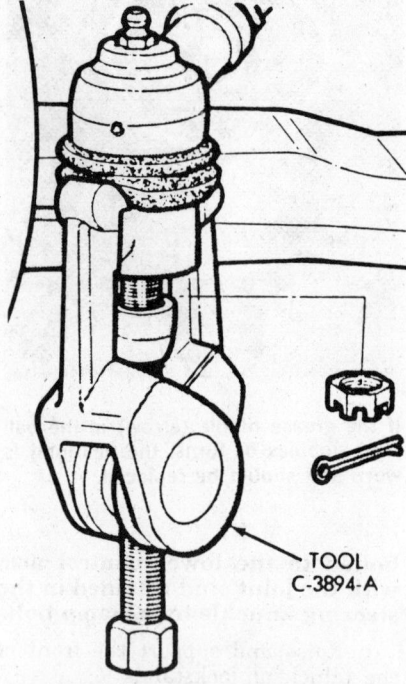

Disconnecting the tie rod from the steering knuckle

Steering Knuckle

REMOVAL & INSTALLATION

1. Remove the grease cap, the cotter pin, the lock nut and the spring washer. Loosen the hub nut with the vehicle on the floor with the brakes applied.

2. Raise and support the front of the vehicle on jackstands. Remove the wheel/tire assembly, the hub nut and the washer.
3. Using a ball joint removal tool No. C-3894-A or equivalent, separate the tie-rod end from the steering knuckle.
4. Disconnect the brake hose retainer from the MacPherson strut.
5. Remove the ball joint-to-steering knuckle bolt. Using pry bar, separate the control arm from the steering knuckle.

NOTE: DO NOT pull out on the steering knuckle, for the inner CV-joint may separate from the transaxle.

6. Remove the brake caliper adaptor bolt and the washer assemblies. Support the caliper on a wire, DO NOT allow it to hang on the brake hose.
7. Remove the brake rotor (disc). Using a soft mallet, pull the steering knuckle assembly outward and tap the halfshaft from the hub. Support the halfshaft on a wire.
8. Mark the location of the strut-to-steering knuckle bolts. Remove the strut-to-steering knuckle bolts and the steering knuckle assembly.
9. To install, reverse the removal procedures. Torque the tie rod-to-steering knuckle nut to 35 ft. lbs., the control arm-to-steering knuckle bolt to 70 ft. lbs., the hub nut bolt to 180 ft. lbs., the brake caliper-to-steering knuckle bolts to 160 ft. lbs. and the strut-to-steering knuckle bolts to 45 ft. lbs. (Omni and Horizon) or 75 ft. lbs. (all other models), plus a ¼ turn.

Sway Bar

REMOVAL & INSTALLATION

1. Raise and support the front of the vehicle on jackstands.
2. Remove the sway bar nuts, bolts and retainers from the control arms.
3. Remove the sway bar-to-crossmember clamp bolts. Remove the clamps and the sway bar.
4. To install, reverse the removal procedures. Torque the sway bar nuts and bolts to 22 ft. lbs.

REAR SUSPENSION

Omni, Horizon, Charger, Turismo, TC3 and 024

The vehicles use an independent

trailing arm assembly, with an integral sway bar. Each wheel spindle is attached to a MacPherson type shock absorber and a trailing arm, that extends rearward from the body mounting point. A crossmember is welded to the trailing arms, which provides lateral location and anti-roll-bar type stabilizing.

All Other Models

The vehicles use a flexible beam axle with trailing links and coil springs. A shock absorber, mounted near each coil spring, is attached to the body and the beam axle. Each wheel spindle is bolted to the outer end of the beam axle.

Shock Absorber

REMOVAL & INSTALLATION

All Models, Except: Omni, Horizon, Charger, Turismo, TC3 and 024

1. Raise and support the rear of the vehicle on jackstands.

NOTE: If equipped with air shocks, disconnect the air lines to the shocks.

2. Disconnect the upper and lower shock absorber bolts.
3. Remove the shock absorber.
4. Purge the new shocks of air, by compressing them while inverted and extending them in their normal position, several times.
5. To install, reverse the removal procedures. Torque the upper shock absorber-to-body bolt to 45 ft. lbs. and the lower shock absorber-to-axle mounting bolt to 40 ft. lbs.

MacPherson Strut

REMOVAL & INSTALLATION

Omni, Horizon, Charger, Turismo, TC3 and 024

1. Remove the protective cap from the upper mounting nut. On the two door models, remove the lower rear quarter panel trim.
2. Remove the upper mounting nut, the isolator retainer and the isolator.
3. Raise and support the rear of the vehicle on jackstands.
4. Remove the lower strut-to-trailing arm bolt.
5. Remove the strut from the vehicle.
6. To install, reverse the removal procedures. Torque the lower strut-to-trailing bolt to 40 ft. lbs. and the upper strut-to-body nut to 20 ft. lbs.

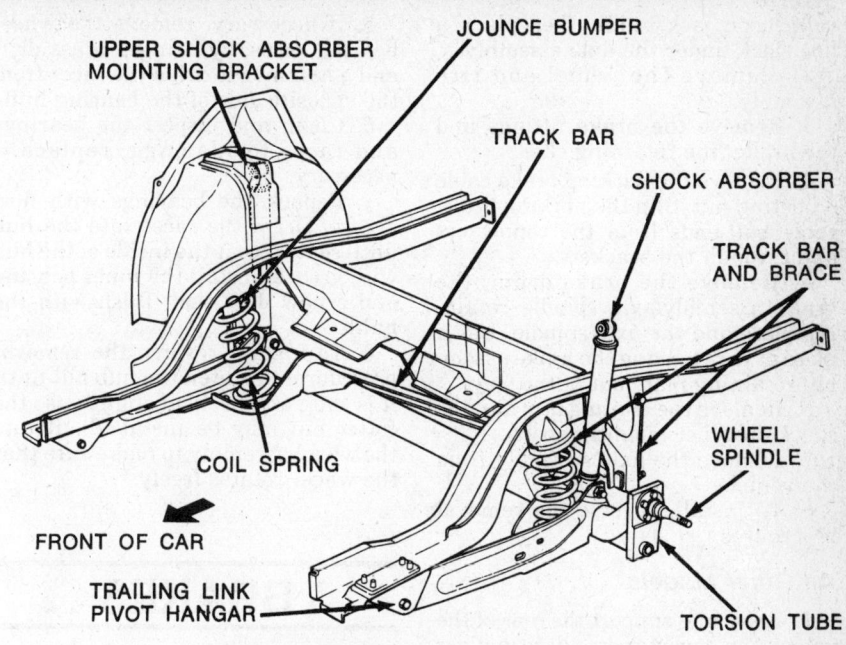

Rear suspension except Omni and Horizon

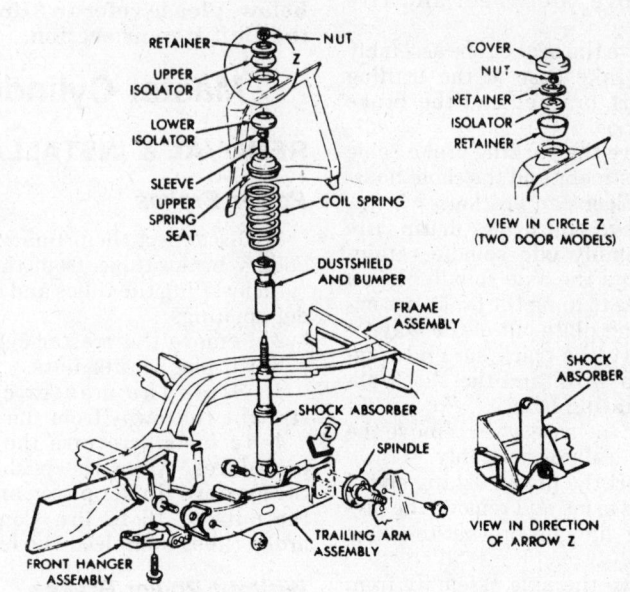

Semi-independent rear suspension system on Omni/Charger and Horizon/Turismo

OVERHAUL

For all strut overhaul procedures, please refer to "Strut Overhaul" in the Unit Repair section.

Rear Springs

REMOVAL & INSTALLATION

All Models, Except: Omni, Horizon, Charger, Turismo, TC3 and 024 – 1981 and Later

1. Raise and support the rear of the

vehicle on jackstands, then place a floor jack under the axle assembly.
2. Remove the shock absorber-to-axle bolts.
3. Lower the axle to remove the spring and insulator.
4. To install, reverse the removal procedures.

Rear Axle Assembly

REMOVAL & INSTALLATION

Omni, Horizon, Charger, Turismo, TC3 and 024

1. Raise and support the rear of the

vehicle on jackstands, then place a floor jack under the axle assembly.

2. Remove the wheel and tire assembly.

3. Remove the brake fittings and the brake line retaining clips.

4. Remove the parking brake cable adjusting nut. Slip the parking brake cable ball-ends from the connectors and through the brackets.

5. Remove the brake drum, the brake assembly/axle spindle retaining plate and the axle spindle. Use a piece of wire to hang the brake assembly retaining plate out of the way.

6. Remove the trailing arm and the shock absorber mounting bolts.

7. Remove the axle assembly from the vehicle.

8. To install, reverse the removal procedures.

All Other Models

1. Raise and support the rear of the vehicle on jackstands. Place a floor jack under the axle assembly.

2. Remove the wheel and tire assembly.

3. Remove the brake tube assembly from the brake hose at the trailing arm support bracket and the brake line lock clips.

4. Remove the parking brake cable at the adjustment and the cable housing at the floor pan bracket.

5. Remove the brake drum, the brake assembly/axle spindle retaining plate and the axle spindle. Use a piece of wire to hang the brake assembly retaining plate out of the way.

6. Remove the track bar end (support with a wire) and the shock absorber mounting bolts.

7. Lower the axle, then remove the spring and isolator assembly.

8. Support the pivot bushing end of the trailing arms and remove the pivot bushing hanger bracket-to-frame bolts.

9. Remove the axle assembly from the vehicle.

10. To install, reverse the removal procedures.

Rear Hub And Bearing

REPLACEMENT & ADJUSTMENT

1. Raise and support the rear of the vehicle.

2. Remove the grease cap, the cotter pin, the lock nut, the hub nut, the washer and the outer wheel bearing.

3. Pull the brake drum and wheel assembly from the axle spindle.

4. From inside the brake drum, remove the dust seal and the inner wheel bearing.

5. If necessary, remove the wheel bearing races, by using a brass drift and a hammer to drive the races from the opposite side of the bearing hub.

6. Clean and inspect the bearings and races for damage, replace if necessary.

7. Repack the bearings with new grease. Drive the races into the hub until seated, coat the inside of the hub with grease, install the inner bearing and a new dust seal (flush with the hub).

8. To install, reverse the removal procedures. Tighten the hub nut until it is snug and back off slightly (so the cotter pin may be installed). Rotate the wheel assembly to make sure that the wheel rotates freely.

BRAKES

For all brake system repair and service procedures not detailed below, please refer to "Brakes" in the Unit Repair section.

Master Cylinder

REMOVAL & INSTALLATION

Power Brakes

1. Disconnect the primary and secondary brake tubes from the master cylinder. Plug the tubes and the cylinder openings.

2. Remove the master cylinder-to-power brake booster nuts.

3. Slide the master cylinder straight out, away from the booster.

4. To install, reverse the removal procedures. Align the pushrod with the master cylinder piston and torque the nuts to 16 ft. lbs. Connect the brake tubes and bleed the brakes.

Without Power Brakes

1. Disconnect the primary and secondary brake tubes from the master cylinder. Plug the tubes and the cylinder openings.

2. Disconnect the stoplight switch mounting bracket from under the instrument panel.

3. Pull the brake pedal backward to disengage the pushrod from the master cylinder piston.

NOTE: This will destroy the grommet.

4. Remove the master cylinder-to-firewall nuts.

5. Slide the master cylinder out and away from the firewall. Remove the pieces of the broken grommet.

6. Install the boot on the pushrod.

7. Install a new grommet on the pushrod.

8. Apply a soap and water solution to the grommet and slide it firmly into position in the primary piston socket. Move the pushrod side-to-side to seat the grommet.

9. From the engine side, press the pushrod through the master cylinder mounting plate and align the mounting studs with the holes in the cylinder.

10. Install the nuts and torque them to 16 ft. lbs.

11. From under the instrument panel, place the pushrod on the pin on the pedal and install a new retaining clip. Lubricate the pin.

12. Install the brake tubes on the master cylinder and bleed the brake system.

Combination Valve

The combination valve combines a pressure warning switch with a dual proportioning valve. It is located below the master cylinder and attached to the frame rail.

REMOVAL & INSTALLATION

1984 and Later

1. Disconnect the elecrical connector from the brake warning switch.

2. Disconnect and plug the brake lines from the combination valve.

3. Remove the combination valve-to-frame rail bolts and the valve from the vehicle.

4. To install, reverse the removal procedures. Bleed the brake system.

Power Brake Booster

REMOVAL & INSTALLATION

1. Remove the master cylinder-to-power brake booster nuts and pull the cylinder away from the booster without disconnecting the brake lines.

NOTE: If equipped with a manual transaxle, remove the clutch cable mounting bracket. Pull the wiring harness away and up from the strut tower.

2. Disconnect the vacuum hose from the booster.

3. Under the instrument panel, pry the retainer clip center tang over the end of the brake pedal pin and pull the retainer clip from the pin. Discard the clip.

NOTE: Except for the Omni, Horizon, Charger, Turismo, TC3 and 024 models, remove the stop lamp switch and the striker plate.

4. Remove the booster-to-firewall nuts and the booster from the vehicle.

5. To install, position the booster on the firewall and reverse the removal procedures. Torque the booster-to-firewall and the master cylinder-to-booster mounting nuts to 21 ft. lbs. Lubricate the bearing surface of the pedal pin. Install a new pushrod-to-pedal pin clip. Check the stoplight operation. With vacuum applied to the power brake unit and pressure applied to the pedal, the master cylinder should vent (force a jet of fluid through the front chamber vent port).

> **CAUTION**
> *Do not attempt to disassemble the power brake unit, since the booster is serviced as a complete assembly ONLY.*

Wheel Cylinder

REMOVAL & INSTALLATION

1. Raise and support the rear of the vehicle on jackstands.

2. Remove the grease cap, the hub nut, the outer wheel bearing and the wheel/tire assembly with the brake drums.

3. Visually inspect the wheel cylinder boots for signs of excessive leakage. Replace any boots that are torn or broken.

NOTE: A slight amount of fluid on the boots may not be a leak but may be preservative fluid used at the factory.

4. If a leak is detected, remove the brake shoes and check for contamination. Replace the linings if they are soaked with grease or brake fluid.

5. Disconnect the brake line from the wheel cylinder.

6. Remove the wheel cylinder attaching bolts, then pull the wheel cylinder out of its support.

7. To install, reverse the removal procedures. Torque the wheel cylinders-to-backing plate bolts to 6 ft. lbs. (8 Nm). Bleed the brake system.

Parking Brake

For the overhaul procedures, please refer to "Brakes" in the Unit Repair Section.

ADJUSTMENT

NOTE: The service brakes must be properly adjusted before adjusting the parking brakes.

1. Raise and support the rear of the vehicle on jackstands. Fully release the parking brake.

2. Locate the cable connector at the rear suspension crossmember. Clean and lubricate the cable assembly.

3. Loosen the adjusting nut until there is slack in the cable.

4. Insert a brake adjusting spoon through the slot in the brake backing plate and rotate the starwheel so there is light shoe-to-drum contact.

5. Back-off the starwheel to allow free drum rotation.

6. Loosen the cable adjusting nut until both rear wheels turn freely, then back-off the nut two full turns.

7. Apply and release the parking brake several times to make sure the wheels rotate freely.

STEERING

Steering Wheel

REMOVAL & INSTALLATION

1. Disconnect the negative battery cable and the wiring connector at the steering column.

2. Remove the horn button or the pad and the horn switch.

3. Remove the steering wheel nut.

4. Using a steering wheel puller tool No. C-3428B or equivalent, remove the steering wheel.

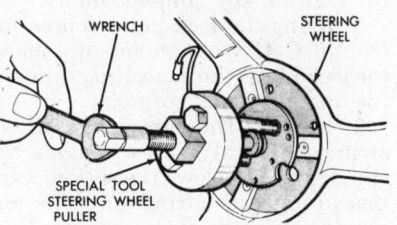

Steering wheel removal

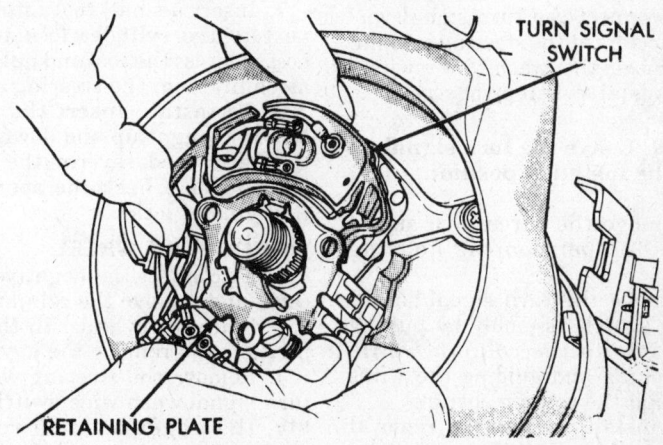

RETAINING PLATE

Turn signal switch removal—Aries and Reliant

5. To install, align the master serration in the wheel hub with the missing tooth on the shaft. Torque the shaft nut to 45 ft. lbs.

> **CAUTION**
> *DO NOT torque the nut against the steering column lock, for damage may occur.*

6. Replace the horn switch and button.

Turn Signal Switch

REMOVAL & INSTALLATION

Omni, Horizon, Charger, Turismo, TC3 and 024

1. Disconnect the electrical connector at the column.

2. Remove the steering wheel and the lower column cover.

3. Remove the wash/wipe switch.

4. Remove the wiring clip, the three turn signal switch screws and the switch from the steering column.

5. To install, reverse the removal procedures.

All Other Models
TILT WHEEL

1. Disconnect the negative battery cable.

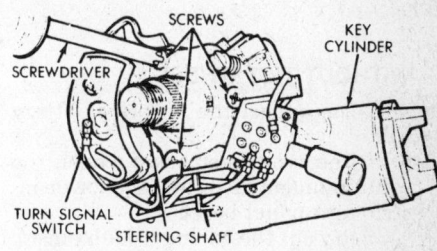

Turn signal switch removal—Omni and Horizon

2. Remove the steering wheel, the column under cover and the lower instrument panel bezel.

3. Pry out the wiring trough plastic retainers and lift out the trough.

4. Position the gearshift lever to the full clockwise position.

5. Disconnect the turn signal wiring connectors at the steering column.

6. If equipped, remove the plastic cover from the lock plate. Using tool No. C-4156 or equivalent, depress the lock plate and pry out the retaining ring.

7. Remove the lock plate, the cancelling cam and the upper bearing spring.

8. Place the turn signal switch in the right turn position and remove the turn signal-to-wash/wiper switch pivot screw.

9. Remove the hazard warning switch knob screw and the three turn signal switch-to-steering column screws.

10. Remove the turn signal/hazard warning switch assembly by pulling the switch up the column, while straightening and guiding the wires up through the column opening.

11. To install, reverse the removal procedures. Lubricate the switch pivot with light grease. Place the switch in the right turn position, when installing. Use tool No. C-4156 or equivalent, to install the upper bearing spring, the cancelling cam and the lock plate.

WITHOUT TILT WHEEL

1. Disconnect the negative battery cable.

2. Remove the steering wheel, the column under cover and the lower instrument panel bezel.

3. Pry out the wiring trough plastic retainers and lift out the trough.

4. Position the gearshift lever to the full clockwise position.

5. Disconnect the turn signal wiring connectors at the steering column.

6. Remove the wash/wipe switch-to-turn signal switch pivot screw.

NOTE: Leave the turn signal lever in the installed position.

7. Remove the three turn signal switch-to-upper bearing housing screws.

8. Remove the turn signal/hazard warning switch assembly by pulling the switch up the column, while straightening and guiding the wires up through the column opening.

9. To install, reverse the removal procedures. Lubricate the switch pivot with light grease.

Ignition And Steering Lock

REMOVAL & INSTALLATION

Omni, Horizon, Charger, Turismo, TC3 and 024

1. Place the cylinder in the LOCK position and remove the key.

2. Remove the steering wheel, the four column cover screws, the covers, the sound deadener panel and the turn signal switch.

3. Using a hacksaw blade, cut the upper $\frac{1}{4}$ inch from the key cylinder retainer pin boss.

4. Using a drift punch, drive the roll pin from the housing and remove the key cylinder.

5. To install, insert the cylinder into the housing, make sure it engages the lug on the ignition switch driver, install the roll pin and reverse the removal procedures. Check the cylinder for free operation.

All Other Models
TILT WHEEL

1. Disconnect the negative battery cable. Remove the column covers and the wiring connectors.

2. Place the cylinder in the LOCK position and remove the key.

3. Remove the tilt lever. Push the hazard warning in and unscrew the knob.

4. Remove the steering wheel and the ignition key lamp assembly.

5. Using the lock plate depressing tool No. C-4156 or equivalent, remove the lock plate, the cancelling arm and the upper bearing spring.

6. Remove the three turn signal switch assembly screws, place the shift bowl in the low (1) position, wrap tape around the wiring connector and pull the switch/wiring connector from the steering column.

7. Insert a small tool into the slot, next to the cylinder lock mounting boss, depress the tool and pull the lock assembly from the steering column.

8. To install, insert the cylinder lock (moving it up and down to align the parts) and reverse the removal procedures. Check the operation of the cylinder lock.

WITHOUT TILT WHEEL

1. Disconnect the negative battery cable and remove the column covers.

2. Place the cylinder in the LOCK position and remove the key.

3. Remove the steering wheel, the turn signal/wash/wipe switch assembly, the upper bearing retaining plate, the lock plate spring and the lock plate from the steering column.

4. Remove the buzzer/chime switch screw and the switch.

5. Remove the two ignition switch-to-column screws and the switch, slide off the actuating rod.

6. Remove the two dimmer switch screws and disengage from the actuating rod.

7. Remove the two bellcrank screws and slide it up the in the lock housing to disengage it from the ignition actuator rod.

8. Insert a small diameter tool into both cylinder lock release holes, push in on the tools and pull the lock cylinder out of the housing.

9. To install, reverse the removal procedures. Check the operation of the cylinder lock.

Ignition Switch

REMOVAL & INSTALLATION

1. Disconnect the negative battery cable. Place the ignition cylinder in the LOCK position and remove the key.

2. Remove the four column cover screws, the covers and the sound deadener panel.

3. Loosen the two ignition switch mounting plate screws and push the switch up to take up the rod system slack.

4. Remove the clutch speed control switch.

5. Remove the mounting bolts and drop the steering column for the switch replacement.

6. Remove the two ignition switch-to-column screws.

7. Rotate the switch 90° and pull-up to disengage the switch from the switch rod.

8. To install, push-up on the switch to take up the rod slack, tighten the lock screws and reverse the removal procedures. Check the switch for proper operation.

Manual Steering Gear

REMOVAL & INSTALLATION

NOTE: On some models, except Omni, Horizon, Charger, Shadow, Sundance, Turismo, TC3 and 024, the steering column must be removed.

1. Raise and support the front of the vehicle on jackstands. Remove front wheel/tire assemblies.

2. Remove the cotter pins, the castle nuts and the tie rod ends (using a ball joint puller) from the steering knuckles.

3. Except for the Omni, Horizon, Charger, Shadow, Sundance,

Turismo, TC3 and 024 models, remove the steering column-to-steering gear coupling pin; follow the procedures in Step 6, this will expose the coupling pin so that it can be driven out. If equipped, remove the anti-rotational link from the crossmember and the air diverter valve bracket, from the left side of the crossmember.

NOTE: The lower universal joint is removed with the steering gear.

4. On the Omni, Horizon, Charger, Shadow, Sundance, Turismo, TC3 and 024 models, drive out the lower roll pin attaching the pinion shaft to the lower universal joint. Use a back-up, to protect the universal joint, while driving the roll pin.

5. On the Omni, Horizon, Charger, Shadow, Sundance, Turismo, TC3 and 024 models, support the front suspension crossmember with a hydraulic jack. Remove the two rear nuts attaching the crossmember to the frame. Loosen the two front bolts attaching the crossmember to the frame and lower the crossmember slightly for access to the boot seal shields.

6. Except for the Omni, Horizon, Charger, Shadow, Sundance, Turismo, TC3 and 024 models, remove the four front suspension crossmember attaching bolts and the lower front suspension crossmember, using a transaxle jack, so that the steering gear can be removed from the crossmember.

7. Remove the splash and the boot seal shields.

8. Remove the steering gear bolts from the front suspension crossmember.

10. Remove the steering gear from the left side of the vehicle.

11. To install, reverse the removal procedures. The right rear crossmember bolt is a pilot bolt that correctly locates the crossmember, tighten it first. Torque the four crossmember bolts to 90 ft. lbs. and the steering gear attaching bolts to 21 ft. lbs.

ADJUSTMENT

1. Loosen the adjuster plug lock nut on the steering gear.

2. Turn the adjuster plug clockwise until it bottoms, then back it off 40–60°.

3. While holding the adjuster plug stationary, torque the lock nut to 50 ft. lbs. (70 Nm).

Power Steering Gear

REMOVAL & INSTALLATION

NOTE: On some models, except Omni, Horizon, Charger, Shadow, Sundance, Turismo, TC3 and 024, the steering column must be removed.

1. Raise and support the front of the vehicle on jackstands. Remove front wheel/tire assemblies.

2. Remove the cotter pins, the castle nuts and the tie-rod ends (using a ball joint puller) from the steering knuckles.

3. Except for the Omni, Horizon, Charger, Shadow, Sundance, Turismo, TC3 and 024 models, remove the steering column-to-steering gear coupling pin; follow the procedures in Step 6, this will expose the coupling pin so that it can be driven out. If equipped, remove the anti-rotational link from the crossmember and the air diverter valve bracket, from the left side of the crossmember.

NOTE: The lower universal joint is removed with the steering gear.

4. Disconnect and plug the oil pressure lines from the power steering pump.

5. On the Omni, Horizon, Charger, Shadow, Sundance, Turismo, TC3 and 024 models, drive out the lower roll pin attaching the pinion shaft to the lower universal joint. Use a back-up, to protect the universal joint, while driving the roll pin.

6. On the Omni, Horizon, Charger, Shadow, Sundance, Turismo, TC3 and 024 models, support the front suspension crossmember with a hydraulic jack. Remove the two rear nuts attaching the crossmember to the frame. Loosen the two front bolts attaching the crossmember to the frame

and lower the crossmember slightly for access to the boot seal shields.

7. Except for the Omni, Horizon, Charger, Shadow, Sundance, Turismo, TC3 and 024 models, remove the four front suspension crossmember attaching bolts and the lower front suspension crossmember, using a transaxle jack, so that the steering gear can be removed from the crossmember.

8. Remove the splash and the boot seal shields.

9. Remove the steering gear bolts from the front suspension crossmember.

10. Remove the steering gear from the left side of the vehicle.

11. To install, reverse the removal procedures. The right rear crossmember bolt is a pilot bolt that correctly locates the crossmember, tighten it first. Torque the four crossmember bolts to 90 ft. lbs. and the steering gear attaching bolts to 21 ft. lbs. Refill the power steering pump. Bleed the power steering system.

ADJUSTMENT

1. Loosen the adjuster plug lock nut on the steering gear.

2. Turn the adjuster plug clockwise until it bottoms, then back it off 40–60°.

3. While holding the adjuster plug stationary, torque the lock nut to 50 ft. lbs. (70 Nm).

Power Steering Pump

REMOVAL & INSTALLATION

1.6L Engine

1. Loosen (DO NOT remove) the power steering pump pressure hose.

2. Remove the belt adjustment nut and loosen the three locking nuts from the rear pump studs.

3. Place a container on the radiator yoke to catch the power steering fluid.

4. Remove the drive belt and the three locking nuts. Lift the pump, the bracket and the rubber isolator (as an assembly) from the engine.

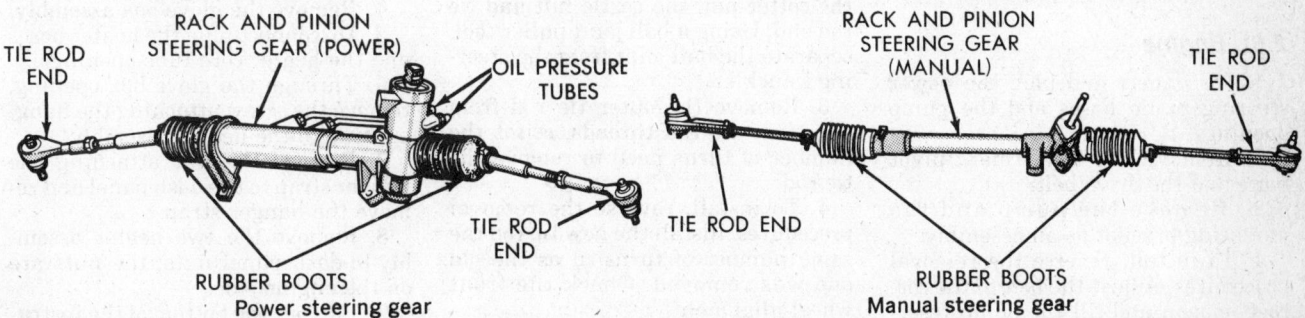

TIE ROD END — RACK AND PINION STEERING GEAR (POWER) — OIL PRESSURE TUBES — TIE ROD END — RUBBER BOOTS — **Power steering gear**

RACK AND PINION STEERING GEAR (MANUAL) — TIE ROD END — TIE ROD END — RUBBER BOOTS — **Manual steering gear**

5. Remove the pump reservoir cap and pour the fluid into the container.

6. Remove and plug the hoses and the pump openings.

7. To install, reverse the removal procedures. Adjust the drive belt tension, refill the reservoir with power steering fluid and torque the locking nuts to 21 ft. lbs. Bleed the power steering system.

1.7L, 2.2L and 2.5L Engines

NOTE: The power steering pump mounting nuts and bolts are metric.

1. Disconnect the vapor separator hose from the carburetor or throttle body and the two wires from the air conditioning clutch cycling switch, if equipped.

2. Loosen the drive belt adjustment bolt and the hose bracket nut, if equipped. Remove the belt from the pump pulley.

3. Raise and support the front vehicle on jackstands.

4. Disconnect the return hose from the gear tube and drain the oil from the pump through the end of the hose.

5. Remove the right side splash shield that protects the drive belts.

6. Disconnect both hoses from the pump. Cap the hoses and the pump openings.

7. Remove the lower stud nut and the pivot bolt from the pump.

8. Lower the vehicle and remove the drive belt from the pulley.

9. Move the pump rearward, to clear the mounting bracket and remove the adjusting bracket.

10. Rotate the pump clockwise, so that the pump pulley faces the rear of the vehicle and pull upwards to remove the pump from the vehicle.

11. To install, use new O-rings on the pressure hose and reverse the removal procedures. Adjust the belt to the correct tension and fill the pump reservoir to the proper level with power steering fluid. Torque the adjustment bolt to 30 ft. lbs., the lower stud nut and the pivot bolt to 40 ft. lbs. Bleed the power steering system.

2.6L Engine

1. Disconnect and plug the power steering pump hoses and the pump openings.

2. Remove the adjustment/pivot bolts and the drive belt.

3. Remove the pump and the mounting bracket as an assembly.

4. To install, reverse the removal procedures. Adjust the belt to the correct tension and fill the pump reser-

voir to the proper level with power steering fluid. Torque the adjustment/pivot bolts to 40 ft. lbs. Bleed the power steering system.

BELT ADJUSTMENT

1. Working on top of the vehicle, loosen the top power steering pump-to-bracket bolt. Working under the vehicle, loosen the bottom power steering pump-to-bracket bolt.

2. Using a 1/2 in. breaker bar, place it into the square hole on the adjusting bracket and turn the adjusting bracket to apply pressure to the belt.

3. Using a straight edge and thumb pressure, at the center of the belt, establish a belt deflection of 1/4 in. (new) or 7/16 in. (used).

4. When the correct deflection is established, torque the top (first) and the bottom (second) power steering pump-to-bracket bolt to 40 ft. lbs. (54 Nm).

SYSTEM BLEEDING

1. Check and/or refill the power steering pump reservoir.

2. Raise and support the front of the vehicle on jackstands.

3. Start the engine, then turn the steering wheel from side-to-side (lock-to-lock) several times, to bleed the air from the system.

4. Check and/or refill the power steering pump reservoir.

NOTE: If air bubbles are still present in the oil, repeat the bleeding procedure until the system is free of air bubbles.

Tie Rod End

REMOVAL & INSTALLATION

1. Loosen the jam nut which connects the outer tie rod end to the inner tie rod. Mark the tie rod position on the threads.

2. At the steering knuckle, remove the cotter pin, the castle nut and tie rod end. Using a ball joint puller tool, separate the ball joint from the steering knuckle.

3. Remove the outer tie rod from the inner tie rod threads, count the number of turns need to remove the tie rod.

4. To install, reverse the removal procedures. Install the new tie rod the same number of turns in as the old one was removed. Check the front wheel alignment.

CHASSIS ELECTRICAL

Heater Assembly Without A/C

REMOVAL & INSTALLATION

Omni, Horizon, Charger, Turismo, TC3 and 024

1. Disconnect the negative battery cable and drain the cooling system.

2. Disconnect the blower motor wiring connector.

3. Remove the ash tray.

4. Depress the temperature control cable tab and pull the cable out of the receiver (on the heater assembly).

5. Remove the glove box and the door assembly.

6. Disconnect/plug the heater hoses and the heater core tube openings.

7. Remove the heater assembly-to-dash mounting nuts.

8. Remove the wire connector at the blower motor heater block.

9. Remove the heater support bracket-to-instrument panel screw and nut.

10. Under the instrument panel, disconnect the plenum stud/heater assembly strap.

11. On the heater assembly, depress the mode door control cable tab and pull the cable out of the receiver.

12. Move the heater assembly to the right side and remove it from the vehicle.

13. To install, reverse the removal procedures.

All Other Models

1. Disconnect the negative battery cable and drain the cooling system.

2. Disconnect the electrical connector from the blower motor.

3. Reach under the unit, depress the tab on the mode door and temperature control cables, pull the flags from the receivers and remove the self-adjust clip from the crank arm.

4. Remove the glove box assembly.

5. Disconnect/plug the heater hoses and the heater core tube openings.

6. Through the glove box opening, remove the screw attaching the hanger strap to the heater assembly.

7. Remove the nut attaching the hanger strap to the dash panel and remove the hanger strap.

8. Remove the two heater assembly-to-dash panel nuts; the nuts are on the engine side.

9. Pull out the bottom of the instru-

ment panel and slide out the heater assembly.

10. To install, reverse the removal procedures.

Heater Blower

REMOVAL & INSTALLATION

NOTE: On A/C equipped vehicles, the blower motor is located inside the Heater/Evaporator Unit; the unit must be removed from the vehicle and disassembled to remove the blower motor.

Omni, Horizon, Charger, Turismo, TC3 and 24

The blower motor is located behind the instrument panel on the left side of the heater assembly.

1. Disconnect the motor electrical connector.
2. Remove the left outlet duct on some models.
3. Remove the motor retaining screws and the motor.
4. To install, reverse the removal procedures.

All Others Models

The blower motor is located under the instrument panel on the left side of the heater assembly.

1. Disconnect the negative battery cable.
2. Disconnect the motor electrical connector.
3. Remove the motor retaining screws and the motor.
4. To install, reverse the removal procedures.

Heater Core (Without A/C)

REMOVAL & INSTALLATION

Omni, Horizon, Charger, Turismo, TC3 and 024

1. Drain the cooling system and remove the heater assembly.
2. Remove the left outlet duct.
3. Remove the blower motor mounting screws and the motor.
4. Remove the outside air and defroster door cover.
5. Remove the defroster door.
6. Remove the defroster door control rod.
7. Remove the core cover.
8. Lift the core out of the unit.
9. To install, reverse the removal procedures.

All Other Models

1. Drain the cooling system and remove the heater assembly.
2. Remove the padding from around the heater core outlets and the upper core mounting screws.
3. Pry loose the retaining snaps from around the outer edge of the housing cover.

NOTE: If a retaining snap should break, the housing cover has provisions for mounting screws.

4. Remove the housing top cover.
5. Remove the bottom heater core mounting screw.
6. Slide the heater core out of the housing.
7. To install, reverse the removal procedures.

Heater Core (With A/C)

REMOVAL & INSTALLATION

Removal of the Heater/Evaporator Unit is required for the core removal. Two people will be required to perform the operation. Discharge, evacuation, recharge and leak testing of the refrigerant system is necessary and should only be performed by someone familiar with and qualified to work on air conditioning systems. During installation, a small can of refrigerant oil will be necessary.

Omni, Horizon, Charger, Turismo, TC3 and 024

1. Disconnect the negative battery cable, discharge the A/C system and drain the cooling system.
2. Disconnect and plug the heater hoses at the heater core and the A/C lines, at the dash.
3. Disconnect the blend air door cable and disengage from the heated air duct clip.
4. Remove the glove box, the center bezel and the bezel.
5. Remove the center distribution duct and the defroster duct adapter.
6. Disconnect the vacuum lines at the engine and the water valve.
7. Remove the assembly retaining nuts at the dash.
8. Remove the right side cowl trim panel.
9. Remove the right instrument panel pivot screw and the steering column-to-panel screws.
10. Remove the top panel cover.
11. Remove all of the panel screws, except the left fenceline screw.
12. Pull the carpet back from under the A/C unit.
13. Remove the hanger strap nut, the blower motor ground cable and the plenum strap nut.
14. Lift the unit rearward, to clear the studs and pull the instrument panel rearward to provide clearance for the unit removal. Drop the unit down and slide it rearward. Remove the assembly from the vehicle and place it on a bench.
15. Remove the $\frac{1}{4}$ in. nut from the top of the mode actuator. Remove the two retaining clips from the front edge of the cover and the actuator door.
16. Remove the assembly cover attaching screws and the cover. Remove the heater core tube retaining bracket screw and lift the core from the unit.
17. To install, reverse the removal procedures. Refill the cooling system and charge the A/C system.

All Other Models

1. Disconnect the negative battery cable, discharge the A/C system and drain the cooling system.
2. Disconnect and plug the heater hoses at the heater core.
3. Disconnect the vacuum lines at the engine intake manifold and the water valve.
4. Remove the right scuff plate and the cowl side trim.
5. Remove the glove box and the A/C control.
6. Remove the console and the forward console mounting bracket (if equipped).
7. Remove the center distribution duct and the demister adapter (if equipped).
8. From under the panel, pull out the defroster duct.
9. Remove the clamp and the condensate drain tube.
10. Disconnect the wiring connectors.
11. On the heater/evaporator assembly, depress the control cable flag and pull the cable out of the retainer.
12. Remove the plenum brace on the right side of the cowl.
13. Pull the carpet back from under the unit.
14. Remove the hanger strap-to-assembly screw.
15. Remove the A/C-to-dash panel mounting nuts from the engine compartment.
16. Pull the unit rearward, to clear the studs, drop it down to the converter tunnel, rotate and slide it to one side. Remove the assembly from the vehicle.
17. Remove the actuator arm by squeezing it off the shaft. Remove the two retaining clips from the front edge of the cover and the actuator door.
18. Remove the assembly cover at-

taching screws and the cover. Remove the heater core tube retaining bracket screw and lift the core from the unit.

19. To install, reverse the removal procedures. Refill the cooling system and charge the A/C system.

Radio

REMOVAL & INSTALLATION

Omni, Horizon, Charger, Turismo, Lancer, LeBaron, TC3 and 024

1. Remove the bezel-to-dash screws and the bezel.
2. Disconnect the radio ground strap. Remove the radio mounting screws.
3. Pull the radio from the panel.
4. Disconnect the wiring and the antenna lead.
5. To install, reverse the removal procedures.

Daytona and Laser

1. Remove the mounting screws from the bottom of the console trim bezel.
2. Remove the bezel from the console.
3. Remove the radio-to-console mounting screws.
4. Pull the radio out and disconnect the wiring connector, the antenna lead and the ground strap.
5. To install, reverse the removal procedures.

Shadow and Sundance

1. Remove the radio bezel from the center of the dash.
2. If equipped with a base console, remove the lower center module cover.
3. If equipped with a full console assembly, remove the right console sidewall.
4. Remove the radio-to-dash screws and pull the radio out of the dash.
5. Disconnect the wiring, the antenna cable and the ground strap from the radio. Remove the radio from the vehicle.
6. To install, reverse the removal procedures.

All Other Models

1. Remove the left (if necessary) and the radio bezel(s).
2. Remove the two screws attaching the radio to the base panel.
3. Pull the radio out and disconnect the wiring connector, the antenna lead and the ground strap.
4. To install, reverse the removal procedures.

Windshield Wiper Switch

The wiper switch is located on the end of the turn signal switch lever.

REMOVAL & INSTALLATION

Refer to the "Turn Signal Removal & Installation" procedures in this section and remove the turn signal/windshield wiper switch from the steering column.

Windshield Wiper Motor

REMOVAL & INSTALLATION

Front Motor

1. Disconnect the linkage from the motor crank arm.
2. Remove the wiper motor plastic cover.
3. Disconnect the wiring harness from the motor.
4. Remove the three mounting bolts and the motor from the motor bracket.
5. To install, reverse the removal procedures.

Rear Motor

OMNI, HORIZON, CHARGER, TURISMO, TC3 AND 024

1. Open the tailgate.
2. Remove the wiper motor plastic cover.
3. Remove the wiper arm assembly.
4. Remove the nut, the ring and the seal from the pivot shaft.
5. From inside the tailgate, disconnect the wiring connector.
6. Remove the motor mounting screws and the motor.
7. To install, reverse the removal procedures.

ALL OTHERS

1. Remove the arm and blade assembly.
2. Open the tailgate.
3. Remove the motor cover (if equipped) and disconnect the wiring connector.
4. Remove the grommet from the glass.
5. Remove the two (Daytona and Laser) or four (all others) bracket retaining screws and the motor from the tailgate.
6. To install, reverse the removal procedures.

Instrument Cluster

For further information on instruments, please refer to "Gauges and Indicators" in the Unit Repair section.

REMOVAL & INSTALLATION

1. Remove the screws and the instrument cluster bezel.
2. Remove the instrument cluster-to-panel screws.
3. Pull the instrument cluster rearward, remove the speedometer cable and the wiring connectors.
4. Remove the cluster assembly.
5. To install, reverse the removal procedures.

Fuses, Fusible Links And Circuit Breakers

LOCATION

Fusible Links

Fusible links are used to prevent major wire harness damage in the event of a short circuit or an overload condition in electrical circuits. Each fusible link is of a fixed value for a specific electrical load. Should a link fail, the cause of the failure must be determined and repaired prior to installing a new link of the same value.

Circuit Breakers

Circuit breakers are used along with the fusible links to protect electrical system components such as headlamps, windshield wipers, electric windows, tailgate front and rear switches. The circuit breakers are located either in the switch or mounted on/near the lower lip of the instrument panel, to the right or left side of the steering column.

Fuse Panels

The fuse panel is used to house the fuses that protect the individual or combined electrical circuits within the vehicle. The turn signal flasher, the hazard warning flasher and the seat belt warning buzzer/timer are located on the fuse panel for quick identification and replacement. The fuses are usually identified by abbreviated circuit names or number, with the number of the rated fuse needed to protect the circuit printed below the fuse holder.

Chrysler Corp.
Rear Wheel Drive
Aspen, Cordoba, Diplomat, Fifth Avenue, Gran Fury, Imperial, LeBaron, Mirada, Newport, New Yorker, St. Regis, Volare

YEAR IDENTIFICATION

1980 Gran Fury

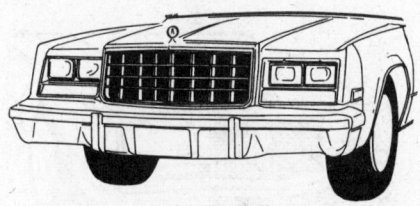

1981 Gran Fury

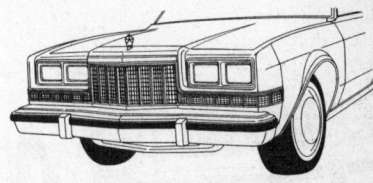

1982–87 Gran Fury

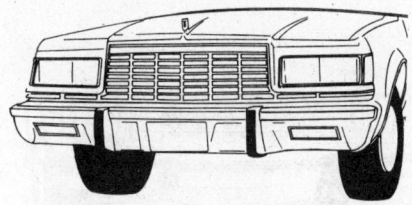

1980 St. Regis

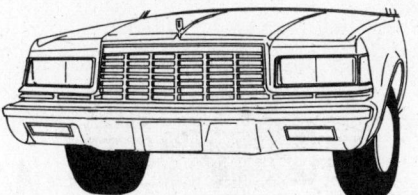

1981 St. Regis

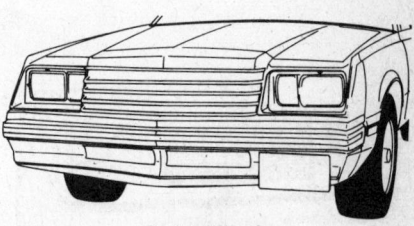

1980 Mirada

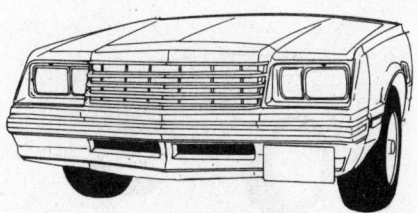

1981 Mirada

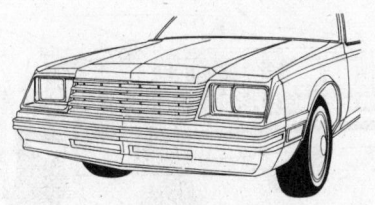

1981-83 Mirada

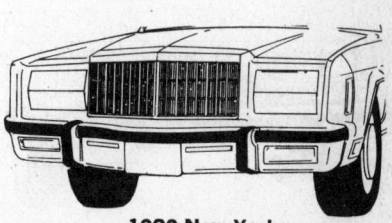

1980 New Yorker

YEAR IDENTIFICATION

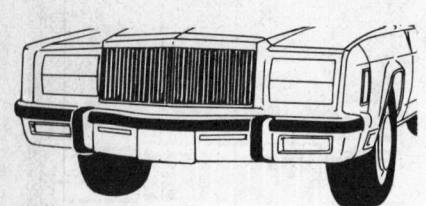

1981 New Yorker

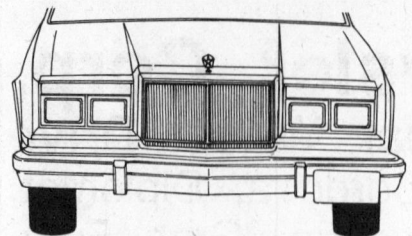

1982-83 New Yorker, 5th Ave.

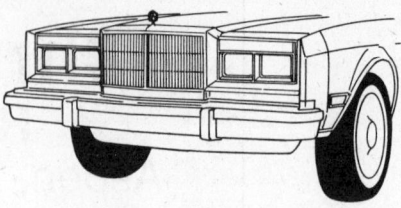

1984—87 Fifth Avenue

1980 LeBaron

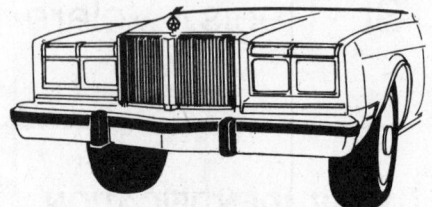

1981 LeBaron

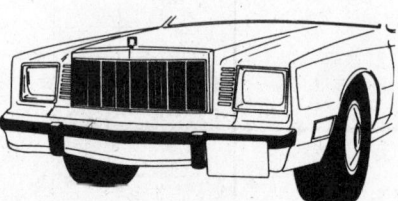

1980 Cordoba

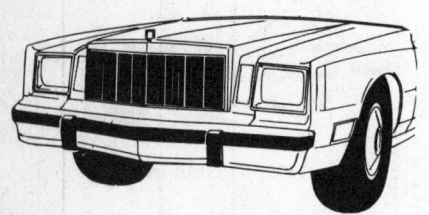

1981-83 Cordoba

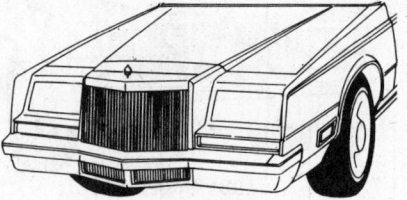

1981-83 Imperial

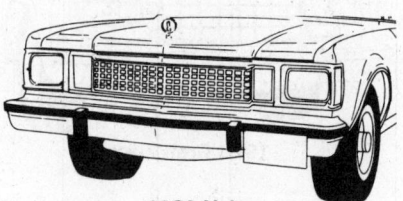

1980 Volare

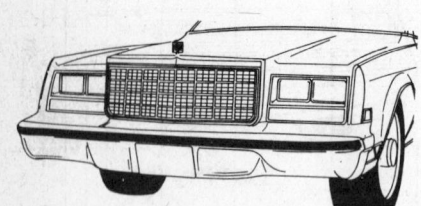

1980 Newport

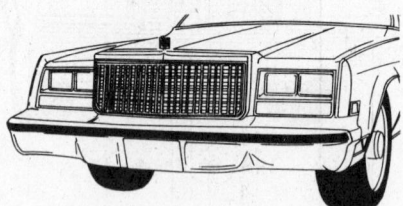

1981 Newport

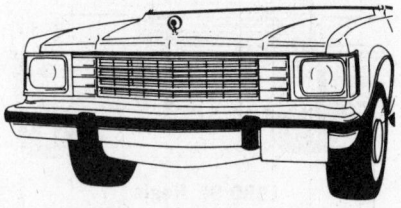

1980 Aspen

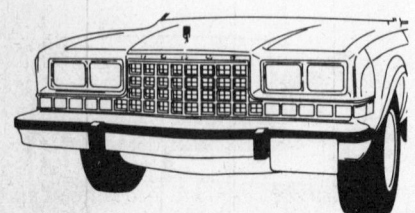

1980 Diplomat

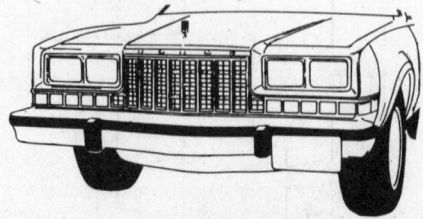

1981—84 Diplomat

1985—87 Diplomat

VEHICLE IDENTIFICATION NUMBER (VIN)

It is important for servicing and ordering parts to be certain of the vehicle and engine identification. The VIN (vehicle identification number) is a 13 or 17 digit number visible through the windshield on the driver's side of the dash and contains the vehicle and engine identification codes. It can be interpreted as follows:

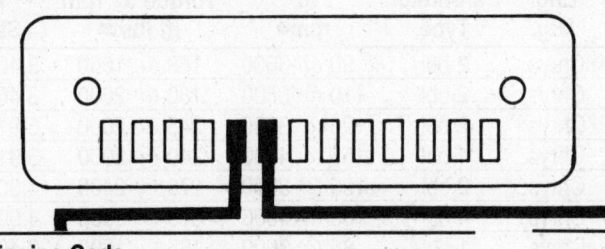

Engine Code						Model Year Code	
Code	Cu. In.	Liters	Cyl.	Carb. (bbl.)	Eng. Mfg.	Code	Year
C	225	3.7	6	1	Chrys.	A	1980
D	225	3.7	6	2	Chrys.		
G	318	5.2	8	2	Chrys.		
H	318	5.2	8	4	Chrys.		
K	360	5.9	8	2	Chrys.		
L	360HP①	5.9	8	4	Chrys.		

The thirteen digit Vehicle Identification Number can be used to determine engine application and model year. The 6th digit indicates the model year, and the 5th digit identifies the factory installed engine.
① High Performance

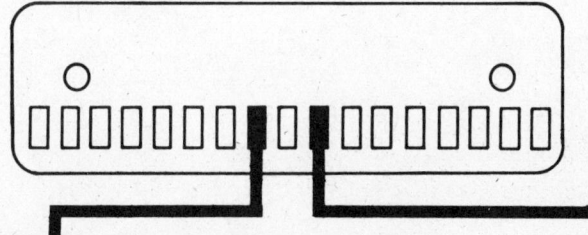

Engine Code						Model Year Code	
Code	Cu. In.	Liters	Cyl.	Carb. (bbl.)	Eng. Mfg.	Code	Year
E	225	3.7	6	1	Chrys.	B	1981
H	225	3.7	6	1	Chrys.	C	1982
F	225	3.7	6	1 H.D.	Chrys.	D	1983
J	225	3.7	6	1 H.D.	Chrys.	E	1984
G	225	3.7	6	2①	Chrys.	F	1985
K	225	3.7	6	2①	Chrys.	G	1986
H	225	3.7	6	2 H.D.①	Chrys.	H	1987
L	225	3.7	6	2 H.D.①	Chrys.		
K	318	5.2	8	2	Chrys.		
P	318	5.2	8	2	Chrys.		
L	318	5.2	8	2 H.D.	Chrys.		
J	318	5.2	8	EFI	Chrys.		
N	318	5.2	8	EFI	Chrys.		
R	318	5.2	8	4	Chrys.		
M	318	5.2	8	4	Chrys.		
N	318	5.2	8	4 H.D.	Chrys.		
S	318	5.2	8	4 H.D.	Chrys.		

H.D. = Heavy Duty
EFI = Electronic Fuel Injection
① = Canada Only
The seventeen digit Vehicle Identification Number can be used to determine engine application and model year. The seventeen digit code supercedes the thirteen digit code which ended in 1980. The 10th digit indicates the model year, and the 8th digit identifies the factory installed engine.
EFI Electronic Fuel Injection

GENERAL ENGINE SPECIFICATIONS

Year	Eng. V.I.N. Code	Engine No. Cyl. Displacement (cu. in.)	Eng. Mfg.	Carburetor Type	Horsepower @ rpm■	Torque @ rpm (ft lbs)■	Bore × Stroke (in.)	Com- pression Ratio	Oil Pressure @ 2000 rpm
'80	C	6-225	Chrys.	2 bbl	90 @ 3600	160 @ 1600	3.406 × 4.125	8.4:1	55
	D	6-225	Chrys.	2 bbl	110 @ 3600	180 @ 2000	3.406 × 4.125	8.4:1	55
	G	8-318	Chrys.	2 bbl	120 @ 3600	245 @ 2000	3.910 × 3.310	8.5:1	55
	H	8-318 Calif.	Chrys.	4 bbl	155 @ 4000	240 @ 2000	3.910 × 3.310	8.5:1	55
	K	8-360 ESC	Chrys.	2 bbl	155 @ 3600	270 @ 2400	4.000 × 3.580	8.4:1	55
	L	8-360 ESC	Chrys.	4 bbl	185 @ 4000	275 @ 2000	4.000 × 3.580	8.0:1	55
'81–'83	E	6-225	Chrys.	1 bbl	85 @ 3600	165 @ 1600	3.406 × 4.125	8.4:1	55
	E	6-225 Calif.	Chrys.	1 bbl	90 @ 3600	165 @ 1200	3.406 × 4.125	8.4:1	55
	K	8-318	Chrys.	2 bbl	130 @ 4000	235 @ 1600	3.910 × 3.310	8.6:1	55
	M	8-318 Calif.	Chrys.	4 bbl	165 @ 4000	240 @ 2000	3.910 × 3.310	8.5:1	55
	J	8-318	Chrys.	EFI	140 @ 4000	245 @ 2000	3.910 × 3.310	8.5:1	55
'84–'85	P	8-318	Chrys.	2 bbl	130 @ 4000	235 @ 1600	3.910 × 3.310	8.6:1 ①	55
	R	8-318	Chrys.	4 bbl	165 @ 4000	240 @ 2000	3.910 × 3.310	8.6:1 ①	55
	S	8-318 HD	Chrys.	4 bbl	165 @ 4000	240 @ 2000	3.910 × 3.310	8.4:1	55
'86–'87	P	8-318	Chrys.	2 bbl	140 @ 3600	265 @ 1600	3.910 × 3.310	9.0:1	55
	R	8-318	Chrys.	4 bbl	165 @ 4000	240 @ 2000	3.910 × 3.310	8.6:1	55
	S	8-318 HD	Chrys.	4 bbl	175 @ 4000	250 @ 3200	3.910 × 3.310	8.0:1	55

■ Horsepower and torque are SAE net figures. They are measured at the rear of the transmission with all accessories installed and operating. Since the figures vary when a given engine is installed in different models, some figures are representative rather than exact.
HP-High Performance
HD Heavy Duty
ESC Electronic Spark Control
EFI Electronic Fuel Injection
EFM Electronic Fuel Metering
① 1985: 9.0:1

TUNE-UP SPECIFICATIONS

(When analyzing compression test results, look for uniformity among cylinders rather than specific pressures.)

Year	Eng. V.I.N. Code	Engine No. Cyl. Displace- ment (cu. in.)	Carb. (bbl.)	Spark Plugs Orig. Type	Gap (in.)	Distributor Point Dwell (deg.)	Point Gap (in.)	Ignition Timing (deg.) ▲ Man. Trans.●	Auto. Trans.	Valves Intake Opens ■ (deg.)	Fuel Pump Pressure (psi)	Idle Speed (rpm) ▲ Man. Trans.●	Auto. Trans.
'80	C	6-225	1	P-560 PR	.035	Electronic		12B	12B	16	3½–5	725	725(750)
	D	6-225	2	P-560 PR	.035	Electronic		—	12B	16	3½–5	725	725②
	G	8-318	2	P-65 PR	.035	Electronic		—	12B	10	5–7	—	700
	H	8-318	4	P-65 PR	.035	Electronic		—	10B②(16B)	10	5–7	—	750②(700)
	K	8-360	2	P-65 PR	.035	Electronic		—	16B	18	5–7	—	750
	L	8-360	4	P-65 PR	.035	Electronic		—	16B	18	5–7	—	750
'81	E	6-225	1	P-560 PR4Y	.048	Electronic		—	12B③	6	4.0–5.5	—	600
	J	8-318	EFI	P-68 ER	.048	Electronic		—	12B	10	15.2–19.4	—	580
	K	8-318	2	P-65 PR4Y	.048④	Electronic		—	16B	10	5.75–7.25	—	600
	M	8-318	4	P-65 PR4Y	.048	Electronic		—	16B	10	5.75–7.25	—	600
'82	E,F,G,H	6-225	1 & 2	560PR	.035	Electronic		—	⑤	6	4.0–5.5	—	750
	J	8-318	EFI	65PR	.048	Electronic		—	⑤	10	5.75–11.5	—	600
	K,L,M,N	8-318	2 & 4	RN-12YC	.035	Electronic		—	⑤	10	5.75–7.25	—	700
'83	H,J,K,L	6-225	1 & 2	RBL-16Y	.035	Electronic		—	⑤	6	4.0–5.5	—	750
	N,P,R,S	8-318	2 & 4	RN-12Y	.035	Electronic		—	⑤	10	5.75–7.25	—	700

TUNE-UP SPECIFICATIONS

(When analyzing compression test results, look for uniformity among cylinders rather than specific pressures.)

Year	Eng. V.I.N. Code	No. Cyl. Displacement (cu. in.)	Carb. (bbl.)	Spark Plugs Orig. Type	Gap (in.)	Distributor Point Dwell (deg.)	Point Gap (in.)	Ignition Timing (deg.) ▲ Man. Trans.●	Auto. Trans.	Valves Intake Opens ■ (deg.)	Fuel Pump Pressure (psi)	Idle Speed (rpm) ▲ Man. Trans.●	Auto. Trans.
'84–'85	P,R,S	8-318	2 & 4	RN-12Y	.035	Electronic		—	⑤	10	5.75–7.25	—	700
'86	P,R,S	8-318	2 & 4	RN-12YC	.035	Electronic		—	⑤	10	5.75–7.25	—	680①
'87				See Underhood Specifications Sticker									

NOTE: The underhood specifications sticker often reflects tune-up specification changes made in production. Sticker figures must be used if they disagree with those in this chart.

Part numbers in this chart are not recommendations by Chilton for any product by brand name.

▲ See text for procedure
■ All figures Before Top Dead Center
● Figure in parentheses indicates California engine
HP High Performance
EFI Electronic Fuel Injection
B Before TDC
① Neutral
② Canada
③ California 16B
④ Late Production .035—See underhood sticker
⑤ See underhood sticker

FIRING ORDERS

NOTE: To avoid confusion, always replace spark plug wires one at a time.

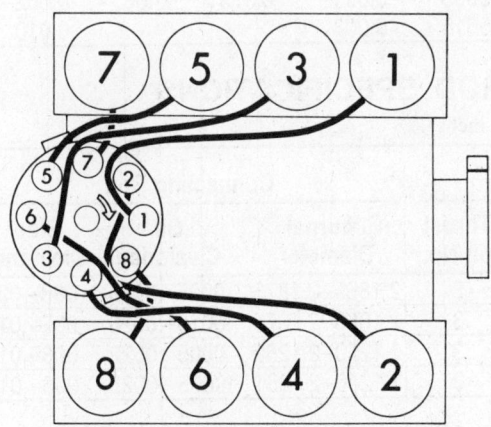

CHRYSLER CORP. 318 V8 Engine
Engine firing order: 1–8–4–3–6–5–7–2
Distributor rotation: clockwise

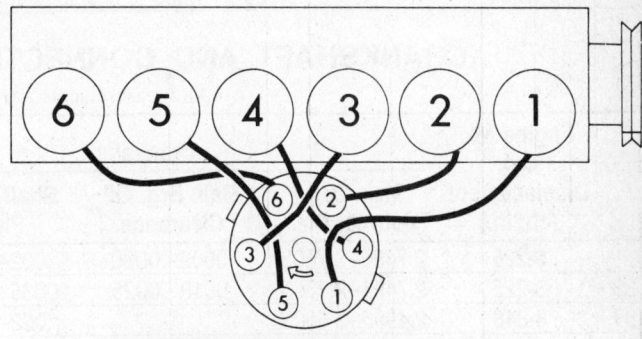

CHRYSLER CORP. 6-cyl.
Engine firing order: 1-5-3-6-2-4
Distributor rotation: clockwise

MECHANICAL VALVE LIFTER CLEARANCE

Year	Engine	Intake (Hot) In.	Exhaust (Hot) In.
'80	Six cylinder	.010	.020

NOTE: 1981 and later slant six engines are equipped with hydraulic valve tappets, which eliminate the need for periodic valve lash adjustment.

CAPACITIES

Year	Engine No. Cyl. Displacement (cu. in.)	Engine Crankcase Add 1 qt. for New Filter	Transmission Pts to Refill After Draining			Drive Axle (pts.)	Gasoline Tank (gals.)	Cooling System (qts.)	
			Manual					With Heater	With A/C
			3-Speed	4-Speed	Automatic				
'80	6-225	4	4.75	7.0①	17⑧⑤	③	18④	11.5	14.5②
	8-318	4	—	—	17⑧⑤	③	18.0⑥	15.0⑦	16.5⑨
	8-360	4	—	—	17⑧⑤	③	18.0⑥	16.0	16.0
'81–'83	6-225	4	4.75	7.0	17	③	18④	11.5	14.5②
	8-318	4	—	—	17	③	18.0⑥	15.0⑦	16.5⑨
'84–'87	8-318	4	—	—	17	②	18.0	15.5	16.5

① Aspen/Volare
② 15 qts on Cordoba, Mirada
③ 7¼″ axle—2.5 pts., 8¼″ axle—4.4 pts., 9¼″ axle—4.5 pts.
④ 19.5 on Aspen/Volare wagon; 21 gal. on 1980 and later Cordoba, Mirada, Gran Fury, St. Regis
⑤ 16.3 pts. on Aspen/Volare with A904 trans.; 15.9 pts. on A727 trans.
⑥ 21 gal. on '80–'81 New Yorker, Dodge/Plymouth full size. 19.5 gal. on 1980 Dodge/Plymouth midsize.
⑦ 15.5 qts. on Imperial
⑧ A904 trans.; 15.9 pts.—A727 trans.
⑨ 17.5 on Imperial and heavy duty cooling systems

CAMSHAFT SPECIFICATIONS

(All measurements in inches.)

Year	V.I.N. Code	Engine	Eng. Mfg.	Journal Diameter					Bearing Clearance	Lobe Lift		Camshaft End Play
				1	2	3	4	5		Intake	Exhaust	
'80–'83	C,D,E,F, G,H,J,K,L	6-225	Chry.	1.998– 1.999	1.982– 1.983	1.967– 1.968	1.951– 1.952		0.001– 0.003	0.406	0.414	0.002– 0.010
'80	K,L	8-360	Chry.	1.998– 1.999	1.982– 1.983	1.967– 1.968	1.951– 1.952	1.5605– 1.5615	0.001– 0.003	0.410	0.410	0.002– 0.010
'80–'87	G,H,J,L,M, N,P,R,S	8-318	Chry.	1.998– 1.999	1.982– 1.983	1.967– 1.968	1.951– 1.952	1.5605– 1.5615	0.001– 0.003	0.373	0.400	0.002– 0.010

CRANKSHAFT AND CONNECTING ROD SPECIFICATIONS

(All measurements are given in inches.)

Year	Engine No. Cyl. Displacement (cu. in.)	Crankshaft				Connecting Rod		
		Main Brg. Journal Dia.	Main Brg. Oil Clearance	Shaft End-Play	Thrust on No.	Journal Diameter	Oil Clearance	Side* Clearance
'80	6-225	2.7495–2.7505	.0005–.0020	.002–.009	3	2.1865–2.1875	.0005–.0025	.006–.025
'81–'83	6-225	2.7495–2.7505	.0010–.0025	.0035–.0095	3	2.1865–2.1875	.0010–.0025	.007–.013
'80–'87	8-318	2.4995–2.5005	①	.002–.010	3	2.1240–2.1250	.0005–.0022	.006–.014
'80–'81	8-360	2.8095–2.8105	.0005–.0020①	.002–.009	3	2.1240–2.1250	.0005–.0025	.006–.014

*Total for two rods on V8s
① No. 1—.0005–.0015; No. 2, No. 3, No. 4, No. 5—.0005–.0020

PISTON AND RING SPECIFICATIONS

(All measurements given in inches.)

Year	Engine No. Cyl. Displacement (cu. in.)	Ring Gap			Ring Side Clearance			Piston Clearance③
		Top Compression	Bottom Compression	Oil Control	Top Compression	Bottom Compression	Oil Control	Piston-to-Bore Clearance
'80	6-225, 8-318, 8-360	.010–.020	.010–.020	.015–.055	.0015–.0030①	.0015–.0030①	.0002–.0050	0.0005–0.0015②
'81–'83	6-225	.010–.020	.010–.020	.015–.055	.0015–.0030	.0015–.0030	.0002–.0050	.0005–.0015
	8-318	.010–.020	.010–.020	.015–.055	.0015–.0030	.0015–.0030	.0002–.0050	.0005–.0015
'84–'87	8-318	.010–.020	.010–.020	.015–.055	.0015–.0030	.0015–.0030	.0002–.0050	.0005–.0015
	8-318HP	.010–.020	.010–.020	.015–.055	.0015–.0030	.0015–.0030	.0002–.0050	.001–.002

① .0015–.0040 in. on 1980 318, 360HP V8s   ② .0010–.0020 in. on 1980 360 V8 w/4 bbl   ③ At top of skirt

VALVE SPECIFICATIONS

Year	Engine No. Cyl. Displacement (cu. in.)	Seat Angle (deg.)	Face Angle (deg.)	Spring Test Pressure (lbs. @ in.)	Spring Installed Height (in.)	Stem-to-Guide Clearance (in.) Intake	Stem-to-Guide Clearance (in.) Exhaust	Stem Diameter (in.) Intake	Stem Diameter (in.) Exhaust
'80	6-225	45	①	143 @ 1.31	1²¹⁄₃₂	.0010–.0030	.0020–.0040	.3725	.3715
	8-318	45	45	177 @ 1.31②	1¹¹⁄₁₆⑦	.0010–.0030③	.0020–.0040④	.3725⑤	.3715⑥
	8-360	45	45	177 @ 1.31	1²¹⁄₃₂	.0010–.0030	.0020–.0040	.3725	.3715
	8-360HP	45	45	193 @ 1.25	1²¹⁄₃₂	.0015–.0035	.0025–.0045	.3720	.3710
'81–'83	6-225	45	①	143 @ 1.31	1²¹⁄₃₂	.0010–.0030	.0020–.0040	.3725	.3715
	8-318	45	45	177 @ 1.31②	1¹¹⁄₁₆⑦	.0010–.0030③	.0020–.0040④	.3725⑤	.3715⑥
'84–'87	8-318	45	45	177 @ 1.31②	1¹¹⁄₁₆	.0010–.0030③	.0020–.0040④	.3725⑤	.3715⑥

HP High Performance
① Intake 45°, Exhaust 43°
② 318 EFM—193 @ 1.25
③ 318 EFM—.0015–.0035
④ 318 EFM—.0025–.0045
⑤ 318 EFM—.3720
⑥ 318 EFM—.3710
⑦ 318 HP—1²¹⁄₃₂

TORQUE SPECIFICATIONS

(All readings in ft. lbs.)

Year	Engine No. Cyl. Displacement (cu. in.)	Cylinder Head Bolts	Rod Bearing Bolts	Main Bearing Bolts	Crankshaft Damper Bolt	Flywheel-to-Crankshaft Bolts	Manifold Intake	Manifold Exhaust
'80–'83	6-225	70	45	85	Press fit	55	①	10
	8-318	95③	45	85	100	55	40	¹⁵⁄₂₀②
'80	8-360	95	45	85	100	55	40	¹⁵⁄₂₀②
'84–'87	8-318	105	45	85	100	55	40	¹⁵⁄₂₀②

① Intake to exhaust manifold bolts—17 ft. lbs., studs—20 ft. lbs.
② Nuts/screws
③ See step No. 15 under "Cylinder Head Removal, V8" in text

WHEEL ALIGNMENT SPECIFICATIONS

Year	Model	Caster Range (deg.)	Caster Pref. Setting (deg.)	Camber Range (deg.)	Camber Pref. Setting (deg.)	Toe-in (in.)	Steering Axis Inclin. (deg.)	Wheel Pivot Ratio Inner Wheel	Wheel Pivot Ratio Outer Wheel
'80–'81	Chrysler	¼N to 2¼P	1P	¼N–1¼P	½P	⅛ ± ¹⁄₁₆	8	20	18.0
'80–'85	Aspen, Volare, Diplomat, Gran Fury, Newport, 5th Avenue	1¼P to 3¾P	2½P ± 1	¼N–1¼P	½P ± ½	⅛ ± ¹⁄₁₆	8	20	18
'86–'87	5th Avenue Diplomat Gran Fury	1¼P to 3¾P	2½P ± 1	¼N–1¼P①	½P ± ½	⅛ ± ¹⁄₁₆②	8	20	18

N Negative
P Positive
① Rear Axle: ¹⁄₁₆N to ³⁄₃₂P
② Rear Axle: ¹⁄₁₆–³⁄₁₆

FRONT END HEIGHT

Year	Model	Front End Height (± ⅛ in.)
'80–'81	Chrysler, St. Regis, Gran Fury	16¾ ② ①
'80–'85	Aspen, Volare, 5th Avenue	12½ ② ①
'86–'87	5th Avenue Diplomat Gran Fury	12½ ② ①

① Measured from the head of the suspension crossmember front isolator bolt to ground
② ± ¼ in.

ENGINE ELECTRICAL

Alternator

For all information on the charging system not detailed below, please refer to the "Charging & Starting" Unit Repair Section.

PRECAUTIONS

To prevent serious damage to the alternator and the rest of the charging system, the following precautions must be observed:

• When installing a battery, make sure that the positive cable is connected to the positive terminal and the negative to the negative.
• When jump-starting the vehicle with another battery, make sure that the like terminals are connected. This also applies when using a battery charger.
• Never operate the alternator with the battery disconnected or on an uncontrolled open circuit. Double-check to see that all connections are tight.
• Do not short across or ground any alternator or regulator terminals.
• Do not apply full battery voltage to the field connector.
• Always disconnect the negative battery terminal before disconnecting the alternator.

BELT TENSION ADJUSTMENT

Tighten the belt so that it can be depressed about ½ in., using moderate thumb pressure, in the center of the longest span between pulleys. Some alternator brackets have a square hole into which a ½ in. square drive socket can be inserted to tension the belt.

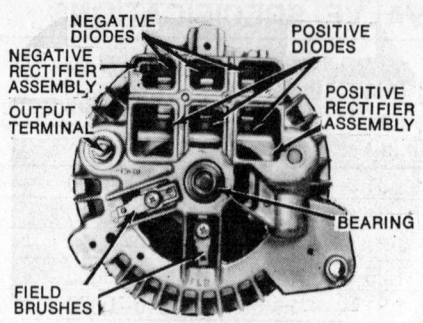

Rear view of the alternator

REMOVAL & INSTALLATION

1. Disconnect the negative battery cable.
2. Disconnect the "BAT", the "FLD" and the ground wire leads from the alternator.
3. Back out the alternator adjusting bolt, push the alternator down and remove the drive belt.

NOTE: If equipped with A/C, remove the alternator pivot bolt.

4. Remove the alternator by removing the mounting bolts and the belt tensioner bracket bolt.
5. To install, reverse the removal procedures. Using a voltmeter, check the current output.

Regulator

REMOVAL & INSTALLATION

All models use a solid-state (silicone transistor) voltage regulator which is nonadjustable. The regulator is mounted in the engine compartment and clearly labeled.

For all information on the regulator testing procedures, please refer to the "Charging & Starting" Unit Repair Section.

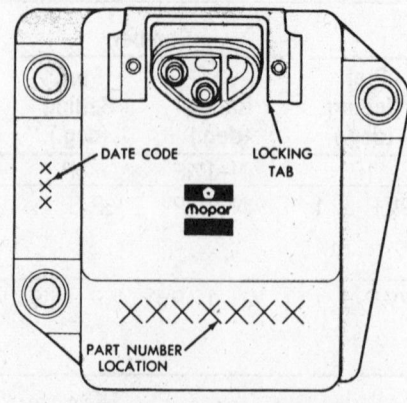

View of the voltage regulator

1. Release the spring clips and pull off the regulator wiring plug.
2. Remove the retaining bolts and regulator.
3. To install, reverse the removal procedures. Be sure that the spring clips engage the wiring plug and that the unit has a good ground.

STARTER

For all information on the starting system not detailed below, please refer to the "Charging & Starting" Unit Repair Section.

REMOVAL & INSTALLATION

1. Disconnect the negative battery cable.
2. Remove the battery cable from the starter.
3. On 1984 and later models, remove the nut and bolt retaining the heat shield to the starter, then remove the heat shield.
4. Disconnect the electrical leads from the solenoid terminals.
5. Remove the starter securing nut/bolt and remove the starter from the flywheel housing. If equipped with an automatic transmission, remove the starter securing nut/bolt, slide the cooler tube bracket off the stud and remove the starter.

NOTE: When removing the starter, be careful not to damage the flywheel housing seal.

6. To install, reverse the removal procedures. Be sure that the starter and flywheel housing mating surfaces are free of dirt and oil. Position the starter to the flywheel housing seal. When tightening the bolt and nut, hold the starter away from the engine to ensure proper alignment.

Distributor

The electronic ignition is standard on all models. The only system maintenance required is to inspect the wiring, replace the spark plugs and check the timing.

NOTE: The tachometer hookup with electronic ignition is the same as with conventional point-type systems. The red lead connects to the negative primary coil terminal and the black lead to a good ground. Some meters will not work with this system.

REMOVAL

1. Disconnect the negative battery cable.

2. Disconnect the vacuum advance line at the distributor and the lead wire(s) at the harness connector.

3. Unfasten the distributor cap retaining clips and lift off the cap.

4. Rotate the engine until the distributor rotor is pointing toward the No. 1 spark plug wire. Matchmark the distributor body and the engine block to indicate the position of the distributor to the block.

5. Place a mark on the edge of the distributor housing to indicate the position of the rotor to the distributor.

6. Remove the distributor hold-down clamp screw and clamp.

7. Carefully lift the distributor from of the block; the shaft will rotate slightly as the distributor is removed.

INSTALLATION

NOTE: Use the reference marks that were made to correctly position the distributor in the block.

Engine Undisturbed

1. On the 6 cyl.. engine, position the rotor just ahead of the No. 1 distributor cap terminal. Lower the distributor into the engine block opening. Mesh the distributor gear with the camshaft drive gear. Be certain that the rubber O-ring seal is in the groove of the distributor shank. Proceed to Step 3.

2. On the V8 engine, clean the top of the engine block around the distributor opening and position the rotor even with the No. 1 distributor cap terminal position. Lower the distributor into the engine block opening and engage the tongue of the distributor shaft with the slot in the distributor oil pump drive gear. Proceed to Step 3.

NOTE: When the distributor is properly seated, the rotor should be directly under the No. 1 distributor cap terminal.

3. Install the distributor hold-down clamp and tighten the retaining screw finger tight.

4. Install the distributor cap. Connect the primary lead wire(s) to the harness connector.

5. Start the engine. Check and adjust the ignition timing.

6. Connect the vacuum advance line to the distributor.

Engine Disturbed

1. Remove the No. 1 spark plug, insert a finger into the hole and rotate

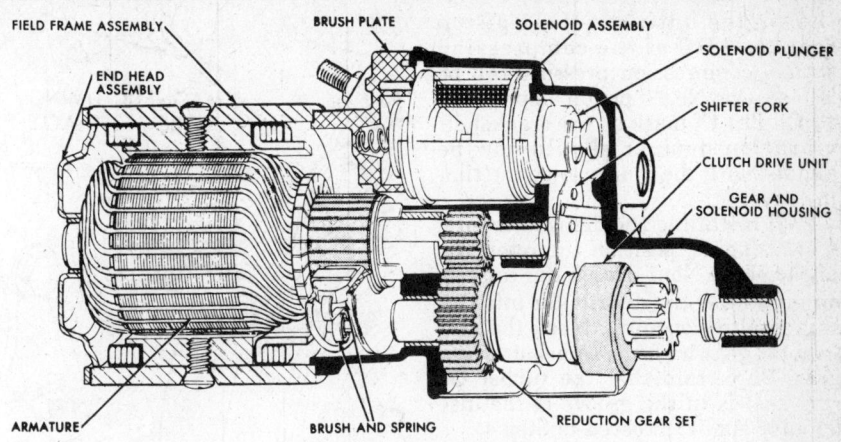

Starter motor details

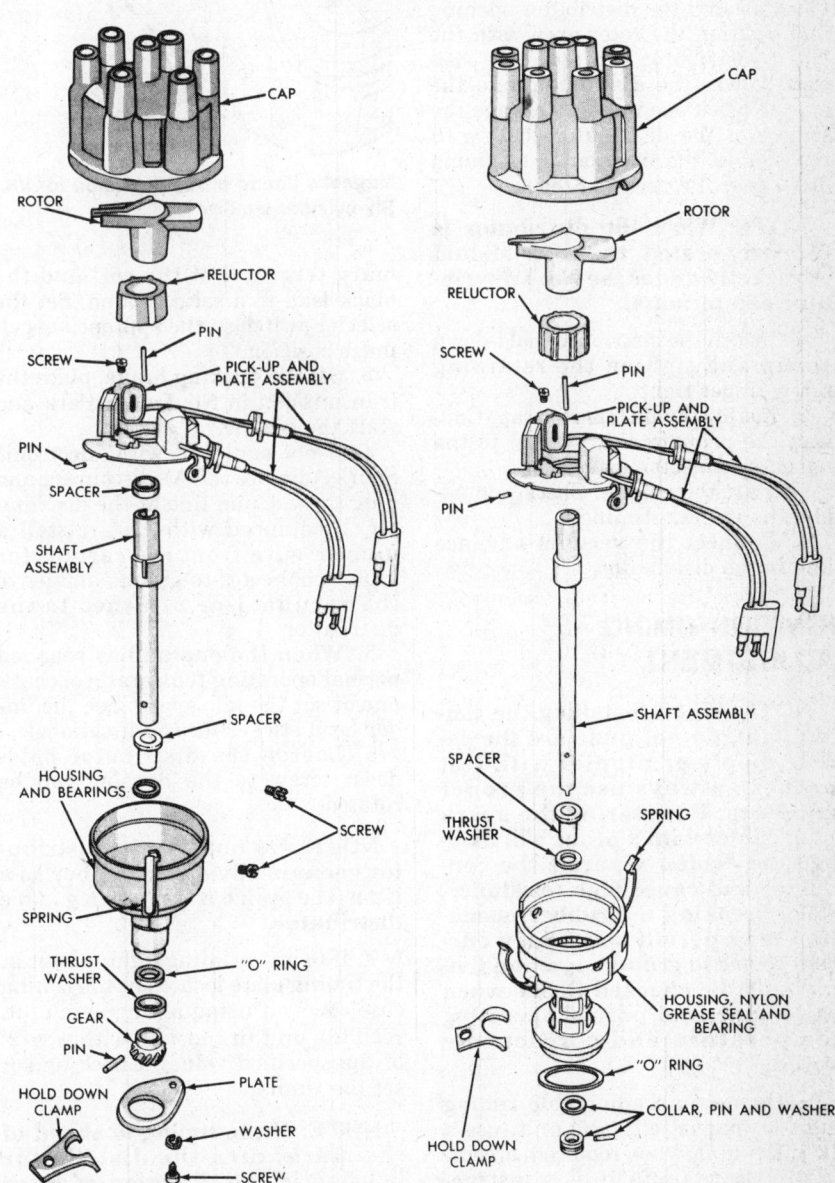

Exploded view of a 6-cyl. ESA dual pick-up distributor. Earlier models use single pick-up

Exploded view of a V8 ESA dual pick-up distributor. Earlier models use single pick-up

the engine until the No. 1 piston reaches TDC of the compression stroke (compression pressure can be felt as the No. 1 piston approaches TDC). The O° mark on the crankshaft vibration damper should now be aligned with the indicator on the timing cover.

2. If installing the distributor on a 6 cyl.. engine, position the rotor just ahead of the No. 1 distributor cap terminal. Lower the distributor into the engine block opening. Mesh the distributor gear with the camshaft drive gear. Be certain that the rubber O-ring seal is in the groove of the distributor shank. Proceed to Step 4.

3. If installing the distributor on a V8 engine, clean the top of the engine block around the distributor opening and position the rotor even with the No. 1 distributor cap terminal position. Lower the distributor into the engine block opening and engage the tongue of the distributor shaft with the slot in the distributor oil pump drive gear. Proceed to Step 4.

NOTE: When the distributor is properly seated, the rotor should be directly under the No. 1 distributor cap terminal.

4. Install the distributor hold-down clamp and tighten the retaining screw finger tight.

5. Install the distributor cap. Connect the primary lead wire(s) to the harness connector.

6. Start the engine. Check and adjust the ignition timing.

7. Connect the vacuum advance line to the distributor.

IGNITION TIMING ADJUSTMENT

NOTE: When installing the timing light, do not puncture the cables, boots or nipples with test probes—always use the proper adaptors. Puncturing the spark plug cables with a probe will damage the cables, separate the conductor and cause high resistance. Also, breaking the rubber insulation may permit secondary current to arc to ground. Ignition timing must be checked ONLY when the engine is at normal operating temperature and correct idle speed.

1. Connect an adjustable timing light or magnetic timing unit (use a 10° offset unit, when required) according to the manufacturer's instructions. Check the underhood sticker for any further instructions.

2. Connect a tachometer to the engine; the red lead to the negative pri-

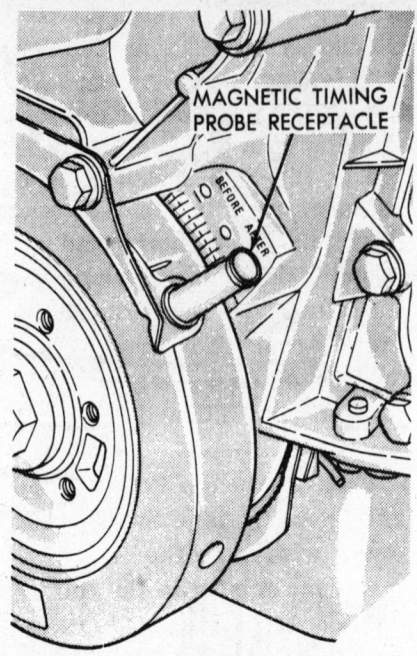

Magnetic timing probe provision for V8. Six-cylinder similiar

mary terminal of the coil and the black lead to a solid ground. Set the selector switch on the appropriate cylinder position.

3. Set the parking brake, place the transmission in Neutral or Park and start the engine.

4. If not equipped with Electronic Spark Advance (ESA), disconnect and plug the vacuum line at the distributor. If equipped with ESA, install a jumper wire from the carburetor switch connector-to-ground and leave the vacuum line attached to the distributor.

5. When the engine has reached normal operating temperature, check and/or set the idle speed (See the under hood sticker for specifications).

6. Loosen the distributor hold-down screw so the housing can be rotated.

NOTE: Do not use the distributor vacuum advance chamber as a handle when turning the distributor.

7. If using a timing light, aim it at the timing plate located at the timing case cover; if using a magnetic unit, read the unit (it must be within ± 2° of the specified value). Check and/or set the timing.

NOTE: If the timing is ahead of the mark, turn the distributor housing in the direction of rotor rotation to retard the timing. If it is past the mark, rotate the distributor against its direction of rotation to advance the timing.

8. When the timing is adjusted to specifications, tighten the distributor lock clamp.

9. If the engine idle speed has changed, readjust the curb idle. Do not reset the timing.

10. Turn off the engine, unplug/reconnect the vacuum line, disconnect the timing equipment and jumper wire.

FUEL SYSTEM

Mechanical Fuel Pump

REMOVAL & INSTALLATION

Mechanical fuel pumps are used on carbureted engines. On the 3.7L 6 cyl.. engine, the pump is driven by a small cam eccentric cast into the main camshaft. On the V8 engines, the pump is driven by a steel eccentric cam pressed on the camshaft gear end. On all engines, the pump is driven directly by the pump rocker arm riding on the cam eccentric.

1. Remove the dirt/oil from the pump exterior and remove the fuel lines.

2. Remove the pump-to-block bolts and the pump.

3. Using a putty knife, clean the gasket mounting surfaces.

4. To install, coat both sides of the pump gasket with sealer and reverse the removal procedures.

NOTE: If the pump is difficult to engage with the drive, rotate the engine slightly.

5. Connect the fuel lines and tighten the pump bolts. Start the engine and check for fuel leaks.

Electric Fuel Pump

The 5.2L V8 (EFI) engine is equipped with an electric fuel pump (with check valve) in the fuel tank, dual fuel filters and a fuel control check valve mounted near the throttle body. Fuel pressure is automatically relieved from the system when the engine is not in operation.

REMOVAL & INSTALLATION

1. Open the trunk and pull back the carpet. Remove the fuel line clamps/hoses from the delivery and return tubes.

2. Disconnect the 4 terminal electrical connector. Using a mallet and a non-metallic drift, drive the sealing

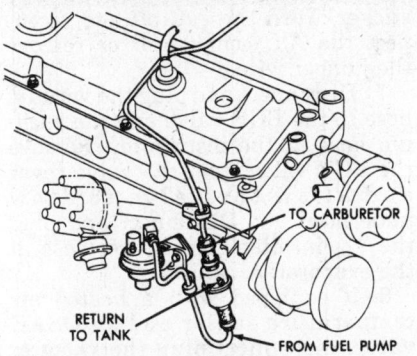

Fuel pump and filter location, six cylinder

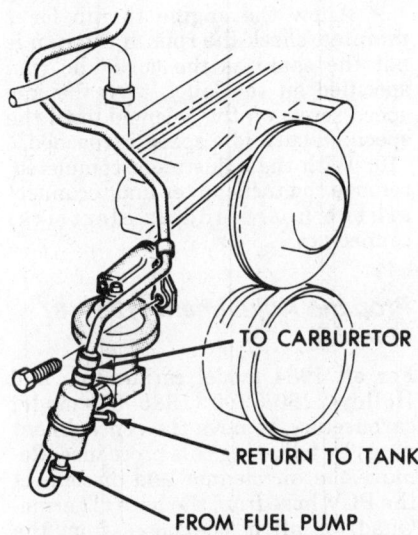

Fuel pump and filter location for V8

ring nut (loosen it) from the fuel tank.

3. Remove the fuel pickup unit from the fuel tank, being careful not to get dirt in the tank. Separate the fuel pump from the in-tank filter and pickup assembly.

4. To install, reverse the removal procedures. Make sure that the sealing ring gasket is secure.

Fuel Filter

REMOVAL & INSTALLATION

Carbureted Engines

Locate the filter in the fuel line between the fuel pump and the carburetor. Using hose clamp pliers, remove the retaining clamps and pull off the filter. Reverse this procedure for installation. Be sure the arrow on the filter is pointing toward the carburetor (direction of fuel flow).

NOTE: Some filters have a third line, to prevent vapor lock by allowing fuel vapors to return to the tank.

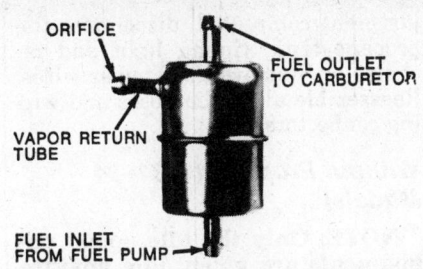

The combination filter and vapor separator found on some models

Fuel Injected Engines

The 5.2L V8 (EFI) engine is equipped with parallel fuel filters mounted in the delivery line between the fuel tank and the throttle body. The filters are mounted side by side on a common bracket. Replace both filters when servicing.

1. Disconnect and plug the fuel lines at the fuel filters.

2. Remove the fuel filter bracket from the vehicle.

3. Remove the fuel filters from the bracket.

4. To install, use new fuel filters and reverse the removal procedures.

NOTE: When installing the fuel filters, be sure to install them in the correct direction of the fuel flow.

Carburetor

Due to the complex nature of modern feedback carburetors and fuel injection systems, comprehensive diagnosis and testing procedures fall outside the confines of this repair manual. For complete information on diagnosis, testing and repair procedures concerning all fuel injection and feedback carburetor systems, please refer to *Chilton's Guide To Fuel Injection And Feedback Carburetors.*

REMOVAL & INSTALLATION

For all information on the carburetor not detailed below, please refer to the "Carburetor" Unit Repair Section.

1. Disconnect the negative battery terminal.

2. Remove the air cleaner.

3. Remove the filler cap from the fuel tank.

4. Place a small fuel container under the fuel inlet fitting.

5. Using 2 wrenches, to avoid twisting the line, disconnect the fuel inlet line and drain the excess fuel into the container.

6. Disconnect all of the vacuum lines, electrical connectors (if equipped), the throttle and choke linkage.

7. Remove the carburetor mounting bolts/nuts and the carburetor from the engine. To avoid spillage of the fuel, hold the carburetor level.

8. To install, reverse the removal procedures. Check to be certain that the choke plate opens/closes fully (when operated) and that full throttle travel is obtained.

NOTE: To prime the carburetor after installation, crank the engine and depress the accelerator several times.

IDLE SPEED & MIXTURE ADJUSTMENTS

NOTE: Federal law prohibits tampering with the idle air mixture adjustment. Only under the provision that an idle defect exists after all normal diagnosis attempts have failed (to cure the faulty condition) can the idle air mixture be adjusted. These procedures require the use of sophisticated testing equipment to ensure that the results are within legal limits. The factory recommended procedures for the idle mixture adjustment requires the addition of propane to the air mixture, in addition to special tools. For the idle speed screw locations, consult the illustrations.

Without Propane (1980 Models)

1. Connect a tachometer and a timing light to the engine.

2. Set the parking brake. Place the transmission in Neutral or Park and operate the engine until the normal operating temperature is reached.

3. Turn off all of the accessories and headlights.

4. Remove and plug the vacuum hose at the distributor. If equipped with an Electronic Spark Control (ESC), use a jumper wire to ground the idle stop carburetor switch.

5. Refer to the Emission Control Label under the hood; then, check and/or adjust the timing.

NOTE: If equipped with an ESC unit, wait one minute after returning to idle to check the timing.

6. Disconnect the PCV hose at the cylinder head cover and the purge hose at the vapor canister (do not plug the hoses).

7. Refer to the Emission Control Label under the hood and adjust the idle speed and idle mixture screws, to obtain the smoothest idle speed.

8. After adjusting the idle speed, reconnect the hoses (the idle speed may increase but DO NOT readjust). Remove the tachometer, the timing light and the ESC jumper wire (if equipped).

Propane Adjustment (1980 Models)

1. Set the parking brake and place the transmission in Neutral or Park.
2. Turn off the accessories and headlights. Connect a tachometer and a timing light to the engine. Start the engine and allow it to reach the operating temperature.
3. Disconnect and plug the vacuum hoses at the distributor or spark control unit and the EGR valve. Disconnect the emission canister purge hose (DO NOT plug).
4. If equipped with an idle stop carburetor switch, connect a jumper wire from it to a ground. Check and/or adjust the timing.

NOTE: If the engine is equipped with ESC, wait one minute after returning to idle to check the timing.

5. Disconnect the vacuum hose from the air cleaner or the choke diaphragm and install the propane hose in its place.
6. With the propane bottle in a safe, upright position, remove the PCV valve from the cylinder head cover and allow it to draw air.
7. Open the propane main valve and slowly open the propane metering valve until maximum engine speed is reached. The engine speed will drop if too much propane is added. To obtain the highest engine speed, fine tune the metering valve.
8. With the propane flowing, adjust the idle speed screw to obtain the correct propane speed. Again, adjust the metering valve to obtain the highest engine speed. Readjust the idle speed screw to obtain the correct propane speed.
9. Turn off the propane and allow the engine speed to stabilize. Slowly adjust the idle mixture screw (pausing between adjustments to allow the engine to stabilize) to obtain the smoothest idle at the correct idle speed.

NOTE: If the idle mixture screw(s) limiter cap(s) interfere with the adjustment procedures, remove the cap(s).

10. Turn on the propane and again fine tune to the highest engine speed. When very little difference in the engine speed is noticed, turn off the propane tank.
11. With the idle and mixture adjustment completed, disconnect the propane tank, timing light and tachometer. Remove the jumper wires. Reassemble all of the hoses and wiring connectors.

Without Propane (1981–84 Models)

NOTE: Only the idle speed adjustments are given; idle mixture adjustments require the use of a special monitoring and adjustment equipment.

1. Set the parking brake, place the transmission in Neutral. Turn off all accessories and run the engine until the operating temperature is reached.
2. Connect a timing light and a tachometer to the engine; then, check and adjust the ignition timing. Allow the engine to stabilize for two minutes before adjusting the timing. Once the timing has been adjusted, remove the timing light.
3. If equipped with a Spark Control Computer (SCC), remove and plug the vacuum line at the transducer; then connect an auxiliary vacuum supply to the transducer and set it for 16 in. Hg.
4. If equipped with a fuel control computer, connect a jumper wire between the carburetor switch and a good ground; disconnect the engine harness lead from the oxygen sensor and ground the lead.

NOTE: To disconnect the oxygen sensor wire lead, a bullet connector is provided approximately 4 in. from the sensor. On the Holley 6145 model, an exhaust manifold heat shield must be removed to gain access to the oxygen sensor. For the Rochester Quadrajet, disconnect the carburetor electrical connector and attach a jumper wire between the violet wire of the harness and a good ground.

5. Turn on the A/C and disconnect the air conditioner compressor by disconnecting its clutch wire. If the vehicle is not equipped with A/C, connect a jumper wire between the positive battery terminal and the Solenoid Idle Stop (SIS) solenoid lead wire to energize the solenoid.

--- **CAUTION** ---
Make sure the jumper wire is connected to the proper wire of the solenoid or the wiring harness could be damaged.

6. With the solenoid energized, remove the screw/spring from the top of the solenoid, insert an ⅛ in. Allen wrench into the solenoid and adjust the rpm to obtain the solenoid idle speed specified on the underhood sticker. Turn the A/C off and reconnect the A/C compressor or remove the jumper wire.
7. Disconnect and plug the vacuum hose to the EGR valve and the vacuum hose at the distributor. Remove the PCV valve from the valve cover and allow the valve to draw underhood air. Disconnect and plug the ³⁄₁₆ in. diameter control hose at the evaporative canister.
8. If equipped with a heated air temperature sensor and an OSAC valve, disconnect/plug the vacuum hoses from the carburetor at the heated air temperature sensor and at the OSAC valve.
9. Allow the engine to run for 2 minutes, check the rpm. If the rpm is not the same as the curb idle rpm specified on the label, turn the idle speed screw on the solenoid until the specified curb idle speed is reached.
10. With the adjustment completed, remove the tachometer and reconnect all the hoses and/or electrical connectors.

Propane Adjustment (1981–87 Models)

For all 1984 model carburetors and Holley 2280/6280 (1985–87) model carburetors, remove the concealment plugs by following this procedure. Remove the air cleaner and disconnect the PCV hose from the base (Thermo-Quad) or all of the hoses from the front of the carburetor (BBD). About ¼ in. from the end of the mixture screw housing, center punch a mark. Using a ³⁄₁₆ in. drill bit, drill through the outer housing. Use a small punch to drive out the concealment plug. For the Rochester Quadrajet (1985–87) remove the concealment plugs by placing the carburetor on a bench with the bottom facing upwards. Using a hacksaw, make 2 cuts down to the plug, at a 45° angle, one on each side (between the indicator marks) of the plug. Using a flat punch, drive the cut segment into the housing, exposing the steel plug. Drive out the steel plug.

1. Remove the carburetor from the engine. Remove the throttle body from the carburetor and clamp it in vise (protect the mounting surfaces of the throttle body from the vise jaws).
2. If a roll pin is present in the idle mixture screw housing, drill a ⁵⁄₆₄ in. pilot hole in the top of the housing (above the roll pin), followed by a ⅛ in. drill; then, using a small punch drive out the roll pin.
3. Place the throttle body in the vise with the bottom facing upwards. Drill a ⁵⁄₆₄ in. pilot hole at a 45° angle towards the concealment plug; then

redrill the pilot hole with a $\frac{1}{8}$ in. drill.

4. Insert a small punch into the hole and drive out the concealment plug.

5. Reinstall the throttle body to the carburetor and reinstall the carburetor to the engine.

6. Set the parking brake and place the transmission in Neutral or Park.

7. Turn off the accessories and headlights. Connect a tachometer and a timing light to the engine. Start the engine and allow it to reach operating temperature.

8. Disconnect and plug the vacuum hoses at the distributor, the EGR valve, the heated air temperature sensor (if equipped), the ESA computer (if equipped) and the OSAC valve (if equipped). Check and/or adjust the timing.

NOTE: If the engine is equipped with ESC, wait one minute after returning to idle to check the timing.

9. If equipped with a fuel control computer, connect a jumper wire between the carburetor switch and a good ground; disconnect the engine harness lead from the oxygen sensor and ground the lead. To disconnect the oxygen sensor wire lead, a bullet connector is provided approximately 4 in. from the sensor.

10. On the 1985–87 2 bbl, disconnect the engine electrical harness connector. From the carburetor side of the connector, remove duty cycle (green) and the idle solenoid (blue) wires, then reconnect the electrical connector.

11. Start the engine and allow it to operate for at least 4 minutes.

NOTE: For the 1984 Rochester Quadrajet, disconnect the carburetor electrical connector and install a jumper wire between violet wire of the harness and a good ground.

12. Turn on the A/C and disconnect the air conditioner compressor by disconnecting its clutch wire. If not equipped with A/C, connect a jumper wire between the positive battery terminal and the Solenoid Idle Stop (SIS) solenoid lead wire (for the 1985 Rochester Quadrajet, use the red wire), this will energize the solenoid.

13. If equipped with a Spark Control Computer (SCC), disconnect and plug the vacuum line at the vacuum transducer; then connect an auxiliary vacuum supply to the transducer and set it for 16 in. Hg.

14. On the 4 bbl carburetors, the propane supply hose must be "T" into the vacuum hose going to the bowl vent solenoid or connected to the choke vacuum nipple of the carburetor. For all others, disconnect the vacuum hose from the choke diaphragm at the tee and install the propane hose in its place (the other tee connectors must remain in place).

15. With the propane bottle in a safe, upright position, remove the PCV valve from the cylinder head cover and allow it to draw air. Disconnect and plug the $\frac{3}{16}$ in. emission canister control hose.

16. Open the propane main valve and slowly open the propane metering valve until maximum engine speed is reached. The engine speed will drop if too much propane is added. To obtain the highest engine speed, fine tune the metering valve.

17. With the propane flowing, adjust the solenoid idle speed screw to obtain the correct propane speed. Again, adjust the metering valve to obtain the highest engine speed. Readjust the idle speed screw to obtain the correct propane speed.

18. Turn off the propane and allow the engine speed to stabilize. Slowly adjust the idle mixture screw (pausing between adjustments to allow the engine to stabilize) to obtain the smoothest idle at the correct idle speed.

19. Turn on the propane and again fine tune to the highest engine speed. When very little difference in the engine speed is noticed, turn off the propane tank.

20. With the idle speed and idle mixture adjusted, disconnect the propane tank and tachometer; then, reconnect all of the hoses and wiring connectors.

21. Install a new concealment plug and drive in the roll pin (if equipped).

SOLENOID IDLE STOP ADJUSTMENT (1981–84)

1. Adjust the ignition timing to specifications. Disconnect and plug the vacuum hose at the EGR valve. Connect a jumper wire between the carburetor switch and a good ground.

2. Disconnect and plug the $\frac{3}{16}$ in. diameter control hose at the canister. Remove or prop up the air cleaner.

3. Remove the PCV valve from the cylinder rocker cover and allow the valve to draw underhood air. Connect a tachometer to the engine.

4. Turn on the air conditioning (if equipped) and set the blower on Low. Disconnect the air conditioning clutch wire.

5. On non-air conditioned models, connect a jumper wire between the battery positive post and the solenoid idle stop lead wire.

CAUTION

Use care in jumping to the proper wire on the solenoid. Applying battery voltage to other than the correct wire will damage the wiring harness.

---

6. Open the throttle slightly to allow the solenoid plunger to extend. Remove the adjusting screw and spring from the solenoid. Insert a $\frac{1}{8}$ in. Allen wrench into the solenoid and adjust to the correct solenoid rpm.

7. Install the solenoid screw and spring. Turn the screw in until it bottoms. To adjust the engine rpm, turn the idle screw on the throttle lever. Turn Off the air conditioning and replace clutch wire or remove jumper wire.

SOLENOID IDLE SPEED & IDLE RPM ADJUSTMENT

1985 2 bbl & 1985–87 4 bbl

1. Adjust ignition timing to specifications. Disconnect and plug the vacuum hose at the EGR valve and the carburetor hose at the heated air temperature sensor.

2. Remove the air cleaner. Disconnect and plug the $\frac{3}{16}$ in. diameter control hose at the canister.

3. Remove the PCV valve from the cylinder rocker cover and allow the valve to draw underhood air. Connect a tachometer to the engine.

4. Start the engine and allow it to reach normal operating temperature. Turn off the engine. For the 2 bbl's, disconnect and reconnect the fusible link at the battery.

5. For the 2 bbl's, connect a ground wire to the carburetor switch; for the 4 bbl's, disconnect the carburetor electrical connector, attach a jumper wire between the ground switch terminal of the electrical harness (violet wire) and a good ground.

6. For the 2 bbl's, disconnect the engine harness lead from the oxygen sensor and ground the lead. Start the engine and operate for 4 minutes to allow the system to stabilize.

7. Connect a jumper wire between the battery positive post and the solenoid idle stop lead (red) wire.

CAUTION

Use care in jumping the proper wire on the solenoid. Applying battery voltage to other than the correct wire will damage the wiring harness.

---

8. Open the throttle slightly to allow the solenoid plunger to extend.

9. Remove the outer screw and spring from the solenoid. Insert a $\frac{1}{8}$ in. Allen wrench into the solenoid and adjust to the correct engine solenoid

speed (900 rpm–2 bbl., 800 rpm–1984 4 bbl. or 850 rpm–1985–87 4 bbl.).

10. Install the screw and spring and tighten the screw until it bottoms. Remove the jumper wire. Turn the solenoid screw to adjust the idle speed (see underhood specification label).

11. Remove the tachometer, install the hoses, the PCV valve and the air cleaner. Remove the jumper wire and connect the electrical connector.

1985–87 2 bbl

1. Adjust ignition timing to specifications. Disconnect and plug the vacuum hose at the EGR valve and the carburetor hose at the heated air temperature sensor.

2. Remove the air cleaner. Disconnect and plug the $3/16$ in. dia. control hose at the canister.

3. Remove the PCV valve from the cylinder rocker cover and allow the valve to draw underhood air. Disconnect and plug the vacuum hose at the Electronic Spark Advance (ESA) unit. Connect a tachometer to the engine.

4. Disconnect the carburetor electrical connector from the engine wiring harness, remove the duty cycle (green) and the idle solenoid (blue) wires from the carburetor electrical connector. With the 2 wires removed from the electrical connector, reconnect the connector to the engine wiring harness.

5. For the 2 bbl's, connect a ground wire to the carburetor switch; for quadrajets, disconnect the carburetor electrical connector, attach a jumper wire between the ground switch terminal of the electrical harness (violet wire) and a good ground.

6. Connect a jumper wire between the battery positive post and the solenoid idle stop lead (blue) wire.

----- CAUTION -----

Use care in jumping the proper wire on the solenoid. Applying battery voltage to other than the correct wire will damage the wiring harness.

----

7. Place the transmission in Neutral, start the engine, place the carburetor on the 2nd step of the fast idle cam and run the engine for 5–10 minutes, then return the engine to idle.

8. Open the throttle slightly to allow the solenoid plunger to extend.

9. Remove the outer screw and spring from the solenoid. Insert a $1/8$ in. Allen wrench into the solenoid and adjust to the correct engine solenoid speed (775 rpm).

10. Install the screw and spring and tighten the screw until it bottoms. Remove the jumper wire. Turn the solenoid screw to adjust the idle speed (see underhood specification label).

11. Remove the tachometer, then install the hoses, the PCV valve, the air cleaner and the engine electrical harness wires. Remove the jumper wire and connect the electrical connector.

FAST IDLE SPEED ADJUSTMENT

2 bbl

1. Check and/or adjust the timing. Disconnect and plug the vacuum hose at the EGR valve.

2. Disconnect and plug the carburetor vacuum hose at the heated air temperature sensor, the hose at the distributor and the $3/16$ dia. control hose at the emissions canister.

3. Connect a jumper wire between the carburetor switch (equipped with Electronic Spark Advance–ESA) and a good ground. Remove or prop up the air cleaner.

4. Remove the PCV valve from the cylinder head cover and allow it to draw underhood air. Connect a tachometer to the engine. Start the engine and allow it to reach normal operating temperature (5–10 minutes).

NOTE: On the 1985–87 models, disconnect the engine electrical harness, then remove the duty cycle (green) and the idle solenoid (blue) wires from the engine wiring harness; reconnect the engine electrical harness (minus the 2 wires).

5. Open the throttle slightly and place the fast idle speed screw on the 2nd step of the fast idle cam.

6. With the choke fully open, adjust the fast idle speed screw to obtain the correct fast idle rpm.

7. Return to idle, then reposition the adjusting screw on the second highest step of the fast idle cam to verify the fast idle speed; readjust, if necessary.

8. Return to idle and turn off the engine. Unplug/reconnect the vacuum hoses and electrical connectors. Remove the tachometer, reinstall the PCV valve and remove the ground wire.

NOTE: The idle speed with the engine in normal operating condition (everything connected) may vary from set speeds. Do not readjust.

4 bbl

1. Check and/or adjust the timing. Disconnect and plug the vacuum hose at the EGR valve.

2. Disconnect and plug the $3/16$ dia. control hose at the emissions canister. Connect a jumper wire between the carburetor switch and a good ground.

3. Remove the PCV valve from the cylinder head cover and allow it to draw underhood air. Connect a tachometer to the engine. Start the engine and allow it to reach normal operating temperature.

NOTE: If the vehicle is equipped with an oxygen feedback system, disconnect the engine harness lead from the oxygen sensor and ground the harness lead. After disconnecting the engine harness lead, allow the engine to operate for 4 minutes for the system to stabilize.

4. Open the throttle slightly and place the fast idle speed screw on the 2nd step of the fast idle cam.

5. With the choke fully open, adjust the fast idle speed screw to obtain the correct fast idle rpm.

6. Return to idle, then reposition the adjusting screw on the second highest step of the fast idle cam to verify the fast idle speed; readjust, if necessary.

7. Return to idle and turn off the engine. Unplug/reconnect the vacuum hoses and electrical connectors. Remove the tachometer, reinstall the PCV valve and remove the ground wire.

NOTE: The idle speed with the engine in normal operating condition (everything connected) may vary from set speeds. Do not readjust.

Electronic Fuel Injection (EFI)

Due to the complex nature of modern fuel injection systems, comprehensive diagnosis and testing procedures fall outside the confines of this repair manual. For complete information on diagnosis, testing and repair procedures concerning all fuel injection systems, please refer to *Chilton's Guide To Fuel Injection And Feedback Carburetors*.

ENGINE COOLING

There are 3 levels of cooling systems: standard, air conditioning and trailer towing packages. The radiator size varies with the engine and cooling level. Other variable items are fan size, fan shrouds, thermostatically

controlled fluid fan drives and external automatic transmission fluid coolers.

NOTE: Chrysler recommends that only ethylene glycol type antifreeze mixed with water be used.

Radiator

REMOVAL & INSTALLATION

1. Place the heater temperature selector to "Full On". Drain the cooling system by opening the drain cock at the bottom of the radiator. When the reserve tank is empty, remove the pressure cap.

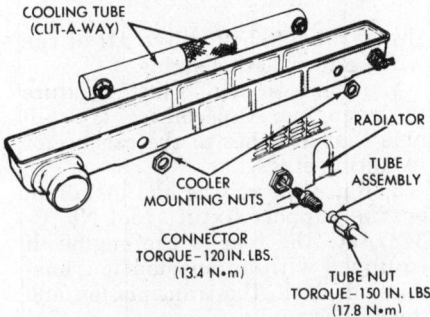

COOLING TUBE (CUT-A-WAY)

RADIATOR

COOLER MOUNTING NUTS

TUBE ASSEMBLY

CONNECTOR TORQUE – 120 IN. LBS. (13.4 N•m)

TUBE NUT TORQUE – 150 IN. LBS. (17.8 N•m)

Internal details of the in-radiator transmission fluid cooler

2. If equipped with an automatic transmission, remove the oil cooler lines from the radiator.

3. Remove the upper and lower hose clamps and hoses. Remove the coolant reserve tank tube.

4. Remove the screws and position the shroud rearward to provide maximum clearance.

5. Loosen the retaining screws at the bottom of the radiator and remove the screws at the top.

6. Lift the radiator out of the engine compartment.

— CAUTION —
Extreme care should be taken during removal not to damage the radiator cooling fins or water tubes.

7. To install, reverse the removal procedures. Fill the radiator to the top of neck and the reserve tank to the "MAX" level. Warm up the engine with the heater on and check the coolant level. On vehicles with automatic transmissions, check the transmission fluid level after warm-up and add fluid as required.

Water Pump

REMOVAL & INSTALLATION

1. Drain the cooling system. Disconnect the negative battery cable.

2. Remove the fan shroud screws and move the shroud out of the way.

3. Remove the upper radiator hose and tie it back over the engine. Remove the lower radiator hose, bypass hose and heater hose from the water pump.

4. Loosen the alternator, power steering pump and air pump (if equipped). Remove all of the accessory belts.

5. Remove the fan, fan shroud, fan spacer, fluid drive and the water pump pulley.

— CAUTION —
With fluid-coupled fan drives, to prevent silicone fluid from leaking, do not set the drive unit with its shaft pointing downward.

6. On the 6 cyl.., remove the air pump pulley. Remove the front bracket of the air compressor, air pump and power steering pump (if equipped). Set aside the air and the power steering pumps.

7. On the V8's, remove and set aside the alternator/brackets, air compressor and the power steering pump.

8. Remove the water pump bolts and the water pump. Discard the gasket and clean the mounting surfaces.

9. To install, use a new gasket and reverse the removal procedures. Torque the water pump bolts to 30 ft. lbs. Fill the radiator to the neck and the reserve tank to the "MAX" level. Warm up the engine with the heater on and inspect the water pump for any leaks. Check the coolant level and add as required.

Thermostat

REMOVAL & INSTALLATION

All engines use a 219° Fahrenheit thermostat. On the V8, it is located at the top-front of the intake manifold. On the 6 cyl.inder, it is located on the left-front side.

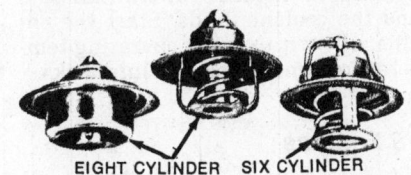

EIGHT CYLINDER   SIX CYLINDER

The spring side (arrow) is inserted into the engine when installing a thermostat

1. Drain the cooling system to below the level of the thermostat.

2. Remove the housing bolts and the thermostat housing and thermostat. Clean the gasket surfaces.

3. To install, use a new gasket

(dipped in water) or sealant and reverse the removal procedures. On the V8, be sure that the pellet end is facing the engine. On the 6 cyl.inder the vent hole must face upwards.

4. Refill the system, allow the engine to warm up with the heater on and check for leaks.

EMISSION CONTROLS

For all emission control maintenance, please refer to "Emission Controls" in the unit repair section. Due to the complex nature of modern electronic engine control systems, comprehensive diagnosis and testing procedures fall outside the confines of this repair manual. For complete information on diagnosis, testing and repair procedures concerning all modern engine and emission control systems, please refer to *Chilton's Guide To Electronic Engine Controls*.

ENGINE MECHANICAL

Engine

REMOVAL & INSTALLATION

6 Cylinder Engine

1. Scribe the hood hinge outlines on the underside of the hood, then remove the hood.

2. Drain the cooling system. Remove the negative battery cable and the air cleaner.

3. Remove the radiator hoses, heater hoses, transmission oil cooler lines (if equipped) and the radiator. Remove the PCV and the evaporative control system from the cylinder head cover.

4. Disconnect the fuel lines, carburetor linkage and wiring from the engine. Disconnect the exhaust pipe at the exhaust manifold and raise the vehicle on a hoist.

5. If equipped with an automatic transmission, drain the transmission/converter fluid. Remove the fluid cooler lines, the filler tube and the shift linkage.

6. Remove the clutch torque shaft

Chilton's TIME SAVER

Although the engine removal procedure requires first removing the transmission, the transmission can be left in the chassis.

If the car is equipped with an automatic transmission, remove the inspection plate from the bellhousing. Attach a remote starter switch to the engine and crank the engine until you have access to the converter-to-driveplate attaching nuts. Remove the nuts, then remove the starter. If the car has a manual transmission, disconnect the clutch torque shaft from the engine block and the clutch linkage from the adjustment rod.

Remove the automatic transmission filler tube. Support the transmission from below and remove the transmission-to-engine or transmission-to clutch bell housing attaching bolts. Place a block of wood on the floor jack lifting pad and position the jack under the transmission. As the engine is lifted out of

the car, raise and lower the jack as necessary so the angle of the transmission follows the angle of the engine as nearly as possible. Use care, as the transmission will probably be tilting on the jack pad.

When installing the engine into a car with an automatic transmission, remember that the crankshaft flange bolt circle, the inner and outer circle of holes in the driveplate, and the four tapped holes in the front face of the converter all have one hole offset. The torque converter must be mounted to the driveplate in its original location to insure proper engine-to-converter balance.

When installing an engine into a car with a manual transmission, it may be necessary to turn the crankshaft pulley slightly, with the transmission in gear, to mesh the transmission input shaft spline with the clutch disc inner hub.

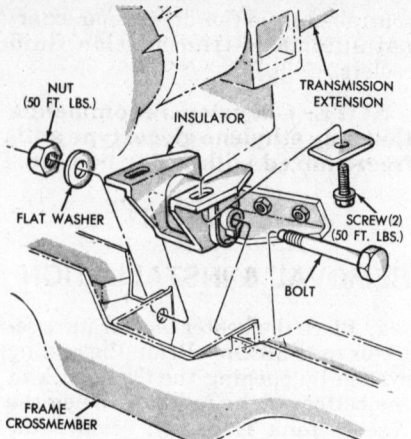

Typical V8 rear engine mount

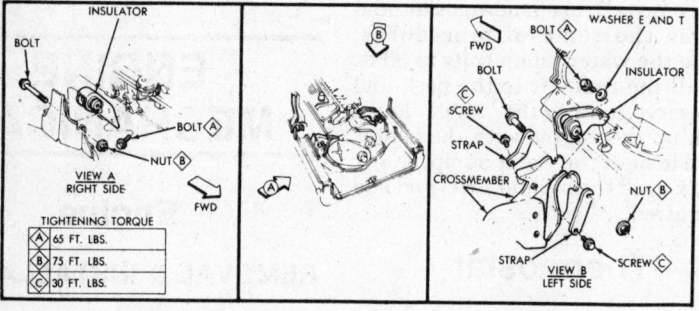

Engine front mounts, Slant Six, all models

TIGHTENING TORQUE	
Ⓐ	65 FT. LBS.
Ⓑ	75 FT. LBS.
Ⓒ	30 FT. LBS.

and the rods. Remove the speedometer cable and the gear shift rods. Disconnect the driveshaft and tie out of the way.

7. Install an engine support fixture to the rear of the engine. Remove the engine rear support crossmember.

8. Remove the transmission mounting bolts from the clutch housing and the transmission.

NOTE: With an automatic transmission, the torque converter must first be unbolted from the flywheel. Matchmark the converter and flywheel for reassembly.

9. Lower the vehicle. Fasten an engine lifting fixture to the engine and attach a vertical hoist to the fixture eyebolt. Remove the bolts from the front engine mounts.

10. Lift the engine out of the engine compartment and attach it to an engine stand.

11. To install, reverse the removal procedures. Replace the transmission and the coolant fluids. Start the engine, allow it to reach operating temperatures and check for fluid leaks.

V8 Engines

1. Scribe the outline of the hood hinge brackets on the hood and remove the hood. Remove the negative battery cable.

2. Drain the cooling system. Remove the radiator hoses, the fan shroud, the transmission oil cooler lines (if equipped) and the radiator.

3. Remove the fuel line(s), the air cleaner, the accelerator linkage and

the carburetor. Remove all of the wires attach to the engine.

4. Attach an engine lifting fixture to the engine. A special tool is available that attaches to the carburetor mounting studs.

5. Raise the vehicle and install an engine support fixture tool No. C-3487A to the rear of the engine. If equipped with an automatic transmission, drain the transmission and the convertor.

6. If equipped with air conditioning and/or power steering, remove the unit(s) from the engine and position it out of the way without disconnecting the lines.

7. Disconnect the exhaust pipe(s) from the exhaust manifold(s). Remove the driveshaft, wires, linkage, speedometer cable and oil cooler lines (if equipped) that are attached to the transmission.

8. If equipped with an automatic transmission, remove the torque converter-to-drive plate bolts. Remove the engine rear support crossmember, the transmission-to-engine bolts and the transmission. See the Time Saver.

9. Lower the vehicle. Attach a vertical hoist to the engine lifting fixture, lift the engine slightly and remove the motor mount bolts.

10. Raise the engine, carefully remove it from the engine compartment and mount it onto an engine stand.

11. To install, reverse the removal procedures. Fill the transmission and cooling system with fluid. Start the engine, allow it to reach operating temperatures and check for leaks.

Intake Manifold

REMOVAL & INSTALLATION

V8 Engines

1. Drain the cooling system. Disconnect the negative battery cable.

2. Remove the alternator, the air cleaner and disconnect the fuel lines(s) from the carburetor or throttle body.

3. Disconnect all vacuum lines and the throttle linkage that attach to the carburetor/throttle body and intake manifold.

4. Disconnect the spark plug wires from the plugs. Remove the distributor cap and wires as an assembly.

5. Disconnect the wires from the coil and the temperature sending unit.

6. Disconnect the heater and by-pass hose from the intake manifold.

7. Remove the intake manifold attaching bolts. Remove the manifold, carburetor/throt-tle body and coil from the engine as an assembly.

8. Clean all gasket mounting surfaces and firmly cement new gaskets to the engine.

NOTE: Do not use sealer on the composition side gasket used on the 5.9L engines.

9. To install, reverse the removal procedures. Torque the bolts to 45 ft. lbs. in 3 steps, in the sequence shown.

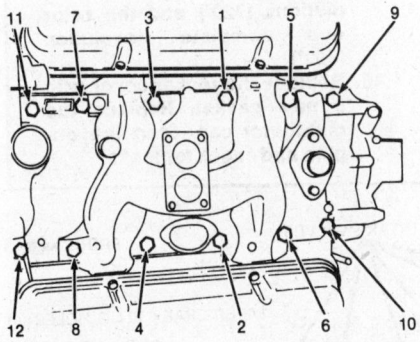

Intake manifold tightening sequence for V8 engines

Exhaust Manifold

REMOVAL & INSTALLATION

V8 Engines

1. Disconnect the exhausts manifold at the exhaust pipe flange. Access to these bolts is from underneath the vehicle. If equipped, disconnect the Air Injection nozzles and the carburetor heated air stove. Disconnect any components of the EGR system which are in the way.

2. Remove the exhaust manifold by removing the securing bolts and washers. To reach these bolts, it may be necessary to jack the engine slightly off its front mounts.

NOTE: When the exhaust manifold is removed, sometimes the securing studs will come out with the nuts. If this occurs, studs must be replaced with the aid of sealing compound on the coarse thread ends. If this is not done, water leaks may develop at the studs.

3. To install the exhaust manifold, reverse the removal procedures. On the center branch of the V8's manifold, no conical washers are used. Torque the exhaust manifold bolts to 20 ft. lbs. and the nuts to 15 ft. lbs.

Combination Manifold

REMOVAL & INSTALLATION

6 Cylinder Engine

1. Disconnect the air cleaner vacuum control tube from the carburetor. Disconnect the flexible connector between the carburetor and the air cleaner.

2. Disconnect the breather cap from the air cleaner line and remove the air cleaner.

3. Disconnect the crankcase ventilation tube, the carbureted bowl vent line and the distributor vacuum line. Remove the carburetor air heater.

4. Disconnect the automatic choke rod, the fuel line, the throttle linkage and remove the carburetor.

5. Disconnect the exhaust pipe at the exhaust manifold.

6. Remove the nuts and washers securing the manifold assembly to the cylinder head. Remove the manifold from the cylinder head.

NOTE: Make note of the location of the different types of washers for installation.

7. Remove the 3 screws securing the intake manifold to the exhaust manifold. Separate the manifolds and discard the gasket.

8. Clean all the gasket surfaces with solvent and blow them dry with compressed air. Check the mating surfaces of the manifolds with a straightedge. Surfaces should be flat within .006 in. per foot of length.

9. To install, place a new gasket between the 2 manifolds. Install the the stud nut and the 2 long screws securing the manifolds, DO NOT tighten the screws. Using a new gasket, position the manifold assembly on the cylinder head.

10. Install the triangular washers/nuts on the upper studs and on the 4 lower studs opposite No. 2 and 5 cylinders. The 8 triangular washers should be positioned squarely on the machined surfaces of both manifold retaining pads. These washers must be installed with the cup side against the manifold. Install the nuts and washers only when the engine is cold.

11. Install the steel conical washers with the cup (concave) side to the manifold, 1 on the center upper stud and 2 on the center lower studs. Install a brass washer on each end, with the flat side to the manifold. Install the nuts with the flat side away from the washers. Snug up the nuts.

12. Tighten the 3 intake-to-exhaust manifold screws to 12 ft. lbs., starting with the inner stud. Tighten the manifold-to-head nuts to 10 ft. lbs.

13. Attach the exhaust pipe-to-man-

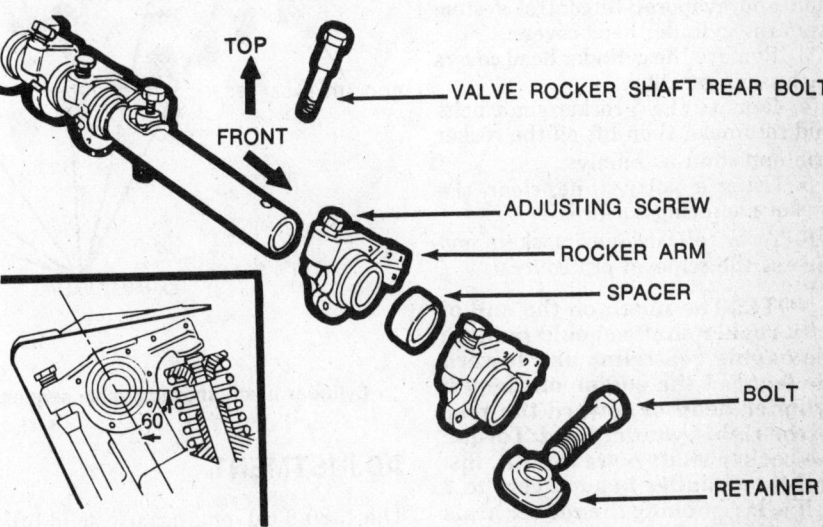

Slant Six rocker shaft details

ifold, using a new gasket and tighten the nuts to 35 ft. lbs. To complete the installation, reverse the removal procedures.

Rocker Arm/Shaft

REMOVAL & INSTALLATION

6 Cylinder Engines

1. Remove the PCV hose, distributor advance hose, vapor canister hose, temperature gauge wire, electric choke wire and the alternator wiring harness clips.

2. Disconnect the fender support bracket. Remove the valve cover with its gasket. If equipped with A/C, hold the A/C hoses up out of the way and then remove the valve cover.

3. Take out the rocker arm/shaft assembly securing bolts and remove the rocker arm/shaft assembly.

4. To install, use a new gasket, reverse the removal procedures and adjust the valves.

NOTE: The shaft oil holes must be installed facing the bottom to provide proper lubrication to the rocker arms. A special bolt goes to the rear of the shaft. Torque the rocker arm bolts to 25 ft. lbs. and be sure to adjust the valves. See Time Saver.

V8 Engines

The stamped steel rocker arms are arranged on one rocker arm shaft per cylinder head. To remove the rocker arms and shaft:

1. Disconnect the spark plug wires.

2. Disconnect the closed ventilation and evaporative control system from the cylinder head covers.

3. Remove the cylinder head covers with their gaskets.

4. Remove the 5 rocker shaft bolts and retainers, then lift off the rocker arm and shaft assembly.

5. Using a putty knife, clean the gasket mounting surfaces.

6. To install, use new gaskets and reverse the removal procedures.

NOTE: The notch on the end of both rocker shafts should point to the engine centerline and toward the front of the engine on the left cylinder head or toward the rear on the right cylinder head. Torque the rocker shaft bolts to 17 ft. lbs. and the cylinder head covers to 7 ft. lbs. If removing the rocker arms from the shaft, be sure to install them in their correct order of removal (right and left directions).

CHILTON'S TIME SAVER

The factory recommends adjusting the valves on six cylinder engines with the engine running, but the amateur mechanic will have better luck with the following procedure:

1. The engine must be at normal operating temperature. Mark the crankshaft pulley into three equal 120° segments, starting at the timing mark.

2. Remove the valve (rocker) cover and the distributor cap.

3. Set the engine at TDC on the No. 1 cylinder by aligning the mark on the crankshaft pulley with the 0° mark on the timing cover pointer. The distributor rotor should point at the position of the No. 1 spark plug wire in the distributor cap. Both rocker arms on No. 1 cylinder should be free to move slightly. If all this isn't the case, you have No. 6 cylinder at TDC and will have to turn the engine 360° in the normal direction of rotation.

4. The cylinders are numbered from front to rear. The intake and exhaust valves are in the following sequence, starting at the front: E-I, E-I, E-I, I-E, I-E, I-E. Note that intake and exhaust valves have different settings.

5. The lash is measured between the rocker arm and the end of the valve.

6. To check the lash, insert the correct size feeler gauge between the rocker arm and the valve. Press down lightly on the other end of the rocker arm. If the gauge cannot be inserted, loosen the self-locking adjustment nut on top of the rocker arm. Tighten the nut until the gauge can just be inserted and withdrawn without buckling.

7. After both valves for the No. 1 cylinder are adjusted, turn the engine so that the pulley turns 120° in the normal direction of rotation (clockwise). The distributor rotor will turn 60°, since it turns at half engine speed.

8. Check that the rocker arms are free and adjust the valves for the next cylinder in the firing order, No. 5. The firing order is 1-5-3-6-2-4.

9. Turn the engine 120° to adjust each of the remaining cylinders in the firing order. When you are done the engine will have made two complete revolutions (720°) and the rotor one complete revolution (360°).

10. Replace the rocker cover with a new gasket. Replace the distributor cap. Start the engine and check for leaks.

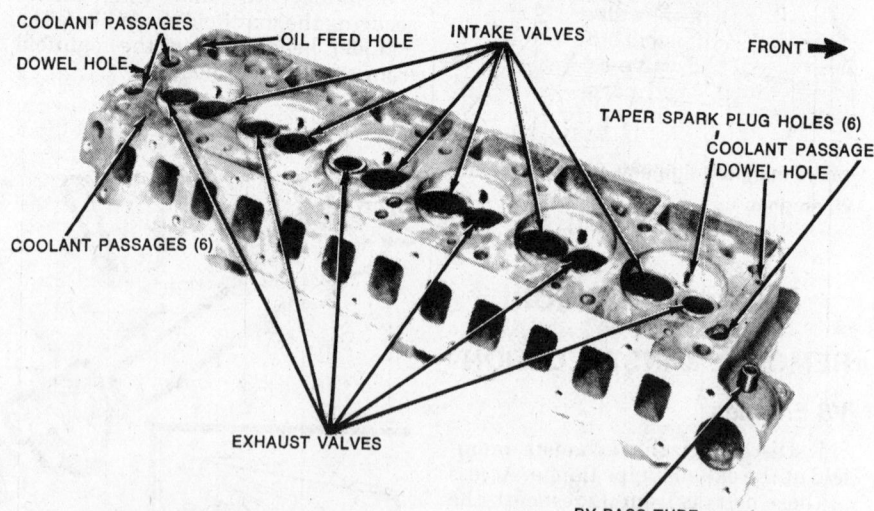

Cylinder head, showing valve sequence—six cylinder engines

ADJUSTMENT

The 1980 6 cyl. engines use solid lifters and adjustable rocker arms, which require valve adjustment. All other engines, including the V8, use hydraulic lifters and non-adjustable rocker arms; no adjustment is possible.

NOTE: After engine reassembly, allow the engine to operate at idle to build up oil pressure.

CAUTION

Do not set the valve lash closer than specified in an attempt to quiet the valve mechanism. This will cause burned valves. It is better to have the valves set slightly loose.

1980 6 Cylinder Engines

1. Set the engine idle speed and operate until the normal operating temperature is reached.

2. Remove the valve cover. Be careful of the hot oil which will splash off the rocker assembly when the cover is removed.

NOTE: Much of the oil splash can be avoided by purchasing a Slant 6 rocker cover from any junkyard and cutting out the center of the cover with aviation shears. Place this modified cover on the head, temporarily fasten it with the rocker cover bolts and proceed with the adjustment through the cut outs. The remaining sides will keep most of the oil splash on the rocker assembly. When the valves are adjusted, remove this cover and replace with the original.

3. Using the proper thickness feeler gauge, measure the clearance between the valve stem tip and the end of the rocker arm adjusting screw at each valve. If necessary, loosen the locknut on each rocker arm and turn the adjusting screw (while the feeler gauge is still inserted) to obtain the correct valve clearance.

4. After all of the valves have been checked and adjusted, stop the engine and replace the valve cover, using a new cylinder head cover gasket. If much oil was lost during the valve adjustment procedures, check the oil level in the crankcase.

Cylinder Head

CAUTION

Do not loosen the head bolts until the engine is thoroughly cool, to prevent warping the head. If the head sticks to the block, operate the starter to loosen it with compression or rap it upward with a soft rubber hammer. Do not force anything between the head and the block. The cylinder head bolts should be retorqued after the first 500 miles, unless a special gasket is used.

REMOVAL & INSTALLATION

6 Cylinder Engines

1. Drain the cooling system and disconnect the negative battery cable.
2. Remove the carburetor air clean-

er and the fuel lines. Disconnect the accelerator linkage.

3. Remove the vacuum lines from the carburetor and the distributor. Disconnect the spark plug wires from the spark plugs.

4. Disconnect the heater hose and the clamp holding the by-pass hose. Disconnect the heat indicator sending unit wire.

5. Disconnect the exhaust pipe at the exhaust manifold flange. If equipped, disconnect the diverter valve vacuum line from the intake manifold; also remove the air injection assembly (if applicable) from the cylinder head cover.

6. Remove the PCV tube, evaporative control system and cylinder head cover.

7. Remove the rocker arm/shaft assembly. Remove the pushrods and keep them in order.

NOTE: It is a good idea to set the pushrods in order (as they are removed from the engine) in a piece of lumber that has been drilled. The twelve pushrods can be labeled by writing the cylinder number and "In" and "Ex" under each hole in the lumber.

8. Remove the 14 head bolts in reverse of the torque sequence and lift off the cylinder head. The cylinder head is removed with the intake and exhaust manifold assembly. Clean the machined mounting surfaces.

NOTE: To ease the installation procedures, remove the intake/exhaust manifold assembly from the cylinder head.

9. To install, use new gaskets and reverse the removal procedures. Check all of the mounting surfaces with a straight edge, flatness should not exceed 0.00075 times the span length in any direction. Torque the cylinder head bolts in sequence to 70 ft. lbs. in 2 steps.

10. Refill the cooling system. Start the engine and operate it (at idle) until the normal operating temperatures have been reached. On the 1980 models, adjust the valve clearance.

V8 Engines

1. Drain the cooling system and disconnect negative battery cable.
2. Remove the alternator, the air cleaner and fuel line(s). Disconnect the accelerator linkage.
3. Remove the vacuum hose(s) from the carburetor/throttle body and the distributor. Remove the distributor cap and wires and the spark plugs.
4. Disconnect the coil wires, the temperature sending wire, the heater and by-pass hoses.

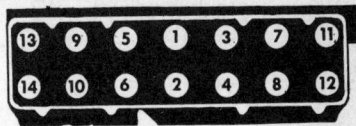

Tightening sequence for 6 cyl. engines

Chilton's TIME SAVER

Frequently valves become bent or warped or their seats become blocked with carbon or other material. Left unattended, this can cause burnt valves, damaged cylinder heads and other expensive troubles. To detect leaking valves early, perform this test whenever the cylinder head is removed.

1. After removing head, replace sparkplugs. Removing sparkplugs before removing heads eliminates breakage.
2. Place head on bench with valves, springs, retainers and keys installed and combustion chambers up.
3. Pour enough safe solvent in each combustion chamber to completely cover both valves. Watch combustion chambers for two minutes for any leakage.

5. Remove the PCV valve, the evaporative control system (if equipped) and the valve covers.

6. Remove the intake manifold, the ignition coil and the carburetor/throttle body as an assembly. Remove the tappet chamber cover, if equipped.

7. Disconnect the exhaust pipes and remove the exhaust manifolds.

8. Remove the rocker arm/shaft assemblies. Remove the pushrods and identify to insure installation in the original locations.

9. Remove the 10 head bolts from each cylinder head and lift off the heads. Clean all of the machine mounting surfaces.

NOTE: If there is any reason to suspect leakage between the cylinder block and head, inspect all surfaces with a straight edge. The out of flatness should not exceed 0.00075 times the span length in any direction.

10. To install, use new head gaskets and reverse the removal procedures. Torque the cylinder head bolts to 105 ft. lbs. in 3 steps.

11. Refill the cooling system. Start the engine and operate it (at idle) until the normal operating temperatures have been reached.

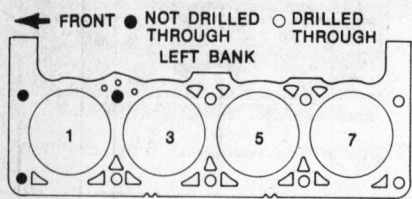
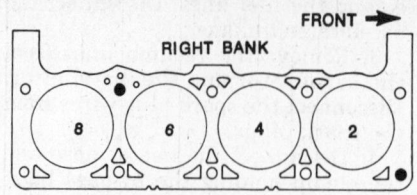

Sealant must be applied to the drilled through head bolt threads on the V8 engines

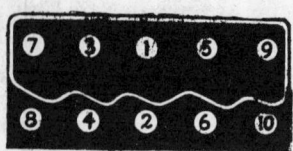

Tightening sequence for V8 engines

OVERHAUL

For all cylinder head overhaul procedures, please refer to "Engine Rebuilding" in the Unit Repair Section.

Front Cover

REMOVAL & INSTALLATION

6 Cylinder Engines

1. Drain the cooling system and disconnect the negative battery cable.
2. Remove the radiator and cooling fan. Remove the alternator, the power steering and the A/C drive belts, if equipped.
3. Remove the damper pulley bolt. Using a suitable puller tool (C-3732A, or equivalent), remove the damper pulley.
4. Remove the front oil pan bolts and the timing cover bolts. Remove the timing cover and gasket.
5. Using a putty knife, clean the gasket mounting surfaces.
6. To install, apply an 1/8 in. bead of sealer to the new gasket and reverse the removal procedures. Torque the oil pan and timing cover bolts to 17 ft. lbs. Refill the cooling system.

V8 Engines

1. Disconnect the negative battery cable and drain the cooling system. Remove the fan shroud bolts.
2. Disconnect the top radiator hose from the radiator and tie it back over the engine. Remove the alternator and the power steering pump (if equipped) belts.
3. Remove the fan, fan shroud, fan spacer, fan drive and the water pump pulley. Disconnect the alternator bracket and it set aside. Remove the

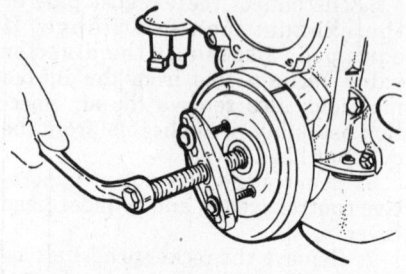

Removing vibration damper assembly, 318 V8. Slant Six similar

air compressor and the power steering pump (if equipped); then, set aside.
4. Disconnect the lower radiator and bypass hose from the water pump. Remove the water pump bolts and the pump.
5. Remove the damper pulley bolt and washer from the crankshaft. Install the bar from the tool set No. C-3688 and the screw from the tool set No. C-3732A, then pull the damper from the crankshaft.
6. Remove the fuel line(s) and fuel pump (carbureted engines). Loosen the oil pan bolts and remove the front bolt from each side.
7. Remove the timing gear cover.
8. Using a putty knife, clean the gasket mounting surfaces.
9. To install, use a new gasket, oil seal (lubricate the seal lips with oil) and reverse the removal procedures. Torque the damper pulley bolt to 135 ft. lbs., the timing chain cover to 30 ft. lbs. and the damper pulley to 17 ft. lbs.

NOTE: Use tool No. C-3688 or equivalent to install the damper pulley onto the crankshaft.

OIL SEAL REPLACEMENT

6 Cylinder Engine

1. Disconnect the negative battery cable and drain the cooling system.
2. Remove the radiator and fan assembly. Remove the drive belt(s) from the damper pulley.
3. Remove the damper from the crankshaft using puller tool C-3732A, or equivalent.
4. Using a small pry bar, pry the seal from the timing cover, being

careful not to damage the sealing surface of the cover.
5. To install the new oil seal, install the threaded shaft of the seal installation tool No. C-4251 into the threads of the crankshaft and place the seal with the spring facing the engine.
6. Using an installation adapter tool No. C-4251-2, place the thrust bearing/nut on the shaft and tighten until the tool is flush with the timing chain cover.
7. Lubricate the seal lip with Lubriplate®, then install the damper using tool No. C-3732A.
8. To complete the installation, reverse the removal procedures.

V8 Engines

1. Disconnect the negative battery cable and drain the cooling system. Remove the fan shroud bolts.
2. Remove the alternator and the power steering pump (if equipped) belts. Remove the fan and the fan shroud.
3. Remove the damper pulley bolt and washer from the crankshaft. Install the bar from tool set No. C-3688 and the screw from tool set No. C-3732A, then pull the vibration damper from the end of the crankshaft.
4. Using a small pry bar, carefully pry the seal from the timing cover without scratching the sealing surface of the cover.
5. To install the seal, lubricate the seal lip with Lubriplate®, place it on the crankshaft with the spring side facing the engine. Mount the threaded shaft of the seal installation tool No. C-4251 into the threads of the crankshaft.

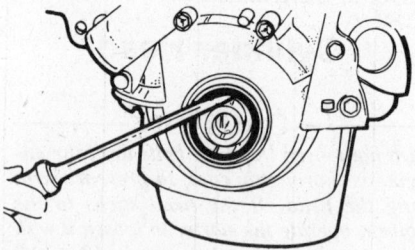

Oil seal removal for V8 6 cyl. similiar

6. Using an installation adapter tool No. C-4251-3, place the thrust bearing and nut on the shaft and tighten until the tool is flush with the timing chain cover.
7. To complete the installation, reverse the removal procedures. Torque the damper bolt to 135 ft. lbs. and the pulley bolts to 17 ft. lbs.

Timing Chain

REMOVAL & INSTALLATION

6 Cylinder Engines

1. Refer to the "Front Cover, Removal and Installation" procedures in this section and remove the front cover.

2. Turn the crankshaft to align (facing each other on the center line of the sprockets) the timing marks, the crankshaft sprocket with the timing mark on the camshaft sprocket.

3. Remove the camshaft sprocket bolt, the camshaft sprocket and crankshaft sprocket with the timing chain.

4. To install the timing chain, place the camshaft and crankshaft sprockets on a flat surface, with the timing indicators (facing each other) on an imaginary centerline through both sprocket bores. Place the timing chain around both sprockets. Be sure the timing marks are in alignment.

5. Align the sprockets with the keyway location in the crankshaft sprocket and the keyway or dowel hole in the camshaft sprocket.

6. Slide both sprockets evenly onto their respective shafts, while keeping the sprockets tight against the chain in the correct position.

NOTE: Using a straightedge, measure the alignment of the sprocket timing marks; they must be perfectly aligned.

7. To complete the installation, reverse the removal procedures. Use tool No. C-3732A to press the damper pulley onto the crankshaft. Torque the camshaft sprocket bolt to 35 ft. lbs. Apply an ⅛ in. bead of sealer to the new gasket. Torque the oil pan and timing cover bolts to 17 ft. lbs. Refill the cooling system.

V8 Engines

1. Refer to the "Front Cover, Removal and Installation" procedures in this section and remove the front cover.

NOTE: Before removing the timing chain, rotate the engine to align the timing mark of the camshaft sprocket with the timing mark of the crankshaft sprocket. The timing marks must face each other on the center line between the two sprockets.

2. Remove the camshaft sprocket lockbolt, cup washer and fuel pump eccentric. Remove the timing chain with both sprockets.

3. To install the timing chain,

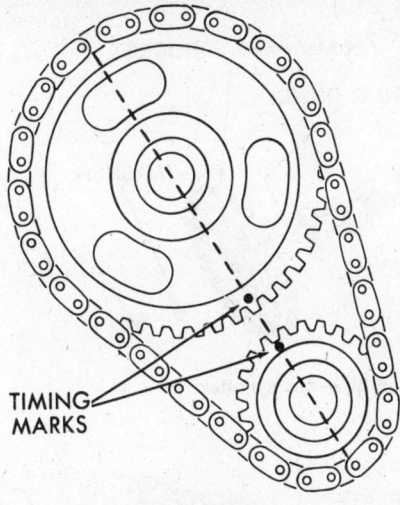

Alignment of timing marks—6 cylinder

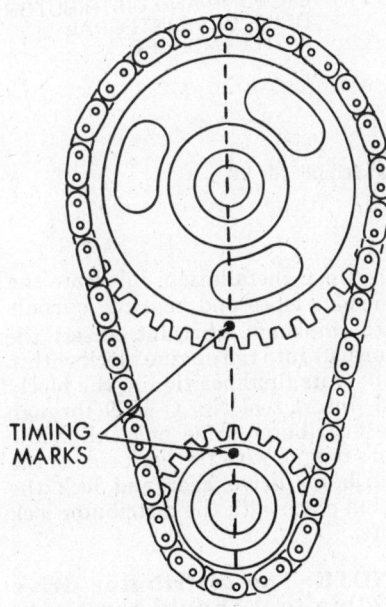

Alignment of timing marks—V8

place the camshaft and crankshaft sprockets on a flat surface, with the timing indicators (facing each other) on an imaginary centerline through both sprocket bores. Place the timing chain around both sprockets. Be sure the timing marks are in alignment.

4. Align the sprockets with the keyway location in the crankshaft sprocket and the keyway or dowel hole in the camshaft sprocket.

5. Slide both sprockets evenly onto their respective shafts, while keeping the sprockets tight against the chain in the correct position.

NOTE: Using a straightedge, measure the alignment of the sprocket timing marks; they must be perfectly aligned.

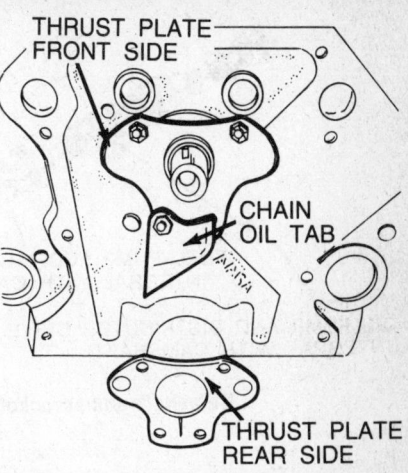

Timing chain oil tab installation, V8s

6. To complete the installation, reverse the removal procedures. To install the damper pulley, use tool No. C-3688. Torque the camshaft bolt to 35 ft. lbs., the damper pulley bolt to 135 ft. lbs., the timing chain cover to 30 ft. lbs. and the damper pulley bolts to 17 ft. lbs.

NOTE: If the camshaft end play exceeds 0.010 in., install a new thrust plate, it should be 0.002–0.006 in. with the new plate.

Camshaft

REMOVAL & INSTALLATION

NOTE: Whenever a new camshaft and/or new tappets are installed, the manufacturer recommends that 1 qt. of crankcase conditioner should be added to the engine oil to aid break-in. This oil mixture should be left in the engine for a minimum of 500 miles. Chrysler recommends that the engine be removed from the vehicle before removing the camshaft. However, in some cases it may be possible to remove the camshaft from the engine, with the engine still in the vehicle, by removing the radiator and grille, then sliding the camshaft out through the front of the vehicle.

6 Cylinder Engines

1. Refer to the "Cylinder Head, Removal and Installation" and the "Timing Chain, Removal and Installation" procedures in this section and remove the cylinder head and the timing chain from the engine.

2. Remove the valve tappets, keeping them in order to insure installation in their original locations.

3. Remove the distributor, the oil pump and the fuel pump.

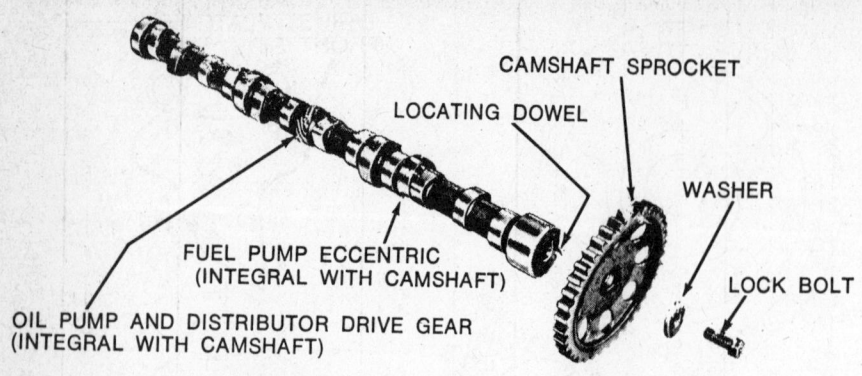

CAMSHAFT SPROCKET

LOCATING DOWEL

WASHER

LOCK BOLT

FUEL PUMP ECCENTRIC
(INTEGRAL WITH CAMSHAFT)

OIL PUMP AND DISTRIBUTOR DRIVE GEAR
(INTEGRAL WITH CAMSHAFT)

Camshaft and sprocket assembly—six cylinder

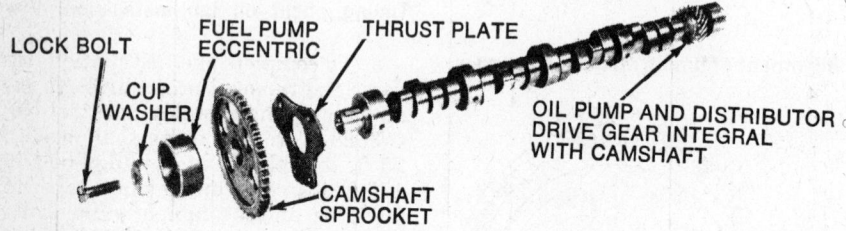

LOCK BOLT

CUP
WASHER

FUEL PUMP
ECCENTRIC

THRUST PLATE

OIL PUMP AND DISTRIBUTOR
DRIVE GEAR INTEGRAL
WITH CAMSHAFT

CAMSHAFT
SPROCKET

Camshaft and sprocket assembly for V8

4. Fit a long bolt into the front of the camshaft to facilitate the camshaft removal. Remove the camshaft, being careful not to damage the cam bearings with the cam lobes.

5. Before installation, lubricate the camshaft lobes and bearing journals. Inspect the crowns of all the tappet faces with a straightedge. Replace any tappets that have dished or worn surfaces.

6. To complete the installation, reverse the removal procedures.

V8 Engines

1. Refer to the "Cylinder Head, Removal and Installation" and the "Timing Chain, Removal and Installation" procedures in this section and remove the cylinder head and the timing chain from the engine.

NOTE: When removing the rocker arm/shaft assemblies, pushrods and valve tappets, keep them in order to ensure reinstallation in their original locations.

2. Remove the distributor and lift out the oil pump driveshaft.

3. Remove the camshaft thrust plate (note the location of the oil tab).

4. Fit a long bolt into the front of the camshaft. Remove the camshaft, being careful not to damage the cam bearings with the cam lobes.

5. Upon installation, lubricate the camshaft lobes and bearing journals with camshaft lubricant. Insert the camshaft into the engine block within 2 in. of its final position in the block.

6. Insert tool No. C-3509 through the distributor drive hole, align the tool's tongue with the back side of the distributor drive gear and lock the tool in place with the distributor lock plate.

NOTE: The distributor drive holding tool should remain in place until the camshaft and crankshaft sprockets and the timing chain have been installed.

7. Install the camshaft thrust plate. Make sure the tang is in the lower right hole of the plate. Torque the thrust plate nuts to 17 ft. lbs. and the camshaft sprocket bolt to 35 ft. lbs. The top edge of the chain oil tab must be flat against the plate. If the camshaft end play exceeds 0.010 in., install a new thrust plate; the new plate clearance should be 0.002–0.006 in.

NOTE: Before installing the lifters, inspect the crowns of the lifter faces with a straightedge; replace any that have dished or worn surfaces.

8. To complete the installation, lubricate all of the parts, reverse the re-moval procedures and adjust the timing.

Piston & Connecting Rod Positioning

For all piston and connecting rod overhaul procedures, refer to "Engine Rebuilding" in the Unit Repair Section.

Upon installation of the piston assemblies, stagger the compression ring gaps, so that they do not line up with the oil ring gaps, as follows:

• On the 6 cyl. engines, rotate the oil ring expander so that the ends are at the right-side of the engine. Rotate the steel rails so the gaps are approximately opposite and positioned above the wrist pin holes.

• On the V8 engines, make sure the oil ring expander ends are butted and the rail gap ends are located as shown in the illustration.

• On all 6 cyl. engines, the squirt hole in the connecting rod should face forward.

• On the V8 engines, a "V" groove is cut into the parting bearing face. When installing, make sure that the groove in the rod aligns with the groove in the cap (the groove provides for lubrication of the cylinder wall in the opposite bank).

• When installing the rod cap to the connecting rod, torque the nuts to 45 ft. lbs.

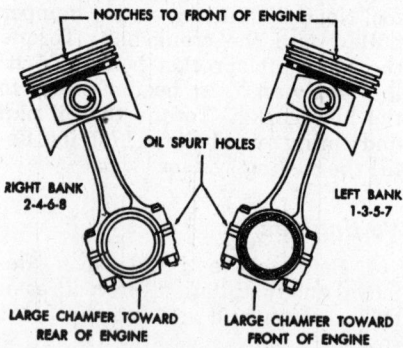

NOTCHES TO FRONT OF ENGINE

OIL SPURT HOLES

RIGHT BANK
2-4-6-8

LEFT BANK
1-3-5-7

LARGE CHAMFER TOWARD
REAR OF ENGINE

LARGE CHAMFER TOWARD
FRONT OF ENGINE

V8 piston and connecting rod assembly

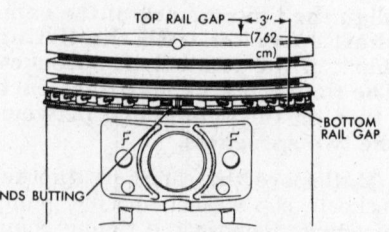

TOP RAIL GAP — 3"
(7.62 cm)

BOTTOM
RAIL GAP

ENDS BUTTING

Proper oil ring installation

INDENT—ASSEMBLY TOWARDS FRONT OF ENGINE

OIL HOLE— ASSEMBLE TOWARDS FRONT OF ENGINE

Piston and connecting rod assembly for 6 cyl.

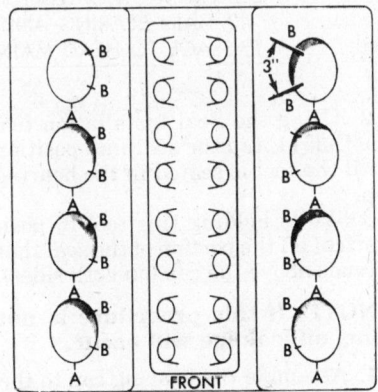

FRONT

TOP VIEW OF BLOCK

A-EXPANDER GAPS   B-RAIL GAPS

Correct piston ring arrangement in cylinder bores, V8s. Slant Six similar

ENGINE LUBRICATION

Oil Pan

REMOVAL & INSTALLATION

6 Cylinder Engines

1. Disconnect the negative battery cable. Remove the engine dipstick.

2. Remove the radiator shroud attaching screws and move the shroud rearward on the engine.

3. Jack up the vehicle and drain the oil. Remove the engine-to-transmission bracket, the exhaust pipe and the torque converter inspection shield, if equipped with an automatic transmission.

4. Remove the steering center link from the steering and idler arm ball joints.

5. Position a jack stand at the right-front corner of the engine oil pan. Be sure not to support the engine at the crankshaft pulley or damper.

6. Remove the front engine mount bolts. Raise the engine about $1\frac{1}{2}$–2 in.

7. Remove the oil pan bolts, rotate the engine crankshaft to clear the counterweights and remove the oil pan.

8. To install, use a new pan gasket set, apply sealer to the 4 junctions of the gaskets, install the oil pan and torque it to 17 ft. lbs. Make sure the pickup screen contacts the bottom of the pan.

9. To complete the installation, reverse the removal procedures. Torque engine mounts to 75 ft. lbs. and the steering/idler arms to 175 ft. lbs.; be sure to install the cotter pins. Replace the oil, then start and run the vehicle for 5 minutes and check for leaks.

V8 Engines

1. Disconnect the negative battery cable and remove the dipstick.

2. Raise and support the front of the vehicle on jackstands, then drain the oil. Remove the torque convertor-to-engine left housing strut, if equipped.

3. Disconnect the steering center link from the steering and idler arms on all vehicles except: 1980–82 Aspen and Volare; 1980–81 LeBaron; 1980–82 Chrysler and Diplomat; 1980–84 Cordoba and Mirada; 1981–83 Imperial; 1982–84 New Yorker, Fifth Avenue and Gran Fury.

4. Disconnect the exhaust pipes from the manifolds and secure them out of the way. On the models mentioned in Step 3, remove the starter, starter mounting stud (if equipped) and the torque converter inspection plate, if equipped with an automatic transmission.

5. Check to see if there is sufficient clearance to reach all of the oil pan bolts; if not, it will be necessary to raise the engine about $1\frac{1}{2}$–2 in. To do this, remove the motor mounts and raise the engine only until the bolts become accessible.

6. Remove the distributor cap for clearance. Remove the oil pan bolts, rotate the engine crankshaft to clear the counterweights and remove the oil pan with a twisting motion.

NOTE: On some models, you may have to unbolt the transmission until the pan clears.

7. When installing the oil pan, be sure the oil strainer is parallel to and in contact with the oil pan bottom. Apply sealer to the corner junctions of the cork and rubber gaskets. The side gaskets should overlap the rear seal. Torque the oil pan bolts to 17 ft. lbs.

8. To complete the installation, reverse the removal procedures. Torque the engine mount bolts to 75 ft. lbs. Install the engine-to-converter housing strut (if equipped).

Rear Main Oil Seal

REMOVAL & INSTALLATION

Split Type

Service replacement seals are of split, rubber-type composition. This type of seal makes it possible to replace the upper half of the rear main oil seal without removing the engine from the vehicle or the crankshaft from the engine. When installing rubber seals, they must be replaced as a set and cannot be combined with the rope type rear main seal. The following procedure is for removing the rope type seal and replacing it with the rubber type seal.

1. Remove the oil pan.

2. Remove the rear seal retainer and the rear main bearing cap.

3. Remove the lower rope seal by carefully prying from the side with a small screwdriver.

4. To remove the upper rope seal, drive the exposed end of the seal with a 6 in. piece of $\frac{3}{16}$ in. diameter brazing rod. When the opposite end of the seal starts to protrude from the block, have an assistant grasp it with pliers and gently pull it from the block while the opposite end is being driven. There are also screw type extractor tools available.

5. Before installing the seal, clean and lightly oil the crankshaft and new seal.

6. To ease the installation, loosen all the main bearing caps slightly to lower the crankshaft.

—— CAUTION ——

DO NOT allow the crankshaft to drop enough to let the main bearings become displaced on the crankshaft.

7. Hold the seal tightly against the crankshaft with thumb pressure (with paint stripe to the rear) and in-

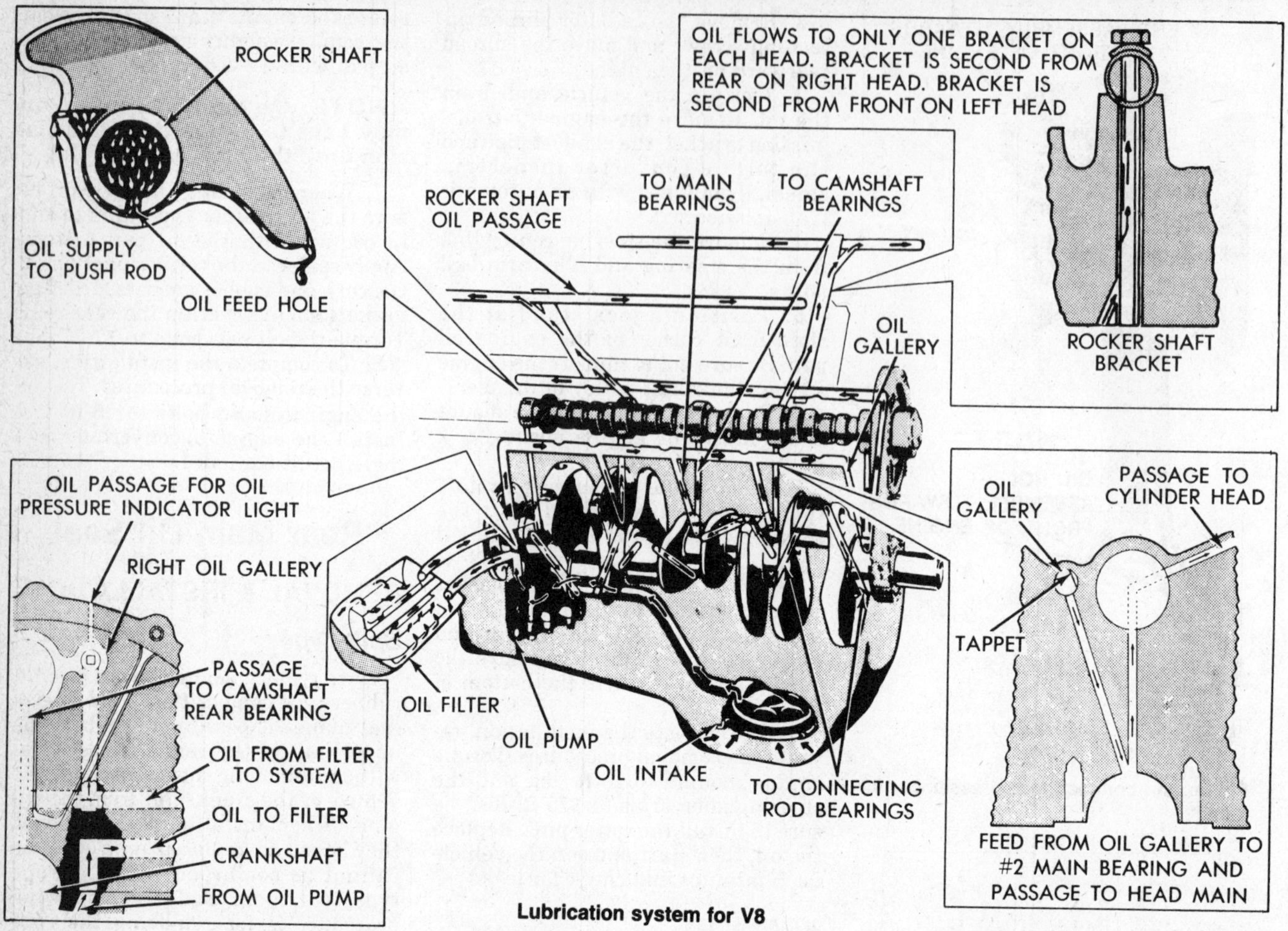

ROCKER SHAFT

OIL SUPPLY TO PUSH ROD

OIL FEED HOLE

OIL FLOWS TO ONLY ONE BRACKET ON EACH HEAD. BRACKET IS SECOND FROM REAR ON RIGHT HEAD. BRACKET IS SECOND FROM FRONT ON LEFT HEAD

ROCKER SHAFT OIL PASSAGE

TO MAIN BEARINGS

TO CAMSHAFT BEARINGS

OIL GALLERY

ROCKER SHAFT BRACKET

OIL PASSAGE FOR OIL PRESSURE INDICATOR LIGHT

RIGHT OIL GALLERY

PASSAGE TO CAMSHAFT REAR BEARING

OIL FROM FILTER TO SYSTEM

OIL TO FILTER

CRANKSHAFT

FROM OIL PUMP

OIL FILTER

OIL PUMP

OIL INTAKE

TO CONNECTING ROD BEARINGS

OIL GALLERY

PASSAGE TO CYLINDER HEAD

TAPPET

FEED FROM OIL GALLERY TO #2 MAIN BEARING AND PASSAGE TO HEAD MAIN

Lubrication system for V8

stall the seal in the block groove. Rotate the crankshaft (if necessary) while installing the seal in the groove. Make sure the sharp edges on the block groove, do not cut or nick the rear of the seal.

8. Install the lower half of the seal (with paint stripe to the rear) into the lower seal retainer. On the 5.2L engine, insert the cap seals into the slots in the bearing cap; the one with the yellow paint goes on the right side. Be sure the narrow edge is facing up. Pull outward on the small end of the seal until its edge lines up with the shoulder. On all engines, lightly oil the seal lips before installation.

9. To complete the installation, install the rear main bearing cap, torque the main bearing caps to 85 ft. lbs. and reverse the removal procedures.

NOTE: Before tightening the main bearing caps, make sure all the main bearings are located in their proper positions.

Rope Type

To perform this procedure, the crankshaft must be removed from the engine.

UPPER SEAL

1. Install the new oil seal in the cylinder block so that the ends of the seal protrude.

2. Using the Seal Installation tool No. C-3511, tap the new seal into position until the tool is seated in the bearing bore.

3. While holding this tool in position, cut off the portion of the seal that extends below the block (on both sides).

LOWER SEAL

1. Install the new oil seal in the bearing cap so that the ends of the seal protrude.

2. Using the Seal Installation tool No. C-3511, tap the seal into position until the tool is seated in the bearing cap.

3. While holding this tool in position, cut off the portion of the seal that extends above the cap (on both sides).

NOTE: If this procedure is not done, oil leakage will occur.

4. Assemble the bearing cap to the

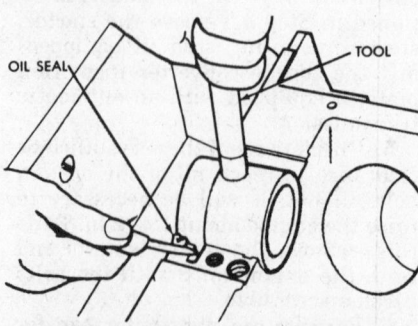

OIL SEAL

TOOL

Installing the seal to the rear main bearing

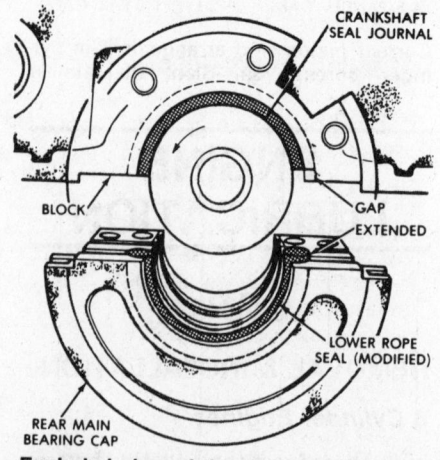

CRANKSHAFT SEAL JOURNAL

BLOCK

GAP

EXTENDED

LOWER ROPE SEAL (MODIFIED)

REAR MAIN BEARING CAP

Exploded view of the rear main bearing oil seal

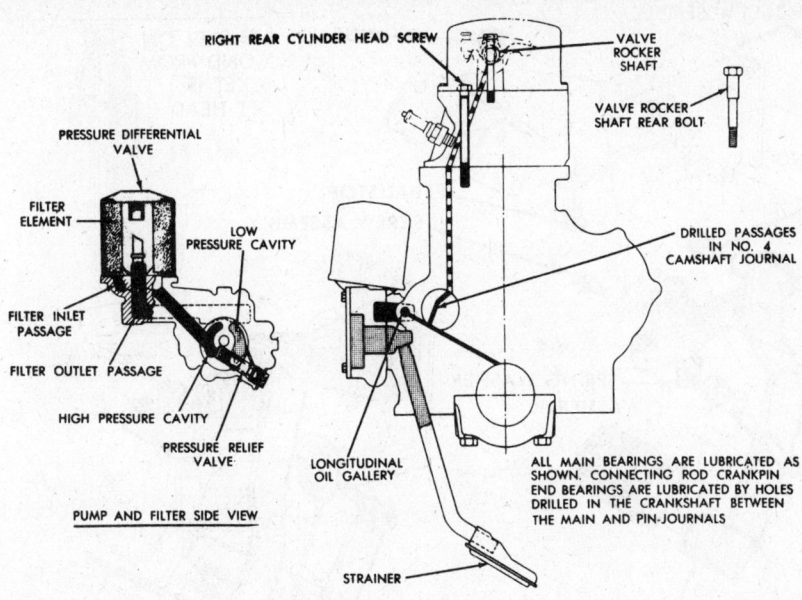

Six-cylinder lubrication system

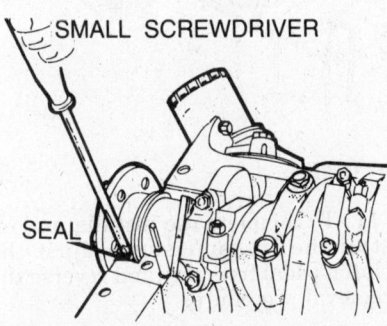

Removing upper main bearing oil seal with screwdriver. View showed with engine out of car.

cylinder block. Torque the bearing cap bolts to 85 ft. lbs.

5. Apply sealer to the bearing cap-to-engine block joints.

6. To complete the installation, reverse the removal procedures.

Oil Pump

REMOVAL & INSTALLATION

6 Cylinder Engines

1. Remove the negative battery cable, the fan shroud screws and the fan shroud, push the fan shroud rearward.

2. Raise vehicle on hoist, support front of engine with jackstand placed under right-front corner of oil pan and remove engine mount bolts. Do not support engine at crankshaft pulley or damper.

3. Raise the engine approximately 1½–2 in.

4. Remove the oil filter, oil pump mounting bolts, oil pump assembly and the gaskets from the mounting surfaces.

5. To install, prime the oil pump and reverse the removal procedures. Torque the oil pump bolts to 17 ft. lbs.

V8 Engines

1. Remove the oil pan.

2. Remove oil pump mounting bolts and the oil pump from the rear main bearing cap.

3. To install, prime the pump with engine oil and reverse the removal procedures. Torque the oil pump bolts to 30 ft. lbs.

CLUTCH

REMOVAL

1. Refer to the "Manual Transmission Removal & Installation" procedures in this section and remove the transmission.

2. Remove the clutch housing pan.

3. Disconnect the fork return spring from the clutch housing and release fork.

4. Remove the spring washer holding the fork rod to the torque shaft lever pin. Remove the fork rod assembly from the torque shaft and the release fork.

5. Remove the sleeve assembly and clutch release bearing from the clutch release fork. Remove the release fork and boot from the clutch housing.

6. Punch-mark the clutch cover and flywheel so they may be installed in the same position.

7. Loosen the clutch cover attaching screws 1–2 turns at a time, in rotation, to avoid bending the cover.

8. Remove the clutch assembly. Be careful not to contaminate the clutch with grease or oil.

INSTALLATION

1. Lightly lubricate the drive pinion bushing in the end of the crankshaft. Use about ½ teaspoon of long-life chassis grease. Lubricant should be inserted in the cavity in front of the bushing. Also coat the inner surface of the bushing with a light film of grease.

2. Thoroughly clean the surfaces of the flywheel and pressure plate with fine sandpaper. All oil or grease must be removed at this time.

3. Position the clutch disc, pressure plate and cover in the mounting position. Springs on disc damper must be facing away from the flywheel. Do not touch the disc face and be careful to avoid disc contact with oil or grease. Insert a clutch disc aligning arbor or suitable substitute (such as a spare transmission input shaft) through the disc hub and into the bushing.

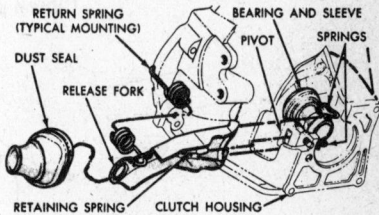

Clutch release fork, bearing and sleeve

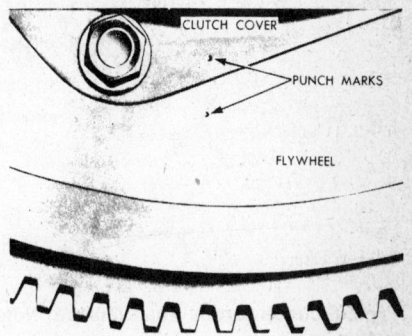

Marking clutch and flywheel

4. Align the punch marks that were made at removal. Install the clutch cover bolts, but do not tighten them.

5. Tighten all the bolts a few turns at a time in an alternate sequence. Torque ⁵⁄₁₆ in. bolts to 17 ft. lbs. and ³⁄₈ in. bolts to 30 ft. lbs. Remove the alignment tool.

6. Apply a light film of high temperature grease to the release fork

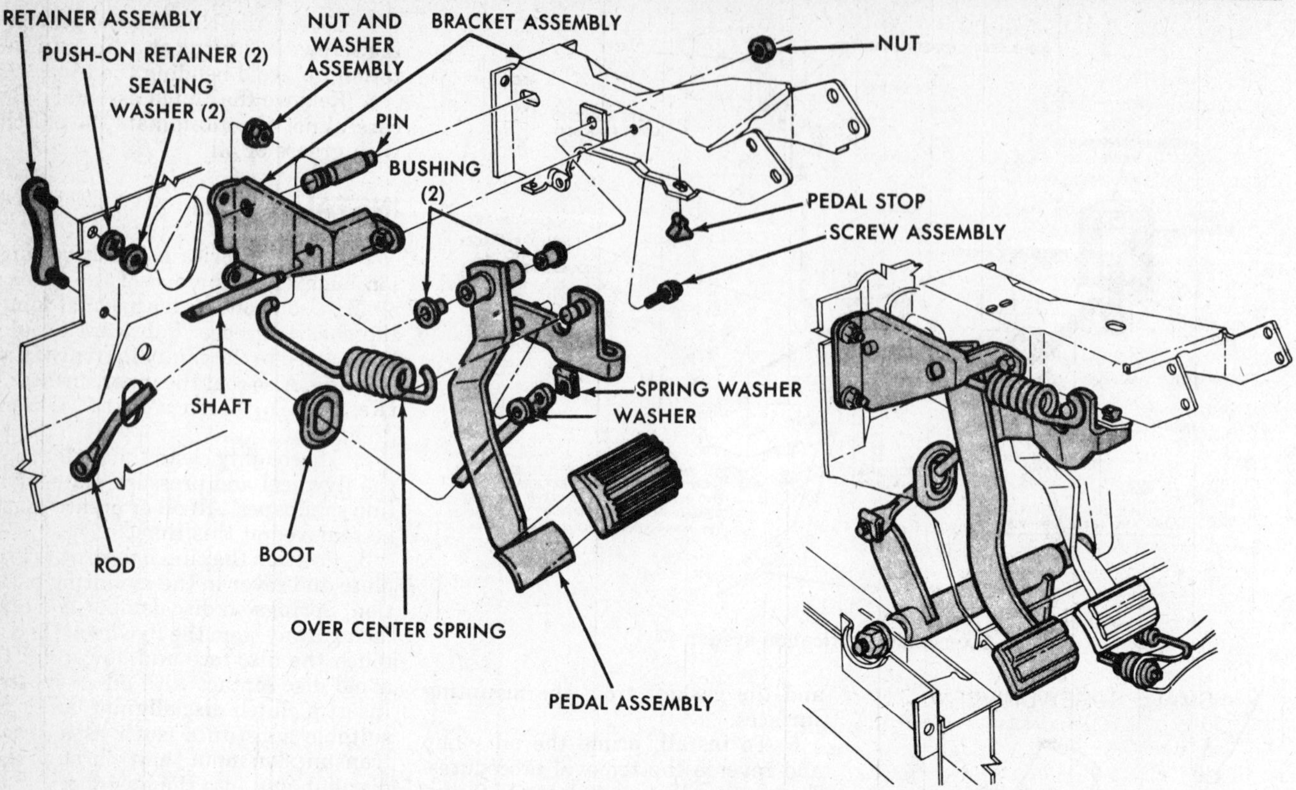

RETAINER ASSEMBLY
PUSH-ON RETAINER (2)
SEALING WASHER (2)
NUT AND WASHER ASSEMBLY
BRACKET ASSEMBLY
NUT
PIN
BUSHING (2)
PEDAL STOP
SCREW ASSEMBLY
SHAFT
SPRING WASHER
WASHER
ROD
BOOT
OVER CENTER SPRING
PEDAL ASSEMBLY

Clutch pedal and linkage for Aspen, Diplomat, LeBaron and Volare

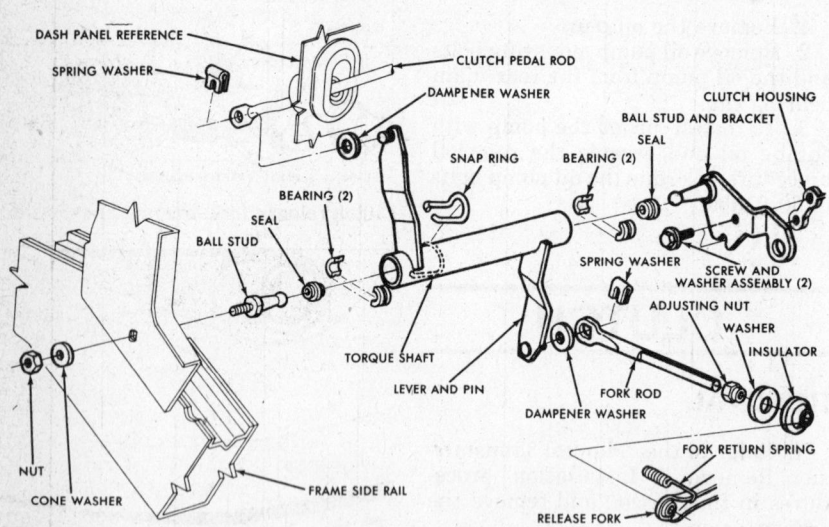

DASH PANEL REFERENCE
SPRING WASHER
CLUTCH PEDAL ROD
DAMPENER WASHER
CLUTCH HOUSING
BALL STUD AND BRACKET
SEAL
SNAP RING
BEARING (2)
BEARING (2)
SEAL
BALL STUD
SPRING WASHER
SCREW AND WASHER ASSEMBLY (2)
ADJUSTING NUT
WASHER
INSULATOR
TORQUE SHAFT
LEVER AND PIN
FORK ROD
DAMPENER WASHER
FORK RETURN SPRING
NUT
CONE WASHER
FRAME SIDE RAIL
RELEASE FORK

Typical clutch torque shaft linkage. Note clutch adjusting nut on right side of illustration

pads on the sleeve. Lightly lubricate the fork fingers and retaining spring.

7. Insert the fork fingers under the clutch sleeve retaining springs, while at the same time engaging the fork retaining spring into the fork pivot. The retaining springs on the sleeve must have lateral freedom.

8. Make sure that the groove in the seal is properly seated in the seal opening flange in the clutch housing.

9. Insert the threaded end of the fork rod assembly into the release fork. Install the washer so that the curved surface will lock the adjusting nut in the tapered hole. Install the fork rod to the torque shaft lever and lock it with a spring washer.

10. Install the fork return spring between the release fork and the clutch housing.

NOTE: When installing the transmission, be sure not to allow grease to get on the splines or pilot end of the transmission drive pinion.

11. To complete the installation, install the transmission, adjust the clutch pedal free-play and reverse the removal procedures.

Linkage Free-Play

ADJUSTMENT

Adjust the fork rod by rotating the self-locking nut to provide $5/32$ in. free-play at the fork end. This adjustment will result in the proper 1 in. free-play at the clutch pedal.

MANUAL TRANSMISSION

A fully synchronized 3-speed A-230 (standard) and a fully synchronized 4-speed A-833/overdrive (optional) transmissions are available for all vehicles. All of the transmissions have a serial number stamped on a pad on the right side of the case.

For the manual transmission overhaul procedures, please refer to "Manual Transmission Overhaul" in the Unit Repair section.

REMOVAL & INSTALLATION

1. Raise the vehicle, support it on jackstands and drain the transmission fluid.

2. Disconnect the negative battery terminal and remove the shift levers from the transmission.

3. On floor shifter models, remove the floor pan retaining screws and the boot. Remove the retaining clips, washers and shift rods from the shift unit levers. Unbolt and remove the shift unit from the extension housing.

4. After marking both parts for re-assembly, detach the driveshaft and the rear universal joint. Carefully pull the shaft yoke out of the transmission extension housing.

--- CAUTION ---

Do not nick or scratch the machined surface on the sliding spline yoke.

5. Remove the back-up light switch and the speedometer pinion retainer with the cable housing attached. When removing the speedometer, carefully work the adapter and pinion out of the extension housing.

6. Install an engine support fixture tool No. C-3487-A, or equivalent, with the support ends against the underside of the oil pan flange. Using the support fixture, raise the engine slightly.

7. Unfasten the extension housing from the center crossmember. Support the transmission on a jack and remove the center crossmember.

8. Remove the bolts which secure the transmission to the clutch housing.

9. Slide the transmission back until the input shaft clears the clutch disc. Lower the transmission and remove it from the vehicle.

10. To install, lubricate the input shaft pilot bearing in the flywheel and the bearing retainer pilot (for the clutch release sleeve). Do not lubricate the clutch splines or clutch release levers.

11. Position the transmission so the drive pinion is centered in the clutch housing bore. Push the transmission forward until the pinion shaft enters the clutch disc. Place the transmission in gear. Twist the output shaft until the splines are in alignment. Push the transmission forward until it is seated against the clutch housing.

--- CAUTION ---

Once the transmission pinion is inside the clutch, support the transmission with a jack, so that the weight is not placed on the input shaft.

12. When the transmission is in po-

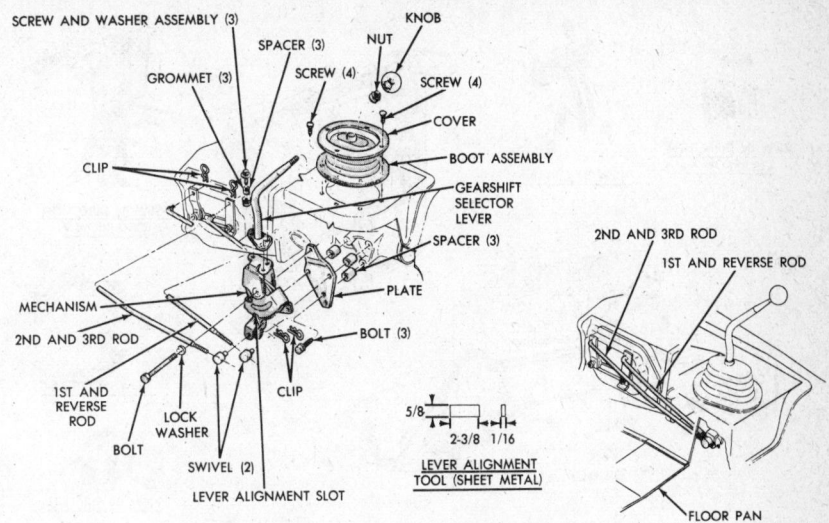

Typical three-speed floor shift linkage

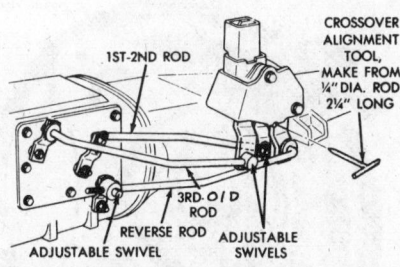

Three speed overdrive shift linkage adjustment

sition, replace the housing bolts and torque to 50 ft. lbs. Using a drift, align the crossmember bolt holes, install them and torque to 50 ft. lbs.

13. Remove the engine support fixture and tighten the engine mount-to-crossmember bolt. Install the gearshift linkage and make adjustments. Connect the driveshaft and universal joints. Fill the transmission with fluid.

14. To complete the installation, reverse the removal procedures.

FLOORSHIFT LINKAGE ADJUSTMENT

1. For the 3-speed, fabricate a $\frac{5}{8}$ in. x $2\frac{3}{8}$ in. x $\frac{1}{16}$ in. alignment tool, out of metal. For the 4-speed, use $2\frac{1}{4}$ in. x $\frac{1}{4}$ in. dia. rod or drill bit.

2. Remove the clips, washers and swivel adjusting ends of the shift rods from the shifter unit levers.

3. From under the vehicle, insert the alignment tool through the slots or holes in the shifter levers and the shifter unit, to hold the levers in the neutral-crossover position.

4. Place the shift levers (on the transmission side cover) in the Neutral or middle position.

5. Adjust the swivels so that they can be installed freely in the shift unit lever holes (on the 4-speed, start with the 1-2 rod). Install the shifting rods with the proper washers and clips.

6. Remove the alignment tool and check the shifting action.

AUTOMATIC TRANSMISSION

Three Torqueflite transmission models are used. The model may be identified by the part number, which is stamped on a pad on the left side of the case pan flange. The A-727 transmission has a more gradual slope to the converter housing than does the A-904. The earlier vehicles use A-904 model, while the later vehicles use the A-904 LA model transmission, which includes slight modifications in the number of clutch plates and discs, added for strength. However, all adjustments are the same as for the A-904.

For pan removal and all maintenance procedures, please refer to "Automatic Transmission" in the Unit Repair section.

REMOVAL

NOTE: The transmission and converter must be removed as an assembly; otherwise, the converter drive plate, pump bushing and/or oil seal may be damaged. The drive plate will not support a load;

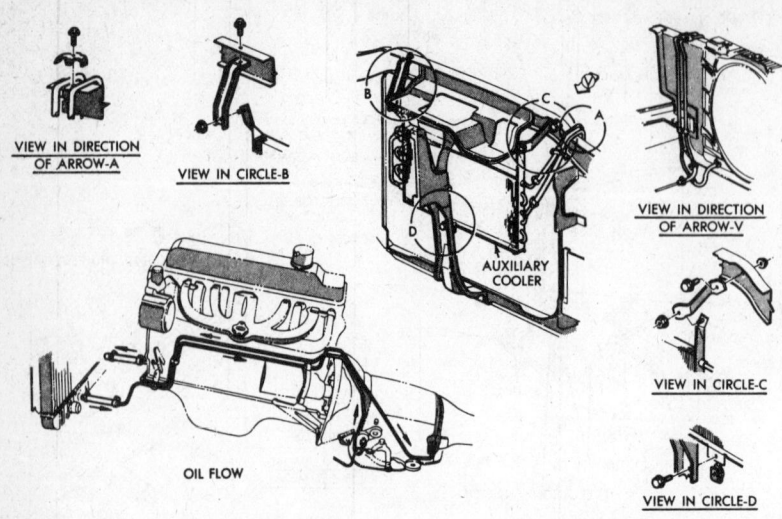

VIEW IN DIRECTION OF ARROW-A

VIEW IN CIRCLE-B

AUXILIARY COOLER

VIEW IN DIRECTION OF ARROW-V

VIEW IN CIRCLE-C

VIEW IN CIRCLE-D

OIL FLOW

Oil flow—transmission oil coolers, Slant Six engine

do not allow the weight of the transmission to rest on the plate during removal. Also, removal and installation will vary slightly from this procedure by vehicle and model.

1. Disconnect the negative battery cable. Raise and support the vehicle on jackstands. A hydraulic floor jack should be used for removal of the transmission unit.

2. Some models may require that the exhaust system be dropped. If necessary, disconnect the exhaust pipe from the exhaust manifold and all of the exhaust system hangers from the vehicle.

3. If equipped, remove the engine-to-transmission struts. Remove and plug the oil cooler lines at the transmission.

4. Remove the starter motor, oil cooler line bracket and the converter access cover.

5. Using a socket wrench on the damper bolt, rotate the engine clockwise to bring the converter drain plug to the bottom. Drain the converter, loosen the pan (break it loose), drain the transmission and reinstall the pan.

6. Match mark the converter and drive plate to aid in later installation. The crankshaft flange bolt circle, the inner/outer circle of holes in the drive plate and the 4 tapped holes in the front face of the converter all have 1 hole offset, so that these parts will be installed in their original positions. This maintains the balance of the engine and converter.

7. Rotate the engine clockwise with the socket wrench on the damper bolt (to position the bolts) and remove the torque converter-to-drive plate bolts.

8. Matchmark the drive shaft at

DRAIN PLUG

1/8 INCH HOLE

Converter and drive plate markings

the rear universal joint for reassembly purposes and disconnect it. Carefully pull the shaft assembly out of the extension housing.

9. Disconnect the wire connector from the back-up light and neutral starting switch.

10. Disconnect the gearshift rod and torque shaft assembly from the transmission

NOTE: When disassembling the linkage rods from the levers, equipped with plastic grommets as retainers, replace the grommets. Use a small pry bar to pry the rod from the grommet; then, cut away the old grommet. Use pliers to snap the new grommet into the lever and the rod into the grommet.

11. Disconnect the throttle rod from the lever at the left side of the transmission. If equipped, remove the linkage bellcrank from the transmission.

12. Remove the oil filer tube and the

speedometer cable. Support the rear of the engine with an engine support fixture tool No. C-3487-A, wooden blocks or jackstands.

13. Using a floor jack, raise the transmission slightly to relieve the load on the support.

14. Remove the transmission-to-crossmember and the crossmember-to-frame bolts, then remove the crossmember.

NOTE: Vehicles with longitudinal torsion bars (in line with the chassis) have a torsion bar anchor crossmember, which remains in place, requiring a careful downward tilt of the front of the transmission as it is being removed. If a dampening weight is bolted to the rear of the extension housing (the long tapered housing on the rear of the transmission), it must be removed to provide additional clearance. These vehicles also have access holes through the crossmembers for the 3 bolts which retain the weight mounting bracket on the extension housing.

15. Remove the bellhousing-to-engine bolts. Carefully work the transmission/con-verter assembly rearward, off the engine block dowels and disengage the converter hub from the crankshaft. Attach a small C-clamp to the edge of the bellhousing to hold converter in place during transmission removal.

16. Lower the transmission and remove it from under the vehicle.

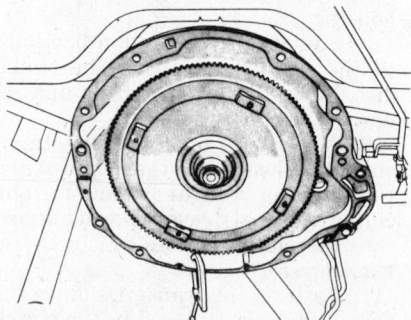

Attach a C-clamp to automatic transmission bell housing to keep torque converter in place when removing engine

INSTALLATION

NOTE: The transmission and converter must be installed as an assembly; otherwise the converter drive plate, pump bushing, and oil seal will be damaged. The drive plate will not support a load, so none of the transmission weight should be allowed to rest on the plate during installation.

1. Coat the crankshaft converter hub hole with multi-purpose grease. Place the transmission/converter assembly on the floor jack and position the assembly under the vehicle. Raise or tilt as necessary until the transmission is aligned with the engine.

2. Rotate the converter so that the mark on the converter (made during removal) will align with the mark on the drive plate. The offset holes in the plate are located next to the $\frac{1}{8}$ in. hole in the inner circle of the drive plate. Carefully work the transmission assembly forward over the engine block dowels, with the converter hub entering the crankshaft opening.

3. After the transmission is in position, install the converter housing bolts and tighten to 30 ft. lbs. If equipped, install the dampener weight on the rear of the extension housing.

4. Install the crossmember to the frame, lower the transmission and fasten the extension housing to the crossmember. The engine support, jackstands or blocks may now be removed.

5. Install the oil filter tube and speedometer cable. Connect the throttle rod to the transmission lever.

6. Connect the gearshift rod and torque shaft assembly to the transmission lever and frame.

7. Place the wire connector on the combination back-up light and Neutral/Park starter switch.

8. Carefully guide the sliding yoke onto the extension housing, output splines. Align the marks made at removal, then connect the driveshaft to the rear axle pinion shaft yoke.

9. Rotate the engine clockwise to install the converter-to-drive plate bolts, matching the marks made at removal. Torque to 22 ft. lbs.

10. Install the converter access cover, the starter motor and the cooler line bracket. Tighten the oil cooler lines to the transmission fittings.

11. Install the engine-to-transmission struts (if equipped) and replace the exhaust system (if it was disturbed).

12. Adjust the shift and throttle linkages. Refill the transmission with Dexron® II type automatic transmission fluid.

DRIVE AXLE

Driveshaft And U-Joints

The driveshaft is a one-piece tubular

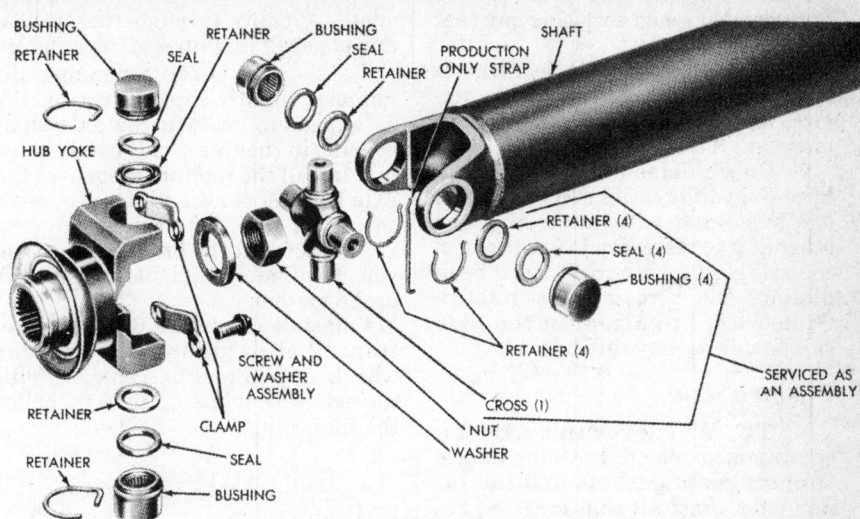

Rear driveshaft universal joint assembly—remove the two clamps to remove driveshaft

shaft with 2 universal joints, one at each end. The front joint yoke serves as a slip yoke on the transmission output shaft. The rear universal joint is of the type that must be unbolted to be removed.

REMOVAL & INSTALLATION

Lubricant loss from the rear of the transmission can be avoided by raising the rear of the vehicle before removing the driveshaft. Keep a clean rag handy to plug the end of the extension housing when the driveshaft is removed.

1. Match-mark the driveshaft, U-joint and pinion flange before disassembly. These marks must be realigned during reassembly to maintain the balance of the driveline. Failure to align them may result in excessive vibration.

2. Remove both clamps from the differential pinion yoke and slide the driveshaft forward slightly to disengage the U-joint from the pinion yoke. Tape the two loose U-joint bearings together to prevent them from falling off.

—— CAUTION ——

Do not disturb the bearing assembly retaining strap. Never allow the driveshaft to hang from either of the U-joints. Always support the unattached end of the shaft to prevent damage to the joints.

3. Lower the rear end of the driveshaft, gently slide the front yoke/driveshaft assembly rearward and disengage the assembly from the transmission output shaft. Be careful not to damage the splines or the surface on which the output shaft seal rides.

4. Check the transmission output shaft seal for signs of leakage.

5. To install, reverse the removal procedures; be sure to align the matchmarks. Torque the U-joint bolts to 14 ft. lbs.

U-JOINT OVERHAUL

For all U-Joint overhaul procedures, please refer to "U-Joint/CV-Joint Overhaul" in the Unit Repair section.

Rear Axle/Axle Shaft

Three different (ring gear diameter) rear axle assemblies are used. These axles can be visually identified as follows:

a. The $7\frac{1}{4}$ in. has 9 bolts and an oval shape.

b. The $8\frac{1}{4}$ in. has a 10 bolt rear cover without a filler plug.

c. The $9\frac{1}{4}$ in. has a 12 bolt cover.

Some vehicles are equipped with the Sure-Trip limited slip differential. Identification of the Sure-Grip rear axle can be made easily by lifting both rear wheels off the ground and turning them. If both rear wheels turn in the same direction simultaneously, the vehicle has the Sure-Grip axle. All axles have a ratio identification tag under one of the cover or carrier bolts.

REMOVAL & INSTALLATION

Because the axle shafts are slightly different from one rear axle assembly to another, individual service procedures are required for each axle shaft assembly. Two very important points to remember when servicing any rear axle assembly are:

1. Always elevate both rear wheels when performing any rear axle service, when using the engine or other means to rotate the axle.

2. On vehicles equipped with a Sure-Grip differential, NEVER rotate one axle shaft without rotating the other. If it is necessary to rotate one of the axle shafts, both shafts must be in position and both must be rotated. Otherwise, alignment of the axle shafts will be very difficult.

7¼ Inch Axle

NOTE: Whenever this axle assembly is serviced, both the brake support plate gaskets and the inner axle shaft oil seal must be renewed. There is no provision for adjusting the axle shaft end-play.

1. Raise and support the rear of the vehicle on jackstands. Remove the rear wheels.

2. Detach the clips securing the brake drum to the axle shaft studs and remove the brake drum.

3. Disconnect and plug the brake lines at the wheel cylinders.

4. Through the access hole in the axle shaft flange, remove the axle shaft retaining nuts.

5. Attach a puller or slide hammer to the axle shaft flange and remove the axle shaft.

6. Remove the brake assembly from the axle housing.

7. Remove the axle shaft oil seal from the axle housing.

— CAUTION —

Never use a torch or other heat source as an aid in removing any axle shaft components as this will result in serious damage to the axle assembly.

8. Place the axle shaft collar in a vise or on an anvil. Using a chisel, cut deeply into the retaining collar at 90° intervals. This will loosen it enough so it can be removed. Using an arbor press, press the bearing from the axle shaft.

9. To assemble, replace the retainer plate, bearing and bearing retainer collar on the axle shaft, use an arbor press to install the bearing retainer on the axle shaft.

10. Lightly grease the outside diameter of a new oil seal and insert it into the axle housing. Using the seal installation tool No. C-3734, or equivalent, seat the seal in the axle housing.

11. Install a new inner gasket on the studs of the axle housing and replace the brake support plate assembly. Refit a new outer gasket onto the axle housing studs.

12. Very carefully slide the axle shaft assembly through the oil seal and engage the splines of the differential side gear. Using a non-metallic hammer, lightly tap the end of the axle shaft to position the axle shaft bearing in the recess of the axle housing. Install the retainer plate over the axle housing studs and torque the securing nuts to 35 ft. lbs.

13. Reconnect the brake lines to the wheel cylinders and bleed the rear brake system.

14. Install the brake drum and retaining clips. Reinstall the rear wheels and lower the vehicle. Refill the axle with lubricant to ⅜ in. below the filler plug.

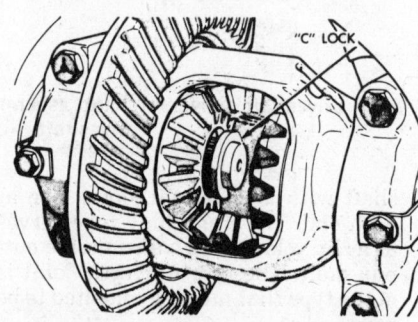

Removing and installing the C-lock on the axle shaft

Removal of differential pinion shaft lock screw on the 8¼ in. rear axle

8¼ Inch & 9¼ Inch Axles

NOTE: There is no provision for adjusting axle shaft end play.

1. Raise and support the rear of the vehicle on jackstands. Remove the wheels.

2. Clean all dirt from the housing cover and remove the cover to drain the lubricant.

3. Remove the brake drum.

4. Rotate the differential case until the differential pinion shaft lockscrew can be removed. Remove the lockscrew and pinion shaft.

5. Push the axle shaft toward the center of the vehicle and remove the C-lock from the groove on the axle shaft.

6. Pull the axle shaft from the housing, being careful not to damage the bearing which remains in the housing.

— CAUTION —

During Sure-Grip axle shaft removal and installation, do not rotate an axle shaft unless both are in position. Rotation of one shaft without the other in place may result in misalignment of the two spline segments with which the axle shaft spline engages and will mean difficult realignment procedures when the shaft is installed.

7. Inspect the axle shaft bearings and replace any doubtful parts. Whenever the axle shaft is replaced, the bearings should also be replaced.

8. Remove the axle shaft seal from the bore in the housing.

9. Using a slide hammer, remove the axle shaft bearing from the housing. DO NOT reuse the seal; always install a new axle shaft seal.

10. Check the bearing shoulder in the axle housing for imperfections. These should be smoothed out with a fine file or polish.

11. Clean the axle shaft bearing cavity.

12. Install the axle shaft bearing in the axle housing. Be sure that the bearing is seated firmly against the shoulder.

13. Install a new axle shaft bearing seal, seat it against the housing flange face.

14. Insert the axle shaft, make sure the splines do not damage the seal. Be sure the splines are properly engaged with the differential side gear splines.

15. Install the C-lock in the grooves on the axle shafts. Pull the shafts outward so the C-locks seat in the counterbore of the differential side gears.

16. Install the differential pinion shaft through the case and pinions. Install the lockscrew and torque to 8 ft. lbs.

17. Clean the housing and gasket surfaces. Install the cover and a new gasket.

NOTE: Replacement gaskets may not be available for differential covers. In this case, use of MOPAR Silicone Rubber Sealant or equivalent. Be sure to replace the rear axle ratio identification tag under one of the cover bolts. Refill the axle with lubricant to ½ in. below the filler plug. MOPAR Hypoid Lubricant & Friction Modifier additive must be used in Sure-Grip limited slip units.

18. Install the brake drum and wheel. Lower the vehicle.

FRONT SUSPENSION

All vehicles in this section utilize the torsion bar type front suspension. Compression type lower ball joints are located in the steering knuckles. When servicing the front suspension, it should be kept in mind that rubber bushings must not be lubricated. Any front suspension part that contains rubber should be tightened with full vehicle weight on the suspension.

Shock Absorber

REMOVAL & INSTALLATION

1. Raise the front of the vehicle (until the wheels clear the ground) and support it on jackstands. Remove the front wheels. Remove the nut and retainer from shock absorber upper end.
2. Grip the shock absorber base, remove the lower attaching nut, retainer and bushing.
3. Fully compress the shock absorber by pushing it upward, disengaging it from the lower control arm. Pull the shock absorber down, firmly and remove it from the vehicle.
4. Check the shock absorber bushings, if they are worn, cracked or scored, replace them. Remove and install the busings with a press or using a drift and a hammer. To ease installation, lubricate with soapy water.

NOTE: Do not use oil to ease the installation.

5. Purge the new shock absorber by repeatedly extending it in the upright position and compressing it in the inverted position. It is normal to have more resistance to extend than to compress.
6. To install, fully compress the new shock absorber, insert the top end through the upper bushing, then install the retainer and nut. Torque the nut to 25 ft. lbs.

NOTE: Be sure all the retainers are installed with the concave side in contact with rubber.

7. Install the shock absorber to the lower control arm mount. Install the bolt (from the rear) or the retainer and nut finger tight. Lower the vehicle and torque the nut to 35 ft. lbs.,

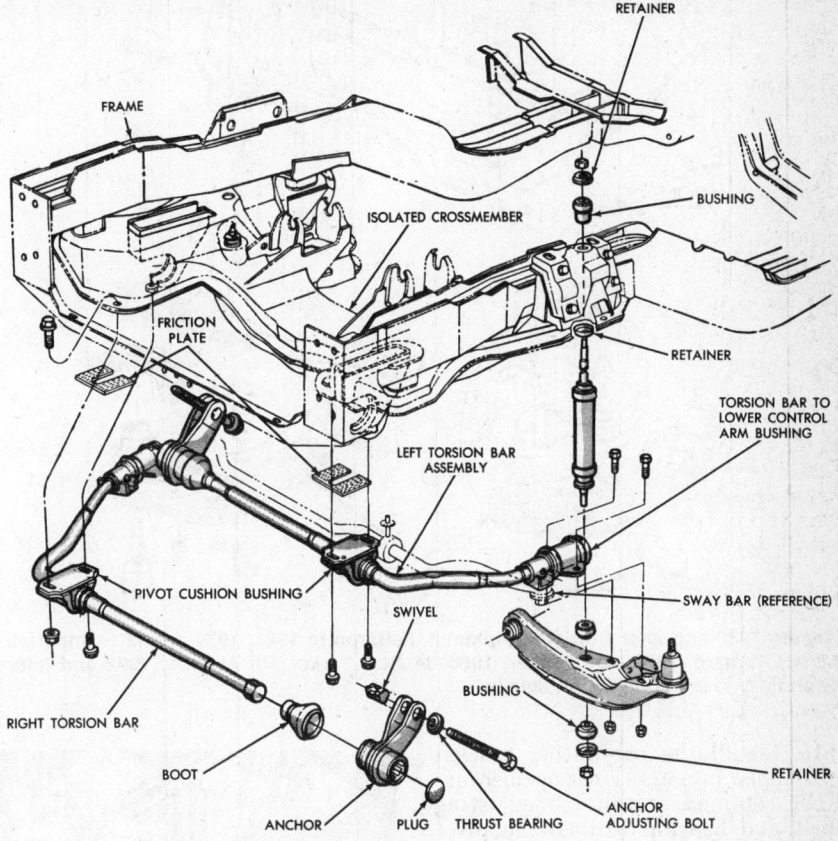

Transverse torsion bar isolated front suspension system

with the full weight of the vehicle on the wheels.

Torsion Bars

REMOVAL & INSTALLATION

The torsion bars are not interchangeable from right to left. Longitudinal bars are marked with an R or an L, according to their location.

1. Raise the vehicle and support it, so that the front suspension is in the rebound position.
2. Release the load on both torsion bars by turning the anchor adjusting bolts counterclockwise.
3. Remove the anchor adjusting bolt on the torsion bar to be removed.
4. Raise the lower control arms until there is $2\frac{7}{8}$ in. clearance between the crossmember ledge, at the jounce bumper and the torsion bar bushing, on the lower control arm.
5. Remove the sway bar-to-lower control arm bolt and retainer. Remove the torsion bar end bushing bolts from the lower control arm.
6. Remove the torsion bar pivot bushing bolts from the crossmember. Remove the bar and anchor assembly from the crossmember.

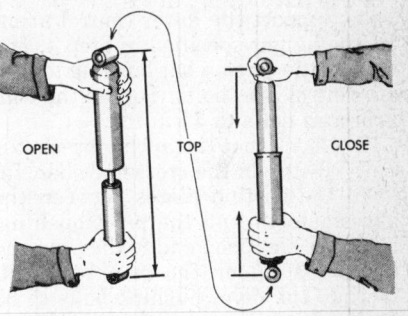

Purging new shock absorbers of air

NOTE: Check the seals on the bar for damage. If corrosion is evident, replace the bar assembly. Touch up any paint nicks or scratches. Check the adjusting bolt and swivel for corrosion or damage. Replace them if necessary.

7. To install, slide the balloon seal over the end of the bar with the cupped end toward the hex.
8. Coat the hex end of the bar with high-temperature waterproof grease.
9. Install the hex end of the bar into the anchor bracket. The ears of the bracket should be nearly straight up. Position the swivel into the anchor bracket ears.

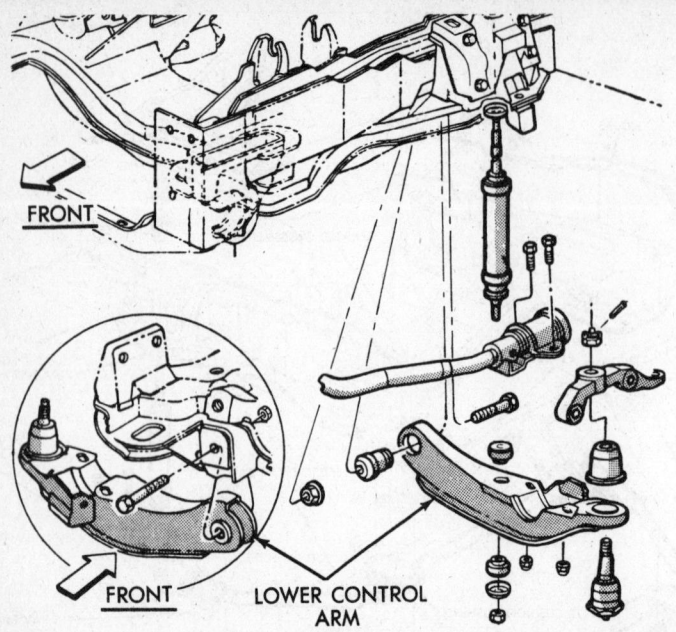

Aspen, 1980 and later Cordoba, Diplomat, LeBaron to 1981, 1981 and later Imperial, Mirada, Volare, 1982 New Yorker, 1983–84 New Yorker 5th Avenue, 1982 and later Gran Fury lower control arm details

10. Install the bar anchor bracket assembly into the crossmember anchor retainer. Install the adjusting bolt and bearing. Attach the pivot bushing to the crossmember, finger-tight.

11. Support the lower control arms, at the height specified in Step 4, and install the torsion bar bushing to lower control arm bolts. Torque the control arm bolts to 70 ft. lbs.

12. Check that the anchor bracket is fully seated in the crossmember. Install the friction plates between the crossmember and the pivot bushing, with the slot open end to the rear and bottomed out on the mounting bolt. Torque the pivot bushing bolts to 85 ft. lbs.

13. Install the balloon seal over the anchor bracket. Install the sway bar end bolt and torque to 50 ft. lbs.

14. Load the torsion bars by turning the adjusting bolts clockwise. Lower the vehicle and adjust the front end height.

Ball Joints

INSPECTION

NOTE: Before making the inspection, verify that the wheel bearings are adjusted correctly and that the control arm bushings are in good condition.

1. Place a jack under the lower control arm as close to the wheel as possible.

2. Raise the vehicle until there is

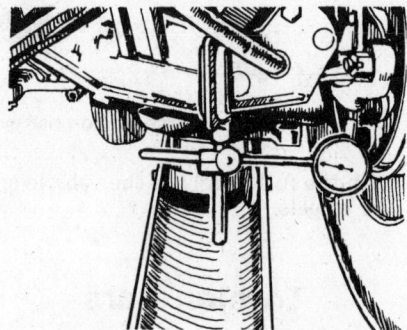

Measuring lower ball joint play; see text for specifications

1–2 in. of clearance under the wheel.

3. Insert a bar under the wheel and pry upward. If the wheel raises noticeably the ball joints are worn. While prying on the wheel, visually check to determine if the upper or lower ball joint is worn.

4. You can make a more accurate measurement by clamping a dial indicator to the lower control arm and measuring the lower ball joint stud movement.

NOTE: Due to the distribution of forces in the suspension, the lower ball joint is usually the one that needs replacing. The manufacturer's limit for lower ball joint play, measured at the joint, is 0.030 in. for all models.

5. Lower the jack enough to let the tire lightly contact the floor. Tighten the wheel bearing adjusting nut enough to remove all play. Have an

assistant try to move the top of the tire in and out while you observe the upper ball joint. If there is any noticeable side play, replace the upper ball joint.

6. Correct the wheel bearing adjustment.

REMOVAL & INSTALLATION

Upper Ball Joint

— CAUTION —
The torsion bar remains under tension during this procedure.

NOTE: Turn the ignition key to the OFF or UNLOCKED position.

1. Raise and support the front of the vehicle. Place a jackstand under the lower control arm as close to the wheel as possible. Remove the wheel. The jackstand should not contact the brake splash shield and the rubber rebound bumper must not contact the frame.

2. Remove the cotter pin and nut that attaches the upper ball joint to the steering knuckle. Remove the cotter pin and nut from the lower ball joint, to enable the removal tool to be used.

3. Slide the ball joint removal tool No. C-3564-A onto the lower ball joint stud, allowing the tool to rest on the knuckle arm. Set the tool securely against the upper stud. Apply pressure to the upper stud by tightening the tool and strike the knuckle sharply to loosen the stud. Never strike the ball joint stud.

NOTE: The brake caliper may have to be removed for clearance.

4. After disengaging the ball joint, support the steering knuckle and brake assembly. Using tool No. C-3560, unscrew the upper ball joint from the upper control arm.

5. Position a new ball joint on the upper control arm and screw the joint into the arm. Be careful not to cross thread the joint in the arm. Torque it to 125 ft. lbs.

6. Position a new seal on the ball joint stud and install the seal, using tool No. C-4039, in the ball joint making sure the seal is fully seated on the ball joint housing.

7. Position the ball joint stud in the steering knuckle and install the retaining nut. Torque the nut to 100 ft. lbs. Install a new cotter pin.

8. To complete the installation, reverse the removal procedures. Lubricate the ball joint. Adjust the wheel alignment.

Remove the lower ball joint with the illustrated tool. Don't try to free the ball joint completely with tool alone: see text

Lower Ball Joint

NOTE: Turn the ignition key to the OFF or UNLOCKED position.

1. Raise the vehicle so that the front suspension drops to the downward limit of its travel. Position jackstands beneath the front frame for extra support.

2. Remove the wheel and tire assembly.

3. Remove the brake caliper and tie it up out of the way, so that there is no strain on the flexible brake hose.

4. Remove the hub/rotor assembly and splash shield. Disconnect shock absorber from the lower control arm.

5. Unload the torsion bar by rotating the adjusting bolt counterclockwise.

6. Remove the cotter pin and nut from the upper and lower ball joints. Slide the ball joint removal tool No. C-3564-A over the upper stud, so that it rests on the steering knuckle. Tighten the tool to place the lower ball joint under pressure. Using a hammer, strike the steering knuckle to loosen the ball joint.

7. Use tool No. C-4212 to press the ball joint out of the lower control arm.

8. To install the new ball joint, use tool No. C-4212 to press it into the lower control arm.

9. Place a new seal over the ball joint. Using the tool No. C-4039, press the retainer portion of the seal down over the ball joint housing, until it locks into position.

10. Insert the ball joint stud through the opening in the steering knuckle and install the stud retaining nut. Torque to 100 ft. lbs. Install the cotter pin and lubricate the ball joint.

11. Load the torsion bar by rotating the adjusting bolt clockwise.

12. Install the shock absorber, the splash shield, hub/rotor assembly and brake caliper. Install the wheel and tire assembly.

13. Adjust the front wheel bearings. Remove the jackstands and lower the vehicle. Adjust the front suspension height.

Control Arms

REMOVAL & INSTALLATION

Upper

1. Refer to the "Upper Ball Joint, Removal and Installation" procedures in this section and separate the upper ball joint from the steering knuckle.

2. Remove the rubber engine splash shield and the pivot shaft nuts. It will be easier to reset the alignment if you mark the original pivot bar location. Remove the control arm and pivot shaft assembly.

3. To install, reverse the removal procedures.

Lower

1. Refer to the "Lower Ball Joint, Removal and Installation" procedures in this section and separate the lower ball joint from the steering knuckle.

CAUTION

Unload both bars even if you are removing only one control arm, to reduce the sway bar reaction.

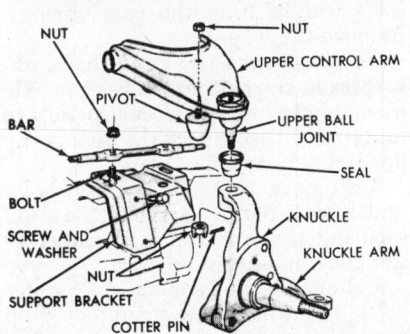

Typical upper control arm and steering knuckle

2. Raise the lower control arm until there is $2\frac{7}{8}$ in. clearance between the crossmember ledge at the jounce bumper and the torsion bar bushing on the lower control arm. Unbolt the torsion bar busing from the control arm.

3. Remove the lower control arm pivot bolt and the control arm.

NOTE: If the control arm shaft bushings indicates wear or deterioration, replace them.

4. To install, reverse the removal procedures. Torque the control arm pivot bolt to 75 ft. lbs. and the torsion bar end bushing-to-lower control arm to 50 ft. lbs.

Sway Bar

REMOVAL & INSTALLATION

1. Raise and support the front of the vehicle on jackstands.

2. Release the load on both torsion bars by turning the anchor adjusting bolts counterclockwise.

3. Raise the lower control arms until there is $2\frac{7}{8}$ in. clearance between the crossmember ledge, at the jounce bumper and the torsion bar bushing, on the lower control arm.

4. Support the lower control arms with a jackstands. Remove the sway bar-to-torsion bar bushing bolts, retainers, cushions and sleeves. Remove the retainer assembly strap and retainer bolts. Remove the sway bar.

NOTE: Inspect the cushions/bushings for wear or deterioration and replace, if necessary.

5. To install, reverse the removal procedures. Torque the sway bar-to-torsion bar to 50 ft. lbs. and the sway bar retainer/strap bolts to 30 ft. lbs. Load the torsion bar by turning the crossmember adjusting bolt clockwise. Lower the vehicle and adjust the torsion bar height.

Steering Knuckle

REMOVAL & INSTALLATION

NOTE: Refer to the "Upper & Lower Ball Joint, Removal & Installation" procedure in this section and remove the steering knuckle from the control arms.

Front Wheel Bearing

ADJUSTMENT

1. Raise and support the front of the vehicle so that the front wheels are off the ground.

2. Remove the hub caps, grease cut, cotter pin and nut lock.

3. Back off on the adjusting nut. Check for free wheel rotation.

4. While rotating the wheel, tighten the wheel bearing adjustment nut to 20–25 ft. lbs.

5. Loosen the adjusting nut, then retighten to finger-tight while rotating the wheel.

6. Position the nut lock so that one pair of slots is in line with the cotter pin hole and install the cotter pin.

This adjustment should give 0.001–0.003 in. end play.

7. To complete the installation, reverse the removal procedures. Repeat the procedures for the other wheel and lower the vehicle.

REAR SUSPENSION

All models utilize rear springs of the semi-elliptical leaf type. They are engineered to operate with little or no camber under conditions of small or no load. Heavy duty springs are offered as an option on most models. They serve to increase the stability of the vehicle under conditions of a heavy load. All vehicles with leaf springs are constructed with zinc interleaves between the normal leaves. They have the purpose of reducing spring corrosion and lengthening the spring life.

Shock Absorber

REMOVAL & INSTALLATION

1. Raise and support the rear of the vehicle with jackstands (under the axle assembly), so as to relieve load from the shock absorbers.

2. Remove the nut, retainer and bushing, attaching the shock to the spring mounting plate. To avoid damage to the shock, grip the base of the shock below the base-to-reservoir weld while loosening the retaining nut.

3. At the upper mount, remove the shock attaching bolt/nut and the shock.

4. Purge the new shock of air by repeatedly extending it in its upright position and compressing it, in an inverted position. It is normal to have more resistance to extend than compress.

5. To install the shock, position it so the upper bolt or nut may be replaced, hand-tighten only. Align the shock with the spring mounting plate and install the bolt or nut, hand-tighten only.

6. Lower the vehicle and tighten the shock absorber mounting bolts. Torque the bottom bolt to 35 ft. lbs. and the top bolt to 70 ft. lbs.

Spring

REMOVAL & INSTALLATION

1. Raise and support the rear of the

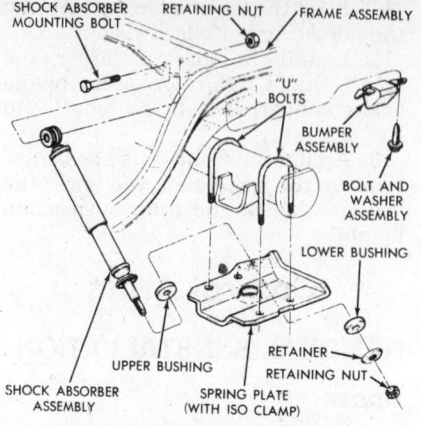

Typical rear shock mounting

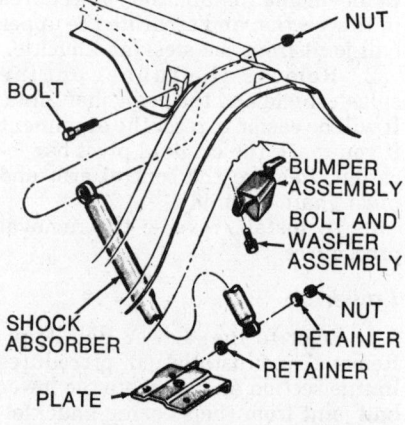

View of the rear shock absorber—Gran Fury, St. Regis, New Yorker and Newport

vehicle on jackstands; placed the jackstands under the axle so as to relieve weight from the rear springs. Remove the wheels.

2. Disconnect the rear shock absorbers at the bottom. Lower the axle assembly to allow the rear springs to hang free. Disconnect the rear sway bar links, if equipped.

3. Remove the U-bolt nuts, bolts and spring plates. Remove the nuts securing the front spring hanger to the body mounting bracket.

4. Remove the rear spring hanger bolts and allow the spring to drop enough to allow the front spring hanger bolts to be removed.

5. Remove the front pivot bolt from the front spring hanger.

6. Remove the shackle nuts and shackle from the rear of the spring.

7. To install, assemble the shackle/bushings in the rear of the spring and hanger. Start the shackle bolt nut. Do not lubricate rubber bushings to ease installation or tighten the bolt nut.

8. Align the front spring hanger with the front spring eye and insert the pivot bolt and nut. Do not tighten.

9. Install the rear spring hanger-to-body bracket and torque the bolts to 35 ft. lbs.

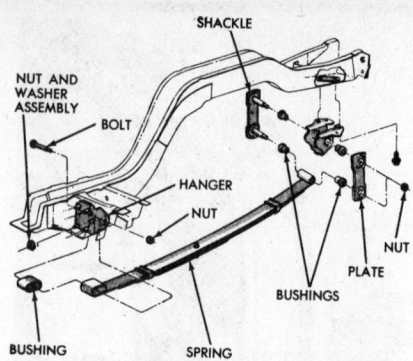

Rear Spring details for Aspen, 1980 and later Cordoba, Diplomat, LeBaron to 1981, 1981 and later Imperial, Mirada, Volare, 1982 New Yorker 5th Avenue, 1983 and later New Yorker 5th Avenue and 1982 and later Gran Fury

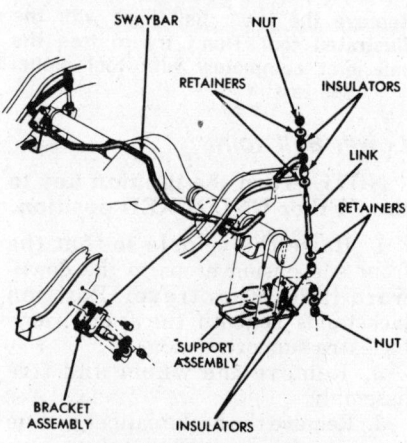

Rear sway bar mounting

10. With the aid of a helper, raise the spring and insert the bolts in the spring hanger mounting bracket holes. Install the nuts and torque them to 35 ft. lbs.

11. Position the axle assembly so it is correctly aligned with the spring center bolt.

12. Position the center bolt over the lower spring plate. Insert the U-bolt and nut. Tighten the U-bolt to 45 ft. lbs. Connect the rear shock absorbers.

13. Lower the vehicle. Torque the pivot bolts to 105 ft. lbs. and the shackle nuts to 35 ft. lbs.

NOTE: Drive the vehicle, then remeasure the front suspension heights and correct, if necessary.

Sway Bar

REMOVAL & INSTALLATION

1. Raise and support the rear of the vehicle on jackstands.

2. Remove the two sway bar link retaining nuts, retainers and insulators from the support.

3. Remove the nuts and retainer from each bracket fastened to each rail.

4. Remove the sway bar.

5. To install, reverse the removal procedures. Torque the retainer-to-bracket bolts to 17 ft. lbs. and the link nuts to 8 ft. lbs.

Rear Wheel Bearings

For information on the rear wheel bearings, refer to the "Rear Axle, Removal and Installation" procedures in this section.

BRAKES

For all brake system repair and service procedures not detailed below, please refer to "Brakes" in the Unit Repair section.

Master Cylinder

REMOVAL & INSTALLATION

1. Disconnect the brake lines from the master cylinder. Plug the brake line outlets to prevent fluid loss.

2. Remove the nuts that attach the master cylinder to the cowl panel or the power brake booster.

3. On non-power brakes models, disconnect the pushrod from the brake pedal and the stop light switch bracket. Pull the brake pedal back hard enough to separate the push rod from the master cylinder piston; the pushrod grommet will be destroyed, replaced it. Upon installation, lubricate the new grommet with a drop of water.

4. Slide the master cylinder straight out and off the cowl panel or power brake booster.

5. To install, reverse the removal procedures. Torque the mounting nuts to 17 ft. lbs. Bleed the brake system.

Combination Valve

The combination valve is located near the master cylinder and attached to the fender splash shield. The valve assembly contains a warning switch (with a hold off valve) and a proportioning valve.

REMOVAL & INSTALLATION

1. Disconnect the electrical connector from the combination valve.

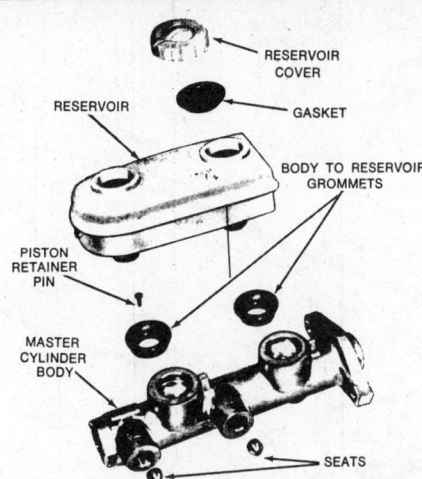

Aluminum master cylinder—primary and secondary system components similar to cast iron cylinder

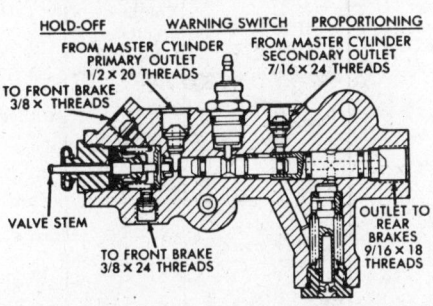

Exploded view of the combination valve

2. Disconnect and plug the brake tubes at the combination valve.

3. Remove the valve-to-fender splash shield bolts and the combination valve from the vehicle.

4. To install, reverse the removal procedures. Bleed the brake system.

Power Brake Booster

REMOVAL & INSTALLATION

1. Remove the master cylinder retaining nuts from the power brake booster and position the master cylinder out of the way without disconnecting the brake lines. Use care not to kink the brake lines.

2. Disconnect the vacuum hose from the power brake booster.

3. Working under the dash, remove the nut and bolt that attaches the power brake booster pushrod to the brake pedal. Use a small screwdriver to expand the retainer clip and remove it from the brake pedal pin. Discard the clip. Unbolt and remove the lower pivot bolt/nut.

4. Remove the 4 power brake booster attaching nuts/washers and the booster assembly from the vehicle.

5. To install, reverse the removal

procedures. Use a new retainer clip. Torque the mounting bolts/nuts to 17 ft. lbs.

Wheel Cylinder

REMOVAL & INSTALLATION

1. Raise and support the rear of the vehicle on jackstands.

2. Remove the wheel assembly and the brake drum.

3. Inspect the wheel cylinder for signs of leakage.

NOTE: If signs of leakage are present, the wheel cylinder must be replaced.

4. Remove the brake shoe assembly from the backing plate.

5. Disconnect and plug the brake line at the wheel cylinder.

6. Remove the wheel cylinder-to-backing plate bolts and separate the wheel cylinder from the backing plate.

NOTE: It is recommended to replace the wheel cylinder instead of rebuilding it.

7. To install, use a new wheel cylinder and reverse the removal procedures. Torque the wheel cylinder-to-backing plate bolts to 6 ft. lbs. Bleed the brake system.

Parking Brake

For overhaul procedures, refer to "Brakes" in the Unit Repair section.

ADJUSTMENT

1. Raise and support the rear of the vehicle on jackstands. Release the parking brake lever. Clean and lubricate the parking brake cable adjusting nut and threads. Loosen the cable adjusting nut.

2. Insert a brake spoon or screwdriver through the brake adjusting hole and rotate the star wheel until a slight drag is felt while rotating the wheels. Back off the star wheel until no drag is felt.

3. Tighten the cable adjusting nut until a slight drag is felt when rotating the rear wheels. Loosen the cable adjusting nut until the rear wheels can be rotated freely. Back off the cable adjusting nut two additional turns.

4. Apply and release the parking brake several times. Check, to verify that the rear wheels rotate freely, without any brake drag.

RIGHT REAR
CABLE ASSEMBLY

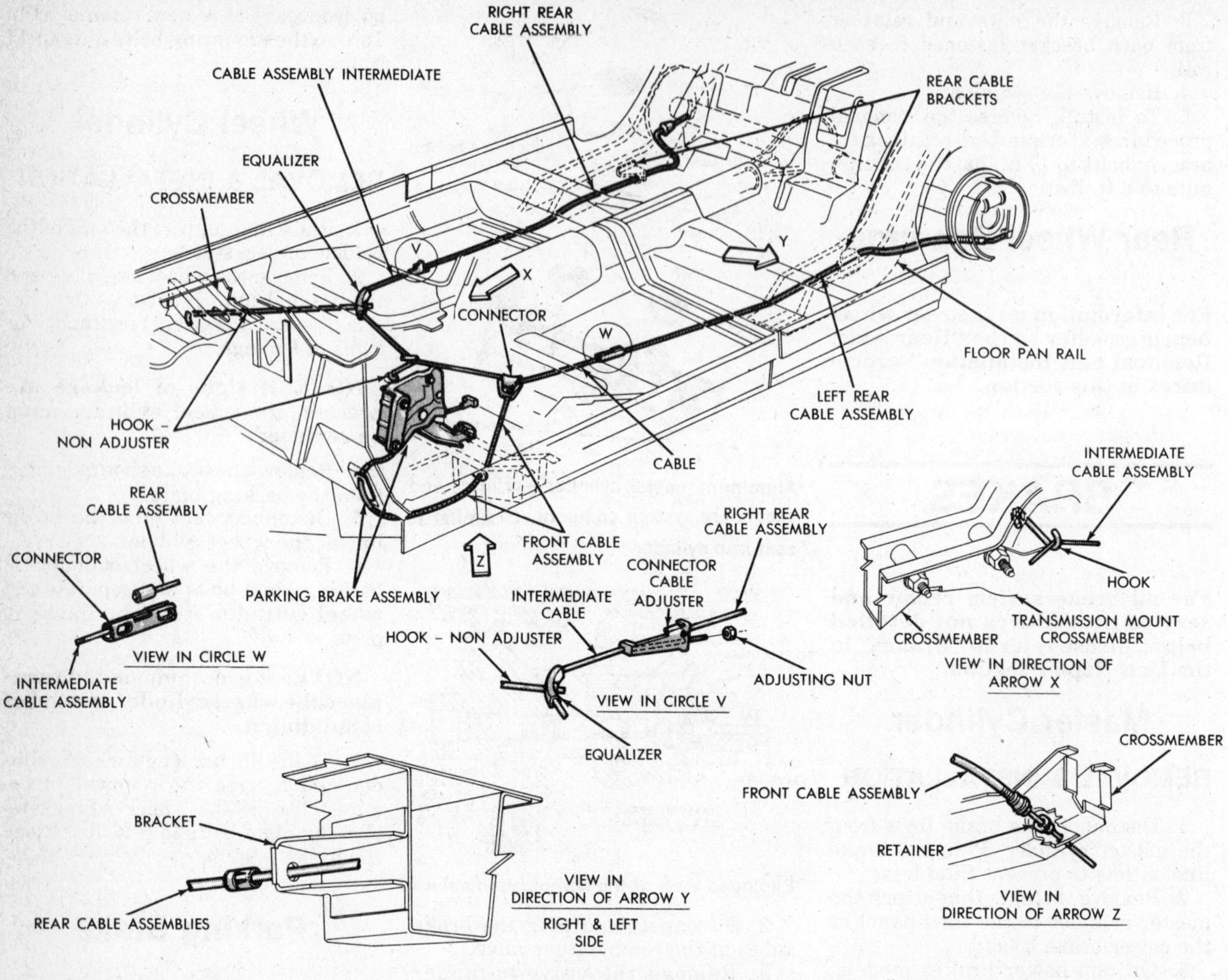

CABLE ASSEMBLY INTERMEDIATE

REAR CABLE
BRACKETS

EQUALIZER

CROSSMEMBER

CONNECTOR

FLOOR PAN RAIL

LEFT REAR
CABLE ASSEMBLY

HOOK –
NON ADJUSTER

CABLE

INTERMEDIATE
CABLE ASSEMBLY

REAR
CABLE ASSEMBLY

RIGHT REAR
CABLE ASSEMBLY

HOOK

CONNECTOR

CONNECTOR
CABLE
ADJUSTER

TRANSMISSION MOUNT
CROSSMEMBER

INTERMEDIATE
CABLE ASSEMBLY

FRONT CABLE
ASSEMBLY

INTERMEDIATE
CABLE

CROSSMEMBER

VIEW IN CIRCLE W

PARKING BRAKE ASSEMBLY

HOOK – NON ADJUSTER

**VIEW IN DIRECTION OF
ARROW X**

ADJUSTING NUT

VIEW IN CIRCLE V

EQUALIZER

CROSSMEMBER

BRACKET

FRONT CABLE ASSEMBLY

RETAINER

REAR CABLE ASSEMBLIES

**VIEW IN
DIRECTION OF ARROW Y**
RIGHT & LEFT
SIDE

**VIEW IN
DIRECTION OF
ARROW Z**

Parking brake cable routing, typical

STEERING

Steering Wheel

REMOVAL & INSTALLATION

———— **CAUTION** ————
All models are equipped with collapsible steering columns. A sharp blow or excessive pressure on the column will cause it to collapse. Do not hammer on the steering wheel or center nut.

1. Disconnect the negative battery cable.

2. Remove the padded center assembly. This center assembly is often held on only by spring clips. There are usually holes in the back of the wheel so the pad can be pushed off. However, on some deluxe interiors it is held on by screws behind the arms of the wheel. Remove the horn wire.

3. Remove the large center nut. Matchmark the steering wheel and steering shaft so that the wheel may be replaced in its original position. In most cases, the wheel can only be installed one way.

4. Using a steering wheel puller tool No. C-3428B, or equivalent, pull the steering wheel from the steering shaft.

5. To install, reverse the removal procedures. When placing the wheel on the shaft, make sure the tires are straight ahead and the matchmarks are aligned. Torque the nut to 45 ft. lbs.

Combination Switch

REMOVAL & INSTALLATION

1. Disconnect the negative battery cable. Remove the steering wheel center cover and the steering wheel.

2. Remove the steering column cover and the lower instrument panel be-zel. Loosen the gearshift housing Allen screw and remove the gearshift indicator.

3. If equipped with a tilt steering wheel, place the steering wheel at its mid-point position. Remove the steering column-to-lower panel reinforcement nuts and the 4 mounting bracket-to-steering column bolts.

———— **CAUTION** ————
Support the steering column to prevent damage.

4. Pry out the plastic buttons holding the wiring trough to the column and remove the trough. Disconnect the turn signal wiring connector and wrap it with a piece of tape.

5. On the standard column, remove the screws holding the turn signal lever assembly to the turn signal switch pivot and the screws/retainers holding the turn signal switch to the upper bearing housing.

6. On the tilt column, remove the plastic cover (if equipped) from the

A steering wheel puller can be made by drilling two holes in a piece of steel the same distance apart as the two threaded holes in the steering wheel. Sometimes an old spring shackle will have the right dimensions. Drill another hole in the center. Place a center bolt with the head against the steering shaft and a nut against the bottom of the homemade puller bar. Thread the two outer bolts into the holes in the wheel. Unscrew the nut on the center bolt to draw the wheel off the shaft.

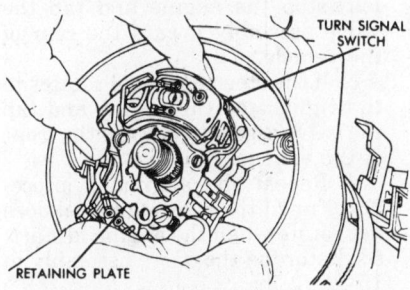

Removing turn signal switch from steering column

lock plate. Using the lock plate depression tool No. C-4156, depress the lock plate and pry the ring out of the groove with a small pry bar.

7. Remove the lock plate, the canceling cam and the canceling cam spring. Place the turn signal switch in the right position. Remove the screw holding the hazard warning switch knob and the screws holding the turn signal switch to the steering column.

8. Pull the switch gently from the column while guiding the wires through the column opening.

9. To install, reverse the removal procedures. Tighten the mounting bracket-to-steering column bolts and the lower reinforcement nuts to 9 ft. lbs.

Ignition Lock Cylinder

REMOVAL & INSTALLATION

Standard Steering Column

1. Refer to the "Steering Wheel Removal & Installation" procedures in this section and remove the steering wheel.

2. Remove the housing cover-to-lock housing screws and the cover.

3. Remove the wash/wipe switch assembly. Pull the hider up the control stalk and remove the control stalk sleeve-to-wash/wipe switch assembly screws. Turn the control stalk shaft clockwise 1 revolution and pull it straight out of the switch.

4. Remove the turn signal/upper bearing retainer screws and retainer. Lift the switch and move out of the way.

5. Disconnect the horn and key light ground wires. Remove the key light assembly screw and lift the assembly out of the way.

6. Remove the 4 bearing housing-to-lock housing screws, the snap ring from the upper end of the steering shaft and the bearing housing.

7. Remove the lock plate and spring from the steering shaft.

8. Remove the ignition key, the buzzer/chime screw and switch.

9. On the steering column housing, remove the ignition switch screws, twist the switch 90° and slide off the actuator rod. Remove the dimmer switch mounting screws and disconnect the switch from the actuator rod.

10. Remove the bellcrank screws and slide it up into the lock housing until it can be disconnected from the ignition actuator rod.

11. Place the ignition cylinder in the LOCK position and remove the key. Install a small screwdriver in both lock cylinder release holes, push inward to release the retaining spring and pull the lock cylinder out of the housing.

12. To install, insert the lock cylinder, insert the key, press inward, rotate the cylinder (this will align the parts), snap the cylinder into place and reverse the removal procedures.

Tilt Steering Column

1. Refer to the "Steering Wheel Removal and Installation" procedures in this section and remove the steering wheel.

2. Remove the tilt lever. Push in and unscrew the hazard knob. Remove the key lamp assembly.

3. Pull off the wash/wipe switch knob. Pull up the control stalk hider, remove the 2 sleeve-to-wash/wipe switch screws and the sleeve.

4. Turn the control stalk shaft clockwise, 1 revolution and pull it straight out of the switch.

5. Remove the plastic cover (if equipped) from the lock plate. Using the lock plate depression tool No. C-4156, depress the lock plate and pry the ring out of the groove with a small pry bar. Remove the lock plate, the canceling cam and the canceling cam spring.

6. Place the shift bowl in the LOW position. Remove the turn signal switch screws, the switch actuator screw, the actuator arm and the key lamp.

7. Place the ignition cylinder in the "LOCK" position. Install a small screwdriver in the lock cylinder (right hand slot) release hole, push inward to release the retaining spring and pull the lock cylinder out of the housing.

8. To install, press the cylinder lock inward (make contact with the drive shaft), move the actuator rod up and down (to align the parts), snap the cylinder into place and reverse the removal procedures.

Ignition Switch

REMOVAL & INSTALLATION

1. Remove the screws that attach the upper steering column housing to the steering column and remove the housing.

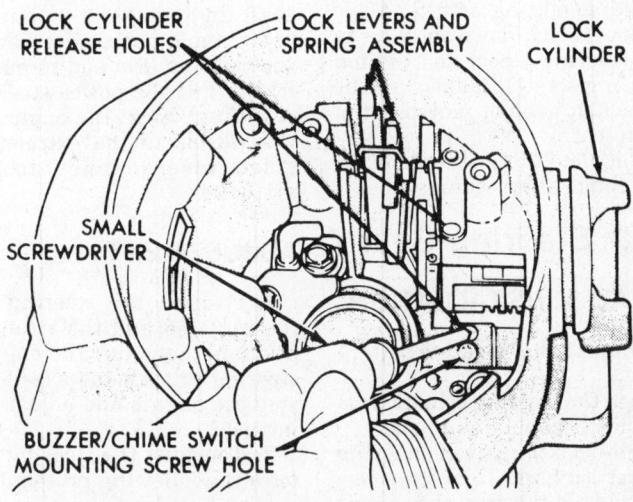

LOCK CYLINDER RELEASE HOLES

LOCK LEVERS AND SPRING ASSEMBLY

LOCK CYLINDER

SMALL SCREWDRIVER

BUZZER/CHIME SWITCH MOUNTING SCREW HOLE

Removal of lock cylinder from steering wheel

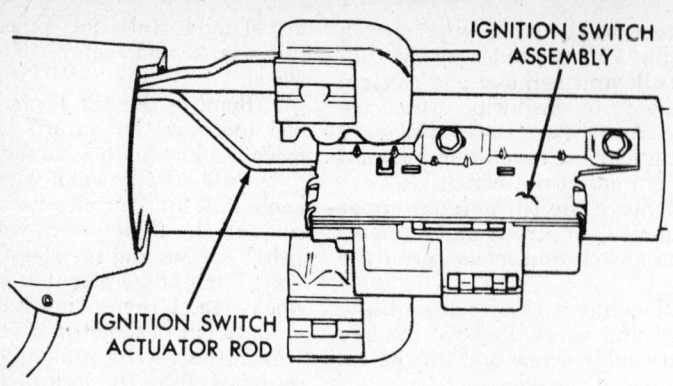

IGNITION SWITCH ASSEMBLY

IGNITION SWITCH ACTUATOR ROD

Ignition switch positioned on steering column

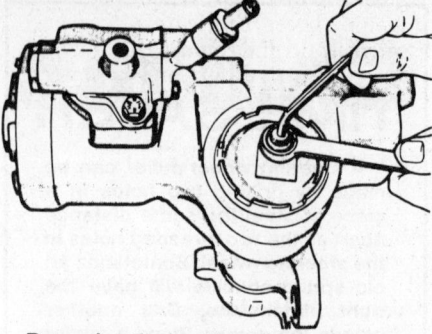

Removing the backlash from the steering gear

2. Move the column to the FULL UP position.

3. Insert a screwdriver into the slot in the spring retainer and press in on the retainer, approximately $\frac{3}{16}$ in. Turn the retainer approximately $\frac{1}{8}$ turn to the left until the ears align with the grooves in the housing. Remove the spring retainer, spring and guide.

4. Make sure the ignition switch is in the ACCESSORY position, then remove the wire connector from the ignition switch and remove the screws that attach the ignition switch to the outside of the steering column.

5. Lift the ignition switch from the column, twisting it to disengage the switch actuating rod from the rack. Remove the switch.

6. To install, place the ignition switch in the ACCESSORY position, insert the actuating rod into the steering column.

7. Twist the switch and rod assembly as required to engage the actuating rod with the rack. Make sure the ignition lock cylinder is in the correct position.

8. Loosely install the ignition switch and mounting screws. Move the ignition switch downward, away from the steering wheel and tighten the mounting screws. Make sure the ignition switch has not moved out of the lock detent.

9. To complete the installation, reverse the removal procedures.

Power Steering Gear

REMOVAL & INSTALLATION

1. Disconnect the negative battery cable.

2. Detach the steering column from the instrument panel and the floor.

3. Disconnect the power steering pressure/return hoses from the steering gear and tie the free ends above the power steering pump to avoid flu-

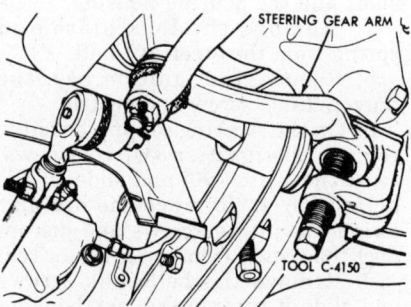

STEERING GEAR ARM

TOOL C-4150

Steering Gear Arm removal

id loss. Cap the steering gear fittings.

4. Remove the steering arm nut and lock washer from under the vehicle. Using the steering arm removal tool No. C-4150, or equivalent, remove the steering arm from the steering gear.

5. Disconnect and drop the exhaust system.

6. Remove the starter heat shield, the steering gear bolts and the steering gear.

7. To install, reverse the removal procedures. Torque the steering gear bolts to 100 ft. lbs. Rotate the worm shaft (by hand) and center the sector shaft at mid-point of its travel. Install the steering arm and torque it to 175 ft. lbs. Fill the power steering pump with fluid. Start the engine and turn the wheels to the extreme opposite sides (several times) to bleed the system.

ADJUSTMENT

1. Position the steering wheel in the centered and the straight ahead position, then start the engine. If the steering wheel rotates (self-steering), stop the engine and adjust the steering gear.

2. To adjust the steering gear, perform the following procedures:

   a. Loosen the valve assembly mounting bolts, then tighten to 7 ft.

lbs. to prevent oil leakage during the centering operation.

   b. Position the steering wheel in the centered and the straight ahead position, then start the engine. If the steering wheel rotates to the left, stop the engine and tap the valve assembly toward the rear of the vehicle.

   c. If the steering wheel rotates to the right, stop the engine and tap the valve assembly toward the front of the vehicle.

   d. Repeat the alignment procedures until the steering wheel does not rotate when the engine is started. Retorque the valve assembly to 15–20 ft. lbs.

3. If backlash is noticed the steering gear, perform the following procedures to remove it:

   a. Using a box-end wrench and an Allen wrench, loosen the sector shaft adjusting screw locknut and turn the adjusting screw to remove the backlash.

NOTE: If the power train has been removed from the vehicle, tighten the adjusting screw $1\frac{1}{4}$ turns and tighten the locknut. Turn the steering wheel several times to align the piston rack and the sector teeth.

   b. Center the steering gear. Tighten the adjusting screw until the backlash has disappeared, then turn the screw in $\frac{3}{8}$–$\frac{1}{2}$ turns and torque the locknut to 28 ft. lbs.

Power Steering Pump

REMOVAL & INSTALLATION

1. Back off the pump mounting adjusting bolts and remove the pump drive belt.

2. Disconnect the pump hoses at the pump.

3. Remove the pump bolts and pump with the bracket.

4. To install, reverse the removal procedures. Torque the pump mount-

ing bolts to 30 ft. lbs. When installing the pump hoses, lubricate with power steering fluid and replace the O-rings. Adjust the drive belt and bleed the system.

BELT ADJUSTMENT

1. Install the pump drive belt and adjust.
2. There should be no more than $\frac{1}{2}$ in. of play, under moderate thumb pressure, on the longest span of belt.
3. If equipped with a $\frac{1}{2}$ in. square hole on the adjusting bracket, use a $\frac{1}{2}$ in. breaker bar to adjust the belt tension.

SYSTEM BLEEDING

1. Fill the pump with power steering fluid.
2. Start the engine and rotate the steering wheel from stop to stop several times, to bleed the system.
3. Make sure that there are no air bubbles in the fluid.
4. Check the pump fluid level and fill as required.

Tie Rod End

REMOVAL & INSTALLATION

1. Loosen the tie rod adjuster sleeve clamp nuts.
2. Remove the tie rod end stud nut and cotter pin.
3. If the outer tie rod end is being removed, remove the stud from the steering knuckle. If the inner tie rod end is being removed, remove the stud from the center link. The studs on all

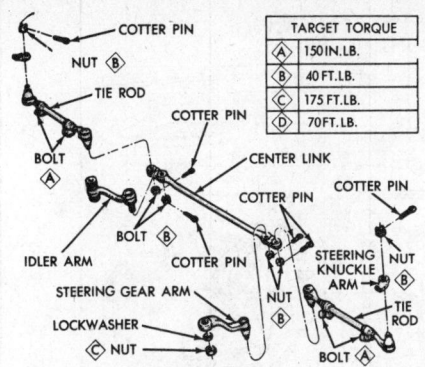

TARGET TORQUE	
Ⓐ	150 IN. LB.
Ⓑ	40 FT. LB.
Ⓒ	175 FT. LB.
Ⓓ	70 FT. LB.

Steering linkage, all models similar

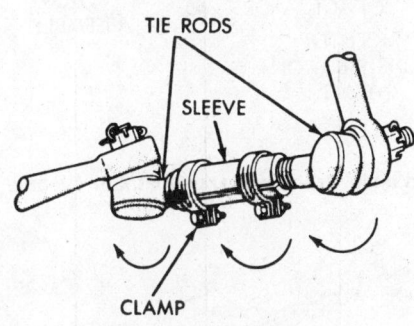

Tie Rod adjustment

the tie rod ends fit in a tapered hole. They can be removed with the ball joint removal tool No. C-3894-A.

NOTE: Use extreme care not to damage the rubber grease seals at the tie-rod ends. If the seals become damaged they must be removed and the tie-rod ends inspected.

4. Unscrew the tie rod end from the threaded sleeve and record the number of turns required to remove it. The threads may be left or right hand threads.
5. To install, grease the tie rod threads and reverse the removal procedures. Screw in the tie rod end as many turns as were needed to remove it. This will give approximately correct toe-in. Torque the stud nuts to 40 ft. lbs. and install new cotter pins. Adjust the toe.

CHASSIS ELECTRICAL

Heater Assembly

REMOVAL & INSTALLATION

NOTE: The heater assembly must be removed in order to service the heater core. If the vehicle is equipped with A/C, discharge the refrigerant system and remove the "H" valve (cap the lines to prevent dirt and moisture entry). Remove the condensation drain tube, the blend air door actuator, the A/C temperature control cable and the compensator hose

1. Disconnect the negative battery cable. Drain the cooling system.
2. Disconnect the heater hoses at the firewall and plug the heater core tubes.

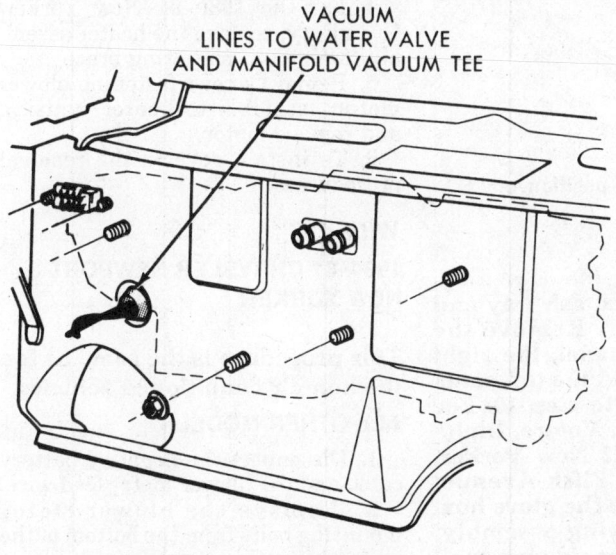

VIEW IN DIRECTION OF ARROW X

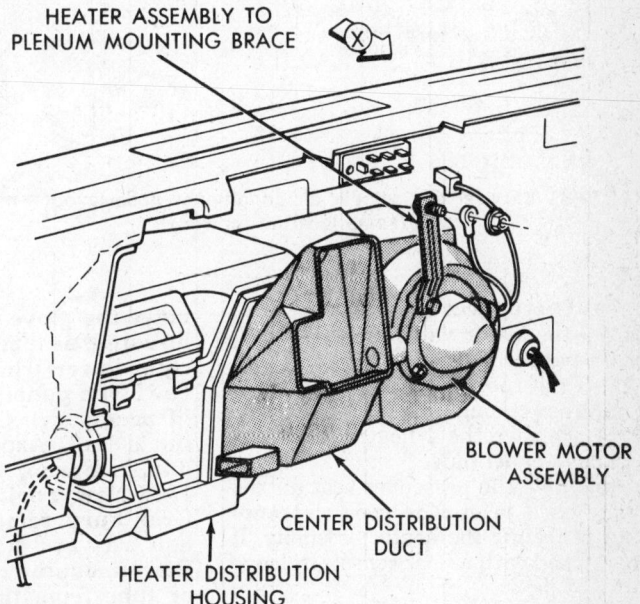

Heater assembly—air conditioned vehicles; 1980–83

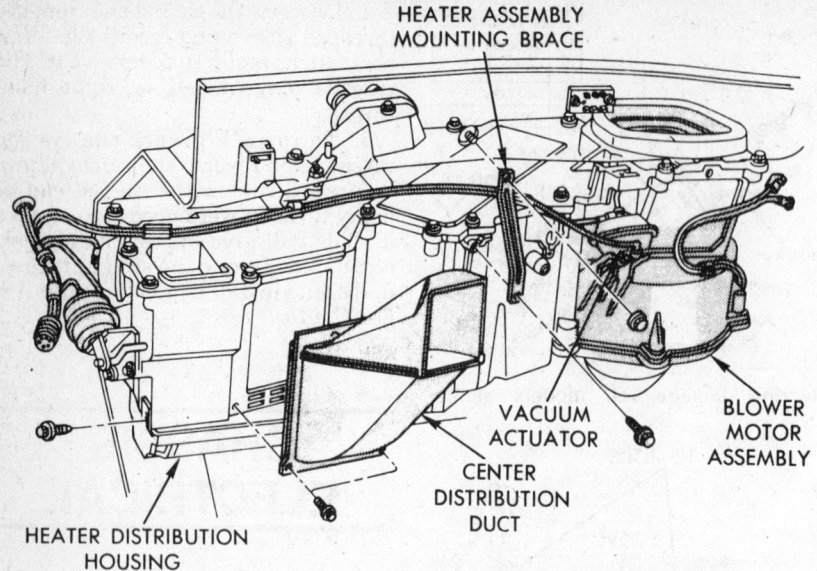

Cordoba, Mirada heater assembly, 1983 and later. Gran Fury, Diplomat, 5th Avenue similar

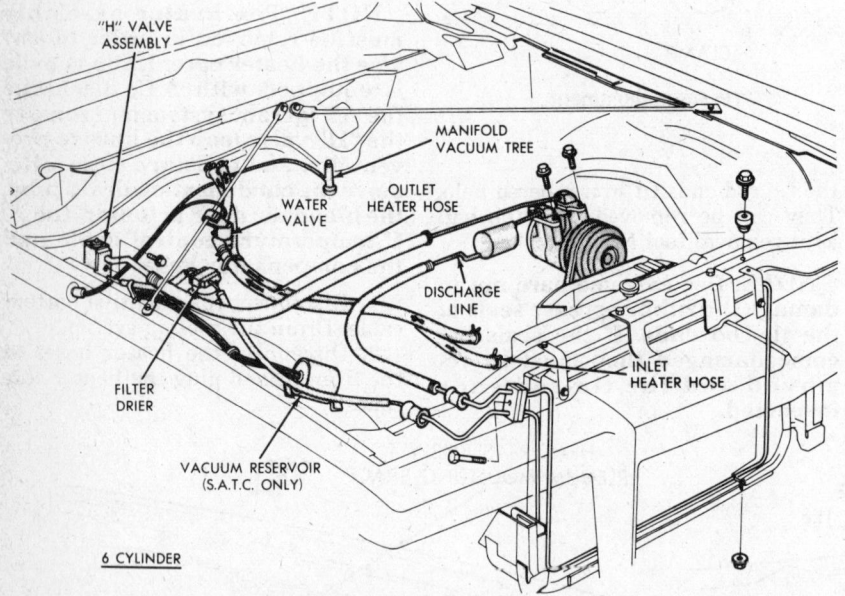

1979-81 Chrysler six cylinder air conditioner hose layout—note position of "H" valve (same location for V8s)

6. Remove the cluster bezel assembly and the upper instrument panel cover.

7. Remove the steering column cover and the right side intermediate cowl trim panel.

8. Remove the lower instrument panel and the center-to-lower reinforcement.

9. Remove the right center air distribution duct.

10. Disconnect the locking tab or defroster duct, the blower feed and ground wires.

11. Depress the flag tab at the temperature control cable and remove the heat distribution housing from the heater housing receiver.

12. While holding up the assembly, remove the heater assembly-to-plenum mounting brace.

13. Pull the assembly out, rotate it to the right and remove it from under the instrument panel.

14. To install, reverse the removal procedures and refill the cooling system.

Heater Blower

REMOVAL & INSTALLATION

NOTE: All service to the blower motor is made under the right-side of the instrument panel.

Without A/C

1. Disconnect the negative battery cable.

2. For the 1980–81 New Yorker/ Gran Fury, remove the glove box assembly.

3. Disconnect motor electrical connections.

4. For the 1980–81 New Yorker/ Gran Fury, remove the heater assembly-to-plenum mounting brace.

5. Remove screws fastening blower motor assembly-to-heater housing and remove motor.

6. To install, reverse the removal procedures.

With A/C

1980–81 CHRYSLER NEWPORT & NEW YORKER

This procedure is the same as for the non-air conditioned vehicles.

ALL OTHER MODELS

1. Disconnect the negative battery cable and the blower motor feed wire.

2. Remove the blower motor mounting bolts from the bottom of the recirculation housing.

3. Separate the lower blower motor housing from the upper housing.

4. Remove the blower motor

3. Disconnect the vacuum lines from the water valve and the manifold vacuum tree (see illustration). Push the rubber grommet and vacuum lines through the dash panel.

4. Remove the 4 heater assembly-to-dash panel nuts.

5. Slide the passenger seat all the way back to provide good clearance for removing the heater assembly. If equipped with a floor console, remove it.

NOTE: For the Mirada, 1980–83 Cordoba and 1981–83 Imperial, re-move the glove box, ash tray and housing assembly. Remove the right lower trim panel, the right cowl trim panel and the top panel (if necessary). Go to step 10. For the 1980–81 Aspen, Volare, Diplomat, LeBaron, 1982 New Yorker, 1983 New Yorker Fifth Avenue/ Gran Fury, remove the glove box, ash tray and housing assembly. Disconnect the right side lap cooler tube from the lap cooler, the right trim bezel and the right cowl trim. Go to step 10.

mounting plate screws, the wire grommet, the mounting plate and the blower motor/wheel assembly.

5. To install, reverse the removal procedures.

Heater Core

REMOVAL & INSTALLATION

1980–81 Aspen, Volare, LeBaron & Diplomat

1. Refer to the "Heater Assembly Removal & Installation" and remove the heater assembly from the vehicle and place it on a bench. If equipped with A/C, remove the blend air door from the shaft.

2. Remove the retainer clips securing the heater housing halves and separate the halves.

3. Remove the screw securing the seal retainer/seal to the heater core tubes.

4. Remove the heater core support clamps and slide out the heater core.

5. To install, reverse the removal procedures.

All Other Models

1. Refer to the "Heater Assembly Removal & Installation" and remove the heater assembly from the vehicle and place it on a bench. If equipped with A/C, remove the blend air door from the shaft.

2. Remove the servo motor from the shaft, the screws from the top of the unit and lift off the cover.

3. Remove the mounting flange screws (if equipped) from behind the core seal on the unit face and lift out the heater core.

4. To install, reverse the removal procedures.

Radio

The following should be observed when working on the radio:

1. Always observe the proper polarity of the connection; positive (+) goes to the power source and negative (–) to ground.

2. NEVER operate the radio without a speaker; damage to the output transistors will result. If a replacement speaker is used, be sure that it is the correct impedance (Ω) for the radio.

3. If a new antenna or antenna cable is used, adjust the antenna trimmer for the best reception of a weak AM station around 1400–1600 KHz; the trimmer screw is located either behind the tuning knob or on the radio case.

4. For best FM reception, the best

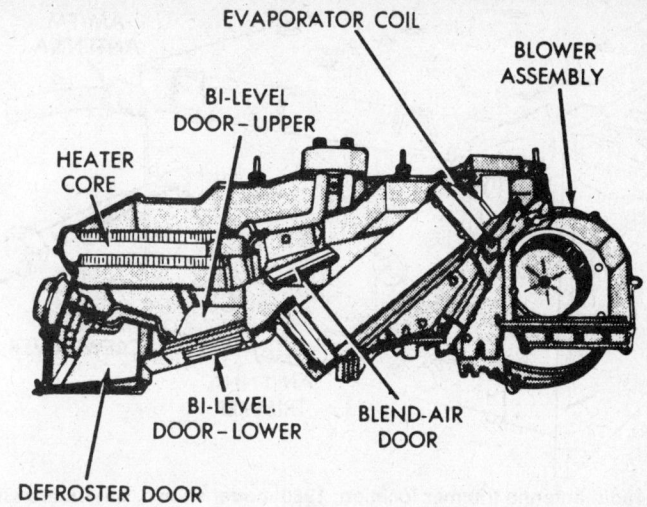

Heater core removal for '83 and later models

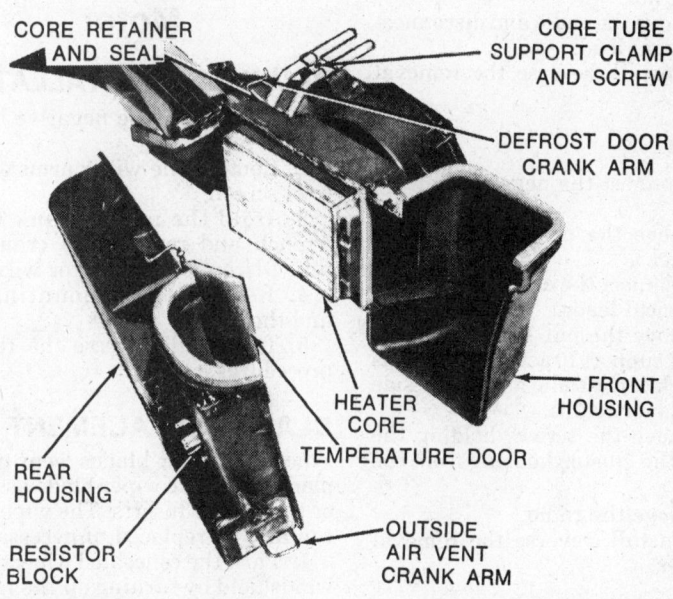

Heater core removal—Aspen, Volare, Diplomat, LeBaron, '82 New Yorker, '83 New Yorker 5th Avenue, '82-'83 Gran Fury

antenna height is 31–33 in.; for best AM reception, the antenna should be at its full length.

REMOVAL & INSTALLATION

Volare, Aspen, Diplomat, 1980–81 LeBaron and Mirada, 1981–83 Imperial, 1982 New Yorker, 1983 New Yorker Fifth Avenue/ Gran Fury

1. Disconnect the negative battery cable.

NOTE: On the Aspen, Volare, Diplomat, LeBaron and 1982 New Yorker/Gran Fury remove the in-strument cluster bezel by remov-ing the 4 screws along the lower edge. Place the automatic trans-mission selector in No.1 position and pull out to detach the top edge clips. Remove the center bezel on 1980–83 Mirada and 1981–83 Imperial.

2. Remove the radio mounting screws.

3. On monaural radios, remove the lamp assembly from the front of the radio.

4. Remove the radio-to-panel mounting screws.

5. Remove the instrument panel upper cover. Work through the access hole in the top of the instrument panel to disconnect the antenna and

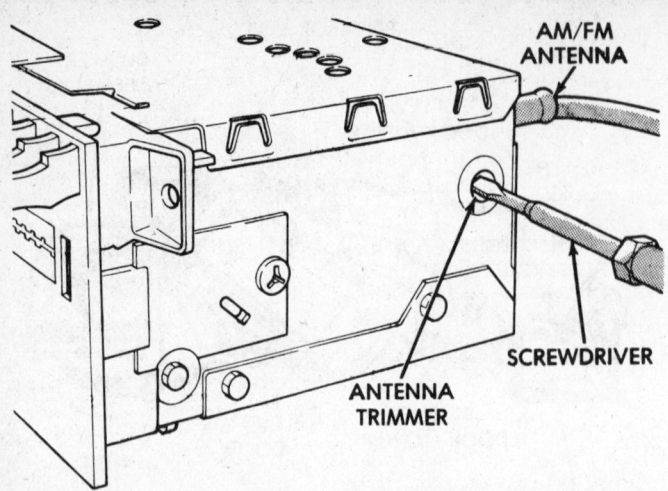

AM/FM radio antenna trimmer location; 1980 model shown, others are similar

speaker leads. Remove the bracket mounting nut.

6. Remove the radio and disconnect the electrical lead.

7. To install, reverse the removal procedures.

Cordoba

1. Disconnect the negative battery cable.

2. Remove the instrument cluster lower bezel.

3. Disconnect the antenna, speaker and electrical leads.

4. Remove the nut holding the radio to the support bracket. The nut is at the back of the radio and on the side of tape player/radios.

5. Remove the screws holding the radio on the cluster housing from the front.

6. Remove the radio.

7. To install, reverse the removal procedures.

All Other Models

1. Disconnect the negative battery cable.

2. Remove the center bezel and the radio-to-panel mounting screws.

3. Pull the radio out through the front face of the panel. Detach the antenna lead, ground strap, power wire and speaker leads.

4. To install, reverse the removal procedures.

Windshield Wiper Switch

REMOVAL & INSTALLATION

Refer to the "Combination Switch Removal & Installation" procedures in this section, to replace the wiper switch.

Windshield Wiper Motor

REMOVAL & INSTALLATION

1. Disconnect the negative battery cable.

2. Remove the wiper arms and the cowl screen.

3. Hold the motor crank with a wrench and remove the crank arm nut. Disconnect the motor wiring.

4. Remove the 3 mounting nuts and the motor.

5. To install, reverse the removal procedures.

BLADE REPLACEMENT

When the wiper blades wear out, replace the entire wiper blade assembly or the rubber inserts. The wiper arms can also be replaced, if necessary.

1. Park the concealed wipers on the windshield by turning off the ignition key while the wipers are running.

2. Lift the wiper arm assembly off the glass.

3. On 1980–83 models, the slot and the pin styles of wiper arm-to-wiper blade connectors are used; to remove them, slide a small screwdriver along side the release lever, depress the lever and pull the wiper blade assembly from the wiper arm.

4. On 1980–83 models, the wiper blade is secured to the wiper blade assembly end bridge by three methods: the bear claw, lock tab and push button. Release the bear claw (pinch the lock) on the end of the blade reinforcement and slide the element off. Release the lock tab (lift the tab) or the push button (depress the button) on the end bridge, separate it from the center bridge and slide the end bridge off the element.

5. On 1984 and later, lift the release tab on the side of the blade as-

sembly and pull the assembly from the wiper arm. To release the element from the blade assembly, pinch the release lock and slide the element off.

6. To install, reverse the removal procedures.

Instrument Cluster

For further information on instruments, please refer to "Gauges and Indicators" in the Unit Repair section.

REMOVAL & INSTALLATION

1. Remove the instrument cluster mounting screws and the cluster bezel.

NOTE: On some models, it will be necessary to remove the instrument panel upper cover and the sub-bezel to gain access to the speedometer cable and the electrical wire connectors that must be disconnected before the cluster can be removed. Care should be exercised not to force the finish panels when removing or installing, for breakage can occur.

2. Place the gearshift indicator on the No. 1 position.

3. The speedometer cable is attached to the speedometer housing by a spring clip, which locks into a groove on the speedometer housing. To release the cable, depress the spring clip arm to disengage it from the groove and pull the cable away from the speedometer housing.

4. Disconnect the wiring connectors and the cluster from the vehicle.

5. To install, reverse the removal procedures.

Fuse Block

LOCATION

The fuse block is located under the lower left-side of the instrument panel, secured by 2 mounting screws. When replacing the fuses or circuit breakers, be sure to replace them with ones of the same amperage.

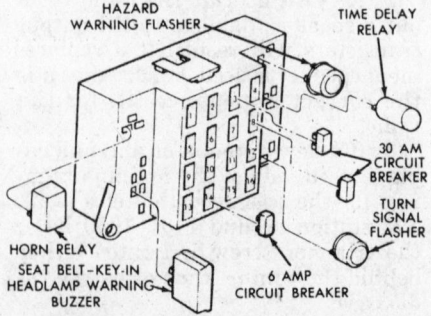

Exploded view of the fuse block

Ford Motor Co.
Front Wheel Drive
Escort, EXP, LN7, Lynx, Tempo, Topaz

YEAR IDENTIFICATION

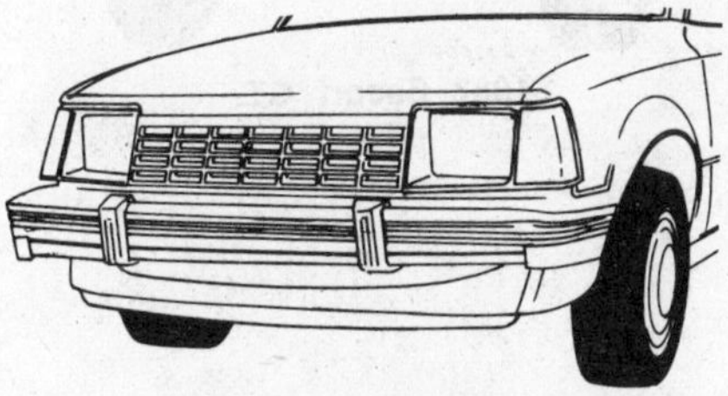

1981–82 Escort

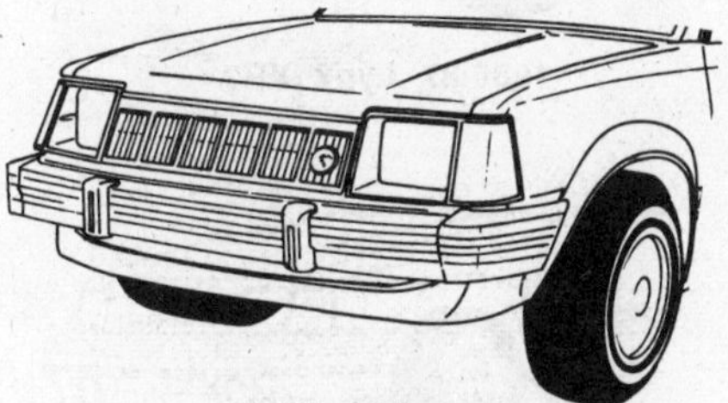

1981–82 Lynx

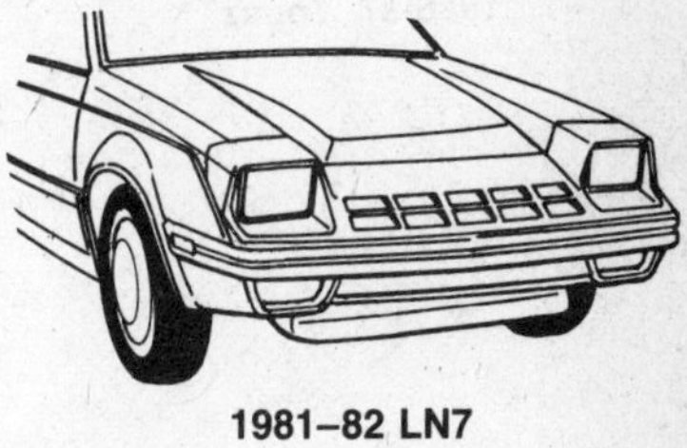

1981–82 LN7

1981–82 EXP

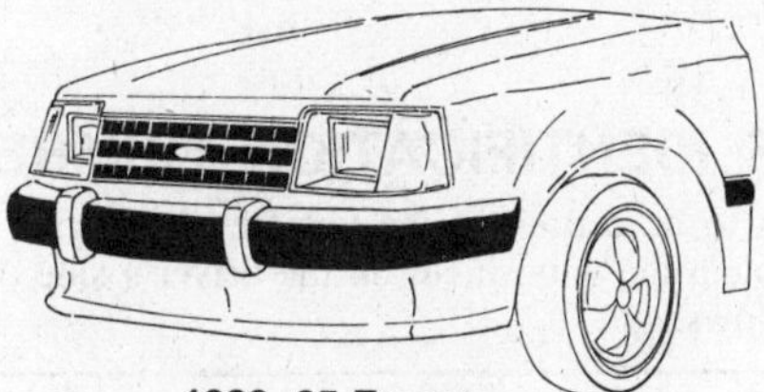

1983–85 Escort

1983–84 Lynx

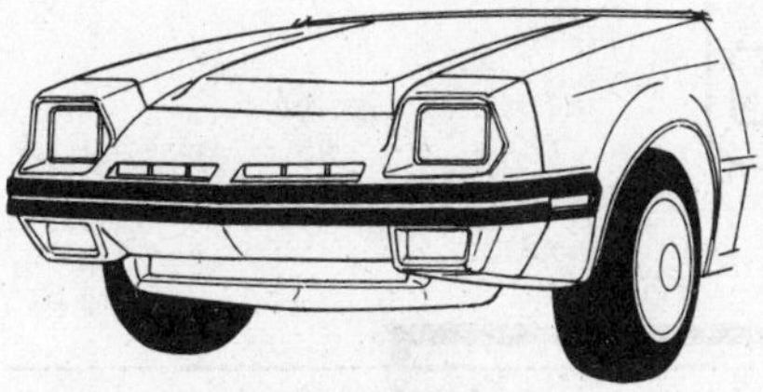

1983 EXP

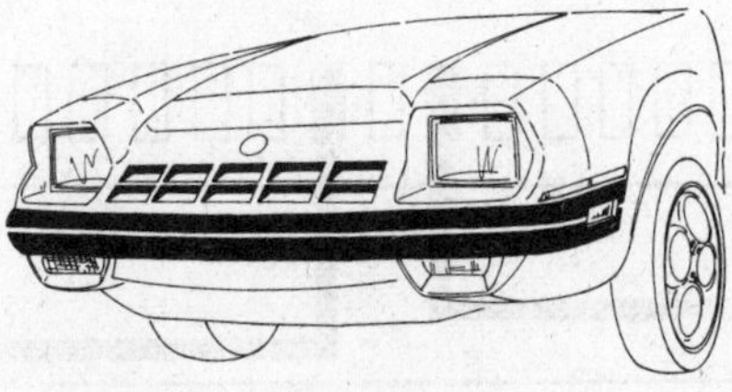

1983 LN7

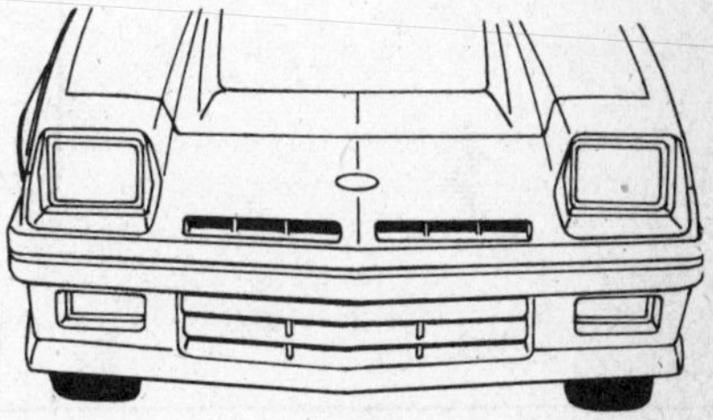

1984–85 EXP

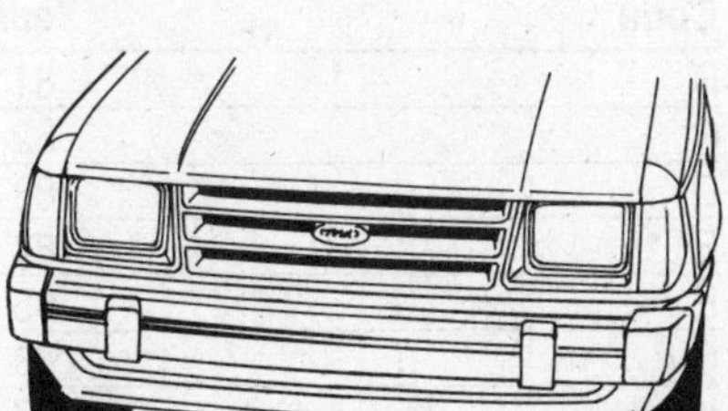

1984 Tempo

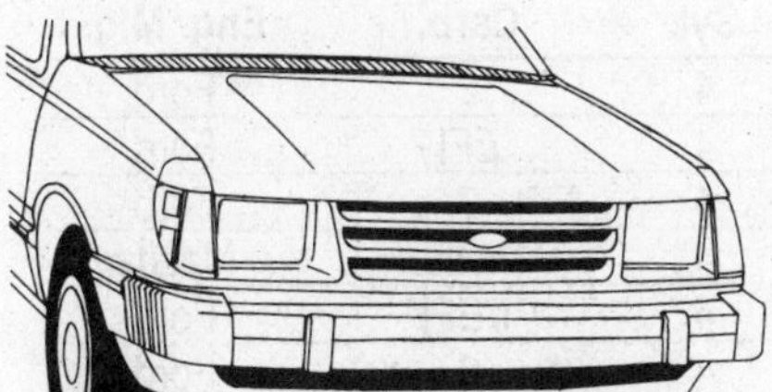

1985 Tempo

1984–85 Topaz

YEAR IDENTIFICATION

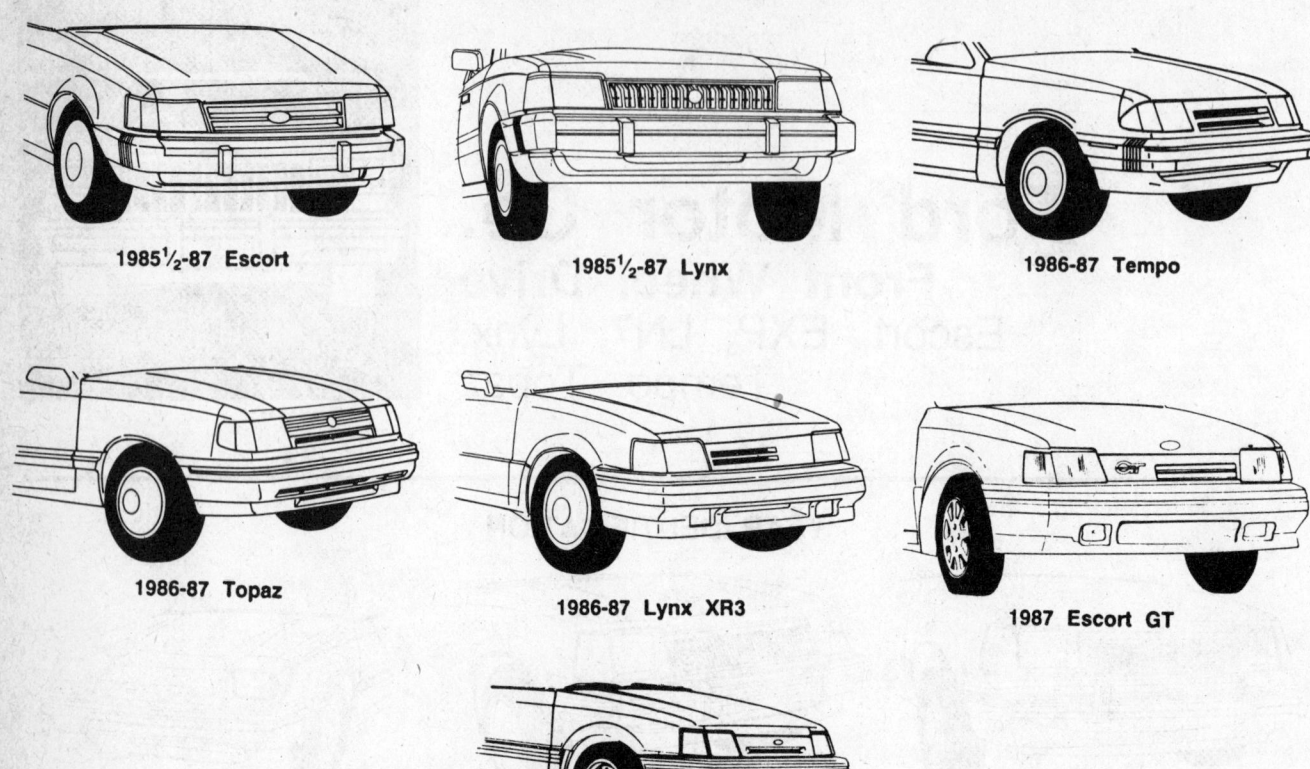

1985½-87 Escort

1985½-87 Lynx

1986-87 Tempo

1986-87 Topaz

1986-87 Lynx XR3

1987 Escort GT

1987 Escort EXP

VEHICLE IDENTIFICATION NUMBER (VIN)

It is important for servicing and ordering parts to be certain of the vehicle and engine identification. The VIN (vehicle identification number) is a 13 or 17 digit number visible through the windshield on the driver's side of the dash and contains the vehicle and engine identification codes. It can be interpreted as follows:

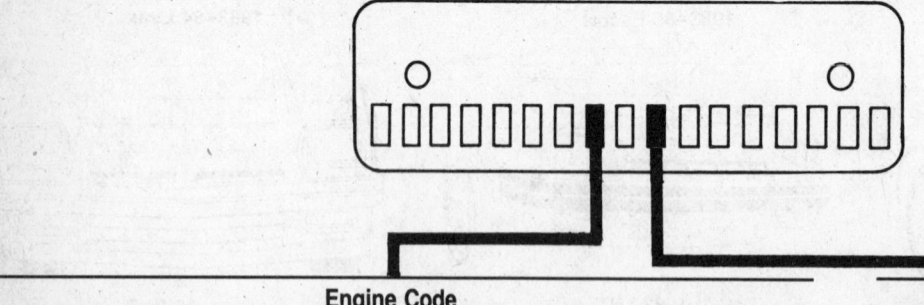

Engine Code						Model Year Code	
Code	Cu. In.	Liters	Cyl.	Carb.	Eng. Mfg.	Code	Year
2	98	1.6	4	2	Ford	B	'81
5	98	1.6	4	EFI	Ford	C	'82
4	98	1.6 HO	4	2	Ford	D	'83
H	121	2.0	4	Diesel	Mazda	E	'84
R	140	2.3 (HSC)	4	1/EFI	Ford	F	'85
9	113.5	1.9	4	2	Ford	G	'86
X	140	2.3 (HSC)	4	CFI	Ford	H	'87

The seventeen digit Vehicle Identification Number can be used to determine engine application and model year. The tenth digit indicates model year, and the eighth digit identifies engine code.

GENERAL ENGINE SPECIFICATIONS

Year	Eng VIN Code	Engine No. Cyl. Displacement (cc)	Eng. Mfg.	Carburetor Type	Horsepower @ rpm ■	Torque @ rpm (ft. lbs.) ■	Bore × Stroke (mm)	Compression Ratio	Oil Pressure @ 2000 rpm
'81	2	4-1597	Ford	2bbl	69 @ 5000	86 @ 3200	80.0 × 79.5①	8.8:1	40
'82	2	4-1597	Ford	2bbl	69 @ 5000	86 @ 3200	80.0 × 79.5①	8.8:1	40
	2	4-1597	Ford	2bbl	80 @ 5800	88 @ 3400	80.0 × 79.5①	9.0:1	40
'83–'85	2	4-1597	Ford	2bbl	70 @ 4600	89 @ 3000	80.0 × 79.5①	8.8:1	40
	4	4-1597	Ford	2bbl	80 @ 5800	88 @ 3400	80.0 × 79.5①	9.0:1	40
	5	4-1597	Ford	EFI	90 @ 5800	89 @ 3000	80.0 × 79.5①	9.0:1	40
	8	4-1597	Ford	Turbo	120 @ 5200	120 @ 3400	80.0 × 79.5①	8.0:1	35–65
	H	4-2000	Mazda	Diesel	52 @ 4000	82 @ 2400	86.0 × 86.0②	22.5:1	55-60
	R	4-2300	Ford	1bbl①	84 @ 4600	118 @ 2600	93.5 × 84⑤	9.0:1	55–70
'86–'87	9	4-1860	Ford	2bbl⑦	86 @ 4800	100 @ 3000	82.0 × 88⑥	8.0:1	35–65
	H	4-2000	Mazda	Diesel	52 @ 4000	82 @ 2400	86.0 × 86.0②	22.5:1	55-60
	R	4-2300	Ford	CFI	86 @ 4000	124 @ 2800	93.5 × 84⑤	9.0:1	55–70
	X	4-2300	Ford	CFI	100 @ 4600	125 @ 3200	93.5 × 84⑤	9.0:1	55–70

■ Horsepower and torque are SAE net figures. They are measured at the rear of the transmission with all accessories installed and operating.
① 3.15 × 3.13 in.
② 3.39 × 3.39 in.
③ Greater than 0.7 KG/CM² @ 700 RPM Oil Temp. 80°C

④ CFI—'84½ and later
⑤ 3.70 × 3.30 in.
⑥ 3.23 × 3.46 in.
⑦ EFI optional

TUNE-UP SPECIFICATIONS

(When analyzing compression test results, look for uniformity among cylinders rather than specific pressures.)

Year	Eng VIN Code	Engine No. Cyl. Displacement cu in. (cc)	Eng. Mfg.	Spark Plugs Orig. Type	Spark Plugs Gap (in.)	Distributor Point Dwell (deg.)	Distributor Point Gap (in.)	Ignition Timing (deg) Man. Trans.	Ignition Timing (deg) Auto. Trans.	Fuel Pump Pressure (psi)	Idle Speed (rpm) Man. Trans.	Idle Speed (rpm) Auto. Trans.
'81	2	4-97.6 (1597)	Ford	AGSP-32④	.042–.046	Electronic		10B①	10B①	4–6	①	①
'82	2	4-97.6 (1597)	Ford	AWSF-32④	.042–.046	Electronic		①	①	4–6	①	①
'83–'84	4	4-97.6 (1597)	Ford	AWSF-34②④	.042–.046	Electronic		①	①	4–6③	①	①
'84	6	4-113.5 (1860)	Ford	AWSF-34C	.044	Electronic		①	①	4–6	①	①
'84–'86	R,X	4-140 (2300)	Ford	AWSF-62	.044	Electronic		10B	15B	5③	①	①
'85–'86	9	4-113.5 (1860)	Ford	⑤	⑤	Electronic		①	①	4–6③	①	①
'87				See Underhood Specifications Sticker								

NOTE: The underhood specifications sticker often reflects tune-up specification changes made in production. Sticker figures must be used if they disagree with those in this chart. Part numbers in this chart are not recommended by Chilton for any product by brand name.
B Before Top Dead Center
① Calibration levels vary from model to model. Always refer to the underhood sticker for your cars requirements.
② EFI Models: AWSF24
③ EFI pressure: 35–45 psi
④ CAUTION: Two different plug designs are used on 1.6L engines. The designs are: gasket equipped and tapered seat (no gasket). All 1981 Escort/Lynx models; and 1982 EXP/LN7 models built before 9/4/81 use gasket equipped plugs. All 1982 and later Escort/Lynx; and EXP/LN7 models built after 9/4/81 are equipped with tapered seat plugs. DO NOT INTERCHANGE TYPES. Tighten gasket equipped plugs to 17–22 ft. lbs. Tapered plugs are tightened to 10–15 ft. lbs. DO NOT OVERTIGHTEN.
⑤ See Underhood Specifications Sticker

FIRING ORDER

NOTE: To avoid confusion, always replace spark plug wires one at a time.

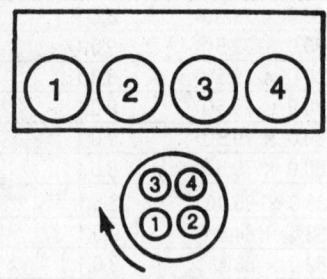

Tempo/Topaz 2300 HSC Engine
Firing Order 1–3–4–2
Distributor Rotation Clockwise

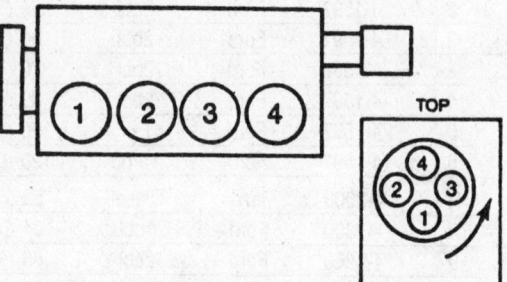

Escort, Lynx 1600cc Engine
Firing Order 1–3–4–2
Distributor Rotation Clockwise

CAPACITIES

Year	Engine No. Cyl. Displacement (cc)	Engine Crankcase Capacity Including Filter (qts.)	Transmission Pts to Refill After Draining		Drive Axle (pts.)	Gasoline Tank (gals.)	Cooling System (qts.)	
			Manual	Automatic (Total Capacity)			With Heater	With A/C
'81–'85	4-1597	4.0	5.0①	②	③	④	6.7	8.1
'84–'87	4-2000	7.2⑤	①	—	③	⑥	8.1	8.1
'84–'87	4-2300	4.5⑦	6.1	②	③	⑥	8.1	8.1
'85–'87	4-160	4.0	5.0①	②	③	⑥	8.1	8.1

① 5 speed: 6.1
② Total dry capacity–converter, cooler and sump drained.
 1981–82: 20 pts, 1983–87: 16.6 pts.
 Partial fluid change (pan sump only), add 8 pts, start engine and check level. Add necessary fluid until correct level is reached.
③ Included in transmission capacity

④ 1981–'82: 10 gal. Standard
  11.3 gal. Extended range
 1983–'85: 10 gal. FE models
  13 gal. Standard
  13 gal. EXP/LN7
⑤ Capacity for complete system-pan capacity is 5.3 qts.

⑥ 1984: 14 gal; 1985–87; 15.2 gal.
⑦ After filter replacement, add 4 qts of oil and run engine. Shut engine off and check oil level. Add ½ qt if necessary.

CAMSHAFT SPECIFICATIONS

(All measurements are given in inches.)

Year	Engine No. Cyl. Displacement (cc)	Lobe Lift	Valve Lift @ Zero Lash		Camshaft End Play	Journal-to-Bearing Clearance	Journal Diameter	Journal Out-of-Round Limit
			Intake	Exhaust				
81–'85	4-1597	.229①	.377②	.377②	.0018–.006	.0008–.0028	③	.008
84–'87	4-2000	—	—	—	.008–.0059	.001–.0026	1.2582–1.2589	—
84–'87	4-2300	④	.392	.377	.009	.001–.003	⑤	.005
85–'87	4-1860	.240	.468②	.468②	.0018–.006	.0013–.0033	1.8017–1.8007	.003

① HO and EFI: .240
② HO and EFI: .396
③ No. 1: 1.761–1.762
  No. 2: 1.771–1.772
  No. 3: 1.781–1.782
  No. 4: 1.791–1.792
  No. 5: 1.801–1.802
④ IN: .249
  EX: .239
⑤ Not Available

CRANKSHAFT AND CONNECTING ROD SPECIFICATIONS

(All measurements are given in inches.)

Year	Engine No. Cyl. Displacement (cc)	Crankshaft Main Brg. Journal Dia.	Crankshaft Main Brg. Oil Clearance	Crankshaft Shaft End-Play	Crankshaft Thrust on No.	Connecting Rod Journal Diameter	Connecting Rod Oil Clearance	Connecting Rod Side Clearance
'81–'85	4-1597	2.2826–2.2834	0.0008–0.0015	0.004–0.008	3	1.885–1.886	0.0002–0.0003	0.004–0.011
'84–'87	4-2000	2.3598–2.3605	0.0012–0.0020	0.0016	3	2.0055–2.0061	0.0031	0.0043–0.0103
'84–'87	4-2300	2.2489–2.2490	0.0008–0.0024	0.004–0.008	3	2.1232–2.1240	0.0008–0.0015	0.0035–0.0105
'85–'87	4-1860	2.2835–2.2827	0.0008–0.0015	0.004–0.008	3	1.8862–1.8854	0.0008–0.0015	0.004–0.011

VALVE SPECIFICATIONS

Year	Engine No. Cyl. Displacement (cc)	Seat Angle (deg.)	Face Angle (deg.)	Spring Test Pressure (lbs @ in.)	Spring Installed Height (in.)	Stem-to-Guide Clearance (in.) Intake	Stem-to-Guide Clearance (in.) Exhaust	Stem Diameter (in.) Intake	Stem Diameter (in.) Exhaust
'81	4-1597	45	45½	180 @ 1.09	1.46	0.0008–0.0027	0.0015–0.0032	0.316	0.315
'82	4-1597	45	45½	180 @ 1.09	1.46	0.0010	0.00210	0.320	0.310
'83–'85	4-1597	45	45½	200 @ 1.09①	1.480②	0.0008–0.0027	0.0018–0.0037	0.316	0.315
'84–'87	4-2000	45	45	—	1.7760	0.0016–0.0029	0.0018–0.0031	0.3138	0.3138
'84–'87	4-2300	45	45½	182 @ 1.107	1.49	0.0018	0.0023	0.3415	0.3411
'85–'87	4-1860	45	45.6	200 @ 1.09③	1.86④	0.0008–0.0027	0.0018–0.0037	0.3167	0.3156

①H.O and EFI Engines: 206 @ 1.09
②H.O and EFI Engines: 1.450–1.480
③EFI Engines 216 @ 1.016
④EFI Engines 1.90

PISTON AND RING SPECIFICATIONS

(All measurements are given in inches.)

Year	Engine Displacement (cc)	Piston Clearance	Ring Gap Top Compression	Ring Gap Bottom Compression	Ring Gap Oil Control	Ring Side Clearance Top Compression	Ring Side Clearance Bottom Compression	Ring Side Clearance Oil Control
'81	1597	0.0008–0.0016	0.012–0.020	0.012–0.020	0.016–0.055	0.001–0.003	0.002–0.003	Snug
'82	1597	0.0012–0.0020	0.012–0.020	0.012–0.020	0.016–0.055	0.001–0.003	0.002–0.003	Snug
'83–'85	1597	0.0018–0.0026	0.012–0.020	0.012–0.020	0.016–0.055	0.001–0.003	0.002–0.003	Snug
'84–'87	2000	0.0013–0.0020	0.0079–0.0157	0.0079–0.0157	0.0079–0.0157	0.0020–0.0035	0.0016–0.0031	Snug
'84–'87	2300	0.0013–0.0021	0.008–0.016	0.008–0.016	0.015–0.055	0.002–0.004	0.002–0.004	Snug
'85–'87	1860	0.0016–0.0024	0.010–0.020	0.010–0.020	0.016–0.055	0.0015–0.0032	0.0015–0.0035	Snug

TORQUE SPECIFICATIONS

(All readings in ft. lbs.)

Year	Engine No. Cyl. Displacement (cc)	Cylinder Head Bolts	Rod Bearing Bolts	Main Bearing Bolts	Crankshaft Bolt	Flywheel-to-Crankshaft Bolts	Manifold Intake	Manifold Exhaust
'81–'85	4-1597	①	19–25	67–80	74–90	59–69	12–15②	15–20
'84–'87	4-2000	①	51–54	61–65	115–123	130–137	12–16	16–19③
'84–'87	4-2300	81③	21–26	60–74	82–103	54–64	15–23	20–30③
'85–'87	4-1860	①	19–25	67–80	74–90	54–64	12–15	15–20

① See head removal procedure for instructions
② Manifold stud nuts: 12–13 ft. lbs.
③ Tighten in two stages

WHEEL ALIGNMENT SPECIFICATIONS

Year	Model	Caster Range (deg.) ■ ▲	Caster Pref. Setting (deg.)	Camber Range (deg.) ■	Camber Pref. Setting (deg.)	Toe (in.)	Steering Axis Inclin. (deg.)
'81–'83	Escort, Lynx, EXP/LN7	9/16P to 2 1/16P	1 5/16P	—	—	1/32 to 7/32	—
		—	—	(Left) 1 13/32P to 2 29/32P	2 5/32 P	—	14 21/32
		—	—		1 23/32P	—	15 3/32
				(Right) 3 1/32P to 2 15/32P			
'84–'87	Escort, Lynx, EXP	5/8P to 2 1/8P	1 2/3P	—	—	1/64 (in) to 7/32 (out)	—
				(Left) 1 3/8P to 2 7/8P	2 1/8P	—	14 21/32
				(Right) 15/16P to 2 7/16P	1 11/16P	—	15 3/32
'84–'87	Tempo, Topaz	9/16P to 2 1/16P	1 5/16P	—	—	1/32 (in) to 7/32 (out)	—
				(Left) 1 1/8P to 2 5/8P	1 7/8P	—	14 5/8
				(Right) 11/16P to 2 3/16P	1 1/2P	—	15 1/8

■ Caster and chamber are pre-set at the factory and cannot be adjusted

▲ Caster measurements must be made on the left side by turning left wheel through the prescribed angle of sweep and on the right side by turning right wheel through prescribed angle of sweep for the equipment being used. When using alignment equipment designed to measure caster on both the right and left side, turning only one wheel will result in a significant error in caster angle for the opposite side.

ENGINE ELECTRICAL

Alternator

REMOVAL & INSTALLATION

For alternator testing and diagnosis, refer to "Charging and Starting" in the Unit Repair section.

1. Disconnect the negative battery cable.
2. If the alternator is equipped with a pulley cover shield, remove the shield at this time.
3. Loosen the alternator pivot bolt. Remove the adjustment bracket to alternator bolt (and nut, if equipped). Pivot the alternator to gain slack in the drive belt and remove the belt.
4. Disconnect and label (for correct installation) the alternator wiring.

NOTE: Some models use a push-on wiring connector on the field and stator connections. Pull or push straight off or on, or damage to the connectors may occur.

5. Remove the pivot bolt and the alternator.
6. Install in the reverse order of removal. Adjust the drive belt tension so that there is approx. 1/4–1/2 in. deflection on the longest belt span between the pulleys. Reinstall the pulley shield, if equipped and connect the negative battery cable.

Voltage Regulator

NOTE: Three different types of regulators are used, depending on model, engine, alternator output and type of dash mounted charging indicator used (light or ammeter). The regulators are 100 percent solid state and are calibrated and preset by the manufacturer. No readjustment is required or possible on these regulators.

SERVICE

Whenever system components are being replaced the following precautions should be followed so that the charging system will work properly and the components will not be damaged.

1. Always use the proper alternator.

2. The electronic regulators are color coded for identification. Never install a different coded regulator for the one being replaced. General coding identification follows, if the regulator removed does not have the color mentioned, identify the output of the alternator and method of charging indication, then consult a parts department to obtain the correct regulator. A black coded regulator is used in systems which use a signal lamp for charging indication. Gray coded regulators are used with an ammeter gauge. Neutral coded regulators are used on models equipped with a diesel engine. The special regulator must be used on vehicles equipped with a diesel engine to prevent glow plug failure.

3. Models using a charging lamp indicator are equipped with a 500 ohm resistor on the back of the instrument panel.

REMOVAL & INSTALLATION

1. Disconnect the negative battery cable.
2. Unplug the wiring harness from the regulator.
3. Remove the regulator mounting bolts.
4. Install in the reverse order of removal.

Starter

For all starter overhaul procedures, please refer to "Charging and Starting" in the Unit Repair section.

REMOVAL & INSTALLATION

1. Disconnect the negative battery cable.
2. Raise and safely support the front of the vehicle on jackstands. Disconnect he starter cable from the starter motor.
3. On models that are equipped with a manual transaxle, remove the three nuts that attach the roll resistor brace to the starter mounting studs at the transaxle. Remove the brace. On models that are equipped with an automatic transaxle, remove the hose bracket mounted on the starter studs.
4. Remove the two bolts attaching the rear starter support bracket, remove the retaining nut from the rear of the starter motor and remove the support bracket.
5. On models equipped with a manual transaxle, remove the three starter mounting studs and the starter motor. On models equipped with an automatic transaxle, remove the two starter mounting studs, mounting bolt and the starter motor.
6. Position the starter motor on the transaxle housing and install in the reverse order of removal. Tighten the mounting bolts or studs to 30–40 ft. lbs.

Distributor

NOTE: Either a conventional Duraspark system, a Thickfilm Integrated (TFI) system, or a Thickfilm Integrated IV (TFI-IV) is used depending on year and model. Trouble-shooting and servicing of the systems are contained in the "Electronic Ignition Systems" Unit Repair section.

REMOVAL & INSTALLATION

1.6L & 1.9L Engine

The camshaft-driven distributor is located at the top left end of the cylinder head. It is retained by two hold down bolts at the base of the distributor shaft housing.

1. Turn engine to No. 1 piston at TDC of the compression stroke. Disconnect negative battery cable. Disconnect the vacuum hose(s), if equipped, from the advance unit. Disconnect the wiring harness at the distributor.

2. Remove the capscrews and remove the distributor cap.
3. Scribe a mark on the distributor body, showing the position of the ignition rotor. Scribe another mark on the distributor body and cylinder head, showing the position of the body in relation to the head. These marks can be used for reference when installing the distributor, as long as the engine remains undisturbed (turned).
4. Remove the two distributor holddown bolts. Pull the distributor out of the head.
5. To install the distributor with the engine undisturbed, place the distributor in the cylinder head, seating the off-set tang of the drive coupling into the groove on the end of the camshaft. Install the two distributor holddown screws and tighten them so that the distributor can just barely be moved. Install the rotor (if removed), the distributor cap and all wiring, then set the ignition timing.
6. If the crankshaft was rotated while the distributor was removed, the engine must be brought to TDC (Top Dead Center) on the compression stroke of the No. 1 cylinder. Remove the No. 1 spark plug. Place your finger over the hole and rotate the crankshaft slowly (use a wrench on the crankshaft pulley bolt) in the direction of normal engine rotation, until engine compression is felt.

— CAUTION —

Turn the engine only in the direction of normal rotation. Backward rotation will cause the cam belt to slip or lose teeth, altering engine timing.

7. When engine compression is felt at the spark plug hole, indicating that the piston is approaching TDC, continue to turn the crankshaft until the timing mark on the pulley is aligned with the "0" mark (timing marking) on the engine front cover. Turn the distributor shaft until the ignition rotor is at the No. 1 firing position. Install the distributor into the cylinder head, as outlined in Step 5 of this procedure.

2.3L Engine

The TFI-IV distributor is mounted on the side of the engine block. Some engines may be equipped with a "security" type distributor hold down bolt which requires a special wrench for removal. The TFI-IV distributor incorporates a "Hall Effect" vane switch stator assembly and an integrally mounted thickfilm module. When the "Hall Effect" device is turned on and a pulse is produced, the EEC-IV electronics computes crankshaft position and engine demand to calibrate spark

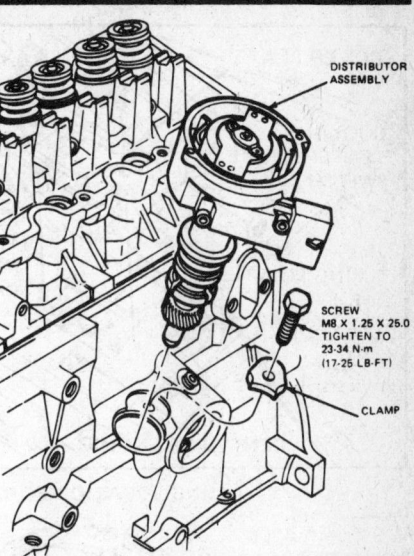

Distributor Installation on 2.3L HSC Engine

advance. Initial ignition timing adjustment/checking is necessary when the distributor has been removed. Repairs to the distributor are accomplished by a distributor replacement.

1. Turn engine to No. 1 piston at TDC of the compression stroke. Disconnect the negative battery cable.
2. Disconnect the wiring harness at the distributor. Mark No. 1 spark plug wire cap terminal location on the distributor body. Remove the coil wire from cap.
3. Remove the distributor cap with plug wires attached and position out of the way. Remove the rotor.
4. Remove the distributor base hold down bolt and clamp. Slowly remove the distributor from the engine. Be careful not to disturb the intermediate driveshaft.
5. Install in reverse order after aligning the center blade of the rotor with the reference mark made on the distributor body for No. 1 plug wire terminal location.
6. If the engine was disturbed (turned) while the distributor was out, the engine will have to be reset at TDC before installation.

IGNITION TIMING ADJUSTMENT

NOTE: If the vehicle (1.6 or 1.9L engines) is equipped with a barometric pressure switch (12A243), disconnect it from the ignition module and place a jumper wire across the pins at the ignition module connector (yellow and black wires). On 1.6L models equipped with EFI (electronic fuel injection), disconnect the single wire connector near the distributor prior to timing operation.

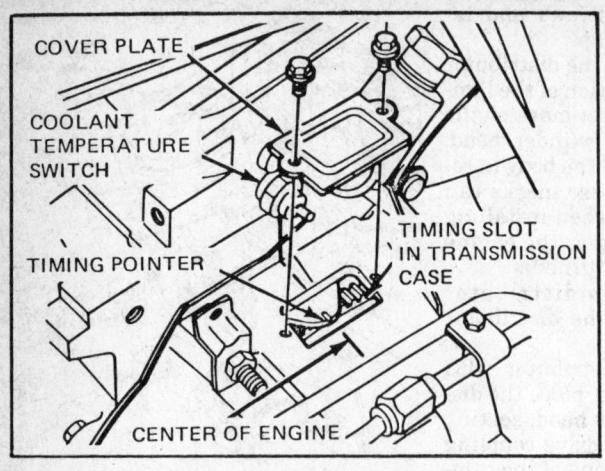

TIMING LOCATION FOR MTX

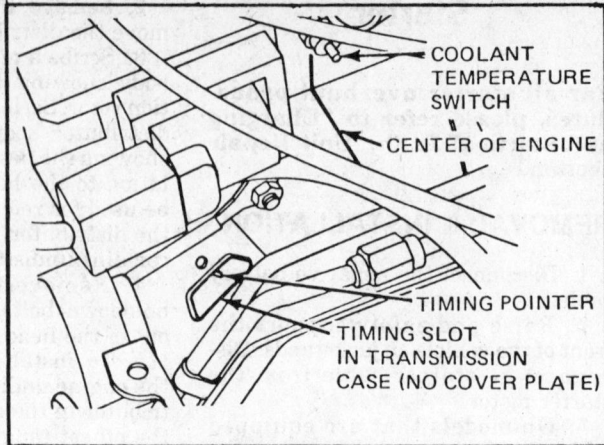

TIMING LOCATION FOR ATX

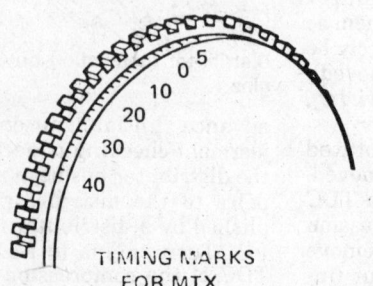

TIMING MARKS FOR MTX

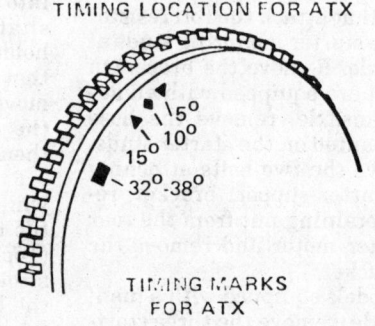

TIMING MARKS FOR ATX

Tempo/Topaz timing marks

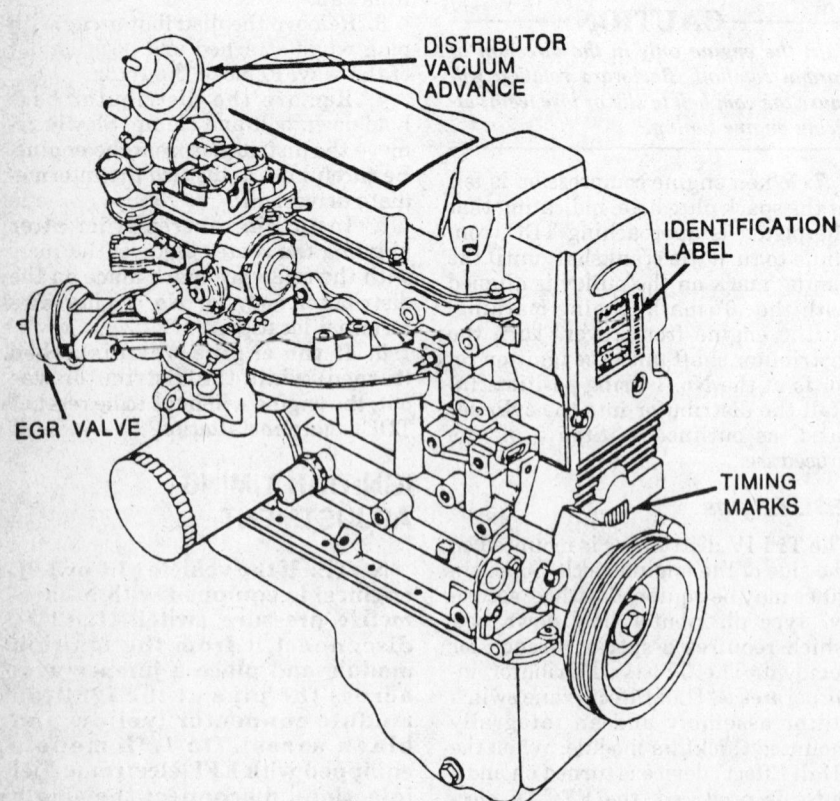

1.6L timing marks are on the front cover

1. Timing marks on 1.6L & 1.9L engines consist of a notch on the crankshaft pulley and a graduated scale molded into the camshaft drive belt cover. The number of degrees before or after TDC (top dead center) represented by each mark can be interpreted according to the decal affixed to the top of the belt cover (emission decal).

2. Timing marks on 2.3L engines are located on the flywheel edge (manual transaxle) or flywheel face (automatic transaxle) and are visible through a slot in the transaxle case at the back of the engine. A cover plate retained by two screws must be removed to view the timing marks on manual cars. Each mark (small graduation) equals two degrees. Early automatic cars have timing marks punched on the flywheel, the marks are 5° apart. The required degree mark should align with the timing slot pointer. Unless the emission decal specifies otherwise, timing for manual transaxle models is 10° BTDC and 15° BTDC for automatic transaxle models.

3. Turn the engine until No. 1 piston is at TDC on the compression stroke. Apply white paint or chalk to the rotating timing mark (notch on pulley or flywheel) after cleaning the metal surface.

4. Attach a timing light and tachometer to the engine. Start the engine and allow to idle until normal operating temperature is reached.

5. Shut off the engine. Refer to the emissions decal for timing, engine rpm vacuum hose (if equipped) status information. Disconnect and plug the distributor vacuum line(s) if required. On models equipped with a 1.6L EFI or 2.3L engine (EEC-IV), disconnect the ignition spout wire (usually black or yellow/light green with dots) from the single wire distributor connector.

6. Be sure the parking brake is applied and wheels blocked. Start the engine and place the transmission in gear specified on emissions decal. Check idle rpm and adjust if necessary.

7. Aim the flashing timing light at the timing marks. If the proper marks are not aligned, loosen the distributor hold down bolt/nut slightly and rotate the distributor body until the marks are aligned. Tighten the hold down.

8. Recheck the ignition timing, readjust if necessary. Shut off the engine and reconnect vacuum hoses or spout connector and barometric pressure switch (if equipped). Start engine and readjust idle rpm if necessary.

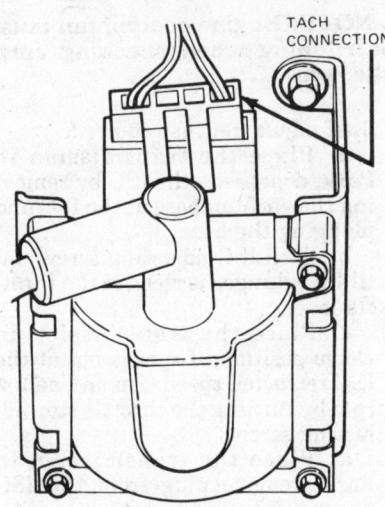

Tach Connection, "E" coil

Tachometer Hookup

Models equipped with a "Conventional" type coil have an adapter on the top of the coil that provides a clip marked "Tach Test". On models (TFI) equipped with an "E" type coil, the tach connection is made at the back of the wire harness connector. A cut-out is provided and the tachometer lead wire alligator clip can be connected to the dark green/yellow dotted wire of the electrical harness plug.

Diesel Glow Plugs

REMOVAL & INSTALLATION

1. Disconnect the battery ground cable from the battery, located in the luggage compartment.

2. Disconnect the glow plug harness from the glow plugs.

3. Using a 12mm deepwell socket, remove the glow plugs.

4. Install the glow plugs, using a 12mm deepwell socket. Tighten the glow plugs to 11–15 ft. lbs.

5. Connect the glow plug harness to the glow plugs. Tighten the nuts to 5–7 ft. lbs.

6. Connect the battery ground cable to the battery.

7. Check the glow plug system operation.

GASOLINE FUEL SYSTEM

NOTE: Many models use push-connect fuel line fittings. Refer to the "Ford Rear Wheel Drive" section for removal and installation procedures.

Mechanical Fuel Pump

REMOVAL & INSTALLATION

1. Loosen the threaded fuel lines at the fuel pump slightly. Have a rag handy to absorb any gasoline spillage.

2. Loosen the fuel pump mounting bolts two turns and free the fuel pump from the engine.

3. Bump the engine around with the starter until reduced pressure (hold pump against the engine) is felt. The fuel pump drive lobe and pushrod are now at their low point.

4. Remove fuel lines and vapor hose (if equipped). Remove the fuel pump mounting bolts and the fuel pump.

5. Remove fuel pump mounting gasket and clean the engine mounting surface.

6. Remove and check the pushrod. Replace is wear is apparent. Install the pushrod, new mounting gasket and fuel pump in the reverse order of removal. Fuel pump mounting bolts are torqued to 11–19 ft. lbs. Pushrod length should be: 1.6L engines = 3.88–3.90 in.; 2.3L engine = 2.43 in.

Electric Fuel Pump

REMOVAL & INSTALLATION

NOTE: The fuel system on in-

jected models is under pressure. Fuel system pressure must be relieved by connecting a service tool to the relief valve on the fuel rail before system service can be performed.

1. Relieve fuel system pressure. Raise and safely support the rear of the car on jackstands.

2. Remove the upper pump mounting bolt after disconnecting the parking brake cable from the mounting clip.

NOTE: Refer to the "Ford Rear Wheel Drive" section for details on disconnecting push type fuel line fitting connectors.

3. Disconnect the electrical connector from the pump. Disconnect the fuel line from the pump to injectors.

4. Disconnect and plug the fuel line from the gas tank. Remove the fuel pump.

5. Install in reverse order. Pressurize the system by turning the ignition key on and off several times, allowing the key to remain on for at least two seconds each time.

6. Start the engine and check for fuel leaks.

Fuel Filter

REMOVAL & INSTALLATION

NOTE: Before removing the fuel filter on models equipped with fuel injection, the fuel system pressure must be relieved.

1. If the fuel filter is located in the carburetor inlet and connected with rubber hose, remove clamps and inlet rubber hose. Unscrew the filter from the carburetor.

2. If the fuel filter is mounted in the carburetor inlet and connected with steel line, hold the filter nut with the proper size wrench and unscrew the inverted fitting nut on the steel line using a flare or suitable wrench. Remove the line and unscrew the filter.

NOTE: See the "Ford Rear Wheel Drive" section for details on disconnecting push type fuel line connectors.

3. If the fuel filter is connected to the carburetor with steel lines, hold the filter nuts with proper wrench and disconnect the steel lines with a flare or suitable wrench.

4. Install the fuel filter in the reverse order of removal.

Carburetors

Due to the complex nature of modern feedback carburetors and fuel injection systems, comprehensive diagnosis and testing procedures fall outside the confines of this repair manual. For complete information on diagnosis, testing and repair procedures concerning all modern feedback carburetors, please refer to *Chilton's Guide To Fuel Injection And Feedback Carburetors.*

REMOVAL & INSTALLATION

1. Remove the air cleaner assembly. Disconnect the throttle control cable and speed control (if equipped).
2. Disconnect the fuel line at the fuel filter or carburetor. Label (for identification and location) and disconnect all vacuum lines, wires, harnesses and linkage attached to the carburetor. Disconnect the TV linkage if the car is equipped with an automatic transaxle.
3. Remove the carburetor base mounting nuts and remove the carburetor.
4. Install in the reverse order. Start engine and allow to reach normal operating temperature. Adjust curb idle.

OVERHAUL

For all overhaul information, please refer to "Carburetors" In the Unit Repair section.

IDLE SPEED AND MIXTURE ADJUSTMENT

NOTE: Most carburetor mixture adjustments are factory set and are designed to reduce engine emissions. Mixture adjustments should be made only if absolutely necessary and should be checked on a machine as soon as possible to see if the emission level is OK. A tachometer must be used while making any idle rpm adjustments.

Refer to the proceeding section for tachometer hook up instructions. Refer to emissions decal for idle speed and specific instructions.

1.6L Engine w/740-2V without Idle Speed Control

1. Place the transmission in Neutral or Park, set the parking brake and block the wheels.
2. Bring the engine to normal operating temperature.
3. Disconnect and plug the vacuum hose at the thermactor air control valve bypass section.
4. Place the fast idle adjustment screw on the second highest step of the fast idle cam. Run engine until cooling fan comes on.
5. Slightly depress the throttle to allow the fast idle cam to rotate. Place the transmission in specified gear, and check/adjust the curb idle rpm to specification.

NOTE: Engine cooling fan must be running when checking curb idle rpm. Use a jumper wire if necessary.

6. Place the transmission in Neutral or Park. Rev the engine momentarily. Place transmission in specified position and recheck curb idle rpm. Readjust if required.
7. If the vehicle is equipped with a dashpot, check/adjust clearance to specification.
8. Remove the plug from the hose at the thermactor air control valve bypass section and reconnect.
9. If the vehicle is equipped with an automatic transmission and curb idle adjustment is more than 50 rpm, an automatic transmission linkage adjustment may be necessary.

1.6L Engine w/740-2V and Mechanical Vacuum Idle Speed Control (ISC)

1. Place the transmission in Neutral or Park, set the parking brake and block the wheels.

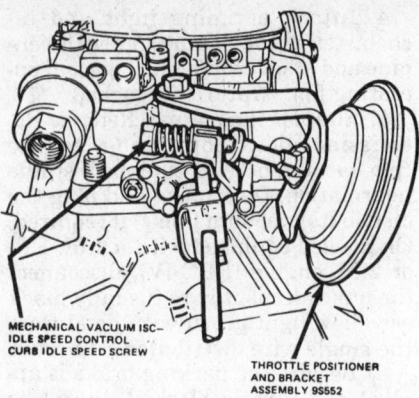

MECHANICAL VACUUM ISC— IDLE SPEED CONTROL CURB IDLE SPEED SCREW

THROTTLE POSITIONER AND BRACKET ASSEMBLY 95552

Fast idle and throttle stop adjustment points

2. Bring the engine to normal operating temperature.
3. Disconnect and plug the vacuum hose at the thermactor air control valve bypass section.
4. Place the fast idle adjustment screw on the second highest step of the fast idle cam. Run the engine until cooling fan comes on.
5. Slightly depress the throttle to allow fast idle cam to rotate. Place the transmission in Drive (fan on) and check curb idle rpm to specification.

NOTE: Engine cooling fan must be running when checking curb idle rpm.

6. If adjustment is required:
   a. Place the transmission in Park, deactivate the ISC by removing the vacuum hose at the ISC and plugging the hose.
   b. Turn ISC adjusting screw until ISC plunger is clear of the throttle lever.
   c. Place the transmission in Drive position, if rpm is not at the ISC retracted speed (fan on), adjust rpm by turning the throttle stop adjusting screw.
   d. Place the transmission in Park, remove plug from the ISC vacuum line and reconnect to ISC.
   e. Place transmission in Drive if rpm is not at the curb idle speed (fan on), adjust by turning the ISC adjustment screw.
7. Place transmission in Neutral or Park. Rev the engine momentarily. Place the transmission in specified position and recheck curb idle rpm. Readjust if required.
8. Remove the plug from the thermactor air control valve bypass section hose and reconnect.
9. If the vehicle is equipped with an automatic transmission and curb idle adjustment is more than 50 rpm, an automatic transmission linkage adjustment may be necessary.

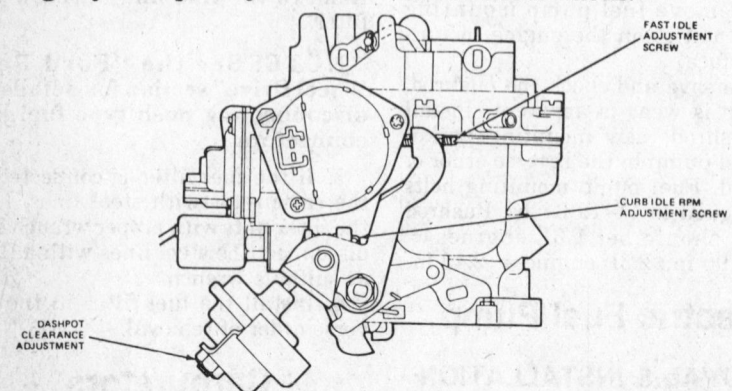

FAST IDLE ADJUSTMENT SCREW

CURB IDLE RPM ADJUSTMENT SCREW

DASHPOT CLEARANCE ADJUSTMENT

1.6L fast idle and curb idle adjustment front points

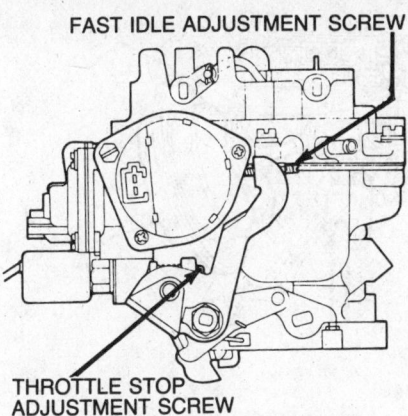

FAST IDLE ADJUSTMENT SCREW

THROTTLE STOP
ADJUSTMENT SCREW

1.6L with idle speed control (ISC) adjustment points

1.6L Engine w/740-2V–Vacuum Operated Throttle Modulator (VOTM)

1. Place the transmission in Neutral or Park, set the parking brake and block the wheels.
2. Bring the engine to normal operating temperature.
3. To check/adjust VOTM rpm:
• Place A/C heat sector in Heat position, blower switch on High.
• Disconnect the vacuum hose from VOTM and plug, install a slave vacuum hose from the intake manifold vacuum to the VOTM.
4. Disconnect and plug the vacuum hose at the thermactor air control valve bypass section.
5. Run the engine until the engine cooling fan comes on.
6. Place the transmission in specified gear, and check/adjust VOTM rpm to specification.

NOTE: Engine cooling fan must be running when checking VOTM rpm. Adjust rpm by turning screw on VOTM.

7. Remove the slave vacuum hose. Remove the plug from the VOTM vacuum hose and reconnect the hose to the VOTM.
8. Return the intake manifold vacuum supply source to original location.

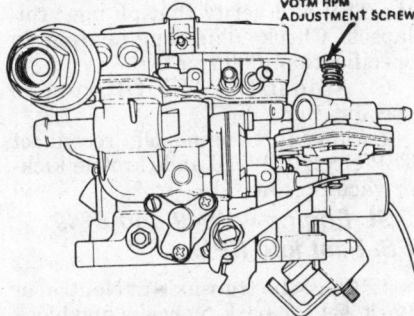

VOTM RPM
ADJUSTMENT SCREW

1.6L VOTM idle adjustment point

9. Remove the plug from the vacuum hose at the thermactor air control valve bypass section and reconnect.

1.9L w/740 2V and Mechanical Vacuum Idle Speed Control (ISC)

1. Connect a tachometer and timing light to the engine. Place the transmission in Neutral or Park, and the A/C-Heater selector in the Off position. Set the parking brake and block the wheels.
2. Run the engine until normal operating temperature is reached.
3. Check the engine ignition timing. Adjust as necessary.
4. Disconnect and plug the vacuum hose at the thermactor air control bypass.
5. Place the fast idle adjusting screw on the second step of the fast idle cam. Run the engine until the cooling fan comes on.
6. Depress the throttle slightly. The engine cooling fan must be on to allow the fast idle cam to rotate. Place the ATX selector in Drive. Check idle rpm, adjust if necessary.
7. If adjustment is required; place the transmission in Park or Neutral (if manual). Deactivate the ISC (idle speed control) by removing the vacuum hose at the ISC and plugging the hose.
8. If the full stroke rpm is not within specifications, adjust the rpm by turning the full stroke speed adjusting screw.
9. Connect a hand vacuum pump to the ISC and supply enough vacuum to retract the ISC plunger clear of the full stroke adjusting screw.
10. Place the transmission in Drive if automatic, Neutral if manual. If the throttle stop rpm is not to specification, adjust by turning the throttle stop adjusting screw.
11. Adjust the dashpot, if necessary at this point; turn the engine off and make sure the ISC plunger is retracted.
12. Depress the dashpot plunger into the dashpot assembly. Measure the distance between the throttle lever pad and the dashpot plunger. Adjust to specification by loosening the dashpot lock nut and turning the dashpot assembly. When adjustment is completed, tighten the locknut. Restart the engine and allow it to reach normal operating temperature with the cooling fan on.
13. Place the transmission in Park or Neutral if manual, remove the vacuum hand pump from the ISC. Remove the plug from the vacuum line and reconnect the ISC.
14. Place the transmission in Drive or Neutral if manual and check idle

rpm. If rpm is not within specification (cooling fan on), adjust by turning the ISC curb idle adjusting screw. The plug must be removed from the back of the unit during adjustment. Reinstall plug after adjustment is completed.
15. Place the transmission in Neutral or Park. Rev the engine momentarily. Place the transmission in Drive or Neutral if manual and recheck idle rpm. Readjust if necessary.
16. Remove the plug from the thermactor air control bypass section hose and reconnect the hose.
17. If the vehicle is equipped with an automatic transmission and the curb idle requires more than a 50 rpm adjustment, refer to the Automatic Transmission Linkage Adjustment.

Dashpot Clearance Adjustment – 1.6L Engine

NOTE: If the carburetor is equipped with a dashpot, it must be adjusted if the curb idle speed is adjusted.

1. With the engine OFF, push the dashpot plunger in as far as possible and check the clearance between the plunger and the throttle lever pad.

NOTE: Refer to the emissions decal for proper dashpot clearance. If not available, set clearance to 0.138 ± 0.020 in.

2. Adjust the dashpot clearance by loosening the mounting locknut and rotating the dashpot.

— **CAUTION** —

If the locknut is very tight, remove the mounting bracket, hold it in a suitable device, so that it will not bend, and loosen the locknut. Reinstall bracket and dashpot.

3. After gaining the required clearance, tighten the locknut and recheck adjustment.

Fast Idle RPM – 1.6L & 1.9L Engine

NOTE: Refer to the emissions decal for the required fast idle speed.

1. Place the transmission in Neutral or Park, set the parking brake and block the wheels.
2. Bring the engine to the normal operating temperature.
3. Disconnect the vacuum hose at the EGR and plug.
4. Place the fast idle adjustment screw on the second highest step of the fast idle cam. Run engine until cooling fan comes on.
5. Check/adjust fast idle rpm to

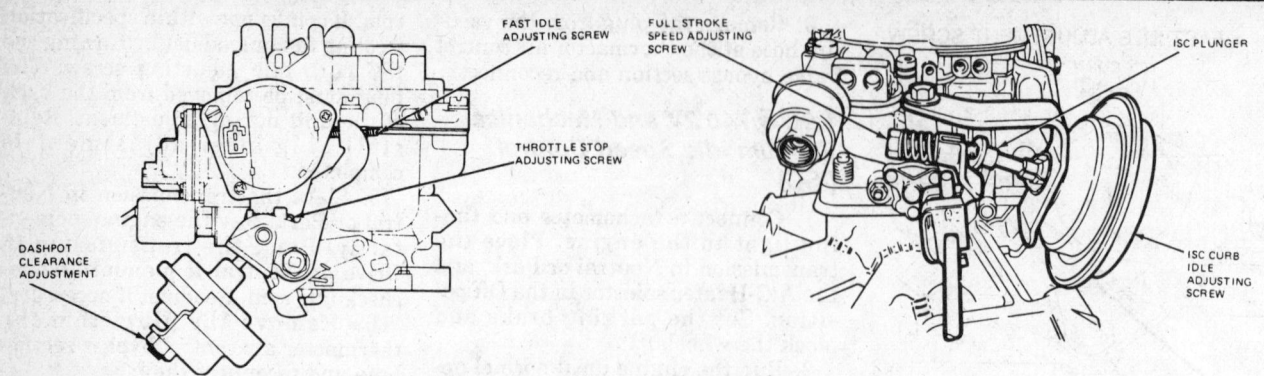

740 carburetor fast idle and curb idle rpm adjustments

specification. If adjustment is required, loosen locknut, adjust and retighten.

NOTE: Engine cooling fan must be running when checking fast idle rpm. Use a jumper wire if necessary.

6. Remove the plug from the EGR hose and reconnect.

Air Conditioning/Throttle Kicker Adjustment – 1.6L Engine

1. Place the transmission in Neutral or Park.
2. Bring engine to normal operating temperature.
3. Identify vacuum source to air by-pass section of air supply control valve. If vacuum hose is connected to carburetor, disconnect and plug hose at air supply control valve. Install slave vacuum hose between intake manifold and air bypass connection on air supply control valve.
4. To check/adjust A/C or throttle kicker rpm:
•If vehicle is equipped with A/C, place selector to maximum cooling, blower switch on High. Disconnect A/C compressor clutch wire.
•If vehicle is equipped with kicker and no A/C, disconnect vacuum hose from kicker and plug, install slave vacuum hose from intake manifold vacuum to kicker.
5. Run engine until engine cooling fan comes on.
6. Place transmission in specified gear and check/adjust A/C or throttle kicker rpm to specification.

NOTE: Engine cooling fan must be running when checking A/C or throttle kicker rpm. Adjust rpm by turning screw on kicker.

7. If slave vacuum hose was installed to check/adjust kicker rpm, remove slave vacuum hose. Remove plug from kicker vacuum hose and reconnect hose to kicker.

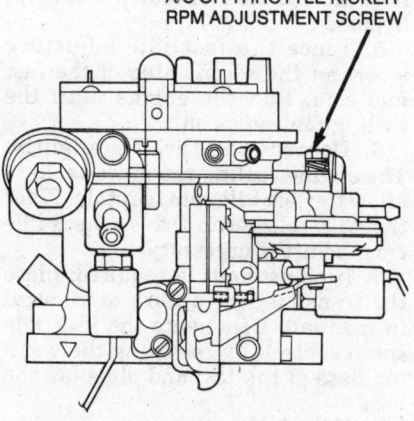

A/C or Throttle Kicker adjustment

8. Remove slave vacuum hose. Return intake manifold supply source to original condition. Remove plug from carburetor vacuum hose and reconnect to air bypass valve.

2.3L Engine w/1949 and 6149 FB–Curb Idle RPM

NOTE: A/C on rpm is non-adjustable. TSP-Off rpm is not required. Verify that TSP plunger extends with ignition key ON.

1. Place the transaxle in Neutral or Park, set the parking brake and block the wheels.
2. Disconnect the throttle kicker vacuum line and plug.
3. Bring the engine to normal operating temperature (cooling fan should cycle).
4. Place the A/C selector in the Off position.
5. Place gear selector in specified position.
6. Activate the cooling fan by grounding the control wire with a jumper wire.
7. Check/adjust curb idle rpm. If adjustment is required, turn curb idle adjusting screw.

8. Place the transaxle in Neutral or Park. Rev the engine momentarily. Place the transaxle in specified position and recheck curb idle rpm. Readjust if required.
9. Reconnect the cooling fan wiring.
10. Turn the ignition key to the OFF position.
11. Reconnect the vacuum line to the throttle kicker.
12. If the vehicle is equipped with an automatic transaxle and curb idle adjustment exceeds 50 rpm, an automatic transaxle linkagae adjustment may be necessary.
13. Remove all test equipment and reinstall the air cleaner assembly.

2.3L Engine w/1949 and 6149 FB–TSP Off RPM

NOTE: This adjustment is not required as part of a normal engine idle RPM check/adjustment. If engine continues to run after ignition key is turned to OFF position.

1. Place the transaxle in Neutral or Park, set the parking brake and block the wheels.
2. Bring the engine to normal operating temperature.
3. Disconnect the throttle kicker vacuum line and plug.
4. Place the A/C selector to Off position.
5. Disconnect the electrical lead to the TSP and verify that plunger collapses. Check/adjust engine rpm to specification (600 rpm).
6. Adjust the TSP Off rpm to specification.
7. Shut the engine off, reconnect TSP electrical lead and throttle kicker vacuum line.

2.3L Engine w/1949 and 6149 FB–Fast Idle RPM

1. Place the transaxle in Neutral or Park, set the parking brake and block the wheels.

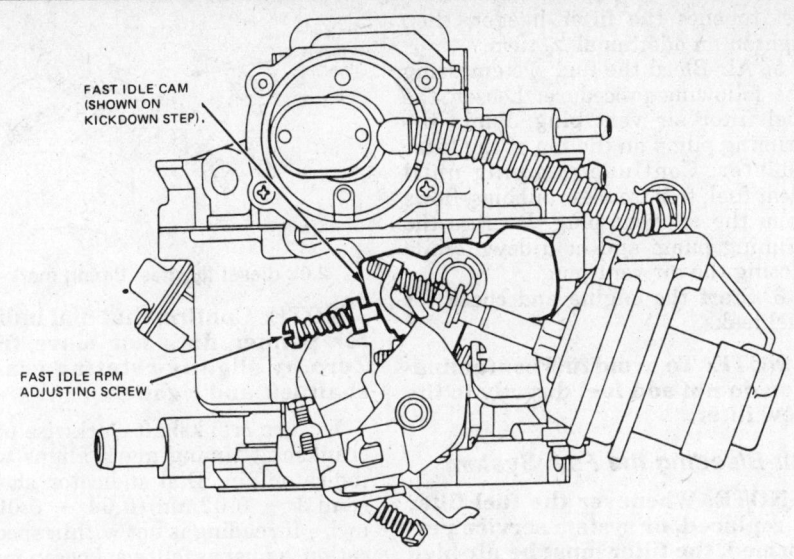

2.3L HSC curb idle adjustment

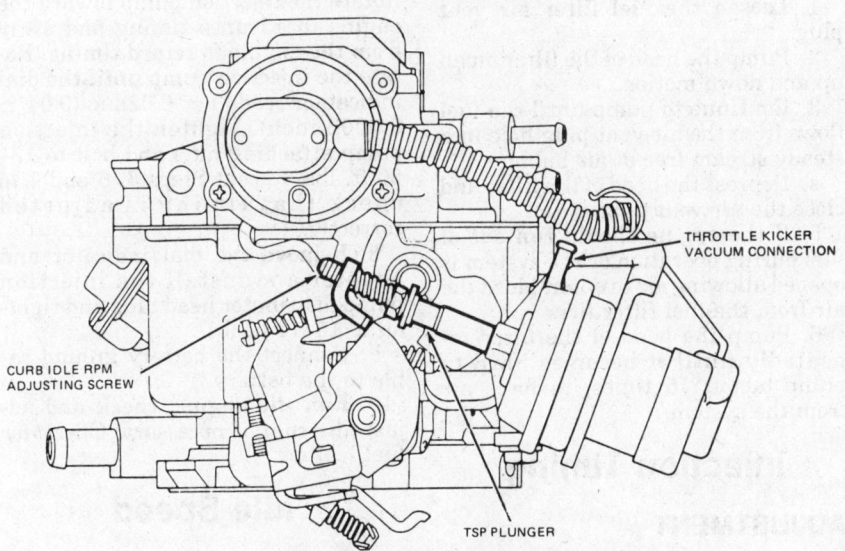

2.3L HSC fast idle adjustment

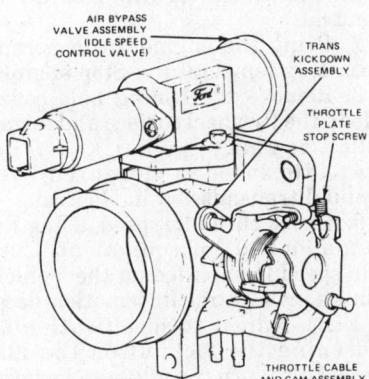

1.6L models with EFI adjustment point

2. Bring the engine to normal operating temperature with the carburetor set on second step of fast idle cam.

3. Return the throttle to normal idle position.

4. Place the A/C selector in the Off position.

5. Disconnect the vacuum hose at the EGR valve and plug.

6. Place the fast idle adjusting screw on the specified step of the fast idle cam.

7. Check/adjust the fast idle rpm to specification.

8. Rev the engine momentarily, allowing engine to return to idle and turn ignition key to OFF position.

9. Remove the plug from the EGR vacuum hose and reconnect.

Fuel Injection

Due to the complex nature of modern carburetor and fuel injection systems, comprehensive diagnosis and testing procedures fall outside the confines of this repair manual. For complete information on diagnosis, testing and repair procedures concerning all modern fuel injection systems, please refer to *Chilton's Guide To Fuel Injection And Feedback Carburetors.*

IDLE SPEED ADJUSTMENT

1.6L Engine w/Electronic Fuel Injection (EFI)–Initial Engine RPM Adjustment (ISC Disconnected)

The purpose of this procedure is to provide a means of verifying the initial engine rpm setting with the ISC disconnected. If engine idle rpm is not within specification after performing this procedure, it will be necessary to have 1.6L EFI EEC IV diagnostics performed.

NOTE: Curb idle rpm is controlled by the EEC IV processor and the Idle Speed Control (ISC) device (part of the fuel charging assembly).

1. Place the transmission in Neutral or Park, set the parking brake and block the wheel.

2. Bring the engine to the normal operating temperature and shut engine off.

3. Disconnect vacuum connector at the EGR solenoid and plug both lines.

4. Disconnect the idle speed control (ISC) power lead.

5. Electric cooling fan must be on during the idle speed setting procedure.

6. Start the engine and operate at 2000 rpm for 60 seconds.

7. Place transmission in Neutral for M/T and Drive for A/T, check/adjust initial engine rpm within 120 seconds by adjusting throttle plate screw.

8. If idle adjustment is not completed with 120 second time limit, shut engine Off, restart and repeat Steps 6 and 7.

9. If the vehicle is equipped with an automatic transmission and initial engine RPM adjustment increases or decreases by more than 50 rpm, an automatic transmission linkage adjustment may be necessary.

10. Turn the engine Off and remove the plugs from the EGR vacuum lines at the EGR solenoid and reconnect.

11. Reconnect the speed control (ISC) power lead.

2.3L Engine w/CFI and Idle Speed Control (ISC)

NOTE: Curb and fast idle

speeds are controlled by the EEC IV processor and the ISC control device. If the control system is operating properly, these speeds are fixed and cannot be changed by traditional adjusting techniques. If a too fast or slow problem exists appropriate EEC IV diagnostic testing will be required.

1. To test the ISC motor for proper operation, start the engine and run for at least 30 seconds, turn the key OFF and visually inspect the ISC motor shaft for retracting and repositioning.

2. To check the curb idle speed; set the parking brake and block the wheels. Connect a tachometer to the engine. If the vehicle is equipped with an automatic parking brake release, place the transmission selector in Reverse when checking the idle speed rpm.

3. Make sure the engine is at normal operating temperature and all accessories are turned off. Check all hoses for vacuum leaks.

4. Idle engine for about 120 seconds and check rpm with the transmission in the proper gear, Drive or Reverse for automatic, Neutral for manual. Place the transmission back into Neutral or Park, engine rpm should increase by approximately 100 rpm.

5. Lightly step on and off the accelerator. The engine rpm should return to specification. If the idle rpm remains high, repeat the sequence. It may take at least 120 seconds for the system to "learn". If correct idle specifications are not present EEC IV system check is required.

DIESEL FUEL SYSTEM

NOTE: Many models use push connect fuel line fittings. Refer to the Ford Rear Wheel Drive section for removal and installation procedures.

Fuel Filter

REMOVAL & INSTALLATION

1. Remove the spin-on filter by turning counterclockwise with hands or suitable tool, and discard filter.

2. Clean the filter mounting surfaces.

3. Coat the gasket of the new filter with clean diesel fuel.

4. Tighten the filter until the gas-

ket touches the filter header, then tighten an additional ½ turn.

5. Air-Bleed the fuel system using the following procedure: Loosen the fuel filter air vent plug. Pump the priming pump on the top of the filter adapter. Continue pumping until clear fuel, free from air bubbles, flows from the air vent plug. Depress the priming pump and hold down while closing the air vent plug.

6. Start the engine and check for fuel leaks.

NOTE: To avoid fuel contamination do not add fuel directly to the new filter.

Air-Bleeding the Fuel System

NOTE: Whenever the fuel filter is replaced, or system service performed, the filter must be air-bled as follows.

1. Loosen the fuel filter air vent plug.

2. Pump the head of the filter an an up and down motion.

3. Continue to pump until the fuel flows from the air vent plug hole in a steady stream free of air bubbles.

4. Depress the head of the filter and close the air vent plug.

5. If the engine should run out of fuel during operation or the system is opened allowing air to enter, bleed the air from the fuel filter first.

6. Pump the head of the filter repeatedly until it becomes hard to pump (about 15 times) to force air from the system.

Injection Timing

ADJUSTMENT

NOTE: Engine coolant temperature must be above 80°C (176°F) before the injection timing can be checked and/or adjusted.

1. Disconnect the battery ground cable from the battery located in luggage compartment.

2. Remove the injection pump distributor head plug bolt and sealing washer.

3. Install Static Timing Gauge Adapter, Rotunda 14-0303 or equivalent with Metric Dial Indicator, so that indicator pointer is in contact with injection pump plunger.

4. Remove timing mark cover from transmission housing. Align timing mark (TDC) with pointer on the rear engine cover plate.

5. Rotate the crankshaft pulley slowly, counterclockwise until the dial indicator pointer stops moving (approximately 30°–50° BTDC).

6. Adjust dial indicator to Zero.

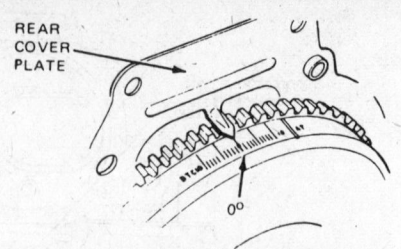

2.0L diesel flywheel timing mark

NOTE: Confirm that dial indicator pointer does not move from Zero by slightly rotating crankshaft left and right.

7. Turn crankshaft clockwise until crankshaft timing mark aligns with indicator pin. Dial indicator should read 1 + 0.02mm (0.04 + 0.0008 inch). If reading is not within specification, adjust as follows: Loosen injection pump attaching bolt and nuts. Rotate the injection pump toward the engine to advance timing and away from the engine to retard timing. Rotate the injection pump until the dial indicataor reads 1 ± 0.02mm (0.04 ± 0.0008 inch). Tighten the injection pump attaching nuts and bolt to 13–20 ft. lbs. Repeat Steps 5, 6 and 7 to check that timing is adjusted correctly.

8. Remove the dial indicator and adapter and install the injection pump distributor head plug and tighten to 10–14 ft. lbs.

9. Connect the battery ground cable to the battery.

10. Run the engine, check and adjust idle rpm, if necessary. Check for fuel leaks.

Idle Speed

ADJUSTMENT

1. Place the transmission in Neutral.

2. Bring the engine up to normal operating temperature. Stop engine.

3. Remove the timing hole cover. Clean the flywheel surface and install reflective tape.

4. Idle speed is measured with manual transmission in Neutral.

5. Check curb idle speed, using Rotunda 99-0001 or equivalent. Curb idle speed is specified on the Vehicle Emissions Control Information decal (VECI). Adjust to specification by loosening the locknut on the idle speed bolt. Turn the idle speed adjusting bolt clockwise to increase, or counterclockwise to decrease engine idle speed. Tighten the locknut.

6. Place transmission in Neutral. Rev engine momentarily and recheck the curb idle rpm. Readjust if necessary.

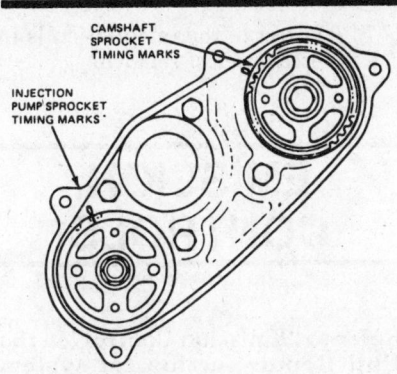

2.0L diesel camshaft and injector pump timing marks

7. Turn A/C On. Check the idle speed. Adjust to specification by loosening nut on the A/C throttle kicker and rotating screw.

Injection Pump

Removal & Installation

1. Disconnect the negative battery cable.
2. Disconnect the air inlet duct from the air cleaner and intake manifold. Cover the intake manifold opening to prevent dirt from entering.
3. Remove the rear timing belt cover and flywheel timing mark cover.
4. Remove the rear timing belt.
5. Disconnect the throttle and speed control cables. Disconnect the vacuum hoses from the altitude compensator and cold start diaphragm. Label the hoses for reinstallation identification.
6. Disconnect the fuel cut-off solenoid connector.
7. Disconnect the fuel supply and fuel return lines at the injection pump.
8. Remove the injector lines from the injection pump and nozzles. Cap the lines to prevent dirt from entering the system.
9. Rotate the injection pump sprocket until the timing marks are aligned. Install two M8 x 1.25mm bolts in the holes to hold the injection pump sprocket in position and remove the sprocket retaining nut.
10. Remove the sprocket using a suitable puller on two M8 x 1.25mm bolts threaded into the holes provided in the sprocket.
11. Remove the bolt attaching the injection pump to the pump front bracket. Remove the two nuts attaching the injection pump to the pump rear bracket and remove the pump.
12. Install the pump in the reverse order of removal. Tighten the rear bracket nuts to 23–24 ft. lbs. Front mounting bolt to 12–16 ft. lbs. Sprock-

et nut to 51–58 ft. lbs. Injector line capnuts to 18–22 ft. lbs.
13. Check and adjust the injection pump timing. Air bleed the fuel system.
14. Run the engine and check for fuel leaks.

Injection Nozzles and Injection Pump Lines

REMOVAL & INSTALLATION

1. Disconnect and remove the injection lines from the injection pump and nozzles. Cap all lines and fitting to prevent dirt contamination.
2. Remove the nuts attaching the fuel return line to the nozzles and remove the return line and seals.
3. Remove the injector nozzles using a 27mm socket. Remove the nozzle gaskets and washers from the nozzle seats using an O-ring pick tool T71P-19703-C or the equivalent.
4. Clean the outside of the nozzles with safety solvent and dry them thoroughly.
5. Position new sealing gaskets in the nozzle seats with the red painted surface facing up.
6. Position new copper gaskets in the nozzles bores. Install the nozzles and tighten to 44–51 ft. lbs.
7. Position the fuel return line on the nozzles using new seals. Install the retaining nuts and tighten to 10 ft. lbs.
8. Install the fuel lines on the injection pump and nozzles. Tignten to 18–22 ft. lbs.
9. Air bleed the fuel system. Run the engine and check for fuel leaks.

ENGINE COOLING

Radiator

REMOVAL & INSTALLATION

1. Disconnect the negative battery cable. Drain the cooling system.
2. On models equipped, remove the carburetor air intake tube and alternator air tube from the radiator support.
3. Remove the upper shroud mountings, disconnect the wire harness to the electric fan motor and remove the shroud and fan as an assembly.
4. Remove the upper and lower ra-

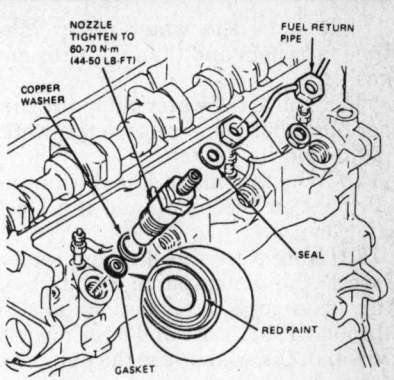

2.0L diesel injection nozzle installation

diator hoses. Disconnect the coolant recovery reservoir.
5. On models equipped with an automatic transaxle, disconnect and plug the transmission cooler lines at the radiator.
6. Remove radiator mountings, tilt radiator toward engine and lift from engine compartment.
7. Install the radiator in the reverse order. Be sure the lower radiator mounts are positioned correctly on the radiator support.

Water Pump

REMOVAL & INSTALLATION

1.6L & 1.9L Engine

1. Disconnect the negative battery cable. Drain the cooling system.
2. Remove the alternator drive belt. If equipped with an air conditioning or power steering, remove the drive belts.
3. Use a wrench on the crankshaft pulley to rotate the engine so No. 1 piston is on TDC of the compression stroke.

— **CAUTION** —
Turn the engine only in the direction of normal rotation. Backward rotation will cause the camshaft belt to slip or lose teeth.

4. Remove the cam belt cover.
5. Loosen the belt tensioner attaching bolts, then secure the tensioner over as far as possible.
6. Pull the belt from the camshaft tensioner, and water pump sprockets. Do not remove it from, or allow it to change its position on, the crankshaft sprocket.

NOTE: Do not rotate the engine with the camshaft belt removed.

7. Remove the camshaft sprocket.
8. Remove the rear timing cover stud. Remove the heater return tube hose connection at the water pump inlet tube.

9. Remove the water pump inlet tube fasteners and the inlet tube and gasket.

10. Remove the water pump to cylinder block bolts and remove the water pump and its gasket.

11. To install, make sure the mating surfaces on the pump and the block are clean.

12. Using a new gasket and sealer, install the water pump and tighten the bolts to 5–7 ft. lbs. on models through 1982. 1983 and later models, 30–40 ft. lbs. Make sure the pump impeller turns freely.

13. Install remaining parts in the reverse order of removal. Use new gaskets and sealer. Install the camshaft sprocket over the cam key. See below for procedure. Install new timing belt and adjust tension. See "Timing Belt Removal and Installation" for procedure.

Thermostat

REMOVAL & INSTALLATION

1. Drain the cooling system to below the thermostat level.

2. It is not necessary to remove the radiator hose from the thermostat housing.

3. Remove the retaining bolts from the thermostat housing and remove the thermostat.

4. Use a new gasket when replacing the thermostat. Coat the gasket with RTV Sealer.

NOTE: Some thermostats use a silicone sealer instead of a gasket.

2.0 Diesel Engine

1. Remove the front timing belt upper cover.

2. Loosen and remove the front timing belt, refer to timing belt in-vehicle services.

3. Drain the cooling system.

4. Raise the vehicle and support safely on jackstands.

5. Disconnect the lower radiator hose and heater hose from the water pump.

6. Disconnect the coolant tube from the thermostat housing and discard gasket.

7. Remove the three bolts attaching the water pump to the crankcase. Remove the water pump. Discard gasket.

8. Clean the water pump and crankshaft gasket mating surfaces.

9. Install the water pump, using a new gasket. Tighten bolts to 23–34 ft. lbs.

10. Connect the coolant tube from the thermostat housing on the water

pump using a new gasket. Tighten bolts to 5–7 ft. lbs.

11. Connect the heat hose and lower radiator hose to the water pump.

12. Lower vehicle.

13. Fill and bleed the cooling system.

14. Install and adjust the front timing belt.

15. Run the engine and check for coolant leaks.

16. Install the front timing belt upper cover.

2.3L Engine

1. Disconnect the negative battery cable. Drain the cooling system.

2. Loosen the thermactor pump mounting and remove the drive belt. Disconnect and remove the hose clamp below the pump. Remove the thermactor pump bracket mounting bolts and remove the thermactor and bracket as an assembly.

3. Loosen the water pump drive belt idler pulley and remove the drive belt.

4. Disconnect the heater hose from the water pump.

5. Remove the water pump mounting bolts and the pump.

6. Clean the engine mounting surface. Apply gasket cement to both sides of the mounting gasket and position the gasket on the engine.

7. Install the pump in reverse order of removal. Torque the mounting bolts to 15–22 ft. lbs.

8. Add the proper coolant mixture, start the engine and check for leaks.

Thermostat

REMOVAL & INSTALLATION

1. Disconnect the negative battery cable. Drain the radiator until the coolant level is below the thermostat.

2. Disconnect the wire connector at the thermostat housing thermoswitch.

3. Loosen the top radiator hose clamp. Remove the thermostat housing mounting bolts and lift up the housing.

4. Remove the thermostat by turning counterclockwise.

5. Clean the thermostat housing and engine gasket mounting surfaces. Install new mounting gasket and fully insert the thermostat to compress the mounting gasket. Turn the thermostat clockwise to secure in housing.

6. Position the housing onto the engine. Install the mounting bolts and torque to 6–8 ft. lbs. on 1.6L & 1.9L engines and 12–18 ft. lbs. on 2.3L engines.

7. The rest of the installation is in the reverse order of removal.

EMISSION CONTROLS

Refer to "Emission Control" in the Unit Repair section for system maintenance procedures. Due to the complex nature of modern electronic engine control systems, comprehensive diagnosis and testing procedures fall outside the confines of this repair manual. For complete information on diagnosis, testing and repair procedures concerning all modern engine and emission control systems, please refer to *Chilton's Guide To Electronic Engine Controls.*

EGR MAINTENANCE REMINDER SYSTEM

Some vehicles are equipped with an EGR Maintenance Reminder System,

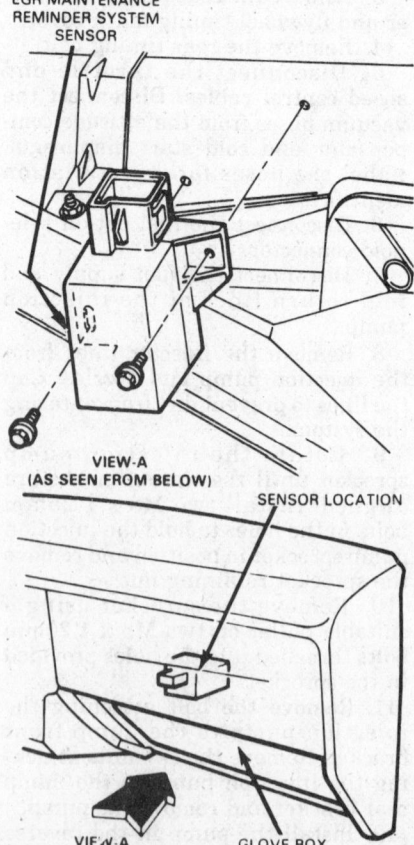

EGR sensor location. Replace the sensor to canal the EGR service light in the dash

that consists of a mileage sensor module, instrument panel warning light, and necessary wiring. This system provides a visual warning to indicate the EGR system needs service at 30,000 miles. The mileage sensor is a blue-plastic box mounted under the dash, behind the glove box. The warning light is in the instrument panel to the left of the steering column.

NOTE: The light will remain on until the sensor module is replaced.

ENGINE MECHANICAL

Gasoline Engines

REMOVAL & INSTALLATION

1.6L & 1.9L Engine

Except Turbocharged Models

NOTE: The following procedure is for engine only removal and installation.

1. Mark the position of the hinges on the hood underside and remove the
2. Remove the air cleaner assembly. Remove the air feed duct and the heat tube. Remove the air duct to the alternator.
3. Disconnect the battery cables from the battery. Remove the battery. If equipped with air conditioning, remove compressor with line still connected and position out of the way.

——— **CAUTION** ———
Never loosen refrigerant lines, as the escaping refrigerant is a deadly poison and can freeze exposed skin instantly.

4. Drain the cooling system. Remove the drive belts from the alternator and thermactor pump. Disconnect the thermactor air supply hose. Disconnect the wiring harness at the alternator. Remove alternator and thermactor.
5. Disconnect and remove the upper and lower radiator hoses. If equipped with an automatic transaxle, disconnect and plug the fluid cooler lines at the radiator.
6. Disconnect the heater hoses from the engine. Unplug the electric cooling fan wiring harness. Remove the fan and radiator shroud as an assembly.
7. Remove the radiator. Label and disconnect all vacuumlines, including power brake booster, from the engine.

Label and disconnect all linkage, including kickdown linkage if automatic, and wiring harness connectors from the engine.

8. If equipped with fuel injection, discharge the system pressure. Remove supply and return fuel lines to the fuel pump. Plug the line from the gas tank.
9. Raise and safely support the car on jackstands. Remove the clamp from the heater supply and return tubes, remove the tubes.
10. Disconnect the battery cable from the starter motor. Remove brace or bracket from the back of the starter and remove the starter.
11. Disconnect the exhaust system from the exhaust manifold. Drain the engine oil.
12. Remove the brace in front of the bell housing (flywheel or converter) inspection cover. Remove the inspection cover.
13. Remove the crankshaft pulley. If equipped with a manual transaxle, remove the timing belt cover lower attaching bolts.
14. If equipped with an automatic transaxle, remove the torque converter to flywheel mounting nuts.
15. Remove the lower engine to transaxle attaching bolts.
16. Loosen the hose clamps on the bypass hose and remove the hose from the intake manifold.
17. Remove the bolt and nut attaching the right front mount insulator to the engine bracket.
18. Lower the car from the jackstands.
19. Attach an engine lifting sling to the engine. Connect a chain hoist to the lifting sling and remove all slack. Remove the through bolt from the right front engine mount and remove the insulator.
20. If the car is equipped with a manual transaxle, remove the timing belt cover upper mounting bolts and remove the cover.
21. Remove the right front insulator attaching bracket from the engine.
22. Position a floor jack under the transaxle. Raise the jack just enough to take the weight of the transaxle.
23. Remove the upper bolts connecting the engine and transaxle.
24. Slowly raise the engine and separate from the transaxle. Be sure the torque converter stays on the transaxle. Remove the engine from the car. On models equipped with manual transaxles, the engine must be separated from the input shaft of transaxle before raising.
25. Install the engine in the reverse order of removal. On manual transaxle models take care when engaging the clutch disc splines. On automatic

transaxle models be sure the converter mounting studs engage the flywheel. Be sure the alignment dowels on the back of the engine engage the transaxle and the engine and transaxle mate together flush.

Turbocharged Models

1. Mark the position of the hood hinges and remove the hood.
2. Disconnect the negative battery cable and drain the cooling system.
3. Remove the air cleaner and vane meter assembly including the air intake tube assembly.
4. Disconnect the secondary wire from the ignition coil. Remove the alternator drive belt, remove the alternator mounting bolts and position the alternator out of the way.
5. Disconnect the radiator hoses at the engine. Remove the radiator guard, fan assembly and the radiator.
6. Disconnect the heater at the metal tube. Label and disconnect all electrical connectors and vacuum lines.
7. Disconnect the fuel supply and return lines at the intake manifold. Disconnect the power brake booster vacuum line (if equipped). Disconnect the throttle cable and bracket at the air horn assembly. Disconnect the carbon canister tube. Disconnect the purge hose at the canister purge solenoid.
8. Raise and support the front of the vehicle on jackstands. Drain the engine oil. Remove the oil cooler.
9. Remove the heater supply and return tubes. Disconnect the battery cable from the starter motor.
10. Remove the knee brace from the front of the starter motor and remove the starter motor.
11. Disconnect the exhaust pipe from the turbocharger. Remove the support bracket located in the front of the bell housing inspection cover and remove the inspection cover. Remove the exhaust pipe support bracket.
12. Remove the crankshaft pulley. Remove the timing cover lower attaching bolts.
13. Remove the flywheel housing lower mounting bolts. Remove the bracket bolt that attaches the negative battery cable to the engine block. Remove the nut and bolt attaching the mounting insulator bracket to the engine bracket located at the front of the engine.
14. Disconnect the EGR tube at the intake manifold. Disconnect the pulse air hose at the check valve and air cleaner assembly.
15. Lower the vehicle from the jackstands.
16. Install suitable lifting brackets on the engine. Attach a chain hoist to

the lifting device and apply slight upward tension.

17. Remove the nuts that attach the casting of the front engine insulator and remove the casting. Remove the remaining timing cover attaching bolts and remove the cover.

18. Position a jack underneath the transaxle and raise the jack enough to support the weight of the transaxle.

19. Remove the flywheel housing upper mounting bolts.

20. Remove the engine from the vehicle. Install the engine in the reverse order of removal.

2.3L Engine

------ **CAUTION** ------

The engine and transaxle assembly are removed together as a unit from underneath the car. Provision must be made to safely raise and support the car for powertrain removal and installation. The air conditioning system (if equipped) must be discharged prior to engine removal. The refrigerant is contained under high pressure and is very dangerous when released. The system should be discharged by a knowledgeable person using the proper equipment.

NOTE: The following procedure is for engine and transaxle removal and installation as an assembly. If services performed while the engine and transaxle are separated include engine oil pan removal, the engine and transaxle must be attached together before the engine oil pan is reinstalled.

1. Mark the position of the hinges on the underside of the hood and remove the hood.

2. Disconnect the battery cables from the battery, negative cable first. Remove the air cleaner assembly.

3. Remove the radiator cap and disconnect the lower radiator hose from the radiator to drain the cooling system.

4. Remove the upper and lower radiator hoses. On models equipped with an automatic transaxle, disconnect and plug the oil cooler lines from the rubber connectors in the radiator.

5. Disconnect and remove the coil from the cylinder head. Disconnect the cooling fan wiring harness. Remove the radiator shroud and electric fan as an assembly.

6. Be sure the air conditioning system is properly and safely discharged. Remove the hoses from the compressor. Label and disconnect all electrical harness connections, linkage and vacuum lines from the engine.

7. On automatic transaxle models disconnect the TV (throttle valve) linkage at the transaxle. On manual transaxle models disconnect the clutch cable from the lever at the transaxle.

8. Disconnect the fuel supply and return lines. Plug the fuel line from the gas tank. Disconnect the thermactor pump discharge hose at the pump.

9. Disconnect the power steering lines at the pump. Remove the hose support bracket from the cylinder head.

10. Install an engine support sling (Ford Tool T79L-5000A, or equivalent) and support the weight of the engine/transaxle assembly.

11. Raise and safely support the car on jackstands.

12. Remove the starter cable from the starter motor terminal. Drain the engine oil and the transaxle lubricant.

13. Disconnect the hose from the catalytic converter. Remove the bolts retaining the exhaust pipe bracket to the oil pan.

14. Remove the exhaust pipe to exhaust manifold mounting nuts. Remove the pipes from the mounting bracket insulators and position out of the way.

15. Disconnect the speedometer cable from the transaxle. Remove the heater hoses from the water pump inlet and intake manifold connector.

16. Remove the water intake tube bracket from the engine block. Remove the two clamp attaching bolts from the bottom of the oil pan. Remove the water pump inlet tube.

17. Remove the bolts attaching the control arms to the body. Remove the stabilizer bar bracket retaining bolts and remove the brackets.

18. Remove the half shafts (drive-axles) from the transaxle.

19. On models equipped with a manual transaxle, remove the roll restrictor nuts from the transaxle and pull the roll restrictor from mounting bracket.

20. On models equipped with a manual transaxle, remove the shift stabilizer bar to transaxle attaching bolts. Remove the shift mechanism to shift shaft attaching nut and bolt at the transaxle.

21. On models equipped with an automatic transaxle, disconnect the shift cable clip from the transaxle lever. Remove the manual shift linkage bracket bolts from the transaxle and remove the bracket.

22. Remove the left rear No. 4 insulator mount bracket from the body by removing the retaining nuts.

23. Remove the left front No. 1 insulator to transaxle mounting bolts.

24. Lower the car and support with stands so that the front wheels are just above the ground. Do not allow the wheels to touch the ground.

25. Connect an engine sling to the lifting brackets provided. Connect a hoist to the sling and apply slight tension. Remove the support sling (Step 10).

26. Remove the right hand insulator intermediate bracket to engine bracket bolts, intermediate bracket to insulator attaching nuts and the nut on the bottom of the double ended stud which attaches the intermediate bracket and engine bracket. Remove the bracket.

27. Lower the engine and transaxle assembly to the ground.

28. Raise and support the car at a height suitable from assembly to be removed.

29. Installation is the reverse of removal procedure.

Diesel Engine

REMOVAL & INSTALLATION

2.0 Diesel Engine

NOTE: Suitable jackstands or hoisting equipment are necessary to remove the engine and transaxle assembly, as the assembly is removed from underneath the vehicle.

------ **CAUTION** ------

The air conditioning system contains refrigerant (R-12) under high pressure. Use extreme care when discharging system. If the tools and know-how are not on hand, have the system discharged prior to start of engine removal.

1. Mark the position of the hood hinges and remove the hood.

2. Remove the negative ground cable from battery that is located in luggage compartment.

3. Remove the air cleaner assembly.

4. Position a drain pan under the lower radiator hose. Remove the hose and drain the engine coolant.

5. Remove the upper radiator hose from the engine.

6. Disconnect the cooling fan at the electrical connector.

7. Remove the radiator shroud and cooling fan as an assembly. Remove the radiator.

8. Remove the starter cable from the starter.

9. Discharge air conditioning system (see opening CAUTION), if so equipped. Remove the pressure and suction lines from the air conditioning compressor.

10. Identify and disconnect all vacuum lines as necessary.

11. Disconnect the engine harness connectors (two) at the dash panel. Disconnect the glow plug relay connectors at the dash panel.

NOTE: Connectors are located under the plastic shield on the dash panel. Remove and save plastic retainer pins. Disconnect the alternator wiring connector on RH fender apron.

12. Disconnect the clutch cable from the shift lever on transaxle.
13. Disconnect the injection pump throttle linkage.
14. Disconnect the fuel supply and return hoses on the engine.
15. Disconnect the power steering pressure and return lines at the power steering pump, if so equipped. Remove the power steering lines bracket at the cylinder head.
16. Install Engine Support Tool D79P-8000-A or equivalent to existing engine lifting eye.
17. Raise vehicle and safely support on jackstands.
18. Remove the bolt attaching the exhaust pipe bracket to the oil pan.
19. Remove the two exhaust pipes to exhaust manifold attaching nuts.
20. Pull the exhaust system out of rubber insulating grommets and set aside.
21. Remove the speedometer cable from the transaxle.
22. Position a drain pan under the heater hoses. Remove one heater hose from the water pump inlet tube. Remove the other heater hose from the oil cooler.
23. Remove the bolts attaching the control arms to the body. Remove the stabilizer bar bracket retaining bolts and remove the brackets.
24. Halfshaft assemblies must be removed from the transaxle at this time.
25. On MTX models, remove the shift stabilizer bar-to-transaxle attaching bolts. Remove the shift mechanism to shift shaft attaching nut and bolt at the transaxle.
26. Remove the LH rear insulator mount bracket from body bracket by removing the two nuts.
27. Remove the LH front insulator to transaxle mounting bolts.
28. Lower vehicle (see CAUTION below). Install lifting equipment to the two existing lifting eyes on engine.

— **CAUTION** —
Do not allow front wheels to touch floor.

29. Remove Engine Support Tool D79L-8000-A or equivalent.
30. Remove RH insulator intermediate bracket to engine bracket bolts, intermediate bracket to insulator attaching nuts and the nut on the bottom of the double ended stud attaching the intermediate bracket to engine bracket. Remove the bracket.
31. Carefully lower the engine and transaxle assembly to the floor.
32. Raise the vehicle and safely support.
33. Position the engine and transaxle assembly directly below the engine compartment.
34. Slowly lower the vehicle over the engine and transaxle assembly.

— **CAUTION** —
Do not allow the front wheels to touch floor.

35. Install the lifting equipment to both existing engine lifting eyes on engine.
36. Raise the engine and transaxle assembly up through engine compartment and position accordingly.
37. Install RH insulator intermediate attaching nuts and intermediate bracket to engine bracket bolts. Install nut on bottom of double ended stud attaching intermediate bracket to engine bracket. Tighten to 75–100 ft. lbs.
38. Install Engine Support Tool D79L-8000-A or equivalent to the engine lifting eye.
39. Remove the lifting equipment.
40. Raise vehicle.
41. Position a suitable floor or transaxle jack under engine. Raise the engine and transaxle assembly into mounted position.
42. Install insulator to bracket nut and tighten to 75–100 lbs.
43. Tighten the LH rear insulator bracket to body bracket nuts to 75–100 ft. lbs.
44. Install the lower radiator hose and install retaining bracket and bolt.
45. Install the shift stabilizer bar to transaxle attaching bolt. Tighten to 23–35 ft. lbs.
46. Install the shift mechanism to input shift shaft (on transaxle) bolt and nut. Tighten to 7–10 ft. lbs.
47. Install the lower radiator hose to the radiator.
48. Install the speedometer cable to the transaxle.
49. Connect the heater hoses to the water pump and oil cooler.
50. Position the exhaust system up and into insulating rubber grommets located at the rear of the vehicle.
51. Install the exhaust pipe to exhaust manifold bolts.
52. Install the exhaust pipe bracket to the oil pan bolt.
53. Place the stabilizer bar and control arm assembly into position. Install control arm to body attaching bolts. Install the stabilizer bar brackets and tighten all fasteners.
54. Halfshaft assemblies must be installed at this time.
55. Lower the vehicle.
56. Remove the Engine Support Tool D79L-6000-A or equivalent.
57. Connect the alternator wiring at RH fender apron.
58. Connect the engine harness to main harness and glow plug relays at dash panel.

NOTE: Reinstall plastic shield.

59. Connect the vacuum lines.
60. Install the air conditioning discharge and suction lines to A/C compressor, if so equipped. Do not charge system at this time.
61. Connect the fuel supply and return lines to the injection pump.
62. Connect the injection pump throttle cable.
63. Install the power steering pressure and return lines. Install bracket.
64. Connect the clutch cable to shift lever on transaxle.
65. Connect the battery cable to starter.
66. Install the radiator shroud and coolant fan assembly. Tighten attaching bolts.
67. Connect the coolant fan electrical connector.
68. Install the upper radiator hose to engine.
69. Fill and bleed the cooling system.
70. Install the negative ground battery cable to battery.
71. Install the air cleaner assembly.
72. Install the hood.
73. Charge air conditioning system, if so equipped. System can be changed at a later time if outside source is used.
74. Check and refill all fluid levels, (power steering, engine, MTX).
75. Start the vehicle. Check for leaks.

Intake Manifold

REMOVAL & INSTALLATION

Gasoline Engine–Except Turbocharged Models

1. Disconnect the negative battery terminal.
2. Remove the air cleaner housing.
3. Partially drain the cooling system and disconnect the heater hose from under the intake manifold.
4. Disconnect and label all vacuum and electrical connections.
5. Disconnect the fuel line and carburetor linkage.
6. Disconnect the EGR vacuum hose and supply tube.
7. On Escort & Lynx models, jack

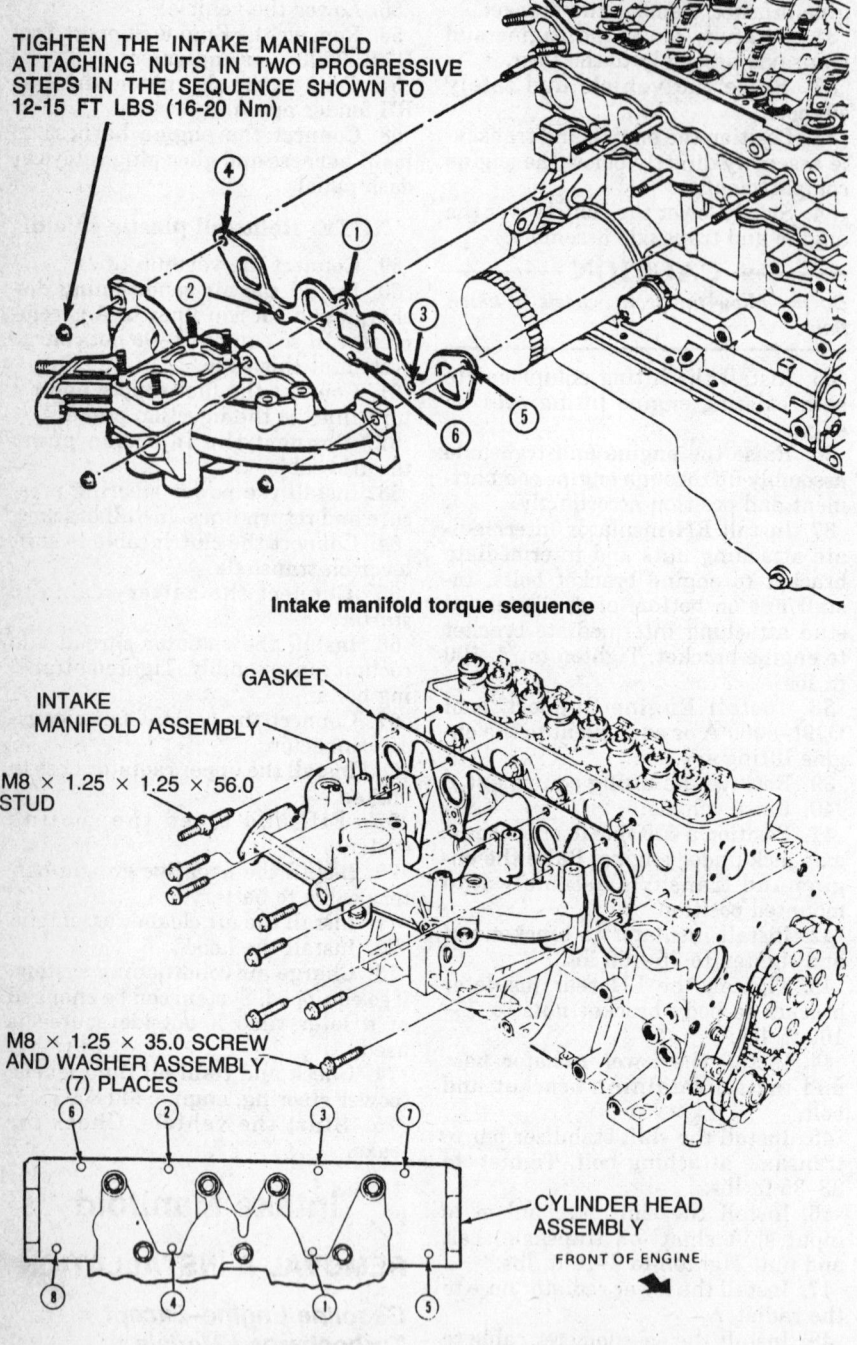

TIGHTEN THE INTAKE MANIFOLD ATTACHING NUTS IN TWO PROGRESSIVE STEPS IN THE SEQUENCE SHOWN TO 12-15 FT LBS (16-20 Nm)

Intake manifold torque sequence

INTAKE MANIFOLD ASSEMBLY

GASKET

M8 × 1.25 × 1.25 × 56.0 STUD

M8 × 1.25 × 35.0 SCREW AND WASHER ASSEMBLY (7) PLACES

CYLINDER HEAD ASSEMBLY

FRONT OF ENGINE

Intake manifold installation—2.3L HSC engine

up the vehicle and support it with jackstands.

8. On Escort & Lynx models, remove the bottom (3) intake manifold nuts.

9. On Escort & Lynx models, remove the vehicle from the jackstands.

10. If equipped with automatic transmission disconnect the throttle valve linkage at the carburetor and remove the cable bracket attaching bolts.

11. If equipped with power steering (Escort/Lynx), remove the thermactor pump drive belt, the pump, the mounting bracket, and the by-pass hose.

12. Remove the fuel pump, (Escort/Lynx). See the fuel pump removal procedure.

13. Remove the intake bolts, the manifold, and gasket.

NOTE: Do not lay the intake

manifold flat as the gasket surfaces may be damaged.

14. Installation is the reverse of removal.

Turbocharged Models

1. Disconnect the negative battery cable.

2. Remove the air supply hose from the air throttle body assembly.

3. Label and disconnect the vacuum hoses. Remove the EGR supply tube. Disconnect the throttle air bypass valve solenoid electrical connector.

4. Raise and support the front of the vehicle on jackstands. Remove the three intake manifold nuts and lower the vehicle.

5. Disconnect the fuel supply and return lines at the intake manifold. Disconnect the accelerator cable from the air throttle body assembly.

6. Disconnect the wiring harness at the shock tower. Disconnect the PCV hose at the rocker arm cover and intake manifold. Remove the PCV valve.

7. Remove the three remaining mounting nuts. The center nut may need tool T81P-9425-A for removal and installation purposes.

8. Install the intake manifold in the reverse order of removal using a new gasket. Tighten the mounting nuts to 12–15 ft. lbs.

Diesel Engine

1. Disconnect the air inlet duct from the intake manifold and install the protective cap in the intake manifold, (part or Protective Cap Set T84P-9395-A or equivalent).

2. Disconnect the flow plug resistor electrical connector.

3. Disconnect the breather hose.

4. Drain the cooling system.

5. Disconnect the upper radiator hose at the thermostat housing.

6. Disconnect the two-coolant hoses at the thermostat housing.

7. Disconnect the connectors to the temperature sensors in the thermostat housing.

8. Remove the bolts attaching the intake manifold to the cylinder head and remove the intake manifold.

9. Clean the intake manifold and cylinder head gasket mating surfaces.

10. Instal the intake manifold, using a new gasket, and tighten the bolts to 12–16 ft. lbs.

11. Connect the temperature sensor connectors.

12. Connect the lower coolant hose to the thermostat housing and tighten the hose clamp.

13. Connect the upper coolant tube, using a new gasket and tighten bolts to 5–7 ft. lbs.

14. Connect the upper radiator hose to the thermostat housing.

15. Connect the breather hose.

16. Connect the glow plug resistor electrical connector.

17. Remove the protective cap and install the air inlet duct.

18. Fill and bleed the cooling system.

19. Run the engine and check for intake air leaks and coolant leaks.

Exhaust Manifold

REMOVAL & INSTALLATION

Gasoline Engine–Except Turbocharged

1. Disconnect the negative battery cable.

2. Remove the air cleaner duct for access to the manifold.

3. Disconnect the Thermactor (air pump) line from the manifold. Disconnect the EGR tube. Remove heat shield. Disconnect sensor wire harness. if equipped. Unbolt the exhaust pipe from the manifold flange.

4. Unbolt and remove the exhaust manifold.

5. Clean the manifold mating surfaces. Place a new gasket on the exhaust pipe-to-manifold flange.

6. Install the manifold. Tighten the bolts in a circular pattern, working from the center to the ends, in three progressive steps.

Turbocharged Models

1. Disconnect the negative battery cable. Remove the guard from the radiator.

2. Loosen the turbo outlet hose clamp at the throttle housing. Remove the hose from the turbo housing and rotate the hose up out of the way.

3. Disconnect the inlet hose to the turbocharger.

4. Remove the alternator and bracket. Disconnect the EGO sensor electrical harness connector.

5. Raise and support the front of the vehicle on jackstands. Disconnect the oil supply line at the coolant outlet and at the turbocharger. Disconnect the oil return line from the bottom of the turbocharger center housing and from the engine block.

6. Lower the vehicle from the jackstands. Remove the exhaust pipe to turbocharger mounting nuts and move the exhaust pipe away from the studs. Remove the bolt that attaches the exhaust shield to the water outlet connector.

7. Remove the nuts attaching the exhaust manifold to the cylinder head. Slide the manifold and turbocharger away from the cylinder head

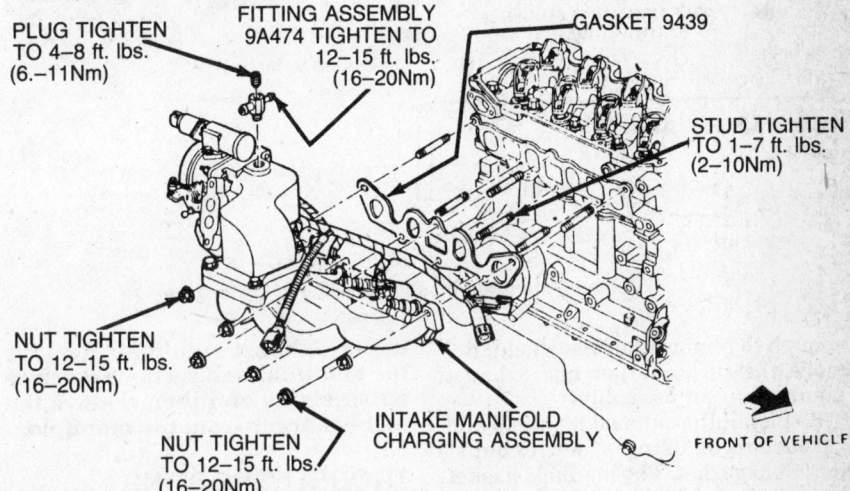

PLUG TIGHTEN TO 4–8 ft. lbs. (6.–11Nm)

FITTING ASSEMBLY 9A474 TIGHTEN TO 12–15 ft. lbs. (16–20Nm)

GASKET 9439

STUD TIGHTEN TO 1–7 ft. lbs. (2–10Nm)

NUT TIGHTEN TO 12–15 ft. lbs. (16–20Nm)

NUT TIGHTEN TO 12–15 ft. lbs. (16–20Nm)

INTAKE MANIFOLD CHARGING ASSEMBLY

FRONT OF VEHICLF

Install the intake manifold and tighten the retaining bolts in the sequence shown on 1.6L engine

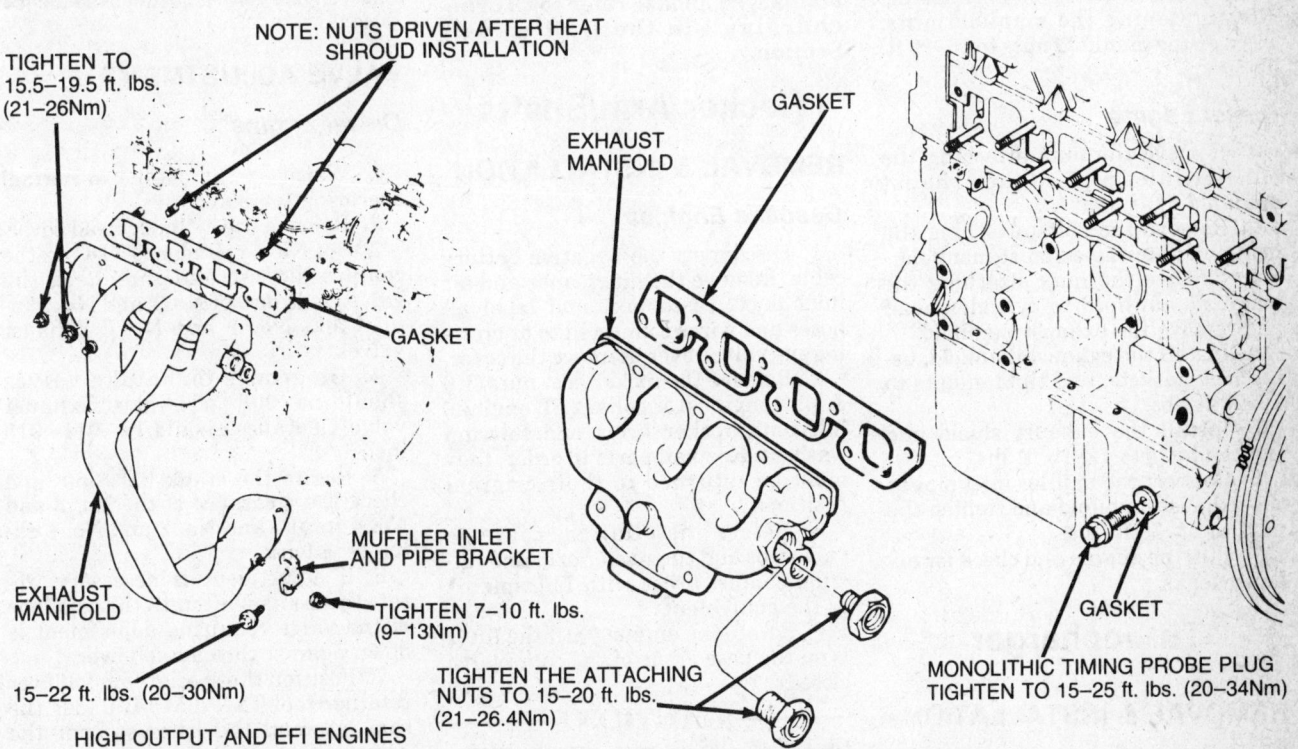

NOTE: NUTS DRIVEN AFTER HEAT SHROUD INSTALLATION

TIGHTEN TO 15.5–19.5 ft. lbs. (21–26Nm)

GASKET

EXHAUST MANIFOLD

GASKET

MUFFLER INLET AND PIPE BRACKET

EXHAUST MANIFOLD

TIGHTEN 7–10 ft. lbs. (9–13Nm)

15–22 ft. lbs. (20–30Nm)

TIGHTEN THE ATTACHING NUTS TO 15–20 ft. lbs. (21–26.4Nm)

GASKET

MONOLITHIC TIMING PROBE PLUG TIGHTEN TO 15–25 ft. lbs. (20–34Nm)

HIGH OUTPUT AND EFI ENGINES

Install the exhaust manifold and tighten the remaining bolts in the sequence shown on 1.6L engine

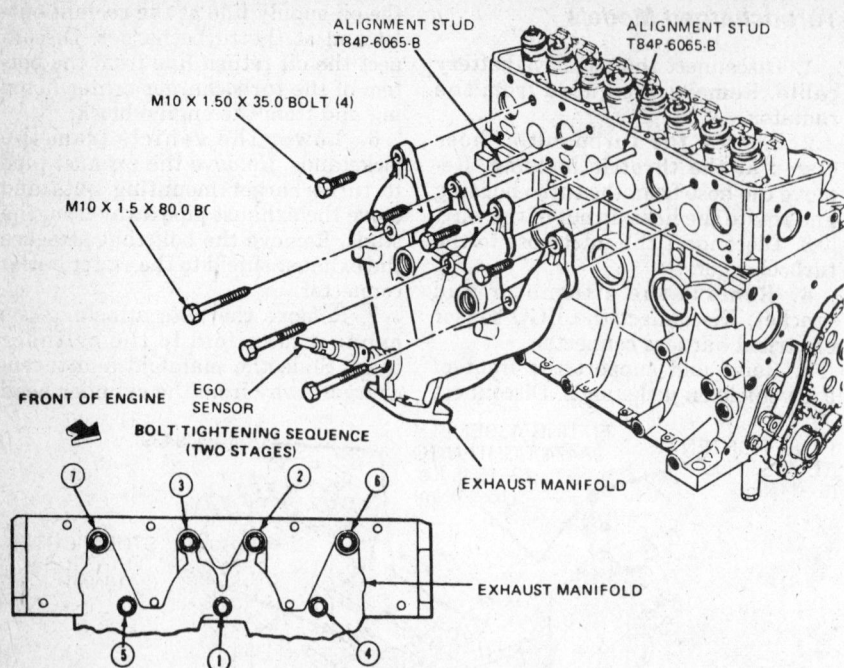

ALIGNMENT STUD
T84P-6065-B

ALIGNMENT STUD
T84P-6065-B

M10 X 1.50 X 35.0 BOLT (4)

M10 X 1.5 X 80.0 BC

FRONT OF ENGINE   EGO
SENSOR

BOLT TIGHTENING SEQUENCE
(TWO STAGES)

EXHAUST MANIFOLD

EXHAUST MANIFOLD

Exhaust manifold installation—2.3L HSC engine

enough to remove the heat shield. Remove the turbocharger and exhaust manifold as an assembly.

8. Install the exhaust manifold and turbocharger using a new exhaust manifold gasket. The manifold gasket has a top and bottom, be sure to install it correctly. After positioning the manifold, loosely install the retaining nuts. Connect the turbo oil lines before tightening the manifold nuts. Tighten the manifold nuts to 16–19 ft. lbs.

Diesel Engine

1. Remove the nuts attaching the muffler inlet pipe to the exhaust manifold.

2. Remove the bolts attaching the heat shield to the exhaust manifold.

3. Remove the nuts attaching the exhaust manifold to cylinder head and remove the exhaust manifold.

4. Install the exhaust manifold, using new gaskets, and tighten nuts to 16–20 ft. lbs.

5. Install the exhaust shield and tighten bolts to 12–16 ft. lbs.

6. Connect the muffler inlet pipe to the exhaust manifold and tighten the nuts to 25–35 ft. lbs.

7. Run the engine and check for exhaust leaks.

Turbocharger

REMOVAL & INSTALLATION

NOTE: Follow the procedure

under exhaust manifold. Remove the manifold and turbocharger as an assembly and then remove the turbocharger from the manifold.

TROUBLESHOOTING

For more information on the turbocharger, please refer to "Turbocharging" in the Unit Repair Section.

Rocker Arm/Shafts

REMOVAL & INSTALLATION

Gasoline Engines

1. Disconnect the negative battery cable. Remove the air cleaner and air inlet duct. Disconnect and label all hoses and wires connected to or crossing the valve cover. Remove the cover.

2. Remove the rocker arm nuts (1.6 & 1.9L engines) or bolts (2.3L engine), fulcrums, rocker arms and fulcrum washers. Keep all parts in order; they must be returned to their original positions.

3. Before installation, coat the valve tips and the rocker arm and fulcrum contact areas with Lubriplate® or the equivalent.

4. Rotate the engine until the lifter is on the base circle of the cam (valve closed).

—— CAUTION ——
On 1.6L & 1.9L engines, turn the engine only in the direction of normal rotation.

Backward rotation will cause the camshaft belt to slip or lose teeth, altering valve timing and causing serious engine damage.

5. Install the rocker arm and components. Be sure the lifter is on the base circle of the cam for each rocker arm as it is installed.

6. Clean the valve cover mating surfaces. Apply a bead of sealer to the cover flange and install the cover. Install all disconnected hoses and wires.

Valve Clearance

The intake and exhaust valves are driven by the camshaft, working through hydraulic lash adjusters and stamped rocker arms (1.6L & 1.9L engine) or through hydraulic lifters, pushrods and rocker arms (2.3L engine). The hydraulic lash adjusters or lifters eliminate the need for periodic valve lash adjustment.

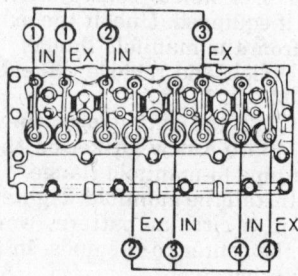

Diesel engine valve adjustment sequence

VALVE ADJUSTMENT

Diesel Engine

1. Warm up the engine to normal operating temperature.

2. Remove the cylinder head cover.

3. Set No. 1 cylinder to TDC on the compression stroke and check the valve clearance of No. 1 and No. 2 intake and No. 1 and No. 3 exhaust valves.

4. Clearance for intake valves should be .008–.011 inch. Exhaust valve clearance should be .011–.015 inch.

5. Rotate the crankshaft 360° and check the clearance of the No. 3 and No. 4 intake and No. 2 and No. 4 exhaust valves.

6. If adjustment is necessary, rotate the crankshaft until the cam lobe of the valve requiring adjustment is down against the cam follower.

7. Position the special cam follower retainer tool T84P-6513-B under the cam between the lobes so that the edge contacts the cam follower needing adjustment.

8. Rotate the camshaft until the cam lobe (of valve needing adjustment) is on its base circle (lobe pointing straight up).

9. Pry the adjusting shim out of the cam follower. Valve shims are available in thicknesses ranging from 3.40 to 4.60mm. If the valve was too tight or loose, install a shim of the appropriate size. Shim thickness is stamped on the valve shim. Install the shim with the numbers down to avoid wearing the numbers off.

10. Rotate the camshaft until the lobe is down and remove the retainer tool.

11. Recheck the valve clearance by repeating the previous steps.

12. Install the engine cover using a new gasket. Tighten the retaining bolts to 5–7 ft. lbs.

Cylinder Head

REMOVAL & INSTALLATION

1.6L & 1.9L Engine

NOTE: The engine must be "overnight" cold before removing the cylinder head, to reduce the possibility of warpage or distortion.

—— **CAUTION** ——

Do not reuse the cylinder head retaining bolts. Use new bolts when installing head.

1. Disconnect the negative battery cable.

2. On non-turbocharged engines, drain the cooling system and disconnect the heater hose at the fitting located under the intake manifold and the upper radiator hose at the engine.

3. Disconnect the wiring terminal from the cooling fan switch.

4. On turbocharged engines, drain the cooling system and disconnect the upper radiator hose at the engine.

5. On non-turbocharged engines, remove the air cleaner assembly, remove the PCV hose, and disconnect all interfering vacuum hoses after marking them for reassembly. On turbocharged engines, remove the air supply hose at the throttle body assembly and remove the PCV oil separator system.

6. Turn the engine until No. 1 piston is at TDC (top dead center) on the compression stroke and the timing marks are aligned.

7. Remove the valve cover and disconnect all accessory drive belts. Remove the crankshaft pulley. (Use crankshaft pulley wrench T81P-6312-A and crankshaft bolt wrench YA-826 or equivalents). Remove the timing belt cover.

8. Remove the distributor cap and spark plug wires as an assembly.

9. Loosen both belt tensioner attaching bolts using special Ford tool T81P-6254-A or the equivalent. Secure the belt tensioner as far left as possible. Remove the timing belt.

10. Disconnect the tube at the EGR valve, then remove the PVS hose connectors using tool T81P-8564-A or equivalent. Label the connectors and set aside.

11. Disconnect the choke wire, the fuel supply and return lines, the accelerator cable and speed control cable (if equipped). Disconnect the altitude compensator, if equipped, from the dash panel and place on the heater/AC air intake.

12. Disconnect and remove the alternator.

13. If equipped with power steering, remove the thermactor pump drive belt, the pump and its bracket. Disconnect the turbocharger inlet hose (if equipped). Disconnect the turbocharger oil supply tube at the turbocharger coolant outlet and the engine block. Remove the supply line.

14. Raise the vehicle and and support on jackstands. Disconnect the exhaust pipe from the manifold or turbocharger. Disconnect the oil return line from the turbocharger.

15. Lower the vehicle and remove the cylinder head bolts and washers. Discard the bolts, they cannot be used again.

16. Remove the cylinder head with the manifolds (and turbocharger) attached. Remove and discard the head gasket. Do not place the cylinder head with combustion chambers down or damage to the spark plugs or gasket surfaces may result.

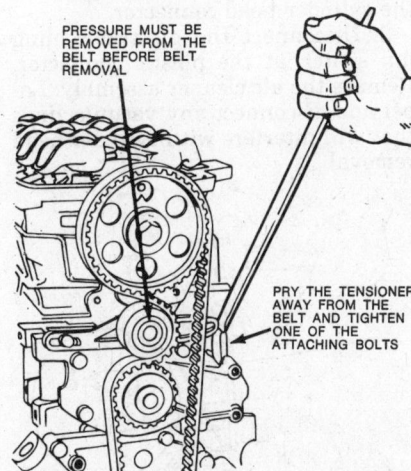

PRESSURE MUST BE REMOVED FROM THE BELT BEFORE BELT REMOVAL

PRY THE TENSIONER AWAY FROM THE BELT AND TIGHTEN ONE OF THE ATTACHING BOLTS

The timing belt tensioner must be released and moved away from the belt before the timing belt can be removed

—— **CAUTION** ——

Before installing the cylinder head on a turbocharged engine, check piston squish height as described following this procedure.

17. To install, clean all gasket material from both the block face and the cylinder head. Prior to installing the cylinder head rotate the crankshaft so that the No. 1 piston is 90° BTDC (before top dead center). Turn the crankshaft back until the crankshaft pulley keyway is at 9 o'clock (this is the 90° BTDC position). This will prevent damage to the valves and pistons. To time the valve train to this piston position, turn the camshaft until the keyway is at the 6 o'clock position. The camshaft and crankshaft must not be turned until after the installation of the timing gears and belt.

18. Position the cylinder head gasket on the engine block. The gasket for the turbocharged engine is different than the non-turbo. Be sure the correct head gasket is installed. Position the cylinder head on the engine block. Lightly oil the threads of the new head bolts and install them. (Do not reuse the cylinder head bolts. Always install new ones.) Follow the tightening torque sequence shown. Tighten the bolts to 44 ft. lbs. Loosen all bolts approximately two turns, then retighten to 44 ft. lbs. again following the torque sequence. After tightening all bolts the second time, turn all bolts an additional 90° in sequence. When the first 90° sequence is completed, repeat an additional 90° turn, once again in sequence.

19. Remaining installation is the reverse of removal. See "Timing Belt Removal and Installation" for timing belt installation procedures and recheck engine timing at No. 1 piston on TDC.

20. Fill the cooling system only with Ford Cooling System Fluid E1FZ-19549-A or Prestone II or the equivalent. Using the wrong type of coolant can damage the engine. Start the engine, check the ignition timing and check for fluid leaks.

INTAKE

9 3 1 5 7

8 6 2 4 10

EXHAUST

Cylinder head tightening sequence. See text for procedure

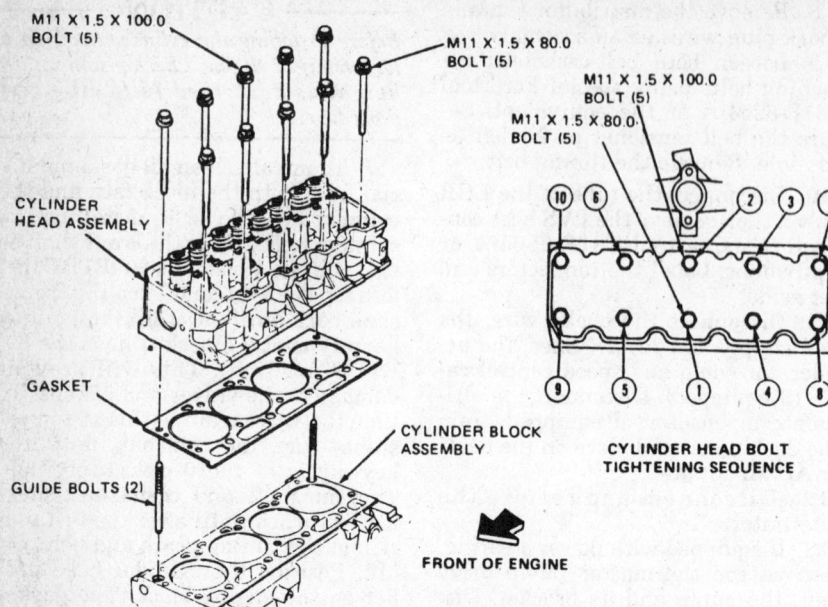

M11 X 1.5 X 100.0
BOLT (5)

M11 X 1.5 X 80.0
BOLT (5)

M11 X 1.5 X 100.0
BOLT (5)

M11 X 1.5 X 80.0
BOLT (5)

CYLINDER
HEAD ASSEMBLY

GASKET

CYLINDER BLOCK
ASSEMBLY

GUIDE BOLTS (2)

CYLINDER HEAD BOLT
TIGHTENING SEQUENCE

FRONT OF ENGINE

Cylinder head torque sequence—2.3L HSC

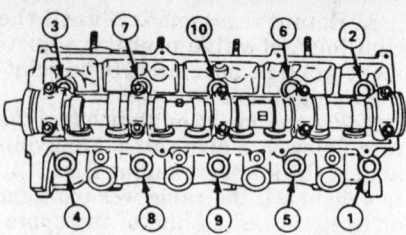

2.0L diesel cylinder head bolt removal

Piston Squish Height—Turbo Engines

NOTE: If no other service or parts other than just cylinder head gasket replacement has been done on the engine, the piston squish height should be within specification. If the cylinder head surface has be reworked or replacement parts such as a crankshaft, connecting rod or piston installed the height clearance between the piston head and cylinder head combustion chamber must be checked.

1. Clean all gasket surfaces, piston head and combustion chamber.
2. Place a small amount of soft lead solder a various points on the high part of the piston head.
3. Rotate the crankshaft to lower the piston in the bore. Install the cylinder head with a used head gasket (a

used gasket is perferred since it has already been compressed, however a new gasket can be used) between the cylinder head and block.
4. Install the used head bolts and tighten them to 44 ft. lbs in the proper torque sequence.
5. Rotate the crankshaft to move the piston through its TDC position. Remove the cylinder head and measure the thickness of the solder. Proper clearance is .039–.070 inch.

2.3L Engine

1. Disconnect the negative battery cable. Drain the cooling system by disconnecting the lower radiator hose.
2. Disconnect the heater hose at the fitting under the intake manifold. Disconnect the upper radiator hose at the cylinder head connector.
3. Disconnect the electric cooling fan switch at the plastic connector. Remove the air cleaner assembly. Label and disconnect any vacuum lines that will interfere with cylinder head removal.

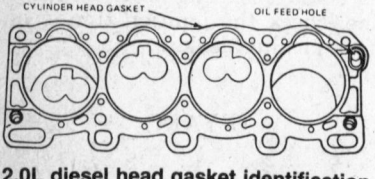

2.0L diesel head gasket identification

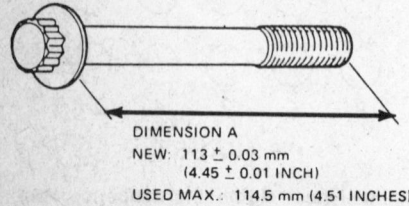

DIMENSION A
NEW: 113 ± 0.03 mm
(4.45 ± 0.01 INCH)
USED MAX.: 114.5 mm (4.51 INCHES)

2.0L diesel head bolt measurement

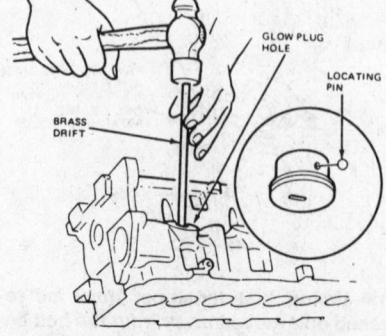

2.0L diesel pre-chamber removal

4. Disconnect all drive belts. Remove rocker arm cover. Remove the distributor cap and spark plug wires as an assembly.
5. Disconnect the EGR tube at EGR valve. Disconnect the choke wire from the choke.
6. Discharge the fuel system pressure if equipped with CFI (fuel injection). Disconnect the fuel supply and return lines. Disconnect the accelerator cable and speed control cable, if equipped. Loosen the bolts retaining the thermactor pump pulley.
7. Raise and safely support the front of the car. Disconnect the exhaust pipe from the exhaust manifold. Lower car.
8. Loosen the rocker arm bolts until the arms can pivot for pushrod removal. Remove the pushrods, keep the pushrods in order for installation in original position.
9. Remove the cylinder head bolts. Remove the cylinder head, gasket, thermactor pump, intake and exhaust manifolds as an assembly. Do not lay the cylinder head down flat before removing the spark plugs; take care not to damage the gasket surface.
10. Clean all gasket material from the head and block surfaces.
11. Position a new head gasket on the block surface, use sealer to retain the gasket.
12. To help with head installation alignment, purchase two head bolts and cut off the heads. Install the "modified" bolts at opposite corners of the block to act as guides.
13. Position the cylinder head over the guide bolts and lower onto the engine block.
14. Install head bolts, remove the guides and replace with regular bolts.
15. Tighten the head bolts to 53–59 ft. lbs. in two stages in the sequence shown.
16. The rest of the cylinder head installation is in the reverse order of removal.

Diesel Engine

1. Disconnect the battery ground cable from the battery, which is located in the luggage compartment.
2. Drain the cooling system.
3. Remove the camshaft cover,

front and rear timing belt covers, and front and rear timing belts.

4. Raise the vehicle and safely support on jackstands.

5. Disconnect the muffler inlet pipe at the exhaust manifold. Lower the vehicle.

6. Disconnect the air inlet duct at the air cleaner and intake manifold. Install a protective cover.

7. Disconnect the electrical connectors and vacuum hoses to the temperature sensors located in the thermostat housing.

8. Disconnect the upper and lower coolant hoses, and the upper radiator hose at the thermostat housing.

9. Disconnect and remove the injection lines at the injection pump and nozzles. Cap all lines and fittings with Cap Protective Set T84P-9395-A or equivalent.

10. Disconnect the glow plug harness from the main engine harness.

11. Remove the cylinder head bolts in the sequence shown. Remove the cylinder head.

12. Remove the glow plugs. Then remove pre-chamber cups from the cylinder head using a brass drift.

13. Clean the pre-chamber cups, pre-chambers in the cylinder heads and all gasket mounting surfaces on the cylinder head and engine block.

14. Install the pre-chambers in the cylinder heads making sure the locating pins are aligned with the slots provided.

15. Install the glow plugs and tighten to 11–15 ft. lbs. Connect glow plug harness to the glow plugs. Tighten the nuts to 5–7 ft. lbs.

—————— CAUTION ——————

Carefully blow out the head bolt threads in the crankcase with compressed air. Failure to thoroughly clean the thread bores can result in incorrect cylinder head torque or possible cracking of the crankcase.

16. Position a new cylinder head gasket on the crankcase making sure the cylinder head oil feed hole is not blocked.

17. Measure each cylinder head bolt dimension A. If the measurement is more than 114.5mm (4.51 inches), replace the head bolt.

—————— CAUTION ——————

Rotate the camshaft in the cylinder head until the cam lobes for No. 1 cylinder are at the base circle (both valves closed). Then, rotate the crankshaft clockwise until No. 1 piston is halfway up in the cylinder bore toward TDC. This is to prevent contact between the pistons and valves.

18. Install the cylinder head on the crankcase.

NOTE: Before installing the cylinder head bolts, paint a white reference dot on each one, and apply a light coat of engine oil on the bolt threads.

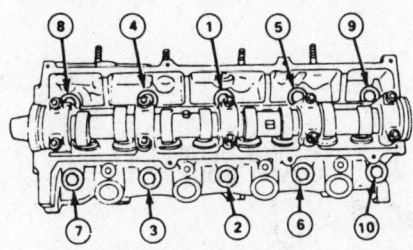

2.0L diesel head bolt tightening sequence

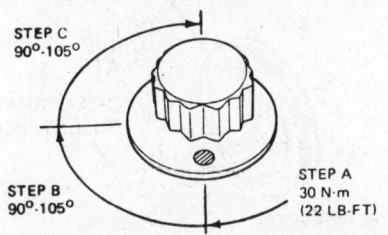

2.0L diesel head bolt tightening steps

19. Tighten cylinder head bolts as follows:

a. Tighten bolts to 22 ft. lbs. in the sequence shown.

b. Using the painted reference marks, tighten each bolt in sequence another 90 degrees to 105 degrees.

c. Repeat Step "b" by turning the bolts another 90–105 degrees.

20. Connect the glow plug harness to main engine harness.

21. Remove the protective caps and install injection lines to the injection pump and nozzles. Tighten capnuts to 18–20 ft. lbs.

22. Air bleed the system.

23. Connect the upper (with a new gasket) and lower coolant hoses, and the upper radiator hose to the thermostat housing. Tighten upper coolant hose bolts to 5–7 ft. lbs.

24. Connect the electrical connectors and the vacuum hoses to the temperature sensors in the thermostat housing.

25. Remove the protective cover and install the air inlet duct to the intake manifold and air cleaner.

26. Raise vehicle and support on jackstands. Connect the muffler inlet pipe to the exhaust manifold. Tighten nuts to 25–35 ft. lbs.

27. Lower the vehicle.

28. Install and adjust the front timing belt.

29. Install and adjust the rear timing belt.

30. Install the front upper timing belt cover and rear timing belt cover. Tighten the bolts to 5–7 ft. lbs.

31. Check and adjust the valves as outlined. Install the valve cover and tighten the bolts to 5–7 ft. lbs.

32. Fill and bleed the cooling system.

33. Check and adjust the injection pump timing.

34. Connect battery ground cable to battery. Run engine and check for oil, fuel and coolant leaks.

OVERHAUL

For all cylinder head overhaul procedures, refer to "Engine Rebuilding" in the Unit Repair Section.

Front Cover

REMOVAL & INSTALLATION

NOTE: The engine must be removed from the car for the following procedure.

1. Remove the bolt and washer retaining the drive pulley. Use a suitable puller and remove the crankshaft pulley.

2. Remove the front cover retaining bolts, pry the top of the cover away from the engine block and remove the cover.

3. Installation is in reverse order of removal.

Timing Belt

REMOVAL & INSTALLATION

1.6L & 1.9L Engine

NOTE: With the timing belt removed and pistons at TDC, do not rotate the camshaft for fear of bending the valves. If the camshaft must be rotated, align the crankshaft pulley 90° BTDC (crankshaft keyway at 9 o'clock).

1. Disconnect the negative battery cable. Remove all accessory drive belts and remove the timing belt cover.

NOTE: Align the timing mark on the camshaft sprocket with the timing mark on the cylinder head.

2. After aligning the camshaft timing marks, reinstall the timing belt cover and confirm that the timing mark on the crankshaft pulley aligns with the TDC mark on the front cover. Remove the timing belt cover.

3. Loosen both timing belt attaching bolts using tool T81P-6254-A or equivalent. Pry the tensioner away from the belt as far as possible and

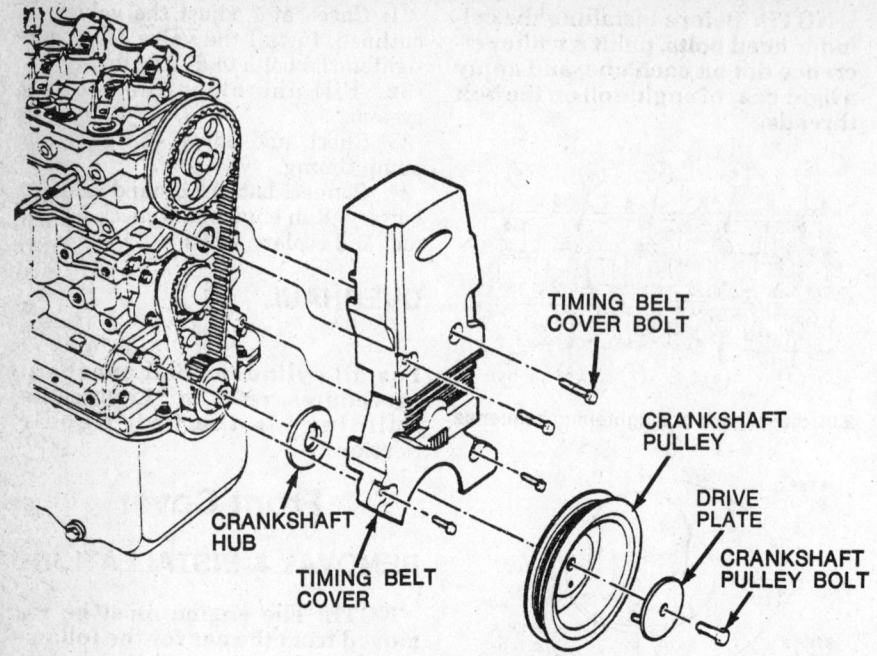

1.6L engine timing belt cover

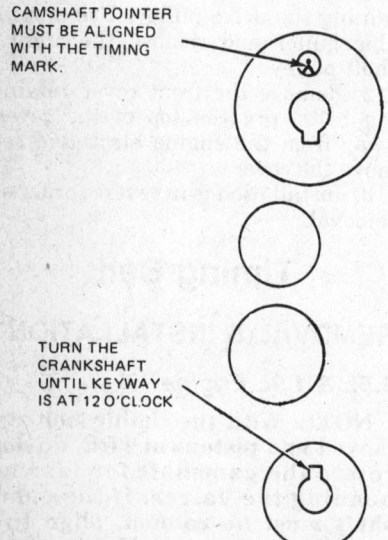

CAMSHAFT POINTER MUST BE ALIGNED WITH THE TIMING MARK.

TURN THE CRANKSHAFT UNTIL KEYWAY IS AT 12 O'CLOCK

When installing the timing belt, the keyway on the crankshaft is at 12 o'clock, the camshaft pointer is aligned with the timing mark and the keyway on the camshaft is at 6 o'clock

hold it in that position by tightening one of the tensioner attaching bolts.

4. Remove the crankshaft pulley and remove and discard the timing belt.

NOTE: Due to limited working space, special tools are required to remove the crankshaft pulley. Crankbelt wrench (Ford) tool number YA826 (to hold the pulley stationary) and a Crankshaft Pulley wrench (Ford) tool number

T81P-6312-A or equivalent tools will make the job easier.

5. To install new belt, fit the timing belt over the gears in a counterclockwise direction starting at the crankshaft. Ensure that belt span between crankshaft and camshaft is kept tight as belt is installed over remaining gears.

6. Loosen belt tensioner attaching bolts and allow tensioner to extend against the belt.

7. Tighten one tensioner attaching bolt using special tool mentioned earlier or its equivalent.

8. Install the crankshaft pulley, drive plate and pulley attaching bolt.

9. Hold the crankshaft pulley stationary using tool YA-826 or equivalent and torque pulley bolt to 74–90 ft. lbs.

10. Disconnect the distributor coil wire and crank the engine for 30 seconds after reconnecting the negative battery cable. Disconnect the negative cable and realign marks. Check that the camshaft sprocket pointer is aligned with the TDC mark, and that the crankshaft is in the TDC position. If the timing marks do not align, remove the belt and reinstall.

11. Loosen belt tensioner attaching bolt (tightened in Step 7) ¼ to ½ turn maximum, while holding the crankshaft stationary with wrench T81P6312A or equivalent.

12. While securing the crankshaft so that it cannot turn, turn the camshaft sprocket counterclockwise using Camshaft Holding Tool D81P6256A or equivalent, and a torque wrench.

Tighten the belt tensioner attaching bolts when the torque wrench measures 27–32 ft. lbs. for a new belt, or 10 ft. lbs, if an old belt was installed.

NOTE: Do not apply torque to the camshaft sprocket attaching bolt. Apply it to the hex on the sprocket.

13. Install the timing belt cover and remaining parts in reverse order of removal.

Diesel Engine
ON CAR SERVICE

NOTE: This procedure is for in-vehicle service of the water pump, camshaft, or cylinder head. The timing belt cannot be replaced with the engine installed in the vehicle.

1. Remove the front timing belt upper cover and the flywheel timing mark cover.

2. Rotate engine clockwise until the timing marks on the flywheel and the front camshaft sprocket are aligned with their pointers.

3. Loosen tensioner pulley lockbolt and slide the timing belt off the water pump and camshaft sprockets.

4. The water pump and/or camshaft can now be serviced.

——— **CAUTION** ———
Unless the camshaft is being removed, DO NOT rotate the crankshaft with the front timing belt removed.

ADJUSTMENTS
Front Belt

1. Remove the flywheel timing mark cover.

2. Remove the front timing belt upper cover.

3. Remove the belt tension spring from the storage pocket in the front cover.

4. Install the tensioner spring in the belt tensioner lever and over the stud mounted on the front of the crankcase.

5. Loosen the tensioner pulley lockbolt.

6. Rotate the crankshaft pulley two revolutions clockwise until the flywheel TDC timing mark aligns with the pointer on the rear cover plate.

7. Check the front camshaft sprocket to see that it is aligned with its timing mark.

8. Tighten the tensioner lockbolt to 23–34 ft. lbs.

9. Check the bolt tension using Rotunda Belt Tension Gauge model 21-0028 or equivalent. Belt tension should be 33–44 lbs.

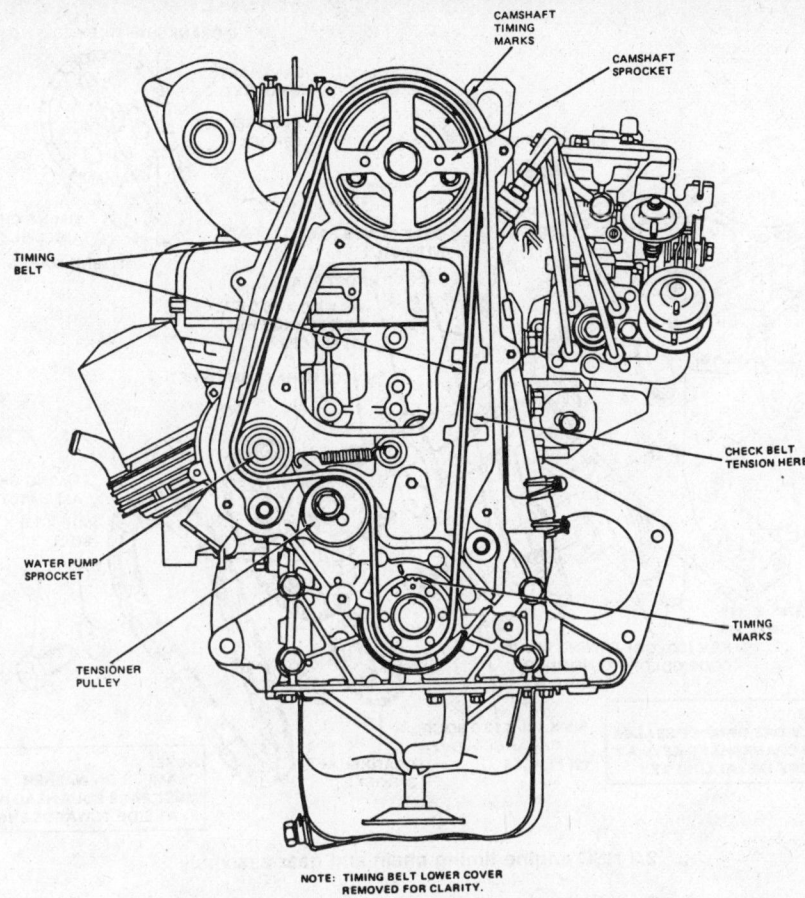

NOTE: TIMING BELT LOWER COVER REMOVED FOR CLARITY.

2.0L diesel timing belt installation

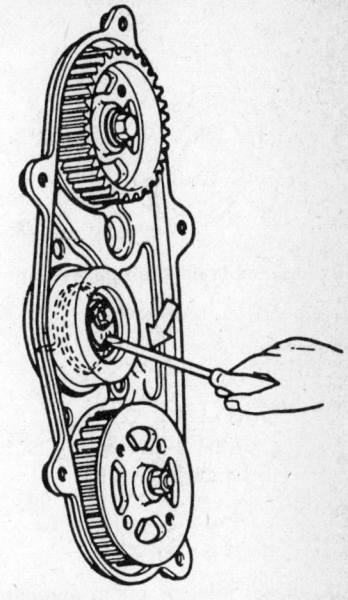

Loosening tensioner pulley on 2.0L diesel

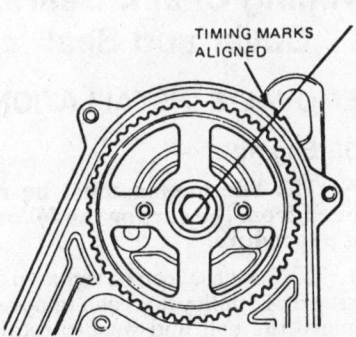

2.0L diesel camshaft timing mark

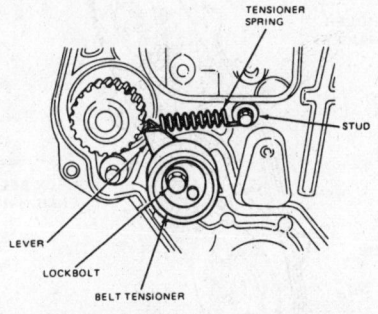

2.0L diesel front timing belt tensioner spring installation

10. Remove the tensioner spring and install it in the storage pocket in the front cover.

11. Install the front cover and tighten the attaching bolts to 5–7 ft. lbs.

12. Install the flywheel timing mark cover.

REAR BELT

1. Remove the flywheel timing mark cover.

2. Remove the rear timing belt cover.

3. Loosen the tensioner pulley locknut.

4. Rotate the crankshaft two revo-lutions until the flywheel TDC timing mark aligns with the pointer on the rear cover plate.

5. Check that the camshaft sprock-et and injection pump sprocket are aligned with their timing marks.

6. Tighten tensioner locknut to 15–20 ft. lbs.

7. Check belt tension using Rotun-da Belt Tension Gauge model 21-0028 or equivalent. Belt tension should be 22–33 lbs.

8. Install the rear timing belt cov-er. Tighten the 6mm bolts to 5–7 ft. lbs. and the 8mm bolt to 12–16 ft. lbs.

9. Install the flywheel timing mark cover.

Replacement
REAR BELT

1. Remove the rear timing belt cover.

2. Remove the flywheel timing mark cover from clutch housing.

3. Rotate the crankshaft until the flywheel timing mark is at TDC on No. 1 cylinder.

4. Check that the injection pump and camshaft sprocket timing marks are aligned.

5. Loosen the tensioner locknut. With a screwdriver, or equivalent tool, inserted in the slot provided, ro-tate the tensioner clockwise to relieve belt tension. Tighten locknut snug.

6. Remove the timing belt.

7. Install the belt.

8. Loosen the tensioner locknut and adjust timing belt as outlined in previous section.

9. Install rear timing belt cover and tighten bolts to 5–7 ft. lbs.

FRONT BELT

NOTE: The engine must be re-moved from the vehicle to replace the front timing belt.

1. With engine removed from the vehicle and installed on an engine stand, remove from timing belt upper cover.

2. Install a Flywheel Holding Tool T84P6375A or equivalent.

3. Remove the six bolts attaching the crankshaft pulley to the crank-shaft sprocket.

4. Install a crankshaft pulley Re-mover T58P63616D or equivalent us-ing Adapter T74P6700B or equiva-lent, and remove crankshaft pulley.

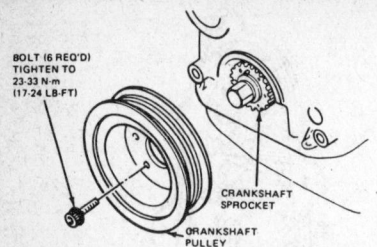

BOLT (6 REQ'D)
TIGHTEN TO
23-33 N·m
(17-24 LB·FT)

CRANKSHAFT
SPROCKET

CRANKSHAFT
PULLEY

2.0L diesel crankshaft pulley removal

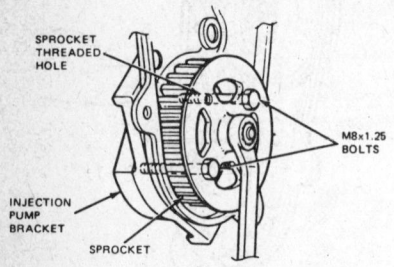

SPROCKET
THREADED
HOLE

M8×1.25
BOLTS

INJECTION
PUMP
BRACKET

SPROCKET

2.0L diesel injector pump sprocket removal

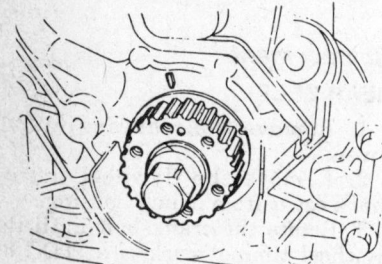

2.0L diesel crankshaft pulley timing marks

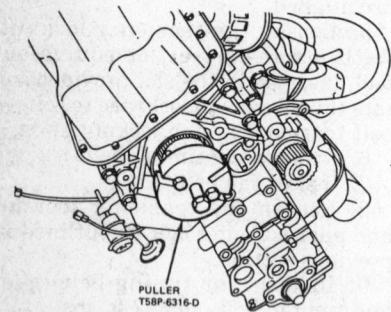

PULLER
T58P-6316-D

2.0L diesel crankshaft sprocket removal

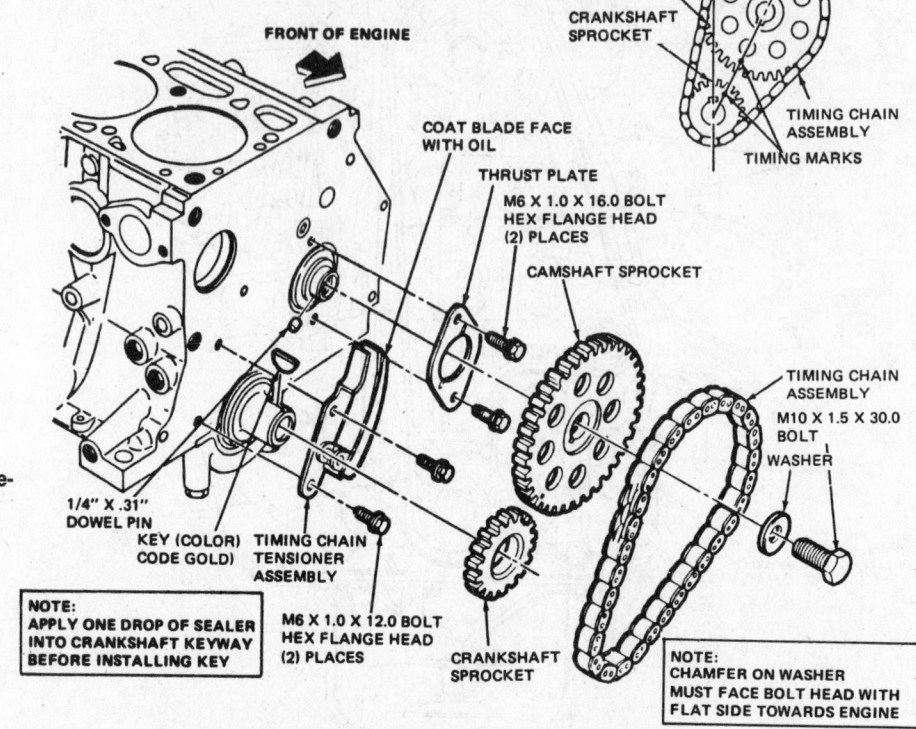

FRONT OF ENGINE

CRANKSHAFT KEY

CAMSHAFT
SPROCKET

CRANKSHAFT
SPROCKET

TIMING CHAIN
ASSEMBLY

TIMING MARKS

COAT BLADE FACE
WITH OIL

THRUST PLATE
M6 X 1.0 X 16.0 BOLT
HEX FLANGE HEAD
(2) PLACES

CAMSHAFT SPROCKET

TIMING CHAIN
ASSEMBLY
M10 X 1.5 X 30.0
BOLT
WASHER

1/4" X .31"
DOWEL PIN

KEY (COLOR)
CODE GOLD)

TIMING CHAIN
TENSIONER
ASSEMBLY

CRANKSHAFT
SPROCKET

NOTE:
APPLY ONE DROP OF SEALER
INTO CRANKSHAFT KEYWAY
BEFORE INSTALLING KEY

M6 X 1.0 X 12.0 BOLT
HEX FLANGE HEAD
(2) PLACES

NOTE:
CHAMFER ON WASHER
MUST FACE BOLT HEAD WITH
FLAT SIDE TOWARDS ENGINE

2.3 HSC engine timing chain and gear assembly

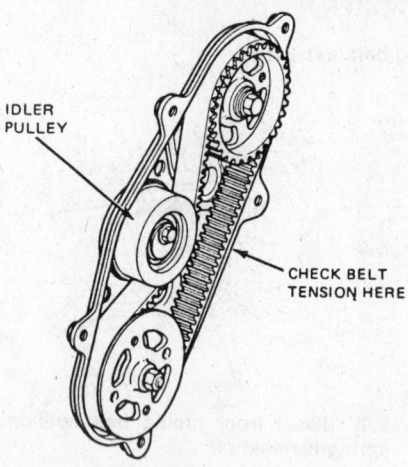

IDLER
PULLEY

CHECK BELT
TENSION HERE

2.0L diesel timing belt tensioner rear belt

5. Remove the front timing belt lower cover.

6. Loosen the tensioning pulley and remove the timing belt.

7. Align the camshaft sprocket with the timing mark.

NOTE: Check the crankshaft sprocket to see that the timing marks are aligned.

8. Remove the tensioner spring from the pocket in the front timing belt upper cover and install it in the slot in the tensioner lever and over the stud in the crankcase.

9. Push the tensioner lever toward the water pump as far as it will travel and tighten lockbolt snug.

10. Install timing belt.

11. Adjust the timing belt tension as outlined in previous section.

12. Install the front timing belt lower cover and tighten bolts to 5–7 ft. lbs.

13. Install the crankshaft pulley and tighten bolts to 17–24 ft. lbs.

14. Install the front timing belt upper cover and tighten bolts to 5–7 ft. lbs.

Timing Chain, Gears, Cover and Seal

REMOVAL & INSTALLATION

2.3L Engine

NOTE: The engine must be removed from the car for the following procedure.

1. The front seal can be replaced after the drive pulley has been removed. Remove the bolt and washer retaining the pulley. Use a suitable puller and remove the crankshaft pulley, install a front seal remover tool (Ford No. T74P6700A or the equivalent) and remove the seal. Coat a new seal with grease and install with suitable tool (Ford No. T83T4676A or the equivalent). Drive the seal in until fully seated. Check the seal after installation to make sure the spring is in proper position around the seal. Install the crankshaft pulley, washer and bolt.

2. To remove the front cover, remove the crankshaft pulley as described above. Remove the front cover retaining bolts, pry the top of the cover away from the engine block and remove the cover.

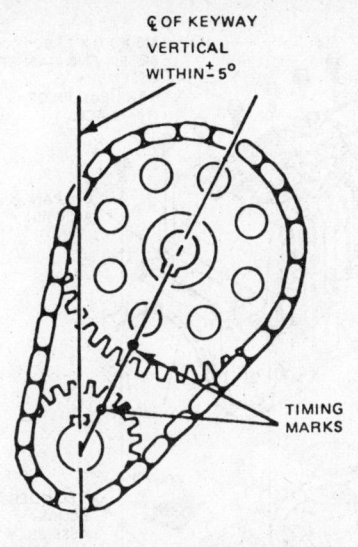

Timing mark alignment 2.3 HSC engine

3. Clean all gasket mounting surfaces. Check the play in the timing chain, replace chain if play is excessive. Check the timing chain tensioner blade for wear, if excessive replace the blade.

4. Turn the engine until the timing marks on the crank and cam gears align.

5. Remove the camshaft gear attaching bolt and washer. Slide the two gears and chain forward and remove as an assembly.

6. Install in the reverse order. Make sure the timing marks on the camshaft and crank gears are in alignment. Check the timing chain damper, located in the front cover, for wear and replace damper if necessary. Lubricate gears, chain, tensioner blade and front cover oil seal before cover installation. Apply an oil resistant sealer to both sides of the front cover gasket.

Camshaft

REMOVAL & INSTALLATION

1.6L & 1.9L Engine

NOTE: The camshaft can be removed with the engine in the car.

1. Remove the fuel pump and plunger. Set the engine to TDC on the compression stroke of No. 1 cylinder. Remove the negative battery cable.

2. Remove the alternator drive belt. Remove the power steering and air conditioning compressor drive belts, if equipped.

3. Remove the camshaft belt cover.

4. Remove the distributor.

5. Remove the rocker arms.

6. Remove the hydraulic valve lash adjusters. Keep the parts in order, as

they must be returned to their original positions.

7. Make sure the engine is set at No. 1 piston at TDC on the compression stroke. Remove the timing belt. DO NOT rotate the crankshaft while the timing belt is removed.

8. Remove the camshaft sprocket and key.

9. Remove the camshaft thrust plate.

10. Remove the ignition coil and coil bracket.

11. Remove the camshaft through the back of of the head towards the transaxle. 12.

Before installing the camshaft, coat the bearing journals, cam lobe surfaces, seal and thrust plate groove with engine oil. Install the camshaft through the rear of the cylinder head. Rotate the camshaft during installation. 13. Install the camshaft thrust plate and tighten the two attaching bolts to 7–11 ft. lbs.

14. Install the cam sprocket and key.

15. Install the timing belt. See timing belt removal and installation procedure.

16. Install the remaining parts in the reverse order of removal.

2.3L Engine

NOTE: The engine must be removed from the car to perform the following procedure.

1. Remove the oil dipstick, all drive belts and pulleys and remove the cylinder head.

2. Use a magnet or suitable tool to remove the hydraulic lifters from the engine. Keep the lifters in order if reusable.

3. Remove the crankshaft pulley and timing case cover.

4. Check camshaft end play, if excessive replace the thrust plate.

5. Remove the fuel pump and pushrod. Remove the timing chain, sprockets and tensioner.

6. Remove the camshaft thrust plate retaining bolts and the plate.

7. Carefully remove the camshaft from the engine. Use caution to avoid damage to the bearings, journals and lobes.

8. Install in reverse order. Apply lubricant to the camshaft lobes and journals and to the bottom of the lifters. Lubricate all assemblies with oil.

Piston and Connecting Rod Positioning

For all piston and connecting rod overhaul procedures, refer to "Engine Rebuilding" in the Unit Repair Section.

Gasoline Engines

1. The "Front" and arrow markings must face the front of the engine when rods and pistons are reinstalled.

2. "Front" and arrow markings must face the front of the engine when the rod and piston assemblies are installed.

3. Position the oil ring gap to the rear of the piston and the top and second compression ring gaps at 180 degree and 90 degree angles.

NOTE: Dip the piston assembly into an oil filled container before compressing the rings. Make sure the cylinder walls and connecting rod journals are clean and oiled before installation of the piston/rod assembly.

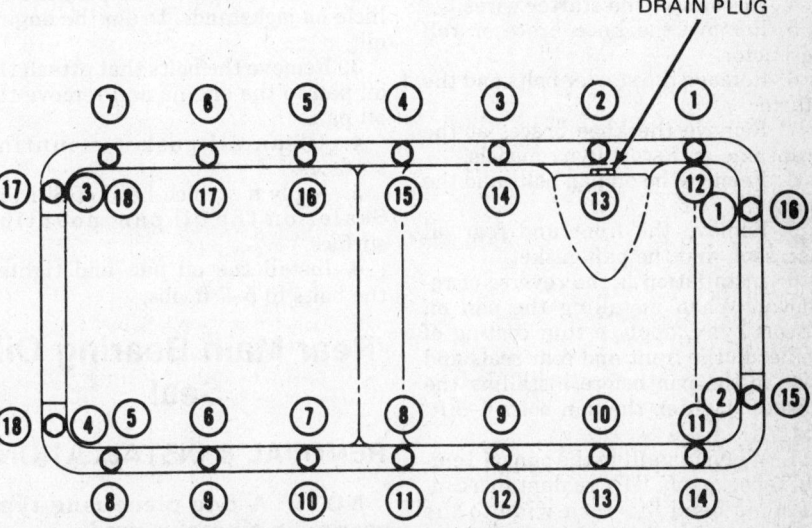

Oil pan removal and installation: tighten the bolts using the sequence inside the diagram, then retighten the bolts using the outside sequence

Diesel Engine

1. When installing rings, make sure that the side with the stamped mark faces upward.

2. Assemble the compression and oil rings. The gap of the top ring and second ring should be positioned on the opposite side from the turbulent flow chamber.

3. The gap of the ring should not be directed toward the thrust side or counter-thrust side and the gap of the top ring should be set opposite (180°) to that of the oil ring.

4. When assembling the connecting rod and bearing cap, make sure the weight marks of the rod and cap are matched correctly.

ENGINE LUBRICATION

Oil Pan

REMOVAL & INSTALLATION

Gasoline Engines

NOTE: The oil pan can be removed with the engine in the car. No suspension or chassis components need be removed.

1. Disconnect the negative battery terminal.

2. Jack up the vehicle and support it with stands.

3. Drain the oil. On Tempo/Topaz, drain cooling system and remove coolant tube (lower hose). Disconnect exhaust pipe. Move A/C line out of the way.

4. Disconnect the starter wires.

5. Remove the knee brace or roll restrictor.

6. Remove the starter bolts and the starter.

7. Remove the knee braces at the transaxle on Escort/Lynx models.

8. Remove the oil pan bolts and the pan.

9. Remove the front and rear oil pan seal, and the pan gasket.

10. Installation is the reverse of removal. When installing the pan on Escort/Lynx, apply a thin coating of sealer to the front and rear seals and also to the pan before installing the gasket. Tighten the pan bolts 6–8 ft. lbs.

11. When installing the pan on Tempo-Topaz, apply RTV sealant in a continuous bead ³/₁₆ inch wide to the groove on the oil pan. Install pan and pan bolts. Tighten bolts enough to

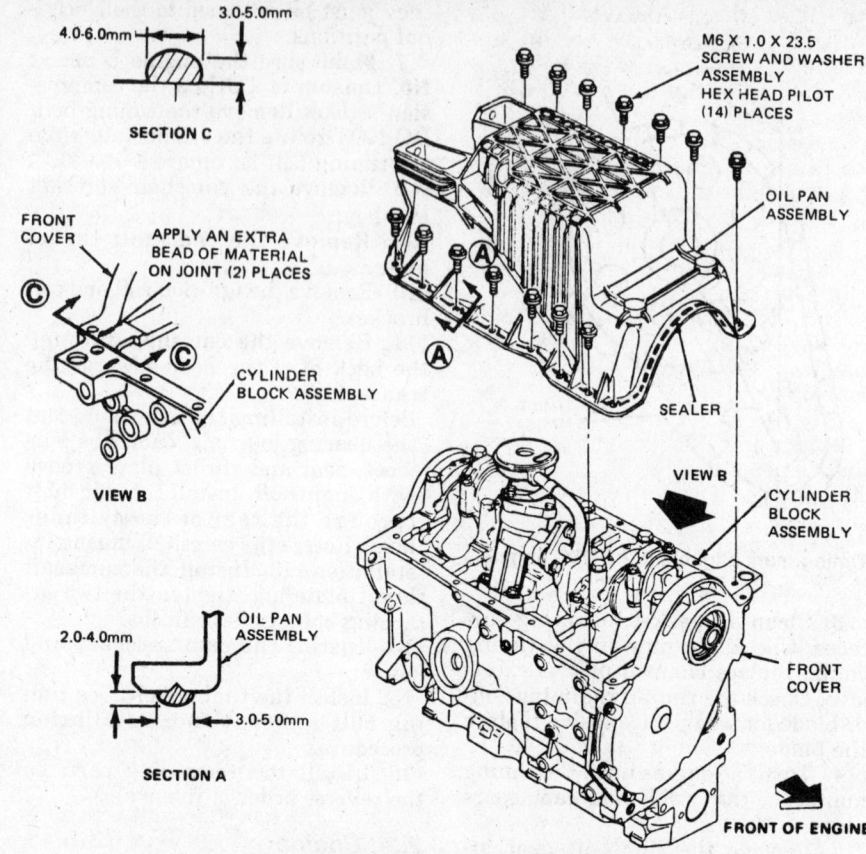

Oil pan installation on 2.3L HSC engine

squeeze RTV sealant to a point where the two transaxle bolts holes align. Install transaxle bolts and tighten to 30–39 ft. lbs. Loosen the bolts one-half turn. Tighten pan mounting bolts to 6–9 ft. lbs.

Diesel Engine

1. Disconnect the negative battery cable.

2. Raise and safely support the vehicle on jackstands. Drain the engine oil.

3. Remove the bolts that attach the oil pan to the engine and remove the oil pan.

4. Clean all gasket mounting surfaces.

5. Apply a ⅛ inch bead of Silicone Sealer on the oil pan mounting surface.

6. Install the oil pan and tighten the bolts to 5–7 ft. lbs.

Rear Main Bearing Oil Seal

REMOVAL & INSTALLATION

NOTE: A one piece ring-type rear main oil seal is used.

1. Remove the transaxle.

2. Remove the rear cover plate.

3. Remove the flywheel or flexplate if so equipped.

4. Remove the rear main seal by carefully punching a hole in the seal and removing with a threaded slide hammer. Be extremely careful not to scratch the crankshaft or seal mating surface.

5. Coat the lips of the new seal with engine oil. Gently tap the seal into place.

6. Installation is the reverse of removal.

Oil Pump

REMOVAL & INSTALLATION

1.6L & 1.9L Engine

1. Disconnect the negative cable at the battery.

2. Loosen the alternator bolt on the alternator adjusting arm. Lower the alternator to remove the accessory drivebelt from the crankshaft pulley.

3. Remove the timing belt cover.

NOTE: Set No. 1 cylinder at TDC prior to timing belt removal.

4. Loosen both belt tensioner attaching bolts using Tool T81P6254A or equivalent on the left bolt. Using a

pry bar or other suitable tool pry the tensioner away from the belt. While holding the tensioner away from the belt, tighten one of the tensioner attaching bolts.

5. Disengage the timing belt from the camshaft sprocket, water pump sprocket and crankshaft sprocket.

6. Raise the vehicle and safely support on jackstands. Drain the crankcase.

7. Using a Crankshaft Pulley Wrench T81P6312A and Crankshaft Bolt Wrench YA-826 or equivalent, remove the crankshaft pulley attaching bolt.

8. Remove the timing belt.

9. Remove the crankshaft drive plate assembly. Remove the crankshaft pulley. Remove the crankshaft sprocket.

10. Disconnect the starter cable at the starter.

11. Remove the knee-brace from the engine.

12. Remove the starter.

13. Remove the rear section of the kneebrace and inspection plate at the transmission.

14. Remove the oil pan retaining bolts and oil pan. Remove the front and rear oil pan seals. Remove the oil pan side gaskets. Remove the oil pump attaching bolts, oil pump and gasket. Remove the oil pump seal.

15. Make sure the mating surfaces on the cylinder block and the oil pump are clean and free of gasket material.

16. Remove the oil pick-up tube and screen assembly from the pump for cleaning.

17. Lubricate the outside diameter of the oil pump seal with engine oil.

18. Install the oil pump seal using Seal Installer T81P6700A or equivalent.

19. Install the pick-up tube and screen assembly on the oil pump. Tighten attaching bolts to 6–9 ft. lbs.

20. Lubricate the oil pump seal lip with light engine oil.

21. Position the oil pump gasket over the locating dowels. Install attaching bolts and tighten to 5–7 ft. lbs.

22. Apply a bead of silicone sealer approximately 3.0mm wide at the corner of the front and rear oil pan seals, and at the seating point of the oil pump to the block retainer joint.

23. Install the front oil pan seal by pressing firmly into the slot cut into the bottom of the pump.

24. Install the rear oil seal by pressing firmly into the slot cut into rear retainer assembly.

NOTE: Install the seal before the sealer has cured (within 10 minutes of application).

25. Apply adhesive sealer evenly to oil pan flange and to the oil pan side of the gaskets. Allow the adhesive to dry past the "wet" stage and then install the gaskets on the oil pan. Position the oil pan on the cylinder block.

26. Install oil pan attaching bolts. Tighten bolts in the proper sequence 6–8 ft. lbs.

27. Position the transmission inspection plate and the rear section of the knee-brace on the transmission. Install the two attaching bolts and tighten to specification.

28. Install the starter.

29. Install the knee-brace.

30. Connect the starter cable.

31. Install the crankshaft gear. Install crankshaft pulley. Install crankshaft drive plate assembly. Install timing belt over the crankshaft pulley.

32. Using the Crankshaft Pulley Wrench T81P6312A and Crankshaft Bolt Wrench YA-826 or equivalent, install the crankshaft pulley attaching bolt. Tighten bolt to specification. (Refer to "Timing Belt" section.)

33. Lower the vehicle.

34. Install the engine front timing cover.

35. Position the accessory drive belts over the alternator and crankshaft pulleys. Tighten the drive belts to specification.

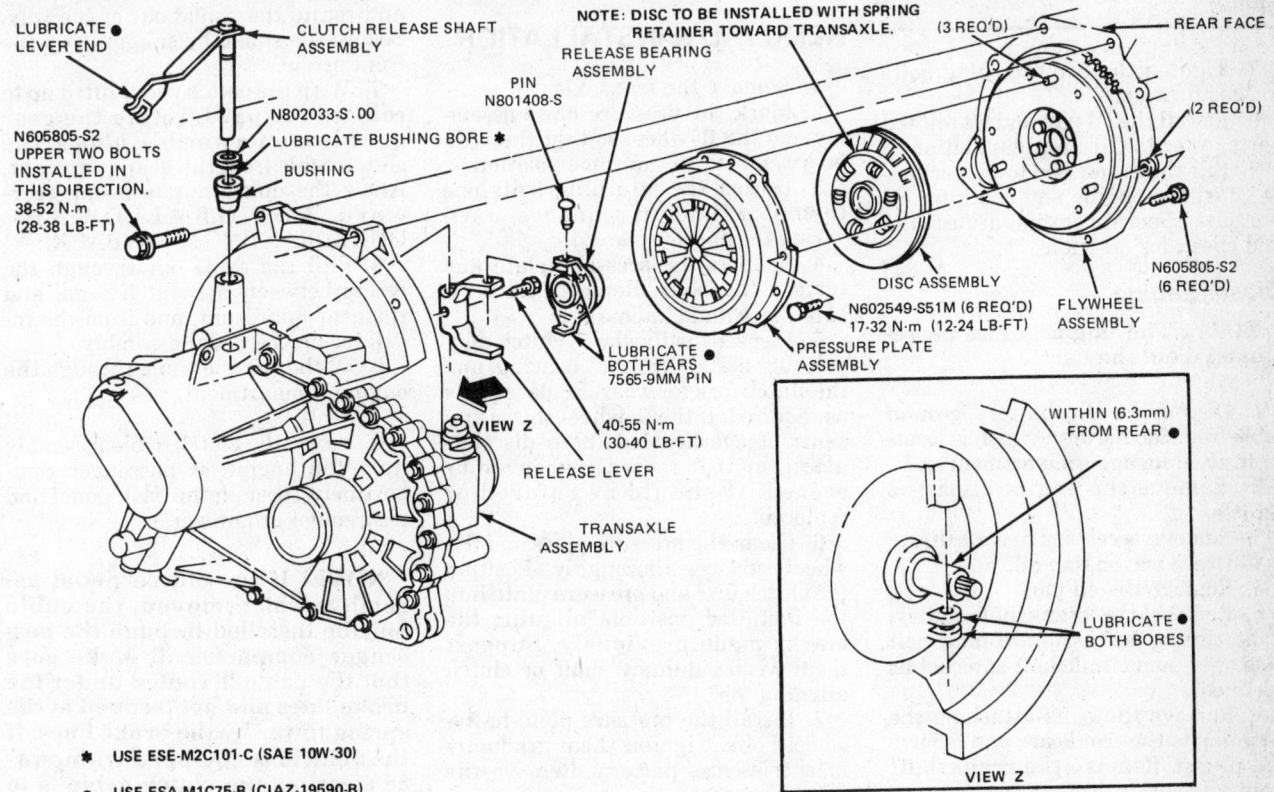

Typical clutch assembly installation

36. Connect the negative cable at the battery. Fill crankcase to the proper level with the specified oil.

37. Start the engine and check for oil leaks. Make sure the oil pressure indicator lamp has gone out. If the lamp remains On, immediately shut off the engine, determine the cause and correct the condition.

2.3L Engine

1. Remove the oil pan.
2. Remove the oil pump attaching bolts and remove the oil pump and intermediate driveshaft.
3. Prime the oil pump by filling inlet port with engine oil. Rotate the pump shaft until oil flows from outlet port.
4. If the screen and cover assembly have been removed, replace the gasket. Clean the screen and reinstall the screen and cover assembly.
5. Position the intermediate driveshaft into the distributor socket.
6. Insert the intermediate driveshaft into the oil pump. Install the pump and shaft as an assembly.

————— CAUTION —————

Do not attempt to force the pump into position if it will not seat. The shaft hex may be misaligned with the distributor shaft. To align, remove the oil pump and rotate the intermediate driveshaft into a new position.

7. Tighten the two attaching bolts to specification.
8. Install the oil pan and all related parts, refer to Oil Pan Installation.
9. Fill the crankcase to proper level. Start the engine and check for oil pressure. Operate engine at fast idle and check for oil leaks.

Diesel Engine

NOTE: The engine must be removed from the car.

1. Disconnect the battery ground cable from the battery, which is located in the luggage compartment.
2. Remove the engine from the vehicle.
3. Remove accessory drive belts.
4. Drain the engine oil.
5. Remove the oil pan.
6. Remove the crankshaft pulley, front timing belt, front timing belt tensioner, and crankshaft sprocket as outlined.
7. Remove the bolts attaching the oil pump to the crankcase and remove the pump. Remove the crankshaft front oil seal.
8. Clean the oil pump and the crankcase gasket mating surfaces.
9. Apply a 1/8 inch bead of silicone

sealer on the oil pump-to-crankcase mating surface.
10. Install a new O-ring.
11. Install the oil pump, making sure the oil pump inner gear engages with the splines on the crankshaft. Tighten the 10mm bolts to 23–34 ft. lbs. and the 8mm bolts to 12–16 ft. lbs.
12. Install a new crankshaft front oil seal.
13. Clean the oil pan-to-crankcase mating surfaces.
14. Apply a 1/8 inch bead of Silicone Sealer on the oil-pan-to-crankcase mating surface.
15. Install the oil pan and tighten the bolts to 5–7 ft. lbs.
16. Install and adjust as necessary the crankshaft sprocket, front timing belt tensioner and front timing belt.
17. Install and adjust the accessory drive belts.
18. Install engine in the vehicle.
19. Fill and bleed the cooling system.
20. Fill the crankcase with the specified quantity and quality of oil.
21. Run the engine and check for oil, fuel and coolant leaks.

————————————————————

CLUTCH

REMOVAL & INSTALLATION

1. Remove the transaxle.
2. Mark the pressure plate assembly and the flywheel so that they can be assembled in the same position.
3. Loosen the attaching bolts one turn at a time, in sequence, until spring tension is relieved.
4. Support the pressure plate and remove the bolts. Remove the pressure plate and clutch disc.
5. Inspect the flywheel, clutch disc, pressure plate, throwout bearing, and the clutch fork for wear. Replace parts as required. If the flywheel shows any signs of overheating (blue discoloration) or if it is badly grooved or scored, it should be refaced or replaced.
6. Clean the pressure plate and flywheel surfaces thoroughly. Position the clutch disc and pressure plate into the installed position, aligning the marks made previously. Support them with a dummy shaft or clutch aligning tool.
7. Install the pressure plate-to-flywheel bolts. Tighten them gradually in a crisscross pattern. Remove the alignment tool.
8. Lubricate the release bearing and install it in the fork.
9. Install the transaxle.

ADJUSTMENT

NOTE: All models are equipped with automatically self-adjusting clutches. No separate free play adjustments are necessary.

Clutch Cable

REMOVAL & INSTALLATION

1. Prop up the clutch pedal to lift the pawl free of the quardrant which is part of the self adjuster mechanism.
2. Remove the air cleaner assembly to gain access to the clutch cable.
3. Grasp the extended end of the clutch cable with a pair of pliers, and unhook the clutch cable from the clutch bearing release lever.

NOTE: Do not grasp wire strand portion of inner cable since this may cut wires and result in cable failure.

4. Disconnect the cable from the insulator that is located on the rib of the transaxle.
5. Remove the panel above the clutch pedal pad (Tempo/Topaz).
6. Position the clutch shield away from the brake pedal support bracket by removing the rear retaining screw (located nearest the instrument panel). Loosen the front retaining screw and rotate the shield out of the way. Secure the shield by snugging up the front screw.
7. With the clutch pedal lifted up to release the pawl, rotate the gear quardrant forward. Unhook the clutch cable from the gear quardrant. Allow the quardrant to swing rearward. DO NOT ALLOW THE QUARDRANT TO SNAP BACK.
8. Pull the cable out through the recess between the clutch pedal and the gear quardrant, and from the insulator on the pedal assembly.
9. Withdraw the cable through engine compartment.
To Install:
10. Insert the clutch cable assembly from the engine or passenger compartment through the dash panel and pash panel groummet.

NOTE: If the clutch pedal assembly was removed, the cable may be installed through the passenger compartment. Make sure that the cable is routed under the brake lines and not trapped at the spring tower by the brake lines. If the vehicle is equipped with power steering, the clutch cable is to be routed inboard of the power steering hose.

11. Push the clutch cable through

insulator on the stop bracket, and through recess between the pedal and gear quardrant.

12. With the clutch pedal lifted up to release the pawl, rotate the gear quardrant forward. Hook the cable into the gear quardrant.

13. Install the clutch shield on the brake pedal support bracket.

14. Install the panel above the clutch pedal.

15. Secure the pedal in the upmost position using a piece of wire, cord or tape.

16. Hook the cable into the clutch release lever in the engine compartment.

17. Remove the device used to temporarily secure the pedal.

18. Adjust the clutch by depressing the pedal several times.

19. Install the air cleaner.

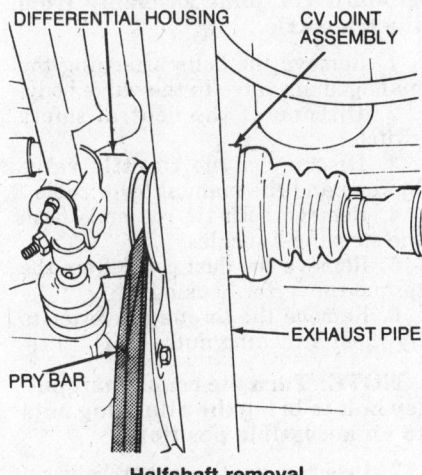

Halfshaft removal

Starter/Clutch Interlock Switch

REMOVAL & INSTALLATION

1. Remove the panel above the clutch pedal (Tempo/Topaz).

2. Disconnect the wiring connector.

3. Remove the clutch interlock attaching screw and hairpin clip, then remove the switch

NOTE: Always install the switch with the self-adjusting clip about 1 inch from the end of the rod. The clutch pedal must be fully up (clutch engaged). Otherwise, the switch may be misadjusted.

To install:

4. Insert eyelet end of the rod over the pin on the clutch pedal and secure with the hairpin clip.

5. Align mounting boss with corresponding hole in bracket. Attach with screw.

6. Reset the clutch interlock switch by pressing the clutch pedal to the floor.

7. Connect the wiring connector.

8. Install the panel (Tempo/Topaz).

MANUAL TRANSAXLE

REMOVAL & INSTALLATION

Escort/Lynx

1. Disconnect the negative battery terminal.

2. Remove the two transaxle to engine top mounting bolts.

3. Remove the clutch cable from the clutch release lever.

4. Raise the vehicle and support it on jack stands.

5. Remove the brake line routing clamps from the front wheels.

6. Remove the bolt that secures the lower control arm ball joint to the steering knuckle assembly, and pry the lower control arm away from the knuckle. When installing, a new nut and bolt must be used.

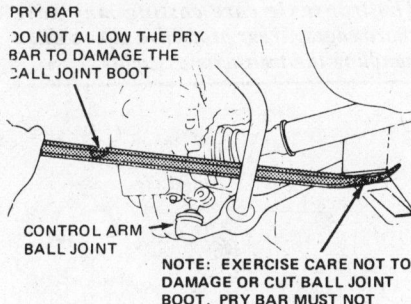

Separating the steering knuckle from the ball joint

NOTE: The plastic shield installed behind the rotor contains a molded pocket for the lower control arm ball joint. When removing the control arm from the knuckle, bend the shield toward the rotor to provide clearance.

7. Pry the right inboard CV-joint from the transaxle, then remove the CV-joint and halfshaft by pulling outward on the steering knuckle; wire the CV-joint/halfshaft assembly out of the way. Wire the joint assembly in a level position to prevent it from expanding.

NOTE: When the CV-joint is pulled out of the transaxle fluid

will leak out. Install shipping plugs T81P-1177-B or their equivalent to prevent the dislocation of the differential side gears.

8. Repeat the procedures and remove the left hand CV-joint/halfshaft from the transaxle.

9. Remove the stabilizer bar.

10. Disconnect the speedometer cable and back-up light.

11. Remove the (3) nuts from the starter mounting studs which hold the engine roll restrictor bracket.

12. Remove the roll restrictor and the starter stud bolts.

13. Remove the stiffner brace.

14. Remove the shift mechanism crossover spring.

15. Remove the shift mechanism stabilizer bar.

16. Remove the shift mechanism.

17. Place a transmission jack under the transaxle.

18. Remove the rear transmission mounts.

19. Remove the front transmission mounts.

20. Lower the transaxle support jack until it clears the rear mount and support the engine with a jack, under the oil pan.

21. Remove the four remaining engine to transaxle bolts.

22. Remove the transaxle.

NOTE: The case may have sharp edges. Wear protective gloves when handling the transaxle.

23. Installation is the reverse of removal.

NOTE: When installing the CV-joint/halfshaft assemblies into the transaxle, install new circlips on the inner stub shafts, carefully install the assemblies into the transaxle to prevent damaging the oil seals, and insure that both joints are fully seated in the transaxle by lightly prying outward to confirm they are seated. If the circlips are not seated, the joints will move out of the transaxle.

Tempo/Topaz

1. Wedge a wood block approximately 7 inches long under the clutch pedal to hold the pedal up slightly beyond its normal position. Grasp the clutch cable and pull forward, disconnecting it from the clutch release shaft assembly. Remove the clutch casing from the rib on the top surface of the transaxle case.

2. Using a 13mm socket, remove the two top transaxle-to-engine mounting bolts. Using a 10mm socket, remove the air cleaner.

3. Raise and safely support the car. Remove the front stabilizer bar to control arm attaching nut and washer (drivers side). Discard the attaching nut. Remove the two front stabilizer bar mounting brackets. Discard the bolts.

4. Using a 15mm socket, remove the nut and bolt that secures the lower control arm ball joint to the steering knuckle assembly. Discard the nut and bolt. Repeat this procedure on the opposite side.

5. Using a large pry bar, pry the lower control arm away from the knuckle.

--- CAUTION ---

Exercise care not to damage or cut the ball joint boot. Pry bar must not contact the lower arm. Repeat this procedure on the opposite side.

6. Using a large pry bar, pry the left inboard CV-joint assembly from the transaxle.

NOTE: Lubricant will drain from the seal at this time. Install shipping plugs (T81P-1177-B or equivalent). Two plugs are required (one for each seal). Remove the inboard CV-joint from the transaxle by grasping the left hand steering knuckle and swinging the knuckle and halfshaft outward from the transaxle.

--- CAUTION ---

Exercise care when using a pry bar to remove the CV-joint assembly. If not careful, damage to the differential oil seal may result.

7. If the CV-joint assembly cannot be pried from the transaxle, insert Differential Rotater Tool (T81P-4026-A or equivalent), through the left side and tap the joint out. Tool can be used from either side of transaxle.

8. Wire the halfshaft assembly in a near level position to prevent damage to the assembly during the remaining operations. Repeat this procedure on the opposite side.

9. Using a small prybar, remove the backup lamp switch connector from the transaxle back-up lamp switch.

10. Using a 15mm socket, remove the three nuts from the starter mounting studs which hold the engine roll restrictor bracket. Remove the engine roll restrictor.

11. Using a 13mm deep well socket, remove the shift mechanism to shift shaft attaching nut and bolt and control selector indicator switch arm. Remove the shift shaft.

13. Using a 15mm socket, remove the shift mechanism stabilizer bar to

transaxle attaching bolt. Remove the $7/32$ in. sheet metal screw and the control selector indicator switch and bracket assembly.

14. Using a 22mm ($7/8$ in.) crows foot wrench, remove the speedometer cable from the transaxle.

15. Using a 13mm universal socket, remove the two stiffener brace attaching bolts from the oil pan to clutch housing.

16. Position a suitable jack under the transaxle. Using an 18mm socket, remove the two nuts that secure the left hand rear No. 4 insulator to the body bracket.

17. Using a 13mm socket, remove the bolts that secure the left hand front No. 1 insulator to the body bracket. Lower the transaxle jack until the transaxle clears the rear insulator. Support the engine with a screw jack stand under the oil pan. Use a 2 x 4 inch piece of wood on top of the screw jack.

18. Using a 13mm socket, remove the four engine to transaxle attaching bolts. One of these bolts holds the ground strap and wiring loom stand off bracket.

19. Remove the transaxle from the rear face of the engine and lower transaxle from the vehicle.

20. Install in reverse order.

--- CAUTION ---

The transaxle case casting may have sharp edges. Wear protective gloves when handling the transaxle assembly.

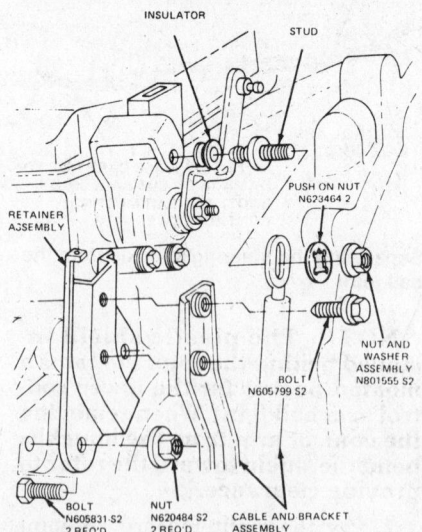

ATX shift control cable and bracket assembly

OVERHAUL

For all overhaul procedures, please refer to "Manual Transmission Overhaul" in the Unit Repair section.

AUTOMATIC TRANSAXLE

For all automatic transaxle adjustment procedures, please refer to "Automatic Transmissions" in the Unit Repair section.

REMOVAL & INSTALLATION

Escort/Lynx

NOTE: Removal of the automatic transaxle is basically the same as the standard transaxle with the following recommendations. Due to the ATX case configuration the right hand halfshaft assembly must be removed first. Special tool T81P-4026-A is then inserted into the transaxle to drive the left hand inboard CV joint assembly from the transaxle.

1. Remove the bolts attaching the managed air valve to the valve body.
2. Disconnect the neutral safety switch.
3. Disconnect the throttle valve linkage and the manual lever cable.
4. Remove both tie rod ends from the steering knuckles.
5. Remove the dust cover from the torque converter housing.
6. Remove the torque converter to flywheel attaching nuts.

NOTE: Turn the crankshaft pulley bolt to bring the attaching nuts to an accessible position.

7. Insert a small pry bar between the flywheel and torque converter, then carefully move the transaxle and converter away from the engine.

Tempo/Topaz

NOTE: The engine and automatic transaxle must be removed as an assembly. Refer to "Engine Removal". Separate the transaxle from the engine after removal from the car. If services performed while the transaxle and engine are separated include engine oil pan removal, the transaxle and engine must be attached together before reinstalling the engine oil pan.

--- CAUTION ---

When removing both the left and right halfshafts special plug T81P-1177-B must be installed. Failure to use these plugs can result in dislocation of the differential side gears. Should these gears become misaligned the differential will have to be removed from the transaxle to re-align the gears.

The halfshaft removal procedure is the same for the ATX and MTX with the following exception. Due to the case configuration on the ATX the right hand halfshaft assembly must be removed first. Driver #T81P-4026-A or equivalent is inserted into the transaxle to drive the left hand inboard CV joint assembly from the transaxle. If only the left hand halfshaft assembly is to be removed for service, remove the right hand halfshaft assembly from the transaxle only. After removal support it with a length of wire, then drive the left hand halfshaft assembly from the transaxle.

DRIVE AXLE

Halfshaft

REMOVAL & INSTALLATION

NOTE: Before attempting this procedure you must be sure to have a new hub nut and a new lower control arm to steering knuckle bolt and nut. Once these parts have been removed they must not be reused.

1. Remove the hub cap and loosen the hub nut.
2. Jack up the vehicle and support it with jack stands.
3. Remove the hub nut and washer.
4. Remove the bolt attaching the brake hose routing clip to the suspension strut.
5. Remove the ball joint to steering knuckle bolt and nut.
6. Separate the ball joint from the steering knuckle using a pry bar.

NOTE: The lower control arm ball joint fits into a pocket formed in the plastic disc brake rotor shield. This shield must be bent away from the ball joint while prying the ball joint out of the steering knuckle.

7. Remove the halfshaft from the differential housing, using a pry bar. Be careful not to damage any seals or boots.
8. Tie the end of the shaft out of the way with a piece of wire.
9. Separate the outboard CV joint from the hub using a puller.
10. Installation is the reverse of removal with the following suggestions. Install a new circlip on the inboard CV joint stub shaft. Stake the new hub nut with a chisel, after torqueing nut to 180–200 ft. lbs.

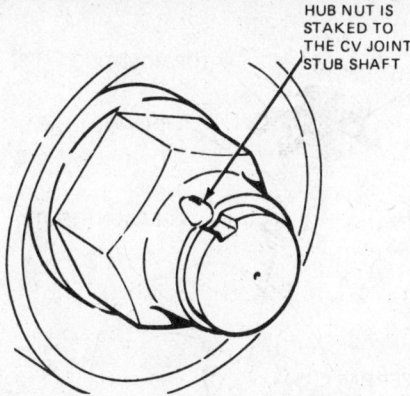

HUB NUT IS STAKED TO THE CV JOINT STUB SHAFT

Stake the front wheel axle nut

Constant Velocity Joint

OVERHAUL

NOTE: The CV joint components are matched during manufacture and cannot be interchanged with components from another CV joint. The joint can be disassembled for inspection and cleaning only; a damaged CV joint must be completely replaced. For disassembly and inspection procedures, please refer to "CV-Joint Overhaul" in the Unit Repair Section.

REAR AXLE

Rear Wheel Bearings

REMOVAL, PACKING, INSTALLATION AND ADJUSTMENT

NOTE: The rear wheel bearings are located in the brake drum hub. The inner wheel bearing is protected by a grease seal. A washer and spindle nut retain the hub/drum assembly and control the bearing endplay.

1. Remove wheel, dust cover, cotter pin nut and drum.
2. The outer bearing will be loose when the drum is removed and may be lifted out by hand. The inner bearing is retained by a grease seal. To remove the inner bearing, insert a wooden dowel or soft drift through the hub from the outer bearing side and carefully drive out the inner bearing and grease seal.

3. Clean the bearings, cups and hubs with a suitable solvent. Inspect the bearings and cups for damage or heat discoloring. Replace as a set if necessary. Always install a new grease seal.
4. If new bearings are to be used, use a three jawed slide-hammer puller to remove the cups from the drumhub. Install the new bearing cups using a suitable driver. Make sure they are fully seated in the hub.
5. Pack the bearings with a multi-purpose grease.
6. Coat the cups with a thin film of grease. Install the inner bearing and grease seal.
7. Coat the bearing surfaces of the spindle with a thin film of grease. Slowly and carefully slide the drum and hub over the spindle and brake shoes. Install the outer bearing over the spindle and into the hub.
8. Install the keyed flat washer and adjusting nut on the spindle.
9. Tighten the adjusting nut to between 17–25 ft. lbs.
10. Back-off the adjusting nut ½ turn. Then retighten it to between 10–15 ft. lbs.
11. Position the nut retainer on the nut and install the cotter pin. Do not tighten the nut to install the cotter pin.
12. Spread the ends of the cotter pin and bend them around the nut retainer. Install the center grease cap.
13. Install the tire and wheel assembly. Lower the car and tighten the wheel lugs.

Front Axle Hub and Bearings

Refer to "Drive Axle" for removal and installation procedures.

FRONT SUSPENSION

All models are equipped with a MacPherson strut front suspension with cast steering knuckles. The shock absorber strut assembly includes a rubber top mount and a coil spring insulator, mounted on the shock strut.

The entire strut assembly is attached to the top by two bolts. The lower end of the assembly is attached to the steering knuckle. A pinch joint is designed into the knuckle. The forged lower arm assembly is attached to the underbody side apron

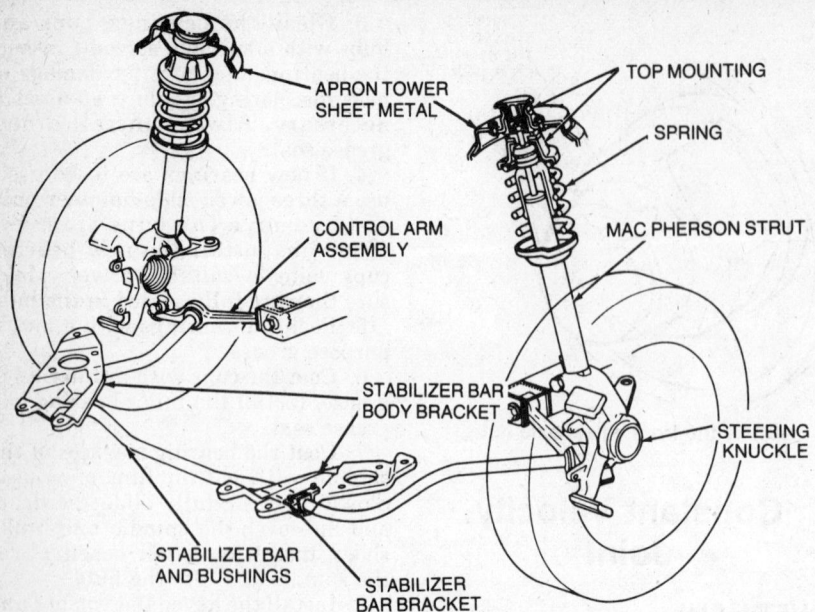

Front suspension components

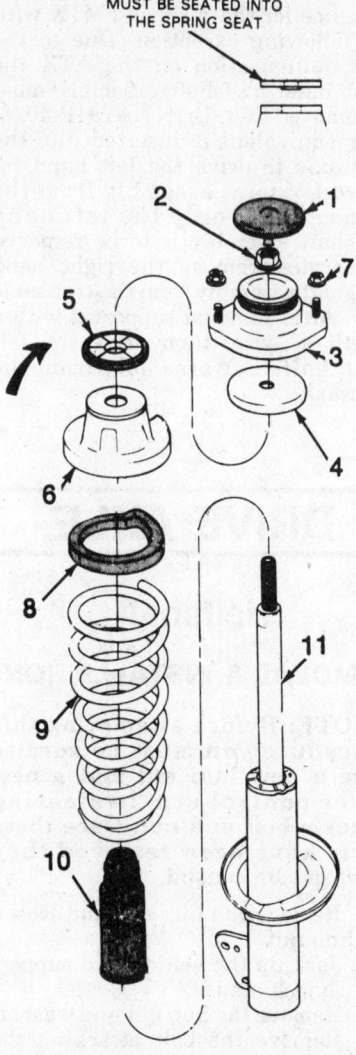

BEARING AND SEAL ASSY. MUST BE SEATED INTO THE SPRING SEAT

Exploded view of strut assembly

and steering knuckle. A stabilizer bar connects the outer end of the lower arm to the engine mount bracket. Caster and camber are preset and non-adjustable. The suspension fittings are "lubed for life"; no grease fittings are provided.

Strut and Shock Absorber

REMOVAL & INSTALLATION

1. Jack up the vehicle and support it with jack stands.
2. Remove the front wheels. Remove caliper and rotor.
3. Remove the brake line flex hose clip from the strut.
4. Jack up the lower control arm and raise the strut as far as possible without lifting the vehicle from the jack stands.
5. Install a spring compressor on the spring.
6. Tighten the spring until there is approximately $\frac{1}{8}$ inch between any two coils.

—————— CAUTION ——————
The spring must be compressed before the strut is removed to insure that excessive force is not applied to the constant velocity joints.

7. Remove the pinch bolt from the steering knuckle.
8. Loosen the two top mounting bolts, but do not remove them.
9. Lower the jack away from the control arm.
10. Use a suitable tool to spread the pinch joint.

11. Place a piece of 2 x 4 wood about $7\frac{1}{2}$ inches long against the shoulder of the knuckle.
12. Insert a pry bar between the wooden block and the strut base, or apron. Separate the strut from the knuckle.
13. Remove the top mounting nuts.
14. Remove the strut and spring assembly.
15. Installation is the reverse of removal. Torque the top mount bolts to 20–30 ft. lbs., and the pinch bolt to 37–44 ft. lbs.

OVERHAUL

For all spring and shock absorber REMOVAL & INSTALLATION procedures, and all strut overhaul procedures, please refer to "Strut Overhaul" in the unit repair section.

Springs

REMOVAL & INSTALLATION

For all spring and shock absorber REMOVAL AND INSTALLATION procedures, please refer to "Strut Overhaul" in the unit repair section.

Front Control Arms and Ball Joints

REMOVAL & INSTALLATION

1. Loosen the wheel nuts, raise and support the car, and remove the wheel and tire.

2. Remove the nut from the stabilizer bar. Pull of the large dished washer.
3. Remove the control arm inner mounting pivot nut and bolt.
4. Remove the ball joint stud pinch bolt from the steering knuckle.
5. Pull the control arm and ball joint down and away from the steering knuckle. Slightly separate the pinch ears with a small pry bar if necessary.
6. Remove the stabilizer spacer bushings. Remove the control arm.
7. Installation is the reverse. Tighten the pinch bolt to 37–44 ft. lbs. Inner control arm mounting bolt and nut to 48–55 ft. lbs. Stabilizer bar nut to 98–115 ft. lbs.

NOTE: Be sure the steering column is unlocked and do not use a hammer to separate the ball joint from the knuckle.

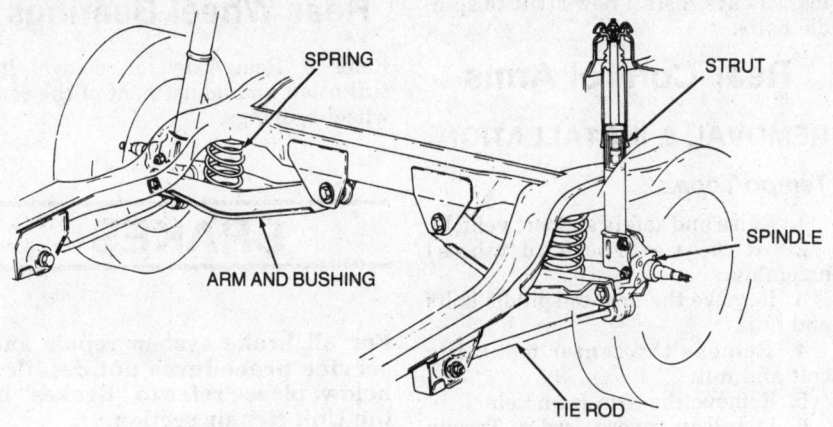

Escort/Lynx rear suspension components

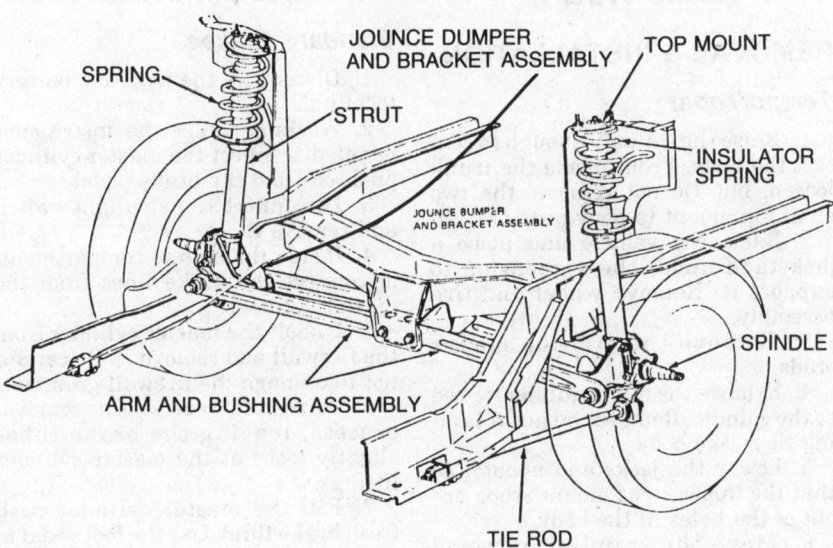

Tempo/Topaz rear suspension

Front Wheel Bearings

NOTE: Sealed front wheel bearings are used, which require no periodic lubrication or adjustment.

REAR SUSPENSION

MacPherson Strut

The Escort, EXP, Lynx and LN7 feature a modified MacPherson strut independent rear suspension. Each side consists of a shock strut, lower control arm, tie rod, forged spindle and a coil spring mounted on the control arm.

The shock strut consists of a rubber insulated top mount, one piece jounce bumper/dust shield and an integral shock absorber. The entire strut assembly is attached to the body side panel by a rubber insulated top mount assembly and nut. The lower end of the assembly is bolted to the spindle. The lower control arm attaches to the crossmember and to the spindle. A coil spring is located on the crossmember. The tie rod attaches to the frame rail and the spindle assembly.

The Tempo/Topaz use a true MacPherson strut independent rear suspension. Each side consists of a shock absorber strut assembly, two parallel control arms per side, tie rod, forged spindle and a jounce bumper and bracket. The shock absorber strut assembly includes a rubber isolated top mount, upper spring seat, coil spring

insulator, coil spring and lower spring seat. The strut assembly is attached at the top by two studs, which retain the top mount of the strut to the inner body side panel. The lower end of the assembly is bolted to the spindle. The two stamped control arms attach to the underbody and spindle with nuts and bolts. A tie rod attaches to the underbody and to the forged spindle. The jounce bumper bracket attaches to the strut with the lower strut attaching bolts.

OVERHAUL

For all strut overhaul procedures, please refer to "Strut Overhaul" in the unit repair section.

Coil Spring

REMOVAL & INSTALLATION

Escort/Lynx

1. Jack up the vehicle and support it with jackstands.
2. Place a jack under the control arm and raise the control arm enough to put tension on the spring.

NOTE: Be careful not to raise the car off the jackstands.

3. Remove the control arm bolt at the spindle.
4. Slowly lower the control arm until the spring can be removed.
5. Installation is the reverse of removal.

Shock Strut

REMOVAL & INSTALLATION

Escort/Lynx

1. Remove the rear compartment access panels. Four door models require the removal of the quarter panel trim.
2. Loosen, but do not remove the top strut nut. If the shock absorber is to be re-used do not grip the shock absorber shaft with pliers, as this will damage the shaft.
3. Jack up the vehicle and support it with jackstands.
4. Remove the rear tire.
5. Support the lower control arm with a jack.
6. Remove the clip retaining the brake hose to the shock and carefully move it out of the way.
7. Loosen the nuts and bolts retaining the shock to the spindle, but do not remove them.
8. Remove the top mounting nut.

9. Remove the bottom bolts and nuts and remove the shock assembly.

10. Installation is the reverse of removal.

Strut and Spring

REMOVAL & INSTALLATION

Tempo/Topaz

1. Raise the jack only enough to contact body.

2. Open the trunk lid and loosen, but do not remove two nuts retaining the upper strut mount to body.

3. Raise the vehicle. Remove the wheel and tire.

4. Place a jackstand under the control arms to support the suspension.

——————— CAUTION ———————

Care should be taken when removing the strut that the rear brake flex hose is not stretched or the steel brake tube is not bent.

5. Remove the bolt attaching the brake hose bracket to the strut and carefully move it out of the way.

6. Remove the two bolts retaining the jounce bumper bracket and strut to the spindle.

7. Remove the jounce bumper bracket from the vehicle.

8. Remove the shock strut from the spindle.

9. Remove the two upper mount-to-body nuts.

10. Remove the strut from vehicle.

NOTE: Refer to the Unit Repair Section for strut servicing.

11. Install in reverse order. Torque the top mount to body bolts 20–30 ft. lbs. and strut to spindle bolts 70–96 ft.

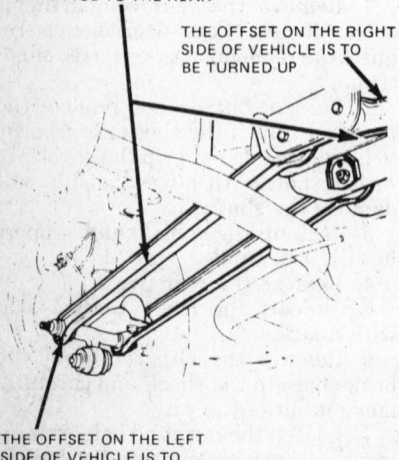

ARMS ARE TO BE INSTALLED WITH THE FLANGE TO THE REAR

THE OFFSET ON THE RIGHT SIDE OF VEHICLE IS TO BE TURNED UP

THE OFFSET ON THE LEFT SIDE OF VEHICLE IS TO BE TURNED DOWN

Tempo/Topaz rear control arm mounting

lbs. Always install new strut to spindle bolts.

Rear Control Arms

REMOVAL & INSTALLATION

Tempo/Topaz

1. Raise and safely support vehicle.

2. Remove tire and wheel assembly.

3. Remove the arm-to-spindle bolts and nut.

4. Remove the center mounting bolt and nut.

5. Remove the arm from vehicle.

6. Install in reverse order. Torque arm to body bolt 40–55 ft. lbs. and arm to spindle bolt 60–86 ft. lbs.

Tie-Rod

REMOVAL & INSTALLATION

Tempo/Topaz

1. Raise the car only enough to contact the body. From inside the trunk loosen, but Do not remove, the two strut top mount-to-body nuts.

2. Raise the vehicle and place a jackstand under the suspension to support it. Remove wheel and tire assembly.

3. Remove the two top mount studs.

4. Remove the nut retaining tie-rod to the spindle. Remove the nut retaining tie-rod to body.

5. Lower the jackstand enough so that the upper strut mount studs are out of the holes in the body.

6. Move the spindle rearward enough so that the tie-rod can be removed.

7. Place the new washers and bushings on both ends of new tie-rod. Bushings at front and rear of tie-rod are different. The rear bushings have indentations in them.

8. Insert one end into the body bracket and install a new bushing, washer and nut. Do not tighten at this time.

9. Pull back on the spindle enough so that the tie rod end can be installed in the spindle.

10. Install a new bushing, washer and nut. Do not tighten at this time.

11. Raise the jackstand enough to hold the two strut mounting studs in place.

12. Install two new strut-to-body mount nuts. Tighten to 20–30 ft. lbs.

13. Raise the suspension to curb height and tighten the two tie-rod nuts to 52–74 ft. lbs.

14. Remove jackstand. Install the tire and wheel assembly. Lower the vehicle.

Rear Wheel Bearings

Refer to "Rear Axle" for removal, installation and adjustment of the rear wheel bearings.

BRADES

For all brake system repair and service procedures not detailed below, please refer to "Brakes" in the Unit Repair section.

Master Cylinder

REMOVAL & INSTALLATION

Standard Brakes

1. Disconnect the negative battery terminal.

2. Working under the instrument panel, disconnect the master cylinder pushrod from the brake pedal.

3. Disconnect the stoplight switch and remove it.

4. Inside the engine compartment, disconnect the brake lines from the master cylinder.

5. Unbolt the master cylinder from the firewall and remove it. Be careful not to damage the firewall grommet.

6. To install, reverse the removal process, leaving the brake tubes slightly loose at the master cylinder fittings.

7. Fill the master cylinder with fresh brake fluid. Use the foot pedal to bleed the master cylinder. Tighten the brake line fittings.

Power Brakes

1. Disconnect the brake lines from the master cylinder.

2. Unbolt the master cylinder from the booster and remove the cylinder.

3. To install, mount the master cylinder on the booster. Attach the brake fluid lines to the master cylinder, but leave the fittings slightly loose.

4. Fill the reservoirs with fresh brake fluid. Use the foot pedal to bleed the master cylinder. Tighten the brake line fittings.

Proportioning Valve

The proportioning valve regulates the rear brake system hydraulic pressure. It is located between the rear brake system inlet and outlet ports. There are no adjustments possible on this valve. If found to be defective it must be replaced.

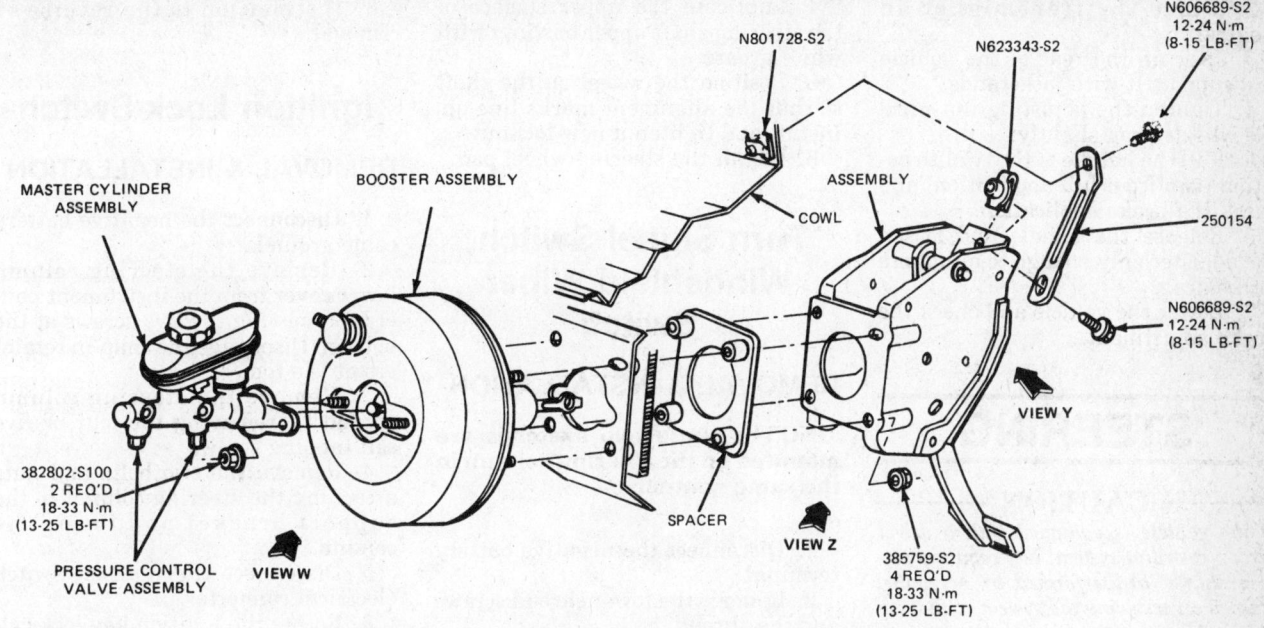

Typical master cylinder and power brake booster mounting

Power Brake Booster

REMOVAL & INSTALLATION

1. Disconnect the battery and remove the tubes from the primary and the secondary outlet ports of the master cylinder.

2. Remove the two nuts attaching the master cylinder to the brake booster assembly and remove the master cylinder.

3. Working inside the vehicle below the instrument panel, remove the wiring connector from the stop lamp switch. Remove the push rod retainer and the outer nylon washer from the pedal pin. Slide the stop lamp switch along the brake pedal pin just far enough for the outer hole to clear the pin. Remove the switch by sliding it upward, being careful not to damage the switch during removal.

4. Remove the booster-to-dash panel attaching nuts. Slide the booster push rod and push rod bushing off the brake pedal pin.

5. Inside the engine compartment, disconnect the manifold vacuum hose from the booster check valve. Move the booster forward until the booster studs clear the dash panel and remove the booster.

To install:

1. Align the pedal support and support spacer inside the vehicle and place the booster in position on the dash panel. Hand start the attaching nuts.

2. Working inside the vehicle, install the pushrod and pushrod bush-ing on the brake pedal pin. Tighten the booster-to-dash panel adjusting nuts to 13–25 ft. lbs.

3. Position the stop lamp switch so that it straddles the booster push rod with the switch slot toward the pedal blade and the hole just clearing the pin. Slide switch down onto pin. Slide the assembly toward the pedal arm, being careful not to damage the switch. Install the nylon washer on pin and secure all parts to pin with hairpin retainer. Make sure that the retainer is fully installed and locked over the pedal pin. Install the stop lamp switch wiring connector on the stop lamp.

4. Connect the manifold vacuum hose to the booster check valve using a hose clamp.

5. Position the master cylinder assembly on the booster assembly studs. Tighten the nuts 13–25 ft. lbs.

6. Install the brake tube fittings into the master cylinder ports and tighten 10–18 ft. lbs.

7. Bleed the brake system.

8. Connect the battery and start the engine. Then check to make sure the power brake system is functioning properly.

Note: On vehicles equipped with speed control, the vacuum dump valve must be adjusted if the brake booster has been removed.

To adjust the vacuum dump valve:

1. Firmly depress and hold the brake pedal.

2. Push in the dump valve until the valve collar bottoms against the retaining clip.

3. Place a 0.050–0.10 in. shim between the white button of the valve and the pad on the brake pedal.

4. Firmly pull the brake pedal rearward to its normal position, allowing the dump valve to ratchet backward in the retaining clip.

Wheel Cylinder
REMOVAL & INSTALLATION

1. Remove the wheel/tire and hub/drum assemblies.

2. Remove the brake shoe assembly.

3. Disconnect the brake tube from the wheel cylinder.

4. Remove the wheel cylinder attaching bolts and remove the wheel cylinder.

NOTE: Use caution to prevent brake fluid from contacting brake linings or they must be replaced.

5. Installation is in the reverse order of the removal procedure.

6. Adjust the brakes and bleed the brake system.

Parking Brake

ADJUSTMENT

1. Apply approximately 100 lbs. pedal effort to the hydraulic service brake three times, before adjusting the parking brake.

NOTE: On cars equipped with power brakes, the engine must be running before completing Step 1.

2. Place the transmission in Neutral.

3. Jack up the rear of the vehicle and support it with jackstands.

4. Tighten the adjusting nut until the wheels drag slightly.

5. Pull the handle to the twelfth position (two from full application) and check the brake application.

6. Release the handle and loosen the adjuster only enough to eliminate brake drag.

7. Lower the vehicle and check the brake adjustment.

STEERING

——— CAUTION ———

If the vehicle is equipped with a driver airbag restraint system, any required service should be performed by personnel trained on servicing the system so that accidental firing of the airbag will not occur.

Steering Wheel

REMOVAL & INSTALLATION

1. Disconnect the negative battery terminal.

2. Remove the steering wheel pad.

3. Remove and discard the nut from the steering shaft. Install a steering wheel puller on the end of the shaft and remove the wheel.

——— CAUTION ———

The use of a knock-off type steering wheel puller or the use of a hammer on the end of the steering shaft will damage the collapsible steering column.

4. Lubricate the upper surface of the steering shaft upper bushing with white grease.

5. Position the wheel on the shaft so that the alignment marks line up. Install and tighten a new locknut.

6. Install the steering wheel pad.

Turn Signal Switch, Windshield Wiper Switch

REMOVAL & INSTALLATION

NOTE: These two switches are mounted on the steering column in the same manner.

1. Disconnect the negative battery terminal.

2. Remove the lower shroud screws and the shroud.

3. Remove the upper shroud.

4. Remove the lever by pulling and twisting straight out (windshield wiper switch only).

5. Peel back the foam cover from the appropriate switch.

6. Disconnect the electrical connectors.

7. Remove the two self tapping screws (hex head screws—wash/wipe switch) that attach the switch to the lock cylinder housing, and remove the switch.

NOTE: On vehicles equipped with cruise control, transfer the ground brush in the turn signal switch cancelling cam to the new switch.

8. Installation is the reverse of removal.

Ignition Lock/Switch

REMOVAL & INSTALLATION

1. Disconnect the negative battery cable ground.

2. Remove the steering column lower cover from the instrument panel by removing the two screws at the bottom. Disengage the snap-in retainers at the top.

3. Remove the steering column shroud by removing the four or five self tapping screws.

4. Remove the two bolts and nuts attaching the steering column to the support bracket and lower the column.

5. Disconnect the ignition switch electrical connector.

6. Rotate the ignition key lock cylinder to the RUN position.

7. Drill out the break-off head bolts that connect the switch to the lock cylinder housing using a $\frac{1}{8}$ in. drill.

8. Remove the two bolts using a screw extractor tool.

9. Disengage the ignition switch from the actuator pin.

To Install:

1. Adjust the ignition switch by sliding the carrier to switch RUN position.

NOTE: A new replacement switch assembly will be set in the RUN position as recieved.

2. Check to ensure that the ignition key lock cylinder is in approximately the RUN position. The RUN position is achieved by rotating the key lock cylinder approximately 90 degrees from the LOCK position.

3. Install the ignition switch onto the actuator pin. It may be necessary to move the switch back and forth slightly to align the switch mounting holes with the column lock housing threaded holes.

4. Install new break-off head bolts and tighten until the heads break off.

5. Connect the electrical connector to the ignition switch.

6. Connect the negative battery ground cable. Check the ignition switch for proper function, including START and ACC positions. Also, make certain the column is locked in the lock position.

7. Align the steering column mounting holes with support bracket. Install two bolts and nuts.

8. Install the steering column trim shrouds.

9. Install the steering column lower cover on the steering panel.

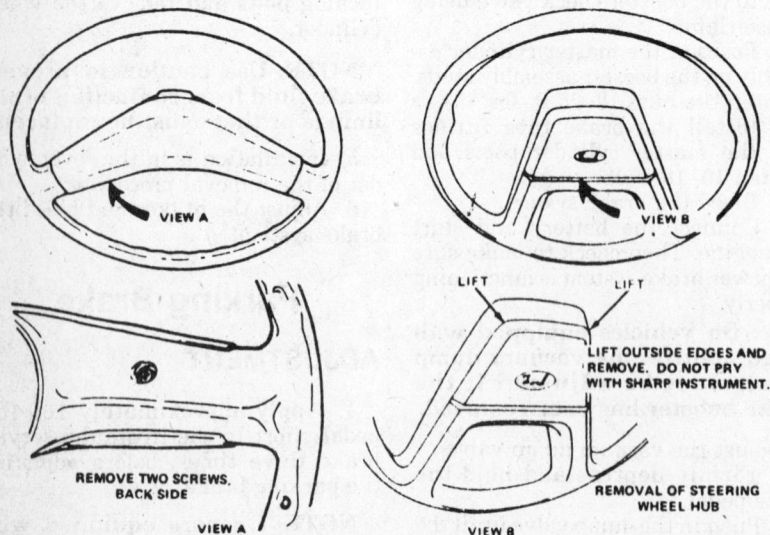

Steering wheel cover removal

Rack and Pinion Steering

REMOVAL & INSTALLATION

Escort/Lynx

1. Disconnect the negative battery cable from the battery. Jack up the front of the car and support it safely on jackstands.
2. Turn the ignition switch to the ON position. Remove the lower access (kick) panel from below the steering wheel.
3. Remove the intermediate shaft bolts at the gear input shaft and at the steering column shaft.
4. Spread the slots of the clamp to loosen the intermediate shaft at both ends. The next steps must be performed before the intermediate shaft and gear input shaft can be separated.
5. Turn the steering wheel full left so the tie-rod will clear the shift linkage. Separate the outer tie rod ends from the steering knuckle by using a tie-rod end remover.
6. Remove the left tie-rod end from the tie-rod (wheel must be at full left position). Disconnect the speedometer cable from the transmission if the car is equipped with an automatic transaxle. Disconnect the secondary air tube at the check valve. Disconnect the exhaust pipe from the exhaust manifold and wire it out of the way to allow enough room to remove the steering gear.
7. Remove the exhaust hanger bracket from below the steering gear. Remove the steering gear mounting brackets and rubber mounting insulators.
8. Have someone help by holding the gear from the inside of the car. Separate the intermediate shaft from the input shaft.
9. Make sure the gear is still in the full left turn position. Rotate the gear forward and down to clear the input shaft through the opening. Move the gear to the right to clear the splash panel and other linkage that interferes with the removal. Lower the gear and remove from under the car.
10. Installation is in the reverse order of removal. Have the toe adjustment checked after installing a new rack and pinion assembly.

NOTE: Removal and installation is basically the same for power rack and pinion steering. However, the pressure and return lines must be disconnected at the intermediate connectors and drained of fluid. It is necessary to remove the pressure switch from the pressure line.

Tempo, Topaz

1. Disconnect the negative battery cable from the battery.
2. Turn the ignition key to the RUN position.
3. Remove the access panel from the dash below the steering column.
4. Remove the intermediate shaft bolts at gear input shaft and at the steering column shaft.
5. With a wide blade tool, spread the slots enough to loosen intermediate shaft at both ends. The intermediate shaft and gear input shaft cannot be separated at this time.
6. From under vehicle, separate the tie-rod ends from steering knuckles, using Tool-3290-C and adapter T81P-3504-W or equivalent. Turn the right wheel to the full left turn position.
7. Disconnect the speedometer cable at transmission (automatic transmission only).

8. Disconnect the secondary air tube at the check valve. Disconnect the exhaust system at the exhaust manifold. Remove the exhaust system.
9. Remove the gear mounting brackets and insulators.

NOTE: Right and left hand brackets and insulators are not interchangeable side to side.

10. Turn the steering wheel full left so the tie-rod will clear the shift linkage during removal.
11. Separate the gear from intermediate shaft, with an assistant pulling up on the shaft from inside the vehicle.

— CAUTION —
Care should be taken during gear removal and installation to prevent tearing or damaging the steering gear bellows.

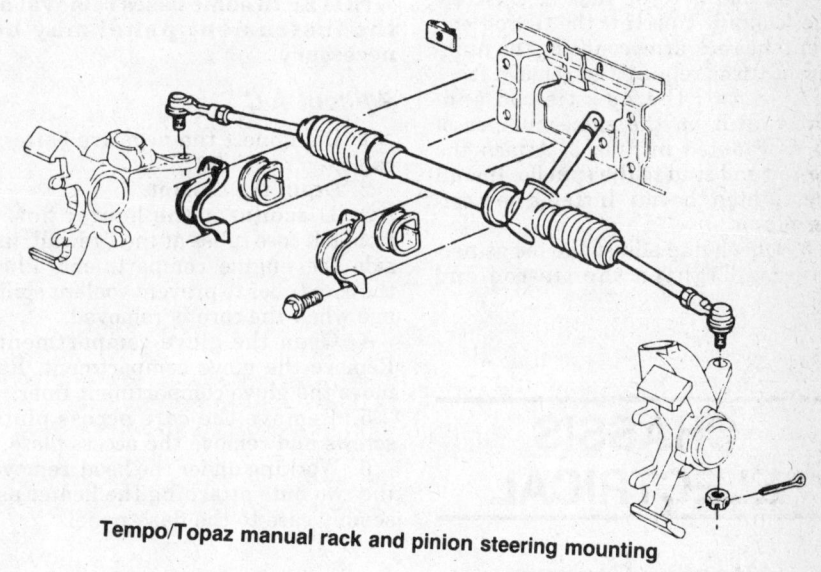

Tempo/Topaz manual rack and pinion steering mounting

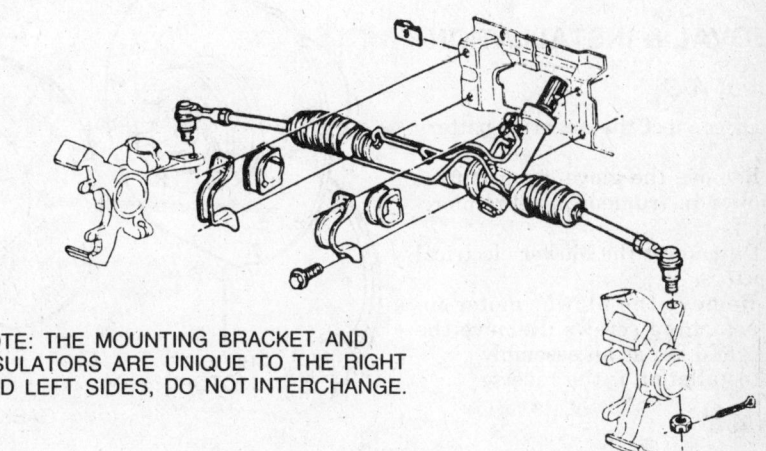

NOTE: THE MOUNTING BRACKET AND INSULATORS ARE UNIQUE TO THE RIGHT AND LEFT SIDES, DO NOT INTERCHANGE.

Tempo/Topaz power rack and pinion steering assembly

12. Rotate the gear forward and down to clear the input shaft through the dash panel opening.

13. Make sure the input shaft is in the full left turn position. Move the gear through the right (passenger) side apron opening until the left tie-rod clears the shift linkage and other parts so it may be lowered.

14. Lower the left side of the gear and remove the gear out of the vehicle.

15. Installation is the reverse of removal. Torque the steering gear mounting bolts to 48–55 ft. lbs.

Tie-Rod End

REMOVAL & INSTALLATION

1. Remove and discard the cotter pin. Remove the nut at the spindle.

2. Separate the tie-rod end stud from the spindle using a puller.

3. Matchmark the position of the locknut with paint on the tie-rod if the tie-rod end is to be reused. Unscrew the locknut. Unscrew the tie-rod end from the rack arm, counting the number of turns required to remove it.

4. Install the new tie-rod end, screwing it on the same number of turns counted in Step 3. Attach the tie-rod end stud to the spindle. Install and tighten the nut. Install a new cotter pin.

5. Check and adjust the toe as necessary. Tighten the tie-rod end locknut.

CHASSIS ELECTRICAL

Heater Blower

REMOVAL & INSTALLATION

Without A/C

1. Disconnect the negative battery cable.

2. Remove the glove compartment and lower instrument panel reinforcing rail.

3. Disconnect the blower electrical connectors.

4. Remove the blower motor-to-case attaching screws. Remove the blower and fan as an assembly.

5. Installation is the reverse.

With A/C

1. Locate and remove two screws at each side of the glove compartment

opening along the lower edge of the instrument panel. Then, remove the glove compartment door and instrumental panel lower reinforcement from the instrument panel.

2. Disconnect the blower motor wires from the wire harness at the hardshell connector.

3. Remove the screws attaching the blower motor and mounting plate to the evaporator case.

4. Rotate the motor until the mounting plate flats clear the edge of the glove compartment opening and remove the motor.

5. Remove the hub clamp spring from the blower wheel hub. Then, remove the blower wheel from the motor shaft.

6. Installation is the reverse of removal.

Heater Core

REMOVAL & INSTALLATION

NOTE: In some cases removal of the instrument panel may be necessary.

Without A/C

1. Disconnect the negative battery cable.

2. Drain the coolant.

3. Disconnect the heater hoses from the core tubes at the firewall, inside the engine compartment. Plug the core tubes to prevent coolant spillage when the core is removed.

4. Open the glove compartment. Remove the glove compartment. Remove the glove compartment liner.

5. Remove the core access plate screws and remove the access plate.

6. Working under the hood, remove the two nuts attaching the heater assembly case to the dash panel.

7. Remove the core through the glove compartment opening. Installation is the reverse of removal procedures.

With A/C

1. Disconnect the negative battery cable and drain the cooling system.

2. Disconnect the heater hoses from the heater core.

3. Working inside the vehicle, remove the floor duct from the plenum (2 screws).

4. Remove the four screws attaching the heater core cover to the plennum, remove the cover and remove the heater core.

5. Installation is the reverse of removal procedures.

Radio

For the best FM reception, adjust the antenna to 31 inches in height. Fading or weak AM reception may be adjusted by means of the antenna trimmer control, located either on the right rear or front side of the radio chassis. See the owner's manual for position. To adjust the trimmer:

1. Extend the antenna to maximum height.

2. Tune the radio to a weak station around 1600 KC. Adjust the volume so that the sound is barely audible.

3. Adjust the trimmer to obtain maximum volume.

REMOVAL & INSTALLATION

1. Disconnect the negative battery cable.

NOTE: Remove the A/C floor duct if so equipped.

2. Remove the ash tray and bracket.

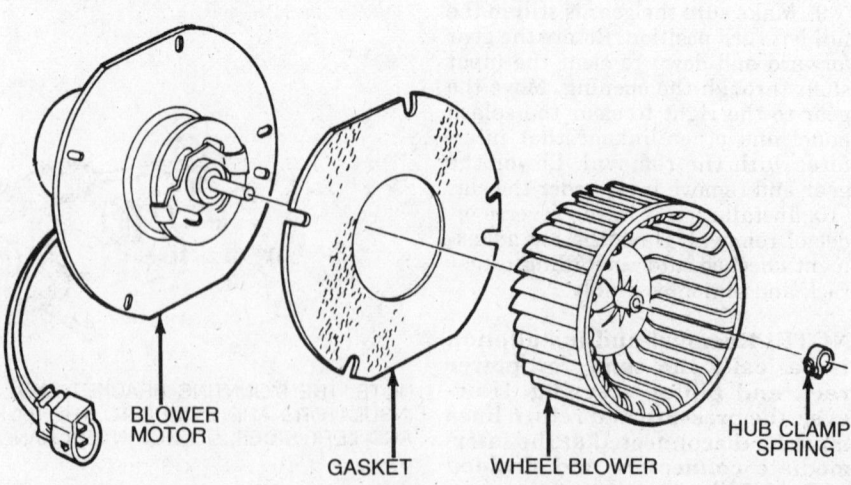

BLOWER MOTOR GASKET WHEEL BLOWER HUB CLAMP SPRING

Blower motor and wheel dissaembly

3. Pull the knobs from the shafts.

4. Working under the instrument panel, remove the support bracket nut from the radio chassis.

5. Remove the shaft nuts and washers.

6. Drop the radio down from behind the instrument panel. Disconnect the power lead, antenna, and speaker wires. Remove the radio.

7. Installation is the reverse.

Windshield Wiper Motor

REMOVAL & INSTALLATION

NOTE: The motor is located in the right rear corner of the engine compartment, in the cowl area above the firewall.

1. Disconnect the negative battery cable.

2. Remove the plastic cowl cover.

3. Disconnect the motor electrical connector.

4. Remove the motor attaching bolts. Disengage the motor from the linkage and remove the motor. Installation is the reverse.

Wiper Linkage

REMOVAL & INSTALLATION

NOTE: The wiper linkage is mounted below the cowl top panel and can be reached by raising the hood.

1. Remove the wiper arm and blade assembly from the pivot shaft. Pry the latch (on the arm) away from the shaft to unlock the arm from the pivot shaft.

2. Raise the hood and disconnect the negative battery cable.

3. Remove the clip and disconnect the linkage drive arm from the motor crank pin.

4. On Tempo/Topaz remove the screws retaining the pivot assemblies to the cowl.

6. On Escort/Lynx, EXP.LN7 remove the large pivot retainer nuts from each pivot shaft.

7. Remove the linkage and pivot assembly from the cowl chamber.

8. Installation is the reverse of removal.

Headlight Switch

REMOVAL & INSTALLATION

1. Disconnect the negative battery terminal.

2. Remove the left hand air vent control cable, and drop the cable and bracket down out of the way (cars without air conditioning only).

3. Remove the fuse panel bracket retaining screws and move the fuse panel assembly out of the way.

4. Pull the headlight knob out, to the on position.

5. Reach behind the dashboard and depress the release button on the switch housing, while at the same time pulling the knob and shaft from the switch.

6. Remove the retaining nut from the dashboard.

7. Pull the switch from the dash and remove the electrical connections.

8. Installation is the reverse of removal.

Instrument Cluster

For further information on instruments, please refer to " Gauges and Indicators" in the Unit Repair Section.

REMOVAL & INSTALLATION

1. Disconnect the negative battery terminal.

2. Remove the bottom steering column cover.

3. Remove the steering column opening cover reinforcement screws.

NOTE: On cars equipped with speed control disconnect the wires from the amplifier assembly.

4. Remove the steering column retaining screws from the steering column support bracket and lower the column.

5. Remove the column trim shrouds.

6. Disconnect all electrical connections from the column.

7. Remove the finish panel screws and the panel.

8. Remove the speedometer cable.

9. Remove the four cluster screws and remove the cluster.

10. Installation is the reverse of removal.

Speedometer Cable

REMOVAL & INSTALLATION

1. Remove the instrument cluster.

2. Pull the speedometer cable from the casing. If the cable is broken, disconnect the casing from the transaxle and remove the broken piece from the transaxle end.

3. Lubricate the new cable with graphite lubricant. Feed the cable into the casing from the instrument panel end.

4. Attach the cable to the speedometer. Install the cluster.

Fuses, Fusible Links and Circuit Breakers

LOCATION

A fuse link is a short length of insulated wire integral with the engine compartment wiring harness. It is several wire gauges smaller than the circuit which it protects and generally located in line directly from the positive terminal of the battery.

Production fuse links are color-coded.

- 12 gauge: Grey.
- 14 gauge: Dark Greeen.
- 16 gauge: Black.
- 18 gauge: Brown.
- 20 gauge: Dark Blue.

NOTE: Replacement fuse link color coding may vary from production fuse link color coding.

When heavy current flows, such as when a booster battery is connected incorrectly or when a short to ground occurs in the wiring harness, the fuse link burns out and protects the alternator or wiring.

Fusible Link

REMOVAL & INSTALLATION

1. Disconnect the negative battery cable.

2. Cut the damaged fuse link from the wiring harness and discard it. If the fuse link is one of three circuits feed by a single feed wire, cut it out of harness at each splice end and discard.

3. Identify and obtain proper fuse link and butt connectors for attaching the fuse link to the harness.

4. Strip away approximately ½ in. of insulation from the two wiring ends. Attach the replacement fuse link to stripped wire ends with two proper size butt connectors.

5. Solder the connectors and wires and insulate with electrical tape.

Fuses and Circuit Breakers

LOCATION

A combination fuse and circuit breaker panel contains most of the fuses and circuit breakers used in the system. The mounting of the fuse panel is

CURCUIT PROTECTION CHART

Circuit	Circuit Protection and Rating	Location
Headlamps and High Beam Indicator	22 Amp CB	Integral with Lighting Switch
Heated Rear Window	16 Gauge Fuse Link	Engine Compartment
Load Circuit	Fuse Link	In Harness
Engine Compartment Lamp (Tempo/Topaz)	15 Amp Fuse	Fuse Panel
Liftgate Wiper (Escort/Lynx)	4.5 Amp CB	Instrument Panel to Left of Radio

CB Circuit Breaker

usually on the left side of the passenger compartment, under the dash, either on the side kick panel or on the firewall to the left of the steering column. Certain models will have the fuse panel exposed while other models will have it covered with a removable trim cover. These fuses and circuit breakers are color coded by amp rating.

- 4 amp: Pink.
- 5 amp: Tan.
- 10 amp: Red.
- 15 amp: Light Blue.
- 20 amp: Yellow.
- 25 amp: Natural.
- 30 amp: Light Green.

The location and values of the fuses and circuit breakers not contained in the panels are given in the circuit protection chart.

REMOVAL & INSTALLATION

1. Remove the malfunctioning fuse or circuit breaker by pulling it out of its cavity.

NOTE: If the fuse or circuit breaker is mounted on the fuse panel, remove the trim cover.

2. Replace the blown fuse or circuit breaker with one of proper amp rating for the circuit, by pushing straight in until the fuse or circuit breaker seats itself fully into the cavity.
3. Replace the trim panel.

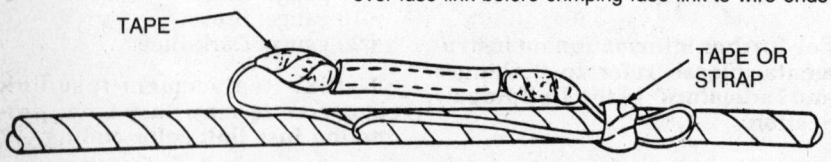

Remove existing vinyl tube shielding and reinstall over fuse link before crimping fuse link to wire ends

Typical repair using the special 17 gauge fuse link

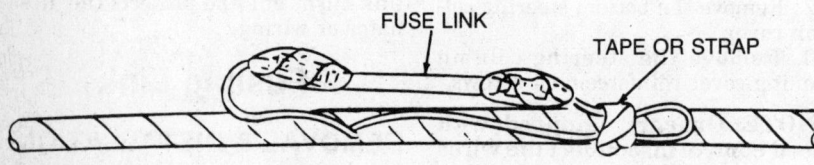

Typical repair for any in-line fuse link using the specified gauge fuse link for the specific circuit

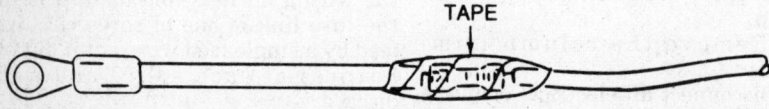

Typical repair using the eyelet terminal fuse link of the specified gauge for attachment to a circuit wire end

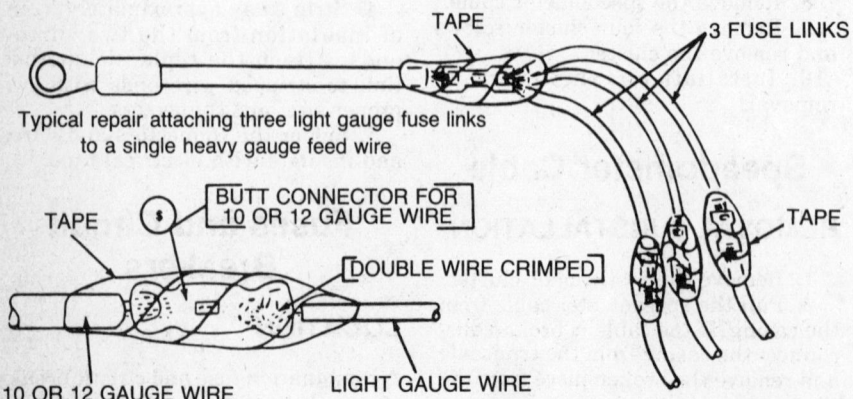

Typical repair attaching three light gauge fuse links to a single heavy gauge feed wire

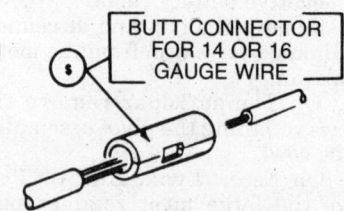

FUSIBLE LINK REPAIR PROCEDURE

General fuse link repair procedures

Ford Motor Co.
Rear Wheel Drive
Ford, Lincoln, Mercury— All Models

YEAR IDENTIFICATION

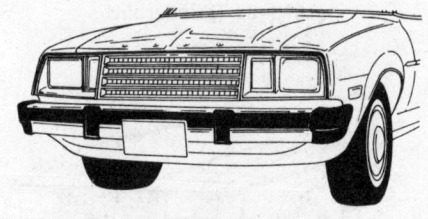

1980 Pinto

1980–81 Mustang

1982 Mustang

1983 Mustang

1984–86 Mustang

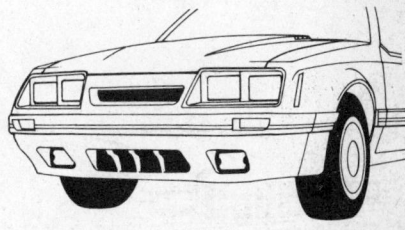

1985–86 Mustang GT

1984–86 Mustang SVO

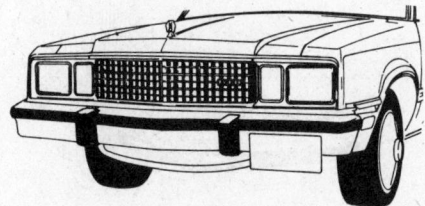

1980–81 Fairmont

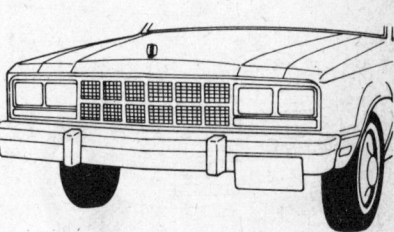

1980 Fairmont Futura

C183

YEAR IDENTIFICATION

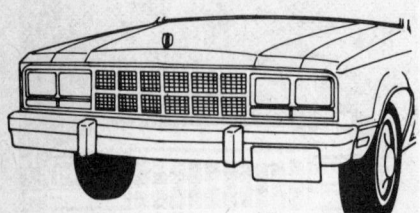

1981-83 Fairmont Futura

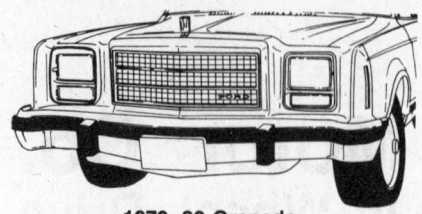

1979–80 Granada

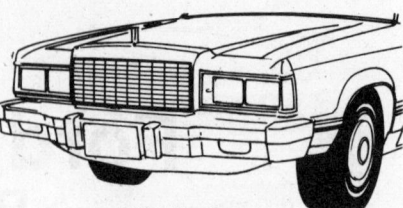

1981-82 Granada

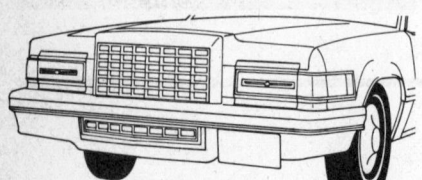

1980 Thunderbird

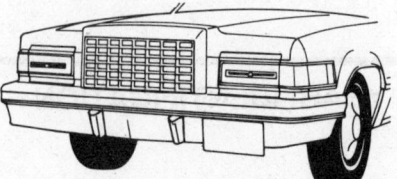

1981-82 Thunderbird

1983–86 Thunderbird

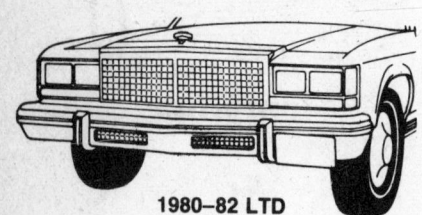

1980–82 LTD

1983 LTD

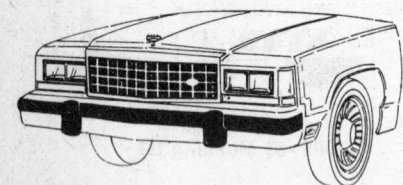

1984–86 LTD

1983–86 Crown Victoria and Country Squire

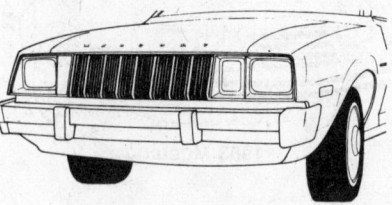

1985–86 Thunderbird Turbo Coupe

1980 Bobcat

1980–82 Capri

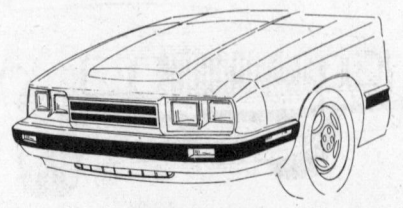

1983–86 Capri

1980 Zephyr

YEAR IDENTIFICATION

1981-83 Zephyr

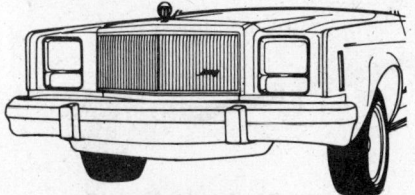

1980 Monarch

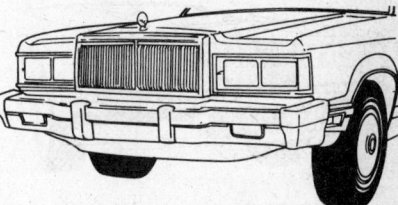

1981-82 Cougar

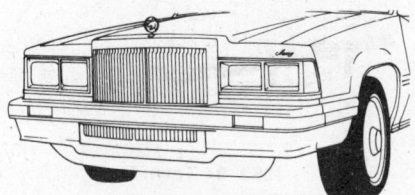

1980 Cougar XR-7

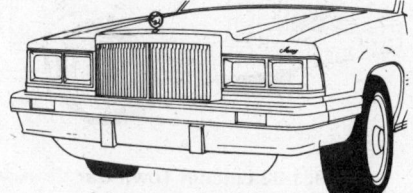

1981-82 Cougar XR-7

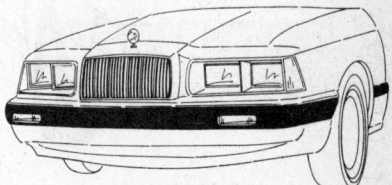

1983—86 Cougar

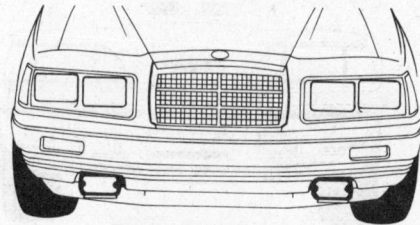

1985 Cougar XR7

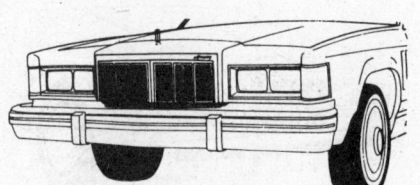

1980 Marquis

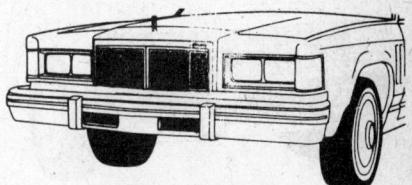

1981-82 Marquis

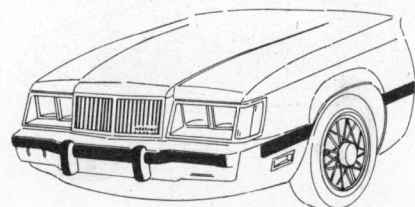

1983—86 Marquis

1983—86 Grand Marquis and Colony Park

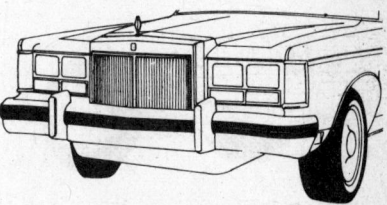

1980 Versailles

1980 Lincoln Continental

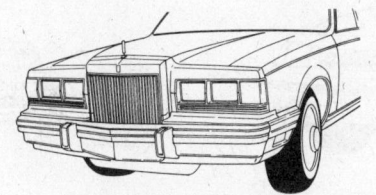

1982 Continental

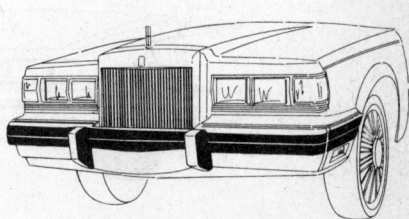

1983 Continental

YEAR IDENTIFICATION

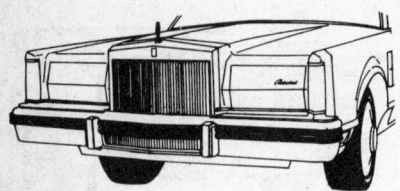

1980–82 Continental Mark VI

1983 Continental Mark VI

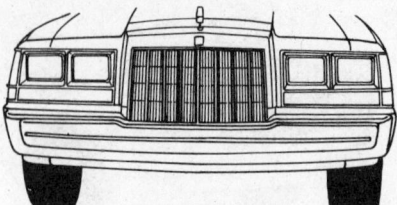

1984-85 Continental

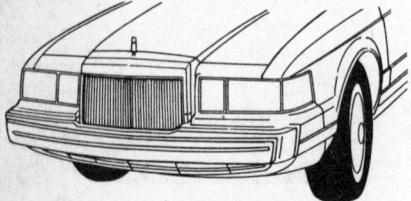

1984–86 Mark VII

1981-82 Lincoln Town Car

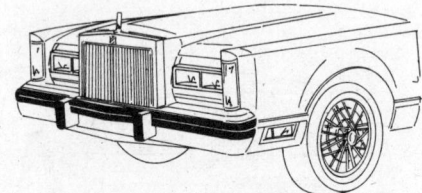

1983–84 Town Car

1985–86 Mark VII LSC

1985–86 Lincoln Town Car

1987 Thunderbird Turbo

1987 Mark VII LSC

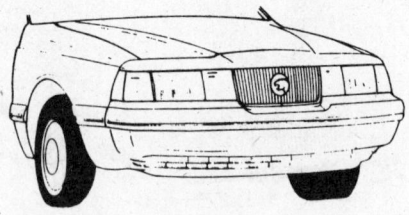

1987 Mustang GT

1987 Cougar

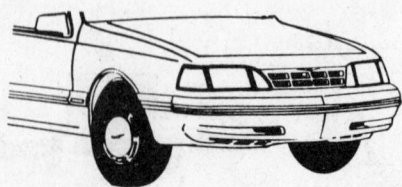

1987 Thunderbird

1987 Lincoln Continental

VEHICLE IDENTIFICATION NUMBER (VIN)

It is important for servicing and ordering parts to be certain of the vehicle and engine identification. The VIN (vehicle identification number) is a 13 or 17 digit number visible through the windshield on the driver's side of the dash and contains the vehicle and engine identification codes. It can be interpreted as follows:

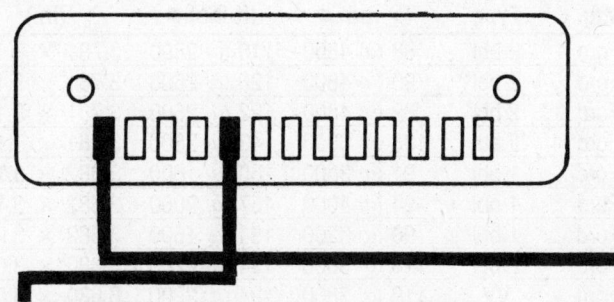

Engine Code						Model Year Code	
Code	Cu. In.	Liters	Cyl.	Carb.	Eng. Mfg.	Code	Year
Y	140	2.3	4	2	Ford	0	1980
T('80)	140	2.3	4	Turbo	Ford		
A('80)	140	2.3	4	2	Ford		
Z	170	2.8	6	2	Ford		
T	200	3.3	6	1	Ford		
B('80)	200	3.3	6	1	Ford		
L	250	4.1	6	1	Ford		
C('80)	250	4.1	6	1	Ford		
D	255	4.2	8	2(W)	Ford		
F	302	5.0	8	2①	Ford		
G,E	351W	5.8	8	2②	Ford		

The thirteen digit Vehicle Identification Number can be used to determine engine application and model year. The first digit indicates model code, and the fifth digit indicates engine application.
① EFI on various models in 1980
② VV carburetor on various models in 1980

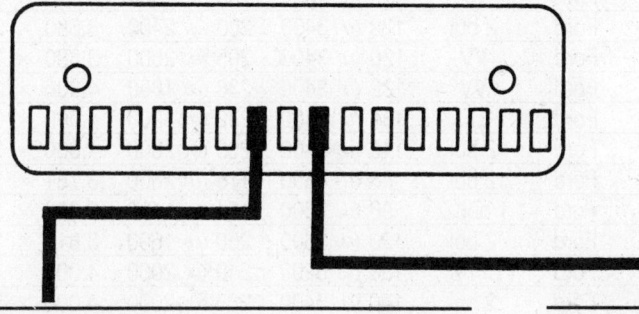

Engine Code						Model Year Code	
Code	Cu. In.	Liters	Cyl.	Carb.	Eng. Mfg.	Code	Year
A	140	2.3	4	2(1)	Ford	B	1981
T	140	2.3	4	Turbo	Ford	C	1982
6	140	2.3	4	Propane	Ford	D	1983
W	140	2.3	4	Turbo	Ford	E	1984
B	200	3.3	6	1	Ford	F	1985
3	232	3.8	V6	①	Ford	G	1986
D	255	4.2	V8	2(VV)	Ford	H	1987
F	302	5.0	V8	①	Ford		
M	302HO	5.0	V8	①	Ford		
G	351W	5.8	V8	①	Ford		
G	351HO	5.8	V8	①	Ford		

The seventeen digit Vehicle Identification Number can be used to determine engine application and model year. The tenth digit indicates the model year, and the eighth digit identifies the engine code.
① EFI, VV, 2 bbl. or 4 bbl. depending on model and year

GENERAL ENGINE SPECIFICATIONS

Year	Eng. V.I.N. Code	Engine No. Cyl. Displacement (Cu. In.)	Eng. Mfg.	Carburetor Type	Horsepower @ rpm ■	Torque @ rpm (ft lbs) ■	Bore × Stroke (in.)	Compression Ratio	Oil Pressure @ 2000 rpm
'80	A	4-140 MT	Ford	2 bbl	88 @ 4800	118 @ 2800	3.781 × 3.126	9.0:1	50
	A	4-140 AT	Ford	2 bbl	90 @ 4800	125 @ 2600	3.781 × 3.126	9.0:1	50
	A	4-140 Cal	Ford	2 bbl	89 @ 4800	122 @ 2600	3.781 × 3.126	9.0:1	50
	T	4-140 T	Ford	2 bbl	135 @ 6000	143 @ 2800	3.781 × 3.126	9.0:1	55
	B	6-200 MT	Ford	1 bbl	91 @ 3800	160 @ 1600	3.682 × 3.126	8.6:1	30–50
	B	6-200 AT	Ford	1 bbl	94 @ 4000	157 @ 2000	3.682 × 3.126	8.6:1	30–50
	C	6-250 All	Ford	1 bbl	90 @ 3200	194 @ 1600	3.682 × 3.910	8.6:1	50
	D	8-255 49	Ford	2 bbl	119 @ 3800	194 @ 2200	3.680 × 3.000	8.8:1	40–60
	D	8-255 Cal	Ford	VV	119 @ 3800	194 @ 2200	3.680 × 3.000	8.8:1	40–60
	F	8-302	Ford	2 bbl	134 @ 3600	232 @ 1600	4.000 × 3.000	8.4:1	40–65
	F	8-302	Ford	VV	131 @ 3600	231 @ 1400	4.000 × 3.000	8.4:1	40–65
	G	8-351 W	Ford	VV	142 @ 3200	286 @ 1400	4.000 × 3.500	8.3:1	40–65
	E	8-351 W	Ford	2 bbl	138 @ 3200	260 @ 2200	4.000 × 3.500	8.3:1	40–65
	G	8-351	Ford	VV	140 @ 3400	265 @ 2000	4.000 × 3.500	8.3:1	40–65
'81	A	4-140	Ford	2 bbl	88 @ 4600	118 @ 2600	3.781 × 3.126	9.0:1	40–60
	B	6-200	Ford	1 bbl	88 @ 3800	154 @ 1400	3.680 × 3.130	8.6:1	30–50
	D	8-255	Ford	2 bbl	115 @ 3400	195 @ 2200	3.680 × 3.000	8.2:1	40–60
	D	8-255	Ford	VV	120 @ 3400	205 @ 2600	3.680 × 3.000	8.2:1	40–60
	F	8-302	Ford	2 bbl	130 @ 3400	235 @ 1600	4.000 × 3.000	8.4:1	40–60
	F	8-302	Ford	VV	130 @ 3400	235 @ 1800	4.000 × 3.000	8.4:1	40–60
	F	8-302	Ford	EFI	130 @ 3400	230 @ 2000	4.000 × 3.000	8.4:1	40–60
	G	8-351	Ford	VV	145 @ 3200	270 @ 1800	4.000 × 3.500	8.3:1	40–60
	G	8-351 HO	Ford	VV	165 @ 3600	285 @ 2200	4.000 × 3.500	8.3:1	40–60
'82	A	4-140	Ford	2 bbl	86 @ 4600	117 @ 2600	3.781 × 3.126	9.0:1	40–60
	B	6-200	Ford	1 bbl	87 @ 3800	154 @ 1400	3.680 × 3.130	8.6:1	30–50
	3	6-232	Ford	2 bbl	112 @ 4000	175 @ 2600	3.810 × 3.390	8.8:1	54–59
	3	6-232	Ford	VV	118 @ 4000	186 @ 2600	3.810 × 3.390	8.8:1	54–59
	D	8-255	Ford	2 bbl	122 @ 3400	209 @ 2400	3.680 × 3.000	8.2:1	40–60
	D	8-255	Ford	VV	120 @ 3400	205 @ 2600	3.680 × 3.000	8.2:1	40–60
	F	8-302	Ford	VV	132 @ 3400	236 @ 1800	4.000 × 3.000	8.4:1	40–60
	F	8-302	Ford	EFI	134 @ 3400	232 @ 3200	4.000 × 3.000	8.4:1	40–60
	G	8-351	Ford	2 bbl	140 @ 3400	265 @ 2000	4.000 × 3.500	8.3:1	40–60
'83	A	4-140	Ford	2 bbl	88 @ 4800	118 @ 2800	3.781 × 3.126	9.0:1	40–60
	X	6-200	Ford	1 bbl①	88 @ 3800	154 @ 1400	3.680 × 3.130	8.6:1	30–50
	3	V6-232	Ford	2 bbl	120 @ 3600	250 @ 1600	3.810 × 3.340	8.7:1	40–60
	F	8-302	Ford	EFI②	130 @ 3200	240 @ 2000	4.000 × 4.000	8.4:1	40–60
	G	8-351	Ford	2 bbl	140 @ 3400	265 @ 2000	4.000 × 3.500	8.3:1	40–60
'84	A	4-140	Ford	1 bbl	88 @ 4000	122 @ 2400	3.781 × 3.126	9.0:1	40–60
	W	4-140T	Ford	EFI	145 @ 4600	180 @ 3600	3.781 × 3.126	8.0:1	40–60
	T	4-140T③	Ford	EFI	175 @ 4400	210 @ 3000	3.781 × 3.126	8.0:1	40–60
	3	V6-232	Ford	CFI①	120 @ 3600	250 @ 1600	3.810 × 3.390	8.7:1	40–60
	F	8-302	Ford	CFI	140 @ 3200	250 @ 1600④	4.000 × 3.000	8.4:1	40–60
	M	8-302 HO	Ford	4 bbl	205 @ 4400	265 @ 3200	4.000 × 3.000	8.3:1	40–60
	G	8-351	Ford	2 bbl	180 @ 3600	285 @ 2400	4.000 × 3.500	8.3:1	40–60
'85	A	4-140	Ford	1 bbl	88 @ 4000	124 @ 2800	3.781 × 3.126	9.0:1	40–60
	W	4-140T	Ford	EFI	145 @ 4600	180 @ 3600	3.781 × 3.126	8.0:1	40–60
	T	4-140T③	Ford	EFI	175 @ 4400	210 @ 3000	3.781 × 3.126	8.0:1	40–60
	R	4-140 HSC	Ford	1/EFI	86 @ 4000	124 @ 2800	3.781 × 3.126	9.0:1	40–60
	3	V6-232	Ford	2 bbl	120 @ 3600	250 @ 1600	3.810 × 3.390	8.7:1	40–60
	C	V6-232	Ford	CFI	120 @ 3600	250 @ 1600	3.810 × 3.390	8.7:1	40–60
	F	8-302	Ford	CFI	165 @ 3200	250 @ 1600	4.000 × 3.000	8.4:1	40–60
	M	8-302	Ford	4 bbl	210 @ 4400	265 @ 3200	4.000 × 3.000	8.3:1	40–60
	G	8-351	Ford	2 bbl	180 @ 3600	285 @ 2400	4.000 × 3.500	8.3:1	40–60

GENERAL ENGINE SPECIFICATIONS

Year	Eng. V.I.N. Code	Engine No. Cyl. Displacement (Cu. In.)	Eng. Mfg.	Carburetor Type	Horsepower @ rpm ■	Torque @ rpm (ft lbs) ■	Bore × Stroke (in.)	Compression Ratio	Oil Pressure @ 2000 rpm
'86–'87	A	4-140	Ford	1 bbl	88 @ 4200	122 @ 2600	3.781 × 3.126	9.0:5	40–60
	T	4-140T	Ford⑥	EFI	145 @ 4400	180 @ 3000	3.781 × 3.126	8.0:1	40–60
	W	4-140T	Ford⑦	EFI	155 @ 4600	190 @ 2800	3.781 × 3.126	8.0:1	40–60
	3	V6-232	Ford	CFI①	120 @ 3600	205 @ 1600	3.810 × 3.390	8.7:1	40–60
	F	8-302	Ford	SEFI	150 @ 3200	270 @ 2000	4.000 × 3.000	8.9:1	40–60
	M	8-302 HO	Ford	4 bbl	210 @ 4400	265 @ 3200	4.000 × 3.000	8.3:1	40–60
	G	8-351	Ford	2 bbl	180 @ 3600	285 @ 2400	4.000 × 3.500	8.3:1	40–60

■ Horsepower and torque are SAE net figures. They are measured at the rear of the transmission with all accessories installed and operating. Since the figures vary when a given engine is installed in different models, some are representative rather than exact.

T Turbocharger
EFI Electronic fuel injection
SEFI Sequential electronic fuel injection
HO High output
HSC High swirl combustion
CFI Central fuel injection
VV Variable Venturi carburetor

① Some models are equipped with a 2-bbl carburetor
② Some Mustang/Capri models are equipped with a 4-bbl carburetor
③ SVO

④ On models equipped with dual exhaust the horsepower is 155 @ 3600 rpm and the torque is 265 @ 2000 rpm.
⑤ Some models are equipped with EFI
⑥ Manual transmission
⑦ Automatic transmission

TUNE-UP SPECIFICATIONS

When analyzing compression test results, look for uniformity among cylinders rather than specific pressures

Year	Eng. V.I.N. Code	No. Cyl. Displacement (cu. in.)	Eng. Mfg.	Spark Plugs Orig. Type	Gap (in.)	Distributor	Ignition Timing (deg) Man. Trans.	Ignition Timing (deg) Auto. Trans.	Valves Intake Opens (deg)	Fuel Pump Pressure (psi)	Idle Speed (rpm) Man. Trans.	Idle Speed (rpm) Auto. Trans.
'80	A	4-140	Ford	AWSF-42	.035	Electronic	6B	20B(12B)	22	5½–6½	850	750
	T	4-140T	Ford	AWSF-32	.050	Electronic	6B(2B)	8B(2B)	22	6½–7½	900	800①
	B	6-200	Ford	BRF-82	.050	Electronic	10B	12B	20	5½–6½	900	700
	C	6-250	Ford	BSF-82	.050	Electronic	8B	10B	18	5½–6½	700	550
	D	8-255	Ford	ASF-42	.050	Electronic	8B	8B(6B)④	16	4–6	500	550(500)④
	F	8-302	Ford	ASF-52	.050	Electronic	—	8B④	16	5½–6½	—	550④
	G	8-351	Ford	ASF-52	.050	Electronic	—	①④	23	6½–8	—	600④
'81	A	4-140	Ford	AWSF-42	.034	Electronic	6B	6B	22	5½–6½	700	700
	T	4-140T	Ford	AWSF-32	.034	Electronic	6B	8B	22	5½–6½	850	750(650)
	B	6-200	Ford	BSF-92	.050	Electronic	10B	10B	20	5½–6½	900	700①
	D	8-255	Ford	ASF-52	.050	Electronic	10B	10B	16	5½–6½	900	700①
	F	8-302	Ford	ASF-52	.050	Electronic	8B	8B	16	5½–6½	800	800
	G	8-351	Ford	ASF-52	.050	Electronic	—	①④	23	6½–8	—	600④
'82	A	4-140	Ford	AWSF-42	.034	Electronic	①	①	22	5½–6½	850	750
	B	6-200	Ford	BSF-92	.050	Electronic	①	①	20	6–8	700①	600(700)①
	3	6-232	Ford	AGSP-52	.044	Electronic	①	①	13	6–8	①	①
	F	8-302	Ford	ASF-52	.050	Electronic	—	①④	16	6–8⑧	①④	①④
	G	8-351	Ford	ASF-52	.050	Electronic	—	①④	23	6½–8	—	600④
'83	A	4-140	Ford	AWSF-44	.044	Electronic	①	①	16	5½–6½	850	800
	X	6-200	Ford	BSF-92	.050	Electronic	①	①	20	6–8	600	600
	3	V-6-232	Ford	AWSF-52	.044	Electronic	①	①	13	39	550	550
	F	8-302	Ford	ASF-52②	.050	Electronic	①	①	16	6–8③	—	550
	G	8-351	Ford	ASF-42	.044	Electronic	①	①	23	6–8	—	700/600
'84	A	4-140	Ford	AWSF-44	.044	Electronic	①	①	16	5–7	850	750
	W	4-140-T	Ford	AWSF-32	.034	Electronic	①	①	—	39	①④	①④
	T	4-140-T⑤	Ford	AWSF-32	.034	Electronic	①	①	—	39	①④	①④
	3	V6-232	Ford	AWSF-54	.044	Electronic	①	①	13	39	—	550
	F	8-302	Ford	ASF-52	.050	Electronic	①	①	16	39	550	550
	M	8-302HO	Ford	ASF-42	.044	Electronic	①	①	—	6–8	700	700
	G	8-351	Ford	ASF-42	.044	Electronic	①	①	23	6–8	—	600⑥

TUNE-UP SPECIFICATIONS

When analyzing compression test results, look for uniformity among cylinders rather than specific pressures

Year	Eng. V.I.N. Code	No. Cyl. Displacement (cu. in.)	Eng. Mfg.	Spark Plugs Orig. Type	Gap (in.)	Distributor	Ignition Timing (deg) Man. Trans.	Auto. Trans.	Valves Intake Opens (deg)	Fuel Pump Pressure (psi)	Idle Speed (rpm) Man. Trans.	Auto. Trans.
'85	A	4-140	Ford	AWSF-44	.044	Electronic	①	①	16	6–8	850	750
	W	4-140T	Ford	AWSF-32	.034	Electronic	①	①	—	39	750	750
	T	4-140T ⑤	Ford	AWSF-32	.034	Electronic	①	①	—	39	①④	①④
	R	4-140-HSC	Ford	AWSF-52	.044	Electronic	①	①	—	39	800	700
	3	V6-232	Ford	AGSP-52	.044	Electronic	①	①	13	6–8	600	600
	F	8-302	Ford	ASF-52	.050	Electronic	①	①	13	39	—	550
	M	8-302HO	Ford	ASF-42	.044	Electronic	①	①	—	6–8	700	700
	G	8-351	Ford	ASF-42	.044	Electronic	①	①	—	6–8	—	600⑥
'86	A	4-140	Ford	AWSF-44C	.044	Electronic	①	①	16	6–8	750	750
	T	4-140T	Ford	AWSF-32C	.034	Electronic	①	①	—	39	825/975	825/975
	W	4-140T	Ford	AWSF-32C	.034	Electronic	①	①	—	39	825/975	825/975
	3	V6-232	Ford	AWSF-54	.044	Electronic	①	①	13	39	—	550
	F	8-302	Ford	ASF-32C	.044	Electronic	①	①	—	39	①④	①④
	M	8-302HO	Ford	ASF-42	.044	Electronic	①	①	—	6–8	700	700
	G	8-351	Ford	ASF-32C	.044	Electronic	①	①	—	6–8	650	650
'87				See Underhood Specifications Sticker								

NOTE: The underhood specifications sticker often reflects tune-up specification changes made in production. Sticker figures must be used if they disagree with those in this chart.

T Turbocharger
B Before Top Dead Center
HO High Output
— Not applicable
HSC High Swirl Combustion

① Calibrations vary depending upon the model; refer to the underhood calibration sticker.
② The carbureted models use spark plug ASF-42 (.044) and the idle speed rpm is 700 rpm.

③ On fuel injected models the pressure is 39 psi.
④ Electronic engine control models the ignition timing, idle speed and idle mixture is not adjustable.
⑤ SVO Mustang
⑥ 700 rpm with the VOTM on.

FIRING ORDERS

NOTE: To avoid confusion, always replace spark plug wires one at a time.

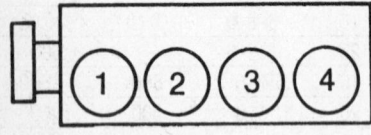

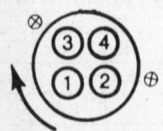

FORD MOTOR CO. 2300 cc 4-cyl.
Engine firing order: 1-3-4-2
Distributor rotation: clockwise

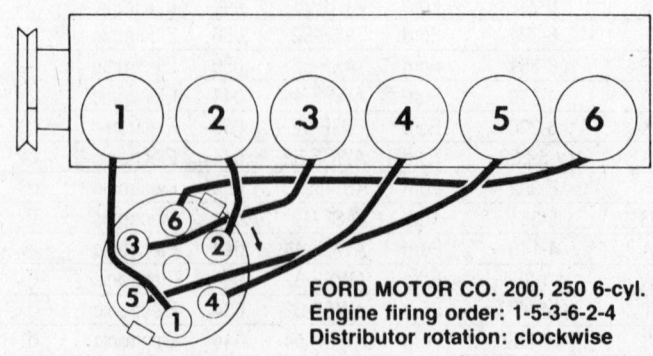

FORD MOTOR CO. 200, 250 6-cyl.
Engine firing order: 1-5-3-6-2-4
Distributor rotation: clockwise

FIRING ORDERS

NOTE: To avoid confusion, always replace spark plug wires one at a time.

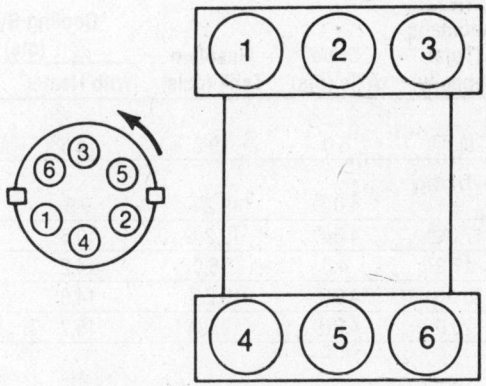

FORD MOTOR CO. 232 V6
Engine firing order: 1-4-2-5-3-6
Distributor rotation: counterclockwise

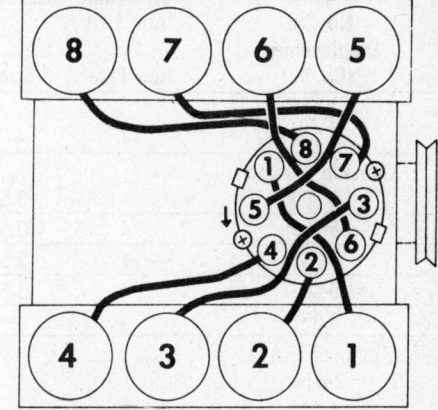

FORD MOTOR CO. 225 and 302 (except HO) engines
Firing Order: 1-5-4-2-6-3-7-8
Distributor rotation: clockwise

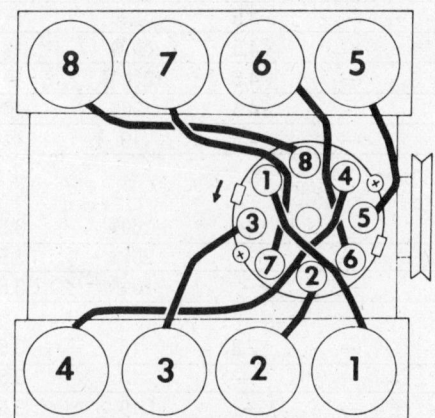

FORD MOTOR CO. 302HO and 351 engines
Firing Order: 1-3-7-2-6-5-4-8
Distributor rotation: counterclockwise

CAPACITIES—PINTO/BOBCAT

Year	Engine No. Cyl. Displacement (Cu. In.)	Engine Crankcase Add 1 Qt For New Filter ■	Transmission Pts. to Refill After Draining		Drive Axle (pts)	Gasoline Tank (gals)	Cooling System (qts)	
			4-Speed Manual	Automatic (Total capacity)			With Heater	With A/C
'80	4-140 (2300cc)	4.0	2.8	①	2.5②	13③④	8.6	9.0
	6-171 (2800cc)	4.5	2.8	①	4.5	13③④	8.5	9.1

■ ½ quart for 2800
— Not applicable
① C3 trans.—16; C4 trans.—14
② 8.00 in. axle—4.5
③ 14 gals on station wagon
④ Optional on some 1980 models—11.7

CAPACITIES—MID-SIZE MODELS

Year	Engine No. Cyl. Displacement (Cu. In.)	Engine Crankcase Add 1 Qt For New Filter	Transmission Pts to Refill After Draining Manual 3-Speed	Transmission Pts to Refill After Draining Manual 4/5-Speed	Automatic (Total Capacity)	Drive Axle (pts)	Gasoline Tank (gals)	Cooling System (qts) With Heater	Cooling System (qts) With A/C
'80	**Versailles**								
	8-302	4	—	—	20.5③	5.0	19.2	14.6	14.6
'80	**Granada, Monarch**								
	6-200	4	3.5	4.0④	—	4.0⑤	19.2⑥	9.9	9.9
	6-250	4	3.5	4.0④	17.0⑦	4.0⑤	19.2⑥	10.5	10.7
	8-255	4	3.5	4.0④	17.2	4.5	18.0	14.2	14.3
	8-302	4	3.5	4.0④	20.0⑧	4.0⑤	19.2⑥	14.6	14.6
	8-351	4	—	—	20.0	4.0⑤	19.2⑥	15.7	16.7
'80–'82	**Fairmont, Futura, Zephyr**								
	4-140	4	—	2.8④	16⑨	⑩	16⑪⑫⑬	8.6⑭	10.2
	6-200	4	3.5	2.8④	19⑨	⑩	16⑪⑬	9.0⑬⑯	9.0⑮⑯
	8-255	4	—	—	16⑨	⑩	16⑪⑬	13.4	13.5
	8-302	4	—	—	20.5⑨	⑩	16⑪	13.9	14.0
'80–'82	**Mustang, Capri**								
	4-140	4	—	2.8	⑰⑨		11.5⑳	8.6⑭	10㉓
	4-140T	4.5	—	3.5	⑰⑨	⑩	11.5⑳	8.6㉒	10.2㉒
	6-170	4.5	—	4.5	⑰⑨	⑩	12.5	9.2	9.2
	6-200	4	—	4.5	12⑰⑲	⑩	16㉑⑳	9⑮	9⑮
	8-255	4	—	4.5	19⑲	⑩	12.5⑳	13.4㉙	13.7㉕
	8-302	4	—	4.5	19	⑩	12.5⑳	13.9	14.2
'80–'81	**Cougar XR-7, Thunderbird**								
	6-200	4	—	—	16⑲	3.5	17.5	13.0	13.2
	8-255	4	—	—	20㉖	3.5	17.5	13.2	13.3
	8-302	4	—	—	20㉖	3.5	17.5	12.7	12.8
'81	**Cougar, Granada**								
	4-140	4	—	2.8	16	3.5	14.7	8.6	8.6
	6-200	4	—	—	16	3.5	16.0	8.1	8.1
	8-255	4	—	—	19	3.5	16.0	13.4	13.5
'82	**Cougar XR-7, Thunderbird, Lincoln Continental**								
	6-200	4	—	—	22	3.25	21	8.4	8.4
	6-232	4	—	—	24	3.25	21㉗	8.3	8.6
	8-255	4	—	—	24	3.25	21	14.9	15
	8-302	4	—	—	24	3.25	22.6	13.3	13.4
'82	**Cougar, Granada**								
	4-140	4	—	—	16	⑩	16.0㉘	10.2	10.2
	6-200	4	—	—	22	3.25	16.0㉘	8.4	8.4
	6-232	4	—	—	22	3.25	16.0㉘	8.3	8.3
'83–'87	**LTD/Marquis**								
	4-140	4	—	2.8	16	3.25㉙	16	8.6	9.4
	4-140P	4	—	—	16	3.25㉙	24	8.6	9.4
	6-200	4	—	—	22	3.25㉙	16	8.4	8.5
	6-232	4	—	—	22㉚	3.25㉙	16	10.7	10.8
'83–'87	**Mustang/Capri**								
	4-140	4㉜	—	2.8㉛	16	3.25㉙	15.4	8.6	9.4㉚
	6-232	4	—	—	22	3.25㉙	15.4	8.4	8.4
	8-302	4	—	4.5	—	3.55	15.4	13.1	13.4
'83	**Fairmont Futura**								
	4-140	4	—	2.8	16	3.25㉙	16	10.2	10.2
	6-200	4	—	—	22	3.25㉙	16	8.4	8.4

CAPACITIES—MID-SIZE MODELS

Year	Engine No. Cyl. Displacement (Cu. In.)	Engine Crankcase Add 1 Qt For New Filter	Transmission Pts to Refill After Draining			Drive Axle (pts)	Gasoline Tank (gals)	Cooling System (qts)	
			Manual		Automatic (Total Capacity)			With Heater	With A/C
			3-Speed	4/5-Speed					
'83–'87	Thunderbird, Cougar Continental								
	4-140 Turbo	4.5㉜	—	4.75	—	3.25㉙	18	8.4	8.7
	6-232	4	—	—	22㉞	3.25㉙	21	10.4	10.7
	8-302	4	—	—	22㉞	3.25㉙	20.7㉝	13.3	13.4

T—Turbocharged
P—Propane
N/A—Specs not available at time of printing
① C4-20 pts; C6-25 pts; FMX-22 pts
② Station wagon; 21.2 gal
③ '80 and later; 20 pts
④ 4 speed overdrive; 4.5 pts
⑤ 8 in.-4.5 pts
  8.7 in.-4.0 pts
  9 in.-5.0 pts
⑥ 1 gal less on certain '76 models; '78–'80; 18 gals
⑦ 16.5 pts with C4
⑧ 17 pts for '76

⑨ '81-13.25 pts with C4; 14.5 pts w/6 cyl; 19 pts w/V8
⑩ 6.75 in. axle-2.5 pts; 7.5 in. axle-3.5 pts
⑪ '80–'81 Station wagon; 14 gals
⑫ '81; 14.7 gals
⑬ '82; 20 gal optional
⑭ '82; 10.2 qts
⑮ '80–'82; 8.1 qts
⑯ '82; 8.4 qts
⑰ C3-16 pts; C4-14 pts
⑱ add ½ qt with filter change
⑲ '82; C5-22 pts
⑳ '82; 15.4 gal
㉑ '80–'81; 12.5 gal

㉒ '80–'81; 9.2 qts
㉓ '80–'81; 9.0 qts-'82; 10.2 qts
㉔ '82; 14.7 qts
㉕ '82; 15 qts
㉖ AOD transmission; 24 pts
㉗ Continental; 20 gals std; 22.6 optional
㉘ 20 gals optional
㉙ Traction-Lok; 3.55 pts
㉚ AOD transmission; 24 pts
㉛ 5 speed transmission; 4.75 pts
㉜ 4.5 Turbo
  add .5 with filter
㉝ 22.3 Continental
㉞ 10.5—Turbo models

CAPACITIES—FULL SIZE MODELS

Year	Engine No. Cyl. Displacement (Cu. In.)	Engine Crankcase Add 1 Qt For New Filter	Automatic Transmission (Total capacity)	Drive Axle (pts)	Gasoline Tank (gals) ■	Cooling System	
						With Heater	With A/C
'80	8-302	4	①	②	19	13.3	13.4
	8-351W	4	①	②	19	14.4	14.5
'81	8-255	4	24	②	20	14.8	15.2
	8-302	4	24	②	20	13.0	13.3
	8-351	4	24	②	20	13.9	14.0
'82–'87	8-225 ④	4	24	②	20	14.8	15.2
	8-302	4	24	②	20③	13.3	13.4
	8-351	4	24	②	20	13.8	13.8

■ Station wagons:
1980—20 gals
PI Police Interceptor
— Not applicable

① See Automatic Trans. Capacities Chart
② 7.5 inch axle—3.5
  8.5 inch axle—4.0

③ Lincoln: 18
④ Discontinued for 1983

AUTOMATIC TRANSMISSION CAPACITIES

(Pts.)

Code	Capacities
X	22
W	20.5
U, Z	25
T	24

NOTE: Total dry capacity.

VALVE SPECIFICATIONS

Year	Engine No. Cyl. Displacement (cu in.)	Seat Angle (deg)	Face Angle (deg)	Spring Test Pressure (lbs @ in.)	Spring Installed Height (in.)	Stem-to-Guide Clearance (in.)		Stem Diameter (in.)	
						Intake	Exhaust	Intake	Exhaust
'80–'84	6-200	45	44	150 @ 122①	1¹⁹⁄₃₂	.0008–.0025	.0010–.0027	.3104	.3102
'80	6-250	45	44	150 @ 122①	1¹⁹⁄₃₂	.0008–.0025	.0010–.0027	.3104	.3102
'80–'86	4-140	45	44	②	1⁹⁄₁₆	.0010–.0027	.0015–.0032	.3420	.3415
'80–'81	8-255	45	44	③	④	.0010–.0027	.0015–.0032	.3420	.3415
'80–'87	8-302	45	44⑧	③	④	.0010–.0027	.0015–.0032	.3420	.3415
'80–'81	8-351W	45	44	③	⑤	.0010–.0027	.0015–.0032	.3420	.3415
'82–'87	8-351	45	45	204 @ 1.33⑪	1⁴⁹⁄₆₄⑫	.0010–.0027	.0015–.0027	.3416–.3423	.3411–.3418
'82–'87	V6-232	⑥	⑥	215 @ 1.79	⑩	.0010–.0027	.0015–.0032	.3420	.3415
'82	8-255	⑥	⑦	⑨	—	.0010–.0027	.0015–.0032	.3420	.3415

① 1980–81 Intake: 51–57 @ 1.59
② 1980–81: Intake 71–79 @ 1.56
    Exhaust: 159–175 @ 1.16
    1982: 167 @ 1.16
    1983: 149 @ 1.12

From 1984: 154 @ 1.12
③ Intake: 196–212 @ 1.36
    Exhaust: 190–210 @ 1.20

④ Intake: 1¹¹⁄₁₆
    Exhaust: 1¹⁹⁄₃₂
⑤ Intake: 1²⁵⁄₃₂
    Exhaust: 1¹⁹⁄₃₂

⑥ 44° 30′–45°
⑦ 45° 30′–45° 45′
⑧ 1982—45
⑨ Intake: 192 @ 1.40

Exhaust: 191 @ 1.23
⑩ 1¾
⑪ Exhaust: 205 @ 1.15
⑫ Exhaust: 1³⁷⁄₆₄

CRANKSHAFT AND CONNECTING ROD SPECIFICATIONS

All measurements are given in inches.

Year	Engine No. Cyl. Displacement (cu in.)	Crankshaft				Connecting Rod		
		Main Brg. Journal Dia	Main Brg. Oil Clearance	Shaft End-Play	Thrust on No.	Journal Diameter	Oil Clearance	Side Clearance
'80–'84	6-200	2.2482–2.2490	.0005–.0022①	.004–.008	5	2.1232–2.1240	.0008–.0015	.0035–.0105
	6-250	2.3982–2.3990	.0005–.0022①	.004–.008	5	2.1232–2.1240	.0008–.0015	.0035–.0105
	8-255, 302	2.2482–2.2490	.0005–.0015②	.004–.008	3	2.1228–2.1236	.0008–.0026③	.010–.020
	8-351W	2.9994–3.0002	.0008–.0015②	.004–.008	3	2.3103–2.3111	.0008–.0026③	.010–.020
'80–'87	4-140	2.3990–2.3982	.0008–.0015	.004–.008	3	2.0464–2.0472	.0008–.0015	.0035–.0105
'82–'87	V6-232	2.5190	.0001–.001	.004–.008	3	2.3103–2.3111	.0008–.0026	.0047–.0114

① .0008–.0015 in. in 1980–81
② .0001–.0015 No. 1 bearing only
③ .0008–.0015 in. in 1980–81

PISTON AND RING SPECIFICATIONS

All measurements in inches

Engine Displacement (cu. in.)	Piston Clearance	Ring Gap			Ring Side Clearance			Wear Limit
		Top Compression	Bottom Compression	Oil Control	Top Compression	Bottom Compression	Oil Control	
4-140(2.3L)	0.0014–0.0022	.010–.020	.010–.020	.015–.055	.002–.004	.002–.004	Snug	.006
4-140(2.3L) ('80–'82 Turbo)	0.0034–0.0042	.010–.020	.010–.020	.015–.055	.002–.004	.002–.004	Snug	.006
4-140(2.3L) ('83–'87 Turbo)	0.0030–0.0038	.010–.020	.010–.020	.015–.055	.002–.004	.002–.004	Snug	.006
6-200(3.3L) 250(4.1L)	0.0013–0.0021	.008–.016	.008–.016	.015–.055	.002–.004	.002–.004	Snug	.006
6-232(3.8L)	0.0014–0.0028	.010–.020	.010–.020	0.15–.055	.002–.004	.002–.004	Snug	.006
8-255(4.2L) 302(5.0L) 351W(5.8L)	0.0018–0.0026	.010–.020	.010–.020	.015–.055	.002–.004	.002–.004	Snug	.006

CAMSHAFT SPECIFICATIONS
All measurements in inches

Engine	Journal Diameter 1	2	3	4	5	Bearing Clearance	Lobe Lift Intake	Exhaust	Endplay
4-140 (2.3L)	1.7713–1.7720	1.7713–1.7720	1.7713–1.7720	1.7713–1.7720	—	.001–.003	.2437①	.2437①	.001–.007
6-200 (3.3L)	1.8095–1.8105	1.8095–1.8105	1.8095–1.8105	1.8095–1.8105	—	.001–.003	.245	.245	.001–.007
6-232 (3.8L)	2.0505–2.0515	2.0505–2.0515	2.0505–2.0515	2.0505–2.0515	—	.001–.003	.240	.241	②
6-250 (4.1L)	1.8095–1.8105	1.8095–1.8105	1.8095–1.8105	1.8095–1.8105	—	.001–.003	.245	.245	.001–.007
8-255 (4.2L)	2.0805–2.0815	2.0655–2.0665	2.0505–2.0515	2.0355–2.0365	2.0205–2.0215	.001–.003	.2375	.2375	.001–.007
8-302 (5.0L)	2.0805–2.0815	2.0655–2.0665	2.0505–2.0515	2.0355–2.0365	2.0205–2.0215	.001–.003	.2375③	.2474③	.001–.003
8-351W (5.8L)	2.0805–2.0815	2.0655–2.0665	2.0505–2.0515	2.0355–2.0365	2.0205–2.0215	.001–.003	.260④	.260④	.001–.007

① '84 and later: .2381
② Endplay controlled by button and spring on camshaft end.
③ HO engine: Intake—:2600; Exhaust—.2780
④ HO engine: Intake—.2780; Exhaust—.2830

TORQUE SPECIFICATIONS
All readings in ft. lbs.

Year	Engine No. Cyl. Displacement (cu in.)	Cylinder Head Bolts	Rod Bearing Bolts	Main Bearing Bolts	Crankshaft Pulley or Damper Bolt	Flywheel-to-Crankshaft Bolts	Manifold Intake	Exhaust
'80–'87	6-200	70–75	21–26	60–70	85–100	75–85	—	13–18②
	6-250	70–75	21–26	60–70	85–100	75–85	—	13–18②
	8-255, 302	65–72	19–24	60–70	70–90	75–85	23–25①⑨	18–24
	8-351W	105–112	40–45	95–105	70–90	75–85	23–25①	18–24
'80–'87	4-140	80–90⑧	30–36	80–90	100–120	54–64	③	16–23
'82–'87	V6-232	⑤	⑥	⑦	85–100	75–85	18.4	15–22

① Retorque with engine hot
② 1977 and later: 18–24
③ Two steps: 5–7, then 14–21, non-Turbo, 13–18 for Turbo
④ Four steps: 3–6, 6–11, 11–15, 15–18; retorque to 15–18; retorque to 15–18 with engine hot

⑤ Soak bolts in oil, torque in sequence to 65–81 ft. lbs., loosen all bolts two complete turns then retorque to 65–81 ft. lbs.
⑥ Soak nuts in oil, torque to 30–36 ft. lbs., loosen two complete turns then retorque to 30–36 ft. lbs.

⑦ Soak bolts in oil, torque to 62–81 ft. lbs., loosen two complete turns then retorque bolts to 62–81 ft. lbs.
⑧ Tighten in two steps: 50–60 then 80–90
⑨ 1981–82-255 V8-18–20

WHEEL ALIGNMENT SPECIFICATIONS—PINTO/BOBCAT

Year	Model	Caster Range (deg)	Caster Pref. Setting (deg)	Camber Range (deg)	Camber Pref. Setting (deg)	Toe-in (in.)	Steering Axis Inclin.	Wheel Pivot Ratio (deg) Inner Wheel	Outer Wheel
'80	Pinto, Bobcat	¼P to 1¾P	1P	¼N to 1¼P	½P	0 to ¼	10.018	20	18.84
'80	Sta. Wag.	½N to 1P	¼P	¼N to 1¼P	½P	0 to ¼	10.018	20	18.84

N Negative   P Positive

WHEEL ALIGNMENT SPECIFICATIONS—MID SIZE MODELS

Year	Model	Caster Range (deg)	Caster Pref Setting (deg)	Camber Range (deg)	Camber Pref Setting (deg)	Toe-in (in.)	Steering Axis Inclin. (deg)	Wheel Pivot Ratio (deg) Inner Wheel	Wheel Pivot Ratio (deg) Outer Wheel
'80	Monarch, Granada, Versailles	1¼N to ¼P	½N	½N to 1P	¼P	0 to ¼	6¾	20	①
'80–'81	Fairmont and Zephyr (exc. Station Wagon)	⅛P to 1⅞P②	1P	⁵⁄₁₆N to 1³⁄₁₆P②	⁷⁄₁₆P	¹⁄₁₆ to ⁵⁄₁₆	15¼	20	19.84
'80–'81	Fairmont and Zephyr (Station Wagon)	⅛N to 1⅝P②	¾P	¼N to 1¼P②	½P	¹⁄₁₆ to ⁵⁄₁₆	15¼	20	19.84
'80–'82	Thunderbird, Cougar XR-7	⅛ to 1⅞P②	1P	½N to 1¼P②	⅜P	¹⁄₁₆ to ⁵⁄₁₆	15⅓	20	24.9③
'80–'82	Mustang, Capri	¼P to 1¾P②	1P	½N to 1P②	¼P	¹⁄₁₆ to ⁵⁄₁₆	15¼	20	19.84
'81–'82	Cougar, Granada	⅛P to 1⅞P	1P	⁵⁄₁₆N to 1³⁄₁₆P②	⁷⁄₁₆P	¹⁄₁₆ to ⁵⁄₁₆	15¼	20	19.84
'82–'83	Fairmont, Futura, Zephyr	⅛P to 1⅞P	1P	⁵⁄₁₆N to 1³⁄₁₆P②	⁷⁄₁₆P	¹⁄₁₆ to ⁵⁄₁₆	15¼	20	19.84
'83–'86	Thunderbird Cougar XR-7	½P to 2P	1¼P	½N to 1P	¼P	¹⁄₁₆ to ⁵⁄₁₆	—	20	19.73
'83–'86	LTD, Marquis (Sedan)	1⅛P to 2⅛P	1⅛P	⁵⁄₁₆N to 1³⁄₁₆P	⁷⁄₁₆P	¹⁄₁₆ to ⁵⁄₁₆	—	20	19.84
'83–'86	LTD, Marquis (Station Wagon)	⅛N to 1⅞P	⅞P	¼N to 1¼P	½P	¹⁄₁₆ to ⁵⁄₁₆	—	20	19.84
'83–'86	Mustang, Capri	½P to 2P	1¼P	¾N to ¾P	0	¹⁄₁₆ to ⁵⁄₁₆	—	20	19.84
'82–'83	Lincoln Continental	⅜P to 2⅛P	1¼P	½N to 1¼P	⅜P	0 to ¼	—	20	19.13
'87	Specifications Not Available At Time Of Publication								

N Negative   P Positive
N.A.—Not Available
① Granada/Monarch, Versailles w/PS—18.20; w/o PS—18.43
② Caster and camber are preset and nonadjustable
③ 1981–82—19.77; 83–85 N.A.

WHEEL ALIGNMENT SPECIFICATIONS—FULL SIZE MODELS

Year	Model	Caster Range (deg)	Caster Pref. Setting (deg)	Camber Range (deg)	Camber Pref. Setting (deg)	Toe-in (in.)	Steering Axis Inclin. (deg)	Wheel Pivot Ratio (deg) Inner Wheel	Wheel Pivot Ratio (deg) Outer Wheel
'80–'86	Lincoln	2¼P to 3¾P	3P	¼N to 1¼P	½P	¹⁄₁₆–³⁄₁₆	10.87	—	18.50
'80–'86	Ford, Mercury	2¼P to 3¾P	3P	¼N to 1¼P	½P	¹⁄₁₆ to ³⁄₁₆	10³¹⁄₃₂	—	18.50
'87	Specifications Not Available At Time Of Publication								

N: Negative P: Positive

DIESEL 2.4L SPECIFICATIONS

ENGINE
Type ....6-cylinder, in-line, 4-cycle, overhead valve, water-cooled
Bore .... 3.150 in. (80mm)
Stroke .... 3.189 in. (81mm)
Displacement .... 149 cu. in. (2442.9cc)
Compression ratio 23:1
Horsepower ....114 at 4800 rpm
Minimum Torque ....150 lb. ft. at 2400 rpm
Compression pressure .... 348 psi (2400 kPa)
Valve clearance (cold engine) .... Intake: 0.010 in. (0.3mm)

DIESEL 2.4L SPECIFICATIONS

ENGINE

Cam Timing   Exhaust: 0.010 in. (0.3mm)

Intake valve opens ....6° BTDC
Intake valve closes ....34° ABDC
Exhaust valve opens ....46° BBDC
Exhaust valve closes ....6° ATDC
Intake valve lift ....0.374 in. (9.5mm)
Exhaust valve lift ....0.376 in. (9.55mm)
Weight ....433 lbs. (196.4 kg) dry

FUEL SYSTEM

Injection firing order 1 5 3 6 2 4
Idle speed ....750 + 50 − 0 rpm
Fast idle (cold-start) speed .... 900–1050 rpm
Injection pump timing ....2.5° BTDC at 750–800 rpm

LUBRICATION SYSTEM

Complete System w/o oil cooler ....7.1 qts. (6.7L)
Complete System ....7.9 qts. (7.5L)
Engine oil pressure ....57–85 psi at 4000 rpm

NOTE: Due to the late introduction of this engine, normal specifications were unavailable at the time of publication.

DIESEL 2.4L TORQUE SPECIFICATIONS

Description	Ft. lbs.	Description	Ft. lbs.
Main bearing caps	43–48	Oil line from turbocharger to crankcase 22mm width across flats hollow bolt	29–36
Engine support straps	28–34		
Valve cover	6–7	Water pump to crankcase	14–17
Oil trap to valve cover	11–14	Fan coupling to water pump nut with left-hand threads	36
Cylinder head bolts			
Step 1	36–43	Fan to fan coupling	6–7
Step 2	65–69	Pulley to water pump	6–7
Step 3 (torque angle)	90 ± 5°	Thermostat housing	6–7
Oil spray bar to cylinder head	14–17	Bleeder screw	4–7
Oil drain plug	24–26	Temperature sensor/temperature switch	12–14
Oil pan to crankcase	6.5–7	Intake manifold to cylinder head	14–17
Front/rear end covers to crankcase	6–7 14–17	Exhaust manifold to cylinder head (upper row of staybolts installed with Loctite 270)	14–17
Flywheel to crankshaft (installed with Loctite No. 270)	71–81	Turbocharger to exhaust manifold	17–20
Vibration damper hub to crankshaft	282–311	Exhaust to turbocharger	31–35
Pulley/vibration damper to vibration damper hub	16–17	Vacuum pump	6–7
		Pulse sensor to engine (holder)	6–7
Connecting rod bolts		Glow plugs	14–22
Step 1	14	Temperature switch to fuel filter housing	22
Step 2 (torque angle)	70°	Wire to glow plug	3–4
Sprocket to camshaft	40–47	Fuel filter housing to holder	31–35
Bearing cap of camshaft	6–7 14–17	Injection pump to holder, rear (nuts and bolts)	14–17
Tensioning roller holder to crankcase	14–17	Injection pump to holder, front	14–17
Clamping bolt in rocker arm	5–6.5	Electric shut-off to injection pump	11–18
Sprocket to auxiliary shaft	40–47	Electric valve for cold start accelerator to injection pump	11–14
Oil pressure switch	22–29		
Oil pump to crankcase	16–17	Injection pump gear to injection pump	33–36
Oil pump cover	6–7	Tensioning torque for tensioning roller holder	33–36
Oil filter housing to crankcase	14–17		

DIESEL 2.4L TORQUE SPECIFICATIONS

Description	Ft. lbs.	Description	Ft. lbs.
Oil filter cover	15–18	Tensioning roller holder to engine (M8 nut and bolt)	18
Oil filter drain plug	7–9	Combination fuel injector in cylinder head	29–33
Oil spray jet to crankcase	6–7		
Oil cooler oil lines to oil filter housing	22–29	Injection line (coupling nut)	14–18
Oil lines to turbocharger	14–17	Nozzle holder to injection pump	33
		Spill valve to injection pump (hollow bolt)	14–22

ENGINE ELECTRICAL

For all charging system test procedures, please refer to the "Charging and Starting Systems" Unit Repair section.

Alternator

REMOVAL & INSTALLATION/ BELT TENSION ADJUSTMENT

1. Disconnect the negative battery ground cable.
2. Loosen the adjustment tensioner bolt (if equipped) and (or) the alternator slotted adjustment and mounting bolt. Remove the drive belt(s). Models equipped with a single drive belt (serpentine): lever the tensioner away from the belt and remove belt from alternator pulley.

NOTE: Various models are equipped with a 5-rib or 6-rib K-section (V-ribbed) belt and an automatic absorber tensioner. A special tool must be fabricated, on some models, to remove the tension from the absorber assembly arm so that the belt can be removed and installed. Loosen the idler pulley pivot and adjustment bolts before using tool to remove the belt.

3. Disconnect the electrical harness connectors from the alternator. Remove the adjustment and mounting bolts and remove the alternator. On some models it is necessary to remove the alternator from the mounting brackets before disconnecting the electrical harness due to clearance restrictions.
4. Install the alternator to the bracket and connect the electrical connectors. Adjust the drive belt tension so that there is approximately ¼

to ½ in. of deflection on the longest span of belt between pulleys. Apply pressure to the front of the alternator housing when adjusting belt tension. A flat is provided, on some models, that will allow an open end wrench to be used for applying tension to the belt.

5. On models equipped with a single drive belt, install the alternator to the bracket, attach the electrical connectors, slide the serpentine belt over the alternator pulley, and release the automatic tensioner. If the vehicle is equipped with AOD transmission and air conditioning, install the belt over the crank and A/C compressor pulleys, then place the absorber arm deflection tool on the arm and push the absorber pulley downward to the bottom of the slot (never push on the ribs of the pulley). Fit the belt over the rest of the pulleys. While holding the absorber pulley down, adjust the idler pulley by hand until it is snug and tighten the adjustment bolt and pivot bolt on the idler pulley assembly. Release the deflection tool; the proper tension will be set automatically.

Voltage Regulator

REMOVAL & INSTALLATION

1. Disconnect the negative battery cable.

2. Disconnect the wire harness connector. Remove the mounting screws and the regulator.
3. Mount the regulator in position and tighten the attaching screws. If equipped with a radio suppression capacitor, mount the capacitor in position.
4. Connect the wiring harness. Connect the negative battery cable. For removal and installation procedures of integral voltage regulators, please refer to "Charging and Starting" in the Unit Repair Section.

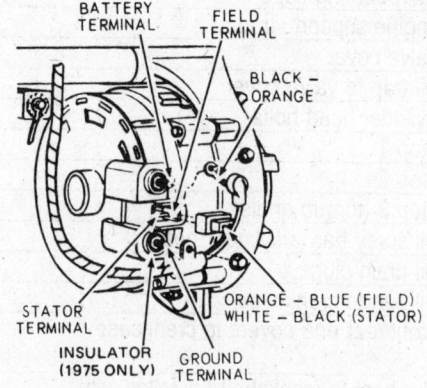

Typical connector details for the side terminal alternator

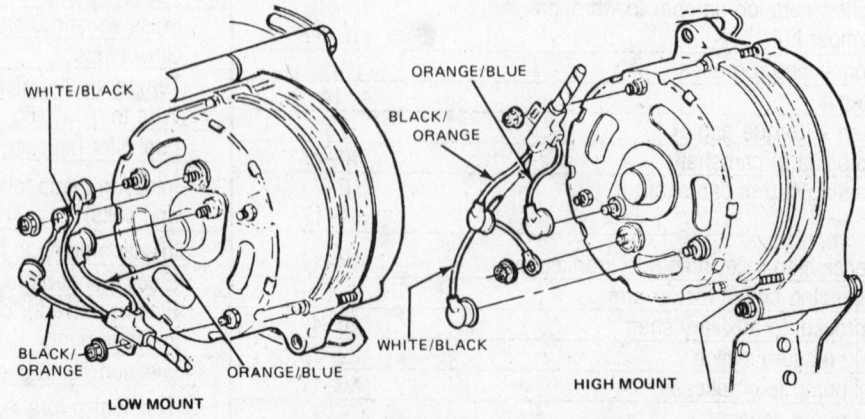

Typical connector details for the rear terminal alternator

Starter

Starter overhaul procedures can be found in the "Charging and Starting" Unit Repair Section.

REMOVAL & INSTALLATION

Gas Engines

1. Disconnect the negative battery cable.
2. Raise the vehicle and support it safely.
3. Disconnect the starter cable from the starter.

NOTE: If clearance is a problem, it may be necessary to remove a brace, raise the engine, etc. Refer to Step 4.

4. On Granada Monarch and Versailles with 302 CID engine, remove the right engine mount and raise the engine. On Fairmont and Zephyr with 200 CID engine, remove the wishbone brace. Pinto, Bobcat and Mustang, remove the crossmember from under the bell housing and remove the steering gear assembly from the side rail. On 1980 and later Thunderbirds and Cougars, XR-7s, LTDs Marquis and Continentals remove the cross brace.
5. Remove the starter housing bolts and crossmember from under the engine. Remove the heat shield, if equipped.
6. Manipulate the starter so that it can be lowered through the steering linkage. On some engine/chassis combinations, this can be accomplished by turning the front wheels either right or left, or by removing the idler arm bracket attaching bolts and lowering the steering linkage away from the engine.
7. The installation of the starter assembly is the reverse of the removal procedure. Tighten the mounting bolts to 15–20 ft. lbs.

2.4L Diesel Engine

1. Disconnect the battery ground (negative) cable.
2. Remove the bolt holding the dipstick tube to the intake manifold.
3. Remove the wires from the starter solenoid. Remove the front starter support bracket.
4. Remove the two starter to torque converter housing mounting bolts.
5. Pull the dipstick tube outward slightly allowing clearance for starter motor removal. Remove the starter motor.
6. Position the starter to torque converter housing and install the two bolts. Tighten to 30–40 ft. lbs.
7. Install the starter support bracket and tighten the attaching bolts to 14–20 ft. lbs.
8. Connect the cables to the starter solenoid. Tighten the red wire to 80–120 inch lbs. Tighten the black wire to 25 inch lbs.
9. Reposition the dipstick to the intake manifold, install the bolt and tighten to 6–7 ft. lbs.
10. Install the battery ground cable.

Distributor

REMOVAL & INSTALLATION

1. Remove the air cleaner on V6 and V8 engines. On 4 and 6 cylinder in-line engines, removal of a thermactor (air) pump mounting bolt and drive belt will allow the pump to be moved to the side and permit access to the distributor. If necessary, disconnect the thermactor air filter and lines as well.
2. Remove the distributor cap and position the cap and ignition wires to the side.
3. Disconnect the wire harness plug from the distributor connector. Disconnect and plug the vacuum hoses from the vacuum diaphragm assembly (if equipped).
4. Rotate the engine (in normal direction of rotation) until No. 1 piston is on TDC (Top Dead Center) of the compression stroke. The TDC mark on the crankshaft pulley and the pointer should align. Rotor tip pointing at No. 1 spark plug wire position on distributor cap.
5. On DuraSpark I or II, turn the engine a slight bit more (if required) to align the stator (pick-up coil) assembly pole with an (the closest) armature pole. On DuraSpark III, the distributor sleeve groove (when looking down from the top) and the cap adaptor alignment slot should align. On models equipped with EEC IV (1984 and later), remove the rotor (2 screws) and note the position of the "polarizing square" and shaft plate for reinstallation reference.
6. Scribe a mark on the distributor body and engine block to indicate the position of the rotor tip and position of the distributor in the engine. DuraSpark III and some EEC IV system distributors are equipped with a notched base and will only locate at one position on the engine.
7. Remove the holddown bolt and clamp located at the base of the distributor. Some DuraSpark III and EEC IV system distributors are equipped with a special holddown bolt that requires a Torx head wrench for removal. Remove the distributor from the engine. Pay attention to the direction the rotor tip points if it moves from the No. 1 position the drive gear disengages. For reinstallation purposes, the rotor should be at this point to insure proper gear mesh and timing.
8. Avoid turning the engine, if possible, while the distributor is removed. If the engine is turned from TDC position, TDC timing marks will have to be reset before the distributor is installed; Steps 4 and 5.

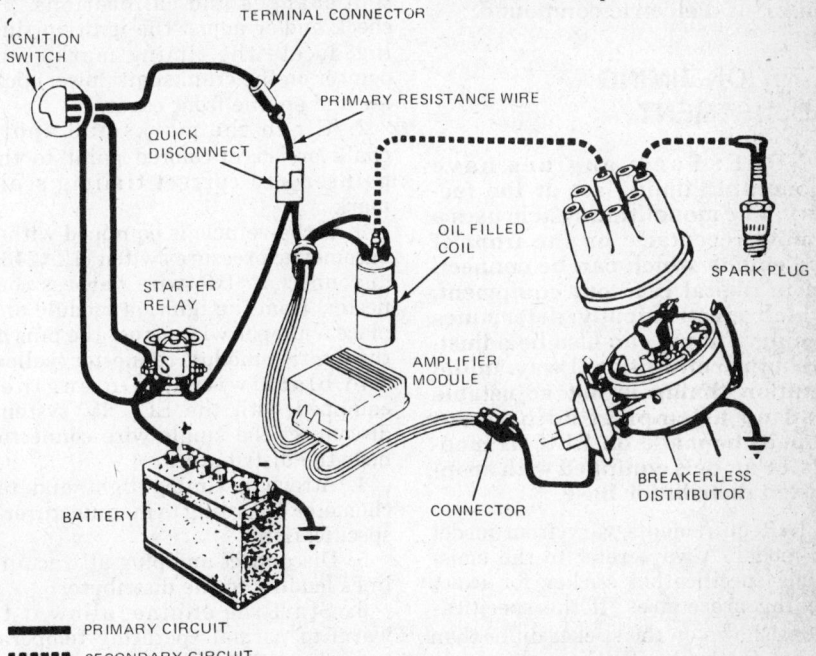

Typical electronic ignition system schematic

9. Position the distributor in the engine with the rotor aligned to the marks made on the distributor, or at the position the rotor pointed when the distributor was removed. The stator and armature or "polarizing square" and shaft plate should also be aligned. Engage the oil pump intermediate shaft and insert the distributor until fully seated on the engine, if the distributor does not fully seat, turn the engine slightly to fully engage the intermediate shaft.

10. Follow the above procedures on models equipped with an indexed distributor base. Make sure when positioning the distributor that the slot in the distributor base will engage the block tab and the sleeve/adaptor slots are aligned.

11. After the distributor has been fully seated onto he block, recheck the timing mark and rotor alignment. Install the holddown bracket and bolt. On models equipped with an indexed base, tighten the mounting bolt. On other models, snug the mounting bolt so the distributor can be turned for ignition timing purposes.

12. The rest of the installation is in the reverse order of removal. Check and reset the ignition timing.

NOTE: A silicone compound is used on rotor tips, distributor cap contacts and on the inside of the connectors on the spark plugs cable and module couplers. Always apply Silicone Dielectric Compound after servicing any component of the ignition system. Various models use a multi-point rotor which does not require the application of dielectric compound.

IGNITION TIMING ADJUSTMENT

NOTE: Some engines have monolithic timing set at the factory. The monolithic system uses a timing receptacle on the front of the engine which can be connected to digital read-out equipment, which electronically determines timing. Timing can also be adjusted in the conventional way. Initial ignition timing is not adjustable and no attempt at adjustment should be made on EEC III models, or models equipped with an indexed distributor base.

1. Requirements vary from model to model. Always refer to the emissions specification sticker for exact timing procedures. If the specifications shown on the sticker differ than those on the FORDRX in this book, follow the sticker procedures and

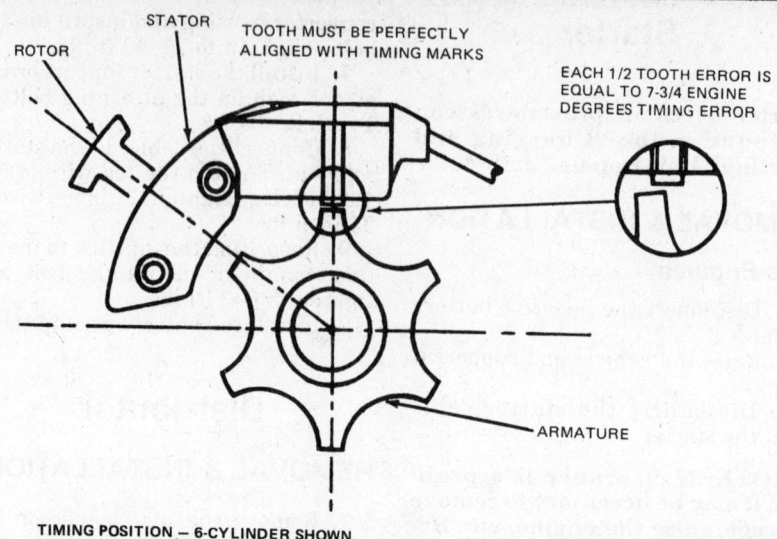

TIMING POSITION — 6-CYLINDER SHOWN, 8-CYLINDER SIMILAR

Distributor firing position with electronic ignition

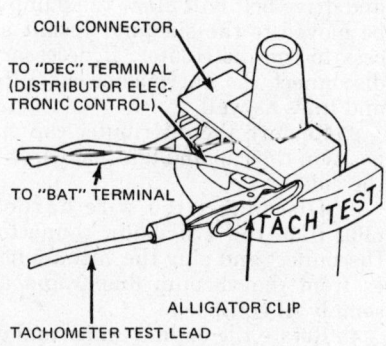

Electronic ignition test tachometer hookup

specifications as they reflect production changes and calibrations. To check and/or adjust the ignition timing; locate the timing marks and pointer on the crankshaft (lower) pulley and engine front cover.

2. Clean the marks and apply chalk or bright-colored paint to the pointer and correct timimg scale mark.

3. If the vehicle is equipped with a barometric pressure switch (12A243), disconnect it (two wire harness connector) from the ignition module and place a jumper wire across the pins at the ignition module connector (yellow and black wires). On engines equipped with the EEC IV system, disconnect the single wire connector near the distributor.

4. Attach a timing light and tachometer according to manufacturer's specifications.

5. Disconnect and plug all vacuum lines leading to the distributor.

6. Start the engine, allow it to warm to normal operating temperature, then set the idle to the specifications given on the underhood sticker.

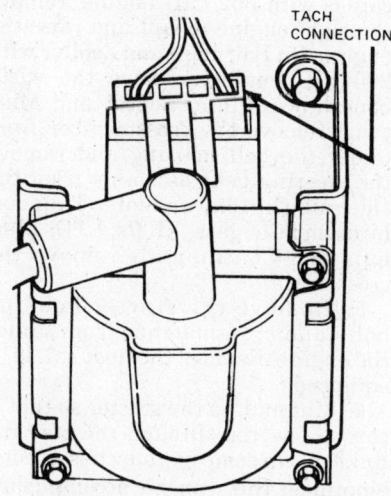

Tach connection, "E" coil

7. Aim the timing light at the timing mark and pointer on the front of the engine. If the timing marks do not align when the light flashes, turn the engine off and loosen the distributor holddown clamp slightly.

8. Start the engine again, and observe the alignment of the timing marks. Turn the distributor counterclockwise or clockwise until the marks are aligned. When altering the timing, it is wise to gently tap the distributor with a wooden hammer handle to move it in the desired direction. Grasping the distributor with your hand may result in a painful electric shock. When the timing marks are aligned, turn the engine off and tighten the distributor hold-down clamp. Remove the test equipment, reconnect the vacuum hoses, single wire connector (EEC IV)and /or the barometric pressure switch (on vehicles equipped).

Tachometer Connection

The coil connector used with DuraSpark is provided with a cavity for connection of a tachometer, so that the connector doesn't have to be removed to check engine rpm. Install the tach lead into the cavity marked TACH TEST and connect the other lead to a good ground. If the coil connector (except vehicles equipped with an "E" coil) must be removed, pull it out horizontally until it is disengaged from the coil terminal.

Diesel Glow Plugs

REMOVAL & INSTALLATION

1. Disconnect the negative battery cable.
2. Unscrew the glow plug electrical connector and remove the wire.
3. Remove the glow plug using a 12mm deepwell socket.
4. Coat the glow plug threads with a copper based, anti-sieze compound.
5. Install the glow plug into the engine block.
6. Tighten the glow plug to 15–22 ft. lbs.
7. Connect the electrical wire to the glow plug with nut and tighten to 3–4 ft. lbs.
8. Connect the negative battery cable.

GASOLINE FUEL SYSTEM

Mechanical Fuel Pump

REMOVAL & INSTALLATION

NOTE: Before removing the pump, rotate the engine so that the low point of the cam lobe is against the pump arm. This can be determined by rotating the engine with a fuel pump mounting bolts loosened slightly; when tension (resistance) is removed from the arm, proceed with removal.

1. Remove the inlet, outlet and return vapor (if equipped) lines from the pump.
2. Remove the fuel pump mounting bolts and remove the pump and gasket. Remove the fuel pump pushrod on 2800 V6 engines.

3. Clean all gasket material from the pump mounting surface on the engine, and apply a coat of oil-resistant sealer to the new gasket.
4. Reinstall the pushrod on models equipped. Position pump on engine and install retaining screws.
5. Reinstall lines, start engine and check for leaks.

NOTE: If resistance is felt while positioning the fuel pump on the block, the camshaft eccentric is in the high position. To ease installation, connect a remote engine starter switch to the engine and tap the remote switch until resistance fades.

Electric Fuel Pump

REMOVAL & INSTALLATION

NOTE: A single internally gas tank mounted pump is used on fuel injected models through 1983 and on 1984 and later Lincoln Town Car, Ford Crown Victoria and Mercury Grand Marquis models (with CFI). Other 1984 and later models equipped with a high output injected or turbocharged injected engine are equipped with two electric pumps. A low-pressure pump is mounted in the tank and a high pressure pump is externally mounted.

——— **CAUTION** ———
Before servicing any part of the fuel injection system it is necessary to depressurize the system. A special tool is available for testing and bleeding the system.

To depressurize the fuel injection system:

a. If the fuel charging assembly is mounted to the engine, remove the fuel tank cap then release pressure from the system by opening the pressure relief valve on the fuel line in the upper right hand corner of the engine compartment. Use Fuel Pressure Gauge T80L-9974-A or equivalent.

NOTE: The cap on the relief valve must be removed.

b. Using an open-end wrench, remove the pressure relief valve from the fuel line.
c. Install the pressure relief valve and cap. Tighten valve to 48–84 inch lbs. Tighten cap to 4–6 inch lbs.

In-Tank Pump

1. Disconnect the negative battery cable.

PRESSURE RELIEF VALVE 9H321 TIGHTEN TO 4-7 FT.LBS.

PRESSURE RESET VALVE CAP 9H323

Fuel pressure relief valve removal and installation

2. Depressurize the system and drain as much gas from the tank by pumping out through the filler neck.
3. Raise the back of the car and safely support on jackstands.
4. Disconnect the fuel supply, return and vent lines at the right and left side of the frame.
5. Disconnect the wiring harness to the fuel pump.
6. Support the gas tank, loosen and remove the mounting straps. Remove the gas tank.
7. Disconnect the lines and harness at the pump flange.
8. Clean the outside of the mounting flange and retaining ring. Turn the fuel pump lock ring counterclockwise and remove.
9. Remove the fuel pump.
10. Clean the mounting surfaces. Put a light coat of grease on the mounting sufaces and on the new sealing ring. Install the new fuel pump.
11. Installation is in the reverse order of removal. If single high pressure pump system, fill the tank with at least 10 gals. of gas. Turn the ignition key ON for three seconds. Repeat 6 or 7 times until the fuel system is pressurized. Check for any fitting leaks. Start the engine and check for leaks.

External Pump

1. Disconnect the negative battery cable.
2. Depressurize the fuel system.
3. Raise and support the rear of the vehicle on jackstands.
4. Disconnect the inlet and outlet fuel lines.
5. Bend down the retaining tab and remove the pump from the mounting bracket ring.
6. Install in reverse order, make sure the pump is indexed correctly in the mounting bracket insulator.

"Quick-Connect" Line Fittings

REMOVAL & INSTALLATION

NOTE: "Quick-Connect" (push) type fittings must be disconnected using proper procedures or the fitting may be damaged. Two types of retainers are used on the push connect fittings. Line sizes of 3/8 in. and 5/16 in. use a "hairpin" clip retainer. 1/4 in. line connectors use a "duck bill" clip retainer.

Hairpin Clip

1. Clean all dirt and/or grease from the fitting. Spread the two clip legs about 1/8 in. each to disengage from the fitting and pull the clip outward from the fitting. Use finger pressure only, do not use any tools.
2. Grasp the fitting and hose assembly and pull away from the steel line. Twist the fitting and hose assembly slightly while pulling, if necessary, when a sticking condition exists.
3. Inspect the hairpin clip for damage, replace the clip if necessary. Reinstall the clip in position on the fitting.
4. Inspect the fitting and inside of the connector to insure freedom of dirt or obstruction. Install fitting into the connector and push together. A click will be heard when the hairpin snaps into proper connection. Pull on the line to insure full engagement.

Duck Bill Clip

1. A special tool is available from Ford for removing the retaining clips (Ford Tool No. T82L-9500-AH). If the tool is not on hand see Step 2. Align the slot on the push connector disconnect tool with either tab on the retaining clip. Pull the line from the connector.
2. If the special clip tool is not available, use a pair of narrow 6 in. channel lock pliers with a jaw width of 0.2 in. or less. Align the jaws of the pliers with the openings of the fitting case and compress the part of the retaining clip that engages the case. Compressing the retaining clip will release the fitting which may be pulled from the connector. Both sides of the clip must be compressed at the same time to disengage.
3. Inspect the retaining clip, fitting end and connector. Replace the clip if any damage is apparent.
4. Push the line into the steel connector until a click is heard, indicting the clip is in place. Pull on the line to check engagement.

Fuel Filter

REMOVAL & INSTALLATION

Carbureted Engines

IN-LINE HOSE CONNECTED FILTERS

1. Remove the air cleaner.
2. Loosen the hose clamps.
3. Unscrew the filter from the carburetor.
4. Disconnect the filter from the hose and discard the filter, hose and clamps. Replacement filters usually come with a length of hose and new clamps, always use the new parts when filter replacement is necessary.
5. Reverse the procedure to install the fuel filter. After installation, start the engine and check for fuel leakage.

INVERTED NUT (STEEL LINE) CONNECTED FILTERS

1. Remove the air cleaner assembly.
2. Position an 11/16 in. open end wrench on the filter hex nut to hold the filter in position and remove the steel fuel line from the filter using a suitable wrench.
3. Unscrew the filter from the carburetor.
4. Install the new filter in reverse order of removal.

IN CARBURETOR FILTERS (VV CARBS)

1. Remove the air cleaner.
2. Disconnect the fuel line from the carburetor inlet fitting, while holding

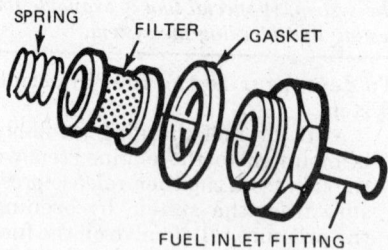

VV carburetor fuel filter

the inlet fitting with a suitable wrench.
3. Remove inlet fitting and fuel filter.
4. Install the spring, filter, gasket and fitting.
5. Connect the fuel line, start engine and check for leaks.

Fuel Injected Engines

NOTE: Models equipped with fuel injection actually have four fuel filters; a nylon mesh "sock" at the fuel pump inlet in the fuel tank; a large paper element filter mounted in the fuel line under the car; a small canister filter mounted in the engine compartment; and individual mesh filters at each injector fuel inlet. Of these, only the undercar paper element filter is scheduled for regular replacement (at 50,000 mile intervals). Filter replacement requires discharging of the fuel injection system pressure prior to filter change. Discharge pressure, disconnect the fuel lines and remove filter retainer. Note the direction of the fuel flow arrow on filter. Install new filter in reverse order.

Carburetors

NOTE: Refer to the Unit Repair section on Carburetors for overhaul and adjustment.

REMOVAL & INSTALLATION

1. Remove the air cleaner.
2. Disconnect the throttle cable or rod at the throttle lever. Disconnect the distributor vacuum line, exhaust gas recirculation line, inline fuel filter, choke heat tube and the positive crankcase ventilation hose at the carburetor.
3. Disconnect the throttle solenoid (if so equipped) and electric choke assist at their connectors.
4. Remove the carburetor retaining nuts. Lift off the carburetor carefully, taking care not to spill any fuel. Remove the carburetor mounting gasket and discard it. Remove the carburetor mounting spacer, if so equipped, from the intake manifold.
5. Prior to installation, clean the gasket mounting surfaces of the intake manifold, spacer (if so equipped), and carburetor. When using a spacer, use two new gaskets, sandwiching the spacer between the gaskets. If a spacer is not used, only one new carburetor mounting gasket is required.
6. Place the new gasket(s) and spacer (if so equipped) on the carburetor mounting studs. Position the carburetor on top of the gasket and hand tighten the retaining nuts. Then tighten the nuts in a crisscross pattern to 10–15 ft. lbs.
7. Connect the throttle linkage, and distributor vacuum line, exhaust gas recirculation line, inline fuel filter, choke heat tube, positive crankcase ventilation hose, throttle solenoid (if so equipped) and electric-choke assist. Adjust to correct idle speed.

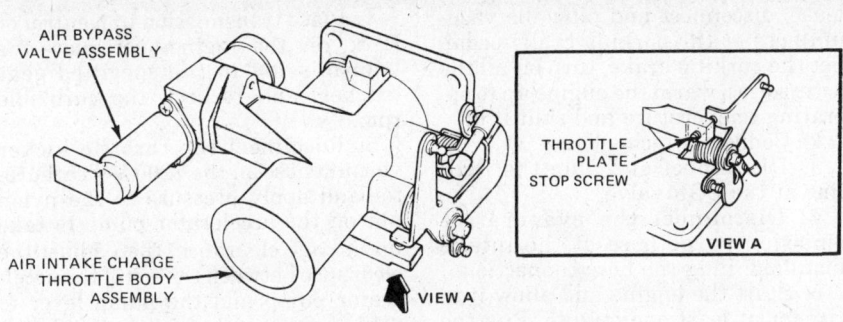

AIR BYPASS VALVE ASSEMBLY

THROTTLE PLATE STOP SCREW

VIEW A

AIR INTAKE CHARGE THROTTLE BODY ASSEMBLY

VIEW A

Multi-point injection adjustment

Fuel Injection

Due to the complex nature of modern fuel injection systems, comprehensive diagnosis and testing procedures fall outside the confines of this repair manual. For complete information on diagnosis, testing and repair procedures concerning all modern fuel injection systems, please refer to *Chiltons Guide To Fuel Injection and Feedback Carburetors.*

Idle Speed

ADJUSTMENTS

NOTE: If the underhood sticker reflects different procedures and/or specifications than those in this book, follow the sticker as it will list the latest production and calibration changes.

1980 2.3L (140) OHC Engine

1. Set the parking brake and block the wheels. Turn off all accessories, bring the engine to normal operating temperature, connect a tachometer and check the ignition timing.
2. Remove or relocate the air cleaner. Remove the plug and molded rubber fitting from the EGR cold weather modulator in the air cleaner (if equipped).
3. On engines with Thermactor systems: Apply vacuum to 1-port dump valves and plug all hoses to 2-port dump valves. Plug all hoses to 2-port dump valves. Disconnect and plug the charcoal canister purge valve vacuum hose, being careful not to damage the purge valve.
4. Check the throttle linkage for freedom of movement.
5. Run the engine at 2500 rpm for 15 seconds before each speed check.
6. There are several different idle speed control devices used. Some models have no speed control devices other than the curb idle screw. Others

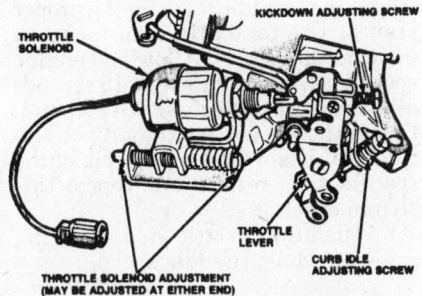

KICKDOWN ADJUSTING SCREW

THROTTLE SOLENOID

THROTTLE LEVER

CURB IDLE ADJUSTING SCREW

THROTTLE SOLENOID ADJUSTMENT (MAY BE ADJUSTED AT EITHER END)

Typical idle speed solenoid adjusting locations

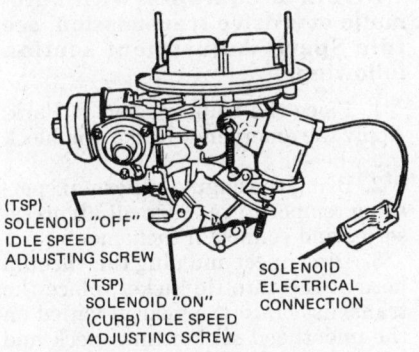

(TSP) SOLENOID "OFF" IDLE SPEED ADJUSTING SCREW

(TSP) SOLENOID "ON" (CURB) IDLE SPEED ADJUSTING SCREW

SOLENOID ELECTRICAL CONNECTION

Model 6500 carburetor idle adjustments— with TSP (Throttle Stop Positioner)

are equipped with a Throttle Stop Positioner (TSP) solenoid which can be accompanied by a throttle modulator (A/C only), or a dashpot. Air conditioned models without the TSP are equipped with the throttle modulator alone.

7. On models with curb idle screw only, adjust the idle speed using the screw. On models with TSP or TSP and A/C, adjust the idle at the TSP-ON adjusting screw, then collapse the TSP plunger with the throttle lever and adjust the TSP-OFF idle speed to specifications. On models with the throttle modulator alone (A/C equipped), loosen the locknut securing the throttle modulator and rotate the modulator until there is clearance between the modulator stem and the throttle pad. Adjust curb idle by turning the throttle stop adjusting screw.

Adjust the throttle modulator by turning it back in until its stem contacts the throttle lever pad, then tighten the locknut.

NOTE: On engines with 2700 VV carburetor, the accelerator pump lever lash must be adjusted every time the idle speed is adjusted. To adjust, apply a slight downward pressure on the top of the nylon nut on the accelerated pump stem and the lever. Lash should be .010 in. Turn the adjusting nut to adjust.

1980 255, 302 and 351W Engines Without EEC

NOTE: If equipped with automatic overdrive transmission, see the "Idle Speed Adjustment" section following.

1. The air cleaner must be installed. If engine speed fluctuates, use the average engine speed. Do not depress the brake pedal on models with hydro-boost brakes. On cars with automatic parking brake release, disconnect and plug the vacuum hose at the parking brake pedal. Set the parking brake, turn off all accessories, warm the engine to operating temperature, and shut off.
2. Disconnect the fuel evaporation purge valve hose by tracing the hose from the charcoal canister to the first fitting. Disconnect and plug the hose: also cap the fitting. Connect a tachometer.
3. On all models except those with the Model 2700 VV (variable venturi) carburetor: Remove the spark delay valve (if equipped) and route the hose directly to the distributor advance fitting. On engines with the VV carburetor: disconnect and plug the distributor vacuum advance hose.
4. Trace the EGR hose to the carburetor. If an EGR/PVS valve is located in the hose, disconnect and plug the hose at the EGR valve.
5. Start the engine (choke fully open, transmission in Park). Place the fast idle lever on the specified step of the cam (see the emission control sticker on the engine for specification). Adjust if not within 100 rpm of specifications. Run the engine to 2500 rpm for 15 seconds and recheck the adjustment.
6. On engines with the VV carburetor only, turn off the engine and disconnect and plug the hose from the throttle modulator. Attach a spare length of vacuum hose from an engine vacuum source to the modulator. Start the engine, open the throttle until the modulator plunger is fully extended. Release the throttle. Check

the auxiliary fast idle rpm (engine sticker). Adjustment is made by loosening the modulator locknut and turning the modulator or, on some models, by turning the adjuster bolt on which the modulator rides. Reconnect the hose after adjustment.

7. After fast idle rpm is set, reconnect the vacuum lines (and spark delay valve, if equipped) removed earlier.

8. Before each idle speed check the following, run the engine at 2500 rpm for 15 seconds (transmission in Neutral), then allow the engine to return to curb idle.

9. The air conditioning must be off, engine warm, choke fully open, parking brake set, and transmission in gear specified on the engine sticker (usually in Drive). If engine rpm in each case is not within 50 rpm of specifications, adjustment is required.

10. If no solenoid is present; turn the throttle stop adjusting screw until specified rpm (engine sticker) is obtained. If equipped with a dashpot, shut off the engine, collapse the dashpot plunger, and measure the clearance between the plunger and the throttle lever pad. Adjust to specifications (sticker) if necessary.

11. On non-air conditioned cars with an anti-diesel TSP (throttle solenoid position): adjust the TSP by rotating the long screw (part of the mounting bracket) until the specified curb idle rpm (engine sticker) is obtained. Then, collapse the TSP plunger by forcing the throttle lever pad against the plunger. Adjust the throttle stop screw until the specified TSP-OFF rpm (sticker) is obtained.

12. On air conditioned cars with an A/C TSP: Turn the A/C on and open the throttle to allow the TSP plunger to extend, then release the throttle. Disconnect the A/C compressor clutch wire at the compressor. Check A/C-ON rpm and adjust, if necessary, by turning the long screw on the TSP BRACKET UNTIL THE SPECIFIED A/C-ON rpm is obtained. Then turn the A/C off, connect the compressor clutch wire, and adjust the throttle stop screw until the specified A/C-OFF rpm is obtained.

1980 Engines With EEC

NOTE: If equipped with automatic overdrive transmission, see the Idle Speed Adjustment Section following.

1. The air cleaner must be installed. If the engine speed fluctuates, use the average engine speed. Do not depress the brake pedal on models equipped with Hydroboost brakes. On cars with automatic parking brake release, disconnect and plug the vacuum hose at the parking brake pedal. Set the parking brake, turn off all the accessories, warm the engine up to operating temperature and shut it off.

2. Connect a tachometer.

3. Disconnect and plug the EGR line at the EGR valve.

4. Disconnect the evaporative emission purge hose at the intake manifold. Plug the hose connection.

5. Start the engine and allow it to run for at least one minute. Run the engine at 2500 rpm for 15 seconds and place the fast idle lever on the proper step of the fast idle lever on the proper step of the fast idle cam (see the underhood sticker). Allow the engine speed to stabilize for about 15 seconds and measure the fast idle speed. Check the sticker for the proper setting. If it is not within 100 rpm of the specification, reset it and repeat this step to check it.

6. Turn the throttle stop adjusting screw to adjust the idle speed.

1981 and Later 225, 302 and 351 Engines (Except CFI)

NOTE: If equipped with automatic overdrive transmission, see Idle Speed Adjustment section following.

1. Place the transmission in Park. Apply the emergency brake and block the wheels.

2. Bring the engine to normal operating temperature. Turn off all accessories and connect a tachometer.

3. Disconnect and plug the vacuum hose at the throttle kicker, place the transmission in the gear specified on the underhood sticker and check and adjust the curb idle rpm. Adjust at the curb idle screw at the throttle valve lever or at the saddle bracket adjusting screw.

4. Place transmission in Neutral or Park, rev the engine once, place the transmission in the specified gear (sticker) and recheck the curb idle rpm.

5. Reconnect the throttle kicker vacuum hose on the 7200 VV carburetor and apply pressure to the nylon nut on the accelerator pump to take up linkage clearance, then adjust the clearance between the top of the accelerator pump and the pump lever to .010 in., using the nylon nut on the pump rod. Turn the pump rod one turn counterclockwise to set the lever lash preload.

6. Reconnect all hoses.

7. To set the throttle kicker speed, set the transmission in Neutral or Park, bring the engine to normal operating temperature and turn off all accessories. Disconnect the vacuum hose at the vacuum Operated Throttle Modulator (kicker) and connect an external vacuum source (10 in. Hg minimum) to the kicker.

8. Place the transmission in the gear specified on the underhood sticker (apply parking brake, block wheels).

9. Disconnect the A/C compressor clutch wire, place the A/C selector to max. blower cooling and check/adjust the VOTM kicker speed. If adjustment is required, turn the saddle bracket adjusting screw.

10. Reconnect all components.

351W With 7200VV Carburetor

1. Follow Steps 1–3 of the "302 CFI" procedure. Additionally, disconnect and plug the EGR vacuum hose from the EGR valve. Disconnect the evaporative emission (charcoal canister) purge hose from the intake manifold; cap the manifold connection.

2. Curb the idle with Cold Start VOTM: Warm the engine. If the rpm

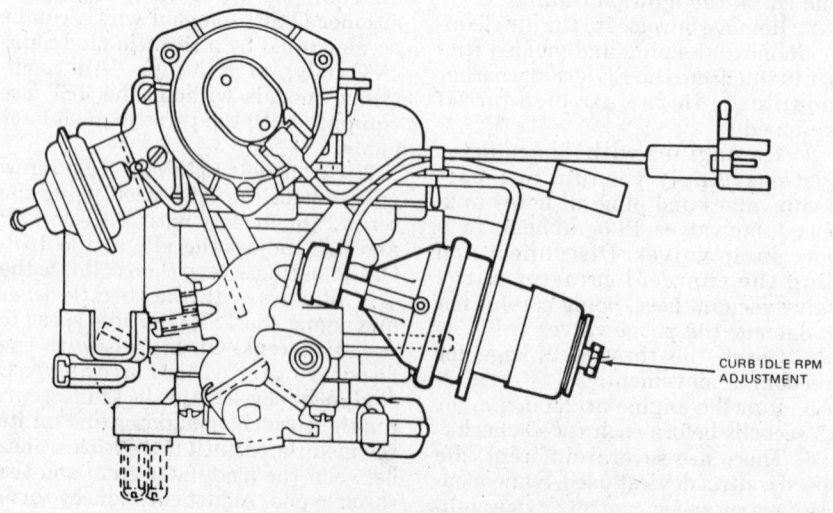

Typical VOTM curb idle adjustment

CURB IDLE RPM ADJUSTMENT

is higher than specified, adjust the throttle stop-screw counterclockwise. If the rpm is low, shut off the engine, turn the throttle stop adjusting screw one turn clockwise, start the engine, and recheck the adjustment. Open and close the throttle and check the speed. See Step 7 of the "302 CFI" procedure.

3. Curb idle with Dashpot: If the car has air conditioning, shut it off. Start the engine and turn the throttle stop adjusting screw until the specified idle speed is reached. Turn the engine off and check the clearance between the dashpot plunger and the throttle lever pad. Adjust if not correct (see the emission control sticker on the car for proper clearance measurement). Start the engine, open and close the throttle and recheck the idle speed; shut off the engine and recheck the dashpot clearance. See Step 7 of the "302 CFI" procedure.

4. Curb Idle without Dashpot: If the car has neither a dashpot nor a VOTM, simply start the engine (A/C off, if equipped) and turn the throttle stop adjusting screw until the specified speed is reached. Open and close the throttle and recheck the adjustment. See Step 7 of the "302 CFI" procedure.

1980 and Later, 302 CFI, V6 (232) CFI

1. Leave all hoses and wires connected to the air cleaner case. The air cleaner assembly can be removed for adjustments, but must be installed when measuring idle speed. If the car has speed control and correct idle speed cannot be achieved, disconnect the accelerator cable at the throttle lever.

2. Apply the parking brake and block the front wheels. If the car has a vacuum-operated parking brake pull-off, disconnect and plug the vacuum hose from the parking brake.

3. Turn off all accessories. Start the engine and allow it to reach normal operating temperature. Check the throttle linkage for freedom of movement and correct as necessary. Connect a tachometer to the engine.

4. Throttle stop screw is not to be adjusted.

5. If the throttle speed is high, adjust the Vacuum Operated Throttle Modulator (VOTM) bracket adjusting screw counterclockwise. When the idle speed is as specified, open and close the throttle and recheck.

6. If the rpm is low, shut off the engine. Turn the VOTM bracket adjusting screw one turn clockwise. Start the engine and run at 2000 rpm for

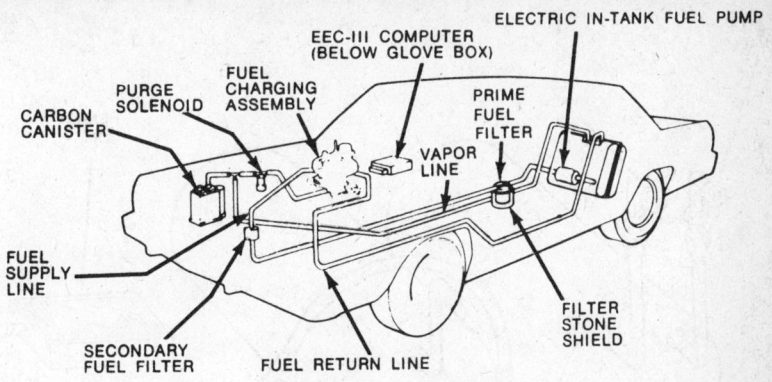

EFI system components

Idle Speed Change	Turns on Linkage Lever Screw
Less than 50 rpm	No change
500–100 rpm increase	1½ turns out
50–100 rpm increase	1½ turns in
100–150 rpm increase	2½ turns out
100–150 rpm decrease	2½ turns in

ten seconds. Let the idle stabilize for one minute (time not to exceed two minutes) and recheck the idle speed. Repeat as necessary.

7. If the idle speed has been altered more than 50 rpm, the Automatic Overdrive Transmission throttle valve control linkage must be adjusted.

AUTOMATIC OVERDRIVE (AOD) IDLE SPEED ADJUSTMENT

If the car is equipped with Ford's automatic overdrive transmission, and the idle speed is adjusted by more than 50 rpm, the adjustment screw on the linkage lever at the carburetor must also be adjusted.

FUEL MIXTURE ADJUSTMENT

The factory recommended procedure for adjusting the idle mixture requires the addition of an artificial mixture enrichment substance (propane) to the air intake. Proprane enrichment adjustment requires special professional equipment, and if not done correctly, will negatively affect emissions calibrations.

PROPANE FUEL SYSTEM

The propane fuel system is a completely closed system which contains a supply of pressurized liquid propane fuel. The liquid propane is delivered by specially approved fuel lines to a fuel lock and a converter/regulator. The converter/regulator changes the pressurized liquid to a low pressure vapor and meters fuel vapor delivery to a simple carburetor. The carburetor, responding the the engine vacuum, mixes fuel vapor with air and regulates delivery to the engine.

—— CAUTION ——

Close the fuel tank manual shut-off valve securely before performing any service, except idle adjustment, on a propane-fueled vehicle. If the fuel system is to be serviced, run the engine out of fuel after shutting the tank valve. If the engine continues to run more than 2–3 minutes, reseat the tank valve. Failure to close the shut-off valve and run the engine out of fuel could result in fuel leakage creating a fire hazard.

NOTE: Open the fuel valve slowly after completing service. Listen for the sound of fuel filling the lines. When the filling sound stops, open the valve fully. If the valve is opened too quickly, the sudden flow will cause the excess flow valve in the tank to block fuel delivery. Should the excess flow valve close, close the manual shut-off valve for 10 seconds. You will hear a faint click from inside the tank when the excess flow valve resets. Slowly reopen the shut-off valve.

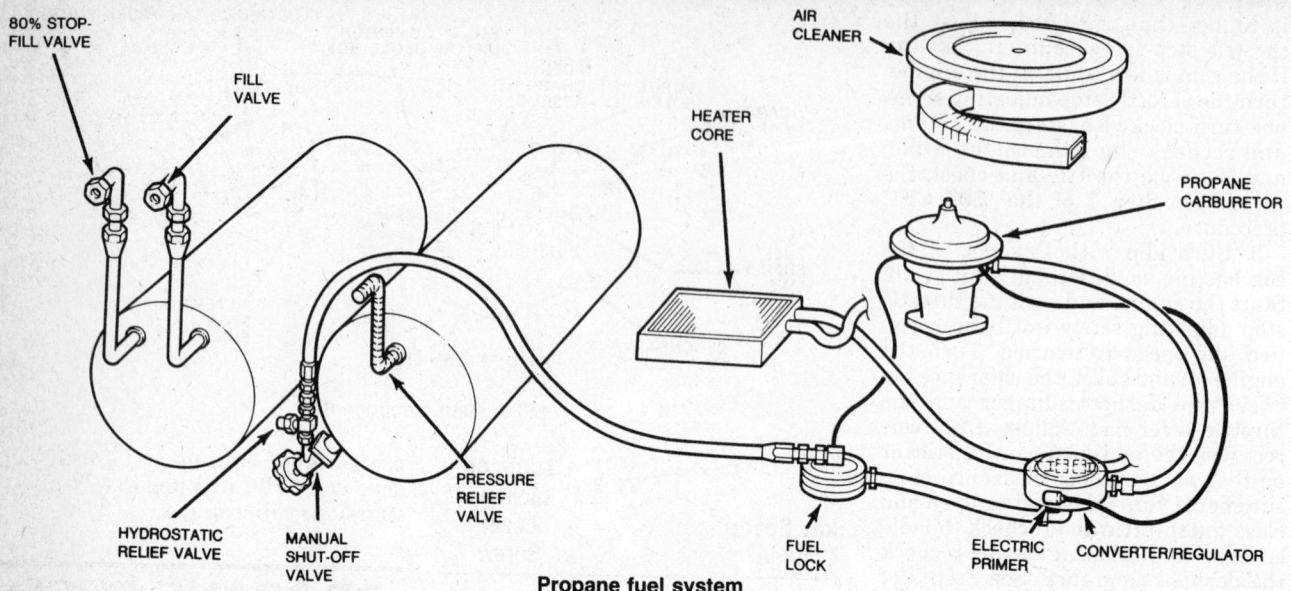

Propane fuel system

DIESEL FUEL SYSTEM

Fuel Filter

REMOVAL & INSTALLATION

1. Drain the fuel from the fuel filter by opening the vent screw on the top of the filter and then depressing the drain valve on the bottom of the filter.
2. Disconnect the Water-in-Fuel sensor connector.
3. Remove the filter cartridge using a standard oil filter wrench, if necessary.
4. Remove the protective cover.
5. Remove the drain valve from the old filter and install on the new filter.
6. Install the protective cover.
7. Coat the surface of the sealing gasket with engine oil and install the filter on the adapter. Turn the filter until the gasket contacts the sealing surface of the filter adapter.
8. Turn the filter an additional one-half turn.
9. Close the vent screw.
10. Start the engine and check for fuel leaks, tightening the filter further, if necessary.

Fuel Heater

REMOVAL & INSTALLATION

1. Disconnect the water-in-fuel sensor connector, fuel temperature sensor and fuel heater connector.
2. Drain the fuel from the fuel filter by opening the vent screw on the top

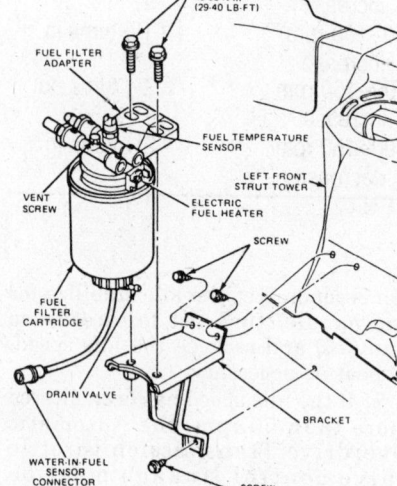

Diesel fuel filter assembly

of the filter and depressing the drain valve on the bottom of the filter.
3. Remove the filter cartridge using a standard oil filter wrench, if necessary.
4. Remove the fuel lines from the fuel filter adapter.
5. Remove the two bolts retaining the fuel heater/filter adapter to the bracket, and remove from vehicle.
6. Unscrew the fuel heater assembly from the fuel filter adapter.
7. Coat the seal with engine oil and install the fuel heater on the fuel filter adapter.
8. Position the fuel filter adapter (with fuel heater attached) to the bracket and install with two bolts. Tighten to 29–40 ft. lbs.
9. Coat the surface of the sealing gasket with engine oil and install the

filter on the adapter. Turn the filter until the gasket contacts the sealing surface of the filter adapter. Turn the filter an additional half turn.
10. Connect the water-in-fuel sensor, temperature sensor and fuel heater connectors.
11. Reconnect the fuel lines to the fuel filter and tighten the vent screw.
12. Start the engine and check for fuel leaks, tightening the filter further, if necessary.

BLEEDING THE FUEL SYSTEM

The fuel system must be bled, before starting the engine, whenever the fuel system has been opened or the vehicle has run out of gas.

1. Make sure that the vent screw is closed and that there is power to the fuel shutoff solenoid.
2. Turn the ignition switch to the ON position and let the electric lift pump run for one or two minutes. Crank the engine. If the engine starts, and runs correctly soon after it starts, STOP. Bleeding is complete. If the engine will not start or is running poorly, procede to the next step.
3. Loosen the coupling nuts on the injectors individually while the engine is running or cranking.

--- **CAUTION** ---
Use extreme care to prevent being struck by high pressure fuel.

4. If the engine runs correctly, Stop, if not, proceed to the next step.
5. If additional bleeding is required it should be done in the following order with a cranking or running engine.

- Fuel return line banjo bolt (labeled out).
- Injection pump distributor head plug bolt.
- Fuel Shut-Off solenoid.
- Injection nozzle fuel lines at injection pump.

6. Tighten all connections and check for leaks.

Injection Timing

NOTE: This procedure requires the use of special tools.

ADJUSTMENT

NOTE: Engine coolant temperature must be above 80°C (176°F) before injection timing can be checked and/or adjusted.

1. Disconnect the negative battery cable, located in the luggage compartment.
2. Remove the injection pump distributor head plug bolt and sealing washer.
3. Install Static Timing Gauge Adapter, Rotunda 014-00303, with Metric Dial Indicator, D82L-4201-A or equivalent so that indicator pointer is in contact with injection pump plunger.
4. Remove the timing mark cover from the transmission housing. Align timing mark (TDC) with pointer on rear engine cover plate.
5. Rotate the crankshaft pulley slowly, counterclockwise until the dial indicator pointer stops moving (approximately 30-50 degrees BTDC).

NOTE: There is a 40° BTDC timing mark on the flywheel.

6. Adjust the dial indicator to zero.

NOTE: Confirm that the dial indicator pointer does not move from zero by slightly rotating the crankshaft left to right.

7. Turn the crankshaft clockwise until the crankshaft timing mark aligns with the indicator pin. Dial indicator should read 0.04 ± 0.0008 in. If reading is not within specification, adjust as follows:
    a. Loosen injection pump bolts and nuts.
    b. Rotate the injection pump toward the engine to advance timing and away from the engine to retard timing. Rotate the injection pump until the dial indicator reads 0.04 ± 0.0008 in.
    c. Tighten the injection pump attaching nuts and bolts to 13-20 ft. lbs.
    d. Repeat Steps 5, 6 and 7 to

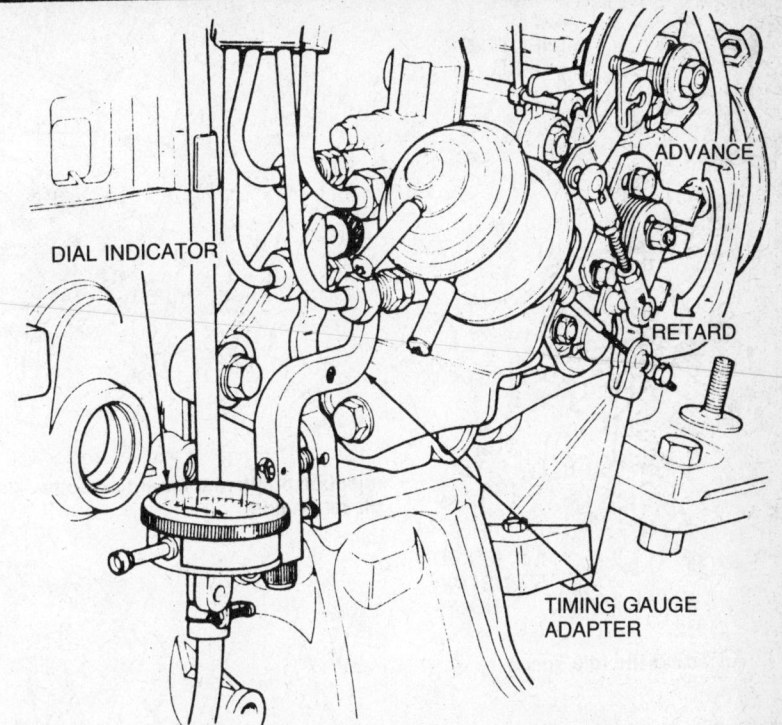

Installing timing gauge on diesel engine

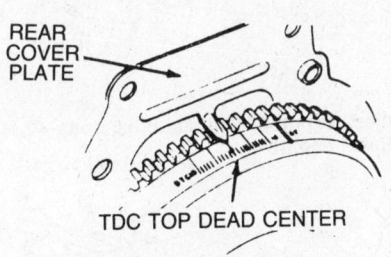

Typical flywheel timing mark

check that the timing is adjusted properly.
8. Remove the dial indicator and adapter and install the injection pump distributor head plug and tighten to 10-15 ft. lbs.
9. Connect the negative battery cable.
10. Run the engine and check and adjust idle rpm, if necessary. Check for fuel leaks.

Idle Speed

ADJUSTMENT

1. Place the transmission in NEUTRAL and apply the parking brake.
2. Bring the engine up to normal operating temperature. Stop the engine.

NOTE: Idle speed is measured with manual transmission in neutral.

3. Remove timing hole cover. Clean the flywheel surface and install reflective tape.
4. Check curb idle speed using Rotunda 099-00001 or equivalent. Curb idle speed is specified on the Vehicle Emissions Control Information (VECI) decal. Adjust to specification by loosening the lock nut on the idle speed bolt. Turn the idle speed adjusting bolt clockwise to increase, or counterclockwise to decrease engine idle speed. Tighten the lock nut.
5. Place the transmission in NEUTRAL. Rev the engine momentarily and recheck curb idle rpm. Readjust if necessary.
6. Turn A/C ON. Check the idle speed. Adjust to specification by loosening the nut on the A/C throttle kicker and rotating the screw.

Injection Pump and Lines

REMOVAL & INSTALLATION

1. Disconnect the battery ground cable. Drain the cooling system.
2. Remove the accessory drive belts.
3. Remove the fan and clutch assembly or electric motor and fan assembly.
4. Remove the camshaft drive belt.
5. Install Injection Pump Sprocket Aligning Pin T84P-9000-A or equiva-

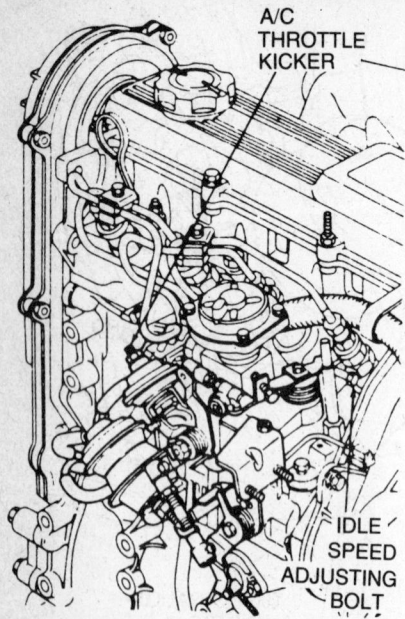

Adjusting the idle speed

lent and remove nut and washer attaching sprocket to the injection pump.

6. Install puller T67L-3600-A or equivalent and remove the sprocket. Remove the woodruff key from pump shaft.

7. Disconnect the clamp attaching the oil dipstick tube to the intake manifold, and position out of the way.

8. Disconnect the turbo pressure indicator switch connector. Remove the diagnostic plug bracket and position out of the way.

9. Loosen the clamp attaching the turbo crossover pipe boot to the intake manifold.

10. Remove the nuts attaching the intake manifold to cylinder head, and remove the intake manifold.

NOTE: To prevent fuel system contamination, cap all fuel lines and fittings.

11. Disconnect and cap the nozzle fuel lines at nozzles.

12. Remove the injection nozzle lines from injection pump using Fuel Line Nut Wrench T84P-9396-A or equivalent. Install caps on each end of each fuel line and pump fitting as it is removed and identify each fuel line accordingly.

13. Disconnect the coolant hoses from the idle speed boost housing.

14. Disconnect the electrical connectors to the fuel shut-off and cold start accelerator valves, micro-switch and fuel pressure switch.

15. Disconnect the nozzle return line at the injection pump.

16. Disconnect the fuel return hose

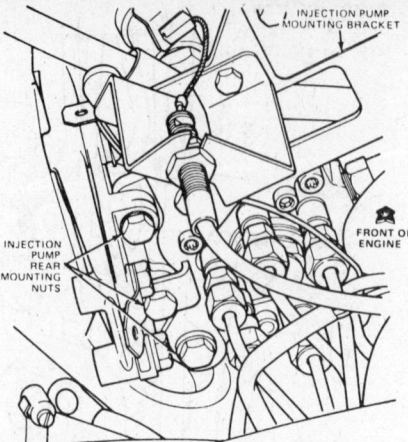

Injection pump rear mounting bolts—2.4L diesel engine

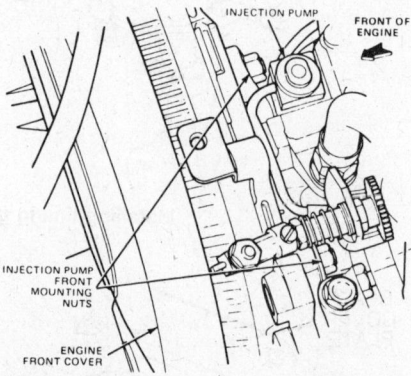

Injection pump front mounting bolts—2.4L diesel engine

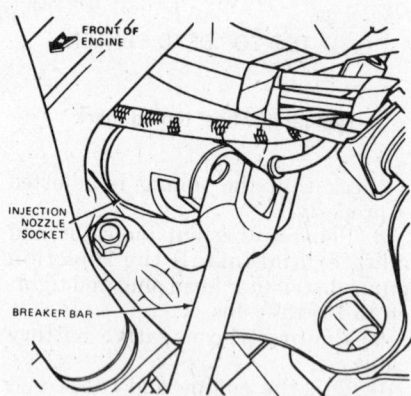

Injector nozzle removal and installation

from the fuel return line on the left fender apron.

17. Disconnect the fuel inlet hose from the fuel inlet line on the left fender apron.

18. Disconnect the vacuum hoses at the altitude compensation valve. Note position of hoses, so they may be returned to the original position.

19. Disconnect the throttle cable and speed control cable, if equipped, from the injection pump.

20. Remove the three nuts attaching the injection pump to mounting bracket.

21. Remove the two nuts attaching the injection pump to the engine front cover, and remove the injection pump.

22. Install the injection pump in position. Line up the mark on the front cover with the mark on the injection pump mounting boss. Install attaching nuts and bolts. Tighten to 14–17 ft. lbs.

23. Connect the throttle cable, and speed control cable, if so equipped.

24. Remove the protective caps and install the fuel inlet hose to the fuel inlet line on left fender apron. Connect the fuel return hose to the fuel return line on the left fender apron.

25. Connect the vacuum hoes to the altitude compensation valve. Refer to the VECI decal.

26. Connect the nozzle return line to the injection pump.

27. Connect the electrical connectors to the fuel pressure sensor, microswitch, cold start accelerator valve and fuel shut-off valve.

28. Connect the coolant hoses to the idle speed boost housing.

29. Install the fuel lines on injection pump, using Tool T84P-9396-A or equivalent, and tighten to 14–17 ft. lbs.

30. Connect the fuel lines to the nozzles and tighten to 14–17 ft. lbs.

31. Clean the intake manifold and cylinder head gasket mating surfaces. Position a new intake manifold gasket on the cylinder head, and install the intake manifold. Be sure the intake manifold inlet port is inserted into the turbo crossover pipe boot. Tighten attaching bolts to 14–17 ft. lbs. Tighten the clamp at the crossover pipe boot.

32. Install the diagnostic plug bracket on the cylinder head, and tighten to 14–17 ft. lbs.

33. Connect the turbo pressure indicator switch connector.

34. Position the oil dipstick tube to the intake manifold and install clamp.

35. Install the woodruff key in injection pump shaft.

36. Install the sprocket on injection pump. Install injection Pump Aligning Pin T84P-9000-A or equivalent, in sprocket. Install the sprocket attaching washer and nut and tighten to 33–36 ft. lbs.

37. Install and adjust camshaft drive belt.

38. Install the camshaft drive belt cover and tighten to 6–7 ft. lbs.

39. Install fan and clutch assembly or electric motor and fan assembly.

40. Install and adjust the accessory drive belts.

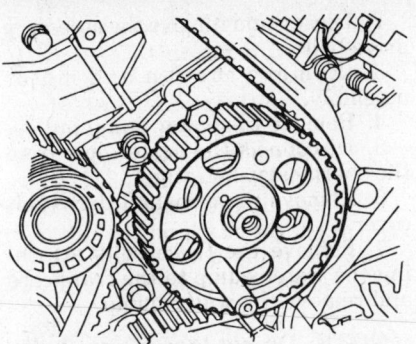

Injector pump aligning tool—2.4L diesel engine

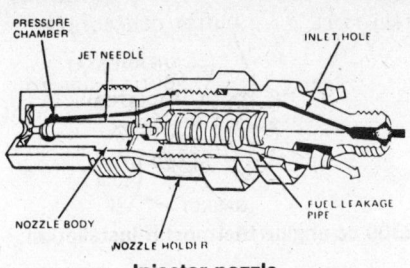

Injector nozzle

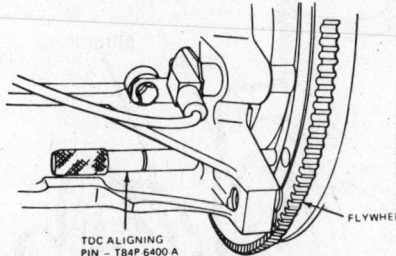

TDC aligning tool installation—2.4L diesel engine

41. Fill and bleed the cooling system.
42. Air bleed the fuel system.
43. Adjust the injection pump timing.
44. Connect the battery ground cable.
45. Start the engine and check for fuel, coolant and oil leaks.
46. Adjust the curb idle, fast idle and injection pump timing.

Fuel Shut-Off Valve

REMOVAL & INSTALLATION

1. Disconnect the battery ground cable.
2. Remove the nut attaching the electrical connector to the shut-off valve and remove the connector.
3. Remove the shut-off valve.

—— **CAUTION** ——
Piston and spring may fall out when removing valve.

4. Replace the O-ring and valve, and install the valve on the injection pump. Tighten to 11–18 ft. lbs.
5. Install the connector on the shut-off valve. Tighten nut to 3-3.5 ft. lbs.
6. Connect the battery ground cable. Run the engine and check for fuel leaks.

Injection Nozzles

REMOVAL & INSTALLATION

1. Pull off the leak oil lines from the injector nozzles.

NOTE: Make sure area around injector is clean.

2. Remove the fuel lines at the injectors and at the fuel injection pump with Fuel Line Wrench T84P-9527-A or equivalent. Cap all fuel lines and openings as the fuel lines are removed.

3. Unscrew the fuel injectors with Injector Nozzle Socket T84P-9527-A or equivalent. Note injector order for installation.

NOTE: On injectors with sensor, disconnect the sensor plug wires and guide sensor wires through Injector Nozzle Socket T84P-9527-A or equivalent, while installing tool on the injector.

4. Plug the cylinder block injector nozzle opening.
5. Clean the injector nozzle opening in the cylinder block.
6. Install new heat shields into the injection nozzle openings.
7. Apply a copper based, anti-sieze compound to the injector nozzle threads. Remove the protective plug in the cylinder block and install injector nozzles in original positions with Injector Nozzle Socket T84P-9527-A or equivalent. Tighten to 30–33 ft. lbs.

NOTE: On injectors with sensors, guide the sensor plug wire through socket before installing the injector nozzle. Reconnect the sensor wire after nozzle installation.

8. Remove the protective caps from the fuel lines, injector pump and injector nozzles and install fuel lines using Fuel Line Wrench T84P-9396-A or equivalent. Tighten to 15–18 ft. lbs.

Injection Nozzle Fuel Lines

REMOVAL & INSTALLATION

1. If all the fuel lines are being re-

moved, remove the intake manifold, and then remove all the fuel lines as an assembly.

NOTE: Do not remove the two clamps holding the fuel lines together.

2. Remove fuel line(s) at the injector nozzles and at the fuel injection pump with Fuel Line Wrench T84P-9395-A or equivalent. Cap all fuel lines and openings as fuel lines are removed.
3. If only one fuel line is being removed, remove the clamps holding fuel lines together and remove the fuel line.
4. If the fuel lines are being installed as an assembly, remove the protective caps and install fuel lines (with clamps installed) to the injector nozzles and injection pump using Fuel Line Wrench T84P-9395-A or equivalent.
5. If only one fuel line is being installed, remove protective caps and position fuel line to the injector nozzle, and injection line using Fuel Line Wrench T84P-9395-A or equivalent. Install clamps holding the fuel lines together.
6. Install the intake manifold if it was previously removed.

Electric Fuel Pump

REMOVAL & INSTALLATION

1. Disconnect the electric fuel pump electrical connector.
2. Remove the hose clamp on the inlet and outlet lines and remove the hoses from the fuel pump.
3. Remove the two fuel pump retaining screws and remove the fuel pump.
4. To install, reverse the removal steps. Tighten attaching screws to 9–11 ft. lbs.

ENGINE COOLING

Radiator

REMOVAL & INSTALLATION

1. Drain the cooling system.
2. Disconnect the upper, lower and overflow hoses at the radiator.
3. On automatic transmission equipped cars, disconnect the fluid cooler lines at radiator.

4. Depending on model; remove the two top mounting bolts and remove radiator and shroud assembly, or remove the shroud mounting bolts and position the shroud out of the way. If the air conditioner condenser is attached to the radiator, remove the retaining bolts and position the condenser out of the way. DO NOT disconnect the refrigerant lines.

5. Remove the radiator attaching bolts or top brackets and lift out the radiator.

6. If a new radiator is to be installed, transfer the petcock from the old radiator to the new one. On cars equipped with automatic transmissions, transfer the fluid cooler line fittings from the old radiator.

7. Position the radiator and install, but do not tighten, the radiator support bolts. On cars equipped with automatic transmissions, connect the fluid cooler lines. Then tighten the radiator support bolts or shroud and mounting bolts.

8. Connect the radiator hoses. Close the radiator petcock. Fill and bleed the cooling system.

9. Start the engine and bring to operating temperature. Check for leaks.

10. On cars equipped with automatic transmission, check the cooler lines for leaks and interference. Check the transmission fluid level.

Water Pump

REMOVAL & INSTALLATION

Gasoline Engines

1. Drain the cooling system.
2. Disconnect the negative battery cable.
3. On cars with power steering, remove the drive belt.
4. If the vehicle is equipped with air conditioning, remove the idler pulley bracket and air conditioner drive belt.
5. On engines with Thermactor, remove the belt.
6. Disconnect the lower radiator hose and heater hose from the water pump.
7. On cars equipped with a fan shroud, remove the retaining screws and position the shroud rearward.
8. Remove the fan, fan clutch and spacer from the engine, and if the car is equipped with an electric motor driven fan, remove the fan as an assembly for working clearance.
9. On 4-cylinders, remove the cam belt outer cover.
10. On cars equipped with "water pump mounted" alternators, loosen the alternator mounting bolts, remove the alternator belt and remove

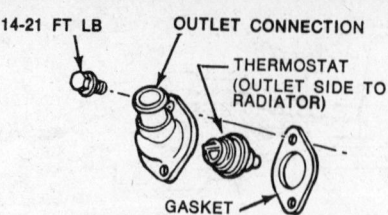

2300 cc engine thermostat installation

14-21 FT LB — OUTLET CONNECTION — THERMOSTAT (OUTLET SIDE TO RADIATOR) — GASKET

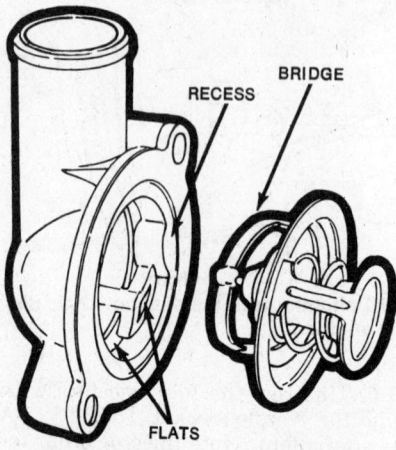

RECESS — BRIDGE — FLATS

V8 engine thermostat installation

the alternator adjusting arm bracket from the water pump. If interference is encountered, remove the air pump pulley and pivot bolt. Remove the air pump adjusting bracket. Swing the upper bracket aside. Detach the air conditioner compressor and lay it aside. Do not disconnect any of the A/C lines. Remove any accessory mounting brackets from the water pump.

11. Loosen the bypass hose clamp at the water pump, if equipped.

12. Remove the water pump mounting screws and remove the pump from the engine.

13. Clean any gasket material from the pump mounting surface. On engines equipped with a water pump backing plate; remove the plate, clean gasket surfaces, install a new gasket and plate on the water pump.

NOTE: The 250 6 cylinder engine originally uses a one-piece gasket for the cylinder front cover and water pump. Trim away the old gasket at the edge of the cylinder cover and replace with service gasket.

14. Remove the heater hose fitting from the old pump and install it on the new pump.

15. Coat both sides of the new gasket with a water-resistant sealer, then install the pump reversing the procedure.

Diesel Engine

1. Drain the cooling system.

2. Loosen and remove the accessory drive belts.

3. Remove the fan and motor assembly.

4. Remove the water pump pulley.

5. Disconnect the heater hose from the thermostat housing.

6. Remove the camshaft drive belt cover.

7. Remove the three bolts attaching the water pump to the crankcase and remove the water pump.

NOTE: Do not loosen cam belt.

8. Clean the gasket mating surfaces of the water pump and crankcase.

9. Install the water pump with a new gasket, on the crankcase and tighten bolts to 14–17 ft. lbs.

10. Install the camshaft drive belt cover and tighten bolts to 6–7 ft. lbs.

11. Connect the heater hose to the thermostat housing.

12. Install the water pump pulley and tighten bolts to 6–7 ft. lbs.

13. Install the fan and motor assembly.

14. Install and adjust the accessory drive belts.

Thermostat

REMOVAL & INSTALLATION

1. Open the drain cock and drain the radiator so the coolant level is below the coolant outlet elbow which houses the thermostat.

NOTE: On some models it will be necessary to remove the distributor cap, rotor and vacuum diaphragm in order to gain access to the thermostat housing mounting bolts.

2. Remove the outlet elbow retaining bolts and position the elbow sufficiently clear of the intake manifold or cylinder head to provide access to the thermostat.

3. Remove the thermostat and the gasket.

4. Clean the mating surfaces of the outlet elbow and the engine to remove all old gasket material and sealer. Coat the new gasket with water-resistant sealer. Install the thermostat in the coolant elbow. The thermostat must be rotated clockwise to lock it in position on all V8s. On 4-cylinders, be sure the full width of the heater outlet tube is visible within the thermostat port.

5. Install the outlet elbow and retaining bolts on the engine. Torque the bolts to 12–15 ft. lbs.

6. Refill the radiator. Run the engine at operating temperature and check for leaks. Recheck the coolant level.

Electro-Drive Cooling Fan

REMOVAL & INSTALLATION

Gasoline Engines

Various models are equipped with a bracket-mounted electric cooling fan that replaces the conventional water pump mounted fan. Operation of the fan motor is dependent on engine coolant temperature and air conditioner compressor clutch engagement. The fan will run only when the coolant temperature is approximately 108°F or higher or when the compressor clutch is engaged. The fan, motor and mount can be removed as an assembly after disconnecting the wiring harnesses and mounting bolts.

——— **CAUTION** ———

The cooling fan is automatic and may come on at any time without warning even if the ignition is switched OFF. To avoid possible injury, always disconnect the negative battery cable when working near the electric cooling fan.

Diesel Engine

1. Disconnect the battery ground cable.
2. Raise the vehicle and support on jackstands.
3. Remove the bolts and nuts attaching the mounting brackets to the radiator support.
4. Disconnect the electrical connector to fan.
5. Remove the bolts securing the hoodlatch to the radiator support and position the latch out of the way.
6. Remove the bolts and nuts attaching the mounting brackets to fan and motor assembly.
7. Remove the fan and motor assembly from the vehicle.
8. Position the fan and motor assembly in the vehicle.
9. Position the mounting brackets on the fan and motor assembly. Tighten the nuts to 4–5 ft. lbs.
10. Install the hood latch.
11. Connect the electrical connector to the fan and motor assembly.
12. Position the mounting brackets in the vehicle. Tighten the mounting bolts to 6–8 ft. lbs. Tighten mounting nuts to 4–5 ft. lbs.
13. Lower the vehicle.
14. Connect the battery ground cable.

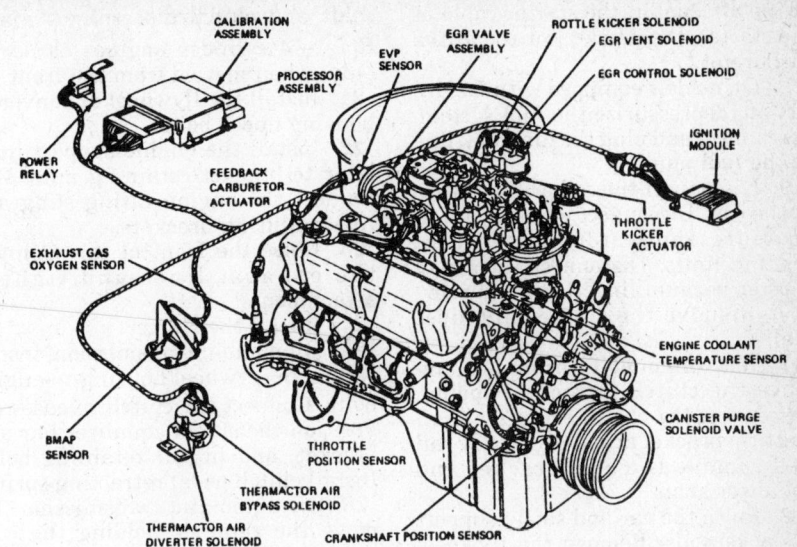

Typical EEC system components (EEC II shown)

EMISSION CONTROLS

For all emission control maintenance information, please refer to "Emission Controls" in the Unit Repair Section. Due to the complex nature of modern electronic engine control systems, comprehensive diagnosis and testing procedures fall outside the confines of this repair manual. For complete information on diagnosis, testing and repair procedures concerning all modern engine and emission control systems, please refer to *Chilton's Guide To Electronic Engine Controls*.

ENGINE MECHANICAL

Engine

REMOVAL & INSTALLATION

NOTE: Disconnect the negative battery cable before beginning any work. Always label all disconnected hoses, vacuum lines and wire harnesses, to prevent incorrect reassembly. Do not disconnect any air conditioning lines unless you are thoroughly familiar with A/C systems and the hazards involved; escaping refrigerant (freon) will freeze any surface it contacts, including skin and eyes. Have the system discharged professionally before required repairs are started.

Gasoline Engine

1. Scribe the hood hinge outline on the under-hood, disconnect the hood and remove.
2. Drain the entire cooling system and crankcase.
3. Remove the air cleaner, disconnect the battery at the cylinder head. On automatic transmission equipped cars, disconnect the fluid cooler lines at the radiator. On the four cylinder, remove the exhaust manifold shroud.
4. Remove the upper and lower radiator hoses and remove the radiator. If equipped with air conditioning, unbolt the compressor and position compressor out of the way with refrigerant lines intact. Unbolt and lay the refrigerant condenser forward without disconnecting refrigerant lines.

——— **CAUTION** ———

If there is not enough slack in the refrigerant lines to position the compressor out of the way, the refrigerant in the system must be evacuated (using proper safety precautions) before the lines can be disconnected from the compressor.

5. Remove the fan, fan belt and upper pulley. On models equipped with an electric cooling fan, disconnect the power lead and remove the fan and shroud as an assembly.
6. Disconnect the heater hoses from the engine. On four cylinder engines, disconnect the heater hose from the water pump and choke fittings.
7. Disconnect the alternator wires

at the alternator, the starter cable at the starter, the accelerator rod at the carburetor.

8. On models equipped with fuel injection, depressurize the fuel system. Disconnect and plug the fuel tank line at the fuel pump.

9. Disconnect the coil primary wire at the coil. Disconnect wires at the oil pressure and water temperature sending units. Disconnect the brake booster vacuum line, if so equipped.

10. Remove the starter and dust seal.

11. With manual transmission, remove the clutch retracting spring. Disconnect the clutch equalizer shaft and arm bracket at the underbody rail and remove the arm bracket and equalizer shaft.

12. Raise the car and safely support on jackstands. Remove the flywheel or converter housing upper retaining bolts.

13. Disconnect the exhaust pipe or pipes at the exhaust manifold or turbocharger. Disconnect the right and left motor mount at the underbody bracket. Remove the flywheel or converter housing cover. On models equipped, disconnect the engine roll dampner on the left front of the engine to frame.

14. On manual shift, remove the lower wheel housing bolts.

15. On automatic transmission, disconnect throttle valve vacuum line at the intake manifold and disconnect the converter from the flywheel. Remove the converter housing lower retaining bolts. On power steering, disconnect power steering pump from cylinder head. Remove the drive belt and wire steering pump out of the way. do not disconnect the hoses.

16. Lower the car. Support the transmission and flywheel or converter housing with a jack.

17. Attach an engine lifting hook. Lift the engine up and out of the compartment and onto workstand.

18. Place a new gasket on exhaust pipe flange.

19. Attach engine sling and lifting device. Lift engine from workstand.

20. Lower the engine into the engine compartment. Be sure the exhaust manifold(s) is/are in proper alignment with the muffler inlet pipe(s), and the dowels in the block engage the holes in the flywheel housing. On a car with automatic transmission, start the converter pilot into the crankshaft make sure converter studs align with flexplate holes. On manual transmissions, start the transmission main drive gear into the clutch disc. If the engine hangs up after the shaft enters, rotate the crankshaft slowly (with transmission in gear) until the shaft and clutch disc splines mesh. Rotate 4 cylinder engines clockwise only, when viewed from the front.

21. Install the flywheel or converter housing upper bolts.

22. Install the engine support insulator to bracket retaining nuts. Disconnect the engine lifting sling and remove lifting brackets.

23. Raise the front of car. Connect the exhaust lines and tighten attachments.

24. Install the starter.

25. On manual transmission, install remaining fywheel housing-to-engine bolts. Connect the clutch release rod. Position the clutch equalizer bar and bracket, and install retaining bolts. Install clutch pedal retracting spring.

26. On automatic transmission, remove the retainer holding the converter in the housing. Attach the converter to the flywheel. Install the converter housing inspection cover and the remaining converter housing retaining bolts.

27. Remove the support from the transmission and lower the car.

28. Connect the engine ground strap and coil primary wire.

29. Connect the water temperature gauge wire and the heater hose at coolant outlet housing. Connect the accelerator rod at the bellcrank.

30. On automatic transmission, connect the transmission filler tube bracket. Connect the throttle valve vacuum line.

31. On power steering, install the drive belt and power steering pump bracket. Install the bracket retaining bolts. Adjust the drive belt to proper tension.

32. Remove the plug from the fuel tank line. Connect the flexible fuel line and the oil pressure sending unit wire.

33. Install the pulley, belt, spacer, and fan. Adjust belt tension.

34. Tighten the alternator adjusting bolts. Connect the wires and the battery ground cable. On the four cylinder, install the exhaust manifold shroud.

35. Install the radiator. Connect radiator hoses. On air conditioned cars, install the compressor and condensor.

36. On automatic transmission, connect fluid cooler lines. On cars with power brakes. connect the brake booster line.

37. Install oil filter. Connect heater hose at water pump and carburetor choke (4 cyl).

38. Bring the crankcase to level with correct grade of oil. Run the engine at fast idle and check for leaks. Install the air cleaner and make final engine adjustments.

39. Install and adjust hood.

Diesel Engine

1. Disconnect the negative battery cable.

2. Disconnect the wiring assembly for the engine underhood light.

3. Scribe hinge mark locations and remove the hood.

4. Drain the cooling system. Drain the engine oil.

5. Remove the air cleaner assembly.

6. Remove the fan shroud attaching bolts and remove the fan shroud. Remove the engine cooling fan assembly.

7. Remove upper and lower radiator hoses.

8. Disconnect the transmission oil cooler tubes from the radiator fittings.

9. Disconnect the muffler inlet pipe.

10. Label and disconnect the vacuum hoses and wiring harnesses.

11. Disconnect the engine oil cooler hoses.

12. Disconnect the accelerator cable at the fuel injection pump.

13. Disconnect the fuel line from the tank to fuel injection pump.

14. Disconnect the transmission gear shift linkage.

15. Disconnect the battery ground cable at engine.

16. Remove the coolant expansion bottle and position it out of the way.

17. Disconnect the heater hoses at the dash panel (firewall).

18. Disconnect the wire to A/C compressor clutch.

19. Disconnect the power steering pump hose(s).

20. Disconnect the fuel line to the injectors.

21. Disconnect the wiring harness to instrument panel. Disconnect engine ground leads.

22. Install an engine support Tool D79F-6000-A or equivalent (bar and "J" hook or chain).

23. Raise the vehicle and safely support on jackstands.

24. Remove the muffler inlet pipe.

25. Remove the lower engine oil cooler bracket and brace.

26. Remove the stabilizer bar, bracket retaining bolts and position forward.

27. Remove the left hand front fender splash shield.

28. Disconnect the steering gear input shaft to steering column shaft coupling.

29. Remove the retainer nuts to the engine insulator supports.

30. Position a jack under the engine. Raise the engine assembly. Position the steering gear out of the way.

31. Lower the engine assembly.

32. Remove the converter housing access cover.

33. Remove the converter assembly retainer nuts.

34. Insert a pair of locking pliers in the converter housing to hold the converter in place during engine removal.

NOTE: Make sure that the upper jaw of the locking pliers contacts the converter while clamped to the converter housing. This will apply adequate pressure on the converter to prevent separation during engine movements and removal.

35. Remove No. 3 crossmember retainer nuts.

36. Remove the transmission gear shift lever bellcrank.

37. Raise the transmission.

38. Remove No. 3 crossmember re-

tainer bolts. Lower the transmission.

39. Remove engine to transmission converter housing retainer bolts.

40. Install crossmember (No. 3) retainer bolts.

41. Lower the vehicle.

42. Install engine lifting equipment.

43. Remove the engine support Tool D79T-6000-A or equivalent.

44. Remove the engine assembly.

45. Position engine and install on engine work stand and service as necessary.

46. Install engine lifting equipment. Raise the engine and install in vehicle.

47. Install engine support Tool D79T-6000-A or equivalent.

48. Remove the engine lifting equipment. Raise vehicle and safely support on jackstands.

49. The remainder of the installation procedure is in reverse order of removal.

Intake Manifold

REMOVAL & INSTALLATION
4 Cylinder Gas Engine
CARBURETOR EQUIPPED

1. Drain the cooling system and remove the air cleaner.

2. Disconnect the accelerator cable.

3. Disconnect and label the vacuum hoses at the carburetor.

4. Remove the engine oil dipstick.

5. Disconnect the heat tube at the EGR valve.

6. Disconnect and plug the fuel line at the carburetor.

7. Remove the bolt attaching the dipstick to the manifold.

8. Remove the PCV valve from the manifold.

9. Remove the two distributor cap screws and the distributor cap.

10. Remove the intake manifold at-

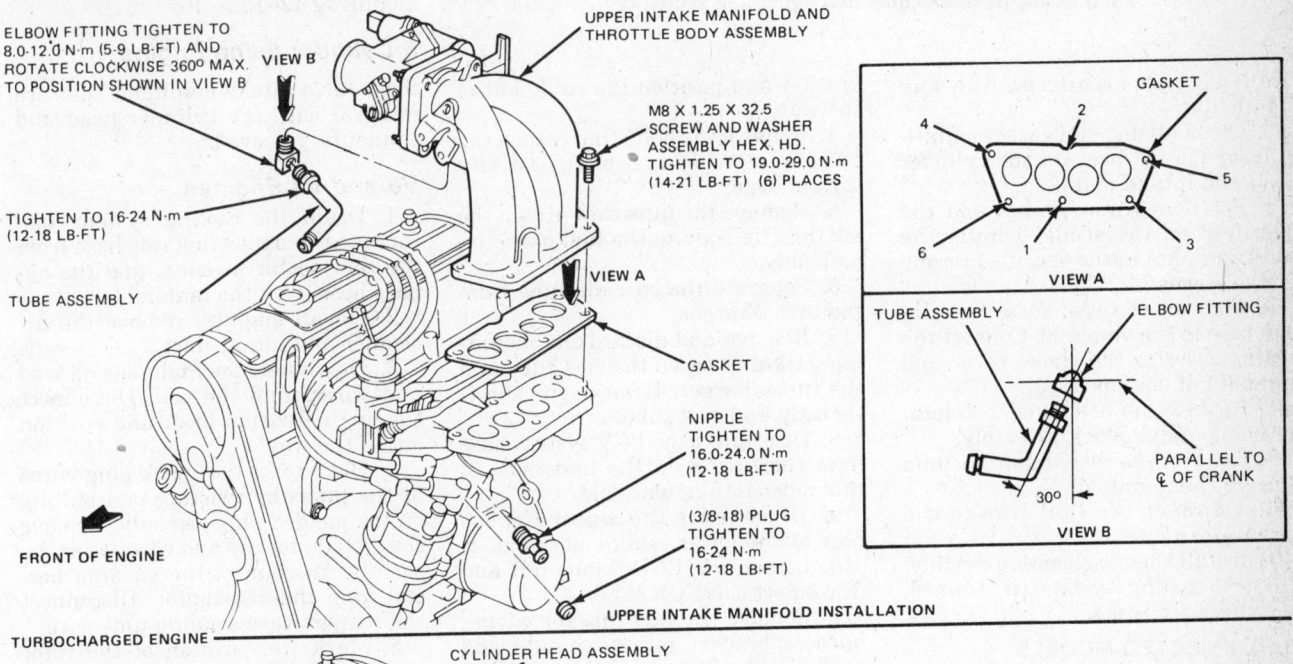

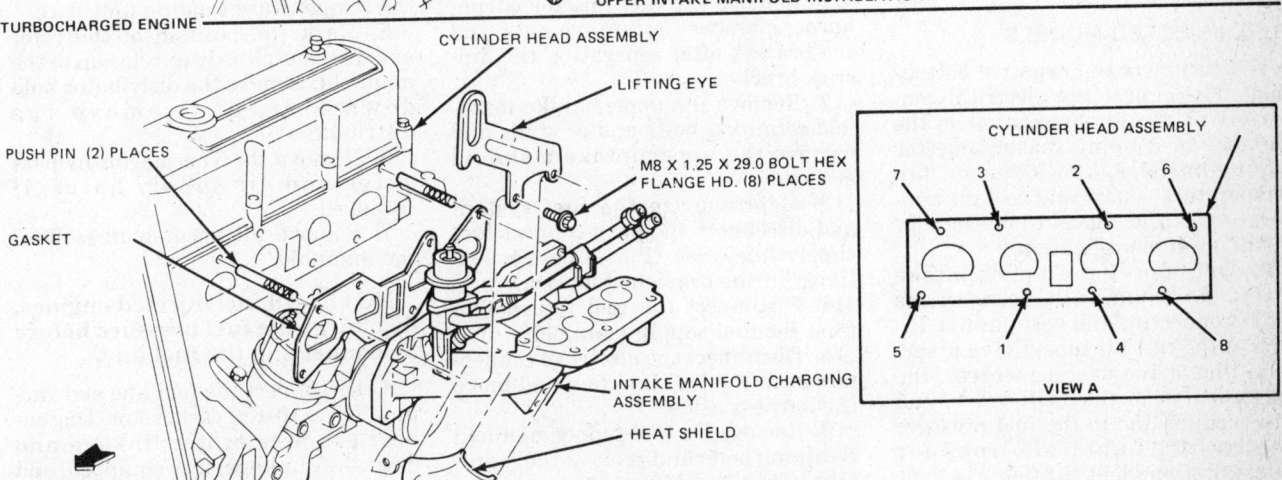

Multi-point injection, 2.3L (140) engine-upper and lower intake manifold removal and installation

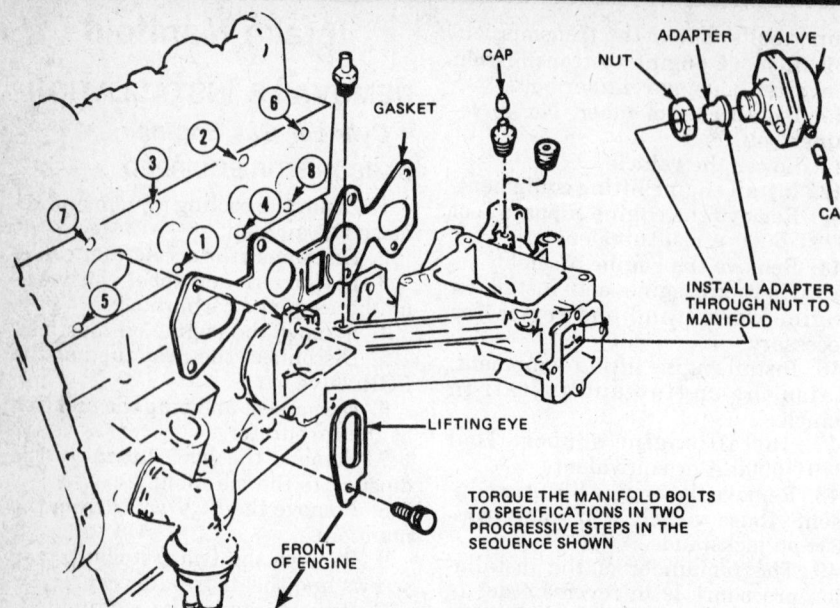

CAP   NUT   ADAPTER   VALVE

GASKET

INSTALL ADAPTER THROUGH NUT TO MANIFOLD

CAP

LIFTING EYE

TORQUE THE MANIFOLD BOLTS TO SPECIFICATIONS IN TWO PROGRESSIVE STEPS IN THE SEQUENCE SHOWN

FRONT OF ENGINE

2300 cc engine intake manifold tightening sequence

17. Remove the bottom and the top retaining bolts from the lower manifold. Remove the manifold.

18. Clean and inspect all mounting surfaces of the fuel charge manifolds and cylinder head.

19. Clean and oil all stud threads. Install a new mounting gasket over the studs.

20. Install the lower manifold to the cylinder head with lift bracket in position. Install the four upper manifold nuts fingertight. Install the four remaining nuts and tighten all nuts to 12–15 ft. lbs. See illustration for torque sequence.

21. Install the remaining components in the reverse order of removal. Fuel supply manifold bolts are tighten to 12–15 ft. lbs. Upper manifold mounting bolts 15–22 ft. lbs. Dipstick and injector wiring harness bolts 15–22 ft. lbs. Cast air tube to turbocharger 14–21 ft. lbs. Air throttle body mounting 12–15 ft. lbs.

6 Cylinder Inline Engine

Sixes have intake manifolds that are integral with the cylinder head and cannot be removed.

V6 and V8 Engines

1. Drain the cooling system, disconnect the upper radiator hose from the thermostat housing, and the by-pass hose from the manifold.

2. On all engines, remove the air cleaner and intake duct.

3. Disconnect the high tension lead and wires from the coil. Disconnect the engine wiring loom and position out of the way.

4. Disconnect the spark plug wires at the plugs by twisting and pulling on the molded plug cap only. Remove the distributor cap and wires as an assembly. Disconnect the vacuum hose(s) from the distributor. Disconnect the temperature sending unit wire.

5. Mark the position of the rotor and distributor body in relation to the manifold, remove the distributor hold down bolt, and remove the distributor.

6. Remove the Thermactor by-pass valve and air supply hoses, if equipped.

7. Remove all vacuum lines from the manifold.

NOTE: On fuel injected engines, discharge the fuel pressure before disconnecting the fuel line.

8. Disconnect the fuel line and vacuum hoses at the carburetor. Disconnect the accelerator linkage and downshift linkage, if so equipped, and position out of the way.

9. Disconnect the crankcase vent hose at the rocker cover.

taching bolts and remove the manifold.

11. Clean all dirt and gasket material from the surfaces on the cylinder head and intake manifold.

12. Position a new gasket and the manifold on the studs. Torque the bolts and nuts to the specified torque in two stages.

13. Connect the crankcase ventilation hose to the manifold. Connect the heater hoses to the choke cover and manifold, if equipped.

14. Replace the heat tube, accelerator cable and dipstick assembly.

15. Connect the distributor vacuum lines to the manifold.

16. Connect the fuel line to the carburetor.

17. Install the air cleaner assembly. Fill the cooling system, if drained, and check for leaks.

FUEL INJECTED MODELS

1. Disconnect the negative battery cable. Disconnect the electrical connectors at the air bypass valve, the throttle positioning sensor, injector wiring harness, knock sensor, fan temperature sensor and coolant temperature sensor. Label connectors for installation identification.

2. Disconnect the upper intake manifold vacuum fitting connections by disconnecting the vacuum line fitting at the cast air tube, the rear vacuum line at the dash panel tree, the vacuum line to the EGR valve, and the vacuum line to the fuel pressure regulator. Label all lines for reinstallation identification.

3. Disconnect the throttle linkage. Unbolt the accelerator cable from the

bracket and position the cable out of the way.

4. Remove the bolts that attach the cast air tube assembly to the turbocharger.

5. Remove the nuts that attach the air throttle body to the fuel charging assembly.

6. Separate the cast air tube from the turbocharger.

7. Remove and discard the mounting gasket between the cast tube and the turbocharger. Remove the throttle body and cast tube.

8. Disconnect the PCV system hose from the fitting on the underside of the upper intake manifold.

9. Disconnect the water bypass hose at the lower intake manifold.

10. Loosen the EGR flange nut and disconnect the EGR tube.

11. Remove the fuel injector wiring harness bracket retaining nuts and the bracket after separating the dipstick bracket.

12. Remove the upper intake manifold retaining bolts and or studs and remove the upper intake manifold assembly.

13. Depressurize the fuel system and disconnect the push-connect fuel supply line. (See "Push-Connect Fittings" in the previous fuel section).

14. Disconnect the fuel return line from the fuel supply manifold.

15. Disconnect the electrical connectors from the fuel injectors and move the harness aside.

16. Remove the fuel supply manifold retaining bolts and remove the manifold carefully. Injectors can be removed at this time by exerting a slight twisting/pulling motion.

10. If the car is air conditioned, re-move the compressor mounting brackets from the manifold and posi-tion the compressor out of the way. Do not disconnect any A/C hoses. Also, on these models, remove the coil.

11. Remove the intake manifold and carburetor as an assembly. Be careful not to damage any gasket sealing surfaces.

12. Clean the mating surfaces of the manifold, block, and heads. Apply a ⅛ in. bead of silicone seal to the four en-gine block-to-cylinder head mating surfaces.

13. Position the new end seals into place on the block, pressing the locat-ing tabs into place. Position new man-ifold gaskets into place on the heads, and apply a ⅛ in. bead of silicone seal to the four end seal-to-manifold gas-ket joints. Do not allow the sealer to fall into the engine "valley".

NOTE: The V6 232 cu. in. engine uses RTV sealant instead of end seals. Be sure to apply an even bead of sealant when installing the manifold.

14. Carefully lower the manifold into place. After it is positioned, run your finger around the seal area to be sure the seals are properly positioned. If they are not, remove the manifold and reposition the seals.

15. Torque the manifold to specifica-tion in three stages, according to the pattern given. The rest of installation is the reverse of removal. After instal-lation, run the engine to operating temperature and retorque the mani-fold bolts.

Diesel Engine

1. Disconnect the battery ground cable.

2. Remove the diagnostic plug bracket and position out of the way.

3. Disconnect the turbo boost pres-sure indicator connector.

4. Disconnect the oil dipstick tube clamp from intake manifold and posi-tion out of the way.

5. Loosen the clamp at the turbo crossover pipe boot.

6. Remove the bolts attaching the intake manifold to the cylinder head and remove the intake manifold.

7. Clean the intake manifold and cylinder head gasket mating surfaces.

8. Install the intake manifold on the cylinder head, with a new gasket, making sure the inlet port is installed in the turbo crossover pipe boot.

9. Tighten the intake manifold bolts to 14–17 ft. lbs., and tighten the crossover pipe boot clamp.

10. Connect the turbo boost pressure indicator switch connector.

11. Install the diagnostic plug

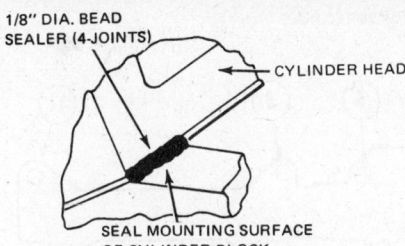

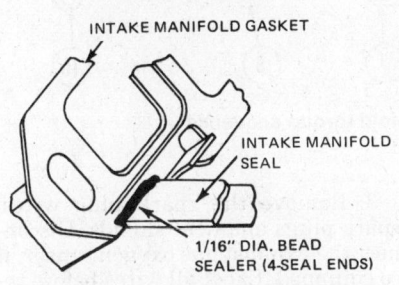

Intake manifold sealer application

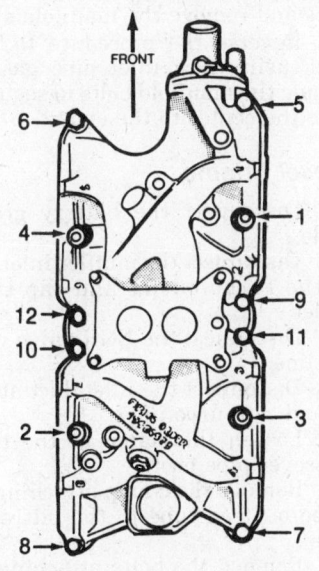

Intake manifold torque sequence—255, 302, 351W; V6 similar

bracket and tighten the bolts to 14–17 ft. lbs.

12. Connect the battery ground ca-ble. Start the engine and check for in-take leaks.

Exhaust Manifold

REMOVAL & INSTALLATION

4 Cylinder Gas Engine

NOTE: Refer to the Turbo-charger Unit Repair Section for reference when servicing high performance models.

NON-TURBOCHARGED MODELS

1. Remove the air cleaner. Remove the heat shroud from the exhaust manifold. Disconnect the hose from the thermactor check valves, if equipped. Disconnect the oxygen sen-sor wiring, on models equipped.

2. Place a block of wood under the exhaust pipe, then disconnect it from the manifold.

3. Remove the attaching nuts and remove the manifold from the head. Clean the mating surfaces.

4. Install a light coat of graphite grease on the exhaust manifold mat-ing surface and position the manifold on the cylinder head.

5. Install the attaching nuts and tighten them to the proper torque.

6. Connect the exhaust pipe to the manifold and remove the wood sup-port from under the pipe.

7. Install the air cleaner, and check valve hose and oxygen sensor wiring if present.

TURBOCHARGED MODELS

1. Remove the air cleaner duct as-sembly. Remove the turbocharger as-sembly. Remove the heat shroud from

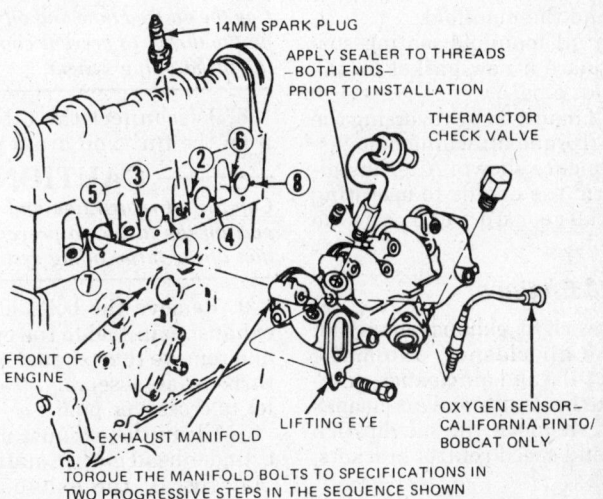

Exhaust manifold installation and tightening sequence— 2300 cc engine

INSTALL 3/8-16 STUD & WASHER ASSEMBLY – HOLES NUMBERED 4 & 5
3/8-16 X 2.62 BOLT – HOLES 3-6-7-8
3/8-16 X 1.12 BOLT – HOLES 1-2-9-10-11

FRONT OF ENGINE ➡

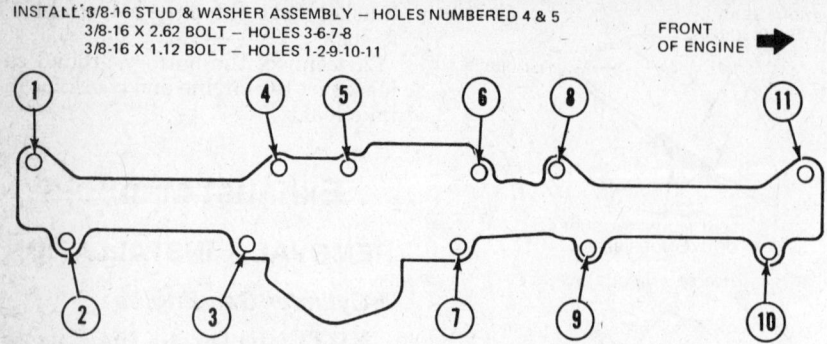

Six cylinder exhaust manifold torque sequence

the exhaust manifold. Disconnect the oxygen sensor wiring.

2. Remove the mounting nuts and bolts retaining the manifold to the cylinder head. Remove the manifold.

3. Install a light coat of graphite grease on the exhaust manifold mating surface and position the manifold on the cylinder head.

4. Install the attaching nuts and tighten them to 16–23 ft. lbs.

5. Complete the installation of the exhaust manifold in the reverse order of removal.

6 Cylinder Inline Engine

1. Remove the air cleaner and heat duct body.

2. Disconnect the muffler inlet pipe and remove the choke hot air tube from the manifold.

3. Remove the EGR tube and any other emission components which will interfere with manifold removal.

NOTE: Some models have a catalytic converter bolted to the manifold; the converter mounts on the four manifold flange studs.

4. Bend the exhaust manifold attaching bolt lock tabs back, remove the bolts and the manifold.

5. Clean all manifold mating surfaces and place a new gasket on the muffler inlet pipe.

6. Install manifold by reversing the procedure. Torque attaching bolts using the sequence shown. After installation, warm the engine to operating temperature and retorque to specifications.

V6 and V8 Engines

1. On the right exhaust manifold, remove the air cleaner, automatic choke heat tube and air cleaner, automatic choke heat tube and air cleaner heat ducts. Remove the oil dipstick and tube, and speed control brackets, if equipped.

2. Disconnect the exhaust manifold(s) from the muffler inlet pipe(s).

3. Remove the spark plug wires, spark plugs and heat shields. Disconnect the exhaust gas oxygen sensor, if so equipped. Label all wires before removal if they are not already marked.

4. Remove the manifold attaching bolts and remove the manifold(s).

5. Reverse the procedure to reinstall, using new inlet pipe gaskets. Torque the manifold bolts in sequence from the center to the ends.

Diesel Engine

1. Disconnect the battery ground cable.

2. Disconnect the muffler inlet pipe at the turbo outlet and cap turbo outlet.

3. Disconnect the EGR valve vacuum line.

4. Disconnect the inlet duct at turbo and cap turbo inlet.

5. Loosen the clamp at the turbo crossover pipe boot.

6. Remove the clamp attaching the turbo oil feed tube to the oil return tube.

7. Remove the bolts attaching the oil feed tube to the turbo.

— CAUTION —

Cap the oil feed tube and oil feed inlet port on the turbo, to prevent contamination of the turbo oiling system.

8. Disconnect the oil return line from the turbo oil drain port.

— CAUTION —

Cap the oil return line and the oil return port on the turbo, to prevent contamination of the turbo oiling system.

9. Remove the bolts attaching the exhaust manifold to the cylinder head and remove the exhaust manifold and turbo as an assembly. Cap turbo outlet to crossover pipe.

10. Clean the exhaust manifold and cylinder head gasket mating surfaces.

11. Install the exhaust manifold, with a new gasket, making sure the turbo outlet is installed in crossover

pipe boot. Tighten bolts to 14–17 ft. lbs., and tighten the crossover pipe boot clamp.

12. Remove the caps and install the oil feed line, with a new gasket, on the turbo oil inlet port. Tighten bolts to 14–17 ft. lbs.

13. Remove the caps and connect the oil return line to the turbo oil return port. Tighten fitting to 29–36 ft. lbs.

14. Install the oil feed tube to the exhaust manifold clamp and tighten to 6.5–7 ft. lbs.

15. Remove the cap and connect the inlet duct to the turbo inlet.

16. Remove the cap and connect the muffler inlet pipe to the turbo exhaust outlet. Tighten bolts to 31–35 ft. lbs.

17. Connect the EGR valve vacuum line.

18. Connect the battery ground cable.

19. Run the engine and check for intake, exhaust and oil leaks.

Turbocharger

NOTE: Refer to "Turbocharging" in the Unit Repair Section for servicing information.

REMOVAL & INSTALLATION

Gasoline Engine

THROUGH 1980

1. Allow engine to cool. Disconnect intake hose between turbocharger and injector unit.

2. Disconnect oil supply lines.

3. Unbolt exhaust pipe. Disconnect sensors.

4. Loosen and remove mounting bolts. Remove the turbocharger.

5. Install in the reverse order.

1983 AND LATER

A blow-through turbocharger system is used. Fuel is introduced downstream of the compressor and provides almost immediate response to accelerator pedal movement due to a reduction in fuel delivery time resulting in smoother performance during all driving situations. Accelerating the engine to top rpm when cold can damage the engine or turbocharger. Immediately shutting off the engine that has been operating at top rpm for an extended period of time can also damage the engine and/or turbocharger. Always permit the engine to idle for a short period of time before shut off. After an oil and filter change, disconnect the distributor wiring harness and crank the engine briefly to build up oil pressure before starting the engine. When installing the turbocharger, or after an oil and filter change, disconnect the distributor

feed harness and crank the engine with the starter motor until the oil pressure light on the dash goes out. Oil pressure must be up before starting the engine.

1. Turbocharger servicing is by replacement only. Prior to starting the removal procedure, clean the turbocharger and area around the turbo with a non-caustic solution. Maintain clean as possible working conditions while removing and installing the turbocharger.

2. When disconnecting lines and feed pipes always cover or plug openings.

3. Disconnect the negative battery cable. Drain the cooling system.

4. Remove the bolts retaining the cast air tube to the turbocharger. Loosen the clamp on the intake hose at the throttle body.

5. Label for identification and disconnect all vacuum hoses and tubes that will interfere with turbocharger removal.

6. Disconnect the PCV tube from the turbocharger air inlet elbow.

7. Remove the cast air tube and hose assembly from between the turbo and throttle body assembly.

8. Disconnect the electrical ground wire from the turbocharger air inlet elbow. Disconnect the water inlet line from the center of the turbocharger housing.

9. Disconnect the oil feed supply line from the turbo.

10. Disconnect the oxygen sensor connector at the turbocharger.

11. Raise and support the front of the vehicle.

12. Disconnect the exhaust pipe from the turbocharger. Disconnect the oil return line from the bottom of the turbocharger. Be careful when handling the oil line, do not kink or damage it. Disconnect the water inlet line from the turbo.

13. Remove the lower turbocharger bracket to engine bolt.

14. Lower the vehicle.

15. Remove the front turbocharger mounting nut.

16. Loosen the other turbocharger mounting nuts a little at a time and slide the turbo on the mounting studs until the nuts can be removed. Remove the turbocharger.

17. Clean all gasket mounting surfaces. Install the turbocharger in the reverse order of removal. Use new mounting gasket on the turbo and oil return line. Use new mounting nuts when installing the turbocharger. Torque is as follows: Lower bracket bolt; 28–40 ft. lbs. Oil return line; 14–21 ft. lbs. Exhaust pipe; 25–35 ft. lbs. Turbo mounting nuts; 28–40 ft. lbs. Cast air pipe to turbo; 15–22 ft. lbs.

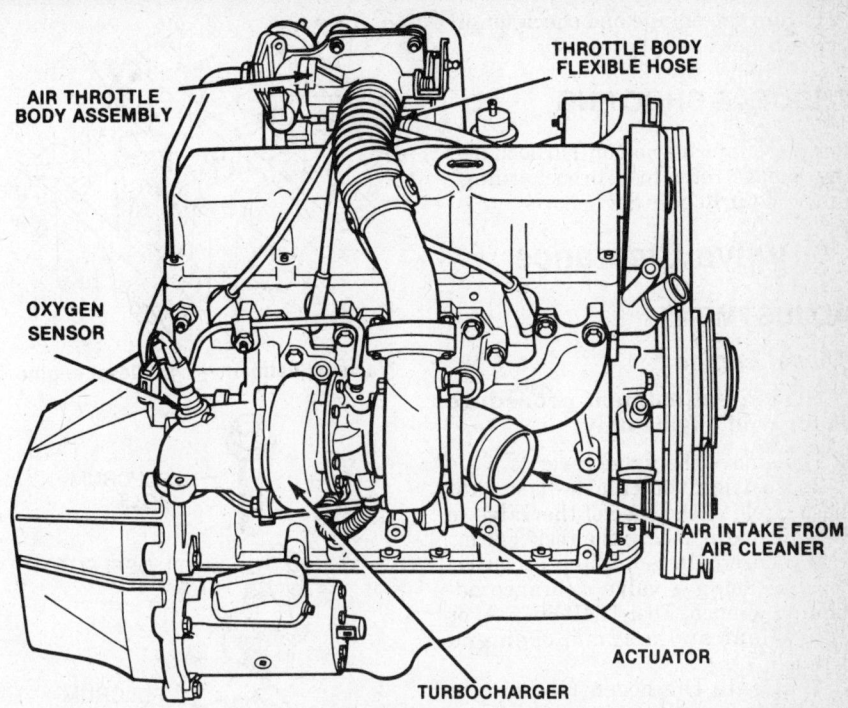

Typical turbocharger mounting

NOTE: When installing the turbocharger, or after an oil and filter change, disconnect the distributor feed harness and crank the engine with the starter motor until the oil pressure light on the dash goes out. Oil pressure must be up before starting the engine.

Diesel Engine

— CAUTION —

Do not accelerate the engine before engine oil pressure has been built up. Also, do not switch off the engine while it is running at high speed; the turbocharger will continue to spin for a long time without oil pressure. These conditions can damage the engine and/or turbocharger.

1. Remove the two bolts attaching the exhaust pipe to the turbocharger.

2. Remove the EGR tube and clamps.

3. Loosen the four hose clamps on the crossover tube and then remove tube.

4. Remove the air cleaner assembly and bellows. Cap turbocharger openings.

5. Remove the two oil supply line bolts on top of the turbocharger center housing.

6. Remove the clamp from oil lines.

7. Remove the oil return line.

8. Remove the bolt and sealing washers attaching the oil supply line to oil filter housing.

9. Disconnect and remove the EGR valve.

Torque Specs	Ft. Lbs. (in. lbs.)
EGR Valve	18
Hose Clamps	(15–22)
Oil Supply Line	15–18
To Turbo	
To Engine Block	26–33
Oil Return Line—To Turbo	15–18
Turbocharger-to-Exhaust Manifold	17–20
Turbocharger-to-Exhaust Pipe	17–20

10. Remove the four bolts attaching the turbocharger to the exhaust manifold and remove the turbocharger.

11. Clean the mating surfaces of the turbocharger and exhaust manifold.

12. Position the turbocharger on the exhaust manifold and install the four mounting bolts. Tighten to 17–20 ft. lbs.

13. Install the EGR valve. Tighten to 18 ft. lbs.

14. Install the oil supply line using new seals. Tighten the bolt to 26–33 ft. lbs.

15. Install the clamp retaining the oil lines.

16. Install the oil supply line bolts to the turbocharger housing and tighten to 15–18 ft. lbs.

17. Remove the protective caps from the turbocharger and install the air cleaner assembly and bellows.

18. Install the crossover tube. Tighten the hose clamps snug.

19. Install the EGR tube clamp.

20. Install the two bolts attaching the exhaust pipe to the turbocharger and tighten to 17–20 ft. lbs.

FORD MOTOR CO. REAR WHEEL DRIVE

21. Run the engine and check for oil and air leaks.

TROUBLE SHOOTING

For more information on Turbocharging, please refer to "Turbocharging" in the Unit Repair Section

Valve Clearance

ADJUSTMENT

Diesel Engine

NOTE: Adjustment procedure is for cold engine only.

1. Remove the valve cover.
2. Position the camshaft so that base circle of the lobe of the valve to be adjusted is facing the rocker arm.
3. Loosen the adjusting eccentric locknut using a valve clearance adjusting wrench, Tool T84P-6575-A, or equivalent and a 12mm open end wrench.
4. Rotate the eccentric using a small punch until the valve clearance is adjusted to specification: Intake: 0.012 in.; Exhaust: 0.016 in. Tighten eccentric locknut.
5. Repeat Steps 2, 3 and 4 for each valve.
6. Install the valve cover.
7. Start the engine and check for oil leaks.

Rocker Arms/Shafts

REMOVAL & INSTALLATION

4 Cylinder Gas Engine

1. Remove the valve cover and associated parts as required.
2. Rotate the camshaft so that the base circle of the cam is against the cam follower you intend to remove.
3. Remove the retaining spring from the cam follower, if so equipped.
4. Using a valve spring compressor tool, collapse the lash adjuster and/or depress the valve spring, as necessary, and slide the cam follower over the lash adjuster and out from under the camshaft.
5. Install the cam follower in the reverse order of removal. Make sure that the lash adjuster is collapsed and released before rotating the camshaft.

6 Cylinder Inline Engine

1. Remove the air cleaner and PCV line, and the accelerator control cable bracket.
2. Remove the rocker arm cover and gasket.
3. Remove the rocker shaft bolts, two turns at a time each, working from the ends toward the center.

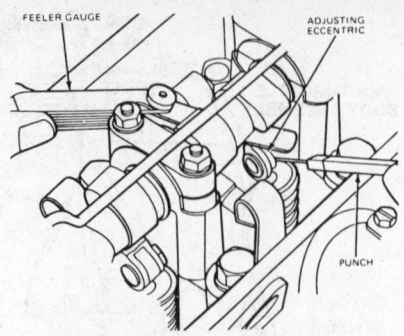

Valve adjustment—2.4L diesel engine

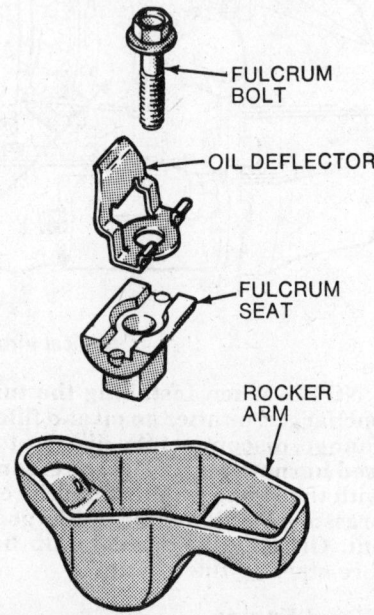

Late model V8 rocker arm mounting—V6 similar

4. Lift off the rocker shaft assembly. Keep the pushrods in order, if removed, for installation in their original positions.
5. Installation is the revers of removal. Torque the rocker shaft bolts, two turns at a time, working from the center toward the ends, to 30–35 ft. lbs.

V6 and V8 Engines

1. Right side: Disconnect the automatic choke heat chamber air inlet hose. Remove the air cleaner and duct. Remove the automatic choke heat tube (232, 302). Remove the PCV fresh air tube from the rocker cover, and disconnect the EGR vacuum amplifier hoses.
2. Remove the Thermactor by-pass valve and air supply hoses.
3. Disconnect the spark plug wires.
4. On the left side: Remove the wiring harness from the clips. Remove the rocker arm cover.
5. Remove the rocker arm stud nut or bolt. fulcrum seat and rocker arm.

6. Lubricate all parts with heavy SE oil before installation. When installing, rotate the crankshaft until the lifter is on the base of the cam circle (low point, no lift) and assemble the rocker arm. Torque the nut or bolt to 17–23 ft. lbs.

NOTE: Some later engines are using RTV sealant instead of valve cover gaskets. Always apply an even $\frac{1}{8}$ in. bead of sealant along the channel of the valve cover after cleaning.

Cylinder Head

REMOVAL & INSTALLATION

NOTE: The engine should be "overnight" cold before removing the cylinder head(s), to prevent warpage or distortion. Always label all disconnected hoses and wires to assure proper assembly.

4 Cylinder Gas Engine

1. Drain the cooling system.
2. Remove the air cleaner and the valve rocker cover.
3. Remove the intake and exhaust manifolds. The intake manifold, installed valves and sensors (if equipped), and carburetor can be removed as an assembly.
4. Remove the camshaft drive belt cover.
5. Loosen the drive belt tensioner and remove the drive belt.
6. Remove the water outlet from the cylinder head.
7. Remove the cylinder head bolts evenly, and remove the cylinder head.
8. Position a new cylinder head gasket on the block. Rotate the camshaft so that the locating pin is at the five o'clock position, to avoid valve damage.
9. Position the cylinder head and camshaft assembly on the block. Install the bolts finger tight, then torque to specifications in two stages.

NOTE: If difficulty in positioning the head on the block is encountered, guide pins may be fabricated by cutting the heads off two extra cylinder head bolts.

10. Set the crankshaft at TDC and be sure that the camshaft drive gear and distributor are positioned correctly at explained under Timing Belt Replacement.
11. Install the camshaft drive belt and release the tensioner. Rotate the crankshaft two full turns clockwise (facing the engine) to remove all slack from the belt. The timing marks

should again be aligned. Tighten the tensioner lockbolt and pivot bolts.

12. Install the camshaft drive belt cover.

TIME SAVER

The following is a method for replacing valve springs, oil seals or spring retainers without removing the cylinder head.

1. Purchase an air chuck with a spark plug hole adapter.
2. Remove the valve rocker cover. Remove the rocker arm from the valve to be worked on.
3. Remove the spark plug from the cylinder to be worked on.
4. Turn the crankshaft to bring the piston of this cylinder down, away from possible contact with the valve head. Sharply tap the valve retainer to loosen the valve lock.
5. Then turn the crankshaft to bring the piston in this cylinder to the Exact Top of its Compression Stroke.
6. Screw the air chuck fitting into the spark plug hole.
7. Hook up an air hose to the chuck and turn on the pressure (about 200 psi).
8. With a strong and constant supply of air holding the valve closed, compress the valve spring and remove the lock and retainer.

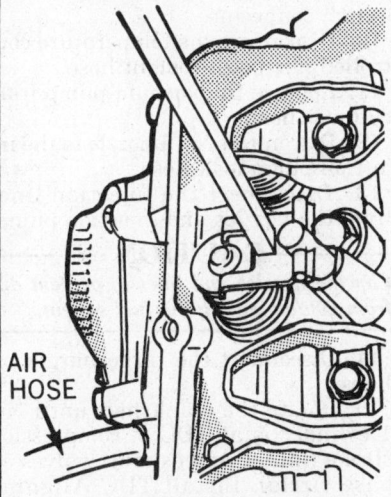

Compressing valve spring

9. Make the necessary replacements and reassemble.

NOTE: It is important that the operation be performed exactly as stated, in this order. The piston in the cylinder must be on exact top-center to prevent air pressure from turning the crankshaft.

13. Apply sealer to the water outlet and new gasket, and install.
14. Install the intake and exhaust manifolds.
15. Adjust the valve clearance.
16. Install a new valve cover gasket and install the valve cover.
17. Install the air cleaner and crankcase ventilation hose.
18. Refill the cooling system.

6 Cylinder Inline Engine

1. Drain cooling system, remove the air cleaner and disconnect the battery cable at the cylinder head.
2. Disconnect the exhaust pipe at the manifold end, swing the exhaust pipe down and remove the flange gasket.
3. Disconnect the fuel and vacuum lines from the carburetor. Disconnect the intake manifold line at the intake manifold.
4. Disconnect the accelerator and

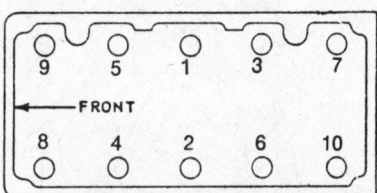

2300 cc head bolt tightening sequence

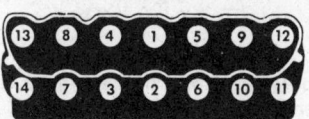

Cylinder head bolt tightening sequence —200, 250 6 cyl.

TIME SAVER

Frequently valves become bent or warped or their seats become blocked with carbon or other material. Left unattended, this can cause burnt valves, damaged cylinder heads and other expensive troubles. To detect leaking valves early, perform this test whenever the cylinder head is removed.

1. After removing head, replace spark plugs. Removing spark plugs before removing heads eliminates breakage.
2. Place head on bench with valves, springs, retainers and keys installed and combustion chambers up.
3. Pour enough gasoline in each combustion chamber to completely cover both valves. Watch combustion chambers for two minutes for any leakage.

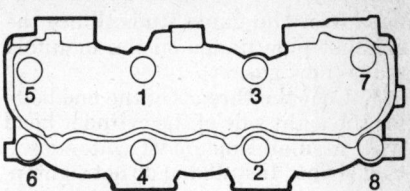

V6 cylinder head bolt torque sequence

retracting spring at the carburetor. Disconnect the transmission kickdown linkage, if equipped.

5. Disconnect the carburator spacer outlet line at the spacer. Disconnect the radiator upper hose and the heater hose at the water outlet elbow. Disconnect the radiator lower hose and the heater hose at the water pump.
6. Disconnect the distributor vacuum control line(s) at the distributor. Disconnect the gas filter line on the inlet side of the filter.
7. Disconnect and label the spark plug wires and remove the plugs. Disconnect the temperature sending unit wire.
8. Remove the rocker arm cover.
9. Loosen the rocker arm shaft attaching bolts and remove the rocker arm and shaft assembly. Remove the valve pushrods, in order, for installation in their original positions.
10. Remove one cylinder head bolt from each end of the head (at opposite corners) and install cylinder head guide studs for lifting the head. Remove the remaining cylinder head bolts and lift off the cylinder head. Do not pry under the cylinder head as damage to the mating surfaces can easily occur.

NOTE: To help in removal and installation of cylinder head, two 6 in. $^{7}/_{16}$ x 14 bolts with heads cut off at the head end slightly tapered and slotted, for installation and removal with a screwdriver, will reduce the possibility of damage during head replacement. These guide studs make a handy tool during head removal and gasket and head replacement.

11. Clean the cylinder head and block surfaces. Check for warpage and surface damage; correct as necessary.
12. Apply cylinder head gasket sealer to both sides of the new gasket and slide the gasket down over the two guide studs in the cylinder block.

NOTE: Apply gasket sealer only to steel shim head gaskets. Steel/asbestos composite head gaskets are to be installed without any sealer.

13. Carefully lower the cylinder

head over the guide studs. Place the exhaust pipe flange on the manifold studs (new gasket).

14. Coat the threads of the end bolts for the right side of the cylinder head with a small amount of water-resistant sealer. Install, but do not tighten, two head bolts at opposite ends to hold the head gasket in place. Remove the guide studs and install the remaining bolts.

15. Cylinder head torquing should proceed in three steps and in prescribed order. Tighten to 55 ft. lbs., then give them a second tightening to 65 ft. lbs. The final step is to 75 ft. lbs., at which they should remain undisturbed.

16. Lubricate both ends of the pushrods and install them in their original locations.

17. Apply lubricant to the rocker arm pads and the valve stem tips and position the rocker arm shaft assembly on the head. Be sure the oil holes in the shaft are in a down position.

18. Tighten all the rocker shaft retaining bolts to 30–35 ft. lbs. Start tightening in the middle and work evenly end to end.

19. Hook up the exhaust pipe.

20. Reconnect the heater and radiator hoses.

21. Connect the distributor vacuum line, the carburetor gas line and the intake manifold vacuum line on the engine.

22. Connect the accelerator rod and retracting spring. Connect the choke wire. Connect the transmission kickdown linkage.

23. Lightly lubricate the spark plug threads and install them. Connect spark plug wires and be sure the wires are all the way down in their sockets. Connect the temperature sending unit wire. Connect the negative battery cable.

24. Coat one side of a new rocker cover gasket with oil-resistant sealer. lay the treated side of the gasket on the cover and install the cover. Be sure the gasket seals evenly all around the cylinder head.

25. Fill the cooling system. Install the PCV system and air cleaner. Start the engine and check for leaks.

V6 and V8 Engines

1. Remove the valve covers and disconnect the negative battery cable.

2. Remove the intake manifold and carburetor assembly.

3. On cars equipped with air conditioning, remove the compressor from the engine and position it on one side, without disconnecting the refrigerant lines.

4. If removing the left cylinder head, on cars equipped with power

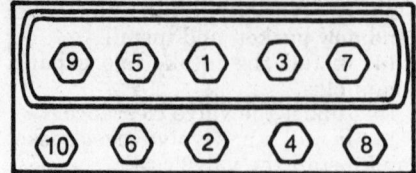

Cylinder head bolt tightening sequence all V8s

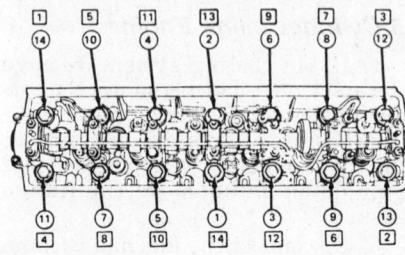

☐ REMOVAL ○ INSTALLATION

2.4L diesel engine cylinder head torque sequence

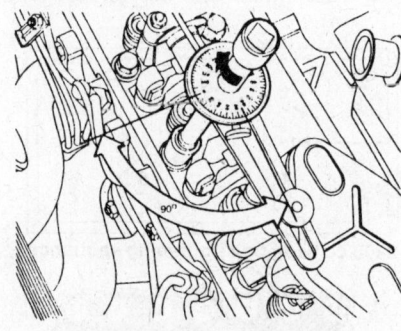

Cylinder head bolt tightening—2.4L diesel engine

steering, remove the pump, bracket, and drive belt and position to one side without disconnecting the lines. On cars with Thermactor emission control system, disconnect the hose from the air manifold on the left cylinder head.

5. If removing the right cylinder head, remove the alternator mounting bracket bolt and spacer, ignition coil, and air cleaner inlet duct. On cars equipped with Thermactor emission control, remove the air pump and bracket. Disconnect the hose from the right cylinder head.

6. Disconnect the exhaust manifolds from the exhaust pipes.

7. Loosen the rocker arm stud nuts or bridge bolts so that the arms can rotate to the side to clear the pushrods. Remove the pushrods. Keep them in order for installation in their original positions.

8. Remove the cylinder head bolts and lift off the cylinder head. On some 351 engines, it may be necessary to remove the exhaust manifold to gain access to the lower cylinder head bolts.

9. Reverse the procedure for installation taking care to follow the specified torque sequence.

Diesel Engine

1. Disconnect the battery ground cable.

2. Drain the cooling system. Disconnect the heater hose(s).

3. Loosen and remove accessory drive belts.

4. Remove the valve cover.

5. Disconnect the diagnostic connectors.

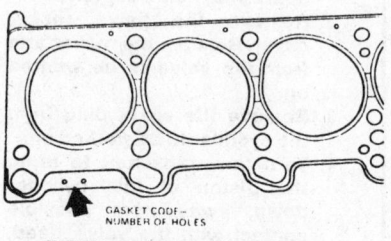

GASKET CODE—
NUMBER OF HOLES

HIGHEST PISTON PROTRUSION OF ALL 6 PISTONS mm	CYL. HEAD GASKET CODE NO OF HOLES	THICKNESS OF CYL HEAD GASKET mm
0.60 – 0.70	1	1.4
0.70 – 0.85	2	1.5
0.85 – 1.00	3	1.6

Cylinder head gasket identification—2.4L diesel engine

6. Disconnect the coolant temperature switch and glow plug connector.

7. Disconnect the breaker hose and bracket.

8. Remove the clamp attaching the oil dipstick tube to the intake manifold and position out of the way.

9. Disconnect the boost pressure switch connector.

10. Disconnect the temperature controlled idle boost coolant hose.

11. Remove the vacuum pump from cylinder head.

12. Disconnect No. 1 nozzle to the injection pump leak hose.

13. Disconnect the injection lines from the nozzles and injection pump.

CAUTION
Cap the nozzles and lines to prevent dirt from contaminating the fuel system.

14. Disconnect the turbocharger oil lines.

15. Rotate the crankshaft until No. 1 cylinder is at TDC of compression stroke (intake and exhaust valves on base circle). Install TDC Aligning Pin, T84P-6400-A or equivalent.

16. Loosen the camshaft drive sprocket retaining bolt.

17. Loosen the camshaft drive belt tensioning roller nut and bolt, and remove drive belt.

18. Loosen the cylinder head bolts in sequence, and remove cylinder head.

19. Clean gasket sealing surfaces on the cylinder head and crankcase.

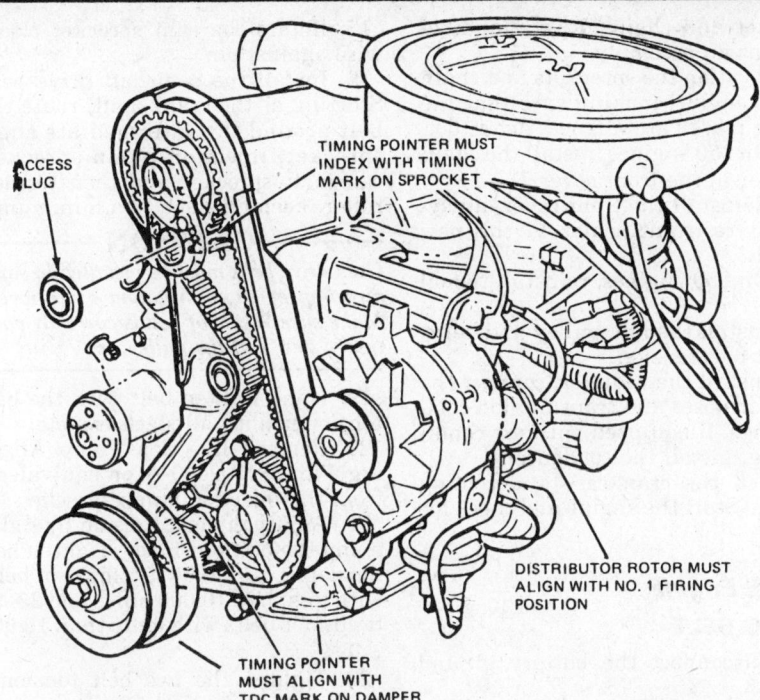

ACCESS PLUG

TIMING POINTER MUST INDEX WITH TIMING MARK ON SPROCKET

DISTRIBUTOR ROTOR MUST ALIGN WITH NO. 1 FIRING POSITION

TIMING POINTER MUST ALIGN WITH TDC MARK ON DAMPER

Crankshaft, camshaft, and distributor timing marks—2300 cc engine

20. Check for cylinder head warpage.

CAUTION

Use care when cleaning gasket surfaces. Slight scoring of these surfaces can cause leakage due to high compression pressures.

21. Clean the top of each piston.
22. Using a dial indicator D82L-4201-A and Piston Height Gauge D84P-6100-A or equivalent, measure the amount the piston top extends above crankcase gasket surface as follows: Mount the dial indicator and bracket with dial indicator tip on piston. Rotate the crankshaft to position piston at TDC, using dial indictor. Zero the dial indicator with tip on crankcase. Move the tip to the front of the piston. Record measurement. Move the tip to the rear of the piston. Record measurement. Repeat this procedure for each cylinder. Average the two readings for each cylinder. Using the measurement of highest piston, refer to the chart provided and select correct cylinder head gasket. Clean carbon and oil deposits from the cylinder head bolts.

CAUTION

Keep oil and/or antifreeze from entering cylinder head bolt holes. If either enters bolt holes, carefully blow out with compressed air. The presence of oil and/or antifreeze in bolt holes could result in insufficient cylinder head bolt tightening, or a cracked crankcase.

24. Position the correct cylinder head gasket on the crankcase.
25. Carefully lower the cylinder head onto the crankcase, using care not to damage the gasket.
26. Install and tighten the cylinder head bolts, in sequence to 36–43 ft. lbs. Wait 15 minutes and tighten the bolts, in sequence to 65–69 ft. lbs.
27. Install and adjust the drive belt.
28. Connect the turbocharger oil lines and tighten to 14–17 ft. lbs.
29. Connect the nozzle high pressure lines to the nozzles and injection pump. Tighten to 14–18 ft. lbs., using fuel line wrench.
30. Connect No. 1 nozzle to the injection pump leak hose.
31. Install the vacuum pump on the cylinder head and tighten to 6–7 ft. lbs.
32. Connect the temperature controlled, idle boost coolant hose.
33. Connect the radiator hoses to the cylinder head.
34. Connect the oil pressure switch connector.
35. Install the oil dipstick tube.
36. Install the breather hose and bracket.
37. Connect the coolant temperature switch and glow plug connectors.
38. Connect the diagnostic connectors.
39. Install the valve cover.
40. Install and adjust accessory drive belts.
41. Connect the heater hose(s).
42. Fill and bleed coolant system. Connect the battery ground cable.

OVERHAUL

For all cylinder head overhaul procedures, please refer to "Engine Rebuilding" in the unit repair section.

Timing Chain/Belt and Tensioner

REMOVAL & INSTALLATION

4 Cylinder Gas Engine

Should the camshaft drive belt jump timing by a tooth or two, the engine could still run; but very poorly. To visually check for correct timing of the crankshaft, auxiliary shaft and the camshaft follow this procedure:

There is an access plug provided in the cam drive belt cover so that the camshaft timing can be checked without removing the drive belt cover. Remove the access plug, turn the crankshaft until the timing mark on the crankshaft damper indicates TDC, and observe that the timing mark on the camshaft drive sprocket is aligned with the pointer on the inner belt cover. Also, the rotor of the distributor must align with the No. 1 cylinder firing position.

NOTE: Never turn the crankshaft in the opposite direction of normal rotation. Backward rotation of the crankshaft may cause the timing belt to slip and alter the timing.

CAUTION

After an procedure requiring removal of the rocker arms, each lash adjuster must be fully collapsed after assembly, then released. This must be done before the camshaft is turned.

TIMING BELT

1. Set the engine TDC as described for checking valve timing. The crankshaft and camshaft timing marks should align with their respective pointers and the distributor rotor should point to the No. 1 plug tower.
2. Loosen the adjustment bolts on the alternator and accessories and remove the drive belts. To provide clearance for removing the camshaft belt, remove the fan and pulley.
3. Remove the belt outer cover.
4. Remove the distributor cap from the distributor and position it out of the way.
5. Loosen the belt tensioner adjustment and pivot bolts. Lever the tensioner away from the belt and retighten the adjustment bolt to hold it away.
6. Remove the crankshaft bolt and

pulley. Remove the belt guide behind the pulley.

7. Remove the camshaft drive belt.

8. Install the new belt over the crankshaft pulley first, then counterclockwise over the auxiliary shaft sprocket and the camshaft sprocket. Adjust the belt fore and aft so that it is centered on the sprockets.

9. Loosen the tensioner adjustment bolt, allowing it to spring back against the belt.

10. Rotate the crankshaft two complete turns in the normal rotation direction to remove any belt slack. Turn the crankshaft until the timing check marks are lined up. If the timing has slipped, remove the belt and repeat the procedure.

11. Tighten the tensioner adjustment bolt to 14–21 ft. lbs., and the pivot bolt to 28–40 ft. lbs.

12. Replace the belt guide and crankshaft pulley, distributor cap, belt outer cover, fan and pulley, drive belts and accessories. Adjust the accessory drive belt tension. Start the engine and check the ignition timing.

6 Cylinder Inline Engines

1. Drain the cooling system and crankcase.

2. Disconnect the upper radiator from the intake manifold and the lower hose from the water pump. On cars with automatic transmission, disconnect the cooler lines from the radiator.

3. Remove the radiator, fan and pulley, and engine drive belts. On models with air conditioning, remove the condenser retaining bolts and position the condenser forward. Do not disconnect the refrigerant lines.

4. Remove the crankshaft pulley bolt and use a puller to remove the vibration damper.

5. On 200 cu. in. engines remove the cylinder front cover retaining bolts and front oil pan bolts and gently pry the cover away from the block. On 250 engines, it is necessary to remove the oil pan before removing the front cover.

6. With a socket wrench of the proper size on the crankshaft pulley bolt, gently rotate the crankshaft in a clockwise direction until all slack is removed from the left side of the timing chain. Next, turn the crankshaft in a counterclockwise direction to remove all the slack from the right side of the chain. Force the left side of the chain outward with the fingers and measure the distance between the reference point and the present position of the chain. If the distance exceeds ½ in., replace the chain and sprockets.

7. Crank the engine until the timing marks are aligned as shown in the illustration. Remove the bolt, slide

sprocket and chain forward and remove as an assembly.

8. Position the sprockets and chain on the engine, making sure that the timing marks are aligned, dot to dot.

9. On 250 engines, install the chain snubber in the front cover.

10. Reinstall the front cover, applying oil resistant sealer to the new gasket.

11. On 250 engines, reinstall the oil pan.

12. Install the fan, pulley and belts. Adjust belt tension.

13. Install the radiator, connect the radiator hoses and transmission cooling lines. If equipped with air conditioning, install the condenser.

14. Fill the crankcase and cooling system. Start the engine and check for leaks.

Diesel Engine

TIMING BELT

1. Disconnect the battery ground cable.

2. Drain the cooling system.

3. Remove the accessory drive belts.

4. Remove the fan assembly and water pump assembly.

5. Remove the vibration damper and pulley.

6. Disconnect the heater hose from the thermostat housing.

7. Remove the four bolts attaching the camshaft drive belt cover to crankcase, and remove the cover.

8. Remove the rocker cover.

9. Rotate the engine until No. 1 cylinder is at TDC on compression stroke (intake and exhaust valves on base circle).

10. Install TDC Aligning Pin T84P-6256-A or equivalent.

NOTE: Flat side of nut or cam position tool should be facing down.

11. Loosen the camshaft sprocket bolt.

12. Using a piece of chalk, or similar marker, mark the direction of engine rotation on drive belt, unless a new belt is to be installed.

13. Loosen the two bolts on the belt tensioner.

14. Remove the camshaft drive belt.

15. Insert a 0.098 in. (2.5mm) thick feeler gauge blade between Cam Positioning Tool T84P-6256-A or equivalent, at the right front corner of the gasket mating surface of the cylinder head if using a new drive belt or a drive belt used with less than 10,000 miles.

16. Install Injection Pump Aligning Pin T84P-9000-A or equivalent, through injection pump sprocket.

17. Rotate the cam sprocket clockwise against pin.

18. Install the camshaft drive belt. Starting at the crankshaft, route the belt around the intermediate shaft sprocket, injection pump sprocket, camshaft sprocket and then tension roller, keeping slack to a minimum.

—— CAUTION ——

Used drive belts must be installed in same direction of engine rotation as removed. Make sure V side of belt is correctly positioned in V's of the pulley.

19. Hand tighten belt with the belt tensioner until all slack is gone.

20. Remove Injection Pump Aligning Pin T84P-9000-A or equivalent, from the injection pump sprocket.

21. Adjust the belt tension by tightening the belt tensioner. Tighten belt tensioner to 34–36 ft. lbs. on belts with less than 10,000 miles and 23–25 ft. lbs. for belts with more than 10,000 miles.

22. Tighten the two belt tensioner holding bolts to 15–18 ft. lbs.

23. Tighten the camshaft sprocket to 41–47 ft. lbs.

24. Remove the Cam Positioning Took, T84P-6265-A or equivalent.

25. Install the camshaft drive belt cover and tighten bolts to 6–7 ft. lbs.

26. Connect the heater hose to the thermostat housing.

27. Install the vibration damper.

28. Install fan and water pump pulley assembly.

29. Install and adjust accessory drive belts.

30. Fill and bleed the cooling system.

31. Connect the battery ground cable.

32. Run the engine and check for oil and coolant leaks.

33. Check the injection pump timing.

Front Cover

REMOVAL & INSTALLATION

1. Disconnect the battery ground cable.

2. Drain the cooling system.

3. Loosen and remove the accessory drive belts.

4. Remove the engine cooling fan.

5. Remove the vibration damper.

6. Disconnect the heater hose from thermostat housing.

7. Remove the four bolts attaching the camshaft drive belt cover to the crankcase and remove the cover.

8. Remove the camshaft drive belt.

9. Remove the bolts attaching the intermediate shaft sprocket using

Holding Tool T84P-6316-A or equivalent.

NOTE: Be sure Allen head screws are aligned with holes in intermediate shaft sprocket.

10. Remove the vibration damper flange and sprocket retaining bolt and remove the flange and sprocket using puller, T67L-3600-A or equivalent.

11. remove the three oil pan-to-front cover attaching bolts. Loosen, but DO NOT REMOVE, the remaining oil pan bolts.

12. Remove the six bolts attaching the front cover to the crankcase, and remove cover.

13. Clean the front cover and crankcase gasket mating surfaces.

14. Inspect and replace the crankshaft and intermediate shaft oil seals, if necessary.

15. If the oil pan gasket is damaged, install a new pan gasket.

16. Install the new front cover gasket.

NOTE: Coat the areas where front cover gasket meets oil pan gasket with a ¼ in. bead of RTV Sealant, D64Z-19562-A or equivalent sealer. RTV Sealant should be applied immediately prior to front cover installation. When applying RTV Sealant always use the bead size specified and join the components within 15 minutes of application. After this amount of tie the sealant begins to "set-up" and its sealing effectiveness may be reduced.

17. Position the front engine cover on the crankcase, and tighten the 6mm bolts to 6.5–7 ft. lbs.

18. Install the three oil pan-to-front cover attaching bolts. Tighten the oil pan bolts to 6.5–7 ft. lbs.

19. Position the vibration damper flange and sprocket on crankshaft, with the shoulder toward front of vehicle.

20. Position the intermediate shaft sprocket on the intermediate shaft, guiding the locating pin into the bore.

21. Install Holding Tool T84P-6316-A or equivalent.

NOTE: Align Allen head screws in tool with holes in intermediate shaft.

22. Install and tighten the vibration damper flange and sprocket bolt to 282–311 ft. lbs.

23. Install and tighten the intermediate shaft sprocket bolt to 40–47 ft. lbs. Remove Tool T84P-6316-A or equivalent.

24. Install and adjust the camshaft drive belt.

25. Install the camshaft drive belt cover and tighten bolts to 6–7 ft. lbs.

26. Connect the heater hose to thermostat housing.

27. Install the vibration damper and pulley. Tighten to 16–17 ft. lbs.

28. Install the fan assembly.

29. Install and adjust the accessory drive belts.

30. Connect the battery ground cable.

31. Start the idle engine. Check for oil leaks.

OIL SEAL REPLACEMENT

1. Remove the engine front cover.

2. Using an arbor press, press old seal(s) out of the front cover.

3. Position the new seals on the front cover and install, using T84P-6019-A for crankcase seal, or T84P-6020-A or equivalent for intermediate shaft seal.

4. Lubricate the sealing lips with engine oil.

5. Install engine front cover.

Timing Gears/ Sprockets

REMOVAL & INSTALLATION

V6 and V8 Engines

1. Drain cooling system, remove air cleaner and disconnect the battery.

2. Disconnect radiator hoses and remove the radiator.

3. Disconnect heater hose at water pump. Slide water pump by-pass hose clamp toward the pump.

4. Loosen alternator mounting bolts at the alternator. Remove the alternator support bolt at the water pump. Remove Thermactor (air) pump on all engines so equipped. If equipped with power steering or air conditioning. unbolt the component, remove the belt, and lay the pump aside with the lines attached.

5. Remove the fan, spacer, pulley and drive belt.

6. Drain the crankcase.

7. Remove pulley from crankshaft

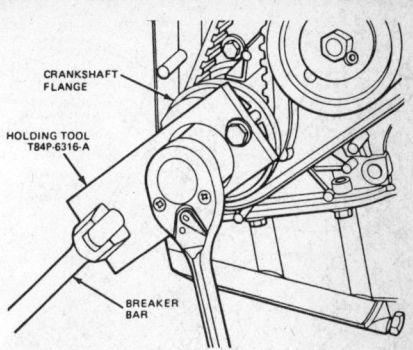

Crankshaft flange removal—2.4L diesel engine

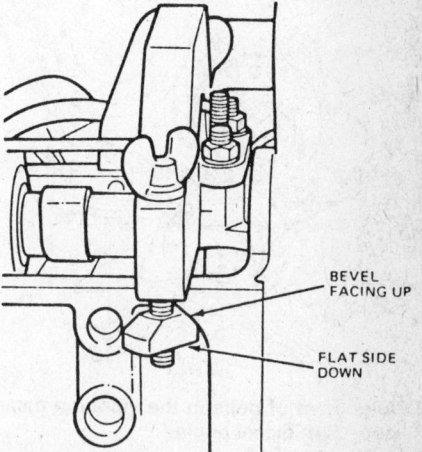

Cam positioning tool nut—2.4L diesel engine

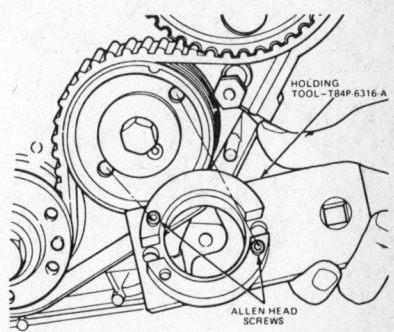

Intermediate shaft allen head screws—2.4L diesel engine

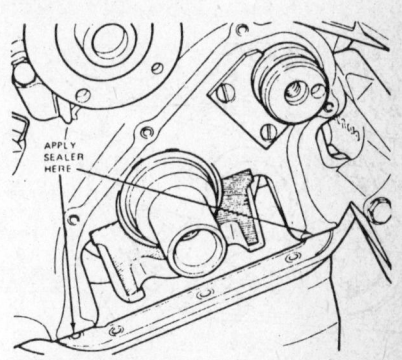

Front cover gasket sealing areas—2.4L diesel engine

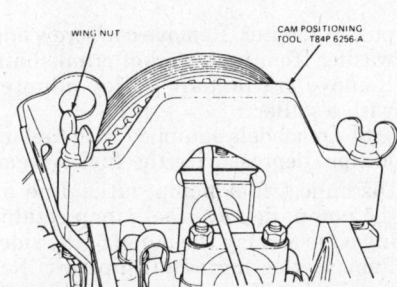

Cam positioning tool—2.4L diesel engine

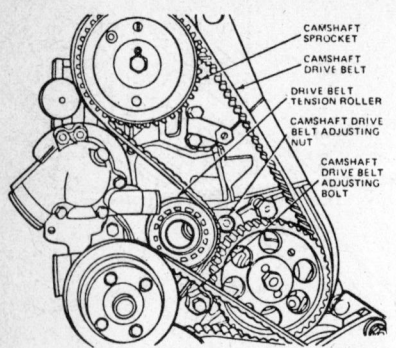

Camshaft drive belt installation—2.4L diesel engine

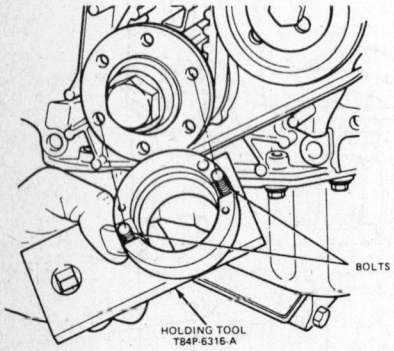

Alignment of bolts in the vibration damper—2.4L diesel engine

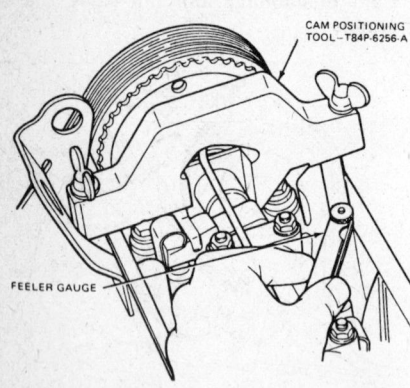

Cam positioning tool with feeler gauge—2.4L diesel engine

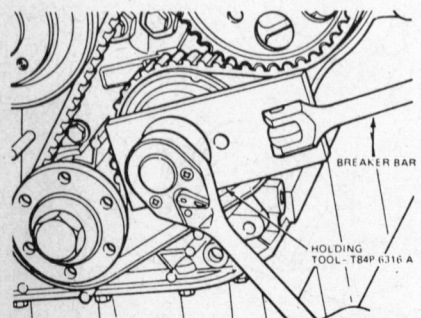

Intermediate shaft sprocket removal—2.4L diesel engine

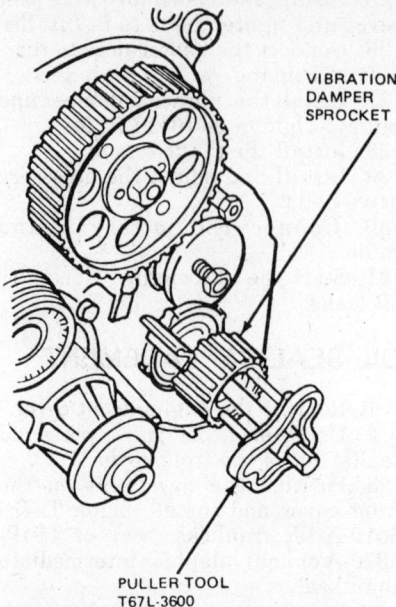

Vibration damper sprocket removal—2.4L diesel engine

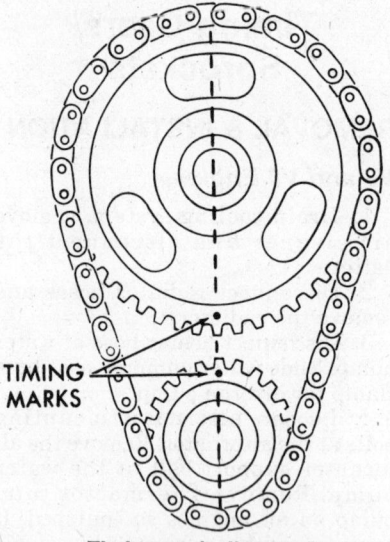

Timing mark alignment

Rear Spring details for Aspen, 1980 and later Cordoba, Diplomat, LeBaron to 1981, 1981 and later Imperial, Mirada, Volare, 1982 New Yorker 5th Avenue, 1983 and later New Yorker 5th Avenue and 1982 and later Gran Fury

pulley adapter. Remove cap screw and washer from front end of crankshaft. Remove crankshaft pulley adapter with a puller.

8. On models equipped with fuel injection, depressurize the fuel system. Disconnect fuel pump outlet line at the pump. Remove fuel pump retaining bolts and lay the pump to the side. Remove the engine oil dipstick. Remove the distributor on V6 232 engines.

NOTE: On the 232 V6 engines, it is necessary to drop the oil pan before the front cover cam be removed.

9. Remove the front cover attaching bolts. On the 232 V6 engine, remove the water pump and front cover as an assembly.

10. Remove the crankshaft oil slinger if so equipped. On the 232 V6 engine, remove the camshaft thrust button and spring.

11. Check timing chain deflection, using the procedure outlines in Step 6 of the six cylinder cover and chain removal.

12. Crank engine until sprocket timing marks are aligned as shown in the valve timing illustration.

13. Remove crankshaft sprocket cap screw, washers, and fuel pump eccentric. Slide both sprockets and chain forward and off as an assembly.

14. Position sprockets and chain on the camshaft and crankshaft with both timing marks dot to dot on a centerline. Install fuel pump eccentric, washers and sprocket attaching bolt. Torque the sprocket attaching bolt to 40–45 ft. lbs.

15. Install the crankshaft front oil slinger.

NOTE: When replacing the front cover on the 232 V6 engine, RTV sealer is used. Apply an even 1/8 in. bead on the cover "gasket" surface.

16. Clean front cover and mating surfaces of old gasket material. Install a new oil seal in the cover. Use a seal driver tool, if available. Oil the lips of the seal to prevent damage.

17. Coat a new cover gasket with sealer and position it on the block.

NOTE: On all engines, trim away the exposed portion of the oil pan gasket flush with the cylinder block. Cut the position and required portion of a new gasket to the oil pan, applying sealer to both sides of it. On 232 V6 engines, after installing the cylinder front cover, install the oil pan using a new gasket.

18. Install front cover, using a crankshaft-to-cover alignment tool. Coat the threads of the attaching bolts with sealer. Torque attaching bolts to 12–15 ft. lbs.

19. Install fuel pump, connect fuel pump outlet tube.

20. Install crankshaft pulley adapter and torque attaching bolt. Install crankshaft pulley.

21. Install water pump pulley, drive belt, spacer and fan.

22. Install alternator support bolt at

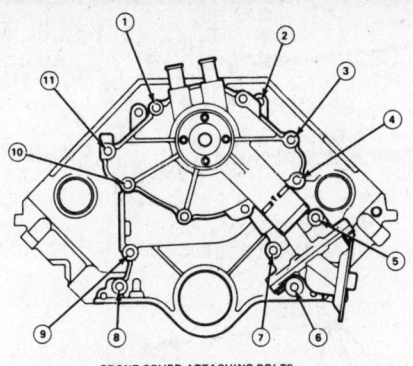

FRONT COVER ATTACHING BOLTS

Front cover attaching bolts—V6 (232) engine

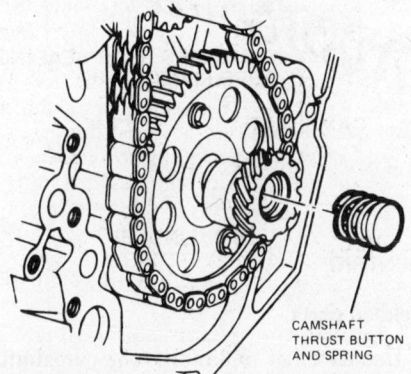

CAMSHAFT THRUST BUTTON AND SPRING

Camshaft thrust button and spring—V6 (232) engine

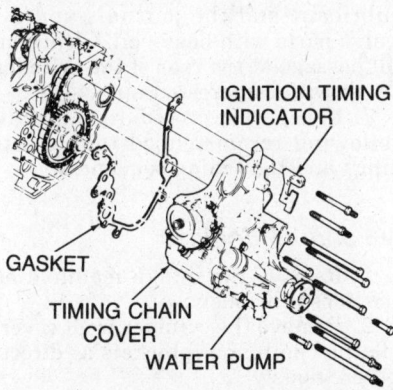

IGNITION TIMING INDICATOR

GASKET

TIMING CHAIN COVER

WATER PUMP

Front cover removal and installation—V6 (232) engine

the water pump. Tighten alternator mounting bolts. Adjust drive belt tension. Install Thermactor pump is so equipped.

23. Install radiator and connect all coolant and heater hoses. Connect battery cables.

24. Refill cooling system and the crankcase. Install the dipstick.

25. Start engine and operate at fast idle.

26. Check for leaks, install air cleaner. Adjust ignition timing and make all final adjustments.

OIL SEAL/COVER SEAL REPLACEMENT

NOTE: It is recommended to replace the cover seal any time the front cover is removed.

1. With the cover removed from the car, drive the old seal from the rear of cover with a pinpunch. Clean out the recess in the cover.

2. Coat the new seal with grease and drive it into the cover until it is fully seated. Check the seal after installation to be sure the spring is properly positioned in the seal.

Camshaft

REMOVAL & INSTALLATION

4 Cylinder Gas Engine

NOTE: The camshaft can be replaced with the cylinder head still mounted on the engine in the vehicle, or with the cylinder head removed from the vehicle.

1. Disconnect the negative battery cable. Drain the cooling system. Remove the air cleaner assembly.

2. Label and remove all wires, electrical harnesses, vacuum lines and cables that will interfere with valve cover removal.

3. On fuel injected models, depressurize the system and remove necessary equipment that interferes with valve cover removal.

4. Remove the alternator and mounting brackets as an assembly and position to the side.

5. Remove the upper and lower radiator hoses. Remove the fan, motor and mounting shroud as an assembly.

6. Remove the valve cover.

7. Set the engine at No. 1 cylinder TDC on the compression stroke. Remove the timing belt.

8. Raise and support the front of the vehicle on jackstands. Remove the right and left engine mount through bolts and joint to bracket retaining bolts.

9. Place a block of wood on a floor jack an raise the engine carefully as high as it will go. Place blocks of wood between the engine mounts and No. 2 crossmember pedestal. Lower the jack and lower the vehicle to the ground.

10. Remove the rocker arms.

11. Remove the camshaft drive gear attaching bolt and washer, and remove the gear and belt guide plate.

12. The camshaft is removed through the front of the cylinder head after removing the front cam bearing seal. Use a new seal during assembly.

13. Reverse the removal procedure to install the camshaft and cylinder head (if removed).

NOTE: Coat the camshaft with oil before sliding it into the cylinder head. Apply a coat of sealer or teflon tape to the cam drive gear bolt before installation.

--- **CAUTION** ---

After any procedure requiring removal of the rocker arms, each lash adjuster must be fully collapsed after assembly, then released. This must be done before the camshaft is turned.

AUXILIARY SHAFT REMOVAL

1. Remove the camshaft drive belt cover.

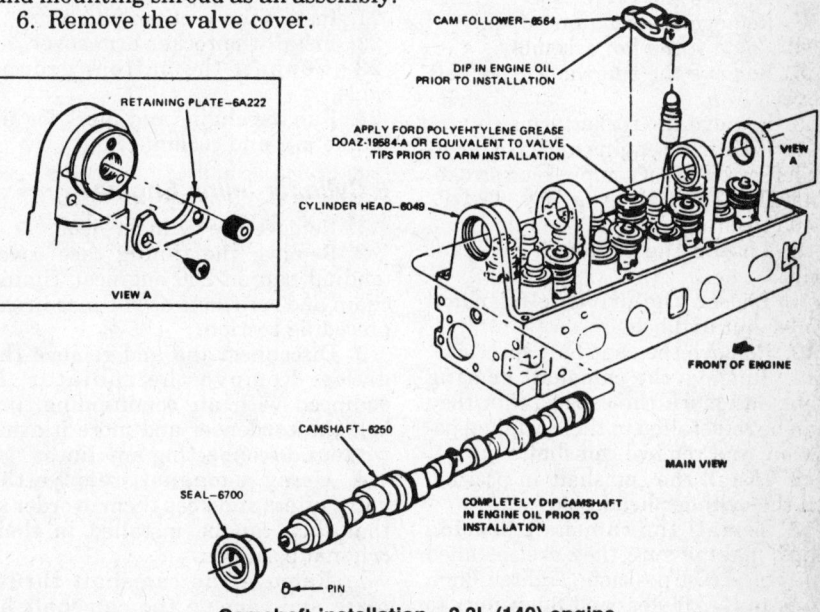

RETAINING PLATE—6A222

VIEW A

CAM FOLLOWER—6564

DIP IN ENGINE OIL PRIOR TO INSTALLATION

APPLY FORD POLYEHTYLENE GREASE DOAZ-19584-A OR EQUIVALENT TO VALVE TIPS PRIOR TO ARM INSTALLATION

CYLINDER HEAD—6049

VIEW A

FRONT OF ENGINE

CAMSHAFT—6250

SEAL—6700

PIN

COMPLETELY DIP CAMSHAFT IN ENGINE OIL PRIOR TO INSTALLATION

MAIN VIEW

Camshaft installation—2.3L (140) engine

2. Remove the drive belt. Remove the auxiliary shaft sprocket. A puller may be necessary to remove the sprocket.

3. Remove the distributor and fuel pump.

4. Remove the auxiliary shaft cover and thrust plate.

5. Withdraw the auxiliary shaft from the block.

NOTE: The distributor drive gear and the fuel pump eccentric on the auxiliary shaft must not be allowed to touch the auxiliary shaft bearings during removal and installation. Completely coat the shaft with oil before sliding it into place.

6. Slide the auxiliary shaft into the housing and insert the thrust plate to hold the shaft.

7. Install a new gasket and auxiliary shaft cover.

NOTE: The auxiliary shaft cover and cylinder front cover share a gasket. Cut off the old gasket around the cylinder cover and use half of the new gasket on the auxiliary shaft cover.

8. Fit a new gasket into the fuel pump and install the pump.

9. Insert the distributor and install the auxiliary shaft sprocket.

10. Align the timing marks and install the drive belt.

11. Install the drive belt cover.

12. Check the ignition timing.

Diesel Engine

1. Disconnect the battery ground cable.

2. Remove the valve cover.

3. Remove the vacuum pump.

4. Remove the fan assembly.

5. Remove the camshaft drive belt cover.

6. Remove the rocker arms.

7. Rotate the engine until No. 1 cylinder is at TDC of compression stroke. Install TDC Aligning Pin, T84-P-6400-A, or equivalent.

8. Loosen the camshaft sprocket bolt.

9. Loosen the drive belt tension roller nut and bolt.

10. Remove the camshaft sprocket.

11. Remove the camshaft bearing caps and mark the caps so that they can be reinstalled in their original position, and remove camshaft.

12. Install the camshaft in position on the cylinder head.

13. Install the camshaft bearing caps, making sure they are installed in the correct position. Tighten 6mm nuts to 6–7 ft. lbs. and 8mm nuts to 14–17 ft. lbs.

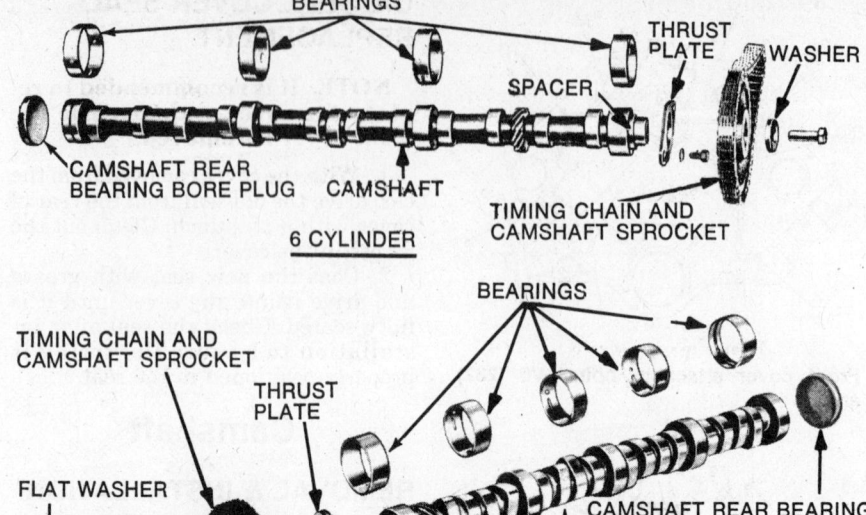

Camshaft and related parts

14. Install the camshaft sprocket but do not tighten at this time.

15. Install and adjust the camshaft drive belt.

16. Adjust the cam and pump timing.

17. Remove the TDC Aligning Pin Tool T84P-6400-A or equivalent.

18. Install the rocker arms.

19. Install the camshaft drive belt cover and tighten the bolts to 6–7 ft. lbs.

20. Install the fan assembly.

21. Install the vacuum pump.

22. Install the rocker arm cover.

23. Connect the battery ground cable.

24. Run the engine and check for oil, intake air, and coolant leaks.

6 Cylinder Inline Engine

1. Remove the cylinder head.

2. Remove the timing case cover (and oil pan on 250 engines), timing chain and sprockets as outlines in the preceding section.

3. Disconnect the and remove the grille. Remove the radiator. If equipped with air conditioning, unbolt the condenser and more it aside without disconnecting any lines.

4. Using a magnet, remove the valve lifters and keep them in order so that they can be installed in their original positions.

5. Remove the camshaft thrust plate and remove the camshaft by pulling it from the front of the engine.

Use care not to damage the camshaft or bearings while removing the cam from the engine.

6. Before installing the camshaft, coat the lobes with engine assembly lubricant and the journals and all valve parts with heavy oil. Clean the oil passage at the rear of the cylinder block with compressed air.

7. Reverse the procedure to install, following recommended torque settings and tightening sequences.

V6 and V8 Engines

1. Remove the intake manifold as outlines previously.

2. Remove the cylinder front cover, timing chain and sprockets as directed previously.

3. Remove the grille and radiator. On models with air conditioning, remove the condenser retaining bolts and position it out of the way. Do not disconnect refrigerant lines. On the Versailles, the hood latch assembly, ambient temperature switch wiring, and the support bracket must be removed.

4. Remove the rocker arm covers.

5. Remove the pushrods and lifters and keep them in order so that they can be installed in their original positions.

6. Remove the camshaft thrust plate and washer if so equipped. Remove the camshaft from the front of the engine. Use care not to damage

camshaft lobes or journals while removing the cam from the engine.

7. Before installing the camshaft, coat the lobes with engine assembly lubricant and the journals and valve parts with heavy oil.

8. Reverse the procedure to install.

ENGINE LUBRICATION

Oil Pan

REMOVAL & INSTALLATION

NOTE: On certain engine-chassis combinations, interference will be encountered between the oil pan and oil pump while attempting to remove the oil pan. If interference occurs, lower the oil pan as far as possible, reach inside and remove the bolts mounting the oil pump or pickup tube. Lower the pump and/or pickup tube into the oil pan. Remove the oil pan. Interference may also occur between the pan and the rear counter balance weight of the crankshaft. Turn the crank to position the weight in an upward position if necessary. Certain late models use "Quick-Connect" transmission oil cooler line fittings. Refer to the "Fuel" section for removal and installation procedures.

4 Cylinder Gas Engine

BOBCAT/PINTO

1. Drain the crankcase.
2. Remove the oil dipstick.
3. Disconnect the steering shaft connection from the rack and pinion.
4. Disconnect the rack and pinion from the crossmember and move it forward to provide clearance.
5. Remove the flywheel housing inspection cover.
6. Remove the oil pan attaching bolts and remove the pan.
7. Clean the gasket mounting surface of the block and the pan.
8. Coat the block surface and the oil pan gasket with oil resistant sealer and position the gasket on the block.
9. Coat the oil pan front oil seal and the front cover with oil resistant sealer and position the seal on the front cover, making sure the ends of the seal contact the oil pan gaskets.
10. Coat the rear oil pan seal with oil resistant sealer and install it in the rear main bearing cap.
11. Position the pan on the block and

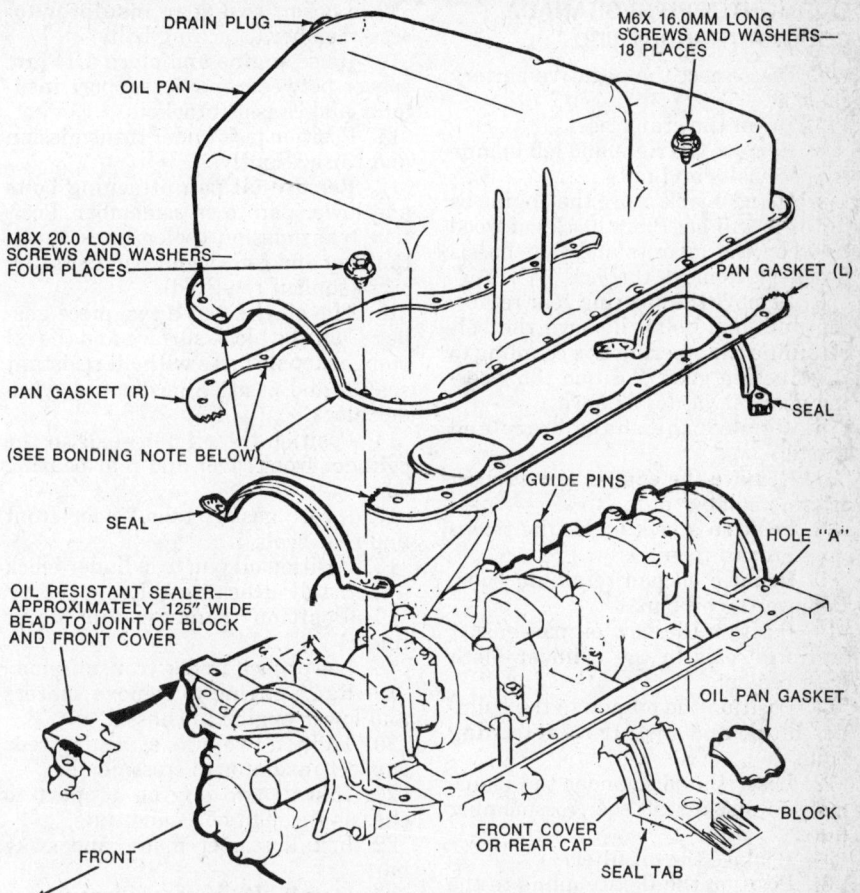

1. APPLY GASKET ADHESIVE EVENLY TO OIL PAN FLANGE AND TO PAN SIDE GASKETS. ALLOW ADHESIVE TO DRY PAST WET STAGE, THEN INSTALL GASKETS TO OIL PAN.
2. APPLY SEALER TO JOINT OF BLOCK AND FRONT COVER. INSTALL SEALS TO FRONT COVER AND REAR BEARING CAP AND PRESS SEAL TABS FIRMLY INTO BLOCK. BE SURE TO INSTALL THE REAR SEAL BEFORE THE REAR MAIN BEARING CAP SEALER HAS CURED.
3. POSITION 2 GUIDE PINS AND INSTALL THE OIL PAN. SECURE THE PAN WITH THE FOUR M8 BOLTS SHOWN ABOVE.
4. REMOVE THE GUIDE PINS AND INSTALL AND TORQUE THE EIGHTEEN M6 BOLTS, BEGINNING AT HOLE "A" AND WORKING CLOCKWISE AROUND THE PAN.

2300 cc engine oil pan torque sequence

tighten the bolts to specification. Tighten all bolts to 7–9 ft. lbs., except 8mm bolts on the 2300. Tighten these to 11–13 ft. lbs.

12. Reverse Steps 1–5 to complete installation.

MUSTANG/CAPRI

1. Disconnect the negative battery cable.
2. Remove the fan shroud or fan shroud and electric fan assembly.
3. Drain the crankcase.
4. Remove the right and left engine support bolts and nuts.
5. Using a jack with a piece of wood between the raising point and jack contact points, raise the engine as high as it will go. Place blocks of wood between the mounts and chassis brackets. Remove the jack. Remove shake brace.
6. Remove the sway bar retaining bolts and lower the sway bar.

7. Remove the starter motor.
8. Remove steering gear retaining bolts and lower the gear.
9. Remove the oil pan retaining bolts. Allow the oil pan to drop the crossmember and remove.
10. Install new oil pan gasket and end seals.
11. Position oil pan to the cylinder block and install retaining bolts.
12. Reposition the steering gear and install bolts and nuts.
13. Install starter.
14. Raise the engine enough to remove the wood blocks, lower the engine and remove jack. Install shake brace.
15. Install the right and left engine support bolts and nuts.
16. Install the sway bar.
17. Install the fan shroud.
18. Fill the crankshaft with oil.
19. Connect battery cable, run engine and check for leaks.

FAIRMONT/ZEPHYR/GRANADA/ COUGAR/THUNDERBIRD

1. Disconnect the negative battery cable.
2. Drain the crankcase.
3. Remove the right and left engine support bolts and nuts.
4. Using a jack, raise the engine as far as it will go. Place blocks of wood between the mounts and the chassis brackets. Remove the jack.
5. Remove the steering gear retaining nuts and bolts. Remove the bolt retaining the steering flex coupling to the steering gear. Position the steering gear forward and down.
6. Remove the shake brace and starter.
7. Remove the engine rear support to crossmember nuts.
8. Position a jack under the transmission and raise.
9. Remove oil pan retaining bolts. Remove the oil pan.
10. Position the new oil pan gasket and end seals to the cylinder block with cement.
11. Position the oil pan to the cylinder block and install its retaining bolts.
12. Loser the jack under the transmission and install the crossmember nuts.
13. Replace the oil filter.
14. Position the flex coupling to the steering gear and install the retaining bolts.
15. Install the steering gear.
16. Install the shake brace. Install the starter.
17. Raise the engine enough to remove the wood blocks. Lower the engine and remove the jack. Install engine support bolts and nuts.
18. Lower the vehicle and fill the crankcase with oil.
19. Connect the battery.
20. Start the engine and check for leaks.

6 Cylinder Inline Engine

1. Disconnect two oil cooler lines at radiator.
2. Remove radiator top support two bolts. Remove or position fan shroud back over fan.
3. Remove oil level dipstick, drain crankcase.
4. Remove four bolts and nuts attaching sway bar to chassis and allow sway bar to hang down.
5. Remove K brace.
6. Lower front steering rack and pinion, or center link and linkage, if necessary for clearance.
7. Remove starter.
8. Remove two nuts attaching engine mounts to support brackets.

9. Loosen two rear insulator-to-crossmember attaching bolts.
10. Raise engine and place a $1\frac{1}{4}$ in. spacer between engine support insulator and chassis brackets.
11. Position jack under transmission and raise slightly.
12. Remove oil pan attaching bolts and lower pan to crossmember. Position transmission cooler lines out of the way and remove oil pan (rotating crankshaft if required).
13. The oil pan has a two piece gasket. Coat the block surface and the oil pan gasket surfaces with oil resistant sealer, and position gaskets to cylinder block.
14. Position the oil pan seals in the cylinder front cover and rear bearing cap.
15. Insert gasket tabs under front and rear seals.
16. Position oil pan to cylinder block and install attaching bolts.
17. Position transmission cooler lines.
18. Lower jack under transmission.
19. Raise engine to remove spacers and lower engine to chassis.
20. Tighten two nuts attaching rear support insulator to crossmember.
21. Install two engine support to chassis through bolts and nuts.
22. Install starter motor and sway bar.
23. Install "K" brace, fill crankcase with oil.
24. Connect oil cooler lines to radiator and install upper radiator support.
25. Lower vehicle, start engine and check for leaks.

V6 Engine

1. Remove the air cleaner assembly including the air intake duct.
2. Remove the fan shroud attaching bolts and position the shroud back over the fan.
3. Remove the oil level dipstick.
4. Remove the screws attaching the vacuum solenoids to the dash panel. Lay the solenoids to the dash panel. Lay the solenoids on the engine without disconnecting the vacuum hoses or electrical connectors.
5. Remove the exhaust manifold to exhaust pipe attaching nuts.
6. Drain the crankcase.
7. Remove the oil filter.
8. Remove the bolts attaching the shift linkage bracket to the transmission bell housing. Remove the starter motor for more clearance if necessary.
9. Disconnect the transmission cooler lines at the radiator. Remove power steering hose retaining clamp from frame.
10. Remove the converter cover.
11. On models equipped with rack and pinion steering vehicles proceed

with the following steps: Remove the engine damper to No. 2 crossmember bracket attaching bolt. The damper must be disconnected from the crossmember. Disconnect steering flex coupling. Remove two bolts attaching steering gear to main crossmember and let steering gear rest on the frame away from oil pan.
12. Remove the nut and washer assembly attaching the front engine insulator to the chassis.
13. Raise the engine 2–3 in. and insert wood blocks between the engine mounts and the vehicle frame.

NOTE: On some models equipped with rack and pinion steering such as Granada/Cougar, Thunderbird/XR-7, LTD/Marquis, it may be necessary to raise the engine as much as 5 in. to provide adequate pan-to-crossmember clearance.

CAUTION

Watch the clearance between the transmission dipstick tube and the thermactor downstream air tube. If the tubes contact before adequate pan-to-crossmember clearance is provided, lower the engine and remove the transmission dipstick tube and the downstream air tube.

14. Remove the oil pan attaching bolts. Work the oil pan loose and remove.
15. On models with limited clearance, lower the oil pan onto the crossmember. Remove the oil pickup tube attaching nut. Lower the pick-up tube/screen assembly into the pan and remove the oil pan through the front of the vehicle.
16. Remove the oil pan seal from the main bearing cap.
17. Clean the gasket surfaces on the cylinder block, oil pan and oil pick-up tube.
18. Apply $\frac{1}{8}$ in. bead of RTV sealer to all matching surfaces of oil pan and engine front cover.
19. Install the oil pan.

NOTE: On models with limited clearance place the oil pick-up tube/screen assembly in the oil pan.

20. Install all other components removed.
21. Fill the crankcase to the correct level with the oil.
22. Start the engine and check the fluid levels in the transmission.
23. Check for engine oil, and transmission fluid leaks.

V8 Engine

NOTE: On vehicles equipped with a dual sump oil pan, both

drain plugs must be removed to thoroughly drain the crankcase. When raising the engine for oil pan removal clearance; drain cooling system, disconnect hoses, check fan to radiator clearance when jacking. Remove radiator if clearance is inadequate.

1. Remove the fan shroud attaching bolts, positioning the fan shroud back over the fan. Remove the dipstick and tube assembly. Disconnect negative battery cable.

2. Drain the crankcase.

3. Remove the stabilizer bar from the chassis (Versailles only). Disconnect the engine stabilizer on models equipped.

4. On rack and pinion models disconnect steering flex coupling. Remove two bolts attaching steering gear to main crossmember and let steering gear rest on frame away from oil pan. Disconnect power steering hose retaining clamp from frame. Remove the starter motor.

6. Remove the idler arm bracket retaining bolts (if equipped) and pull the linkage down and out of the way.

7. Disconnect and plug the fuel line from the gas tank at the fuel pump. Disconnect and lower the exhaust pipe/converter assemblies if they will interfere with pan removal/installation. Raise the engine and place two wood blocks between the engine mounts and the vehicle frame. Remove converter inspection cover.

NOTE: On fuel injected model, depressurize the fuel system prior to line disconnection.

8. Remove the K braces (four bolts).

9. Remove the oil pan attaching bolts and lower oil pan to the frame.

10. Remove oil pump attaching bolts and the inset tube attaching nut from the No. 3 main bearing cap stud and lower the oil pump into the oil pan.

11. Remove the oil pan, rotating the crankshaft as necessary to clear the counterweights.

12. Clean the gasket mounting surfaces thoroughly. Coat the surfaces on the block and pan with sealer. Position the pan side gaskets on the engine block. Install the rear main cap seal with the tabs over the pan side gaskets.

13. Position oil pump and inlet tube into oil pan. Slide oil pan into position under the engine. With the oil pump intermediate shaft in position in the oil pump, position the oil pump to the cylinder block, and the inlet tube to the stud on No. 3 main bearing cap attaching bolt. Install the attaching bolts and tighten to specification. Position the oil pan on the engine and

install the attaching bolts. Tighten the bolts (working from the center toward the ends) 9-11 ft. lbs. for $^5/_{16}$ in. bolts and 7-9 ft. lbs. for $^1/_4$ in. bolts.

14. Position the steering gear to the main crossmember. Install the two attaching bolts and tighten to specification. Connect steering flex coupling.

15. Position the rear K braces and install the four attaching bolts.

16. Raise the engine and remove the wood blocks.

17. Install the stabilizer bar (Versailles only).

18. Lower the engine and install the engine mount attaching bolts. Tighten to specification. Install the converter inspection cover.

19. Install the oil dipstick and tube assembly, and fill crankcase with the specified engine oil. Install the idler arm.

20. Connect the transmission oil cooler lines. Connect the battery cable.

21. Position the shroud to the radiator and install the two attaching bolts. Start the engine and check for leaks.

Piston and Connecting Rod

POSITIONING

Four and six cylinder in-line and V-type engines, should have their piston and rod assemblies installed with the notch on the piston crown toward the front and the oil squirt hole in the rod toward the right side. V8 pistons are assembled with the notch or arrow on the piston crown toward the front and the numbered side of the rod toward the outside.

Oil Pan and Pump

REMOVAL & INSTALLATION

Diesel Engine

1. With the engine removed from vehicle and placed on an engine stand, remove the bolts attaching the oil pan to the crankcase.

2. Remove the two bolts attaching the oil pump pickup to the crankcase.

3. Remove the three bolts attaching the oil pump to the crankcase, and remove the oil pump.

4. Remove the oil pump driveshaft, if necessary.

5. Install the oil pump driveshaft, if removed, making sure it is fully engaged with intermediate shaft.

6. Install the oil pump on the crankcase, making sure driveshaft is fully engaged in the oil pump. Tight-

en the oil pump and oil pick-up bolts to 16-17 ft. lbs.

Oil Pump

REMOVAL & INSTALLATION

Except V6 Engine

1. Remove the oil pan.

2. Remove the oil pump inlet tube and screen assembly.

3. Remove the oil pump attaching bolts and remove the oil pump gasket and intermediate shaft.

4. Prime the oil pump by filling inlet and outlet port with engine oil and rotating shaft of pump to distribute it.

5. Position the intermediate drive shaft into the distributor socket.

6. Position new gasket on pump body and insert intermediate drive shaft into pump body.

7. Install the pump and intermediate shaft as an assembly.

NOTE: Do not force pump if it does not seal readily. The drive shaft may be misaligned with the distributor shaft. To align, rotate the intermediate drive shaft into a new position.

8. Install and torque the oil pump attaching screws to 12-15 ft. lbs. on in line six cylinder 20-25 ft. lbs. on V8s.

9. Install oil pan.

V6 Engine

NOTE: The oil pump is mounted in the front cover assembly. Oil pan removal is necessary for pick-up tube/screen replacement or service.

1. Raise and safely support the vehicle on jackstands.

2. Remove the oil filter.

3. Remove the cover/filter mount assembly.

4. Lift the two pump gears from their mounting pocket in the front cover.

5. Clean all gasket mounting surfaces.

6. Inspect the mounting pocket for wear. If excessive wear is present, complete timing cover assembly replacement is necessary.

7. Inspect the cover/filter mount gasket to timing cover surface for flatness. Place a straight edge across the flat and check clearance with a feeler gauge. If the measured clearance exceeds .004 in., replace the cover/filter mount.

8. Replace the pump gears if wear is excessive.

9. Remove the plug from the end of the pressure relief valve passage us-

ing a small drill and slide hammer. Use caution when drilling.

10. Remove the spring and valve from the bore. Clean all dirt, gum and metal chips from the bore and valve. Inspect all parts for wear. Replace as necessary.

11. Install the valve and spring after lubricating them with engine oil. Install cover/filter mount using a new mounting gasket. Tighten the mounting bolts to 18–22 ft. lbs. Install the oil filter, add necessary oil for correct level.

Rear Main Oil Seal

REMOVAL & INSTALLATION

NOTE: Refer to the "build" dates listed below to determine if the engine is equipped with a split-type or one piece rear main oil seal. Engines after the date indicated have a one-piece oil seal. 2.3L (140) OHC: after 9/28/81; 232 V6: after 4/1/83; 302 V8: after 12/1/82; 351W-V8: after 7/11/83.

Split-Type Seal – Gas Engines

1. Remove the oil pan, and, if required, the oil pump.

2. Loosen all main bearing caps allowing the crankshaft to lower slightly.

NOTE: The crankshaft should not be allowed to drop more than $\frac{1}{32}$ in.

3. Remove the rear main bearing cap and remove the seal from the cap and block. Be very careful not to scratch the sealing surface. Remove the oil seal retaining pin from the cap, if equipped. It is not used with the replacement seal.

4. Carefully clean the seal grooves in the cap and block with solvent.

5. Soak the new seal halves in clean engine oil.

6. Install the upper half of the seal in the block with the undercut side of the seal toward the front of the engine. Slide the seal around the crankshaft journal until $\frac{3}{8}$ in. protrudes beyond the base of the block.

7. Tighten all the main bearing caps (except the rear main bearings) to specifications.

8. Install the lower seal into the rear cap, with the undercut side facing the front of the engine. Allow $\frac{3}{8}$ in. of the seal to protrude above the surface, at the opposite end from the block seal.

9. Squeeze a $\frac{1}{16}$ in. bead of silicone sealant onto the outer center edges of the bearing cap.

10. Install the rear cap and torque to specifications.

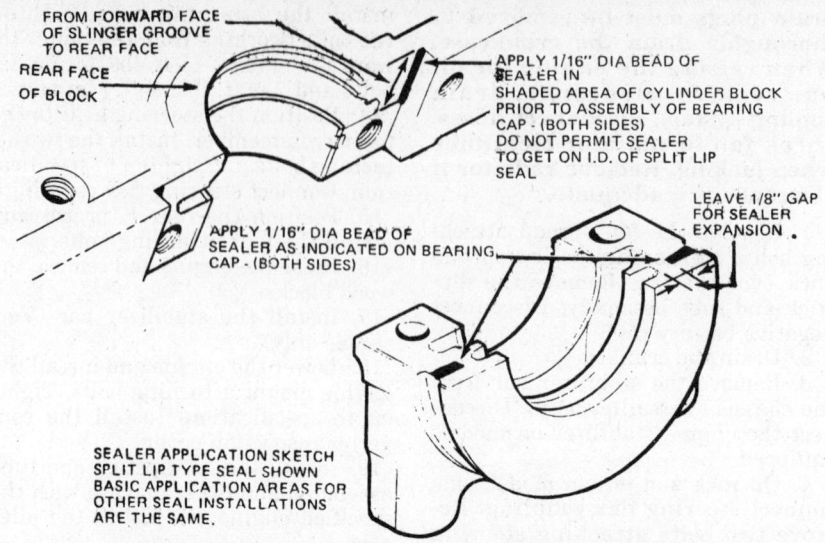

Rear main bearing cap sealer application—with split seal

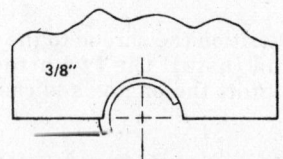

SEAL HALVES TO PROTRUDE BEYOND PARTING FACES THIS DISTANCE TO ALLOW FOR CAP TO BLOCK ALIGNMENT

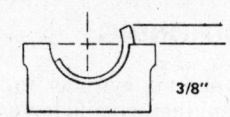

REAR FACE OF REAR MAIN BEARING CAP AND CYLINDER BLOCK
INSTALL SEAL WITH LIP TOWARDS FRONT OF ENGINE

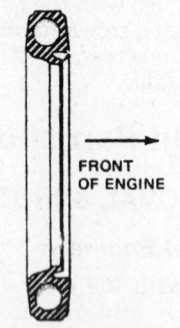

VIEW LOOKING AT PARTING FACE OF SPLIT, LIP-TYPE CRANKSHAFT SEAL

Rear main bearing seal installation—with split seal

11. Install the oil pump and pan. Fill the crankcase with oil, start the engine, and check for leaks.

One-Piece Seal – Gas Engines

1. Remove the transmission, clutch and flywheel or driveplate after referring to the appropriate section of this text for instructions.

2. Punch two holes in the crankshaft rear oil seal on opposite sides of the crankshaft just above the bearing cap to cylinder block split line. Install a sheet metal screw in each of the holes or use a small slide hammer, and pry the crankshaft rear main oil seal from the block.

NOTE: Use extreme caution not to scratch the crankshaft oil seal surface.

3. Clean the oil seal recess in the cylinder block and main bearing cap.

4. Coat the seal and all of the seal mounting surfaces with oil and install the seal in the recess, driving it in place with an oil seat installation tool or a large socket.

5. Install the driveplate or flywheel and clutch and transmission in the reverse order of removal.

Diesel Engine

1. Raise the vehicle and safely support on jackstands.

2. Remove the transmission.

3. Remove the flywheel.

4. Remove the four oil pan to rear engine cover attaching bolts.

5. Loosen, but DO NOT REMOVE the remaining oil pan bolts.

6. Remove the six engine rear cover bolts and remove the cover.

7. Clean the crankcase and the engine rear cover gasket mating surfaces.

8. Using an arbor press, press the old seal out of the cover.

9. Position a new seal on the cover and press in using Crankshaft Rear Seal Replacer T84P-6701-A, or equivalent.

10. Lubricate the sealing lips on the seal with engine oil.

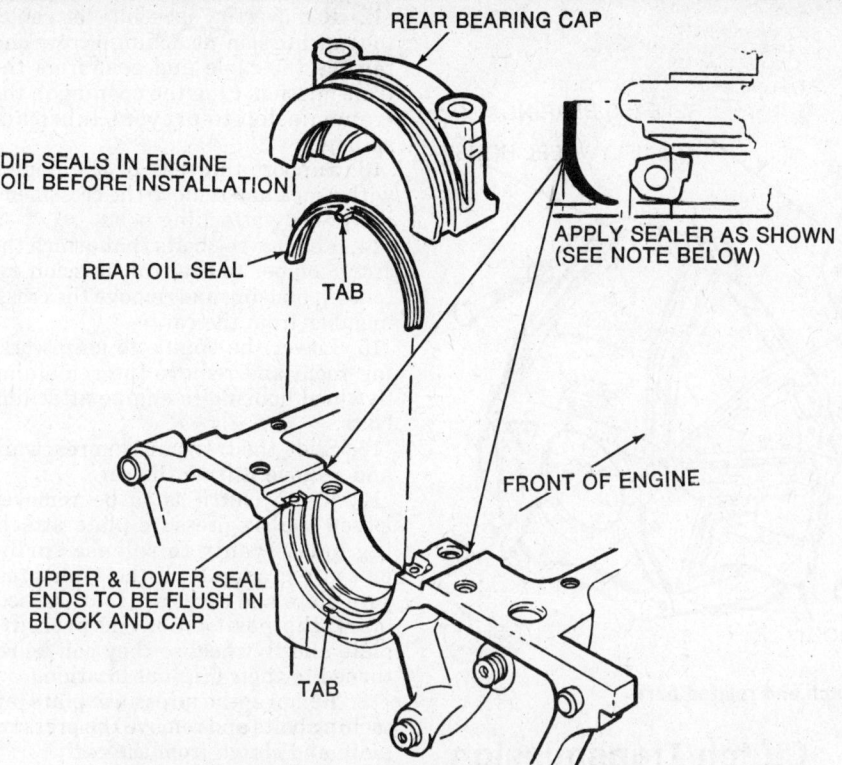

REAR BEARING CAP

DIP SEALS IN ENGINE
OIL BEFORE INSTALLATION

REAR OIL SEAL

TAB

APPLY SEALER AS SHOWN
(SEE NOTE BELOW)

FRONT OF ENGINE

UPPER & LOWER SEAL
ENDS TO BE FLUSH IN
BLOCK AND CAP

TAB

NOTE: CLEAN THE AREA WHERE SEALER IS TO BE APPLIED BEFORE INSTALLING THE SEALS. AFTER THE SEALS ARE IN PLACE, APPLY A 1/16 INCH BEAD OF SEALER AS SHOWN. *SEALER MUST NOT TOUCH SEALS*

Replacement of the rear main bearing oil seal—2.3L (140) engine with split seal

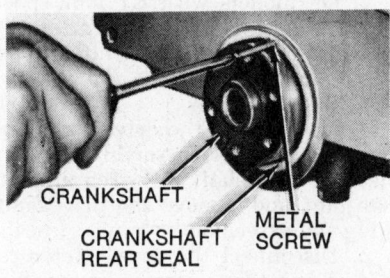

CRANKSHAFT

CRANKSHAFT
REAR SEAL

METAL
SCREW

Typical one piece rear main bearing seal removal

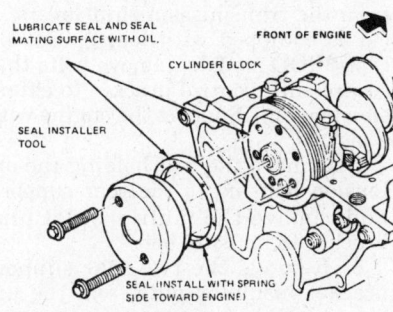

LUBRICATE SEAL AND SEAL
MATING SURFACE WITH OIL.

FRONT OF ENGINE

CYLINDER BLOCK

SEAL INSTALLER
TOOL

SEAL (INSTALL WITH SPRING
SIDE TOWARD ENGINE)

NOTE: REAR FACE OF SEAL MUST BE WITHIN
0.127mm (0.005 INCH) OF THE REAR FACE OF THE BLOCK

One piece rear main bearing oil seal installation

11. Position new rear cover gasket on the crankcase.

12. Apply gasket sealer at points where the rear cover gasket meets the oil pan gasket.

13. Position the rear cover on the crankshaft.

14. Install the rear cover bolts and tighten 6mm bolts to 6–7 ft. lbs. and 8mm bolts to 14–17 ft. lbs.

15. Install the four oil pan to rear cover attaching bolts. Tighten all oil pan bolts to 6.5–7 ft. lbs.

16. Install the flywheel.

17. Install the transmission.

18. Lower the vehicle.

19. Run the engine and check for oil leaks.

CLUTCH

NOTE: All 1981 and later models have self-adjusting clutches. No adjustments are necessary.

SELF-ADJUSTING CLUTCH

The free play in the clutch is adjusted by a built in mechanism that allows the clutch controls to be self-adjusted during normal operation. The self-adjusting feature should be checked every 5000 miles. This is accomplished by insuring that the clutch pedal travels to the top of its upward position. Grasp the clutch pedal with your hand or put your foot under the clutch pedal, pull up on the pedal until it stops. Very little effort is required (about 10 lbs.). During the application of upward pressure, a click may be heard which means an adjustment was necessary and has been accomplished.

ADJUSTMENT

Pinto and Bobcat

1. Loosen the cable locknut on the transmission side of the flywheel housing.

2. Pull the cable toward the front of the car until the tabs on the adjuster nut are clear of the housing. Rotate the nut toward the front of the car about $\frac{1}{4}$ in.

3. Release the cable. Then pull the cable forward again until there is no release lever free movement. Rotate the adjusting nut toward the housing until the tabs touch the housing, then drop the tabs into the nearest groove.

4. Tighten the locknut.

All Except Fairmont, Zephyr, Mustang and Capri

1. Disconnect clutch return spring from release lever.

2. Loosen release lever rod locknut and adjusting nut. On models through 1980, remove the release lever rod locking pin and loosen the adjusting nut.

3. Move clutch release lever rearward until release bearing lightly contacts clutch pressure plate release fingers.

4. Adjust rod length until rod seats in release lever pocket.

5. Insert specified feeler gauge between adjusting nut and swivel sleeve. Tighten adjusting nut against gauge.

6. Tighten locknut against adjusting nut, taking care not to disturb adjustment. On models through 1980, rotate the rod to align the flat with the pin hole in the adjusting nut and install the pin. Remove feeler gauge.

7. Install clutch return spring.

8. Check free travel at pedal. Readjust if necessary to obtain specified travel. Moving adjusting nut away from swivel sleeve increases travel. Moving adjusting nut toward swivel sleeve decreases travel.

9. As final check, measure pedal free travel with transmission in neu-

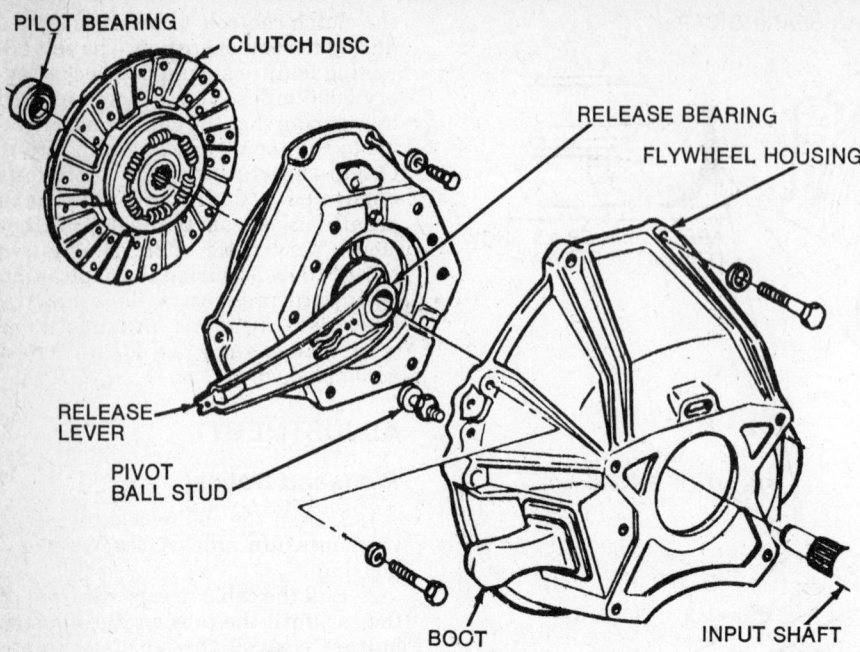

Exploded view of clutch and related parts

tral and engine running at 3000 rpm. If pedal travel is not minimum of $\frac{1}{2}$ in., readjust free travel.

1980 Fairmont, Zephyr, Mustang and Capri

NOTE: These models no longer have free-play adjustments. Pedal height is adjusted instead.

4 CYLINDER, 255 AND 302 V8

1. Working under the car, remove the dust shield.
2. Loosen the clutch cable locknut. To raise the pedal, turn the adjusting nut clockwise; to lower the pedal, turn it counterclockwise.
3. On the four cylinder engine, adjust the pedal height to 5.3 in.; on the 255 and 302 V8 adjust the height to 6.5 in.
4. Tighten the locknut. When the pedal is adjusted properly, the pedal can be raised about $2\frac{1}{2}$ in. on the V8 to reach the pedal stop.
5. Install the dust shield.

INLINE SIX CYLINDER

1. Pull the clutch cable toward the front of the car until the adjusting nut can be rotated. In order to free the nut from the rubber insulator, it may be necessary to block the clutch release forward so the clutch is partially disengaged.
2. Rotate the adjusting nut to obtain a 5.3 in. pedal height. Depress the pedal a few times and recheck the adjustment. When the pedal is properly adjusted, it can be raised about $2\frac{1}{4}$ in. to reach the pedal stop.

Clutch/Transmission

REMOVAL & INSTALLATION

Pinto and Bobcat

1. Place the gearshift lever in the Neutral position. Raise the care and remove the back-up light switch from the transmission extension housing.
2. Loosen the shift lever locknut. Remove the knob and the locknut from the shift lever. Remove the four rubber boot attaching screws and remove the boot.
3. Compress the corrugated rubber spring, then remove the retaining snap-ring and slide the spring upward on the lever.
4. Bend the shift lever locktabs up, then thread the plastic dome nut from the extension housing.
5. Lift the shift lever from the extension housing.
6. Working from under the hood, remove the upper flywheel housing-to-engine attaching bolts.
7. Raise the vehicle and matchmark the driveshaft and the rear axle pinion flange.
8. Disconnect and remove the driveshaft. Place rags in the extension housing to prevent loss of lubricant.
9. Remove the clutch release lever dust cover.
10. Disconnect the clutch cable from the clutch release lever.
11. Remove the starter motor attaching bolts and position the motor out of the way.

12. Remove the speedometer cable-to-transmission attaching screw and remove the cable and gear from the transmission. Plug the opening in the transmission to prevent lubricant spillage.
13. Support the rear of the engine with a jack and remove the crossmember-to-body attaching bolts.
14. Remove the bolts that attach the crossmember to the transmission extension housing and remove the crossmember from the car.
15. Lower the engine to gain working room and remove the remaining flywheel housing-to-engine attaching bolts.
16. Slide the transmission rearward and remove it from the car.
17. If the clutch is to be removed loosen the six pressure plate attaching bolts evenly to release spring pressure gradually. If the same pressure plate and cover are to be reused, mark the position of the pressure plate and flywheel so they can be returned to their original location.
18. Remove the pressure plate attaching bolts and remove the pressure plate and clutch from the car.
19. Installation is in reverse order.

Except Pinto/Bobcat

1. Disconnect and remove starter and dust ring, if the clutch is to be removed. On floorshift models, remove the boot retainer and shift lever.
2. On models with ET four speed transmission: working under the hood, remove the upper clutch housing-to-engine bolts.
3. Raise the car.
4. Matchmark the drive shaft and axle flange for reassembly. Disconnect the driveshaft at the rear universal joint and remove the driveshaft. Plug the extension housing.
5. Disconnect the speedometer cable at the transmission extension. Disconnect the seat belt sensor wires. Remove the clutch lever boot and cable on models so equipped.
6. Disconnect the gear shift rods from the transmission shift levers. If car is equipped with four speed, except SROD models, remove bolts that secure shift control bracket to extension housing. Support the engine with a jack.
7. Remove the bolt holding the extension housing to the rear support, and remove the muffler inlet pipe bracket to housing bolt.
8. Remove the two rear support bracket insulator nuts from the underside of the crossmember. Remove crossmember.
9. Place a jack (equipped with a protective piece of wood) under the rear of the engine oil pan. Raise or

lower the engine slightly as necessary to provide access to the bolts.

10. Remove transmission-to-flywheel housing bolts.

11. Slide the transmission back and out of the car. It may be necessary to slide the catalytic converter bracket forward to provide clearance on some models.

12. To remove the clutch, remove release lever retracting spring. Disconnect pedal at the equalizer bar, or the clutch cable from the housing, as applicable.

13. Remove bolts that secure engine rear plate to front lower part of bellhousing.

14. Remove bolts that attach bell housing to cylinder block and remove housing and release lever as a unit. Remove the clutch release lever by pulling it through the window in the housing until the retainer spring disengages from the pivot.

15. Loosen six pressure plate cover attaching bolts evenly to release spring pressure. Mark cover and flywheel to facilitate reassembly in same position.

16. Remove six attaching bolts while holding pressure plate cover. Remove pressure plate and clutch disc.

— **CAUTION** —

Do not depress the clutch pedal while the transmission is removed.

17. Before installing the clutch, clean the flywheel surface. Inspect the flywheel and pressure plate for wear, scoring, or burn marks (blue color). Light scoring and wear may require refacing of the flywheel or replacement of the damaged parts.

18. Attach the clutch disc and pressure plate assembly to the flywheel. The three dowl pins on the flywheel. The three dowl pins on the flywheel, if so equipped, must be properly aligned. Damaged pins must be replaced. Avoid touching the clutch plate surface. Tighten the bolts finger tight.

19. Align the clutch disc with the pilot bushing. Torque cover bolts to 12–24 ft. lbs. with the four cylinder, 12–20 ft. lbs. for all others.

20. Lightly lubricate the release lever fulcrum ends. Install the release lever in the flywheel housing and install the dust shield.

21. Apply very little lubricant on the release bearing retainer journal. Fill the groove in the release bearing hub with grease. Clean all excess grease from the inside bore of the hub to prevent clutch disc contamination. Attach the release bearing and hub on the release lever.

22. Make sure the flywheel housing and engine block are clean. Any missing or damaged mounting dowels must be replaced. Install the flywheel housing and torque the attaching bolts to 38–61 ft. lbs. on all V8s and 250 sixes, and 28–38 ft. lbs. on fours. Install the dust cover and torque the bolts to 17–20 ft. lbs.

23. Connect the release rod or cable and the retracting spring. Connect the pedal-to-equalizer rod at the equalizer bar.

24. Install starter and dust ring.

25. After moving the transmission back just far enough for the pilot shaft to clear the clutch housing, move it upward and into position on the flywheel housing. It may be necessary to put the transmission in gear and rotate the output shaft to align the input shaft and clutch splines.

26. Move the transmission forward and into place against the flywheel housing, and install the transmission attaching bolts finger-tight.

27. Tighten the transmission bolts to 37–42 ft. lbs. on all cars.

28. Install the crossmember and torque the mounting bolts to 20–30 ft. lbs. Slowly lower the engine onto the crossmember.

29. Torque the rear mount to 30–50 ft. lbs.

30. Connect gear shift rods and the speedometer cable.

31. Remove the plug from the extension housing and install the driveshaft, aligning the marks made previously.

32. Refill the transmission to proper level. On floorshift models, install the boot retainer and shift lever.

NOTE: See the Manual Transmission Application Chart in the Unit Repair section for all applications and overhaul information.

Shift Linkage

ADJUSTMENT

Column Shift

NOTE: With the transmission in Neutral, the shift lever should be in a horizontal plane the parallel to the instrument panel line. Corrective adjustments should be made at the gear shift rods.

1. Place shift lever in Neutral.
2. Loosen two gear shift rod adjustment nuts.
3. Insert $3/16$ in. diameter alignment pin through first and reverse gear shift lever and second and third gear shift lever. Align levers to insert pin.
4. Tighten gear shift rod adjustment nuts, and remove pin.

5. Check gear level for smooth crossover.

3 SPEED FLOOR AND CONSOLE SHIFT

1. Loosen three shift linkage adjustment nuts.
2. Install a ¼ in. diameter alignment pin through control bracket and levers.
3. Tighten three shift linkage adjustment nuts and remove alignment pin.
4. Check gear lever for smooth crossover.

AUTOMATIC TRANSMISSION

NOTE: Refer to the "Automatic Transmission" Unit Repair section for linkage and band adjustments and fluid and filter changes. Transmissions may be identified by the code on the vehicle certification label.

TRANSMISSION CODES

C.	C5 automatic
S.	JATCO automatic
T.	AOD (automatic overdrive)
U.	C6 automatic
V.	C3 automatic
W.	C4 automatic
X.	FMX automatic
Z.	C6 police automatic

REMOVAL & INSTALLATION

Except ZF Transmission

1. Disconnect the negative battery cable. Raise and safely support the vehicle.

2. Place the drain pan under the transmission fluid pan. Remove the fluid filler tube, if pan mounted, and drain the transmission fluid. On models that do not have a pan mounted filler tube, loosen the pan attaching bolts and allow the fluid to drain. Start loosening the bolts at the rear of the pan and work toward the front. Finally remove all of the pan attaching bolts except two at the front, to allow the fluid to further drain. After the fluid has drained, install two bolts on the rear side of the pan to temporarily hold in place.

3. Remove the converter drain plug access cover from the lower end of the converter housing.

4. Remove the converter-to-flywheel attaching nuts. Place a wrench on the crankshaft pulley attaching bolt to turn the engine to gain access to the nuts. DO NOT turn OHC (overhead cam) engines opposite the normal direction or damage to the timing belt may occur.

5. With the wrench on the crankshaft pulley attaching bolt, turn the engine to gain access to the converter drain plug. then, remove the plug. Place a drain pan under the converter to catch the fluid. After the fluid has been drained from the converter, reinstall the plug. Tighten to 20–25 ft. lbs.

6. Remove the drive shaft and plug the back of the transmission extension housing to prevent dirt from entering.

7. Label and remove all vacuum lines and wiring harnesses connected to the transmission. Remove the filler tube from the transmission after removing the engine mounting bolt. Disconnect the speedometer cable. Disconnect the shift linkage and kickdown cable.

8. Remove the transmission support to crossmember bolts or nuts. Disconnect the starter cable and remove the starter motor. Remove any exhaust system parts (pipes, converters, brackets etc.) that will interfere with transmission removal.

9. Disconnect the oil cooler lines from the transmission case.

10. Position a suitable transmission jack to support the transmission and secure the transmission to the jack with a safety chain.
Raise the transmission slightly and remove the crossmember attaching bolts and remove the crossmember.

── CAUTION ──
The engine will lower slightly when the transmission is removed. Check clearance between the front of the engine and radiator shroud. Remove the shroud and position it over the fan if necessary. Remove the air cleaner assembly. Check the upper radiator hose, do not permit it to be stretched, or damage to the radiator connection may occur.

11. Remove the converter housing-to-engine attaching bolts. Pull the transmission back and away from the engine after prying the converter away from the drive plate. After the transmission is separated from the engine, attach a small C-clamp on the housing in a position that will not permit the converter from fall out of the housing. Lower the transmission slowly and remove it from under the vehicle.

12. Apply white lube to the converter hub. Install the transmission in the reverse order of removal. Make sure the converter holes are aligned correctly with the drive pale and the converter pilot is flat against the crank pilot. Torque the Converter to Drive Plate; 25–30 ft. lbs. Transmission to Engine; 23–28 ft. lbs. on 4 and in-line 6 cylinder engines. 40–50 ft. lbs. On V6 (232) and V8 engines except with the AOD transmission. AOD transmission; 35–40 ft. lbs.

13. If the converter has been completely drained, add four quarts of the proper transmission fluid, start the engine and move the selector through the gears. Recheck and add fluid as necessary until the correct level is reached.

NOTE; Dexron®II fluid is used in all transmissions except FMX models which use Type F and C5 models which require Type H.

ZF Transmission

1. Remove the kickdown (TV) cable and insert from the injection pump side lever and cable bracket in the engine compartment.

2. Place the transmission selector lever in N (Neutral). Raise the vehicle on a hoist.

3. Remove the outer manual lever and nut from the transmission selector shaft.

4. Remove position sensor from converter housing.

5. Remove the engine brace from the lower end of the converter housing.

6. Place a transmission jack under the transmission.

7. Place a wrench on the crankshaft pulley attaching bolt and turn the converter to gain access to the converter-to-flywheel attaching nuts. Remove the converter-to-flywheel attaching nuts.

NOTE: The converter studs are installed in the converter with Loc-Tite®. During disassembly the nuts may override the Loc-Tite® and the nut and stud come out as a "bolt". This poses no concern. The stud and converter threads should be cleaned, Loc-Tite® applied, and the "bolt" reinstalled and tightened to specification.

8. Disconnect the driveshaft from the rear axle and slide shaft rearward from the transmission.

NOTE: To maintain driveshaft balance, mark the rear driveshaft yoke and axle companion flange so the driveshaft can be installed in its original position. Install a seal installation tool in the extension housing to prevent fluid leakage.

9. Disconnect the neutral start switch electrical connector.

10. Remove the extension housing damper.

11. Remove the rear support-to-crossmember attaching nuts and the two crossmember-to-side support attaching bolts.

12. Remove the two engine rear support-to-extension housing attaching bolts and remove the rear mount from the exhaust system.

13. On Continental with column shift, remove the two bolts securing the bellcrank bracket to the engine-to-transmission brace.

14. Disconnect each oil line from the fittings on the transmission using push connect service Tool T82L-9500-AH or equivalent.

15. Disconnect the speedometer wiring harness from the extension housing.

16. Remove the two converter housing to starter motor bolts.

17. Secure the transmission to the jack with a safety chain and lower the jack slightly.

18. Remove the four converter housing-to-cylinder block attaching bolts.

19. Remove the filler tube and dipstick.

20. Carefully move the transmission and converter assembly away from the engine and, at the same time, lower the jack to clear the underside of the vehicle.

21. Mount the transmission in a holding fixture.

22. Place the transmission on the jack. Secure the transmission to the jack with a safety chain.

23. Rotate the converter until the studs are in alignment with the holes in the flywheel and flexplate.

24. Move the converter and transmission assembly forward into position, using care not to damage the flywheel, flexplate and the converter pilot. The converter face must seat squarely against the flexplate (This indicates that the converter pilot is not binding in the engine crankshaft).

25. Install the filler tube and dipstick, position bracket over the upper right housing to engine bolt holes.

26. Install and tighten the four converter housing-to-engine attaching bolts to 38–48 ft. lbs.

27. Remove the safety chain from around the transmission.

28. Connect the oil cooler lines by pushing them into the fittings on the transmission (located on the intermediate plate).

29. Connect the speedometer wiring harness to the extension housing.

30. Install the extension housing damper with three bolts. Tighten bolts to 18–25 ft. lbs.

31. Install the rear support on the exhaust system.

32. Install the crossmember on the side supports and install the attaching bolts and nuts. Position the rear support on the crossmember and tighten the nuts to specification.

33. Secure the engine rear support to the extension housing and tighten the bolts to specification.

34. If removed, install exhaust system hardware.

35. Lower the transmission and remove the jack.

36. On the Continental equipped with column shift, position the bellcrank to the engine-to-transmission brace and install the two attaching bolts. Tighten the bolts to 10–20 ft. lbs.

37. Guide the kickdown (TV) cable up into the engine compartment.

38. Install the outer manual lever on the transmission selector shaft. Tighten the attaching nut to 10–20 ft. lbs.

39. Install the converter to flywheel attaching nuts (or bolts) and tighten to 20–34 ft. lbs.

40. Install the engine brace on the lower end of the converter housing and engine block. Tighten the bolts to 15–18 ft. lbs.

41. Connect the neutral start switch harness at the transmission.

42. Install position sensor to converter housing.

43. Connect the driveshaft to the rear axle. Install the driveshaft so the index marks, made during removal, are correctly aligned. Lubricate the yoke splines with C1AZ-19590-B or equivalent.

44. Lower the vehicle and adjust the kickdown (TV) cable.

45. Fill the transmission to the correct level with the specified fluid. Start the engine and shift the transmission to all positions, then recheck the fluid level.

TV CABLE ADJUSTMENT

1. Set the injection pump lever at the full throttle position.

2. Tighten the rear adjusting nut on the threaded barrel until a gap of (1.54–1.57 in.) exists between the edge of the crimped bead on the cable closest to the barrel and the end of the threaded barrel.

3. Tighten the forward adjusting nut to lock the cable assembly to the bracket to 80–106 inch lbs.

4. Recheck the gap and readjust as necessary.

NOTE: Kickdown on this transmission is controlled by the injection pump linkage adjustments.

DRIVE AXLE

Driveshaft and U-Joints

REMOVAL & INSTALLATION

NOTE: Universal joints are retained at the rear by either U-bolts or a coupling flange that is bolted to the pinion (differential) flange. Various models are equipped with a double Cardan-type universal joint at the rear. Service for the front U-joint on these models is the same as for other models.

1. Matchmark the rear driveshaft yoke and the companion flange so that the parts may be reassembled in the same way to maintain balance.

2. Remove the U-bolts and straps or coupling flange nuts and bolts at the rear of the driveshaft, and tape the loose bearing caps to the spider.

3. Allow the rear of the driveshaft to drop down slightly. Pull the driveshaft and slip yoke out of the transmission extension housing.

4. Plug the transmission to prevent fluid leakage.

5. To install, lubricate the yoke splines and install the yoke into the transmission extension housing, aligning the splines. Be careful not to bottom the slip yoke hard against the transmission seal.

6. Rotate the pinion flange as necessary to align the matchmarks made earlier. Install the matchmarks made earlier. Install the U-bolts and tighten to 8–15 ft. lbs. On the Versailles, tighten the coupling-to-pinion flange bolts to 70–90 ft. lbs. Various models use special wax-dipped coupling-to-pinion flange bolts which may not be reused. They must be replaced with special new bolts, torqued to 71–96 ft. lbs.

Universal Joint

OVERHAUL

NOTE: Universal joint and Double-Cardan joint overhaul procedures are given in the "U-Joints and CV-Joints" Unit Repair section.

Rear Axle, Bearing and Seal

NOTE: Both integral and removable carrier type axles are used. Traction-Lok (limited slip)

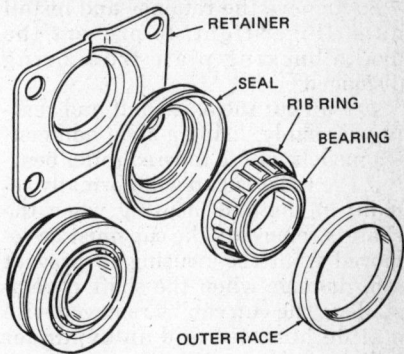

Tapered bearing and retainer–removable carrier axle

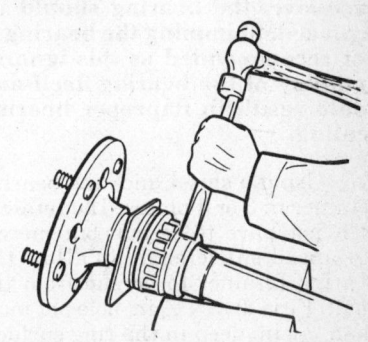

Axle shaft bearing retainer removal —removable carrier axle

axles are available only as removable carrier types. The axle type and ratio are stamped on a plate attached to a rear housing cover bolt. Axle types also indicate whether the axle shafts are retained by C-locks. To properly identify a C-lock axle, drain the lubricant, remove the rear cover and look for the C-lock on the end of the axle shaft in the differential side gear bore. If the second letter of the axle model code is F, it is a Traction-Loc axle. Always refer to the axle tag code and ratio when ordering parts.

REMOVAL & INSTALLATION

Except C-Lock Type

NOTE: Bearings must be pressed on and off the shaft with an arbor press. Unless you have access to one, it is unadvisable to attempt any repair work on the axle shaft bearing assemblies.

1. Remove the wheel, tire, and brake drum. With disc brakes, remove the caliper, retaining nuts, and rotor. New anchor plate bolts will be needed for reassembly.

2. Remove the nuts holding the retainer plate to the backing plate, or axle shaft retainer bolts from the housing. Disconnect the brake line with drum brakes.

3. Remove the retainer and install nuts, finger-tight, to prevent the brake backing plate from being dislodged.

4. Pull out the axle shaft and bearing assembly, using a slide hammer. On models with a tapered roller bearing, the tapered cup will normally remain in the axle housing when the shaft is removed. The cup must be removed from the housing to prevent seal damage when the shaft is reinstalled. The cup can be removed with a slide hammer and an expander puller.

NOTE: If end-play is found to be excessive, the bearing should be replaced. Shimming the bearing is not recommended as this ignores end-play of the bearing itself and could result in improper bearing seating.

5. Using a chisel, nick the bearing retainer in 3 or 4 places. The retainer does not have to be cut, but merely collapsed sufficiently to allow the bearing retainer to be slid from the shaft. First drill a $\frac{1}{4}$ in. hole not more than $\frac{5}{16}$ in. deep in the ring surface.

6. Press off the bearing and install the new one by pressing it into position. With tapered bearings, place the lubricated seal and bearing on the axle shaft (cup rib ring facing the flange). Make sure that the seal is the correct length. Disc brake seal rims are black, drum brake seal rims are grey. Press the bearing and seal onto the shaft.

7. Press on the new retainer.

NOTE: Do not attempt to press the bearing and the retainer on at the same time.

8. On ball bearing models, to replace the seal: remove the seal from the housing with an expanded cone type puller and a slide hammer. The seal must be replaced whenever the shaft is removed. Wipe a small amount of sealer onto the outer edge of the new seal before installation; do not put sealer on the sealing lip. Press the seal into the housing with a seal installation tool.

9. Assemble the shaft and bearing in the housing, being sure that the bearing is seated properly in the housing. On ball bearing models, be careful not to damage the seal with the shaft. With tapered bearings, first install the tapered cup on the bearing, and lubricate the outer diameter of the cup and the seal with axle lube. Then install the shaft and bearing assembly into the housing.

10. Install the retainer, drum or rotor and caliper, wheel and tire. Bleed the brakes.

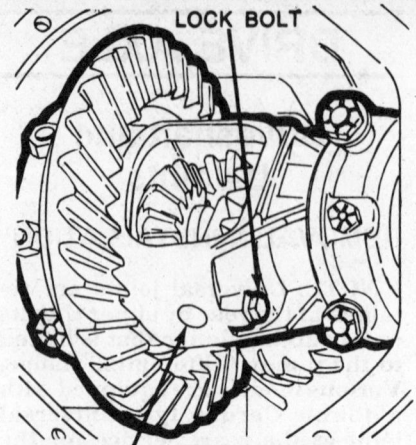
DIFFERENTIAL PINION SHAFT
Removing the differential pinion shaft lockbolt

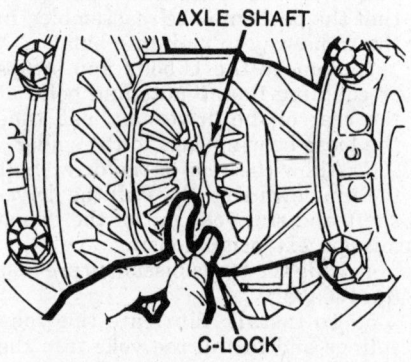
Removing the axle shaft C-locks

C-Lock Type

1. Jack up and support the rear of the car.

2. Remove the wheels and tires from the brake drums.

3. Place a drain pan under the housing and drain the lubricant by loosening the housing cover.

4. Remove the locks securing the brake drums to the axle shaft flanges and remove the drums.

5. Remove the housing cover and gasket, if used.

6. Position jackstands under the rear frame member and lower the axle housing. This is done to give easy access to the inside of the differential.

7. Working through the opening in the differential case, remove the side gear pinion shaft lockbolt and the side gear pinion shaft.

8. Push the axle shafts inward and remove the C-locks from the inner end of the axle shafts. Temporarily replace the shaft and lockbolt to retain the differential gears in position.

9. Remove the axle shafts with a slide hammer. Be sure the seal is not damaged by the splines on the axle shaft.

10. Remove the bearing and oil seal from the housing. Both the seal and bearing can be removed with a slide hammer. Two types of bearings are used on some axles, one requiring a press fit and the other a loose fit. A loose fitting bearing does not necessarily indicate excessive wear.

11. Inspect the axle shaft housing and axle shafts for burrs or other irregularities. Replace any work or damaged parts. A light yellow color on the bearing journal of the axle shaft is normal, and does not require replacement of the axle shaft. Slight pitting and wear is also normal.

12. Lightly coat the wheel bearing rollers with axle lubricant. Install the bearings in the axle housing until the bearing seats firmly against the shoulder.

13. Wipe all lubricant from the oil seal bore, before installing the seal.

14. Inspect the original seals for wear. If necessary, these may be replaced with new seals, which are prepacked with lubricant and do not require soaking.

15. Install the oil seal.

— **CAUTION** —

Installation of the seal without the proper tool can cause distortion and seal leakage. Seals may be colored coded for side indentification. Do not interchange seals from side to side, if they are coded.

16. Remove the lockbolt and pinion shaft. Carefully slide the axle shafts into place. Be careful that you do not damage the seal with the splined end of the axle shaft. Engage the splined end of the shaft with the differential side gears.

17. Install the axle shaft C-locks on the inner end of the axle shafts and seat the C-locks in the counterbore of the differential side gears.

18. Rotate the differential pinion gears until the differential pinion shaft can be installed. Install the differential pinion shaft lockbolt. Tighten to 15–22 ft. lbs.

19. Install the brake drum on the axle shaft flange.

20. Install the wheel and tire on the brake drum and tighten the attaching nuts.

21. Clean the gasket surface of the rear housing and install a new cover gasket and the housing cover. Some models do not use a "paper" gasket. On these models, apply a bead of silicone sealer on the gasket surface. The bead should run inside of the bolt holes.

22. Raise the rear axle so that it is in the running position. Add the amount of specified lubricant to bring the lubricant level to $\frac{1}{2}$ in. below the filler hole.

JACKING

──── CAUTION ────

The electrical power supply to the air suspension system on vehicles equipped, must be shut off prior to hoisting, jacking or towing an air suspension vehicle. This can be accomplished by disconnecting the battery or turning off the power switch located in the trunk on the LH side. Failure to do so may result in unexpected inflation or deflation of the air springs, which may result in shifting of the vehicle during these operations. Before lowering a vehicle on to its tires after lifting, hoisting and/ or towing, it is necessary to check the air springs to insure that they are not creased or deflated but properly folded and in place.

When it becomes necessary to raise the car for service, proper safety precautions must be taken. Depending on year and model a bumper jack or screw-type jack is provided. Bumper slots or frame rail notches are provided for jack placement. Never crawl under the car when it is supported only by the jack. If the jack should slip or tip over, as jacks sometimes do, you would be pinned under two tons of automobile.

When raising the car with a jack to change a tire, follow these precautions: Fully apply the parking brake, block the wheel diagonally opposite the wheel to be raised, stop the engine, place the gear lever in Park (automatic), and make sure that the jack is firmly planted on a level, solid surface. If you are going to work beneath the car, always install jackstands beneath an adjacent frame member, or at either front lower arm strut connection.

FRONT SUSPENSION

Shock Absorber/Type 1: Coil Spring on Upper Arm

NOTE: For models equipped with air suspension refer to section following "Front and Rear Suspension." Different front suspension designs are used depending on year and model. Type 1: Coil Spring on Upper Control Arm; Type 2: Coil Spring on Lower Control Arm; Type 3: Strut Suspension.

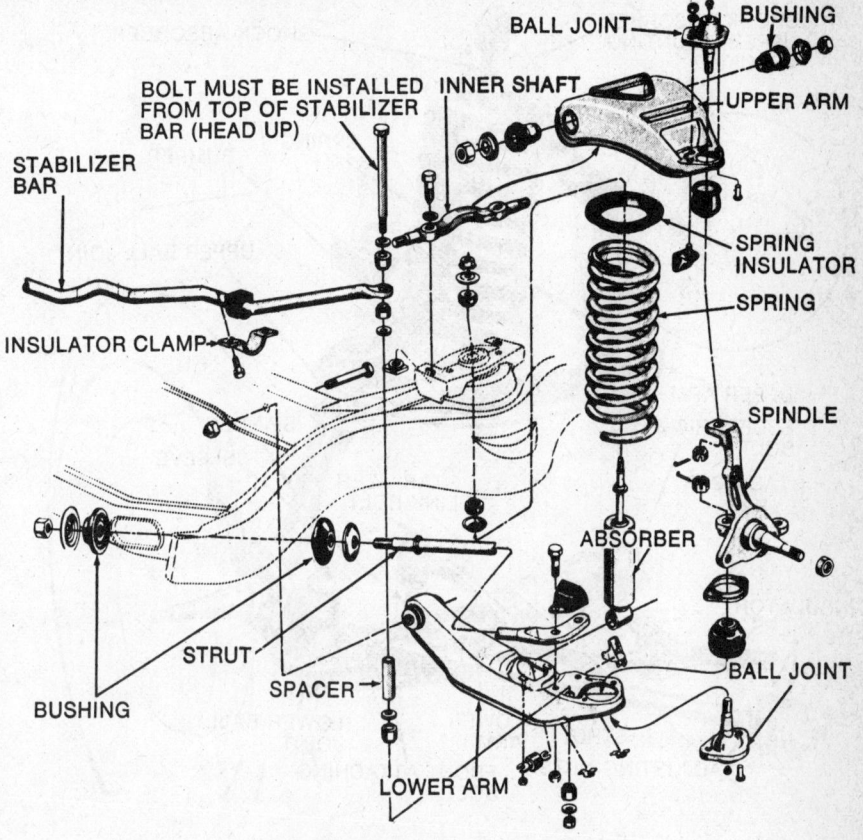

Front suspension—spring on lower arm

REMOVAL & INSTALLATION

NOTE: Purge a new shock of air by repeatedly extending it in its normal position and compressing it while inverted.

1. Raise the hood and remove the three shock absorber-to-spring tower attaching bolts.
2. Raise the front of the vehicle and place jackstands under the lower control arms.
3. Remove the shock absorber lower attaching nuts, washers, and insulators.
4. Lift the shock absorber and upper bracket from the spring tower and remove the bracket from the shock absorber. Remove the insulators from the lower attaching studs.
5. Install the upper mounting bracket on the shock absorber. Torque to 22–25 ft. lbs. Install the insulators on the lower attaching studs.
6. Place the shock absorber and upper bracket assembly in the spring tower, making sure that the shock absorber lower studs are in the pivot plate holes.
7. Install the two washers and attaching nuts on the lower studs of the shock absorbers. Torque to 8–12 ft. lbs.

8. Install the shock absorbers upper mounting bracket attaching nuts. Torque to 32–48 ft. lbs.
9. Remove the jackstands and lower the vehicle.

Coil Spring

──── CAUTION ────

Do not attempt to remove or install coil springs without a spring compressor. If a coil spring is removed without the use of a spring compressor, severe injury may result.

1. Remove the shock absorber and mounting bracket.
2. Jack up the car and install a jackstand beneath the inboard end of the lower control arm. Remove the wheel. Place a wooden block between the control arm and frame.
3. Remove the grease cap, cotter pin, lock nut and outer bearing from the hub.
4. Remove the disc brake caliper and rotor.
5. Install a spring compressor(s) on the coil spring and compress the spring until tension is released from the control arm.
6. Remove the two nuts retaining the upper control arm to the spring

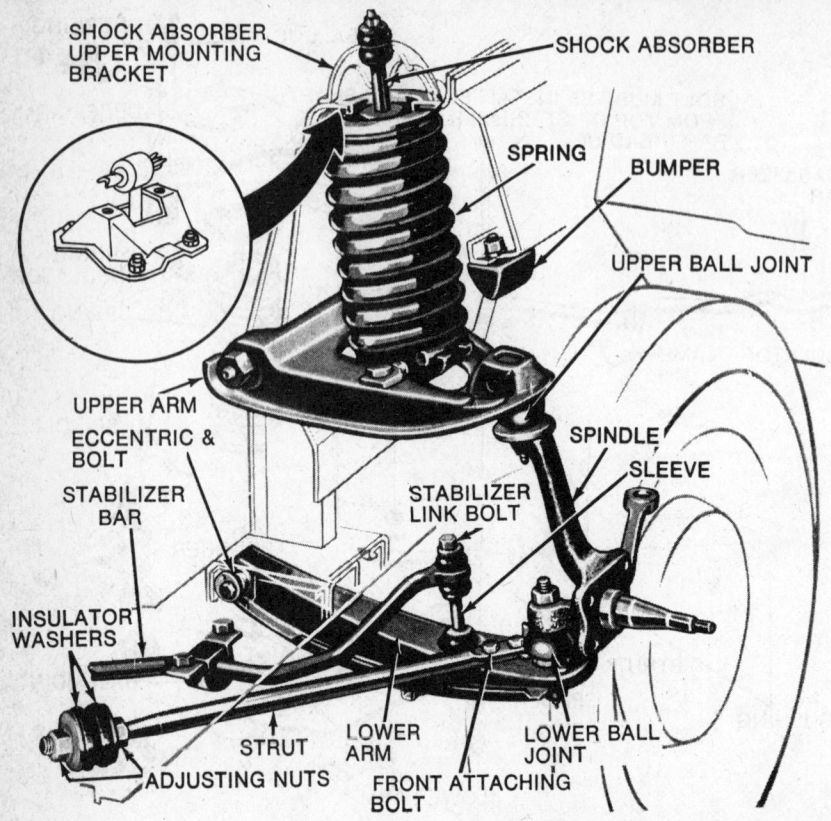

Front suspension—spring on upper arm

8. Remove the link bolt from the lower arm.

9. Remove the strut bar to lower control arm attaching nuts and bolts.

10. Remove the lower ball joint cotter pin and back off the nut. Using a ball joint removal tool, loosen the ball joint stud in the spindle.

11. Remove the nut from the lower ball joint stud and lower the arm.

12. Remove the lower arm to underbody cam attaching parts and remove the arm.

13. To install, position the lower arm in the underbody and install the ball joint and cam attaching parts loosely.

14. Raise the lower arm, install the ball joint stud into place and loosely install the stud nut.

15. Install the stabilizer and strut and tighten the stabilizer nuts to 6–12 ft. lbs. Tighten the strut-to-arm nuts to 90–115 ft. lbs.

16. Tighten the ball joint stud to 75 ft. lbs., then continue to tighten until the cotter pin holes align. Install a new cotter pin. Tighten the lower arm bolts to 85–100 ft. lbs.

17. Lower the car and remove the upper arm support.

18. Front end alignment must be rechecked.

tower and swing the arms outboard from the tower.

7. Slowly release the spring compressor from the spring, remove the tool and remove the spring from the vehicle.

8. To install, tape the spring upper insulator on the top of the spring. Place the spring in the tower. Install the spring compressor and compress the spring.

9. Swing the upper control arm inboard and insert the retaining bolts through the bolt holes in the spring tower. Install the retaining nuts and torque them to 110–130 ft. lbs.

10. Slowly release the spring compressor, while guiding the lower end of the spring into the upper control arm spring seat. The end of the spring must be no more than $\frac{1}{2}$ in. from the tab on the spring seat.

11. Remove the spring compressor.

12. Reinstall the rotor and the caliper. Install the outer bearing, washer and adjusting nut on the spindle. Adjust the wheel bearing. Install the locknut, cotter pin and grease cap.

13. Install the wheel, remove the jackstand and lower the vehicle. Install the shock absorber and upper mounting bracket.

Lower Ball Joint

NOTE: On cars which have the coil springs mounted on the upper control arm, the lower ball joint is an integral part of the lower control arm. If the lower ball joint is defective the entire lower control arm must be replaced.

1. Raise the vehicle with a floor jack so that the front wheel falls to the full down position.

2. Have an assistant grasp the bottom of the tire and move the wheel in and out.

3. As the wheel is being moved, observe the lower control arm where the spindle attaches to it.

4. Any movement between the lower part of the spindle and the lower control arm indicates a worn ball joint which must be replaced.

NOTE: During this check, the upper ball joint will be unloaded and may move; this is normal and not an indication of a bad ball joint. Also, do not mistake a loose wheel bearing for a worn ball joint.

5. Position a support between the upper arm and side rail.

6. Raise the vehicle, position jack stands and remove the wheel and tire.

7. Remove the stabilizer bar to link attaching nut and disconnect the bar from the link.

When upper control arm bushings become low on lubrication, they become very noisy. This can often be corrected by lubrication; it is not necessary to replace the bushings. On early models that do not contain grease plugs it is necessary to drill and tap the bushing to accept a grease fitting. On later models with grease plugs it is difficult to remove the plug and grease the bushing with conventional tools. Ford Motor Co. has available an upper A-arm lubrication kit which greatly eases the performance of this operation.

Upper Ball Joint

1. Raise the vehicle on a hoist or floor jack so that the front wheels hang in full down position.

2. Have an assistant grasp the wheel top and bottom and apply alternate in and out pressure to the top and bottom of the wheel.

3. Radial play of $\frac{1}{4}$ in. is acceptable measured at the inside of the wheel adjacent to the upper arm on all models except the Granada, Monarch, and Versailles; on those models only, any

detectable play indicates worn ball joints.

NOTE: This radial play measurement is multiplied at the outer circumference of the tire and should not be measured here. Measure only at the inside of the wheel. The factory procedure for ball joint replacement is to install a new upper control arm. The factory does not recommend installation of a new ball joint. However, ball joint replacements are available from auto parts dealers, and may be installed using the following procedure.

4. Position a support between the upper arm and frame rail.
5. Raise the vehicle and remove the tire and wheel.
6. Remove the upper ball joint cotter pin and loosen the nut.
7. Using a ball joint removal tool, loosen the ball joint in the spindle.
8. Remove the three ball joint retaining rivets using a large chisel.
9. Remove the nut from the ball joint stud and remove the ball joint.
10. Clean and remove all burrs from the ball joint mounting area of the control arm before installing a new ball joint to the arm.
12. Install and tighten the ball joint stud nut and install the cotter pin.
13. Lubricate the new joint with a hand type grease gun only; using an air pressure gun may loosen the ball joint seal.
15. Check front end alignment.

Upper Control Arm

NOTE: The upper arm shaft and bushings may not be replaced separately from the upper arm.

1. Remove the shock absorber and upper mounting bracket from the car as an assembly. Install a wood block as support between the upper arm and the body.
2. Raise the vehicle and remove the wheel and tire as an assembly.
3. Install spring compressor tool.
4. Place a safety stand under the lower arm.
5. Remove the cotter pin from the upper ball joint and loosen the nut.
6. Using a ball joint removal tool, loosen the ball joint in the spindle, then remove the nut and lift the stud from the spindle.
7. Remove the upper arm attaching nuts from the engine compartment and remove the upper arm.
8. To install the arm, position it on the mounting bracket and install the attaching nuts on the inner shaft attaching nuts on the inner shaft at-

taching bolts. Torque to 110–130 ft. lbs.
9. Install the upper ball joint stud in the spindle and tighten the nut according to the procedure in Step 12 of the lower ball joint procedure. Install a new cotter pin.
10. Remove spring compressor and position spring on upper arm. Install wheel and check front end alignment.

Shock Absorber/Type 2: Coil Spring on Lower Arm

REMOVAL & INSTALLATION

NOTE: Purge a new shock of air by repeatedly extending it in its normal position and compressing it while inverted.

1. Remove the nut, washer, and bushing from the upper end of the shock absorber.
2. Raise the vehicle and install jackstands under the frame rails.
3. Remove the two bolts securing the shock absorber to the lower control arm and remove the shock absorber.
4. Install a new bushing and washer on the top of the shock absorber and position the unit inside the front spring. Install the two lower attaching bolts and torque them to 8–15 ft. lbs.
5. Remove the jackstands and lower the vehicle.
6. Place a new bushing and washer on the shock absorber top stud and install a new attaching nut. Torque to 22–30 ft. lbs.

Coil Spring and Lower Control Arm

1980 AND LATER

1. Raise the car and support it with jackstands. Remove the tire and wheel.
2. Disconnect the stabilizer link from the lower arm.
3. Remove the lower shock absorber attaching bolts.
4. Remove the shock absorber upper nut and remove the shock.
5. Remove the steering center link from the pitman arm.
6. Install a spring compressor tool. Insert the securing pin through the upper ball nut and the compression rod. This pin can only be inserted one way. With the upper ball nut secured, turn the upper plate so it walks up the coil and contacts the upper spring seat. Back the nut off ½ turn.
7. Install the lower ball nut and the thrust washer on the compression rod

and tighten the forcing nut until the spring is free in the seat.
8. Remove the two lower control arm pivot bolts.
9. Disengage the arm from the frame and remove the spring assembly.
10. If a new spring is being installed, mark the position of the upper and lower plates on the old spring. Also, measure the length of the spring and the amount of curvature in order to simplify the compressing and installation of the new spring.
11. Loosen the forcing nut and remove the spring from the tool.

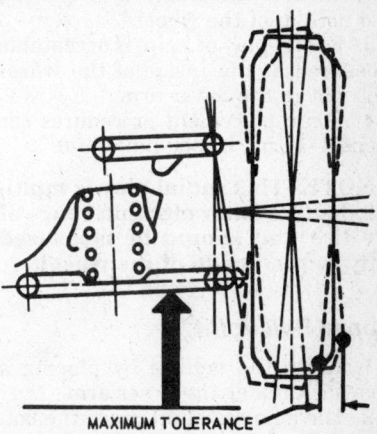

MAXIMUM TOLERANCE

Measuring lower ball joint radial play –spring on lower arm

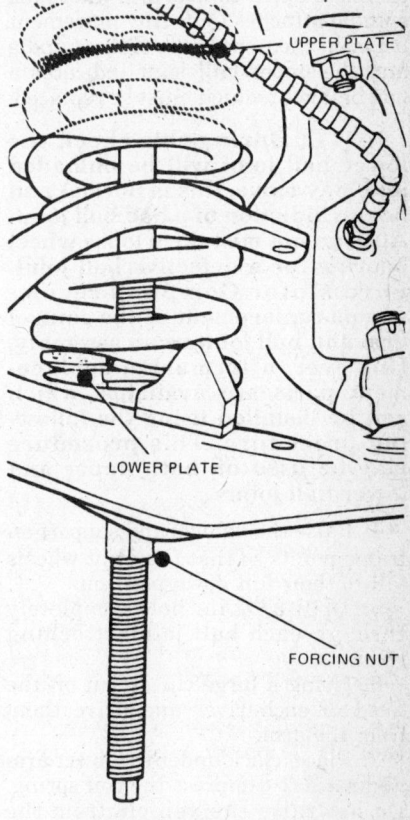

UPPER PLATE

LOWER PLATE

FORCING NUT

Spring compressor installed

12. Assemble the spring compressor tool on the new spring in the same position as the old spring was removed.

13. Position the spring in the lower arm.

14. Reverse the removal procedure to reinstall.

Lower Ball Joint

1. Raise the vehicle by placing a floor jack under the lower arm to remove the preload from the lower ball joint.

2. Have an assistant grasp the wheel top and bottom and apply alternate in and out pressure to the top and bottom of the wheel.

3. Radial play of ¼ in. is acceptable measured at the inside of the wheel adjacent to the lower arm.

4. For replacement procedures see Steps 4–11 of "Upper Ball Joint".

NOTE: This radial play is multiplied at the outer circumference of the tire and should be measured only at the inside of the wheel.

Upper Ball Joint

1. Raise the vehicle by placing a floor jack under the lower arm.

2. Have an assistant grasp the bottom of the tire and move the wheel in and out.

3. As the wheel is being moved, observe the upper control arm where the spindle attaches to it. Any movement between the upper part of the spindle and the upper ball joint indicates a bad ball joint which must be replaced.

NOTE: During this check the lower ball joint will be unloaded and may move; this is normal and not an indiction of a bad ball joint. Also, do not mistake a loose wheel bearing for a defective ball joint. Ford Motor Company recommends replacement of the control arm and ball joint as an assembly. However, aftermarket replacement parts are available, which can be installed using the following procedure. This procedure may be used on both upper and lower ball joints.

4. Raise the vehicle and support on frame points so that the front wheels fall to their full down position.

5. Drill a ⅛ in. hole completely through each ball joint attaching rivet.

6. Using a large chisel, cut off the head of each rivet and drive them from the arm.

7. Place a jack under the lower arm and raise to compress the coil spring. Do not raise the vehicle from the jackstands.

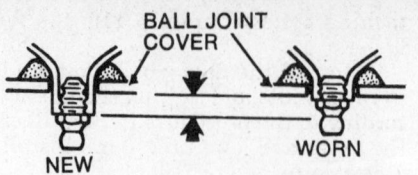

BALL JOINT COVER

NEW   WORN

CHECKING SURFACE
Ball joint wear indicator

8. Remove the cotter pin and attaching nut from the ball joint stud.

9. Using a ball joint removal tool, loosen the ball joint stud from the spindle and remove the ball joint from the arm.

10. Clean all metal burrs from the arm and install the new ball joint, using the service part nuts and bolts to attach the ball joint. Do not attempt to rivet the ball joint once it has been removed.

11. Check front end alignment.

Upper Control Arm

1. Raise the car and support the frame with jackstands placed just behind the lower arm rear pivot point. Remove the wheel.

2. Remove the cotter pin from the upper ball joint stud nut. Loosen the nut a few turns but do not remove.

3. Install a ball joint removal tool between the upper and lower ball joint studs. Expand the tool until it places the upper stud under compression. Tap the spindle near the stud with a hammer to loosen the stud.

4. Remove the tool. Raise the lower arm with a jack until pressure is relieved from the upper stud. Remove the upper stud nut.

5. Remove the upper shaft attaching bolts and the upper arm.

6. To install, position the arm to the frame, install the attaching nuts, and torque to 120–140 ft. lbs. Connect the upper stud to the spindle. Install the attaching nuts, and tighten to 75

ft. lbs., then continue to tighten until the cotter pin holes align. Install a new cotter pin. Install the wheel, adjust the wheel bearings, and lower the car. Caster, camber and toe must be adjusted after installation.

Type 3: Strut Suspension

REMOVAL & INSTALLATION

Strut and Upper Mount Assembly

1. Raise the front of the car and place jackstands under the jacking pads just behind the lower arms.

2. Remove the wheel and tire. Remove the brake caliper and position out of the way, do not allow the caliper to hang from the brake hose. Raise the lower arm with a floor jack to compress the spring.

3. Remove the three upper strut mounting nuts from the top of the shock tower. (if the upper mount is to be replaced on Thunderbird/Cougar models, loosen the 16mm strut rod nut at this time).

4. Remove the two lower strut nuts that attach the strut to the spindle bracket. Leave the bolts in place.

5. Compress the strut to clear the upper mount. With the strut compressed, remove the lower shock strut through bolts. Push the mounting bracket free of the spindle and remove the strut.

NOTE: On models equipped with gas pressurized struts, the strut will remain fully extended. Carefully remove both lower strut to spindle bolts, push the bracket free of the spindle and remove the strut.

7. To install, place the upper

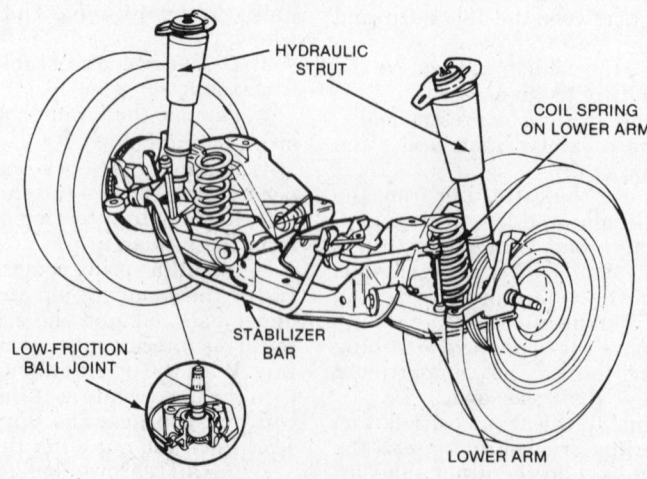

HYDRAULIC STRUT

COIL SPRING ON LOWER ARM

LOW-FRICTION BALL JOINT

STABILIZER BAR

LOWER ARM

Typical strut suspension

mount in position on the shock tower. Loosely install new upper mounting nuts. Extend the strut and position in the spindle bracket. Install the two lower mounting bolts and nuts. Tighten the nuts to 140–170 ft. lbs.

8. Raise the control arm with a floorjack and tighten the upper mount nuts to 50–70 ft. lbs.

9. Install the remaining parts in the reverse order of removal.

Ball Joint

Only one ball joint is used on each side, located in the lower arm. It is provided with a grease fitting, which projects beyond the ball joint cover. When the checking surface (the round boss into which the grease fitting is threaded) is flush with the cover, the ball joint is due for replacement. The ball joint and lower arm must be replaced as an assembly. Follow the instructions for arm replacement.

Coil Spring

1. Raise and support the front of the vehicle on jackstands placed on the pads just behind the lower arms. Remove the wheel and brake caliper. Suspend the caliper with hose connected, out of the way with wire.

2. Disconnect the tie rod end from the steering spindle.

3. Disconnect the stabilizer bar from the arm. Remove the steering gear bolts and lower the gear out of the way to provide clearance (if necessary) for suspension arm bolt removal.

4. Install a spring compressor. Turn the tightening nut on the tool so the spring is free in the seat.

5. Remove the two lower control arm pivot bolts and disengage the arm from the frame. Remove the spring.

6. Reverse to install. Be sure the lower end of the spring is properly positioned between the two holes in the lower arm spring pocket.

Lower Arm

1. Raise and support the front of the vehicle on jackstands. Position the stands on the lifting pads that are just behind the lower control arms.

2. Remove the wheels. Remove the brake caliper and rotor.

3. Disconnect the tie rod end from the steering spindle. Remove the steering gear mounting bolts and position the steering gear to give clearance for suspension mounting bolt removal.

4. Disconnect the stabilizer bar from the lower arm. Remove the cotter pin from the ball joint stud nut, and loosen the nut two turns. Rap the

spindle boss to loosen the stud in the spindle.

5. Compress the coil spring with a suitable spring compressor. Remove the ball joint nut and raise the entire strut and spindle assembly. Wire in position, up and out of the way.

6. Remove the suspension arm to crossmember mounting bolts and nuts. Remove the control arm and coil spring.

7. Install in the reverse order of removal. Tighten the stud nut to 80 ft. lbs., then continue to tighten to align the cotter pin holes. Install a new cotter pin. Torque the tie rod end nut to 35–47 ft. lbs. Stabilizer bolt torque is 9–12 ft. lbs. Steering gear to crossmember is 90–100 ft. lbs.

8. Check front end alignment.

Front Suspension Components
—— CAUTION ——

Power to the air system must be shut off by turning the air suspension switch (in luggage compartment) OFF or by disconnecting the battery when servicing any suspension components.

REMOVAL & INSTALLATION

Stabilizer Bar Link Insulators

1. Turn the air suspension switch OFF.

2. Raise the vehicle and support on jackstands.

3. Remove the nut, washer, and insulator from the end of the stabilizer bar link attaching bolt.

4. Remove the bolt and the remaining washers, insulators and spacer.

5. Install the stabilizer bar link insulators by reversing the removal procedure.

6. Tighten the attaching nut.

7. Lower the vehicle. Turn air suspension system ON.

Stabilizer Bar and/or Bushing

1. Turn the air suspension switch OFF.

2. Raise the vehicle and support on jackstands.

3. Disconnect the stabilizer bar from each link and bushing U-clamps. Remove the stabilizer bar assembly.

4. Remove the adapter brackets and U-clamps.

5. Cut the worn bushings from the stabilizer bar.

6. Coat the necessary parts of the stabilizer bar with Ford Rubber Suspension Insulator Lubricant, E25Y-19533-A or equivalent, and slide bushings onto the stabilizer bar. Reinstall the U-clamps.

7. Reinstall the adapter brackets on the U-clamps.

8. Using a new nut and bolt, secure each end of the stabilizer bar to the lower suspension arm.

9. Using new bolts, clamp the stabilizer bar to the attaching brackets on the side rail.

10. Lower the vehicle. Turn air suspension switch ON.

Shock Strut Replacement

1. Turn the air suspension switch OFF.

2. Turn the ignition key to the unlocked position to allow free movement of the front wheels.

3. From the engine compartment, loosen but do not remove the one 16mm strut-to-upper mount attaching nut. A suitable tapered tool inserted in the slot will hold the rod stationary while loosening the nut. The vehicle should be in place to be raised and must not be driven with the nut loosened or removed.

4. Raise and support the vehicle. Position safety stands under the lower control arms as far outboard as possible being sure that the lower sensor mounting bracket is clear. Lower until vehicle weight is supported by the lower arms.

5. remove tire and wheel assembly.

6. Remove the brake caliper and wire out of the way.

7. Remove the strut-to-upper mount attaching nut and then the two lower nuts and bolts attaching the strut to the spindle.

NOTE: The strut should be held firmly during the removal of the last bolt since the gas pressure will cause the strut to fully extend when removed.

8. Lift the strut up from the spindle to compress the rod and then remove the strut.

9. Prime the new strut by extending and compressing the strut rod five times.

10. Place the strut rod through the upper mount, hand start and secure a new 16mm nut.

11. Compress the strut, and position onto the spindle.

12. Install two new lower mounting bolts, and hand start the nuts.

13. Raise the vehicle to remove load from the lower control arms, and tighten the lower mounting nuts.

14. Install the brake caliper. Install the tire and wheel assembly.

15. Remove safety stands and lower the vehicle to the ground.

16. Turn air suspension switch ON.

NOTE: Front wheel alignment should be checked and adjusted, if out of specification.

Upper Mount Assembly

NOTE: Upper mounts are one piece units and cannot be disassembled.

1. Turn the air suspension system OFF.

2. Turn the ignition key to the unlocked position to allow free movement of the front wheels.

3. From the engine compartment, loosen but do not remove the three 12mm upper mount retaining nuts. Vehicle should be in place over a hoist and must not be driven with these nuts removed. Do not remove the pop rivet holding the camber plate in position.

4. Loosen 16mm strut rod nut at this time.

5. Raise the vehicle and position safety stands under the lower control arms as far outboard as possible being sure that the lower sensor mounting bracket is clear. Lower until the vehicle weight is supported by the lower arms.

6. Remove the tire and wheel assembly.

7. Remove brake caliper and rotate out of position and wire securely out of the way.

8. Remove the upper mount retaining nuts and the two lower nuts and bolts that attach the strut to the spindle.

NOTE: The strut should be held firmly during the removal of the last bolt since the gas pressure will cause the strut to fully extend when removed.

9. Lift the strut up from the spindle to compress the rod, and then remove the strut.

10. Remove the upper mount from the strut.

11. Install a new upper mount on the strut and hand start a new 16mm nut.

12. Position the upper mount studs into the body and start and secure three new nuts. Secure the strut rod 16mm nut.

13. Compress the strut and position onto the spindle.

14. Install two new lower mounting bolts, and hand start nuts.

15. Raise the vehicle to remove load from the lower control arms and tighten the lower mounting nuts to 126–179 ft. lbs.

16. Install the brake caliper. Install the tire and wheel assembly.

17. Remove safety stands and lower vehicle to the ground.

18. Turn air suspension switch ON.

19. Front wheel alignment should be checked and adjusted if out of specification.

Spindle Assembly

1. Turn the air suspension switch OFF.

2. Raise and support the vehicle on jackstands.

3. Remove the wheel and tire assembly.

4. Remove the brake caliper, rotor and dust shield.

5. Remove the stabilizer link from the lower arm assembly.

6. Remove the tie rod end from the spindle.

7. Remove the cotter pin from the ball joint stud nut, and loosen the nut one or two turns.

— **CAUTION** —

DO NOT remove the nut from the ball joint stud at this time.

8. Tap the spindle boss smartly to relieve stud pressure.

9. Place a floor jack under the lower arm, compress the air spring and remove the stud nut.

10. Remove the two bolts and nuts attaching the spindle to the shock strut. Compress the shock strut until working clearance is obtained.

11. Remove the spindle assembly.

12. Place the spindle on the ball joint stud, and install the new stud nut. DO NOT tighten at this time.

13. Lower the shock strut until the attaching holes are in line with the holes in the spindle. Install two new bolts and nuts.

14. Tighten ball joint stud nut and install cotter pin.

15. Lower the floor jack from under the suspension arm, and remove jack.

16. Tighten the shock strut to spindle attaching nuts.

17. Install stabilizer bar link and tighten attaching nut.

18. Attach the tie rod end, and tighten the retaining nut.

19. Install the disc brake dust shield, rotor, and caliper.

20. Install the wheel and tire assembly.

21. Remove the safety stands, and lower the vehicle.

22. Turn air suspension switch ON.

23. Front wheel alignment should be checked and adjusted if out of specification.

Suspension Control Arm

1. Turn the air suspension switch OFF.

2. Raise the vehicle and support on jackstands, so the control arms hang free (full rebound).

3. Remove the wheel and tire assembly.

4. Disconnect the tie rod assembly from the steering spindle.

5. Remove the steering gear bolts, if necessary, and position the gear so that the suspension arm bolt may be removed.

6. Disconnect the stabilizer bar link from the lower arm.

7. Disconnect the lower end of the height sensor from the lower control arm sensor mounting stud. Remove the sensor mounting stud and unscrew from lower arm, noting the position of stud on the lower arm bracket.

8. Remove the cotter pin from the ball joint stud nut, and loosen the ball joint nut one or two turns. DO NOT remove the nut at this time. Tap spindle boss smartly to relieve stud pressure.

9. Vent the air spring(s) to atmospheric pressure. Then reinstall the solenoid.

10. Remove the air spring to lower arm fastener clip.

11. Remove the ball joint nut, and raise the entire strut and spindle assembly (strut, rotor, caliper and spindle). Wire it out of the way to obtain working room.

12. Remove the suspension arm to crossmember nuts and bolts, and remove the arm from the spindle.

13. Position the arm into the crossmember and install new arm to crossmember bolts and nuts. DO NOT tighten at this time.

14. Remove the wire from the strut and spindle assembly and attach to the ball joint stud. Install a new ball joint stud nut. DO NOT tighten at this time.

15. Position the air spring in the arm and install a new fastener.

16. Attach the sensor mounting stud and screw to lower arm in the same position as original arm location. Connect the lower end of sensor to the lower arm mounting stud.

17. With a suitable jack, raise the suspension arm to curb height.

18. With the jack still in place, tighten the lower arm to crossmember attaching nut to 150–180 ft. lbs.

19. Tighten ball joint stud nut to 100–120 ft. lbs., and install a new cotter pin. Remove jack.

20. Install the steering gear to crossmember bolts and nuts (if removed). Hold the bolts, and tighten nuts to 90–100 ft. lbs.

21. Position the tie rod assembly into the steering spindle, and install the retaining nut. Tighten the nut to 35 ft. lbs., and continue tightening the nut to align the next castellation with cotter pin hole in the stud. Install a new cotter pin.

22. Connect the stabilizer bar link to the lower suspension arm, and tighten the attaching nut to 9–12 ft. lbs.

23. Install the wheel and tire assem-

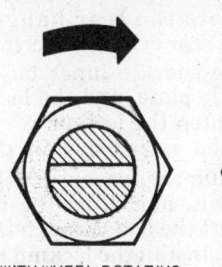

| WITH WHEEL ROTATING, TORQUE ADJUSTING NUT, TO 17-25 FT. LBS. | BACK ADJUSTING NUT OFF 1/2 TURN | TIGHTEN ADJUSTING NUT TO 10-15 IN.-LBS. | INSTALL THE LOCK AND A NEW COTTER PIN |

Front wheel bearing adjustment

bly, and lower the vehicle but DO NOT allow tires to touch the ground.

24. Turn the air suspension switch ON.

25. Refill the air spring(s).

26. Front wheel alignment should be checked and adjusted if out of specification.

Front Wheel Bearing

ADJUSTMENT

1. Raise and support the front of the vehicle on jackstands.

2. Remove the wheel cover and grease cap.

3. Remove the cotter pin and nut lock.

4. Loosen the adjusting nut three turns and rock the wheel back and forth a few times to release the brake pads from the rotor.

5. While rotating the wheel and hub assembly, tighten the adjusting nut to 17–25 ft. lbs.

6. Back off the adjusting nut ½ turn, then retighten to 10-15 inch lbs.

7. Install the locknut and a new cotter pin. Check the wheel rotation. If it is noisy or rough, the bearings either need to be cleaned or repacked, or readjusted. After adjustments are complete, replace the grease cap.

REAR SUSPENSION

NOTE: For models equipped with air suspension refer to the section following "Rear Suspension". Rear suspension differs depending on year and model. The basic designs are leaf spring and coil spring (either between the axle housing and frame, or control arm and frame).

Shock Absorbers

REMOVAL & INSTALLATION

NOTE: Purge a new shock of air by repeatedly extending it in its normal position and compressing it while inverted. Models equipped with axle dampers are serviced by supporting the rear of the vehicle, removing the wheel and disconnecting the front and rear mounting nuts and removing the damper.

1. Remove the lower end of the shock absorber from the spring plate.

2. Remove the nut retaining the upper end of the shock absorber to the mounting bracket underneath the car.

3. Compress and remove the shock absorber. Discard the nuts.

4. Transfer the washers and bushings to the new shock absorber. Insert the upper stud through the mounting bracket, and install a new attaching nut finger-tight.

5. Compress and install the shock absorber to the spring plate. Install the washer, bushings, and attaching nuts.

6. Tighten the upper and lower attaching nuts.

SPRING BETWEEN AXLE HOUSING AND FRAME

1. Raise the vehicle and install jackstands.

2. Remove the shock absorber outer attaching nut, washer and insulator from the stud at the top side of the spring upper seat. Compress the shock sufficiently to clear the spring seat hole, and remove the inner insulator and washer from the upper attaching stud.

3. Remove the locknut and disconnect the shock absorber lower stud at the mounting bracket on the axle housing. Remove the shock absorber.

4. Position a new inner washer and insulator on the upper spring seat. While maintaining the shock in this position, install a new outer insulator, washer, and nut on the stud from the top side of the spring upper seat.

5. Extend the shock absorber. Locate the lower stud in the mounting bracket hole on the axle housing and install the locknut.

SPRING BETWEEN LOWER CONTROL ARM AND FRAME

1. Remove the upper attaching nut,

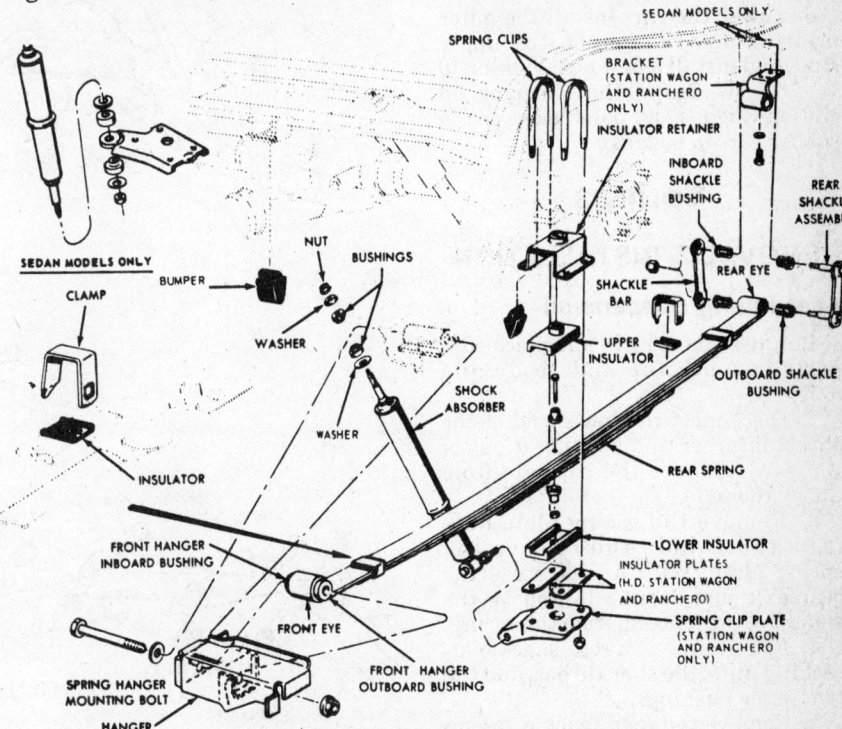

Exploded view of leaf spring rear suspension

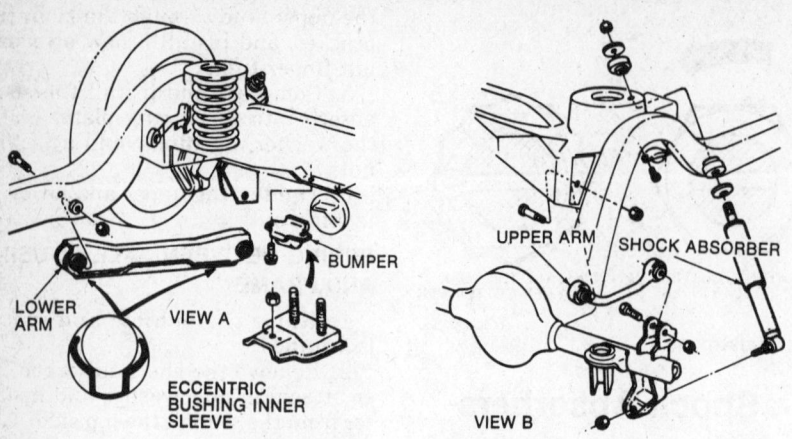

Coil spring rear suspension

washer, and insulator. Access is through the trunk on sedans or side panel trim covers on station wagons and hatchbacks. Sedan studs have rubber caps.

2. Raise the car. Compress the shock to clear the upper tower. Remove the lower nut and washer; remove the shock.

3. Purge the shock of air and compress. Place the lower mounting eye over the lower stud and install the washer and a new locking nut. Do not tighten the nut yet.

4. Place the insulator and washer on the upper stud. Extend the shock, installing the stud through the upper mounting hole.

5. Torque the lower mounting nut to 40–55 ft. lbs.

6. Lower the car. Install the outer insulator and washer on the upper stud, and install a new nut. Tighten to 14–26 ft. lbs. Install the trim panel on station wagons and hatchbacks or the rubber cap on sedans.

Springs

REMOVAL & INSTALLATION

Leaf Spring Suspension

1. Raise the vehicle and place supports beneath the underbody and axle.

2. Disconnect the lower end of the shock absorber and position it out of the way. Remove the supports from under the axle.

3. Remove the spring plate nuts from the U-bolt and remove the spring plate. With a jack, raise the rear axle just enough to remove the weight of the housing from the spring.

4. Remove the two rear shackle attaching nuts, the shackle bar, and the two inner bushings.

5. Remove the rear shackle assembly and the two outer bushings.

6. Remove the nut from the spring mounting bolt and tap the bolt out of the bushing at the front hanger. Lift out the spring assembly.

NOTE: All used attaching components (nuts, bolts, etc.) must be discarded and replaced with new ones prior to assembly. Bushings may be lubricated with soap and water to ease bolt installation; do not use grease or oil.

7. Position the leaf spring under the axle housing and insert the shackle assembly into the rear hanger bracket and the rear eye of the spring.

8. Install the shackle inner bushings, the shackle plate, and the locknuts. Hand tighten the locknuts.

9. Position the spring eye in the front hanger, slip the washer on the front hanger bolt, and, from the inboard side, insert the bolt through the hanger and eye. Install the locknut on the hanger bolt finger-tight.

10. Lower the rear axle housing so that it rests on the spring. Place the spring plate on the U-bolt and tighten the nuts.

11. Attach the lower end of the shock absorber to the spring plate using a new nut.

12. Place jackstands under the rear axle. Lower the vehicle until the spring is in the approximate curb load position, and tighten the front hanger locknut.

13. Tighten the rear shackle locknuts.

14. Remove the jackstands and lower the vehicle.

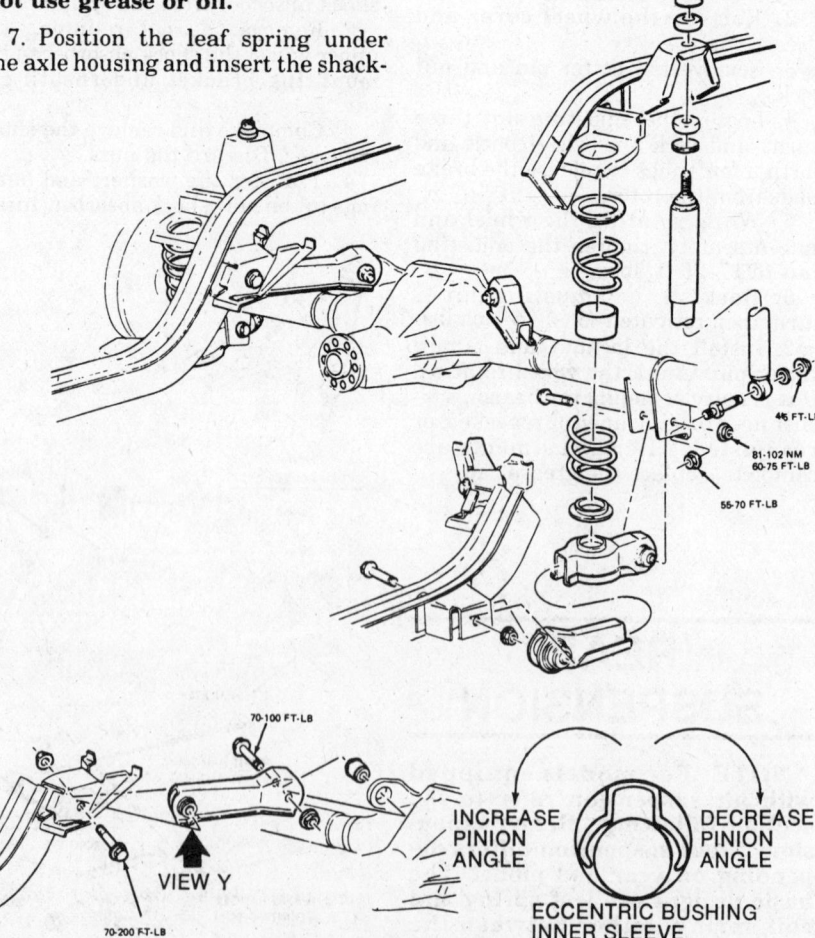

Four-bar link coil spring suspension

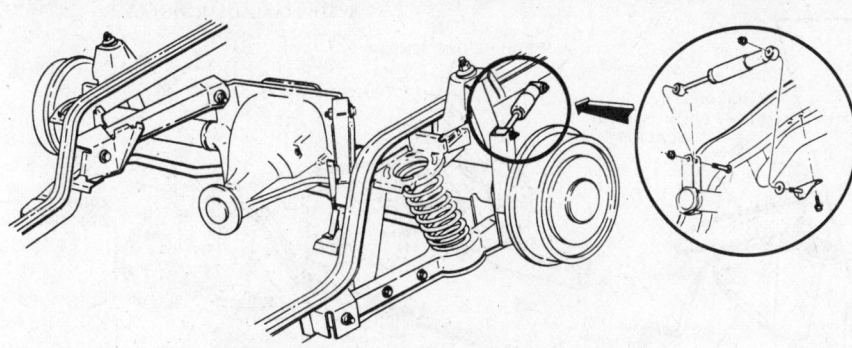

Hydra–trac rear suspension showing dampers

Coil Spring Suspension

REMOVAL & INSTALLATION

Coil Spring

SPRING BETWEEN AXLE HOUSING AND FRAME.

1. Place a jack under the rear axle housing. Raise the vehicle and place jackstands under the frame side rails.
2. Disconnect the lower studs of the shock absorbers from the mounting brackets on the axle housing.
3. Lower the axle housing until the springs are fully released.
4. Remove the springs and insulators from the vehicle.
5. Place the insulators in each upper seat and position the springs between the upper and lower seats.
6. With the springs in position, raise the axle housing until the lower studs of the rear shock absorbers reach the mounting brackets on the axle housing. Connect the lower studs and install the attaching nuts.
7. Remove the jackstands and lower the vehicle.

SPRING BETWEEN LOWER CONTROL ARM AND FRAME

NOTE: If one spring must be replaced, the other should be replaced also. If the car has a stabilizer bar, the bar must be removed first.

1. Raise and support the car at the rear crossmember, while supporting the axle with a jack.
2. Lower the axle until the shocks are fully extended.
3. Place a jack under the lower control arm pivot bolt. Remove the pivot bolt and nut. Carefully and slowly lower the arm until the spring load is relieved.
4. Remove the spring and insulators.

5. To install, tape the insulator in place in the frame, and place the lower insulator in place on the arm. Install the internal damper in the spring.
6. Position the spring in place and slowly raise the jack under the lower arm. Install the pivot bolt and nut, with the nut facing outwards. Do not tighten the nut.
7. Raise the axle to curb height, and tighten the lower pivot bolt to 70–100 ft. lbs. Remove the crossmember stands and lower the car.

Rear Control Arm

1. Raise and support the vehicle on jackstands positioned on the rear frame pads.
2. Position a floor jack under the rear axle and raise slightly. Position jackstands at both ends of the axle to support the axle weight.
3. Position a jack under the lower arm pivot bolt and raise to support. Remove the pivot bolt and nut.
4. Lower the jack slowly and remove the coil spring. Remove the control arm.
5. Install the control arm in the reverse order of removal.

AIR SUSPENSION

Components

REMOVAL & INSTALLATION

——————— **CAUTION** ———————

DO NOT remove an air spring under any circumstances when there is pressure in the air spring. Do not remove any components supporting an air spring without either exhausting the air or providing support for the air spring.

Suspension Fasteners

Suspension fasteners are important attaching parts in that they could affect performance of vital components and systems and/or could result in major service expense. They must be replaced with fasteners of the same part number, or with an equivalent part, if replacement becomes necessary. DO NOT use a replacement part of lesser quality or substitute design. Torque values must be used, as specified, during assembly to assure proper retention of parts. New fasteners must be used whenever old fasteners are loosened or removed and when new component parts are installed.

Jacking and Supporting

——————— **CAUTION** ———————

The electrical power supply to the air suspension system must be shut off prior to hoisting, jacking or towing an air suspension vehicle. This can be accomplished by disconnecting the battery or turning off the power switch located in the trunk on the LH side. Failure to do so may result in unexpected inflation or deflation of the air springs which may result in shifting of the vehicle during these operations.

Raise the front of the vehicle at the No. 2 crossmember until the tires are above the floor. Support the vehicle body with jackstands at each front corner and then lower the floor jack so that the front suspension is in full rebound. Repeat this procedure for the rear suspension, except raise the body at the rear jacking location.

——————— **CAUTION** ———————

Power to the air system must be shut off by turning the air suspension switch (in luggage compartment) Off or by disconnecting the battery when servicing any air suspension components. Do not attempt to install or inflate any air spring that has become unfolded. Any spring which has unfolded must be refolded, prior to being installed in a vehicle. Do not attempt to inflate any air spring which has been collapsed while uninflated from the rebound hanging position to the jounce stop. After inflating an air spring in hanging position, it must be inspected for proper shape. Failure to follow the above may result in a sudden failure of the air spring or suspension system.

Air Spring Solenoid

The air spring solenoid valve has a two stage solenoid pressure relief fitting similar to a radiator cap. A clip is first removed, and rotation of the solenoid out of the spring will release air from the assembly before the solenoid can be removed.

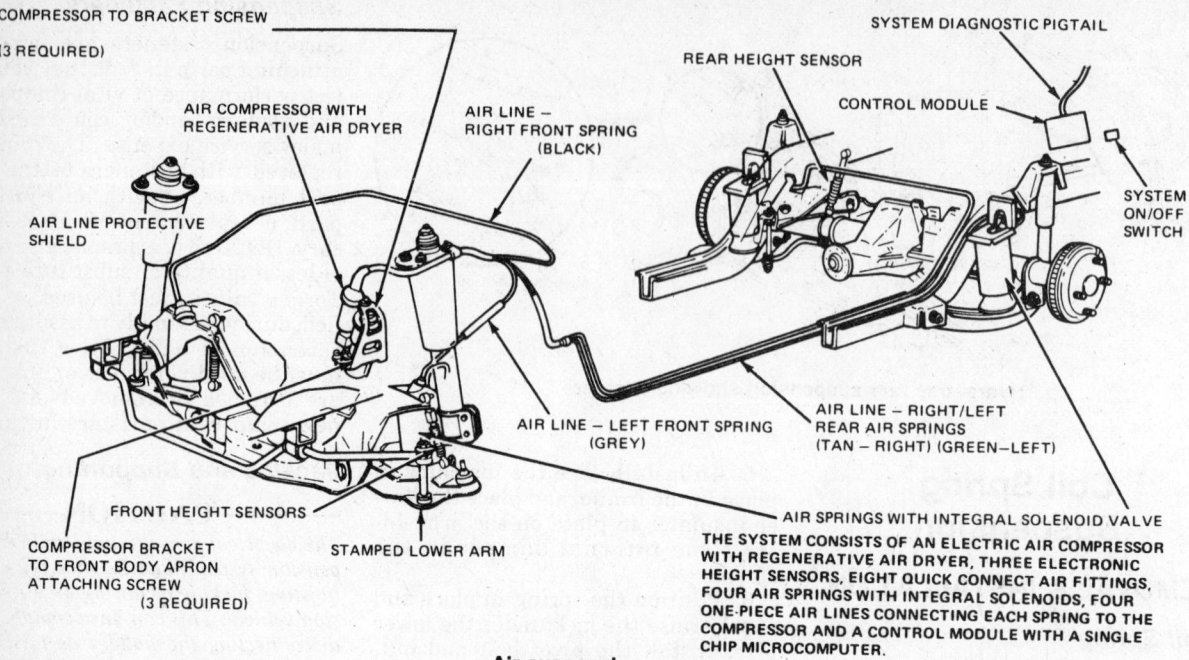

COMPRESSOR TO BRACKET SCREW
(3 REQUIRED)

AIR COMPRESSOR WITH REGENERATIVE AIR DRYER

AIR LINE PROTECTIVE SHIELD

AIR LINE – RIGHT FRONT SPRING (BLACK)

SYSTEM DIAGNOSTIC PIGTAIL

REAR HEIGHT SENSOR

CONTROL MODULE

SYSTEM ON/OFF SWITCH

FRONT HEIGHT SENSORS

COMPRESSOR BRACKET TO FRONT BODY APRON ATTACHING SCREW (3 REQUIRED)

STAMPED LOWER ARM

AIR LINE – LEFT FRONT SPRING (GREY)

AIR LINE – RIGHT/LEFT REAR AIR SPRINGS (TAN – RIGHT) (GREEN–LEFT)

AIR SPRINGS WITH INTEGRAL SOLENOID VALVE

THE SYSTEM CONSISTS OF AN ELECTRIC AIR COMPRESSOR WITH REGENERATIVE AIR DRYER, THREE ELECTRONIC HEIGHT SENSORS, EIGHT QUICK CONNECT AIR FITTINGS, FOUR AIR SPRINGS WITH INTEGRAL SOLENOIDS, FOUR ONE-PIECE AIR LINES CONNECTING EACH SPRING TO THE COMPRESSOR AND A CONTROL MODULE WITH A SINGLE CHIP MICROCOMPUTER.

Air suspension

1. Turn the air suspension switch OFF.

2. Raise the vehicle. Remove wheel and tire assembly.

3. Disconnect the electrical connector and then disconnect the air line.

4. Remove the solenoid clip. Rotate the solenoid counterclockwise to the first stop.

5. Pull the solenoid straight out slowly to the second stop to bleed air from the system.

CAUTION

Do not fully release solenoid until air is completely bled from the air spring.

6. After the air is fully bled from the system, rotate the solenoid counterclockwise to the third stop, and remove the solenoid from the air spring assembly.

7. Check the solenoid O-ring for abrasion or cuts. Replace O-ring as required. Lightly grease the O-ring area of solenoid with silicone dielectric compound WA-10 D7AZ-19AA331-A or equivalent.

8. Insert the solenoid into the air spring end cap and rotate clockwise to the third stop, push into the second stop, then rotate clockwise to the first stop.

9. Install solenoid clip. Connect the air line and the electrical connector.

10. Refill the air spring(s). Install the wheel and tire assembly.

Air Spring Fill

1. Turn ON the air suspension switch. Diagnostic pigtail is to be ungounded.

2. Connect a battery charger to reduce battery drain.

3. Cycle the ignition from the OFF to RUN position, hold in the RUN position for a minimum of five seconds, then return to the OFF position. Driver's door is open with all other doors shut.

4. Change the diagnostic pigtail from an ungrounded state to a grounded state by attaching a lead from the diagnostic pigtail to vehicle ground. The pigtail must remain grounded during the spring fill sequence.

5. While applying the brakes, turn the ignition switch to the RUN position. (The door must be open. Do not start the vehicle). The warming lamp will blink continuously once every two seconds to indicate the spring pump sequence has been entered.

6. To fill a rear spring(s), close and open the door twice. After a 6 second delay, the rear spring will be filled for 60 seconds.

7. To fill a front spring(s), close and open the door twice. After a 6 second delay, the rear spring will be filled for 60 seconds.

8. To fill rear and front springs, fill the rear springs first (Step 6). When the rear fill has finished, close and open the door once to initiate the front spring fill.

9. Terminate the air spring fill by turning the ignition switch to OFF, actuating the brake, or ungrounding the diagnostic pigtail. The diagnostic pigtail must be ungrounded at the end of spring fill.

10. Lower vehicle and start engine.

Allow the vehicle to level with doors closed.

Air Spring (Front or Rear)

1. Turn the air suspension switch OFF.

2. Raise and support the vehicle. Suspension must be at full rebound.

3. Remove tire and wheel assembly.

4. Remove the air spring solenoid.

5. Remove the spring to lower arm fasteners. Remove the clip for front spring and/or remove bolts for rear spring.

6. Push down on the spring clip on the collar of the air spring and rotate collar counterclockwise to release the spring from the body spring seat. Remove the air spring.

7. Install the air spring solenoid. Correctly position the solenoid. For LH installation (front or rear spring), the notch on the collar is to be in line with the centerline of the solenoid. For RH installation (front or rear), the flat on the collar is to be in line with the centerline of the solenoid.

8. Install the air spring into the body spring seat, taking care to keep the solenoid air and electrical connections clean and free of damage. Rotate the air spring collar until the spring clip snaps into place. Be sure that the air spring collar is retained by the three rolled tabs on the body spring seat.

9. Attach the air line and electrical connector to the solenoid assembly.

10. Align and secure the lower arm to spring attachment with suspension

at full rebound and supported by the shock absorbers.

CAUTION

The air springs may be damaged if suspension is allowed to compress before spring is inflated.

---

11. Replace the tire and wheel assembly.
12. Lower the vehicle until the tire and wheel assembly are 1–3 in. above floor. Refill the air spring(s).

Air Compressor and Dryer Assembly

1. Turn the air suspension switch OFF.
2. Disconnect the electrical connector located on the compressor.
3. Remove the air line protector cap from the dryer by releasing the two latching pins located on the bottom of the cap 180 degrees apart.
4. Disconnect the four air lines from dryer.
5. Remove the three screws retaining the air compressor to mounting bracket.
6. Position the air compressor and dryer assembly to the mounting bracket and install the three mounting screws.
7. Connect the four air lines into the dryer.
8. Connect the electrical connection. Install the air line protector cap onto the dryer.
9. Turn the air suspension switch ON.

Dryer, Air Compressor

1. Turn the air suspension switch OFF.
2. Remove the air line protector cap from the dryer by releasing the two latching pins located on the bottom of the cap 180 degrees apart.
3. Disconnect the four air lines from the dryer.
4. Remove the dryer retainer clip and screw.
5. Remove from the head assembly.
6. Check to ensure the old O-ring is not in the head assembly.
7. Check the dryer end to ensure new O-ring is in proper position.
8. Insert the dryer into the head assembly and install the retainer clip and screw.
9. Connect the four air lines into the dryer.
10. Install the air protector cap onto the dryer.
11. Turn the air suspension switch ON.

Mounting Bracket, Air Compressor

1. Turn the air suspension switch OFF.
2. Remove the air compressor and dryer assembly.
3. Raise and support the vehicle on jackstands.
4. Remove the left front tire and wheel assembly.
5. Remove the left front inner fender liner.
6. Remove the three bolts attaching the mounting bracket to body side apron.
7. Position the mounting bracket to the body side apron with the two locating tabs.
8. Secure the three bolts attaching the bracket to the body side apron.
9. Install the left front inner fender liner.
10. Install the tire and wheel assembly.
11. Lower the vehicle.
12. Install compressor and dryer assembly. Turn the air suspension switch On.

Height Sensors—Front

1. Turn the air suspension switch OFF.
2. Disconnect the sensor electrical connector. The front sensor connectors are located in the engine compartment behind the shock towers.
3. Push the front sensor connector through the access hole in the rear of the shock tower.
4. Raise and support the vehicle on jackstands. Suspension must be at full rebound.
5. Disconnect the bottom and then the top end of the sensor from the attaching studs.
6. Disconnect the sensor wire harness from plastic clips on the shock tower and remove sensor.
7. Connect the top and then the bottom end of the sensor to the attaching studs. Route the sensor electrical connector as required to connect to the vehicle wire harness.
8. Lower the vehicle. Connect the sensor connector. Turn the air suspension switch ON.

Height Sensor—Rear

1. Turn the air suspension switch OFF.
2. Disconnect the sensor electrical connector located in the luggage compartment in front of the forward trim panel. also pull the luggage compartment carpet back for access to the sensor sealing grommet located on the floor pan.
3. Raise and support the vehicle on jackstands. Suspension must be at full rebound.
4. Disconnect the bottom and then the top end of the sensor from the attaching studs.
5. Push upwards on the sealing grommet to unseat and then push sensor through the floor pan hole into the luggage compartment.
6. Lower the vehicle.
7. Connect the sensor connector and then push sensor through the floor pan hole being sure to seat the sealing grommet. Replace the luggage compartment carpet.
8. Raise and support the vehicle on jackstands.
9. Connect the top and then the bottom end of the sensor. Lower the vehicle.
10. Turn the air suspension switch ON.

Control Module

1. Turn the air suspension switch OFF. Ignition switch is also to be OFF.
2. Remove the LH luggage compartment trim panel.
3. Disconnect the wire harness from the module.
4. Remove the three attaching nuts.
5. Remove the module.
6. Position the module and secure it with the three attaching nuts.
7. Connect the wire harness to the module.
8. Attach the LH luggage compartment trim panel. Turn the air suspension switch ON.

Nylon Air Line

If a leak is detected in an air line, it can be serviced by carefully cutting the line with a sharp knife to ensure a good, clean, straight cut. Then, install a service fitting. If more tube is required, it can be obtained in bulk. The four air lines are color coded to show which spring they are connecting, but do not require orientation at the air compressor dryer. A protective plastic cap and convoluted tube protect the air lines from the dryer rearward over the left shock tower in the engine compartment. Routing of the lines after exiting the protective tube follows:

Left Front/Grey: Down and through the rear wall of the left shock tower to the air spring solenoid.

Right Front/Back: To cowl and along cowl on the right side of the vehicle, forward and down through the rear wall of the right shock tower to the air spring solenoid.

Left Rear/Green, Right Rear/Tan: Through the left side apron into the fender well, through the left upper dash panel (sealing grommet) into the passenger compartment, down the dash panel to the left rocker, along the rocker to the left rear fender well, over the fender well into the luggage compartment. The left air line goes down through the floor pan (sealing grommet) in front of the left rear shock tower. The right air line goes across the rear seat support and then down through the floor pan (sealing grommet) in front of the right rear shock tower.

Quick Connect Fittings

If a leak is detected in any of the eight quick connect fittings, it can be serviced using a repair kit containing a new O-ring, collet, release ring, and O-ring removal tool. The outer housing of the fitting cannot be serviced.

To remove the collet and O-ring, insert a scrap piece of air line, grasp the air line firmly (do not use pliers) and pull straight out (DO NOT use the release button). A force of 30–50 lbs. is required to remove the collet. After the retainer is removed, use the repair tool to remove the old O-ring.

To service, insert the new O-ring and seat it in the bottom of the fitting housing. Then, insert the new collet, being sure the end with four prongs is inserted. Press the collet into position with finger pressure. Install the new release button.

O-ring Seals

The areas that have O-ring seals that can be serviced are: Air compressor head to dryer. One O-ring. Air spring solenoid to end cap: Two O-rings each solenoid. Quick connect fitting: Four O-rings at dryer, one O-ring at each spring. If air leaks are detected in these areas, the components can be removed, following the procedures outlined in this section, and new O-rings can be installed.

Air Suspension Switch

1. Disconnect the electrical connector.
2. Depress the retaining clips that retain the switch to the brace, and remove switch.
3. Push the switch into position in the brace, making sure retaining clips are fully seated.
4. Connect electrical connector.

Compressor Relay

1. Disconnect the electrical connector.
2. Remove the screw retaining the relay to the left front shock tower and remove the relay.
3. Position the relay on the shock

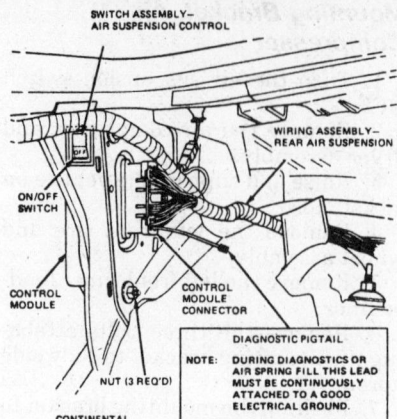

Air suspension cut-off switch—Continental

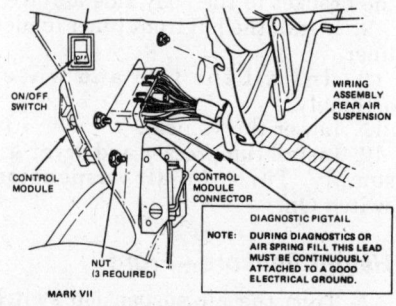

Air suspension cut-off switch—Mark VII

tower and install the retaining screw.
4. Connect the electrical connector.

Rear Suspension Components

REMOVAL & INSTALLATION

Shock Absorber

————— CAUTION —————

Power to the air system must be shut off by turning the air suspension switch (in luggage compartment) OFF or by disconnecting the batter when servicing any suspension component.

1. Turn the air suspension switch OFF.
2. Open the luggage compartment and remove inside trim panels to gain access to the upper shock stud.
3. Loosen but do not remove the shock rod attaching nut.
4. Raise the vehicle and position two safety stands under the rear axle. Lower the vehicle until weight is supported by the rear axle.
5. Remove the upper attaching nut, washer and insulator and then remove the lower shock protective cover (right shock only) and lower shock absorber cross bolt and nut from the lower shock brackets.

6. From under the vehicle, compress the shock absorber to clear it from the hold in the upper shock tower.

————— CAUTION —————

Shock absorbers will extend unassisted. Do not apply heat or flame to the shock absorber tube during removal.

7. Remove the shock absorber.
8. Prime the new shock absorber by extending and compressing shock absorber five times.
9. Place the inner washer and insulator on the upper attaching stud. Position stud through shock tower mounting hole and position an insulator, washer on stud from the luggage compartment. Hand start the attaching nut and then secure.
10. Place the shock absorber's lower mounting eye between the ears of the lower shock mounting bracket, compressing shock as required. Insert the bolt, (bolt head must seat on the inboard side of the shock bracket), through the shock bracket and the shock absorber mounting eye. Hand start and then secure the original attaching nut.
11. Install the protective cover, to the RH shock absorber. This is done by inserting the bolt point and nut into the cover's open end, sliding the cover over the shock bracket, and snapping the closed end of the cover over the bolt head. Properly installed, the cover will completely conceal the bolt point, nut, and bolt head. The rounded or closed end of the cover should be pointing inboard.
12. Raise the vehicle and remove safety stands from under axle, then lower the vehicle.
13. Reinstall the inside trim panels.
14. Turn air suspension switch ON.

Lower Control Arm

NOTE: If one arm requires replacement, replace the other arm also.

1. Turn the air suspension switch OFF.
2. Raise and support the vehicle so that the suspension will be at full rebound.
3. Remove tire and wheel assembly.
4. Vent air spring(s) to atmospheric pressure. Then, reinstall the solenoid.
5. Remove the two air spring-to-lower arm bolts and remove the air spring from the lower arm.
6. Remove the frame-to-arm and the axle-to-arm bolts and remove the arm from the vehicle.
7. Position the lower arm assembly into the front arm brackets, and in-

sert a new arm-to-frame pivot bolt and nut with nut facing outwards. DO NOT tighten at this time.

8. Position the rear bushing in the axle bracket and install a new arm-to-axle pivot bolt and nut with nut facing outwards. DO NOT tighten at this time.

9. Install two new air spring-to-arm bolts. DO NOT tighten at this time.

10. Using a suitable jack, raise the axle to curb height. Tighten the lower arm front bolt, the rear pivot bolt, and the air spring to arm bolt being sure that the air spring piston is flat on the lower arm. Remove the jack.

11. Replace tire and wheel assembly.

12. Lower the vehicle.

13. Turn the air suspension switch ON.

14. Refill the air spring(s).

Upper Control Arm and Axle Bushings

NOTE: If one arm requires replacement, replace the other arm also.

1. Turn the air suspension switch OFF.

2. Raise and support the vehicle so that the suspension will be at full rebound.

3. On the RH side detach rear height sensor from side arm. Note position of the sensor adjustment bracket on the upper arm.

4. Remove the upper arm-to-axle pivot bolt and nut.

5. Remove the upper arm-to-frame pivot bolt and nut. Remove upper arm from vehicle.

6. Place the upper arm axle bushing remover tool in position and remove the bushing assembly.

7. Using the installer tool, install the bushing assembly into the bushing ear of the rear axle.

8. Place the upper arm into the bracket of body side rail. Insert a new upper arm-to-frame pivot bolt and nut (nut facing outboard). DO NOT tighten at this time.

9. Align the upper arm-to-axle pivot hole with the hole in the axle bushing. If required, raise the axle using a suitable jack to align. Install a new pivot bolt and nut (nut inboard). DO NOT tighten at this time.

10. On the RH side, reattach rear height sensor to the arm. Set the adjustment bracket to the same position as on the replaced arm and tighten nut.

11. Using a suitable jack, raise the axle to curb height, and tighten the front upper arm bolt, and the rear upper arm bolt.

12. Remove the jackstands supporting the axle.

13. Lower the vehicle.

14. Turn the air suspension switch ON.

Stabilizer Bar Link Insulators

1. Turn the air suspension switch OFF.

2. Raise and support the vehicle on jackstands.

3. Remove the nut, washer and insulator from the end of the stabilizer bar link attaching bolt.

4. Remove the bolt and the remaining spacer, washer and insulators.

5. Install the stabilizer bar link insulators by reversing the removal procedure. A new bolt and nut must be used.

6. Tighten the attaching nut.

7. Lower the vehicle.

8. Turn the air suspension switch ON.

Stabilizer Bar Bushings

1. Turn the air suspension switch OFF.

2. Raise and support the vehicle on jackstands.

3. Disconnect the stabilizer bar from each link and bushing U-clamp. Remove the stabilizer bar assembly.

4. Remove the U-clamps.

5. Cut the worn bushings from the stabilizer bar.

6. Coat the necessary parts of the stabilizer bar with Ford Rubber Suspension Insulator Lubricant E25Y-19553-A or equivalent and slide new bushings onto the stabilizer bar. Reinstall U-clamps.

7. Using new bolts and nuts, attach stabilizer bar to the axle. Do not tighten bolts at this time.

8. Using new bolts and nuts, attach the link end of the stabilizer bar to the body. Tighten the link attaching nut and then the axle attaching bolts.

9. Lower the vehicle.

10. Turn the air suspension switch ON.

Rear Wheel Bearings

Refer to the rear axle removal and installation section for rear wheel bearing removal, packing and installation procedures.

BRAKES

For all brake system repair and service procedures not detailed below,

please refer to "Brakes" in the Unit Repair section.

NOTE: An independent parking brake operates the rear wheel brake shoes or pads through a mechanical cable linkage. Front disc brakes are used. Rear disc brakes are standard on Versailles and available on various models when equipped with the hydraulically assisted Hydro-Boost System. Complete service procedures are in the Unit Repair Section.

Master Cylinder

REMOVAL & INSTALLATION
Standard Brakes

1. Working under the dash, disconnect the master cylinder pushrod from the brake pedal. The pushrod cannot be removed from the master cylinder.

2. Disconnect the stoplight switch wires and remove the switch from the brake pedal, using care not to damage the switch.

3. Disconnect the brake lines from the master cylinder.

4. Remove the attaching screws from the firewall and remove the master cylinder from the car.

5. Reinstall in reverse order, leaving the brake line fittings loose at the master cylinder.

6. Fill the master cylinder, and with the brake lines loose, slowly bleed the air from the master cylinder using the foot pedal.

Power Brakes

1. Disconnect the brake lines from the master cylinder.

2. Remove the two nuts and lockwashers that attach the master cylinder to the brake booster.

3. Remove the master cylinder from the booster.

4. Reverse the procedure to reinstall.

5. Fill master cylinder and bleed entire brake system.

6. Refill master cylinder.

Brake Control Valve Assembly (Proportioning, Metering, and Pressure Differential Valve)

REMOVAL & INSTALLATION

1. Disconnect the brake warning

lamp switch wire harness connector from the warning lamp switch.

2. Disconnect the front brake system inlet tube and rear system inlet tube from the brake control valve assembly.

3. Disconnect the left and right front brake outlet tubes from the brake control valve assembly.

4. Disconnect the rear system outlet tube from the brake control valve assembly.

5. Remove the screw that retains the brake control valve assembly on the frame. Remove the assembly from the vehicle.

NOTE: The brake control valve assembly is serviced only as an assembly.

6. Position the brake valve assembly on the frame. Install the mounting screw for frame mounting, and tighten to 7–11 ft. lbs.

7. Install the inlet and outlet tubes in the reverse order of the removal procedure, and torque tube nuts to 10–18 ft. lbs.

8. Connect the brake warning lamp switch wiring harness connector to the brake warning lamp switch. Verify the connection by turning the ignition switch to the ON position; lamp must go on. Also confirm that the two locking fingers on the connector are locked into the switch.

9. Bleed the brake system and centralize the the pressure differential valve by:

   a. Turn the ignition switch to the ON or ACC position.

   b. Depress the brake pedal and the piston will center itself, causing the brake warning lamp to go out (if it was illuminated).

   c. Turn the ignition switch to the OFF position.

   d. Before driving the vehicle, check the operation of the brakes and be sure that a firm pedal is obtained.

NOTE: During the brake system bleeding operation on vehicles equipped with a metering valve, the metering valve bleeder rod must be pushed in (pressure bleeding).

Power Brake Vacuum Unit

REMOVAL & INSTALLATION

1. Working inside the car below the instrument panel, disconnect booster valve operating rod from the brake pedal assembly. To do this, disconnect the stop light switch wires at the con-

nector. Remove the hairpin retainer and nylon washer from the pedal pin. Slide the switch off just enough for the outer arm to clear the pin. Remove the switch. Slide the boost push rod, bushing and inner nylon washer off the pedal pin.

2. Remove the air cleaner for working clearance if necessary. On four cylinder models, disconnect the accelerator cable at the carburetor. Remove the securing screw from the accelerator shaft bracket and remove the cable from the bracket. Remove the two screws attaching the bracket to the manifold; rotate the bracket toward the engine.

3. Disconnect the brake lines at the master cylinder outlet fittings.

4. Disconnect manifold vacuum hose from the booster unit. On cars equipped with speed control, remove the left cowl screen in the engine compartment. Remove three nuts retaining the speed control servo to the firewall and move the servo out of the way.

5. Remove the four bracket-to-firewall attaching bolts.

6. Remove the booster and bracket assembly from the firewall, sliding the valve operating rod out from the engine side.

7. Installation is the reverse of removal. Bleed the brakes after installation is complete.

Hydro-Boost Power Unit

REMOVAL & INSTALLATION

1. Open the hood and remove the two nuts attaching the master cylinder to the brake booster.

2. Remove the master cylinder from the Hydro-Boost accumulator.

3. Set the master cylinder aside without disturbing the hydraulic lines.

4. Disconnect the pressure, steering and return lines from the accumulator.

5. Plug the lines and ports.

6. Working below the instrument panel, disconnect the Hydro-Boost pushrod from the brake pedal. To do this, disconnect the stoplight switch at the connector. Remove the hairpin retainer. Slide the stoplight switch from the brake pedal pin far enough to clear the switch outer pin hole. Remove the switch from the pin.

7. Loosen the Hydro-Boost attaching nuts and remove the pushrod, washers and bushings from the brake pedal pin.

8. Remove the accumulator.

9. Installation is the reverse of re-

moval. Leave the Hydro-Boost mounting nuts loose until the pushrod and stoplight switch are connected to the brake pedal. After installation, remove the coil wire from the distributor. Fill the power steering reservoir, and while cranking the engine, pump the brake pedal. Do not move the steering wheel until all the air has been pumped out of the system. Check the power steering fluid level, install the coil wire, start the engine and pump the brakes while steering from lock to lock. Check for leaks.

Wheel Cylinder

REMOVAL & INSTALLATION

1. Remove the wheel and brake drum.

2. Remove the brake shoe assemblies.

3. Disconnect the brake tube from the brake cylinder at the backing plate.

4. Remove the wheel cylinder attaching bolts and remove the wheel cylinder.

5. Installation is in the reverse order of the removal procedure.

6. Torque the wheel cylinder attaching bolts to 10–20 ft. lbs. Torque the brake tube fitting nut to 10–18 ft. lbs. using a tube nut wrench.

7. Install the links in the ends of the wheel cylinder, and install the shoes and adjuster assemblies.

8. Adjust the brakes. Install the brake drum and wheel. Bleed the brakes.

Parking Brake

ADJUSTMENT

NOTE: If a new cable is installed, prestretch it by applying and releasing five times before making any adjustments.

Rear Drum Brakes

NOTE: In most cases, a rear brake shoe adjustment will provide satisfactory parking brake action. However, if parking brake cables are excessively loose after releasing the handbrake, proceed as follows.

HAND OPERATED LEVER

1. Fully release the parking brake.

2. Place the transmission in Neutral and raise the rear axle until the rear wheels clear the floor.

3. Pry the handle cover up inside the car. The rear of the cover is held by two screws. tighten the adjusting nut until the rear brakes drag when the rear wheels are turned.

4. Loosen the adjusting nut until the rear wheels can be turned without the rear brakes dragging. Apply the parking brake, release it, and repeat Steps 3 and 4 one time.

5. Lower the rear of the vehicle and check the operation of the parking brake.

FOOT OPERATED LEVER

1. Fully release the parking brake.
2. Loosen locknut on equalizer rod under the car. Then loosen the nut in front of the equalizer, several turns.
3. Turn the locknut forward against the equalizer until the cables are tight enough so that the rear wheels cannot be turned by hand. Then, back off the adjustment until the rear wheels turn freely.
4. When cables are properly adjusted, tighten both nuts against the equalizer.
5. Apply and release the brake and feel for freeness of rear wheels.

Disc Brakes

1. Fully release the parking brake.
2. Place the transmission in Neutral. If it is necessary to raise the car to reach the adjusting nut and observe the parking brake levers, use an axle hoist or a floor jack positioned beneath the differential. This is necessary so that the rear axle remains at the curb attitude, not stretching the parking brake cables.

——— CAUTION ———

If you are raising the rear of the car only, block the front wheels.

3. Locate the adjusting nut beneath the car on the driver's side. While observing the parking brake actuating levers on the rear calipers, tighten the adjusting nut until the levers just begin to move. The, loosen the nut sufficiently for the levers to fully return to the stop position. The levers are in the stop position. the levers are in the stop position when a ¼ in. pin can be inserted past the side of the lever into the holes in the cast iron housing.

4. Check the operation of the parking brake. Make sure the actuating levers return to the stop position by attempting to pull them rearward. If the lever moves rearward, the cable adjustment is too tight, which will cause a drag in rear brake and consequent brake overheating and fade.

STEERING

Steering Wheel

REMOVAL & INSTALLATION

1. Disconnect the negative battery cable.
2. If the vehicle is equipped with a horn ring, remove it by rotating it clockwise. If equipped with a steering wheel crash pad, remove the retaining screws from the underside of the steering wheel and then remove the crash pad. Disconnect the horn and speed control (if so equipped) wires from the inside of the steering wheel center. On 1980 and later models, remove the steering wheel hub cover by pushing the cover retaining posts out with a rod through the two holes provided on the back side of the hub.
3. Remove and discard the steering wheel nut, install a steering wheel puller on the end of the shaft, and remove the steering wheel.

——— CAUTION ———

The use of a knockoff type steering wheel puller and a hammer may damage the steering column bearing or (in the case of the collapsible-type steering wheel) the column itself.

4. With the front wheels positioned straight ahead, line up the marks on the steering wheel and column and install the steering wheel and a new locknut. Tighten the nut to 30–40 ft. lbs.
5. Connect the horn and speed control wires and install the horn ring and the crash pad and retaining screws. On 1980 and later models, locate the hub cover posts in the holes and push the cover into place.
6. Connect the negative battery cable.

Turn Signal Switch

REMOVAL & INSTALLATION

1. On standard steering columns, remove the upper extension shroud (below the steering wheel) by snapping the shroud from the retaining clip. On tilt columns, remove the trim shroud by removing the five self-tapping screws.
2. Use a pulling and twisting motion, while pulling straight out, to remove the turn signal switch lever.
3. Peel back the piece of foam rubber from around the switch.
4. Disconnect the two switch electrical connectors.
5. Remove the two self-tapping screws which secure the switch to the lock cylinder housing, and disengage the switch from the housing.
6. To install, align the switch mounting holes with the corresponding holes in the lock cylinder housing. Install the two screws.
7. Stick the foam back into place.
8. Align the key on the turn signal lever with the keyway in the switch and push the lever into place.
9. Install the two electrical connectors and the trim shrouds.

Ignition Switch

REMOVAL & INSTALLATION

1. Disconnect the negative battery cable.
2. Remove the upper shroud below the steering wheel by unsnapping the retaining clips. On the tilt column it will be necessary to remove the five attaching screws.
3. Disconnect the electrical connector from the ignition switch.
4. Drill out the bolts holding the switch to the lock cylinder using a ⅛ in. drill bit.
5. Remove the bolts using an Easy-Out® bolt extractor.
6. Disengage the switch from the actuator pin.
7. Adjust the new ignition switch by sliding the carrier to the Lock position. Insert a small drill bit through the switch housing and into the carrier to restrict movement of the carrier with respect to the switch housing. A new replacement comes with an adjusting pin already installed.
8. Turn the ignition key to the LOCK position.
9. Install the ignition switch on the actuator pin.
10. Install new "break-off head" bolts and tighten them until the heads break off. Tighten bolts evenly.
11. Remove the drill bit or adjusting pin.
12. Connect all electrical connections and the negative battery cable.
13. Start the car and check for proper operation of the switch.
14. Install the steering column shroud.

Manual Steering Gear

REMOVAL & INSTALLATION
Worm and Recirculating Ball

1. Position the steering wheel in the straight ahead position.
2. Remove the bolt(s) retaining the flex coupling to the steering shaft. Match mark the coupling and steering shaft and separate.

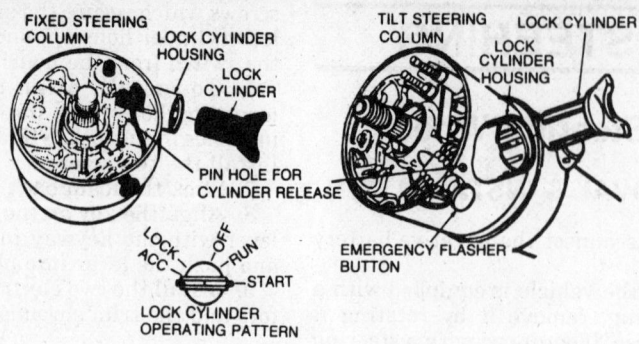

Lock cylinder replacement with locking column

NOTE: Separation can be accomplished when the steering gear box is lowered, if necessary.

3. Remove the retaining nut and washer from the pitman arm to sector shaft. Matchmark the components. Using a puller, separate the pitman arm from the sector shaft.

4. It may be necessary to disconnect the clutch linkage on vehicles equipped with manual transmissions. V8 models may have to have the exhaust system lowered to provide clearance for the removal of the steering gear.

5. Remove the steering gear to side rail retaining bolts and remove the steering gear from vehicle.

6. The installation is the reverse of the removal procedure. Be sure to align the reference marks when reinstalling.

Rack and Pinion

1. Disconnect the battery negative cable.

2. Remove the retaining bolt from the flexible coupling to the steering shaft.

3. Place the ignition switch in the ON position and raise the vehicle and support safely.

4. Remove the right and left tie-rod end retaining nuts and separate the studs from the spindle arms, using a separator tool.

5. Support the steering rack and pinion assembly and remove the retaining nuts, bolts and washers insulators.

NOTE: On certain models such as the Mustang, it is necessary to remove the crossmember to allow clearance for the removal of the rack and pinion.

6. Remove the rack and pinion from the vehicle.

7. The installation is the reverse of the removal procedure.

Power Steering Gear

REMOVAL & INSTALLATION

Non-Integral Control Valve Assembly

1. Raise the vehicle and support safely.

2. Remove the clamp retaining fluid lines to the outside of the control valve.

3. Disconnect the fluid lines after marking each for reassembly. Allow the fluid to drain.

4. Turn the wheels to the left and right several times to force the fluid from the system.

5. Loosen the clamping bolt at the end of the control valve sleeve.

6. Remove the roll-pin from the steering arm to center link through the slot in the control valve sleeve.

7. Remove the cotter pin and the control valve ball stud nut. Remove the ball stud from the sector shaft arm (pitman arm).

8. Remove the valve assembly from the centerlink by turning the control valve counterclockwise until the valve assembly separates from the centerlink.

9. The installation should follow the outline procedure: Thread the control valve on the centerlink until approximately four threads are visible on the centerline. Position the ball stud in the sector shaft arm and measure the distance between the center of the left spindle connecting rod hole in the centerlink to the end of the control valve. This distance must be 2.55–2.65 in. If the distance is not correct, disengage the ball stud from the sector shaft arm and turn the control valve on the centerlink to increase or decrease the distance. When the ball stud is correctly positioned, align the hole in the steering arm to centerlink with the slot near the end of the valve sleeve and install the roll pin. Com-

plete the assembly in the reverse of the removal procedure, fill with fluid and bleed the system.

Power Cylinder

1. Raise the vehicle and support safely.

2. Disconnect the fluid lines from the power cylinder and allow to drain.

3. Remove the PAL crimp nut, washer and the insulator form the end of the power cylinder rod.

4. Remove the cotter pin and nut that retains the power cylinder to the center link.

5. Disconnect the power cylinder stud from the center link by using a steering arm remover tool.

6. Remove the insulator sleeve and washer from the end of the power cylinder rod. Remove the cylinder rod boot and discard the clamp.

7. The installation of the power cylinder is the reverse of the removal procedure. Fill with fluid and bleed the system.

Integral Gear Assembly

1. Remove the stone shield, if equipped.

2. Tag the fluid lines and remove from the steering gear. Allow to drain.

3. Plug the lines and ports to avoid entry of dirt.

4. Remove the bolts that retain the flexible coupling to the steering column and gear.

5. Raise the vehicle and support safely. Remove the sector shaft nut and washer.

6. Remove the pitman arm with a special pulling tool to avoid damage to the shaft.

7. Support the steering gear assembly and remove the gear box retaining bolts from the side rail or bracket.

8. Remove the clamping bolt from the flexible coupling and work the steering gear from the flex coupling and remove from the vehicle.

9. The installation of the steering gear is the reverse of the removal procedure. Fill with fluid and bleed the system.

Rack and Pinion Power Steering

1. Disconnect the negative battery cable.

2. Remove the bolt retaining the flexible coupling to the steering input shaft.

3. Place the ignition key in the ON position and raise the vehicle and support safely.

4. Remove the two tie rod end retaining nuts and cotter pins. Separate the tie rod stud from the spindle arms with the use of a separator tool.

5. Support the rack and pinion and remove the retaining nuts, washers and bolts from the rack and pinion to the crossmember.

6. Lower the gear assembly slightly to gain access to the pressure and return line fittings. Disconnect the fittings and plug the openings to prevent the entry of dirt.

7. Remove the rack and pinion gear assembly from the vehicle.

8. The installation of the rack and pinion assembly is the reverse of the removal procedure. Fill with fluid and bleed the system.

Power Steering Pump

REMOVAL & INSTALLATION

1. Drain the fluid from the pump reservoir by disconnecting the fluid return hose at the pump. Disconnect the pressure hose from the pump.

2. Remove the drive belt. Remove the mounting bolts from the mounting bracket(s) and remove the pump. In some cases, depending on model, it is necessary to remove the pulley (using a special puller) before the pump can be removed from the mounting bracket.

3. To reinstall the pump, position on mounting bracket and loosely install the mounting bolts and nuts. Put the drive belt over the pulley and move the pump outward against the belt until the proper belt tension is obtained. Do not pry against the pump body. Measure the belt tension with a belt tension gauge for the proper adjustment. Only in cases where a belt tension gauge is not available should the belt deflection method be used.

4. Tighten the mounting bolts and nuts.

Tie Rod End

REMOVAL & INSTALLATION

Except Rack and Pinion

1. Raise and support the front end.
2. Remove the cotter pin and nut from the rod end ball stud.
3. Loosen the sleeve clamp bolts and remove the rod end from the spindle arm center link using a ball joint separator.
4. Remove the rod end from the sleeve, counting the exact number of turns required.
5. Install the new end using the exact number of turns it took to remove the old one.
6. Install all parts. Torque the stud to 40–43 ft. lbs., and the clamp to 20–22 ft. lbs.
7. Check the toe-in.

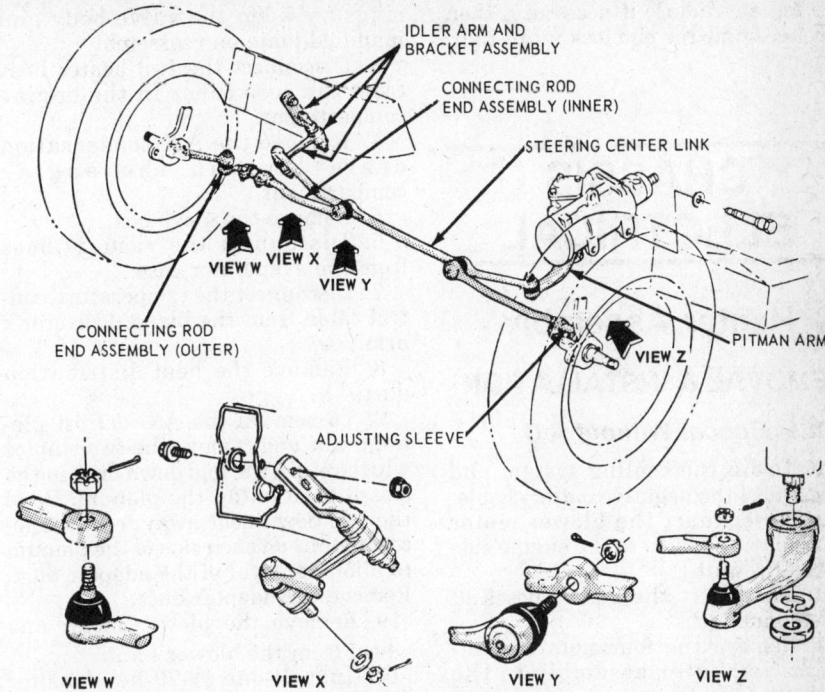

Typical steering linkage

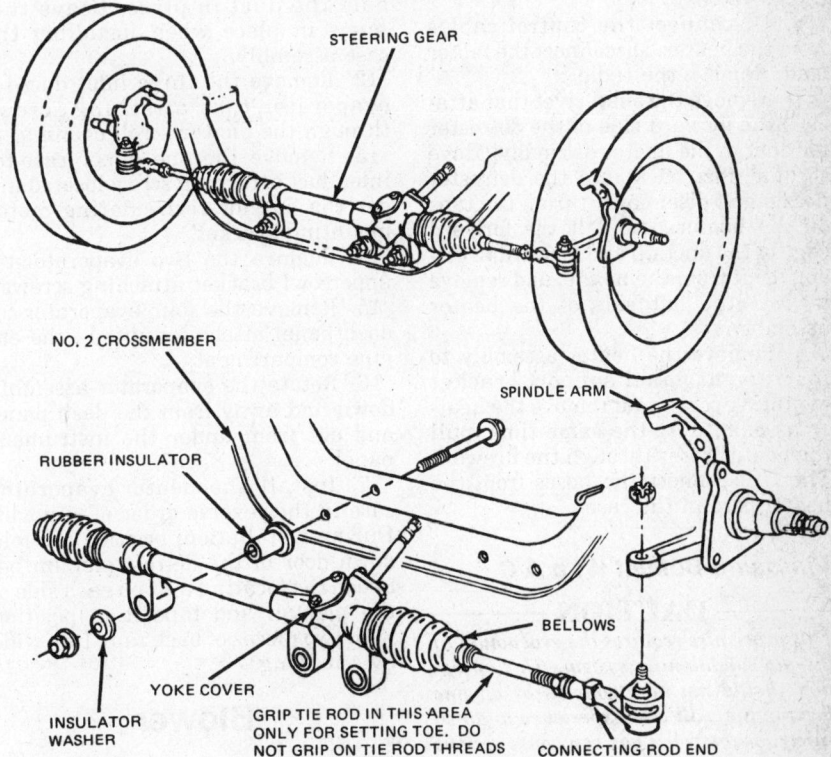

Rack and pinion steering linkage

Rack and Pinion Models

1. Remove the cotter pin and nut at the spindle. Separate the tie-rod end stud from the spindle with a puller.
2. Matchmark the position of the locknut with paint on the tie rod. Unscrew the locknut. Unscrew the tie rod end, counting the number of turns required to remove.
3. Install the new end the same number of turns. Attach the tie rod end stud to the spindle. Install the nut and torque to 35 ft. lbs., then continue to tighten until the cotter pin holes align. Install a new cotter pin. Check

the toe and adjust if necessary, then torque the tie rod end locknut to 35 ft. lbs.

CHASSIS ELECTRICAL

Heater Assembly

REMOVAL & INSTALLATION

Pinto, Bobcat Without A/C

1. Drain the cooling system and disconnect the negative battery cable.
2. Disconnect the blower motor ground wire (black) at the engine side of the firewall.
3. Disconnect the heater hoses at the engine block.
4. Remove the four nuts that attach the heater assembly to the firewall, from the engine side.
5. Working inside the car, remove the glove box.
6. Disconnect the control cables from the heater. Disconnect the motor lead. Remove the radio.
7. Remove the snap-rivet that attaches the forward side of the defroster air duct to the heater assembly. Move the air duct back into the defroster nozzle and disengage it from the tabs on the heater box. Tilt the forward edge of the duct up and forward to disengage it from the nozzle, and remove it from the left side of the heater assembly.
8. Remove the heater assembly to instrument panel support bracket mounting screw and remove the heater assembly. At the same time, pull the heater hoses through the firewall. Then, disconnect the hoses from the heater core in the case.

Pinto and Bobcat With A/C

——— CAUTION ———

This procedure requires the evacuation of the air conditioning system. This operation should not be attempted by anyone lacking the skill and experience to do so safely, as the freon gas can cause serious injury on contact. Have the system evacuated by a professional if in doubt.

1. Drain the engine coolant, discharge the air conditioning system and disconnect the battery.
2. Remove the two hex screws attaching the evaporator manifold plate to the expansion valve body and the STV housing manifold from the evaporator manifold plate. Use new O-

rings between the valve body and manifold plate on reassembly.
3. Disconnect the two heater hoses from the core tubes in the engine compartment.
4. Remove the A/C condensation drain hose in the engine compartment.
5. Remove the glove box.
6. Disconnect the vacuum hoses from the evaporator case.
7. Disconnect the temperature control cable from the blend door crank arm.
8. Remove the heat distribution duct.
9. To remove the A/C defrost plenum: Cut and remove the two staples which retain the fold down door in the closed position on the plenum. Bend the fold down door away from the locating tabs on each side of the plenum to allow removal of the adaptor duct. Remove the adapter duct.
10. Remove the blower motor and wheel from the blower scroll.
11. Install one ¼-20 hex-washer head screw to the mounting tab on the inlet duct to upper cowl bracket to hold the duct in place. Leave this screw in place when installing the case assembly.
12. Remove the three inlet duct-to-evaporator case attaching screws through the blower scroll opening.
13. Remove the one upper case-to-inlet duct attaching screw located under the outside-recirculating motor mounting bracket.
14. Remove the two evaporator-to-uppercowl bracket attaching screws.
15. Remove the four evaporator-to-dash panel attaching nuts in the engine compartment.
16. Rotate the evaporator assembly down and away from the dash panel and out from under the instrument panel.
17. Install the heater/evaporator case in the reverse order of removal. During installation, position the fold down door of the defrost plenum between the locating tabs on each side of the plenum and tape it in position with two pieces of back tape 1 in. wide by 4 in. long.

Blower

REMOVAL & INSTALLATION

1980 and Later Full Size Ford, Mercury and Lincoln

1. Disconnect the negative battery cable.
2. Disconnect the blower motor wiring.
3. Remove the blower motor cooling tube from the motor.

4. Remove the four blower motor attaching screws and remove the motor.
5. Reverse to install.

1980 Granada, Monarch Without A/C

1. Drain radiator coolant.
2. Disconnect heater hoses at core connections.
3. Disconnect battery ground cable.
4. Remove glove box.
5. Remove register air duct.
6. Remove floor discharge duct assembly and defroster nozzle retaining clip.
7. Disconnect the two air door control cables from heater case and doors. Remove right vent cable bracket from instrument panel.
8. Disconnect resistor electrical connector.
9. Remove vent duct to upper cowl mounting bolt.
10. Remove three heater case to dash panel mounting stud nuts and remove heater case and vent duct as an assembly.
11. Remove four mounting screws and slide blower motor assembly from blower scroll in case.
12. Reverse procedure for installation.

Fairmont, Zephyr, Mustang, Capri, Granada 1981 and Later, Cougar 1981 and Later Without A/C

1. Remove right ventilator assembly.
2. Disconnect blower motor lead wire from resistor assembly, push wire back thru the hole in case.
3. Remove right side cowl trim panel for access to blower motor ground wire connector and remove retaining screw.
4. Remove blower motor flange retaining screws from inside of housing.

NOTE: Blower wheel may be removed to improve access.

5. Remove blower motor from housing.
6. Reverse procedure for installation.

Continental With A/C

1. Disconnect the ground cable from the battery.
2. Disconnect the blower motor lead connector from wiring harness connector.
3. Remove the four (4) retaining screws.
4. Turn the motor and wheel assembly slightly to the right so that the bottom edge of the mounting plate

follows the contour of the wheel well splash panel. Lift up on the blower and remove it from the blower housing.

5. Installation is the reverse of removal.

1980 Granada, Monarch With A/C

1. Disconnect the negative battery cable.

2. Loosen the passenger side door sill scuff plate and the right "A" pillar trim cover. Remove the right cowl trim panel.

3. Remove the bolt retaining the lower side of the instrument panel to the cowl. Remove the right cowl side brace bolts.

4. Disconnect the wiring harness connectors at the blower motor.

5. If so equipped, remove the cooling tube from the blower motor.

6. Remove the 4 screws retaining the blower motor and wheel assembly to the scroll. To remove the motor, pull rearward on the lower edge of the instrument panel to provide clearance.

7. Installation is the reverse of removal. If necessary, cement the cooling tube to the blower motor.

Fairmont, Zephyr, Mustang, Capri, Thunderbird, XR7, Granada 1981 and Later, Cougar 1981 and Later, LTD and Marquis 1983 and Later Models With A/C

1. Remove glove box, disconnect outside-recirc door vacuum motor hose.

2. Remove instrument panel lower right-to-side cowl attaching bolt.

3. Remove support brace to top of air inlet duct retaining screw.

4. Disconnect blower motor lead wire at connector.

5. Remove blower housing lower support bracket to evaporator case retaining nut.

6. Remove side cowl trim panel and remove blower motor ground wire screw.

7. Remove screw securing top of air inlet duct to evaporator case.

8. Remove the air inlet duct and blower housing down and away from evaporator case.

9. Remove four blower motor mounting plate screws and remove blower motor assembly from blower housing.

NOTE: DO NOT remove the mounting plate from blower motor.

10. Reverse procedure for installation.

Versailles With A/C

1. Disconnect battery ground cable.

2. Remove instrument panel lower finish plate.

3. Remove glove box.

4. Loosen right door sill plate and remove right side cowl trim panel.

5. Remove instrument panel to-side cowl lower right attaching bolt.

6. Remove bolt securing the brace to the lower edge of the instrument panel below glove box.

7. Disconnect blower motor wires.

8. Remove four blower motor attaching screws and remove blower motor assembly.

NOTE: DO NOT remove the mounting plate from the blower motor.

Pinto, Bobcat

WITHOUT A/C

1. Remove the heater assembly.

2. Disconnect the blower motor lead wire from the resistor.

3. Remove the four blower motor mounting plate attaching nuts and remove the motor and wheel.

4. Install in the reverse order.

WITH A/C

NOTE: The blower motor and wheel is integrally located within the scroll portion of the evaporator assembly on the right-side of the evaporator case. To remove the blower motor and wheel, remove the glove box and remove the four screws retaining the blower motor and wheel in the blower scroll. It may be necessary to remove the instrument panel to right side cowl bolt, to allow the panel to be pulled rearward for clearance. Install the blower motor and wheel in the reverse order of removal.

Heater Core Without A/C

REMOVAL & INSTALLATION

Full Size Ford and Mercury

1. Drain the coolant and save for reuse.

2. Disconnect the negative cable from the battery.

3. Remove the heater hoses from the heater core.

4. Plug the heater core tubes to prevent coolant from spilling under the dash during plenum removal.

5. Remove the plenum to dash bolt, located under the windshield wiper motor at the left end of the plenum.

6. Remove the one nut from the heater case (engine side).

7. Disconnect the vacuum supply hose from the vacuum fitting and push the grommet and hose into the passenger compartment.

8. Remove the glove box assembly.

9. Loosen the right door sill plate and remove the right side cowl trim panel.

10. Remove the lower right instrument panel to side cowl bolt.

11. Remove the instrument panel pad.

12. Remove the temperature control cable from the top of the plenum. Then, disconnect the temperature control cable from the blend door crank arm.

13. Remove the push clip attaching the center register duct bracket to the plenum and rotate the bracket up to the right.

14. Disconnect the vacuum jumper harness at the multiple vacuum connector near the floor air distribution duct.

15. Disconnect the white vacuum hose from the outside air door vacuum motor.

16. Remove the two (2) screws attaching the seat side of the floor air distribution duct to the plenum.

NOTE: It may be necessary to remove the two (2) screws attaching the lower panel door vacuum motor to the mounting bracket to gain access to the floor air distribution duct screw.

17. Remove the plastic push-pin fastener from the floor air distribution duct and remove the duct.

18. Remove the remaining two (2) plenum retaining nuts from the lower flange of the plenum.

19. Move the plenum toward the seat to allow the heater core tubes to clear the holes in the dash panel.

20. Rotate the top of the plenum down and out from under the instrument panel.

21. Remove the four (4) heater core cover retaining screws and lift off the cover.

22. Remove the heater core tube bracket retaining screw.

23. Pull the heater core and seal from the plenum assembly.

24. Installation is the reverse of removal. Connect the negative battery cable and refill the cooling system. Check heater operation.

1980 Granada and Monarch

1. Drain radiator coolant.

2. Disconnect heater hoses at core connections.

3. Remove glove box.

4. Remove right register air duct.

5. Remove floor discharge duct assembly and clip holding defroster nozzle to the heater.

6. Disconnect air door control cables from heater case and doors. Remove right vent cable bracket from instrument panel.

7. Disconnect resistor assembly electrical connector.

8. Remove vent duct-to-upper-cowl mounting bolt.

9. Remove three heater case to dash panel mounting stud nuts and remove heater case and vent duct as an assembly.

10. Remove heater core cover and pad and slide core out of the case.

11. Reverse procedure for installation.

Fairmont, Zephyr, Capri, Mustang, Granada 1981 and Later, Cougar 1981 and Later

1. Drain radiator coolant.

2. Disconnect heater hoses at core connections.

3. Remove glove box.

4. Remove instrument panel-to-cowl brace retaining screws and brace.

5. Move temperature control lever to warm position.

6. Remove four heater core cover retaining screws.

7. Remove heater core cover through glove box opening.

8. In engine compartment, loosen heater case assembly mounting stud nuts.

9. Push heater core tubes and seal toward passenger compartment to loosen heater core assembly from heater case assembly.

10. Remove heater core from heater case assembly through the glove box opening.

11. Reverse procedure for installation.

Pinto, Bobcat

1. Remove heater assembly.

2. Remove the compression gasket from the cowl air inlet and remove the eleven clips from the case. Separate the case and remove the heater core.

Heater Core With A/C Including Automatic Temperature Control
—— CAUTION ——

Removal of the heater-air conditioner housing requires evacuation of the air conditioner refrigerant. This operation requires special tools and training. Failure to follow proper safety precautions may cause personal injury. It is recommended

that discharging the charging of the A/C system be performed by an experienced professional mechanic.

REMOVAL & INSTALLATION

Ford and Mercury (Full Size), Lincoln Continental

1. Disconnect negative battery cable.

2. Remove the heater hoses from the core tubes and plug the ends to prevent coolant loss.

3. Plug the heater core tubes to prevent coolant loss during plenum and core removal.

4. In the engine compartment, remove the bolt located under the windshield wiper motor. Remove the nut at the upper left corner (engine side) of the evaporator case.

5. Disconnect the control system vacuum supply hose from the vacuum source and push the grommet and vacuum supply hose into the passenger compartment.

6. Remove the glove box assembly.

7. Loosen the right door sill plate and remove the right side cowl trim panel.

8. Remove the lower right instrument panel to side cowl bolt.

9. Remove the instrument panel pad.

10. Vehicles W/O Automatic Temperature Control; Remove the bracket from the temperature control cable housing at the top of the plenum assembly. Disconnect the temperature control cable from the blend door crank arm. Remove the push clip attaching the center register duct bracket to the plenum and rotate the bracket up to the right. Disconnect the vacuum jumper vacuum harness at the multiple vacuum connector near the floor air distribution duct.

11. Vehicles With Automatic Temperature Control; Disconnect the temperature control cable from the ATC sensor. Disconnect the vacuum harness connector from the ATC sensor. Disconnect the ATC sensor tube from the sensor and evaporator case connector. Also, disconnect the wire connector from the top end of the electric-vacuum relay, located on the right side of the plenum case. Disconnect the white vacuum hose from the outside recirculating door vacuum motor.

12. Remove the two attaching screws from the floor air distribution duct, at the seat side of the air distribution duct.

13. Remove the plastic push fastener, holding the air distribution duct to the left of the plenum and remove the air distribution duct.

14. Remove the final two retaining

nuts from the lower flange of the plenum assembly.

15. Move the plenum assembly toward the seat to allow the heater core tubes to clear the holes in the dash panel. Rotate the plenum assembly down and out from under the dash panel.

16. Installation is the reverse of removal. Refill the cooling system and check the heater operation.

1980 Granada, Monarch

NOTE: The refrigerant system components and charge do not have to be disturbed when removing and installing the heater core.

1. Drain the coolant and disconnect the battery.

2. Disconnect two heater hose clamps at the dash panel in the engine compartment. Plug the core tubes to prevent coolant leakage during removal.

3. Remove the heat distribution duct from the instrument panel.

4. Remove the glove box liner.

5. Loosen the right door sill cuff plate, right "A" pillar trim cover and remove the right cowl side trim panel.

6. Loosen the instrument panel-to-right cowl side bolt and remove the instrument panel brace bolt at the lower rail, below the glove box. Remove the tunnel to cowl brace at the left end of the plenum assembly, if equipped.

7. Remove the instrument panel crash pad.

8. Remove the radio speaker or panel cowl brace.

9. Remove the 4 defroster nozzle-to-cowl bracket mounting screws.

10. Lift the defroster nozzle upward through the crash pad opening.

NOTE: On models with automatic temperature control; remove the retaining nut and remove the vacuum harness from the electro-vacuum relay.

11. Disconnect the vacuum hoses from the A/C-Defrost and Heat/Defrost door motors. Remove the screw from the clip holding the vacuum harness to the plenum.

12. Remove two Heat/Defrost door mounting nuts and swing the motor rearward on the door crankarm.

13. Remove two screws attaching the plenum to the left mounting bracket. Then remove the two screws and three clips securing the plenum to the evaporator case.

14. Swing the bottom of the plenum away from the evaporator case to disengage the S-clip on the forward flange of the plenum. Raise the ple-

num to clear the tabs on the top of the evaporator case.

15. Move the plenum to the left as far as possible, (about 4 in.) pulling rearward on the instrument panel to gain clearance. Take care when pulling back on the instrument panel to avoid cracking the plastic panel.

NOTE: There is very little clearance between the plenum and the wiper motor assembly.

16. Pull the heater core to the left using the tab molded into the rear heater core seal. As the rear surface of the heater core clears the evaporator case, pull the core rearward and downward to clear the instrument panel.

17. Reverse the above procedure to install.

NOTE: Before installing the core, make sure that the heater core tube to dash panel seal is in place between the evaporator case and the dash panel.

Thunderbird, Cougar, XR7, 1983 and Later LTD and Marquis

1. Remove the instrument panel and lay it on the front seat.

2. Drain the coolant from the cooling system. Disconnect the heater hoses from the core tubes and plug the tubes to prevent spillage.

3. From the engine compartment side remove the two nuts attaching the evaporator case to the dash panel.

4. Under the dash area remove the screws attaching the evaporator case support bracket and the air inlet duct support bracket to the cowl top panel.

5. Remove the retaining nut from the bracket at the left side of the evaporator case and the nut attaching the heater core access cover to the evaporator case.

6. Carefully pull the evaporator case assembly away from the dash panel to gain access to the screws retaining the heater core access cover to the evaporator case.

7. Remove the heater core cover attaching screws and remove the cover.

8. Lift the heater core and seals from the evaporator case. Remove the two seals from the core tubes.

9. Installation is the reverse of removal. Refill the cooling system and check heater operation.

NOTE: AUTOMATIC TEMPERATURE CONTROL: Removal and installation of the heater core are the same for manual and automatic temperature control systems, on the 1980–81 Thunderbird, Cougar and XR7.

Fairmont, Zephyr, Mustang, Capri, 1981 and Later Granada, Cougar

1. Remove the instrument panel.
2. Drain radiator coolant.
3. Disconnect heater hoses at core connections.
4. In engine compartment, remove two evaporator case to dash panel retaining nuts.
5. Under dash, remove screws securing evaporator case support bracket and air inlet duct support bracket to the cowl top panel.
6. Remove one nut retaining the bracket at the left end of evaporator case to the dash panel, and one nut securing the bracket below the case to the dash panel.
7. Remove five heater core access cover screws and the cover from the evaporator case.
8. Remove heater core and seals from evaporator case.
9. Reverse procedure for installation.

Versailles

1. Drain radiator coolant.
2. Disconnect heater hoses at core connections.
3. Disconnect battery ground cable.
4. Remove instrument panel lower finish panel and support brace bolt under glove box.
5. Remove instrument panel pad.
6. Loosen right door sill plate retaining screws, remove side cowl trim panel attaching screws and remove trim panel.
7. At right end of instrument panel, remove the instrument panel lower attaching screw.
8. Remove two screws and the floor air distribution duct from the plenum assembly.
9. Remove radio speaker or instrument panel to cowl brace.
10. Remove four screws securing defroster nozzle to the cowl mounting tabs. Lift defroster nozzle up and out from behind instrument panel.
11. Remove cowl to floor brace at left side of plenum assembly.
12. Disconnect vacuum from A/C defrost and heat-defrost door vacuum motors.
13. Remove retaining nut and vacuum harness connector from the electro-vacuum relay, located on the right side of the plenum case.
14. Remove screw securing vacuum harness clamp to plenum assembly.
15. Remove two heat-defrost door motor retaining nuts and swing motor to rear.
16. Remove two screws attaching front corner of plenum assembly to

the plenum-to-cowl mounting bracket.
17. Remove two screws attaching the lower edge of the plenum to the evaporator case.
18. Remove two clips securing the rear edge of the plenum assembly to the evaporator case.
19. Raise plenum assembly until top edge of plenum is disengaged from evaporator case. Then move plenum to the left as far as possible.
20. Pull rearward on lower edge of instrument panel carefully and remove plenum assembly from behind instrument panel.
21. Pull the heater core from the evaporator housing carefully, using the tab molded into heater core seal.
22. Disengage heater core tubes from dash panel seal as core is removed. As heater core clears evaporator case, pull heater core rearward and down to clear instrument panel.
23. Reverse procedure for installation.

Pinto, Bobcat

1. Remove the evaporator case assembly from the vehicle.
2. Remove the upper-to-lower case attaching screws. Remove the blower motor and wheel.
3. Remove the rubber seal from the heater core tubes.
4. Remove the upper half of the evaporator case.
5. Move the rubber seal on the evaporator core forward to clear the case mounting stud and pull the core out of the lower case.
6. Install in the reverse order of removal. Be sure to install new rope sealer around the flange of the lower case before installing the upper half of the case. Install new O-rings on the manifold plate. Dip the new O-rings in refrigerant oil before installing them.

Radio

NOTE: For the best FM reception, adjust the antenna, if adjustable, to 32 in. height. Fading or weak AM reception may be corrected by adjusting the trimmer control The trimmer control is located either on the right rear or front side of the radio. See the owner's manual for position if you are in doubt. To adjust the trimmer:

1. Extend the antenna to maximum height.
2. Tune the radio to a weak station around 1600 KC. Adjust the volume so that the sound is barely audible.
3. Adjust the trimmer to obtain maximum volume.

REMOVAL & INSTALLATION

Lincoln

1. Disconnect the negative battery cable.
2. Remove the four radio plate-to-panel screws. Pull the radio with the front plate attached rearward until the rear bracket is clear.
3. Disconnect the wires from the chassis. If equipped with premium sound, remove the control assembly attaching nut and washer, remove the switch, and remove the illumination lamp socket from the front bracket.
4. Remove the radio with the front plate attached. Remove the four screws and remove the plate. Installation is the reverse of removal.

Full Size Ford and Mercury

1. Disconnect the battery ground cable.
2. On all-electronic radios, remove the radio-to-mounting plate screws and remove the mounting plate.
3. Remove the radio knobs, the screws that attach the bezel to the instrument panel, and remove the bezel.
4. Remove the radio mounting plate attaching screws (standard radios), and disengage the radio by pulling it from the lower rear support bracket.
5. Disconnect all the leads from the radio.
6. Remove the radio mounting plate and the rear upper support; remove the radio from the instrument panel.
7. Reverse the procedure to install.

Granada, Monarch, Versailles

1. Disconnect the negative battery cable.
2. Remove the headlight switch from the instrument panel. Remove the heater, air conditioner, windshield wiper/washer knobs, and radio knobs and discs.
3. Remove the six screws which attach the applique to the instrument panel and remove the applique. Disconnect the antenna lead-in cable from the radio.
4. Remove the four screws which attach the radio bezel to the instrument panel. Slide the radio and bezel out of the lower rear support bracket and instrument panel opening toward the interior far enough to disconnect the electrical connections, and remove the radio.
5. Remove the nut attaching the rear support bracket to the radio and remove the bracket. Remove the nuts

and washer from the radio control shafts and remove the bezel.
6. To install, attach the rear support bracket to the radio. Install the bezel, washers and nuts.
7. Insert the radio with rear support bracket and bezel through the instrument panel opening far enough to connect the electrical leads and antenna lead-in cable. Install the radio upper rear support bracket into the lower rear support bracket.
8. Center the radio and bezel in the opening and install the four bezel attaching screws.
9. Install the instrument panel applique with its six attaching screws. Install all knobs removed from the instrument panel and radio. Install the headlight switch. Connect the negative battery cable.

Fairmont, Zephyr, Mustang, Capri, 1981 and Later Granada and Cougar, Futura

1. Disconnect the negative battery cable.
2. Disconnect the electrical, speaker, and antenna leads from the radio.
3. Remove the knobs, discs, and control shaft nuts and washers from the radio shafts.
4. Remove the ash tray receptacle and bracket.
5. Remove the rear support nut from the radio.
6. Remove the instrument panel lower reinforcement and the heater or air conditioning floor ducts.
7. Remove the radio from the rear support, and drop the radio down and out from behind the instrument panel.
8. To install, reverse the removal procedure.

Thunderbird and Cougar XR-7, 1982 and Later Continental, 1983 and Later LTD, Marquis

1. Disconnect the negative battery cable.
2. Remove the radio knobs (pull off). Remove the center trim panel.
3. Remove the radio mounting plate screws. Pull the radio towards the front seat to disengage it from the lower bracket.
4. Disconnect the radio and antenna connections.
5. Remove the radio. Remove the nuts and washers (conventional radios) as necessary.
6. On electronic radios, install the mounting plates before installing the retaining nuts and washers or screws. The rest of installation is the reverse of removal.

Headlight Switch

REMOVAL & INSTALLATION

Ford and Mercury

1. Disconnect the negative battery cable.
2. Underneath the instrument panel, depress the shaft retaining knob and pull the knob straight out.
3. Unscrew the trim bezel and remove the locknut.
4. Underneath the instrument panel, move the switch toward the front of the car while tilting it downward.
5. Disconnect the wiring from the switch and remove the switch from the car.
6. Installation is the reverse of removal.

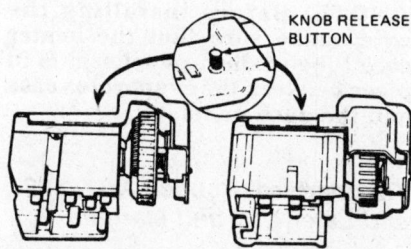

Headlight switch and release button location

Lincoln and Mark VI

1. Disconnect the battery ground cable.
2. Remove the switch knob and shaft.
3. Carefully pull the two control bezels out with pliers. There will be only one bezel if the car doesn't have automatic dimmer control.
4. Unscrew the threaded headlight switch bezel. On Lincoln, remove the screw at the rear corner of the bracket. On Mark V, remove the cluster opening finish panel. On 1980 and later models, remove the steering column lower shroud and the lower left instrument panel trim bezel.
5. On Mark V, remove the four screws from the bracket on the front of the switch. On 1980 and later models, remove the five switch mounting bracket-to-instrument panel screws.
6. Disconnect the wires and remove the switch. Note their location and detach any vacuum lines. Remove the bracket from the switch on the Mark V and all 1980 and later models.
7. Reverse the procedure for installation.

Pinto and Bobcat

1. Disconnect the negative battery cable.

2. Pull the headlight switch out to the ON position.

3. Reach under the dashboard and depress the knob and shaft retainer button on the headlight switch.

4. Disconnect the wiring and remove the switch.

5. Reverse to install.

Windshield Wiper Switch

REMOVAL & INSTALLATION

1980 Versailles, Granada, Monarch

1. Disconnect the negative battery cable.

2. Separate the wiring connector after releasing the locking tabs.

3. Remove the two lower instrument panel screws and remove the lower instrument panel shield.

4. Remove the two steering column cover screws and separate the steering column cover halves.

5. Remove the wiring shield.

6. Using a number T-20 internal driver bit or its equivalent, remove the retaining screw and remove the windshield wiper/washer-turn signal arm assembly.

NOTE: The wiper/washer switch is an integral part of the turn signal switch arm and cannot be repaired separately.

Pinto and Bobcat

1. Disconnect the battery ground cable.

2. Remove the instrument cluster.

3. Remove the wiper switch knob and the bezel nut.

4. Disconnect the wiring connector on the switch and remove the switch from the dash.

5. Reverse the removal procedure to install.

Ford, Mercury, Fairmont, Zephyr, Mustang, Capri, Lincoln, Mark VI, Thunderbird, XR-7, Cougar

1. Disconnect the negative battery cable.

2. Remove the split steering column cover retaining screws.

3. Separate the two halves and remove the wiper switch retaining screws.

4. Disconnect the wire connector and remove the wiper switch.

5. The installation of the wiper switch is the reverse of the removal procedure.

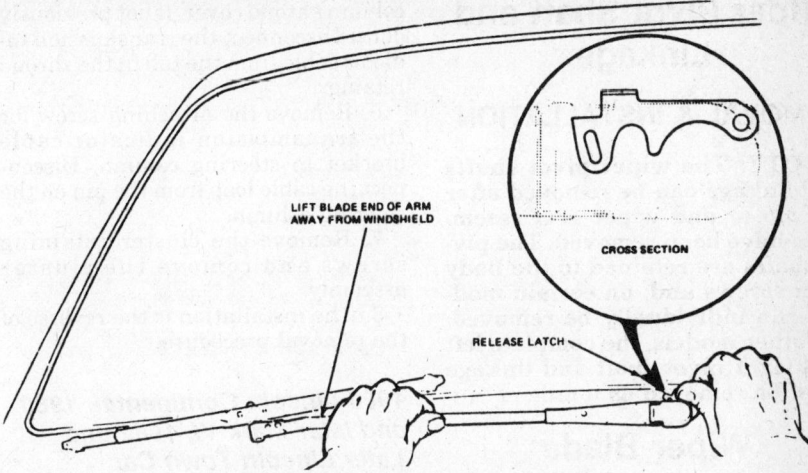

Windshield wiper arm removal typical

Wiper Motor

REMOVAL & INSTALLATION

Ford, Mercury, Mark VI, 1980-81 Lincoln Continental, 1982 and Later Lincoln Town Car

1. Disconnect the battery ground cable.

2. 1982 and later models, remove the hood seal.

3. Disconnect the right washer nozzle hose and remove the right wiper arm and blade assembly from the pivot shaft.

4. Remove the windshield wiper motor and linkage cover by removing the two attaching screws.

5. Disconnect the linkage drive arm from the motor output arm crankpin by removing the retaining clip.

6. Disconnect the two push-on wire connectors from the motor.

7. Remove the three bolts that retain the motor to the dash panel extension and remove the motor.

8. To install, be sure the output arm is in the park position and reverse the removal procedure.

Fairmont, Zephyr, Thunderbird, Cougar, XR-7, 1981 and Later Mustang, Capri, Granada

1. Disconnect the battery ground cable.

2. Remove the right wiper and blade assembly.

NOTE: On Fairmont, Zephyr, Granada and Cougar models, also remove the left wiper arm and blade.

3. Remove the grille on the top of the cowl.

4. Disconnect the linkage drive arm from the motor crankpin after removing the clip.

5. Disconnect the wiper motor electrical connector and remove the three attaching screws from the motor. Pull the motor from the opening.

6. Be sure the motor crank arm is in the park position and reverse the removal procedure to install.

Pinto, Bobcat, 1980 Mustang, Capri

1. Loosen the two nuts and disconnect the pivot shaft and link assembly from the motor drive arm ball on the Pinto and Bobcat models. remove the link retaining clip on the Mustang and Capri models.

2. Remove the three motor attaching screws and lower the motor away from under the left side of the instrument panel.

3. Disconnect the wiper motor electrical wires and remove the motor.

4. Be sure the motor is in the park position and reverse the removal procedure to install.

1980 Granada, Monarch, Versailles

1. Disconnect the battery ground cable.

2. Remove the instrument panel pad.

3. Remove the radio speaker mounting bracket and remove the speaker.

4. Remove the defroster nozzle and the air distribution duct.

5. Remove the interlock module from the bracket and disconnect the multiple connector.

6. Remove the motor bracket to cowl bolts and drive arm clip.

7. Remove motor from vehicle.

8. Reverse the removal procedure to install.

Wiper Pivot Shaft and Linkage

REMOVAL & INSTALLATION

NOTE: The wiper pivot shafts and linkage can be removed after the motor and wiper arm assemblies have been removed. The pivot shafts are retained to the body with screws and, on certain models, can individually be removed. On other models, the complete left and right pivot shaft and linkage must be removed as a unit.

Wiper Blade

REPLACEMENT

Wiper blades used are from either Trico or Anco companies. With a bayonet type blade, the blade saddle slides over the end of the arm and is engaged by a locking stud. with a side saddle in type, a pin on the arm enters the side of the blade saddle and engages a loaded spring (Trico) or a loaded clip (Anco) in the saddle.

Instrument Cluster

REMOVAL & INSTALLATION

─── **CAUTION** ───

Extreme care must be exercised during the removal and installation of the instrument cluster and dash components to avoid damage or breakage. Wooden paddles should be used to separate dash components, if required. Tape or cover dash areas that may be damaged to the removal and installation of the dash components.

NOTE: During the removal and installation procedures, slight variations may be required from the general outline, to facilitate the removal and installation of the instrument panel an cluster components, due to slight changes from model year to model year.

All Full-Size Models (Except Below)

1. Disconnect the battery ground cable.
2. Remove the lower steering column cover.
3. Remove the instrument cluster trim cover and the bottom half of the steering column shroud.
4. Reach behind the cluster and disconnect the cluster electrical feed plug and the speedometer cable.
5. Unsnap and remove the steering column shroud cover, if not previously done. Disconnect the transmission indictor cable from the tab in the shroud retainer.
6. Remove the attaching screw for the transmission indicator cable bracket to steering column. Disconnect the cable loop from the pin on the steering column.
7. Remove the cluster retaining screws and remove the cluster assembly.
8. The installation is the reverse of the removal procedure.

1980 Lincoln Continental, 1980 and later Mark VI, 1982 and Later Lincoln Town Car

1. Disconnect the battery ground cable.
2. Remove the lower steering column cover.
3. Remove the instrument cluster trim cover. Remove the bottom half of the steering column shroud.
4. Reach behind the cluster and disconnect the cluster electrical feed plug and the speedometer cable.
5. Unsnap and remove the steering column shroud cover, if not previously done. Disconnect the transmission indictor cable from the tab in the shroud retainer.
6. Remove the attaching screw for the transmission indicator cable bracket to steering column. Disconnect the cable loop from the pin on the steering column.
7. Remove the cluster retaining screws and remove the cluster assembly.
8. The installation is the reverse of the removal procedure.

AUXILIARY INSTRUMENT CLUSTER

1. Remove the main instrument cluster as previously outlined.
2. Remove the auxiliary cluster housing trim from the instrument panel.
3. Disconnect the electrical connection from the rear of the cluster.
4. Remove the auxiliary instrument cluster from the rear, through the opening of the removed main instrument cluster.
5. The installation is the reverse of the removal procedure.

Thunderbird, Cougar XR-7 and Continental

1. Disconnect the negative battery cable.
2. Disconnect the speedometer cable (Standard cluster).
3. Remove the instrument panel trim cover and steering column lower shroud.

4. Remove the cluster retaining screws (Electronic cluster).
5. Remove the attaching screw from the transmission indictor quadrant cable bracket to the steering column. Disconnect the cable loop from the pin on the steering column.
6. Remove the cluster retaining screws (Standard cluster).
7. Pull the cluster away from the instrument panel and disconnect the speedometer cable (Electronic cluster).
8. Remove the cluster from the instrument panel.
10. Reverse the removal procedure to install.

Fairmont, Zephyr, 1981 and Later Granada, Cougar

NOTE: Certain special ordered cluster assemblies have two printed circuits.

1. Disconnect the battery negative cable.
2. remove the steering column shroud and the cluster trim cover.
3. Remove one screw from the shift quadrant control cable bracket to steering column and disconnect the cable loop from the pin n the shift cane lever. Remove the plastic clamp from around the steering column.
4. Remove the retaining screws holding the cluster to the instrument panel.
5. Pull the cluster away from the instrument panel and disconnect the speedometer cable. disconnect the electrical connectors and remove the cluster from the dash.
6. To install the cluster, reverse the removal procedure.

1980 Granada, Monarch, Versailles

1. Disconnect the negative battery cable.
2. Remove the retaining screws from the lower cluster applique cover, below the steering column.
3. Remove the steering column shroud.
4. From under the instrument panel, release the headlamp switch control knob and shaft assembly.
5. Remove the threaded headlight switch bezel.
6. Remove the retaining screws from the cluster front cover.
7. Insert a right angle standard tip screwdriver along the edges of the finish panel, withdrawing the studs in sequence gradually around the outer edge of the panel.
8. Remove the cluster front cover.
9. If the vehicle is equipped with automatic transmission, remove one screw attaching the shift quadrant

control cable bracket to the steering column. Disconnect the cable loop from the pin on the steering column.

10. From under the instrument panel, disconnect the speedometer cable.

11. Remove the retaining screws and pull the cluster away from the instrument panel.

12. Disconnect the electrical connectors from the rear of the cluster.

13. Remove the cluster from the instrument panel.

14. To install the cluster, reverse the removal procedure.

Mustang, Capri

1. Disconnect the battery ground cable.

2. Remove the instrument trim cover.

3. From under the dash, reach up and disconnect the speedometer cable.

4. Remove the cluster retaining screws and pull the cluster away from the dash. Disconnect the tachometer and wiring connectors. Remove the cluster assembly.

5. The installation is the reverse of the removal procedure.

1983 and Later LTD, Marquis

STANDARD CLUSTER

1. Disconnect the negative battery cable.

2. Disconnect the speedometer cable. Remove the screws retaining the cluster trim panel and remove the panel.

3. Remove the steering wheel shroud. Remove the screw retaining the shift indictor control cable to the steering column. Disconnect the indicator cable loop from the shift lever pin. Remove the plastic clamp from the steering column.

4. Remove the cluster retaining screws. Disconnect the cluster feed plug from the printed circuit. Disconnect the engine warning lamp.

5. Remove the instrument cluster.

6. Install the cluster in the reverse order of removal.

ELECTRONIC CLUSTER

1. Disconnect the negative battery cable.

2. Remove the screws retaining the lower instrument cluster trim panel. Remove the steering column cover.

3. Remove the screws retaining the instrument cluster to the instrument panel.

4. Remove the screw attaching the transmission indictor cable bracket to the steering column. Disconnect the cable loop from the pin on the steering column.

5. Carefully pull the instrument cluster away from the panel and disconnect the speedometer cable. Disconnect the cluster feed plug and ground receptacle from the cluster back plate.

6. Remove the cluster assembly.

7. Install the cluster in the reverse order of removal.

Instrument Panel

REMOVAL & INSTALLATION

1983 and Later Mustang and Capri (Manual A/C)

1. Disconnect the ground (negative) cable from the battery.

2. Remove the instrument panel pad.

3. Remove the two screws attaching the steering column lower cover to the instrument panel and remove the cover.

4. Remove the steering column trim shrouds by removing the screws on the underside of the shroud.

5. Remove the four nuts attaching the steering column to the brake pedal support and carefully lower the steering column only enough for access to the transmission gear shift selector lever and cable assembly (automatic transmission vehicles only).

NOTE: Care must be used to assure that the column is not lowered too far to prevent damage to the selector lever and/or cable.

6. Reach between the steering column and the instrument panel and gently lift the selector lever cable off the shift selector lever. Then, remove the cable clamp from the steering column tube.

7. Lay the steering column to rest on the front seat.

8. Remove the one screw attaching the instrument panel to the brake pedal support at the steering column opening.

9. Remove the one screw attaching the lower brace to the lower edge of the instrument panel below the radio.

10. Remove the one screw attaching the brace to the lower edge of the instrument panel below the glovebox.

11. Disconnect the temperature control cable from the temperature blend door and the evaporator case bracket.

12. Disconnect the 7-port vacuum hose connector at the evaporator case.

13. Disconnect the blower resistor wire connector from the resistor on the evaporator housing, and the blower motor feed wire at the in-line connector near the blower resistor wire connector.

14. Support the instrument panel and, with an angle philips screwdriver, remove three screws attaching the

top of the instrument panel to the cowl.

15. Remove the screws attaching each end of the instrument panel to the cowl side panels.

16. Move the instrument panel rearward and disconnect the speedometer cable from the speedometer and any wires that will not allow the instrument panel to lay on the front seat. Use care not to scratch the instrument panel or the steering column.

17. Place the instrument panel near the installed position and connect any wires or connectors that were disconnected during removal.

18. Connect the speedometer cable to the speedometer.

19. Place the instrument panel is position and install one screw at each end of the instrument panel.

20. Install the three screws along the top front edge of the instrument panel with an angle philips screwdriver.

21. Connect the two support braces to the lower edge of the instrument panel with one screw each.

22. Connect the temperature control cable to the temperature blend door crank arm and the bracket.

23. Connect the vacuum hoes at the 7-port connector and the blower motor wires at the resistor and in-line connector near the resistor.

24. Install the screw attaching the instrument panel to the brake pedal support.

25. Position the steering column near the brake panel support.

26. Connect the transmission gear shift selector lever cable to the shift selector lever. Then, connect the cable clamp to the steering column tube.

27. Position the steering column against the brake pedal support and install the four attaching nuts.

28. Adjust the transmission selector indicator.

29. Install the steering column shroud.

30. Position the steering column opening cover to the instrument panel and install the two attaching screws.

31. Install the instrument panel pad and connect the ground cable to the battery.

32. Connect the temperature control cable to the temperature blend door crank arm and adjust as necessary.

1983 and Later Thunderbird/ Cougar (Manual A/C)

1. Disconnect the ground cable from battery.

2. Remove the three screws attaching steering column opening cover to instrument panel. Then, pull the panel rearward to disengage the clips and remove the cover.

3. Remove the three screws attaching the reinforcement to the instrument panel below the steering column opening and remove the reinforcement.

4. Remove the two nuts retaining the hood latch release handle mounting bracket to the brake pedal support below the steering column.

5. Remove the sound insulator from under the left side of the instrument panel.

6. Remove the two nuts retaining the steering column clamp to the brake pedal support and allow the steering column to rest on the seat.

7. Remove the instrumental panel pad.

8. Disconnect the speedometer cable from the rear of the speedometer.

9. Remove the two screws attaching the console tray (automatic transmission) or console cover around the gear shift lever (manual transmission) to the console. Then, remove the tray or cover from the console.

10. Remove the four console switch panel cover attaching screws. Disconnect the switch wires and remove the switch panel.

11. Open the console box cover and remove the two screws from the bottom of the console box and the two screws from the top front of the box.

12. Remove the two screws attaching front end of the console to the lower edge of the instrument panel.

13. Remove the two screws attaching the bracket at the front end of console and lower the edge of the instrument panel to the floor pan. The, lift the rear of the console and pull the console rearward to disengage it from the instrument panel. Position the console out of the way.

14. Remove the two plastic push pins attaching the glove box door straps to the glove box. Allow the glove box and door to hang by the hinge.

15. Remove the one bolt attaching instrument panel brace to the instrument panel at the glove box opening.

16. Remove the one nut attaching the lower edge of the instrument panel to the brake pedal support.

17. Remove the one bolt attaching a second instrument panel brace to the lower edge of instrument panel just to the left of the console extension.

18. Remove the one bolt attaching each end of the instrument panel to the cowl side panel.

19. Support the instrument panel and remove the three screws attaching top edge of instrument panel to cowl top panel.

20. Cover the steering column and seats. Then, carefully position the instrument panel toward seat disconnecting wires and vacuum harness as

necessary. Allow instrument panel to rest on front seat.

21. Place the instrument panel in position and connect any wires and vacuum harness that were disconnected during instrument panel removal. Then, install the three screws to attach top edge of instrument panel to cowl top panel.

22. Install one bolt to attach each end of instrument panel to cowl side panel.

23. Install one bolt each to attach the instrument panel left and right braces to the lower edge of instrument panel near the console extension.

24. Install a nut to attach lower edge of instrument panel to the brake pedal support.

25. Position the console to vehicle and install two screws to attach the front bracket to the floor.

26. Install two screws to attach the console box to the floor and two screws to attach the top front of box to the support bracket.

27. Position the console switch panel to the console and connect wires. Then, install four panel cover attaching screws.

28. Position the console tray or cover around gear shift lever to console and install the two attaching screws.

29. Connect the speedometer cable to the speedometer.

30. Install the instrument panel pad.

31. Position the steering column to brake pedal support and install retaining clamp (two nuts).

32. Position the reinforcement across the steering column opening of instrument panel and install the three attaching screws.

33. Position the steering column opening cover to the instrument panel and install three attaching screws.

34. Install the sound insulator under left side of instrument panel.

35. Connect the support straps to the glove box.

36. Connect the ground cable to the battery.

37. Check for proper operation of all instruments and controls.

1983 and Later LTD/Marquis (Manual A/C)

1. Remove the instrument panel upper finish panel and pad assemblies.

2. Loosen the steering column and carefully lower only enough for access to transmission gear shaft selector lever and cable assembly.

NOTE: Care must be used to assure that the column is not lowered too far to prevent damage to the selector lever and/or cable.

3. Reach between the steering column and the instrument panel and gently lift the selector lever cable off the shift selector lever. Then, remove the cable clamp from the steering column tube.

4. Lay the steering column to rest on the front seat.

5. Remove one screw attaching the instrument panel to the brake pedal support at the steering column opening.

6. Disconnect the temperature control cable from the temperature blend door and the evaporator case bracket.

7. Disconnect the vacuum hose connectors at the evaporator case.

8. Disconnect the blower resistor wire connector from the resistor on the evaporator housing, and the blower motor feed wire at the in-line connector near the blower resistor wire connector.

9. Support the instrument panel and remove three screws attaching the top of the instrument panel to the cowl.

10. Remove one screw attaching each end of the instrument panel to the cowl side panels.

11. Remove the two screws that hold the instrument panel to floor.

12. Move the instrument panel rearward and disconnect the speedometer cable from the speedometer and any wires that will not allow the instrument panel to lay on the front seat. Use care not to scratch the instrument panel or the steering column.

13. Place the instrument panel near the installed position and connect any wires or connectors that were disconnected during removal.

14. Connect the speedometer cable to the speedometer.

15. Place the instrument panel in position and install one screw at each end of the instrument panel.

16. Install three screws along the top front edge of the instrument panel.

17. Install two screws retaining instrument panel to floor.

18. Connect the temperature control cable to the temperature blend door crank arm and the bracket.

1983 and Later Thunderbird/ Cougar (Automatic A/C)

1. Disconnect the ground cable from the battery.

2. Remove three screws attaching the steering column opening cover to the instrument panel. Then, pull the panel below the steering column opening and remove the reinforcement.

3. Remove the three screws attaching the reinforcement to the instrument panel below the steering column

opening and remove the reinforcement.

4. Remove two nuts retaining the hood latch release handle mounting bracket to the brake pedal support below the steering column.

5. Remove the sound insulator from under the left side of the instrument panel.

6. Remove two nuts retaining the steering column clamp to the brake pedal support and allow the steering column to rest on the seat.

7. Remove the instrument panel pad.

8. Disconnect the speedometer cable from the speedometer head.

9. Remove two screws attaching the console tray (automatic transmission) or the console cover around the hear shift lever (standard transmission) to the console. The, remove the tray or cover from the console.

10. Remove four console switch panel cover attaching screws. Disconnect the switch wires and remove the switch panel.

11. Open the console box cover and remove two screws from the bottom of the console box and two screws from the top front of the box.

12. Remove two screws attaching the front end of the console to the lower edge of the instrument panel.

13. Remove two screws attaching the bracket at the front end of the console and lower edge of the instrument panel to the floor pan. The, lift the rear end of the console and pull the console rearward to disengage it from the instrument panel. Position the console out of the way.

14. Remove two plastic push pins attaching the glove box door straps to the glove box. Allow the glove box and door to hang by the hinge.

15. Remove one bolt attaching the instrument panel brace to the instrument panel at the glove box opening.

16. Remove one nut attaching the lower edge of the instrument panel to the brake pedal support.

17. Remove one bolt attaching a second instrument panel brace to the lower edge of the instrument panel just to the left of the console extension.

18. Remove one bolt attaching each end of the instrument panel to the cowl side panel.

19. Support the instrument panel and remove three screws attaching the top edge of the instrument panel to the cowl top panel.

20. Cover the steering column and seats. Then, carefully position the instrument panel toward the seat disconnecting wires and vacuum harness as necessary. Allow the instrument panel to rest on the front seat.

21. Place the instrument panel in position and connect any wires and vacuum harness that were disconnected during the instrument panel removal. Then, install the three screws to attach the top edge of the instrument panel to the cowl top panel.

22. Install one bolt to attach each end of the instrument panel to the cowl side panel.

23. Install one bolt each to attach the instrument panel left and right braces to the lower edge of the instrument panel near the console extension.

24. Install the nut to attach the lower edge of the instrument panel to the brake pedal support.

25. Position the console to the vehicle and install two screws to attach the front bracket to the floor.

26. Install two screws to attach the console box to the floor and two screws to attach the top front of the box to the support bracket.

27. Position the console switch panel to the console and connect the wires. Then, install the four panel cover attaching screws.

28. Position the console tray or the cover around the gear shift lever to the console and install the two attaching screws.

29. Connect the speedometer cable to the speedometer.

30. Install the instrument panel pad.

31. Position the steering column to the brake pedal support and install the retaining clamp (two nuts).

32. Position the reinforcement across the steering column opening of the instrument panel and install the three attaching screws.

33. Position the steering column opening cover to the instrument panel and install the three attaching screws.

34. Install the sound insulator under the left side of the instrument panel.

35. Connect the support straps to the glove box.

36. Connect the ground cable to the battery.

37. Check for proper operation of all instruments and controls.

1983 and Later LTD/Marquis (Automatic A/C)

1. Remove the instrument panel upper finish panel and pad assemblies.

2. Loosen the steering column and carefully lower only enough for access to transmission gear shift selector lever and cable assembly.

NOTE: Care must be used to assure that the column is not lowered too far to prevent damage to the selector lever and/or cable.

3. Reach between the steering column and the instrument panel and gently lift the selector lever cable off the shift selector lever. Then, remove the cable clamp from the steering column tube.

4. Lay the steering column to rest on the front seat.

5. Remove one screw attaching the instrument panel to the brake pedal support at the steering column opening.

6. Disconnect the servo motor from the temperature blend door.

7. Disconnect the vacuum hose connectors at the evaporator case.

8. Disconnect the blower resistor wire connector from the resistor on the evaporator housing, and the blower motor feed wire at the in-line connector near the blower resistor wire connector.

9. Support the instrument panel and remove three screws attaching the top of the instrument panel to the cowl.

10. Remove one screw attaching each end of the instrument panel to the cowl side panels.

11. Remove two screws holding instrument panel to floor.

12. Move the instrument panel rearward and disconnect the speedometer cable from the speedometer and any wires that will not allow the instrument panel to lay on the front seat. Use care not to scratch the instrument panel or the steering column.

13. Place the instrument panel near the installed position and connect any wires or connectors that were disconnected during removal.

14. Connect the servo motor to the temperature blend door crank arm and the bracket.

15. Connect the vacuum hoses at the 7-port connectors and the blower motor wires at the resistor and in-line connector near the resistor.

16. Connect the speedometer cable to the speedometer.

17. Place the instrument panel in position and install one screw at each end of the instrument panel.

18. Install three screws along the top front edge of the instrument panel.

19. Install two screws retaining instrument panel to floor.

20. Install the screw attaching the instrument panel to the brake pedal support.

21. Position the steering column near the brake pedal support.

22. Connect the transmission gear shift selector lever cable to the shift lever. Then, connect the cable clamp to the steering column tube.

23. Position the steering column against the brake pedal support and install the four attaching nuts.

24. Adjust the transmission selector indicator as necessary.

25. Install the steering column shroud.

26. Position the steering column opening cover to the instrument panel and install the two attaching screws.

27. Install the instrument panel pad and connect the ground cable to the battery.

Speedometer Cable

REMOVAL & INSTALLATION

1. Reach up behind the speedometer and depress the flat, quick-disconnect tab, while pulling back on the cable.

2. If the inner cable is broken, raise and support the car and remove the cable-to-transmission clamp and pull the cable from the transmission.

3. Pull the core from the cable.

4. Installation is the reverse of removal. Lubricate the core with speedometer cable lubricant prior to installation.

Fuses, Fusible Links and Circuit Breakers

Fusible links are used to protect the main wiring harness and selected branches from complete burn-out, should a short circuit or electrical overload occur.

Circuit breakers are used on certain electrical components requiring high amperage, such as the headlamp circuit, electrical seats and/or windows to name a few. The advantage of the circuit breaker is its ability to open and close the electrical circuit as the lead demands, rather than the necessity of a part replacement, should the circuit be opened with another protective device in line.

A fuse panel is used to house the numerous fuses protecting the various branches of the electrical system and is normally the most accessible. the mounting of the fuse panel is usually on the left side of the passenger compartment, under the dash, either on the side kick panel or on the firewall to the left of the steering column. Certain models will have the fuse panel exposed while other models will have it covered with a removable trim cover.

Ford Motor Co.
Taurus, Sable

YEAR IDENTIFICATION

1987 Taurus

1987 Sable

VEHICLE IDENTIFICATION NUMBER (VIN)

It is important for servicing and ordering parts to be certain of the vehicle and engine identification. The VIN (vehicle identification number) is a 13 or 17 digit number visible through the windshield on the driver's side of the dash and contains the vehicle and engine identification codes. It can be interpreted as follows:

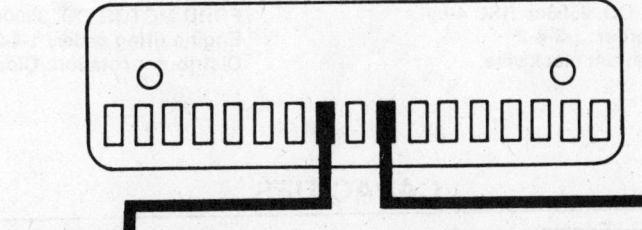

Engine Code						Model Year Code	
Code	Cu. In.	Liters	Cyl.	Fuel Delivery	Eng. Mfg.	Code	Year
D	153	2.5	4	CFI	Ford	G	'86
U	182	3.0	6	EFI	Ford	H	'87

The seventeen digit Vehicle Identification Number can be used to determine engine application and model year. The 10th digit indicates the model year and the 8th digit identifies the factory installed engine.
CFI Central Fuel Injection
EFI Electronic Fuel Injection

C265

GENERAL ENGINE SPECIFICATIONS

Year	Eng. VIN Code	Engine No. Cyl. Displacement (cu. in.)	Eng. Mfg.	Fuel Delivery	Horsepower @ rpm	Torque @ rpm (ft lbs)	Bore × Stroke (in.)	Compression Ratio	Oil Pressure @ 2000 rpm
'86–'87	D	4-153	Ford	CFI	88 @ 4600	130 @ 2800	3.7 × 3.6	9.0	55–70
	U	6-182	Ford	EFI	140 @ 4800	160 @ 3000	3.5 × 3.1	9.3	55–70

CFI Central Fuel Injection
EFI Electronic Fuel Injection

TUNE-UP SPECIFICATIONS

When analyzing compression test results, look for uniformity among cylinders rather than specific pressures.

Year	Eng. VIN Code	Engine No. Cyl. Displacement (cu. in.)	Eng. Mfg.	hp	Spark Plugs Orig Type	Spark Plugs Gap (in.)	Ignition Timing (deg) Man Trans	Ignition Timing (deg) Auto Trans	Fuel Pump Pressure (psi)	Idle Speed (rpm) Man Trans	Idle Speed (rpm) Auto Trans
'86	D	4-153	Ford	88	AWSF-32C	0.044	①	①	15–16	①	①
	U	6-182	Ford	140	AWSF-32C	0.044	—	①	36–39	—	①
'87					See Underhood Specifications Sticker						

① See underhood sticker

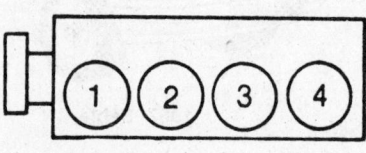

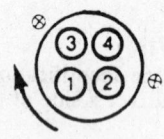

FORD MOTOR CO. 2500cc HSC 4-cyl.
Engine firing order: 1-3-4-2
Distributor rotation: clockwise

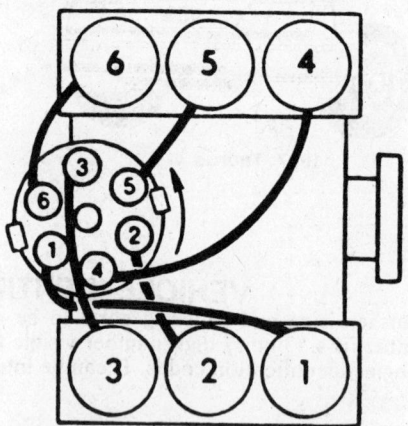

FORD MOTOR CO, 3000cc V6
Engine firing order: 1-4-2-5-3-6
Distributor rotation: Clockwise

CAPACITIES

Year	Engine No. Cyl. Displacement (cu. in.)	Engine Crankcase Add 1 Qt For New Filter	Transmission (Pts To Refill After Draining) 4sp/5sp	Transmission (Pts To Refill After Draining) Automatic	Gasoline Tank (gals)	Cooling System (qts) With Heater	Cooling System (qts) With A/C
'86–'87	4-135	4.0	12.4	16.6	16.0②	8.3	8.3
	6-182	3.5	—	21.8	16.0②	11.0	11.0①

① 11.8 qts—station wagon with A/C
② 18.6 gal.—optional

CAMSHAFT SPECIFICATIONS
All measurements are given in inches.

Year	Engine No. Cyl. Displacement (ca. in.)	Lobe Lift	Valve Lift @ Zero Lash Intake	Exhaust	Camshaft End Play	Journal-to-Bearing Clearance	Journal Diameter	Journal Out-of-Round Limit
'86–'87	4-153	0.239	0.392	0.377	0.009	0.001–0.003	2.010–2.009	0.005
	6-182	0.260	0.419	0.419	①	0.001–0.003	2.0074–2.0084	0.001

① No end-play, camshaft is retained by a spring.

VALVE SPECIFICATIONS

Year	Engine No. Cyl. Displacement (cu. in.)	Seat Angle (deg)	Face Angle (deg)	Spring Test Pressure (lbs @ in.)	Spring Installed Height (in.)	Stem-to-Guide Clearance Intake	Exhaust	Stem Diameter Intake	Exhaust
'86–'87	4-153	45	44	70–78 @ 1.5①	1.49	0.0018	0.0023	0.3415–0.3422	0.3411–0.3418
	6-182	45	44	73 @ 1.5①	1.85	0.001–0.0027	0.0015–0.0032	0.3126–0.3134	0.3121–0.3129

① Unloaded

CRANKSHAFT AND CONNECTING ROD SPECIFICATIONS
All measurements are given in inches.

Year	Engine No. Cyl. Displacement (cu. in.)	Main Brg. Journal Dia	Main Brg. Oil Clearance	Shaft End-Play	Thrust on No.	Journal Diameter	Oil Clearance	Side Clearance
'86–'87	4-153	2.2489–2.2490	0.0008–0.0015	0.004–0.008	3	2.2388–2.2396	0.0008–0.0015	0.0035–0.0105
	6-182	2.5190–2.5198	0.001–0.0014	0.004–0.008	3	2.250–2.251	0.001–0.0014	0.006–0.014

PISTON AND RING SPECIFICATIONS
All measurements are given in inches.

Year	V.I.N. Code	Engine Type Disp. (cu. in.)	Eng. Mfg.	Piston-to-Bore Clearance	Top Compression	Bottom Compression	Oil Control	Top Compression	Bottom Compression	Oil Control
'86–'87	D	4-153	Ford	0.0012–0.0022	0.008–0.016	0.008–0.016	0.015–0.055	0.002–0.004	0.002–0.004	—
	U	6-182	Ford	0.0012–0.0023	0.010–0.020	0.010–0.020	0.010–0.049	0.0016–0.0037	0.0016–0.0037	—

TORQUE SPECIFICATIONS
All readings in ft. lbs.

Year	Engine	Cylinder Head Bolts	Rod Bearing Bolts	Main Bearing Bolts	Crankshaft Bolt	Flywheel to Crankshaft Bolts	Manifold Intake	Manifold Exhaust
'86–'87	4-153	①	21–26	51–66	140–170	54–64	15–23	②
	6-182	③	④	65–81	141–169	54–64	⑤	15–22

① Tighten in 2 steps: 51.6 ft. lbs., 70–76 ft. lbs.
② Tighten in 2 steps: 5–7 ft. lbs., 20–30 ft. lbs.
③ Tighten in 2 steps: 48–54 ft. lbs., 63–80 ft. lbs.
④ Tighten to 20–28 ft. lbs., back off 2 turns, tighten to 20–25 ft. lbs.
⑤ Tighten in 3 steps: 11 ft. lbs., 18 ft. lbs., 24 ft. lbs.

FRONT WHEEL ALIGNMENT SPECIFICATIONS

Year	Model	Caster① Range (deg)	Camber Range (deg)	Toe-in (in.)
'86–'87	Taurus/Sable Wagon	2⅞P–5⅞P	1¹⁄₃₂N–⁵⁄₃₂P	⁷⁄₃₂N–¹⁄₆₄P
	Taurus Sedan	3P–6P	1³⁄₃₂N–³⁄₃₂P	⁷⁄₃₂N–¹⁄₆₄P
	Sable Sedan	2⅞P–5⅞P	1¹⁄₃₂N–³⁄₃₂P	⁷⁄₃₂N–¹⁄₆₄P

① The caster measurements are made by turning each individual wheel through the prescribed angle of sweep.

REAR WHEEL ALIGNMENT SPECIFICATIONS

Year	Model	Camber Range (deg)	Toe-in (in.)
'86–'87	Taurus	1⅝N–¼N	13⁄₆₄N–① 19⁄₆₄P
	Sable	1⁹⁄₁₆N–³⁄₁₆N	11⁄₆₄N–① 19⁄₆₄P

① Individual sides

ENGINE ELECTRICAL

Alternator

PRECAUTIONS

To prevent serious damage to the alternator and the rest of the charging system, the following precautions must be observed:

1. When installing a battery, make sure that the positive cable is connected to the positive terminal and the negative to the negative.

2. When jump-starting the vehicle with another battery, make sure that like terminals are connected. This also applies when using a battery charger.

3. Never operate the alternator with the battery disconnected or otherwise on an uncontrolled open circuit. Double-check to see that all connections are tight.

4. Do not short across or ground any alternator or regulator terminals.

5. Do not try to polarize the alternator.

6. Do not apply full battery voltage to the field connector.

7. Always disconnect the battery ground cable before disconnecting the alternator lead.

BELT TENSION ADJUSTMENT

2.5L Engine

The drive belt tension is maintained by an automatic tensioner which does not require adjustment.

3.0L Engine

1. Loosen the alternator adjusting arm bolt.

2. Adjust the drive belt tension using one of the following methods:

a. Using the Belt Tension Gauge tool No. 021-00028, install it onto the drive belt, on the longest belt span between the pulleys and adjust the drive belt tension to 100–140 lbs. (new) or 80–100 lbs. (used).

b. Apply thumb pressure on the longest belt span between the pulleys so that there is approximately $1/4$–$1/2$ in. of deflection.

3. Torque the adjusting arm bolt 22–32 ft. lbs. and recheck the belt tension.

REMOVAL & INSTALLATION

The engines are equipped with V-ribbed belts. To increase the belt life, make sure that the V-grooves make proper contact on the pulleys.

2.5L Engine

1. Place a $1/2$ in. flex handle in the square hole of the belt tensioner or an 18mm socket on the tensioner pulley nut.

2. Turn the tensioner counterclockwise and remove the drive belt.

3. At the back of the alternator, label and disconnect the electrical connectors.

NOTE: The alternator uses a push-on wiring connector on the field and stator connections. Depress the locking tab when removing the electrical connector from the alternator.

4. Remove the mounting bolts and the alternator from the vehicle.

5. To install, reverse the removal procedures. Torque the alternator-to-engine bolts to 45–57 ft. lbs.

6. To install the drive belt, place it on the pulleys (except the alternator), turn the tensioner counterclockwise and position the belt on the alternator pulley so that the V-grooves are aligned correctly.

3.0L Engine

1. Disconnect the negative battery cable.

2. Remove the alternator adjusting arm bolt and the drive belt.

3. At the back of the alternator, label and disconnect the electrical connectors.

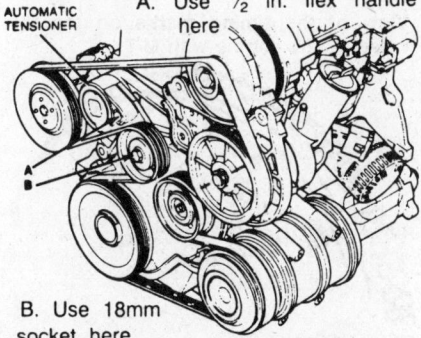

A. Use $1/2$ in. flex handle here

B. Use 18mm socket here

View of the drive belts—2.5L engine.

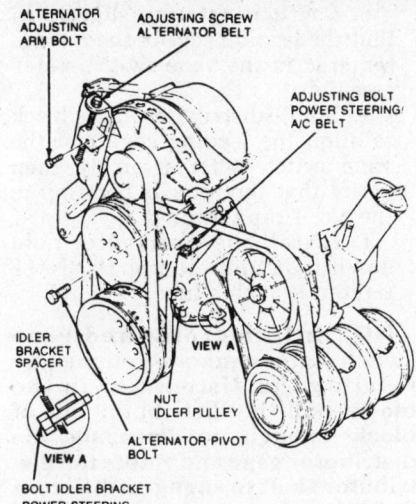

View of the drive belts—3.0L engine

NOTE: The alternator uses a push-on wiring connector on the field and stator connections. Depress the locking tab when removing the electrical connector from the alternator.

4. Remove the pivot bolt and the alternator from the vehicle.

5. To install, reverse the removal procedures. Adjust the drive belt tension. Torque the pivot bolt to 45–57 ft. lbs. and the adjusting arm bolt to 22–32 ft. lbs.

Voltage Regulator

NOTE: Two types of regulators are used, depending on the engine, the alternator output and type of dash mounted charging indicator used (light or ammeter). The regulators are 100% solid state, calibrated and preset by the manufacturer. No readjustments are required or possible.

REMOVAL & INSTALLATION

External

1. Remove the negative battery cable.

2. Disconnect the electrical connectors from the regulator.

3. Remove the regulator mounting screws and the regulator.

4. To install, reverse the removal procedures.

5. Test the system for proper voltage regulation.

Internal

1. Disconnect the negative battery cable.

2. At the rear of the alternator, disconnect the electrical connector from the regulator.

3. Remove the regulator-to-alternator screws and the regulator. If necessary, remove the brush holder-to-regulator screws and the brush holder from the regulator.

4. If the brush holder was removed from the regulator, install it using the following procedure:

a. Push the brushes into the brush holder and install a stiff wire (in the brush holder pin hole) to hold the brushes in place.

b. Align the brush holder with the regulator and torque the screws to 20–30 inch lbs.

c. When the regulator/brush holder assembly is installed on the alternator, remove the wire retainer from the brush holder.

NOTE: If the wire is not removed from the brush holder, a short circuit will result and destroy the regulator.

5. To install, reverse the removal procedures. Torque the regulator-to-alternator screws to 25–35 inch lbs.

Starter

REMOVAL & INSTALLATION

1. Disconnect the negative battery cable and the cable from the starter.

2. Raise and support the front of the vehicle safely and block the rear wheels.

3. From the upper starter stud bolt, remove the cable support and ground cable connection.

4. Remove the starter brace and the starter.

5. Remove the three starter-to-bell housing bolts (2.5L engine) or the two starter-to-bell housing bolts (3.0L engine).

6. If equipped with an ATX, remove the starter between the subframe and radiator. If equipped with a MTX, remove the starter between the sub-frame and the engine.

7. To install, reverse the removal procedures. Torque the starter bolts to 30–40 ft. lbs.

Distributor

The distributor is a new universal gear driven design with a die cast base that incorporates an integrally mounted TFI-IV (Thick Film Ignition) ignition module, a "Hall Effect" vane switch stator assembly and provision for a fixed octane adjustment. The new design eliminates the conventional centrifugal and vacuum advance mechanisms.

NOTE: No distributor calibration is required; initial timing is the normal adjustment.

REMOVAL

1. Disconnect the primary wiring connector from distributor.

NOTE: Before removing the distributor cap, mark the relationship of the No. 1 wire tower on the distributor base.

2. Using a screwdriver, remove distributor cap (with the wires attached) and position it aside.

3. Turn the crankshaft to align the rotor with the No. 1 tower position, then remove the rotor.

4. Remove the TFI-IV harness connector.

NOTE: Some engines may be equipped with a security type distributor hold down bolt. Using the tool No. T82L-12270-A, remove the distributor hold down bolt.

5. Remove the distributor hold down bolt/clamp and the distributor; be careful not to disturb the intermediate driveshaft.

INSTALLATION

1. If the engine has been disturbed (crankshaft rotated), perform the following procedures:

a. Remove the No. 1 spark plug.

b. Rotate the crankshaft until the No. 1 piston is on the compression stroke. Align timing marks for correct initial timing.

c. Position the distributor shaft so that the center of the rotor is pointing toward the mark previously made on distributor base.

d. Continue rotating slightly so that the leading edge of the rotor is centered in the vane switch stator assembly.

e. Rotate distributor in the block to align the leading edge and the vane switch stator assembly, then verify that the rotor is pointing to the No. 1 cap terminal.

f. Install the distributor hold down bolt and clamp; DO NOT tighten it at this time.

NOTE: If the rotor and vane switch stator cannot be aligned by rotating the distributor in the block, pull the distributor out of block (enough) to disengage the distributor gear and rotate the distributor shaft to engage a different distributor gear tooth. Repeat Step 1 as necessary.

2. Reinstall the electrical harness connector to the distributor.

3. Install the distributor cap/ignition wire assembly. Check that the ignition wires are securely connected to the distributor cap and spark plugs. Torque the distributor cap screws to 18–23 inch lbs. (2.0–2.6 Nm).

4. Using a timing light, set the initial timing by referring to the Vehicle Emission Control Information Decal.

5. Torque the distributor hold down bolt to 17–25 ft. lbs. (23–34 Nm).

6. Check and/or adjust the initial timing (if necessary).

IGNITION TIMING ADJUSTMENT

The locations of the timing marks on the 2.5L engine are as follows:

a. Manual transaxles – the timing marks are located on the flywheel and visible through a hole in the top of the transaxle case. To view the timing marks, a cover plate on top of the transaxle must be removed.

b. Automatic transaxles – the timing marks are visible through a hole in the transmission case. There in no cover plate.

The 3.0L engine employs timing marks on the crankshaft pulley and a timing pointer near the pulley.

1. Place the transaxle in the Park (ATX) or Neutral (MTX) position.

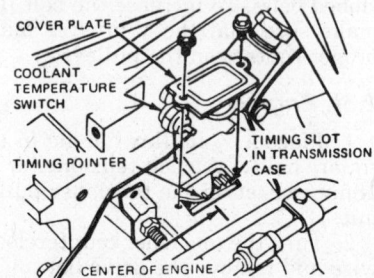

Location of the timing marks—2.5L engine with M/T

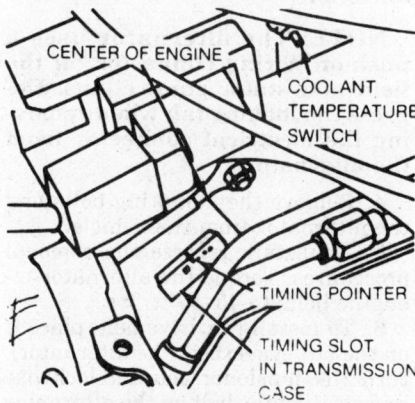

Location of the timing marks—2.5L engine with A/T

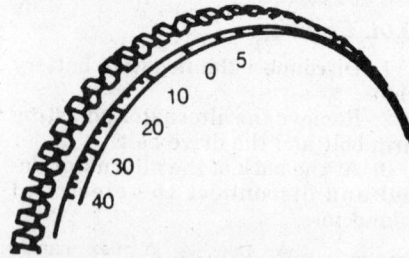

View of the timing marks on the flywheel—2.5L engine with M/T

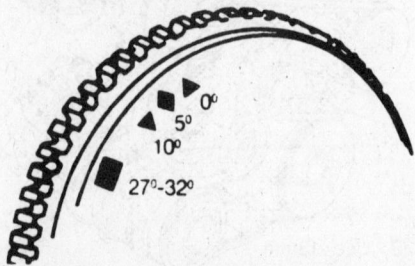

View of the timing marks on the flywheel—2.5L engine with A/T

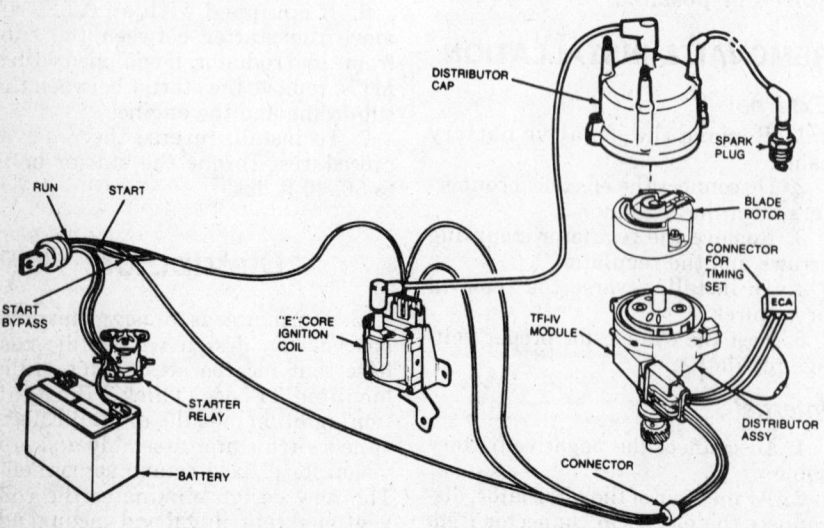

Exploded view of the distributor and ignition system

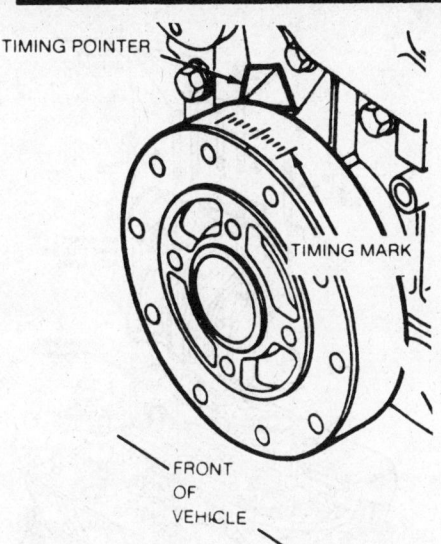

View of the timing marks on the 3.0L engine

2. Open the hood and clean the timing marks with a stiff brush or solvent. On the 2.5L, MTX models, it will be necessary to remove the transaxle cover plate which allows access to the timing marks.

3. Using a white chalk or paint mark, mark the specified timing mark and pointer.

4. Near the distributor, disconnect the in-line spout connector. The spout connector is the center wire between the Electronic Control Assembly (ECA) connector and the Thick Film Ignition (TFI) module.

5. Connect an inductive type timing light (Rotunda tool No. 059-00006) to the No. 1 spark plug wire. DO NOT puncture the ignition wire with any type of probing device.

NOTE: The high ignition coil currents generated in the EEC-IV ignition system may falsely trigger the timing lights with capacitive or direct connect pick-ups. It is necessary that an inductive type timing light be used in this procedure.

6. Connect a tachometer (Rotunda tool No. 099-00003) to the ignition coil.

NOTE: The ignition coil electrical connector allows a test lead with an alligator clip to be connected to it's dark green/yellow dotted wire terminal without removing the connector. Be careful not to ground the alligator clip, for permanent damage will result to the coil.

7. Start the engine and allow it run until the normal operating temperature is reached.

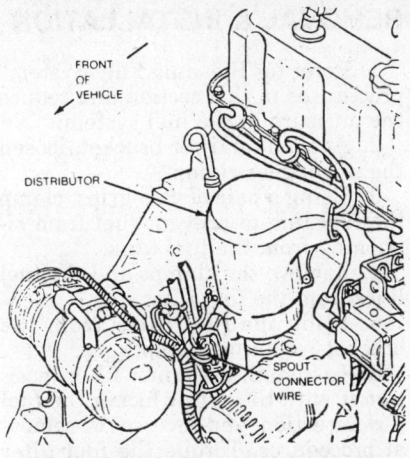

View of the in-line spout connector

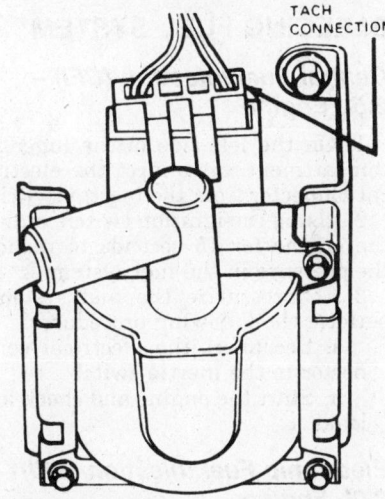

View of the tachometer connecting point of the "E" coil

8. Check the engine idle rpm, if it is not within specifications, adjust as necessary. After the rpm has been adjusted or checked, aim the timing light at the timing marks. If they are not aligned, loosen the distributor clamp bolts slightly and rotate the distributor body until the marks are aligned under the timing light illumination.

9. Tighten the distributor clamp bolts and recheck the ignition timing. Re-adjust the idle speed (if necessary).

10. Turn the engine OFF, remove the test equipment, reconnect the in-line spout connector to the distributor and reinstall the cover plate on the MTX models.

FUEL SYSTEM

This book contains simple testing and service procedures for your vehicle's fuel injection system.

Due to the complex nature of modern fuel injection systems, comprehensive diagnosis, testing and repair procedures fall outside the confines of this repair manual. For complete information on diagnosis, testing and repair procedures concerning all modern fuel injection systems, please refer to *Chilton's Guide To Fuel Injection And Feedback Carburetors.*

Mechanical Fuel Pump

REMOVAL & INSTALLATION

2.5L Engine

1. Loosen the threaded fuel lines at the fuel pump slightly. Have a rag handy to absorb any gasoline spillage.

2. Loosen the fuel pump mounting bolts two turns and free the fuel pump from the engine.

3. Bump the engine around (with the starter) until the fuel pressure (hold the pump against the engine) is reduced. The fuel pump drive lobe and pushrod are now at their low point.

4. Remove the fuel lines and vapor hose (if equipped). Remove the fuel pump mounting bolts and the fuel pump.

5. Remove the fuel pump mounting gasket and clean the engine mounting surface.

6. Remove and check the pushrod; replace it, if wear is apparent.

7. To install, use a new gasket and reverse the removal procedures. Torque the fuel pump-to-engine bolts to 11–19 ft. lbs. The push rod length should be 2.43 in.

Electric Fuel Pump

The electric fuel pump is located in the fuel tank and is a part of the fuel gauge sending unit.

REMOVAL & INSTALLATION

3.0L Engine

1. Position the vehicle on a level surface.

2. Refer to "Bleeding Fuel System" procedures in this section and reduce the pressure in the fuel system.

3. Remove the fuel from the fuel tank by pumping it out through the filler neck. Take precautions to avoid the risk of fire when handling gasoline.

4. Raise and support the rear of the vehicle on jackstands. Remove the fuel filler tube (neck).

5. Support the fuel tank and remove the fuel tank straps. Lower the fuel tank slightly, then remove the fuel lines, the electrical connectors and the vent lines.

6. Remove the fuel tank and place it on a workbench. Clean any dirt from around the fuel pump attaching flange.

7. Using a brass drift and a hammer, turn the fuel pump locking ring counterclockwise and remove it.

8. Remove the fuel pump assembly from the fuel tank and discard the flange gasket.

9. To install, perform the following procedures:

a. Using Multipurpose Long Life Lubricant C1AZ-19590 or equivalent, coat the new O-ring and install it in the fuel ring groove.

b. Refill the fuel tank.

c. Install a fuel pressure gauge and turn the ignition switch ON and OFF, 5–10 times, for three second intervals, until the pressure gauge reads 13 psi (CFI) or 30 psi (EFI).

d. Remove the pressure gauge, start the engine and check for fuel leaks.

Fuel Filter

The fuel filter is mounted under the vehicle, next to the right front corner of the fuel tank.

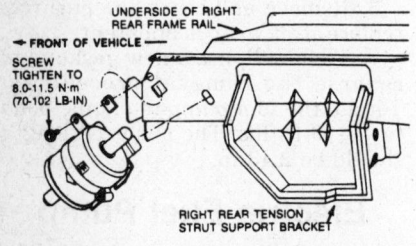

Exploded view of the fuel filter

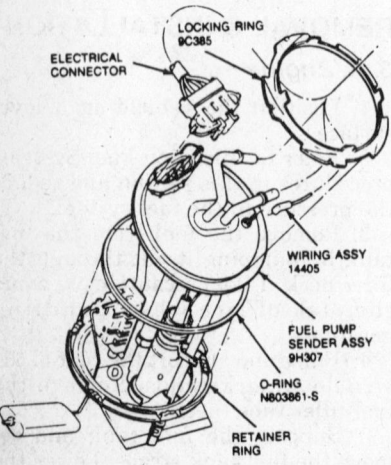

Exploded view of the fuel sender assembly

REMOVAL & INSTALLATION

1. Refer to "Bleeding Fuel System" procedures in this section and reduce the pressure in the fuel system.

2. At the fuel filter bracket, loosen the worm gear clamp.

3. Using a pair of vise-grips, clamp the fuel line to prevent fuel from siphoning from the fuel tank.

4. Remove the clamps and the fuel lines from the fuel filter.

5. Slide the fuel filter from the bracket retaining clamp.

6. To install, use a new filter (position it with the arrow facing the fuel flow direction) and reverse the removal procedures. Torque the fuel filter clamp to 15–25 inch lbs. (1.7–2.8 Nm).

BLEEDING FUEL SYSTEM

Central Fuel Injection (CFI)— 2.5L Engine

1. On the left side of the luggage compartment, disconnect the electrical connector from the inertia switch.

2. Using the ignition switch, crank the engine for 15 seconds to reduce the pressure in the fuel system.

3. To pressurize the fuel system, perform the following procedures:

a. Reconnect the electrical connector to the inertia switch.

b. Start the engine and check for leaks.

Electronic Fuel Injection (EFI)— 3.0L Engine

1. Remove the fuel tank cap and the air filter.

2. Disconnect the negative battery cable.

3. Using the Fuel Pressure Gauge tool No. T80L-9974-A, connect it to the pressure relief valve (remove the valve cap) on the fuel injection manifold.

4. Open the pressure relief valve and reduce the fuel pressure.

5. To pressurize the fuel system, perform the following:

a. Tighten the pressure relief valve and remove the pressure gauge.

b. Reinstall the negative battery cable.

c. Start the engine and check for leaks.

Throttle Body

REMOVAL & INSTALLATION

Central Fuel Injection (CFI)— 2.5L Engine

The Central Fuel Injection (CFI) system is a single-point, time pulse mod-

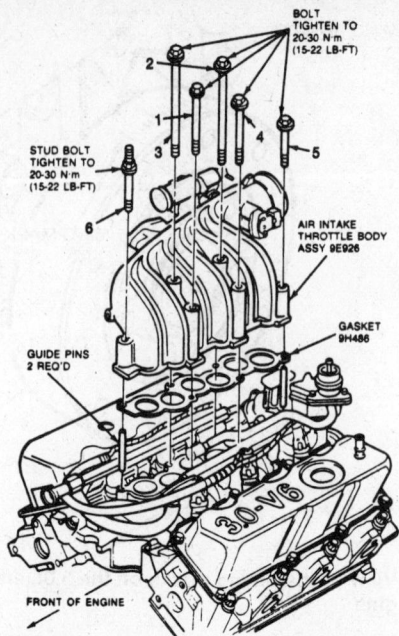

Exploded view of the throttle body—3.0L engine

ulated injection system. Fuel is injected into the intake manifold by a single solenoid injection valve mounted on the throttle body.

1. Remove the air duct-to-throttle body clamp and separate the air duct from the throttle body.

2. At the left side of the luggage compartment, disconnect the electrical connector from the inertia switch.

3. Using the ignition switch, crank the engine for 15 seconds to reduce the pressure in the fuel system.

4. At the throttle body, disconnect the throttle cable and the throttle valve lever (if equipped with ATX).

5. Disconnect the electrical connectors form the Idle Speed Control (ISC), the Throttle Position Sensor (TPS) and the fuel injector.

6. At the throttle body, disconnect the fuel inlet, the fuel outlet and the PCV lines.

7. Remove the throttle body-to-manifold nuts and the throttle body from the vehicle; discard the gasket.

8. To install, use a new gasket and reverse the removal procedures. Torque the throttle body-to-intake manifold nuts to 14–16 ft. lbs. (18–21 Nm). Start the engine and check for fuel leaks. Check and/or adjust the idle speed.

Electronic Fuel Injection (EFI)— 3.0L Engine

The Electronic Fuel Injection (EFI) system is a multi-point, time pulse, mass air flow injection system. Metered fuel is injected into the cylinders by 6 fuel injectors, located on the tuned intake manifold.

1. Remove the air intake duct between the air cleaner and the throttle body.

2. Disconnect the electrical connectors from the throttle position sensor (TPS), the air by-pass valve and the air charge temperature (ACT) sensor.

3. Disconnect the vacuum hoses from the intake manifold.

4. Disconnect and remove the accelerator and speed control cables from the accelerator mounting bracket and throttle lever.

5. If equipped with an ATX, remove the transmission valve (TV) linkage from the throttle lever.

6. Remove the throttle body-to-intake manifold bolts, the throttle body and the gasket from the vehicle.

7. Using a putty knife, clean the gasket mounting surfaces.

8. To install, use a new gasket and reverse the removal procedures. Torque the throttle body-to-intake manifold bolts (in sequence) to 15–22 ft. lbs. (20–30 Nm).

IDLE SPEED AND MIXTURE ADJUSTMENT

2.5L CFI Engine

The curb idle and fast idle speeds are controlled by the EEC-IV computer and the idle speed control (ISC) device. If the control system is operating correctly, the speeds are fixed and should not be changed.

1. Apply the parking brake and block the wheels.

NOTE: If equipped with an ATX and an automatic parking brake, ALWAYS place the transaxle in Reverse (not Drive) when checking the idle speed in gear.

2. Start the engine and allow it run until normal operating temperatures are reached; make sure that all of the accessories are turned Off. Connect a tachometer to the ignition coil connector.

3. Check the vacuum lines for leaks.

4. Place the transaxle in Drive (or Reverse) for ATX or Neutral for MTX and allow the engine to operate for two minutes. The idle speed should be within specifications listed on the underhood decal.

NOTE: If the electric cooling fan turns ON during this procedure, wait for it to turn OFF before proceeding.

5. If equipped with an ATX, place it in Neutral, the idle speed should increase about 100 rpm.

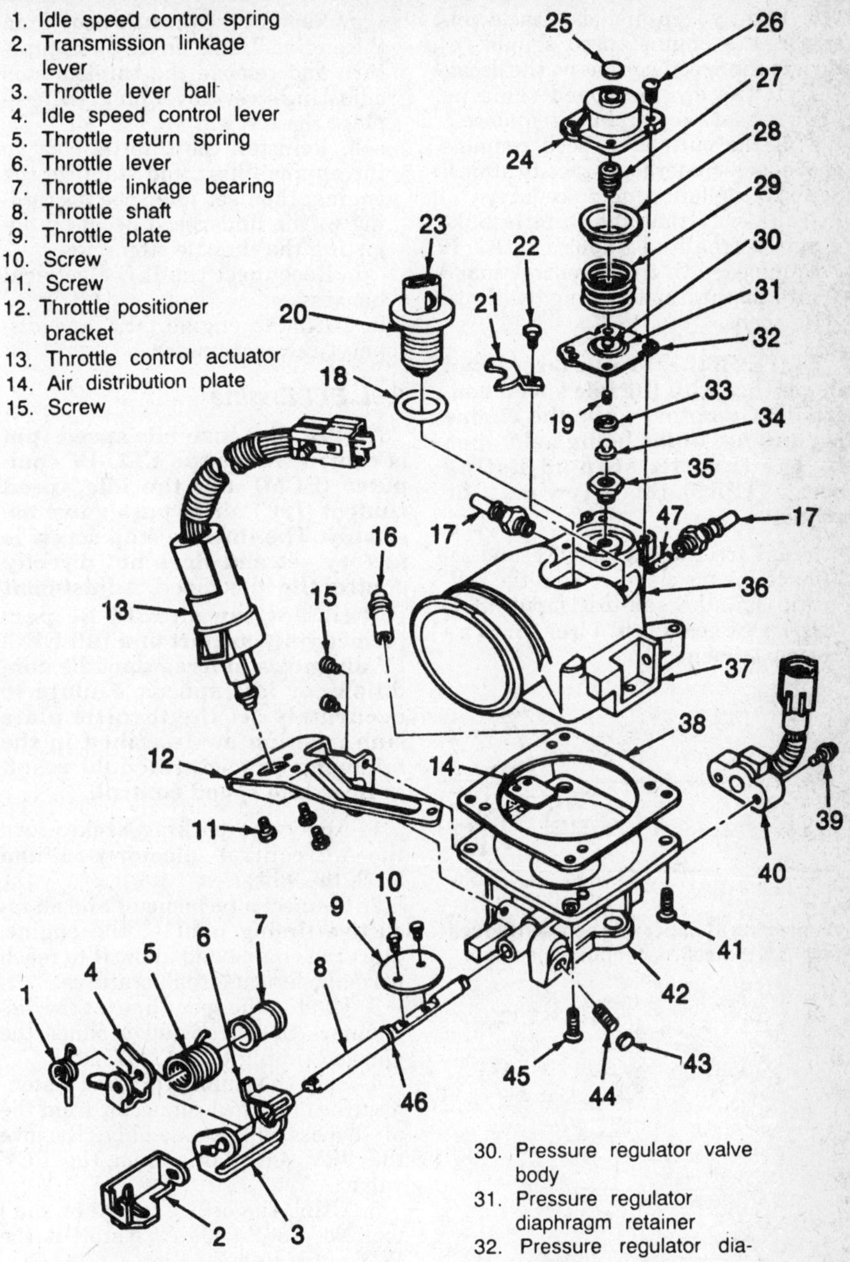

1. Idle speed control spring
2. Transmission linkage lever
3. Throttle lever ball
4. Idle speed control lever
5. Throttle return spring
6. Throttle lever
7. Throttle linkage bearing
8. Throttle shaft
9. Throttle plate
10. Screw
11. Screw
12. Throttle positioner bracket
13. Throttle control actuator
14. Air distribution plate
15. Screw
16. Emission inlet tube
17. Fuel injection quick connector
18. O-ring
19. Fuel pressure regulator spring
20. O-ring
21. Fuel injector retainer
22. Screw
23. Fuel injector
24. Fuel pressure regulator cover
25. Expansion plug
26. Screw
27. Pressure regulator adjusting screw
28. Pressure regulator diaphragm cup
29. Pressure regulator diaphragm spring
30. Pressure regulator valve body
31. Pressure regulator diaphragm retainer
32. Pressure regulator diaphragm
33. Pressure regulator valve retainer
34. Pressure regulator valve
35. Pressure regulator outlet tube
36. Fuel charging body assembly
37. Fuel charging main body
38. Gasket
39. Screw
40. Potentiometer assembly
41. Screw
42. Fuel charging throttle body
43. Extension plug
44. Screw
45. Screw
46. Fuel charging shaft seal
47. Fuel inlet screen

Exploded view of the throttle body—2.5L engine

6. Lightly step on and off the accelerator, the engine speed should return to the specifications on the decal.

7. If the engine speed remains high, repeat the checking sequence.

8. If the curb idle speed remains above the underhood specifications, perform the following procedures:

a. Verify that the throttle linkage is free and unobstructed. If equipped with cruise control, make sure that it is not holding the throttle open.

NOTE: If the throttle lever is not in contact with the idle speed control (ISC) motor, while the engine is running, but is being held open by the throttle stop adjusting screw (TSAS), this screw must be adjusted.

b. Turn the engine OFF and remove the air cleaner. Find the self-test and the self-test input (STI) connectors in the engine compartment.

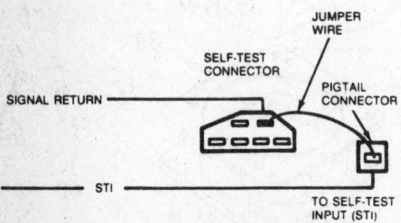

Connecting a jumper wire between the self-test and the self-test input connectors

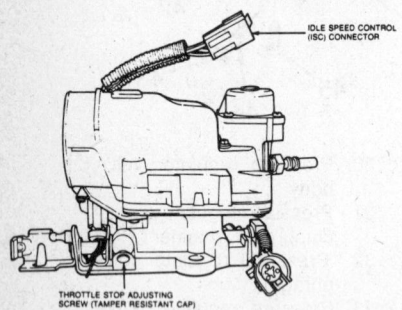

View of the throttle body—2.5L CFI engine

c. Using a jumper wire, connect it between the self-test input (STI) connector and the signal return pin on the self-test connector.

d. Turn the ignition switch ON but do not start the engine.

e. Wait for 10–15 seconds, the idle speed control (ISC) plunger will fully contract; if not, inspect the ISC.

f. Turn the ignition switch OFF and wait for 10–15 seconds, then remove the jumper wire and disconnect the ISC electrical connector from the electrical harness.

g. Remove the throttle body from the engine. Using an ice pick, puncture and remove the throttle stop adjusting screw cover plug, then replace the screw.

h. Reinstall the throttle body to the engine. Start and stabilize the engine, then set idle speed according to the underhood decal by adjusting the throttle stop screw.

i. Reconnect the ISC electrical harness.

9. Turn the engine OFF and disconnect the tachometer.

3.0L EFI Engine

NOTE: The curb idle speed rpm is controlled by the EEC-IV computer (ECM) and the idle speed control (ISC) air bypass valve assembly. The throttle stop screw is factory set and does not directly control the idle speed. Adjustment to this setting should be performed only as part of a full EEC-IV diagnosis of irregular idle conditions or idle speeds. Failure to accurately set the throttle plate stop position as described in the following procedure could result in false idle speed control.

1. Apply the parking brake, turn the A/C control selector OFF and block the wheels.

2. Connect a tachometer and an inductive timing light to the engine. Start the engine and allow it to reach normal operating temperatures.

3. Unplug the spout line (at the distributor), the check and/or adjust the ignition timing to 8–12° BTDC.

4. Shut the engine OFF and disconnect the electrical connector from the air bypass valve assembly. Remove the PCV entry line from the PCV valve.

5. Using the orifice (0.200 in. dia.) tool No. T86P-9600-A, install it the PCV entry line.

6. Start the engine. Place the transaxle in Drive (ATX) or Neutral (MTX). Disconnect the electrical connector from the electric cooling fan.

7. Check and/or adjust (if necessary) the idle speed to 595–655 rpm by turning the throttle plate stop screw.

8. After adjusting the idle speed, turn Off the engine and wait for 3–5 minutes.

9. Start the engine and confirm that the idle speed is now adjusted to specifications, if not, readjust as necessary.

10. Turn the engine OFF and remove the orifice. Reconnect the PCV entry line, the spout line, the ISC motor and the electric cooling fan.

11. Make sure that the throttle plate is not stuck in the bore and that the throttle plate stop screw is setting on the rest pad with the throttle closed. Correct any condition that will not allow the throttle to close to the stop set position.

12. Restart the engine. After 3–5 minutes of operation, the engine idle speed should be at specifications.

Fuel Rail

REMOVAL & INSTALLATION

3.0L EFI Engine

1. Refer to "Bleeding Fuel System" procedures in this section and reduce the pressure in the fuel system.

2. Refer to the "Throttle Body Removal & Installation" procedures in this section and remove the throttle body from the engine.

3. Using the Fuel Line Coupling Disconnect tool No. T81P-19623-G, disconnect the fuel lines from the fuel rail.

4. Disconnect the electrical harness connectors from the fuel injectors.

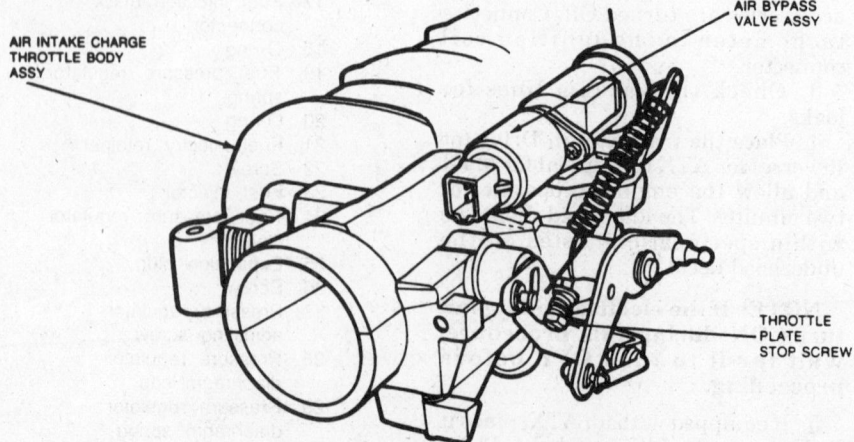

View of the throttle body—3.0L EFI engine

5. Remove the fuel rail-to-intake manifold screws, then carefully disengage and lift the fuel rail from the fuel injectors.

6. Remove and discard the O-rings from the top of the fuel injectors.

7. To install, use new fuel injector O-rings and reverse the removal procedures. Torque the fuel rail-to-intake manifold screws to 6–8 ft. lbs. (8–12 Nm) and the throttle body-to-intake manifold bolts to 15–22 ft. lbs. (20–30 Nm). Start the engine and then check for fuel leaks.

Fuel Injector(s)

REMOVAL & INSTALLATION

2.5L CFI Engine

1. Refer to "Bleeding Fuel System" procedures in this section and reduce the pressure in the fuel system.

2. At the throttle body, disconnect the electrical connector from the fuel injector.

3. Remove the fuel injector retaining screw, the retainer and the fuel injector from the throttle body.

4. In the throttle body injector cavity, remove and discard the lower O-ring.

5. To install, use a new O-ring and reverse the removal procedures. Start the engine and check for fuel leaks.

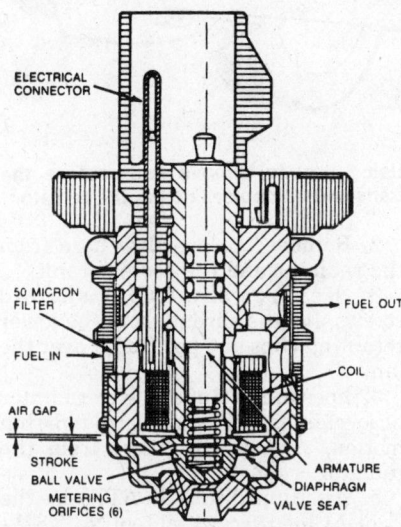

Cross-sectional view of the fuel injector—2.5L CFI engine

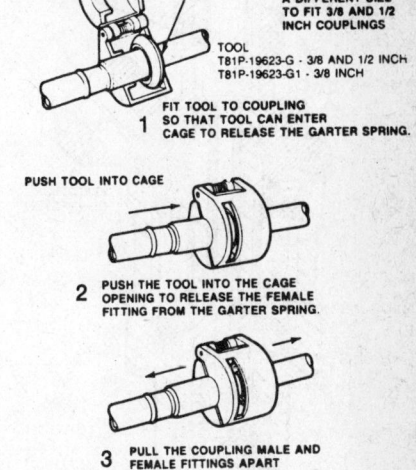

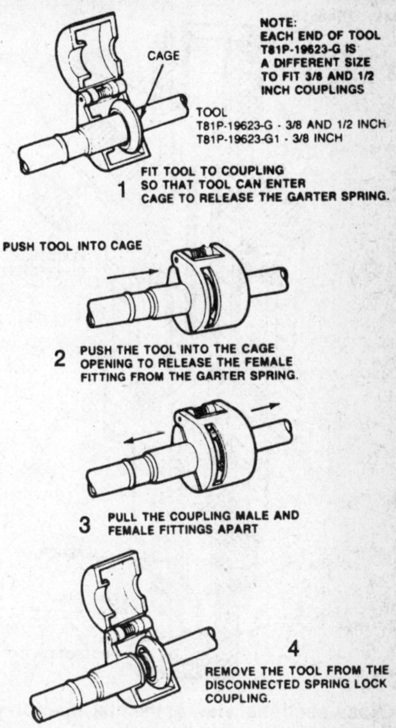

Disconnecting the fuel lines from the fuel rail

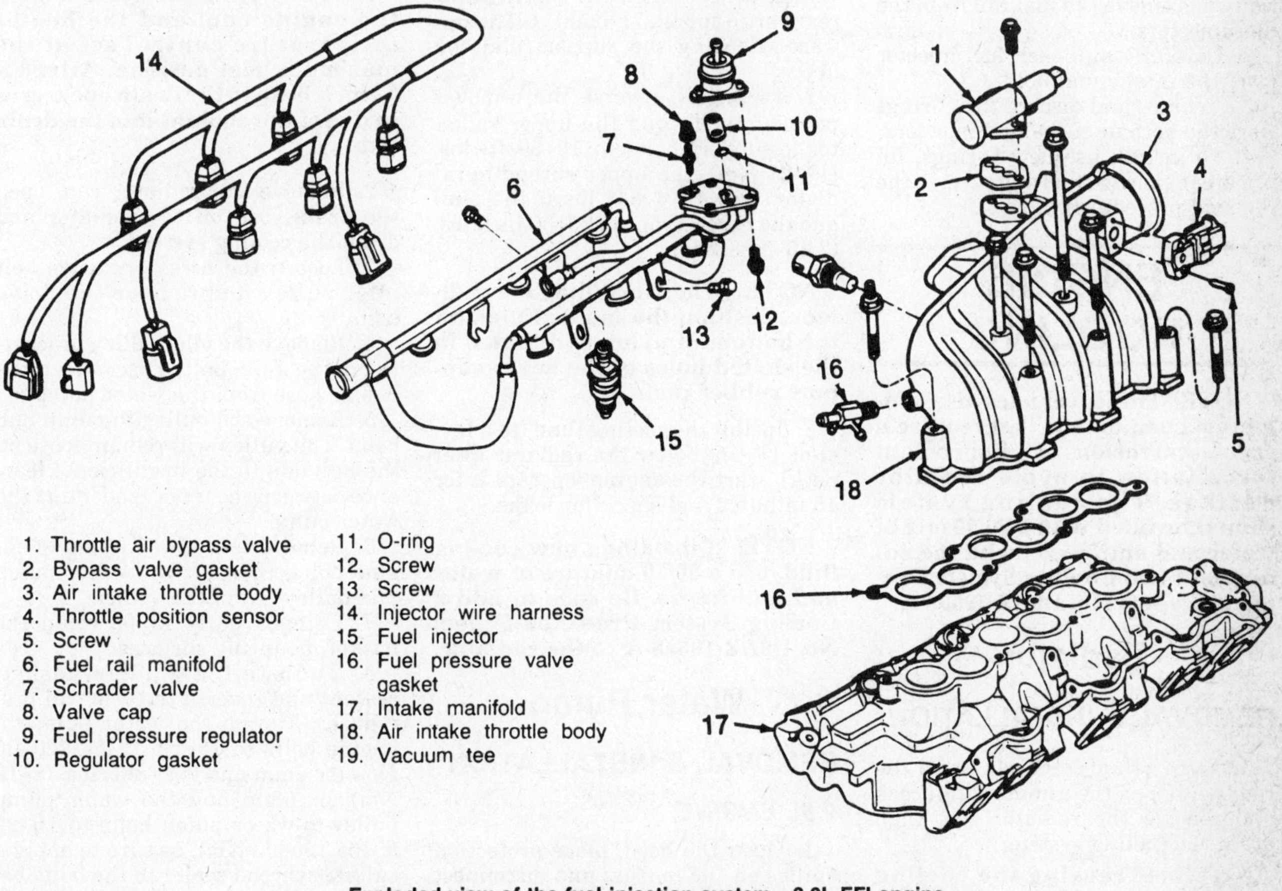

Exploded view of the fuel injection system—3.0L EFI engine

1. Throttle air bypass valve
2. Bypass valve gasket
3. Air intake throttle body
4. Throttle position sensor
5. Screw
6. Fuel rail manifold
7. Schrader valve
8. Valve cap
9. Fuel pressure regulator
10. Regulator gasket
11. O-ring
12. Screw
13. Screw
14. Injector wire harness
15. Fuel injector
16. Fuel pressure valve gasket
17. Intake manifold
18. Air intake throttle body
19. Vacuum tee

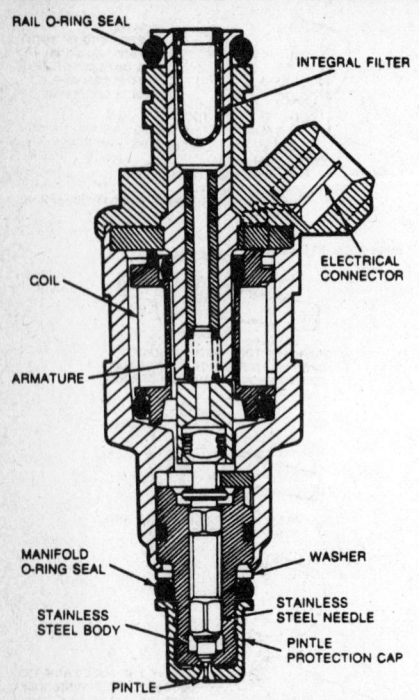

Cross-sectional view of the fuel injector—3.0L EFI engine

3.0L EFI Engine

1. Refer to the "Fuel Rail Removal & Installation" procedures in this section and remove the fuel rail from the fuel injectors.

2. Carefully pull each fuel injector from the intake manifold.

3. Remove and discard the O-rings from the bottom of the fuel injectors.

4. To install, use new O-rings, lubricate them with fuel and reverse the removal procedures.

ENGINE COOLING

NOTE: These vehicles use aluminum components that require a special corrosion inhibitor coolant formulation to avoid radiator damage. The cooling system should be filled with a 50/50 mix of water and antifreeze, with the addition of two Cooling System Protector Pellets No. D9AZ-19558-A.

Radiator

REMOVAL & INSTALLATION

1. Place a fluid catch pan under the radiator. Open the radiator draincock and remove the radiator cap, then drain the cooling system.

NOTE: If reusing the cooling fluid, be sure to keep it dirt free.

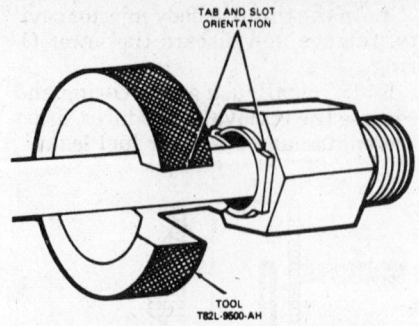

Using tool to remove and replace the transaxle oil cooler lines at the radiator

2. Remove the overflow hose from the radiator and the coolant tank.

3. Remove the upper shroud screws, lift the shroud from the lower retaining clips and position it over the fan.

4. Loosen the upper/lower radiator hose clamps, then using a twisting motion, remove the hoses from the radiator.

5. If equipped with an ATX, use the Cooler Line Disconnect tool No. T82L-9500-AH to disconnect the oil cooling lines from the radiator. Be sure to plug the cooling lines to prevent fluid draining from the transaxle.

6. Remove the upper radiator-to-vehicle screws, then tilt the radiator rearward (approx. 1 inch), lift it upward (clearing the support and the fan).

7. To install, reverse the removal procedures. Torque the upper radiator-to-support bolts to 13–20 ft. lbs. (17–27 Nm), the upper shroud-to-radiator screws to 4–6 ft. lbs. (5.5–8 Nm) and the hose clamps to 20–30 inch lbs. (2.25–3.38 Nm).

NOTE: When installing the radiator, position the molded pins (at the bottom of nylon end tanks) in the slotted holes of the lower support rubber pads.

8. Install the cooling fluid (to a level of $1\frac{1}{2}$ in. below the radiator filler neck), start the engine, operate it for 15 minutes and check for leaks.

NOTE: If installing new cooling fluid, use a 50/50 mixture of water and anti-freeze. Be sure to add 2 Cooling System Protector Pellets No. D9AZ-19558-A to the radiator.

Water Pump

REMOVAL & INSTALLATION

2.5L ENGINE

1. Open the hood, place protection aprons on the fenders and disconnect the negative battery cable.

2. Remove the radiator cap and position a drain pan under the radiator.

3. Raise and support the front of the vehicle on jackstands. Remove the lower radiator hose from the radiator and drain the coolant into the drain pan.

4. Remove the water pump inlet tube. Loosen the belt tensioner by inserting a $\frac{1}{2}$ inch flex handle in the square hole of the tensioner, then rotate the tensioner counterclockwise and remove the drive belt from the vehicle.

5. Disconnect the heater hose from the water pump. Remove the three water pump-to-engine block bolts and the pump from the engine.

6. Using a putty knife, clean the gasket mounting surfaces.

7. To install, use a new gasket, sealant and reverse the removal procedures. Torque the water pump-to-engine bolts to 15–22 ft. lbs. (20–30 Nm). Check and/or adjust the drive belt tension. Refill the cooling system, start the engine and check for leaks.

3.0L Engine

1. Disconnect the negative battery cable and place a suitable drain pan under the radiator draincock.

NOTE: Drain the system with the engine cool and the heater temperature control set at the maximum heat position. Attach a $\frac{3}{8}$ inch hose to the drain cock so as to direct the coolant into the drain pan.

2. Remove the radiator cap, open the drain cock on the radiator and drain the cooling system.

3. Loosen the accessory drive belt idler pulley and remove the drive belts.

4. Remove the idler pulley bracket-to-engine nuts/bolt. Disconnect the heater hose from the water pump.

5. Remove the pulley-to-pump hub bolts. The pulley will remain loose on the hub due to the insufficient clearance between the inner fender and the water pump.

6. Remove the water pump-to-engine bolts, then lift the water pump and pulley out of the vehicle.

7. Using a putty knife, clean the gasket mounting surfaces.

8. To install, use a new gasket, sealant and reverse the removal procedures. Torque the water pump-to-engine bolts to 15–22 ft. lbs. (20–30 Nm) for 8mm and to 6–8 ft. lbs. (8–12 Nm) for 6mm and the water pump pulley-to-water pump bolts to 15–22 ft. lbs. (20–30 Nm); be sure to apply a suitable thread sealer to the bolts before installing them. Check and/or ad-

just the drive belt tension. Refill the cooling system, start the engine and check for leaks.

Thermostat

The thermostat is located in the water outlet housing at the top rear of the engine.

REMOVAL & INSTALLATION

1. Raise the hood and place protective aprons on the fenders.
2. Disconnect the negative battery cable.
3. Position a drain pan under the radiator, remove the radiator cap, open the draincock and drain the coolant into the drain pan.

NOTE: Drain the cooling system to a level below the water outlet housing.

4. On the 2.5L engine, remove the vent plug from the water outlet housing.
5. Loosen the upper hose clamp at the radiator. Remove the water outlet housing-to-engine bolts, lift the outlet clear of the engine and remove the thermostat from the housing.
6. Using a putty knife, clean the gasket mounting surfaces.
7. To install, use a new gasket, sealant, thermostat and reverse the removal procedures. Torque the water outlet housing-to-engine bolts to 12–18 ft. lbs. (16–24 Nm) for 2.5L engine and 6–8 ft. lbs. (8–12 Nm) for 3.0L engine.

NOTE: When installing the thermostat, rotate it clockwise into the water outlet housing. On the 3.0L engine, position the thermostat ball check valve at the top.

EMISSION CONTROLS

Please refer to the "Emission Control" Unit Repair section for system maintenance and servicing procedures. Due to the complex nature of modern electronic engine control systems, comprehensive diagnosis and testing procedures fall outside the confines of this repair manual. For complete information on diagnosis, testing and repair procedures concerning all modern engine and emission control systems, please refer to *Chilton's Guide To Electronic Engine Controls*.

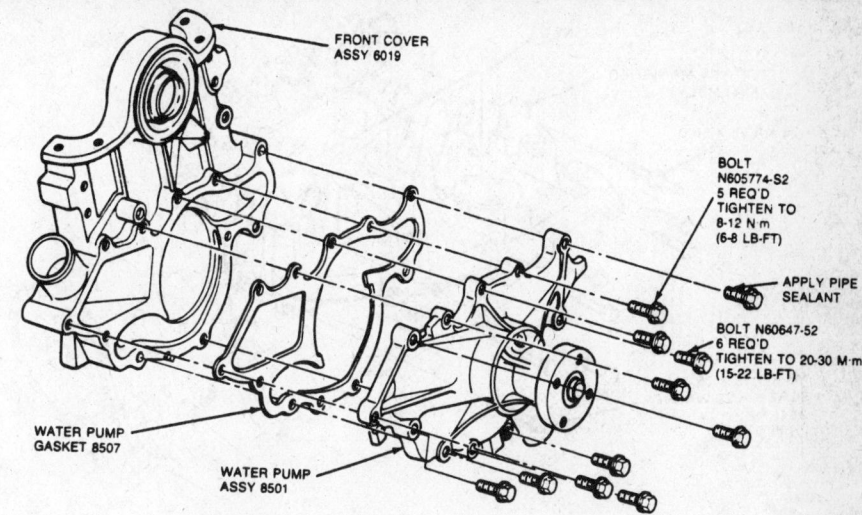

Exploded view of the water pump assembly—3.0L engine

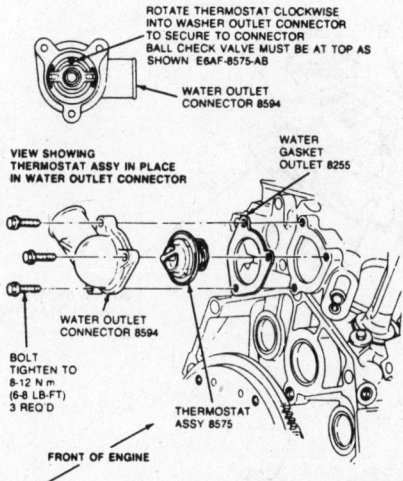

Exploded view of the thermostat and housing—3.0L engine

ENGINE MECHANICAL

Engine

REMOVAL & INSTALLATION

2.5L Engine

1. Using a scribing tool, mark the position of the hood hinges and remove hood.
2. Remove the negative battery cable. Remove the air cleaner and the air duct.
3. Place a drain pan under the radiator, open the draincock and drain the cooling system. Remove upper/lower radiator-to-engine cooling hoses.
4. If equipped with an ATX, use the Cooler Line Disconnect tool No. T82L-

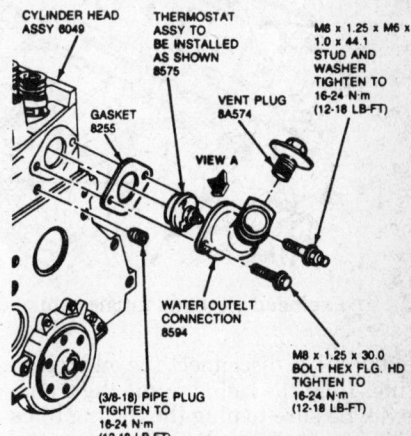

Position of water pump bolts on 2.0L engine—one M8 and two M6

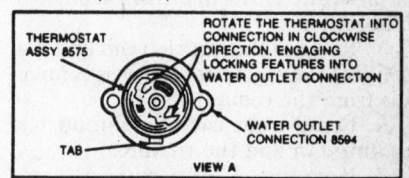

Exploded view of the thermostat and housing—2.5L engine

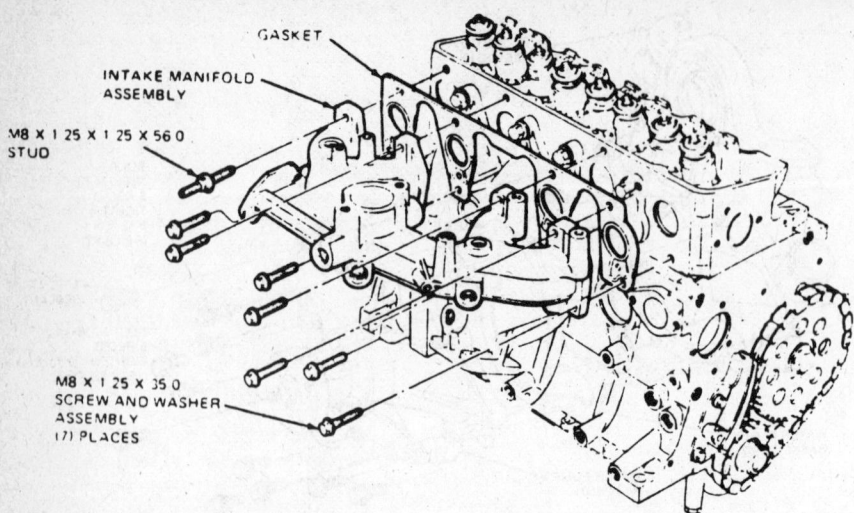

Exploded view of the intake manifold—2.5L engine

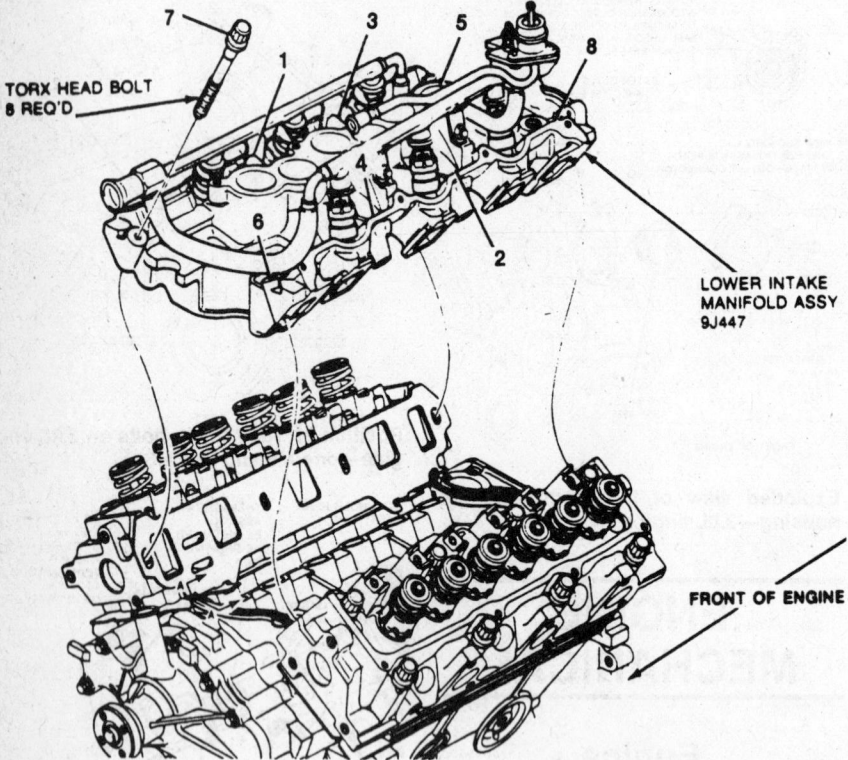

Exploded view and torque sequence of the intake manifold—3.0L engine

9500-AH to disconnect the oil cooling lines from the radiator and the transaxle. Be sure to plug the cooling lines to prevent fluid draining from the transaxle.

5. Remove the ignition coil and disconnect the electrical harness connector from the coolant fan.

6. Remove the radiator shroud, the cooling fan and the radiator.

7. If equipped with with A/C, remove the compressor and move it aside, DO NOT disconnect the refrigerant lines.

8. Label and disconnect all of the electrical connectors and vacuum lines, as necessary.

9. If equipped with a MTX, disconnect the TV linkage or clutch cable from the transaxle.

10. Disconnect the accelerator linkage and the fuel lines.

— CAUTION —

Before disconnecting the fuel lines, relieve the pressure in the fuel system.

11. Disconnect the thermactor pump discharge hose from the pump.

12. If equipped with power steering, disconnect lines from the pump.

13. Install an engine support tool to the engine lifting eye.

14. Raise and support the front of the vehicle on jackstands.

15. Remove the electrical connectors from the starter and the hose from the catalytic converter.

16. Remove the exhaust pipe bracket-to-oil pan bolt and the exhaust pipe-to-exhaust manifold nuts.

17. Remove the exhaust system from the rubber insulating grommets and set it aside.

18. Remove the speedometer cable from the transaxle.

19. Remove the heater hoses from the water pump inlet tube and the intake manifold.

20. Remove the water pump inlet tube clamp bolt from the engine and the clamp bolts from the underside of oil pan, then the inlet tube.

21. Remove the control arm-to-body bolts. Remove the stabilizer bar bracket bolts and the brackets.

22. Remove the halfshaft assemblies from the transaxle.

23. If equipped with a MTX, perform the following procedures:

   a. Remove the roll restrictor-to-transaxle nuts.

   b. Pull the roll restrictor from the mounting bracket.

   c. Remove the shift stabilizer bar-to-transaxle bolts.

   d. Remove the shift mechanism-to-shift shaft nut/bolt.

24. If equipped with an ATX, disconnect the shift cable-to-lever clip from the transaxle, the shift linkage bracket-to-transaxle bolts and the bracket.

25. Remove the left-hand rear insulator mount bracket-to-body nuts and the bracket.

26. Remove the left-hand front insulator-to-transaxle bolts.

27. Lower the vehicle but DO NOT allow front wheels to touch floor. Install a vertical hoist to the engine lifting eyes.

28. Remove the Engine Support tool.

29. Remove the right-hand No. 3A insulator intermediate bracket-to-engine bracket bolts, the intermediate bracket-to-insulator nuts, the intermediate bracket-to-engine bracket double ended stud bottom nut and the bracket.

30. Carefully lower the engine/transaxle assembly to the floor.

31. Remove the engine-to-transaxle bolts and separate the engine from the transaxle.

32. To install, reverse the removal procedures. Refill the cooling system. Check and/or refill the engine oil and the transaxle fluid. Start the engine and check for leaks.

3.0L Engine

1. Disconnect the battery cables from the battery. Place a drain pan under the radiator and drain the cooling system. Using a scribing tool, mark the hood hinge location and remove the hood.

2. If equipped with A/C, remove the compressor and move it aside; DO NOT discharge the A/C system.

3. Remove the air cleaner assembly, the battery and the battery tray.

4. Remove the integrated relay controller, the cooling fan and the radiator with fan shroud. Remove the engine bounce damper bracket from the shock tower.

5. Remove the evaporative emission line, the upper/lower radiator hose and the starter brace.

6. Remove the exhaust pipes from both exhaust manifolds. Remove and plug the power steering pump lines.

7. Bleed the pressure from the fuel system and remove the fuel lines. Remove and tag all of the necessary vacuum lines.

8. Disconnect the ground strap, the heater hoses, the accelerator cable linkage, the throttle valve linkage and the speed control cable (if equipped).

9. Disconnect the electrical cable connectors from the following items: the alternator, the A/C clutch, the oxygen sensor, the ignition coil, the radio frequency suppressor, the cooling fan voltage resistor, the engine coolant temperature sensor, the thick film ignition module, the fuel injectors, the ISC motor wire, the throttle position sensor, the oil pressure sending switch, the ground wire, the block heater (if equipped), the knock sensor, the EGR sensor and the oil level sensor.

10. Remove the engine mounting bolts and engine mounts. Remove the transaxle-to-engine mounting bolts and the transaxle brace assembly.

11. Connect an engine lifting plate and a vertical hoist to the engine, then remove the engine from the vehicle. Remove the main wire harness from the engine.

12. To install, reverse the removal procedures. Torque the transaxle brace assembly bolts to 40–55 ft. lbs., the engine mount nuts to 55–75 ft. lbs. and the engine mount bolts to 40–55 ft. lbs. Refill the cooling system. Check and/or refill the crankcase and the transaxle. Start the engine and check for leaks.

NOTE: On the engine mount assembly 6F063 and 6F065 torque the engine mount nuts and engine mount bolts to 70–96 ft. lbs.

Intake Manifold

REMOVAL & INSTALLATION

2.5L Engine

1. Raise and secure the hood. Disconnect the negative battery cable.

2. Place a drain pan under the radiator, remove the radiator cap, open the draincock and drain the cooling system.

3. Remove the accelerator cable, the air cleaner assembly and the heat stove tube from the heat shield.

4. Label and remove the necessary vacuum lines.

5. Remove the thermactor belt, the hose below thermactor pump and the thermactor pump.

6. Remove the heat shield from the exhaust manifold.

7. Disconnect the thermactor check valve hose from the tube assembly. Remove the bracket-to-EGR valve nuts.

8. Disconnect the water inlet tube from the intake manifold.

9. Disconnect the EGR tube from the EGR valve.

10. Remove the intake manifold-to-engine bolts and the manifold.

11. Using a putty knife, clean the gasket mounting surfaces.

12. To install, use a new gasket and reverse the removal procedures. Torque the intake manifold-to-engine bolts to 15–23 ft. lbs. (20–30 Nm). Refill the cooling system. Start engine and check for leaks.

3.0L Engine

1. Disconnect the negative battery cable. Drain the cooling system to a level below the intake manifold.

2. Refer to the "Throttle Body Removal & Installation" procedures in this section and remove the throttle body from the engine.

3. Reduce the pressure in the fuel system and disconnect the fuel lines.

4. Disconnect and remove the fuel injector electrical harness from the engine.

5. Label and disconnect the spark plug wires (for easy installation). Mark and remove the distributor from the engine.

6. Disconnect the upper radiator hose, the water outlet heater hose and the thermostat housing from the engine.

7. Remove the intake manifold-to-engine bolts/studs, the intake manifold; discard the side gaskets and end seals.

NOTE: The intake manifold assembly can be removed with the fuel rails and injectors in place.

8. Using a putty knife, clean the gasket/seal mounting surfaces.

9. To install, use new gaskets, sealant and reverse the removal procedures. Torque the intake manifold-to-engine bolts (in 2 steps) to 18 ft. lbs. (24 Nm), the throttle body-to-engine bolts to 15–22 ft. lbs. (20–30 Nm) and the thermostat housing-to-engine bolts to 6–8 ft. lbs. Refill the cooling system. Start the engine and check for leaks. Check and/or adjust the idle speed, the transmission throttle linkage and the speed control.

NOTE: Lightly oil all of the bolts/stud threads before installation. When using a silicone rubber sealer, assembly must occur within 15 minutes after the sealer has been applied. After this time, the sealer may start to set-up and its sealing quality may be reduced. In high temperature/humidity conditions the sealant will start to set up in approximately 5 minutes.

Exhaust Manifold

REMOVAL & INSTALLATION

2.5L Engine

1. Refer to the "Intake Manifold Removal & Installation" procedures in this section and remove the intake manifold.

2. Disconnect the exhaust pipe-to-exhaust manifold nuts.

3. Remove the heat shield from the exhaust manifold.

4. Disconnect the electrical connector from the EGO sensor.

5. Disconnect the thermactor check valve hose from the tube assembly. Remove the bracket-to-EGR valve nuts.

6. Remove the exhaust manifold-to-engine bolts and the manifold from the vehicle.

7. Using a putty knife, clean the gasket mounting surfaces.

NOTE: When installing the exhaust manifold, use the Alignment Stud tools No. T84P-6065-B to align the manifold with the block.

8. To install, use new gaskets and reverse the removal procedures. Torque the exhaust manifold-to-engine bolts (in two steps) to 20–30 ft. lbs. (27–41 Nm), the exhaust manifold-to-exhaust pipe nuts to 25–34 ft. lbs. (34–47 Nm) and the intake manifold-to-engine bolts to 15–23 ft. lbs. (20–30 Nm). Refill the cooling system. Start engine and check for leaks.

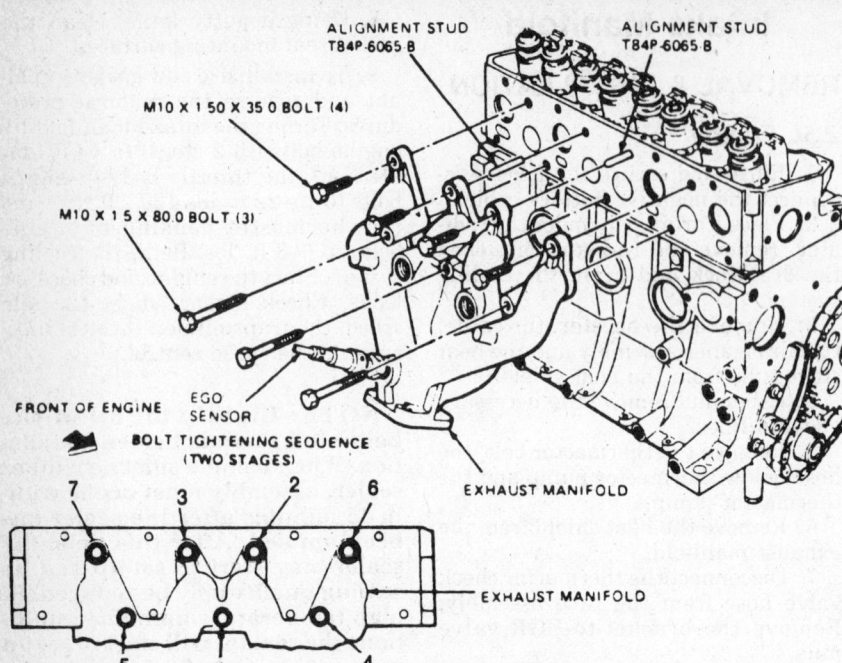

Exploded view and torque sequence of the exhaust manifold—2.5L engine

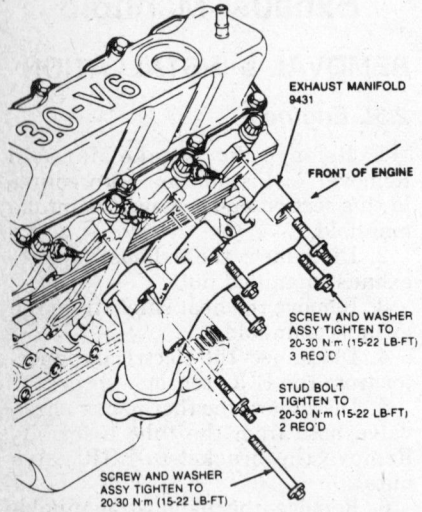

Exploded view of the left-side exhaust manifold—3.0L engine

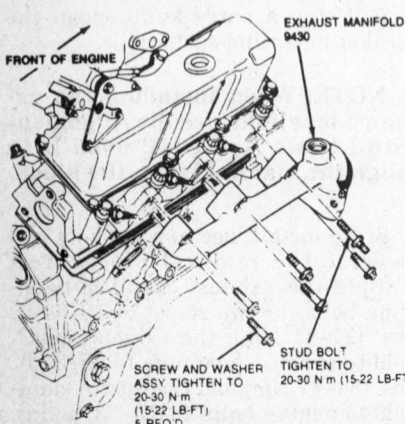

Exploded view of the right-side exhaust manifold—3.0L engine

3.0L Engine
LEFT SIDE

1. Disconnect the negative battery cable. Remove the oil level indicator support bracket.
2. Disconnect and plug the power steering hoses at the power steering pump.
3. Remove the exhaust manifold-to-exhaust pipe nuts and separate the exhaust pipe from the exhaust manifold.
4. Remove the exhaust manifold-to-cylinder head bolts and the exhaust manifold from the engine.
5. Using a putty knife, clean gasket mounting surfaces. Lightly oil all of the bolt/stud threads prior to installation.
6. To install, use new gaskets and reverse the removal procedures. Torque the exhaust manifold-to-engine bolts to 15–22 ft. lbs. (20–30 Nm) and the exhaust manifold-to-exhaust pipe nuts to 16–24 ft. lbs. (21–32 Nm). Refill and bleed the power steering system.

3.0L Engine
RIGHT SIDE

1. Disconnect the negative battery cable. Remove the heater hose support bracket.
2. Disconnect and plug the heater hoses. Remove the EGR tube from the exhaust manifold; use a back-up wrench on the lower adapter.
3. Remove the exhaust manifold-to-exhaust pipe nuts and separate the pipe from the manifold.

4. Remove the exhaust manifold-to-engine bolts and the exhaust manifold from the engine.
5. Using a putty knife, clean the gasket mounting surfaces.

NOTE: Lightly oil all of the bolt and stud threads prior to installation.

6. To install, use new gaskets and reverse the removal procedures. Torque the exhaust manifold-to-engine bolts to 15–22 ft. lbs. (20–30 Nm), the exhaust pipe-to-exhaust manifold nuts to 16–24 ft. lbs. (21–32 Nm) and the EGR tube-to-exhaust manifold to 25–36 ft. lbs. (35–50 Nm).

Rocker Arms/Shafts

REMOVAL & INSTALLATION

2.5L Engine

1. Raise the hood and place protective aprons on the fenders.
2. Remove the oil filler cap. Disconnect the PCV hose, the throttle linkage and the speed control cable from the top of the rocker arm cover (if equipped).
3. Remove the rocker arm cover-to-cylinder head bolts and the cover.
4. Remove the rocker arm fulcrum bolts, the fulcrums and the rocker arms. If necessary, remove the push rods.

NOTE: When removing the rocker arm assemblies, be sure to keep all of the parts in order for installation purposes.

5. Using a putty knife, clean the gasket mounting surfaces. Inspect and/or replace any damaged parts.

NOTE: The following procedure may require the removal of the front timing cover.

6. To install the rocker arm components, perform the following procedures:
   a. Position the push rods in the hydraulic lifter.
   b. Install the rocker arms, the fulcrums and the bolts onto the valves.
   c. Position the crankshaft so that the timing marks, on the crankshaft and the camshaft sprockets, are facing each other (on the center line). Torque the rocker arm fulcrum bolts of the intake valves of cylinders No. 1 and 2 and the exhaust valves of cylinders No. 1 and 3 to 4.5–7.5 ft. lbs. (6–10 Nm).
   d. Turn the crankshaft one complete revolution so that the timing marks, on the crankshaft and the camshaft sprockets, are facing op-

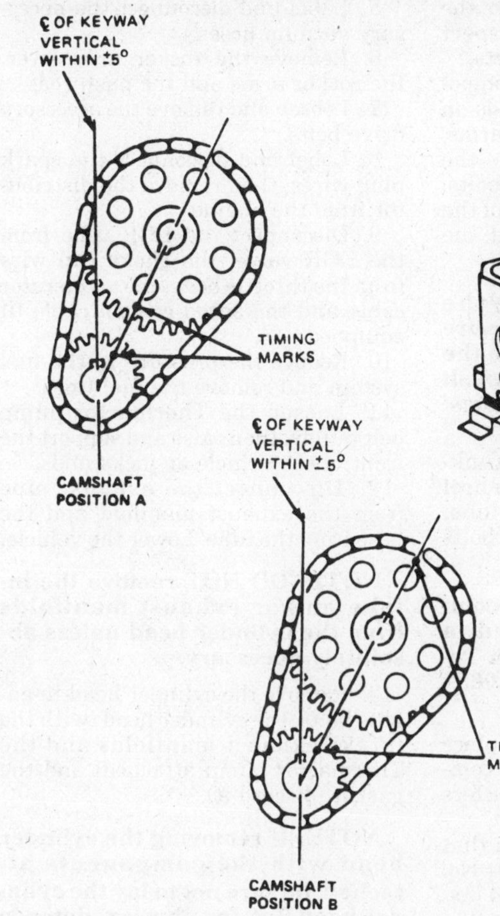

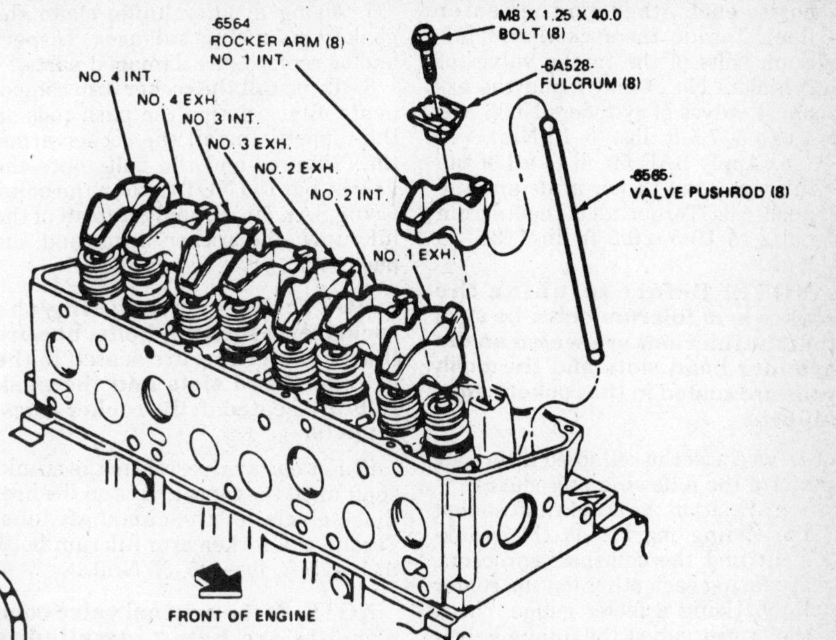

Ȼ OF KEYWAY
VERTICAL
WITHIN ±5°

TIMING MARKS

CAMSHAFT POSITION A

Ȼ OF KEYWAY
VERTICAL
WITHIN ±5°

TIMING MARKS

CAMSHAFT POSITION B

6564 ROCKER ARM (8)
NO. 1 INT.

NO. 4 INT.
NO. 4 EXH.
NO. 3 INT.
NO. 3 EXH.
NO. 2 EXH.
NO. 2 INT.
NO. 1 EXH.

M8 X 1.25 X 40.0
BOLT (8)

6A528 FULCRUM (8)

6565 VALVE PUSHROD (8)

FRONT OF ENGINE

CYL. NO.	CAMSHAFT POSITION	
	A	B
	TIGHTEN FULCRUM BOLTS AS NOTED	
1	INTAKE-EXHAUST	—
2	INTAKE	EXHAUST
3	EXHAUST	INTAKE
4	—	INTAKE-EXHAUST

Exploded view of the rocker arm assemblies and valve procedures—2.5L engine

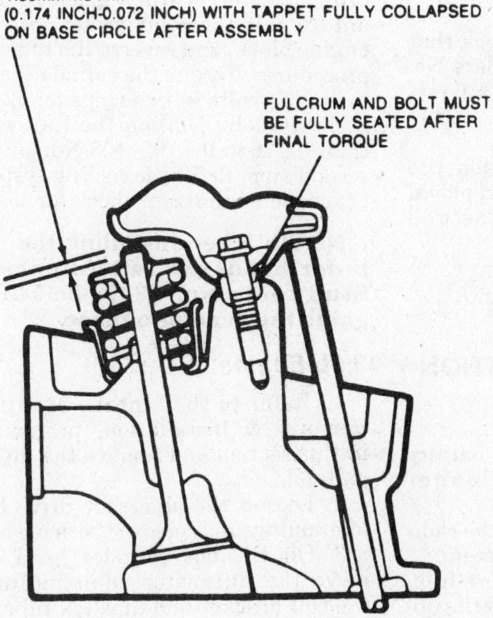

1.80mm-4.34mm
(0.174 INCH-0.072 INCH) WITH TAPPET FULLY COLLAPSED
ON BASE CIRCLE AFTER ASSEMBLY

FULCRUM AND BOLT MUST
BE FULLY SEATED AFTER
FINAL TORQUE

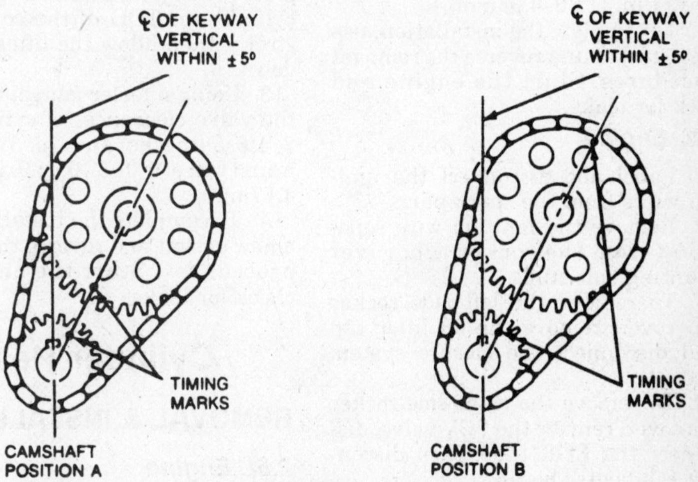

Ȼ OF KEYWAY
VERTICAL
WITHIN ±5°

TIMING MARKS

CAMSHAFT POSITION A

Ȼ OF KEYWAY
VERTICAL
WITHIN ±5°

TIMING MARKS

CAMSHAFT POSITION B

CYL. NO.	CAMSHAFT POSITION	
	A	B
	TIGHTEN FULCRUM BOLTS AS NOTED	
1	INTAKE-EXHAUST	—
2	INTAKE	EXHAUST
3	EXHAUST	INTAKE
4	—	INTAKE-EXHAUST

Checking the tappet gap—2.5L engine

C281

posite each other (on the center line). Torque the rocker arm fulcrum bolts of the intake valves of cylinders No. 3 and 4 and the exhaust valves of cylinders No. 2 and 4 to 4.5–7.5 ft. lbs. (6–10 Nm).

e. Apply SAE 50 oil to all of the fulcrums, the rocker arms and the push rods. Torque all of the fulcrum bolts to 19.5–26.5 ft. lbs. (26–38 Nm).

NOTE: Before torquing the rocker arm fulcrum bolts, be sure that the fulcrums are seated on the cylinder head slots and the push rods are seated in the rocker arms/tappets.

7. To check the collapsed lifter gap, perform the following procedures:

a. Position the crankshaft so that the timing marks, on the crankshaft and the camshaft sprockets, are facing each other (on the center line). Using a feeler gauge, check the tappet gap of the intake valves of cylinders No. 1 and 2 and the exhaust valves of cylinders No. 1 and 3; the tappet gap should be 0.072–0.174 in. (1.80–4.34mm).

b. Turn the crankshaft one complete revolution so that the timing marks, on the crankshaft and the camshaft sprockets, are facing opposite each other (on the center line). Using a feeler gauge, check the tappet gap of the intake valves of cylinders No. 3 and 4 and the exhaust valves of cylinders No. 2 and 4; the tappet gap should be 0.072–0.174 in. (1.80–4.34mm).

8. To complete the installation, use a new gasket and reverse the removal procedures. Start the engine and check for leaks.

3.0L Engine

1. Label and disconnect the ignition wires from the spark plugs.

2. Remove the ignition wire separators from the rocker arm cover mounting bolt studs.

3. To remove the left side rocker arm cover, remove the oil filler cap then disconnect the closure system hose.

4. To remove the right side rocker arm cover, remove the PCV valve, disconnect the EGR tube, then disconnect the heater hoses.

5. Remove the rocker arm covers-to-cylinder head screws and the covers.

6. Remove the rocker arm fulcrum bolts, the fulcrums and the rocker arms. If necessary, remove the push rods.

NOTE: When removing the rocker arm assemblies, be sure to keep all of the parts in order for installation purposes.

7. Using a putty knife, clean the gasket mounting surfaces. Inspect and/or replace any damaged parts.

8. To install the rocker arm components, first position the push rods on the tappets. Install the rocker arms, the fulcrums and the bolts onto the valves, but DO NOT tighten the bolts. Using SAE 50 oil, apply it to all of the fulcrums, the rocker arms and the push rods.

NOTE: Before torquing the rocker arm fulcrum bolts, be sure that the fulcrums are seated in the cylinder head slots and the push rods are seated in the rocker arms/tappets.

9. For each valve, rotate the crankshaft until the tappet rests on the heel (base circle) of the camshaft lobe. Torque the rocker arm fulcrum bolts to 19–25 ft. lbs. (25–35 Nm).

NOTE: If the original valve components are being installed, a valve clearance check is not required. Valve Clearance: 0.088–0.189 in. (2.23–4.77mm).

10. Rotate the crankshaft to place the No. 1 cylinder on TDC of the compression stroke, then allow the lifters to bleed down.

11. Using a feeler gauge, check that the valve clearances of cylinders No. 1, 3 & 6 (intake) and No. 1, 2 & 4 (exhaust) are 0.088–0.189 in. (2.23–4.77mm).

12. Rotate the crankshaft one complete revolution, positioning the No. 2 cylinder on TDC of the compression stroke, then allow the lifters to bleed down.

13. Using a feeler gauge, check that the valve clearances of cylinders No. 2, 4 & 5 (intake) and No. 3, 5 & 6 (exhaust) are 0.088–0.189 in. (2.23–4.77mm).

14. To complete the installation, use a new gasket and reverse the removal procedures. Start the engine and check for leaks.

Cylinder Head

REMOVAL & INSTALLATION

2.5L Engine

1. Disconnect the negative battery cable. Remove the air cleaner assembly.

2. Place a drain pan under the radiator and drain the cooling system.

3. Disconnect the heater hose from under the intake manifold and the upper radiator hose from the cylinder head.

4. Disconnect the electrical harness connector from the cooling fan.

5. Label and disconnect the necessary vacuum hoses.

6. Remove the rocker arm cover, the rocker arms and the push rods.

7. Loosen and remove the accessory drive belts.

8. Label and disconnect the spark plug wires, then remove the distributor from the engine.

9. Disconnect the EGR tube from the EGR valve, the choke cap wire from the throttle body, the accelerator cable and the speed control cable (if equipped).

10. Reduce the pressure in the fuel system and remove the fuel lines.

11. Loosen the Thermactor pump belt pulley, then raise and support the front of the vehicle on jackstands.

12. Disconnect the exhaust pipe from the exhaust manifold and the hose from the tube. Lower the vehicle.

NOTE: DO NOT remove the intake and/or exhaust manifolds from the cylinder head unless absolutely necessary.

13. Remove the cylinder head-to-engine bolts, the cylinder head (with the intake/exhaust manifolds and the Thermactor pump attached) and the gasket (discard it).

NOTE: If removing the cylinder head with the components attached, be sure not to lay the cylinder head flat, for physical damage may occur to the gasket surface or spark plugs.

14. Using a putty knife, clean the gasket mounting surfaces.

15. To install, use new gaskets, sealant (to hold the head gasket onto the engine block) and reverse the removal procedures. Torque the cylinder head-to-engine bolts in two steps to: 52–59 ft. lbs. (70–80 Nm) on the first step, then 70–76 ft. lbs. (95–103 Nm) on the second step. Refill the cooling system. Start the engine and check for leaks.

NOTE: When installing the cylinder head, use two Alignment Stud tools No. T84P-6065-A to guide the head into place.

3.0L Engine

1. Refer to the "Intake Manifold Removal & Installation" procedures in this section and remove the intake mainfold.

2. Loosen the accessory drive belt idler pulley and remove the drive belt.

3. On the left cylinder head, remove the alternator adjusting arm, the coil bracket and dipstick tube.

4. On the right cylinder head, remove the accessory drive belt idler pulley and the grounding strap throttle cable support bracket.

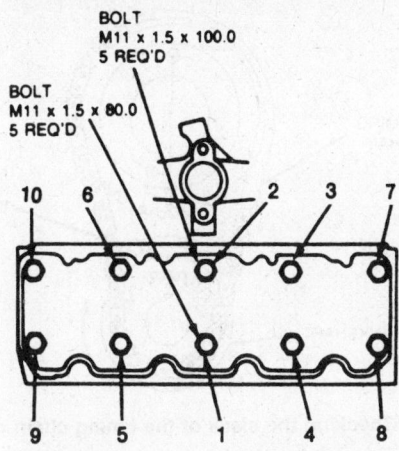

BOLT
M11 x 1.5 x 100.0
5 REQ'D

BOLT
M11 x 1.5 x 80.0
5 REQ'D

Cylinder head bolt torquing sequence—2.5L engine

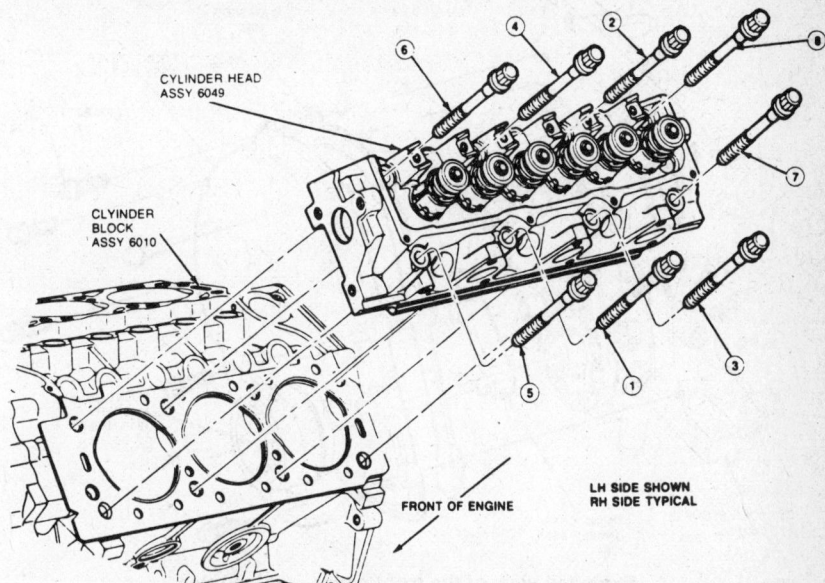

CYLINDER HEAD
ASSY 6049

CLYINDER
BLOCK
ASSY 6010

FRONT OF ENGINE

LH SIDE SHOWN
RH SIDE TYPICAL

Cylinder head bolt torquing sequence— 3.0L engine

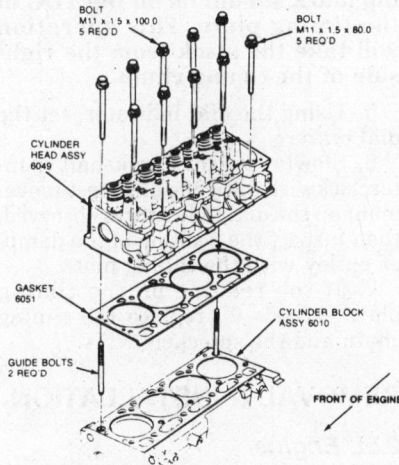

BOLT
M11 x 1.5 x 100.0
5 REQ D

BOLT
M11 x 1.5 x 80.0
5 REQ D

CYLINDER
HEAD ASSY
6049

GASKET
6051

GUIDE BOLTS
2 REQ D

CYLINDER BLOCK
ASSY 6010

FRONT OF ENGINE

Installing the cylinder head—2.5L engine

5. If equipped with power steering, remove the pump mounting bracket-to-engine bolts. Leave the pump hoses connected and position the pump out of the way.

6. From both sides of the engine, remove the exhaust pipe-to-exhaust manifold bolts, the exhaust manifolds-to-cylinder head bolts and the exhaust manifolds from the engine.

7. Remove the PCV valve and the rocker arm covers. Loosen the rocker arm fulcrum bolts enough to allow the rocker arm to be lifted from the push rod and rotated to one side.

8. Remove the push rods, keeping them in order for installation purposes.

9. Remove the cylinder head-to-engine bolts and the cylinder heads from the engine. Discard the cylinder head gaskets.

10. Using a putty knife, clean and inspect the gasket mounting surfaces.

11. To install, use new gaskets and reverse the removal procedures. Torque the cylinder head-to-engine bolts (in two steps) to 48–54 ft. lbs.

(1st step) and to 63–80 ft. lbs. (2nd step), the rocker arm fulcrum bolts to 19–29 ft. lbs. Start the engine and check for coolant, fuel, oil and exhaust leaks. Check and/or adjust the transmission throttle linkage and the speed control.

NOTE: Before installation, lightly oil the bolt and stud threads except for those specifying that a special sealant be applied. If the flat surface of the cylinder head is warp, do not plane or grind off more than 0.010 in. (0.254mm). If the head is machined past the resurface limit, it will have to be replaced with a new one.

OVERHAUL

For all cylinder head overhaul procedures, please refer to "Engine Rebuilding" in the Unit Repair section.

Front Cover

REMOVAL & INSTALLATION

2.5L Engine

1. Refer to the "Engine Removal & Installation" procedures in this section and remove the engine and the transmission from the vehicle as an assembly.

2. Remove the dipstick and the accessory drive pulley (if equipped). Remove the crankshaft pulley-to-crankshaft bolt, the washer and the pulley.

3. Remove front cover-to-engine bolts and pry the top of the front cover away from the block.

NOTE: The front cover oil seal must be removed in order to use the Front Cover Aligner tool No. T84P-6019-C to install the front cover.

4. Using a putty knife, clean the gasket mounting surfaces.

5. To install, use a new gasket, sealant, a new oil seal and reverse the removal procedures. Torque the front cover-to-engine bolts to 6–9 ft. lbs. and the crankshaft pulley bolt to 140–170 ft. lbs.

3.0L Engine

1. Refer to the "Water Pump Removal & Installation" procedures in this section and remove the water pump.

2. Remove the crankshaft pulley and the damper from the crankshaft.

3. Remove the lower radiator hose from the front cover.

4. Remove the oil pan-to-front cover bolts, the front cover-to-engine bolts and the front cover.

5. Using a putty knife, clean the gasket mounting surfaces.

NOTE: When the front cover is removed, the oil seal should be replaced.

6. To install, use new gaskets, sealant and reverse the removal procedures. Torque the front cover-to-engine bolts to 15–22 ft. lbs. (20–30 Nm), the front cover-to-oil pan bolts to 80–106 inch lbs. (9–12 Nm), the water pump-to-front cover bolts to 6–8 ft.

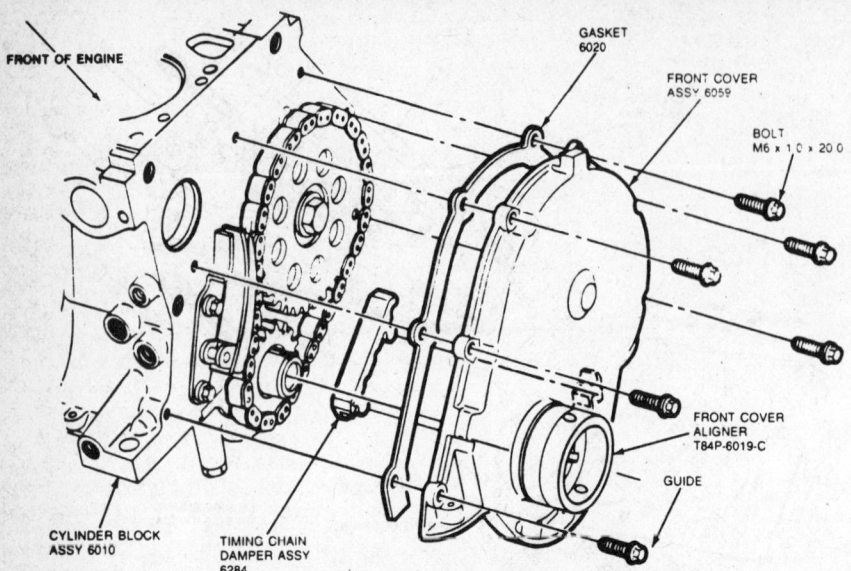

Exploded view of the front cover—2.5L engine

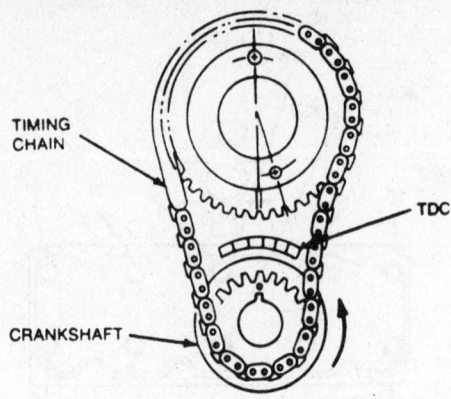

Checking the slack of the timing chain

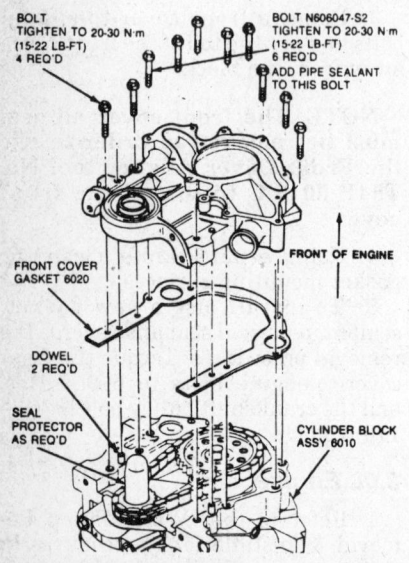

Exploded view of the front cover assembly—3.0L engine

lbs. (8–12 Nm), the crankshaft damper-to-crankshaft bolt to 141–169 ft. lbs. (190–230 Nm), the crankshaft pulley-to-crankshaft damper bolts to 20–28 ft. lbs. (26–38 Nm), the water pump pulley-to-water pump bolts to 15–22 ft. lbs. (20–30 Nm).

Timing Chain And Tensioner

ADJUSTMENT

NOTE: No adjustments of the timing chain are necessary or possible. The ONLY check to be make is the deflection measurement. If

the deflection is beyond specifications, replace the timing chain, the sprockets and/or the tensioners.

2.5L Engine

1. To check the timing chain deflection, first rotate the crankshaft in the counterclockwise direction (view from the front) to take up the slack.

2. Mark a reference point on the block, then measure the mid-section distance to the timing chain on the left side.

3. Rotate the crankshaft clockwise to take up the slack on the right side of the chain.

4. If the deflection exceeds 1/2 in., replace the timing chain and/or the sprockets.

NOTE: The deflection measurement is the difference between the two measurements.

5. Check the timing chain tensioner blade for wear depth. If the wear depth exceeds specification, replace the tensioner.

3.0L Engine

1. Refer to the "Rocker Arm Removal & Installation" procedures in this section and remove the left side valve cover.

2. Loosen the exhaust valve fulcrum bolt of the No. 5 cylinder and rotate the rocker to one side.

3. Using a dial indicator, install it onto the push rod.

4. Rotate the crankshaft until the No. 1 cylinder is at the TDC of the compression stroke.

NOTE: The damper pulley timing mark should be on the TDC of the timing plate. This operation will take the slack from the right side of the timing chain.

5. Using the dial indicator, set the dial on zero.

6. Slowly turn the crankshaft counterclockwise until the slightest movement on the dial indicator is observed, then inspect the position of the damper pulley with the timing plate.

7. If the reading on the timing plate exceeds 6°, replace the timing chain and the sprockets.

REMOVAL & INSTALLATION

2.5L Engine

1. Refer to the "Engine Removal & Installation" procedures in this section and remove the engine and secure it to a workstand.

2. Refer to the "Front Cover Removal & Installation" procedures in this section and remove the front cover.

3. Rotate the crankshaft until the timing marks of the crankshaft and the camshaft sprockets are aligned.

4. Remove the camshaft sprocket bolt and washer, then slide the camshaft sprocket, the timing chain and the crankshaft sprocket off as an assembly.

5. Inspect and/or replace the parts as necessary.

6. To install, slide the sprocket/timing chain assembly onto the camshaft and crankshaft with timing marks aligned. Oil the timing chain, the sprockets and the tensioner after installation.

7. Apply oil resistant sealer to a new front cover gasket and position gasket onto the front cover.

8. Using the Front Cover Aligner tool No. T84P-6019-C, position it onto the end of the crankshaft, ensuring that the crank key is aligned with the

keyway in the tool. Bolt the front cover to the engine. Tighten all the attaching bolts to specification. Remove the front cover aligner tool.

9. Lubricate the hub of the crankshaft pulley with Polyethylene Grease to prevent damage to the seal during installation and initial engine start. Install crankshaft pulley.

10. To complete the installation, reverse the removal procedures. Torque the camshaft sprocket-to-camshaft bolt to 41–56 ft. lbs. (55–75 Nm), the front cover-to-engine bolts to 6–9 ft. lbs. (8–12 Nm), the crankshaft pulley bolt to 140–170 ft. lbs. (190–230 Nm). Refill the cooling system. Start the engine and check for leaks.

3.0L Engine

1. Refer to the "Engine Removal & Installation" procedures in this section and remove the engine and secure it to a workstand.

2. Refer to the "Front Cover Removal & Installation" procedures in this section and remove the front cover.

3. Rotate the crankshaft until the No. 1 piston is at the TDC of it's compression stroke and the timing marks are aligned.

4. Remove the camshaft sprocket-to-camshaft bolt and washer, then slide the sprockets and timing chain forward to remove them as an assembly.

5. Inspect the timing chain and sprockets for excessive wear; replace them, if necessary.

6. Using a putty knife, clean the gasket mounting surfaces.

7. Apply oil to the timing chain and sprockets after installation.

NOTE: The camshaft bolt has a drilled oil passage in it for timing chain lubrication. If the bolt is damaged do not replace it with a standard bolt.

8. Apply a bead of RTV sealant on the gap at the cylinder block.

9. Apply an oil resistant sealer No. B5A-19554-A, or equivalent, to a new front gasket and position the gasket onto the front cover.

10. Position the front cover on the engine taking care not to damage the front seal. Make sure the cover is installed over the alignment dowels.

NOTE: When installing the front cover onto the engine, make sure that the oil pan seal is not dislodged.

11. If necessary, replace the front cover seal using the Seal Installation tool No. T70P-6B070-A.

12. To complete the installation, reverse the removal procedures. Torque the camshaft sprocket-to-camshaft bolt to 40–51 ft. lbs. (55–70 Nm), the front cover-to-engine bolts to 15–22 ft. lbs. (20–30 Nm), the water pump-to-front cover bolts to 6–8 ft. lbs. (8–12 Nm), the damper-to-crankshaft bolt to 141–169 ft. lbs. (190–230 Nm) and the damper pulley-to-damper bolts to 20–28 ft. lbs. (26–38 Nm). Refill the crankcase and the cooling system. Start the engine and check for leaks.

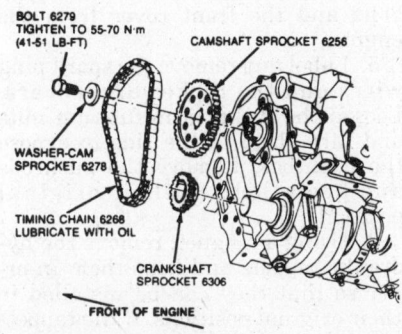

Exploded view of the timing chain and sprockets—3.0L engine

OIL SEAL REPLACEMENT

2.5L Engine

1. Refer to the "Engine Removal & Installation" procedures in this section, then remove the engine from the vehicle and place it on a workstand.

2. Remove the drive belt(s), the crankshaft pulley bolt/washer and the crankshaft pulley.

3. Using the Damper Removal tool No. T77F-4220-B1, remove the crankshaft pulley.

4. Using the Front Seal Removal tool No. T74P-6700-A, remove the front cover oil seal.

5. Coat the new seal with grease. Using the Front Seal Replacer tool No. T83T-4676-A, install the seal into the cover; drive the seal in until it is fully seated. Check the seal after installation to ensure that the seal spring is properly positioned.

6. To complete the installation, reverse the removal procedures. Torque the crankshaft pulley-to-crankshaft bolt to 140–230 ft. lbs. (190–230 Nm).

3.0L Engine

1. Disconnect the negative battery cable and loosen the accessory drive belts.

2. Raise and support the front of the vehicle on jackstands, then remove the right side front wheel.

3. Remove the crankshaft pulley-to-damper bolts. Disengage the acces-

sory drive belts and remove the crankshaft pulley.

4. Using the Crankshaft Damper Removal tool No. T58P-6316-D and the Vibration Damper Removal Adapter tool No. T82L-6316-B, remove the crankshaft damper from the crankshaft.

5. Using a small pry bar, pry the oil seal from the front cover; be careful not to damage the front cover and the crankshaft.

NOTE: Before installation; inspect the front cover and shaft seal surface of the crankshaft damper for damage, nicks, burrs or other roughness which may cause the new seal to fail. Service or replace the components as necessary.

6. To install, lubricate the new seal lip with clean engine oil, then install the seal using the seal installer tool No. T82L-6316-A and the front cover seal replacer tool No. T70P-6B070-A, or their compatible equivalents.

7. Coat the crankshaft damper sealing surface with clean engine oil. Apply RTV to the keyway of the damper prior to installation. Install the damper using the vibration damper seal installer tool No. T82L-6316-A.

8. To complete the installation, reverse the removal procedures. Torque the crankshaft damper-to-crankshaft bolt to 141–169 ft. lbs. (190–230 Nm), the crankshaft pulley-to-damper bolts to 20–28 ft. lbs. (26–38 Nm).

Camshaft

REMOVAL & INSTALLATION

2.5L Engine

1. Refer to the "Engine Removal & Installation" procedures in this section, then remove the engine and place it on a workstand.

2. Refer to the "Timing Chain Removal & Installation" procedures in this section and remove the timing chain with the camshaft sprocket.

3. Remove the cylinder head.

4. Using a magnet, remove the hydraulic tappets and keep them in order so that they can be installed in their original positions. If the tappets are stuck in the bores by excessive varnish, etc., use a Hydraulic Tappet Puller to remove the tappets.

6. Remove the oil pan.

7. To check the camshaft end play, install the camshaft sprocket to the camshaft and perform the following procedures:

   a. Push the camshaft toward the rear of the engine and install a dial

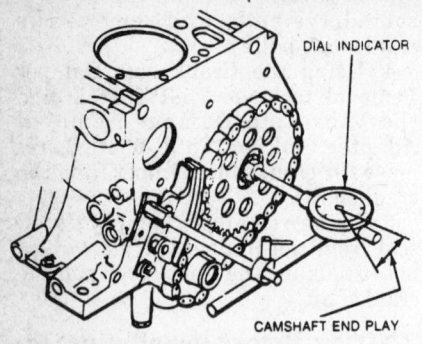

Using a dial micrometer to check the camshaft end play

indicator tool No. 4201-C, so that the indicator point is on the camshaft sprocket mounting bolt.

b. Zero the dial indicator. Position a large screwdriver between the camshaft sprocket and the block.

c. Pull the camshaft forward and release it. Compare the dial indicator reading with the camshaft end play specification of 0.009 in. (0.229mm).

d. If the camshaft end play is over the amount specified, replace the thrust plate.

8. Remove the camshaft sprocket and the camshaft thrust plate.

9. Carefully remove the camshaft by pulling it toward the front of the engine. Use caution to avoid damaging the bearings, journals and lobes.

10. Using a putty knife, clean the gasket mounting surfaces.

11. Clean the oil pump inlet tube screen, the oil pan and the cylinder block gasket surfaces. Prime the oil pump by filling the inlet opening with oil and rotate the pump shaft until oil emerges from the outlet tube. Install the oil pump, the oil pump inlet tube screen and the oil pan.

12. To install, use new gaskets, sealant, lubricate the internal parts with SAE 50 weight oil and reverse the removal procedures. Torque the camshaft thrust plate bolts to 6–9 ft. lbs. (8–12 Nm), the camshaft sprocket-to-camshaft bolt to 41–56 ft. lbs. (55–75 Nm), the front cover-to-engine bolts to 6–9 ft. lbs. (8–12 Nm), the damper pulley-to-crankshaft bolt to 140–170 ft. lbs. (190–230 Nm), the cylinder head bolts-to-engine bolts (in 2 steps) to 70–76 ft. lbs. (95–103 Nm) and the oil pan-to-engine bolts to 15–23 ft. lbs. (20–30 Nm). Adjust the valves. Refill the cooling system and the crankcase. Start the engine and check for leaks. Check and/or adjust the ignition timing and the engine idle speed.

3.0L Engine

1. Refer to the "Engine Removal &

Installation" procedures in this section, then remove the engine and place it on a workstand.

2. Ensure the cooling system, the fuel system and the crankcase have been drained.

3. Remove the idler pulley and bracket assembly. Remove the drive and accessory belts. Remove the water pump.

4. Remove the crankshaft pulley and damper. Remove the lower radiator hose. Remove the oil pan-to-front cover bolts, the front cover-to-engine bolts and the front cover from the engine.

5. Label and remove the spark plug wires and the rocker arm covers. Loosen the rocker arm fulcrum nuts and turn them to the side to expose the pushrods. Remove the pushrods and keep them in their original position.

6. Using a magnet, remove the hydraulic tappets and keep them in order so that they can be installed in their original positions. If the tappets are stuck in the bores by excessive varnish, use the Hydraulic Tappet Puller tool No. T70L-6500-A, to remove the tappets.

7. To check the camshaft end play, perform the following procedures:

a. Push the camshaft toward the rear of the engine and install a Dial Indicator tool No. 4201-C, so that the indicator point is on the camshaft sprocket bolt. Zero the dial indicator.

b. Using a medium pry bar, position it between the camshaft sprocket and the block.

NOTE: When applying pressure to the camshaft sprocket, be careful not to break the powdered metal camshaft sprocket.

c. Pry the camshaft forward and release it. Compare the dial indicator reading with the camshaft end play specification of 0.009 in.

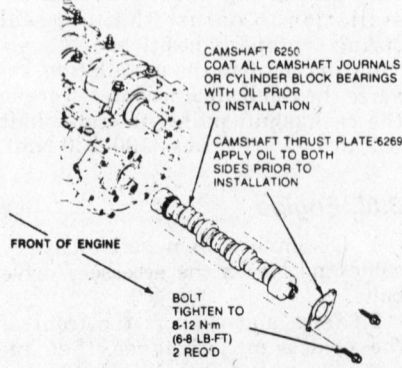

CAMSHAFT 6250
COAT ALL CAMSHAFT JOURNALS OR CYLINDER BLOCK BEARINGS WITH OIL PRIOR TO INSTALLATION

CAMSHAFT THRUST PLATE-6269 APPLY OIL TO BOTH SIDES PRIOR TO INSTALLATION

FRONT OF ENGINE

BOLT TIGHTEN TO 8–12 N·m (6–8 LB-FT) 2 REQ'D

Exploded view of the camshaft—3.0L engine

d. If the camshaft end play is over the amount specified, replace the thrust plate.

8. Remove the timing chain and sprockets.

9. Remove the camshaft thrust plate. Carefully remove the camshaft by pulling it toward the front of the engine. Remove it slowly to avoid damaging the bearings, journals and lobes.

10. Using a putty knife, clean the gasket mounting surfaces.

11. To install, use new gaskets, sealant, lubricate the internal parts with SAE 50 weight oil and reverse the removal procedures. Torque the camshaft thrust plate-to-engine bolts to 6–8 ft. lbs. (8–12 Nm), the front cover-to-engine bolts to 15–22 ft. lbs. (20–30 Nm), the water pump-to-front cover bolts to 6–8 ft. lbs. (8–12 Nm), the crankshaft damper-to-crankshaft bolt to 141–169 ft. lbs. (190–230 Nm) and the crankshaft pulley-to-damper bolts to 20–28 ft. lbs. (26–38 Nm). Refill the cooling system and the crankcase. Start the engine and check for leaks. Check and/or adjust the ignition timing and the engine idle speed.

Piston And Connecting Rod Positioning

For all piston and connecting rod overhaul procedures, please refer to "Engine Rebuilding" in the Unit Repair section.

ENGINE LUBRICATION

Oil Pan

REMOVAL & INSTALLATION

2.5L Engine

1. Disconnect negative ground cable at battery.

2. Raise and support the front of the vehicle of jackstands.

3. Remove the crankcase oil plug and drain the fluid. Remove the lower radiator hose and drain the cooling system.

4. If equipped with a manual transaxle, remove the roll restrictor.

5. Disconnect the starter electrical connectors and the starter.

6. Disconnect the exhaust pipe from the oil pan.

7. Remove the engine coolant tube located near the lower radiator hose, the water pump and the oil pan tabs.

8. If equipped with A/C, position the air conditioner line off to the side.

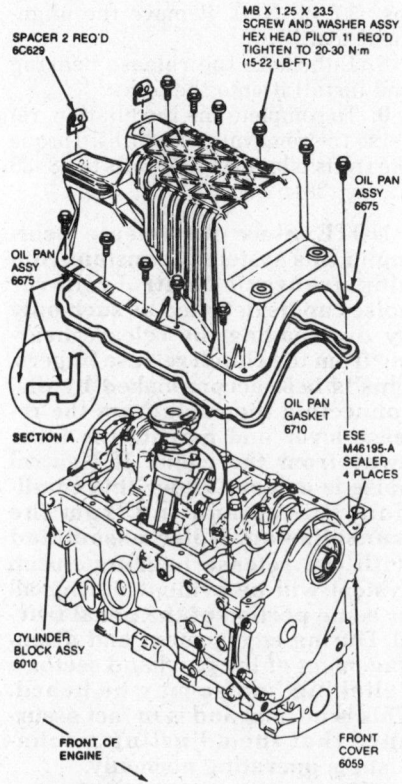

Exploded view of the oil pan assembly— 2.5L engine

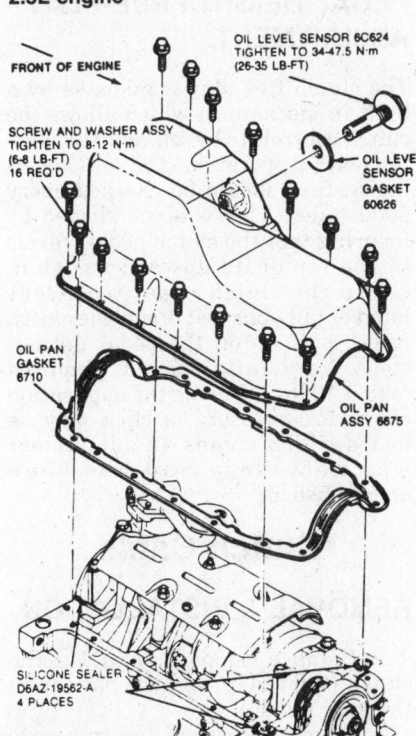

Exploded view of the oil pan assembly— 3.0L engine

9. Remove the oil pan-to-engine bolts and the pan.

10. Using a putty knife, clean the gasket mounting surfaces. Remove and clean oil pump pickup tube and screen assembly. After cleaning, install tube and screen assembly.

NOTE: Before proceeding, a trial installation of the pan to cylinder block must be performed to insure smooth pan installation, thus preventing smearing of the sealant. Check again for any residual oil that may have leaked down (particularly the rear of the engine) and clean as necessary.

11. To install, use new gaskets, sealant and reverse the removal procedures. Torque the oil pan-to-engine bolts to 15–23 ft. lbs. (20–30 Nm). Refill the cooling system and the crankcase. Start the engine and check for leaks.

3.0L Engine

1. Disconnect the negative battery cable and remove the oil level dipstick.

2. Raise and support the front of the vehicle on jackstands. If equipped with a low oil level sensor, remove the retainer clip and the electrical connector from the sensor.

3. Drain the engine oil into an oil catch pan.

4. Disconnect the electrical connectors from the starter and remove the starter.

5. In the exhaust manifold, disconnect the electrical connector from the oxygen sensor.

6. Remove the catalytic converter and the exhaust pipe assembly.

7. From the torque converter housing, remove the lower engine/flywheel dust cover.

8. Remove the oil pan-to-engine mounting bolts and the pan from the engine.

9. Using a putty knife, clean the gasket mounting surfaces.

10. To install, use new gaskets, sealant and reverse the removal procedures. Torque the oil pan-to-engine bolts to 6–8 ft. lbs. (8–12 Nm). Refill the cooling system and the crankcase. Start the engine and check for leaks.

Rear Main Bearing Oil Seal

REMOVAL & INSTALLATION

1. Refer to the "Transaxle Removal & Installation" procedures in this section and remove the transaxle from the vehicle.

2. If equipped with a MTX, remove the pressure plate, the clutch and the flywheel. If equipped with an ATX, remove the flexplate.

3. Using a sharp awl, punch a hole into the seal metal surface between the seal lip and the block.

4. Using the threaded end of slide hammer tool No. T77L-9533-B, screw it into the hole in the seal. Using a slide hammer, pull the seal from the block.

NOTE: Use caution to avoid damaging the oil seal surface.

5. Inspect the crankshaft seal area for damage which may cause the seal to leak. If damage is evident, service or replace the crankshaft as necessary.

6. Coat the crankshaft seal area and the seal lip with engine oil.

7. Using the seal installer tool No. T82L-6701-A, or equivalent, press the new seal into the block.

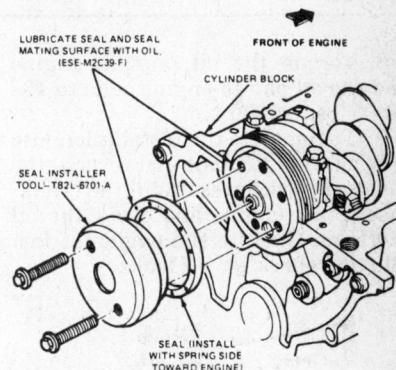

Installing the rear main oil seal—typical

NOTE: When using the installer tool, tighten the bolts evenly so that the seal is straight and seats without misalignment.

8. To complete the installation, reverse the removal procedures. Torque the flywheel-to-crankshaft bolts to 54–64 ft. lbs. (73–87 Nm); the pressure plate-to-flywheel bolts to 12–24 ft. lbs. (17–32 Nm); the torque converter-to-flexplate bolts to 23–39 ft. lbs. (31–53 Nm); the transaxle-to-engine bolts to 25–33 ft. lbs. (34–45 Nm). Install the shift control linkage.

Oil Pump

REMOVAL & INSTALLATION

2.5L Engine

1. Refer to the "Oil Pan Removal & Installation" procedures in this section and remove the oil pan.

2. Remove oil pump-to-engine bolts, the oil pump and the intermediate driveshaft.

3. Using a putty knife, clean the gasket mounting surfaces.

4. Prime the oil pump by filling the inlet port with engine oil. Rotate the pump shaft until oil flows from the outlet port.

5. If the screen and cover assembly have been removed, replace the gasket. Clean the screen, then reinstall the screen and cover assembly.

6. Position the intermediate driveshaft into the distributor socket.

7. Insert the intermediate driveshaft into the oil pump. Install the oil pump and the shaft as an assembly.

--- **CAUTION** ---

DO NOT attempt to force the pump into position if it will not seat. The shaft hex may be misaligned with the distributor shaft. To align, remove the oil pump and rotate the intermediate driveshaft into a new position.

8. Torque the oil pump-to-engine and the oil pan-to-engine bolts to 15–23 ft. lbs. (20–30 Nm).

9. To complete the installation, use new gaskets, sealant and reverse the removal procedures. Refill the crankcase. Start engine and check the oil pressure. Operate the engine at fast idle and check for oil leaks.

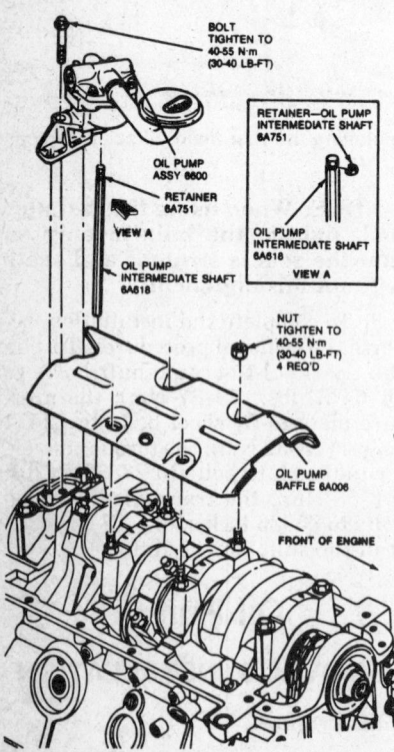

Exploded view of the oil pump assembly—3.0L engine

3.0L Engine

1. Refer to the "Oil Pan Removal & Installation" procedures in this section and remove the oil pan.

2. Remove the oil pump-to-engine bolts, then lift the pump from the engine and withdraw the oil pump driveshaft.

3. Using a putty knife, clean the gasket mounting surfaces.

4. Prime the oil pump by filling either the inlet or the outlet port with engine oil. Rotate the pump shaft to distribute the oil within the oil pump body.

5. Insert the oil pump driveshaft into the pump with the retainer end facing inward. Place the oil pump in the proper position with a new gasket and install the mounting bolts.

6. Torque the oil pump-to-engine bolts to 30–40 ft. lbs. (40–55 Nm). Clean and install the oil pump inlet tube and screen assembly with a new gasket.

7. To complete the installation, use new gaskets, sealant and reverse the removal procedures. Torque the oil pan-to-engine bolts to 6–8 ft. lbs. (8–12 Nm). Refill the crankcase to the proper level with recommended engine oil. Start the engine and check the oil pressure. Operate the engine at fast idle and check for oil leaks.

CLUTCH

REMOVAL & INSTALLATION

1. Refer to the "Transaxle Removal & Installation" procedures in this section and remove the transaxle.

2. Make alignment marks on the pressure plate assembly and the flywheel for reassembly purposes.

3. Loosen the pressure plate-to-flywheel bolts one turn at a time, in sequence, until spring tension is relieved.

4. Support the pressure plate/clutch disc assembly and remove the bolts. Remove the pressure plate and the clutch disc.

5. Inspect the flywheel, the clutch disc, the pressure plate, the throwout bearing and the clutch fork for wear; replace the parts as necessary. If the flywheel shows signs of overheating (blue discoloration) or is badly grooved/scored, it should be refaced or replaced.

6. Clean the pressure plate and flywheel surfaces thoroughly. Position the clutch disc and pressure plate into the assembled position, aligning the

match-marks. Support the assembly with the clutch arbor tool No. T81P-7550-A, or equivalent.

7. Install the pressure plate-to-flywheel bolts and torque them gradually in a criss-cross pattern to 12–24 ft. lbs. (17–32 Nm). Remove the alignment tool.

8. Lubricate the release bearing and install it onto the fork.

9. To complete the installation, reverse the removal procedures. Torque the transaxle-to-engine bolts to 28–38 ft. lbs. (38–52 Nm).

NOTE: Since the release bearing in this system is constant-running, transaxle Neutral rollover noise can be detected as such only by disengaging the release bearing from the clutch release fingers. This is best accomplished by disconnecting the cable from the release lever and moving the lever away from the cable. If Neutral noise is evident under this condition, it is emanating from the transmission. Noise associated with the release bearing/clutch system will be evident during all or some portion of the pedal travel. During engagement and disengagement of the pawl and sector a "clicking" noise may be heard. This is normal and is in fact assurance that the adjusting mechanism is operating normally.

PEDAL HEIGHT/FREE-PLAY ADJUSTMENT

The clutch free play is adjusted by a built in mechanism which allows the clutch controls to be self-adjusted during normal operation. The self-adjusting feature should be checked every 5000 miles. This is accomplished by insuring that the clutch pedal travels to the top of its upward position. Grasp the clutch pedal with your hand or put your foot under the clutch pedal, pull up on the pedal until it stops. Very little effort is required (about 10 lbs.). During the application of upward pressure, a click may be heard which means an adjustment was necessary and has been accomplished.

Clutch Cable

REMOVAL & INSTALLATION

1. Using a support, prop Up the clutch pedal to release the pawl from the gear quadrant.

2. To gain access to the transaxle end of the clutch cable, remove the air cleaner.

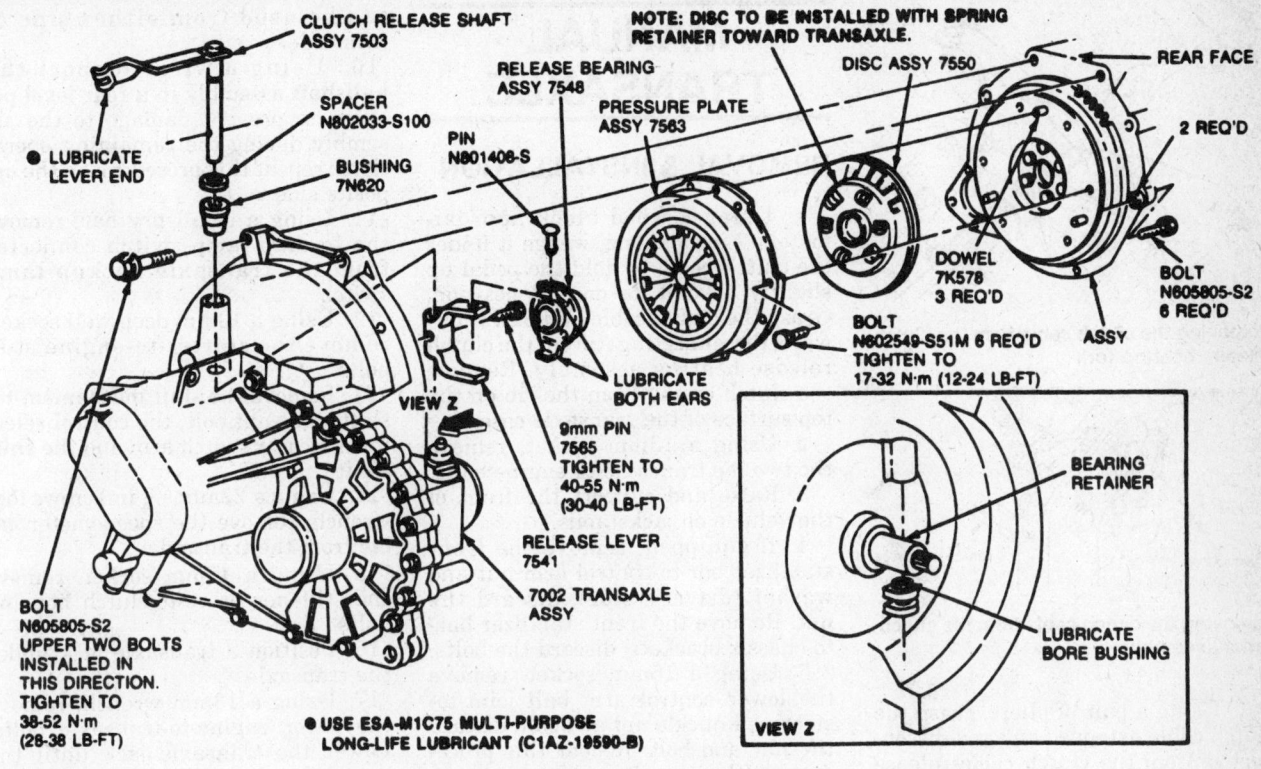

CLUTCH RELEASE SHAFT
ASSY 7503

NOTE: DISC TO BE INSTALLED WITH SPRING
RETAINER TOWARD TRANSAXLE.

RELEASE BEARING
ASSY 7548

SPACER
N802033-S100

PRESSURE PLATE
ASSY 7563

DISC ASSY 7550

REAR FACE

2 REQ'D

● LUBRICATE
LEVER END

PIN
N801406-S

BUSHING
7N620

● LUBRICATE
BOTH EARS

DOWEL
7K578
3 REQ'D

BOLT
N605805-S2
6 REQ'D

ASSY

BOLT
N602549-S51M 6 REQ'D
TIGHTEN TO
17-32 N·m (12-24 LB-FT)

VIEW Z

9mm PIN
7565
TIGHTEN TO
40-55 N·m
(30-40 LB-FT)

RELEASE LEVER
7541

7002 TRANSAXLE
ASSY

BOLT
N605805-S2
UPPER TWO BOLTS
INSTALLED IN
THIS DIRECTION.
TIGHTEN TO
38-52 N·m
(28-38 LB-FT)

● USE ESA-M1C75 MULTI-PURPOSE
LONG-LIFE LUBRICANT (C1AZ-19590-B)

BEARING
RETAINER

LUBRICATE
BORE BUSHING

VIEW Z

Exploded view of the clutch assembly

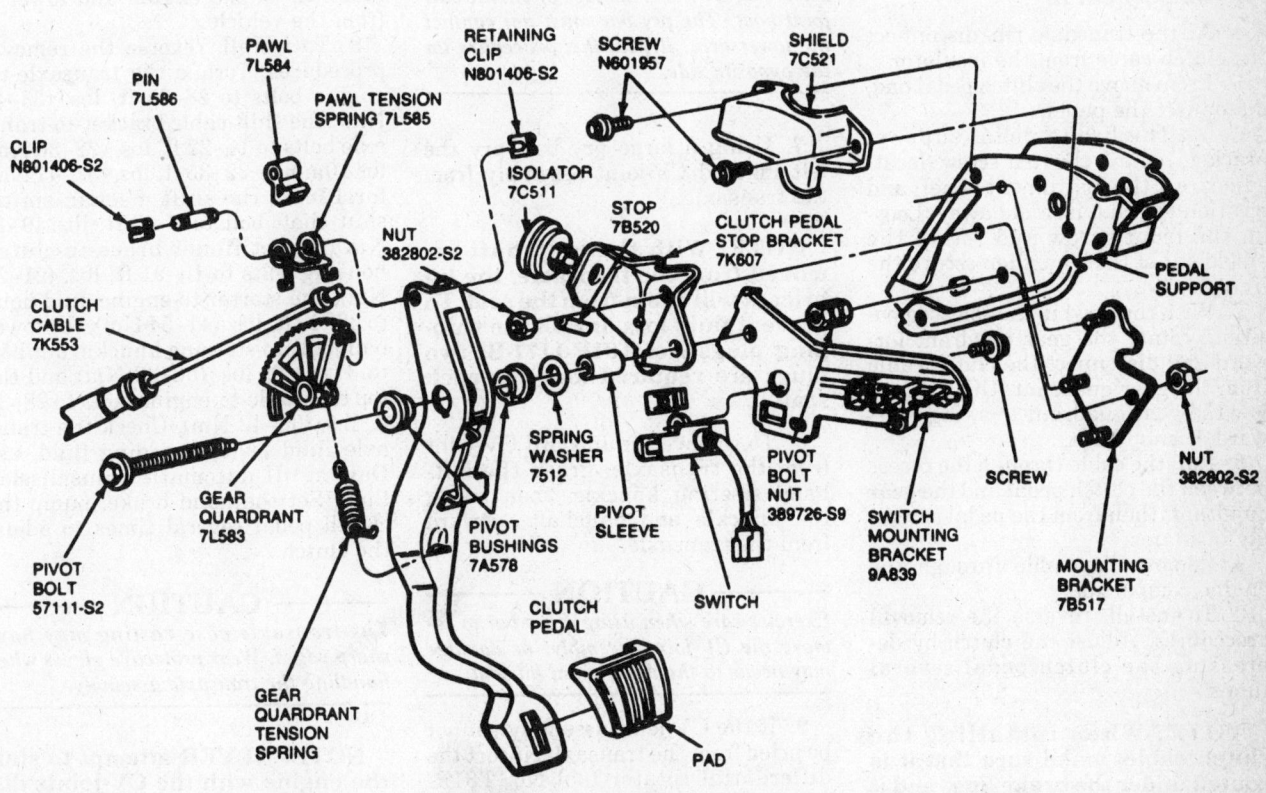

PIN
7L586

PAWL
7L584

PAWL TENSION
SPRING 7L585

RETAINING
CLIP
N801406-S2

SCREW
N601957

SHIELD
7C521

CLIP
N801406-S2

ISOLATOR
7C511

STOP
7B520

CLUTCH PEDAL
STOP BRACKET
7K607

PEDAL
SUPPORT

NUT
382802-S2

CLUTCH
CABLE
7K553

SPRING
WASHER
7512

PIVOT
BOLT NUT
389726-S9

SWITCH
MOUNTING
BRACKET
9A839

SCREW

NUT
382802-S2

GEAR
QUARDRANT
7L583

PIVOT
BUSHINGS
7A578

PIVOT
SLEEVE

SWITCH

MOUNTING
BRACKET
7B517

PIVOT
BOLT
57111-S2

CLUTCH
PEDAL

GEAR
QUARDRANT
TENSION
SPRING

PAD

Exploded view of the clutch pedal assembly

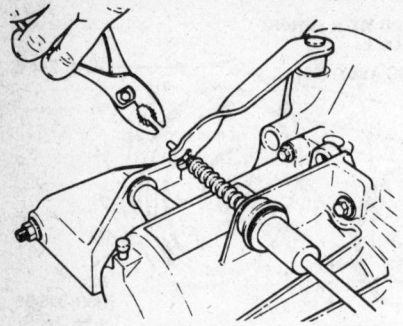

Removing the clutch cable from the clutch release bearing fork

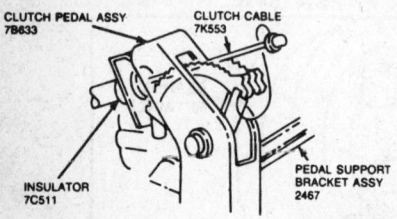

Removing the clutch cable from the clutch pedal assembly

3. Using a pair of pliers, grasp the clutch cable extended end and disconnect it from the clutch cable release bearing fork.

NOTE: When releasing the clutch cable from the release bearing fork, DO NOT grasp the wire strand portion of the inner cable, for you may cut it.

4. At the transaxle rib, disconnect the clutch cable from the insulator.

5. From above the clutch pedal pad, disconnect the panel.

6. At the brake pedal support bracket, remove the rear screw (located nearest the instrument panel) and position the clutch shield away. Loosen the front screw and rotate the shield out of the way, then secure the front screw.

7. With the pawl in the released position, rotate the gear quadrant forward and disconnect the clutch cable from the gear quadrant. DO NOT allow the gear quadrant to swing rearward or snap back.

8. Pull the cable through the recess between the clutch pedal and the gear quadrant, then from the pedal assembly insulator.

9. Remove the cable through the engine compartment.

10. To install, reverse the removal procedures. Adjust the clutch by depressing the clutch pedal several times.

NOTE: When installing the clutch cable, make sure that it is routed under the brake lines and it is not trapped at the spring tower near the brake lines.

MANUAL TRANSAXLE

REMOVAL & INSTALLATION

1. Using a wood block approximately 7 inches long, wedge it under the clutch pedal to hold the pedal up slightly beyond its normal position. Grasp the clutch cable and pull it forward, disconnecting it from the clutch release bearing assembly. Remove the clutch casing from the rib on the top surface of the transaxle case.

2. Using a 13mm socket, remove the two top transaxle-to-engine bolts.

3. Raise and support the front of the vehicle on jackstands.

4. If equipped, remove the front stabilizer bar-to-control arm nut and washer (driver's side); discard the nut. Remove the front stabilizer bar-to-chassis brackets; discard the bolts.

5. Using a 15mm socket, remove the lower control arm ball joint-to-steering knuckle nut and bolt; discard the nut and bolt. Repeat this procedure on the opposite side.

6. Using a large pry bar, separate the lower control arm from the steering knuckle.

— **CAUTION** —
Exercise care not to damage or cut the ball joint boot. The pry bar must not contact the lower arm. Repeat this procedure on the opposite side.

7. Using a large pry bar, pry the left inboard CV-joint assembly from the transaxle.

NOTE: With the halfshaft removed from the transaxle, the lubricant will drain from the seal. To prevent fluid loss, install the shipping plugs No. T81P-1177-B; two plugs are required (one for each seal).

8. To remove the inboard CV-joint from the transaxle, grasp the left-hand steering knuckle, then swing the knuckle and halfshaft outward from the transaxle.

— **CAUTION** —
Exercise care when using a pry bar to remove the CV-joint assembly, or damage may occur to the differential oil seal.

9. If the CV-joint assembly cannot be pried from the transaxle, insert the differential rotator tool No. T81P-4026-A, or equivalent, through the left side and tap the joint out; the tool

can be used from either side of transaxle.

10. Using a wire, support the halfshaft assembly in a rear level position to prevent damage to the assembly during the remaining operations; repeat this procedure on the opposite side.

11. Using a small pry bar, remove the backup lamp switch connector from the transaxle backup lamp switch.

12. Using a 13mm deep well socket, remove the starter-to-engine stud bolts.

13. Remove the shift mechanism-to-shift shaft nut/bolt, the control selector indicator switch arm and the shift shaft.

14. Using a 22mm ($^7/_8$ in.) crows foot wrench, remove the speedometer cable from the transaxle.

15. Using a 13mm socket, remove the stiffener brace-to-clutch housing bolts.

16. Position a transaxle jack under the transaxle.

17. Using a 13mm wrench, remove the lower engine-to-transaxle bolts. Lower the transaxle jack until the transaxle clears the rear insulator. Support the engine with a screw jack stand under the oil pan; use a 2 x 4 inch piece of wood on top of the screw jack.

18. Remove the transaxle from the rear face of the engine and lower it from the vehicle.

19. To install, reverse the removal procedures. Torque the transaxle-to-engine bolts to 28–31 ft. lbs. (38–42 Nm), the shift cable/bracket-to-transaxle bolts to 16–22 ft. lbs. (22–30 Nm) for 10mm or 22–35 ft. lbs. (31–47 Nm) for 12mm; the shift mechanism-to-shift shaft bolt to 7–10 ft. lbs. (9–13 Nm); the stiffener brace-to-clutch housing bolts to 15–21 ft. lbs. (21–28 Nm); the starter-to-engine stud bolts to 30–40 ft. lbs. (41–54 Nm); the lower ball joint-to-steering knuckle nut/bolt to 37–44 ft. lbs. (50–60 Nm) and the top transaxle-to-engine bolts to 28–31 ft. lbs. (38–42 Nm). Check the transaxle fluid level; if adding fluid, use Dexron® II automatic transmission fluid. Set the hand brake, pump the clutch pedal several times to adjust the clutch.

— **CAUTION** —
The transaxle case casting may have sharp edges. Wear protective gloves when handling the transaxle assembly.

NOTE: NEVER attempt to start the engine with the CV-joints disconnected from the transaxle or side gear dislocation may occur.

OVERHAUL

For all manual transaxle overhaul procedures, please refer to "Manual Transmission Overhaul" in the Unit Repair section.

AUTOMATIC TRANSAXLE

For all linkage and band adjustments, plus the fluid and filter change procedures, please refer to "Automatic Transmission" in the Unit Repair section.

REMOVAL & INSTALLATION

1. Disconnect the battery negative cable. Remove the air cleaner, the hoses and the tubes.

2. Using a screwdriver, place it in the shift cable/bracket assembly slot, to keep the assembly from moving, then remove the assembly-to-transaxle bolt and the assembly from the transaxle.

3. Disconnect the electrical connector from the neutral safety switch and the electrical bulkhead connector from the rear of the transaxle.

4. Disconnect the throttle valve cable from the throttle body lever and the throttle valve-to-transaxle bolt. Pull the throttle valve cable Up and disconnect it from the TV link.

NOTE: Pulling too hard on the throttle valve cable may bend the internal TV bracket.

5. On the left side, remove the engine support-to-strut nut/bolt. Remove the top torque converter housing-to-engine bolts.

6. Attach the engine support (3-bar system) hooks to the engine lifting points, tighten the hooks to slowly lift the engine.

7. Raise and support the front of the vehicle on jackstands. Remove the front wheel/tire assemblies.

8. Remove the tie-rod-to-steering knuckle cotter pin and castle nut, then separate the tie-rod end from the steering knuckle.

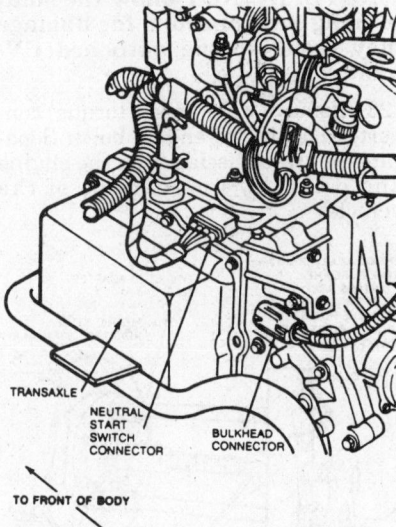

View of the transaxle electrical connectors—automatic overdrive

9. Remove the lower control arm ball joint-to-steering knuckle cotter pin and castle nut, then separate the ball joint from the steering knuckle.

10. Remove the stabilizer bar-to-control arm nuts.

11. Remove the rack/pinion-to-subframe nuts/bolts and the engine mount-to-subframe nuts. Using a piece of wire, support the steering gear from the tie-rod end to the coil spring to hold the steering gear in position.

12. Disconnect the electrical connector from the oxygen sensor.

13. Remove the exhaust pipe-to-exhaust manifold nuts and the rear portion of the convertor pipe-to-exhaust pipe.

14. Using an assistant, lower the adjustable jacks and allow the subframe to lower. Rotate the front of the subframe down and pick up the rear of the subframe off the exhaust pipe. Work the subframe rearward until it can be lowered past the exhaust pipe.

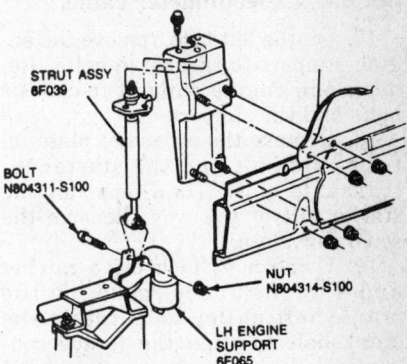

Exploded view of the vehicle speed sensor—automatic overdrive

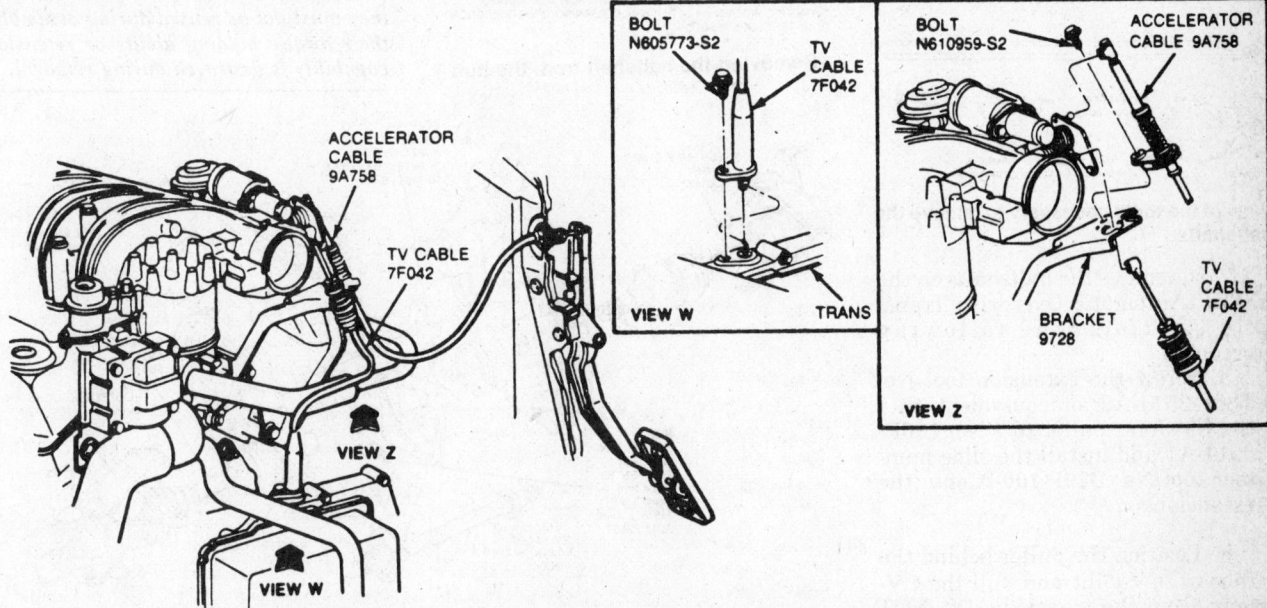

Exploded view of the TV cable assembly—automatic overdrive

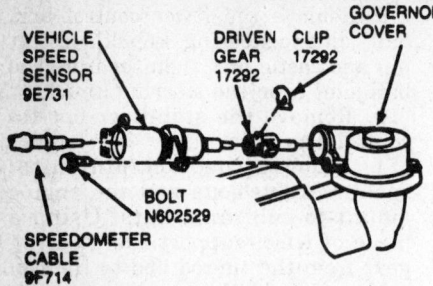

Exploded view of the engine stabilizer—automatic overdrive

15. Remove the subframe-to-chassis bolts. Remove the left side engine support mount-to-subframe nuts/bolts and lower the subframe.

16. Position a transmission jack under the transaxle oil pan. Remove the vehicle speed sensor from the transaxle.

NOTE: Vehicles equipped with electronic instrument clusters do not use a speedometer cable.

17. On the left side, remove the engine support-to-transaxle bolts. Remove the engine support-to-chassis bolts and the support.

18. Remove the separator plate-to-transaxle bolt and the starter-to-transaxle bolts, then position the starter out of the way. Remove the separator plate.

19. Using a ½ inch drive ratchet and a ⅞ inch deep socket on the crankshaft pulley bolt, rotate the crankshaft to align the torque converter bolts with the starter drive hole. As the torque converter-to-flywheel nuts are exposed, remove them.

20. Disconnect and plug the oil cooler lines from the transaxle.

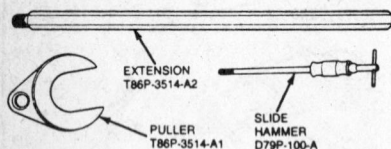

View of the tools necessary to remove the halfshafts

21. To remove the halfshafts on the AXOD (Automatic Overdrive Transaxle), perform the following procedures:

a. Screw the extension tool No. T86P-3514-A2, or equivalent, into the CV-Joint puller tool No. T86P-3514-A1 and install the slide hammer tool No. D79P-100-A onto the extension.

b. Position the puller behind the inboard CV-joint and pull the CV-joint from the transaxle; DO NOT pry against the case.

22. To remove the halfshafts on the ATX (Automatic Transaxle), perform the following procedures:

a. On the right side, remove the link shaft bearing support-to-bracket bolts.

b. While supporting the bearing support, slide the link shaft from the transaxle.

c. Using a wire, support the link shaft.

d. Using the driver tool No. T81P-4026-A, or equivalent, insert it into the transaxle to drive the left-hand inboard CV-joint assembly from the transaxle.

e. Using a length of wire, support the halfshaft.

NOTE: DO NOT allow the shaft to hang unsupported, for damage may result to the outboard CV-joint.

23. Remove the lower torque converter housing-to-engine bolts. Separate the transmission from the engine and carefully lower it out of the vehicle.

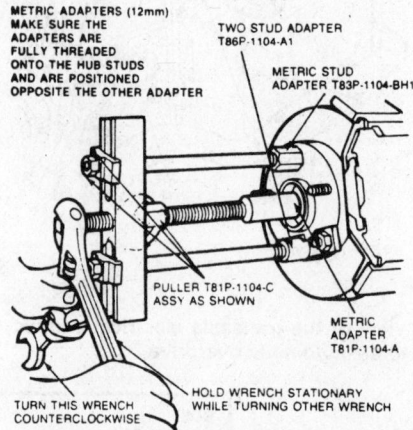

Removing the halfshaft from the hub

24. To install, reverse the removal procedures. Torque the transaxle-to-engine bolts to 41–50 ft. lbs. (55–68 Nm); the control arm ball joint-to-steering knuckle nut to 36–44 ft. lbs. (50–60 Nm); the tie-rod end-to-steering knuckle nut to 23–35 ft. lbs. (31–47 Nm); the starter-to-transaxle bolts to 30–40 ft. lbs. (41–54 Nm); the engine support-to-transaxle bolts to 25–33 ft. lbs. (34–45 Nm); the engine mount-to-subframe bolts to 55–70 ft. lbs. (75–90 Nm); the subframe-to-chassis bolts to 40–50 ft. lbs. (55–70 Nm); the stabilizer bar-to-control arm nuts to 98–125 ft. lbs. (133–169 Nm) and the stabilizer U-clamp-to-chassis bolts to 60–70 ft. lbs. (81–95 Nm). Check and add fluid to the transaxle.

DRIVE AXLE

Halfshafts

When removing both the left and right halfshafts, shipping plug tools No. T81P-1177-B must be installed. Failure to use these tools can result in dislocation of the differential side gears. Should the gears become misaligned, the differential will have to be removed from the transaxle to realign the side gears.

— CAUTION —

DO NOT start this procedure unless the following parts are to known to be available: A new hub nut assembly, a new lower control arm-to-steering knuckle nut/bolt and a new inboard CV-joint stub shaft circlip. Once these parts are removed, they must not be reused during assembly; their torque holding ability or retension capability is destroyed during removal.

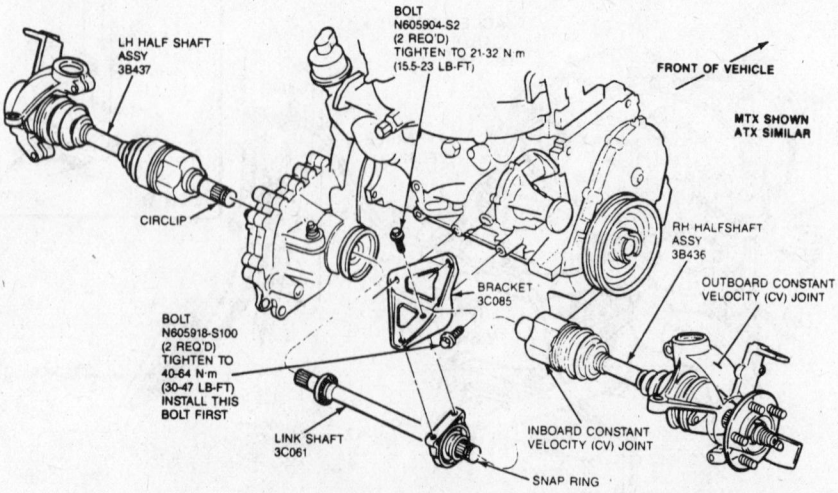

Exploded view of the halfshaft assembly—used with the MTX and ATX

REMOVAL & INSTALLATION

1. Remove the wheel cover/hub cover and loosen the wheel nuts. Remove the hub retainer and washer; the nut must be discarded after removal.

2. Raise and support the vehicle on jackstands, then remove the wheel/tire assembly.

3. Remove the ball joint-to-steering knuckle nut. Using a punch and a hammer, drive the bolt from the steering knuckle; discard the bolt/nut.

4. Using a medium pry bar, separate the ball joint from the steering knuckle.

NOTE: Position the end of the pry bar outside of the bushing pocket to avoid damaging the bushing. Use care to prevent damage to the ball joint boot.

5. Remove the stabilizer bar link from the stabilizer bar.

6. If removing the right side halfshaft/link shaft from the ATX or the MTX, perform the following procedures:

   a. Remove the bearing support-to-bracket bolts, then slide the shaft out of the transaxle. Using a piece of wire, support the end of the shaft from a convenient underbody component.

NOTE: DO NOT allow the shaft to hang unsupported, for damage to the outboard CV-joint may occur.

   b. Separate the outboard CV-joint from the hub using the front hub remover tool No. T81P-1104-C, the meteric adapters tools No. T83-P-1104-BH, T86P-1104-Al and T81P-1104-A.

NOTE: NEVER use a hammer to separate the outboard CV-joint stub shaft from the hub; damage to the CV-joint threads and internal components may result. The right side link shaft and halfshaft assembly is removed as a complete unit.

7. If removing the right side halfshaft from the AXOD (overdrive) or the left side halfshaft from the MTX, perform the following procedures:

   a. Position the CV-joint puller tool No. T86P-3514-A1 between the CV-joint and transaxle case. Turn the steering hub and/or wire the strut assembly out of the way.

   b. Assemble the screw extension tool No. T86P-3514-A2 into the CV-Joint puller and hand tighten. Assemble the screw impact slide ham-

mer tool No. D79-100-A onto the extension and remove the CV-joint.

   c. Support the end of the shaft by suspending it from a convenient underbody component with a piece of wire. DO NOT allow the shaft to hang unsupported, damage to the outboard CV-joint may occur.

   d. Separate the outboard CV-joint from the hub using the front hub remover tool No. T81P-1104-C, the meteric adapters tools No. T83-P-1104-BH, T86P-1104-Al and T81P-1104-A.

NOTE: Never use a hammer to separate the outboard CV-joint stub shaft from the hub. Damage to the CV-joint threads and internal components may result.

   e. Remove the halfshaft assembly from the vehicle.

8. To remove the left side halfshaft from an ATX, perform the following procedures:

NOTE: Due to the ATX case configuration, the right side halfshaft assembly MUST BE removed first.

   a. Remove the right hand halfshaft assembly (from the transaxle) and support it on a wire.

   b. Insert the differential rotator tool No. T81P-4026-A into the transaxle and drive the left side inboard CV-joint assembly from the transaxle.

   c. Support the end of the shaft by suspending it from a convenient underbody component with a piece of wire. DO NOT allow the shaft to hang unsupported, for damage to the outboard CV-joint may occur.

   d. Using the front hub removal tool No. T81P-1104-C, the meteric adapter tools No. T83-P-1104-BH, T86P-1104-Al and T81P-1104-A, separate the outboard CV-joint from the hub.

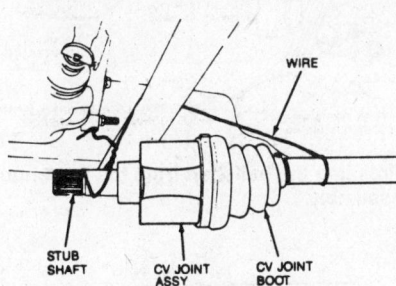

Supporting the halfshaft with a wire

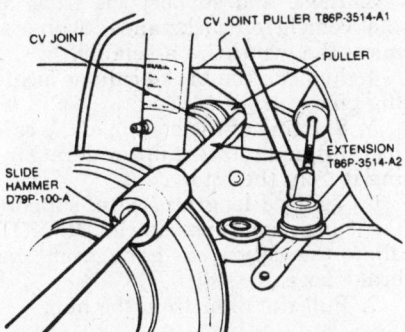

Removing the halfshaft assemblies from AXOD (right-side) or MTX (left-side)

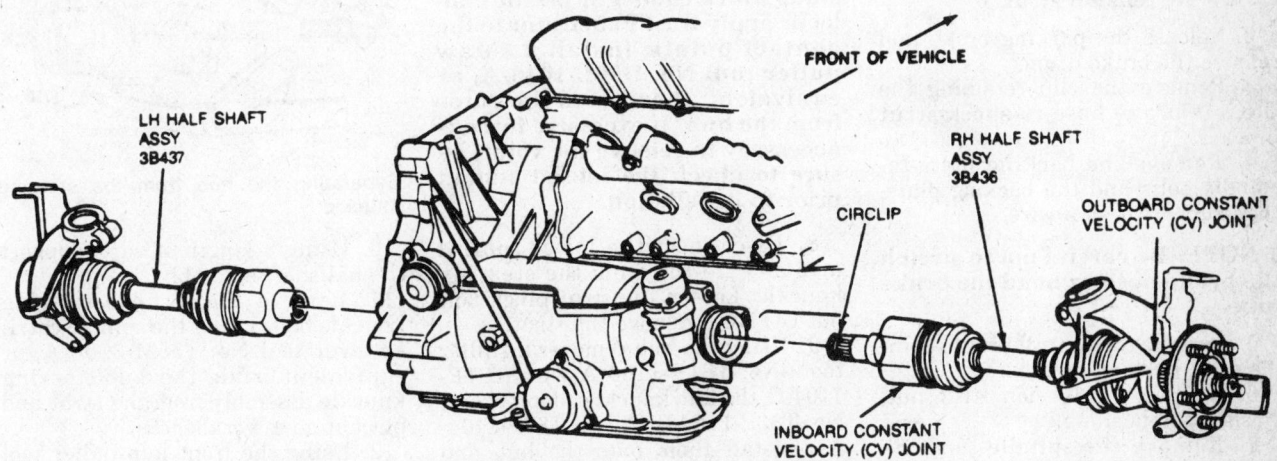

Exploded view of the halfshaft assembly—AXOD

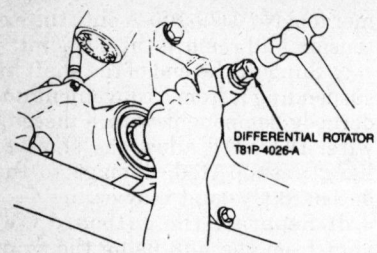

Removing the left-side halfshaft from the transaxle—ATX

NOTE: Never use a hammer to separate the outboard CV-joint halfshaft from the hub. Damage to the CV-joint threads and internal components may result.

    e. Remove the halfshaft assembly from the vehicle.

9. To install, use a new circlip (on the inboard CV-joint), oil seal and reverse the removal procedures. Torque the control arm-to-steering knuckle nut/bolt to 40–55 ft. lbs.; the stabilizer bar-to-stabilizer link nut to 35–50 ft. lbs.; the wheel lug nuts to 80–105 ft. lbs. and the hub nut to 180–200 ft. lbs. Refill the transaxle to the proper level with the specified fluid.

CV-JOINT OVERHAUL

For all CV-joint overhaul procedures, please refer to "CV-Joint Overhaul" in the Unit Repair section.

Rear Axle Spindle

REMOVAL & INSTALLATION

1. Raise and support the rear of the vehicle on jackstands. Remove the wheel/tire assembly.

NOTE: DO NOT raise the vehicle by the tension strut.

2. Release the parking brake and remove the brake drum.
3. Remove the clip retaining the flexible brake hose-to-shock strut bracket.
4. Remove the backing plate-to-spindle bolts and the backing plate, then support it on a wire.

NOTE: Be careful not to stretch the brake hose or bend the brake tube.

5. Remove the control arm-to-spindle nuts, bolts and washers.
6. Remove the tension strut nut, washer and bushing.
7. Remove the spindle-to-shock strut pinch bolt (discard it) and the spindle from the vehicle.

8. Use a new strut-to-spindle nut/bolt (do not tighten) and new control arm-to-spindle nuts/bolts.
9. Position a jackstand under the rear suspension and support it at curb height.
10. Torque the spindle-to-strut bolt to 55–81 ft. lbs. (70–110 Nm); the control arm-to-spindle nuts to 52–74 ft. lbs. (70–100 Nm).
11. To complete the installation, reverse the removal procedures. Check and/or adjust the rear wheel alignment.

Front Axle Hub And Bearings

REMOVAL & INSTALLATION

1. Remove the front wheel cover and the hub cover, then loosen the wheel lug nuts. Remove the hub nut (discard it) and the washer.

NOTE: The hub nut features a crimped collar which requires sufficient torque to remove it. DO NOT use an impact type tool to remove the hub nut.

2. Loosen the strut top-to-apron nuts.
3. Raise and support the front of the vehicle on jackstands, then remove the wheel/tire assembly.
4. Remove the brake caliper locating pins.
5. Starting at the bottom of the caliper, lift and rotate it upward, removing it from the rotor.
6. Using a length of wire, support the caliper from the vehicle; DO NOT allow the caliper to hang from the brake hose.
7. Pull the rotor from the hub.

NOTE: If the rotor is difficult to remove, strike it between the studs with a rubber or plastic mallet or apply Rust Penetrator to the contact points, install a 3-Jaw puller tool No. D80L-1013-A, or equivalent, then press the rotor from the hub. If excessive force is necessary to remove the rotor, be sure to check the lateral runout prior to installation.

8. Disconnect the lower control arm and tie-rod from the steering knuckle. Loosen the strut pinch bolt but DO NOT remove the strut.
9. Using the hub remover/installer tools No. T81P-1104-A with T81P-1104-C, the hub knuckle adapter tools No. T83P-1104-BH1 and T86P-1104-A1, install them onto the hub and press the halfshaft from the hub/bearing assembly.

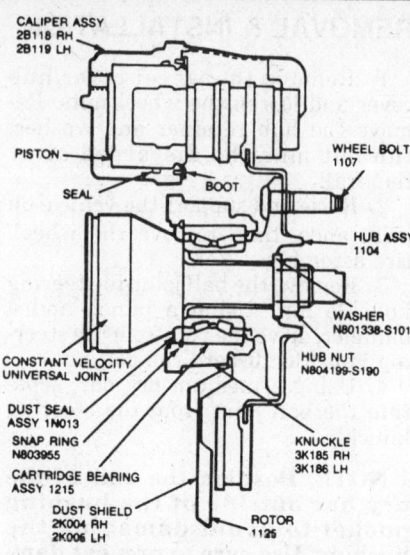

Cross-sectional view of the front hub/bearing assembly

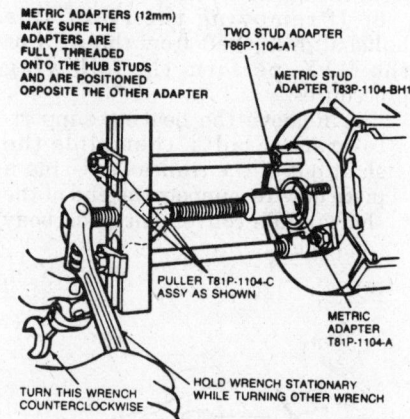

Pressing the halfshaft from the front hub assembly

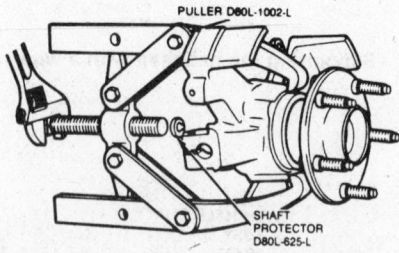

Separating the hub from the steering knuckle

10. Using a length of wire, support the halfshaft assembly.
11. Remove the strut-to-steering knuckle bolt. Using the spindle carrier lever tool No. T85M-3206-A, or equivalent, slide the hub/steering knuckle assembly from the strut and place it on a workbench.
12. Using the front hub puller tool No. D80L-1002-L and the shaft protector tool No. D80L-625-1, place the

puller jaws on the steering knuckle bosses and remove the hub from the steering knuckle assembly.

NOTE: When pressing the hub from the steering knuckle, be sure that the shaft protector is centered, clears the bearing internal diameter and rests on the end face of the hub journal.

13. Using snapring pliers, remove and discard the bearing-to-steering knuckle snapring.

14. Using a hydraulic press, the front bearing spacer tool No. T86P-1104-A2 and the bearing remover tool No. T83P-1104-AH2, press the bearing from the steering knuckle; discard the bearing.

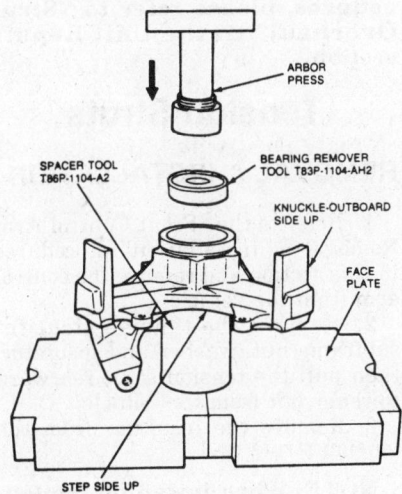

Removing the bearing from the steering knuckle

15. Clean and inspect the hub and steering knuckle bearing surfaces for damage.

16. To install the new bearing into the steering knuckle, perform the following procedures:

   a. Using an arbor press, position the front bearing spacer tool No. T86P-1104-A2, with the step-side down, on the arbor plate.

   b. Position the outboard edge of the steering knuckle on the front bearing spacer tool No. T86P-1104-A2.

   c. Position the new bearing on the inboard side of the steering knuckle.

   d. Using the bearing installer tool No. T86P-1104-A3, position it on the new bearing. Press the new bearing into the steering knuckle until it seats against the seat of the knuckle bore.

17. To install the hub into the steering knuckle, perform the following procedures:

   a. Using an arbor press, position the front bearing spacer tool No. T86P-1104-A2, or equivalent, on the arbor plate.

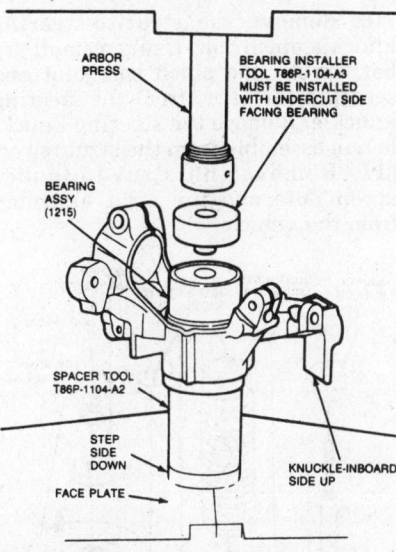

Installing the new bearing into the steering knuckle

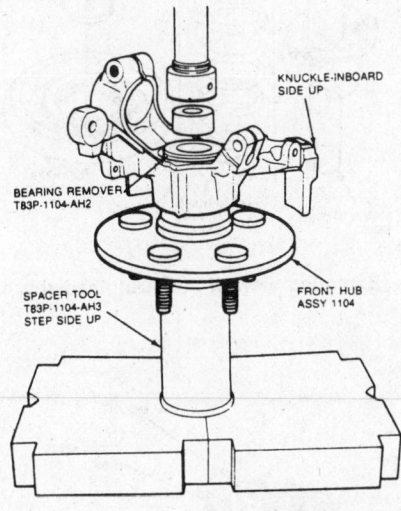

Pressing the hub into the steering knuckle

   b. Position the hub (with the lugs facing down) on the front bearing spacer tool No. T86P-1104-A2.

   c. Position the steering knuckle (outboard side facing down) onto the hub barrel.

   d. Position the bearing remover tool No. T83P-1104-AH2 (flat-side facing down) onto the bearing inner race.

   e. Press the hub into the steering knuckle assembly together until it is fully seated. Check that the hub rotates freely in the knuckle.

18. To replace the dust seal, perform the following procedures:

NOTE: TAP UNIFORMLY TO REMOVE DUST SEAL, USING LIGHT DUTY HAMMER AND SCREWDRIVER.

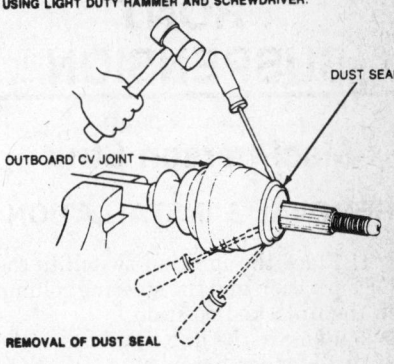

Removing the dust seal from the halfshaft

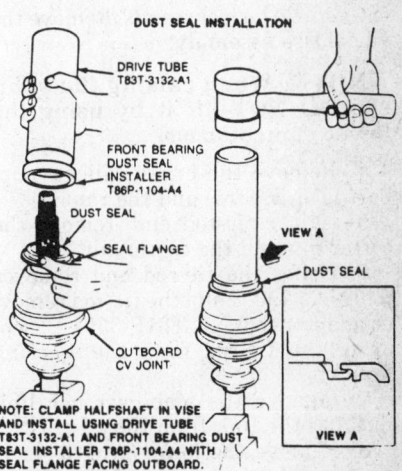

Installing the dust seal onto the halfshaft

   a. Remove the halfshaft from the vehicle and place it in a vise.

   b. Using a punch and a mallet, drive the dust seal from the halfshaft.

   c. Using a new dust seal, position it on the halfshaft with the flange outboard edge facing the bearing.

   d. Using the drive tube tool No. T83T-3132-A1 and the front bearing dust seal installer tool No. T86P-1104-A4, drive the new dust seal onto the halfshaft.

NOTE: When installing the hub nut, DO NOT use an impact type tools to tighten the nut.

19. To install, lubricate the halfshaft end with engine oil and reverse the removal procedures. Torque the hub nut to 180–200 ft. lbs. (245–270 Nm); the strut-to-apron nuts to 22–32 ft. lbs.; the strut-to-steering knuckle bolt to 70–95 ft. lbs. (95–129 Nm); the control arm-to-steering knuckle nuts to 40–55 ft. lbs. (54–74 Nm); the tie-rod end-to-steering knuckle nut to 23–25 ft. lbs. (31–47 Nm) and the wheel lug nuts to 80–105 ft. lbs. (109–142 Nm). Check and/or adjust the front wheel alignment.

FRONT SUSPENSION

MacPherson Strut

REMOVAL & INSTALLATION

1. Place the ignition switch to the OFF position and the steering column in the unlocked position.

2. Remove the hub nut. Loosen the strut-to-fender apron nuts; DO NOT remove the nuts.

3. Raise and support the front of the vehicle on jackstands. Remove the wheel/tire assembly.

NOTE: When raising the vehicle, DO NOT lift it by using the lower control arms.

4. Remove the brake caliper (support it on a wire) and the rotor.

5. At the tie-rod end, remove the cotter pin and the castle nut.

6. Using the tie-rod end remover tool No. 3290-C and the tie-rod remover adapter tool No. T81P-3504-C, separate the tie-rod from the steering knuckle.

7. Remove the stabilizer bar link nut and the link from the strut.

8. Remove the lower arm-to-steering knuckle pinch bolt and nut; it may be necessary to use a drift punch to remove the bolt. Using a screwdriver, spread the knuckle-to-lower arm pinch joint and remove the lower arm from the steering knuckle.

9. Remove the halfshaft from the hub and support it on a wire.

NOTE: When removing the halfshaft, DO NOT allow it to move outward for the tripod CV-joint could separate from the internal parts, causing failure of the joint.

10. Remove the strut-to-steering knuckle pinch bolt. Using a small pry bar, spread the pinch bolt joint and separate the strut from the steering knuckle. Remove the steering knuckle/hub assembly from the strut tower.

11. Remove the strut-to-fender apron nuts and the strut assembly from the vehicle.

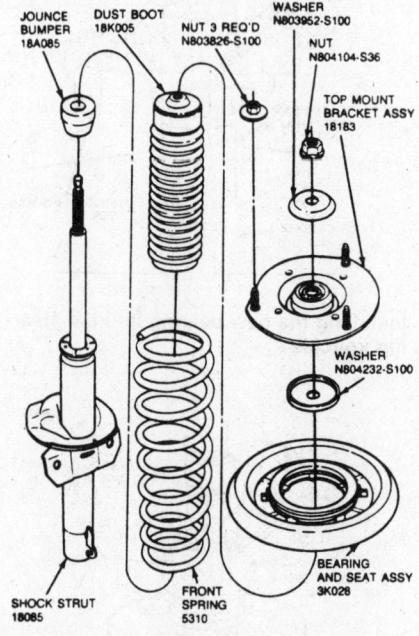

Exploded view of the strut assembly

12. To install, reverse the removal procedures. Torque the strut-to-fender apron nuts to 22–32 ft. lbs. (30–40 Nm); the strut-to-steering knuckle bolt to 70–95 ft. lbs. (95–129 Nm); the control arm-to-steering knuckle bolt to 40–55 ft. lbs. (54–74 Nm); the stabilizer bar-to-link assembly nut to 35–48 ft. lbs. (47–65 Nm); the tie-rod end-to-steering knuckle nut to 23–25 ft. lbs. (31–47 Nm) and the hub nut to 180–200 ft. lbs. (244–271 Nm). Check the front end alignment.

OVERHAUL

For all spring and shock absorber removal and installation procedures, and all strut overhaul procedures, please refer to "Strut Overhaul" in the Unit Repair section.

Tension Struts

REMOVAL & INSTALLATION

1. Refer to the "Front Control Arm Removal & Installation" procedures in this section and remove the control arm from the vehicle.

2. Remove the tension strut-to-subframe nut, washer and insulator, then pull the tension strut rearward to remove it from the vehicle.

3. Remove the insulator from the tension strut.

NOTE: When installing the tension strut, use a new insulator, washer and nut.

4. To install, reverse the removal procedures. Torque the tension strut-to-subframe nut to 70–95 ft. lbs. (95–129 Nm); the control arm-to-frame nut/bolt to 70–95 ft. lbs. (95–129 Nm); the control arm-to-steering knuckle nut/bolt to 40–55 ft. lbs. (54–74 Nm); the tension strut-to-control arm nut to 70–95 ft. lbs. (95–129 Nm) and the wheel lug nuts to 80–105 ft. lbs. (109–142 Nm). Check and/or adjust the front wheel alignment.

Ball Joints

The ball joint is a part of the front control arm and cannot be replaced. If the ball joint is bad, the control arm must be replaced.

INSPECTION

1. Raise and support the front of the vehicle on jackstands.

NOTE: DO NOT raise or support the vehicle on the front control arms.

View of the front suspension system

2. Have an assistant grasp the lower edge of the wheel/tire assembly and move the assembly in and out.

3. As the wheel is being moved, observe the control arm-to-steering knuckle ball joint for movement; movement indicates a worn out ball joint.

4. If ball joint movement is present, replace the lower control arm.

REMOVAL & INSTALLATION

1. Refer to the "Front Control Arm Removal & Installation" procedures in this section and remove the control arm from the vehicle.

2. To install, use a new control arm and reverse the removal procedures.

Front Control Arms

REMOVAL & INSTALLATION

1. Raise and support the front of the vehicle on jackstands. Remove the tire assembly. Position the steering column in the unlocked position.

2. Remove the tension strut-to-control arm nut and the dished washer.

3. Remove and discard the control arm-to-steering knuckle pinch bolt. Using a small pry bar, spread the pinch joint and separate the control arm from the steering knuckle.

NOTE: When separating the control arm from the steering knuckle, DO NOT use a hammer. Be careful not to damage the bolt seal.

4. Remove the control arm-to-frame nut/bolt, then the control arm from the frame and the tension strut.

NOTE: DO NOT allow the halfshaft to move outward or the tripod CV-joint could separate from the internal parts, causing failure of the joint.

5. To install, use a new pinch nut/bolt and reverse the removal procedures. Torque the control arm-to-frame nut/bolt to 70–95 ft. lbs. (95–129 Nm); the control arm-to-steering knuckle nut/bolt to 40–55 ft. lbs. (54–74 Nm); the tension strut-to-control arm nut to 70–95 ft. lbs. (95–129 Nm) and the wheel lug nuts to 80–105 ft. lbs. (109–142 Nm). Check the front end alignment.

NOTE: When installing a new control arm, be sure to saturate the new bushing with vegetable oil; DO NOT use brake fluid, petroleum-based oil or mineral oil.

Stabilizer Bar

REMOVAL & INSTALLATION

1. Raise and support the front of the vehicle on jackstands.

NOTE: DO NOT raise or support the vehicle on the front control arms.

2. Remove the stabilizer bar link-to-stabilizer bar nut, the stabilizer bar link-to-strut nut and the link from the vehicle.

3. Remove the steering gear-to-subframe nuts and the gear from the subframe.

4. Position a set of jackstands under the subframe and remove the rear subframe-to-frame bolts. Lower the subframe rear to obtain access to the stabilizer bar brackets.

5. Remove the stabilizer bar U-bracket bolts and the stabilizer bar from the vehicle.

NOTE: When removing the stabilizer bar, replace the insulators and the U-bracket bolts with new ones.

6. To install, reverse the removal procedures. Torque the U-bracket-to-subframe bolts to 21–32 ft. lbs. (28–43 Nm); the subframe-to-steering gear bolts to 85–100 ft. lbs. (115–135 Nm); the stabilizer bar-to-stabilizer bar link nut to 35–48 ft. lbs. (47–65 Nm) and the stabilizer bar-to-strut nut to 35–48 ft. lbs. (47–65 Nm).

Front Wheel Bearing

The front wheel bearing is a sealed cartridge type that requires no adjustment; if the bearing shows signs of wear, replace it.

REMOVAL & INSTALLATION

1. Refer to the "Front Axle Hub and Bearings Removal & Installation" procedures in this section and remove the bearings from the hub.

2. To install, reverse the removal procedures.

REAR SUSPENSION

MacPherson Strut

REMOVAL & INSTALLATION

1. Raise and support the rear of the vehicle on jackstands. Remove the rear tires.

NOTE: DO NOT raise or support the vehicle using the tension struts.

2. Raise the rear lid and loosen (do not remove) the upper strut-to-body nuts.

3. Remove the brake differential control valve-to-control arm bolt. Using a wire, secure the control arm to the body to ensure proper support leaving six inches clearance to aid in the strut removal.

4. Remove the brake hose-to-strut bracket clip and move the hose out of the way.

5. If equipped, remove the stabilizer bar U-bracket from the vehicle.

6. If equipped, remove the stabilizer bar-to-stabilizer link nut, washer and insulator, then separate the stabilizer bar from the link.

NOTE: When removing the strut, be sure that the rear brake flex hose is not stretched or the steel brake tube is not bent.

7. Remove the tension strut-to-spindle nut, washer and insulator. Move the spindle rearward to separate it from the tension strut.

8. Remove the shock strut-to-spindle pinch bolt. If necessary, use a medium pry bar, spread the strut-to-spindle pinch joint to remove the strut.

9. Lower the jackstand and separate the shock strut from the spindle.

10. Support the shock strut, then remove the top strut-to-body nuts and the strut from the vehicle.

11. To install, reverse the removal procedures. Torque the shock strut-to-body nuts to 19–26 ft. lbs. (26–35 Nm); the shock strut-to-spindle bolt to 55–81 ft. lbs. (75–110 Nm); the control arm-to-spindle bolt to 52–74 ft. lbs. (70–100 Nm); the control arm-to-body bolt to 52–74 ft. lbs. (70–100 Nm); the tension strut-to-spindle nut to 52–74 ft. lbs. (70–100 Nm); the stabilizer bar link-to-stabilizer bar nut to 6–12 ft. lbs. (8–16 Nm) and the stabilizer U-bracket-to-body bolts to 15–25 ft. lbs. (20–34 Nm).

OVERHAUL

For all spring and shock absorber removal and installation procedures, and all strut overhaul procedures, please refer to "Strut Overhaul" in the Unit Repair section.

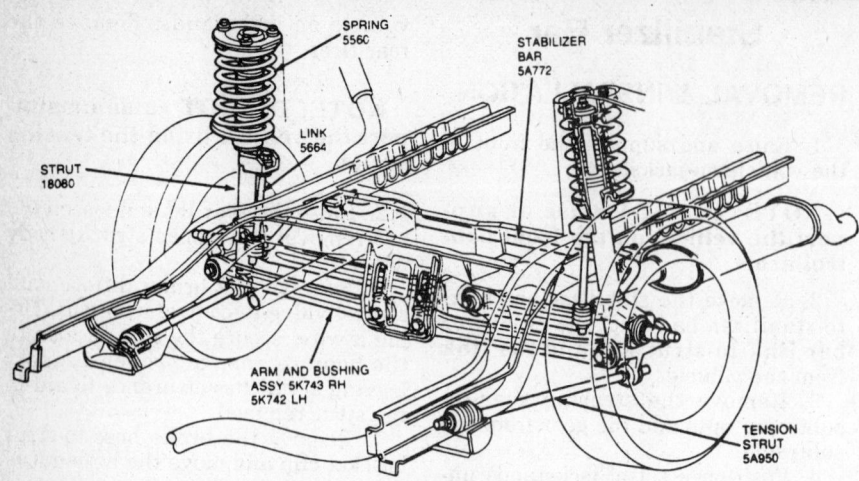

View of the rear suspension components

Rear Control Arms

REMOVAL & INSTALLATION

1. Raise and support the rear of the vehicle on jackstands. Remove the rear tires.

NOTE: DO NOT raise or support the vehicle using the tension struts.

2. From the left side of the front control arm, disconnect the brake proportioning valve.

3. From the front of the control arms, disconnect the parking brake cable.

4. Remove the control arm-to-spindle bolt, washer and nut.

5. Remove the control arm-to-body nut/bolt and the arm from the vehicle.

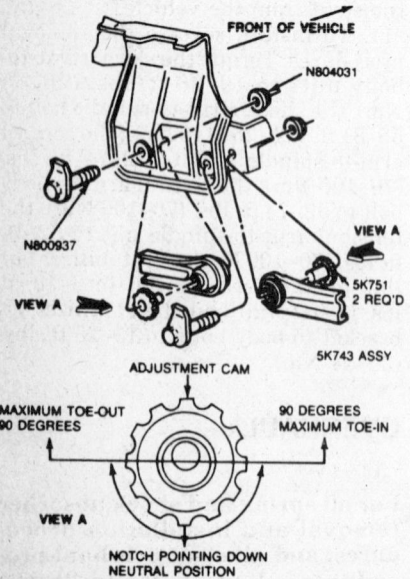

Exploded view of the rear control arm adjusting cam

NOTE: When installing new control arms, be sure that the offset is facing upwards; the arms are stamped with "Bottom" on the lower edge. The flange edge of the right side rear arm stamping MUST face the front of the vehicle; the other three MUST face the rear of the vehicle.

6. To install, position the arm (cam where required) at the center of the vehicle (insert the bolts but do not tighten) and reverse the removal procedures. Torque the control arm-to-spindle bolts to 52–74 ft. lbs. (70–100 Nm) and the control arm-to-body nuts to 52–74 ft. lbs. (70–100 Nm). Check the rear wheel alignment.

Spindle

REMOVAL & INSTALLATION

1. Raise and support the rear of the vehicle on jackstands. Remove the rear tires.

NOTE: DO NOT raise or support the vehicle using the tension struts.

2. Remove the brake drum.

3. Remove the brake flex hose-to-shock strut bracket retaining clip and the backing plate-to-spindle bolts. Remove the backing plate and suspend it on a wire.

NOTE: Be sure that the brake hose and/or the brake tube does not become stretched or bent.

4. Remove the control arm-to-spindle bolts, washers and nuts.

5. Remove the spindle-to-shock strut pinch bolt (discard it) and the spindle from the vehicle.

6. To install, reverse the removal procedures. Torque the spindle-to-shock strut pinch bolt to 55–81 ft. lbs. (70–110 Nm); the tension strut-to-spindle nut to 52–74 ft. lbs. (70–100 Nm) and the control arm-to-spindle nuts to 52–74 ft. lbs. (70–100 Nm). Check the rear wheel alignment.

Stabilizer Bar/Link

REMOVAL & INSTALLATION

1. Raise and support the rear of the vehicle on jackstands. Remove the rear tires.

NOTE: DO NOT raise or support the vehicle using the tension struts.

2. Remove the stabilizer bar-to-link (both sides) nuts, washers and insulators.

3. Remove the stabilizer bar U-bracket-to-body bolts and the stabilizer bar from the vehicle.

4. Remove the stabilizer link-to-shock strut bracket nut, washer and insulator; inspect the condition of the insulators and replace them if necessary.

5. To install, reverse the removal procedures. Torque the stabilizer link-to-shock strut bracket nut to 6–12 ft. lbs. (8–16 Nm); the stabilizer bar U-bracket-to-body bolts to 15–25 ft. lbs. (20–34 Nm) and the stabilizer bar-to-link nuts to 6–12 ft. lbs. (8–16 Nm).

Rear Tension Strut

REMOVAL & INSTALLATION

1. Raise and support the rear of the vehicle on jackstands. Remove the rear tires.

NOTE: DO NOT raise or support the vehicle using the tension struts.

2. Loosen but DO NOT remove the shock strut-to-body nuts.

3. Remove the tension strut-to-spindle nut and the tension strut-to-body nut.

4. Move the spindle rearward and remove the tension strut from the vehicle.

NOTE: The tension strut bushings at the front and the rear are different; the rear bushings have indentations in them.

5. To install, use new tension strut washers/bushings and reverse the removal procedures. Torque the tension strut-to-spindle nut to 52–74 ft. lbs. (70–100 Nm) and the tension strut-to-

body bracket nut 52–74 ft. lbs. (70–100 Nm). Check the rear wheel alignment.

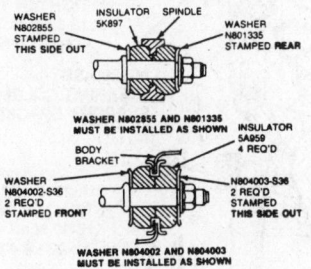

Cross-sectional view of the rear tension strut washers and bushings

Rear Wheel Bearings

ADJUSTMENT

The following procedure should be performed whenever the wheel is excessively loose on the spindle or it does not rotate freely.

NOTE: The rear wheel uses a tapered roller bearing which may feel loose when properly adjusted; this feel should be considered normal.

1. Raise and support the rear of the vehicle on jackstands.
2. Remove the wheelcover or the ornament and nut covers. Remove the hub grease cap.

NOTE: If the vehicle is equipped with styled steel or aluminum wheels, the wheel/tire assembly must be removed to remove the dust cover.

3. Remove the cotter pin and the nut retainer.
4. Back off the hub nut one full turn.
5. While rotating the hub/drum assembly, tighten the adjusting nut to 17–25 ft. lbs. (23–24 Nm). Back off the adjusting nut ½ turn, then retighten it to 10–15 inch lbs. (1.1–1.7 Nm).
6. Position the nut retainer over the adjusting nut so that the slots are in line with cotter pin hole (without rotating the adjusting nut).
7. Install the cotter pin and bend the ends around the retainer flange.
8. To complete the installation, reverse the removal procedures.

BRAKES

For all brake system repair and service procedures not detailed below, please refer to "Brakes" in the Unit Repair section.

Master Cylinder

REMOVAL & INSTALLATION

1. Disconnect and plug the brake lines from the master cylinder.
2. Disconnect the electrical connector (brake warning lamp) from the master cylinder.
3. Remove the master cylinder-to-power booster nuts and the master cylinder from the vehicle.
4. To install, reverse the removal procedures. Torque the master cylinder-to-power booster nuts to 13–25 ft. lbs. (18–33 Nm). Bleed the brake system.

Control Valve

REMOVAL & INSTALLATION

Sedan

The control valve is mounted to the floorpan near the left-rear wheel. It utilizes a mechanical linkage to the lower suspension arm to vary the valve performance based on the rear weight of the vehicle.

1. Raise and support the rear of the vehicle on jackstands.

NOTE: DO NOT raise or support the vehicle using the tension struts.

2. Label and disconnect the brake tubes from the control valve assembly.
3. Remove the valve bracket-to-underbody screws and the control valve assembly from the vehicle.

NOTE: The service replacement control valve will have a red plastic gauge clip on it, which MUST NOT BE removed until installation.

4. To install, make sure that the rear suspension is in the Full rebound position and the control valve operating screw is loose, then reverse the removal procedures. Torque the control valve-to-underbody screws to 8–10 ft. lbs. (11–13 Nm). Perform the control valve assembly adjustment procedures. Bleed the rear brake system.

Station Wagon

The control valves are screwed into the bottom of the master cylinder. They control the braking pressure to the rear wheels to minimize rear wheel skidding during hard braking.

1. Disconnect and plug the primary and/or secondary brake tube from the master cylinder.
2. Remove the pressure control valve(s) from the master cylinder.
3. To install, reverse the removal procedures. Torque the control valve-to-master cylinder to 10–18 ft. lbs. (13–24 Nm). Bleed the brake system.

ADJUSTMENT

Sedan

1. Place the vehicle on a hoist or an alignment machine, so that the vehicle is at the curb load level and the wheels are on a flat surface.
2. At the control valve, loosen the adjuster screw.
3. Using a piece of rubber or plastic tubing $\frac{5}{8}$–$\frac{21}{64}$ inch (16–16.5mm), $\frac{3}{8}$ inch (OD) x $\frac{1}{4}$ inch (ID), slice it lengthwise (on one edge), then install it on the operating rod.

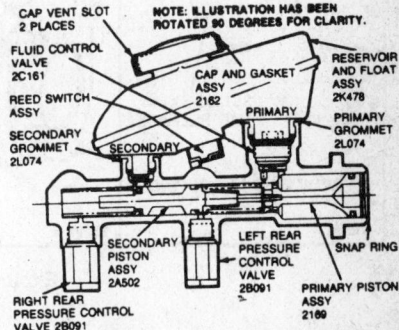

Cross-sectional view of the master cylinder—station wagon model

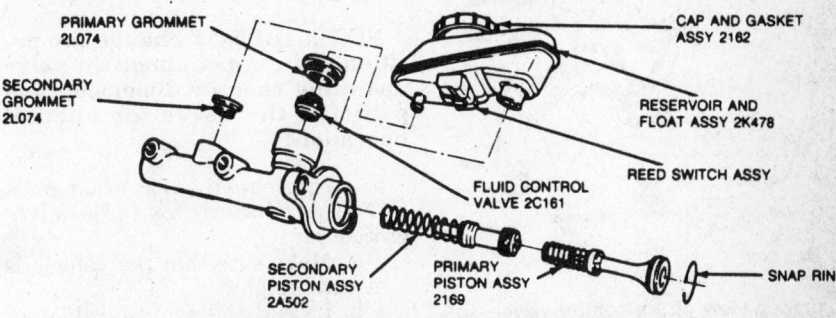

Exploded view of the master cylinder—sedan model

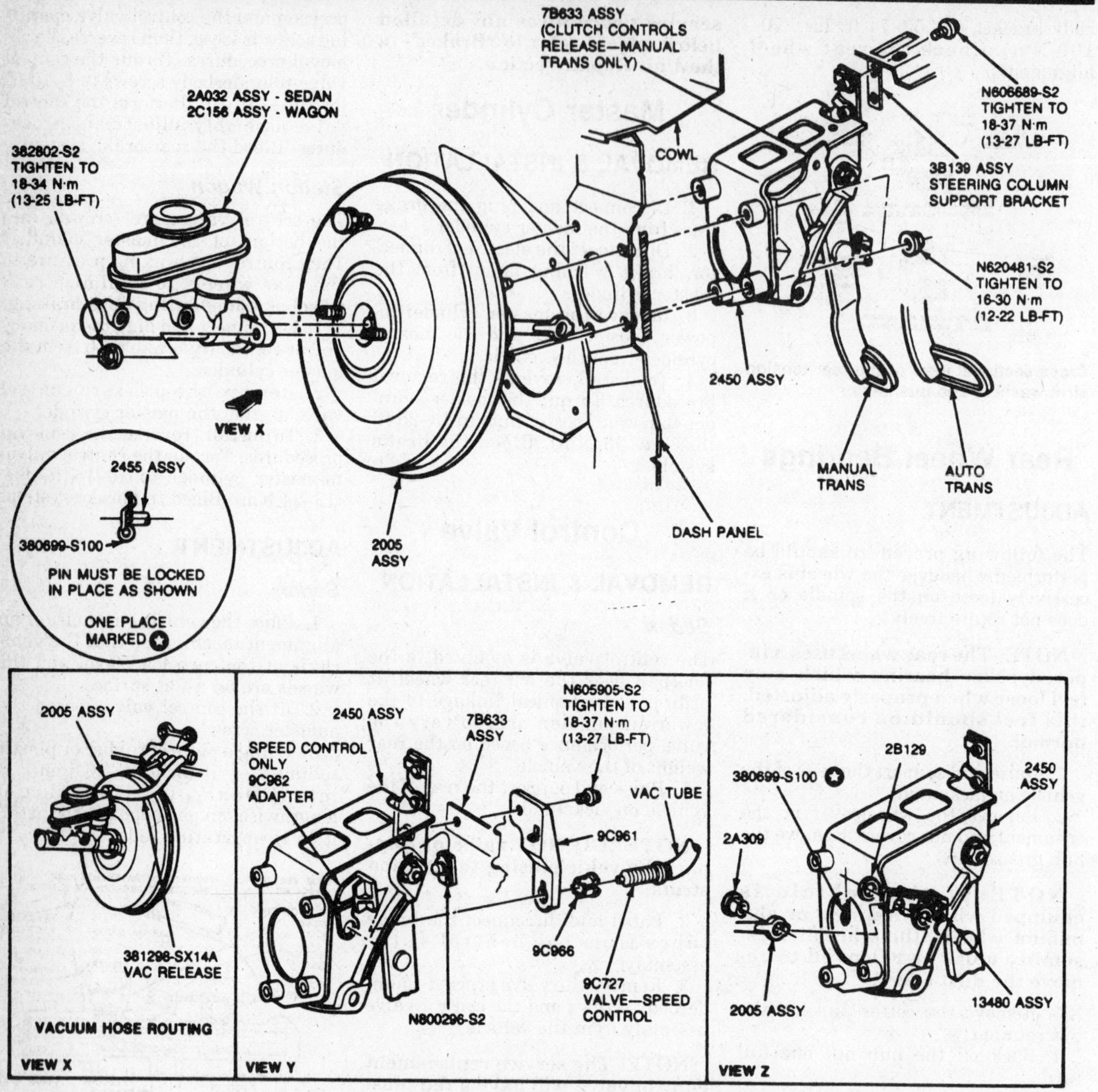

2A032 ASSY - SEDAN
2C156 ASSY - WAGON

382802-S2
TIGHTEN TO
18-34 N·m
(13-25 LB-FT)

7B633 ASSY
(CLUTCH CONTROLS
RELEASE—MANUAL
TRANS ONLY)

COWL

N606689-S2
TIGHTEN TO
18-37 N·m
(13-27 LB-FT)

3B139 ASSY
STEERING COLUMN
SUPPORT BRACKET

N620481-S2
TIGHTEN TO
16-30 N·m
(12-22 LB-FT)

2450 ASSY

MANUAL
TRANS

AUTO
TRANS

VIEW X

2455 ASSY

380699-S100

PIN MUST BE LOCKED
IN PLACE AS SHOWN

ONE PLACE
MARKED ✪

2005
ASSY

DASH PANEL

2005 ASSY

381298-SX14A
VAC RELEASE

VACUUM HOSE ROUTING

VIEW X

2450 ASSY

SPEED CONTROL
ONLY
9C962
ADAPTER

7B633
ASSY

N605905-S2
TIGHTEN TO
18-37 N·m
(13-27 LB-FT)

VAC TUBE

9C961

9C966

9C727
VALVE—SPEED
CONTROL

N800296-S2

VIEW Y

2B129

380699-S100 ✪

2450
ASSY

2A309

2005 ASSY

13480 ASSY

VIEW Z

Exploded view of the power brake booster/master cylinder

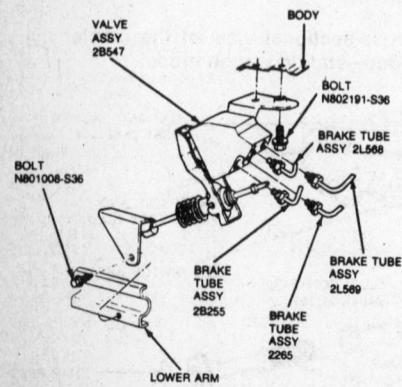

VALVE
ASSY
2B547

BODY

BOLT
N802191-S36

BRAKE TUBE
ASSY 2L568

BOLT
N801008-S36

BRAKE
TUBE
ASSY
2B255

BRAKE TUBE
ASSY
2L589

BRAKE
TUBE
ASSY
2265

LOWER ARM

Exploded view of the control valve—sedan model

---

4. Make sure that the adjuster is resting on the lower mounting bracket; tighten the set screw.

NOTE: DO NOT change the position of the upper nut on the valve operating rod; the dimension will position the valve for normal operation.

5. To decrease the rear brake pressure, perform the following procedures:

a. Make sure that the vehicle is at curb height.

b. Loosen the control valve set screw.

c. Move the piston Down the operating rod 1mm for each 60 psi pressure decrease.

d. Tighten the set screw in the desired position.

6. To increase the rear brake pressure, perform the following procedures:

a. Make sure that the vehicle is at curb height.

b. Loosen the control valve set screw.

c. Move the piston Up the operating rod 1mm for each 60 psi pressure increase.

d. Tighten the set screw in the desired position.

Power Brake Booster

REMOVAL & INSTALLATION

1. Refer to the "Master Cylinder Removal & Installation" procedures in this section and remove the master cylinder from the power brake booster; it is not necessary to remove the brake tubes from the master cylinder.

2. Remove the vacuum hose from the power brake booster.

3. From under the instrument panel, remove the pushrod retainer and outer nylon washer from the brake pin.

4. Remove the power brake booster-to-pedal support nuts. Slide the booster pushrod and pushrod bushing off of the brake pedal pin.

5. Move the booster forward until the studs clear the dash panel and remove the it from the vehicle.

6. To install, reverse the removal procedures. Torque the power brake booster-to-brake assembly bracket nuts to 12–22 ft. lbs. (16–30 Nm) and the master cylinder-to-power brake booster nuts to 13–25 ft. lbs. (18–34 Nm). If the brake tubes were disconnected, bleed the brake system.

Wheel Cylinder

REMOVAL & INSTALLATION

1. Raise and support the rear of the vehicle on jackstands.

NOTE: DO NOT raise or support the vehicle using the tension struts.

2. Remove the wheelcover or the ornament and nut covers. Remove the hub grease cap.

NOTE: If the vehicle is equipped with styled steel or aluminum wheels, the wheel/tire assembly must be removed to remove the dust cover.

3. Remove the cotter pin and the nut retainer.

4. Remove the hub nut, the thrust washer, the outer bearing and the brake drum assembly.

5. Remove the brake shoes, the retainers and the springs from the backing plate.

6. Disconnect and plug the brake tube at the rear-side of the wheel cylinder.

7. Remove the wheel cylinder-to-backing plate bolts and the wheel cylinder from the vehicle.

8. To install, reverse the removal procedures. Torque the wheel cylin-

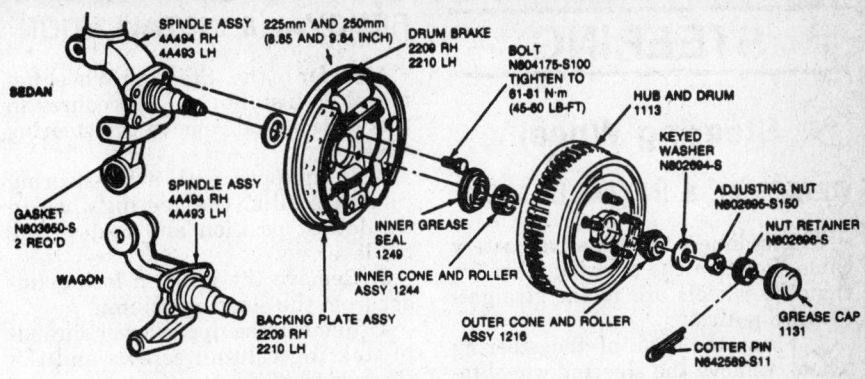

Exploded view of the rear wheel assembly

der-to-backing plate bolts to 7.5 ft. lbs. (10–14 Nm). Adjust the rear wheel bearing. Bleed the rear brake system.

HYDRAULIC SYSTEM BLEEDING

1. Clean all the dirt from around the master cylinder filler cap. If the master cylinder is known or suspected of containing air, it must be bleed before the wheel cylinders and/or calipers.

2. To bleed the master cylinder, loosen the upper secondary left-front outlet fitting aprroximately ¾ of a turn. Have an assistant depress the brake pedal through the Full travel. Close the outlet fitting and return the pedal (slowly) to the fully released position. Wait five seconds, then repeat the operation until the air bubbles cease to appear.

3. Loosen the upper primary right-front outlet fitting approximately ¾ of a turn and repeat step two.

4. To continue to bleed the system, remove the rubber cap dust cap from the wheel cylinder bleeder fitting or caliper fitting. Check to make sure the bleeder fitting is positioned at the upper-half of the caliper front; if not, the caliper is located on the wrong side.

5. Place a box wrench on the bleeder fitting and attach the rubber drain tube to the fitting. Submerge the free end of the tube in a container partially filled with clean brake fluid and loosen the bleeder fitting approximately ¾ of a turn.

6. Have the assistant depress the brake pedal (slowly) through the Full travel. Close the bleeder fitting, then return the pedal to the Full released position. Wait 5 seconds, then repeat this operation until the air bubbles cease to appear at the submerged end of the bleeder tube.

7. When the fluid is completely free of air bubbles, secure the bleeder screw and install the rubber dust cap on the bleeder fitting. Repeat this process on the opposite diagonal system. Refill the master cylinder reservoir after each wheel cylinder or caliper is bled and reinstall the master cylinder cap.

8. With the bleeding operation completed, fill the reservoir to the maximum fill level, indicated on the reservoir. Always ensure the disc brake pistons are returned to their normal positions by depressing the brake pedal several times until the normal pedal travel is established. Check the brake pedal, if the pedal feels spongy, repeat the bleeding procedure.

Parking Brake

ADJUSTMENT

1. Raise and support the rear of the vehicle on jackstands.

NOTE: DO NOT raise or support the vehicle using the tension struts.

2. Make sure that the parking brake is fully released and the transmission is in Neutral.

3. At the parking brake cable equalizer, turn the adjusting nut until drag is felt at the rear wheel.

4. Loosen the adjusting nut until no brake drag is felt.

NOTE: If the parking brake cables have been replaced, in a foot-operated control assembly, set the parking brake lever with approx. 100 lbs. of pressure, then release the control and repeat the procedure.

5. Lower the vehicle and check the operation of the parking brake.

STEERING

Steering Wheel

REMOVAL & INSTALLATION

1. Disconnect the negative battery cable. Position the steering wheel so that the wheels are in the straight-forward position.

2. From the rear of the steering wheel, remove the steering wheel-to-horn pad screws.

3. Disconnect the horn pad electrical connectors. If equipped with cruise control, disconnect the electrical connector from the slip ring terminal.

4. Remove and discard the steering wheel-to-steering column nut.

5. Firmly grasp the steering wheel and pull it from the steering column; DO NOT use a wheel puller.

6. To install, reverse the removal procedures. Torque the steering wheel-to-steering column nut to 50–62 ft. lbs. (68–84 Nm).

Combination Switch

The combination switch is mounted on the steering column and consists of the following switches: turn signal, cornering lights, hazard warning, headlight dimmer, headlight flash-to-pass, windshield washer and windshield wiper.

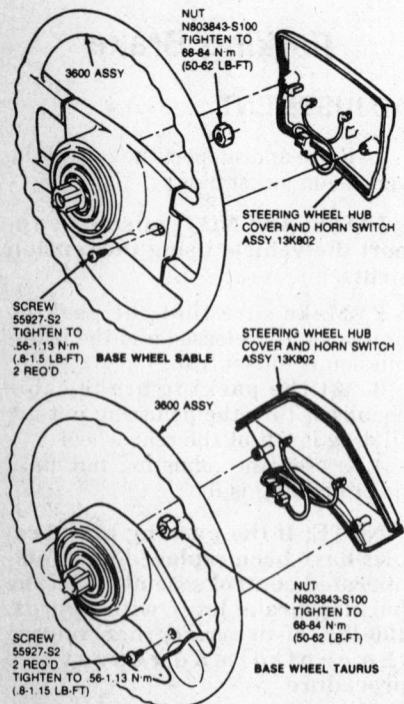

Exploded view of the steering wheel assembly

REMOVAL & INSTALLATION

1. Refer to the "Steering Wheel Removal & Installation" procedures in this section and remove the steering wheel.

2. If equipped with a tilt-steering column, position the steering wheel in the lowest position and remove the tilt lever.

3. Remove the ignition lock cylinder from the steering column.

4. Remove the upper/lower shroud-to-steering column screws and the shrouds from the steering column.

5. Remove the electrical harness-to-steering column retainer and disconnect the electrical connectors from the steering column.

6. Remove the combination switch-to-steering column screws and the switch from the steering column.

7. To install, reverse the removal procedures. Torque the combination switch-to-steering column screws to 18–27 inch lbs. (2–3 Nm), the shroud-to-steering column screws to 6–10 inch lbs. (0.7–1.0 Nm), the steering wheel-to-steering column nut to 50–62 ft. lbs. (68–84 Nm) and the tilt lever-to-steering column screw to 6–8.5 inch lbs. (0.7–1.0 Nm).

Ignition Lock/Switch

REMOVAL & INSTALLATION

1. Disconnect the negative battery cable.

2. Turn the ignition lock cylinder to the Run position.

3. Using a ⅛ inch diameter punch, insert it through the access hole in the bottom of the lower steering column shroud, depress the retaining pin and remove the ignition lock cylinder.

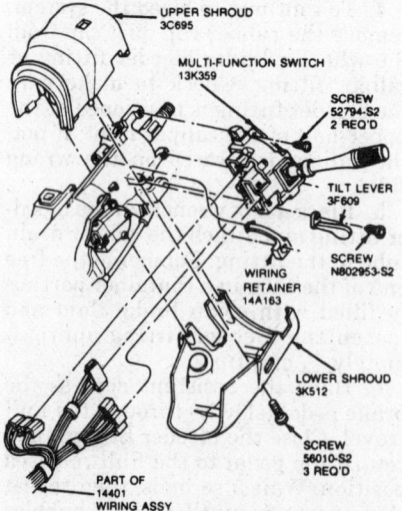

Exploded view of the upper steering column—combination switch

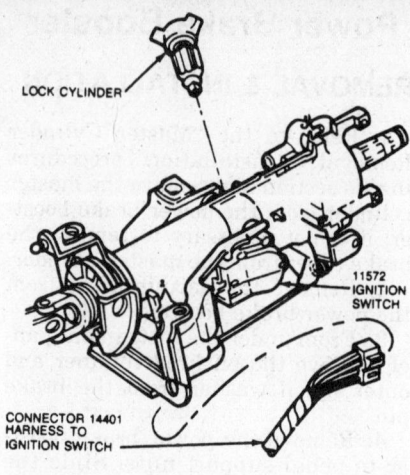

Removing the lock cylinder from the ignition switch assembly

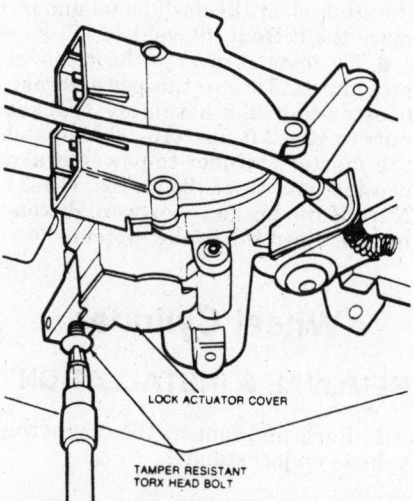

Removing the actuator cover from the ignition switch assembly

4. If equipped with a tilt steering column, remove the tilt lever-to-steering column screw and the tilt lever.

5. Remove the lower instrument panel cover screws and the cover.

6. Using a Phillips head screw driver, remove the upper/lower steering column shrouds.

7. Remove the steering column-to-support bracket nuts/bolts and lower the steering column.

8. Disconnect the electrical harness connector from the ignition switch.

9. Using the Torx Driver tool No. D83L-2100-A, remove the ignition lock actuator cover plate Torx® head bolt and the cover plate.

10. Using the driver tool No. D83L-2100-A, or equivalent, remove the ignition switch-to-actuator assembly Torx® head bolts and the cover assembly from the actuator assembly, then slide the lock actuator from the actuator assembly.

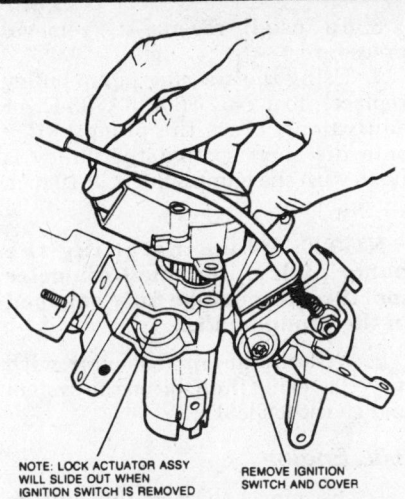

NOTE: LOCK ACTUATOR ASSY
WILL SLIDE OUT WHEN
IGNITION SWITCH IS REMOVED

REMOVE IGNITION
SWITCH AND COVER

Removing the ignition switch from the lock actuator assembly

11. To install the ignition switch-to-actuator assembly, perform the following procedures:

a. Position the ignition switch in the RUN position.

NOTE: To position the ignition switch in the RUN position, turn the switch drive shaft fully clockwise to the START position and release it.

b. Using a small ruler, insert the lock actuator into the actuator assembly to a depth of 0.46–0.54 inch (11.75–13.25mm).

c. While holding the lock actuator at the proper depth, install the ignition switch.

d. Using new tamper-resistant Torx® head bolts, torque the ignition switch-to-actuator assembly bolts to 30–48 inch lbs. (3.4–5.4 Nm).

12. To install the lock cylinder, perform the following procedures:

a. While measuring the lock actuator, turn the ignition switch to the Lock position and install it, the depth should be 0.92–1.0 inch (23.5–25.5mm); if this specification is not met, repeat the lock actuator installation.

b. Using a new tamper-resistant Torx® head bolt, torque the cover-to-lock actuator assembly bolt to 30–48 inch lbs. (3.4–5.4 Nm).

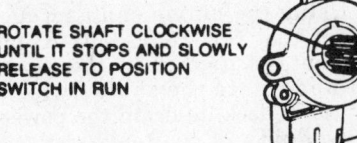

ROTATE SHAFT CLOCKWISE
UNTIL IT STOPS AND SLOWLY
RELEASE TO POSITION
SWITCH IN RUN

Positioning the ignition switch drive shaft

13. Check the operation of the ignition lock cylinder; if the operation is OK, remove the ignition lock cylinder.

14. To complete the installation, reverse the removal procedures and reinstall the ignition lock cylinder. Torque the steering column support bracket-to-instrument panel nuts/bolts to 15–25 ft. lbs. (20–34 Nm), the shroud-to-steering column screws to 6–10 inch lbs. (0.7–1.1 Nm) and the tilt release lever-to-steering column screw to 6.5–8.5 ft. lbs. (9–11 Nm).

Power Steering Gear

REMOVAL & INSTALLATION

1. From inside the vehicle, remove the steering shaft weather boot-to-dash panel nuts.

2. Remove the intermediate shaft-to-steering column shaft bolts and set the weather boot aside. Remove the steering gear input shaft pinch bolt and the intermediate shaft.

3. Raise and support the front of the vehicle on jackstands.

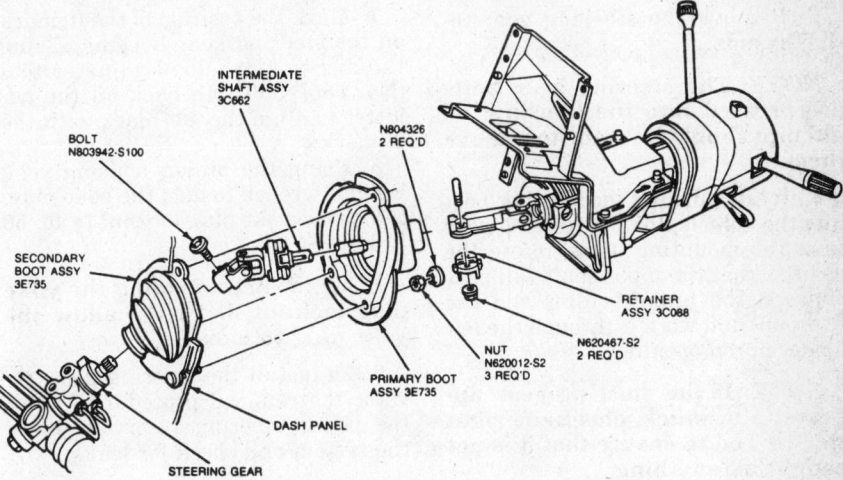

Exploded view of the steering column assembly

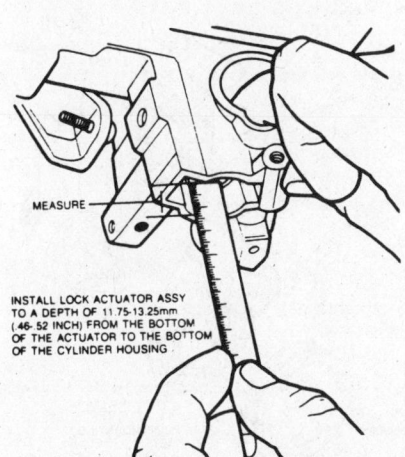

MEASURE

INSTALL LOCK ACTUATOR ASSY
TO A DEPTH OF 11.75-13.25mm
(.46-.52 INCH) FROM THE BOTTOM
OF THE ACTUATOR TO THE BOTTOM
OF THE CYLINDER HOUSING .

Installing the lock actuator into the actuator housing

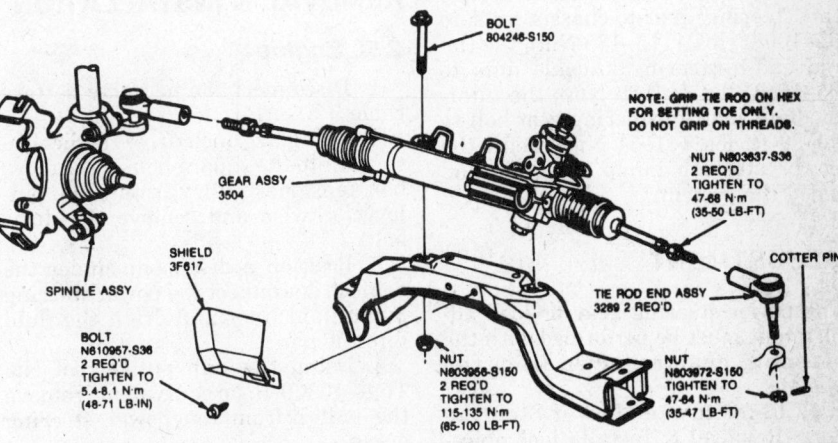

Exploded view of the power steering gear assembly

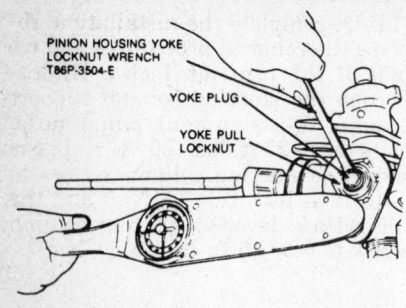

Adjusting the yoke plug of the power steering gear

4. Remove the left front wheel and the heat shield, then cut the bundling strap lines from the steering gear.

5. Remove the tie-rod ends from the steering knuckles.

6. Position a drain pan under the vehicle, then disconnect the pressure and return hoses from the steering gear and drain the fluid.

NOTE: The pressure and return hoses are located on the front of the valve housing.

7. Remove the steering gear-to-chassis nuts.

NOTE: The steering gear bolts are pressed into the housing; no attempt should be made to remove them.

8. While pushing the weather boot into the vehicle, lift the steering gear from the mounting holes, rotate the gear (so that the input shaft will pass between the brake booster and the floorpan) and work it through the left fender apron opening.

NOTE: If the steering gear appears to be stuck, check the right side tie-rod to ensure that it is not caught on anything.

9. To install, use new plastic seals on the hydraulic line fittings and reverse the removal procedures. Torque the steering gear-to-chassis nuts to 85–100 ft. lbs. (115–135 Nm); the tie-rod end-to-steering knuckle nuts to 35–50 ft. lbs. (48–68 Nm); the intermediate shaft-to-steering gear bolt to 30–38 ft. lbs. (41–51 Nm). Refill the power steering pump reservoir and bleed the system.

ADJUSTMENT

The power steering gear preload adjustment must be performed with the steering gear removed from the vehicle.

1. Refer to the "Power Steering Gear Removal & Installation" procedures in this section and remove the power steering gear from the vehicle.

2. Mount the steering gear in a holding fixture (tool No. T57L-500-B, or equivalent).

NOTE: If the steering gear mounting holes in the holding fixture are too small, drill the holes larger using a 9/16 inch drill bit.

3. DO NOT remove the external pressure lines from the steering gear unless they are leaking or damaged. If they are removed, they must be replaced with new ones.

4. Using the pinion shaft torque adjuster tool No. T86P-3504-K, or equivalent, position it on the input shaft and rotate the shaft (twice), from lock-to-lock, to drain the power steering fluid.

5. Using the pinion housing yoke locknut wrench (tool No. T86P-3504-E, or equivalent), loosen the yoke plug locknut and the yoke plug.

6. Position the steering gear in the center of it's travel and torque the yoke plug to 45–50 inch lbs. (5–5.6 Nm); be sure to clean the yoke plug threads before torquing.

7. Mark the position of the 0° mark on the steering gear housing. Using the disc-yoke preload adjuster tool (No. T86P-3504-H), back off the adjuster to align the 48° mark with the 0° mark.

8. Using the pinion housing yoke locknut wrench to hold the yoke plug, torque the yoke plug locknut to 40–50 .ft. lbs. (54–68 Nm).

NOTE: When torquing the yoke plug locknut, DO NOT allow the yoke plug to move.

9. To install the steering gear, reverse the removal procedures. Refill the power steering reservoir, bleed the system and check for leaks.

Power Steering Pump

REMOVAL & INSTALLATION

2.5L Engine

1. Disconnect the negative battery cable.

2. Using a 1/2 inch drive ratchet, insert it into the square hole of the drive belt tensioner pulley, rotate the pulley clockwise and remove the drive belt.

3. Position a drain pan under the vehicle, disconnect the power steering pump fluid lines and drain the fluid into the pan.

4. Using the hub puller tool No. T69L-10300-B, or equivalent, remove the pulley from the power steering pump.

5. Remove the pump-to-bracket bolts and the pump from the vehicle.

6. To install, reverse the removal procedures.

7. Using the steering pump pulley replacer tool No. T65P-3A733-C, or equivalent, press the pump pulley onto the shaft so that the pulley is flush with the pump shaft ± 0.10 inch (± 0.25mm).

NOTE: When installing the pump pulley, the small diameter tool threads must be fully engaged in the pump shaft.

8. Fill the pump reservoir with power steering fluid, bleed the system and check for leaks.

3.0L Engine

1. Disconnect the negative battery cable.

2. Loosen the idler pulley and remove the power steering belt.

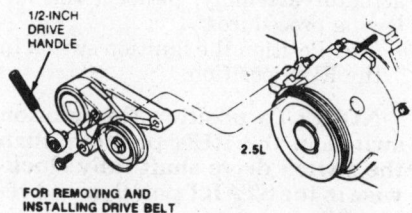

FOR REMOVING AND INSTALLING DRIVE BELT

Removing the drive belt—2.5L engine

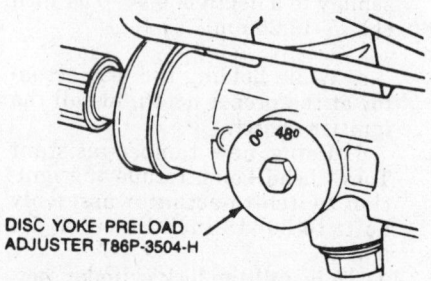

DISC YOKE PRELOAD ADJUSTER T86P-3504-H

View of the power steering gear adjuster disc

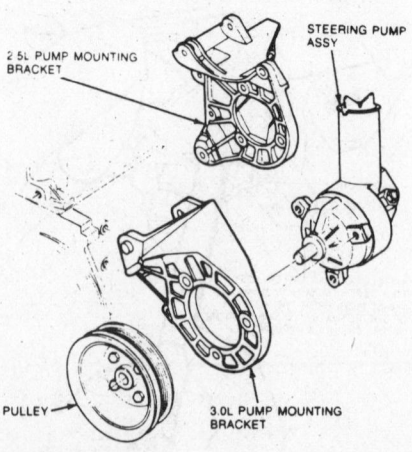

Exploded view of the power steering pump

3. Remove the pulley from the pump hub and the return line from the pump.

4. Back off the pressure line nut until the line separates from the pump.

5. Remove the pump-to-bracket bolts and the pump from the vehicle.

6. To install, reverse the removal procedures. Torque the pump-to-bracket bolts to 30–45 ft. lbs. (40–62 Nm). Refill the power steering pump reservoir, bleed the system and check for leaks.

BELT ADJUSTMENT

Refer to the "Alternator, Belt Tension Adjustment" procedures in this section and adjust the drive belt.

SYSTEM BLEEDING

1. Using power steering pump fluid, fill the power steering reservoir to the maximum fill line.

2. Run the engine until it reaches normal operating temperature.

3. Turn the steering wheel from the left-to-right (all the way) several times.

NOTE: When turning the steering wheel, do not hold it in the far left or right positions too long.

4. Check and/or refill the power steering system.

5. If air remains in the system, it must be purged by performing the following procedures:

   a. Remove the power steering pump dipstick cap assembly.

   b. Using type F automatic transmission fluid, fill the reservoir to the COLD FULL mark on the pump dipstick.

   c. Disconnect the ignition coil wire. Raise and support the front of the vehicle on jackstands.

   d. Using the starter motor, crank the engine, then check the fluid level; do not turn the steering wheel at this time.

   e. Check and/or refill the pump reservoir. Using the starter motor, crank the engine and cycle the steering wheel from lock-to-lock, then recheck the fluid level.

   f. Using the vacuum tester tool No. 021-00014, or equivalent, press the rubber stopper into the pump reservoir. Install the coil wire and start the engine.

   g. Apply 15 in. Hg to the pump reservoir for at least three minutes (engine idling).

NOTE: As the air is being purged from the system, the vacuum will fall off; be sure to maintain adequate vacuum with the vacuum source.

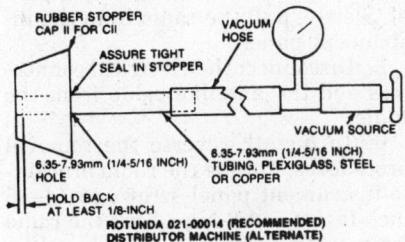

View of the Rotunda Vacuum Tester tool No. 021-00014

   h. Release and remove the vacuum source. Check and/or refill the power steering pump reservoir to the Cold Full mark.

   i. With the engine idling, connect 15 in. Hg to the pump reservoir. Every 30 seconds for approx. 5 min., turn the steering wheel from lock-to-lock; DO NOT hold the steering wheel in the lock position.

NOTE: When bleeding the power steering system, be sure to maintain adequate vacuum.

   j. Release the vacuum, remove the vacuum equipment and add fluid, if necessary.

   k. Start the engine, cycle the steering wheel and check for oil leaks.

Tie-Rod End

REMOVAL & INSTALLATION

1. Raise and support the front of the vehicle on jackstands.

2. Remove the discard the cotter pin and the nut from the tie-rod end ball stud.

3. Using the tie-rod remover tool No. TOOL-3290-C, or equivalent, separate the tie-rod end from the steering knuckle.

4. While holding the tie-rod end, loosen the tie-rod jam nut.

5. Note the depth of the tie-rod end-to-tie-rod, then remove the tie-rod end from the tie-rod.

6. To install, reverse the removal procedures. Torque the tie-rod-to-steering knuckle nut to 36 ft. lbs. (48 Nm) and the tie-rod end-to-tie-rod nut to 35–50 ft. lbs. (47–68 Nm). Check and/or adjust the toe-in.

CHASSIS ELECTRICAL

Heater Blower

REMOVAL & INSTALLATION

1. Open the glove box door, release the retainers and lower the door.

2. Remove the recirc duct support bracket-to-cowl screw, the vacuum line-to-vacuum motor hose and the recirc duct-to-heater assembly screws.

3. Remove the recirc duct-to-heater assembly duct, then lower the recirc duct from between the instrument panel and the heater case.

4. Disconnect the heater motor electrical connector.

5. Remove the heater motor wheel clip and the heater wheel.

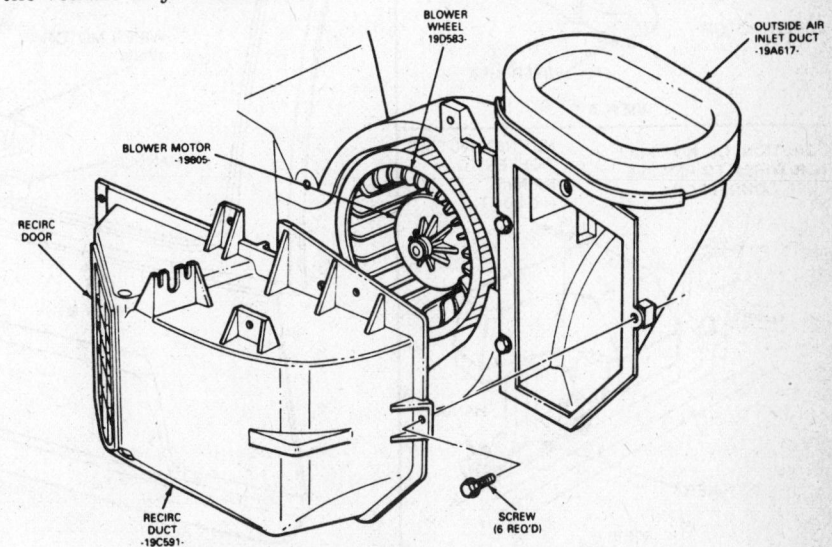

Exploded view of the heater motor and duct assembly

6. Remove the heater motor-to-mounting plate screws and heater motor from the evaporator case.

7. To install, reverse the removal procedures.

Heater Core

REMOVAL & INSTALLATION

1. Refer to the "Instrument Panel Removal & Installation" procedures in this section and remove the instrument panel and lay it on the front seat.

2. Remove the heater case from the vehicle.

3. Remove the vacuum line from the heater core tube seal.

4. Remove the seal from the heater core tubes.

5. Remove the heater core access cover screws and the cover from the heater case.

6. Remove the heater core and the seals from the heater case.

7. To install, reverse the removal procedures.

Radio

REMOVAL & INSTALLATION

1. Disconnect the negative battery cable.

2. Remove the trim panel-to-center instrument panel.

3. Remove the radio/bracket-to-instrument panel screws.

4. Push the radio toward the front, then raise the rear of the radio slightly so that the rear support bracket clears the clip in the instrument panel. Slowly, pull the radio from the instrument panel.

5. Disconnect the electrical connectors and the antenna cable from the radio.

6. To install, reverse the removal procedures. Torque the radio/bracket-to-instrument panel screws to 14–16 inch lbs. (1.5–1.9 Nm). Test the radio for operation.

Windshield Wiper Switch

REMOVAL & INSTALLATION

Front

The front windshield wiper switch is a part of the combination switch, which is mounted to the steering column. Refer to the "Combination Switch Removal & Installation" procedures in this section and remove the combination switch from the steering column. To install, reverse the removal procedures.

Rear—Station Wagon

1. Disconnect the negative battery cable.

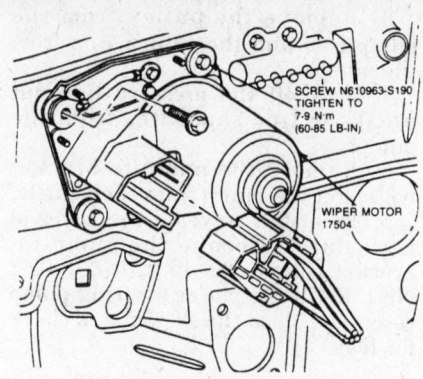

View of the front wiper motor assembly

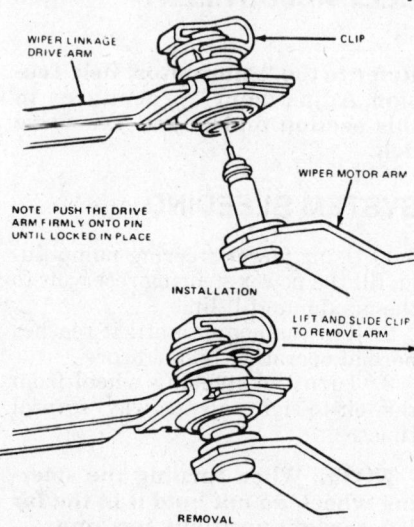

Separating the wiper linkage from the motor linkage

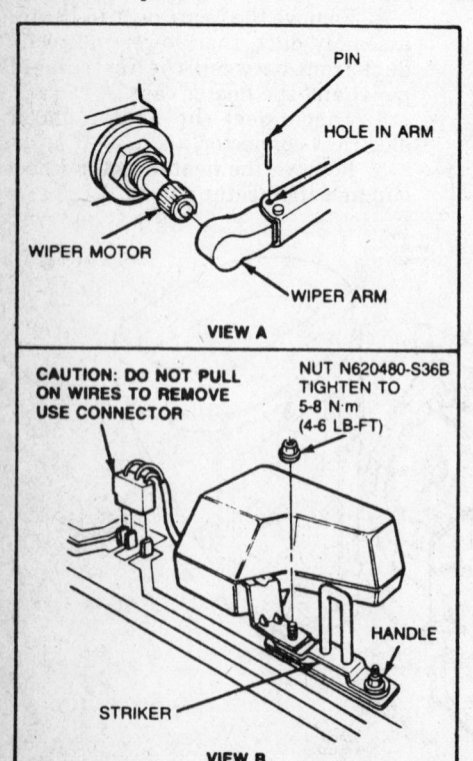

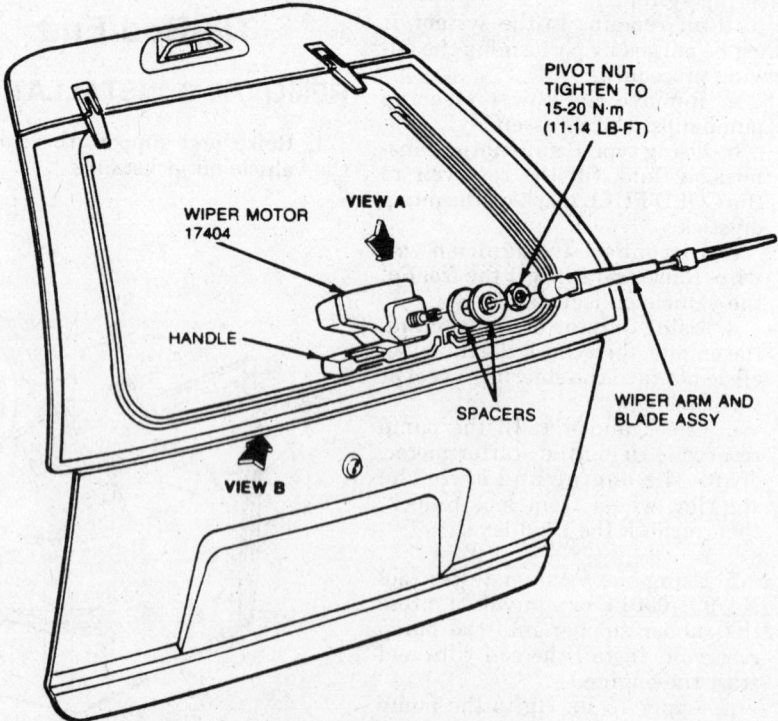

Exploded view of the rear wiper motor assembly—Station Wagon model

2. Remove the finish panel-to-instrument panel screws, then rock the upper edge toward the driver seat.

3. Disconnect the electrical connector from the rear washer switch.

4. Remove the washer switch from the instrument panel.

NOTE: On the Sable model, the switch is retained by two screws.

5. To install, reverse the removal procedures.

Windshield Wiper Motor

REMOVAL & INSTALLATION

Front

1. Disconnect the negative battery cable.

2. Disconnect the electrical connector from the motor.

3. Remove the left side wiper arm.

4. On the passenger side, lift the water shield cover from the cowl.

5. Remove the linkage-to-operating arm clip.

NOTE: When removing the retaining clip, lift up the locking tab and pull the clip away from the pin.

6. Remove the motor/bracket-to-cowl bolts and the assembly from the vehicle.

7. To install, reverse the removal procedures. Torque the motor/bracket-to-cowl bolts to 60–85 inch lbs. (7–9 Nm).

Rear—Station Wagon

1. Disconnect the negative battery cable.

2. Remove the wiper arm/blade assembly from the rear wiper motor.

3. Remove the rear motor pivot shaft-to-glass nut/spacers.

4. Disconnect the electrical connector from the rear wiper motor.

NOTE: When removing the electrical connector, pull on the connector and not the wire.

5. Remove the motor-to-handle nut and the motor from the vehicle.

6. To install, reverse the removal procedures. Torque the motor-to-handle nut to 4–6 ft. lbs. (5–8 Nm) and the wiper motor-to-glass nut to 11–14 ft. lbs. (15–20 Nm).

Instrument Cluster

For further information on instruments, please refer to "Gauges and Indicators" in the Unit Repair section.

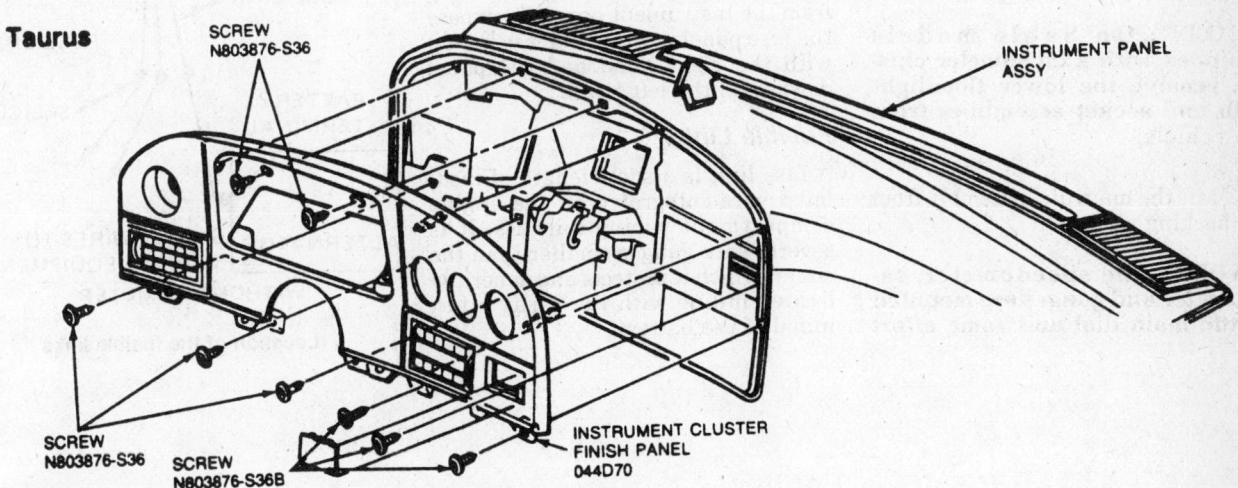

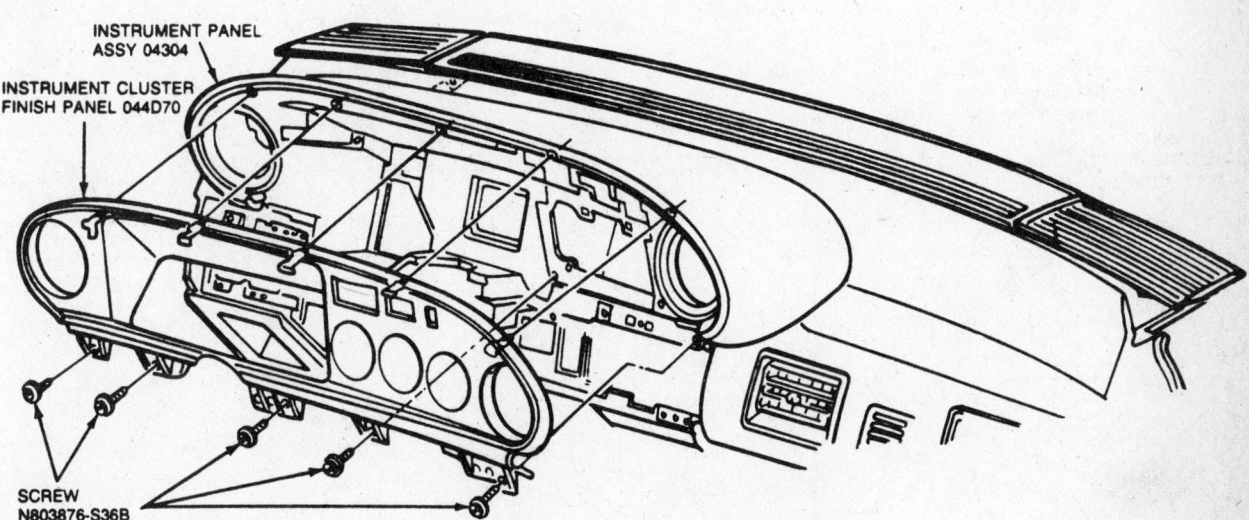

Exploded view of the instrument clusters

REMOVAL & INSTALLATION

1. Disconnect the negative battery cable.

2. Drop the fuse panel (on its hinges), to provide access to the speedometer cable latch attachment. Disconnect the speedometer cable by disengaging the cable latch from the speedometer head and pulling the cable away from the speedometer.

3. Remove the instrument panel finish screws and the instrument finish panel.

4. Remove the steering column shroud.

NOTE: On Sable models equipped with a tachometer cluster, remove the lower trim panel screws and the trim panel from the vehicle.

5. Remove the mask/lens-to-instrument panel screws and the mask/lens from the vehicle.

NOTE: On Sable models equipped with a tachometer cluster, remove the lower floodlight bulb and socket assemblies from the vehicle.

6. Lift the main dial assembly from the backing plate.

NOTE: The speedometer, tachometer and gauges are mounted to the main dial and some effort

may be required to pull the quick-connect electrical terminals from the clip.

7. On column shift vehicles, remove the transmission selector indicator-to-main dial (PRNDL or PRN D D1) screws and the indicator from the vehicle.

8. Remove the instrument cluster-to-instrument panel screws and the instrument cluster from the vehicle.

9. To install, reverse the removal procedures.

Fuses And Fusible Links

LOCATION

Fuses

The fuses are installed on the fuse panel which is located to the left side of the steering column and is hung from the instrument panel. To expose the fuse panel, pull the release bar up with the right hand and the panel down with the left hand.

Fusible Links

A fuse link is a short length of insulated wire integral with the engine compartment wiring harness. It is several wire gauges smaller than the circuit which it protects and generally located in-line with the positive terminal of the battery.

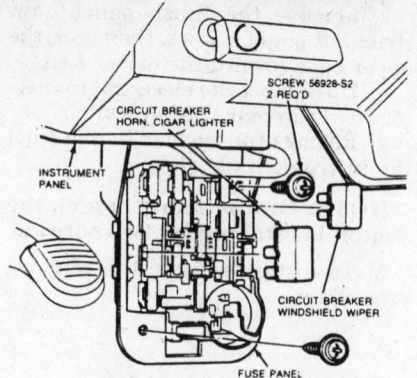

Location of the fuse panel

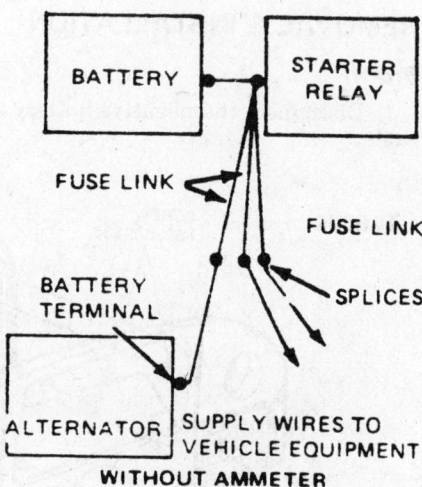

Location of the fusible links

Buick

Rear Wheel Drive
Electra, Century, Regal, LeSabre

YEAR IDENTIFICATION

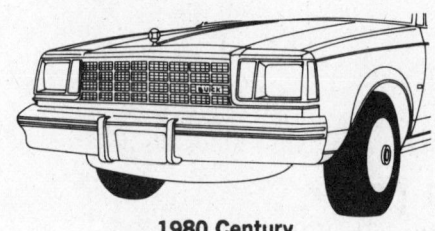

1980 Century

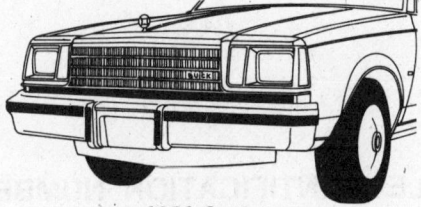

1981 Century

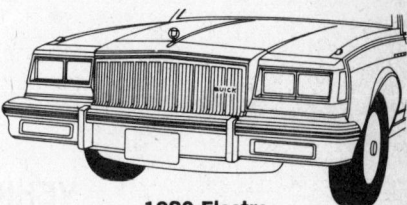

1980 Electra

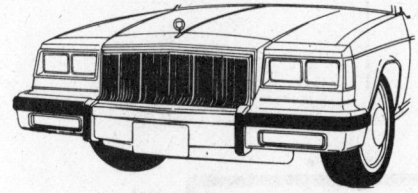

1981 Electra

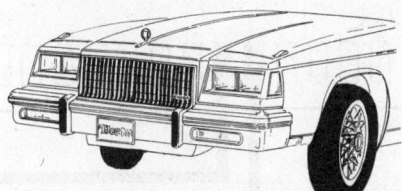

1982–83 Electra

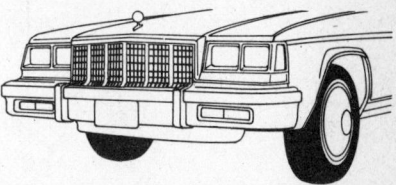

1984 Electra

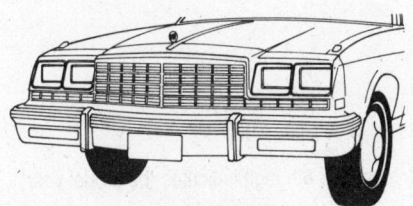

1980 LeSabre

1981 LeSabre

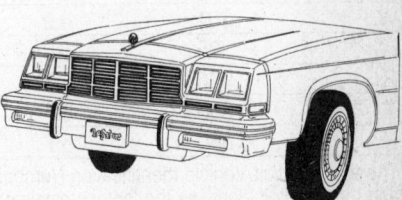

1982–83 LeSabre

YEAR IDENTIFICATION

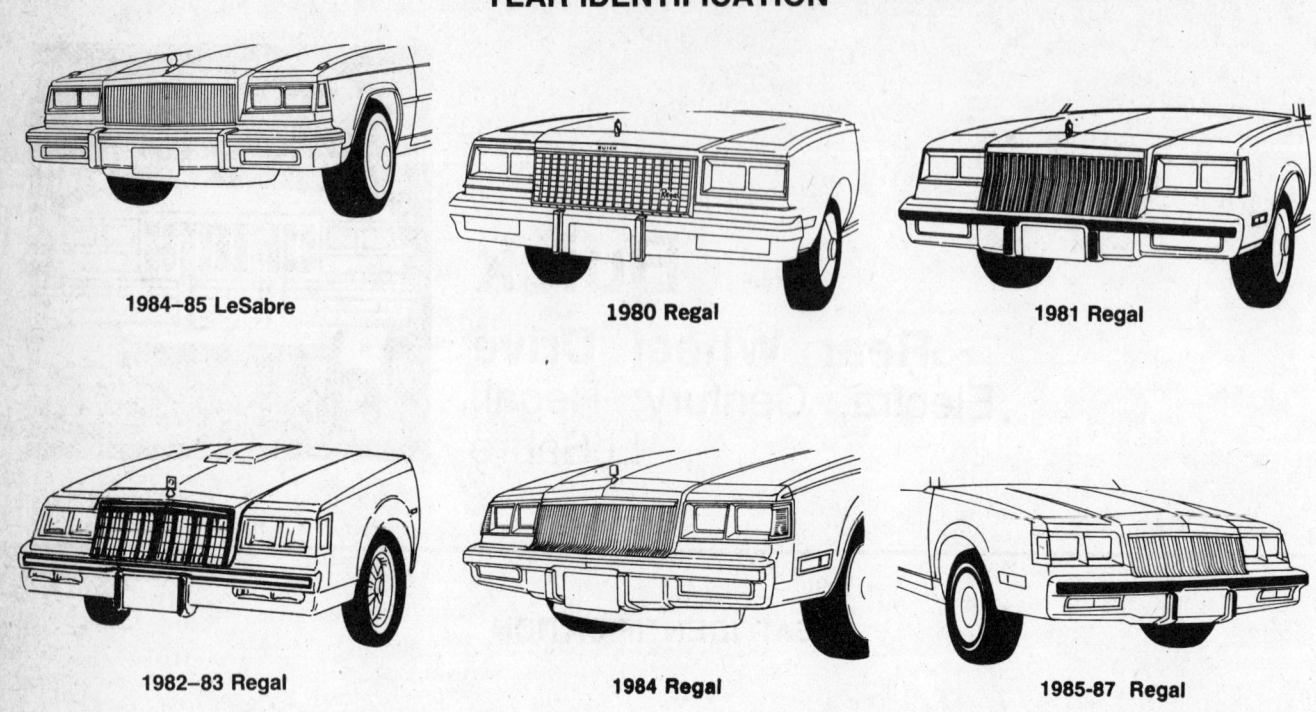

1984–85 LeSabre

1980 Regal

1981 Regal

1982–83 Regal

1984 Regal

1985-87 Regal

VEHICLE IDENTIFICATION NUMBER (VIN)

It is important for servicing and ordering parts to be certain of the vehicle and engine identification. The VIN (vehicle identification number) is a 13 or 17 digit number visible through the windshield on the driver's side of the dash and contains the vehicle and engine identification codes. It can be interpreted as follows:

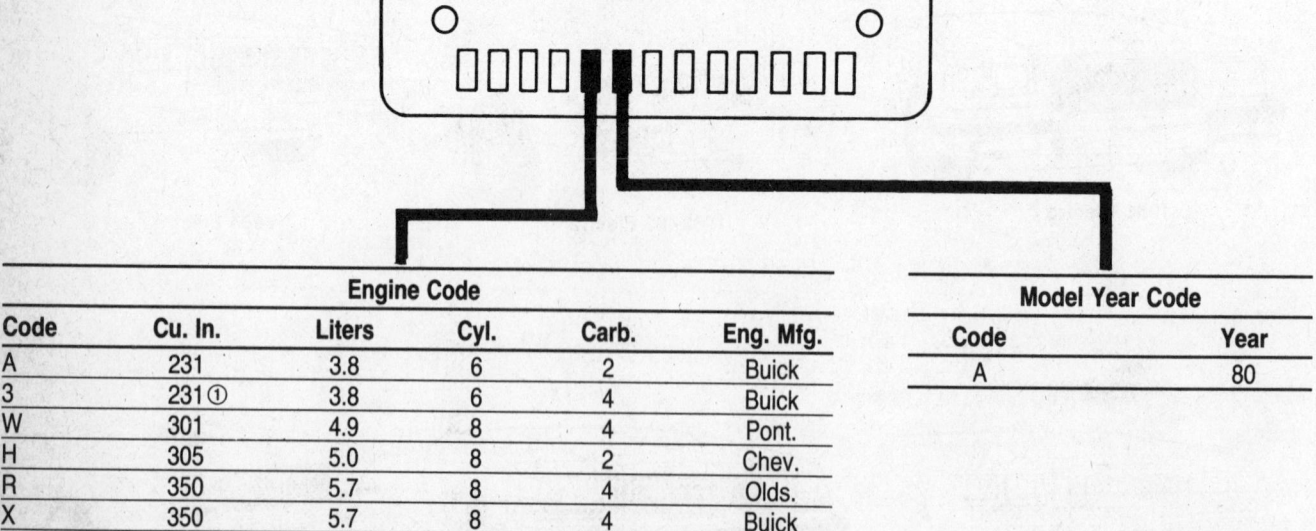

Engine Code							Model Year Code	
Code	Cu. In.	Liters	Cyl.	Carb.	Eng. Mfg.		Code	Year
A	231	3.8	6	2	Buick		A	80
3	231①	3.8	6	4	Buick			
W	301	4.9	8	4	Pont.			
H	305	5.0	8	2	Chev.			
R	350	5.7	8	4	Olds.			
X	350	5.7	8	4	Buick			
N	350	5.7	8	Diesel	Olds.			

The thirteen digit Vehicle Identification Number can be used to determine engine application and model year. The 6th digit indicates the model year, and the 5th digit identifies the factory-installed engine.
① Turbocharged

VEHICLE IDENTIFICATION NUMBER (VIN)

It is important for servicing and ordering parts to be certain of the vehicle and engine identification. The VIN (vehicle identification number) is a 13 or 17 digit number through the windshield on the driver's side of the dash and contains the vehicle and engine identification codes. It can be interpreted as follows:

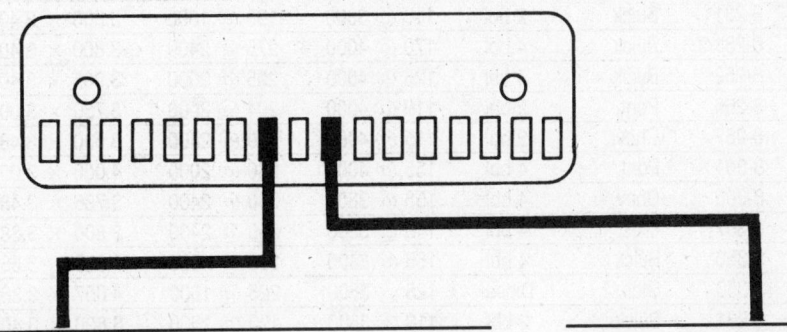

Engine Code

Code	Cu. In.	Liters	Cyl.	Carb.	Eng. Mfg.
A	231	3.8	6	2	Buick
3	231 ①	3.8	6	4	Buick
7	231	3.8	6	SFI	Buick
8	231 ①	3.8	6	4	Buick
9	231 ①	3.8	6	SFI	Buick
4	252	4.1	6	4	Buick
V	263	4.3	6	Diesel	Olds.
S	265	4.3	8	2	Pont.
J	267	4.4	8	2	Chev.
W	301	4.9	8	4	Pont.
H	305	5.0	8	4	Chev.
Y	307	5.0	8	4	Olds.
X	350	5.7	8	4	Buick
N	350	5.7	8	Diesel	Olds.

Model Year Code

Code	Year
B	81
C	82
D	83
E	84
F	85
G	86
H	87

The seventeen digit Vehicle Identification Number can be used to determine engine application and model year. The 10th digit indicates the model year, and the 8th digit identifies the factory-installed engine.
SFI—Sequential Fuel Injection
① Turbocharged

GENERAL ENGINE SPECIFICATIONS

Year	Engine V.I.N. Code	Engine Type	Engine Manufac- turer	Fuel Delivery	Horsepower @ rpm①	Torque @ rpm (ft. lbs.)①	Bore × Stroke (in.)	Compression Ratio	Oil Pressure (psi @ rpm)
'80	A	6-231	Buick	2 bbl	115 @ 3800	190 @ 2000	3.800 × 3.400	8.0:1	37 @ 2400
	3	6-231②	Buick	4 bbl	165 @ 4000	265 @ 2800	3.800 × 3.400	8.0:1	37 @ 2400
	4	6-252	Buick	4 bbl	125 @ 4000	205 @ 2000	3.965 × 3.400	8.0:1	37 @ 2400
	S	8-265	Pont.	2 bbl	120 @ 3600	210 @ 1600	3.750 × 3.000	8.0:1	40 @ 2600
	W	8-301	Pont.	4 bbl	150 @ 4000	240 @ 2000	4.000 × 3.000	8.2:1	40 @ 2600
	H	8-305	Chev.	4 bbl	155 @ 4000	235 @ 2400	3.736 × 3.480	8.5:1	35-40 @ 2400
	N	8-350	Olds.	Diesel	120 @ 3600	220 @ 2200	4.057 × 3.385	22.5:1	37 @ 1500
	R	8-350	Olds.	4 bbl	170 @ 3800	275 @ 2000	4.057 × 3.385	8.5:1	37 @ 1500
	X	8-350	Buick	4 bbl	155 @ 3400	280 @ 1800	3.800 × 3.850	8.0:1	37 @ 2400

GENERAL ENGINE SPECIFICATIONS

Year	Engine V.I.N. Code	Engine Type	Engine Manufacturer	Fuel Delivery	Horsepower @ rpm①	Torque @ rpm (ft. lbs.)①	Bore × Stroke (in.)	Compression Ratio	Oil Pressure (psi @ rpm)
'81	A	6-231	Buick	2 bbl	110 @ 3800	190 @ 1600	3.800 × 3.400	8.0:1	37 @ 2400
	3	6-231②	Buick	4 bbl	170 @ 4000	275 @ 2400	3.800 × 3.400	8.0:1	37 @ 2400
	4	6-252	Buick	4 bbl	125 @ 4000	205 @ 2000	3.965 × 3.400	8.0:1	37 @ 2400
	S	8-265	Pont.	2 bbl	119 @ 4000	204 @ 2000	3.750 × 3.000	8.0:1	35-40 @ 2600
	J	8-267	Chev.	2 bbl	115 @ 4000	200 @ 2400	3.500 × 3.480	8.3:1	45 @ 2000
	W	8-301	Pont.	4 bbl	155 @ 4000	240 @ 2000	4.000 × 3.000	8.2:1	40 @ 2600
	H	8-305	Chev.	4 bbl	155 @ 3800	230 @ 2400	3.736 × 3.480	8.6:1	35-40 @ 2400
	Y	8-307	Olds.	4 bbl	148 @ 3800	250 @ 2400	3.800 × 3.385	8.0:1	40 @ 1500
	X	8-350	Buick	4 bbl	155 @ 3400	280 @ 1800	3.800 × 3.850	8.0:1	37 @ 2400
	N	8-350	Olds.	Diesel	125 @ 3600	225 @ 1600	4.057 × 3.385	22.5:1	40 @ 1500
'82	A	6-231	Buick	2 bbl	110 @ 3800	190 @ 1600	3.800 × 3.400	8.0:1	37 @ 2400
	3	6-231②	Buick	4 bbl	170 @ 3800	275 @ 2600	3.800 × 3.400	8.0:1	37 @ 2400
	4	6-252	Buick	4 bbl	125 @ 4000	205 @ 2000	3.965 × 3.400	8.0:1	37 @ 2400
	V	6-263	Olds.	Diesel	85 @ 3600	165 @ 1600	4.057 × 3.385	21.6:1	30-45 @ 1500
	J	8-267	Chev.	2 bbl	115 @ NA	205 @ NA	3.500 × 3.480	8.3:1	35-40 @ 2600
	H	8-305	Chev.	4 bbl	140 @ 3600	240 @ 1600	3.736 × 3.480	8.0:1	35-40 @ 2600
	Y	8-307	Olds.	4 bbl	150 @ 3800	260 @ 2400	3.800 × 3.385	8.5:1	30-45 @ 1500
	N	8-350	Olds.	Diesel	105 @ 3200	200 @ 1600	4.057 × 3.385	22.5:1	30-45 @ 1500
'83–'84	A	6-231	Buick	2 bbl	110 @ 3800	190 @ 1600	3.800 × 3.400	8.0:1	37 @ 2400
	8	6-231②	Buick	4 bbl	170 @ 3800	275 @ 2600	3.800 × 3.400	8.0:1	37 @ 2400
	9	6-231②	Buick	SFI	190 @ 4000	300 @ 2480	3.800 × 3.400	8.0:1	37 @ 2400
	4	6-252	Buick	4 bbl	125 @ 4000	205 @ 2000	3.965 × 3.400	8.0:1	37 @ 2400
	V	6-263	Olds.	Diesel	85 @ 3600	165 @ 1600	4.057 × 3.385	21.6:1③	30-45 @ 1500
	Y	8-307	Olds.	4 bbl	150 @ 3800	260 @ 2400	3.800 × 3.385	8.5:1	30-45 @ 1500
	N	8-350	Olds.	Diesel	105 @ 3200	200 @ 1600	4.057 × 3.385	22.5:1	30-45 @ 1500
'85	A	6-231	Buick	2 bbl	110 @ 3800	190 @ 1600	3.800 × 3.400	8.0:1	37 @ 2400
	9	6-231②	Buick	SFI	190 @ 4000	300 @ 2400	3.800 × 3.400	8.0:1	37 @ 2400
	V	6-263	Olds.	Diesel	85 @ 3600	165 @ 1600	4.057 × 3.385	22.5:1	30-45 @ 1500
	Y	8-307	Olds.	4 bbl	140 @ 3600	240 @ 1600	3.800 × 3.385	8.0:1	30-45 @ 1500
	N	8-350	Olds.	Diesel	105 @ 3200	200 @ 1600	4.057 × 3.385	22.5:1	30-45 @ 1500
'86–'87	A	6-231	Buick	2 bbl	110 @ 3800	190 @ 1600	3.800 × 3.400	8.0:1	37 @ 2400
	7	6-231	Buick	SFI	235 @ 4400	330 @ 2800	3.800 × 3.400	8.0:1	37 @ 2400
	Y	8-307	Olds.	4 bbl	140 @ 3200	255 @ 2000	3.800 × 3.385	7.99:1	30 @ 1500

NA—Not available at time of publication.
SFI—Sequential Fuel Injection
① Horsepower and torque are SAE net figures. They are measured at the rear of the transmission with all accessories installed and operating. Since the figures vary when a given engine is installed in different models, some are representative rather than exact.
② Turbocharged engine
③ 1984—22.5:1

GASOLINE ENGINE TUNE-UP SPECIFICATIONS

(When analyzing compression test results, look for uniformity among cylinders rather than specific pressures.)

Year	Engine V.I.N. Code	Engine Type	Engine Manufacturer	Spark Plugs Type①	Gap (in.)	Ignition Timing (deg. B.T.D.C.)② Manual Transmission③	Automatic Transmission③	Intake Valve Opens (°B.T.D.C.)	Fuel Pump Pressure (psi)	Idle Speed (rpm)④ Manual Transmission③	Automatic Transmission③
'80	A	6-231	Buick	R45TSX	.060	15 @ 550	15 @ 550	16	5½–6½	800/600	670/550⑤ (620/550)⑥ 550⑦
	3	6-231	Buick	R45TS	.040	—	15 @ 650	16	5½–6½	—	650
	4	6-252	Buick	R45TSX	.060	—	15 @ 550	—	5½–6½	—	680/550⑧ 550⑦
	S	8-265	Pont.	R45TSX	.060	—	10 @ 700	27	7–8½	—	650/550⑧ 550⑦
	W	8-301	Pont.	R45TSX	.060	—	12 @ 500	27	7–8½	—	650/500⑧ 550⑦
	H	8-305	Chev.	R45TS	.035	—	4 @ 550	28	7½–9	—	650/550⑧ 550⑦
	R	8-350	Olds.	R46SX or R47SX	.080 .080	—	18 @ 1100	16	5½–6½	—	650/550
	X	8-350	Buick	R45TSX	.060	—	15 @ 550	13½	6–7½	—	550
'81	A	6-231	Buick	R45TS8	.080	—⑨—		16	5½–6½	—⑨—	
	3	6-231	Buick	R45TS	.040	—⑨—		16	5½–6½	—⑨—	
	4	6-252	Buick	R45TS8	.080	—⑨—		16	5½–6½	—⑨—	
	S	8-265	Pont.	R45TSX	.060	—⑨—		27	5½–6½	—⑨—	
	J	8-267	Chev.	R45TS	.045	—⑨—		28	5½–6½	—⑨—	
	W	8-301	Pont.	R45TSX	.060	—⑨—		27	5½–6½	—⑨—	
	H	8-305	Chev.	R45TS	.045	—⑨—		28	5½–6½	—⑨—	
	Y	8-307	Olds.	R45TS4	.060	—⑨—		20	5½–6½	—⑨—	
	X	8-350	Buick	R45TSX	.060	—⑨—		13½	5½–6½	—⑨—	
'82	A	6-231	Buick	R45TS8	.080	—⑨—		16	5½–6½	—⑨—	
	3	6-231	Buick	R45TSX	.060	—⑨—		16	5½–6½	—⑨—	
	4	6-252	Buick	R45TS8	.080	—⑨—		16	5½–6½	—⑨—	
	J	8-267	Chev.	R45TS	.045	—⑨—		28	5½–6½	—⑨—	
	H	8-305	Chev.	R45TS	.045	—⑨—		28	5½–6½	—⑨—	
	Y	8-307	Olds.	R46SX	.080	—⑨—		20	5½–6½	—⑨—	
'83	A	6-231	Buick	R45TS8	.080	—⑨—		16	5½–6½	—⑨—	
	8	6-231	Buick	R45TSX	.060	—⑨—		16	5½–6½	—⑨—	
	4	6-252	Buick	R45TS8	.080	—⑨—		16	5½–6½	—⑨—	
	Y	8-307	Olds.	R46SX	.080	—⑨—		20	5½–6½	—⑨—	
'84	A	6-231	Buick	R45TSX	.060	—⑨—		16	5½–6½	—⑨—	
	9	6-231	Buick	R44TS	.045	—⑨—		16	26–51	—⑨—	
	4	6-252	Buick	R45TSX	.060	—⑨—		16	5½–6½	—⑨—	
	Y	8-307	Olds.	R46SX	.080	—⑨—		20	5½–6½	—⑨—	
'85–'87	A	6-231	Buick	R45TSX	.060	—⑨—		16	5½–6½	—⑨—	
	7,9	6-231	Buick	R44TS	.045	—⑨—		16	26–51	—⑨—	
	Y	8-307	Olds	R45TS	.060	—⑨—		20	5½–6½	—⑨—	

NOTE: The underhood specifications sticker often reflects tune-up specification changes made in production. Sticker figures must be used if they disagree with those in this chart.

① All models use electronic ignition systems.
B.T.D.C.—Before top dead center (No. 1 cylinder)
C.I.D.—Cubic inch displacement
Min.—Minimum
Part numbers in this chart are not recommendations by Chilton for any product by brand name.

② On some models, the engine must be held at a specific rpm to accurately check and adjust the ignition timing. See the text for specific procedures.

③ Figure in parenthesis ( ) indicates a special figure for California models; figure in brackets

[ ] indicates a special figure for high-altitude models.

④ Most 1979 and later carburetors have idle mixture screws concealed with hardened steel plugs. Normal idle mixture adjustments are not required on these carburetors.

⑤ With air conditioning, 49 states models only
⑥ With air conditioning, California models only
⑦ All models without air conditioning
⑧ All models with air conditioning

⑨ On vehicles equipped with computerized emissions systems (which have no distributor vacuum advance unit), the idle speed and ignition timing are controlled by the emissions computer.

DIESEL ENGINE TUNE-UP SPECIFICATIONS

Year	Engine V.I.N. Code	Engine No. of cyl.- Displacement- Manufacturer	Fuel Pump Pressure (psi)①	Injection Nozzle Pressure (psi)	Compression Pressure (psi)②	Injection Pump Setting (deg)	Intake Valve Opens (°B.T.D.C.)	Idle Speed (rpm)
'80–'85	N	8-350 Olds	5.5–6.5	1225	275 minimum	④	16	⑤
'82–'85	V	6-260 Olds	5.8–8.7	③	275 minimum	6ATDC	16	⑤

NOTE: The underhood specifications sticker often reflects tune-up specification changes made in production. Sticker figures must be used if they disagree with those in this chart.

B.T.D.C.—Before top dead center (No. 1 cylinder)

① Fuel transfer pump pressure given. Injection pump reading should be 8–12 psi @ 1000 rpm taken at injection pump pressure tap

② The lowest cylinder reading must not be less than 70% of the highest cylinder reading.

③ Injector opening pressure:
   New nozzle (green band/no color)—1000 psi
   New nozzle (red band)—800 psi
   Used nozzle—200 psi less than new

④ Align the marks on injection pump and drive flange on 1980–81 models. 4ATDC on 1982 and later models

⑤ See the underhood specifications sticker.

FIRING ORDERS

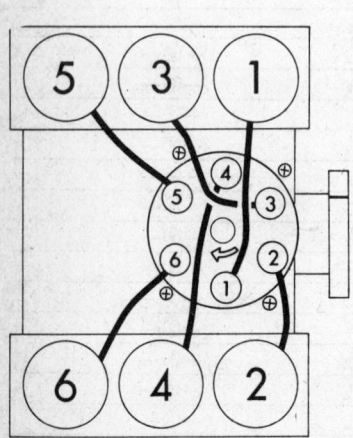

GM (Buick) 231, 252 V6
Engine firing order: 1-6-5-4-3-2
Distributor rotation: clockwise

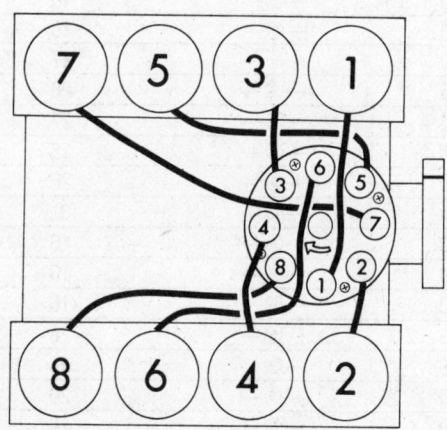

Buick-manufactured V8 engines
Engine firing order: 1-8-4-3-6-5-7-2
Distributor rotation: clockwise

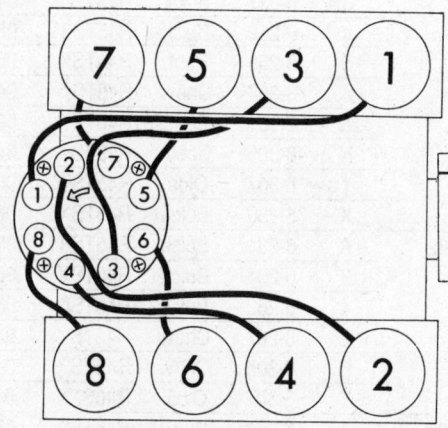

Oldsmobile and Pontiac-manufactured V8 engines Engine firing order: 1-8-4-3-6-5-7-2 Distributor rotation: counterclockwise.

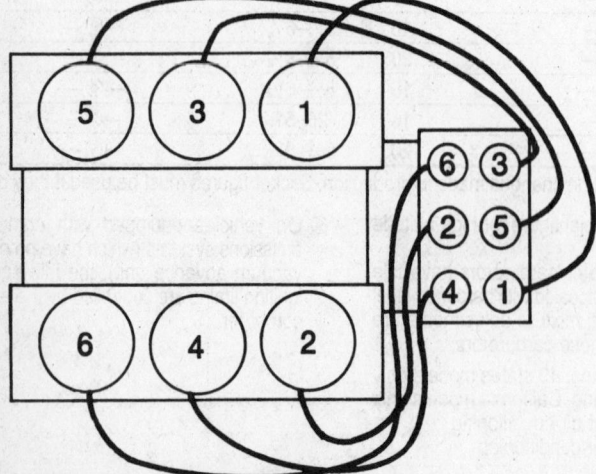

Buick V6 with C3I ignition system
Engine firing order: 1-6-5-4-3-2

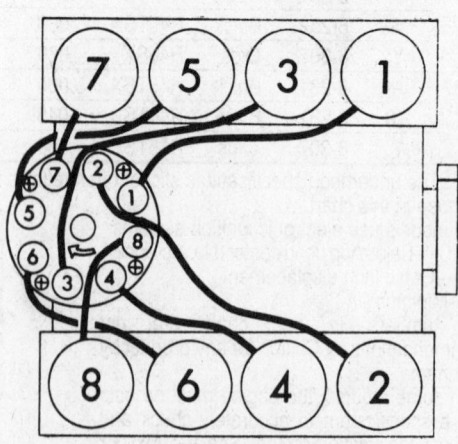

Chevrolet-manufactured V8 engines
Engine firing order: 1-8-4-3-6-5-7-2
Distributor rotation: clockwise

CAPACITIES
Electra and LeSabre

Year	Engine No. Cyl. Displacement (cu. in.)	Engine Crankcase Add 1 qt. for New Filter	Transmission Pts to Refill After Draining — Manual 3-Speed	4-Speed	Automatic ●	Drive Axle (pts.)	Gasoline Tank (gals.)	Cooling System (qts.) With Heater	With A/C	Heavy Duty
'80–'81	6-231 Buick	4	—	—	③	①	25.0②	13.0	13.0	13.0
	6-252 Buick	4	—	—	③	①	25.0②	13.0	13.0	13.0
	8-301 Pont.	4	—	—	③	①	25.0②	18.9	18.9	18.9
	8-307 Olds.	4	—	—	③	①	25.0②	15.6	16.3	16.0
	8-350 Buick	4	—	—	③	①	25.0②	14.3	14.2	14.7
	8-350 Olds.	4	—	—	③	①	25.0②	—	14.5	15.2
	8-350 Diesel	6	—	—	⑦	①	23.0④	18.3	18.0	18.0
'82–'84	6-231, 6-252	4	—	—	③	①	25	13	13.1	—
	8-307 Olds.	4	—	—	③	①	25	15.4	16.2	16.1
	8-350 Diesel	6	—	—	③	①	25	—	17.9	—
'85–'86	6-231	4	—	—	③	①	25	13	13.1	—
	8-307 Olds.	4	—	—	③	①	25	15.4	16.2	16.1
	8-350 Diesel	6	—	—	③	①	25	—	17.9	—

● Specifications do not include torque converter
—: Not applicable
① 7.5 in. ring gear—3.5; 8.5 in. ring gear—4.25; 8.75 in. ring gear—5.4
② Estate wagon—22 gal.
③ Add 6 pts, start engine and allow to warm up, fill as required
④ Wagon—27 gal.

CAPACITIES
Century and Regal

Year	Engine No. Cyl. Displacement (cu. in.)	Engine Crankcase Add 1 qt. for New Filter	Transmission Pts to Refill After Draining — Manual 3-Speed	4-Speed	Automatic ●	Drive Axle (pts.)	Gasoline Tank (gals.)	Cooling System (qts.) With Heater	With A/C	Heavy Duty
'80	6-231 Buick②	4	3.5	—	③	⑤	18.1①①	13.4	13.4	—
	8-265 Pont.	4①	—	—	③	⑤	18.0	20.3	21.0	—
	8-301 Pont.	4①	—	—	③	⑤	18.0	20.3	21.0	20.8
	8-305 Chev.	4	—	—	③	⑤	18.0	17.6	—	18.1
'81	6-231 Buick	4	3.5	—	③	⑤	18.1④	13.4	13.4	—
	6-252 Buick	4	3.5	—	③	⑤	18.1④	13.0	13.0	—
	8-265 Pont.	4	—	—	③	⑤	25⑦	20.3	21.0	—
	8-350 Diesel	7	—	—	③	⑤	—	—	17.3	—
'82	6-231 Buick	4	—	—	⑧	⑤	18.1	13	13.1	—
	6-252 Buick	4	—	—	⑧	⑤	18.1	13	13.1	—
	6-263 Diesel	6	—	—	⑧	⑤	18.1⑥	—	14.8	—
	8-267 Chev.	4	—	—	⑧	⑤	18.1⑥	—	21	—
	8-305 Chev.	4	—	—	⑧	⑤	18.1⑥	—	19	—
'83–'84	6-231 Buick	4	—	—	⑧	⑤	19	13	13.1	—
	6-252 Buick	4	—	—	⑧	⑤	19	13	13.1	—
	6-263 Diesel	6	—	—	⑧	⑤	19	—	14.8	—

CAPACITIES
Century and Regal

Year	Engine No. Cyl. Displacement (cu. in.)	Engine Crankcase Add 1 qt. for New Filter	Transmission Pts to Refill After Draining			Drive Axle (pts.)	Gasoline Tank (gals.)	Cooling System (qts.)		
			Manual		Automatic ●			With Heater	With A/C	Heavy Duty
			3-Speed	4-Speed						
'85	6-231 Buick	4	—	—	⑧	⑤	19	13	13.1	—
	6-263 Diesel	6	—	—	⑧	⑤	19	—	14.8	—
'86–'87	6-231 Buick	4	—	—	⑧	⑤	18	13	13	—
	6-231 Buick (SFI)	5	—	—	⑧	⑤	18	12	12	—
	8-307 Olds	4	—	—	⑧	⑤	18	15.5	15	16

● Specifications do not include torque convertor
— Not applicable
SFI Sequential Fuel Inj.
① 4 quarts total
② Century and Regal
③ THM 200—6; THM 350—3

④ Wagon—18.2
⑤ 7.5 in. ring gear—3.5; 8.5 in. ring gear—4.25; 8.75 in. ring gear—5.4
⑥ Station wagon—18.2 gal.
⑦ Station wagon—22 gal.
⑧ THM 200 & 200R-4—7 pts.; THM 250C—8 pts.; THM 350C—6.3 pts.

VALVE SPECIFICATIONS

Year	Engine No. Cyl. Displacement (cu. in.)	Seat Angle (deg.)	Face Angle (deg.)	Spring Test Pressure (lbs @ in.)	Spring Installed Height (in.)	Stem-to-Guide Clearance (in.)		Stem Diameter (in.)	
						Intake	Exhaust	Intake	Exhaust
'80–'81	6-196 Buick	45	45	164 @ 1.340	1⁴⁷⁄₆₄	.0015–.0032	.0015–.0032	.3405–.3412	.3405–.3412
	6-231 Buick	45	45	164 @ 1.34④	1⁴⁷⁄₆₄	.0015–.0035	.0015–.0032	.3401–.3412	.3405–.3412
	6-252 Buick	45	45	164 @ 1.34④	1⁴⁷⁄₆₄	.0015–.0035	.0015–.0032	.3401–.3412	.3405–.3412
	8-265 Pont.	46	45	170 @ 1.260	1⁴⁷⁄₆₄	.0017–.0020	.0017–.0020	.3400	.3400
	8-301 Pont.	46	45	170 @ 1.260	1⁴⁷⁄₆₄	.0017–.0020	.0017–.0020	.3400	.3400
	8-305 Chev.	46	45	200 @ 1.250	1²³⁄₃₂	.0010–.0037	.0010–.0047	.3410	.3410
	8-307 Olds.	45②	46③	187 @ 1.27	1⁴⁷⁄₆₄	.0010–.0027	.0015–.0032	.3428	.3424
	8-350 Buick	45	45	180 @ 1.340	1⁴⁷⁄₆₄	.0015–.0035	.0015–.0032	.3720–.3730	.3723–.3730
	8-350 Olds.	45①	44①	187 @ 1.27	1⁴⁷⁄₆₄	.0010–.0027	.0015–.0032	.3425–.3432	.3420–.3427
	8-350 Olds. Diesel	45①⑦	44①⑦	151 @ 130⑤	1⁴⁷⁄₆₄	.0010–.0027	.0015–.0032	.3425–.3432	.3420–.3427
	8-403 Olds.	45①	44①	187 @ 1.27	1⁴⁷⁄₆₄	.0010–.0027	.0015–.0032	.3425–.3432	.3420–.3427
'82	6-231 Buick	45	45	182 @ 1.340	1⁴⁷⁄₆₄	.0015–.0035	.0015–.0032	.3407	.3409
	6-252 Buick	45	45	164 @ 1.34④	1⁴⁷⁄₆₄	.0015–.0035	.0015–.0032	.3401–.3412	.3405–.3412
	6-263 Olds. Diesel	45①	44①	217 @ 1.220 in.	—	.0010–.0027	.0015–.0032	.3425–.3432	.3420–.3427
	8-267 Chev.	46	45	180 @ 1.25	1²³⁄₃₂	.0010–.0027	.0010–.0027	.3414	.3414
	8-305 Chev.	46	45	180 @ 1.25	1²³⁄₃₂⑥	.0010–.0027	.0010–.0027	.3414	.3414
	8-307 Olds.	45①	44①	187 @ 1.27	1⁴⁷⁄₆₄	.0010–.0027	.0015–.0032	.3425–.3432	.3400–.3427
	8-350 Diesel	45①	44①	210 @ 1.23	1⁴⁷⁄₆₄	.0010–.0027	.0015–.0032	.3425–.3432	.3400–.3427
'83	6-231 Buick	45	45	182 @ 1.34⑧	1⁴⁷⁄₆₄	.0015–.0035	.0015–.0032	.3401–.3412	.3405–.3412
	6-252 Buick	45	45	182 @ 1.34	1⁴⁷⁄₆₄	.0015–.0035	.0015–.0032	.3401–.3412	.3405–.3412
	8-307 Olds.	45①	44①	187 @ 1.27	1⁴⁷⁄₆₄	.0010–.0027	.0015–.0032	.3425–.3432	.3420–.3427
	6-263 Diesel	45①	44①	209 @ 1.22	—	.0010–.0027	.0015–.0032	.3425–.3432	.3420–.3427
	8-350 Diesel	45①	44①	209 @ 1.22	1⁴⁷⁄₆₄	.0010–.0027	.0015–.0032	.3425–.3432	.3420–.3427
'84	6-231 Buick	45	45	182 @ 1.34⑧	1⁴⁷⁄₆₄	.0015–.0035	.0015–.0032	.3401–.3412	.3405–.3412
	6-252 Buick	45	45	182 @ 1.34	1⁴⁷⁄₆₄	.0015–.0035	.0015–.0032	.3401–.3412	.3405–.3412
	8-307 Olds.	45①	44①	187 @ 1.27	1⁴⁷⁄₆₄	.0010–.0027	.0015–.0032	.3425–.3432	.3420–.3427
	6-263 Diesel	45①	44①	209 @ 1.22	—	.0010–.0027	.0015–.0032	.3425–.3432	.3420–.3427
	8-350 Diesel	45①	44①	209 @ 1.22	1⁴⁷⁄₆₄	.0010–.0027	.0015–.0032	.3425–.3432	.3420–.3427

VALVE SPECIFICATIONS

Year	Engine No. Cyl. Displacement (cu. in.)	Seat Angle (deg.)	Face Angle (deg.)	Spring Test Pressure (lbs @ in.)	Spring Installed Height (in.)	Stem-to-Guide Clearance (in.) Intake	Stem-to-Guide Clearance (in.) Exhaust	Stem Diameter (in.) Intake	Stem Diameter (in.) Exhaust
'85	6-231 Buick	45	45	182 @ 1.34⑧	1⁴⁷⁄₆₄	.0015–.0035	.0015–.0032	.3401–.3412	.3405–.3412
	8-307 Olds.	45①	44①	187 @ 1.27	1⁴⁷⁄₆₄	.0010–.0027	.0015–.0032	.3425–.3432	.3420–.3427
	6-263 Diesel	45①	44①	209 @ 1.22	—	.0010–.0027	.0015–.0032	.3425–.3432	.3420–.3427
	8-350 Diesel	45①	44①	209 @ 1.22	1⁴⁷⁄₆₄	.0010–.0027	.0015–.0032	.3425–.3432	.3420–.3427
'86–'87	6-231 Buick	45	45	182 @ 1.34⑧	1⁴⁷⁄₆₄	.0015–.0035	.0015–.0032	.3412–.3401	.3412–.3405
	8-307 Olds	45②	46③	187 @ 1.27	1⁴⁷⁄₆₄	.0010–.0027	.0015–.0032	.3425–.3432	.3420–.3427

① Exhaust valve seat angle—31°, exhaust valve face angle—30°
② Exhaust—59°
③ Exhaust—60°
④ Exhaust: 182 @ 1.34
⑤ 1981 210 @ 1.23
⑥ Exhaust—1¹⁹⁄₃₂
⑦ Seat: Intake—45°, exhaust—59°
Face: Intake—46°, exhaust—60°
⑧ 6-231 with Sequential Fuel Injection— 185 @ 1.34

TORQUE SPECIFICATIONS

(All readings in ft. lbs.)

Year	Engine No. Cyl. Displacement (cu in.)	Cylinder Head Bolts	Rod Bearing Bolts	Main Bearing Bolts	Crankshaft Bolt	Flywheel-to-Crankshaft Bolts	Manifold Intake	Manifold Exhaust
'80–'81	6-231 Buick	80	40	100	225②	60	45	25
	6-252 Buick	80	40	100	225②	60	45	25
	6-265 Pont.	95	35	70①	160	95	40	35
	8-301 Pont.	90	35	60①	160	95	40	35
	8-305 Chev.	65	45	70	60	60	30	20
	8-350 Buick	80	40	100	225②	60	45	25
	8-350 Chev.	65	45	70	60	60	30	20
	8-307, 350 Olds.	130	42	80③	255②	60④	40	25
	8-350 Diesel	130	42	120	200–310②	60	40	25
'82	6-231 Buick	80	40	100	225	60	45	25
	6-252 Buick	80	40	100	225②	60	45	25
	6-263 Diesel	142⑤	42	107	160–350	48	41	29
	8-267 Chev.	65	45	70	60	60	30	20
	8-301 Pont.	90	35	60①	160	95	40	35
	8-305 Chev.	65	45	70	60	60	30	20
	8-350 Buick	80	40	100	225②	60	45	25
	8-350 Chev.	65	45	70	60	60	30	20
	8-307, 350 Olds.	130	42	80③	255②	60④	40	25
	8-350 Diesel	130	42	120	200–310②	60	40	25
'83–'84	6-231 Buick	80	40	100	200	60	47	25
	6-252 Buick	80	40	100	225	60	45	25
	6-263 Diesel	142⑤	42	89⑥	203–350	⑦	41	28
	8-307 Olds.	125	42	80③	200–310	60	40	25
	8-350 Diesel	130	42	120	200–310	60	40	25
'85	6-231 Buick	80	40	100	200	60	47	25
	8-307 Olds.	125	42	80③	200–310	60	40	25
	6-263 Diesel	142⑤	42	89⑥	203–350	⑦	41	28
	8-350 Diesel	130	42	120	200–310	60	40	25
'86–'87	6-231 Buick	⑨	40	100	200	60	45	20
	8-307 Olds	125⑧	42	80③	200–310	60	40⑧	25

① Rear main—100 ft. lbs.
② Fan pulley to balancer—20 ft. lbs.
③ Rear main—120 ft. lbs.
④ Manual transmission—90 ft. lbs.
⑤ No. 5, 6, 11, 12, 13 and 14—59 ft. lbs.
⑥ No. 2 and 3 outer—52 ft. lbs. Type II bolts
⑦ VIN T—76 ft. lbs.
VIN V—57 ft. lbs.
Torque converter—46 ft. lbs.
⑧ Clean and dip entire bolt in engine oil before tightening
⑨ See procedure in Text

PISTON AND RING SPECIFICATIONS

(All measurements are given in inches. To convert inches to metric units, refer to the Metric Conversion Chart.)

Year	V.I.N. Code	Engine Type/ Disp. cu. in.	Eng. Mfg.	Piston-to-Bore Clearance	Ring Gap			Ring Side Clearance		
					Top Compression	Bottom Compression	Oil Control	Top Clearance	Bottom Clearance	Oil Control
'80	A,3	6-231	Buick	.0008–.0020	.013–.023	.013–.023	.015–.035	.0030–.0050	.0030–.0050	.0035 max
'81–'87	A,3, 7,8,9	6-231	Buick	.0008–.0020 ①	.010–.020	.010–.020	.015–.055	.0030–.0050	.0030–.0050	.0035 max
'81–'84	4	6-252	Buick	.0008–.0020	.010–.020	.010–.020	.015–.055	.0030–.0050	.0030–.0050	.0035 max
'82–'85	V	6-263	Olds Diesel	.0035–.0045	.019–.027	.013–.021	.015–.055	.0050–.0070	.0030–.0050	.001–.005
'79–'80	S	8-265	Pont.	.0025–.0033	.010–.020	.010–.020	.035 max	.0015–.0035	.0015–.0035	.0015–.0035
'81	S	8-265	Pont.	.0025–.0033	.010–.028	.010–.028	.015–.055	.0015–.0035	.0015–.0035	.0015–.0035
'81–'82	J	8-267	Chev.	.0025–.0033	.010–.020	.010–.025	.015–.055	.0012–.0032	.0012–.0032	.0020–.0080
'80–'81	W	8-301	Pont.	.0025–.0033	.010–.020	.010–.020	.035 max	.0015–.0035	.0015–.0035	.0015–.0035
'80–'82	H	8-305	Chev.	.0027 max	.010–.030	.010–.035	.015–.065	.0012–.0032	.0012–.0032	.0020–.0080
'81–'87	Y	8-307	Olds.	.0008–.0018	.009–.019 ②	.009–.019 ②	.015–.055 ③	.0020–.0040	.0020–.0040	.0010–.0050
'80–'81	X	8-350	Buick	.0008–.0020	.010–.020 ④	.010–.020 ④	.015–.035	.0030–.0050	.0030–.0050	.0035 max
'80	R	8-350	Olds.	.0010–.0020	.010–.023 ⑤	.010–.023 ⑤	.015–.055	.0020–.0040	.0020–.0040	.0010–.0050
'80–'85	N	8-350	Olds Diesel	.0035–.0045	.015–.025	.015–.025	.015–.055	.0040–.0060 ⑥	.0018–.0038 ⑦	.0010–.0050

① 1985–86 6-231 Turbo: .0022–.0034 in.
② With TRW rings: .010–.025 in.
③ With TRW rings: .010–.025 in.
④ 1980: .103–.023 in.
⑤ With Sealed Power rings: .010–.020 in.
⑥ 1982–85: .0050–.0070 in.
⑦ 1982–85: .0030–.0050 in.

CRANKSHAFT AND CONNECTION ROD SPECIFICATIONS

(All measurements are given in inches.)

Year	Engine Displacement (cu in.)	Crankshaft				Connecting Rod		
		Main Brg. Journal Dia	Main Brg. Oil Clearance	Shaft End-Play	Thrust on No.	Journal Diameter	Oil Clearance	Side Clearance
'80–'81	6-231 Buick	2.4995	.0003–.0018	.003–.009 ⑤	2	2.2487–2.2495	.0005–.0026	.006–.023 ⑪
	265 Pont.	3.000	.0004–.0020 ⑫	.006–.022 ⑦	4	2.250 ⑧	.0005–.0025	.006–.022 ⑪
	6-252 Buick	2.4995	.0003–.0018	.003–.009	2	2.2487–2.2495	.0005–.0026	.006–.023
	8-301 Pont.	3.000	.0004–.0020	.006–.022	4	2.250	.0005–.0025	.006–.022
	8-305 Chev.	③	④	.002–.006	5	2.099–2.100	.0035 max	.006–.014
	8-307 Olds.	2.49793–2.4998 ⑥	.0005–.0021 ①	.0035–.0135	3	2.1238–2.1248	.0004–.0033	.006–.020
	8-350 Buick	3.000	.0004–.0015	.003–.009	3	1.991–2.000	.0005–.0026	.006–.023
	8-350 Olds	2.4985–2.4995 ②	.0005–.0021 ①	.0035–.0135	3	2.1238–2.1248	.0004–.0033	.006–.020
	8-350 Diesel	2.9993–3.0003	.0005–.0021 ①	.0035–.0135	3	2.2495–2.2500	.0005–.0026	.006–.020

CRANKSHAFT AND CONNECTION ROD SPECIFICATIONS

(All measurements are given in inches.)

Year	Engine Displacement (cu in.)	Crankshaft				Connecting Rod		
		Main Brg. Journal Dia	Main Brg. Oil Clearance	Shaft End-Play	Thrust on No.	Journal Diameter	Oil Clearance	Side Clearance
'82	6-231 Buick	2.4995	.0003–.0018	.003–.009	2	2.2487–2.2495	.0005–.0026	.006–.023
	6-252 Buick	2.4995	.0003–.0018	.003–.009	2	2.2487–2.2495	.0005–.0026	.006–.023
	6-263 Diesel	2.9993–3.0003	.0005–.0021	.0035–.0135	3	2.1238–2.2148	.0005–.0026	.006–.020
	8-267 Chev.	③	.0008–.0020⑨ -	.002–.006	5	2.0986–2.0998	.0013–.0035	.006–.014
	8-305 Chev.	③	.0008–.0020	.002–.006	5	2.0986–2.0998	.0013–.0035	.006–.014
	8-307 Olds	2.4973–2.4998⑥	.0005–.0021①	.0035–.0135	3	2.1238–2.1248	.0004–.0033	.006–.020
	8-350 Diesel	2.9993–3.0003	.0005–.0021	.0035–.0135	3	2.2495–2.2500	.0005–.0026	.006–.020
'83–'84	6-231 Buick	2.4995	.0003–.0018	.003–.011	2	2.2487–2.2495	.0005–.0026	.006–.023⑪
	6-252 Buick	2.4995	.0003–.0018	.003–.011	2	2.2487–2.2495	.0005–.0026	.006–.023
	6-263 Diesel	2.9993–3.0003	.0005–.0021	.0035–.0135	3	2.1238–2.2148	.0005–.0026	.006–.020
	8-307 Olds	2.4973–2.4998⑥	.0005–.0021①	.0035–.0135	3	2.1238–2.1248	.0004–.0033	.006–.020
	8-350 Diesel	2.9993–3.0003	.0005–.0021⑩	.0035–.0135	3	2.2495–2.2500	.0005–.0026	.006–.020
'85	6-231 Buick	2.4995	.0003–.0018	.003–.011	2	2.2487–2.2495	.0005–.0026	.003–.015
	6-263 Diesel	2.9993–3.0003	.0005–.0021	.0035–.0135	3	2.2498–2.249	.0004–.0026	.008–.021
	8-307	2.4985–2.4995⑥	.0005–.0021①	.003–.013	3	2.1238–2.1248	.0005–.0026	.006–.020
	8-350 Diesel	2.9993–3.0003	.0005–.0021⑩	.0035–.0135	3	2.2495–2.5000	.0005–.0026	.006–.020
'86–'87	6-231 Buick	2.4995	.0003–.0018	.003–.011②		2.2487–2.2495	.0005–.0026	.003–.015
	8-307 Olds	2.4988–2.4998⑬	.0005–.0021①	.0035–.0135	3	2.1238–2.1248	.0004–.0033	.006–.020

① Number five main bearing clearance—.0015–.0031
② No. 1:2.4988–2.4998
③ No. 1:2.4484–2.4493 2,3,4:2.4481–2.4490 No. 5:2.4479–2.4488
④ 1980: No. 1:.0015, No. 2,3,4:.0025, No. 5:.0035
⑤ 1981—.003–.011
⑥ 2.4990–2.4995 (Nos. 2,3,4,5)
⑦ 1981—.0035–.0085
⑧ 1981—2.000
⑨ Intermediate—.0011–.0023 Rear—.0017–.0033
⑩ .0020–.0034 No. 5
⑪ Total for both rods per journal
⑫ 1981—.0002–.0018
⑬ 2.4985–2.4995 (Nos. 2,3,4,5)

WHEEL ALIGNMENT SPECIFICATIONS
Electra and LeSabre

Year	Model	Caster		Camber		Toe-in (in.)
		Range (deg)	Pref. Setting (deg)	Range (deg)	Pref. Setting (deg)	
'80–'82	All	2P–4P	3P	0–1⅝P	13/16P	1/16–1/4
'83–'85	All	2P–4P	3P	0–1⅝P	—	1/16–1/4

P Positive

WHEEL ALIGNMENT SPECIFICATIONS
Century and Regal

Year	Model	Caster		Camber		Toe-in (in.)
		Range (deg)	Pref. Setting (deg)	Range (deg)	Pref. Setting (deg)	
'80–'81	All w/manual steer.	½P–1½P	1P	0–1P	½P	1/16–3/16
	All w/power steer.	2½P–3½P	3P	0–1P	½P	1/16–3/16
'82	All	2½P–3½P	3P	0–1P	½P	1/16–3/16
'83–'87	All	2P–4P	3P	⅓N–1⅓P	①	1/16–1/4

① Not available　　N Negative　　P Positive

ENGINE ELECTRICAL

The Delco SI alternator system is standard on all models. For further information on the charging system, please refer to "Charging and Starting" in the Unit Repair section.

For all information on the diesel charging system, see the "Oldsmobile Rear Wheel Drive" section.

Alternator

REMOVAL & INSTALLATION

Disconnect the negative battery cable at the battery. Disconnect and label the electrical connections. Remove the bolt holding the slotted adjusting bracket to the unit. Release the drive belt. Remove the thru-bolt to release the alternator from the engine. When reinstalling, adjust the drive belt to allow $\frac{1}{2}$ in. play on the longest run between pulleys.

NOTE: On some models, it may be necessary to loosen and rotate the fan shroud. On models with air conditioning, it may be necessary to remove the compressor bracket. Do not discharge the A/C.

Voltage Regulator

REMOVAL & INSTALLATION

The voltage regulator is in the alternator, and requires no adjustment. The alternator must be disassembled to remove the regulator.

Starter

For further information on the starting system, please refer to "Charging and Starting" in the Unit Repair section. For all information on the diesel starting system, see the "Oldsmobile Rear Wheel Drive" section.

REMOVAL & INSTALLATION

1. Disconnect the negative battery cable.
2. Raise the vehicle and support it safely. Remove the starter brace and any shields which are in the way. On some automatic transmission models, it may be necessary to remove the exhaust crossover pipe. On manual transmission models, loosen the en-gine crossmember by removing the six crossmember bolts and the two stabilizer shaft bolts from the passenger's side and loosening the four crossmember bolts on the driver's side.
3. Label and disconnect the wires from the solenoid.
4. Remove the starter bolts, taking note of any shims and their placement.
5. Remove the starter.
6. Installation is the reverse of removal.

Distributor

A solid-state, High Energy Ignition (HEI) system is standard equipment. 1981 and later models use an (EST) Electronic Spark Timing distributor. The EST distributor uses no mechanical or vacuum advance and is easily identified by the absence of a vacuum advance and the presence of a four terminal connector. On both distributors there are no contact points or condenser to replace, nor any cam or rubbing block to wear out, thus eliminating distributor maintenance.

1984 and later turbocharged V6 engines with fuel injection are equipped with the Computer Controlled Coil Ignition (C3I) system, which eliminates the distributor. The C3I system uses a rectangular coil which fires the spark plugs on command from the on-board computer. The C3I is also equipped with an electronic spark control (ESC) system, in which the computer uses reference signals from a camshaft and/or crankshaft sensor to determine the degree of engine rotation and uses this information to compute the optimum spark timing in response to engine load and operating conditions.

See the "Oldsmobile Rear Wheel Drive" section for a description of the diesel engine compression-ignition process.

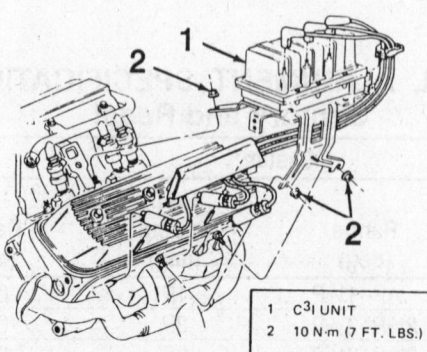

1 C3I UNIT
2 10 N·m (7 FT. LBS.)

Computer Controlled Coil Ignition (C3I) assembly removal.

REMOVAL & INSTALLATION

1. Remove the distributor cap, primary wire and vacuum line at the distributor. Unplug the V6 and V8 distributor cap HEI connectors.
2. Scribe a mark on the distributor body, to locate the position of the rotor. Then scribe another mark on the engine block to show the position of the distributor body in the block.
3. Remove the hold-down clamp. Mark the position of the rotor, then lift the distributor out of the block until the rotor stops turning. Mark the position of the rotor again and remove the distributor.

NOTE: For firing order and cylinder numbering, see the specifications at the beginning of this section.

4. If the engine has not been disturbed, insert the distributor into the engine, making sure the tip of the rotor is aligned with the marks that were scribed on the distributor housing and the engine block.
5. If the engine has been cranked with the distributor out, remove the No.1 spark plug and place a finger over the hole. Slowly turn the engine until compression is felt. Align the timing marks so No.1 cylinder is in firing position. Position the distributor in the block with the rotor at No. 1 firing position. Make sure the oil pump intermediate drive shaft is properly seated in the oil pump.
6. Install the distributor lock but do not tighten.
7. Rotate the distributor body clockwise. Tighten the retaining screw.
8. Connect the primary wire and the vacuum line to the distributor, then install distributor cap.
9. Start the engine and check the ignition timing.

IGNITION TIMING ADJUSTMENT

NOTE: On all models with computer-controlled ignition systems (no vacuum advance unit), refer to the specific timing adjustment procedure shown on the underhood emissions sticker. Timing marks are located on the front engine cover and on the harmonic balancer or pulley.

1. Disconnect the distributor vacuum advance hose from the distributor and plug the hose.
2. On 1981 and later models with the EST distributor, disconnect the four-termi-nal connector from the wiring harness. Some models may re-

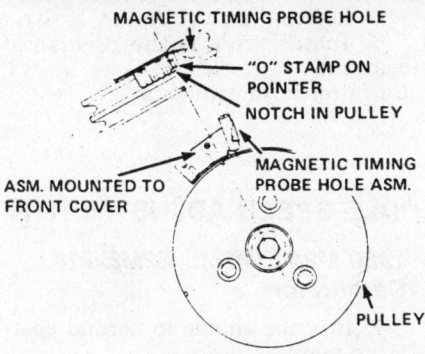

Location of hole for magnetic timing probe.

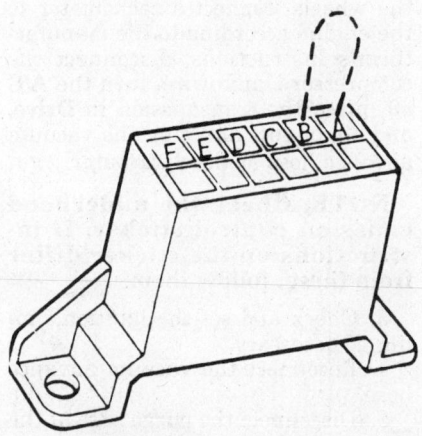

Ground the diagnostic lead with a jumper wire as shown, if necessary.

quire that the diagnostic terminal located under the dashboard be grounded to set the computerized engine control system in the base timing mode. Check the underhood sticker for timing instructions before continuing.

3. Make sure the timing marks are clean and readable. The engine must be at normal operating temperature.

NOTE: It may be necessary to put a small amount of white paint or chalk on the timing marks to make them more visible.

4. Connect a timing light to the No. 1 cylinder, using an adaptor or inductive pickup. Do not pierce the high energy ignition wires with any sharp instrument. Connect a tachometer according to the manufacturer's instructions.

5. Loosen the distributor hold down clamp.

6. Start the engine and run it at the rpm specified in the Tune-up Specifications chart. Aim the timing light at the marks on the engine, then rotate the distributor until the correct marks line up. Tighten the distributor clamp and recheck the timing.

7. Reconnect the vacuum hose or the four-terminal connector.

FUEL SYSTEM

For diesel engine fuel injection adjustments, timing, and removal and installation procedures, see "Oldsmobile Rear Wheel Drive" section. For additional information on the fuel system, please refer to "Carburetors" or "Diesel Maintenance" in the Unit Repair section. For further information on the fuel injection system, please refer to *Chilton's Guide To Fuel Injection And Feedback Carburetors.*

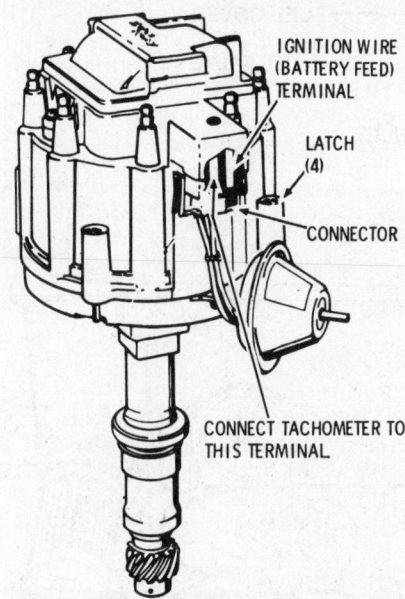

Tachometer connection for the HEI system

Fuel Pump

All air conditioned cars with carbureted V8 engines have a special fuel pump. This pump has a vapor return line which returns hot fuel and fuel vapor to the fuel tank. The possibility of vapor lock is thus greatly reduced by keeping cool fuel circulating through the pump.

The fuel pump used on engines equipped with fuel injection is an electric, high-pressure unit that maintains a constant fuel pressure of 28–50 psi. It is attached to the fuel sending unit located in the fuel tank.

REMOVAL & INSTALLATION

Mechanical Fuel Pump

1. Disconnect the fuel inlet hose from the pump. Disconnect the vapor return hose, if so equipped. Disconnect the inlet hose.

2. Remove the two bolts.

3. Remove the fuel pump.
4. Install a new gasket.
5. Install a new pump and bolts.
6. Tighten the bolts alternately and evenly.
7. Reconnect the hoses, start the engine, and check for leaks.

Electric Fuel Pump

——— **CAUTION** ———

The fuel injection system is constantly under pressure. Fuel pressure must be relieved before disconnecting any fuel lines. To relieve the fuel pressure, remove the fuel pump fuse and start the engine. Allow the engine to run until it stalls, then crank the engine an additional three seconds to make sure all fuel is exhausted from the lines. Replace the fuel pump fuse with the key OFF.

1. Relieve the fuel system pressure as described above.
2. Disconnect the negative battery cable.
3. Raise the vehicle and support it safely.
4. Remove the fuel tank as outlined under "Fuel Tank Removal".
5. Remove the fuel tank sending unit and pump assembly by turning the cam lock ring counterclockwise. Lift the assembly from the fuel tank and remove the fuel pump from the sending unit.
6. Pull the fuel pump up into the attaching hose while pulling outward away from the bottom support. Take care to prevent damage to the rubber insulator and strainer during removal. After the pump assembly is clear of the bottom support, pull the pump assembly out of the rubber connector for removal.
7. Inspect the pump attaching hose for any signs of deterioration and replace as necessary. Check the rubber sound insulator at the bottom of the pump and replace as required.
8. Push the fuel pump into the attaching hose.
9. Install the tank sending unit and pump assembly into the fuel tank. Use a new O-ring during assembly.
10. Install the cam lock over the assembly and lock into place by turning clockwise.
11. Install the fuel tank.

Fuel Filter

REMOVAL & INSTALLATION

Carbureted Models

NOTE: When purchasing a new filter, be sure that it matches the old one exactly.

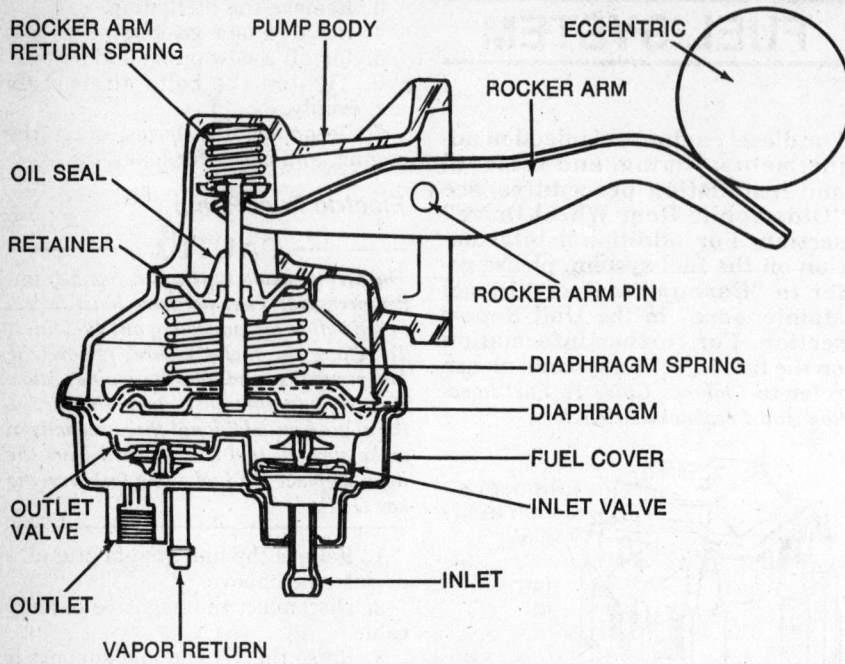

Vapor return type fuel pump

1. Disconnect the fuel line connection at the inlet of the carburetor.
2. Remove the inlet fuel filter nut from the carburetor with a box wrench.
3. Remove the filter element and spring.
4. If it is a bronze element, blow through the cone end; the element should allow air to pass freely.
5. Install the element spring and a new element into the carburetor. Bronze elements are installed with the small section of the cone facing outward.
6. Install a new gasket on the fitting nut and install the nut.
7. Install the fuel line and tighten it securely. Start the engine and check for leaks.

Fuel Injection Models

━━━━━━ **CAUTION** ━━━━━━

Fuel system is under pressure. See the Caution under "Fuel Pump Removal" for procedure to relieve fuel pressure before attempting to remove any fuel lines.

The fuel injection system uses an inline filter located in the fuel feed line under the hood, attached to the frame rail, or on the rear crossmember of the vehicle. Always use a back-up wrench on the fittings any time a fuel filter is removed or installed, and never replace a metal fuel line with a rubber insert. The high pressure fuel system used with all fuel injection systems requires metal fuel lines to contain the pressure. Replace the O-

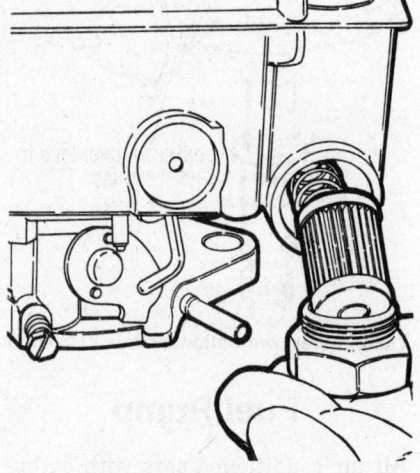

Typical fuel filter

ring at the connection and torque the fuel fitting to 22 ft. lbs. (30 Nm).

Carburetors

For further information on feedback carburetors, please refer to *Chilton's Guide To Fuel Injection And Feedback Carburetors*.

REMOVAL & INSTALLATION

1. Remove the air cleaner.
2. Disconnect and cap the fuel line.
3. Disconnect and label all vacuum lines and/or electrical connectors to the carburetor.
4. Remove the carburetor mounting bolts.

5. Remove the carburetor.
6. Installation is the reverse of removal
Carburetor Adjustments

IDLE SPEED ADJUSTMENTS

1980 M2ME/M2MC/E2ME-210 Carburetor

1. Run the engine to normal operating temperature.
2. Make sure that the choke is fully opened, set the parking brake, block the wheels, connect a tachometer to the engine according to the manufacturer's instructions, disconnect the compressor clutch wire, turn the A/C off, place the transmission in Drive, and disconnect and plug the vacuum advance hose at the distributor.

NOTE: Check the underhood emission control sticker. If instructions on the sticker differ from these, follow them.

3. Check and set the ignition timing, if necessary.
4. Reconnect the vacuum advance hose.
5. Disconnect the purge hose at the vapor canister.
6. On cars without A/C, set the idle speed by turning the idle screw to obtain the specified rpm. On cars with A/C, set the idle speed screw to the specified rpm. Turn the A/C on. Open the throttle momentarily to extend the solenoid plunger, then adjust the solenoid screw to obtain the solenoid idle speed shown on the underhood sticker. Turn the A/C off.
7. Connect all hoses, and remove the tachometer.

1980 and Later M4MC/M4ME Carburetor with Idle Speed Solenoid

1. Run the engine to normal operating temperature.
2. Make sure that the choke is fully opened, turn the A/C off, set the parking brake and block the wheels.
3. Connect a tachometer to the engine according to the manufacturer's instructions.
4. Disconnect the purge hose from the vapor canister. On the 350 engine, plug the purge hose.
5. Disconnect and plug the EGR vacuum hose at the valve. Disconnect and plug the vacuum advance hose.
6. Place the transmission in PARK.
7. Check and adjust the timing, if necessary.

8. Reconnect the vacuum advance hose.

9. Place the transmission in Drive.

NOTE: If instructions on the underhood sticker differ from these, follow underhood sticker.

10. On cars without A/C: Turn the idle speed screw to obtain the specified rpm. On cars with A/C: Turn the idle speed screw to set the specified curb idle speed. Turn the A/C on and disconnect the compressor clutch wire. Open the throttle momentarily to extend the solenoid plunger. Adjust the solenoid screw to obtain the solenoid idle speed shown on the underhood sticker. Reconnect the compressor clutch and turn the A/C off.

11. Reconnect all hoses and remove the tachometer.

1981 and Later M4MC/M4ME Carburetor without Idle Speed Solenoid

Most 1981 and later models are equipped with an Idle Speed Control (ISC) mounted on the float bowl. Idle speeds are computer controlled and the ICS should not be adjusted.

On some V8 models an Idle Load Compensator (ILC) is mounted on the float bowl to control the curb idle speed. The ILC is adjusted at the factory and capped to prevent readjustment.

On cars that do not include either an ISC or ILC, but are quipped with air conditioning, an idle speed solenoid is used to maintain idle speed. For adjustment of these models refer to the previous 1980 and Later adjustment procedures.

NOTE: The underhood sticker specifies which idle system the car is equipped with.

IDLE MIXTURE ADJUSTMENT

1980 Models

Adjustments of idle mixture in 1980 car carburetors are impossible without using a propane enrichment system. Backing out the mixture screw alone will have little or no effect on the mixture. All carburetors have mixture screws concealed by staked-in-plugs. Mixture adjustments are possible only during carburetor overhaul.

1981 and Later

On these models the air/fuel mixture is controlled by the electronic control module of the computer command control system. No adjustment should be attempted.

Fuel Injection Models

The fuel mixture is controlled by the on-board computer. No adjustments are possible.

For further information on the fuel injection system, please refer to *Chilton's Guide To Fuel Injection And Feedback Carburetors.*

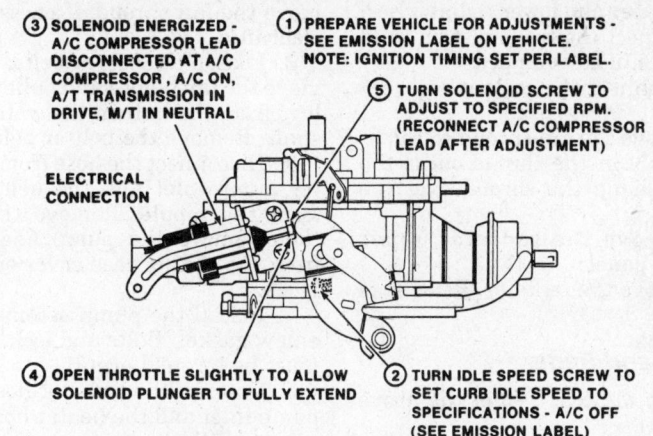

③ SOLENOID ENERGIZED - A/C COMPRESSOR LEAD DISCONNECTED AT A/C COMPRESSOR , A/C ON, A/T TRANSMISSION IN DRIVE, M/T IN NEUTRAL

ELECTRICAL CONNECTION

① PREPARE VEHICLE FOR ADJUSTMENTS - SEE EMISSION LABEL ON VEHICLE. NOTE: IGNITION TIMING SET PER LABEL.

⑤ TURN SOLENOID SCREW TO ADJUST TO SPECIFIED RPM. (RECONNECT A/C COMPRESSOR LEAD AFTER ADJUSTMENT)

④ OPEN THROTTLE SLIGHTLY TO ALLOW SOLENOID PLUNGER TO FULLY EXTEND

② TURN IDLE SPEED SCREW TO SET CURB IDLE SPEED TO SPECIFICATIONS - A/C OFF (SEE EMISSION LABEL)

Idle speed solenoid adjustment

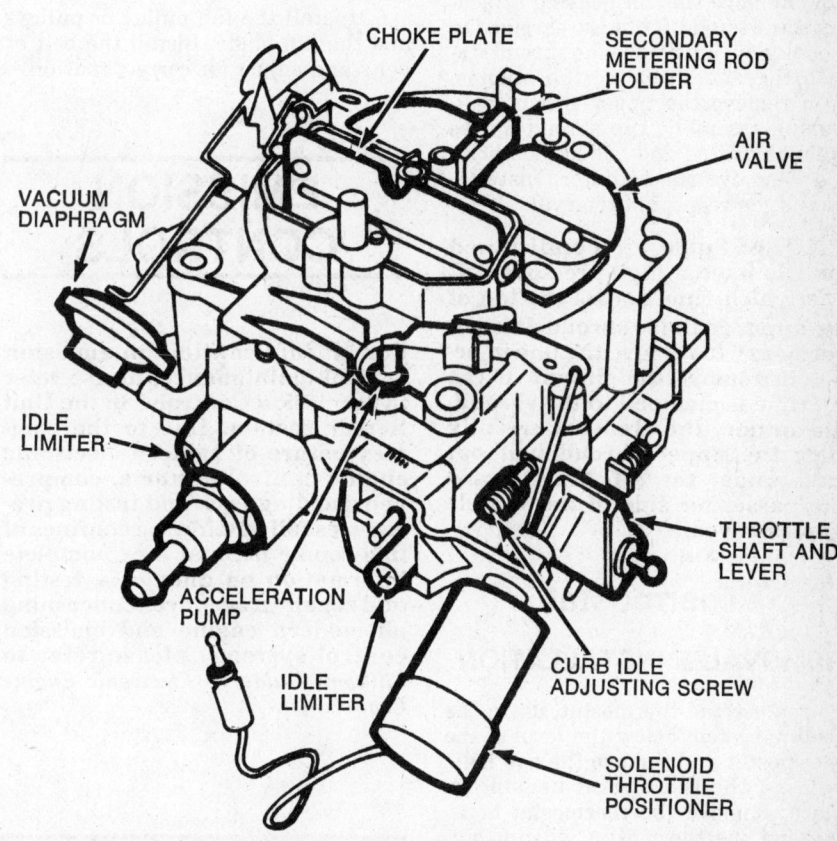

CHOKE PLATE

SECONDARY METERING ROD HOLDER

AIR VALVE

VACUUM DIAPHRAGM

IDLE LIMITER

ACCELERATION PUMP

IDLE LIMITER

THROTTLE SHAFT AND LEVER

CURB IDLE ADJUSTING SCREW

SOLENOID THROTTLE POSITIONER

4 bbl carburetor

ENGINE COOLING

Radiator

REMOVAL & INSTALLATION

LeSabre and Electra

1. Drain the radiator and discon-

nect the upper and lower radiator hoses. Disconnect the transmission fluid cooler lines, if so equipped.

2. Disconnect the coolant recovery hose.

3. Remove the fan-shroud-to-radiator screws. Lift the shroud out of the clips and hang the shroud over the fan.

4. Remove the radiator upper mounting panel.

5. Remove the radiator. Reverse to install.

Century and Regal

1. Refer to Steps 1–2 of the above procedure.

2. Remove the fan blade and the fan clutch.

3. Remove the fan housing attaching screws and lift out the shroud. On models which have the fan shroud stapled together, remove the staples, then remove the upper shroud half. During assembly, the shroud halves must be drilled and bolted together.

4. Remove the radiator. Installation is the reverse of removal.

NOTE: Some air conditioned models have a high-pressure A/C line which runs across the top of the upper radiator shroud. It is not necessary to remove the line in order to remove the shroud. If the A/C line is clamped to the shroud, disconnect the clamp. Carefully slide the upper shroud half out from under the A/C line, toward the passenger side of the vehicle (fan removed).

Thermostat

REMOVAL & INSTALLATION

To replace the thermostat, drain the cooling system below the level of the thermostat and remove the two bolts holding the thermostat housing in place. Remove the thermostat housing and the thermostat will lift out. Clean the mating surfaces of both the intake manifold and the thermostat housing. Use a new gasket when installing a new thermostat. If only silicone sealer was used from the factory, use only sealer during assembly.

Be sure the thermostat is not reversed in its installed position. The spring should be installed toward the engine.

Water Pump

REMOVAL & INSTALLATION

1. Drain the cooling system. Re-

move the fan shroud, if necessary for clearance.

2. Loosen the belt or belts, then remove the fan blades and pulley or pulleys from the hub on the water pump shaft. Remove the belt or belts.

3. Disconnect the hose from the water pump inlet and the heater hose from the nipple. Remove the bolts, then remove the pump and gasket from the timing case cover or engine block.

4. Install the pump assembly with a new gasket. Bolts and lock washers must be torqued evenly.

5. Connect the radiator hose to the pump inlet and the heater hose to the nipple. Fill the cooling system and check all points of possible coolant leaks.

6. Install the fan pulley or pulleys and the fan blade. Install the belt or belts and adjust for correct tension.

EMISSION CONTROLS

For all information on emission control maintenance, please refer to "Emission Controls" in the Unit Repair section. Due to the complex nature of modern electronic engine control systems, comprehensive diagnosis and testing procedures fall outside the confines of this repair manual. For complete information on diagnosis, testing and repair procedures concerning all modern engine and emission control systems, please refer to Chilton's Guide To Electronic Engine Controls.

ENGINE MECHANICAL

NOTE: Refer to the charts in the beginning of this section to determine the type and manufacturer of the engine used in your vehicle. Specific repair information for engines other than those manufactured by Buick may be found in the car section of the engine manufacturer (e.g. Chevrolet 267 and 305 engines will be found in the "Chevrolet Rear Wheel Drive" section of this book).

Engine

REMOVAL & INSTALLATION

1. Scribe marks at the hood hinges and the hinge brackets. Remove the hood.

2. Disconnect the battery and drain the coolant.

3. Remove the air cleaner.

4. On cars with air conditioning (A/C), disconnect the compressor ground wire from the bracket. Remove the electrical connector from the compressor. Remove the compressor and position the compressor out of the way. Do not disconnect any hoses.

—— CAUTION ——
If the compressor refrigerant lines do not have enough slack to position the compressor out of the way without disconnecting the refrigerant lines, the air conditioning system will have to be discharged. Do not attempt to bleed the freon unless you are familiar with air conditioning systems. Compressed refrigerant will freeze any surface it contacts (including skin and eyes) and forms a poisonous gas in the presence of flame.

5. Remove the fan blades, pulley, and belts.

6. Disconnect the radiator and heater hoses. Remove the radiator and shroud assembly.

7. Remove the power steering pump and move it out of the way. Do not disconnect any hoses.

8. Remove the fuel pump hoses and plug them.

—— CAUTION ——
On engines equipped with fuel injection, the fuel system must be depressurized before removing any fuel lines. To depressurize the fuel system, remove the fuel pump fuse and run the engine until it stalls. Crank the engine for three seconds after it stalls to make sure all fuel is exhausted from the fuel lines, then replace the fuel pump fuse with the key OFF.

9. Disconnect the vapor emission lines from the carburetor, the vacuum supply hose from the carburetor to the vacuum manifold, and the power brake vacuum hoses, if so equipped.

10. Disconnect the throttle linkage.

11. Disconnect the oil and coolant switches.

12. Disconnect the engine-to-body ground strap.

13. Raise the car and disconnect the starter wires.

14. Disconnect the pipes from the exhaust manifold and support the exhaust system.

15. On Pontiac-built engines:

a. On models with automatic transmission, remove the converter cover, the converter retaining bolts and slide the converter retaining bolts and slide the converter to the rear.

b. On models with manual transmission, disconnect the clutch linkage and remove the clutch cross-shaft, starter motor and the lower flywheel cover.

c. Remove two bellhousing bolts from each side.

d. On automatic transmission models, disconnect the transmission filler tube.

e. Remove the two front motor mount nuts.

f. Lower the car and support the transmission.

g. Remove the remaining bellhousing bolts and raise the transmission slightly.

h. Remove the engine with a suitable lifting device.

16. On Oldsmobile-built engines:

a. Remove the torque converter cover and the converter-to-flywheel retaining bolts.

b. Remove the engine mounting bolts.

c. Remove the three engine-to-transmission bolts from the left side and remove the engine.

17. On Chevrolet and Buick-built engines:

a. Remove the flywheel and converter cover.

b. On cars with automatic transmission, remove the flywheel-to-converter attaching bolts. Matchmark the converter to the flywheel. On all automatic transmission models, remove the engine-to-transmission attaching bolts. On manual transmission models, disconnect the driveshaft, the shaft linkage, the clutch equalizer shaft and the transmission mount.

c. Remove the motor mount fasteners and the cruise control bracket, if so equipped.

d. Lower the car and support the transmission, except for models with manual transmissions.

e. Raise the engine slightly so the engine mount through bolts can be removed. On models with manual transmissions, remove the engine and transmission as a unit.

18. Install the engine in the reverse order of removal. Note that there are dowel pins in the block that have matching holes in the bellhousing. These dowel pins must be in almost perfect alignment before the engine will go together with the transmission. See "Manual Transmission, Removal and Installation" for clutch alignment procedures.

Intake Manifold

REMOVAL & INSTALLATION

V6 and V8

1. Disconnect the negative battery cable and drain the radiator.

2. Remove the air cleaner. Remove the mass air flow sensor on fuel injection models.

3. Disconnect the upper radiator hose and the heater hose at the manifold. Remove the serpentine drive belt, if equipped.

4. Disconnect the accelerator linkage and the linkage bracket at the manifold. Remove the cruise control chain, if so equipped.

5. Remove the fuel line from the carburetor and the booster vacuum pipe from the manifold. Remove turbocharger, if so equipped.

— CAUTION —
On fuel injected models, the fuel system must be depressurized before disconnecting any fuel lines.

6. Disconnect and label the transmission vacuum modulator line, idle stop solenoid wire (if so equipped), distributor wires and the temperature sending unit wire.

7. Disconnect and mark the vacuum hoses at the distributor and the carburetor.

8. Disconnect the coolant bypass hose at the manifold.

9. On six cylinder models, remove the distributor cap and wires to gain access to the Torx® head bolt. Remove the bolt. On fuel injected models, remove the C3I ignition coil assembly.

10. Remove the throttle linkage springs.

11. Remove the A/C compressor top bracket, if so equipped.

12. Remove the manifold.

13. Use a new gasket to install. Use sealer on the ends of the rubber gasket seals. Carefully guide the manifold onto the engine block dowel pin. Observe "Turbocharger Precautions" given with the previous Turbocharger information. Tighten the bolts in the proper order. Installation is the reverse of the removal steps.

Exhaust Manifold

REMOVAL & INSTALLATION

All Models, Both Sides

1. Raise the vehicle and support on jack stands. Disconnect and tag the spark plug wires.

2. Disconnect the exhaust crossover pipe from the manifolds on both sides of the engine and lower it. On

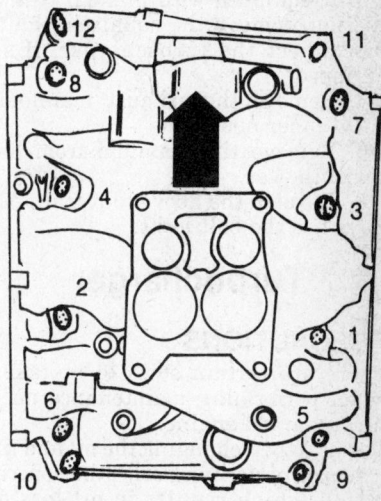

Intake manifold torque sequence for the Buick-manufactured V8 engine

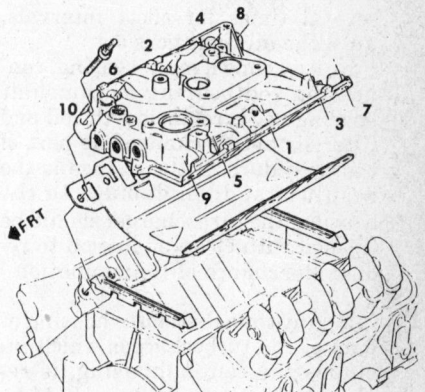

Intake manifold bolt tightening sequence—Buick built V6 engines with carburetor

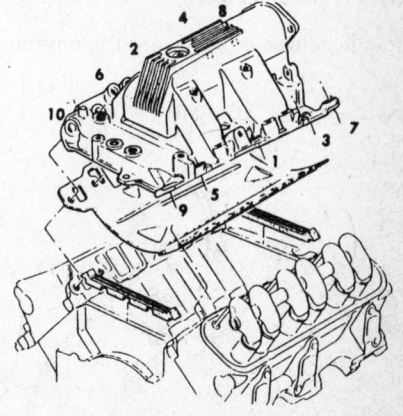

Intake manifold bolt tightening sequence-V6 engine with fuel injection.

the V6, disconnect the choke pipe if you are working on the right side, the Early Fuel Evaporation (EFE) line if you are working on the left side. Disconnect the oxygen sensor wire. Remove the heat shield, if equipped.

3. If equipped with manual transmission, remove the equalizer shaft, Disconnect the turbocharger, if so equipped.

4. Remove the exhaust manifold-to-cyl-inder head bolts.

5. Remove the manifold from beneath the car.

6. Reverse the above to install. Always use the bolt locks.

Turbocharger

PRECAUTIONS

There are certain steps to be taken when performing maintenance on a turbocharged engine:

a. When changing the oil and filter, or performing any other operation which results in oil loss or drainage, before re-starting the engine, disconnect the pink wire for the distributor, crank the engine several times for short intervals, until the oil light goes out.

b. Any time a main bearing, connecting rod bearing or camshaft bearing needs replacing, the oil and filter should be changed as part of the procedure. If the change is the result of sudden damage to the bearing, the turbocharger should be flushed with clean engine oil to reduce the chance of contamination.

c. Any time the center housing or part of the turbocharger which includes the center housing, is replaced, the oil and filter should be changed as part of the procedure.

COMPONENT REMOVAL & INSTALLATION

In the course of servicing the engine, component parts of the turbocharger assembly, including the unit itself, piping, hoses and lines, and electrical connections may have to be removed or disconnected. If removal and installation of turbocharger components becomes necessary, refer to the proper service procedure below.

If the turbocharger unit has to be removed, first clean around the unit thoroughly with a non-caustic solution. When removing the turbocharger, take great care to avoid bending, nicking or in ANY WAY damaging the compressor or turbine blades. Any damage to the blades will result in imbalance, failure of the center housing bearing, damage to the unit and possible personal injury or damage to other engine parts.

ESC Detonation Sensor

1. Squeeze the side of the connector and carefully pull it straight up.

2. Using a deep socket, unscrew the sensor.

3. To install, reverse the removal procedure. Torque the sensor to 14 ft. lbs. Do not over-torque the sensor or apply a side-load when installing.

Wastegate Actuator Assembly

1. Disconnect the two hoses from the actuator.

2. Remove the wastegate linkage-to-ac-tuator rod clip.

3. Remove the two bolts attaching the actuator to the compressor housing.

4. Installation is the reverse of removal.

Center Housing Assembly

1. Disconnect the exhaust outlet pipe from the elbow assembly.

2. Raise and support the car.

3. Disconnect the exhaust outlet pipe from the catalytic converter.

4. Lower the car.

5. Disconnect the exhaust inlet pipe from the turbine housing.

6. Disconnect the exhaust inlet pipe from the right exhaust manifold.

7. Remove the two turbine housing-to-intake manifold bolts.

8. Disconnect the oil feed pipe from the center housing rotating assembly.

9. Remove the oil drain hose from the oil drain pipe.

10. Remove the wastegate linkage-to-ac-tuator rod clip.

11. Remove the six bolts and three clamps attaching the center housing to the compressor housing.

12. Installation is the reverse of removal.

Plenum

1. Remove the turbocharger and actuator assembly.

2. Remove the four bolts attaching the carburetor to the plenum.

3. Installation is the reverse of removal. Torque the bolts to 20 ft. lbs.

Turbocharger Assembly

REMOVAL & INSTALLATION

1. Disconnect the exhaust inlet and outlet pipes from the turbocharger.

2. Disconnect the oil feed pipe from the center housing.

3. Remove the nut attaching the air intake elbow to the carburetor and remove the elbow and the flex tube from the carburetor.

4. Disconnect the accelerator, cruise and detent linkages from the carburetor. Disconnect the plenum linkage bracket.

5. Remove the two bolts attaching the plenum to the side bracket.

6. Disconnect the fuel line and all vacuum lines from the carburetor.

7. Drain the cooling system.

8. Disconnect the coolant lines from the front and rear of the plenum.

9. Disconnect the power brake vacuum line from the plenum.

10. Remove the two bolts attaching the turbine housing to the intake manifold bracket.

11. Remove the two bolts attaching the EGR valve manifold to the plenum. Loosen the two bolts attaching the EGR valve to the intake manifold.

12. Remove the AIR bypass hose from the check valve.

13. Remove the three bolts attaching the compressor housing to the intake manifold.

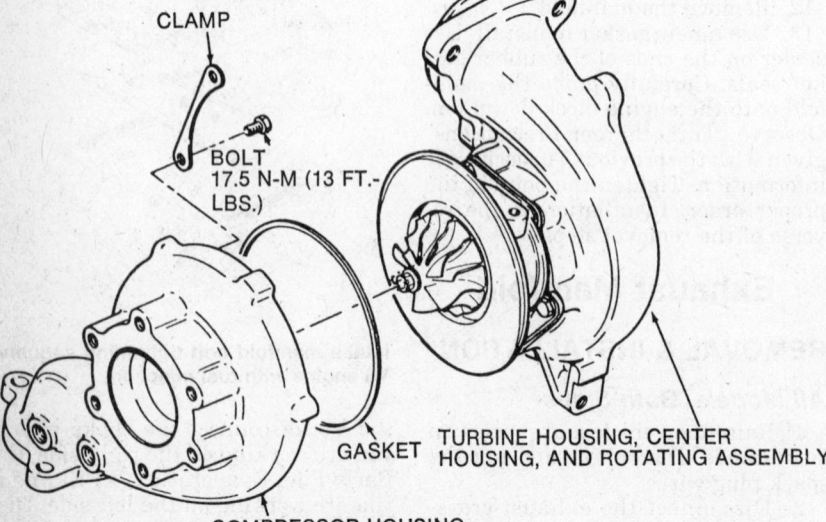

CLAMP

BOLT 17.5 N-M (13 FT.-LBS.)

GASKET

TURBINE HOUSING, CENTER HOUSING, AND ROTATING ASSEMBLY

COMPRESSOR HOUSING

Compressor Housing

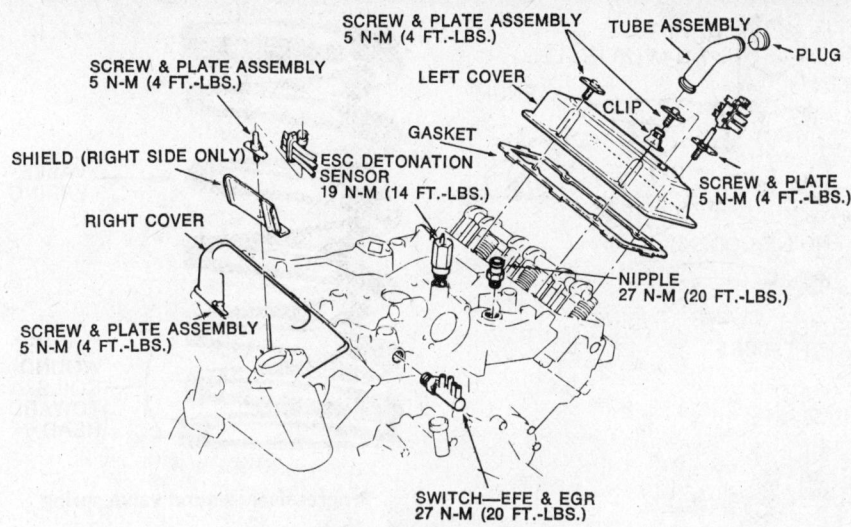

ECS Detonation Sensor

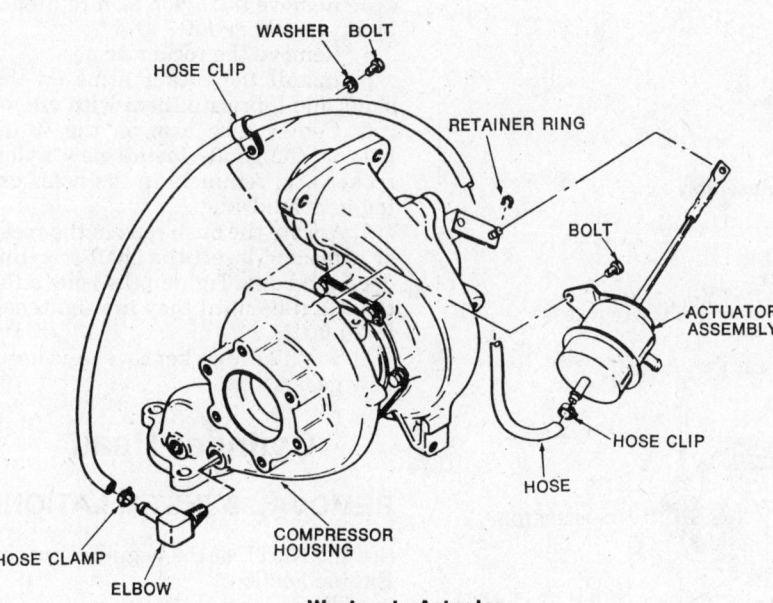

Wastegate Actuator

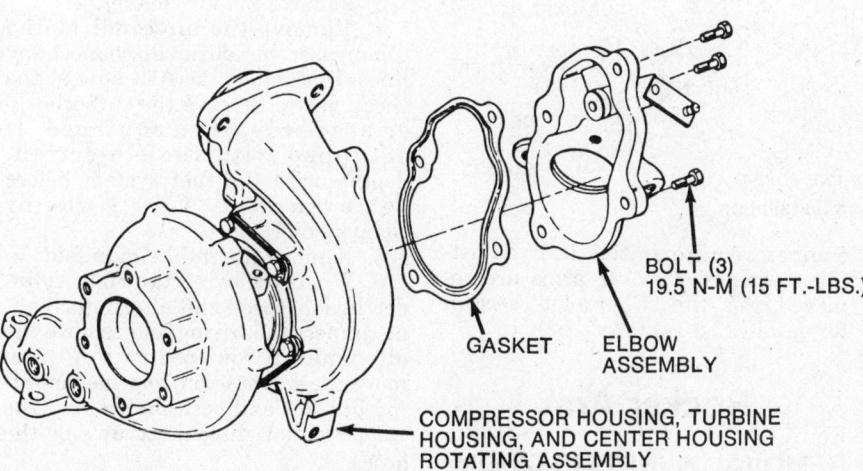

Elbow Assembly

14. Remove the turbocharger, actuator, carburetor and plenum from the engine.

15. Remove the six bolts attaching the carburetor and plenum to the turbocharger and actuator.

16. Remove the oil drain from the center housing.

To install:

1. Install the oil drain on the center housing. Torque to 15 ft. lbs.

2. Install the six turbocharger/actuator-to-carburetor/plenum bolts.

3. Place the assembly on the engine and connect all vacuum hoses.

4. Install the three bolts attaching the compressor housing to the intake manifold. Torque to 35 ft. lbs.

5. Install the AIR bypass hose.

6. Loosely install the two bolts attaching the EGR valve manifold to the plenum. Tighten the two bolts attaching the EGR valve to 15 ft. lbs. Tighten the EGR manifold-to-plenum bolts to 15 ft. lbs.

7. Install the two bolts attaching the turbine housing to the intake manifold bracket. Torque to 20 ft. lbs.

8. Connect the power brake vacuum line at the plenum. Torque to 10 ft. lbs.

9. Connect the plenum front bracket and install one bolt attaching the bracket to the manifold. Torque to 20 ft. lbs.

10. Connect the coolant hoses to the plenum.

11. Refill the cooling system.

12. Connect the carburetor fuel line and remaining vacuum hoses.

13. Install the two bolts attaching the plenum to the side bracket. Torque to 20 ft. lbs.

14. Connect the linkage bracket to the plenum. Torque to 20 ft. lbs.

15. Connect the accelerator, detent and cruise linkages to the carburetor.

16. Install the nut attaching the air intake elbow to the carburetor. Torque to 15 ft. lbs.

17. Connect the oil feed pipe to the center housing. Torque to 7 ft. lbs.

18. Connect the inlet and outlet pipes to the turbocharger. Torque to 14 ft. lbs.

Valve System

All Buick engines use rocker arm shafts, while the engines from other GM Divisions use separate rocker arms mounted on studs. All lifters are the hydraulic type.

NOTE: Some of the engines use progressively wound valve springs. The coils are closer together at one end than the other. The close wound end must go

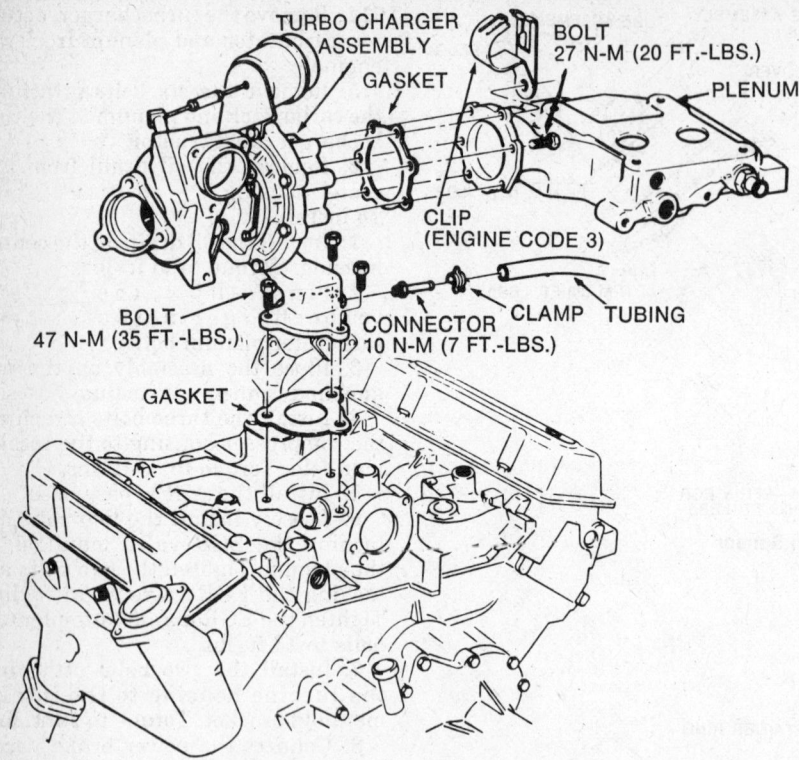

Turbocharger and Plenum Assembly

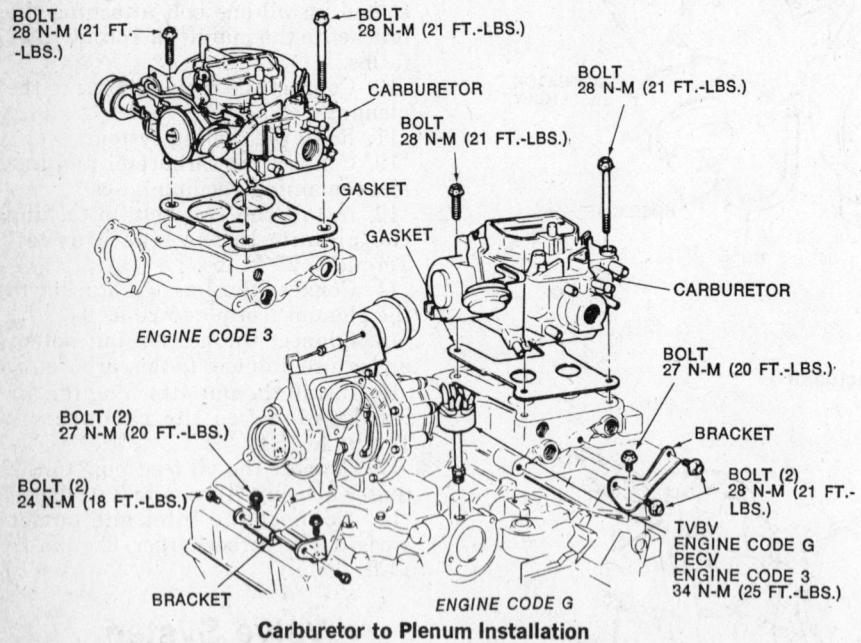

Carburetor to Plenum Installation

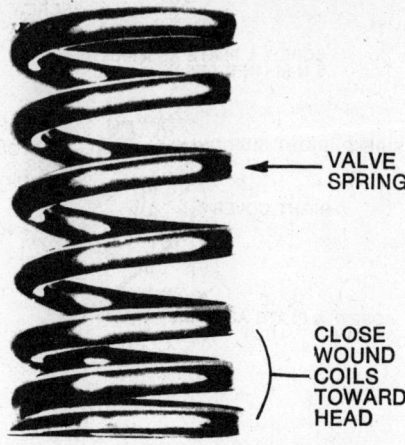

Progressively wound valve spring

against the cylinder head. See the NOTE at the beginning of the Engine Section.

VALVE ADJUSTMENT

The valves on Buick engines cannot be adjusted. If there is excessive clearance in the valve train, look for worn push rods, rocker arms, valve springs or collapsed or stuck lifters. Chevrolet engines require initial lash adjustment whenever rocker arms are removed. See the "Chevrolet" section for details.

Rocker Arm

REMOVAL & INSTALLATION

1. Remove the rocker arm cover.

2. Remove the rocker arm shaft assembly bolts and the assembly.

3. Remove the nylon arm retainers by prying them out.

4. Remove the rocker arms.

5. Install the rocker arms on the shaft and lubricate them with oil.

6. Center each arm on the ¼ in. hole in the shaft. Install new nylon rocker arm retainers in the holes using a ½ in. drift.

7. Locate the push rods in the rocker arms and insert the shaft-to-cylinder head bolts. Tighten the bolts a little at a time until they are tightened to 30 ft. lbs.

8. Install the rocker cover and use a new gasket.

Cylinder Head

REMOVAL & INSTALLATION

See the NOTE at the beginning of the Engine Section.

1. Disconnect the battery.

2. Drain the coolant.

3. Remove the air cleaner.

4. Remove the air conditioning compressor, but do not disconnect any lines. Disconnect the AIR hose at the check valve. Remove the turbocharger assembly, if so equipped. If equipped with fuel injection, depressurize the fuel system before removing any fuel lines or components.

5. Remove the intake manifold.

6. When removing the right cylinder head, loosen the alternator belt, disconnect the wiring and remove the alternator. If equipped with A/C, remove the compressor from the mounting bracket and position it out of the way. Do not disconnect any of the hoses.

7. When removing the left cylinder head, remove the dipstick, power

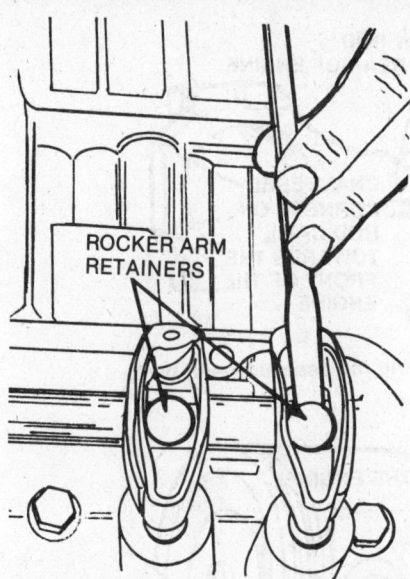

Removing nylon rocker arm retainer

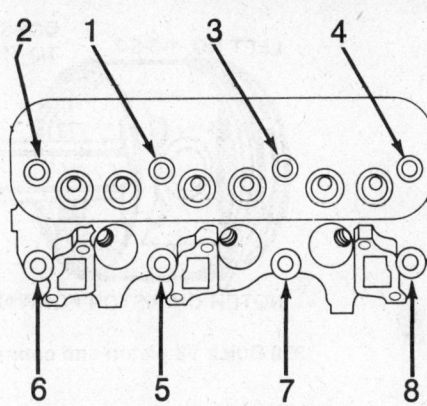

V6 231,252 cylinder head torque sequence

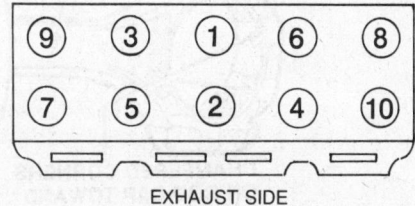

EXHAUST SIDE

Cylinder head bolt torque sequence for the Buick-manufactured V8 engine

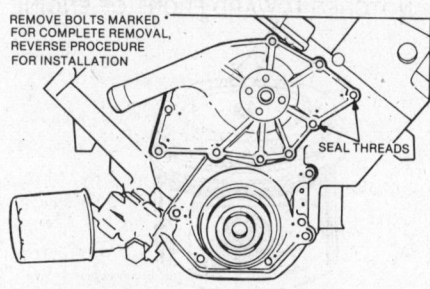

REMOVE BOLTS MARKED FOR COMPLETE REMOVAL, REVERSE PROCEDURE FOR INSTALLATION

SEAL THREADS

Timing chain cover removal and installation Buick-manufactured V8 engine

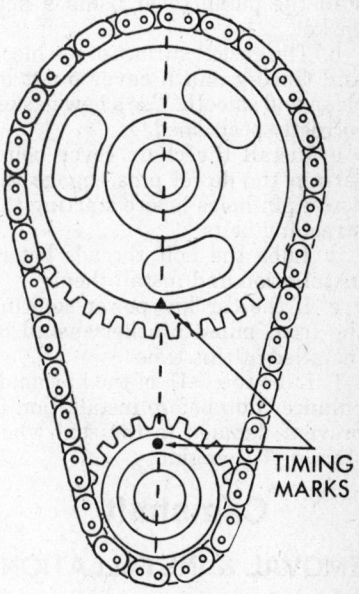

TIMING MARKS

Valve timing marks

steering pump and air pump if so equipped.

8. Disconnect and label the plug wires.

9. Disconnect exhaust manifold from the head being removed.

10. Remove the rocker arm cover and rocker shaft assembly. Lift out the push rods. Be extremely careful to avoid getting dirt into the valve lifters. Keep the pushrods in order; they must be returned to their original positions.

11. Remove the cylinder head bolts.

12. Remove the cylinder head and gasket.

13. Reverse the above steps to install. Torque the head bolts to specifications in three steps. On the 1986 and later 231 cu. in. V6 engine torque the head bolts in the following matter:

a. Use a heavy duty thread sealer on the head bolts.

b. Torque the head bolts to 25 ft. lbs. in the sequence shown in the illustration.

NOTE: Should you reach 60 ft. lbs. at any time in the next two steps, you should stop at this point. Do not complete the balance of the 90°turn.

c. Tighten each bolt ¼ turn (90°) in sequence.

d. Tighten each bolt an additional ¼ turn (90°) in sequence.

Timing Cover, Chain and Seal

REMOVAL & INSTALLATION

1. Drain the cooling system.

2. Remove the radiator, fan, pulley and belt.

3. Remove the fuel pump and alternator, if necessary to remove cover.

4. Remove the distributor, if located on the front of the engine. If the timing chain and sprockets will not be disturbed, note the position of the distributor for installation in the same position.

5. Remove the thermostat bypass hose.

6. Remove the harmonic balancer.

7. Remove the timing chain-to-crankcase bolts.

8. Remove the oil pan-to-timing chain cover bolts and remove the timing chain cover.

9. Using a punch, drive out the old seal and the shedder toward the rear of the seal.

10. Coil the new packing around the opening so the ends are at the top. Drive in the shedder using a punch. Properly size the packing by rotating a hammer handle around the packing until the balancer hub can be inserted through the opening.

11. Align the timing marks on the sprockets.

12. Remove the camshaft sprocket bolt without changing the position of the sprocket. On the V6, remove the oil pan.

13. Remove the front crankshaft oil slinger.

14. On the 350, remove the crankshaft distributor drive gear retaining bolt and washer. Remove the drive gear and the fuel pump eccentric. On the V6, remove the camshaft sprocket bolts.

15. Using two large screwdrivers, carefully pry the camshaft sprocket and the crankshaft sprocket forward until they are free. Remove the sprockets and the chain.
To install:

1. Make sure, with sprockets temporarily installed, that No. 1 piston is at top dead center and the camshaft sprocket O-mark is straight down and on the centerline of both shafts.

2. Remove the camshaft sprocket and assemble the timing chain on both sprockets. Then slide the sprockets-and-chain assembly on the shafts with the O-marks in their closest together position and on a centerline with sprocket hubs.

3. Assemble the slinger on the crankshaft with I.D. against the sprocket, (concave side toward the

front of engine). Install the oil pan, if removed.

4. On the 350, slide the fuel pump eccentric on the camshaft and the Woodruff key with the oil groove forward. On the V6 install the camshaft sprocket bolts.

5. Install the distributor drive gear.

6. Install the drive gear and eccentric bolt and retaining washer. Torque to 40–55 ft. lbs.

7. Install the timing case cover. Install a new seal by lightly tapping it in place. The lip of the seal faces inward. Pay particular attention to the following points.

　a. Remove the oil pump cover and pack the space around the oil pump gears completely full of petroleum jelly. There must be no air space left inside the pump. Reinstall the pump cover using a new gasket.

　b. The gasket surface of the block and timing chain cover must be clean and smooth. Use a new gasket correctly positioned.

　c. Install the chain cover being certain the dowel pins engage the dowel pin holes before starting the attaching bolts.

　d. Lube the bolt threads before installation and install them.

　e. If the car has power steering the front pump bracket should be installed at this time.

　f. Lube the O.D. of the harmonic balancer hub before installation to prevent damage to the seal when starting the engine.

Camshaft

REMOVAL & INSTALLATION

See the NOTE at the beginning of the Engine Section.

1. Complete Steps 1–8 under "Timing Cover, Chain and Seal, Removal and Installation" above. Skip Steps 9 and 10, complete Steps 11–15.

NOTE: If equipped with air conditioning, unbolt the condenser and position it out of the way. If this is not possible, the A/C system will have to be discharged. Do not attempt to discharge the freon unless familiar with air conditioning systems. Refrigerant will freeze any surface it touches (like skin and eyes) and gives off a poisonous vapor when exposed to an open flame.

2. Remove the hydraulic lifters, keeping them in order for installation.

3. Slide the camshaft forward, out of the bearing bores. Do this carefully,

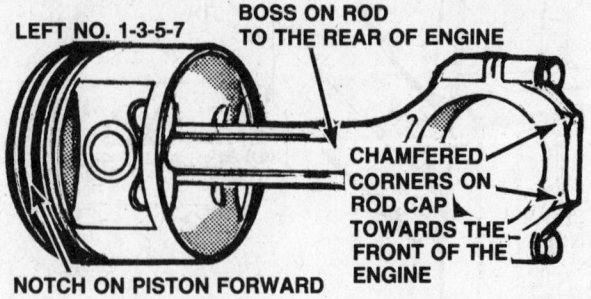

LEFT NO. 1-3-5-7

BOSS ON ROD TO THE REAR OF ENGINE

CHAMFERED CORNERS ON ROD CAP TOWARDS THE FRONT OF THE ENGINE

NOTCH ON PISTON FORWARD

350 Buick V8 piston and connecting rod assembly—left bank

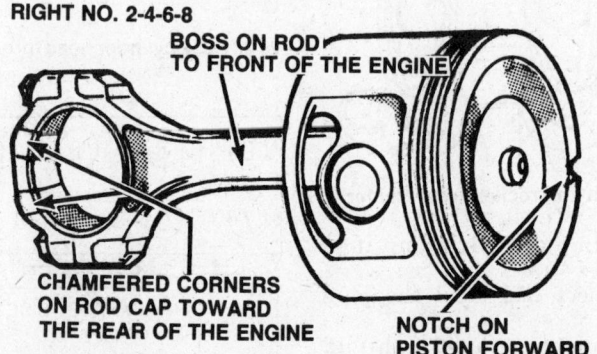

RIGHT NO. 2-4-6-8

BOSS ON ROD TO FRONT OF THE ENGINE

CHAMFERED CORNERS ON ROD CAP TOWARD THE REAR OF THE ENGINE

NOTCH ON PISTON FORWARD

350 Buick V8 piston and connecting rod assembly—right bank

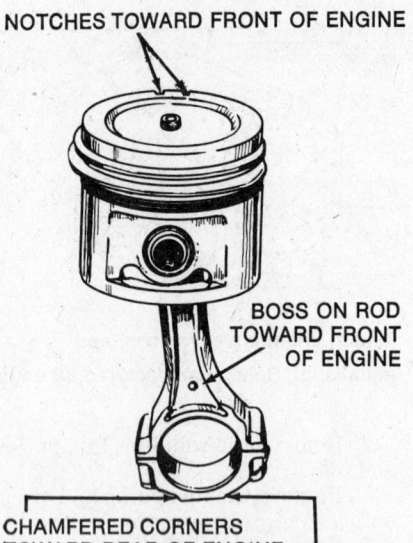

NOTCHES TOWARD FRONT OF ENGINE

BOSS ON ROD TOWARD FRONT OF ENGINE

CHAMFERED CORNERS TOWARD REAR OF ENGINE

RIGHT NO. 2-4-6

Right bank piston and rod positioning—231 and 252

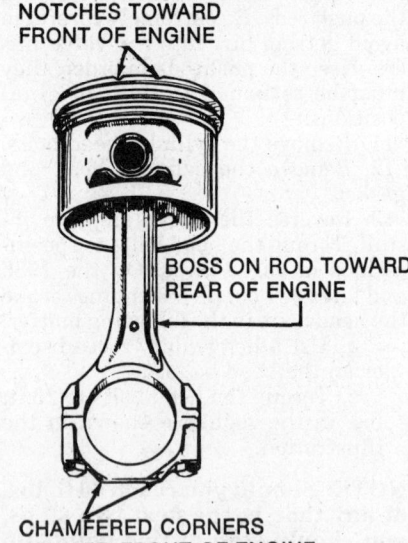

NOTCHES TOWARD FRONT OF ENGINE

BOSS ON ROD TOWARD REAR OF ENGINE

CHAMFERED CORNERS TOWARD FRONT OF ENGINE

LEFT NO. 1-3-5

Left bank piston and rod positioning—231 and 252

to avoid damage to the bearing surfaces and bearings.

4. Reverse to install. Clean all gasket surfaces thoroughly and use new gaskets. Lubricate the camshaft lobes with heavy oil before installation, and be careful not to contact any of the bearings with the cam lobes. Make sure that the camshaft timing marks are aligned with the crankshaft marks. See installation steps under "Timing Cover, Chain and Seal" above.

Pistons and Connecting Rods

On the V6, starting at the front, the cylinders in the right bank are numbered 2-4-6 and in the left bank are

numbered 1–3–5. On the V8, starting at the front, the cylinders on the right are 2–4–6–8 and the cylinders on the left are numbered 1–3–5–7.

All compression rings are marked with a dimple, a letter "T", a letter "O", or the word "TOP" to identify the side of the ring which must face toward the top of the piston.

When the piston and connecting rod assembly is properly installed, the oil spurt hole in the connecting rod will face the camshaft. The notch on the piston will face the front of the engine. On all engines, the chamfered corners of the bearing caps should face toward the front of the left bank and toward the rear of the right bank. The boss on the connecting rod should face toward the front of the engine for the right bank and to the rear of the engine on the left bank.

ENGINE LUBRICATION

Oil Pan

REMOVAL & INSTALLATION

V8 Engines

1. Disconnect the battery ground cable.
2. Remove the fan shroud-to-radiator screws.
3. Remove the air cleaner and disconnect the throttle linkage.
4. Raise the front end and support it on jackstands.
5. Drain the oil.
6. Disconnect the exhaust crossover pipe at the engine.
7. Remove the lower flywheel housing cover.
8. Remove the shift linkage bolt and swing it out of the way.
9. Remove the front engine mount bolts.
10. Raise the front of the engine, either by placing a block of wood and a jack under the crankshaft pulley mounting or by lifting it with a hoist.

— CAUTION —

On air conditioned cars, place a support under the right-side of the transmission before raising the engine. If you don't do this, the weight of the air conditioner will flip the engine and transmission to the right.

11. Unbolt and remove the pan. It may be necessary to turn the crank-

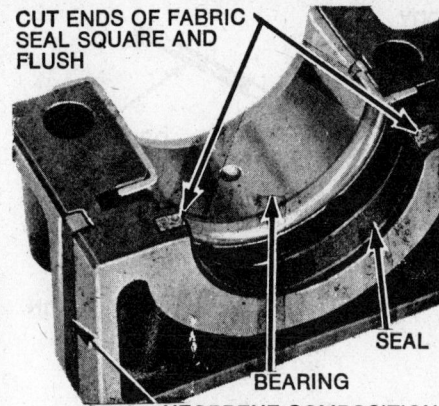

CUT ENDS OF FABRIC SEAL SQUARE AND FLUSH

SEAL

BEARING

SEAL—NEOPRENE COMPOSITION

Rear main bearing cap

shaft so that it doesn't interfere with the front of the pan.

12. Reverse the procedure for installation.

V6 Engines

1. Raise the car and drain the oil.
2. Remove the flywheel cover.
3. Remove the exhaust crossover pipe.
4. Remove the oil pan attaching bolts and remove the oil pan.

Rear Main Seal

REMOVAL & INSTALLATION

Braided fabric seals are used. The upper seal half cannot be replaced without removing the crankshaft, unless the Time Saver in this section is used.

1. Remove the oil pan and rear main bearing cap.
2. Remove the old seal from the bearing cap and place a new seal in the groove with both ends projecting above the parting surface of the cap.
3. Force the seal into the groove by rubbing down with a hammer handle or smooth tool, until the seal projects above the groove not more than $1/16$ in. Cut the ends off flush with the surface of the cap with a razor blade.
4. On the 231, 252, and 350, place new neoprene seals in the grooves in the sides of the bearing cap after soaking the seals in kerosene for a minute or two.

NOTE: The neoprene composition seals will swell up once exposed to the oil and heat. It is normal for the seals to leak for a short time, until they become properly seated. The seals must not be cut to fit.

5. To install, reverse the above. Use a small amount of sealer on the bearing cap mating surface. The engine must

TOP HALF, REAR MAIN BEARING OIL SEAL REPLACEMENT

Although the factory recommends removing the crankshaft to replace the top half of the oil seal, the following procedure can be used without removing the crankshaft.

1. Remove the oil pan and rear main bearing cap.
2. Loosen the rest of the crankshaft main bearings and allow the crankshaft to drop about 1/16 in.
3. Remove the old upper half of the oil seal.
4. Wrap some soft copper wire around the end of the new seal and leave about 12 in. on the end. Generously lubricate the new seal with oil.
5. Slip the free end of the copper wire into the oil seal groove and around the crankshaft. Pull the wire until the seal protrudes an equal amount on each side. Rotate the crankshaft as the seal is pulled into place.
6. Remove the wire. Push any excess seal that may be protruding back into the groove.
7. Before tightening the crankshaft bearing caps, visually check the bearings to make sure they are in place. Torque the bearing cap bolts to specifications. Make sure there is no oil on the parting surfaces.
8. Replace the oil pan. Run the engine slowly for the first few minutes of operation.

be operated at low rpm when first started, after a new seal is installed.

Oil Pump

REMOVAL & INSTALLATION

See the NOTE at the beginning of the Engine Section. On the V6 and V8, the oil pump is located on the left side of the timing chain cover, where it is connected by a drilled passage in the cylinder crankcase to an oil screen housing and standpipe assembly.

1. Remove the oil filter.
2. Unbolt the pump cover assembly from the timing chain cover.
3. Remove the cover assembly and slide out the pump gears.

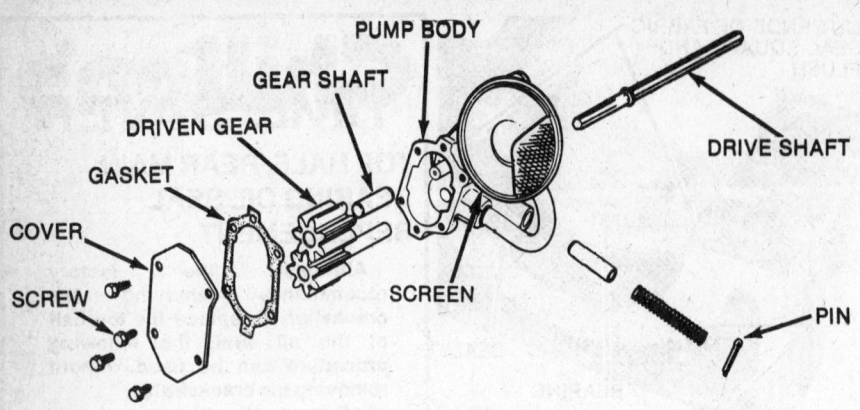

PUMP BODY
GEAR SHAFT
DRIVEN GEAR
GASKET
COVER
SCREW
SCREEN
DRIVE SHAFT
PIN

Typical oil pump assembly

4. Remove the oil pressure relief valve cap, spring, and valve. Do not remove the oil filter by-pass valve and spring.

5. Check that the relief valve spring isn't worn on its side or collapsed. Check that the relief valve is no more than an easy slip fit in its bore in the cover. If there is any perceptible sideplay, replace the valve. If there is still sideplay, replace the cover.

6. Check the filter by-pass valve for the good condition.

To assemble the pump:

7. Lubricate and install the pressure relief valve and spring in the cover bore. Install the gasket and cap, torquing the cap to 35 ft. lbs.

8. Install the gears and check that gear-to-cover end clearance is between 0.002–0.006 in. If the clearance is less, check the timing cover gear pocket for wear.

9. Remove the gears and pack the gear pocket full of petroleum jelly. Don't use grease.

CAUTION

Unless the pump is primed this way, it won't produce any oil pressure when the engine is started.

---

10. Install the gears. Install a new gasket and the cover. Torque the bolts evenly to 10 ft. lbs. Replace the filter.

CLUTCH

The only service adjustment necessary on the clutch is to maintain the correct pedal free-play (see Linkage Adjustment, below).

REMOVAL & INSTALLATION

1. Remove the pedal return spring from the clutch fork. Remove the transmission as outlined under Manual Transmission.

2. Remove the flywheel housing.

3. Remove the throw-out bearing from the clutch fork.

4. Disconnect the clutch fork from the ball stud.

5. Mark the clutch cover and the flywheel to assure proper balance on reassembly.

6. Loosen the clutch cover to flywheel bolts one turn at a time until the spring pressure is released.

7. Support the pressure plate and cover assembly while removing the last bolts, then remove the cover assembly and the driven plate.

8. Inspect the flywheel for scoring, grooves, or signs of overheating (discoloration). Reface or replace the flywheel as necessary.

9. Install the clutch by reversing the removal procedure. Use a clutch aligning pilot or a spare transmission input shaft through the hub of the driven plate and into the pilot bushing. Be sure to align the clutch cover-to-flywheel index marks.

LINKAGE ADJUSTMENT

Century and Regal

1. Remove the return spring.

2. Turn the clutch lever and shaft assembly until the pedal is firmly against the stop.

3. Push the outer end of the clutch fork to the rear until the throwout bearing touches the spring fingers.

4. Install the lower pushrod in the fork and the swivel in the gauge hole. Turn the rod clockwise as viewed from the front to remove all play from the linkage.

5. Remove the swivel from the gauge hole and install it in the hole furthest from the centerline of the lever and shaft assembly. Install the washers and retainer.

6. Tighten the locknut against the swivel, being careful not to change the rod length.

7. Install the clutch retainer spring. The above procedure should produce ⅔–1⅓ in. of free play when measured at the pedal pad center.

MANUAL TRANSMISSION

A fully-synchronized Saginaw three-speed transmission has been available in these cars. It can be identified by the single bolt at the top of the side cover. The production code and transmission serial number are on the right side of the transmission case. The only four-speed used is a Saginaw unit. The production code and transmission serial number are stamped on the right side of the transmission case.

For overhaul procedures, please refer to "Manual Transmission Overhaul" in the Unit Repair section.

REMOVAL & INSTALLATION

1. Raise the vehicle on a hoist and drain the transmission fluid.

2. Mark the universal joint and transmission shaft companion flange to aid proper alignment at the time of installation. Remove the two U-bolts and disconnect the driveshaft at the rear joint. Slide the driveshaft rearward as far as possible and remove it.

3. Disconnect the shift linkage from the transmission.

4. Disconnect the speedometer cable and the back-up light switch at the transmission.

5. Remove the crossmember-to-transmission mounting bolts, the catalytic converter-to-transmission bracket (if so equipped) and remove the cross-member-to-frame bolts. Raise the transmission slightly and remove the crossmember.

6. Remove the two upper transmission-to-flywheel housing bolts.

NOTE: If guide pins are not used, damage to the clutch driven plate can result.

8. Slide the transmission back until the drive gear shaft disengages the clutch disc and clears the flywheel housing. Lower the transmission.

9. On installation, install the guide pins in the upper and lower right side bolt holes for alignment. If the guide

pins aren't used, the clutch plate might be damaged.

LINKAGE ADJUSTMENT

Three-Speed Floorshift

1. Place the transmission levers into neutral.
2. Loosen the shift rod adjusting clamp bolts.
3. Place a rod in the notch in the rear portion of the shift bracket assembly.
4. Move both shift levers back against the rod.
5. Tighten the shift rod adjusting bolts.

Four-Speed Floorshift

1. Place the transmission levers in neutral positions.
2. Place a $5/16$ in. diameter rod in the rear lower portion of the shift bracket assembly.
3. Adjust all three shift levers back against the rod.
4. Tighten the adjusting clamp bolts.

AUTOMATIC TRANSMISSION

Many different automatic transmissions have been used in rear wheel drive Buick models. Basically, all of the transmissions used fall into 3 basic groups, as follows:

For further information on automatic transmissions, please refer to "Automatic Transmissions" in the Unit Repair section.

REMOVAL & INSTALLATION

1. Disconnect the negative battery cable at the battery.
2. If so equipped, disconnect the detent/downshift cable at its upper end (accelerator pedal or carburetor).
3. Raise the vehicle and support it safely with jackstands. Preferably, the front AND rear of the vehicle should be raised to provide adequate clearance for transmission removal.
4. Disconnect the exhaust crossover pipe at the manifolds, if exhaust system-to-transmission interference is obvious. It may be necessary to remove the catalytic converter, exhaust pipe, or just the brackets in order to clear the transmission.

TRANSMISSION TYPE

THM-200 GROUP	
THM-200	3S-standard duty
THM-200-C	3S-LTC
THM-200-4R	4S-LTC

THM-350 GROUP	
THM-250	3S-light duty
THM-350	3S-standard duty
THM-350-C	3S-LTC
THM-375-B	3S-heavy duty

THM-400 GROUP	
THM-400	3S-heavy duty

3S-3 speed, 4S-4 speed
LTC-Locking torque convertor

— **CAUTION** —

Exhaust system services must be performed while all components are COLD.

5. Remove the transmission inspection cover.
6. Remove the torque converter-to-flywheel bolts. The relationship between the flywheel and converter must be marked so that proper balance is maintained after installation.
7. Matchmark the drive shaft and the rear yoke (for reinstallation purposes). With a drain pan positioned under the front yoke, unbolt and remove the drive shaft.
8. Mark and disconnect the vacuum lines, wiring, and speedometer cable from the transmission, as required.
9. Place a transmission jack (carefully) up against the transmission oil pan, then secure the transmission to the jack.
10. Remove the transmission mounting pad bolt(s), then carefully raise the transmission just enough to take the weight of the transmission off of the supporting crossmember.

— **CAUTION** —

Exercise extreme care to avoid damage to underhood components while raising or lowering the transmission.

11. Unbolt and remove the transmission crossmember, complete with the mount. It may be necessary to raise or lower the transmission a small amount to remove the crossmember.
12. Remove the transmission dipstick, then unbolt and remove the filler tube.
13. Disconnect the shift linkage (or cable on floor shift equipped models) and oil cooler lines from the transmission.
14. Support the engine using a jackstand placed beneath the engine oil pan. Be sure to put a block of wood

between the jackstand and the oil pan, to prevent damage to the pan.
15. Securely wire the torque converter to the transmission case.
16. Remove the transmission-to-engine mounting bolts, then carefully move the transmission rearward, downward and out from beneath the vehicle.

— **CAUTION** —

If interference is encountered with the cable(s), cooler lines, etc., remove the component(s) before finally lowering the transmission. Refer to the Automatic Transmission segment of the Unit Repair section for further information.

Installation is the reverse of the previous steps. Note the following points during and after installation:
1. Torque the transmission-to-engine mounting bolts to 30–40 ft. lbs.
2. Align the matchmarks of the drive shaft with the marks of the rear yoke before installing the joint straps and bolts.
3. Align the converter and flywheel markings before installing the converter bolts.
4. Add the proper type and quantity of transmission fluid. If the converter was replaced, an additional 4 pints (approx.) should be added. NEVER overfill the transmission.
5. Adjust the shift linkage (or cable) and the detent/downshift cable.
6. Make sure that all vacuum lines, electrical connections, and oil cooler line connections are secure before driving the vehicle.
7. Check for fluid leakage, then after the transmission is hot, recheck the fluid level.

DRIVE AXLE

Driveshaft

REMOVAL & INSTALLATION

1. Mark the driveshaft rear yoke and the differential flange to assure correct alignment upon reassembly.
2. Remove the bolts and straps from the differential flange.
3. Remove the driveshaft assembly by first sliding the driveshaft sufficiently forward to disengage the differential flange and then slide the shaft downward and rearward to disengage the front splined yoke from the transmission output shaft.
4. Installation is the reverse of removal. Be sure to align the match marks made before disassembly.

U-Joint

REMOVAL & INSTALLATION

For all U-Joint repair procedures, please refer to "U-joints/CV - Joints" in the Unit Repair section.

Axle Shaft, Bearing and Seal

REMOVAL & INSTALLATION

These cars use two different types of drive axle, the C-lock and the non C-lock type. Axle shafts in the C-lock type are retained by C-shaped locks. which fit grooves at the inner end of the shaft. Axle shafts in the non C-lock type are retained by the brake backing plate, which is bolted to the axle housing. Bearings in the C-lock type axle consist of an outer race, bearing rollers and a roller, cage, retained by snaprings. The non C-lock type axle uses a unit roller bearing (inner race, rollers and outer race), which is pressed onto the shaft up to a shoulder. It is imperative to determine the axle type before attempting any service.

The axle identification number is stamped on the front of the passenger side axle tube next to the differential carrier on all models except those with an $8\frac{1}{2}$ in. ring gear. These models have the I.D. on a tag under one of the differential rear cover bolts.

Non C-Lock Type

Design allows for maximum axle shaft end-play of 0.022 in., which can be measured with a dial indicator. If end-play is found to be excessive, the bearing should be replaced. Shimming the bearing is not recommended as this ignores end-play of the bearing itself and could result in improper seating of the bearing.

Before attempting any service to the drive axle or axle shafts, remove the differential carrier cover and visually determine if the axle shafts are retained by C-shaped locks at the inner end, or by the brake backing plate at the outer end. If the shafts are not retained by C-locks, proceed as follows:

1. Remove the wheel, tire and brake drum.
2. Remove the nuts holding the retainer plate to the backing plate. Disconnect the brake line.
3. Remove the retainer and install the two lower nuts finger tight, to prevent the brake backing plate from being dislodged.

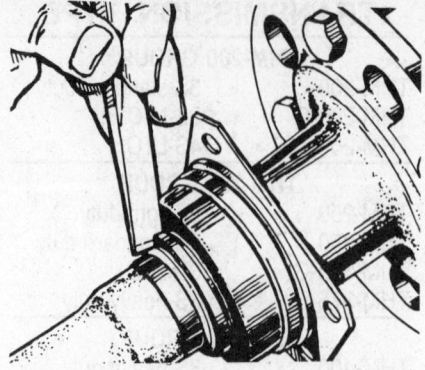

Breaking the bearing retainer with a chisel

4. Pull out the axle shaft and bearing assembly, using a slide hammer.
5. Using a chisel, nick the bearing retainer in three or four places. The retainer does not have to be cut, merely collapsed sufficiently to allow the bearing retainer to be slid from the shaft.
6. Press off the bearing and install the new one by pressing it into position.

NOTE: Do not attempt to press the bearing and the retainer on at the same time.

7. Assemble the shaft and bearing in the housing, being sure that the bearing is seated properly in the housing.
8. Install the retainer, drum, wheel and tire. Bleed the brakes.

C-Lock Type

Before attempting any service to the drive axle or axle shafts, remove the carrier cover and visually determine if the axle shaft(s) are retained by C-shaped locks at the inner ends or by a brake backing plate at the outer end. If they are retained by C-shaped locks, proceed as follows:

1. Raise the vehicle and remove the wheels.
2. The differential cover has already been removed (see Caution above). Remove the differential pinion shaft lockscrew and the differential pinion shaft.
3. Push the flanged end of the axle shaft toward the center of the vehicle and remove the C-lock from the end of the shaft.
4. Remove the axle shaft from the housing, being careful not to damage the oil seal.
5. Remove the oil seal by inserting the button end of the axle shaft behind the steel case of the oil seal. Pry the seal loose from the bore.
6. Seat the legs of a bearing puller behind the bearing. Seat a washer against the bearing and hold it in

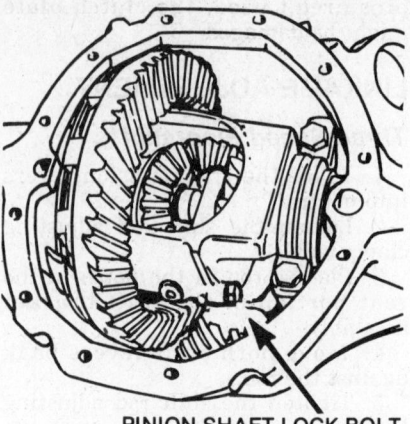

PINION SHAFT LOCK BOLT

Removing pinion shaft lock bolt from differential

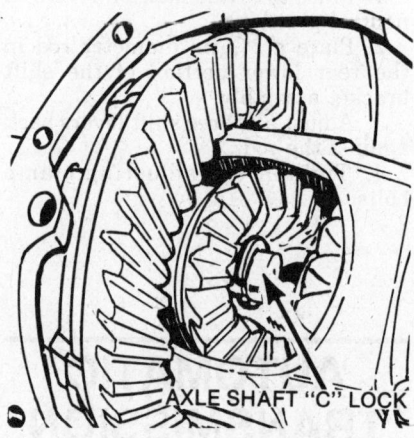

AXLE SHAFT "C" LOCK

Removing the axle shaft C lock

place with a nut. Use a slide hammer to pull the bearing.

7. Pack the cavity between the seal lips with a wheel bearing lubricant and lubricate a new wheel bearing with same.
8. Use a suitable driver and install the bearing until it bottoms against a tube. Install the oil seal.
9. Slide the axle shaft into place. Be sure that the splines on the shaft do not damage the oil seal. Make sure that the splines engage the differential side gear.
10. Install the axle shaft C-lock on the inner end of the axle shaft and push the shaft outward so that the C-lock seats in the differential side gear counterbore.
11. Position the differential pinion shaft through the case and pinions, aligning the hole in the case with the hole for the lockscrew.
12. Install the pinion shaft lockscrew.
13. Use a new gasket and install the carrier cover. Be sure that the gasket surfaces are clean before installing the gasket and cover.

14. Fill the axle with lubricant to the bottom of the filler hole.

15. Install the brake drum and wheels and lower the car. Check for leaks and road test the car.

FRONT SUSPENSION

Shock Absorber

REMOVAL & INSTALLATION

1. Remove the upper shock absorber attaching nut, grommet retainer, and grommet.

2. Remove the lower retaining screw. Lower the shock through the hole in the lower control arm.

NOTE: Purge new shocks of air by repeatedly extending them in their normal position and compressing them while inverted.

3. Reverse the above steps to install. Tighten the upper nut to 8 ft. lbs; the lower bolts to 20 ft. lbs.

Ball Joint

INSPECTION

Lower Ball Joint

All cars have visual wear indicators on the lower ball joints. The lower ball joint grease plug screws into the wear indicator which protrudes from the bottom of the ball joint housing. As long as the wear indicator extends out of the ball joint housing, the ball joint is not worn. If the tip of the wear indicator is parallel with, or recessed into the ball joint housing, the ball joint is defective.

Upper Ball Joint

Place a jack under each lower control arm between the suspension sprint pocket and the ball joint and raise the car. Grasp the wheel at the 6 and 12 o'clock position and shake the top of the wheel in and out. Observe the steering knuckle for any movement relative to the control arm. If the ball joint is loose, it must be replaced.

Upper Control Arm and/or Ball Joint

REMOVAL & INSTALLATION

1. Raise the car and place a jack

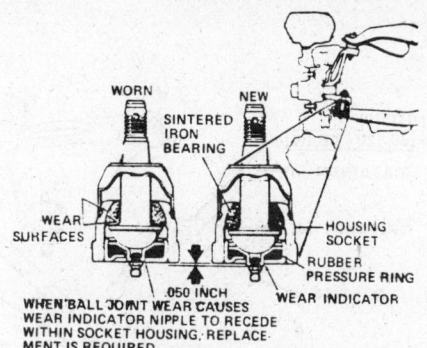

Lower ball joint wear indicator

under the frame. Remove the wheel and tire.

2. With another jack, support the car weight under the outer edge of the lower control arm. Raise the jack enough to free the upper control arm for the upper ball stud.

3. Remove the cotter pin from the upper ball joint stud.

4. Loosen, but do not remove, the nut.

——— CAUTION ———
If the nut is removed, the full force of the coil spring could be released. Use a ball joint removal tool to free the stud from the knuckle.

5. Wire the brake and knuckle in place to prevent brake hose damage, then lift the upper arm from the knuckle.

NOTE: If only the ball joints are to be replaced, stop at this point. Center punch and drill out the four rivets, then chisel off their heads. Remove the old ball joint. The new joint comes with four specially hardened bolts which must be torqued to 8 ft. lbs. The nut goes on top.

6. Remove the upper control arm shaft-to-bracket nuts and lock washers. Carefully note the number, thickness, and location of the adjusting shims. Remove the control arm assembly.

7. Reverse the above steps to install. Observe the following torque figures: Upper control are-to-frame nuts: 46 ft. lbs. Ball joint stud nut: 60–65 ft. lbs. Upper control arm bushing nuts: 85 ft. lbs. The upper control arm bushing nuts must be torqued with the weight of the car on the wheels.

——— CAUTION ———
When installing the cotter pin, never loosen the nut to align the cotter pin holes. Always tighten the nut to the next slot that lines up with the hole.

Lower Control Arm Or Spring

REMOVAL & INSTALLATION

1. Raise the front of the car and remove the wheel.

2. Disconnect and remove the shock absorber.

3. Remove the front stabilizer rod link from the lower control arm.

4. Disconnect the brake reaction rod from the lower control arm.

5. As a safety precaution and to gain maximum leverage, place a jack about ½ in. below the lower ball joint stud. Now, remove the ball stud cotter pin and loosen the nut about ⅛ in. Do not remove the nut.

——— CAUTION ———
If the nut is removed, the full force of the coil spring could be released.

6. Rap the steering knuckle in the area of the stud or use a ball joint removal tool to separate the stud from the knuckle.

7. After the stud has broken loose from the knuckle, raise the jack against the control arm. Remove the nut and separate the steering knuckle from the tapered stud.

8. Carefully lower the jack under the control arm and release the spring. With the jack entirely lowered, it may be necessary to pry the spring off its seat on the lower control arm with a pry bar.

9. After the spring is removed, the lower control arm may be removed by removing the lock nut which attaches the control arm to the frame.

10. Reverse to install. Torque the control arm to frame bolts, with the car on the ground, to 85 ft. lbs.

Lower Ball Joint

REMOVAL & INSTALLATION

1. Refer to Steps 1–7 of the "Lower Control Arm" procedure.

2. Install a ball joint remover and tighten the tool to force the ball joint out of the lower control arm.

3. Reverse the above to install. Tighten the castellated nut to 85–90 ft. lbs. Always tighten the castellated nut to the next slot if necessary to align the cotter pin.

Front Wheel Bearing

ADJUSTMENT

1. Lift the wheel off the ground by jacking under the lower control arm.

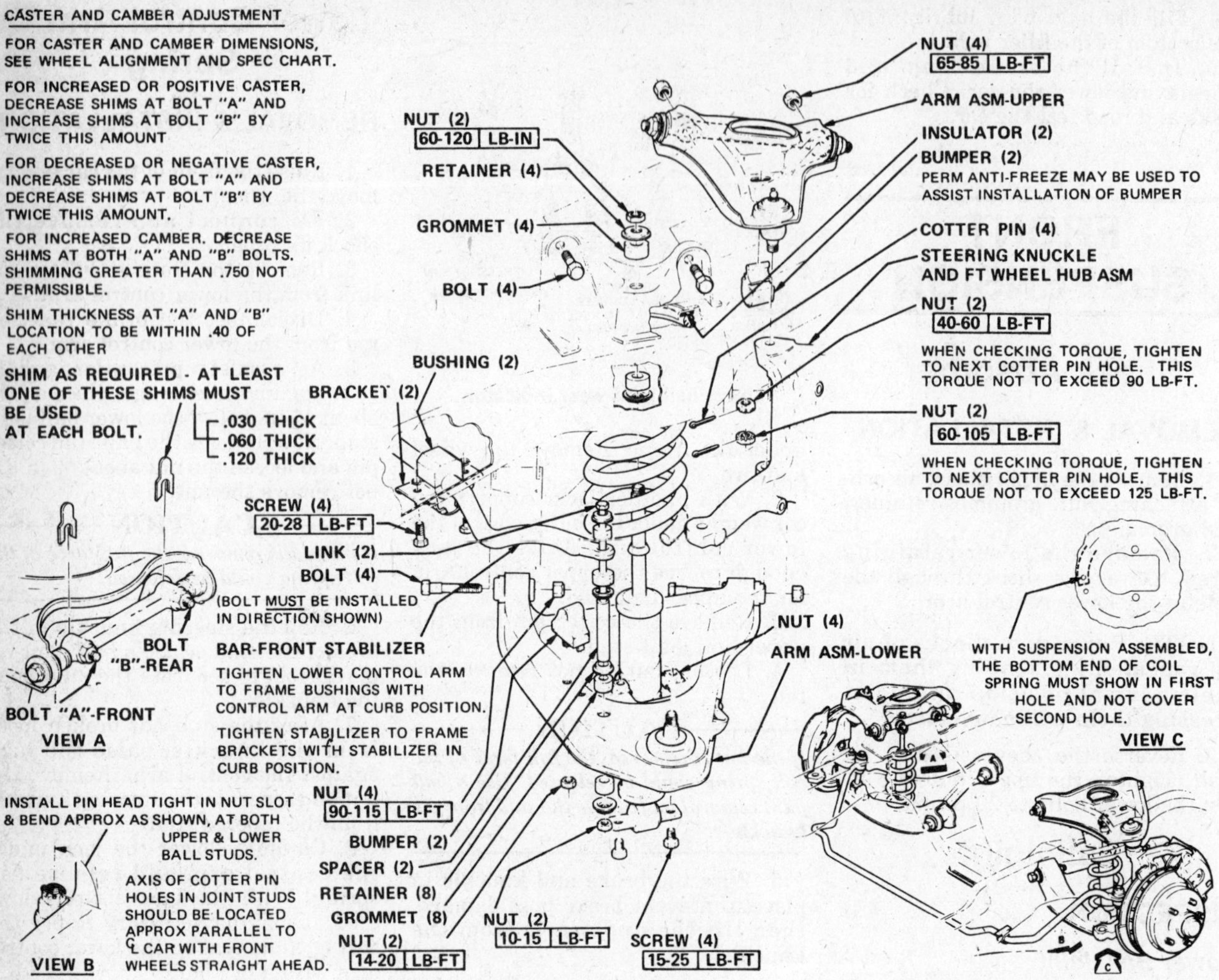

CASTER AND CAMBER ADJUSTMENT

FOR CASTER AND CAMBER DIMENSIONS, SEE WHEEL ALIGNMENT AND SPEC CHART.

FOR INCREASED OR POSITIVE CASTER, DECREASE SHIMS AT BOLT "A" AND INCREASE SHIMS AT BOLT "B" BY TWICE THIS AMOUNT

FOR DECREASED OR NEGATIVE CASTER, INCREASE SHIMS AT BOLT "A" AND DECREASE SHIMS AT BOLT "B" BY TWICE THIS AMOUNT

FOR INCREASED CAMBER, DECREASE SHIMS AT BOTH "A" AND "B" BOLTS. SHIMMING GREATER THAN .750 NOT PERMISSIBLE.

SHIM THICKNESS AT "A" AND "B" LOCATION TO BE WITHIN .40 OF EACH OTHER

SHIM AS REQUIRED - AT LEAST ONE OF THESE SHIMS MUST BE USED AT EACH BOLT.
- .030 THICK
- .060 THICK
- .120 THICK

BOLT "B"-REAR

BOLT "A"-FRONT

VIEW A

INSTALL PIN HEAD TIGHT IN NUT SLOT & BEND APPROX AS SHOWN, AT BOTH UPPER & LOWER BALL STUDS.

AXIS OF COTTER PIN HOLES IN JOINT STUDS SHOULD BE LOCATED APPROX PARALLEL TO ℄ CAR WITH FRONT WHEELS STRAIGHT AHEAD.

VIEW B

NUT (2) | 60-120 | LB-IN
RETAINER (4)
GROMMET (4)
BOLT (4)
BUSHING (2)
BRACKET (2)
SCREW (4) | 20-28 | LB-FT
LINK (2)
BOLT (4)
(BOLT MUST BE INSTALLED IN DIRECTION SHOWN)

BAR-FRONT STABILIZER
TIGHTEN LOWER CONTROL ARM TO FRAME BUSHINGS WITH CONTROL ARM AT CURB POSITION.
TIGHTEN STABILIZER TO FRAME BRACKETS WITH STABILIZER IN CURB POSITION.

NUT (4) | 90-115 | LB-FT
BUMPER (2)
SPACER (2)
RETAINER (8)
GROMMET (8)
NUT (2) | 14-20 | LB-FT
NUT (2) | 10-15 | LB-FT
SCREW (4) | 15-25 | LB-FT

NUT (4) | 65-85 | LB-FT
ARM ASM-UPPER
INSULATOR (2)
BUMPER (2)
PERM ANTI-FREEZE MAY BE USED TO ASSIST INSTALLATION OF BUMPER
COTTER PIN (4)
STEERING KNUCKLE AND FT WHEEL HUB ASM
NUT (2) | 40-60 | LB-FT
WHEN CHECKING TORQUE, TIGHTEN TO NEXT COTTER PIN HOLE. THIS TORQUE NOT TO EXCEED 90 LB-FT.
NUT (2) | 60-105 | LB-FT
WHEN CHECKING TORQUE, TIGHTEN TO NEXT COTTER PIN HOLE. THIS TORQUE NOT TO EXCEED 125 LB-FT.

NUT (4)
ARM ASM-LOWER

WITH SUSPENSION ASSEMBLED, THE BOTTOM END OF COIL SPRING MUST SHOW IN FIRST HOLE AND NOT COVER SECOND HOLE.

VIEW C

Typical front suspension

FRAME
BUSHING (8)
SLEEVE (4)
SHACKLE (4)
NUT (4) | 100-125 | LB-FT
BOLT (4)
VIEW B

UNDERBODY BUMPER ASM
BRACKET
NUT (4) | 120-180 | LB-IN
NUT (2) | 120-180 | LB-IN
BOLT (4)
BOLT (2)
NUT | 15-25 | LB-FT
VIEW D

SHOCK ABSORBER
L-WASHER (2)
NUT (2) | 55-75 | LB-FT
VIEW E

U-BOLT (4)
NUT (2) | 65-80 | LB-FT
CUSHION (2)
BOLT (2)
CUSHION (2)
PLATE
LUBRICATE AREAS INDICATED BEFORE ASSEMBLY
VIEW A
SPRING ASM
NUT (8) | 30-45 | LB-FT

BUMPER ASM (2)
SCREW (4) | 72 MIN | LB-IN
MUST BE TIGHT BUT NOT STRIPPED
VIEW C

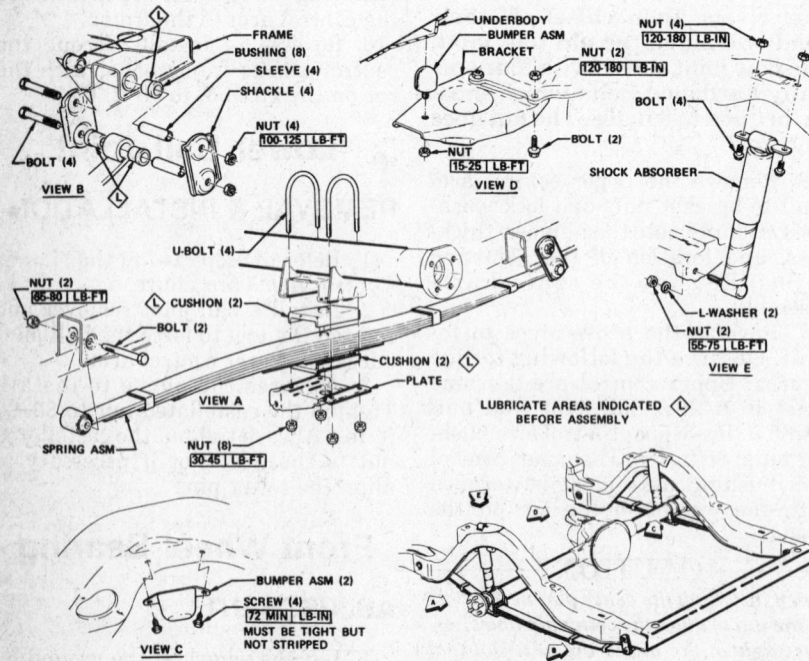

Rear leaf spring suspension details

2. Remove the dust cap from the hub.

3. Remove the cotter pin and discard it.

4. Tighten the spindle nut to 12 ft. lbs. while turning the wheel. Then back off the nut ¼–½ turn.

5. Retighten the nut by hand until it is finger-tight.

6. Loosen the nut no more than ¹⁄₆ of a turn until the nearest hole in the spindle lines up with the slot in the spindle nut, and insert a new cotter pin.

7. Feel the looseness in the hub assembly. There should be 0.001–0.005 in. end-play.

8. Replace the dust cover and lower the car.

REAR SUSPENSION

Shock Absorber

REMOVAL & INSTALLATION

NOTE: Purge new shocks of air

by repeatedly extending them in their normal position and compressing them while inverted.

1. Raise the car at the axle housing.

2. Remove the nut, retainer, and grommet, or nut and lockwasher, as equipped which attaches the lower end of the shock absorber to its mounting.

3. Remove the two shock absorber upper attaching screws and remove the shock absorber.

4. Reverse the removal procedures to install. Tighten the upper bolts to 18–20 ft. lbs. for all models. The nuts that lock the upper bolts on some models are torqued to 12 ft. lbs. Tighten the lower nut to 65 ft. lbs.

Leaf Spring

REPLACEMENT

1. Raise the rear of the car on stands.

2. Support the rear axle to take its weight off the springs.

3. Disconnect the bottom of the shock absorber.

4. Loosen the front spring eye bolt.

5. Unbolt the spring front bracket from the underbody.

6. Lower the axle slightly and remove the front bracket from the spring.

7. Pry the parking brake cable out of its retainer bracket on the axle spring mounting plate.

8. Unbolt the spring from the axle.

9. Remove the spring plate and cushion from the bottom of the spring. There should also be a cushion between the axle and the spring.

10. Remove the upper bolt from the rear spring shackle. Lower the spring and remove the bottom bolt.

11. On installation, attach the front bracket to the spring eye. The head of the bolt should be toward the center of the car.

12. Assemble the shackle loosely to the rear spring eye.

13. Raise the rear end of the spring and install the upper shackle bolt loosely, making sure that the parking brake cable goes under the spring.

14. Raise the front end of the spring and loosely attach the front bracket to the underbody. Make sure that the bracket tab goes into its slot.

15. Make sure that the upper and lower spring cushions are aligned properly. The upper one has locating ribs and the lower one, a locating dowel.

16. Install the spring lower mounting plate over the locating dowel and

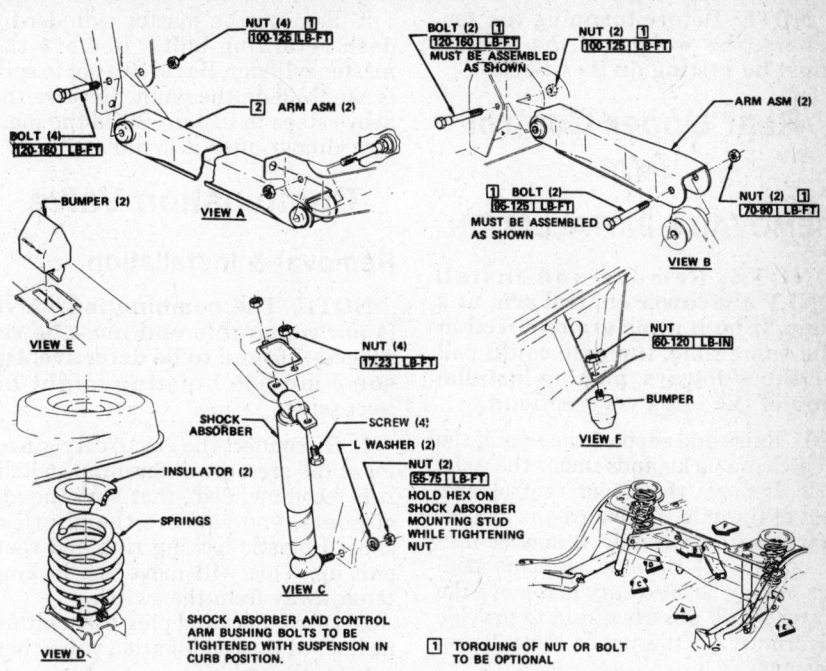

Rear suspension details—typical with coil springs

loosely install the nuts. Don't forget the parking brake cable bracket.

17. Attach the bottom of the shock absorber.

18. Attach the parking brake cable to the bracket on the lower spring plate.

19. Let the vehicle weight down on the springs. Tighten all the bolts. Torques are: rear shackle bolts — 40–60 ft. lbs.; front eye bolt — 65–80 ft. lbs., and axle bolts — 35–50 ft. lbs.

Coil Spring

REMOVAL & INSTALLATION

1. Jack up the back of the car and support both sides on jack stands on the frame, in front of the rear axle. Support the rear axle with an adjustable lifting device. Disconnect the shock absorber.

2. Detach the upper control arm at the differential.

3. Disconnect the stabilizer bar, if so equipped.

4. Remove any brake hose supports but disconnect the brake hose, only if necessary.

5. Carefully lower the axle until the tension is released from the coil spring. Be careful not to stretch the brake hose. Remove the spring. Note the direction in which the end of the last coil is pointing. Install the spring in the same position.

6. When installing a new coil spring, make certain that the bottom

of the coil is properly inserted into the socket in the frame and into the form plate on the trailing arm.

7. Jack the axle into place and reinstall the control arm bolt. Tighten the bolts with the car's weight on the springs.

Rear Lower Control Arm

REMOVAL & INSTALLATION

NOTE: Remove and install ONLY one lower control arm at a time. If both arms are removed at the same time, the axle could roll or slip sideways, making installation of the arms very difficult.

1. Raise and support the rear of the vehicle on jackstands under the rear axle. If equipped with a stabilizer bar, remove it.

2. Remove the control arm attaching fasteners and the control arm.

3. To install, reverse the removal procedures. Torque the control arm-to-frame nut to 92 ft. lbs. (LeSabre and Electra) or 70 ft. lbs. (Century and Regal), the control arm-to-axle nut to 92 ft. lbs. (Impala and Caprice) and the control arm-to-axle bolt to 125 ft. lbs. (LeSabre and Electra) or 79 ft. lbs. (Century and Regal). If equipped with a stabilizer bar, torque the mounting fasteners to 52 ft. lbs. (LeSabre and Electra) or 35 ft. lbs. (Century and Regal).

NOTE: Before torquing the fasteners, the weight of the vehicle must be resting on its wheels.

Rear Upper Control Arm

REMOVAL & INSTALLATION

NOTE: Remove and install ONLY one lower control arm at a time. If both arms are removed at the same time, the axle could roll or slip sideways, making installation of the arms very difficult.

1. Raise and support the rear of the vehicle on jackstands under the axle.
2. Remove the upper control arm nut at the axle. To remove the mounting bolt from the axle, it may be necessary to rock the axle. On some models, it may be necessary to remove the lower shock absorber stud to provide clearance for the upper control arm removal.
3. Remove the upper control arm-to-frame nut and bolt, then the control arm.
4. To install, reverse the removal procedures. Torque the upper control arm-to-axle nut to 70 ft. lbs., the upper control arm-to-axle bolt to 79 ft. lbs. and the upper control arm-to-frame bolt to 92 ft. lbs. (LeSabre and Electra) or 70 ft. lbs. (Century and Regal).

Rear Wheel Bearings

For rear wheel bearing removal and installation, please refer to the "Drive Axle" section.

BRAKES

For information of the brake system not detailed below, please refer to "Brakes" in the Unit Repair section.

Master Cylinder

REMOVAL & INSTALLATION

1. Disconnect the brake lines from the master cylinder and tape the end of the lines to keep dirt out.
2. Disconnect the brake pedal from the master cylinder at the pushrod.

NOTE: This step isn't required with power brakes

3. Remove the master cylinder-to-dash retaining bolts. Remove the master cylinder. Be careful not to spill brake fluid on the paint. Reverse the above steps to install. Bleed the master cylinder after it is reinstalled.

Combination Valve

Removal & Installation

NOTE: The combination valve is not repairable and must be replaced if found to be defective. On some models hoisting might be necessary.

1. Disconnect the electrical connector at the pressure differential switch. It is recommended, that with the aid of pliers, you squeeze the eliptical shaped plastic locking ring and then pull up. This will move the locking tangs away from the switch.
2. Disconnect and plug the hydraulic lines at the combination valve then remove the valve.
3. Installation is the reverse of removal.
4. Bleed the entire brake system.

——————— CAUTION ———————
Do not move the car until a firm brake pedal is obtained.

Power Brake Booster

REMOVAL & INSTALLATION

1. Unbolt the master cylinder from the power unit. Being careful not to kink or bend the brake lines, pull the master cylinder away from the power unit without disconnecting the brake lines.
2. Disconnect and plug the vacuum hose.
3. Disconnect the power brake pushrod from the brake pedal.
5. Remove the unit.
6. To install, mount the unit to the firewall.
7. Install the master cylinder to the power unit and torque the nuts to 15 ft. lbs. on the Century (through 1981) and Regal, and 25 ft. lbs. on all others.
8. Connect the vacuum hose.
9. Connect the power brake pushrod to the brake pedal.

Wheel Cylinder

Removal & Installation

LeSabre and Electra

1. Raise and safely support the car.
2. Mark the relationship of the wheel to the axle flange.
3. Remove the wheel, drum and brake shoes.
4. Clean all dirt around around the wheel cylinder at the brake line and disconnect the brake line.
5. Remove the wheel cylinder from the backing plate.
6. Installation is the reverse of removal. Torque the rear wheel brake pipe to wheel cylinder to 12 ft. lbs.

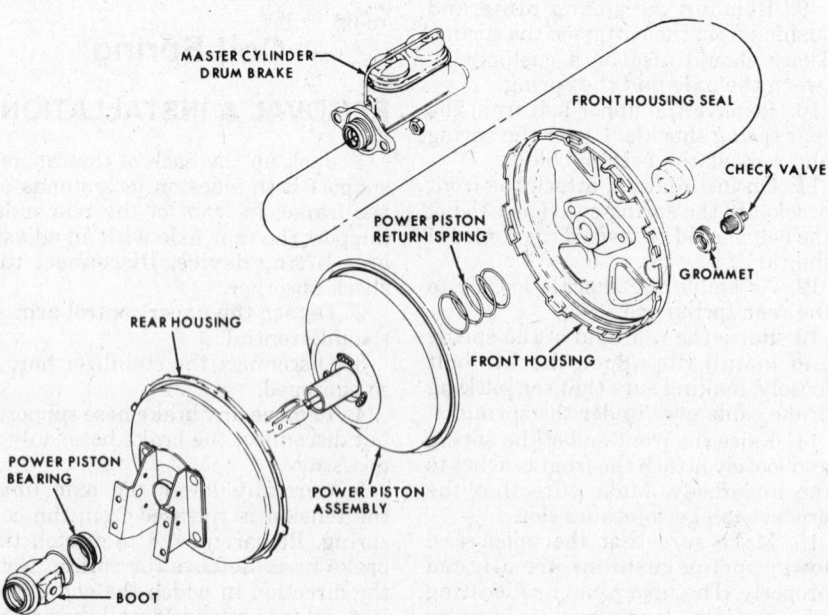

Power brake unit and master cylinder

7. Bleed the system.

Century and Regal

1. Insert awls or pins, $^1/_8$ in. diameter or less, into the access slots between the wheel cylinder pilot and retainer locking tabs.

2. Bend both tabs away simultanously until they spring over the abutment shoulder releasing the wheel cylinder. Discard the old retaining clip.

3. For ease of installation hold the wheel cylinder against the backing plate by inserting a block betwen the wheel cylinder and the axle shaft flange.

4. Position the wheel cylinder retainer clip so the tabs will be away from and in a horizontal position with the backing plate when installing.

5. Press the new retaining clip over the wheel cylinder abutment and into position using a $1^1/_8$ in. 12 point socket. Make sure the retainer tabs are properly snapped under the abutment shoulder.

6. Install the brake shoes, drum and wheel.

7. Flush and bleed the hydraulic system.

Parking Brake

ADJUSTMENT

NOTE: Be sure that the parking brake does not drag. An overtightened, dragging parking brake on a car with automatic brake adjusters will result in an extremely short life for rear brake linings.

Adjustment of the parking brake is necessary whenever the rear brake cables have been disconnected or the parking brake pedal can be depressed more than eight rachet clicks under heavy foot pressure. The car should first be raised on a lift.

1. Make sure that the service brakes are properly adjusted.

2. Depress the parking brake pedal two rachet clicks.

3. Loosen the jam nut on the equalizer adjusting nut. Tighten the adjusting nut until the left rear wheel can just be turned rearward by hand, but not forward.

4. Release the rachet one click; the rear wheel should rotate rearward freely and forward with a slight drag.

5. Release the rachet fully; the rear wheel should turn freely in either direction.

STEERING

Steering Wheel

REMOVAL & INSTALLATION

Except Tilt and Telescope Column

1. Disconnect the battery ground and unplug the horn wire connector from the steering column.

2. On cars with a standard wheel or optional wood-rim wheel, pull off the cap, remove the three screws and the contact, insulator, and spring. On cars with the bar-type horn actuator, remove the screws securing the actuator from the underside of the steering wheel, unhook the lead connector plug, and remove the actuator assembly.

3. Loosen the steering wheel nut.

4. Apply the steering wheel puller and pull the wheel up to the nut. Now remove the puller, nut and steering wheel.

—————— CAUTION ——————
Don't pound on the steering wheel in either direction or the collapsible steering columns will collapse, requiring replacement.

5. Location marks are provided on the steering wheel and shaft to simplify proper indexing at the time of installation. Install wheel with the location mark aligned with that of the shaft.

6. Install the wheel nut and torque to 30 ft. lbs.

7. Reinstall horn button or actuator assembly.

Tilt and Telescope Column

1. Disconnect the battery ground.

2. Remove the attaching screws and lift the pad from the column.

3. Disconnect the horn wire by pushing in the connector and turning it counterclockwise.

4. Push the locking lever counterclockwise until full release is obtained.

5. Mark the lock plate-to-locking lever position and remove the plate and lever.

6. Remove the steering wheel retaining nut and remove the wheel with a puller.

7. Install a $^5/_{16}$ in. × 18 set screw into the upper shaft at the fully extended position and lock it.

8. Install the steering wheel, observing the aligning mark on the hub and the slash mark on the end of the shaft. Make certain that the unat-

tached end of the horn upper contact assembly is seated flush against the top of the horn contact carrier button.

9. Install the nut on the upper steering shaft and torque to 30 ft. lbs.

10. Remove the set screw installed in Step 7.

11. Install the plate assembly finger tight.

12. Position the locking lever in the vertical position and move it counterclockwise until the holes in the plate align with the holes in the lever. Install the attaching screws.

13. Align the pad assembly with the holes in the steering wheel and install the retaining screws.

14. Connect the battery.

15. Make certain that the locking lever securely locks the wheel travel and that the wheel travel is free in the unlocked position.

Turn Signal Switch

REMOVAL & INSTALLATION

Except Tilt and Telescope Column

NOTE: The steering wheel must always be supported. Use extreme care not to bend the steering column.

1. Remove the steering wheel.

2. Remove the three cover screws and the cover. Steering columns have a lock plate which is removed by inserting a screwdriver in the cover slot and prying out. This is done in at least two of the slots to avoid breaking the plate.

3. Depress the lock plate and remove the snap-ring. Remove the lock plate.

4. Remove the spring and horn contact signal canceling cam. Remove the thrust washer.

5. Place the turn signal lever in the right turn position, remove the attaching screw and remove the turn signal lever. On models with the dimmer switch mounted on the column, remove the actuator arm screw and the actuator arm. Pull the turn signal lever straight out to remove. Depress the hazard warning knob, and remove the knob. Some models have a screw in the end of the knob which must be removed.

6. Remove the three turn signal switch mounting screws.

7. Remove the instrument panel lower trim panel and disconnect the turn signal connector from the harness.

8. Remove the four bracket attaching screws and remove the bracket.

9. On models with automatic transmissions, loosen the shift indicator

needle attaching screw and remove the needle.

10. Remove the two steering column supporting bolts while supporting the column. Do not allow the column to drop suddenly.

11. Remove the bracket and wiring from the column. Loosely reinstall the column supporting bolts, if removed.

12. Pull the switch straight up with the wire protector and wire harness.

13. Reverse the above steps to install.

Tilt and Telescope Column

1. Disconnect the battery ground.

2. Remove the steering wheel and lock plate as previously described.

3. Remove the upper bearing pre-load spring.

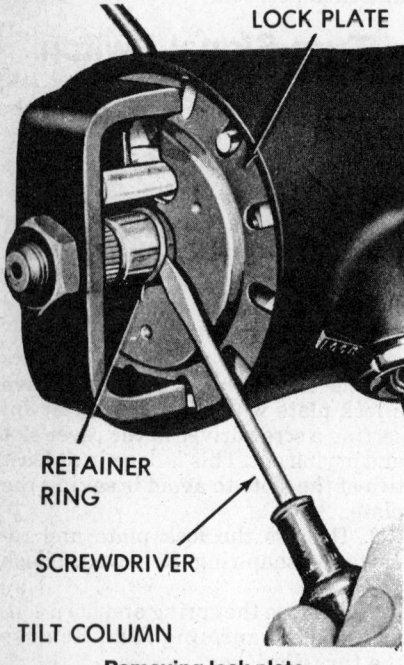

LOCK PLATE

RETAINER RING

SCREWDRIVER

TILT COLUMN

Removing lock plate

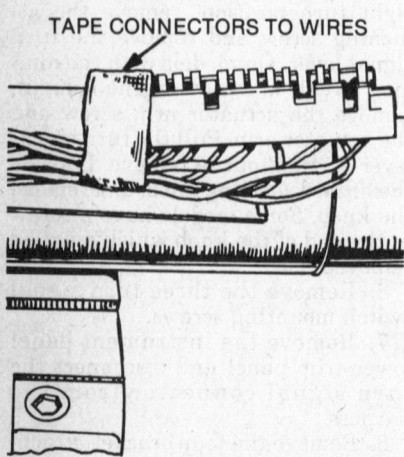

TAPE CONNECTORS TO WIRES

Tape the connector to the wires so that it will slip easily up the steering column

4. Position the turn signal lever in the right turn position and remove the lever and screw.

5. With column mounted dimmer switches, remove the actuator arm and screw, then remove the turn signal arm by pulling it straight out.

6. Push in on the warning hazard knob, then remove the retaining screw and knob.

7. Position the column in the center position and remove the three turn signal switch attaching screws.

8. Remove the instrument panel lower trim pad and disconnect the turn signal harness connector. Lift the connector from the mounting bracket on the right side of the jacket.

9. Remove the toe pan bolts.

10. Remove the four bolts attaching the bracket assembly to the jacket.

11. Remove the shift indicator retaining clip.

12. Support the column and remove the bracket assembly. Remove the wire protector from the turn signal wiring. Pull the turn signal switch and wiring from the column.

13. Prior to installation, coat all moving parts with lithium-base grease.

14. Insert switch wiring into the column.

15. Place the switch in the right turn position and push it straight down until seated.

NOTE: Angling or cocking of the switch can cause damage to the buzzer terminal or tangs.

16. Install the switch attaching screws and torque them to 25 inch lbs.

17. Position the turn signal in the center.

18. Connect the wiring to the harness.

19. Install the hazard warning knob and turn signal level.

20. Install the lock plate and carrier and the steering wheel.

21. Install the wiring protector and bracket. Torque the bracket bolts to 18 ft. lbs. and the nuts to 24 ft. lbs.

22. Install the shift indicator needle or clip.

23. Position the harness connector in the bracket on the right side of the jacket.

24. Install the instrument panel lower trim pad and connect the battery ground.

Ignition Switch and Lock Cylinder

REMOVAL & INSTALLATION

Standard Column

1. Refer to the "Turn Signal Switch

Replacement" procedure, Steps 1–6

2. Disconnect the turn signal connector from the harness and pull out the turn signal switch. Allow it to hang.

3. With the lock cylinder in the RUN position, insert a small screwdriver into the slot next to the turn signal switch mounting screw boss (right-hand slot), depress the spring latch and remove the key lock.

4. Pull the buzzer switch straight out, depressing the switch clip with pliers.

5. Place the ignition switch in the OFF-UNLOCKED position by pulling up on the connecting rod until there is a definite stop or detent felt.

6. Remove the two attaching screws and the ignition switch.

7. Assembly is the reverse of the above. However, note the following steps before proceeding with the reassembly.

8. To install the steering lock, hold the lock cylinder sleeve and rotate the knob clockwise against the stop. Insert the cylinder into the cover bore with the key on the cylinder sleeve aligned with the keyway in the housing. Then push the cylinder until it bottoms. Maintaining a light inward pressure, rotate the knob counterclockwise until the drive section of the cylinder mates with the drive shaft. Push in until the snapring pops into the groove and the lock cylinder is secured in the cover. Check for free rotation.

9. Move the switch slider to the extreme left position (ACC), then two detents to the right, to the OFF-UNLOCKED position. Fit the actuator rod into the hole and attach the switch to the column.

10. The neutral start switch is adjusted with he shift lever in the Drive position.

Tilt Column

1. Refer to the "Turn Signal Switch Replacement" procedure for tilt and telescopic columns, Steps 1–6.

2. Position the tilt column in the center position and remove the three turn signal switch screws. Tape the wires to the wire connector at the upper end and place the shift bowl in Low. Pull the switch straight up and out, allowing it to hang.

3. Insert a small screwdriver into the slot next to the turn signal switch mounting screw boss (right-hand slot), depress the spring latch and remove the key lock. Remove the retaining screw and the lock cylinder.

4. Remove the buzzer switch straight out, depressing the switch clip with pliers.

5. Remove the three housing cover screws and cover.

6. Install the tilt release lever and place column in full UP position.

7. Place a screwdriver in the slot of the tilt spring retainer, press in about $3/16$ in. and turn counterclockwise. Remove the spring and guide.

NOTE: The spring is very strong; be careful.

8. Place the column in neutral position, push in on the upper steering shaft, remove the inner race seat and race.

9. Remove the upper flange pinch bolt, place the ignition switch in the accessory position, remove the two switch mounting screws and switch.

NOTE: The neutral start switch can be removed at this time, if necessary.

10. Assembly is the reverse of the above. However, note the following steps before proceeding with the reassembly.

11. To install the steering lock, hold the lock cylinder sleeve and rotate the knob clockwise against the stop. Insert the cylinder into the cover bore with the key on the cylinder sleeve aligned with the keyway in the housing. Push the cylinder in until it bottoms. Maintaining a light inward pressure, rotate the knob counterclockwise until the drive section of the cylinder mates with the drive shaft. Push in until the snap-ring pops into the groove and the lock cylinder is secured in the cover. Check for free rotation.

12. When installing the ignition switch, be sure the lock cylinder is in the LOCK position. Put the shift bowl or shroud in the PARK position. Make sure the ignition switch is in the LOCK position. Insert the actuator rod into the switch and assemble the switch to the column.

13. The neutral-start switch is adjusted with the shift lever in the DRIVE position.

Steering Gear

REMOVAL & INSTALLATION

——— **CAUTION** ———

Failure to disconnect the flexible coupling from the steering gear stub shaft can result in damage to the steering gear and/or intermediate shaft. This damage can cause loss of steering control which could result in a vehicle crash and bodily injury.

1. Remove flexible coupling shield.
2. Disconnect the hoses from the gear and cap the hose fittings.

3. Hoist the car.
4. Remove the Pitman shaft nut, then disconnect the Pitman arm from the pitman shaft using Puller J-29107 or a similar puller.
5. Remove the three bolts attaching the gear to the frame side rail and remove the gear with the hoses attached.

NOTE: If Mounting threads are stripped, do not repair. Replace housing.

6. Installation is the reverse of removal. Tighten the steering gear to frame bolts 80 ft. lbs. and the Pitman shaft nut 180 ft. lbs.

Power Steering Pump

REMOVAL & INSTALLATION

1. Remove the hoses at the pump and tape the openings shut to prevent contamination. Position the disconnected lines in a raised position to prevent leakage.
2. Remove the pump belt.
3. Loosen the retaining bolts and any braces, and remove the pump.
4. Install the pump on the engine with the retaining bolts hand-tight.
5. Connect and tighten the hose fittings.
6. Refill the pump with fluid and bleed by turning the pulley and adjust the tension.

Tie Rod End

REMOVAL & INSTALLATION

1. Raise and support the car. Loosen the tie rod adjuster sleeve clamp nuts.
2. Remove the tie rod stud nut cotter pin and nut.
3. Remove the tie rod stud from the steering arm or intermediate rod. This is a taper fit. Removal is accomplished using a ball joint removal tool or by hitting the steering arm sharply with a hammer, while using a heavy hammer as a backup.
4. Unthread the tie rod from the adjusted sleeve. Outer tie rods have right-hand threads and inner tie rods have left-hand threads. Count the number of turns the tie rod must be rotated to remove it from the adjusting sleeve. This will allow a reasonably accurate realignment upon reassembly.

NOTE: If a turning force of more than 7 ft. lbs. is needed for removal, after breakaway, the nuts and bolts should be replaced.

5. Reverse the removal procedures

to install. Clean rust and dirt from the threads. Observe the following torque specifications: steering arm-to-tie rod end nut, 35 ft. lbs.; tie rod clamp nuts, 11–14 ft. lbs.; tie rod-to-intermediate nut, 40 ft. lbs. Check the alignment and adjust as necessary.

CHASSIS ELECTRICAL

Blower Motor

NOTE: Vacuum hose routing clips, electrical wires and relays, weather seals, and other items, may be attached to the heater housing, and will have to be re-located during removal and replacement of the heater core and/or the blower motor. Always tag any disconnected hoses or wires for installation.

REMOVAL & INSTALLATION

Century and Regal

1. Disconnect the blower motor wire.
2. Remove the blower motor attaching screws and the motor.

Electra and LeSabre

1. Disconnect the blower motor wires.
2. On A/C equipped cars, disconnect the cooling tube from the case.
3. Remove the motor attaching screws and lift the motor from the case.
4. Installation is the reverse of removal. Replace any damaged sealer.

Heater Core

REMOVAL & INSTALLATION

Without A/C

CENTURY AND REGAL

1. Disconnect the heater hoses at the core tubes. Place the hoses in an up position to prevent excess coolant loss.
2. Disconnect all electrical connectors at the module case.
3. Remove the front case from the module on 1979 models and the top module cover on 1980 and later models.
4. Remove the core. On later models, remove core bracket and ground screws to gain access.

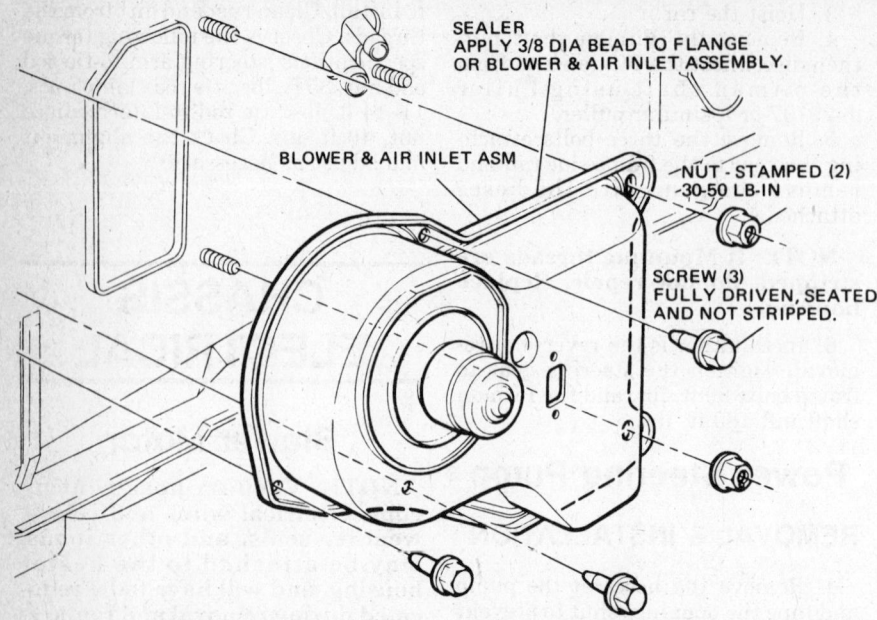

SEALER
APPLY 3/8 DIA BEAD TO FLANGE
OR BLOWER & AIR INLET ASSEMBLY.

BLOWER & AIR INLET ASM

NUT - STAMPED (2)
-30-50 LB-IN

SCREW (3)
FULLY DRIVEN, SEATED
AND NOT STRIPPED.

Blower motor and air inlet assembly—Century, Regal

5. Reverse the above for installation. Replace any damaged sealer.

ELECTRA AND LESABRE

1. Drain the radiator.
2. Disconnect the heater core inlet and outlet hoses at the firewall.
3. Detach the electrical connections.
4. Remove the screws holding the front case to the heater module assembly.
5. Remove the front case, then the heater core.
6. On installation, reseal the case.

Heater Core

REMOVAL & INSTALLATION

With A/C

CENTURY AND REGAL

1. Engage the right-hand wiper arm so it is in the UP position.
2. Drain the radiator enough so you can disconnect the heater core hoses, then disconnect the hoses and plug them. Disconnect the battery ground cable.
3. Pull off the trim seal and remove the screens from the assembly. Mark and remove any electrical connections in the way.
4. Loosen and move up the lower windshield trim. Remove the windshield molding cowl brackets.
5. Tape a strip of wood below the lower edge of the windshield glass near the module for protection. Remove all module cover screws.
6. Cut through the sealing material along the cowl with a knife.

7. Pry the module cover off from the side, not down from the top, to insure the windshield isn't damaged.
8. Lift the cover off and away from the flange of the fender-cowl brace.
9. Remove the core.
10. Reverse to install. Use new strip-caulk sealer.

ELECTRA AND LESABRE

1. Disconnect the battery ground cable.
2. Drain the coolant. Disconnect the heater hoses at the firewall.
3. Disconnect the electrical connections. Remove the diagnostic connector.
4. Remove the thermostatic switch from the heater/conditioning module cover.
5. Remove the weather seal on top of the module cover.
6. Remove the cowl screen and windshield washer nozzle.
7. Remove the screws and take off the module cover.
8. Remove the core retaining clip, twist the heater core, and pull it up and out.
9. On installation, reseal the module cover.

Wiper Motor

REMOVAL & INSTALLATION

1. Disconnect the battery.
2. Loosen the two nuts on the adjustable motor drive link at the crank arm and slip the drive link off.
3. Remove the electrical connectors from the washer motor and pump.

4. Disconnect the washer pump hoses.
5. Remove the three bolts securing the motor to the cowl and carefully lift the motor away from the cowl.
5. Reverse the above steps for installation.

Wiper Blade

REMOVAL & INSTALLATION

Any one of three methods of blade attachment may be used on these models. If there is a small tab on top of the blade, depress it and slide off the blade. If there is a small spring visible in the top of the blade, insert a screwdriver in the opening, press down and slide the blade off. If there is a clip on the under side of the arm, press down on the clip and slide the blade off. Wiper blade element replacement is covered in the Maintenance Unit Repair Section.

Radio

The antenna trim must be adjusted on AM radios, when major repair has been done to the unit or the antenna changed. The trimmer screw is located behind the right side knob. Raise the antenna to its full height. Tune to a weak station around 1400 and turn the volume down until barely audible. Turn the trimmer screw until the maximum volume is achieved.

REMOVAL & INSTALLATION

--- CAUTION ---
Don't turn on the radio without first connecting the speaker. The output transistors may be damaged.

Century and Regal

1. Disconnect the negative battery cable and remove the radio knobs.
2. Remove the center trim plate.
3. Remove the glove box to gain access to the radio.
4. Disconnect the radio mounting bracket.

Antenna trim screw

5. Disconnect the radio wiring.

6. Remove the radio with the bracket attached. Reverse to install.

Electra and LeSabre

1. Disconnect the battery ground cable.

2. Remove the ashtray and bracket.

3. Pull off the radio knobs and trim washers.

4. Remove the lower left air duct.

5. Remove the two retaining nuts from the control shafts.

6. Unplug the power lead, speaker wire and antenna lead.

7. Remove the rear radio mounting nut.

8. Reverse the procedure for installation.

NOTE: On 1981 and later models, remove the headlight switch and place the gear shift lever in the LOW position to remove the left hand instrument panel.

Speedometer Cable

REMOVAL & INSTALLATION

1. Reach up underneath the instrument panel and disconnect the cable housing from the cluster housing. On some models you might first have to remove the left air conditioning duct.

2. Carefully pull the cable housing from the cluster housing down and pull out the cable.

3. Hold the cable vertically and turn it slowly between your fingers. If it is kinked, you will notice it flopping around. Replace any kinked cable.

4. If the cable is broken, raise and support the car. Disconnect the cable housing from the transmission, remove the gear and pull the cable from the cable housing.

5. Install the new cable in the cable housing after thoroughly lubricating it.

Instrument Cluster

REMOVAL & INSTALLATION

1. Pull steering column collar filler out.

2. Remove headlamp switch knob and bezel:

   a. To remove knob use an awl and push forward in slot on knob to release clip while pulling knob out.

   b. Bezel unscrews unless vehicle is equipped with twilight sentinel. Then it simply comes off by pulling straight out.

3. Grasp trim plate on both sides and gently pull out.

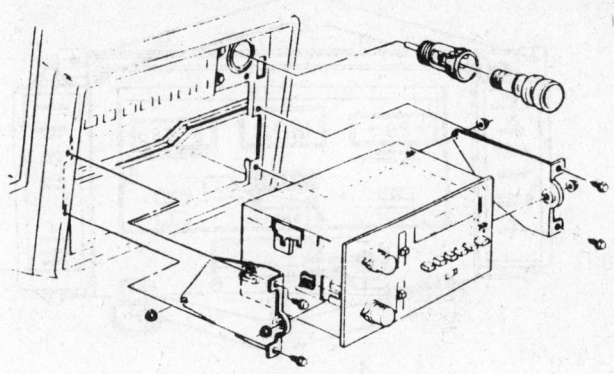

Typical radio mounting

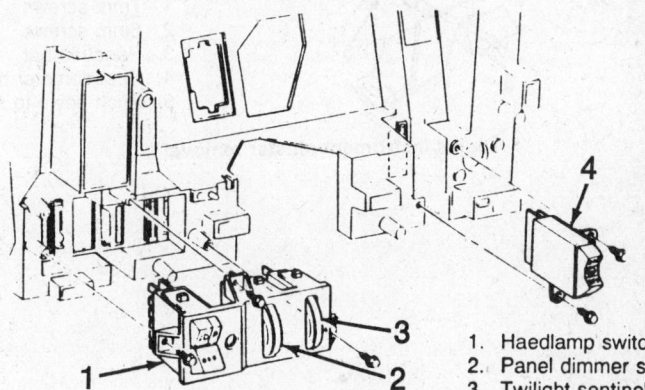

1. Haedlamp switch
2. Panel dimmer switch
3. Twilight sentinel control
4. Rear defogger switch

Typical switch mounting

NOTE: The gauges can be removed after removing the instrument panel cover.

4. Installation is the reverse of removal.

Light Switch

REPLACEMENT

1. Disconnect the battery.

2. Disconnect the multiple connector from the switch.

3. Pull the switch knob to the last notch and depress the spring loaded latch button on top of the switch while pulling the knob and rod out of the switch. Depress the retainer tab behind the knob and remove the knob.

NOTE: On cars with air conditioning, remove the left duct.

4. Remove the escutcheon and the switch.

5. Install in the reverse of the above.

Fuses

The fuse block is located beneath the instrument panel above the headlight dimmer floor switch. Fuse holders are labeled as to their service and the correct amperage. Always replace blown fuses with new ones of the correct amperage. Otherwise electrical overloads and possible wiring damage will result.

Fusible Links

Fusible links are sections of wire, with special insulation, designed to melt under electrical overload. Replacements are simply spliced into the wire. The fusible links are all two wire gauge sizes smaller than the wires they protect. There may be as many as five of these in the engine compartment wiring harnesses. These are:

1. Horn relay-to-fuse panel circuit—one link.

2. Charging circuit, from the starter solenoid to the horn relay—two links.

3. Starter solenoid-to-ammeter circuit—one link.

4. Horn relay-to-rear window defroster circuit—one link.

NOTE: Most models have fusible links at these locations.

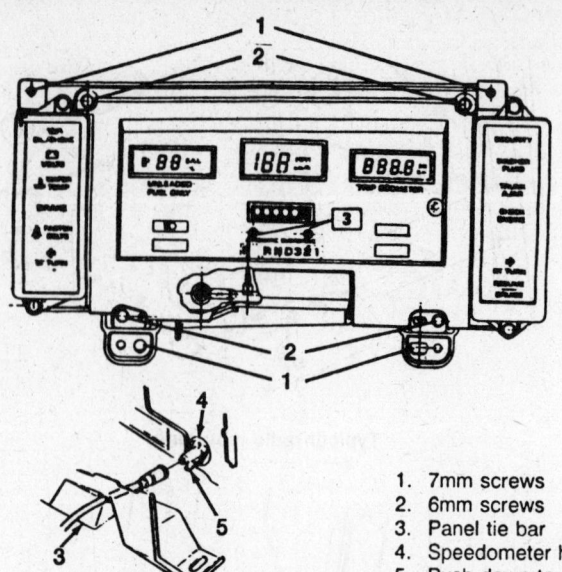

1. 7mm screws
2. 6mm screws
3. Panel tie bar
4. Speedometer head
5. Push down to release

Digital instrument cluster removal

REPLACEMENT

1. Disconnect the battery ground cable.

2. Disconnect the fusible link from the junction block or starter solenoid.

3. Cut the harness directly behind the connector to remove the damaged fusible link.

4. Strip the harness wire approximately ½ in.

5. Connect the new fusible link to the harness wire using a crimp-on connector. Solder the connection using rosin-core solder.

6. Tape all exposed wires with plastic electrical tape.

7. Connect the fusible link to the junction block or starter solenoid and reconnect the battery ground cable.

Cadillac
Rear Wheel Drive
Deville, Fleetwood

YEAR IDENTIFICATION

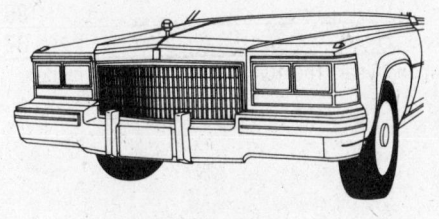

1980 Cadillac

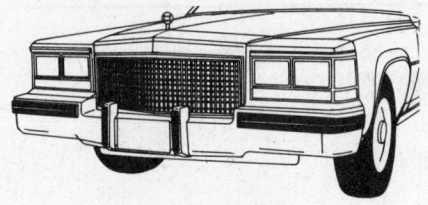

1981 Cadillac

1982-87 Cadillac

1980 VEHICLE IDENTIFICATION NUMBER (VIN)

It is important for servicing and ordering parts to be certain of the vehicle and engine identification. The VIN (vehicle identification number) is a 13 or 17 digit number visible through the windshield on the driver's side of the dash and contains the vehicle and engine identification codes. It can be interpreted as follows:

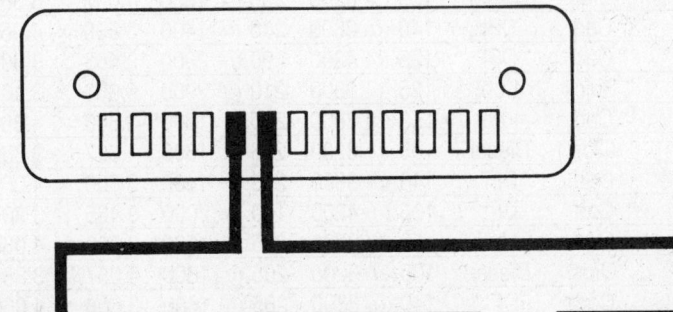

Engine Code						Model Year Code	
Code	Cu. In.	Liters	Cyl.	Carb.	Eng. Mfg.	Code	Year
N	350	5.7	8	Diesel	Olds.	A	'80
6	368	6.0	8	4bbl	Cad.		

The thirteen digit Vehicle Identification Number can be used to determine engine application and model year. The 6th digit indicates the model year, and the fifth digit identifies the factoy installed engine.

1981–87 VEHICLE IDENTIFICATION NUMBER (VIN)

It is important for servicing and ordering parts to be certain of the vehicle and engine identification. The VIN (vehicle identification number) is a 13 or 17 digit number visible through the windshield on the driver's side of the dash and contains the vehicle and engine identification codes. It can be interpreted as follows:

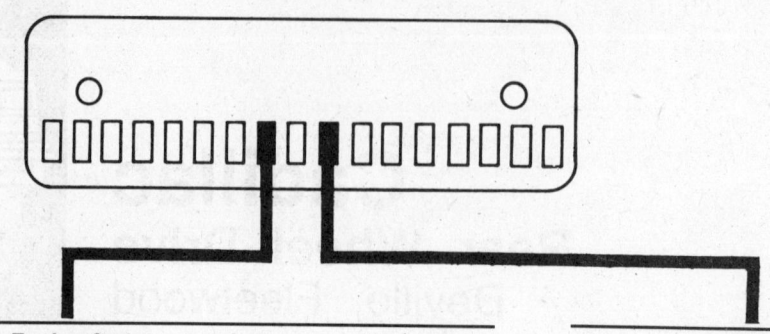

Engine Code

Code	Cu. In.	Liters	Cyl.	Carb.	Eng. Mfg.
8	250	4.1	8	DFI	Cad.
4	252	4.1	6	4bbl	Buick
N	350	5.7	8	Diesel	Olds.
9	368	6.0	8	DFI	Cad.
6	368	6.0	8	4bbl	Cad.
Y	307	5.0	8	4bbl	Olds.

Model Year Code

Code	Year
B	'81
C	'82
D	'83
E	'84
F	'85
G	'86
H	'87

The seventeen digit Vehicle Identification Number can be used to determine engine application and model year. The 10th digit indicates the model year, and the 8th digit identifies the factory installed engine.

GENERAL ENGINE SPECIFICATIONS

Year	Eng. V.I.N. Code	Engine Displacement cu. in.	Eng. Mfg.	Fuel Delivery	Horsepower @ rpm ■	Torque @ rpm (ft. lbs.) ■	Bore × Stroke (in.)	Compression Ratio	Oil Pressure @ 2000 rpm
'80	6	368	Cad.	4 bbl	150 @ 3800	265 @ 1600	3.800 × 4.060	8.2:1	35
	N	350	Olds.	Diesel	105 @ 3200	205 @ 1600	4.057 × 3.385	22.5:1	40
	9	368	Cad.	DFI	140 @ 3800	265 @ 1400	3.800 × 4.060	8.2:1	35
'81	4	252	Buick	4 bbl	125 @ 3800	210 @ 2000	3.965 × 3.400	8.0:1	35
	6	368	Cad.	4 bbl	150 @ 3800	265 @ 1600	3.800 × 4.060	8.2:1	35
	N	350	Olds.	Diesel	105 @ 3200	205 @ 1600	4.057 × 3.385	22.5:1	40
	9	368	Cad.	DFI	140 @ 3800	265 @ 1400	3.800 × 4.060	8.2:1	35
'82	8	250	Cad.	DFI	125 @ 4200	190 @ 2000	3.465 × 3.307	8.5:1	30
	4	252	Buick	4 bbl	125 @ 3800	210 @ 2000	3.965 × 3.400	8.0:1	35
	6	368	Cad.	4 bbl	150 @ 3800	265 @ 1600	3.800 × 4.060	8.2:1	35
	N	350	Olds.	Diesel	105 @ 3200	205 @ 1600	4.057 × 3.385	22.5:1	40
	9	368	Cad.	DFI	140 @ 3800	265 @ 1400	3.800 × 4.060	8.2:1	35
'83–'84	8	250	Cad.	DFI	135 @ 4200	190 @ 2000	3.465 × 3.307	8.5:1	30
	6	368	Cad.	4 bbl	150 @ 3800	265 @ 1600	3.800 × 4.060	8.2:1	35
	N	350	Olds.	Diesel	105 @ 3200	205 @ 1600	4.057 × 3.385	22.5:1	40
	9	368	Cad.	DFI	140 @ 3800	265 @ 1400	3.800 × 4.060	8.2:1	35
'85–'87	8	250	Cad.	DFI	135 @ 4200	265 @ 1400	3.465 × 3.307	8.5:1	35
	N	350	Olds.	Diesel	105 @ 3200	205 @ 1600	4.057 × 3.385	22.5:1	40
	Y	307	Olds.	4 bbl	140 @ 3600	240 @ 1600	3.800 × 3.385	7.9:1	30

■ Horsepower and torque are SAE net figures. They are measured at the rear of the transmission with all accessories installed and operating. Since the figures vary when a given engine is installed in different models, some are representative rather than exact.
DFI Digital Fuel Injection

TUNE-UP SPECIFICATIONS

(When analyzing compression test results, look for uniformity among cylinders rather than specific pressures.)

Year	V.I.N. Code	Engine Displacement (cu. in.)	Mfg.	Fuel Delivery	Spark Plugs Orig. Type	Gap (in.)	Distributor	Ignition Timing (deg.) ▲	Valves Intake Opens ■ (deg.)	Fuel Pump Pressure (psi)	Idle Speed (rpm) ▲
'80	N	350	Olds.	Diesel	—	—	—	5B①	16	5.5–6.5②	650/575
	6	368	Cad.	4 bbl	R-45NSX	.060	Electronic	18B	11	5.5–6.5	575
	9	368	Cad.	DFI	R-45NSX	.060	Electronic	10B	11	12–14	450④
'81	6	368	Cad.	4 bbl	R-45NSX	.060	Electronic	18B	11	5.5–6.5	575
	9	368	Cad.	DFI	R-45NSX	.060	Electronic	10B	11	12–14	450④
	N	350	Olds.	Diesel	—	—	—	—	16	5½–6½②	⑤
	4	252	Buick	4 bbl	R-45TSX	.060	Electronic	15B	16	4¼–5¾	550③
'82	9	368	Cad.	DFI	R-45NAX	.060	Electronic	⑥	11	12–14	450④
	N	350	Olds.	Diesel	—	—	—	⑥	16	5½–6½	⑤
	8	250	Cad.	DFI	R-43NT66	.060	Electronic	⑥	37	12–14	450
	4	252	Buick	4 bbl	R-45TSX	.060	Electronic	⑥	16	4¼–5¾	550③
	6	368	Cad.	4 bbl	R-45NSX	.060	Electronic	⑥	11	5.5–6.5	575
'83–84	N	350	Olds.	Diesel	—	—	—	⑥	16	5½–6½	⑥
	8	250	Cad.	DFI	R42CLTS6	.060	Electronic	⑥	20	12–14	⑥
	9	368	Cad.	DFI	R-45NAX	.060	Electronic	⑥	11	12–14	⑥
	6	368	Cad.	4 bbl	R-45NSX	.060	Electronic	⑥	11	5.5–6.5	⑥
'85–'86	8	250	Cad.	DFI	R-42CLTS6	.060	Electronic	⑥	20	12–14	⑥
	N	350	Olds.	Diesel	—	—	—	⑥	16	5.5–6.5	⑥
	Y	307	Olds	4 bbl	4-45TS	.060	Electronic	⑥	20	5.5–6.5	⑥
'87	All				See Underhood Specifications Sticker						

NOTE: The underhood specifications sticker often reflects tune-up specification changes made in production. Sticker figures must be used if they disagree with those in this chart. Part numbers in this chart are not recommendations by Chilton for any product by brand name.

▲ See text for procedure
■ All figures Before Top Dead Center
B Before Top Dead Center
EFI Electronic Fuel Injection
DFI Digital Fuel Injection
—Not applicable

① Static
② Injector opening pressure: 1800 psi
③ In drive
④ Drive or neutral
⑤ 600 RPM in drive, warm engine; 750 RPM in drive, cold engine
⑥ See underhood decal

FIRING ORDERS

NOTE: To avoid confusion, always replace spark plugs and wires one at a time.

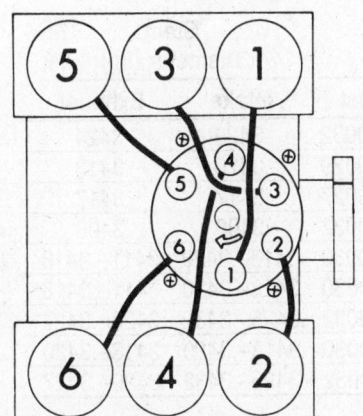

GM (Buick) 252 (4.1 L) V6
Engine firing order: 1-6-5-4-3-2
Distributor rotation: clockwise

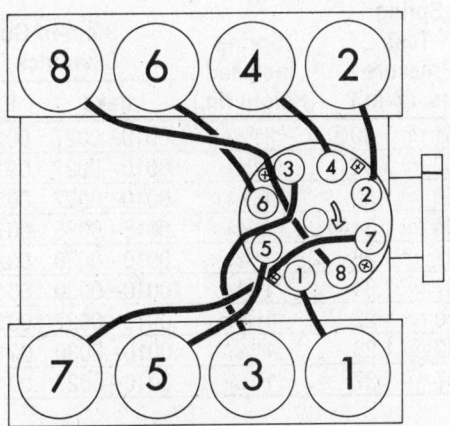

GM (Cadillac) 8-368 (6.0L)
Engine firing order: 1–5–6–3–4–2–7–8
Distributor rotation: clockwise

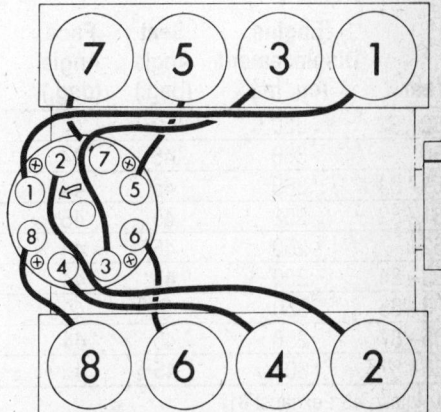

GM (Cadillac) 8-250 (4.1L)
GM (Oldsmobile) 8-307 (5.0L)
Engine firing order: 1–8–4–3–6–5–7–2
Distributor rotation: counterclockwise

CAPACITIES

Year	Engine Displacement (cu. In.)	Engine Crankcase Add 1 qt. for New Filter	Transmission Pts. to Refill After Draining ●	Drive Axle (pts.)	Gasoline Tank (gals.)	Cooling System (qts.)
'80–'82	350 Diesel	6	6	4.25	27	23.7
'80–'84	368	4	8	4.25	25	21.4
'81–'82	252	4	8	4.25	25	18.2
'82	250	4	8	4.25	25	10.8
'83–'87	250	4	10.6	3.5	24.5	11.0
'83–'85	350 Diesel	6	10.6	3.5	26	23.7
'86–'87	307	4	10.6	3.5	24.5	16.2

● Specifications do not include torque converter

CAMSHAFT SPECIFICATIONS
(All specifications in inches.)

Years	Engine Displ. (cu. in.)	Journal Diameter 1	2	3	4	5	Bearing Clearance	Lobe Lift Int.	Exh.	End Play
'82–'87	250	N.A.	N.A.	N.A.	N.A.	N.A.	.0018–.0037	.384	.396	N.A.
'81–'82	252	1.7850–1.7860	1.7850–1.7860	1.7850–1.7860	1.7850–1.7860	—	①	N.A.	N.A.	N.A.
'80–'83	350	2.0357–2.0365	2.0157–2.0165	1.9957–1.9965	1.9757–1.9765	1.9557–1.9565	.0020–.0058	.252	.279	.011–.077
'84–'85	350	2.0352–2.0365	2.0154–2.0161	1.9956–1.9964	1.9756–1.9963	1.9547–1.9559	.0020–.0058	.252	.279	0
'80–'84	368	N.A.	N.A.	N.A.	N.A.	N.A.	.0010–.0020	.457	.473	N.A.
'86–'87	307	2.0352–2.0365	2.0152–2.0166	1.9952–1.9965	1.9752–1.9765	1.9552–1.9565	.0020–.0058	.247–.251	.006–.022	

① No. 1: .0005–.0025
Nos. 2, 3, 4: .0005–.0035
N.A.: Information not available

VALVE SPECIFICATIONS

Year	Engine Displacement (cu. in.)	Seat Angle (deg.)	Face Angle (deg.)	Spring Test Pressure (lbs. @ in.)	Spring Installed Height (in.)	Stem-to-Guide Clearance (in.) Intake	Exhaust	Stem Diameter (in.) Intake	Exhaust
'80–'82	350	①	②	151 @ 1.30④	1⁴³/₆₄	.0010–.0027	.0015–.0032	.3429	.3424
'80	368	45	44	160 @ 1.50	1⁶¹/₆₄	.0010–.0027	.0012–.0029	.3420	.3418
'81–'84	368	45	44	160 @ 1.50	1⁶¹/₆₄	.0010–.0027	.0010–.0027	.3417	.3417
'81–'82	252	45	45	164 @ 1.34③	1⁴⁷/₆₄	.0015–.0035	.0015–.0032	.3406	.3408
'82	250	45	44	182 @ 1.28	1⁴³/₆₄	.0010–.0030	.0010–.0030	.3413–.3420	.3411–.3418
'83–'84	250	45	44	167 @ 1.34	1⁴³/₆₄	.0010–.0030	.0010–.0030	.3413–.3420	.3411–.3418
'83–'85	350	①	②	210 @ 1.22	1⁴³/₆₄	.0010–.0027	.0015–.0032	.3425–.3432	.3420–.3427
'85–'87	250	45	44	182 @ 1.28	1⁴⁷/₆₄	.0010–.0030	.0010–.0030	.3413–.3420	.3413–.3420
'86–'87	307	45⑤	46⑤	187 @ 1.27	1⁴⁷/₆₄	.0010–.0027	.0015–.0032	.3425–.3432	.3420–.3427

① Intake 45°; exhaust 31°
② Intake 44°; exhaust 30°
③ Exhaust 182 @ 1.34
④ 210 @ 1.30—1981–82
⑤ Exhaust: 59°
⑥ Exhaust: 60°

CRANKSHAFT AND CONNECTING ROD SPECIFICATIONS

(All measurements are given in inches.)

| Year | Engine Displacement (cu. in.) | Crankshaft | | | | Connecting Rod | | |
		Main Brg. Journal Dia.	Main Brg. Oil Clearance	Shaft End-Play	Thrust on No.	Journal Diameter	Oil Clearance	Side Clearance
'80–'82	368	3.250	.0001–.0026	.002–.012	3	2.5000	.0005–.0028	.008–.020
'81–'82	252	2.4995	.0003–.0018	.003–.009	2	2.2487–2.2495	.0005–.0026	.006–.023
'82–'87	250	2.6377	.0004–.0027	.001–.007	3	1.9291	.0005–.0028	.008–.020
'80–'85	350	2.9993–3.0003	.0005–.0021①	.0035–.0135	3	2.1238–2.1248②	.0005–.0026	.006–.020
'86–'87	307	2.4988–2.4998③	.0005–.0021④	.0035–.0135	3	2.1238–2.1248	.0004–.0033	.006–.020

① No. 5: 1980—.0015–.0031
      1981–86—.0020–.0034
② 1980: 1.8770–1.8870
③ Nos. 2, 3, 4, 5: 2.4985–2.4995
④ No. 5: .0015–.0031

TORQUE SPECIFICATIONS

(All readings in ft. lbs.)

| Year | Engine Displacement (cu. in.) | Cylinder Head Bolts | Rod Bearing Bolts | Main Bearing Bolts | Crankshaft Bolt | Flywheel-to-Crankshaft Bolts | Manifold | |
							Intake	Exhaust
'80–'84	368	95②	40	90	Press fit	75	30	①
'81–'82	252	80	40	100	225	60	45	25
'82–'87	250	③	20	85	18	37	④	18
'80–'85	350	130②	42	120	200–310	60	40	25
'86–'87	307	125②	42	80⑤	200–310	60	40②	25

① Long bolt—35, Short bolt—12
② Dip bolt in oil before tightening
③ Tighten all bolts in sequence to 45 ft. lbs.,
    then in sequence to 90 ft. lbs.
④ Tighten bolts 1, 2, 3, 4 in sequence to
    11–15 ft. lbs.; tighten bolts 5 thru 16 to
    18–22 ft. lbs., Retighten all bolts in
    sequence to 18–22 ft. lbs.
⑤ Rear Main: 120

PISTON AND RING SPECIFICATIONS

(All measurements are given in inches.)

| Year | Engine Displ. (cu. in.) | Piston-Bore Clearance | Ring Side Clearance | | | Ring Gap | | |
			Top Compression	Bottom Compression	Oil Control	Top Compression	Bottom Compression	Oil Control
'81–'82	252	.0013–.0035③	.0030–.0050	.0030–.0050	.0035 max.	.013–.023	.013–.023	.015–.035
'82–'83	250	.0010–.0018	.0016–.0037	.0016–.0037	None①	.009–.020	.009–.020	.010–.050
'80–'82	350	.0050–.0060	.005–.007	.0018–.0038	.0010–.0050	.015–.025	.015–.025	.015–.055
'80–'84	368	.0006–.0014	.0017–.0040	.0017–.0040	None①	.013–.023	.013–.023	.015–.055
'84–'87	250	.0010–.0018	.0016–.0037	.0016–.0037	None①	.015–.025	.015–.025	.010–.050
	350	.0035–.0045	.0050–.0070	.0030–.0050	.0010–.0050	.019–.027	.013–.021	.010–.022②
'86–'87	307	.0008–.0018	.0020–.0040	.0020–.0040	.0010–.0050	.009–.019④	.009–.019④	.015–.055

① Side sealing
② 1983: .015–.055
③ Measured at skirt bottom
④ With TRW rings: .010–.025

RIDE HEIGHT ADJUSTMENT SPECIFICATIONS

Models	Years		Front Suspension		Rear Suspension	
			Inches	mm	Inches	mm
DeVille w/o ALC	'80–'81	(Gas)	2.185	55.5	5.508	139.9
		(Diesel)	2.240	56.9	5.519	140.2
	'82–'84	(Gas)	2.244	57.0	5.512	140.0(1)
		(Diesel)	2.283	58.0	5.551	141.0
DeVille w/ALC	'80–'81	(Gas)	2.106	53.5	5.149	130.8
		(Diesel)	2.165	55.0	5.161	131.1
	'82–'84	(Gas)	2.165	55.0	5.157	131.0(2)
		(Diesel)	2.204	56.0	5.169	132.0
Brougham w/o ALC	'80–'81	(Gas)	2.185	55.5	5.508	139.9
		(Diesel)	2.240	56.9	5.519	140.2
	'82–'84	(Gas)	2.244	57.0	5.551	141.0
		(Diesel)	2.283	58.0	5.551	141.0
Brougham w/ALC	'80–'81	(Gas)	2.106	53.5	5.149	130.8
		(Diesel)	2.165	55.0	5.161	131.1
	'82–'84	(Gas)	2.165	55.0	5.196	132.0
		(Diesel)	2.204	56.0	5.196	132.0
Brougham	'85–'87	(Gas)	2.165	55.0	5.196	132.0
		(Diesel)	2.204	56.0	5.196	132.0

ALC: Automatic Level Control
(1) Sedan DeVille: 141mm
(2) Sedan DeVille: 132mm

WHEEL ALIGNMENT SPECIFICATIONS

Year	Model	Caster		Camber		Toe-in (in.)
		Range (deg.)	Pref. Setting (deg.)	Range (deg.)	Pref. Setting (deg.)	
'80–'84	All	2½P to 3½P	3P	$^3/_{32}$P to $^{57}/_{64}$P	½P	⅛
'85–'87	All	3½P to 4½P	4P	$^3/_{32}$P to $^{57}/_{64}$P	½P	⅛

N: Negative
P: Positive

ENGINE ELECTRICAL

For further information on the charging system, please refer to Charging and Starting in the Unit Repair section.

Alternator

REMOVAL & INSTALLATION

1. Disconnect the negative battery cable.
2. Disconnect and tag the electrical leads from the alternator.
3. Remove the screw from the alternator adjusting bracket.
4. Remove the screw from the rear of the alternator, retaining the shims for reinstallation.
5. Loosen the alternator pivot bolt and remove the drive belt.
6. Remove the air pump pulley for access to the pump bolt behind the pulley.
7. Loosen the two screws securing the front bracket to the engine.
8. Remove the alternator, spacer and lower through bolt by twisting the alternator toward the fender for clearance.
9. Install the alternator in the reverse order of removal.

NOTE: On Heavy Duty Alternator ONLY (100/145 amp with external voltage adjuster): after connecting the negative cable, momentarily connect a jumper wire between the BAT and R alternator terminals to polarize the charging system. Start the engine and run it at fast idle for ten seconds; the charge light should go out.

Starter

REMOVAL & INSTALLATION

1. Disconnect the negative battery terminal. Raise the car and support it with jackstands.
2. Disconnect and tag the battery lead and the wires from the solenoid.
3. Remove the bolt holding the support bracket to the starter.
4. Remove the two starter-to-engine bolts.
5. Remove the starter by pulling it forward and down.
6. Installation is the reverse of removal.

NOTE: On some models it may be necessary to remove the exhaust crossover pipe to complete this procedure.

Distributor

The High Energy Ignition system is standard equipment on all models. The HEI system consists of an ignition coil, electronic module and a magnetic pick-up assembly all within the distributor. A terminal in the top of the distributor cap is provided for the connection of a tachometer. The terminal is marked TACH.

For more information on the ignition system, please refer to "Electronic Ignition Systems" in the Unit Repair section. See the Oldsmobile Rear Wheel Drive section for a description of the diesel engine compression ignition process.

REMOVAL

Unplug and remove the distributor cap, leaving the plug wires installed. Disconnect the vacuum line. Disconnect the primary lead at the distributor. Using a scribe mark, index the distributor body to the cylinder block, and the tip of the rotor to the distributor housing so that the distributor body will be correctly replaced at reassembly. Remove the clamp bolt or nut and distributor.

On the 8-250 a special tool No. J-29791 is used to loosen the hold down nut. A thrust washer is used between the distributor drive gear and the crankcase. This washer may stick to the bottom of the distributor as it is removed. Make sure this washer is at the bottom of the distributor bore before installation. The malfunction trouble codes must be cleared after removal or adjustment of the distributor. This is accomplished by removing battery voltage to terminal R for 10 seconds.

INSTALLATION

If the engine has been rotated with the distributor removed, see the following procedure for installation with the engine disturbed. If the engine has not been moved, install the distributor so that the vacuum advance unit aligns with the match-mark made at removal. Turn the rotor slightly left of center so that as the gear engages the camshaft it will revolve into the proper position, pointing to the No. 1 contact in the cap. Install the hold-down clamp. Connect the primary lead and install the cap.

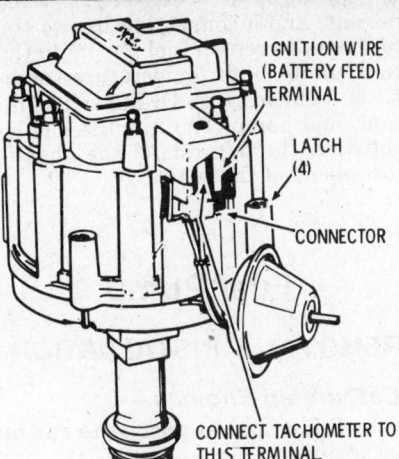

IGNITION WIRE (BATTERY FEED) TERMINAL

LATCH (4)

CONNECTOR

CONNECT TACHOMETER TO THIS TERMINAL

HEI tachometer hookup

Rotate the lubricator. Plug the distributor vacuum line to the carburetor. Connect a timing light to the No. 1 spark plug wire. Clean the crankshaft pulley markings and the pointer. Set the timing to specifications. Tighten the clamp bolt or nut. Remove the plug and adapter pin and reconnect the vacuum line to the advance unit.

INSTALLATION (ENGINE DISTURBED)

If the engine has been cranked after removing the distributor, perform the following procedure for installation:

1. Crank the engine until No. 1 piston is at the top of its compression stroke. The compression stroke can be determined by removing the spark plug from No. 1 cylinder and placing your thumb over the hole while an assistant slowly cranks the engine. Crank until compression is felt at the hole and then continue cranking slowly until the timing mark on the crankshaft pulley lines up with the zero degrees (0°) timing mark located on the timing chain cover.

2. Position the distributor in the block but do not, at this time, allow it to engage with its drive gear at the base of the mounting hole. Observe the position of the vacuum control unit on the distributor. If the distributor is located correctly, the vacuum unit will be positioned normally so that the vacuum hose can easily connect to it.

3. Rotate the distributor shaft so that the rotor points between No. 1 and No. 8 spark plug towers and push the distributor down to engage the camshaft. It may be necessary to turn the rotor a small amount in either direction in order to achieve this engagement. The rotor will rotate slightly as the distributor gear engages. If installed correctly, the rotor should point toward the No. 1 spark plug terminal in the distributor cap.

4. Press down firmly on the distributor housing. This will ensure that the distributor shaft engages the oil pump shaft, thereby allowing the distributor to fully contact the engine block.

5. Install the holddown clamp and tighten the bolt until it is snug.

6. Install the distributor cap, making sure that the rotor points to No. 1 terminal in the cap.

7. Attach all wires and the vacuum advance hose.

8. Start the engine. If it fails to start, or runs roughly, the distributor may be 180° out of time. Lift up on the distributor, turn the rotor one-half revolution, and install the distributor. Repeat Steps 1–8 if the engine continues to run poorly.

9. Check the ignition timing and adjust as necessary.

IGNITION TIMING ADJUSTMENT

1. Loosen the distributor holddown bolt so that the distributor can be turned without being too loose. On the 8-250 a special tool No. J-29791 is used to loosen the holddown nut.

2. Follow the instructions on the tune-up label located in the engine compartment. Some models may require the diagnostic terminal to be grounded.

3. Connect the timing light. Make certain that the timing marks are visible.

4. Connect a tachometer to the engine following the manufacturer's hook-up directions. Secure the parking brake and block the wheels. Start the engine and place the selector in Drive. It is best to also have an assistant in the car with the brakes firmly applied.

NOTE: Do not stand in front of the vehicle while performing the next step.

5. Adjust the idle speed to the specified rpm, then place the transmission in Park or Neutral.

6. Point the timing light at the pulley and observe the notch in the pulley in relation to the notches on the front cover. Check the specification chart for the correct timing setting.

7. If the setting is not correct, rotate the distributor until the correct timing is obtained. Tighten the distributor clamp nut and recheck the timing.

8. Reconnect all vacuum hoses and

remove the jumper wire from the diagnostic terminal, if grounded for timing procedure.

passage and a connecting line to the fuel tank to return fuel vapors to the tank under high temperature conditions. Vehicles with DFI have one intank fuel pump. The fuel filter is located on the left side of the chassis just ahead of the rear axle.

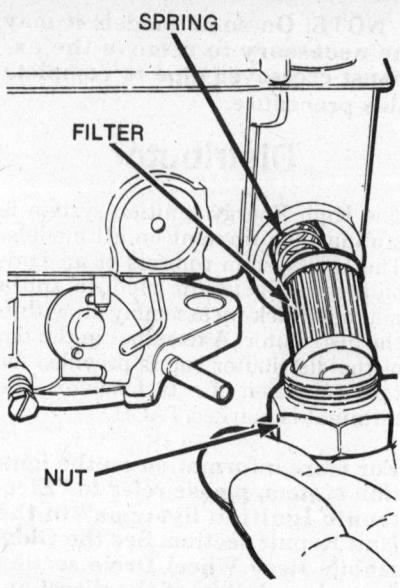

Fuel filter—typical of all carbureted models

FUEL SYSTEM

Carbureted Gasoline Engines

The carbureted fuel system includes the fuel pump, fuel filter, lines, carburetor and intake manifold.

Diesel Engine

The diesel engine is produced by General Motor's Oldsmobile division. Details on the diesel fuel system can be found in the "Oldsmobile Rear Wheel Drive" car section.

Fuel Injected Gasoline Engines

The Digital Fuel Injection (DFI) system is optional on all 1981 and later 6.0 liter 368 cu. in. and standard on all 4.1 liter, 250 cu. in. engines. Not only is the air/fuel mixture monitored and controlled by the DFI, but also the electronic spark timing, idle speed and EGR control. The system also incorporates a diagnostic readout of problems or system malfunctions.

The DFI system consists of four basic systems: the fuel delivery system, air induction system, the network of sensors, and the electronic control unit (ECU). The DFI system also includes a detailed digital electronic control module (ECM) and four subsystems: the electronic spark timing system (EST), idle speed control system (ISC), EGR control system, and the failure operation circuit and diagnostics readout system.

For further information on the fuel injection system, please refer to *Chilton's Guide To Fuel Injection And Feedback Carburetors*.

Fuel Pump and Filter

The fuel pump on carbureted engines is a mechanical unit, operated by an eccentric on the camshaft. The fuel filter is mounted in the carburetor behind the fuel inlet nut. A check valve is included in the fuel filter. On air conditioned cars, the fuel filter has a

Fuel Pump

REMOVAL & INSTALLATION

Carbureted Engines

1. Raise and support the car on jackstands.
2. Disconnect the fuel line and the vapor return hose from the fuel pump. Have a towel handy to catch any fuel that leaks out. Plug the hoses.
3. Disconnect the fuel pump-to-carburetor line and position it out of the way.
4. Remove the pump mounting screws and, tipping the pump upward, remove it from the car. Clean the old gasket material from the fuel pump mounting surface.
5. Installation is the reverse of removal. Use a new gasket.

Fuel Injected Engines

1. Disconnect the negative battery terminal, open the fuel tank filler door and disconnect the sending unit feed wire.
2. Siphon the fuel from the fuel tank. If the rear of the car is raised one foot higher than the front, more fuel can be taken out.
3. Raise the rear of the car and remove the screw securing the ground wire to the cross member.
4. Disconnect the fuel line, evaporative emission lines and the fuel return lines at the front of the tank.
5. Support the tank with a jack and wooden block and remove one screw on each side securing the fuel tank support straps to the body at the front of the tank.
6. Lower the jack and tank enough so that the fuel pump electrical lead can be disconnected. Disconnect the wire.
7. Remove the fuel tank from the car.
8. Remove the locknuts securing the fuel gauge tank unit and fuel pump feed wires to the tank unit.
9. Turn the cam locking ring counterclockwise with a soft non-ferrous punch and hammer. When the lock ring is disengaged, remove it and lift the gauge/pump unit from the tank.
10. Install in the reverse order of removal. Tighten the fuel tank retaining strap screws to 25 ft. lbs.

Fuel Filter

REMOVAL & INSTALLATION

Carbureted Engines

1. Disconnect the fuel line at the carburetor inlet.
2. Remove the fuel inlet nut from the carburetor using a box wrench.
3. Remove the fuel filter element and spring.
4. Install the filter spring and new fuel filter element into the carburetor.
5. Install a new gasket on the fuel inlet nut and install the nut.
6. Connect the fuel line to the fuel inlet nut and tighten securely. Start the engine and check for leaks.

Fuel Injected Engines

NOTE: The fuel filter element can be replaced by unscrewing the bottom cover and removing it.

1. Bleed the pressure from the fuel delivery system and remove the fuel inlet and outlet hoses from the fuel filter.
2. Remove the two screws retaining the fuel filter to the bracket and remove the filter from the engine or frame.
3. Remove the inlet and outlet fittings from the filter assembly if they are needed for the new filter.
4. Install the fittings to the new filter, using a sealer on the threads.
5. Attach the filter to the bracket and tighten the retaining screws to 12 ft. lbs.
6. Connect the inlet and outlet line, using new clamps.

NOTE: It may require considerable cranking before the engine starts, due to the drained fuel lines.

Throttle Body

REMOVAL & INSTALLATION

1. Remove the air cleaner assembly.
2. Disconnect the ISC motor, IPS, both injectors, and position the electrical connections out of the way.
3. Remove both throttle return springs, cruise control, throttle linkage, and downshift cable.
4. Disconnect the fuel inlet and return line, brake booster line, MAP hose and AIR hose from the rear of the throttle body.
5. Remove the PCV, EVAP, and EGR hoses from the front of the throttle body.
6. Remove the three throttle body mounting bolts and remove the throttle body and gasket.
7. Installation is the reverse of removal.

After installation, check and adjust the throttle position sensor (TPS) and the idle speed control (ISC) motor as necessary.

Fuel Injector

REMOVAL & INSTALLATION

1. Disconnect the negative battery terminal.
2. Remove the air cleaner assembly.
3. Disconnect the Idle Speed Control (ISC) motor and the Throttle Position Sensor (TPS).
4. Remove the screws securing the pressure regulator assembly and remove the regulator.
5. Use a small pair of pliers to gently grasp the center collar of the injector (between the electrical terminals) and carefully remove the injectors with a lifting-twisting motion.
6. Discard the upper and lower O-rings. Note the presence of the backup washer under the upper O-ring.
7. Installation is the reverse of removal. Lubricate the new O-rings with oil prior to installing them on the injector.

NOTE: Do not attempt to make any adjustments to the pressure regulator during this procedure as all adjustments are preset at the factory.

Idle Speed and Mixture Adjustments

CARBURETED ENGINE

Changes have been made in the carburetors for these model years that make idle speed and mixture adjustments impossible without the use of a propane enrichment system. Most carburetors have mixture needles concealed under staked-in screws. Mixture adjustments are possible only during carburetor overhaul.

NOTE: Vehicles equipped with the Computer Command Control System can't use the propane enrichment or lean drop methods of idle mixture adjustment.

FUEL INJECTED ENGINES

NOTE: No idle speed adjustments are possible on the DFI system.

Throttle Position Sensor

ADJUSTMENT

1. Remove the air cleaner and run the engine to normal operating temperature.

2. Connect a tachometer and a high impedance voltmeter as follows:
   a. Plus (+) lead to the TPS harness test point which connects to pin A (0.8 dark blue wire).
   b. Negative (-) lead to the TPS harness test point which connects to pin B (0.8 black/white wire).
   c. Select the 2V DC scale.
3. Open the set timing connector.
4. Retract the ISC motor by pressing the plunger (switch activated) in while the throttle is opened to approximately 1500 rpm. When the ISC motor fully retracts, disconnect the ISC connector before releasing the throttle.
5. Jump the ISC harness connector pins A and B together.
6. The ISC plunger should not be touching the throttle lever. If contact is noted, adjust the plunger (turn in) with pliers.
7. The idle speed should now be approximately 375–400 rpm. Adjust the throttle stop screw to the proper rpm if necessary.
8. The digital voltmeter should indicate 0.50 volts. If necessary adjust the TPS as outlined in Steps 9–11. If the voltmeter is correct, proceed to step twelve.
9. Remove the throttle body assembly from the intake manifold. Invert the throttle body assembly to gain access to the spot welds that hold the TPS screws in place. Use a $\frac{5}{16}$ inch

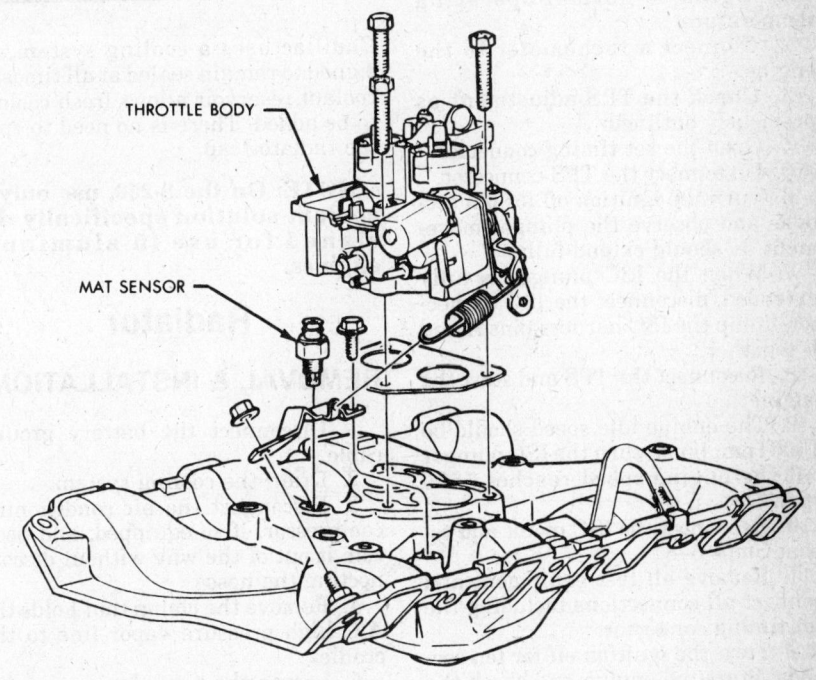

Throttle body Installation-DFI System

drill bit to drill through the spot welds to gain access to the screws. Loosen the screws enough to permit rotation of the sensor.

10. With the engine idling 375–400 rpm loosen the TPS mounting screws and position the TPS lever so the voltmeter reads 0.50 volts.

11. Tighten the TPS mounting screws with the sensor in this position. Recheck the voltmeter to make sure the adjustment hasn't changed.

12. Remove all test equipment and reconnect all connections including the set timing connector.

13. Turn off the ignition for ten seconds. The ISC motor should move to the extended position.

14. The above procedure may have turned on the CHECK ENGINE light, and may have set a trouble code. Refer to the procedure at the end of Idle Speed Control (ISC) Motor Adjustment, to clear the trouble code from the system.

Idle Speed Control

ADJUSTMENT

Adjustment of the ISC motor is necessary to establish the initial position of the motor after it has been replaced. It may be necessary if the throttle pedal ratchets when the ignition is turned off or on.

1. Remove the air cleaner, and run the engine to normal operating temperature.

2. Connect a tachometer to the engine.

3. Check the TPS adjustment as previously outlined.

4. Open the set timing connector.

5. Disconnect the TPS connector.

6. Turn the ignition off for ten seconds and observe the plunger movement. It should extend fully.

7. When the ISC plunger is fully extended, disconnect the ISC connector. Jump the ISC harness pins A and B together.

8. Reconnect the TPS and start the engine.

9. The engine idle speed should be 1500 rpm. If not, turn the ISC plunger till the engine speed reaches 1500 rpm.

10. Reconnect the ISC motor and repeat Steps 5–8.

11. Remove all test equipment and connect all connections including the set timing connector.

12. Turn the ignition off for ten seconds. Start the engine and check the ISC for proper operation.

13. Turn the ignition off for ten seconds. The ISC motor should move to the full extended position.

14. This procedure may have turned on the check engine light, and may have set a trouble code. To clear the code from the system, turn the key on, and simultaneously press and hold the OFF and WARMER buttons in the climate control panel until 88 appears in the readout. To clear the codes, depress the OFF and HI buttons simultaneously.

Carburetor

For further information on feedback carburetors, please refer to *Chilton's Guide To Fuel Injection And Feedback Carburetors.*

REMOVAL & INSTALLATION

1. Remove the air cleaner.
2. Disconnect the fuel line.
3. Disconnect the throttle linkage.
4. Disconnect and label all vacuum hoses.
5. Remove the retaining bolts.
6. Remove the carburetor.
7. Installation is the reverse of removal.

COOLING SYSTEM

Cadillac uses a cooling system designed to remain sealed at all times. A coolant reservoir allows fresh coolant to be added. There is no need to open the radiator cap.

NOTE: On the 8-250, use only a coolant solution specifically designed for use in aluminum engines.

Radiator

REMOVAL & INSTALLATION

1. Disconnect the battery ground cable.
2. Drain the cooling system.
3. Disconnect the air conditioning compressor, if so equipped, and position it out of the way without disconnecting the hoses.
4. Remove the clamp that holds the A/C high pressure vapor line to the cradle.
5. Loosen the hose clamps and disconnect the upper and lower radiator hoses.
6. Disconnect the two transmission cooler lines and plug them.

NOTE: Disconnect the heater return hose, if so equipped.

7. Remove the two top radiator cradle clamps, straps or sheet metal cover and the fan shroud. Disconnect the reservoir hose from the filler neck.

8. Remove the vacuum hoses, if so equipped. Mark them for proper installation.

9. Pull the radiator straight up and out of the car.

10. Installation is the reverse of removal.

Water Pump

REMOVAL & INSTALLATION

1. Disconnect the negative battery cable and drain the radiator.

2. Remove two screws from the radiator support rods on each side. Loosen one screw on each side and move the support rod out of the way.

3. Remove the two screws from the upper fan shroud and remove the radiator hose brace-to-shroud screw.

4. Drill out the upper fan shroud attaching rivets and remove the upper shroud.

5. Loosen the alternator bracket and remove the pulley so the fan can be rotated. Remove the four screws attaching the fan hub to the water pump.

6. Loosen the power steering pump bracket and remove the belt.

7. Disconnect the hose at the water pump.

8. Disconnect the fuel line at the carburetor and the fuel pump and remove the line.

9. Loosen the four crankshaft pulley-to-hub screws.

10. Remove the water pump attaching screws and remove the pump.

11. Reverse to install, using a new gasket.

Thermostat

REMOVAL & INSTALLATION

1. Drain the cooling system until the coolant level is below the level of the thermostat.

2. Remove the upper radiator hose at the thermostat housing.

3. Remove the two thermostat housing attaching bolts. Remove the housing.

4. Pull the thermostat from the engine block.

5. Position the thermostat in the block with the valve up.

6. Install a new gasket coated with sealer onto the engine block.

7. Position and secure the thermostat housing; tighten the screws.

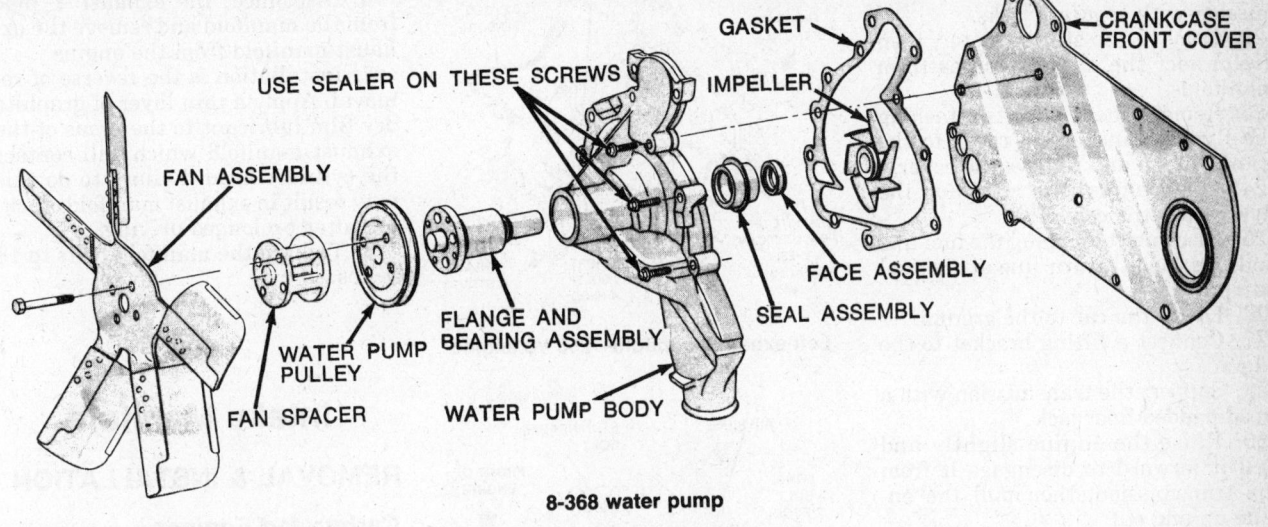

8-368 water pump

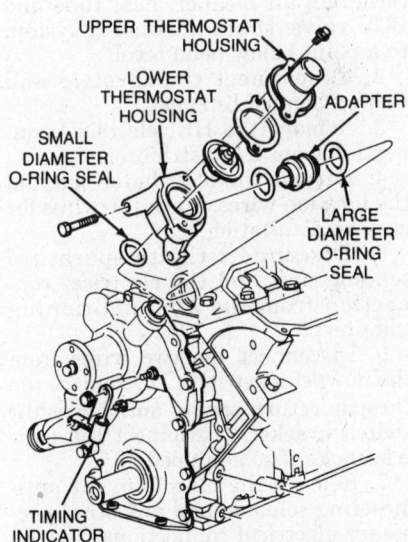

Thermostat and housing installation—250 V8 engine

8. Connect the radiator hose and refill the system to the proper level.

EMISSION CONTROLS

For all information concerning emission control maintenance, please refer to "Emission Controls" in the Unit Repair section.

Due to the complex nature of modern electronic engine control systems, comprehensive diagnosis and testing procedures fall outside the confines of this repair manual. For complete information on diagnosis, testing and repair procedures concerning all modern engine and emission control systems, please refer to Chilton's Guide To Electronic Engine Controls.

ENGINE MECHANICAL

NOTE: The 8-350 diesel engine and the 8-307 are produced by General Motor's Oldsmobile division. For all engine service procedures, please refer to the "Oldsmobile Rear Wheel Drive" car section. The 6-252 (4.1 liter) engine is produced by the Buick division. Service procedures are contained in the "Buick Rear Wheel Drive" car section.

Engine

REMOVAL & INSTALLATION

1. Disconnect the negative battery cable.
2. Remove the hood, after scribing the hood hinge outline for proper alignment.
3. Remove the air cleaner and heat shroud.
4. Drain the cooling system. Unfasten the fender struts from the radiator shroud.
5. Remove the radiator hose bracket, radiator cover and fan.
6. Remove the upper radiator hose.
7. Disconnect the throttle and Cruise Control linkage at the carburetor.

8. Disconnect the brake vacuum hose from the vacuum pipe. Remove the Cruise Control power unit on cars so equipped.
9. Disconnect the power steering pump bracket and swing the pump out of the way with the hoses still connected. Position the power steering fluid cooler out of the way.
10. Remove the A/C compressor bracket bolts and swing the compressor out of the way with hoses still connected. Do not discharge the system.
11. Disconnect the temperature sender wire, idle speed-up wire (if so equipped), ignition primary wire, downshift switch wire, S.C.S. solenoid (if so equipped) and anti-dieseling solenoid wires, electronic ignition connector block temperature sender lead, and all ground straps.
12. Bend back the clips and position the wiring harness out of the way.
13. Disconnect all vacuum hoses, and purge hose from the charcoal canister. Disconnect the automatic level control line, on models so equipped.
14. Disconnect the alternator, heater switch and oil pressure sender wires.
15. Remove the wiring harness from the clips.
16. Remove the water hose from the fitting at rear of the right cylinder head.
17. Loosen and remove the alternator and A.I.R. pumps and remove the belts.
18. Disconnect the struts and swing them out of the way.
19. Raise the car and support on it jackstands. Remove the six engine-to-transmission bolts and remove each engine mount through bolt.
20. Relieve the fuel pressure on fuel injected cars.

21. Support the engine and transmission with separate jacks.

22. Remove the starter motor, then disconnect the exhaust pipes from manifolds.

23. Remove the four bolts attaching the flywheel inspection cover to the transmission and remove the cover.

24. Remove the bolts attaching the flywheel to the converter.

25. Disconnect and plug the fuel line and the vapor return line at the fuel pump.

26. Lower the car to the ground.

27. Connect a lifting bracket to the engine.

28. Support the transmission with a wood-padded floor jack.

29. Raise the engine slightly and pull it forward to disengage it from the transmission, then pull the engine up and out.

30. Installation is the reverse of removal.

Exhaust Manifold

REMOVAL & INSTALLATION

8-368

NOTE: Before attempting this procedure it may be easier to remove the crossover pipe.

1. In order to remove the left exhaust manifold, remove the air cleaner assembly, then remove the air cleaner bracket and heat stove from the manifold.

2. Unfasten the nuts which secure the downpipes to either manifold. Remove the two studs retaining the EFE valve to the right-side manifold and remove the EFE valve.

3. Remove the bolts which secure the manifold to the cylinder heads.

NOTE: It may not be possible to remove the fifth bolt from the front of the cylinder head completely. Back the bolt all the way out and remove it with the manifolds.

4. Lift the manifold out of the engine compartment.

5. Installation is the reverse of removal. Lubricate the cylinder head installation surface with moly grease. Install the fifth screw from the front prior to installing the manifold. Tighten the bolts to specifications. On the right manifold, position the EFE valve on the manifold with the actuator toward the engine block. Tighten the two stud bolts.

8-250

LEFT SIDE

1. Remove the air cleaner and tube

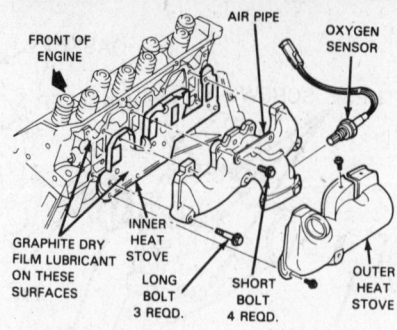

Left exhaust manifold—250 V8 engine

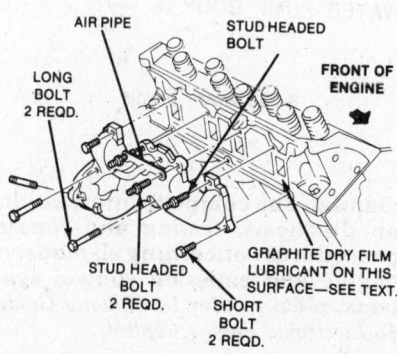

Right exhaust manifold installation—250 V8 engine

assembly from the preheat stove.

2. Remove the one screw securing the oil dipstick tube to the preheat stove.

3. Remove the preheat stove from the manifold.

4. Disconnect the Y pipe from the exhaust manifold.

5. Remove the oxygen sensor using a 1 inch box-end wrench.

6. Remove the exhaust manifold retaining bolts and remove the exhaust manifold from the car.

7. Installation is the reverse of removal. Apply a thin layer of graphite dry film lubricant to the areas of the exhaust manifold which will contact the cylinder head and heat stove. Failure to do this may result in exhaust manifold cracking after prolonged driving.

8. Tighten the manifold bolts to 18 ft. lbs.

RIGHT SIDE

1. Remove the nut securing the transmission cooler line bracket to the exhaust manifold.

2. Remove the two nuts retaining the air valve bracket to the manifold.

3. Remove the upper exhaust manifold to head bolts.

4. Jack up the car and safely support with jack studs.

5. Remove the lower exhaust manifold to head bolts.

6. Disconnect the exhaust Y pipe from the manifold and remove the exhaust manifold from the engine.

7. Installation is the reverse of removal. Apply a thin layer of graphite dry film lubricant to the areas of the exhaust manifold which will contact the cylinder head. Failure to do this may result in exhaust manifold cracking after prolonged driving.

8. Tighten the manifold bolts to 18 ft. lbs.

Intake Manifold

REMOVAL & INSTALLATION

Carbureted Engines

1. Remove the negative battery terminal, air cleaner, heat tube and PCV valve. Drain the cooling system to a point below head level.

2. Disconnect the throttle and Cruise Control linkages.

3. Remove the HEI electrical connection from the distributor.

4. Remove the distributor cap and the ignition wires. Mark the wires for easy reinstallation.

5. Disconnect the temperature sending unit and the electrical connection from the air conditioning compressor.

6. Disconnect the two wires from the downshift switch. Disconnect the throttle return spring and downshift switch bracket. Disconnect the electric choke if so equipped.

7. Remove the plug from the anti-dieseling solenoid and any other necessary electrical connections. Tag all connectors for installation.

8. Disconnect the power brake booster vacuum and vacuum modulator lines. Remove the cruise control mechanism if so equipped. Disconnect the A/C vacuum hose from the rear of the manifold.

9. Disconnect the fuel line from the carburetor.

10. Disconnect the vacuum advance line (if so equipped) and the canister purge hoses and position them out of the way.

11. Remove the air conditioning compressor and position it out of the way. Do not disconnect the refrigerant lines.

12. Disconnect the coolant by-pass hose at the manifold, if so equipped.

13. Remove the carburetor.

14. Remove the manifold bolts and the manifold.

15. Installation is the reverse of removal. See Steps 19–21 under Fuel Injected Engines.

Fuel Injected Engines

8-368

1. Disconnect the negative battery cable and remove the air cleaner and crankcase filter.

2. Disconnect the throttle cable and cruise control linkage at the throttle body. Remove the cable from the bracket and move it aside.

3. Disconnect the coolant temperature switch wire, the HEI wire, speed sensor wire, downshift switch wire, and the injector wiring harness from the fuel rail brackets and move the harness out of the way.

4. Disconnect the two vacuum hoses from the throttle body to the thermal vacuum switch (TVS).

5. Disconnect the vacuum hoses and power brake pipe from the rear of the throttle body.

6. Bleed the pressure from the fuel delivery system. Using a back-up wrench to avoid kinking the fuel line, disconnect the fuel line from the fuel rail.

7. Disconnect the EGR solenoid wires, air temperature sensor wire and the MAP sensor vacuum hose.

8. Remove the PCV valve from the rocker cover and move it out of the way.

9. Remove the spark plug wires and the distributor cap.

10. Drain the radiator, disconnect the upper radiator hose, the thermostat bypass hose and the heater hose at the rear of the manifold.

11. Remove the air conditioning compressor and tie it out of the way. Do not disconnect the refrigerant lines.

12. Remove the fuel feed and return hose from the fuel pressure regulator.

13. Remove the intake manifold retaining screws and remove the manifold. Do not pry or lift the manifold by the fuel rails or their mounting brackets.

14. Clean all gasket material from the mating surfaces of the manifold, cylinder heads and block.

15. Place new rubber intake manifold seals over the rails at the front and rear of the cylinder block. The tabs on the gasket should be positioned in the holes in the rails and the beveled ends of the gasket tucked into the slot at the mating of the head and rail.

16. Coat the ends of the new rubber end seals with RTV sealer. Place the seals on the front and rear of the cylinder block. Position the seal tabs in the holes provided. Tuck the beveled ends of the seals under the edge of the cylinder head.

17. Apply a thin layer of graphite along both sides of the top fiber part of

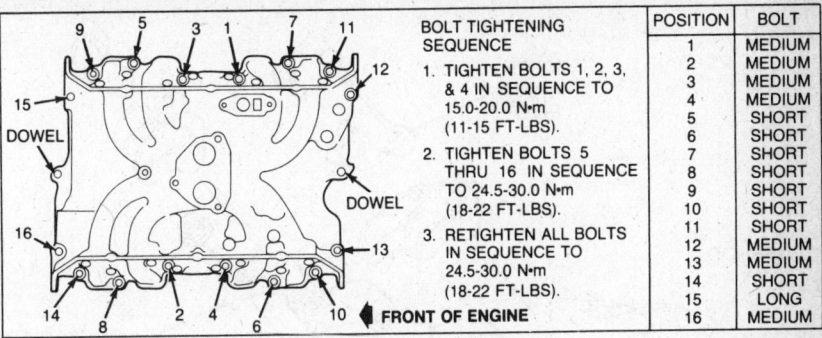

BOLT TIGHTENING SEQUENCE	POSITION	BOLT
1. TIGHTEN BOLTS 1, 2, 3, & 4 IN SEQUENCE TO 15.0-20.0 N·m (11-15 FT-LBS).	1	MEDIUM
	2	MEDIUM
	3	MEDIUM
	4	MEDIUM
2. TIGHTEN BOLTS 5 THRU 16 IN SEQUENCE TO 24.5-30.0 N·m (18-22 FT-LBS).	5	SHORT
	6	SHORT
	7	SHORT
	8	SHORT
	9	SHORT
	10	SHORT
3. RETIGHTEN ALL BOLTS IN SEQUENCE TO 24.5-30.0 N·m (18-22 FT-LBS).	11	SHORT
	12	MEDIUM
	13	MEDIUM
	14	SHORT
	15	LONG
	16	MEDIUM

Intake manifold installation—250 V8 engine

the intake gasket. Install intake gasket.

18. Position the intake manifold on the cylinder heads by lowering straight down. The manifold is centered evenly by tightening the third bolt from each side. Tighten all mounting bolts to 30 ft. lbs.

8-250

1. Drain the coolant from the radiator and disconnect the upper radiator hose from the upper thermostat housing.

2. Disconnect the electrical connections from the following: Coolant sensor, MAT sensor, throttle position sensor, the 12 volt feed wire and the 4 way connector, ISC motor and the injectors.

3. Disconnect the heater hose from the rear of the intake manifold.

4. Disconnect the fuel lines from the throttle body.

5. Remove the distributor.

6. Remove the rocker arm covers then remove the rocker arm support with the rocker arms intact by removing the five nuts which attach the support to the studed head bolts. Keep the pushrods in sequence so they may be reassembled in their original position.

7. Partially disconnect the A/C compressor and move to one side without discharging the A/C system.

8. Remove the vacuum harness connections from the TVS at the rear of the intake manifold.

9. Remove the 16 intake manifold bolts and remove the two bolts securing the lower thermostat housing to the front cover.

10. Bend the front and rear engine lift brackets out of the way then remove the intake manifold and lower thermostat housing as as assembly by lifting straight up.

11. Installation is the reverse of removal. Make sure the pushrods are properly seated and tighten the retaining nuts alternately and evenly until the nuts are all the way down, then tighten to 35 ft. lbs. Use new gas-

kets and O-rings. With the new intake gaskets in position, apply RTV sealant to the four corners where the end seals meet the side gaskets. Refer to the intake manifold torque sequence illustration and tighten the bolts to the torque values given in the Torque Specifications chart. Check the ignition timing.

Valve System

All Cadillac engines use hydraulic lifters. Valve systems with hydraulic lifters operate with zero clearance in the valve train. The rocker arms are nonadjustable. The lifter itself will compensate if there is slack in the system but if there is excessive play, the entire system should be examined.

If the valve guides are found to be worn past allowable limits, they will have to be rebored and valves with oversize stems installed. Three oversize valves of different stem diameters are available for each engine.

Sometimes a valve guide bore is made oversize at the factory. Oversize valve guide bores from the factory are marked on the inboard side of the cylinder heads on the cylinder head gasket surface in line with the oversize valve.

Rocker Arm

REMOVAL & INSTALLATION

8-368

These engines feature a modulated displacement design that can operate eight, six or four cylinders depending on driving requirements. The selective operation of the number of cylinders is controlled by a microprocessor that operates four engine valve selector units. The selector units are electromechanical devices which can deactivate both the intake and exhaust valves of a cylinder.

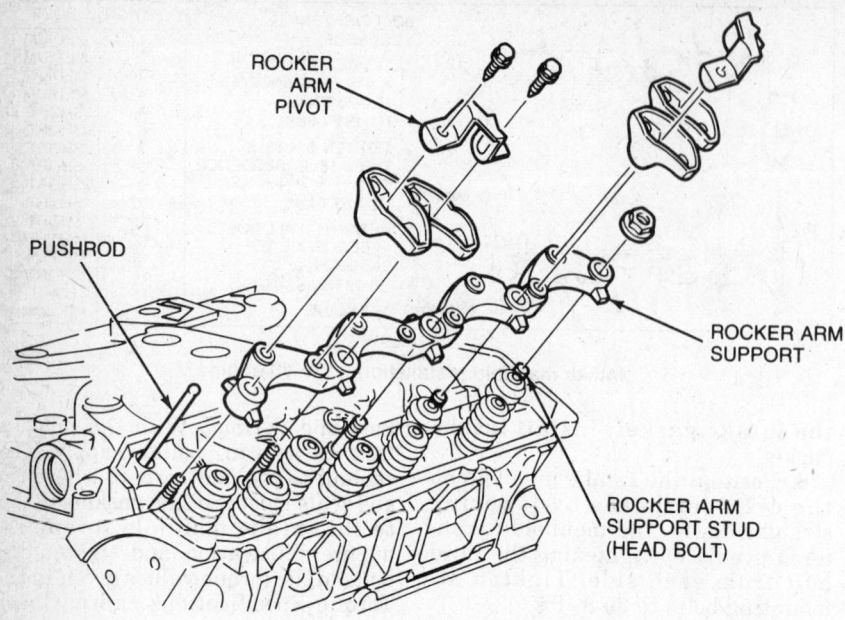

Rocker arm assembly—250 V8 engine

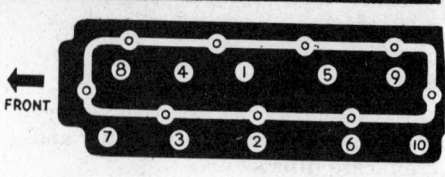

8-368 head bolt torque sequence

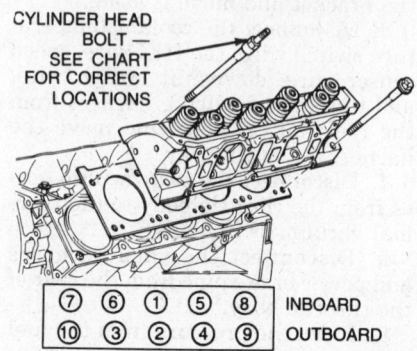

Cylinder head removal and installation—250 V8 engine

Rocker Arm or Rocker Arm Pivot

REMOVAL & INSTALLATION

8-250

1. With the valve cover removed, remove the five nuts from the stud headed head bolts and remove the valve train support with the rocker arms and pivots attached as an assembly.

NOTE: This method is preferred as the pivot assemblies may be damaged if pivot bolt torque is not removed evenly against valve spring pressure.

2. Place the support in a vise and individually remove the rocker arms and pivot.

3. With the valve train support secured in a vise, position the rocker arms and pivots to the valve train support and loosely install the pivot bolts. Torque the pivot bolts to 20 ft. lbs.

NOTE: When installing new parts, thoroughly lubricate all parts with an EP lubricant such as an axle lube.

4. Position the valve train support with the rocker arms and pivots installed over the five stud headed head bolts.

5. Position the pushrod into the seat of each rocker arm and loosely install the five retaining nuts. Tighten the live nuts alternately and evenly while checking the positioning of the push rods from time to time. When the nuts are all the way down, tighten to 35 ft. lbs.

6. Install the rocker cover.

Cylinder Head

REMOVAL & INSTALLATION

8-368

1. Drain the cooling system.
2. Remove the intake and exhaust manifolds.
3. Remove the rocker arm covers.
4. Disconnect the electrical and ground connections from the engine. Tag the wires.
5. For left head removal, dismount the power steering pump and position it out of the way.
6. For the right head, disconnect the heater hose at the head. Dismount the alternator and AIR pump, and position them out of the way.
7. Remove the rocker arm support bolts and lift off the rocker arm assemblies. Identify them so they can be installed in their original positions.
8. Remove the pushrods.
9. Install two $^7/_{16}$ x 14 x 6 in. bolts in two of the rocker arm support bolt holes. These bolts can be used as lifting handles.
10. Remove the head bolts. The bolts are different sizes, so identify them, somehow, to make sure they are installed in their original positions. The lower left bolt on each head will be trapped, due to clearance problems. To facilitate head removal, pull the bolt up as high as possible and hold it there with a spring clamp or spring-type clothes pin.
11. Thoroughly clean the gasket surfaces of the head and block. Check the head for flatness with a straight-edge. Always use a new head gasket.
12. Position the new hasket on the block. Place the lower left bolt in the head and lower the head into position on the block.
13. Place the remaining bolts in the head, making sure that they are installed in their original positions. Tighten the bolts finger tight, only.
14. Tighten the bolts, in three sequential passes, according to the illustration provided, to 30, then 60, then 95 ft. lbs.
15. Install the rocker arm assemblies, in their original positions, and torque the bolts to 60 ft. lbs.
16. Install all remaining parts in reverse of their removal procedures.

8-250

RIGHT SIDE

1. Remove both rocker arm covers.
2. Remove the intake manifold.
3. Remove the generator and AIR pump.
4. Remove the ground strap from the right front of the cylinder head.
5. Remove the exhaust manifold from the cylinder head.
6. Remove the screw retaining the AIR pipe to the cylinder head.
7. Remove the head bolts and remove the cylinder head.
8. Installation is the reverse of re-

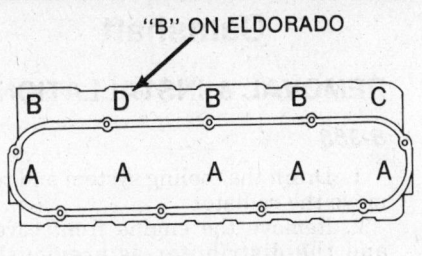

"B" ON ELDORADO

← FRONT OF ENGINE

8-368 head bolt location and length

BOLT LOCATION	LENGTH
A—BOLT	4.36"
B—BOLT	4.77"
C—BOLT	3.02"
D—BOLT/STUD	3.02"

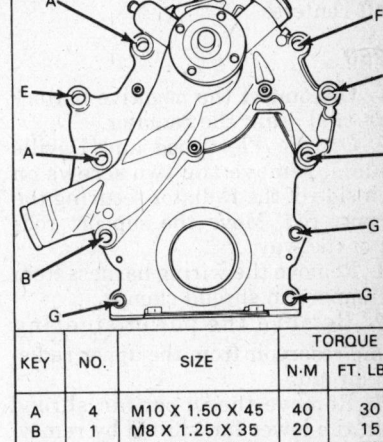

Front cover screw location and torque specifications—250 V8 engine

KEY	NO.	SIZE	TORQUE	
			N·M	FT. LBS.
A	4	M10 X 1.50 X 45	40	30
B	1	M8 X 1.25 X 35	20	15
E	1	M10 X 1.50 (NUT)	40	30
F	1	M10 X 1.50 X 50	40	30
G	3	M8 X 1.25 X 20	20	15

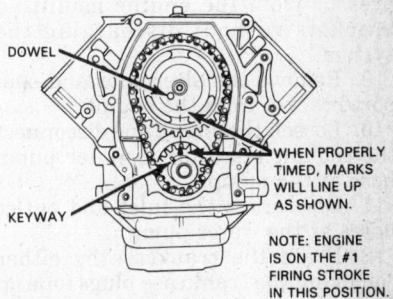

Camshaft timing marks—250 V8 engine

moval with the exception of the following precautions:

a. Apply graphite lubricant to the exhaust faces of the cylinder head.

b. Install the cylinder head bolts finger tight with the studed head bolts in the upper row and the conventional bolts in the lower row, then tighten all bolts in sequence to 45 ft. lbs. and again in sequence to 90 ft. lbs. as shown in illustration.

LEFT SIDE

1. Remove both rocker arm covers.
2. Remove the intake manifold.
3. Remove the vacuum pump and mounting bracket.
4. Remove the upper bolt and loosen the lower bolt securing the power steering pump to the engine.
5. Remove the exhaust manifold and heat stove from the cylinder head.
6. Remove the AIR pipe from the rear of the cylinder head.
7. Remove the head bolts and remove the cylinder head and gasket.
8. Installation is the reverse of removal. Perform Steps 8a, and 8b, of the Right Side procedure.

Timing Chain Cover, Chain, and Sprocket

REMOVAL

8-368

1. Disconnect negative battery cable and drain cooling system.
2. Detach upper radiator hose retainer from cradle and position hose out of the way.
3. Remove the fan, alternator and power steering belts.
4. Remove four capscrews that se-

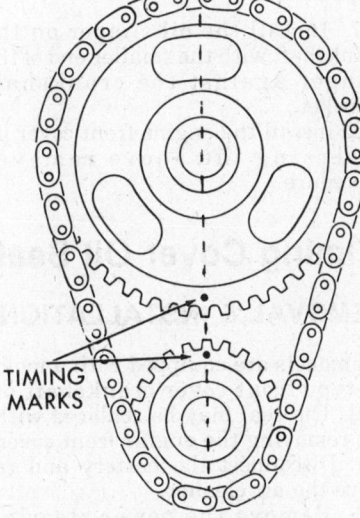

TIMING MARKS

8-368 timing mark alignment

cure crank pulley to harmonic balancer, then remove the pulley.

5. Remove the plug from the end of the crankshaft. Install the puller and remove the harmonic balancer.
6. Drain the engine oil and remove the oil pan.

NOTE: It may be necessary to remove the starter to gain access to the bolts that are directly behind it.

7. Disconnect lower radiator hose from water pump, then remove the screws that hold front cover to engine. Remove cover with water pump attached.
8. Remove distributor and fuel pump.
9. Remove oil slinger and fuel pump eccentric.

10. Remove capscrews that secure camshaft sprocket.
11. Remove camshaft sprocket along with timing chain.
12. To install, reverse removal procedure. Mount the timing chain over the camshaft and the crankshaft sprocket and start the camshaft sprocket over the shaft, being certain the aligning dowel is in a position where it will enter the hole in the camshaft freely. Make certain that the timing marks on the sprockets are in line between shaft centers. Camshaft sprockets are a tight fit. However, a comparatively easy way to install a tight-fitting sprocket is to draw it on carefully with two bolts somewhat longer then the regular mounting bolts. By drawing alternately against each bolt, and tapping gently with a plastic hammer, even a very tight camshaft gear sprocket can be installed.
13. When the camshaft is secured, turn the engine two full revolutions until the timing marks again assume the original position. Check to make certain that the punch marks, which are stamped into the front face of the

sprockets, are in line between the shaft centers.

8-250

1. Disconnect the negative battery cable and drain the radiator.
2. On the Fleetwood and Deville models, remove the two screws on each side of the radiator securing the support rod. Move the support rods out of the way.
3. Remove the wiring harness from the upper fan shroud clamps.
4. Remove the power steering pump reservoir from the upper radiator shroud.
5. Remove the upper fan shroud from the lower fan shroud by removing the staples.
6. Remove the clutch fan assembly.
7. Remove the generator, A.I.R. Pump, vacuum pump, and A/C pump drive belts.
8. Partially remove the A/C compressor from the engine mounting brackets without discharging the system.
9. Remove the alternator and support bracket from the engine.
10. Loosen the clamp and disconnect the coolant reservoir to water pump hose at the pump.
11. Disconnect the inlet and outlet hoses at the water pump.
12. Drain the crankcase by either removing the crankcase plugs (one on each side) or by elevating the rear wheels. This will prevent coolant from draining into the oil pan as the front cover is removed.
13. Remove the water pump and crankcase pulleys.
14. Remove the A/C bracket at the water pump.
15. Remove the timing mark tab from the front cover.
16. Remove the crankcase pulley to hub bolts and separate the pulley from the hub.
17. Remove the plug from the end of the crankshaft. Install a puller and remove the hub or balancer.
18. Remove the remaining front cover attaching screws and remove the cover with the water pump and lower thermostat housing as an assembly.
19. The timing chain and sprocket may now be removed as follows:
20. Remove the oil slinger from the crankshaft.
21. Rotate the engine and line up the timing marks as shown in the illustration.
22. Remove the screw securing the camshaft sprocket to the camshaft, then remove the camshaft and crankshaft sprocket with the chain attached.
23. Installation is the reverse of removal. After installing the timing

chain over the camshaft sprocket rotate the crankshaft until the timing mark on the crank sprocket is positioned straight up.
24. Install the cam sprocket and timing chain over the crankshaft so that the timing marks are aligned as the illustration.
25. Hold the camshaft sprocket in position against the end of the camshaft and press the sprocket on the camshaft by hand. Make sure the index pin in the camshaft is lined up with the index hole in the sprocket.
26. If necessary, keep the engine from rotating while torquing the camshaft sprocket screw to 37 ft. lbs.

NOTE: Engine timing has been set so that the No. 1 cylinder is in the TDC firing position. If for some reason the distributor was removed make sure the rotor is set so that cylinder No. 1 is in the firing position.

27. Install the oil slinger on the crankshaft with the smaller end of the slinger against the crankshaft sprocket.
28. Install the engine front cover by reversing the above removal procedure.

Timing Cover Oil Seal

REMOVAL & INSTALLATION

All models are equipped with a molded-type front cover crankshaft oil seal. The seal may be replaced without removing the engine front cover.
1. Disconnect the battery and remove the air cleaner.
2. Remove the power steering pump drive belt.
3. Remove the alternator drive belt.
4. On air conditioned cars, and cars equipped with the A.I.R. system, remove the pump drive belts.
5. Reuse and support the front of the car on jackstands. Remove the fan.
6. Remove pulley and harmonic balancer, as outlined in "Timing Chain and Sprocket Removal".
7. With a suitable tool, pry out front cover oil seal.
8. Lubricate new oil seal with wheel bearing grease. Position the seal on the end of the crankshaft with the garter spring side toward the engine.
9. Using a seal installer, drive the front seal into the front cover until it bottoms.
10. Assemble and install the remaining parts in reverse order of disassembly.

Camshaft

REMOVAL & INSTALLATION

8-368

1. Drain the cooling system and remove the radiator.
2. Remove the engine front cover and the distributor as previously outlined.
3. Remove the oil pump and the oil slinger from the crankshaft.
4. Remove the fuel pump and the fuel pump eccentric from the camshaft.
5. Remove the camshaft sprocket and the timing chain.

NOTE: Make certain that the marks on the two sprockets are correctly aligned before removing the timing chain.

6. Remove the lifters and slide the camshaft carefully out of the engine block. Do not allow the camshaft lobes to scratch the camshaft bearings.
7. To install the camshaft, reverse the procedure. Before installation, the camshaft should be lubricated with a thin coat of engine oil and then carefully inserted to avoid bearing damage.
8. The camshaft sprocket screws should be torqued to 18 ft. lbs. while the fuel pump eccentric screw is tightened to 35 ft. lbs.

8-250

1. Remove the radiator, timing chain and valve lifters as described in this section.
2. Temporarily reinstall the cam sprocket or a long bolt to use as a handle and slide the camshaft forward until it is out of the engine. Do not allow the camshaft lobes to scratch the camshaft bearings.
3. Installation is the reverse of removal. Apply a thin coat of rear axle lubricant or equivalent to the camshaft lobes, distributor gear teeth and bearing journals.

Piston and Rod Installation

The numbers on the connecting rods face away from the camshaft; that is, the numbers on the left bank (even) face to the left; the numbers on the right bank (odd) face to the right.

On the 8-368, the word rear, (or R), stamped on the piston, faces the rear of the engine on both banks and an arrow on the piston top points to the front of the engine.

On the 8-250, the piston is placed in the cylinder with the notch in the top

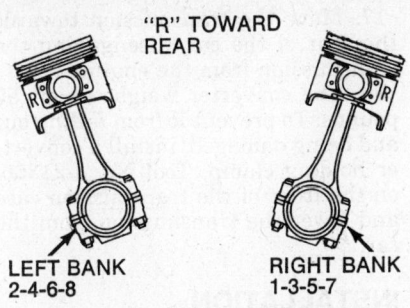

"R" TOWARD
REAR

LEFT BANK
2-4-6-8

RIGHT BANK
1-3-5-7

Typical piston-to-rod relationship

of the piston or the F on the side of the piston facing toward the front of the engine. The oil spurt hole in the connecting rod faces toward the camshaft.

LUBRICATION

Oil Pan

REMOVAL & INSTALLATION

8-250

1. Disconnect the negative battery cable.
2. Jack the car up and support it with jackstands.
3. Drain the engine oil and remove the oil filter.
4. Remove the flywheel inspection cover and support struts.
5. Disconnect the exhaust Y-pipe at the exhaust manifolds and remove the one bolt at the catalytic converter bracket.
6. Remove the oil pan screws, lower the exhaust pipe, and remove the oil pan.

NOTE: If the pan is difficult to remove, lightly tap the edges of the pan with a plastic hammer.

7. Seal the oil pan to the block with RTV sealant.
8. Tighten the oil pan retaining screws and nuts to 11 ft. lbs.
9. The remainder of the installation is the reverse of removal.

8-368

1. Remove the wheel housing struts from the fenders. Disconnect the negative battery terminal.
2. Remove the 3 screws from the upper radiator shroud, two securing the shroud, and one securing the top radiator hose. Drill out the rivets securing the upper shroud to the lower one and remove the shroud. Use bolts

and nuts to replace the rivets when reinstalling the shroud.
3. Loosen the drive belts and remove the crankshaft pulley.
4. Jack up your car and support it with jackstands.
5. Remove the through-bolt from each motor mount.
6. Remove the crossover pipe and the converter as an assembly.
7. Remove the starter.
8. Remove the torque converter cover.
9. Drain the oil pan.
10. Using a jack, with a block of wood on top, place it under the crankshaft hub. Jack up the engine, remove the pan bolts and the pan.
11. Clean all the gasket material from the pan and the block mating surfaces. Use a new gasket kit and sealer. Make sure the seals are firmly positioned on the flange surfaces with each seal properly located in the cutout notches of the pan gasket.
12. Installation is the reverse of removal. Torque the pan bolts to 10 ft. lbs.

Oil Pump

REMOVAL & INSTALLATION

8-250

1. Raise the car and support it with jackstands.
2. Remove the oil pan.
3. Remove the two screws and one nut securing the oil pump to the engine.
4. To disassemble, remove the four screws holding the oil pump cover to the housing, then slide the drive shaft, drive gear and driven gear out of the pump housing.
5. Remove the oil pressure regulator valve and spring from the bore in the housing assembly.
6. Inspect the oil pressure regulator valve for nicks and burrs.
7. Measure the free length of the regulator valve spring. It should be 2.57–2.69 in.
8. Inspect the drive gear and driven gear for nicks and burrs.
9. Assemble the pump drive gear over the drive shaft so that the retaining ring is inside the gear. Position the drive gear over the pump housing shaft closest to the pressure regulator bore.
10. Slide the driven gear over the remaining shaft in the pump housing, meshing the driven gear with the drive gear.
11. Install the oil pressure regulator spring and valve in the bore of the pump housing assembly.

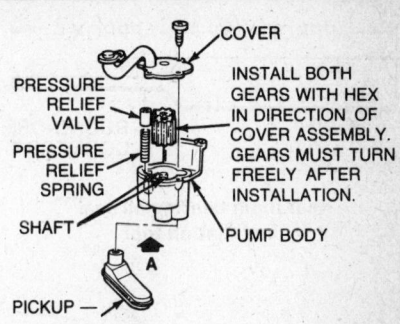

COVER

PRESSURE
RELIEF
VALVE

PRESSURE
RELIEF
SPRING

SHAFT

PICKUP

INSTALL BOTH
GEARS WITH HEX
IN DIRECTION OF
COVER ASSEMBLY.
GEARS MUST TURN
FREELY AFTER
INSTALLATION.

PUMP BODY

A

Oil pump disassembled—250 V8 engine

12. Install the pump cover and four retaining screws.
13. Install the oil pump assembly to the block, engaging the drive shaft to the distributor gear. Tighten the nut to 22 ft. lbs. and the two screws to 15 ft. lbs.
14. Install the oil pan and lower the car.

8-368

1. Jack up the car, support it with jackstands, and remove the oil filter.
2. Remove five capscrews that secure oil pump to engine.

NOTE: Remove screw nearest pressure regulator last.

3. Slide drive shaft drive gear and driven gear out of housing.
4. Remove plug from housing cover, using $\frac{5}{16}$ in. wrench. Remove pressure regulator valve and spring.
5. Check free length of regulator spring—it should be 2.57–2.69 in.
6. Inspect gears and housing for burrs or scoring.
7. Check pump clearance limits.
8. On installation, pack the pump with petroleum jelly. Use a new gasket, engage the pump driveshaft with the distributor drive, and install screw nearest pressure regulator first. Install remaining screws and tighten all five screws to 15 ft. lbs. Install oil filter, add one quart oil to engine, run engine and check for leaks.

Rear Main Seal

REMOVAL & INSTALLATION

1. Remove the oil pan.
2. Remove the rear main bearing cap and loosen the bolts holding the other four bearings about three turns each. Remove the old rear main bearing seals.
3. Clean the groove in the cap and in the block. Lubricate seals with engine oil.
4. Make an installation tool.

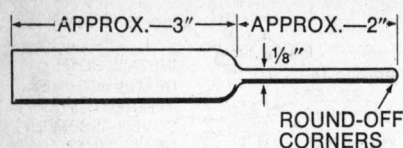

Rear main bearing oil seal
installation tool

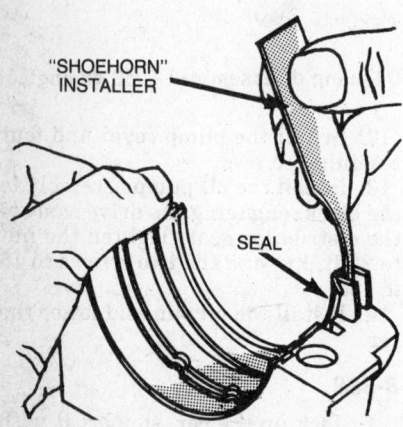

Installing rear main bearing oil seal

5. Start the upper half into the groove in the block with the lip facing forward and rotate it into position, using the tool as a guide. Press firmly on both ends to be sure it is protruding uniformly on each side.

6. Install the lower half of the seal into the bearing cap with the lip facing forward and one end of the seal over the ridge and flush with the split line. Hold one finger over this end to prevent it from slipping, and push the seal into seated position by applying pressure to the other end. Be sure the seal is firmly seated and protrudes evenly on each side. Do not apply pressure to the lip. This may damage the effectiveness of the seal.

NOTE: On vehicles equipped with neoprene type seals, make sure that the seal is flush at the split line to avoid leaks.

7. Apply rubber cement to the mating surfaces of the block and cap being careful not to get any cement on the bearing, the crankshaft or the seal. The cement coating should be about 0.010 in. thick.

8. Tighten the bearing bolts to 89 ft. lbs. for the 8-250, and 90–100 ft. lbs. for the 8-368. Be sure to tighten the bolts of the other four bearings also. Rotate the crankshaft one full turn to check for binding.

9. Reinstall the oil pan.

AUTOMATIC TRANSMISSION

All cars use a Turbo Hydra-Matic transmission. For all automatic transmission service procedures, please refer to Automatic Transmissions in the Unit Repair section.

REMOVAL

1. Disconnect the negative battery cable.

2. Raise the car and make sure it is supported securely.

3. Disconnect the transmission linkage by removing the one nut from the shaft on the left side of the transmission.

4. Remove the speedometer drive cable.

5. Disconnect and cap the oil cooler pipes at the transmission. Plug the connector holes in the transmission.

6. Disconnect the vacuum modulator hose.

7. Remove the propeller shaft.

8. Remove the lower flex plate housing cover.

9. Remove the three converter to flex plate attaching bolts. Rotate the converter and flexplate until the bolts can be reached for removal. A bolt in the end of the crankshaft balancer can be used to rotate the flexplate.

NOTE: Do not pry on the flexplate ring gear or transmission case to rotate the converter, as the flexplate or case may be damaged.

10. Place a jack or other suitable device under the rear of the engine.

11. Remove the two nuts from the tunnel strap and remove the strap.

12. Remove the two rear engine mount to extension housing bolts.

13. Position a transmission jack under the transmission and raise it just enough to take the load off the rear engine support and remove the shim.

14. Remove the four bolts from the rear engine support and swivel the support out of the way. Allow the support to hang by the parking brake cable.

15. Disconnect the exhaust pipe from the exhaust manifold and remove the rear engine support cross member.

16. Remove the six transmission case-to-engine attaching bolts. If necessary, lower the engine and transmission slightly to gain access to the upper attaching bolts.

17. Move the transmission towards the rear of the car, disengaging the transmission from the engine.

18. The converter weighs about 50 pounds. To prevent it from falling out and being damaged, install a converter holding clamp, Tool No. J-21366, on the front of the transmission case and lower the transmission from the car.

INSTALLATION

1. Install the converter on the turbine shaft making certain that the converter drive hub is fully engaged with the pump gear tangs, and install the converter holding clamp on the front of the transmission case.

2. Align the front of the transmission case dowel holes with the dowels on the engine. Install the six transmission case-to-engine attaching bolts and tighten the bolts to 35 ft. lbs.

3. Rotate the converter until two of the three weld nuts on the converter line up with the bolt holes in the flexplate.

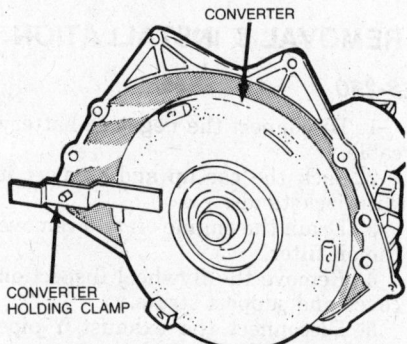

Installation of the converter holding clamp

NOTE: Make certain the converter rotates freely in this position and is not cocked, and that the pilot in the center of the converter is properly seated in the crankshaft.

4. Install the two flexplate-to-converter attaching bolts through the access holes in the flexplate and tighten them finger tight.

NOTE: The bolts must not be tightened at this time to assure proper alignment of the converter.

5. Rotate the converter and install the third attaching bolt. Tighten all bolts to 35 ft. lbs.

6. Lower the transmission carefully and install the two rear engine mount-to-exten-sion housing bolts. Tighten the bolts to 55 ft. lbs.

7. Install the exhaust pipe, rear engine support crossmember and the tunnel strap.

8. The remainder of installation is the reverse of removal. Add transmission fluid as required.

DRIVESHAFT AND U-JOINTS

Universal joints and driveshafts can be divided into two groups: single piece shaft models and two piece shaft models.

For all U-joint removal and repair procedures, please refer to "U-Joints & CV-Joint Overhaul" in the Unit Repair Section.

Single Piece Shaft

REMOVAL & INSTALLATION

1. Put the transmission in Neutral, then raise the car and support it with jackstands.
2. Remove the two accessible rear U-joint flange capscrews.
3. Rotate the driveshaft and remove the other two capscrews, while supporting the shaft. Never let the full weight of the driveshaft be supported only by the front universal joint.
4. Push shaft forward to clear pinion flange, then pull rearward to disengage slip yoke from transmission. Plug transmission to prevent oil leakage or entry of dirt.
5. Lubricate slip yoke inside diameter with gear lube, outside of splines with ATF.
6. To install, reverse the removal procedure, tightening the rear U-joint fasteners to 70 ft. lbs. Place the transmission in Park to hold the shaft while tightening the capscrews.

Two Piece Shaft

REMOVAL & INSTALLATION

1. Follow Steps 1–6 of Single Piece Shaft Removal and Installation, with the addition of the following step.
2. Remove center bearing support after matchmarking it and crossmember. When installing, tighten the bolts to 16 ft. lbs.

REAR AXLE

Axle Shaft, Bearing, and Seal

REMOVAL & INSTALLATION

1. Raise the car on a hoist and remove the wheel and brake drum.
2. Clean any dirt from the differential cover and loosen the cover attaching bolts, allowing the lubricant to drain out into a suitable container.
3. Remove the pinion cross shaft lockscrew and remove the cross shaft.
4. Push in on the flanged end of the axle shaft and remove the C-lock from the splined end of the axle shaft.
5. Remove the axle shaft from the housing, being cautious not to damage the oil seal.
6. Use a suitable tool to pry the oil seal out of the bore. Use an axle shaft bearing puller on a slide hammer to remove the axle bearing from the bearing bore.
7. Install the new bearing in the bearing bore until it is 0.550 in. from the end of the axle tube. Use a block of wood and a hammer to tap the bearing in place. Install the axle shaft bearing seal until it is flush with the end of the axle tube.
8. Slide the axle shaft into the housing until the splines on the end of the shaft engage the splines of the differential side gear. Handle the shaft gently when trying to engage the splines.
9. Install the axle shaft C-lock on the splined end of the axle shaft in the differential. Push the shaft outward so that the shaft lock seats in the counterbore of the differential side gear.
10. Install the pinion cross shaft through the differential case and pinion gears. Align the lockscrew hole and install the lockscrew, tightening it to 25 ft. lbs.
11. Clean the differential housing and cover mating surfaces and install the cover with a new gasket.
12. Fill the differential with lubricant, install the brake drum and wheel, and lower the car.

FRONT SUSPENSION

All cars use the same front suspension system. The system is a coil spring suspension which consists of two upper and two lower control arm assemblies, shock absorbers, a stabilizer bar, two steering knuckles, and a pair of coil springs.

Shock Absorber

REMOVAL & INSTALLATION

NOTE: Purge a new shock of air by repeatedly extending it in its normal position and compressing it while inverted.

1. Open the hood. Remove the retaining nut from the frame spring tower. Use a pair of locking type pliers, to prevent the shock stem from turning while the nut is being unfastened.
2. Remove the bottom shock absorber bolts.
3. Remove the shock through the bottom of the lower arm.
4. Install the retainer and the lower grommet.
5. Extend the shock rod as far as it will go.
6. Install the shock up through the coil spring and install the top grommet, retainer and nut.
7. Position the lower end of the shock on the lower control arm. Install the bolt, lockwasher, and nut. Tighten the bolt to 22 ft. lbs.
8. Tighten the retaining nut on the upper stem to 15 ft. lbs., while holding the stem with a pair of locking type pliers keep it from turning.

NOTE: Hold the shock absorber on the square tip with locking pliers to prevent damaging the threads when removing or installing the top nut.

Lower Control Arm and Coil Spring

REMOVAL & INSTALLATION

1. Raise the car and support it by the frame so the control arms hang freely.
2. Remove the lower shock absorber mounting bolts.
3. Attach a special supporting tool (No. J-23028-01) to a floor jack. Position the tool and the jack so as to cradle the inner bushings.
4. Remove the stabilizer-to-lower control arm attaching bolt.
5. Raise the jack to relieve the tension on the lower control arm pivot bolts. As a safety measure, install a chain around the spring and through the lower control arm.
6. Lower the jack slowly.
7. When all the spring pressure is relieved, remove the safety chain and the spring.
8. Remove the lower ball joint stud cotter pin.
9. Loosen, but do not remove, the ball joint nut.
10. Install a ball stud remover between the studs and screw the threaded end of the tool until the stud is freed.
11. Remove the lower stud nut.

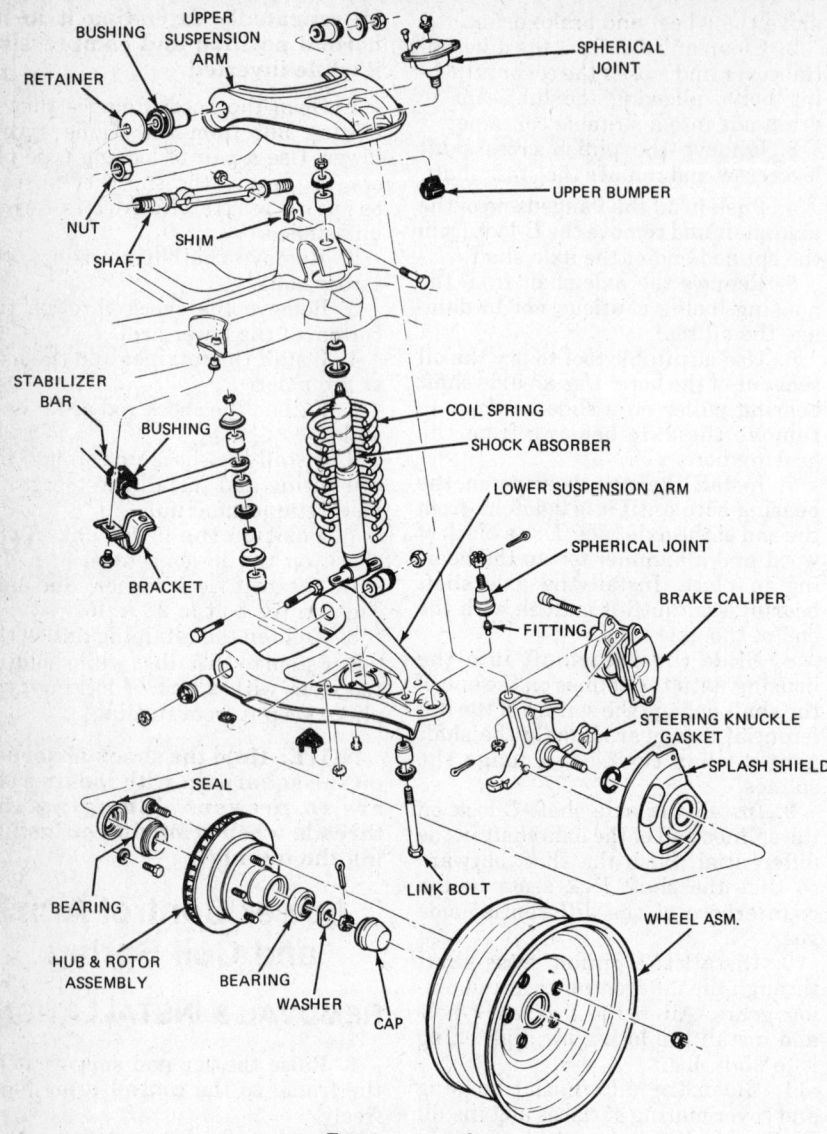

Front suspension

12. Pull outward on the bottom of the tire while at the same time pushing the tire upward to free the steering knuckle from the ball joint stud.

13. Remove the lower control arm from the car.

14. Installation is the reverse of removal. Tighten the attaching bolts to the following values: lower control arm ball joint stud-to-steering knuckle boss, 80 ft. lbs. (tighten to align the cotter pin hole); control arm pivot bolts, 90 ft. lbs.

BALL JOINT INSPECTION

NOTE: Before performing this inspection, make sure the wheel bearings are adjusted correctly and that the control arm bushings are in good condition.

1. Jack the car up under the front lower control arm at the spring seat.

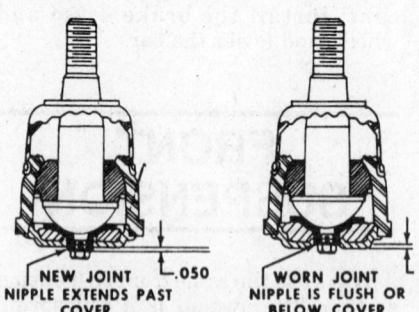

NEW JOINT L.050
NIPPLE EXTENDS PAST COVER

WORN JOINT
NIPPLE IS FLUSH OR BELOW COVER

Lower ball joint wear indicator

2. Raise the car until there is 1–2 in. of clearance under the wheel.

3. Insert a bar under the wheel and pry upward. If the wheel rises more than $\frac{1}{8}$ in., the ball joints are worn. Determine if the upper or lower ball joint is worn by visual inspection while prying on the wheel.

NOTE: Due to the distribution of forces in the suspension, the lower ball joint is usually the defective joint. Cadillacs are equipped with wear indicators on the lower ball joint. As long as the wear indicator neck extends below the ball stud seat, replacement is unnecessary.

Lower Ball Joint
REMOVAL & INSTALLATION

1. Jack up the car and support it with jackstands. Remove the wheel.

2. Remove the lower ball joint stud cotter pin. Loosen (not more than one turn), but do not remove, the stud nut.

3. Install a ball joint removal tool between the studs and turn the threaded end of the tool until the stud is free of the steering knuckle.

----- **CAUTION** -----
The lower control arm must be supported so that the spring cannot force the arm down.
----

4. Remove the lower stud nut. Pull out on bottom of the brake disc and simultaneously push up to free the steering knuckle from the ball joint stud. If additional leverage is needed, it may be necessary to reinstall the tire for the above procedure.

5. Lift up on the upper control arm (with the steering knuckle and hub attached), and place a block of wood between the frame and the upper arm. Be careful not to pull on the brake hose when lifting the knuckle and hub. Remove the tie rod end from the steering knuckle only if necessary.

6. Use a ball joint removal tool to push the ball joint from the lower control arm.

7. To install, place the lower ball joint in the lower control arm and seat it. Position the bleed vent in the rubber boot of the new ball joint facing inward.

8. Turn the ball joint stud cotter pin hole fore and aft. Remove the wood block holding the upper control arm.

NOTE: Examine the tapered hole in the steering knuckle. Clean the area. The knuckle MUST be replaced if any out-of-roundness, deformation, or damage is found.

9. Attach the ball joint stud to the steering knuckle and install the stud nut. Torque the nut to 80 ft. lbs. and install a new cotter pin.

NOTE: 125 ft. lbs. or $\frac{1}{6}$ turn maximum is allowed to align the cotter pin slot. Do not back off the nut to install the cotter pin.

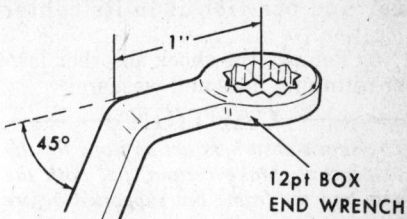

Rear shock absorber wrench

10. Lubricate the ball joint. If removed, install the tie rod end and torque the nut to 35 ft. lbs. Install the cotter pin.

11. Install the wheel and tire and lower the car. Have the front wheel alignment checked and adjusted as necessary.

Upper Ball Joint

REMOVAL & INSTALLATION

1. Raise the car and support it on jackstands.

2. Remove the wheel and tire.

3. Remove the disc brake caliper assembly and support it with a length of wire.

NOTE: Never let the caliper hang by the brake hose.

4. Remove the cotter pin from the upper ball joint stud. Loosen the stud nut but do not remove it.

5. Use a ball joint stud removing fork, or a screw press, to free the stud from control arm with a jack.

6. Support the lower control arm with a jack.

7. Remove the upper ball joint stud nut; remove the joint from the steering knuckle; allow the knuckle to swing out of the way.

8. Lift the upper control arm and place a block of wood between it and the frame as a support.

9. Remove the rivets from the upper control arm with either a chisel or a grinding wheel. Drive them out with a punch after removing the heads. Do not damage the ball joint seat.

10. Install the new ball joint in the upper control arm and attach it with the nuts and bolts provided. Insert the bolts from the bottom and tighten them to 25 ft. lbs.

11. Turn the ball joint stud so the cotter pin hole runs front-to-rear.

12. Remove the block of wood from between the frame and the upper control arm.

13. Before installing the ball joint stud in the steering knuckle, check the tapered hole and remove any dirt or debris. If the hole is distorted or damaged, the steering knuckle must be replaced.

14. Install the ball joint stud in the hole in the top of the steering knuckle. Install the castellated nut and tighten it to 60 ft. lbs. Tighten the nut to a maximum of 100 ft. lbs. to install the cotter pin. Do not back the nut off in order to install the cotter pin.

15. Install the brake caliper assembly.

16. Grease the ball joint.

17. Install the wheel and lower the car.

Upper Control Arm

REMOVAL & INSTALLATION

1. Raise and support the car.

2. Place a jackstand under the lower control arm.

3. Remove the wheel.

4. Remove the upper ball joint stud from the steering knuckle.

5. Remove the two nuts securing the upper arm shaft to the frame bracket and remove the arm.

6. Note the number and position of shims for reassembly.

7. Install the suspension arm cross shaft on the attaching bolts.

8. Using a free running nut instead of a locknut, tighten both nuts until the serrated bolts are reseated.

9. Remove the free running nuts and install the locknuts.

10. Install the shims as removed.

11. Torque the mounting nuts to 75 ft. lbs.

NOTE: Tighten the nut on the thinner shim pack first.

12. Install the ball joint stud through the knuckle and tighten the nut to 60 ft. lbs. Install the cotter pin.

13. Install the wheel and torque the lug nuts to 100 ft. lbs.

Wheel Bearing

ADJUSTMENT

1. Raise the front of the car. Remove the dust cap from the wheel bearing and remove the cotter pin.

2. While spinning the wheel, tighten the adjusting nut to 12 ft. lbs. Stop spinning the wheel.

3. Back off the nut until it is free and then tighten it finger tight.

4. Insert the cotter pin. If the pin cannot be installed in this position, back off the nut until the holes align. Make certain that the pin fits tightly.

REAR SUSPENSION

Except Commercial Chassis: A four link rear suspension system, consisting of upper and lower control arms, coil springs and shock absorbers is used. The coil springs are placed on brackets on the rear axle housing at their lower ends, the upper ends being seated in the frame crossmember. Some vehicles are equipped with Electronic Level Control.

Commercial Chassis: The Commercial Chassis uses semi-elliptic leaf springs. Electronic Level Control is optional.

Shock Absorber

REMOVAL & INSTALLATION

NOTE: Purge a new shock of air by repeatedly extending it in its normal position and compressing it while inverted.

1. Raise the rear of the vehicle and support both the frame and the axle with separate jackstands.

2. If the vehicle is equipped with Electronic Level Control, remove the air lines at the shocks.

——— CAUTION ———
The shocks act as rebound stops for the rear suspension and under no circumstances should the rear end be raised excessively high while disconnecting the shocks, unless both the rear axle and the frame are supported.

3. Remove the upper retaining bolts and nuts. To do this, bend a ½ in. box end wrench, as illustrated, to form a 45° angle at a point one inch from the center of the box diameter. This is used to hold the upper mounting nut.

4. Remove the lower retaining nut while holding the stem by the grommet to keep the stem from turning. Pull the shock off.

5. Installation is the reverse of removal.

Coil Spring

REMOVAL & INSTALLATION

1. Raise and support the car.

2. Place a jack under the differential housing.

3. Remove the wheels.

4. If the car has level control, disconnect the link at the overtravel le-

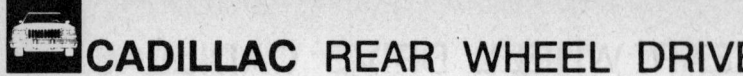

ver and position it in its center location.

5. Remove the shock absorber lower retaining nuts and washers.

——————— **CAUTION** ———————

The shock absorbers act as stops for the suspension. Make certain that both the axle and the frame are supported before continuing.

6. Disconnect the brake line retaining clip from the axle and frame, but do not disconnect the brake line. This should allow enough slack, as the axle is lowered, to eliminate the need for disconnecting and reconnecting the brake line. If enough slack cannot be obtained, disconnect the brake line from the hose and plug both openings. Be sure to bleed the brakes after installation.

7. Disconnect rear U-joint and wire the driveshaft out of the way. Do not allow the driveshaft to hang unsupported.

8. Remove nuts and bolts that secure both upper control arms to the axle brackets.

9. Lower rear axle assembly slowly until the springs are free and remove the springs.

——————— **CAUTION** ———————

Do not allow the differential to wind up as it is lowered as the spring may fly out.

10. To install, reverse the removal procedure. Tighten the upper and lower control arm bolts to 75 ft. lbs.

Leaf Spring

REMOVAL & INSTALLATION

1. Raise the rear of the car and support it so the axle can be raised or lowered. Raise the axle so that all tension is relieved from the spring.

2. Disconnect the rear automatic leveling valve over-travel lever from its link and hold the lever in the exhaust position (down) to deflate the shock absorbers.

3. Disconnect the lower half of the shock absorbers and move them out of the way.

4. Loosen the parking brake adjustment at the equalizer and remove the parking brake cable clip from the front retaining bracket on the spring. Remove the cable clamps from the under side of the springs.

5. Loosen the spring front eye busing-to-retaining bracket bolt.

6. Remove the bolts retaining the front spring bracket to the underbody.

7. Lower the axle enough to permit access to the front eye bolt and remove the bracket from the spring.

NOTE: The front eye bushing can be replaced at this time.

8. Remove the U-bolt and T-bolt nuts retaining the lower spring plate to the axle and stabilizer bar brackets.

9. Remove the upper and lower spring pads and spring plate.

10. Support the spring with a jackstand and remove the two nuts from the rear shackle.

11. Separate the shackle and remove the spring from the vehicle.

12. If the spring is being replaced, remove the spring damper for installation on the new spring by removing the clamp bolt and bending the bottom half of the clamp down about 2 in. Slide the clamp rearward over the damper and remove the damper from the spring.

13. Position the spring damper on the new spring and position it $\frac{1}{8}$ in. from the front spring eye. Slide the clamp forward over the damper and position the clamp at the second leaf of the spring. The clamp must face upward and the nut must be on the outside of the spring. Install the clamp bolt pointing up and tighten to 20 ft. lbs.

NOTE: Do not tighten any of the attaching hardware to specifications until Step 25. Allow the retaining nuts and bolts to remain only finger tight.

14. Position the front eye of the spring to the front mounting bracket and install the attaching bolt and washer with the bolt head on the inside. Bolt torque is 105 ft. lbs.

15. Install the upper shackle bushings in the frame. Position the shackles to the bushings and install the bolt and nut. Torque is 50 ft. lbs.

16. Install the bushing halves in the rear spring eye and install the spring to the shackle. Lower shackle bolt and nut torque is 50 ft. lbs.

17. Raise the front end of the spring and position the bracket to the underbody. Make sure the tab on the bracket is aligned in the slot in the underbody.

18. Install the screws retaining the front spring bracket to the underbody. Torque is 30 ft. lbs.

19. Position the spring upper cushion between the spring and the axle bracket so the cushion ribs align with the bracket locating ribs.

20. Position the lower mounting plate over the locating dowel on the lower spring pad and install the retaining nuts. Torque is 45 ft. lbs.

21. Position the stabilizer brackets on the lower spring plate. Retaining bolts and nut torque is 30 ft. lbs.

22. Connect the lower shock absorb-

er mount to the lower spring bracket. Torque is 45 ft. lbs.

22. Connect the lower shock absorber mount to the lower spring bracket. Torque is 45 ft. lbs.

23. Install the parking brake cable under the leaf spring and secure it at the front of the spring with the wire clip and clamp. Adjust the parking brake cable.

24. Connect the rear leveling overtravel lever to its link.

25. Tighten all of the attaching hardware to the specified torques.

26. Lower the vehicle.

BRAKES

Hydro-boost is installed on all diesel-engined cars. Hydro-boost is a hydraulically-assisted power brake booster. The power steering pump provides the hydraulic fluid pressure to operate both the power brake booster and the power steering gear.

For all brake system removal, installation and adjustment procedures not detailed below, please refer to "Brakes" in the Unit Repair section.

Master Cylinder

REMOVAL & INSTALLATION

NOTE: It is possible to remove the master cylinder unit without removing the power booster from the vehicle.

1. Disconnect and plug the front and rear brake lines at the master cylinder.

2. Remove the two securing nuts which hold the master cylinder to the power booster.

3. Remove the master cylinder.

4. To install, reverse the removal procedure. Bleed the hydraulic system.

Combination Valve

REMOVAL & INSTALLATION

The valve is non-serviceable and is located on the frame extension on the left side of the car. To remove it, disconnect the brake lines at the valve and unbolt and remove the valve. Installation is the reverse of removal. Bleed the brakes.

Vacuum Power Brake Unit

REMOVAL & INSTALLATION

1. Disconnect and cap hydraulic lines from master cylinder.
2. Disconnect vacuum line from vacuum check valve on unit.
3. Remove steering column lower cover.
4. Remove cotter pin, washer and spring spacer that secure power unit pushrod to brake pedal arm.
5. Remove the four nuts that secure power unit to firewall, then remove power unit.
6. To install, reverse removal procedure. Bleed the hydraulic system.

Hydro-Boost Power Brake Unit

REMOVAL & INSTALLATION

— CAUTION —

Power steering fluid and brake fluid are incompatible. If the brake seals contact the steering fluid or the steering seals contact the brake fluid, the seals will be ruined.

1. With the engine off, pump the brake pedal four or five times to empty the accumulator of pressurized fluid.
2. Disconnect the brake lines from the master cylinder and cap the lines.
3. Remove and plug the three hydraulic lines from the booster. Remove the washer and retainer that secures the booster pedal rod to the brake pedal arm.

NOTE: To avoid booster damage, do not pry the pedal rod off the pedal arm.

4. Remove the four nuts holding the booster to the firewall.
5. Loosen the booster from the firewall and move the booster pedal rod inboard until it disconnects from the brake pedal arm. Remove the spring washer from the brake pedal arm and remove the booster.
6. To install, reverse the removal procedure. Tighten the booster mounting nuts to 30 ft. lbs. and the master cylinder-to-booster mounting nuts to 20 ft. lbs. Bleed the Hydro-boost system as explained in the Brakes Unit Repair section.

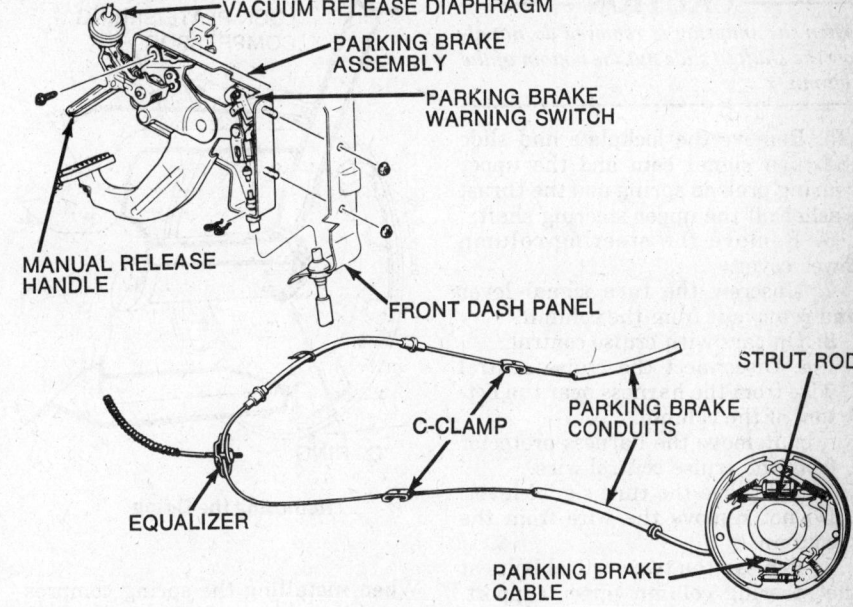

Parking brake system

Parking Brake

ADJUSTMENT

NOTE: Make certain that the rear brakes are properly adjusted before adjusting the parking brake.

1. Make a check of the parking brake linkage for the free movement of all the cables. Lubricate, if necessary.
2. Depress the parking brake pedal 1½ in.
3. Raise the rear wheels off the ground.
4. While holding the cable stud to keep it from turning, tighten the equalizer nut until a light drag is felt on either wheel when they are spun in the forward direction.
5. When the parking brake is released there should be no brake shoe drag.

STEERING

Steering Wheel

REMOVAL & INSTALLATION

— CAUTION —

Do not strike the end of the steering column in an effort to remove the steering wheel. Delicate parts of the column may be damaged.

1. Disconnect the negative battery cable.
2. Remove the screws on the underside of the steering wheel spokes near the center and remove the pad assembly.
3. Remove the horn contact wire from the plastic tower by pushing in on the wire and turning it counterclockwise. Turning the ignition on will facilitate removal.
4. Remove the nut holding the steering wheel to the steering shaft.
5. On tilt wheels, remove locking lever and flange and screw assembly.
6. Matchmark the shaft and wheel for installation in the original position and use a puller to remove the steering wheel.
7. On installation, tighten the steering shaft nut to 30 ft. lbs. (1980-81); 35 ft. lbs. (1982 and later).

Turn Signal Switch

REMOVAL & INSTALLATION

Standard Column

1. Disconnect the negative battery cable.
2. Remove the steering wheel.
3. Insert a thin screwdriver into the lockplate and remove the lockplate cover assembly.
4. Install a spring compressor onto the steering shaft. Tighten the tool to compress the lockplate and the spring. Remove the snapring from the groove in the shaft.

CAUTION

When the snapring is removed do not allow the shaft to slide out the bottom of the column.

5. Remove the lockplate and slide the turn signal cam and the upper bearing preload spring and the thrust washer off the upper steering shaft.

6. Remove the steering column lower cover.

7. Unscrew the turn signal lever and remove it from the column.

8. On cars with cruise control:

a. Disconnect the cruise control wire from the harness near the bottom of the column.

b. Remove the harness protector from the cruise control wire.

c. Remove the turn signal lever. Do not remove the wire from the column.

9. Remove the two vertical bolts at the steering column upper support. Remove the shim packs. Keep the shims in order for reinstallation.

10. Remove the four screws securing the column upper mounting bracket to the column and remove the bracket.

11. Disconnect the turn signal wiring and remove the wires from the plastic protector.

12. Remove the turn signal switch mounting screws.

13. Slide the switch connector out of the bracket on the steering column.

14. If the switch is known to be bad, cut the wires and discard the switch. Tape the connector of the new switch to the old wires, and pull the new harness down through the steering column while removing the old wires.

15. If the original switch is to be reused, wrap tape around the wire and connector and pull the harness up through the column. It may be helpful to attach a length of wire or string to the harness connector before pulling it up through the column to facilitate installation.

16. After freeing the switch wiring protector from its mounting, pull the turn signal switch straight up and remove the switch, switch harness, and the connector from the column.

17. To reassemble reverse the removal procedure.

Tilt and Telescopic Columns

1. Disconnect the battery and remove the steering wheel.

2. Remove the rubber sleeve bumper from the steering shaft.

3. Remove the plastic retainer with a screwdriver, disengaging the tabs on the retainer from the C-ring.

4. Compress the upper steering shaft preload spring with a spring compressor and remove the C-ring.

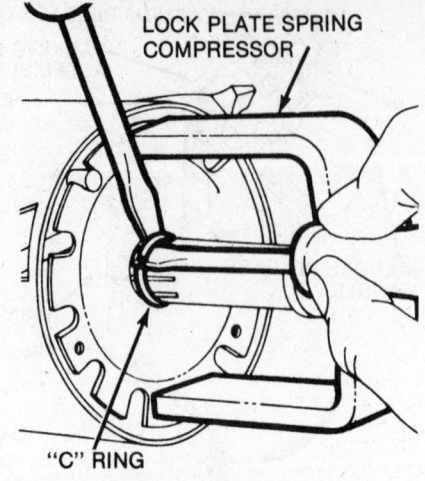

LOCK PLATE SPRING COMPRESSOR

"C" RING

Removing the C-ring

When installing the spring compressor, pull the upper shaft up about 1 in. and turn the ignition to the LOCK position to hold the shaft in place.

5. Remove the spring compressor and remove the upper steering shaft lockplate, horn contact carrier and the preload spring.

6. Remove the steering column lower cover.

7. Unscrew and remove the turn signal lever. If equipped with cruise control:

a. Disconnect the cruise control wire from the harness near the bottom of the steering column.

b. Remove the lever attaching screw and carefully pull the lever out enough to allow the removal of the turn signal switch.

8. Remove the two nuts and shim packs from the upper column support. Keep the shims together as a unit for reinstallation.

9. Remove the bracket from the steering column by removing the two attaching screws from each side.

10. Disconnect the turn signal wiring harness from the car harness and remove the wires from the plastic protector.

11. Remove the turn signal switch retaining screws and pull the switch up out of the steering column.

12. If the switch is to be replaced, cut the wires from the switch and tape the new switch connector to the old wires. Carefully pull the new harness down through the column as the old wires are removed.

13. If the old switch is to be reused, tape the connector to the wires and carefully pull the harness up out of the column.

14. Feed the wiring harness down through the steering column to replace the old switch.

15. Secure the switch in the steering column.

16. Install the upper shaft preload spring.

17. Install the lock plate and carrier assembly. Make sure that the flat on the lower end of the steering shaft is pointing up and that the small plastic tab on the carrier is up or nearest the top of the column. The flat surface of the lock plate must be installed facing down against the turn signal switch.

18. Install the spring compressor, compress the preload spring and lock plate and install the C-ring with the wide side toward the keyway.

19. Remove the spring compressor and install the plastic retainer on the C-ring.

20. Install the rubber sleeve bumper over the steering shaft and install the steering wheel.

21. Install the turn signal lever. If the vehicle is equipped with cruise control, secure the lever to the switch with the retaining screw and install the wiring harness.

22. Remove the tape from the end of the harness and connect the switch and cruise control, if so equipped, to the car harness.

23. Cover both harnesses with the plastic protector and position it to the column. The turn signal connector slides on the tabs of the column.

24. Position the steering column upper bracket over the turn signal switch harness plastic protector.

25. Install the mounting bracket nuts and shims in their original positions.

26. Install the steering column lower cover.

Steering Linkage

REMOVAL & INSTALLATION

1. Remove the steering shock damper, if so equipped, from the frame bracket. Remove the cotter pins and nuts from the outer tie rod pivots.

2. Remove the outer tie rod pivots from the steering knuckles using a tie rod end puller.

3. Remove the idler arm screws and lockwashers from the side member.

4. Remove the pitman arm cotter pin, nut and washer at the steering linkage.

5. Remove the steering linkage from the pitman arm.

6. Remove the intermediate rod with the tie rods and idler arm attached.

7. Remove the cotter pins and nuts from the idler arm pivot and inner tie rod pivots.

8. Remove the tie rod.

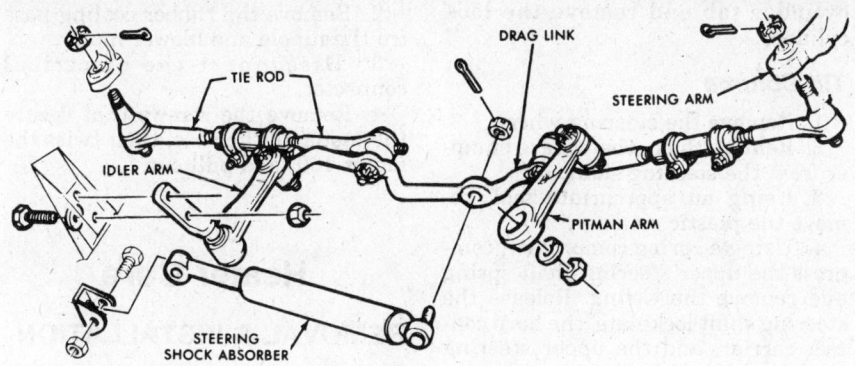

Cadillac steering linkage

9. Remove the idler arm from the intermediate rod.

10. Remove the dust seals from the pitman arm and idler arm pivot studs.

11. Remove the outer tie rod pivots by loosening the nuts on the outer clamp bolts and unscrewing the pivot from adjuster tubes.

12. To install, reverse the removal procedure. Tighten the idler arm nuts to 35 ft. lbs. Install the cotter pin. Do not tighten more than 10 ft. lbs. over specification to align the cotter pin.

Power Steering Pump

REMOVAL & INSTALLATION

8-368

1. Disconnect and plug the fluid lines at the pump.

2. Remove the nut securing the pump mounting bracket to the cylinder head stud.

3. Remove the belt adjustment bolt.

4. Remove the steering pump bracket attaching bolt from the front of the cylinder block.

5. Remove the drive belts.

6. Remove the bottom pivot bolt and remove the pump with the bracket and filter attached.

7. Installation is the reverse of removal. Tighten the fluid line fittings to 20 ft. lbs. Tighten all mounting and adjusting bols to 25 ft. lbs.

8-350 Diesel

1. On 1981 and later models, dismount the cruise control servo and position it out of the way.

2. Dismount the alternator and position it out of the way. Leave the through-bolt in place.

3. Remove the alternator adjustment bracket.

4. Disconnect and plug the pressure and return lines.

5. Loosen the pump adjusting bolt, pivot bolt and pivot nut. Remove the belt.

6. Remove the two nuts and spacer securing the pump mounting bracket to the water pump and timing chain cover.

7. Remove the pump bracket bolt and lift out the pump and bracket.

8. Installation is the reverse of removal. Torque the lower nut at the spacer to 30 ft. lbs.; the upper nut and bracket bolt to 20 ft. lbs., each; the adjusting bolt to 30 ft. lbs.; the hose connections to 35 ft. lbs.

6-252

1. Disconnect and plug the fluid lines at the pump.

2. Loosen the two adjusting bolts on the front bracket, and the nut on the rear bracket. Remove the belt.

3. Remove the two front mounting bracket bolts.

4. Remove the nut securing the pump to the rear bracket.

5. Remove the pivot bolt and lift out the pump and bracket.

6. Installation is the reverse of removal. Torque all mounting and adjusting bolts to 34 ft. lbs. Tighten the hose fitting to 21 ft. lbs.

8-250

1. Disconnect and plug the fluid lines at the pump.

2. Loosen the air conditioning compressor and AIR pump and remove all of the drive belts.

3. Using a puller, such as tool J-25034-B, remove the power steering pump pulley. It is not necessary to remove the fan shroud.

4. Remove the pump-to-block mounting bolts and lift out the pump.

5. Installation is the reverse of removal. Torque the rear mounting bolt to 18 ft. lbs.; the front mounting bolt to 29 ft. lbs.

Steering Gear

REMOVAL & INSTALLATION

1. Position a container under the gear to catch dripping fluid.

2. Disconnect the pressure and return lines at the steering gear. Plug all openings to prevent the loss of fluid and the entrance of dirt into the system.

3. Disconnect the stone shield from the return pipe.

4. Remove the pinch bolt and disconnect the flex coupling from the gear.

—————— **CAUTION** ——————

Failure to disconnect the flexible coupling from the steering gear stub shaft can result in damage to the steering control which could result in a vehicle crash and bodily injury.

5. Raise the car and support it with jackstands.

6. Remove one nut and lockwasher from the pitman shaft. Using a pitman arm puller, remove the pitman arm from the steering gear.

7. Remove the three screws and flat washers that hold the gear to the frame side rail, and lower the gear assembly down and out of the car.

8. Installation is the reverse of removal. The following torques are necessary: pinch bolt, 30 ft. lbs.; steering gear mounting bolts, 70 ft. lbs.; pitman arm 185 ft. lbs.

Ignition Switch

REPLACEMENT

1. Disconnect the negative battery terminal.

2. Position the lock cylinder in the LOCK position.

3. Remove the steering column lower cover.

4. Loosen the two nuts on upper steering column, allowing the column to drop.

—————— **CAUTION** ——————

Do not remove the nuts, as the column may bend under its own weight.

5. Disconnect the ignition switch connector at the switch.

6. Remove the two screws securing the ignition switch to the steering column. Remove the switch.

7. To install, first assemble the ignition switch on the actuator rod and adjust it to the LOCK position, as follows:

a. Standard Column — Hold the switch actuating rod stationary with one hand while moving the switch toward the bottom of the column until the switch reaches the end of its travel (ACC position). Back off one detent, then, with the key also in the LOCK position, tighten the two switch mounting screws to 35 inch lbs.

b. Tilt column — Hold the switch actuating rod stationary with one hand while moving the switch toward the upper end of the column until the switch reaches the end of its travel (ACC position). Back off one detent, then, with the key also in the LOCK position, tighten the two switch mounting screws to 35 inch lbs.

8. Connect the wires, tighten the two steering column nuts, install the lower cover and reconnect the battery.

Lock Cylinder

REPLACEMENT

Standard Steering Column

1. Remove the steering wheel.
2. Remove the lockplate cover assembly.
3. After compressing the lockplate spring, remove the snapring from the groove in the shaft.

——————— CAUTION ———————

When the snapring is removed, do not allow the shaft to slide out the bottom of the column.

4. Remove the lockplate and slide the turn signal cam and the upper bearing preload spring off the upper steering shaft.
5. Remove the thrust washer from the shaft.
6. Remove the hazard warning switch from the column along with the turn signal lever.
7. Use the following procedure if the car is equipped with cruise control:

a. Attach a piece of stiff wire to the connector on the cruise control switch harness.
b. Gently pull the harness up and out of the column.

8. Remove the turn signal switch mounting screws.
9. Slide the switch connector out of the bracket on the steering column.
10. After freeing the switch wiring protector from its mounting, pull the turn signal switch straight up and remove the switch, switch harness and the connector from the column.
11. Turn the ignition switch to ON or RUN and then insert a small drift pin into the slot next to the switch mounting screw boss. Push the lock

cylinder tab and remove the lock cylinder.

Tilt Column

1. Remove the steering wheel.
2. Remove the rubber sleeve bumper from the steering shaft.
3. Using an appropriate tool, remove the plastic retainer.
4. Using a spring compressor, compress the upper steering shaft spring and remove the C-ring. Release the steering shaft lockplate, the horn contact carrier, and the upper steering shaft preload spring.
5. Remove the four screws which hold the upper mounting bracket and then remove the bracket.
6. Slide the harness connector out of the bracket on the steering column. Tape the upper part of the harness and connector.
7. Disconnect the hazard button and position the shift bowl in Park. Remove the turn signal lever from the column.
8. Use the following procedure for cars with cruise control:

a. Remove the harness protector from the harness.
b. Attach a piece of piano wire to the switch harness connector.
c. Before removing the turn signal lever, loop a piece of piano wire and insert it into the turn signal lever opening. Using the wire, pull the cruise control harness out through the opening.
d. Pull the rest of the harness up through and out of the column.
e. Remove the guide wire from the connector and secure the wire to the column.
f. Remove the turn signal lever.

9. Pull the turn signal switch up until the end connector is within the shift bowl. Remove the hazard flasher lever. Allow the switch to hang.
10. Place the ignition key in the RUN position.
11. Depress the center of the lock cylinder retaining tab with a screwdriver and then remove the lock cylinder.
12. To install, reverse the removal procedure.

CHASSIS ELECTRICAL

Heater Blower
REMOVAL & INSTALLATION

1. Disconnect the negative battery cable.

2. Remove the rubber cooling hose fro the nipple and blower motor.
3. Disconnect the electrical connector.
4. Remove the screws that secure the motor to the case, then twist the motor 180° and pull out.

Heater Core

REMOVAL & INSTALLATION

1. Disconnect wiring from the blower, resistors, and thermostatic cylinder switch.
2. Drain the cooling system.
3. Remove the right windshield washer nozzle.
4. Remove the primary and secondary right air inlet screens and moldings from the plenum.
5. Remove the two screws securing the thermostatic compressor cycling switch to the module and carefully reposition the switch off the module cover.
6. Remove the 16 fasteners securing the module cover and remove the cover.
7. Remove the hoses from the core nipples.
8. Remove one screw and retainer holding the core to the frame at the top.
9. Place the temperature door in the MAX HOT position and reach through the temperature housing and push the lower forward corner of the heater core away from the housing. This causes the core to snap out of the lower clamp. The core may not be removed in a vertical direction.
10. Installation is the reverse of removal. Always use new sealer wherever sealer was removed during core removal.

Radio

REMOVAL & INSTALLATION

1. Remove the radio knobs and antirattle springs. Disconnect the negative battery terminal.
2. Remove the two hex nuts securing the bezel to the radio.
3. Remove the two center air conditioning outlet grilles. Remove the one screw in each outlet.
4. Remove the maplights and remove the center panel insert.
5. Unbolt and remove the radio from the panel.
6. Disconnect the wiring.
7. Installation is the reverse of removal.

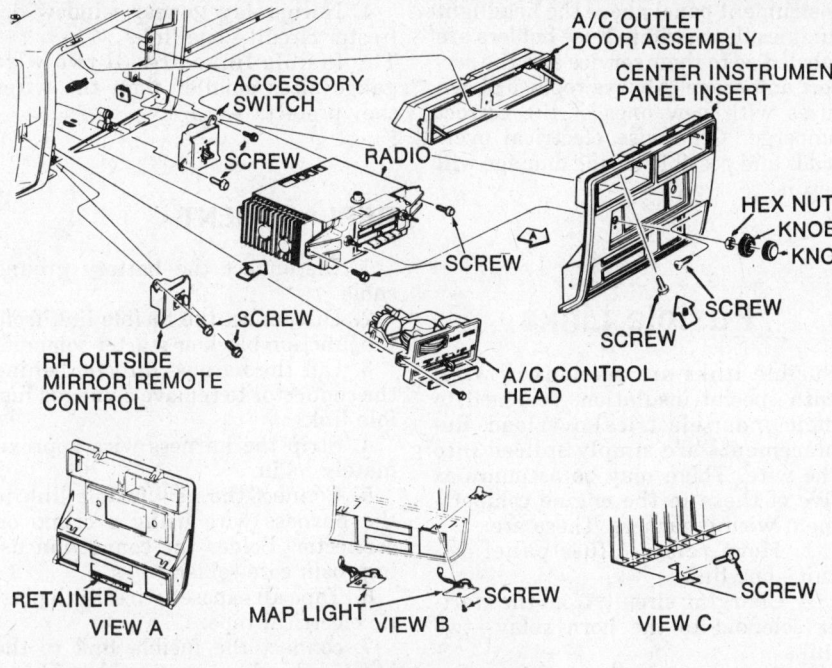

Typical radio installation

Windshield Wiper Switch

REMOVAL & INSTALLATION

1. Remove the left climate control door knob.
2. Remove the left climate control oultet grille.
3. Remove the screws securing the left instrument panel trim plate bezel.
4. Remove the steering column cover screws.
5. Break loose the steering column seal at the lower trim panel and remove the trim panel.
6. Disconnect the wiring from the switch, remove the attaching screws and pull out the switch.
7. Installation is the reverse of removal.

Wiper Motor

REMOVAL & INSTALLATION

1. Disconnect the negative battery cable.
2. Remove the cowl screen.
3. Reach through the opening and disengage the transmission drive link from the wiper crank arm by loosening two nuts.
4. Disconnect the wiring and washer hoses.
5. Remove the bolts that secure the wiper/washer unit to firewall.
6. Remove the entire assembly.

7. to install, reverse the removal procedure, making sure the wiper crank arm is in the Park position.

Wiper Blade

REPLACEMENT

Two methods are used to retain the blades to the arms. One method uses a press type tab. When the tab is depressed, the blade assembly can be slid off the arm. The other method uses a spring retainer. A suitable tool must be inserted on top of the spring and the spring pushed downward. The blade assembly can then be slid off the pin. The rubber element can be replaced separately from the blade. Replacement procedures are given in the Maintenance Section at the rear of this book.

Headlight Switch

REMOVAL & INSTALLATION

1. Disconnect the battery ground.
2. Remove the left instrument panel insert.
3. Remove the three screws securing the switch to the instrument panel.
4. On cars equipped with cruise control and Twilight Sentinel, remove the two screws securing the cruise control switch to the instrument panel.
5. Disconnect the Twilight Sentinel and/or Guidematic switches.

6. Disconnect the switch 2-piece connector.
7. Remove the switch rod.
8. Slide the cruise control switch forward.
9. Disconnect the headlight switch wires.
10. Remove the nut from the lens housing and remove the switch.
11. Installation is the reverse of removal.

Speedometer Cable

REMOVAL & INSTALLATION

1. Remove the left instrument panel insert.
2. Disconnect the battery ground.
3. Place the shift lever in Park and remove the screw securing the shift indicator cable to the column.
4. Remove the two upper screws securing the cluster assembly to the panel horizontal support.
5. Remove the two lower inside screws securing the cluster to the horizontal support.
6. Remove the screw located directly above the steering column securing the cluster to the speedometer mounting plate.
7. Pull the cluster outward to disengage the cable and remove the cluster. Placing the shift lever in the low range and tilting the steering wheel all the way down will help during removal.
8. Disconnect the cable housing from the locking spring on the mounting plate and pull it through the firewall.
9. Pull the core from the cable. If the core is broken or frayed on the transmission end, raise and support the car and disconnect the cable from the transmission. Be sure the entire cable has been removed.
10. Installation is the reverse of removal.

Instrument Cluster

REMOVAL & INSTALLATION

1. Loosen the set screw in the left climate control outlet door knob and remove the knob.
2. Remove the left climate control air outlet grille using tool J24612-01 or its equivalent.
3. Remove the screws securing the bezel to the instrument panel. One is located inside of the left A/C outlet grille opening.
4. Remove the screws in the lower steering column cover.
5. Disconnect steering column seal

on lower surface and remove trimplate.

6. Disconnect negative battery cable.

7. With shift lever in PARK, remove shift indicator clip from steering column.

8. Remove upper screws securing cluster assembly to instrument panel horizontal support.

9. Pull cluster outward and release the speedometer cable. Disconnect the speed control sensor electrical connector if so equipped.

10. Disconnect the printed circuit connector from cluster.

11. Remove the cluster.

12. Installation is the reverse of removal.

NOTE: Removal or installation can be done by placing shift lever in NEUTRAL and tilting the wheel to the lowest position of so equipped.

Fuses

The fuse block is located beneath the instrument panel above the headlight dimmer floor switch. Fuse holders are labeled as to their service and the correct amperage. Always replace blown fuses with new ones of the correct amperge. Otherwise electrical overloads and possible wiring damage will result.

Fusible Links

Fusible links are sections of wire, with special insulation, designed to melt under electrical overload. Replacements are simply spliced into the wire. There may be as many as five of these in the engine compartment wiring harness. These are:

1. Horn relay to fuse panel circuit—one link.

2. Charging circuit, from the starter solenoid to the horn relay—two links.

3. Starter solenoid to ammeter circuit—one link.

4. Horn relay to rear window defroster circuit—one link.

The fusible links are all two wire gauge sizes smaller than the wires they protect.

REPLACEMENT

1. Disconnect the battery ground cable.

2. Disconnect the fusible link from the junction block or starter solenoid.

3. Cut the harness directly behind the connector to remove damaged fusible link.

4. Strip the harness wire approximately $1/2$ in.

5. Connect the new fusible link to the harness wire using a crimp on connector. Solder the connection using rosin core solder.

6. Tape all exposed wires with plastic electrical tape.

7. connect the fusible link to the junction block or starter solenoid and reconnect the battery ground cable.

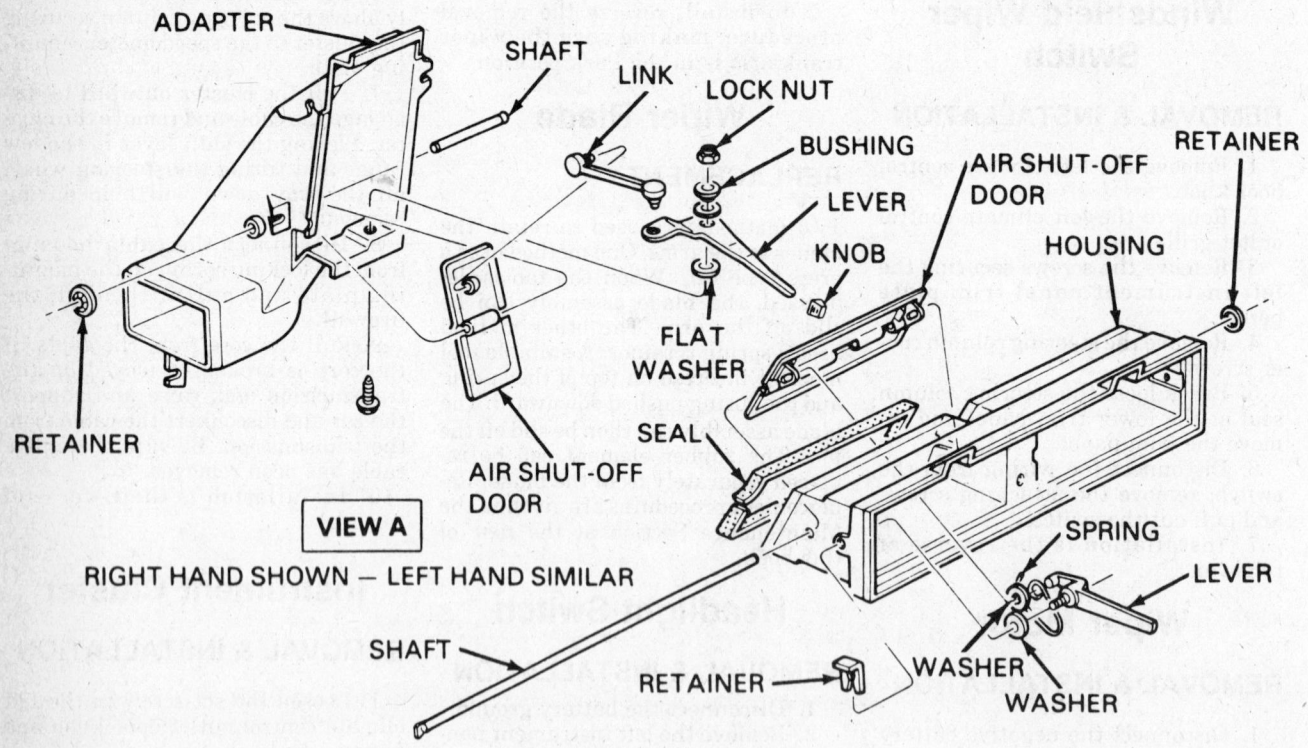

VIEW A

RIGHT HAND SHOWN — LEFT HAND SIMILAR

A/C ducts (DeVille and Brougham)

Chevrolet
Corvette

YEAR IDENTIFICATION

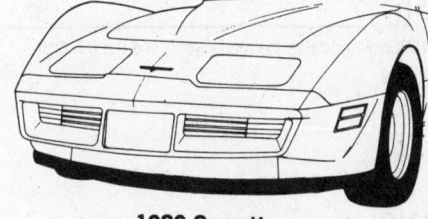

1980 Corvette

1984–86 Corvette

1981-82 Corvette

VEHICLE IDENTIFICATION NUMBER (VIN)

It is important for servicing and ordering parts to be certain of the vehicle and engine identification. The (VIN) (vehicle identification number) is a 13 or 17 digit number visible through the windshield on the driver's side of the dash and contains the vehicle and engine identification codes. It can be interpreted as follows:

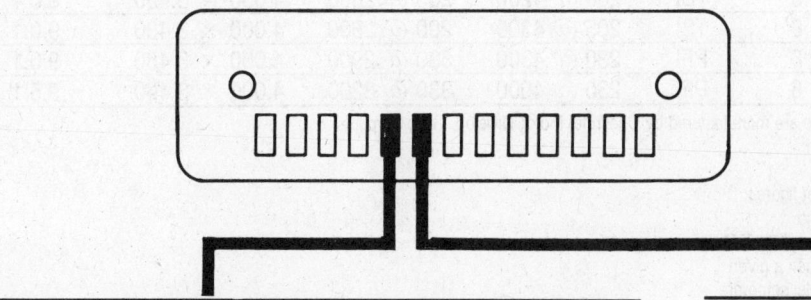

Engine Code						Model Year Code	
Code	Cu. In.	Liters	Cyl.	Carb.	Eng. Mfg.	Code	Year
8①	350	5.7	8	4 bbl.	Chev.	A	'80
H	305	5.7	8	4 bbl.	Chev.		
6②	350	5.7	8	4 bbl.	Chev.		

The thirteen digit Vehicle Identification Number can be used to determine engine application and model year. The 6th digit indicates the model year, and the 5th digit identifies the factory installed engine.
① Standard performance L48 engine
② High performance L82 engine

C373

VEHICLE IDENTIFICATION NUMBER (VIN)

It is important for servicing and ordering parts to be certain of the vehicle and engine identification. The VIN (vehicle identification number) is a 13 or 17 digit number visible through the windshield on the driver's side of the dash and contains the vehicle and engine identification codes. It can be interpreted as follows:

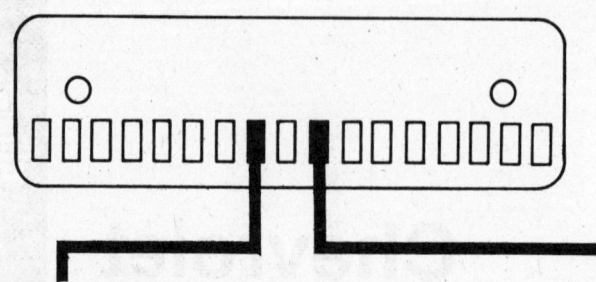

Engine Code

Code	Cu. In.	Liters	Cyl.	Carb.	Eng. Mfg.
6	350	5.7	8	4 bbl.	Chev.
8	350	5.7	8	①	Chev.

Model Year Code

Code	Year
B	'81
C	'82
E	'84
F	'85
G	'86
H	'87

The seventeen digit Vehicle Identification Number can be used to determine engine application and model year. The 10th digit indicates the model year, and the 8th digit identifies the factory installed engine. There is no 1983 Corvette model.
① Throttle body fuel injection (TBI) on 1984 models; port fuel injection (PFI) on 1985 and later.

GENERAL ENGINE SPECIFICATIONS

Year	Engine No. of Cyl. Displacement (cu. in.)	Engine VIN Code	Fuel Delivery	Horsepower @ rpm①	Torque @ rpm (ft. lbs.)①	Bore × Stroke (in.)	Compression Ratio	Oil Pressure @ 2000 rpm
'80	8-305	H	4 bbl.	180 @ 4200	255 @ 2000	3.736 × 3.480	8.6:1	45
	8-350	8	4 bbl.	190 @ 4400	280 @ 2400	4.000 × 3.480	8.2:1	45
	8-350	6	4 bbl.	230 @ 5200	275 @ 3600	4.000 × 3.480	9.0:1	45
'81	8-350	6	4 bbl.	190 @ 4200	280 @ 1600	4.000 × 3.480	8.2:1	45
'82	8-350	8	TBI	200 @ 4200	285 @ 2800	4.000 × 3.480	9.0:1	45
'84	8-350	8	TBI	205 @ 4300	290 @ 2800	4.000 × 3.480	9.0:1	50–65
'85	8-350	8	PFI	230 @ 4300	330 @ 2900	4.000 × 3.480	9.0:1	50–65
'86–'87	8-350	8	PFI	230 @ 4000	330 @ 3200	4.000 × 3.480	9.5:1	50–65

NOTE: All engines used in the Corvette are manufactured by Chevrolet Motor Division, G.M. Corp.
TBI—Throttle body fuel injection system
PFI—Port fuel injection system
① Horsepower and torque are SAE net figures. They are measured at the rear of the transmission with all accessories installed and operating. Since the figures vary when a given engine is installed in different models, some are representative rather than exact.

TUNE UP SPECIFICATIONS

(When analyzing compression test results, look for uniformity among cylinders rather than specific pressures.)

Year	No. of Cyl. Displacement (cu. in.)	VIN Code	Option Code	hp	Type (A.C.)	Gap (in.)	Man. Trans.	Auto. Trans.	Valves Intake Opens ⑤(deg.)	Fuel Pump Pressure (psi)	Man. Trans.	Auto. Trans.
	Engine				**Spark Plugs**		**Ignition Timing (deg.)④**		**Valves Intake Opens**	**Fuel Pump**	**Idle Speed (rpm)④**	
'80	8-305	H	LG4	180	R45TS	0.045	4B	4B	28	7½–9	②	②
	8-350	8	L48	190	R45TS	0.045	6B③	6B	28	7½–9	②	②
	8-350	6	L82	230	R45TS	0.045	12B	12B	52	7½–9	②	②
'81	8-350	6	L81	190	R45TS	0.045	6B	6B	38	7½–9	②	②
'82	8-350	8	L83	200	R45TS	0.045	①	②	32	9–13	①	②
'84	8-350	8	L83	205	R45TS	0.045	②	②	32	9–13	②	②
'85–'86	8-350	8	L98	230	R45TS	0.045	②	②	NA	NA	②	②
'87	See Underhood Specifications Sticker											

NOTE: All models use electronic ignition systems. No adjustments are necessary. The underhood specifications sticker often reflects tuneup specification changes made in production. Sticker figures must be used if they disagree with those in this chart. Part numbers in this chart are not recommendations by Chilton for any product by brand name.

B—Before Top Dead Center
① Manual transmission not available
② See Underhood Sticker
③ Except Calif. and High Altitude: 6B Calif. and
  High Altitude: 8B
④ See text for procedure
⑤ All figures Before Top Dead Center

FIRING ORDER

NOTE: To avoid confusion, always replace spark plug wires one at a time.

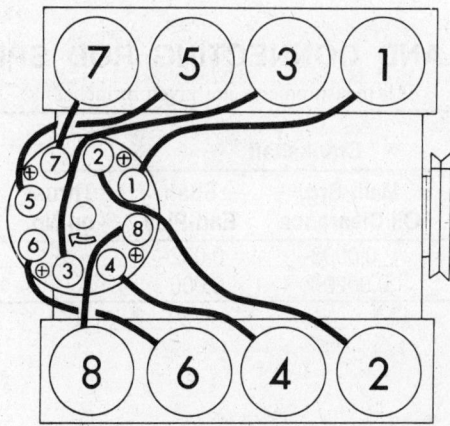

GM (Chevrolet) V8
Engine firing order: 1-8-4-3-6-5-7-2
Distributor rotation: clockwise

CAPACITIES

Year	Engine No. of Cyl. Displacement (cu. in.)	Engine Crankcase Add 1 qt. for New Filter	Transmission (Pts. to Refill After Draining)			Gasoline Tank (gals.)	Cooling System (qts.)	
			Manual 4-Speed	Automatic ①	Drive Axle (pts.)		With Heater	With A/C
'80–'81	8-305, 350	4	3	8	4	24	21	22
'82	8-350	4	—	10	4	24	21	22
'84–'87	8-350	4	3.5②	10	3.75	20	14	14

① For pan removal and filter change only
— Not applicable
② Four speed overdrive uses Dexron® II in the overdrive section and 80W GL 5 Lube in the trans. section

VALVE SPECIFICATIONS

Year	Engine No. of Cyl. Displacement (cu. in.)	Seat Angle (deg.)	Face Angle (deg.)	Spring Test Pressure (lbs. @ in.)	Spring Installed Height (in.)	Stem-to-Guide Clearance (in.)		Stem Diameter (in.)	
						Intake	Exhaust	Intake	Exhaust
'80	8-305, 350 (std. perf.)	46	45	180–188 @ 1.25	1²³⁄₃₂	0.0010–0.0027	0.0010–0.0027	0.3410–0.3417	0.3410–0.3417
'80	8-350 (high perf.)	46	45	196–204 @ 1.25	1²³⁄₃₂ ①	0.0010–0.0027	0.0010–0.0027	0.3410–0.3417	0.3410–0.3417
'81–'82	All	46	45	196–204 @ 1.25	1²³⁄₃₂ ①	0.0010–0.0027	0.0010–0.0027	0.3410–0.3417	0.3410–0.3417
'84–'87	8-350	46	45	194–206 @ 1.25②	1²³⁄₃₂ ①	0.0010–0.0027	0.0010–0.0027	0.3410–0.3417	0.3410–0.3417

① 1¹⁹⁄₃₂ for the exhaust valve spring
② 1.16 exhaust valve

CRANKSHAFT AND CONNECTING ROD SPECIFICATIONS
(All measurements are given in inches)

Year	Engine No. of Cyl. Displacement (cu. in.)	Crankshaft				Connecting Rod		
		Main Brg. Journal Dia.	Main Brg. Oil Clearance	Shaft End-Play	Thrust on No.	Journal Diameter	Oil Clearance	Side Clearance
'80–'87	8-305, 350	2.4484–2.4493①	0.0008–0.0020②	0.002–0.006	5	2.0988–2.0998	0.0013–0.0035	0.008–0.014

① Nos. 2, 3, 4—2.4481–2.4490; No. 5—2.4479–2.4488
② Nos. 2, 3, 4—.0011–.0023; No. 5—.0017–.0033
③ 1986 and later: .006–.014

TORQUE SPECIFICATIONS
(All readings in ft. lbs.)

Year	Engine No. of Cyl. Displacement (cu. in.)	Cylinder Head Bolts	Rod Bearing Bolts	Main Bearing Bolts	Crankshaft Balancer Bolt	Flywheel-to-Crankshaft Bolts	Manifold	
							Intake	Exhaust
'80–'83	8-305, 350	65	45	75①	60	60	30③	②
'84–'87	8-350	65	45	80	60	60	35	20

① Engines with 4-bolt mains—Outer bolts: 70 ft. lbs.
② Center bolts—30, end bolts 20
③ Also torque the throttle body plate bolts to 20–34 ft. lbs. on 1982 models.

PISTON AND RING SPECIFICATIONS

(All measurements are given in inches)

Year	Engine No. of Cyl. Displacement (cu. in.)	Ring Gap			Ring Side Clearance			Piston-to-Bore Clearance (in.)
		Top Compression	Bottom Compression	Oil Control	Top Compression	Bottom Compression	Oil Control	
'80	8-305, 350 (exc. L82)	.010–.020	.010–.023①	.015–.065	.0012–.0032	.0012–.0032	.0020–.0080	.0007–.0027
'80	8-350 (L82)	.010–.020	.010–.023	.015–.065	.0012–.0032	.0012–.0032	.0020–.0080	.0036–.0061
'81	8-350	.010–.020	.010–.023	.015–.065	.0012–.0032	.0012–.0032	.0020–.0080	.0046–.0061
'82	8-350	.010–.020	.010–.023	.015–.065	.0012–.0032	.0012–.0032	.0020–.0080	.0025–.0045
'84–'87	8-350	.010–.020	.010–.025	.015–.055	.0012–.0032	.0012–.0032	.002–.007	.0025–.0035

① 185 and 195 horsepower engines—.010–.025

WHEEL ALIGNMENT SPECIFICATIONS

(All measurements stated in degrees, unless noted.)

Year	Front Wheel Caster		Front Wheel Camber		Rear Wheel Camber		Toe (in.)	
	Range	Preferred	Range	Preferred	Range	Preferred	Front Wheel	Rear Wheel
'80	2P–2½P	2¼P	¼P–1P	¾P	½N ± ½	—	3/32–5/32	3/32 ± 1/32
'81–'82	1¾P–2¾P	2¼P	¼P–1¼P	¾P	0 ± ½	—	¼ ± 1/16	1/16 ± 1/16
'84	2½P–3½P	3P	5/16P–15/16P	13/16P	½N–½	0	0–¼P	3/32P–7/13P
'85–'86	2½P–3½P	3P	5/16P–15/16P	13/16P	1/32N–29/52P	13/32	0–¼P	3/32P–7/32P

N—Negative
P—Positive

ENGINE ELECTRICAL

For further information on the charging system, please refer to "Charging and Starting" in the Unit Repair Section.

Alternator

REMOVAL & INSTALLATION

1. Disconnect the negative battery terminal.
2. Disconnect and identify the wire leads from the alternator.
3. Remove the alternator brace bolt, then remove the drive belt.
4. Remove the alternator pivot attaching bolt and remove the alternator.
5. To install, reverse the above procedure and adjust the belt tension.

NOTE: On all 1984 and later engines, a single serpentine belt is used to drive all accessories formerly driven with V-belts. Belt tension is maintained by a spring loaded tensioner which has the ability to maintain belt tension over a broad range of belt lengths. There is an indicator to make sure the tensioner is adjusted to within its operating range. The belt tension is adjusted with a ½ inch breaker bar inserted into the square hole in the tensioner arm and a belt tension gauge (BT 7825 or equivalent) to 120–140 lbs., as read on the tension gauge installed between the alternator and the A.I.R. pump.

Starter

For further information on the starting system, please refer to "Charging and Starting" in the Unit Repair section.

REMOVAL & INSTALLATION

1. Disconnect the battery cables at the battery.
2. Raise the front of the vehicle to a convenient working height and support with jackstands.

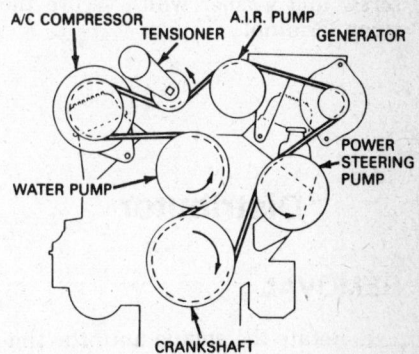

Serpentine drive belt tensioner—1984 and later

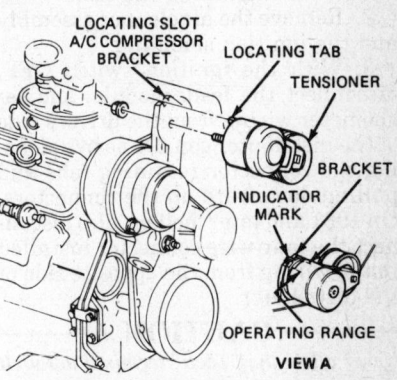

Serpentine drive belt installation

3. Disconnect the wiring from the starter solenoid. Replace each connector nut as the terminals are removed as the thread sizes differ between connectors. Tag the wiring positions to avoid improper connections during installation.

4. Remove the front starter support bracket and the heat shield (if so equipped).

5. Loosen the two main starter mounting bolts, support the starter and remove the bolts. Lower the starter front end first and remove the starter.

6. Reverse the previous steps to install the starter. Torque the two main starter mounting bolts to 25–35 ft. lbs.

Solenoid Replacement

1. Remove the screw and washer from the motor connector strap terminal.

2. Remove the two solenoid retaining screws.

3. Twist the solenoid clockwise to remove the solenoid flange key from the keyway in the housing. Remove the solenoid.

4. To reinstall the unit, place the return spring on the plunger and place the solenoid body on the drive housing. Push the solenoid inward and turn counterclockwise to engage the flange key. Install and tighten the solenoid retaining screws and the screw and washer which secure the strap terminal.

Distributor

REMOVAL

1. Rotate the engine until the timing mark on the balancer is aligned with the top dead center mark (TDC or "O") on the timing tab scale.

2. Remove the air cleaner assembly and the ignition shielding.

3. With the ignition switch OFF, disconnect the feed, module, and tachometer wiring from the drivers side of the distributor cap. Do so by releasing the connector retaining tabs and pulling downward on the connectors. On 1981 and later models, also disconnect the four-wire connector installed in the wiring from the opposite side of the distributor.

--------- CAUTION ---------
Never allow the "Tach" terminal to touch ground.

4. Locate the locking tabs of the spark plug wire retaining ring on the distributor cap. Move each of the two locking tabs outward to release the retaining ring. With the plug wires still attached to the ring, carefully remove the retaining ring from the distributor cap. Pull the plug wires out of the locating looms (if so equipped) and move the ring (with the wires still attached) out of the way.

5. Release the distributor cap holddown screws (push downward and turn counterclockwise) and lift the cap assembly off of the distributor. Check that the firing tip of the rotor is pointed at the No. 1 terminal of the distributor cap; if it is not, rotate the engine one full revolution and again align the timing marks. Recheck the position of the rotor.

6. Disconnect the vacuum line from the distributor vacuum advance unit, if so equipped.

7. Disconnect the battery cables at the battery.

8. Mark the relationships between the following items:

   a. Rotor firing tip and distributor body

   b. distributor body and either the firewall or the intake manifold.

The combination of these marks will assure that the distributor gear is properly meshed with the camshaft and the ignition timing will be close enough to start the engine after the distributor is reinstalled.

9. Remove the distributor holddown bolt and plate.

10. Carefully pull the distributor from the engine. Note the rotation of the rotor as the distributor is pulled upward, caused by the angled distributor drive teeth.

11. Installation is the reverse of removal.

--------- CAUTION ---------
Do not rotate the engine while the distributor is removed.

INSTALLATION—ENGINE DISTURBED

1. Turn the crankshaft until the No. 1 cylinder is at the top of its compression stroke. Remove the No. 1 spark plug to feel the compression.

2. Align the timing mark on the vibration damper with the TDC indicator or O mark on the timing scale.

3. With distributor body oriented in its normal position hold the rotor pointing toward the No. 1 plug wire location, then turn the rotor approximately $\frac{1}{8}$ turn counterclockwise and push the distributor down until it engages the camshaft, rotating the shaft slightly if necessary.

4. Press down on the distributor and crank the engine to make sure the oil pump shaft is engaged.

5. Return the crankshaft to No. 1 cylinder compression stroke with the timing marks aligned, then tighten the distributor clamp bolt.

6. Install the distributor cap, checking that the rotor points to the No. 1 terminal. Make sure that the spark plug wires are in their supports and are securely connected.

7. Connect distributor vacuum line (if so equipped) and primary wire.

8. Start engine and set the timing.

9. Install the ignition shielding.

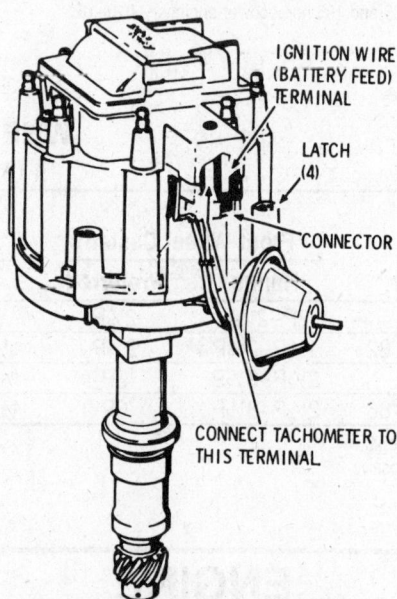

IGNITION WIRE (BATTERY FEED) TERMINAL

LATCH (4)

CONNECTOR

CONNECT TACHOMETER TO THIS TERMINAL

HEI system tachometer hookup

HEI SYSTEM TACHOMETER HOOKUP

Connect one dwell/tach lead to the TACH terminal on the side of the distributor and the other to ground. Some tachometers must be connected to the TACH terminal and the battery positive terminal. Not all tachometers will operate correctly with the HEI system. Check with the manufacturer if there is any doubt.

--------- CAUTION ---------
The TACH terminal should never be connected to ground. When hooking up a remote starter switch, disconnect the BATT terminal.

IGNITION TIMING ADJUSTMENT

NOTE: Before using the timing light, wipe the dirt and grease from the scale and mark the notch on the harmonic balancer with white paint or chalk. Refer to the underhood emissions label for the proper timing adjustment procedure.

The use of an inductive pick-up timing light is recommended for HEI systems. Follow the timing light manufacturers instructions to attach the timing light. Due to the battery location of Corvettes, 12 volt DC timing lights may be connected as follows: positive lead—alternator BAT terminal; negative lead—ground on engine.

—————— CAUTION ——————
DO NOT use an older style timing light which requires piercing of the spark plug lead.

Disconnect the distributor spark advance hose and plug the vacuum opening. Start the engine and run it at idle speed. Aim the timing light at the degree scale just over the harmonic balancer. Adjust the timing by loosening the securing clamp and rotating the distributor until the desired ignition advance is achieved, then tighten the clamp. To advance the timing, rotate the distributor opposite to the normal direction of rotor rotation. Retard the timing by rotating the distributor in the normal direction of rotor rotation. On 1980 and later models with computer controlled emissions systems, follow the information on the underhood decal to set the timing.

The ignition timing on 1982 and later models is adjusted in basically the same manner as previously described. The exception to this is that the EST BYPASS wire from the distributor must be disconnected prior to the timing adjustment. Trace the four wires from the distributor housing which join at a common multi-connector, close to the distributor. Follow the tan wire with a black stripe (EST BYPASS wire) from the multiconnector. Past the multiconnector, the EST BYPASS wire has its own, single connector. Separate this connector before adjusting the timing. While the EST BYPASS wire is disconnected, the CHECK ENGINE light on the instrument panel will illuminate. After adjusting the timing, reconnect the EST BYPASS connector; the CHECK ENGINE light will go out.

NOTE: It is not necessary to adjust the idle speed on 1982 and later models prior to the timing adjustment, though the engine must be at normal operating temperature.

1984 and later models incorporate an Electronic Spark Control (ESC) into the distributor which retards the spark advance when engine detonation occurs. If the controller fails, the result could be no ignition, no retard or full retard. Some engines will also have a magnetic timing probe hole for use with electronic timing equipment. Consult the manufacturer's instructions for the use of this equipment.

FUEL SYSTEM

Mechanical Fuel Pump

The 1980–81 Corvette fuel pump is a mechanically operated diaphragm-type pump. The camshaft of the engine has an eccentric (similar to a cam lobe, but more rounded) cast as part of the camshaft. As the camshaft rotates, the eccentric actuates a pushrod which pushes the fuel pump rocker arm to activate the pump. The fuel pump is attached to the right front of the cylinder block. A fuel pump mounting plate is used, with two gaskets; one between the pump and the plate; the other between the plate and the block.

The inlet, or suction line of the pump, draws fuel from the tank. The outlet, or pressure line of the pump, supplies pressurized fuel to the carburetor. Some models use a third line, which is a vapor return. The purpose of the vapor return is to route the hot fuel and fuel vapor from the pump back to the fuel tank, which considerably reduces the chance of vapor lock.

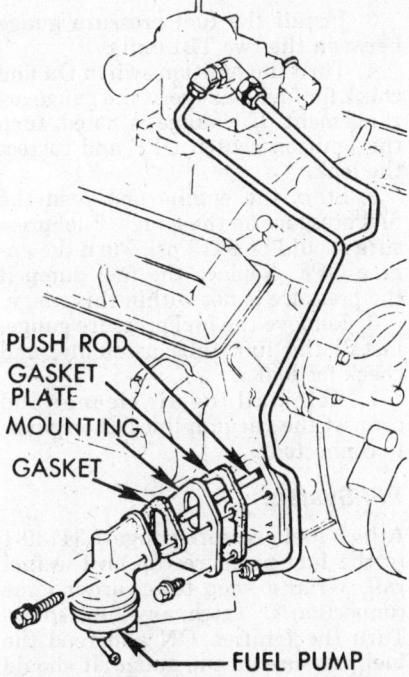

PUSH ROD
GASKET
PLATE
MOUNTING
GASKET

FUEL PUMP
Typical mechanical fuel pump mounting

NOTE: The fuel pump is not rebuildable—if defective, it must be replaced.

REMOVAL & INSTALLATION

1. Disconnect the fuel inlet, outlet, and vapor return (if equipped) lines from the pump.
2. Remove the bolt from the front right face of the engine block which is almost opposite the forward pump mounting bolt. Insert a longer bolt ($\frac{3}{8}$ x 16 x 2 in.) in this hole and snug down the bolt. This will hold the fuel pump pushrod in place.
3. Remove the fuel pump mounting bolts and remove the fuel pump.
4. Installation of the pump is the revere of the previous steps. Replace the fuel pump gasket(s) during installation. Start the engine and check the pump for proper operation, and check for leaks.

TESTING

If the engine exhibits a tendency to "starve-out", never assume that the fuel pump is defective until you test the pump. In most cases, a "starve-out" condition is caused by a weak ignition system, plugged fuel filter, or restricted fuel line.

1. Disconnect the fuel line from the carburetor. While this line is disconnected, check the fuel filter.
2. Run a piece of fuel-resistant rubber hose from the line to a graduated container.
3. Disconnect the BAT connector from the coil terminal.
4. Crank the engine. Fuel should be pumped into the container at a rate of 1 pint in 30 seconds.
5. Remove the added hose and container and connect a fuel pressure gauge to the fuel line. Crank the engine and read the highest pressure obtained on the gauge. See the Tune-Up Specifications Chart.
6. If the pump failed the tests of either Step 4 or 5, replace the pump as previously outlined. If the pump checked okay, remove the pressure gauge and reconnect the fuel line to the carburetor. Reconnect the ignition wiring as originally connected.

Electric Fuel Pump

An electric, impeller-type fuel pump is used on all 1982 and later Corvettes. The pump is designed to deliver a constant flow of fuel to the fuel injectors. The fuel pressure is regulated at the pressure regulator and/or compensator units. The pump is mounted inside the fuel tank as part of the fuel gauge sending unit. The pump can be replaced independently of the sending unit.

FUEL SYSTEM BLEEDING

—— CAUTION ——
This procedure must be performed to relieve the fuel system pressure before ANY work is done to the fuel system which requires that a fuel line be disconnected.

1982–84 TBI (Throttle Body Injection) System

With the engine OFF, remove the fuse from the fuse block designated "F.P." (fuel pump). Start the engine and allow it to run until it dies due to fuel starvation. Turn the ignition OFF, replace the fuel pump fuse, and service the fuel system as required.

1985 and later PFI (Port Fuel Injection)System

Connect fuel gauge J34730-1 or equivalent to the fuel pressure tap. Wrap a shop towel around the fitting while connecting the gauge to catch any fuel spray. Install a bleed hose into a suitable container, then open the valve to bleed the fuel system pressure.

REMOVAL & INSTALLATION

1982 and later

1. Disconnect the battery cables at the battery.
2. Remove the fuel filler door and bezel.

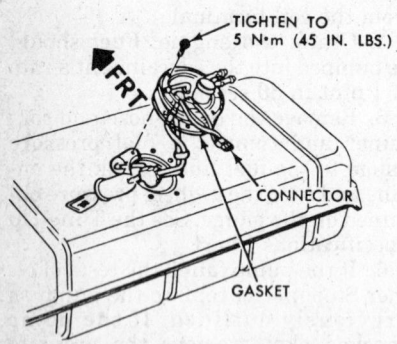

Electric fuel pump and gauge sender assembly, used with T.B.I.-equipped models

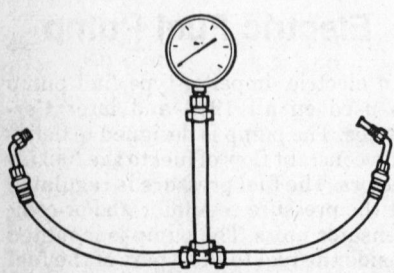

Fuel pressure gauge used to test the fuel pressure on T.B.I.-equipped models

3. Remove the fuel filler neck seal and drain hose.
4. Disconnect the lines and electrical connectors from the sending unit/pump assembly, and remove the screws which retain the assembly.
5. Remove the sending unit/pump assembly and the gasket.
6. Separate the pump from the sending unit.
7. Installation is the reveres of the previous steps. DO NOT connect the battery until all other steps have been completed.

TESTING

NOTE: A special fuel pressure gauge is required to safely perform this test. On engines equipped with TBI, use gauge J-29658; on engines equipped with PFI, use gauge J-34730-1.

TBI System

1. Remove the air cleaner assembly and plug the vacuum connection(s) at the TBI unit.
2. Remove the fuel tube which connects between both TBI units.

NOTE: Use two line wrenches of the appropriate sizes to disconnect each fitting, one wrench to hold the large fitting, the other to loosen the smaller fitting.

—— CAUTION ——
A small amount of fuel will be released from the connections.

3. Install the fuel pressure guage between the two TBI units.
4. Turn the ignition switch On and check for fuel leakage at the gauge arrangement. If leakage is noted, turn the ignition switch OFF and correct the leak.
5. Start the engine and read the fuel pressure on the gauge. Fuel pressure should be 9–13 psi. Turn the engine OFF. Replace the fuel pump if the pressure is not within this range.
6. Remove the fuel pressure gauge, install the fuel tube assembly, and check for leaks.
7. Reinstall the air cleaner and connect the vacuum lines as originally connected.

PFI System

Attach fuel pressure gauge J-34730-1 to the fuel pressure tap on the fuel rail. Wrap a shop towel around the connection to catch any fuel spray. Turn the ignition ON and read the fuel pressure on the gauge. It should be 34–39 psi.

Fuel Filter

REMOVAL & INSTALLATION

1980–81

Fuel filters are integral with the carburetor body. The filter element can be replaced as follows:
1. Disconnect the fuel line.
2. Remove the fuel filter nut from the carburetor.
3. Remove the filter element and spring. Blow through the filter end. If the air does not flow freely, replace the element. Do not attempt to clean the filter element.
4. Install the spring, then the element.
5. Install the inlet fitting using a new gasket.
6. Install the fuel line.

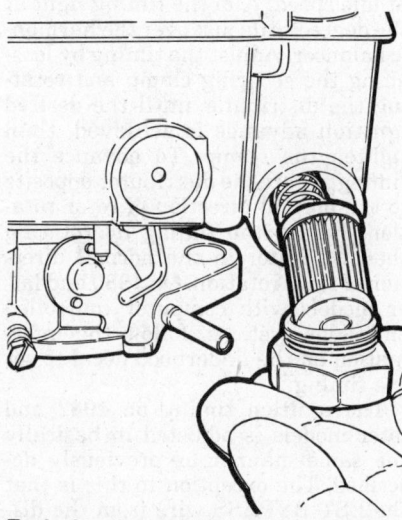

Typical carburetor-mounted fuel filter installation

NOTE: A check valve is installed in the fuel filter to meet roll over safety standards. New service replacement filters (paper) include the check valve. Install the check valve end of the filter toward the fuel line.

1982 and later

On these models, an inline fuel filter is used. The filter is mounted on the passenger-side frame rail, beneath the vehicle. Replacement is a simple matter of bleeding the system (as previously outlined), disconnecting the hoses, and dismounting the filter. Check for leaks after installing the new filter

NOTE: A woven plastic filter is located on the lower end of the fuel pick-up tube in the tank. This filter is self cleaning and normally requires no maintenance.

Carburetor

For further information on feedback carburetors, please refer to *Chilton's Guide To Fuel Injection And Feedback Carburetors.*

REMOVAL & INSTALLATION

1. Remove the air cleaner.
2. Disconnect the fuel line.
3. Disconnect the throttle linkage.
4. Disconnect and label all vacuum hoses.
5. Remove the retaining bolts.
6. Remove the carburetor.
7. Installation is the reverse of removal.

IDLE SPEED AND MIXTURE ADJUSTMENTS

When adjusting a carburetor with two idle mixture screws, adjust them alternately and evenly, unless otherwise stated. In the following adjustment procedures the term "lean roll" means turning the mixture adjusting screws in (clockwise) from optimum setting to obtain an obvious drop in engine speed (usually 20 rpm).

For all adjustments and specifications not detailed below, please refer to "Carburetors" in the Unit Repair section.

Idle Speed

Refer to the underhood tune-up decal for specific procedures concerning your vehicle.

1980 Idle Mixture

Changes in the idle systems of these models make it impossible to adjust the mixture without the aid of a propane enrichment system, not available to the general public. Blacking out the mixture screw, of itself, will have little or no effect. All carburetors have mixture screws concealed under staked-in plugs. Mixture adjustments are possible only during carburetor overhaul.

1981 and Later Idle Mixture

The previously used propane enrichment or lean drop methods should not be used when adjusting carburetors used on Computer Command Control equipped vehicles. Because of the sensitivity of the CCC system any adjustments to the carburetor can impair the ability of the system to maintain correct control of the air/fuel mixture. The only time adjustments should be made is when the carburetor is being overhauled.

Fuel Injection

Refer to the GM "F" Body ("Camaro/Firebird") section for throttle body removal and installation procedures. For further information on fuel injection, please refer to *Chilton's Guide To Fuel Injection And Feedback Carburetors.*

ENGINE COOLING

Most 1980–82 Corvettes equipped with the heavy duty cooling system option use an auxiliary electric cooling fan. The auxiliary fan is used to supplement the engine-mounted fan during conditions of very high engine temperatures. The auxiliary fan circuit is energized anytime the ignition switch is in the RUN position, though the fan itself will not operate until the engine coolant temperature reaches 238°F, as sensed by the cylinder head-mounted coolant temperature sensor. When the coolant temperature decreases to approximately 201°F, the fan will turn off.

--- CAUTION ---
Keep hands, tools, clothing, etc. clear of the auxiliary fan. The fan can come on automatically even when the engine is not running.

1984 and later models use an electric, single-speed cooling fan mounted directly to the radiator shroud. Fan operation is determined by coolant temperature, operating only enough to maintain engine coolant at or below a preset maximum temperature. The cooling fan is energized through a relay controlled by a sending unit located in the right cylinder bank. The fan only operates when road speed is below 35 mph.

Radiator

REMOVAL & INSTALLATION

1980–82

1. Disconnect the negative battery cable at the battery.
2. Drain the cooling system.
3. Remove the air cleaner snorkel.
4. Raise the front of the vehicle and support it with jack stands.
5. Disconnect the fan shroud from the radiator support bracket.
6. If so equipped, disconnect the au-

tomatic transmission cooler lines from the radiator.
7. Remove the radiator support brackets.
8. Disconnect the radiator upper and lower hoses and the overflow tube from the radiator.
9. Remove the radiator.
10. Installation is the reverse of removal. When installing the radiator make sure it is seated in the mounting pads. When replacing the radiator cap make sure the arrows line up with the overflow tube.

NOTE: It may be necessary to remove the fan shroud when removing the radiator.

1984 and Later

1. Disconnect the negative battery cable.
2. Drain the cooling system.
3. Remove the upper and lower radiator hoses.
4. Remove overflow hose at radiator.
5. Remove A/C accumulator and move aside.
6. Remove transmission cooler line.
7. Remove fan wires from fan and shroud.
8. Remove fan to gain access to lower cooler line.
9. Remove transmission cooler line at fitting.
10. Remove upper shroud bolts.
11. Remove upper shroud.
12. Remove radiator.
13. Installation is the reverse of removal.

Water Pump

REMOVAL & INSTALLATION

1980–82

1. Drain the radiator and loosen the fan pulley bolts.
2. Disconnect the heater hose, lower radiator hose and, if applicable, the bypass hose at the water pump.
3. Remove the alternator upper brace. Loosen the swivel bolt and remove the fan belt.
4. Remove the fan blade and pulley. Replace a bent or damaged fan.

NOTE: Thermostatic fan clutches must be kept in an "in-car" position. When removed from the car the assembly should be supported so that the clutch disc remains in a vertical plane to prevent silicone fluid leakage.

5. Remove the water pump attaching bolts and, if applicable, the power

steering-to-pump bolts and remove the pump and gasket.

6. Install the pump assembly using a new gasket. Coat the gasket on both sides with sealer. Tighten the ³⁄₈ in. bolts to 30 ft. lbs.

7. Install the pulley and fan.

8. Connect the hoses and fill the cooling system.

9. Install the alternator upper brace and fan belt. Install the power steering pump bolt.

10. Adjust the belts, then start the engine and check for leaks.

1984 and Later

NOTE: If the compressor lines do not have enough slack to move the compressor out of the way without disconnecting the refrigerant lines, the air conditioning system must be evacuated (using the required tools) before the refrigerant lines can be disconnected.

―――――― CAUTION ――――――

Do not disconnect any refrigerant lines unless experienced with air conditioning systems. Escaping refrigerant will freeze any surface it contacts, including skin and eyes.

1. Disconnect the negative battery cable.

2. Drain cooling system.

3. Remove serpentine drive belt.

4. Remove water pump pulley.

5. Remove AIR pump pulley.

6. Remove air management valve adapter.

7. Remove AIR pump.

8. Disconnect fuel inlet and return lines.

9. Remove rear A/C compressor braces.

10. Remove lower A/C compressor mounting bolt.

11. Remove A/C compressor and idler pulley bracket nuts.

12. Disconnect A/C compressor wires.

13. Slide mounting bracket forward and rear A/C compressor bolt.

14. Remove A/C compressor.

15. Remove right and left AIR hoses at check valve.

16. Remove AIR pipe at intake and power steering reservoir bracket.

17. Remove power steering reservoir bracket including top alternator bolt.

18. Remove lower AIR bracket on water pump.

19. Remove lower radiator and heater hose at water pump.

20. Remove water pump.

21. Installation is the reverse of removal. Tighten the water pump bolts 100–125 inch lbs.

Thermostat

REMOVAL & INSTALLATION

The thermostat is located on the front of the intake manifold directly in the center. It is not necessary to remove the top radiator hose to remove the thermostat.

1. Drain the cooling system about halfway.

2. Remove the two retaining bolts from the thermostat housing and lift up the housing with the hose attached. Remove the thermostat.

3. Insert the new thermostat, spring end down, and install the housing with a new gasket.

EMISSION CONTROLS

For all information on emission controls, please refer to "Emission Controls" in the Unit Repair section. Due to the complex nature of modern electronic engine control systems, comprehensive diagnosis and testing procedures fall outside the confines of this repair manual. For complete information on diagnosis, testing and repair procedures concerning all modern engine and emission control systems, please refer to *Chilton's Guide To Electronic Engine Controls*.

ENGINE MECHANICAL

NOTE: Engine application and specification tables may be found at the beginning of this section. With the exception of some engine procedures outlined in this section please refer to the "Chevrolet Rear Wheel Drive" section for the following removal and installation procedures: Intake and exhaust manifolds, cylinder head and valve system, timing case cover and pulley, camshaft, pistons and connecting rods.

REMOVAL

This procedure is basically the same for all models regardless of which engine is used. Certain pieces of optional equipment require minor specific changes, but overall, the operation remains the same.

1. Mark the relationship between each hood hinge, and the hood. Remove the hood.

2. Disconnect the battery cables at the battery.

3. Remove the air cleaner assembly and cover the carburetor or throttle body assembly. Mark any disconnected hoses so that they may be reinstalled properly.

4. Raise the front of the vehicle and support it with jackstands.

5. Locate and remove the engine coolant drain plugs. There is a drain plug on each side of the engine block, just above the top of the oil pan.

6. Loosen the radiator drain petcock and allow the coolant to drain from the radiator.

7. Remove the radiator hoses and the heater hoses.

8. On models through 1982 remove the radiator fan shroud, radiator, engine cooling fan(s) and fan clutch (if so equipped).

9. On 1984 and later models, remove the serpentine belt.

10. Drain the engine oil.

11. Remove the ignition shielding and release the distributor cap holddown screws. Move the distributor cap (with wires still intact) out of the way (away from the firewall).

12. On models so equipped, disconnect the four wire connector at the distributor.

13. On 1984 and later models, remove the distributor.

14. During this step, mark the location and/or connection point of each item so that these items may be properly reinstalled/reconnected.

  a. Disconnect the wiring from the starter and distributor.

  b. Disconnect the wiring from the alternator.

  c. Disconnect the wires from both the water temperature sender and oil pressure sender.

  d. Disconnect the engine ground wires.

  e. Disconnect the wiring from the idle solenoid, if so equipped.

  f. Disconnect the wiring from the various emission control items, as applicable (e.g. oxygen sensor, barometric sensor, air control valve, etc.).

  g. Disconnect the accelerator and transmission linkage (or cables, on late models) at the carburetor or throttle body injection (TBI) unit. If equipped with cables, unbolt the cable brackets from the engine.

  h. Disconnect the fuel supply and evaporative emission lines at the fuel pump. On TBI equipped models, disconnect the flexible hoses which connect the frame-mounted lines to the engine-mounted lines. In either case, plug the fuel supply line to prevent fuel siphoning from the tank. See the procedure for depressurizing the fuel system before disconnecting any fuel lines in the Fuel Injection section of Unit Repair.

i. On PFI equipped models, disconnect the PFI harness at the engine. When moving or disconnecting fuel lines please refer to the Fuel System Section under Fuel System Bleeding for relieving the fuel system pressure.

j. Disconnect any vacuum lines which run from a body-mounted item to an engine-mounted item (e.g. power brake unit, cruise control, etc.).

15. Remove the drive belt(s) from both the power steering pump and the air conditioning compressor, if equipped with these items. Unbolt the pump and compressor from their respective mounting brackets and tie these units out of the way (with lines still attached—DO NOT disconnect the refrigerant lines from the A/C compressor.

16. Disconnect the cruise control chain or cable from the engine, if so equipped.

17. Disconnect the exhaust pipes from the exhaust manifold flanges.

18. Remove the starter and solenoid as an assembly.

19. Remove the flywheel splash shield or convertor underpan, as applicable.

20. On automatic transmission equipped vehicles, remove the torque convertor-to-flywheel attaching bolts. Also, on these models, removes the transmission dipstick and tube.

21. On manual transmission equipped models, disconnect the linkage from each of the two levers of the clutch cross shaft. Loosen the outer ball stud nut and slide the stud out of the bracket slot. Move the cross shaft as required to clear the inboard ball stud. Remove the cross shaft from the vehicle.

22. Unless you have a suitable plug to prevent the transmission from draining after the driveshaft is removed, drain the transmission. On automatics without drain plugs it will be necessary to carefully remove the transmission pan, drain the fluid, and reinstall the pan. Using chain or heavy wire, secure the torque convertor to the transmission so that the convertor will not fall out as the engine is removed (auto. trans. only).

23. Matchmark the driveshaft to the rear axle flange. Unbolt the universal joint straps from the flange and re-

move the driveshaft assembly.

24. Support the transmission using a floor jack and remove the transmission-to-engine mounting bolts (auto. trans.) or the bellhousing-to-engine mounting bolts (man. trans.).

25. Remove the engine mount thrubolts (one per side, positioned front-to-back).

26. Attach the engine lifting devices to the engine lifting brackets. Most engines are equipped with these brackets bolted to the intake manifold. If the engine does not have these brackets, remove the valve covers and the center head bolt from each cylinder head. Attach the lifting apparatus to the cylinder heads and secure with the cylinder head bolts.

— CAUTION —

Be absolutely sure that the chain which is used has a weight rating greater than the weight of the engine. If possible, use chain rated at least at 1000 lbs. Avoid using chains with a lesser rating; serious injury could result if an inferior chain is used.

27. Move the engine forward, enough to disengage the engine from the transmission. Raise the engine enough to clear the front of the car and carefully move the engine over and away from the nose of the vehicle.

25. Service the existing engine as necessary, or install a replacement.

— CAUTION —

Do not allow the engine to hang from the engine hoist for an extended period of time. Never work on the engine when it is attached to the hoist. Support the engine safely on the floor or on an engine stand.

INSTALLATION

Installation of the engine is the reverse of the removal procedure. Make note of the following points before installing the engine:

1. Be sure that all wires, lines, etc., are connected as they originally were.

2. Be absolutely sure that the fuel lines are tightened properly and the throttle return springs are installed properly before attempting to start a new or rebuilt engine.

3. Follow all available bolt torque specifications.

4. Be sure to fill the engine, transmission, and cooling system with the correct quantities and qualities of fluids.

5. If a new camshaft was installed in the engine, the engine should be run for at least one hour after started at a minimum of 1500 rpm to properly "break-in" a new cam. If the cam manufacturers instructions differ, follow their recommendations.

6. Even though most head gasket manufacturers state that their gaskets require no "hot retorque", it is good practice to retorque the head bolts after the engine has been run for a couple of hours.

7. During engine installation, it is wise to replace "disposable" items such as radiator hoses, heater hoses, belts, and flexible fuel lines to prevent annoying (and possibly dangerous) post-installation problems associated with these items (e.g. coolant leaks, fuel leaks, overheating, etc.).

8. Adjust all belt(s) to the proper tension. If the belts are new, recheck their tension after about a $1/2$ hour of running time (with a new cam, do this after the cam "break-in" period). Note that a belt is considered "used" after just 5 minutes of running time, and additional belt stretch will usually occur.

9. ALWAYS check for coolant, fuel, and oil leaks after the engine is started. If there is leakage, turn off the engine, determine the source of leakage and fix the problem before restarting the engine.

10. Adjust the ignition timing after the engine is started.

Intake Manifold

REMOVAL & INSTALLATION

1985 and Later

1. Disconnect the negative battery cable.

2. Drain the cooling system.

3. Relieve the fuel system pressure, then remove the PFI sub-assembly consisting of the mass air flow sensor, plenum, runners, and fuel rail.

4. Disconnect all necessary vacuum hoses. Be sure and tag for reassembly.

5. Disconnect all necessary electrical connections.

6. Remove the distributor cap.

7. Mark and remove the distributor.

8. Disconnect the upper radiator hose at the thermostat housing.

9. Remove the AIR pump brace.

10. Disconnect the EGR pipe at the intake manifold.

11. Dioconnect the heater control vacuum line at the intake.

12. Remove the thermostat outlet.

13. Disconnect the coolant temperature sensor.

14. Remove the intake bolts and remove the manifold.

15. Installation is the reverse of removal. Clean all sealing surfaces then install the gaskets on the cylinder heads. The blocked openings in the gaskets are to be positioned at the

rear of the engine. Bend the gasket tab flush with the rear face of the cylinder head and place a $^3/_{16}$ inch bead of RTV sealent on the front and rear ridges of the cylinder case. Apply Loctite or equivalent to the attaching bolts.

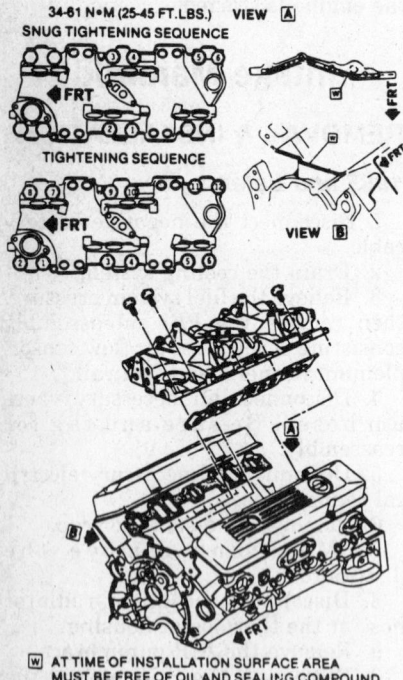

1. PLENUM
2. FUEL RAIL
3. INTAKE MANIFOLD
4. RUNNER

1985 and later PFI assembly

34-61 N•M (25-45 FT.LBS.)   VIEW Ⓐ
SNUG TIGHTENING SEQUENCE

TIGHTENING SEQUENCE

VIEW Ⓑ

Ⓦ AT TIME OF INSTALLATION SURFACE AREA MUST BE FREE OF OIL AND SEALING COMPOUND MUST BE WET TO TOUCH WHEN BOLT/SCREWS ARE TORQUED. APPLY SEALING COMPOUND .12 THICK.

Intake manifold assembly—1985 and later

Valve Clearance Adjustment

Engines equipped with hydraulic lifters VERY RARELY need adjustment of the valve lash. If the vehicle runs

well and there is no audible "clicking" in the valve train, leave it alone. This is because removal of the valve covers on vehicles equipped with air conditioning, various emission controls, cruise control, etc., can be a major project in itself.

On air conditioned models, the A/C compressor must be moved out of the way to gain access to one of the valve covers. Do not disconnect the refrigerant lines to move the compressor.

ENGINE LUBRICATION

Oil Pan

REMOVAL & INSTALLATION

NOTE: Before attempting this procedure, it may be necessary to remove the exhaust crossover pipe from the exhaust manifolds.

1. Disconnect the negative battery terminal.
2. Jack up your car and support it with jack stands.
3. Drain the oil and remove the filter.
4. Remove the starter and flywheel splash shield.
5. Disconnect the idler arm and lower steering linkage.
6. Remove the oil pan bolts and the pan.
7. Discard the old seals and gaskets.
8. On high performance engines the oil baffle must be removed before additional operations can be performed.
9. Installation is the reverse of removal. Remember to always install new gaskets.

Oil Pump

REPLACEMENT

The oil pump is located in the oil pan, and it is driven by a tang from the distributor shaft. The oil pump is bolted to the rear, main bearing cap. Oil is fed from the pump up through the rear main bearing cap. The pump may be removed after removing the oil pan as previously outlined.

Rear Main Bearing Oil Seal

NOTE: Removal and installa-

tion procedures for Corvette engines are given in the GM "F" BODY ("Camaro/Firebird") section.

CLUTCH

— CAUTION —

When servicing clutch parts, do not create dust by grinding or sanding clutch disc or by cleaning parts with a dry brush or with compressed air. A water dampened cloth—NOT SOAKED—should be used. The clutch disc contains asbestos fibers which can become airborne if dust is created during servicing. Breathing dust containing asbestos fibers may cause serious bodily harm.

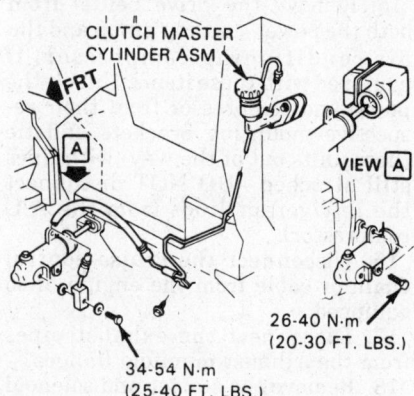

CLUTCH MASTER CYLINDER ASM

FRT

VIEW Ⓐ

26-40 N·m (20-30 FT. LBS.)

34-54 N·m (25-40 FT. LBS.)

Exploded view of hydraulic clutch components

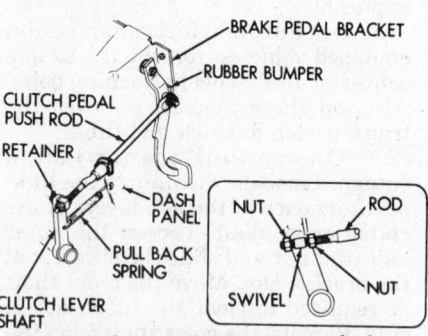

BRAKE PEDAL BRACKET
RUBBER BUMPER
CLUTCH PEDAL PUSH ROD
RETAINER
DASH PANEL
PULL BACK SPRING
CLUTCH LEVER SHAFT
NUT   ROD
SWIVEL   NUT

Clutch linkage adjustment—typical

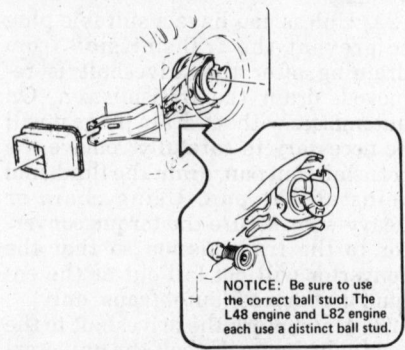

NOTICE: Be sure to use the correct ball stud. The L48 engine and L82 engine each use a distinct ball stud.

Ball stud attachment

Clutches on 1980–82 models are of the diaphragm spring type. The throwout bearing is a ball bearing with no provision for lubrication. The throwout fork pivots on a ball stud which is mounted in the rear face of the bellhousing.

1984 and later models use a hydraulic clutch system. The hydraulic system consists of a master cylinder and a slave cylinder. When pressure is applied to the clutch pedal (pedal depressed), the push rod contacts the plunger and pushes it up the bore of the master cylinder. In the first $\frac{1}{32}$ in. of movement, the center valve seal closes the port to the fluid reservoir tank and as the plunger continues to move up the bore of the cylinder, the fluid is forced through the outlet line to the slave cylinder mounted on the clutch housing. As fluid is pushed down the pipe from the master cylinder, this in turn forces the piston in the slave cylinder outward. A push rod is connected to the slave cylinder and rides in the pocket of the clutch fork. As the slave cylinder piston moves rearward the push rod forces the clutch fork and release bearing to disengage the pressure plate from the clutch disc. On the return stroke (pedal released), the plunger moves back as a result of the return pressure of the clutch. Fluid returns to the master cylinder and the final movement

of the plunger lifts the valve seal off the seat, allowing an unrestricted flow of fluid between system and reservoir. A piston return spring in the slave cylinder preloads the clutch linkage and assures contact of the release bearing with the clutch release fingers at all times. As the driven disc wears, the diaphragm spring fingers move rearward forcing the release bearing, fork and push rod to move. This movement forces the slave cylinder piston forward in its bore, displacing hydraulic fluid up into the mast cylinder reservoir, thereby proving the "self-adjusting" feature of the hydraulic clutch linkage system.

REMOVAL & INSTALLATION

1. Support the engine and remove the transmission as described in the Manual Transmission section.
2. Disconnect the clutch fork pushrod and spring. On 1984 and later models, remove the slave cylinder attaching bolts.
3. Remove the flywheel housing.
4. Slide the clutch fork from the ball stud and remove the fork from the dust boot. The ball stud is threaded into the clutch housing and is easily replaced, if necessary.
5. Install a clutch pilot tool.

NOTE: Look for the assembly markings "X" on the flywheel and the clutch cover (pressure plate assembly). If there are none, scribe marks to identify the position of the clutch cover relative to the flywheel.

6. Loosen the clutch cover bolts evenly until the spring pressure is relieved, then remove the bolts and clutch assembly.
7. Before installing, clean the pressure plate and the flywheel face.
8. Position the disc and pressure plate assembly on the flywheel and install a pilot tool.

NOTE: On single-disc models, the clutch disc is installed with the damper springs and slinger toward the transmission. On dual disc models, the discs are installed with the springs away from the flywheel.

9. Install the pressure plate assembly bolts. Make sure the mark on the cover is aligned with the mark on the flywheel. Tighten the bolts alternately and evenly to 35 ft. lbs.
10. Remove the pilot tool.
11. Remove the release fork and lubricate the ball socket and the fork fingers at the throwout bearing with graphite or Moly Grease. Reinstall the release fork.
12. Lubricate the inside recess and the fork groove of the throwout bearing with a light coat of graphite or moly grease.
13. Install the clutch release fork and dust boot in the clutch housing and the throwout bearing on the fork, then install the flywheel housing. Tighten flywheel housing bolts to 30 ft. lbs. Reinstall the slave cylinder.
14. Connect the fork pushrod and spring.
15. Adjust the shift linkage.
16. Adjust the clutch pedal free play. Bleed the hydraulic clutch system on 1984 and later models.

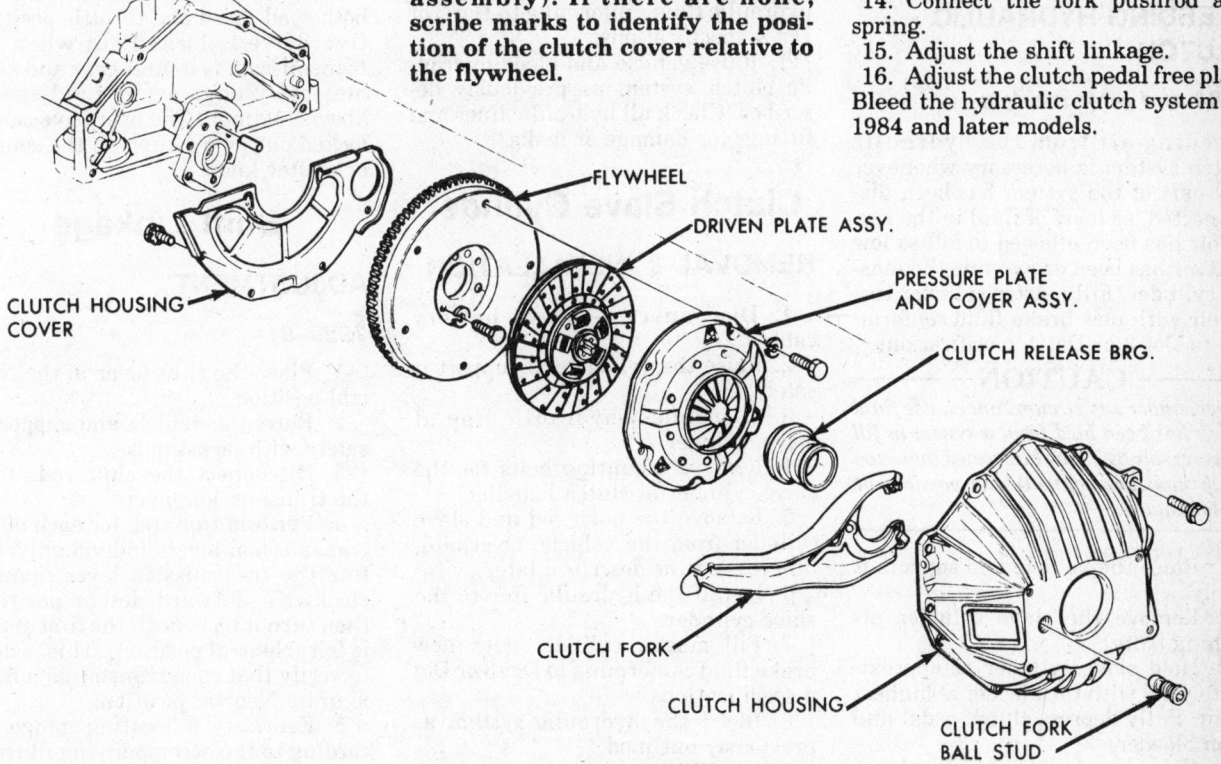

FLYWHEEL

DRIVEN PLATE ASSY.

PRESSURE PLATE AND COVER ASSY.

CLUTCH RELEASE BRG.

CLUTCH HOUSING COVER

CLUTCH FORK

CLUTCH HOUSING

CLUTCH FORK BALL STUD

Exploded view of the clutch assembly

ADJUSTMENT

1. Disconnect the return spring between the floor and the cross shaft.
2. Push the clutch lever and shaft assembly until the clutch pedal is tightly against the rubber stop under the dash.
3. Loosen the two locknuts on the shaft.
4. Push the shaft until the throwout bearing just touches the pressure plate spring.
5. Tighten the top locknut towards the swivel until the distance between it and the swivel is 0.4 in.
6. Tighten the bottom locknut against the swivel.
7. Check pedal free travel. It should be 1–1½ in.

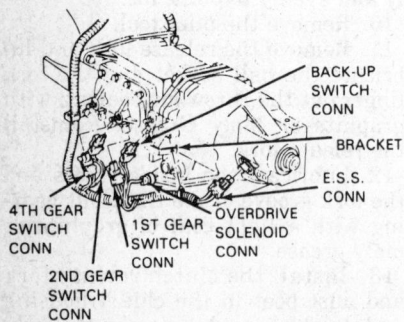

Electrical connectors on 4–speed overdrive transmission

BLEEDING HYDRAULIC CLUTCH

1984 and later models

Bleeding air from the hydraulic clutch system is necessary whenever any part of the system has been disconnected, or level of fluid in the reservoir has been allowed to fall so low that air has been drawn into the master cylinder. Fill master cylinder reservoir with new brake fluid conforming to Dot 3 or Dot 4 specifications.

— CAUTION —

Never, under any circumstances, use fluid which has been bled from a system to fill the reservoir as it may be aerated (have too much moisture content) and possibly be contaminated.

1. Raise the vehicle and support it safely.
2. Remove the slave cylinder attaching bolts.
3. Hold slave cylinder at approximately 45° with the bleeder at highest point. Fully depress clutch pedal and open bleeder.
4. Close bleeder valve and release clutch pedal.

5. Repeat Steps 3 and 4 until all air is evacuated from the system. Check and refill master cylinder reservoir as required to prevent air from being drawn through the master cylinder.

NOTE: Never release a depressed clutch pedal with the bleeder screw open, or air will be drawn into the system.

Clutch Master Cylinder

REMOVAL & INSTALLATION

1984 and Later Models

1. Disconnect negative battery cable.
2. Remove hush panel from under dash.
3. Disconnect push rod from clutch pedal.
4. Disconnect hydraulic line at the clutch master cylinder.
5. Remove the mounting bolts for the master cylinder at the front of dash assembly. Remove master cylinder and overhaul, if necessary, as described later.
6. Install mounting bolts for master cylinder at front of dash. Torque bolts to 15–22 ft. lbs. (20–30 Nm).
7. Connect hydraulic line at master cylinder.
8. Connect push rod at clutch pedal. Lubricate pivot point.
9. Install hush panel.
10. Fill master cylinder with new hydraulic fluid conforming to Dot 3 or Dot 4 specifications.
11. Raise vehicle and bleed hydraulic clutch system as previously described. Check all hydraulic lines and fittings for damage or leaks.

Clutch Slave Cylinder

REMOVAL & INSTALLATION

1. Disconnect negative battery cable.
2. Raise the vehicle and support it safely.
3. Disconnect hydraulic line at slave cylinder.
4. Remove mounting bolts for the slave cylinder at clutch housing.
5. Remove the push rod and slave cylinder from the vehicle. Overhaul, if necessary, as described later.
6. Install the hydraulic line to the slave cylinder.
7. Fill master cylinder with new brake fluid conforming to Dot 3 or Dot 4 specifications.
8. Bleed the hydraulic system as previously outlined.
9. Install slave cylinder to clutch housing. Lubricate leading end of cyl-

inder with Girling Rubber Lube or equivalent. Torque mounting bolts to 20–30 ft. lbs. (26–40 Nm).

MANUAL TRANSMISSION

Transmission refill capacities are in the Capacities table of this section. The Warner is used on all 1980–82 models. Identification is determined by side cover design and linkage. The Warner 4-speeds have the reverse fork mounted in the tailshaft.

1984 and later models use a computer-controlled 4-speed overdrive manual transmission which is essentially a combination of two separate transmissions. The first is a conventional 4-speed (83mm) manual system. The second is a three-speed overdrive system that is electronically controlled by the Electronic Control Module (ECM). By combining these two transmissions, the complete unit is actually capable of operating with seven separate gear ratios. The overdrive unit performs its function using a planetary gear system in combination with two sets of clutch packs controlled by hydraulic pressure just like the automatic transmission. The ECM is programmed to control the entire overdrive unit by monitoring both road speed and throttle position. Overdrive is locked out when the transmission is in first gear and automatically engages at road speeds above 110 mph. The overdrive can be locked out by a switch on the console or shifter knob.

Shift Linkage

ADJUSTMENT

1980–81

1. Place the shift lever in the Neutral position.
2. Raise the vehicle and support it safely with jackstands.
3. Disconnect the shift rods from the transmission levers.
4. Perform this step for each of the transmission levers individually: Rotate the transmission lever counterclockwise (forward detent position) then turn it back until the first detent is felt (Neutral position). This is done to verify that each transmission lever is in its Neutral position.
5. Fabricate a locating gauge according to the accompanying illustration. Insert the locating gauge into the notch of the shifter housing and

through the shift levers to properly align the levers. It may be necessary to move the shift lever(s) to install the locating gauge completely.

6. Loosen the locknuts of the 3–4 shift rod swivel (front of transmission side cover) and turn the swivel as necessary to allow the swivel to easily enter the hole of the 3–4 transmission lever. Apply a slight rearward pressure to the transmission lever and tighten the swivel locknuts. Attach the shift rod to the transmission lever with the retaining clip (and washer, if used).

7. Repeat Step 6 for the 1–2 and reverse shift rods.

8. Remove the locating gauge and lower the vehicle.

NOTE: After the adjustments have been made, the centerlines of the shifter levers must be aligned to prevent rubbing.

1984 and Later

1. Disconnect the negative battery cable.

2. Remove the left seat from the vehicle. If equipped with power seats, disconnect the electrical leads.

3. Remove the shift knob.

4. Remove the console cover.

5. Remove the glove box lock.

6. Remove the left side panel from the console.

7. Remove the shifter cover.

8. Loosen the adjusting nuts on the shifter rods.

9. With the transmission and shifter in neutral, install the alignment pin in shifter as illustrated.

10. Equalize the swivels on all three shift levers. Hand tighten the forward and rear adjusting nuts at the same time with equal force. Do this for all three shifter rods and then torque the forward and rear adjusting nuts at the same time to specifications.

11. Reintall the components removed by reversing Steps 1–7. Lubricate all linkage pivot points. If, after adjusting the linkage, it is found that high shift effort still exists, an anti-chatter lubricant (positraction additive) may be used. The lubricant is available in a small plastic bottle and can be squirted into the transmission through the filler plug.

REMOVAL & INSTALLATION

1980–82

1. Disconnect the battery ground cable.

2. Remove the shifter ball and "T" handle.

3. Remove the console trim plate.

4. Raise the vehicle on a hoist.

5. Remove the right and left ex-

haust pipes. It may be necessary to remove the catalytic converter and its mounting bracket to gain sufficient clearance to remove the transmission.

6. Disconnect the driveshaft at the transmission, lower the driveshaft and remove the slip yoke from the transmission.

7. Remove the rear mount to bracket bolts, then jack the engine enough to raise the transmission from the mount.

8. Remove the transmission linkage mounting to frame bolts.

9. Disconnect the shift levers at the transmission.

10. Remove the bolts attaching the gearshift assembly to mounting bracket and remove the mounting bracket. Remove the shifter mechanism with the rods and levers attached.

11. Disconnect the speedometer cable.

12. Remove the transmission mount bracket.

13. Remove the transmission to clutch housing retaining bolts and the lower left extension bolt.

14. Pull the transmission rearward until it is clear of the clutch housing, then rotate it clockwise while pulling to the rear.

15. To allow room for the transmission removal slowly lower the rear of the engine until the distributor gently touches the fire wall.

NOTE: Do not allow the engine to rest against the distributor as damage may result. Place two blocks of wood directly behind the heads to keep the engine weight off the distributor.

16. Installation is the reverse of removal. Adjust the shift linkage. Torque the transmission-to-clutch housing bolts to 52 ft. lbs. Torque the crossmember bolts to 25 ft. lbs.

1984 and Later

1. Disconnect the negative battery cable.

2. Remove the air cleaner assembly.

3. Disconnect the throttle valve (T.V.) cable at the left TBI unit.

4. Remove the distributor cap and lay aside.

5. Raise the vehicle and support it safely.

6. Remove the complete exhaust system assembly as follows:

   a. Disconnect A.I.R. pipe at the catalytic converter.

   b. Disconnect A.I.R. pipe clamps at exhaust manifold.

   c. Disconnect oxygen sensor electrical lead.

   d. Remove the bolts attaching the mufflers to the hangers.

   e. Remove hanger bracket at the converter.

   f. Disconnect the exhaust pipes from the exhaust manifolds and remove the exhaust system.

7. Remove the exhaust hanger at the transmission.

8. Support the transmission with a jack.

9. Remove the bolts attaching the driveline beam at the axle and transmission. Remove the driveline beam from the vehicle.

10. Mark the relationship of the propeller shaft to the axle companion flange. Remove the trunnion bearing straps and disengage the rear universal joint from the axle. Slide the propeller shaft slip yoke out from the overdrive unit and remove shaft from the vehicle.

11. Disconnect the cooler lines at the overdrive unit.

12. Disconnect the T.V. cable at the overdrive unit.

13. Disconnect the shift linkage at the side cover.

14. Disconnect the electrical connectors at the side cover switches, backup light switch, overdrive unit and speedometer sensor.

15. Lower the transmission and support the engine.

16. Remove the bolts attaching the transmission to the bellhousing. Slide the transmission rearward to disengage the input shaft from the clutch. Remove the transmission from the vehicle.

17. Inspect the clutch components for signs of wear or heat damage. See the Clutch Section, if necessary.

18. Installation is the reverse of removal. Clean and repack the clutch release bearing.

19. Refer to the Rear Suspension Section for installation and specifications for the driveline beam.

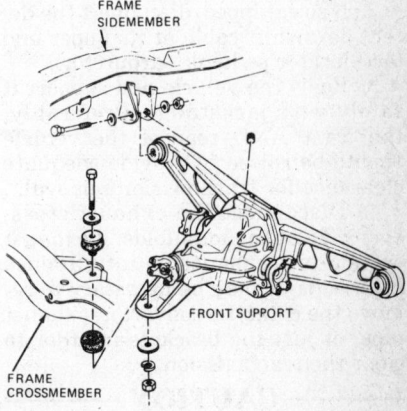

Differential carrier installation—1980 and later

20. Torque all fasteners to specifications. Do not overtighten.
21. Adjust the throttle valve (T.V.) cable.
22. Refill the transmission with fluid (4 Speed) SAE-80W or SAE-80W-90 GL-5 gear lube (Overdrive Unit) Dexron® II Automatic Transmission Fluid.

——— CAUTION ———
Do not overtorque the bolts attaching the driveline beam to the transmission. Overtorqueing can damage the bushing and seal in the overdrive unit and result in fluid leakage. Inadequate fluid level will damage the transmission.

OVERHAUL

For all overhaul information, please refer to "Manual Transmission Overhaul" in the Unit Repair section.

AUTOMATIC TRANSMISSION

Two Turbo HydraMatics have been available. The 700R-4 is the only automatic available in 1982 and later models. Identification can be made by the shape of the pan.

For automatic transmission adjustments, fluid and filter changing and identification details, please refer to "Automatic Transmissions" in the Unit Repair Section.

REMOVAL & INSTALLATION

1. Disconnect the negative battery cable at the battery.
2. If so equipped, disconnect the detent/downshift cable at its upper end (accelerator pedal or carburetor).
3. Raise the vehicle and support it safely with jackstands. Preferably, the front AND rear of the vehicle should be raised to provide adequate clearance for transmission removal.
4. Disconnect the exhaust crossover pipe at the manifolds, if exhaust system-to-transmission interference is obvious. It may be necessary to remove the catalytic convertor, exhaust pipe, or just the brackets in order to clear the transmission.

——— CAUTION ———
Exhaust system services must be performed while all components are COLD.

5. Remove the transmission inspection cover.
6. Remove the torque convertor-to-flywheel bolts. The relationship between the flywheel and convertor must be marked so that proper balance is maintained after installation.
7. Matchmark the drive shaft and the rear yoke (for reinstallation purposes). With a drain pan positioned under the front yoke, unbolt and remove the drive shaft.
8. Mark and disconnect vacuum lines, wiring, and the speedometer cable from the transmission as required.
9. Place a transmission jack (carefully) up against the transmission oil pan, then secure the transmission to the jack.
10. Remove the transmission mounting pad bolt(s), then carefully raise the transmission just enough to take the weight of the transmission off of the supporting crossmember. Remove the transmission mounting pad.

——— CAUTION ———
Exercise extreme care to avoid damage to underhood components while raising or lowering the transmission.

11. Remove the transmission dipstick, then unbolt and remove the filler tube.
12. Disconnect the shift linkage (or cable on floor shift-equipped models) and oil cooler lines from the transmission.
13. Support the engine using a jack placed beneath the engine oil pan. Be sure to put a block of wood between the jack and the oil pan, to prevent damage to the pan.
14. Securely wire the torque convertor to the transmission case.
15. Remove the transmission-to-engine mounting bolts, then carefully move the transmission rearward, downward, and out from beneath the vehicle.

——— CAUTION ———
If interference is encountered with the cable(s), cooler lines, etc., remove the component(s) before finally lowering the transmission.

16. Installation is basically the reverse of the previous steps. Note the following points during and after installation:
   a. Torque the transmission-to-engine mounting bolts to 30–40 ft. lbs.
   b. Align the matchmarks of the drive shaft with the marks of the rear yoke before installing the joint straps and bolts.
   c. Align the converter and flywheel markings before installing the converter bolts.
   d. Add the proper type and quantity of transmission fluid. If the converter was replaced, an additional 4 pints (approx.) should be added. NEVER overfill the transmission.
   e. Adjust the shift linkage (or cable) and the detent/downshift cable.
   f. Make sure that all vacuum lines, electrical connections, and oil cooler line connections are secure before driving the vehicle.
   g. Check for fluid leakage, then after the transmission is hot, recheck the fluid level.

DRIVE AXLE

NOTE: For driveshaft removal, refer to the procedure listed in the Chevrolet Rear Wheel Drive car section. For all universal joint information refer to "U-Jonts/CV-Joints" in the Unit Repair section. The 1984 and later Corvette uses two types of driveshafts, one shaft uses an aluminum tube and the other uses a steel tube.

Differential Carrier and Cover

REMOVAL & INSTALLATION

1980–82

1. Raise the vehicle on a hoist.
2. Remove the spare tire.
3. Remove the support hooks attached to the carrier cover and remove the spare tire cover.
4. Remove the exhaust system.
5. Place jackstands under the front control arms to support the vehicle.
6. Remove the heat shield.
7. Using an adjustable floor jack and a C-clamp raise the spring to relieve load and disconnect the spring.
8. Remove the transverse spring at the cover plate.
9. Mark the cam bolt and remove from the bracket.
10. Remove the two bolts attaching the strut bracket to the carrier and lower the strut rods by pushing away on the tire and wheel assembly.
11. Mark the driveshaft and disconnect it at the companion flange in order to gain access to the insulator attaching bolt.

NOTE: It may be necessary to support the driveshaft to gain access to the insulator attaching bolt.

12. Place a jackstand under the carrier, and remove the carrier-to-body attaching bolts.

13. Lower the differential in order to gain access to all cover bolts.

14. Drain the differential and remove the cover.

15. To remove the carrier disconnect the driveshaft at the spindle companion flange.

16. Lower and remove the differential assembly and remove the driveshafts from the side yokes.

17. Installation is the reverse of removal.

1984 and Later

1. Remove air cleaner.

2. Disconnect distributor cap from distributor.

3. Raise vehicle and support it with jackstands.

4. Remove spare tire.

5. Remove spare tire cover by removing support hooks.

6. Removes the complete exhaust system as an assembly by:

   a. Disconnect A.I.R. pipe at the converter.

   b. Disconnect A.I.R. pipe clamps at exhaust manifold.

   c. Disconnect oxygen sensor electrical lead.

   d. Remove the bolts attaching the mufflers to the hangers.

   e. Remove hanger bracket at converter.

   f. Disconnect the exhaust from the exhaust manifold and remove the exhaust system.

7. Disconnect leaf spring at the knuckles and remove attaching bolts

at the cover. Remove leaf spring from vehicle.

8. Scribe mark on cam bolts and mounting bracket so they can be realigned in the same position. Remove cam bolts and then mounting bracket from carrier.

9. Disconnect both tie rod ends from the knuckles.

10. Remove the axle shaft union straps from the side gear yokes. Push wheel and tire assemblies outboard to disengage the trunnions from the side gear yokes.

11. Remove prop shaft trunnion straps at pinion flange. Push prop shaft forward into transmission and tie shaft to the support beam.

12. Support transmission.

13. Remove differential cover/beam attaching bolts at frame brackets.

14. Remove support beam attaching bolts at the front of the differential carrier. Remove differential carrier assembly from the vehicle.

15. Installation is the reverse of removal.

Rear Axle Shaft

NOTE: For the rear axle shaft removal and installation procedure, please refer to the Rear Hub And Bearing procedure in the Rear Suspension section.

FRONT SUSPENSION

Corvette utilizes conventional short-long arm suspension, with coil springs and tube shocks. A stabilizer bar is used between the lower arms to reduce body lean during cornering, thereby keeping more tire surface on the ground. 1984 and later models use a transverse leaf spring on the front suspension, along with forged aluminum components.

Shock Absorber

REMOVAL & INSTALLATION

1. Remove the upper stem nut while holding the stem to keep it from turning.

2. Remove the two bolts holding the shock absorber to the lower control arm and pull the shock through the arm.

NOTE: Pulling the shock through the lower control arm applies only to 1980–82 models.

3. Purge the new shock of air by repeatedly extending it in its normal position and compressing it while inverted. Extend the shock absorber and insert it up through the lower control arm. Make sure that the upper stem goes through the hole in the upper control arm frame bracket.

4. Install the grommet, retainer cup, and nut to the shock absorber upper stem.

5. Hold the shock absorber stem and tighten the upper nut to 8 ft. lbs. 1980–82, 18 ft. lb. 1984 and later models.

6. Install the lower control arm retaining bolts and tighten to 13 ft. lbs. 1980–82, 22 ft. lbs. 1984 and later models.

DISPOSAL OF PRESSURIZED SHOCK ABSORBERS

1984 and Later Models

Due to the high pressure of gas it is advised that, upon scrapping or disposal of these shock absorbers, the pressure be released. This is carried out as follows:

1. Clamp shock in vise with piston rod pointing down.

2. Measure approx. 0.5 in. (10–15mm) from bottom of shock and drill an approx. 5mm hole so the gas can escape.

3. Measure approx. 5.5–6.0 in. (140–150mm) from first hole and drill an approx. 5mm hole to facilitate drainage of oil. Drain all oil from shock absorber.

NOTE Hold stud at this point to obtain torque.

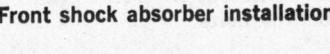

Front shock absorber installation

NOTE Spring to be installed with tape at lowest position. Bottom of spring is coiled helical, and the top is coiled flat with a gripper notch near end of wire.

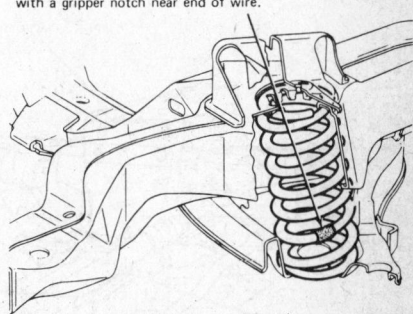

NOTE After assembly, end of spring coil must cover all or part of one inspection drain hole. The other hole must be partly exposed or completely uncovered.

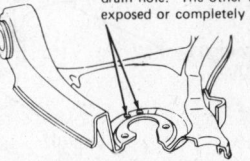

Front spring positioning

Spring

REMOVAL & INSTALLATION

1980–82

1. Raise the car on hoist and remove nut, retainer and grommet from the top of the shock absorber. Support car so that the control arms swing free.

2. Disconnect stabilizer bar from lower control arm and remove shock absorber.

3. Bolt a spring remover tool to a suitable jack and place it under the lower control arm bushings so that the bushings seat in the grooves of the tool.

NOTE: This tool is a cradle which, when fastened to a hydraulic jack, allows the lowering of the control arm and slow decompression of the spring. A similar tool can be fabricated in the shop. Always safety chain the spring and control arm when using this method.

4. Remove the cross shaft rear retaining nut and the two front retaining bolts.

5. Slowly release jack, swing control arm forward, then remove spring.

6. Install by reversing procedure above. Torque the retaining nut to 92 ft. lb. Torque the retaining bolts 59–75 ft. lbs.

NOTE: Chevrolet recommends this cradle spring removal tool for all models. Other methods may be used, depending on the availability of tools.

1984 and Later

1. Raise the vehicle on a hoist.
2. Remove the wheels and tires, left side.

3. Remove the left front caliper bracket and rotor.

4. Remove spring protectors both sides.

5. Install spring compressor J-33432-4 or its equivalent.

6. Disconnect outer tie rod left side.

7. Remove the stabilizer link, left side.

8. Remove the lower shock mount, left side.

9. Compress special tool J-33432-4.

10. Remove the lower ball joint, left side.

11. Remove the spring hold down brackets.

12. Remove the transverse spring from the left side.

13. Installation is the reverse of removal. The following torques are needed during reinstallation: Spring protector to crossmember 18 ft. lbs. Stabilizer link 35 ft. lbs.

BALL JOINT INSPECTION

NOTE: Before performing this inspection, make sure the wheel bearings are adjusted correctly and that the control arm bushings are in good condition.

1. Jack the car up under the front lower control arm at the spring seat.

2. Raise the car until there is 1–2 in. of clearance under the wheel.

3. Insert a bar under the wheel and pry upward. If the wheel raises more then $\frac{1}{8}$ in. the ball joints are worn. Determine if the upper or lower ball joint is worn by visual inspection while prying on the wheel.

NOTE: Due to the distribution of forces in the suspension, the lower ball joint is usually the defective joint.

Upper & Lower Ball Joint

REMOVAL & INSTALLATION

1. Raise the car on a hoist.
2. Remove the tire and wheel assembly.
3. Support the lower control arm with a jack.
4. Loosen the upper ball stud nut.
5. Install a ball joint remover tool and unseat the upper joint from the steering knuckle. Remove the upper stud nut and install a block of wood under the upper control arm.
6. Chisel or grind off the ball joint mounting rivets.
7. Drill out the ball stud attaching holes to accept the service ball joint attaching bolts.

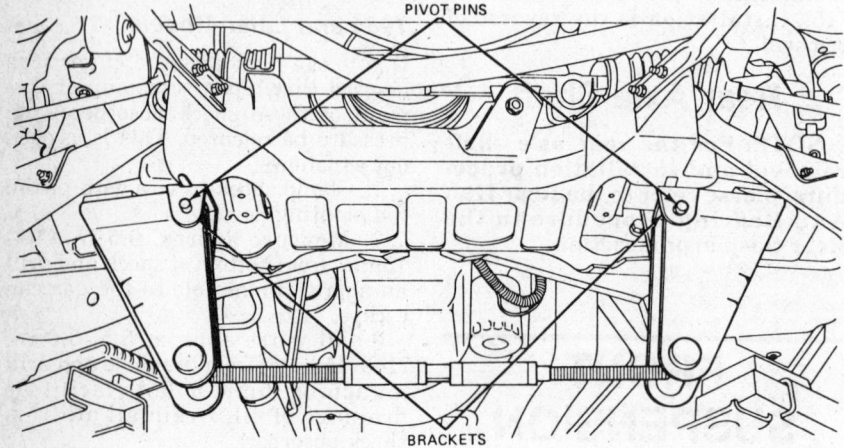

NOTE: PIVOT PINS ARE REMOVED SO THAT THE BRACKET MAY BE PLACED OVER THE TOP OF THE SPRING.

Front spring removal/installation

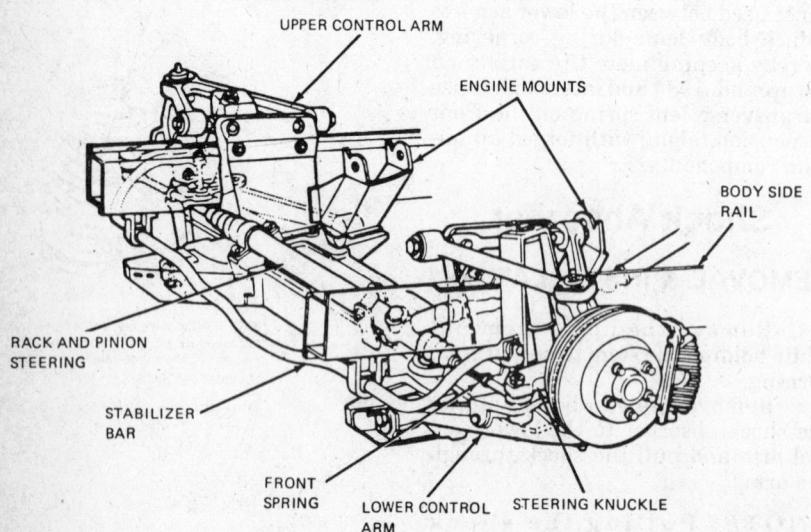

1984 front suspension components

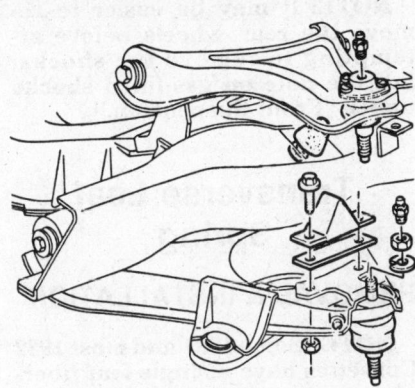

Upper and lower ball joints

3. Remove the wheel.

4. Separate the upper ball joint from the steering knuckle as described above under "Upper Ball Joint Removal & Installation".

5. Remove the control arm shaft to frame nuts.

NOTE: Tape the shims together and identify them so that they can be installed in the positions from which they were removed.

6. Remove the bolts which attach the control arm shaft to the frame and remove the control arm. Note the positions of the bolts.

7. Install in the reverse order of removal. Make sure the shaft to frame bolts are installed in the same position they were in before removal and that the shims are in their original positions. Tighten the shaft to frame bolts to 55 ft. lbs. The control arm shaft nuts are torqued to 60 ft. lbs.

8. Install the ball joint with the nuts and bolts supplied with the new joint.

9. Install the lube fitting in the new joint.

10. Mate the upper control arm to the steering knuckle and install the ball stud through the knuckle boss.

11. Tighten the ball stud nut to 50 ft. lbs. plus whatever is necessary to align the cotter pin holes. Install the cotter pin. Never loosen the nut to align the cotter pin holes.

12. Install the wheel. Lower the car.

Lower Control Arm

REMOVAL & INSTALLATION

1980–82

1. Remove the spring.

2. Remove the ball stud from the steering knuckle as described above.

3. Remove the control arm pivot bolts and remove the control arm. On some Corvettes, the pivot bolt is secured to the frame with two bolts.

4. Installation is the reverse.

1984 and later

1. Raise vehicle and remove wheel and tire.

2. Remove spring protector.

3. Using tool J-33432 or its equal compress spring.

4. Remove lower shock bracket.

5. Using tool J-33436 or its equal disconnect lower ball joint.

6. Remove lower control arm.

7. Installation is the reverse of removal. Torque the ball joint nut to 48 ft. lbs.

Upper Control Arm

REMOVAL & INSTALLATION

1980–82

1. Raise the vehicle on a hoist.

2. Support the outer end of the lower control arm, with a jack.

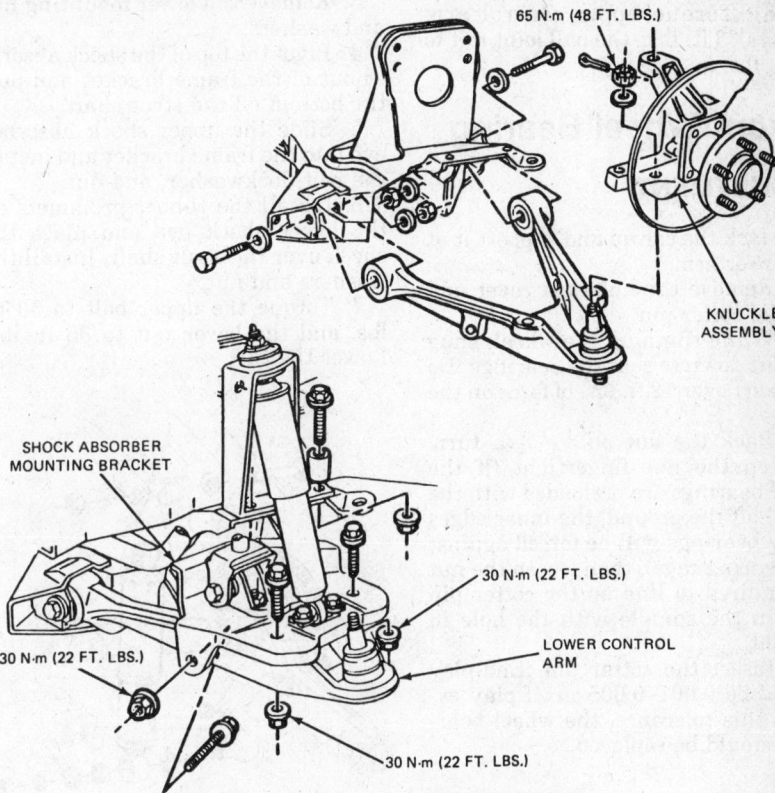

Lower control arm removal/installation

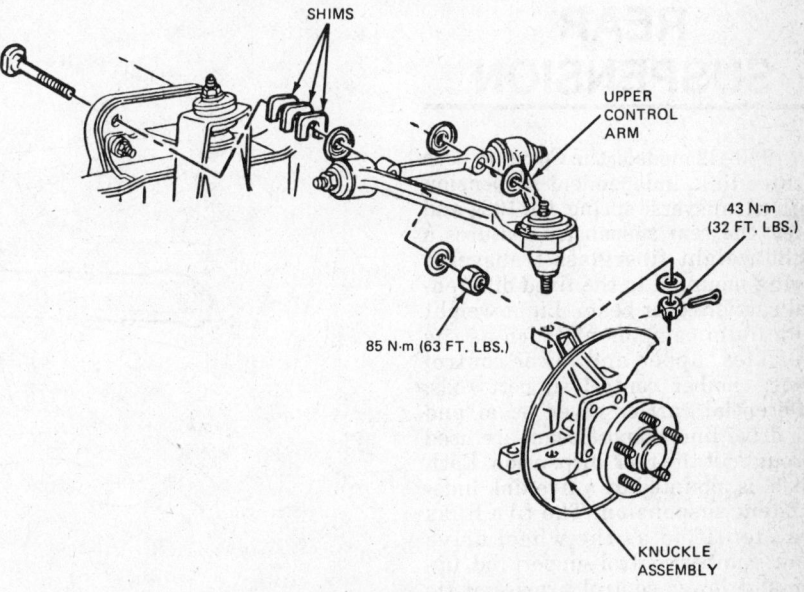

Upper control arm removal/installation

1984 and Later

1. Raise vehicle and remove wheel and tire.
2. Remove spring protector.
3. Using spring compressor J-33432 or its equivalent, compress and loosen the spring.
4. Use tool J-33436 or its equivalent to disconnect the upper ball joint from the knuckle.
5. Remove the upper control arm.
6. Installation is the reverse of removal. Torque upper control arm bolts to 63 ft. lbs., the ball joint nut to 32 ft. lbs.

Front Wheel Bearing

ADJUSTMENT

1. Jack the car up and support it at the lower arm.
2. Remove the hub dust cover and spindle cotter pin.
3. While spinning the wheel, snug the nut down to seat the bearings. Do not exert over 12 ft. lbs. of force on the nut.
4. Back the nut off $\frac{1}{4}$–$\frac{1}{2}$ a turn. Tighten the nut fingertight (if the roller bearings are preloaded with the wheel off the ground, the inner edges of the bearings will be forced against the bearing cage), then loosen the nut as required to line up the cotter pin hole in the spindle with the hole in the nut.
5. Insert the cotter pin. End-play should be 0.001–0.005 in. If play exceeds this tolerance, the wheel bearings should be replaced.

REAR SUSPENSION

On 1980–82 models, the Corvette uses a three-link, independent suspension with a transverse spring. On 1984 and later, the rear suspension features a light weight fiberglass transverse spring mounted to the fixed differential carrier cover beam. Light weight aluminum components such as the knuckles, upper and lower control arms, camber control support rods, differential carrier cover beam and the drive line support beam are used throughout the rear suspension. Each wheel is mounted by a five link independent suspension. The five links are identified as the wheel drive shaft, camber control support rod, upper and lower control arms and tie rod.

Shock Absorber

REMOVAL & INSTALLATION

NOTE: Purge new shocks of air by repeatedly extending them in their normal position and compressing them while inverted.

1. Jack the car to a convenient working height.
2. Remove the upper bolt and nut.
3. Remove the lower mounting nut and washers.
4. Pivot the top of the shock absorber out of the frame bracket and pull the bottom off the strut shaft.
5. Slide the upper shock absorber eye into the frame bracket and install the bolt, lockwasher, and nut.
6. Install the rubber grommets on the lower shock eye and place the shock over the strut shaft. Install the washers and nut.
7. Torque the upper bolt to 50 ft. lbs. and the lower nut to 35 ft. lbs. Lower the car.

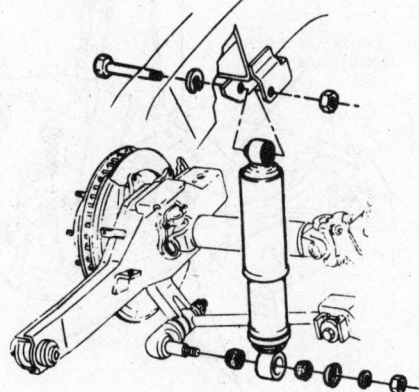

Rear shock absorber mounting

NOTE

NOTE: It may be easier to remove the rear wheels before attempting to remove the shocks. See the note on gas-filled shocks under "Front Suspension."

Transverse Leaf Spring

REMOVAL & INSTALLATION

NOTE: Some 1981 and most 1982 Corvettes have a single leaf fiberglass rear spring. All 1984 and later models use a single leaf rear spring.

1. Raise car and support it by the frame, slightly forward of torque control pivot points. Remove wheel assemblies.
2. Place a floor jack under the spring near the link bolt, and raise the spring until it is nearly flat.
3. Tie the end of the spring to the suspension crossmember to hold this flat attitude, with a $\frac{1}{4}$ in. or $\frac{5}{16}$ in. chain and grab hook wrapped around the spring and crossmember. To prevent chain slipping, use a C-clamp on the spring adjacent to the chain.
4. Remove link bolt and rubber bushings.
5. Support and raise the spring end, as before, and remove chain.
6. Carefully lower jack to completely relax spring.
7. Repeat the procedure on the other side of the car.
8. Remove bolts and washers attaching the springs at the center.
9. Remove the spring by sliding it over the exhaust pipes and out one side of the car.

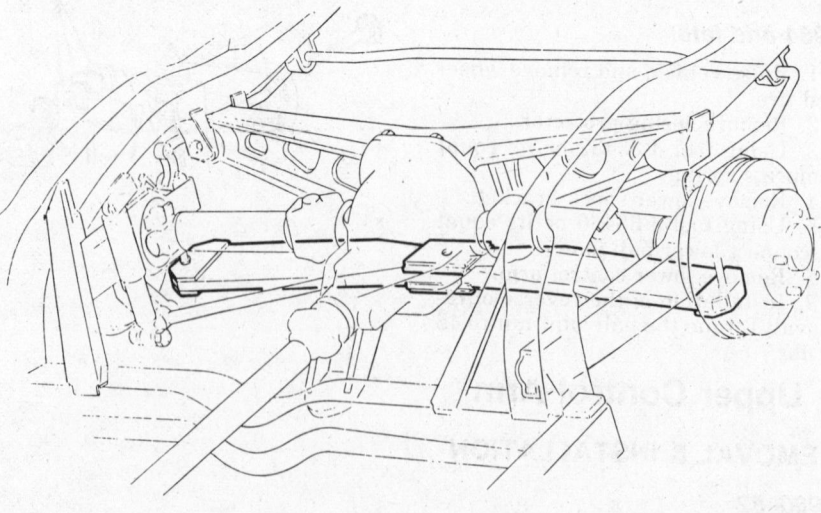

Fiberglass-reinforced plastic (F.R.P.) single leaf rear spring—most 1981 and later models

10. Install by reversing removal procedure. Always use new link bolts and cushions. Torque the rear spring to carrier bolts to 33 ft. lbs. through 1980 and 50 ft. lbs. 1981 and later. Install the nut on the link bolt just far enough to expose the cotter pin hole, then insert the pin.

Strut Rod & Bracket

REMOVAL & INSTALLATION

—— CAUTION ——

The strut rod shaft is often very hard to remove; take care not to distort either the shaft or the spindle support in the removal process.

1. Raise car on a hoist.
2. Disconnect shock absorber lower eye from strut rod shaft.
3. Remove strut rod shaft cotter pin and nut. Withdraw shaft by pulling toward the front of the car.
4. Mark related position of camber adjustment, so that adjustment is maintained upon reassembly.
5. Loosen camber bolt and nut. Remove four bolts holding strut rod bracket to carrier and lower the bracket.
6. Remove cam bolt and cam bolt assembly. Pull strut down out of bracket and remove bushing caps.
7. Inspect strut rod bushings for wear and replace where necessary. Replace strut rod if it is bent or damaged in any way.

NOTE: The strut rod shaft has a flat side which should line up with the matching flat in the spindle support.

8. Install by reversing removal procedure. Torque the strut rod-to-spindle support to 75 ft. lbs. plus as needed to align cotter pin hole. Torque the bracket-to-carrier to 35 ft. lbs., 20 ft. lbs. on 1981 and later.
9. Check rear wheel camber and adjust to specifications.

Spindle Support Rod

REMOVAL & INSTALLATION

1984 and Later

1. Raise and support vehicle.
2. Scribe mark on cam bolt and mounting bracket so they can be realigned in the same position.
3. Remove cam bolt and separate spindle support rod from the mounting bracket.
4. Remove the spindle support rod bolt at the knuckle and remove rod.

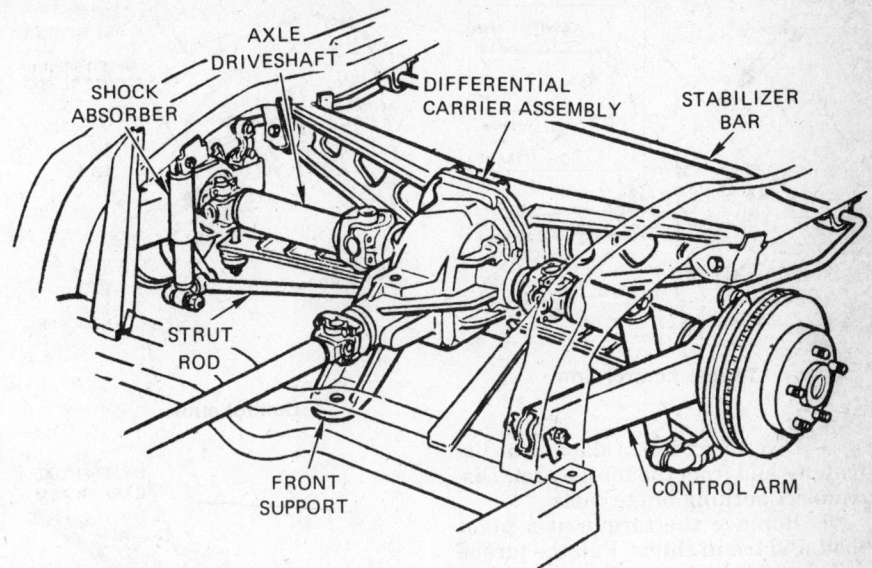

Independent rear suspension—1980 and later

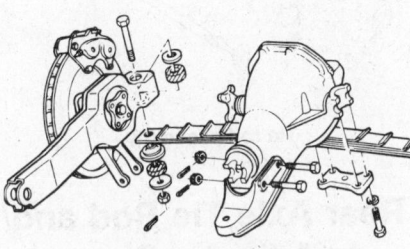

Transverse spring mounting

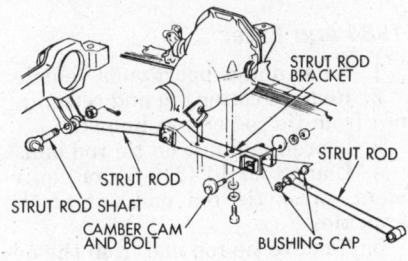

Strut rod mounting

5. Installation is the reverse of removal. Torque the spindle end of support rod 95–188 ft. lbs; differential end 47–62 ft. lbs.

Torque Control Arm

REMOVAL & INSTALLATION

1. Disconnect spring on the side from which the torque arm is to be removed. Follow procedure for "Spring Removal & Installation".

NOTE: If so equipped, disconnect stabilizer rod from torque arm.

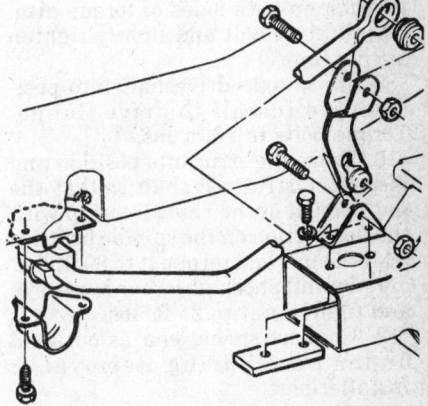

Stabilizer shaft installation, Corvette

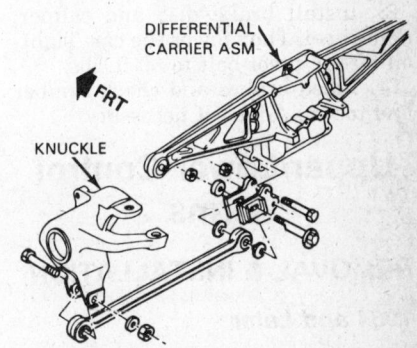

Spindle support rod

2. Remove shock absorber lower eye from strut rod shaft.
3. Disconnect and remove strut rod shaft and swing strut rod down.
4. Remove four bolts holding the axle driveshaft to spindle flange and disconnect drive shaft.

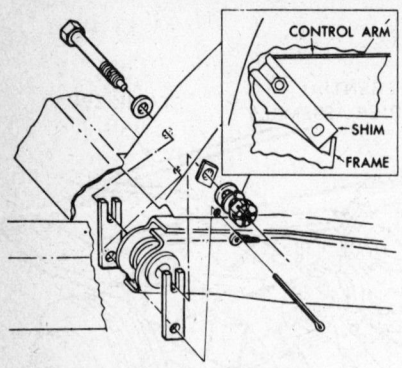

Torque control arm

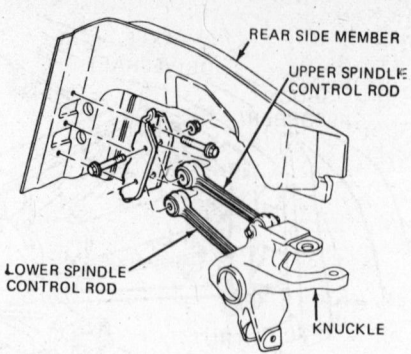

Control arms.

5. Disconnect the brake line at the caliper and from the torque arm. Disconnect parking brake cable.

6. Remove the torque arm pivot bolt and toe-in shims. Pull the torque arm out of the frame. Tape the shims together to assure proper reassembly.

7. To install, place torque arm in frame opening.

8. Position toe-in shims in original location on both sides of torque arm. Install pivot bolt and lightly tighten at this time.

9. Raise axle driveshaft into position and install to drive flange. Torque bolts to 75 ft. lbs.

10. Raise the strut into position and insert the strut rod shaft so that the flat portion of the shaft lines up with the flat portion on the spindle fork. Install the nut and torque it to 80 ft. lbs.

11. Install shock absorber lower eye and tighten nut to 35 ft. lbs.

12. Connect spring end as outlined under Leaf Spring Removal & Installation.

NOTE: If car is so equipped, connect stabilizer shaft.

13. Install brake disc and caliper, and wheel. Then lower the car. Tighten torque pivot bolt to 50 ft. lbs.

14. Bleed brakes and check camber and toe-in. Adjust if necessary.

Upper/Lower Control Arms

REMOVAL & INSTALLATION

1984 and Later

1. Raise and support the vehicle.
2. Remove shock absorber.
3. Remove control arm bolt at the knuckle.
4. Remove control arm bolt at the body bracket and remove the arm.
5. Installation is the reverse of removal. Torque the control arm-to-body bolts 55–70 ft. lbs; control arm-to-knuckle 125–154 ft. lbs.

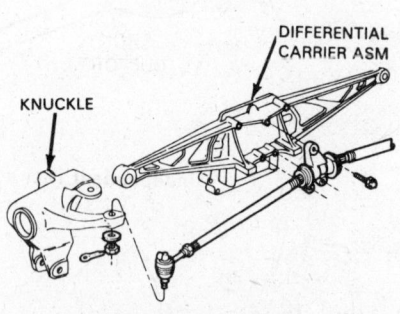

Tie rod assembly

Rear Axle Tie Rod and/ or Adjuster Sleeve

REMOVAL & INSTALLATION

1984 and Later

1. Raise and support your vehicle.
2. Remove cotter pin and retaining nut from tie rod end at knuckle.
3. Loosen jam nut on tie rod end.
4. Using tool J-24319-01, or equivalent, press tie rod end out of the knuckle.
5. Remove tie rod end from the adjusting sleeve. Count the number of turns.
6. Remove adjuster sleeve.
7. Installation is the reverse of removal. Install the same number of turns as removal. Torque tie rod end-to-knuckle: 29–36 ft. lbs; tie rod housing to cover beam: 35–44 ft. lbs. Realign suspension if necessary.

Support Beam

REMOVAL & INSTALLATION

1984 and Later

1. Raise vehicle and support with jackstands. Allow exhaust system to cool.
2. Remove the complete exhaust system as an assembly by:

a. Disconnect A.I.R. pipe at the converter.
b. Disconnect A.I.R. pipe clamps at exhaust pipe.
c. Disconnect oxygen sensor electrical lead.
d. Remove the bolts attaching the mufflers to the hangers.
e. Remove hanger bracket at the converter.
f. Disconnect the exhaust from the exhaust manifold and remove the exhaust system.

3. Support transmission, using a suitable jack.

4. Remove support beam attaching bolts at the differential carrier and transmission extension housing.

5. Remove the propeller shaft.

6. Remove the support beam by prying transmission to the driver side of vehicle. Remove support beam from the vehicle.

7. Installation is the reverse of removal. Torque the rear support beam bolts: 51–66 ft. lbs; front beam bolts: 47–55 ft. lbs. Apply sealant as illustrated during installation.

Stabilizer Bar

REMOVAL & INSTALLATION

1984 and Later

1. Raise and support the vehicle on jackstands.
2. Remove spare tire and tire carrier.
3. Disconnect stabilizer bar from knuckles.
4. Remove stabilizer bar bushing retainers, bushings and bar from the vehicle.
5. Installation is the reverse of removal. Torque the stabilizer bar to body bolts and link bracket bolts 14–22 ft. lbs.; Stabilizer link to bracket and bar bolts 25–35 ft. lbs.

Rear Wheel Bearing

REMOVAL & INSTALLATION

1980–82

The Corvette rear wheel spindle is mounted on two tapered roller bearings contained in the spindle support arm, which is bolted to the torque control arm. The flanged end of the spindle is riveted to the brake disc assembly. These rivets are not to be removed for the following service procedures. Bearing end-play is controlled by a solid tubular spacer and a shim.

1. Jack up the vehicle and support it with jackstands.
2. Remove the wheel and tire assembly.

3. Remove the axle drive shaft.

4. Apply the parking brake to prevent the rotors from turning.

5. Remove the cotter pin, nut and flange.

NOTE: It may be necessary to use special tool J08614-01, or its equivalent, to remove the flange.

6. Install tool J-21859-1, or equivalent, over the spindle threads, then remove the drive spindle from its support using tool J-22602 or equivalent. When using this tool make sure the puller plate is positioned vertically in the torque control arm before applying pressure to the puller screw.

7. When the spindle is removed, the outer bearing will remain on the spindle. The inner bearing, tubular spacer, end-play adjustment shim and both outer races will remain in the spindle support.

8. Remove the bearing, spacer and shim. Record the shim thickness for later use.

9. With the spindle assembly on the bench, position tool J24489-1, or equivalent, between the outer bearing and the seal.

10. Using puller J-8433-1, or equivalent, draw the bearing off the spindle.

11. Remove the outer seal from the spindle shaft and inspect it for damage. Replace if necessary.

12. Remove the outer races from the spindle shaft and install new ones, using tool J-7817, or equivalent.

13. Pack the new wheel bearing with grease.

14. Installation is the reverse of removal.

Rear Hub & Bearing

REMOVAL & INSTALLATION

1984 and Later

1. Remove center cap from wheel.

2. Remove cotter pin, spindle nut and washer.

3. Raise vehicle and support with jack stands.

4. Remove wheel and tire.

5. Remove brake caliper and support.

6. Remove brake rotor.

7. Disconnect tie rod end from the knuckle.

8. Disconnect transverse spring from the knuckle.

9. Scribe mark on cam bolt and mounting bracket so they can be re-aligned in the same position.

10. Remove cam bolt and separate spindle support rod from the mounting bracket.

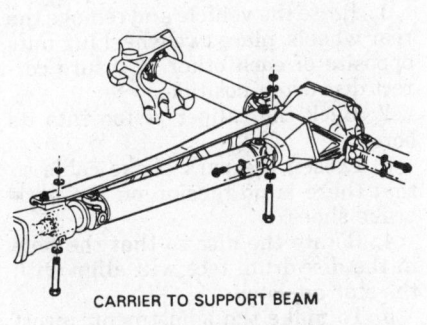

CARRIER TO SUPPORT BEAM

Support beam attachments

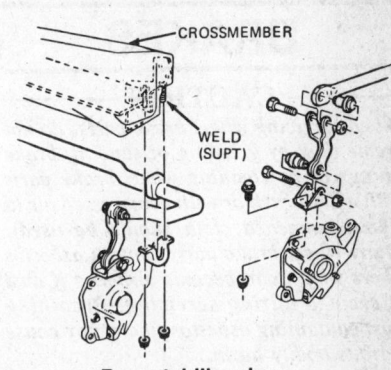

CROSSMEMBER

WELD (SUPT)

Rear stabilizer bar

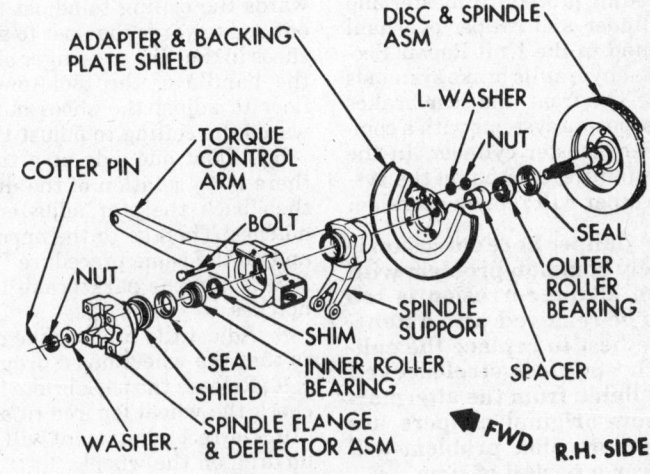

ADAPTER & BACKING PLATE SHIELD

DISC & SPINDLE ASM

L. WASHER

NUT

TORQUE CONTROL ARM

COTTER PIN

BOLT

SEAL

OUTER ROLLER BEARING

NUT

SPINDLE SUPPORT

SHIM

SPACER

SEAL

SHIELD

INNER ROLLER BEARING

SPINDLE FLANGE & DEFLECTOR ASM

WASHER

FWD   R.H. SIDE

Exploded view of spindle

11. Remove the trunnion straps at the side yoke shaft. Push out on the knuckle and separate axle shaft from the side gear yoke shaft. Remove the axle shaft from the vehicle.

12. Using J-34161 (Torx® No. 45) or its equivalent, remove the hub and bearing mounting bolts.

13. Remove hub and bearing from the vehicle and support the parking brake backing plate.

14. Installation is the reverse of removal. Check suspension alignment and adjust as needed.

WHEEL BEARING END PLAY CHECK

The rear wheel bearings should have end play of 0.001–0.008 inches. When necessary, adjust them using the following procedure.

1. Jack up your vehicle and support it with jackstands.

2. Remove the tire and wheel assembly.

3. Remove the axle driveshaft.

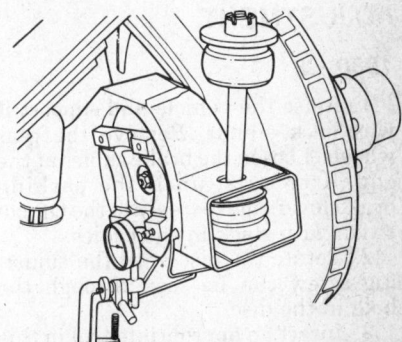

Checking spindle bearing end play

4. Mark the camber cam in relation to the bracket. Loosen and turn the camber bolt until the strut rod forces the torque control arm outward.

5. Mount a dial indicator on the torque control surface and rest the pointer on the flange end.

6. Grasp the rotor and move it in and out. If the bearing movement is with specifications no adjustment is necessary. If the adjustment is not within these limits you must add or subtract shims accordingly.

BRAKES

CAUTION

When servicing wheel brake parts, do not create dust by grinding or sanding brake linings or by cleaning wheel brake parts with a dry brush or with compressed air (a water dampened cloth should be used). Many wheel brake parts contain asbestos fibers which can become airborne if dust is created during servicing. Breathing dust containing asbestos fibers may cause serious bodily harm.

Brake adjustments, lining replacement, bleeding procedure, master and wheel cylinder and caliper overhaul can be found in the Unit Repair Section. A dual hydraulic brake system is employed. The front and rear brakes are each separate systems with a common tandem master cylinder. In the event of a failure in either of the systems, the other will remain operable.

NOTE: Caliper bore corrosion is a relatively common problem with Corvettes. If the corrosion is too severe to be removed with a hone, it may be best to replace the calipers with stainless steel-sleeved units available from the aftermarket. If new original calipers are used, the corrosion problem will return over a period of time.

Parking Brake

ADJUSTMENT

1980–82

1. Raise the vehicle and support it with jack stands. Remove the rear wheels. Loosen the brake cables at the equalizer nuts, until the parking brake levers move freely to the Off position with slack in the cables.
2. Rotate the disc until the adjusting screw can be seen through the hole in the disc.
3. Insert an appropriate tool in this hole and adjust with an up-and-down motion.
4. Tighten the adjuster until the disc cannot move, then back off 6–8 notches.
5. Install the rear wheels.
6. Apply the parking brake to the 13th notch.
7. Tighten the check nuts until an 80 lb. pull is obtained while pulling into the 14th notch.
8. Torque the check nuts to 70 inch lbs.
9. Release the parking brake and check for a no drag condition.

1984 and Later

1. Raise the vehicle and remove the rear wheels, place two wheel lug nuts opposite of each other to insure correct disc/drum position.
2. Back the caliper piston into its bore.
3. Loosen the park brake cable so that there is no tension on the park brake shoes.
4. Rotate the disc so that the hole in the disc/drum face will align with the star adjuster.
5. To make the adjustment, insert a brake adjusting spoon through the hole in the disc face. For the driver's side, move the handle of the tool towards the ceiling to adjust the shoes out and towards the floor to adjust the shoes in. For the passenger side, move the handle of the tool towards the floor to adjust the shoes out and towards the ceiling to adjust them in.
6. Adjust one side at a time until there is no rotation of the disc/drum, then back the star adjuster off 5–7 notches. Then go to the opposite side and do the same procedure.
7. Apply the park brake lever two notches.
8. Adjust the cable at the equalizer so that the wheel has a drag.
9. Release the park brake lever and check the wheel for free rotation.
10. Correct adjustment will result in no drag on the wheel.

Power Brake Unit

REMOVAL & INSTALLATION

1. Remove the vacuum hose from the brake booster.
2. Disconnect the hydraulic brake lines from the master cylinder.

NOTE: Do not spill brake fluid on painted surfaces.

3. Remove the master cylinder from the brake booster.
4. Disconnect the push rod at the brake pedal.
5. Remove the nuts and lockwashers that secure the unit to the firewall.

6. Installation is the reverse of removal. Torque the mounting bolts to 24 ft. lbs. Remember to bleed the brake system.

Master Cylinder

REMOVAL & INSTALLATION

1. Disconnect the brake lines at the master cylinder.
2. Remove the two mounting nuts and lift off the cylinder.
3. Installation is the reverse of removal. Torque the nuts to 24 ft. lbs. and bleed the system.

CAUTION

The use of silicone brake fluid such as Delco Supreme No. 24 can damage the seals and rubber components in the brake system. Use brake fluid that meets or exceeds DOT 3 specifications, such as Delco Supreme No. 11 or equivalent.

Combination Valve

NOTE: The combination valve is not repairable and must be replaced if found to be defective. 1984 and later models use a proportioning valve which is an integral part of the master cylinder.

REMOVAL & INSTALLATION

1980–82

1. Disconnect and plug the hydraulic lines at the combination valve.
2. Disconnect the warning switch wiring harness from the valve switch terminal.
3. Remove the combination valve.
4. Installation is the reverse of removal. Make sure to bleed the entire brake system and that a firm brake pedal is obtained before moving the vehicle.

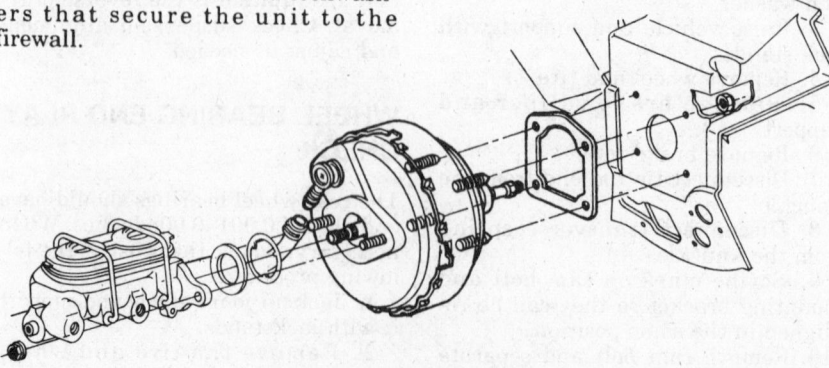

Power brake booster installation

STEERING

The manual steering gear on the Corvette is of the recirculating ball type. Relay-type steering linkage is used on all models, with a pitman arm connected to one end of a relay rod and a frame-mounted idler arm at the other end. Two tie rod assemblies connect the relay rod to the steering arms. The tie rod ends are threaded into sleeves to provide adjustment.

The Corvette uses a linkage assist power steering system. A valve attached to the linkage modulates pressure according to power requirements. A power cylinder supplies the actual assist. 1984 and later models use a power rack and pinion system. This system has a rotary control valve which directs hydraulic fluid, coming from the hydraulic pump, to either side of the rack piston. The integral rack piston is attached to the rack and converts hydraulic pressure to a linear force which moves the rack left or right. The force is then transmitted through the inner and outer tie rods to the steering knuckles which turn the wheels.

NOTE: Procedures for removal and installation of the ignition switch, key warning buzzer switch, and ignition lock cylinder can be found in the "Chevrolet" car section.

Tie Rod

REMOVAL & INSTALLATION

NOTE: Before attempting this procedure mark the tie rod threads with paint or chalk for easy reinstallation.

1. Remove the cotter pins and nuts from the tie rod end studs.
2. Tap on the steering arm near the tie rod end (use another hammer as backing) and pull down on the tie rod, if necessary, to free it.
3. Remove the inner stud in the same manner as the outer.
4. Loosen the clamp bolts and unscrew the ends if they are being replaced.
5. Lubricate the tie rod end threads with chassis grease if they were removed. Install each end assembly an equal distance from the sleeve.
6. Insure that the tie rod end stud threads and nut are clean. Install new seals and install the studs into the steering arms and relay rod.
7. Install the stud nuts. Tighten the nuts to 35 ft. lbs. plus as needed to align the cotter pin hole.

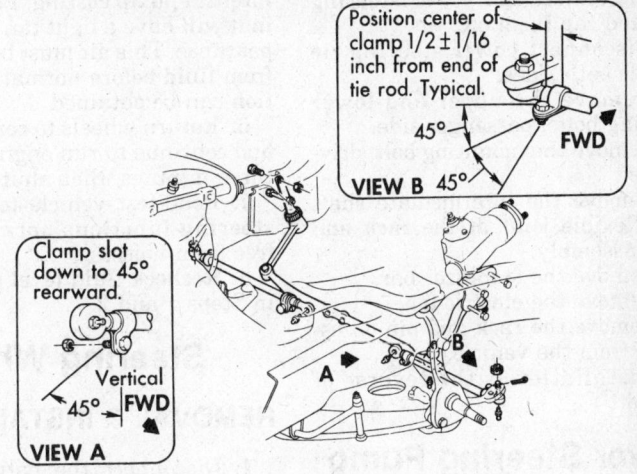

Steering linkage

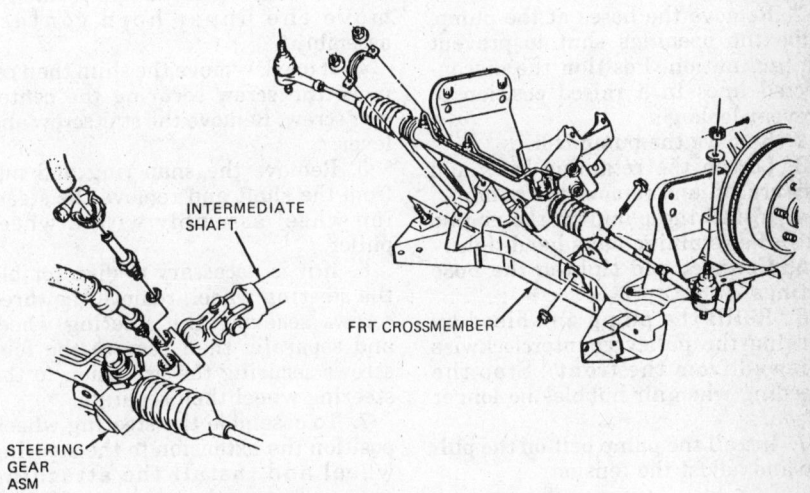

INTERMEDIATE SHAFT

FRT CROSSMEMBER

STEERING GEAR ASM

Power rack and pinion assembly

8. Adjust the toe-in.

NOTE: Before tightening the sleeve clamps, ensure that the clamps are positioned so that the adjusting sleeve slot is covered by the clamp.

Steering Gear

REMOVAL & INSTALLATION

1980–82

1. Disconnect battery ground cable.
2. Remove retaining nuts, lockwashers, and bolts at steering coupling to steering shaft flange.
3. Remove pitman arm nut and washer from pitman shaft and mark relation of arm position to shaft.
4. Remove pitman arm with Tool J-6632, or its equivalent.
5. Remove screws securing steering gear to frame and remove gear from vehicle.

Steering gear attachment

6. Installation is the reverse of removal. Torque the gear-to-frame nuts to 30 ft. lbs.; flange-to-flex coupling bolt to 30 ft. lbs.

1984 and Later

1. Raise the vehicle on a hoist.
2. Remove the drivers side wheel and tire.

3. Disconnect the power steering hoses and cap them off.

4. Disconnect power steering tie rod ends both sides.

5. Remove the upper and lower mounting bolts, passenger side.

6. Remove the mounting bolt, drivers side.

7. Remove the intermediate shaft lower flexible joint at the rack and pinion assembly.

8. Remove the stabilizer bar.

9. Remove the electric fan.

10. Remove the rack and pinion assembly from the vehicle.

11. Installation is the reverse of removal.

Power Steering Pump

REMOVAL & INSTALLATION

1. Remove the hoses at the pump. Tape the openings shut to prevent contamination. Position the disconnected lines in a raised position to prevent leakage.

2. Remove the pump belt.

3. Loosen the retaining bolts and any braces, and remove the pump.

4. Install the pump on the engine with the retaining bolts hand-tight.

5. Connect and tighten the hose fittings.

6. Refill the pump and bleed by turning the pulley counterclockwise (viewed from the front). Stop the bleeding when air bubbles no longer appear.

7. Install the pump belt on the pulley and adjust the tension.

BLEEDING POWER STEERING SYSTEM

1. Run engine until power steering fluid reaches normal operating temperature, approximately 170°F (80°C), then shut engine off. Remove reservoir filler cap and check oil level.

2. If oil level is low, add power steering fluid to proper level and replace filler cap. When adding or making a complete fluid change, always use GM No. 1050017, or equivalent, power steering fluid. Do not use transmission fluid.

3. When checking fluid level after the steering system has been serviced, air must be bled from the system. With wheels turned all the way to the left, add power steering fluid to level indicated on reservoir.

4. Start engine, and running at fast idle, recheck fluid level. Add fluid if necessary.

5. Bleed system by turning wheels from side to side without hitting stops. Maintain fluid level just above

internal pump casting. Fluid with air in it will have a light tan or milky appearance. This air must be eliminated from fluid before normal steering action can be obtained.

6. Return wheels to center position and continue to run engine for two or three minutes, then shut engine off.

7. Road-test vehicle to make sure steering functions normally and is free from noise.

8. Recheck fluid level as described in Steps 1 and 2.

Steering Wheel

REMOVAL & INSTALLATION

1. Disconnect the battery ground cable.

2. Pry off the horn button cap.

3. Remove the three screws and remove the upper horn contact assembly.

4. If used, remove the shim then remove the screw securing the center star screw. Remove the star screw and lever.

5. Remove the snap ring and nut from the shaft and remove the steering wheel assembly with a wheel puller.

6. If it is necessary to disassemble the steering wheel, remove the three screws securing the steering wheel and separate; then remove the four screws securing the extension to the steering wheel then separate.

7. To assemble the steering wheel, position the extension to the steering wheel and install the attaching screws. Torque the screws to 20 inch lbs.

8. Position the spring, eyelet and insulator to the lower contact assembly. Position the assembly to the steering wheel and install the three screws.

9. Position the steering wheel assembly to the steering column and torque the nut to 30 ft. lbs.

10. Install the snap ring.

11. Position the lever to the steering column and install the star screw. Install the screws. Remove the switch by pulling it.

12. Position the upper contact assembly and shim if used and install the three retaining screws.

13. Install the retainer and horn button cap.

14. Connect the battery ground cable.

Turn Signal Switch

REMOVAL & INSTALLATION

1. Remove the steering wheel as

previously outlined and press of the hub with a puller.

2. Remove the steering column/dash trim cover.

3. Remove the C-ring plastic retainer, if so equipped.

4. Install the special lockplate compressing tool (J-23653 1980–81 and 84; J-23063 1982) over the steering shaft. Position a $^5/_{16}$ in. nut under each tool leg and reinstall the star screw to prevent the shaft from moving.

5. Compress the lockplate by turning the shaft nut clockwise until the C-ring can be removed.

6. Remove the tool and lift out the lockplate, horn contact carrier, and the upper bearing preload spring.

7. Pull the switch connector out of the mast jacket and tape the upper part to facilitate switch removal.

8. Remove the turn signal lever. Push the flasher in and unscrew it.

9. Position the turn signal and shifter housing in Low position. Remove the switch by pulling it straight up while guiding the wiring harness out of the housing.

10. Install the replacement switch by working the harness connector down through the housing and under the mounting bracket.

11. Install the harness cover and clip the connector to the mast jacket.

12. Install the switch mounting screws, signal lever, and the flasher knob.

13. With the turn signal lever in neutral and the flasher knob out, install the upper bearing pre-load spring, horn contact carrier, and lockplate onto the shaft.

14. Position the tool as in Step 4 and compress the plate far enough to allow the C-ring to be installed.

15. Remove the tool. Install the plastic C-ring retainer.

16. Install the column/dash trim cover. Install the steering wheel.

CHASSIS ELECTRICAL

Headlight Switch

REMOVAL & INSTALLATION

1. Disconnect the negative battery terminal.

2. Remove the left air distribution duct.

3. Remove the instrument cluster attaching screws and pull the cluster rearward.

4. Disconnect the speedometer cable, electrical connectors and remove the cluster.

5. Remove the instrument panel to left door pillar attaching screws and pull the left side of the instrument panel slightly forward for access.

6. Depress the shaft retainer, pull the knob and shaft assembly out and remove the switch bezel.

7. Disconnect the vacuum hoses from the switch, tagging them for installation.

8. Pry the connector from the switch and remove the switch from the panel.

9. Installation is the reverse of removal.

Instrument Cluster

REMOVAL & INSTALLATION

1980–82

1. Disconnect negative battery cable.

2. Remove left air distribution duct.

3. Remove lens to bezel attaching screws and remove lens.

4. Remove cluster to instrument panel attaching screws.

5. Pull cluster assembly slightly forward to obtain clearance for removal of speedometer cable housing, headlamp switch connectors and panel illuminating lamps.

6. Install by reversing removal procedure, being careful not to kink the speedometer cable casing.

1984 and Later

1. Disconnect battery ground cable.

2. Remove light switch knob (spring loaded), and light switch nut.

3. Remove steering column trim cover.

4. Remove 2 steering column attaching bolts and lower steering column for access.

5. Remove cluster bezel front and left side attaching screws.

6. Remove cluster bezel from instrument panel.

7. Remove 4 cluster to instrument panel attaching screws.

8. Pull cluster rearward for access to disconnect cluster electrical connectors. Metal retaining clips are located at back side of connectors.

9. Remove cluster from instrument panel. Odometer may be removed for service or replacement.

10. Reverse above procedure to install.

Speedometer Cable

REMOVAL & INSTALLATION

Reach behind the speedometer and depress the retaining clip. Pull the cable from the casing. If the cable is broken, raise the car and disconnect the cable at the transmission. Lubricate only the bottom $\frac{3}{4}$ of the cable with speedometer cable lubricant. Reconnect all parts.

NOTE: This procedure is easier to complete if the instrument panel cluster has previously been removed.

Windshield Wiper Motor

REMOVAL & INSTALLATION

1980–82

1. With wiper motor in park position and hood open, disconnect the washer hoses and all wiring from the motor assembly.

2. Remove the plenum chamber grill.

3. Remove the nut which retains the crank arm to the motor assembly.

4. Remove the ignition shield, if used, and distributor cap. Remove and identify the left bank spark plug leads.

5. Remove the motor mounting screws or nuts and remove the motor.

6. To install, reverse the above procedure.

1984 and Later

1. Open hood and install fender covers.

2. Remove wiper arms.

3. Remove air inlet leaf screen.

4. Turn ignition ON, and activate motor with wiper switch. Allow motor crank arm to rotate to point to a position between 4 and 5 o'clock as viewed from passenger compartment. Stop crank arm in this position by turning off ignition switch.

5. Disconnect battery ground cable.

6. Disconnect upper motor electrical connectors.

7. Remove motor mounting bolts.

8. With crank arm in position described in Step 4 above, motor may now be removed from vehicle. Lower electrical connector may be disconnected as motor is partially removed.

9. To install, reverse the above procedure.

Wiper Transmission

REMOVAL & INSTALLATION

1. Make sure the wiper is in the park position.

2. Disconnect the battery ground cable.

3. Open the hood and remove the plenum chamber screen.

4. Loosen the nuts retaining the ball sockets to the crank arm and detach the drive rod from the crank arm.

5. Remove the transmission nuts, then lift the rod assemblies from the plenum chamber.

6. Remove the transmission linkage from the plenum chamber.

7. To install, reverse the removal procedure. Make sure the wipers are in the park position.

Radio

REMOVAL & INSTALLATION

1. Disconnect the battery ground cable.

2. Remove the console tunnel side panels.

3. Pull the radio control knobs from the shaft.

4. Remove the two screws that secure the console trim plate to the instrument cluster.

5. Remove the rear defogger switch if so equipped.

6. Remove the five screws from around the upper perimeter of the instrument cluster.

7. Pull the instrument cluster enough to disconnect the electrical connector from the rear of the cluster.

NOTE: The center instrument cluster trim panel is designed to collapse under impact. Do not deflect the panel to gain access to the radio.

8. Remove the screw holding the radio bracket reinforcement to the floor pan.

9. Pull the radio outward and disconnect the wiring from the back.

10. Installation is the reverse of removal. If a new radio is being installed, save the mounting bracket from the rear of the old one.

NOTE: The radio heat sink must be removed when radio service is required. It is located behind the passenger side dash panel.

Fuses

The fuse block is located beneath the instrument panel above the headlight dimmer floor switch. Fuse holders are labeled as to their service and the correct amperage. Always replace blown fuses with new ones of the correct amperage. Otherwise electrical overloads and possible wiring damage will result.

The fuse block on some models is a swing-down unit located in the underside of the instrument panel adjacent to the steering column. Access to the fuse block on some models is gained through the glove box opening. The Convenience Center on some models is a swing-down unit located on the underside of the instrument panel. The swing-down feature provides center location and easy access to buzzers, relays and flasher units. All units are serviced by plug-in replacement. Location of Convenience Center on specific models may vary.

Circuit Breaker

A circuit breaker is an electrical switch which breaks the circuit during an electrical overload. The circuit breaker will remain open until the short or overload condition in the circuit is corrected.

Fusible Links

Fusible links are sections of wire, with special insulation, designed to melt under electrical overload. Replacements are simply spliced into the wire. There may be as many as five of these in the engine compartment wiring harnesses. These are:

1. Horn relay to fuse panel circuit—one link.

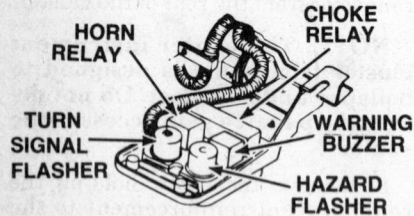

Swing–down convenience center—typical

2. Charging circuit, from the starter solenoid to the horn relay—two links.

3. Starter solenoid to ammeter circuit—one link.

4. Horn relay to rear window defroster circuit—one link.

5. Connect the new fusible link to the harness wire using a crimp on connector. Solder the connection using rosin core solder.

6. Tape all exposed wires with plastic electrical tape.

7. Connect the fusible link to the junction block or starter solenoid and reconnect the battery ground cable.

Heater Blower

REMOVAL & INSTALLATION

1980–82

1. Disconnect the negative battery cable at the battery.

2. On 1981–82 models, unbolt the A/C compressor and move the compressor out of the way. DO NOT disconnect the refrigerant lines from the compressor.

3. Remove the coolant recover jar, if so equipped.

4. Disconnect the wiring from the motor and the cooling tube from the motor case, if so equipped.

5. Remove the mounting screws from the blower motor and remove the motor. If the motor sticks to the case due to the sealer, pry the motor GENTLY away from the case.

6. Installation is the reverse of the previous steps.

1984 and Later

1. Open the hood and disconnect the battery ground cable.

2. Remove the front wheel house rear panel and move wheel house seal aside.

3. Remove the motor cooling tube.

4. Remove the relay.

5. Remove the blower motor assembly to case attaching screws.

6. Remove the motor and impeller.

7. To install, reverse the replacement procedure and check the operation.

Heater Core

REMOVAL & INSTALLATION

1980

1. Disconnect the battery ground cable.

2. Drain the cooling system. It is not necessary to evacuate the A/C refrigerant.

3. Disconnect the heater hoses at the firewall and plug the pipes.

4. Remove the nuts from the distributor studs protruding through the firewall.

5. Remove the right side dash pad and center dash cluster.

6. Disconnect the right dash outlet from the center duct.

7. Remove the center duct from the selector duct.

8. Remove the selector duct to the dash panel and pull it to the right and to the rear.

9. Remove the cables and wiring connectors from the selector and remove it from the car.

10. Remove the temperature door cam plate from the selector duct.

11. Remove the heater core and housing from the selector.

12. Reverse the removal procedure to install.

1981–82

1. Disconnect the negative cable at the battery.

2. Drain the coolant from the radiator.

3. Raise the right front of the vehicle and support it safely.

4. Disconnect the heater hoses at the heater core connections.

5. Remove the heater case retaining nut which is located on the top of the blower case.

6. Remove the glove box.

7. Remove the console side panels.

8. Remove the knobs and nuts from the radio shafts.

9. Remove the two screws which secure the console trim plate to the instrument cluster.

10. Remove the instrument cluster attaching screws.

11. Pull the cluster out slightly and disconnect the electrical connector from the rear of the cluster.

12. Remove the radio as previously outlined.

13. Remove the right side windshield pillar trim panel.

14. Remove the right side dash panel retaining screws and pull the panel rearward to release the upper retaining clip.

15. Remove the following ducts.
  a. Right side vent
  b. Main vent distribution
  c. Lower heater deflector
  d. Heater-defroster distribution duct assembly (Disconnect the vacuum line).

16. Disconnect both the temperature cable and the vacuum line at the heater housing.

17. Remove the heater core from the housing.

18. Installation is the reverse of the previous steps.

1984 and Later

1. Disconnect the negative battery cable. Drain the cooling system.

2. Remove the instrument cluster bezel including the tilt wheel lever and instrument panel pad.

3. Remove the A/C distributor duct and disconnect the flex hose.

4. Remove the right side hush panel.

5. Remove the side window defroster flex hose.

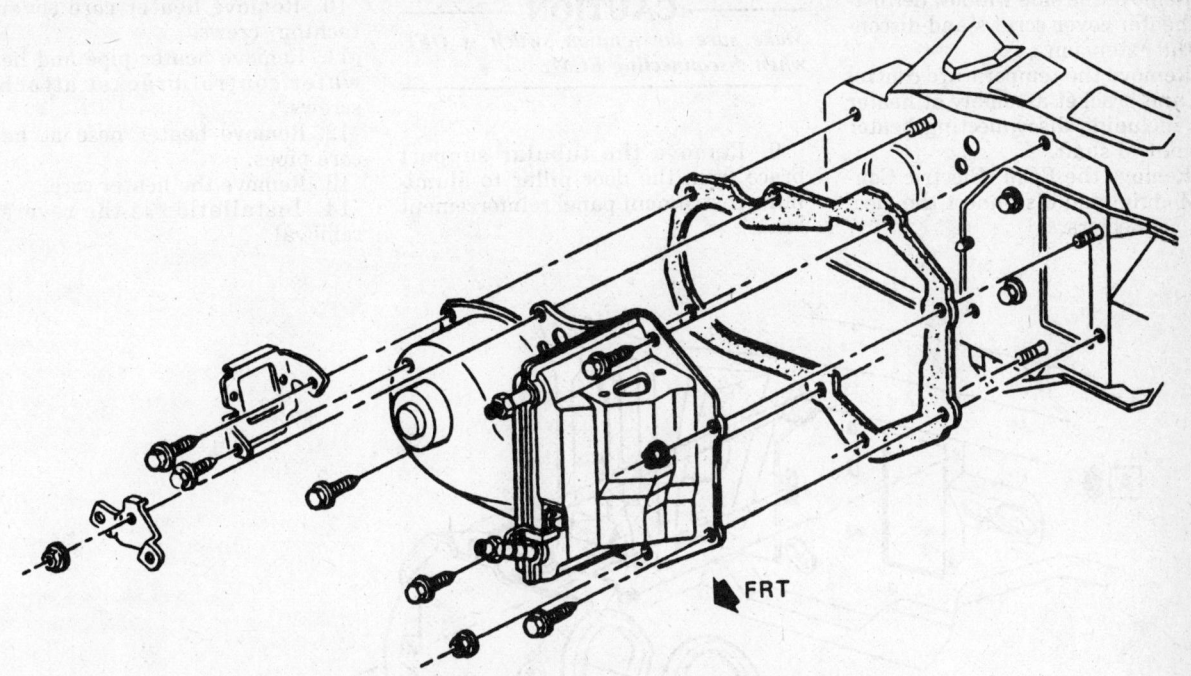

Blower motor and evaporator module—1984 and later

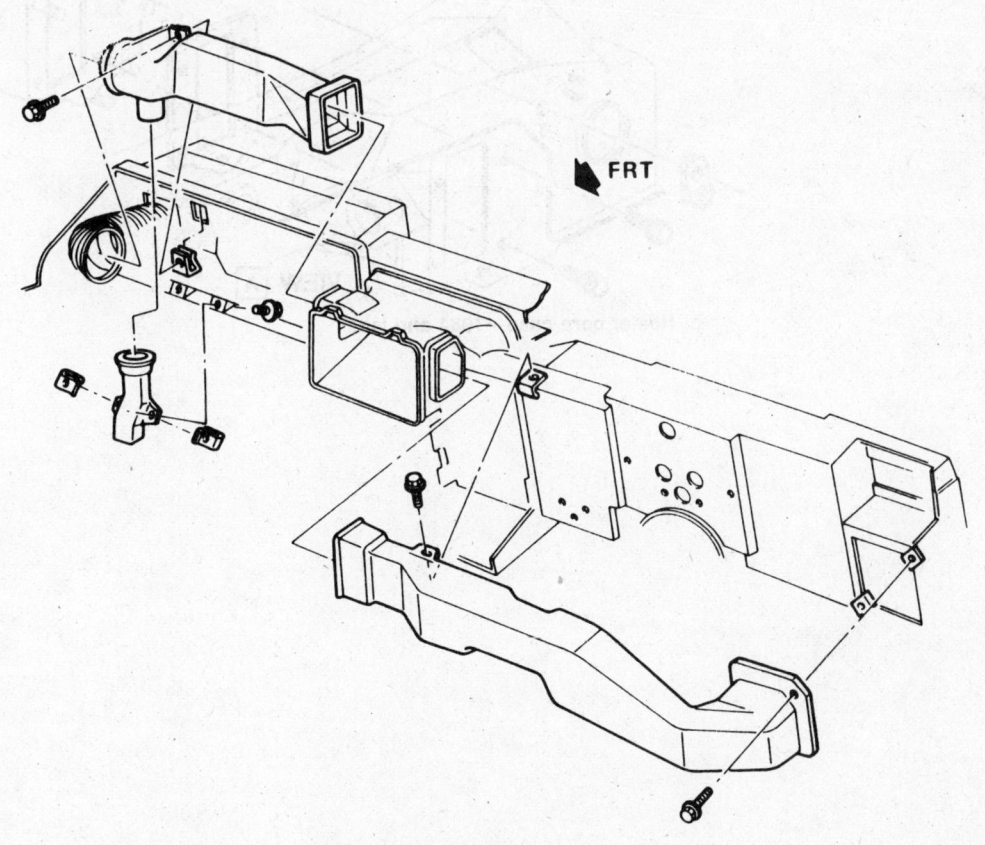

Duct assembly—1984 and later

6. Remove the side window defroster to heater cover screws and disconnect the extension.

7. Remove the temperature control cable and bracket assembly at heater cover including disconnecting heater door control shaft.

8. Remove the ECM (Electric Control Module) and disconnect the electrical connectors.

— CAUTION —
Make sure the ignition switch is OFF when disconnecting ECM.

9. Remove the tubular support brace from the door pillar to aluminum, instrument panel reinforcement brace.

10. Remove heater core cover attaching screws.

11. Remove heater pipe and heater water control bracket attaching screws.

12. Remove heater hose at heater core pipes.

13. Remove the heater core.

14. Installation is the reverse of removal.

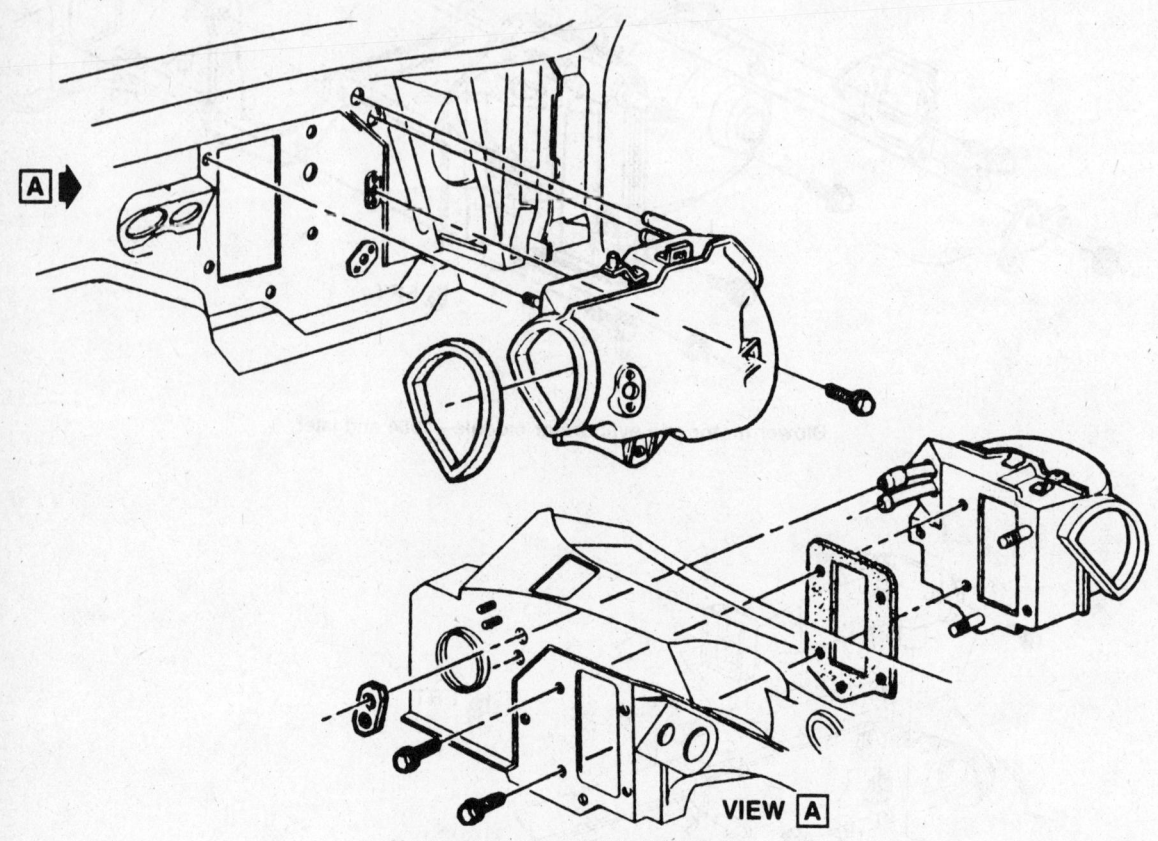

Heater core case—1984 and later

Chevrolet
Rear Wheel Drive
Caprice, Impala, Malibu, Monte Carlo

YEAR IDENTIFICATION

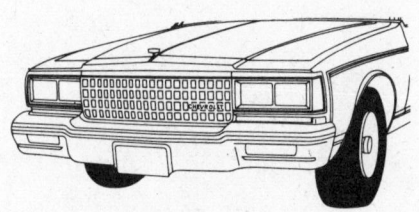

1980 Caprice

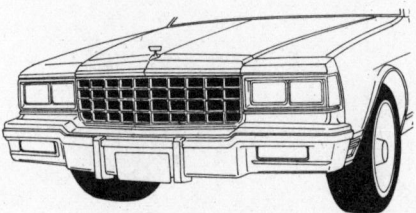

1981-87 Caprice

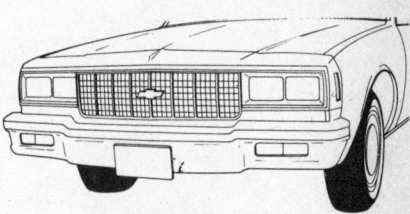

1980 Impala

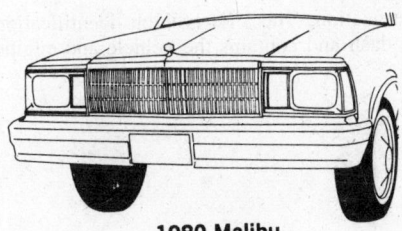

1980 Malibu

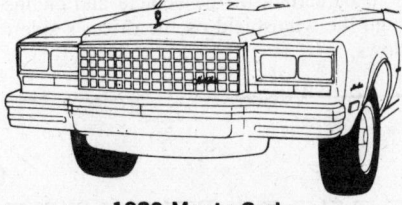

1980 Monte Carlo

1981 Malibu

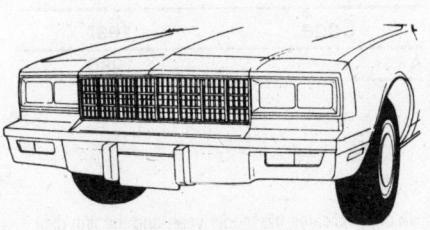

1981 Monte Carlo

1981–84 Impala

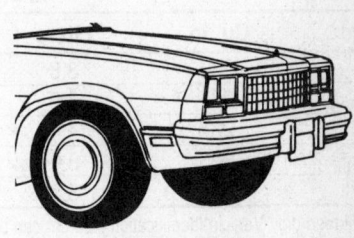

1982–83 Malibu

C403

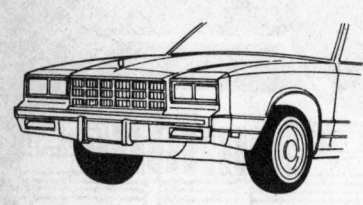

YEAR IDENTIFICATION

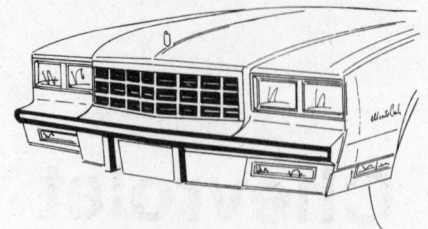

1982 Monte Carlo   **1983–86 Monte Carlo**   **1984-87 Monte Carlo SS**

VEHICLE IDENTIFICATION NUMBER (VIN)

It is important for servicing and ordering parts to be certain of the vehicle and engine identification. The VIN (vehicle identification number) is a 13 or 17 digit number visible though the windshield on the driver's side of the dash and contains the vehicle and engine identification codes. It can be interpreted as follows:

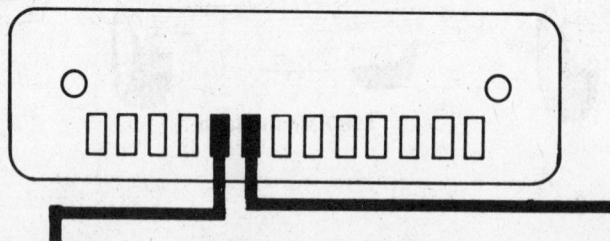

	Engine Code						Model Year Code	
Code	Cu. In.	Liters	Cyl.	Carb.	Eng. Mfg.		Code	Year
K	229	3.8	6	2	Chev.		A	1980
A	231	3.8	6	2	Buick			
J	267	4.4	8	2	Chev.			
H	305	5.0	8	4	Chev.			
N	350	5.7	8	Diesel	Olds.			

The thirteen digit Vehicle Identification Number can be used to determine engine application and model year. The sixth digit indicates the model year, and the fifth digit identifies the factory installed engine.

VEHICLE IDENTIFICATION NUMBER (VIN)

It is important for servicing and ordering parts to be certain of the vehicle and engine identification. The VIN (vehicle identification number) is a 13 or 17 digit number visible though the windshield on the driver's side of the dash and contains the vehicle and engine identification codes. It can be interpreted as follows:

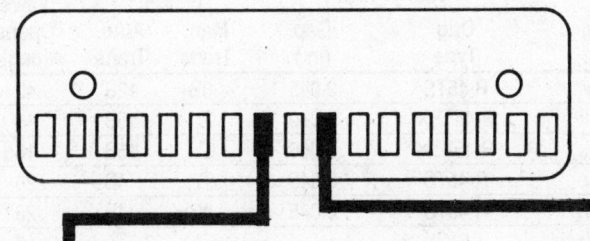

Engine Code						Model Year Code	
Code	Cu. In.	Liters	Cyl.	Carb.	Eng. Mfg.	Code	Year
K	229	3.8	6	2	Chev.	B	1981
9	229	3.8	6	2	Chev.	C	1982
A	231	3.8	6	2	Buick	D	1983
3	231	3.8	6	Turbo	Buick	E	1984
Z	262	4.3	6	TBI	Chev.	F	1985
V	263	4.3	6	Diesel	Olds.	G	1986
J	267	4.4	8	2	Chev.	H	1987
G	305	5.0	8	4	Chev.		
H	305	5.0	8	4	Chev.		
7	305	5.0	8	4	Chev.		
Y	307	5.0	8	4	Olds.		
N	350	5.7	8	Diesel	Chev.		

The seventeen digit Vehicle Identification Number can be used to determine engine application and model year. The tenth digit indicates the model year, and the eighth digit identifies the factory installed engine.

GENERAL ENGINE SPECIFICATIONS

Year	VIN Code	Engine No. Cyl. Displ. (cu. in.)	Eng. Mfg.	Fuel Delivery System	Horsepower @ rpm	Torque @ rpm ft. lb.	Bore × Stroke	Compression Ratio	Oil Pressure 2000 rpm
'80–81	K	6-229	Chev.	2 bbl	110 @ 4200	170 @ 2000	3.736 × 3.480	8.2:1	37
'82	K	6-229	Chev.	2 bbl	115 @ 4000	170 @ 2000	3.736 × 3.480	8.6:1	45
'83–'84	9	6-229	Chev.	2 bbl	115 @ 4000	170 @ 2000	3.736 × 3.480	8.6:1	45
'80–'81	3	6-231	Buick	Turbo	170 @ 4000	275 @ 2400	3.800 × 3.400	8.0:1	37
'80–'84	A	6-231	Buick	2 bbl	110 @ 3800	190 @ 1600	3.800 × 3.400	8.0:1	45
'85	Z	6-262	Chev.	TBI	130 @ 3600	218 @ 2000	4.000 × 3.480	9.3:1	45
'86–'87	Z	6-262	Chev.	TBI	140 @ 3800	225 @ 2200	4.000 × 3.480	9.3:1	45
'82–'83	V	6-263	Olds.	Diesel	85 @ 3200	165 @ 1600	4.057 × 3.385	21.6:1	45①
'80–'82	J	8-267	Chev.	2 bbl	115 @ 4000	200 @ 2400	3.500 × 3.480	8.3:1	45
'80–'87	H	8-305	Chev.	4 bbl	150 @ 3800	240 @ 2400	3.736 × 3.480	8.6:1	45
'85–'87	H	8-305	Chev.	4 bbl②	165 @ 4200	245 @ 2400	3.736 × 3.480	9.5:1	45
'84	G	8-305	Chev.	4 bbl	134 @ 4800	319 @ 3200	3.736 × 3.480	9.5:1	45
'85–'87	G	8-305	Chev.	4 bbl	180 @ 4800	235 @ 3200	3.736 × 3.480	9.5:1	45
'86–'87	Y	8-307	Olds.	4 bbl	148 @ 3800	250 @ 2400	3.800 × 3.385	8.0:1	40①
'80–'86	N	8-350	Olds.	Diesel	105 @ 3200	205 @ 1600	4.057 × 3.385	22.5:1	30–45①

The seventeen digit Vehicle Identification Number can be used to determine engine application and model year. The tenth digit indicates the model year, and the eighth digit identifies the factory installed engine.

■ Horsepower and torque are SAE net figures. They are measured at the rear of the transmission with all accessories in stalled and operating. Since the figures vary when a given engine is installed in different models, some are representative rather than exact.
① @ 1500
① Caprice only

TUNE-UP SPECIFICATIONS

(When analyzing compression test results, look for uniformity among cylinders rather then specific pressures.)

Year	V.I.N. Code	Eng. No. Cyl. Displ. Cu. In.	Eng. Mfg.	Spark Plugs Orig Type	Spark Plugs Gap (in.)	Ignition Timing (deg)▲● Man. Trans	Ignition Timing (deg)▲● Auto. Trans	Intake Valve Opens ■(deg)●	Fuel Pump Pressure (psi)	Idle Speed (rpm)▲* Man.● Trans	Idle Speed (rpm)▲* Auto. Trans
'80–'81	K	6-229	Chev.	R-45TS	0.045	8B	12B	42	4.5–6.0	700	600
	A	6-231	Buick	R-45TSX	0.060	①	15B	16	4.25–5.75	①	560(600)
	3	6-231	Buick	R-45TSX	0.060	①	15B	16	4.25–5.75	①	550(600)
	J	8-267	Chev.	R-45TS	0.045	①	4B	28	7.5–9.0	①	500
	H	8-305	Chev.	R-45TS	0.045	4B	4B	28	7.5–9.0	700	500(550)
	N	8-350	Olds. Diesel	—	—	①	①	16	5.5–6.5	①	①
'82	K	6-229	Chev.	R-45TS	0.045	—	6B	42	4.5–6.0	—	600
'82–'83	A	6-231	Buick	R-45TS	0.045	—	15B	16	4.25–5.75	—	500
	V	6-263	Olds. Diesel	—	—	—	①	16	5.5–6.5	—	①
'82	J	8-267	Chev.	R-45TS	0.045	—	6B	44	5.5–7.0	—	500
'82–'83	H	8-305	Chev.	R-45TS	0.045	—	6B	44	5.5–7.0	—	500
'82–'84	N	8-350	Olds. Diesel	—	—	—	①	16	5.5–6.5	—	①
'83–'84	9	6-229	Chev.	R-45TS	0.045	—	6B	42	4.5–6.0	—	600
'84	A	6-231	Buick	R-45TS	0.045	—	①	16	4.25–5.75	—	①
'84–'86	G	8-305	Chev.	R-45TS③	0.045③	—	①	—	7.5–9.0	—	①
	H	8-305	Chev.	R-45TS③	0.045③	—	①	44	5.5–7.0	—	①
'86–'87	Z	6-262	Chev.	R-43CTS②	0.035	—	①	—	—	—	①
'86	Y	8-307	Olds.	FR3LS6	0.060	—	①	—	6–7.5	—	①
'87	All			See Underhood Specifications Sticker							

NOTE: The underhood specifications sticker often reflects tune-up specifications changes made in production. Sticker figures must be used if they disagree with those in this chart.

▲ See text for procedure
● Figure in parentheses indicates California engine
■ All figures Before Top Dead Center
* When two idle speed figures are spearated by a
  slash, the lower figure is with the idle speed
  solenoid disconnected
B Before Top Dead Center

TDC Top Dead Center
— Not available
① Refer to underhood specifications sticker
② '86 Monte Carlo: R-43TS w/.035 gap
③ '86 Caprice: R-44TS w/.035 gap

FIRING ORDERS

NOTE: To avoid confusion, always replace spark plug wires one at a time.

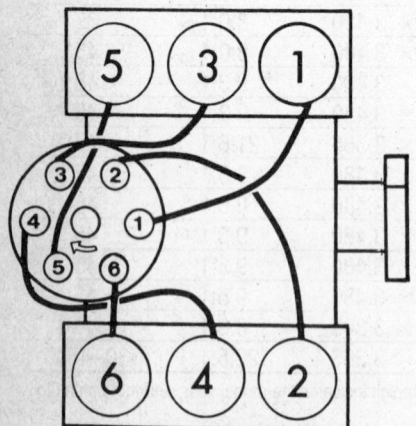

Chevrolet-built V6 engine
Engine firing order: 1-6-5-4-3-2
Distributor rotation: clockwise

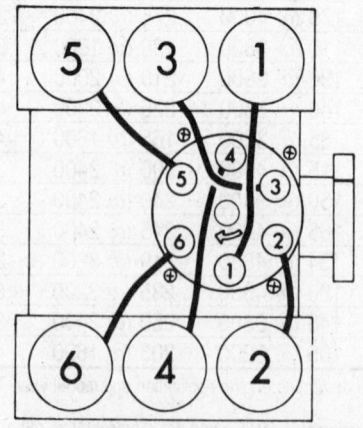

GM (Buick) 231 V6
Engine firing order: 1-6-5-4-3-2
Distributor rotation: clockwise

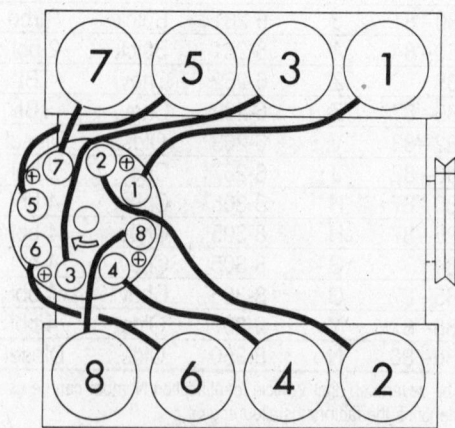

GM (Chevrolet) V8
Engine firing order: 1-8-4-3-6-5-7-2
Distributor rotation: clockwise

CAPACITIES
Monte Carlo, Malibu

Year	Engine No. Cyl. Displacement (cu. in.)	Engine Crankcase Add 1 qt For New Filter ■	Transmission (Pts to Refill After Draining)			Drive Axle (pts.)	Gasoline Tank (gals.)	Cooling System (qts.)	
			Manual		Automatic ●			With Heater	With A/C
			3-Speed	4-Speed					
'80–'81	6-229 Chev.	4	3.0	3.4	②	3.25	18.1	18.8⑦	18.8⑦
	6-231 Buick	4	3.0	3.4	②	3.25	18.1	15.4④	15.4④
	8-267 Chev.	4	3.0	3.4	②	3.25	18.1	20.6	20.6
	8-305 Chev.	4	3.0	3.4	②	⑤	18.1	①	①
	8-350 Chev.	4	3.0	3.4	②	4.25	18.1	16.4	16.4
'82	6-229 Chev.	4	—	—	②	3.5	18.1	15.0	15.0
	6-231 Buick	4	—	—	②	3.5	18.1	12.5	12.2
	6-263 Diesel	6③	—	—	②	3.5	18.1	15.0	15.0
	8-267 Chev.	4	—	—	6.0	3.5	18.1	18.9	18.0
	8-305 Chev.	4	—	—	6.0	3.5	18.1	16.5	16.5
	8-350 Diesel	7③	—	—	6.0	3.5	18.1	18.0	18.0
'83	6-229 Chev.	4	—	—	6.0	3.5	18.1	15.0	15.0
	6-231 Buick	4	—	—	6.0	3.5	18.1	15.0	15.0
	6-263 Diesel	6③	—	—	6.0	3.5	18.1	15.0	15.0
	8-305 Chev.	4	—	—	6.0	3.5	18.1	15.0	15.0
	8-350 Diesel	7③	—	—	6.0	3.5	18.1	18.0	18.0
'84	6-229 Chev.	4	—	—	6.0	3.5	18.1	15.0	15.0
	6-231 Buick	4	—	—	6.0	3.5	18.1	15.0	15.0
	8-305 Chev. ⑧	4	—	—	6.0	3.5	18.1	16.6	16.6
	8-305 Chev.	4	—	—	6.0	3.5	18.1	16.3	16.3
	8-350 Diesel	7③	—	—	6.0	3.5	19.8	17.3	17.3
'85–'87	6-262 Chev.	4③	—	—	7.0	⑤	17.6	12.0	12.0
	8-305 Chev.	4③	—	—	7.0	⑤	18.1	16.3	16.3
	8-305 Chev. ⑧	4③	—	—	7.0	⑤	18.1	16.6	16.6

● Specifications do not include torque converter
Add just enough fluid to fill the transmission to the proper level. It takes only one pint to raise the level from the "ADD" to "FULL" with a hot transmission. Do not overfill.
■ On models with micro oil filters, capacity is the same with or without new filter
① 19.2 (thru 1980), 16.5 (1981)
② 1981–82: 7.0 pts. w/200, 200C, 200-4R; 8.0 pts w/250, 250C; 6.3 pts w/350, 350C

③ Includes mandatory filter change
④ 1981: 12.5 w/heater, 12.2 w/A/C
⑤ With 7.5 inch ring gear: 3.5
   With 8.75 inch ring gear: 5.4
⑥ 1981: 15.2 w/heater, 15 w/A/C
⑦ 1981: 16.61 w/heater, 16.63 w/A/C—Not applicable
⑧ Eng. Code G

CAPACITIES
Impala, Caprice

Year	Engine No. Cyl. Displacement (cu. in.)	Engine Crankcase Add 1 qt. For New Filter	Transmission Pts to Refill After Draining			Drive Axle (pts.)	Gasoline Tank (gals.)	Cooling System (qts.)	
			Manual		Automatic ●			With Heater	With A/C
			3-Speed	4-Speed					
'80–'82	6-229	4③	—	—	7.0	4.0	18.5⑨	—	14¼④
	6-231	4③	—	—	7.0	4.0①	18.5⑨	—	11¾④
	8-267	4	—	—	6.0②	4.0①	18.5⑨	—	16¾
	8-305	4	—	—	6.0	4.0①	18.5⑨	—	15½
	8-350⑤	4	—	—	6.0	4.0①	18.5⑨	—	16¼
	8-350 Diesel	7	—	—	6.0	4.0①	18.5⑨	—	16¼

CAPACITIES
Impala, Caprice

Year	Engine No. Cyl. (cu. in.) Displacement	Engine Crankcase Add 1 qt. For New Filter	Transmission Pts to Refill After Draining Manual 3-Speed	4-Speed	Automatic ●	Drive Axle (pts.)	Gasoline Tank (gals.)	Cooling System (qts.) With Heater	With A/C
'83	6-229	4	—	—	6.0	⑥	⑦	—	14¼ ④
	6-231	4	—	—	6.0	⑥	⑦	—	11¾ ④
	8-305	4	—	—	6.0 ⑧	⑥	⑦	—	15½
	8-350 Diesel	6	—	—	6.0	⑥	⑦	—	18.3
'84	6-229	4	—	—	6.0	⑥	⑦	—	14¼
	6-231	4	—	—	6.0	⑥	⑦	—	11¾
	8-305	4	—	—	6.0 ⑧	⑥	⑦	—	15½
	8-350 Diesel	6	—	—	6.0 ⑧	⑥	⑦	—	18.3
'85	6-262	4 ③	—	—	7.0 ⑧	⑥	⑦	—	14.0 ⑩
	8-305	4	—	—	7.0 ⑧	⑥	⑦	—	15.3 ⑪
	8-350 Diesel	6	—	—	7.0 ⑧	⑥	⑦	—	18.3
'86-'87	6-262	4 ③	—	—	7.0 ⑧	⑥	⑦	12.2	12.5
	8-305	4	—	—	7.0 ⑧	⑥	⑦	16.8	17.5
	8-307	4	—	—	7.0 ⑧	⑥	⑦	17.1	17.6

● Specifications do not include torque converter Add just enough fluid to fill the transmission to the proper level. It takes only one pint to raise the level from "ADD" to "FULL" with a hot transmission. Do not overfill.
—Not applicable
① with 7.5 inch ring gear: 3.25

② 7.5 pt. w/200 T.H. Trans.
③ 4 qt. with filter change
④ Cooling system capacity, Station wagon heavy duty capacity 16¾ qts.
⑤ Not available after 1980.
⑥ 7.5" ring gear: 3.5 pts
   8.5" ring gear: 4.25 pts
   8.75" ring gear: 5.0 pts

⑦ Gasoline coupe and sedan—25 gal; Diesel coupe and sedan—26 gal All station wagons—22 gal.
⑧ Automatic Overdrive: 10 pts
⑨ Station wagons: 22 gal.
⑩ With H.D. cooling: 14.6
⑪ With H.D. cooling: 16.1

VALVE SPECIFICATIONS

Year	VIN Code	Engine No. Cyl. Displacement (cu. in.)	Seat Angle (deg.)	Face Angle (deg.)	Spring Test Pressure (lbs. @ in.)	Spring Installed Height (in.)	Stem-to-Guide Clearance (in.) Intake	Exhaust	Stem Diameter (in.) Intake	Exhaust
'80-'82	K	6-229	46	45	200 @ 1.25	1.70	0.0010–0.0027	0.0010–0.0027	0.3414	0.3414
'80-'84	A	6-231	45	45	168 @ 1.32	1.72	0.0015–0.0032	0.0015–0.0032	0.3407	0.3409
'83-'84	9	6-229	46	45	200 @ 1.25	1.70	0.0010–0.0027	0.0010–0.0027	0.3414	0.3414
'80-'81	3	6-231	45	45	168 @ 1.32	1.72	0.0015–0.0032	0.0015–0.0032	0.3407	0.3409
'82-'83	V	6-263 Diesel	45 ③	44 ③	210 @ 1.22	1.67	0.0015–0.0032	0.0015–0.0032	0.3429	0.3424
'80-'82	J	8-267	46	45	200 @ 1.25	1.70	0.0010–0.0027	0.0010–0.0027	0.3414	0.3414
'80-'87	H	8-305	46	45	200 @ 1.25	1.70	0.0010–0.0027	0.0010–0.0027	0.3414	0.3414
'84-'87	G	8-305	46	45	200 @ 1.25	1.70	0.0010–0.0027	0.0010–0.0027	0.3414	0.3414
'85-'87	Z	6-262	46	45	200 @ 1.25	1.70	0.0010–0.0027	0.0010–0.0027	0.3414	0.3414
'86-'87	Y	8-307	45	44	187 @ 1.27	1.67	0.0010–0.0027	0.0015–0.0032	0.3429	0.3429
'80-'85	N	8-350 Diesel	45 ①	44 ②	205 @ 1.300	1.67	0.0010–0.0027	0.0015–0.0032	0.3429	0.3424

① Exhaust: 31°
② Exhaust: 30°
③ Exhaust: Face 30°; Seat 31°

CRANKSHAFT AND CONNECTING ROD SPECIFICATIONS

(All measurements are given in inches.)

Year	VIN Code	Engine No. Cyl. Displacement (cu. in.)	Crankshaft				Connecting Rod		
			Main Brg. Journal Dia.	Main Brg. Oil Clearance	Shaft End-Play	Thrust on No.	Diameter Journal	Clearance Oil	Clearance Side
'84–'87	G	8-305	2.4484–2.4493⑤	.0008–.0020④	.002–.006	5	2.0986–2.0998	.0013–.0035	.006–.014
'80–'81	3	6-231	2.4995	.0004–.0015	.004–.008	2	2.2495–2.2487	.0005–.0026	.006–.027
'83–'84	9	6-229	2.4484–2.4493④	.0008–.0020③	.002–.006	4	2.0986–2.0998	.0013–.0035	.006–.014
'85–'87	Z	6-262	2.4484–2.4493④	.0008–.0020③	.002–.006	4	2.0986–2.0998	.0013–.0035	.006–.014
'80–'82	K	6-229	2.4484–2.4493④	.0008–.0020③	.002–.006	4	2.0986–2.0998	.0013–.0035	.006–.014
'80–'84	A	6-231	2.4995	.0004–.0015	.004–.008	2	2.2495–2.2487	.0005–.0026	.006–.027
'82–'83	V	6-263 Diesel	2.9993–3.0003	.0005–.0021①	.0035–.0135	3	2.2490–2.2510	.0005–.0026	.006–.020
'80–'82	J	8-267	2.4484–2.4493④	.0008–.0020③	.002–.006	5	2.0986–2.0998	.0013–.0035	.006–.014
'80–'87	H	8-305	2.4484–2.4493④	.0008–.0020③	.002–.006	5	2.0986–2.0998	.0013–.0035	.006–.014
'86–'87	Y	8-307	2.4990–2.4995⑤	.0005–.0021②	.0035–.0135	3	2.1238–2.1248	.0004–.0033	.006–.020
'80–'85	N	8-350 Diesel	2.9993–3.0003	.0005–.0021②	.0035–.0135	3	2.1238–2.1248	.0005–.0026	.006–.020

① No. 4: .0020–.0034
② No. 5: .0015–.0031
③ Intermediate—.0011–.0023
　Rear—.0017–.0032
④ Intermediate—2.4481–2.4490
　Rear—2.4479–2.4488
⑤ No. 1: 2.4993–2.4998

PISTON AND RING SPECIFICATIONS

(All measurements are given in inches.)

Year	V.I.N. Code	Engine Type/ Disp. cu. in.	Piston-to-Bore Clearance	Ring Gap			Ring Side Clearance		
				Top Compression	Bottom Compression	Oil Control	Top Compression	Bottom Compression	Oil Control
'80–'82	A	6-231	0.0008–0.0012	0.010–0.020	0.010–0.020	0.015–0.035	0.003–0.005	0.003–0.005	0.0035 Max.
'80–'82	K	6-229	0.0012	0.010–0.020	0.010–0.025	0.010–0.035	0.0012–0.0032	0.0012–0.0032	0.0020–0.0070
'80–'81	3	6-231	0.0008–0.0012	0.010–0.020	0.010–0.020	0.015–0.035	0.003–0.005	0.003–0.005	0.0035 Max.
'80–'82	J	8-267	0.0012	0.010–0.020	0.010–0.025	0.015–0.055	0.0012–0.0032	0.0012–0.0032	0.0020–0.0070
'83–'84	9	6-229	0.0012	0.010–0.020	0.010–0.025	0.010–0.055	0.0012–0.0032	0.0012–0.0032	0.0020–0.0070
'85–'87	Z	6-262	0.0012	0.010–0.020	0.010–0.025	0.015–0.055	0.0012–0.0032	0.0012–0.0032	0.0020–0.0070
'82–'83	V	6-263	0.0030–0.0040	0.015–0.025	0.015–0.025	0.015–0.035	0.005–0.007	0.003–0.005	0.001–0.005
'80–'87	H	8-305	0.0012	0.010–0.020	0.010–0.025	0.015–0.055	0.0012–0.0032	0.0012–0.0032	0.0020–0.0070
'84–'87	G	8-305	0.0012	0.010–0.020	0.010–0.025	0.015–0.055	0.0012–0.0032	0.0012–0.0032	0.0020–0.0070
'86–'87	Y	8-307	0.0008–0.0018	0.009–0.019	0.009–0.019	0.015–0.055	0.0020–0.0040	0.0020–0.0040	0.000–0.0035
'80–'85	N	8-350	0.0050–0.0060	0.015–0.025	0.015–0.025	0.015–0.055	0.005–0.007	0.003–0.005	0.0010–0.0050
'83–'84	A	6-231	0.0012	0.010–0.020	0.010–0.025	0.015–0.055	0.0012–0.0032	0.0012–0.0032	0.0020–0.0070

TORQUE SPECIFICATIONS

(All readings in ft. lbs.)

Year	Engine No. Cyl. Displacement (cu. in.)	Cylinder Head Bolts	Rod Bearing Bolts	Main Bearing Bolts	Crankshaft Bolt	Flywheel to Crankshaft Bolts	Manifold	
							Intake	Exhaust
'80–'87	6-200, 6-229, 8-267, 8-305	65	45	70⑥	60	60	30	20④
'80–'84	6-231	80	40	100	175②	60	45	25
'85–'87	6-262	60–75	45	80	70	70	45	20
'86–'87	8-307	125①	42	⑤	200–310	46	40①	25
'82–'85	8-350 Diesel	130①	42	120	310	60	40①	25
'82–'83	6-263 Diesel	142③	42	107	160–350	48	41	29

① Dip bolt in oil before tightening
② '83–'85: 225 ft. lbs.
③ Bolts No. 5, 6, 11, 12, 13, 14: 59 ft. lbs.
④ 8-305 inside bolts: 25 ft. lbs.
⑤ No. 1 thru 4: 80 ft. lbs., No. 5: 120 ft. lbs.
⑥ '85–'87: 70–85

WHEEL ALIGNMENT SPECIFICATIONS

Year	Model	Caster		Camber		Toe In (in.)	Steering Axis (deg.) Inclination
		Range (deg.)	Pref. Setting (deg.)	Range (deg.)	Pref. Setting (deg.)		
'80–'87	Malibu, Monte Carlo, Man. Steer.	0 to 2P	1P	³⁄₁₀N to 1³⁄₁₀P	½P	¹⁄₁₆ to ¼	7⅞
	Malibu, Monte Carlo, Pow. Steer.	2P to 4P	3P	³⁄₁₀N to 1³⁄₁₀P	½P	¹⁄₁₆ to ¼	7⅞
'80–'87	Impala, Caprice	2P to 4P	3P	0 to 1⅗P	⅘P	¹⁄₁₆ to ¼	—

N Negative   P Positive

ENGINE ELECTRICAL

The voltage regulator is a solid-state, non-adjustable unit integral with the alternator. The alternator must be disassembled to remove the regulator.

For further information on the charging system, please refer to the "Charging and Starting" Unit Repair section.

Alternator

REMOVAL & INSTALLATION

1. Disconnect negative battery cable to prevent diode damage.
2. Disconnect and label the alternator wiring.
3. Remove the brace bolt. If equipped with power steering, loosen pump brace and mount nuts, then detach the drive belt(s).
4. Support the alternator and remove the mounting bolt(s). Remove the unit from the vehicle.
5. To install, reverse the removal procedures. Adjust the drive belt to have ¼–¹/₂ in. play on longest run of belt.

Integral Voltage Regulator

An alternator with an integral voltage regulator is standard equipment. There are no adjustments possible with this unit.

For the testing procedures, please refer to the "Charging and Starting" Unit Repair section.

Starter

For further information on the starting system, please refer to the "Charging and Starting" Unit Repair section.

REMOVAL & INSTALLATION

1. Disconnect negative battery cable.
2. Raise and support vehicle on jackstands.
3. Disconnect all wires at solenoid terminals.

NOTE: When removing the electrical connectors, make a note of the wiring color codes for reinstallation purposes.

4. Remove the starter support bracket mount bolts. On engines with a solenoid heat shield, remove the front bracket upper bolt and the bracket from the starter motor.
5. Loosen the front bracket bolt or nut, then rotate the bracket clear. Lower and remove the starter.

NOTE: If shims are present,

note their location and remove them, so that they may be replaced in the same positions upon installation.

6. To install, reverse the removal procedures.

Distributor

A solid-state, High Energy Ignition (HEI) system is standard equipment on all models. Most 1981 and later models use an Electronic Spark Timing (EST) distributor. The EST distributor uses no mechanical or vacuum advance and is easily identified by the absence of a vacuum advance and the presence of a four-terminal connector.

REMOVAL

1. Disconnect the negative battery cable.

NOTE: On the V6 and V8 systems, disconnect the ignition switch battery feed wire from the distributor cap.

2. Tag and disconnect the electrical connectors from the distributor cap.
3. Disconnect the hose at the vacuum advance unit, if equipped.
4. Turn the 4 distributor cap-to-housing screws counterclockwise and lift off the cap assembly.

NOTE: It may be necessary to remove the secondary wires from the distributor cap.

5. Using crayon or chalk, make locating marks on the rotor, the distributor housing and the engine for installation purposes.
6. Remove the distributor clamp bolt and clamp, then lift the distributor (turning the rotor counterclockwise) from the engine.

INSTALLATION

Undisturbed Engine

1. Install a new O-ring on the distributor housing.
2. Install the distributor into the engine opening, aligning the marks made during the removal procedure.
3. Install the distributor clamp, the hold down bolt and the distributor cap.
4. Check the ignition timing.

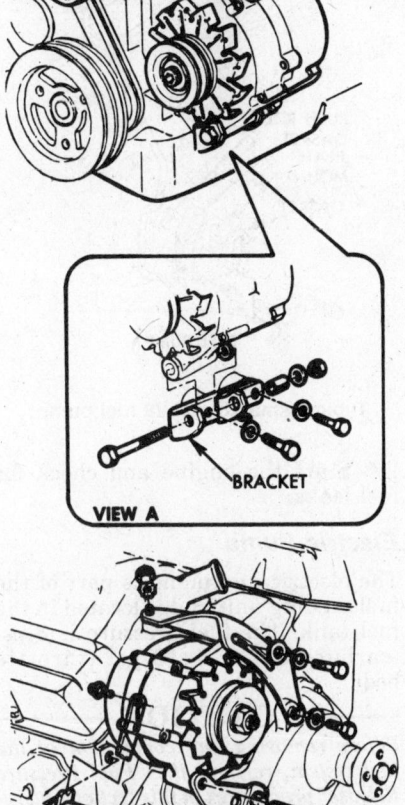

VIEW A

BRACKET

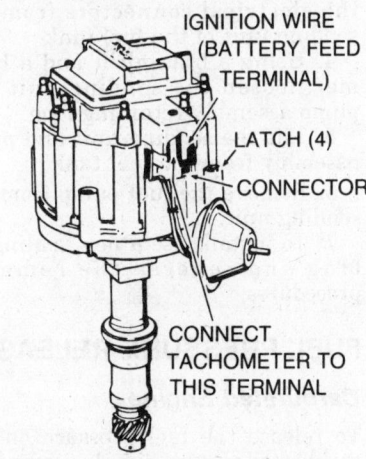

Typical alternator mounting on the Chevrolet V6 and V8 engines

IGNITION WIRE (BATTERY FEED TERMINAL)

LATCH (4)

CONNECTOR

CONNECT TACHOMETER TO THIS TERMINAL

HEI system tachometer hookup

Disturbed Engine

1. Remove the No. 1 spark plug.
2. Place a finger over the No. 1 spark plug hole and rotate the engine by hand until the compression can be felt.

3. Align the timing mark on the crankshaft pulley with the "0" mark on the timing plate.
4. Align the distributor rotor near the No. 1 spark plug tower.
5. Install the distributor, the hold down clamp, the bolt and the cap. It may be necessary to turn the rotor a little in either direction in order to engage the gears.

NOTE: With the distributor installed, make sure that the rotor is aligned with the No. 1 spark plug tower of the cap.

6. Check the ignition timing.

IGNITION TIMING ADJUSTMENT

NOTE: The underhood timing instructions on the Vehicle Emissions Control Information label may differ from these procedures; always follow the underhood label's directions.

1. If equipped, disconnect and plug the vacuum advance hose from the distributor. If equipped with electronic spark timing (EST), disconnect the 4-wire connector to the distributor.
2. Make sure the timing marks are clean and readable. The engine must be at normal operating temperature.

NOTE: It may be necessary to put a small amount of white paint or chalk on the timing marks to make them more visible.

3. Connect a timing light to No. 1 cylinder.
4. Loosen the distributor clamp.
5. Start the engine and run it at the rpm specified in the Tune-Up chart. Rotate the distributor to align the timing marks. Tighten the distributor clamp and recheck the timing.
6. Reconnect the vacuum hose or the 4-wire connector.

HEI SYSTEM TACHOMETER HOOKUP

There is a terminal marked TACH on the distributor cap. Connect one tachometer lead to this terminal and the other lead to a ground. On some tachometers, the leads must be connected to the TACH terminal and to the battery positive terminal.

CAUTION

Never ground the TACH terminal; serious module and ignition coil damage will result. If there is any doubt as to the correct tachometer hookup, check with the tachometer manufacturer.

Diesel Glow Plugs

For information on diesel glow plugs please refer to the Oldsmobile Rear Wheel Drive section. For glow plug testing procedures, please refer to the "Diesel Maintenance" Unit Repair section.

FUEL SYSTEM

The fuel pump is the single action AC diaphragm type. The pump is actuated by an eccentric located on the engine camshaft. On inline engines, the eccentric actuates the pump rocket arm. On V6 and V8 engines, a pushrod between the camshaft eccentric and the fuel pump actuates the pump rocker arm.

NOTE: Refer to the Oldsmobile Rear Wheel Drive section for all diesel engine fuel system service procedures

Fuel Pump

REMOVAL & INSTALLATION

Mechanical Pump

The mechanical fuel pumps are diaphragm operated and are located on either side of the engine. The fuel pressure should be 4–6.5 psi.

NOTE: Some vehicles have a special fuel pump, which has a metering outlet, for vapor return; this system reduces the possibility of vapor lock. Before working on the fuel system, release the fuel pressure. Remove and replace the fuel tank cap.

1. Disconnect the fuel inlet, outlet and return (if equipped) hoses from the fuel pump.
2. Remove the retaining bolts.
3. Remove the fuel pump, pushrod, gasket and mounting plate (if equipped).

NOTE: On the Chevrolet engines (V6 and V8), if the pushrod is not to be removed, remove the upper bolt from the right front mounting boss. Insert a longer bolt ($3/8$ x 16 x 2 in.) in this hole to hold the fuel pump pushrod.

4. Clean the gasket mounting surfaces.
5. To install, use new gasket(s) and reverse the removal procedures. Torque the mounting plate bolts to 3 ft. lbs. and the fuel pump bolts to 27 ft.

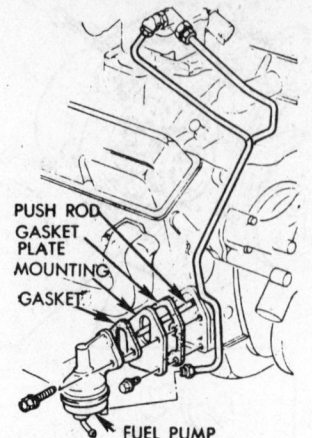

Typical small block V8 fuel pump

lbs. Start the engine and check for fuel leaks.

Electric Pump

The electric fuel pump is part of the fuel sending unit and is located in the fuel tank. The fuel pressure is 3 psi (carbureted) or 4–13 psi (throttle body).

--- CAUTION ---
Before removing any component of the fuel system, refer to the "Fuel Pressure Release" procedures in this section and release the fuel pressure.

1. Disconnect the negative battery cable.
2. Remove the fuel tank.
3. Disconnect the fuel tubes and the electrical connectors from the sending unit of the fuel tank.
4. Using a brass drift and a hammer, loosen the sending unit and pump assembly retaining ring.
5. Lift the sending unit and pump assembly from the fuel tank.
6. Remove the fuel pump from the sending unit.
7. To install, use a new sealing O-ring and reverse the removal procedures.

FUEL PRESSURE RELEASE

Carbureted Engine

To release the fuel pressure on the carbureted system, simply remove the fuel tank cap. All tank pressure will vent out the fuel filler pipe.

Throttle Body Fuel Injection (TBI)

To release the fuel pressure on the TBI system, remove the fuel pump fuse from the fuse panel, then start and operate the engine until it stalls.

Crank the engine for three seconds to make sure all fuel pressure is removed from the fuel lines, then replace the fuel pump fuse.

Fuel Filter

There are three types of fuel filters; internal (in the carburetor fitting), inline (in the fuel line) and in-tank (the sock on the fuel pick-up tube).

Before removing any component of the fuel system, refer to the "Fuel Pressure Release" procedures in this section and release the fuel pressure.

REMOVAL & INSTALLATION

Internal Filter

1. Disconnect the fuel line connection at the fuel inlet filter nut on the carburetor.
2. Remove the fuel inlet filter nut from the carburetor.
3. Remove the filter and the spring.

NOTE: If a check valve is not present with the filter, one must be installed when the filter is replaced.

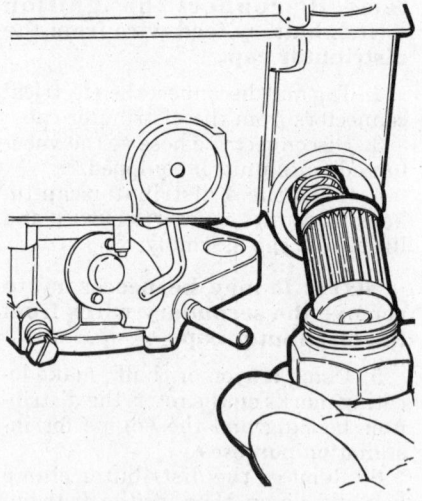

Fuel filter—typical

4. Install the spring, filter and check valve (must face the fuel line), then reverse the removal procedures.
5. Start the engine and check for leaks.

Inline Filter

1. Disconnect the fuel lines.
2. Remove the fuel filter from the retainer or mounting bolt.
3. To install, reverse the removal procedures. Start the engine and check for leaks.

NOTE: The filter has an arrow (fuel flow direction) on the side of the case, be sure to install it correctly in the system, with the arrow facing away from the fuel tank.

In-Tank Filter

To service the in-tank fuel filter, refer to the "Electric Fuel Pump Removal & Installation" procedures in this section.

Fuel Tank

-------- CAUTION --------

Before removing any component of the fuel system, refer to the "Fuel Pressure Release" procedures in this section and release the fuel pressure.

REMOVAL & INSTALLATION

1. Release the pressure from the fuel system.
2. Raise and support the vehicle on jackstands.
3. Remove the fuel tank cap and drain the fuel.
4. Disconnect the fuel pump and the sending unit electrical connector at the body electrical harness connector.

-------- CAUTION --------

DO NOT pry up on the cover connector. The pump and the sending unit wiring harness are an integral part of the sending unit.

5. Disconnect the flexible fuel lines from the metal fuel pipes at the tank.
6. Remove the fuel tank strap bolts and lower the fuel tank from the vehicle.
7. To install, reverse the removal procedures.

Carburetors

For all carburetor adjustments and specifications, please refer to the "Carburetor" Unit Repair section. For further information on feedback carburetors, please refer to *Chilton's Guide To Fuel Injection And Feedback Carburetors.*

-------- CAUTION --------

Fuel system is under pressure. Before removing any component of the fuel system, refer to the "Fuel Pressure Release" procedures in this section and relieve the fuel pressure.

REMOVAL & INSTALLATION

1. Disconnect the negative battery cable.

2. Remove the air cleaner and gasket.
3. Disconnect the fuel and vacuum lines.
4. Disconnect the choke electrical connector.
5. Disconnect the throttle linkage.
6. If equipped with an automatic transmission, disconnect the throttle valve linkage.
7. Remove the carburetor attaching nuts or bolts, gasket and the carburetor.
8. If equipped, remove the electric EFE heater and the insulator.
9. Clean the gasket mounting surfaces.
10. To install, use a new gasket and reverse the removal procedures. Torque the short bolts to 7 ft. lbs. and the long bolts or nuts to 12 ft. lbs. Check the idle and the fast idle speeds.

IDLE SPEED ADJUSTMENT

1980 Models

NOTE: An idle speed control system is used on some engines to control the idle speed. No adjustments are necessary with this system.

V6-229, V6-231, V8-267 AND V8-305

1. Run the engine to normal operating temperature.
2. Make sure that the choke is fully opened, turn the A/C off, set the parking brake, block the drive wheels and connect a tachometer to the engine according to the manufacturer's instructions.
3. Disconnect and plug the vacuum hoses at the EGR valve and the vapor canister.
4. Place the transmission in Park (A/T) or Neutral (M/T).
5. Disconnect and plug the vacuum advance hose at the distributor. Check and adjust the timing.
6. Connect the distributor vacuum line.
7. If the vehicle is equipped with a M/T and NOT equipped with A/C or a solenoid, place the idle speed screw on the low step of the fast idle cam and turn the screw to achieve the specified idle speed.

NOTE: If equipped with A/C, turn the idle speed screw to the specified rpm, disconnect the compressor clutch wire, turn the A/C On. Open the throttle momentarily to extend the solenoid plunger, then adjust the solenoid screw to obtain the specified rpm. If NOT equipped with A/C, momentarily open the throttle to extend the so-

lenoid plunger, turn the solenoid screw to obtain the specified rpm, disconnect the solenoid wire and turn the idle speed screw to obtain the slow engine idle speed.

1981 And Later Rochester E2ME Or E4ME Models

NOTE: No idle speed adjustment is necessary; the idle speed is controlled by the ECM.

IDLE MIXTURE ADJUSTMENT

1980 Rochester M2ME/M2MC Models

1. Set the parking brake and block the drive wheels.

NOTE: If the vehicle is equipped with a vacuum parking brake release system, disconnect and plug the vacuum hose at the brake.

2. Refer to the "Emission Control Label" in the engine compartment, then disconnect and plug the necessary hoses.
3. Connect a tachometer to the engine, disconnect and plug the vacuum advance hose at the distributor, then set the timing.
4. Operate the engine to normal operating temperatures; make sure that the choke is open and the A/C is turned OFF.

NOTE: If the vehicle is equipped with Electronic Spark Timing (EST), check and/or adjust the timing and the idle speed according to the "Emission Control Label."

5. Disconnect the crankcase ventilation tube from the air cleaner.
6. Using tool J-26911, connect a hose from a propane tank to the crankcase tube opening at the air cleaner.

-------- CAUTION --------

When using a propane tank, make sure it is secured in a safe place in the vertical position.

7. Place the transmission in Drive (A/T) or Neutral (M/T).
8. Slowly OPEN the propane valve until the maximum engine speed is reached.

NOTE: The addition of too much propane will cause the engine speed to drop.

9. If the idle mixture speed does not meet specifications, remove the idle mixture screw plug covers.
10. Turn the idle mixture screws

clockwise until lightly seated, then back them OUT (equally) until the lean best idle point at the enriched idle speed is reached.

11. Turn the propane tank OFF and turn the idle mixture screws clockwise (equally) until the curb idle speed is reached.

12. Turn ON the propane tank and recheck the engine speed, if not within specifications, repeat the propane setting.

13. To complete the adjustment, turn OFF the engine, remove the propane tank, reconnect the crankcase ventilation tube to the air cleaner and install new idle mixture screw plugs.

1980 And Later Rochester E2ME And E4ME

1. Remove the carburetor from the engine and invert it. Using a hacksaw, make 2 parallel cuts ($^1/_8$ in. deep) in the throttle body on either side of the idle mixture screws locator points; the cuts should reach to the steel plug.

2. Using a flat center punch, hold it at a 45° angle to the cut segiments, drive the segiments into the throttle body. Using a center punch, drive the steel plugs from the housing.

3. Using tool J-29030, turn the idle mixture screws inward until they seat, then back them out 2 turns.

4. Reinstall the carburetor to the engine but NOT the gasket and air cleaner.

5. Set the parking brake, block the drive wheels, start the engine and run it to normal operating temperatures.

6. Disconnect and plug the vacuum hoses according to the Emission Control Label in the engine compartment.

7. Connect a dwell meter, a tachometer and a timing light to the engine, then check and/or adjust the timing.

8. Place the transmission in Drive (A/T) or in Neutral (M/T), then check and/or adjust the idle speed.

NOTE: If equipped with an Idle Speed Control (ISC) or an Idle Load Compensator (ILC), DO NOT adjust the curb idle speed.

9. The dwell should be within the 10°–50° range, if NOT, perform the following procedures:

   a. Turn the engine OFF. Cover the internal bowl vents and the Air Bleed Valve inlets with masking tape, then the primary and the secondary carburetor air intakes with a shop cloth.

   b. Using a No. 35 (0.110 in.) drill bit, drill the rivet heads from the Idle Air Bleed Valve cover. Using a small drift punch, drive the remaining rivets from the air horn tower.

   c. Remove and discard the cover. Using compressed air, blow the metal chips from the air horn.

   d. Remove the cloth and the masking tape from the carburetor, start the engine and allow it to idle with the A/T in Drive or the M/T in Neutral. Slowly turn the Idle Air Bleed Valve until the dwell reading varies between 25°–35°; adjust it as close to 30° as possible.

NOTE: The Idle Air Bleed Valve is very sensitive and should ONLY be turned in $^1/_8$ in. increments. If the dwell remains below 25°, turn the idle mixture screws OUT one full turn. If the dwell is above 35°, turn the idle mixture screws IN one full turn. Readjust the Idle Air Bleed Valve to obtain the dwell limits.

10. After adjusting the idle mixture, fill the idle mixture screw holes with silicone sealant.

11. Check and/or adjust the Fast Idle Speed according the Emissions Control Label.

12. Remove the test equipment, unplug and reconnect the vacuum hoses, then install the air cleaner and gasket.

FAST IDLE ADJUSTMENT

1980 Rochester M2ME, 1980 And Later Rochester E2ME

1. Refer to the emission label and prepare the vehicle for adjustment.

2. Place the transmission in Park (A/T) or Neutral (M/T).

3. Place the fast idle screw on the highest step of the fast idle cam.

4. Turn the fast idle screw to obtain the fast idle speed.

1982 And Later Rochester E4ME Models

NOTE: The fast idle speed adjustment must be performed according to the emission control label instructions. See the underhood sticker.

1982 AND LATER IDLE MIXTURE ADJUSTMENT (ALL MODELS)

All models have sealed idle mixture screws; in most cases these are concealed under staked-in plugs. Idle mixture is adjustable only during carburetor overhaul and requires the addition of propane as an artificial enrichener, along with the use of an emissions tester or infrared analyzer.

Electronic Fuel Injection

Before removing any component of the fuel injection system, refer to the "Fuel Pressure Release" procedures in this section and release the fuel pressure. For all fuel injection information, removal and service procedures, please refer to *Chilton's Guide To Fuel Injection And Feedback Carburetors.*

IDLE SPEED ADJUSTMENT

NOTE: The throttle body injection (TBI) unit is set at the factory and no further adjustment is necessary. Only if the TBI unit has been replaced should an idle speed adjustment be performed.

1985 And Later

1. Run the engine until it reaches normal operating temperatures.

2. Remove the air cleaner and the gasket.

3. Using an awl, puncture idle stop screw cover plug and pry the plug from the TBI.

4. Connect a jumper lead from the Idle Air Control (IAC) motor diagnostic lead to ground.

5. Connect a tachometer to the engine.

6. Turn the ignition ON (DO NOT start the engine) and wait 30 seconds.

7. Disconnect the IAC electrical connector from the TBI unit and remove the diagnostic lead jumper wire.

8. Start the engine, engage the parking brake, place the Automatic Transmission in Drive and allow the engine speed to stabilize.

9. Adjust the idle speed screw to 475–625 rpm.

10. Stop the engine and reconnect the IAC motor electrical connector.

11. Using a voltmeter, adjust the Throttle Position Sensor (TPS) to 0.450–0.600 volts.

12. Recheck the setting.

13. Start the engine and check for proper idle operation.

14. To install, seal the throttle stop screw with silicone sealant and reverse the removal procedures.

ENGINE COOLING

Radiator

REMOVAL & INSTALLATION

1. Disconnect the negative battery

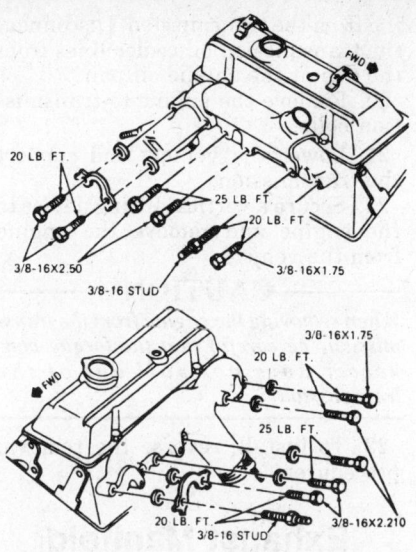

Exhaust manifold installation for the Chevrolet 200 and 229 V6 engine

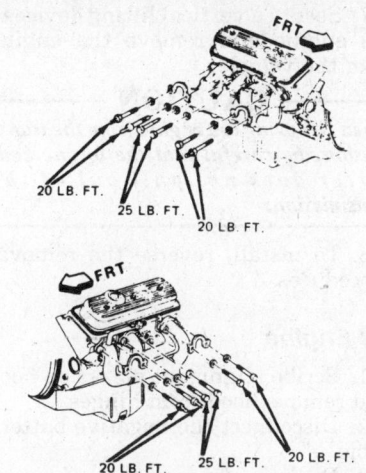

Exhaust manifold installation for the Chevrolet 262 V6 engine

cable and drain cooling system.

2. If necessary, remove the fan, the upper fan shroud or the upper support.

3. Disconnect upper and lower hoses.

4. Disconnect and plug the oil cooler lines, if equipped with an A/T.

5. Lift radiator and shroud straight up and out of vehicle.

NOTE: If equipped with a clutch type fan, keep it in an upright position to prevent the fluid from leaking.

6. To install, reverse the removal procedures. Make sure the lower cradles are properly located and A/T is full.

Water Pump

The water pump is a die cast, centrifugal-type with sealed bearings. Since it is pressed together, it must be serviced as a unit.

REMOVAL & INSTALLATION

All Engines Except Diesel

1. Disconnect the negative battery cable and drain the cooling system.

2. If necessary, remove the fan shroud or the upper radiator support.

3. Remove the necessary drive belts.

4. Remove the fan and water pump pulley.

5. Remove the alternator and the power steering pump (if equipped) brackets, then move the units aside.

6. Remove the heater hose and the lower radiator hose from the pump.

7. Remove the water pump retaining bolts and the pump. Clean the gasket mounting surfaces.

NOTE: Use an anti-seize compound on the water pump bolt threads.

8. To install, use new gaskets and reverse the removal procedures. Torque the water pump, the alternator and the power steering (if equipped) mounting bolts to 30 ft. lbs. Adjust the drive belts and fill the cooling system.

NOTE: If a belt tensioning gauge is available, adjust the belts to 100–130 lbs. of tension on new belts or to 70 lbs. on used belts. If the gauge is not available, adjust the belts so that a $1/4$–$1/2$ inch deflection can be made on the longest span of the belt under moderate thumb pressure.

Diesel Engine Only

1. Disconnect the negative battery cable and drain the cooling system.

2. Disconnect the lower radiator hose, the heater hose and the by-pass hose from the water pump.

3. Remove the fan assembly, the drive belts and the water pump pulley.

4. Remove the alternator, the power steering pump and the A/C compressor (if equipped) brackets, then move the units aside.

5. Remove the water pump mounting bolts and the pump. Clean the gasket mounting surfaces.

6. To install, use new gaskets, sealant and reverse the removal procedures. Torque the water pump bolts to 22 ft. lbs. Adjust the belts and refill the cooling system.

NOTE: Apply sealer to the lower water pump bolts.

Thermostat

REMOVAL & INSTALLATION

1. Disconnect the negative battery cable and drain cooling system to below thermostat level.

2. Remove the air cleaner. Disconnect upper radiator hose.

3. Remove the thermostat housing bolts, the housing and the thermostat.

4. Clean the gasket mounting surfaces.

NOTE: When installing the thermostat, place the pin side facing upwards.

5. To install, use new gaskets, sealant and reverse the removal procedures. Torque the thermostat housing mounting bolts to 30 ft. lbs. Refill the cooling system.

EMISSION CONTROLS

Due to the complex nature of modern electronic engine control systems, comprehensive diagnosis and testing procedures fall outside the confines of this repair manual. For complete information on diagnosis, testing and repair procedures concerning all modern engine and emissions control systems, please refer to *Chilton's Guide To Electronic Engine Controls.*

ENGINE MECHANICAL

The V8 engines include the 267, 305, 307 and 350 cu. in. engines. All of the V8's are Chevrolet built except the Oldsmobile built 307. For all service procedures for the 307, see the Oldsmobile Rear Wheel Drive section. The V6–200, 229 and 262 are cut down small block Chevrolet V8's which share common parts with the V8.

The V6 231 engine is built by Buick and is not covered in this section. For all service procedures for the 231, see the Buick Rear Wheel Drive section. The 350 V8 and 4.3L

V6 diesel engines are built by Oldsmobile. Service procedures for these engines are given in the Oldsmobile Rear Wheel Drive section.

Engine

REMOVAL & INSTALLATION

———— CAUTION ————

Do not discharge the compressor or disconnect the A/C lines. Damage to the A/C system or personal injury could result.

V6 Engine

1. Scribe alignment marks at the hood hinges and remove the hood.
2. Disconnect the negative battery cable.
3. Disconnect the exhaust pipe from the exhaust manifold.
4. Remove the bell housing cover and drain the transmission oil cooler lines at the oil pan.
5. Remove the left engine mount through bolt and loosen the right engine mount through bolt.
6. If equipped with an A/T, remove the torque converter cover, the converter-to-flex plate bolts and the engine-to-transmission bolts.

NOTE: Before removing the torque converter bolts, scribe a mark to ensure the relationship between the torque converter and the flex plate.

7. Disconnect the CCC wiring harness from the transmission and the knock sensor from the engine.
8. Disconnect the fuel hoses from the frame and the lower fan shroud.
9. Lower the vehicle and remove the windshield washer bottle.
10. Disconnect and label the CCC wiring harness, other necessary wiring connectors and the vacuum hoses from the engine.
11. Remove the air cleaner, the upper fan shroud, the accelerator and the T.V. cables.
12. Drain the cooling system, then remove the heater and the radiator hoses.
13. If equipped with A/C and power steering, remove the compressor and the power steering pump, then move them aside.
14. Disconnect the transmission oil cooler lines and the overflow tube from the radiator, then remove the radiator.
15. Remove the A/C hose and the adjusting bracket from the alternator.
16. Disconnect the battery cables from the frame and the heater hose from the bracket.

17. Secure a vertical lifting device to the engine and remove the engine from the vehicle.

———— CAUTION ————

When removing the engine from the transmission, be careful that the torque converter does not pull out of the transmission.

18. To install, reverse the removal procedures.

V8 Engine

1. Scribe alignment marks on hood and remove hood from hinges.
2. Disconnect the negative battery cable.
3. Drain cooling system, then remove the heater hoses and the radiator hoses from the engine.
4. Remove the upper fan shroud and the fan assembly.
5. If equipped with A/C and power steering, remove the compressor and the power steering pump, then move them aside.
6. Disconnect the accelerator and the T.V. cables.
7. Remove the transmission oil cooler lines (if equipped) from the radiator and the radiator.
8. Disconnect and label the vacuum hoses and the CCC wiring harness connector(s) from the engine.
9. Remove the AIR pipe from the converter.
10. Remove the windshield washer bottle.
11. Disconnect and mark the wiring harness at the bulkhead and related engine wiring.
12. Remove the distributor cap and the cruise control cable (if equipped).
13. Disconnect the positive battery cable from the battery and the frame. Disconnect the negative battery cable from the A/C hose/alternator bracket.
14. Raise and support the vehicle on jackstands.
15. Remove the crossover pipe and the catalytic converter as an assembly.
16. If equipped with an A/T, remove the torque converter cover and the torque converter bolts.

NOTE: Before removing the torque converter bolts, scribe a mark to ensure the relationship between the torque converter and the flex plate.

17. Remove the engine-to-mount bolts.
18. Disconnect the fuel line from the fuel pump. Be sure to relieve fuel system pressure as previously described before disconnecting any fuel lines.
19. If equipped with an A/T, disconnect the torque converter clutch wir-

ing from the transmission. Disconnect the transmission oil cooler lines from the clip at the engine oil pan.
20. Remove the engine-to-transmission bolts.
21. Lower the vehicle and support the transmission.
22. Secure a vertical lifting device to the engine and remove the engine from the vehicle.

———— CAUTION ————

When removing the engine from the transmission, be careful that the torque converter does not pull out of the transmission.

23. To install, reverse the removal procedures.

Exhaust Manifold

REMOVAL & INSTALLATION

V6 Engine

RIGHT SIDE

1. Disconnect the negative battery cable.
2. Raise and support the vehicle on jackstands.
3. Remove the exhaust pipe from the exhaust manifold.
4. Lower the vehicle.
5. Disconnect the air management valve bracket, the AIR hoses and the AIR pipe at the converter, the cylinder heads and the exhaust manifold.
6. Disconnect the spark plug wires.
7. Remove the exhaust manifold bolts and the manifold.
8. Using a putty knife, clean the gasket mounting surfaces.
9. To install, use a new gasket and reverse the removal procedures. Torque the exhaust manifold mounting bolts to 20 ft. lbs. (outside) and 25 ft. lbs. (center).

LEFT SIDE

1. Disconnect the negative battery cable.
2. Raise and support the vehicle on jackstands.
3. Remove the exhaust pipe from the exhaust manifold.
4. If equipped with A/C, remove the compressor and the rear adjusting bracket, then move it aside.
5. If equipped with power steering, remove the power steering pump and lower the rear adjusting bracket, then move the pump aside.
6. Disconnect the spark plug wires from the plugs.
7. Remove the exhaust manifold bolts and the manifold.
8. Using a putty knife, clean the gasket mounting surfaces.
9. To install, use a new gasket and

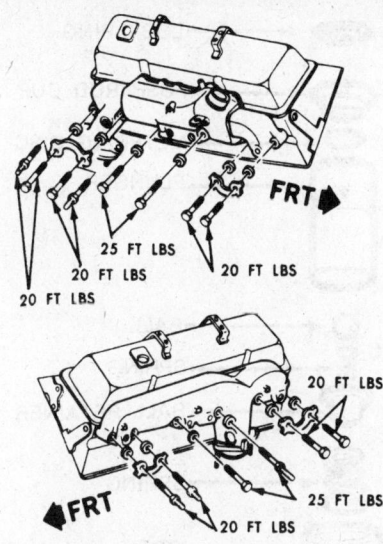

Exhaust manifold installation for the Chevrolet built V8 engines

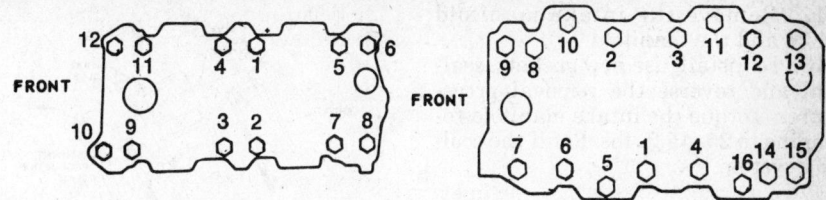

Intake manifold torque sequence for the Chevrolet V8 engines

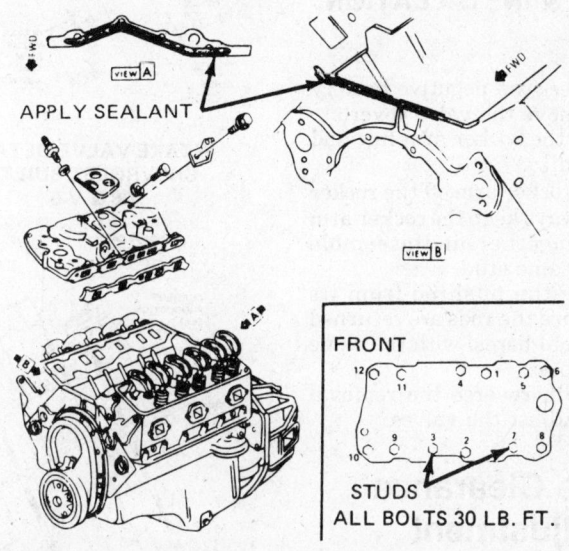

Intake manifold seal installation and torquing sequence for the Chevrolet—200 and 229 V6 engine

Intake manifold installation and torque sequence for the Chevrolet 262 V6 engine

reverse the removal procedures. Torque the exhaust manifold mounting bolts to 20 ft. lbs. (outside) and 25 ft. lbs. (center).

V8 Engine

RIGHT SIDE

1. Disconnect the negative battery cable.
2. Raise and support the vehicle on jackstands.
3. Disconnect the exhaust pipe from the exhaust manifold.
4. Lower the vehicle.
5. Remove the air cleaner.
6. Disconnect the spark plug wires on the right side, the EFE vacuum hose and the AIR hose.
7. Loosen the alternator bracket adjustment, then remove the drive belt and the lower alternator bracket.
8. Remove the AIR valve and disconnect the converter AIR pipe from the back of the manifold.
9. Remove the exhaust manifold bolts and the manifold.
10. To install, use a new gasket and reverse the removal procedures. Torque the exhaust manifold-to-engine bolts to 20 ft. lbs. (outside) and 25 ft. lbs. (center).

LEFT SIDE

1. Disconnect the negative battery cable.
2. Raise and support the vehicle on jackstands.
3. Remove the exhaust pipe from the exhaust manifold and the electrical connector from the oxygen sensor.
4. Lower the vehicle.
5. Disconnect the AIR hose.
6. If equipped, remove the power

steering pump, the bracket and move the pump aside.
7. If equipped with A/C, loosen the bracket at the front of the cylinder head, then remove the compressor and the rear bracket.
8. Remove the exhaust manifold bolts, the wire loom holder from the valve cover and the manifold. Using a putty knife, clean the gasket mounting surfaces.
9. To install, use a new gasket and reverse the removal procedures.

Intake Manifold

REMOVAL & INSTALLATION

All Engines

1. Disconnect the negative battery cable.
2. Remove the air cleaner and drain the cooling system.
3. Disconnect the CCC harness and move aside.
4. Remove the heater and the radiator hoses from the engine.
5. Remove the upper alternator bracket, the electrical wiring connectors and the vacuum hoses.
6. On the V8, disconnect the fuel line and the brake pipes from the carburetor. On the V6, disconnect the

clips and the fuel lines from the TBI unit.
7. Disconnect the accelerator and the T.V. cables from the engine.
8. On the V8, remove the spark plug wires and the exhaust manifold from the right cylinder head.
9. Remove the distributor cap, then mark the position of the rotor with the distributor and the engine. Remove the distributor.
10. On the V6, remove the coil. On the V8, remove the carburetor.
11. If equipped with A/C, remove the compressor and the bracket, then move aside.

12. Remove the intake manifold bolts and the manifold.

13. To install, use new gaskets, sealant and reverse the removal procedures. Torque the intake manifold-to-engine to 25–45 ft. lbs. Refill the cooling system.

Rocker Arm

REMOVAL & INSTALLATION

All Engines

1. Disconnect the negative battery cable and remove the valve cover(s).

2. Remove the rocker arm nut and rocker arm ball.

3. Lift the rocker arm off the rocker arm stud. Always keep the rocker arm assemblies together and assemble them on the same stud.

4. Remove the pushrod from its bore. Make sure the rods are returned to their original bores, with the same end in the block.

5. To install, reverse the removal procedures. Adjust the valves.

Valve Clearance Adjustment

All Engines

1. Disconnect the negative battery cable and remove the valve covers.

2. Turn the crankshaft until the mark on the damper pulley aligns with the 0° mark on the timing plate at the front of the engine and the No. 1 cylinder is at the TDC of the compression stroke.

NOTE: If the valves of the No. 1 cylinder DO NOT move when checked, the engine is at the TDC of the compression stroke. If the valves DO move when checked, the engine is at the TDC of the No. 6 cyl. (V8) or No. 4 cyl. (V6); turn the crankshaft one complete revolution.

3. With the engine at the No. 1 firing position, adjust the exhaust valves of No. 1, 3, 4 & 8 on the V8 or No. 1, 5 & 6 on the V6 and the intake valves of No. 1, 2, 5 & 7 on the V8 or No. 1, 2 & 3 on the V6.

NOTE: To adjust the valves, back out the adjusting nut until lash is felt at the pushrod, then turn the nut in until the lash disappears. Once the play has be removed, turn the adjusting IN one full turn (to center the lifter plunger).

4. After adjusting the indicated valves, turn the crankshaft one full

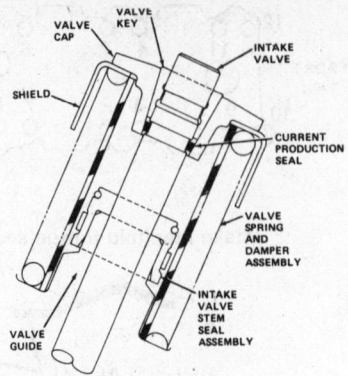

**INTAKE VALVE DETAIL
CHEVROLET BUILT
V-8 & V-6**

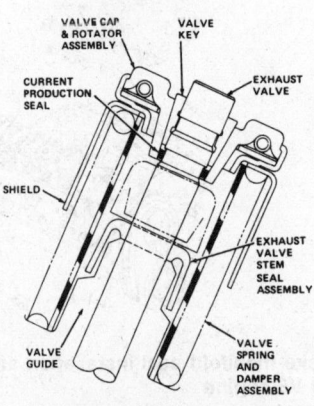

**EXHAUST VALVE DETAIL
CHEVROLET BUILT
V-8 & V-6**

Valve seal and retainer details for
Chevrolet V6 and V8 engines

rotation and align the mark on the damper pulley with the 0° mark on the timing plate.

5. With the engine at the No. 6 (V8) or No. 4 (V6) firing position, adjust the exhaust valves of No. 2, 5, 6 & 7 on the V8 or No. 2, 3 & 4 on the V6 and the intake valves of No. 3, 4, 6 & 8 on the V8 or No. 4, 5 & 6 on the V6.

6. To install, reverse the removal procedures. Start the engine, then check and/or adjust the idle speed.

Cylinder Head

REMOVAL & INSTALLATION

NOTE: The engine should be "overnight" cold before the cylinder head is removed to prevent warpage.

—————— **CAUTION** ——————
DO NOT discharge the compressor or disconnect the A/C lines. Personal injury could result.

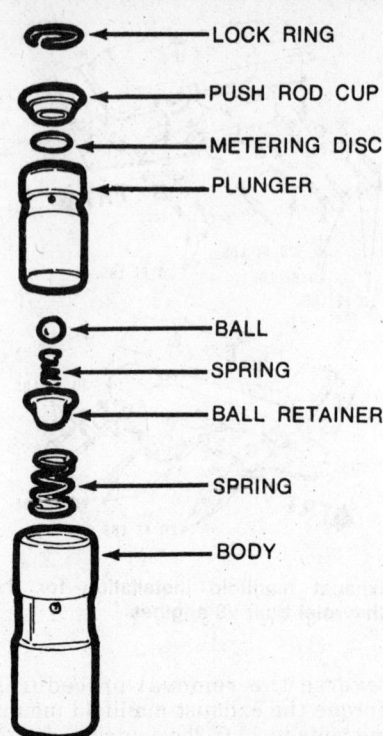

Hydraulic lifter plunger and body are fitted pairs and must not be mismated

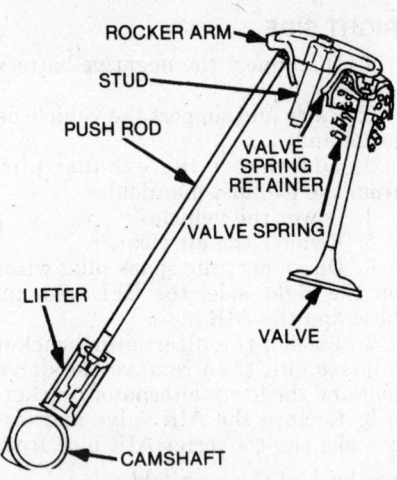

V8 engine valve system

All Engines

1. Refer to the "Intake Manifold Removal & Installation" procedures in this section and remove the intake manifold.

2. Remove the alternator's lower mounting bolt and move the unit aside.

3. Remove the exhaust manifold(s), the rocker arm cover(s) and the rocker arm assemblies.

4. Drain the cooling system.

5. Remove the diverter valve, the cylinder head bolts and the cylinder head(s).

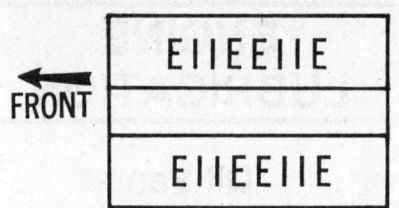

Chevrolet intake(I) and exhaust(E) valve arrangements (except the V6 engine)

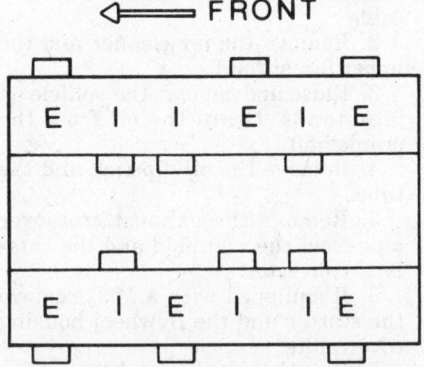

Valve arrangement of the Chevrolet-built V6 engines (E-exhaust; I-intake)

6. Using a putty knife, clean the gasket mounting surfaces.

7. To install, use new gaskets and reverse the removal procedures. Torque the cylinder head bolts in sequence to specifications. Check and/or adjust the valve clearances, the timing and the idle speed.

Timing Chain Cover

REMOVAL & INSTALLATION

All Engines

1. Disconnect the negative battery cable and drain the cooling system.

2. Remove the fan assembly, the drive belts and the fan pulley.

3. Raise and support the vehicle on jackstands.

4. Remove the crankshaft pulley and the damper pulley bolt.

5. Using tool J-23523, remove the damper pulley.

6. Remove the alternator and the brackets. If equipped with power steering, remove the lower pump bracket and swing aside.

7. Remove the heater and the lower radiator hoses from the water pump.

8. Remove the water pump bolts and the pump from the engine.

9. Remove the timing cover bolts and the timing cover.

10. Using a putty knife, clean the gasket mounting surfaces.

11. To install, use new gaskets, sealant and reverse the removal proce-

dures. Torque the timing cover bolts to 8 ft. lbs., the damper bolts to 65–75 ft. lbs. and the water pump bolts to 25–35 ft. lbs.

Timing Chain Cover Oil Seal

REMOVAL & INSTALLATION

All Engines
COVER REMOVED

1. Refer to the "Timing Cover Removal & Installation" procedures in this section and remove the timing cover.

2. Using a small pry bar, pry the oil seal from the timing cover.

3. Using tool J-23042, drive the new oil seal into the timing cover.

NOTE: When installing the new oil seal, be sure to support the rear side of the timing cover.

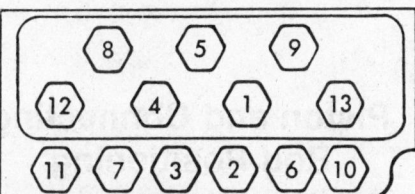

Cylinder head torque sequence-200, 229, and 262 V6 engines

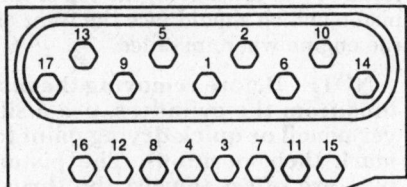

Cylinder head torque sequence for the Chevrolet V8 engines

4. To complete the installation, reverse the removal procedures.

COVER INSTALLED

1. Refer to the "Timing Cover Removal & Installation" procedures in this section and remove the balancer from the crankshaft.

2. Using a small pry bar, pry the oil seal from the timing cover.

3. Place the new seal (open end toward the engine) on the timing cover and drive it into the cover using tool J-23042.

4. To complete the installation, reverse the removal procedures. Torque the balancer bolt to 65–75 ft. lbs.

Timing Chain And Sprocket

REMOVAL & INSTALLATION

All Engines

1. Refer the "Timing Cover Remov-

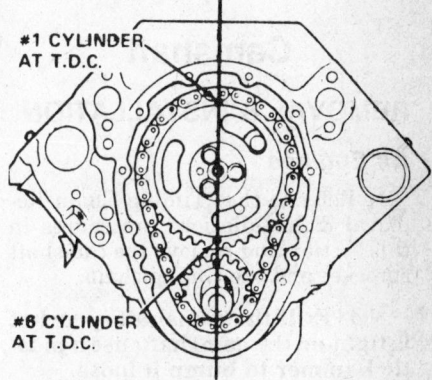

Timing mark alignment on 1980 and later Chevrolet V6 and V8 engines

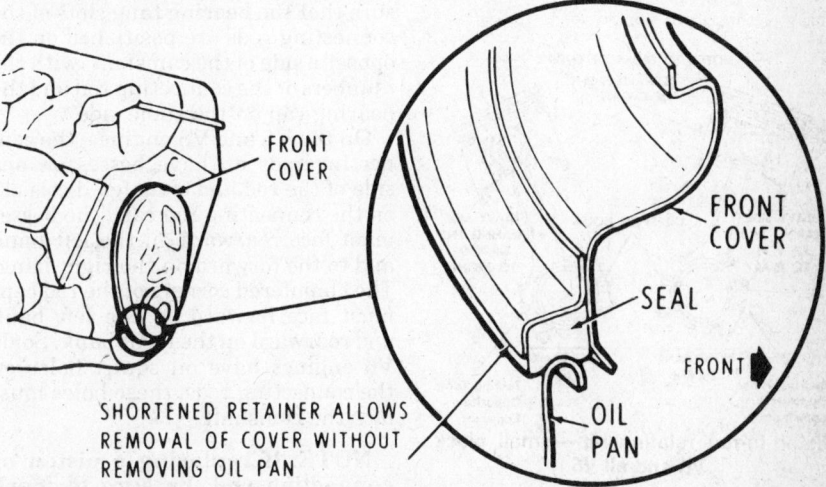

On Chevrolet V6 and V8 engines, it is not necessary to lower or remove the oil pan in order to remove the timing cover. The seal retainer is short enough to clear the pan

al & Installation" procedures in this section and remove the water pump and the timing cover.

2. Turn the crankshaft until the mark on the camshaft sprocket aligns with the mark on the crankshaft sprocket.

3. Remove the camshaft sprocket bolts, the camshaft sprocket, the timing chain and the crankshaft sprocket (if necessary).

NOTE: When installing the timing chain, install the sprockets with the timing marks facing each another; this position is TDC of the No. 6 cyl. (V8) or No. 4 cyl. (V6). To locate the TDC of the No. 1 cyl., turn the crankshaft one full revolution, the camshaft timing mark will now be at the top of the sprocket.

4. To install, use new gaskets, sealant and reverse the removal procedures. Torque the camshaft sprocket bolts to 13–23 ft. lbs. Check and/or adjust the engine timing.

Camshaft

REMOVAL & INSTALLATION

All Engines

1. Refer to the "Timing Chain Removal & Installation" procedures in this section and remove the camshaft sprocket and the timing chain.

NOTE: If the camshaft sprocket is tight on the camshaft, use a plastic hammer to bump it loose.

2. On the V8, remove the oil cooler lines and the hoses from the radiator, then the radiator.

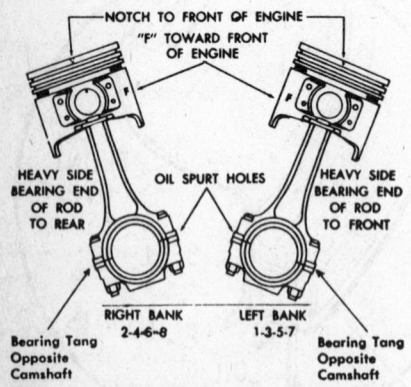

Piston-to-rod relationship—small block V8 and all V6

Piston-to-rod relationship for Chevrolet V8 engines

---

3. Remove the intake manifold and the rocker arm covers.

4. On the V8, remove the AIR pump bracket and disconnect the fuel lines at the fuel pump, then remove the fuel pump.

5. If equipped with A/C on the V8, remove the compressor and the condenser, then move them aside.

6. Remove the rocker arm assemblies, the push rods and the valve lifters.

7. Install two $5/16$ x 18 x 4 in. bolts in the camshaft and carefully pull the it from the front of the engine.

NOTE: When removing or replacing the camshaft, be careful not to damage the camshaft bearings.

8. To install, reverse the removal procedures. Torque the camshaft mounting bolts to 13–23 ft. lbs. Check and/or adjust the engine timing. Refill the cooling system.

Piston and Connecting Rod Positioning

The pistons have a machined hole, the letter "F" or a notch on the top of the piston, which should face the front of the engine when installed.

NOTE: Before removing the pistons from the cylinders, use a silver pencil or quick drying paint to mark their positions. The piston pins are offset toward the thrust side (right hand side).

On the inline 6 cyl. engine, make sure that the bearing tang slots of the connecting rods are positioned on the opposite side of the camshaft, with the numbers of the connecting rod and the bearing cap on the same side.

On the V6 and V8 engines, the connecting rods on have bosses on one side of the rod and chamfered corners on the connecting rod cap. The bosses must face rearward on the left bank and to the forward on the right bank. The chamfered corners of the rod caps must face forward on the left bank and rearward on the right bank. Some V6 engines have oil squirt holes on the connecting rods; these holes must face the camshaft.

NOTE: If replacing a piston or connecting rod, be sure to mark the position of the new part and replace the connecting rod bearing with a new one.

ENGINE LUBRICATION

Oil Pan

REMOVAL & INSTALLATION

V6 Engine

1. Disconnect the negative battery cable.

2. Remove the air cleaner and the upper fan shroud.

3. Raise and support the vehicle on jackstands. Drain the oil from the crankshaft.

4. Remove the oil dipstick and the tube.

5. Remove the exhaust crossover pipe from the manifold and the catalytic converter.

6. If equipped with a M/T, remove the starter and the flywheel housing cover plate.

7. If equipped with an A/T, remove the torque converter housing cover plate and disconnect the transmission oil cooler lines at the oil pan.

8. On the 1980 and later models, remove the left engine mount through bolt and loosen the right engine mount through bolt.

9. Using a floor jack, raise the engine and install the left engine mount through bolt.

NOTE: If, when removing the oil pan, the crankshaft throw or the counter balance weight blocks the pan removal, turn the crankshaft to put the throw in the horizontal plane.

10. Remove the oil pan bolts and lower the pan.

11. Using a putty knife, clean the gasket mounting surfaces.

12. To install, use a new gasket, sealant and reverse the removal procedures. Torque the oil pan bolts to 13 ft. lbs. (231 cu. in.) or 7 ft. lbs. (all others) and the engine mount bolts to 50 ft. lbs.

V8 Engine

NOTE: 1986 and later V8 models use a new one-piece molded silicone oil pan gasket to eliminate the joints.

Malibu And Monte Carlo

1. Disconnect the negative battery cable.

2. Remove the air cleaner, the upper radiator mounting panel and the fan shroud.

3. Raise and support the vehicle on

jackstands. Drain the oil from the engine.

4. Remove the distributor cap and the fan assembly.

5. Disconnect the AIR hose from the converter pipe and the AIR pipe from the exhaust manifold.

6. Remove the exhaust crossover pipe from the manifold and the catalytic converter.

7. If equipped with an A/T, remove torque converter housing cover plate and disconnect the transmission oil cooler lines at the oil pan. If equipped with a M/T, remove the starter and the flywheel housing cover plate.

8. Rotate crankshaft until timing mark on torsional damper is at 6 o'clock position, this positions the crankshaft throw in the horizontal place.

9. Remove front engine mount through bolts.

10. Raise engine and insert blocks under engine mounts.

NOTE: The block thickness should be 3 in. for Malibu and Monte Carlo.

11. Remove the oil pan bolts and lower the pan.

12. Using a putty knife, clean the gasket mounting surfaces.

13. To install, use a new gasket, sealant and reverse the removal procedures. Torque the oil pan bolts to 7 ft. lbs. and the engine mount bolts to 50 ft. lbs.

CAPRICE AND IMPALA

1. Disconnect the negative battery cable.

2. Remove the distributor cap and the upper fan shroud (if necessary).

3. Drain the cooling system. Remove the radiator hoses, then the oil dipstick and the tube (if necessary).

4. Remove the fan blade assembly. If equipped with A/C, remove the vacuum reservoir.

5. Raise and support the vehicle on jackstands. Drain the oil from the engine.

6. Remove the front engine mount through bolts and the starter.

NOTE: If equipped, disconnect the AIR hose from the converter pipe and the AIR pipe from the exhaust manifold.

7. If equipped with an A/T, remove the torque converter housing cover plate and the oil cooler lines from the oil pan (if necessary). If equipped with a M/T, remove the flywheel housing cover plate.

8. Disconnect the exhaust "Y" pipe from the exhaust manifolds.

9. Rotate the crankshaft until the timing mark on the damper is at the 6

o'clock position, this will position the crankcase throw in the horizontal plane.

10. Using wooden blocks (2 x 4 in.) and a floor jack, raise the engine enough to insert the blocks under the engine mounts, then lower the engine onto the blocks.

11. Remove the oil pan bolts and lower the pan.

12. Using a putty knife, clean the gasket mounting surfaces.

13. To install, use a new gasket, sealant and reverse the removal procedures. Torque the oil pan bolts to 7 ft. lbs. and the engine mount bolts to 50 ft. lbs.

Oil Pump

REMOVAL & INSTALLATION

All Engines

1. Refer to the "Oil Pan Removal & Installation" procedures in this section and remove the oil pan.

2. Remove the oil pump bolts from the rear main bearing cap.

3. Remove the oil pump and the extension shaft (if necessary).

4. Clean the gasket mounting surfaces.

5. To install, use new gaskets and reverse the removal procedures. Torque the oil pump bolts to 65 ft. lbs. (1980–81) or 7 ft. lbs. (1982 and later) and the oil pan bolts to 7 ft. lbs. Refill the crankcase.

Rear Main Oil Seal

REMOVAL & INSTALLATION

All Engines

1. Refer to the "Oil Pan Removal & Installation" procedures in this section and remove the oil pan.

2. Remove the oil pump and the rear main bearing cap.

3. Using a small pry bar, pry the oil seal from the rear main bearing cap.

4. Using a small hammer and a brass pin punch, drive the top half of the oil seal from the rear main bearing. Drive it out far enough, so it may be removed with a pair of pliers.

5. Using a non-abrasive cleaner, clean the rear main bearing cap and the crankshaft.

6. Fabricate an oil seal installation tool from 0.004 in. shim stock, shape the end to $^1/_2$ in. long by $^{11}/_{64}$ in. wide.

7. Coat the new oil seal with engine oil; DO NOT coat the ends of the seal.

8. Position the fabricated tool between the crankshaft and the seal seat in the cylinder case.

9. Position the new half seal be-

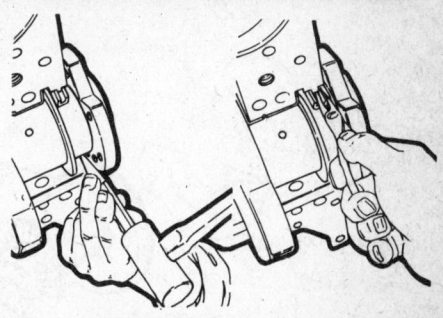

Removing the rear main seal

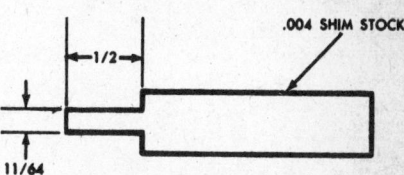

.004 SHIM STOCK

Rear main seal installation tool

tween the crankshaft and the tip of the tool, so that the seal bead contacts the tip of the tool.

NOTE: Make sure that the seal lip is positioned toward the front of the engine.

10. Using the fabricated tool as a shoe horn, to protect the seal's bead from the sharp edge of the seal seat surface in the cylinder case, roll the seal around the crankshaft. When the seal's ends are flush with the engine block, remove the installation.

11. Using the same manner of installation, install the lower half onto the lower half of the rear main bearing cap.

12. Apply sealant to the cap-to-case mating surfaces and install the lower rear main bearing half to the engine; keep the sealant off of the seal's mating line.

13. Install the rear main bearing cap bolts and torque to 10–12 ft. lbs. Using a lead hammer, tap the crankshaft forward and rearward, to line up the thrust surfaces. Torque the main bearing bolts to 70–85 ft. lbs. and reverse the removal procedures. Refill the crankcase.

CLUTCH

The only service adjustment necessary on the clutch is to maintain the correct pedal free play. The clutch pedal free play or throwout bearing lash decreases with disc wear.

REMOVAL & INSTALLATION

1. Support the engine and remove the transmission.

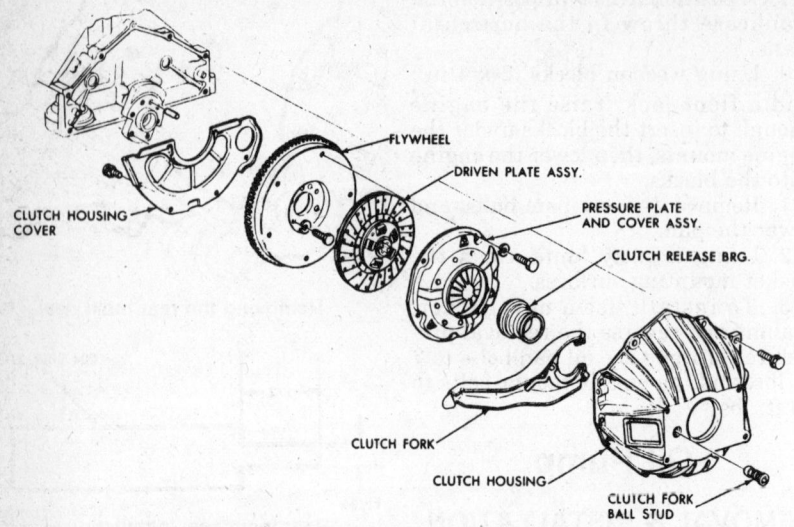

Exploded view of a typical clutch assembly

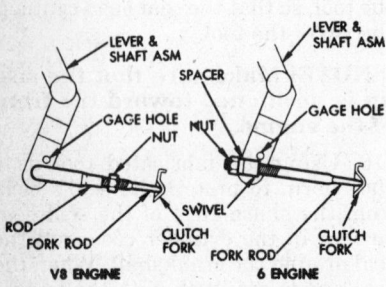

Clutch pedal free-play adjustment

2. Disconnect the clutch fork push rod and the spring.

3. Remove the flywheel housing.

4. Slide the clutch fork from the ball stud and remove the fork from the dust boot. The ball stud is threaded into the clutch housing and may be replaced, if necessary.

5. Install a clutch alignment tool J-5824 to support the clutch assembly during removal. Mark the relationship between the flywheel and the clutch cover for reinstallation, if they do not already have X marks.

6. Loosen the clutch-to-flywheel bolts evenly, one turn at a time, until the spring pressure is released, then remove bolts and the clutch assembly.

7. To install, reverse the removal procedures. Adjust the shift linkage and the clutch pedal free-play.

FREE PLAY ADJUSTMENT

This adjustment must be made under the vehicle on the clutch operating linkage. Free play is measured at the clutch pedal.

1. Disconnect the return spring at the clutch operating fork.

2. Rotate the clutch lever until the pedal is firmly against the bumper.

3. Push the outer end of the clutch operating fork to the rear until the release bearing can just be felt to contact the pressure plate fingers.

4. Detach the front end of the operating rod from the clutch pivot shaft arm and place it in the gauge hole on the arm.

5. Loosen the locknut and lengthen the rod just enough to take all the play out of the linkage, then tighten the locknut.

6. Replace the operating rod in its original location.

7. Replace the return spring and check the free play at the pedal pad, it should be about $^3/_4$–$1^1/_4$ inches.

MANUAL TRANSMISSION

For further identification and overhaul information, please refer to the "Manual Transmission" Unit Repair section.

REMOVAL & INSTALLATION

1. On the 3 speed floorshift models, remove the shift knob; on the 4 speed floorshift models, remove the spring and the "T" handle.

2. Raise and support the vehicle on jackstands.

3. Disconnect the speedometer cable at the transmission.

4. Remove the driveshaft.

5. Support the rear of the engine and remove the crossmember.

6. Detach the shift rods from the transmission levers.

NOTE: On the floorshift models, disconnect the back drive rod at the bell crank (if equipped).

7. On floorshift models, remove the shift control assembly from the transmission; pull down until the shift lever clears the rubber boot.

8. Remove the upper transmission-to-clutch housing bolts and replace them with headless guide pins, then remove the lower bolts.

9. Slide the transmission back along the guide pins until the input shaft clears the clutch and remove the transmission.

10. To install, reverse the removal procedures. If the input shaft won't engage the clutch splines, put the transmission in gear and turn the output shaft slightly. Torque the transmission-to-clutch housing bolts to 52 ft. lbs. (4 speed–83mm) or 75 ft. lbs. (all others).

SHIFT LINKAGE ADJUSTMENT

Floorshift

1. Turn the ignition switch OFF.

2. Raise and support the vehicle on jackstands.

3. Loosen the locknuts on the shift rods; the rods should pass freely through the swivels.

4. Set the transmission levers in the Neutral position.

5. Set the floorshift lever in the Neutral position. Install a locating gauge (3.0 x $^1/_8$ in. dia.) into the alignment slot of the control lever bracket assembly.

6. Adjust the length of the shift rods at the swivels and tighten the locknuts, then remove the locating gauge.

7. Shift the control lever into Reverse and LOCK the ignition switch.

NOTE: If equipped with a back drive rod, pull down slightly on the rod at the steering column, to remove any slack and tighten the locknut. The ignition switch must move freely to the LOCK position and it must not be possible to turn the key to LOCK when in any other position. If the interlock binds, leave the switch in the Lock position and readjust the back drive rod.

8. Check shifting operation and re-adjust (if necessary).

AUTOMATIC TRANSMISSION

Identification can be made by the shape of the pan. See the Unit Repair Section for further visual differences and service procedures.

REMOVAL & INSTALLATION

1. Disconnect the negative battery cable.
2. Remove the air cleaner, the T.V. cable at the upper end and the transmission fluid level indicator.
3. Raise and support the vehicle on jackstands.
4. Remove the drive shaft and the floor pan reinforcement (if necessary).
5. At the transmission, disconnect the speedometer cable, the shift linkage, the electrical leads and retainers.
6. Remove the torque converter cover and the torque converter-to-flexplate nuts.

NOTE: Before removing the torque converter from the flex plate, mark the flex plate to the torque converter.

7. If equipped with a gas engine, remove the catalytic converter bracket.
8. Remove the transmission support mounting bolts at the transmission mount and at the frame, then raise the transmission slightly and slide the support rearward.
9. Lower the transmission, then remove the oil cooler lines and the T.V. cable from the transmission.

NOTE: After removing the oil cooler lines, be sure to cap them to prevent dirt from entering the system.

10. Support the engine and remove the transmission-to-engine bolts.
11. Install the torque converter holding tool J-21366 to support the torque converter, then remove the transmission from the vehicle.
12. To install, reverse the removal procedures. Torque the transmission-to-engine bolts to 35 ft. lbs., the torque converter-to-flex plate to 46 ft. lbs., the transmission support-to-frame bolts to 40 ft. lbs., the transmission support-to-transmission mount to 25 ft. lbs. and the drive shaft bolts to 16 ft. lbs. Adjust the shift linkage and the T.V. cable.

SHIFT LINKAGE ADJUSTMENT, BAND ADJUSTMENT, FLUID AND FILTER CHANGE

For all automatic transmission service procedures, please refer to the "Automatic Transmission" Unit Repair section.

DRIVE AXLE

The universal joints are lubricated and sealed at the factory, they require no periodic maintenance. Two basic universal joints are used: the Cleveland and the Saginaw. The Cleveland type uses snapring bearing cap retainers. The Saginaw uses injection molded plastic to retain the bearing caps. On the Saginaw type there is a snapring groove in the bearing housing inboard of the yoke to facilitate installation with a repair kit.

Driveshaft

REMOVAL & INSTALLATION

1. Raise and support the vehicle on jackstands.
2. Mark the relationship of the driveshaft to the pinion flange.
3. Remove the rear universal joint retainers and separate it from the pinion flange.

NOTE: If the universal joint cups are loose, tape the cups to the universal joint to keep them from falling off the joint.

4. Support the driveshaft and remove it from the transmission. When removing or installing the driveshaft, DO NOT allow the universal joints to bend to extreme angles, this might rupture the internally injected seals.
5. To install, reverse the removal procedures. Torque the universal joint fasteners to 15 ft. lbs.

Universal Joint

For all universal joint service information refer to the "U-Joints/ CV-Joints" Unit Repair section.

REMOVAL & INSTALLATION

1980–81 Cleveland Type

1. Refer to the "Driveshaft Removal & Installation" procedures in this section and remove the driveshaft.

NOTE: NEVER clamp the driveshaft tube in a vise; this may dent the tube. Support the driveshaft horizontally and clamp on the yokes of the universal joints.

2. Remove the lock rings from the ends of the trunion yoke.
3. Support the driveshaft in the horizontal position with the base plate of a press, so that the lower ear of the yoke is supported on a piece of $1^1/_4$ in. I.D. pipe.
4. Place a socket on the upper bearing cup and press the lower bearing cup out of the yoke ear.

NOTE: Since the bearing cup cannot be fully pressed from the yoke ear, grasp the cup in the jaws of a vice and work it from the yoke.

5. Rotate the driveshaft to the opposite bearing cup and press the bearing cup from the yoke, using the same removal procedure.
6. With both bearing cups removed from the yoke, separate the yoke from the driveshaft.
7. Repeat the removal procedures for the other bearing cups.
8. Clean and inspect all of the parts.

NOTE: If the used universal joints are going to be reinstalled, repack with new grease.

9. To install, use new universal joints or repack the old ones. Place a bearing cup part way into 1 side of the yoke (place the yoke ear to the bottom).
10. Insert the cross into the yoke so that the trunion seats freely into the bearing cup.
11. Insert the opposite bearing cup part way into the yoke ear. Install the cross into the cup, making sure that both trunions are straight and true with the bearing cups.
12. Using an arbor press, press the bearing cups into the yoke, making sure that the cross trunions are free to turn. Install the bearing retainers.
13. Assemble the other side of the yoke in the same manner.
14. To complete the installation, reverse the removal procedures.

1980 And Later Saginaw Type

1. Refer to the "Driveshaft Removal & Installation" procedures in this section and remove the driveshaft.

NOTE: Never clamp the drive-shaft tube in a vise, for this may dent the tube. Support the driveshaft horizontally and clamp on the yokes of the universal joints.

2. Support the driveshaft in the horizontal position with the base plate of a press, so that the lower ear of the yoke is supported on a $1^1/_8$ in. socket.

3. Place the cross press tool J-9522-3 on the open horizontal bearing cups and press (shear the plastic retaining ring) the lower bearing cup out of the yoke ear.

NOTE: If the bearing cup is not completely removed, lift the cross tool and place a spacer tool J-9522-5 between the seal and the bearing cup. Repeat the pressing procedure to drive the bearing cup from the yoke.

4. Rotate the driveshaft to the opposite bearing cup and press the bearing cup from the yoke.

5. With both bearing cups removed from the yoke, separate the yoke from the driveshaft.

6. Repeat the removal procedures for the other bearing cups.

NOTE: Since there are no bearing retainer grooves in the production bearing cups, the universal cannot be reused.

7. Remove the remains of the sheared bearing cups and check for nicks in the yoke ears.

8. To install, use new universal joints and place a bearing cup part way into one side of the yoke (place the yoke ear to the bottom).

9. Insert the cross into the yoke so that the trunion seat freely into the bearing cup.

10. Insert the opposite bearing cup part way into the yoke ear. Install the cross into the cup, making sure that both trunions are straight and true with the bearing cups.

11. Using the press, press the bearing cups into the yoke, making sure that the cross trunions are free to turn.

12. As soon as the bearing retainer groove(s) clears the yoke, stop pressing and install the bearing retainer(s) onto the groove(s).

NOTE: It may be necessary to strike the yoke with a hammer to align the seating of the bearing retainers.

13. Assemble the other side of the yoke in the same manner.

14. To complete the installation, reverse the removal procedures.

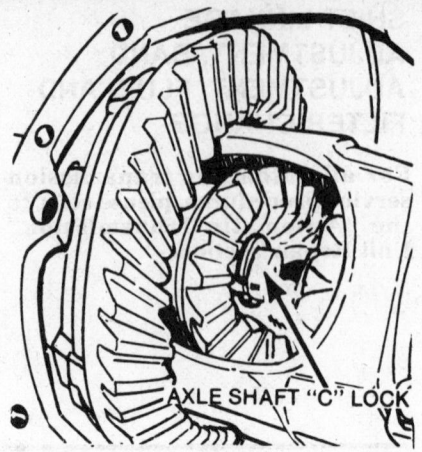

Axle shaft C-clips inside the differential

AXLE SHAFTS

The axle shafts are retained by C-shaped locks, which fit into grooves at the inner end of the shaft.

REMOVAL & INSTALLATION

1. Raise and support the rear of the vehicle on jackstands.

2. Remove the wheel and the brake drum assembly.

3. Clean the dirt from around the carrier cover.

4. Remove the carrier cover and drain the oil from the housing.

5. Remove the rear axle pinion shaft lock screw and the shaft.

6. Push the flanged end of the axle shaft toward the center of the vehicle and remove the C-lock from the end of the shaft.

7. Pull the axle shaft from the housing, being careful not to damage the oil seal.

8. Using a pry bar, pry the oil seal from the axle housing.

9. Install the bearing removal tool J-23689 ($8^3/_4$ in. axle) or J-22813 (all other axles) to the axle bearing, connect it to a slide hammer and pull the bearing from the housing.

10. To install, lubricate a new bearing and drive it into the housing (until it seats) using the bearing installation tool J-23690.

11. Lubricate the lips of a new oil seal and drive it into the housing (until it is flush with the housing), using the seal installation tool J-21128.

12. To complete the installation, slide the axle shaft into the housing (making sure that it engages the splines of the side gear) and reverse

the removal procedures. Torque the pinion lock screw to 20 ft. lbs. Install a new carrier cover and torque the cover bolts to 20 ft. lbs. Fill the axle with lubricant to $^3/_8$ in. below the filler hole.

FRONT SUSPENSION

NOTE: Many 1980 and later vehicles have been gradually switched over to metric fasteners. Most models use metric prevailing torque nuts to fasten the upper and lower ball joint studs to the steering knuckle. American standard inch calibrated wrenches will not fit metric nuts and bolts.

Coil Spring

REMOVAL & INSTALLATION

1. Raise and support the vehicle on jackstands.

2. Remove shock absorber.

3. Secure tool J-23028 to a jack and position the assembly under the control arm, supporting the inner bushing.

4. Disconnect stabilizer bar at lower control arm.

5. Raise the jack to take the tension off of the control arm pivots. Install a chain around the spring and through the control arm as a safety measure, then remove the inner control arm-to-crossmember pivot bolts.

6. Carefully lower the control arm, allowing the spring to relax.

— **CAUTION** —
Allow the spring to completely expand before attempting to remove it.

7. Remove the chain and the spring.

8. To install, reverse the removal procedures. Torque the lower control arm pivot nuts to 65 ft. lbs. (Malibu and Monte Carlo) or 90 ft. lbs. (Impala and Caprice) with the weight of the vehicle on the springs.

Shock Absorber

REMOVAL & INSTALLATION

New shock absorbers must be purged of air before installation. This is done by repeatedly extending the shock in its normal mounted position, inverting it and compressing it.

1. Remove the nut, retainer and

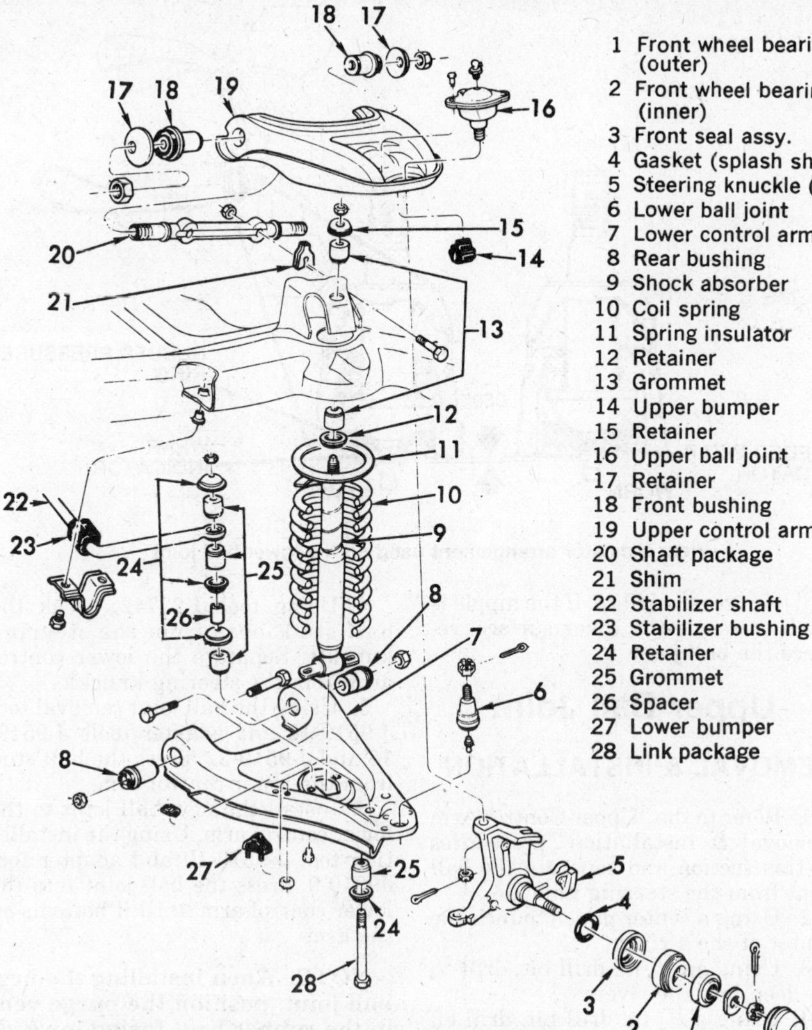

1 Front wheel bearing (outer)
2 Front wheel bearing (inner)
3 Front seal assy.
4 Gasket (splash shield)
5 Steering knuckle (r.h.)
6 Lower ball joint
7 Lower control arm
8 Rear bushing
9 Shock absorber
10 Coil spring
11 Spring insulator
12 Retainer
13 Grommet
14 Upper bumper
15 Retainer
16 Upper ball joint
17 Retainer
18 Front bushing
19 Upper control arm
20 Shaft package
21 Shim
22 Stabilizer shaft
23 Stabilizer bushing
24 Retainer
25 Grommet
26 Spacer
27 Lower bumper
28 Link package

Exploded view of the front suspension assembly

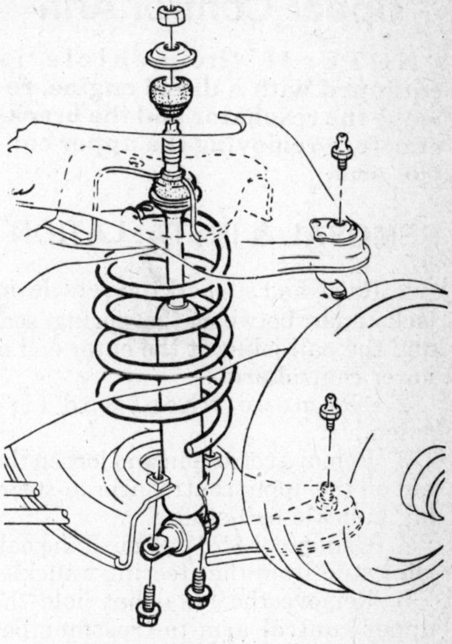

NOTE After assembly, end of spring coil must cover all or part of one inspection drain hole. The other hole must be partly exposed or completely uncovered.

NOTE Spring to be installed with tape at lowest position. Bottom of spring is coiled helical, and the top is coiled flat with a gripper notch near end of wire.

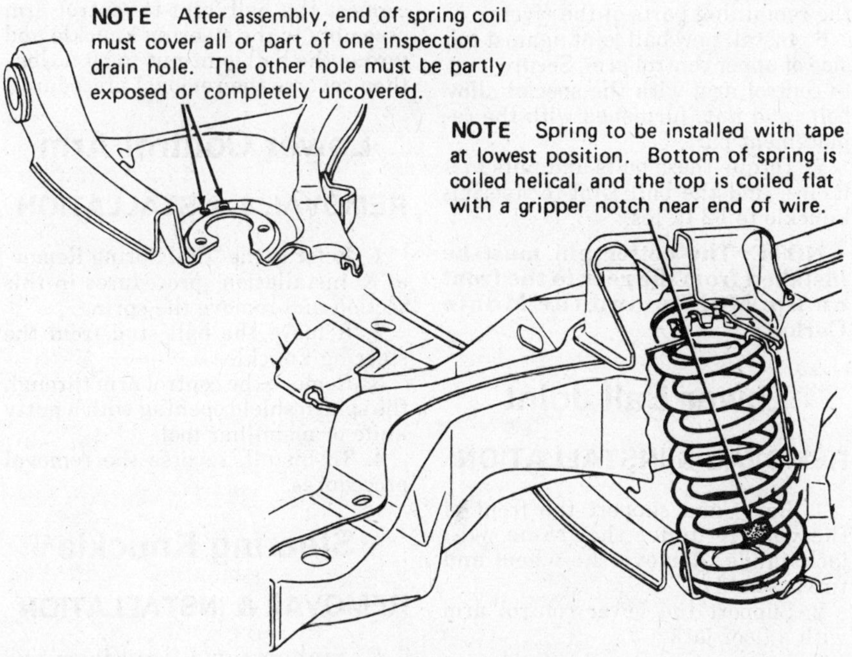

Positioning the coil spring

Installing the shock absorbers

grommet, which are attached to the upper end of the shock absorber and seat against the frame bracket.

NOTE: It may be necessary to hold the shock absorber shaft to remove the nut. This may be done with a wrench on the end of the shaft.

2. Raise and support the vehicle on jackstands to allow the shock to be dropped from the lower control arm.

3. Remove the shock absorber lower attaching screws and lower the shock from the control arm.

4. To install, reverse the removal procedures. Make sure all grommets are in the correct position and tighten the upper nut.

Front Wheel Bearing

ADJUSTMENT

1. Lift the wheel off the ground by jacking under the lower control arm.

2. Remove the dust cap from the hub.

3. Remove the cotter pin and discard it.

4. Snug up the spindle nut while spinning the wheel to seat the bearings (12 ft. lbs.). Then back off the nut $1/4$–$1/2$ turn.

5. Retighten the nut by hand until it is finger tight.

6. Loosen the nut until the nearest hole in the spindle lines up with a slot in the spindle nut, then insert a new cotter pin. When the bearing is properly adjusted, there will be 0.001–0.005 in. end play.

7. Replace the dust cover and lower the vehicle.

Upper Control Arm

NOTE: If the vehicle is equipped with a diesel engine, remove the resonator and the bracket before removing the upper control arm.

REMOVAL & INSTALLATION

1. Raise and support the vehicle on jackstands between the spring seat and the ball joint, at the outer end of lower control arm.
2. Remove wheel and tire assembly.
3. Remove cotter pin and loosen the nut on the upper control arm-to-steering knuckle ball stud.
4. Using tool J-23742, push the ball joint stud from the steering knuckle.
5. Remove the nuts that hold the upper control arm-to-crossmember and the control arm. Count number of shims at each bolt.
6. To install, reverse the removal procedures. Install same number of shims as removed at each bolt. Torque the control arm-to-frame to 48 ft. lbs. (Malibu and Monte Carlo) or 70 ft. lbs. (all others) and the upper ball joint-to-steering knuckle to 52 ft. lbs. Insert cotter pin. Check caster and camber.

Ball Joint

INSPECTION

NOTE: Before performing this inspection, make sure the wheel bearings are adjusted correctly and that the control arm bushings are in good condition.

1. Raise and support the vehicle on jackstands under the front lower control arm at the spring seat. Raise the vehicle until there is 1–2 in. of clearance under the wheel.
2. Insert a bar under the wheel and pry upward. If the wheel raises more than $^1/_8$ in., the ball joints are worn. While prying on the wheel, determine by visual inspection whether the upper or lower ball joint is worn.

NOTE: Due to the distribution of forces in the suspension, the lower ball joint is usually the defective joint.

LOWER BALL JOINT WEAR INDICATORS–ALL MODELS

These vehicles have a visual wear indicator on the lower ball joint. Wear is indicated by the position of the $^1/_2$ in. nipple into which the grease fitting is screwed. On a new joint, the nipple should project 0.050 in. beyond the

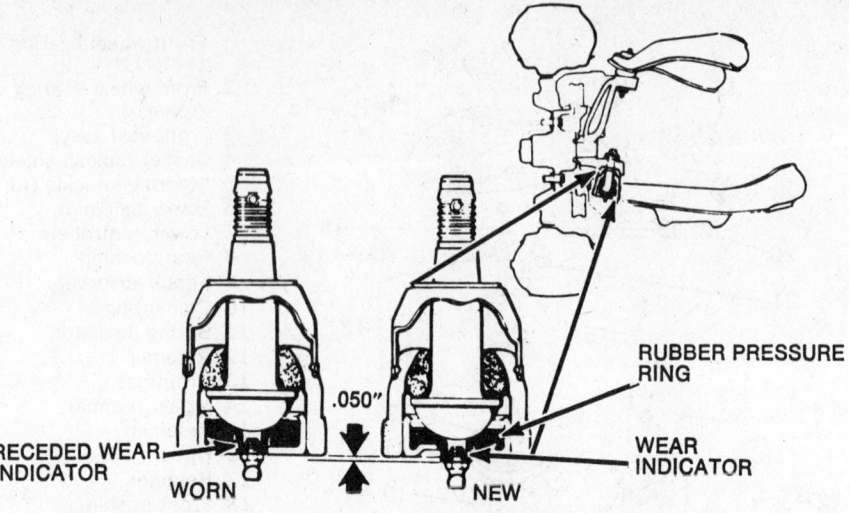

Wear indicator arrangement used on the lower ball joint

ball joint cover surface. If the nipple is flush or inside the cover surface, replace the ball joint.

Upper Ball Joint

REMOVAL & INSTALLATION

1. Refer to the "Upper Control Arm Removal & Installation" procedures in this section and separate the ball joint from the steering knuckle.
2. Using a center punch, punch the center of the 4 rivets.
3. Using an $^1/_8$ in. drill bit, drill $^1/_4$ in. deep into the rivets.
4. Using a $^1/_2$ in. drill bit, drill off the heads of the rivets.
5. Using a drift punch, drive out the remaining parts of the rivets.
6. Install new ball joint against top side of upper control arm. Secure joint to control arm with the special alloy bolts and nuts furnished with the replacement part.
7. Torque these bolts and nuts to 9 ft. lbs. and the ball joint-to-steering knuckle to 52 ft. lbs.

NOTE: The cotter pin must be installed from the rear to the front on the Malibu and the Monte Carlo.

Lower Ball Joint

REMOVAL & INSTALLATION

1. Raise and support the front of the vehicle under the frame with jackstands. Remove the wheel and tire assembly.
2. Support the lower control arm with a floor jack.
3. Remove the cotter pin and loosen the lower ball stud nut.

4. Using tool J-23742, break the ball stud loose from the steering knuckle. Separate the lower control arm from the steering knuckle.
5. Using the ball joint removal tool J-9519-10 and adapter tools J-9519-16 and J-9519-22, press the ball stud from the lower control arm.
6. Install the new ball joint to the lower control arm. Using the installation tool J-9519-10 and adapter tool J9519-9, press the ball joint into the lower control arm until it bottoms on the arm.

NOTE: When installing the new ball joint, position the purge vent in the rubber boot facing inward.

7. To complete the installation, connect the ball joint-to-control arm assembly to the steering knuckle and torque the ball joint nut to 90 ft. lbs., then reverse the removal procedures.

Lower Control Arm

REMOVAL & INSTALLATION

1. Refer to the "Coil Spring Removal & Installation" procedures in this section and remove the spring.
2. Remove the ball stud from the steering knuckle.
3. Remove the control arm through the splash shield opening with a putty knife or a similiar tool.
4. To install, reverse the removal procedures.

Steering Knuckle

REMOVAL & INSTALLATION

1. Siphon some fluid from the brake master cylinder.

2. Raise and support the vehicle on jackstands.

3. Remove the wheel and tire assembly.

4. Remove the caliper from the steering knuckle and support on a wire.

5. Remove the grease cup, the cotter pin, the castle nut and the hub assembly.

6. Remove the 3 bolts holding the shield to the steering knuckle.

7. Using the ball joint removal tool J-6627, disconnect the tie rod from the steering knuckle.

8. Using ball joint removal tool J-23742, disconnect the ball joints from the steering knuckle.

9. Place a floor jack under the lower control arm (near the spring seat) and disconnect the ball joint from the steering knuckle.

10. Raise the upper control arm and disconnect the ball joint from the steering knuckle.

11. Remove the steering knuckle from the vehicle.

12. To install, reverse the removal procedures. Torque the upper ball joint-to-steering knuckle nut to 65 ft. lbs., the lower ball joint-to-steering knuckle nut to 90 ft. lbs. and the tie rod-to-steering knuckle nut to 40 ft. lbs. Adjust the wheel bearing and refill the master cylinder.

Stabilizer Bar

REMOVAL & INSTALLATION

1. Raise and support the front of the vehicle on jackstands.

2. Disconnect the stabilizer link bolts at the lower control arms.

3. Remove the stabilizer-to-frame clamps.

4. Remove the stabilizer bar.

5. To install, reverse the removal procedures. Torque the stabilizer-to-lower control arm bolts to 13 ft. lbs. and the stabilizer-to-frame bolts to 24 ft. lbs.

REAR SUSPENSION

All models use a coil spring rear suspension located by two lower control arms and two diagonally mounted upper control arms. The fore and aft axle movements are controlled by the lower control arms, while the lateral movement is controlled by the upper control arms.

Shock Absorber

REMOVAL & INSTALLATION

New shock absorbers must be purged of air before installation. This is done by repeatedly extending the shock in its normal mounted position, inverting and compressing it.

1. Raise and support the vehicle on jackstands at the axle housing, to prevent stretching the brake hose.

NOTE: If equipped with super lift shock absorbers, disconnect the air line snap on connector at the shock absorber.

2. On some models, it may be necessary to remove the wheel and tire assembly.

3. Remove the nut, the retainer, the grommet and lock washer, which attach the lower end of the shock absorber to its mounting.

NOTE: On some models, it may be necessary to remove the upper shock absorber bracket by reaching between the tire and the frame to remove the mounting nuts.

4. Remove the bolts, nuts and lock washers from the upper end of the shock absorber and the shock absorber.

5. To install, reverse the removal procedures. Torque the lower shock-to-frame bolt to 65 ft. lbs. and the upper shock-to-frame nut to 12 ft. lbs. or bolt to 20 ft. lbs.

NOTE: If equipped with super lift shocks, add 10 psi of air pressure to the shock (to prevent damage) and torque the upper nut to 20 ft. lbs.

Rear Spring

REMOVAL & INSTALLATION

1. Raise and support the rear of the vehicle on jackstands at the frame rails.

2. Remove the clip that attaches the brake hose to the mounting bracket on the frame crossmember.

3. Support the rear axle with a floor jack.

4. Disconnect the upper control arms from the axle housing.

5. If equipped with a stabilizer bar, remove the bar from the control arms.

6. Remove the nut and the lock washer from the shock absorber, then disconnect the shock from the axle. It may be necessary to adjust the height of the jack to disconnect the shock.

7. Carefully lower the jack until the spring is free, then remove the spring.

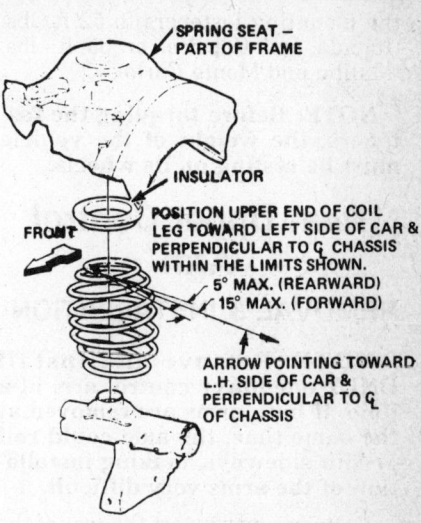

Coil spring positioning

NOTE: When removing the spring, mark its position so that it may be installed in the same position.

8. To install, reverse the removal procedures. Torque the upper control arm-to-front bushing nut to 92 ft. lbs. (Impala and Caprice) or 70 ft. lbs. (Malibu and Monte Carlo), the upper control arm-to-rear bushing nut to 70 ft. lbs., the upper control arm-to-rear bushing bolt to 80 ft. lbs., the stabilizer bar mounts to 52 ft. lbs. (Impala and Caprice) or 35 ft. lbs. (Malibu and Monte Carlo).

Rear Lower Control Arm

REMOVAL & INSTALLATION

NOTE: Remove and install ONLY one lower control arm at a time. If both arms are removed at the same time, the axle could roll or slip sideways, making installation of the arms very difficult.

1. Raise and support the rear of the vehicle on jackstands under the rear axle.

NOTE: If equipped the a stabilizer bar, remove it.

2. Remove the control arm attaching fasteners and the control arm.

3. To install, reverse the removal procedures. Torque the control arm-to-frame nut to 92 ft. lbs. (Impala and Caprice) or 70 ft. lbs. (Malibu and Monte Carlo), the control arm-to-axle nut to 92 ft. lbs. (Impala and Caprice) and the control arm-to-axle bolt to 125 ft. lbs. (Impala and Caprice) or 79 ft. lbs. (Malibu and Monte Carlo). If equipped with a stabilizer bar, torque

the mounting fasteners to 52 ft. lbs. (Impala and Caprice) or 35 ft. lbs. (Malibu and Monte Carlo).

NOTE: Before torquing the fasteners, the weight of the vehicle must be resting on its wheels.

Rear Upper Control Arm

REMOVAL & INSTALLATION

NOTE: Remove and install ONLY one lower control arm at a time. If both arms are removed at the same time, the axle could roll or slip sideways, making installation of the arms very difficult.

1. Raise and support the rear of the vehicle on jackstands under the axle.
2. Remove the upper control arm nut at the axle.

NOTE: To remove the mounting bolt from the axle, it may be necessary to rock the axle. On some models, it may be necessary to remove the lower shock absorber stud to provide clearance for the upper control arm removal.

3. Remove the upper control arm-to-frame nut and bolt, then the control arm.
4. To install, reverse the removal procedures. Torque the upper control arm-to-axle nut to 70 ft. lbs., the upper control arm-to-axle bolt to 79 ft. lbs. and the upper control arm-to-frame bolt to 92 ft. lbs. (Impala and Caprice) or 70 ft. lbs. (Malibu and Monte Carlo).

Stabilizer Bar

REMOVAL & INSTALLATION

1. Raise and support the rear of the vehicle on jackstands under the frame.
2. Support the axle assembly with a floor jack.
3. Remove the stabilizer bar-to-lower control arm bolts and the stabilizer bar.
4. To install, reverse the removal procedures. Torque the stabilizer bar-to-lower control arm to 52 ft. lbs. (Impala and Caprice) or 35 ft. lbs. (Malibu and Monte Carlo).

BRAKES

For all information on the brake system not detailed below, please refer to the "Brakes" Unit Repair section.

Master Cylinder

NOTE: Vehicles with disc brakes do not have a check valve in the front outlet port of the master cylinder. If one is installed, the front discs will quickly wear out due to residual hydraulic pressure holding the pads against the rotor.

REMOVAL & INSTALLATION

1. Disconnect the hydraulic lines from master cylinder.
2. Remove the retaining nuts and the lockwashers holding the cylinder to the cowl or the brake booster.

NOTE: If equipped with non-power brakes, disconnect the pushrod at brake pedal.

3. Remove the master cylinder, the gasket and the rubber boot.
4. To install, reverse the removal procedures. Torque the master cylinder mounting nuts to 22 ft. lbs. and the hydraulic lines to 18 ft. lbs. Refill the master cylinder, bleed the brake system and check the brake pedal free-play.

NOTE: On non-powered brakes, position the master cylinder on the cowl, making sure that the pushrod goes through the rubber boot into the piston. Reconnect the pushrod clevis to the brake pedal. If equipped with power brakes, install the master cylinder on the power booster.

Power Brake Booster

REMOVAL & INSTALLATION

1. Disconnect the vacuum hose from the vacuum check valve.
2. Unbolt the master cylinder and carefully move it aside without disconnecting the hydraulic lines.

NOTE: If sufficient booster clearance cannot be obtained, it will be necessary to disconnect the hydraulic lines from the master cylinder, then remove the master cylinder.

3. Disconnect the pushrod at the brake pedal assembly.

NOTE: Some brake boosters may be held on with sealant; this can be easily removed with tar remover.

4. Remove the booster-to-cowl nuts

and lockwashers and the booster from engine compartment.

5. To install, reverse the removal procedures. Torque the booster-to-cowl and the master cylinder-to-booster mounting nuts to 28 ft. lbs.

NOTE: Make sure to check the operation of the stop lights. Allow the engine vacuum to build before applying the brakes. Bleed the hydraulic system if the lines were disconnected from the master cylinder.

Hydro-Boost Brake Booster

REMOVAL & INSTALLATION

For an explanation, troubleshooting and bleeding of the Hydro-Boost brake system, please refer to the "Brakes" Unit Repair section.

1. Turn the engine off and pump the brake pedal 4 or 5 times to deplete the accumulator.
2. Remove the nuts from the master cylinder, then move the master cylinder away from the booster, with brake lines still attached.
3. Remove the hydraulic lines from the booster.
4. Remove the retainer and washer at the brake pedal.
5. Remove the attaching nuts retaining the booster fastened to the cowl and the booster.
6. To install, reverse the removal procedures. Torque the booster-to-cowl nuts to 15 ft. lbs. and the master cylinder-to-booster nuts to 20 ft. lbs. Bleed the power steering and hydro booster system.

Powermaster Power Brake Assembly

The unit is a complete, integral power brake system. The assembly consists of an Electro-Hydraulic (E-H) pump, a fluid accumulator, a pressure switch, a fluid reservoir and a hydraulic booster with a dual master cylinder.

For further information on the Powermaster system, please refer to the "Brakes" Unit Repair section.

REMOVAL & INSTALLATION

—— CAUTION ——
Before performing any service to the brake system, depressurize the powermaster assembly.

1. Turn the ignition OFF, then depress the brake pedal 10 times using at least 50 lbs. of force.

NOTE: Before removing the master cylinder form the cowl, remove some of the brake fluid from the reservoir.

2. Remove the electrical connectors from the pressure switch and the E-H pump.
3. Disconnect and plug the brake lines at the master cylinder.
4. Remove the master cylinder-to-cowl mounting nuts.
5. Disconnect the master cylinder push rod from the brake pedal.
6. Remove the powermaster unit from the vehicle.
7. To install, reverse the removal procedures. Torque the master cylinder-to-cowl nuts to 22–30 ft. lbs. Refill the master cylinder and bleed the brake system.

BLEEDING POWERMASTER UNIT

1. Fill the reservoir to the full marks.
2. Remove the brake lines at the outlet ports of the master cylinder and allow the cylinder to bleed until the fluid runs out, the reconnect the brake lines. ONLY tighten the line closest to the cowl.
3. Have an assistant depress the brake pedal slowly (exerting 50 lbs. of pressure) until the full pedal travel is accomplished, then tighten the forward brake line. WAIT for 5 seconds.
4. Reapply the brake pedal and hold it, then open the forward connector $1/2$ turn to release the air and retighten the connector. Repeat the procedure until the air is bled from the master cylinder.
5. Refill the master cylinder.

Combination Valve

REMOVAL & INSTALLATION

NOTE: The combination valve is not repairable and must be replaced if found to be defective. On some models hoisting might be necessary.

1. Disconnect the electrical connector at the pressure differential switch. It is recommended , that with the aid of pliers, you squeeze the eliptical shaped plastic locking ring and then pull up. This will move the locking tangs away from the switch.
2. Disconnect and plug the hydraulic lines at the combination valve then remove the valve.

3. Installation is the reverse of removal.
4. Bleed the entire brake system.

─────── **CAUTION** ───────
Do not move the car until a firm brake pedal is obtained.

Wheel Cylinder

REMOVAL & INSTALLATION

Impala and Caprice

1. Raise and safely support the car.
2. Mark the relationship of the wheel to the axle flange.
3. Remove the wheel, drum and brake shoes.
4. Clean all dirt around around the wheel cylinder at the brake line and disconnect the brake line.
5. Remove the wheel cylinder from the backing plate.
6. Installation is the reverse of removal. Torque the rear wheel brake pipe to wheel cylinder to 12 ft. lbs.
7. Bleed the system.

Malibu and Monte Carlo

1. Insert awls or pins, $1/8$ in. diameter or less, into the access slots between the wheel cylinder pilot and retainer locking tabs.
2. Bend both tabs away simultanously until they spring over the abutment shoulder releasing the wheel cylinder. Discard the old retaining clip.
3. For ease of installation hold the wheel cylinder against the backing plate by inserting a block betwen the wheel cylinder and the axle shaft flange.
4. Position the wheel cylinder retainer clip so the tabs will be away from and in a horizontal position with the backing plate when installing.
5. Press the new retaining clip over the wheel cylinder abutment and into position using a $1^1/8$ in. 12 point socket. Make sure the retainer tabs are properly snapped under the abutment shoulder.
6. Install the brake shoes, drum and wheel.
7. Flush and bleed the hydraulic system.

Parking Brake

ADJUSTMENT

The automatic self-adjusting feature incorporated in the rear brake mechanism normally maintains proper parking brake adjustment. For this reason, the rear brake adjustment

must be checked before any adjustment of the parking brake cables is made. Check the parking brake mechanism and cables for free movement and lubricate all working surfaces before proceeding.

─────── **CAUTION** ───────
It is very important that the parking brake cables not be too tight. If the cables are too tight, they create a drag and position the secondary shoes so that the self-adjusters continue to operate to compensate for drag wear. The result is rapidly worn rear brake linings.

1. Raise and support the rear of the vehicle on jackstands.
2. Set the parking brake at 2 clicks.
3. Loosen the equalizer locknut. Tighten the adjusting nut until the left wheel can be turned backward with two hands, but is locked in the forward rotation.
4. Tighten the locknut.
5. Fully release the parking brake and rotate the rear wheels; no drag should be felt in either direction.

STEERING

Tie Rod End

REMOVAL & INSTALLATION

1. Raise and support the vehicle on jackstands.
2. Loosen the tie rod adjuster sleeve clamp nut.
3. Remove the tie rod cotter pin and nut from the ball joint stud.

NOTE: If the torque required to remove the nuts and bolts exceeds 7 ft. lbs., it's best to discard them and use new fasteners of equal grade quality.

4. Using tool J-6627 or BT-7101, remove the tie rod stud from the steering knuckle.

NOTE: The outer tie rods have right hand threads and the inner tie rods have left hand threads.

5. Unscrew the tie rod from the adjuster sleeve. Count the number of turns the tie rod must be rotated to remove it from the adjusting sleeve; this will allow a reasonably accurate realignment upon reassembly.
6. To install, reverse the removal procedures. Clean all rust and dirt from the threads. Torque the tie rod-to-steering knuckle to 30 ft. lbs., the tie rod-to-intermediate rod nut to 40 ft. lbs. and the tie rod clamp nuts to 15

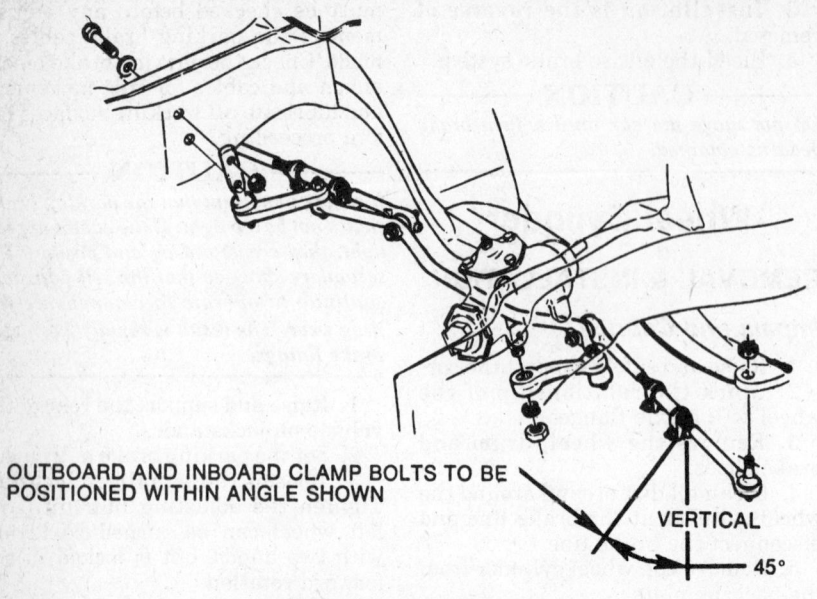

OUTBOARD AND INBOARD CLAMP BOLTS TO BE POSITIONED WITHIN ANGLE SHOWN

VERTICAL

45°

Steering linkage-typical all models

ft. lbs. Check and/or adjust the alignment, if necessary.

Steering Gear

REMOVAL & INSTALLATION

1. Disconnect the pressure and return hoses from the steering gear housing.
2. Disconnect the negative battery cable and remove coupling shield.
3. Remove the steering coupling-to-steering shaft flange retaining nuts, bolts and lock washers.
4. Remove the pitman arm nut and washer from the pitman shaft, then mark the relationship of the arm to the shaft.
5. Using tool J-6632, remove the pitman arm from the steering gear.
6. Remove the steering gear-to-frame screws and the steering gear from vehicle.
7. To install, reverse the removal procedures. Torque the steering gear-to-frame to 70 ft. lbs., the coupling shaft pinch bolt to 30 ft. lbs. and the pitman arm shaft nut to 185 ft. lbs. Bleed the power steering system.

Power Steering Pump

REMOVAL & INSTALLATION

1. Disconnect and cap the pressure hoses at the pump or the steering gear.

── **CAUTION** ──
On some engines, the power steering pump is located low on the engine. Do not attempt to check the fluid level with the engine running or personal injury can result.

2. Loosen the bracket-to-pump mounting bolts and remove the drive belt.
3. Remove the bracket-to-pump mounting bolts and remove the pump.
4. To install, reverse the removal procedures. Bleed the pump of air by turning the pulley counterclockwise until no bubbles appear in the reservoir. Bleed the power steering system.

BLEEDING

1. Fill the reservoir with power steering fluid.

NOTE: The use of automatic transmission fluid in the power steering system is NOT recommended; use power steering fluid ONLY.

2. Allow the reservoir and fluid to be undisturbed for a few minutes.
3. Start the engine, allow it to run for approximately 3–5 minutes to warm up the fluid, then turn it off.
4. Check the reservoir fluid level and add fluid, if necessary.
5. Repeat the above steps until the fluid level stabilizes.
6. Raise the front of the vehicle so that the wheels are off of the ground and set the parking brake.
7. Start the engine and increase the engine speed to about 1500 rpm.
8. Turn the front wheels right to left (and back) several times, lightly contacting the wheel stops at the ends of travel.

9. Check the reservoir fluid level. Add fluid as required.
10. Repeat Step 8 until the fluid level in the reservoir stabilizes.
11. Lower the vehicle and repeat Steps 8 and 9.

Steering Wheel

REMOVAL & INSTALLATION

Standard Wheel

1. Disconnect the negative battery cable.
2. Remove the shroud screws from the underside of the steering column, then lift the shroud and the horn contact lead assembly from the steering wheel.
3. Remove the snap ring and the steering wheel nut.
4. Using the steering wheel puller tool J-2927, pull the steering wheel from the steering column shaft.
5. To install, center the steering wheel and reverse the removal procedures. Torque the steering wheel nut to 31 ft. lbs.

Padded Rim Wheel

1. Disconnect the negative battery cable.
2. Pry out the center cap and retainer, then remove the shaft snapring.

NOTE: On the tilt-telescope wheel, remove the three upper contact retaining screws, the contact and the shim (if used), then the center star screw and lever.

3. Remove the steering wheel nut and washer.
4. Remove the receiving cup screws, then the cup belleville spring, bushing and pivot ring.
5. Mark the wheel-to-shaft relationship and remove the wheel with a puller.
6. To install, center the steering wheel on the shaft and reverse the removal procedures. Torque the steering wheel nut to 30 ft. lbs.

Turn Signal Switch

REMOVAL & INSTALLATION

1. Refer to the "Steering Wheel Removal & Installation" procedures in this section and remove the steering wheel.
2. Remove the column-to-instrument panel trim cover.
3. Using a small screwdriver, carefully pry the cover up and out of the steering column.
4. Position the lockplate compress-

ing tool J-23653 or equivalent on the end of the steering shaft and compress the lock plate by turning the shaft nut clockwise. Pry the wire snapring out of the shaft groove and discard the ring.

5. Remove the tool and lift the lockplate off the shaft.

6. Slide the cancelling cam, upper bearing preload spring and thrust washer off the shaft.

7. Remove the retaining screw and the turn signal lever. Push the flasher knob in and unscrew it.

NOTE: If equipped with a button and a knob, remove the button retaining screw, then remove the button, spring and knob.

8. Remove the mounting screws. Pull the switch connector out of the jacket bracket, wrap it with tape, lift up on the switch and pull the connector through the column support bracket.

NOTE: On tilt wheels, place the turn signal and shifter housing in the low position and remove the harness cover.

9. To install, attach a long piece of wire to the turn signal switch connector, feed the wire through the column housing and under the bracket, then pull the wire, the switch connector and the cover into position.

NOTE: On tilt wheel models, pull the connector down through the housing under the bracket, then install the cover over the harness.

10. Install the switch mounting screws and the connector on the jacket bracket. Install the column-to-dash trim plate.

11. Install the flasher knob and the turn signal lever.

12. With the turn signal lever in the neutral or off position and the flasher knob out, slide the thrust washer, upper bearing preload spring and cancelling cam onto the shaft.

13. Using tool J-23653 or equivalent, press the lock plate down on the shaft and install a NEW snapring in the shaft groove. Do not reuse the old snap ring.

14. Install the cover and the steering wheel. Torque the turn signal switch screws to 3 ft. lbs. and the steering wheel-to-shaft nut to 35 ft. lbs.

Ignition Switch

REMOVAL & INSTALLATION

The switch is located inside the chan-

nel section of the brake pedal support and is completely inaccessible without first lowering the steering column. The switch is actuated by a rod and rack assembly. A gear on the end of the lock cylinder engages the toothed upper end of the rod.

1. Support and lower the steering column.

2. Place the ignition switch in the OFF-UNLOCKED position and move the actuating rod two detents from the top.

3. Remove the two mounting screws and the ignition switch assembly.

4. Before installing, place the new switch in the OFF-UNLOCKED position and make sure the ignition lock cylinder and the actuating rod are in the OFF-UNLOCKED (2nd detent from the top) position.

5. Install the actuating rod into the switch, mount the switch to the column and torque the mounting screws to 3 ft. lbs.

NOTE: Use only the specified screws since over length screws could impair the collapsibility of the column.

6. Install the steering column and torque the steering column-to-bracket nuts to 25 ft. lbs.

Ignition Lock Cylinder

REMOVAL & INSTALLATION

1. Disconnect the battery ground cable, then refer to the "Turn Signal Switch Removal & Installation" procedure in this section and lift the turn signal switch to allow access to the lock cylinder.

NOTE: When lifting the turn signal switch, pull it rearward far enough to slip it over the end of the steering shaft. DO NOT pull the wiring harness from the column.

2. Place the ignition lock in the ON position.

3. Remove the buzzer switch, the lock cylinder screw and lock cylinder.

——————— CAUTION ———————

If the screw is dropped during removal and falls into the column, complete disassembly will be necessary to retrieve the screw.

4. To install, rotate the cylinder clockwise to align cylinder key with the keyway in the housing. Push the lock in all the way, then install the screw and torque it to 22 inch lbs. for adjustable columns and 40 inch lbs. for standard.

CHASSIS ELECTRICAL

Headlight Switch

REMOVAL & INSTALLATION

1980–84 Caprice And Impala

1. Disconnect the negative battery cable.

2. Pull the knob out to ON position.

3. Reach under the instrument panel and depress the switch shaft retainer. Remove the knob and shaft assembly, then the windshield wiper switch.

4. Remove the retaining ferrule nut.

5. Remove the switch from instrument panel.

6. Disconnect the multi-plug connector from the switch.

7. To install, reverse the removal reverse the removal procedure.

1985 and Later Caprice And Impala

1. Disconnect the negative battery cable.

2. Remove the steering column trim cover.

3. Remove the headlight switch mounting screws.

4. Pull the switch and disconnect the lighting connector.

5. To install, reverse the removal procedures.

Malibu and Monte Carlo

1. Disconnect negative battery cable.

2. Remove the screws and the instrument panel pad.

3. Remove the three windshield wiper/light switch mounting screws.

4. Pull the knob to the ON position.

5. Reach behind the instrument panel and depress the switch shaft retainer, then remove the knob and shaft assembly.

6. Remove the ferrule nut and the switch assembly from the instrument panel.

7. To install, reverse the removal procedures.

Speedometer Cable

REMOVAL & INSTALLATION

Caprice And Impala

1. Disconnect the negative battery cable.

2. Remove the attaching screws and lower the steering column bottom cover.

3. Disconnect the shift lever indicator from the steering column.

4. Unbolt the column from the instrument panel.

5. Remove the screws and the plastic snap retainers, then lift off the lens.

6. Remove the screws from the upper surface of the grey sheet metal trim plate.

7. Remove the nuts from two studs in the lower corner of the cluster.

8. Reach behind the cluster and depress the cable retaining clip, then remove the speedometer cable.

9. Pull the core from the casing.

NOTE: If the core is broken, raise the vehicle and disconnect the cable from the transmission.

10. To install, lubricate the new cable core with speedometer cable lubricant and reverse the removal procedures.

Malibu And Monte Carlo

1. Disconnect the negative battery cable.

2. Remove the radio knobs and the clock stem.

3. Remove the instrument bezel retaining screws.

4. Disconnect the tailgate release or defogger switch.

5. Remove the instrument cluster bezel.

6. Remove the speedometer head.

7. Disconnect the cable from the head by depressing the clip.

8. Pull the core from the casing. If the core is broken, raise the vehicle and remove the lower cable end from the transmission.

9. To install, lubricate the core with cable lubricant and reverse the removal procedures.

Instrument Cluster

REMOVAL & INSTALLATION

1. Disconnect the negative battery cable.

2. Remove the screws securing the steering column lower cover and remove the cover.

3. Disconnect the shift indicator cable from the steering column.

4. Remove the screws securing the steering column to the instrument panel and lower the steering column.

5. Remove the screws and the snap-in-plastic fasteners from the perimeter of the instrument cluster lens.

6. Remove the 2 screws from the

upper surface of the grey sheet metal trim plate.

7. Remove the nuts from the two studs in the lower corner of the cluster.

8. Reach behind the instrument cluster and disconnect the speedometer cable, then remove the cluster by pulling outward.

9. To install, reverse the removal procedure.

Wiper Motor

REMOVAL & INSTALLATION

All Models

1. Disconnect the negative battery cable.

2. Remove the cowl vent screen.

3. Remove the transmission drive link(s) from the motor crank arm.

4. Disconnect the electrical connectors and the washer hoses.

5. Remove the 3 motor mounting screws and the motor.

NOTE: When removing the motor, guide the crank arm through the hole.

6. To install, reverse the removal procedures. The motor must be in the "Park" position before assembling the crank arm to the transmission drive link(s).

Wiper Arms And Blades

REMOVAL & INSTALLATION

If the wiper assembly has a press type release tab at the center, simply depress the tab and remove the blade. If the blade had no release tab, use a screwdriver to depress the spring at the center. This will release the assembly. To install the assembly, position the blade over the pin at the tip of the arm and press until the spring retainer engages the groove in the pin. To remove the element, either depress the release button or squeeze the spring type retainer clip at the outer end together and slide the blade element out. Just slide the new element in until it latches.

1. Insert tool J-8966 or equivalent under the wiper arm and lever the arm off the shaft.

2. If equipped, disconnect the washer hose from the arm. Remove the arm.

3. To install, reverse the removal procedures.

NOTE: Be sure that the motor is in the "Park" position before installing the wiper arms.

Blower Motor

REMOVAL & INSTALLATION

Caprice And Impala

1. Disconnect the negative battery cable.

2. Disonnect the blower lead wire.

3. Remove the attaching screws and gently pry the blower from the case.

NOTE: When removing the blower motor, the sealer may act as an adhesive.

4. To install, reverse the removal procedures. Replace the sealer if it was damaged.

Malibu And Monte Carlo

1. Disconnect the negative battery cable.

2. Disconnect the motor lead wire.

NOTE: If equipped with A/C, disconnect the cooling tube.

3. Remove the blower-to-case screws.

4. Remove the retaining nut, then separate the motor from the wheel.

5. To install, place the open end of the blower away from the motor and reverse the removal procedures.

Heater Core

REMOVAL & INSTALLATION

Caprice And Impala

WITHOUT A/C

1. Disconnect the negative battery cable.

2. Drain the cooling system.

3. Disconnect and plug the heater hoses at the core and the core tubes.

NOTE: It may be necessary to remove the inner fender to remove the heater hoses.

4. Remove the screws from the perimeter of the core cover on the engine side of the cowl.

5. Pull the core cover from the cowl mounting.

6. Pull the core assembly from the module.

7. To install, reverse the removal procedures.

WITH A/C

1. Disconnect the negative battery cable and drain the cooling system.

2. Disconnect and plug the heater hoses at the core tubes.

NOTE: It may be necessary to remove the inner fender to remove the heater hoses.

3. Remove the module retaining bracket and the ground strap.

4. Remove the module rubber seal and the screen.

5. Remove the right windshield wiper arm.

6. Remove the diagnostic connector, the high blower relay and the thermal switch mounting screws.

7. Remove the electrical connectors from the top of the module.

8. Remove the module top cover and lift out the core.

9. To install, reverse the removal procedures.

Malibu And Monte Carlo

WITHOUT A/C

1. Disconnect the negative battery cable and drain the cooling system.

2. Disconnect and plug the heater hoses from the core tubes.

3. Disconnect the electrical connectors from the module case.

4. Unbolt and remove the module's front cover and lift out the core.

5. To install, use sealant and reverse the removal procedures.

WITH A/C

1. Disconnect the negative battery cable and drain the cooling system.

2. Disconnect and plug the heater hoses at the core tubes.

3. Remove the retaining bracket and the ground strap.

4. Remove the module's rubber seal and screen.

5. Remove the right windshield wiper arm.

6. Remove the diagnostic connector, the high blower relay and the thermostatic switch.

7. Disconnect the electrical connectors from the module.

8. Remove the module's top cover and the screen.

9. To install, use sealant and reverse the removal procedures.

Radio

REMOVAL & INSTALLATION

1980–84 Caprice And Impala

1. Disconnect the negative battery cable.

2. Pull the control knobs off of the radio.

3. Remove the 3 mounting screws and the trim plate.

4. Remove the 2 receiver bracket-to-instrument panel screws and bottom nut.

5. Detach the wiring and the antenna from the rear of the receiver.

6. Remove the radio and the mounting bracket.

7. To install, reverse the removal procedures.

NOTE: To prevent damage to the radio, always connect the speaker wiring harness before applying power to the radio.

1980–83 Malibu, 1980–84 Monte Carlo

1. Disconnect the negative battery cable.

2. Pull the control knobs from the shafts.

3. Remove the trim plate screws and the trim plate.

4. Remove the wiring and antenna cable from the rear of the radio.

5. Remove the receiver stud nut at the right side bracket.

6. Remove the control knob nuts.

7. Remove the instrument panel bracket screws and the bracket.

8. Remove the radio through the panel opening.

9. To install, reverse the removal procedures.

NOTE: To prevent damage to the radio, always connect the speaker wiring harness before applying power to the radio.

1985 And Later Models

1. Disconnect the negative battery cable.

2. Remove the glove box, then remove the temperature control cable from the temperature door.

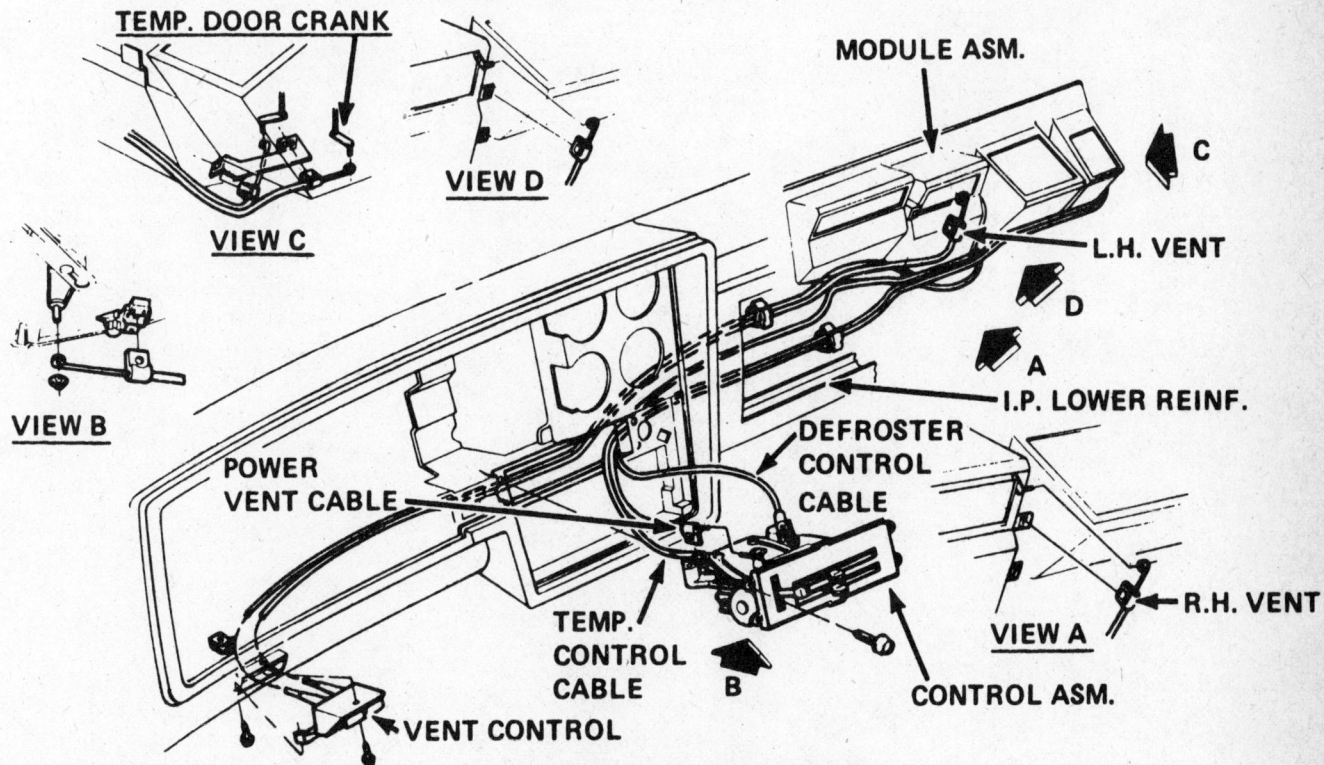

Chevrolet heater control cable adjustments

3. Remove the radio, heater and A/C assembly control panel-to-dash fasteners.

4. Pull the panel from the dash, then remove the vacuum and the electrical connectors.

5. Remove the A/C control trim plate and knobs.

6. Remove the radio from the bracket.

7. To install, reverse the removal procedures.

NOTE: To prevent damage to the radio, always connect the speaker wiring harness before applying power to the radio.

FUSES

The fuse block is located on a swing down unit beneath the instrument panel next to the steering column or through the glove box opening (on some models). Fuse holders are labeled as to their service and the correct amperage. Always replace blown fuses with new ones of the correct amperage. Otherwise electrical overloads and possible wiring damage will result.

Fusible Links

Fusible links are sections of wire, with the special insulation, designed to melt under electrical overload. Replacements are simply spliced into the wires. There may be as many as five of these in the engine compartment wiring harness. These are:

1. Horn relay-to-fuse panel circuit — 1 link.

2. Charging circuit, from the starter solenoid-to-horn relay — 2 links.

3. Starter solenoid-to-ammeter circuit — 1 link.

4. Horn relay-to-rear window defroster circuit — 1 link.

NOTE: The fusible links are all two wire gauge sizes smaller than the wires they protect.

REPLACEMENT

1. Disconnect the negative battery cable.

2. Disconnect the fusible link from the junction block or the starter solenoid.

3. Cut the harness directly behind the connector to remove the damaged fusible link.

4. Strip the harness wire approximately $1/2$ in.

5. Connect the new fusible link to the harness wire using a crimp on connector. Solder the connection using rosin core solder.

6. Tape all of the exposed wires with plastic electrical tape.

7. To complete the installation, reverse the removal procedures.

Oldsmobile
Rear Wheel Drive
Cutlass, 88, 98

YEAR IDENTIFICATION

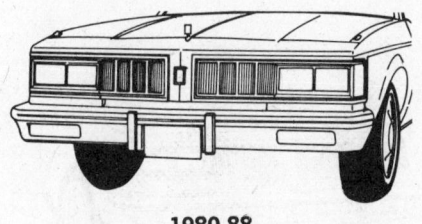

1980 88

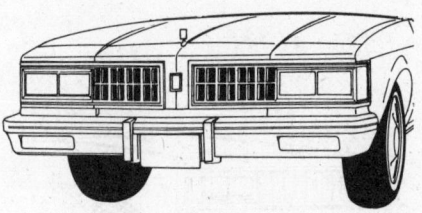

1981 88

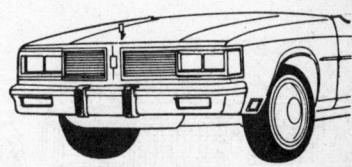

1982 Delta 88

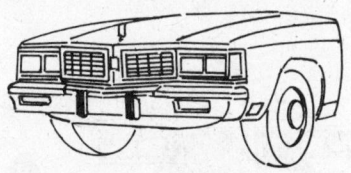

1983–84 Delta 88

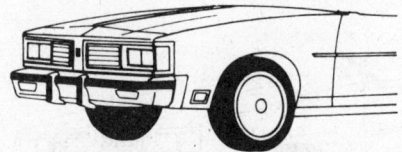

1985 Delta 88 Royale Coupe

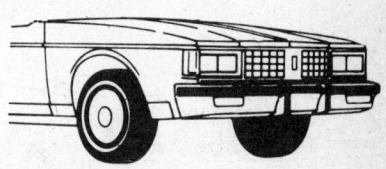

1985 Delta 88 Royale Brougham LS Sedan

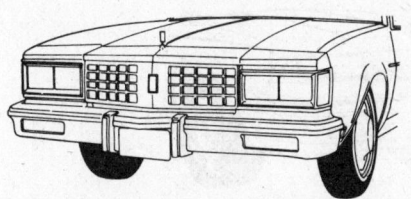

1980 98

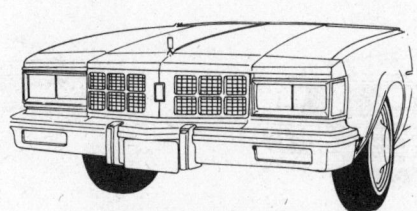

1981-82 98

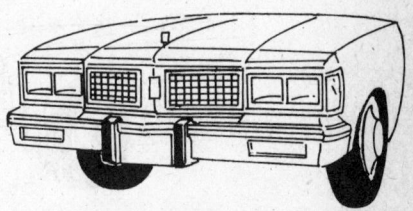

1983–84 98

C435

YEAR IDENTIFICATION

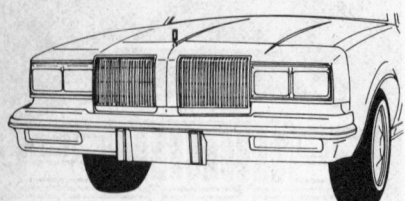

1980 Cutlass Supreme

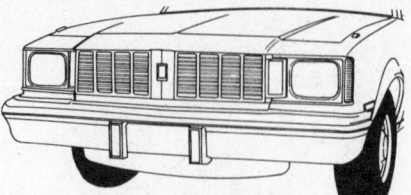

1980 Cutlass Salon

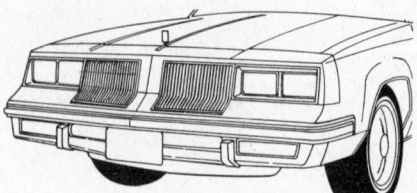

1981-82 Cutlass Supreme

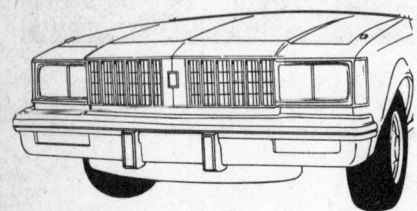

1981-82 Cutlass Salon

1981 Cutlass Supreme Brougham

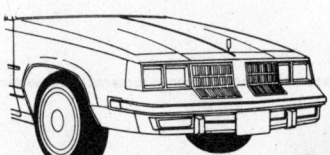

1982-83 Cutlass Supreme

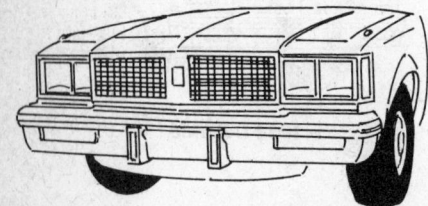

1983 Cutlass Supreme Brougham Sedan

1985-87 Cutlass Supreme Brougham

1984 Cutlass Supreme Coupe

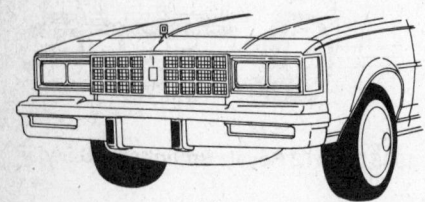

1984 Cutlass Supreme Sedan

1985-87 Cutlass Supreme

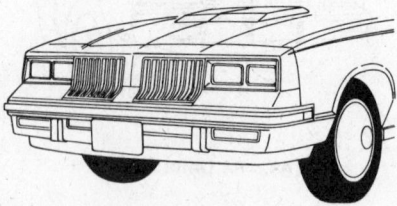

1984-87 Hurst Olds

1986 Cutlass Salon

1987 Custom Cruiser

VEHICLE IDENTIFICATION NUMBER (VIN)

It is important for servicing and ordering parts to be certain of the vehicle and engine identification. The VIN (vehicle identification number) is a 13 or 17 digit number visible through the windshield on the driver's side of the dash and contains the vehicle and engine identification codes. It can be interpreted as follows:

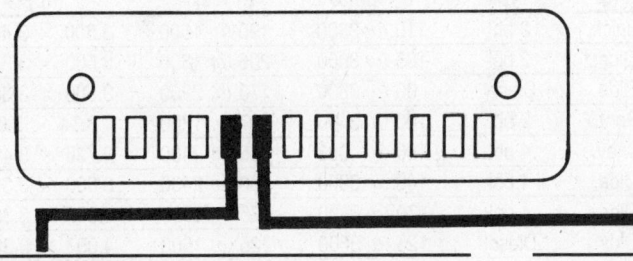

Engine Code						Model Year Code	
Code	Cu. In.	Liters	Cyl.	Carb.	Eng. Mfg.	Code	Year
C	231	3.8	6	2bbl	Buick	A	1980
A	231	3.8	6	2bbl	Buick		
F	260	4.3	6	2bbl	Olds.		
P	260	4.3	8	Diesel	Olds.		
S	265	4.3	8	2bbl	Pont.		
U	305	5.0	8	2bbl	Chev.		
H	305	5.0	8	4bbl	Chev.		
Y	307	5.0	8	4bbl	Olds.		
9	307	5.0	8	4bbl	Olds.		
R	350	5.7	8	4bbl	Olds.		
N	350	5.7	8	Diesel	Olds.		

The thirteen digit Vehicle Identification Number can be used to determine engine application and model year. The 6th digit indicates the model year, and the 5th digit identifies the factory installed engine.

VEHICLE IDENTIFICATION NUMBER (VIN)

It is important for servicing and ordering parts to be certain of the vehicle and engine identification. The VIN (vehicle identification number) is a 13 or 17 digit number visible through the windshield on the driver's side of the dash and contains the vehicle and engine identification codes. It can be interpreted as follows:

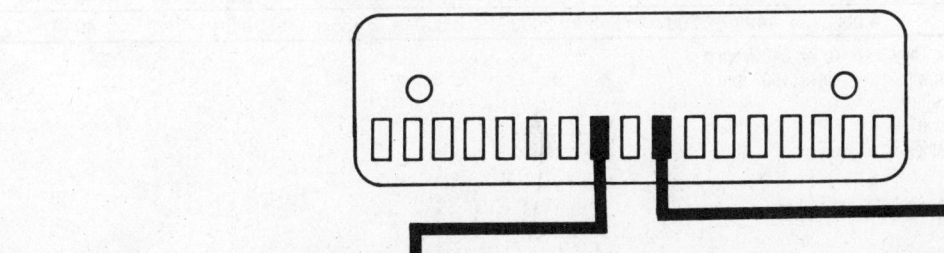

Engine Code						Model Year Code	
Code	Cu. Inc.	Liters	Cyl.	Carb.	Eng. Mfg.	Code	Year
A	231	3.8	6	2bbl	Buick	B	1981
4	252	4.1	6	4bbl	Buick	C	1982
F	260	4.3	8	2bbl	Olds.	D	1983
V	263	4.3	6	Diesel	Olds.	E	1984
Y	307	5.0	8	4bbl	Olds.	F	1985
9	307	5.0	8	4bbl	Olds	G	1986
N	350	5.7	8	Diesel	Olds.	H	1987

The seventeen digit Vehicle Identification Number can be used to determine engine application and model year. The 10th digit indicates the model year and the 8th digit identifies the factory installed engine.

GENERAL ENGINE SPECIFICATIONS

Year	Eng. VIN Code	Engine No. Cyl. Displacement (cu. in.)	Eng. Mfg.	Carburetor Type	Horsepower @ rpm ■	Torque @ rpm (ft lbs) ■	Bore × Stroke (in.)	Compression Ratio	Oil Pressure @ 2000 rpm
'80	A	6-231	Buick	2 bbl	110 @ 3800	190 @ 1600	3.800 × 3.400	8.0:1	37
	F	8-260	Olds.	2 bbl	105 @ 3600	205 @ 1800	3.500 × 3.385	7.5:1	40
	P	8-260	Olds.	Diesel	90 @ 3600	170 @ 2200	3.500 × 3.385	22.5:1	40
	S	8-265	Pont.	2 bbl	100 @ 3400	207 @ 1800	3.74 × 3.00	8.2:1	40
	H	8-305	Chev.	4 bbl	160 @ 4000	235 @ 2400	3.736 × 3.480	8.6:1	45
	Y	8-307	Olds.	4 bbl	148 @ 3800	250 @ 2400	3.800 × 3.385	7.9:1	40
	R	8-350	Olds.	4 bbl	170 @ 3800	275 @ 2000	4.057 × 3.385	8.0:1	40
	N	8-350	Olds.	Diesel	125 @ 3600	225 @ 1600	4.057 × 3.385	22.5:1	40
'81	A	6-231	Buick	2 bbl	110 @ 3800	190 @ 1600	3.800 × 3.400	8.0:1	37
	4	6-252	Buick	4 bbl	125 @ 4000	205 @ 2000	3.965 × 3.400	8.0:1	37①
	F	8-260	Olds.	2 bbl	105 @ 3600	205 @ 1800	3.500 × 3.385	7.5:1	40②
	Y	8-307	Olds.	4 bbl	148 @ 3800	250 @ 2400	3.800 × 3.385	8.0:1	40②
	N	8-350	Olds.	Diesel	125 @ 3600	225 @ 1600	4.057 × 3.385	22.5:1	40②
'82	A	6-231	Buick	2 bbl	110 @ 3800	190 @ 1600	3.800 × 3.400	8.0:1	37
	4	6-252	Buick	4 bbl	125 @ 4000	205 @ 2000	3.965 × 3.400	8.0:1	37①
	F	8-260	Olds.	2 bbl	105 @ 3600	205 @ 1800	3.500 × 3.385	7.5:1	40②
	V	6-263	Olds.	Diesel	85 @ 3600	165 @ 1600	4.057 × 3.385	21.6:1	40
	Y	8-307	Olds.	4 bbl	148 @ 3800	250 @ 2400	3.800 × 3.385	8.0:1	40②
	N	8-350	Olds.	Diesel	125 @ 3600	225 @ 1600	4.057 × 3.385	22.5:1	40②
'83	A	6-231	Buick	2 bbl	110 @ 3800	190 @ 1600	3.800 × 3.400	8.0:1	37
	4	6-252	Buick	4 bbl	125 @ 4000	205 @ 2000	3.965 × 3.400	8.0:1	37①
	F	8-260	Olds.	2 bbl	105 @ 3600	205 @ 1800	3.500 × 3.385	7.5:1	40②
	V	6-263	Olds.	Diesel	85 @ 3600	165 @ 1600	4.057 × 3.385	21.6:1	40
	Y	8-307	Olds.	4 bbl	148 @ 3800	250 @ 2400	3.800 × 3.385	8.0:1	40②
	N	8-350	Olds.	Diesel	125 @ 3600	225 @ 1600	4.057 × 3.385	22.5:1	40②
'84–'85	A	6-231	Buick	2 bbl	110 @ 3800	190 @ 1600	3.800 × 3.400	8.0:1	37
	4	6-252	Buick	4 bbl	125 @ 4000	205 @ 2000	3.965 × 3.400	8.0:1	37①
	V	6-263	Olds.	Diesel	85 @ 3600	165 @ 1600	4.057 × 3.385	21.6:1	40
	Y	8-307	Olds.	4 bbl	148 @ 3800	250 @ 2400	3.800 × 3.385	8.0:1	40②
	N	8-350	Olds.	Diesel	125 @ 3600	225 @ 1600	4.057 × 3.385	22.5:1	40②
'86–'87	A	6-231	Buick	2 bbl	110 @ 3800	190 @ 1600	3.800 × 3.400	8.0:1	37
	Y	8-307	Olds.	4 bbl	140 @ 3200	255 @ 2000	3.800 × 3.385	8.0:1	40②
	9	8-307	Olds.	4 bbl	140 @ 3200	255 @ 2000	3.800 × 3.385	8.0:1	40②

■ Horsepower and torque are SAE net figures. They are measured at the rear of the transmission with all accessories installed and operating. Since the figures vary when a given engine is installed in different models, some are representative rather than exact.

① @ 2400 rpm
② @ 1500 rpm

TUNE-UP SPECIFICATIONS
Cutlass, Omega

(When analyzing compression test results, look for uniformity among cylinders rather than specific pressures.)

Year	Eng. VIN Code	No. Cyl Displacement (cu in.)	Eng. Mfg.	hp	Spark Plugs Orig. Type	Spark Plugs Gap ● (in.)	Distributor Point Dwell (deg.)	Distributor Point Gap (in.)	Ignition Timing (deg.)● Man. Trans.	Ignition Timing (deg.)● Auto. Trans.	Valves Intake Opens ■(deg.)●	Fuel Pump Pressure (psi)	Idle Speed (rpm) ▲ Man. Trans. ●	Idle Speed (rpm) ▲ Auto. Trans.
'80	A	6-231	Buick	110	R-45TS (R-45TSX)	0.040 0.060	Electronic		15B	15B	16	3–4.5	800/600	670/550 (620/550)
	F	8-260	Olds.	All	R-46SX	0.080	Electronic		—	20B⑦	—	5.5–6.5	—	625/500

TUNE-UP SPECIFICATIONS
Cutlass, Omega

(When analyzing compression test results, look for uniformity among cylinders rather than specific pressures.)

Year	Eng. VIN Code	No. Cyl Displacement (cu in.)	Eng. Mfg.	hp	Spark Plugs Orig. Type	Gap ● (in.)	Distributor Point Dwell (deg.)	Point Gap (in.)	Ignition Timing (deg.)● Man. Trans.	Auto. Trans.	Valves Intake Opens ■(deg.)●	Fuel Pump Pressure (psi)	Idle Speed (rpm) ▲ Man. Trans. ●	Auto. Trans.
'80	H	8-305	Chev.	All	R-45TS	(0.045)	Electronic		—	4B	28	7.5–9	—	600(650)/500(550)
	R	8-350	Olds.	All	R-46SX	0.080	Electronic		—	18B	16	5.5–6.5	—	600(650)/500(550)
'81	A	6-231	Buick	All	R-45TSX	0.080	Electronic		15B	15B	16	4.25–5.75	①	①
	F	8-260	Olds.	All	R-46SX	0.080	Electronic		—	20B②	14	5.5–6.5	—	①
	Y	8-307	Olds.	All	R-46SX	0.080	Electronic		—	15B	14	5.5–6.5	—	①
'82	A	6-231	Buick	All	R45TX	0.040	Electronic		—	①	16	4.25–5.75	—	①
	F	8-260	Olds.	All	R46SX	0.080	Electronic		—	①	—	5.5–6.5	—	①
	Y	8-307	Olds.	All	R46SX	0.080	Electronic		—	①	—	6–7.5	—	①
'83	A	6-231	Buick	All	R45TX	0.040	Electronic		—	①	16	4.25–5.75	—	①
	Y	8-307	Olds	All	R46SX	0.080	Electronic		—	①	—	6–7.5	—	①
'84–'85	A	6-231	Buick	All	R45TX	0.040	Electronic		—	①	16	4.25–5.75	—	①
	Y	8-307	Olds.	All	R46SX	0.080	Electronic		—	①	—	6–7.5	—	①
'86	A	6-231	Buick	110	R45TSX	0.060	Electronic		—	15B	—	5.5–6.5	—	①
	Y	8-307	Olds.	140	FR3LS6	0.060	Electronic		—	20B	—	5.5–6.5	—	600
	9	8-307	Olds.	140	FR3LS	.060	Electronic		—	20B	—	5.5–6.5	—	600
'87	SEE UNDERHOOD SPECIFICATIONS STICKER													

NOTE: The underhood specifications sticker often reflects tune-up specification changes made in production. Sticker figures must be used if they disagree with those in this chart. Part numbers in this chart are not recommendations by Chilton for any product by brand name.

■ All figures Before Top Dead Center
● Figure in parentheses indicates California engine. Where two idle speed figures appear separated by a slash, the second is with the idle speed solenoid disconnected.
B Before Top Dead Center
— Not applicable
① See underhood sticker
② Station wagon—18B @ 1100

DIESEL TUNE-UP SPECIFICATIONS

Year	Eng. V.I.N. Code	Engine No. Cyl. Displacement (Cu. in.)	Eng. Mfg.	Fuel Pump Pressure (psi)	Compression (lbs)	Intake Valve Opens (deg)	Idle Speed ● (rpm)
'80	N	8-350	Olds.	5.5-6.5	275 min.	16	750/600
'81	N	8-350	Olds.	5.5-6.5	275 min.	16	①
'82	V	6-263	Olds.	5-6	②	②	①
	N	8-350	Olds.	5.5-6.5	275 min.	16	①
'83	V	6-263	Olds.	5-6	②	②	①
	N	8-350	Olds.	5.5-6.5	275 min.	16	①
'84–'85	V	6-263	Olds.	5-6	②	—	①
	N	8-350	Olds.	5.5-6.5	275 min.	—	①

NOTE: The underhood specifications sticker often reflects tuneup specification changes made in production. Sticker figures must be used if they disagree with those in this chart.
① See underhood specifications sticker
② Not available
● Where two idle speed figures appear separated by a slash, the first is idle speed with solenoid energized, the second is idle speed with solenoid disconnected.

FIRING ORDER

NOTE: To avoid confusion, always replace spark plug wires one at a time.

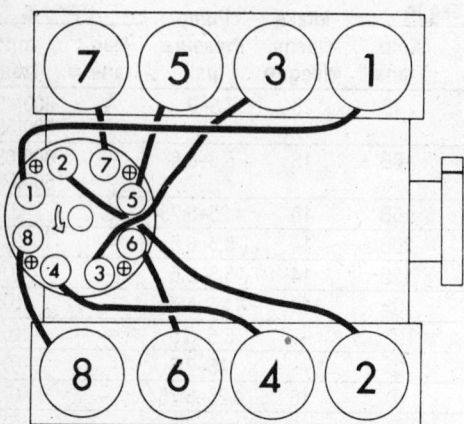

GM (Oldsmobile) 260 V8
Engine firing order: 1-8-4-3-6-5-7-2
Distributor rotation: counterclockwise

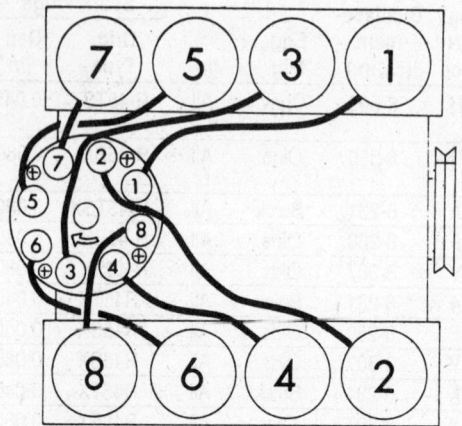

GM (Chevrolet) V8
Engine firing order: 1-8-4-3-6-5-7-2
Distributor rotation: clockwise

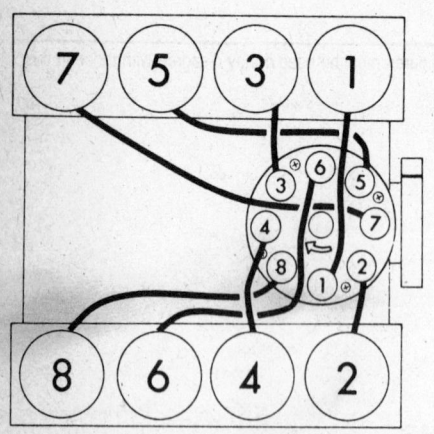

GM (Buick) Omega 350 V8
Engine firing order: 1-8-4-3-6-5-7-2
Distributor rotation: clockwise

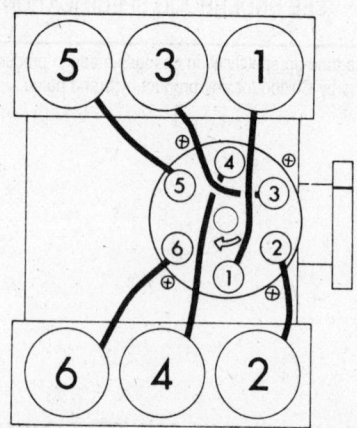

GM (Buick) 231, 252 V6
(3.2 L, 3.8 L, 4.1 L)
Engine firing order: 1-6-5-4-3-2
Distributor rotation: clockwise

V6 harmonic balancers have two timing marks: one is 1/8 in. wide, and one is 1/16 in. wide. Use the 1/16 in. mark for timing with a hand held light. The 1/8 in. mark is used only with a magnetic timing pick-up probe.

CAPACITIES
Cutlass, Omega

Year	Engine No. Cyl. Displacement (cu. in.)	Engine Crankcase Add 1 qt. for New Filter	Transmission (Pts. to Refill after Draining)			Drive Axle (pts.)	Gasoline Tank (gals.)	Cooling System (qts.)		
			3 sp	4sp/5sp	Automatic●			With Heater	With A/C	Heavy Duty Cooling
'80	6-231 Buick	4	—	3	6	3.5	18.5	11.9	12.4	—
	6-231 Buick	4	3	3	6	3.5	18	13	13	—
	8-260 Olds.	4	—	3	6	3.5	18	16	16.5	—
	8-350 Chev.	4	—	3	6	3.5	18	15.25	15.25	16
	8-350 Olds.	4	—	—	6	3.5	18	15	15	—
	8-350 Diesel	7①	—	—	6	3.5	18	17.25	17.25	—

CAPACITIES
Cutlass, Omega

Year	Engine No. Cyl. Displacement (cu. in.)	Engine Crankcase Add 1 qt. for New Filter	Transmission (Pts. to Refill after Draining)			Drive Axle (pts.)	Gasoline Tank (gals.)	Cooling System (qts.)		Heavy Duty Cooling
			3 sp	4sp/5sp	Automatic●			With Heater	With A/C	
'81	6-231 Buick	4	3	—	6	3.5	18.1	N.A.	N.A.	N.A.
	8-260 Olds.	4	—	—	6	3.5	18.1	15.9	15.6	15.5
	8-307 Olds.	4	—	—	6	3.5	18.1	14.9	15.6	15.5
	8-350 Diesel	7①	—	—	6	3.5	19.8	17.4	17.3	17.3
'82	6-231 Buick	4	3	—	6	3.5	18.2	13.3	13.3	—
	6-263 Diesel	6①	—	—	6	—	19.8	—	—	—
	8-260 Olds.	4	—	—	6	3.5	18.2	20.0	20.0	—
	8-305 Olds.	4	—	—	6	3.5	18.2	15.5	15.5	—
	8-350 Diesel	7①	—	—	6	3.5	19.8	18	18	—
'83	6-231 Buick	4	—	—	6	3.5	18.2	13.3	13.3	—
	6-263 Diesel	6①	—	—	6	3.5	19.8	12.9	12.9	—
	8-307 Olds	4	—	—	6	3.5	18.2	15.5	15.5	—
	8-350 Diesel	7①	—	—	6	3.5	19.8	18	18	—
'84–'85	6-231 Buick	4	—	—	6	3.5	19.8	13.3	13.3	—
	6-263 Diesel	6①	—	—	6	3.5	19.8	12.9	12.9	—
	8-307 Olds.	4	—	—	6	3.5	18.2	15.5	15.5	—
	8-350 Diesel	7①	—	—	6	3.5	19.8	18	18	—
'86–'87	6-231 Buick	4	—	—	7	3.5	18.1	13	13	13.5
	8-307 Olds.	4	—	—	7	3.5	18.1	14.9	15.6	15.5
	8-307 Olds.	4	—	—	7	3.5	22.0	—	15.5	16.0

● Check dip stick and gradually fill to correct level.
  See the General Maintenance section of Unit
  Repair.
① Includes mandatory filter change

CAPACITIES
Oldsmobile 88, 98

Year	Engine No. Cyl. Displacement (cu. in.)	Engine Crankcase Add 1 qt. for New Filter	Transmission (Pts. to Refill after Draining) Automatic●	Drive Axle (pts.)	Gasoline Tank (gals.)	Cooling System (qts.)		Heavy Duty Cooling
						With Heater	With A/C	
'80	6-231 Buick	4	6	④	20.75	13.0	13.0	—
	8-307 Olds.	4	6	④	25①	15.5	15.25	16.25
	8-350 Olds.	4	6	④	25②	14.5	14.5	15.5
	8-350 Diesel	7⑤	6	④	27③	18.25	18.0	—
'81	6-231 Buick	4	6	4	25③	13.7	13.7	—
	6-252 Buick	4	6	4	25③	13.7	13.7	—
	8-260 Pont.	4	6	4	25③	15.9	15.5	16.6
	8-307 Olds.	4	6	4	25③	14.9	15.6	15.6
	8-350 Diesel	7⑤	6	4	27③	18.0	18.0	18.0
'82	6-231 Buick	4	6	4	25③	13.7	13.7	—
	6-252 Buick	4	6	4	25③	13.7	13.7	—
	8-260 Olds.	4	6	4	25③	16	16.5	—
	8-263 Diesel.	6⑤	6	4	27	⑥	⑥	—
	8-307 Olds.	4	6	4	25③	17.5	17.5	—
	8-350 Diesel	7⑤	6	4	27③	18.0	18.0	18.0

CAPACITIES
Oldsmobile 88, 98

Year	Engine No. Cyl. Displacement (cu. in.)	Engine Crankcase Add 1 qt. for New Filter	Transmission (Pts. to Refill after Draining) Automatic●	Drive Axle (pts.)	Gasoline Tank (gals.)	Cooling System (qts.)		Heavy Duty Cooling
						With Heater	With A/C	
'83	6-231 Buick	4	6	4	25③	13.7	13.7	—
	6-252 Buick	4	6	4	25③	13.7	13.7	—
	8-307 Olds.	4	6	4	25③	17.5	17.5	—
	8-350 Diesel	7⑤	6	4	27③	18	18	—
'84–'85	6-231 Buick	4	6	4	25③	13.7	13.7	—
	8-307 Olds.	4	6	4	25③	17.5	17.5	—
	8-350 Diesel	7⑤	6	4	27③	18	18	—

● Check dipstick and gradually fill to the correct level. See the General Maintenance section of the Unit Repair.
— Not applicable
① Royale, Royal Brougham Coupe and Sedan: 20.75

② 20.75 for Calif. 350 or w/power seats
③ 22 gals on station wagon
④ 7.5 inch ring gear: 3.5
   8.5 and 8.75 inch ring gear: 4.25
⑤ Includes mandatory filter change
⑥ Not available at time of publication

VALVE SPECIFICATIONS

Year	Engine No. Cyl. Displacement (cu. in.)	Seat Angle (deg.)	Face Angle (deg.)	Spring Test ■ Pressure (lbs @ in.)	Spring Installed Height (in.)	Stem-to-Guide Clearance (in.)		Stem Diameter (in.)	
						Intake	Exhaust	Intake	Exhaust
'80	6-231 Buick	45	45	168 @ 1.340	1⁴⁷⁄₆₄	.0015–.0035	.0015–.0032	.3402–.3412	.3405–.3412
	8-260 Olds.	①	④	187 @ 1.270	1⁴³⁄₆₄	.0010–.0027	.0015–.0032	.3425–.3432	.3420–.3427–
	8-260 Diesel	①	④	151 @ 1.300	1⁴³⁄₆₄	.0010–.0027	.0015–.0032	.3425–.3432	.3420–.3427
	8-305 Chev.	46	45	200 @ 1.160	③	.0010–.0037	.0010–.0037	.3414	.3414
	8-307 Olds.	①	④	187 @ 1.270	1⁴³⁄₆₄	.0010–.0027	.0015–.0032	.3429	.3424
	8-350 Chev.	46	45	200 @ 1.160	③	.0010–.0037	.0010–.0037	.3414	.3414
	8-350 Olds.	①	④	187 @ 1.270	1⁴³⁄₆₄	.0010–.0027	.0015–.0032	.3425–.3432	.3420–.3427
	8-350 Diesel	①	④	151 @ 1.300	1⁴³⁄₆₄	.0010–.0027	.0015–.0032	.3425–.3432	.3420–.3427
'81	6-231 Buick	45	45	182 @ 1.340	1⁴⁷⁄₆₄	.0015–.0035	.0015–.0032	.3407	.3409
	6-252 Buick	45	45	182 @ 1.340	1⁴⁷⁄₆₄	.0015–.0035	.0015–.0032	.3407	.3409
	8-260 Olds.	①	④	187 @ 1.270	1⁴³⁄₆₄	.0010–.0027	.0015–.0032	.3429	.3424
	8-307 Olds.	①	④	187 @ 1.270	1⁴³⁄₆₄	.0010–.0027	.0015–.0032	.3429	.3424
	8-350 Diesel	①	④	210 @ 1.22	1⁴³⁄₆₄	.0010–.0027	.0015–.0032	.3429	.3424
'82	6-231 Buick	45	45	182 @ 1.340	1⁴⁷⁄₆₄	.0015–.0035	.0015–.0032	.3407	.3409
	6-252 Buick	45	45	182 @ 1.340	1⁴⁷⁄₆₄	.0015–.0035	.0015–.0032	.3407	.3409
	6-263 Diesel	①	④	210 @ 1.220	—	.0010–.0027	.0015–.0032	.3429	.3429
	8-260 Olds.	①	④	187 @ 1.270	1⁴³⁄₆₄	.0010–.0027	.0015–.0032	.3429	.3424
	8-307 Olds.	①	④	187 @ 1.270	1⁴³⁄₆₄	.0010–.0027	.0010–.0032	.3429	.3429
	8-350 Diesel	①	④	210 @ 1.22	1⁴³⁄₆₄	.0010–.0027	.0015–.0032	.3429	.3429
'83	6-231 Buick	45	45	182 @ 1.340	1⁴⁷⁄₆₄	.0015–.0035	.0015–.0032	.3407	.3409
	6-252 Buick	45	45	182 @ 1.340	1⁴⁷⁄₆₄	.0015–.0035	.0015–.0032	.3407	.3409
	6-263 Diesel	①	④	210 @ 1.220	—	.0010–.0027	.0015–.0032	.3429	.3429
	8-307 Olds	①	④	187 @ 1.270	1⁴³⁄₆₄	.0010–.0027	.0015–.0032	.3429	.3429
	8-350 Diesel	①	④	210 @ 1.22	1⁴³⁄₆₄	.0010–.0027	.0015–.0032	.3429	.3429

VALVE SPECIFICATIONS

Year	Engine No. Cyl. Displacement (cu. in.)	Seat Angle (deg.)	Face Angle (deg.)	Spring Test ■ Pressure (lbs @ in.)	Spring Installed Height (in.)	Stem-to-Guide Clearance (in.)		Stem Diameter (in.)	
						Intake	Exhaust	Intake	Exhaust
'84–'85	6-231 Buick	45	45	182 @ 1.340	1⁴⁷⁄₆₄	.0015–.0035	.0015–.0032	.3407	.3409
	6-263 Diesel	①	④	210 @ 1.220	—	.0010–.0027	.0015–.0032	.3429	.3429
	8-307 Olds	①	④	187 @ 1.270	1⁴³⁄₆₄	.0010–.0027	.0010–.0032	.3429	.3429
	8-350 Diesel	①	④	210 @ 1.22	1⁴³⁄₆₄	.0010–.0027	.0015–.0032	.3429	.3429
'86–'87	6-231 Buick	46	45	63–71 @ 1.727	1.697–1.757	.0015–.0035	.0015–.0032	3.407	.3409
	8-307(4) Olds.	45⑤	46⑤	76–84 @ 1.67	⑥	.0010–.0027	.0015–.0032	3.428	3.424
	8-307(9) Olds.	45⑤	46⑤	85–95 @ 1.67	⑥	.0010–.0027	.0015–.0032	3.428	3.424

■ Valve open
① Intake 45°, exhaust 31°
② Intake 200 @ 1.25
③ Intake: 1⁴⁵⁄₆₄
   Exhaust: 1³⁹⁄₆₄
④ Intake 44°, exhaust 30°
⑤ Applies to intake. Exhaust 59, 60
⑥ Special tool GM BT-6428 or J-25289 must be used

CRANKSHAFT AND CONNECTING ROD SPECIFICATIONS

(All measurements are given in inches.)

Year	Engine No. Cyl. Displacement (cu. in.)	Crankshaft			Thrust on No.	Connecting Rod		
		Main Brg. Journal Dia	Main Brg. Oil Clearance	Shaft End-Play		Journal Diameter	Oil Clearance	Side Clearance
'80	6-231 Buick	2.4995	0.0003–0.0018	0.004–0.008	2	2.2487–2.2495	0.0005–0.0026	0.006–0.027
	8-260 Olds.	2.4985–2.4995①	0.0005–0.0021②	0.0035–0.0135	3	2.1238–2.1248	0.0004–0.0033	0.006–0.020
	8-305 Chev.	③	④	0.002–0.006	3	2.0986–2.0998	0.003 max	0.006–0.014
	8-307 Olds.	2.4985–2.4995①	0.0005–0.0021②	0.0035–0.0135	3	2.1238–2.1248	0.0004–0.0033	0.006–0.020
	8-350 Chev.	③	④	0.002–0.006	3	2.0986–2.0998	0.003 max	0.006–0.014
	8-350 Olds.	2.4985–2.4995①	0.0005–0.0021②	0.0035–0.0135	3	2.1238–2.1248	0.0004–0.0033	0.006–0.020
	8-260, 350 Diesel	2.9993–3.0003	0.0005–0.0021②	0.0035–0.0135	3	2.1238–2.1248	0.0005–0.0026	0.006–0.020
'81	6-231 Buick	2.4995	0.0003–0.0018	0.011–0.003	2	2.2487–2.2495	0.0005–0.0026	0.006–0.023
	6-252 Buick	2.4955	0.0003–0.0018	0.011–0.003	2	2.2487–2.2495	0.0005–0.0026	0.006–0.023
	8-260 Olds.	2.5000	0.0005–0.0021②	0.0035–0.0135	3	2.1238–2.1248	0.0004–0.0033	0.006–0.020
	8-307 Olds.	2.4990–2.4995	0.0005–0.0021②	0.0035–0.0135	3	2.1238–2.1248	0.0004–0.0033	0.006–0.020
	8-350 Diesel	2.9993–3.0003	0.0005–0.0021②	0.0035–0.0135	3	2.24995–2.500	0.0005–0.0026	0.006–0.020
'82	6-231 Buick	2.4955	0.0003–0.0018	0.011–0.003	2	2.2487–2.2495	0.0005–0.0026	0.006–0.023
	6-252 Buick	2.4955	0.0003–0.0018	0.011–0.003	2	2.2487–2.2495	0.0005–0.0026	0.006–0.023
	6-263 Diesel	2.9993–3.0003	0.0005–0.0021②	0.0035–0.0135	3	2.2490–2.2510	0.0005–0.0026	0.006–0.020
	8-260 Olds.	2.4990–2.4995⑦	0.0005–0.0021②	0.0035–0.0135	3	2.1238–2.1248	0.0004–0.0033	0.006–0.020
	8-307 Olds	2.4990–2.4995⑦	0.0005–0.0021②	0.0035–0.0135	3	2.1238–2.1248	0.0004–0.0033	0.006–0.020
	8-350 Diesel	2.9993–3.0003	0.0005–0.0021②	0.0035–0.0135	3	2.2495–2.2500	0.0005–0.0026	0.006–0.020
'83	6-231 Buick	2.4995	0.0003–0.0018	0.011–0.003	2	2.2487–2.2495	0.0005–0.0026	0.006–0.015
	6-252 Buick	2.4995	0.0003–0.0018	0.011–0.003	2	2.2487–2.2495	0.0005–0.0026	0.006–0.015
	6-263 Diesel	2.9993–3.0003	0.0005–0.0021②	0.0035–0.0135	3	2.2490–2.2510	0.0005–0.0026	0.006–0.020
	8-307 Olds	2.4990–2.4995⑦	0.0005–0.0021②	0.0035–0.0135	3	2.1238–2.1248	0.0004–0.0033	0.006–0.020
	8-350 Diesel	2.9993–3.0003	0.0005–0.0021②	0.0035–0.0135	3	2.2495–2.2500	0.0005–0.0026	0.006–0.020

CRANKSHAFT AND CONNECTING ROD SPECIFICATIONS

(All measurements are given in inches.)

Year	Engine No. Cyl. Displacement (cu. in.)	Crankshaft			Thrust on No.	Connecting Rod		
		Main Brg. Journal Dia	Main Brg. Oil Clearance	Shaft End-Play		Journal Diameter	Oil Clearance	Side Clearance
'84–'86	6-231 Buick	2.4995	0.0003–0.0018	0.011–0.003	2	2.2487–2.2495	0.0005–0.0026	0.006–0.015
	6-263 Diesel	2.9993–3.003	0.0005–0.0021②	0.0035–0.0135	3	2.2490–2.2510	0.0005–0.0026	0.006–0.020
	8-307 Olds.	2.4990–2.4995⑦	0.0005–0.0021②	0.0035–0.0135	3	2.1238–2.1248	0.0004–0.0033	0.006–0.020
	8-350 Diesel	2.9993–3.0003	0.0005–0.0021②	0.0035–0.0135	3	2.2495–2.2500	0.0005–0.0026	0.006–0.020
'86–'87	6-231	2.2487–2.2495	0.0003–0.0018	0.003–0.011	3	2.2487–2.2495	0.0005–0.0026	0.003–0.015
	8-307	2.4985–2.4995⑧	0.005–0.002⑨	0.0035–0.0135	3	2.1238–2.1248	0.0004–0.0033	0.006–0.020

① #1: 2.4988-2.4998
② #5: .0015-.0031
③ #1: 2.4484-2.4493
   #2,3,4: 2.4481-2.4490
   #5: 2.4479-2.4488
④ Front—.001-.0015;
   Intermediate—.001-.0025;
   Rear—.0025-.0035

⑤ #1: .0020 max
⑥ Diameter may also be 2.240
⑦ #2,3,4,5—#1 2.4993-2.4998
⑧ Applies to #2, 3, 4, 5—#1 2.4988-2.4998
⑨ Applies to 1, 2, 3, 4—#5—0.0015-0.0031

PISTON AND RING SPECIFICATIONS

(All measurements are given in inches. To convert inches to metric units, refer to the Metric Information section.)

Year	VIN Code	Engine Type/ Disp. (cu. in.)	Eng. Mfg.	Piston-to-Bore Clearance	Ring Gap			Ring Side Clearance		
					Top Compression	Bottom Compression	Oil Control	Top Compression	Bottom Compression	Oil Control
'81	F	8-260	Olds.	0.0008–0.0018	0.010–0.020①	0.010–0.020①	0.015–0.035	0.0020–0.0040	0.0020–0.0040	0.005–0.011
'80	R	8-350	Olds.	0.0008–0.0018	0.013–0.023②	0.013–0.023②	0.015–0.055	0.0015–0.0035	0.0015–0.0035	0.001–0.005
'80–'85	A,4	6-231, 252	Buick	0.0016–0.0038	0.013–0.023	0.013–0.023	0.015–0.055	0.0030–0.0050	0.0030–0.0050	0.0015–0.0035
'80	L,H	8-305, 350	Chevy.	0.0027 max.	0.010–0.030	0.010–0.035	—	0.012–0.0032	0.012–0.0032	0.005–0.011
'86	N	8-350	Olds. Diesel	0.005–0.006③	0.015–0.025	0.015–0.025	—	0.005–0.007	0.0018–0.0038	0.001–0.005
'80–'86	Y	8-307	Olds.	0.0008–0.0018	0.009–0.019	0.009–0.019	—	0.0020–0.0040	0.0020–0.0040	0.000–0.0035
'82–'86	V	6-263	Olds. Diesel	0.003–0.004	0.015–0.025	0.015–0.025	—	0.005–0.007	0.0018–0.0038	0.001–0.005
'86–'87	A	C-231	Buick	0.0013–0.0035	0.010–0.020	0.010–0.020	0.015–0.055	0.003–0.005	0.003–0.005	0.005–0.009
	Y	8-307	Olds.	0.00075–0.00175	0.009–0.019	0.009–0.019	0.015–0.055	0.0018–0.0038	0.0018–0.0038	0.001–0.005

① w/Sealed Power rings—.009-.009
② w/Sealed Power rings—.010-.020
③ At bottom of skirt

TORQUE SPECIFICATIONS

(All readings in ft. lbs.)

Year	Engine	Cylinder Head Bolts	Rod Bearing Bolts	Main Bearing Bolts	Crankshaft Bolt	Flywheel to Crankshaft Bolts	Manifold	
							Intake	Exhaust
'81	6-231 Buick	80	40	100	225	60	45	25
	6-252 Buick	80	40	100	225	60	45	25
	8-260 Olds.	85③	42	①	200–310	②	40③	25
	8-305 Chev.	65	45	70	60	60	30	20
	8-307 Olds.	130③	42	①	200–310	60	40③	25
	8-350 Buick	80	40	100	225	60	45	25
	8-350 Chev.	65	45	70	60	60	30	③
	8-350 Olds.	130③	42	①	200–310	60	40③	25
	8-260, 350 Diesel	130③	42	120	200–310	60	40③	25

TORQUE SPECIFICATIONS

(All readings in ft. lbs.)

Year	Engine	Cylinder Head Bolts	Rod Bearing Bolts	Main Bearing Bolts	Crankshaft Bolt	Flywheel to Crankshaft Bolts	Manifold Intake	Manifold Exhaust
'82	6-231 Buick	80	40	100	225	60	45	25
	6-252 Buick	80	40	100	225	60	45	25
	6-263 Diesel	142	42	107	160–350	48	41	29
	8-260 Olds.	85③	42	①	200–310	②	40③	25
	8-307 Olds.	130③	42	①	200–310	60	40③	25
	8-350 Diesel	130③	42	120	200–310	60	40③	25
'83–'85	6-231 Buick	80	40	100	225	60	45	25
	6-252 Buick	80	40	100	225	60	45	25
	6-263 Diesel	142	42	107	160–350	57	41	29
	8-307 Olds.	130③	42	①	200–310	60	40③	25
	8-350 Diesel	130③	42	120	200–310	60	40③	25
'86–'87	6-231 Buick	80	45	100	200	60	45	20
	8-307 Olds.	125③	42	①	200–310	60	40	25

① 80 on No. 1-4, 120 on No. 5
② A.T. 60 ft lbs.; M.T. 90 ft lbs.
③ Dip bolt in oil before tightening
④ 70 on No. 1-4, 100 on No. 5

WHEEL ALIGNMENT

Year	Model	Caster Range (deg.)	Caster Pref. Setting (deg.)	Camber Range (deg.)	Camber Pref. Setting (deg.)	Toe-in (in.)	Steering Axis Inclin. (deg.)	Wheel Pivot Ratio (deg.) Inner Wheel	Wheel Pivot Ratio (deg.) Outer Wheel
'80–'85	Cutlass Pwr. str.	2P to 4P	3P	5/16N to 1 5/16	1/2P	1/16 to 1/4	—	—	—
	Man. str.	0P to 2P	1P	5/16N to 1 5/16	1/2P	1/16 to 1/4	—	—	—
	88-98	2P to 4P	3P	0 to 1 5/8P	3/4P	0 to 1/4	—	—	—
'86–'87	Cutlass	1 1/5P–2 1/5P	1 1/5P	1/3N–1 1/3P	1/2P	1/2P–2 1/2P	—	—	—
	Custom Cruiser	1 1/5P–2 1/5P	1 1/5P	0–1 3/5P	4/5P	1/2P–2 1/2P	—	—	—

— Not specified
N Negative   P Positive

ENGINE ELECTRICAL

Alternator

The Delco SI alternator with integral, non-adjustable regulator is standard on all models. The alternator has integral capacitors to supress radio interference.

In the diesel engine models, a single, standard Delcotron supplies two parallel-connected 12 volt batteries. The two batteries are needed to cope with the load imposed by the eight glow plugs and the larger starter. There are no special switches or relays in the charging system.

NOTE: See Charging and Starting Systems in the Unit Repair Section for charging system test procedures.

REMOVAL & INSTALLATION

1. Disconnect the battery ground cable and the wiring from the alternator.
2. Remove the mounting bolt, adjusting bolt, and drive belt.
3. Lift out the alternator.
4. To install, reverse the removal procedure, connect the battery ground cable and tighten the alternator belt. Determine belt tension at a point halfway between the pulleys by press-

ing on the belt with moderate thumb pressure. If the distance between the pulleys (measured at the pulley center) is 13–16 in., the belt should deflect $\frac{1}{2}$ in. at the halfway point or $\frac{1}{4}$ in. if the distance is 7–10 in.

Voltage Regulator

REMOVAL & INSTALLATION

This is a completely sealed unit that cannot be adjusted or disassembled. Replacement of the voltage regulator requires the disassembly of the alternator.

Starter

See Charging and Starting Systems in the Unit Repair Section for starter motor service procedures. The diesel engine starter is of conventional design, but somewhat larger and with a greater output to turn the engine at 100 rpm for starting. The diesel's high compression ratio makes this necessary.

REMOVAL & INSTALLATION

1. Disconnect the battery and carefully raise the car.
2. Remove upper support attaching bolts and the brace and wire guide tube bolt, if so equipped.
3. Remove the flywheel housing cover.
4. Remove the two starter mounting bolts.
5. Lower starter, disconnect wiring, and remove starter.
6. Install by reversing the procedure. If shims were removed, they must be installed in their original location to assure proper drive pinion-to-flywheel engagement.

Distributor

A high energy ignition (HEI) system is standard equipment. The HEI distributor replaces the points and condenser with a timing wheel, magnetic pick-up and control module. On V6 and V8 engines, the coil is built into the distributor cap. For further description, as well as service procedures for HEI, see the Electronic Ignition unit repair section.

Unlike the gasoline engine, which is a spark-ignition design, the diesel engine is a compression-ignition type. When air is compressed to an extreme, high temperatures are produced. At the moment of extreme compression a small quantity of fuel

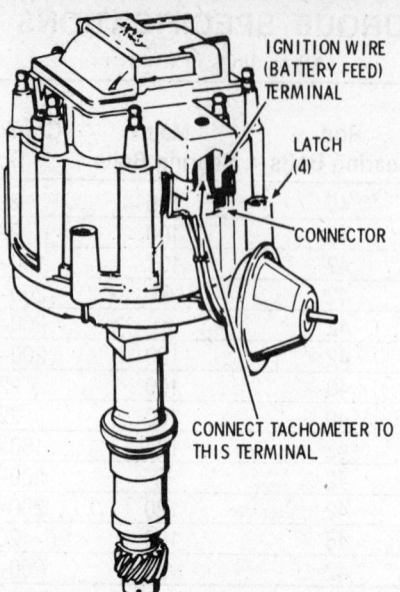

IGNITION WIRE
(BATTERY FEED)
TERMINAL

LATCH
(4)

CONNECTOR

CONNECT TACHOMETER TO
THIS TERMINAL

HEI system tachometer hook-up

is sprayed, under high pressure, into the compression chambers. The temperature of compression ignites the tiny fuel droplets. A temperature of about 1000°F is needed for the fuel ignition. Glow plugs are used to start combustion because the combustion chambers are cold prior to an initial startup, and the first few revolutions of the engine could not produce sufficiently high combustion chamber temperatures for fuel ignition. The glow plugs warm the chambers for a few seconds, bringing them up to the required temperatures to aid in starting combustion. Then glow plugs are automatically shut off.

REMOVAL & INSTALLATION

1. Remove distributor cap, primary (or feed) wire and vacuum line at the distributor.
2. Scribe a mark on the distributor body, locating the position of the rotor, and scribe another mark on the distributor body and engine block, showing the position of the body in the block.
3. Remove the hold-down screw and lift the distributor out of the block.

NOTE: Do not crank the engine with the distributor removed; this will change the timing.

If engine has not been disturbed (cranked) after removing the distributor, perform the following operations for installation:

1. Turn the rotor until it is about $\frac{1}{8}$ turn past the locating mark previously made on the distributor housing.
2. Push the distributor down into

the block. It may be necessary to turn the rotor slightly until the shaft engages in the block. The mark on the distributor housing must line up with the mark made on the engine block.

3. Tighten the hold-down bolt until it is snug, then connect the vacuum advance line.
4. Connect the primary wire to the coil or, on HEI, connect the feed wire and install the distributor cap.
5. Check the timing and adjust it as necessary. Tighten the holddown bolt. If engine has been disturbed (cranked) after removing distributor, perform the following operations for installation:

1. Crank the engine until No. 1 piston is at the top of its compression stroke. The compression stroke can be determined by removing the spark plug from the No. 1 cylinder and placing your thumb over the hole while an assistant slowly cranks the engine. Crank until compression is felt at the hole, then continue cranking slowly until the timing mark on the crankshaft pulley lines up with the 0° mark.
2. Position the distributor in the block but do not allow it to engage with its drive gear. Observe the position of the vacuum control unit on the distributor. If the distributor is located correctly, the vacuum unit will be positioned normally so that the vacuum hose can be easily connected to it.
3. Position the distributor rotor so that it is between terminal No. 1 and the last spark plug tower of the firing order on the distributor cap.
4. Install the distributor, making sure the distributor shaft engages the oil pump shaft, thereby allowing the distributor to fully contact the engine block.
5. Install the hold-down clamp and tighten the bolt until it is snug.
6. Install the distributor cap.
7. Attach all wires and the vacuum advance hose.
8. Check the timing and adjust it as necessary.

IGNITION TIMING ADJUSTMENT

1. Use an adapter between the No. 1 spark plug and No. 1 spark plug lead when connecting a timing light. Connect the timing light to the adapter; DO NOT pierce the spark plug lead. Because of the higher voltage used in the HEI system, any break in the insulation will cause electricity to jump to the nearest ground, making the No. 1 plug misfire.
2. The tachometer terminal is next to the ignition switch connector on the cap of four cylinder, V6 and V8 distributors.
3. Most new tachometers can be

used. Tachometers without a relay can't be used. Check the tach's instructions if you aren't sure. If you don't have the instructions, hook up the tach and check the readings on both the high and low rpm scales. If they agree, the tach is OK; if they don't, use another tach.

4. There is no way of adjusting dwell, since this is controlled by the electronic module.

5. To crank the engine without starting it, disconnect the ignition switch wire at the distributor cap.

NOTE: Always consult the underhood sticker before adjusting timing. If the sticker differs from these procedures, follow the sticker.

1. Disconnect the vacuum advance hose from the distributor and plug it.

NOTE: On 1981 and later models the 4 terminal EST connector at the distributor must be disconnected before timing the engine. Check the underhood sticker for ignition timing instructions.

2. Remove the air cleaner and tape over the vacuum hose fitting.

3. Connect the tachometer and adjust the engine speed to specifications.

4. Connect a timing light, loosen the distributor mounting bolt, and turn the distributor until the specified timing is obtained.

5. Tighten the mounting bolt and recheck timing to see if it changed during 6.Unplug the vacuum advance hose and connect it to the distributor.

7.Remove the tape from the vacuum hose fitting and install and connect the hose, if so equipped.

8.Install the air cleaner.

NOTE: All V6 engine harmonic balancers have two timing marks, one measuring $1/8$ in. wide and one measuring the normal $1/16$ in. wide. The smaller mark is used for setting the timing with a hand held timing light. The $1/8$ in. wide mark is required when using magnetic timing equipment. All engines have a mounting bracket on the front cover which will accept the magnetic timing pickup probe.

Diesel Glow Plugs

There are two types of glow plugs used on General Motors Corp. diesels; the "fast glow" type and the "slow glow" type. The fast glow type uses pulsing current applied to 6 volt glow plugs while the slow glow type use continuous current applied to 12 volt glow plugs.

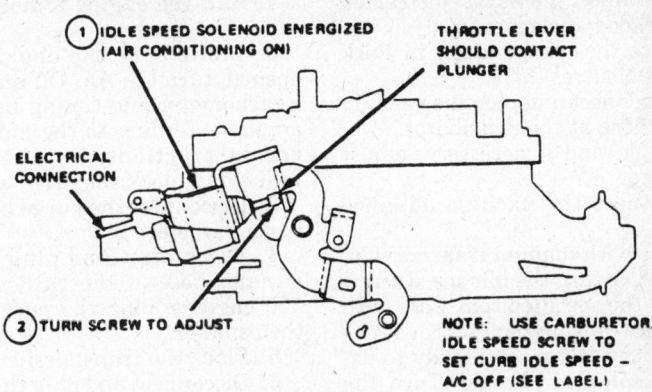

2MC, M2MC unloader adjustment

An easy way to tell the plugs apart is that the fast glow (6 volt) plugs have a $5/16$ in. wide electrical connector plug while the slow glow (12 volt) connector plug is $1/4$ in. wide. Do not attempt to interchange any parts of these two glow plug systems.

GASOLINE FUEL SYSTEM

Fuel Pump

REMOVAL & INSTALLATION

GASOLINE ENGINES

NOTE: On models equipped with an air pump, this pump must be removed to reach the fuel pump. On A/C equipped models the compressor (with lines attached) must be removed first and put aside.

1. Disconnect the fuel lines at the fuel pump.

2. Remove the two mounting bolts or nuts.

3. Remove the shields and oil filter on V6 engines.

4. Remove the pump and gasket. Installation is in the reverse order of removal.

Fuel Filter

REMOVAL & INSTALLATION

All carburetors have a fuel filter in the carburetor body. To replace the filter element, remove the fuel inlet line, then remove the inlet fitting and pull out the filter element. Be careful when tightening the brass fitting because the threads are easily stripped.

Carburetor

REMOVAL & INSTALLATION

1. Disconnect the negative battery cable.

2. Remove the air cleaner.

3. Disconnect the accelerator linkage.

4. Disconnect the transmission detent cable.

5. Disconnect the cruise control, if so equipped.

6. Disconnect the fuel line at the carburetor.

7. Disconnect all necessary vacuum lines. Number them for easy reinstallation.

8. Remove the attaching bolts, then the carburetor.

9. Installation is in the reverse order of removal.

IDLE SPEED AND MIXTURE ADJUSTMENTS

NOTE: When adjusting the idle speed and mixture, always check the underhood specifications sticker. If the sticker gives different instructions from the procedures given here, follow the underhood sticker's instructions.

Idle Speed Adjustment

NOTE: On vehicles equipped with Idle Speed Control (ISC) no adjustments are possible.

2GC AND 2GE 2BBL CARBURETORS

1. Run the engine to normal operating temperature. Make sure that the choke is fully opened, turn the A/C Off and connect a tachometer and timing light to the engine according to the manufacturers' instructions.

2. Set the parking brake and block the drive wheels.

3. Disconnect hoses as instructed on underhood sticker.

4. Place the transmission in Park (AT) and Neutral (MT).

5. Disconnect and plug the vacuum advance hose at the distributor.

6. Check, and if necessary, adjust the timing.

7. Connect the vacuum advance hose.

8. Cars with manual transmission, without A/C: turn the idle speed screw to obtain the specified rpm. Cars with automatic transmission, without A/C: open the throttle momentarily to extend the solenoid plunger. Turn the solenoid screw to adjust the speed to the curb idle rpm listed on the underhood sticker. Turn the idle speed screw to the specified rpm. Cars with A/C: Turn the idle speed screw to obtain the specified rpm. Disconnect the A/C compressor clutch wire. Turn the A/C On. Open the throttle momentarily to extend the solenoid plunger. Turn the solenoid screw to obtain the rpm specified on the underhood sticker. Connect the compressor clutch wire.

9. Connect all hoses. Remove the tachometer and timing light.

M2MC-210 AND M2ME 2BBL CARBURETOR

NOTE: See the Note at beginning of "Idle Speed and Mixture Adjustments."

1. Run the engine to normal operating temperature. ◊ 2. Disconnect the A/C compressor clutch wire, turn the A/C Off, make sure the choke is fully opened, place the manual transmission in Neutral, and the automatic transmission in Drive. Set the parking brake and block the drive wheels.

3. Disconnect and plug the vacuum advance hose at the distributor.

4. Check and adjust the timing.

5. Connect the vacuum advance hose.

6. Disconnect the purge hose at the vapor canister.

7. Cars without A/C: turn the idle speed screw to obtain the specified rpm. Cars with A/C: turn the idle speed screw to obtain the specified rpm, turn the A/C On, open throttle momentarily to extend the solenoid plunger and set the solenoid screw to obtain the rpm specified on the underhood sticker. Turn the A/C off.

8. Connect all hoses and remove the tachometer and timing light. Connect the compressor clutch wire.

M4MC 4BBL CARBURETOR

NOTE: See the Note at beginning of "Idle Speed and Mixture Adjustments."

1. Run the engine to normal operating temperature.

2. Make sure the choke is fully opened, turn the A/C Off and connect a tachometer and timing light to the engine according to the manufacturers' instructions. Set the parking brake and block the drive wheels.

3. Disconnect the purge hose at the vapor canister.

4. Disconnect and plug the EGR vacuum hose at the EGR valve. On 350 engines, plug the purge hose at the canister.

5. Place the transmission in Park.

6. Disconnect and plug the vacuum advance line at the distributor.

7. Check and adjust the timing.

8. Connect the vacuum advance line.

9. Place the transmission in Drive.

10. On cars without A/C: adjust the idle speed screw to obtain the specified rpm. On cars with A/C: disconnect the compressor clutch wire. Open the throttle momentarily to extend the solenoid plunger. Turn the A/C ON and adjust the solenoid screw to obtain the rpm specified on the underhood sticker. Connect the compressor clutch wire and turn the A/C off.

11. Connect all hoses and remove the tachometer and timing light.

2SE, E2SE, E2ME 2-BBL, E4ME AND E4MC 4-BBL CARBURETORS

1. Run the engine until it reaches normal operating temperature.

2. Prepare the vehicle for adjustment as indicated on the emission label under the hood.

3. Check the ignition timing and adjust as necessary.

4. Reconnect the vacuum advance line.

5. With the A/C Off, turn the idle speed screw to obtain the curb idle as specified on the emissions label.

6. With the automatic transmission in Drive or the manual transmission in Neutral, disconnect the A/C compressor wire at the compressor and turn the A/C On.

7. Open the throttle slightly to extend the solenoid plunger.

8. Turn the solenoid screw to obtain the correct rpm.

9. Turn the engine off and reconnect the A/C compressor line and all hoses.

1980 Idle Mixture Adjustment

Changes in the idle system have made adjustment of the fuel mixture impossible without the aid of a propane enrichment system not available to the general public. Backing out the mixture screws will have little or no effect on the mixture. All models have their mixture screws concealed by staked-in plugs; mixture is set during manufacture and is not adjustable.

1981 and Later Idle Mixture Adjustment.

On 1981 and later models equipped with Computer Control Command no mixture adjustments are possible. Adjustments are controlled by the ECM.

For further information on Carburetors, please refer to "Carburetors" in the Unit Repair Section.

DIESEL FUEL SYSTEM

The fuel system is the heart of the diesel engine. The main components are the injection pump, injection lines and fuel injectors. The fuel injection pump is a small, high pressure rotary pump which delivers a small, metered amount of fuel to the injection nozzles at the proper time. The high pressure lines are all of equal length to avoid differences in timing. The nozzles project into the combustion chambers and inject the fuel into the chambers in a finely atomized, precisely controlled spray. A small, low pressure transfer pump is employed in the inlet line to the injection pump to keep the injection pump supplied. Engine rpm is controlled by a rotary fuel metering valve operated by the accelerator linkage. A fuel filter is located between the transfer pump and the injection pump. On all engines, the fuel pump is of the mechanical diaphragm type, mounted on the engine.

For further information on the Diesel Fuel System, please refer to "Diesel Maintenance" in the Unit Repair Section.

Fuel Pump

REMOVAL & INSTALLATION

V8 Models

The fuel supply pump on the V8 diesel engine is serviced in the same manner as the fuel pump on the gasoline engine. See the procedure above under "Gasoline Fuel System."

V6 MODELS

NOTE: The fuel pump used on the V6 diesel engine is located at the front of the engine, next to the fuel heater.

1. Disconnect negative battery cable, remove the air cleaner, and unplug all electrical connectors from the pump.

2. Place a rag under the pump inlet and outlet fittings, and carefully unscrew the inlet and the outlet fittings. Cap all fittings to keep dirt out.

3. Remove the pump mounting bracket nut, then the fuel pump.

4. To install fuel pump, reverse above procedure, and tighten the nut of the pump mounting bracket to 18 ft. lbs. Then torque inlet and outlet line fittings to 19 ft. lbs.

NOTE: In some cases you may have to adjust pump position slightly to align pump fittings with the fuel lines.

5. After installing the fuel pump, position a catch basin and disconnect the fuel line at the filter and turn on the ignition switch to prime and bleed the lines. If after torquing the fuel line, the pump runs with a click-like sound, or the fuel bubbles, check for leaks in the fuel lines. When the pump quiets down, tighten the fuel line at the filter.

Diesel Fuel Filter

REMOVAL & INSTALLATION

The fuel filter is a square assembly located at the back of the engine above the intake manifold. Disconnect the fuel lines and remove the filter. Install the lines to the new filter. Start the engine and check for leaks.

Tachometer Hook-Up—Diesel Engine

A magnetic pickup tachometer is necessary because of the lack of an ignition system. The tachometer probe is inserted into the hole in the timing indicator.

Idle Speed Adjustment

1. Run the engine until it reaches operating temperature.

2. Insert the probe of a magnetic pickup tachometer into the timing indicator hole.

3. Set the parking brake and block the drive wheels.

4. Put the transmission in Drive and, if necessary, turn the A/C off.

5. Turn the slow idle screw on the injection pump to obtain the idle speed specified on the emisson control label.

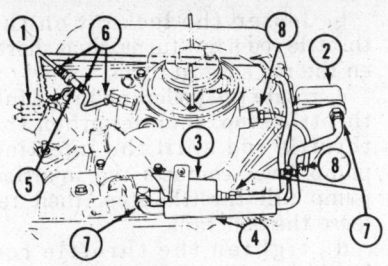

1. RETURN LINE
2. FUEL FILTER
3. FUEL PUMP
4. FUEL LINE HEATER (OPTIONAL)
5. HOUSING PRESSURE ALTITUDE ADVANCE
6. 10 FT. LBS.
7. 19 FT. LBS.
8. 11 FT. LBS.

V6 diesel fuel lines

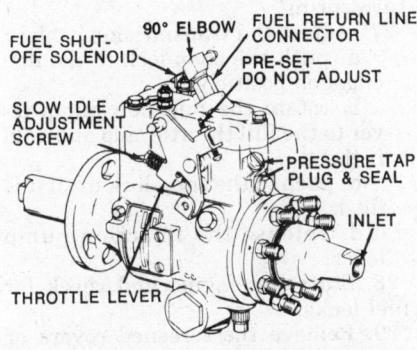

Injection pump slow idle screw

Fast Idle Solenoid Adjustment

1. With the ignition off, disconnect the single green wire from the fast idle relay located on the front of the firewall. This will energize the solenoid so the adjustment you will be making will effect fast idle speed only.

2. Set the parking brake and block the drive wheels.

3. Start the engine and adjust the solenoid to the specifications on the emission control label.

4. Turn the ignition switch off and reconnect the green wire to the fast idle relay.

Fast Idle Solenoid

ADJUSTMENT

1. With the ignition OFF, disconnect the single green wire from the fast idle relay located on the front of the firewall.

2. Set the parking brake and block the drive wheels.

3. Start the engine and adjust the solenoid (energized) to the specifications on the underhood emission control label.

4. Turn the ignition switch OFF and reconnect the green wire.

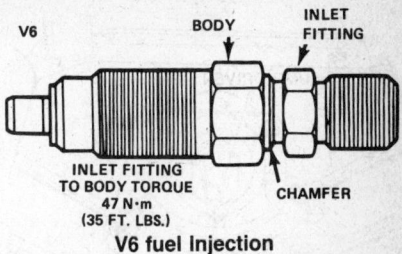

V6

INLET FITTING TO BODY TORQUE 47 N·m (35 FT. LBS.)

V6 fuel Injection

Cruise Control Servo Relay Rod

ADJUSTMENT

1. Turn the engine OFF.

2. Adjust the rod to minimum slack then put the clip in the first free hole closest to the bellcrank, but within the servo ball.

Diesel Injection Pump and Lines

REMOVAL & INSTALLATION

NOTE: This procedure contains throttle rod and transmission cable adjustments.

1. Remove the air cleaner.

2. Remove the filters and pipes from the valve covers and air crossover.

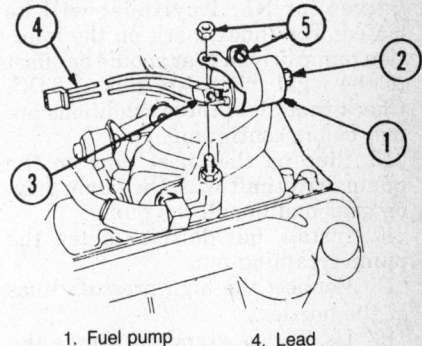

1. Fuel pump
2. Inlet
3. Outlet
4. Lead
5. 18 ft. lbs.

V6 diesel fuel pump

3. Remove the air crossover and cap the intake manifold with screened covers (J-26996-1 on V8's or 29657 V6), or tape.

4. Disconnect the throttle rod and return spring.

5. Remove the bellcrank.

6. Remove the throttle and transmission cables from the intake manifold brackets.

7. Disconnect the fuel lines from the filter and remove the filter.

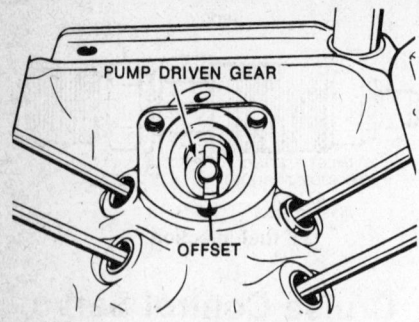

Offset on pump driven gear

8. Disconnect the fuel inlet line at the pump.

9. Remove the rear A/C compressor brace and remove the fuel line.

10. Disconnect the fuel return line from the injection pump.

11. Remove the clamps and pull the fuel return lines from each injection nozzle.

12. Using two wrenches, disconnect the high pressure lines at the nozzles.

13. Remove the three injection pump retaining nuts with tool J-26987 or its equivalent.

14. Remove the pump and cap all lines and nozzles.

To install:

15. Remove the protective caps from all lines and nozzles. Place the engine on TDC for the No. 1 cylinder. The mark on the harmonic balancer on the crankshaft will be aligned with the zero mark on the timing tab, and both valves for No. 1 cylinder will be closed. The index mark on the injection pump driven gear should be offset to the right when No. 1 is at TDC. Check that all of these conditions are met before continuing.

16. Line up the offset tang on the pump driveshaft with the pump driven gear and install the pump.

17. Install, but do not tighten the pump retaining nuts.

18. Connect the high pressure lines at the nozzles.

19. Using two wrenches, torque the high pressure line nuts to 25 ft. lbs.

20. Connect the fuel return lines to the nozzles and pump.

21. Align the timing mark on the injection pump with the line on the timing mark adapter and torque the mounting nuts to 35 ft. lbs. V6, 18 ft. lbs. V8.

NOTE: A ¾ in. open end wrench on the boss at the front of the injection pump will aid in rotating the pump to align the marks.

22. Adjust the throttle rod:

a. remove the clip from the cruise control rod and remove the rod from the bellcrank.

b. loosen the locknut on the throttle rod a few turns, then shorten the rod several turns.

c. rotate the bellcrank to the full throttle stop, then lengthen the throttle rod until the injection pump lever contacts the injection pump full throttle stop, then release the bellcrank.

d. tighten the throttle rod locknut.

23. Install the fuel inlet line between the transfer pump and the filter.

24. Install the rear A/C compressor brace.

25. Install the bellcrank and clip.

26. Connect the throttle rod and return spring.

27. Adjust the transmission cable:

a. push the snap-lock to the disengaged position.

b. rotate the injection pump lever to the full throttle stop and hold it there.

c. push in the snap-lock until it is flush.

d. release the injection pump lever.

28. Start the engine and check for fuel leaks.

29. Remove the screened covers or tape and install the air crossover.

30. Install the tubes in the air flow control valve in the air crossover and install the ventilation filters in the valve covers.

31. Install the air cleaner.

32. Start the engine and allow it to run for two minutes. Stop the engine, let it stand for two minutes, then restart. This permits the air to bleed off within the pump.

Injection Timing

ADJUSTMENT

For the engine to be properly timed, the lines on the top of the injection pump adapter and the flange of the injection pump must be aligned.

1. The engine must be off for resetting the timing.

2. Loosen the three pump retaining nuts with J-26987 on V8's or J-25304 on V6's, an injection pump intake manifold wrench, or its equivalent.

3. Align the timing marks and torque the pump retaining nuts to 35 ft. lbs.

NOTE: The use of a ¾ in. open end wrench on the boss at the front of the pump will aid in rotating the pump to align the marks.

4. Adjust the throttle rod. See "Fuel Injection Pump Removal and Installation," Step 22.

INLET FITTING TO BODY TORQUE DIESEL EQUIPMENT — 45 FT. LBS. C.A.V. LUCAS — 25 FT. LBS.

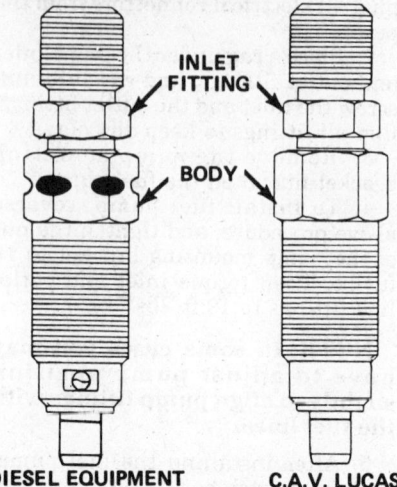

V8 diesel fuel injector identification—1980 and later

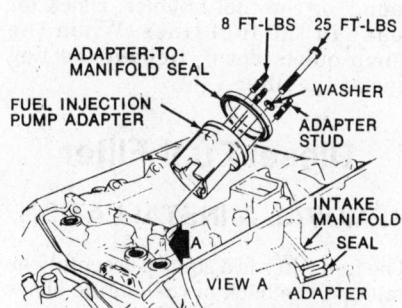

Injection pump adapter bolts

Injection Nozzle

REMOVAL & INSTALLATION

The injection nozzles on these engines are simply unbolted from the cylinder head, after the fuel lines are removed, in similar fashion to a spark plug. Be careful not to damage the nozzle end and make sure you remove the copper nozzle gasket from the cylinder head if it does not come off with the nozzle.

Clean the carbon off the tip of the nozzle with a soft brass wire brush and install the nozzles, with gaskets.

NOTE: 1981 and later models use two types of injectors, CAV Lucas and Diesel Equipment. When installing the inlet fittings, torque the Diesel Equipment injector fitting to 45 ft. lbs. and the CAV Lucas to 25 ft. lbs.

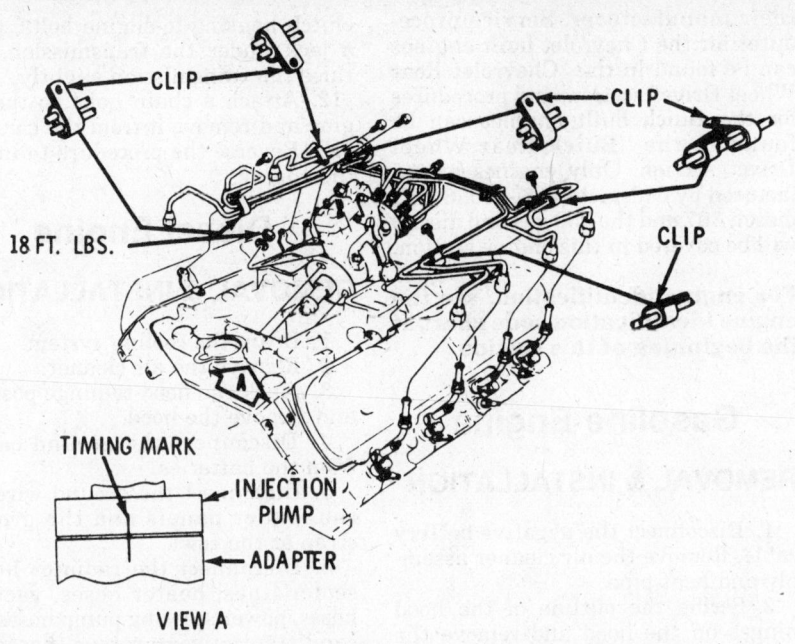

V8 diesel engine injection pump timing marks—V6 similar

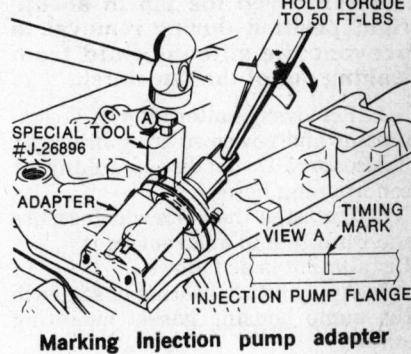

Marking Injection pump adapter

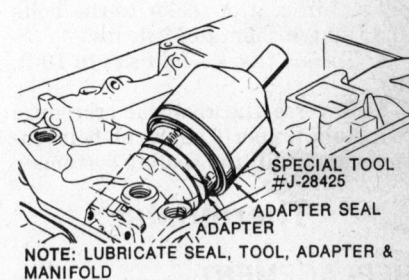

NOTE: LUBRICATE SEAL, TOOL, ADAPTER & MANIFOLD

Installing adapter seal

Injection Pump Adapter, Seal, Timing Mark

REMOVAL & INSTALLATION

NOTE: Skip Steps 4 and 9 if a new adapter is not being installed.

1. Remove injection pump and lines as described earlier.
2. Remove the injection pump adapter.
3. Remove the seal from the adapter.
4. File the timing mark from the adapter. Do not file the mark off the pump.
5. Position the engine at TDC of No. 1 cylinder. Align the mark on the balancer with the zero mark on the indicator. The index is offset to the right when No. 1 is at TDC.
6. Apply chassis lube to the seal areas. Install, but do not tighten the injection pump.
7. Install the new seal on the adapter using tool J-28425, or its equivalent.
8. Torque the adapter bolts to 25 ft. lbs.
9. Install timing tool J-26896 into the injection pump adapter. Torque the tool, toward No. 1 cylinder, to 50 ft. lbs. Mark the injection pump adapter. Remove the tool.
10. Install the injection pump.

COOLING SYSTEM

The diesel engine cooling system is the same as that used on the gasoline engine except that the radiator tank has two oil coolers. One is connected to the transmission, the other to the oil filter base.

Radiator

REMOVAL & INSTALLATION

1. Drain the cooling system.
2. Remove the upper radiator baffle and slide the shroud back over the fan.
3. Unfasten the upper and lower hoses from the radiator.
4. Disconnect the overflow hose or the optional coolant recovery system hose.
5. On models equipped with an automatic transmission, disconnect and cap the lines which run to the fluid cooler. On vehicles with diesel engines remove the engine oil cooler lines from the radiator.
6. Unfasten the radiator's securing bolts and move the radiator upward to disengage it from its supports. Remove the radiator from the car.

NOTE: It may be necessary to rotate the fan blades in order to keep them out of the way.

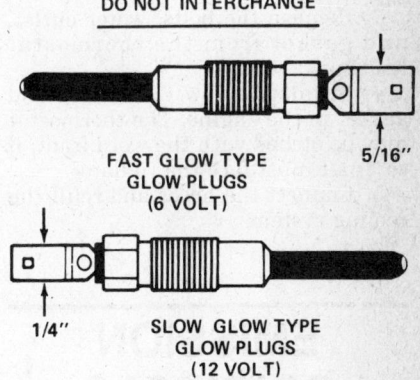

DO NOT INTERCHANGE

FAST GLOW TYPE GLOW PLUGS (6 VOLT)   5/16"

1/4"   SLOW GLOW TYPE GLOW PLUGS (12 VOLT)

Glow plug identification

7. Installation is in the reverse order of removal. Refill the cooling system.

Water Pump

REMOVAL & INSTALLATION

1. Drain the cooling system.
2. Unfasten the heater, bypass, and lower radiator hoses from the pump.
3. Loosen the drive belts. Remove the fan assembly and the four spacer bolts. On cars with A/C, remove the fan and clutch assembly.

NOTE: Keep the fan in an upright position during removal to prevent the silicone fluid from leaking out of the fan clutch.

4. Remove the alternator, A/C compressor and power steering brackets, if necessary. Do not disconnect any air conditioning hoses.

5. Unfasten the bolts which secure the water pump and remove it. Installation is as follows:

1. Apply a thin coating of sealer to the pump housing gasket mounting surface.

2. Place a new gasket on the housing.

3. Install the pump assembly. Apply a thin coat of sealer to the bolts and tighten them to 13 ft. lbs.

4. Torque the $5/16$ in. bolts to 10 ft. lbs.

5. Reverse the removal procedure to install. Properly adjust all belt tensions and refill the cooling system.

Thermostat

REPLACEMENT

1. Drain the coolant level below the thermostat.

2. Remove the hoses from the thermostat housing.

3. Remove the bolts, water outlet, and gasket from the thermostat housing.

4. Install the new thermostat and gasket in the engine. The thermostat may be etched with the word front; if so, front must face the radiator.

5. Connect the hoses and refill the cooling system.

EMISSION CONTROLS

Please refer to the "Emission Control" section of the Unit Repair Section for a description and service of the emission control systems.

ENGINE MECHANICAL

The Chevrolet, V8-305, 350 Buick, 231, 252 and V8-350; have been used by Oldsmobile in various models. Service procedures for these engines will be found in car sections dealing with

their manufacturer. Service procedures for the Chevrolet built engines can be found in the "Chevrolet Rear Wheel Drive" section, and procedures for the Buick built engines can be found in the "Buick Rear Wheel Drive" section. Only engines manufactured by Oldsmobile (8–260, 6–263 diesel, 307 and the 350 gas and diesel) will be covered in this engine section.

For engine identification, see the engine identification code chart at the beginning of this section.

Gasoline Engine

REMOVAL & INSTALLATION

1. Disconnect the negative battery cable. Remove the air cleaner assembly and heat pipe.

2. Scribe the outline of the hood hinges on the hood and remove the hood.

3. Drain the cooling system and disconnect the radiator and heater hoses from the engine. Remove the fan blade, pulleys, and belts. On 1986 307 engines, remove the radiator.

4. Disconnect the engine ground strap from the cylinder head. Remove the fan shroud.

5. Disconnect and tag all vacuum lines and electrical leads from the engine.

6. Disconnect the throttle linkage. Disconnect the fuel line from the fuel pump. Remove the clutch equalizer on manual transmission cars.

7. If the car is equipped with an automatic transmission, disconnect the cooler lines from the radiator. If equipped with power steering or air conditioning, remove the pump and bracket or compressor and bracket from the engine without disconnecting the lines.

CAUTION
Disconnecting the air conditioner lines could result in personal injury.

---

8. Remove the radiator. Remove the fan, if necessary to gain working clearance. Raise the car and drain the engine oil.

9. Disconnect the exhaust pipes and/or crossover pipes from the exhaust manifolds. Remove the motor mount throughbolts. Remove the starter.

10. On models equipped with an automatic transmission, remove the torque converter cover. Matchmark the flywheel and converter. Turn the crankshaft pulley to gain access to the three torque converter-to-flywheel attaching bolts and remove the bolts.

11. Remove the transmission or

clutch housing-to-engine bolts, place a jack under the transmission, and raise the transmission slightly.

12. Attach a chain hoist to the engine and remove it from the car.

13. Reverse the procedure to install the engine.

Diesel Engine

REMOVAL & INSTALLATION

1. Drain the cooling system.

2. Remove the air cleaner.

3. Mark the hood-to-hinge position and remove the hood.

4. Disconnect the ground cables from the batteries.

5. Disconnect the ground wires at the fender panels and the ground strap at the cowl.

6. Disconnect the radiator hoses, cooler lines, heater hoses, vacuum hoses, power steering pump hoses, air conditioning compressor (hoses attached), fuel inlet hose and all attached wiring.

7. Remove the bellcrank clip.

8. Disconnect the throttle and transmission cables.

9. Remove the upper radiator support and the radiator.

10. Raise and support the car.

11. Disconnect the exhaust pipes at the manifold.

12. Remove the torque converter cover and the three bolts holding the converter to the flywheel.

13. Remove the engine mount bolts or nuts.

14. Remove the three right side transmission-to-engine bolts. Remove the starter.

15. Lower the car and attach a hoist to the engine.

16. Slightly raise the transmission with a jack.

17. Remove the three left side transmission-to-engine bolts and lift out the engine.

18. Installation is in the reverse order of removal. Converter cover bolts are torqued to 40 ft. lbs. on the 350 V8 and 35 ft. lbs. on the 263 V6.

Intake Manifold

REMOVAL & INSTALLATION

Gasoline Engine

1. Remove the air cleaner, drain the radiator, and disconnect the negative battery terminal.

2. Disconnect the upper radiator hose, by-pass hose, and heater hose from the manifold.

3. Disconnect the throttle linkage, vacuum and gas lines, and the brake booster line from the carburetor.

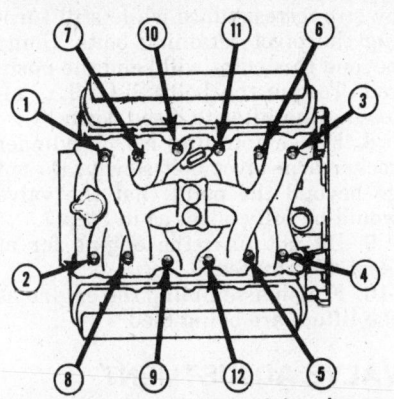

V8 intake manifold bolt tightening sequence—gasoline and diesel engines

V8 diesel engine intake manifold and gasoline

4. Remove the alternator and air conditioning compressor brackets if necessary (the rear brackets must be removed on late model 307 engines).

— **CAUTION** —

Do not disconnect the A/C lines. Personal injury could result.

5. Disconnect the temperature gauge wire and, on later models, the CCC wire.

6. Remove the intake manifold bolts and remove the manifold with the carburetor attached.

7. Install in the reverse order of removal, tightening all bolts first to 15 ft. lbs., then to the figure specified in the torque chart, in the sequence illustrated. Coat all gasket surfaces with sealer.

Diesel Engine

1. Remove the air cleaner.

2. Drain the radiator. Loosen the upper bypass hose clamp, remove the thermostat housing bolts, and remove the housing and the thermostat from the intake manifold.

3. Remove the breather pipes from the rocker covers and the air crossover. Remove the air crossover.

4. Disconnect the throttle rod and the return spring. If equipped with cruise control, remove the servo.

5. Remove the hairpin clip at the bellcrank and disconnect the cables. Remove the throttle cable from the bracket on the manifold; position the cable away from the engine. Disconnect and label any wiring as necessary.

6. Remove the alternator bracket if necessary. On the 350 cu. in. engine, if equipped with air conditioning, remove the compressor mounting bolts and move the compressor aside, without disconnecting any of the hoses or wiring. Remove the compressor

mounting bracket from the intake manifold.

7. Disconnect the fuel line from the pump and the fuel filter. Remove the fuel filter and bracket.

8. Remove the fuel injection pump and lines. See above for procedures.

9. Disconnect and remove the vacuum pump or oil pump drive assembly from the rear of the engine.

10. Remove the intake manifold drain tube.

11. Remove the intake manifold bolts and remove the manifold. Remove the adapter seal. Remove the injection pump adapter.

12. Clean the mating surfaces of the cylinder heads and the intake manifold using a putty knife.

13. Coat both sides of the gasket surface that seal the intake manifold to the cylinder heads with GM sealer 1050026 or the equivalent. Position the intake manifold gaskets on the cylinder heads. To install the front and rear end seals, apply 1052915, 22521437, G.E. 1673 RTV sealer or equivalent to the end seals only. Then install the end seals, making sure that the ends are positioned under the cylinder heads.

14. Carefully lower the intake manifold into place on the engine.

15. Clean the intake manifold bolts thoroughly, then dip them in clean engine oil. Install the bolts and on the 350 V8 tighten to 15 ft. lbs. in the sequence shown. Next, tighten all the bolts to 30 ft. lbs., in sequence, and finally tighten to 40 ft. lbs. in sequence. On the 263 V6 engine tighten to 15 ft. lbs. in the sequence shown, then retorque to 41 ft. lbs.

16. Install the intake manifold drain tube and clamp.

17. Install injection pump adapter. See under "Fuel System," above: "Diesel Engine, Injection Pump Adapter, Adapter Seal and New Adapter Timing Mark Removal & Installation." If a new adapter is not being used, skip Steps 4 and 9.

18. Install the fuel injection pump. See "Diesel Engine," under "Fuel System," above for procedures.

19. Install the vacuum pump or coil pump drive assembly.

— **CAUTION** —

Do not operate the engine without vacuum pump/oil pump assembly in place as this assembly drives the engine oil pump.

20. Install the remaining components in the reverse sequence of their removal. For throttle rod and transmission cable adjustments, see "Diesel Engine, Fuel Injection Pump" removal and installation, Steps 22 and 27, under "Fuel System," above.

Exhaust Manifold

REMOVAL & INSTALLATION

Right Side Except Diesel

1. Disconnect the negative battery cable.

2. Raise the car and remove the right front wheel, if necessary, the exhaust and crossover pipe, and the manifold bolts. If so equipped, disconnect the oxygen sensor lead.

3. Remove the lower engine mounting bolt and raise the engine slightly, if necessary for clearance.

4. Remove the manifold from below.

88 AND 98—Left Side

1. Remove the air cleaner.

2. Remove the hot air shroud.

3. Remove the lower alternator bracket.

4. Raise the car and remove the crossover pipe.

5. Lower the car and remove the manifold.

Cutlass—Left Side

1. Raise the car and disconnect the left side crossover pipe.

2. Lower the car and disconnect the intermediate steering column shaft.

3. Remove the hot air shroud.

4. Remove the exhaust manifold.

Diesel Engine—Left Side

1. Remove the air cleaner.

2. Remove the alternator lower bracket.

3. Raise and support the car.

4. Remove the crossover pipe.

5. Lower the car.

6. Remove the exhaust manifold.

7. Installation is in the reverse order of removal.

Diesel Engine—Right Side

1. Raise and support the car.

2. Remove the crossover pipe.

3. Disconnect the exhaust pipe.

4. Remove the right front wheel.

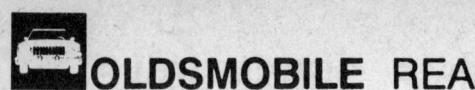
5. Remove the exhaust manifold from under the car.

6. Installation is in the reverse order of removal.

Valve System

Hydraulic lifters are used on all engines. Valve guides are not replaceable, but may be reamed oversize. Occasionally a valve guide bore will be oversize as manufactured. These are marked on the inboard side of the cylinder heads on the machined surface just above the intake manifold. Valve lifters and pushrods should be kept in order when removed, so they can be reinstalled in their original locations. Valve lifters used in diesel engines are not the same as those used in gasoline engines. See the NOTE at the beginning of the Engine section.

Rocker Arms

REMOVAL & INSTALLATION

Gasoline Engine

Remove the valve covers. Remove the two bolts that attach the rocker arm pivot to the cylinder head. Remove the rocker arms in pairs. Install the rocker arms for each cylinder only when the lifters are off the cam lobe and the valves are closed. Lubricate all pivot and rocker arm wear points with white grease. Torque the hardened flanged retaining bolts to 25 ft. lbs.

V8 Diesel Engine

NOTE: When the diesel engine rocker arms are removed or loosened, the lifters must be bled down to prevent oil pressure buildup inside each lifter, which could cause it to raise up higher than normal and bring the valves within striking distance of the pistons.

1. Remove the valve cover.
2. Remove the rocker arm pivot bolts, the bridged pivot and rocker arms.
3. Remove each rocker set as a unit.
4. To install, lubricate the pivot wear points and position each set of rocker arms in its proper location. Do not tighten the pivot bolts, to prevent bending the valves when the engine is turned.
5. The lifters can be bled down for six cylinders at once with the crankshaft in either of the following two positions:
   a. For cylinders number 3, 5, 7, 2, 4 and 8, turn the crankshaft so the

saw slot on the harmonic balancer is at 0° on the timing indicator.
   b. For cylinders 1, 3, 7, 2, 4 and 6, turn the crankshaft so the saw slot on the harmonic balancer is at 4 O'clock.
6. Tighten the rocker arm pivot bolts to 28 ft. lbs. It will take 45 minutes to completely bleed down the lifters in this position. If additional lifters must be bled, rotate the engine to the other position, tighten the rocker arm pivot bolts, and again wait 45 minutes before rotating the crankshaft.
7. Assemble the remaining components in the reverse order of disassembly. The rocker covers do not use gaskets, but instead are sealed with a bead of RTV (room temperature vulcanizing) silicone sealer.

V6 Diesel Engine

NOTE: When the diesel engine rocker arms are removed or loosened, the lifters must be bled down to prevent oil pressure buildup inside each lifter, which could cause it to raise up higher than normal and bring the valves within striking distance of the pistons.

1. Remove the valve cover.
2. Remove the rocker arm pivot bolts, the bridged pivot and rocker arms.
3. Remove each rocker set as a unit.
4. Before installing any removed rocker arms, rotate the engine crankshaft so that No. 1 cylinder is 32° before top dead center. This is 2 in. counterclockwise from the 0° pointer. To verify that No. 1 cylinder TDC is coming up, if only the right valve cover was removed, remove the No. 1 cylinder glow plug, then turn the engine: compression pressure will force air out the glow plug hole. If the left valve cover was removed, rotate the crankshaft until the No. 5 cylinder intake valve pushrod ball is 0.28 in. above the No. 5 cylinder exhaust valve pushrod ball.

NOTE: Use only hand wrenches to torque the rocker arm pivot bolts to avoid engine damage.

5. If removed, install the No. 5 cylinder pivot and rocker arms, then torque the bolts alternately between the intake and exhaust valves until the intake valve begins to open, then stop.
6. Install the remaining rocker arms, except No. 3 exhaust (if this rocker was removed).
7. If removed, install the No. 3 cylinder exhaust valve pivot, but do not torque beyond the point that the valve would be fully open. This is indicated

by strong resistance while still turning the pivot retaining bolts. Going beyond this point will bend the pushrod. Torque the bolts SLOWLY, allowing the lifter to bleed down.
8. Finish torquing No. 5 cylinder rocker arm pivot bolt slowly. Do not go beyond the point that the valve would be fully open, as in Step 7.
9. Do not turn the engine for at least 45 minutes.
10. Finish assembling the engine as the lifters are being bled.

VALVE ADJUSTMENT

These valves cannot be adjusted. If there is excessive clearance in the valve train, look for worn pushrods, rocker arms, valve springs, or collapsed or stuck valve lifters.

Cylinder Head

See the NOTE at the beginning of the Engine section.

REMOVAL & INSTALLATION

——— CAUTION ———
Do not disconnect the A/C lines. Severe personal injury could result.

Gasoline Engine

1. Drain the cooling system.
2. Remove the intake manifold and carburetor as an assembly.
3. Remove exhaust manifolds.
4. Loosen or remove any accessory brackets which interfere.
5. Remove the valve cover. Loosen any accessory brackets which are in the way.
6. Remove the battery ground strap from the cylinder head.
7. Remove rocker arm bolts, pivots, rocker arms and pushrods. Scribe the pivots and identify the rocker arms and pushrods so that they may be installed in their original locations.
8. Remove cylinder head bolts and cylinder head(s).
9. Install in the reverse order of removal. It is recommended that the head gasket be coated on both sides with sealer. Dip head bolts in oil before installing and allow them to drain so oil will not fill threads and interfere with proper torquing. On the 231 V 6, apply a heavy bolt sealer to the bolt threads instead of oil. Tighten all head bolts in the correct sequence to 60–70 ft. lbs., then again in sequence to the specified torque. See Specifications at the beginning of this section for correct head bolt torque. Re-torque the bolts after engine is warmed up. On the 1985-87

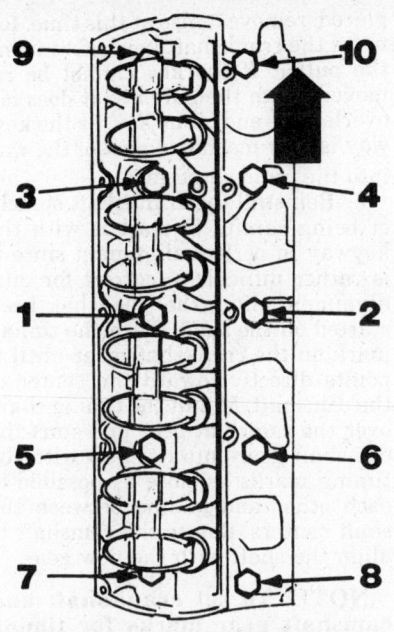

9 — 10
3 — 4
1 — 2
5 — 6
7 — 8

V8 head bolt torque sequence

APPLY A 6 MM (1/4") BEAD OF 1052915, GE 1673 OR EQUIVALENT RTV SEALER ON THE VALVE COVER AS SHOWN.

Application of RTV sealer to valve covers on the 307 V8

models, apply a ¼ in. bead of rtv sealer along the horizontal surface of the cover, running inside the boltholes.

NOTE: In 1981 and later models the head gaskets must be installed without sealer. The gaskets for the 260 cu. in. V8 are to be installed with the stripe facing up. The 307 cu. in. V8 gaskets do not have a stripe.

Diesel Engine

1. Remove the intake manifold, using the procedure outlined above.
2. Remove the rocker arm cover(s), after removing any accessory brackets which interfere with cover removal.
3. Disconnect and label the glow plug wiring.
4. If the right cylinder head is being removed, remove the ground strap from the head.
5. Remove the rocker arm bolts, the bridged pivots, the rocker arms, and the pushrods, keeping all the parts in order so that they can be returned to their original locations. It is a good practice to number or mark the parts to avoid interchanging them.
6. Remove the fuel return lines from the nozzles.

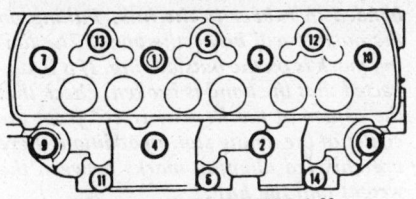

TORQUE ALL BOLTS (EXCEPT 5, 6, 11, 12, 13 & 14) TO 193 N·m (142 FT. LBS.). NUMBERS 5, 6, 11, 12, 13 & 14 TORQUE TO 80 N·m (59 FT. LBS.).

V6 diesel engine cylinder head torque sequence

7. Remove the exhaust manifold(s), using the procedure outlined above.
8. Remove the engine block drain plug on the side of the engine from which the cylinder head is being removed. On V6s, remove the pipe-thread plugs covering the upper cylinder head bolts.
9. Remove the head bolts. Remove the cylinder head.
10. To install, first clean the mating surfaces thoroughly. Install new head gaskets on the engine block. Do not coat the gaskets with any sealer. The gaskets have a special coating that eliminates the need for sealer. The use of sealer will interfere with this coating and cause leaks. Install the cylinder head onto the block.
11. Clean the head bolts (and pipe-thread plugs-V6s) thoroughly. On the V8, dip the bolts in clean engine oil and install into the cylinder block until the heads of the bolts lightly contact the cylinder head. On V6s, coat the plug threads, bolt threads and the area under the bolt threads with sealer/lu-bricant part No. 1052080 or equivalent.

NOTE: The correct sealer must be used or coolant leaks and bolt torque loss will result.

12. On the V8, tighten the bolts, in the sequence illustrated, to 100 ft. lbs. When all bolts have been tightened to this figure, begin the tightening sequence again, and torque all bolts to 130 ft. lbs.
13. On V6s, tighten all head bolts in sequence to the following torques: all except bolts 5, 6, 11, 12, 13 and 14—100 ft. lbs.; bolts 5, 6, 11, 12, 13 and 14—41 ft. lbs. Finally, tighten all bolts except 5, 6, 11, 12, 13 and 14 to 142 ft. lbs., and bolts 5, 6, 11, 12, 13 and 14 to 59 ft. lbs. in the proper sequence. Install the pipe thread plugs.
14. Install the engine block drain plug(s), the exhaust manifold(s), the fuel return lines, the glow plug wiring, and the ground strap for the right cylinder head.
15. Install the valve train assembly. Refer to "Diesel Engine, Rocker

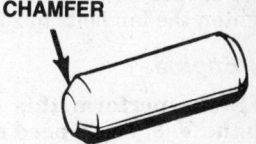

CHAMFER

Chamfer the alignment pin

Arm Replacement," above, for valve lifter bleeding procedures.
16. Install the intake manifold.
17. Install the rocker cover(s). The valve covers are sealed with RTV (room temperature vulcanizing) silicone sealer instead of a gasket. Use GM No. 1052434 or its equivalent. Install the cover to the head within 10 minutes (while the sealer is still wet).

OVERHAUL

For all cylinder head overhaul procedures, please refer to "Engine Rebuilding" in the unit repair section.

Timing Case and Camshaft

See the NOTE at the beginning of the Engine section.

V8 FRONT COVER REMOVAL & INSTALLATION

Gasoline Engine

1. Drain the coolant. Disconnect the radiator hose and the bypass hose. Remove the fan, belts and pulley.
2. Remove the vibration damper and crankshaft pulley.
3. Drain the oil and remove the oil pan.
4. Remove the front cover attaching bolts and remove the cover, timing indicator and water pump from the front of the engine.
5. Grind a chamfer on the end of each dowel pin as illustrated. When installing the dowel pins, they must be inserted chamfered end first. Trim about ⅛ in. from each end of the new front pan seal and trim any excess material from the front edge of the oil pan gasket. Be sure all mating surfaces are clean.
6. Install in the reverse order of removal using a new gasket with sealing compound. Tighten water pump attaching screws to 13 ft. lbs., 5/16 in. front cover attaching bolts to 25 ft. lbs. and the four bottom bolts (cover plate) to 35 ft. lbs. Torque the pulley hub bolt to 310 ft. lbs. Crankshaft pul-

ley bolts should be torqued to 10 ft. lbs. Tighten the fan bolts to 20 ft. lbs.

Diesel Engine

NOTE: To perform this operation on the V-8, you'll need a set of special tools designed to pull the crankshaft pulley off the crankshaft without damaging the rubber insert separating inner and outer pulley halves. Use tools equivalent to GM No. J-8614-3, J-8614-2, J-8614-1, and J-7583-3.

1. Drain the cooling system and disconnect the radiator hoses.
2. Remove all belts, fan and pulley. Remove the crankshaft pulley and balancer, utilizing the special tools described in the note above on the V-

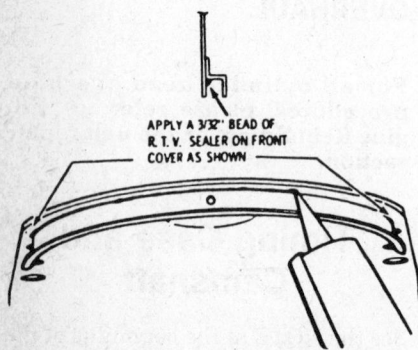

APPLY A 3/32" BEAD OF R.T.V. SEALER ON FRONT COVER AS SHOWN

V6 diesel engine front cover installation—apply R.T.V. sealer on the front cover oil pan seal retainer as shown

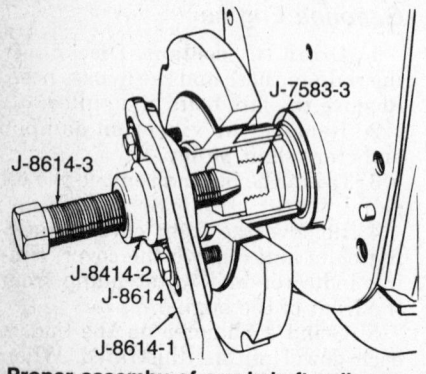

J-7583-3
J-8614-3
J-8414-2
J-8614
J-8614-1

Proper assemby of crankshaft pulley removal tools (307 V8)

8. See the illustration for proper assembly of these tools. On the V-6, make sure you use a puller that will bolt to the outside of the balancer and pull it off by applying pressure to a pilot inserted into the center of the crankshaft.

——— CAUTION ———
The use of any other type of puller, such as a universal claw type which pulls on the outside of the hub, can destroy the balancer. The outside ring of the balancer is

bonded in rubber to the hub. Pulling on the outside will break the bond. The timing mark is on the outside ring. If it is suspected that the bond is broken, check that the center of the keyway is 16° from the center of the timing slot. In addition, there are chiseled aligning marks between the weight and the hub.

3. Unbolt and remove the cover, timing indicator and water pump.
4. It may be necessary to grind a flat on the cover for gripping purposes.
5. Grind a chamfer on one end of each dowel pin.
6. Cut the excess material from the front end of the oil pan gasket on each side of the block.
7. Clean the block, oil pan and front cover mating surfaces with solvent.
8. Trim about $1/8$ in. off each end of a new front pan seal.
9. Install a new front cover gasket on the block and a new seal in the front cover.
10. Apply sealer to the gasket around the coolant holes.
11. Apply sealer to the block at the junction of the pan and front cover. On V6, apply RTV sealer on the front cover oil pan seal retainer.
12. Place the cover on the block and press down to compress the seal. Rotate the cover left and right and guide the pan seal into the cavity using a small screwdriver. Oil bolt threads and heads, install two to hold the cover in place, then install both dowel pins (chamfered end first). Install remaining front cover bolts.
13. Apply a lubricant, compatible with rubber, on the balancer seal surface.
14. Install the balancer and bolt. Torque the bolt to 200–300 ft. lbs. on V8, 160–350 ft. lbs. on V6.
15. Install all other parts in the reverse order of removal.

V8 Timing Chain

REPLACEMENT & VALVE TIMING

1. Remove the timing case cover, turn crankshaft to line up timing marks, and take off the camshaft sprocket.

NOTE: The fuel pump operating cam is bolted to the front of the camshaft sprocket and the sprocket is mounted on the camshaft by means of a dowel.

2. Remove the oil slinger, timing chain, and the camshaft sprocket. If the crankshaft sprocket is to be re-

placed, remove it also at this time. Remove the crankshaft key before using the puller. If the key cannot be removed, align the puller so it does not overlap the end of the key, as the keyway is only machined part of the way into the crankshaft gear.

3. Reinstall the crankshaft sprocket being careful to start it with the keyway in perfect alignment since it is rather difficult to correct for misalignment after the gear has been started on the shaft. Turn the timing mark on the crankshaft gear until it points directly toward the center of the camshaft. Mount the timing chain over the camshaft gear and start the camshaft gear onto its shaft with the timing marks as close as possible to each other and in line between the shaft centers. Rotate the camshaft to align the shaft with the new gear.

NOTE: To set crankshaft and camshaft gear marks for timing chain installation; position the engine with the No. 6 piston at top dead center. Slowly rotate the crankshaft one revolution until the camshaft gear mark is at 12 o'clock. No. 1 piston will now be at TDC on the compression stroke.

4. Install the fuel pump eccentric with the flat side toward the rear.
5. Drive the key in with a brass hammer until it bottoms.
6. Install the oil slinger.

NOTE: Whenever the timing chain and gears are replaced on the diesel engine it will be necessary to retime the engine. Refer to the paragraph on "Diesel Engine Injection Timing."

V6 Diesel Timing Chain and Sprocket

REPLACEMENT

1. Remove the front cover. See above for procedure. Remove the valve covers.
2. Loosen all rocker arm pivot bolts evenly so that lash exists between the rocker arms and valves. It is not necessary to completely remove the rocker arms unless related service is being performed.
3. Remove the crankshaft oil slinger and the camshaft sprocket bolt and washer.
4. Remove the timing chain, camshaft and crankshaft sprockets. If the crankshaft sprocket is a tight fit on the crankshaft use an appropriate puller to remove it.
5. If the camshaft sprocket-to-cam key comes out with the camshaft

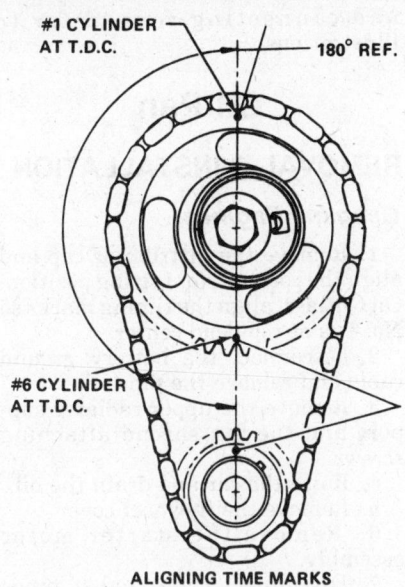

#1 CYLINDER AT T.D.C.   180° REF.

#6 CYLINDER AT T.D.C.

ALIGNING TIME MARKS

V8 engine timing mark alignment

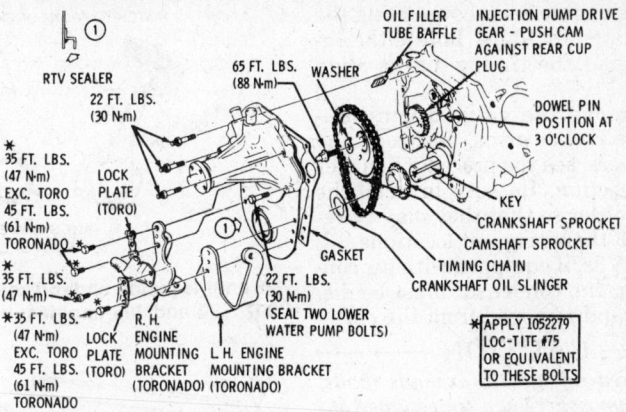

OIL FILLER TUBE BAFFLE   INJECTION PUMP DRIVE GEAR - PUSH CAM AGAINST REAR CUP PLUG

RTV SEALER
22 FT. LBS. (30 N·m)

65 FT. LBS. (88 N·m)   WASHER

DOWEL PIN POSITION AT 3 O'CLOCK

* 35 FT. LBS. (47 N·m) EXC. TORO 45 FT. LBS. (61 N·m) TORONADO

LOCK PLATE (TORO)

KEY
CRANKSHAFT SPROCKET
CAMSHAFT SPROCKET
TIMING CHAIN
CRANKSHAFT OIL SLINGER

GASKET

35 FT. LBS. (47 N·m)

* 35 FT. LBS. (47 N·m) EXC. TORO 45 FT. LBS. (61 N·m) TORONADO

R. H. ENGINE MOUNTING BRACKET (TORONADO)

L. H. ENGINE MOUNTING BRACKET (TORONADO)

22 FT. LBS. (30 N·m) (SEAL TWO LOWER WATER PUMP BOLTS)

LOCK PLATE (TORO)

*APPLY 1052279 LOC-TITE #75 OR EQUIVALENT TO THESE BOLTS

V8 diesel engine front cover and timing chain assembly

sprocket, remove the front camshaft bearing retainer and install the key into the injection pump drive gear. Install the bearing retainer.

6. Install the key in the crankshaft, if removed.

7. Install the camshaft sprocket, crankshaft sprocket and the timing chain together, align the timing marks on the camshaft and the crankshaft. Tighten the camshaft sprocket bolt to 70 ft. lbs.

8. Install the oil slinger and the remaining parts of the front cover assembly.

9. After installing the front cover, bleed down the valve lifters as instructed in "Diesel Engine Rocker Arm Replacement", above.

10. Remaining installation is in the reverse order of removal. Sealant is used in place of valve cover gaskets.

V8 Camshaft

REMOVAL & INSTALLATION

—— CAUTION ——

All Oldsmobile V8s require discharging of the air conditioning for camshaft removal. This should not be attempted by anyone who lacks the skill and experience to do so, as contact with the refrigerant can cause serious personal injury.

Gasoline Engine

1. Disconnect the battery.
2. Drain and remove the radiator.
3. Disconnect the fuel line at the fuel pump and remove the pump.
4. Disconnect the throttle cable and the air cleaner.

5. Remove the alternator belt, loosen the alternator bolts, and move the alternator to one side.

6. Remove the power steering pump from its brackets and move it out of the way.

7. Remove the air conditioning compressor from its brackets and move the compressor out of the way without disconnecting the lines.

8. Disconnect the hoses from the water pump.

9. Disconnect the electrical and vacuum connections.

10. Mark the distributor as to location in the block. Remove the distributor.

11. Raise the car and drain the oil pan.

12. Remove the exhaust crossover pipe and starter motor.

13. Disconnect the exhaust pipe at the manifold.

14. Remove the harmonic balancer and pulley.

15. Support the engine and remove the front motor mounts.

16. Remove the flywheel inspection cover.

17. Remove the engine oil pan.

18. Support the engine by placing wooden blocks between the exhaust manifolds and the front crossmember.

19. Remove the engine front cover.

20. Remove the valve covers.

21. Remove the intake manifold, oil filler pipe, and temperature sending switch.

22. Mark the lifters, pushrods, and rocker arms as to location so that they may be installed in the same position. Remove these parts.

23. If the car is equipped with air conditioning, discharge the A/C system and remove the condenser. See CAUTION above.

24. Remove the fuel pump eccentric, camshaft gear, oil slinger, and timing chain.

25. Carefully remove the camshaft from the engine.

26. Inspect the shaft for signs of excessive wear or damage.

27. Liberally coat camshaft and bearings with heavy engine oil or engine assembly lubricant and insert the cam into the engine.

28. Align the timing marks on the camshaft and crankshaft gears. See Timing Chain Replacement and Valve Timing for details.

29. Install the distributor using the locating marks made during removal. If any problems are encountered, see "Distributor Installation."

30. To install, reverse the removal procedure but pay attention to the following points:

a. Install the timing indicator before installing the power steering pump bracket.

b. Install the flywheel inspection cover after installing the starter.

c. Replace the engine oil and radiator coolant.

Diesel Engine

NOTE: If the camshaft is to be removed the air conditioning system must be discharged by a professional and the condenser removed. Removal of the camshaft also requires removal of the injection pump drive and driven gears, removal of the intake manifold, disassembly of the valve lifters, and re-timing of the injection pump.

1. Disconnect the negative battery cables. Drain the coolant. Remove the radiator.

2. Remove the intake manifold and gasket and the front and rear intake manifold seals. Refer to the intake manifold removal and installation procedure. Remove the oil pump drive assembly on the V6.

3. Remove the balancer pulley and the balancer. See "Caution" under V8 diesel engine front cover removal and installation, above, for V8 engine. Re-

move the engine front cover using the appropriate procedure. Rotate the engine so that the timing marks align on V6s.

4. Remove the valve covers. Remove the rocker arms, pushrods and valve lifters; see the procedure earlier in this section. Be sure to keep the parts in order so that they may be returned to their original locations.

5. On V8s, if equipped with air conditioning, the condenser must be discharged and removed from the car.

--- CAUTION ---

Compressed refrigerant expands (boils) into the atmosphere at a temperature of -26°F . It will freeze any surface it contacts, including your skin or eyes.

6. Remove the camshaft sprocket retaining bolt, and remove the timing chain and sprockets, using the procedure outlined earlier.

7. On V6s, remove the front camshaft bearing retainer bolt and the retainer, then remove the camshaft sprocket key and the injection pump drive gear.

8. Position the camshaft dowel pin at the 3 o'clock position on the V8.

9. On V8s, push the camshaft rearward and hold it there, being careful not to dislodge the oil plug at the rear of the engine. Remove the fuel injection pump drive gear by sliding it from the camshaft while rocking the pump driven gear.

10. To remove the fuel injection pump driven gear, remove the injection pump intermediate pump adapter (V6s) and the pump adapter (All), remove the snap ring, and remove the selective washer. Remove the driven gear and spring.

11. Remove the camshaft by sliding it out the front of the engine. Be extremely careful not to allow the cam lobes to contact any of the bearings, or the journals to dislodge the bearings during camshaft removal. Do not force the camshaft, or bearing damage will result.

12. If either the injection pump drive or driven gears are to be replaced, replace both gears. Make certain the marks are in alignment on both gears before inserting the cam gear key on the V6.

13. Coat the camshaft and the cam bearings with GM lubricant No.1052365 or the equivalent.

14. Carefully slide the camshaft into position in the engine.

15. Fit the crankshaft and camshaft sprockets, aligning the timing marks as shown in the timing chain removal and installation procedure, above. Remove the sprockets without disturbing the timing.

16. Install the injection pump driven

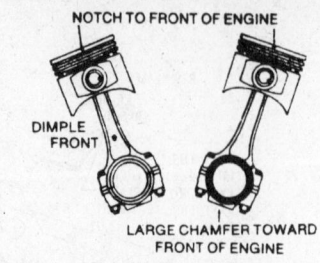

V8 engine piston and rod assembly—260, 307, 350 and 403 engines

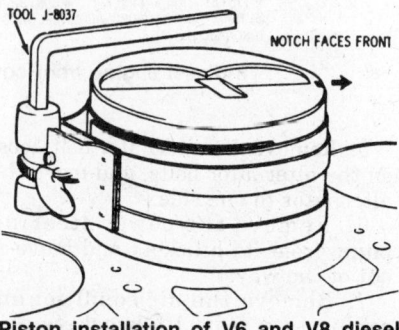

Piston installation of V6 and V8 diesel engines

gear, spring, shim, and snap ring. Check the gear end play. If the end play is not within .002–.015 in. on V8s, or .002–.006 on V6s, replace the shim to obtain the specified clearance. Shims are available in 0.003 in. increments, from 0.080–0.115 in.

17. On V8s, bring the camshaft dowel pin to the 3 o'clock position. Align the zero marks on the pump drive gear and pump driven gear. Hold the camshaft in the rearward position and slide the pump drive gear onto the camshaft. On the V6, align the zero marks on the injection pump drive and driven gears, then install the camshaft sprocket key. Install the camshaft bearing retainer.

18. Install the timing chain and sprockets, making sure the timing marks are aligned.

19. Install the lifters, pushrods and rocker arms. See "Rocker Arm Replacement, Diesel Engine" for lifter bleed down procedures. Failure to bleed down the lifters could bend valves when the engine is turned over.

20. Install the injection pump adapter and injection pump. See the appropriate sections under "Fuel System" above for procedures.

21. Install the remaining components in the reverse order of removal.

Piston and Connecting Rod Positioning

For the correct positioning of pistons and connecting rods, refer to illustrations.

Oil Pan

REMOVAL & INSTALLATION

Gasoline Engines

1. Remove the distributor cap and align the rotor to No. 1 firing position. On Cutlass, align the timing marks so No. 1 is at top dead center.

2. Disconnect the battery ground cable and remove the dipstick.

3. Remove the upper radiator support and the fan shroud attaching screws.

4. Raise the car and drain the oil.

5. Remove the flywheel cover.

6. Remove the starter motor assembly.

7. Disconnect the exhaust pipes and the crossover pipe.

8. Disconnect the engine mounts and raise the front of the engine as far as possible.

9. Remove the oil pan attaching bolts and remove the pan.

10. Coat both sides of the new gasket with sealer when installing. Installation is in the reverse order of removal. Torque the attaching bolts to 10 ft. lbs.

Diesel Engines

1. On V8s, remove the vacuum pump and drive (with A/C) or the oil pump drive (without A/C). On V6s, remove the oil pump drive and vacuum pump.

2. Disconnect the batteries and remove the dipstick.

3. Remove the upper radiator support and fan shroud.

4. Raise and support the car. Drain the oil.

5. Remove the flywheel cover.

6. Disconnect the exhaust and crossover pipes.

7. Remove the oil cooler lines at the filter base.

8. Remove the starter assembly. Support the engine with a jack.

9. Remove the engine mounts from the block.

10. Raise the front of the engine and remove the oil pan.

11. Installation is in the reverse order of removal.

Oil Pump

REMOVAL & INSTALLATION

Gasoline and Diesel Except 231 V6

The oil pump is mounted on the bottom of the block and is accessible only

by removing the oil pan. On V8 engines, including diesel, and the V6 diesel, remove the oil pan, then unbolt and remove the oil pump and screen as an assembly.

231 V6

1. Place a drain pan under the area of the pump and filter and then remove the filter.

2. Remove the bolts attaching the oil pump cover to the timing chain cover. Then, remove the cover, and oil pump drive and driven gears.

3. Remove the oil pressure relief valve cap, spring, and the valve itself.

NOTE: Leave the oil filter bypass valve and spring (which are staked in place) undisturbed.

4. Install in reverse order. Just before installing the cover assembly, carefully and completely pack the oil pump with petroleum jelly. **Failure to do this will prevent the oil pump from priming itself, which will cause extensive engine damage.** Torque the cover assembly screws alternately and evenly to 12 ft. lbs.

Rear Main Seal

REPLACEMENT

Gasoline and Diesel Except 231 V6

The crankshaft need not be removed to replace the rear main bearing upper oil seal.

1. Drain the crankcase and remove the oil pan and rear main bearing cap.

2. Using a blunt-ended tool, drive the upper seal into its groove on each side until it is tightly packed. This is usually $1/4$–$3/4$ in.

3. Cut pieces of new seal $1/16$ in. longer than required to fill the grooves and install, packing into place.

4. Carefully trim any protruding seal, being sure not to scratch or damage the bearing surface.

5. Install a new seal in the bearing cap. Now, seat the seal with a tool such as BT–7923 or J–2528A. Rock the tool back and forth slightly and then cut off either end flush with the cap, holding the seal in position with the tool. Apply a thin film of chassis grease to the rope seal. Apply a sealer such as 1050026 or the equivalent to the area of the cap around the ends of the seal. Use the sealer sparingly and keep it out of the bolt threads. Install cap, tightening bolts to 120 ft. lbs. (107 ft. lbs. on V6 diesel). Install the oil pan.

231 V6

NOTE: This procedure requires the use of a special set of seal guide and packing tools J–21526–1 and J–21526–2 and BT–7923 or J–2528A.

1. Drain the crankcase and remove the oil pan and rear main bearing cap.

2. Insert the packing tool into the seal groove and against one end of the seal in the block. Then, tap the tool to pack the seal tightly into the block $1/4$–$3/4$ in. Repeat the same step on the other side.

3. Measure the amount the seal was driven up on one side by measuring the insertion of the packing tool. Then, with a single edged razor blade, cut that amount from the old seal removed from the bearing cap, using the rear cap as a holding fixture for the seal. Repeat this procedure for the other side (or end) of the seal. Retain both these short pieces.

4. Install the guide tool J–21526–1 onto the block. Using the packing tool, work the short pieces cut in the last step into the guide tool and then pack them into the block, one on either side. Oil the pieces to make them easier to insert. The guide and packing tools have been machined to provide a built in stop so the pieces will be inserted just the right amount.

4. Remove the guide tool. Then, apply a sealer such as LOCTITE 414® or Fel-Pro Mighty Grip® to the seal groove. Then, promptly (within one minute), insert the seal into the groove and roll it into place with a wooden dowel or similar device until it projects above the groove no more than $1/16$ in. Apply a thin film of chassis grease to the rope seal. Apply a sealer such as 1052756 or Fel-Pro Set and Seal® or the equivalent to the bearing cap mating surface. Make sure to keep the sealer out of bolt threads and to use it sparingly. Soak sealing strips in light oil for five minutes and then install them into the cap. Install service composition strips into grooves along the sides of the main cap.

5. Install the main bearing cap and torque to specifications. Perform the remaining steps in reverse of the removal procedure.

CLUTCH

Clutch Pedal

ADJUSTMENT

1980–82 Cutlass

1. Loosen the locknut on the pushrod swivel.

2. Detach the pedal return spring.

3. Turn the clutch lever and shaft assembly until the clutch pedal seats against the rubber bumper on the dash brace.

4. Push the outer end of the clutch fork rearward, so that the throwout bearing just contacts the clutch plate.

5. Remove the retaining clip from the lower pushrod swivel and install the swivel in the upper gauge hole. Install the retaining clip.

6. Lengthen the pushrod until there is no lash.

7. Remove the retaining clip and reinstall the swivel in the lower hole on the lever and shaft assembly.

8. Tighten the locknut against the swivel. Be sure the rod length remains unchanged.

9. Install the pedal return spring and check pedal free-play.

REPLACEMENT

Cutlass

1. Remove the transmission.

2. Detach the clutch return spring and clutch release rod assembly.

3. Remove the throwout bearing.

4. Without removing the starter from the engine, remove the flywheel housing.

NOTE: The release yoke, boot and ball stud will remain in the housing.

5. Scribe a mark opposite the X mark on the flywheel cover. This mark is for proper flywheel balancing.

6. Loosen the pressure plate evenly, one turn at a time.

—————— **CAUTION** ——————

Do not lubricate the splines as the lubricant will be forced onto the damper, resulting in clutch rattle.

Clutch installation is performed in the following order:

1. Install the clutch disc/cover assembly and finger-tighten its securing bolts.

NOTE: Align the mark made during removal with the X mark on the flywheel cover.

2. Use a clutch arbor or an old input shaft to align the disc by inserting it through the disc and into the pilot bearing.

3. Tighten every other bolt until the cover assembly is within $1/4$ in. of the flywheel.

4. Repeat Step 3 for the three remaining bolts.

5. Tighten the first three bolts to 30 ft. lbs. and then tighten the remaining three bolts to the same torque.

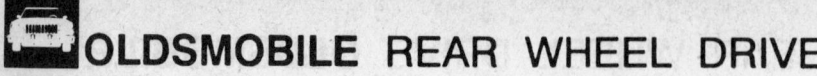

6. Remove the arbor. Lubricate the inside groove of the throwout bearing and the release yoke ball stud with wheel bearing grease.

7. Install the throwout bearing.

8. Install the flywheel housing and the transmission. Adjust clutch freeplay as outlined above.

MANUAL TRANSMISSION

The 3-speed transmission is the Saginaw unit. The standard 4-speed transmission in all models is also a Saginaw unit. On the Saginaw, all three shift rods go to levers on the side cover. See the Capacities table at the beginning of this section for manual transmission refill capacities. For manual transmission overhaul procedures, see the Unit Repair Section.

Transmission

REMOVAL & INSTALLATION

1. Disconnect throttle linkage and raise car. If applicable, disconnect T.C.S. switch.

2. Remove the driveshaft.

3. Support the rear of the engine. Remove the catalytic converter and/or brackets, if they are in the way.

4. On console equipped floorshifts, disconnect shifter assembly at transmission, allowing this unit to remain in car.

5. Disconnect parking brake cables and remove the cross member.

6. Disconnect speedometer cable and back-up light switch.

7. Remove transmission upper and lower bolts.

--- CAUTION ---

During removal, use aligning studs to support the transmission, otherwise distortion of the clutch driven plate will result.

8. Slide transmission rearward and remove. On models equipped with dual exhaust, it may be necessary to disconnect left exhaust pipe at the manifold.

9. Install by reversing the removal

Transmission to Clutch	Ft. Lbs.
Housing	53
Crossmember to frame	25
Crossmember to transmission	35
U-joint strap bolt	15

procedure. Observe the following torque figures:

3-Speed Linkage

ADJUSTMENT

Cutlass

1. Turn the ignition switch to OFF.

2. Raise and support the car.

3. Remove the retainer from the shift rods.

4. Place the transmission levers in Neutral.

5. Align the control levers and place a ¼ in. gauge pin into the levers and brackets, with the shift handle in Neutral.

6. Loosen the nuts on the shift rods and adjust the trunnion and pin assembly on First/Reverse, then tighten the nuts and install the shift rod and retainer.

7. Loosen the shift rod nut and adjust the trunnion and pin assembly on Second/Third, then tighten the nuts and install the shift rod and retainer.

8. Remove the gauge pin from the control lever assembly and check the operation of the control lever. Readjust as required.

9. Lower the car.

1980–82 4-Speed Linkage

ADJUSTMENT

1. Turn the ignition switch to the OFF position.

2. Raise and support the car.

3. Loosen the lock nuts at the swivels on the shift rods.

4. Set the transmission levers in Neutral.

5. Place the shifter in Neutral.

6. Align the control levers and place a ¼ in. gauge pin into the levers and bracket.

7. Tighten the First/Second shift rod nut against the swivel. Torque to 10 ft. lbs.

8. Tighten the Third/Fourth shift rod nut against the swivel. Torque to 10 ft. lbs.

9. Tighten the reverse shift control rod nut to 10 ft. lbs.

10. Remove the gauge pin, check for proper operation of the levers and lower the car.

OVERHAUL

For all manual transmissions or transaxle overhaul procedures, please refer to "Manual Transmission Overhaul" in the unit repair section.

AUTOMATIC TRANSMISSION

All Oldsmobile models use the Turbo Hydra-Matic automatic transmission. The transmission can be identified visually: The 200, 250, 350, and 375B have a square or oblong pan with the right rear corner cut off; the 375 and 400 have an irregular pan shape. Some 200s have the word METRIC embossed in the pan. The 200 has ten pan bolts; the 350 and 375B have thirteen. The 250 has an intermediate band adjusting screw on the right side of the case. The 200, 250, 350, and 375B have a downshift cable between the carburetor linkage and the transmission; the 375 and 400 have an electrical downshift switch on the accelerator pedal linkage. The 200-R4 four speed automatic overdrive transmission was introduced in 1981.

For automatic transmission service procedures, refer to the Unit Repair Section.

REMOVAL & INSTALLATION

1. Remove the air cleaner assembly.

2. Disconnect the throttle valve detent cable.

3. Remove the dipstick (and the bolt holding the dipstick tube if accessible).

4. Jack up the car and support it with jack-stands.

5. Remove the driveshaft.

6. Disconnect the speedometer cable and shift linkage.

7. Disconnect any electrical leads.

8. Remove the flywheel cover; matchmark the flywheel and converter to maintain original balance.

9. Remove the torque converter to flywheel bolts and/or nuts.

NOTE: It may be necessary to disconnect the catalytic converter support bracket.

10. Remove the transmission support bracket (rear crossmember).

11. Remove the oil cooler lines.

12. Support the transmission with a transmission jack and remove the bellhousing bolts.

13. Support the engine with a jack and remove the transmission.

NOTE: Carefully remove the transmission to prevent the torque converter from falling off the mainshaft.

14. Installation is in the reverse order of removal.

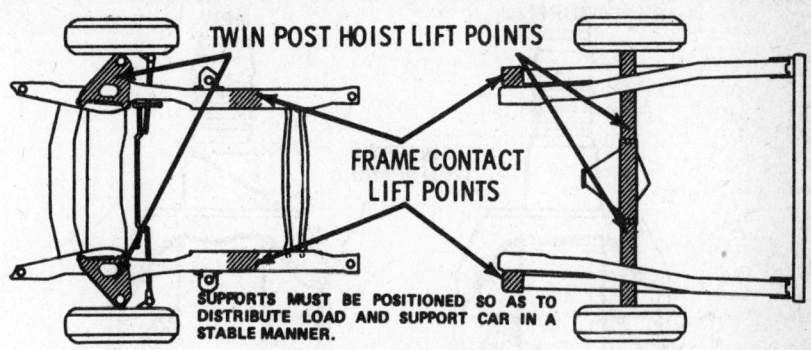

Omega hoisting points through 1979

DRIVESHAFT AND U-JOINTS

Driveshaft

REMOVAL & INSTALLATION

1. Matchmark the relationship of the driveshaft to the differential flange.

2. Unbolt the straps or flange. Tape the bearing caps in place to prevent losing the bearing rollers. Support the driveshaft to prevent excessive strain on the universal joint.

3. Pull the shaft back and remove it. Be careful not to damage the splines at the transmission end.

4. If the transmission splined slip yoke does not have a vent hole at the center, it should be lubricated with engine oil. If it does have a vent hole, it should be lubricated with grease. Slide the slip yoke into place.

5. Align the matchmarks and tighten the bolts. Strap bolts should be tightened to 16 ft. lbs.

UNIVERSAL JOINT OVERHAUL

See the "U-Joints and CV-Joint" Unit Repair section for overhaul procedures.

REAR AXLE

Axle, Shaft, Bearing and Seal

REMOVAL & INSTALLATION

C-lock Type

These cars use the C-lock type rear axles. The axle shafts are retained by C-shaped locks, which fit grooves at the inner end of the shaft. Bearings in the C-lock type axle consist of an outer race, bearing rollers and a roller cage, retained by snaprings.

1. Raise the vehicle and remove the wheels and brake drums.

2. Clean the area of the cover and drain the fluid from the carrier by removing the cover. Remove the differential pinion shaft.

3. Push the flanged end of the axle shaft toward the center of the vehicle and remove the C-lock from the end of the shaft.

4. Remove the axle shaft from the housing, being careful not to damage the oil seal.

5. Remove the oil seal with a pry bar inserted behind the steel case of the oil seal. Pry the seal loose from the bore.

6. Seat the legs of the bearing puller behind the bearing. Seat a washer against the bearing and hold it in place with a nut. Use a slide hammer to pull the bearing.

7. Pack the cavity between the seal lips with wheel bearing lubricant and lubricate a new wheel bearing with same.

8. Use a suitable driver and install the bearing until it bottoms. Lubricate the lips of the oil seal and tap it into place so it is flush with the axle tube.

9. Slide the axle shaft into place. Be sure that the splines on the shaft do not damage the oil seal. Make sure that the splines engage the differential side gear.

10. Install the axle shaft C-lock on the inner end of the axle shaft and push the shaft outward so that the C-lock seats in the differential side gear counterbore.

11. Position the differential pinion shaft through the case and pinions, aligning the hole in the case with the hole for the lockscrew.

12. Install the pinion shaft lockscrew.

13. Use a new gasket and install the carrier cover. Be sure that the gasket surfaces are clean before installing the gasket and cover.

14. Fill the axle with lubricant to the bottom of the filler hole.

15. Install the brake drum and wheels and lower the car. Check for leaks and road test the car.

FRONT SUSPENSION

Shock Absorber

REPLACEMENT

1. Remove the two bolts and lockwashers securing the shock to the lower control arm.

2. Remove the upper nut, retainer, and grommet from the shock.

3. To install, reverse the removal procedure.

NOTE: Purge new shock absorbers of air by repeatedly extending in their normal position and compressing while inverted.

Lower Ball Joint

INSPECTION

All Models

These lower ball joints contain a visual wear indicator. The lower ball joint grease plug screws into the wear indicator which protrudes from the bottom of the ball joint housing. As long as the wear indicator extends out of the ball joint housing, the ball joint is not worn. If the tip of the wear indicator is parallel with, or recessed into the ball joint housing, the ball joint is defective.

Lower Ball Joint

REMOVAL & INSTALLATION

1. Raise the car and support the frame with jack stands.

2. Remove the tire and wheel.

3. Place a floor jack under the control arm spring seat.

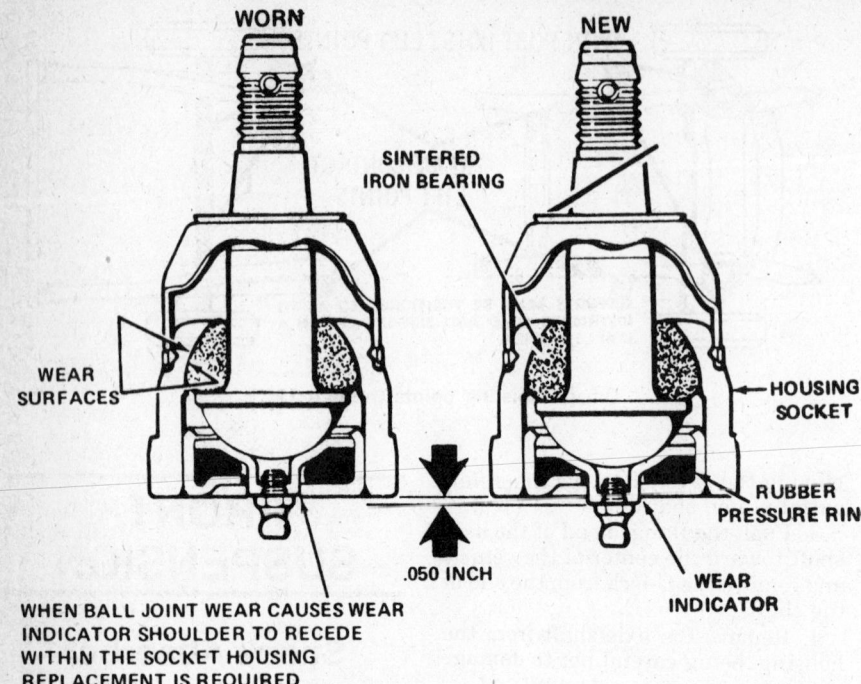

WORN

NEW

SINTERED
IRON BEARING

HOUSING
SOCKET

RUBBER
PRESSURE RING

WEAR
SURFACES

.050 INCH

WEAR
INDICATOR

WHEN BALL JOINT WEAR CAUSES WEAR
INDICATOR SHOULDER TO RECEDE
WITHIN THE SOCKET HOUSING
REPLACEMENT IS REQUIRED

Lower ball joint wear indicator

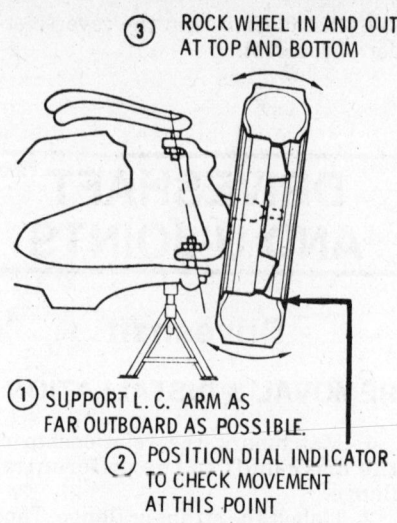

③ ROCK WHEEL IN AND OUT
AT TOP AND BOTTOM

① SUPPORT L. C. ARM AS
FAR OUTBOARD AS POSSIBLE.

② POSITION DIAL INDICATOR
TO CHECK MOVEMENT
AT THIS POINT

Checking upper ball joint

CAUTION

Leave the jack under the spring seat during removal and installation, in order to keep the spring and control arm positioned.

4. Remove the cotter pin from the ball joint stud and, using a ball joint stud removal tool, separate the ball joint from the steering knuckle.
5. When the stud comes loose, remove the stud nut.
6. Guide the lower control arm through the opening in the splash shield using a screwdriver.
7. Block the steering knuckle out of the way by using a block of wood between the frame and the upper control arm.
8. Pry the retainer off the ball joint seal with a driftpin and remove the seal.
9. Using a ball joint remover, remove the lower ball joint from the control arm.
10. Press in a new balljoint until it bottoms on the lower control arm.

NOTE: On disc brake cars, make sure the grease purge on the seal faces away from the brakes.

11. Assemble the suspension and torque the nut to 95 ft. lbs. Install the cotter pin and bend it to the side, not over the top of the nut. The cotter pin on the Cutlass must be installed parallel to the center line of the car.
12. Install the ball joint fitting and lube until grease appears at the seal.

13. Install the tire and wheel assembly.

Upper Ball Joint

INSPECTION

1. Jack up the car and place jack stands under the left and right control arms as near as possible to the lower ball joints. Make sure the car sits steadily on the floor stands.
2. Position a dial indicator so that its button contacts the inside lip of the wheel trim.
3. Grasp the wheel at the 6 and 12 o'clock positions. Push in on the bottom of the wheel while pulling on the top. Read the gauge and reverse the push/pull procedure. If the total deflection on the gauge reads more than 0.125 in., the ball joint is worn and must be replaced.

Upper Ball Joint

REMOVAL & INSTALLATION

1. Raise the front of car and place floor stands under the lower control arm between the spring seats and the ball joints.

CAUTION

Leave the jack under the spring seat during removal and installation, in order to keep the spring and control arm positioned.

2. Remove the wheel.
3. Remove the cotter pin from the upper ball joint stud and loosen the upper ball joint nut.
4. Using a ball joint remover tool, break the stud loose and remove the nut and pull the stud out of the knuckle. Support the steering knuckle to prevent damage to the brake line.
5. Using a $1/8$ in. diameter drill bit, drill into each of the four rivet heads to a depth of $1/2$ in.
6. Drill off the rivet heads with a $1/2$ in. diameter bit.
7. Punch out the rivets and remove the ball joint.
8. To install, place the new ball joint in the upper control arm and secure it with four bolts and nuts in place of rivets. Tighten the nuts to 8 ft. lbs.
9. Connect the ball joint to steering knuckle. Torque the nut to 65 ft. lbs.

NOTE: When replacing ball joints, use only high-quality replacement parts and bolts and nuts specified to be strong enough to endure the stress. Always advance the ball stud nut to align the cotter pin hole.

10. Install the grease fitting and lubricate until grease appears at the seal.
11. Install the wheel.

Upper Control Arm

REMOVAL & INSTALLATION

1. Raise the car and place jack stands between the spring seats and the ball joints of the lower control arms.
2. Remove tire and wheel.

3. Place floor jack under lower control arm spring seat.

—— CAUTION ——

Leave the jack under the spring seat during removal and installation, in order to keep the spring and control arm positioned.

4. Remove ball joint stud from steering knuckle, by removing cotter pin and nut and pressing joint loose from knuckle with a ball joint remover. Support hub assembly to prevent damage to the brake line.

5. Loosen the pivot shaft-to-frame nuts and remove the alignment shims. Support hub assembly and remove upper arms by sliding shaft off end of bolts.

NOTE: Mark alignment shims for reassembly in their original locations.

6. It is necessary to remove upper control arm attaching bolts to gain clearance to remove arm assembly.

7. Remove control arm from car.

8. To reinstall, position bolts loosely in frame and install pivot shaft on bolts.

9. Install alignment shims placing them in position from which they were removed. Torque the nuts to 73 ft. lbs. for 88 and 98; 45 ft. lbs. for 1980–82 Cutlass.

10. Connect the ball joint stud to the steering knuckle and torque to 65 ft. lbs. minimum for Cutlass, 88 and 98. Install the cotter pin.

11. Install the wheel and check the alignment on all models.

Lower Control Arm and/or Spring

REMOVAL & INSTALLATION

1. Place the transmission in Neutral so the steering wheel is unlocked.

2. Raise the car and remove the wheel. Support the car with stands.

3. Remove the shock absorber.

4. Insert a spring removal tool into the shock hole. Rotate the tool so the plate is well seated in the lower control arm spring seat.

5. Rotate the nut on the tool to compress the spring slightly, just enough so it is free in the seat.

6. On all models remove the two lower control arm pivot bolts and disengage the arm from the frame.

7. Rotate the arm and remove the spring.

8. Loosen the lower ball joint stud nut a few turns. Using a ball joint re-

Lower front suspension—88 and 98, 1978 and later Cutlass

mover, expand the tool to snap the ball joint loose from the knuckle.

9. Remove the stud nut and the control arm.

10. Installation is in the reverse order of removal. Torque the lower ball joint stud nut to 95 ft. lbs.

Wheel Bearing

ADJUSTMENT

1. Raise the car so the wheel can spin freely. Remove the dust cap.

2. Tighten the adjusting nut to 12 ft. lbs. while turning the wheel.

3. Back off on the nut $\frac{1}{2}$ turn.

4. Finger-tighten the nut and install the cotter pin or the retaining ring.

NOTE: If the cotter pin cannot be installed, back off on the nut until the slot aligns with the serrations on the nut. Do not back off on the nut more than $\frac{1}{24}$ of a turn.

5. Once adjusted, the front wheel bearings should have 0.001–0.005 in. end play.

REAR SUSPENSION

Shock Absorber

REMOVAL & INSTALLATION

NOTE: Purge new shock absorbers of air by repeatedly extending in their normal position and compressing while inverted.

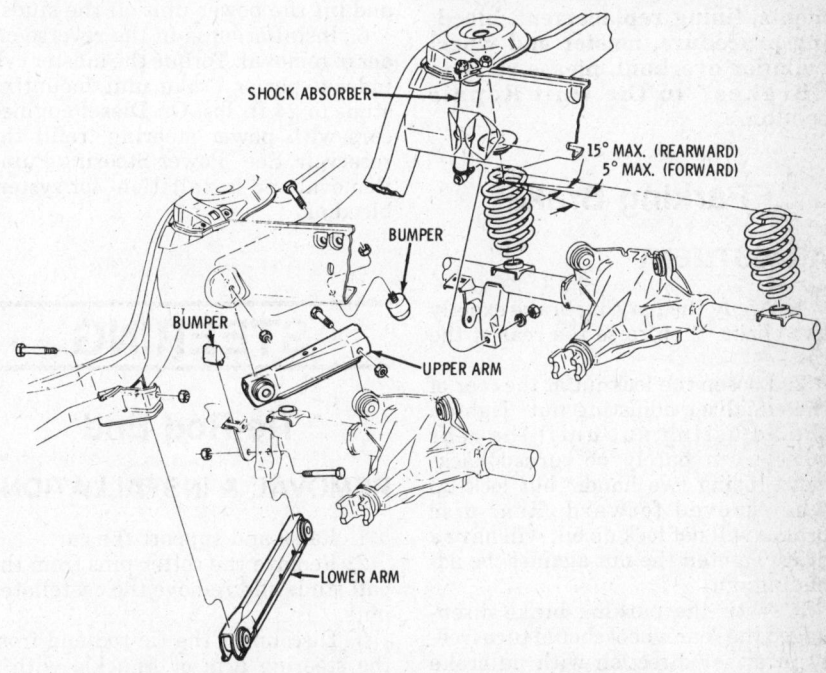

Cutlass, 88 and 98 rear suspension (except wagon)

1. Raise the vehicle and support the rear axle housing.

2. Remove the lower shock mounting bolt from the shock absorber eye.

3. Unfasten the upper mounting bracket bolts and remove the shock.

4. Installation is in the reverse order of removal, except that the upper attaching bolts should remain loose while the lower (eye) is being tightened.

Coil Spring

REMOVAL & INSTALLATION

1. Raise the rear of the car on the axle housing and place jack stands under the frame. Do not lower the jack.◊ 2. Disconnect the brake line at the axle housing and at the differential housing.

3. Disconnect the upper control arms at the differential.

4. Remove the shock absorber lower mount and lower the jack. Be careful not to stretch the brake hose.

5. Remove the spring.

6. Installation is in the reverse order of removal.

BRAKES

For information on brake adjustments, lining replacement, bleeding procedure, master and wheel cylinder overhaul, please refer to "Brakes" in the Unit Repair section.

Parking Brake

ADJUSTMENT

1. Apply the parking brake exactly two clicks, then raise the rear of the car.

2. Loosen the locknut at the rear of the equalizer adjusting nut. Tighten the adjusting nut until the rear wheels can barely be turned backward (using two hands) but lock up when moved forward. Rear disc brakes will not lock up but will have a drag. Tighten the nut against the adjusting nut.

3. With the parking brake disengaged the rear wheel should turn freely in either direction with no brake drag.

Master Cylinder

REMOVAL & INSTALLATION

NOTE: Be sure the area where the master cylinder is mounted is clean, before beginning removal.

1. Disconnect and cap or plug hydraulic lines. Disconnect the electrical lead, if so equipped.

2. On non-power brakes, disconnect the pushrod at the brake pedal.

3. Remove the attaching bolts and master cylinder.

4. Install in the reverse order of removal. Fill with fluid and bleed.

Power Brake Unit

REMOVAL & INSTALLATION

1. Remove the two nuts holding the master cylinder fastened to the power unit. Carefully position the master cylinder out of the way, being careful not to kink any of the hydraulic lines. It is not necessary to disconnect the brake lines.

2. Disconnect the vacuum hose from the vacuum check valve on the front housing. Plug the hose. On cars with diesel engine, disconnect the three hydraulic lines from the power cylinder. Plug the lines immediately.

3. Loosen the four nuts that hold the power unit mounted on the firewall.

4. Disconnect the pushrod from the brake pedal. Do not force the pushrod to the side when disconnecting.

5. Remove the four mounting nuts and lift the power unit off the studs.

6. Installation is in the reverse order of removal. Torque the master cylinder-to-power brake unit mounting studs to 24 ft. lbs. On Diesel engined cars with power steering, refill the reservoir. See "Power Steering Pump Removal and Installation" for system bleeding.

STEERING

Tie-Rod End

REMOVAL & INSTALLATION

1. Raise and support the car.

2. Remove the cotter pins from the ball studs and remove the castellated nuts.

3. Disconnect the tie-rod end from the steering arm or knuckle with a ball joint separator.

4. Remove the inner ball stud from the intermediate rod with a puller. Mark the tie-rod end position before removal.

5. Loosen the clamp bolts and unscrew the ends from the adjuster tubes. If a force of more than 7 ft. lbs. is required to remove the ends after breakaway, the fasteners should be replaced.

6. Clean and inspect all parts. When installing, run the tie-rod end to the position marked. Torque the ball stud nuts to 30 ft. lbs.

Steering Gear

REMOVAL & INSTALLATION

1. Remove the flexible coupling shield.

NOTE: Failure to disconnect the flexible coupling from the steering gear stub shaft can result in damage to the steering gear and/or immediate shaft. This damage can cause loss of steering control which could result in a vehicle crash and bodily injury.

2. Disconnect the hoses from the gear and cap the hose fittings.

3. Jack up the car. Support it with jackstands.

4. Remove the pitman shaft nut. Then disconnect the pitman arm from the pitman shaft. Special puller J29107 or its equivalent must be used.

5. Remove the three bolts attaching the gear to the frame side rail. Remove the gear.

6. Installation is in the reverse order of removal.

NOTE: If the threads are stripped, do not repair, replace the housing. Tighten the gear-to-frame bolts 70 ft. lbs., pitman shaft to 85 ft. lbs. and the coupling flange hub bolts to 30 ft. lbs.

Steering Wheel

REMOVAL & INSTALLATION

Except Tilt and Telescope Models

1. Disconnect the battery ground cable.

2. On the stock wheel, remove the two screws attaching the horn pad assembly to the wheel. Disconnect the horn contact from the pad assembly. On the deluxe wheel, remove the pad attaching screws, lift up the pad, and disconnect the horn wire by pushing on the insulator and turning counterclockwise. On the sport steering

wheel, pull up on the emblem to remove it. Remove the contact assembly attaching screws and the contact assembly.

3. On all models remove the steering wheel nut retainer.

4. Remove the retaining nut and the steering wheel, using a suitable puller.

5. Installation is in the reverse order of removal. Align the marks on the wheel hub and the steering shaft. If the spokes of the wheel are not horizontal, it is necessary to adjust the tie-rod ends. Torque the attaching bolt 30 ft. lbs.

——— CAUTION ———

Do not hammer on the steering shaft. The energy-absorbing column will be damaged and require replacement.

Tilt and Telescope Models

1. Disconnect the battery ground.

2. Remove the three pad attaching screws, lift off the pad assembly and disconnect the horn wire.

3. Push the locking lever counterclockwise to full release.

4. Mark the plate assembly where the two attaching screws attach the plate assembly tot he locking lever and remove the two screws.

5. Unscrew and remove the plate assembly. Remove the steering wheel nut.

6. Using a puller, remove the steering wheel.

7. Install a $5/16$ in. x 18 set screw into the upper shaft at the full extended position and lock.

8. Install the steering wheel, aligning the scribe mark on the hub with the slash mark on the end of the shaft. Make sure that the attached end of the upper horn contact assembly is seated flush against the top of the horn contact assembly.

9. Install the steering wheel nut and torque to 30 ft. lbs. The remainder of the installation is in the reverse order of removal. Remove the set screw after steering wheel installation.

Turn Signal Switch

REMOVAL & INSTALLATION

1. Disconnect the negative battery cable.

2. Remove the steering wheel.

3. Pry the lockplate cover off with a screwdriver.

4. Place a lock plate removal tool over the steering shaft and tighten the nut to depress the lockplate. Remove the snap ring retainer.

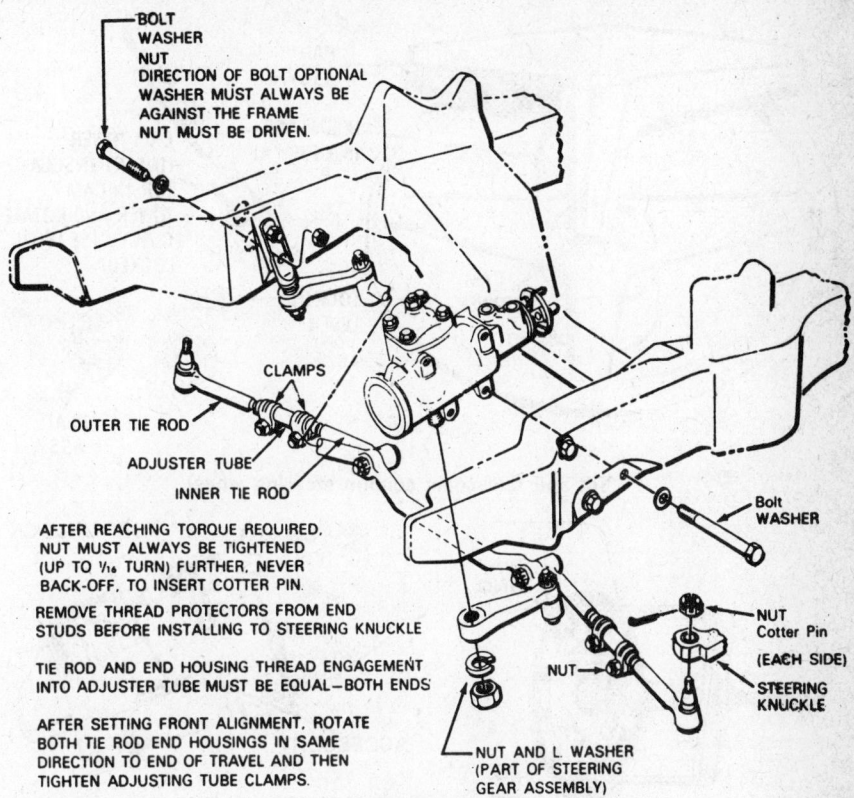

BOLT
WASHER
NUT
DIRECTION OF BOLT OPTIONAL
WASHER MUST ALWAYS BE
AGAINST THE FRAME
NUT MUST BE DRIVEN.

CLAMPS

OUTER TIE ROD

ADJUSTER TUBE

INNER TIE ROD

Bolt WASHER

NUT Cotter Pin (EACH SIDE)

STEERING KNUCKLE

NUT

AFTER REACHING TORQUE REQUIRED,
NUT MUST ALWAYS BE TIGHTENED
(UP TO 1/16 TURN) FURTHER, NEVER
BACK-OFF, TO INSERT COTTER PIN.

REMOVE THREAD PROTECTORS FROM END
STUDS BEFORE INSTALLING TO STEERING KNUCKLE

TIE ROD AND END HOUSING THREAD ENGAGEMENT
INTO ADJUSTER TUBE MUST BE EQUAL—BOTH ENDS

AFTER SETTING FRONT ALIGNMENT, ROTATE
BOTH TIE ROD END HOUSINGS IN SAME
DIRECTION TO END OF TRAVEL AND THEN
TIGHTEN ADJUSTING TUBE CLAMPS.

NUT AND L. WASHER
(PART OF STEERING
GEAR ASSEMBLY)

Steering linkage

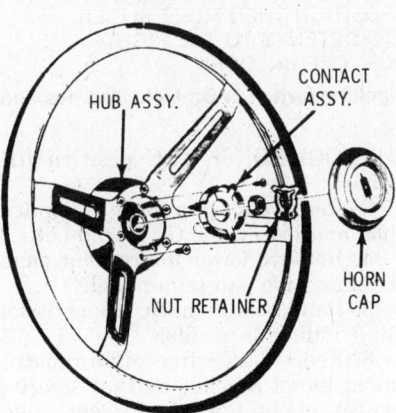

HUB ASSY.

CONTACT ASSY.

HORN CAP

NUT RETAINER

Sport steering wheel

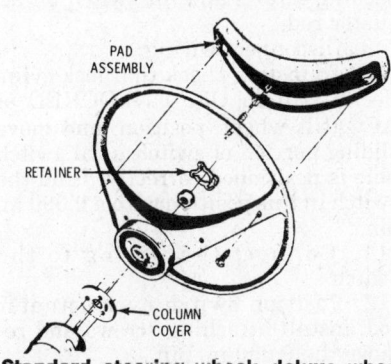

PAD ASSEMBLY

RETAINER

COLUMN COVER

Standard steering wheel; deluxe wheel similar

5. Remove the lock plate and the cancelling cam.

6. Remove the upper bearing preload spring. With the turn signal lever in the right turn position, remove the lever attaching screw and the lever. On models with the dimmer switch in the turn signal lever, remove the actuator arm screw and the arm. Remove the turn signal lever. Remove the three turn signal switch screws.

7. Push in the hazard switch knob and remove the retaining screw and the knob. On tilt columns, position the housing in the center position.

8. Remove the lower trim panel from the instrument panel and disconnect the turn signal connector from the wiring harness. Remove the connector.

9. Remove the bolts attaching the surrounding bracket assembly to the jacket. On all column shift automatics remove the shift indicator needle attaching screw and remove or disconnect the needle.

10. Hold the steering column in place and remove the two attaching nuts from below. Remove the bracket assembly and the wire protector. Loosely reinstall the nuts to hold the column in place.

11. Carefully remove the turn signal switch and the wiring.

12. To install, place the switch in the right turn position and push the

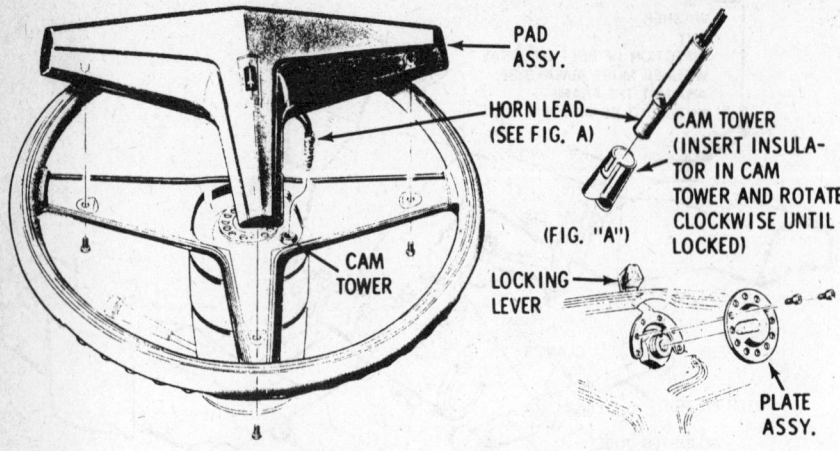

PAD ASSY.

HORN LEAD (SEE FIG. A)

CAM TOWER (INSERT INSULATOR IN CAM TOWER AND ROTATE CLOCKWISE UNTIL LOCKED)

(FIG. "A")

LOCKING LEVER

CAM TOWER

PLATE ASSY.

Tilt and telescopic column steering wheel

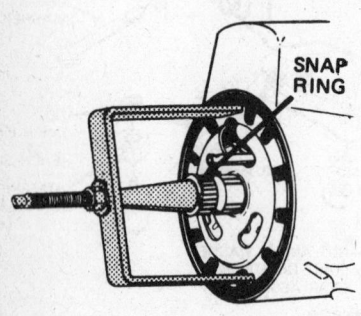

SNAP RING

Lock plate removal tool

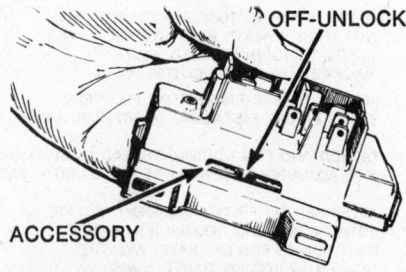

OFF-UNLOCK

ACCESSORY

MOVE SWITCH SLIDER TO EXTREME LEFT (ACCESSORY) POSITION THEN MOVE SLIDER TO DETENTS TO THE RIGHT OF "OFF-UNLOCK"

Ignition switch in Off-Unlocked position

switch in until it is properly seated. Torque the three attaching nuts to 35 inch lbs. Return the switch to the neutral position and reverse the removal procedure.

Power Steering Pump

REMOVAL & INSTALLATION

1. Disconnect negative battery cable, and remove the drive belt.
2. Use a puller to remove the pump pulley.
3. Detach and cap the hoses.
4. Remove the pump and mounting bracket.
5. Reverse the procedure for installation. Bleed the system of air by turning the wheels from side to side without hitting the stops, with the wheels off the floor and the engine running.

Ignition Switch and/or Lock Cylinder

REMOVAL & INSTALLATION

Ignition Switch

1. Disconnect negative battery cable.
2. Place ignition switch in OFF-

UNLOCKED, or ACC position (tilt wheel).
3. Remove toe pan cover (if applicable) and loosen the toe clamp bolts.
4. Remove lower instrument panel trim and toe pan trim panel.
5. Remove automatic transmission shift indicator needle.
6. Remove steering column instrument panel bracket and let steering wheel rest on the driver's seat.
7. Remove the two dimmer switch retaining screws and remove the switch.
8. Remove two ignition switch attaching screws and lift switch off actuator rod.
9. Disconnect wiring.
10. To install, check that lock cylinder is still in OFF-UNLOCKED or ACC (tilt wheel) position, and move sliding portion of switch until switch hole is positioned correctly. Hold the switch in this position with a 0.090 in. pin.
11. Connect the wiring to the switch.
12. Position switch over actuator rod, install attaching screws and remove the 0.090 in. pin.
13. Reverse Steps 1–6 to complete installation.

Lock Cylinder

1. Refer to the turn signal removal procedure, Steps 1–7.
2. Disconnect the turn signal connector from the harness. Remove the connector from the mounting bracket.
3. Carefully pull the turn signal switch from the column, allowing it to hang.
4. Position the lock assembly in the RUN position, and remove the retaining screw and the lock.
5. To install the lock cylinder, hold the lock cylinder and rotate the tabs clockwise until they stop. Insert the cylinder into the housing, aligning the keyway groove. Push the cylinder and rotate the tabs counterclockwise while lightly pushing inward on the cylinder until the drive section of the cylinder mates with drive shaft. Reverse the removal procedure.

INSTRUMENT PANEL

Headlight Switch

REMOVAL & INSTALLATION

Cutlass Without A/C

1. Disconnect the negative battery cable.
2. Remove the left hand control panel to gain access to the electrical connector.
3. Remove the connector from the switch.
4. Pull the switch to the ON position, depress the spring-loaded release button on the switch body and pull the knob and stem from the switch.
5. Remove the shaft mounting bushing.
6. Remove the switch. Reverse the above to install.

Cutlass Models With A/C

1. Disconnect the negative battery cable.
2. Remove the instrument cluster pad.
3. Remove the two switch mounting screws and remove the switch.

88 and 98 Models

1. Disconnect the negative battery cable.
2. Rotate the headlight switch so the notch is on the bottom. Bend a small hook in the end of a paper clip

and use it to release the knob retaining clip and remove the knob.

3. Remove the left hand trim cover.

4. Remove the switch mounting plate screws and pull the switch through the opening.

5. Remove the electrical connector and remove the switch.

Speedometer Cable

REMOVAL & INSTALLATION

The speedometer cable is retained at the rear of the speedometer head by quick release clip. To remove the cable, reach up behind the speedometer and depress the clip while pushing in, then pull back on the cable. The cable may then be pulled from the firewall and into the engine compartment. Raise the car and support it on jack stands. Disconnect the cable from the transmission and remove the core. When replacing the core, coat all but the top $\frac{1}{3}$ with speedometer cable lubricant. Cable replacement is in the reverse order of removal.

Instrument Cluster

REMOVAL & INSTALLATION

Cutlass Models

1. Disconnect the speedometer cable at the cruise control transducer, if so equipped.

2. Remove the instrument cluster pad assembly.

3. Remove the steering column trim cover.

4. Disconnect the shift indicator clip from the steering column shift bowl.

5. Remove the instrument cluster screws.

6. Pull the cluster rearward. Disconnect the speedometer cable.

7. Disconnect the speed sensor, if so equipped.

8. Remove the cluster.

9. Installation is in the reverse order of removal.

88 and 98 Models

1. Remove the steering column trim cover.

2. Remove the screws attaching the gauge cluster to the left hand trim cover.

3. Pull the gauge cluster rearward. Disconnect gauges wiring, lamp sockets, and speedometer cable.

4. Remove the gauge cluster.

5. Installation is in the reverse order of removal.

WINDSHIELD WIPERS

Wiper Motor

REMOVAL & INSTALLATION

1. Remove the cowl screen or grille.

2. Loosen the linkage drive link-to-crankarm attaching nuts, and remove the link from the arm.

3. Disconnect the wiring and washer hoses.

4. Remove the three motor attaching screws, guide the crankarm through the hole in the dash, and remove the motor.

5. Reverse the above steps to install.

WIPER BLADE

REPLACEMENT

Depending on model and availability, one of three methods is used:

1. A tab on the arm saddle is depressed.

2. A spring type blade clip is depressed.

3. A coil spring retainer is depressed with a screwdriver.

Details can be found in the Maintenance Unit Repair Section.

RADIO

REMOVAL & INSTALLATION

88 and 98 Models

1. Disconnect the negative battery cable.

2. Remove the knobs from the radio and pull out the cigarette lighter.

3. Remove the two trim cover attaching screws and remove the cover.

4. Remove the radio bracket attaching screw from the lower tie bar.

5. Remove the four mounting plate screws and pull the radio out to obtain access to the electrical connections. Detach the wiring harness and the antenna lead.

6. Remove the mounting plate nuts and remove the radio. Installation is in the reverse order of removal.

Cutlass Models

1. Disconnect the negative battery cable.

2. Remove the radio knobs. Pull the lower trim cover outward, off the retaining clips.

3. Remove the four mounting plate screws and the screw from the radio support bracket on the lower tie bar.

4. Pull the radio out and detach the wiring and the antenna lead.

CIRCUIT PROTECTION

Fuses

The fuse block is located beneath the instrument panel, drivers side of firewall. Fuse holders are labeled as to their service and the correct amperage. Always replace blown fuses with new ones of the correct amperage; otherwise, electrical overloads and possible wiring damage will result.

Fusible Links

Fusible links are sections of wire, with special insulation, designed to melt under electrical overload. Replacements are simply spliced into the wire. There may be as many as five of these in the engine compartment wiring harnesses. These are:

1. Horn relay-to-fuse panel circuit—one link.

2. Charging circuit, from the starter solenoid to the horn relay—two links.

3. Starter solenoid-to-ammeter circuit—one link.

4. Horn relay-to-rear window defroster circuit—one link.

NOTE: The fusible links are all two wire gauge sizes smaller than the wires they protect. Most models have fusible links at these locations.

Replacement

1. Disconnect the battery ground cable.

2. Disconnect the fusible link from the junction block or starter solenoid.

3. Cut the harness directly behind the connector to remove the damaged fusible link.

4. Strip the harness wire approximately $\frac{1}{2}$ in.

5. Connect the new fusible link to the harness wire using a crimp-on connector. Solder the connection using rosin-core solder.

6. Tape all exposed wires with plastic electrical tape.

7. Connect the fusible link to the junction block or starter solenoid and reconnect the battery ground cable.

HEATER

Blower Motor and Heater Core

REMOVAL & INSTALLATION

Without Air Conditioning

88 and 98 Blower Motor

1. Disconnect the negative battery and the blower motor wiring.
2. Remove the retaining screws and remove the motor. Use sealer as needed upon installation for a watertight seal.

88 and 98 Heater Core

1. Disconnect the negative battery cable, the blower motor wiring and the heater core ground strap.
2. Drain the cooling system and disconnect the heat hoses.
3. It may be necessary to move the temperature air valve by disconnecting the cable and tapping the hinge pin down to clear the upper pivot.
4. Remove the screws attaching the blower case to the heater case. Remove the heater core shroud screws and remove the shroud and core.

Cutlass Blower Motor and Heater Core

1. Remove the glove box, the heater air distribution outlet, the upper level vent duct, and the defroster outlet attaching screw.
2. Disconnect the blower motor wiring and the cables at the blower motor.
3. Drain the cooling system.
4. Remove the right hand windshield wiper arm.
5. Remove the leaf screen.
6. Disconnect the heater hoses at the heater assembly. Remove the heater assembly-to-cowl screws and remove the assembly.
7. Remove blower motor mounting screws and remove the motor from the assembly.
8. Remove the front cover screws and remove the heater core. Reverse above procedures to install.

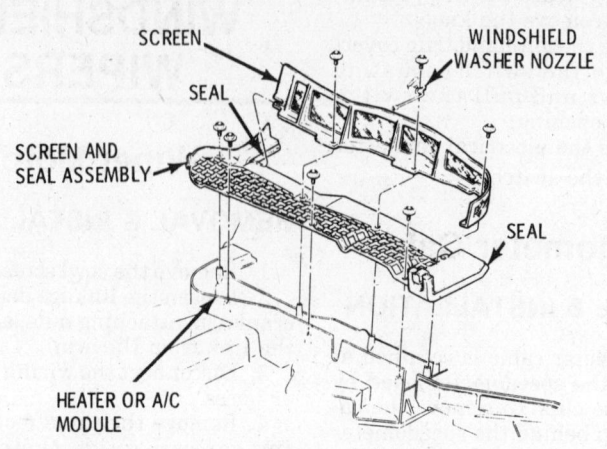

Air inlet screen-88. 98

Blower Motor

REMOVAL & INSTALLATION

88 & 98 with A/C

The blower motor is mounted in the upper evaporator and blower case, by 6 screws (7 with noise suppressor). Disconnect the electrical connectors and remove the screws. Lift the blower straight up to remove.

Cutlass with A/C

1. Disconnect the battery ground.
2. Disconnect the blower wiring.
3. Unbolt and remove the motor.
4. Installation is in the reverse order of removal. Replace any damaged sealer.

Heater Core

REMOVAL & INSTALLATION

88 and 98 with A/C

1. Disconnect the battery ground.
2. Disconnect the blower wiring.
3. Remove the thermostatic switch and diagnostic connector.
4. Remove the right end of the hood seal and the air inlet screen screws.

5. Remove the 5 case-to-firewall screws at the top, 9 upper case-to-lower case screws at the flange and two more at the plenum.
6. Lift the upper case straight up and off. Remove the pipe bracket screws from the case. Disconnect the hoses and position them to prevent spillage.
7. Disconnect and lift out the heater core.
8. Installation is in the reverse order of removal. Replace any damaged sealer.

Cutlass with A/C

1. Drain the cooling system.
2. Disconnect the hoses at the core pipes.
3. Remove the retaining bracket and ground strap.
4. Remove the module rubber seal.
5. Remove the module screen.
6. Remove the right windshield wiper arm.
7. Remove the diagnostic connector, high blower relay and thermostatic switch mounting screws.
8. Disconnect all electrical connections at the module.
9. Remove the module top cover.
10. Lift out the core.
11. Installation is in the reverse order of removal. Replace all insulation.

Pontiac
Rear Wheel Drive
Bonneville, Catalina, Parisienne, Grand Prix, LeMans, Grand Am

YEAR IDENTIFICATION

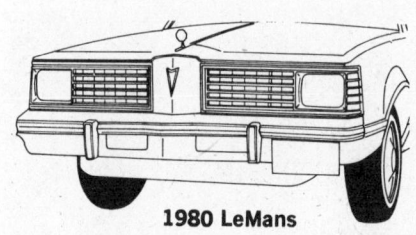

1980 LeMans

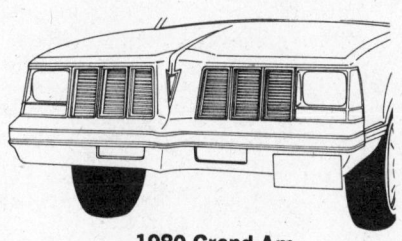

1980 Grand Am

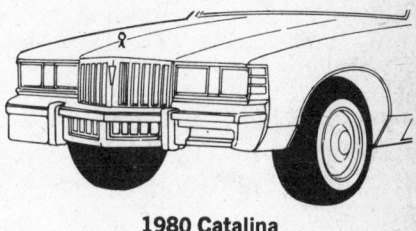

1980 Catalina

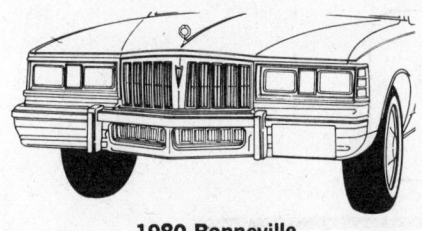

1980 Bonneville

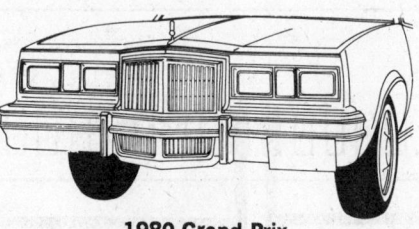

1980 Grand Prix

1981 LeMans

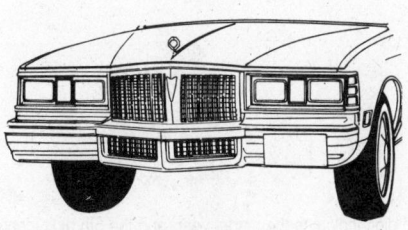

1981 Bonneville

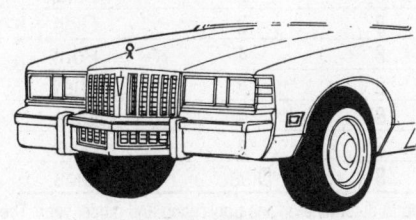

1981 Catalina

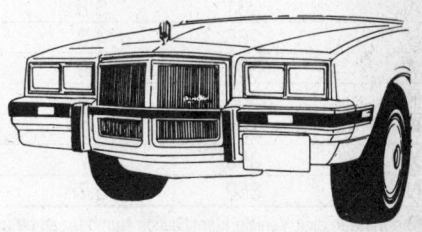

1981-84 Grand Prix

YEAR IDENTIFICATION

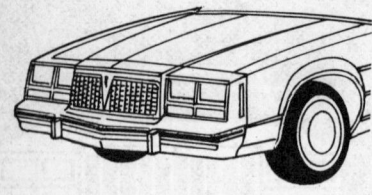
1982–84 Bonneville

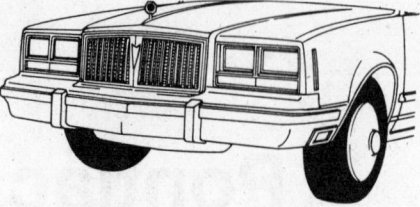
1984–87 Parisienne

1985–86 Grand Prix

1985-87 Bonneville

1987 Grand Prix

VEHICLE IDENTIFICATION NUMBER (VIN)

It is important for servicing and ordering parts to be certain of the vehicle and engine identification. The VIN (vehicle identification number) is a 13 or 17 digit number visible through the windshield on the driver's side of the dash and contains the vehicle and engine identification codes. It can be interpreted as follows:

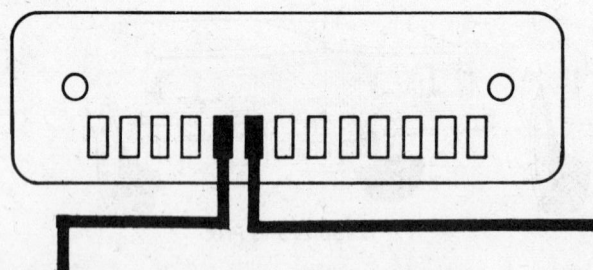

	Engine Code						Model Year Code	
Code	Cu. In.	Liters	Cyl.	Carb.	Eng. Mfg.		Code	Year
K	229	3.8	6	2	Chev.		A	80
A	231	3.8	6	2	Buick			
S	265	4.3	8	2	Olds.			
W	301	4.9	8	4	Pont.			
H	305	5.0	8	4	Chev.			
R	350	5.7	8	4	Olds.			
X	350	5.7	8	4	Buick			
N	350	5.7	8	Diesel	Olds.			

The thirteen digit Vehicle Identification Number can be used to determine engine application and model year. The 6th digit indicates the model year, and the 5th digit identifies the factory installed engine.

VEHICLE IDENTIFICATION NUMBER (VIN)

It is important for servicing and ordering parts to be certain of the vehicle and engine identification. The VIN (vehicle identification number) is a 13 or 17 digit number visible through the windshield on the driver's side of the dash and contains the vehicle and engine identification codes. It can be interpreted as follows:

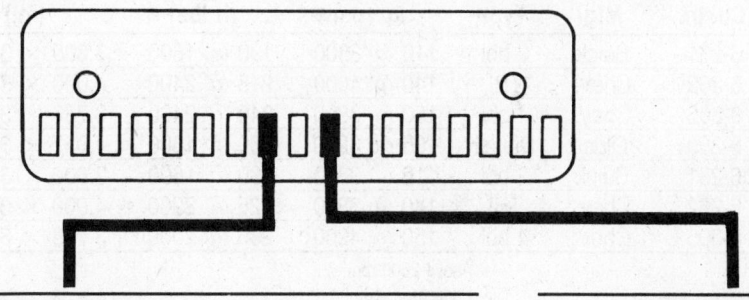

Engine Code						Model Year Code	
Code	Cu. In.	Liters	Cyl.	Carb.	Eng. Mfg.	Code	Year
A	231	3.8	6	2	Buick	B	81
4	252	4.1	6	2	Buick	C	82
Z	262	4.3	6	TBI	Chev.	D	83
S	265	4.3	8	2	Pont.	E	84
W	301	4.9	8	4	Pont.	F	85
H	305	5.0	8	4	Chev.	G	86
Y	307	5.0	8	4	Olds.	H	87
N	350	5.7	8	Diesel	Olds.		

The seventeen digit Vehicle Identification Number can be used to determine engine application and model year. The 10th digit indicates the model year, and the 8th digit identifies the factory installed engine.
TBI Throttle Body Injection

GENERAL ENGINE SPECIFICATIONS

Year	Eng. V.I.N. Code	Engine No. Cyl. Displacement Cu. In.	Eng. Mfg.	Carburetor Type	Horsepower @ rpm ■	Torque @ rpm (ft lbs) ■	Bore × Stroke (in.)	Compression Ratio	Oil Pressure @ 2000 rpm
'80	K	6-229	Chev.	2 bbl	110 @ 4200	170 @ 2000	3.736 × 3.480	8.6:1	45
	A	6-231	Buick	2 bbl	115 @ 3800	188 @ 2000	3.800 × 3.400	8.0:1	37
	S	8-265	Pont.	2 bbl	120 @ 3600	210 @ 1600	3.750 × 3.000	8.0:1	40②
	W	8-301	Pont.	4 bbl	150 @ 4000	240 @ 2000	4.000 × 3.000	8.2:1	40②
	H	8-305	Chev.	4 bbl	150 @ 3800	230 @ 2400	3.736 × 3.480	8.5:1	40
	X	8-350	Buick	4 bbl	155 @ 3400	280 @ 1800	3.800 × 3.850	8.0:1	35
	R	8-350	Olds.	4 bbl	160 @ 3800	260 @ 2400	4.000 × 3.480	8.5:1	40
	N	8-350	Olds.	Diesel	125 @ 3600	225 @ 1600	4.057 × 3.385	22.5:1	40③
'81	A	6-231	Buick	2 bbl	115 @ 3800	188 @ 2000	3.800 × 3.400	8.0:1	37
	S	8-265	Pont.	2 bbl	120 @ 3600	210 @ 1600	3.750 × 3.000	8.0:1	40④
	W	8-301	Pont.	4 bbl	155 @ 4000	240 @ 2000	4.000 × 3.000	8.2:1	40
	Y	8-307	Olds.	4 bbl	148 @ 3800	250 @ 2400	3.800 × 3.385	8.0:1	40③
	N	8-350	Olds.	Diesel	105 @ 3200	205 @ 1600	4.057 × 3.385	22.5:1	40③
'82	A	6-231	Buick	2 bbl	110 @ 3800	190 @ 1600	3.800 × 3.400	8.0:1	37
	4	6-252	Buick	4 bbl	125 @ 4000	205 @ 2000	3.965 × 3.400	8.0:1	37
	N	8-350	Olds.	Diesel	105 @ 3200	200 @ 1600	4.057 × 3.385	22.5:1	40
'83–'84	A	6-231	Buick	2 bbl	110 @ 3800	190 @ 1600	3.800 × 3.400	8.0:1	37
	H	8-305	Chev.	4 bbl	150 @ 3800	240 @ 2400	3.736 × 3.480	8.6:1	45
	N	8-350	Olds.	Diesel	105 @ 3200	200 @ 1600	4.057 × 3.385	22.5:1	40③

GENERAL ENGINE SPECIFICATIONS

Year	Eng. V.I.N. Code	Engine No. Cyl. Displacement Cu. In.	Eng. Mfg.	Carburetor Type	Horsepower @ rpm ■	Torque @ rpm (ft lbs) ■	Bore × Stroke (in.)	Compression Ratio	Oil Pressure @ 2000 rpm
'85	A	6-231	Buick	2 bbl	110 @ 3800	190 @ 1600	3.800 × 3.400	8.0:1	37
	Z	6-262	Chev.	TBI	140 @ 4000	218 @ 2400	4.000 × 3.480	9.3:1	50–65 ①
	H	8-305	Chev.	4 bbl	150 @ 3800	240 @ 2400	3.736 × 3.480	9.5:1	45
	N	8-350	Olds.	Diesel	105 @ 3200	200 @ 1600	4.057 × 3.385	22.7:1	40 ③
'86–'87	A	6-231	Buick	2 bbl	110 @ 3800	190 @ 1600	3.800 × 3.400	8.0:1	37
	Z	6-262	Chev.	TBI	140 @ 3800	225 @ 2200	4.000 × 3.480	9.3:1	45
	H	8-305	Chev.	4 bbl	150 @ 4200 ⑤	235 @ 2000 ⑥	3.736 × 3.480	9.5:1	45

TBI Throttle Body Injection
① Pressure at 2000 prm
② Pressure at 2600 rpm
③ Pressure at 1500 rpm
④ Above 2600 rpm
⑤ Parisienne: 165 hp
⑥ Parisienne: 245 @ 2400

TUNE-UP SPECIFICATIONS

When analyzing compression test results, look for uniformity among cylinders rather than specific pressures.

Year	Eng. V.I.N. Code	No. Cyl. Displacement	Eng. Mfg.	Spark Plugs Orig. Type	Spark Plugs Gap (in.)	Distributor Point Dwell (deg)	Distributor Point Gap (in.)	Ignition Timing (deg)	Valves Intake Opens (deg)	Fuel Pump Pressure (psi)	Idle Speed (rpm) ●
'80	K	6-229	Chev.	R-45TS	.045	Electronic		12B	42	4.5–6.0	600
	A	6-231	Buick	R-45TSX ③	.060 ③	Electronic		15B	16	3–4½	620/550
	S	8-265	Pont.	R-45TSX	.060	Electronic		10B	27	7–8½	650/550
	W	8-301	Pont.	R-45TSX	.060	Electronic		12B	16	7–8½	650/500
	H	8-305	Chev.	R-45TS	.045	Electronic		4B	28	7½–9	650/550
	X	8-350	Buick	R-46TSX	.060	Electronic		15B	16	4.5–5.5	550
	R	8-350	Olds.	R-43TS	.045	Electronic		6B	28	7½–9	650/550
	N	8-350 ②	Olds.	—	—	—		5B	16	5½–6½	600
'81	A	6-231	Buick	R-45TS8	.080	Electronic		①	16	4.25–5.75	①
	S	8-265	Pont.	R-45TSX	.060	Electronic		12B	16	7–8.5	①
	W	8-301	Pont.	R-45TSX	.060	Electronic		12B	16	7.0–8.5	650/500
	Y	8-307	Olds.	R-46SX	.080	Electronic		15B	14	5.5–6.5	①
	N	8-350 ②	Olds.	—	—	—		①	16	5.5–6.5	①
'82	A	6-231	Buick	R-45TS8	.080	Electronic		15B	16	4.25–5.75	①
	4	6-252	Buick	R-45TS8	.080	Electronic		15B	16	4.25–5.75	①
	N	8-350 ②	Olds.	—	—	—		①	16	5.5–6.5	①
'83–'84	A	6-231	Buick	R-45TS8	.080	Electronic		15B	16	4.25–5.75	①
	H	8-305	Chev.	R-45TS	.045	Electronic		44		5.5–7.0	①
	N	8-350 ②	Olds.	—	—	—		①	16	5.5–6.5	①
'85	Z	6-262	Chev.	R-43CTS	.035	Electronic		①	NA	12.0	①
	A	6-231	Buick	R-45TSX	.060	Electronic		①	16	4.25–5.75	①
	H	8-305	Chev.	R-45TS	.045	Electronic		①	44	5.5–7.0	①
	N	8-350 ③	Olds.	—	—	—		①	16	5.5–6.5	①
'86–'87	A	6-231	Buick	R-45TSX	.060	Electronic		①	NA	4.25–5.75	①
	Z	6-262	Chev.	R-43TS	.035	Electronic		①	NA	4.13	①
	H	8-305	Chev.	R-45TS	.045	Electronic		①	NA	7.5–9.0	①
	H	8-305	Chev.	R-45TS	.045	Electronic		①	NA	5.5–7.0	①

NOTE: The underhood specifications sticker often reflects tune-up specification changes made in production. Sticker figures must be used if they disagree with those in this chart. Part numbers in this chart are not recommendations by Chilton for any product by brand name.

● Where two idle speeds appear separated by a slash, the second is with the solenoid disconnected.

B Before Top Dead Center
① See the underhood sticker
② Diesel
③ Low Altitude without C-4: R-45TS; gap: .040

FIRING ORDERS

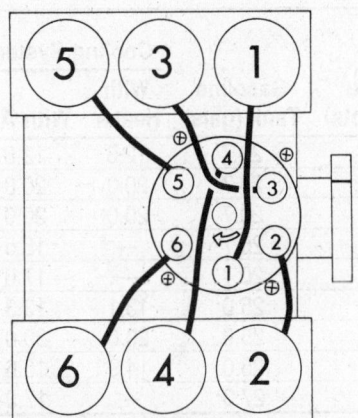

GM (Buick) 231, 252 V6
Engine firing order: 1-6-5-4-3-2
Distributor rotation: clockwise

V6 harmonic balancers have two timing marks: one is 1/8 in. wide, and one is 1/16 in. wide. Use the 1/16 in. mark for timing with a hand held light. The 1/8 in. mark is used only with a magnetic timing pick-up probe.

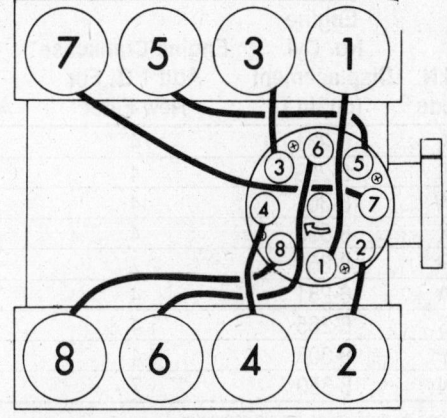

GM (Buick) 350 V8
Engine firing order: 1-8-4-3-6-5-7-2
Distributor rotation: clockwise

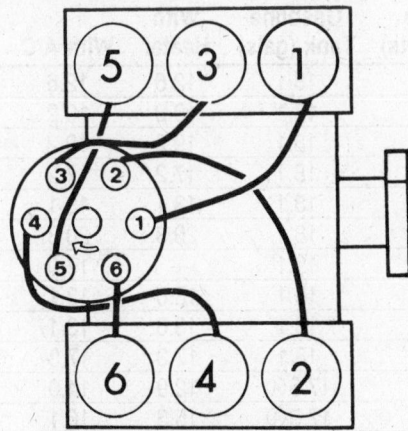

GM (Chevrolet) 229, V6
Engine firing order: 1-6-5-4-3-2
Distributor rotation: clockwise

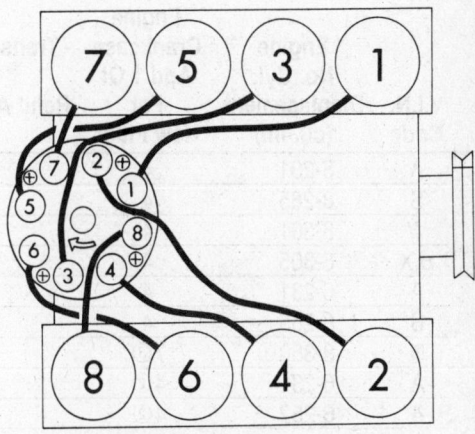

GM (Chevrolet) 305 V8
Engine firing order: 1-8-4-3-6-5-7-2
Distributor rotation: clockwise

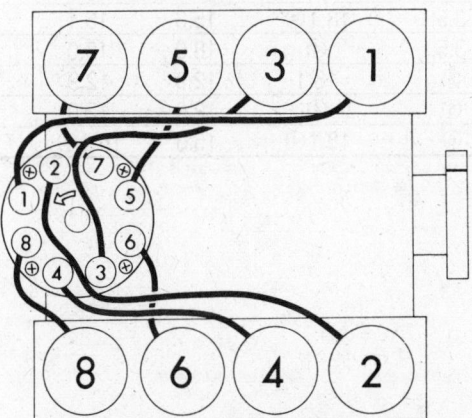

GM (Oldsmobile) 307,350 V8
Engine firing order: 1-8-4-3-6-5-7-2
Distributor rotation< counterclockwise

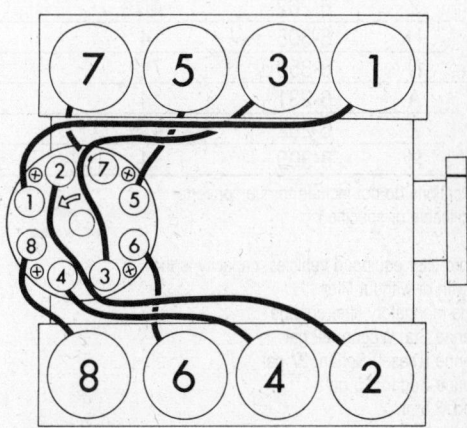

GM (Pontiac) 265 V8
Engine firing order: 1-8-4-3-6-5-7-2
Distributor rotation: counterclockwise

CAPACITIES
1980-81 Bonneville, Catalina

Year	V.I.N. Code	Engine No. Cyl. Displacement (cu. in.)	Engine Crankcase Add 1 Qt For New Filter	Transmission Pts to Refill After Draining ①	Drive Axle (pts)	Gasoline Tank (gals)	Cooling System With Heater	With A/C	With Super Cooling
'80	A	6-231	4	8	3.4	20.7	12.6	12.6	—
	S	8-265	4	8	3.4	20.7	20.0	20.0	20.0
	W	8-301	4	6	3.4	20.7	20.0	20.0	20.0
	X	8-350	4	6	3.4	20.7	—	15.5	15.5
	R	8-350	7	6	3.4	20.7	—	17.0	17.0
'81	A	6-231	4	8	3.4	25.0	13.1	13.3	13.3
	S	8-265	4	8	3.4	25.0	20.0	20.0	20.0
	Y	8-307	4	8	3.4	25.0	14.9	15.6	15.6
	N	8-350	7	8	3.4	27.0	—	17.0	17.0

① Specifications do not include torque converter
— Not applicable

CAPACITIES
All Other Models

Year	V.I.N. Code	Engine No. Cyl. Displacement (cu. in.)	Engine Crankcase Add 1 Qt For New Filter	Transmission Pts to Refill After Draining ●	Drive Axle (pts)	Gasoline Tank (gals)	Cooling System (qts) With Heater	With A/C	With Super Cooling
'80	A	8-231	4	6	3.4	18.1	12.6	12.6	—
	S	8-265	4	6	3.4	18.1	19.2	19.2	19.2
	W	8-301	4	6	3.4	18.1	19.2	19.2	19.2
	H,X	8-305	4	6	3.4	18.1	17.2	17.2	17.2
'81	A	6-231	4	8	3.4	18.1	13.1	13.1	—
	S	8-265	4	8	3.4	18.1	20.3	20.3	20.3
	N	8-350①	7③	8	3.4	19.1	—	17.0	—
'82	A	6-231	4②	8	3.5	18.1	13.0	13.1	13.1
	4	6-252	4②	8	3.5	18.1	13.0	13.1	13.1
	N	8-350①	7③	8	3.5	18.1	17.3	17.3	—
'83–'84	A	6-231	4	6⑤	3.5	17.5④	12.9	14.0	—
	H	8-305	4	6⑤	3.5	17.5④	15.3	16.1	—
	N	8-350①	7③	6⑤	3.5	19.8④	17.2	18.0	—
'85	A	6-231	4	⑤	3.5	18.1	12.9	12.9	—
	Z	6-262	4②	⑤	3.5	25.0④	12.2	12.0	12.6
	H	8-305	4	⑤	3.5	18.1④	15.3	16.1	16.1
	N	8-350①	7③	⑤	3.5	④	18.0	18.0	—
'86–'87	A	6-231	4	⑤	⑥	18.1	12.9	12.9	
	Z	6-262	4③	⑤	⑥	④	12.0	12.0	
	H	8-305	4	⑤	⑥	18.1④	15.3	16.1⑦	

● Specifications do not include torque converter
— Not applicable or specified
① Diesel
② On micro-filter equipped vehicles, capacity is the same with or without filter
③ Includes mandatory filter change
④ Parisienne Sta. Wagon: 22 gal
  Parisienne (Diesel) Sedan: 27 gal
  Parisienne Sedan: 25 gal
⑤ 3-speed: 8.5 pt.
  Overdrive: 10.1 pt.
⑥ 7.5" ring gear: 3.5, 8.5" ring gear: 4.3 (262 eng.)
  5.4 (305 eng.)
⑦ Parisienne: 15.0

CRANKSHAFT AND CONNECTING ROD SPECIFICATIONS

All measurements are given in inches.

Year	V.I.N. Code	Engine No. Cyl. Displacement (cu. in.)	Crankshaft				Connecting Rod		
			Main Brg. Journal Dia.	Main Brg. Oil Clearance	Shaft End-Play	Thrust on No.	Journal Diameter	Oil Clearance	Side Clearance*
'80	A	6-231	2.4995–2.5000	.0003–.0017	.003–.009	2	2.2487–2.2495	.0005–.0025	.006–.023
	K	6-229	②	③	.002–.007	5	2.0990–2.1000	.0013–.0035	.006–.016
	S	8-265	3.0000	.0002–.0018	.0035–.0085	4	2.0000	.0005–.0026	.006–.022
	W	8-301	3.0000	.0004–.0020	.003–.009	4	2.2500	.0005–.0025	.006–.022
	H	8-305	②	③	.002–.007	5	2.0990–2.1000	.0013–.0035	.006–.016
	R	8-350	2.4985–2.4995④	.0005–.0021①	.0035–.0135	3	2.1238–2.1248	.0004–.0033	.006–.020
	X	8-350	3.0000–3.0005	.0004–.0015	.003–.009	3	1.9910–2.0000	.0005–.0026	.006–.023
	N	8-350⑤	2.9993–3.0003	.0005–.0021①	.0035–.0135	3	2.1238–2.1248	.0005–.0026	.006–.020
'81	A	6-231	2.4995–2.5000	.0003–.0018	.003–.009	2	2.2487–2.2495	.0005–.0026	.006–.023
	S	8-265	3.0000	.0002–.0018	.0035–.0085	4	2.0000	.0005–.0026	.006–.022
	W	8-301	3.0000	.0004–.0020	.003–.009	4	2.2500	.0005–.0025	.006–.022
	Y	8-307	2.4985–2.4995④	.0005–.0021①	.0035–.0135	3	2.1238–2.1248	.0004–.0033	.006–.020
	N	8-350⑤	2.9993–3.0003	.0005–.0021①	.0035–.0135	3	2.1238–2.1248	.0005–.0026	.006–.020
'82	A	6-231	3.4955	.0003–.0018	.003–.011	2	2.2491	.0005–.0026	.006–.023
	4	6-252	2.4955	.0003–.0018	.003–.009	2	2.2487–2.2495	.0005–.0026	.006–.023
	N	8-350⑤	3.000	—	.0035–.0135	3	2.1243	.0005–.0026	.006–.020
'83–'84	A	6-231	3.4955	.0003–.0018	.003–.011	2	2.2491	.0005–.0026	.006–.020
	H	8-305	②	③	.002–.007	5	2.0995	.0013–.0035	.006–.016
	N	8-350⑤	2.9998	.0005–.0021	.0035–.0135	3	2.1243	.0005–.0026	.006–.020
'85–'87	Z	6-262	②	③	.002–.006	5	2.2487–2.2498	.0013–.0035	.006–.014
	A	6-231	3.4995	.0003–.0018	.003–.011	2	2.1243	.0005–.0026	.006–.020
	H	8-305	②	③	.002–.006	5	2.0995	.0013–.0035	.006–.016
	N	8-350⑤	2.9998	.0005–.0021	.0035–.0135	3	2.1243	.0005–.0026	.006–.020

* Total for two rods
① No. 5—.0015–.0031
② No. 1: 2.4484–2.4493
  No. 2, 3, 4: 2.4481–2.4490
  No. 5: 2.4479–2.4488
③ No. 1: .0008–.0020
  No. 2: .0011–.0023
  No. 3: .0017–.0033
④ No. 1: 2.4988–2.4998
⑤ Diesel

TORQUE SPECIFICATIONS

All readings in ft. lbs.

Year	V.I.N. Code	Engine No. Cyl. Displacement (cu. in.)	Cylinder Head Bolts	Rod Bearing Bolts	Main Bearing Bolts	Crankshaft Bolt	Flywheel to Crankshaft Bolts	Manifold	
								Intake	Exhaust
'80	K	6-229	65	45	70	60	60	30	20
	A	6-231	80	40	100	225	60	45	25
	S	8-265	95	30	③	160	95	35	40
	W	8-301	95	30	③	160	95	35	40
	H	8-305	65	45	70	60	60	30	20
	X	8-305	65	45	70	60	60	30	20
	R	8-350	130①	42	120	200–310	60	40	25
	N	8-350⑤	130①	42	120	200–310	60	40	25
'81	A	6-231	80	40	100	225	60	45	25
	S	8-265	95	30	③	160	95	35	40
	W	8-301	95	30	③	160	95	35	40
	Y	8-307	130①	42	120④	200–310	60②①	40	25
	N	8-350⑤	130①	42	120	200–310	60②	40	25

TORQUE SPECIFICATIONS

All readings in ft. lbs.

Year	V.I.N. Code	Engine No. Cyl. Displacement (cu. in.)	Cylinder Head Bolts	Rod Bearing Bolts	Main Bearing Bolts	Crankshaft Bolt	Flywheel to Crankshaft Bolts	Manifold Intake	Manifold Exhaust
'82	A	6-231	80	40	100	225	60	45	25
	4	6-252	80	40	100	225	60	45	25
	N	8-350⑤	130①	42	120	200-310	60②	40	25
'83-'84	A	6-231	80	40	100	225	60	45	25
	H	8-305	65	45	70	60	60	30	20
	N	8-350⑤	130①	42	120	200-310	60②	40	25
'85-'87	Z	6-262	65	45	70-85	70	60	30	20
	A	6-231	80⑦	40⑥	100	225	60	45	25
	H	8-305	65	45	70-85	70	60	30	20
	N	8-350⑤	130①	42	120	200-310	60②	40	25

① Dip bolts in oil before tightening
② Flywheel-to-converter: 40
③ 7/16" bolt—70 ft. lbs., 1/2" bolt—100 ft. lbs., rear main bearing—100 ft. lbs.
④ 80 on Nos. 1-4, 120 on No. 5
⑤ Diesel
⑥ '86-'87; 45
⑦ For Models through 1985 only. For 1986 and later 231 engs. follow the torque procedure in the Buick (RWD) section.

WHEEL ALIGNMENT SPECIFICATIONS

Year	Model	Caster Range (deg)	Caster Pref. Setting (deg)	Camber Range (deg)	Camber Pref. Setting (deg)	Toe-in (in.)	Steering Axis Inclin. (deg)
'80	Bonneville, Catalina	1/2P to 1 1/2P	1P	1/4P to 1 1/4P	3/4P	1/16 to 3/16	10 5/16
	Grand AM, Grand Prix, LeMans, Man. Str.	1/2P to 1 1/2P	1P	0 to 1P	1/2P	1/16 to 3/16	8
	Pow. Str.	2 1/2P to 3 1/2P	3P	0 to 1P	1/2P	1/16 to 3/16	8
'81	Bonneville, Catalina	1/2P to 1 1/2P	1P	1/4P to 1 1/4P	3/4P	1/16 to 3/16	10 5/16
	Grand Prix, LeMans, Man. Str.	1/2P to 1 1/2P	1P	1/2P to 1 1/2P	1P	1/16 to 3/16	8
	Pow. Str.	2 1/2P to 3 1/2P	3P	1/2P to 1 1/2P	1P	1/16 to 3/16	8
'82-'87	Bonneville, Grand Prix, Man. Str.	1/2P to 1 1/2P	1P	0 to 1P	1/2P	1/16 to 3/16	—
	Pow. Str.	2 1/2P to 3 1/2P	3P	0 to 1P	1/2P	1/16 to 3/16	—
	Parisenne	2 1/2P to 3 1/2P	3P	1/3P to 1 1/3P	4/5P	1/16 to 3/16	—

— Not specified
N Negative
P Positive

PISTON AND RING SPECIFICATIONS

(All measurements are given in inches.)

Year	V.I.N. Code	Engine Type/ Disp. cu. in.	Piston-to-Bore Clearance	Ring Gap Top Compression	Ring Gap Bottom Compression	Ring Gap Oil Control	Ring Side Clearance Top Compression	Ring Side Clearance Bottom Compression	Ring Side Clearance Oil Control
'80	K	6-229	.0007-.0017	.010-.020	.010-.025	.010-.035	.0012-.0032	.0012-.0032	.002-.007
'80-'87	A	6-231	.0013-.0035	.013-.023	.013-.023	.015-.035	.003-.005	.003-.005	.001-.0035
'82	4	6-252	.0013-.0035	.010-.020	.010-.020	.015-.055	.003-.005	.003-.005	.001-.0035
'85-'87	Z	6-262	.0007-.0017	.010-.020	.010-.025	.015-.055	.0012-.0032	.0012-.0032	.002-.007
'80-'81	S	8-265	.0025-.0033	.010-.022	.010-.028	.010-.050	.0015-.0035	.0015-.0035	.0015-.0035
'80-'81	W	8-301	.0025-.0033	.010-.028	.010-.028	.015-.055	.0015-.0035	.0015-.0035	.0015-.0035
'80	H	8-305	.0007-.0017	.010-.020	.010-.025	.015-.055	.0012-.0032	.0012-.0032	.002-.007
'83-'87	H	8-305	.0007-.0017	.010-.020	.010-.025	.015-.055	.0012-.0032	.0012-.0032	.002-.007
'81	Y	8-307	.0007-.0017	.009-.019	.009-.019	.015-.055	.002-.004	.002-.004	.001-.005
'80	X	8-350	.0013-.0035	.010-.020	.010-.020	.015-.035	.003-.005	.003-.005	.001-.0035
'80	R	8-350	.0007-.0017	.010-.020	.010-.020	.015-.035	.002-.004	.002-.004	.001-.005
'80-'85	N	8-350①	.00035-.00045	.015-.025	.015-.025	.015-.055	.005-.007	.005-.007	.001-.005

① Diesel

ENGINE ELECTRICAL

The charging system is of the SI integral regulator type. The generator utilizes a solid state voltage regulator, which is attached to the slip ring end frame of the alternator.

Alternator

REMOVAL & INSTALLATION

NOTE: For alternator testing and diagnosis procedures, refer to "Charging and Starting" in the Unit Repair section.

1. Disconnect the battery ground cable to prevent diode damage.
2. Tag and disconnect the alternator wiring.
3. Remove the alternator brace bolt. If the vehicle is equipped with power steering, loosen the pump brace and mount nuts. Detach the drive belt(s).
4. Support the alternator and remove the mounting bolt(s). Remove the unit from the vehicle.
5. To install, reverse the removal procedures. Adjust the belt to to allow approximately $\frac{1}{2}$ in. of play, on the longest run between the pulleys.

Voltage Regulator

The voltage regulator is electronic and is housed within the alternator. Adjustments to the regulator are not possible. Should replacement of the regulator become necessary, the alternator must be disassembled.

Starter

NOTE: For all starter overhaul procedures, please refer to "Charging and Starting" in the Unit Repair section.

REMOVAL & INSTALLATION

1. Disconnect the negative battery cable.
2. Raise and support the vehicle on jack stands.
3. Remove the starter braces, shields and other items, which may be in the way.

NOTE: On certain models, it may necessary to remove the frame support. The support runs from the corner of the frame to the front crossmember. To remove it,
loosen the support-to-frame mounting bolt(s), remove the support-to-crossmember mounting bolt(s) and swing the support out of the way.

4. Remove the two starter-to-engine bolts. Lower the starter and remove the electrical connections.
5. To install, reverse the removal procedures. Torque the mounting bolts to 25–35 ft. lbs.

NOTE: If shims were inserted between the starter and engine block, they must be replaced in their original locations.

Distributor

A solid-state, High Energy Ignition system (HEI) is standard equipment. Most 1981 and later models use an (EST) Electronic Spark Timing distributor. The EST distributor uses no mechanical or vacuum advance and is easily identified by the absence of a vacuum advance and the presence of a 4 terminal connector.

For all electronic ignition system description, testing and service, please refer to "Electronic Ignition Systems" in the Unit Repair section.

REMOVAL

1. Disconnect the negative battery cable.

NOTE: On the V6 and V8 systems, disconnect the ignition switch battery feed wire from the distributor cap.

2. Tag and disconnect the electrical connectors from the distributor cap.
3. Disconnect the hose at the vacuum advance unit, if equipped.
4. Turn the four distributor cap-to-housing screws counterclockwise and lift off the cap assembly.

NOTE: It may be necessary to remove the secondary wires from the distributor cap.

5. Using crayon or chalk, make locating marks on the rotor, the distributor housing and the engine for installation purposes.
6. Remove the distributor clamp bolt and clamp, then lift the distributor (turning the rotor counterclockwise) from the engine.

INSTALLATION
Undisturbed Engine

1. Install a new O-ring on the distributor housing.
2. Install the distributor into the engine opening, aligning the marks made during the removal procedure.
3. Install the distributor clamp, the hold down bolt and the distributor cap.
4. Check the ignition timing.

Disturbed Engine

1. Remove the No. 1 spark plug.
2. Place a finger over the No. 1 spark plug hole and rotate the engine until the compression can be felt.
3. Align the timing mark on the crankshaft pulley with the "0" mark on the timing plate.
4. Align the distributor rotor near the No. 1 spark plug tower.
5. Install the distributor, the hold down clamp, the bolt and the cap. It may be necessary to turn the rotor a little in either direction in order to engage the gears.

NOTE: With the distributor installed, make sure that the rotor is aligned with the No. 1 spark plug tower of the cap.

6. Reverse the removal procedures and check the ignition timing.

Ignition Timing

NOTE: The underhood timing instructions on the Vehicle Emissions Control Information label may differ from these procedures; always follow the underhood label's directions.

1. If equipped, disconnect and plug the vacuum advance hose from the distributor. If equipped with electronic spark timing (EST), disconnect the 4-wire connector to the distributor.
2. Make sure the timing marks are clean and readable. The engine must be at normal operating temperature.

NOTE: It may be necessary to put a small amount of white paint or chalk on the timing marks to make them more visible.

3. Connect a timing light to No. 1 cylinder.
4. Loosen the distributor clamp.
5. Start the engine and run it at the rpm specified in the Tune-Up chart. Rotate the distributor to align the timing marks. Tighten the distributor clamp and recheck the timing.
6. Reconnect the vacuum hose or the 4-wire connector.

HEI System Tachometer Hookup

There is a terminal marked TACH on the distributor cap. Connect one ta-

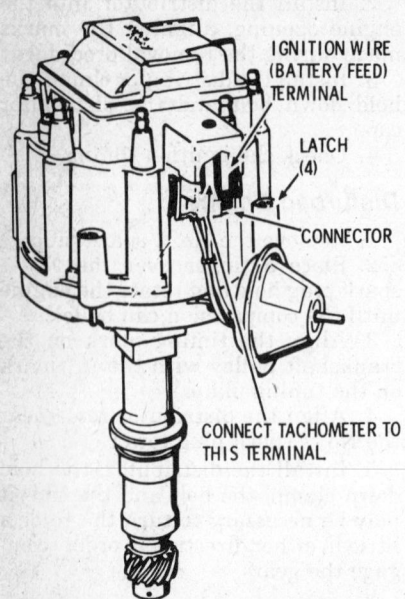

Tachometer hookup—V8 HEI and V6

chometer lead to this terminal and the other lead to a ground. On some tachometers, the leads must be connected to the TACH terminal and to the battery positive terminal.

--------- CAUTION ---------

Never ground the TACH terminal; serious module and ignition coil damage will result. If there is any doubt as to the correct tachometer hookup, check with the tachometer manufacturer.

GASOLINE FUEL SYSTEM

For all carburetor service information, please refer to "Carburetors" in the Unit Repair section.

Fuel Pump

REMOVAL & INSTALLATION

Mechanical Pump

The mechanical fuel pumps are diaphragm operated and are located on the either side of the engine. The fuel pressure should be 4–6.5 psi.

NOTE: **Some vehicles have a special fuel pump, which has a metering outlet for vapor return; this system reduces the possibility of vapor lock. Before working on the fuel system, release the fuel pressure. Remove and replace the fuel tank cap.**

When replacing a fuel pump on a Chevrolet 305, 307 or 350 cu. in. engine, considerable time can be saved as follows:
1. **Before removing the old pump, remove the upper bolt from the engine's right front mounting boss. This bolt hole is in direct alignment with the fuel pump pushrod. The threaded bolt hole continues into the pump pushrod bore. The bolt acts as an oil plug.**
2. **Temporarily insert a longer bolt, (about 3/8—16 x 2 in.) into the hole. Screw the bolt into the bore until it bottoms against the pump pushrod. (Don't tighten the bolt with a wrench or the rod can be damaged.)**
3. **The mechanic is now free to remove and install the fuel pump without worrying about fuel pump pushrod misalignment.**

1. Disconnect the fuel inlet, outlet and return (if equipped) hoses from the fuel pump.
2. Remove the retaining bolts.
3. Remove the fuel pump, pushrod, gasket and mounting plate (if equipped).

NOTE: **On the Chevrolet engines (V6 and V8), if the pushrod is not to be removed, remove the upper bolt from the right front mounting boss. Insert a longer bolt ($\frac{3}{8}$ x 16 x 2 in.) in this hole to hold the fuel pump pushrod.**

4. Clean the gasket mounting surfaces.
5. To install, use new gasket(s) and reverse the removal procedures. Torque the mounting plate bolts to 3 ft. lbs. and the fuel pump bolts to 27 ft. lbs. Start the engine and check for fuel leaks.

Electric Pump

The electric fuel pump is part of the fuel sending unit and is located in the fuel tank. The fuel pressure is 3 psi (carbureted), 4–13 psi (throttle body) or minimum 50 psi (port injection).

--------- CAUTION ---------

Before removing any component of the fuel system, refer to the "Fuel Pressure Release" procedures in this section and relieve the fuel pressure.

1. Disconnect the negative battery cable.
2. Remove the fuel tank.

3. Disconnect the fuel tubes and the electrical connectors from the sending unit of the fuel tank.
4. Using a brass drift and a hammer, loosen the sending unit and pump assembly retaining ring.
5. Lift the sending unit and pump assembly from the fuel tank.
6. Remove the fuel pump from the sending unit.
7. To install, use a new sealing O-ring and reverse the removal procedures.

FUEL PRESSURE RELEASE

Carbureted

To release the fuel pressure on the carbureted system, remove and replace the fuel tank cap.

Throttle Body Injection (TBI)

To release the fuel pressure on the TBI system, remove the fuel pump fuse from the fuse panel, start and operate the engine until it stalls, then replace the fuse.

Tuned Port Injection (TPI)

To release the fuel pressure on the TPI system, connect a tube to the pressure release fitting on the fuel rail, place the other end of the tube in a container. Open the pressure fitting, bleed off the fuel and close the fitting.

Fuel Filter

There are three types of fuel filters; internal (in the carburetor fitting), in-line (in the fuel line) and in-tank (the sock on the fuel pick-up tube). Before removing any component of the fuel system, refer to the "Fuel Pressure Release" procedures in this section and release the fuel pressure.

REMOVAL & INSTALLATION

Internal Filter

1. Disconnect the fuel line connection at the fuel inlet filter nut on the carburetor.
2. Remove the fuel inlet filter nut from the carburetor.
3. Remove the filter and the spring.

NOTE: **If a check valve is not present with the filter, one must be installed when the filter is replaced.**

4. Install the spring, filter and check valve (must face the fuel line), then reverse the removal procedures.
5. Start the engine and check for leaks.

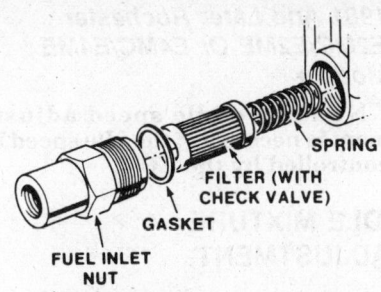

View of the carburetor fuel filter

In-Line Filter

1. Disconnect the fuel lines.
2. Remove the fuel filter from the retainer or mounting bolt.
3. To install, reverse the removal procedures. Start the engine and check for leaks.

NOTE: The filter has an arrow (fuel flow direction) on the side of the case, be sure to install it correctly in the system, with the arrow facing away from the fuel tank.

In-Tank Filter

To service the in-tank fuel filter, refer to the "Electric Fuel Pump Removal & Installation" procedures in this section.

Fuel Tank
CAUTION
Before removing any component of the fuel system, refer to the "Fuel Pressure Release" procedures in this section and release the fuel pressure.

REMOVAL & INSTALLATION

1. Release the pressure from the fuel system.
2. Raise and support the vehicle on jack stands.
3. Remove the fuel tank cap and drain the fuel.
4. Disconnect the fuel pump and the sending unit electrical connector at the body electrical harness connector.

CAUTION
DO NOT pry up on the cover connector. The pump and the sending unit wiring harness are an integral part of the sending unit.

5. Disconnect the flexible fuel lines from the metal fuel pipes at the tank.
6. Remove the fuel tank strap bolts and lower the fuel tank from the vehicle.
7. To install, reverse the removal procedures.

Carburetors
For all carburetor adjustments and specifications, refer to the "Carburetor" Unit Repair section. For further information on feedback carburetors, please refer to *Chilton's Guide To Fuel Injection And Feedback Carburetors.*

CAUTION
Before removing any component of the fuel system, refer to the "Fuel Pressure Release" procedures in this section and release the fuel pressure.

REMOVAL & INSTALLATION

1. Disconnect the negative battery cable.
2. Remove the air cleaner and gasket.
3. Disconnect the fuel and vacuum lines.
4. Disconnect the choke electrical connector.
5. Disconnect the throttle linkage.
6. If equipped with an automatic transmission, disconnect the throttle valve linkage.
7. Remove the carburetor attaching nuts or bolts, gasket and the carburetor.
8. If equipped, remove the electric EFE heater and the insulator.
9. Clean the gasket mounting surfaces.
10. To install, use a new gasket and reverse the removal procedures. Torque the short bolts to 7 ft. lbs. and the long bolts or nuts to 12 ft. lbs. Check the idle and the fast idle speeds.

IDLE SPEED ADJUSTMENT

1980 V6 With 2GE And V8–305 With 2GC Carburetors

1. Connect a tachometer to the engine according to the manufacturer's instructions.
2. Run the engine to normal operating temperatures. If equipped with A/C, turn it off and make sure that the choke is fully open. Set the parking brake and block the wheels. Disconnect and plug the hoses from the vapor canister at the canister and the EGR valve at the valve.
3. Place the transmission in Park (A/T) or Neutral (M/T).
4. Disconnect and plug the vacuum advance line at the distributor.
5. Check and/or adjust the timing.
6. Unplug and reconnect the vacuum advance line.
7. Turn the idle screw to the specified rpm.
8. Unplug and reconnect the hoses.

1980 V8–301 With M2MC; 1980 V6–231 And V8–265 With M2ME Carburetors

1. Connect a tachometer to the engine according to the manufacturer's instructions. Run the engine to normal operating temperatures. If equipped with A/C, disconnect the clutch wire connector at the compressor. Set the parking brake and block the wheels. Make sure that the choke is fully opened and place the transmission in Drive (A/T) or Neutral (M/T).
2. Disconnect and plug the vacuum advance line at the distributor.
3. Check and/or adjust the timing.
4. Unplug and reconnect the vacuum advance line.
5. Disconnect the purge hose from the canister.

NOTE: If equipped with A/C, turn the idle screw to obtain the specified rpm, turn the A/C switch on, open the throttle momentarily to extend the idle solenoid plunger, adjust the idle solenoid to the specified rpm and turn the A/C off. If NOT equipped with A/C, turn the idle screw to obtain the specified rpm.

6. Place the transmission in Park (A/T) or Neutral (M/T).
7. Disconnect and plug the vacuum hose at the EGR valve.
8. Place the fast idle screw on the 2nd step of the cam and adjust to the specified rpm.
9. Unplug and reconnect the hose. Reconnect the purge hose to the canister and the A/C compressor clutch wire.

1980 V8–305, 350 With M4MC Or M4ME Carburetors

1. Connect a tachometer to the engine following the manufacturer's specifications.
2. Run the engine to normal operating temperature.
3. Make sure that the choke is fully open, turn off the A/C, set the parking brake and block the wheels.
4. On the 350 engine, plug the disconnected purge hose.
5. Place the transmission in Drive (A/T) or Neutral (M/T).
6. Disconnect and plug the vacuum advance hose at the distributor.
7. Check and/or adjust the timing.
8. If equipped with a manual transmission, reconnect the vacuum advance line. On the 1980 models, disconnect the purge hose at the canister.
9. Turn the idle screw on all models to obtain the specified rpm.

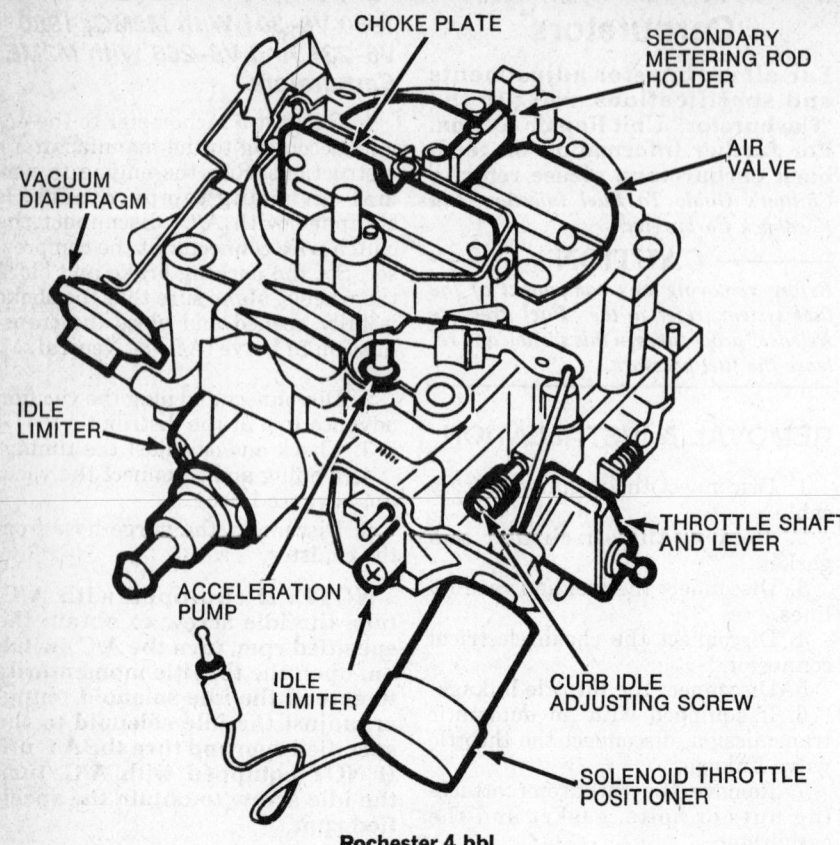

CHOKE PLATE

SECONDARY
METERING ROD
HOLDER

AIR
VALVE

VACUUM
DIAPHRAGM

IDLE
LIMITER

THROTTLE SHAFT
AND LEVER

ACCELERATION
PUMP

IDLE
LIMITER

CURB IDLE
ADJUSTING SCREW

SOLENOID THROTTLE
POSITIONER

Rochester 4 bbl

10. If equipped with A/C, turn it on and disconnect the compressor clutch wire. Open the throttle momentarily to fully extend the solenoid plunger. Adjust the solenoid screw to the rpm specified on the underhood sticker. Reconnect the compressor clutch and turn the A/C off. If NOT equipped with A/C, turn the idle screw to obtain the specified rpm.

11. Unplug and reconnect the canister and the EGR hoses.

1983–84 Rochester E2SE Models

For all overhaul and service adjustment procedures, refer to "Carburetors" in the Unit Repair section. For further information on feedback carburetors, refer to *Chilton's Guide To Fuel Injection And Feedback Carburetors*.

—— **CAUTION** ——

Before performing the idle adjustments, block the drive wheels and set the parking brake.

WITH A/C

1. Refer to the emission label in the engine compartment and follow the instructions to prepare the vehicle for adjustment.

2. Place the A/T in Drive or the M/T in Neutral and turn the throttle slightly to allow the solenoid plunger to fully extend.

3. Turn the solenoid screw to adjust the curb idle rpm, then disconnect the solenoid lead.

4. Turn the idle speed screw to set the basic idle speed. Reconnect the solenoid electrical lead after adjustment.

WITHOUT A/C

1. Refer to the emission label in the engine compartment and follow the instructions to prepare the vehicle for adjustment.

2. With the A/C off, turn the idle speed screw to set the curb idle speed.

3. Turn the A/C on, disconnect the A/C compressor lead at the A/C compressor and place the A/T in Drive or the M/T in Neutral.

4. Open the throttle slightly to allow the solenoid plunger to extend.

5. Turn the solenoid screw to adjust to the specified rpm. After adjustment, reconnect the A/C compressor lead.

6. Turn the A/C off. Set the curb idle speed by turning the idle speed screw.

1981 And Later Rochester E2MC/E2ME Or E4MC/E4ME Models

NOTE: No idle speed adjustment is necessary; the idle speed is controlled by the ECM.

IDLE MIXTURE ADJUSTMENT

1980 Rochester M2ME/M2MC Models

1. Set the parking brake and block the drive wheels.

NOTE: If the vehicle is equipped with a vacuum parking brake release system, disconnect and plug the vacuum hose at the brake.

2. Refer to the emission control label in the engine compartment, then disconnect and plug the necessary hoses.

3. Connect a tachometer to the engine, disconnect and plug the vacuum advance hose at the distributor, then set the timing.

4. Operate the engine to normal operating temperatures; make sure that the choke is open and the A/C is turned OFF.

NOTE: If the vehicle is equipped with Electronic Spark Timing (EST), check and/or adjust the timing and the idle speed according to the emission control label.

5. Disconnect the crankcase ventilation tube from the air cleaner.

6. Using tool J-26911, connect a hose from a propane tank to the crankcase tube opening at the air cleaner.

—— **CAUTION** ——

When using a propane tank, make sure it is secured in a safe place in the vertical position.

7. Place the transmission in Drive (A/T) or Neutral (M/T).

8. Slowly OPEN the propane valve until the maximum engine speed is reached.

NOTE: The addition of too much propane will cause the engine speed to drop.

9. If the idle mixture speed does not meet specifications, remove the idle mixture screw plug covers.

10. Turn the idle mixture screws clockwise until lightly seated, then back them OUT (equally) until the lean best idle point at the enriched idle speed is reached.

11. Turn the propane tank OFF and turn the idle mixture screws clockwise (equally) until the curb idle speed is reached.

12. Turn ON the propane tank and recheck the engine speed, if not within specifications, repeat the propane setting.

13. To complete the adjustment, turn OFF the engine, remove the propane tank, reconnect the crankcase ventilation tube to the air cleaner and install new idle mixture screw plugs.

1980 And Later Rochester E2ME/E2MC, E4ME/E4MC And 1984 6510-C (Canada)

1. Remove the carburetor from the engine and invert it. Using a hacksaw, make two parallel cuts ($^1/_8$ in. deep) in the throttle body on either side of the idle mixture screws locator points; the cuts should reach to the steel plug.

2. Using a flat center punch, hold it at a 45° angle to the cut segments, drive the segments into the throttle body. Using a center punch, drive the steel plugs from the housing.

3. Using tool J-29030, turn the idle mixture screws inward until they seat, then back them out two turns.

4. Reinstall the carburetor to the engine but NOT the gasket and air cleaner.

5. Set the parking brake, block the drive wheels, start the engine and run it to normal operating temperatures.

6. Disconnect and plug the vacuum hoses according to the Emission Control Label in the engine compartment.

7. Connect a dwell meter, a tachometer and a timing light to the engine, then check and/or adjust the timing.

8. Place the transmission in Drive (A/T) or in Neutral (M/T), then check and/or adjust the idle speed.

NOTE: If equipped with an Idle Speed Control (ISC) or an Idle Load Compensator (ILC), DO NOT adjust the curb idle speed.

9. The dwell should be within the 10–50° range, if NOT, perform the following procedures:

   a. Turn the engine OFF. Cover the internal bowl vents and the Air Bleed Valve inlets with masking tape, then the primary and the secondary carburetor air intakes with a shop cloth.

   b. Using a No. 35 (0.110 in.) drill bit, drill the rivet heads from the Idle Air Bleed Valve cover. Using a small drift punch, drive the remaining rivets from the air horn tower.

   c. Remove and discard the cover.

Using compressed air, blow the metal chips from the air horn.

   d. Remove the cloth and the masking tape from the carburetor, start the engine and allow it to idle with the A/T in Drive or the M/T in Neutral. Slowly turn the Idle Air Bleed Valve until the dwell reading varies between 25–35°; adjust it as close to 30° as possible.

NOTE: The Idle Air Bleed Valve is very sensitive and should ONLY be turned in $^1/_8$ in. increments. If the dwell remains below 25°, turn the idle mixture screws OUT one full turn. If the dwell is above 35°, turn the idle mixture screws IN one full turn. Readjust the Idle Air Bleed Valve to obtain the dwell limits.

10. After adjusting the idle mixture, fill the idle mixture screw holes with silicone sealant.

11. Check and/or adjust the Fast Idle Speed according to the Emissions Control Label.

12. Remove the test equipment, unplug and reconnect the vacuum hoses, then install the air cleaner and gasket.

1983–84 Rochester E2SE Models

1. Remove the carburetor from the engine and invert it. Using a hacksaw, make two parallel cuts ($^1/_8$ in. deep) in the throttle body on either side of the idle mixture screws locator points; the cuts should reach to the steel plug.

2. Using a flat center punch, hold it at a 45° angle to the cut segments, drive the segments into the throttle body. Using a center punch, drive the steel plugs from the housing.

3. Using tool J-29030, turn the idle mixture screws inward until they seat, then back them out 4 turns.

4. Reinstall the carburetor to the engine but NOT the gasket and air cleaner.

5. Remove the vent stack screen assembly from the air horn to gain access to the lean mixture screw. Turn the lean mixture screw IN until lightly seated and back it OUT $2^1/_2$ turns.

6. Set the parking brake, block the drive wheels, start the engine and run it to normal operating temperatures.

7. Disconnect the bowl vent, the EGR valve and the canister purge hoses at the carburetor, then cap the ports. Disconnect and plug the temperature sensor hose at the air cleaner.

8. Connect the positive lead of a dwell meter to the mixture control solenoid test lead (green connector) and

the other lead to ground. Connect a tachometer to the distributor lead (brown connector).

NOTE: If the vehicle is equipped with a tachometer, connect the new tachometer to the distributor side of the tach filter.

9. Place the transmission in Park (A/T) or in Neutral (M/T), then run the engine (on the highest step of the fast idle cam) for 3 minutes, until the cooling fan starts to operate.

10. Operate the engine at 3000 rpm, then adjust the lean mixture screw to establish a dwell of 35°, using tool J-28696-10.

NOTE: When adjusting the lean mixture screw, turn it in small increments and allow time for the engine to stabilize between adjustments. If the dwell remains low, turn the screw OUT; if the dwell is high, turn the screw IN.

11. Return the engine to curb idle speed.

12. Using tool J-29030, adjust the idle mixture screw to establish a dwell of 25° with the cooling fan OFF.

NOTE: Allow time for the reading to adjust after each adjustment. If the reading is low, back the screw OUT; if the reading is high, turn the screw IN.

13. When the cooling fan is in the OFF cycle, disconnect the mixture control solenoid and check for an idle speed change of more than 50 rpm. If the rpm change is NOT great enough, check the idle air bleed circuit for restrictions or leaks.

14. Operate the engine at 3000 rpm for a few moments and check for a dwell varying around 35°; if NOT, reset the lean mixture and the idle mixture screws.

15. When the adjustment is completed, remove the test equipment and replace the removed hoses and electrical connectors.

FAST IDLE ADJUSTMENT

1980 Rochester M2ME, 1980 And Later Rochester E2ME And 1982–84 Rochester E2SE Models

1. Refer to the emission label and prepare the vehicle for adjustment.

2. Place the transmission in Park (A/T) or Neutral (M/T).

3. Place the fast idle screw on the highest step of the fast idle cam.

4. Turn the fast idle screw to obtain the fast idle speed.

1982 and Later Rochester E4ME Models

NOTE: The fast idle speed adjustment must be performed according to the emission control label instructions. See the underhood sticker.

1982 AND LATER IDLE MIXTURE ADJUSTMENT (ALL MODELS)

All models have sealed idle mixture screws; in most cases these are concealed under staked-in plugs. Idle mixture is adjustable only during carburetor overhaul and requires the addition of propane as an artificial enrichener.

Electronic Fuel Injection

Before removing any component of the fuel system, refer to the "Fuel Pressure Release" procedures in this section and release the fuel pressure. For all information, removal and service procedures, please refer to *Chilton's Guide To Fuel Injection And Feedback Carburetors*.

IDLE SPEED ADJUSTMENT

NOTE: The TBI unit is adjusted at the factory and no further adjustment is necessary. Only if the TBI unit has been replaced should an adjustment be performed.

1985 and Later

1. Run the engine until it reaches normal operating temperatures.
2. Remove the air cleaner and the gasket.
3. Using an awl, puncture idle stop screw cover plug and pry the plug from the TBI.
4. Connect a jumper lead from the Idle Air Control (IAC) motor diagnostic lead to ground.
5. Connect a tachometer to the engine.
6. Turn the ignition ON (DO NOT start the engine) and wait 30 seconds.
7. Disconnect the IAC electrical connector from the TBI unit and remove the diagnostic lead jumper wire.
8. Start the engine, engage the parking brake, place the automatic transmission in Drive and allow the engine speed to stabilize.
9. Adjust the idle speed screw to 475–625 rpm.
10. Stop the engine and reconnect the IAC motor electrical connector.

11. Using a voltmeter, adjust the Throttle Position Sensor (TPS) to 0.450–0.600 volts.
12. Recheck the setting.
13. Start the engine and check for proper idle operation.
14. To install, seal the throttle stop screw with silicone sealant and reverse the removal procedures.

DIESEL FUEL SYSTEM

For information and service procedures, refer to "Diesel Maintenance" in the Unit Repair section. See the Oldsmobile Rear Wheel Drive section for complete diesel engine fuel system procedures.

ENGINE COOLING

Radiator

REMOVAL & INSTALLATION

1. Disconnect the negative battery cable and drain cooling system.
2. If necessary, remove the fan, the upper fan shroud or the upper support.
3. Disconnect upper and lower hoses.
4. Disconnect and plug the oil cooler lines, if equipped with an A/T.
5. Lift radiator and shroud straight up and out of vehicle.

NOTE: If equipped with a clutch type fan, keep it in an upright position to prevent the fluid from leaking.

6. To install, reverse the removal procedures. Make sure the lower cradles are properly located and A/T is full.

Water Pump

The water pump is a die cast, centrifugal-type with sealed bearings. Since it is pressed together, it must be serviced as a unit.

REMOVAL & INSTALLATION

All Engines Except Diesel

1. Disconnect the negative battery cable and drain the cooling system.
2. If necessary, remove the fan shroud or the upper radiator support.
3. Remove the necessary drive belts.
4. Remove the fan and water pump pulley.
5. Remove the alternator and the power steering pump (if equipped) brackets, then move the units aside.
6. Remove the heater hose and the lower radiator hose from the pump.
7. Remove the water pump retaining bolts and the pump. Clean the gasket mounting surfaces.

NOTE: Use an anti-seize compound on the water pump bolt threads.

8. To install, use new gaskets and reverse the removal procedures. Torque the water pump, the alternator and the power steering (if equipped) mounting bolts to 30 ft. lbs. Adjust the drive belts and fill the cooling system.

NOTE: If a belt tensioning gauge is available, adjust the belts to 100–130 lbs. of tension on new belts or to 70 lbs. on used belts. If the gauge is not available, adjust the belts so that a $1/4$–$1/2$ inch deflection can be made on the longest span of the belt under moderate thumb pressure.

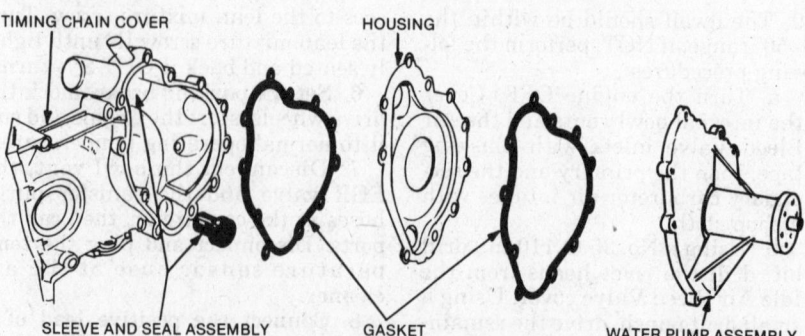

Pontiac V8 water pump assembly

Diesel

1. Disconnect the negative battery cable and drain the cooling system.
2. Disconnect the lower radiator hose, the heater hose and the by-pass hose from the water pump.
3. Remove the fan assembly, the drive belts and the water pump pulley.
4. Remove the alternator, the power steering pump and the A/C compressor (if equipped) brackets, then move the units aside.
5. Remove the water pump mounting bolts and the pump. Clean the gasket mounting surfaces.
6. To install, use new gaskets, sealant and reverse the removal procedures. Torque the water pump bolts to 22 ft. lbs. Adjust the belts and refill the cooling system.

NOTE: Apply sealer to the lower water pump bolts.

Thermostat

REMOVAL & INSTALLATION

1. Disconnect the negative battery cable and drain cooling system to below thermostat level.
2. Remove the air cleaner. Disconnect upper radiator hose.
3. Remove the thermostat housing bolts, the housing and the thermostat.
4. Clean the gasket mounting surfaces.

NOTE: When installing the thermostat, place the pin side facing upwards.

5. To install, use new gaskets, sealant and reverse the removal procedures. Torque the thermostat housing mounting bolts to 30 ft. lbs. Refill the cooling system.

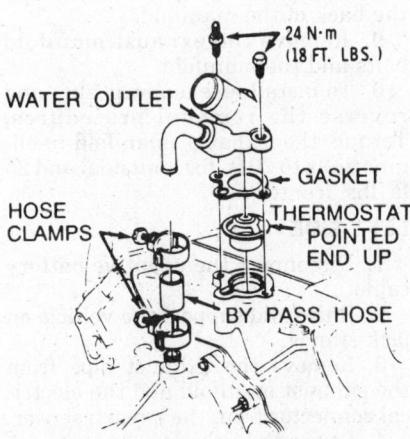

View of the thermostat—diesel engine

EMISSION CONTROLS

For a description and service procedures, please refer to "Emission Controls" in the Unit Repair section. Due to the complex nature of modern electronic engine control systems, comprehensive diagnosis and testing procedures fall outside the confines of this repair manual. For complete information on diagnosis, testing and repair procedures concerning all modern engine and emission control systems, please refer to *Chilton's Guide To Electronic Engine Controls*.

ENGINE MECHANICAL

NOTE: Pontiac uses engines produced by several other GM divisions. If the following engine procedures do not apply, identify the engine by using the VIN code as explained at the beginning of this section, then determine the engine builder from the Engine Identification Chart. When the engine has been identified, refer to the appropriate vehicle section of this book. For all 3.8L V6 procedures, for example, refer to the Buick section. For all V6 and V8 diesel information, refer to the Oldsmobile Rear Wheel Drive section.

Engine

REMOVAL & INSTALLATION

V6 Engine

1. Scribe alignment marks at the hood hinges and remove the hood.
2. Disconnect the negative battery cable.
3. Disconnect the exhaust pipe from the exhaust manifold.
4. Remove the bell housing cover and drain the transmission oil cooler lines at the oil pan.
5. Remove the left engine mount through bolt and loosen the right engine mount through bolt.
6. If equipped with an A/T, remove the torque converter cover, the converter-to-flex plate bolts and the engine-to-transmission bolts.

NOTE: Before removing the torque converter bolts, scribe a mark to ensure the relationship between the torque converter and the flex plate.

7. Disconnect the CCC wiring harness from the transmission and the knock sensor from the engine.
8. Disconnect the fuel hoses from the frame and the lower fan shroud.
9. Lower the vehicle and remove the windshield washer bottle.
10. Disconnect and label the CCC wiring harness, other necessary wiring connectors and the vacuum hoses from the engine.
11. Remove the air cleaner, the upper fan shroud, the accelerator and the T.V. cables.
12. Drain the cooling system, then remove the heater and the radiator hoses.
13. If equipped with A/C and power steering, remove the compressor and the power steering pump, then move them aside.
14. Disconnect the transmission oil cooler lines and the overflow tube from the radiator, then remove the radiator.
15. Remove the A/C hose and the adjusting bracket from the alternator.
16. Disconnect the battery cables from the frame and the heater hose from the bracket.
17. Secure a vertical lifting device to the engine and remove the engine from the vehicle.

— **CAUTION** —
When removing the engine from the transmission, be careful that the torque converter does not pull out of the transmission.

18. To install, reverse the removal procedures.

V8 Engine

1. Scribe alignment marks on hood and remove hood from hinges.
2. Disconnect the negative battery cable.
3. Drain cooling system, then remove the heater hoses and the radiator hoses from the engine.
4. Remove the upper fan shroud and the fan assembly.
5. If equipped with A/C and power steering, remove the compressor and the power steering pump, then move them aside.
6. Disconnect the accelerator and the T.V. cables
7. Remove the transmission oil cooler lines (if equipped) from the radiator and the radiator.
8. Disconnect and label the vacuum hoses and the CCC wiring harness connector(s) from the engine.

9. Remove the AIR pipe from the converter.

10. Remove the windshield washer bottle.

11. Disconnect and mark the wiring harness at the bulkhead and related engine wiring.

12. Remove the distributor cap and the cruise control cable (if equipped).

13. Disconnect the positive battery cable from the battery and the frame. Disconnect the negative battery cable from the A/C hose/alternator bracket.

14. Raise and support the vehicle on jack stands.

15. Remove the crossover pipe and the catalytic converter as an assembly.

16. If equipped with an A/T, remove the torque converter cover and the torque converter bolts.

NOTE: Before removing the torque converter bolts, scribe a mark to ensure the relationship between the torque converter and the flex plate.

17. Remove the engine-to-mount bolts.

18. Disconnect the fuel line from the fuel pump.

19. If equipped with an A/T, disconnect the torque converter clutch wiring from the transmission. Disconnect the transmission oil cooler lines from the clip at the engine oil pan.

20. Remove the engine-to-transmission bolts.

21. Lower the vehicle and support the transmission.

22. Secure a vertical lifting device to the engine and remove the engine from the vehicle.

— CAUTION —
When removing the engine from the transmission, be careful that the torque converter does not pull out of the transmission.

23. To install, reverse the removal procedures.

Intake Manifold

REMOVAL & INSTALLATION

1. Disconnect the negative battery cable.

2. Remove the air cleaner and drain the cooling system.

3. Disconnect the CCC harness and move aside.

4. Remove the heater and the radiator hoses from the engine.

5. Remove the upper alternator bracket, the electrical wiring connectors and the vacuum hoses.

6. On the V8, disconnect the fuel line and the brake pipes from the car-

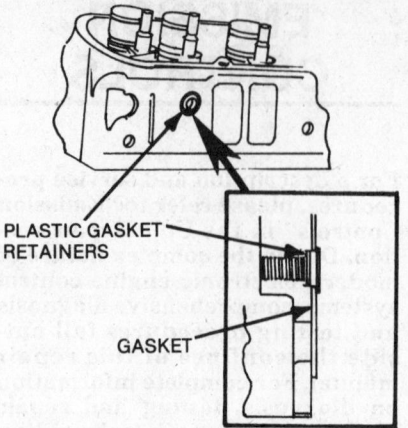

PLASTIC GASKET RETAINERS

GASKET

Pontiac V8 intake manifold gaskets can be held in place by using plastic retainers, available at Pontiac dealers

buretor. On the V6, disconnect the clips and the fuel lines from the TBI unit.

7. Disconnect the accelerator and the T.V. cables from the engine.

8. On the V8, remove the spark plug wires and the exhaust manifold from the right cylinder head.

9. Remove the distributor cap, then mark the position of the rotor with the distributor and the engine. Remove the distributor.

10. On the V6, remove the coil. On the V8, remove the carburetor.

11. If equipped with A/C, remove the compressor and the bracket, then move aside.

12. Remove the intake manifold bolts and the manifold.

13. To install, use new gaskets, sealant and reverse the removal procedures. Torque the intake manifold-to-engine to 25–45 ft. lbs. Refill the cooling system.

Exhaust Manifold

REMOVAL & INSTALLATION

V6 Engine
RIGHT SIDE

1. Disconnect the negative battery cable.

2. Raise and support the vehicle on jack stands.

3. Remove the exhaust pipe from the exhaust manifold.

4. Lower the vehicle.

5. Disconnect the air management valve bracket, the AIR hoses and the AIR pipe at the converter, the cylinder heads and the exhaust manifold.

6. Disconnect the spark plug wires.

7. Remove the exhaust manifold bolts and the manifold.

8. Using a putty knife, clean the gasket mounting surfaces.

9. To install, use a new gasket and reverse the removal procedures. Torque the exhaust manifold mounting bolts to 20 ft. lbs. (outside) and 25 ft. lbs. (center).

LEFT SIDE

1. Disconnect the negative battery cable.

2. Raise and support the vehicle on jack stands.

3. Remove the exhaust pipe from the exhaust manifold.

4. If equipped with A/C, remove the compressor and the rear adjusting bracket, then move it aside.

5. If equipped with power steering, remove the power steering pump and lower the rear adjusting bracket, then move the pump aside.

6. Disconnect the spark plug wires from the plugs.

7. Remove the exhaust manifold bolts and the manifold.

8. Using a putty knife, clean the gasket mounting surfaces.

9. To install, use a new gasket and reverse the removal procedures. Torque the exhaust manifold mounting bolts to 20 ft. lbs. (outside) and 25 ft. lbs. (center).

V8 Engine
RIGHT SIDE

1. Disconnect the negative battery cable.

2. Raise and support the vehicle on jack stands.

3. Disconnect the exhaust pipe from the exhaust manifold.

4. Lower the vehicle.

5. Remove the air cleaner.

6. Disconnect the spark plug wires on the right side, the EFE vacuum hose and the AIR hose.

7. Loosen the alternator bracket adjustment, then remove the drive belt and the lower alternator bracket.

8. Remove the AIR valve and disconnect the converter AIR pipe from the back of the manifold.

9. Remove the exhaust manifold bolts and the manifold.

10. To install, use a new gasket and reverse the removal procedures. Torque the exhaust manifold-to-engine bolts to 20 ft. lbs. (outside) and 25 ft. lbs. (center).

LEFT SIDE

1. Disconnect the negative battery cable.

2. Raise and support the vehicle on jack stands.

3. Remove the exhaust pipe from the exhaust manifold and the electrical connector from the oxygen sensor.

4. Lower the vehicle.

5. Disconnect the AIR hose

6. If equipped, remove the power

steering pump, the bracket and move the pump aside.

7. If equipped with A/C, loosen the bracket at the front of the cylinder head, then remove the compressor and the rear bracket.

8. Remove the exhaust manifold bolts, the wire loom holder from the valve cover and the manifold. Using a putty knife, clean the gasket mounting surfaces.

9. To install, use a new gasket and reverse the removal procedures.

Rocker Arm

REMOVAL & INSTALLATION

1. Disconnect the negative battery cable and remove the valve covers.

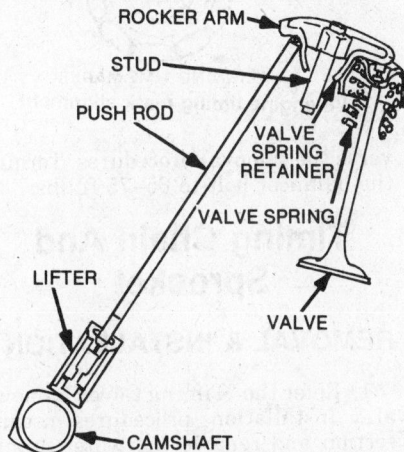

Pontiac V8 valve train assembly

2. Remove the rocker arm nut and rocker arm ball.

3. Lift the rocker arm off the rocker arm stud. Always keep the rocker arm assemblies together and assemble them on the same stud.

4. Remove the pushrod from its bore. Make sure the rods are returned to their original bores, with the same end in the block.

5. To install, reverse the removal procedures. Adjust the valves.

VALVE CLEARANCE ADJUSTMENT

1. Disconnect the negative battery cable and remove the valve covers.

2. Turn the crankshaft until the mark on the damper pulley aligns with the 0° mark on the timing plate at the front of the engine and the No. 1 cylinder is at the TDC of the compression stroke.

NOTE: If the valves of the No. 1 cylinder DO NOT move when checked, the engine is at the TDC

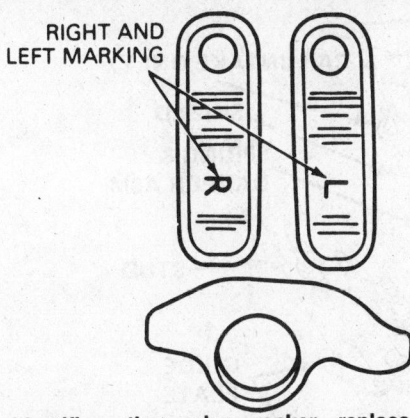

Identifing the valve rocker replacements—V6 engine

TORQUE SEQUENCES

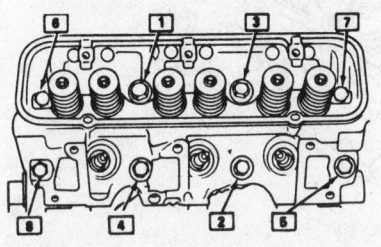

Cylinder head bolt torquing sequence—3.8L engine

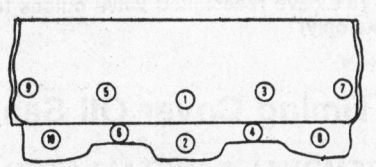

Cylinder head torque sequence—265 and 301 V8's

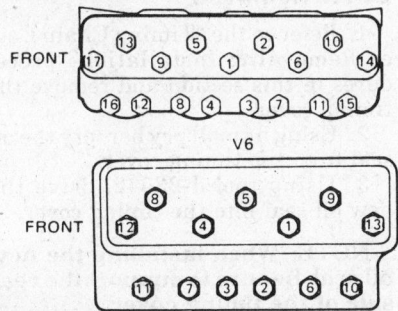

Cylinder head bolt torquing sequence—except 3.8L engine

of the compression stroke. If the valves DO move when checked, the engine is at the TDC of the No. 6 cylinder (V8) or No. 4 cylinder (V6); turn the crankshaft one complete revolution.

3. With the engine at the No. 1 firing position, adjust the exhaust valves of No. 1, 3, 4 & 8 on the V8 or No. 1, 5 & 6 on the V6 and the intake valves of No. 1, 2, 5 & 7 on the V8 or No. 1, 2 & 3 on the V6.

NOTE: To adjust the valves, back out the adjusting nut until lash is felt at the pushrod, then turn the nut in until the lash disappears. Once the play has been removed, turn the adjusting IN one full turn (to center the lifter plunger).

4. After adjusting the indicated valves, turn the crankshaft one full rotation and align the mark on the damper pulley with the 0° mark on the timing plate.

5. With the engine at the No. 6 (V8) or No. 4 (V6) firing position, adjust the exhaust valves of No. 2, 5, 6 & 7 on the V8 or No. 2, 3 & 4 on the V6 and the intake valves of No. 3, 4, 6 & 8 on the V8 or No. 4, 5 & 6 on the V6.

6. To install, reverse the removal procedures. Start the engine, then check and/or adjust the idle speed.

Cylinder Head

REMOVAL & INSTALLATION

1. Refer to the "Intake Manifold Removal & Installation" procedures in this section and remove the intake manifold.

2. Remove the alternator lower mounting bolt and move the unit aside.

3. Remove the exhaust manifold(s), the rocker arm cover(s) and the rocker arm assemblies.

4. Drain the cooling system.

5. Remove the diverter valve, the cylinder head bolts and the cylinder head(s).

6. Using a putty knife, clean the gasket mounting surfaces.

7. To install, use new gaskets and reverse the removal procedures. Torque the cylinder head bolts in sequence to 20–35 ft. lbs. Check and/or adjust the valve clearances, the timing and the idle speed.

Timing Chain Cover

REMOVAL & INSTALLATION

1. Disconnect the negative battery cable and drain the cooling system.

2. Remove the fan assembly, the drive belts and the fan pulley.

3. Raise and support the vehicle on jack stands.

4. Remove the crankshaft pulley and the damper pulley bolt.

5. Using tool J-23523, remove the damper pulley.

6. Remove the alternator and the brackets. If equipped with power steering, remove the lower pump bracket and swing aside.

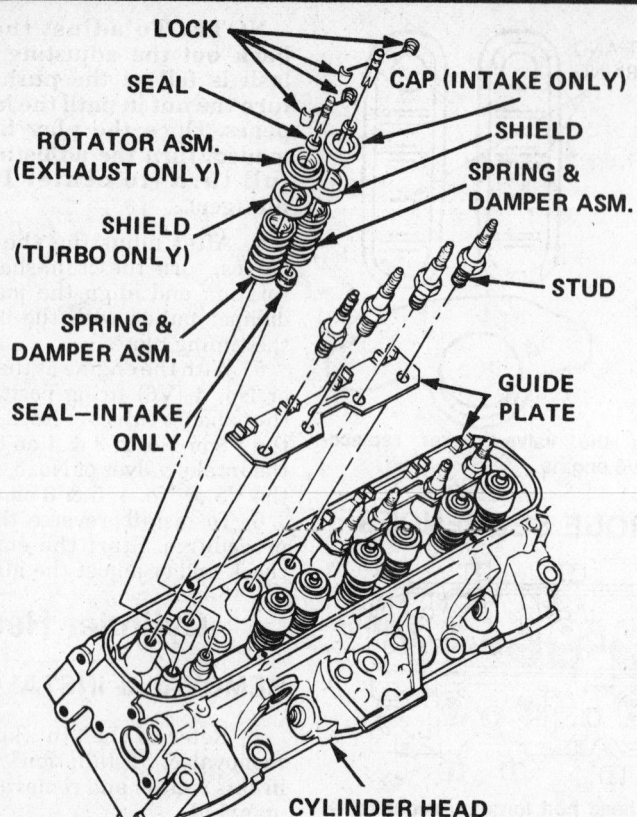

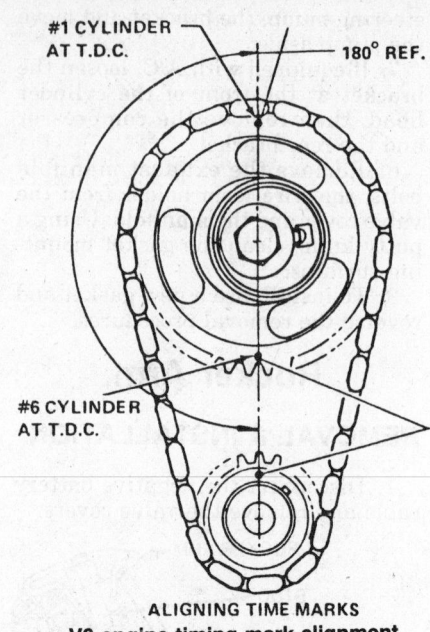

ALIGNING TIME MARKS
V8 engine timing mark alignment

The 1981 and later 265 (4.3L) and 301 (4.9L) V8's have redesigned valve guides to allow the use of valve stem seals (on the intake only)

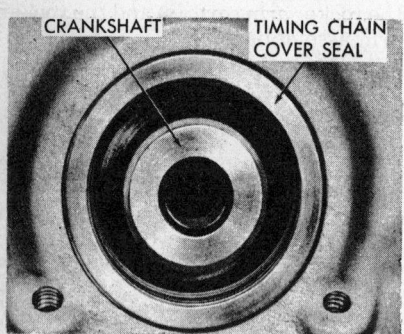

Timing chain cover oil seal
(© Pontiac Div., G.M. Corp)

Timing Cover Oil Seal

REMOVAL & INSTALLATION

Cover Removed

1. Refer to the "Timing Chain Cover Removal & Installation" procedures in this section and remove the timing cover.
2. Using a small pry bar, pry the oil seal from the timing cover.
3. Using tool J-23042, drive the new oil seal into the timing cover.

NOTE: When installing the new oil seal, be sure to support the rear side of the timing cover.

4. To complete the installation, reverse the removal procedures.

Cover Installed

1. Refer to the "Timing Cover Removal & Installation" procedures in this section and remove the balancer from the crankshaft.
2. Using a small pry bar, pry the oil seal from the timing cover.
3. Place the new seal (open end toward the engine) on the timing cover and drive it into the cover using tool J-23042.
4. To complete the installation, re-

verse the removal procedures. Torque the balancer bolt to 65–75 ft. lbs.

Timing Chain And Sprocket

REMOVAL & INSTALLATION

1. Refer the "Timing Cover Removal & Installation" procedures in this section and remove the water pump and the timing cover.
2. Turn the crankshaft until the mark on the camshaft sprocket aligns with the mark on the crankshaft sprocket.
3. Remove the camshaft sprocket bolts, the camshaft sprocket, the timing chain and the crankshaft sprocket (if necessary).

NOTE: When installing the timing chain, install the sprockets with the timing marks facing each another; this position is TDC of the No. 6 cyl (V8) or No. 4 cyl (V6). To

7. Remove the heater and the lower radiator hoses from the water pump.
8. Remove the water pump bolts and the pump from the engine.
9. Remove the timing chain cover bolts and the cover.
10. Using a putty knife, clean the gasket mounting surfaces.
11. To install, use new gaskets, sealant and reverse the removal procedures. Torque the timing cover bolts to 8 ft. lbs., the damper bolts to 65–75 ft. lbs. and the water pump bolts to 25–35 ft. lbs.

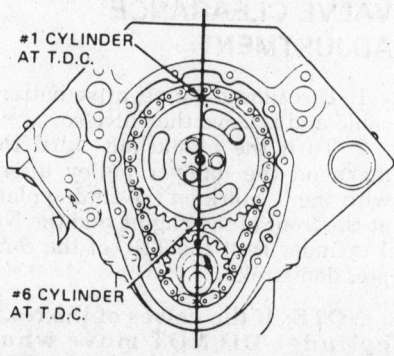

Aligning the timing marks—gas engine

locate the TDC of the No. 1 cylinder, turn the crankshaft one full revolution, the camshaft timing mark will now be at the top of the sprocket.

4. To install, use new gaskets, sealant and reverse the removal procedures. Torque the camshaft sprocket bolts to 13–23 ft. lbs. Check and/or adjust the engine timing.

Camshaft

REMOVAL & INSTALLATION

1. Refer to the "Timing Chain Removal & Installation" procedures in this section and remove the camshaft sprocket and the timing chain.

NOTE: If the camshaft sprocket is tight on the camshaft, use a plastic hammer to bump it loose.

2. On the V8, remove the oil cooler lines and the hoses from the radiator, then the radiator.

3. Remove the intake manifold and the rocker arm covers.

4. On the V8, remove the AIR pump bracket and disconnect the fuel lines at the fuel pump, then remove the fuel pump.

5. If equipped with A/C on the V8, remove the compressor and the condenser, then move them aside.

6. Remove the rocker arm assemblies, the push rods and the valve lifters.

7. Install two $5/16$ x 18 x 4 in. bolts in the camshaft and carefully pull the it from the front of the engine.

NOTE: When removing or replacing the camshaft, be careful not to damage the camshaft bearings.

8. To install, reverse the removal procedures. Torque the camshaft mounting bolts to 13–23 ft. lbs. Check and/or adjust the engine timing. Refill the cooling system.

Piston and Connecting Rod

The letter "F" or the notch on the edge of each piston, faces the front of the engine. The connecting rods have bosses on one side of the rod and chamfered corners on the connecting rod cap. The bosses must face rearward on the left bank and to the forward on the right bank. The chamfered corners of the rod caps must face forward on the left bank and rearward on the right bank. Some V6 engines have oil squirt holes on the connecting rods; these holes must face the camshaft.

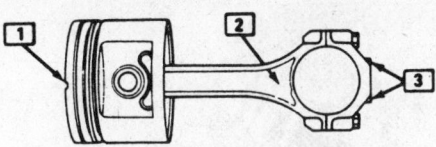

Piston and connecting rod assembly—3.8L engine

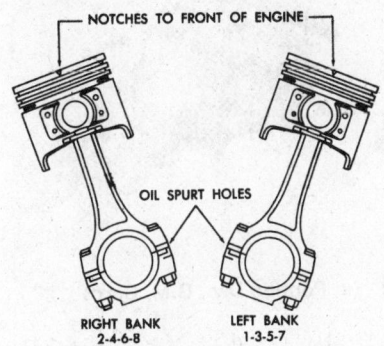

Pontiac V8 piston and rod assembly

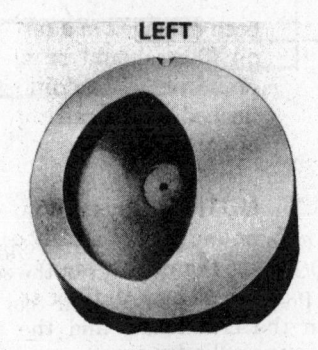

Turbocharged engine piston indentification

ENGINE LUBRICATION

Oil Pan

REMOVAL & INSTALLATION

1. Disconnect the negative battery cable.

2. Remove the air cleaner and the upper fan shroud.

3. Raise and support the vehicle on jack stands. Drain the oil from the engine.

4. On the V8, disconnect the AIR hose from the converter pipe and the AIR pipe from the exhaust manifold.

5. Remove the exhaust crossover pipe from the manifold and the converter.

6. Remove the starter and the flywheel cover.

7. Disconnect the transmission oil cooler lines at the oil pan.

8. On the V8, remove the engine through bolts. On the V6, loosen the right engine mount bolt and remove the left engine mount bolt.

NOTE: If, when removing the oil pan, the crankshaft throw or the counter balance weight block the pan removal, turn the crankshaft to put the throw in the horizontal plane.

9. Remove the oil pan bolts and lower the pan.

10. Using a putty knife, clean the gasket mounting surfaces.

11. To install, use a new gasket, sealant and reverse the removal procedures. Torque the oil pan bolts to 13 ft. lbs. (3.8L) or 7 ft. lbs. (all others) and the engine mount bolts to 50 ft. lbs.

Oil Pump

REMOVAL & INSTALLATION

1. Refer to the "Oil Pan Removal & Installation" procedures and remove the oil pan.

2. Remove the oil pump-to-rear bearing bolts, the pump and the extension shaft.

3. Using a putty knife, clean the gasket mounting surfaces.

4. To install, use new gaskets, sealant and reverse the removal procedures. Torque the oil pump and the oil pan bolts to 7 ft. lbs. Refill the crankcase.

Rear Main Seal

REMOVAL & INSTALLATION

1. Refer to the "Oil Pan Removal & Installation" procedures in this section and remove the oil pan.

2. Remove the oil pump and the rear main bearing cap.

3. Using a small screwdriver, pry the oil seal from the rear main bearing cap.

4. Using a small hammer and a brass pin punch, drive the top half of

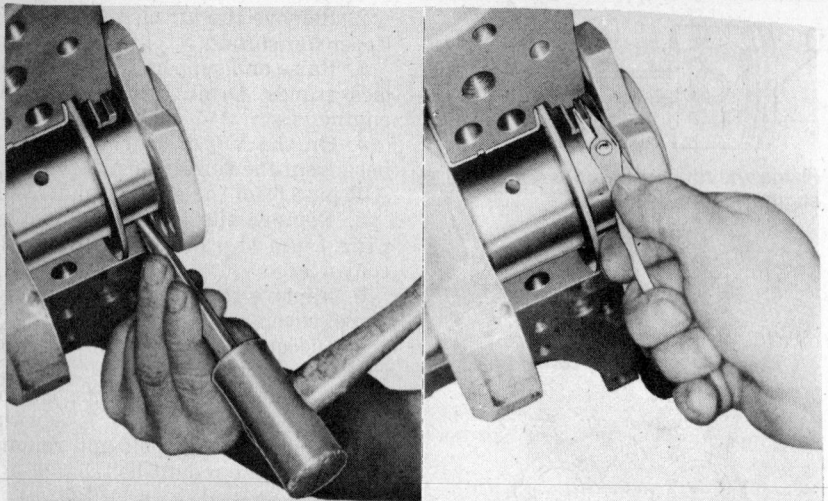

Rear main oil seal removal—upper half (© Pontiac Div., G.M. Corp)

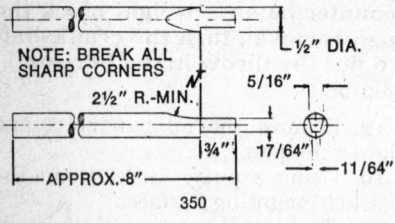

NOTE: BREAK ALL SHARP CORNERS
2½" R.-MIN.
5/16"
½" DIA.
¾"
17/64"
11/64"
APPROX.-8"
350

Pontiac V8 rear main bearing upper seal tool

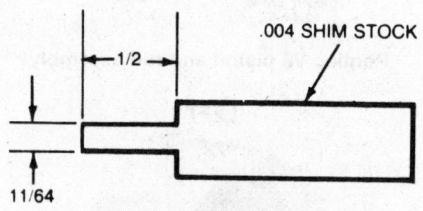

.004 SHIM STOCK
1/2
11/64

Oil seal installation tool

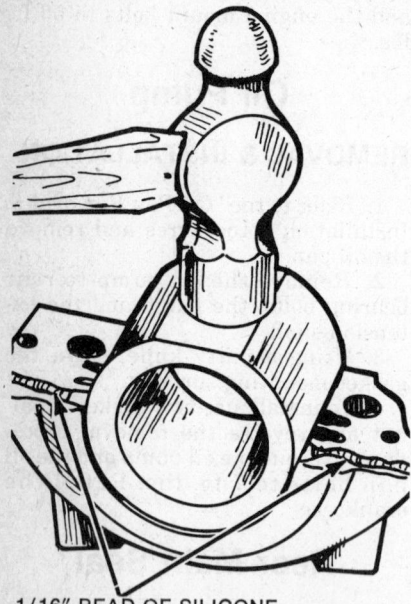

1/16" BEAD OF SILICONE RUBBER SEALER
Forming a new crankshaft seal

the oil seal from the rear main bearing. Drive it out far enough, so it may be removed with a pair of pliers.

5. Using a non-abrasive cleaner, clean the rear main bearing cap and the crankshaft.

6. Fabricate an oil seal installation tool from 0.004 in. shim stock, shape the end to ½ in. long by 11/64 in. wide.

7. Coat the new oil seal with engine oil; DO NOT coat the ends of the seal.

8. Position the fabricated tool between the crankshaft and the seal seat in the cylinder case.

9. Position the new half seal between the crankshaft and the tip of the tool, so that the seal bead contacts the tip of the tool.

NOTE: Make sure that the seal lip is positioned toward the front of the engine.

10. Using the fabricated tool as a shoe horn, to protect the seal's bead from the sharp edge of the seal seat surface in the cylinder case, roll the seal around the crankshaft. When the seal's ends are flush with the engine block, remove the installation.

11. Using the same manner of installation, install the lower half onto the lower half of the rear main bearing cap.

12. Apply sealant to the cap-to-case mating surfaces and install the lower rear main bearing half to the engine; keep the sealant off of the seal's mating line.

13. Install the rear main bearing cap bolts and torque to 10–12 ft. lbs. Using a lead hammer, tap the crankshaft forward and rearward, to line up the thrust surfaces. Torque the main bearing bolts to 70–85 ft. lbs. and reverse the removal procedures.

CLUTCH

REMOVAL & INSTALLATION

1. Raise and support the vehicle on jack stands.

2. Disconnect the negative battery cable.

3. Support rear of engine and remove the driveshaft.

4. Remove rear crossmember bolts from frame and transmission mounts, then remove the crossmember.

5. Disconnect the transmission shift linkage, the speedometer cable and the clutch return spring.

NOTE: After removing the transmission components, the clutch fork pushrod should hang free.

6. Remove flywheel cover screws and the cover plate.

7. Lower the engine to gain access to the bell housing-to-engine bolts, then remove all but uppermost bolt.

8. Hold the transmission and bell housing assembly against block over dowel pins, while removing last bolt. Remove the transmission and bell housing as an assembly.

9. Matchmark the pressure plate to the flywheel with paint to make sure that correct balance is maintained.

10. Loosen the pressure plate screws (a little at a time) until the clutch diaphragm spring tension is released, remove the bolts and the clutch assembly.

NOTE: The pilot bearing is an oil-impregnated type bearing pressed into the crankshaft; inspect and renew it, if necessary.

11. Install the clutch disc with the long hub facing the flywheel.

12. Install the pressure plate and cover assembly, then align the clutch disc by inserting a pilot tool or old an transmission input shaft into the

clutch splines. Align the mark on the clutch cover with the mark on the flywheel, then align nearest bolt holes.

13. Install the bolts in the cover and tighten them alternately.

14. Remove the clutch pilot tool and ensure that it can be reinserted and moved freely.

15. Install the clutch fork and dust boot into the clutch housing. Lubricate the throwout bearing with a high melting point grease.

16. To complete the installation, reverse the removal procedures. Torque the housing bolts. Adjust the shifter and the clutch release linkage.

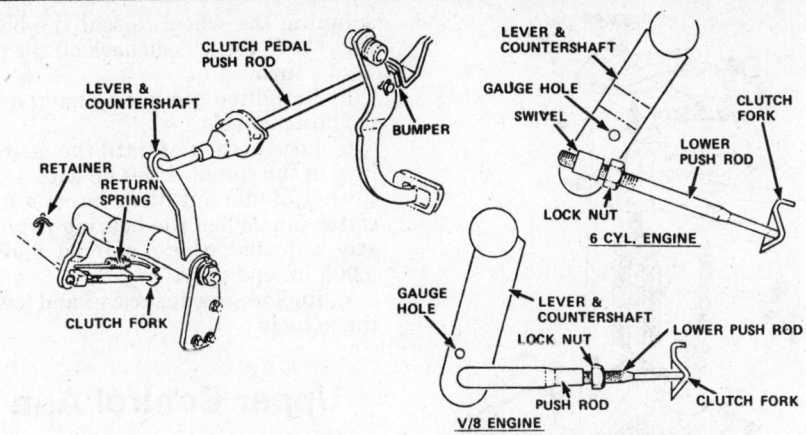

Typical clutch linkage and adjustment points

ADJUSTMENT

1. Disconnect the clutch fork return spring.
2. Loosen the pushrod locknut.
3. Detach the swivel or pushrod from the countershaft lever.
4. Install the swivel or pushrod into the gauge hole of the countershaft lever.
5. Push on the countershaft lever so that the clutch pedal is up against the stop.
6. Hold the clutch fork to the rear so that the release bearing lightly contacts the release levers.
7. Adjust the pushrod length to remove all lash from the linkage.
8. Reinstall the swivel or pushrod in the original hole on the countershaft lever. Tighten the locknut.
9. Replace the spring. The pedal free travel should now be $3/4$–$1^1/4$ in.

AUTOMATIC TRANSMISSION

NOTE: For information on transmission adjustments and service procedures, please refer to "Automatic Transmissions" in the Unit Repair section.

REMOVAL & INSTALLATION

1. Disconnect the negative battery cable.
2. Remove the air cleaner, the T.V. cable at the upper end and the transmission fluid level indicator.
3. Raise and support the vehicle on jack stands.
4. Remove the drive shaft and the floor pan reinforcement (if necessary).
5. At the transmission, disconnect the speedometer cable, the shift linkage, the electrical leads and retainers.

6. Remove the torque converter cover and the torque converter-to-flexplate nuts.

NOTE: Before removing the torque converter from the flex plate, mark the flex plate to the torque converter.

7. If equipped with a gas engine, remove the catalytic converter bracket.
8. Remove the transmission support mounting bolts at the transmission mount and at the frame, then raise the transmission slightly and slide the support rearward.
9. Lower the transmission, then remove the oil cooler lines and the T.V. cable from the transmission.

NOTE: After removing the oil cooler lines, be sure to cap them to prevent dirt from entering the system.

10. Support the engine and remove the transmission-to-engine bolts.
11. Install the torque converter holding tool J-21366 to support the torque converter, then remove the transmission from the vehicle.
12. To install, reverse the removal procedures. Torque the transmission-to-engine bolts to 35 ft. lbs., the torque converter-to-flex plate to 46 ft. lbs., the transmission support-to-frame bolts to 40 ft. lbs., the transmission support-to-transmission mount to 25 ft. lbs. and the drive shaft bolts to 16 ft. lbs. Adjust the shift linkage and the T.V. cable.

DRIVE AXLE

Driveshaft

REMOVAL & INSTALLATION

1. Raise and support the vehicle on jack stands.

2. Mark the driveshaft rear yoke and the differential flange to assure correct alignment upon reassembly.
3. Remove the U-bolts and nuts from the differential flange.
4. Remove the driveshaft assembly by first sliding the driveshaft forward to disengage the differential flange, then slide the shaft downward and rearward to disengage the front splined yoke from the transmission output shaft.
5. To install, reverse the removal procedures. Be sure to align the matchmark made before assembly. Torque the drive shaft mounting bolts to 15 ft. lbs.

U-Joint

REMOVAL & INSTALLATION

For information on universal joint removal, installation and overhaul procedures, please refer to "U-Joints/CV-Joints" in the Unit Repair section.

Axle Shaft Bearing and Seal

The axle shafts are retained by C-shaped locks, which fit into grooves at the inner end of the shaft.

REMOVAL & INSTALLATION

1. Raise and support the rear of the vehicle on jack stands.
2. Remove the wheel and the brake drum assembly.
3. Clean the dirt from around the carrier cover.
4. Remove the carrier cover and drain the oil from the housing.
5. Remove the rear axle pinion shaft lock screw and the shaft.
6. Push the flanged end of the axle shaft toward the center of the vehicle and remove the "C" lock from the end of the shaft.
7. Pull the axle shaft from the housing, being careful not to damage the oil seal.
8. Using a pry bar, pry the oil seal from the axle housing.
9. Install the bearing removal tool J-23689 ($8^3/4$ in. axle) or J-22813-01 (all other axles) to the axle bearing, connect it to a slide hammer and pull the bearing from the housing.
10. To install, lubricate a new bearing and drive it into the housing (until it seats) using the bearing installation tool J-23690.

11. Lubricate the lips of a new oil seal and drive it into the housing (until it is flush with the housing), using the seal installation tool J-21128 ($8^3/_4$ in. axle) or J-23771 (all other axles).

12. To complete the installation, slide the axle shaft into the housing (making sure that it engages the splines of the side gear) and reverse the removal procedures. Torque the pinion lock screw to 20 ft. lbs. Install a new carrier cover and torque the cover bolts to 20 ft. lbs. Fill the axle with lubricant to $^3/_8$ in. below the filler hole.

FRONT SUSPENSION

NOTE: Many 1980 and later Pontiacs have been gradually switched over to metric fasteners. Most models use metric prevailing torque nuts to fasten the upper and lower ball joint studs to the steering knuckle. American standard inch calibrated wrenches will not fit metric nuts and bolts.

Coil Spring

REMOVAL & INSTALLATION

1. Raise and support the vehicle on jack stands.
2. Remove shock absorber.
3. Secure tool J-23028 to a jack and position the assembly under the control arm, supporting the inner bushing.
4. Disconnect stabilizer bar at lower control arm.
5. Raise the jack to take the tension off of the control arm pivots. Install a chain around the spring and through the control arm as a safety measure, then remove the 2 inner control arm-to-crossmember pivot bolts.
6. Carefully lower the control arm, allowing the spring to relax.

─── **CAUTION** ───
Allow the spring to completely expand before attempting to remove it.

6. Remove the chain and the spring.
7. To install, reverse the removal procedures. Torque the lower control arm pivot nuts to 90 ft. lbs. (Parisienne) or 65 ft. lbs. (all others) with the weight of the vehicle on the springs.

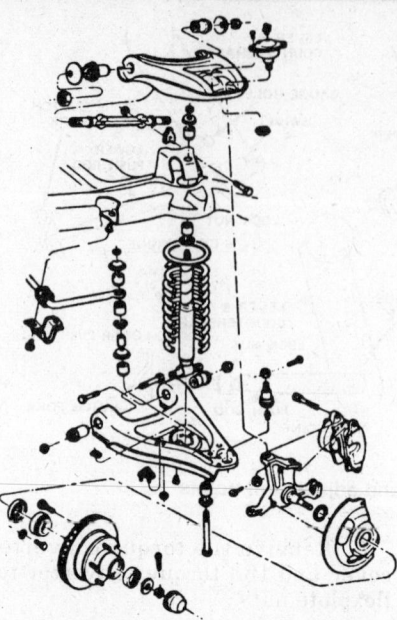

Exploded view of typical front suspension

Shock Absorber

REMOVAL & INSTALLATION

New shock absorbers must be purged of air before installation. This is done by repeatedly extending the shock in its normal mounted position, inverting it and compressing it.

1. Remove the nut, retainer and grommet, which are attached to the upper end of the shock absorber and seat against the frame bracket.

NOTE: It may be necessary to hold the shock absorber shaft to remove the nut. This may be done with a wrench on the end of the shaft.

2. Raise and support the vehicle on jack stands to allow the shock to be dropped from the lower control arm.
3. Remove the two shock absorber lower attaching screws and lower the shock from the control arm.
4. To install, reverse the removal procedures. Make sure all grommets are in the correct position and tighten the upper nut.

Front Wheel Bearing

ADJUSTMENT

1. Lift the wheel off the ground by jacking under the lower control arm.
2. Remove the dust cap from the hub.
3. Remove the cotter pin and discard it.
4. Snug up the spindle nut while spinning the wheel to seat the bearings (12 ft. lbs.). Then back off the nut $^1/_4$–$^1/_2$ turn.

5. Retighten the nut by hand until it is finger tight.
6. Loosen the nut until the nearest hole in the spindle lines up with a slot in the spindle nut, then insert a new cotter pin. When the bearing is properly adjusted, there will be 0.001–0.005 in. end play.
7. Replace the dust cover and lower the vehicle.

Upper Control Arm

REMOVAL & INSTALLATION

1. Raise and support the vehicle on jack stands between the spring seat and the ball joint, at the outer end of lower control arm.
2. Remove wheel and tire assembly.
3. Remove cotter pin and loosen the nut on the upper control arm-to-steering knuckle ball stud.
4. Using tool J-23742, push the ball joint stud from the steering knuckle.
5. Remove the two nuts that hold the upper control arm-to-crossmember and the control arm. Count number of shims at each bolt.
6. To install, reverse the removal procedures. Install same number of shims as removed at each bolt. Torque the control arm-to-frame to 70 ft. lbs. (Parisienne) or 48 ft. lbs. (all others) and the upper ball joint-to-steering knuckle to 52 ft. lbs. Insert cotter pin. Check caster and camber.

Ball Joint

INSPECTION

NOTE: Before performing this inspection, make sure the wheel bearings are adjusted correctly and that the control arm bushings are in good condition.

1. Raise and support the vehicle on jack stands under the front lower control arm at the spring seat. Raise the vehicle until there is 1–2 in. of clearance under the wheel.
2. Insert a bar under the wheel and pry upward. If the wheel raises more than $^1/_8$ in., the ball joints are worn. While prying on the wheel, determine by visual inspection whether the upper or lower ball joint is worn.

NOTE: Due to the distribution of forces in the suspension, the lower ball joint is usually the defective joint.

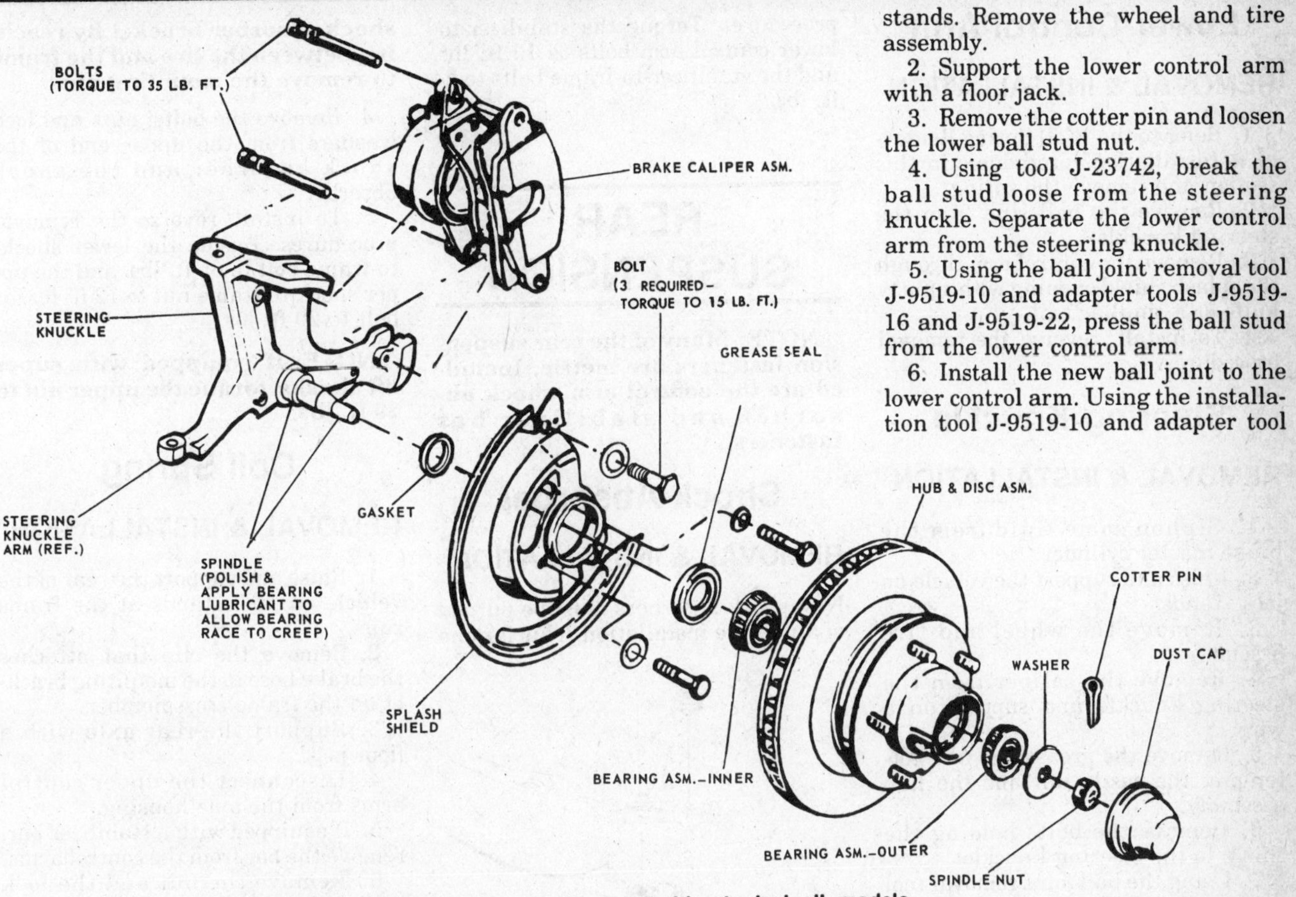

BOLTS
(TORQUE TO 35 LB. FT.)

BRAKE CALIPER ASM.

STEERING
KNUCKLE

BOLT
(3 REQUIRED—
TORQUE TO 15 LB. FT.)

GREASE SEAL

STEERING
KNUCKLE
ARM (REF.)

GASKET

SPINDLE
(POLISH &
APPLY BEARING
LUBRICANT TO
ALLOW BEARING
RACE TO CREEP)

HUB & DISC ASM.

COTTER PIN

WASHER

DUST CAP

SPLASH
SHIELD

BEARING ASM.—INNER

BEARING ASM.—OUTER

SPINDLE NUT

Steering knuckle, hub and disc assembly—typical all models

LOWER BALL JOINT WEAR INDICATORS

These vehicles have a visual wear indicator on the lower ball joint. Wear is indicated by the position of the $1/2$ in. nipple into which the grease fitting is screwed. On a new joint, the nipple should project 0.050 in. beyond the ball joint cover surface. If the nipple is flush or inside the cover surface, replace the ball joint.

Upper Ball Joint

REMOVAL & INSTALLATION

1. Refer to the "Upper Control Arm Removal & Installation" procedures in this section and separate the ball joint from the steering knuckle.
2. Using a center punch, punch the center of the four rivets.
3. Using an $1/8$ in. drill bit, drill $1/4$ in. deep into the rivets.
4. Using a $1/2$ in. drill bit, drill off the heads of the rivets.
5. Using a drift punch, drive out the remaining parts of the rivets.
6. Install new ball joint against top side of upper control arm. Secure joint to control arm with the special alloy

stands. Remove the wheel and tire assembly.

2. Support the lower control arm with a floor jack.
3. Remove the cotter pin and loosen the lower ball stud nut.
4. Using tool J-23742, break the ball stud loose from the steering knuckle. Separate the lower control arm from the steering knuckle.
5. Using the ball joint removal tool J-9519-10 and adapter tools J-9519-16 and J-9519-22, press the ball stud from the lower control arm.
6. Install the new ball joint to the lower control arm. Using the installation tool J-9519-10 and adapter tool

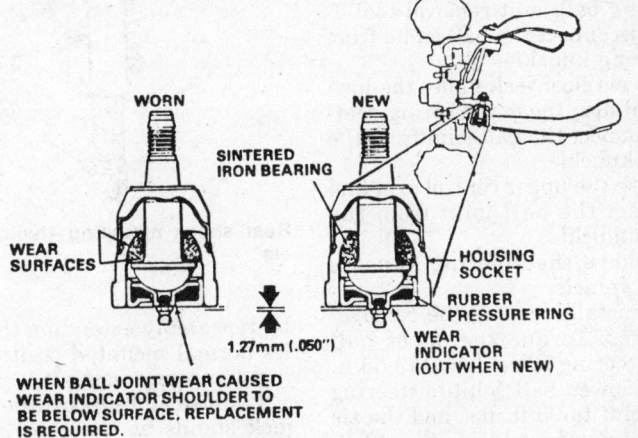

WORN

NEW

SINTERED
IRON BEARING

WEAR
SURFACES

1.27mm (.050")

WHEN BALL JOINT WEAR CAUSED
WEAR INDICATOR SHOULDER TO
BE BELOW SURFACE, REPLACEMENT
IS REQUIRED.

HOUSING
SOCKET

RUBBER
PRESSURE RING

WEAR
INDICATOR
(OUT WHEN NEW)

Lower ball joint wear indicator

bolts and nuts furnished with the replacement part.

7. Torque these bolts and nuts to 9 ft. lbs. and the ball joint-to-steering knuckle to 52 ft. lbs.

Lower Ball Joint

REMOVAL & INSTALLATION

1. Raise and support the front of the vehicle under the frame with jack

J9519-9, press the ball joint into the lower control arm until it bottoms on the arm.

NOTE: When installing the new ball joint, position the purge vent in the rubber boot facing inward.

7. To complete the installation, connect the ball joint-to-control arm assembly to the steering knuckle and torque the ball joint nut to 90 ft. lbs., then reverse the removal procedures.

Lower Control Arm

REMOVAL & INSTALLATION

1. Refer to the "Coil Spring Removal & Installation" procedures in this section and remove the spring.
2. Remove the ball stud from the steering knuckle.
3. Remove the control arm through the splash shield opening with a putty knife or a similiar tool.
4. To install, reverse the removal procedures.

Steering Knuckle

REMOVAL & INSTALLATION

1. Siphon some fluid from the brake master cylinder.
2. Raise and support the vehicle on jack stands.
3. Remove the wheel and tire assembly.
4. Remove the caliper from the steering knuckle and support on a wire.
5. Remove the grease cup, the cotter pin, the castle nut and the hub assembly.
6. Remove the bolts holding the shield to the steering knuckle.
7. Using the ball joint removal tool J-6627, disconnect the tie rod from the steering knuckle.
8. Using ball joint removal tool J-23742, disconnect the ball joints from the steering knuckle.
9. Place a floor jack under the lower control arm (near the spring seat) and disconnect the ball joint from the steering knuckle.
10. Raise the upper control arm and disconnect the ball joint from the steering knuckle.
11. Remove the steering knuckle from the vehicle.
12. To install, reverse the removal procedures. Torque the upper ball joint-to-steering knuckle nut to 65 ft. lbs., the lower ball joint-to-steering knuckle nut to 90 ft. lbs. and the tie rod-to-steering knuckle nut to 40 ft. lbs. Adjust the wheel bearing and refill the master cylinder.

Stabilizer Bar

REMOVAL & INSTALLATION

1. Raise and support the front of the vehicle on jack stands.
2. Disconnect the stabilizer link bolts at the lower control arms.
3. Remove the stabilizer-to-frame clamps.
4. Remove the stabilizer bar.
5. To install, reverse the removal procedures. Torque the stabilizer-to-lower control arm bolts to 13 ft. lbs. and the stabilizer-to-frame bolts to 24 ft. lbs.

REAR SUSPENSION

NOTE: Many of the rear suspension fasteners are metric. Included are the control arm, shock absorber and stabilizer bar fasteners.

Shock Absorber

REMOVAL & INSTALLATION

New shock absorbers must be purged of air before installation. This is done

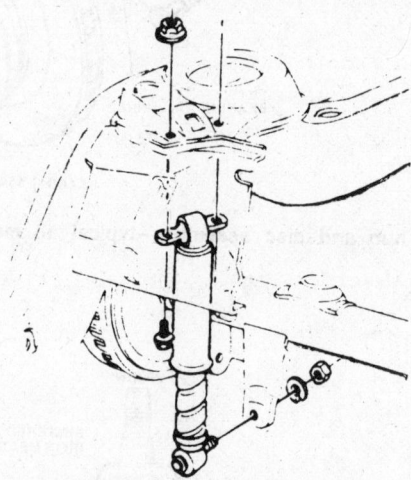

Rear shock mounting—typical all models

by repeatedly extending the shock in its normal mounted position, inverting and compressing it.

1. Raise and support the vehicle on jack stands at the axle housing, to prevent stretching the brake hose.

NOTE: If equipped with super lift shock absorbers, disconnect the air line snap on connector at the shock absorber.

2. On some models, it may be necessary to remove the wheel and tire assembly.
3. Remove the nut, the retainer, the grommet and lock washer, which attach the lower end of the shock absorber to its mounting.

NOTE: On some models, it may be necessary to remove the upper

shock absorber bracket by reaching between the tire and the frame to remove the mounting nuts.

4. Remove the bolts, nuts and lock washers from the upper end of the shock absorber and the shock absorber.
5. To install, reverse the removal procedures. Torque the lower shock-to-frame bolt to 65 ft. lbs. and the upper shock-to-frame nut to 12 ft. lbs. or bolt to 20 ft. lbs.

NOTE: If equipped with super lift shocks, torque the upper nut to 20 ft. lbs.

Coil Spring

REMOVAL & INSTALLATION

1. Raise and support the rear of the vehicle on jack stands at the frame rails.
2. Remove the clip that attaches the brake hose to the mounting bracket on the frame crossmember.
3. Support the rear axle with a floor jack.
4. Disconnect the upper control arms from the axle housing.
5. If equipped with a stabilizer bar, remove the bar from the control arms.
6. Remove the nut and the lock washer from the shock absorber, then disconnect the shock from the axle. It may be necessary to adjust the height of the jack to disconnect the shock.
5. Carefully lower the jack until the spring is free, then remove the spring.

NOTE: When removing the spring, mark it's position so that it may be installed in the same position.

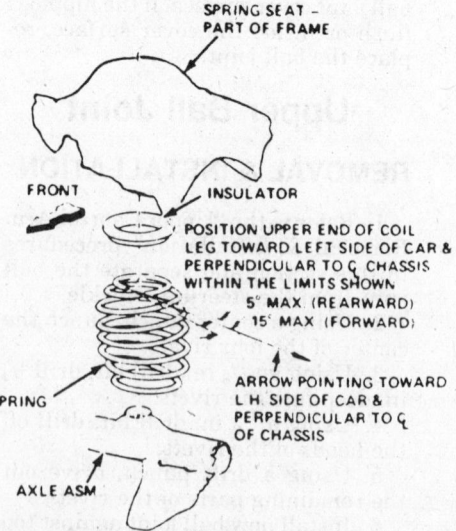

Coil spring installation—all full size models

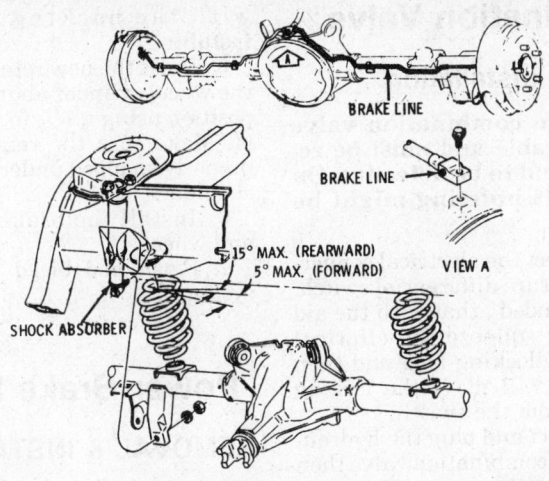

Rear coil spring installation—all except full size models

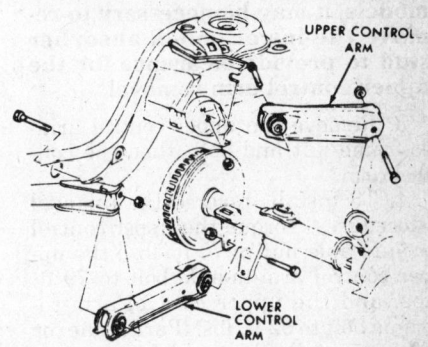

Control arms—all full size models

the same time, the axle could roll or slip sideways, making installation of the arms very difficult.

1. Raise and support the rear of the vehicle on jack stands under the axle.

6. To install, reverse the removal procedures. Torque the upper control arm-to-front bushing nut to 92 ft. lbs. on the Parisienne, or 70 ft. lbs. on all other models; the upper control arm-to-rear bushing nut to 70 ft. lbs. and the upper control arm-to-rear bushing nut to 80 ft. lbs.

Rear Lower Control Arm

REMOVAL & INSTALLATION

NOTE: Remove and install ONLY one lower control arm at a time. If both arms are removed at the same time, the axle could roll or slip sideways, making installation of the arms very difficult.

1. Raise and support the rear of the vehicle on jack stands under the rear axle.

NOTE: If equipped the a stabilizer bar, remove it.

2. Remove the control arm attaching fasteners and the control arm.

3. To install, reverse the removal procedures. Torque the control arm-to-frame nut to 92 ft. lbs. (Parisienne) or 70 ft. lbs. (all other models), the control arm-to-axle nut to 92 ft. lbs. (Parisienne) and the control arm-to-axle bolt to 125 ft. lbs. (Parisienne) or 79 ft. lbs. (all other models).

Rear Upper Control Arm

REMOVAL & INSTALLATION

NOTE: Remove and install ONLY one lower control arm at a time. If both arms are removed at

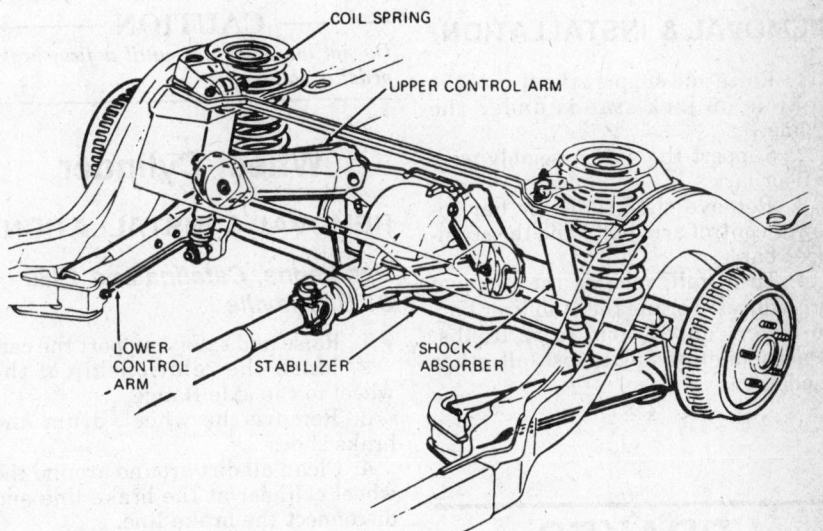

Rear suspension assembly—all full size models

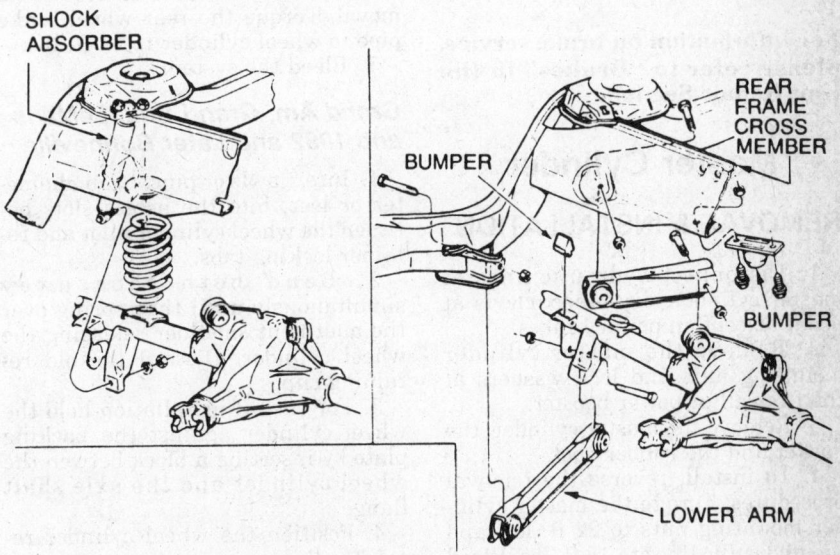

Rear suspension assembly—all except full size models

2. Remove the upper control arm nut at the axle.

NOTE: To remove the mounting bolt from the axle, it may be necessary to rock the axle. On some models, it may be necessary to remove the lower shock absorber stud to provide clearance for the upper control arm removal.

3. Remove the upper control arm-to-frame nut and bolt, then the control arm.

4. To install, reverse the removal procedures. Torque the upper control arm-to-axle nut to 70 ft. lbs., the upper control arm-to-axle bolt to 79 ft. lbs. and the upper control arm-to-frame bolt to 92 ft. lbs. (Parisienne) or 70 ft. lbs. (all other models).

Stabilizer Bar

REMOVAL & INSTALLATION

1. Raise and support the rear of the vehicle on jack stands under the frame.

2. Support the axle assembly with a floor jack.

3. Remove the stabilizer bar-to-lower control arm bolts and the stabilizer bar.

4. To install, reverse the removal procedures. Torque the stabilizer bar-to-lower control arm to 52 ft. lbs. (Parisienne) or 35 ft. lbs. (all other models).

BRAKES

For information on brake service, please refer to "Brakes" in the Unit Repair Section.

Master Cylinder

REMOVAL & INSTALLATION

1. Disconnect hydraulic lines at master cylinder; disconnect clevis at pedal (except on power brakes).

2. Remove the master cylinder mounting nuts and lock washers at the firewall or master booster.

3. Remove the master cylinder, the gasket and the rubber boot.

4. To install, reverse the removal procedures. Torque the master cylinder mounting nuts to 22 ft. lbs. and the hydraulic lines to 18 ft. lbs. Bleed the brake system.

Combination Valve

Removal & Installation

NOTE: The combination valve is not repairable and must be replaced if found to be defective. On some models hoisting might be necessary.

1. Disconnect the electrical connector at the pressure differential switch. It is recommended , that with the aid of pliers, you squeeze the eliptical shaped plastic locking ring and then pull up. This will move the locking tangs away from the switch.

2. Disconnect and plug the hydraulic lines at the combination valve then remove the valve.

3. Installation is the reverse of removal.

4. Bleed the entire brake system.

——— CAUTION ———
Do not move the car until a firm brake pedal is obtained.

Wheel Cylinder

REMOVAL & INSTALLATION

Parisienne, Catalina and 1980–81 Bonneville

1. Raise and safely support the car.
2. Mark the relationship of the wheel to the axle flange.
3. Remove the wheel, drum and brake shoes.
4. Clean all dirt around around the wheel cylinder at the brake line and disconnect the brake line.
5. Remove the wheel cylinder from the backing plate.
6. Installation is the reverse of removal. Torque the rear wheel brake pipe to wheel cylinder to 12 ft. lbs.
7. Bleed the system.

Grand Am, Grand Prix, LeMans and 1982 and Later Bonneville

1. Insert awls or pins, $1/8$ in. diameter or less, into the access slots between the wheel cylinder pilot and retainer locking tabs.
2. Bend both tabs away simultanously until they spring over the abutment shoulder releasing the wheel cylinder. Discard the old retaining clip.
3. For ease of installation hold the wheel cylinder against the backing plate by inserting a block betwen the wheel cylinder and the axle shaft flange.
4. Position the wheel cylinder retainer clip so the tabs will be away from and in a horizontial position

with the backing plate when installing.

5. Press the new retaining clip over the wheel cylinder abutment and into position using a $1^1/8$ in. 12 point socket. Make sure the retainer tabs are properly snapped under the abutment shoulder.

6. Install the brake shoes, drum and wheel.

7. Flush and bleed the hydraulic system.

Power Brake Booster

REMOVAL & INSTALLATION

1. Remove the vacuum hose from the front housing and discard the grommet. Remove the master cylinder and position away from the booster. It is not necessary to disconnect the lines from the master cylinder, if it is not to be repaired.

2. Remove the clevis pin retainer from the brake pedal inside the vehicle.

3. Remove the nuts from the vacuum cylinder studs under the dash and remove the vacuum power section.

4. To install, reverse the removal procedures. Torque the booster-to-cowl and the master cylinder-to-booster mounting nuts to 28 ft. lbs.

Hydro-Boost Brake Booster

REMOVAL & INSTALLATION

For an explanation, troubleshooting and bleeding of the Hydro-Boost brake system, please refer to "Brakes" in the Unit Repair section.

1. Turn the engine off and pump the brake pedal 4 or 5 times to deplete the accumulator.

2. Remove the nuts from the master cylinder, then move the master cylinder away from the booster, with brake lines still attached.

3. Remove the hydraulic lines from the booster.

4. Remove the retainer and washer at the brake pedal.

5. Remove the 4 attaching nuts retaining the booster fastened to the firewall and the booster.

6. To install, reverse the removal procedures. Torque the booster-to-cowl nuts to 15 ft. lbs. and the master cylinder-to-booster nuts to 20 ft. lbs. Bleed the power steering and hydro booster system.

Parking Brake

ADJUSTMENT

The automatic self-adjusting feature incorporated in the rear brake mechanism normally maintains proper parking brake adjustment. For this reason, the rear brake adjustment must be checked before any adjustment of the parking brake cables is made. Check the parking brake mechanism and cables for free movement and lubricate all working surfaces before proceeding.

— CAUTION —

It is very important that the parking brake cables not be too tight. If the cables are too tight, they create a drag and position the secondary shoes so that the self-adjusters continue to operate to compensate for drag wear. The result is rapidly worn rear brake linings.

1. Raise and support the rear of the vehicle on jack stands.
2. Set the parking brake at 2 clicks.
3. Loosen the equalizer locknut. Tighten the adjusting nut until the left wheel can be turned backward, but is locked in forward rotation.
4. Tighten the locknut.
5. Fully release the parking brake and rotate the rear wheels; no drag should be felt in either direction.

STEERING

Tie Rod End

REMOVAL & INSTALLATION

1. Raise and support the vehicle on jack stands.
2. Loosen the tie rod adjuster sleeve clamp nut.
3. Remove the tie rod cotter pin and nut from the ball joint stud.

NOTE: If the torque required to remove the nuts and bolts exceeds 7 ft. lbs., it's best to discard them and use new fasteners of equal grade quality.

3. Using tool J-6627 or BT-7101, remove the tie rod stud from the steering knuckle.

NOTE: The outer tie rods have right-hand threads and the inner tie rods have left-hand threads.

4. Unscrew the tie rod from the ad-

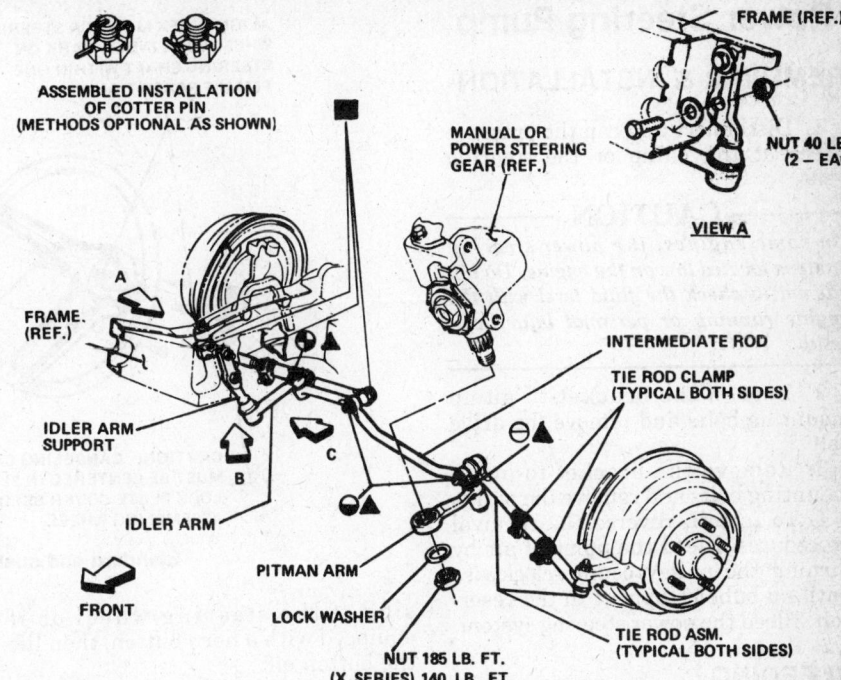

ASSEMBLED INSTALLATION OF COTTER PIN
(METHODS OPTIONAL AS SHOWN)

VIEW A

FRAME (REF.)

MANUAL OR POWER STEERING GEAR (REF.)

NUT 40 LB. FT (2 – EACH)

FRAME. (REF.)

IDLER ARM SUPPORT

IDLER ARM

PITMAN ARM

LOCK WASHER

FRONT

NUT 185 LB. FT. (X SERIES) 140 LB. FT.

INTERMEDIATE ROD

TIE ROD CLAMP (TYPICAL BOTH SIDES)

TIE ROD ASM. (TYPICAL BOTH SIDES)

Steering linkage

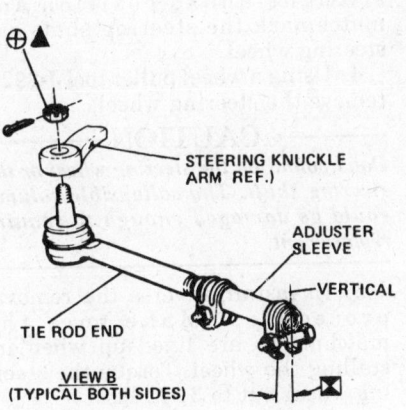

STEERING KNUCKLE ARM REF.)

ADJUSTER SLEEVE

VERTICAL

TIE ROD END

VIEW B
(TYPICAL BOTH SIDES)

Tie rod assembly—typical

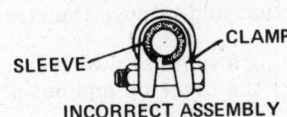

POSITION OF TIE ROD ADJUSTER SLEEVE & CLAMP

CLAMP

SLEEVE

INCORRECT ASSEMBLY

CORRECT ASSEMBLY

NOTE: SLOT IN TIE ROD ADJUSTER SLEEVE MAY BE IN ANY POSITION EXCEPT AT EDGES OF CLAMP JAWS.

Tie rod clamp installation

juster sleeve. Count the number of turns the tie rod must be rotated to remove it from the adjusting sleeve; this will allow a reasonably accurate realignment upon reassembly.

5. To install, reverse the removal procedures. Clean all rust and dirt from the threads. Check and/or adjust the alignment, if necessary.

Steering Gear

REMOVAL & INSTALLATION

1. Disconnect the pressure and return hoses from the steering gear housing.
2. Disconnect the negative battery cable and remove coupling shield.
3. Remove the steering coupling-to-steering shaft flange retaining nuts, bolts and lock washers.
4. Remove the pitman arm nut and washer from the pitman shaft, then mark the relationship of the arm to the shaft.
5. Using tool J-6632, remove the pitman arm from the steering gear.
6. Remove the steering gear-to-frame screws and the steering gear from vehicle.
7. To install, reverse the removal procedures. Torque the steering gear-to-frame to 70 ft. lbs., the coupling shaft pinch bolt to 30 ft. lbs. and the pitman arm shaft nut to 185 ft. lbs. Bleed the power steering system.

C495

Power Steering Pump

REMOVAL & INSTALLATION

1. Disconnect and cap the pressure hoses at the pump or the steering gear.

—————— CAUTION ——————

On some engines, the power steering pump is located low on the engine. Do not attempt to check the fluid level with the engine running or personal injury can result.

2. Loosen the bracket-to-pump mounting bolts and remove the drive belt.

3. Remove the bracket-to-pump mounting bolts and remove the pump.

4. To install, reverse the removal procedures. Bleed the pump of air by turning the pulley counterclockwise until no bubbles appear in the reservoir. Bleed the power steering system.

BLEEDING

1. Fill the reservoir with power steering fluid.

NOTE: The use of automatic transmission fluid in the power steering system is NOT recommended; use power steering fluid ONLY.

2. Allow the reservoir and fluid to be undisturbed for a few minutes.

3. Start the engine, allow it to run for approximately 3–5 minutes to warm up the fluid, then turn it off.

4. Check the reservoir fluid level and add fluid, if necessary.

5. Repeat the above steps until the fluid level stabilizes.

6. Raise the front of the vehicle so that the wheels are off of the ground and set the parking brake.

7. Start the engine and increase the engine speed to about 1500 rpm.

8. Turn the front wheels right to left (and back) several times, lightly contacting the wheel stops at the ends of travel.

9. Check the reservoir fluid level. Add fluid as required.

10. Repeat Step 8 until the fluid level in the reservoir stabilizes.

11. Lower the vehicle and repeat Steps 8 and 9.

Steering Wheel

REMOVAL & INSTALLATION

1. Disconnect the negative battery cable.

2. On the deluxe models, remove the trim cover screws on the under-

ALIGN INDEX MARK ON STEERING WHEEL WITH INDEX MARK ON STEERING SHAFT WITHIN ONE FEMALE SERRATION.

PUSH INSULATOR INTO CAM TOWER & ROTATE CLOCKWISE TO LOCK IN POSITION.

PAD ASM.

SHAFT NUT (SEE VIEW A)   RETAINER

35 LB. FT.   STEERING COLUMN SHAFT

STEERING WHEEL

CAUTION: CANCELING CAM TOWER MUST BE CENTERED IN SLOT OF LOCK PLATE COVER BEFORE ASSEMBLING WHEEL.

VIEW A

Standard and cushion steering wheel

side of the steering wheel or if equipped with a horn button, then lift the button off.

2. Remove the snap-ring and steering wheel nut from the steering shaft.

3. Position the wheels in the straight-ahead position and matchmark the steering shaft and steering wheel.

4. Using a wheel puller tool J-2927, remove the steering wheel.

—————— CAUTION ——————

Don't pound on the steering wheel or the steering shaft. The collapsible column could be damaged enough to require replacement.

5. To install, reverse the removal procedures. Make sure the matchmarks are lined up when installing the wheel. Torque the steering wheel nut to 31 ft. lbs.

Turn Signal Switch

REMOVAL & INSTALLATION

1. Refer to the "Steering Wheel Removal & Installation" procedures in this section and remove the steering wheel.

2. Using a small screwdriver, carefully pry the cover up and out of the steering column.

3. Position the lockplate compressing tool J-23653 or equivalent on the end of the steering shaft and compress the lock plate by turning the shaft nut clockwise. Pry the wire snap-ring out of the shaft groove and discard the ring.

4. Remove the tool and lift the lockplate off the shaft.

5. Slide the cancelling cam, upper bearing preload spring and thrust washer off the shaft.

6. Remove the retaining screw and the turn signal lever. Push the flasher knob in and unscrew it.

NOTE: If equipped with a button and a knob, remove the button retaining screw, then remove the button, spring and knob.

7. Remove the three mounting screws. Pull the switch connector out of the jacket bracket, wrap it with tape, lift up on the switch and pull the connector through the column support bracket.

NOTE: On tilt wheels, place the turn signal and shifter housing in the low position and remove the harness cover.

8. To install, attach a long piece of wire to the turn signal switch connector, feed the wire through the column housing and under the bracket, then pull the wire, the switch connector and the cover into position.

NOTE: On tilt wheel models, pull the connector down through the housing under the bracket, then install the cover over the harness.

9. Install the switch mounting screws and the connector on the jacket bracket. Install the column-to-dash trim plate.

10. Install the flasher knob and the turn signal lever.

11. With the turn signal lever in the neutral or off position and the flasher knob out, slide the thrust washer, upper bearing preload spring and cancelling cam onto the shaft.

12. Using tool J-23653 or equivalent, press the lock plate down on the shaft and install a new snap-ring in

the shaft groove. Do not re-use the old snap ring.

13. Install the cover and the steering wheel. Torque the turn signal switch screws to 3 ft. lbs. and the steering wheel-to-shaft nut to 31 ft. lbs.

Ignition Switch

REMOVAL & INSTALLATION

The switch is located inside the channel section of the brake pedal support and is completely inaccessible without first lowering the steering column. The switch is actuated by a rod and rack assembly. A gear on the end of the lock cylinder engages the toothed upper end of the rod.

1. Support and lower the steering column.

2. Place the ignition switch in the OFF-UNLOCKED position and move the actuating rod two detents from the top.

3. Remove the two mounting screws and the ignition switch assembly.

4. Before installing, place the new switch in the OFF-UNLOCKED position and make sure the ignition lock cylinder and the actuating rod are in the OFF-UNLOCKED (2nd detent from the top) position.

5. Install the actuating rod into the switch, mount the switch to the column and torque the mounting screws to 3 ft. lbs.

NOTE: Use only the specified screws since over length screws could impair the collapsibility of the column.

6. Install the steering column and torque the steering column-to-bracket nuts to 25 ft. lbs.

ADJUSTMENT

Standard Column

1. Place the switch in the OFF position.

2. Position the switch on the column, then move the slider to the extreme left (toward the wheel).

3. Move the slider back two positions to the right of ACCESSORY position.

4. Place the key in any run position and shift the transmission into any position but Park for automatics. Put it in Reverse for manual.

5. Position the lock toward ACCESSORY with a light finger pressure and secure the switch.

Tilt Column

1. Place the key in ACCESSORY position and leave the key in the lock.

2. Loosen the switch mounting screws.

3. Push the switch upward toward the wheel to make certain it is in ACCESSORY detent.

4. Hold the key in full counterclockwise ACCESSORY position and tighten the switch mounting screws.

5. If the switch is properly adjusted, it will go into ACCESSORY position, the key can be removed when in

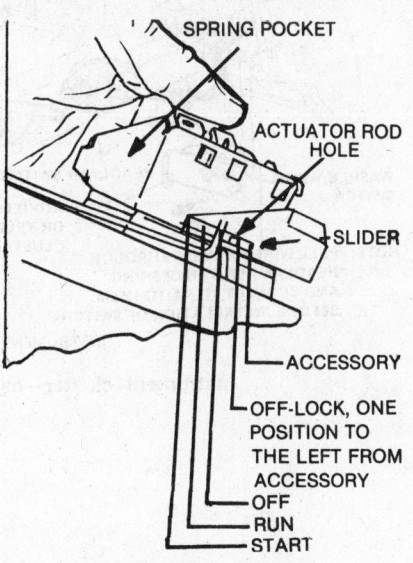

Installing ignition switch

the LOCK position and the switch will go into the START position.

Ignition Cylinder Lock

REMOVAL & INSTALLATION

1. Disconnect the battery ground cable, then refer to the "Turn Signal Switch Removal & Installation" procedure in this section and lift the turn signal switch to allow access to the lock cylinder.

NOTE: When lifting the turn signal switch, pull it rearward far enough to slip it over the end of the steering shaft. DO NOT pull the wiring harness from the column.

2. Place the ignition lock in the ON position.

3. Remove the buzzer switch, the lock cylinder screw and lock cylinder.

— CAUTION —
If the screw is dropped during removal and falls into the column, complete disassembly will be necessary to retrieve the screw.

4. To install, rotate the cylinder clockwise to align cylinder key with the keyway in the housing. Push the lock in all the way, then install the screw and torque it to 22 inch lbs. for adjustable columns and 40 inch lbs. for standard columns.

CHASSIS ELECTRICAL

Headlamp Switch

REMOVAL & INSTALLATION

1. Disconnect the negative battery cable.

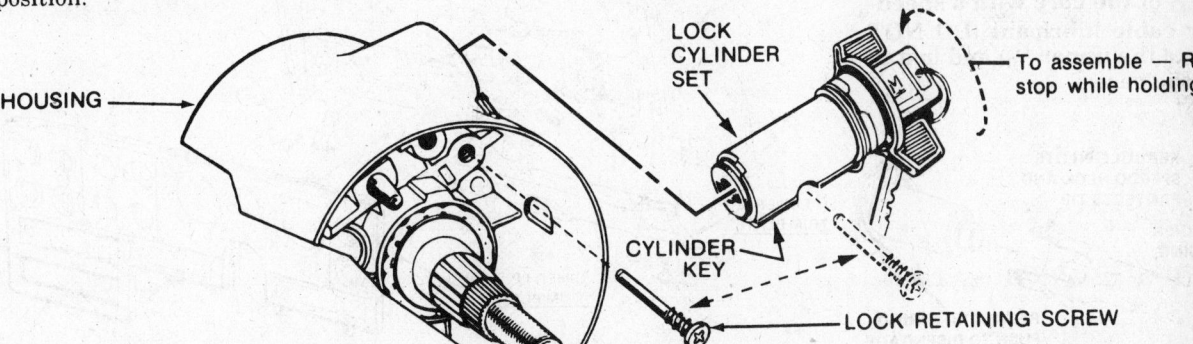

Lock cylinder installation details

2. Pull the control knob to ON position.

3. Reach under the instrument panel and depress the switch shaft retainer, then remove the knob and the shaft assembly.

NOTE: Disconnect vacuum hose on vacuum-operated headlamp models.

4. Remove the windshield wiper switch and the retaining ferrule nut.

5. Remove switch from instrument panel.

6. Disconnect multi-plug connector from switch.

NOTE: To remove the electrical connector from the switch, insert a small pry bar into the side of the switch and pry the switch away from the connector.

7. To install, reverse the removal procedures.

Speedometer Cable

REMOVAL & INSTALLATION

1. Remove the upper and the lower instrument panel trim plates.

2. If equipped with cruise control, disconnect the speedometer cable at the cruise control transducer. If NOT equipped with cruise control, disconnect the speedometer cable strap at the power brake booster.

3. Reach up behind or pull out the speedometer and find where the cable attaches to the speedometer head. Press the retaining clip downward and slide the cable from the head.

4. Slide the old core from the casing. If the core is broken, raise the vehicle and remove the cable retaining clip from the transmission. Pull out the remaining piece of the core.

5. To install, reverse the removal procedures.

NOTE: Prior to installing the cable core, wipe the core clean, flush the casing with solvent, coat the lower $2/3$ of the core with a speedometer cable lubricant (DO NOT lubricate the upper $1/3$) and install it into the case.

Instrument Cluster

REMOVAL & INSTALLATION

All Except Parisienne

1. Disconnect the negative battery cable.

2. Disconnect the shift indicator cable from the column jacket.

3. Remove the 2 steering column nuts and lower the steering column.

4. Remove the lower trim panel.

5. Remove the cigar lighter retaining nut.

6. Remove the upper trim panel screws and the panel.

7. Remove the gauges as needed.

8. To install, reverse the removal procedures.

Parisienne

1. Disconnect the negative battery cable.

2. Remove the steering column lower cover screws and cover.

3. If equipped with an automatic transmission, disconnect the shift indicator cable from the steering column.

4. Remove the two steering column-to-instrument panel screws and lower the steering column.

— **CAUTION** —

Use extreme care when lowering the steering column to prevent damage to the column assembly.

5. Remove the six screws and the three snap-in fasteners from the pe-

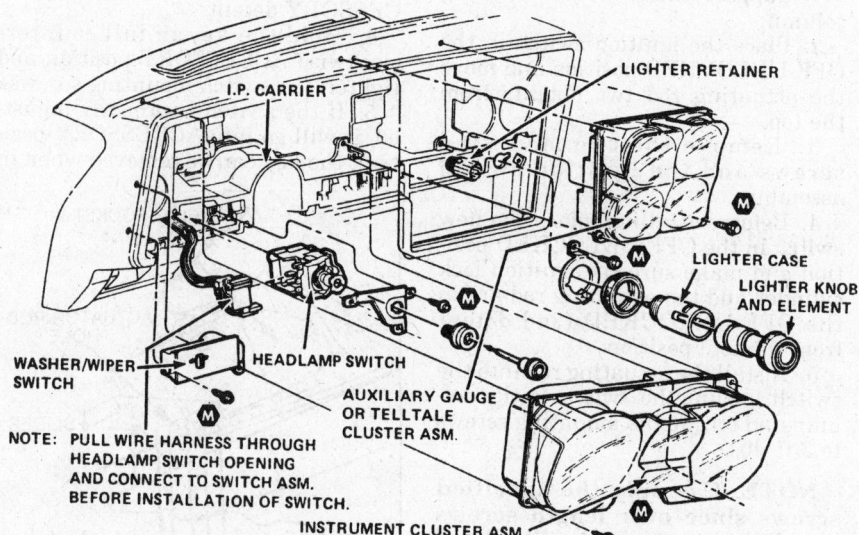

Instrument cluster—all except full size models

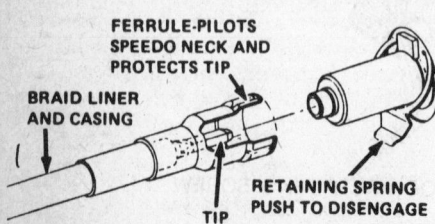

Speedometer cable attachment details

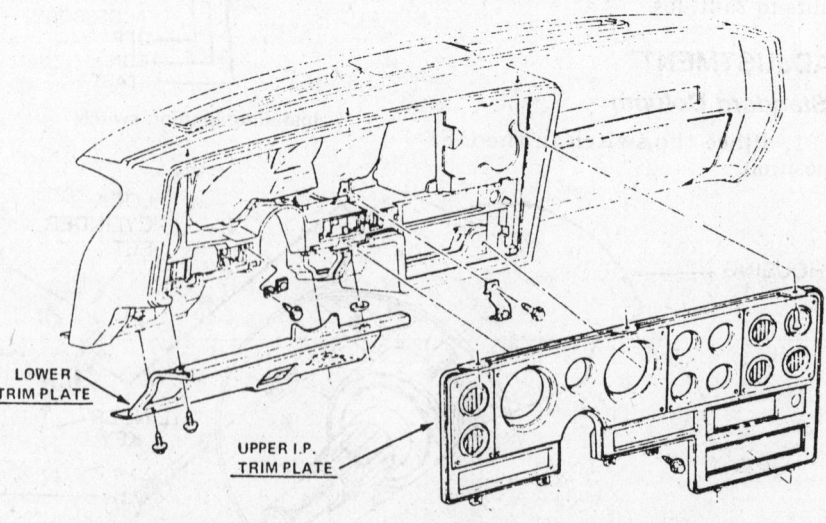

Instrument panel trim plate—all except full size models

rimeter of the instrument cluster lens.

6. Remove the two screws from the upper surface of the grey sheet metal trim plate.

7. Remove the two stud nuts from the lower corner of the cluster.

8. Disconnect the speedometer cable and pull the cluster from the instrument panel.

9. Disconnect the electrical connectors from the cluster and remove the cluster from the vehicle.

10. To install, reverse the removal procedures.

Windshield Wiper Motor

REMOVAL & INSTALLATION

1. Remove the cowl screen.

2. Remove the wiper motor transmission drive link nuts and the drive link from the motor.

3. Disconnect the wiring connectors and the washer hoses from the motor.

4. Remove the 3 motor mounting screws.

5. Remove the motor by guiding the crank arm through the hole.

6. To install, reverse the removal procedures. Make sure that the motor is in the Park position.

Wiper Blade

REMOVAL & INSTALLATION

1. At the end of the wiper element, squeeze the clip, pull down and out to release the element from the wiper arm.

2. Slide the element up and out of the housing retaining tabs.

3. To install, reverse the removal procedures.

Heater Blower Motor

REMOVAL & INSTALLATION

The blower motor is located in the engine compartment on the right side of the cowl.

With A/C

This procedure is the same as for non air-conditioned vehicles.

Without A/C

1. Disconnect the negative battery cable.

2. Disconnect the blower motor feed wire and the ground wire.

3. Remove the blower motor retaining screws and the motor.

4. To install, reverse the removal procedures.

Heater Core

REMOVAL & INSTALLATION

Without A/C

1980–81 LEMANS AND 1982 AND LATER BONNEVILLE

1. Disconnect and plug the heater hoses at the heater core tubes.

2. On the engine side of the firewall, remove the heater core cover from the case.

3. Remove the core bracket and the ground screw.

4. Lift out the core.

5. To install, reverse the removal procedures.

1980–81 BONNEVILLE AND CATALINA

1. Drain the cooling system.

2. Remove and plug the heater hoses at the heater core tubes.

3. Disconnect the electrical connectors.

4. Remove the front module cover screws and the module assembly.

5. Remove the heater core from the module.

6. To install, reverse the removal procedures. Use a strip caulk type sealer when installing the module to the cowl.

1984 AND LATER PARISIENNE AND 1980 AND LATER GRAND PRIX

1. Disconnect and plug the heater hoses at the core tubes.

2. Remove the core cover from the module.

3. Remove the core bracket and the ground screw.

4. Lift out the heater core.

5. To install, reverse the removal procedures. Replace the damaged sealer, if necessary.

With A/C

1980 AND LATER GRAND PRIX, 1980–81 LEMANS AND GRAND AM AND 1982 AND LATER BONNEVILLE

1. Position the wipers in the UP position.

2. Disconnect and unplug the heater hoses at the heater case.

3. Remove the module top cover seals.

4. Remove the module top screens.

5. Disconnect the electrical connectors from the heater case.

6. Move the lower windshield reverse molding aside.

7. Remove the cowl brackets.

8. Tape a strip of wood to the lower edge of the windshield glass, to protect the glass.

9. Remove the top cover screws.

10. Cut the sealing material along the cowl with a knife.

11. Pry the cover from the side, NOT from the top.

12. Remove the heater core and the seal.

13. To install, reverse the removal procedures. Use new sealing material.

1980–81 BONNEVILLE AND CATALINA AND 1984 AND LATER PARISIENNE

1. Drain the cooling system.

2. Disconnect and plug the heater hoses at the heater case.

3. Remove the retaining bracket and the ground strap.

4. Disconnect the module's rubber seal and screen.

5. Remove the right windshield wiper arm.

6. Remove the diaphragm connections, the hi-blower relay, the thermal switch mounting screws and the electrical connectors from the module top.

7. Remove the module top cover and pull out the heater core.

8. To install, reverse the removal procedures. Apply a strip of caulk type sealer when installing the module top.

Radio

REMOVAL & INSTALLATION

LeMans, 1982 and Later Bonneville

1. Disconnect the negative battery cable.

2. Remove the radio knobs and the bezels.

3. Remove the upper and the lower instrument panel trim plates.

4. Remove the front radio retaining screws.

5. Open the glove box door and lower it by releasing spring clips. Pull the radio out after loosening the rear right side nut. Disconnect all of the wiring and remove the radio.

6. To install, reverse the removal procedures. If the radio is to be replaced, remove the bushing from the rear of the radio and install it on the replacement radio.

Grand Am and Grand LeMans

1. Disconnect the negative battery cable.

2. Remove the radio knobs, the be-

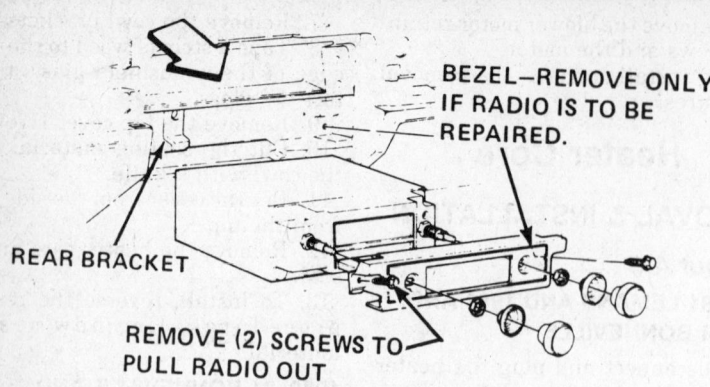

BEZEL—REMOVE ONLY IF RADIO IS TO BE REPAIRED.

REAR BRACKET

REMOVE (2) SCREWS TO PULL RADIO OUT

Pontiac radio removal

zels and the retaining hex nut from the right hand radio tuning shaft.

3. Remove the 4 retaining screws and the trim plate.

4. Remove the front retaining screw and the mounting bracket screw.

5. Remove the radio and the mounting bracket from the dash, disconnecting the electrical connections and the antenna lead.

6. To install, reverse the removal procedures.

1980–81 Bonneville, Catalina and 1984 and Later Parisienne

1. Disconnect the negative battery cable.

2. Remove the upper trimplate. Remove the radio trimplate by removing the two top screws, the ashtray assembly, the lighter and the ashtray bracket.

3. Remove the two radio screws.

4. Remove the radio through the instrument panel and detach all of the electrical connectors.

5. To install, reverse the removal procedures.

Grand Prix

1. Disconnect the negative battery cable.

2. Remove the knobs, the bezels and right hand hex nut from the radio. Remove the upper and the lower instrument panel trimplates.

3. Remove the 4 retaining screws and the radio trim plate.

4. Remove the front retaining screw and the radio mounting bracket retaining screw (below the radio). Open the glove box and loosen the rear nut at the right side of the radio.

5. Remove the radio and the bracket as an assembly, then disconnect the radio connections and the antenna lead-in while the radio is pulled out.

6. To install, reverse the removal procedures.

Fuses

The fuse block is located beneath the left side of the instrument panel. The fuse holders are labeled as to their service and the correct amperage. Always replace the blown fuses with new ones of the correct amperage oth-

erwise electrical overloads and possible wiring damage will result.

Fusible Links

Fusible links are sections of wire, with a special insulation, designed to melt under electrical overload. Replacements are simply spliced into the wire. There may be as many as five of these in the engine compartment wiring harnesses. These are:

1. Horn relay-to-fuse panel circuit—one link.

2. Charging circuit—from the starter solenoid-to-horn relay—two links.

3. Starter solenoid-to-ammeter circuit—one link.

4. Horn relay-to-rear window defroster circuit—one link.

NOTE: The fusible links are all two wire gauge sizes smaller than the wires they protect.

REPLACEMENT

1. Disconnect the negative battery cable.

2. Disconnect the fusible link from the junction block or the starter solenoid.

3. Cut the harness directly behind the connector to remove the damaged fusible link.

4. Strip the harness wire approximately $1/2$ in.

5. Connect the new fusible link to the harness wire using a crimp on connector. Solder the connection using rosin core solder.

6. Tape all of the exposed wires with plastic electrical tape.

7. Connect the fusible link to the junction block or the starter solenoid and reconnect the battery ground cable.

GM "A" & "X" Body

Celebrity, Century, Cutlass
Ciera, 6000, Citation, Omega,
Phoenix, Skylark

YEAR IDENTIFICATION

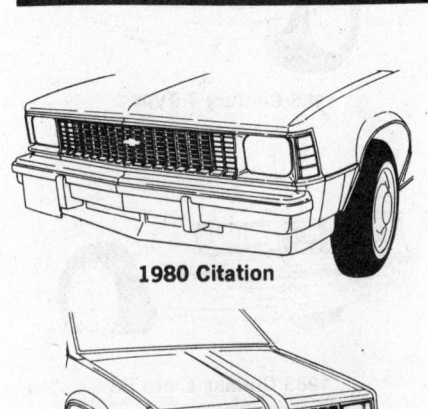

1980 Citation

1981–85 Citation

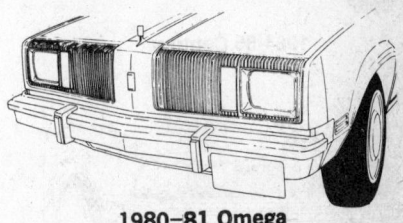

1980–81 Omega

1982-83 Omega

1984 Omega ES

1984 Omega Sedan

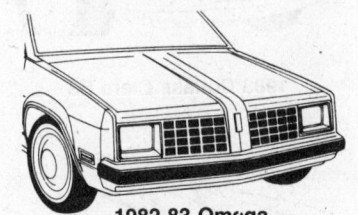

1980 Phoenix

1981 Phoenix

1982 Phoenix

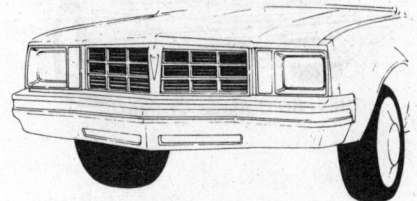

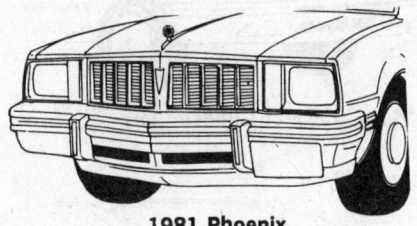

1983 Phoenix

1984 Phoenix LE, SE

1980 Skylark

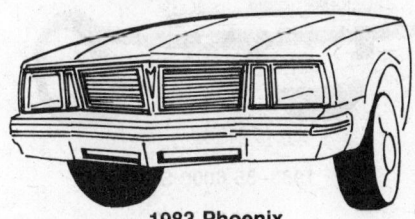

1981 Skylark

1984–85 Skylark T Type

1981–85 Skylark Sport Coupe

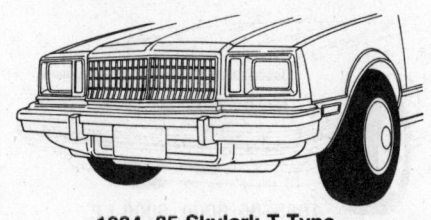

C501

YEAR IDENTIFICATION

1982-83 Celebrity

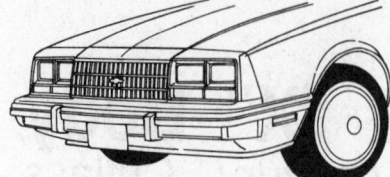

1984—85 Celebrity

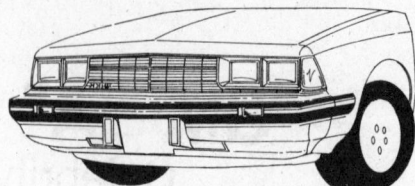

1982-83 Century

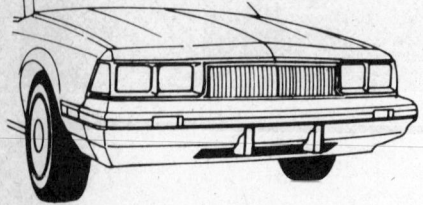

1984-85 Century

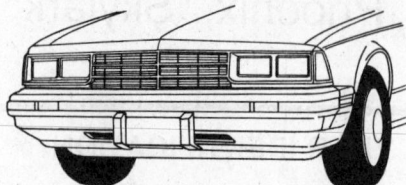

1984 Century T Type

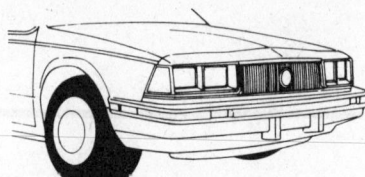

1985 Century T Type

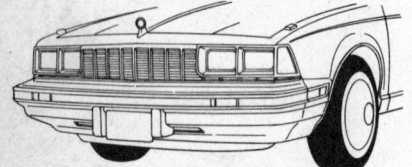

1984 Century Custom, Limited

1982-83 Cutlass Ciera

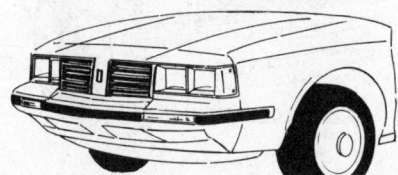

1983 Cutlass Ciera ES

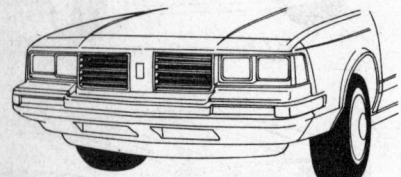

1984 Ciera

1985—86 Cutlass Ciera

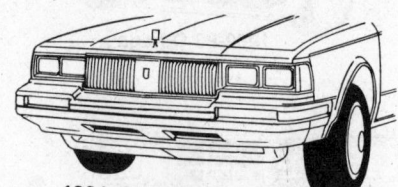

1984—86 Cutlass Cruiser

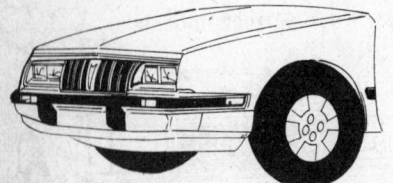

1982-83 6000

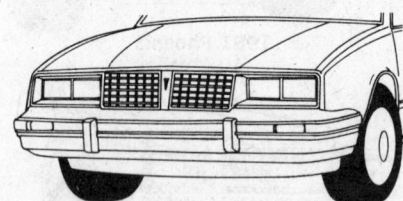

1984 6000, 6000 LE

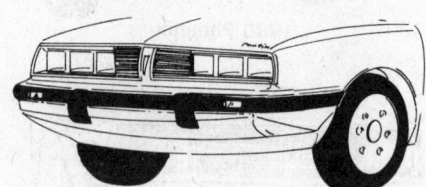

1983—85 6000 STE

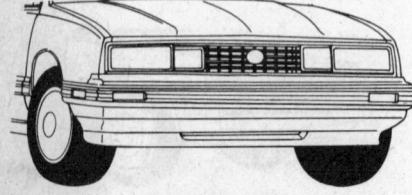

1986 6000 STE

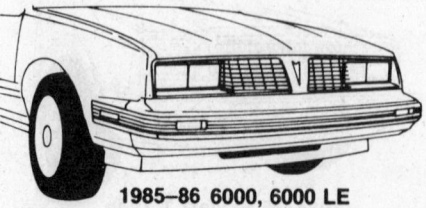

1985—86 6000, 6000 LE

1982—85 Citation

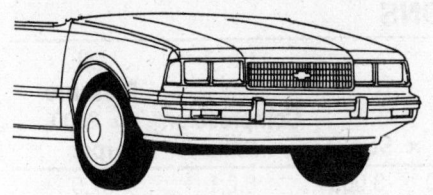

1986-87 Celebrity

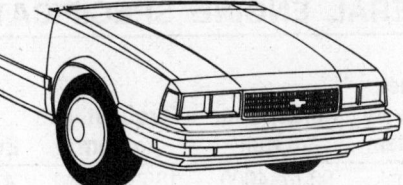

1985-87 Celebrity Eurosport

1986-87 Century

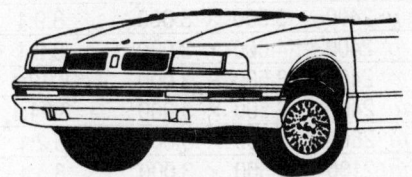

1987 Cutlass Ciera GT

1987 Pontiac 6000 SE

VEHICLE IDENTIFICATION NUMBER (VIN)

It is important for servicing and ordering parts to be certain of the vehicle and engine identification. The VIN (vehicle identification number) is a 13 or 17 digit number visible through the windshield on the driver's side of the dash and contains the vehicle and engine identification codes. It can be interpreted as follows:

Engine Code						Model Year Code	
Code	Cu. In.	Liters	Cyl.	Carb.	Eng. Mfg.	Code	Year
5	151	2.5	4	2	Pont.	A	1980
7	173	2.8	V6	2	Chev.		

The thirteen digit Vehicle Identification Number can be used to determine engine application and model year. The 6th digit indicates the model year, and the 5th digit identifies the factory installed engine.

VEHICLE IDENTIFICATION NUMBER (VIN)

It is important for servicing and ordering parts to be certain of the vehicle and engine identification. The VIN (vehicle identification number) is a 13 or 17 digit number visible through the windshield on the driver's side of the dash and contains the vehicle and engine identification codes. It can be interpreted as follows:

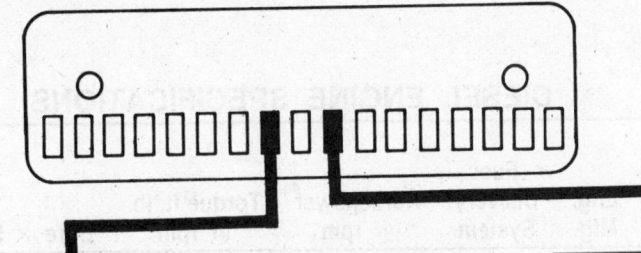

Engine Code						Model Year Code	
Code	Cu. In.	Liters	Cyl.	Carb.	Eng. Mfg.	Code	Year
5	151	2.5	4	2	Pont.	B	1981
R	151	2.5	4	TBI	Pont.	C	1982
X	173	2.8	V6	2	Chev.	D	1983
Z	173(HO)	2.8	V6	2	Chev.	E	1984
E	181	3.0	V6	2	Buick	F	1985
3	231	3.8	V6	MFI	Buick	G	1986
T	263	4.3	V6	Diesel	Olds	H	1987
W	173	2.8	V6	MFI	Chev.		
L	181	3.0	V6	MFI	Buick		
B	231	3.8	V6	SFI	Buick		

The seventeen digit Vehicle Identification Number can be used to determine engine application and model year. The 10th digit indicates the model year and the 8th digit identifies the factory installed engine.

TBI: Throttle Body Injection
MFI: Multi-Point Fuel Injection

GENERAL ENGINE SPECIFICATIONS

Year	VIN Code	Engine No. Cyl. Displ. (cu. in.)	Eng. Mfg.	Fuel Delivery System	Horsepower @ rpm	Torque ft. lb. @ rpm	Bore × Stroke	Compression Ratio	Oil Pressure @ 2000 rpm
1980–81	5	4-151	Pont.	2-bbl	90 @ 4000	135 @ 2400	4.000 × 3.000	8.2:1	37.5
	5	4-151 Calif.	Pont.	2-bbl	90 @ 4400	128 @ 2400	4.000 × 3.000	8.2:1	37.5
	X	6-173	Chev.	2-bbl	115 @ 4800	150 @ 2000	3.500 × 3.000	8.5:1	30–45
	X	6-173 Calif.	Chev.	2-bbl	110 @ 4800	140 @ 2000	3.500 × 3.000	8.5:1	30–45
	Z	6-173 HO	Chev.	2-bbl	135 @ 4800	165 @ 2400	3.500 × 3.000	8.9:1	30–45
1982	R	4-151	Pont.	TBI	90 @ 4000	134 @ 2400	4.000 × 3.000	8.2:1	37.5
	X	6-173	Chev.	2-bbl	112 @ 5100	148 @ 2400	3.500 × 3.000	8.42:1	30–45
	Z	6-173 HO	Chev.	2-bbl	135 @ 5400	142 @ 2400	3.500 × 3.000	8.94:1	30–45
1983	R	4-151	Pont.	TBI	92 @ 4000	134 @ 2800	4.000 × 3.000	8.2:1	37.5
	X	6-173	Chev.	2-bbl	112 @ 4800	145 @ 2100	3.500 × 3.000	8.5:1	50–65
	Z	6-173 HO	Chev.	2-bbl	135 @ 5400	145 @ 2400	3.500 × 3.000	8.9:1	①
	E	6-181	Buick	2-bbl	110 @ 4800	145 @ 2600	3.800 × 2.660	8.45:1	35–42
1984	R	4-151	Pont.	TBI	92 @ 4000	134 @ 2800	4.000 × 3.000	9.0:1	37.5
	X	6-173	Chev.	2-bbl	112 @ 4800	145 @ 2100	3.500 × 2.990	8.5:1	①
	Z	6-173 HO	Chev.	2-bbl	130 @ 5400	145 @ 2400	3.500 × 2.900	8.9:1	①
	E	6-181	Buick	2-bbl	110 @ 4800	145 @ 2600	3.800 × 2.660	8.45:1	35–42
1985	R	4-151	Pont.	TBI	92 @ 4400	134 @ 2800	4.000 × 3.000	9.0:1	37.5
	X	6-173	Chev.	2-bbl	112 @ 4800	145 @ 2100	3.500 × 2.990	8.5:1	50–65
	W	6-173	Chev.	MFI	130 @ 4800	155 @ 3600	3.500 × 2.990	8.9:1	50–65
	Z	6-173 HO	Chev.	MFI	125 @ 4500	165 @ 3600	3.500 × 2.990	8.9:1	50–65
	E	6-181	Buick	2-bbl	110 @ 4800	145 @ 2600	3.800 × 2.660	8.45:1	35–42
	3	6-231	Buick	MFI	125 @ 4400	195 @ 2000	3.800 × 3.400	8.0:1	35–42
1986–87	R	4-151	Pont.	TBI	92 @ 4400	134 @ 2800	4.000 × 3.000	9.0:1	37.5
	X	6-173	Chev.	2-bbl	112 @ 4800	145 @ 2100	3.500 × 2.990	8.0:1	①
	W	6-173	Chev.	MFI	125 @ 4800	160 @ 3600	3.500 × 2.990	8.5:1	①
	L,E	6-181	Buick	MFI	125 @ 4900	150 @ 2400	3.800 × 2.660	8.45:1	②
	B,3	6-231	Buick	SFI	125 @ 4400	195 @ 2000	3.800 × 3.400	8.0:1	②

① 50–65 psi @ 1200 rpm
② 37 psi @ 2400 rpm

DIESEL ENGINE SPECIFICATIONS

Year	VIN Code	Engine No. Cyl. (cu. in.)	Eng. Mfg.	Fuel Delivery System	Horsepower @ rpm	Torque ft. lb. @ rpm	Bore × Stroke	Compression Ratio	Oil Pressure @ 2000 rpm
1983–85	T	6-263	Olds.	Diesel	85 @ 3600	165 @ 1600	4.057 × 3.385	21.6:1	40–45

TUNE-UP SPECIFICATIONS

When analyzing compression test results, look for uniformity among cylinders rather than specific pressures.

Year	VIN Code	Eng. No. Cyl. Displ. (cu. in.)	Eng. Mfg.	hp	Spark Plugs Orig. Type	Spark Plugs Gap (in.)	Ignition Timing (deg) ▲ Man. Trans.	Ignition Timing (deg) ▲ Auto. Trans.	Intake Valve Opens (deg)■	Fuel Pump Pressure (psi)	Idle Speed (rpm) ▲ Man. Trans.	Idle Speed (rpm) ▲ Auto. Trans.
'80	5	4-151	Pont.	90	R-43TSX	0.060	10B③	10B	33	6.5–8.0	1000	650
	X	6-173	Chev.	110	R-44TS	0.045	2B④	6B⑤	25	6.0–7.5	750⑥	750⑥
'81	5	4-151	Pont.	90	R-44TSX	0.060	4B	4B	33	6.5–8.0	1000	675
	X	6-173	Chev.	110	R-43TS	0.045	6B	10B	25	6.0–7.5	850	850⑦
	Z	6-173 HO	Chev.	135	R-42TS	0.045	10B	10B	31	6.0–7.5	700	700

TUNE-UP SPECIFICATIONS

When analyzing compression test results, look for uniformity among cylinders rather than specific pressures.

Year	VIN Code	Eng. No. Cyl. Displ. (cu. in.)	Eng. Mfg.	hp	Spark Plugs		Ignition Timing (deg) ▲		Intake Valve Opens (deg)■	Fuel Pump Pressure (psi)	Idle Speed (rpm) ▲	
					Orig. Type	Gap (in.)	Man. Trans.	Auto. Trans.			Man. Trans.	Auto. Trans.
'82–'84	5,R	4-151	Pont.	90	R-44TSX	0.060	8B	8B	33	6.0–8.0	950①	750②
	X	6-173	Chev.	112	R-43CTS	0.045	10B	10B	25	6.0–7.5	800	600
	Z,W	6-173 HO	Chev.	135⑩	R-42CTS	0.045	6B	10B	31	6.0–7.5	850⑧	750
	E	6-181	Buick	110	R-44TS8	0.080	—	15B	16	6.0–8.0	—	see text
	3	6-231	Buick	125	R-44TS8	0.080	⑨	⑨	4.0–6.5	⑨	⑨	
	T	6-263	Olds.	85	—	—	—	6A	N.A.	5.8–8.7	—	650
'85–'86	5	4-151	Pont.	92	R-43TXS	0.045	⑨	⑨	33	6.0–7.0	⑨	⑨
	R	4-151	Pont.	92	R-43TXS	0.60	⑨	⑨	33	12.0	⑨	⑨
	X	6-173	Chev.	112	R-43CTS	.045	⑨	⑨	25	6.0–7.5	⑨	⑨
	Z	6-173 HO	Chev.	—	R-42CTS	.045	⑨	⑨	—	—	⑨	⑨
	W	6-173	Chev.	125	R-42CTS	.045	⑨	⑨	—	24.0–37.0	⑨	⑨
	B,3	6-231	Buick	125	R-44TS8	.080	⑨	⑨	—	—	⑨	⑨
	T	6-263	Diesel Olds	—	—	—	⑨	⑨	—	5.5–6.5	⑨	⑨
	L,E	6-181	Buick	110	R-44TS	.060	⑨	⑨	—	3.9–6.5	⑨	⑨
'87					See Underhood Specifications Sticker							

NOTE: The underhood specifications sticker often reflects tune-up specification changes made in production. Sticker figures must be used if they disagree with those in this chart.

▲ See text for procedure
■ All figures Before Top Dead Center
B: Before Top Dead Center
A: After Top Dead Center
Part numbers in this chart are not recommendations by Chilton for any product by brand name.
N.A.: Information not available
① Without air conditioning: 850
② Without air conditioning: 680

③ Calif.: 12B
④ Calif.: 6B
⑤ Calif.: 10B
⑥ Calif.: 700
⑦ With A/C: 900
⑧ Calif.: 750
⑨ See underhood specifications sticker
⑩ 130 on code W engine, 1982–84 only

FIRING ORDERS

NOTE: To avoid confusion, always replace sparkplug wires one at a time.

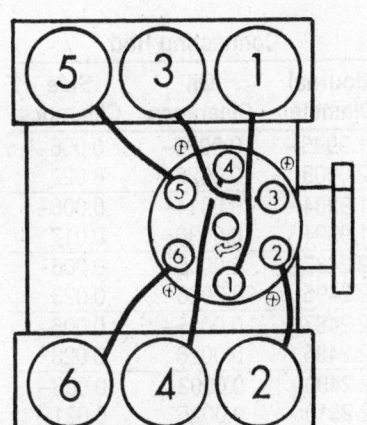

GM (Buick) 181 V6 (3.0L)
GM (Buick) 231 V6 (3.8 L)
Engine firing order: 1-6-5-4-3-2
Distributor rotation: clockwise

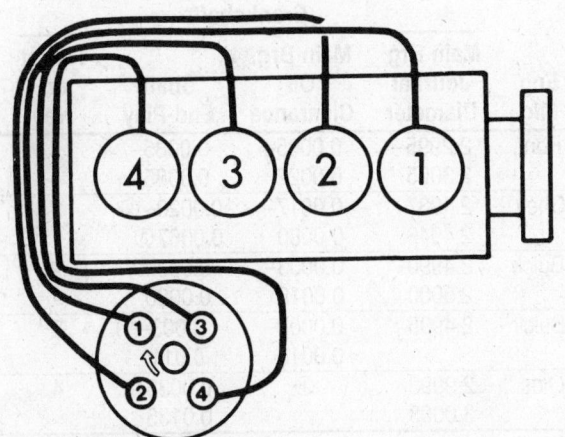

GM (Pontiac) 151-4
Engine firing order: 1-3-4-2
Distributor rotation: clockwise

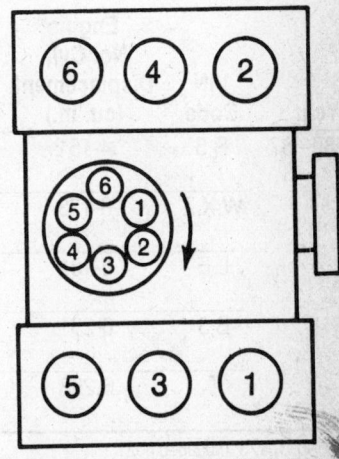

GM (Chevrolet) 173 V6 (2.8 L)
Engine firing order: 1-2-3-4-5-6
Distributor rotation: clockwise

CAPACITIES
A-Body

Year	VIN Code	Engine Displacement (cu. in.)	Eng. Mfg.	Crankcase (qts)		Transaxle Pints		Gas Tank (gals)	Cooling System (qts)	
				w/filter	wo/filter	Manual	Auto		w/heater	w/AC
'82–'85	R	4-151	Pont.	3.0	2.8	6.0	10.0	16.0	9.5	9.75
	X,W	6-173	Chev.	4.0	3.0	6.0	10.0	16.0	11.5	11.75
	E	6-181	Buick	4.0	3.0	—	10.0②	16.0	13.5	14.25
	3	6-231	Buick	4.0①	4.0	—	13.0	16.0	12.25	12.75
	T	6-263	Olds.	6.0	5.5	—	10.0②	16.0	13.25	13.75
'86–'87	R	4-151	Pont.	3.0	3.0	6.0	18	15.7	9.8	9.6
	X,W	6-173	Chev.	4.0	4.0	6.0	18	16.4	12.5	12.6
	L,E	6-181	Buick	4.0	4.0	6.0	18	16.0	14.4	14.0
	B,3	6-231	Buick	4.0	4.0	6.0	18	16.0	11.4	12.0
	T	6-263	Olds.	6.0	6.0	6.0	18	16.6	13.2	13.9

① Add as necessary to bring to appropriate level.
② 13.0 pts w/440T4 transaxle.

CAPACITIES
X-Body

Year	VIN Code	Engine No. Cyl. Displacement (cu. in.)	Engine Crankcase Add 1 qt For New Filter	Transmission (Pts-to-Refill After Draining)		Drive Axle (pts)	Gasoline Tank (gals)	Cooling System (qts)	
				Manual	Automatic			w/Heater	w/AC
'80–'81	5,R	4-151	3	5.9	10.5	①	14	8.3	8.6
	7	6-173	4	5.9	10.5	①	14	10.2	10.6
'82	5,R	4-151	3	5.9	10.5	①	14	8.3	8.6
	X	6-173	4	5.9	10.5	①	14	10.6	10.8
'83–'85	5,R	4-151	3	5.9	10.5	①	14.6	8.3	8.6
	X	6-173	4	5.9	10.5	①	15.1	10.6	10.8
	Z,W	6-173 HO	4	5.9	10.5	①	15.5	10.6	10.8

① Transaxle refill given with transmission capacity

CRANKSHAFT AND CONNECTING ROD SPECIFICATIONS
All measurements are given in inches

Year	VIN Code	Engine No. Cyl. Displacement (cu. in.)	Eng. Mfg.	Crankshaft				Connecting Rod		
				Main Brg. Journal Diameter	Main Brg. Oil Clearance	Shaft End-Play	Thrust on No.	Journal Diameter	Oil Clearance	Side Clearance
'80–'87	R,5	4-151	Pont.	2.2995–2.3005	0.0005–0.0022	0.0035–0.0085	5	1.9995–2.0005	0.0005–0.0026	0.006–0.022
	W,X,Z	6-173	Chev.	2.4937–2.4946	0.0017–0.0030	0.0020–②0.0067③	3	1.9984–1.9994	0.0014–0.0036	0.006–0.017
	L,E	6-181	Buick	2.4990–2.5000	0.0003–0.0018	0.0030–0.0090	2	2.2487–2.2495	0.0005–0.0026	0.006–0.023
	B,3	6-231	Buick	2.4995	0.0003–0.0018	0.003–0.011	2	2.2487–2.2495	0.0005–0.0026	0.006–0.023
	T	6-263	Olds.	2.9993–3.0003	①	0.0035–0.0135	4	2.2490–2.2510	0.0003–0.0025	0.008–0.021

① No. 1, 2, 3: 0.0005–0.0021
   No. 4: 0.0020–0.0034
② 1980: 0.0020–0.0079
③ 1986–87: 0.0020–0.0033

VALVE SPECIFICATIONS

Year	VIN Code	Engine No. Cyl. Displacement (cu. in.)	Eng. Mfg.	Seat Angle (deg)	Face Angle (deg)	Spring Test Pressure (lbs. @ In.)	Spring Installed Height (in.)	Stem-to-Guide Clearance (in.)		Stem Diameter (in.)	
								Intake	Exhaust	Intake	Exhaust
'80–'85	R,5	4-151	Pont.	46	45	176 @ 1.254	1.660	0.0010–0.0027	0.0010–0.0027	0.3418–0.3425	0.3418–0.3425
	W,X,Z	6-173	Chev.	46	45	155 @ 1.160	1.610	0.0010–0.0027	0.0010–0.0027	0.3410–0.3416	0.3410–0.3416
	E	6-181	Buick	45	45	220 @ 1.340	1.727	0.0015–0.0035	0.0015–0.0032	0.3401–0.3412	0.3402–0.3415
	3	6-231	Buick	45	45	220 @ 1.340	1.727	0.0015–0.0035	0.0015–0.0032	0.3401–0.3412	0.3405–0.3412
	T	6-263	Olds.	①	②	210 @ 1.220	1.670	0.0010–0.0027	0.0015–0.0032	0.3425–0.3432	0.3420–0.3427
'86–'87	R	4-151	Pont.	46	45	170–180 @ 1.260	1.690	—	—	0.3420–0.3430	0.3420–0.3430
	W,X	6-173	Chev.	46	45	155 @ 1.160	1.610	0.0260–0.0268	0.0260–0.0268	—	—
	L,E	6-181	Buick	45	45	220 @ 1.340	1.727	0.0015–0.0032	0.0015–0.0032	0.3405–0.3412	0.3405–0.3412
	B,3	6-231	Buick	45	45	220 @ 1.340	1.727	0.0015–0.0032	0.3405–0.3412	0.3405–0.3412	
	T	6-263	Olds.	①	②	210 @ 1.220	1.670	0.0010–0.0027	0.0015–0.0027	0.3425–0.3432	0.3420–0.3427

① Intake: 45
   Exhaust: 32
② Intake: 44
   Exhaust: 30

CAMSHAFT SPECIFICATIONS

All measurements are given in inches

Year	VIN Code	Engine	Eng. Mfg.	Journal Diameter					Bearing Clearance	Lobe Lift		Camshaft End Play
				1	2	3	4	5		Intake	Exhaust	
'80–'85	R,5	4-151	Pont.	1.869	1.869	1.869	—	—	0.0007–0.0027	0.398	0.398	0.0015–0.0050
	W,X,Z	6-173	Chev.	1.869	1.869	1.869	1.869	—	0.0010–0.0040	0.231	0.263	
	E	6-181	Buick	1.786	1.786	1.786	1.786	1.786	①	0.406	0.406	—
	3	6-231	Buick	1.786	1.786	1.786	1.786	1.786	①	N.A.	N.A.	—
	T	6-263	Olds.	②	2.205	2.185	2.165	—	0.0020–0.0059	N.A.	N.A.	0.0008–0.0228
'86–'87	R	4-151	Pont.	1.869	1.869	1.869	—	—	0.0007–0.0027	0.398	0.398	0.0015–0.0050
	W	6-173	Chev.	1.8678	1.8678	1.8678	1.8678	—	.001–.004	.2626	.2732	—
	X	6-173	Chev.	1.8678	1.8678	1.8678	—	—	.001–.004	.231	.2626	—
	L,E	6-181	Buick	1.786	1.786	1.786	1.786	①	.358	.384	—	
	B	6-231	Buick	1.786	1.786	1.786	1.786	①	.392	.392	—	
	T	6-263	Olds.	②	2.205	2.185	2.165	—	0.0020–0.0059	N.A.	N.A.	0.0008–0.0228
	3	6-231	Buick	1.786	1.786	1.786	1.786	①	.368	.384	—	

① No. 1: 0.0005–0.0025
   No. 2–5: 0.0005–0.0035
② No. 1 bearing is not borable, but must be
   replaced separately.
N.A. Not available.

PISTON AND RING SPECIFICATIONS

All measurements are given in inches.

Year	VIN Code	Engine Type/ Disp. (cu. in.)	Eng. Mfg.	Piston-to- Bore Clearance	Ring Gap			Ring Side Clearance		
					Top Compression	Bottom Compression	Oil Control	Top Compression	Bottom Compression	Oil Control
'80–'85	R,5	4-151	Pont.	0.0025–0.0033	0.010–① 0.022	0.020–② 0.027	0.015–0.055	0.0015–0.0030	0.0015–0.0030	snug
	W,X,Z	6-173	Chev.	0.0017–0.0027	0.0098–0.0197	0.0098–0.0197	0.020–③ 0.055	0.0012–④ 0.0028	0.0016–④ 0.0037	0.008 max.
	L,E	6-181	Buick	0.0008–0.0020	0.013–0.023	0.013–0.023	0.015–0.035	0.0030–0.0050	0.0030–0.0050	0.0035 max.
	B,3	6-231	Buick	0.0008–0.0020	0.010–0.020	0.010–0.020	0.015–0.055	0.0030–0.0050	0.0030–0.0050	0.0035 max.
	T	6-263	Olds.	0.0030–0.0040	0.015–0.025	0.015–0.025	0.015–0.055	0.0050–0.0070	0.0030–0.0070	0.001–0.005
'86–'87	R	4-151	Pont.	0.0014–0.0022	0.010–0.020	0.010–0.020	0.020–0.060–	0.002–0.003	0.001–0.003	0.015–0.055
	W	6-173	Chev.	0.001–0.002	0.0012–0.0027	0.0016–0.0037	0.020–0.055	0.0098–0.0197	0.0098–0.0197	0.020–0.055
	L,E	6-181	Buick	0.0008–0.0020	0.010–0.020	0.010–0.020	0.015–0.055	0.0030–0.0050	0.0030–0.0050	0.0035 max
	B,3	6-231	Buick	0.0008–0.0020	0.010–0.020	0.010–0.020	0.015–0.055	0.0030–0.0050	0.0030–0.0050	0.0035 max
	X	6-173	Chev.	0.0007–0.0017	0.0012–0.0027	0.0016–0.0037	0.020–0.055	0.0098–0.0197	0.0098–0.0197	0.020–0.055

① 1980: 0.015–0.025
② 1980: 0.009–0.019
③ 1980: 0.015–0.055
④ 1980: 0.012–0.032

TORQUE SPECIFICATIONS

All readings in ft. lbs.

Year	VIN Code	Engine No. Cyl. Displacement (cu. in.)	Eng. Mfg.	Cylinder Head Bolts	Rod Bearing Bolts	Main Bearing Bolts	Crankshaft Bolt	Flywheel-to- Crankshaft Bolts	Manifold	
									Intake	Exhaust
'80–'87	R,5	4-151	Pont.	85③	32	70	200	44④	29	44
	W,X,Z	6-173	Chev.	70⑤	37	68	75	50	23	25
	L,E	6-181	Buick	80	40–45	100	225	60	32	25–37
	B,3	6-231	Buick	80	40–45	100	225	60	32	25–37
	T	6-263	Olds.	①	42	107	255②	76	41	29

① All exc. No. 5, 6, 11, 12, 13, 14: 142
No. 5, 6, 11, 12, 13, 14: 59
② Range: 160–350 ft. lb.
③ 1980–81: 75
1984–87: 92
④ 1986–87: 55
⑤ 1986–87: 65–90

WHEEL ALIGNMENT SPECIFICATIONS

Year	Model	Caster*		Camber		Toe-In (in.)	Steering Axis (deg) Inclination
		Range (deg)	Pref. Setting (deg)	Range (deg)	Pref. Setting (deg)		
'80–'81	All	2N–2P	0	0–1P	½P	0–³⁄₁₆	14.5
'82–'87	All	0–4P	2P	½N–½P	0	¹³⁄₆₄–¹³⁄₆₄	14.5

* Caster is not adjustable

ENGINE ELECTRICAL

Alternator

Alternator and regulator troubleshooting are covered in the "Charging and Starting Systems" Unit Repair section.

REMOVAL, INSTALLATION & BELT TENSION ADJUSTMENT

1. Disconnect battery ground cable to prevent diode damage.
2. Disconnect and label the alternator wiring.
3. Remove the brace bolt. If power steering equipped, loosen pump brace and mount nuts. Detach drive belt(s).
4. On 4 cylinder engines, remove the upper bracket.
5. Support the alternator and remove mount bolt(s). Remove unit from vehicle.
6. Reverse procedure to install. Adjust drive belt to have $\frac{1}{4}$–$\frac{1}{2}$ in. play on longest run of belt.

Integral Voltage Regulator

An alternator with an integral voltage regulator is standard equipment. There are no adjustments possible with this unit; testing procedures will be found in the Charging and Starting Systems Unit Repair Section.

Starter

Starter motor troubleshooting and repairs are covered in the "Charging and Starting Systems" Unit Repair section.

REMOVAL & INSTALLATION

All Except Diesel

NOTE: On some models it may be necessary to move the fuel lines out of the way. Remove the fuel lines from the retaining clamp and loosen at the regulator. If equipped with fuel injection, relieve the pressure from the fuel system before disconnecting fuel lines. See "Electric Fuel Pump Removal" for procedure.

1. Disconnect battery ground cable.

2. Raise and support vehicle.
3. Disconnect all wires at solenoid terminals. Note color coding of wires for reinstallation.
4. Remove starter support bracket mount bolts (4 cylinder engines use two nuts; V6 engines use one nut). On engines with solenoid heat shield, remove front bracket upper bolt and detach bracket from starter motor.
5. Loosen the front bracket bolt or nut and rotate bracket clear. Lower and remove starter. Note the location of any shims so that they may be replaced in the same positions upon installation.
6. Reverse procedure to install.

Diesel

1. Disconnect the negative cable at the battery(s).
2. Raise and support the car on jackstands.
3. Remove the lower starter shield nut and bend the shield out of the way.
4. Disconnect the wires from the starter. It's a good idea to tag the wires.
5. Remove the front starter bolt. Loosen the rear starter bolt and remove the starter with the rear bolt remaining in the starter housing.
6. Installation is the reverse of removal.

TO DECREASE STARTER NOISE INSTALL SHIMS AS REQUIRED (0.38MM AT A TIME NOT TO EXCEED 1.14MM THICKNESS). RETORQUE FASTENERS.

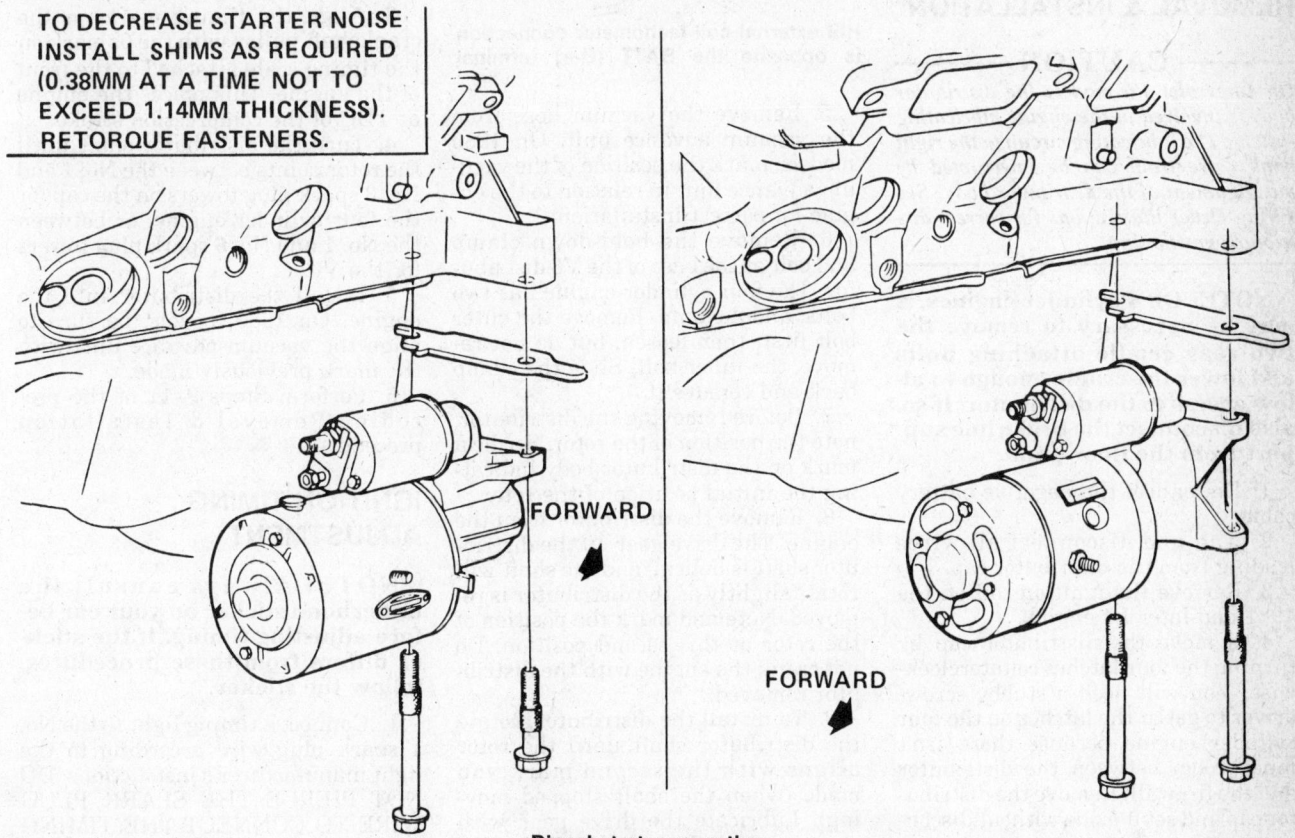

FORWARD

FORWARD

Diesel starter mounting

Distributor

All models are equipped with the HEI distributor and ignition system. When using an auxiliary starter switch on HEI systems, the distributor BATT lead must be disconnected. Failure to do this may cause damage to the grounding circuit in the ignition switch.

HEI SYSTEM TACHOMETER HOOKUP

On all 1980 models and all 1981 and later models with the V6 engine, there is a terminal on the distributor cap marked TACH. On all 1981 and later models with the L4 engine, there is a terminal on the ignition coil where the brown wire is connected. Connect one tachometer lead to this terminal and the other lead to a suitable ground. On some tachometers, the leads must be connected to the TACH terminal and then to the positive battery terminal.

Never ground the TACH terminal; serious module and ignition coil damage will result. If there is any doubt as to the correct tachometer hookup, check with the tachometer manufacturer.

REMOVAL & INSTALLATION

-------- CAUTION --------

On Chevrolet V6 models the distributor body is involved in the engine lubricating system. The lubricating circuit to the right bank valve train can be interrupted by misalignment of the distributor body. See Firing Order illustrations for correct distributor positioning.

NOTE: On 4 cylinder engines, it may be necessary to remove the two rear cradle attaching bolts and lower the cradle enough to allow access to the distributor. If so, also disconnect the brake line support from the floor pan.

1. Disconnect the negative battery cable.
2. Tag and disconnect all wires leading from the distributor cap.
3. Remove the ignition coil on the 1981 and later L4 engine.
4. Remove the distributor cap by turning the four latches counterclockwise. You will need a stubby screwdriver to get at the latches on the four cylinder engine because there isn't much room between the distributor and the firewall. Remove the distributor cap and set it aside without disconnecting any of the wires.

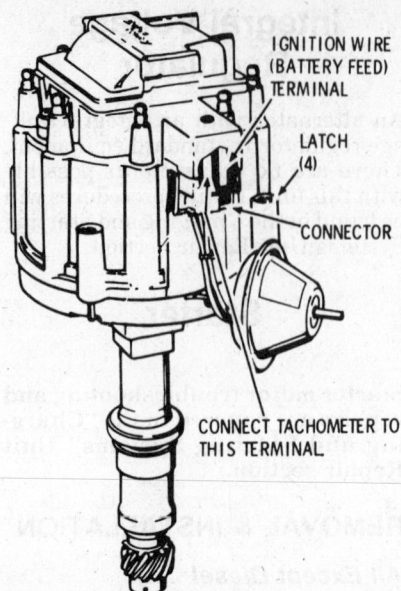

IGNITION WIRE (BATTERY FEED) TERMINAL

LATCH (4)

CONNECTOR

CONNECT TACHOMETER TO THIS TERMINAL

HEI coil-in-cap distributor tachometer hookup

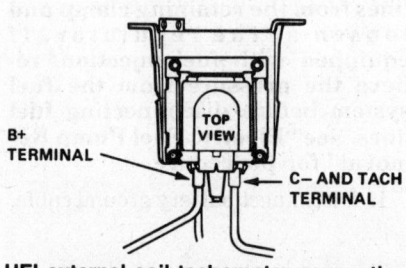

B+ TERMINAL

TOP VIEW

C– AND TACH TERMINAL

HEI external coil tachometer connection is opposite the BATT (B+) terminal

5. Remove the vacuum hose from the vacuum advance unit. On 1980 models, mark the position of the vacuum advance unit in relation to the engine for correct installation.
6. Remove the hold-down clamp and bolt at the base of the V6 distributor. The four cylinder engine has two bolts, and a clamp. Remove the outer bolt first, then loosen, but do not remove, the inner bolt. Slide the clamp back and remove it.
7. Before removing the distributor, note the position of the rotor. Scribe a mark on the distributor body indicating the initial position of the rotor.
8. Remove the distributor from the engine. The drive gear on the distributor shaft is helical, and the shaft will rotate slightly as the distributor is removed. Note and mark the position of the rotor at this second position. Do not crank the engine with the distributor removed.
9. To install the distributor, rotate the distributor shaft until the rotor aligns with the second mark you made (when the shaft stopped moving). Lubricate the drive gear with clean engine oil, then install the dis-

tributor into the engine. On 1980 engines, align the vacuum advance unit with the mark previously made. As the distributor is installed, the rotor should move to the mark you made first, indicating rotor position before the distributor was removed. This will ensure proper timing. If the marks do not align properly, remove the distributor and try again.

10. Install the clamp and hold-down bolt. Tighten them until the distributor can just be moved with a little effort.
11. Connect the ignition wire and tachometer wire, and install the distributor cap. Plug the vacuum advance hose (if so equipped). Set the ignition timing. Connect the vacuum hose.

INSTALLATION IF THE ENGINE WAS DISTURBED

If the engine was cranked while the distributor was removed, you will have to place the engine on TDC of the compression stroke to obtain proper ignition timing.

1. Remove the No. 1 spark plug.
2. Place your thumb over the spark plug hole. Crank the engine slowly until compression is felt. It will be easier if you have someone rotate the engine by hand, using a wrench on the crankshaft pulley.
3. Align the timing mark on the crankshaft pulley with the 0° mark on the timing scale attached to the front of the engine. This places the engine at TDC of the compression stroke.
4. Turn the distributor shaft until the rotor points between the No. 1 and No. 3 spark plug towers on the cap for the four cylinder engine, or between the No. 1 and No. 6 spark plug towers for the V6.
5. Install the distributor into the engine. On 1980 models, be sure to align the vacuum advance unit with the mark previously made.
6. Perform Steps 9–11 of the preceding Removal & Installation procedure.

IGNITION TIMING ADJUSTMENT

NOTE: Always consult the underhood sticker on your car before adjusting timing. If the sticker differs from these procedures, follow the sticker.

1. Connect a timing light to the No. 1 spark plug wire according to the light manufacturer's instructions. DO NOT PIERCE THE SPARK PLUG WIRE TO CONNECT THE TIMING LIGHT.

2. Disconnect the distributor spark advance hose (if equipped) and plug the vacuum opening.

3. On models with Electronic Spark Timing (EST) distributor, disconnect the 4 terminal plug at the distributor. The EST distributor uses no mechanical or vacuum advance and is easily identified by the absence of a vacuum advance and the presence of a four terminal connector.

4. Start the engine and run it at idle speed.

5. Aim the timing light at the degree scale just over the harmonic balancer.

6. Adjust the timing by loosening the securing clamp and rotating the distributor until the desired ignition advance is achieved, then tighten the clamp.

7. On the four cylinder, loosen the distributor clamp outer bolt, then slide the clamp back slightly. Do not remove the retaining bolt.

8. Adjust the timing, then replace and tighten the clamp. To advance the timing, rotate the distributor opposite the normal direction of rotor rotation. Retard the timing by rotating the distributor in the normal direction of rotor rotation.

Diesel Glow Plugs

NOTE: A burned out Fast Glow glow plug may bulge then break off and drop into the pre-chamber when the glow plug is removed. When this occurs the cylinder head must be removed and the pre-chamber removed from the head to remove the broken tip. When installing a glow plug, apply lubricant 1052771 or equavalent to the threads only when the engine is equipped with aluminum cylinder heads.

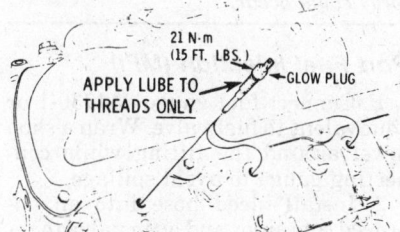

21 N·m
(15 FT. LBS.)
APPLY LUBE TO THREADS ONLY
GLOW PLUG

Diesel glow plug installation

—— CAUTION ——

It is important that the pre-chamber be fully installed flush to the surface of the cylinder head. If this is not done, cylinder head gasket or piston damage can occur.

NOTE: When servicing the right

rear glow plugs, it may be necessary to follow the procedures below.

REMOVAL AND INSTALLATION

1. Remove the engine support strut (see above figure).

2. Rotate the intermediate steering shaft so that the steering gear stub shaft clamp bolt is in the up position and remove the clamp bolt. Then disconnect the intermediate shaft from the stub shaft.

—— CAUTION ——

Failure to disconnect the intermediate shaft from the rack and pinion stub shaft can result in damage to the steering gear and/or intermediate shaft. This damage can cause loss of steering control which could result in a vehicle crash with possible bodily injury.

3. Place a floor jack under the front crossmember of the cradle and raise the jack until the jack just starts to raise the car.

4. Remove the front two body mount bolts with the lower cushions and retainers. Then remove the cushions from the bolts.

The removal of any one body mount requires the loosening of the adjacent body mounts to permit the cradle to seperate from the body. Take care to prevent breaking the plastic fan shroud, or damaging frame attachments such as steering hoses and brake pipes, during replacement of body mounts.

When installing a body mount, take care to ensure that the body is seated properly in the frame mounting hole; otherwise, direct metal to metal contact will result between the frame and the body. The tube spacer should be in all bolt-in-body mounts. The insulator and metal washer should be positioned to prevent contact with the frame rail. Do not overtighten the body mount; a collasped tube spacer or stripped bolt may result.

Proper clamping by the mount depends on clean dry surfaces. If the body mount bolt doesn't screw in smoothly, it may be necessary to run a tap through the cage nut in the body to remove foreign material. Take care to ensure that the tap does not punch through the underbody.

Whenever the body is going to be moved in relation to the cradle, the intermediate shaft should be disconnected from the rack and pinion steering gear stub shaft.

—— CAUTION ——

Failure to disconnect the intermediate

shaft from the rack and pinion steering gear stub shaft can result in damage to the steering gear and/or intermediate shaft. This damage can cause loss of steering control which could result in a vehicle crash with possible bodily injury.

5. Thread the body mount bolts with retainers a minimum of three turns into the cage nuts so that the bolts retain cradle movement.

6. Release the floor jack slowly until the crossmember contacts the body mount bolt retainers. As the jack is being lowered, watch and correct any interference with hoses, lines, pipes and cables.

—— CAUTION ——

Do not place your hands between the crossmember and body mount to remove objects or correct interference while the jack is lowering.

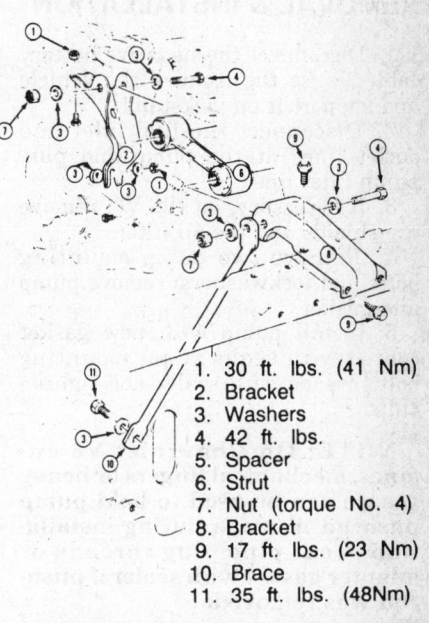

1. 30 ft. lbs. (41 Nm)
2. Bracket
3. Washers
4. 42 ft. lbs.
5. Spacer
6. Strut
7. Nut (torque No. 4)
8. Bracket
9. 17 ft. lbs. (23 Nm)
10. Brace
11. 35 ft. lbs. (48Nm)

Engine mounting strut and bracket

NOTE: Do not lower the cradle without it being restrained as possible damage can occur to the body and underhood items.

7. Reverse the procedure for installation. Torque the intermediate steering shaft clamp bolt to 46 ft. lbs. and the body mount bolts to 77 ft. lbs.

TESTING

For complete diagnosis, testing and repair procedures of the diesel glow plug system, please refer to "Diesel Maintenance" in the Unit Repair section.

GASOLINE FUEL SYSTEM

The fuel pump on the 1980–81 4–151 and on the V6 is the single action AC diaphragm type. The pump is actuated by an eccentric located on the engine camshaft. On the V6 a pushrod between the camshaft eccentric and the fuel pump actuates the pump rocker arm. 1982 and later 4 cylinder engines with the TBI (Throttle Body Injection) system use an electric fuel pump located in the fuel tank. The pump is activated by signals from the ECM (Electronic Control Module) through fuel pump relay.

Mechanical Fuel Pump

REMOVAL & INSTALLATION

1. Disconnect the negative battery cable. Raise the front of the vehicle and support it on jackstands.
2. Disconnect the fuel inlet and outlet lines at the pump and plug pump inlet line.
3. If necessary, on the V6, remove the shields and the oil filter.
4. Remove two pump mounting bolts and lockwashers; remove pump and gasket.
5. Install pump with new gasket coated with sealer. Coat mounting bolt threads with sealer and tighten bolts.

NOTE: On Chevrolet V6 engines, mechanical fingers or heavy grease can be used to hold pump pushrod in place during installation. Coat pipe plug threads or adapter gasket with sealer if pushrod was removed.

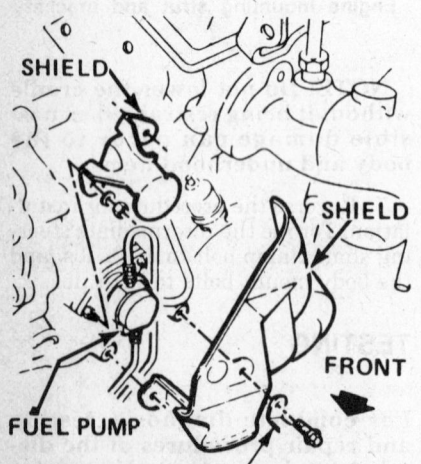

173 V6 fuel pump installation details

6. Install the shields and oil filter on the V6, if removed.
7. Connect inlet and outlet lines, start engine and check for leaks.

Electric Fuel Pump

REMOVAL & INSTALLATION

— CAUTION —

Before opening any part of the fuel system, the pressure must be relieved. Follow the procedure below to relieve the pressure:

1. Remove the fuel pump fuse from the fuse panel.
2. Start the engine and let it run until all fuel in the line is used.
3. Crank the starter an additional three seconds to relieve any residual pressure.
4. With the ignition OFF, replace the fuse.
5. Drain the fuel tank.
6. Disconnect wiring from the tank.
7. Remove the ground wire retaining screw from under the body.
8. Disconnect all hoses from the tank.
9. Support the tank on a jack and remove the retaining strap nuts.
10. Lower the tank and remove it.
11. Remove the fuel gauge/pump retaining ring using a spanner wrench such as tool J-24187.
12. Remove the gauge unit and the pump.
13. Installation is the revere of removal. Always replace the O-ring under the gauge/pump retaining ring.

Fuel Filter

REMOVAL & INSTALLATION

Carbureted Engines

1. An in-carburetor filter is used. With the engine cold, disconnect the fuel inlet line at the carburetor. Hold the large nut with a wrench while turning the smaller nut.
2. Remove the large nut from the carburetor inlet.
3. Remove the filter and spring from the carburetor.
4. New filters usually come with a new inlet nut gasket. If not, obtain one. Never reuse the old gasket. Install the new filter, gasket and nut. Do not overtighten the nut.
5. Install the fuel line. Use two wrenches to avoid distorting the fuel pipe.

Fuel Injected Models

Before servicing any part of the fuel

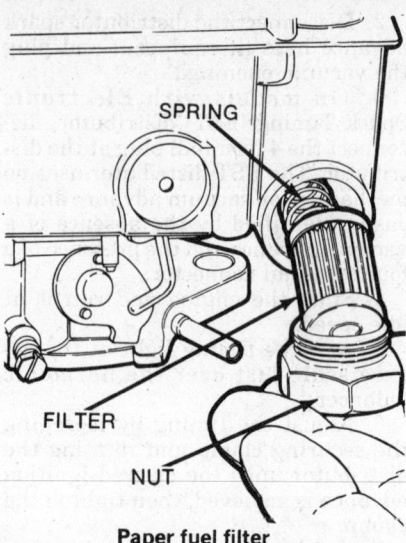

Paper fuel filter

system, it is necessary to releave the fuel system pressure. This will reduce the risk of fire or personal injury. The system contains an orafice, in the pressure regulator, which allows the fuel pressure to bleed off after the engine has been shut off. To relieve the fuel pressure:

Throttle Body Injection (TBI)

1. Remove the fuse marked "Fuel Pump" from the fuse block in the passenger compartment.
2. Crank the engine, engine will start and run until the fuel supply remaining in the fuel lines is exhausted. When the engine stops, engage the starter again for 3.0 seconds to assure dissapation of any remaining pressure.
3. With the ignition OFF, replace the "Fuel pump" fuse.

— CAUTION —

Unless this procedure is followed before servicing fuel lines or connections, fuel spray could occur.

Port Fuel Injection (MFI)

1. Connect fuel gauge J34730-1 or equavalent to fuel valve. Wrap a shop towel around the fitting while connecting gauge to avoid spillage.
2. Install bleed hose into an approved container and open valve to relieve system pressure.

FILTER REMOVAL

The filter is an inline unit ahead of the TBI unit. To remove the filter, make sure the engine is cold, unclamp and remove the fuel hose, then unscrew the filter from the steel fuel line. Installation is the reverse of removal.

Carburetor

REMOVAL & INSTALLATION

All Models

1. Disconnect the battery.
2. Remove the air cleaner.
3. Disconnect the accelerator linkage.
4. Disconnect the transmission detent cable.
5. If equipped, disconnect the cruise control cable.
6. Disconnect all electrical connectors at the carburetor, or which might interfere with carburetor removal. Tag the wires for installation.
7. Disconnect and tag all vacuum lines at the carburetor.
8. Disconnect the fuel line at the carburetor inlet.
9. Installation is the reverse of removal. Torque the long bolts to 7 ft. lb.; the short bolts to 11 ft. lb.

IDLE SPEED AND MIXTURE ADJUSTMENT

Carbureted Models

1980

1. Run the engine to normal operating temperature.
2. Make sure that the choke is fully opened, turn the A/C Off, set the parking brake, block the drive wheels and connect a tachometer to the engine according to the manufacturer's instructions.
3. Disconnect and plug the vacuum hoses at the EGR valve and the vapor canister.
4. Place the transmission in Park (AT) or Neutral (MT).
5. Disconnect and plug the vacuum advance hose at the distributor. Check and adjust the timing.
6. Connect the distributor vacuum line.
7. Manual transmission cars with A/C and without solenoid; place the idle speed screw on the low step of the fast idle cam and turn the screw to achieve the specified idle speed.
Cars with A/C: set the idle speed screw to the specified rpm. Disconnect the compressor clutch A/C On. Open the throttle momentarily to extend the solenoid plunger. Turn the solenoid screw to obtain the specified rpm.
Automatic transmission cars without A/C; manual transmission cars without A/C, solenoid-equipped carburetor: momentarily open the throttle to extend the solenoid plunger. Turn the solenoid screw to obtain the specified rpm. Disconnect the solenoid wire and turn the idle speed screw to obtain the slow engine idle speed.

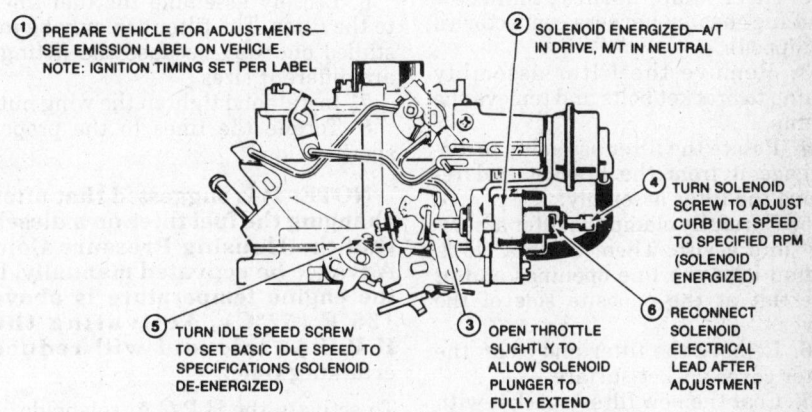

1980 idle speed adjustment w A/C

1980 Idle speed adjustment—w/o A/C

1981–87

Mixture adjustments are a function of the Computer Command Control (CCC) system. The idle speed on models equipped with an Idle Speed Control (ISC) motor is also automatically adjusted by the Computer Command Control System, making manual adjustment unnecessary. The underhood specifications sticker will indicate ISC motor use.

On non-A/C models not equipped with ISC, the idle speed is adjusted at the idle speed screw on the carburetor. Before adjusting, check the underhood sticker for any preparations required. On A/C equipped models which do not have an ISC motor, an idle speed solenoid similar to the ones on earlier models is used. This solenoid is adjusted at the solenoid screw, using the same procedures as on earlier models. Consult the underhood specifications sticker for special instructions.

FUEL INJECTED MODELS

No idle speed or mixture adjustments are possible on 1982 and later fuel injected engines.

Fuel Injection

Due to the complex nature of modern fuel injection systems, comprehensive diagnosis and testing procedures fall outside the confines of this repair manual. For complete information on diagnosis, testing and repair procedures concerning all modern fuel injection systems, please refer to Chiltons Guide To Fuel Injection and Feedback Carburetors.

DIESEL FUEL SYSTEM

Fuel Filter

REMOVAL AND INSTALLATION

Models with NB9

NB9 is an assembly which performs several functions:
   a. Filtration.
   b. Water seperation.
   c. Water detection.
   d. Water drainage and fuel heating (fuel heating is optional).

To remove the fuel filter:

1. Disconnect the fuel lines from the inlet and outlet ports.
2. Disconnect the drain hose, drain fuel filter lamp, harness connector and fuel heater harness connector (if equipped).
3. Remove the filter assembly clamp to bracket bolts and remove the clamp.
4. Rotate the filter assembly to disengage it from the bracket and remove the filter assembly.
5. Carefully clamp the filter assembly into a vise. Then, using a cloth, clamp it at the line openings and at the flat at the opposite side of the cover.
6. Remove the filter and clean the filter cover gasket surface.
7. Coat the new filter's gasket with engine oil or diesel fuel.
8. Install the filter on the cover and tighten the filter ⅔ turn beyond intial gasket contact.
9. Loosely assemble the filter assembly in the bracket. Engage the bracket lock tab into the bracket. Using new "O" rings, loosely assemble the fuel lines to the filter assembly.
10. Install the clamps and tighten the bolts to specification.
11. Install the drain hose and torque the clamp.
12. Torque the fuel lines to specification.
13. Connect the harness connectors to the drain fuel filter lamp module and fuel heater.
14. Carefully expand the clamps that retain the fuel hoses to the in-line filter (sight glass).
15. Disconnect the hoses and remove the in-line filter.
16. Install a new in-line filter on the hoses with the arrow pointing to the engine. Carefully expand the clamps and position them in the orioginal location.

17. Loosen the air bleed screw.
18. Turn the ignition switch to the "RUN" position. This will energize the fuel pump.
19. Close the air bleed screw when fuel is present at the air bleed ports. Clean any spilled fuel.
20. Start the engine and inspect for leaks.

BLEEDING FUEL SYSTEM

Refer to the last three steps of the above procedure.

Models Without NB9

1. Remove the air cleaner and install J-26996-1 air crossover cover.
2. Place a rag under the fuel filter to catch any fuel that will drop out.
3. Disconnect the fuel lines from the filter.
4. Remove the wing nut and remove the filter assembly.
5. Place a new filter into position.
6. Loosely assemble the fuel lines to the filter. The filter can only be installed one way because the fittings are different sizes.
7. Install and tighten the wing nut.
8. Torque the lines to the proper specification.

NOTE: It is suggested that after changing the fuel filter on a diesel, that the Housing Pressure Cold Advance be activated manually, if the engine temperature is above 125°F (52°C). Activating the H.P.C.A. solenoid will reduce cranking time.

To activate the H.P.C.A. solenoid:
   a. Disconnect the two lead connector at the engine temperature switch.
   b. Bridge the connector with a jumper and start the engine.
   c. After the engine is running, remove the jumper and reconnect the connector to the engine temperature switch.

9. Start the engine and check for leaks in the fuel lines and fittings.

Injection Timing

CHECKING AND/OR ADJUSTING TIMING USING J-33075 TIMING METER

The timing meter picks up the engine speed and crankshaft position from the crankshaft balancer. It uses a luminosity signal through a glow plug probe to determine combustion timing. Certain engine malfunctions may cause incorrect timing readings. Engine malfunctions should be corrected before timing adjustment is made.

The marks on the pump and adapter flange will normally be aligned within .050 in. (1.27mm).

NOTE: Alignment of timing marks may be used in emergency situations (i.e. timing meter not available). However for optimum engine operation, the timing should be adjusted with the timing meter as soon as possible.

1. Place the transmission selector lever in Park, apply the parking brake and block the drive wheels.
2. Start the engine and let it run at idle until fully warmed up. Then shut off the engine.

NOTE: Failure to have the engine fully warmed up will result in incorrect timing reading and adjustments.

3. Remove the air cleaner assembly and install cover J-26996-1. The EGR valve hose must be disconnected.
4. Clean any dirt from the engine probe holder (rpm counter) and crankshaft balancer rim.
5. Clean the lens on both ends of the glow plug probe and clean the lens in the photo-electric pick-up. Use a dulled tooth pick to scrape the carbon from the combustion chamber side of the glow plug probe. Look through the probe to be sure it's clean. Retarded readings will result if the probe is not clean.
6. Install the rpm probe into the crankshaft rpm counter (probe holder).
7. Remove the glow plug from No. 1 cylinder. Install the glow plug from No. 1 cylinder. Install the glow plug probe in the glow opening. Torque the probe to 9 ft. lbs.
8. Set the timing meter offset selector to A (20).
9. Connect the battery leads; red to positive, black to negative.
10. Start the engine and adjust the engine rpm to the speed specified on the underhood emission control sticker.
11. Observe the timing reading, then at two minute intervals, again observe the reading. When the readings stabilize over the two minute interval, compare that reading to the one specified on the underhood sticker. The timing reading when set to specification will be ATDC (after top dead center).
12. Disconnect the timing meter.
13. Lubricate only the threads of the removed glow plug with lubricant 9985462 or equivalent.

NOTE: Failure to apply the correct lubricant can cause engine damage.

14. Install the removed glow plug. Torque the glow plug to 15 ft. lbs.

15. Install the air cleaner being certain to reconnect the EGR valve hose.

ADJUSTMENT

1. Shut off the engine.

2. Note the relative position of the marks on the pump flange and pump intermediate adapter.

3. Loosen the bolts holding the pump to the adapter to a point where the pump can be rotated. Use a 1 in. open end wrench. Tool J-25304 has the proper offset on the handle to clear the fuel return line.

4. Rotate the pump to the left to advance the timing and to the right to retard the timing. The width of the mark on the intermediate adapter is about $\frac{2}{3}$ degree. Move the pump the amount that is needed and tighten the pump retaining bolts to 35 ft. lbs.

5. Start the engine and recheck the timing reading as outlined previously. Reset and recheck the timing if needed.

6. Reset the idle speed. Please note the following:

a. Sooty or dirty probes will result in retarded readings.

b. The luminosity probe will soot up very fast when used in a cold engine.

c. Wild needle fluctuations on the timing meter indicate a cylinder not firing properly. Correction of this condition must be made prior to adjusting the timing.

Idle Speed

ADJUSTMENT

1. Apply the parking brake, place the transmission selector lever in Park and block the drive wheels.

2. Start engine and allow it to run until warm, usually 10–15 minutes.

3. Shut off the engine, remove the air cleaner assembly.

4. Clean the front cover rpm counter (probe holder) and the crankshaft balancer rim.

5. Install the magnetic pick-up probe of tool J-26925 fully into the rpm counter. Connect the battery leads; red to positive (+) and black to negative (−).

6. Disconnect the two-lead connector at the generator.

7. Turn off all electrical accessories.

8. Allow no one to touch either the steering wheel or service brake pedal.

9. Start the engine and place the transmission selector lever in Drive.

10. Check the slow idle speed read-

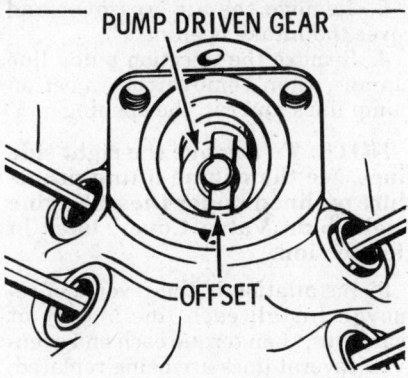

PUMP DRIVEN GEAR

OFFSET

Injection pump offset

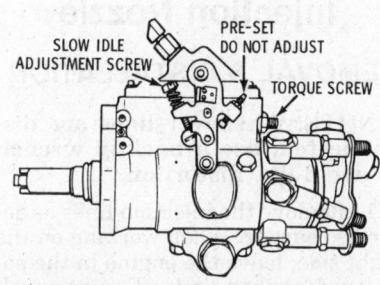

SLOW IDLE ADJUSTMENT SCREW — PRE-SET DO NOT ADJUST — TORQUE SCREW

Idle speed adjustment points, CAV pump shown

ing against the one given on the underhood emission control sticker. Reset if required.

11. Unplug the connector from the fast idle cold advance (engine temp.) switch and install a jumper between the connector terminals. Do not allow the jumper to touch ground.

12. Check the fast idle solenoid speed against the one given on the underhood sticker and reset if required.

13. Remove the jumper and reconnect the connector to the temperature switch.

14. Recheck and reset the slow idle speed if necessary.

15. Shut off the engine.

16. Reconnect the lead at the generator.

17. Disconnect and remove the tachometer.

18. If equipped with cruise control adjust the servo throttle cable to minimum slack then install the clip on the servo stud.

Injection Pump

REMOVAL

1. Remove the air cleaner assembly.

2. Remove the crankcase ventilation filter and pipes from the valve cover and air crossover.

3. Remove the air crossover and in-

stall intake manifold screened covers J-29657. Remove the fuel lines, filter and fuel pump as an assembly.

4. Disconnect the throttle cable and T.V. cable from the pump throttle lever. Disconnect the throttle return spring.

5. Remove the throttle and T.V./ Detent cables from the intake manifold brackets. Position the cables away from the engine.

6. Disconnect the fuel return line from the injection pump.

7. Disconnect the injection line clamps that are closest to the pump.

8. Disconnect the injection lines from the pump and cap all openings. Carefully reposition the lines to gain enough clearance for pump removal.

9. Remove the two bolts retaining the injection pump.

10. Remove the pump and discard the pump to adapter O-ring.

INSTALLATION

1. Position engine No. 1 cylinder to firing position by aligning the mark on the balancer with zero mark on the indicator located on the front of the engine (the index is offset to the right when No. 1 cylinder is at TDC).

2. Line up offset tang on pump driveshaft with the pump driven gear. Install a new pump to adapter O-ring, then install the pump fully, seating the pump by hand.

3. If a new or intermediate adapter is installed, set the injection pump at the center of the slots in the pump mounting flange. If the original intermediate adapter is being retained, align the pump timing mark with the mark on the intermediate adapter. Install the two bolts and washers retaining the pump and torque to 35 ft. lbs.

4. Remove the caps from the openings and connect the injection lines to the pump. Install the disconnected injection line clamps.

5. Connect the fuel return line.

6. Install the throttle and T.V. cables into the intake manifold bracket.

7. Connect the throttle cable and T.V. cable to the pump throttle lever. Connect the throttle return spring. Adjust the T.V. cable. See the "Automatic Transmission" Unit Repair section for adjustment procedures.

8. Install all remaining fuel lines and fuel filter.

9. Start the engine and check for leaks.

10. Check and if necessary reset the pump timing, see "Checking and/or Adjusting Pump Timing."

11. Adjust the vacuum regulator valve. See the Unit Repair Section.

12. Adjust the idle speeds.

13. Remove the screened covers

from intake manifold, then install the air crossover.

14. Install tubes and hoses in the air crossover and ventilation filters in the valve cover.

15. Install the air cleaner being certain to reconnect the EGR valve hose.

Electric Fuel Transfer Pump

The pump is located at the front of the engine next to the fuel heater.

REMOVAL & INSTALLATION

1. Place the ignition in the OFF position.
2. Remove the air cleaner.
3. Disconnect the pump electrical supply wire.
4. Using a $\frac{1}{4}$ in. wrench on the inlet fitting, unscrew the fuel pipe inlet tube.
5. Using the same method, unscrew the outlet pipe.
6. Unbolt the pump mounting bracket and the pump.
7. Installation is the reverse of removal. Using the two wrench method, torque the fuel lines to 19 ft. lbs. The pump bracket is torqued to 18 ft. lbs. In some cases, it may be necessary to adjust the pump location slightly to get a good alignment on the fuel lines. When the pump is installed, disconnect the fuel line at the filter, and run the pump with the key ON to bleed the lines.

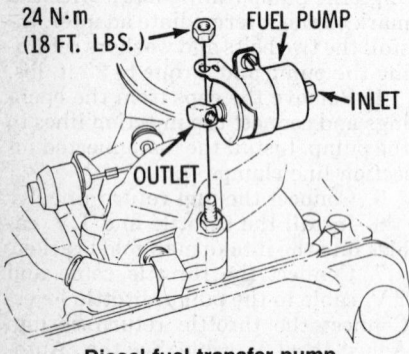

Diesel fuel transfer pump

Fuel Injection Pump Lines

REMOVAL & INSTALLATION

1. Remove the air cleaner.
2. Remove the filters and pipes from the valve covers and air crossover.

3. Remove the air crossover and cover the intake opening.
4. Remove the injection pump line clamps, then remove the injection pump lines and cap the opening.

NOTE: To remove the right side lines, see the engine lifting procedure outlined under Diesel Engine Right Side Valve Cover, later in this section.

5. Installation is the reverse of removal. Install each line loosely at both ends, then torque each end evenly. If several lines are being replaced, start with the lower lines.

Injection Nozzle

REMOVAL & INSTALLATION

NOTE: Whenever lines are disconnected, use a backup wrench to avoid line distortion.

1. Remove the injection lines as described earlier. When working on the right side, leave the engine in the position described under line removal, for injection removal.
2. Remove the nozzle by applying the wrench to the larger hex on the nozzle.
3. Remove the copper gasket from the head if it did not remain on the nozzle.
4. Installation is the revere of removal. Use lubricant 9985462 or its equivalent on the nozzle threads. Always use a new copper gasket on each nozzle. Torque the nozzle and the lines to 25 ft. lbs. each.

ENGINE COOLING

Radiator

REMOVAL & INSTALLATION

1. Disconnect the negative battery cable.
2. Drain the cooling system.
3. Remove the forward strut bracket for the engine at the radiator. Loosen the bolt to prevent shearing the rubber bushing, then swing the strut rearward.
4. Disconnect the headlamp wiring harness from the fan frame. Unplug the fan electrical connector.
5. Remove the attaching bolts for the fan.
6. Scribe the hood latch location on the radiator support, then remove the latch.

7. Disconnect the coolant hoses from the radiator. Remove the coolant recover tank hose from the radiator neck. Disconnect and plug the automatic transmission fluid cooler lines from the radiator, if so equipped.
8. Remove the radiator attaching bolts and remove the radiator. If the car has air conditioning, it first may be necessary to raise the left side of the radiator so that the radiator neck will clear the compressor.

To install:
1. Install the radiator in the car, tightening the mounting bolts to 7 in. lbs. Connect the transmission cooler lines and hoses. Install the coolant recovery hose.
2. Install the hood latch. Tighten to 6 ft. lbs.
3. Install the fan, making sure the bottom leg of the frame fits into the rubber grommet at the lower support. Install the fan wires and the headlamp wiring harness. Swing the strut and brace forward, tightening to 11 ft. lbs. Connect the engine ground strap to the strut brace. Install the negative battery cable, fill the cooling system, and check for leaks.

Water Pump

REMOVAL & INSTALLATION

4 Cyl Engine

1. Disconnect battery negative cable.
2. Remove accessory drive belts.
3. Remove water pump attaching bolts and remove pump.
4. If installing a new water pump, transfer pulley from old unit. With sealing surfaces cleaned, place a $\frac{1}{8}$ in. (3mm) bead of sealant No. 1052289 or equivalent on the water pump sealing surface. While sealer is still wet, install pump and torque bolts to 6 ft. lbs.
5. Install accessory drive belts.
6. Connect battery negative cable.

V6–173 Engine

1. Disconnect battery negative cable.
2. Drain cooling system and remove heater hose.
3. Remove water pump attaching bolts and nut and remove pump.
4. With the sealant surfaces cleaned, place a $\frac{3}{32}$ in. (2mm) bead of sealant No. 1052357 or equivalent on the water pump sealing surface.
5. Clean old sealant from pump.
6. Coat bolt threads with pipe sealant No. 1052080 or equivalent.
7. Install pump and torque bolts to 10 ft. lbs.
8. Connect battery negative battery cable.

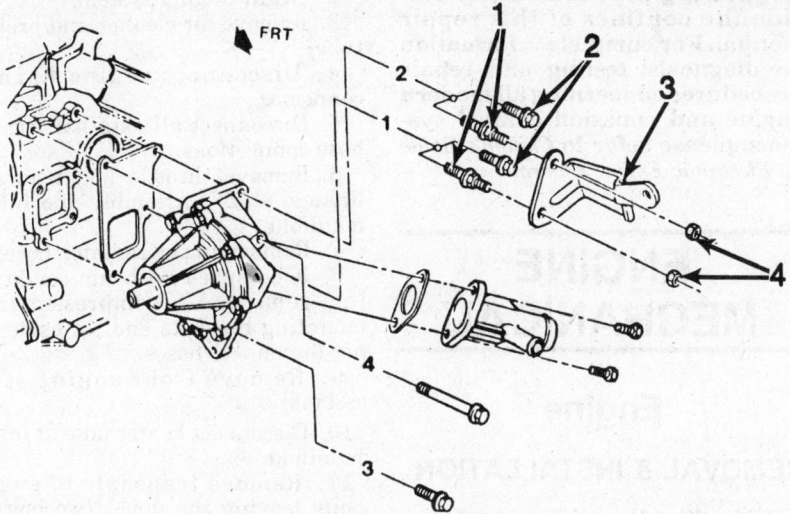

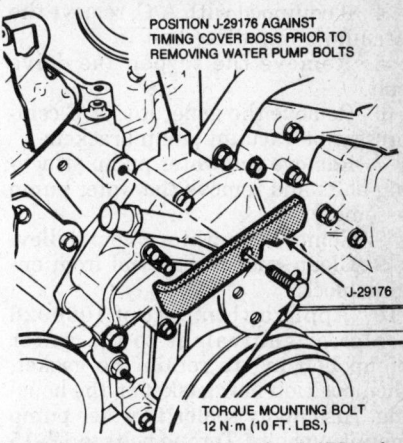

POSITION J-29176 AGAINST TIMING COVER BOSS PRIOR TO REMOVING WATER PUMP BOLTS

J-29176

TORQUE MOUNTING BOLT 12 N·m (10 FT. LBS.)

Install the special tool on the 6-173 to insure that the front cover does not separate from the crankcase when the pump is removed

4-151 water pump installation

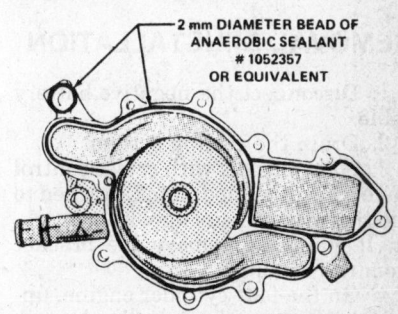

2 mm DIAMETER BEAD OF ANAEROBIC SEALANT # 1052357 OR EQUIVALENT

Applying sealer to the 6-173 water pump

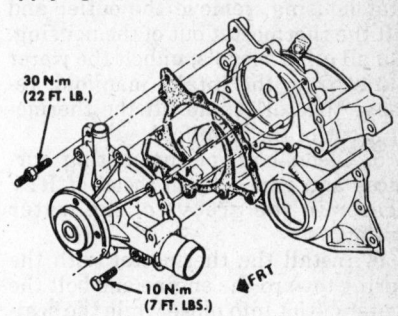

30 N·m (22 FT. LB.)

10 N·m (7 FT. LBS.)

FRT

Water pump installation—6–181 and 6–231

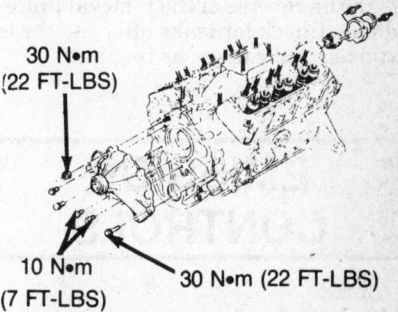

30 N·m (22 FT-LBS)

10 N·m (7 FT-LBS)

30 N·m (22 FT-LBS)

6-173 water pump installation

NOTE: When replacing the water pump on a car equipped with the V6 engine, the timing cover must be clamped to the cylinder block prior to removing the water pump bolts. Certain bolts holding the water pump pass through the front cover, and when removed, may allow the front cover to pull away from the cylinder block, breaking the seal. This may or may not be readily apparent and if left undetected, could allow coolant to enter the crankcase. To prevent this possible separation during water pump removal, Special Tool No. J29176 will have to be installed.

1982–83 V6–181 Engine

1. Disconnect the negative battery cable.
2. Remove accessory drive belts.
3. Remove water pump attaching bolts.
4. Remove the engine support strut.
5. Place a floor jack under the front crossmember of the cradle and raise the jack until the jack just starts to raise the car.
6. Remove the front two body mount bolts with the lower cushions and retainers.
7. Thread the body mount bolts with retainers a minimum of three turns into the cage so that the bolts restrain cradle movement.
8. Release the floor jack slowly until the crossmember contacts the body mount bolt retainers. As the jack is being lowered watch and correct any interference with hoses, lines, pipes and cables.

NOTE: Do not lower the cradle without its being restrained as possible damage can occur to the body and underhood items.

9. Remove water pump from engine.
10. Reverse removal procedure.
11. Install pump and torque to 25 ft. lbs.
12. Connect negative battery cable.
13. Fill with coolant and check for leaks.

1984 AND LATER V6–181, 231 Engines

1. Disconnect the negative cable at the battery.
2. Drain the cooling system.
3. Remove the accessory drive belts.
4. Disconnect the radiator and heater hoses at the water pump.
5. Remove the water pump pulley bolts (long bolt removed through access hole provided in the body side rail), then remove the pulley.
6. Remove the water pump attaching bolts, then remove the water pump.
7. Clean all gasket mating surfaces.
8. Using a new gasket, install the water pump on the engine. Torque the bolts to specifications.
9. Install the water pump pulley, then torque the bolts to specifications. (See illustration).
10. The remainder of the installation is the reverse of removal.

V6–263 Engine

1. Drain radiator.
2. Disconnect lower radiator hose at water pump.

3. Disconnect the heater return hose at the water pump, remove the bolt retaining the heater water return pipe to the intake manifold and position the pipe out of the way.

4. If equipped with A/C, remove the vacuum pump drive belt.

5. Remove the serpentine drive belt.

6. Remove the generator. A/C compressor or vacuum pump brackets.

7. Remove the water pump attaching bolts and remove the water pump assembly.

8. Remove the water pump pulley.

9. Clean gasket material from engine block.

10. Apply a thin coat of 1050026 sealer or equivalent to the water pump housing to retain the gasket, then position new gasket on the housing. Also apply sealer to water pump mounting bolts. Torque bolts to 12–15 ft. lbs.

Thermostat

REMOVAL & INSTALLATION

1. Disconnect the negative battery cable.

2. Drain the cooling system.

3. Some models with cruise control have a vacuum modulator attached to the thermostat housing with a bracket. If equipped, remove the bracket from the housing.

4. On the four cylinder engine, unbolt the water outlet from the thermostat housing, remove the outlet and lift the thermostat out of the housing. On all other models, unbolt the water outlet from the intake manifold, remove the outlet and lift the thermostat out of the manifold.

5. Clean both of the mating surfaces and run a $\frac{1}{8}$ in. bead of RTV sealer in the groove of the water outlet.

6. Install the thermostat with the spring toward the engine and bolt the water outlet into place while the sealer is still wet. Torque the bolts to 21 ft. lbs. The remainder of the installation is in the reverse of the removal procedures. Check for leaks after the car is started and correct as required.

EMISSION CONTROLS

For all routine maintenance and service information on emission control systems, please refer to "Emission Controls" in the Unit Repair section. Due to the complex nature of modern electronic engine and emission control systems, comprehensive diagnosis and testing procedures fall outside the confines of this repair manual. For complete information on diagnosis, testing and repair procedures concerning all modern engine and emission control systems, please refer to *Chilton's Guide To Electronic Engine Controls*.

ENGINE MECHANICAL

Engine

REMOVAL & INSTALLATION

4–151 with Manual Transaxle

NOTE: Before attempting this procedure, relieve the pressure in the fuel system as described under "Fuel Pump."

1. Disconnect battery cables at battery.

2. Raise the car and support it safely.

3. Remove front mount-to-cradle nuts.

4. Remove forward exhaust pipe.

5. Remove starter assembly (wires attached and swing to side).

6. Remove flywheel inspection cover.

7. Lower the car.

8. Remove air cleaner.

9. Remove all bellhousing bolts.

10. Remove forward torque reaction rod from engine and core support.

11. If equipped with A/C, remove A/C belt and compressor and swing to side.

12. Remove emission hoses at canister.

13. Remove power steering hose (if so equipped).

14. Remove vacuum hoses and electrical connectors at solenoid.

15. Remove heater blower motor.

16. Disconnect throttle cable.

17. Drain cooling system.

18. Disconnect heater hose.

20. Disconnect engine harness at bulkhead connector.

21. With engine lifting tool, hoist engine (remove heater hose at intake manifold and disconnect fuel line).

22. Installation is the reverse of removal.

4–151 with Automatic Transaxle

NOTE: Relieve the pressure in the fuel system as described under Fuel Pump.

1. Disconnect battery cables at battery.

2. Drain cooling system.

3. Remove air cleaner and preheat tube.

4. Disconnect engine harness connector.

5. Disconnect all external vacuum hose connections.

6. Remove throttle and transaxle linkage at EFI assembly and intake manifold.

7. Remove upper radiator hose.

8. If equipped with air conditioning, remove A/C compressor from mounting brackets and set aside. Do not disconnect hoses.

9. Remove front engine strut assembly.

10. Disconnect heater hose at intake manifold.

11. Remove transaxle to engine bolts leaving the upper two bolts in place.

12. Remove front mount-to-cradle nuts.

13. Remove forward exhaust pipe.

14. Remove flywheel inspection cover and remove starter motor.

15. Remove torque converter to flywheel bolts.

16. Remove power steering pump and bracket and move to one side.

17. Remove heater hose and lower radiator hose.

18. Remove two rear transaxle support bracket bolts.

19. Remove fuel supply line at fuel filter.

20. Using a floor jack and a block of wood placed under the transaxle, raise engine and transaxle until engine front mount studs clear cradle.

21. Connect engine lift equipment and put tension on engine.

22. Remove two remaining transaxle bolts.

23. Slide engine forward and left from car. Install engine on stand.

24. Installation is the reverse of removal. Do not completely lower the engine with a jack supporting the transaxle.

6–173 with Manual Transaxle

1. Disconnect cables from battery.

2. Remove air cleaner.

3. Drain cooling system.

4. Disconnect vacuum hosing to all nonengine mounted components.

5. Disconnect accelerator linkage from carburetor.

6. Disconnect engine harness connector.

7. Disconnect radiator hoses from radiator.

8. Disconnect heater hoses from engine.

9. If equipped, remove power steering pump and bracket assembly from engine.

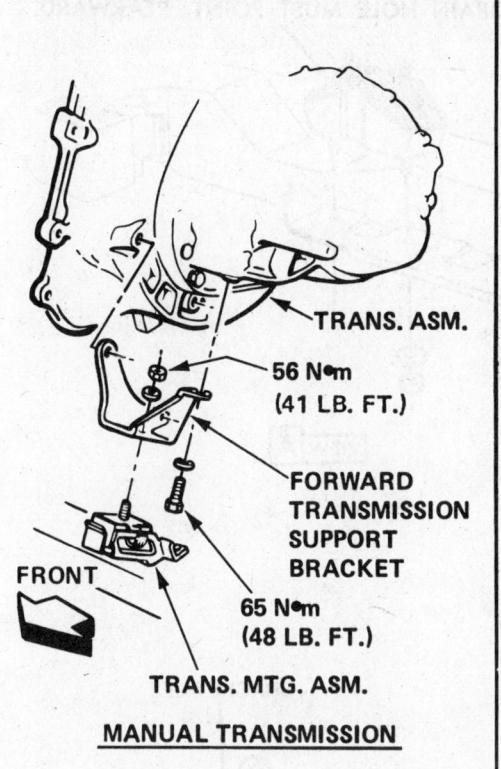

TRANS. ASM.

56 N•m
(41 LB. FT.)

FORWARD
TRANSMISSION
SUPPORT
BRACKET

FRONT

65 N•m
(48 LB. FT.)

TRANS. MTG. ASM.

MANUAL TRANSMISSION

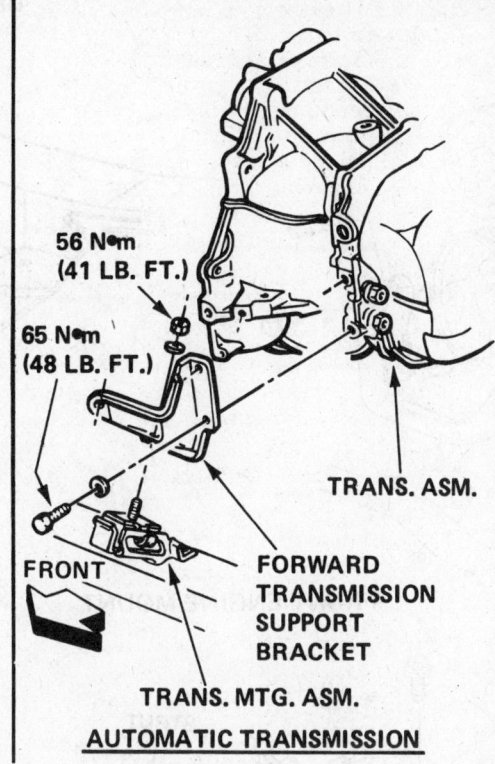

56 N•m
(41 LB. FT.)

65 N•m
(48 LB. FT.)

TRANS. ASM.

FRONT

FORWARD
TRANSMISSION
SUPPORT
BRACKET

TRANS. MTG. ASM.

AUTOMATIC TRANSMISSION

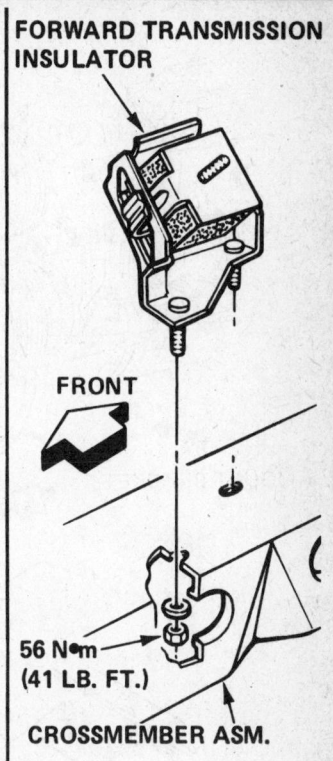

FORWARD TRANSMISSION
INSULATOR

FRONT

56 N•m
(41 LB. FT.)

CROSSMEMBER ASM.

4-151 front mounts

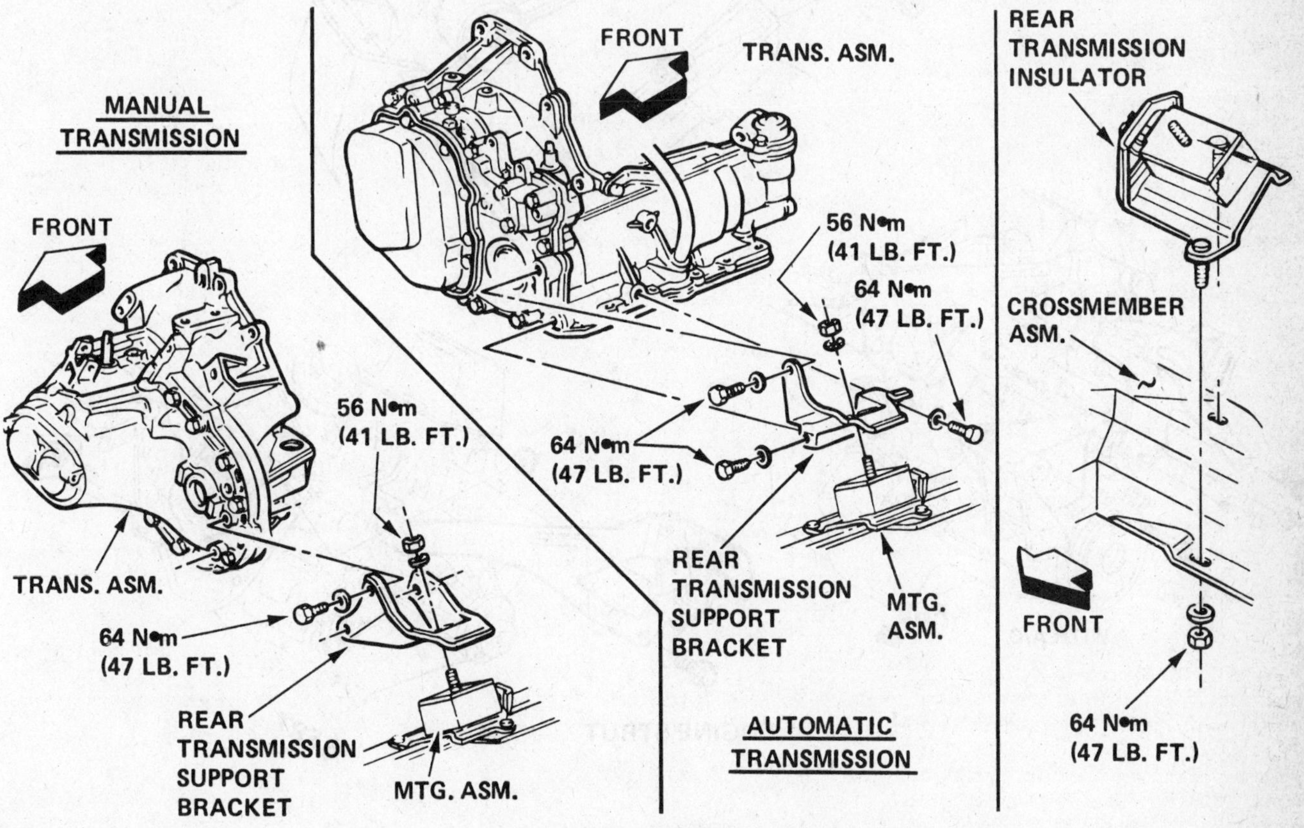

MANUAL
TRANSMISSION

FRONT

FRONT   TRANS. ASM.

56 N•m
(41 LB. FT.)

64 N•m
(47 LB. FT.)

TRANS. ASM.

56 N•m
(41 LB. FT.)

64 N•m
(47 LB. FT.)

64 N•m
(47 LB. FT.)

REAR
TRANSMISSION
SUPPORT
BRACKET

MTG.
ASM.

REAR
TRANSMISSION
SUPPORT
BRACKET

MTG. ASM.

**AUTOMATIC
TRANSMISSION**

REAR
TRANSMISSION
INSULATOR

CROSSMEMBER
ASM.

FRONT

64 N•m
(47 LB. FT.)

4-151 rear mounts

C519

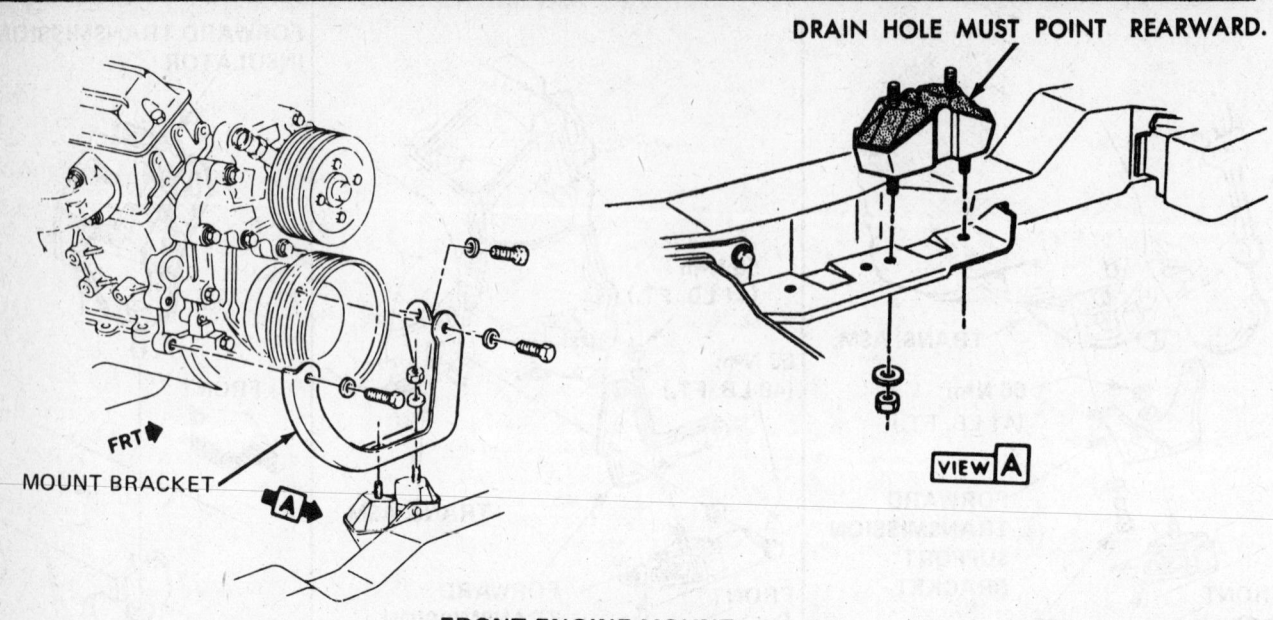

DRAIN HOLE MUST POINT REARWARD.

FRT

MOUNT BRACKET

VIEW **A**

FRONT ENGINE MOUNT

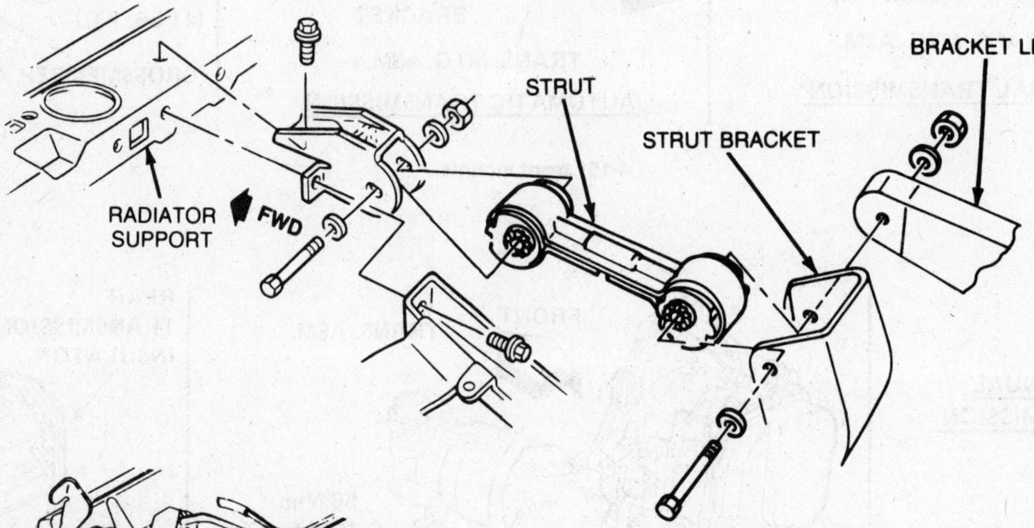

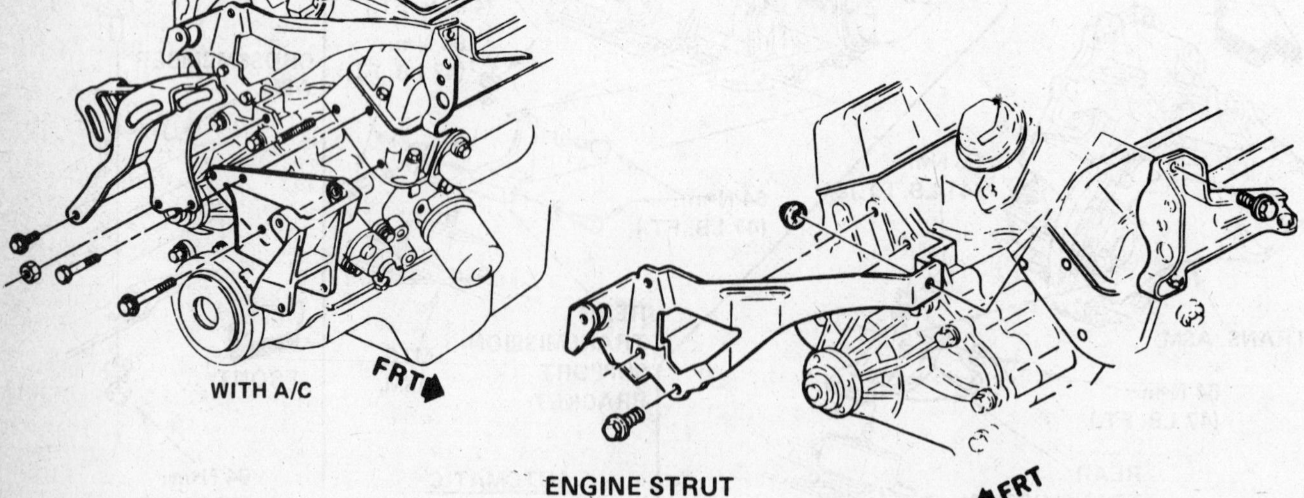

RADIATOR SUPPORT

FWD

STRUT

STRUT BRACKET

BRACKET LE2

WITH A/C

FRT

ENGINE STRUT

FRT

6-173 mounts

10. Disconnect clutch cable from transaxle.

11. Disconnect shift linkage from transaxle shift levers. Remove cables from transaxle bosses.

12. Disconnect speedometer cable from transaxle.

13. Install engine support fixture. Raise engine until weight is relieved from mount assemblies.

14. Remove exhaust crossover.

15. Remove all but the transaxle to engine retaining bolts.

16. Remove side and crossmember assembly.

17. Disconnect exhaust pipe.

18. Remove all powertrain mount to cradle attachments.

19. Using tool J-28468 or J-33008, pull both axle drive shafts from transaxle assembly.

20. Lower vehicle.

21. Lower left side of engine/transaxle assembly by loosening tool J-22825.

22. Place jack under transaxle.

23. Remove the final transaxle to engine attaching bolt and separate transaxle from engine and lower.

24. Lower vehicle.

25. Install engine lifting fixture.

26. If A/C equipped, remove compressor from mounting bracket and swing aside.

27. Disconnect forward strut bracket from radiator support. Swing aside.

28. Lift engine out of vehicle.

29. Installation is the reverse of removal.

6–173 with Automatic Transaxle

1. Disconnect battery cables from battery.

2. Remove air cleaner.

3. Drain cooling system.

4. Disconnect vacuum hosing to all non-engine mounted components.

5. Disconnect detent cable from carburetor lever.

6. Disconnect accelerator linkage.

7. Disconnect engine harness connector.

8. Disconnect ground strap from engine at engine forward strut.

9. Disconnect radiator hoses from radiator.

10. Disconnect heater hoses from engine.

11. Remove power steering pump and bracket assembly from engine, if equipped.

12. Raise the vehicle and support it safely.

13. Disconnect exhaust pipe.

14. Disconnect fuel lines at rubber hose connections at right side of engine.

15. Remove engine front mount to cradle retaining nuts (right side of vehicle).

16. Disconnect battery cables from engine (Starter and transaxle housing bolt).

17. Remove flex plate cover and disconnect torque convertor from flex plate.

18. Remove transaxle case to cylinder case support bracket bolts.

19. Lower vehicle. Place a support under the transaxle rear extension.

20. Remove engine strut bracket from radiator support and swing rearward.

21. Remove exhaust crossover pipe.

22. Remove transaxle to cylinder case retaining bolts. Make note of ground stud location.

23. If A/C equipped, remove compressor from mounting bracket and lay aside.

24. Install lift fixture to engine and remove engine from vehicle.

25. Installation is the reverse of removal.

6–181 and 231 Engines

1. Disconnect battery cables from battery.

2. Remove air cleaner.

3. Drain cooling system.

4. Disconnect vacuumhosing to all nonengine mounted components.

5. Disconnect detent cable from carburetor lever.

6. Disconnect accelerator linkage.

7. Disconnect engine harness connector.

8. Disconnect ground strap from engine at engine forward strut.

9. Disconnect radiator hoses from radiator.

10. Disconnect heater hoses from engine.

11. Remove power steering pump and bracket assembly from engine.

12. Raise the vehicle and support it safely.

13. Disconnect exhaust pipe at manifold.

14. Disconnect fuel lines at rubber hose connections.

15. Remove engine front mount to cradle retaining nuts (right side of vehicle).

16. Disconnect battery cables from engine (Starter and transaxle housing bolt).

17. Remove flex plate cover and disconnect torque converter from flex plate.

18. Remove transaxle case to cylinder case support bracket bolts.

19. Lower vehicle. Place a support under the transaxle rear extension.

20. Remove engine strut bracket from radiator support and swing rearward.

21. Remove transaxle to cylinder case retaining bolts. Make note of ground stud location.

22. If A/C equipped, remove compressor from mounting bracket and lay aside.

23. Install lift fixture to engine and remove engine from vehicle.

24. Installation is the reverse of removal.

6–263 Engine

1. Drain the cooling system. Remove the serpentine drive belt (and vacuum pump drive belt, if A/C equipped).

2. Remove air cleaner and install cover J-26996.

3. Disconnect battery negative cable(s) at batteries and ground wires at inner fender panel. Disconnect engine ground strap, rear (right) head to cowl.

4. Raise the car and support it safely.

5. Remove the flywheel cover.

6. Remove the flywheel to torque converter bolts.

7. Disconnect the exhaust pipe from the rear exhaust manifold.

8. Remove the engine to transaxle brace.

9. Remove the engine mount to cradle retaining nuts and washers.

10. Disconnect the leads to the starter motor, No. 2 cylinder glow plug and battery ground cable to engine bolt.

11. Disconnect the lower oil cooler hose and cap the openings.

12. Remove the accessible power steering pump bracket fasteners.

13. Lower the car.

14. Remove the remaining power steering pump bracket/brace fasteners and lower the power steering pump with hoses out of the way.

15. Remove heater water return pipe.

16. Disconnect all remaining glow plug leads at the glow plugs.

17. Disconnect all other leads at the engine, disconnect the engine harness at the cowl connector and body mounted relays and position the engine harness aside.

18. If A/C equipped, disconnect the compressor with brackets and lines attached and position aside.

19. Disconnect the fuel and vacuum hoses, cap all fuel line openings.

20. Disconnect the throttle and T.V. cables at the injection pump and cable bracket. Position cables aside.

21. Disconnect the upper oil cooler hose and cap the openings.

22. Remove the exhaust crossover pipe heat shield.

23. Disconnect and move aside the transaxle filler tube.

24. Remove the exhaust crossover pipe.

25. Remove the engine mounting strut and strut brackets.

26. Install a suitable engine lifting device. Make certain that when installing chains to the cylinder heads that washers are used under the chains and bolt heads and that the bolts are torqued to 20 ft. lbs.

CAUTION

Failure to properly secure the engine lift to the aluminum cylinder heads can result in personal injury.

27. Position a support under the transaxle rear extension. It may be necessary to raise the support as the engine is being removed.

28. Remove the engine to transaxle bolts and remove the engine.

29. Installation is the reverse of removal. Note the following:

a. Before installing the flex plate-to-converter bolts, make sure that the weld nuts on the converter are flush with the flex plate, and the converter rotates freely by hand.

b. Use only new O-rings at all connections.

c. Adjust the throttle valve cable as outlined in the "Automatic Transmission" Unit Repair section.

Intake Manifold

REMOVAL & INSTALLATION

4-151 Engine

CAUTION

Bleed pressure from the fuel system, if equipped with fuel injection, before attempting this procedure.

1. Remove the air cleaner and the PCV valve.
2. Drain the cooling system into a clean container.

3. Disconnect the fuel and vacuum lines and the electrical connections at the carburetor and manifold.

4. Disconnect the throttle linkage at the EFI unit and disconnect the transaxle downshift linkage and cruise control linkage.

5. Remove the carburetor and the spacer.

6. Remove the bell crank and the throttle linkage. Position to the side for clearance.

7. Remove the heater hose at the intake manifold.

8. Remove the pulse air check valve bracket from the manifold.

9. Remove the manifold attaching bolts and remove the manifold.

6-173 Engine

1. Remove the rocker covers.
2. Drain the cooling system.
3. If equipped, remove the AIR pump and bracket.
4. Remove the distributor cap. Mark the position of the ignition rotor in relation to the distributor body, and remove the distributor. Do not crank the engine with the distributor removed.
5. Remove the heater and radiator hoses from the intake manifold.
6. Remove the power brake vacuum hose.
7. Disconnect and label the vacuum hoses. Remove the EFE pipe from the rear of the manifold.
8. Remove the carburetor linkage. Disconnect and plug the fuel line.
9. Remove the manifold retaining bolts and nuts.
10. Remove the intake manifold. Remove and discard the gaskets, and scrape off the old silicone seal for the front and rear ridges.

To install:

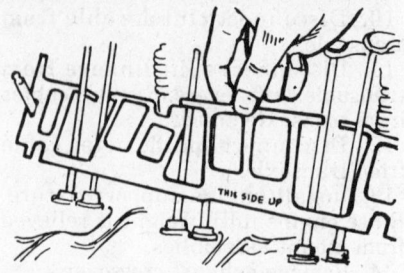

Cut the 6-173 intake manifold gasket as necessary

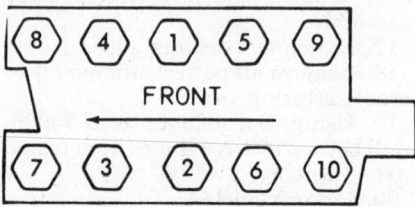

6-173 intake manifold torque sequence

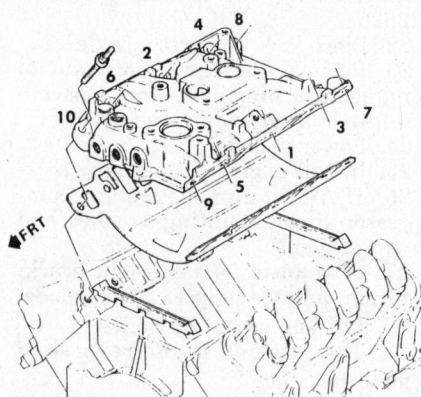

Intake manifold torque sequence—6-181 and 6-231

1. The gaskets are marked for right and left side installation; do not interchange them. Clean the sealing surface of the engine block and apply a $\frac{3}{16}$ in. bead of silicone sealer to each ridge.

2. Install the new gaskets onto the heads. The gaskets will have to be cut slightly to fit past the center push-rods. Do not cut any more material than necessary. Hold the gaskets in place by extending the ridge bead of sealer $\frac{1}{4}$ in. onto the gasket ends.

3. Install the intake manifold. The area between the ridges and the manifold should be completely sealed.

4. Install the retaining bolts and nuts, and tighten in sequence to 23 ft. lbs. Do not overtighten; the manifold is made from aluminum, and can be warped or cracked with excessive force.

5. The rest of installation is the reverse of removal. Adjust the ignition timing after installation, and check the coolant level after the engine has warmed up.

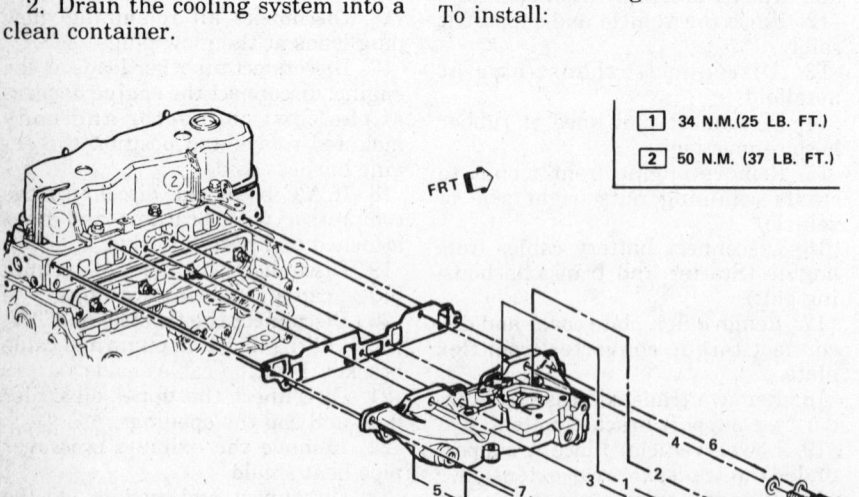

| 1 | 34 N.M. (25 LB. FT.) |
| 2 | 50 N.M. (37 LB. FT.) |

4-151 intake manifold torque sequence

6-181 and 6-231 Engines

RIGHT SIDE

1. Disconnect the battery ground.
2. Drain the cooling system.
3. Remove the air cleaner.
4. Disconnect all hoses and wiring from the manifold.
5. Disconnect the accelerator linkage and cruise control chain.
6. Disconnect the fuel line at carburetor.
7. Remove the distributor cap and rotor and remove the Torx® head bolt from the left side of the manifold.
8. Unbolt and remove the manifold.
9. Installation is the reverse of removal. When installing the front and rear seals, make sure that the ends of the seals fit snugly against the block and head. Install Nos. 1 & 2 bolts first and tighten them until snug, then install the other bolts in order.

6-263 Engine

NOTE: This procedure requires the removal, disassembly draining and reassembly of the valve lifters. Read that procedure (later on) before continuing.

1. Remove the air cleaner assembly.
2. Drain the radiator, then disconnect the upper radiator hose from the water outlet.
3. Disconnect the heater inlet hose from the outlet on the intake manifold and disconnect the heater outlet pipe from the intake manifold attachments and move it aside.
4. Remove air crossover and the fuel injection pump.
5. Disconnect wiring as necessary at the generator, A/C compressor and switches, if so equipped.
6. Remove the cruise control servo if so equipped.
7. Remove the A/C compressor bracket and brace bolts and position the compressor (if so equipped) with lines attached out of the way.
8. Remove the generator assembly.
9. Disconnect the engine mounting strut.
10. Remove the fuel lines, filter and brackets. Cap all openings.
11. Disconnect the electrical leads to the glow plug controller and sending units.
12. Disconnect the exhaust crossover pipe head shield.
13. Remove the left (forward) injection lines and cap all openings. Use a backup wrench on the nozzles.
14. Disconnect the throttle and T.V. cables from the bracket.
15. Remove the drain tube.
16. Remove the intermediate pump adapter.

17. Remove pump adapter and seal.
18. Remove the intake manifold.
19. Clean the machined surfaces of cylinder head and intake manifold with a putty knife. Use care not to gouge or scratch the machined surfaces. Clean all bolts and bolt holes.
20. Coat both sides of gasket sealing surface that seal the intake manifold to the head with 1050026 sealer or equivalent and position intake manifold gasket. Install end seals, making sure that ends are positioned under the cylinder heads. The seals and mating surfaces must be dry. Any liquid, including sealer will act as a lubricant and cause the seal to move during assembly. Use RTV sealer only on each end of the seal.
21. Position intake manifold on engine. Lubricate the entire intake manifold bolt (all) with lubricant 1052080 or equivalent.
22. Torque the bolts in sequence shown to 20 ft. lb. Then retorque to 41 ft. lbs.
23. Install the drain tube.
24. Install the pump adapter.
25. Apply chassis lube to seal area of intake manifold and pump adapter.
26. Apply chassis lube to inside and outside diameter of seal and seal area of tool J-28425.
27. Install seal on tool and install the seal.
28. Install intermediate pump adapter.
29. Reverse the order of removal and install all other removed parts except the air crossover.
30. Fill the cooling system.
31. Install manifold covers, J-29657.
32. Start engine and check for leaks.
33. Check and reset the injection pump timing, if necessary.
34. Remove screen covers from manifold.
35. Install air crossover.
36. Install the air cleaner.
37. Road test car and inspect for leaks.

Exhaust Manifold

REMOVAL & INSTALLATION

4-151 Engine

1. Remove the air cleaner and the EFI preheat tube.
2. Remove the manifold strut bolts from the radiator support panel and the cylinder head.
3. Remove the A/C compressor bracket to one side. Do not disconnect any of the refrigerant lines.
4. If necessary, remove the dipstick tube attaching bolt, and the engine mount bracket from the cylinder head.

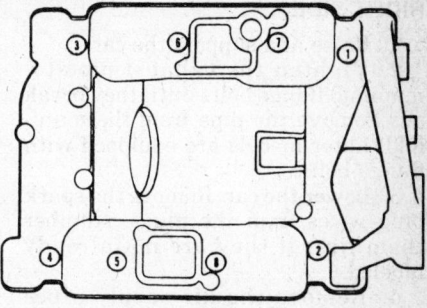

6-263 intake manifold torque sequence

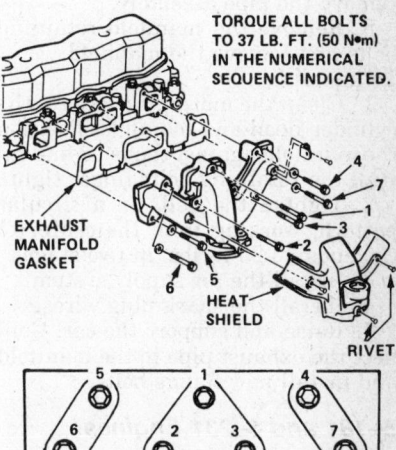

TORQUE ALL BOLTS TO 37 LB. FT. (50 N•m) IN THE NUMERICAL SEQUENCE INDICATED.

EXHAUST MANIFOLD GASKET

HEAT SHIELD

RIVET

BOLT LOCATIONS

4-151 exhaust manifold torque sequence

5. Raise the car and disconnect the exhaust pipe from the manifold.
6. Remove the manifold attaching bolts and remove the manifold.
7. Reverse to install.

6-173 Engine

LEFT SIDE

1. Remove the air cleaner. Remove the carburetor heat stove pipe.
2. Remove the air supply plumbing from the exhaust manifold.
3. Raise and support the car. Unbolt and remove the exhaust pipe at the manifold.
4. Unbolt and remove the manifold.

To install:

1. Clean the mating surfaces of the cylinder head and manifold. Install the manifold onto the head, and install the retaining bolts finger tight.
2. Tighten the manifold bolts in a circular pattern, working from the center to the ends, to 25 ft. lbs. in two stages.
3. Connect the exhaust pipe to the manifold.
4. The remainder of installation is the reverse of removal.

RIGHT SIDE

1. Raise and support the car.
2. Tighten the exhaust pipe-to-manifold flange bolts until they break off. Remove the pipe from the manifold. Later models are equipped with flange bolts.
3. Lower the car. Remove the spark plug wires from the plugs. Number them first if they are not already labeled.
4. Remove the air supply pipes from the manifold. Remove the PULSAIR bracket bolt from the rocker cover, on models so equipped, then remove the pipe assembly.
5. Remove the manifold retaining bolts and remove the manifold.

To install:
1. Clean the mating surfaces of the cylinder head and manifold. Position the manifold against the head and install the retaining bolts finger tight.
2. Tighten the bolts in a circular pattern, working from the center to the ends, to 25 ft. lbs. in two stages.
3. Install the air supply system.
4. Install the spark plug wires.
5. Raise and support the car. Connect the exhaust pipe to the manifold and install new flange bolts.

6–181 and 6–231 Engines

1. Disconnect the battery ground cable.
2. Remove the pinch bolt at the steering gear intermediate shaft and separate the intermediate shaft from the stub shaft.

--- CAUTION ---

Failure to disconnect the intermediate shaft from the rack and pinion stub shaft can result in damage to the steering gear and/or intermediate shaft. This damage can cause loss of steering control which could result in a vehicle crash with possible bodily injury.

3. Raise and support the car on jackstands.
4. Unbolt the exhaust pipe from the manifold.
5. Lower the car.
6. Remove the upper engine support strut.
7. Place a floor jack under the front crossmember and take up the weight of the car.
8. Remove the two front body mount bolts along with their cushions and retainers.
9. Remove the cushions from the bolts and thread the bolts and their retainers a minimum of three turns into the cradle cage nuts so that the bolts serve to hold the cradle and prevent movement.
10. Lower the floor jack so that the crossmember contacts the body mount

bolt retainers. Check for any hose or wire interference problems.
11. Remove the alternator, disconnect the power steering pump and remove its bracket.
12. Disconnect the manifold from the crossover pipe.
13. Unbolt and remove the manifold.
14. Installation is the reverse of removal.

LEFT SIDE

1. Disconnect the battery ground.
2. Unbolt and remove the crossover pipe.
3. Remove the upper engine support strut.
4. Unbolt and remove the manifold.
5. Installation is the reverse of removal.

6–263 Engine

LEFT SIDE

1. Remove the crossover pipe from the manifolds.
2. Raise and support the car on jackstands.
3. Unbolt and remove the manifold.
4. Installation is the reverse of removal. Lubricate the entire length of each manifold bolt with lubricant 1052080 or its equivalent.

RIGHT SIDE

1. Remove the engine support strut.
2. Place a floor jack under the front crossmember and take up the weight of the car.

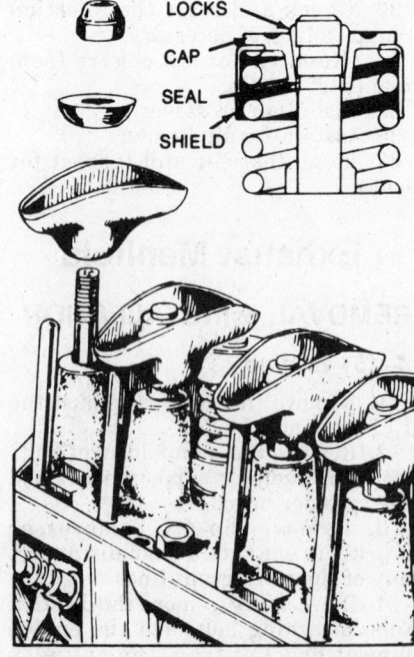

4-151 and 6-173 rocker arm, pivot and nut

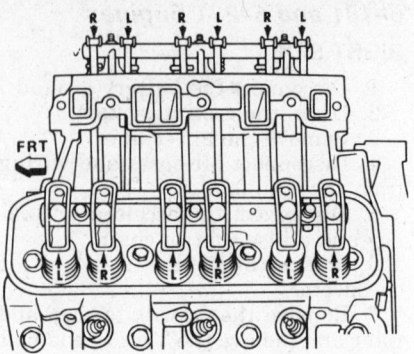

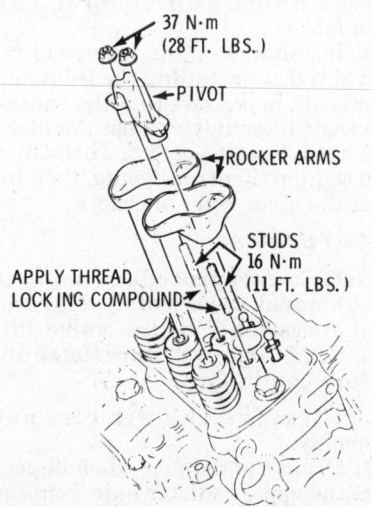

Position of rocker arms on shaft—6-181

3. Remove the two front body mount bolts. Remove the cushions from the bolts.
4. Thread the body mount bolts with their retainers into the cage nuts so that the bolts restrict movement of the engine cradle.
5. Lower the jack until the crossmember contacts the body mount bolt retainers. Check for any hose or wire interference.
6. Remove the crossover pipe.
7. Raise and support the car on jackstands.
8. Disconnect the exhaust pipe from the manifold.
9. Lower the car.
10. Unbolt and remove the manifold.
11. Installation is the reverse of removal. Lubricate the entire length of each manifold bolt with lubricant 1052080 or its equivalent.

Valve Guides

RECONDITIONING

Valve guides on all engines are integral with the cylinder head. Worn guides can be reamed to accept oversized valve stems, or knurled to retain original sized stems. See the chart at the beginning of this section to determine the original stem diameter.

Rocker Arm, Shaft and Pushrod

REMOVAL & INSTALLATION

4–151 Engine

1.
Remove the valve cover.

2. On fuel injected engines, see the fuel pump section to relieve pressure in the fuel system before disconnecting any fuel lines.

3. If only the pushrod is being removed, loosen the rocker arm bolt and swing the rocker arm aside.

4. Remove the rocker arm nut and ball.

5. Lift the rocker arm off the stud, keeping rocker arms in order for installation.

6–173 Engine

NOTE: Some engines are assembled using RTV (Room Temperature Vulcanizing) silicone sealant in place of rocker arm cover gasket. If the engine was assembled using RTV, never use a gasket when reassembling. Conversely, if the engine was assembled using a rocker arm cover gasket, never replace it with RTV. When using RTV, a 1/8 in. bead is sufficient. Always run the bead on the inside of the bolt holes.

Rocker arms are removed by removing the adjusting nut. Be sure to adjust valve lash after replacing rocker arms. When replacing an exhaust rocker, move an old intake rocker arm to the exhaust rocker arm stud and install the new rocker arm on the intake stud. Cylinder heads use threaded rocker arm studs. If the threads in the head are damaged or stripped, the head can be retapped and a helical type insert installed. If engine is equipped with the A.I.R. (air pump) exhaust emission control system, the interfering components of the system must be removed. Disconnect the lines at the air injection nozzles in the exhaust manifolds.

6–181 and 6–231 Engines

1. Remove the rocker arm cover(s).
2. Remove the rocker arm shaft(s).
3. Place the shaft on a clean surface.
4. Remove the nylon rocker arm retainers. A pair of slip joint pliers is good for this.

5. Slide the rocker arms off the shaft and inspect them for wear or damage. Keep them in order.

6. Installation is the reverse of removal. If new rocker arms are being installed, note that they are stamped R (right) or L (left). Each rocker arm must be centered over its oil hole. New nylon retainers must be used.

6–263 Engine

NOTE: This procedure requires that the valve lifters be bled.

1. Remove the valve cover(s) as outlined below.

2. Remove the rocker arm nuts, pivot and rocker arms.

3. If rocker arms are being replaced, they must be replaced in cylinder sets. Never replace just one rocker arm per cylinder. If a stud was replaced, coat the threads with locking compound and torque it to 11 ft. lbs.

4. Installation is the reverse of removal. See the section on Valve lifter bleed-down. This is absolutely necessary; if lifters are not bled, engine damage will be unavoidable. Torque the rocker arm nuts to 28 ft. lbs.; the cover to 5 ft. lbs.

Diesel Engine Right Side Valve Cover

REMOVAL & INSTALLATION

1. Remove the injection lines. See the procedure outlined earlier in this section.

2. Disconnect the crankcase ventilation pipes, grommets, filter and crankcase vacuum regulator valve.

3. Remove the engine support strut.

4. Place a floor jack under the front crossmember and take up the weight of the engine.

5. Remove the two front body mount bolts. Remove the cushions from the bolts and thread the bolts and retainers into the cage nuts to restrict engine cradle movement.

6. Lower the jack until the crossmember contacts the body mount bolt retainers.

7. Remove the valve cover.

8. Installation is the reverse of removal. These covers are installed with a 1/8 in. bead of RTV gasket material in place of a gasket. Coat each bolt with No. 1052080 lubricant or equivalent.

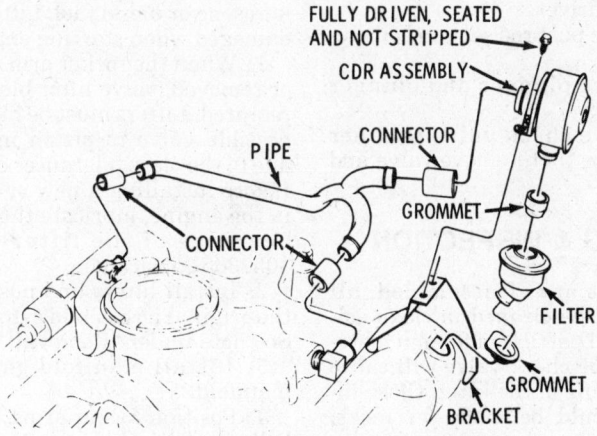

Diesel crankcase ventilation system

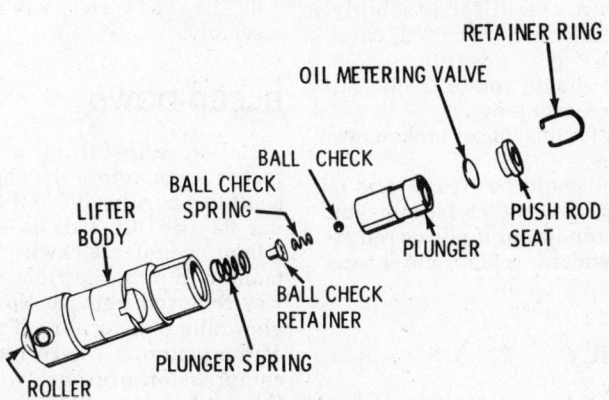

Valve lifter disassembled

Diesel Engine Valve Lifter Bleed-Down

If the intake manifold and valve rocker arms have been removed, it will be necessary to remove, disassemble, drain and reassemble the lifters on that side. If the rocker arms have been loosened or removed, but the intake manifold was not removed, skip down to the Bleed-Down procedure.

REMOVAL

Keep lifters and pushrods in order. This is absolutely necessary for installation, since these parts have differences which could result in engine damage if not installed in their original positions.

1. Remove intake manifold. Refer to "Intake Manifold".
2. Remove valve covers, rocker arm assemblies and pushrods.
3. Remove the valve lifter guide retainer bolts.
4. Remove the retainer guides and valve lifters.

DISASSEMBLY

1. Remove the retainer ring with a small screwdriver.
2. Remove pushrod seat and oil metering valve.
3. Remove plunger and plunger spring.
4. Remove check valve retainer from plunger, then remove valve and spring.

CLEANING & INSPECTION

After lifters are disassembled, all parts should be cleaned in clean solvent. A small particle of foreign material under the check valve will cause malfunctioning of the lifter. Close inspection should be made for nicks, burrs or scoring of parts. If either the roller body or plunger is defective, replace with a new lifter assembly. Whenever lifters are removed, check as follows:

1. Roller should rotate freely, but without excessive play.
2. Check for missing or broken needle bearings.
3. Roller should be free of pits or roughness. If present, check camshaft for similar condition. If pits or roughness are evident replace lifter and camshaft.

ASSEMBLY

1. Coat all lifter parts with a coating of clean kerosene or diesel fuel.

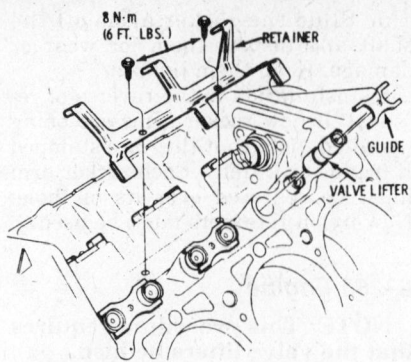

Diesel valve lifters, guides and retainers

2. Assemble the ball check, spring and retainer into the plunger.
3. Install plunger spring over check retainer.
4. Hold plunger with spring up and insert into lifter body. Hold plunger vertically to prevent cocking spring.
5. Submerge the lifter in clean kerosene or diesel fuel.
6. Install oil metering valve and push rod seat into lifter and install retaining ring.

INSTALLATION

Prime new lifters by working lifter plunger while submerged in clean kerosene or diesel fuel. Lifter could be damaged when starting engine if dry.

1. When the rocker arm is loosened or removed, valve lifter bleed down is required. Lifters must be bled down as possible valve to piston interference due to the close tolerances could exist. Before installing a new or used lifter in the engine, lubricate the roller and bearings of the lifter with No. 1052365 lubricant.
2. Install lifters and pushrods into the original position in cylinder block. See note under "Removal."
3. Install manifold gaskets and manifold.
4. Position rocker arms, pivots and bolts on cylinder head.
5. Install valve covers.
6. Install intake manifold assembly.

BLEED-DOWN

1. Before installing any removed rocker arms, rotate the engine crankshaft to a position of No. 1 cylinder being 32° BTDC. This is about 2 in. (50mm) counterclockwise from the 0° pointer. If only the right valve cover was removed, remove No. 1 cylinder glow plug to determine if the position of the piston is the correct one. The compression pressure will indicate the right position. If the left valve cover was removed, rotate the crankshaft

until the No. 5 cylinder intake valve pushrod ball is 0.28 in. (7.0mm) above the No. 5 cylinder exhaust valve pushrod ball.

NOTE: Use only hand wrenches to torque the rocker arm pivot nuts to avoid engine damage.

2. If removed, install the No. 5 cylinder pivot and rocker arms. Torque the nuts alternately between the intake and exhaust valves until the intake valve begins to open, then stop.
3. Install remaining rocker arms except No. 3 exhaust valve (if this rocker arm was removed).
4. If removed, install but do not torque No. 3 valve pivots beyond the point that the valve would be fully open. This is indicated by strong resistance while still turning the pivot retaining bolts. Going beyond this would bend the pushrod. Torque the nuts SLOWLY allowing the lifter to bleed down.
5. Finish torquing No. 5 cylinder rocker arm pivot nut SLOWLY. Do not go beyond the point that the valve would be fully open. This is indicated by strong resistance while still turning the pivot retaining bolts. Going beyond this would bend the pushrod.
6. DO NOT turn the engine crankshaft for at least 45 minutes.
7. Finish reassembling the engine as the lifters are being bled.

NOTE: Do not rotate the engine until the valve lifters have been bled down, or damage to the engine will occur.

Valve Lash

ADJUSTMENT

4–151, 6–181, 6–231 and 6–263 Engines

No routine adjustment is necessary.

6–173 Engine

Anytime the V6 valve train is disturbed, the valve lash must be adjusted. Crank the engine until the timing mark aligns with the "O" mark on the timing scale, and both valves in the No. 1 cylinder are closed. If the valves are moving as the timing marks align, the engine is in the No. 4 firing position. Turn the crankshaft one more revolution. With the engine in the No. 1 firing position, adjust the following valves:

- exhaust – 1,2,3
- intake – 1,5,6

Rotate the crankshaft one full revolution, until it is in the No. 4 firing position. Adjust the following valves:

- Exhaust—4, 5, 6
- Intake—2, 3, 4

Adjustment is made by backing off the rocker arm adjusting nut until there is play in the pushrod. Tighten the nut to remove the pushrod clearance (this can be determined by rotating the pushrod with your fingers while tightening the adjusting nut). When the pushrod cannot be freely turned, tighten the nut 1½ additional turns to place the hydraulic lifter in the center of its travel. No further adjustment is required.

Cylinder Head

REMOVAL & INSTALLATION

4–151 Engine

— CAUTION —

On fuel injected engines, relieve the pressure in the fuel system before disconnecting any fuel line connections. The engine should be overnight cold.

1. Drain the cooling system into a clean container.
2. Remove the air cleaner.
3. Remove the intake and exhaust manifolds as previously outlined.
4. Remove the alternator bracket bolts.
5. Remove the A/C compressor bracket bolts and position the compressor to one side. Do not disconnect any of the refrigerant lines.
6. Disconnect all vacuum and electrical connections from the cylinder head.
7. Disconnect the upper radiator hose.
8. Disconnect the spark plug wires and remove the plugs.
9. Remove the rocker arm cover, rocker arms, and pushrods.
10. Unbolt and remove the cylinder head.
11. Clean the gasket surfaces thoroughly.
12. Install a new gasket over the dowels and position the cylinder head.
13. Coat the head bolt threads with sealer and install finger tight.
14. Tighten the bolts in sequence, in three equal steps to the specified torque.
15. Install all parts in the revere of removal.

6–173 Engine

LEFT SIDE

1. Raise the vehicle and support it safely.
2. Drain the coolant from the block and lower the car.
3. Remove the intake manifold.
4. Remove the crossover.

5. Remove the alternator and AIR pump brackets.
6. Remove the dipstick tube.
7. Loosen the rocker arm bolts and remove the pushrods. Keep the pushrods in the same order as removed.
8. Remove the cylinder head bolts in stages and in the reverse order of the tightening sequence.

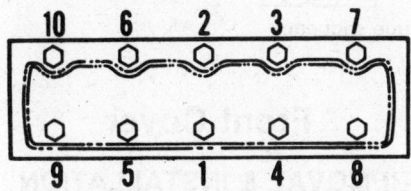

4-151 head bolt torque sequence

9. Remove the cylinder head. Do not pry on the head to loosen it.
10. Installation is the reverse of removal. The words "This Side Up" on the new cylinder head gasket should face upward. Coat the cylinder head bolts with sealer and torque to specifications in the sequence shown. Make sure the pushrods seat in the lifter seats and adjust the valves.

RIGHT SIDE

1. Raise the vehicle and drain the coolant from the block.
2. Disconnect the exhaust pipe and lower the vehicle.
3. If equipped, removes the cruise control servo bracket.
4. Remove the air management valve and hose.
5. Remove the intake manifold.

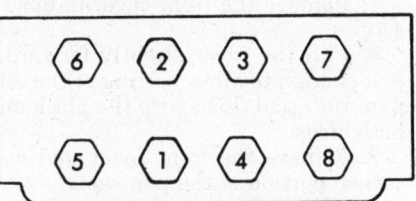

6-173 head bolt torque sequence

6. Remove the exhaust crossover.
7. Loosen the rocker arm nuts and remove the pushrods. Keep the pushrods in the order in which they were removed.
8. Remove the cylinder head bolts in stages and in the reverse order of the tightening sequence.
9. Remove the cylinder head. Do not pry on the cylinder head to loosen it.
10. Installation is the reverse of removal. The words "This Side Up" on the new cylinder head gasket should face upwards. Coat the cylinder head bolts with sealer and tighten them to specifications in the sequence shown. Make sure the lower ends of the pushrods seat in the lifter seats and adjust the valves.

6–181 and 6–231 Engine

1. Disconnect negative battery cable.
2. Remove intake manifold.
3. Loosen and remove belt(s).
4. When removing LEFT cylinder head:
   a. Remove oil dipstick.
   b. Remove air and vacuum pumps with mounting bracket if present, and move out of the way with hoses attached.
5. When removing RIGHT cylinder head:
   a. Remove alternator.
   b. Disconnect power steering gear pump and brackets attached to cylinder head.
6. Disconnect wires from spark plugs, and remove the spark plug wire clips from the rocker arm cover studs.

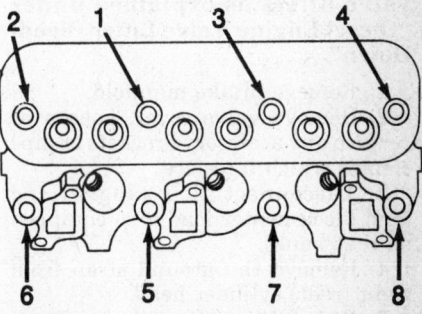

Head bolt torque sequence—6–181 and 6–231

7. Remove exhaust manifold bolts from head being removed.
8. With air hose and cloths, clean dirt off cylinder head and adjacent area to avoid getting dirt into engine. It is extremely important to avoid getting dirt into the hydraulic valve lifters.
9. Remove rocker arm cover and rocker arm and shaft assembly from cylinder head. Lift out pushrods. If lifters are to be serviced, remove them at this time and place them in a container with numbered holes or a similar device, to keep them identified as to engine position. If they are not to be removed, protect lifters and camshaft from dirt by covering area with a clean cloth.
10. Loosen all cylinder head bolts, then remove bolts and lift off the cylinder head.
11. With cylinder head on bench, remove all spark plugs for cleaning and to avoid damaging them during work on the head.
12. Installation is the reverse of removal. Clean all gasket surfaces thoroughly. Always use a new head gasket. The head gasket is installed with the bead downward. Coat the head bolt threads with thread sealer.

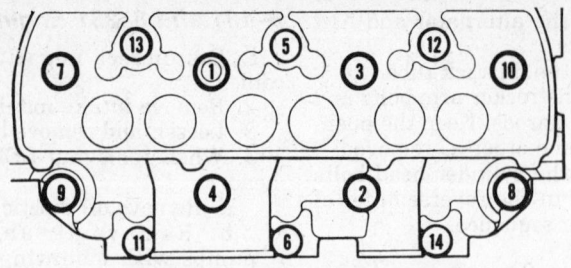

6-263 head bolt torque sequence

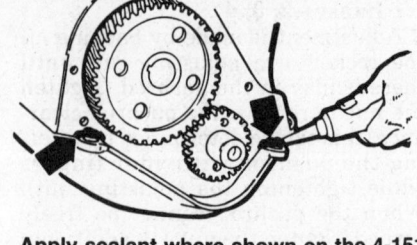

Apply sealant where shown on the 4-151

Torque the head bolts in three equal stages. Recheck head bolt torque after the engine has been warmed to operating temperature.

6–263 Engine

NOTE: This procedure requires the complete disassembly of the valve lifters as explained under "Diesel Engine Valve Lifter Bleed-Down".

1. Remove intake manifold.
2. Remove valve cover. Loosen or remove any accessory brackets or pipe clamps which interfere.
3. Disconnect glow plug wiring (and block heater lead if so equipped on rear bank).
4. Remove the ground strap from right (rear) cylinder head.
5. Remove rocker arm nuts, pivots, rocker arms and pushrods. Scribe pivots and keep rocker arms separated so they can be installed in their original locations.
6. Disconnect the exhaust crossover pipe from the exhaust manifold on the side being worked on and loosen it on the other.
7. Remove engine block drain plug, from side of the block where head is being removed.
8. Remove the pipe plugs covering the upper cylinder head bolts.
9. Remove all the cylinder head bolts and remove the cylinder head.
10. If necessary to remove the prechamber, remove the glow plug and injection nozzle, then tap out with a small blunt 1/8 in. drift. DO NOT use a tapered drift.
11. Installation is the reverse of removal. Do not use sealer on the head gasket. If a pre-chamber was replaced, measure the chamber height and grind the new one to within 0.001 in. of the old chamber's height, using No. 80 grit wet sandpaper to polish it. Coat the head bolts with sealer.

OVERHAUL

For all cylinder head overhaul procedures, please refer to "Engine Rebuilding" in the Unit Repair Section.

Front Cover

REMOVAL & INSTALLATION

4–151 Engine

—— CAUTION ——

On fuel injected engines, relieve the pressure in the fuel system before disconnecting the fuel line connections.

1. Remove the crankshaft hub. It is necessary to remove the inner fender splash shield.
2. Remove the alternator lower bracket.
3. Remove the front engine mounts.
4. Using a floor jack, raise the engine.
5. Remove the engine mount mounting bracket-to-cylinder block bolts. Remove the bracket and mount as an assembly.
6. Remove the oil pan-to-front cover screws.
7. Remove the front cover-to-block screws.
8. Pull the cover slightly forward, just enough to allow cutting of the oil pan front seal flush with the block on both sides.
9. Remove the front cover and attached portion of the pan seal.
10. Clean the gasket surfaces thoroughly.
11. Cut the tabs from the new oil pan front seal.
12. Install the seal on the front cover, pressing the tips into the holes provided.
13. Coat the new gasket with sealer and position it on the front cover.
14. Apply a 1/8 in. bead of silicone sealer to the joint formed at the oil pan and block.
15. Align the front cover seal with a centering tool and install the front cover. Tighten the screws and install the hub.

6–173 Engine

—— CAUTION ——

The outer ring (weight) of the harmonic balancer is bonded to the hub with rubber.

Breakage may occur if the balancer is hammered back onto the crankshaft. A press or special installation tool is necessary.

1. Remove the water pump.
2. Remove the compressor without disconnecting any A/C lines and lay it aside.
3. Remove harmonic balancer, using a puller.

NOTE: The outer ring (weight) of the harmonic balancer is bonded to the hub with rubber. The balancer must be removed with a puller which acts on the inner hub only. Pulling on the outer portion of the balancer will break the rubber bond or destroy the tuning of the torsional damper.

4. Disconnect the lower radiator hose and heater hose.
5. Remove timing gear cover attaching screws, and cover and gasket.
6. Clean all the gasket mounting surfaces on the front cover and block. Apply a continuous 3/22 in. bead of sealer (GM No. 1052357 or equivalent) to front cover sealing surface and around coolant passage ports and central bolt holes.
7. Apply a bead of silicone sealer to the oil pan-to-cylinder block joint.
8. Install a centering tool in the crankshaft snout hole in the front cover and install the cover.
9. Install the front cover bolts finger tight, remove the centering tool and tighten the cover bolts. Install the harmonic balancer, pulley, water pump, belts, radiator, and all other parts.

6–181 and 6–231 Engines

1. Drain the cooling system.
2. Disconnect the lower radiator hose and the heater hose at the water pump.
3. Remove the two nuts from the front engine mount at the cradle and raise the engine using a suitable lifting device.
4. Remove the water pump pulley and all drive belts.

5. Remove the alternator and brackets.

6. Remove the distributor.

NOTE: If the timing chain and sprockets are not going to be disturbed, note the position of the distributor rotor for reinstallation in the same position.

7. Remove the balancer bolt and washer, and using a puller, remove the balancer.

8. Remove the cover-to-block bolts. Remove the two oil pan-to-cover bolts.

9. Remove the cover and gasket.

10. Installation is the reverse of removal. Always use a new gasket coated with sealer. Remove the oil pump cover and pack the area around the gears with petroleum jelly so that no air space is left within the pump. Apply sealer to the cover bolt threads.

6–263 Engine

1. Drain the cooling system.

2. Disconnect the lower radiator hose and the heater hose at the water pump. Disconnect the heater outlet pipe at the manifold.

3. Disconnect the power steering pump, vacuum pump, belt tensioner, air conditioning compressor and alternator brackets.

— CAUTION —

Do not disconnect any refrigerant lines.

4. Remove the crankshaft balancer using a puller.

5. Unbolt and remove the front cover and gasket.

6. Installation is the reverse of removal. Grind a chamfer on the end of each dowel pin to aid in cover installation. Trim $\frac{1}{8}$ inch from the ends of the new front pan seal. Apply RTV sealer

to the oil pan seal retainer. After the cover gasket is in place, apply sealer to the junction of the pan, gasket and block. When installing the cover, rotate it right and left while guiding the pan seal into place with a small screwdriver.

Oil Seal

REMOVAL & INSTALLATION

1. After removing the timing cover, pry oil seal out of front of cover.

2. Install new lip seal with lip (open side of seal) inside and drive or press seal carefully into place.

NOTE: The timing cover oil seal can be replaced without removing the cover. Remove the fan belts, crankshaft pulley and harmonic balancer. Pry the oil seal out the cover working carefully to prevent damage to the seal mating surface. Lubricate the new seal and drive it into place with the open side toward the engine. Use a seal installer to avoid damaging or cocking the seal.

Timing Gear and/or Chain

REMOVAL & INSTALLATION

4–151 Engine

See the Camshaft Removal & Installation procedure.

6–173 Engine

To replace the chain, remove the crankcase front cover. This will allow

access to the timing chain. Crank the engine until the marks punched on both sprockets are closest to one another and in line between the shaft centers. Take out the three bolts that hold the camshaft sprocket to the camshaft. This sprocket is a light press fit on the camshaft and will come off readily. It is located by a dowel. The chain comes off with the

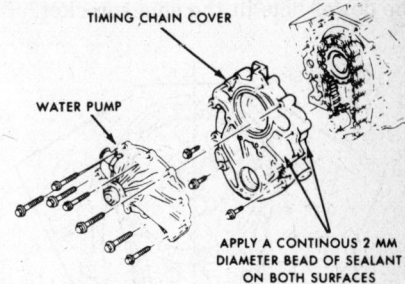

6-173 timing cover removal

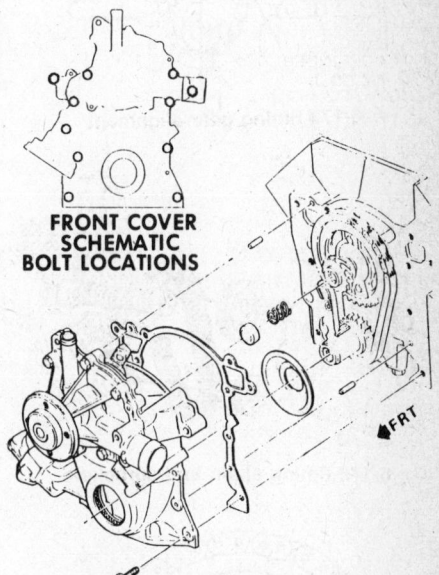

Timing cover removal—6–181 and 6–231

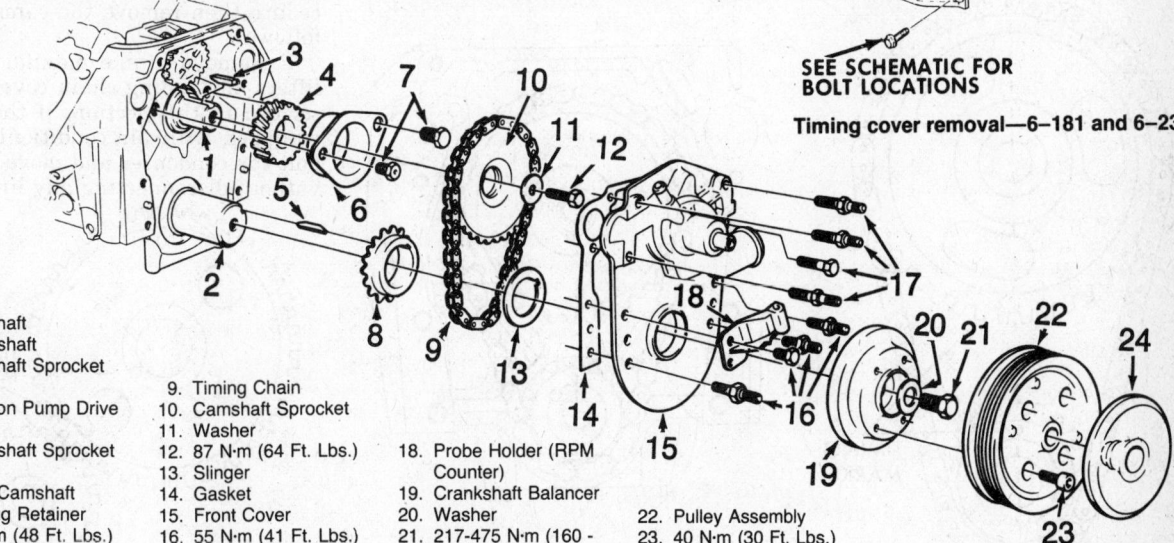

1. Camshaft	
2. Crankshaft	
3. Camshaft Sprocket Key	
4. Injection Pump Drive Gear	9. Timing Chain
5. Crankshaft Sprocket Key	10. Camshaft Sprocket
	11. Washer
6. Front Camshaft Bearing Retainer	12. 87 N·m (64 Ft. Lbs.)
7. 65 N·m (48 Ft. Lbs.)	13. Slinger
8. Crankshaft Sprocket	14. Gasket
	15. Front Cover

16. 55 N·m (41 Ft. Lbs.)	
17. 28 N·m (21 Ft. Lbs.)	
18. Probe Holder (RPM Counter)	
19. Crankshaft Balancer	
20. Washer	22. Pulley Assembly
21. 217-475 N·m (160 - 350 Ft. Lbs.)	23. 40 N·m (30 Ft. Lbs.)
	24. Cover

6-263 timing cover and chain removal

camshaft sprocket. A gear puller will be required to remove the crankshaft sprocket.

Without disturbing the position of the engine, mount the new crank sprocket on the shaft, then mount the chain over the camshaft sprocket. Arrange the camshaft sprocket in such a way that the timing marks will line up between the shaft centers and the camshaft locating dowel will enter the dowel hole in the cam sprocket.

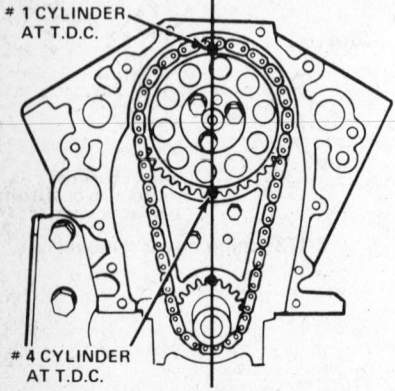

6-173 timing gear alignment

6-173 timing chain and sprockets

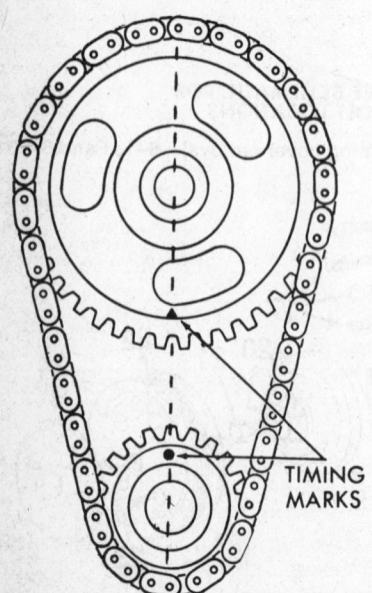

Timing gear alignment—6–181 and 6–231

Place the cam sprocket, with its chain mounted over it, in position on the front of the camshaft and pull up with the three bolts that hold it to the camshaft. After the sprockets are in place, turn the engine two full revolutions to make certain that the timing marks are in correct alignment between the shaft centers.

6–181 and 6–231 Engines

1. Remove the timing chain cover as outlined earlier.
2. Turn the crankshaft so that the timing marks are aligned.
3. Remove the crankshaft oil slinger.
4. Remove the camshaft sprocket bolts.
5. Use two prybars to alternately pry the camshaft and crankshaft sprocket free along with the chain.
6. Installation is the reverse of removal. If the engine was turned, make sure that the No. 1 cylinder is at TDC.

6–263 Engine

NOTE: The following procedure requires the bleed-down of the valve lifters. Read that procedure before proceeding.

1. Remove the front cover.
2. Loosen all the rocker arms. See Rocker Arm Removal & Installation.
3. Remove the crankshaft oil slinger.
4. Remove the camshaft sprocket bolt.
5. Using two prybars, work the camshaft and crankshaft sprockets alternately off their shafts along with the chain. It may be necessary to remove the crankshaft sprocket with a puller.

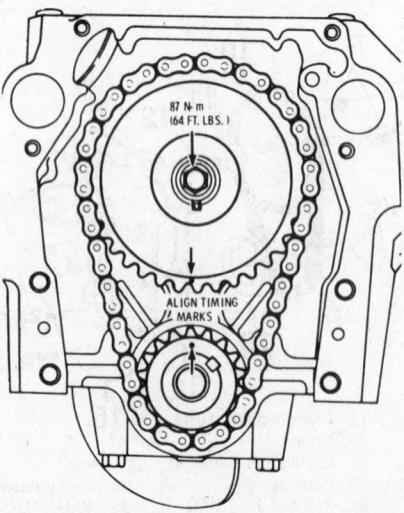

6-263 timing gear alignment

6. Installation is the reverse of removal. If the engine was turned, make sure that the No. 1 piston is at TDC. Bleed the lifters following the procedure under "Diesel Engine Valve Lifter Bleed-Down."

Camshaft

REMOVAL & INSTALLATION

4–151 Engine

CAUTION

Relieve the pressure in the EFI system on fuel injected engines before disconnecting the fuel line connections.

---

1. Remove the engine as previously outlined.
2. Remove the rocker cover, rocker arms, and pushrods.
3. Remove the distributor, spark plugs, and fuel pump.
4. Remove the pushrod cover and gasket. Remove the lifters.
5. Remove the alternator, the alternator lower bracket and the front engine mount bracket assembly.
6. Remove the oil pump driveshaft and gear assembly.
7. Remove the crankshaft hub and timing gear cover.
8. Remove the two camshaft thrust plate screws by working through the holes in the gear.
9. Remove the camshaft and gear assembly by pulling it through the front of the block. Take care not to damage the bearings.
10. Install in the reverse order. Torque the thrust plate screws to 75 inch lbs.

6–173 Engine

Follow the 6–173 engine removal procedure then remove the camshaft as follows:

1. Remove intake manifold, valve lifters and timing chain cover as described in this section. If the car is equipped with air conditioning, unbolt the condenser and move it aside without disconnecting any lines.

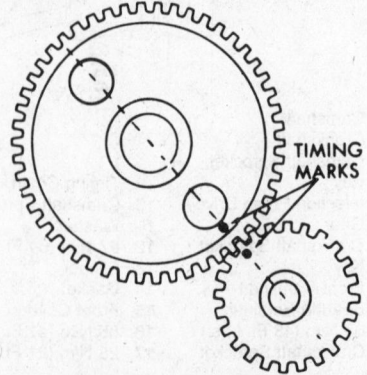

4-151 timing gear alignment

2. Remove fuel pump and pump pushrod.

3. Remove camshaft sprocket bolts, sprocket and timing chain. A light blow to the lower edge of a tight sprocket should free it (use a plastic mallet).

4. Install two bolts in cam bolt holes and pull cam from block.

5. To install, reverse removal procedure aligning the sprocket timing marks.

6–181 and 6–231 Engines

1. Remove the engine as described earlier.
2. Remove the intake manifold.
3. Remove the rocker arm covers.
4. Remove the rocker arm assemblies, pushrods and lifters.
5. Remove the timing chain cover.

NOTE: Align the timing marks of the camshaft and crankshaft sprockets to avoid burring the camshaft journals by the crankshaft.

6. Remove the timing chain and camshaft sprocket as described earlier.

7. Installation is the reverse of removal.

6–263 Engine

NOTE: This procedure requires the removal, disassembly, cleaning, reassembly and bleed-down of all the valve lifters. Read that procedure, described earlier, before proceeding.

1. Remove the engine as described earlier.
2. Remove the intake manifold.
3. Remove the oil pump drive assembly.
4. Remove the timing chain cover.
5. Align the timing marks.
6. Remove the rocker arms, pushrods and lifters, keeping them in order for reassembly.
7. Remove the timing chain and camshaft sprocket as described earlier.
8. Remove the camshaft bearing retainer.
9. Remove the cam sprocket key.
10. Remove the injection pump drive gear.
11. Remove the injection pump driven gear, intermediate pump adapter and pump adapter. Remove the snap ring and selective washer. Remove the driven gear and spring.
12. Carefully slide the camshaft out of the block.
13. If the camshaft bearings are being replaced, you'll have to remove the oil pan.

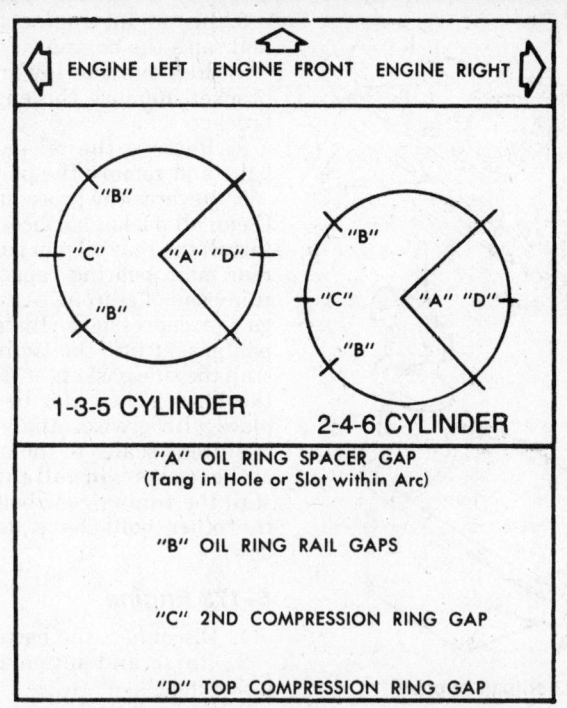

"A" OIL RING SPACER GAP
(Tang in Hole or Slot within Arc)

"B" OIL RING RAIL GAPS

"C" 2ND COMPRESSION RING GAP

"D" TOP COMPRESSION RING GAP

Arrange the piston rings on all V6 engines as shown

On all engines, the piston assemblies are installed with the notch facing forward

14. Installation is the reverse of removal. Perform the complete valve lifter bleed-down procedure mentioned earlier.

Piston and Connecting Rod Positioning

For all piston and connecting rod overhaul procedures, please refer to "Engine Rebuilding" in the Unit Repair Section. See the accompanying illustrations to properly install piston and connecting rod assemblies.

ENGINE LUBRICATION

Oil Pan

REMOVAL & INSTALLATION

4–151 Engine

1. Raise and support the vehicle safely. Drain the oil.
2. Remove the engine cradle-to-front engine mounts.
3. Disconnect the exhaust pipe at both the exhaust manifold and at the rear transaxle mount.
4. Disconnect and remove the starter. Remove the flywheel housing or torque converter cover.
5. Remove the alternator upper bracket.

6. Install an engine lifting chain and raise the engine.

7. Remove the lower alternator bracket. Remove the engine support bracket.

8. Remove the oil pan retaining bolts and remove the pan.

9. Reverse the procedure to install. Clean all gasket surfaces thoroughly. Install the rear oil pan gasket into the rear main bearing cap, then apply a thin bead of silicone sealer to the pan gasket depressions. Install the front pan gasket into the timing cover. Install the side gaskets onto the pan, not the block. They can be retained in place with grease. Apply a thin bead of silicone sealer to the mating joints of the gaskets. Install the oil pan; install the timing gear bolts last, after the other bolts have been snugged down.

6-173 Engine

1. Disconnect the battery ground.
2. Raise and support the car on jackstands.
3. Drain the oil.
4. Remove the bellhousing cover.
5. Remove the starter.
6. Support the engine.
7. Unbolt the engine from its mounts.
8. Remove the oil pan bolts.
9. Raise the engine with a jack, just enough to remove the oil pan.
10. Installation is the reverse of removal. The pan is installed using RTV gasket material in place of a gasket. Make sure that the sealing surfaces are free of old RTV material. Use a ⅛ in. bead of RTV material on the pan sealing flange. Torque the pan bolts to 8–10 ft. lbs.

6-181 and 6-231 Engines

1. Disconnect the battery ground cable.
2. Raise and support the car on jackstands.
3. Drain the oil.
4. Remove the bellhousing cover.

5. Unbolt and remove the oil pan.
6. Installation is the reverse of removal. RTV gasket material is used in place of a gasket. Make sure that the sealing surfaces are free of all old RTV material. Use a ⅛ in. bead of RTV material on the oil pan sealing flange. Torque the pan bolts to 10–14 ft. lbs.

6-263 Engine

--------- CAUTION ---------

The following procedure will be personally hazardous unless the procedures are followed exactly.

1. Install the engine support fixture assembly shown in the accompanying illustration. Be certain to arrange washers on the fixture so that the bolt securing the chain to the cylinder head can be torqued to 20 ft. lbs. THIS IS ABSOLUTELY NECESSARY.

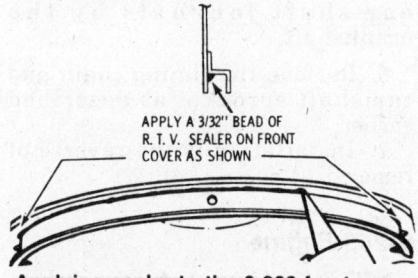

APPLY A 3/32" BEAD OF R. T. V. SEALER ON FRONT COVER AS SHOWN

Applying sealer to the 6-263 front cover

2. Raise the front and rear of the car and support it on jackstands with the rear slightly lower than the front. The front jackstands should be located at the front lift points.
3. Drain the oil.
4. Remove the left side steering gear cradle bolt and loosen the right side cradle bolts.
5. Remove the front stabilizer bar.
6. Using a ½ in. drill bit, drill through the spot weld located between the rear holes at the left front stabilizer bar mounting.
7. Remove the nuts securing the engine and transaxle to its cradle.
8. Disconnect the left lower ball joint from the knuckle.
9. Place a wood block on a floor jack and raise the transaxle under the pan until the mount studs clear the cradle.
10. Remove the bolts securing the front crossmember to the right side of the cradle.
11. Remove the bolts from the left side front body mounts.
12. Remove the left side and front crossmember assemblies. It will be necessary to lower the rear crossmember below the left side of the body through the careful use of a large prybar.

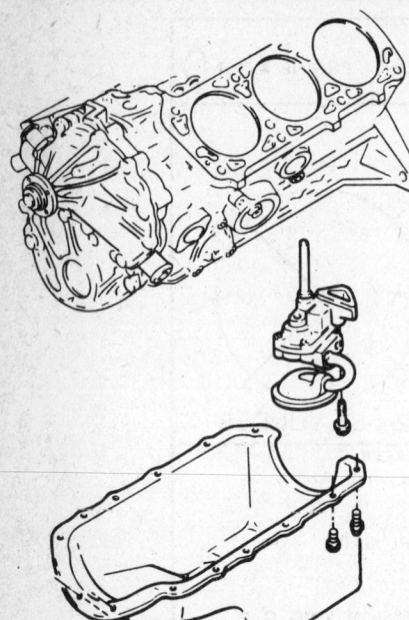

6-173 oil pan removal

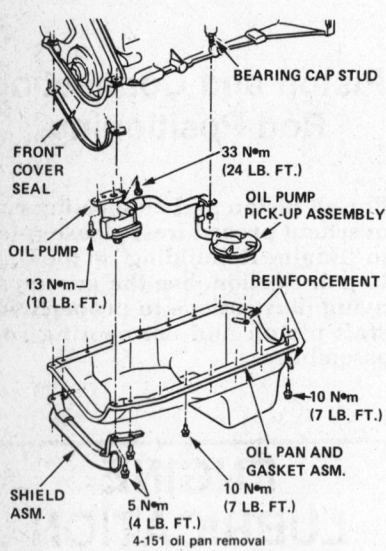

BEARING CAP STUD

FRONT COVER SEAL

OIL PUMP
33 N•m (24 LB. FT.)

OIL PUMP PICK-UP ASSEMBLY

13 N•m (10 LB. FT.)

REINFORCEMENT

10 N•m (7 LB. FT.)

OIL PAN AND GASKET ASM.

SHIELD ASM.

5 N•m (4 LB. FT.)

10 N•m (7 LB. FT.)

4-151 oil pan removal

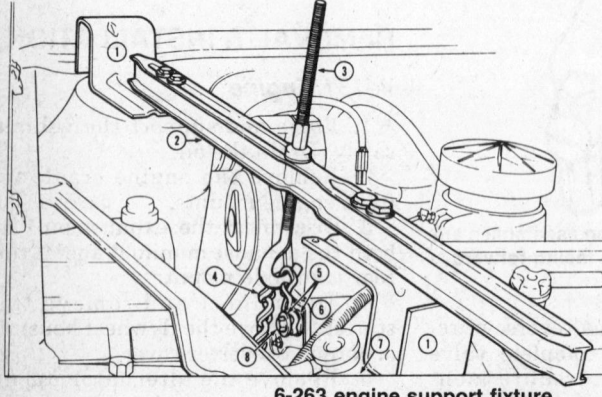

1. J-22825-45 Supports
2. BT-6603 or J-22825-1 Bar
3. J-22825-48 Hook
4. Chain
5. Washers
6. Right Cylinder Head
7. Radiator Support
8. 27 N•m (20 Ft. Lbs.)

6-263 engine support fixture

13. Remove the bellhousing cover.
14. Remove the starter.
15. Remove the engine front mount bracket.
16. Unbolt and remove the oil pan.
17. Installation is the reverse of removal. Apply sealer to both sides of the oil pan gasket and make sure that the tabs on the gaskets are installed in the seal notches. Apply RTV sealer to the front cover oil pan seal retainer, and to each seal where it contacts the block. Wipe the seal area of the pan with clean engine oil before installing the pan. Torque the pan bolts to 10 ft. lbs.

Rear Main Bearing Oil Seal

REMOVAL & INSTALLATION

4–151 Engine

1. Remove the transaxle and flywheel.
2. Being careful not to scratch the crankshaft, pry out the old seal with a screwdriver.
3. Coat the new seal with clean engine oil, and install it by hand onto the crankshaft. The seal backing must be flush with the block opening.
4. Install all other parts in reverse of removal.

6–173 Engine

1. Remove the oil pan and pump.
2. Remove the rear main bearing cap.
3. Gently pack the upper seal into the groove approximately $\frac{1}{4}$ in. on each side.
4. Measure the amount the seal was driven in on one side and add $\frac{1}{16}$ in. Cut this length from the old lower cap seal. Be sure to get a sharp cut. Repeat for the other side.
5. Place the piece of cut seal into the groove and pack the seal into the block. Do this for each side.

NOTE: GM makes a guide tool (J-29114-1) which bolts to the block via an oil pan bolt hole, and a packing tool (J-29114-2) which are machined to provide a built-in stop for the installation of the short cut pieces. Using the packing tool, work the short pieces of seal onto the guide tool, then pack them into the block with the packing tool.

6. Install a new lower seal in the rear main cap.
7. Install a piece of Plastigage or the equivalent on the bearing journal. Install the rear cap and tighten to 70 ft. lbs. Remove the cap and check the

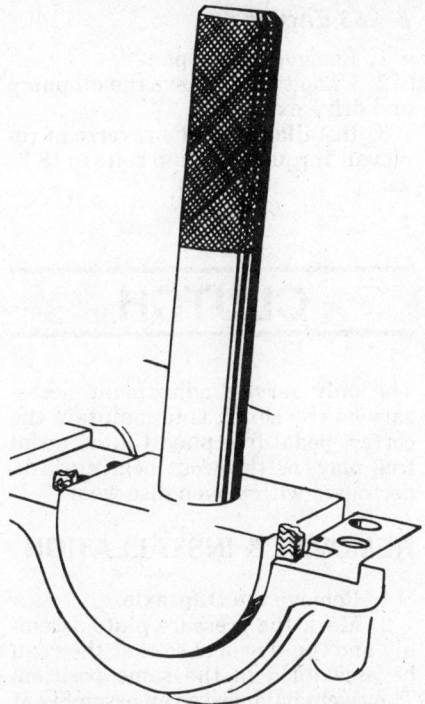

AFTER CORRECTLY POSITIONING SEAL, ROTATE TOOL SLIGHTLY AND CUT OFF EACH END OF SEAL FLUSH WITH BLOCK

Installing the lower seal half

gauge for bearing clearance. If out of specification, the ends of the seal may be frayed or not flush, preventing the cap from proper sealing. Correct as required.
8. Clean the journal, and apply a thin film of sealer to the mating surfaces of the cap and block. Do not allow any sealer to get onto the journal or bearing. Install the bearing cap and tighten to 70 ft. lbs. Install the pan and pump.

6–181, 6–231 and 6–263 Engines

Braided fabric seals are pressed into grooves formed in crankcase and rear bearing cap to rear of the oil collecting groove, to seal against leakage of oil around the crankshaft.

A new braided fabric seal can be installed in crankcase only when crankshaft is removed, but it can be repaired while crankshaft is installed, as outlined under "Rear Main Bearing Upper Oil Seal Repair". The seal can be replaced in cap whenever the cap is removed. Remove old seal and place new seal in groove with both ends projecting above parting surface of cap. Force seal into groove rubbing down with hammer handle or smooth stick until seal projects above the groove not more than $\frac{1}{16}$ in. Cut ends off flush with surface of cap, using sharp knife or razor blade.

The engine must be operated at

slow speed when first started after a new braided seal is installed. Neoprene composition seals are placed in grooves in the sides of bearing cap to seal against leakage in the joints between cap and crankcase. The neoprene composition swells in the presence of oil and heat. The seals are undersize when newly installed and may even leak for a short time until the seals have had time to swell and seal the opening.

The neoprene seals are slightly longer than the grooves in the bearing cap. The seals must not be cut to length. Before installation of seals, soak for 1 to 2 minutes in light oil or kerosene. After installation of bearing cap in crankcase, install seal in bearing cap.

To help eliminate oil leakage at the joint where the cap meets the crankcase, apply silicone sealer, or equivalent, to the rear main bearing cap split line. When applying sealer, use only a thin coat as an over abundance will not allow the cap to seat properly. After seal is installed, force seals up into the cap with a blunt instrument to be sure of a seal at the upper parting line between the cap and case.

REAR MAIN BEARING UPPER OIL SEAL REPAIR

1. Remove oil pan.
2. Insert packing tool (J-21526-2) against one end of the seal in the cylinder block. Drive the old seal gently into the groove until it is packed tight. This varies from $\frac{1}{4}$–$\frac{3}{4}$ in. depending on the amount of pack required.
3. Repeat Step 2 on the other end of the seal in the cylinder block.
4. Measure the amount the seal was driven up on one side and add $\frac{1}{16}$ in. Using a single edge razor blade, cut that length from the old seal removed from the rear main bearing cap. Repeat the procedure for the other side. Use the rear main bearing cap as a holding fixture when cutting the seal.
5. Install Guide Tool (J-21526-1) onto cylinder block.
6. Using packing tool, work the short pieces cut in Step 4 into the guide tool and then pack into cylinder block. The guide tool and packing tool have been machined to provide a built-in stop. Use this procedure for both sides. It may help to use oil on the short pieces of the rope seal when packing into the cylinder block.
7. Remove the guide tool.
8. Install a new fabric seal in the rear main bearing cap. Install cap and torque to specifications.
9. Install oil pan.

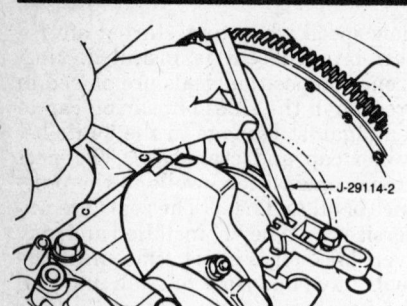

Installing the upper rear main seal on 6-173 engines

Oil Pump

REMOVAL & INSTALLATION

4–151 and 6–173 Engines

1. Remove the oil pan as described earlier.
2. Unbolt and remove the oil pump and pickup.
3. Installation is the reverse of removal. Torque the 4–151 pump to 22 ft. lbs. and the 6–173 pump bolts to 26–35 ft. lbs.

6–181 and 6–231 Engines

1. Remove the oil filter.
2. Unbolt the oil pump cover from the timing chain cover.
3. Slide out the oil pump gears. Clean all parts thoroughly in solvent and check for wear. Remove the oil pressure relief valve cap, spring and valve.

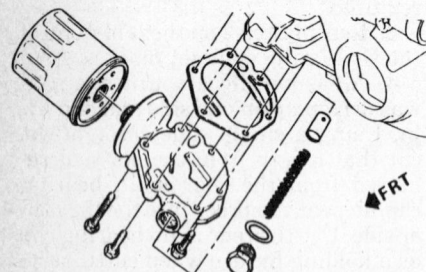

Oil pump—6–181 and 6–231

4. Installation is the reverse of removal. Torque the pressure relief valve cap to 35 ft. lbs. Install the pump gears and check their clearances:
    a. End clearance: 0.002–0.006 in.
    b. Side clearance: 0.002–0.005 in.

Place a straightedge across the face of the pump cover and check that it is flat to within 0.001 in. Pack the oil pump cavity with petroleum jelly so that there is no air space. Install the cover and torque the bolts to 10 ft. lbs.

6–263 Engine

1. Remove the oil pan.
2. Unbolt and remove the oil pump and drive extension.
3. Installation is the reverse of removal. Torque the pump bolts to 18 ft. lbs.

CLUTCH

The only service adjustment necessary on the clutch is to maintain the correct pedal free play. Clutch pedal free play, or throwout bearing lash, decreases with driven disc wear.

REMOVAL & INSTALLATION

1. Remove the transaxle.
2. Mark the pressure plate assembly and the flywheel so that they can be assembled in the same position. They were balanced as an assembly at the factory.
3. Loosen the attaching bolts one turn at a time until spring tension is relieved.
4. Support the pressure plate and remove the bolts. Remove the pressure plate and clutch disc. Do not disassemble the pressure plate assembly; replace it if defective.
5. Inspect the flywheel, clutch disc, pressure plate, throwout bearing and the clutch fork and pivot shaft assembly for wear. Replace the parts as required. If the flywheel shows any signs of overheating, or if it is badly grooved or scored, it should be replaced.
6. Clean the pressure plate and flywheel mating surfaces thoroughly. Position the clutch disc and pressure plate into the installed position, and support with a dummy shaft or clutch

aligning tool. The clutch plate is assembled with the damper springs offset toward the transaxle. One side of the factory-supplied clutch disc is stamped "Flywheel side".
7. Install the pressure plate-to-flywheel bolts. Tighten them gradually in a crisscross pattern.
8. Lubricate the outside groove and the inside recess of the release bearing with high temperature grease. Wipe off any excess. Install the release bearing.
9. Install the transaxle.

CLUTCH LINKAGE AND PEDAL HEIGHT/FREE–PLAY ADJUSTMENT

All cars use a self-adjusting clutch mechanism which may be checked as follows. As the clutch friction material wears, the cable must be lengthened. This is accomplished by simply pulling the clutch pedal up to its rubber bumper. This action forces the pawl against its stop and rotates it out of mesh with the quadrant teeth, allowing the cable to play out until the quadrant spring load is balanced against the load applied by the release bearing. This adjustment procedure is required every 5000 miles or less.

1. With engine running and brake on, hold the clutch pedal approximately $\frac{1}{2}$ in. from floor mat and move shift lever between first and reverse several times. If this can be done smoothly without clashing into reverse, the clutch is fully releasing. If shift is not smooth, clutch is not fully releasing and linkage should be inspected and corrected as necessary.
2. Check clutch pedal bushings for sticking or excessive wear.
3. Have an assistant sit in the driver's seat and fully apply the clutch pedal to the floor. Observe the clutch

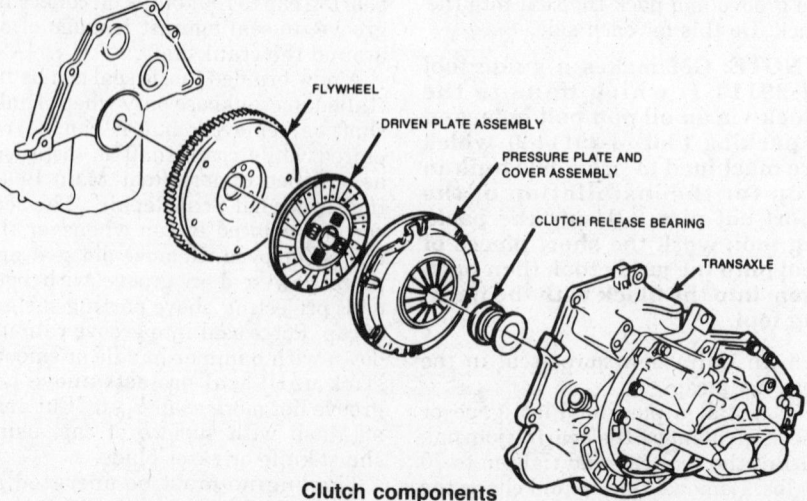

FLYWHEEL

DRIVEN PLATE ASSEMBLY

PRESSURE PLATE AND COVER ASSEMBLY

CLUTCH RELEASE BEARING

TRANSAXLE

Clutch components

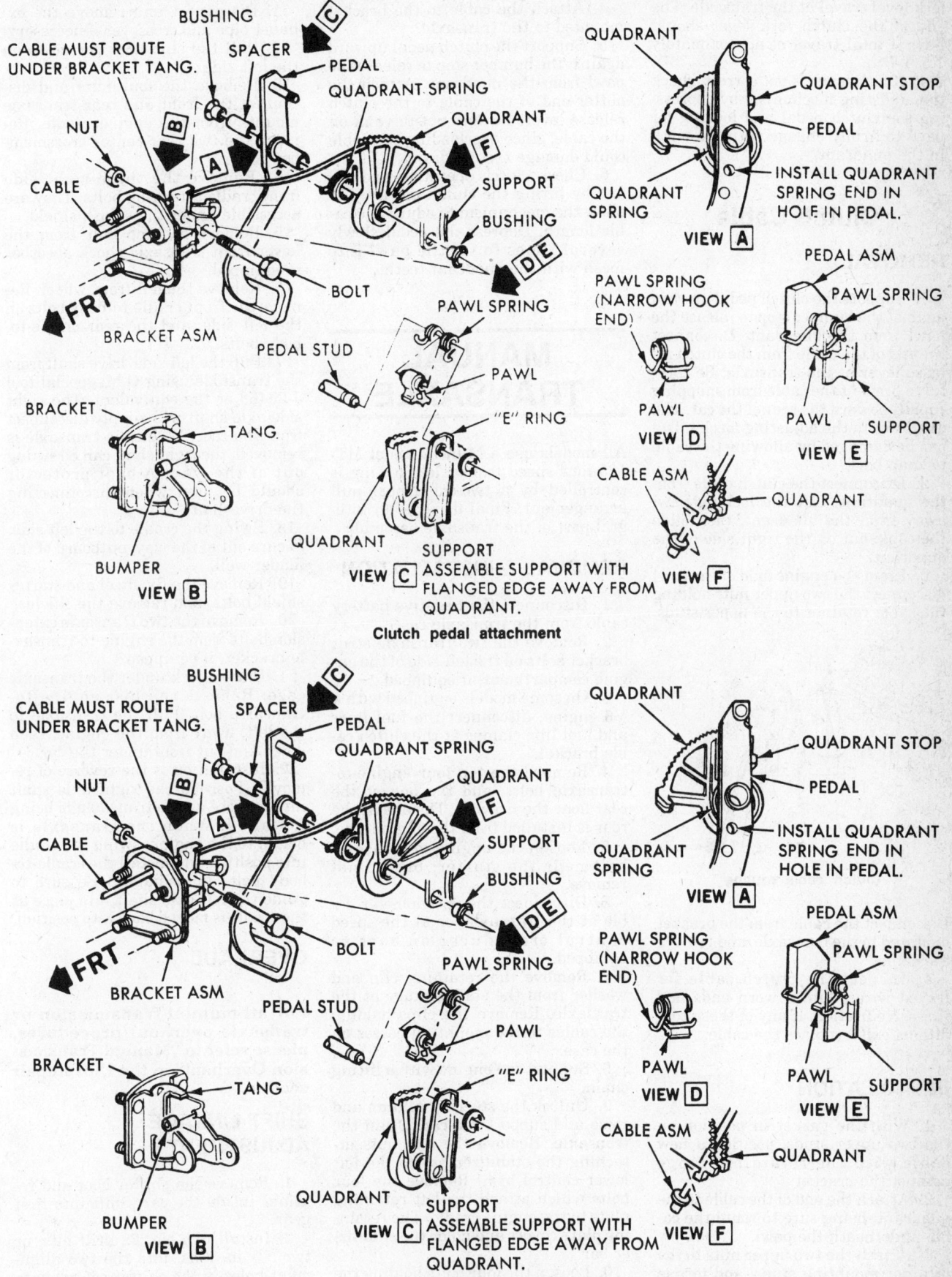

Clutch pedal attachment

Clutch cable and pedal

fork level travel at the transaxle. The end of the clutch fork lever should have a total travel of approximately 1.5–1.7 in.

4. If fork lever is not correct, check the adjusting mechanism by depressing the clutch pedal and looking for pawl to firmly engage with the teeth in the quadrant.

Clutch Cable

REMOVAL

1. Support the clutch pedal upward against the bumper stop to release the pawl from the quadrant. Disconnect the end of the cable from the clutch release lever at the transaxle. Be careful to prevent the cable from snapping rapidly toward the rear of the car. The quadrant in the adjusting mechanism can be damaged by allowing the cable to snap back.

2. Disconnect the clutch cable from the quadrant. Lift the locking pawl away from the quadrant, then slide the cable out on the right side of the quadrant.

3. From the engine side of the cowl disconnect the two upper nuts holding the cable retainer to the upper studs.

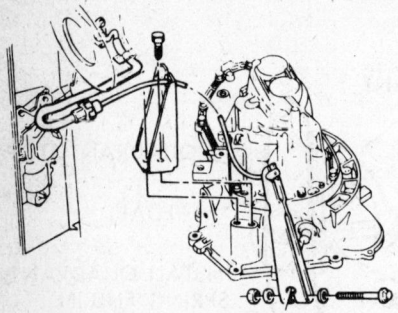

Clutch cable routing

Disconnect the cable from the bracket mounted to the transaxle, and remove the cable.

4. Inspect the clutch cable for frayed wires, kinks, worn ends and excessive friction. If any of these conditions exist, replace the cable.

INSTALLATION

1. With the gasket in position on the two upper studs, position a new cable with the retaining flange against the bracket.

2. Attach the end of the cable to the quadrant, being sure to route the cable underneath the pawl.

3. Attach the two upper nuts to the retainer mounting studs, and torque to specifications.

4. Attach the cable to the bracket mounted to the transaxle.

5. Support the clutch pedal upward against the bumper stop to release the pawl from the quadrant. Attach the outter end of the cable to the clutch release lever. Be sure not to yank on the cable, since overloading the cable could damage the quadrant.

6. Check clutch operation and adjust by lifting the clutch pedal up to allow the mechanism to adjust the cable length. Depress the pedal slowly several times to set the pawl into mesh with the quadrant teeth.

MANUAL TRANSAXLE

All models use a Muncie model MT-125 four speed transaxle. Shifting is controlled by a two-cable push-pull arrangement. Final drive is an integral part of the transaxle assembly.

REMOVAL & INSTALLATION

1. Disconnect the negative battery cable from the transaxle case.

2. Remove the two transaxle strut bracket bolts on the left side of the engine compartment, if equipped.

3. On some models equipped with a V6 engine, disconnect the fuel lines and fuel line clamps at the clutch cable bracket.

4. Remove the top four engine-to-transaxle bolts, and the one at the rear near the firewall. The one at the rear is installed from the engine side.

5. Loosen the engine-to-transaxle bolt near the starter, but do not remove.

6. Disconnect the speedometer cable at the transaxle, or at the speed control transducer on cars so equipped.

7. Remove the retaining clip and washer from the shift linkage at the transaxle. Remove the clips holding the cables to the mounting bosses on the case.

8. Support the engine with a lifting chain.

9. Unlock the steering column and raise and support the car. Drain the transaxle. Remove the two nuts attaching the stabilizer bar to the left lower control arm. Remove the four bolts which attach the left retaining plate to the engine cradle. The retaining plate covers and holds the stabilizer bar.

10. Loosen the four bolts holding the right stabilizer bracket.

11. Disconnect and remove the exhaust pipe and crossover if necessary.

12. Pull the stabilizer bar down on the left side.

13. Remove the four nuts and disconnect the front and rear transaxle mounts from the engine cradle. Remove the two rear center crossmember bolts.

14. Remove the three right side front cradle attaching bolts. They are accessible under the splash shield.

15. Remove the top bolt from the lower front transaxle shock absorber if equipped.

16. Remove the left front wheel. Remove the front cradle-to-body bolts on the left side, and the rear cradle-to-body bolts.

17. Pull the left side drive shaft from the transaxle using G.M. special tool J-28468 or the equivalent. The right side axle shaft will simply disconnect from the case. When the transaxle is removed, the right shaft can be swung out of the way. A boot protector should be used when disconnecting the driveshafts.

18. Swing the cradle to the left side. Secure out of the way, outboard of the fender well.

19. Remove the flywheel and starter shield bolts, and remove the shields.

20. Remove the two transaxle extension bolts from the engine-to-transaxle bracket, if equipped.

21. Place a jack under the transaxle case. Remove the last engine-to-transaxle bolt. Pull the transaxle to the left, away from the engine, then down and out from under the car.

22. Installation is the reverse of removal. Position the right axle shaft into its bore as the transaxle is being installed. When the transaxle is bolted to the engine, swing the cradle into position and install the cradle-to-body bolts immediately. Be sure to guide the left axle shaft into place as the cradle is moved back into position.

OVERHAUL

For all manual transmission or transaxle overhaul procedures, please refer to "Manual Transmission Overhaul" in the Unit Repair section.

SHIFT LINKAGE ADJUSTMENT

1. Remove the shifter boot and retainer inside the car. Shift into first gear.

2. Install two No. 22 drill bits, or two $5/32$ in. rods, into the two alignment holes in the shifter assembly to hold it in first gear.

3. Place the transaxle into first gear by pushing the rail selector shaft down just to the point of feeling the resistance of the inhibitor spring. Then rotate the shift lever all the way counterclockwise.

4. Install the stud, with the cable attached, into the slotted area of the select lever, while gently pulling on the lever to remove all lash.

5. Remove the two drill bits or pins from the shifter.

6. Check the shifter for proper operation. It may be necessary to fine tune the adjustment after road testing.

AUTOMATIC TRANSMISSION

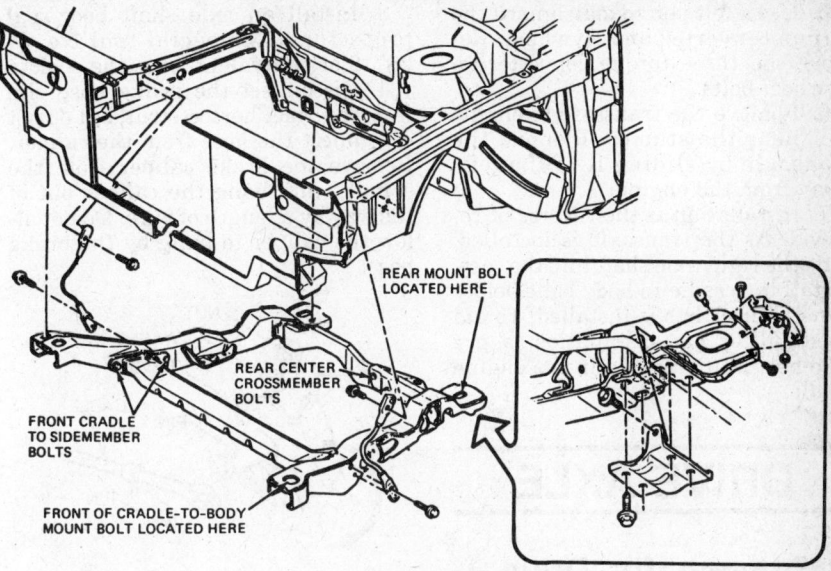

Typical engine/transaxle cradle

All models use a Turbo Hydro-Matic 125 or 125C automatic transmission. The 125C is equipped with a torque converter clutch (TCC) which under certain conditions mechanically couples the engine to the transaxle for greater power transfer efficiency and increased fuel mileage. A cable operated throttle valve linkage is used. Automatic transaxle operation is provided through a conventional three element torque converter, a compound planetary gear set, and a dual sprocket and drive link assembly.

SHIFT LINKAGE ADJUSTMENT, BAND ADJUSTMENT, FLUID AND FILTER CHANGE

All automatic transmission service procedures are contained in the "Automatic Transmission" Unit Repair section of this book.

─────── CAUTION ───────

Any inaccuracies in shift linkage adjustments may result in premature failure of the transmission due to operation without the controls in full detent. Such operation results in reduced fluid pressure and in turn, partial engagement of the affected clutches. Partial engagement of the clutches, with sufficient pressure to permit apparently normal vehicle operation will result in failure of the clutches and/or other internal parts after only a few miles of operation.

REMOVAL & INSTALLATION

1. Disconnect the negative battery cable from the transaxle. Tape the wire to the upper radiator hose to keep it out of the way.

2. Remove the air cleaner and disconnect the detent cable. Slide the detent cable in the opposite direction of the cable to remove it from the carburetor.

3. Unbolt the detent cable attaching bracket at the transaxles.

4. Pull up on the detent cable cover at the transaxle until the cable is exposed. Disconnect the cable from the rod.

5. Remove the two transaxle strut bracket bolts at the transaxle, if equipped.

6. Remove all the engine-to-transaxle bolts except the one near the starter. The one nearest the firewall is installed from the engine side; you will need a short handled box wrench or ratchet to reach it.

7. Loosen, but do not remove the engine-to-transaxle bolt near the starter.

8. Disconnect the speedometer cable at the upper and lower coupling. On cars with cruise control, remove the speedometer cable at the transducer.

9. Remove the retaining clip and washer from the shift linkage at the transaxle. Remove the two shift linkage at the transaxle. Remove the two shift linkage bracket bolts.

10. Disconnect and plug the two fluid cooler lines at the transaxle. These are inch-size fittings ($\frac{1}{2}$ and $\frac{11}{16}$); use a back-up wrench to avoid twisting the lines.

11. Install an engine holding chain or hoist. Raise the engine enough to take its weight off the mounts.

12. Unlock the steering column and raise the car.

13. Remove the two nuts holding the anti-sway (stabilizer) bar to the left lower control arm (driver's side).

14. Remove the four bolts attaching the covering plate over the stabilizer bar to the engine cradle on the left side (driver's side).

15. Loosen but do not remove the four bolts holding the stabilizer bar bracket to the right side (passenger's side) of the engine cradle. Pull the bar down on the driver's side.

16. Disconnect the front and rear transaxle mounts at the engine cradle.

17. Remove the two rear center crossmember bolts.

18. Remove the three right (passenger) side front engine cradle attaching bolts. The nuts are accessible under the splash shield next to the frame rail.

19. Remove the top bolt from the lower front transaxle shock absorber, if equipped (V6 engine only).

20. Remove the left (driver) side front and rear cradle-to-body bolts.

21. Remove the left front wheel. Attach an axle shaft removing tool (GM No. J-28468 or equivalent) to a slide hammer. Place the tool behind the axle shaft cones and pull the cones out away from the transaxle. Remove the right shaft in the same manner. Set the shafts out of the way. Plug the openings in the transaxle to prevent fluid leakage and the entry of dirt.

22. Swing the partial engine cradle to the left (driver) side and wire it out of the way outboard of the fender well.

23. Remove the four torque converter and starter shield bolts. Remove the two transaxle extension bolts from the engine-to-transaxle bracket.

24. Attach a transaxle jack to the case.

25. Use a felt pen to matchmark the torque converter and flywheel. Remove the three torque converter-to-flywheel bolts.

26. Remove the transaxle-to-engine bolt near the starter. Remove the transaxle by sliding it to the left, away from the engine.

27. Installation is the reverse of removal. As the transaxle is installed, slide the right axle shaft into the case. Install the cradle-to-body bolts before the stabilizer bar is installed. To aid in stabilizer bar installation, a pry hole has been provided in the engine cradle.

DRIVE AXLE

Driveshafts, Drive Axles and U-Joints

REMOVAL & INSTALLATION

CAUTION

Use care when removing the drive axle. Tri-pots can be damaged if the drive axle is overextended.

1. Remove the hub nut.
2. Raise the front of the car. Remove the wheel and tire.

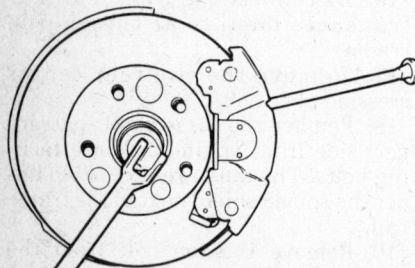

Insert a drift into the caliper when tightening the hub nut

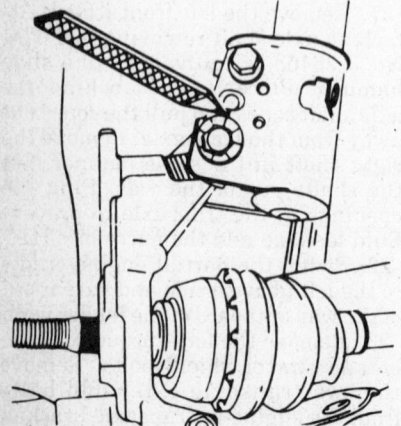

Mark the camber eccentric before removal

3. Install an axle shaft boot seal protector, GM special tool No. J-28712 or equivalent, onto the seal.

4. Disconnect the brake hose clip from the MacPherson strut, but do not disconnect the hose from the caliper. Remove the brake caliper from the spindle, and hang the caliper out of the way by a length of wire. Do not allow the caliper to hang by the brake hose.

5. Mark the camber alignment cam bolt for reassembly. Remove the cam bolt and the upper attaching bolt from the strut and spindle.

6. Pull the steering knuckle assembly from the strut bracket.

7. Using GM special tool J-28733 or the equivalent spindle remover, remove the axle shaft from the hub and bearing assembly.

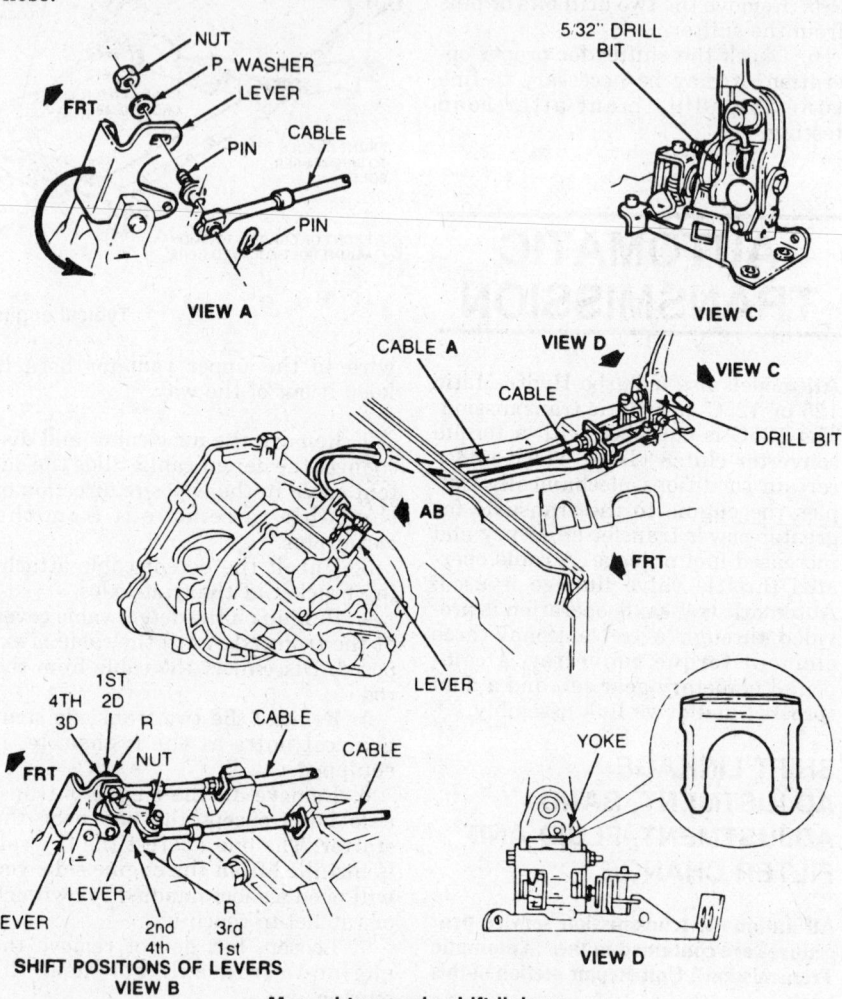

Manual transaxle shift linkage

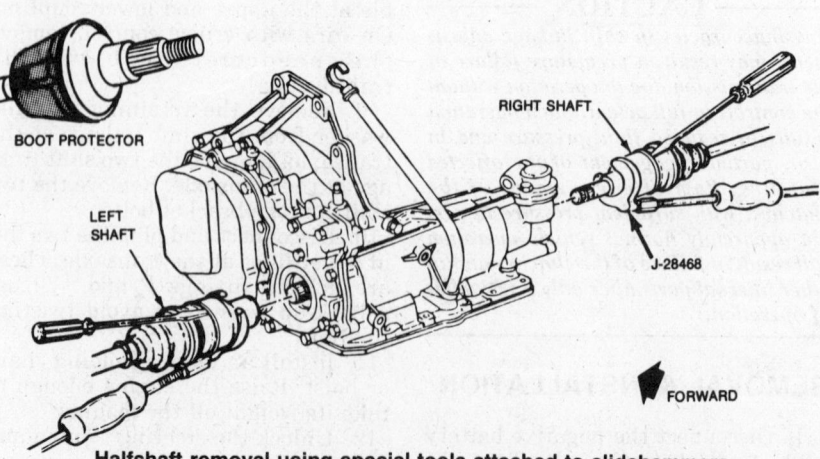

Halfshaft removal using special tools attached to slidehammers

8. If a new drive axle is to be installed, a new knuckle seal should be installed first.

9. Loosely install the drive axle into the transaxle and steering knuckle.

10. Loosely attach the steering knuckle to the suspension strut.

11. The drive axle is an interference fit in the steering knuckle. Press the axle into place, then install the hub nut. When the shaft begins to turn with the hub, insert a drift through the caliper into one of the cooling slots in the rotor to keep it from turning. Insert a long bolt in the hub flange to prevent the shaft from turning. Tighten the hub nut to 70 ft. lbs. to completely seat the shaft.

12. Install the brake caliper. Tighten the bolts to 30 ft. lbs.

13. Load the hub assembly by lowering it onto a jackstand. Align the camber cam bolt marks made during removal, install the bolt and tighten to 140 ft. lbs. Tighten the upper nut to the same value.

14. Install the axle shaft all the way into the transaxle using a screwdriver inserted into the groove provided on the inner retainer. Tap the screwdriver until the shaft seats in the transaxle. Remove the boot seal protector.

15. Connect the brake hose clip to the strut. Install the tire and wheel, lower the car, and tighten the hub nut to 225 ft. lbs. (1980–82); 185 ft. lbs. (1983 and later).

Constant Velocity Joint

OVERHAUL

Please refer to the "U-Joint/CV-Joint" Unit Repair section.

REAR AXLE

Hub and Bearing

REMOVAL & INSTALLATION

A single unit hub and bearing assembly is bolted to both ends of the rear axle assembly. These take the place of "rear axles" used on rear wheel drive cars. The hub and bearing assembly is a sealed unit which requires no maintenance. The unit must be replaced as an assembly and cannot be disassembled or adjusted.

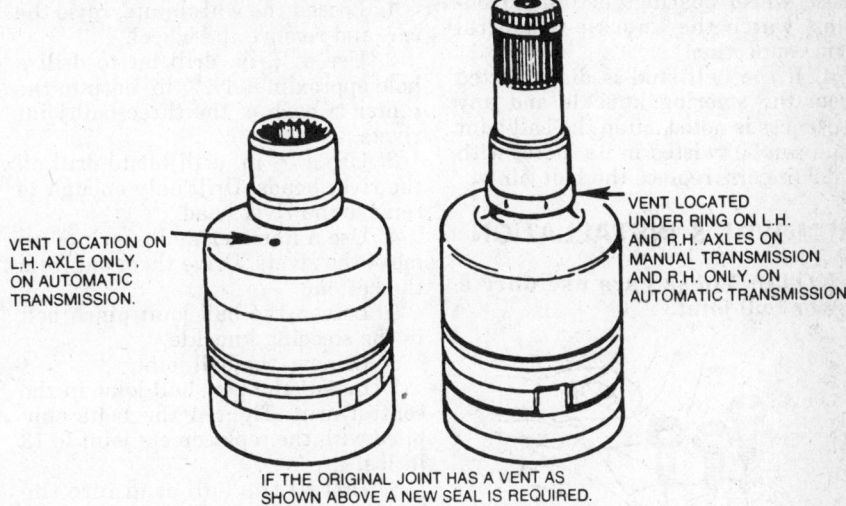

IF THE ORIGINAL JOINT HAS A VENT AS SHOWN ABOVE A NEW SEAL IS REQUIRED.

Comparison of old and new CV joints

The hub and bearing can be removed by removing the rear brake drum, removing the four hub and bearing-to-axle assembly attaching bolts and pulling the unit out. Installation is the reverse of removal. Tighten the bolts to 35–39 ft. lbs.

FRONT SUSPENSION

MacPherson Strut

REMOVAL & INSTALLATION

The MacPherson strut is a combination coil spring and shock absorber (damper) unit. The strut is removed as an assembly from the car. A special strut compressor must be used to disassemble the strut and coil spring.

For all spring and shock absorber Removal & Installation procedures, and all strut overhaul procedures, please refer to "Strut Overhaul" in the Unit Repair section.

1. Loosen the wheel nuts, raise the car, and remove the wheel and tire.

2. Remove the brake hose clip-to-strut bolt (if equipped). Do not disconnect the hose from the caliper. Install a drive axle cover to protect the axle boot.

3. Mark the camber cam eccentric adjuster for assembly.

4. Remove the two lower strut-to-steering knuckle bolts and the three upper strut-to-body nuts. Remove the strut.

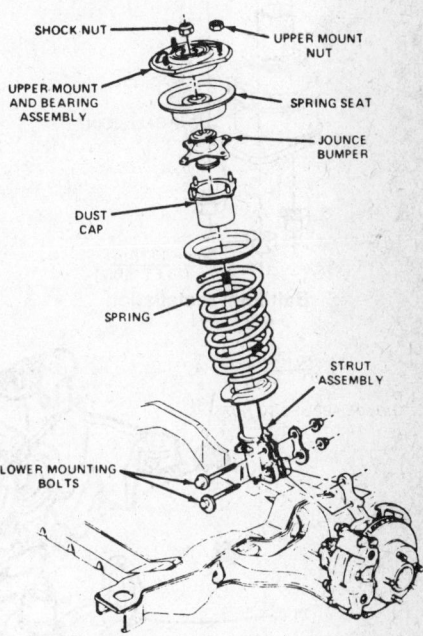

Front suspension components

Ball Joints

INSPECTION

1. Raise the front of the car with a lift placed under the engine cradle. The front wheels should be clear of the ground.

2. Grasp the wheel at the top and bottom and shake the wheel in and out.

3. If any movement is seen of the steering knuckle relative to the control arm, the ball joints are defective and must be replaced. Note that movement elsewhere may be due to

loose wheel bearings or other troubles; watch the knuckle-to-control arm connection.

4. If the ball stud is disconnected from the steering knuckle and any looseness is noted, often the ball joint stud can be twisted in its socket with your fingers, replace the ball joints.

REMOVAL & INSTALLATION

NOTE: These cars use only a lower ball joint.

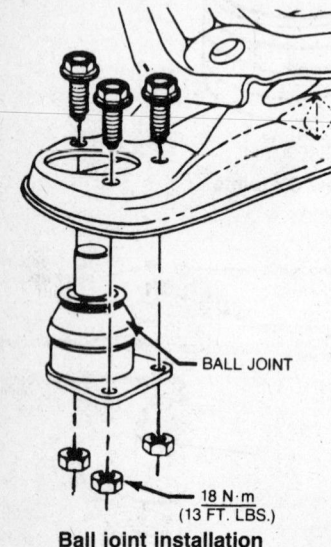

Ball joint installation

18 N·m (13 FT. LBS.)

BALL JOINT

1. Loosen the wheel nuts, raise the car, and remove the wheel.

2. Use a $\frac{1}{8}$ in. drill bit to drill a hole approximately $\frac{1}{4}$ in. deep in the center of each of the three ball joint rivets.

3. Use a $\frac{1}{2}$ in. drill bit to drill off the rivet heads. Drill only enough to remove the rivet head.

4. Use a hammer and punch to remove the rivets. Drive them out from the bottom.

5. Loosen the ball joint pinch bolt in the steering knuckle.

6. Remove the ball joint.

7. Install the new ball joint in the control arm. Tighten the bolts supplied with the replacement joint to 13 ft. lbs.

8. Install the ball stud into the steering knuckle pinch bolt fitting. It should go in easily; if not, check the stud alignment. Install the pinch bolt from the rear to the front. Tighten to 45 ft. lbs.

9. Install the wheel and lower the car.

Lower Control Arm

REMOVAL & INSTALLATION

1. Loosen the wheel nuts, raise the car, and remove the wheel.

2. Remove the stabilizer bar from the control arm.

3. Remove the ball joint from the steering knuckle.

4. Remove the control arm pivot bolts and the control arm.

5. To install, insert the control arm into its fittings. Install the pivot bolts from the rear to the front. Tighten the bolts to 48 ft. lbs. on 1980 models and 50 ft. lbs. on 1981–85 models.

6. Insert the ball stud into the pinch bolt fitting. It should go in easily; if not, check the ball joint stud alignment.

7. Install the pinch bolt from the rear to the front. Tighten to 45 ft. lbs. on 1980 models and 40 ft. lbs. on 1981–85 models.

8. Install the stabilizer bar attachment. Tighten to 35 ft. lbs.

9. Install the wheel and lower the car.

Front Wheel Bearing

ADJUSTMENT

These models use a permanently sealed and lubricated front wheel bearing assembly. No adjustments are necessary or possible.

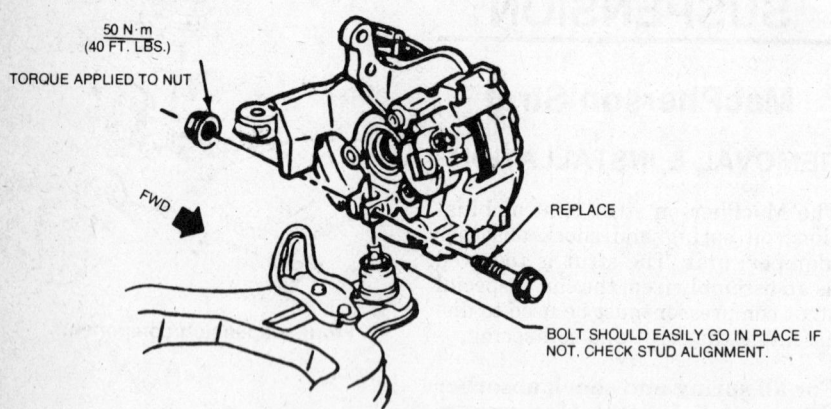

50 N·m (40 FT. LBS.)
TORQUE APPLIED TO NUT

FWD

REPLACE

BOLT SHOULD EASILY GO IN PLACE IF NOT, CHECK STUD ALIGNMENT.

Ball joint stud should go in easily

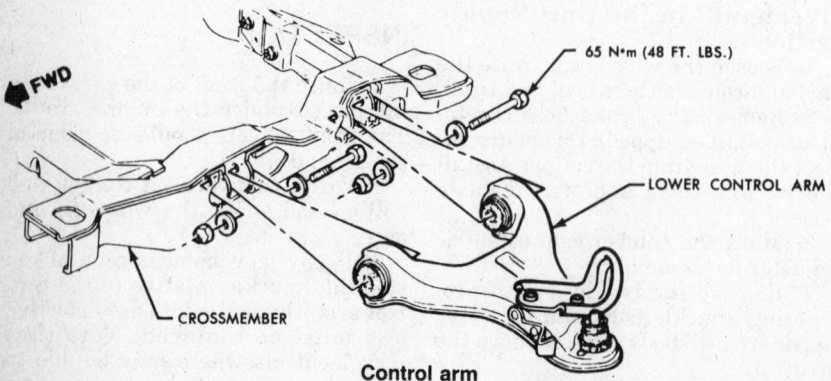

FWD

65 N·m (48 FT. LBS.)

LOWER CONTROL ARM

CROSSMEMBER

Control arm

REAR SUSPENSION

Shock Absorber

REMOVAL & INSTALLATION

1. Open the deck or trunk lid, remove the trim cover, and remove the upper shock nut. Remove and replace one shock at a time when replacing both shocks.

2. Jack the car to a convenient working height. Support the rear axle assembly.

3. Remove the lower attaching bolt and remove the shock absorber. On cars equipped with air shocks, disconnect the air line.

NOTE: Purge new shocks of air by repeatedly compressing them while inverted and extending them in their normal installed position.

4. Install the shock absorber in a reverse of the removal procedure. Torque the lower nuts to 43 ft. lbs.; the upper nut to 13 ft. lbs.

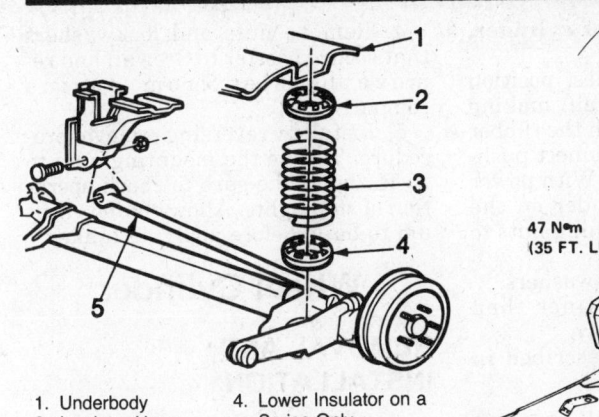

1. Underbody
2. Insulator Upper
3. Spring
4. Lower Insulator on a Series Only
5. Track Bar

A-Body Rear suspension

Spring

REMOVAL & INSTALLATION

1. Raise and support the car on a hoist. Do not use twin-post hoist. The swing arc of the axle may cause it to slip from the hoist when the bolts are removed. If a suitable hoist is not available, raise and support the car on jackstands, and use a jack under the axle.

2. Support the axle with a jack that can be raised and lowered.

3. Remove the brake hose attaching brackets (right and left), allowing the hoses to hang freely. Do not disconnect the hoses.

4. Remove the track bar attaching bolts from the rear axle.

5. Remove both shock absorber lower attaching bolts from the axle.

6. Lower the axle. Remove the coil spring and insulator.

NOTE: Do not suspend rear axle by brake hoses.

7. To install, position the spring and insulator on the axle. The leg on the upper coil of the spring must be parallel to the axle, facing the left hand side of the car.

8. Install the shock absorber bolts. Tighten to 43 ft. lbs. Install the track bar, if equipped, tightening to 33 ft. lbs. Install the brake line brackets. Tighten to 8 ft. lbs.

Rear Wheel Hub And Bearing Assembly

REMOVAL AND INSTALLATION

1. Raise and support the car on a hoist.

2. Remove the wheel and brake drum.

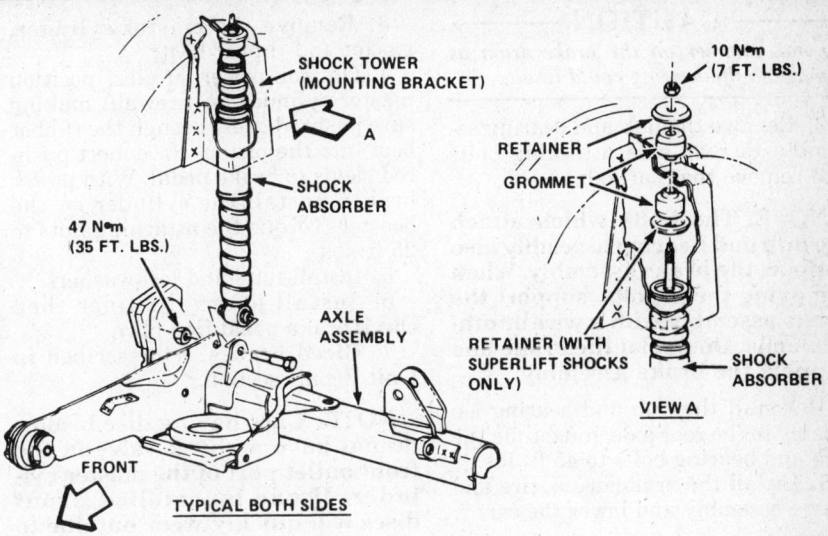

Shock absorber installation

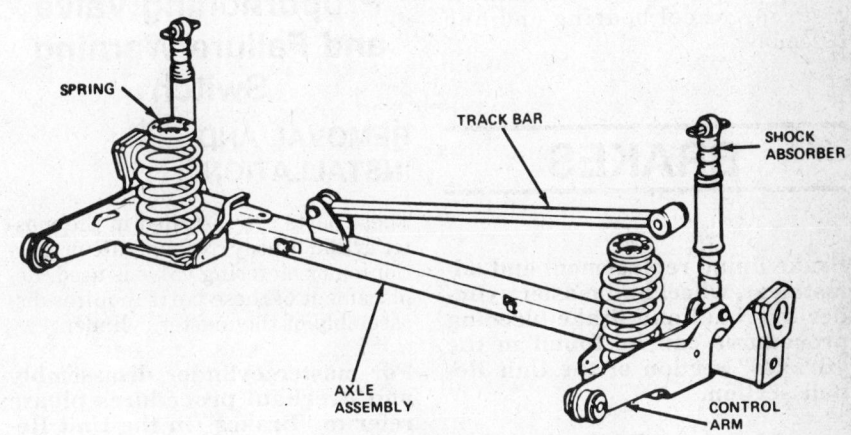

X-Body rear suspension

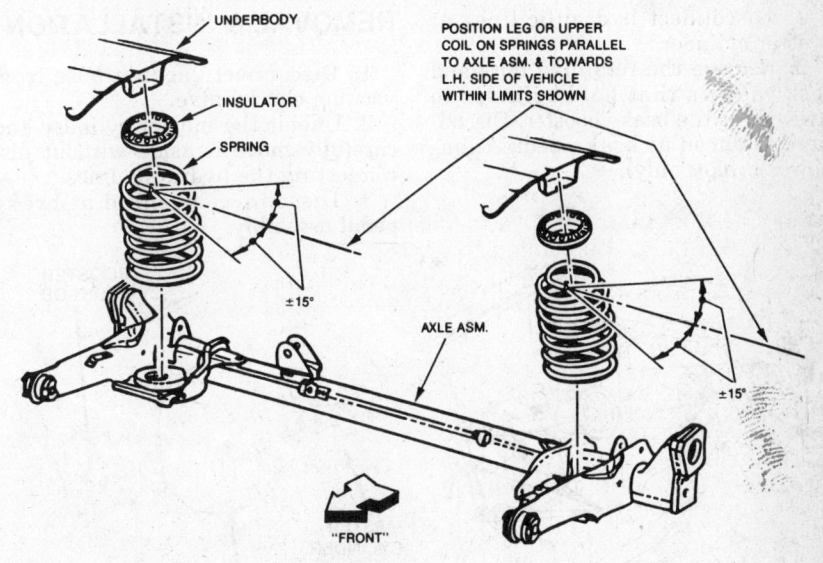

Spring installation

CAUTION

Do not hammer on the brake drum as damage to the bearing could result.

3. Remove the hub and bearing assembly to rear axle attaching bolts and remove the rear axle.

NOTE: The bolts which attach the hub and bearing assembly also support the brake assembly. When removing these bolts, support the brake assembly with a wire or other means. Do not let the brake line support the brake assembly.

4. Install the hub and bearing assembly to the rear axle and torque the hub and bearing bolts to 45 ft. lbs.

5. Install the brake drum, tire and wheel assembly and lower the car.

ADJUSTMENT

There is no necessary adjustment to the rear wheel bearing and hub assembly.

BRAKES

Brake lining replacement and adjustment, wheel and master cylinder overhaul and brake bleeding procedures can be found in the "Brakes" section of the Unit Repair section.

Master Cylinder

REMOVAL & INSTALLATION

1. Disconnect hydraulic lines at master cylinder.

2. Remove the retaining nuts and lockwashers that hold cylinder to firewall or the brake booster. Disconnect pushrod at brake pedal (non-power brakes only).

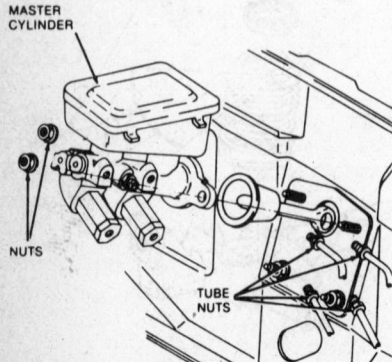

MASTER CYLINDER

NUTS

TUBE NUTS

Typical master cylinder installation

3. Remove the master cylinder, gasket and rubber boot.

4. On non-power brakes, position master cylinder on firewall, making sure pushrod goes through the rubber boot into the piston. Reconnect pushrod clevis to brake pedal. With power brakes, install the cylinder on the booster. Torque the attaching nuts to 25 ft. lbs.

5. Install nuts and lockwashers.

6. Install hydraulic lines then check brake pedal free play.

7. Bleed brakes, as described in Unit Repair section.

NOTE: Cars having disc brakes do not have a check valve in the front outlet port of the master cylinder. If one is installed, front discs will quickly wear out due to residual pressure holding pads against rotor.

Proportioning Valve and Failure Warning Switch

REMOVAL AND INSTALLATION

These parts are installed in the master cylinder body. No seperate proportioning or metering valve is used. Replacement of these parts requires disassembly of the master cylinder.

For master cylinder disassembly and overhaul procedures please refer to "Brakes" in the Unit Repair Section.

Power Brake Booster

REMOVAL & INSTALLATION

1. Disconnect vacuum hose from vacuum check valve.

2. Unbolt the master cylinder and carefully move it aside without disconnecting the hydraulic lines.

3. Disconnect pushrod at brake pedal assembly.

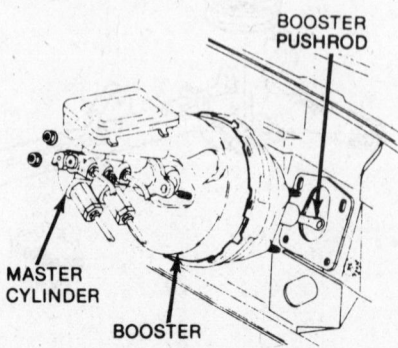

BOOSTER PUSHROD

MASTER CYLINDER

BOOSTER

Power booster removal

4. Remove nuts and lockwashers that secure booster to firewall and remove booster from engine compartment.

5. Install by reversing removal procedure. Torque the mounting nuts to 25 ft. lbs. Make sure to check operation of stop lights. Allow engine vacuum to build before applying brakes.

Wheel Cylinder

REMOVAL AND INSTALLATION

1. Loosen the wheel lug nuts, raise and support the vehicle properly, and remove the wheel. Remove the drum and brake shoes. Leave the hub and wheel bearing assembly in place.

2. Remove any dirt from around the brake line fitting. Disconnect the brake line.

3. Remove the wheel cylinder retainer by using two awls or punches with a tip diameter of $\frac{1}{8}$ in. or less. Insert the awls or punches into the access slots between the wheel cylinder pilot and retainer locking tabs. Bend both tabs away simultaneously. Remove the wheel cylinder from the backing plate.

4. To install, position the wheel cylinder against the backing plate and holt it in place with a wooden block between the wheel cylinder and the hub and bearing assembly.

5. Install a new retainer over the wheel cylinder abutment on the rear of the backing plate by pressing it into place with a $1\frac{1}{8}$ in. 12 point socket and an extension.

6. Install a new bleeder screw into the wheel cylinder. Install the brake line and tighten to 10–15 ft. lbs.

7. The rest of installation is the reverse of the removal procedure. After installing the brake drum, bleed the system.

Parking Brake

ADJUSTMENT

1. Raise the rear of the car and support it safely with jackstands, with both rear wheels off the ground.

2. Apply parking brake two ratchet clicks (1980–81), or three ratchet clicks (1982 and later) from fully released position.

3. Loosen the equalizer locknut, then tighten the adjusting nut until a light to moderate drag is felt when the rear wheels are rotated.

4. Tighten the locknut.

5. Fully release parking brake and rotate rear wheels—no drag should be felt.

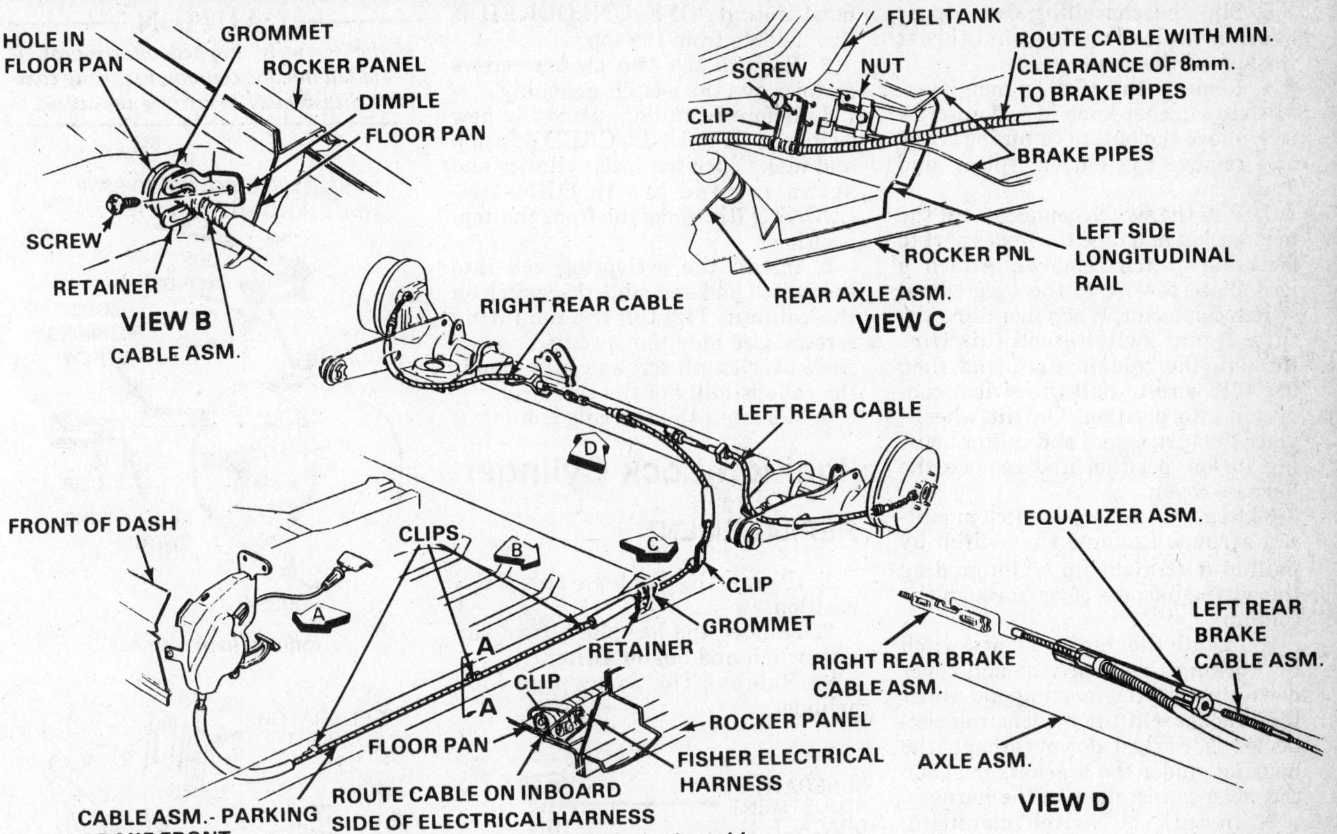

Parking brake cable diagram

HOLE IN FLOOR PAN
GROMMET
ROCKER PANEL
DIMPLE
FLOOR PAN
SCREW
RETAINER
VIEW B
CABLE ASM.

FUEL TANK
SCREW
NUT
CLIP
ROUTE CABLE WITH MIN. CLEARANCE OF 8mm. TO BRAKE PIPES
BRAKE PIPES
ROCKER PNL
REAR AXLE ASM.
LEFT SIDE LONGITUDINAL RAIL
VIEW C

RIGHT REAR CABLE
LEFT REAR CABLE
FRONT OF DASH
CLIPS
CLIP
GROMMET
RETAINER
CLIP
FLOOR PAN
ROUTE CABLE ON INBOARD SIDE OF ELECTRICAL HARNESS
CABLE ASM.- PARKING BRAKE FRONT
ROCKER PANEL
FISHER ELECTRICAL HARNESS
EQUALIZER ASM.
RIGHT REAR BRAKE CABLE ASM.
LEFT REAR BRAKE CABLE ASM.
AXLE ASM.
VIEW D

Parking brake cable

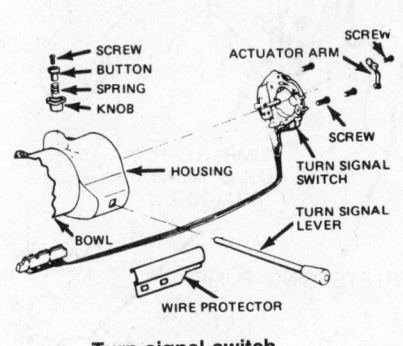

SCREW
BUTTON
SPRING
KNOB
ACTUATOR ARM
SCREW
SCREW
HOUSING
BOWL
WIRE PROTECTOR
TURN SIGNAL SWITCH
TURN SIGNAL LEVER

Turn signal switch

SNAP RING

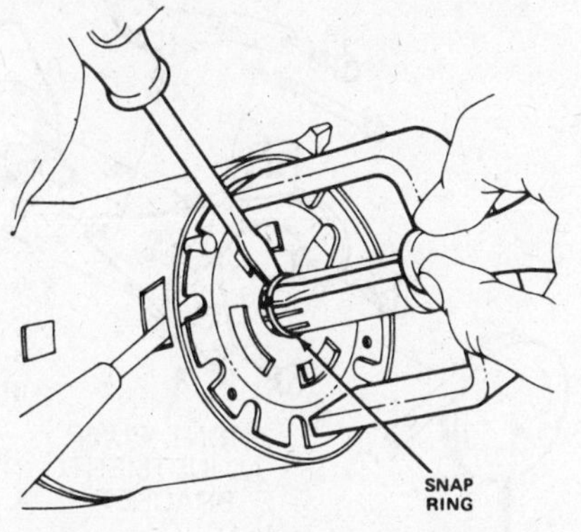

Depress the lockplate and remove the snapring

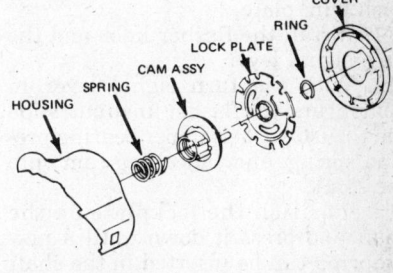

COVER
RING
LOCK PLATE
CAM ASSY
SPRING
HOUSING

Remove these parts for access to the turn signal switch

STEERING

Steering Wheel

REMOVAL & INSTALLATION

—— **CAUTION** ——

Disconnect the battery ground cable before removing the steering wheel. When installing a steering wheel, always make sure that the turn signal lever is in the neutral position.

1. Remove the trim retaining screws from behind the wheel. On wheels with a center cap, pull off the cap.

2. Lift the trim off and pull the horn wires from the turn signal cancelling cam.

3. Remove the retainer and the steering wheel nut.

4. Mark the wheel-to-shaft relationship, and then remove the wheel with a puller.

5. Install the wheel on the shaft aligning the previously made marks. Tighten the nut to 30 ft. lbs.

6. Insert the horn wires into the cancelling cam.

7. Install the center trim and reconnect the battery cable.

Turn Signal Switch

REMOVAL & INSTALLATION

1. Remove the steering wheel as previously outlined. Remove the trim cover.

2. Loosen the cover screws. Pry the cover off with a screwdriver, and lift the cover off the shaft.

3. Position the U-shaped lockplate compressing tool on the end of the steering shaft and compress the lockplate by turning the shaft nut clockwise. Pry the wire snapring out of the shaft groove.

4. Remove the tool and lift the lockplate off the shaft.

5. Slip the cancelling cam, upper bearing preload spring, and thrust washer off the shaft.

6. Remove the turn signal lever. Push the flasher knob in and unscrew it. Remove the button retaining screw and remove the button, spring and knob.

7. Pull the switch connector out the mast jacket and tape the upper part to facilitate switch removal. Attach a long piece of wire to the turn signal switch connector. When installing the turn signal switch, feed this wire through the column first, and then use this wire to pull the switch connector into position. On tilt wheels, place the turn signal and shifter housing in low position and remove the harness cover.

8. Remove the three switch mounting screws. Remove the switch by pulling it straight up while guiding the wiring harness cover through the column.

9. Install the replacement switch by working the connector and cover down through the housing and under the bracket. On tilt models, the connector is worked down through the housing, under the bracket, and then the cover is installed on the harness.

10. Install the switch mounting screws and the connector on the mast jacket bracket. Install the column-to-dash trim plate.

11. Install the flasher knob and the turn signal lever.

12. With the turn signal lever in neutral and the flasher knob out, slide the thrust washer, upper bearing preload spring, and cancelling cam onto the shaft.

13. Position the lockplate on the shaft and press it down until a new snapring can be inserted in the shaft groove. Always use a new snapring when assembling.

14. Install the cover and the steering wheel.

Ignition Switch

REPLACEMENT

The switch is located inside the channel section of the brake pedal support and is completely inaccessible without first lowering the steering column. The switch is actuated by a rod and rack assembly. A gear on the end of the lock cylinder engages the toothed upper end of the rod.

1. Lower the steering column; be sure to properly support it.

2. Put the ignition switch in the OFF–UNLOCKED position. With the cylinder removed, the rod is in LOCK when it is in the next to the upper-most detent. OFF–UNLOCKED is two detents from the top.

3. Remove the two switch screws and remove the switch assembly.

4. Before installing, place the new switch in OFF–UNLOCKED position and make sure the lock cylinder and actuating rod are in OFF–UN-LOCKED (third detent from the top) position.

5. Install the activating rod into the switch and assemble the switch on the column. Tighten the mounting screws. Use only the specified screws since overlength screws could impair the collapsibility of the column.

6. Reinstall the steering column.

Ignition Lock Cylinder

REPLACEMENT

1. Place the lock in the RUN position.

2. Remove the lockplate, turn signal switch and buzzer switch.

3. Remove the screw and lock cylinder.

CAUTION

If the screw is dropped on removal, it could fall into the column, requiring complete disassembly to retrieve the screw.

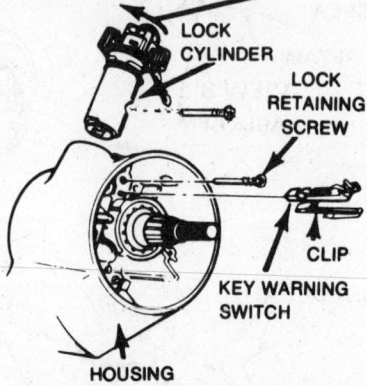

Ignition lock cylinder

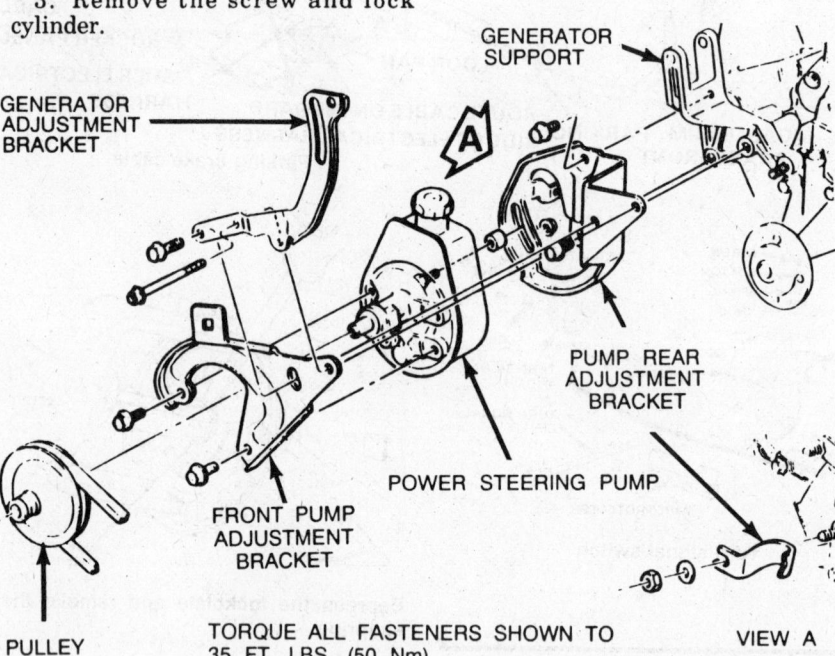

TORQUE ALL FASTENERS SHOWN TO 35 FT. LBS. (50 Nm)

Power steering pump removal—6-181

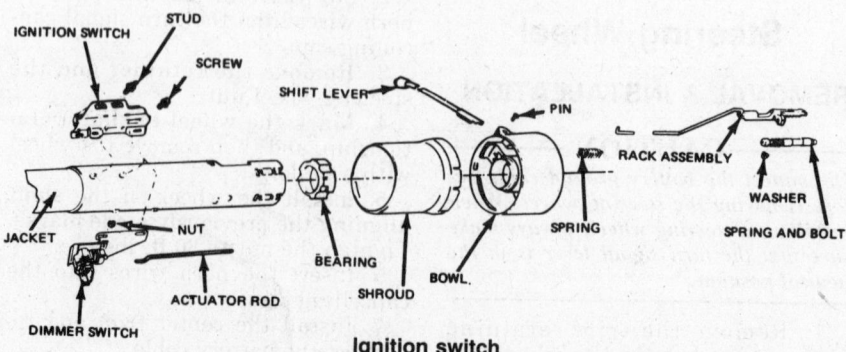

Ignition switch

4. Rotate the cylinder clockwise to align cylinder key with the keyway in the housing.

5. Push the lock all the way in.

6. Install the screw. Tighten the screw to 14 inch lbs. for adjustable columns and 25 inch lbs. for standard columns.

Steering Gear

REMOVAL & INSTALLATION

1. Raise and support the front end of the car with jackstands under the frame members. Allow the front suspension to hang freely. Disconnect the power steering hoses from the gear, where equipped.

2. Move the intermediate shaft seal upward and remove the intermediate shaft-to-stub shaft pinch bolt.

3. Remove both front wheels.

4. Remove the cotter pins and nut from both tie rod ends. Disconnect the tie rod ends from the steering knuckles.

5. Remove the air management system pipe bracket bolt from the crossmember.

6. Support the engine cradle with a floor jack. Remove the two rear cradle mount bolts and, using a jack, lower the rear of the engine cradle about 4–5 inches. DON'T LOWER IT TOO FAR OR DAMAGE TO SURROUNDING COMPONENTS WILL RESULT.

7. Remove the rack and pinion heat shield.

8. Remove the two rack and pinion mount bolts.

9. Remove the rack and pinion assembly through the left wheel opening.

10. Installation is the reverse of removal. Torque the mount bolts to 70 ft. lbs.; the tie rod end nuts to 30 ft. lbs.; the pinch bolt to 45 ft. lbs.

Power Steering Pump

REMOVAL & INSTALLATION

Gasoline Engines

All models use integral rack and pinion power steering. A pump delivers hydraulic pressure through two hoses to the steering gear itself.

1. Remove the hoses at the pump and tape the openings shut to prevent contamination. Position the disconnected lines in a raised position to prevent leakage.

2. Remove the pump belt.

3. On the four cylinder, remove the radiator hose clamp bolt. On the 6–173, disconnect the negative battery

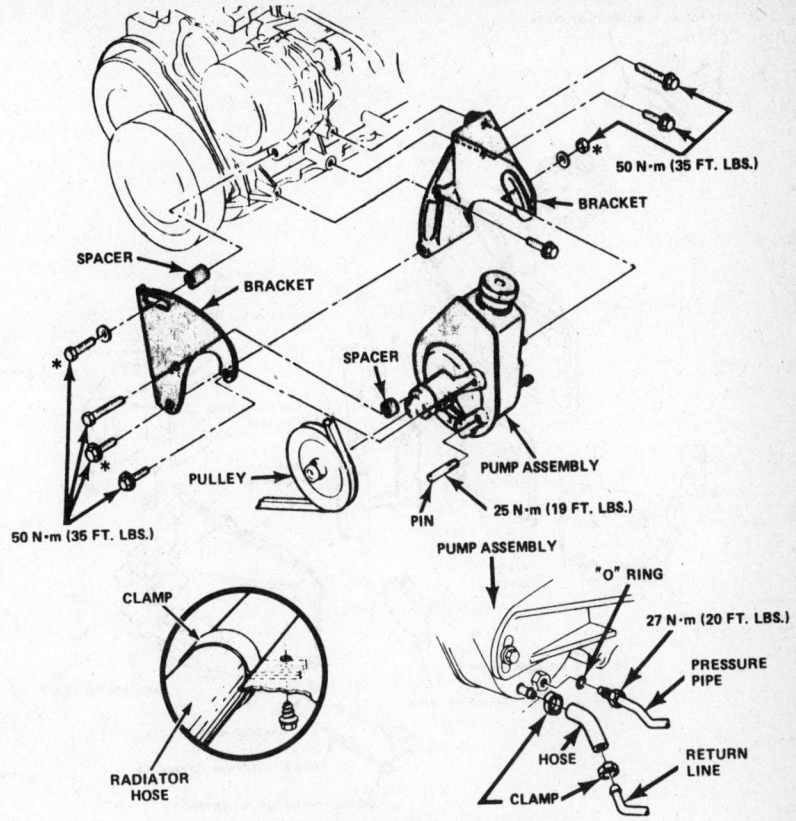

Power steering pump removal, 4-151

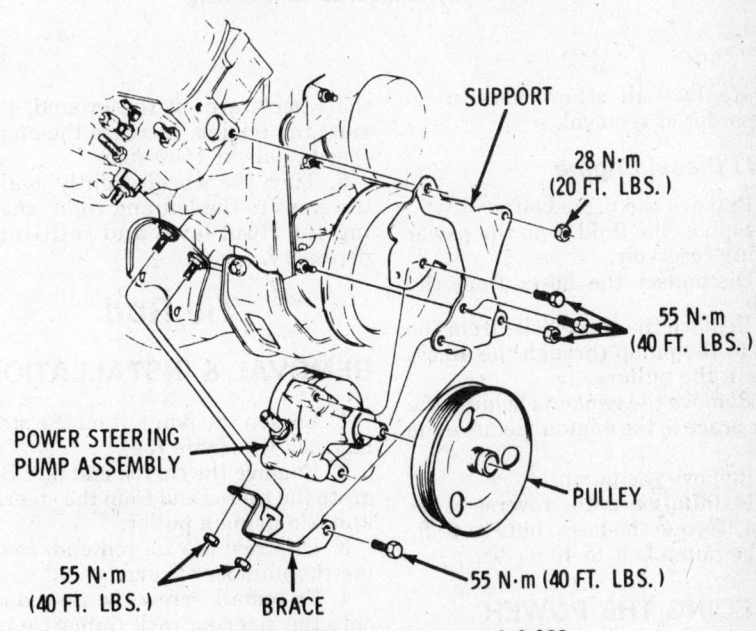

Power steering pump removal, 6-263

cable and the electrical connector at the blower motor, drain the cooling system, and remove the heater hose at the water pump. On the 6–183, remove the alternator.

4. Loosen the retaining bolts and any braces, and remove the pump.

5. Install the pump on the engine with the retaining bolts hand tight.

6. Connect and tighten the hose fittings.

7. Refill the pump with fluid and bleed by turning the pulley counterclockwise (viewed from the front). Stop the bleeding when air bubbles no longer appear.

8. Install the pump belt on the pulley and adjust the tension.

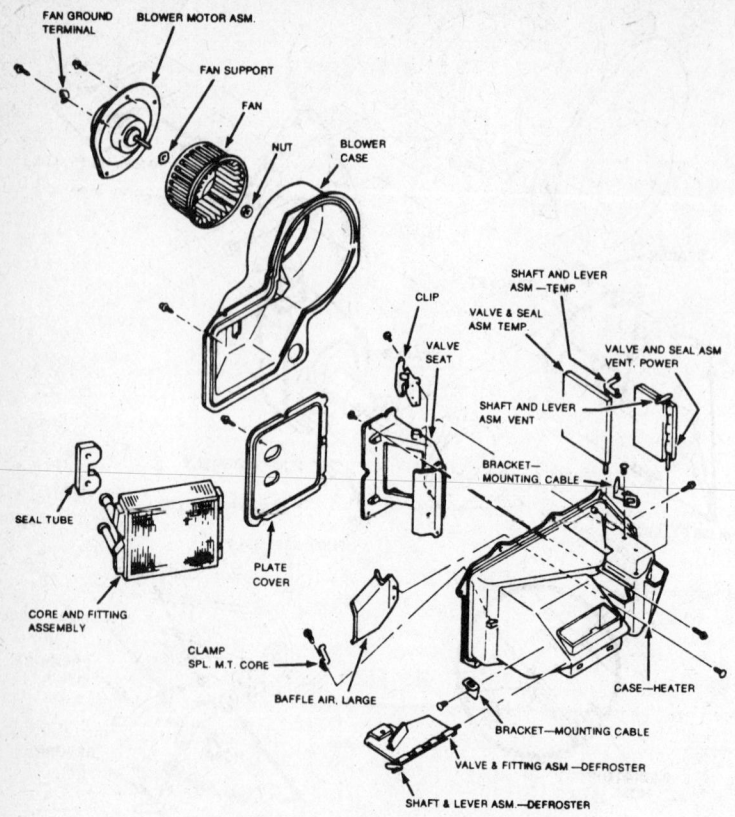

Heater assembly, without air conditioning

9. Replace all other parts in reverse order of removal.

6–263 Diesel Engine

1. Remove the drive belt.
2. Siphon the fluid from the power steering reservoir.
3. Disconnect the hoses from the pump.
4. Remove the three bolts from the front of the pump through the access holes in the pulley.
5. Remove the two nuts holding the lower brace to the engine. Remove the brace.
6. Remove the pump.
7. Installation is the reverse of removal. Torque the brace nuts to 40 ft. lbs; the pump bolt to 40 ft. lbs.

BLEEDING THE POWER STEERING SYSTEM

1. Fill the fluid reservoir.
2. Let the fluid stand undisturbed for two minutes, then crank the engine for about two seconds. Refill reservoir if necessary.
3. Repeat Steps 1 and 2 above until the fluid level remains constant after cranking the engine.
4. Raise the front of the car until the wheels are off the ground, then start the engine. Increase the engine speed to about 1500 rpm.
5. Turn the wheels lightly against the stops to the left and right, checking the fluid level and refilling if necessary.

Tie Rod

REMOVAL & INSTALLATION

1. Loosen the jam nut on the steering rack (inner tie rod).
2. Remove the tie rod end nut. Separate the tie rod end from the steering knuckle using a puller.
3. Unscrew the tie rod end, counting the number of turns.
4. To install, screw the tie rod end onto the steering rack (inner tie rod) the same number of turns as counted for removal. This will give approximately correct toe.
5. Install the tie rod end into the knuckle. Install the nut and tighten to 40 ft. lbs.
6. If the toe must be adjusted, use pliers to expand the boot clamp. Turn the inner tierod to adjust. Replace the clamp.
7. Tighten the jamnut to 50 ft. lbs.

CHASSIS ELECTRICAL

Heater Blower

REMOVAL & INSTALLATION

This procedure is for all cars, with or without air conditioning.

1. Disconnect the negative cable at the battery.
2. Working inside the engine compartment, disconnect the blower motor electrical leads.
3. Remove the motor retaining screws, and remove the blower motor.
4. Reverse to install.

Heater Core

REMOVAL & INSTALLATION

Cars Without Air Conditioning

1. Drain the cooling system.
2. Remove the heater inlet and outlet hoses at the firewall, inside the engine compartment.
3. Remove the radio noise suppression strap.
4. Remove the heater core cover retaining screws. Remove the cover.
5. Remove the core. Reverse to install.

Cars With Air Conditioning
A–BODY CARS

1. Drain the cooling system.
2. On the diesel, raise and support the car on jackstands.
3. Disconnect the hoses at the core.
4. On the diesel, remove the instrument panel lower sound absorber.
5. Remove the heater duct and lower side covers.
6. Remove the lower heater outlet.
7. Remove the two housing cover-to-air valve housing clips.
8. Remove the housing cover.
9. Remove the core restraining straps.
10. Remove the core tubing retainers and lift out the core.
11. Installation is the reverse of removal.

X–BODY CARS

1. Drain the cooling system.
2. Remove the heater hoses from the core tubes at the firewall.
3. Remove the heater duct and heater case side cover from under the instrument panel.
4. Remove the core retaining clamps. Remove the inlet and outlet tube support clamps.

5. Remove the core. Reverse to install.

Radio

REMOVAL & INSTALLATION

Celebrity

1. Disconnect the battery ground.
2. Remove the steering column trim panel including hush panel.
3. Remove the ashtray and ashtray assembly fuse block, seperate ashtray assembly from fuse block. Move both for access.
4. Disconnect the cigarette lighter and rear defogger switch connectors.
5. Remove the cigarette lighter.
6. Remove the glove box.
7. Remove the instrument panel center trim panel attaching nuts.
8. Pull the trim panel away from the instrument panel (enough to remove the radio).
9. Remove the radio.
10. Installation is the reverse of removal.

Century

1. Disconnect the battery ground.
2. Remove the instrument panel trim plate.
3. Remove the right side instrument panel rocker switch trim panel by removing the three screws and gently rocking it out.
4. Remove the four radio mounting screws.
5. Unplug the antenna and all other wires.
6. Remove the radio.
7. Installation is the reverse of removal.

Ciera

1. Disconnect the battery ground.
2. Remove the left instrument panel trim pad.
3. Remove the instrument panel cover.
4. Unbolt the radio from the upper and lower mounting brackets.
5. Pull the radio out to disconnect the wires, then remove it.
6. Installation is the reverse of removal.

6000

1. Disconnect the battery ground.
2. Remove the lower center instrument panel trim plate.
3. Unbolt and remove the radio.
4. Installation is the reverse of removal.

Citation

1. Disconnect the negative battery cable.

2. Remove the radio knobs, the shaft nuts, and the clock knob, if equipped.
3. Remove the instrument cluster trim bezel attaching screws and pull the bezel rearward.
4. Remove the headlamp shaft and knob. Reach behind the instrument panel bezel with a long screwdriver and push the headlamp shaft release button to release the knob.
5. Disconnect the wiring and remove the bezel.
6. Remove the two screws attaching the radio bracket to the instrument panel.
7. Pull the radio rearward while at the same time twisting it slightly to the left, and disconnect the electrical connectors and antenna lead. Remove the lamp socket.
8. Remove the radio.
9. Installation is the reverse of removal.

Omega

1. Remove the instrument panel molding.
2. Remove the ash tray receiver.
3. Remove the four screws attaching the ash tray assembly and remove the ash tray light bulb and socket assembly.
4. Pull the radio and ash tray retainer assembly out far enough to disconnect the radio wiring and remove the radio.
5. Installation is the reverse of removal.

Skylark

1. Disconnect the negative battery cable.
2. Remove the center instrument panel trim plate.

3. Remove the radio attaching screws and pull the radio out to gain access to the wiring. You may have to remove the ashtray retainer assembly to gain access to the radio wiring.
4. Disconnect the wiring. Remove the knobs and separate the face plate from the radio.
5. Installation is the reverse of removal.

Phoenix

1. Disconnect the negative battery cable.
2. Remove the center instrument panel trim plate.
3. Remove the radio attaching screws and pull the radio out to gain access to the wiring.

Windshield Wiper Switch

REMOVAL AND INSTALLATION

1. Disconnect the negative battery cable.
2. Remove the steering wheel, the cover and the lock plate assembly.
3. Remove the turn signal actuator arm, the lever and the hazard flasher button.
4. Remove the turn signal switch screws, the lower steering column trim panel and the steering column bracket bolts.
5. Disconnect the the turn signal switch and the wiper switch connectors.
6. Pull the turn signal switch rearward 6-8 inches, then remove the key buzzer switch and cylinder lock.

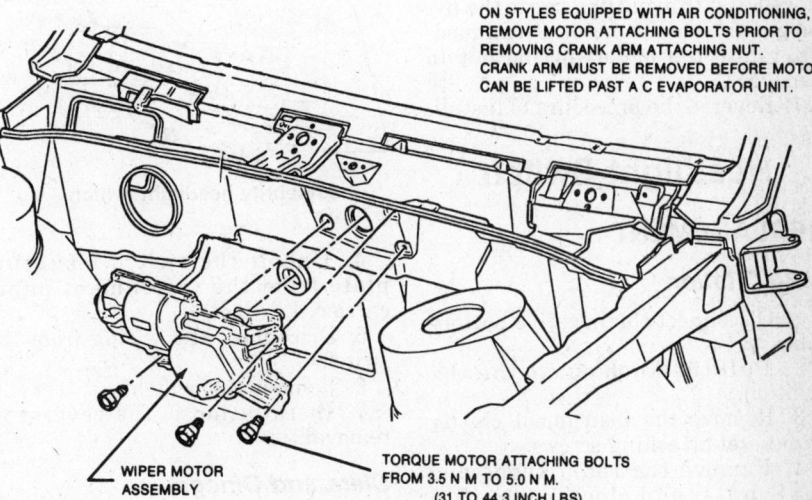

ON STYLES EQUIPPED WITH AIR CONDITIONING, REMOVE MOTOR ATTACHING BOLTS PRIOR TO REMOVING CRANK ARM ATTACHING NUT. CRANK ARM MUST BE REMOVED BEFORE MOTOR CAN BE LIFTED PAST A C EVAPORATOR UNIT.

WIPER MOTOR ASSEMBLY

TORQUE MOTOR ATTACHING BOLTS FROM 3.5 N M TO 5.0 N M. (31 TO 44.3 INCH LBS)

Wiper motor removal

7. Remove and pull the steering column housing rearward, then remove the housing cover screw.

8. Remove the wiper switch pivot and the switch.

9. To install, reverse the removal procedure.

Windshield Wiper Motor

REMOVAL & INSTALLATION

A–Body Cars

1. Raise the hood.
2. Remove the grille.
3. Loosen the wiper linkage to drive arm attaching nuts.
4. Remove the transmission link from the drive arm.
5. Disconnect the wiring and hoses from the motor.
6. Unbolt and remove the motor.
7. Installation is the reverse of removal.

X–Body Cars

1. Remove the wiper arms.
2. Remove the lower windshield reveal molding, the front cowl panel and the cowl screen. Disconnect the washer hose under the screen.
3. Disconnect the motor electrical leads.
4. Loosen, but do not remove, the transmission drive link attaching nuts to the motor crank arm.
5. Disconnect the drive link from the motor crank arm.
6. Remove the three motor attaching bolts. On models with air conditioning, remove the bolts and while supporting the motor, remove the motor crank arm nut using lock-ring type pliers and a closed end wrench. The motor attaching bolts must be removed first to avoid damage to the nylon gear inside the motor. On all models, rotate the motor up and out to remove.
7. Reverse the procedure to install.

Headlight Switch

REPLACEMENT

1980 Citation

1. Disconnect the negative battery cable.
2. Pull the knob out to the ON position.
3. Remove the instrument cluster trim bezel attaching screws.
4. Remove the radio knobs and shaft nuts, and clock knob, if so equipped.
5. Pull the bezel rearward slightly and depress the shaft retaining button. Pull the knob and shaft from the switch.
6. Disconnect the accessory electrical connectors.
7. Remove the bezel.
8. Remove the switch retaining nut and push the switch out from its mounting hole.
9. Disconnect the electrical connector and remove the switch.

1981–85 Citation

1. Disconnect the negative battery cable.
2. Pull the headlamp switch knob out to the last detent.
3. Remove the spring clip retainer on the knob shaft and remove the shaft.
4. Disconnect all accessory switch connectors.
5. Remove the headlamp switch ferrule nut and push switch forward out of the mounting hole.
6. Lift the switch up and out through the opening above the switch mounting and disconnect the switch electrical connector.
7. Remove the switch from the instrument panel.
8. Installation is the reverse of removal.

Celebrity

1. Disconnect the battery ground.
2. Remove the headlamp switch knob.
3. Remove the instrument panel trim pad.

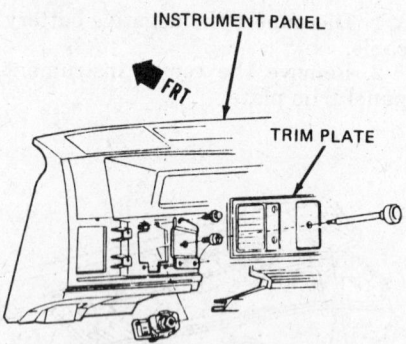

Celebrity headlight switch

4. Unbolt the switch mounting plate from the instrument panel carrier.
5. Disconnect the wiring from the switch.
6. Remove the switch.
7. Installation is the reverse of removal.

Ciera and Omega

1. Remove the left side instrument panel trim pad.

2. Remove the three screws that attach the switch to the instrument panel.
3. Pull the switch rearward and remove it.
4. Installation is the reverse of removal.

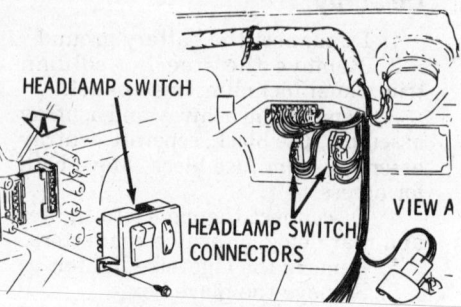

Ciera headlight switch

6000 and Phoenix

1. Disconnect the battery ground.
2. Remove the steering column trim cover and headlight rod and knob by reaching behind the instrument panel and depressing the lock tab with a screwdriver.
3. Remove the left instrument panel trim plate.
4. Unbolt and remove the switch and bracket assembly from the instrument panel.
5. Loosen the bezel and remove the switch from the bracket.
6. Installation is the reverse of removal.

Skylark

1. Disconnect the negative battery cable.
2. Remove the light switch knob by depressing the retaining clip behind the knob and removing the knob from the shaft.
3. Turn the sleeve counterclockwise and spin the knob off the shaft.
4. Remove the instrument panel trim plate.
5. Remove the mounting screws and unplug the switch.
6. Installation is the reverse of removal.

Century

1. Disconnect the battery ground.
2. Remove the instrument panel trim plate.
3. Remove the left side instrument panel switch trim panel by removing the three screws and gently rocking the panel out.
4. Remove the three screws and pull the switch straight out.
5. Installation is the reverse of removal.

Instrument Cluster

For further information on instruments, please refer to "Gauges and Indicators" in the Unit Repair Section.

REMOVAL & INSTALLATION

Citation

1. Disconnect the negative battery cable.
2. Remove the radio knobs (pull off), the shaft nuts, and the clock knob.
3. Remove the instrument cluster bezel (trim plate) attaching screws; there are three at the top and one each in the two lower corners. Pull the bezel slightly rearward.
4. Remove the headlamp shaft and knob.
5. Disconnect the accessory switch wiring.
6. Remove the bezel.
7. Remove the four screws holding the instrument cluster to the instrument panel.
8. Disconnect the shift indicator cable from the steering column shift bowl on models with automatic transaxle.
9. Pull the cluster towards you and disconnect the speedometer cable and instrument electrical connections.
10. Remove the instrument cluster. Installation is the reverse of removal.

Omega

1. Remove the steering column trim cover.
2. Lower the steering column.
3. Remove the four screws holding the instrument panel trim cover to the panel.
4. Pull the trim cover rearward and disconnect the switch wiring, and the remote control mirror cable if your car has one. Remove the trim panel.
5. Remove the four screws holding the instrument cluster to the panel.
6. Disconnect the shift indicator cable from the steering column shift bowl, if your Omega has an automatic transaxle.
7. Pull the cluster towards you and disconnect the speedometer cable and electrical wiring.
8. Remove the instrument cluster. Installation is the reverse.

Phoenix

1. Disconnect the negative battery cable.
2. Remove the speedometer cluster trim plate. There is one screw in each corner.

3. Remove the screws attaching the steering column trim cover to the instrument panel and remove the trim cover.
4. Remove the four cluster attaching screws.
5. With automatic transaxle, disconnect the shift indicator cable, marking the cable location on the steering column shift bowl prior to disconnecting.
6. Disconnect the speedometer cable and pull the cluster toward you. Disconnect the electrical wiring from the back of the cluster and remove the cluster. Installation is the reverse of removal.

Skylark

1. Disconnect the negative battery cable.
2. Remove the radio and accessory switch knobs.
3. Remove the instrument panel trim plate.
4. With automatic transaxle, disconnect the shift indicator cable from the steering column shift bowl.
5. Remove the four cluster attaching screws.
6. Disconnect the speedometer cable and electrical wiring from the back of the cluster. Remove the cluster. Installation is the reverse of removal.

Century

1. Disconnect the battery ground.
2. Disconnect the speedometer cable and pull it through the firewall.
3. Remove the left side hush panel by removing the three 7mm screws and one 11mm nut.
4. Remove the right side hush panel by removing the five 7mm screws and two 11mm nuts.
5. Remove the shift indicator cable clip.
6. Remove the steering column trim plate.
7. Put the gear selector in LOW, remove the nine retaining screws and gently pull out the instrument panel trim plate.
8. Disconnect the parking brake cable at the lever by pushing it forward and sliding it out of its slot.
9. Unbolt and lower the steering column (3 bolts and 1 nut).
10. Remove the gauge cluster by removing the four screws and pulling the cluster out far enough to disconnect any wires, then pull the cluster out.
11. Installation is the reverse of removal.

Celebrity

1. Disconnect battery ground cable.

2. Remove instrument panel hush panel.
3. Remove vent control housing (heater only vehicles).
4. On non A/C cars remove steering column trim cover screws and lower cover with vent cables attached. On A/C equipped vehicles, remove trim cover attaching screws (6) and remove cover.
5. Remove instrument cluster trim pad as outlined in this section.
6. Remove ash try, retainer and fuse block, disconnect wires as necessary.
7. Remove headlamp switch knob and instrument panel trim plate and disconnect electrical connectors of any accessory switches in trim plate.
8. Remove cluster assembly and disconnect speedometer cable. PRNDL and cluster electrical connectors.
9. Installation is the reverse of removal.

Ciera

1. Remove left instrument panel trim pad.
2. Remove instrument panel cluster trim cover.
3. Disconnect speedometer cable at transmission or cruise control transducer if equipped.
4. Remove steering column trim cover.
5. Disconnect shift indicator clip from steering column shift bowl.
6. Remove 4 screws attaching cluster assembly to instrument panel.
7. Pull assembly out far enough to reach behind cluster and disconnect speedometer cable.
8. Remove cluster assembly.
9. Installation is the reverse of removal.

6000

1. Remove the center and left-hand lower instrument panel trim plates.
2. Remove 6–8 screws holding instrument cluster to instrument panel carrier.
3. Remove instrument cluster lens to gain access to speedometer head and instrument/gauges.
4. Installation is the reverse of removal.

Speedometer Cable

REPLACEMENT

1. Remove the instrument cluster.
2. Slide the cable out from the casing. If the cable is broken, the casing will have to be unscrewed from the transaxle and the broken piece removed from that end.

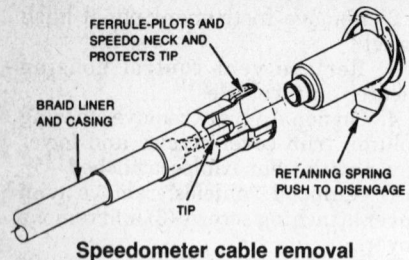

FERRULE-PILOTS AND SPEEDO NECK AND PROTECTS TIP

BRAID LINER AND CASING

TIP

RETAINING SPRING PUSH TO DISENGAGE

Speedometer cable removal

3. Before installing a new cable, slip a piece of cable into the speedometer and spin it between your fingers in the direction of normal rotation. If the mechanism sticks or binds, the speedometer should be repaired or replaced.

4. Inspect the casing; if it is cracked, kinked, or broken, the casing should be replaced.

5. Slide a new cable into the casing, engaging the transaxle end securely. Sometimes it is easier to unscrew the casing at the transaxle end, install the cable into the transaxle fitting, and screw the casing back into place. Install the instrument cluster.

Fuses, Fusible Links & Circuit Breakers

On some models, the fuse block is a swingdown unit located in the underside of the instrument panel, adjacent to the steering column. On other models, access to the fuse block is gained through the glove box. All models use miniaturized plug type fuses which are color-coded and stamped with the amperage rating.

Fusible links are provided in all circuits and fed directly from the battery. Fusible links are lengths of copper wire, about 4 in. long and four gauge sizes smaller than the wire that they protect. Burned out fusible links should be replaced with the same gauge wire for continued circuit protection.

The head lights are protected by a circuit breaker in the headlamp switch. If the circuit breaker trips, the headlights will either flash on and off, or stay off all toeghter. The circuit breaker resets automatically after the overload is removed.

The windshield wipers are also protected by a circuit breaker. If the motor overheats, the circuit breaker will trip, remaining off until the motor cools and the overload is removed.

The circuit breakers for the power door locks and power windows are located in the fuse box.

GM "C" Body

Buick Electra Limited, Park Avenue, T-Type, Cadillac Deville, Fleetwood Brougham, Oldsmobile 98 Regency, 98 Regency Brougham

YEAR IDENTIFICATION

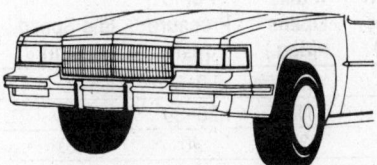

1985–86 Cadillac DeVille, Fleetwood

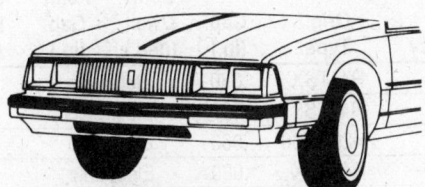

1985–86 Oldsmobile Ninety-Eight Regency Brougham

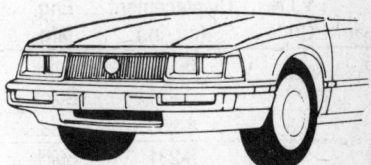

1985–86 Buick Electra, Park Avenue

1986-87 Touring Sedan

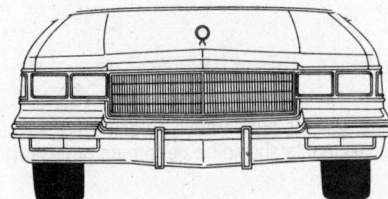

1986-87 Coupe DeVille

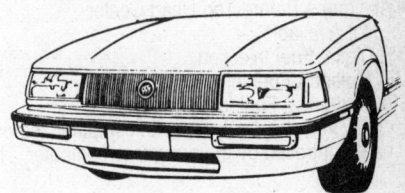

1987 Electra

VEHICLE IDENTIFICATION NUMBER (VIN)

It is important for servicing and ordering parts to be certain of the vehicle and engine identification. The VIN (vehicle identification number) is a 17 digit number visable through the windshield on the driver's side of the dash and contains the vehicle and engine identification codes. It can be interpreted as follows:

Engine Code						Model Year Code	
Code	Cu. In.	Liters	Cyl.	Carb.	Eng. Mfg.	Code	Year
E	181	3.0	6	2bbl	Buick	F	85
3	231	3.8	6	SFI	Buick	G	86
B	231	3.8	6	SFI	Buick	H	87
T	263	4.3	6	Diesel	Olds.		
8	250	4.1	8	DFI	Cad.		

The seventeen digit Vehicle Identification Number can be used to determine engine application and model year. The 10th digit indicates the model year, and the 8th digit identifies the factory installed engine.

DFI Digital Fuel Injection

SFI Sequential Fuel Injection

GENERAL ENGINE SPECIFICATIONS

Year	Eng. V.I.N. Code	Engine Displacement (cu. in.)	Eng. Mfg.	Fuel Delivery	Horsepower @ rpm■	Torque @ rpm (ft lbs)■	Bore × Stroke (in.)	Compression Ratio	Oil Pressure @ 2000 rpm
'85	E	6-181	Buick	2bbl	110 @ 4800	145 @ 2600	3.800 × 2.660	8.4:1	35–42
	3	6-231	Buick	MFI	125 @ 4400	195 @ 2000	3.800 × 3.400	8.0:1	35–40
	T	6-263	Olds.	Diesel	85 @ 3600	165 @ 1600	4.057 × 3.385	21.6:1	30–45
	8	8-250	Cad.	DFI	135 @ 4200	190 @ 2000	3.465 × 3.307	8.5:1	30
'86–'87	3	6-231	Buick	SFI	150 @ 4400	200 @ 2000	3.800 × 3.400	8.5:1	37
	B	6-231	Buick	SFI	140 @ 4400	200 @ 2000	3.800 × 3.400	8.5:1	37
	8	8-250	Cad.	DFI	130 @ 4200	200 @ 2000	3.465 × 3.307	8.5:1	30

■ Horsepower and torque are SAE net figures. They are measured at the rear of the transmission with all accessories installed and operating. Since the figures vary when a given engine is installed in different models, some are representative rather than exact.
MFI Multi-Point Fuel Injection
DFI Digital Fuel Injection
SFI Sequential Fuel Injection

GASOLINE ENGINE TUNE-UP SPECIFICATIONS

(When analyzing compression test results, look for uniformity among cylinders rather than specific pressures.)

Year	Eng. V.I.N. Code	No. Cyl. Displacement (cu. in.)	Eng. Mfg.	Fuel Delivery	Orig. Type	Gap (in.)	Point Dwell (deg.)	Point Gap (in.)	Ignition Timing (deg.)	Valves Intake Opens ■(deg.)	Fuel Pump Pressure (psi)	Idle Speed (rpm)
'85	E	6-181	Buick	2 bbl	R44TSX	.060	Electronic		①	16	3.9–6.5	①
	3	6-231	Buick	MFI	R44TS8	.080	Electronic		①	NA	28–36	①
	8	8-250	Cad.	DFI	R42CLTS6	.060	Electronic		①	37	40	①
'86	3	6-231	Buick	SFI	R44TSX	.080	Electronic		①	NA	28–36	①
	B	6-231	Buick	SFI	R44TSX	.080	Electronic		①	NA	38–36	①
	8	8-250	Cad.	DFI	R42CLTS6	.060	Electronic		①	37	40	①
'87					See Underhood Specifications Sticker							

NOTE: The underhood specifications sticker often reflects tune-up specification changes made in production. Sticker figures must be used if they disagree with those in this chart. Part numbers in this chart are not recommendations by Chilton for any product by brand name.
■ All figures Before Top Dead Center
NA Not Available
DFI Digital Fuel Injection
MFI Multiport Fuel Injection
SFI Sequential Fuel Injection
① Only vehicles equipped with computerized emissions systems (which have no distributor vacuum advance unit), the idle speed and ignition timing are controlled by the emissions computer. These adjustments should be performed professionally on models so equipped.

FIRING ORDER

NOTE: To avoid confusion, always replace spark plug wires one at a time.

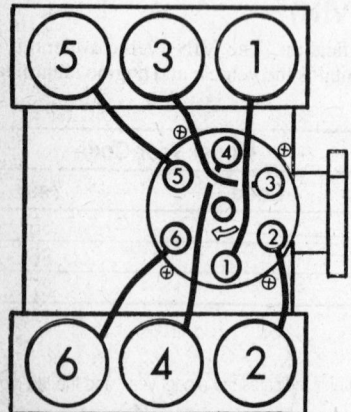

V 6 harmonic balancers have two timing marks: one is ⅛ in. wide. Use the 1/16 in. mark for timing with a hand held light. The ⅛ in. mark is used only with a magnetic timing pick-up probe.

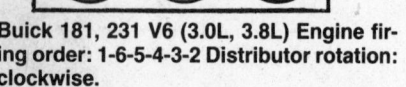

Buick 181, 231 V6 (3.0L, 3.8L) Engine firing order: 1-6-5-4-3-2 Distributor rotation: clockwise.

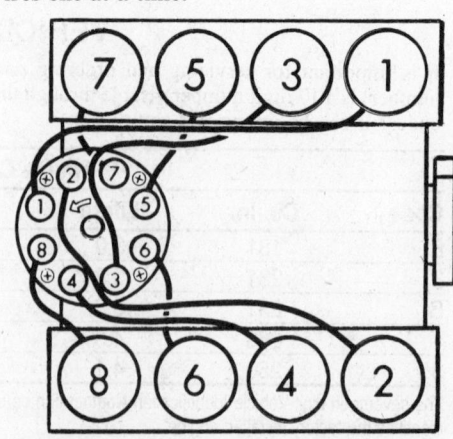

Cadillac 250 V8 (4.1L) Engine firing order: 1-8-4-3-6-5-7-2 Distributor rotation: counterclockwise

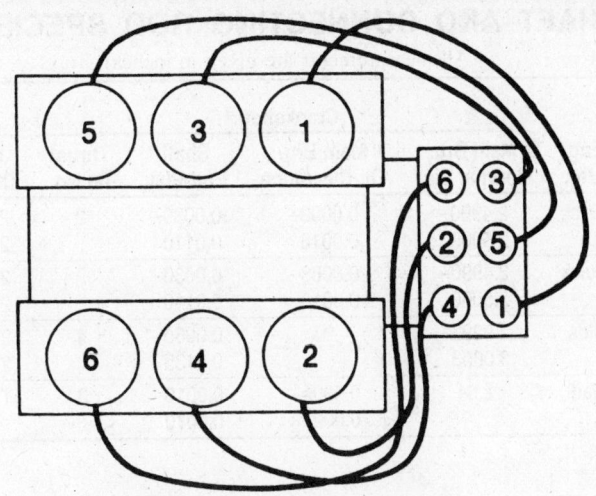

Buick V6 with C3I ignition system
Engine firing order: 1-6-5-4-3-2

DIESEL ENGINE TUNE-UP SPECIFICATIONS

Year	Engine No. of Cyl. Displacement Manufacturer	Fuel Pump Pressure (psi)	Compression Pressure (psi)②	Intake Valve Opens (°B.T.D.C.)	Idle Speed (rpm)
'85	6-263-Olds.	5.5–6.5	275 minimum	16	①

① See the Underhood Specifications Sticker

CAPACITIES

Year	Engine No. Cyl. Displacement (cu. in.)	Engine Crankcase Add 1 qt. for New Filter	Transaxle Automatic Pts. to Refill After Draining ●	Gasoline Tank (gals.)	Cooling System (qts) With Heater	With A/C
'85	6-181	4.0	13	18	13.3	13.6
'85–'87	6-231	4.0	13	18	13.1	13.2
'85	6-263	5.5	13	18	13.3	13.3
'85–'87	8-250	4.0	13	18	NA①	NA①

● Specifications do not include torque converter
NA Not available at time of publication
① The 4.1L V8 uses a coolant solution specifically designed for use in aluminum engines. Be sure that the coolant you choose meets GM spec. #1825M or is labeled for use in aluminum engines.

VALVE SPECIFICATIONS

Year	Engine No. Cyl. Displacement (cu. in.)	Seat Angle (deg.)	Face Angle (deg.)	Spring Test Pressure (lbs. @ in.)	Spring Installed Height (in.)	Stem-to-Guide Clearance (in.) Intake	Exhaust	Stem Diameter (in.) Intake	Exhaust
'85	6-181	45	45	220 @ 1.340	1.727	0.0015–0.0035	0.0015–0.0032	0.3401–0.3412	0.3405–0.3412
'85–'87	6-231	45	45	220 @ 1.340	1.727	0.0015–0.0035	0.0015–0.0032	0.3401–0.3412	0.3405–0.3412
'85	6-263	①	②	210 @ 1.220	1.670	0.0010–0.0027	0.0015–0.0032	0.3425–0.3432	0.3420–0.3427
'85–'87	8-250	45	44	182 @ 1.280	1.730	0.001–0.003	0.001–0.003	0.3413–0.3420	0.3411–0.3418

① Intake: 45   ② Intake: 44
   Exhaust: 31   Exhaust: 30

CRANKSHAFT AND CONNECTING ROD SPECIFICATIONS

(All measurements are given in inches)

Year	V.I.N. Code	Engine No. Cyl. Displacement (cu. in.)	Eng. Mfg.	Crankshaft				Connecting Rod		
				Main Brg. Journal Dia.	Main Brg. Oil Clearance	Shaft End-Play	Thrust on No.	Journal Diameter	Oil Clearance	Side Clearance
'85	E	6-181	Buick	2.4990–2.5000	0.0003–0.0018	0.0030–0.0110	2	2.2487–2.2495	0.0005–0.0026	0.006–0.023
'85–'87	3, B	6-231	Buick	2.4990–2.5000	0.0003–0.0018	0.0030–0.0110	2	2.2487–2.2495	0.0005–0.0026	0.006–0.023
'85	T	6-263	Olds.	2.9993–3.0003	①	0.0035–0.0135	4	2.2490–2.2500	0.0005–0.0025	0.008–0.018
'85–'87	8	8-250	Cad.	2.64	0.0004–0.0027	0.0010–0.0070	3	1.9291	0.0005–0.0028	0.008–0.020

① No. 1, 2, & 3: 0.0005–0.0020
No. 4: 0.0020–0.0034

CAMSHAFT SPECIFICATIONS

(All measurements in inches.)

Year	V.I.N. Code	Engine	Eng. Mfg.	Journal Diameter					Bearing Clearance	Lobe Lift		Camshaft End Play
				1	2	3	4	5		Intake	Exhaust	
'85	E	6-181	Buick	1.786	1.786	1.786	1.786	1.786	①	0.406	0.406	NA
'85–'87	3, B	6-231	Buick	1.786	1.786	1.786	1.786	1.786	①	0.406	0.406	NA
'85	T	6-263	Olds.	②	2.015–2.016	1.995–1.996	1.975–1.976		0.0020–0.0059	0.252	0.279	0.0008–0.0228
'85–'87	8	8-250	Cad.	NA	NA	NA	NA	NA	0.0018–0.0037	0.384	0.396	NA

NA Not available at time of publication
① No. 1: 0.0005–0.0025
No. 2–5: 0.0005–0.0035
② No. 1 bearing is not borable, but must be replaced separately.

PISTON AND RING SPECIFICATIONS

(All measurements are given in inches.)

Year	V.I.N. Code	Engine Type/ Disp. (cu. in.)	Eng. Mfg.	Piston-to-Bore Clearance	Ring Gap			Ring Side Clearance		
					Top Compression	Bottom Compression	Oil Control	Top Compression	Bottom Compression	Oil Control
'85	E	6-181	Buick	0.0008–0.0020	0.010–0.020	0.010–0.020	0.015–0.055	0.0030–0.0050	0.0030–0.0050	0.0035 max.
'85–'87	3,B	6-231	Buick	0.0008–0.0020	0.010–0.020	0.010–0.020	0.015–0.055	0.0030–0.0050	0.0030–0.0050	0.0035 max.
'85	T	6-263	Olds.	0.0035–0.0045	0.019–0.027	0.013–0.021	0.010–0.022	0.005–0.007	0.003–0.005	0.001–0.005
'85–'87	8	8-250	Cad.	0.0010–0.0018	0.023–0.025	0.023–0.025	0.010–0.050	0.0016–0.0037	0.0016–0.0037	None (side sealing)

TORQUE SPECIFICATIONS

(All readings in ft. lbs.)

Year	V.I.N. Code	Engine No. Cyl. Displacement (cu. in.)	Eng. Mfg.	Cylinder Head Bolts	Rod Bearing Bolts	Main Bearing Bolts	Crankshaft Bolt	Flywheel-to-Crankshaft Bolts	Manifold	
									Intake	Exhaust
'85	E	6-181	Buick	80	40	100	200	60	47	25
'85–'87	3, B	6-231	Buick	80	40	100	200	60	47	25
'85	T	6-263	Olds.	①	42	89	②	76	41	31
'85–'87	8	8-250	Cad.	90③	22	85	18	63	④	18

① All exc. bolt No. 5, 6, 11, 12, 13, 14: 142 } See text
   Bolt No. 5, 6, 11, 12, 13, 14: 59

② Crankshaft balancer to crankshaft bolt: 203–350
   Crankshaft pulley to balancer bolts: 30
③ See text for proper tightening sequence
④ Bolts No. 1, 2, 3, 4: 15
   Bolts No. 5–16: 22
   All bolts: 22 } See text
   Repeat 3rd step

WHEEL ALIGNMENT SPECIFICATIONS

Year	Model	Caster Range (deg.)	Caster Pref. Setting (deg.)	Canber Range (deg.)	Canber Pref. Setting (deg.)	Toe (in.)
'85–'87	Buick	2–3	2 1/2	L −1–0 R 0–1	1/2	7/32 ①
'85	Cad.	1 1/2– 3 1/2	2 1/2	−5/16– 1 1/4	1/2	3/32 ①
'86–'87	Cad.	1 1/2– 3 1/2	2 1/2	L −1–0 R 0–1	−1/2 1/2	3/32 ①
'85–'87	Olds	2–3	2 1/2	L −1–0 R 0–1	−1/2 1/2	3/32 ①

L Left
R Right
① In or out. Pref: 0

ENGINE ELECTRICAL

For all information on the charging system not detailed below, please refer to "Charging and Starting" in the Unit Repair section.

Alternator

REMOVAL & INSTALLATION

1. Disconnect the negative battery cable.
2. Tag and disconnect the battery charge wire, 3-prong connector and the ground wire at the back of the alternator.

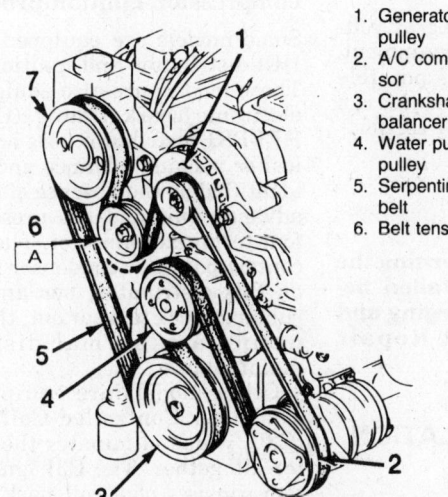

1. Generator pulley
2. A/C compressor
3. Crankshaft balancer
4. Water pump pulley
5. Serpentine belt
6. Belt tensioner
7. P/S pump pulley
A. Rotate the drive belt tensioner in direction of arrow in order to install or remove the drive belt

Serpentine drive belt routing on V6 engine

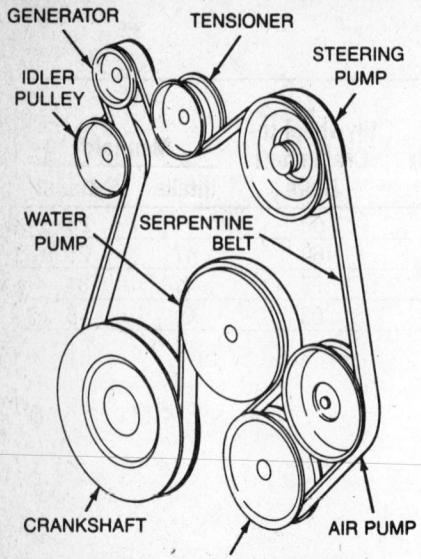

GENERATOR
TENSIONER
STEERING PUMP
IDLER PULLEY
WATER PUMP
SERPENTINE BELT
CRANKSHAFT
AIR PUMP
AIR CONDITIONER COMPRESSOR

Serpentine drive belt routing on V8 engine.

3. Remove the brace at the back of the alternator (if so equipped).

4. Loosen the adjusting bolt, swivel the alternator in and remove the drive belt. If a serpentine drive belt is used, loosen the belt tensioner and rotate it counterclockwise to remove the drive belt.

5. If necessary, loosen the power steering pump brace mounting bolts.

6. Support the alternator, remove the mounting bolts and then remove the alternator.

7. Installation is in the reverse order of removal. Adjust the drive belt to have ¼–½ in. play midway along the longest freespan of the belt. If a serpentine drive belt is used, tighten the tensioner pulley.

Voltage Regulator

An alternator with an integral voltage regulator is standard equipment. There are no adjustments possible with this unit. Testing procedures can be found in the Unit Repair section.

Starter

For all information concerning the starter which is not detailed below, please refer to "Charging and Starting" in the Unit Repair section.

REMOVAL & INSTALLATION

All Except Diesel

1. Disconnect the negative battery cable.

2. Raise and support the vehicle on jackstands.

3. Tag and disconnect all wires at the solenoid. Note color coding of wires for reinstallation.

NOTE: On some models it may be necessary to remove the cross-over pipe to complete this procedure.

4. Remove starter support bracket mount bolts. On engines with solenoid heat shield, remove front bracket upper bolt and detach bracket from starter motor.

5. Loosen the front bracket bolt or nut and rotate bracket clear. Lower and remove starter. Note the location of any shims so that they may be replaced in the same positions upon installation.

6. Installation is in the reverse order of removal. Don't forget any shims.

Diesel

1. Disconnect the negative battery cable.

2. Raise and support the vehicle on jackstands.

3. Remove the lower starter shield nut and then carefully bend the shield out of the way.

4. Tag and disconnect the starter leads at the starter.

5. Remove the front starter bolt. Loosen the rear starter mounting bolt and then remove the starter with the rear bolt still in housing.

6. Installation is in the reverse order of removal.

Distributor

NOTE: See the "Oldsmobile Rear Wheel Drive" section for a description of the diesel engine compression-ignition process.

Some models are equipped with the HEI distributor and ignition system. These models are also equipped with electronic spark timing (EST). The HEI-EST distributor uses no mechanical or vacuum advance and is easily identified by the absence of a vacuum advance unit and the presence of a four terminal connector. There are no contact points or condensor to replace on this distributor, nor any cam or rubbing block to wear out, thus eliminating any normal distributor maintenance.

Other models are equipped with Computer Controlled Coil Ignition (C3I), which eliminates the distributor altogether. The C3I ignition system consists of a coil pack, ignition module and crankshaft sensor. There are two types of C3I coils used. Type 1

coils have three plug wires on each side of the coil assembly; Type 2 coils have all six wires connected on one side of the coil. When troubleshooting or replacing components, it is important to determine which C3I system is installed on the engine.

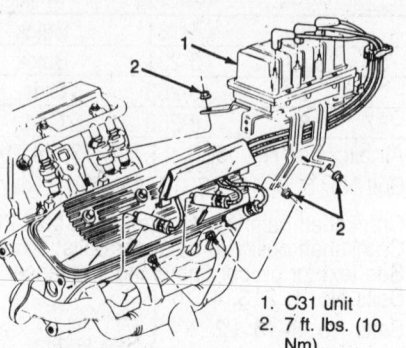

1. C3I unit
2. 7 ft. lbs. (10 Nm)

Computer Controlled Coil Ignition (C3I) mounting

NOTE: When using an auxiliary starter switch on HEI systems, the distributor BAT lead must be disconnected. Failure to do this may cause damage to the grounding circuit in the ignition switch. For further information on the ignition systems, please refer to "Electronic Ignition Systems" in the Unit Repair section.

HEI SYSTEM TACHOMETER HOOKUP

On all models, there is a terminal on the distributor cap marked TACH (usually next to the BAT terminal). Connect one tachometer lead to this terminal and the other lead to a suitable ground. On some tachometers, the leads must be connected to the TACH terminal and then to the positive battery terminal.

——— CAUTION ———

Never ground the TACH terminal; serious module and ignition coil damage will result. If there is any doubt as to the correct tachometer hookup, check with the tachometer manufacturer.

REMOVAL & INSTALLATION

HEI Distributor

1. Disconnect the negative battery cable.

2. Tag and disconnect all wires leading from the distributor cap. DO NOT use a screwdriver or other tool to release the locking tabs.

3. Remove the distributor cap by turning the four latches counterclockwise. Lift off the distributor cap and carefully set it aside.

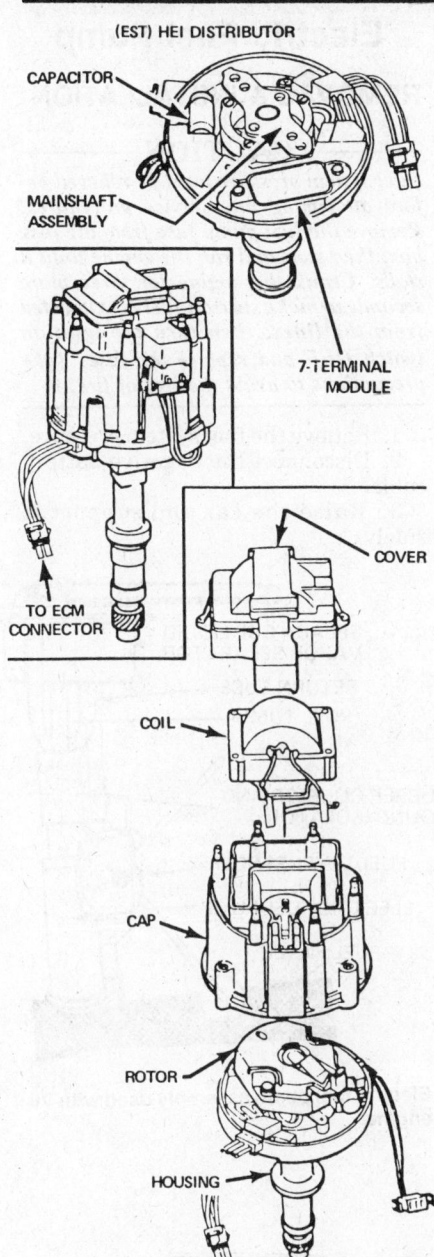

(EST) HEI DISTRIBUTOR

CAPACITOR

MAINSHAFT ASSEMBLY

7-TERMINAL MODULE

TO ECM CONNECTOR

COVER

COIL

CAP

ROTOR

HOUSING

Exploded view of the HEI distributor

NOTE: The location of the distributor cap "doghouse" must be in the same position on reinstallation in order to provide sufficient clearance for adjustment.

4. Disconnect the four terminal ECM connector harness from the distributor if not already done.

5. Loosen, but do not remove, the distributor hold-down clamp.

NOTE: On the V8-250 engine, a special tool (No. J-29791) will be required to loosen the hold-down clamp.

6. Scribe a mark on the distributor body to note the initial position of the rotor. Pull the distributor upward until the rotor just stops turning (counterclockwise); note the position of the rotor once again. Remove the distributor.

NOTE: Do not crank the engine with the distributor removed.

7. On certain models, a thrust washer is used between the distributor drive gear and the crankcase. This washer may stick to the bottom of the distributor as it is removed. Always make sure that this washer is at the bottom of the distributor bore before installation. On DFI systems (Digital Fuel Injection), the malfunction trouble codes must be cleared after removal or adjustment of the distributor. This is accomplished by removing battery voltage to terminal "R" for 10 seconds.

8. To install the distributor, rotate the distributor shaft until the rotor aligns with the second mark you made (when the shaft stopped moving). Lubricate the drive gear with clean engine oil, and install the distributor into the engine. As the distributor is installed, the rotor should rotate to the first mark you made in Step 6. This will ensure proper timing. If the marks do not align properly, remove the distributor and try again. Don't forget the thrust washer when installing the distributor, if so equipped.

9. Install the clamp and hold-down bolt. Tighten them until the distributor can just be moved with a little effort.

10. Connect all wires and hoses. Install the distributor cap. Check the ignition timing and adjust if necessary.

INSTALLATION (ENGINE DISTURBED)

If the engine has been disturbed (cranked) after removing the distributor, perform the following procedure for installation:

1. Crank the engine until No. 1 piston is at the top of its compression stroke (TDC). The compression stroke can be determined by removing the spark plug from No. 1 cylinder and placing your thumb over the hole while an assistant slowly cranks the engine. Crank until compression is felt at the hole and then continue cranking slowly until the timing mark on the crankshaft pulley lines up with the zero degrees (0°) timing mark located on the timing chain cover.

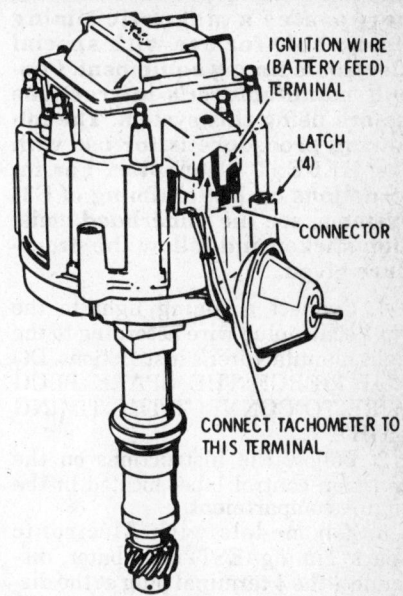

IGNITION WIRE (BATTERY FEED) TERMINAL

LATCH (4)

CONNECTOR

CONNECT TACHOMETER TO THIS TERMINAL

HEI coil-in-cap distributor tachometer hook-up

2. Position the distributor in the block but do not, at this time, allow it to engage with its drive gear at the base of the mounting hole.

3. Rotate the distributor shaft so that the rotor points between No. 1 and No. 8 spark plug towers on the V8, No. 1 and No. 6 on the V6 and push the distributor down to engage the camshaft. It may be necessary to turn the rotor a small amount in either direction in order to achieve this engagement. The rotor will rotate slightly as the distributor gear engages. If installed correctly, the rotor should point toward the No. 1 spark plug terminal in the distributor cap.

4. Press down firmly on the distributor housing. This will ensure that the distributor shaft engages the oil pump shaft, thereby allowing the distributor to fully contact the engine block.

5. Install the hold-down clamp and tighten the bolt until it is snug.

6. Install the distributor cap, making sure that the rotor points to No. 1 terminal in the cap.

7. Attach all wires and hoses.

8. Start the engine. If it fails to start, or runs roughly, the distributor may be 180° out of time. Lift up on the distributor, turn the rotor one-half revolution, and install the distributor. Repeat Steps 1–8 if the engine continues to run poorly.

9. Check the timing and adjust it as necessary.

IGNITION TIMING ADJUSTMENT

NOTE: The 4.1L V8 engine in-

corporates a magnetic timing probe hole for use with special electronic timing equipment. Consult manufacturer's instructions before using this system. The following procedure is for use with the HEI-EST distributor. For instructions on initial timing of C3I systems, see the underhood emission sticker and follow the procedure given.

1. Connect a timing light to the No.1 spark plug wire according to the light manufacturer's instructions. DO NOT PIERCE THE SPARK PLUG WIRE TO CONNECT THE TIMING LIGHT.

2. Follow the instructions on the emission control label located in the engine compartment.

3. On models with Electronic Spark Timing (EST) distributor, disconnect the 4 terminal plug at the distributor. Some models may require grounding the diagnostic connector (ALCL) located under the left side of the dash.

4. Start the engine and run it at idle speed.

5. Aim the timing light at the degree scale just over the harmonic balancer.

6. Adjust the timing by loosening the securing clamp and rotating the distributor until the desired ignition advance is achieved. When the correct timing is set, tighten the clamp.

NOTE: On the 4.1-250 cu. in. V8 engine, a special tool No. J-29791 is used to loosen the hold down nut.

7. Adjust the timing, then replace and tighten the clamp. To advance the timing, rotate the distributor opposite the normal direction of rotor rotation. Retard the timing by rotating the distributor in the normal direction of rotor rotation.

NOTE: On DFI systems (Digital Fuel Injection), the malfunction trouble codes must be cleared after removal or adjustment of the distributor. This is accomplished by removing battery voltage to terminal "R" for 10 seconds.

FUEL SYSTEM

NOTE: For further information on fuel injection, please refer to "Chilton's Guide to Fuel Injection and Feedback Carburetors". For further information on the diesel engine, please refer to "Diesel Maintenance" in the Unit Repair section.

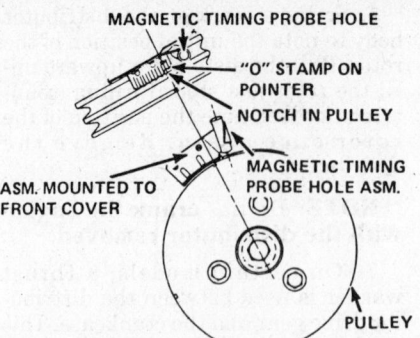

The 4·1L V8 incorporates a special magnetic timing probe hole

Mechanical Fuel Pump

REMOVAL & INSTALLATION

1. The mechanical fuel pump is located on either the right or left front of the engine. Disconnect the fuel inlet hose from the pump. Disconnect the vapor return hose, if equipped.

2. Disconnect the fuel outlet line. Always use a backup wrench on fuel fittings.

3. Remove the two mounting bolts holding the fuel pump to the block.

4. Remove the fuel pump, pushrod, gasket and mounting plate, if used.

5. Installation is the reverse of removal. Clean all gasket mating surfaces and make sure the pushrod is installed properly. When installing the mounting bolts, turn them both evenly, a little at a time.

Electric Fuel Pump

REMOVAL & INSTALLATION

CAUTION

Fuel system pressure must be relieved before attempting any service procedures. Remove the fuel pump fuse from the fuse box, then start and run the engine until it stalls. Crank the engine for three more seconds to make sure all fuel is exhausted from the lines, then turn the ignition switch OFF and replace the fuse. Take precautions to avoid the risk of fire.

---

1. Relieve the fuel system pressure.
2. Disconnect the negative battery cable.
3. Raise the car and support it safely.

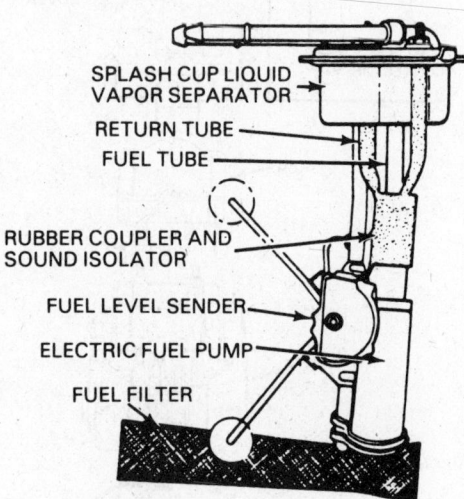

Electric fuel pump assembly used with V8 engine

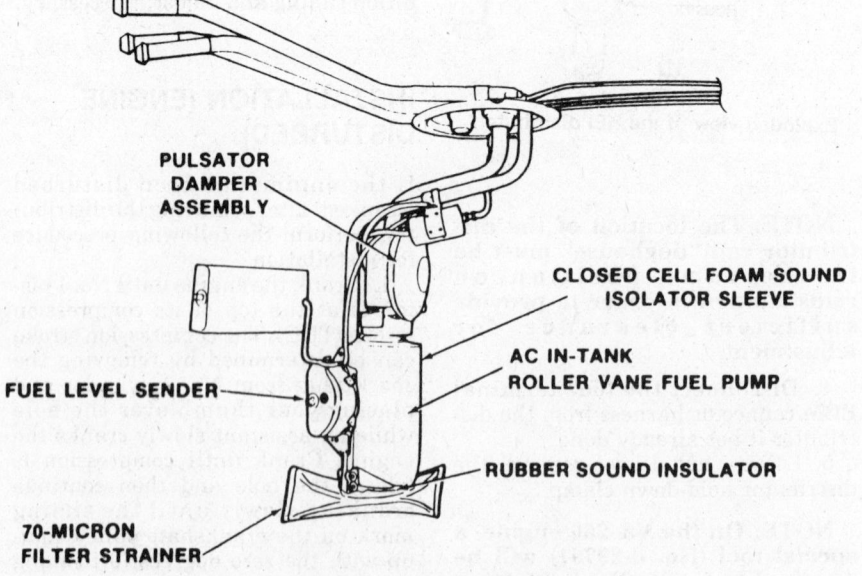

Electric fuel pump assembly used with V6 engine

4. Drain and remove the fuel tank. Make sure the car is supported in such a way that it won't tip forward when the fuel tank is removed.

5. Remove the fuel lever sending unit and pump assembly by turning the cam lock ring counterclockwise and lifting the assembly from the fuel tank. There is a special tool made for turning the lock ring, but careful tapping with a brass drift will work.

6. Pull the fuel pump up into the attaching hose while pulling outward away from the bottom support. Take care to prevent damage to the rubber sound insulator and strainer during removal. Once the pump assembly is clear of the bottom support, pull it out of the rubber connector.

7. Installation is the reverse of removal. Use a new O-ring when installing the assembly into the fuel tank. When installing the fuel tank, make sure all rubber sound isolators or anti-squeak spacers are replaced in their original locations.

Fuel Filter

REMOVAL & INSTALLATION

Carburetor Models

Internal fuel filters are used in the inlet fittings on all carburetors. Filter elements are placed in the inlet hole with the gasket surface facing outward. A spring seals the element by pressing the gasket against the inlet fitting. To replace the filter, disconnect the fuel feed line from the inlet fitting, then remove the fitting from the carburetor. Ribs on the closed end of the fuel filter prevent the element from being installed incorrectly unless forced.

Fuel Injection Models

—— **CAUTION** ——
Fuel system is under pressure. See the Caution under "Fuel Pump Removal" for procedure to relieve fuel pressure before attempting to remove any fuel lines.

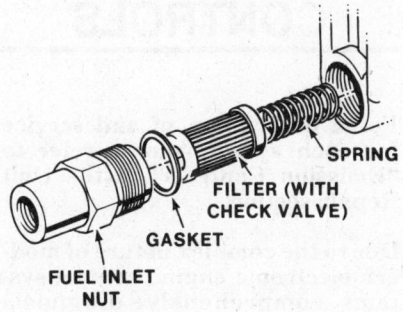

SPRING
FILTER (WITH CHECK VALVE)
GASKET
FUEL INLET NUT

Typical fuel filter

The fuel injection system uses an inline filter located in the fuel feed line under the hood, attached to the frame rail, or on the rear crossmember of the vehicle. Always use a backup wrench on the fittings any time a fuel filter is removed or installed, and never replace a metal fuel line with a rubber insert. The high pressure fuel system used with all fuel injection systems requires metal fuel lines to contain the pressure. Replace the O-ring at the connection and torque the fuel fitting to 22 ft. lbs. (30 Nm).

Carburetor

For all carburetor adjustments and specifications, please refer to "Carburetors" in the unit repair section. For further information on feedback carburetors, please refer to *Chilton's Guide To Fuel Injection And Feedback Carburetors*.

REMOVAL & INSTALLATION

1. Remove the air cleaner.
2. Disconnect and cap the fuel line.
3. Disconnect and label all vacuum lines and/or electrical connectors to the carburetor.
4. Remove the carburetor mounting bolts.
5. Remove the carburetor.
6. Installation is the reverse of removal

IDLE SPEED AND MIXTURE ADJUSTMENT

On all models, the idle speed and fuel mixture is computer-controlled and not adjustable. No periodic adjustment or maintenance is required. For adjustments after carburetor overhaul, please refer to "Carburetors" in the unit repair section.

COOLING SYSTEM

Radiator

REMOVAL & INSTALLATION

1. Disconnect the negative battery cable. The engine should be cool for this procedure.
2. Drain the coolant into a suitable container. Unless contaminated, the coolant may be reused.
3. On the 4.1L V8, detach the electrical connectors, remove the mounting bolts and then remove the left and right cooling fans.

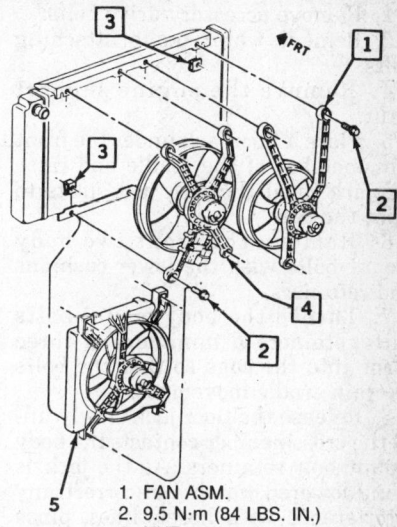

1. FAN ASM.
2. 9.5 N·m (84 LBS. IN.)
3. NUT
4. FAN ASM.—DIESEL
5. FAN ASM.—GAS

Certain models utilize two cooling fans

4. Loosen the clamp-screws and remove the coolant reservoir and upper radiator hoses.
5. Disconnect the engine, transaxle and auxiliary oil cooler lines at the radiator. Wire the lines out of the way.
6. Disconnect the lower radiator hose at the radiator.
7. Remove the mounting bolts and lift out the radiator.
8. On all other engines, remove the upper fan mounting bolts (the 4.3L diesel engine has two cooling fans like the 4.1L V8).
9. Disconnect the upper air cleaner duct and/or silencer on the Ninety Eight.
10. Unscrew the mounting bolts and remove the upper radiator valance panel.
11. Unscrew the clamp-screws and disconnect the coolant recovery tank hose and the upper radiator hose from the radiator.
12. Disconnect the transaxle and engine (diesel only) oil cooler lines from the radiator side tank. Wire them out of the way.
13. Unscrew the mounting bolts and then lift the radiator from the engine compartment.
14. Installation is in the reverse order of removal.

NOTE: When installing the engine oil cooler lines on the diesel, always use new O-rings. Tighten to 26 ft. lbs. (35 Nm).

Water Pump

REMOVAL & INSTALLATION

3.0L V6

1. Disconnect the negative battery cable.

2. Remove accessory drive belts.

3. Remove water pump attaching bolts.

4. Remove the engine support strut.

5. Place a floor jack under the front crossmember of the cradle and raise the jack until the jack just starts to raise the car.

6. Remove the front two body mount bolts with the lower cushions and retainers.

7. Thread the body mount bolts with retainers a minimum of three turns into the cage so that the bolts restrain cradle movement.

8. Release the floor jack slowly until the crossmember contacts the body mount bolt retainers. As the jack is being lowered watch and correct any interference with hoses, lines, pipes and cables.

NOTE: Do not lower the cradle without its being restrained as possible damage can occur to to the body and underhood items.

9. Remove water pump from engine.

10. Installation is the reverse of the removal procedure.

11. Install the pump and torque the mounting bolts to 25 ft. lbs.

12. Connect negative battery cable.

13. Fill with coolant and check for leaks.

3.8L V6

1. Drain the cooling system. Remove the fan shroud, if necessary for clearance.

2. Loosen the belt or belts, then remove the fan blades and pulley or pulleys from the hub on the water pump shaft. Remove the belt or belts.

3. Disconnect the hose from the water pump inlet and the heater hose from the nipple. Remove the bolts, then remove the pump and gasket from the timing case cover or engine block.

4. Install the pump assembly with a new gasket. Bolts and lock washers must be torqued evenly.

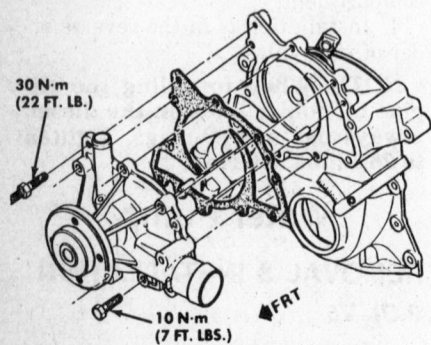

Water pump installation—3.0L V6

30 N·m
(22 FT. LB.)

10 N·m
(7 FT. LBS.)

FRT

5. Connect the radiator hose to the pump inlet and the heater hose to the nipple. Fill the cooling system and check all points of possible coolant leaks.

6. Install the fan pulley or pulleys and the fan blade. Install the belt or belts and adjust for correct tension.

4.3L V6 (Diesel)

1. Drain the radiator.

2. Disconnect lower radiator hose and water pump.

3. Disconnect the heater return hose at the water pump, remove the bolt retaining the heater water return pipe to the intake manifold and position the pipe out of the way.

4. If equipped with A/C, remove the vacuum pump drive belt.

5. Remove the serpentine drive belt.

6. Remove the generator, A/C compressor vacuum pump brackets.

7. Remove the water pump attaching bolts and remove the water pump assembly.

8. Remove the water pump pulley.

9. Clean gasket material from engine block.

10. Apply a thin coat of 1050026 sealer or equivalent to the water pump housing to retain the gasket, then position new gasket on the housing. Also apply sealer to water pump mounting bolts. Torque bolts to 12–15 ft. lb.

4.1L V8

1. Disconnect the negative battery terminal.

2. Drain the coolant.

3. Disconnect the A/C accumulator from the bracket and then position it out of the way. Disconnect the bracket from the wheel arch.

4. Remove the right side cross-car brace.

5. Remove the drive belt, the idler pulley and the bracket.

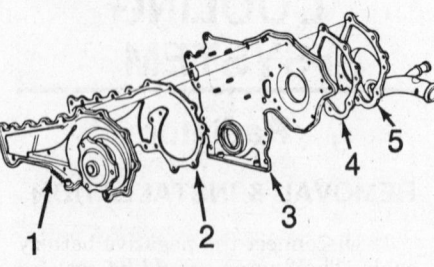

1. WATER PUMP ASSEMBLY
2. WATER PUMP GASKET
3. FRONT COVER
4. WATER PUMP INLET GASKET
5. WATER PUMP INLET

Water pump installation—4.1L V8

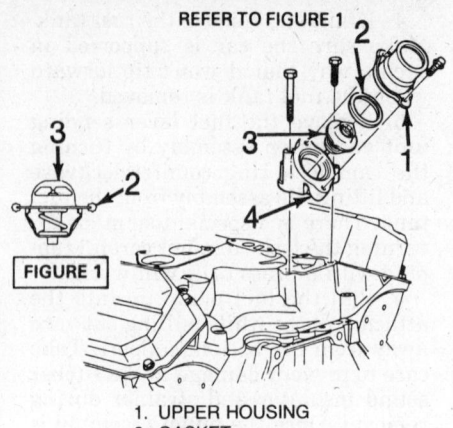

REFER TO FIGURE 1

FIGURE 1

1. UPPER HOUSING
2. GASKET
3. THERMOSTAT ASSEMBLY
4. LOWER HOUSING

Thermostat installation—4.L V8

6. Unscrew the three mounting bolts and remove the water pump pulley.

7. Remove the water pump and gasket.

8. Installation is in the reverse order of removal. Always use a new water pump gasket.

Thermostat

REMOVAL & INSTALLATION

To replace the thermostat, drain the cooling system below the level of the thermostat and remove the two bolts holding the water neck in place. Remove the water neck and the thermostat will lift out. Clean the mating surfaces of both the intake manifold and the water neck. Use a new gasket when installing a new thermostat. If only silicone sealer was used from the factory, use only silicone sealer during assembly.

— **CAUTION** —
Be sure the thermostat is not reversed in its installed position. The spring should be installed toward the engine.

EMISSION CONTROLS

For a description of and service for each system, please refer to "Emission Controls" in the Unit Repair section.

Due to the complex nature of modern electronic engine control systems, comprehensive diagnosis and testing procedures fall outside the confines of this repair manual. For complete information on diagnosis, testing and repair procedures concerning all modern engine and emission control systems, please refer to *Chilton's Guide To Electronic Engine Controls.*

ENGINE

NOTE: For all engine service procedures not detailed below, please refer to the "Buick Rear Wheel Drive" section (3.8L V6), "Cadillac Rear Wheel Drive" section (4.1L V8) or the "GM A & X Body" section (3.0L V6, 4.3L V6).

Engine

REMOVAL & INSTALLATION

3.0L V6 and 3.8L V6

1. Disconnect the negative battery cable. Matchmark the hood hinges and remove the hood.

2. Tag and disconnect the air flow sensor wiring, if equipped with fuel injection. Depressurize the fuel system as described under "Fuel Pump Removal."

3. Disconnect the air intake duct. Drain the engine coolant.

4. Raise the front of the vehicle and support it on jackstands.

5. Unscrew the retaining bolts and separate the exhaust pipe from the manifold.

6. Loosen and remove the engine mount bolts.

7. Remove the bolts and then disconnect the driveline vibration absorber.

8. Tag and disconnect the starter wiring and then remove the starter.

9. Disconnect the A/C compressor and position it out of the way. DO NOT disconnect the refrigerant lines.

10. Disconnect the hydraulic lines at the power steering pump and wire them out of the way.

11. Loosen and remove the lower transaxle-to-engine bolts.

NOTE: One bolt is situated between the transaxle case and the engine block. It is installed in the opposite direction of the other bolts.

12. Remove the flexplate cover. Matchmark the flexplate-to-torque converter relationship to insure proper alignment upon installation. Remove the flexplate-to-torque converter bolts.

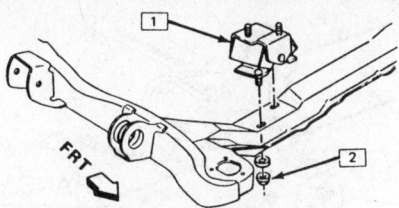

1. ENGINE MOUNT
2. NUT 41 N·m (30 FT. LBS.)

Right side engine mounts—3.0L V6 and 3.8L V6

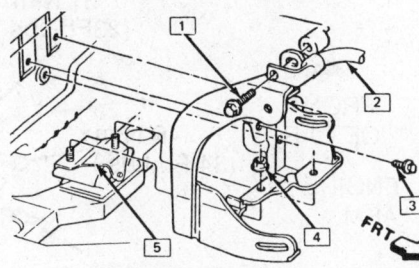

1. BOLT 50 N·m (37 FT. LBS.)
2. NEGATIVE BATTERY CABLE
3. BOLT 95 N·m (70 FT. LBS.)
4. NUT 35 N·m (25 FT. LBS.)
5. ENGINE MOUNT

Left side engine mounts—3.0L V6 and 3.8L V6

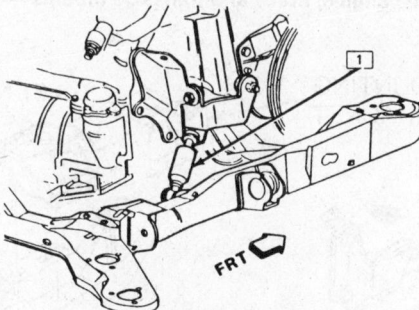

1. DRIVELINE VIBRATION ABSORBER

Typical driveline vibration absorber

13. Disconnect the engine support bracket at the transaxle and then lower the vehicle.

14. Disconnect the Radiator and heater hoses at the engine and position them out of the way.

15. Remove the alternator and rotate to the cowl.

16. Disconnect the engine wiring harness at the electronic control unit, then feed the main connector through the firewall and lay it across the engine.

17. Remove the remaining upper transaxle-to-engine bolts.

18. Install a lifting fixture to the engine and remove the engine from the vehicle.

19. Installation is the reverse of the removal procedure.

4.1L V8

1. Disconnect the negative battery cable. Drain the radiator coolant.

2. Remove the air cleaner. Matchmark the hood to the support brackets and remove the hood.

3. Disconnect the A/C hose strap from the strut tower. Disconnect The A/C accumulator from its bracket and position it out of the way. Do not disconnect any refrigerant lines.

4. Tag and disconnect the canister hoses and ground wire from the accumulator bracket and then remove the bracket itself from the inner strut tower.

5. Disconnect or remove the cooling fans, the drive belt and the radiator and heater hoses.

6. Tag and disconnect the following:
   a. Oil pressure switch
   b. Coolant temperature sensor
   c. Distributor wires
   d. EGR solenoid
   e. Engine temperature switch
   f. Accelerator cable
   g. Cruise control linkage
   h. Transmission TV cable

7. Remove the cruise control diaphragm and its bracket.

8. Remove the vacuum supply hose and the exhaust crossover pipe.

9. Disconnect the oil cooler lines at the oil filter adapter, unscrew their mounting bracket at the transaxle and position them out of the way.

10. Remove the air cleaner mounting bracket.

11. CAREFULLY bleed the fuel pressure at the Schraeder valve and then disconnect the fuel lines at the throttle body.

— CAUTION —
When bleeding the fuel system, be sure to have a container or rags on hand to catch excess fuel. Take precautions to avoid the risk of fire.

12. Unscrew the fuel line bracket at the transaxle and wire the fuel lines out of the way.

13. Tag and disconnect the small vacuum line at the brake booster.

14. Tag and disconnect the AIR solenoid electrical and hose connections. Remove the AIR valves and bracket.

15. Tag and disconnect the wires at the following:
   a. Idle Speed Control (ISC) motor
   b. Throttle Position Switch (TPS)
   c. Fuel injectors
   d. Manifold Air Temperature (MAT) sensor
   e. Oxygen sensor
   f. Throttle body base warmer
   g. Alternator

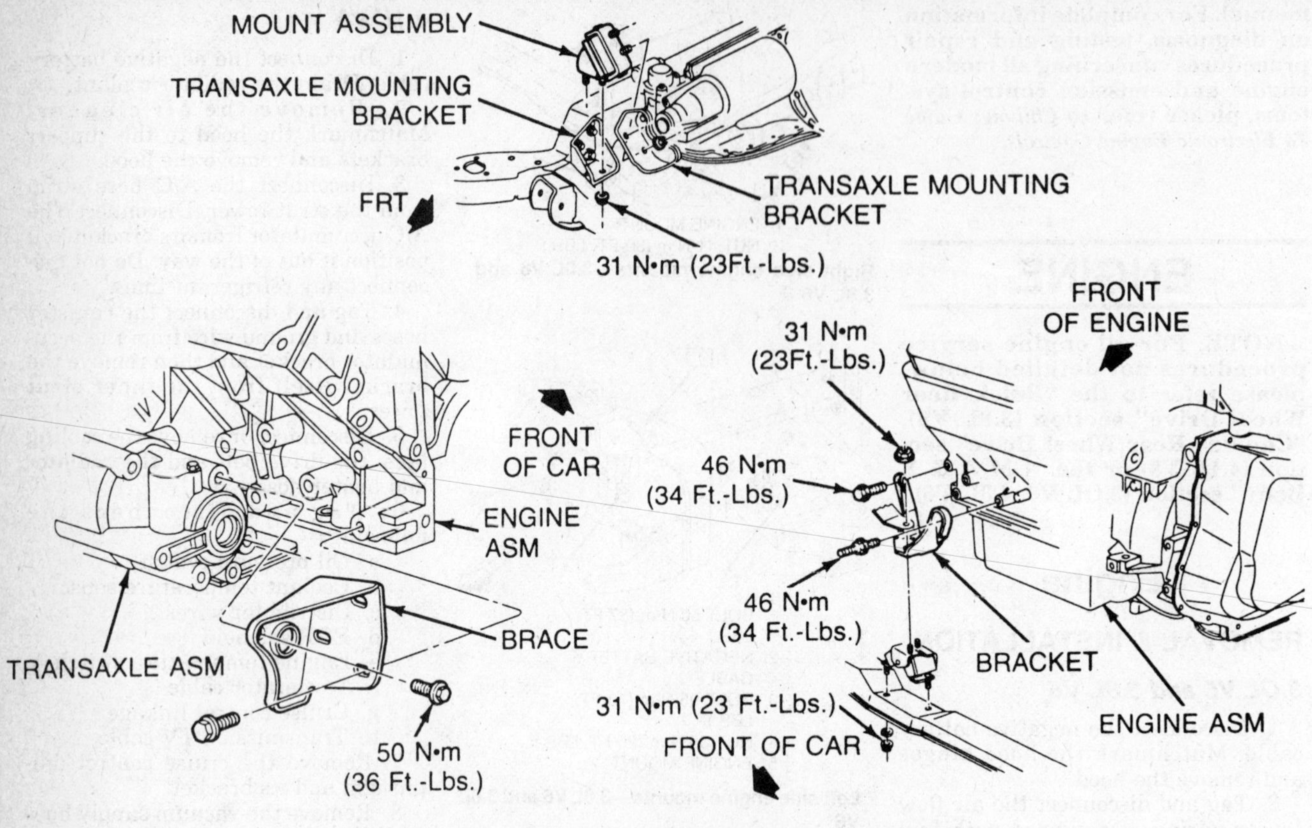

MOUNT ASSEMBLY

TRANSAXLE MOUNTING BRACKET

FRT

TRANSAXLE MOUNTING BRACKET

31 N•m (23Ft.-Lbs.)

FRONT OF ENGINE

31 N•m (23Ft.-Lbs.)

FRONT OF CAR

46 N•m (34 Ft.-Lbs.)

ENGINE ASM

BRACE

46 N•m (34 Ft.-Lbs.)

TRANSAXLE ASM

BRACKET

31 N•m (23 Ft.-Lbs.)

FRONT OF CAR

ENGINE ASM

50 N•m (36 Ft.-Lbs.)

Right side engine, brace and transaxle mounts—4.1L V8

TRANSAXLE MOUNTING BRACKET

31 N•m (23 FT-LBS)

B

A

TRANSAXLE MOUNTING BRACKET

FRT

31 N•m (23 FT-LBS)

FRAME ASM

52 N•m (38 FT-LBS)

FRT

GUIDE — OIL COOLER PIPES

52 N•m (38 FT-LBS)   VIEW A

FRT

VIEW B

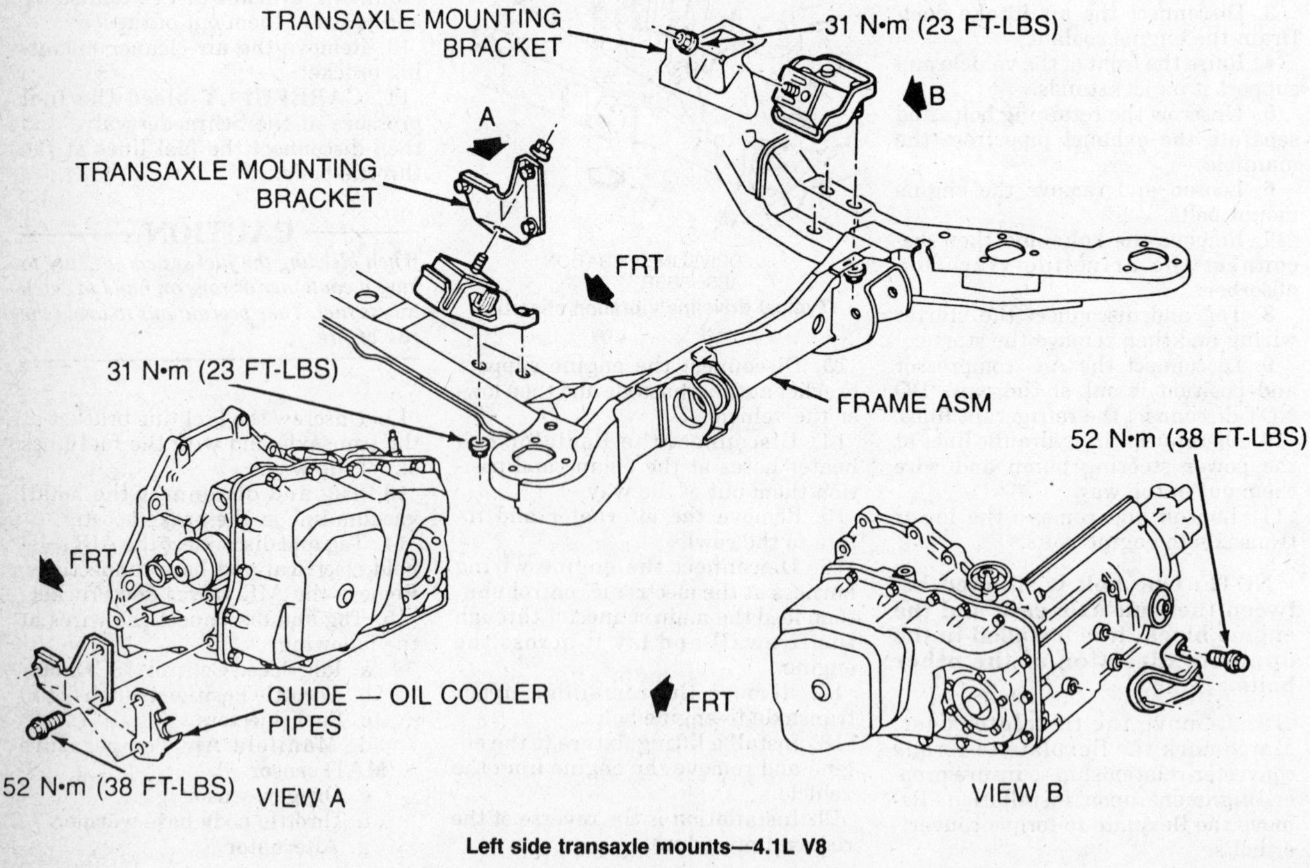

Left side transaxle mounts—4.1L V8

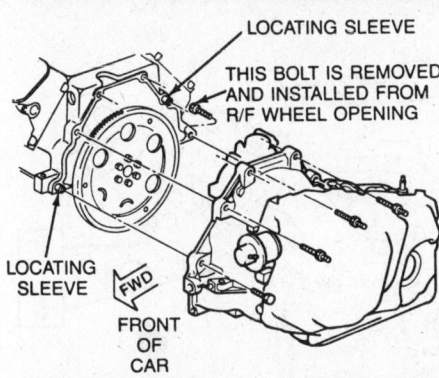

LOCATING SLEEVE

THIS BOLT IS REMOVED AND INSTALLED FROM R/F WHEEL OPENING

LOCATING SLEEVE

FWD

FRONT OF CAR

Transaxle-to-engine attaching bolts—4.1L V8

16. Remove the idler pulley. Remove the power steering pump hose strap from the stud-headed bolt in front of the right cylinder head. Remove the stud-headed bolt.

17. Remove the AIR pipe clip near the No. 2 spark plug.

18. Remove the power steering pump and belt tensioner (with bracket). Wire them out of the way.

19. Raise the vehicle and support it on jack stands.

20. Tag and disconnect the starter wires and the ground wire at the cylinder block.

21. Remove the two flex plate covers. Remove the starter. Remove the three flexplate-to-converter bolts.

22. Remove the A/C compressor lower dust shield.

23. Remove the right front wheel. Remove the outer wheelhouse plastic shield.

24. Remove the A/C compressor mounting bolt and lower the compressor out of the way.

25. Remove the lower radiator hose.

26. Remove the driveline vibration dampener and its brackets from the lower right front of the engine and cradle.

27. Remove the three right front engine-to-transaxle bracket bolts.

28. Disconnect the exhaust pipe at the manifold. Remove the AIR pipe-to-con-verter bracket from the exhaust manifold stud.

NOTE: Be careful not to lose the springs when detaching the exhaust pipe.

29. Remove the lower right hand bell housing-to-engine bolt. Support the engine with a jack.

30. Remove the five upper bell housing-to-engine bolts. Remove the three left front engine mount bracket-to-engine bolts.

31. Attach a suitable lifting fixture and remove the engine.

32. Installation is in the reverse order of removal.

4.3L V6 (Diesel)

1. Disconnect the negative battery cable. Matchmark the hood to the support brackets and then remove the hood. Drain the cooling system.

2. Remove the serpentine drive belt. Remove the vacuum drive belt.

3. Remove the air cleaner. Install an air crossover screen cover (No. J-26996-1) or equivalent.

4. Tag and disconnect the ground wires at the inner fender panel and the engine ground strap.

5. Raise the vehicle and support it on jack stands.

6. Remove the engine-to-transaxle brace.

7. Remove the flywheel cover and then remove the flywheel-to-torque converter bolts.

8. Disconnect the exhaust pipe from the rear exhaust manifold.

9. Remove the engine mount-to-cradle retaining nuts and washers.

10. Remove the engine absorbers assembly from the frame bracket.

11. Tag and disconnect the following:
   a. Starter motor wires
   b. Glow plug wire at No. 2 cylinder
   c. Battery ground cable

12. Disconnect the lower oil cooler hose and cap the opening.

13. Remove the accessible power steering pump bracket fasteners. Lower the vehicle.

14. Remove the remaining power steering pump bracket/brace fasteners and lower the pump (with hoses connected) out of the way.

15. Disconnect the heater water return pipe.

16. Tag and disconnect the remaining glow plug leads and all other electrical leads connected to the engine.

17. Disconnect the engine harness at the cowl connector and body-mounted relays.

18. Remove the A/C compressor with the lines and brackets attached. Wire the compressor out of the way.

19. Disconnect all fuel and vacuum lines. Exercise caution as some fuel may spray from the fittings when loosened. Use a rag to catch any fuel spray.

NOTE: Cap all open fuel lines.

20. Disconnect the throttle and TV cables at the injection pump and cable brackets.

21. Disconnect the upper oil cooler line. Cap the openings.

22. Remove the crossover pipe heat shield and the transaxle filler tube.

23. Remove the exhaust crossover pipe.

24. Install a suitable engine lifting device to the lift hooks on the bloc.

25. Use a floor jack to support the transaxle under the rear extension housing.

26. Remove the engine-to-transaxle bolts and remove the engine.

27. Installation is in the reverse order of removal.

Exhaust Manifold

REMOVAL & INSTALLATION

3.0L V6 and 3.8L V6

LEFT SIDE

1. Disconnect the negative battery cable.

2. Remove the mass air flow sensor, air intake duct and crankcase ventilation pipe.

3. Remove the two bolts attaching the exhaust crossover pipe to the manifold.

4. Tag and disconnect the spark plug wires.

5. Remove the mounting bolts and remove the manifold.

NOTE: The oil dipstick tube may need to be removed to provide access to the manifold bolts.

6. Installation is in the reverse order of removal.

RIGHT SIDE

1. Disconnect the negative battery cable.

2. Repeat Step 2 of the "Left Side" procedure.

3. Disconnect the IAC connector at the throttle body (3.8L only).

4. Tag and disconnect the spark plug wires and the oxygen sensor lead.

5. Disconnect the heater inlet pipe from the manifold studs.

6. Remove the exhaust crossover pipe.

7. Remove the front alternator support bracket.

8. Remove the exhaust manifold mounting bolts. Raise and support the front of the vehicle.

9. Disconnect the exhaust pipe from the manifold.

10. Remove the front exhaust pipe. Remove the manifold.

11. Installation is in the reverse order of removal.

4.1L V8

RIGHT SIDE

1. Disconnect the negative battery cable. Remove the air cleaner.

2. Remove the exhaust crossover pipe. Disconnect the oxygen and coolant temperature sensors.

3. Remove the catalytic converter-to-AIR pipe clip bolt.

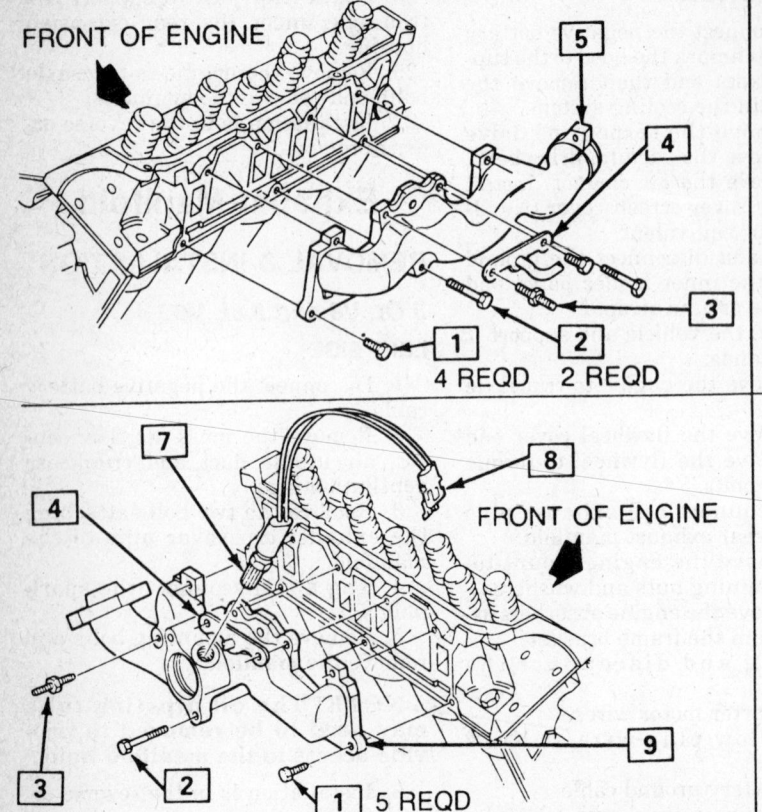

FRONT OF ENGINE

5

4

3

1
4 REQD

2
2 REQD

FRONT OF ENGINE

9

6

5

7

8

FRONT OF ENGINE

4

3

2

1
5 REQD

9

1. SHORT BOLT
2. LONG BOLT
3. STUD HEADED BOLT
4. AIR PIPE
5. LEFT EXHAUST
   MANIFOLD
6. EXHAUST CROSSOVER
   PIPE
7. OXYGEN SENSOR
8. OXYGEN SENSOR
   CONNECTOR
9. RIGHT EXHAUST
   MANIFOLD

Exhaust manifold removal—4.1L V8

4. Remove the two front manifold mounting bolts. Raise and support the front of the car.

5. Disconnect the converter air pipe bracket from the stud and then remove the converter-to-manifold exhaust pipe.

6. Remove the remaining manifold mounting bolts, disconnect the AIR pipe and remove the manifold.

7. Installation is in the reverse order of removal.

LEFT SIDE

1. Disconnect the negative battery cable.

2. Remove both cooling fans and the exhaust crossover pipe.

3. Remove the drive belt and the AIR pump pivot bolt.

4. Remove the belt tensioner and the power steering pump brace.

5. Remove the manifold mounting bolts. Disconnect the air pipe and remove the manifold.

6. Installation is in the reverse order of removal.

4.3L V6 (Diesel)
LEFT SIDE

1. Disconnect the negative battery cable.

APPLY LUBRICANT
TO ENTIRE BOLT

2
1
3
4
5

1. EXHAUST MANIFOLD
2. GASKET
3. WASHER (3)
4. LOCK (3)
5. BOLT-38 N·m (28 LB. FT.)

Left (front) exhaust manifold removal—4.3L V6 (diesel).

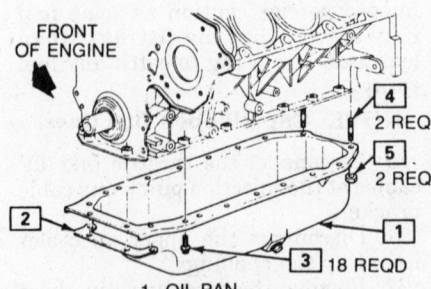

FRONT
OF ENGINE

4
2 REQD

5
2 REQD

2

3
18 REQD

1

1. OIL PAN
2. REINFORCEMENT
3. BOLT
4. STUD
5. NUT

Oil pan installation—4.1L V8

2. Remove the exhaust crossover pipe.

3. Raise and support the front of the car.

4. Remove the right engine splash shield.

5. Remove the vacuum pump-to-exhaust manifold brace.

6. Remove the mounting bolts and remove the manifold.

7. Installation is in the reverse order of removal.

RIGHT SIDE

1. Disconnect the negative battery cable.

2. Remove the exhaust crossover pipe.

3. Raise and support the front of the car.

4. Disconnect the exhaust pipe from the manifold.

5. Remove the mounting bolts and remove the manifold.

6. Installation is in the reverse order of removal.

Oil Pan

REMOVAL & INSTALLATION

4.1L V8

1. Disconnect the negative battery cable.

2. Remove the two flywheel covers.

3. Drain the oil.

4. Remove the mounting bolts and nuts and then remove the oil pan.

NOTE: If the pan is difficult to remove, try tapping the edges lightly with a rubber mallet.

5. Seal the oil pan to the block with RTV sealant.

6. Install the mounting bolts and nuts and tighten to 12 ft. lbs. (16 Nm).

7. Installation of the remaining components is in the reverse order of removal.

AUTOMATIC TRANSMISSION

For all automatic transmission band and linkage adjustments, and any other service procedures, please refer to "Automatic Transmissions" in the Unit Repair Section.

CAUTION

Any inaccuracies in shift linkage adjustments may result in premature failure of the transmission due to operation without the controls in full detent. Such operation results in reduced fluid pressure and in turn, partial engagement of the affected clutches. Partial engagement of the clutches, with sufficient pressure to permit apparently normal vehicle operation will result in failure of the clutches and/or other internal parts after only a few miles of operation.

REMOVAL & INSTALLATION

3.0L V6 and 3.8L V6

1. Disconnect the negative battery cable. Disconnect the wire connector at the mass air flow sensor (3.8L only).

2. Remove the air intake duct and the mass air flow sensor as an assembly.

3. Disconnect the cruise control assembly. Disconnect the shift control linkage.

4. Tag and disconnect the following:
   a. Park/Neutral switch
   b. Torque converter clutch
   c. Vehicle speed sensor
   d. Vacuum modulator hose at the modulator.

NOTE: Care must be exercised on reassembly of the Park/Neutral

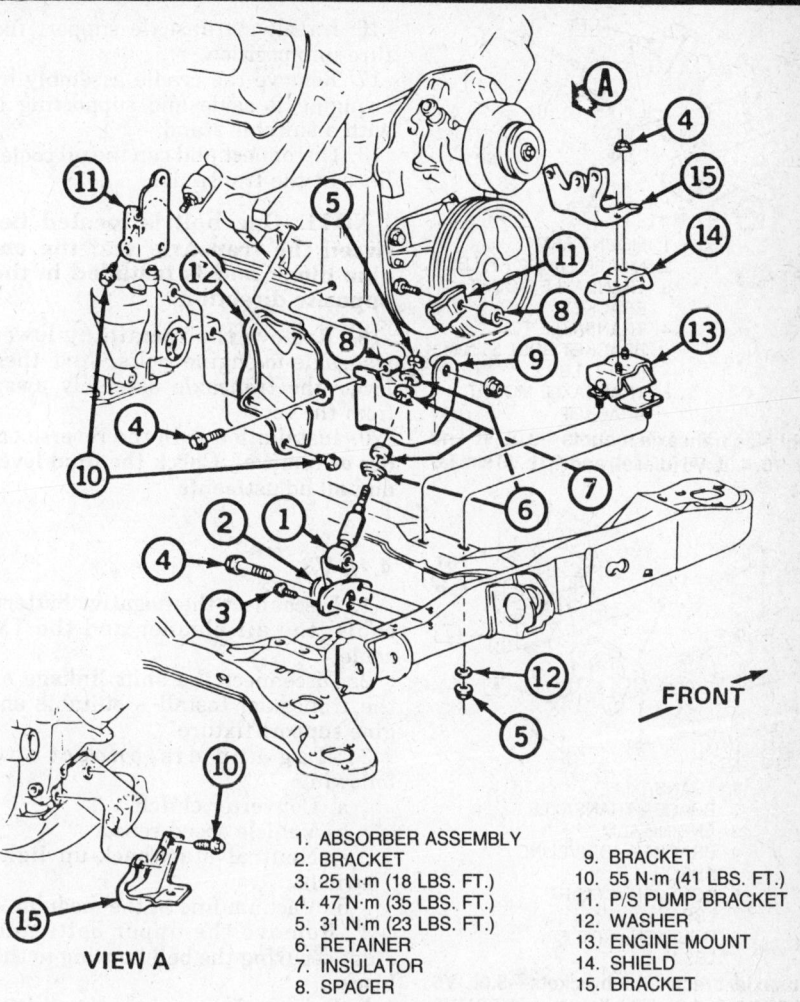

1. ABSORBER ASSEMBLY
2. BRACKET
3. 25 N·m (18 LBS. FT.)
4. 47 N·m (35 LBS. FT.)
5. 31 N·m (23 LBS. FT.)
6. RETAINER
7. INSULATOR
8. SPACER
9. BRACKET
10. 55 N·m (41 LBS. FT.)
11. P/S PUMP BRACKET
12. WASHER
13. ENGINE MOUNT
14. SHIELD
15. BRACKET

Engine mounting—4.3L V6 (diesel)

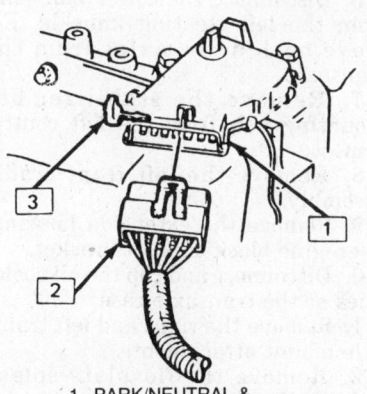

1. PARK/NEUTRAL & BACKUP LAMP SWITCH
2. SWITCH CONN.
3. "T" LATCH
T-latch connector

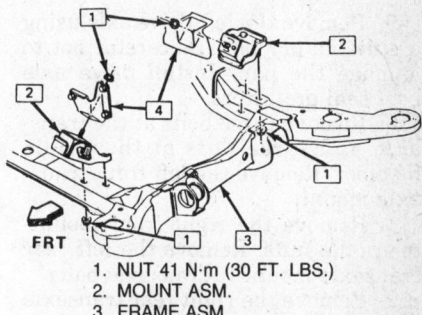

1. NUT 41 N·m (30 FT. LBS.)
2. MOUNT ASM.
3. FRAME ASM.
4. TRANSAXLE MOUNTING BRACKET

Left side transaxle mounts—3.0L V6 and 3.8L V6. 4.3L V6 (diesel) and 4.1L V8 similar

switch to ensure a proper fit of both the connector and the T-latch. Failure to do so may result in intermittent loss of switch functions.

5. Remove the three top transaxle-to-engine block bolts. Install an engine support fixture.

6. Remove both front wheels and then turn the steering wheel to the full left position.

7. Remove the right front ball joint nut and separate the control arm from the steering knuckle.

8. Remove the right drive axle as detailed later in this section.

NOTE: Be careful not to allow the drive axle splines to contact any portion of the lip seal.

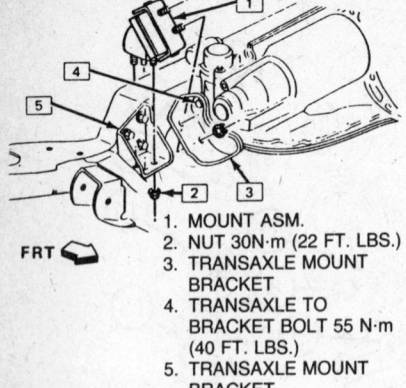

1. MOUNT ASM.
2. NUT 30N·m (22 FT. LBS.)
3. TRANSAXLE MOUNT BRACKET
4. TRANSAXLE TO BRACKET BOLT 55 N·m (40 FT. LBS.)
5. TRANSAXLE MOUNT BRACKET

Right side transaxle mounts—3.0L V6 and 3.8L V6. 4.3L V6 (diesel) and 4.1L V8 similar

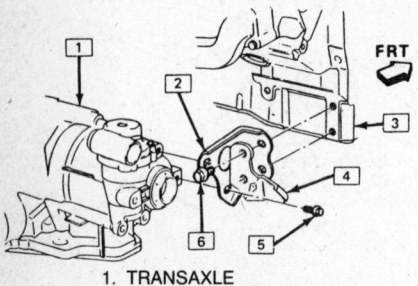

1. TRANSAXLE
2. BRACE—TRANSAXLE
3. ENGINE ASM.
4. BRACKET—DRIVELINE ABSORBER
5. BOLT 45N·m (33 FT. LBS.)
6. BOLT 95N·m (70 FT. LBS.)

Transaxle brace and brackets—3.0L V6 and 3.8L V6. 4.3L V6 (diesel) and 4.3L V8 similar

9. Remove the left drive axle using a suitable pry bar. Be careful not to damage the pan. Install drive axle boot seal protectors.

10. Remove three bolts at the transaxle and three nuts at the cradle member. Remove the left front transaxle mount.

11. Remove the right front mount-to-cradle nuts. Remove the left rear transaxle mount-to-transaxle bolts.

12. Remove the right rear transaxle mount as in Step 10. Remove the engine support bracket-to-transaxle case bolts.

13. Remove the flywheel cover. Remove the flywheel-to-converter bolts.

NOTE: Be sure to matchmark the flywheel-to-converter relationship for proper alignment upon reassembly.

14. Remove the bolts attaching the rear cradle member to the front cradle dog leg.

15. Remove the front left cradle-to-body bolt. Remove the front cradle dog leg-to-right cradle member bolts.

16. Install a transaxle support fixture into position.

17. Remove the cradle assembly by swinging it aside and supporting it with a suitable stand.

18. Disconnect and cap the oil cooler lines at the transaxle.

NOTE: One bolt is located between the transaxle and the engine block and is installed in the opposite direction.

19. Remove the remaining lower transaxle-to-engine bolts. And then lower the transaxle assembly away from the car.

20. Installation is in the reverse order or removal. Check the fluid level and all adjustments.

4.1L V8

1. Disconnect the negative battery cable, the air cleaner and the TV cable.

2. Disconnect the shift linkage at the transaxle. Install a suitable engine support fixture.

3. Tag and disconnect the following:
   a. Converter clutch
   b. Vehicle speed sensor
   c. Neutral start/back-up light switch
   d. Vacuum line at the modulator

4. Remove the upper bolts and studs securing the bell housing to the block.

5. Raise and support the car and remove both front wheels.

6. Disconnect the lower ball joint from the left steering knuckle. Remove both drive axles from the transaxle.

7. Remove the stabilizer bar mounting bolt from the left control arm.

8. Remove the left front cradle assembly.

9. Remove the extension housing-to-engine block support bracket.

10. Disconnect and cap the oil cooler lines at the transaxle case.

11. Remove the right and left transaxle mount attachments.

12. Remove the flexplate splash shield. Remove the converter-to-flexplate bolts.

13. Remove all the lower bell housing bolts except the lower rear on (No. 6).

14. Position a jack under the transaxle and then remove the last bell housing bolt.

NOTE: To reach the last bell housing bolt, you will need a 3 in. socket wrench extension and you must come through the right wheel arch opening.

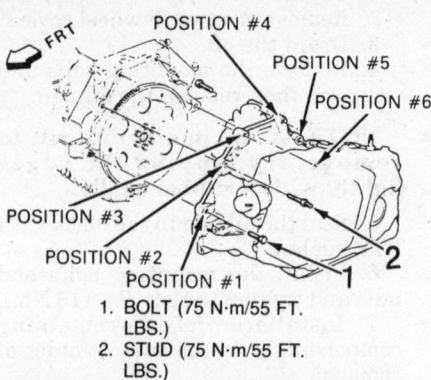

1. BOLT (75 N·m/55 FT. LBS.)
2. STUD (75 N·m/55 FT. LBS.)

Transaxle-to-engine mounting bolts. Remove No. 6 last—4.1L V8

15. Remove the transaxle assembly.

16. Installation is in the reverse order of removal. Check the fluid level and all adjustments.

4.3L V6 (Diesel)

1. Disconnect the negative battery cable. Disconnect the TV cable at the injection pump and transaxle.

2. Remove the crossover pipe shield and disconnect the shift control linkage.

3. Tag and disconnect the following:
   a. Park/Neutral switch
   b. Torque converter clutch
   c. Vehicle speed sensor
   d. Vacuum hose at the modulator.

4. Remove the three upper engine-to-transaxle bolts.

5. Loosen, but do not remove, the engine-to-transaxle bolt at the starter.

6. Install a suitable engine support fixture.

7. Raise and support the car. Remove both front wheels and then turn the steering wheel to the full left position.

8. Disconnect the right front ball joint from the steering knuckle. Remove the right drive axle from the transaxle.

9. Remove the left front and rear transaxle-to-cradle mounts.

10. Remove the transaxle brace and its bracket.

11. Disconnect the speedometer cable.

12. Remove the right rear transaxle mount and disconnect the left stabilizer link.

13. Remove the flywheel cover and then remove the flywheel-to-torque converter bolts.

14. Remove the bolts attaching the rear cradle member to the front cradle dog leg. Remove one stabilizer brace and loosen the other.

15. Remove the front cradle-to-body bolt and the right front motor mount.

16. Remove the wiring harness cover on the cradle and position it out of the way.

17. Install a suitable transmission support fixture.

18. Slide the cradle assembly to one side and support it.

19. Disconnect and cap the oil cooler lines at the transaxle.

20. Remove the exhaust connector pipe and the rear exhaust manifold.

21. Remove the remaining engine-to-transaxle bolts.

NOTE: One engine-to-transaxle bolt is installed in the opposite direction.

22. Lower the transaxle and remove it.

23. Installation is in the reverse order of removal. Check the fluid level and all adjustments.

DRIVESHAFTS, DRIVE AXLES AND U-JOINTS

NOTE: Removal and installation procedures for the front drive axles (halfshafts) and the CV-joints may be found in the "GM A & X Body" section.

For further information on CV-Joint overhaul, please refer to "U-Joints and CV-Joints" in the Unit Repair section.

REAR AXLE

Hub

REMOVAL & INSTALLATION

A single unit hub and bearing assembly is bolted to both ends of the rear axle assembly. These take the place of rear axles used on rear wheel drive cars. The hub and bearing assembly is a sealed unit which requires no maintenance. The unit must be replaced as an assembly and cannot be disassembled or adjusted.

The hub and bearing can be removed by removing the rear brake drum, removing the four hub and bearing-to-axle assembly attaching bolts and pulling the unit out. Installation is the reverse of removal. Tighten the bolts to 35–39 ft. lbs.

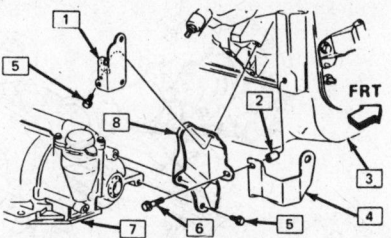

1. POWER STEERING PUMP BRACKET
2. SPACER
3. ENGINE
4. SHOCK ABSORBER BRACKET
5. BOLT 55N·m (40 FT. LBS.)
6. BOLT 45N·m (33 FT LBS.)
7. TRANSAXLE
8. BRACE

Transaxle brace and brackets—4.3L V6 (diesel)

FRONT SUSPENSION

MacPherson Strut

REMOVAL & INSTALLATION

For spring and shock absorber removal and installation, and any other strut overhaul procedures, please refer to "Strut Overhaul" in the Unit Repair section.

1. Remove the three nuts attaching the top of the strut assembly to the body.

2. Raise the car and support it with jack stands under the engine cradle.

3. Lower the car slightly so that the weight rests on the jack stands.

4. Remove the wheels and tires.

NOTE: Always install drive axle boot seal protectors. Care must be taken to prevent overextension of the inner Tri-Pot joints.

5. Remove the brake line bracket bolt from the strut assembly. Do not disconnect the brake line from the caliper.

6. Remove the strut-to-steering knuckle bolts and then carefully remove the strut assembly.

7. Installation is in the reverse order of removal. Please note the following:

a. Check wheel alignment

b. Tighten the strut-to-body bolts to 18 ft. lbs. (24 Nm)

c. Tighten the strut-to-steering knuckle bolts to 144 ft. lbs. (195 Nm).

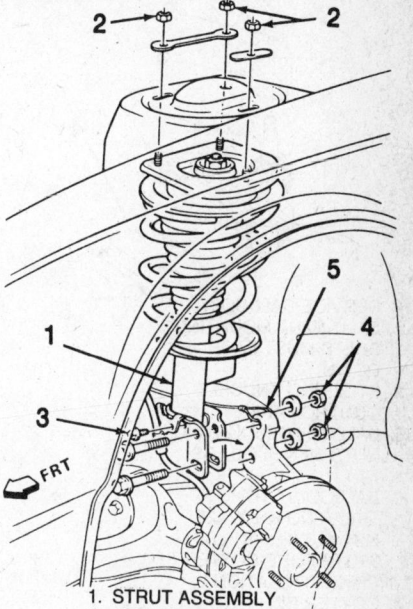

1. STRUT ASSEMBLY
2. STRUT TO BODY NUTS 24 N·m (18 LBS. FT.)
3. BRAKE LINE BRACKET BOLT 17 N·m (13 LBS. FT.)
4. STRUT TO STEERING KNUCKLE NUTS 195 N·m (144 LBS. FT.)
5. RETAIN STEERING KNUCKLE WITH WIRE ONCE STRUT ASSEMBLY IS REMOVED

Front MacPherson strut assemlby

Front Wheel Bearing

ADJUSTMENT

All models covered in this section utilize a permanently sealed and lubricated front wheel bearing assembly. No adjustments are either necessary or possible.

Ball Joints

INSPECTION

1. Raise the front of the car with a lift placed under the engine cradle. The front wheel should be clear of the ground.

2. Grasp the wheel at the top and bottom and shake the wheel in and out.

3. If any movement is seen of the steering knuckle relative to the control arm, the ball joints are defective and must be replaced. Note that movement elsewhere may be due to loose wheel bearings or other troubles; watch the knuckle-to-control arm connection.

4. If the ball stud is disconnected from the steering knuckle and any

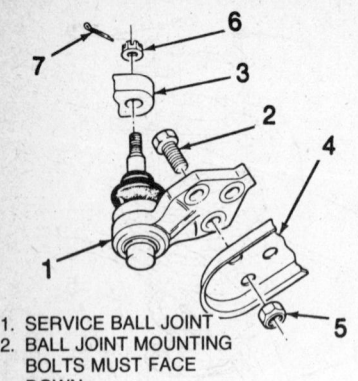

1. SERVICE BALL JOINT
2. BALL JOINT MOUNTING BOLTS MUST FACE DOWN
3. STEERING KNUCKLE
4. CONTROL ARM
5. BALL JOINT MOUNTING NUTS 68 N·m (50 LBS. FT.)
6. BALL JOINT TO STEERING KNUCKLE NUT 110 N·m (81 LBS. FT.) BEFORE COTTER PIN INSTALLATION
7. COTTER PIN

Ball joint installation

looseness is noted, often the ball joint stud can be twisted in its socket with your fingers, replace the ball joints.

REMOVAL & INSTALLATION

1. Raise the front of the car and support it with jackstands underneath the engine cradle. Lower the car slightly so that the weight rests primarily on the jack stands.
2. Remove the wheel and tire assemblies.
3. Install drive axle covers to protect the drive axle boot seals.
4. Pull the cotter pin from the ball joint and install a ball joint separator tool. Turn the castellated nut counterclockwise to separate the ball joint from the steering knuckle.
5. Use a $\frac{1}{8}$ in. drill bit to drill a hole approximately $\frac{1}{4}$ in. deep in the center of each of the three ball joint rivets.
6. Use a $\frac{1}{2}$ in. drill bit to drill off the rivet heads. Drill only enough to remove the rivet head.
7. Use a hammer and punch to remove the rivets. Drive them out from the bottom.
8. Loosen the stabilizer bar bushing assembly nut.
9. Pull down on the control arm and remove the ball joint from the steering knuckle and control arm.
10. Install the new ball joint in the steering knuckle and line up the holes with those in the control arm.
11. Install the three ball joint nuts facing down and tighten the nuts to 50 ft. lbs. (68 Nm).
12. Install the castellated nut and tighten to 81 ft. lbs. (110 Nm).

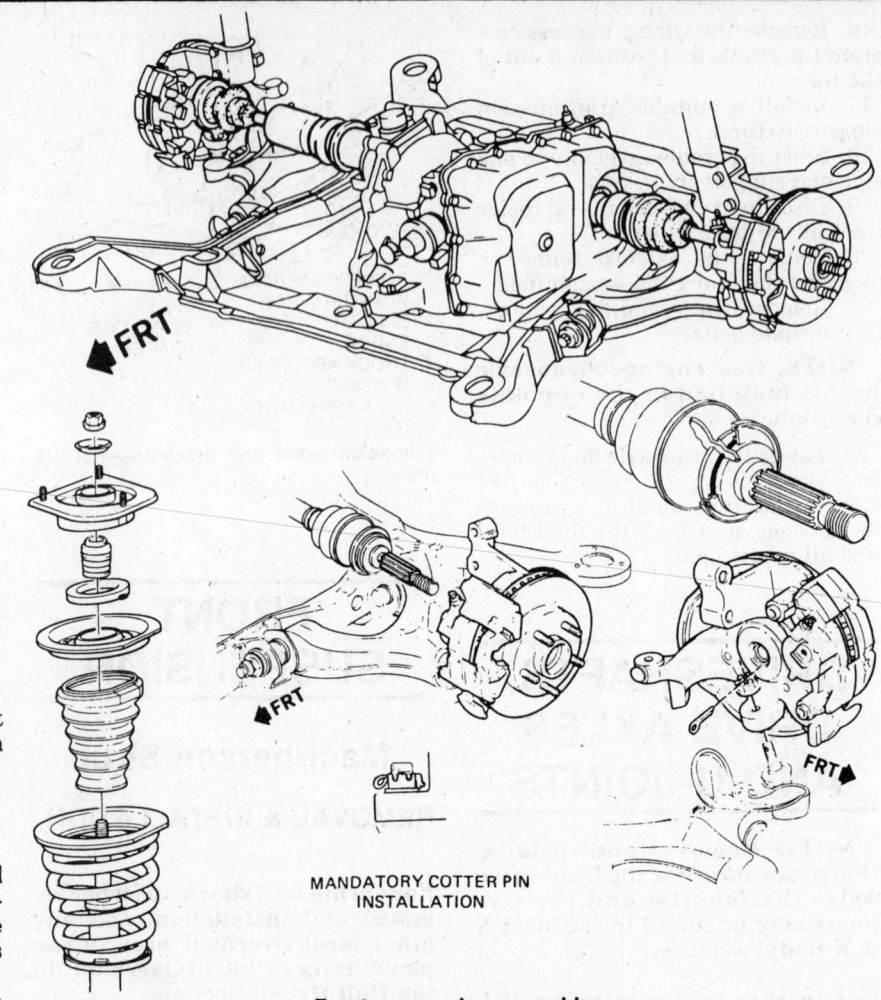

MANDATORY COTTER PIN INSTALLATION

Front suspension assembly

NOTE: Tightening the nut for cotter pin alignment is allowed, but do not loosen it once the torque value has been reached.

13. Install the cotter pin.
14. Installation of the remaining components is in the reverse order of removal.

Lower Control Arm

REMOVAL & INSTALLATION

1. Perform Steps 1–3 of the "Ball Joint Removal and Installation" procedure.
2. Remove the stabilizer bar bushing-to-control arm bolt.
3. Pull the cotter pin from the ball joint and install a ball joint separator tool. Turn the castellated nut counterclockwise to separate the ball joint from the steering knuckle.
4. Remove the remaining control arm bolts and remove the control arm from the vehicle.

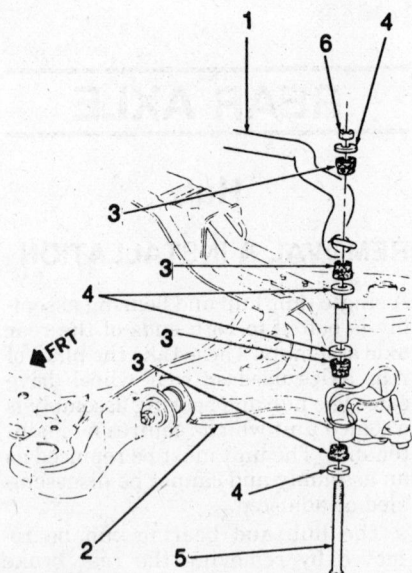

1. Stabilizer Bar
2. Control Arm
3. Insulator (4)
4. Retainer (4)
5. Bolt
6. Nut 17 N·m (13 LBS. FT.)

Stabilizer bar bushing assembly

5. Position the control arm and install the mounting bolts, but DO NOT tighten.

6. Install the stabilizer bar bushing assembly. Reconnect the ball joint to the steering knuckle.

7. Hoist the vehicle slightly so the weight of the vehicle is supported by the control arms.

NOTE: The weight of the vehicle MUST be supported by the control arms when tightening the mounting nuts.

8. Tighten the:
  a. Stabilizer bar bushing nut to 13 ft. lbs. (17 Nm)
  b. Rear control arm mounting nut to 90 ft. lbs. (123 Nm)
  c. Front control arm mounting nut to 140 ft. lbs. (190 Nm)
  d. Ball joint nut to 81 ft. lbs. (110 Nm).

9. Installation of the remaining components is in the reverse order of removal.

REAR SUSPENSION

Superlift Strut

REMOVAL & INSTALLATION

1. Remove the inner trunk side cover.

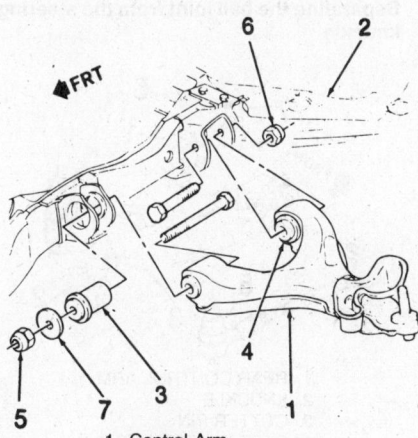

1. Control Arm
2. Cradle
3. Cradle Mounted Bushing
4. Control Arm Mounted Bushing
5. Cradle Mounted Bushing Nut 190 N·m (140 LBS. FT.)
6. Control Arm Mounted Bushing Nut 123 N·m (90 LBS. FT.)
7. Washer

Lower control arm assembly

2. Raise and support the rear of the vehicle. Remove the wheels and tires.

3. Disconnect and plug the ELC air line.

4. Remove the strut tower mounting nuts from inside the trunk.

5. Remove the strut anchor bolts, washers and nuts from the rear knuckle and knuckle bracket.

6. Remove the strut.

7. Installation is in the reverse order of removal. Please note the following:
  a. Tighten the strut tower mounting nuts to 19 ft. lbs. (25 Nm)
  b. Tighten the strut anchor nuts to 144 ft. lbs. (195 Nm)
  c. Lightly pressurize the ELC system by momentarily grounding the compressor test lead in the engine compartment.
  d. Check rear wheel alignment.

Coil Springs

REMOVAL & INSTALLATION

1. Raise the rear of the vehicle and support it so that the control arms hang free. Remove the rear wheels.

2. Separate the rear stabilizer bar from the knuckle bracket and remove it.

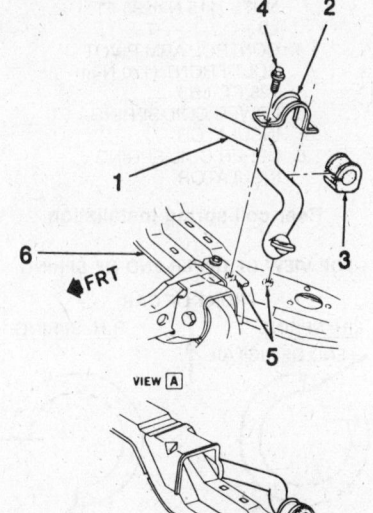

VIEW A

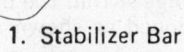

1. Stabilizer Bar
2. Stabilizer Bar Mounting Bracket
3. Stabilizer Bar Mounting Bushing
4. 50 N·m (37 LBS. FT.)
5. Frame Welded Nuts
6. Cradle

Stabilizer bar installation

3. Disconnect the ELC height sensor link (right control arm) and/or the parking brake cable retaining clip (left control arm).

4. Position the special tool J-23028-01 or its equivalent, so as to cradle the control arm bushings.

NOTE: Special tool J-23028-01 should be secured to a suitable jack.

5. Raise the jack to remove the tension from the control arm pivot bolts.

——— **CAUTION** ———
Secure a chain around the spring and through the control arm as a safety precaution.

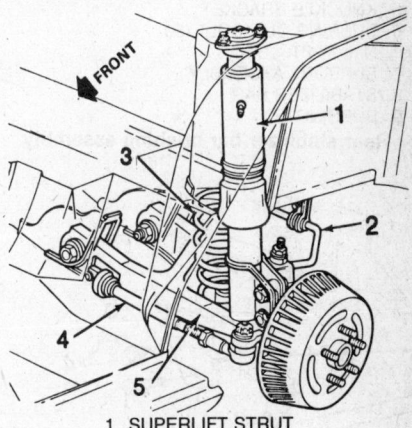

1. SUPERLIFT STRUT
2. STABILIZER BAR
3. COIL SPRING
4. SUSPENSION ADJUSTMENT LINK
5. LOWER CONTROL ARM
Rear suspension

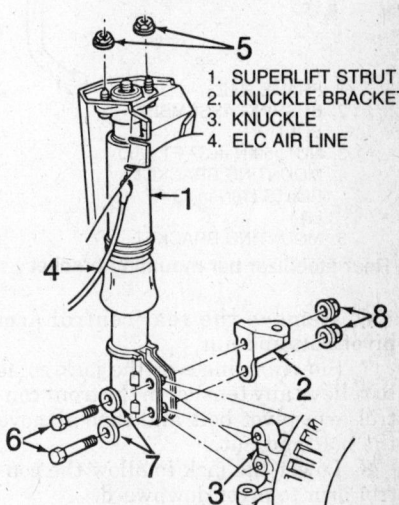

1. SUPERLIFT STRUT
2. KNUCKLE BRACKET
3. KNUCKLE
4. ELC AIR LINE
5. STRUT MOUNTING NUTS (25 N·m/19 FT. LBS.)
6. STRUT ANCHOR BOLTS
7. STRUT ANCHOR WASHERS
8. STRUT ANCHOR NUTS (195 N·m/144 FT. LBS.)

Rear strut installation

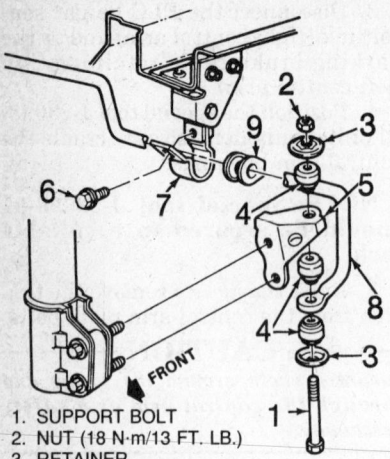

1. SUPPORT BOLT
2. NUT (18 N·m/13 FT. LB.)
3. RETAINER
4. INSULATORS
5. KNUCKLE BRACKET
6. BUSHING CLIP BOLT (50 N·m/37 FT. LB.)
7. SUPPORT ASSEMBLY
8. STABILIZER BAR
9. BUSHING

Rear stabilizer bar bushing assembly

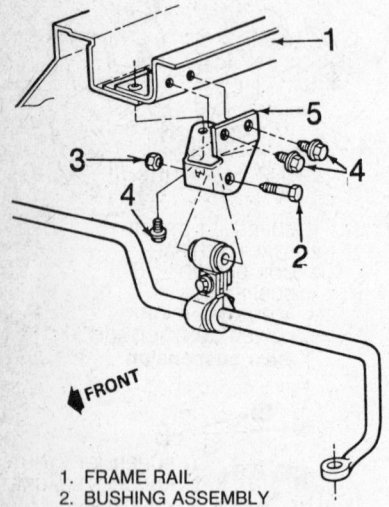

1. FRAME RAIL
2. BUSHING ASSEMBLY BOLT
3. NUT (50 N·m/37 FT. LB.)
4. MOUNTING BRACKET BOLTS (18N·m/13 FT. LB.)
5. MOUNTING BRACKET

Rear stabilizer bar mounting bracket

6. Remove the rear control arm pivot bolt and nut.

7. Slowly maneuver the jack so as to relieve any tension in the front control arm pivot bolt and then remove the bolt and nut.

8. Lower the jack to allow the control arm to pivot downward.

9. When all pressure is removed from the coil spring, remove the safety chain, spring and insulators.

NOTE: The spring insulators should be inspected for cuts or tears. They should be replaced if the vehicle has over 50,000 miles.

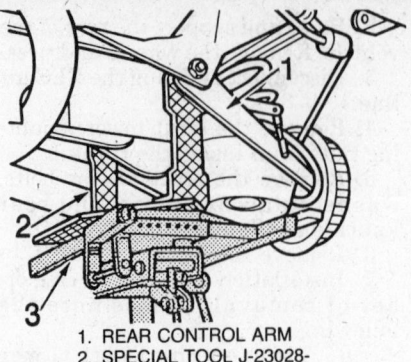

1. REAR CONTROL ARM
2. SPECIAL TOOL J-23028-01
3. TRANSMISSION JACK

Use the special tool and a transmission jack to cradle the control arm

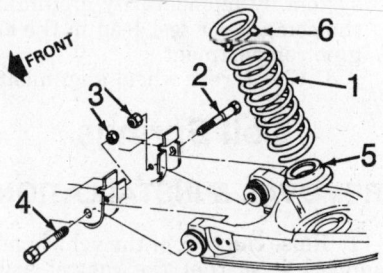

1. COIL SPRING
2. CONTROL ARM PIVOT BOLT-REAR (170 N·m/125 FT. LB.)
3. CONTROL ARM PIVOT NUTS (115 N·m/85 FT. LB.)
4. CONTROL ARM PIVOT BOLT-FRONT (170 N·m/125 FT. LB.)
5. LOWER COIL SPRING INSULATOR
6. UPPER COIL SPRING INSULATOR

Rear coil spring installation

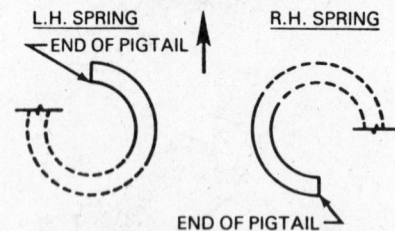

TOP VIEW OF UPPER END OF SPRING

FRONT OF CAR

L.H. SPRING — END OF PIGTAIL

R.H. SPRING

END OF PIGTAIL

Rear coil spring positioning

10. Snap the upper insulator onto the spring. Position the lower insulator and the spring in the control arm. Install the coil springs so that the upper ends are positioned as shown in the illustration.

11. Installation of the remaining components is in the reverse order of removal. Control arm mounting nuts should not be tightened until the vehicle is unsupported and resting on its wheels at normal trim height.

Ball Joint

REMOVAL & INSTALLATION

1. Raise and support the rear of the vehicle and remove the wheels.

2. Disconnect the ELC height sensor link (right control arm) and/or the parking brake cable retaining link (left control arm).

3. Remove the cotter pin and castellated nut from the outer suspension adjustment link.

4. Separate the outer suspension link from the knuckle.

5. Support the control arm with a suitable jack. The lower control arm MUST be supported to prevent the coil spring from forcing the control arm downward.

6. Remove the ball stud cotter pin.

7. Remove the castellated nut and then reinstall it with the flat side facing upward. DO NOT tighten the nut.

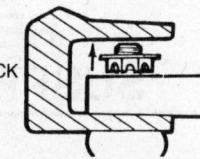

REMOVE CASTELLATED NUT AND REINSTALL WITH FLAT SIDE FACING UPWARD.
PLACE J-34505 INTO POSITION AS SHOWN. LOOSEN NUT AND BACK OFF UNTIL. . .

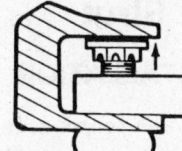

. . .THE NUT CONTACTS THE TOOL. CONTINUE BACKING OFF THE NUT UNTIL THE NUT FORCES THE BALL STUD OUT OF THE KNUCKLE.

Separating the ball joint from the steering knuckle

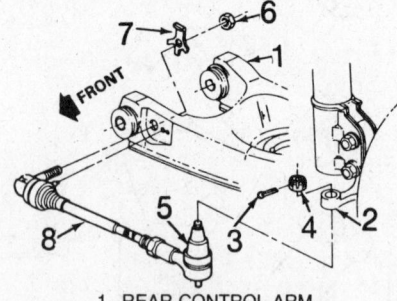

1. REAR CONTROL ARM
2. KNUCKLE
3. COTTER PIN
4. CASTELLATED NUT (50 N·m/37 FT. LB.)
5. OUTER SUSPENSION ADJUSTMENT LINK
6. LINK RETAINING NUT (85 N·m/63 FT. LB.)
7. LINK RETAINER
8. SUSPENSION ADJUSTMENT LINK ASSEMBLY

Rear suspension adjustment link

8. Install a ball joint separator tool and separate the knuckle from the ball stud by backing off the inverted nut against the tool.

9. Separate the ball joint from the control arm.

10. Installation is in the reverse order of removal. Please note following:

   a. Tighten a NEW castellated nut to 7.5 ft. lbs. (10 Nm). Tighten the nut an additional 2/3 of a turn.

   b. Align the slot in the nut to the cotter pin hole by tightening only. Do not loosen the nut to align the holes.

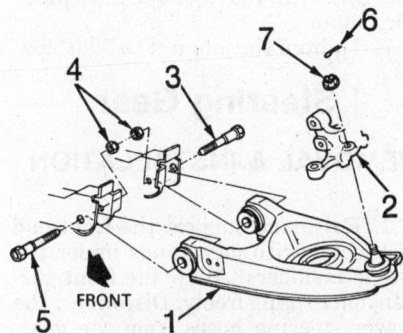

1. REAR CONTROL ARM
2. KNUCKLE
3. CONTROL ARM PIVOT BOLT—REAR
4. CONTROL ARM PIVOT NUTS (115 N·m/85 FT-LB)
5. CONTROL ARM PIVOT BOLT—FRONT
6. COTTER PIN
7. CASTELLATED NUT

Rear control arm

Control Arm

REMOVAL & INSTALLATION

1. Perform Steps 1–2 of the "Ball Joint Removal & Installation" procedures.

2. Remove the suspension adjustment link retaining nut and retainer.

3. Separate the link assembly from the control arm.

4. Remove the coil spring as detailed previously.

5. Perform Steps 6–9 of the "Ball Joint Removal & Installation" procedure.

6. Remove the control arm.

7. Installation is in the reverse order of removal.

BRAKES

For all brake system removal, installation and adjustment procedures, please refer to "Brakes" in the Unit Repair section.

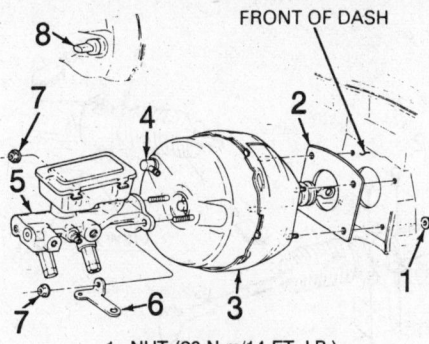

FRONT OF DASH

1. NUT (20 N·m/14 FT. LB.)
2. SEAL
3. POWER BOOSTER
4. CHECK VALVE
5. MASTER CYLINDER
6. VACUUM SWITCH BRACKET (DIESEL)
7. NUT (30 N·m/22 FT. LB.)
8. VACUUM SWITCH (GAS)

Typical master cylinder and power booster mounting

Master Cylinder

REMOVAL & INSTALLATION

1. Disconnect and plug hydraulic lines, and drain the cylinder.

2. Remove the attaching nuts and remove the master cylinder from the power booster unit.

3. Reverse to install. Bleed the system.

Power Booster

REMOVAL & INSTALLATION

1. From inside the car, detach the brake pushrod from the brake pedal.

2. Disconnect the hydraulic lines from the front of the master cylinder.

3. Remove the nuts from the mounting studs which hold the unit to the dash panel. Remove the unit and clean it prior to installation.

4. Install in reverse order of removal. Bleed system.

Parking Brake

ADJUSTMENT

1. Depress the parking brake pedal 1½ in. (35mm).

2. Raise the vehicle and support it with jack stands.

3. Tighten the adjusting nut until the left rear wheel can just be turned to the rear with both hands, but is locked when forward rotation is attempted.

4. With the mechanisms totally disengaged, both rear wheels should turn freely in either direction with no brake drag.

— CAUTION —
Do not adjust the parking brake cable so tight as to cause brake drag.

5. Lower the vehicle.

STEERING

Tie-Rod

REMOVAL & INSTALLATION

1. Loosen the jam nut on the steering rack (inner tie-rod).

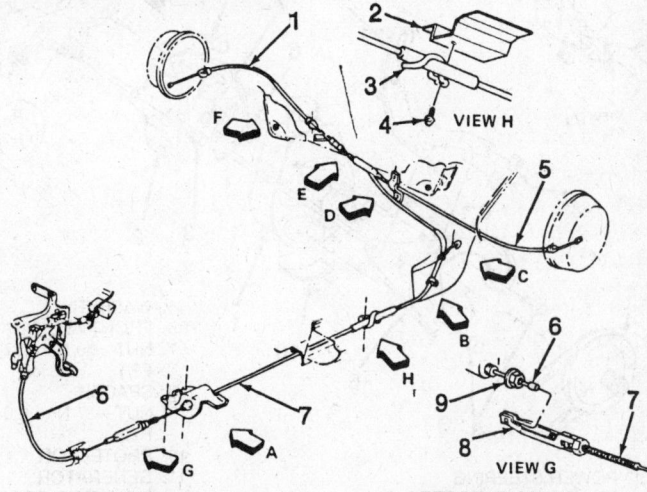

1. RIGHT REAR CABLE
2. UNDERBODY
3. GUIDE
4. BOLT/SCREW 38 N·m (28 FT. LB.)
5. LEFT REAR CABLE
6. CABLE ASM—FRONT
7. CABLE ASM—INTERMEDIATE.
8. EQUALIZER ASM
9. NUT

Parking brake cable routing

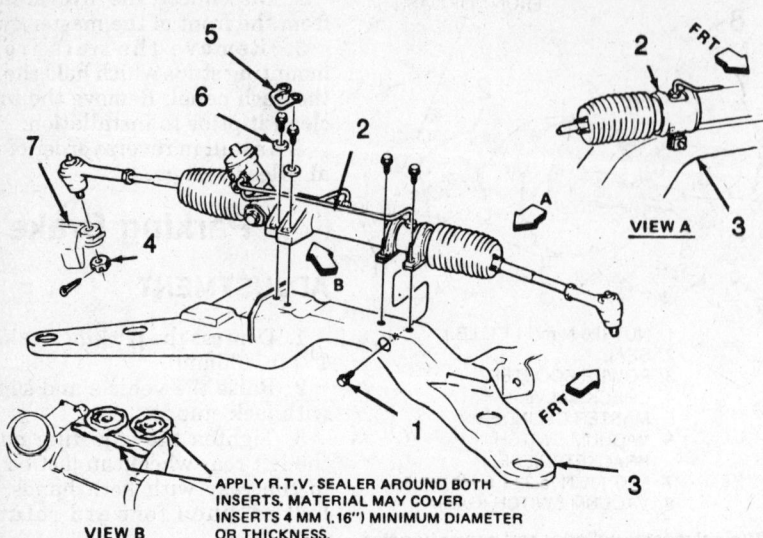

1. BOLT 68 N·m (50 LBS. FT.) AFTER SECOND REUSE OF BOLT, LOCTITE THREAD LOCKING KIT, #1052624 MUST BE USED
2. STEERING GEAR
3. FRAME
4. 50 N·m (35 LBS. FT.), 70 N·m (52 LBS. FT.) MAXIMUM PERMISSIBLE TORQUE TO ALIGN COTTER PIN SLOT. (⅙

TURN MAXIMUM) DO NOT BACK OFF FOR COTTER PIN INSERTION
5. RETAINER
6. WASHER
7. STEERING KNUCKLE

APPLY R.T.V. SEALER AROUND BOTH INSERTS. MATERIAL MAY COVER INSERTS 4 MM (.16") MINIMUM DIAMETER OR THICKNESS.

Rack and pinion assembly

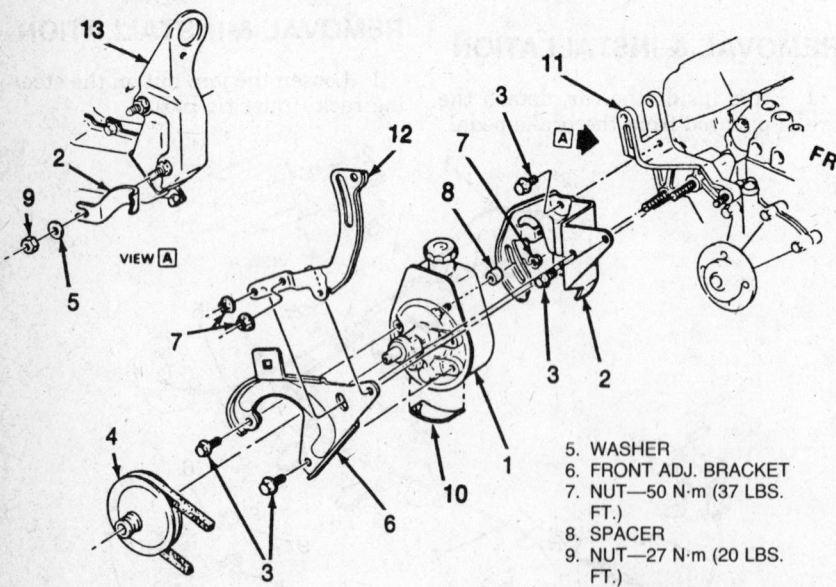

1. POWER STEERING PUMP
2. REAR ADJ. BRACKET
3. BOLT—50 N·m (37 LBS. FT.)
4. PULLEY

5. WASHER
6. FRONT ADJ. BRACKET
7. NUT—50 N·m (37 LBS. FT.)
8. SPACER
9. NUT—27 N·m (20 LBS. FT.)
10. PROTECTOR
11. GENERATOR MOUNTING BRACKET
12. GENERATOR ADJ. BRACKET
13. ENGINE LIFT BRACKET & SHIELD

Power steering pump mounting—3.0L V6 and 3.8L V6

2. Remove the tie-rod end nut. Separate the tie-rod end from the steering knuckle using a puller.

3. Unscrew the tie-rod end, counting the number of turns.

4. To install, screw the tie-rod end onto the steering rack (inner tie-rod) the same number of turns as counted for removal. This will give approximately correct toe.

5. Install the tie-rod end into the knuckle. Install the nut and tighten to 40 ft. lbs.

6. If the toe must be adjusted, use pliers to expand the boot clamp. Turn the inner tie-rod to adjust. Replace the clamp.

7. Tighten the jam nut to 50 ft. lbs.

Steering Gear

REMOVAL & INSTALLATION

1. Raise and support the front end of the car with jackstands under the frame members. Allow the front suspension to hang freely. Disconnect the power steering hoses from the gear, where equipped.

2. Move the intermediate shaft seal upward and remove the intermediate shaft-to-stub shaft pinch bolt.

3. Remove both front wheels.

4. Remove the cotter pins and nut from both tie-rod ends. Disconnect the tie-rod ends from the steering knuckles.

5. Remove the line retainer.

6. Remove the outlet and pressure hose.

7. Remove the five rack and pinion assembly mounting bolts.

8. Loosen the front engine cradle mounting bolts. Install jack stands and the lower the rear of the cradle about 3 in. (76mm).

— **CAUTION** —
Do not lower the rear of the engine cradle too far.

9. Remove the rack and pinion assembly.

10. Installation is in the reverse order of removal. Tighten the rack mounting bolts to 50 ft. lbs. (68 Nm). Tighten the tie-rod end nut to 35–52 ft. lbs. (50–70 Nm). Bleed the power steering system and check for leaks.

Power Steering Pump

REMOVAL & INSTALLATION

3.0L V6 and 3.8L V6

1. Disconnect the negative battery cable.

2. Remove the air cleaner assembly on the 3.0L.

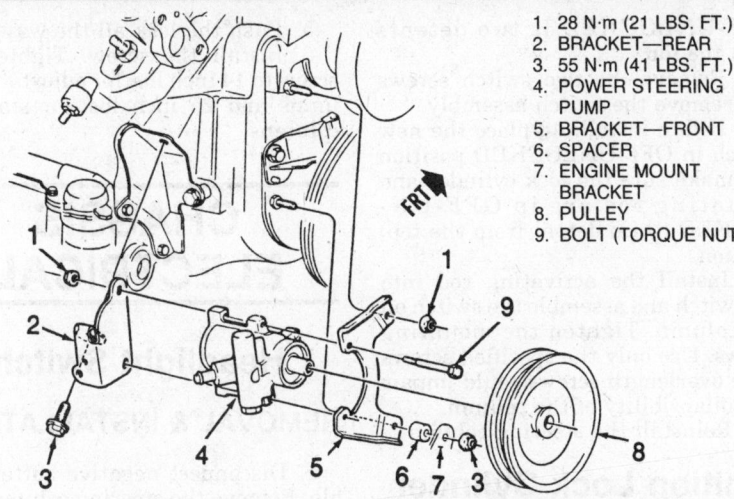

1. 28 N·m (21 LBS. FT.)
2. BRACKET—REAR
3. 55 N·m (41 LBS. FT.)
4. POWER STEERING PUMP
5. BRACKET—FRONT
6. SPACER
7. ENGINE MOUNT BRACKET
8. PULLEY
9. BOLT (TORQUE NUT)

Power steering pump mounting—4.3L V6 (diesel)

3. Remove the drive belt and then the alternator itself.

4. Raise the front of the vehicle and support it on jack stands.

5. Disconnect and plug the pressure and return lines at the pump.

6. Remove the rear pump adjustment bracket-to-pump nut. Remove the power steering belt and lower the vehicle.

7. Remove the alternator adjustment bracket and support brace.

8. Remove the rear pump adjustment bracket and then remove the pump assembly.

9. Remove the front pump adjustment bracket and then remove the pulley.

10. Installation is in the reverse order of removal. Adjust the drive belts and bleed the power steering system.

4.1L V8

1. Disconnect the negative battery cable.

2. Remove the drive belt and the pulley.

3. Disconnect and plus the high pressure and pump feed lines.

4. Remove the two pump mounting bolts. Remove the power steering pump.

5. Installation is in the reverse order of removal. Tighten the pump mounting bolts to 25 Nm. Adjust the drive belt tension and bleed the power steering system.

4.3L V6 (Diesel)

1. Disconnect the negative battery cable. Raise and support the front of the vehicle on jack stands.

2. Remove the engine splash shield. Remove the crankshaft pulley and the engine shock absorber.

3. Disconnect the reservoir hose from the power steering pump and drain the reservoir.

4. Remove the high pressure hose support. Disconnect and cap the high pressure hose.

5. Remove the three bracket bolts. Remove the pump assembly along with its brackets.

6. Installation is in the reverse order of removal. Adjust the drive belt tension. Bleed the power steering system and check for leaks.

BLEEDING THE POWER STEERING SYSTEM

1. Fill the fluid reservoir.

2. Let the fluid stand undisturbed for two minutes, then crank the engine for about two seconds. Refill reservoir if necessary.

3. Repeat Steps 1 and 2 above until the fluid level remains constant after cranking the engine.

4. Raise the front of the car until the wheels are off the ground, then start the engine. Increase the engine speed to about 1500 rpm.

5. Turn the wheels lightly against the stops to the left and right, checking the fluid level and refilling if necessary.

Steering Wheel

REMOVAL & INSTALLATION

—— CAUTION ——
Disconnect the battery ground cable before removing the steering wheel. When installing a steering wheel, always make sure that the turn signal lever is in the neutral position.

1. Remove the trim retaining screws from behind the wheel. On wheels with a center cap, pull off the cap.

2. Lift the trim off and pull the horn wires from the turn signal cancelling cam.

3. Remove the retainer and the steering wheel nut.

4. Mark the wheel-to-shaft relationship, and then remove the wheel with a puller.

5. Install the wheel on the shaft aligning the previously made marks. Tighten the nut.

6. Insert the horn wires into the cancelling cam.

7. Install the center trim and reconnect the battery cable.

Turn Signal Switch

REMOVAL & INSTALLATION

1. Remove the steering wheel as previously outlined. Remove the trim cover.

2. Loosen the cover screws. Pry the cover off with a screwdriver, and lift the cover off the shaft.

3. Position the U-shaped lockplate compressing tool on the end of the steering shaft and compress the lock plate by turning the shaft nut clockwise. Pry the wire snapring out of the shaft groove.

4. Remove the tool and lift the lock plate off the shaft.

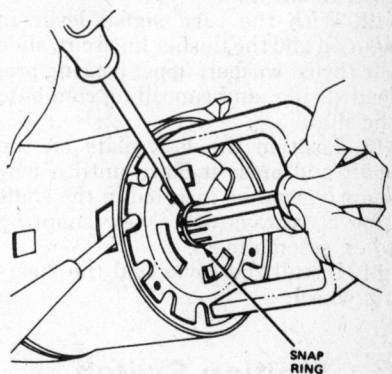

Depress the lockplate and remove the snapring

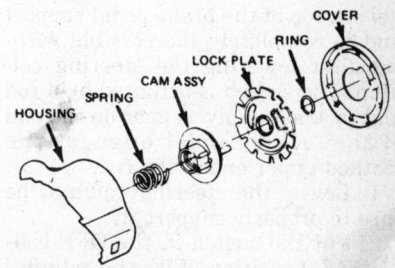

Remove these parts for access to the turn signal switch

5. Slip the cancelling cam, upper bearing preload spring, and thrust washer off the shaft.

6. Remove the turn signal lever. Push the flasher knob in and unscrew it. Remove the button retaining screw and remove the button, spring and knob.

7. Pull the switch connector out the mast jacket and tape the upper part to facilitate switch removal. Attach a long piece of wire to the turn signal switch connector. When installing the turn signal switch, feed this wire through the column first, and then use this wire to pull the switch connector into position. On tilt wheels, place the turn signal and shifter housing in low position and remove the harness cover.

8. Remove the three switch mounting screws. Remove the switch pulling it straight up while guiding the wiring harness cover through the column.

9. Install the replacement switch by working the connector and cover down through the housing and under the bracket. On tilt models, the connector is worked down through the housing, under the bracket, and then the cover is installed on the harness.

10. Install the switch mounting screws and the connector on the mast jacket bracket. Install the column-to-dash trim plate.

11. Install the flasher knob and the turn signal lever.

12. With the turn signal lever in neutral and the flasher knob out, slide the thrust washer, upper bearing preload spring, and cancelling cam onto the shaft.

13. Position the lock plate on the shaft and press it down until a new snapring can be inserted in the shaft groove. Always use a new snapring when assembling.

14. Install the cover and the steering wheel.

Ignition Switch

REMOVAL & INSTALLATION

The switch is located inside the channel section of the brake pedal support and is completely inaccessible without first lowering the steering column. The switch is actuated by a rod and rack assembly. A gear on the end of the lock cylinder engages the toothed upper end of the rod.

1. Lower the steering column; be sure to properly support it.

2. Put the switch in the OFF-UN-LOCKED position. With the cylinder removed, the rod is in LOCK when it is in the next to the uppermost detent.

OFF-UNLOCKED is two detents from the top.

3. Remove the two switch screws and remove the switch assembly.

4. Before installing, place the new switch in OFF-UNLOCKED position and make sure the lock cylinder and actuating rod are in OFF-UN-LOCKED (third detent from the top) position.

5. Install the activating rod into the switch and assemble the switch on the column. Tighten the mounting screws. Use only the specified screws since overlength screws could impair the collapsibility of the column.

6. Reinstall the steering column.

Ignition Lock Cylinder

REMOVAL & INSTALLATION

1. Place the lock in the RUN position.

2. Remove the lock plate, turn signal switch and buzzer switch.

3. Remove the screw and lock cylinder.

——— CAUTION ———
If the screw is dropped on removal, it could fall into the column, requiring complete disassembly to retrieve the screw.

4. Rotate the cylinder clockwise to align cylinder key with the keyway in the housing.

5. Push the lock all the way in.

6. Install the screw. Tighten the screw to 14 inch lbs. for adjustable columns and 25 inch lbs. for standard columns.

CHASSIS ELECTRICAL

Headlight Switch

REMOVAL & INSTALLATION

1. Disconnect negative battery cable. Remove the steering column lower cover or the instrument panel trim plate covering the headlamp switch, if a rocker-type headlamp switch is used.

2. Disconnect wiring harness retainer below headlight switch assembly. On Buick models, the switch connector is integral with the instrument panel; simply pull the switch outward to disconnect it.

3. On knob-type switches, depress spring loaded release button on top of headlight switch and remove switch, knob and rod assembly (switch ON).

4. Remove screw with ground wire at bottom of switch housing and any other mounting screws.

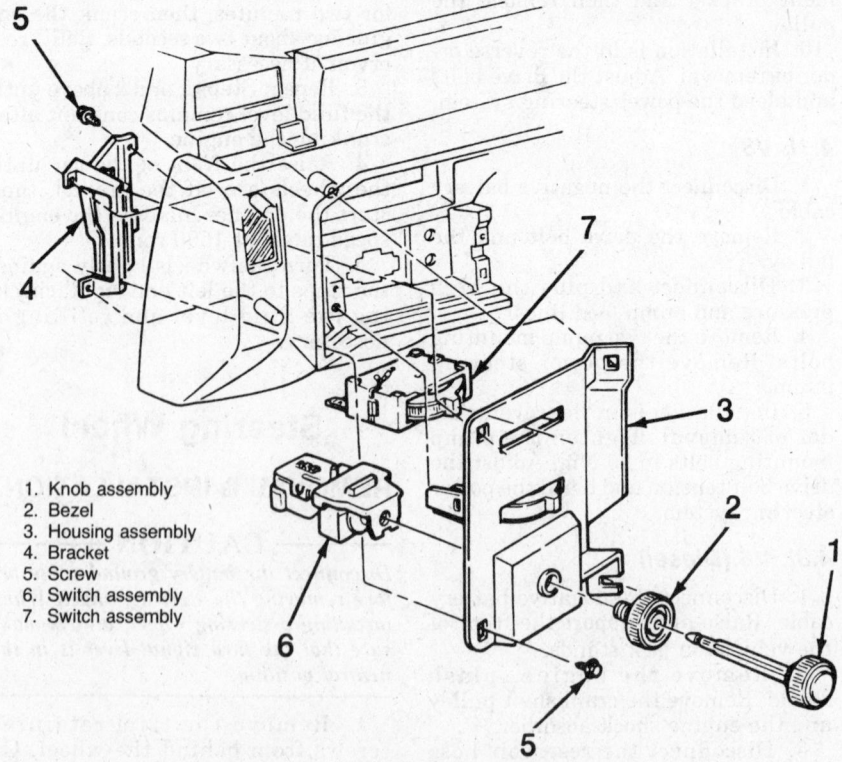

1. Knob assembly
2. Bezel
3. Housing assembly
4. Bracket
5. Screw
6. Switch assembly
7. Switch assembly

Cadillac knob type headlamp switch mounting

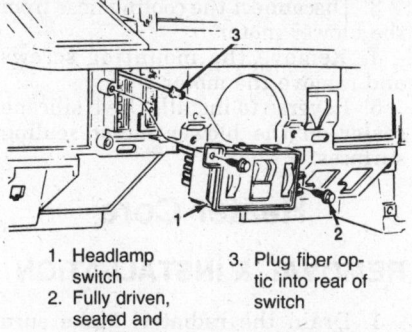

1. Headlamp switch
2. Fully driven, seated and not stripped
3. Plug fiber optic into rear of switch

Oldsmobile rocker type headlamp switch mounting—Buick similar

5. Pull assembly down and rearward, disconnect wiring harness connectors, bulb(s) and remove assembly.
6. Installation is the reverse of removal.

Instrument Cluster

REMOVAL & INSTALLATION

1. Disconnect the negative battery cable.
2. Remove the left sound insulator.
3. Remove the instrument panel insert and applique trim from the instrument panel.
4. Place the shift lever in the Park position and remove the shift indicator clip from the steering column.
5. Remove the nuts securing the steering column to the upper mounting bracket and lower the steering column.
6. Remove the screw securing the upper steering column mounting bracket to the cowl and lower the bracket.
7. Remove the cluster retaining screws, disconnect the speedometer cable, printed circuit connector and remove the cluster.
8. The installation is the reverse of the removal procedure. Be sure the shift indicator is properly aligned.

Emissions Indicator

An emissions indicator flag may appear in the odometer window of the speedometer on some 1980 and later General Motors vehicles. The flag could say "Sensor", "Emissions" or "Catalyst" depending on the part or assembly that is scheduled for regular emissions maintenance replacement. The word "Sensor" indicates a need for oxygen sensor replacement and the words "Emissions" or "Catalyst" indicate the need for catalytic converter replacement.

RESET PROCEDURE

1. Remove the instrument panel trim plate.
2. Remove the instrument cluster lens.
3. Locate the flag indicator reset notches at the drivers side of the odometer.
4. Use a pointed tool to apply light downward pressure on the notches, until the indicator is reset.

Speedometer Cable

REMOVAL & INSTALLATION

1. Disconnect the negative battery cable.
2. Remove the screws holding the instrument cluster trim plate, left and right telltale lens, instrument cluster lens and the transaxle shift indicator assembly. Remove the lens and retainers.
3. Remove the temperature indicator and the fuel gauge.
4. Remove the screws holding the speedometer assembly to the housing. Pull the speedometer head out and disconnect the screw holding the vehicle speed sensor.
5. Remove the speedometer cable. Remove the speedometer.
6. Installation is the reverse of the removal procedure.

Windshield Wiper Motor

REMOVAL & INSTALLATION

1. Remove the cowl screen or grille.
2. Loosen the linkage drive link-to-crankarm attaching nuts, and remove the link from the arm.
3. Disconnect the wiring and washer hoses.
4. Remove the three motor attaching screws, guide the crankarm through the hole in the dash, and remove the motor.
5. Reverse the above steps to install.

WIPER BLADE REPLACEMENT

Depending on model and availability, one of three methods is used:
1. A tab on the arm saddle is depressed.
2. A spring type blade clip is depressed.
3. A coil spring retainer is depressed with a screwdriver.
Details can be found in the Maintenance Unit Repair Section.

Radio

REMOVAL & INSTALLATION

1. Disconnect the negative battery cable.
2. Remove the screws from the top of the instrument panel center insert.
3. Remove the radio knobs and remove the insert.
4. Remove the rear window defogger switch to gain access to the left side mounting screw if so equipped.
5. Remove the mounting screws.
6. Remove the radio and disconnect the wiring. Reverse to install.

Fuses

The fuse block is located beneath the instrument panel above the headlight dimmer floor switch. Fuse holders are labeled as to their service and the correct amperage. Always replace blown fuses with new ones of the correct amperage. Otherwise electrical overloads and possible wiring damage will result.

Electrical Component Location

FUSE PANEL

The fuse panel is located on the left side of the vehicle. It is under the instrument panel assembly. In order to gain access to the fuse panel, it may be necessary to first remove the under dash padding.

ELECTRONIC CONTROL MODULE

The electronic control module is located on the right side of the vehicle. It is positioned in front of the right hand kick panel, or up behind the glovebox. In order to gain access to the assembly you must first remove the trim panel.

TURN SIGNAL FLASHER

The turn signal flasher is located directly under the steering column of the vehicle. It is secured in place with a plastic retainer. In order to gain access to the component, it may first be necessary to remove the under dash padding.

CONVENIENCE CENTER

The convenience center is located on the right side of the vehicle. It is positioned under the dash panel. In order to gain access to the convenience center it may be necessary to remove instrument panel sound absorber.

CIRCUIT BREAKER

A circuit breaker is an electrical switch which breaks the circuit during an electrical overload. The circuit breaker will remain open until the short or overload condition in the circuit is corrected. Circuit breakers are located in the fuse box.

FUSIBLE LINKS

Fusible links are sections of wire, with special insulation, designed to melt under electrical overload. Replacements are simply spliced into the wire. There may be as many as five of these in the engine compartment wiring harnesses. These are:

1. Horn relay to fuse panel circuit — one link.
2. Charging circuit, from the starter solenoid to the horn relay — two links.
3. Starter solenoid to ammeter circuit — one link.
4. Horn relay to rear window defroster circuit — one link.
The fusible links are all two wire gauge sizes smaller than the wires they protect.

NOTE: Most models have fusible links at these locations.

REPLACEMENT

1. Disconnect the battery ground cable.
2. Disconnect the fusible link from the junction block or starter solenoid.
3. Cut the harness directly behind the connector to remove the damaged fusible link.
4. Strip the harness wire approximately $\frac{1}{2}$ in.
5. Connect the new fusible link to the harness wire using a crimp on connector. Solder the connection using resin core solder.
6. Tape all exposed wires with plastic electrical tape. Silicone sealer may be used to weatherproof the connection.
7. Connect the fusible link to the junction block or starter solenoid and reconnect the battery ground cable.

Heater Blower Motor

REMOVAL & INSTALLATION

1. Disconnect the negative battery cable.
2. Disconnect the electrical connections at the blower motor.

3. Disconnect the cooling hose from the blower motor.
4. Remove the mounting screws and remove the motor.
5. Reverse to install. Use a silicone sealer on the blower motor sealing surfaces.

Heater Core

REMOVAL & INSTALLATION

1. Drain the radiator. Make sure the engine is cool for this procedure.
2. Remove the heater hoses from the core and plug the hoses and the nipples to prevent spillage.
3. Remove the instrument panel.
4. Remove the four defroster nozzle attaching screws at the cowl and the screw on the case and remove the nozzle.
5. Disconnect the vacuum hoses.
6. Disconnect the electrical connector at the programmer.
7. Under the hood, remove the heater case-to-cowl attaching screws.
8. Under the instrument panel, remove the heater case-to-cowl attaching screw.
9. Remove the heater case.
10. Remove the four case-to-core screws and remove the core.
11. Installation is the reverse of removal.

GM "E" & "K" Body
Riviera, Eldorado, Seville, Toronado

YEAR IDENTIFICATION

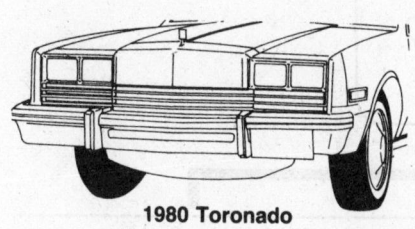

1980 Toronado

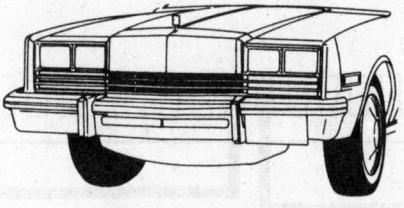

1981–84 Toronado

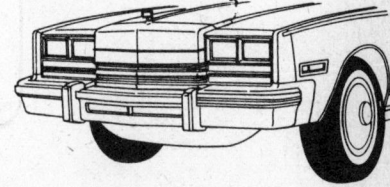

1984–85 Toronado Calienta

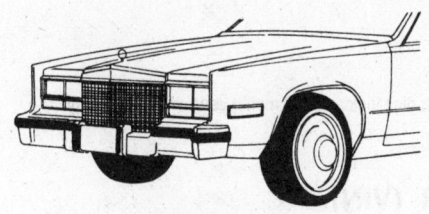

1981 Eldorado

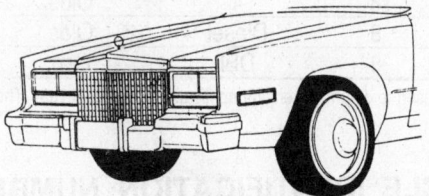

1980 Eldorado

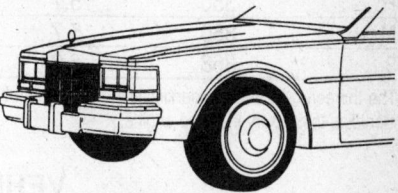

1981 Seville

1982–84 Eldorado

1985 Eldorado

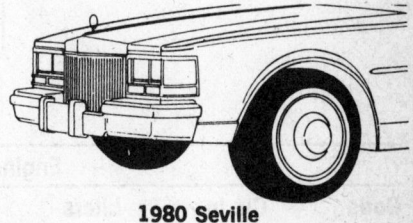

1980 Seville

1980 Riviera

1986-87 Riviera

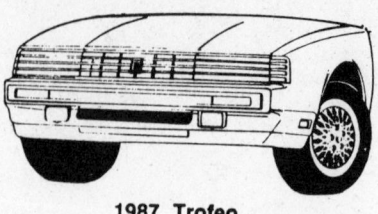

1987 Trofeo

1986-87 Toronado

VEHICLE IDENTIFICATION NUMBER (VIN)

It is important for servicing and ordering parts to be certain of the vehicle and engine identification. The VIN (vehicle identification number) is a 13 or 17 digit number visible through the windshield on the driver's side of the dash and contains the vehicle and engine identification codes. It can be interpreted as follows:

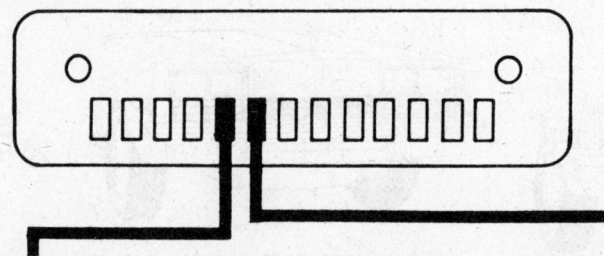

Engine Code						Model Year Code	
Code	Cu. In.	Liters	Cyl.	Carb.	Eng. Mfg.	Code	Year
3	231	3.8	6	4	Buick	A	80
Y	307	5.0	8	4	Olds.		
R	350	5.7	8	4	Olds.		
N	350	5.7	8	Diesel	Olds.		
9	368	6.0	8	DFI	Cad.		

The thirteen digit Vehicle Identification Number can be used to determine engine application and model year. The 6th digit indicates the model year, and the 5th digit identifies the factory installed engine.

VEHICLE IDENTIFICATION NUMBER (VIN)

It is important for servicing and ordering parts to be certain of the vehicle and engine identification. The VIN (vehicle identification number) is a 13 or 17 digit number visible through the windshield on the driver's side of the dash and contains the vehicle and engine identification codes. It can be interpreted as follows:

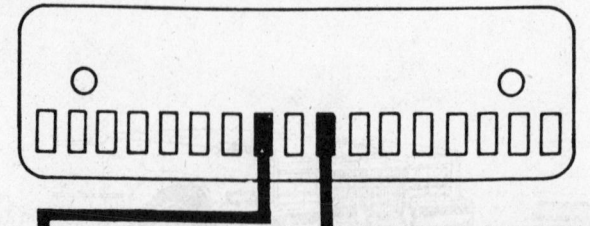

Engine Code						Model Year Code	
Code	Cu. In.	Liters	Cyl.	Carb.	Eng. Mfg.	Code	Year
8	231	3.8	6	4	Buick	B	81

VEHICLE IDENTIFICATION NUMBER (VIN)

It is important for servicing and ordering parts to be certain of the vehicle and engine identification. The VIN (vehicle identification number) is a 13 or 17 digit number visible through the windshield on the driver's side of the dash and contains the vehicle and engine identification codes. It can be interpreted as follows:

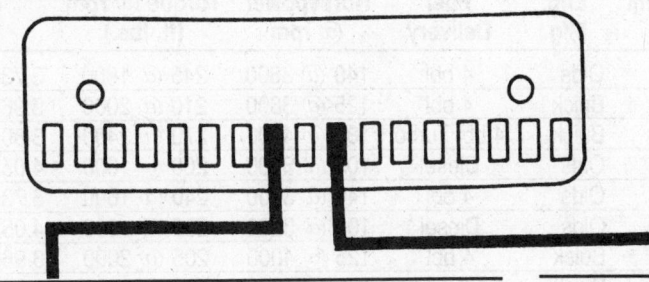

Engine Code

Code	Cu. In.	Liters	Cyl.	Carb.	Eng. Mfg.
3	231	3.8	6	4	Buick
9	231	3.8	6	SFI-Turbo	Buick
B	231	3.8	6	SFI	Buick
A	231	3.8	6	SFI	Buick
7	231	3.8	6	SFI-Turbo	Buick
8	250	4.1	8	DFI	Cad.
4	252	4.1	6	4	Buick
4	252	4.1	6	DFI	Buick
Y	307	5.0	8	4	Olds.
N	350	5.7	8	Diesel	Olds.
9	368	6.0	8	DFI	Cad.

Model Year Code

Code	Year
C	82
D	83
E	84
F	85
G	86
H	87

The seventeen digit Vehicle Identification Number can be used to determine engine application and model year. The 10th digit indicates the model year, and the 8th digit identifies the factory installed engine.

DFI Digital Fuel Injection
SFI Sequential Fuel Injection

GENERAL ENGINE SPECIFICATIONS

Year	Eng. Code	No. Cyl.– Displacement (cu. in.)	Eng. Mfg.	Fuel Delivery	Horsepower @ rpm	Torque @ rpm (ft. lbs.)	Bore × Stroke (in.)	Compression Ratio	Oil Pressure @ 1500 rpm
TORONADO									
'80	Y	8-307	Olds	4 bbl	148 @ 3800	250 @ 2400	3.800 × 3.385	7.9:1	40
	R	8-350	Olds	4 bbl	165 @ 3600	275 @ 2400	4.057 × 3.385	8.0:1	38
	N	8-350	Olds	Diesel	125 @ 3600	225 @ 1600	4.057 × 3.385	22.5:1	38
'81–'82	4	6-252	Buick	4 bbl	125 @ 4000	205 @ 2000	3.965 × 3.400	8.0:1	37①
	Y	8-307	Olds	4 bbl	148 @ 3800	250 @ 3800	3.800 × 3.385	8.0:1	40
	N	8-350	Olds	Diesel	125 @ 3600	225 @ 1600	4.057 × 3.385	22.5:1	38
'83–'84	4	6-252	Buick	4 bbl	125 @ 4000	205 @ 2000	3.965 × 3.400	8.0:1	37①
	Y	8-307	Olds	4 bbl	140 @ 3600	240 @ 1600	3.800 × 3.385	8.0:1	40
	N	8-350	Olds	Diesel	105 @ 3200	200 @ 1600	4.057 × 3.385	22.5:1	38
'85	Y	8-307	Olds	4 bbl	140 @ 3600	240 @ 1600	3.800 × 3.385	8.0:1	40
'86–'87	B	6-231	Buick	SFI	140 @ 4400	200 @ 2000	3.800 × 3.400	8.0:1	37
RIVIERA									
'80	R	8-350	Olds	4 bbl	170 @ 3800	275 @ 2000	4.057 × 3.385	8.5:1	37
	3	6-231	Buick	4 bbl Turbo	165 @ 4000	265 @ 2800	3.800 × 3.400	8.0:1	37②
'81	Y	8-307	Olds	4 bbl	140 @ 3600	245 @ 1600	3.736 × 3.385	8.0:1	37
	4	6-252	Buick	4 bbl	125 @ 4000	205 @ 2000	3.965 × 3.400	8.0:1	37②
	3	6-231	Buick	4 bbl Turbo	180 @ 4000	270 @ 2400	3.800 × 3.400	8.0:1	37②
	N	8-350	Olds	Diesel	105 @ 3200	200 @ 1600	4.057 × 3.385	22.5:1	38

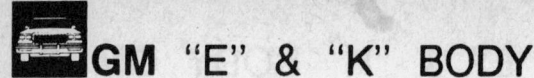

GENERAL ENGINE SPECIFICATIONS

Year	Eng. Code	No. Cyl.– Displacement (cu. in.)	Eng. Mfg.	Fuel Delivery	Horsepower @ rpm	Torque @ rpm (ft. lbs.)	Bore × Stroke (In.)	Com- pression Ratio	Oil Pres- sure @ 1500 rpm
'82	Y	8-307	Olds	4 bbl	140 @ 3600	245 @ 1600	3.736 × 3.385	8.0:1	37
	4	6-252	Buick	4 bbl	125 @ 3800	210 @ 2000	3.965 × 3.400	8.0:1	35③
	3	6-231	Buick	4 bbl Turbo	180 @ 4000	270 @ 2400	3.800 × 3.400	8.0:1	37②
	N	8-350	Olds	Diesel	105 @ 3200	200 @ 1600	4.057 × 3.385	22.5:1	38
'83–'84	Y	8-307	Olds	4 bbl	140 @ 3600	240 @ 1600	3.736 × 3.385	8.0:1	37
	N	8-350	Olds	Diesel	105 @ 3200	200 @ 1600	4.057 × 3.385	22.5:1	38
	4	6-252	Buick	4 bbl	125 @ 4000	205 @ 2000	3.965 × 3.400	8.0:1	35③
	8	6-231	Buick	4 bbl Turbo	180 @ 4000	290 @ 2400	3.800 × 3.400	8.0:1	37②
'85	8	6-231	Buick	SFI Turbo	190 @ 4000	300 @ 2400	3.800 × 3.400	8.0:1	37②
	Y	8-307	Olds	4 bbl	140 @ 3600	240 @ 1600	3.736 × 3.385	8.0:1	37
'86–'87	A	6-231	Buick	SFI	140 @ 4400	200 @ 2000	3.800 × 3.400	8.0:1	37
SEVILLE									
'80	9	8-368	Cadillac	DFI	145 @ 3800	265 @ 1400	3.800 × 4.060	8.2:1	35③
	N	8-350	Olds	Diesel	105 @ 3200	205 @ 1600	4.057 × 3.385	22.5:1	40③
'81	9	8-368	Cadillac	DFI	140 @ 3800	265 @ 1400	3.800 × 4.060	8.2:1	35③
	N	8-350	Olds	Diesel	105 @ 3200	204 @ 1600	4.057 × 3.385	22.5:1	40③
	4	6-252	Buick	4 bbl	125 @ 3800	210 @ 2000	3.965 × 3.400	8.0:1	35③
'82	8	8-250	Cadillac	DFI	135 @ 4200	190 @ 2000	3.465 × 3.307	8.5:1	40
	4	6-252	Buick	4 bbl	125 @ 3800	210 @ 2000	3.965 × 3.400	8.0:1	35③
	N	8-350	Olds	Diesel	105 @ 3200	205 @ 1600	4.057 × 3.385	22.5:1	40③
'83–'84	8	8-250	Cadillac	DFI	135 @ 4200	190 @ 2000	3.465 × 3.307	8.5:1	40
	N	8-350	Olds	Diesel	105 @ 3200	205 @ 1600	4.057 × 3.385	22.5:1	35③
'85	8	8-250	Cadillac	DFI	135 @ 4200	200 @ 2000	3.465 × 3.307	8.5:1	40
'86–'87	8	8-250	Cadillac	DFI	130 @ 4200	200 @ 2000	3.465 × 3.307	9:1	33–40
ELDORADO									
'80	9	8-368	Cadillac	DFI	145 @ 3800	265 @ 1600	3.800 × 4.060	8.2:1	35③
	N	8-350	Olds.	Diesel	105 @ 3200	205 @ 1600	4.057 × 3.385	22.5:1	40③
'81–'82	9	8-368	Cadillac	DFI	140 @ 3800	265 @ 1400	3.800 × 4.060	8.2:1	35③
	N	8-350	Olds	Diesel	105 @ 3200	205 @ 1600	4.057 × 3.385	22.5:1	40③
	8	8-250	Cadillac	DFI	135 @ 4200	190 @ 2000	3.465 × 3.307	8.5:1	40
	4	6-252	Buick	4 bbl	125 @ 3800	210 @ 2000	3.965 × 3.400	8.0:1	35③
'83–'84	8	8-250	Cadillac	DFI	135 @ 4200	190 @ 2000	3.465 × 3.307	8.5:1	40
	N	8-350	Olds	Diesel	105 @ 3200	205 @ 1600	4.057 × 3.385	22.5:1	35③
'85	8	8-250	Cadillac	DFI	135 @ 4200	200 @ 2000	3.465 × 3.307	8.5:1	40
'86–'87	8	8-250	Cadillac	DFI	130 @ 4200	200 @ 2000	3.465 × 3.307	9:1	33–40

Note: Horsepower and torque are SAE net figures.
They are measured at the rear of the transmission
with all accessories installed and operating. Since
the figures vary when a given engine is installed in
different models, some are representative rather than
exact.
① @ 2400 rpm
② @ 2600 rpm
③ @ 2000 rpm
DFI Digital Fuel Injection
SFI Sequential Fuel Injection

FIRING ORDERS

NOTE: To avoid confusion, always replace spark wires one at a time.

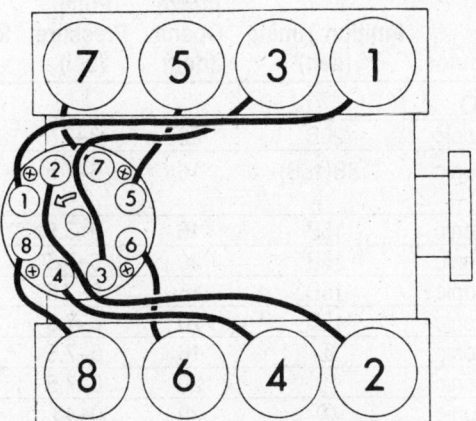

GM 250, 307, 350 V8s, including diesel
Engine firing order: 1-8-4-3-6-5-7-2 Distributor rotation: counterclockwise

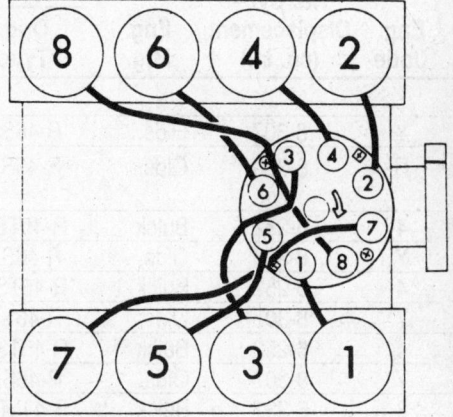

GM 368, 425 V8s
Engine firing order: 1-5-6-3-4-2-7-8 Distributor rotation: clockwise

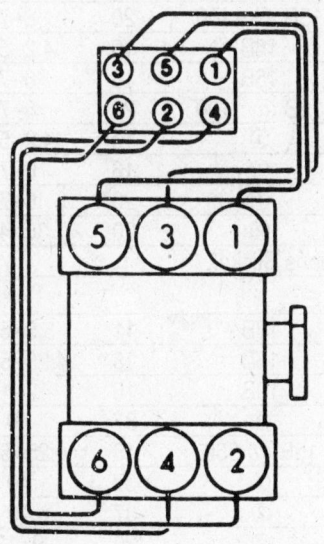

GM Buick 231 V6 firing order with the C3 coilless ignition system

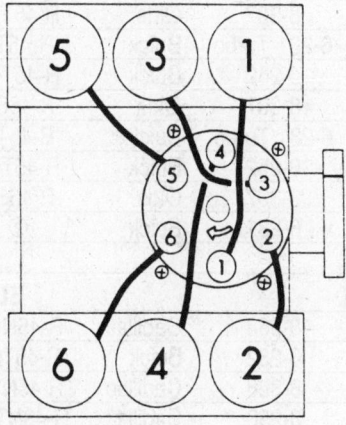

GM (Buick) 231 and 252 V6 (3.8 and 4.1L)
Engine firing order: 1-6-5-4-3-2
Distributor rotation: clockwise
V6 Harmonic balancers have two timing marks: one is 1/8 in. wide, and one is 1/16 in. wide. Use the 1/16 in. mark for timing with a hand held light. The 1/8 in. mark is used only with a magnetic timing pick-up probe.

GASOLINE ENGINE TUNE-UP SPECIFICATIONS

When analyzing compression test results, look for uniformity among cylinders rather than specific pressures.

Year	Eng. Code	No. Cyl.– Displacement (cu. in.)	Eng. Mfg.	Spark Plugs Orig. Type	Gap (in.)	Distributor	Ignition Timing (deg)	Valves Intake Opens (deg)	Fuel Pump Pressure (psi)	Idle Speed (rpm)
						TORONADO				
'80	Y	8-307	Olds.	R-46SX	.080	Electronic	20B	20	6–7.5	600/500
	R	8-350	Olds.	R-46SX	.080	Electronic	18B(16B)	16	6–7.5	600/500 (650/550)
'81–'82	4	6-252	Buick	R-45TS8	.080	Electronic	15B	16	6–7.5	①
	Y	8-307	Olds.	R-46SX	.080	Electronic	15B	20	6–7.5	①
'83–'84	4	6-252	Buick	R-45TS8	.080	Electronic	15B	16	6–7.5	①
	Y	8-307	Olds.	R-46SX	.080	Electronic	15B	20	6–7.5	①
'85	4	6-252	Buick	R-45TS8	.080	Electronic	①	16	6–7.5	①
	Y	8-307	Olds.	R-46SX	.080	Electronic	①	20	6–7.5	①
'86	8	6-231	Buick	R-44LTS	.045	Electronic	②	20	30–40	⑤
'87	All			See Underhood Specifications Sticker						
						RIVIERA				
'80	3	6-231 Turbo	Buick	R-45TS	.040	Electronic	15B	16	5.0	600
	R	8-350	Olds.	R-46SX	.080	Electronic	18B @ 1100	16	5.5–6.5	500
'81–'82	3	6-231 Turbo	Buick	R-45TS	.040	Electronic	15B	16	4.2–5.8	①
	4	6-252	Buick	R-45TS8	.080	Electronic	15B	16	4.2–5.9	①
	Y	8-307	Olds.	R-45TS4	.060	Electronic	15B @ 1100	20	6–7.5	①
'83	8	6-231 Turbo	Buick	R-45TS	.040	Electronic	15B	16	4.2–5.8	①
	4	6-252	Buick	R-45TS8	.080	Electronic	15B	16	6–7.5	①
	Y	8-307	Olds.	R-46SX	.080	Electronic	15B @ 1100	20	6–7.5	①
'84–'85	8,9	6-231 Turbo	Buick	R-45TS	.040	Electronic	①	16	4.2–5.8	①
	4	6-252	Buick	R-45TS8	.080	Electronic	①	16	6–7.5	①
	Y	8-307	Olds.	R-46SX	.080	Electronic	①	20	6–7.5	①
'86	A	6-231	Buick	②	②	Electronic	②	20	30–40	⑤
'87	All			See Underhood Specifications Sticker						
						ELDORADO AND SEVILLE				
'80	9	8-368	Cadillac	R-45NSX	.060	Electronic	18B	11	5.5–6.5	575
'81	4	6-252	Buick	R-45TS8	.060	Electronic	15B	16	4.25–5.75	550③
	9	8-368	Cadillac	R-45NSX	.060	Electronic	10B	16	12–14	470④
'82	8	8-250	Cadillac	R-43NTS6	.060	Electronic	②	37	40	⑤
	4	6-252	Buick	R-45TSV	.060	Electronic	15B @ 550	16	4.25–5.75	550 ③
'83	8	8-250	Cadillac	R-43NTS6	.060	Electronic	②	37	40	⑤
'84–'85	8	8-250	Cadillac	R-43NTS6⑥	.060	Electronic	②	37	40	⑤
'86	8	8-250	Cadillac	②	②	Electronic	②	20	9–12	⑤
'87	All			See Underhood Specifications Sticker						

Note: Check the underhood emission control sticker for correct timing procedure. Some engines require grounding 2 connector or disconnecting a distributor plug to set base timing. Use the sticker specifications if different from above.

① Where two figures appear separated by a slash, the first is idle speed with solenoid energized, the second is idle speed with solenoid disconnected. Figure in parentheses indicates California engine. Solenoid energized (higher) idle speed is set with A/C on and compressor clutch wires disconnected.
② See Underhood Sticker
③ In Drive; A/C 680
④ Drive or Neutral
⑤ Electronic controlled idle, no adjustment is possible
⑥ R42CLTS6
B Before Top Dead Center
Part numbers in this chart are not recommendations by Chilton for any product by brand name.

DIESEL ENGINE TUNE-UP SPECIFICATIONS

Year	Eng. Mfg.	No. Cyl–Displacement (cu in.)	Fuel Pump Pressure (psi)	Compression (lbs)	Intake Valve Opens (deg)	Idle Speed (rpm) ①
'80	Olds.	8-350	5.5–6.5	275 min.	16	750/600
'81–'85	Olds.	8-350	5.5–7.0	275 min.	16	②

NOTE: The underhood specifications sticker often reflects tune-up specification changes made in production. Sticker figures must be used if they disagree with those in this chart.

① Where two idle speed figures appear separated by a slash, the first is idle speed with solenoid energized, the second is idle speed with solenoid disconnected.

② See underhood specifications sticker

CAPACITIES

Year/Model	No. Cyl.–Displacement (cu. in.)	Engine Crankcase Add 1 Qt for New Filter	Transmission Pts to Refill After Draining Automatic ① ③	Drive Axle (pts)	Gasoline Tank (gals)	Cooling System (qts) With A/C	Cooling System (qts) With Heavy Duty
'80	8-368	4	6	3.25	20.7	22.4	—
	8-350	4	6	3.25	20.5	15.2	—
	8-350 Diesel	6②	6	3.25	23	18.5	—
	8-307	4	6	3.25	21	16.5	—
	6-231	4	6	3.25	21	13.6	14.1
'81	8-368	4	6	3.25	20.3	22.4	
	8-307	4	6	3.25	21	16.25	16.25
	8-350 Diesel	6②	6	3.25	④	⑤	—
	6-252	4	6	3.25	21	13.1	—
	6-231	4	6	3.25	21	13.6	14.1
'82	8-350 Diesel	6②	6	3.25	23	⑦	⑦
	8-250	4	6	3.25	20.3	11.8	12.5
	8-307	4	6	3.25	21	⑥	—
	6-252	4	6	3.25	21	13.1	—
	6-231	4	6	3.25	21	13.6	14.1
'83	8-307	4	6	3.25	21	16.2	—
	8-250	4	6	3.25	20.3	11.8	—
	8-350 Diesel	6②	6	3.25	22.8	18.2	—
	8-252	4	6	3.25	21.1	13.1	—
	6-231	4	6	3.25	21.1	13.6	14.1
'84–'85	8-307	4	6	3.25	21.1	16.2	—
	8-250	4	6	3.25	20.3	11.8	12.5
	8-350 Diesel	6②	6	3.25	22.8	18.2	—
	6-252	4	6	3.25	21.1	12.5	—
	6-231	5	6	3.25	21.1	12.9	13.7
'86–'87	6-231	4	13⑧	—	18	12	12
	8-250	4	13	—	18	12.6	—

① Does not include torque converter
② Filter change is mandatory
③ Start engine and allow to warm up. Add fluid as necessary to mark on dipstick
④ Riviera: 27, Eldorado/Seville: 22.8, Toronado: 23
⑤ Riviera: 18.2, Eldorado/Seville: 18.4, Toronado: 18
⑥ Rivivera: 18.9, Toronado: 16.5
⑦ Riviera: 23, Eldorado/Seville: 22.8, Toronado: 23
⑧ Riviera: 8

VALVE SPECIFICATIONS

Year	No. Cyl.–Displacement (cu. in.)	Seat Angle (deg)	Face Angle (deg)	Spring Test Pressure (lbs. @ in.)	Spring Installed Height (in.)	Stem-to-Guide Clearance (in.) Intake	Stem-to-Guide Clearance (in.) Exhaust	Stem Diameter (in.) Intake	Stem Diameter (in.) Exhaust
'80	8-368	45	44	160 @ 1.50	1¹⁵⁄₃₂	.0010–.0027	.0012–.0029	.3420	.3418
	8-350 Diesel	45①	44②	152 @ 1.30	1⁴³⁄₆₄	.0010–.0027	.0015–.0032	.3429	.3424
	8-350	45①	44②	187 @ 1.27	1⁴³⁄₆₄	.0010–.0027	.0015–.0032	.3429	.3424
	8-307	45①	44②	187 @ 1.27	1⁴³⁄₆₄	.0010–.0027	.0015–.0032	.3429	.3424
	6-231	45	45	164 @ 1.34③	1⁴⁷⁄₆₄	.0015–.0035	.0015–.0032	.3406	.3408
'81	8-368	45	44	160 @ 1.50	1¹⁵⁄₃₂	.0010–.0027	.0012–.0029	.3420	.3418
	8-350 Diesel	45①	44②	152 @ 1.30④	1⁴³⁄₆₄	.0010–.0027	.0015–.0032	.3429	.3424
	8-307	45①	44②	187 @ 1.270	1⁴³⁄₆₄	.0010–.0027	.0015–.0032	.3429	.3424
	6-252	45	45	182 @ 1.34	1⁴⁷⁄₆₄	.0015–.0035	.0015–.0032	.3407	.3409
	6-231	45	45	164 @ 1.34③	1⁴⁷⁄₆₄	.0015–.0035	.0015–.0032	.3406	.3408
'82	8-350 Diesel	45①	44②	152 @ 1.30④	1⁴³⁄₆₄	.0010–.0027	.0015–.0032	.3429	.3424
	8-307	45①	44②	187 @ 1.270	1⁴³⁄₆₄	.0010–.0027	.0015–.0032	.3429	.3424
	8-250	45	44	182 @ 1.28	1⁴⁷⁄₆₄	.0010–.0027	.0012–.0029	.3420	.3408
	6-252	45	45	220 @ 1.34	1⁴⁷⁄₆₄	.0015–.0035	.0015–.0032	.3407	.3409
	6-231	45	45	164 @ 1.34③	1⁴⁷⁄₆₄	.0015–.0035	.0015–.0032	.3406	.3408
'83	6-231	45	45	182 @ 1.34	1⁴⁷⁄₆₄	.0015–.0035	.0015–.0032	.3401–.3412	.3405–.3412
	6-252	45	45	182 @ 1.34	1⁴⁷⁄₆₄	.0015–.0035	.0015–.0032	.3401–.3412	.3405–.3412
	8-250	45	44	182 @ 1.28	1⁴³⁄₆₄	.001–.003	.001–.003	.3413–.3420	.3411–.3418
	8-307	45	44	187 @ 1.27	1⁴³⁄₆₄	.0010–.0027	.0015–.0032	.3425–.3432	.3420–.3427
	8-350 Diesel	45①	44②	210 @ 1.22	1⁴³⁄₆₄	.0010–.0027	.0015–.0032	.3425–.3432	.3420–.3427
'84–'85	6-231	45	45	220 @ 1.34	1⁴⁷⁄₆₄	.0015–.0035	.0015–.0032	.3401–.3412	.3405–.3412
	6-252	45	45	182 @ 1.34	1⁴⁷⁄₆₄	.0015–.0035	.0015–.0032	.3401–.3412	.3405–.3412
	8-250	45	44	182 @ 1.28	1⁴³⁄₆₄	.001–.003	.001–.003	.3413–.3420	.3411–.3418
	8-307	45	44	187 @ 1.27	1⁴³⁄₆₄	.0010–.0027	.0015–.0032	.3425–.3432	.3420–.3427
	8-350 Diesel	45①	44②	210 @ 1.22	1⁴³⁄₆₄	.0010–.0027	.0015–.0032	.3425–.3432	.3420–.3427
'86–'87	6-231	45	—	185 @ 1.34	1⁴⁷⁄₆₄	.0015–.0035	.0015–.0032	.3401–.3412	.3405–.3412
	8-250	45	44	182 @ 1.28	1⁴⁷⁄₆₄	.001–.003	.001–.003	.3420–.3413	.3420–.3413

① Exhaust valve seat 31°
② Exhaust valve face 30°
③ Exhaust 182 @ 1.34
④ 210 @ 1.30 1981–82

PISTON AND RING SPECIFICATIONS

All measurements given in inches

Year	VIN Code	Engine Type/Disp. cu. in.	Eng. Mfg.	Piston-to-Bore Clearance	Ring Gap Top Compression	Ring Gap Bottom Compression	Ring Gap Oil Control	Ring Side Clearance Top Compression	Ring Side Clearance Bottom Compression	Ring Side Clearance Oil Control
'80	R	8-350	Olds.	.0010–.0020	.010–.023	.010–.023	.015–.055	.0020–.0040	.0020–.0040	.0006–.0096

PISTON AND RING SPECIFICATIONS
All measurements given in inches

Year	VIN Code	Engine Type/ Disp. cu. in.	Eng. Mfg.	Piston-to-Bore Clearance	Ring Gap			Ring Side Clearance		
					Top Compression	Bottom Compression	Oil Control	Top Compression	Bottom Compression	Oil Control
'80–'84	N	8-350 Diesel	Olds.	.0005– .0006	.015– .025①	.015– .025②	.015– .055	.0040– .0060③	.0018– .0038④	.0010– .0050
'80–'86	Y	8-370	Olds.	.0007– .0017⑤	.009– .019	.009– .019	.015– .055	.0020– .0040	.0020– .0040	.0150– .0550
'80–'81	9	8-368	Cad.	.0006– .0014	.013– .023	.013– .023	.015– .055	.0017– .0040	.0017– .0040	None ⑥
'80–'82	3	6-231	Buick	.0008– .0020⑦	.013– .023	.013– .023	.015– .035	.0030– .0050	.0030– .0050	.0035 max
'83–'85	9	6-231	Buick	.0022– .0034⑧	.010– .020	.010– .020	.015– .055	.0030– .0050	.0030– .0050	.0035 max
'86–'87	A,8	8-231	Buick	.0008– .0020⑦	.010– .020	.010– .020	.015– .055	.077– .078⑨	.077– .078⑨	.183– .189⑨
'81–'82	4	6-252	Buick	.0008– .0020	.013– .023	.013– .023	.015– .035	.0030– .0050	.0030– .0050	.0035 max
'83–'86	4	6-252	Buick	.0008– .0020	.010– .020	.010– .020	.015– .055	.0030– .0050	.0030– .0050	.0035 max
'82–'85	8	8-250	Cad.	.0010– .0018	.009– .020	.009– .020	.010– .050	.0016– .0037	.0016– .0037	None ⑥
'85–'87	8	8-250	Cad.	.0010– .0018	.015– .024	.015– .024	.010– .050	.0016– .0037	.0016– .0037	None ⑥

① 1983–84: .019–.027
② 1983–84: .013–.021
③ 1983–84: .005–.007
④ 1983–84: .003–.005
⑤ 1980: .0005–.0015
⑥ Side sealing
⑦ Measured at top of skirt
⑧ Measured at piston pin centerline
⑨ Ring Width

CRANKSHAFT AND CONNECTING ROD SPECIFICATIONS
All Models
All measurements are given in inches

Year	Engine No. Cyl. Displacement (cu. in.)	Crankshaft				Connecting Rod		
		Main Brg. Journal Dia.	Main Brg. Oil Clearance	Shaft End-Play	Thrust on No.	Journal Diameter	Oil Clearance	Side Clearance
'80	8-368	3.2500	.0001–.0026	.002–.012	3	2.5000	.0005–.0028	.008–.020
	8-350	2.4995–2.4985③	.0005–.0021②	.0035–.0135	3	2.1243	.0004–.0033	.006–.020
	8-350 Diesel	2.9998	.0005–.0021②	.0035–.0135	3	2.1243	.0004–.0033	.006–.020
	8-307	2.4995–2.4990④	.0005–.0021②	.0035–.0135	3	2.1243	.0004–.0033	.006–.020
	6-231	2.4995	.0003–.0017	.004–.008	2	2.2491	.0005–.0026	.006–.027
'81	8-368	3.2500	.0001–.0026	.002–.012	3	2.5000	.0005–.0028	.008–.020
	8-350 Diesel	2.9998	.0005–.0021②	.0035–.0135	3	2.1243	.0004–.0033	.006–.020
	8-307	2.4995–2.4990④	.0005–.0021②	.0035–.0135	3	2.1243	.0004–.0033	.006–.020
	6-252	2.4995	.0003–.0018	.003–.009	2	2.2491	.0005–.0026	.006–.023
	6-231	2.4995	.0003–.0017	.004–.008	2	2.2491	.0005–.0026	.006–.027
'82	8-350 Diesel	2.9998	.0005–.0021②	.0035–.0135	3	2.1243	.0004–.0033	.006–.020
	8-307	2.4995–2.4990④	.0005–.0021②	.0035–.0135	3	2.1243	.0004–.0033	.006–.020
	8-250	2.64	.0004–.0030	.001–.007	3	1.929	.0005–.0028	.008–.020
	6-252	2.4995	.0003–.0018	.003–.009	2	2.2491	.0005–.0026	.006–.023
	6-231	2.4995	.0003–.0017	.004–.008	2	2.2491	.0005–.0026	.006–.027

CRANKSHAFT AND CONNECTING ROD SPECIFICATIONS
All Models
All measurements are given in inches

| Year | Engine No. Cyl. Displacement (cu. in.) | Crankshaft | | | | Connecting Rod | | |
		Main Brg. Journal Dia.	Main Brg. Oil Clearance	Shaft End-Play	Thrust on No.	Journal Diameter	Oil Clearance	Side Clearance
'83–'84	6-231	2.4995	.0003–.0018	.003–.011	2	2.2491	.0005–.0026	.006–.027
	6-252	2.4995	.0003–.0018	.003–.011	2	2.2491	.0005–.0026	.006–.023
	8-250	2.64	.0004–.0027	.001–.007	3	1.929	.0005–.0028	.008–.020
	8-307	2.4995–2.4990④	.0005–.0021②	.0035–.0135	3	2.1238–2.1248	.0004–.0033	.006–.020
	8-350 Diesel	2.9993–3.0003	.0005–.0021①	.0035–.0135	3	2.1238–2.1248	.0005–.0026	.006–.020
'85	6-231	2.4995	.0003–.0018	.003–.011	2	2.2491	.0005–.0026	.006–.027
	6-252	2.4995	.0003–.0018	.003–.011	2	2.2491	.0005–.0026	.006–.023
	8-250	2.64	.0004–.0027	.001–.007	3	1.929	.0005–.0028	.008–.020
	8-307	2.4995–2.4990④	.0005–.0021②	.0035–.0135	3	2.1238–2.1248	.0004–.0033	.006–.020
'86–'87	6-231	2.4995	.003–.0018	.003–.011	2	2.2487–2.2495	.0009–.0026	.003–.015
	8-250	2.6374–2.6384	.0016–.0039⑤	.0010–.007	3	1.929	.0005–.0028	.008–.020

① No. 5—.0020–.0034
② No. 5—.0015–.0031
③ No. 1—2.4998–2.4988; 1980 Cadillac 2.4990
④ No. 1—2.4998–2.4993
⑤ Applies to No. 2, 3, 4, 5. No. 1—.0008–.0031

TORQUE SPECIFICATIONS
All readings in ft. lbs.

| Year | Engine No. Cyl. Displacement (cu. in.) | Cylinder Head Bolts | Bearing Bolts Rod | Bearing Bolts Main | Crankshaft Bolt | Flywheel to Crankshaft Bolts | Manifold | |
							Intake	Exhaust
'80	8-368	95③	40	90	Press fit	75	30	35①
	8-350	130③	42	80②	200–310	60	40③	25
	8-350 Diesel	130③	42	120	200–310	60	40③	25
	8-307	130③	42	80②	200–310	60	40③	25
	6-231	80	40	100	225	60	45	25
'81	8-368	95③	40	90	Press fit	75	30	35①
	8-350 Diesel	130③	42	120	200–310	60	40③	25
	8-307	130③	42	80②	200–310	60	40③	25
	6-252	80	40	100	225	60	45	25
	6-231	80	40	100	225	60	45	25
'82	8-350 Diesel	130③	42	120	200–310	60	40③	25
	8-307	130③	42	80②	200–310	60	40③	25
	8-250	④	22	85	225	75	20	18
	6-252	80	40	100	225	60	45	25
	6-231	80	40	100	225	60	45	25
'83	6-231	80	40	100	225	60	45	25
	6-252	80	40	100	225	60	45	25
	8-250	④	20	85	20	35	20	20
	8-307	130③	42	80②	200–310	60	40③	25
	8-350 Diesel	130③	42	120	200–310	60	40③	25
'84–'85	6-231	80	40	100	225	60	45	25
	6-252	80	40	100	225	60	45	25
	8-250	④	20	85	20	35	20	20
	8-307	125③	42	80②	200–310	60	40③	25
	8-350 Diesel	130③	42	120	200–310	60	40③	25

TORQUE SPECIFICATIONS
All readings in ft. lbs.

Year	Engine No. Cyl. Displacement (cu. in.)	Cylinder Head Bolts	Bearing Bolts Rod	Bearing Bolts Main	Crankshaft Bolt	Flywheel to Crankshaft Bolts	Manifold Intake	Manifold Exhaust
'86–'87	6-231	⑦	40	100	200	60	32	37
	8-250	⑤	22	85	18	37	⑥	18

① 12 ft. lbs. for short bolt
② 120 on No. 5
③ Bolts must be oiled before tightening
④ Pull first to 45 ft. lbs. in sequence, then tighten to 90 ft. lbs. in sequence

⑤ See text. Torque in sequence first to 38, then to 68, and finally to 90.
⑥ Refer to text. Final torque is 22.
⑦ See text. You must measure torque and angle and follow a sequence.

WHEEL ALIGNMENT SPECIFICATIONS

Year	Caster Range (deg)	Caster Pref. Setting (deg)	Camber Range (deg)	Camber Pref. Setting (deg)	Toe-in (in.)	Steering Axis Inclin. (deg)
TORONADO						
'80–'82	2P to 3P	2½P	½N to ½P	0	0 ± 1/16	11
'83–'85	1½P to 3½P	2½P	½N to ½P	0	1/8 in to 1/8 out	—
'86–'87	1½P–3½P	2½P	½N–½P	0	± 1/10 ③	—
ELDORADO AND SEVILLE						
'80–'82	1½P to 3½P	2½P	3/16N to 3/16P	0	1/8P to 1/8P	11
'83–'84	2P to 3P	2½P	½N to ½P	0	1/8 in to 1/8 out	—
'85	2P to 2½P	2P	0 to ½P	0	0 to ½P	—
'86–'87	1½P–3½P	2½P	½N–½P	0	1/5 in–1/5 out ③	—
RIVIERA						
'80–'82	1½P to 3½P	2½P	13/16N to 13/16P	0	1/8 in to 1/8 out	—
'83–'85	2P to 3P	2½P	½N to ½P	0	1/8 in to 1/8 out	—
'86–'87	1½P–3½P	2½P	½N–½P	0	± 1/10 ③	—

① Left side
② Right side
③ Toe-in expressed in degrees

NOTE: Service procedures for the Charging System, Starting System, Ignition System, Fuel System, Cooling System, and Emission Controls on the Toronado up through 1985 can be found in the Oldsmobile section.

ENGINE ELECTRICAL

Alternator

For all alternator and regulator troubleshooting and overhaul procedures, please refer to the "Charging and Starting Systems" Unit Repair Section.

REMOVAL & INSTALLATION

1. Disconnect battery ground cable to prevent diode damage.
2. Disconnect and label the alternator wiring.
3. Remove the brace bolt. If power steering equipped, loosen pump brace and mount nuts. Detach drive belt(s).
4. On Rivieras, first remove the air conditioner compressor bracket that is in the way without disturbing the refrigerant lines. On Eldorado and Seville, disconnect air pump hose at check valve and remove heater hose clip from adjusting link (if so equipped). Support the alternator and remove the mounting bolt(s), then remove the unit from the vehicle.
5. Reverse the procedure to install. On Heavy Duty Alternator ONLY (100 amp with external voltage ad-

justers): after connecting the negative ground cable, momentarily connect a jumper wire between "Bat" and "R" alternator terminals to re-establish residual magnetism in the rotor. Start the engine and run it at a fast idle for ten seconds; the charge light should go out. Adjust drive belt to have ¼–½ in. play on longest run of belt.

Integral Voltage Regulator

An alternator with an integral voltage regulator is standard equipment. There are no adjustments possible with this unit; testing procedures will be found in the "Charging and Starting Systems" Unit Repair Section.

Starter

For all starter motor troubleshooting and overhaul procedures, please refer to the "Charging and Starting Systems" Unit repair section.

REMOVAL & INSTALLATION

All Except Diesel

NOTE: On some models it may be necessary to move the fuel lines out of the way. Remove the fuel lines from the retaining clamp and loosen the regulator.

——— CAUTION ———

If equipped with fuel injection, relieve the pressure from the fuel system before disconnecting the fuel lines. See "Electric Fuel Pump Removal."

1. Disconnect the battery ground cable.
2. Raise and support the vehicle on jackstands.
3. Disconnect all wires at solenoid terminals. Note color coding of wires for reinstallation.
4. Remove starter support bracket mount bolts. On engines with solenoid heat shields, remove front bracket upper bolt and detach bracket from starter motor.
5. Loosen the front bracket bolt or nut and rotate bracket clear. Lower and remove starter. Note the location of any shims so that they may be replaced in the same position upon installation.
6. Installation is the reverse of removal procedures.

Diesel Engine

1. Disconnect the negative cable at the battery(s).
2. Raise and support the car on jackstands.
3. Remove the lower starter shield nut and bend the shield out of the way.
4. Disconnect the wires from the starter. It's a good idea to tag the wires.
5. Remove the front starter bolt. Loosen the rear starter bolt and remove the starter with the rear bolt remaining in the starter housing.
6. Installation is the reverse of removal procedures.

Distributor

Most 1981 and later models use an Electronic Spark Timing (EST) distributor. The EST distributor uses no mechanical or vacuum advance and is easily identified by the absence of a vacuum advance and the presence of a four terminal connector.

REMOVAL & INSTALLATION

1. Label all hoses and wires before removal. Remove spark plug cables and wire connectors from cap.
2. Remove the cap. Remove the vacuum hose from the vacuum unit.
3. Crank the engine until the rotor points toward the rear of the engine and the No. 1 piston is almost at TDC.
4. Turn the engine until the crankshaft pulley timing mark is at "0". The white mark on the side of the rotor will be aligned with the white pointer in the distributor.
5. Remove the distributor clamp and pull the distributor up until the rotor stops turning and note the position of the rotor. Remove the distributor.
6. When installing, make sure that the timing marks are aligned and the rotor and pointer marks are aligned.

FUEL SYSTEM

For diesel engine fuel injection adjustments, timing, and removal and installation procedures, see the Oldsmobile Rear Wheel Drive section. For additional information on the fuel system, please refer to "Carburetors" or "Diesel Maintenance" in the Unit Repair section.

IDLE SPEED AND MIXTURE ADJUSTMENT

1980–85 M4MC/M4ME Carburetors With Idle Speed Solenoid

1. Run the engine to normal operating temperature.
2. Make sure that the choke is fully opened, turn the A/C OFF, set the parking brake and block the wheels.
3. Connect a tachometer to the engine according to the manufacturer's instructions.
4. Disconnect the purge hose from the vapor canister. On the 350, plug the purge hose.
5. Disconnect and plug the EGR vacuum hose at the valve. Disconnect and plug the vacuum advance hose.
6. Place the transmission in PARK.
7. Check and adjust the timing.
8. Reconnect the vacuum advance hose.

9. Place the transmission in DRIVE.

NOTE: If instructions on the underhood sticker differ from these, follow the underhood sticker.

10. On cars without A/C: Turn the idle speed screw to attain specified rpm. On cars with A/C: Turn the idle speed screw to set the specified curb idle. Turn the A/C ON and disconnect the compressor clutch wire. Open the trottle momentarily to extend the solenoid plunger. Adjust the solenoid screw to obtain the solenoid idle speed shown on the underhood sticker. Reconnect the compressor clutch and turn the A/C OFF.
11. Reconnect all hoses and remove the tachometer.

1981–85 M4MC/M4ME Carburetors Without Idle Speed Solenoid

Most 1981 and later models are equipped with an Idle Speed Control (ISC) mounted on the float bowl. Idle speeds are computer controlled and the ICS should not be adjusted.

On some V8 models an Idle Load Compensator (ILC) is mounted on the float bowl to control the curb idle speed. The ILC is adjusted at the factory and capped to prevent readjustment.

On cars that do not include either an ISC or ILC, but are equipped with A/C, an idle speed solenoid is used to maintain idle speed. For adjustment of these refer to the previous 1980–86 adjustment procedures.

NOTE: The underhood sticker specifies which idle system your car is equipped with.

Idle Air Control Valve

The Idle Air Control valve controls engine idle speed by bypassing air around the throttle valve. It responds to a number of electronic signals, actually compensating for changes in engine load. It is not adjustable.

IDLE MIXTURE ADJUSTMENT

1980 Models

1980 model carburetors have a propane enrichment system which cannot be adjusted, and mixture screws that are concealed by staked in plugs, and cannot be adjusted unless the carburetor is being completely overhauled.

1981–87 Models

On these models the air/fuel mixture

is controlled by the electronic control module of the Computer Command Control (CCC) system. No adjustment should be attempted.

Mechanical Fuel Pump

All air conditioned cars with V8 engines have a special fuel pump. This pump has a vapor return line which returns hot fuel and fuel vapor to the fuel tank. The possibility of vapor lock is thus greatly reduced by keeping cool fuel circulating through the pump.

REMOVAL & INSTALLATION

1. Disconnect the fuel inlet hose from the pump. Disconnect the vapor return hose, if so equipped. Disconnect the inlet hose.
2. Remove the two bolts.
3. Remove the fuel pump.
4. Install a new gasket.
5. Install a new pump and bolts.
6. Tighten the bolts alternately and evenly.
7. Reconnect the hoses, start the engine, and check for leaks.

Electric Fuel Pump

REMOVAL & INSTALLATION

——— CAUTION ———

Because of the large amount of fuel under pressure in the injection system, it is dangerous to disassemble system parts unless fuel system presssure is first relieved. There is considerable risk of fire unless you follow the proper procedures.

1. Remove the fuse labeled "Fuel Pump" from the fuse block.
2. Crank the engine and allow it to run until it stalls. Then, crank the starter for at least three full seconds to ensure that there is no fuel pressure in the system.
3. Disconnect the negative battery cable.
4. Raise the car on a hoist and support it securely. Make sure the front end is supported securely so that in the next step, weight distribution changes will not cause it to become unbalanced.
5. Remove the fuel tank as described later in this section.
6. Turn the cam locking ring counterclockwise to release it, and then lift the pump and sending unit out of the tank. Remove the sending unit from the pump.
7. Pull the fuel pump up into the attaching hose while pulling it outward, away from the support on the

bottom of the tank. Make sure you do not damage the the rubber insulator and the strainer as you work. When the pump is entirely clear of the bottom support, pull it out of the rubber connector to remove it.
8. Inspect the attaching hose and the rubber sound insulator at the bottom of the pump for signs of deterioration and replace parts as necessary.
9. Push the fuel pump into the attaching hose. Install a new O-ring. Reattach the level sending unit to the pump and install both into the tank.
10. Install the cam locking ring over the assembly and turn it clockwise to lock it. Reverse the remaining removal proceudres.

Fuel Filter

REMOVAL & INSTALLATION

Carbureted Engines

NOTE: When purchasing a new filter, be sure that it matches the old one EXACTLY.

1. Disconnect the fuel line connection at the inlet of the carburetor.
2. Remove the inlet fuel filter nut from the carburetor with a box wrench.
3. Remove the filter element and spring.
4. If it's a bronze element, blow through the cone end; the element should allow air to pass through freely.

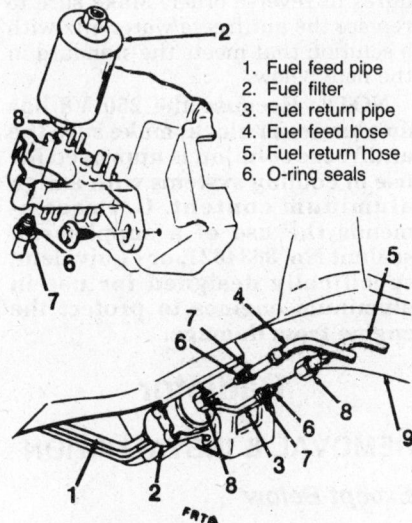

1. Fuel feed pipe
2. Fuel filter
3. Fuel return pipe
4. Fuel feed hose
5. Fuel return hose
6. O-ring seals

7. Torque these fittings to 26 ft. lbs.
8. Flats for use of a backup wrench
9. Left hand frame rail in the engine compartment

Replacing fuel filter on 1985–'86 fuel injected engines

5. Install the spring and a new element into the carburetor. Bronze elements are installed with the small section of the cone facing outward.
6. Install a new gasket on the fitting nut and install the nut.
7. Install the fuel line and tighten it securely. Start the engine and check for leaks.

Fuel Injected Engines

1. Relieve fuel system pressure as described in Steps 1 and 2 of the Electric Fuel Pump Removal & Installation procedure above. Raise the vehicle and support it securely.
2. Using a backup wrench, disconnect both fuel lines at the filter.
3. Unclamp and remove the filter.
4. Install the filter in reverse order, using new O-rings. Operate the engine to check for leaks, and repair as necessary.

Carburetor

REMOVAL & INSTALLATION

1. Remove the air cleaner.
2. Disconnect the fuel line.
3. Disconnect and label all vacuum lines to the carburetor.
4. Remove the carburetor mounting bolts.
5. Remove the carburetor.
6. Installation is the reverse of removal.

Fuel Injection

Due to the complex nature of modern fuel injection systems, comprehensive diagnostic and test procedures fall outside the confines of this repair manual. For complete information on diagnosis, testing, and repair procedures concerning all modern fuel injection systems, please refer to *Chilton's Guide to Fuel Injection and Feedback Carburetors.*

COOLING SYSTEM

Thermostat

REMOVAL & INSTALLATION

To replace the thermostat, drain the cooling system below the level of the thermostat and remove the two bolts holding the thermostat housing in place. Remove the thermostat housing and lift out the thermostat. Clean the mating surfaces of the intake manifold and the thermostat housing. Use a new gasket when installing a new thermostat. If only silicone sealer was used from the factory, use only sealer during assembly.

Note that on the Cadillac 250 cu. in. V8, there are both an upper and a lower thermostat housing. You need not remove the lower housing to replace the thermostat. Make sure you replace the ring seal after the thermostat is in place.

On the fuel-injected 231 V6, the thermostat is located in the front of the intake manifold. To remove it, first remove the bolt fastening the housing mounting clamp. Then pull the housing off the manifold. Remove the sealing ring and thermostat. Install in reverse order.

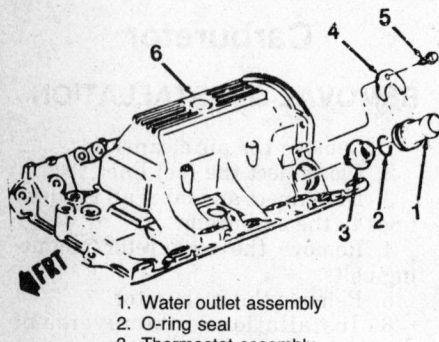

1. Water outlet assembly
2. O-ring seal
3. Thermostat assembly
4. Clamp
5. Torque to 20 ft. lbs.
6. Intake manifold

Replacing the thermostat on the 231 V6 with injection

--- **CAUTION** ---

Be sure not to install the thermostat upside down. The spring and wax pellet must always be positioned downward or on the cylinder block side.

Water Pump

REMOVAL & INSTALLATION

All Models Except Cadillac 250 V8

1. Drain the cooling system. Remove the fan shroud, if necessary for clearance.
2. Loosen the belts, then remove the fan blades and pulleys from the hub on the water pump shaft. Remove the belts.
3. Disconnect the hose from the water pump inlet and the hester hose from the nipple. Remove the bolts, then remove the pump and gasket from the timing case cover or engine block.

4. Install the pump assembly with a new gasket. Bolts and lock washers must be torqued evenly.
5. Connect the radiator hose to the pump inlet and the heater hose to the nipple. Fill the cooling system and check all points of possible coolant leaks.
6. Install the fan pulleys and the fan blade. Install the belts and adjust for correct tension.

Cadillac 250 V8

1. Drain the cooling system. Disconnect the negative battery cable. Remove the filter for the A.I.R. pump. Disconnect and remove the coolant recovery tank.
2. Disconnect and remove the body brace on the water pump side.
3. Remove the belt which drives all the accessories. Then, remove the drive belt idler pulley and its bracket.
4. Remove the water pump pulley. Then, remove the pump mounting bolts, and remove the pump and gasket.
5. Clean all the sealing surfaces. Then, thread the three pump pulley bolts into the hub squarely to thread the holes.
6. Install the gasket, coated on both sides with sealer onto the front cover studs. Then, install the pump. Install the Torx® screws, nuts, and hex bolts as shown in the illustration. Torque the Torx® screws and stud nuts to 30 ft. lbs. Torque the remaining fasteners to 5 ft. lbs.
7. Perform the remaining procedures in reverse order. Make sure to replace the antifreeze/water mix with a solution that meets the standard in the note below.

NOTE: Because the 250 V8 has an aluminum block, make sure the antifreeze solution is approved for use in cooling systems with a high aluminum content. GM recommends the use of a supplement/sealant No. 3634621, or equivalent, specifically designed for use in aluminum engines to protect the engine from damage.

Radiator

REMOVAL & INSTALLATION

Except Below

1. Drain the radiator and disconnect the upper and lower hoses and the transmission coolant lines.
2. Disconnect the coolant recovery hose.
3. Remove the fan shroud-to-radiator screws. Lift the shroud out of the clips and hang the shroud over the fan.

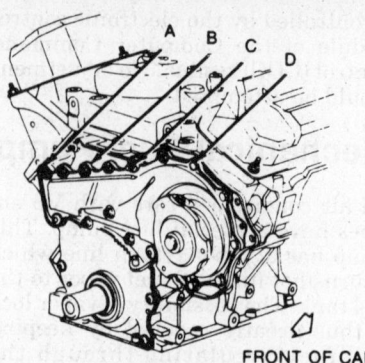

FRONT OF CAR

Locations for fasteners on the 250 V8 water pump. A fasteners are Torx [RG] screws; B fasteners are nuts; C fasteners are studs; and D fasteners are hex screws. Torque A and C to 30 ft. lbs. and B and D to 5 ft. lbs.

4. Remove the radiator upper mounting panel.
5. Remove the radiator. Reverse to install.

1986–87 231 V6

1. Disconnect the negative battery cable and drain the engine coolant. Remove the electric fan upper mounting bolts.
2. Remove the upper air cleaner duct/silencer.
3. Disconnect the radiator hoses and coolant recovery hose from the radiator.
4. Remove the three mounting bolts from the cowl and remove the upper radiator mounting panel. The radiator will now be free to tilt to the rear.
5. Disconnect the fan connector. Disconnect the transaxle oil cooler lines at the side tank. Remove the radiator.
6. Installation is the reverse of removal. Observe the following torque figures:
   a. Radiator support attaching clamp bolts—88 inch lbs.
   b. Transaxle oil cooler lines—20 ft. lbs.
   c. Fan assembly attaching bolts—85 inch lbs.

1986–87 250 V8

1. Disconnect the negative battery cable. Drain the engine coolant.
2. Remove the air cleaner duct and air conditioner hose bracket, disconnect the electrical connector, and remove the cooling fan from behind the radiator.
3. Disconnect the coolant reservoir hose at the radiator filler neck. Disconnect the upper radiator hose.
4. Disconnect the engine and transaxle oil cooler lines at the radiator. Disconnect the lower radiator hose at the radiator.

5. Unbolt and remove the upper radiator support bracket, and then remove the radiator.

6. Install in reverse order. Make sure you use a 50/50 antifreeze mix to replace any that is lost and that you heed the warnings about antifreeze and additives in the note below.

NOTE: Because the 250 V8 has an aluminum block, make sure the antifreeze solution is approved for use in cooling systems with a high aluminum content. GM recommends the use of a supplement/sealant No. 3634621, or equivalent, specifically designed for use in aluminum engines to protect the engine from damage.

EMISSION CONTROLS

For information on emission controls, please refer to "Emission Controls" in the Unit Repair Section. Due to the complex nature of modern electronic engine control systems, comprehensive diagnosis and testing procedures fall outside the confines of this repair manual. For complete information on diagnosis, testing and repair procedures concerning all modern engine and emission control systems, please refer to *Chilton's Guide To Electronic Engine Controls*.

ENGINE

NOTE: Please refer to the "Oldsmobile Rear Wheel Drive" or "Cadillac" sections for any engine procedures not included in this section. Service procedures to the Buick built 231 and 252 cu. in. (3.8 and 4.1L) V6 are in the "Buick Rear Wheel Drive" car section. The 350 cu. in. diesel engine is produced by the General Motors Oldsmobile division; diesel engine service procedures will therefore be found in the "Oldsmobile Rear Wheel Drive" car section. On the Eldorado and 1980 and later Seville, special mounting brackets are welded to the frame to provide the front attaching points and a special crossmember is used for the rear mount.

Engine

REMOVAL & INSTALLATION

1980–85 Toronado

1. Drain the cooling system.
2. Remove hood, marking hinge for reassesmbly.
3. If equipped with a fan shroud, unhook the strap and remove the clips holding the seal to the venturi ring. Move the seal toward the radiator.
4. Disconnect battery.
5. Disconnect radiator hoses, oil cooler lines, heater hoses, vacuum hoses, engine-to-body ground strap, fuel hose from fuel pump, wiring and accelerator cable. Remove the air cleaner, hot air pipe, air conditioner compressor and power steering pump without disconnecting lines and set them aside.
6. Remove the throttle control switch bracket, the radiator support and the radiator.
7. Raise the car.
8. Disconnect exhaust pipes at manifold. Loosen, but do not remove, upper left flywheel cover attaching bolt.
9. Disconnect wires and remove starter.
10. Remove torque converter cover and remove three bolts securing the converter to flywheel. Scribe marks on converter and flywheel for reassembly. Remove the splash shield.
11. Support the final drive assembly.
12. Remove two attaching bolts from right output shaft support bracket and one thru-bolt attaching final drive to engine block on the left side. Scribe around the washers for correct reassembly.
13. Remove engine mount to crossmember nuts and front engine mount nuts. Remove the lower right engine-to-transmission attaching bolt.
14. Support the final drive assembly with a chain stretched under and across the final drive assembly and attached to holes in the frame members.
15. Lower the car.
16. Support engine by using a lifting fixture.
17. Remove the remaining transmission-to-engine bolts.
18. Lift the engine from the car.

--- **CAUTION** ---
If car is to be moved, install a torque converter holding tool. If it is necessary to reposition the air conditioner compressor, do not disconnect the refrigerant lines.

19. To install, reverse removal procedure.

1980–85 Eldorado and Seville

1. Matchmark the hood hinges for reassembly.
2. Drain the cooling system.
3. Disconnect the battery cables and remove the battery.
4. Remove the air cleaner.
5. Loosely install a special valve compressor tool on the EFI line pressure fitting. Place a towel around the fitting to catch any spray. Slowly tighten the tool to relieve fuel pressure.
6. Raise the car on a hoist and remove the exhaust pipe flange bolts from the manifolds. Separate the left side pipe from the Y-pipe and remove the exhaust pipe from the car.
7. Disconnect the shift linkage from the transmission.
8. Disconnect the flexible fuel line from the main fuel pipe. Use a new clamp on installation.
9. Remove the six drive axle-to-output shaft attaching screws from each side.
10. Remove the nuts from the engine and transmission mounts.
11. Remove the lower fan shroud attaching screws and disconnect the lower radiator hose.
12. Lower the car and disconnect the upper radiator hose and the transmission cooler lines.
13. Remove the radiator upper cover and remove the radiator.
14. Remove the four clutch fan nuts and the fan shroud.
15. Disconnect the power steering hoses at the steering gear. Cap the ends to prevent entry of dirt.
16. Disconnect the flexible fuel line from the pressure regulator fuel return pipe. Use a new clamp on installation.
17. If equipped with cruise control, disconnect the vacuum lines from the power unit. Pull the hoses out of the tie-down straps and position them out of the way.
18. Disconnect:
   a. the canister hose.
   b. canister vacuum supply hose.
   c. throttle cable from the throttle body.
   d. heater hoses at the water valve and the water pump.
   e. brake vacuum line at the brake pipe.
   f. speedometer at the transmission.
   g. engine wiring harness at the center bulkhead connector.
   h. distributor wiring.
   i. heater wire from the water valve.
   j. Wiring at the windshield wiper motor and the washer bottle.
   k. engine ground strap from the cowl.

l. wiring at the A/C compressor.

19. Remove the coolant reservoir tank.

20. Loosen the A.I.R. pump and remove the belt from the A/C compressor.

21. Remove the compressor-to-bracket screws and position the compressor out of the way. Do not disconnect any of the lines.

22. Install a lifting chain.

23. Remove the engine, transmission and final drives as a unit.

24. Reverse to install. Torque the engine mounts to 65 ft. lbs.; transmission mounts to 48 ft. lbs.; output shaft-to-drive axle attaching screws to 60 ft. lbs.

1980–85 Riviera

1. Refer to Steps 1–4, 6 and 11–12 of the above procedure.

2. Remove the radiator.

3. Refer to Steps 8–13 of the above procedure.

4. Disconnect the turbo outlet exhaust pipe at the turbocharger assembly.

5. Raise the car and remove the starter motor.

6. Drain the engine oil.

7. Remove the torque convertor cover and remove the converter to flywheel bolts.

8. Remove the transmission to engine bolts.

9. Remove the right output shaft support bolts.

10. Remove the front engine mounts.

11. Install a support chain to the final drive.

12. Remove the final drive to engine bracket.

13. Install a lifting device to the engine and remove it from the car. Installation is the reverse of removal.

1986–87 Toronado and Riviera with 231 V6

1. Disconnect the negative battery cable. Drain the coolant from both engine and radiator by removing all drain plugs. Remove the hood bolts and remove the hood.

2. Remove the air inlet and radiator hoses.

3. Disconnect the following electrical connectors:

a. Fuel rail and other injection system connectors.

b. Engine grounds.

c. Oil pressure sending unit.

d. EGR solenoid.

e. Coolant temperature sending units.

f. Throttle body electrical connections.

g. Crankshaft and camshaft sensors.

h. Generator.

4. Remove the serpentine belt. Remove the power steering pump.

5. Remove the generator. Disconnect and remove the heater hoses.

6. Remove the throttle cable bracket and the accelerator, trasnmission, and cruise control cables at the throttle lever.

7. Disconnect both fuel lines.

8. Remove the cooling fan and radiator as described above.

9. Disconnect the exhaust Y-pipe and then remove the exhaust manifold on the forward side of the engine. Label and then disconnect the vacuum lines running between the engine and components mounted on the firewall or fender wells.

10. Remove the three engine-to-transaxle bolts.

11. Remove the vibration damper and bracket from the engine.

12. Remove the ground strap and wiring harness bolts. Remove the engine-to-transaxle bracket. Then, raise the car and support it securely.

13. Unbolt the A/C compressor from the bracket, move it to one side, and support it without disturbing the refrigerant hoses. Remove the exhaust pipe.

14. Remove the engine mount nuts. Then, remove the torque converter lower cover and matchmark the relationship between the converter and flywheel.

15. Remove the three converter-to-flywheel bolts, rotating the crankshaft as necessary for access.

16. Remove the left/front wheel. Then, remove the remaining engine-to-transaxle bolts. Make sure you remove the one bolt that faces in the opposite direction.

17. Remove the engine-to-transaxle bracket. Then, lower the car and pull the engine out the top of the engine compartment.

18. Install in reverse order, noting the following points:

a. When reattaching the converter to the flexplate, line up the matchmarks for balance. Make sure the weld nuts on the converter are flush with the flywheel. Tighten the bolts first finger tight. Then, tighten them completely. Repeat tightening the first bolt tightened as it may loosen as the others are torqued.

b. Observe the following torques: Balancer assembly to crankshaft—200 ft. lbs.; exhaust manifold to cylinder head—37 ft. lbs.; transaxle to cylinder block—55 ft. lbs.; engine mount to cylinder block—70 ft. lbs.

c. Replenish all fluids as necessary. Operate the engine and check for leaks.

1986–87 Eldorado and Seville

1. Disconnect the negative battery cable and drain the cooling system through both the radiator drain and the plugs in the block.

2. Remove the air cleaner. Remove the hood from the engine compartment.

3. Remove the cooling fan and the accessory drive belt.

4. Remove the upper radiator hose and disconnect the heater hose at the thermostat housing.

5. Disconnect the following electrical connectors, positioning the wires out of the way:

a. Oil pressure sending unit

b. Coolant temperature sensor

c. Distributor

d. EGR solenoid

e. Engine temperature switch

f. Idle Speed Control

g. Throttle position sensor

h. Injector electrical connections

i. MAT sensor

j. Oxygen sensor

k. Throttle body base warmer

l. Alternator

m. Ground wires at the alternator mounting bracket.

6. Disconnect the accelerator, cruise control, and transmission throttle valve cables at the throttle lever.

7. Disconnect the cruise control diaphragm and bracket and move them out of the way.

8. Disconnect the oil cooler and transmission oil cooler lines at the radiator. Then, remove the radiator as described previously.

9. Disconnect the oil cooler lines ar the oil filter adapter and remove them.

10. Remove the bracket which fastens the oil cooler lines at the transmission.

11. Remove the air cleaner mounting bracket. Remove the oil filter housing adapter.

12. Disconnect the air injection tubes at the diverter valve.

13. Remove the right/front and right/rear body braces.

14. Remove the right/front heater hose and the coolant reservoir.

15. Remove the Air Injection Reactor filter box and bracket. Remove the idler pulley for the accessory drive belt.

16. Remove the power steering line brace from the right cylinder head. Then, remove the pump and belt tensioner as an assembly and position them forward of the engine for working room.

17. Have the air conditioner system discharged of refrigerant by someone trained and experienced in refrigeration repair. Then, remove the A/C lines to the accumulator and condenser.

18. Position a metal container and a rag so as to catch the fuel, and then carefully depress the center-pin at the Schrader valve on the fuel line until all fuel pressure is exhausted. Then, disconnect supply and return fuel lines at the throttle body. Remove the fuel line bracket at the transmission and move the fuel lines out of the way.

19. Remove the EGR lines and brackets. Remove the vacuum modulator line and the fuel filter and reposition them out of the way.

20. Raise the car and support it in a secure manner.

21. Remove the starter heat shield. Label and remove the electrical connectors from the starter. Disconnect any ground wires still connected at the block.

22. Disconnect and remove the exhaust crossover pipe. Remove the starter.

23. Remove the two flexplate covers. Then, remove the three flexplate-to-torque converter bolts.

24. Remove the air conditioner compressor lower dust shield. Remove the right/front tire.

25. Remove the outer wheelhouse plastic shield.

26. Remove the right/rear transmission-to-engine mounting bolt. Then, remove the nut from the lower engine mounting damper.

27. Remove the nuts from the front engine mount and the bolts from the right/rear transmission mounts.

28. Remove the alternator. Remove the oxygen sensor wires. Remove the heater bypass bracket from the right side of the car.

29. Remove the right side engine brace. Then, lower the car to the ground.

30. Remove the five remaining engine-to-transmssion mounting bolts accessible from the top.

31. Run a chain from a lifting crane down to the two lift points on top of the engine and ensure it is secure. Lift the engine out of the car.

32. To install, first situate a floorjack under the transaxle and raise it slightly so it will line up with the engine. Then, lower the engine into the engine compartment, being careful not to damage accessories that are still in position. Change the engine and transaxle angles as necessary to get good alignment and then engage the dowels that are on the engine block with the corresponding holes in the transaxle.

32. Install the five transmission-to-engine bolts that are accessible from above into the bell housing. Then, lower and remove the floorjack that is under the transaxle. Then, lower the engine, directing it carefully and squarely onto its mounts. Remove the lifting equipment.

33. Install the remaining accesories, mounting bolts and nuts, etc. in reverse order. Make sure to replenish all fluids with the required type and quantity. Have the air conditioning system charged. Make sure to run the engine and check for leaks.

Exhaust Manifold

REMOVAL & INSTALLATION

V8 – Left Side

1. Remove the air cleaner and the carburetor heat shroud on the manifold on gasoline engine. Disconnect shift linkage and remove heat shield on diesel engine.

2. Remove the lower alternator bracket; raise the front of the car and support it safely.

3. Disconnect the exhaust pipe.

4. Lower the car and remove the manifold attaching bolts. Remove the manifold from above.

5. To install, reverse the removal procedure using the correct torque for the manifold attaching bolts.

V8 – Right Side

1. Raise the car and support it safely.

2. Disconnect the exhaust pipe and then remove the right front wheel.

3. Remove the attaching bolts and lower the manifold down and out from under the vehicle.

4. To install, reverse the removal procedure.

V6 – Left Side

1. Raise the front of the car and support it safely.

2. Disconnect the exhaust crossover pipe.

3. Remove the left front engine mount thru-bolt and loosen the thru-bolt on the right mount.

4. Raise the engine slightly, unscrew the manifold mounting bolts and remove the manifold.

5. Installation is in the reverse order of removal.

V6 – Right Side

1. Raise the front of the car and support it safely.

2. Disconnect the exhaust pipe from both manifolds and lower it.

3. Unscrew the manifold mounting bolts and remove the manifold from underneath the car.

4. Installation is in the reverse order of removal.

Eldorado and Seville

For exhaust manifold procedures for the 350 EFI Cadillac and diesel engines, refer to the "Cadillac" section.

Transversely Mounted 231 V6

1. Disconnect the negative battery cable and the oxygen sensor.

2. Raise the car and support it securely. Then, disconnect the exhaust pipe on both sides.

3. Remove the six manifold bolts on each side and remove each manifold.

4. Clean all sealing surfaces and provide new gaskets. Install in reverse order. Make sure to reconnect the oxygen sensor.

Turbocharger

REMOVAL & INSTALLATION

1. Disconnect the turbocharger exhaust inlet pipe and the exhaust outlet pipe at the turbocharger.

2. Disconnect the oil feed pipe from the center housing.

3. Remove the nut attaching the air intake elbow to the carburetor and remove the elbow, still attached to the flex tube, from the carburetor.

4. Disconnect the accelerator, cruise, and detent linkages at the carburetor. Disconnect the linkage bracket from the plenum.

5. Remove the two bolts attaching the plenum to the side bracket.

6. Disconnect the carburetor fuel line and necessary vacuum hoses.

7. Drain the cooling system.

8. Disconnect the coolant hoses from the front and rear of the plenum.

9. Disconnect the power brake vacuum line at the plenum.

10. Remove the two bolts attaching the turbo housing to the bracket on the intake manifold.

11. Remove the two bolts attaching the EGR valve manifold to the plenum. Loosen the two bolts attaching the EGR valve manifold to the intake manifold.

12. Loosen the clamp attaching the hose from the AIR by-pass to the pipe

to the check valve. Remove the hose from the pipe.

13. Remove the three bolts attaching the compressor housing to the intake manifold.

14. Remove the turbocharger and actuator, still attached to the carburetor and plenum assembly, from the engine. Disconnect any vacuum hoses as necessary (after labeling them for later assembly).

15. Remove the six bolts attaching the turbo and actuator unit to the plenum and carburetor assembly.

16. Remove the oil drain from the turbo center housing.

17. Installation is the reverse of removal. Be sure to refill the cooling system.

NOTE: Before installing the turbo unit, make certain that all parts and connections are clean. Serious damage to the turbo unit and engine will result if dirt and/or foreign matter enters into the engine.

Intake Manifold

REMOVAL & INSTALLATION

V8 with Carburetor

1. Remove the negative battery terminal, air cleaner, heat tube and PCV valve.

2. Disconnect the throttle and Cruise Control linkages.

3. Remove the HEI electrical connection from the distributor.

4. Remove the distributor cap and the ignition wires. Mark the wires for easy reinstallation.

5. Disconnect the temperature sending unit and the electrical connection from the air conditioning compressor.

6. Disconnect the two wires from the downshift switch. Disconnect the throttle return spring and downshift switch bracket. Disconnect the electric choke if so equipped.

7. Remove the plug from the anti dieseling solenoid and any other necessary electrical connections.

8. Disconnect the power brake booster vacuum and vacuum modulator lines. Remove the cruise control mechanism if so equipped. Disconnect the A/C vacuum hose from the rear of the manifold.

9. Disconnect the fuel line from the carburetor.

10. Disconnect the vacuum advance line (if so equipped) and the canister purge hoses and position them out of the way.

11. Remove the air conditioning compressor and position it out of the way. Do not disconnect the refrigerant lines.

12. Disconnect the coolant bypass hose at the manifold, if so equipped.

13. Remove the carburetor.

14. Remove the manifold bolts and the manifold.

15. Installation is the reverse of removal.

Diesel Engine Or Electronic Fuel Injection (EFI)

For intake manifold removal and installation on the Cadillac V8 with electronic fuel injection see the "Cadillac" section. For diesel engine servicing refer to the "Oldsmobile" section.

V6 Engines with Carburetors

Refer to the "Buick Rear Wheel Drive" section for 231 and 252 V6 intake manifold procedures for carbureted engines, built prior to 1986.

1986–87 231 V6 with EFI

1. Disconnect the negative battery cable. Drain the cooling system. Remove the air intake duct and mass airflow sensor.

2. Remove the accessory drive belt, generator, and generator bracket.

3. Remove the ignition module and associated wiring. Disconnect wiring and vacuum lines that will interfere with removal of the manifold.

4. Disconnect and remove the throttle, cruise control, and transmission valve cables at the throttle body.

5. Remove the upper radiator hose. Disconnect the heater hoses at the throttle body.

6. Relieve fuel system pressure as described under the Electric Fuel Pump Removal & Installation procedure above. Then, disconnect the electrical connections for the injectors.

7. Remove the attaching bolts and remove the fuel rail. Replace all O-rings on injectors that are to be reused.

8. Label and remove the spark plug wires.

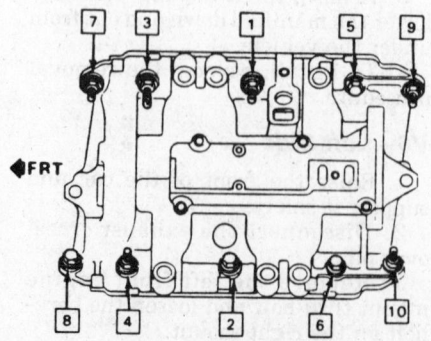

Torque sequence for the intake manifold bolts—fuel injected 231 V6

9. Remove the intake manifold bolts. Remove the manifold and gasket. Clean all mating surfaces. Replace the gasket. Use a sealer such as GM Part No. 1050026 for a steel gasket. Pipe thread fittings must be sealed with a sealer and lubricant such as GM Part No. 1052080.

10. Place the manifold in position squarely over the gaskets. Torque the bolts to 32 ft. lbs. in the sequence shown.

11. Perform the remaining steps of the installation in reverse order. Replenish all fluids, operate the engine and check for leaks.

Timing Cover

REMOVAL & INSTALLATION

Toronado V8 and Buick V6

NOTE: Refer to the "Cadillac" section for Cadillac V8 Procedures.

1. Drain the cooling system. Disconnect the upper and lower radiator, heater and bypass hoses.

2. Remove the radiator, belts, fan and fan pulley, crankshaft pulley and the harmonic balancer. Remove the fuel lines and pump on the V6.

3. Remove the alternator and brackets on the V6.

4. Remove the distributor on the V6. If timing chain and sprockets are not going to be disturbed, matchmark the distributor rotor and housing to aid installation. On V6, loosen and slide the front clamp on the thermostat bypass hose rearward.

5. Remove the timing cover attaching bolts and pull off the cover. On the V8, remove the timing pointer and water pump.

6. Remove both front cover dowel pins. Grind a chamfer on one end of each dowel pin. See Step 5 and illustration under "V8 Front Cover Removal and Installation" in the Oldsmobile Rear Wheel Drive section.

7. Before assembly, remove all old gaskets and install a new timing cover gasket. Use sealer around the coolant holes and at the junction of the block pan and front cover. On the V6, remove the oil pump cover and pack the space around the oil pump gears completely full of petroleum jelly to prime it.

8. Position the front cover, timing pointer and the water pump.

9. Lubricate the attaching bolts and install. Install the fuel pump on V6.

10. Install the harmonic balancer on the crankshaft after lubrication. Replace the engine if removed.

11. Connect all cooling hoses.

12. Install the crankshaft pulley.
13. Install the fan and the fan pulley.
14. Install the drive belts and adjust.
15. Fill the crankcase, if drained, and the radiator.
16. Run the engine and check for leaks.

Timing Chain

REMOVAL & INSTALLATION

V8 Engine

Refer to the "Cadillac" or "Oldsmobile Rear Wheel Drive" section for procedures.

V6 Engine

Refer to "Buick Rear Wheel Drive" section for procedures.

Camshaft

REMOVAL & INSTALLATION

Refer to the "Cadillac" section, and the "Oldsmobile Rear Wheel Drive Section" for V8 procedures and "Buick Rear Wheel Drive" section for V6 procedures.

ENGINE LUBRICATION

Oil Pan

REMOVAL & INSTALLATION

1980–85 Toronado V8 (Gas and Diesel)

1. Disconnect the negative battery cable.
2. Remove the three final drive-to-transmission bolts.
3. Raise the front of the car and support it with jackstands.
4. Disconnect the two lower frame braces, if necessary for working clearance.
5. Disconnect the idler and the pitman arms from the relay rod.
6. Disconnect the right and left side drive axles from their respective output shafts.
7. Disconnect the battery cable bracket from the output shaft support and then disconnect the support itself from the engine block.
8. Remove the remaining final drive-to-transmission bolts, position a

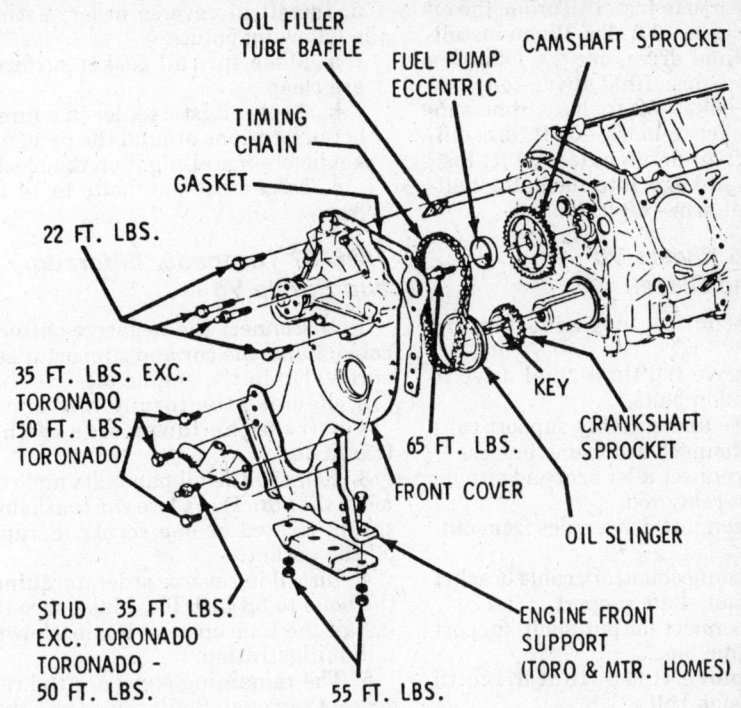

transmission jack under the final drive and remove the final drive.
9. Disconnect the starter wiring and remove the starter, remove the splash shield.
10. Remove the two front motor mount nuts and loosen the front motor mount bolts from the side of the engine block. Jack up the engine 1–1½ inches to gain working clearance. Drain the oil, remove the oil pan bolts and remove the oil pan.
11. Installation is in the reverse order of removal. Use sealer on both sides of the new gasket and tighten the oil pan bolts to 10 ft. lbs.

1980–85 Eldorado And Seville V8 (Gas and Diesel)

1. Disconnect the negative battery cable.
2. Raise the car and support it safely.
3. Remove the frame brace front attaching bolts from both sides and pivot the braces outward.
4. Remove the six securing bolts from the drive axle to the output shaft on both sides. Separate the flanges of the output shafts and drive axles to gain clearance for removal with the shafts attached.
5. Remove the battery cable-to-output shaft retaining screws and remove the two screws securing the support to the engine block.

6. Remove the final drive-to-transmission screw that holds the front of the shield. Remove the shield.
7. Remove the remaining final drive-to-transmission bolts.
8. Remove the final drive support bracket-to-engine block screw.
9. Using a puller, separate the steering linkage intermediate shaft from the pitman arm and the idler arm. Push the linkage toward the front of the car.
10. With the aid of a helper, slide the final drive assembly forward, off the transmission splined shaft, and remove the unit with the output shaft attached. Do not use the shafts as handles, as damage to the seals will occur.
11. Remove the battery cable and the wiring harness connectors from the starter solenoid BAT terminal.
12. Remove the harness connector from the solenoid S terminal.
13. Remove the harness from the clip on the solenoid and position it out of the way.
14. Remove the starter motor attaching bolts and remove the starter.
15. Drain the engine oil.
16. Remove the oil pan attaching screws and remove the oil pan.

NOTE: On cars equipped with diesel engines it is necessary to loosen the motor mounts and jack up the engine slightly to remove the oil pan.

17. Reverse to install. Torque the oil pan screws to 10 ft. lbs. When installing the final drive, use the following torque values: final drive-to-transmission bolts—30 ft. lbs.; front support bracket-to-block—50 ft. lbs.; output shaft-to-drive axle—60 ft. lbs.; steering linkage intermediate shaft-to-pitman arm—60 ft. lbs.

1980–85 Riviera V8 (Gas and Diesel)

1. Disconnect negative battery cable.
2. Remove top three final drive to transmission bolts.
3. Raise and suitably support car.
4. Disconnect two frame braces.
5. Disconnect idler arm and pitman arm from relay rod.
6. Disconnect drive axles from output shafts.
7. Disconnect battery cable bracket from output shaft support.
8. Disconnect output shaft support from engine block.
9. Remove three final drive to transmission bolts.
10. Install transmission jack and remove final drive.
11. Clean gasket material from mating surfaces.
12. Remove splash shield.
13. Disconnect starter wires.
14. Remove starter.
15. Drain oil pan.
16. Remove oil pan bolts and pan.
17. Installation is the reverse of removal. Tighten pan bolts to 10 ft. lbs.

1981–84 Toronado V6

1. Disconnect the negative battery cable. Raise the car and support it securely.
2. Drain the engine oil. Remove the flywheel cover.
3. Remove the crossover pipe.
4. Disconnect the engine mounts from the frame brackets. Then, use a suitable means to raise the engine by the vibration damper for clearance. Remove the oil pan retaining bolts and remove the pan.
5. Install in reverse order. While original production uses silastic sealer, it is best to reinstall the pan with a new gasket. Torque the oil pan mounting bolts to 14 ft. lbs.

1981–82 Eldorado and Seville V6

1. Disconnect the negative battery cable. Raise the vehicle and support it securely.
2. Drain the engine oil.
3. Remove the final drive assembly as described later in this section.
4. Remove the exhaust crossover pipe. Remove the oil pan.

5. Install in reverse order, noting the following points:
   a. Make sure all gasket surfaces are clean.
   b. Apply silastic sealer in a number of locations around the pan gasket before installing it on the block.
   c. Torque the pan bolts to 14 ft. lbs.

1986–87 Toronado, Eldorado, and Seville V6

1. Disconnect the negative battery cable. Raise the car and support it securely. Drain the engine oil.
2. Remove the torque converter cover from the lower side of the transaxle.
3. Remove the oil pan bolts and remove the pan. Don't lose the tensioner spring located at one corner. Scrape off the gasket.
4. Install in reverse order, torquing the bolts to 88 inch lbs. Make sure to install the tensioner spring as shown in the illustration.
5. The remaining steps are the reverse of removal. Refill the oil pan the proper level, operate the engine and check for leaks.

1986–87 Riviera V6

1. Disconnect the negative battery cable. Raise the car and support it securely. Drain the oil pan.
2. Remove the flywheel or torque converter cover. Remove the crossover pipe.
3. Securely support the engine from above. Then, unbolt the engine mounts at the frame brackets.
4. Raise the engine until there will be clearance to remove the oil pan from underneath. Then, remove the oil pan bolts and remove the pan.
5. Remove the old gasket and clean the oil pan and cylinder block mating surfaces thoroughly.
6. Install a new gasket, and the pan and bolts. Torque the bolts to 6–9 ft. lbs. Complete the installation in reverse order. Refill the pan with the specified quantity and type of oil. Operate the engine and check for leaks.

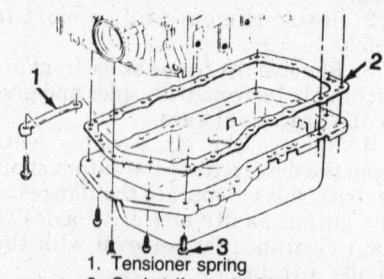

Replacing the oil pan on the 1986–87 231 V6. Note the location of the tensioner spring (1)

1. Tensioner spring
2. Gasket (formed rubber)
3. Torque to 88 in. lbs.

Oil Pump

REMOVAL & INSTALLATION

1980–85 Toronado V8

Remove the oil pan. Remove the oil baffle. Remove the oil pump to rear main bearing cap attaching bolts, then remove the pump and drive shaft extension. When installing the pump, make sure the drive shaft extension is properly inserted in the drive gear.

1980–85 Toronado V6

1. Remove oil pan.
2. Remove screws attaching oil pump pipe and screen assembly to the cylinder block. Remove the oil filter and pump cover, then remove the oil pump gears.
3. On installation, pack the oil pump gears and assembly with petroleum jelly.

1980–85 Eldorado and Seville

ALL EXCEPT 350 CU. IN. ENGINE

1. Remove the oil pan as previously outlined.
2. Remove the oil filter.
3. Remove the bolts securing the oil pump to the engine. The screw nearest the pressure regulator should be removed last, allowing the pump to come down with the screw. Always discard the oil pump-to-crankcase gasket.
4. Remove the oil pump drive shaft.
5. Installation is the reverse of removal.

350 CU. IN. ENGINE

1. Remove the oil pan as previously outlined.
2. Remove the bolts attaching the pump to the rear main bearing cap and remove the pump and drive shaft extension.
3. Installation is the reverse of removal. Be sure the shaft is properly mated with the distributor drive gear. Torque the bolts to 35 ft. lbs.

Eldorado and Seville 250 V8

1. Remove the oil pan as described above. Remove the two screw and single nut securing the pump to the engine and remove the pump.
2. Remove the old O-ring from the oil pump outlet pipe and replace it with a new one.
3. To install, first position the pump assembly and drive shaft against the block, engaging the driveshaft with the distributor gear. Then, install the two screws and one nut that mount the pump. Torque the nut to 22 ft. lbs. and the screws to 15 ft. lbs.

4. Complete the installation in reverse order. Make sure to refill the pan with the correct amount of the proper oil, start the car, and verify that oil pressure is established promptly. Run the engine and check for leaks.

Riviera except 231 V6

For oil pump removal & installation procedures, please refer to the "Buick Rear Wheel Drive" section.

Oil Pump Cover and Gears

REMOVAL & INSTALLATION

231 V6 Engine

1. Disconnect the negative battery cable. Raise the car and support it securely. Remove the oil filter.
2. Remove the bolts that fasten the oil pump cover to the timing chain cover. Then, remove the oil pump cover assembly, drive gear and the driven gear.
3. Remove the oil pump pressure relief valve cap, spring, and relief valve.
4. Clean all the parts in solvent and blow them dry with compressed air.
5. Install in reverse order. Make sure you lubricate all the relief valve parts with clean engine oil. After reassembling the gears into the pump housing, thoroughly pack all the voids between gears and the housing with petroleum jelly to ensure that the pump will prime itself.

NOTE: Failure to pack the oil pump with petrolum jelly before reassembling the pump and starting the engine will cause engine damage.

6. Complete the assembly in reverse order. Torque the cap for the pressure regulator valve to 35 ft. lbs.. Torque the bolts for the pump cover alternately and in several stages to 12 ft. lbs.
7. Refill the oil pan, operate the engine, and check for leaks.

Rear Main Seal

REPLACEMENT

Toronado

See "Oil Pan Removal and Installation". Remove the oil pan and rear main bearing cap. Using a blunt-ended tool, drive the upper seal into its groove on each side until it is tightly packed. This is usually $\frac{1}{4}$–$\frac{3}{4}$ in. Cut pieces of the old bearing cap seal $\frac{1}{16}$ in. longer than the distance each side of the upper seal was compressed. Install these pieces into each side of the upper seal seat, packing them into place. Carefully trim any protruding seal, being sure not to scratch or damage the bearing surface. Install a new seal in the bearing cap and install the cap, tightening bolts to the specified torque. Install the oil pan.

Eldorado and Seville
EXCEPT 250 V8

In order to replace the upper main bearing seal, the crankshaft must be removed from the engine. Only the lower rear main oil seal is covered here.

1. Remove the oil pan as previously outlined.
2. Remove the rear main bearing cap.
3. Remove the rear main bearing insert and the old seal. Thoroughly clean the grooves and inspect it for cracks.
4. Install the new seal into the cap.
5. Cut the seal flush with the mating surface.
6. Clean the bearing insert and install it in the bearing cap.
7. Clean the bearing cap mating surface and apply sealer to the cap.
8. Lubricate the threads of the cap bolts and install the cap. Torque to 120 ft. lbs.
9. Install the oil pan.

250 V8 ENGINE

NOTE: To perform this procedure, you will need a seal removal tool J-26868 or equivalent and a seal installer J-34604.

1. Remove the transaxle as described below. Unbolt and remove the flexplate from the rear end of the crankshaft.
2. Pry the old seal out using the seal removing tool. Thoroughly clean the seal bore of any leftover seal material with a clean rag.
3. Lubricate the lip of a new seal with wheel bearing grease. Position it over the crankshaft and into the seal bore with the spring facing inside the engine.
4. Then, press the seal into place using the seal installer. The seal must be square (this is the purpose of the installer) and flush with the block to 1 mm indented. Install the flexplate and torque the bolts to 37 ft. lbs.
5. Install the transaxle in reverse order. Operate the engine and check for leaks.

Riviera

For rear main bearing oil seal procedures, please refer to the "Buick Rear Wheel Drive" section. For the 231 V6, refer to the procedure for Toronado above.

AUTOMATIC TRANSMISSION

All automatic transmission service procedures are covered in the "Automatic Transmission" Unit Repair section.

REMOVAL & INSTALLATION

NOTE: When performing this procedure on the turbocharged Riviera, the turbo must be removed first.

1. Open hood and disconnect negative battery cable (two cables on diesels).
2. Disconnect the speedometer cable at the transmission. Remove transmission oil dipstick tube.
3. Remove the air cleaner.
4. Disconnect the transmission throttle valve (T.V./detent) cable at its upper end. Disconnect the linkage by removing one nut from shaft on left side of transmission, if so equipped.
5. Safely support the engine unit from underneath, or install an engine holding fixture between the cowl and radiator support.
6. Remove the top and two upper left final drive to transmission bolts.
7. Remove the remaining accessible engine-to-transmission bolts (5).
8. Jack up the car and safely support it with jackstands.
9. Remove the starter assembly.
10. Disconnect the transmission converter clutch connector.
11. Disconnect the transmission oil cooler lines and plug the openings.
12. Remove the flywheel inspection cover (loosen top left bolt). Matchmark the flywheel-to-converter relationship for later assembly.
13. On V8s, disconnect the exhaust Y-pipe connection to the left exhaust pipe. On all, disconnect the right exhaust pipe at the manifold. On gasoline cars, disconnect the catalytic converter hanger bolts (2). On all, lower the exhaust system about 5 inches and support it.
14. Remove the four bolts holding the second frame crossmember.

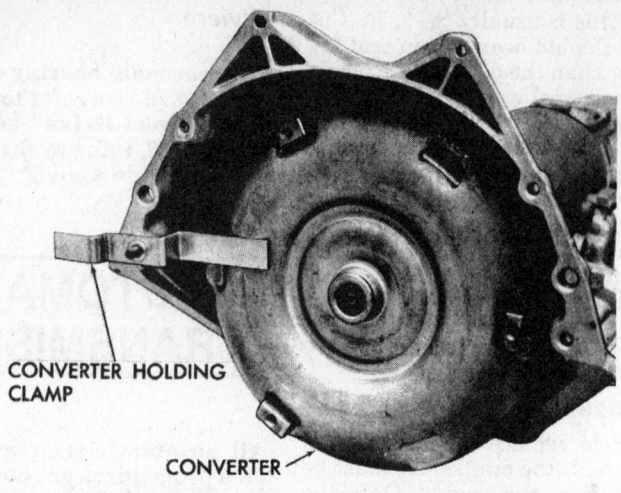

CONVERTER HOLDING CLAMP

CONVERTER

Converter holding clamp. A C-clamp will also work

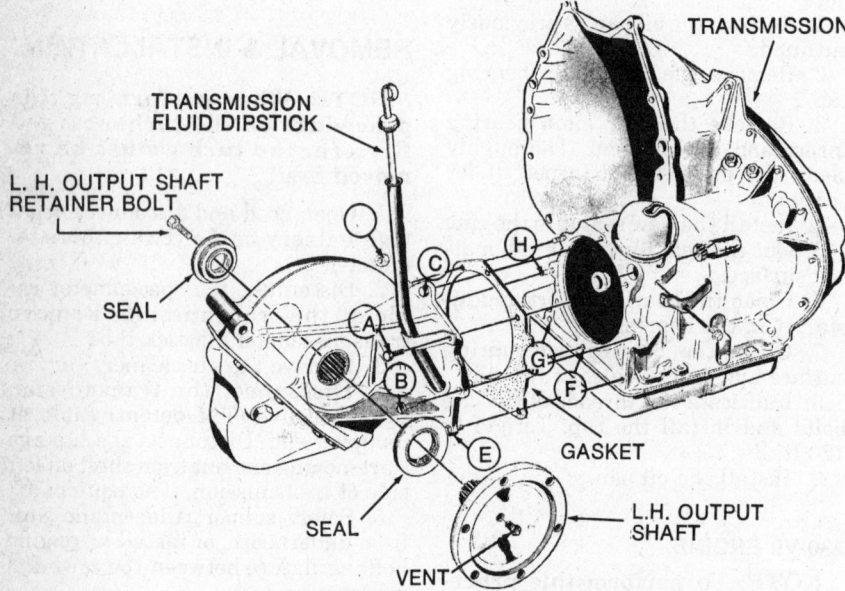

TRANSMISSION

TRANSMISSION FLUID DIPSTICK

L. H. OUTPUT SHAFT RETAINER BOLT

SEAL

H

C

A

B

G

F

E

GASKET

SEAL

VENT

L.H. OUTPUT SHAFT

Typical transmission attachment points

15. Position a hydraulic floor jack underneath the transmission, with a wooden block on the jack pad to protect the transmission case. Jack up the transmission slightly.

16. Remove the three remaining final drive to transmission bolts.

17. Remove the converter-to-flywheel bolts.

18. Disconnect the shift linkage at the transmissions.

19. Remove the final drive support bracket bolt.

20. Remove the right transmission mount (through bolt and three bracket bolts).

21. Remove the left transmission mount through bolt. Remove the lower bracket-to-transmission bolt. Raise the transmission assembly about two inches for access to the two remaining upper bracket-to-transmission bolts.

Remove the remaining transmission-to-engine bolts.

22. Carefully lower the transmission unit while disengaging the final drive.

23. Install a C-clamp or converter holding clamp in front of the torque converter (attached to the bell housing) to hold the converter in place. Remove the transmission from the car.

24. To install the transmission, reverse the removal procedure. Always replace the final drive-to-transmission gasket. Use care when engaging the final drive-to-transmission splines and make sure the final drive-to-transmission mounting faces are in alignment with each other. After the splines are engaged, loosely install the two final drive-to-transmission lower attaching bolts. You can save time here by installing two engine-to-

transmission bolts from above first to aid alignment.

25. After the final drive and transmission are mated align the transmission with the engine and install the remaining attaching bolts. Before the flywheel-to-converter bolts, make sure the weld nuts on the converter are flush with the flywheel and that the converter rotates freely by hand in this position. Then hand start all three bolts and tighten finger tight. This will insure proper converter alignment. Torque the transmission-to-engine bolts to 35 ft. lbs., the final drive-to-transmission bolts to 30 ft. lbs., and the final drive support bracket to final drive bolts to 35 ft. lbs.

1986–87 Riviera and Toronado

NOTE: To perform this procedure, you'll need an engine support J-28467 or equivalent and a Drive Axle Remover J-33008 or equivalent.

1. Disconnect the negative battery cable. Install the engine support fixture.

2. Disconnect the: vacuum line at the modulator; electrical connections involved with the transmission; transmission valve cable at the throttle body and at the transaxle; the cruise control servo.

3. Disconnect the shift selector bracket and cable from the transaxle. Disconnect the neutral start switch.

4. Remove the top three transaxle mounting bolts.

5. Remove the bolts that fasten the wiring harness to the transaxle. Remove the driveline dampener bracket.

6. Raise the car and support it securely.

7. Disconnect and drain the transaxle oil cooler lines at the transaxle.

8. Remove the torque converter cover. Scribe the relationship between the flexplate and the converter so the same relationship may be established on reinstallation for balance. Then remove the converter-to-flexplate bolts, turning the crankshaft as necessary to do so.

9. Remove the left side transaxle mounting bolts. Remove the engine mounting nuts.

10. Disconnect the sway bar links. Disconnect the left side balljoint from the knuckle.

11. Disconnect the left side driveshaft from the transaxle using a special tool such as J-33008 or equivalent. See the procedure below.

12. Disconnect the left side of the frame by removing the bolts.

13. Position a floor jack under the transaxle and support it securely.

14. Remove the two remaining en-

gine-to-transaxle bolts. Note that one of the bolts is located between the transaxle case and the block and is installed in the direction opposite to the others.

15. Remove the engine-to-transaxle bracket.

16. Remove the right drive axle from the transaxle and hang it securely.

17. Remove the transaxle.

18. To install the transaxle, first slide it into position and then install the two lower engine-to-transaxle bolts, torquing to 55 ft. lbs.

19. Install the engine-to-transaxle bracket. Install the left side frame assembly bolts.

20. Install the engine mounting nuts. Install the left side transaxle mounting bolts.

21. Reverse the remaining removal procedures to install. Note that transaxle mounting bolts should be torqued to 55 ft. lbs. Converter-to-flexplate mounting bolts should be torqued to 46 ft. lbs. Make sure that the scribe marks are aligned.

1986–87 Eldorado and Seville

1. Disconnect the negative battery cable. Remove the air cleaner assembly. Disconnect the transmission throttle valve cable.

2. Remove the cruise control servo and bracket assembly. Disconnect the electrical connectors going to the distributor, oil pressure sending unit, and transaxle.

3. Remove the bracket for the engine oil cooler lines.

4. Remove the shift linkage bracket from the transaxle and the manual shift lever from the manual shift shaft. You can leave the cable attached to the lever and bracket.

5. Remove the fuel line bracket. Remove the neutral safety switch connector.

6. Remove the vacuum modulator.

7. Remove the throttle valve cable support bracket and engine oil cooler line bracket. Then, remove the bell housing bolts which are in positions 2,3,4, and 5 in the illustration.

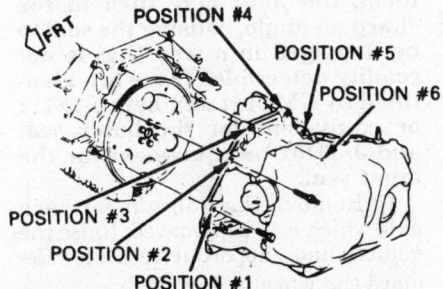

Removing the transaxle-to-engine mounting bolts on the 1986 and later Eldorado and Seville

POSITION #4
POSITION #5
POSITION #6
POSITION #3
POSITION #2
POSITION #1
FRT

8. Remove the air injection reactor crossover pipe fitting and reposition the pipe. Remove the radiator hose bracket and transaxle mount-to-bracket nuts.

9. Install an engine support fixture, noting the positions of the hooks.

10. Raise the vehicle so you can work underneath and support it in a secure manner.

11. Remove both front wheels. Remove right and left stabilizer link bolts. Remove the ball joint cotter pins and nuts and then press the ball joints out of the steering knuckles.

12. Remove the air conditioner splash shield, and the mount cover for the forwardmost cradle insulator.

13. Remove the hose connections from the ends of the air injection reactor pipes. Remove the vacuum hoses and the wire loom from the clips at the front of the cradle.

14. Remove the engine mount and dampener-to-cradle attachments. Remove the transaxle mount-to-cradle attachments. Remove the wire loom clip from the transmission mount bracket. Then, lower the vehicle so you can work from above.

15. Use the two left side support hooks on the engine support fixture to raise the transaxle two inches from its normal position. Then, again raise the vehicle and support it safely so you can work underneath.

16. Remove the right/front and left/rear transaxle-to-cradle bolts, and the left stabilizer mount bolts. Remove the foremost cradle mount insulator bolt and the left cradle member, separating the right/front corner first.

17. Remove the air injection reactor management valve and its bracket assembly from the transaxle mount bracket and reposition the bracket to the transaxle stud bolts.

18. Lower the transaxle to its normal position so you can gain access to the transaxle mounting bracket. Remove the mounting bracket.

19. Raise the vehicle again and support it safely. Remove the right/rear transaxle mount-to-transaxle bracket Remove the engine-to-transaxle brace bolts that pass into the transaxle VSS connector.

20. Mark the relationship between torque converter and flexplate for reassembly in the same position. Remove the flywheel covers and then remove the torque converter bolts, rotating the crankshaft with a socket wrench as necessary to gain access. Position a jack under the transaxle to support it.

21. Remove the bell housing bolts in positions No. 1 and No. 6. Note that you must gain access through the right wheelhouse opening to remove

the bolt at position No. 6. You need to use a 3 foot long socket extension to reach it.

22. Disconnect the oil cooler lines at the transaxle, drain them, and plug the openings. Then, install drive axle boot seal protectors and disconnect the driveshafts at the transaxle, as described below. Suspend the drive axles out of the way. Then, remove the transaxle.

23. Put the transaxle into position. Then, install the lower bell housing bolts at positions No. 1 and No. 2.

24. Install the remaining bell housing bolts and torque all of them to 55 ft. lbs. Note that studs go in positions No. 2,3,4 and bolts go in No. 1,5,6. Note also that No. 6 is longer than Nos. 1 and 5.

25. Turn the converter until it is aligned with the flexplate as originally installed. Install the converter-to-flexplate bolts, and torque them to 46 ft. lbs. Install the splach shield under the converter. Unplug and reconnect the oil cooler lines to the transmission case. Torque the fittings to 15 ft. lbs.

26. To cmplete the installation, reverse the remaining steps of the removal procedure, observing the following torque figures, all given in ft. lbs.:

    a. Forward most insulator mount bolt — 74

    b. Cradle to cradle mounting bolts — 74

    c. Upper transaxle mount bracket stud bolts — 74

    d. Side transaxle mount bracket stud bolts — 50

    e. Left or rear transaxle mount nuts — 35

    f. Engine mount-to-cradle attachments — 35

    g. Right rear mount bracket-to-transaxle bolts — 50

    h. Right rear mount bracket nuts — 35

    i. Stabilizer mount bolts — 38

    j. Ball joint nuts — 81

    k. Shift cable bracket-to-transaxle bolts — 18

    l. Lug nuts — 100

27. Adjust the transmission valve cable and the shift linkage. Fill the transmission to the proper level, using the capacities chart. Then, run it until hot and adjust the level until it is correct.

DRIVE AXLE

Drive axles consist of an axle shaft with an inner and outer constant velocity joint. The right axle shaft has a torsional damper mounted in the center. The inner constant velocity joint has complete flexibility, plus inward and outward movement. The outer constant velocity joint has complete flexibility but doesn't allow for inward and outward movement. Refer to the "U-Joint/CV-Joint" Unit Repair section for constant velocity joint service.

REMOVAL & INSTALLATION

1980–85 Toronado, Riviera, Eldorado and Seville

RIGHT SIDE

1. Hoist car under lower control arms and remove the wheel.
2. Remove drive axle cotter pin, retainer, nut and washer from the wheel hub.
3. Remove oil filter on V8.
4. Remove inner constant velocity joint attaching bolts.
5. Push inner constant velocity joint outward enough to disengage the right hand final drive output shaft, then remove rearward.
6. Remove right hand output shaft bracket bolts to engine and final drive.
7. Remove right hand output shaft and drive axle assembly.

CAUTION

Care must be exercised so that constant velocity joints do not turn to full extremes, and that seals are not damaged against shock absorber or stabilizer bar.

8. Carefully place the right hand drive axle assembly into the lower control arm and insert the outer race splines into the steering knuckle.
9. Lubricate final drive output shaft seal with special seal lubricant.
10. Install the right hand output shaft into the final drive and attach the support bolts to the engine and brace. Torque the bolts to 50 ft. lbs.
11. Move right hand drive axle assembly toward front of car and align with right hand output shaft. Install attaching bolts and torque to 60 ft. lbs.
12. Install oil filter on V8.
13. Install washer and nut on drive axle. Torque to and 175 ft. lbs., then install the retainer and cotter pin.
14. Remove floor stands and lower hoist.
15. Check engine oil level on V8.

LEFT SIDE

1. Hoist car under lower control arms.
2. Remove wheel. Remove disc.
3. Remove drive axle cotter pin, nut and washer.
4. Remove tie rod end cotter pin and nut.
5. Remove the tie rod end from the knuckle with a puller.
6. Remove bolts from drive axle assembly and left output shaft. Insert a spacer between the axle shaft and lower control arm.
7. Remove the upper control arm ball joint cotter pin and nut.
8. Using a hammer and brass drift, drive on the knuckle until the upper ball joint stud is free.
9. Remove the steering knuckle and support, then support the knuckle with wire or string.
10. Carefully remove the drive axle assembly.

NOTE: Care must be exercised so that constant velocity joints do not turn to full extremes and that seals are not damaged against shock absorber or stabilizer bar.

11. Carefully guide left hand drive axle assembly onto lower control arm and into position on spacer.
12. Insert lower control ball joint stud into knuckle and attach nut on 1979 models. Do not torque.
13. Center left hand drive axle assembly in opening of knuckle and insert upper ball joint stud.
14. Place brake hose clip over upper ball joint stud and install nut. Do not torque.
15. Insert tie rod end stud into knuckle and attach nut. Torque to 35 ft. lbs. Install cotter pin and crimp.
16. Align inner constant velocity joint with output shaft and install attaching bolts. Torque to 60 ft. lbs.
17. Torque the lower ball joint stud nuts to 65 ft. lbs., and the upper to 90 ft. lbs. for 1980 and 55 ft. lbs for 1981–82. Install cotter pins and crimp.

NOTE: Upper ball joint cotter pin must be crimped toward upper control arm to prevent interference with outer constant velocity joint seal.

18. Install drive axle washer and nut. Torque to 175 ft. lbs. Install cotter pin and crimp.
19. Install wheel.
20. Remove floor stands and lower hoist.
21. Check camber, caster and toe in and adjust if necessary. Refer to Front End Alignment specifications.

1986–87 Riviera and Toronado

NOTE: To accomplish this procedure, you will need a drive axle spindle remover set, J-33008 and J-28733 or equivalent. Also, if the car uses silicone boot seals, drive axle boot seal protectors must be installed before the driveshaft is disconnected. If the car uses thermoplastic seals, these are not required. They are needed with the silicone seals because, without them, the joint may turn to too sharp an angle, causing the seal to be damaged in a way that is not readily detectible. These are identified by GM part numbers J-28712 or equivalent for the outer seal and J-33162 or equivalent for the inner seal.

1. Remove the hub nut for each axle which is to be removed. Raise the vehicle and support it securely. Remove the wheel(s) involved.
2. If the car has silicone seals, install the protectors described in the note above.

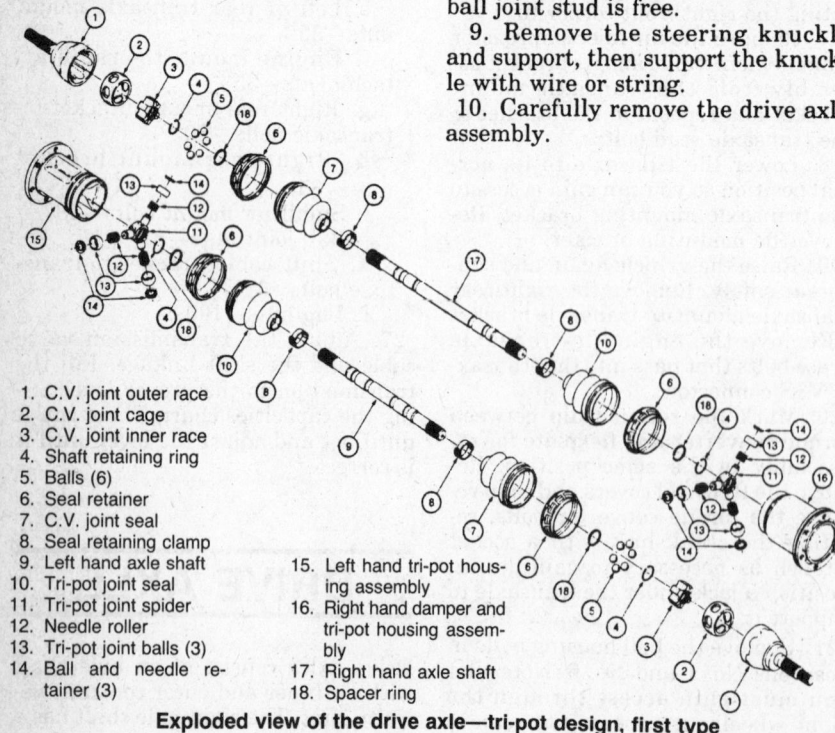

1. C.V. joint outer race
2. C.V. joint cage
3. C.V. joint inner race
4. Shaft retaining ring
5. Balls (6)
6. Seal retainer
7. C.V. joint seal
8. Seal retaining clamp
9. Left hand axle shaft
10. Tri-pot joint seal
11. Tri-pot joint spider
12. Needle roller
13. Tri-pot joint balls (3)
14. Ball and needle retainer (3)
15. Left hand tri-pot housing assembly
16. Right hand damper and tri-pot housing assembly
17. Right hand axle shaft
18. Spacer ring

Exploded view of the drive axle—tri-pot design, first type

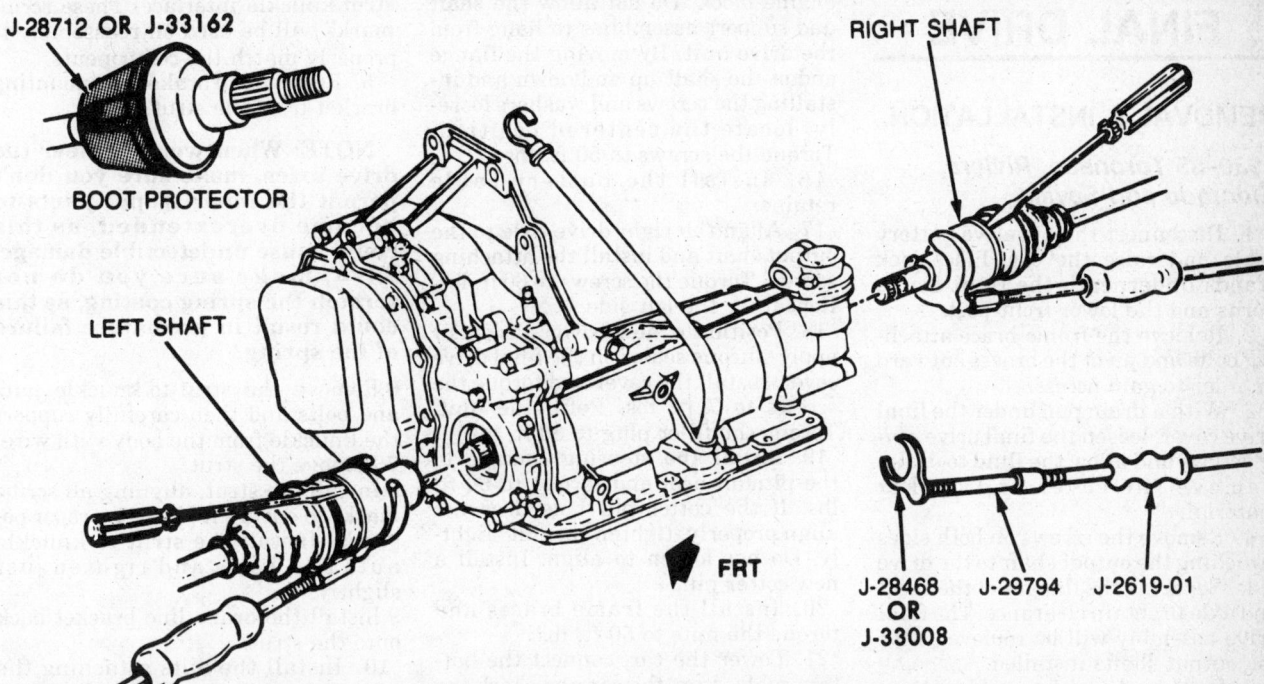

Use a screwdriver and the special tool shown to pull the drive axles out of the transaxle. Make sure to support the axles at the center to avoid putting downward force on the outer joint

3. Remove the brake caliper, caliper support, and rotor as described later in this section.

4. Remove the bolts that attach the steering knuckle to the strut. Then, pull the knuckle out of the strut bracket.

5. Pull the driveshafts out of the transaxle using the special tool as shown.

NOTE: Support the shafts at the center so there will be no downward force on the outer joints.

6. Remove the axles from the hub and bearing assembly with the spindle remove J-28733 or equivalent, and remove them from the vehicle. Do not remove the boot seal protectors unless complete disassembly is necessary.

7. To install, first loosely position the drive axle into the steering knuckle and transaxle.

8. Put the the steering knuckle into position in the strut bracket and install the bolts. Torque to 144 ft. lbs.

9. Install the rotor and brake caliper as desicribed below. Install a drift or screwdriver into one rotor slot to keep it from turning.

─── **CAUTION** ───

If the vehicle uses a prevailing torque hub nut, you MUST use a new one, torque it to specification, and make sure the threads are undamaged and free of oil and grease. Otherwise, the drive axle may not be retained safely.

10. Install a new prevailing torque hub nut and washer and torque them to 74 ft. lbs. Then remove the object used to hold the rotor stationary.

11. Seat the drive axle into the transaxle with a screwdriver resting against the groove provided on the inner retainer, as shown. Tap the screwdriver lightly to seat the snapring and lock the driveshaft into the transaxle. Then, verify that the snapring has been seated by grasping the housing (NOT the shaft itself) and pulling it outboard. If the shaft is locked, you will not be able to pull it free.

12. Remove the boot seal protector. Install the wheel and tire, tightening the lugnuts partway. Lower the vehicle and torque the lugnuts to 100 ft. lbs.

1986–87 Eldorado and Seville

NOTE: To perform this procedure, you will need a special puller J-28733 or equivalent and new prevailing torque nut for each axle you need to remove.

1. Remove the prevailing torque hub nut and washer. Then, raise the vehicle and support it securely.

2. Remove the wheel and tire.

3. Remove the brake caliper and rotor (see the brake section below).

4. Disconnect: the stabilizer bar at the control arm; the tie rod end at the steering knuckle; the lower ball joint stud at the steering knuckle. Then, use a prypar and screwdriver against

a wood block (to protect the case) to pry the drive axle out of the transaxle case.

5. Then, use the special puller to force the drive axle out of the hub, and remove the axle from the car. Inspect the boot seals for damage and replace as necessary.

6. Position the drive axle ends into the steering knuckle and transaxle without fully seating them.

7. Refasten the lower ball joint stud to the steering knuckle and torque the nut as described later under "Front Suspension". Reconnect the stabilizer bar to the lower control arm and the tie rod end to the steering knuckle, again referring to "Front Suspension".

8. Reinstall the brake caliper with new mounting bolts as described below.

9. Install a washer and new prevailing torque nut; torque the nut to 74 ft. lbs. You can prevent the axle from turning when you torque the nut by inserting a screwdriver into one of the slots in the brake rotor.

10. Seat the drive axle in the transaxle by positioning a screwdriver into the groove on the CV-joint housing and tapping the screwdriver with a hammer until the axle is seated. Grab the housing of the drive axle (not the driveshaft) and pull it outward to make sure the axle is properly seated.

11. Install the wheel and tire. Lower the vehicle. Then, final-torque the hub nut to 183 ft. lbs.

FINAL DRIVE

REMOVAL & INSTALLATION

1980–85 Toronado, Riviera, Eldorado and Seville

1. Disconnect the negative battery cable and raise the car. Place jack stands underneath the front frame horns and the lower front post.
2. Remove the frame brace attaching bolts and pivot the braces outward in order to gain access.
3. With a drain pan under the final drive cover, loosen the final drive cover screws and allow the fluid to drain. Remove the cover and gasket material.
4. Remove the screws on both sides attaching the output shaft to the drive axle. Separate the flanges of the shaft and axle to obtain clearance. The final drive assembly will be removed with the output shafts installed.
5. Remove the battery cable retaining screws from the right output shaft and the screws securing the support to the engine block. Rotate the support downward for clearance.
6. Remove the screws which attach the final drive shield to the transmission and the support bracket. Remove the shield.
7. Remove the remaining final drive screws.
8. Remove the final drive support-to-engine block attaching screws.
9. Using a puller, separate the steering linkage from the pitman arm. Push the linkage toward the front of the car.
10. Slide the final drive assembly forward, off the transmission shaft and remove the unit. Do not hold the unit by the output shafts as the seals or splines could easily be damaged.
11. To install, thoroughly clean all the gasket surfaces and position a new gasket on the final drive. Do not use a sealer on the gasket.
12. Align the final drive assembly, with the output shafts attached, to the transmission and install all the attaching screws except the one used to hold the shield. Torque in rotation to 30 ft. lbs. in two steps.
13. Loosen the front support bracket screws and install the bracket to the engine block while holding the bracket flush on the housing pad. Torque to 50 ft. lbs.
14. Install the final drive shield. Torque the drive-to-transmission screw to 30 ft. lbs. and the bracket-to-housing screws to 34 ft. lbs.
15. Align the right output shaft support with the attaching holes in the engine block. Do not allow the shaft and support assemblies to hang from the drive unit. By moving the flange end of the shaft up and down and installing the screws and washers loosely, locate the centered position. Torque the screws to 50 ft. lbs.
16. Install the battery cable retainer.
17. Align the right drive axle to the output shaft and install the attaching screws. Torque the screws to 60 ft. lbs. Repeat for the left side.
18. Position a new cover gasket or apply silicone sealer on the final drive cover. Install the cover and torque the screws to 7 ft. lbs. Refill the unit. Torque the filler plug to 30 ft. lbs.
19. Install the steering linkage to the pitman arm and torque to 60 ft. lbs. If the cotter pin hole does not align properly, tighten the nut slightly. Do not loosen to align. Install a new cotter pin.
20. Install the frame braces and torque the nuts to 50 ft. lbs.
21. Lower the car, connect the battery cable, start the car and check the transmission fluid. When the final drive has reached operating temperature, check it for leaks.

FRONT SUSPENSION

MacPherson Strut

REMOVAL & INSTALLATION

1. Open the hood and remove the nuts attaching the top of the strut to the body.
2. Raise the car and suitably support it, on Riviera and Tornado. On the Seville and Eldorado, hoist the vehicle with a twin hoist and install jackstands under the cradle. Then, lower the vehicle so it rests on the jackstands and not on the control arms.

CAUTION

Support all cars at the rear so that as components are removed, the weight will not shift, causing the car to fall off the supports.

---

3. Remove the tire and wheel related to the strut(s) involved.
4. Using a sharp tool, scribe the knuckle along the lower/outboard radius of the strut. Then, scribe the strut flange on the inboard side, right along the curve of the knuckle. Finally, make a scribe mark across the strut/knuckle interface. These scribe marks will be used on reassembly to properly match the components.
5. Remove the brake line mounting bracket from the strut.

NOTE: When working near the drive axles, make sure you don't permit the inner tri-pot joints to become overextended, as this could cause undetectible damage. Also, make sure you do not scratch the spring coating, as this could result in premature failure of the spring.

6. Remove the strut-to-knuckle nuts and bolts and then carefully support the knuckle from the body with wire.
7. Remove the strut.
8. Install the strut, aligning all scribe marks to ensure it is in the proper position. Install the strut-to-knuckle nuts and bolts and tighten just slightly.
9. Install the brake line bracket back onto the strut.
10. Install the nuts attaching the top of the strut to the body. Now, torque those nuts to 18 ft. lbs. Torque the strut-to-knuckle bolts/nuts to 145 ft. lbs.
11. Install the tire and wheel, and torque the nuts to 100 ft. lbs. Lower the car.

OVERHAUL

For all spring and shock absorber removal & installation procedures, and all strut overhaul procedures, please refer to "Strut Overhaul" in the Unit Repair section.

Wheel Hub and Bearing Assembly

REMOVAL & INSTALLATION

1980–85 Toronado

1. Remove drive axle cotter pin, nut and washer. Remove the lug nuts and wheel. Remove the brake disc.
2. Position access slot in hub assembly so each of the attaching bolts can be removed.
3. Install a front hub puller and slide hammer.
4. Remove hub and bearing assembly.
5. To install, reverse removal procedure. Tighten the axle nut to 175 ft. lbs.

NOTE: O.D. of bearing must be lubricated with E.P. chassis lubricant. Use care when installing hub assembly over drive axle splines.

1986—87 Toronado

NOTE: You will need a front hub spindle remover, J-28733 or equivalent to perform this operation.

1. Raise the car and support it securely. Remove the front wheel.

2. Insert a drift through slots in the rotor and caliper to keep it from turning; then, remove the hub nut and washer.

3. Remove the brake caliper and its support as described later in this section.

4. Separate the drive axle from the hub using the spindle remover.

5. Remove the hub and bearing retaining bolts, and then remove the hub and bearing assembly.

6. To install, reverse the above procedure, noting the following points:

   a. Torque the hub and bearing retaining bolts to 70 ft. lbs.

   b. Use only new caliper mounting bolts and remount the caliper as described later in this section.

   c. Partially tighten the hub nut, holding the rotor with a drift as during removal. Then, after the wheel is installed and the car lowered to the ground, torque the hub nut to 180 ft. lbs. and the wheel nuts to 100 ft. lbs.

1980—85 Eldorado, Riviera and Seville

1. Remove hub cap, loosen wheel nuts, remove drive axle cotter pin and loosen drive axle nut.

2. Jack up car and place jack stands under lower control arms.

3. Remove axle nut and wheel and tire assembly.

4. Remove brake hose and caliper.

NOTE: Match-mark disc and hub, then remove the disc.

5. Remove the upper ball joint cotter pin and loosen steering stud nut.

6. Strike steering knuckle near upper joint to separate it from taper.

7. Cover the lower control arm torsion bar connector with a short piece of rubber hose to avoid damaging the inboard tripod joint seal when the hub and knuckle are removed.

8. Remove tie rod end cotter pin and nut.

9. Separate tie rod end from steering knuckle using a tie rod splitter.

10. Remove lower ball joint cotter pin and stud nut.

11. Disconnect lower ball joint.

12. Remove hub, backing plate and steering knuckle as an assembly.

13. To install, reverse removal procedure. Tighten upper ball joint stud to 60 ft. lbs.; tighten lower ball joint

stud to 80 ft. lbs. Tighten drive axle nut to 110 ft. lbs. Tighten wheel lug nuts to 130 ft. lbs.

1986—87 Eldorado, Riviera and Seville

NOTE: To perform this procedure you will need a front hub spindle remover GM tool No. J-28733 or equivalent and a hub seal installer GM tool No. J-34657 or equivalent.

1. Raise the vehicle and then position jackstands under its cradle. Lower the vehicle so the jackstands support it and the control arms are free. Remove the front wheel.

2. Use a drift inserted through one of the slots in the rotor and a caliper slot to keep the rotor from turning. Then, remove the hub nut and washer.

3. Remove the caliper and caliper support as described later in this section. Remove the rotor as also described later.

4. Use the spindle remover to force the drive axle out of the hub.

5. Remove the hub and bearing retaining bolts and remove the hub and bearing assembly. Remove the seal by driving it toward the center of the car and then cutting it off the drive axle.

6. Lubricate the lip of a new seal with wheel bearing grease and install it with a seal driver.

7. The remainder of installation is the reverse of removal. Keep in mind the following points:

   a. Torque the hub and bearing retaining bolts to 70 ft. lbs.

   b. Install the caliper with new mounting bolts as described later in this section.

   c. Install the hub nut and washer part way with the vehicle in the air. Install the wheel, and then raise the vehicle, remove the jackstands, and lower it to the floor. Final torque the hub nut to 180 ft. lbs.

Wheel Bearing

REPLACEMENT

All Models

1. Remove the front brake caliper.

2. Remove the hub and outer bearing assembly.

3. Carefully pry the seal from the hub, then remove the inner bearing assembly.

4. If necessary, remove the outer bearing brace.

5. Wash all parts in clean solvent and either blow dry with compressed air or let air dry. Check bearings for cracked cages and worn or pitted roll-

ers. Check the bearing races for cracks, scores, or brinelling.

6. To install the new bearings, drive or press the outer braces into the hub (if removed). A large socket (preferably one that's already beat up) can be used with a rubber hammer to drive the bearing in.

7. Thoroughly clean the hub and spindle with clean solvent.

8. Apply a thin wipe of quality, high temperature wheel bearing grease to the spindle at the outer bearing seat and at the inner bearing seat, shoulder and seal seat.

9. Apply a small wipe of grease inboard of each bearing cup in the hub. Pack the bearing cone and roller assemblies full of grease by hand, working the grease thoroughly into the bearings between the rollers, cone, and cage.

10. Place the inner bearing cone and roller assembly in the hub. Then, using your finger, put an additional wipe of grease outboard on the bearing.

11. Install a new grease seal, using a flat place to seat the seal flush, into the hub. Lubricate the seal lip with a thin layer of grease.

12. Carefully install the hub and rotor assembly. Place the outer bearing cone and roller assembly in the outer bearing cup. Install the washer and nut and initially tighten the nut to 12 ft. lbs. while turning the wheel assembly forward by hand. Put another wipe of grease outboard of the bearing. This will give the bearings extra grease.

13. Install the brake caliper, then the wheel assembly.

ADJUSTMENT

Tapered Roller Bearings

The proper functioning of the front suspension cannot be maintained unless the front wheel tapered roller bearings are correctly adjusted. Cones must be a slip fit on the spindle and the inside diameter of cones should be lubricated to insure that the cones will creep. Spindle nut must be a free-running fit on threads.

1. Remove cotter pin from spindle and spindle nut.

2. Tighten the spindle nut to 12 ft. lbs. while turning the wheel assembly forward by hand to fully seat the bearings. This will remove any grease or burrs which could cause excessive wheel bearing play later.

3. Back off the nut to the "just loose" position.

4. Hand tighten the spindle nut. Loosen spindle nut until either hole in the spindle lines up with a slot in the

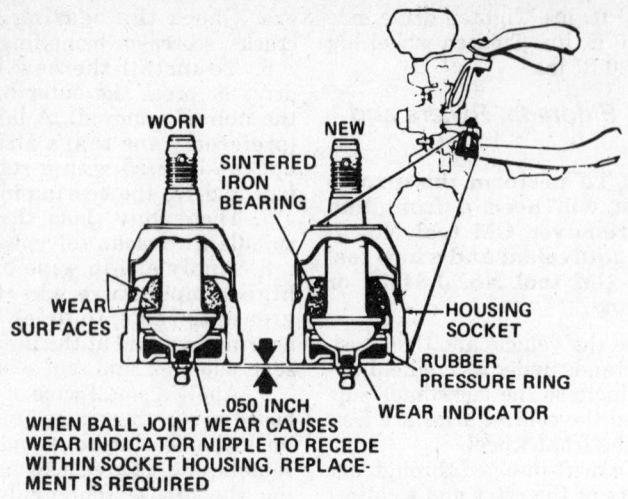

WHEN BALL JOINT WEAR CAUSES WEAR INDICATOR NIPPLE TO RECEDE WITHIN SOCKET HOUSING, REPLACEMENT IS REQUIRED

Lower ball joint wear indicator, typical

nut (not more than ½ flat).

5. Install new cotter pin. Bend the ends of the cotter pin against nut. Cut off extra length to ensure ends will not interfere with the dust cap.

6. Measure the looseness in the hub assembly. There will be from .001–005 inches end play when properly adjusted.

7. Install dust cap on hub.

Bolt-On Type Bearings

All E and K Series, 1981 and later, have front and rear sealed wheel bearings. The bearings are preadjusted and require no lubrication maintenance or adjustment. There are darkened areas on the bearing assembly. These darkened areas are from a heat treatment process and do not indicate a need for bearing replacement.

Torsion Bar

REMOVAL & INSTALLATION

All Models

1. Raise the car and support the frame.

2. Install a torsion bar remover tool, remove the torsion bar adjusting bolt and nut, noting the number of turns to remove, and relax the torsion bar. Do the same on the other torsion bar.

3. Remove the bolts and retainer from the torsion bar crossmember. Move the crossmember back until the bars are free and the adjusting arms can be removed. You may have to slide the torsion bars forward.

4. Reverse the procedure for installation.

Upper Control Arm

REMOVAL & INSTALLATION

All Models

NOTE: The upper control arm is serviced as an assembly, less bushings.

1. Hoist the car under the lower control arm and remove the wheel.

2. Remove the upper shock attaching bolt. It is not necessary, but it does allow more working room.

3. Remove cotter pin and nut on upper ball joint.

4. Disconnect brake hose clamp from ball joint stud.

5. Separate upper ball joint stud from steering knuckle using a hammer and drift.

6. Guide upper control arm over shock absorber and install bushing ends into frame horns.

7. Install cam assemblies.

8. Install ball joint stud into knuckle.

9. Install brake hose clip onto ball joint stud.

10. Install ball joint nut. Torque to the following values and insert cotter pin and crimp: 1980—90 ft. lbs. 1981 and later—55 ft. lbs.

NOTE: Cotter pin must be crimped toward upper shock attaching bolt and nut. Torque to 95 ft. lbs.

11. Install upper shock attaching bolt, if removed and tighten to 95 ft. lbs.

12. Install wheel.

13. Lower the vehicle.

14. Check camber, caster and toe-in and adjust if necessary.

Lower Control Arm

REMOVAL & INSTALLATION

All 1980–85 Models

1. Hoist car and support at lift points. Remove wheel assembly.

2. Place torsion bar remover and installer over crossmember so that center screw is seated in dimple of torsion adjusting arm.

3. Remove torsion bar adjusting bolt and nut, counting the number of turns necessary.

NOTE: This number of turns will be used when installing, to obtain initial ride height.

4. Turn center screw of tool until torsion bar is completely relaxed.

5. Disconnect shock absorber and stabilizer link from lower control arm.

6. Remove drive axle nut. Remove the bolt and nut from the front of the frame brace. Loosen the rear bolt and move the brace out.

7. Remove cotter pin and nut from lower ball joint stud.

8. Remove ball joint stud from knuckle, using puller.

9. Push drive axle in and pull knuckle outward to gain clearance, then remove lower control arm from knuckle and torsion bar.

10. Install by reversing removal procedure. Check and adjust ride height if necessary.

All 1986–87 Models

NOTE: Throughout this procedure, take care not to overextend the tri-pot joints. Overextension could result in separation of internal components, resulting in eventual failure of the joint. The damage done would not be readily detectible. You will need a 90° angle torque wrench J-35551 or equivalent.

1. Raise the car and support it by the cradle on jackstands. Remove the tire and wheel.

2. Disconnect the stabilizer shaft insulator, retainers, spacer, and bolt from the control arm.

3. Disconnect the lower ball joint from the knuckle.

4. Remove the control arm bushing bolt and front nut, retainer, and insulator. Then, remove the control arm from the frame.

5. Install the control arm back onto the frame with the bushing bolt and front nut, retainer, and insulator.

6. Install the control arm bushing bolt and front nut, the retainer, and

the insulator, but do not tighten them.

7. Connect the lower ball joint to the knuckle.

8. Connect the stabilizer shaft insulator, retainers, spacer, and bolt. Tighten the shaft nut/bolt to 13 ft. lbs.

9. Insert the ball joint stud into the steering knuckle, and install the nut. Torque the nut to 88 inch lbs. (7 ft. lbs. on Cadillac); then turn it an additional 120° (180° on Cadillac) while watching the required torque. It must reach at least 37 ft. lbs. torque (48 ft. lbs. on Cadillac) in that 120°. Install the cotter pin.

10. Install the tire and wheel, and then lower the car to the ground, leaving it unsupported by any jacking equipment. Torque the wheel nuts (100 ft. lbs.). Then, torque the control arm bushing bolt to 100 ft. lbs., or if it is a nut, to 91 ft. lbs. Torque the retaining nut to 52 ft. lbs.

BALL JOINT CHECK

All Models

1. Raise the car and position jack stands under the left and right lower control arm, as near as possible to each lower ball joint. Car must be stable and should not rock on jack stands. The upper control arm bumper must not contact the frame. The wheel bearing must be correctly adjusted.

2. Position the dial indicator to register vertical movement at the base of the tire rim for upper ball joint and at center of hub for lower ball joint.

3. Grasp the tire at the 12 o'clock and 6 o'clock positions and rock it in and out for upper ball joint. Pry with a pry bar between the lower control arm and the outer race of the CV-joint for lower ball joint. The vertical reading must not exceed .125 in. in either case.

REMOVAL & INSTALLATION

All 1980–85 Models

NOTE: Although not absolutely necessary, removal of the individual control arm will facilitate ball joint removal.

1. Remove the steering knuckle.
2. Drill the top rivet head off.
3. Drill the rivets just deep enough to remove the rivet head.
4. Using a hammer and punch, drive the rivets out of the control arm.
5. Install service ball joint into control arm and torque bolts and nut. Torque the bolts to 8 ft. lbs. Stake the upper nut.
6. Install knuckle.

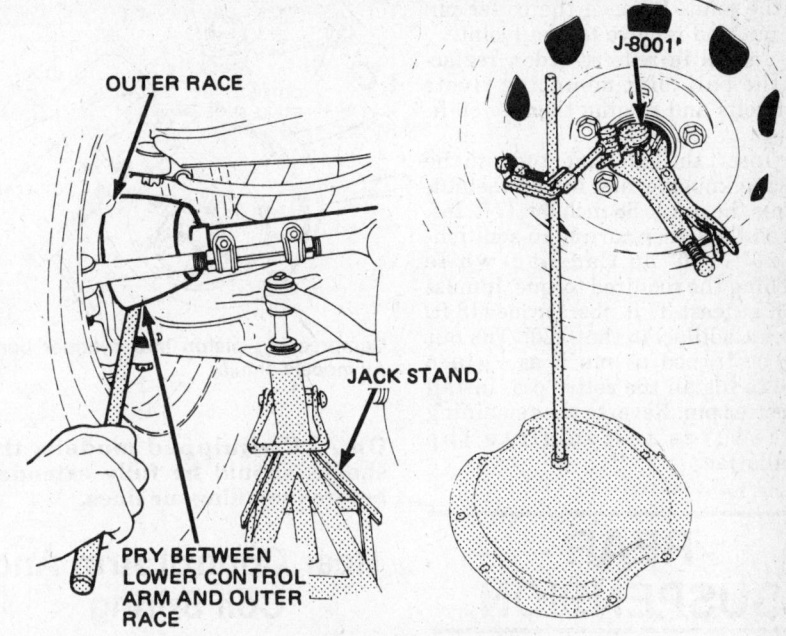

Lower ball joint check

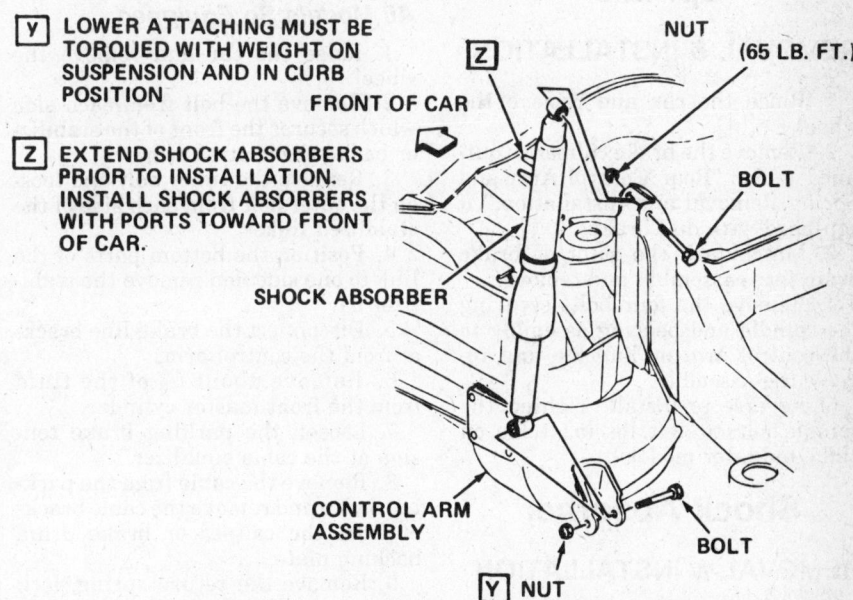

Y LOWER ATTACHING MUST BE TORQUED WITH WEIGHT ON SUSPENSION AND IN CURB POSITION

Z EXTEND SHOCK ABSORBERS PRIOR TO INSTALLATION. INSTALL SHOCK ABSORBERS WITH PORTS TOWARD FRONT OF CAR.

Typical 1979 and later rear shock absorber mounting

All 1986–87 Models

NOTE: To perform this job, you will need a ball joint separator J-35315 or equivalent, and a ball joint nut wrench J-35551.

1. Raise the vehicle and support it under the cradle in a secure manner, so the control arms will hang free.
2. Remove the wheel.
3. Remove the stabilizer bar insulators, retainers, spacer, and bolt.
4. Using a ¼ in. drill bit for the first pass, and a ½ in. bit for the second, drill out the three rivets retain-

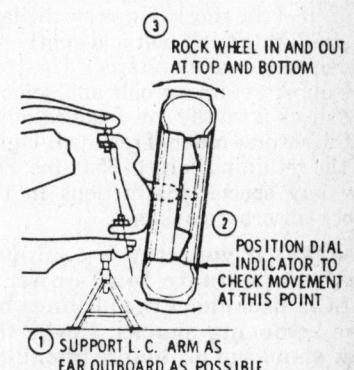

③ ROCK WHEEL IN AND OUT AT TOP AND BOTTOM

② POSITION DIAL INDICATOR TO CHECK MOVEMENT AT THIS POINT

① SUPPORT L.C. ARM AS FAR OUTBOARD AS POSSIBLE.

Upper ball joint check

ing the joint. Remove the cotter pin and nut and remove the ball joint.

5. Install in reverse order, replacing the ball joint mounting rivets with bolts and torqing them to 50 ft. lbs.

6. Insert the ball joint stud into the steering knuckle, and install the nut. Torque the nut to 88 inch lbs. (7 ft. lbs. on Cadillac); then turn it an additional 120° (180° on Cadillac) while watching the required torque. It must reach at least 37 ft. lbs. torque (48 ft. lbs. on Cadillac) in that 120°. The nut may be turned as much as $\frac{1}{6}$ turn more to install the cotter pin. Install the cotter pin. Reverse the remaining procedures to complete the installation.

REAR SUSPENSION

Spindle

REMOVAL & INSTALLATION

1. Raise the car and remove the wheel.

2. Remove the brake caliper as outlined under "Rear Control Arm and Spring Removal and Installation," if equipped with disc brakes.

3. Matchmark the rotor or brake drum for reassembly and remove.

4. Remove the four bolts securing the spindle and bearing assembly to the control arm or knuckle and remove the assembly.

5. Reverse to install. Tighten the spindle bolts to 32 ft. lbs. (52 ft. lbs. on 1986 and later models).

Shock Absorber

REMOVAL & INSTALLATION

Raise the rear of the car and support the control arm with a jack to take the load off of the shock. Unscrew the lower shock retaining bolt and gently tap the shock out of its retainer. Unscrew the upper retaining bolt and remove the shock from the car. Installation is in the reverse order of removal. Tighten the retaining bolts to 65 ft. lbs. Follow any special instructions in the shock absorber packages.

NOTE: If your car is equipped with automatic load control (ALC), disconnect the air lines before removing shocks. Purge the new shocks of air before installing (on all models) by repeatedly extending and compressing them.

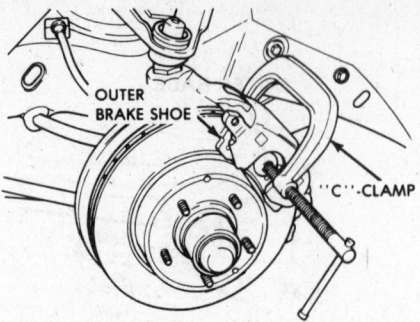

Compressing piston into cylinder bore. All models similar

On ALC-equipped models, the shocks should be fully extended before installing air lines.

Rear Control Arm And Coil Spring

REMOVAL & INSTALLATION

All Models So Equipped

1. Raise the car and remove the wheel.

2. Remove the bolt from each side which secures the front of the stabilizer bar to the control arm.

3. Remove the inner bolt and loosen the outer bolt from each side of the stabilized link.

4. Position the bottom parts of the link to one side and remove the stabilizer bar.

5. Disconnect the brake line bracket from the control arm.

6. Remove about $\frac{2}{3}$ of the fluid from the front master cylinder.

7. Loosen the parking brake tension at the cable equalizer.

8. Remove the cable from the parking brake and remove the cable bracket from the caliper or brake drum backing plate.

9. Remove the return spring, lock nut, lever and the anti-friction washer on disc brakes. The lever must be held while removing the nut.

10. Install and tighten a 7 in. C-clamp on the caliper as shown to bottom the cylinder pistons.

11. Disconnect the brake line from the brake and plug the openings to prevent the entrance of dirt.

12. With a $\frac{3}{8}$ in. allen wrench, remove the two caliper mounting bolts and remove the caliper, pads and rotor. On drum brakes, remove the hub and bearing assembly and remove the brake backing plate, along with the brake shoes.

13. If working on the left side, snap the Electronic Level Control link off the control arm.

14. Support the bottom of the control arm with a floor jack.

15. Remove the ELC line at the shock.

16. Remove the shock absorber.

17. Lower the control arm to relieve tension on the spring. Remove the spring and the insulators.

18. Remove the two control arm mounting bolts and remove the control arm.

19. Reverse to install. Use the following torque values: control arm-to-frame bolts—98 ft. lbs; shock absorbers—65 ft. lbs.; brake caliper mounting bolts—30 ft. lbs.; brake lines—15 ft. lbs; wheel lug nuts—100 ft. lbs.

Transverse Rear Spring

REMOVAL & INSTALLATION

1986–87 Models

1. Raise the car and support it securely by the frame. Remove the tire and wheel from the side from which you wish to draw out the spring.

2. If you are working on the left side, and the car has Electronic Level Control, disconnect the ELC height sensor link.

3. If the car has a stabilizer bar, disconnect the mounting bolt at the strut.

4. Reinstall two wheel nuts opposite each other to hold the rotor onto the hub/bearing assembly.

5. Remove and suspend the brake caliper as described in the brake unit repair section.

6. Loosen but DO NOT REMOVE the knuckle pivot bolt on the outboard end of the control arm.

7. Remove the strut rod cap, mounting nut, retainer, and upper insulator. Then, compress the strut by hand and remove the lower insulator.

8. Remove the inner control arm nuts. Support the knuckle and control arm with a floorjack and then remove the inner control arm bolts. Remove the control arm, knuckle, strut, hub/bearing, and rotor as an assembly.

9. Using a jackstand capable of suspending the entire weight of the car, suspend the outer end of the spring securely.

——— CAUTION ———
Make sure the jackstand is square under the spring so that the stand will not shift or personal injury could result
————————————————

10. Gradually and cautiously lower the car until its weight compresses the spring so there is no weight on the

spring retainer. Then, remove the retainer mounting bolts, the retainer, and the lower insulator from that side of the car. Raise the car slowly until the jackstand is free of downward pressure from the spring and remove it.

11. Draw the spring out of the rear suspension. Remove the upper spring insulators as necessary.

12. Install any insulators that required replacement. Upper/outboard insulators must be installed so that the molded arrow points toward the vehicle centerline. Torque the center and outboard insulator nuts to 21 ft. lbs.

13. Position the spring into the crossmember. Make sure the outboard and center insulator locating bands are centered on the insulators.

14. Follow Step 9 and the caution above to support the outer end of the spring with a jackstand. Then, carefully and gradually lower the car until its weight will permit easy installation of the spring retainer.

15. Install the lower insulator and spring retainer and torque the bolts to 21 ft. lbs. Raise the vehicle carefully and when the spring is clear, remove the jackstand.

16. Position the assembled control arm, knuckle, strut, hub and bearing and rotor assembly into the crossmember assembly and install the inner control arm bolts and nuts JUST HAND TIGHT.

17. Install the lower strut insulator and position the strut rod into the suspension support assembly.

18. Install the upper strut insulator, retainer, and nut. Torque the upper strut nut to 65 ft. lbs., the knuckle pivot bolt to 59 ft. lbs. and the inner control arm bolts to 66 ft. lbs.

19. Install the strut rod cap. Install the stabilizer mounting bolt if the car has a stabilizer bar. Torque this bolt to 43 ft. lbs.

20. Remove the two wheel nuts retaining the brake rotor. Install the remaining parts in reverse of the removal procedure. Have the rear end alignment checked and adjusted, if necessary.

BRAKES

Brake adjustment, brake lining replacement, hydraulic cylinder overhaul and bleeding procedures for all models covered here can be found in the "Brakes" Unit Repair Section.

Parking Brake

ADJUSTMENT

All 1980 Models

1. Depress the parking brake pedal exactly 1 click.

2. Tighten the adjusting nut at the cable equalizer until the left rear wheel can just be turned rearward using 2 hands, but is locked in forward rotation.

3. With the parking brake off, the rear wheels should rotate freely in either direction with no drag.

All 1981–87 Models

1. Lube the cables at the underbody rub points and at the equalizer hooks. Check for free movement of all cables.

2. Set the parking brake pedal in the fully released position, raise, and support the rear of the car.

3. Hold the brake cable stud and tighten the equalizer nut until all cable slack is removed. Make sure the caliper levers are against the stops on the caliper housing; if they are not, loosen the cable until they are.

4. Operate the parking brake pedal several times to check the adjustment, it should travel approximately 4–5 $\frac{1}{2}$ in. on 1980–85 models. On 1986–87 models, note that the pedal should become firm after $3\frac{1}{2}$ strokes.

5. Lower the car and check that the caliper levers are still on their stops. If not, back off the parking brake adjuster until they are.

Master Cylinder

REMOVAL & INSTALLATION

All Models

1. Disconnect and plug hydraulic lines, and drain the cylinder.

2. Remove the attaching nuts and remove the master cylinder from the power booster unit.

3. Reverse to install. Bleed the system.

Power Booster

REMOVAL & INSTALLATION

All Models

1. From inside the car, detach the brake pushrod from the brake pedal.

2. Detach the vacuum hose at the vacuum cylinder.

3. Remove the nuts from the mounting studs which hold the unit to the dash panel. Remove the unit and clean it prior to installation.

4. Install in reverse order of removal. Bleed system.

Brake Caliper

REMOVAL & INSTALLATION

Please refer to the preceeding "Rear Control Arm and Coil Spring Removal and Installation" procedure.

Brake Disc

REMOVAL & INSTALLATION

All Models

NOTE: Brake disc removal and installation for all models is covered in the Brake Unit Repair Section.

STEERING

All steering system procedures are the same as those given in the Oldsmobile, Buick and Cadillac Rear Wheel Drive sections.

CHASSIS ELECTRICAL

Headlight Switch

REMOVAL & INSTALLATION

1980–87 Toronado

1. Remove the left hand trim cover.
   a. Remove the headlight switch knob and the radio knobs.
   b. Remove the steering column trim cover and the four screws beneath the cover.
   c. Remove the left hand sound absorber and carefully pull the trim cover rearward to remove.

2. Remove the two screws attaching the switch to the dash frame.

3. Pull the switch rearward to remove.

Riviera
1980–82

1. Disconnect the negative battery cable.

2. Pull the switch knob to the last notch and depress the spring loaded latch button on top of the switch, while pulling the knob and rod out of the switch.

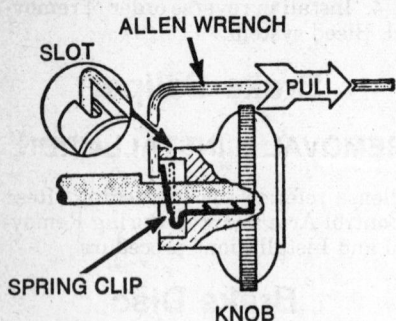

Radio knob removal, most models

3. Remove the escutcheon, trim plate and the retaining nut or screws. Remove the switch from the cluster.

4. Disconnect the multiple connector.

5. Installation is the reverse of removal.

1983–85

1. Disconnect the negative battery cable.

2. Remove the switch trim cover and the left hand trim cover.

3. Remove the headlight switch screws from the instrument panel.

4. Disconnect any electrical connector and pull the switch forward out of the dash panel.

1986–87

1. Disconnect the negative battery cable.

2. Remove the left and center trim covers.

3. Remove the two screws from the left switch panel assembly. Disconnect the electrical connector from the switch panel assembly. Remove the switch.

4. Installation is the reverse of removal.

1980–87 Eldorado and Seville

1. Disconnect negative battery cable. Remove the steering column lower cover.

2. Disconnect wiring harness retainer below headlight switch assembly.

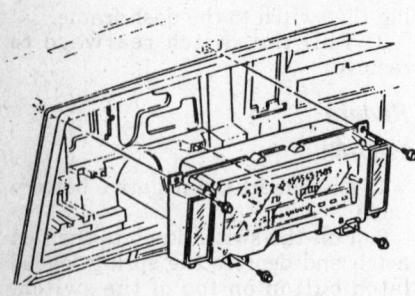

Riviera instrument cluster

3. Depress spring loaded release button on top of headlight switch and remove switch, knob and rod assembly (switch ON).

4. Remove screw with ground wire at bottom of switch housing and any other mounting screws.

5. Pull assembly down and rearward, disconnect wiring harness connectors, bulb(s) and remove assembly.

6. Installation is the reverse of removal.

Instrument Cluster

REMOVAL & INSTALLATION

Riviera

1980–85

1. Disconnect the negative battery cable.

2. Slide the steering column collar upward on the steering column.

3. Remove the headlight knob, escutcheon assembly and all remaining knobs.

4. Depending on the car model, pry either the right or left (or both) trim plates from the instrument panel.

NOTE: Remove the center trim plate by first removing the right and left trim plates. Then remove the radio knobs and screws securing the center trim plate.

5. Disconnect the seelite and remove.

6. Remove all cluster retaining screws. Pull the cluster out slightly and disconnect the speedometer cable. Remove the instrument cluster assembly.

7. To install, attach the cluster assembly and speedometer cable. Secure the center trim plate. Hold the left and right trim plates in position, press into place and reposition the rubber filler ring and headlight knob assembly.

1986–87 Digital Cluster

1. Disconnect the negative battery cable.

2. Remove the center, left, and right trim covers.

3. Remove the four mounting bolts from the instrument cluster. Then, pull the cluster straight out of the housing.

4. Installation is the reverse of removal.

1980–81 Seville and Eldorado

1. Disconnect the negative battery cable.

2. Remove the top and bottom applique retaining screws.

3. Unsnap the portion of the appli-

que trim below the steering column.

4. If equipped with tilt wheel, place in its lowest position and remove the applique panel.

5. Remove the cluster retaining screws, pull the cluster outward, disconnect the speedometer and circuit board connectors and remove the cluster.

6. The installation is the reverse of the cluster removal.

1982–85 Seville and Eldorado

1. Disconnect the negative battery cable.

2. Remove the left sound insulator.

3. Remove the instrument panel insert and applique trim from the instrument panel.

4. Place the shift lever in the Park position and remove the shift indicator clip from the steering column.

5. Remove the nuts securing the steering column to the upper mounting bracket and lower the steering column.

6. Remove the screw securing the upper steering column mounting bracket to the cowl and lower the bracket.

7. Remove the cluster retaining screws, disconnect the speedometer cable, printed circuit connector and remove the cluster.

8. The installation is the reverse of the removal procedure. Be sure the shift indicator is properly aligned.

1986–87 Seville and Eldorado

1. Remove the seven screws located along the top and remove the instrument panel trim plate.

2. Remove the four mounting screws and remove the filter lens.

3. Remove the two mounting screws and remove the warning light lens. Remove the trip odometer reset button.

4. Remove the two screws retaining the instrument panel cluster. Pull the cluster off the electrical connections, and remove it. Use pliers to hold the retaining tabs at either end of the cluster board, and remove the board.

5. To install the cluster, first align it with the electrical connectors and then push it into the instrument panel. The remainder of the installation procedure is the reverse of removal.

1980–85 Toronado

NOTE: To remove the left side sound absorber from under the dash take out two screws and one nut. Pull the absorber down the slide from the steering column.

1. Disconnect the ALCL computer connector, if equipped.

CRUISE CONTROL
1
2

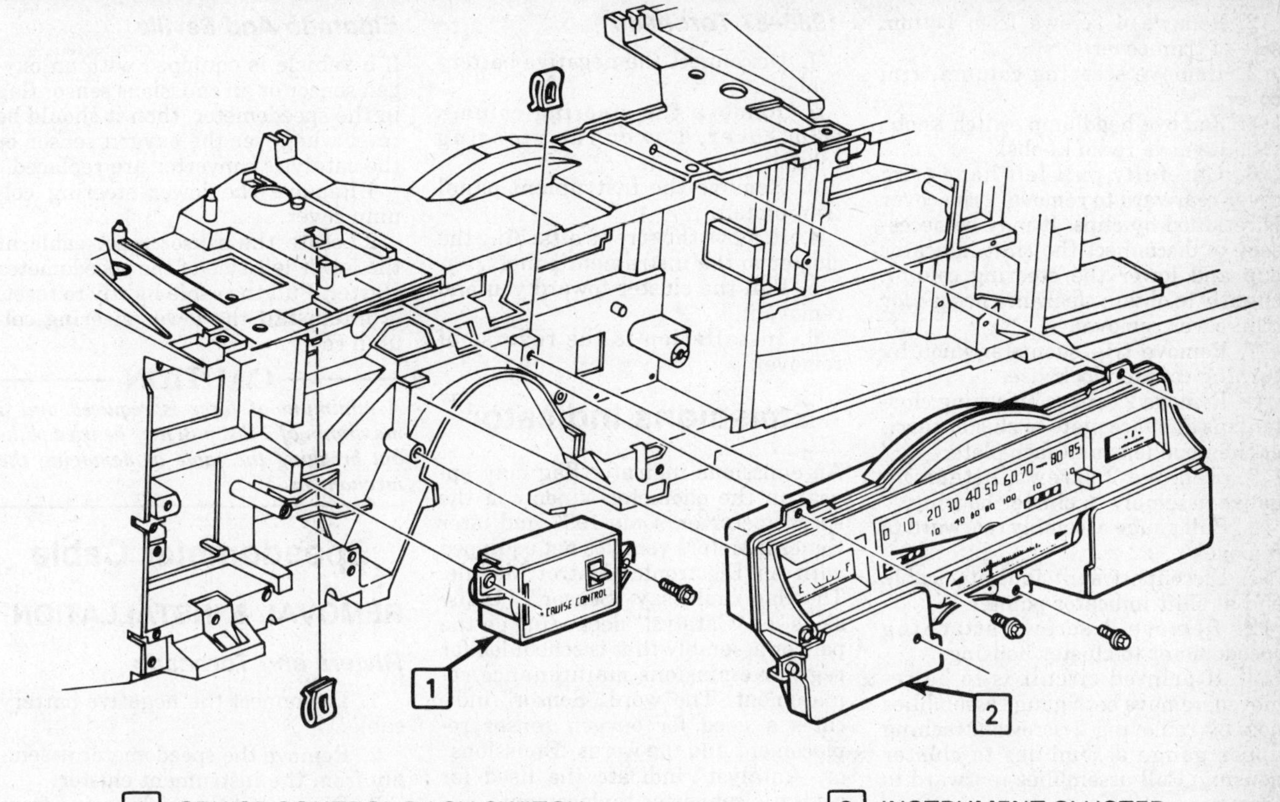

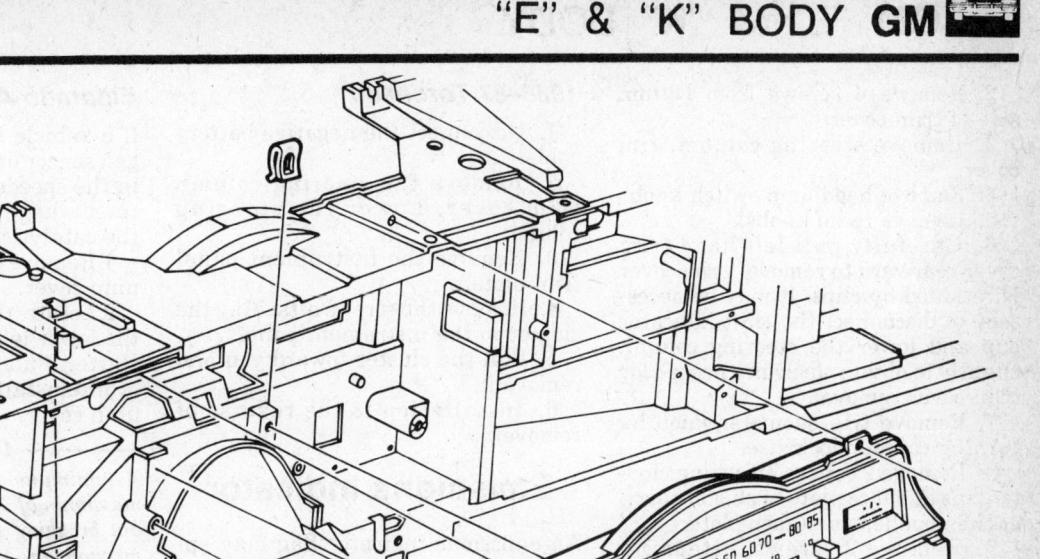

1 CRUISE CONTROL DASH SWITCH   2 INSTRUMENT CLUSTER

Typical instrument cluster (Eldorado and Seville)

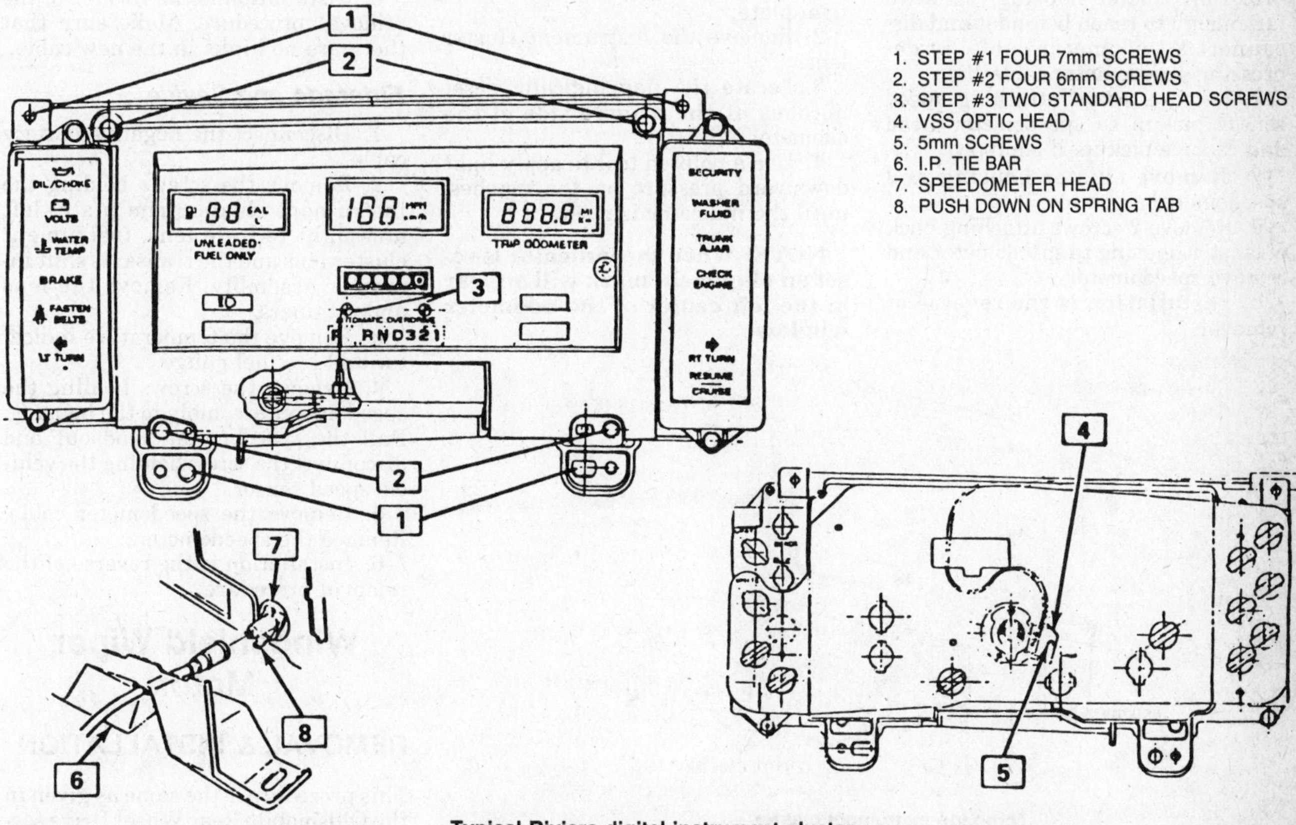

1. STEP #1 FOUR 7mm SCREWS
2. STEP #2 FOUR 6mm SCREWS
3. STEP #3 TWO STANDARD HEAD SCREWS
4. VSS OPTIC HEAD
5. 5mm SCREWS
6. I.P. TIE BAR
7. SPEEDOMETER HEAD
8. PUSH DOWN ON SPRING TAB
OIL/CHOKE
VOLTS
WATER TEMP
BRAKE
FASTEN BELTS
LT TURN
UNLEADED FUEL ONLY
TRIP ODOMETER
RND321
SECURITY
WASHER FLUID
TRUNK AJAR
CHECK ENGINE
RT TURN
RESUME
CRUISE

Typical Riviera digital instrument cluster

2. Remove 4 screws from bottom side of trim cover.

3. Remove steering column trim cover.

4. Remove headlamp switch knob.

5. Remove radio knobs.

6. Carefully pull left hand trim cover rearward to remove. Trim cover is retained by clips. It may be necessary to disconnect the shift indicator clip and lower the steering column slightly to obtain clearance needed for trim cover removal.

7. Remove trip odometer knob by turning counterclockwise.

8. Remove 2 screws attaching cluster lens and face plate to cluster carrier. Remove lens and face plate.

9. Remove 2 screws, attaching gauge assembly to cluster housing.

10. Pull gauge assembly rearward to remove.

11. Disconnect shift indicator cable end at shift indicator pointer.

12. Remove 3 screws attaching speedometer to cluster housing.

13. If printed circuit is to be removed, remove both gauge assemblies now by removing 2 screws attaching upper gauge assemblies to cluster housing. Pull assemblies rearward to remove.

14. Remove 2 screws attaching lower cluster housing to cluster carrier.

15. Disconnect speedometer cable at transmission or at transducer on cars equipped with cruise control.

16. Pull cluster housing rearward far enough to reach behind it and disconnect speedometer cable by depressing speedometer cable clip.

17. Remove screw attaching speed sensor pickup to speedometer head and remove pickup, if equipped.

18. Remove cluster housing and speedometer assembly.

19. Remove 2 screws attaching back of cluster housing to speedometer and remove speedometer.

20. Installation is the reverse of removal.

1986–87 Toronado

1. Disconnect the negative battery cable.

2. Remove the steering column trim cover. Lower the steering column.

3. Remove the instrument panel trim plate.

4. Remove the screws attaching the cluster to the instrument panel.

5. Pull the cluster toward you and remove it.

6. Installation is the reverse of removal.

Emissions Indicator

An emissions indicator flag may appear in the odometer window of the speedometer on some 1980 and later General Motors vehicles not equipped with an Electronic Control Module. The flag could say "Sensor", "Emissions" or "Catalyst" depending on the part or assembly that is scheduled for regular emissions maintenance replacement. The word "Sensor" indicates a need for oxygen sensor replacement and the words "Emissions" or "Catalyst" indicate the need for catalytic converter replacement.

RESET PROCEDURE

All Except Eldorado and Seville

1. Remove the instrument panel trim plate.

2. Remove the instrument cluster lens.

3. Locate the flag indicator reset notches at the drivers side of the odometer.

4. Use a pointed tool to apply light downward pressure on the notches, until the indicator is reset.

NOTE: When the indicator is reset an alignment mark will appear in the left center of the odometer window.

Eldorado And Seville

If a vehicle is equipped with an oxygen sensor or an emissions sensor flag in the speedometer, then it should be reset whenever the oxygen sensor or the catalytic converter are replaced.

1.Remove the lower steering column cover.

2.Locate the sensor reset cable at the lower left side of the speedometer cluster. Pull the cable lightly to reset.

3.Reinstall the lower steering column cover.

— CAUTION —

A minimum of force is required and a maximum of 2 lbs. pull may be used without breaking the cable or damaging the mechanism.

Speedometer Cable

REMOVAL & INSTALLATION

Riviera and Toronado

1. Disconnect the negative battery cable.

2. Remove the speedometer assembly from the instrument cluster.

3. Disconnect the cable casing from the speedometer head.

4. Pull the inner cable from the upper casing.

5. Disconnect the cable from the transaxle. Remove the cable.

6. Installation is the reverse of the removal procedure. Make sure that there are no kinks in the new cable.

Eldorado and Seville

1. Disconnect the negative battery cable.

2. Remove the screws holding the instrument cluster trim plate, left and right telltale lens, instrument cluster lens and the transaxle shift indicator assembly. Remove the lens and retainers.

3. Remove the temperature indicator and the fuel gauge.

4. Remove the screws holding the speedometer assembly to the housing. Pull the speedometer head out and disconnect the screw holding the vehicle speed sensor.

5. Remove the speedometer cable. Remove the speedometer.

6. Installation is the reverse of the removal procedure.

Windshield Wiper Motor

REMOVAL & INSTALLATION

This procedure is the same as given in the Oldsmobile Rear Wheel Drive sec-

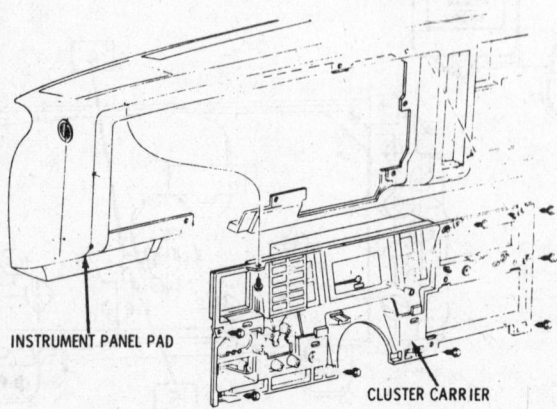

INSTRUMENT PANEL PAD

CLUSTER CARRIER

Toronado instrument cluster

tion. Replacement procedures for the rubber wiper element for all models are given in the Maintenance Section.

Radio

REMOVAL & INSTALLATION

Toronado

This procedure is the same as given in the "Oldsmobile Rear Wheel Drive" section.

1980–85 Eldorado and Seville

1. Disconnect the negative battery cable.
2. Remove the screws from the top of the instrument panel center insert.
3. Remove the radio knobs and remove the insert.
4. Remove the rear window defogger switch to gain access to the left side mounting screw if so equipped.
5. Remove the mounting screws.
6. Remove the radio and disconnect the wiring. Reverse to install.

1986–87 Eldorado and Seville

1. Disconnect the negative battery cable. Remove the radio trim plate.
2. Remove the seven screws attaching the instrument panel trim plate to the instrument panel and remove it.
3. Remove the mounting nuts under the radio. Disconnect the electrical connectors and aerial. Remove the radio.
4. Installation is the reverse of removal.

1980–85 Riviera

1. Disconnect the battery ground cable.

NOTE: On 1981 and later Rivieras remove the center trim plate by grasping it firmly and pulling out. Be careful not to lose the spring retaining clips.

2. Remove the ashtray and bracket.
3. Pull off the radio knobs and trim washers.
4. Remove the lower left air duct.
5. Remove the two retaining nuts from the control shafts.
6. Unplug the power lead, speaker wire, and antenna lead.
7. Remove the rear radio mounting nut.

1986–87 Riviera

1. Remove the left side trim cover from the instrument panel.
2. Remove the right hand switch trim plate.

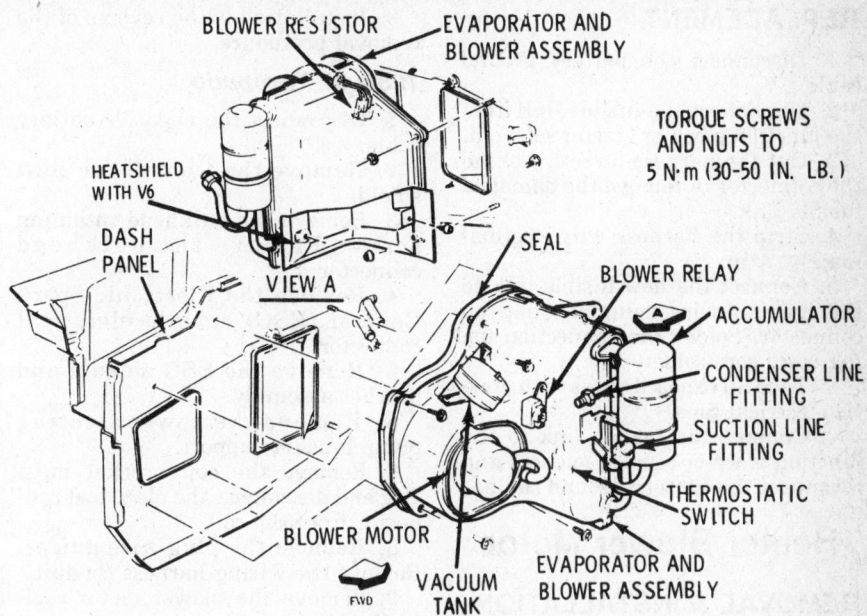

TORQUE SCREWS AND NUTS TO 5 N·m (30-50 IN. LB.)

Blower assembly

3. Remove the 4 screws that hold the radio in place. Then, disconnect the three electrical connectors, antenna lead, and the clock connector (if the car has a clock). Remove the radio.
4. Installation is the reverse of removal.

Fuse Panel

The fuse panel is located on the left side of the vehicle on 1980–85 models. It is under the instrument panel assembly. In order to gain access to the fuse panel, it may be necessary to first remove the under dash padding. On 1986–87 models, it is in the underside of the instrument panel, usually near the steering column.

Electronic Control Module (ECM)

The electronic control module is located on the right side of the vehicle. It is positioned in front of the right hand kick panel, or up behind the glove box. In order to gain access to the assembly you must first remove the trim panel or glove box assembly.

Turn Signal Flasher

The turn signal flasher is located directly under the steering column of the vehicle. It is secured in place with a plastic retainer. In order to gain access to the component, it may first be necessary to remove the under dash padding.

Convenience Center

The convenience center is located on the right side of the vehicle. It is positioned under the dash panel. In order to gain access to the convenience center it may be necessary to remove instrument panel sound absorber.

Circuit Breaker

A circuit breaker is an electrical switch which breaks the circuit during an electrical overload. The circuit breaker will remain open until the short or overload condition in the circuit is corrected.

Fusible Links

Fusible links are sections of wire, with special insulation, designed to melt under electrical overload. Replacements are simply spliced into the wire. There may be as many as five of these in the engine compartment wiring harnesses. These are:
1. Horn relay to fuse panel circuit—one link.
2. Charging circuit, from the starter solenoid to the horn relay—two links.
3. Starter solenoid to ammeter circuit—one link.
4. Horn relay to rear window defroster circuit—one link.
The fusible links are all two wire gauge sizes smaller than the wires they protect.

NOTE: Most models have fusible links at these locations.

REPLACEMENT

1. Disconnect the battery ground cable.

2. Disconnect the fusible link from the junction block or starter solenoid.

3. Cut the harness directly behind the connector to remove the damaged fusible link.

4. Strip the harness wire approximately ½ in.

5. Connect the new fusible link to the harness wire using a crimp on connector. Solder the connection using resin core solder.

6. Tape all exposed wires with plastic electrical tape.

7. Connect the fusible link to the junction block or starter solenoid and reconnect the battery ground cable.

Heater Blower Motor

REMOVAL & INSTALLATION

Riviera

1. Disconnect the blower motor wires.

2. On A/C equipped cars, disconnect the cooling tube from the case.

3. Remove the motor attaching screws and lift the motor from the case.

4. Installation is the reverse of removal. Replace any damaged sealer.

1980–85 Eldorado And Seville

1. Disconnect the negative battery cable.

2. Disconnect the electrical connections at the blower motor.

3. Disconnect the cooling hose from the blower motor.

4. Remove the mounting screws and remove the motor.

5. Reverse to install. Use a silicone sealer on the blower motor sealing surfaces.

1986–87 Seville and Eldorado

1. Disconnect the negative battery cable.

2. Remove the air cleaner assembly.

3. Remove the cross tower brace.

4. Disconnect the wiring harness support bracket.

5. Tag and disconnect the electrical connector. Remove the cooling hose and mounting screws.

6. Tilt the blower motor in the case and remove the fan from the blower motor.

NOTE: Be careful not to bend the fan upon removal as a fan imbalance will result after reassembly.

7. Remove the blower motor and fan assembly from the vehicle.

8. Installation is the reverse of the removal procedure.

1986–87 Toronado

1. Disconnect the negative battery cable.

2. Remove the front of the cowl shield.

3. Remove the bulkhead retaining screw. Remove the bulkhead connector.

4. Remove the Electronic Spark Control (ESC) module electrical connector.

5. Remove the ESC module and bracket assembly.

6. Remove the power steering pump bracket support.

7. Remove the coil bracket nuts. Tag and disconnect the electrical connector from the coil.

8. Remove the plug wire guides. Remove the wiring harness conduit.

9. Remove the blower motor cooling tube.

10. Tag and disconnect the electrical connectors from the blower motor. Remove the blower motor mounting screws.

11. Remove the blower motor mounting screws. Remove the blower motor.

12. Installation is the reverse of the removal procedure.

Heater Core

REMOVAL & INSTALLATION
Toronado and Riviera

NOTE: This procedure involves removing the dashboard.

1. Disconnect the negative battery cable.

2. Drain the radiator. Remove the heater hoses from the heater core.

3. Remove the instrument panel sound absorbers which cover the underside of the dash area.

4. Loosen and lower the steering column and remove the left hand trim cover. See Step 1 of "Headlight Switch Replacement," above.

5. Remove the instrument cluster:

   a. Remove headlight switch (see above for procedures).

   b. Remove the windshield switch, the radio and heater/AC control.

   c. Remove all cluster electrical connections and disconnect the speedometer cable.

   d. Remove the nine attaching screws and remove the cluster.

6. Remove the front speakers, the three screws attaching the manifold to the heater case, the four upper and three lower instrument panel retaining screws, and disconnect the brake release cable.

7. Disconnect the instrument panel wiring harness from the dash wiring assembly and disconnect the right hand remote control mirror cable from the instrument panel.

8. Disconnect the speedometer cable from its clip and the heater control cable at the heater case.

9. Disconnect all vacuum lines and wiring necessary to remove the instrument panel. If car is equipped with pulse wipers remove the wiper switch, unlock the connector from the cluster carrier and separate the pulse jumper harness from the connector.

10. Remove the instrument panel and harness assembly.

11. Remove defroster ducts, disconnect vacuum hoses and temperature cable; remove blower resistor and the three heater assembly retaining nuts.

12. Remove the heater assembly-to-dash screw and clip from inside the car.

13. Remove the heater assembly.

14. Remove the heater core. Reverse procedures to install.

1980–85 Eldorado and Seville

1. Drain the radiator.

2. Remove the heater hoses from the core and plug the hoses and the nipples to prevent spillage.

3. Remove the instrument panel.

4. Remove the four defroster nozzle attaching screws at the cowl and the screw on the case and remove the nozzle.

5. Disconnect the vacuum hoses.

6. Disconnect the electrical connector at the programmer.

7. Under the hood, remove the heater case-to-cowl attaching screws.

8. Under the instrument panel, remove the heater case-to-cowl attaching screw.

9. Remove the heater case.

10. Remove the four case-to-core screws and remove the core.

11. Reverse to install.

1986–87 Eldorado and Seville

1. Disconnect the negative battery cable.

2. Drain the engine coolant.

3. Remove the screws retaining the glove box. Tag and disconnect the electrical connectors from the glove box.

4. Remove the glove box assembly from the vehicle.

5. Remove the lower sound insulator to gain working clearance.

6. Remove the screws retaining the Electronic Control Module (ECM), and remove the ECM.

7. Remove the heater core cover. Remove the hoses to the heater core.

8. Remove the heater core retaining screws. Remove the heater core.

9. Installation is the reverse of the removal procedure.

GM "F" Body
Camaro, Firebird

YEAR IDENTIFICATION

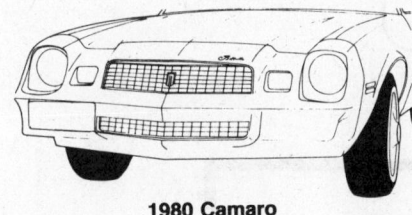

1980 Camaro

1981 Camaro

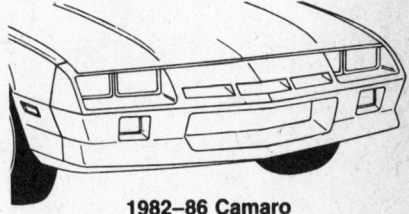

1982–86 Camaro

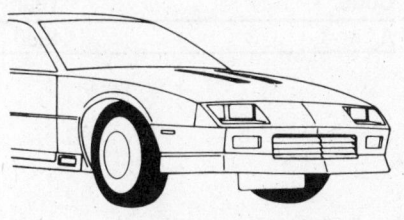

1982-87 Camaro Z-28

1985-87 Camaro IROC

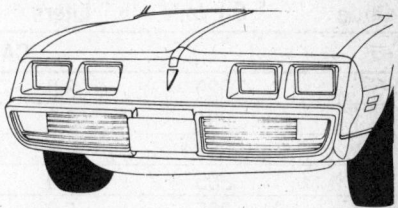

1980 Firebird

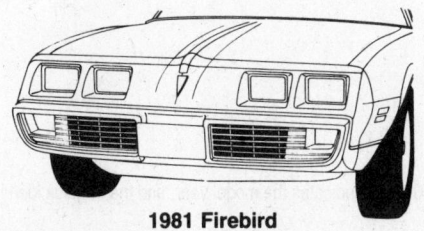

1981 Firebird

1982–86 Firebird

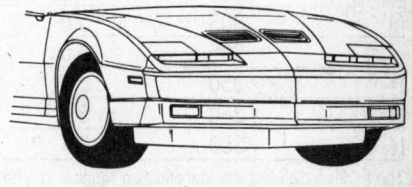

1985-87 Trans Am

YEAR IDENTIFICATION

1987 Firebird Formula

VEHICLE IDENTIFICATION NUMBER (VIN)

It is important for servicing and ordering parts to be certain of the vehicle and engine identification. The VIN (vehicle identification number) is a 13 or 17 digit number visible through the windshield on the driver's side of the dash and contains the vehicle and engine identification codes. It can be interpreted as follows:

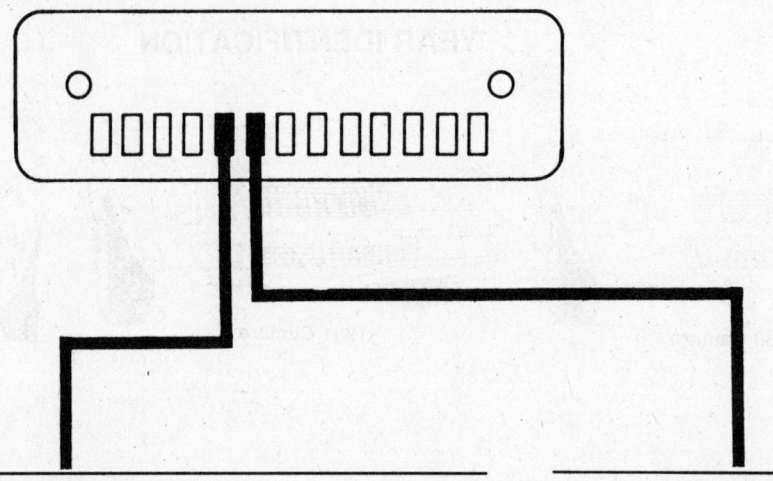

Engine Code

Code	Cu. In.	Liters	Cyl.	Carb.	Eng. Mfg.
CAMARO					
K	229	3.8	6	2	Chev.
A	231	3.8	6	2	Buick
J	267	4.4	8	2	Chev.
U	305	5.0	8	2	Chev.
G	305	5.0	8	2	Chev.
H	305	5.0	8	4	Chev.
L	350	5.7	8	4	Chev.
FIREBIRD					
A	231	3.8	V6	2	Buick
S	265	4.3	8	2	Pont.
W	301	4.9	8	4	Pont.
U	305	5.0	8	2	Chev.
G	305	5.0	8	2	Chev.
H	305	5.0	8	4	Chev.
P	350	5.7	8	4	Pont.
R	350	5.7	8	4	Olds.
L	350	5.7	8	4	Chev.

Model Year Code

Code	Year
A	1980

The thirteen digit Vehicle Identification Number can be used to determine engine application and model year. The 6th digit indicates the model year, and the 5th digit identifies the factory installed engine.

VEHICLE IDENTIFICATION NUMBER (VIN)

It is important for servicing and ordering parts to be certain of the vehicle and engine identification. The VIN (vehicle identification number) is a 13 or 17 digit number visible through the windshield on the driver's side of the dash and contains the vehicle and engine identification codes. It can be interpreted as follows:

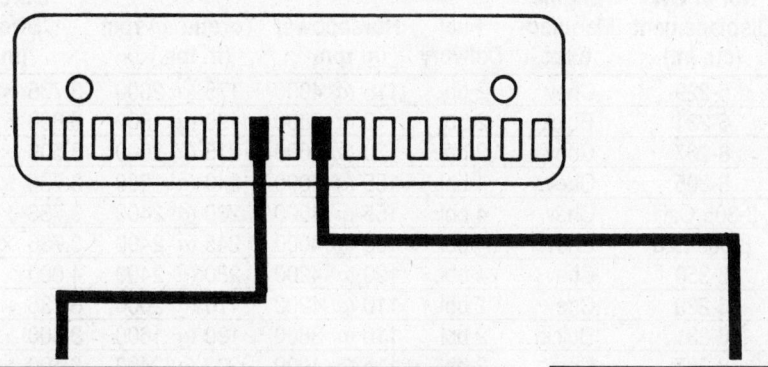

Engine Code					
Code	Cu. In.	Liters	Cyl.	Carb.	Eng. Mfg.
CAMARO					
2	151	2.5	4	TBI	Pont.
F③	151	2.5	4	2	Pont.
1	173	2.8	V6	2	Chev.
L	173	2.8	V6	2	Chev.
S②	173	2.8	6	MFI	Chev.
K	229	3.8	6	2	Chev.
A	231	3.8	V6	2	Buick
J	267	4.4	8	2	Chev.
H	305	5.0	8	4	Chev.
7③	305	5.0	8	TBI	Chev.
G	305	5.0	8	4	Chev.
S③	305	5.0	8	TBI	Chev.
F②	305	5.0	8	TPI	Chev.
L④	350	5.7	8	4	Chev.
8	350	5.7	8	TPI	Chev.
FIREBIRD					
2	151	2.5	4	TBI	Pont.
F③	151	2.5	4	2	Pont.
1	173	2.8	V6	2	Chev.
L	173	2.8	V6	2	Chev.
S②	173	2.8	6	MFI	Chev.
A	231	3.8	V6	2	Buick
S④	265	4.3	8	2	Pont.
W	301	4.9	8	4	Pont.
T	301①	4.9	8	4	Pont.
H	305	5.0	8	4	Chev.
7③	305	5.0	8	TBI	Chev.
G	305	5.0	8	4	Chev.
F②	305	5.0	8	TPI	Chev.

Model Year Code	
Code	Year
B	1981
C	1982
D	1983
E	1984
F	1985
G	1986
H	1987

The seventeen digit Vehicle Identification Number can be used to determine engine application and model year. The 10th digit indicates the model year, and the 8th digit identifies the factory installed engine.

① Turbocharged engine
② 1985 and later
③ 1982–83
④ 1981 only
TBI—Throttle body (fuel) injection
TPI—Tuned Port Injection
MFI—Multi-Port Fuel Injection

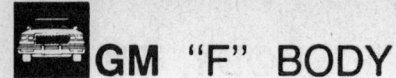

GENERAL ENGINE SPECIFICATIONS
Camaro

Year	Engine VIN Code	Engine No. of Cyl. Displacement (cu. in.)	Engine Manufacturer	Fuel Delivery	Horsepower @ rpm①	Torque @ rpm (ft. lbs.)①	Bore × Stroke (in.)	Compression Ratio	Oil Pressure @ 2000 rpm
'80	K	6-229	Chev.	2 bbl.	115 @ 4000	175 @ 2000	3.736 × 3.480	8.6:1	45
	A	6-231	Buick	2 bbl.	110 @ 3800	190 @ 1600	3.800 × 3.400	8.0:1	45
	J	8-267	Chev.	2 bbl.	120 @ 3600	215 @ 2000	3.500 × 3.480	8.3:1	45
	H	8-305	Chev.	4 bbl.	155 @ 4000	240 @ 1600	3.736 × 3.480	8.6:1	45
	H	8-305 Calif.	Chev.	4 bbl.	155 @ 4000	230 @ 2400	3.736 × 3.480	8.6:1	45
	H	8-305/Z28	Chev.	4 bbl.	165 @ 4000	245 @ 2400	3.736 × 3.480	8.6:1	45
	L	8-350	Chev.	4 bbl.	190 @ 4200	280 @ 2400	4.000 × 3.480	8.2:1	45
'81	K	6-229	Chev.	2 bbl.	110 @ 4200	170 @ 2000	3.736 × 3.480	8.6:1	45
	A	6-231	Buick	2 bbl.	110 @ 3800	190 @ 1600	3.800 × 3.400	8.0:1	45
	J	8-267	Chev.	2 bbl.	115 @ 4000	200 @ 2400	3.500 × 3.480	8.3:1	45
	H	8-305	Chev.	4 bbl.	150 @ 3800	240 @ 2400	3.736 × 3.480	8.6:1	45
	H	8-305/Z28	Chev.	4 bbl.	165 @ 4000	245 @ 2400	3.736 × 3.480	8.6:1	45
	L	8-350	Chev.	4 bbl.	175 @ 4000	275 @ 2400	4.000 × 3.480	8.2:1	45
'82	2	4-151	Pont.	TBI	90 @ 4000	134 @ 2400	4.000 × 3.000	8.2:1	36–41
	1	6-173	Chev.	2 bbl.	102 @ 4800	145 @ 2400	3.503 × 2.992	8.5:1	40
	H	8-305	Chev.	4 bbl.	145 @ 4000	240 @ 2400	3.736 × 3.480	8.6:1	40
	7	8-305	Chev.	TBI	165 @ 4200	240 @ 2400	3.736 × 3.480	9.5:1	40
'83	2	4-151	Pont.	TBI	92 @ 4000	134 @ 2400	4.000 × 3.000	8.2:1	36–41
	1	6-173	Chev.	2 bbl.	102 @ 4800	145 @ 2400	3.503 × 2.992	8.5:1	40
	H	8-305	Chev.	4 bbl.	145 @ 4000	240 @ 2400	3.736 × 3.480	8.6:1	40
	S	8-305	Chev.	TBI	175 @ 4200	250 @ 2800	3.736 × 3.480	9.5:1	50–65
'84	2	4-151	Pont.	TBI	90 @ 4000	134 @ 2400	4.000 × 3.000	8.2:1	36–41
	1	6-173	Chev.	2 bbl.	102 @ 4800	145 @ 2400	3.503 × 2.992	8.5:1	40
	H	8-305	Chev.	4 bbl.	145 @ 4000	240 @ 2400	3.736 × 3.480	8.6:1	40
	G	8-305	Chev.	4 bbl.	190 @ 4800	240 @ 3200	3.736 × 3.480	9.5:1	40
'85–'87	2	4-151	Pont.	TBI	90 @ 4000	134 @ 2400	4.000 × 3.000	9.0:1	36–41
	F	8-305	Chev.	TPI	265 @ 4400	275 @ 3200	3.736 × 3.480	9.5:1	40
	S	6-173	Chev.	MFI	135 @ 5100	165 @ 3600	3.503 × 2.992	8.9:1	40
	H	8-305	Chev.	4 bbl.	155 @ 4200	245 @ 2000	3.736 × 3.480	9.5:1	40
	G	8-305	Chev.	4 bbl.	190 @ 4800	240 @ 3200	3.736 × 3.480	9.5:1	40
	8	8-350	Chev.	TPI	230 @ 4000	330 @ 3200	4.000 × 3.480	9.5:1	50–65

① Horsepower and torque are SAE net figures. They are measured at the rear of the transmission with all accessories installed and operating. Since the figures vary when a given engine is installed in different models, some are representative, rather than exact.

GENERAL ENGINE SPECIFICATIONS
Firebird

Year	Engine VIN Code	Engine No. of Cyl. Displacement (cu. in.)	Engine Manufacturer	Fuel Delivery	Horsepower @ rpm①	Torque @ rpm (ft. lbs.)①	Bore × Stroke (in.)	Compression Ratio	Oil Pressure @ 2050 rpm
'80	A	6-231	Buick	2 bbl.	115 @ 3800	188 @ 2000	3.800 × 3.400	8.0:1	37
	S	8-265	Pont.	2 bbl.	120 @ 3600	210 @ 1600	3.750 × 3.000	8.3:1	37③
	W	8-301	Pont.	4 bbl.	150 @ 4000	240 @ 2000	4.000 × 3.000	8.1:1	38③
	H	8-305	Chev.	4 bbl.	150 @ 3800	230 @ 2400	3.736 × 3.480	8.4:1	40

GENERAL ENGINE SPECIFICATIONS
Firebird

Year	Engine VIN Code	Engine No. of Cyl. Displacement (cu. in.)	Engine Manufacturer	Fuel Delivery	Horsepower @ rpm①	Torque @ rpm (ft. lbs.)①	Bore × Stroke (in.)	Compression Ratio	Oil Pressure ⑴ 2050 rpm
'81	A	6-231	Buick	2 bbl.	115 @ 3800	190 @ 1600	3.800 × 3.400	8.0:1	37
	S	8-265	Pont.	2 bbl.	119 @ 4000	205 @ 2000	3.750 × 3.000	8.3:1	38③
	W	8-301	Pont.	4 bbl.	155 @ 4000	245 @ 2000	4.000 × 3.000	8.1:1	38③
	T	8-301②	Pont.	4 bbl.	210 @ 4000	340 @ 2000	4.000 × 3.000	7.5:1	58③
	H	8-305	Chev.	4 bbl.	155 @ 3800	240 @ 2400	3.736 × 3.480	8.6:1	40
'82	2	4-151	Pont.	TBI	90 @ 4000	134 @ 2400	4.000 × 3.000	8.2:1	36–41
	1	6-173	Chev.	2 bbl.	102 @ 4800	145 @ 2400	3.503 × 2.992	8.5:1	40
	H	8-305	Chev.	4 bbl.	145 @ 4000	240 @ 2400	3.736 × 3.480	8.6:1	40
	7	8-305	Chev.	TBI	175 @ 4200	240 @ 2400	3.736 × 3.480	9.5:1	40
'83	2	4-151	Pont.	TBI	90 @ 4000	134 @ 2400	4.000 × 3.000	8.2:1	36–41
	1	6-173	Chev.	2 bbl.	102 @ 4800	145 @ 2400	3.503 × 2.992	8.5:1	40
	L	6-173	Chev.	2 bbl.	125 @ 5400	145 @ 2400	3.503 × 2.992	8.9:1	50–65
	H	8-305	Chev.	4 bbl.	145 @ 4000	240 @ 2400	3.736 × 3.480	8.6:1	40
	S	8-305	Chev.	TBI	175 @ 4200	250 @ 2800	3.736 × 3.480	9.5:1	50–65
'84	2	4-151	Pont.	TBI	90 @ 4000	134 @ 2400	4.000 × 3.000	8.2:1	36–41
	1	6-173	Chev.	2 bbl.	102 @ 4800	145 @ 2400	3.503 × 2.992	8.5:1	40
	L	6-173HO	Chev.	2 bbl.	125 @ 5400	145 @ 2400	3.503 × 2.992	8.9:1	50–65
	H	8-305	Chev.	4 bbl.	145 @ 4000	240 @ 2400	3.736 × 3.480	8.6:1	40
	G	8-305	Chev.	4 bbl.	190 @ 4800	240 @ 3200	3.736 × 3.480	9.5:1	40
'85–'87	2	4-151	Pont.	TBI	90 @ 4000	134 @ 2400	4.000 × 3.000	8.2:1	36–41
	S	6-173	Chev.	MFI	135 @ 5100	165 @ 3600	3.503 × 2.992	8.9:1	40
	H	8-305	Chev.	4 bbl.	155 @ 4200	245 @ 2000	3.736 × 3.480	8.6:1	40
	G	8-305	Chev.	4 bbl.	180 @ 4800	240 @ 3200	3.736 × 3.480	9.5:1	40
	F	8-305	Chev.	TPI	215 @ 4400	275 @ 3200	3.736 × 3.480	9.5:1	50–65

TBI—Throttle Body Injection
MFI—Multi-Port Fuel Injection
TPI—Tuned Port Injection
① Horsepower and torque are SAE net figures.
They are measured at the rear of the
transmission with all accessories installed and
operating. Since the figures vary when a given
engine is installed in different models, some are
representative, rather than exact.
② Turbo charged engine
③ Oil pressure above 2600 rpm

TUNE-UP SPECIFICATIONS
Camaro

Year	Engine VIN Code	Engine No. of Cyl. Displacement (cu. in.)	Engine Manufacturer	Spark Plugs Type	Spark Plugs Gap (in.)	Ignition Timing (deg)①② Man. Trans.	Ignition Timing (deg)①② Auto. Trans.	Intake Valve Opens (deg)③	Fuel Pump Pressure (psi)	Idle Speed (rpm)①② Man. Trans.	Idle Speed (rpm)①② Auto. Trans.
'80	K	6-229	Chev.	R-45TS⑤	0.045	8B	12B	42	4½–6	700	600
	A	6-231	Buick	R-45TSX	0.060	—	15B	16	4¼–5¾	—	600
	J	8-267	Chev.	R-45TS	0.045	—	4B	28	7½–9	—	500
	H	8-305	Chev.	R-45TS	0.045	4B	4B	28	7½–9	700	500(550)
	L	8-350	Chev.	R-45TS	0.045	6B	6B	28	7½–9	700	500
'81	K	6-229	Chev.	R-45TS	0.045	6B	6B	42	4½–6	700⑥	600⑥
	A	6-231	Buick	R-45TS8	0.080	—	15B	16	4¼–5¾	—	500⑥
	J	8-267	Chev.	R-45TS	0.045	—	6B	44	7½–9	—	500⑥
	H	8-305	Chev.	R-45TS	0.045	6B	6B	44	7½–9	700	500
	L	8-350	Chev.	R-45TS	0.045	—	6B	38	7½–9	—	500⑥

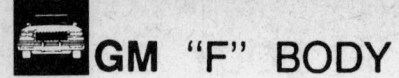

TUNE-UP SPECIFICATIONS
Camaro

Year	Engine VIN Code	Engine No. of Cyl. Displacement (cu. in.)	Engine Manufacturer	Spark Plugs Type	Gap (in.)	Ignition Timing (deg)① ② Man. Trans.	Auto. Trans.	Intake Valve Opens (deg)③	Fuel Pump Pressure (psi)	Idle Speed (rpm)① ② Man. Trans.	Auto. Trans.
'82	2	4-151	Pont.	R-44TSX	0.060	⑦	⑦	—	9–13	⑦	⑦
	1	6-173	Chev.	R-43TS	0.045	⑦	⑦	—	5½–6½	⑦	⑦
	H	8-305	Chev.	R-45TS	0.045	⑦	⑦	—	5½–6½	⑦	⑦
	7	8-305	Chev.	R-45TS④	0.045	⑦	⑦	—	9–13	⑦	⑦
'83	2	4-151	Pont.	R-44TSX	0.060	⑦	⑦	—	9–13	⑦	⑦
	1	6-173	Chev.	R-43CTS	0.045	⑦	⑦	—	5½–6½	⑦	⑦
	H	8-305	Chev.	R-45TS	0.045	⑦	⑦	—	5½–6½	⑦	⑦
	S	8-305	Chev.	R-45TS	0.045	⑦	⑦	—	9–13	⑦	⑦
'84	2	4-151	Pont.	R-44TSX	0.060	⑦	⑦	—	9–13	⑦	⑦
	1	6-173	Chev.	R-43CTS	0.045	⑦	⑦	—	5½–6½	⑦	⑦
	H	8-305	Chev.	R-45TS	0.045	⑦	⑦	—	5½–6½	⑦	⑦
	G	8-305	Chev.	R-45TS	0.045	⑦	⑦	—	9–13	⑦	⑦
'85	2	4-151	Pont.	R-43TSX	0.060	⑦	⑦	—	9–13	⑦	⑦
	S	6-173	Chev.	R-42CTS	0.045	⑦	⑦	—	40.5–47	⑦	⑦
	F	8-305	Chev.	R-43CTS	0.045	⑦	⑦	—	40.5–47	⑦	⑦
	H	8-305	Chev.	R-45TS	0.045	⑦	⑦	—	5½–6½	⑦	⑦
	G	8-305	Chev.	R-44TS	0.045	⑦	⑦	—	9–13	⑦	⑦
'86	2	4-151	Pont.	R-43CTS6	0.060	⑦	⑦	—	9–13	⑦	⑦
	S	6-173	Chev.	R-42CTS	0.045	⑦	⑦	—	40½–47	⑦	⑦
	F	8-305	Chev.	R-43TS	0.035	⑦	⑦	—	40½–47	⑦	⑦
	H	8-305	Chev.	R-45TS	0.045	⑦	⑦	—	5½–6½	⑦	⑦
	G	8-305	Chev.	R-43TS	0.035	⑦	⑦	—	9–13	⑦	⑦
'87	ALL			See Underhood Specifications Sticker							

NOTE: The underhood specifications sticker often reflects tune-up specification changes made during the production run. Sticker figures must always be used if they disagree with those in this chart. Part numbers in this chart are not recommendations by Chilton for any product by brand name.
All models use electronic ignition systems.
B Before Top Dead Center
TDC Top Dead Center
—Not applicable
① See text for procedure
② Figure in parenthesis indicates California engine
③ All figures Before Top Dead Center (BTDC)
④ R-44TS if a colder plug is needed
⑤ With automatic trans.—R-45TS

⑥ Equipped with Idle Speed Control (I.S.C.)
⑦ These functions are controlled by the emissions computer. In rare instances when adjustment is necessary, refer to the underhood emissions sticker for specifications.

TUNE-UP SPECIFICATIONS
Firebird

Year	Engine VIN Code	Engine No. of Cyl. Displacement (cu. in.)	Engine Manufacturer	Spark Plugs Type	Gap (in.)	Ignition Timing (deg)③ ④ Man. Trans.	Auto. Trans.	Intake Valve Opens (deg)⑤	Fuel Pump Pressure (psi)	Idle Speed (rpm)③ ④ Man. Trans.	Auto. Trans.
'80	A	6-231	Buick	R-45TSX⑥	0.060⑥	15B	15B	16	3–4.5	800/600①	620/550①
	S	8-265	Pont.	R-45TSX	0.060	—	10B	27	7.5–9	—	650/550①
	W	8-301	Pont.	R-45TSX	0.060	—	12B	16	7.5–9	—	650/500①
	W	8-301⑦	Pont.	R-45TSX	0.060	—	12B	17	7.5–9	700	550
	H	8-305	Chev.	R-45TS	0.045	—	4B	28	7.5–9	—	650/550①
'81	A	6-231	Buick	R-45TS8	0.080	15B	15B	16	4.25–5.75	800	500
	S	8-265	Pont.	R-45TSX	0.060	—	12B	16	7.5–9	—	450 ± 24
	W	8-301	Pont.	R-45TSX	0.060	—	12B	16	7.5–9	—	450 ± 32
	T	8-301⑧	Pont.	R-45TSX	0.060	—	6B	16	7.5–9	—	450 ± 32
	H	8-305	Chev.	R-45TS	0.045	6B	6B	44	7.5–9	800	800

TUNE-UP SPECIFICATIONS
Firebird

Year	Engine VIN Code	Engine No. of Cyl. Displacement (cu. in.)	Engine Manufac-turer	Spark Plugs Type	Gap (in.)	Ignition Timing (deg)③④ Man. Trans.	Auto. Trans.	Intake Valve Opens (deg)⑤	Fuel Pump Pressure (psi)	Idle Speed (rpm)③④ Man. Trans.	Auto. Trans.
'82	2	4-151	Pont.	R-44TSX	0.060	⑨	⑨	—	9–13	⑨	⑨
	1	6-173	Chev.	R-43TS	0.045	⑨	⑨	—	5.5–6.5	⑨	⑨
	H	8-305	Chev.	R-45TS	0.045	⑨	⑨	—	5.5–6.5	⑨	⑨
	7	8-305	Chev.	R-45TS②	0.045	⑨	⑨	—	9–13	⑨	⑨
'83	2	4-151	Pont.	R-44TSX	0.060	⑨	⑨	—	9–13	⑨	⑨
	1	6-173	Chev.	R-43CTS	0.045	⑨	⑨	—	5.5–6.5	⑨	⑨
	L	6-173HO	Chev.	R-42CTS	0.045	⑨	⑨	—	6–7½	⑨	⑨
	H	8-305	Chev.	R-45TS	0.045	⑨	⑨	—	5.5–6.5	⑨	⑨
	S	8-305	Chev.	R-45TS	0.045	⑨	⑨	—	9–13	⑨	⑨
'84	2	4-151	Pont.	R-44TSX	0.060	⑨	⑨	—	9–13	⑨	⑨
	1	6-173	Chev.	R-43CTS	0.045	⑨	⑨	—	5½–6½	⑨	⑨
	L	6-173	Chev.	R-42CTS	0.045	⑨	⑨	—	6–7½	⑨	⑨
	H	8-305	Chev.	R-45TS	0.045	⑨	⑨	NA	5½–6½	⑨	⑨
	G	8-305	Chev.	R-45TS	0.045	⑨	⑨	—	9–13	⑨	⑨
'85	2	4-151	Pont.	R-43TSX	0.060	⑨	⑨	—	9–13	⑨	⑨
	S	6-173	Chev.	R-42CTS	0.045	⑨	⑨	—	40.5–47	⑨	⑨
	F	8-305	Chev.	R-43CTS	0.045	⑨	⑨	—	40.5–47	⑨	⑨
	H	8-305	Chev.	R-45TS	0.045	⑨	⑨	—	5½–6½	⑨	⑨
	G	8-305	Chev.	R-44TS	0.045	⑨	⑨	—	9–13	⑨	⑨
'86	2	4-151	Pont.	R-43CTS6	0.060	⑨	⑨	—	9–13	⑨	⑨
	S	6-173	Chev.	R-42CTS	0.045	⑨	⑨	—	40½–47	⑨	⑨
	F	8-305	Chev.	R-43TS	0.035	⑨	⑨	—	40½–47	⑨	⑨
	H	8-305	Chev.	R-45TS	0.045	⑨	⑨	—	5½–6½	⑨	⑨
	G	8-305	Chev.	R-43TS	0.035	⑨	⑨	—	9–13	⑨	⑨
'87	All			See Underhood Specifications Sticker							

NOTE: The underhood specifications sticker often reflects tune-up specification changes made during the production run. Sticker figures must always be used if they disagree with those in this chart. Part numbers in this chart are not recommendations by Chilton for any product by brand name.

All models use electronic ignition systems.
B Before Top Dead Center
TDC Top Dead Center
—— Not applicable
NA—Not available
① Lower figure indicates idle speed with solenoid disconnected
② R-44TS if a colder plug is needed
③ See text for procedure
④ Figure in parentheses indicates California engine

⑤ All figures are in degrees Before Top Dead Center. Where two figures appear, the first represents timing with manual transmission, the second with automatic transmission
Auto—29
Trans Am—16
⑥ All M/T and low altitude A/T—R-45TS, gap 0.040
⑦ With performance package
⑧ Turbocharged engine
⑨ These functions are controlled by the emissions computer. In rare instances when adjustment is necessary, refer to the underhood emissions sticker for specifications.

FIRING ORDER

NOTE: To avoid confusion, replace spark plugs and wires one at a time.

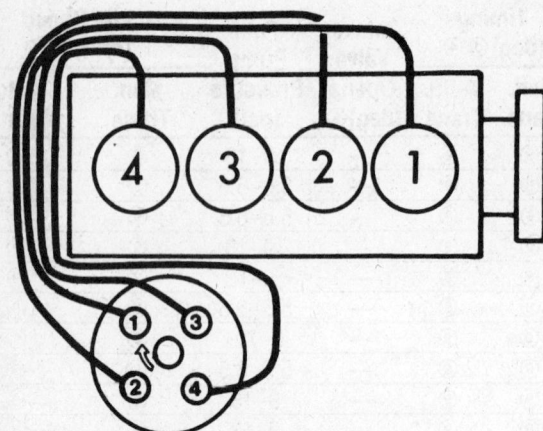

Pontiac-built 151-4 cylinder engine
Engine firing order: 1-3-4-2
Distributor rotation: clockwise

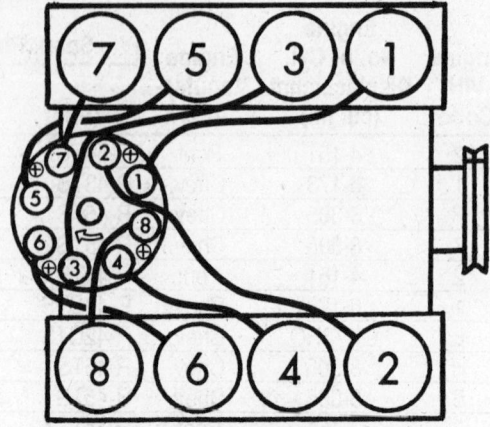

Chevrolet-built V8 engines
Engine firing order: 1-8-4-3-6-5-7-2
Distributor rotation: clockwise

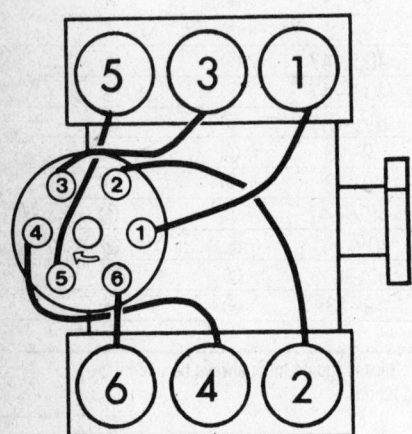

Chevrolet-built 229-V6 engine
Engine firing order: 1-6-5-4-3-2
Distributor rotation: clockwise

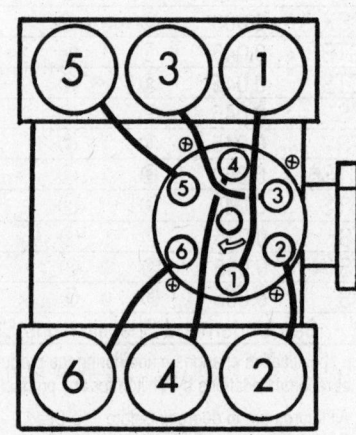

Buick-built 231-V6 engine
Engine firing order: 1-6-5-4-3-2
Distributor rotation: clockwise

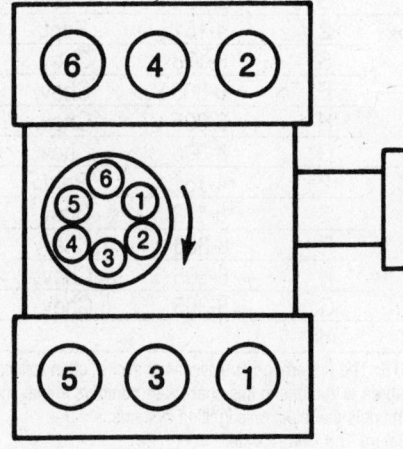

Chevrolet-built 173-V6 engine
Engine firing order: 1-2-3-4-5-6
Distributor rotation: clockwise

Pontiac-built V8 engines
Engine firing order: 1-8-4-3-6-5-7-2
Distributor rotation: counterclockwise

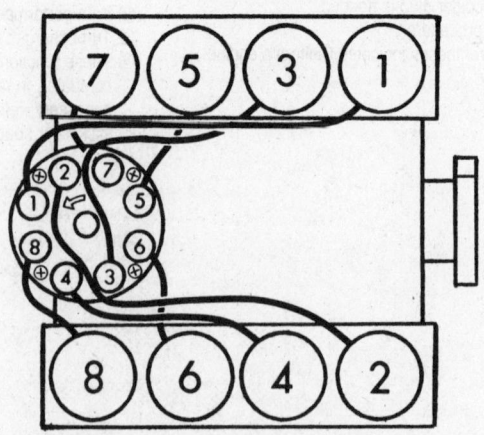

Oldsmobile-built V8 engines
Engine firing order: 1-8-4-3-6-5-7-2
Distributor rotation: counterclockwise

CAPACITIES
Camaro

Year	Engine VIN Code	Engine No. Cyl. Displacement (cu. in.)	Engine Crankcase Add 1 qt for New Filter	Transmission Pts to Refill After Draining			Drive Axle (pts)	Gasoline Tank (gals)	Cooling System (qts)	
				Manual		Automatic ①			With Heater	With A/C
				3-Speed	4-Speed					
'80–'81	K	6-229	4②	3	—	7③	4.25④	21	14.5	15.5
	A	6-231	4②	—	—	7③	4.25④	21	12.0	13.0
	J	8-267	4	—	—	7③	4.25④	21	15.0	16.0
	H	8-305	4	—	3.4	7③	4.25④	21	15.0	16.0
	L	8-350	4	—	—	7④	4.25④	21	16.0	17.0
'82	2	4-151	3②	—	4.3	8.5	3.5	16	12.8	13.0
	1	6-173	4②	—	4.3	8.5	3.5	16	12.8	12.8
	H	8-305⑤	4	—	4.3	8.5	3.5	16	17.2	17.2
	7	8-305⑥	4	—	4.3	8.5	3.5	16	15.9	15.9
'83	2	4-151	3②	—	4.3⑦	8.5⑧	3.5	16	12.8	13.0
	1	6-173	4②	—	4.3⑦	8.5⑧	3.5	16	12.8	12.8
	H	8-305⑤	4	—	4.3⑦	8.5⑧	3.5	16	17.2	17.2
	S	8-305⑥	4	—	4.3⑦	8.5⑧	3.5	16	15.9	15.9
'84	2	4-151	3②	—	3.5⑦	8.5⑧	3.5	16	8.8	9.1
	1	6-173	4②	—	3.5⑦	8.5⑧	3.5	16	12.5	12.5
	H	8-305⑤	4	—	3.5⑦	8.5⑧	3.5	16	15.0	15.0
	G	8-305⑥	4	—	3.5⑦	8.5⑧	3.5	16	15.0	15.0
'85–'87	2	4-151	3②	—	2.5⑦	9.5	3.5	15.5	12.75	13.0
	S	6-173	4②	—	2.5⑦	9.5	3.5	15.5	12.5	12.5
	H	8-305	4	—	2.5⑦	9.5	3.5	16	15.5	16.0
	G	8-305	4	—	2.5⑦	9.5	3.5	16	17.0	17.0
	F	8-305	4	—	2.5⑦	9.5⑧	3.5	15.5	17.0	17.0
	8	8-350	4	—	—	9.5⑧	3.5	20	17.0	—

—Not applicable
① Drain and refill only—does not include torque convertor
② Capacity same with or without filter change
③ With 350—6 pints
④ With 7½" ring gear—3.5 pints; with 8¾" ring gear—5.4 pints
⑤ With 4 bbl. carburetor
⑥ With throttle body fuel injection
⑦ 5-speed—6.87 pints
⑧ Overdrive transmission—9.9 pints: Add 4 pints, run engine and check dipstick; fill as necessary

CAPACITIES
Firebird

Year	Engine VIN Code	Engine No. Cyl. Displacement (cu. in.)	Engine Crankcase (Add 1 Qt For New Filter)	Transmission (Pts-to-Refill After Draining)			Drive Axle (pts)	Gasoline Tank (gals)	Cooling System (qts)	
				Manual		Automatic (Pts.) ①			With Heater	With A/C
				3 Spd.	4 Spd.					
'80	A	6-231	4	3.5	3.5	6	4.25	20.8	13.2	13.2
	S	8-265	4②	—	—	6	4.25	20.8	20.4	20.4
	W	8-301	4②	3.5	3.5	6	4.25	20.8	20.4	20.4
	H	8-305	4	—	—	6	4.25	20.8	—	16.4
'81	A	6-231	4	3.5	—	6	4.25	21.0	20.9	—
	S	8-265	4	—	—	6	4.25	21.0	20.9	—
	W	8-301	4	—	—	6	4.25	21.0	20.9	—
	T	8-301	4	—	—	6	4.25	21.0	20.9	—
	H	8-305	4	—	—	6	4.25	21.0	20.9	—

CAPACITIES
Firebird

Year	Engine VIN Code	Engine No. Cyl. Displacement (cu. in.)	Engine Crankcase (Add 1 Qt For New Filter)	Transmission (Pts-to-Refill After Draining)		Automatic (Pts.)①	Drive Axle (pts)	Gasoline Tank (gals)	Cooling System (qts)	
				Manual					With Heater	With A/C
				3 Spd.	4 Spd.					
'82	2	4-151	3②	—	4.3	8.5	3.5	16.0	12.8	13.0
	1	6-173	4②	—	4.3	8.5	3.5	16.0	12.8	12.8
	H	8-305	4	—	4.3	8.5	3.5	16.0	17.2	17.2
	7	8-305	4	—	4.3	8.5	3.5·	16.0	15.9	15.9
'83	2	4-151	3②	—	4.3③	8.5④	3.5	16.0	12.8	13.0
	1,L	6-173	4②	—	4.3③	8.5④	3.5	16.0	12.8	12.8
	H	8-305	4	—	4.3③	8.5④	3.5	16.0	17.2	17.2
	S	8-305	4	—	4.3③	8.5④	3.5	16.0	15.9	15.9
'84	2	4-151	3②	—	3.5③	8.5④	3.5	16.0	8.8	9.1
	1,L	6-173	4②	—	3.5③	8.5④	3.5	16.0	12.5	12.5
	H	8-305	4	—	3.5③	8.5④	3.5	16.0	15.0	15.0
	G	8-305	4	—	3.5③	8.5④	3.5	16.0	15.0	15.0
'85–'87	2	4-151	3②	—	2.5③	9.5④	3.5	15.5	13.0	12.4
	S	6-173	4②	—	2.5③	9.5④	3.5	15.5	12.5	12.5
	H	8-305	4	—	2.5③	9.5④	3.5	16.0	17.2	17.2
	G	8-305	4	—	2.5③	9.5④	3.5	16.0	17.2	17.2
	F	8-305	4	—	2.5③	9.5④	3.5	15.5	17.0	16.9

—Not applicable
① Drain and refill only—does not include torque convertor.
② Capacity same with or without filter change.
③ 5-speed—6.87 pints
④ Overdrive transmission—9.9 pints. Add 4 pints, run engine and check dipstick-fill as necessary

VALVE SPECIFICATIONS
Camaro

Year	Engine VIN Code	Engine No. Cyl. Displacement (cu. in.)	Seat Angle (deg)	Face Angle (deg)	Spring Test Pressure (lbs. @ in.)	Spring Installed Height (in.)	Stem-to-Guide Clearance (in.)		Stem Diameter (in.)	
							Intake	Exhaust	Intake	Exhaust
'80–'81	K	6-229	46	45	200 @ 1.25	1²³⁄₃₂	.0010–.0027	.0010–.0027	.3414	.3414
	A	6-231	45	45	168 @ 1.33	1⁴⁷⁄₆₄	.0015–.0032	.0015–.0032	.3407	.3409
	J	8-267	46	45	200 @ 1.25	1²³⁄₃₂	.0010–.0027	.0010–.0027	.3414	.3414
	H	8-305	46	45	200 @ 1.25	1²³⁄₃₂	.0010–.0027	.0010–.0027	.3414	.3414
	L	8-350	46	45	200 @ 1.25	1²³⁄₃₂	.0010–.0027	.0010–.0027	.3414	.3414
'82–'84	2	4-151	46	45	122–180 @ 1.25	1.69	.0010–.0027	.0010–.0027 ①	.3418–.3425	.3418–.3425
	1	6-173	46	45	194 @ 1.18	1.57	.0010–.0026	.0010–.0026	.3410–.3420	.3410–.3420
	H,7, S,G	8-305	46	45	194–206 @ 1.25	1²³⁄₃₂	.0010–.0027	.0010–.0027	.3410–.3420	.3410–.3420
'85–'87	2	4-151	46	45	170–180 @ 1.25	1.69	.0010–.0027	.0010–.0027 ①	.3420–.3430	.3420–.3430
	S	6-173	46	45	194 @ 1.18	1.57	.0010–.0026	.0010–.0026	.3410–.3420	.3410–.3420
	F,G,H	8-305	46	45	194–206 @ 1.25	1²³⁄₃₂	.0010–.0027	.0010–.0027	.3410–.3420	.3410–.3420
	8	8-350	46	45	194–206 @ 1.25	1²³⁄₃₂	.0010–.0027	.0010–.0027	.3410–.3420	.3410–.3420

① Figure given is measured at the top of the guide; .0020–.0037 is measured at the bottom of the guide.

VALVE SPECIFICATIONS
Firebird

Year	Engine VIN Code	Engine No. Cyl. Displacement (cu. in.)	Seat Angle (deg)	Face Angle (deg)	Spring Test Pressure (lbs. @ in.)	Spring Installed Height (in.)	Stem-to-Guide Clearance (in.)		Stem Diameter (in.)	
							Intake	Exhaust	Intake	Exhaust
'80	A	6-231	45	45	182 @ 1.34	1⁴⁷/₆₄	.0015–.0032	.0015–.0032	.3402–.3412	.3405–.3412
	S	8-265	46	45	175 @ 1.29	1⁴³/₆₄	.0010–.0027	.0010–.0027②	.3425	.3425
	W	8-301	46	45	175 @ 1.29	1⁴³/₆₄	.0010–.0027	.0010–.0027②	.3425	.3425
	H	8-305	46	45	199 @ 1.25	1²³/₃₂	.0010–.0037	.0010–.0047	.3414	.3414
'81	A	6-231	45	44	182 @ 1.34	1⁴⁷/₆₄	.0015–.0035	.0015–.0032	.3401–.3412	.3405–.3412
	S	8-265	46	45	175 @ 1.29	1⁴³/₆₄	.0010–.0027	.0010–.0027②	.3425	.3425
	W,T	8-301	46	45	175 @ 1.29	1⁴³/₆₄	.0010–.0027	.0010–.0027②	.3425	.3425
	H	8-305	46	45	194 @ 1.25	1²³/₃₂	.0010–.0027	.0010–.0027	.3425–.3432	.3420–.3427
'82–'84	2	4-151	46	45	122–180 @ 1.25	1.69	.0010–.0027	.0010–.0027①	.3418–.3425	.3418–.3425
	1,L	6-173	46	45	194 @ 1.18	1.57	.0010–.0026	.0010–.0026	.3410–.3420	.3410–.3420
	7,G,H,S	8-305	46	45	194–206 @ 1.25	1²³/₃₂	.0010–.0027	.0010–.0027	.3410–.3420	.3410–.3420
'85–'87	2	4-151	46	45	170–180 @ 1.25	1.69	.0010–.0027	.0010–.0027①	.3420–.3430	.3420–.3430
	S	6-173	46	45	194 @ 1.18	1.57	.0010–.0026	.0010–.0026	.3410–.3420	.3410–.3420
	F,G,H	8-305	46	45	194–206 @ 1.25	1²³/₃₂	.0010–.0027	.0010–.0027	.3410–.3420	.3410–.3420

NA: Not available
① Figure given is measured at the top of the guide; .0020–.0037 is measured at the bottom of the guide.
② Bottom exhaust: .0020–.0037

TORQUE SPECIFICATIONS
Camaro
All readings in ft. lbs.

Year	Engine VIN Code	Engine No. Cyl. Displacement (cu. in.)	Cylinder Head Bolts	Rod Bearing Bolts	Main Bearing Bolts	Crankshaft Pulley Bolts	Flywheel-to-Crankshaft Bolts	Manifold	
								Intake	Exhaust
'82–'84	2	4-151	85	32	70	160	44	29	44
'82–'84	1	6-173	70	37	69	75	50	23	25
'80–'81	K	6-229	65	45	70	60	60	30	20
'80–'81	A	6-231	80	40	100	175	60	45	25
'80–'81	J	8-267	65	45	70	60	60	30①	20
'85	2	4-151	92	32	70	200	44	④	④
'80–'81	L	8-350	65	45	70	60	60	30①	20
'85–'87	S	6-173	65–90	34–45	63–83	75	50	13–25	19–31
'85–'87	F,G,H	8-305	60–75	45	③	60	60	25–45	②
'80–'84	H,7,S,G	8-305	65	45	70	60	60	30①	20
'86–'87	2	4-151	92	32	70	162	55	④	④
'87	8	8-350	60–75	42–47	③	60	63–85	25–45	20–32

① 20–34 for T.B.I. plate bolts
② Inner: 20–32 ft. lbs.
   Outer: 14–26 ft. lbs.
③ Inner: 70–85 ft. lbs.
   outer: 60–75 ft. lbs.
④ See text and illustration

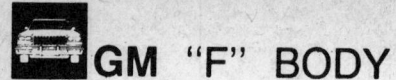
TORQUE SPECIFICATIONS
Firebird
All readings in ft. lbs.

Year	Engine VIN Code	Engine No. Cyl. Displacement (cu. in.)	Cylinder Head Bolts	Rod Bearing Bolts	Main Bearing Bolts	Crankshaft Bolt	Flywheel to Crankshaft Bolts	Manifold Intake	Manifold Exhaust
'80–'81	A	6-231	80	40	100	225	60	45	25
	S	8-265	95	30	①	160	95	35	40
	W	8-301	95	30	①	160	95	35	40
	T	8-301 Turbo	93	28	100	163	—	37	40
	H	8-305	65	45	70	60	60	30	20
'82–'84	2	4-151	85	32	70	160	44	29	44
	1,L	6-173	65–75	34–40	63–74	66–84	45–55	20–25	22–28
	7,G,H,S	8-305	65	45	70	60	60	30②	20
'85	2	4-151	92	32	70	200	44	29④	④
'85–'87	S	6-173	65–90	34–45	63–83	75	50	13–25	19–31
'85–'87	F,G,H	8-305	60–75	45	⑤	60	60	25–45	③
'86–'87	2	4-151	92	32	70	162	55	④	④

① 7/16" bolt—70; 1/2" bolt—100; rear main bearing— 100.
② 20-34 for TBI plate bolts
③ Inner: 20-32 ft. lbs.
Outer: 14-26 ft. lbs.
④ See text and illustration
⑤ Inner: 70-85 ft. lbs.
⑥ Outer: 60-75 ft. lbs.

CRANKSHAFT AND CONNECTING ROD SPECIFICATIONS
Camaro
All measurements are given in inches

Year	Engine VIN Code	Engine No. Cyl. Displacement (cu. in.)	Crankshaft Main Brg. Journal Dia.	Crankshaft Main Brg. Oil Clearance	Crankshaft Shaft End-Play	Crankshaft Thrust on No.	Connecting Rod Journal Diameter	Connecting Rod Oil Clearance	Connecting Rod Side Clearance
'80–'81	K	6-229	①	②	.0020–.0060	4	2.0986–2.0998	.0013–.0035	.0060–.0140
	A	6-231	2.4995	.0004–.0015	.0040–.0080	2	2.2495–2.2487	.0005–.0026	.0060–.0270
	J	8-267	①	②	.0020–.0060	5	2.0986–2.0998	.0013–.0035	.0060–.0140
	H	8-305	①	②	.0020–.0060	5	2.0986–2.0998	.0013–.0035	.0060–.0140
	L	8-350	①	②	.0020–.0060	5	2.0986–2.0998	.0013–.0035	.0060–.0140
'82–'84	2	4-151	2.300	.0005–.0022	.0035–.0085	5	2.000	.0005–.0026	.0060–.0220
	1	6-173	2.493–2.494	.0017–.0029	.0019–.0066	3	1.998–1.999	.0014–.0035	.0060–.0170
	H,7,S,G	8-305	①	②	.0020–.0060	5	2.098–2.099	.0018–.0039	.0080–.0140
'85–'87	2	4-151	2.300	.0005–.0022	.0035–.0085	5	2.000	.0005–.0026	.0060–.0220
	S	6-173	2.647–2.648	.0016–.0032	.0019–.0067	3	1.998–1.999	.0014–.0035	.0060–.0170
	F,G,H	8-305	①	②	.0020–.0060	5	2.098–2.099	.0018–.0039	.0080–.0140
'87	8	8-350	①	②	.0020–.0060	5	2.098–2.099	.0018–.0039	.0080–.0140

① No. 1—2.4484–2.4493
Nos. 2, 3, 4—2.4481–2.4490
No. 5—2.4479–2.4488
② No. 1—0.0008–0.0020
Nos. 2, 3, 4—0.0011–0.0023
No. 5—0.0017–0.0032

CRANKSHAFT AND CONNECTING ROD SPECIFICATIONS
Firebird
All measurements are given in inches

Year	Engine VIN Code	Engine Displacement (cu. in.)	Crankshaft Main Brg. Journal Dia.	Crankshaft Main Brg. Oil Clearance	Crankshaft Shaft End-Play	Crankshaft Thrust on No.	Connecting Rod Journal Diameter	Connecting Rod Oil Clearance	Connecting Rod Side Clearance
'80–'81	A	6-231	2.499	.0003–.0018	.0030–.0090	2	2.000	.0005–.0026	.0060–.0230
	S	8-265	3.000	.0002–.0018	.0030–.0090	4	2.250	.0005–.0025	.0060–.0220①
	W,T	8-301	3.000	.0002–.0018	.0030–.0090	4	2.250	.0005–.0025	.0060–.0220①

CRANKSHAFT AND CONNECTING ROD SPECIFICATIONS
Firebird
All measurements are given in inches

Year	Engine VIN Code	Engine Displacement (cu. in.)	Crankshaft				Connecting Rod		
			Main Brg. Journal Dia.	Main Brg. Oil Clearance	Shaft End-Play	Thrust on No.	Journal Diameter	Oil Clearance	Side Clearance
'80–'81	H	8-305	④	②	.0020–.0060	5	2.098–2.099	.0030	.0060–.0140
'82–'84	2	4-151	2.300	.0005–.0022	.0035–.0085	5	2.000	.0005–.0026	.0060–.0220
	1,L	6-173	2.493–2.494	.0017–.0029	.0019–.0066	3	1.998–1.999	.0014–.0035	.0060–.0170
	7,G,H,S	8-305	④	③	.0020–.0060	5	2.098–2.099	.0018–.0039	.0080–.0140
'85–'87	2	4-151	2.300	.0005–.0022	.0035–.0085	5	2.000	.0005–.0026	.0060–.0220
	S	6-173	2.647–2.648	.0016–.0032	.0019–.0067	3	1.998–1.999	.0014–.0035	.0060–.0170
	F,G,H	8-305	④	③	.0020–.0060	5	2.098–2.099	.0018–.0039	.0080–.0140

① Total for two connecting rods
② No. 1: .001–.0015
No.'s 2, 3, 4: .001–.0025
No. 5: .0025–.0035
③ No. 1: .0008–.0020
No.'s 2, 3, 4: .0011–.0023
No. 5: .0017–.0033
④ No. 1: 2.4484–2.4493
No.'s 2, 3, 4: 2.4481–2.4490
No. 5: 2.4479–2.4488

CAMSHAFT SPECIFICATIONS
Camaro
All measurements in inches. To convert inches to metric units, refer to Metric Information section.

Year	Engine VIN Code	Engine Type/Disp. L(cu in.)	Journal Diameter					Lobe Lift		Camshaft End Play
			1	2	3	4	5	Intake	Exhaust	
'80–'81	K	3.8(6-229)	All 1.8682–1.8692					0.3570	0.3900	0.004–0.012
	A	3.8(6-231)	All 1.7850–1.7860					—	—	—
	J	4.4(8-267)	All 1.8682–1.8692					0.3570	0.3900	0.004–0.012
	H	5.0(8-305)	All 1.8682–1.8692					0.2484	0.2667	0.004–0.012
	L	5.7(8-350)	All 1.8682–1.8692					0.2600	0.2733	0.004–0.012
'82	2	2.5(4-151)	All 1.8690					0.3980	0.3980	0.0015–0.0050
	1	2.8(6-173)	All 1.8976–1.8996					0.2350	0.2660	—
	H	5.0(8-305)	All 1.8682–1.8692					0.2380	0.2600	0.004–0.012
	7	5.0(8-305)	All 1.8682–1.8692					0.2600	0.2730	0.004–0.012
'83	2	2.5(4-151)	All 1.8690					0.3980	0.3980	0.0015–0.0050
	1	2.8(6-173)	All 1.8976–1.8996					0.2350	0.2660	—
	H	5.0(8-305)	All 1.8682–1.8692					0.2340	0.2570	0.004–0.012
	G	5.0(8-305)	All 1.8682–1.8692					0.2690	0.2760	0.004–0.012
	S	5.0(8-305)	All 1.8682–1.8692					0.2570	0.2690	0.004–0.012
'84	2	2.5(4-151)	All 1.8690					0.3980	0.3980	0.0015–0.0050
	1	2.8(6-173)	All 1.8976–1.8996					0.2350	0.2660	—
	H	5.0(8-305)	All 1.8682–1.8692					0.2340	0.2570	0.004–0.012
'85–'87	2	2.5(4-151)	All 1.8690					0.3980	0.3980	0.0015–0.0050
	S	2.8(6-173)	All 1.8976–1.8996					0.2625	0.2732	—
	H	5.0(8-305)	All 1.8682–1.8692					0.2340	0.2570	0.004–0.012
	F,G	5.0(8-305)	All 1.8682–1.8692					0.2690	0.2760	0.004–0.012
	8	5.7(8-350)	All 1.8682–1.8692					0.273	0.282	.0040–.0120

— Not available

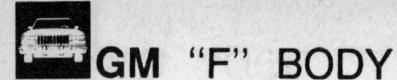

CAMSHAFT SPECIFICATIONS
Firebird

All measurements in inches. To convert inches to metric units, refer to Metric Information section.

Year	Engine VIN Code	Engine Type/Disp. L(cu in.)	Journal Diameter					Lobe Lift		Camshaft End Play
			1	2	3	4	5	Intake	Exhaust	
'80–'81	A	3.8(6-231)	All 1.7850–1.7860					—	—	—
	S	4.3(8-265)	All 1.9000					—	—	—
	W,T	4.9(8-301)	All 1.9000					—	—	—
	H	5.0(8-305)	All 1.8682–1.8692					0.2484	0.2667	0.004–0.012
'82	2	2.5(4-151)	All 1.8690					0.3980	0.3980	0.0015–0.0050
	1,L	2.8(6-173)	All 1.8976–1.8996					0.2350	0.2660	—
	H	5.0(8-305)	All 1.8682–1.8692					0.2380	0.2600	0.004–0.012
	7	5.0(8-305)	All 1.8682–1.8692					0.2600	0.2730	0.004–0.012
'83	2	2.5(4-151)	All 1.8690					0.3980	0.3980	0.0015–0.0050
	1,L	2.8(6-173)	All 1.8976–1.8996					0.2350	0.2660	—
	H	5.0(8-305)	All 1.8682–1.8692					0.2340	0.2570	0.004–0.012
	G	5.0(8-305)	All 1.8682–1.8692					0.2690	0.2760	0.004–0.012
	S	5.0(8-305)	All 1.8682–1.8692					0.2570	0.2690	0.004–0.012
'84	2	2.5(4-151)	All 1.8690					0.3980	0.3980	0.0015–0.0050
	1,L	2.8(6-173)	All 1.8976–1.8996					0.2350	0.2660	—
	H	5.0(8-305)	All 1.8682–1.8692					0.2340	0.2570	0.004–0.012
	G	5.0(8-305)	All 1.862–1.8692					0.2690	0.2760	0.004–0.012
'85–'87	2	2.5(4-151)	All 1.8690					0.3980	0.3980	0.0015–0.0050
	S	2.8(6-173)	All 1.8976–1.8996					0.2625	0.2732	—
	H	5.0(8-305)	All 1.8682–1.8692					0.2340	0.2570	0.004–0.012
	F,G	5.0(8-305)	All 1.8682–1.8692					0.2690	0.2760	0.004–0.012

— Not available

RING SPECIFICATIONS
Camaro

All measurements are given in inches

Year	Engine VIN Code	Engine No. of Cyl. Displacement (cu. in.)	Ring Gap			Ring Side Clearance		
			Top Compression	Bottom Compression	Oil Control	Top Compression	Bottom Compression	Oil Control
'82–'87	2	4-151	.0100–.0220	.0100–.0270	.0150–.0550	.0015–.0030	.0015–.0030	.0010–.0050
'82–'84	1	6-173	.0098–.0196	.0098–.0196	.0020–.0550	.0011–.0027	.0015–.0037	.0078 max.
'80–'81	K	6-229	.0100–.0200	.0100–.0250	.0150–.0550	.0012–.0032	.0012–.0032	.0020–.0070
'80–'81	A	6-231	.0100–.0200	.0100–.0200	.0150–.0350	.0030–.0050	.0030–.0050	.0035 max.
'80–'81	J	8-267	.0100–.0200	.0100–.0250	.0150–.0550	.0012–.0032	.0012–.0032	.0020–.0070
'80–'81	L	8-350	.0100–.0200	.0100–.0250	.0150–.0550	.0012–.0032	.0012–.0032	.0020–.0070
'80–'87	G	8-305	.0100–.0200	.0100–.0250	.0150–.0550	.0012–.0032	.0012–.0032	.0020–.0070
'85–'87	S	6-173	.0098–.0196	.0098–.0196	.0020–.0550	.0011–.0027	.0015–.0037	.0078 max.

RING SPECIFICATIONS
Camaro

All measurements are given in inches

Year	Engine VIN Code	Engine No. of Cyl. Displacement (cu. in.)	Ring Gap			Ring Side Clearance		
			Top Compression	Bottom Compression	Oil Control	Top Compression	Bottom Compression	Oil Control
'85–'87	F	8-305	.0100–.0200	.0100–.0250	.0150–.0550	.0012–.0032	.0012–.0032	.0020–.0070
'80–'87	H	8-305	.0100–.0200	.0100–.0250	.0150–.0550	.0012–.0032	.0012–.0032	.0020–.0070
'83	S	8-305	.0100–.0200	.0100–.0250	.0150–.0550	.0012–.0032	.0012–.0032	.0020–.0070
'82	7	8-305	.0100–.0200	.0100–.0250	.0150–.0550	.0012–.0032	.0012–.0032	.0020–.0070
'87	8	8-350	.0100–.0200	.0100–.0250	.0150–.0500	.0012–.0032	.0012–.0032	.0020–.0070

RING SPECIFICATIONS
Firebird

All measurements are given in inches

Year	Engine VIN Code	Engine No. of Cyl. Displacement (cu. in.)	Ring Gap			Ring Side Clearance		
			Top Compression	Bottom Compression	Oil Control	Top Compression	Bottom Compression	Oil Control
'80–'87	G	8-305	.0100–.0200	.0100–.0250	.0150–.0550	.0012–.0032	.0012–.0032	.0020–.0070
'80–'81	W	8-301	.0100–.0200	.0100–.0200	.0150–.0550	.0015–.0035	.0015–.0035	.0015–.0035
'80–'81	S	8-265	.0100–.0200	.0100–.0200	.0150–.0550	.0015–.0035	.0015–.0035	.0015–.0035
'81	T	8-301	.0100–.0200	.0100–.0200	.0150–.0550	.0015–.0035	.0015–.0035	.0015–.0035
'80–'81	A	6-231	.0130–.0230	.0130–.0230	.0150–.0350	.0030–.0050	.0030–.0050	.0035 max.
'82–'86	2	4-151	.0100–.0220	.0100–.0270	.0150–.0550	.0015–.0030	.0015–.0030	.0010–.0050
'82–'84	1,L	6-173	.0098–.0196	.0098–.0196	.0200–.0550	.0011–.0027	.0015–.0037	.0078 max.
'85–'87	S	6-173	.0098–.0196	.0098–.0196	.0020–.0550	.0011–.0027	.0015–.0037	.0078 max.
'80–'86	H	8-305	.0100–.0200	.0100–.0250	.0150–.0550	.0012–.0032	.0012–.0032	.0020–.0070
'85–'87	F	8-305	.0100–.0200	.0100–.0250	.0150–.0550	.0012–.0032	.0012–.0032	.0020–.0070
'83	S	8-305	.0100–.0200	.0100–.0250	.0150–.0550	.0012–.0032	.0012–.0032	.0020–.0070
'82	7	8-305	.0100–.0200	.0100–.0250	.0150–.0550	.0012–.0032	.0012–.0032	.0020–.0070

PISTON CLEARANCE
Camaro

Year	Engine VIN Code	Engine No. Cyl. Displacement (cu. in.)	Piston-to-Bore Clearance (in.)
'80–'81	K	6-229	.0007–.0017①
'80–'81	A	6-231	.0008–.0020②
'80–'81	J	8-267	.0007–.0017①
'80–'87	H	8-305	.0007–.0017①
'80–'81	L	8-350	.0007–.0017①
'82	7	8-305	.0007–.0017①
'82–'87	2	4-151	.0017–.0033③
'82–'84	1	6-173	.0007–.0016
'83	S	8-305	.0007–.0017①
'84–'87	G	8-305	.0007–.0017①
'85–'87	S	6-173	.0007–.0017①
'85–'87	F	8-305	.0007–.0017①
'87	8	8-350	.0025–.0035

① .75" below piston pin centerline
② 2.5" from top of cylinder (bore); across piston pin centerline (piston)
③ 2.25" from top of cylinder (bore); 1¹³⁄₁₆" from top of piston

PISTON CLEARANCE
Firebird

Year	Engine VIN Code	Engine No. Cyl. Displacement (cu. in.)	Piston-to-Bore Clearance (in.)
'81	T	8-301	.0017–.0025
'80–'81	W	8-301	.0025–.0033②
'80–'81	A	6-231	.0008–.0020②
'80–'81	S	8-265	.0017–.0025②
'82	7	8-305	.0007–.0017①
'82–'87	2	4-151	.0017–.0033③
'82–'84	1,L	6-173	.0007–.0016
'81–'87	H	8-305	.0007–.0017①
'83	S	8-305	.0007–.0017①
'84–'87	G	8-305	.0007–.0017①
'85–'87	S	6-173	.0007–.0017
'85–'87	F	8-305	.0007–.0017①

① .75 in. below piston pin C/L
② 2.5 in. from top of cylinder (bore); across piston pin centerline (piston)
③ 2.25 in. from top of cylinder (bore); 1¹³⁄₁₆ in. from top of piston

WHEEL ALIGNMENT SPECIFICATIONS
Firebird

Year	Caster Range (deg)	Caster Pref. Setting (deg)	Camber Range (deg)	Camber Pref. Setting (deg)	Toe-in (in.)	Steering Axis Inclin.
'80–'81	½P to 1½P	1P	½P to 1½P	1P	¹⁄₁₆ to ³⁄₁₆	10.35
'82–'87	2½P to 3½P	3P	½P to 1½P	1P	①	—

N—Negative
P—Positive
—Not available
① Except Trans Am—.2° ± .05°; Trans Am—.15° ± .05°

WHEEL ALIGNMENT SPECIFICATIONS
Camaro

Year	Caster Range(deg)	Caster Pref. Setting (deg)	Camber Range (deg)	Camber Pref. Setting (deg)	Toe-in (in.)	Steering Axis Inclin (deg)
'80–'81	0 to 2P	1P	⅕N to 1⅘P	1P	¹⁄₁₆–¼	10.35①
'82–'87	2½P to 3½P③	3P	½P to 1½P	1P	②	NA

N—Negative
P—Positive
—Not available
① At 1° camber
② Except Z28—.2° ± 0.5°; Z-28—.15° ± .05°
③ Z28: 3P to 4P

ENGINE ELECTRICAL

The charging system is of the SI integral regulator type. The generator utilizes a solid molded, solid state voltage regulator, which is attached to the slip ring end frame of the alternator.

Alternator

REMOVAL & INSTALLATION

NOTE: For alternator testing and diagnosis procedures, refer to "Charging and Starting" in the Unit Repair section.

1. Disconnect the battery ground cable to prevent diode damage.
2. Tag and disconnect the alternator wiring.
3. Remove the alternator brace bolt. If the vehicle is equipped with power steering, loosen the pump brace and mount nuts. Detach the drive belt(s).
4. Support the alternator and remove the mounting bolt(s). Remove the unit from the vehicle.
5. To install, reverse the removal procedures. Adjust the belt to to allow approximately $\frac{1}{2}$ in. of play, on the longest run between the pulleys.

Voltage Regulator

The voltage regulator is electronic and is housed within the alternator. Adjustments to the regulator are not possible. Should replacement of the regulator become necessary, the alternator must be disassembled.

Starter

NOTE: For the starter overhaul procedures, refer to "Charging and Starting" in the Unit Repair section.

REMOVAL & INSTALLATION

1. Disconnect the negative battery cable.
2. Raise and support the vehicle on jackstands.
3. Remove the starter braces, shields and other items, which may be in the way.

NOTE: On certain models, it may necessary to remove of the frame support. The support runs from the corner of the frame to the front crossmember. To remove it:

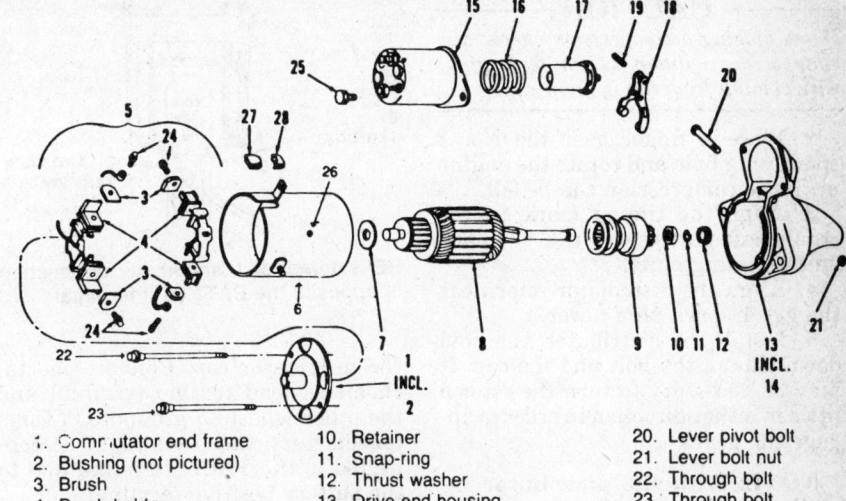

1. Commutator end frame
2. Bushing (not pictured)
3. Brush
4. Brush holder
5. Brush and holder package
6. Field and frame assembly
7. Washer
8. Armature
9. Drive assembly
10. Retainer
11. Snap-ring
12. Thrust washer
13. Drive end housing
14. Bushing (not pictured)
15. Solenoid assembly
16. Plunger spring
17. Plunger
18. Lever
19. Roll pin
20. Lever pivot bolt
21. Lever bolt nut
22. Through bolt
23. Through bolt
24. Brush screw
25. Bolt
26. Field retaining screw
27. Plug
28. Grommet

Exploded view of a typical starter

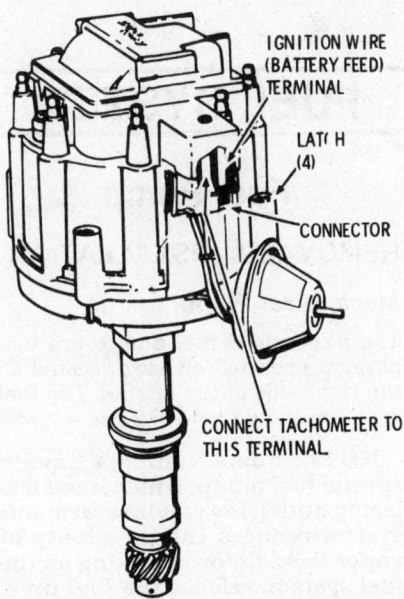

IGNITION WIRE (BATTERY FEED) TERMINAL

LATCH (4)

CONNECTOR

CONNECT TACHOMETER TO THIS TERMINAL

HEI coil-in-cap distributor tachometer hookup

Loosen the support-to-frame mounting bolt(s), remove the support-to-crossmember mounting bolt(s) and swing the support out of the way.

4. Remove the two starter-to-engine bolts. Lower the starter and remove the electrical connections.
5. To install, reverse the removal procedures. Torque the mounting bolts to 25–35 ft. lbs.

NOTE: If shims were inserted between the starter and engine block, they must be replaced in their original locations.

Distributor

REMOVAL

1. Disconnect the negative battery cable.
2. Tag and disconnect the electrical connectors from the distributor cap.
3. Disconnect the hose at the vacuum advance unit, if equipped.
4. Turn the four distributor cap-to-housing screws counterclockwise and lift off the cap assembly.

NOTE: It may be necessary to remove the secondary wires from the distributor cap.

5. Using crayon or chalk, make locating marks on the rotor, the distributor housing and the engine for installation purposes.
6. Remove the distributor clamp bolt and clamp, then lift the distributor (turning the rotor counterclockwise) from the engine.

INSTALLATION

Undisturbed Engine

1. Install a new O-ring on the distributor housing.
2. Install the distributor into the engine opening, aligning the marks made during the removal procedure.
3. Install the distributor clamp, the hold down bolt and the distributor cap.
4. Check the ignition timing.

Disturbed Engine

1. Remove the No. 1 spark plug.

2. Place a finger over the No. 1 spark plug hole and rotate the engine until the compression can be felt.

3. Align the timing mark on the crankshaft pulley with the "0" mark on the timing plate.

4. Align the distributor rotor near the No. 1 spark plug tower.

5. Install the distributor, the hold down clamp, the bolt and the cap. It may be necessary to turn the rotor a little in either direction in order to engage the gears.

NOTE: With the distributor installed, make sure that the rotor is aligned with the No. 1 spark plug tower of the cap.

6. Reverse the removal procedures and check the ignition timing.

Ignition Timing

NOTE: The underhood timing instructions on the Vehicle Emissions Control Information label may differ from these procedures; always follow the underhood label's directions.

1. If equipped, disconnect and plug the vacuum advance hose from the distributor. If equipped with electronic spark timing (EST), disconnect the 4-wire connector to the distributor. If equipped with the 5.0L TBI engine, DO NOT disconnect the 4-wire connector to the distributor. Disconnect the tan/black striped wire near the rear of the right-side cylinder head cover.

2. Make sure the timing marks are clean and readable. The engine must be at normal operating temperature.

NOTE: It may be necessary to put a small amount of white paint or chalk on the timing marks to make them more visible.

3. Connect a timing light to No. 1 cylinder.

4. Loosen the distributor clamp.

5. Start the engine and run it at the rpm specified in the Tune-Up chart. Rotate the distributor to align the timing marks. Tighten the distributor clamp and recheck the timing.

6. Reconnect the vacuum hose or the 4-wire connector.

HEI System Tachometer Hookup

There is a terminal marked TACH on

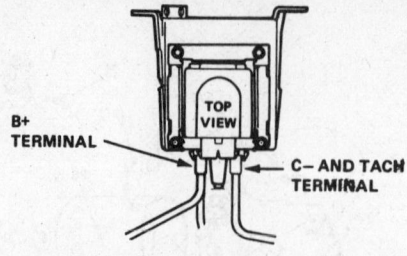

HEI external coil tachometer connection is opposite the BATT (B⁺) terminal

the distributor cap. Connect one tachometer lead to this terminal and the other lead to a ground. On some tachometers, the leads must be connected to the TACH terminal and to the battery positive terminal.

FUEL SYSTEM

Fuel Pump

REMOVAL & INSTALLATION

Mechanical Pump

The mechanical fuel pumps are diaphragm operated and are located on the right-side of the engine. The fuel pressure should be 4–6.5 psi.

NOTE: Some vehicles have a special fuel pump, which has a metering outlet, for vapor return; this system reduces the possibility of vapor lock. Before working on the fuel system, release the fuel pressure. Remove and replace the fuel tank cap.

1. Disconnect the fuel inlet, outlet and return (if equipped) hoses from the fuel pump.

2. Remove the retaining bolts.

3. Remove the fuel pump, push rod, gasket and mounting plate (if equipped).

NOTE: On the Chevrolet engines (V6 and V8), if the push rod is not to be removed, remove the upper bolt from the right front mounting boss. Insert a longer bolt (³⁄₈–16 x 2 in.) in this hole to hold the fuel pump pushrod.

4. Clean the gasket mounting surfaces.

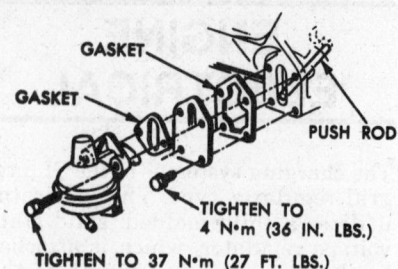

Typical mechanical fuel pump installation on a V8 engine

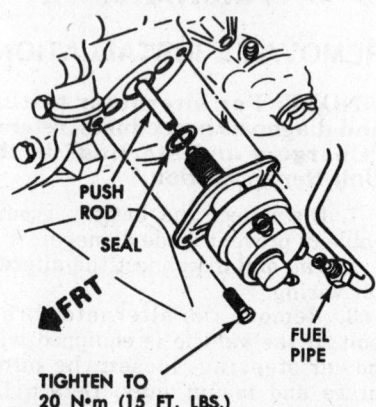

Mechanical fuel pump installation on the 173 V6 engine

5. To install, use new gasket(s) and reverse the removal procedures. Torque the mounting plate bolts to 3 ft. lbs. and the fuel pump bolts to 27 ft. lbs. Start the engine and check for fuel leaks.

Electric Pump

The electric fuel pump is part of the fuel sending unit and is located in the fuel tank. The fuel pressure is 3 psi (carbureted), 4–13 psi (throttle body) or minimum 50 psi (port injection).

1. Disconnect the negative battery cable.

2. Remove the fuel tank.

3. Disconnect the fuel tubes and the electrical connectors from the sending unit of the fuel tank.

4. Using a brass drift and a hammer, loosen the sending unit and pump assembly retaining ring.

5. Lift the sending unit and pump assembly from the fuel tank.

6. Remove the fuel pump from the sending unit.

7. To install, use a new sealing O-ring and reverse the removal procedures.

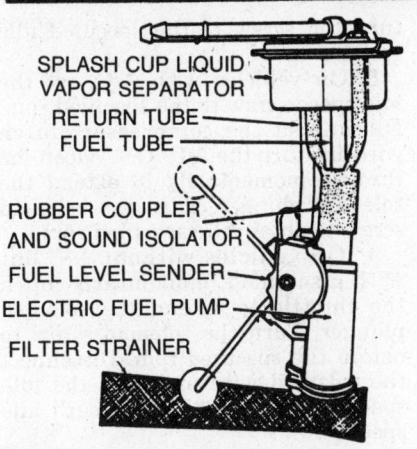

SPLASH CUP LIQUID
VAPOR SEPARATOR
RETURN TUBE
FUEL TUBE

RUBBER COUPLER
AND SOUND ISOLATOR
FUEL LEVEL SENDER
ELECTRIC FUEL PUMP
FILTER STRAINER

Electric fuel pump and fuel gauge meter assembly of a TBI-equipped engine

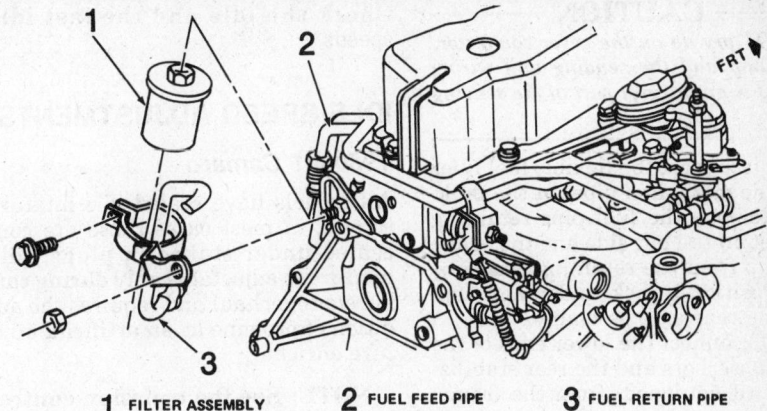

1 FILTER ASSEMBLY   2 FUEL FEED PIPE   3 FUEL RETURN PIPE

Fuel filter mounting on a TBI-equipped 151 4 cylinder engine

Fuel Pressure Release

Carbureted

To release the fuel pressure on the carbureted system, remove and replace the fuel tank cap.

Throttle Body Injection (TBI)

To release the fuel pressure on the TBI system, remove the fuel pump fuse from the fuse panel, start and operate the engine until it stalls, then replace the fuse.

Tuned Port Injection (TPI)

To release the fuel pressure on the TPI system, connect a tube to the pressure release fitting on the fuel rail, place the other end of the tube in a container. Open the pressure fitting, bleed off the fuel and close the fitting.

Fuel Filter

There are three types of fuel filters; internal (in the carburetor fitting), in-line (in the fuel line) and in-tank (the sock on the fuel pick-up tube). Before removing any component of the fuel system, refer to the "Fuel Pressure Release" procedures in this section and release the fuel pressure.

REMOVAL & INSTALLATION

Internal Filter

1. Disconnect the fuel line connection at the fuel inlet filter nut on the carburetor.
2. Remove the fuel inlet filter nut from the carburetor.
3. Remove the filter and the spring.

NOTE: If a check valve is not present with the filter, one must be

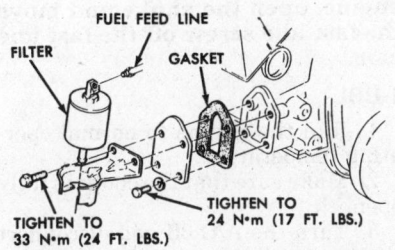

FUEL FEED LINE

FILTER   GASKET

TIGHTEN TO
33 N·m (24 FT. LBS.)   TIGHTEN TO
24 N·m (17 FT. LBS.)

Fuel filter mounting on a TBI-equipped V8 engine. Note that the mounting location is the same as that for the fuel pump on carbureted engines

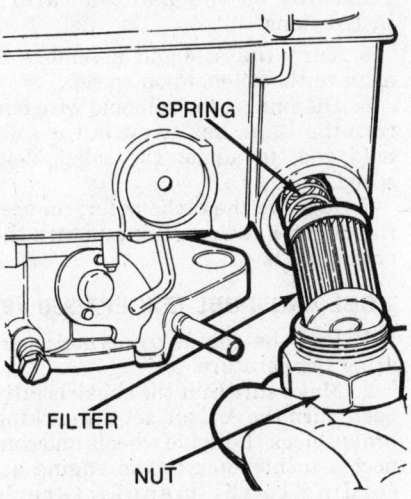

SPRING

FILTER

NUT

Typical fuel filter of carbureted engines

installed when the filter is replaced.

4. Install the spring, filter and check valve (must face the fuel line), then reverse the removal procedures.
5. Start the engine and check for leaks.

In-Line Filter

1. Disconnect the fuel lines.

2. Remove the fuel filter from the retainer or mounting bolt.
3. To install, reverse the removal procedures. Start the engine and check for leaks.

NOTE: The filter has an arrow (fuel flow direction) on the side of the case, be sure to install it correctly in the system, with the arrow facing away from the fuel tank.

In-Tank Filter

To service the in-tank fuel filter, refer to the "Electric Fuel Pump Removal & Installation" procedures in this section.

Fuel Tank

——— CAUTION ———

Before removing any component of the fuel system, refer to the "Fuel Pressure Release" procedures in this section and release the fuel pressure.

REMOVAL & INSTALLATION

1. Release the pressure from the fuel system.
2. Raise and support the vehicle on jackstands.
3. Remove the fuel tank cap and drain the fuel.
4. Disconnect the exhaust pipe at the catalytic convertor and the rear hanger. Allow the exhaust system to hang over the rear axle assembly.
5. Remove the tailpipe and the muffler heat shields.
6. Remove the fuel filler neck shield from behind the left rear wheel.
7. Remove the rear suspension track bar and the brace.
8. Disconnect the fuel pump and the sending unit electrical connector at the body electrical harness connector.

9. Disconnect the flexible fuel lines from the metal fuel pipes at the tank.

10. Remove the fuel pipe retaining bracket on the left side and the brake line clip from the retaining bracket.

11. Position a jack under and support the rear axle assembly.

12. Disconnect the lower ends of the shock absorbers and the rear stabilizer bar (if equipped) from the frame. Lower the axle assembly and remove the coil springs.

13. Lower the rear axle assembly as far as possible without causing damage to the brake lines and cables.

14. Remove the fuel tank strap bolts.

15. Remove the fuel tank by rotating the front of the tank downward and slide it to the right side.

16. To install, reverse the removal procedures.

Carburetors

For all carburetor adjustments and specifications, please refer to the "Carburetor" Unit Repair section. For further information on feedback carburetors, please refer to *Chilton's Guide To Fuel Injection And Feedback Carburetors.*

REMOVAL & INSTALLATION

1. Disconnect the negative battery cable.

2. Remove the air cleaner and gasket.

3. Disconnect the fuel and vacuum lines.

4. Disconnect the choke electrical connector.

5. Disconnect the throttle linkage.

6. If equipped with an automatic transmission, disconnect the throttle valve linkage.

7. If equipped, disconnect the EGR line and the idle stop solenoid.

8. Remove the carburetor attaching nuts or bolts, gasket and the carburetor.

9. Clean the gasket mounting surfaces.

10. To install, use a new gasket and reverse the removal procedures.

Check the idle and the fast idle speeds.

IDLE SPEED ADJUSTMENTS

1980–81 Camaro

All models have sealed idle mixture screws; in most cases these are concealed under staked-in plugs. Idle mixture is adjustable only during carburetor overhaul and requires the addition of propane as an artificial mixture enricher.

NOTE: See the emission control label in the engine compartment for procedures and specifications not supplied here. To prepare the engine for adjustment, warm the engine, open the choke and move the fast idle screw off the fast idle cam.

1 BBL

1. Run the engine to normal operating temperature.

2. Make sure that the choke is fully open.

3. Turn the A/C off and disconnect the vacuum line at the vapor canister. Plug the line.

4. Set the parking brake, block the drive wheels and place the transmission in Drive (A/T) or Neutral (M/T). Connect a tachometer to the engine according to the manufacturer's instructions.

5. Turn the solenoid assembly to achieve the solenoid-on speed.

6. Disconnect the solenoid wire and turn the $\frac{1}{8}$ in. hex screw in the solenoid end, to adjust the solenoid-off speed.

7. Remove the tachometer, connect the canister vacuum line and turn the engine OFF.

2 BBL AND 4 BBL (EXCEPT 350 V8)

1. Run the engine to normal operating temperature.

2. Make sure that the choke is fully open, turn the A/C off, set the parking brake, block the drive wheels and connect a tachometer to the engine according to the manufacturer's instructions.

3. Disconnect and plug the vacuum hoses at the EGR valve and the vapor canister.

4. Place the transmission in Park (A/T) or Neutral (M/T).

5. Disconnect and plug the vacuum advance hose at the distributor. Check and adjust the timing.

6. Connect the distributor vacuum line.

7. On vehicles without A/C or a solenoid, place the idle speed screw on the low step of the fast idle cam and

turn the screw to the specified idle speed.

8. On vehicles with A/C, set the idle speed screw to the specified rpm. Disconnect the compressor clutch wire and turn the A/C "ON." Open the throttle momentarily to extend the solenoid plunger. Turn the solenoid screw to obtain the specified rpm.

9. On vehicles without A/C but with a solenoid, momentarily open the throttle to extend the solenoid plunger. Turn the solenoid screw to obtain the specified rpm. Disconnect the solenoid wire and turn the idle speed screw to the correct curb idle speed.

350 V8 ONLY

1. Run the engine to normal operating temperature.

2. Set the parking brake and block the drive wheels.

3. Connect a tachometer to the engine according to the manufacturer's instructions.

4. Disconnect and plug the purge hose at the vapor canister. Disconnect and plug the EGR vacuum hose at the EGR valve.

5. Turn the A/C off.

6. Place the transmission in Park (A/T) or Neutral (M/T).

7. Disconnect and plug the vacuum advance line at the distributor. Check and adjust the timing.

8. Connect the vacuum advance line. Place the automatic transmission in Drive.

9. On manual transmission vehicles without A/C, adjust the idle stop screw to obtain the specified rpm.

10. On vehicles with A/C, turn the A/C off, adjust the idle stop screw to obtain the specified rpm. Disconnect the compressor clutch wire and turn the A/C on. Open the throttle slightly to allow the solenoid plunger to extend. Turn the solenoid screw to obtain the solenoid rpm listed on the underhood emission sticker.

11. Connect all hoses and remove the tachometer.

1980–81 Firebird

231 V6 and 305 V8

1. Run the engine until it reaches normal operating temperature. Make sure that the choke is open and the A/C is off.

2. Connect a tachometer and a timing light as detailed earlier in this chapter.

3. Set the parking brake and block the wheels.

4. Tag, disconnect and plug all vapor canister and EGR vacuum hoses.

5. Start the engine and place the

transmission in Drive (A/T) or Neutral (M/T).

6. Disconnect the vacuum advance hose and then set the timing to specifications.

7. Reconnect the vacuum advance hose and set the idle speed to specifications.

8. Connect all hoses and stop the engine.

301 V8

1. Run the engine until it reaches normal operating temperature. Make sure that the choke is fully open and the A/C is off.

2. Connect a tachometer and a timing light as detailed earlier in this chapter.

3. Set the parking brake and block the wheels.

4. Disconnect the A/C compressor clutch lead wire.

5. Start the engine and place the transmission in Drive (A/T) or Neutral (M/T).

6. Check and adjust the ignition timing.

7. Disconnect the purge hose at the vapor canister.

8. If equipped with A/C, set the idle speed screw to the specified rpm. Turn the A/C on. Open the throttle momentarily to ensure that the solenoid plunger is fully extended. Adjust the idle speed solenoid to the speed given on the underhood sticker. Turn the A/C off.

9. On vehicles without A/C, turn the idle speed screw until the specified rpm is reached.

10. If equipped, place the automatic transmission in Park.

11. Disconnect and plug the vacuum hose at the EGR valve.

12. Set the fast idle screw on the 2nd step of the fast idle cam and adjust the screw to the specified rpm.

13. Stop the engine and reconnect the EGR vacuum hose, the vapor canister hose and the A/C compressor clutch connector.

350, 400 and 403 V8

1. Follow Steps 1–3 of the preceeding "301 V8" procedure.

2. Disconnect the plug the hose from the vapor canister.

3. Disconnect and plug the vacuum hose leading from the EGR valve.

4. Start the engine and place the transmission in Park (A/T) or Neutral (M/T).

5. Check and adjust the ignition timing.

NOTE: After checking the timing on the 400 V8 with a manual transmission, DO NOT reconnect the vacuum advance line.

6. Place the automatic transmission in Drive.

7. If equipped with A/C, turn the A/C on and disconnect the electrical connector at the compressor clutch. Open the throttle momentarily to ensure that the solenoid plunger is fully extended. Adjust the solenoid screw to the specified rpm. Reconnect the compressor clutch connector and turn the A/C off.

8. If NOT equipped with A/C, turn the idle screw to obtain the specified rpm.

9. Reconnect the disconnected hoses and turn off the engine.

1982–84 Rochester E2SE Models

For all overhaul and service adjustment procedures, refer to "Carburetors" in the Unit Repair section.

—— CAUTION ——

Before performing the idle adjustments, block the drive wheels and set the parking brake.

WITH A/C

1. Refer to the emission label in the engine compartment and follow the instructions to prepare the vehicle for adjustment.

2. Place the A/T in Drive or the M/T in Neutral and turn the throttle slightly to allow the solenoid plunger to fully extend.

3. Turn the solenoid screw to adjust the curb idle rpm, then disconnect the solenoid lead.

4. Turn the idle speed screw to set the basic idle speed. Reconnect the solenoid electrical leak after adjustment.

WITHOUT A/C

1. Refer to the emission label in the engine compartment and follow the instructions to prepare the vehicle for adjustment.

2. With the A/C off, turn the idle speed screw to set the curb idle speed.

3. Turn the A/C on, disconnect the A/C compressor lead at the A/C compressor and place the A/T in Drive or the M/T in Neutral.

4. Open the throttle slightly to allow the solenoid plunger to extend.

5. Turn the solenoid screw to adjust to the specified rpm. After adjustment, reconnect the A/C compressor lead.

6. Turn the A/C off. Set the curb idle speed by turning the idle speed screw.

1982 and Later Rochester E4ME Models

NOTE: No idle speed adjustment is necessary; the idle speed is controlled by the ECM.

FAST IDLE ADJUSTMENT

1982–84 Rochester E2SE Models

1. Refer to the emission label and prepare the vehicle for adjustment.

2. Place the transmission in Park (A/T) or Neutral (M/T).

3. Place the fast idle screw on the highest step of the fast idle cam.

4. Turn the fast idle screw to obtain the fast idle speed.

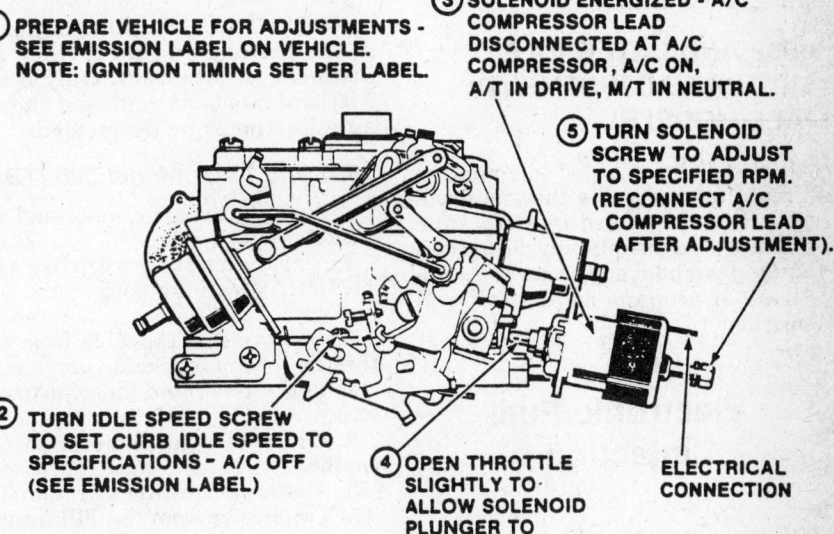

① PREPARE VEHICLE FOR ADJUSTMENTS - SEE EMISSION LABEL ON VEHICLE. NOTE: IGNITION TIMING SET PER LABEL.

③ SOLENOID ENERGIZED - A/C COMPRESSOR LEAD DISCONNECTED AT A/C COMPRESSOR, A/C ON, A/T IN DRIVE, M/T IN NEUTRAL.

⑤ TURN SOLENOID SCREW TO ADJUST TO SPECIFIED RPM. (RECONNECT A/C COMPRESSOR LEAD AFTER ADJUSTMENT).

② TURN IDLE SPEED SCREW TO SET CURB IDLE SPEED TO SPECIFICATIONS - A/C OFF (SEE EMISSION LABEL)

④ OPEN THROTTLE SLIGHTLY TO ALLOW SOLENOID PLUNGER TO FULLY EXTEND.

ELECTRICAL CONNECTION

Idle speed adjustment—E2SE model (with A/C)

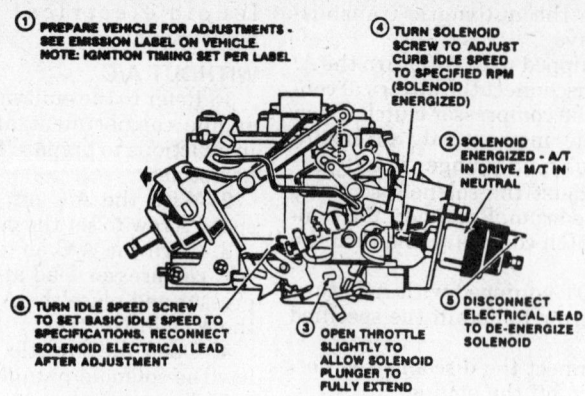

Idle speed adjustment—E2SE model (without A/C)

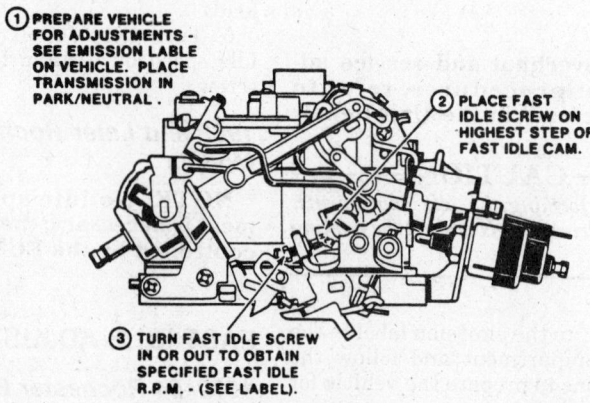

Fast Idle speed adjustment

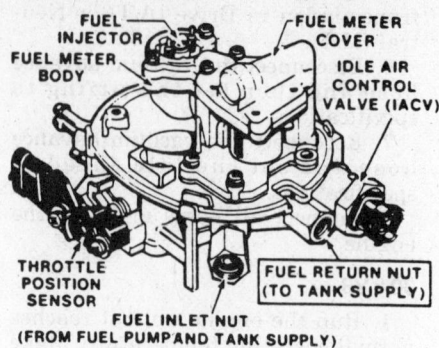

Throttle body—model 300

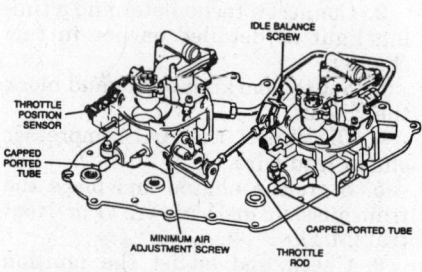

Cross fire injection system—model 400

1982 and Later Rochester E4ME Models

NOTE: The fast idle speed adjustment must be performed according to the emission control label instructions. See the underhood sticker.

1982 AND LATER IDLE MIXTURE ADJUSTMENT (ALL MODELS)

All models have sealed idle mixture screws; in most cases these are concealed under staked-in plugs. Idle mixture is adjustable only during carburetor overhaul and requires the addition of propane as an artificial enrichener.

Electronic Fuel Injection

Before removing any component of the fuel system, refer to the "Fuel Pressure Release" procedures in this section and release the fuel pressure. For further information on the fuel injection system, please refer to *Chilton's Guide To Fuel Injection And Feedback Carburetors*.

IDLE SPEED ADJUSTMENT

NOTE: The TBI unit is adjusted at the factory and no further adjustment is necessary. Only if the TBI unit has been replaced should an adjustment be performed.

1982 and Later Model 300 (TBI)

1. Remove the air cleaner and the gasket.
2. Disconnect and plug the THERMAC vacuum port on the TBI unit.
3. Remove the TV cable from the throttle control bracket to provide access to the minimum air adjustment screw.
4. Connect a tachometer to the engine.
5. Remove the Idle Air Control (IAC) connector from the TBI unit.
6. Start the engine, place the transmission in Park (A/T) or Neutral (M/T) and allow the engine speed to stabilize.

7. Install tool J-33047 into the idle air passage of the TBI unit. Make sure that the tool is seated and no air leaks exist.
8. Using a No. 20 Torx Bit tool, turn the idle stop screw until the engine speed is 475–525 rpm (A/T) or 750–800 rpm (M/T).
9. Stop the engine and remove the tool from the TBI unit.
10. To install, seal the throttle stop screw with silicone sealant and reverse the removal procedures.

1982–83 Model 400 (CFI)

1. Remove the air cleaner and the gasket.
2. Disconnect and plug the THERMAC vacuum port at the rear TBI unit.
3. If necessary, remove the plug covering the minimum air adjusting screw.
4. Block the wheels, set the parking brake, connect a tachometer to the engine, start the engine and allow the engine speed to stabilize.
5. Place the automatic transmission in Drive.
6. Using 2 tools J-33047, plug the idle air passages of each throttle body. Make sure that the tools are seated and no air leaks exist.

NOTE: When the plugs are installed, the rpm should drop below the curb idle speed. If the speed does not drop, check for an air leak.

7. At the rear TBI unit, remove the cap from the ported tube and connect a water manometer J-23951.

8. Adjust the minimum air adjustment screw to obtain 6 inches of water on the manometer. Remove the manometer and install the cap on the ported tube.

9. At the front TBI unit, remove the cap from the ported tube and connect the water manometer J-23951. The reading should be 6 inches of water on the manometer.

NOTE: If the manometer reading is not correct, locate the idle balance screw on the throttle linkage. If the screw is welded, break the weld and install a new screw with thread sealing compound. Adjust the screw to obtain 6 inches of water on the manometer.

10. Remove the manometer and install the cap on the ported tube.

11. At the rear TBI unit, adjust the minimum air adjustment screw to obtain 475 rpm.

12. Stop the engine and remove the idle air passage plugs.

13. Place the transmission in Neutral and start the engine.

NOTE: The engine will run at a high rpm but will decrease when the IAC motors close the air passages. When the rpm drops, stop the engine.

14. Check the Throttle Position Sensor (TPS) voltage and adjust, if necessary.

15. To install, reverse the removal procedures. Reset the IAC motors.

NOTE: To reset the IAC motors, drive the vehicle at 30 mph or if equipped with cruise control, disconnect the speedometer cable at the transducer, turn the key ON and rotate the cable to 30 mph.

1985 and Later Tuned Port Injection (TPI) — V6 and V8

NOTE: The idle stop screw of the throttle body, used to regulate the minimum idle speed, is adjusted and sealed at the factory. If it is necessary to adjust the idle speed, perform the following procedures.

1. Using an awl, pierce the idle stop screw plug and pry it from the throttle body housing.

2. With Idle Air Control (IAC) valve connected to the throttle body, ground the diagnostic lead.

3. Turn the ignition ON (DO NOT start the engine), wait 30 seconds, disconnect the IAC electrical connector, remove the IAC grounded lead and start the engine.

4. If equipped with an automatic transmission, place it in Drive and adjust the idle speed to 450–550 rpm by turning the idle stop screw. If equipped with a manual transmission, place it in Neutral and adjust the idle speed to 550–650 rpm (V6).

5. Turn the ignition OFF and reinstall the electrical connector to the IAC valve.

6. At the Throttle Position Sensor (TPS), install 3 jumper wires between the TPS and the harness connector.

7. Connect a digital voltmeter to terminals "A" and "B" of the TPS. Turn the ignition ON and check for voltages of 0.50–0.60 volts (V6) or 0.465–0.615 volts (V8).

NOTE: If the voltages of the TPS are not correct, loosen the mounting screws and adjust the TPS. When the voltages meet specifications, tighten the mounting screws.

8. If adjustment of the TPS has been performed, recheck the readings.

9. Turn the ignition OFF, remove the jumper wires and reconnect the electrical connector to the TPS.

10. Start the engine and check for proper idle operation. Seal the idle stop screw housing with silicone sealant.

ENGINE COOLING

Radiator

REMOVAL & INSTALLATION

1980–81 Models

1. Disconnect the negative battery cable.

2. Drain the cooling system.

3. Disconnect the upper and lower hoses.

4. If equipped with an automatic transmission, disconnect and plug the oil cooler lines.

5. Remove the upper fan shield (6 cyl.) or the upper shroud bracket (V8).

6. Remove the radiator hold-down bolts, the radiator and shroud assembly from the vehicle.

7. To install, reverse the removal procedures. Torque the hold-down bolts to 12 ft. lbs. Refill the cooling system. If equipped with an automatic transmission, make sure that the fluid level is correct.

1982 and Later Models

1. Disconnect the negative battery cable.

2. Drain the cooling system.

3. Remove the engine cooling fan. If equipped with a fan clutch, the clutch MUST be set aside in an upright position to prevent seal leakage.

4. Disconnect the radiator hoses from the radiator.

5. If equipped with an automatic transmission, disconnect and plug the transmission cooler lines at the radiator.

6. If equipped, remove the fan shield assembly.

7. Remove the radiator and shroud assembly, then lift the assembly straight up.

NOTE: The radiator assembly is held at the bottom by 2 cradles which are secured to the radiator support. If installing a new radiator, transfer the fittings from the old radiator to the new one.

8. To install, reverse the removal procedures. Refill the cooling system, operate the engine to normal operating temperatures and check for leaks. If equipped with an automatic transmission, check the transmission fluid level and adjust the level as required.

Water Pump

REMOVAL & INSTALLATION

NOTE: When servicing the V6 or V8 engines, disconnect the negative battery cable.

1. Drain the cooling system.

2. If equipped, remove the fan shroud and the upper radiator support.

3. Remove the drive belts.

4. Remove the fan and the pulley from the water pump.

NOTE: Viscous drive fans should not be stored horizontally. The silicone fluid can leak out of the fan assemble if it is not kept upright.

5. On the V6 and V8 engines, remove the upper and lower brackets, the air brace, the bracket and the lower power steering pump bracket (if equipped), swing the bracket aside.

6. Disconnect the heater and lower radiator hoses, from the water pump.

7. Remove the water pump-to-cylinder block bolts and the water pump.

NOTE: On the 4 cyl. engines, remove the pump by pulling it straight out of the block.

8. Clean the gasket mounting surfaces.

9. To install, use a new gasket and reverse the removal procedures. Torque the water pump bolts to 15 ft. lbs. and the accessories mounting bolts to 30 ft. lbs. Adjust the drive belt deflection to $\frac{1}{2}$ in. between the longest span of two pulleys. Refill the cooling system. Start the engine and check for leaks.

Thermostat

REMOVAL & INSTALLATION

1. Drain the cooling system until the coolant level is below the thermostat.

2. Remove the air cleaner assembly (except on 4 cyl).

3. Remove the upper radiator hose.

4. Remove the thermostat housing bolts, the thermostat housing and the thermostat.

5. Clean the gasket mounting surfaces.

6. To install, use a new gasket(s) and RTV sealant, the thermostat and reverse the removal procedures. Torque the housing bolts to 20 ft. lbs. (4 cyl.) or 30 ft. lbs. (all others). Fill the cooling system. Start the engine and check for leaks.

NOTE: Make sure that the thermostat spring is installed towards the engine; NOT the radiator.

—————— CAUTION ——————
DO NOT use RTV sealant with the gasket on 4 cylinder engines.

EMISSION CONTROLS

For all information on emission control systems please refer to "Emission Controls" in the Unit Repair section. Due to the complex nature of modern electronic engine control systems, comprehensive diagnosis and testing procedures fall outside the confines of this repair manual. For complete information on diagnosis, testing and repair procedures concerning all modern engine and emission control systems, please refer to *Chilton's Guide To Electronic Engine Controls*.

ENGINE MECHANICAL

Engine

REMOVAL & INSTALLATION

NOTE: In the middle of the 1986 model year, General Motors introduced a limited addition (1000) Camaro IROC, equipped with the 350 5.7L engine similar to the Corvette engine. There is no information on that particular vehicle at the time of publication. If information is needed on the that particular engine, please refer to the Corvette section in this book.

Four Cylinder Engine

1. Disconnect the negative battery cable.

2. Mark the location of the hood on the hood hinges and remove the hood.

3. Drain the cooling system.

4. Remove the A/C compressor and any necessary brackets to gain working clearance.

—————— CAUTION ——————
DO NOT disconnect any air conditioning refrigerant lines unless you are familiar with discharge procedures. Escaping refrigerant can freeze any surface it contacts, including skin and eyes.

5. Remove the radiator hoses from the engine. Remove the fan assembly. Remove the radiator shroud and radiator.

6. If the vehicle is equipped with power steering, remove the power steering pump.

7. Tag and disconnect the electrical connector at the bulkhead connector.

8. Disconnect the fuel lines at the carburetor.

9. Remove the brake hoses from the filter and the ground strap from the rear of the cylinder head.

10. Working from inside the vehicle, remove the right-hand hush panel and the ECM harness at the main ECM connector. Remove the right-hand splash shield from the right fender and feed the ECM harness out from inside the vehicle.

11. Disconnect the heater hoses from the heater core. Remove the canister hose and the throttle cable from the electronic fuel injection if equipped.

12. Raise the vehicle and support safely. Disconnect the electrical connections from the transmission.

13. Remove the flywheel dustcover. If the vehicle is an automatic, remove

the torque converter to flywheel holding bolts.

14. Remove the bolts holding the bellhousing to the engine. Remove the bellhousing to engine exhaust pipe support.

15. Remove the exhaust pipe at the manifold. Remove the catalytic converter assembly.

16. Remove the starter assembly.

17. Remove the clutch fork return spring if vehicle is equipped with a manual transmission.

18. Remove the motor mount bolts.

19. Lower the vehicle and install a suitable engine lifting device.

20. Position a floor jack under the transmission to support the transmission.

21. Lift the engine from the vehicle and place in a suitable engine holding fixture.

22. Installation is the reverse of the removal procedures.

V6 and V8 Engines

1. Disconnect the negative battery cable.

2. Mark the location of the hood and remove the hood from the vehicle.

3. Drain the cooling system. Remove the lower raditor hose and the upper fan shroud. Remove the fan assembly.

4. Remove the upper radiator hose and the coolant recovery hose. Remove the radiator.

5. Remove the transmission cooler lines.

6. Remove the heater hoses.

7. Disconnect the carburetor linkage. If the vehicle is equipped with cruise control, disconnect the detent cable.

8. Remove the vacuum brake booster line.

9. Remove the distributor cap and lay aside with the wiring to gain working clearance.

10. Disconnect all necessary wires and hoses.

11. Remove the power steering pump and lay aside.

12. Raise the vehicle support safely.

13. Remove the exhaust pipes from the manifold. Remove the dust cover from the vehicle. Remove the converter bolts.

14. Disconnect the starter wires and remove the starter assembly.

15. Remove the bellhousing bolts. Remove the motor mount through bolts.

16. Disconnect the fuel lines at the fuel pump.

17. Lower the vehicle and support the transmission using a suitable fixture.

18. Remove the air injection reaction system if equipped.

19. Attach a suitable engine lifting device and remove the engine from the vehicle.

20. Installation is the reverse of the removal procedure.

Exhaust Manifold

REMOVAL & INSTALLATION

1980–81 Models

V6 and V8 ENGINES

1. If equipped with the AIR system, remove the air injector manifold assembly.

NOTE: The $^1/_4$ in. pipe threads in the manifold are straight threads.

2. Disconnect the negative battery cable.

3. If equipped, remove the air cleaner pre-heater shroud.

4. Remove the spark plug wire heat shields.

5. At the left exhaust manifold, disconnect and remove the alternator.

6. Disconnect the exhaust pipe from the manifold and allow it to hang.

7. Bend the locktabs. Remove the end bolts, the center bolts and the exhaust manifold.

NOTE: A $^9/_{16}$ in. thin wall 6-point socket, sharpened at the leading edge and tapped onto the head of the bolt, simplifies bending the

locktabs. If installing a new manifold on the right side of the 1980–81 V8, transfer the heat stove from the old manifold to the new one.

8. Clean the gasket mounting surfaces.

9. To install, use new gaskets and reverse the removal procedures. Torque the bolts to specifications from the inside working outwards.

1982 and Later Models

4 CYL ENGINE

1. Disconnect the negative battery cable.

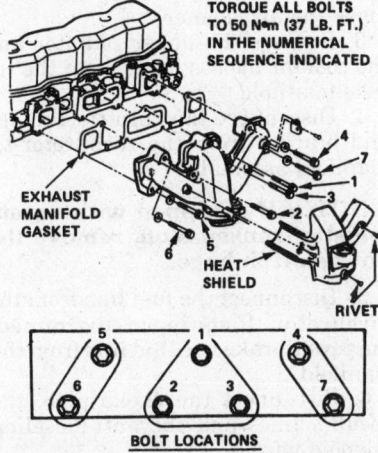

Exhaust manifold bolt tightening sequence—151 4 cylinder engine through 1984

2. Remove the air cleaner assembly, mark the hoses for proper identification.

3. Remove the EFI preheat tube.

4. Remove the oxygen sensor and disconnect the exhaust pipe from the exhaust manifold.

5. Remove the engine oil level dipstick and tube.

6. Remove the exhaust manifold attaching bolts and the manifold.

7. To install, use a new gasket and reverse the removal procedures. Torque the bolts to 44 ft. lbs.

V6 ENGINE—LEFT SIDE

1. Disconnect the negative battery cable.

2. Raise and support the vehicle on jackstands.

3. Disconnect the exhaust pipe from the exhaust manifold.

4. Remove the 4 rear manifold bolts and the nut, then lower the vehicle.

5. Disconnect the air management hoses and wires.

6. If equipped, remove the power steering bracket and move aside.

7. Remove the manifold attaching bolts and the manifold.

8. To install, use a new gasket and reverse the removal procedures. Torque the manifold bolts to 25 ft. lbs., from the inside working outwards.

V6 ENGINE—RIGHT SIDE

1. Disconnect the negative battery cable.

2. Raise and support the vehicle on jackstands.

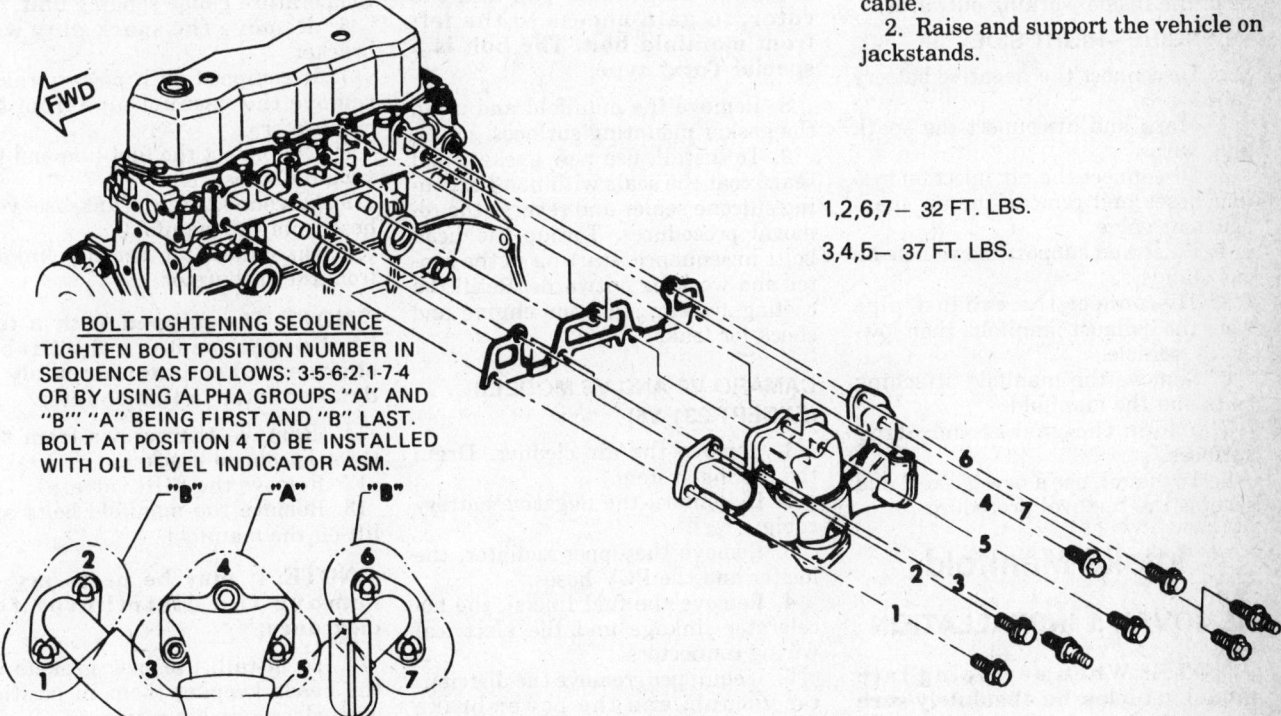

1,2,6,7 – 32 FT. LBS.

3,4,5 – 37 FT. LBS.

Exhaust manifold bolt tightening sequence—1985 and later 151 4 cylinder engine

3. Disconnect the exhaust pipe from the exhaust manifold.

4. Lower the vehicle and remove the exhaust manifold bolts.

5. Disconnect the air management hose and remove the manifold.

6. To install, reverse the removal procedures. Torque the manifold bolts, from the inside working outwards.

V8 ENGINE—LEFT SIDE

1. Disconnect the negative battery cable.

2. Mark and disconnect the spark plug wires.

3. Disconnect the air injection system hoses.

4. If equipped with A/C, unbolt the A/C compressor (DO NOT disconnect the refrigerant hoses) and move aside.

5. If equipped with power steering, remove the power steering pump (DO NOT disconnect the hydraulic lines) and move aside.

6. Remove the rear A/C and power steering adjusting brackets.

7. Raise and support the vehicle on jackstands.

8. Disconnect the exhaust pipe from the exhaust manifold, then lower the vehicle.

9. Remove the manifold attaching bolts and the manifold.

10. Clean the gasket mounting surfaces.

11. To install, use a new gasket and reverse the removal procedures. Torque the manifold bolts to 20 ft. lbs. from the inside working outwards.

V8 ENGINE—RIGHT SIDE

1. Disconnect the negative battery cable.

2. Mark and disconnect the spark plug wires.

3. Disconnect the air injection system hoses and remove the air management valve.

4. Raise and support the vehicle on jackstands.

5. Disconnect the exhaust pipe from the exhaust manifold, then lower the vehicle.

6. Remove the manifold attaching bolts and the manifold.

7. Clean the gasket mounting surfaces.

8. To install, use a new gaskets and reverse the removal procedures.

Intake Manifold

REMOVAL & INSTALLATION

NOTE: When servicing late model vehicles, be absolutely sure to mark the vacuum hoses and wiring, so that these items may be properly reconnected during in-stallation. Also, when disconnecting fittings or metal lines (fuel, power brake vacuum), always use two flare nut wrenches. Hold the wrench as if tightening the fitting (clockwise), THEN loosen and disconnect the smaller fitting from the larger fitting. If this is not done, damage to the line will result.

1980–81 Models
231 V6

1. Disconnect the negative battery cable.

2. Drain the cooling system. Remove the air cleaner.

3. Remove the upper radiator and the coolant bypass hoses from the intake manifold.

4. Disconnect the throttle linkage and bracket from the carburetor-to-manifold assembly.

NOTE: If equipped with an automatic transmission, remove the downshift linkage.

5. Disconnect the fuel line from the carburetor. If equipped, disconnect the power brake vacuum line from the manifold.

6. Disconnect the choke pipe, the vacuum lines and the anti-dieseling solenoid wire.

7. Remove the manifold bolts.

NOTE: It will be necessary to remove the distributor cap and the rotor, to gain access to the left front manifold bolt. The bolt is a special Torx® type.

8. Remove the manifold and clean the gasket mounting surfaces.

9. To install, use new gaskets and seals, coat the seals with non-hardening silicone sealer and reverse the removal procedures. Torque the head bolts in sequence, starting in the center and working outwards. Refill the cooling system, start the engine and check for leaks.

CAMARO V6 AND V8 MODELS (EXCEPT 231 V6)

1. Remove the air cleaner. Drain the cooling system.

2. Disconnect the negative battery cable.

3. Remove the upper radiator, the heater and the PCV hoses.

4. Remove the fuel line(s), the accelerator linkage and the electrical wiring connectors.

5. If equipped, remove the distributor vacuum and the power brake hoses.

6. Remove the distributor cap and the distributor.

NOTE: Before removing the distributor, mark the rotor's position, relative to the distributor housing and the engine.

7. If necessary, remove the alternator upper bracket, the air cleaner bracket and the accelerator bellcrank.

8. If equipped, remove the A/C compressor and bracket, set the compressor aside.

9. If equipped, remove the crusie control servo and bracket.

10. Remove the manifold-to-head attaching bolts, then remove the manifold and carburetor assembly.

NOTE: If the mainfold is to be replaced, transfer the manifold equipment to the new manifold.

11. Clean the gasket mounting surfaces.

12. To install, use new gaskets and sealant, then reverse the removal procedures. Torque the manifold bolts to 25–45 ft. lbs. Refill the cooling system, then start the engine and check for leaks. Check the ignition timing.

FIREBIRD V8 MODELS

1. Drain the cooling system.

2. Remove the air cleaner assembly.

3. Remove the water outlet fitting allowing the radiator hose to remain attached.

4. If necessary, disconnect the heater hose from its fitting.

5. Disconnect the wire from the temperature gauge sending unit.

6. Remove the spark plug wire bracket.

7. If equipped with power brakes, remove the vacuum pipe from the carburetor.

8. Disconnect the fuel line and the vacuum hoses.

9. Disconnect the crankcase vent hose from the manifold.

10. Disconnect the throttle linkage from the carburetor.

NOTE: If equipped with a turbocharger, remove the turbocharger and actuator assembly as detailed later in this chapter.

11. Remove the screws from the throttle control bracket.

12. Remove the EGR valve.

13. Remove the manifold bolts and lift off the manifold.

NOTE: It may be necessary to remove the distributor for clearance.

14. To install, use new gaskets on the heads keeping them in position with plastic gasket retainers.

15. Lower the intake manifold onto the engine and then install the O-ring seal.

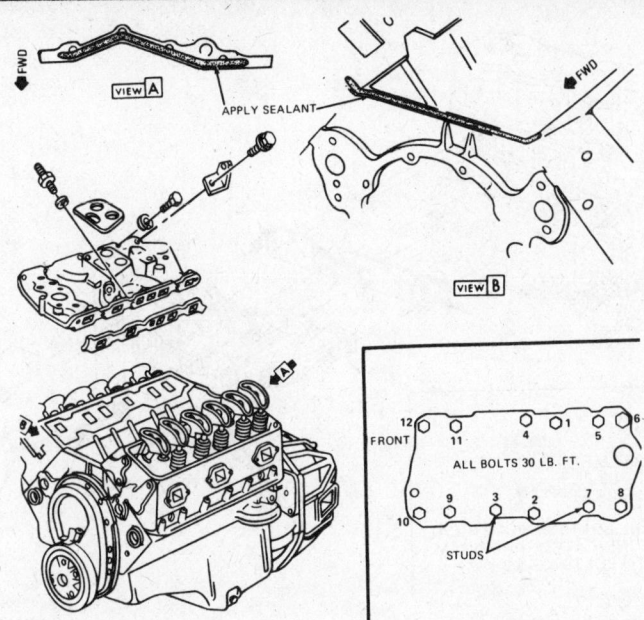

Intake manifold bolt tightening sequence for the Chevrolet-built 229 V6 engine

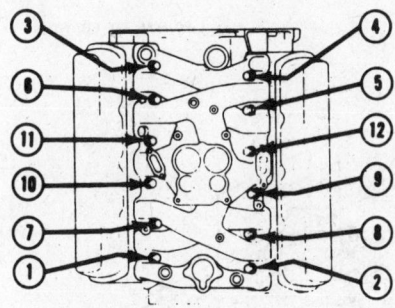

Intake manifold bolt tightening sequence for the Oldsmobile-built V8 engines

16. Loosely install the bolts and nuts.

17. Install the throttle control bracket assembly.

18. Install a new O-ring seal between the timing chain cover and the intake manifold and tighten the bolt to 15 ft. lbs.

19. Tighten all bolts and nuts evenly to 40 ft. lbs. starting from the center and working out.

20. To complete the installation, reverse the removal procedures.

1982 and Later Carbureted Engines

173 V6 ENGINE

1. Disconnect the negative battery cable.

2. Remove the air cleaner assembly.

3. Drain the cooling system.

4. Mark and disconnect the wiring and hoses from the carburetor.

5. Disconnect the fuel line at the carburetor.

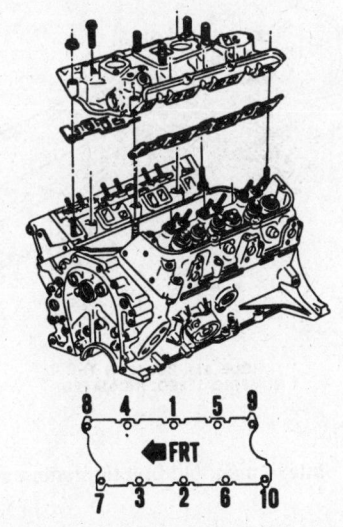

Intake manifold bolt tightening sequence for the 173 V6 engine

6. Disconnect the throttle, the transmission downshift and the cruise control (if equipped) cables from the carburetor.

7. Mark and disconnect the wiring from the ignition coil.

8. Remove the distributor cap and the distributor.

NOTE: Before removing the distributor, mark the rotor position relative to the distributor housing and the engine.

9. Remove the air management hose.

10. Mark and disconnect the hoses from the emission canister. Remove

the pipe bracket from the left valve cover.

11. Remove the left valve cover.

12. Remove the air management bracket from the right valve cover.

13. Remove the right valve cover.

14. Remove the upper radiator hose and the heater hose from the intake manifold.

15. Disconnect the wiring from the coolant switches.

16. Remove the manifold attaching bolts and lift the manifold from the engine.

17. Clean the gasket mounting surfaces.

18. To install, use new gaskets, sealant and reverse the removal procedures. Torque the manifold bolts to 23 ft. lbs. in sequence. Refill the cooling system, start the engine and check for leaks. Check the idle speed.

NOTE: The manifold gaskets are marked for the right and left sides of the engine.

V8 ENGINES

1. Disconnect the negative battery cable.

2. Remove the air cleaner assembly.

3. Drain the cooling system.

4. Disconnect the upper radiator and the heater hoses from the intake manifold.

5. Disconnect the throttle, the transmission downshift and the cruise control (if equipped) linkages from the carburetor.

6. Disconnect the fuel line at the carburetor.

7. Disconnect the spark plug wires from the right side of the engine.

8. Mark and disconnect the wiring and hoses from the intake manifold and carburetor.

9. Remove the distributor as previously outlined.

NOTE: Before removing the distributor, mark the rotor's position, relative to the distributor housing and the engine.

10. If equipped, remove the A/C compressor from the brackets and move aside. DO NOT disconnect the refrigerant hoses.

11. If equipped, remove the A/C compressor brackets and the cruise control servo and bracket, then move aside.

12. Remove the upper alternator mounting bracket.

13. Remove the EGR solenoid and the bracket.

14. Remove the power brake booster vacuum line from the intake manifold-to-carburetor connection.

15. Remove the manifold mounting

bolts and lift the manifold from the engine.

16. Clean the gasket mounting surfaces.

17. To install, use new gaskets, sealant and reverse the removal procedures. Refill the cooling system, start the engine and check for leaks. Check the engine speed.

TBI Equipped Engines

151 4 CYL

1. Disconnect the negative battery cable.

2. Remove the air cleaner assembly and the heat stove pipe.

3. Remove the PCV valve and hose.

4. Drain the cooling system.

5. Disconnect the fuel lines from the Throttle Body Injection (TBI) unit.

--- **CAUTION** ---

Before removing any component of the fuel system, refer to the "Fuel Pressure Release" procedures in this section and release the fuel pressure.

6. Mark and disconnect the vacuum lines and the electrical connections from the TBI unit.

7. Disconnect the throttle and the bell crank linkages from the TBI unit, then move aside.

8. Disconnect the transaxle downshift and the cruise control (if equipped).

9. Remove the ignition coil and the alternator brace.

10. Disconnect the heater hose from the intake manifold.

11. Remove the A/C compressor support brackets and the compressor, move aside. DO NOT disconnect the refrigerant lines from the compressor.

12. Remove the manifold attaching bolts and the manifold.

13. Clean the gasket mounting surfaces.

14. To install, use new gaskets and sealant, then reverse the removal procedures. Torque the manifold bolts to 29 ft. lbs. Refill the cooling system, start the engine and check for leaks. Adjust the idle speed.

305 V8 TBI

1. Disconnect the negative battery cable.

2. Remove the air cleaner assembly. Drain the cooling system.

3. Disconnect the fuel inlet line at the front Throttle Body Injection (TBI) unit.

--- **CAUTION** ---

Before removing any component of the fuel system, refer to the "Fuel Pressure Release" procedures in this section and release the fuel pressure.

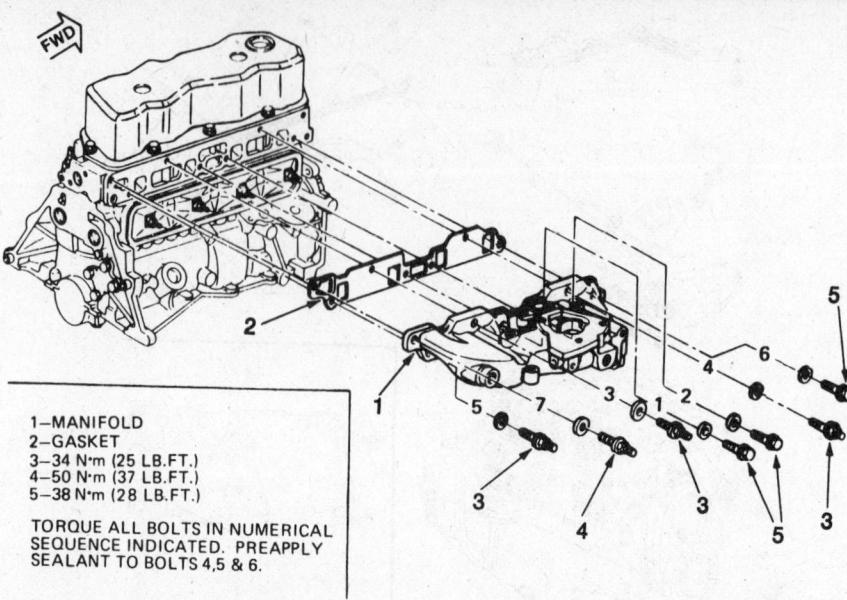

1—MANIFOLD
2—GASKET
3—34 N·m (25 LB. FT.)
4—50 N·m (37 LB. FT.)
5—38 N·m (28 LB. FT.)

TORQUE ALL BOLTS IN NUMERICAL SEQUENCE INDICATED. PREAPPLY SEALANT TO BOLTS 4, 5 & 6.

Intake manifold bolt tightening sequence—1986 and later 151 4 cylinder engine

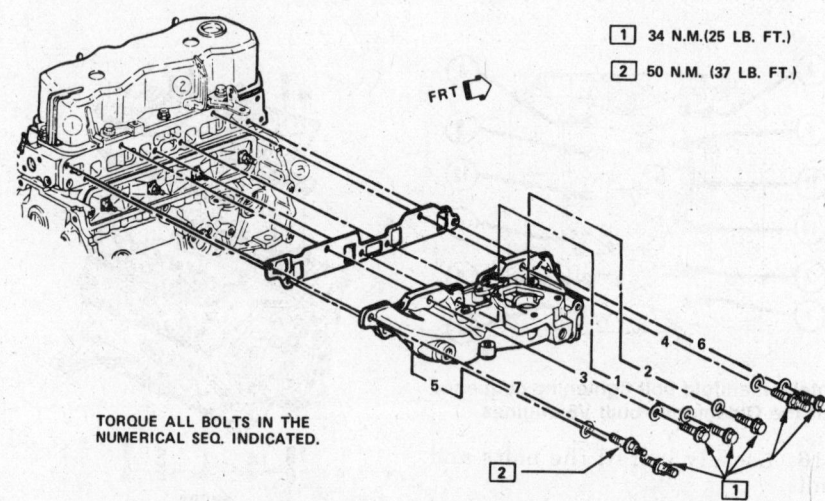

1️⃣ 34 N.M. (25 LB. FT.)
2️⃣ 50 N.M. (37 LB. FT.)

TORQUE ALL BOLTS IN THE NUMERICAL SEQ. INDICATED.

Intake manifold bolt tightening sequence—1982–85 151 4 cylinder engine

4. Remove the exhaust gas recirculation (EGR) solenoid.

5. Remove the alternator adjusting bracket.

6. Disconnect the wiring from the idle air motors, the injectors and the throttle position sensor (TPS).

7. Disconnect the fuel return line at the rear TBI unit.

8. Remove the power brake booster line.

9. Disconnect the accelerator and cruise control cables, tie the cable and bracket assembly aside.

NOTE: If equipped with A/C, remove the A/C strut.

10. Disconnect the positive crankcase ventilation (PCV) hose at the manifold and move the hose aside.

11. Mark and disconnect any vacuum hoses which will interfere with the removal of the manifold.

12. If you plan on removing the TBI units from the upper manifold plate, remove the fuel balance tube (connecting the units).

13. Remove the TBI plate-to-intake manifold bolts, then lift the TBI and plate assembly from the intake manifold.

14. Remove the distributor cap and the distributor.

NOTE: Before removing the distributor, mark the rotor's position, relative to the distributor housing and the engine.

15. Disconnect the upper radiator hose from the thermostat housing.

16. Disconnect the heater hose from the intake manifold.

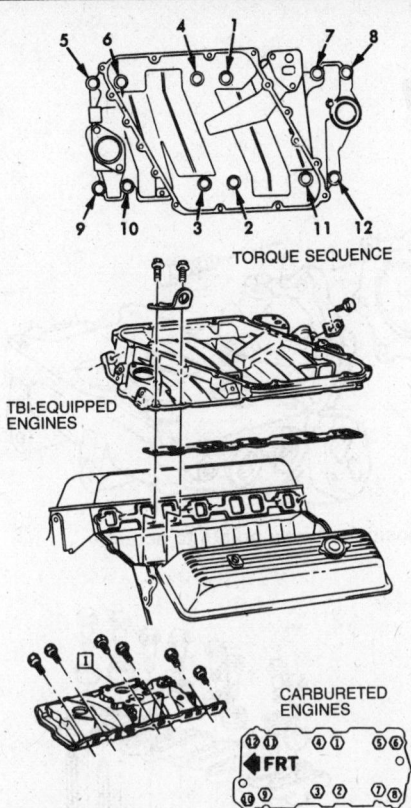

TORQUE SEQUENCE

TBI-EQUIPPED ENGINES

CARBURETED ENGINES

Intake manifold bolt tightening sequence of all Chevrolet-built V8 engines. Note that the lower sequence is used for all carbureted engines, whereas the upper sequence is used for all TBI-equipped engines

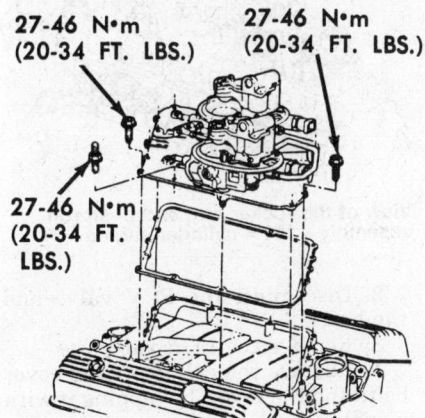

27-46 N•m (20-34 FT. LBS.)

27-46 N•m (20-34 FT. LBS.)

27-46 N•m (20-34 FT. LBS.)

TBI plate and gasket installation on V8 engines, if equipped

17. Remove the intake manifold-to-cylinder head bolts and lift the intake manifold assembly from the engine.

18. Clean the gasket mounting surfaces.

19. To install, use new gaskets and sealant, then reverse the removal procedures. Torque the intake manifold bolts to 25–45 ft. lbs., the TBI and

plate assembly bolts to 20–34 ft. lbs. Refill the cooling system. Adjust the throttle body linkages. Check the idle speed.

173 V6 MFI

1. Disconnect the negative battery cable.

2. Remove the air cleaner and drain the cooling system.

3. Remove the air intake plenum and the fuel rail assembly.

— **CAUTION** —

Before removing any component of the fuel system, refer to the "Fuel Pressure Release" procedures in this section and release the fuel pressure.

4. Disconnect the wires from the spark plugs and the coil.

5. Remove the distributor cap and the spark plug wires.

NOTE: Mark the rotor's position, relative to the distributor housing and the engine.

6. Remove the distributor.

7. If equipped with a manual transmission, remove the air management hose and bracket.

8. Disconnect the emission canister hoses.

9. Remove the pipe bracket from the left valve cover and the left valve cover.

10. Remove the right valve cover, the upper radiator hose and the heater hose.

11. Disconnect the coolant switches.

12. Remove the bolts and the manifold.

13. Clean the gasket mounting surfaces.

14. To install, use new gaskets and sealant, then reverse the removal procedures. Torque the manifold bolts to 13–25 ft. lbs. Refill the cooling system. Start the engine and check for leaks. Check the timing.

305 V8 PFI

1. Disconnect the negative battery cable.

2. Drain the cooling system.

3. Disconnect the accelerator T.V. and the cruise control cables.

4. Remove the air intake duct.

5. Disconnect the coolant hoses and the electrical wiring connectors from the throttle body.

6. Disconnect the vacuum and the breather hoses from the throttle body.

7. Remove the throttle body from the plenum.

8. Remove the shield from the distributor and the distributor.

NOTE: Before removing the distributor, mark the rotor's position, relative to the distributor housing and the engine.

NOTE APPLY A SMOOTH CONTINUOUS BEAD APPROX 2.0-3.0 WIDE AND 3.0-5.0 THICK ON BOTH SURFACES. BEAD CONFIGURATION MUST INSURE COMPLETE SEALING OF WATER AND OIL. SURFACE MUST BE FREE OF OIL AND DIRT TO INSURE ADEQUATE SEAL.

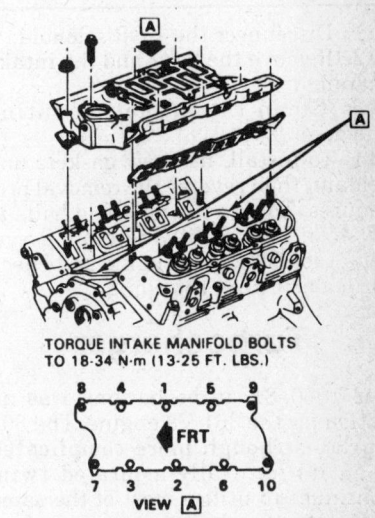

TORQUE INTAKE MANIFOLD BOLTS TO 18-34 N·m (13-25 FT. LBS.)

VIEW A

View of the PFI intake manifold

34-61 N·M (25-45 FT.LBS.) SNUG TIGHTENING SEQUENCE VIEW A

TIGHTENING SEQUENCE

VIEW B

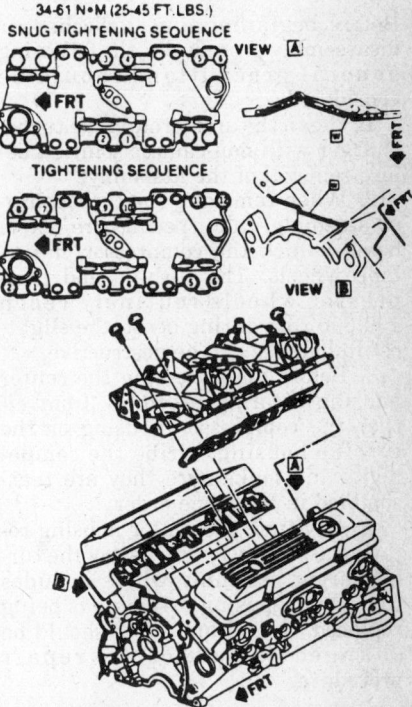

W AT TIME OF INSTALLATION SURFACE AREA MUST BE FREE OF OIL AND SEALING COMPOUND MUST BE WET TO TOUCH WHEN BOLT/SCREWS ARE TORQUED. APPLY SEALING COMPOUND .12 THICK.

View of the MFI intake manifold

9. Remove the power brake and the vacuum hoses from the plenum.

10. Remove the plenum, the fuel rail, the cold start injector and the runners.

CAUTION

Before removing any component of the fuel system, refer to the "Fuel Pressure Release" procedures in this section and release the fuel pressure.

---

11. Disconnect the EGR solenoid.

12. Remove the bolts and the intake manifold.

13. Clean the gasket mounting surfaces.

14. To install, use new gaskets and sealant, then reverse the removal procedures. Torque the manifold bolts to 25–45 ft. lbs. Refill the cooling system, then start the engine and check for leaks. Check the timing.

Turbocharger

For 1980–81, turbocharging was an option for the 301 V8 engine. The 301 Turbo, although more complicated than its naturally-aspirated twin, continues to utilize most of the same design features.

PRECAUTIONS

Before beginning any turbocharger disassembly procedures, the following general precautions should be considered.

1. Clean the area around the turbocharger with non-caustic solution before removal of the assembly.

2. When removing the turbocharger assembly, take special care not to bend or nick the compressor or turbine wheels. The turbine and compressor wheels routinely reach 130,000 rpm during boost, the slightest imbalance can be destructive.

3. Before disconnecting the center housing rotating assembly from either the compressor housing or the turbine housing, scribe the components and make sure they are reassembled in the same order.

4. Any time the center housing rotating assembly or any part of the turbocharger assembly which includes the center housing assembly is being replaced, the oil and filter should be changed as part of the repair procedure.

Wastegate Actuator

REMOVAL & INSTALLATION

1. Disconnect the hoses from the actuator.

2. Remove the wastegate linkage-to-actuator rod clip.

3. Remove the bolts attaching the actuator to the compressor housing.

4. To install, reverse the removal procedures.

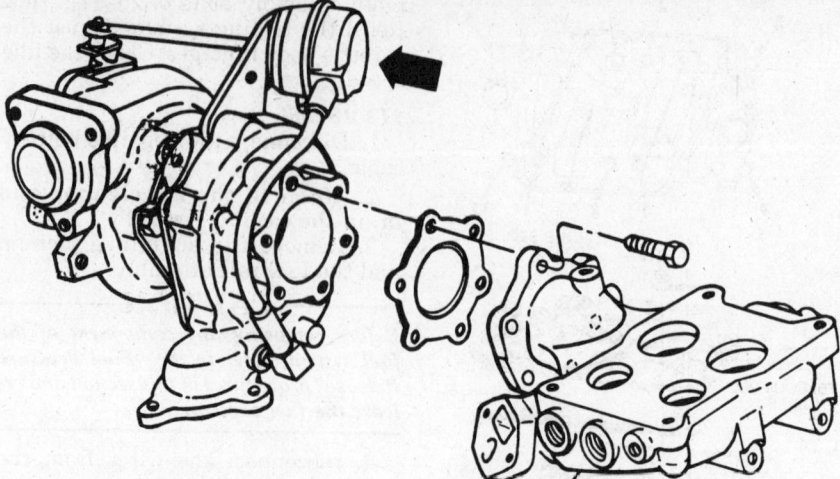

Turbocharger to plenum mounting position. Arrow points to actuator

Turbocharger and Actuator Assembly

REMOVAL & INSTALLATION

The carburetor and plenum are removed as a unit with the turbocharger.

1. Disconnect the turbocharger exhaust inlet and outlet pipes at the turbocharger. Remove the air cleaner.

2. Disconnect the carburetor and transmission control linkages at the carburetor. Disconnect and plug the carburetor fuel line and the necessary vacuum lines.

3. Drain about 3 qts. of coolant from the radiator. Disconnect the coolant hoses from the front and rear of the plenum.

4. Disconnect the EGR pipe at the intake manifold fitting. Remove the two turbine housing-to-bracket bolts on the intake manifold.

5. Remove the three compressor housing-to-intake manifold bolts.

6. Remove the turbocharger, actuator, carburetor and plenum as an assembly. Disconnect the vacuum hoses, as necessary.

7. Remove the six turbocharger-to-carburetor and plenum assembly bolts, then separate the components.

8. To install, use new gaskets and reverse the removal procedures.

Rocker Arm and Pushrod

REMOVAL & INSTALLATION

4 Cyl Engine

1. Remove the air cleaner, the spark plug wires and clips.

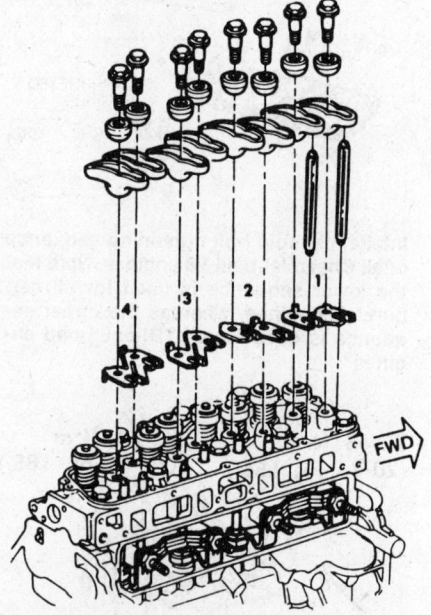

View of the rocker arm and push rod assembly—151 4 cylinder engine

2. Disconnect the PCV valve and the hose.

3. Remove the EGR valve.

4. Remove the cylinder head cover bolts and the cover, by tapping it with a rubber mallet.

5. Remove the rocker arm bolt, the ball washer and the rocker arm.

NOTE: If removing the push rod ONLY, loosen the rocker arm bolt, swing it aside and remove the push rod. If removing all of the push rods, mark them for reinstallation purposes.

6. Clean the gasket mounting surfaces.

7. To install, use a new head gasket and reverse the removal procedures.

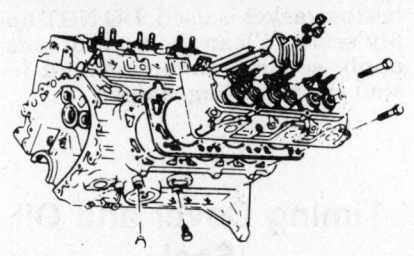

View of the rocker arm and pushrod assembly—V6 shown, V8 is similiar

Torque the rocker arm bolt to 20 ft. lbs. Start the engine, check the timing and the idle speed.

V6 Engine

1. Disconnect the negative battery cable.
2. Disconnect the air management hose, the vacuum hoses, the electrical wiring, the coil and the pipe brackets.
3. If equipped with a carburetor, remove the fuel lines, the throttle controls and the bracket from the carburetor.
4. Remove the cylinder head cover nuts and the covers, by tapping on them with a rubber mallet.
5. Remove the rocker arm nuts, the ball washers, the rocker arms and the push rods.

NOTE: If removing the push rod ONLY, loosen the rocker arm nut, swing it aside and remove the push rod. If removing all of the push rods, mark them for reinstallation purposes.

6. Clean the gasket mounting surfaces.
7. To install, use new cylinder head cover gaskets and sealant, then reverse the removal procedures. Torque the rocker arms nuts to 10–15 ft. lbs. and the cylinder head cover nuts to 7–15 ft. lbs. Adjust the valves. Start the engine, check the timing and the idle speed.

V8 Engine

1. Disconnect the negative battery cable.
2. Remove the air cleaner.
3. Disconnect the AIR, the EGR solenoid, the air management, the power brake booster hoses and/or tubes.
4. Remove the EGR solenoid.
5. Remove the PCV valve and the air management valve bracket, then move aside.
6. Move the air management bracket and the wiring harness aside.
7. Disconnect the alternator wiring.
8. Remove the cylinder head cover

nuts and the covers, by tapping on them with a rubber mallet.
9. Remove the rocker arm nuts, the ball washers, the rocker arms and the push rods.

NOTE: If removing the push rod ONLY, loosen the rocker arm nut, swing it aside and remove the push rod. If removing all of the push rods, mark them for reinstallation purposes.

10. Clean the gasket mounting surfaces.
11. To install, lubricate the parts, use new cylinder head cover gaskets and sealant, then reverse the removal procedures. Torque the rocker arm nuts to 5–11 ft. lbs. and the cylinder head cover nuts to 4–6 ft. lbs. Adjust the valves. Start the engine, check the timing and the idle speed.

VALVE CLEARANCE ADJUSTMENT

4 Cyl Engine

All engines use hydraulic valve lifters, which eliminate the need for periodic valve adjustments. If the rocker arms and/or cylinder heads have been removed or replaced, the rocker arms must be adjusted for zero lash.

V6 Engine

1. Refer to the "Rocker Arm and Pushrod Removal & Installation" procedures in this section and remove the cylinder head covers.
2. Turn the crankshaft, so that the No. 1 cyl. is at TDC of the compression stroke and the timing mark of the crankshaft pulley is on "0" of the timing plate.

NOTE: With the No. 1 cyl. on TDC, adjust the intake valves of cylinders No. 1, 5 and 6 and the exhaust valves of cylinders No. 1, 2 and 3.

3. Loosen the rocker arm nut until valve lash is felt. Tighten the nut until the lash disappears, then tighten the nut down 1½ more turns.
4. Turn the crankshaft 1 revolution, so that the No. 4 cyl. is at TDC and the crankshaft pulley is on the "0" mark of the timing plate.

NOTE: With the No. 4 cyl. on TDC, adjust the intake valves of cylinders No. 2, 3 and 4 and the exhaust valves of cylinders No. 4, 5 and 6.

5. Adjust the remaining valves in a similar manner.
6. Clean the gasket mounting surfaces.

7. To install, use new cylinder head cover gaskets and sealant, then reverse the removal procedures. Torque the cylinder head cover nuts to 7–15 ft. lbs. Start the engine, check the timing and the idle speed.

V8 Engine

1. Refer to the "Rocker Arm and Pushrod Removal & Installation" procedures in this section and remove the cylinder head covers.
2. Turn the crankshaft, so that the No. 1 cyl. is at TDC of the compression stroke and the timing mark of the crankshaft pulley is on "0" of the timing plate.

NOTE: With the No. 1 cyl. on TDC, adjust the intake valves of cylinders No. 1, 2, 5 and 7 and the exhaust valves of cylinders No. 1, 3, 4 and 8.

3. Loosen the rocker arm nut until valve lash is felt. Tighten the nut until the lash disappears, then turn the nut down 1 full rotation.
4. Turn the crankshaft 1 revolution, so that the No. 6 cyl. is at TDC and the crankshaft pulley is on the "0" mark of the timing plate.

NOTE: With the No. 6 cyl. on TDC, adjust the intake valves of cylinders No. 3, 4, 6 and 8 and the exhaust valves of cylinders No. 2, 5, 6 and 7.

5. Adjust the remaining valves in the similar manner.
6. Clean the gasket mounting surfaces.
7. To install, use new cylinder head cover gaskets and sealant, then reverse the removal procedures. Torque the cylinder head cover nuts to 4–6 ft. lbs. Start the engine, check the timing and the idle speed.

Cylinder Head

REMOVAL & INSTALLATION

NOTE: When servicing late model vehicles, be absolutely sure to mark vacuum hoses and wiring so that these items may be properly reconnected during installation. Also, when disconnecting fittings of metal lines (fuel, power brake vacuum), always use two flare nut (or line) wrenches. Hold the wrench on the large fitting with pressure on the wrench as if you were tightening the fitting (clockwise), THEN loosen and disconnect the smaller fitting from the larger fitting. If this is not done, damage to the line will result.

1. Refer to the "Intake Manifold Removal & Installation" procedures in this section and remove the intake manifold.

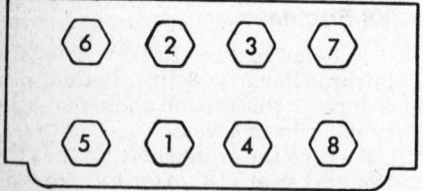

173 V6 cylinder head torque sequence

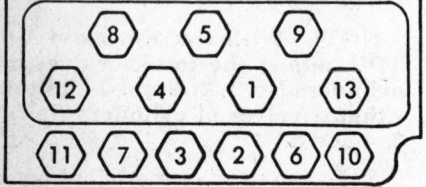

Chevrolet-built 229-V6 cylinder head bolt torque sequence

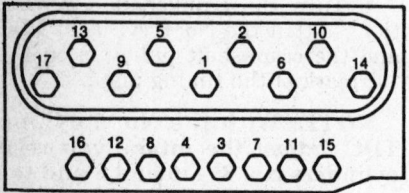

Chevrolet-built V8 engine cylinder head bolt torque sequence

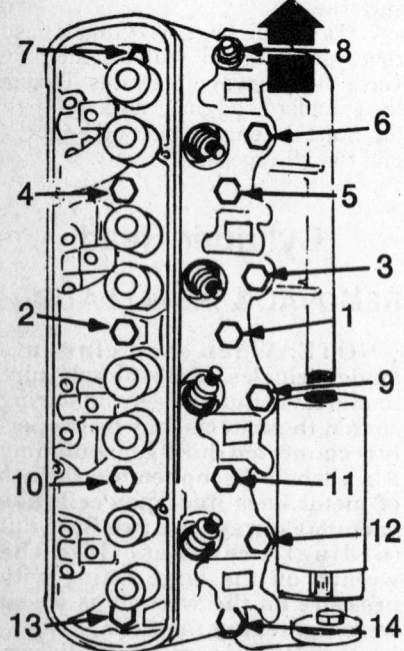

Pontiac-built V8 engine cylinder head bolt torque sequence

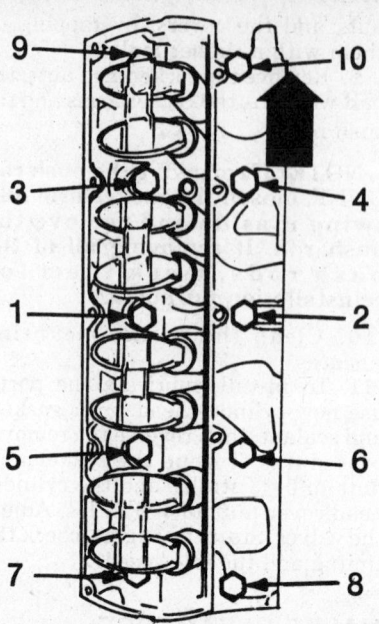

Oldsmobile-built V8 engine cylinder head bolt torque sequence

2. Remove the alternator and lay the unit aside. If necessary, remove the alternator brackets.

3. Disconnect the exhaust pipe(s) and remove the exhaust manifold(s). If equipped with A/C, dismount the A/C compressor and position it aside.

NOTE: On the 4 cyl. engine, remove the power steering pump, if it is top mounted. On the V6 engine, it may be necessary to remove the dipstick tube. On the V8 engine, remove the diverter valve, if equipped.

4. Remove the cylinder head cover(s).

5. Back off the rocker arm nuts and pivot the rocker arms so that the pushrods can be removed. Identify the pushrods so that they can be reinstalled in their original locations.

6. Remove the cylinder head bolts and the cylinder head(s).

7. Clean the gasket mounting surfaces.

8. To install, use new gaskets and reverse the removal procedures. The head gasket is installed with the bead up. Torque the head bolts to 60–75 ft. lbs. (V8), 65–90 ft. lbs. (V6), and 92 ft. lbs. (4 cyl.) a little at a time, in sequence, starting in the center and working towards both ends. Adjust the valves. Refill the cooling system, start the engine and check for leaks. Check the idle speed.

NOTE: On engines using a steel gasket, thinly and evenly, coat both sides with sealer. If a steel as-bestos gasket is used, DO NOT apply sealer. Clean the bolt threads, apply sealing compound and install the bolts finger tight.

Timing Cover and Oil Seal

REMOVAL & INSTALLATION

4 Cyl Engine

1. Remove the drive belts from the crankshaft pulley.

2. Remove the damper pulley bolt and the pulley from the crankshaft.

3. Remove the oil pan-to-timing cover and the timing cover-to-engine bolts.

4. Remove the timing cover.

5. Clean the gasket mounting surfaces.

6. Using a small pry bar, pry the oil seal from the timing cover.

NOTE: The oil seal may be removed from the timing cover without removing the cover. To do this, remove the damper pulley and pry the oil seal from the timing cover, using a small pry bar.

7. To install, use a new oil seal, a new cover gasket and sealant, then reverse the removal procedures. Torque the timing cover bolts to 7 ft. lbs. and the damper pulley bolt to 160 ft. lbs.

NOTE: Place a Seal Installation Tool J-34995 on the crankshaft (to prevent damaging the seal) when installing the new oil seal or the front cover. To install the new oil seal, place the seal's open end toward the inside of the cover and drive it into the cover.

V6 and V8 Engines

1. Refer to the "Water Pump Removal & Installation" procedures in this section and remove the water pump.

2. If equipped with A/C, remove the compressor and the mounting bracket, then move the compressor aside.

3. Remove the crankshaft center bolt. Using the wheel puller tool J-23523, pull the damper pulley from the crankshaft.

NOTE: On the V6 engine, disconnect the lower radiator hose from the timing cover and the heater hose from the water pump.

4. Remove the timing cover bolts and the cover.

5. Using a small pry bar, pry the oil seal from the timing cover.

NOTE: The oil seal may be removed from the timing cover without removing the cover. To do this, remove the damper pulley and pry the oil seal from the timing cover, using a small pry bar.

6. Clean the gasket mounting surfaces.

7. To install, use a new oil seal, a new timing cover gasket and sealant, then reverse the removal procedures. Torque the timing cover bolts to 20 (V6) or 6–8 (V8) ft. lbs., the water pump bolts to 22 (V6) or 25–35 (V8) ft. lbs. and the damper pulley bolt to 67–85 ft. lbs.

NOTE: To install the new oil seal, place the seal's open end toward the inside of the cover and drive it into the cover using tool J-23042. To install the damper pulley, lubricate with engine oil and use tool J-29113 (V6) or J-23523 (V8) to press it onto the crankshaft.

Timing Gear

REMOVAL & INSTALLATION

4 Cyl Engine

NOTE: The timing gear is pressed onto the camshaft. To remove or install the timing gear an arbor must be used.

1. Refer to the "Camshaft Removal & Installation" procedures in this section and remove the camshaft from the engine.

2. Using an arbor press, a press plate and a gear removal tool J-971, press the timing gear from the camshaft.

NOTE: When pressing the timing gear from the camshaft, be certain that the position of the press plate does not contact the woodruff key.

3. To assemble, position the press plate to support the camshaft at the back of the front journal. Place the gear spacer ring and the thrust plate over the end of the camshaft, then install the woodruff key. Press the timing gear onto the camshaft, until it bottoms against the gear spacer ring.

NOTE: The end clearance of the thrust plate should be 0.0015–0.005 in. If less than 0.0015 in., replace the spacer ring; if more than 0.005 in., replace the thrust plate.

4. To complete the installation,

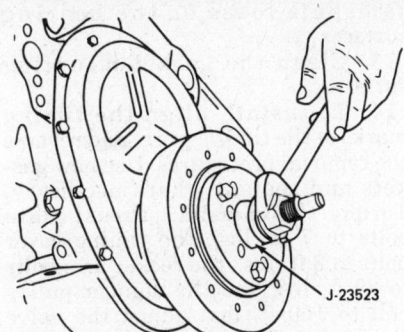

Removing and installing the damper pulley—V8, V6 is similiar

Removing the camshaft thrust plate screws—151 4 cylinder engine

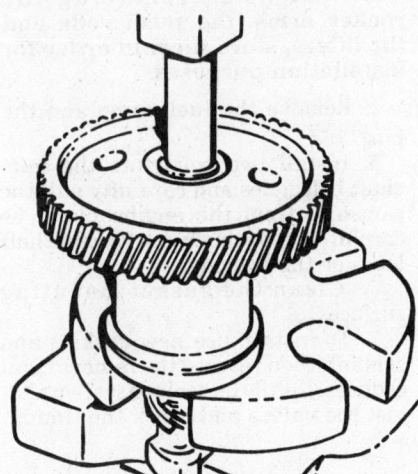

Removing the timing gear from the camshaft—151 4 cylinder engine

align the marks on the timing gears and reverse the removal procedures.

Timing Chain and Sprocket

REMOVAL & INSTALLATION

V6 and V8 Engines

1. Refer to the "Timing Cover Removal & Installation" procedures in this section and remove the timing cover.

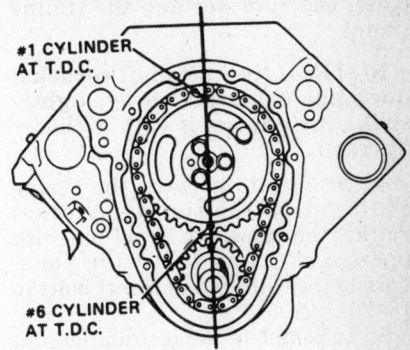

#1 CYLINDER AT T.D.C.
#6 CYLINDER AT T.D.C.

Installing the timing chain and sprocket—V8 engine

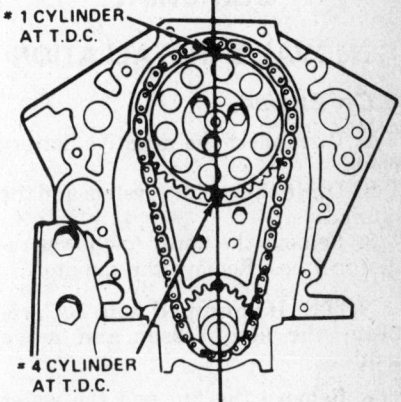

1 CYLINDER AT T.D.C.
4 CYLINDER AT T.D.C.

Installing the timing chain and sprocket—V6 engine

1—ARBOR PRESS

2—J-21474-13 OR J-21795-1

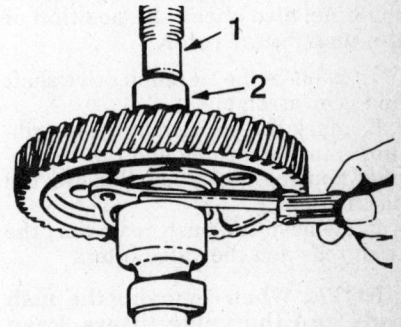

Checking the thrust plate end clearance—151 4 cylinder engine

2. Turn the crankshaft to place the No. 1 cylinder on TDC and so that the camshaft sprocket the No. 4 cyl (V6) or the No. 6 cyl (V8) TDC mark aligns with the crankshaft sprocket mark.

3. Remove the camshaft sprocket

C645

bolts, the sprocket and the timing chain.

NOTE: If the camshaft sprocket does not come off easily, lightly strike the edge of it with a plastic hammer.

4. To install, lubricate the timing chain, align the timing sprocket marks, the camshaft dowel pin with the sprocket and insert the bolts. Torque the camshaft sprocket bolts to 15–20 ft. lbs.

5. To complete the installation, reverse the removal procedures.

Camshaft

REMOVAL & INSTALLATION

4 Cyl Engine

1. Disconnect the negative battery cable.

2. Drain the cooling system and the crankcase.

3. Remove the upper and lower radiator hoses. Remove the radiator.

NOTE: If equipped with A/C, remove the compressor and move aside.

4. Remove the fan and the water pump pulley.

5. Remove the cylinder head cover. Loosen the valve rocker bolts and pivot the rockers out of the way.

6. Remove the spark plugs. Turn the crankshaft to place the No. 4 cyl on TDC.

NOTE: To determine TDC, place a finger over No. 4 cyl spark plug hole. Turn the crankshaft until the pressure blows your finger from the hole; also check the position of the distributor rotor.

7. Remove the oil pump drive shaft and gear assembly.

8. Mark the position of the distributor rotor-to-housing and the distributor housing-to-engine. Remove the distributor.

9. Remove the push rod cover, the push rods and the valve lifters.

NOTE: When removing the push rods and the valve lifters, keep them in order for reinstallation purposes.

10. Remove the damper pulley and the timing gear cover.

11. Working through the holes in the camshaft gear, remove the 2 thrust plate bolts.

12. Pull the camshaft and gear assembly out through the front of the engine, be careful not to damage the

camshaft lobes or the bearing surfaces.

13. Clean the gasket mounting surfaces.

14. To install, align the timing marks of the timing gears and reverse the removal procedures. Use new gaskets and sealant where necessary. Torque the camshaft thrust plate bolts to 7 ft. lbs., the timing cover bolts to 8 ft. lbs., the rocker arm bolts to 20 ft. lbs. and the damper pulley bolt to 160 ft. lbs. Adjust the valve clearances. Replace the cooling fluid and the engine oil. Check and/or adjust the timing.

V6 and V8 Engines

1. Refer to the "Intake Manifold Removal & Installation" and the "Timing Chain and Sprockets Removal & Installation" procedures in this section, then remove the intake manifold and the timing cover.

2. Remove the grille and the radiator. Remove the valve covers.

3. Remove the rocker arm nuts, the ball washers, the rocker arms, the push rods and the lifters.

NOTE: When removing the rocker arms, the push rods and the lifters, store them in order for installation purposes.

4. Remove the fuel pump and the push rod.

5. Install two bolts into the camshaft bolt holes and carefully pull the camshaft from the engine block, be careful not to damage the camshaft lobes or the bearings.

6. Clean the gasket mounting surfaces.

7. To install, use new gaskets and sealant, then reverse the removal procedures. Refill the cooling system. Adjust the valves and check the timing.

Piston and Connecting Rod Positioning

The pistons have a machined hole or a cast notch in the top of the piston, which should face the front of the engine when installed. Before removing the pistons from the cylinders, use a silver pencil or quick drying paint to mark their positions. The piston pins are offset toward the thrust side (right hand side).

NOTE: On the 4 cylinder engine, make sure that the raised notch side of the connecting rod (at the bearing end) is installed opposite the notch on the piston head.

Make sure that the bearing tang slots of the connecting rods are positioned on the opposite side of the camshaft, with the numbers of the connecting rod and the bearing cap on the same side. When installing the piston rings make sure that the marked side of the ring faces upwards. Position the rings according to the illustration.

NOTE: If replacing a piston or connecting rod, be sure to mark the position of the new part and replace the connecting rod bearing with a new one.

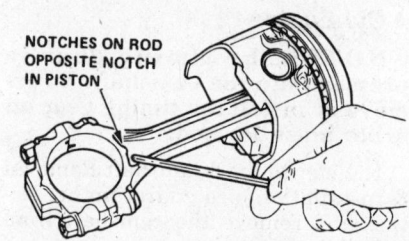

NOTCHES ON ROD OPPOSITE NOTCH IN PISTON

Connecting rod identification—4 cylinder engine

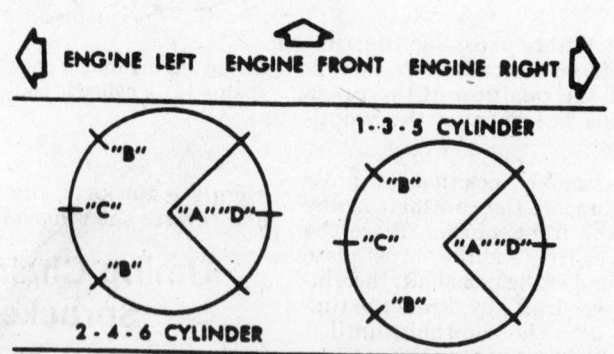

ENGINE LEFT ENGINE FRONT ENGINE RIGHT

1-3-5 CYLINDER

2-4-6 CYLINDER

"A" OIL RING SPACER GAP
(Tang in Hole or Slot within Arc)

"B" OIL RING RAIL GAPS

"C" 2ND COMPRESSION RING GAP

"D" TOP COMPRESSION RING GAP

Ring gap location for a V6 engine—V8 is similiar

ENGINE LUBRICATION

Oil Pan

REMOVAL & INSTALLATION

1980–81 V6 ENGINE

1. Disconnect the negative battery cable.
2. Raise the front of the vehicle and drain the oil.
3. Remove the exhaust pipe crossover tube mounting nuts at the manifold. Lower the cross-over tube.
4. If equipped with an automatic transmission, remove the torque converter cover.
5. Remove the starter brace and the inboard starter bolt, then swing the starter assembly aside.
6. Remove the mounting bolts and the oil pan.
7. Clean the gasket mounting surfaces.
8. To install, use new gaskets, seals and sealant, then reverse the removal procedures. Torque the oil pan $1/4$ in. bolts to 7 ft. lbs. and the $5/16$ in. bolts to 22 ft. lbs. Refill the crankcase.

1980–81 V8 ENGINE

1. Disconnect the negative battery cable.
2. Remove the distributor cap and move aside.
3. Remove the air cleaner and the snorkel.
4. Remove the upper half of the fan shroud.
5. Raise the vehicle on a hoist and drain the engine oil.
6. Disconnect the exhaust crossover pipe from the manifold and the converter.
7. Remove the torque converter cover.

NOTE: If equipped with a manual transmission, remove the starter before the flywheel cover.

8. Remove the front engine mount through-bolts.
9. Raise the engine and reinstall the engine mount through bolts, DO NOT torque.
10. Remove the oil pan bolts and the oil pan. Torque the oil pan $1/4$ in. bolts to 7 ft. lbs. and the $5/16$ in. bolts to 22 ft. lbs. Refill the crankcase.

NOTE: If the front crankshaft throw prohibits removal of the pan, turn the crankshaft to position the throw horizontally.

11. Clean the gasket mounting surfaces and the oil pan.
12. To install, use new gaskets, seals and sealant, then reverse the removal procedures.

1982 and Later 4 Cyl Engine

1. Disconnect the negative battery cable.
2. Raise and support the vehicle on jackstands.
3. Drain the engine oil.
4. Remove the exhaust pipe at the manifold and loosen the exhaust pipe hanger bracket, then allow the exhaust system to hang down.
5. Remove the front engine mount through bolts.
6. Raise the engine to provide sufficient clearance to lower the oil pan.
7. Remove the oil pan bolts and the oil pan.
8. Clean the gasket mounting surfaces and the oil pan.
9. To install, use new gaskets, seals and apply sealant where necessary.

NOTE: When installing the new oil pan gaskets, apply a small amount of RTV where the gasket engages the engine block. Install the side gaskets, using grease as a sealant. Install the oil pan-to-timing cover bolts last, for these holes will not align until the other pan bolts are snug.

10. To complete the installation, reverse the removal procedures. Torque the oil pan bolts to 6 ft. lbs.

V6 and V8 Engines

1. Disconnect the negative battery cable.
2. Remove the air cleaner, the distributor cap and the fan shroud.

NOTE: On the V6 engine, remove the upper half of the fan shroud.

3. Raise and support the vehicle on jackstands.
4. Drain the engine oil.

——————— CAUTION ———————

Be sure that the catalytic converter is cool before proceeding.

5. Remove the exhaust pipe at the manifold and the Air Injection Pipe (AIR) clamp (if equipped).
6. Remove the catalytic converter dust cover (if equipped), the converter hanger bolts and allow the exhaust system to hang down.
7. Remove the starter brace, the mounting bolts and then lay the starter aside.

NOTE: If equipped with a manual transmission, it may be neces-
sary to remove the oil filter in order to remove the inspection cover.

8. Remove the inspection cover and the front engine mount through bolts.
9. Raise the engine enough to provide sufficient clearance for oil pan removal.
10. Remove the oil pan bolts.

NOTE: If the front crankshaft throw prohibits removal of the pan, turn the crankshaft to position the throw horizontally.

11. Remove the oil pan.
12. Clean the gasket mounting surfaces and the oil pan.
13. To install, use new gaskets, seals and sealant, then reverse the removal procedures. On the V8, torque the oil pan $1/4$ in. bolts to 6–8 ft. lbs. and the $5/16$ in. bolts to 13–15 ft. lbs. On the V6, torque the oil pan 6mm bolts to 6–15 ft. lbs. and the 8mm bolts to 15–30 ft. lbs.

Oil Pump

REMOVAL & INSTALLATION

1980–81 V6 (Except 3.8L) and V8 Engines

1. Refer to the "Oil Pan Removal & Installation" procedures in this section and remove the oil pan.
2. Remove the oil pump bolts from the rear main bearing cap.
3. Remove the oil pump and the extension shaft (if necessary).
4. Clean the gasket mounting surfaces.
5. To install, use new gaskets and reverse the removal procedures. Torque the oil pump bolts to 65 ft. lbs. Refill the crankcase.

3.8L V6 Engine

NOTE: The oil pump is located in the timing chain cover and is connected by a drilled passage to the oil screen housing and pipe assembly in the oil pan. The oil is discharged from the pump to the oil pump cover assembly, where the oil filter is mounted.

1. Remove the oil filter.
2. Remove the oil pump cover assembly-to-timing chain cover screws.
3. Remove the cover assembly and slide out the oil pump gears. Clean and inspect the gears for wear or damage.
4. Remove the oil pressure relief valve cap, the spring and the valve. Clean and inspect for wear or damage. Check the relief valve spring for wear and strength, replace if questionable.

NOTE: **The relief valve should slip easily into the bore. If any noticeable shake can be felt, replace the valve and/or the cover.**

5. To install, lubricate the pressure relief valve and the spring, then place them in the cover. Install the cap and the gasket. Torque the cap to 35 ft. lbs.

NOTE: **Pack the pump cavity with petroleum jelly, DO NOT use gear lube. Reinstall the oil pump gears so that the petroleum jelly is forced into every air pocket. There must be no air spaces. Unless the pump is packed, it may not pump oil when the engine is started.**

6. To complete the installation, reverse the removal procedures. Torque the oil pump-to-timing gear cover screws to 10 ft. lbs.

1982 and Later 4 Cyl Engine

1. Refer to the "Oil Pan Removal & Installation" procedures in this section and remove the oil pan.
2. Remove the mounting bolts, the pickup tube nut and the pump assembly.
3. To install, reverse the removal procedures. Torque the pump bolts to 22 ft. lbs.

V6 and V8 Engines

1. Refer to the "Oil Pan Removal & Installation" procedures in this section and remove the oil pan.
2. Remove the oil pump bolts from the rear main bearing cap.
3. Remove the oil pump and the extension shaft (if necessary).
4. Clean the gasket mounting surfaces.
5. To install, use new gaskets and reverse the removal procedures. Torque the oil pump bolts to 25–35 ft. lbs. (2.8L V6) or 65 ft. lbs. (all others). Refill the crankcase.

Rear Main Seal

REMOVAL & INSTALLATION

4 Cyl Engine

1. Refer to the "Transmission Removal & Installation" procedures in this section and remove the transmission.

NOTE: **If equipped with a manual transmission, remove the pressure plate and the clutch.**

2. Remove the mounting bolts and the flywheel.
3. Using a small pry bar, pry the oil seal from the engine.

4. Clean the seal mating surfaces.
5. To install the new seal, lubricate it with engine oil and press it into the engine housing with finger pressure.
6. To complete the installation, reverse the removal procedures. Torque the flywheel to 44–68 ft. lbs.

2.8L V6

1982–84 (One Piece Type)

1. Refer to the "Transmission Removal & Installation" procedures in this section and remove the transmission.

NOTE: **If equipped with a manual transmission, remove the pressure plate and the clutch.**

2. Remove the flywheel from the crankshaft.
3. Using a small pry bar, pry the rear oil seal from the housing.

NOTE: **When prying the oil seal from the housing, be careful not to damage the machined surfaces.**

4. Using the seal installation tool J-34686, lubricate the new oil seal lip and slide it onto the installation (lip side against the tool) until it seats.
5. Align the installation tool dowel

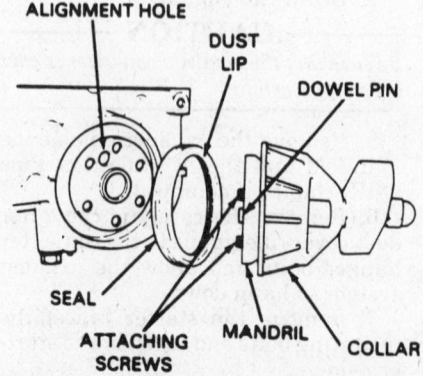

Installing the oil seal on the installation tool—1982–84 2.8L engine

Installing the oil seal in the engine—1982–84 2.8L engine

pin with the dowel pin hole in the crankshaft. Torque the mounting screws to 2–5 ft. lbs.
6. Turn the "T" handle and push the seal into the housing, until it bottoms out against the housing.
7. Loosen the "T" handle until it comes to a stop. Remove the mounting screws of the installation tool.
8. Check the seal and make sure that it is squarely seated in the bore.
9. To complete the installation, reverse the removal procedures. Torque the fylwheel bolts to 50 ft. lbs.

1985 and Later (Two Piece Type)

1. Remove the oil pan and oil pump.
2 Remove the rear main bearing cap.
3. Remove the upper and lower rope seal. Clean the seal channel to remove any rope pieces and oil.

NOTE: **Loosing No. 2 and No. 3 main bearing bolts may be necessary in both removing the upper rope seal and installing the new upper seal.**

4. Apply a very thin coat of GM Gasket Sealing Compound 1050026, or equivalent, to the O.D. of the rubber seal. Keep the sealing compound off of the seal lips.
5. Roll the seal into position in the cylinder case by turning the crankshaft as it is being installed. A piece of shim stock must be used as a shoe horn between the seal outside diameter and the edge of the block seal channel to prevent damaging the seal during installation.

NOTE: **The seal lip must be positioned inboard of the engine and the small dust lip to the flywheel side.**

6. Apply sealing compound to the other half of the new seal as described in Step 4.
7. Apply approximately $1/32$ in. of RTV sealant GM 1052357, or equivalent, to the cap between the rear main seal and the oil pan rear seal groove.

NOTE: **Keep the sealant off of the rear main seal, bearing, and out of the drain slot.**

8. Just prior to assembly, apply a light coat of engine oil on the crankshaft surface that will contact the seal.
9. Install the rear main bearing cap and torque to 70 ft. lbs.

3.8L V6 Engine

1. Refer to the "Oil Pan Removal & Installation" procedures in this section and remove the oil pan.

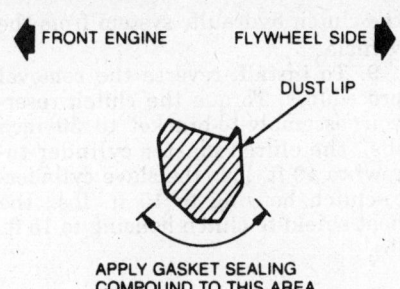

Installing the new rear main seal—1985 and later 2.8L engine

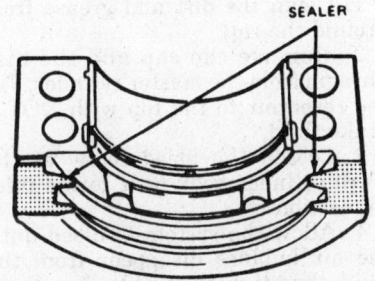

Applying sealant to the rear main bearing cap—1985 and later 2.8L engine

2. Remove the rear main bearing cap from the engine.

3. Using the packing tool J-21526-2, gently drive one end of the old seal into the engine block until it is lightly packed. Repeat the procedure on the other side.

4. Measure the amount the seal was driven up on 1 side and add $^1/_{16}$ in.

5. Using a single edge razor blade, cut the measured amount from the old seal of the lower half (use the rear main bearing cap as a holding fixture). Repeat the procedure for the other side.

6. Bolt the guide tool J-21526-1 to the upper half of the rear main bearing.

7. Using the packing and the guide tools, drive the short pieces of cut seal into the cylinder block until they are equal with the parting line. Perform this procedure on both sides.

NOTE: The guide and the packing tools have been machined to provide built in stops. Apply oil to the short pieces of seal before driving them into the engine block.

8. Install a new fabric seal to the rear main bearing cap and trim the ends flush the parting line.

9. To complete the installation, reverse the removal procedures. Torque

the rear main bearing cap to 100 ft. lbs.

V8 Engine

1. Refer to the "Oil Pan Removal & Installation" procedures in this section and remove the oil pan.

2. Remove the oil pump and the rear main bearing cap.

3. Using a small pry bar, pry the oil seal from the rear main bearing cap.

4. Using a small hammer and a brass pin punch, drive the top half of the oil seal from the rear main bearing. Drive it out far enough, so it may be removed with a pair of pliers.

5. Using a non-abrasive cleaner, clean the rear main bearing cap and the crankshaft.

6. Fabricate an oil seal installation tool from 0.004 in. shim stock, shape the end to $^1/_2$ in. long by $^{11}/_{64}$ in. wide.

7. Coat the new oil seal with engine oil; DO NOT coat the ends of the seal.

8. Position the fabricated tool between the crankshaft and the seal seat in the cylinder case.

9. Position the new half seal be-

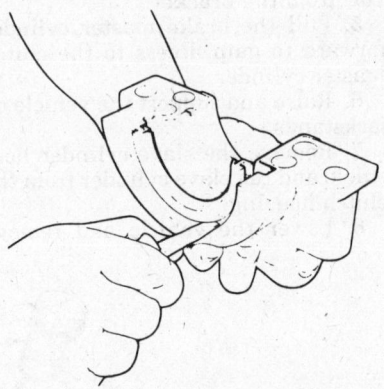

Removing the oil seal from the lower half—V8 engine

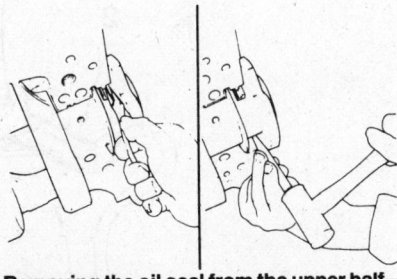

Removing the oil seal from the upper half—V8 engine

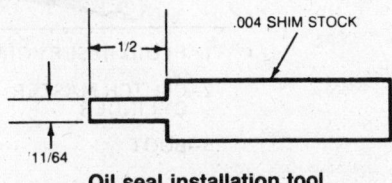

Oil seal installation tool

tween the crankshaft and the tip of the tool, so that the seal bead contacts the tip of the tool.

NOTE: Make sure that the seal lip is positioned toward the front of the engine.

10. Using the fabricated tool as a shoe horn, to protect the seal's bead from the sharp edge of the seal seat surface in the cylinder case, roll the seal around the crankshaft. When the seal's ends are flush with the engine block, remove the installation.

11. Using the same manner of installation, install the lower half onto the lower half of the rear main bearing cap.

12. Apply sealant to the cap-to-case mating surfaces and install the lower rear main bearing half to the engine; keep the sealant off of the seal's mating line.

13. Install the rear main bearing cap bolts and torque to 10–12 ft. lbs. Using a lead hammer, tap the crankshaft forward and rearward, to line up the thrust surfaces. Torque the main bearing bolts to 70–85 ft. lbs. and reverse the removal procedures. Refill the crankcase.

CLUTCH

REMOVAL & INSTALLATION

—— **CAUTION** ——
If equipped with a hydraulic clutch system, disconnect the master cylinder from the clutch pedal BEFORE removing the slave cylinder from the clutch lever. This procedure is to prevent any possible damage to the slave cylinder.

1. Refer to the "Transmission Removal & Installation" procedures in this section and remove the transmission.

NOTE: On 1984 and later vehicles, a hydraulic operated clutch is used. When removing the clutch, remove the slave cylinder heat shield and the cylinder from the clutch housing.

2. Disconnect the clutch fork push rod and spring.

3. Remove the clutch housing.

4. Slide the clutch fork from the ball stud and remove the fork from the dust boot. The ball stud is threaded into the clutch housing and may be replaced, if necessary.

5. Install an alignment tool J-5824 (4-speed) or J-33169 (5-speed) to support the clutch assembly during re-

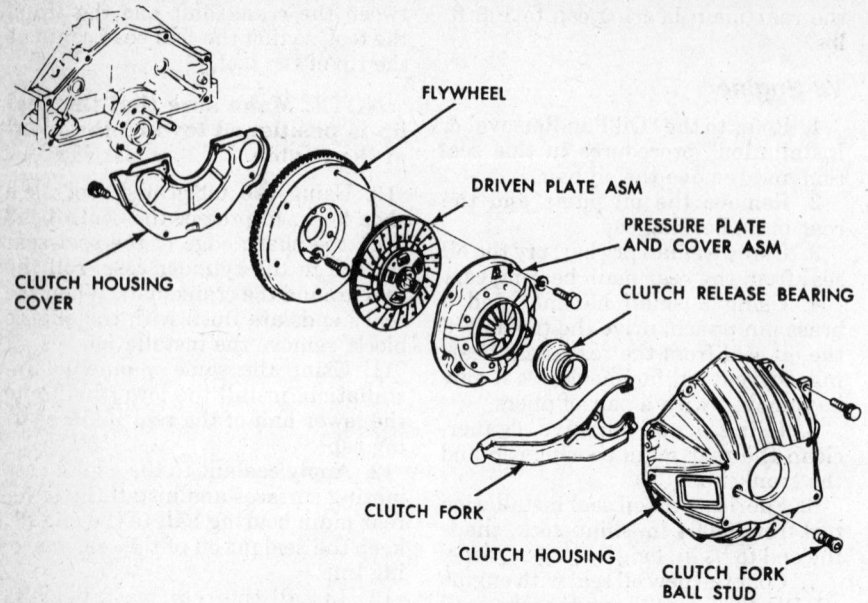

FLYWHEEL

DRIVEN PLATE ASM

PRESSURE PLATE AND COVER ASM

CLUTCH RELEASE BEARING

CLUTCH HOUSING COVER

CLUTCH FORK

CLUTCH HOUSING

CLUTCH FORK BALL STUD

Exploded view of the clutch assembly

moval. Mark the flywheel and the pressure plate for reinstallation, if they do not already have "X" marks.

6. Loosen the pressure plate-to-flywheel attaching bolts evenly, 1 turn at a time, until spring pressure is released. Remove the bolts and clutch assembly.

7. To install, place the clutch disc with the damper springs toward the transmission and reverse the removal procedures. Torque the clutch housing-to-engine to 55 ft. lbs., the slave cylinder-to-clutch housing to 15 ft. lbs. and the heat shield-to-clutch housing to 15 ft. lbs.

Clutch Hydraulic System

The clutch hydraulic system was introduced in 1984 and is to be serviced as a complete unit.

REMOVAL & INSTALLATION

——— **CAUTION** ———

Disconnect the master cylinder from the clutch pedal BEFORE removing the slave cylinder from the clutch lever; this procedure is to prevent any possible damage to the slave cylinder.

1. Disconnect the negative battery cable.
2. Remove the hush panel and the steering column trim cover.
3. Disconnect the master cylinder push rod from the clutch pedal.
4. Remove the clutch master cylinder-to-cowl nuts, the brake booster-to-

cowl nuts and the clutch fluid reservoir from the bracket.
5. Pull the brake master cylinder forward to gain access to the clutch master cylinder.
6. Raise and support the vehicle on jackstands.
7. Remove the slave cylinder heat shield and the slave cylinder from the clutch housing.
8. Lower the vehicle and remove

the clutch hydraulic system from the vehicle.
9. To install, reverse the removal procedures. Torque the clutch reservoir assembly-to-bracket to 30 inch lbs., the clutch master cylinder-to-cowl to 10 ft. lbs., the slave cylinder-to-clutch housing to 15 ft. lbs., the heat shield-to-clutch housing to 15 ft. lbs.

BLEEDING

——— **CAUTION** ———

When adding fluid to the hydraulic system, NEVER use fluid that has been bled from the system.

1. Clean the dirt and grease from around the cap.
2. Remove the cap and the diaphragm from the master cylinder. Fill the reservoir to the top with DOT 3 brake fluid.
3. Loosen the bleed screw at the slave cylinder body next to the inlet connection.
4. Allow the system to bleed until the air bubbles disappear from the fluid, then tighten the bleeder screw.
5. Refill the reservoir, then install the diaphragm and the cap.
6. To expell any air trapped in the system, exert 20 lbs. of force on the clutch release lever and open the bleeder screw. Maintain pressure until a steady stream of fluid flows from the bleeder screw, then tighten the screw.

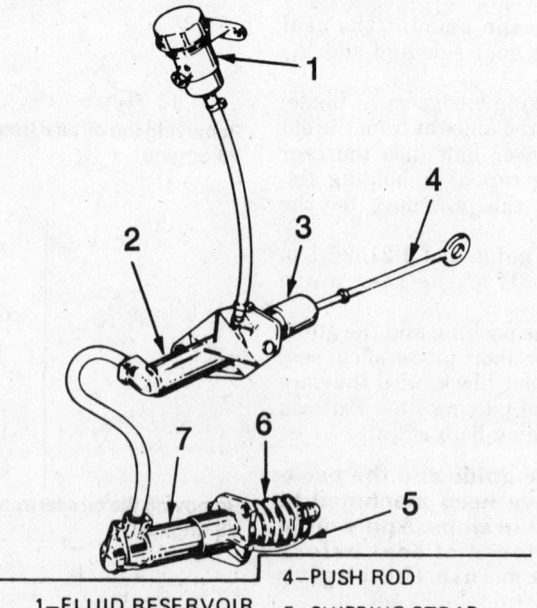

1—FLUID RESERVOIR

2—CLUTCH MASTER CYLINDER

3—BOOT

4—PUSH ROD

5—SHIPPING STRAP

6—BOOT

7—CLUTCH SLAVE CYLINDER

Hydraulic clutch assembly

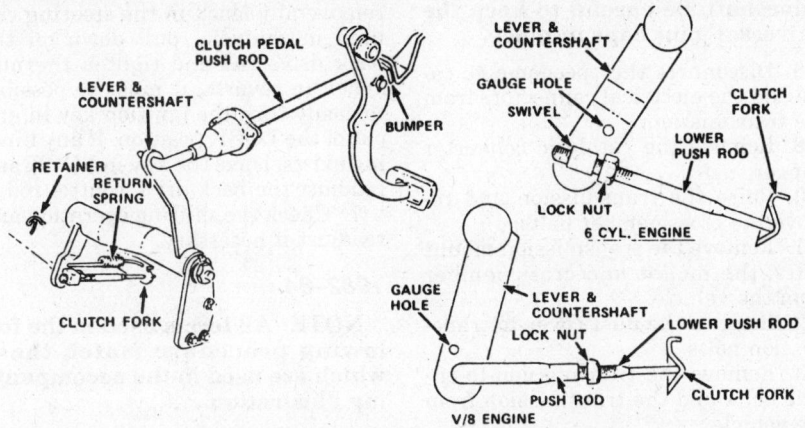

Typical clutch linkage and adjustment points

NOTE: It may be necessary to repeat the bleeding procedure several times, until all of the air is bleed from the system.

7. Refill the reservoir.
8. Check the system, by starting the engine, operating the clutch and shifting the transmission into several gears; there should be no grinding of the gears.

FREE-PLAY ADJUSTMENT— 1980–83

1. Disconnect the return spring at the clutch fork.
2. Hold the pedal against the rubber bumper on the dash brace.
3. Push the clutch fork so that the throwout bearing lightly contacts the pressure plate fingers.
4. Loosen the locknut and adjust the length of the rod so that the swivel or rod can slip freely into the gauge hole of the lever. Increase the length of the rod until the free-play is removed.
5. Remove the rod or swivel from the gauge hole and insert it in the other (original) hole on the lever. Install the retainer and tighten the locknut.
6. Install the return spring and check free-play measurement from the floor mat to top of the pedal pad. It should measure: $^{7}/_{8}$ to $1^{1}/_{2}$ in. (1980–81) or $^{7}/_{8}$ to $1^{1}/_{8}$ in. (1982 and later).

MANUAL TRANSMISSION

For all identification and overhaul information, please refer to the "Manual Transmission" Unit Repair section.

REMOVAL & INSTALLATION

1980–81 Models

1. On the floor-shift models, remove the shift knob and console trim plate.
2. Raise and support the vehicle on jackstands.
3. Disconnect the speedometer cable and the TCS switch wiring, if equipped.
4. Remove the driveshaft.
5. Remove the transmission mounts-to-crossmember bolts, the crossmember-to-frame bolts and the crossmember.
6. Remove the shift levers from the transmission.

7. Disconnect the back drive rod from the bellcrank.
8. Remove the shift control assembly bolts and lower the assembly until the shift lever clears the rubber shift boot. Remove the assembly from the vehicle.
9. Remove the transmission-to-clutch housing bolts and lower the transmission from the vehicle.
10. To install, reverse the removal procedures. Refill the transmission and adjust the shift linkage.

1982–84 4-Speed

1. Disconnect the negative battery cable.
2. Raise and support the vehicle on jackstands.
3. Drain the lubricant from the transmission.
4. Remove the torque arm from the vehicle.
5. Mark the driveshaft and the rear

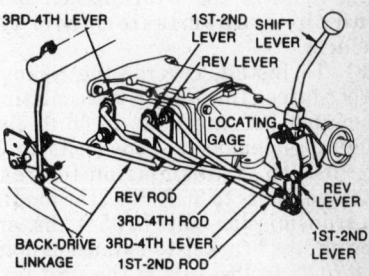

Typical 4 speed transmission shift linkage—through 1981

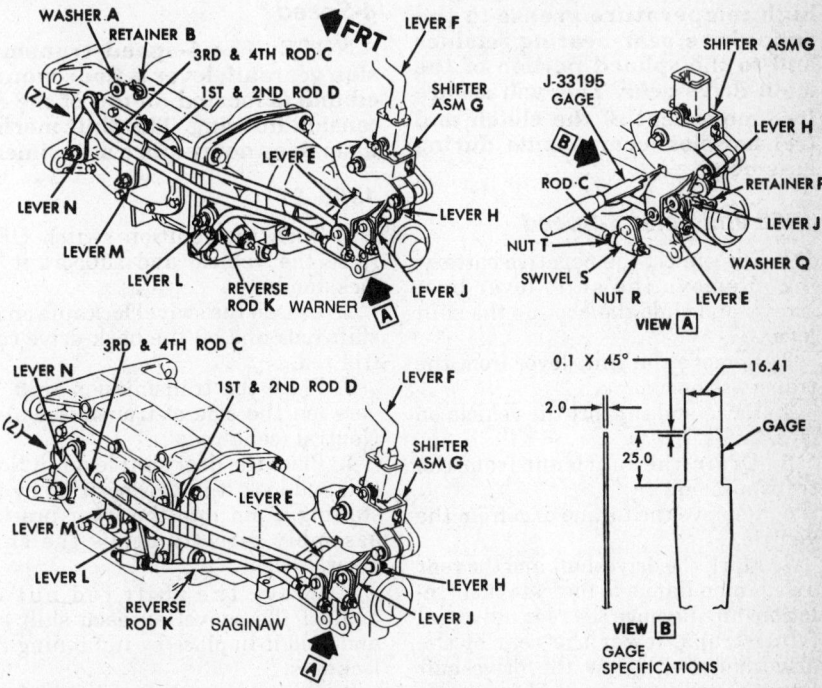

Transmission shift linkage adjustment on 1982–84 models. Note that all component references in the illustration match those in the text and that the dimensions are expressed in millimeters.

axle pinion flange to indicate their relationship. Remove the rear universal joint straps, lower the rear of the driveshaft, withdraw the it from the transmission and remove it from the vehicle.

NOTE: When removing the driveshaft, be careful to keep the universal joint caps in place.

6. Disconnect the speedometer cable and the electrical connectors from the transmission.
7. Remove the exhaust pipe brace.
8. Remove the transmission shifter support-to-transmission bolts.
9. Disconnect the shift linkage at the shifter.
10. Raise the transmission and remove the crossmember attaching bolts.
11. Remove the transmission mount bolts, the mount and the crossmember from the vehicle.
12. Remove the transmission bolts and the transmission from the vehicle.
13. To install, reverse the removal procedures. Torque the transmission-to-clutch housing bolts to 55 ft. lbs., the crossmember-to-body bolts to 35 ft. lbs., the transmission-to-crossmember bolts to 35 ft. lbs., the mount-to-transmission bolts to 35 ft. lbs. and the shifter bracket-to-extension housing to 25 ft. lbs. Adjust the shift linkage and refill the transmission with lubricant.

NOTE: Apply a light coating of high temperature grease to the main drive gear bearing retainer and to the splined portion of the main drive gear. This will assure free movement of the clutch and transmission components during assembly.

1982 and Later 5-Speed

1. Disconnect the negative battery.
2. Remove the shift lever boot screws and slide the boot up the shift lever.
3. Remove the shift lever from the transmission.
4. Raise and support the vehicle on jackstands.
5. Drain the lubricant from the transmission.
6. Remove the torque arm from the vehicle.
7. Mark the driveshaft and the rear axle pinion flange to indicate their relationship. Remove the rear universal joint straps, lower the rear of the driveshaft, withdraw the driveshaft from the transmission and remove it from the vehicle.

NOTE: When removing the

driveshaft, be careful to keep the universal joint caps in place.

8. Disconnect the speedometer cable and the electrical connectors from the transmission.
9. Remove the catalytic converter hanger.
10. Raise the transmission and remove the crossmember bolts.
11. Remove the transmission mount bolts, the mount and crossmember from the vehicle.
12. Remove the dust cover-to-transmission bolts.
13. Remove the transmission-to-engine bolts and the transmission from the vehicle.
14. To install, reverse the removal procedures. Torque the transmission-to-clutch housing bolts to 55 ft. lbs., the crossmember-to-body bolts to 35 ft. lbs., the mount-to-crossmember bolts to 35 ft. lbs. and the mount-to-transmission bolts to 35 ft. lbs. Adjust the shift linkage and refill the transmission with lubricant.

NOTE: Apply a light coating of high temperature grease to the main drive gear bearing retainer and to the splined portion of the main drive gear. This will assure free movement of the clutch and transmission components during assembly.

SHIFT LINKAGE ADJUSTMENT

4-Speed

NOTE: The 5-speed transmission gearshift lever is floor-mounted and is located on top of the extension housing. The shift mechanism does not require adjustment.

1980–81

1. Turn the ignition switch OFF, raise the vehicle and support it on jackstands.
2. Loosen the swivel locknuts on all shift rods and on the back drive control rod.
3. Place the transmission shift levers (on the side of transmission) in Neutral (centered).
4. Place the floor shift lever in Neutral and lock it in this position by installing a pin into the lever bracket assembly directly below the shift lever.
5. Move the shift rod nut up against the swivel on each shift rod and hold it in place by tightening the locknuts.
6. Remove the locating pin from the control bracket assembly and shift the transmission into Reverse. Place the ignition key in LOCK. To

remove any slack in the steering column mechanism, pull down on the back drive rod and tighten the nut. When in reverse, it must be possible to easily turn the ignition key in and out of the LOCK position. If any binding exists, leave the key in LOCK and readjust the back drive control rod.
7. Check the shifting operation and readjust if necessary.

1982–84

NOTE: All terms used in the following procedure match those which are used in the accompanying illustration.

1. Disconnect the negative battery cable.
2. Place the shift control lever (F) in Neutral.
3. Raise and support the vehicle on jackstands.
4. Remove the swivel retainers (P) from the levers (E, H and J).
5. Remove the swivels (S) from the the shifter assembly (G) and loosen the swivel locknuts (R and T).
6. Make sure that levers L, M and N are in their Neutral positions (center detents).
7. Align the holes of levers E, H and J with the notch in the shifter assembly (G). Insert an alignment gauge (J-33195) to hold the levers in this position.
8. Insert swivel S into lever E and install washer Q. Secure with retainer P.
9. Apply rearward pressure (Z) to lever N. Tighten locknuts R and T (at the same time) against swivel S to 25 ft. lbs.
10. Repeat Steps 8 and 9 for rod D and levers J and M.
11. Repeat Steps 8 and 9 for rod K and levers H and L.
12. Remove the alignment gauge, lower the vehicle and check the operation of the shifting mechanism.
13. Reconnect the negative battery cable.

AUTOMATIC TRANSMISSION

For all service information, please refer to the "Automatic Transmission" Unit Repair section.

REMOVAL & INSTALLATION

1. Disconnect the negative battery cable.

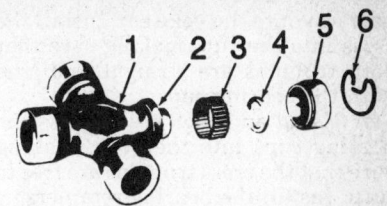

1. Trunnion 4. Washer
2. Seal 5. Cap
3. Bearings 6. Snapring

Universal joint assembly—Cleveland type

2. Remove the air cleaner assembly.
3. Disconnect the throttle valve (TV) control cable at the carburetor.
4. Remove the transmission dipstick and the tube.
5. Raise and support the vehicle on jackstands.

NOTE: In order to provide adequate clearance for transmission removal, it may be necessary to raise both the front and the rear of the vehicle.

6. Mark the relationship between the driveshaft and the rear pinion flange, so that the driveshaft may be reinstalled in its original position.
7. Unbolt the universal joint straps from the pinion flange (use care to keep the universal joint caps in place), lower and remove the driveshaft from the vehicle.
8. Disconnect the catalytic convertor support bracket at the transmission.

NOTE: If equipped with a PM generator, instead of a speedometer cable, disconnect the electrical connector at the generator.

9. Disconnect the speedometer cable, electrical connectors and the shift control cable from the transmission.

——— **CAUTION** ———
During the next step, rear spring force will cause the torque arm to move toward the floor pan. When disconnecting the arm from the transmission, carefully place a piece of wood between the floor pan and the torque arm. This will prevent possible personal injury and/or floor pan damage.

10. Remove the torque arm-to-transmission bolts.
11. Remove the flywheel cover, then mark the relationship between the torque convertor and the flywheel, so that these parts may be reassembled in the same relationship.

NOTE: On the 1980–81 models, remove the tunnel strap.

12. Remove the torque convertor-to-flywheel bolts.
13. Support the transmission with a jack and remove the transmission mount bolt.
14. Unbolt and remove the transmission crossmember.
15. Lower the transmission slightly. Disconnect the TV cable and oil cooler lines from the transmission.
16. Fasten the support tool BT-6424 to the engine, then remove the transmission-to-engine mounting bolts.

——— **CAUTION** ———
The transmission must be secured to the transmission jack.

17. Remove the transmission from the vehicle. Be careful not to damage the oil cooler lines, TV cable or the shift control cable.

NOTE: When removing the transmission, install the torque converter holding tool J-21366 to keep the torque convertor from falling out of the transmission.

18. To install, reverse the removal procedures. Torque the transmission-to-engine bolts to 35 ft. lbs., the torque converter-to-flywheel to 35 ft. lbs., the transmission-to-frame bolts to 40 ft. lbs. and the transmission-to-mount to 25 ft. lbs. Adjust the shift linkages, the TV cable and add fluid to the transmission (if necessary).

NOTE: Before installing the convertor-to-flywheel bolts, be sure that the weld nuts on the convertor are flush with the flywheel and that the convertor rotates freely by hand in this position. Install a O-ring to the dipstick tube.

DRIVE AXLE

Driveshaft

REMOVAL & INSTALLATION

1. Raise and support the vehicle on jackstands.
2. Mark the relationship of the driveshaft to the pinion flange.
3. Remove the rear universal joint retainers and separate it from the pinion flange.

NOTE: If the universal joint cups are loose, tape the cups to the universal joint to keep them from falling off the joint.

4. Support the driveshaft and remove it from the transmission. When removing or installing the driveshaft, DO NOT allow the universal joints to bend to extreme angles, this might rupture the internally injected seals.
5. To install, reverse the removal procedures. Torque the universal joint fasteners to 15 ft. lbs.

Universal Joint

For all universal joint service information, please refer to "U-

Joints/CV-Joints" in the Unit Repair section.

REMOVAL & INSTALLATION

1980–81 Cleveland Type

1. Refer to the "Driveshaft Removal & Installation" procedures in this section and remove the driveshaft.

NOTE: NEVER clamp the driveshaft tube in a vise, for this may dent the tube. Support the driveshaft horizontally and clamp on the yokes of the universal joints.

2. Remove the lock rings from the ends of the trunion yoke.
3. Support the driveshaft in the horizontal position with the base plate of a press, so that the lower ear of the yoke is supported on a piece of 1¼ in. I.D. pipe.
4. Place a socket on the upper bearing cup and press the lower bearing cup out of the yoke ear.

NOTE: Since the bearing cup cannot be fully pressed from the yoke ear, grasp the cup in the jaws of a vice and work it from the yoke.

5. Rotate the driveshaft to the opposite bearing cup and press the bearing cup from the yoke, using the same removal procedure.
6. With both bearing cups removed from the yoke, separate the yoke from the driveshaft.
7. Repeat the removal procedures for the other bearing cups.
8. Clean and inspect all of the parts.

NOTE: If the used universal joints are going to be reinstalled, repack with new grease.

9. To install, use new universal joints or repack the old ones. Place a bearing cup part way into 1 side of the yoke (place the yoke ear to the bottom).
10. Insert the cross into the yoke so that the trunion seats freely into the bearing cup.
11. Insert the opposite bearing cup

part way into the yoke ear. Install the cross into the cup, making sure that both trunions are straight and true with the bearing cups.

12. Using an arbor press, press the bearing cups into the yoke, making sure that the cross trunions are free to turn. Install the bearing retainers.

13. Assemble the other side of the yoke in the same manner.

14. To complete the installation, reverse the removal procedures.

1980 and Later Saginaw Type

1. Refer to the "Driveshaft Removal & Installation" procedures in this section and remove the driveshaft.

NOTE: NEVER clamp the driveshaft tube in a vise, for this may dent the tube. Support the driveshaft horizontally and clamp on the yokes of the universal joints.

2. Support the driveshaft in the horizontal position with the base plate of a press, so that the lower ear of the yoke is supported on a $1^1/_8$ in. socket.

3. Place the cross press tool J-9522-3 on the open horizontal bearing cups and press (shear the plastic retaining ring) the lower bearing cup out of the yoke ear.

NOTE: If the bearing cup is not completely removed, lift the cross tool and place a spacer tool J-9522-5 between the seal and the bearing cup. Repeat the pressing procedure to drive the bearing cup from the yoke.

4. Rotate the driveshaft to the opposite bearing cup and press the bearing cup from the yoke.

5. With both bearing cups removed from the yoke, separate the yoke from the driveshaft.

6. Repeat the removal procedures for the other bearing cups.

NOTE: Since there are no bearing retainer grooves in the production bearing cups, the universal cannot be reused.

7. Remove the remains of the sheared bearing cups and check for nicks in the yoke ears.

8. To install, use new universal joints and place a bearing cup part way into one side of the yoke (place the yoke ear to the bottom).

9. Insert the cross into the yoke so that the trunion seat freely into the bearing cup.

10. Insert the opposite bearing cup part way into the yoke ear. Install the cross into the cup, making sure that both trunions are straight and true with the bearing cups.

11. Using the press, press the bear-

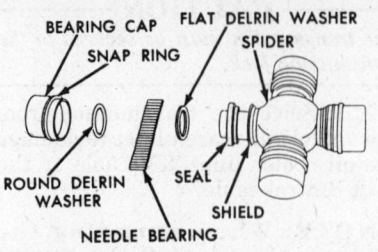

Universal joint assembly—Saginaw type

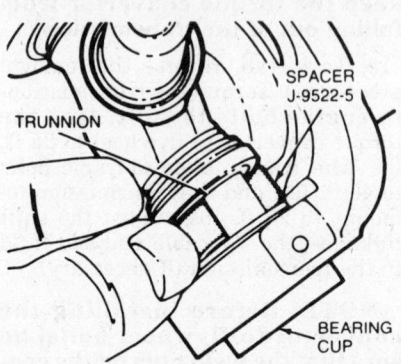

Pressing the bearing cup from the yoke

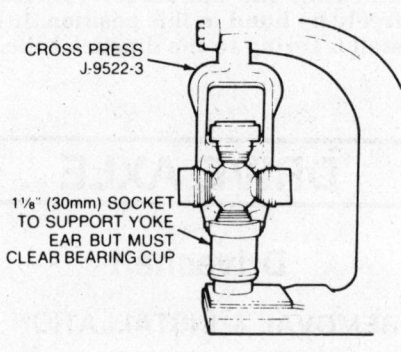

Pressing the universal joint from the yoke

ing cups into the yoke, making sure that the cross trunions are free to turn.

12. As soon as the bearing retainer groove(s) clears the yoke, stop pressing and install the bearing retainer(s) onto the groove(s).

NOTE: It may be necessary to strike the yoke with a hammer to align the seating of the bearing retainers.

13. Assemble the other side of the yoke in the same manner.

14. To complete the installation, reverse the removal procedures.

Axle Shaft and Bearing

REMOVAL & INSTALLATION

1. Raise and support the rear of the vehicle on jackstands.

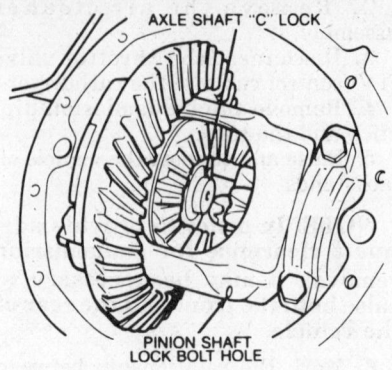

Removing the "C" lock from the axle shaft

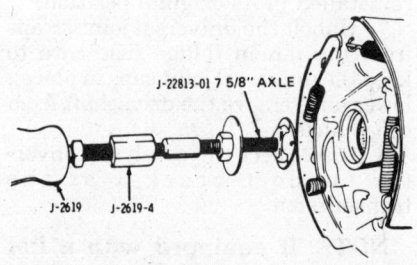

Removing the bearing from the axle housing

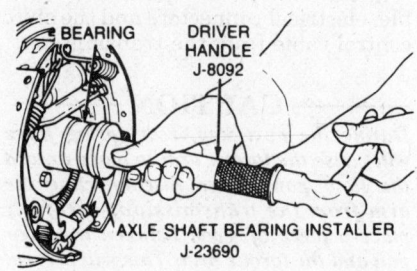

Installing the axle bearing

2. Remove the wheel and the brake drum assembly.

3. Clean the dirt from around the carrier cover.

4. Remove the carrier cover and drain the oil from the housing.

5. Remove the rear axle pinion shaft lock screw and the shaft.

6. Push the flanged end of the axle shaft toward the center of the vehicle and remove the "C" lock from the end of the shaft.

7. Pull the axle shaft from the housing, being careful not to damage the oil seal.

8. Using a pry bar, pry the oil seal from the axle housing.

9. Install the bearing removal tool J-23689 ($8^3/_4$ in. axle) or J-22813 (all other axles) to the axle bearing, connect it to a slide hammer and pull the bearing from the housing.

10. To install, lubricate a new bearing and drive it into the housing (un-

til it seats) using the bearing installation tool J-23690.

11. Lubricate the lips of a new oil seal and drive it into the housing (until it is flush with the housing), using the seal installation tool J-21128.

12. To complete the installation, slide the axle shaft into the housing (making sure that it engages the splines of the side gear) and reverse the removal procedures. Torque the pinion lock screw to 20 ft. lbs. Install a new carrier cover and torque the cover bolts to 20 ft. lbs. Fill the axle with lubricant to $\frac{3}{8}$ in. below the filler hole.

FRONT SUSPENSION

Coil Spring

REMOVAL & INSTALLATION

1980–81

1. Remove the shock absorber and disconnect the stabilizer bar.

2. Support the front of the vehicle at the frame so the control arms hang free.

3. Support the inner end of the control arm with a floor jack; dealers have a device that cradles the inner bushings.

4. Raise the jack to take the tension off the lower control arm pivot blots.

5. Chain the spring to the lower control arm, for safety purposes.

6. Remove the rear and then the front pivot bolt.

7. Lower the jack until all the spring tension is released.

8. Note the way in which the spring is installed to the control arm and remove it.

9. To install, position the spring to the control arm and raise it into place, then reverse the removal procedures. Install the pivot bolts and torque the nuts to 90 ft. lbs.

1982 and Later Models

1. Raise and support the front of the vehicle on jackstands.

2. Remove the front wheel(s).

3. Disconnect the stabilizer link from the lower control arm.

NOTE: If the steering gear hinders removal procedures, detach the unit and move it aside.

4. Disconnect the tie-rod from the steering knuckle using a ball joint removal tool J-24292A.

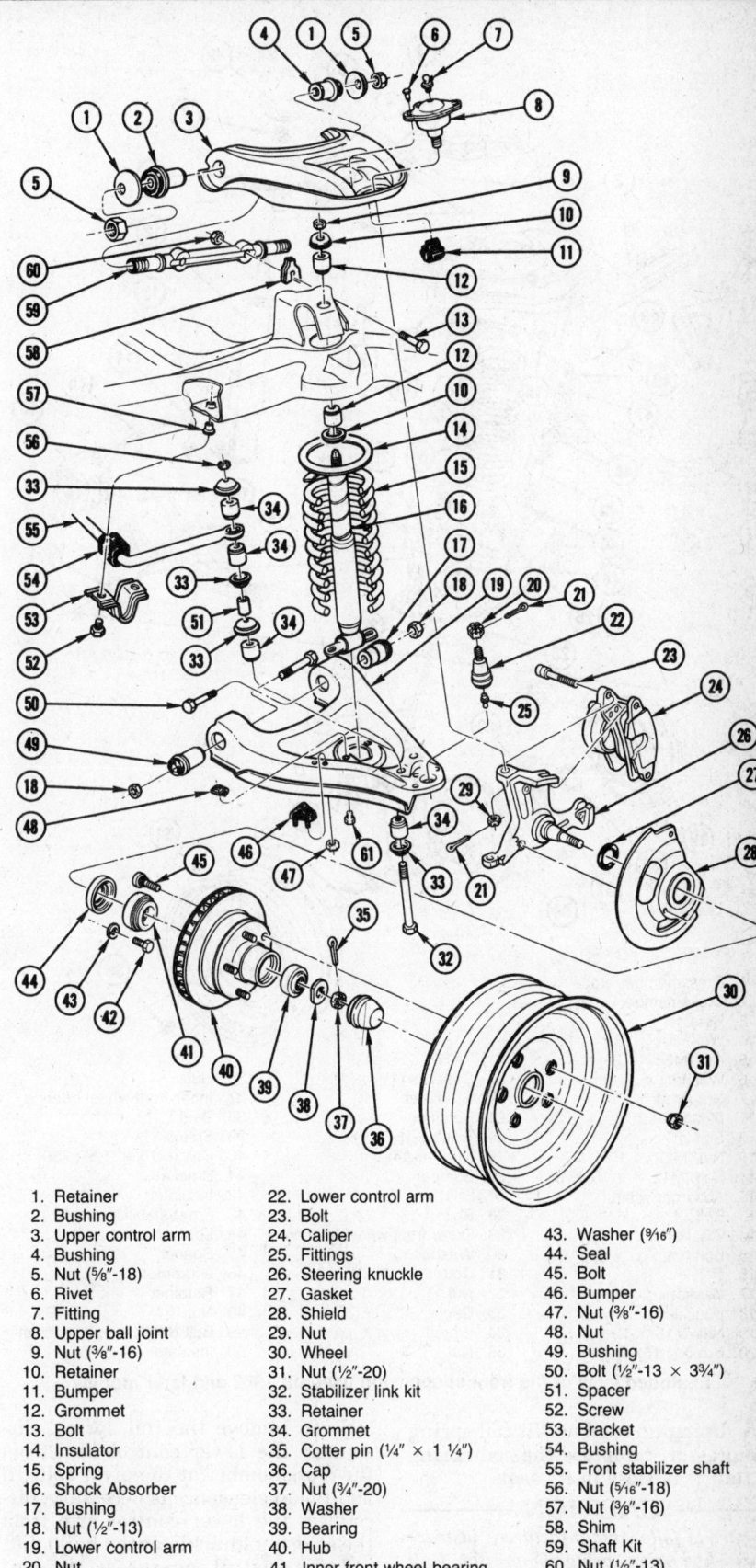

1. Retainer	22. Lower control arm	
2. Bushing	23. Bolt	
3. Upper control arm	24. Caliper	43. Washer (9/16")
4. Bushing	25. Fittings	44. Seal
5. Nut (5/8"-18)	26. Steering knuckle	45. Bolt
6. Rivet	27. Gasket	46. Bumper
7. Fitting	28. Shield	47. Nut (3/8"-16)
8. Upper ball joint	29. Nut	48. Nut
9. Nut (3/8"-16)	30. Wheel	49. Bushing
10. Retainer	31. Nut (1/2"-20)	50. Bolt (1/2"-13 × 3¾")
11. Bumper	32. Stabilizer link kit	51. Spacer
12. Grommet	33. Retainer	52. Screw
13. Bolt	34. Grommet	53. Bracket
14. Insulator	35. Cotter pin (1/4" × 1 1/4")	54. Bushing
15. Spring	36. Cap	55. Front stabilizer shaft
16. Shock Absorber	37. Nut (3/4"-20)	56. Nut (5/16"-18)
17. Bushing	38. Washer	57. Nut (3/8"-16)
18. Nut (1/2"-13)	39. Bearing	58. Shim
19. Lower control arm	40. Hub	59. Shaft Kit
20. Nut	41. Inner front wheel bearing	60. Nut (1/2"-13)
21. Cotter pin (1/8" × 1 1/4")	42. Bolt	61. Bolt

Exploded view of the front suspension used on models through 1981

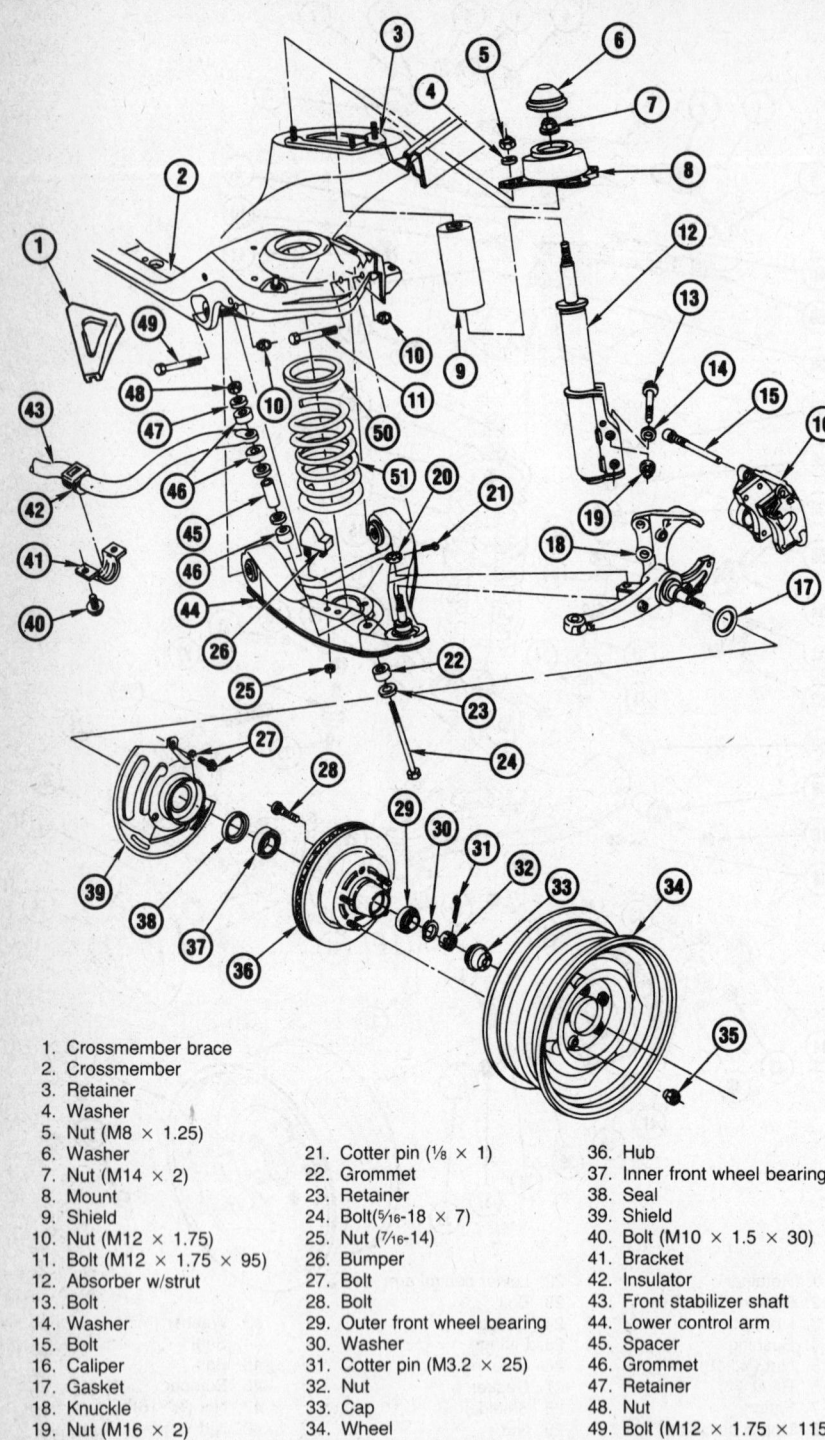

1. Crossmember brace
2. Crossmember
3. Retainer
4. Washer
5. Nut (M8 × 1.25)
6. Washer
7. Nut (M14 × 2)
8. Mount
9. Shield
10. Nut (M12 × 1.75)
11. Bolt (M12 × 1.75 × 95)
12. Absorber w/strut
13. Bolt
14. Washer
15. Bolt
16. Caliper
17. Gasket
18. Knuckle
19. Nut (M16 × 2)
20. Nut (⁹⁄₁₆-18)
21. Cotter pin (⅛ × 1)
22. Grommet
23. Retainer
24. Bolt(⁵⁄₁₆-18 × 7)
25. Nut (⁷⁄₁₆-14)
26. Bumper
27. Bolt
28. Bolt
29. Outer front wheel bearing
30. Washer
31. Cotter pin (M3.2 × 25)
32. Nut
33. Cap
34. Wheel
35. Nut
36. Hub
37. Inner front wheel bearing
38. Seal
39. Shield
40. Bolt (M10 × 1.5 × 30)
41. Bracket
42. Insulator
43. Front stabilizer shaft
44. Lower control arm
45. Spacer
46. Grommet
47. Retainer
48. Nut
49. Bolt (M12 × 1.75 × 115)
50. Insulator

Exploded view of the front suspension used on 1982 and later models

5. Using an internal-fit coil spring compressor, compress the coil spring so that it is loose in its seat.

— **CAUTION** —
Be sure to follow manufacturer's instructions when using spring compressor. Coil springs in a compressed state contain enormous energy which, if released accidentally, could cause serious injury.

6. To remove the coil spring, disconnect the lower control arm from the crossmember at the pivot bolts. If additional clearance is necessary, disconnect the lower control arm from the steering knuckle at the ball joint.

7. To install, compress the coil spring until spring height is the same as when removed, then position the spring on the control arm. Make sure

the lower end of the coil spring is properly positioned in the lower control arm and that the upper end fits correctly in its pad.

8. To complete the installation, reverse the removal procedures. Torque the lower control arm-to-steering knuckle to 78 ft. lbs., the pivot bolt nuts to 63 ft. lbs., the tie rod-to-steering knuckle to 35 ft. lbs. and the stabilizer-to-control arm to 13 ft. lbs.

Shock Absorber

REMOVAL & INSTALLATION

1980–81 Models

1. Remove the upper stem nut while holding the stem to keep it from turning.
2. Remove the bolts holding the shock absorber-to-lower control arm and pull the shock through the arm.
3. Extend the new shock absorber and insert it up through the lower control arm. Make sure that the upper stem goes through the hole in the upper control arm frame bracket.

NOTE: **Purge new shocks of air by repeatedly compressing them while inverted and extending them in their normal installed position.**

4. Install the grommet, retainer cup and nut to the shock absorber upper stem.
5. Hold the shock absorber stem and tighten the upper nut to 8 ft. lbs.
6. Install the lower control arm retaining bolts and tighten to 20 ft. lbs.

Strut

REMOVAL & INSTALLATION

1982 and Later Models

1. Place the ignition key in the unlocked position so that the front wheels can be moved.
2. At the front wheelhouse reinforcement, remove the strut-to-upper mount cover and nut.

— **CAUTION** —
DO NOT attempt to move the vehicle with the upper strut fastener disconnected.

3. Raise and support the front of the vehicle, place jackstands under the lower control arms.
4. Remove the wheel and tire assembly.
5. Remove the brake hose from the strut bracket.
6. Remove the bolts attaching the strut-to-steering knuckle.
7. Lift the strut up from the steer-

ing knuckle to compress the rod, then pull down and remove the strut.

8. To install, extend the rod through the upper mount and start the upper fastener, then reverse the removal procedures. Torque the strut-to-wheelhouse nut to 44 ft. lbs., the strut-to-steering knuckle bolts to 202 ft. lbs.

OVERHAUL

NOTE: For all overhaul information, please refer to "Strut Overhaul" in the Unit Repair section.

FRONT WHEEL BEARING ADJUSTMENT

1. Raise and support the front of the vehicle on jackstands.
2. Remove the hub dust cover, the cotter pin and loosen the hub nut.
3. Spin the wheel and tighten the nut to seat the bearings. DO NOT exert over 12 ft. lbs. of force on the nut.
4. Back the nut off until it is just loose. Line up the cotter pin hole in the spindle with the hole in the nut.
5. Insert a new cotter pin and bend the ends of the pin.

NOTE: The end play should be between 0.001 and 0.005 in. If the play exceeds this tolerance, the wheel bearings should be replaced.

6. To complete the installation, reverse the removal procedures.

Ball Joints

INSPECTION

NOTE: Before performing this inspection, make sure the wheel bearings are adjusted and that the control arm bushings are in good condition.

1. Raise and support the front of the vehicle on jackstands, until there is 1–2 in. of clearance under the wheels.
2. Insert a bar under the wheel and pry upward. If the wheel raises more than $^1/_8$ in., the ball joints are worn. Determine if the upper or lower ball joint is worn by visual inspection while prying on the wheel.
3. The upper ball joint can be further inspected after partial suspension disassembly. If the stud has any detectable side-to-side movement or if it can be twisted with your fingers, it should be replaced.

NOTE: Due to the distribution of forces in the suspension, the

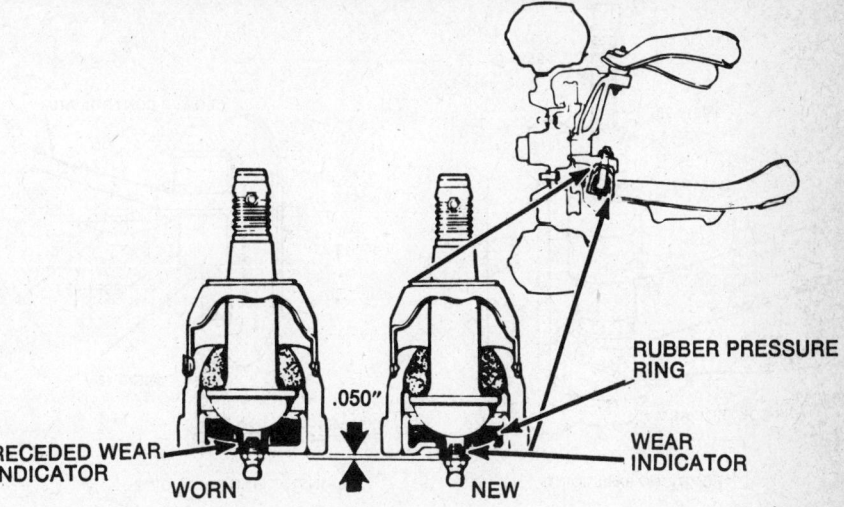

Wear indicator used on the lower ball joints—all models

lower ball joint is usually the defective joint. Because of this, most models are equipped with wear indicators on the lower ball joint as long as the indicator extends below the ball stud seat, replacement is unnecessary.

REMOVAL & INSTALLATION

Upper Ball Joint

NOTE: On 1982 and later vehicles, an upper ball joint is not used due to the strut design.

1. Raise and support the vehicle on jackstands.
2. Remove the tire and wheel assembly.
3. Support the lower control arm with a jack.
4. Remove the upper ball stud nut.
5. Using the ball joint removal tool J-23742, press the the ball joint from the steering knuckle.
6. Using a $^1/_8$ in. drill bit, drill heads of the 4 ball joint rivets on the upper control arm to $^1/_4$ in. deep.
7. Using a $^1/_2$ in. drill bit, drill the remaining heads flush with the control arm, then use a small punch to drive the rivets out of holes.
8. Install the ball joint with the nuts and bolts supplied with the new joint (nuts on top). Torque the nuts and bolts to 8 ft. lbs.
9. Install the lube fitting in the new joint.
10. Mate the upper control arm to the steering knuckle and install the ball stud through the knuckle boss. Torque the ball stud to 65 ft. lbs. (1980–81). Install the cotter pin.

CAUTION

DO NOT back off on the nut to align the cotter pin.

12. To complete the installation, reverse the removal procedures.

Lower Ball Joint

1. Raise and support the front of the vehicle under the frame with jackstands. Remove the wheel and tire assembly.
2. Support the lower control arm with a floor jack.
3. Remove the cotter pin and loosen the lower ball stud nut.
4. Using tool J-24292A, break the ball stud loose from the steering knuckle. Separate the lower control arm from the steering knuckle.
5. Using the ball joint removal tool J-9519-10 (1980–81) or J-9519-23 (1982 and later) and adapter tool J-9519-7, press the ball stud from the lower control arm.
6. Install the new ball joint to the lower control arm. Using the installation tool J-9510-10 (1980–81) or J-9519-23 (1982 and later) and adapter tool J9519-9, press the ball joint into the lower control arm until it bottoms on the arm.

NOTE: When installing the new ball joint, position the purge vent in the rubber boot facing inward.

7. To complete the installation, connect the ball joint-to-control arm assembly to the steering knuckle and torque the ball joint nut to 83 ft. lbs. (1980–81) or 77 ft. lbs. (1982 and later), then reverse the removal procedures.

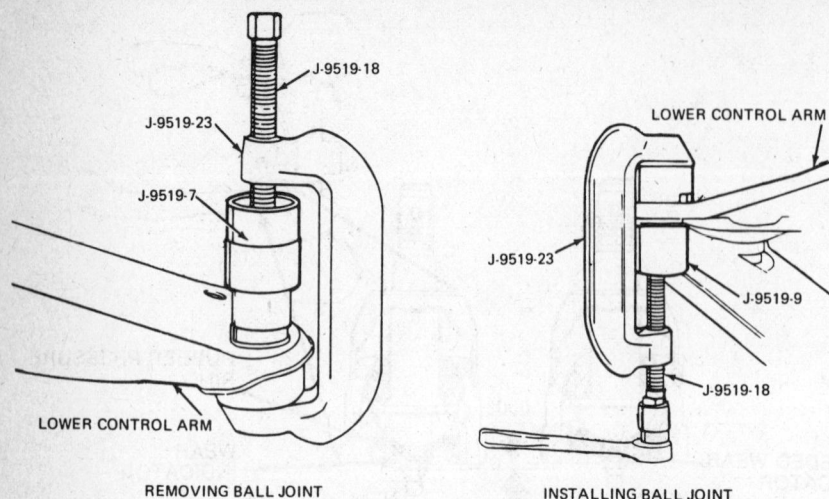

Removing and installing the lower ball joint

Lower Control Arm

REMOVAL & INSTALLATION

1. Refer to the "Coil Spring Removal & Installation" procedures in this section and remove the coil spring.
2. Remove the ball stud from the steering knuckle.
3. Remove the pivot bolts and the lower control arm.
4. To install, reverse the removal procedures. Torque the control arm pivot bolts to 90 ft. lbs. (1980–81) or 63 ft. lbs. (1982 and later).

Upper Control Arm

REMOVAL & INSTALLATION

1. Refer to the "Upper Ball Joint Removal & Installation" procedures in this section and separate the upper ball joint from the steering knuckle.
2. Remove the upper control arm shaft pivot nuts.

NOTE: Tape the shims together and identify them so that they can be installed in the position from which they were removed.

3. Support the hub assembly to prevent damage to the brake line.
4. Remove the upper control arm from the vehicle.
5. To install, reverse the removal procedures. Make sure the shaft to frame bolts are installed in the same position they were in before removal and that the shims are in their original positions. Torque the control arm pivot bolt nuts to 85 ft. lbs.

Steering Knuckle

REMOVAL & INSTALLATION
1980–81 Models

1. Siphon some fluid from the brake master cylinder.
2. Raise and support the vehicle on jackstands.
3. Remove the wheel and tire assembly.
4. Remove the caliper from the steering knuckle and support on a wire.
5. Remove the grease cup, the cotter pin, the castle nut and the hub assembly.
6. Remove the three bolts holding the shield to the steering knuckle.
7. Using the ball joint removal tool J-6627, disconnect the tie rod from the steering knuckle.
8. Using ball joint removal tool J-23742, disconnect the ball joints from the steering knuckle.
9. Place a floor jack under the lower control arm (near the spring seat) and disconnect the ball joint from the steering knuckle.
10. Raise the upper control arm and disconnect the ball joint from the steering knuckle.
11. Remove the steering knuckle from the vehicle.
12. To install, reverse the removal procedures. Torque the upper ball joint-to-steering knuckle nut to 65 ft. lbs., the lower ball joint-to-steering knuckle nut to 90 ft. lbs. and the tie rod-to-steering knuckle nut to 40 ft. lbs. Adjust the wheel bearing and refill the master cylinder.

1982 and Later Models

1. Siphon some fluid from the brake master cylinder.
2. Raise and support the vehicle on jackstands.
3. Remove the wheel and tire assembly.
4. Remove the brake hose from the strut.
5. Remove the caliper from the steering knuckle and support on a wire.
6. Remove the grease cup, the cotter pin, the castle nut and the hub assembly.
7. Remove the splash shield.
8. Disconnect the tie rod from the steering knuckle.
9. Support the lower control arm on a jackstand. Using ball joint removal tool J-24292A, disconnect the ball joint from the steering knuckle.
10. Remove the strut-to-steering knuckle bolts and remove the steering knuckle.
11. To install, reverse the removal procedures. Torque the strut-to-steering knuckle bolts to 202 ft. lbs., the ball joint-to-steering knuckle nut to 78 ft. lbs. and the tie rod-to-steering knuckle nut to 35 ft. lbs. Adjust the wheel bearing and refill the master cylinder.

Stabilizer Bar

REMOVAL & INSTALLATION

1. Raise and support the front of the vehicle on jackstands.
2. Disconnect the stabilizer link bolts at the lower control arms.
3. Remove the stabilizer-to-frame clamps.
4. Remove the stabilizer bar.
5. To install, reverse the removal procedures. Torque the stabilizer-to-lower control arm bolts to 13 ft. lbs. and the stabilizer-to-frame bolts to 24 ft. lbs. (1980–81) or 37 ft. lbs. (1982 and later).

REAR SUSPENSION

1980–81 vehicles use a leaf spring rear suspension; 1982 and later vehicles use a coil spring suspension having a torque arm and a track bar to stabilize the axle assembly. Antisway (stabilizer) bars are optional equipment on all models.

Shock Absorber

REMOVAL & INSTALLATION

1. Raise and support the rear of the vehicle on jackstands.

2. If the vehicle is equipped with Superlift shock absorbers, disconnect the air line.

3. On 1980–81 models, remove the lower shock absorber nut, the retainer and the grommet. Remove the upper bolts and the shock.

4. On the 1982 and later models, pull back the carpet, disconnect the upper shock attaching nut, remove the lower shock-to-axle mounting bolt and the shock absorber.

5. To install, reverse the removal procedures. On 1980–81 models, torque the upper fasteners to 18 ft. lbs. and the lower to 7 ft. lbs. On the 1982 and later models, torque the upper fasteners to 13 ft. lbs. and the lower to 70 ft. lbs.

Springs

REMOVAL & INSTALLATION

1980–81 Models

1. Raise and support the rear of the vehicle by the frame, so that the rear axle can be independently raised and lowered.

2. Support the rear axle with a floor jack.

3. Disconnect the shock absorber lower mount.

4. Loosen the retaining bolt through the front spring eye. Unbolt the front bracket from the body.

5. Lower the axle, then remove the bracket and retaining bolt from the front spring eye.

6. Pry the parking brake cable from the spring mounting plate retainer.

7. Remove the U-bolt nuts, the spring plate, the upper and lower spring pads.

8. Support the spring, then remove the lower rear shackle bolt and the spring.

9. When installing, install the front bracket to the spring eye, install the rear shackle, bolt the front bracket in place, install the U-bolts and the shock absorber. Torque the bolts with the weight of the vehicle on the springs. Torque the front bracket mounting bolt to 25 ft. lbs., the U-bolts to 40 ft. lbs. and the rear shackle bolts to 50 ft. lbs.

1982 and Later Models

1. Raise and support the vehicle with jackstands under the frame, so

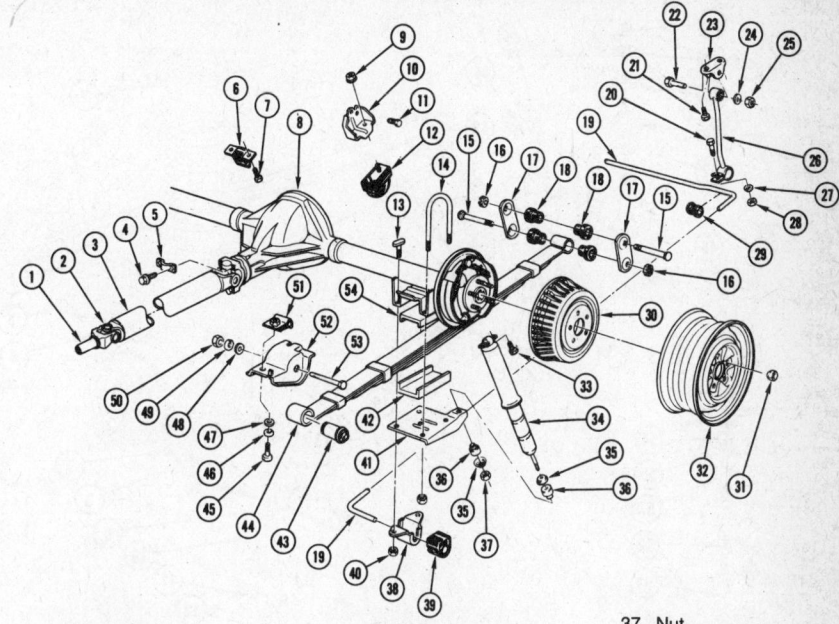

1. Universal joint yoke	19. Rear stabilizer shaft	37. Nut
2. Universal joint	20. Bolt (5⁄16″-18 × 3⁄8″)	38. Bracket
3. Driveshaft	21. Screw (3⁄8″-16 × 1 1⁄8″)	39. Bushing
4. Bolt (5⁄16″-24 × 1 7⁄16″)	22. Bolt (3⁄8″-16 × 2 1⁄4″)	40. Nut
5. Strap	23. Bracket	41. Plate
6. Bumper	24. Washer (1″ × 25⁄64″)	42. Cushion
7. Bolt (5⁄16″-12 × 3⁄4″)	25. Nut (3⁄8″-16)	43. Bushing ASM
8. Housing	26. Support	44. Rear Leaf Spring
9. Nut	27. Washer (1⁄4″)	45. Bolt
10. Bracket	28. Nut (5⁄16″-18)	46. Washer (3⁄8″)
11. Screw (3⁄8″-16 × 1 1⁄8″)	29. Bushing	47. Washer
12. Bumper ASM	30. Brake Drum	48. Washer
13. Bolt (7⁄16″-20 × 1 5⁄16″)	31. Nut	49. Washer (1⁄2″)
14. Bolt (7⁄16″-20 U-shape)	32. Wheel	50. Nut (1⁄2″-20)
15. Pin	33. Screw (5⁄16″-18 × 1″)	51. Nut (3⁄8″-16)
16. Nut (7⁄16″-20)	34. Shock Absorber	52. Bracket
17. Shackle unit	35. Retainer	53. Bolt (1⁄2″-20 × 4 7⁄8″)
18. Bushing	36. Grommet	54. Cushion

Leaf spring rear suspension used on models through 1981

that the rear axle can be independently raised and lowered.

2. Support the rear axle with a floor jack.

3. If equipped with brake hose attachment brackets, disconnect the brackets allowing the hoses to hang free. DO NOT disconnect the hoses. Perform this step only if the hoses will be unduly stretched when the axle is lowered.

4. Disconnect the track bar from the axle.

5. Remove the lower shock absorber bolts and lower the axle. Make sure the axle is supported securely on the floor jack and that there is no chance of the axle slipping after the shock absorbers are disconnected.

6. Lower the axle and remove the coil spring. DO NOT lower the axle past the limits of the brake lines or the lines will be damaged.

7. To install, reverse the removal

procedures. Make sure the spring is seated in the same position as before removal. Torque the track bar bolt-to-axle to 93 ft. lbs., the track bar-to-body to 58 ft. lbs. and the shock absorber-to-axle bolts to 70 ft. lbs.

Track Bar

REMOVAL & INSTALLATION

1982 and Later Models

1. Raise and support the rear of the vehicle on jackstands at the curb height position.

2. Remove the track bar mounting fasteners and the track bar.

3. To install, clean the track bar fasteners and reverse the removal procedures. Torque the track bar-to-axle assembly to 93 ft. lbs. and the track bar-to-body bracket to 58 ft. lbs.

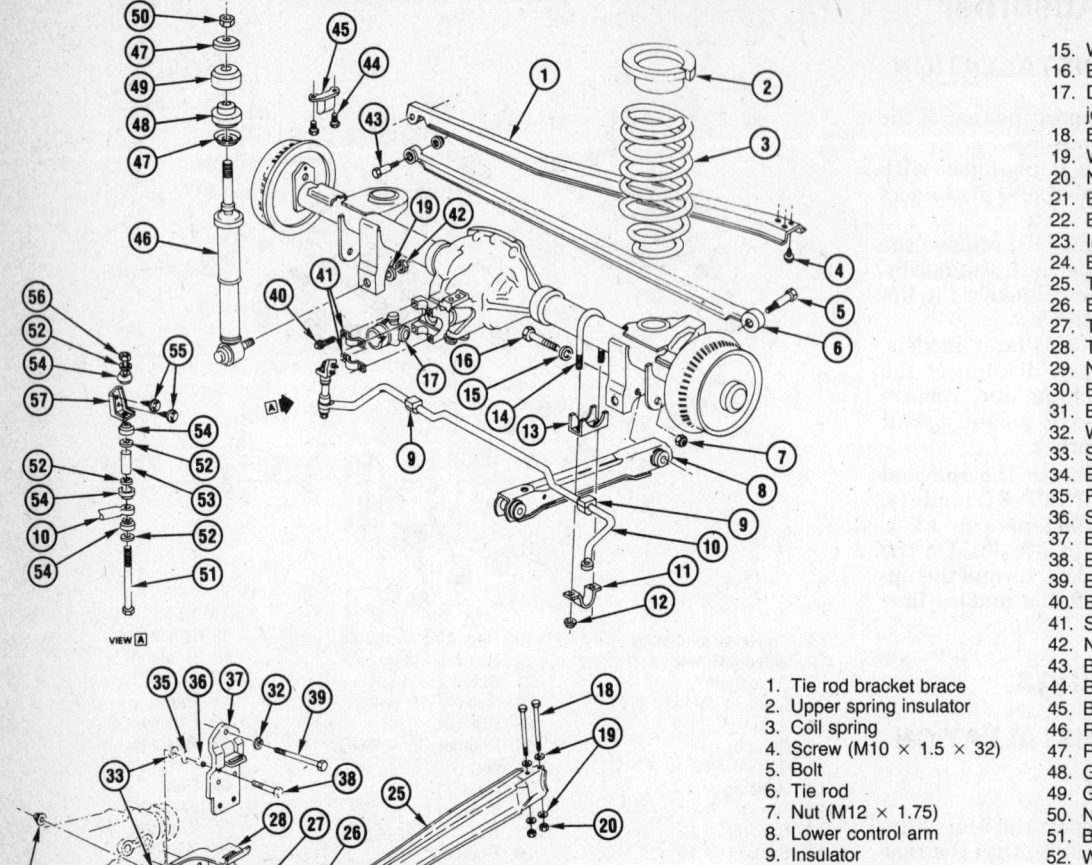

15. Washer
16. Bolt (M12 × 1.75 × 95)
17. Driveshaft w/universal joint
18. Bolt (M14 × 2 × 185)
19. Washer
20. Nut (M14 × 2)
21. Bolt (M8 × 1.25 × 25)
22. Bracket
23. Insulator
24. Bolt (M5 × 0.8 × 10)
25. Torque arm
26. Bolt (M5 × 0.8 × 10)
27. Torque arm insulator
28. Torque arm bracket
29. Nut, "U" (M8 × 1.25)
30. Bolt (M10 × 1.5 × 20)
31. Bolt (M10 × 1.5 × 70)
32. Washer (M10 × 18.3)
33. Spacer
34. Bolt (M4 × 0.7 × 20)
35. Push-nut
36. Spacer
37. Bracket
38. Bolt
39. Bolt (M10 × 1.5 × 110)
40. Bolt
41. Strap
42. Nut (M 14.0 × 2)
43. Bolt w/screw
44. Bolt (M8 × 1.25 × 16)
45. Bumper
46. Rear shock absorber
47. Retainer
48. Grommet
49. Grommet
50. Nut (M10 × 1.5)
51. Bolt (M8 × 1.25 × 180)
52. Washer
53. Spacer
54. Grommet
55. Screw (M10 × 1.5 × 32)
56. Nut (M8 × 1.25)(*2)
57. Bracket

1. Tie rod bracket brace
2. Upper spring insulator
3. Coil spring
4. Screw (M10 × 1.5 × 32)
5. Bolt
6. Tie rod
7. Nut (M12 × 1.75)
8. Lower control arm
9. Insulator
10. Rear stabilizer shaft
11. Clamp
12. Nut (M8 × 1.25)
13. Bracket
14. Bolt

Coil spring rear suspension used on 1982 and later models

Track Bar Brace

REMOVAL & INSTALLATION

1982 and Later Models

1. Raise and support the rear of the vehicle on jackstands under the rear axle.
2. Remove the heat shield screws from the track bar brace.
3. Remove the track bar brace-to-body brace screws.
4. Remove the track bar-to-body bracket fasteners and remove the track bar brace.
5. To install, reverse the removal procedures. Torque the track bar nut-to-body brace to 58 ft. lbs. and the track bar brace-to-body bracket screws to 34 ft. lbs."

Rear Lower Control Arm

REMOVAL & INSTALLATION

1982 and Later Models

NOTE: Remove and install ONLY one lower control arm at a time. If both arms are removed at the same time, the axle could roll or slip sideways, making installation of the arms very difficult.

1. Raise and support the rear of the vehicle on jackstands under the rear axle.
2. Remove the control arm attaching fasteners and the control arm.
3. To install, reverse the removal

procedures. Torque the control arm bolts to 68 ft. lbs.

Torque Arm

REMOVAL & INSTALLATION

1982 and Later Models

NOTE: The coil springs must be removed BEFORE the torque arm. If the torque arm is removed first, damage will result.

1. Raise and support the rear of the vehicle on jackstands under the frame. Place a floor jack under the rear axle.
2. Remove the track bar mounting bolt at the axle assembly, then loosen the track bar bolt at the body brace.
3. Disconnect the rear brake hose

clip at the axle assembly, which will allow additional drop of the axle.

4. Remove the lower attaching nuts from both rear shock absorbers and disconnect the shock absorbers from their lower attaching points.

5. If equipped with a 4 cylinder engine, remove the driveshaft.

6. Carefully lower the rear axle assembly and remove the rear coil springs.

——— CAUTION ———

DO NOT over stress the brake hose when lowering the axle—damage will result.

7. Remove the torque arm rear attaching bolts.

8. Remove the front torque arm outer bracket.

9. Remove the torque arm from the vehicle.

10. To install; place the torque arm in position and loosely install the rear torque arm bolts, then reverse the removal procedures. Torque the front torque arm bracket nuts to 20 ft. lbs. and the rear torque arm nuts to 100 ft. lbs.

11. Place the rear springs and insulators in position, then raise the rear axle assembly until all of the weight is supported by the spring. Torque the shocks-to-axle nuts to 70 ft. lbs., the track bar-to-axle bolt to 93 ft. lbs. and the track bar-to-bracket nut to 58 ft. lbs.

NOTE: On the 4 cylinder models, reinstall the driveshaft.

BRAKES

For all overhaul and service information not covered below, please refer to "Brakes" in the Unit Repair section.

Master Cylinder

REMOVAL & INSTALLATION

1. Disconnect the brake tubes from the master cylinder. If NOT equipped with a power brake booster, disconnect the brake pedal from the master cylinder push rod.

2. Remove the nuts securing the master cylinder to the cowl or the power brake booster.

3. Remove the master cylinder from the vehicle.

——— CAUTION ———

Be careful not to spill brake fluid on the painted surfaces. It will lift the paint.

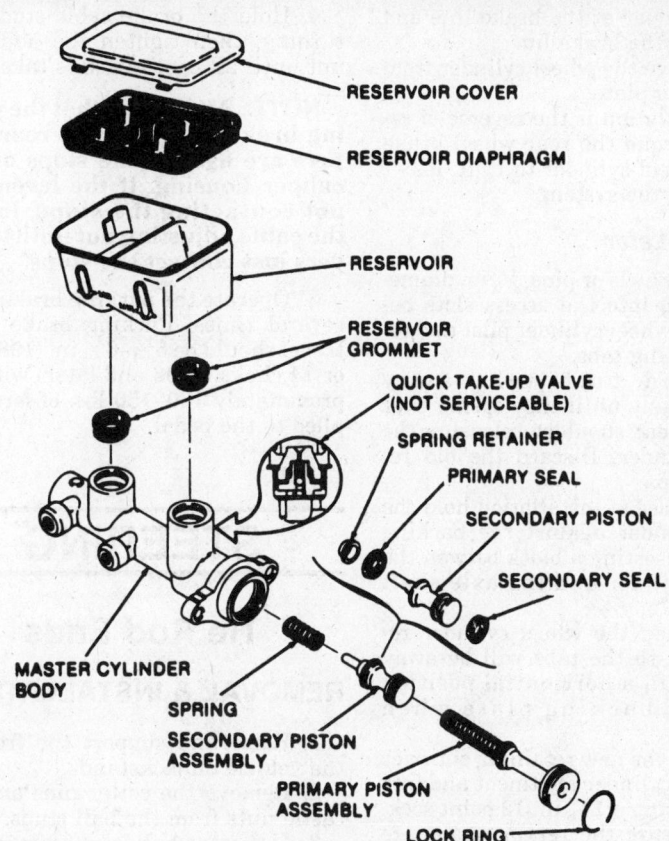

Exploded view of the master cylinder

4. To install, reverse the removal procedures. Torque the master cylinder mounting nuts to 20–30 ft. lbs. Bleed the brake system.

Power Brake Booster

REMOVAL & INSTALLATION

1. Refer to the "Master Cylinder Removal & Installation" procedures in this section and remove the master cylinder.

2. Disconnect the push rod from the brake pedal. Remove the vacuum hose from the power brake booster.

3. From under the dash, remove the 4 mounting nuts from the power brake booster unit.

4. Remove the power brake booster unit.

5. To install, reverse the removal procedures. Torque the power brake booster-to-cowl and the master cylinder-to-booster mounting nuts to 22–30 ft. lbs.

Combination Valve

REMOVAL & INSTALLATION

NOTE: The combination valve is not repairable and must be re-

placed if found to be defective. On some models hoisting might be necessary.

1. Disconnect the electrical connector at the pressure differential switch. It is recommended, that with the aid of pliers, you squeeze the eliptical shaped plastic locking ring and then pull up. This will move the locking tangs away from the switch.

2. Disconnect and plug the hydraulic lines at the combination valve then remove the valve.

3. Installation is the reverse of removal.

4. Bleed the entire brake system.

——— CAUTION ———

Do not move the car until a firm brake pedal is obtained.

Wheel Cylinder

REMOVAL & INSTALLATION

1980–81

1. Raise and safely support the car.

2. Mark the relationship of the wheel to the axle flange.

3. Remove the wheel, drum and brake shoes.

4. Clean all dirt around around the

wheel cylinder at the brake line and disconnect the brake line.

5. Remove the wheel cylinder from the backing plate.

6. Installation is the reverse of removal. Torque the rear wheel brake pipe to wheel cylinder to 12 ft. lbs.

7. Bleed the system.

1982 and Later

1. Insert awls or pins, $1/8$ in. diameter or less, into the access slots between the wheel cylinder pilot and retainer locking tabs.

2. Bend both tabs away simultanously until they spring over the abutment shoulder releasing the wheel cylinder. Discard the old retaining clip.

3. For ease of installation hold the wheel cylinder against the backing plate by inserting a block betwen the wheel cylinder and the axle shaft flange.

4. Position the wheel cylinder retainer clip so the tabs will be away from and in a horizontial position with the backing plate when installing.

5. Press the new retaining clip over the wheel cylinder abutment and into position using a $1^1/8$ in. 12 point socket. Make sure the retainer tabs are properly snapped under the abutment shoulder.

6. Install the brake shoes, drum and wheel.

7. Flush and bleed the hydraulic system.

Parking Brake

ADJUSTMENT

Rear Drum Brakes

1. Depress the parking brake pedal exactly two ratchet clicks.

2. Raise and support the rear of the vehicle on jackstands.

3. Tighten the brake cable adjusting nut until the left rear wheel can be turned rearward with both hands, but locks when forward rotation is attempted.

4. Release the parking brake pedal; both rear wheels must turn freely in either direction without brake drag.

Rear Disc Brakes

1. Check for free movement of the parking brake cables and lubricate the underbody rub points of the cables. Also lubricate the equalizer hooks.

2. Release the parking brake pedal completely.

3. Raise and support the rear of the vehicle on jackstands.

4. Hold the brake cable stud from turning, then tighten the adjusting nut until all cable slack is taken up.

NOTE: Make sure that the parking brake levers on the rear calipers are against the stops on the caliper housing. If the levers are not contacting the stops, loosen the cable adjusting nut until the levers just contact the stops.

5. Operate the parking brake cable several times. Parking brake pedal travel should be $5^1/4$–$6^3/4$ in. (1980–81) or 14 clicks (1982 and later) with approximately 130–150 lbs. of force applied to the pedal.

STEERING

Tie Rod Ends

REMOVAL & INSTALLATION

1. Raise and support the front of the vehicle on jackstands.

2. Remove the cotter pins and the castle nuts from the ball studs.

3. Using the ball joint removal tool J-24319-01 or J-6627, remove the ball joint. If necessary, pull downward on the tie rod to disconnect it from the steering arm.

4. Using the same removal tool, remove the inner ball stud from the relay rod using a similar procedure.

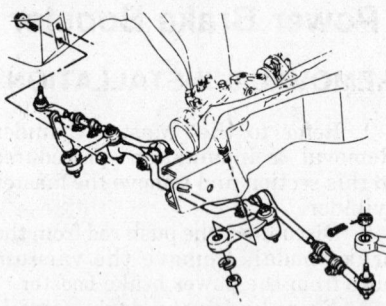

Steering linkage—typical of all models

NOTE: When removing the tie rod ends from the tie rod, be sure to mark their positions or count the number of turns necessary to remove them.

5. To remove the tie rod end or ends from the tie rod, loosen the clamp bolt and unscrew the ends.

NOTE: Lubricate tie rod threads with chassis grease and install new tie rod(s). Make sure both ends are an equal distance from the tie rod and tighten clamp

bolts. **Make sure ball studs, tapered surfaces and the threaded surfaces are clean and smooth and free of grease. Install new seals on ball studs, then position them in the steering knuckle and the relay rod.**

6. Make sure clamp slots and sleeve slots are aligned before tightening clamps. Make sure tightening bolts will be in a horizontal position to 50° upward (in the forward direction) when the tie rod is in its normal position.

7. To install, reverse the removal procedures. Torque the ball joint nuts to 35 ft. lbs. and the tie rod clamps to 14 ft. lbs. Install the new cotter pins. Lubricate new tie rod ends. Check the wheel alignment.

Steering Gear (Manual or Power)

REMOVAL & INSTALLATION

1. Disconnect the negative battery cable. Remove the coupling shield. If equipped with power steering, remove the fluid hoses at the steering gear and cap them to prevent foreign material from entering the system.

2. Remove the mounting bolts, lock washers and nuts at the steering coupling-to-steering shaft flange.

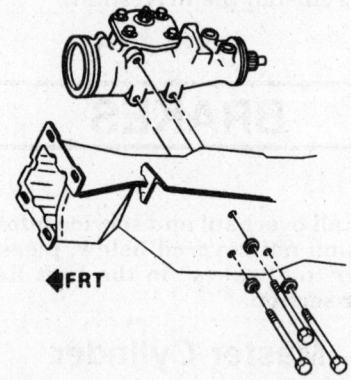

Power steering gear mounting—typical of all models

3. Remove the pitman arm nut and washer. Mark the relation of the arm position-to-shaft.

4. Using removal tool J-6632, remove the pitman arm from the steering gear.

5. Remove the steering gear-to-frame bolts. Remove the steering gear from the vehicle.

6. To Install, reverse the removal procedures.

Power Steering Pump

REMOVAL & INSTALLATION

1. Remove the hoses at the pump and tape the openings shut to prevent contamination. Position the disconnected lines in a raised position to prevent leakage.
2. Remove the pump drive belt.
3. Loosen the retaining bolts, the braces (if equipped) and the pump.
4. To install, reverse the removal procedures. Torque the bracket-to-engine bolts to 24 ft. lbs. and the pump-to-mounting bolts to 22 ft. lbs. Adjust the drive belt tension. Refill the pump with fluid and bleed the system.

SYSTEM BLEEDING

1. Fill the reservoir with power steering fluid.

NOTE: The use of automatic transmission fluid in the power steering system is NOT recommended. Use power steering fluid only.

2. Allow the reservoir and fluid to be undisturbed for a few minutes.
3. Start the engine, allow it to run for approximately 3–5 minutes to warm up the fluid, then turn it off.
4. Check the reservoir fluid level and add fluid, if necessary.
5. Repeat the above steps until the fluid level stabilizes.
6. Raise the front of the vehicle so that the wheels are off of the ground and set the parking brake.
7. Start the engine and increase the engine speed to about 1500 rpm.
8. Turn the front wheels right to left (and back) several times, lightly contacting the wheel stops at the ends of travel.
9. Check the reservoir fluid level. Add fluid as required.
10. Repeat Step 8 until the fluid level in the reservoir stabilizes.
11. Lower the vehicle and repeat Steps 8 and 9.

Steering Wheel

REMOVAL & INSTALLATION
——— CAUTION ———

Disconnect the battery ground cable before removing the steering wheel. When installing a steering wheel, always make sure that the turn signal lever is in the neutral position.

Standard Wheel

1. Remove the shroud screws from behind the steering wheel.

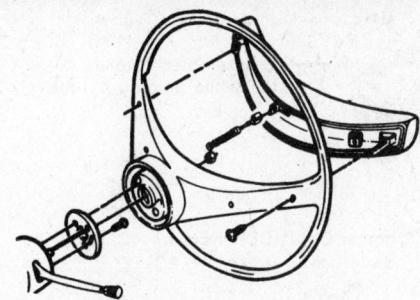

Standard steering wheel

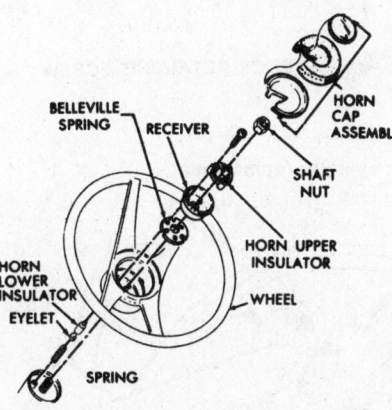

Cushioned rim steering wheel assembly

2. Lift the shroud and the horn lead assembly from the steering wheel.
3. Remove the shaft snap-ring and the steering wheel nut.
4. Mark the wheel-to-shaft relationship. Using the steering wheel puller J-2927 or equivalent, pull the steering wheel from the steering shaft.
5. To install, align the match marks, slide the steering wheel onto the steering shaft and reverse the removal procedures. Torque the steering wheel nut to 35 ft. lbs. (1980), 30 ft. lbs. (1981 and later) and the steering wheel shroud screws to 8 ft. lbs.

1980 Padded Rim Wheel

1. Pry out the center cap and retainer carefully with a small screwdriver. Remove the shaft snap-ring.
2. Remove the steering wheel nut and washer.
3. Remove the three upper horn insulator screws, the insulator, receiver, belleville spring and the shim.
4. Mark the wheel-to-shaft relationship. Using the puller J-2927 or equivalent, pull the steering wheel from the steering column.
5. To install, align the matchmarks and slide the steering wheel onto the steering shaft. Torque the steering wheel nut to 35 ft. lbs.

Turn Signal Switch

REMOVAL & INSTALLATION

1. Refer to the "Steering Wheel Removal & Installation" procedures in this section and remove the steering wheel.
2. Remove the column-to-instrument panel trim cover.
3. Using a small screwdriver, carefully pry the cover up and out of the steering column.
4. Position the lockplate compressing tool J-23653 or equivalent on the end of the steering shaft and compress the lock plate by turning the shaft nut clockwise. Pry the wire snap-ring out of the shaft groove and discard the ring.
5. Remove the tool and lift the lockplate off the shaft.
6. Slide the cancelling cam, upper bearing preload spring and thrust washer off the shaft.
7. Remove the retaining screw and the turn signal lever. Push the flasher knob in and unscrew it.

NOTE: If equipped with a button and a knob, remove the button retaining screw, then remove the button, spring and knob.

8. Remove the three mounting screws. Pull the switch connector out of the jacket bracket, wrap it with tape, lift up on the switch and pull the connector through the column support bracket.

NOTE: On tilt wheels, place the turn signal and shifter housing in the low position and remove the harness cover.

9. To install, attach a long piece of wire to the turn signal switch connector, feed the wire through the column housing and under the bracket, then pull the wire, the switch connector and the cover into position.

NOTE: On tilt wheel models, pull the connector down through the housing under the bracket, then install the cover over the harness.

10. Install the switch mounting screws and the connector on the jacket bracket. Install the column-to-dash trim plate.
11. Install the flasher knob and the turn signal lever.
12. With the turn signal lever in the neutral or off position and the flasher knob out, slide the thrust washer, upper bearing preload spring and cancelling cam onto the shaft.
13. Using tool J-23653 or equivalent, press the lock plate down on the

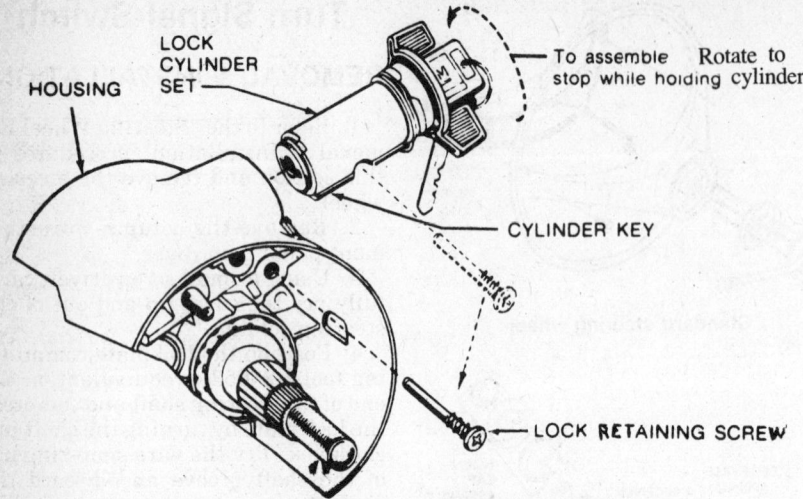

HOUSING

LOCK CYLINDER SET

To assemble Rotate to stop while holding cylinder

CYLINDER KEY

LOCK RETAINING SCREW

1979 and later ignition lock cylinder replacement

shaft and install a NEW snap-ring in the shaft groove. Do not re-use the old snap ring.

14. Install the cover and the steering wheel. Torque the turn signal switch screws to 3 ft. lbs. and the steering wheel-to-shaft nut to 35 ft. lbs.

Ignition Switch

REMOVAL & INSTALLATION

The switch is located inside the channel section of the brake pedal support and is completely inaccessible without first lowering the steering column. The switch is actuated by a rod and rack assembly. A gear on the end of the lock cylinder engages the toothed upper end of the rod.

1. Support and lower the steering column.

2. Place the ignition switch in the OFF-UNLOCKED position and move the actuating rod two detents from the top.

3. Remove the two mounting screws and the ignition switch assembly.

4. Before installing, place the new switch in the OFF-UNLOCKED position and make sure the ignition lock cylinder and the actuating rod are in the OFF-UNLOCKED (2nd detent from the top) position.

5. Install the actuating rod into the switch, mount the switch to the column and torque the mounting screws to 3 ft. lbs.

NOTE: Use only the specified screws since overlength screws could impair the collapsibility of the column.

6. Install the steering column and

Depressing the lock cylinder spring latch

torque the steering column-to-bracket nuts to 25 ft. lbs.

Ignition Lock Cylinder

REMOVAL & INSTALLATION

1. Disconnect the battery ground cable, then refer to the "Turn Signal Switch Removal & Installation" procedure in this section and lift the turn signal switch to allow access to the lock cylinder.

NOTE: When lifting the turn signal switch, pull it rearward far enough to slip it over the end of the steering shaft. DO NOT pull the wiring harness from the column.

2. Place the ignition lock in the ON position.

3. Remove the buzzer switch, the lock cylinder screw and lock cylinder.

──── CAUTION ────

If the screw is dropped during removal and falls into the column, complete disas-

sembly will be necessary to retrieve the screw.

4. To install, rotate the cylinder clockwise to align cylinder key with the keyway in the housing. Push the lock in all the way, then install the screw and torque it to 22 inch lbs. for adjustable columns and 40 inch lbs. for standard columns.

CHASSIS ELECTRICAL

Headlamp Switch

REMOVAL & INSTALLATION

1980–81 Camaro

1. Disconnect negative battery cable and pull the light switch to the ON position.

2. Remove steering column lower cover (6 screws).

3. Reach up under cluster on the left side and depress the light switch shaft retainer (located on the switch), while pulling gently on the switch knob.

4. Remove the nut that secures the switch to the instrument cluster.

5. Remove four cluster carrier screws in front and two from the rear, then tilt the right side of the cluster out.

6. Disconnect the wiring harness connector and remove the switch.

7. To install, reverse the removal procedures.

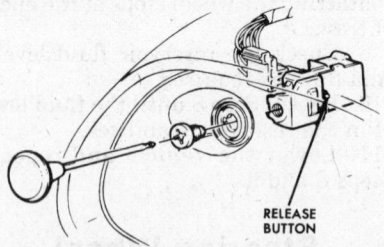

RELEASE BUTTON

Typical headlamp switch. Note the position of the knob and shaft release button

1980–81 Firebird

1. Disconnect the negative battery cable.

2. Pull the headlamp switch knob to the ON position.

3. Reach under the instrument panel and depress the locking button for the knob and shaft (located on the switch), then remove the knob and shaft.

4. Remove the switch retaining nut.

5. Remove the switch from the instrument panel and disconnect the electrical connector.

6. To install, reverse the removal procedures.

1982 and Later (All Models)

1. Disconnect the negative battery cable.

2. Remove the right and left lower trim plates; DO NOT remove the lower instrument panel cover.

3. Remove the instrument panel cluster trim plate.

4. Remove the two screws retaining the switch assembly.

5. Depress the side tangs and pull out the switch assembly.

6. Disconnect the wiring harness connector and remove the headlight switch from the assembly.

7. Reverse the removal procedures to install.

Speedometer Cable

REMOVAL & INSTALLATION

1980–81 Models

1. Disconnect the battery ground cable.

2. Reach up behind the speedometer and depress the retaining tab while pushing in, then out on the cable end.

3. Remove the firewall panel sealing plug to allow movement of the cable.

4. Pull the speedometer cable core from the casing. If the core is broken, it will be necessary to raise the car and disconnect the cable from the transmission to remove the other end.

5. To install, lubricate the core with cable lubricant and reverse the removal procedures. Make sure the core engages the transmission drive unit.

1982 and Later Models

1. Disconnect the negative battery cable.

2. On Firebird models, remove the upper and lower instrument panel trim plates.

3. If equipped with cruise control, disconnect the speedometer cable at the cruise control transducer. If NOT equipped with cruise control, disconnect the speedometer cable strap at the power brake booster.

4. On Camaro models, remove the instrument cluster bezel.

5. Remove the six instrument cluster screws and pull the cluster out far

enough to gain access to the rear of the speedometer head.

6. Reach beneath the cable connection at the speedometer head, push in on the cable retaining spring and disconnect the cable from the speedometer.

7. Slide the old cable out of the speedometer cable case. If the cable is broken, remove the cable from both ends of the casing.

NOTE: Using a short piece of the old cable to fit the speedometer connection, turn the speedometer to increase the speed indicated on the dial and check for any binding during rotation. If binding is noted, the speedometer must be removed for repair or replacement. Check the entire cable casing for extreme bends, chafing, breaks, etc. and replace if necessary.

8. To install, wipe the cable clean using a lint-free cloth. Flush the casing with petroleum spirits and blow dry with compressed air, then lubricate the speedometer cable with an appropriate lubricant (be sure to cover the lower $2/3$ of the cable). Insert the cable into the case and reverse the removal procedures.

Instrument Cluster

REMOVAL & INSTALLATION

1980–81 Camaro

1. Disconnect the negative battery cable.

2. Remove the six cluster bezel screws.

3. Reach behind the left cluster and

press on the retainer button of the headlamp switch shaft while pulling on the switch knob.

4. Remove the retaining nut from the switch.

5. Disconnect the electrical connectors from the headlight and the windshield wiper switches.

6. Remove the headlight switch from the bezel mounting hole.

7. Remove the two wiper switch screws and the switch from the bezel.

8. Disconnect the cigarette lighter electrical connector and unscrew the retainer from the bezel.

9. Pull the bezel rearward and remove the cluster retaining screws.

10. Pull the cluster rearward and disconnect the printed circuit connector.

11. Disconnect the speedometer cable and the wiring harness clips, then remove the cluster.

12. To install, reverse the removal procedures.

Optional Instrument Cluster

To remove all bulbs, instruments (except speedometer) and printed circuits, it is not necessary (unless air-conditioned) to remove the cluster. The instruments are installed and removed from the rear of the cluster. On vehicles with air conditioning, it will be necessary to remove the cluster to remove the ammeter, fuel and temperature gauges.

NOTE: When performing any operation behind the cluster, disconnect the battery ground cable.

1980–81 Firebird

1. Disconnect the battery ground cable.

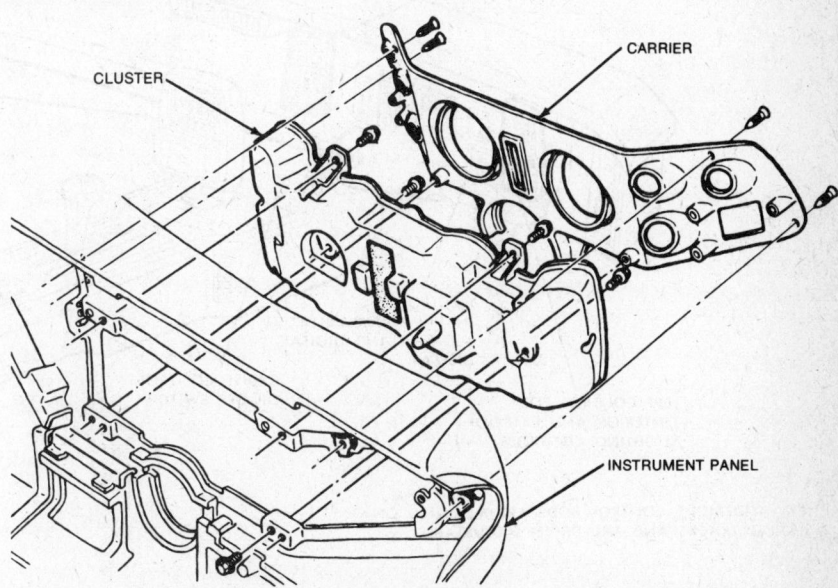

Typical instrument cluster mounting—models through 1981

2. Remove the upper instrument panel trim plate.

3. Remove the lower instrument panel trim and bracket at the steering column.

4. Loosen the two steering column nuts and carefully lower the column.

5. Remove the cluster screws and pull out the cluster. Disconnect the speedometer cable and wiring harness for the printed circuit.

6. Remove the cluster.

7. To install, reverse the removal procedures.

1982 and Later Camaro

SPORT COUPE MODEL

1. Disconnect the negative battery cable.

2. Remove the instrument cluster bezel.

3. Remove the six cluster attachment screws.

4. Pull the cluster out, then disconnect the speedometer cable and electrical connections.

5. Remove the cluster lens.

6. To install, reverse the removal procedures.

BERLINETTA MODEL

1. Disconnect the negative battery cable.

2. Remove the instrument cluster bezel.

3. Remove the eight steering column trim cover screws and the trim cover.

4. Remove the right and the left hand pod attaching screws at the bot-

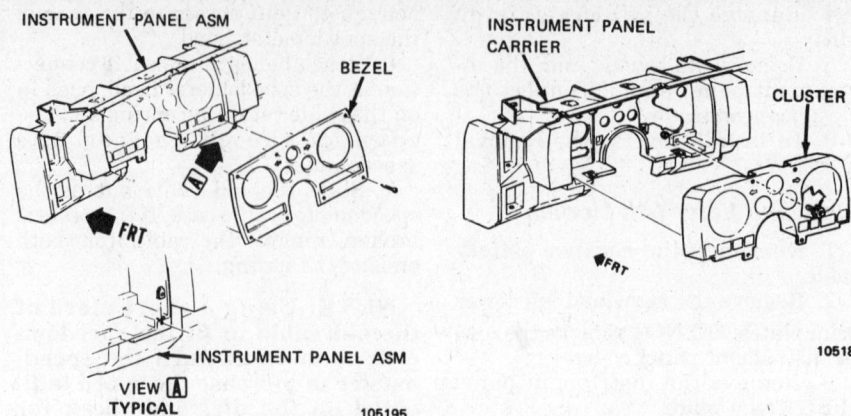

Typical instrument panel cluster and bezel—1982 and later

tom front of each pod. Pull the pods rearward and disconnect the electrical connection.

5. Remove the five cluster lens screws and the lens.

6. Remove the two steering column bolts and lower the column.

7. Pull the instrument cluster rearward and disconnect the electrical connection. Remove the instrument cluster.

8. To install, reverse the removal procedures.

1982 and Later Firebird

1. Disconnect the negative battery cable.

2. Remove the right and left lower trim plates.

3. Remove the instrument cluster trim plate.

4. Remove the cluster attachment screws, pull the cluster back, then disconnect the speedometer cable and the electrical connections.

5. Remove the trip odometer, reset knob (if equipped) and the cluster lens.

6. To install, reverse the removal procedures.

Emissions Indicator

An emissions indicator flag may appear in the odometer window of the speedometer, on some vehicles (1980 and later). The flag could say "Sensor" "Emissions" or "Catalyst" depending on the part or assembly that is scheduled for regular emissions maintenance replacement. The word "Sensor" indicates a need for oxygen sen-

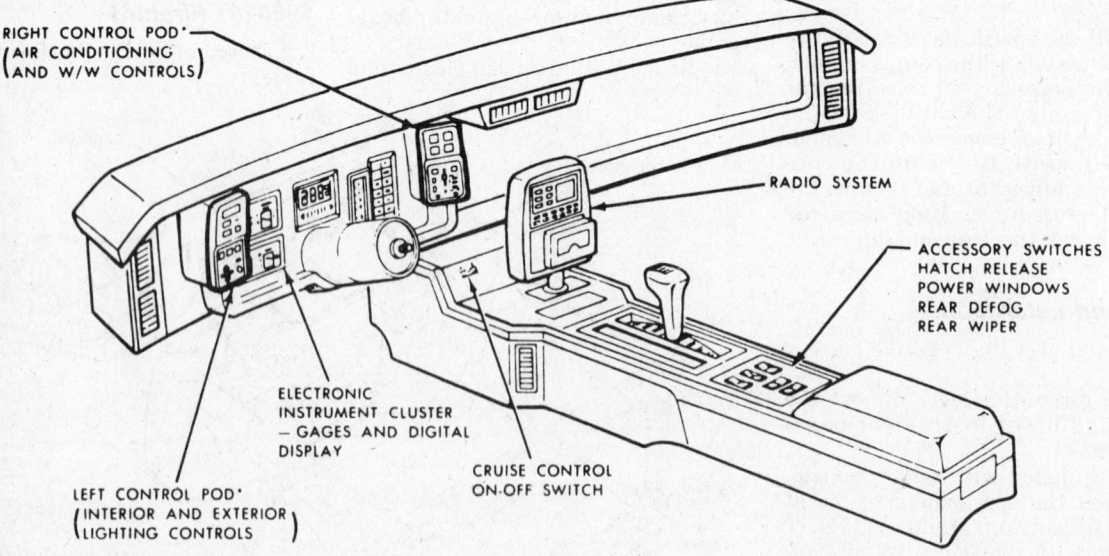

View of the Berlinetta style dash and console

sor replacement and the words "Emissions" or "Catalyst" indicate the need for catalytic converter replacement.

Reset Procedure

1. Remove the instrument panel trim plate.
2. Remove the instrument cluster lens.
3. Locate the flag indicator reset notches at the driver's side of the odometer.
4. Use a pointed tool to apply light downward pressure on the notches, until the indicator is reset.

NOTE: When the indicator is reset an alignment mark will appear in the left center of the odometer window.

Wiper Motor

REMOVAL & INSTALLATION

1. Disconnect the negative battery cable.
2. Remove the cowl vent screen.
3. Remove the transmission drive link(s) from the motor crank arm.
4. Disconnect the electrical connectors and the washer hoses.
5. Remove the motor mounting screws and the motor.

NOTE: When removing the motor, guide the crank arm through the hole.

6. To install, reverse the removal procedures. The motor must be in the "Park" position before assembling the crank arm to the transmission drive link(s).

Wiper Arms and Blades

REMOVAL & INSTALLATION

If the wiper assembly has a press type release tab at the center, simply depress the tab and remove the blade. If the blade had no release tab, use a screwdriver to depress the spring at the center. This will release the assembly. To install the assembly, position the blade over the pin at the tip of the arm and press until the spring retainer engages the groove in the pin. To remove the element, either depress the release button or squeeze the spring type retainer clip at the outer end together and slide the blade element out. Just slide the new element in until it latches.

1. Insert tool J-8966 or equivalent

under the wiper arm and lever the arm off the shaft.
2. If equipped, disconnect the washer hose from the arm. Remove the arm.
3. To install, reverse the removal procedures.

NOTE: Be sure that the motor is in the "Park" position before installing the wiper arms.

Radio

REMOVAL & INSTALLATION

1980–81 Models

1. Disconnect the battery ground cable.
2. Pull off the knobs and bezels.
3. Remove the control shaft nuts and washers, using a deep well socket. If equipped, remove the center air duct and hose.
4. Remove the mounting bracket screws or nuts.
5. Move the radio back until the shafts clear the instrument panel. Lower it and disconnect the antenna, speaker and power wires.
6. Remove the radio.
7. To install, reverse the removal procedures.

NOTE: Make sure to connect the speaker leads before turning the radio ON. Operating the radio without a speaker will damage the transistors.

1982 and Later Models

SPORT COUPE MODEL

1. Disconnect the negative battery cable.
2. Remove the console bezel screws and the console bezel.
3. Remove the radio-to-console mounting screws.
4. Remove the radio and disconnect the electrical connector.
5. To install, reverse the removal procedures.

— CAUTION —
Never apply power to the radio until the speaker wiring is connected; radio damage could result.

BERLINETTA MODEL

1. Disconnect the negative battery cable. Remove the four screws at the console trim plate.
2. Lift the receiver, with the connector attached and turn to one side.
3. Remove the four control head mounting bracket screws. Remove the control head by pulling back on the

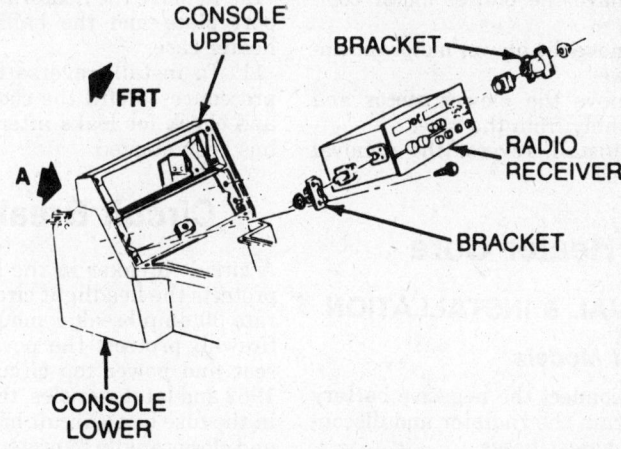

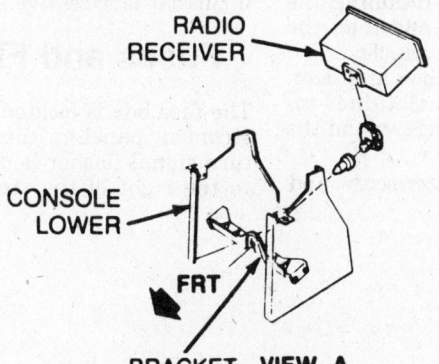

Removing the radio—Sport Coupe model

pawl spring and pulling up on the control head.

4. Disconnect the electrical connectors from the control head.

5. Remove the four screws at the receiver bracket and the slotted screw at the receiver.

6. Disconnect the electrical connector and remove the receiver.

7. To install, reverse the removal procedures.

Blower Motor

REMOVAL & INSTALLATION

1980-81 Models

1. Disconnect the negative battery cable.

2. Tag and disconnect any electrical connections at the motor.

3. Remove the heater front module screws and nuts.

4. Lift off the front module and the motor.

5. Installation is in the reverse order of removal. Replace all sealer.

1982 and Later Models

1. Disconnect the negative battery cable.

2. Tag and disconnect the electrical connectors from the blower motor.

3. Remove the blower motor cooling tube.

4. Remove the blower motor retaining screws.

5. Remove the blower motor and fan assembly from the case.

6. To install, reverse the removal procedures.

Heater Core

REMOVAL & INSTALLATION

1980-81 Models

1. Disconnect the negative battery cable. Drain the radiator and disconnect the heater hoses.

2. Remove the nuts retaining the heater case-to-firewall and then the screws from inside the vehicle.

3. Remove the glovebox and door. Remove the heater outlet duct-to-heater case mounting screws and the duct.

4. Remove the defroster screw and

pull the heater case out. The core may now be pulled from the case.

5. To install, reverse the removal procedures.

NOTE: Use new seals between the heater case and the firewall.

1982 and Later Models

1. Disconnect the negative battery cable.

2. Drain the cooling system.

3. Disconnect the heater hoses from the heater core.

4. Remove the right side lower hush panel.

5. Remove the right side lower instrument panel trim panel and the electronic spark control (ESC) module, if necessary.

6. Remove the right side lower instrument panel carrier-to-cowl screw.

7. Remove the four heater case cover screws.

NOTE: The upper left heater case cover screw may be reached with a long socket extension. Carefully lift the lower right corner of the instrument panel to align the extension.

8. Remove the heater case cover.

9. Remove the heater core support plate and the baffle screws.

10. Remove the heater core, the support plate and the baffle from the heater case.

11. To install, reverse the removal procedures. Refill the cooling system and check for leaks after the engine has been started.

Circuit Breakers

A circuit breaker in the light switch protects the headlight circuit. A separate 30 amp breaker mounted on the firewall protects the power window, seat and power top circuits. On the 1982 and later vehicles, this is located in the fuse box. Circuit breakers open and close rapidly to protect the circuit if current is excessive.

Fuses and Flashers

The fuse box is located under the instrument panel on the left side. The turn signal flasher is under the dash to the right of the steering column.

The hazard flasher is under the dash, to the left of the steering column. On the 1980–81 models, both the turn signal flasher and the hazard flasher are located at the lower left hand and the upper right hand corners of the fuse box respectively. On 1982 and later vehicles, the hazard flasher is located in the convenience center which is on the underside of the instrument panel to the right of the steering column. The turn signal flasher is located in a clip behind the instrument panel to the right of the steering column. There is an inline fuse for the underhood and spotlamp circuit. The fuse box is marked to indicate fuse size and the circuit(s) protected.

Fusible Links

In addition to circuit breakers and fuses, the wiring harness incorporates fusible links to protect the wiring. Links are used rather than a fuse, in wiring circuits that are not normally fused, such as the ignition circuit. Fusible links are color coded red in the charging and load circuits they protect. Each link is 4 gauges smaller than the cable it protects and is marked on the insulation with the gauge size because the insulation makes it appear heavier than it really is.

The engine compartment wiring harness has several fusible links. The same size wire with a special hypalon insulation must be used when replacing a fusible link. The links are typically located in the following areas:

1. A molded splice at the starter solenoid "Bat" terminal, a 14 gauge red wire.

2. A 16 gauge red fusible link at the junction block to protect the unfused wiring of 12 gauge or larger wire. This link stops at the bulkhead connector.

3. The alternator warning light and field circuitry is protected by a 20 gauge red wire fusible link used in the "battery feed-to-voltage regulator No. 3 terminal." The link is installed as a molded splice in the circuit at the junction block.

4. The ammeter circuit is protected by two 20 gauge fusible links installed as molded splices in the circuit at the junction block and battery to starter circuit.

GM "H" Body
Rear Wheel Drive
Monza, Skyhawk, Starfire, Sunbird

YEAR IDENTIFICATION

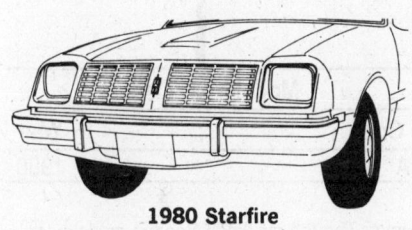

1980 Starfire

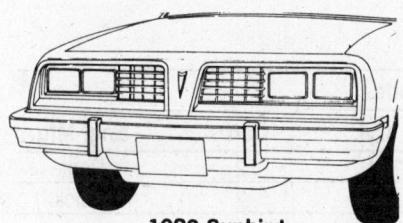

1980 Sunbird

1980 Skyhawk

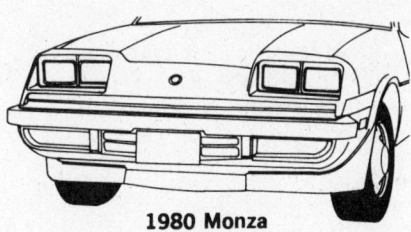

1980 Monza

GENERAL ENGINE SPECIFICATIONS

Year	Eng. V.I.N. Code	Engine No. Cyl. Displacement (cu. in.)	Eng. Mfg.	Carburetor Type	Horsepower (@ rpm)■	Torque @ rpm (ft lbs)■	Bore × Stroke (in.)	Compression Ratio	Oil Pressure @ rpm (psi)
'80	V	4-151	Pont.	2 bbl	90 (@ 4400)①	128 (@ 2400)②	4.000 × 3.000	8.2:1	37–41
	A	6-231	Buick	2 bbl	110 (@ 3800)	190 (@ 1600)	3.800 × 3.400	8.0:1	37

■ Horsepower and torque are SAE net figures. They are measured at the rear of the transmission with all accessories installed and operating. Since the figures vary when a given engine is installed in different models, some are representative rather than exact.
① Calif.—85 (@ 4400)
② Calif.—123 (@ 2800)

VEHICLE IDENTIFICATION NUMBER (VIN)

It is important for servicing and ordering parts to be certain of the vehicle and engine identification. The VIN (vehicle identification number) is a 13 or 17 digit number visible through the windshield on the driver's side of the dash and contains the vehicle and engine identification codes. It can be interpreted as follows:

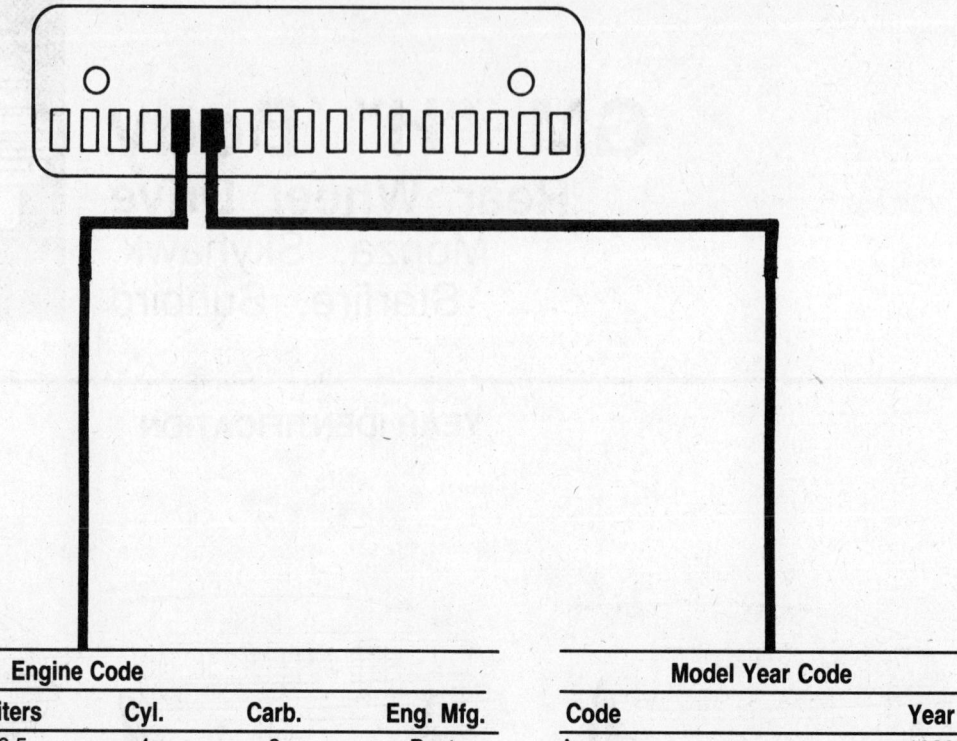

Engine Code

Code	Cu. in.	Liters	Cyl.	Carb.	Eng. Mfg.
V	151	2.5	4	2	Pont.
A	231	3.8	V-6	2	Buick

Model Year Code

Code	Year
A	1980

The thirteen digit Vehicle Identification Number can be used to determine engine application and model year. The 6th digit indicates the model year, and the 5th digit identifies the factory installed engine.

TUNE-UP SPECIFICATIONS

Year	Model	Eng. V.I.N. Code	Engine No. Cyl. Displacement (cu. in.)	Eng. Mfg.	Spark Plugs Orig. Type	Gap (in.)	Distributor (deg)	(in.)	Ignition Timing (deg) Man.	Auto.	Intake Valve Opens (deg)	Fuel Pump Pressure (psi)	Idle Speed (rpm) ▲ Man.	Auto.
'80	All	V	4-151	Pont.	R43TSX ⑥	0.060	Electronic		12B	12B	33	4.5–5	1000/550①②	650/550③④
	All	A	6-231	Buick	R46TSX	0.060	Electronic		15B	15B	16	4.5–7.5	800/600	650/550⑤

NOTE: The underhood specifications sticker often reflects tune-up specification changes made in production. Sticker figures must be used if they disagree with those in this chart. Where two figures are separated by a slash, the first figure is for idle speed with the solenoid connected, while the second is for the idle speed with the solenoid disconnected.

④ 49 States with a/c—1250/1000
⑤ Calif. without a/c—1000/500, Calif. with a/c—1200/1000
⑥ 49 State with a/c—850/650
⑦ Calif. without a/c—650/550, Calif. with a/c—850/650
⑧ Calif. with a/c—670/620
⑨ Calif. Monza—R44TSX, Sunbird, Starfire—R44TSX
▲ Lines separated by a slash show solenoid on/off

FIRING ORDERS

NOTE: To avoid confusion, always replace spark wires one at a time.

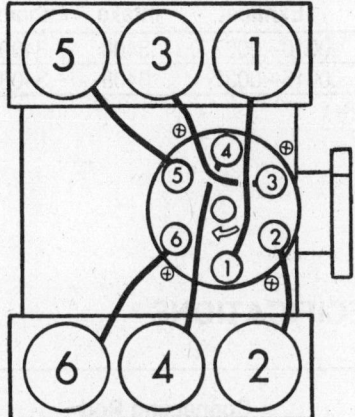

GM (Buick) 196, 231 V6
Engine firing order: 1-6-5-4-3-2
Distributor rotation: clockwise

V6 harmonic balancers have two timing marks: one is 1/8 in. wide, and one is 1/16 in. wide. Use the 1/16 in. mark for timing with a hand held light. The 1/8 in. mark is used only with a magnetic timing pick-up probe.

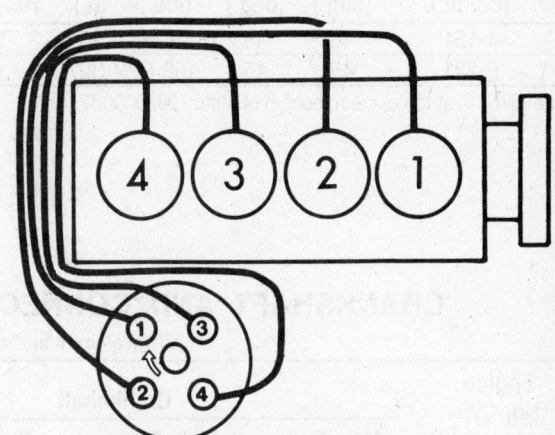

GM Pontiac 151 4-cyl. (1979-80)
Engine firing order: 1-3-4-2
Distributor rotation: clockwise

CAPACITIES

Year	Model	Engine No. Cyl. Displacement (cu. in.)	Engine Crankcase (qts.)		Transmission (pts.) ▲			Drive Axle (pts.)	Gasoline Tank (gals.)	Cooling System (qts.)	
			With Filter	Without Filter	Manual		Automatic ●			W/ AC	W/O AC
					3-spd	4-5 spd					
'80	Monza	4-151	4	3	3.4	—	7.5	3.5	18.5	11.5	11.6
		6-231	5	4	3.4	—	7.0	3.5	18.5	11.9	11.9
	Sunbird	4-151	4	3	3.0	3.5	8.0	3.5	18.5①	11.5	11.0
		6-231	5	4	3.4	—	7.0	3.5	18.5①	11.9	11.9
	Starfire	4-151	4	3	3.0	—	6.0	3.5	18.5	11.5	11.0
		6-231	5	4	3.0	—	6.0	3.5	18.5	12.4	11.9
	Skyhawk	6-231	5	4	3.0	—	6.0	3.5	18.5	12.2	12.3

● Specifications do not include torque converter
▲ Pints to refill after draining
① Sta. wag.—15.0

VALVE SPECIFICATIONS

Year	Engine No. Cyl. Displacement (cu. in.)	Seat Angle (deg.)	Face Angle (deg.)	Spring Test Pressure (lbs. @ in.)	Spring Installed Height (in.)	Stem-to-Guide Clearance (in.)		Stem Diameter (in.)	
						Intake	Exhaust	Intake	Exhaust
'80	4-151	46	45	176 @ 1.254	1.69	.0010–.0027	.0010–.0027①	.3400	.3400
'80	6-231	45	45	168 @ 1.327	1.727	.0015–.0032	.0015–.0032	.3408	.3408

① Figure given is at top of stem; bottom of stem: .0020-.0037

CRANKSHAFT AND CONNECTING ROD SPECIFICATIONS

(All measurements are given in inches.)

Year	Engine No. Cyl. Displacement (cu. in.)	Crankshaft				Connecting Rod		
		Main Brg. Journal Dia.	Main Brg. Oil Clearance	Shaft End-Play	Thrust on No.	Journal Diameter	Oil Clearance	Side Clearance
'80	4-151	2.2988	.0002–.0022	.0035–.0085	5	2.0000	.0005–.0026	.006–.022
'80	6-231	2.4995	.0003–.0017	.004–.008	2	2.2487–2.2495	.0005–.0026	.006–.027

PISTON AND RING SPECIFICATIONS

(All measurements are given in inches.)

Year	V.I.N. Code	Engine Type/ Disp. (cu. in.)	Eng. Mfg.	Piston-to-Bore Clearance	Ring Gap			Ring Side Clearance		
					Top Compression	Bottom Compression	Oil Control	Top Compression	Bottom Compression	Oil Control
'80	V	151	Pont.	.0025–.0033①	.016–.026	.009–.019	.015–.055	.0015–.0035	.0015–.0035	.0015–.0035
'80	A	231	Buick	.0008–.0020	.013–.023	.013–.023	.015–.035	.0030–.0050	.0030–.0050	.0035 Max.

① Measured 1.11 in. from top of piston

TORQUE SPECIFICATIONS

(All readings in ft. lbs.)

Year	Engine No. Cyl. Displacement (cu. in.)	Cylinder Head Bolts	Rod Bearing Bolts	Main Bearing Bolts	Crankshaft Pulley Bolt	Flywheel to Crankshaft Bolts	Manifold	
							Intake	Exhaust
'80	4-151	85③	30	65	160	55	①	①
'80	6-231	80	40	100	225②	60	45	25

① Bolt—40; Nut—30
② Harmonic balancer; not the pulley
③ 1980 engines require that the head bolt threads be coated with non-hardening sealer

WHEEL ALIGNMENT SPECIFICATIONS

Year	Model	Caster Range (deg.)	Caster Pref. Setting (deg.)	Camber Range (deg.)	Camber Pref. Setting (deg.)	Toe-out (in.)	Steering Axis Inclin. (deg.)
'80	All	⅓N to 1⅓N	⅘N	³⁄₁₀N to ⁷⁄₁₀P	⅕P	0 to ¹⁄₁₆	8.55

ENGINE ELECTRICAL

Alternator

A 10-SI Series Delcotron alternator is used on all models. This unit features a nonadjustable, integral solid state regulator mounted inside the slipring end frame. The alternator must be disassembled to replace the voltage regulator.

For further information on the charging system, please refer to "Charging and Starting" in the Unit Repair section.

REMOVAL & INSTALLATION

1. Disconnect the battery cables.
2. Tag and disconnect the alternator wiring.
3. Remove the alternator brace bolt and V-belt.
4. Remove the pivot mount bolt and the alternator.
5. Installation is the reverse of the removal procedure.
6. Adjust the belt to have ¼–½ in. play on the longest span of the belt. If a tensioning gauge is available, adjust the belt to 80 lbs.

Starter

The starter is a solenoid actuated Delco-Remy unit. The starter has no R terminal. The HEI system does not use the solenoid-to-coil wire.

For further information on the starting system, please refer to "Charging and Starting" in the Unit Repair section.

REMOVAL & INSTALLATION

1. Disconnect the battery ground cable and all the wiring at the sole-

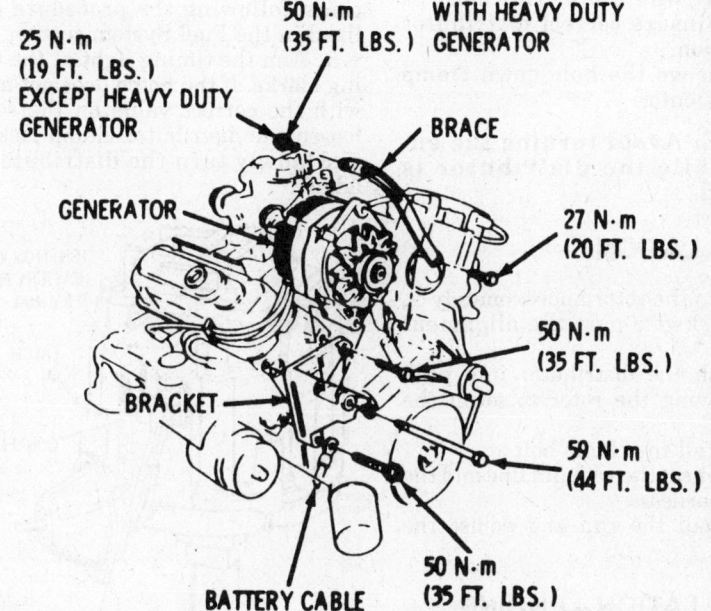

25 N·m (19 FT. LBS.) EXCEPT HEAVY DUTY GENERATOR

50 N·m (35 FT. LBS.) WITH HEAVY DUTY GENERATOR

BRACE

GENERATOR

27 N·m (20 FT. LBS.)

50 N·m (35 FT. LBS.)

59 N·m (44 FT. LBS.)

BRACKET

50 N·m (35 FT. LBS.)

BATTERY CABLE

Typical alternator mounting on V6 engine

noid terminals. Install each nut on the terminal from which it was removed, as these nuts are not interchangeable.
2. Loosen the front starter bracket and remove the two mounting bolts.
3. Remove the front bracket bolt and rotate the bracket out of the way.
4. Remove the starter from the car, lowering the front end first and being careful to catch any shims that may be used.
5. To install, reverse the removal procedure. Tighten the mounting bolts, and then install the brace.

Distributor

The V6 distributor is camshaft-driven and mounted at the front of the engine. The 151 cu. in. distributor is mounted at the right rear. Electronic ignition is standard equipment on all models, eliminating the points and condensor. Two types of HEI distributors are used. 4-151 and V6 distribu-

tors combine all ignition components in one unit. The coil is in the distributor cap and connects directly to the rotor.

TIMING LIGHT CONNECTIONS

Timing light connections should be made in parallel using an adapter at the distributor No. 1 terminal. Some models incorporate a magnetic timing probe hole near the timing scale for use with a special magnetic timing device.

TACHOMETER CONNECTIONS

NOTE: There is a tachometer connecting terminal next to the ignition switch connector on the 151-4 and V6 HEI distributor cap. Most tachometers will work when connected to this terminal and to the positive battery terminal.

Some tachometers won't work at all with this system or may require a special hookup. Never ground the tachometer terminal; the system will be damaged.

REMOVAL

1. Disconnect the wiring harness connectors at the side of the cap and remove the cap.
2. Disconnect the vacuum line and the primary lead.
3. Mark the distributor housing and the engine in line with the rotor centerline with chalk. This must be done to insure correct distributor installation.
4. Remove the hold-down clamp and distributor.

NOTE: Avoid turning the engine while the distributor is removed.

INSTALLATION

1. Turn the rotor approximately 1/8 turn clockwise past the alignment mark.
2. Push the distributor into position, moving the rotor to mesh the gears.
3. Install the clamp bolt.
4. Connect the vacuum line and the wiring harness.
5. Install the cap and adjust the timing.

INSTALLATION – ENGINE DISTURBED

1. Remove No. 1 spark plug and place a finger over the plug hole. Remove the center coil wire and crank the engine until compression is felt in No. 1 cylinder. Rotate the engine until the timing pointer is aligned with the proper mark.
2. Line up the rotor and the mark made on the distributor housing with the mark made on the engine, then turn the rotor clockwise about 1/8 turn past the marks and install the distributor. As the rotor gear engages the drive gear the rotor should rotate back into line with the marks. If not, repeat procedure until it does. Tighten the clamp bolt.
3. Install the rotor, cap and vacuum line.
4. Connect the wiring harness.
5. Check and adjust the ignition timing.

IGNITION TIMING ADJUSTMENT

The timing marks are on a plate mounted on the front of the block and the timing notch is on the crankshaft pulley. Timing is set as follows:

1. Bring the engine to normal operating temperature, then shut the engine off and connect a timing light according to the manufacturer's instructions. Clean the timing plate and mark the notch in the pulley with chalk.
2. Disconnect and plug the vacuum line to the distributor.
3. See the underhood sticker for information on preparing the engine for ignition timing.
4. Set the idle speed to specifications, following the procedure outlined in the Fuel System section.
5. Aim the timing light at the timing marks. If the notch does not align with the correct value on the scale, loosen the distributor clamp locknut and slowly turn the distributor to adjust.

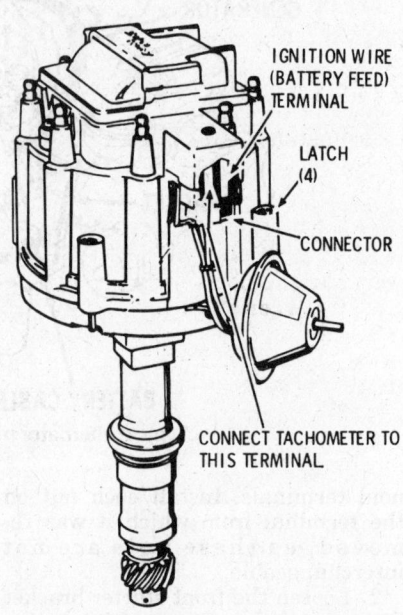

HEI distributor tachometer connecion

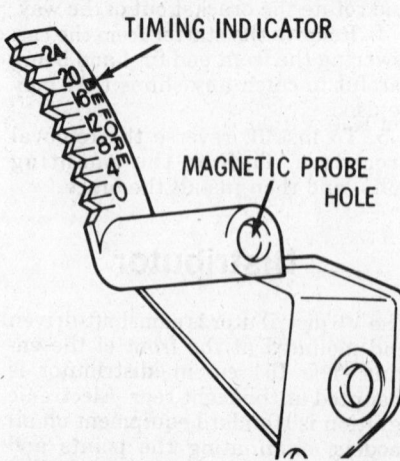

Timing scale indicator showing magnetic timing probe hole

6. Tighten the clamp locknut. Adjust the carburetor idle speed screw to give the specified idle speed with the solenoid disconnected.
7. Reconnect the idle stop solenoid lead. Increase the engine speed to allow the solenoid to extend and then adjust the solenoid plunger screw to obtain the idle speed specified with the solenoid connected.
8. Shut the engine off and connect the vacuum and evaporative emission line.

FUEL SYSTEM

The Holley 5210-C, 6510-C and Rochester 2SE carburetors are used on the 4-151. The Rochester M2ME and M2MC are used on the 6-231. An electric fuel pump is used with V6 engines. It is an integral part of the fuel tank unit assembly, which includes the fuel gauge metering unit. The fuel pump is energized by the ignition switch when the key is in the START or ON position. After the engine starts, the pump receives current through the oil pressure safety switch as long as there is approximately 2 psi oil pressure.

Electric Fuel Pump

REMOVAL & INSTALLATION

NOTE: It is not necessary to raise the car for fuel tank removal.

1. Disconnect the battery ground cable and siphon the fuel from the tank.
2. Disconnect the gauge sending-unit and pump wires at the rear harness connector.
3. Thoroughly clean and disconnect the fuel line and tank vent line at the tank. These connectors are short pieces of rubber hose secured by squeeze clamps. They are located adjacent to the top of the right rear tire. Wear eye protective goggles when working in this area as the lines and connectors are covered with road dirt. A pair of angled slip-joint pliers will be necessary to reach the clamps.
4. Thoroughly clean the area around the filler neck where it enters the rubber connector pipe. Remove the clamp. Remove the three filler pipe-to-body screws.
5. Place a floor jack under the fuel tank to take up the weight. Remove the nuts from the tank straps and slowly lower the tank until the wire connectors in the top of the tank are

visible. Reach up and disconnect the wires. At this point it will probably be necessary to pull the filler tube from the rubber connector pipe. This requires considerable twisting and maneuvering. Take care to avoid getting dirt in the tank.

6. Lower the tank the rest of the way and remove it. The tank may stick to the straps which are coated with a sticky anti-squeak compound.

7. A special wrench is available to remove the lock ring from the gauge pickup unit. If this is not available, use a brass drift. Always use a plastic or hardwood mallet.

8. Spray the area with penetrating oil and very carefully tap the ears of the lockring around, alternating ears, until it is free.

9. Carefully remove the pump and sending unit from the tank. Remove the rubber gasket from the lock ring.

10. Remove the nut or screw securing the pump ground wire to the bracket. Remove the two wire connector nuts from the pump, making sure you know which wire is which for replacement. Slide the pump and hose connector from the pickup tube.

11. Install the new pump. A filter screen should be supplied with the new pump. Do not overtighten the nuts securing the wires to the pump motor.

12. Carefully lower the assembly into position in the tank. The pickup screen should lie flat along the bottom of the tank facing forward.

13. The rubber gasket is reusable if not damaged. Coat it with lithium-based grease prior to installation to avoid twisting and to ease installation.

14. If the special removal tool was not used, carefully tap the lock ring around until around until it contacts the tab stops. The lock ring will tend to move to one side when tapping it. Be certain that the gasket is centered at all times or leakage will result. Alternating the points at which you tap will help prevent cocking, but time and care are necessary.

15. The tank is installed by reversing the removal procedure. Care should be taken to avoid getting dirt in the tank when installing the filler pipe. Refill the tank and check the pump operation and for leaks.

Mechanical Fuel Pump

REMOVAL & INSTALLATION

1. Disconnect the negative battery cable.
2. Disconnect the fuel inlet hose from the pump.

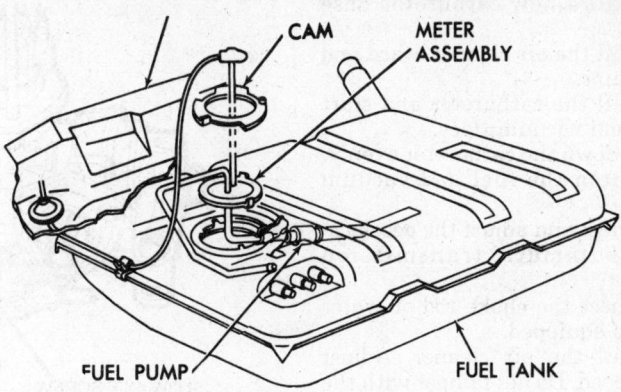

Electric fuel pump installation

3. Disconnect the vapor return hose, if so equipped.
4. Disconnect the fuel outlet pipe.
5. Remove the pump mounting bolts.
6. Remove the fuel pump.
7. To install, position the new pump using a new gasket.
8. Install the bolts and tighten evenly.
9. Install the fuel outlet pipe.

NOTE: If difficulty is encountered starting the fitting on the outlet pipe, disconnect the upper end of the pipe from the carburetor and hold the fuel pump nut with a wrench while tightening the fitting securely. Reconnect and tighten the fitting at the carburetor.

10. Install the fuel inlet hose.
11. Install the vapor return hose, if so equipped.
12. Connect the negative batter cable.
13. Start the engine and check for leaks.

Fuel Filter

REMOVAL & INSTALLATION

The fuel filter is located in the carburetor body, directly behind the fuel inlet line. Either a paper or a bonze filter may be used, depending on the carburetor model.

1. Disconnect the fuel line at the intake fuel filter nut on the carburetor.

——— **CAUTION** ———

Two wrenches, one on the line nut and one on the filter nut are necessary to avoid damage to the line and/or threads.

2. Remove the intake fuel filter nut.
3. Remove the filter element and spring.

4. Install the element spring and element. Bronze filters are installed with the conical section facing out and with a gasket between the filter element and the fuel intake nut.
5. Install the nut using a new gasket and tighten. Do not overtighten this nut, as it is easily stripped.
6. Install fuel line and tighten the connector.

Carburetor

For carburetor overhaul and adjustments not detailed below, please refer to "Carburetors" in the unit repair section. For further information on feedback carburetors, please refer to *Chilton's Guide To Fuel Injection And Feedback Carburetors.*

REMOVAL & INSTALLATION

1. Remove the air cleaner and gasket.
2. Disconnect the fuel and vacuum lines.
3. Disconnect the choke rod, or on an early model two-barrel, remove the choke water cover by removing the three attaching screws. Disconnect all electrical connections.
4. Disconnect the accelerator linkage.
5. Disconnect the Powerglide throttle valve linkage or Turbo Hydra-Matic detent cable.
6. Unbolt the carburetor and remove the carburetor and solenoid assembly.
7. Remove the insulator gasket, air cleaner bracket, and flange gasket.
8. Before installation, make sure that the carburetor and manifold sealing surfaces are clean.

NOTE: To reduce the possibility of backfiring fill the carburetor bowl with a small amount of fuel.

9. Install a new carburetor base gasket.

10. Install the air cleaner brace and the insulator.

11. Install the carburetor and start the fuel and vacuum lines.

12. Bolt down the carburetor evenly.

13. Tighten the fuel and vacuum lines.

14. Connect and adjust the accelerator and automatic transmission linkage.

15. Connect the choke rod or water cover if so equipped.

16. Install the air cleaner. Adjust the idle speed. Do not tamper with the idle mixture screw unless the carburetor has been repaired, overhauled or replaced.

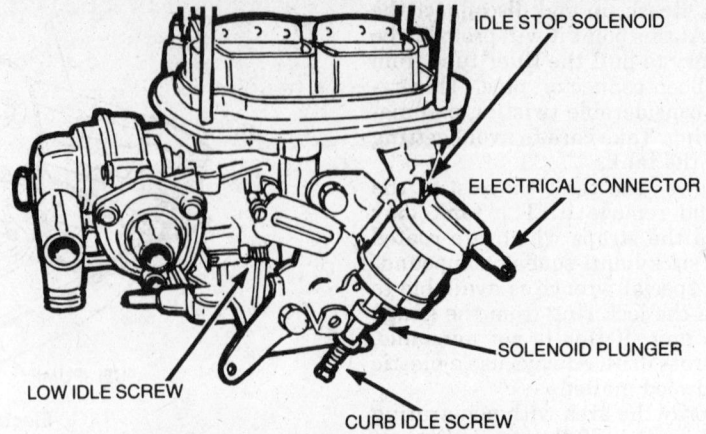

Holley 5210-C, 6510-C idle speed adjustment

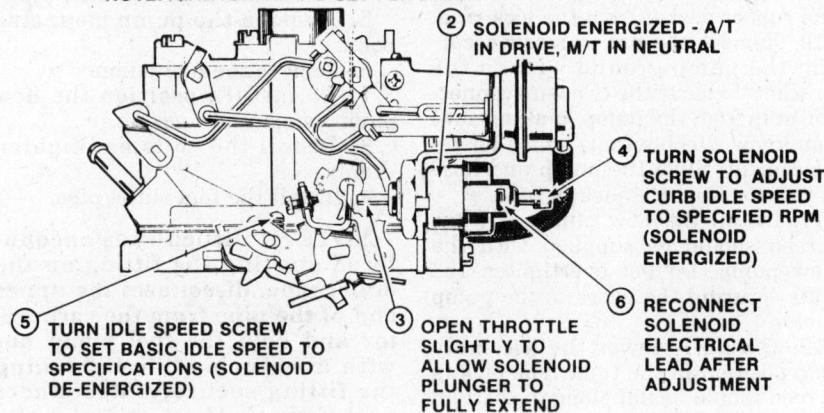

Rochester 2SE idle speed adjustment

IDLE SPEED ADJUSTMENT

4-151 Engine

Check the Vehicle Emission Control Information label for the latest certified information or special instructions.

1. Run the engine to normal operating temperature with the choke fully open. The air conditioning should be off.

2. Connect a tachometer and a timing light according to the manufacturer's instructions.

3. Set the parking brake firmly and block the drive wheels.

4. Disconnect and plug the PCV hose at the canister and the vacuum hose at the distributor.

5. Start the engine and place the transmission in Drive (AT) or Neutral (MT).

6. Check and, if necessary, adjust the timing.

7. Unplug and reconnect the vacuum hose at the distributor. Adjust the idle speed screw to obtain the specified rpm.

8. On cars with automatic transmission or manual transmission with A/C, turn the idle speed screw to obtain the specified rpm, then:

    a. disconnect the electrical line at the wide open throttle A/C override switch located on the accelerator linkage bracket.

    b. turn the A/C on.

    c. momentarily open the throttle to allow the solenoid plunger to extend.

    d. adjust the solenoid screw to the rpm specified in the Tune-Up table.

    e. reconnect the electrical connector.

    f. turn the A/C off.

    g. reconnect the PCV hose at the canister.

V6 Engine

1. Prepare the vehicle according to the instructions found on the underhood emission label.

2. Turn the idle speed screw to obtain the rpm specified in the Tune-Up table. Follow Steps 3–6 if the car is equipped with air conditioning.

3. Disconnect the A/C lead at the compressor clutch.

4. Turn the A/C on. Open the throttle slightly to extend the solenoid plunger.

5. Turn the solenoid screw to obtain the rpm specified in the Tune-Up table.

6. Reconnect the A/C lead.

IDLE MIXTURE ADJUSTMENT

NOTE: Changes in the mixture system have made the adjustment of the air/fuel mixture impossible without a propane enrichment system.

1. Set the parking brake firmly and block the drive wheels.

2. Refer to the emission control label in the engine compartment, then disconnect and plug the necessary hoses.

3. Connect a tachometer to the engine, disconnect and plug the vacuum advance hose at the distributor, then check and/or set the ignition timing.

4. Operate the engine to normal operating temperatures; make sure that the choke is open and the A/C is turned OFF.

NOTE: If the vehicle is equipped with Electronic Spark Timing (EST), check and/or adjust the timing and the idle speed according to the emission control label.

5. Disconnect the crankcase ventilation tube from the air cleaner.

6. Using tool J-26911, connect a hose from a propane tank to the crankcase tube opening at the air cleaner.

7. Place the transmission in Drive (A/T) or Neutral (M/T).

8. Slowly OPEN the propane valve until the maximum engine speed is reached.

NOTE: The addition of too much propane will cause the engine speed to drop.

9. If the idle mixture speed does not meet specifications, remove the idle mixture screw plug covers. This involves removing the carburetor to knock out the mixture adjustment plugs.

10. Turn the idle mixture screws clockwise until lightly seated, then back them OUT (equally) until the lean best idle point at the enriched idle speed is reached.

11. Turn the propane tank OFF and turn the idle mixture screws clockwise (equally) until the curb idle speed is reached.

12. Turn ON the propane tank and recheck the engine speed, if not within specifications, repeat the propane setting.

13. To complete the adjustment, turn OFF the engine, remove the propane tank, reconnect the crankcase ventilation tube to the air cleaner and install new idle mixture screw plugs.

ENGINE COOLING

The intake manifold is water heated to provide an even intake temperature. All models have a radiator drain petcock. All models are equipped with a coolant recovery system reservoir. A translucent plastic reservoir allows for hot coolant expansion. When the engine cools, coolant is drawn into the radiator by vacuum. Additional coolant should be added to the reservoir, not the radiator.

Radiator

REMOVAL & INSTALLATION

1. Drain the radiator. Save the coolant for reuse if not contaminated.

2. On models with the heavy duty radiator, remove the fan shroud.

3. Disconnect the intake and outlet hoses.

4. Remove the front lighting wiring harness from the clips on the fan guard. Remove the two screws which secure the fan guard to the radiator support, then remove the support and the two radiator pads.

NOTE: On vehicles with the heavy duty radiator, remove the two upper brackets (instead of the single support).

5. Lift the radiator up and out of the lower brackets.

6. To install, reverse the removal procedure.

Water Pump

REMOVAL & INSTALLATION

1. Drain the coolant from the radiator.

2. Loosen the fan pulley bolts.

3. If necessary, remove the alternator with the drive belt and brackets.

4. If necessary, remove the air pump with the drive belt and brackets.

5. Disconnect the lower radiator hose and the heater hose at the water pump.

6. Remove the fan and pulley.

7. Remove the pump-to-cylinder block and power steering-to-pump bolts and remove the water pump and old gasket.

8. Installation is the reverse of removal. Use a new gasket coated with sealer. Adjust the alternator and air pump drive belt tension. Fill the cooling system, run the engine and check for leaks.

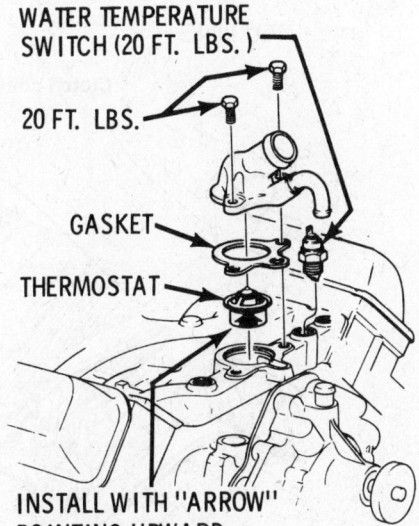

WATER TEMPERATURE SWITCH (20 FT. LBS.)

20 FT. LBS.

GASKET

THERMOSTAT

INSTALL WITH "ARROW" POINTING UPWARD.

Typical thermostat installation

Thermostat

REMOVAL & INSTALLATION

The 4-151 thermostat is located in a housing at the cylinder head water outlet adjacent to the intake manifold. On the V6 engine, the thermostat is in the water outlet housing in the front of the intake manifold.

1. Drain the coolant to a level below that of the water outlet housing.

2. Remove the radiator upper hose.

3. Remove the housing bolts and remove the water outlet housing and gasket.

4. Remove the thermostat.

5. Installation is the reverse of removal. Clean all gasket mating surfaces and use a new gasket.

EMISSION CONTROLS

Please refer to the "Emission Control" section of the Unit Repair Section for a description and service of the emission control systems.

ENGINE MECHANICAL

NOTE: Service procedures for the Pontiac built 151 cu. in. four cylinder engine may be found in the "Pontiac Rear Wheel Drive" section of this book. The use of anti-seize compound is recommended on all bolts installed in aluminum engine blocks. Service procedures for the Buick built 231 cu. in. engine may be found in the "Buick Rear Wheel Drive" section of this book.

Engine

REMOVAL & INSTALLATION

1. Raise and support the hood.

2. Disconnect the battery cables.

3. Raise and support the car.

4. Drain the coolant, engine and transmission.

5. Disconnect the exhaust pipe(s) at the manifold.

6. Remove the flywheel or converter underpan.

7. On automatic transmissions, remove the converter-to-flywheel retaining bolts and install a converter retaining strap.

8. Remove the accessible converter housing or flywheel housing-to-engine bolts.

9. Remove the transmission cooler lines from the retaining clips on the side of the engine.

10. Remove the engine front mounting bolts at the frame brackets and lower the car.

11. Remove the radiator panel or shroud.

12. Remove the radiator and fan.

13. Disconnect the heater hose from the water pump and manifold.

14. Remove the air cleaner.

15. Disconnect the electrical leads from:

  a. alternator
  b. distributor
  c. starter solenoid
  d. oil pressure switch
  e. engine temperature switch
  f. temperature gauge switch
  g. choke secondary pull-off solenoid

16. Unclip the wiring harness from the rocker cover and position it out of the way.

17. Disconnect the automatic transmission vacuum modulator and air conditioning vacuum line from the manifold.

18. Disconnect the rubber fuel line at the rear of the engine.

19. Disconnect the following:

  a. canister vacuum hose at the carburetor
  b. accelerator at the carburetor and manifold bracket
  c. air conditioning blower delay lead at the rear of the engine.

20. On air conditioned cars, remove the compressor from its mount. Do not disconnect any fittings. Secure the compressor to the fender.

21. Disconnect the power steering pump and lay it aside.

22. Install a floor jack under the transmission.

23. Install a hoist on the engine and raise the engine slightly to take the weight off the engine mounts. Remove the remaining engine to transmission bolts.

24. Remove the engine from the car.
To install the engine:

25. Install transmission-to-engine guide pins, made from ³/₈ in. bolts with the heads cut off, into the engine.

26. Install the engine, aligning the engine with the transmission housing.

27. Align the engine mounts with the frame brackets and lower the engine onto the brackets. Loosely install

the engine mount bolts.

28. Remove the guide pins and install the engine-to-housing bolts. Remove the lifting equipment.

29. Remove the support from the transmission and raise and support the car.

30. Remove the converter retaining strap and install and tighten the engine-to-hous-ing bolts.

31. Tighten the engine front mount bolts.

32. Install the converter to the flywheel.

33. Install the flywheel cover or converter underpan.

34. Install the transmission cooler lines in the clips on the side of the block.

35. Connect the exhaust pipe(s) at the manifold and lower the car.

36. Install the air conditioning compressor and power steering pump. Adjust the drive belts.

37. Connect the following:

  a. canister vacuum hose to carburetor.

  b. accelerator cable at carburetor and manifold bracket
  c. air conditioning blower delay lead at side of engine.
  d. fuel line to rubber hose at rear of engine.
  e. air conditioning vacuum line.

38. Install the electrical harness in the clip in the rocker cover and connect the following:

  a. alternator
  b. distributor
  c. starter solenoid
  d. oil pressure switch
  e. engine temperature switch
  f. temperature gauge switch
  g. choke secondary pull-off solenoid.

39. Connect the heater hose at the water pump and at the manifold.

40. Install the radiator, fan, radiator panel or shroud, fill the cooling system, add engine oil and fill the transmission.

41. Install the air cleaner.

42. Connect the battery cables, start the engine and check for leaks.

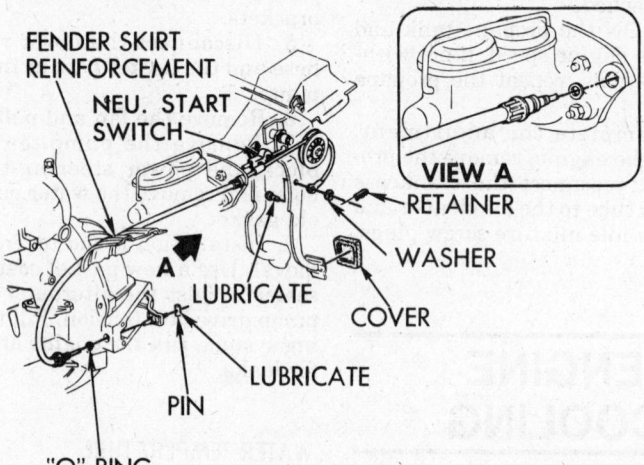

Clutch control cable

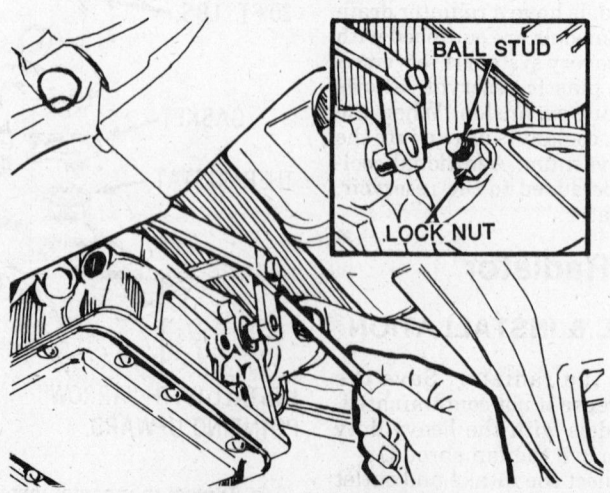

Clutch ball stud adjustment

CLUTCH

The clutch assembly consists of a driven plate, a pressure plate, and a release bearing, and is connected to the clutch pedal by a cable.

Clutch Pedal Free Travel

ADJUSTMENT

1. Remove the ball stud cap and loosen the locknut on the ball stud end, located to the left of the transmission, on the clutch housing.
2. Adjust the ball stud to obtain ⅛ inch clearance between the release bearing face and the pressure plate release fingers.
3. Tighten the ball stud locknut to 25 ft. lbs., being careful not to change the adjustment, and install the ball stud cap.
4. Pull the cable at the clutch fork until the clutch pedal is firmly against the rubber bumper.
5. Push the clutch fork forward until the release bearing contacts the pressure plate fingers, and screw the pin on the cable forward until it contacts the fork. Turn the pin ¼ turn clockwise and seat the pin in its seat on the clutch fork.

6. Attach the cable return spring and install the clutch fork cover.
7. Check the clutch pedal free play. This procedure should provide 0.90 ± 0.25 inch lash at the clutch pedal.

NOTE: When the adjustment of the ball stud and the cable have been completed, verify the clearance between the pressure plate fingers and the release bearing. The release bearing should not be in constant contact with the pressure plate fingers.

Clutch Disc

REMOVAL & INSTALLATION

1. Raise the vehicle on a hoist.
2. Remove the transmission as outlined in this section.
3. Remove the clutch fork cover, then disconnect the clutch return spring and control cable from the clutch fork.
4. Remove the input shaft oil seal from the clutch release bearing sleeve.
5. Remove the flywheel housing lower cover.
6. Remove the flywheel housing from the engine.
7. To remove the release bearing from the clutch fork and sleeve, slide the lever off the ball stud against the spring action. If necessary to replace

the ball stud, removes the cap, locknut and stud from the housing.
8. If assembly marks on the clutch assembly and flywheel are not distinguishable, remark with paint or center-punch.
9. Loosen the clutch cover to flywheel attaching bolts one turn at a time until the spring pressure is released, to avoid bending the clutch cover flange.
10. Support the pressure plate and cover assembly, then remove the bolts and clutch assembly.

--- **CAUTION** ---
Do not disassemble the clutch cover, spring and pressure plate for repair. If defective replace the complete assembly.

11. Index the alignment marks on the clutch assembly and the flywheel. Place the driven plate with the long end of the splined end facing forward, the plate damper springs facing the pressure plate, and insert a dummy input shaft or aligning tool through the cover and the driven plate.
12. Position the complete assembly against the flywheel and insert the dummy shaft or aligning tool into the pilot bearing in the crankshaft.
13. Index the alignment marks and install clutch cover to flywheel bolts finger tight.

--- **CAUTION** ---
Tighten all bolts evenly and gradually until tight to avoid possible clutch distortion.

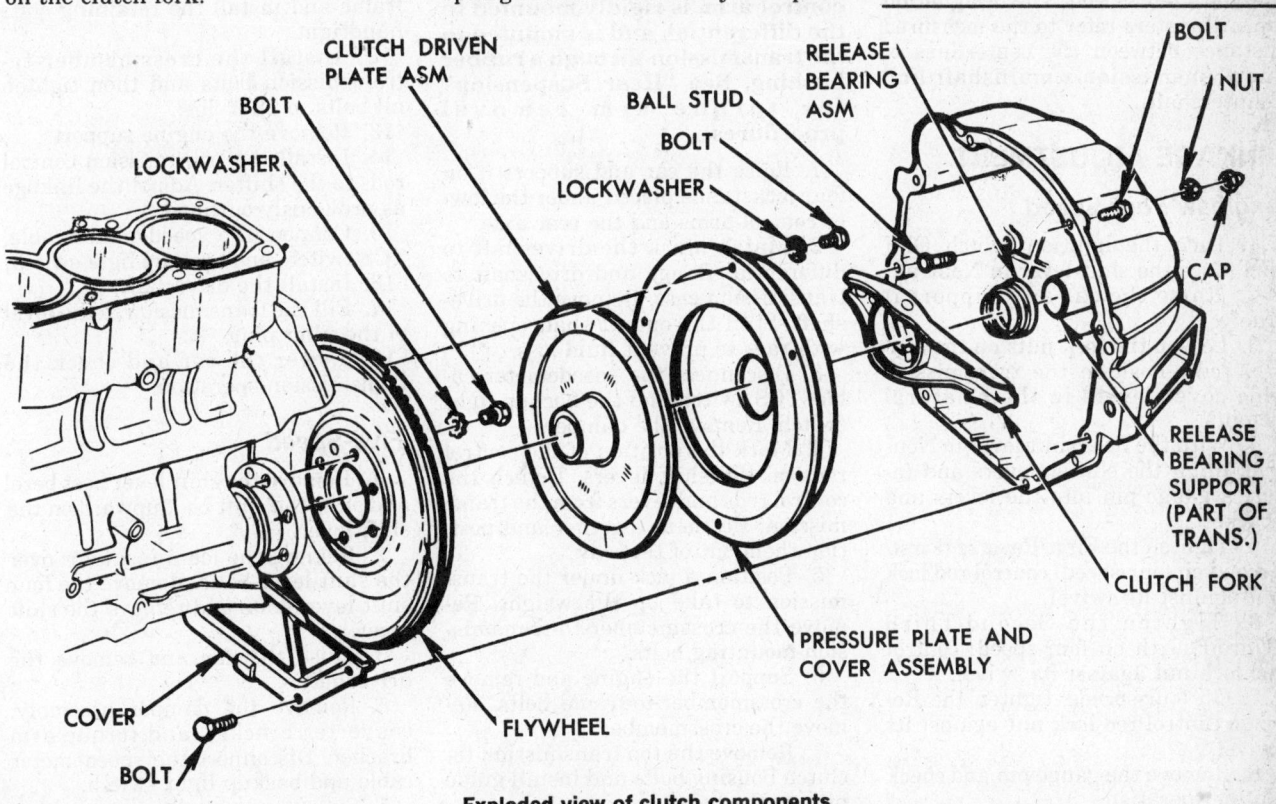

Exploded view of clutch components

14. Lubricate the clutch fork ball socket and the fingers at the release bearing with a high melting point grease such as graphite grease.

15. Lubricate the recess on the inside of the throwout bearing collar and the fork groove with a light coat of graphite grease. Install the fork in the housing but not on the stud.

16. Install the bearing on the sleeve, then position the clutch fork over the bearing in the housing and slide the fork onto the ball stud.

17. Install the flywheel housing and the lower cover. Tighten the bolts.

18. Install the transmission as outlined previously.

19. Adjust the clutch as previously outlined.

20. Lower and remove the vehicle from the hoist.

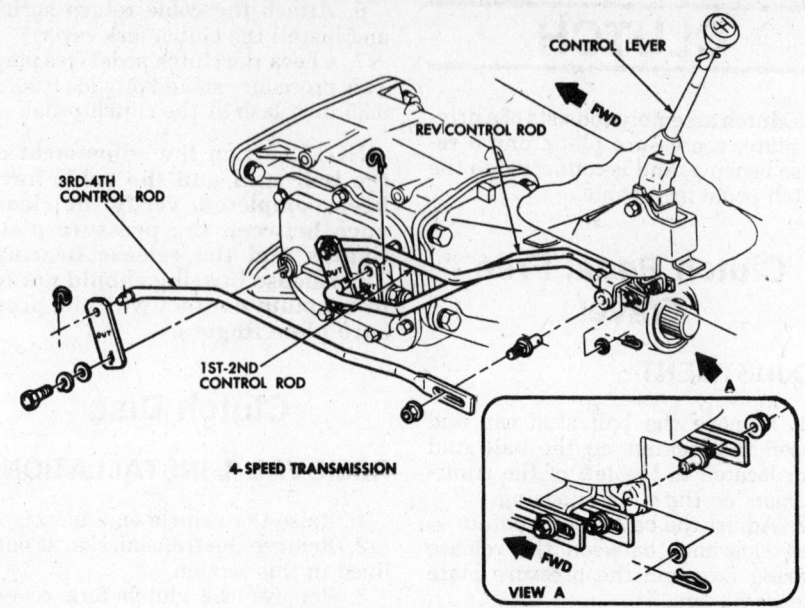

Saginaw four speed inkage

MANUAL TRANSMISSION

A five-speed Borg Warner T-50 transmission (also called the 77mm transmission) is optional. Fourth gear is direct drive with fifth gear an overdrive. The transmission is shifted by a single shift rail enclosed within the transmission. The Saginaw four-speed, now called the 76mm, is the standard four-speed. The designation by millimeters refer to the measured distance between the centerlines of the transmission's mainshaft and countershaft.

LINKAGE ADJUSTMENT

Saginaw Four-Speed

1. Turn the ignition switch OFF and place the shift lever in Neutral.

2. Raise the car and support it safely.

3. Loosen the lock nuts on the control rods. Position the transmission side cover levers in their natural detents.

4. With the floor shift lever in Neutral, align the shifter levers and insert a gauge pin into the levers and bracket.

5. Tighten the First/Reverse (First/Second on four-speed) control rod lock nut against its swivel.

6. Tighten the Second/Third (Third/Fourth on four-speed) control rod lock nut against its swivel.

7. On four-speeds, tighten the Reverse control rod lock nut against its swivel.

8. Remove the gauge pin and check shifter operation.

Transmission

REMOVAL & INSTALLATION

Four-Speed

NOTE: Transmission removal will require additional work due to the torque arm rear suspension. The torque arm serves as an upper control arm, is rigidly mounted to the differential, and is mounted to the transmission through a rubber bushing. See "Rear Suspension" for torque arm removal procedures.

1. Raise the car and support it on four jackstands placed under the lower control arms and the rear axle.

2. Match-mark the driveshaft to differential flange and driveshaft to transmission case. Remove the driveshaft. Stuff the output shaft opening with rags to prevent fluid loss.

3. Disconnect the speedometer cable, TCS switch and the backup light switch. Remove the damper.

4. Mark the position of the control rods on the shift levers. Detach the control rods and levers from the transmission, tie them together and position them out of the way.

5. Position a jack under the transmission to take up the weight. Remove the crossmember-to-transmission mounting bolts.

6. Support the engine and remove the crossmember-to-frame bolts. Remove the crossmember.

7. Remove the top transmission-to-clutch housing bolts and install guide pins in the holes.

8. Remove the lower bolts and pull the transmission back and out of the car.

9. Guide the input shaft through the throwout bearing and into the pilot bearing.

10. Install the transmission retaining bolts and lockwashers. Tighten the bolts to 40 ft. lbs.

11. Position the crossmember on the frame and install the retaining bolts handtight.

12. Install the crossmember-to-transmission bolts and then tighten all bolts to 28 ft. lbs.

13. Remove the engine support.

14. Install the transmission control rods to the shifter. Adjust the linkage as previously outlined.

15. Connect the speedometer cable, TCS switch, and back-up light switch.

16. Install the driveshaft.

17. Fill the transmission to the level of the filler plug.

18. Lower the car and check transmission operation.

Five-Speed

1. Remove the shift lever boot bezel and slide the shift boot upward on the shift lever.

2. Remove the foam insulator over the shift lever bolts. Remove the four shift lever bolts and remove the shift lever.

3. Raise the car and remove the driveshaft.

4. Remove the damper assembly, converter bracket, and torque arm bracket. Disconnect the speedometer cable and backup light switch.

5. Support the transmission with a

jack and remove the transmission support.

6. Remove the transmission-to-clutch housing bolts and slide the exhaust bracket forward. Slide the transmission to the rear and remove it.

7. To install, make sure that the main drive gear splines are clean and dry. Position the transmission to the clutch housing and slide it forward.

8. Slide the exhaust bracket into place and install the transmission-to-clutch housing attaching bolts.

9. Install the rear transmission mount and transmission support. Install the converter bracket, damper, and torque arm.

10. Install the driveshaft, connect the speedometer cable and back-up light switch.

11. Fill the transmission with 3 pints of Dexron® II automatic transmission fluid.

12. Lower the car and install the shift lever and foam insulator. Install the shift lever boot and bezel.

13. Check the transmission for proper operation.

OVERHAUL

For all overhaul procedures, please refer to "Manual Transmission Overhaul" in the Unit Repair section.

AUTOMATIC TRANSMISSION

For all adjustments and service procedures, please refer to "Automatic Transmissions" in the Unit Repair section.

REMOVAL & INSTALLATION

1. Before raising the car, disconnect the negative battery cable and the detent cable at the bracket and carburetor.

2. Remove the air cleaner and dipstick.

3. On vehicles with air conditioning, remove the (5) five heater core cover screws from the heater assembly. Disconnect the wire connector and with hoses attached, place the heater core cover out of the way.

4. Raise vehicle on hoist and remove propeller shafts as outlined in Section 4.

5. Disconnect speedometer cable, electrical lead to case connector and oil cooler pipes.

6. Disconnect shift control linkage.

7. Support transmission with suitable transmission jack and remove the four (4) rear transmission support bolts.

8. Remove the nuts holding the converter bracket to the support.

9. Disconnect the exhaust pipe at the rear of the catalytic converter.

10. Disconnect exhaust pipe at manifold and remove the exhaust pipe, catalytic converter and converter bracket as an assembly.

11. Remove the torque converter under pan.

12. Remove converter to flywheel bolts.

13. Lower transmission until jack is barely supporting it and remove transmission to engine mounting bolts.

14. Raise transmission to its normal position, support engine with jack and slide transmission rearward from engine and lower it away from vehicle.

——— CAUTION ———

Use converter holding Tool J-5384 when lowering transmission or keep rear of transmission lower than front so not to lose converter.

The installation of the transmission is the reverse of the removal with the following added steps.

1. Place a 2 in. block between the rack and pinion housing assembly and the engine oil pan. This will permit correct alignment of the engine and transmission.

2. Before installing the flexplate to converter bolts, make certain that the welded brackets on the converter are flush with the flexplate and the converter rotates freely by hand in this position. Then, hand start all bolts and tighten finger tight before torquing to 20–30 ft. lbs. (27–41 Nm). This insures proper converter alignment.

3. After installation of transmission, remove car from hoist. Check linkage for proper adjustment.

DRIVESHAFT AND U-JOINTS

Driveshaft

REMOVAL & INSTALLATION

1. Raise and support the car with jack stands. Mark the relationship of

the shaft to the companion flange and disconnect the rear universal joint by removing the trunnion bearing U-bolts. Tape the bearing cups to the trunnion to prevent loss of the bearing rollers.

2. Withdraw the driveshaft front yoke from the transmission by moving the shaft rearward and passing it under the axle housing. Plug the transmission opening with rags to prevent fluid or oil loss.

3. Inspect the yoke seal in the transmission extension; replace if necessary.

4. Insert the driveshaft front yoke into transmission extension, making sure that the output shaft splines mate with the driveshaft yoke splines.

5. Align the driveshaft with the companion flange using the reference marks established in the removal procedure. Remove the tape from the U-joint, install the U-bolt to the rear axle flange, and torque them to 15 ft. lbs.

Universal Joint

OVERHAUL

For all U-joint overhaul procedures, please refer to "U-Joint/CV-Joint Overhaul" in the Unit Repair section.

REAR AXLE

All axles are the C-lock type with C-locks retaining the axle shafts. All axles are hypoid type, semi-floating with an integral gear carrier and a removable cover plate.

Axle Shaft, Bearing and Seal

REMOVAL & INSTALLATION

See the "Chevrolet Rear Wheel Drive" car section for the proper procedures.

FRONT SUSPENSION

The independent front suspension, like the rear is very similar to that used on some larger GM vehicles.

Each wheel is suspended by unequal length control arms. The steering knuckle, which supports the wheel is attached to the upper and lower control arms by ball joints. The coil spring is located between the lower control arm and the frame. The tubular shock absorber is mounted inside the spring. Vehicles equipped with the optional ride and handling package (standard on GT) have a stabilizer bar mounted in rubber bushings to the body and connecting the lower control arms.

Shock Absorber

REMOVAL & INSTALLATION

NOTE: To purge air from the shock absorber before installation, extend the shock fully and invert it. Compress the shock, and return it to its upright position. Repeat this operation several times. Do not extend the shock absorber while it is inverted.

1. Pry out the access plug in the engine compartment so that the upper mount is visible.
2. Raise the front of the car and support it with jack stands.
3. Turn the wheels for clearance.
4. Hold the upper shock stud with a wrench. Loosen and remove the locknut.
5. Unbolt the lower end and pull the shock down and out.
6. Place the lower retainer and rubber grommet on the shock stud.
7. Put the shock in place and tighten the lower bolts. Torque to 20 ft. lbs.
8. Place the upper grommet, retainer, and nut on the shock stud.
9. Hold the stud with a wrench and tighten the nut to 10 ft. lbs., or just enough to avoid distorting the rubber grommets.

Ball Joint

INSPECTION

Upper

To check the ball joints for excessive wear:
1. Place jackstands under the lower control arms.
2. Turn the wheels straight ahead.
3. Lift and shake the wheel up and down vertically. If there is noticeable looseness, the ball joints are worn.
4. Grasp the top and bottom of the tire and rock it by pushing in on the top and pulling out on the bottom, then pulling out on the top and pushing in on the bottom. A 1/4 in. play indicates worn ball joints. Make sure

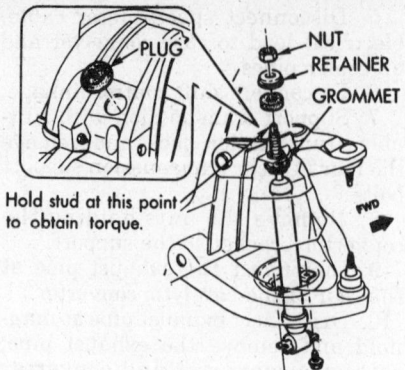

Front shock absorber mounting

that loose wheel bearings do not result in a false reading.

Lower

The lower ball joints incorporate wear indicators. They can be inspected visually; when the 1/2 in. diameter grease fitting is flush with, or inside the cover surface, replace the ball joint. Inspect the grease fitting with the car supported on its wheels so that the lower ball joint is in a loaded condition. Normal protrusion on the grease fitting is 0.050 in. beyond the cover surface. Ball joint tightness can also be checked using the preceding procedure.

REMOVAL & INSTALLATION

Upper

1. Jack up the front of the car and support it under the crossmember braces. Remove the wheel.
2. Place a hydraulic jack under the lower control arm.
3. Remove the cotter pin from the ball joint stud. Loosen, but do not remove the nut.
4. The stud may now be pressed out of the steering knuckle. There is a special tool available to do this.
5. Remove the ball joint by grinding off the rivets, or removing the heads of the rivets with a chisel.
6. Bolt the new ball joint on, using the nuts and bolts supplied with the replacement joint.
7. Install the stud to the steering knuckle and torque the nut to 30 ft. lbs. If the cotter pin hole does not align, tighten the nut 1/2 of a turn further to line it up. Install a new cotter pin.
8. Install the wheel and lower the car.

Lower

1. Repeat steps one through three of the upper ball joint procedure.
2. The stud may now be pressed out of the steering knuckle.

3. The old ball joint must be pressed out of the control arm. A special tool is available for this purpose.
4. Press in the new joint, positioning it so that the grease bleed vent in the rubber boot is facing inward.
5. Install a lubrication fitting in the new joint.
6. Install the stud to the steering knuckle and torque the nut to 60 ft. lbs. If the cotter pin hole does not align, tighten it 1/6 of a turn further. Do not loosen the nut to install the cotter pin.
7. Install the wheel and lower the car.

Spring

REMOVAL & INSTALLATION

1. Raise the front of the car and support it with jackstands placed under the front crossmember braces.
2. Remove the wheel, shock absorbers, and stabilizer bar.
3. Support the lower control arm outer end with a hydraulic floor jack and a block of wood.
4. Securely fasten the spring to the lower control arm with a heavy chain.
5. To detach the tie rod, remove the cotter pin and nut, and tap on the steering arm (not the tie rod end) with a hammer. Hold another hammer behind the steering arm to take the force of the tapping. The tie rod should then fall free.
6. Remove the lower ball joint stud from the steering knuckle as described in the "Lower Ball Joint Removal and Installation" procedure.
7. Very cautiously lower the jack until the spring is fully expanded.
8. Place the spring in its pads on the lower control arm and shock tower. Spring insulators are used on all models; make sure that the insulator is indexed with its closed end located at the high point in the spring seat. Secure the spring with a safety chain as in Step 4.
9. Carefully raise the jack.
10. Place the lower ball joint stud in the steering knuckle. Torque the stud nut to 60 ft. lbs. If the cotter pin does not align, tighten it 1/16 of a turn further and insert a new cotter pin.
11. Install the tie rod end to the steering arm. Torque the nut to 35 ft. lbs. If the cotter pin hole does not align, tighten further up to a maximum of 50 ft. lbs. Insert a new cotter pin.
12. Replace the shock absorber as described in "Shock Absorber Removal and Installation." Do not attach the top end of the shock at this point.
13. Install the stabilizer bar. Tighten the bracket bolts to 30 ft. lbs. and

the control arm bolts to 10 ft. lbs.

14. Replace the wheel and lower the car. Install the upper end of the shock absorber.

Lower Control Arm

REMOVAL & INSTALLATION

1. Raise the vehicle on a hoist and remove the wheel.
2. Support the lower control arm with a floor jack.
3. Remove the upper ball stud nut and remove the ball stud from the steering knuckle.
4. Remove the control arm pivot bolts and remove the control arm from the vehicle.
5. Install the upper control arm to the vehicle at the inner pivot.

NOTE: The inner pivot bolts must be installed with the bolt heads to the front (on the front bushing) and to the rear (on the rear bushing).

6. Install the inner pivot nuts.
7. Position the control arm in a horizontal plane and tighten the inner pivot nuts to 48 ft. lbs.
8. Install the ball stud to the steering knuckle. Torque the nut to 30 ft. lbs. and install a cotter pin.
9. Install the tire and wheel assembly and lower the vehicle.

Wheel Bearings

ADJUSTMENT

1. Jack up the front of the car and support it with jackstands.
2. Remove the dust cap with a pair of slip-joint pliers.

3. Remove and discard the cotter pin. Loosen the spindle nut.
4. Rotate the wheel and tighten the spindle nut to 12 ft. lbs.
5. Back the nut off one flat and insert a new cotter pin. If the hole does not line up, back the nut off $\frac{1}{2}$ flat or less to align the hole.
6. Check that the wheel turns freely, and then lock the cotter pin.
7. Bearing end-play should be between 0.001–0.005 in. Tap the dust cap back on and lower the car.

REAR SUSPENSION

A torque arm rear suspension is used on all models. The suspension consist of lower control arms and a track bar to control lateral movement. A torque arm is used to control rear axle wind-up. A stabilizer bar is standard and the upper control arms have been eliminated.

Shock Absorber

REMOVAL & INSTALLATION

NOTE: To purge air from the shock absorber before installation, extend the shock fully and invert it. Compress the shock, and return it to its upright position. Repeat this operation several times. Do not extend the shock absorber while it is inverted.

1. Raise the vehicle and support the rear axle.

2. Remove the upper attaching bolts and lower through-bolt.
3. Remove the shock absorber.
4. Install the retainer and the rubber grommet onto the shock.
5. Place the shock absorber into the installed position and install the upper retaining bolts. Torque to 18 ft. lbs.
6. Coat the through-bolt shank with chassis lube and install it and a rubber grommet on each side of the shock eye. Torque the nut to 42 ft. lbs.
7. Lower the car.

Rear Spring

REMOVAL & INSTALLATION

1. Raise the vehicle and support the rear axle, with a hydraulic jack.
2. Disconnect the shock absorber lower bolt, only on one side at a time.
3. Mark the position of the spring ends on their pads, if the springs are being reused. Lower the axle and remove the spring and spring insulators.

—— CAUTION ——

When lowering the axle, do not stretch the brake hose running from frame to axle.

4. Install the insulators on the top and bottom of the spring and position it on the axle.
5. Raise the axle and reconnect the shock absorber. Torque the bottom stud or bolt nuts to 42 ft. lbs.
6. Lower the vehicle.

Lower Control Arm

REMOVAL & INSTALLATION

—— CAUTION ——

If both control arms are to be replaced, remove and replace one control arm at a time to prevent the axle from rolling or slipping sideways.

1. Raise the vehicle on a hoist.
2. Support the rear axle.
3. Disconnect the stabilizer bar if so equipped.
4. Remove the control arm front and rear attaching bolts and remove the control arm.
5. Replacement of these bushings is the same procedure as that described for the upper control arm.
6. Place the control arm into position and install the front and rear bolts. Torque to 80 ft. lbs. with the weight of the car on the suspension.
7. Attach the stabilizer bar and the restraint cable, if so equipped.
8. Remove the support from the axle.
9. Lower the vehicle.

Closed end of Insulator must be located to high point in spring seat.

UPPER CONTROL ARM

LOWER CONTROL ARM

ANTI-ROTATION TAB

Position spring insulators as shown

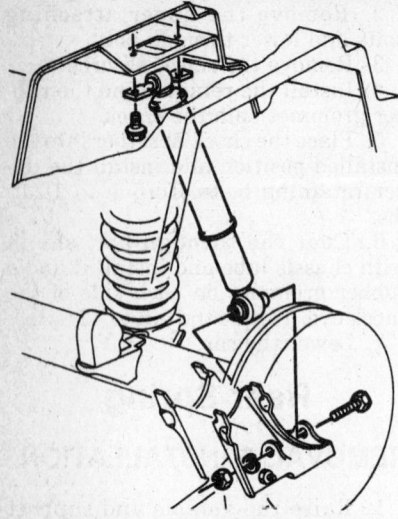

Rear shock absorber mounting

Torque Arm

REMOVAL & INSTALLATION

1. Raise and support the car at the rear axle.
2. Remove the torque arm mounting bracket from the transmission, then remove the through bolt at the bracket.
3. Remove the mounting bolts at the rear axle.
4. Installation is the reverse of removal. See the torque figures in the accompanying illustration.

Track Rod (Tie Rod)

REMOVAL & INSTALLATION

1. Raise and support the car at the rear axle.

2. Remove the track rod mounting bolt at the underbody point, then remove the mounting bolt at the rear axle.
3. Installation is the reverse of removal. Lubricate the track rod bushings, prior to installation, with clean brake fluid. This will prevent squeaking and cracking. Note the torque figures in the accompanying illustration.

NOTE: It is important to use shims as shown to position the axle assembly so that equal clearance exists between the tire and wheelhouse on both sides of the vehicle.

BRAKES

Front disc brakes are standard equipment on all models, with power brakes optional. Hub and disc are one-piece and the assembly is mounted to a one-piece steering knuckle and steering arm. The disc caliper design is similar to the single-piston Delco-Moraine disc brake used on other Chevrolet vehicles. Rear brakes are drum-type. Adjustment occurs automatically when the brakes are applied during a reverse stop.

To service brake shoes, drums and wheel cylinders or brake pads and calipers, please refer to "Brakes" in the Unit Repair section.

Master Cylinder

REMOVAL & INSTALLATION

1. On non-power brakes, disconnect the master cylinder from the brake pedal by detaching the clip and pin.
2. Disconnect the two hydraulic lines at the master cylinder, plugging or covering the ends of the lines.
3. Remove the master cylinder attaching nuts and remove the master cylinder.

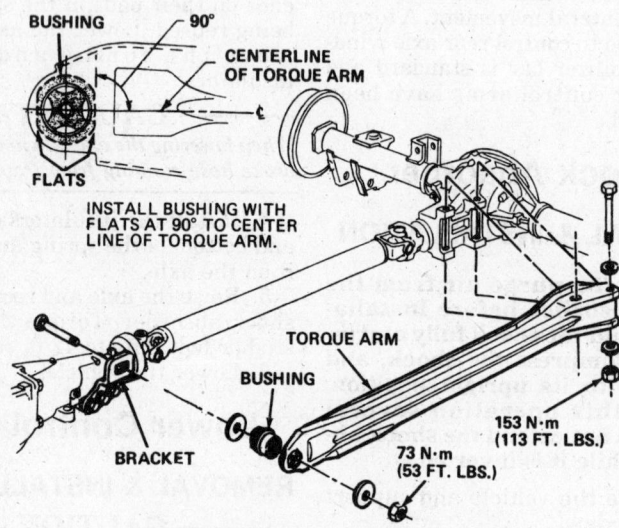

Torque arm removal and installation

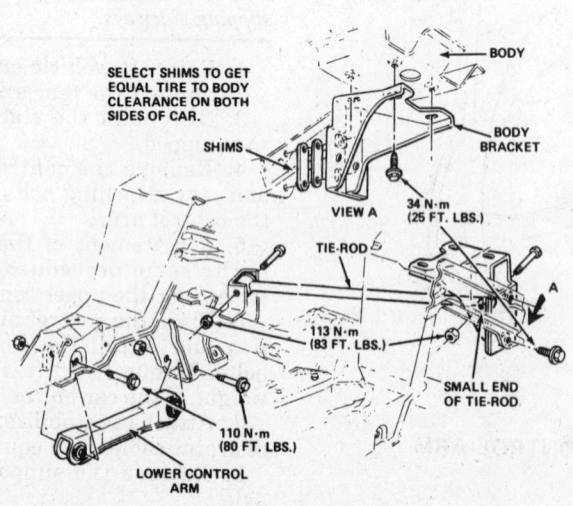

Track rod and lower control arm installation

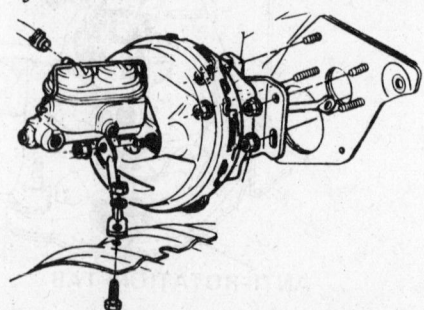

Power brake booster installation details

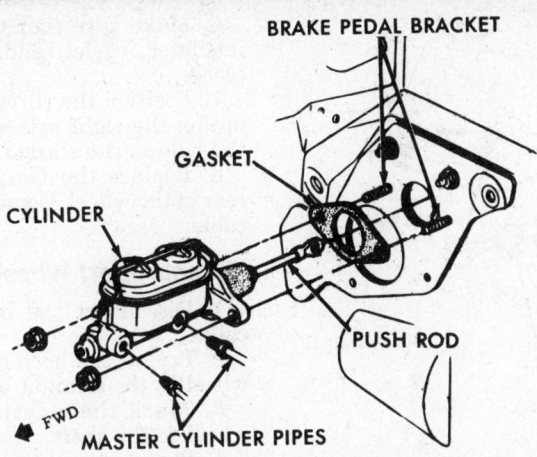

Master cylinder installation details

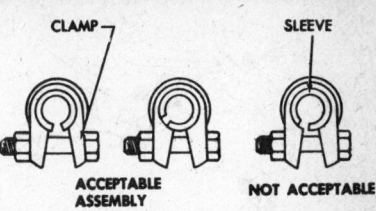

Tie-rod clamp installation

4. Reverse the removal procedure to install. Torque the mounting nuts to 24 ft. lbs.

5. Bleed the hydraulic system.

Power Booster

REMOVAL & INSTALLATION

1. Remove the vacuum hose from the check valve.

2. Remove the master cylinder-to-power booster nuts.

3. Remove the brake line distribution and switch mounting bolt from the fender skirt.

4. Pull forward on the master cylinder until the cylinder clears the power booster.

5. Carefully remove the mast cylinder with the brake lines attached and set the master cylinder aside. Support the cylinder so that there is no stress on the brake lines. The master cylinder should be moved the minimum distance necessary.

6. Unbolt the power booster from the firewall.

7. Remove the brake pedal pushrod from the pedal pin.

8. Remove the power brake booster.

9. Installation is the reverse of removal. Be sure the brake lines are properly routed to provide sufficient clearance.

Parking Brake

ADJUSTMENT

1. Raise and support the rear of the car.

2. Apply the parking brake one notch from the fully released position.

NOTE: It may be necessary to remove the driveshaft to gain access to the parking brake equalizer.

3. Loosen the adjusting locknut and tighten the adjusting nut until a slight drag is felt when the rear wheels are rotated.

4. Tighten the locknut securely.

5. The rear wheels should rotate freely when the parking brake is fully released.

6. Lower the vehicle.

STEERING

Tie Rod

REMOVAL & INSTALLATION

1. Place the vehicle on a hoist.

2. Remove the cotter pins from the ball studs and remove the special nuts.

3. To remove the outer ball stud, tap on the steering arm at the tie rod end with a hammer while using a heavy hammer or similar tool as a backing.

4. Remove the inner ball stud from the relay rod using the same procedure as described in Step 3.

5. To remove the tie rod ends from the tie rod, loosen the clamp bolts and unscrew the end assemblies.

6. If the tie rod ends were removed, lubricate the tie rod threads with chassis lube and install the ends on the tie rod making sure that both ends are threaded an equal distance from the tie rod.

7. Make sure that the threads on the ball studs and in the ball stud nuts are perfectly clean and smooth. Check the condition of the ball stud seals; replace if necessary.

NOTE: If threads are not clean and smooth, the ball studs may turn in the tie rod ends when attempting to tighten nut.

8. Install the ball studs in the steering arms and the relay rod.

9. Install the ball stud nut, tighten and install new cotter pins. Lubricate the tie rod ends.

10. Remove the vehicle from the hoist.

11. Adjust toe-in.

Manual Steering Gear

REMOVAL & INSTALLATION

1. Remove the pot joint coupling clamp bolt at the steering gear wormshaft.

2. Raise the vehicle and support it with jack stands.

3. Remove the "K" brace.

4. Remove the pitman arm nut and washer from the pitman shaft and mark the relation of the arm position to the shaft.

5. Remove the pitman arm. GM recommends the use of tool No. J-6632 for this procedure.

6. Remove the bolts retaining the steering gear to the frame and remove the gear from the vehicle.

7. Installation is the reverse of removal.

Power Steering Gear

REMOVAL

1. To provide access to the steering gear remove the battery.

2. Remove the clamp which secures the intermediate shaft to the steering shaft.

3. Disconnect both hydraulic lines at the steering gear and allow to drain. Position the lines out of the way.

4. Raise the vehicle and support it with jack stands.

5. Remove the clamps at the stabilizer bar.

6. Remove the left front crossmember and bolts.

7. Remove the pitman shaft nut.

8. Remove the pitman arm from the pitman shaft. G.M. recommends

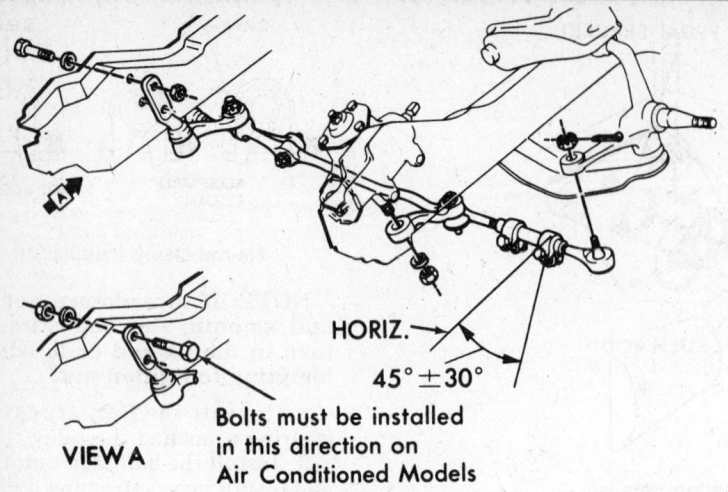

VIEW A

HORIZ.

45° ± 30°

Bolts must be installed
in this direction on
Air Conditioned Models

Steering linkage

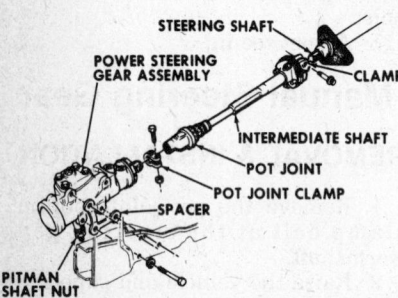

STEERING SHAFT

POWER STEERING
GEAR ASSEMBLY

CLAMP

INTERMEDIATE SHAFT

POT JOINT

POT JOINT CLAMP

SPACER

PITMAN
SHAFT NUT

Power steering gear

using tool J-6632 or J-5504 for this procedure.

9. Remove the three steering gear mounting bolts and lock washers.

10. Lift the steering gar, remove the pitman arm, and position to provide clearance for removal.

11. Remove the steering gear and intermediate shaft as an assembly from the vehicle.

12. Remove the plastic shield.

13. Remove the pot joint clamp and separate the steering gear from the pot joint.

INSTALLATION

1. Install the pot joint to the intermediate shaft.

2. Position the pot joint to the steering gear and install the pot joint clamp and torque to 30 ft. lbs.

3. Install the plastic shield.

4. Install the steering gear and intermediate shaft assembly to the steering column. Torque to 30 ft. lbs.

NOTE: Make sure the flat on the steering shaft is properly aligned. Also the steering gar can rotate three full turns from lock to lock, therefore, it is important for proper installation that the steering gear be centrally positioned (1 ½ turns) when installing the pot joint and intermediate shaft assembly.

5. Install the hydraulic lines to the steering gear.

6. Raise the vehicle and support it with jack stands.

7. Lift the steering gear and position the pitman arm to the pitman shaft.

8. Install the spacer and the three steering gear mounting bolts and lock washers to 70 ft. lbs.

9. Install the pitman shaft nut and lock washer and torque to 185 ft. lbs.

10. Install the left front crossmember and bolts.

11. Install the stabilizer bar clamps.

12. Lower the vehicle, install the battery and check the level of power steering fluid.

Steering Wheel

REMOVAL & INSTALLATION

Standard Wheel

1. Disconnect the battery ground cable.

2. Remove the two screws from the back of the wheel, allowing the shroud (horn actuator bar) to be removed.

3. Set the wheel straight ahead. Mark the relationship of the wheel to the shaft and remove the snap-ring and nut.

4. Remove the steering wheel with a puller, using the two threaded holes in the wheel.

5. Install the wheel, aligning the previously made marks. Make sure that the turn signal switch is in the neutral position. Torque the nut to 30 ft. lbs.

NOTE: Steering wheel rub may be encountered by over-torquing the steering wheel nut by as little as 2 ft. lb.

6. Make sure that the lower horn insulator, eyelet, and spring are in place.

7. Position the shroud, seating the pin on the right side of the wheel in the hole in the shroud.

8. Replace the two screws in the rear of the wheel. Connect the battery cable.

GT and Sport Wheel

1. Disconnect the battery ground cable.

2. Pry off the horn button. Set the wheel in the straight ahead position.

3. Mark the relationship of the wheel to the shaft.

4. Remove the three screws and the upper horn insulator, receiver, and round belleville spring. Remove the snap-ring and nut.

5. Remove the steering wheel with a puller, utilizing the two threaded holes in the wheel.

6. Replace the wheel, aligning the marks previously made. Make sure that the turn signal switch is in the neutral position. Torque the nut to 30 ft. lbs.

NOTE: Steering wheel rub may be encountered by over-torquing of the steering wheel nut by as little as 2 ft. lb.

7. Make sure that the lower horn insulator, eyelet, and spring are in place.

8. Install the belleville spring, receiver, upper horn insulator, and three screws.

Turn Signal Switch

REMOVAL & INSTALLATION

Standard Column

1. Remove the steering wheel as outlined above.

2. On models the cover can be pried off with a screwdriver.

3. The lockplate must be depressed with a special tool. Depress the lockplate and remove the wire snapring from the shaft.

4. Remove the cancelling cam, upper bearing pre-load spring, and thrust washer from the shaft.

5. Remove the turn signal lever screw and the lever.

6. Push the hazard knob in and unscrew it.

7. Unplug the switch connector from the column and wrap the upper part of the connector with tape.

8. Remove the three switch mounting screws and pull the switch straight up. Guide the wiring connector through the column.

9. Tape the new switch connector. Feed the connector down through the column housing and under the mounting bracket.

10. Install the three switch mounting screws.

CAUTION

It is extremely important that only the specified length fasteners be used. Use of overlength fasteners could prevent designed collapse of the steering column during impact.

11. Replace the hazard flasher knob and the turn signal lever. The turn signal switch should be in Neutral and the hazard flasher knob out.

12. Place the thrust washer, upper bearing preload spring, and canceling cam on the shaft.

13. Place the lockplate and a new snap-ring on the shaft. Press the lockplate down as in Step 3 and install the new snap-ring.

14. Replace the cover and its three screws.

15. Install the steering wheel.

Tilt Column

1. Remove the steering wheel.

2. Remove the cover from the steering shaft. The screws have plastic retainers on the back of the cover. It is not necessary to completely remove the screws.

3. Remove the turn signal lever screw and lever.

4. Push the hazard warning knob in and remove the knob.

5. Depress the shaft lockplate and remove the retaining snap-ring. Remove the lockplate.

6. Slide the turn signal cancelling cam and upper bearing preload spring off the end of the shaft.

7. Remove the column mounting bracket and gently lower the column. Support the column.

8. Remove the signal switch wire protective cover and strip the wires from the protector. Do not damage the wires. Disconnect the switch connector from the bracket. Tape the wires close to the connectors to facilitate removal.

9. Remove the switch mounting screws and pull the switch straight up, guiding the wiring harness through the column.

10. Tape a new turn signal switch wiring harness and connector and feed the harness through the housing. Push the hazard warning switch in to aid in installation.

11. Reinstall the protective signal switch wire cover.

12. Install the column bracket and raise the column into position.

13. Install the mounting screws and clip the connector to the bracket on the steering column jacket.

14. Install the hazard warning knob and turn signal lever.

15. Be sure the switch is in the neutral position and the hazard warning knob is out. Slide the upper bearing preload spring and canceling cam onto the shaft.

16. Install the lockplate on the end of the shaft. Compress the lockplate and install a new snap-ring.

17. Reinstall the cover on the end of the shaft.

18. Install the steering wheel.

Ignition Switch

REMOVAL & INSTALLATION

The ignition switch is mounted on top of the column jacket under the dashboard, completely inaccessible unless the steering column is lowered. The energy absorbing column is fragile when disconnected and should not be subjected to any shock or excess pressure. Since the column will distort under its own weight, make sure that it is fully supported along its entire length while it is disconnected from the dashboard.

1. Disconnect the battery ground cable.

2. Remove the steering wheel.

3. On manual steering columns, remove the pot joint coupling clamp bolt.

4. On power steering columns, remove the flexible coupling pinch bolt.

5. Move the front seat back out of the way.

6. Remove the three floor pan bracket screws.

7. Remove the two column-to-instrument panel nuts and carefully lower the column far enough to allow the harness plugs to be disconnected.

8. Disconnect the turn signal and ignition switch harnesses.

9. Place the ignition switch in LOCK position.

10. Remove the two switch screws and the switch assembly.

11. When installing, make sure that the switch is in LOCK position.

12. Install the rod to the switch and the switch to the column. Do not use mounting screws longer than the original ones because they could interfere with the ability of the column to collapse.

NOTE: The following is a mandatory column installation procedure, and must be followed exactly to prevent severe column damage.

13. On power steering models, place the pot joint clamp over the lower end of the pot joint and assemble the intermediate shaft assembly (pot joint, intermediate shaft and flex coupling) to the steering gear stub shaft, aligning the flat on the stub shaft with the flat in the pot joint.

14. Position the column in the vehicle.

15. On manual steering models, place the pot joint clamp over the lower end of the pot joint and assemble the pot joint to the steering gear wormshaft with the flat in the pot joint. On power steering models, align the steering shaft flat with the flat in the flex coupling. When the shaft is bottomed against the coupling reinforcement, install and tighten bolt to 30 ft. lbs.

16. Connect the turn signal and ignition switch wiring harnesses.

17. Loosely install the steering column bracket to instrument panel stud nuts.

18. Align the pot joint clamp with the groove across the end of the pot joint. Install bolt and nut, tightening nut to 55 ft. lbs.

NOTE: The bolt must pass through the shaft undercut.

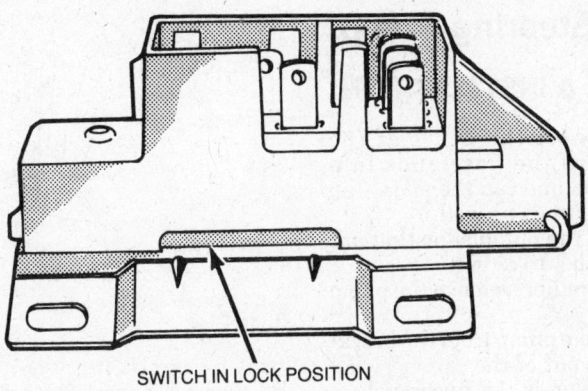

SWITCH IN LOCK POSITION

Ignition switch in lock position

19. With the vehicle on the ground, tighten instrument panel nuts to 19 ft. lbs.

20. Slide the toe plate down the column to the floorboard and install the three screws.

NOTE: On power steering models, alignment flange on the toe plate must be engaged with the front of the toe pan before driving screws. On manual steering models, no side load is allowed during installation of the attaching screws. A side load could cause misalignment.

21. On manual steering models, remove the alignment spacers. The minimum allowable clearance between the O.D. of the steering shaft and the I.D. of the column jacket lower plastic bushing after installation is 0.18 in.

22. Install the steering wheel.

23. Connect the battery ground cable.

Ignition Lock Cylinder

REMOVAL & INSTALLATION

1. Place the lock in the Run position.

2. Remove the lock plate, turn signal switch and buzzer switch.

3. Remove the screw and lock cylinder.

—————— CAUTION ——————

If the screw is dropped on removal, it could fall into the column, requiring complete disassembly to retrieve the screw.

4. Rotate the cylinder clockwise to align the cylinder key with the keyway in the housing.

5. Push the lock all the way in.

6. Install the screw. Tighten the screw to 14 in. lb. for standard columns.

Power Steering Pump

REMOVAL & INSTALLATION

1. Disconnect the hoses at the pump. Secure the hose ends in a raised position and cap the ends. Cap the pump openings as well.

2. Remove the pump adjusting nut and remove the drive belt.

3. Using a puller, remove the pump pulley.

4. Unbolt the pump from the brackets and lift it out of the car.

5. Installation is the reverse of removal. Adjust the drive belt, fill the reservoir and bleed the system.

CHASSIS ELECTRICAL

Speedometer Cable and Instrument Cluster

REMOVAL & INSTALLATION

The speedometer and the instruments are removed from the front of the panel by removing the bezel and the lens.

Lift the speedometer away from the panel and disconnect the speedometer cable and wiring from the rear of the cluster. The cable core can then be removed with the aid of a pair of needle nose pliers. If the cable core is broken, it may be necessary to remove the broken piece from the transmission end of the cable. The cable is easily unscrewed from the left side of the transmission case. When installing a new cable, liberally lubricate it with graphite base speedometer cable lubricant.

All of the indicator bulbs are of the quarter twist type and are removed from the rear of the instrument cluster.

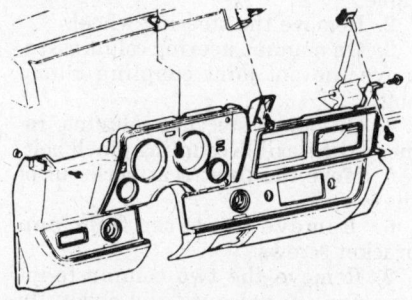

Standard instrument cluster, except station wagons—typical

Headlight Switch

REMOVAL & INSTALLATION

1. Disconnect the battery ground cable.

2. Pull the light switch to ON position.

3. Reach up under the instrument panel and depress the switch retainer button while pulling on the knob.

4. Remove the knob and shaft, then remove the ferrule nut with a large screwdriver.

5. Disconnect the multi-contact connector, prying gently with a small screwdriver.

6. Connect the new switch and reverse the removal procedure to complete the replacement.

Wiper Blade

REMOVAL & INSTALLATION

Three methods of blade attachment may be used. If there is a small tab on top of the blade, depress it and slide the blade off. If there is a small spring visible in the top of the blade, insert a screwdriver in the opening, press down and slide the blade off. If there is a clip on the underside of the arm, press down on the clip and slide the blade off.

Wiper Motor

REMOVAL & INSTALLATION

1. Raise the hood.

2. Reaching through cowl opening, loosen the two transmission drive link attaching nuts to the motor crankarm.

3. Remove the transmission drive link from the motor crankarm.

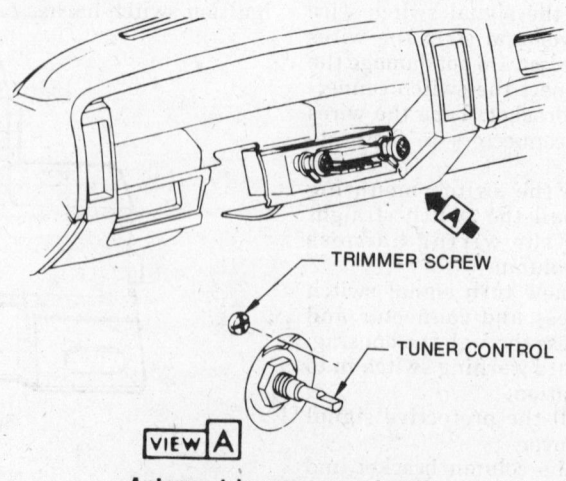

— TRIMMER SCREW

— TUNER CONTROL

VIEW A

Antenna trimmer screw location

4. Disconnect the wiring, and washer hoses.

5. Remove the three motor attaching screws.

6. Remove the motor while guiding the crankarm through the hole.

7. To install, reverse the removal procedure.

Radio

ANTENNA TRIM ADJUSTMENT

1. Remove the right knob and bezel, and locate the trimmer screw above and to the left of the shaft.

2. Temporarily reinstall the knob and tune the radio to a weak station near 1400 KC on the AM dial. Remove the knob.

3. Adjust the trimmer screw until the maximum volume has been reached.

4. Replace the knob and bezel on the radio shaft.

REMOVAL & INSTALLATION

1. Remove the battery ground cable.

2. Remove the knobs, controls, washers and nuts from the radio bushings.

3. On some models it may be necessary to remove the heater outlet duct on cars with air conditioning.

4. Disconnect the antenna lead, power connector, and speaker connectors from the rear of the receiver.

5. Remove the two screws securing the radio mounting bracket to the instrument panel lower reinforcement and lift out the radio receiver.

6. To install, reverse the removal procedure.

Heater Blower Motor

REMOVAL & INSTALLATION

1. Disconnect the battery ground cable.

2. Remove the coolant recovery tank attaching screws and move the tank aside; draining the tank is unnecessary.

3. Disconnect the blower motor lead wire. Disconnect the motor cooling tube on air conditioned models.

4. Scribe the blower motor flange to case position.

5. Remove the blower to case attaching screws and remove the blower wheel and motor assembly. Pry the flange gently if the sealer is retaining the assembly.

6. Remove the blower wheel retain-

ing nut and separate the motor and wheel.

7. To install, reverse Steps 1–5, lining up the match-marks on the motor flange and case which were made at removal.

NOTE: Assemble the blower wheel to the motor with the open end of the blower away from the motor. Reseal the motor flange, if necessary.

Heater Core

REMOVAL & INSTALLATION

All Models Without Air Conditioning

1. Disconnect the battery ground cable.

2. Disconnect the blower motor lead wire.

3. Place a pan under the vehicle. Disconnect the heater hoses at the core connections and secure the ends of the hoses in a raised position.

4. Remove the coil bracket to firewall stud nut and move the coil out of the way.

5. Remove the blower intake to firewall screws and nuts and remove the blower intake, blower motor and wheel as an assembly.

6. Remove the core retaining strap screws and remove the core from the vehicle.

7. To install, reverse Steps 1–6.

NOTE: Be sure that the blower intake sealer is intact, replace if necessary.

Monza "S" Hatchback and Station Wagon With Air Conditioning

1. Disconnect the battery ground.

2. Disconnect the hoses at the core tubes and place in a raised position.

3. Remove the nuts from the selector duct studs in the engine compartment.

4. Remove the glove box and door.

5. Remove the right outlet to instrument panel screws and remove the outlet and hose.

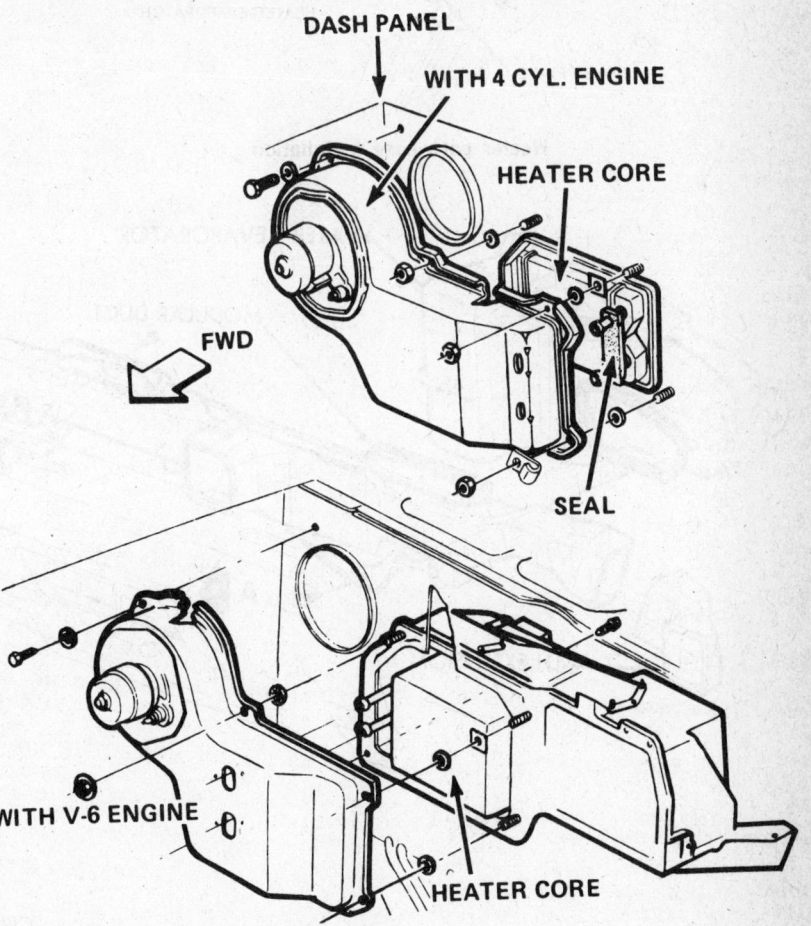

Blower motor and case assembly without A/C

6. Remove the intermediate duct leading to the left outlet.

7. Lower the steering column as described in Ignition Switch Removal and Installation.

8. Remove the instrument panel bezel. Remove the ashtray and retainer.

9. Remove the screws securing the A/C control head to the instrument panel.

10. Disconnect the radio leads and antenna wire.

11. Remove the instrument cluster screws and allow the entire cluster, including the radio, to rest on the steering column.

12. Disconnect the speedometer cable and remove the A/C control head.

13. Remove the center duct screws, then slide it first to the left, then to the right then remove it.

14. Remove the defroster duct and remaining selector ducts.

15. Disconnect all electrical and vacuum lines from the evaporator.

16. Disconnect the temperature door cables.

17. Pry off or punch out the temperature door bell crank.

18. Remove the temperature door.

19. Remove the screws securing the temperature door cable retainer and backing plate.

20. Remove the heater core and backing plate assembly and remove the straps from the core.

21. Installation is the reverse of removal. When installing the ducts, make sure the firewall seals are positioned correctly. When installing the cluster, position the A/C control head and connect the speedometer cable before the cluster is secured. Adjust the temperature door at the selector duct attachment. With the temperature lever and door in the Off position, tighten the cable attaching screw.

Monza (Except Monza "S" and Station Wagon) With Air Conditioning

1. Disconnect the battery ground.

2. Remove the floor outlet duct.

3. Remove the left and right dash outlets.

5. Remove the instrument panel pad.

6. Disconnect the vacuum hoses and electrical wires from the heater-evaporator case.

7. Remove the insulation tray below the instrument cluster and loosen the console and slide it rearward.

8. Lower the steering column assembly, following the instructions in Ignition Switch Removal and Installation.

9. Remove the instrument panel attaching screws and allow the in-

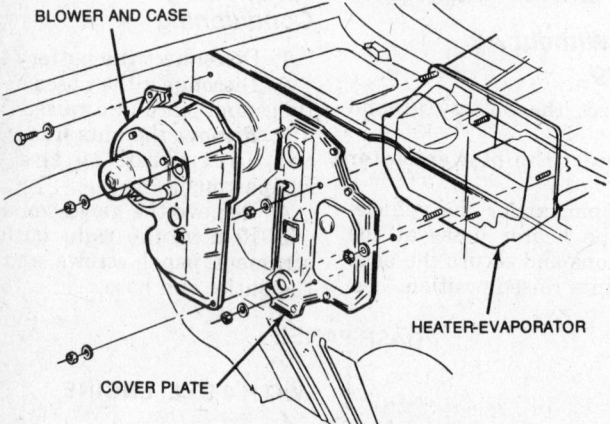

Heater core case installation

BLOWER AND CASE

HEATER-EVAPORATOR

COVER PLATE

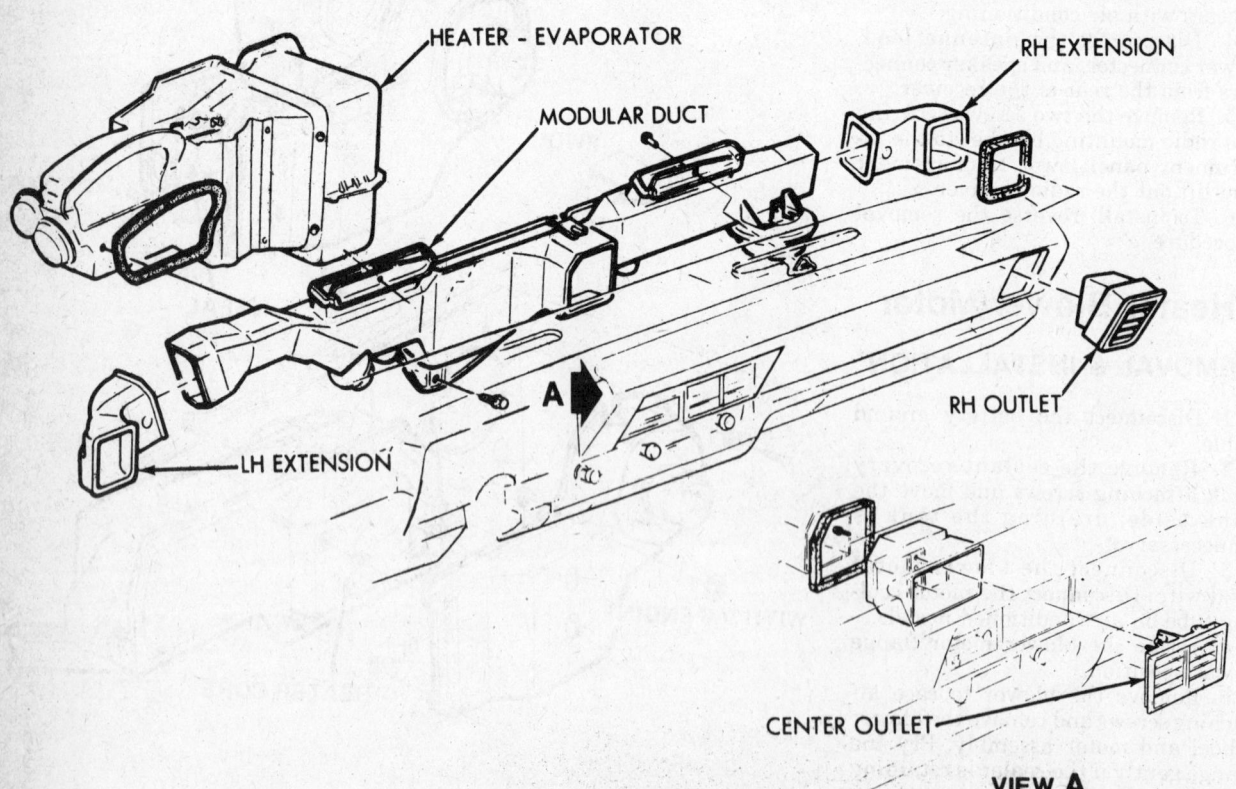

HEATER - EVAPORATOR

MODULAR DUCT

RH EXTENSION

RH OUTLET

LH EXTENSION

A

CENTER OUTLET

VIEW A

A/C Air Distribution Ducts

strument panel to rest on the steering column.

10. Disconnect the speedometer cable, radio wiring and control head wiring.

11. Remove the right side instrument panel and lap duct.

12. Remove the modular duct from the case.

13. Disconnect the temperature door cable and the wiring harness.

14. Remove the heater hoses from the core tubes and position them upright to avoid coolant loss.

15. Remove the three heater case stud nuts.

16. Remove the heater core case-to-evaporator core case screws.

17. Hammer on the studs, carefully, to break loose the heater core case.

18. Unbolt the core from the case.

19. Installation is the reverse of removal. Replace any damaged sealer.

Sunbird with Air Conditioning.

1. Have the air conditioning system purged of refrigerant.

2. Disconnect the negative battery cable.

3. Remove the blower and case assembly.

4. Remove and plug the heater hoses at the core tubes and then hang them out of the way.

5. Remove the evaporator to firewall cover plate screws and remove the plate.

6. Remove (from inside the car), the floor outlet duct, the glove compartment assembly and the dash outlets on both sides. To remove the dash outlets, use a putty knife and pry them out.

7. Remove the eleven instrument panel pad screws and pry the pad off.

8. Remove the right side instrument panel to dash and kick pad screws, then loosen the left side instrument cluster to instrument panel screws.

9. Pull out on the right side of the instrument cluster to gain the necessary clearance to remove the right side instrument panel and lower duct.

10. Disconnect the vacuum hoses on the left side of the heater unit and tag them for later reinstallation.

11. Remove the modulator duct to heater unit screw, then pull the carpet and pad to the rear to make room for the heater unit removal.

12. Pull the heater unit toward you

until the core tubes clear the firewall, then pull it to the right until there is enough clearance to disconnect the control cable.

13. After disconnecting the control cable, disconnect the wiring harness and remove the heater assembly.

14. Remove the screws and separate the heater case, then remove the core to case screws and remvove the core.

15. Installation is the reverse of the above procedure, but before assembly, add 3 oz. of refrigerant oil to the evaporator core.

16. When installing the refrigerant lines, coat all the O-rings with refrigerant oil.

Skyhawk with Air Conditioning

NOTE: This procedure requires purging the air conditioning system of refrigerant. Do not attempt this unless you are thoroughly familiar with air conditioning systems. Freon will freeze any surface it contacts (including eyeballs) and turns into a poisonous gas in the presence of an open flame.

1. Have the air conditioning system purged of refrigerant.

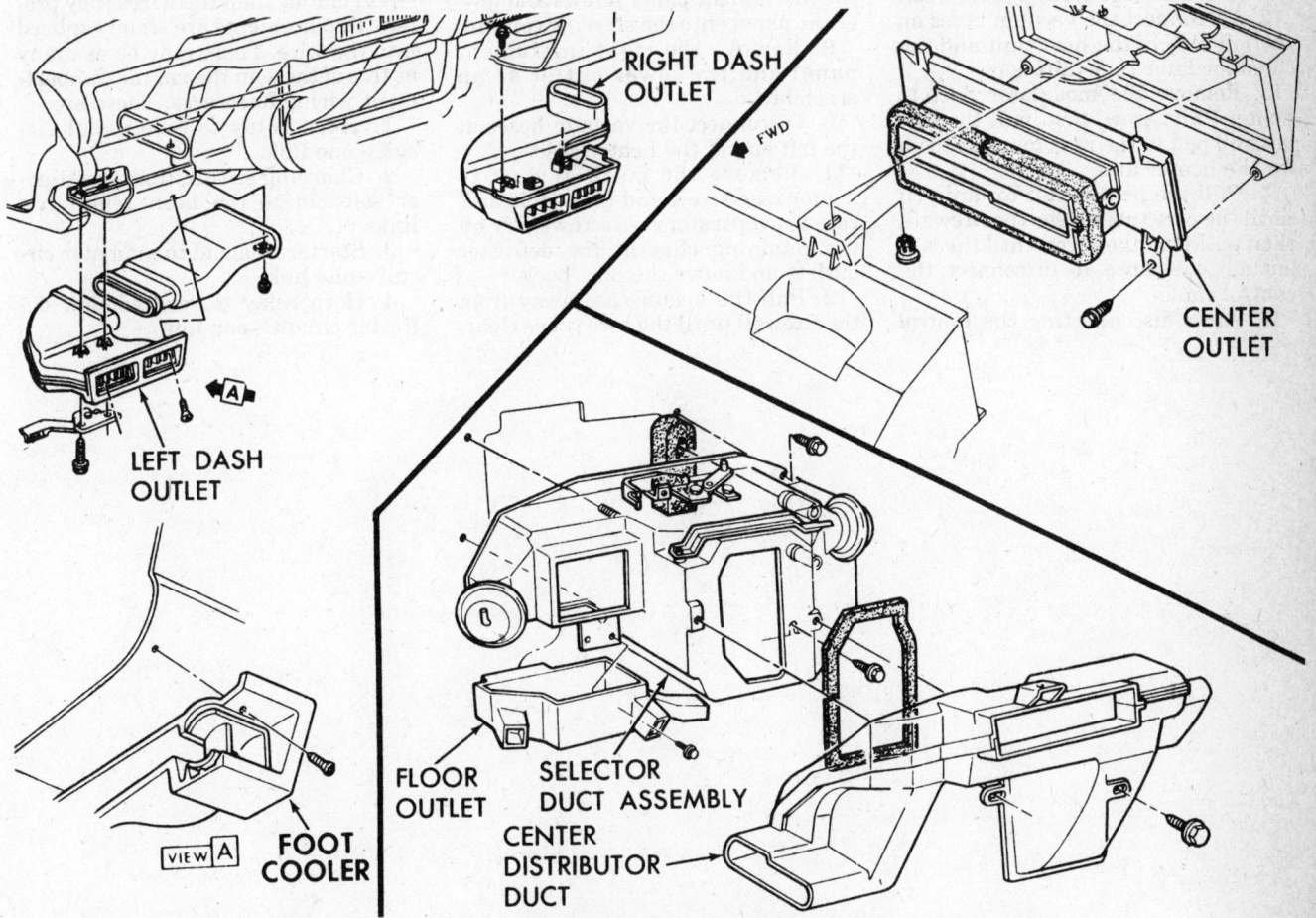

Air distributor and ducts

2. Disconnect the negative battery cable.

3. Disconnect the inlet and outlet lines and the oil bleed line from the accumulator assembly.

4. Remove the accumulator to blower case strap screw, and remove the accumulator unit. Cap all the open connections immediately.

5. Remove the blower and case assembly.

6. Remove and plug the heater hoses at the core tubes and then hang them out of the way.

7. Remove the evaporator to firewall cover plate screws and remove the plate.

8. Remove (from inside the car), the floor outlet duct, the glove compartment assembly and the dash outlets on both sides. Use a putty knife to pry out the dash outlets.

9. Remove the eleven instrument panel pad screws and pry the pad off.

10. Remove the right side instrument panel to dash and kick pad screws, then loosen the left side instrument cluster to instrument panel screws.

11. Pull out on the right side of the instrument cluster to gain the necessary clearance to remove the right side instrument panel and lower duct.

12. Disconnect the vacuum hoses on the left side of the hear unit and tag them for later reinstallation.

13. Remove the modulator duct to heater unit screw, then pull the carpet and pad to the rear to make room for the heater unit.

14. Pull the heater unit toward you until the core tubes clear the firewall, then pull it to the right until there is enough clearance to disconnect the control cable.

15. After disconnecting the control cable, disconnect the wiring harness and remove the heater assembly.

16. Remove the screws and separate the heater case, then remove the core to case screws and remove the core.

17. Installation is the reverse of the above procedure, but before assembly, add 3 oz. of refrigerant oil to the evaporator core.

18. When installing the refrigerant lines, coat all the O-rings with refrigerant oil.

Starfire with Air Conditioning

1. Disconnect the negative battery cable.

2. Remove the three nuts from the engine compartment side of the cover plate.

3. Disconnect the heater hoses and fasten them in a raised position to prevent coolant loss. Plug the core tubes.

4. Remove the heater floor outlet.

5. Remove the glove box and door.

6. Remove the right and left air outlets.

7. Unscrew and move the console back.

8. Remove the column nuts and let the wheel rest on the seat. Remove the instrument panel screws and lower the panel onto the steering column.

9. Remove the right instrument panel and the lower outlet as an assembly.

10. Disconnect the vacuum hoses at the left end of the heater case.

11. Remove the modular duct to heater case screw and the two heater case to evaporator case screws. Pry off the retaining clips at the defroster outlets and move the duct back.

12. Pull the heater case away from the firewall until the core tubes clear, then disconnect the temperature cable.

13. Remove the core to case screws and remove the core.

14. Reverse the procedures for installation. Torque the steering column nuts to 25 ft. lbs.

Fuses, Fusible Links and Circuit Breakers

The fuse block is located beneath the instrument panel above the headlight dimmer floor switch. Fuse holders are labeled as to their service and the correct amperage. Always replace blown fuses with new ones of the correct amperage. Otherwise electrical overloads and possible wiring damage will result.

A circuit breaker is an electrical switch which breaks the circuit during an electrical overload. The circuit breaker will remain open until the short or overload condition in the circuit is corrected.

Fusible links are sections of wire, with special insulation, designed to melt under electrical overload. The fusible links are all two wire gauge sizes smaller than the wires they protect. Replacements are simply spliced into the wire. There may be as many as five of these in the engine compartment wiring harnesses. These are:

1. Horn relay to fuse panel circuit—one link.

2. Charging circuit, from the starter solenoid to the horn relay—two links.

3. Starter solenoid to ammeter circuit—one link.

4. Horn relay to rear window defroster circuit—one link.

GM "H" Body
Front Wheel Drive
Buick LeSabre,
Oldsmobile Delta 88 Royale

YEAR IDENTIFICATION

1986 LeSabre Limited **1987 Oldsmobile Delta 88** **1987 LeSabre Sedan**

VEHICLE IDENTIFICATION NUMBER (VIN)

It is important for servicing and ordering parts to be certain of the vehicle and engine identification. The VIN (vehicle identification number) is a 13 or 17 digit number visible through the windshield on the driver's side of the dash and contains the vehicle and engine identification codes. It can be interpreted as follows:

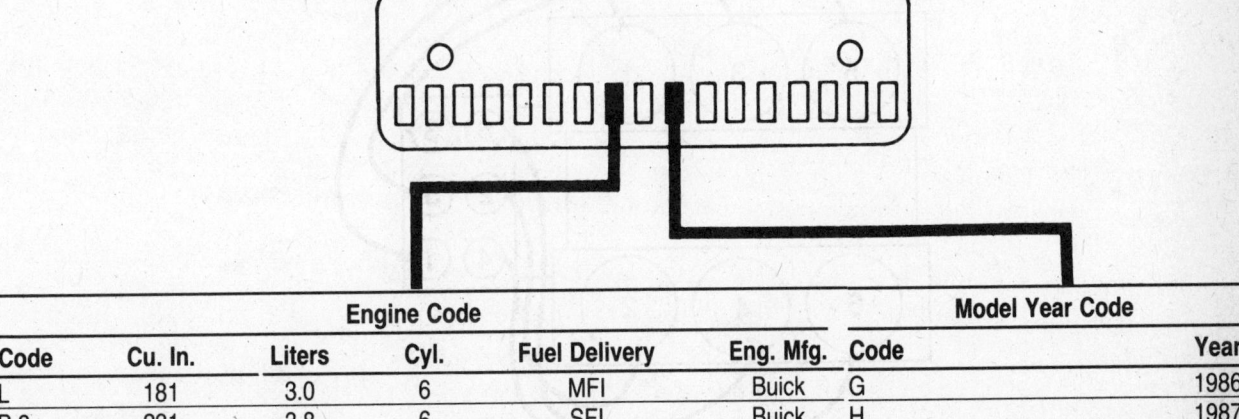

Engine Code						Model Year Code	
Code	Cu. In.	Liters	Cyl.	Fuel Delivery	Eng. Mfg.	Code	Year
L	181	3.0	6	MFI	Buick	G	1986
B,3	231	3.8	6	SFI	Buick	H	1987

The seventeen digit Vehicle Identification Number can be used to determine engine application and model year. The tenth digit indicates model year, and the eighth digit identifies engine code.

MFI Multiport Fuel Injection
SFI Sequential Fuel Injection

GENERAL ENGINE SPECIFICATIONS

Year	V.I.N. Code	Engine No. Cyl. Displ. (cu. in.)	Eng. Mfg.	Fuel Delivery System	Horsepower @ rpm	Torque ft. lb. @ rpm	Bore × Stroke	Compression Ratio	Oil Pressure @ 2400 rpm
'86–'87	L	6-181	Buick	MFI	125 @ 4900	150 @ 2400	3.80 × 2.66	9.0:1	37
	B,3	6-231	Buick	SFI	150 @ 4400	200 @ 2000	3.80 × 3.40	8.5:1	37

MFI Multiport Fuel Injection
SFI Sequential Fuel Injection

TUNE-UP SPECIFICATIONS

(When analyzing compression test results, look for uniformity among cylinders rather than specific pressures.)

Year	V.I.N. Code	Eng. No. Cyl. Displ. (cu. in.)	Eng. Mfg.	hp	Spark Plugs Orig. Type	Gap (in.)	Ignition Timing (deg.)①	Fuel Pump Pressure (psi)	Idle Speed (rpm)
'86	L	6-181	Buick	125	R44LTS	.045	①	34–44	①
	B,3	6-231	Buick	150	R44LTS	.045	①	26–36	①
'87	All				See Underhood Specifications Sticker				

NOTE: The underhood specifications sticker often reflects tune-up specification changes made in production. Sticker figures must be used if they disagree with those in this chart. Part numbers in this chart are not recommendations by Chilton for any product by brand name.

① Idle speed and ignition timing are computer-controlled and not adjustable

FIRING ORDERS

NOTE: To avoid confusion, replace spark plugs and wires one at a time

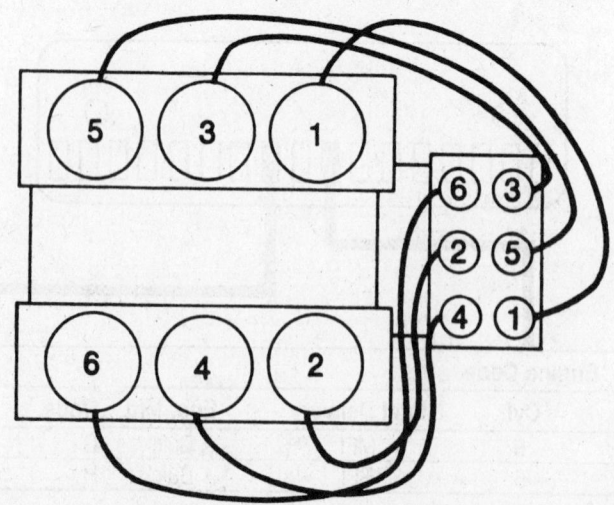

183 cu. in. (3.0L) and 231 cu. in. (3.8L) engines
Firing Order: 1–6–5–4–3–2

CAPACITIES

Year	V.I.N. Code	Engine Displacement (cu. in.)	Eng. Mfg.	Crankcase Quarts w/filter	wo/filter	Transaxle Pints	Gas Tank (gals.)	Cooling System (qts.) Standard	HD
'86–'87	L	181	Buick	4.0①	4.0	13.0	18	12.1	12.2
	B,3	231	Buick	4.0①	4.0	13.0	18	12.5	12.6

① Add as necessary to bring to appropriate
level on dipstick

VALVE SPECIFICATIONS

Year	V.I.N. Code	Engine No. Cyl. Displacement (cu. in.)	Eng. Mfg.	Seat Angle (deg.)	Face Angle (deg.)	Spring Test Pressure (lbs. @ in.)	Spring Installed Height (in.)	Stem-to-Guide CLearance (in.) Intake	Exhaust	Stem Diameter (in.) Intake	Exhaust
'86–'87	L	6-181	Buick	45	45	220 @ 1.340	1.727	.0015–.0035	.0015–.0032	.3401–.3412	.3405–.3412
	B,3	6-231	Buick	45	45	185 @ 1.340	1.727	.0015–.0035	.0015–.0032	.3401–.3412	.3405–.3412

CRANKSHAFT AND CONNECTING ROD SPECIFICATIONS

(All measurements are given in inches.)

Year	V.I.N. Code	Engine No. Cyl. Displacement (cu. in.)	Eng. Mfg.	Crankshaft Main Brg. Journal Diameter	Main Brg. Oil Clearance	Shaft End-Play	Thrust on No.	Connecting Rod Journal Diameter	Oil Clearnace	Side Clearance
'86–'87	L	6-181	Buick	2.4995	.0003–.0018	.003–.011	2	2.2487–2.2495	.0005–.0026	.006–.023
	B,3	6-231	Buick	2.4995	.0003–.0018	.003–.011	2	2.2487–2.2495	.0005–.0026	.006–.023

CAMSHAFT SPECIFICATIONS

(All measurements are given in inches)

Year	V.I.N. Code	Engine (cu. in.)	Eng. Mfg.	Journal Diameter 1	2	3	4	5	Bearing Clearance	Lobe Lift Intake	Exhaust	Camshaft End Play
'86–'87	L	181	Buick	—1.785–1.786—					①	.358	.384	—
	B,3	231	Buick	—1.785–1.786—					①	.368②	.384②	—

① No. 1: 0.0005–0.0025
No. 2–5: 0.0005–0.0035
② VIN B—.392 in.

PISTON AND RING SPECIFICATIONS

(All measurements are given in inches.)

Year	V.I.N. Code	Engine Type/ Disp. (cu. in.)	Eng. Mfg.	Piston-to-Bore Clearance	Ring Gap			Ring Side Clearance		
					Top Compression	Bottom Compression	Oil Control	Top Compression	Bottom Compression	Oil Control
'86–'87	L	6-181	Buick	①	.010–.020	.010–.020	.015–.055	.0030–.0050	.0030–.0050	.0035 max
	B,3	6-231	Buick	①	.010–.020	.010–.020	.015–.055	.0030–.0050	.0030–.0050	.0035 max

① Top land—.046–.056 in.
Skirt top—.0008–.0020 in.
Skirt bottom—.0013–.0035 in.

TORQUE SPECIFICATIONS

(All readings in ft. lbs.)

Year	V.I.N. Code	Engine No. Cyl. Displacement (cu. in.)	Eng. Mfg.	Cylinder Head Bolts	Rod Bearing Bolts	Main Bearing Bolts	Crankshaft Bolt	Flywheel-to-Crankshaft Bolts	Manifold	
									Intake	Exhaust
'86–'87	L	6-181	Buick	①	40	100	200	60	32	37
	B,3	6-231	Buick	①	40	100	200	60	32	37

① See text for angle torque procedure

WHEEL ALIGNMENT SPECIFICATIONS

Year	Model	Caster		Camber		Toe (in.)
		Range (deg.)	Pref. Setting (deg.)	Range (deg.)	Pref. Setting (deg.)	
'86–'87	Buick	2–3	2 1/2	L –1–0 R 0–1	1/2	7/32 ①
'86–'87	Olds	2–3	2 1/2	L –1–0 R 0–1	–1/2 1/2	3/32 ①

L Left
R Right
① In or out. Pref: 0

ENGINE ELECTRICAL

For all information on the charging system not detailed below, please refer to "Charging and Starting" in the Unit Repair section.

Alternator

REMOVAL & INSTALLATION

1. Disconnect the negative battery cable.
2. Tag and disconnect the battery charge wire, 3-prong connector and the ground wire at the back of the alternator.
3. Remove the brace at the back of the alternator (if so equipped).
4. Loosen the serpentine belt tensioner and rotate it counterclockwise to remove the drive belt.
5. Support the alternator, remove the mounting bolts and then remove the alternator.
8. Installation is in the reverse order of removal. Tighten the serpentine belt tensioner pulley.

Voltage Regulator

An alternator with an integral voltage regulator is standard equipment. There are no adjustments possible with this unit. Testing procedures can be found in the Unit Repair section.

Starter

For all information concerning the starter which is not detailed below, please refer to "Charging and Starting" in the Unit Repair section.

REMOVAL & INSTALLATION

1. Disconnect the negative battery cable.
2. Raise and support the vehicle on jackstands. Remove the starter splash shield and any braces which may be in the way.
3. Remove the two starter mounting bolts and lower the starter slightly to gain access to the wires.
4. Tag and disconnect all wires at the solenoid. Note color coding of wires for reinstallation.
5. Installation is in the reverse order of removal. Don't forget any shims, if present.

Computer Controlled Coil Ignition (C3I) System

All models are equipped with Computer Controlled Coil Ignition (C3I), which eliminates the distributor. The C3I ignition system consists of a coil

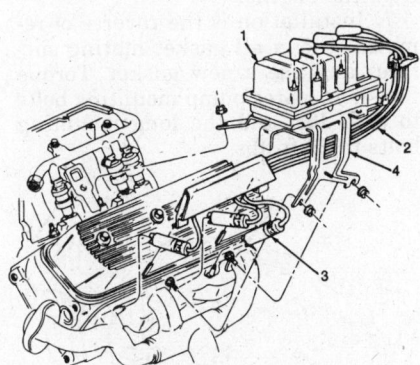

1. C3I coil and module assembly
2. Rear spark plug wire harness
3. Spark plug heat shield
4. Coil and module bracket

C3I ignition system

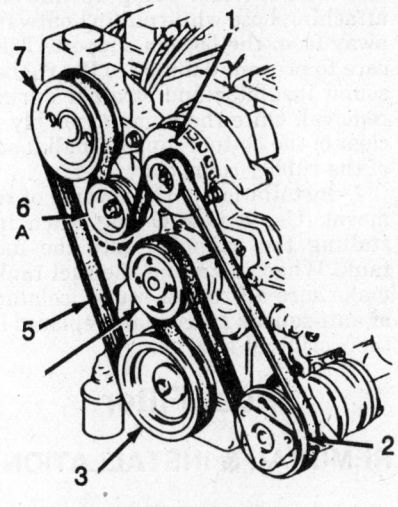

1. Alternator pulley
2. A/C compressor
3. Crankshaft balancer
4. Water pump pulley
5. Serpentine belt
6. Belt tensioner
7. Power steering pump
A. Rotate drive belt tensioner in direction of arrow to remove or install belt

Serpentine drive belt routing

pack, ignition module, camshaft and crankshaft sensor. There are two types of C3I coils used. Type 1 coils have three plug wires on each side of the coil assembly; Type 2 coils have all six wires connected on one side of the coil. When troubleshooting or replacing components, it is important to determine which C3I system is installed on the engine.

1. Crankshaft sensor
2. Mounting bolt
3. Camshaft sensor
4. Mounting bolts

Crankshaft and camshaft sensor location

The C3I system consists of the coil pack, ignition module, crankshaft sensor, interruptor rings and electronic control module (ECM). All components are serviced as complete assemblies, although individual coils are available for Type 2 coil packs. Since the ECM controls the ignition timing, no timing adjustments are necessary or possible.

Crankshaft Sensor

REMOVAL & INSTALLATION

1. Disconnect the negative battery cable.
2. Remove the serpentine drive belt.
3. Raise the car and support it safely.

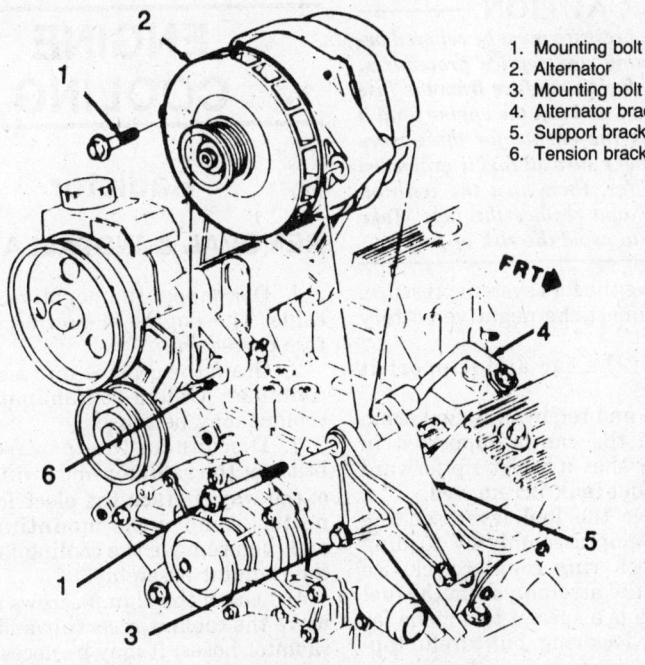

1. Mounting bolt
2. Alternator
3. Mounting bolt
4. Alternator brace
5. Support bracket
6. Tension bracket

Typical V6 alternator mounting

4. Remove the right front tire.

5. Remove the inner fender splash shield.

6. Remove the crankshaft balancer bolt and balancer.

7. Remove the mounting bolts and remove the crankshaft sensor from the front cover. Disconnect the electrical connector and remove the sensor from the vehicle.

8. Installation is the reverse of removal. Make sure the electrical T-latch connector is assembled properly or an intermittent loss of operation may occur. The sensor must be carefully aligned with the interruptor rings to avoid damage when the engine is cranked. Tighten the crankshaft sensor mounting bolts to 22 ft. lbs. (30 Nm) and the crankshaft balancer bolt to 200 ft. lbs. (270 Nm).

For all testing and diagnosis procedures on the C3I ignition system, please refer to *Chilton's Guide To Electronic Engine Controls.*

FUEL SYSTEM

For further information on fuel injection, please refer to *"Chilton's Guide to Fuel Injection and Feedback Carburetors".*

Electric Fuel Pump

REMOVAL & INSTALLATION

CAUTION
Fuel system pressure must be relieved before attempting any service procedures. Remove the fuel pump fuse from the fuse box, then start and run the engine until it stalls. Crank the engine for three more seconds to make sure all fuel is exhausted from the lines, then turn the ignition switch OFF and replace the fuse. Take precautions to avoid the risk of fire.

1. Relieve the fuel system pressure.

2. Disconnect the negative battery cable.

3. Raise the car and support it safely.

4. Drain and remove the fuel tank. Make sure the car is supported in such a way that it won't tip forward when the fuel tank is removed.

5. Remove the fuel lever sending unit and pump assembly by turning the cam lock ring counterclockwise and lifting the assembly from the fuel tank. There is a special tool made for turning the lock ring, but careful tapping with a brass drift will work.

6. Pull the fuel pump up into the attaching hose while pulling outward away from the bottom support. Take care to prevent damage to the rubber sound insulator and strainer during removal. Once the pump assembly is clear of the bottom support, pull it out of the rubber connector.

7. Installation is the reverse of removal. Use a new O-ring when installing the assembly into the fuel tank. When installing the fuel tank, make sure all rubber sound isolators or anti-squeak spacers are replaced in their original locations.

Fuel Filter

REMOVAL & INSTALLATION

CAUTION
Fuel system is under pressure. See the Caution under "Fuel Pump Removal" for procedure to relieve fuel pressure before attempting to remove any fuel lines.

The fuel injection system uses an inline filter located in the fuel feed line under the hood, attached to the frame rail, or on the rear crossmember of the vehicle. Always use a back-up wrench on the fittings any time a fuel filter is removed or installed, and never replace a metal fuel line with a rubber insert. The high pressure fuel system used with all fuel injection systems requires metal fuel lines to contain the pressure. Replace the O-ring at the connection and torque the fuel fitting to 22 ft. lbs. (30 Nm).

ENGINE COOLING

Radiator

REMOVAL & INSTALLATION

1. Disconnect the negative battery cable. The engine should be cool for this procedure.

2. Drain the coolant into a suitable container. Unless contaminated, the coolant may be reused.

3. Disconnect the forward strut brace at the radiator and swing it out of the way. Detach the electrical connector, remove the mounting bolts and then remove the cooling fan from the radiator assembly.

4. Loosen the clamp-screws and remove the coolant reservoir and upper radiator hoses. It may be necessary to remove the hood latch; if so, scribe alignment marks around the latch assembly before removing it from the radiator support.

5. Disconnect the transaxle and auxiliary oil cooler lines at the radiator. Wire the lines out of the way.

6. Disconnect the lower radiator hose at the radiator.

7. Remove the mounting bolts and lift out the radiator.

8. Installation is in the reverse order of removal.

Water Pump

REMOVAL & INSTALLATION

1. Disconnect the negative battery cable.

2. Drain the cooling system.

3. Remove the serpentine drive belt.

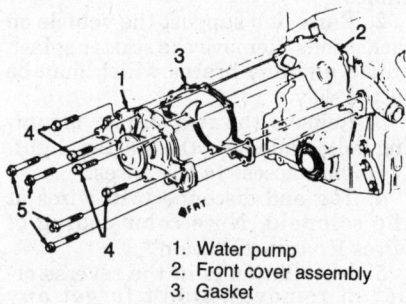

1. Water pump
2. Front cover assembly
3. Gasket
4. 8 ft. lbs.
5. 29 ft. lbs.

Typical water pump mounting

4. Disconnect the coolant hoses at the water pump.

5. Remove the water pump pulley bolts. The long bolt is removed through the access hole provided in the body side rail. Remove the pulley.

6. Remove the water pump attaching bolts and remove the water pump from the engine.

7. Installation is the reverse of removal. Clean all gasket mating surfaces and use a new gasket. Torque the short water pump mounting bolts to 8 ft. lbs. and the long mounting bolts to 29 ft. lbs.

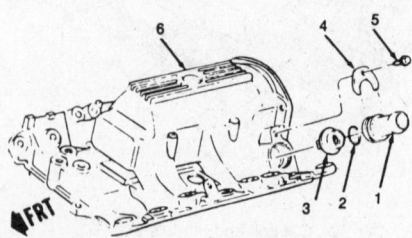

1. Water outlet
2. O-ring
3. Thermostat
4. Clamp
5. 20 ft. lbs.
6. Intake manifold

3.8L V6 thermostat and housing assembly

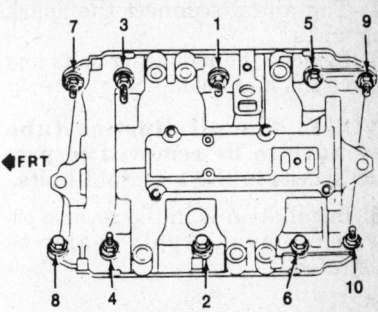

Thermostat

REMOVAL & INSTALLATION

To replace the thermostat, drain the cooling system below the level of the thermostat and remove the bolt(s) holding the water neck in place. Remove the water neck and the thermostat will lift out. Clean the mating surfaces of both the intake manifold and the water neck. Use a new gasket when installing a new thermostat. If only silicone sealer was used from the factory, use only silicone sealer during assembly. Some late model engines only use an O-ring to seal the thermostat water neck.

— **CAUTION** —

Be sure the thermostat is not reversed in its installed position. The spring should be installed toward the engine.

EMISSION CONTROLS

For a description of and service for each system, please refer to "Emission Controls" in the Unit Repair section.

Due to the complex nature of modern electronic engine control systems, comprehensive diagnosis and testing procedures fall outside the confines of this repair manual. For complete information on diagnosis, testing and repair procedures concerning all modern engine and emission control systems, please refer to *Chilton's Guide To Electronic Engine Controls*.

ENGINE MECHANICAL

Engine

REMOVAL & INSTALLATION

1. Disconnect the negative battery cable. Matchmark the hood hinges and remove the hood.
2. Tag and disconnect the air flow sensor wiring. Depressurize the fuel system as described under "Fuel Pump Removal."

3. Disconnect the air intake duct. Drain the engine coolant.
4. Raise the front of the vehicle and support it on jackstands.
5. Unscrew the retaining bolts and separate the exhaust pipe from the manifold.
6. Loosen and remove the engine mount bolts.
7. Remove the bolts and then disconnect the driveline vibration absorber.
8. Tag and disconnect the starter wiring and then remove the starter.
9. Disconnect the A/C compressor and position it out of the way. DO NOT disconnect the refrigerant lines.
10. Disconnect the hydraulic lines at the power steering pump and wire them out of the way.
11. Loosen and remove the lower transaxle-to-engine bolts.

NOTE: One bolt is situated between the transaxle case and the engine block. It is installed in the opposite direction of the other bolts.

12. Remove the flexplate cover. Matchmark the flexplate-to-torque converter relationship to insure proper alignment upon installation. Remove the flexplate-to-torque converter bolts.
13. Disconnect the engine support bracket at the transaxle and then lower the vehicle.
14. Disconnect the radiator and heater hoses at the engine and position them out of the way.
15. Remove the alternator.
16. Either disconnect the engine wiring harness at the electronic control unit, then feed the main connector through the firewall and lay it across the engine, or tag and disconnect all engine sensor connectors from the wiring harness.
17. Remove the remaining upper transaxle-to-engine bolts.
18. Install a lifting fixture to the engine and remove the engine from the vehicle. Lift the engine slowly and make sure no wiring or hoses are snagged as the engine is removed.
19. Installation is the reverse of the removal procedure.

Intake Manifold

REMOVAL & INSTALLATION

1. Disconnect the negative battery cable.
2. Remove the mass air flow sensor and air intake duct.
3. Remove the serpentine drive belt, alternator and bracket.
4. Remove the C3I ignition module and wiring.

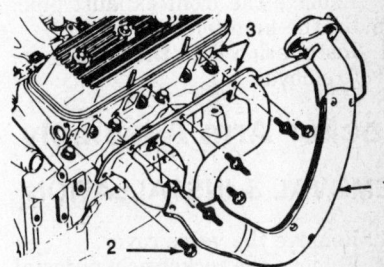

Intake manifold bolt torque sequence

5. Tag and disconnect all vacuum hoses and wiring connectors as necessary.
6. Disconnect the throttle, cruise control and throttle valve cables from the throttle body.
7. Drain the cooling system.
8. Disconnect the heater hoses from the throttle body.
9. Disconnect the upper radiator hose from the intake manifold.
10. Depressurize the fuel system and remove the fuel lines, fuel rail and injectors as an assembly.
11. Remove the intake manifold bolts in reverse of the torque sequence and lift off the intake manifold.
12. Installation is the reverse of removal. Clean all gasket mating surfaces and apply sealer if a steel gasket is used. Torque the intake manifold bolts in the sequence shown to 32 ft. lbs.

Exhaust Manifold

REMOVAL & INSTALLATION

Left Side

1. Disconnect the negative battery cable.
2. Remove the mass air flow sensor, air intake duct and crankcase ventilation pipe.
3. Remove the two bolts attaching the exhaust crossover pipe to the manifold.

1. Exhaust manifold
2. 20 ft. lbs.
3. Apply sealant between manifold and cylinder head

Right exhaust manifold mounting

4. Tag and disconnect the spark plug wires.

5. Remove the mounting bolts and remove the manifold.

NOTE: The oil dipstick tube may have to be removed to provide access to the manifold bolts.

6. Installation is in the reverse order of removal. Apply sealer as illustrated.

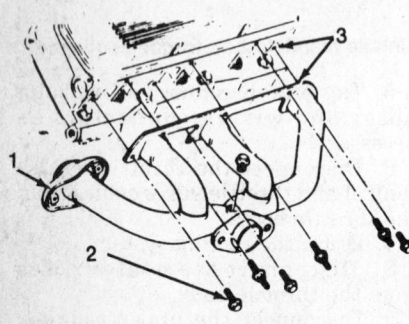

1. Exhaust manifold
2. 20 ft. lbs.
3. Apply sealant between manifold and cylinder head

Left exhaust manifold mounting

Right Side

1. Disconnect the negative battery cable.

2. Repeat Step 2 of the "Left Side" procedure.

3. Disconnect the IAC connector at the throttle body (3.8L only).

4. Tag and disconnect the spark plug wires and the oxygen sensor lead.

5. Disconnect the heater inlet pipe from the manifold studs.

6. Remove the exhaust crossover pipe.

7. Remove the front alternator support bracket.

8. Remove the exhaust manifold mounting bolts. Raise and support the front of the vehicle.

9. Disconnect the exhaust pipe from the manifold.

10. Remove the front exhaust pipe. Remove the manifold.

11. Installation is in the reverse order of removal.

Rocker Arm Assembly

REMOVAL & INSTALLATION

1. Remove the valve cover.

2. Remove the rocker arm pedestal retaining bolts. Note the positions of the double-ended bolts for reassembly.

3. Remove the pedestal and rocker arm assembly. Place the assemblies in order on a clean workbench so they can be reassembled in their original locations.

4. Installation is the reverse of removal. Torque the rocker arm pedestal bolts to 45 ft. lbs. and the valve cover bolts to 7 ft. lbs.

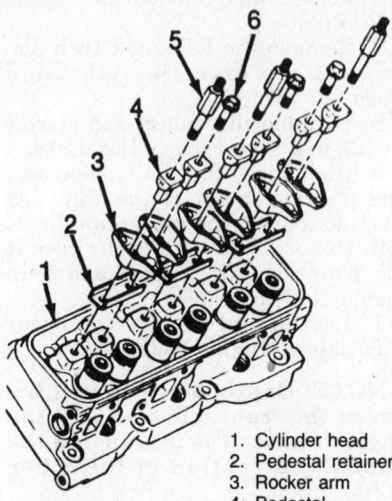

1. Cylinder head
2. Pedestal retainer
3. Rocker arm
4. Pedestal
5. 45 ft. lbs.
6. 45 ft. lbs.

Exploded view of rocker arm assembly

Cylinder Head

REMOVAL & INSTALLATION

1. Drain the cooling system and disconnect the negative battery cable.

2. Remove the intake manifold.

3. Disconnect the exhaust crossover pipe.

4. Remove the exhaust manifold.

5. Remove the valve covers, rocker arms and pushrods. Keep all parts in order so they may be reassembled in their original locations.

6. Loosen the cylinder head bolts in reverse of the torque sequence, then remove the bolts and lift off the cylinder head.

7. Clean all gasket mating surfaces and the cylinder head bolt holes in the block.

8. Installation is the reverse of removal. Torque the cylinder head bolts in the sequence shown to 25 ft. lbs.

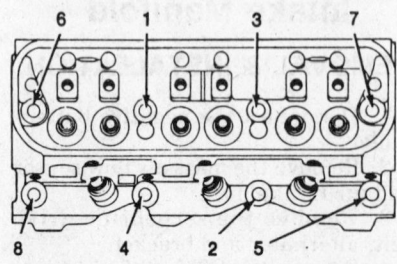

Cylinder head bolt torque sequence

9. Tighten each cylinder head bolt ¼ turn (90°) in sequence.

10. Tighten each cylinder head bolt an additional ¼ turn in sequence.

CAUTION

Should you reach 60 ft. lbs. of torque at any time in Steps 9 and 10, stop tightening the bolt at this point. Do not complete the balance of the 90° turn.

OVERHAUL

For all cylinder head overhaul procedures, please refer to "Engine Rebuilding" in the unit repair section.

Front Cover and Seal

REMOVAL & INSTALLATION

1. Drain the cooling system and disconnect the upper and lower radiator hoses.

2. Disconnect the heater return hose.

3. Remove the two nuts from the front engine mount a the cradle and raise the engine slightly with a suitable lifting device.

4. Remove the serpentine drive belt and the water pump pulley.

5. Remove the alternator and mounting bracket.

6. Remove the front clamp on the coolant bypass hose.

7. Remove the right front tire and the inner fender splash shield, then remove the crankshaft balancer and pulley assembly.

8. Remove the timing chain cover mounting bolts at the block and oil pan.

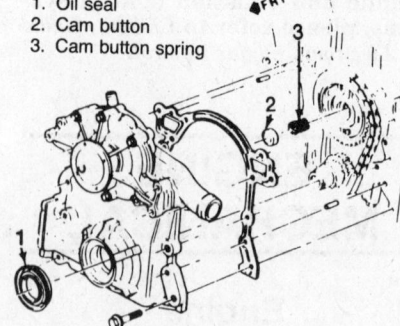

1. Oil seal
2. Cam button
3. Cam button spring

Timing chain cover

9. Remove the timing chain cover. Clean all gasket mating surfaces and pry out the old oil seal with a suitable tool. Install a new oil seal using tool J-35354 or equivalent.

10. Installation is the reverse of removal.

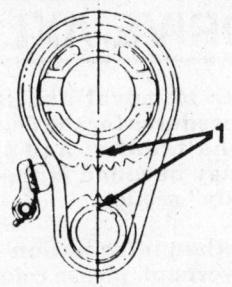

ASSEMBLED VIEW

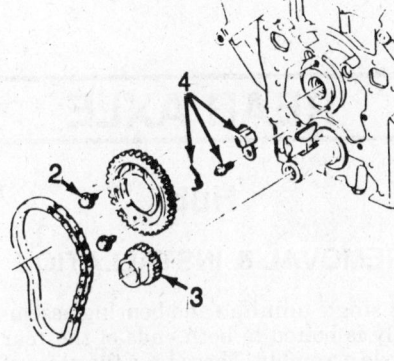

1. Timing mark alignment
2. 19 ft. lbs.
3. Crankshaft sprocket
4. Dampener assembly

Timing chain and sprockets

Timing Chain and Sprockets

REMOVAL & INSTALLATION

1. Rotate the engine until the No. 1 cylinder is at TDC/compression.
2. Remove the front cover as described above.
3. Remove the timing chain dampener.
4. Remove the camshaft sprocket bolts, then remove the camshaft sprocket, timing chain and crankshaft sprocket.
5. Installation is the reverse of removal. Align the timing marks as illustrated and install the sprockets and timing chain together. Torque the camshaft sprocket bolts to 19 ft. lbs.

Camshaft

REMOVAL & INSTALLATION

Camshaft replacement requires removal and disassembly of the engine. Remove the engine and refer to "Engine Rebuilding" in the unit repair section.

Piston and Connecting Rod Positioning

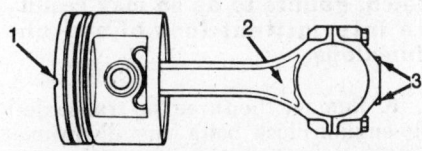

1 **NOTCH ON PISTON TOWARDS FRONT OF ENGINE**
2 **LEFT BANK**
NO. 1, 3 & 5 TWO BOSSES ON ROD TOWARDS REAR OF ENGINE (NOT SHOWN)
3 **CHAMFERED CORNERS ON ROD CAP TOWARDS FRONT OF ENGINE**
4 **RIGHT BANK**
NO. 2, 4 & 6 TWO BOSSES ON ROD TOWARDS FRONT OF ENGINE (NOT SHOWN)
CHAMFERED CORNERS ON ROD CAP TOWARDS REAR OF ENGINE

Piston and connecting rod positioning

ENGINE LUBRICATION

Oil Pan

REMOVAL & INSTALLATION

1. Disconnect the negative battery cable.
2. Raise the car and support it safely.
3. Drain the engine oil into a suitable container and discard.
4. Remove the transmission converter cover.
5. Remove the oil pan mounting bolts and lower the oil pan. Clean all gasket mating surfaces.
6. Installation is the reverse of removal. Use a new oil pan gasket and tighten the retaining bolts to 7 ft. lbs. Do not overtighten or leakage may occur.

Rear Main Seal

REMOVAL & INSTALLATION

1. Raise the vehicle and support it safely.
2. Drain the engine oil and remove the oil pan.
3. Remove the rear main bearing cap.
4. Remove the old seal from the bearing cap.
5. Insert packing tool J-21526-2 or equivalent against one end of the seal in the cylinder block. Pack the old seal into the groove until it is packed

tight, then repeat the procedure on the other end of the seal.
6. Measure the amount the seal was driven up, then add approximately $\frac{1}{16}$ in. Cut this length from the old seal removed from the lower bearing cap, then repeat for the other side. Use the lower bearing cap as a holding fixture when cutting the short lengths with a razor blade.
7. Install seal packer guide J-21526-1 or equivalent onto the cylinder block.
8. Using the packing tool, work the short pieces into the guide tool and pack into the cylinder block until the tool hits the built-in stop.

NOTE: It may help to use oil on the short seal pieces when packing into the block.

9. Repeat Steps 7 and 8 for the other side.
10. Remove the guide tool.
11. Install a new rope seal into the lower bearing cap.
12. Install the lower main bearing cap and torque the bolts to 100 ft. lbs. (135 Nm).
13. Install the remaining components in the reverse order of removal. Fill the crankcase with engine oil, start the engine and check for leaks.

Oil Pump

REMOVAL & INSTALLATION

1. Disconnect the negative battery cable.
2. Remove the front cover from the engine as outlined under "Front Cover and Oil Seal."
3. Remove the oil filter adapter, pressure regulator valve and valve spring.
4. Remove the oil pump cover retaining screws and cover.
5. Remove the oil pump gears.
6. Installation is the reverse of removal.

1. 7 ft. lbs.
2. Oil pump cover
3. Pump outer gear
4. Pump inner gear
5. Front cover

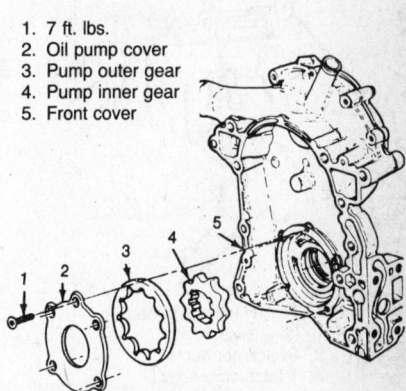

Oil pump and housing

AUTOMATIC TRANSMISSION

For all automatic transmission band and linkage adjustments, and any other service procedures, please refer to "Automatic Transmissions" in the Unit Repair Section.

CAUTION

Any inaccuracies in shift linkage adjustments may result in premature failure of the transmission due to operation without the controls in full detent. Such operation results in reduced fluid pressure and in turn, partial engagement of the affected clutches. Partial engagement of the clutches, with sufficient pressure to permit apparently normal vehicle operation will result in failure of the clutches and/or other internal parts after only a few miles of operation.

REMOVAL & INSTALLATION

1. Disconnect the negative battery cable. Disconnect the wire connector at the mass air flow sensor.
2. Remove the air intake duct and the mass air flow sensor as an assembly.
3. Disconnect the cruise control assembly. Disconnect the shift control linkage.
4. Tag and disconnect the following:
   a. Park/Neutral switch
   b. Torque converter clutch
   c. Vehicle speed sensor
   d. Vacuum modulator hose at the modulator.

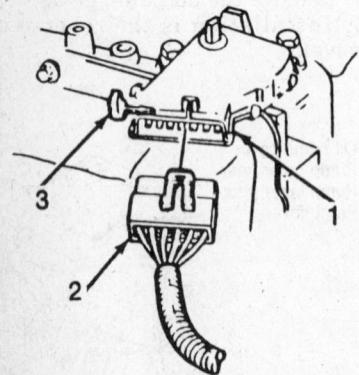

1. Park/Neutral and backup lamp switch
2. Switch connector
3. T-latch connector

T-latch connector assembly

NOTE: Care must be exercised on reassembly of the Park/Neutral switch to ensure a proper fit of both the connector and the T-latch. Failure to do so may result in intermittent loss of switch functions.

5. Remove the three top transaxle-to-engine block bolts. Install an engine support fixture.
6. Remove both front wheels and then turn the steering wheel to the full left position.
7. Remove the right front ball joint nut and separate the control arm from the steering knuckle.
8. Remove the right drive axle as detailed later in this section.

NOTE: Be careful not to allow the drive axle splines to contact any portion of the lip seal.

9. Remove the left drive axle using a suitable pry bar. Be careful not to damage the pan. Install drive axle boot seal protectors.
10. Remove three bolts at the transaxle and three nuts at the cradle member. Remove the left front transaxle mount.
11. Remove the right front mount-to-cra-dle nuts. Remove the left rear transaxle mount-to-transaxle bolts.
12. Remove the right rear transaxle mount as in Step 10. Remove the engine support bracket-to-transaxle case bolts.
13. Remove the flywheel cover. Remove the flywheel-to-converter bolts.

NOTE: Be sure to matchmark the flywheel-to-converter relationship for proper alignment upon reassembly.

14. Remove the bolts attaching the rear cradle member to the front cradle dog leg.
15. Remove the front left cradle-to-body bolt. Remove the front cradle dog leg-to-right cradle member bolts.
16. Install a transaxle support fixture into position.
17. Remove the cradle assembly by swinging it aside and supporting it with a suitable stand.
18. Disconnect and cap the oil cooler lines at the transaxle.

NOTE: One bolt is located between the transaxle and the engine block and is installed in the opposite direction.

19. Remove the remaining lower transaxle-to-engine bolts. And then lower the transaxle assembly away from the car.
20. Installation is in the reverse order or removal. Check the fluid level and all adjustments.

DRIVE AXLE

NOTE: Removal and installation procedures for the front drive axles (halfshafts) and the CV-joints may be found in the "GM A & X Body" section.

For further information on CV-Joint overhaul, please refer to "U-Joints and CV-Joints" in the Unit Repair section.

REAR AXLE

Hub

REMOVAL & INSTALLATION

A single unit hub and bearing assembly is bolted to both ends of the rear axle assembly. These take the place of rear axles used on rear wheel drive cars. The hub and bearing assembly is a sealed unit which requires no maintenance. The unit must be replaced as an assembly and cannot be disassembled or adjusted.

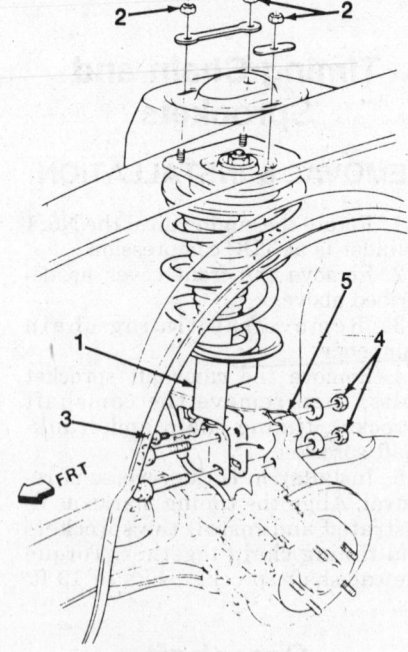

1. Strut assembly
2. 18 ft. lbs.
3. 13 ft. lbs.
4. 144 ft. lbs.
5. Retain knuckle with wire once strut assembly is removed

Strut assembly mounting

The hub and bearing can be removed by removing the rear brake drum, removing the four hub and bearing-to-axle assembly attaching bolts and pulling the unit out. Installation is the reverse of removal. Tighten the bolts to 35–39 ft. lbs.

FRONT SUSPENSION

MacPherson Strut

REMOVAL & INSTALLATION

For spring and shock absorber removal and installation, and any other strut overhaul procedures, please refer to "Strut Overhaul" in the Unit Repair section.

1. Remove the three nuts attaching the top of the strut assembly to the body.
2. Raise the car and support it with jack stands under the engine cradle.
3. Lower the car slightly so that the weight rests on the jack stands.
4. Remove the wheels and tires.

NOTE: Always install drive axle boot seal protectors. Care must be taken to prevent overextension of the inner Tri-Pot joints.

5. Remove the brake line bracket bolt from the strut assembly. Do not disconnect the brake line from the caliper.
6. Remove the strut-to-steering knuckle bolts and then carefully remove the strut assembly.
7. Installation is in the reverse order of removal. Please note the following:
   a. Check wheel alignment
   b. Tighten the strut-to-body bolts to 18 ft. lbs. (24 Nm)
   c. Tighten the strut-to-steering knuckle bolts to 144 ft. lbs. (195 Nm).

Ball Joints

INSPECTION

1. Raise the front of the car with a lift placed under the engine cradle. The front wheel should be clear of the ground.
2. Grasp the wheel at the top and bottom and shake the wheel in and out.
3. If any movement is seen of the steering knuckle relative to the con-

trol arm, the ball joints are defective and must be replaced. Note that movement elsewhere may be due to loose wheel bearings or other troubles; watch the knuckle-to-control arm connection.
4. If the ball stud is disconnected from the steering knuckle and any looseness is noted, often the ball joint stud can be twisted in its socket with your fingers, replace the ball joints.

REMOVAL & INSTALLATION

1. Raise the front of the car and support it with jackstands underneath the engine cradle. Lower the car slightly so that the weight rests primarily on the jack stands.
2. Remove the wheel and tire assemblies.
3. Install drive axle covers to protect the drive axle boot seals.
4. Pull the cotter pin from the ball joint and install a ball joint separator tool. Turn the castellated nut counterclockwise to separate the ball joint from the steering knuckle.
5. Use a ⅛ in. drill bit to drill a hole approximately ¼ in. deep in the center of each of the three ball joint rivets.

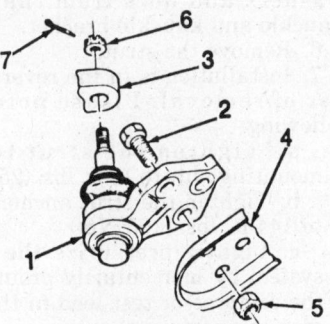

1. Ball joint assembly
2. Mounting bolt
3. Steering knuckle
4. Control arm
5. 50 ft. lbs.
6. 81 ft. lbs.
7. Cotter pin

Ball joint replacement

6. Use a ½ in. drill bit to drill off the rivet heads. Drill only enough to remove the rivet head.
7. Use a hammer and punch to remove the rivets. Drive them out from the bottom.
8. Loosen the stabilizer bar bushing assembly nut.
9. Pull down on the control arm and remove the ball joint from the steering knuckle and control arm.
10. Install the new ball joint in the steering knuckle and line up the holes with those in the control arm.
11. Install the three ball joint nuts facing down and tighten the nuts to 50 ft. lbs. (68 Nm).

12. Install the castellated nut and tighten to 81 ft. lbs. (110 Nm).

NOTE: Tightening the nut for cotter pin alignment is allowed, but do not loosen it once the torque value has been reached.

13. Install the cotter pin.
14. Installation of the remaining components is in the reverse order of removal.

Lower Control Arm

REMOVAL & INSTALLATION

1. Perform Steps 1–3 of the "Ball Joint Removal and Installation" procedure.
2. Remove the stabilizer bar bushing-to-control arm bolt.
3. Pull the cotter pin from the ball joint and install a ball joint separator tool. Turn the castellated nut counterclockwise to separate the ball joint from the steering knuckle.
4. Remove the remaining control arm bolts and remove the control arm from the vehicle.
5. Position the control arm and install the mounting bolts, but DO NOT tighten.
6. Install the stabilizer bar bushing assembly. Reconnect the ball joint to the steering knuckle.
7. Hoist the vehicle slightly so the weight of the vehicle is supported by the control arms.

NOTE: The weight of the vehicle MUST be supported by the control arms when tightening the mounting nuts.

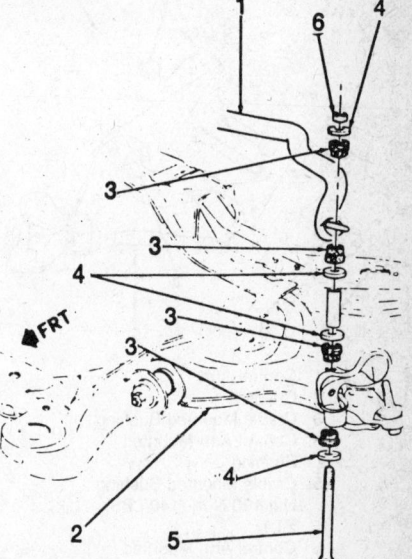

1. Stabilizer Bar
2. Control Arm
3. Insulator (4)
4. Retainer (4)
5. Bolt
6. Nut 17 N·m (13 LBS. FT.)

Stabilizer bar bushing assembly

8. Tighten the:
   a. Stabilizer bar bushing nut to 13 ft. lbs. (17 Nm)

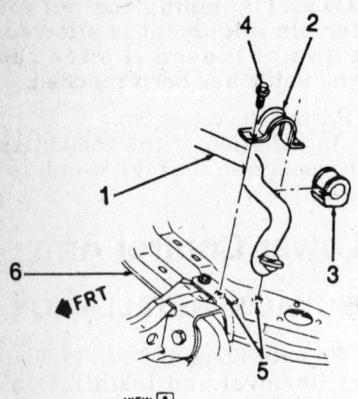

VIEW Ⓐ

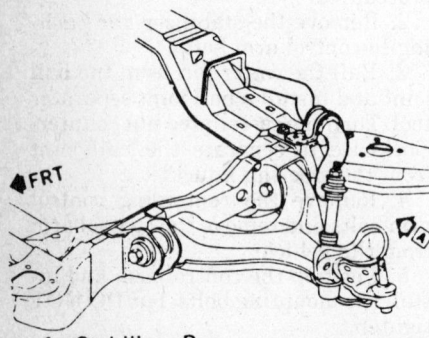

1. Stabilizer Bar
2. Stabilizer Bar Mounting Bracket
3. Stabilizer Bar Mounting Bushing
4. 50 N·m (37 LBS. FT.)
5. Frame Welded Nuts
6. Cradle

Stabilizer bar installation

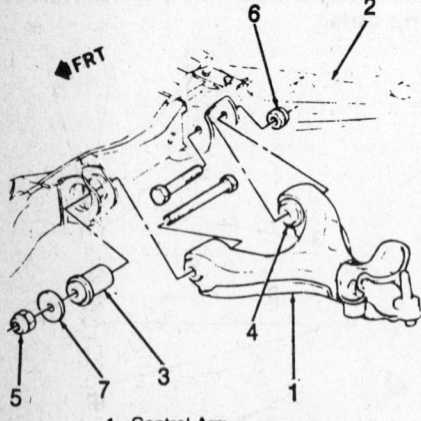

1. Control Arm
2. Cradle
3. Cradle Mounted Bushing
4. Control Arm Mounted Bushing
5. Cradle Mounted Bushing Nut 190 N·m (140 LBS. FT.)
6. Control Arm Mounted Bushing Nut 123 N·m (90 LBS. FT.)
7. Washer

Lower control arm assembly

b. Rear control arm mounting nut to 90 ft. lbs. (123 Nm)
   c. Front control arm mounting nut to 140 ft. lbs. (190 Nm)
   d. Ball joint nut to 81 ft. lbs. (110 Nm).

9. Installation of the remaining components is in the reverse order of removal.

REAR SUSPENSION

Superlift Strut

REMOVAL & INSTALLATION

1. Remove the inner trunk side cover.
2. Raise and support the rear of the vehicle. Remove the wheels and tires.
3. Disconnect and plug the ELC air line.
4. Remove the strut tower mounting nuts from inside the trunk.
5. Remove the strut anchor bolts, washers and nuts from the rear knuckle and knuckle bracket.
6. Remove the strut.
7. Installation is in the reverse order of removal. Please note the following:
   a. Tighten the strut tower mounting nuts to 19 ft. lbs. (25 Nm)
   b. Tighten the strut anchor nuts to 144 ft. lbs. (195 Nm)
   c. Lightly pressurize the ELC system by momentarily grounding the compressor test lead in the engine compartment.
   d. Check rear wheel alignment.

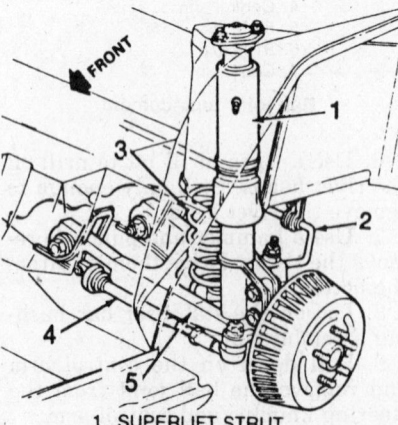

1. SUPERLIFT STRUT
2. STABILIZER BAR
3. COIL SPRING
4. SUSPENSION ADJUSTMENT LINK
5. LOWER CONTROL ARM

Rear suspension

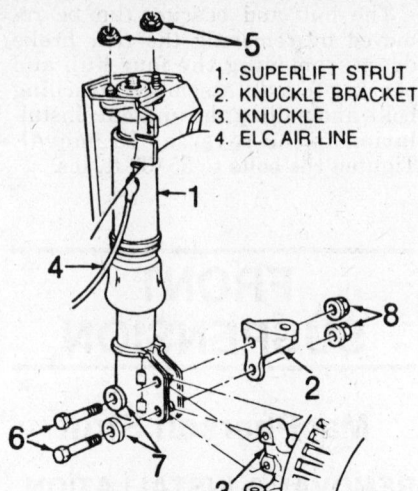

1. SUPERLIFT STRUT
2. KNUCKLE BRACKET
3. KNUCKLE
4. ELC AIR LINE

5. STRUT MOUNTING NUTS (25 N·m/19 FT. LBS.)
6. STRUT ANCHOR BOLTS
7. STRUT ANCHOR WASHERS
8. STRUT ANCHOR NUTS (195 N·m/144 FT. LBS.)

Rear strut installation

Coil Springs

REMOVAL & INSTALLATION

1. Raise the rear of the vehicle and support it so that the control arms hang free. Remove the rear wheels.
2. Separate the rear stabilizer bar from the knuckle bracket and remove it.
3. Disconnect the ELC height sensor link (right control arm) and/or the parking brake cable retaining clip (left control arm).
4. Position the special tool J-23028-01 or its equivalent, so as to cradle the control arm bushings.

NOTE: Special tool J-23028-01 should be secured to a suitable jack.

5. Raise the jack to remove the tension from the control arm pivot bolts.

— CAUTION —
Secure a chain around the spring and through the control arm as a safety precaution.

6. Remove the rear control arm pivot bolt and nut.
7. Slowly maneuver the jack so as to relieve any tension in the front control arm pivot bolt and then remove the bolt and nut.
8. Lower the jack to allow the control arm to pivot downward.
9. When all pressure is removed from the coil spring, remove the safety chain, spring and insulators.

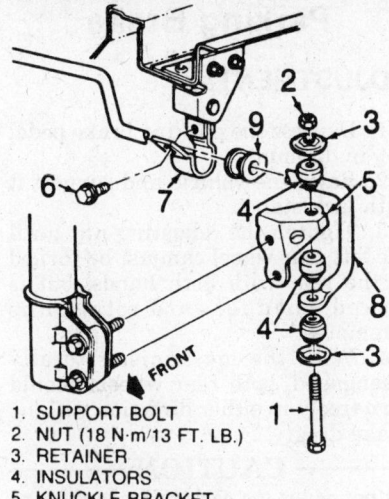

1. SUPPORT BOLT
2. NUT (18 N·m/13 FT. LB.)
3. RETAINER
4. INSULATORS
5. KNUCKLE BRACKET
6. BUSHING CLIP BOLT (50 N·m/37 FT. LB.)
7. SUPPORT ASSEMBLY
8. STABILIZER BAR
9. BUSHING

Rear stabilizer bar bushing assembly

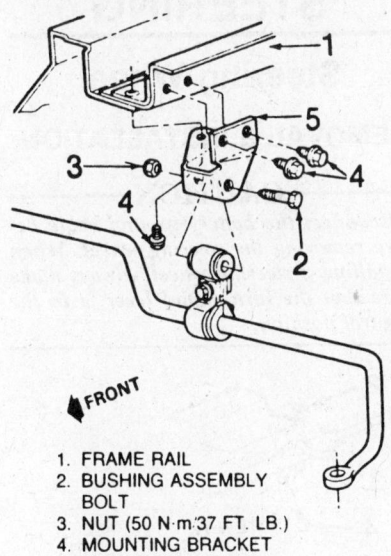

1. FRAME RAIL
2. BUSHING ASSEMBLY BOLT
3. NUT (50 N·m/37 FT. LB.)
4. MOUNTING BRACKET BOLTS (18N·m/13 FT. LB.)
5. MOUNTING BRACKET

Rear stabilizer bar mounting bracket

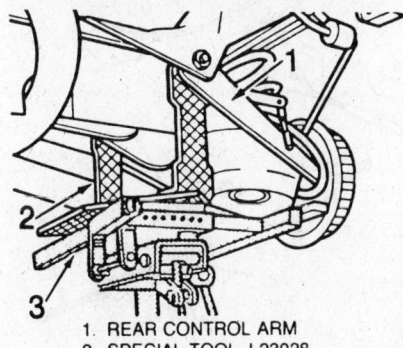

1. REAR CONTROL ARM
2. SPECIAL TOOL J-23028-01
3. TRANSMISSION JACK

Use the special tool and a transmission jack to cradle the control arm

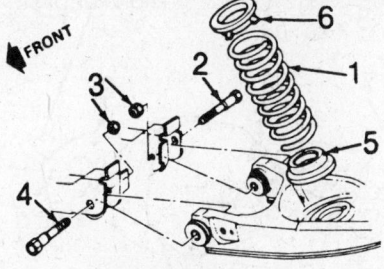

1. COIL SPRING
2. CONTROL ARM PIVOT BOLT-REAR (170 N·m 125 FT. LB.)
3. CONTROL ARM PIVOT NUTS (115 N·m 85 FT. LB.)
4. CONTROL ARM PIVOT BOLT-FRONT (170 N·m/ 125 FT. LB.)
5. LOWER COIL SPRING INSULATOR
6. UPPER COIL SPRING INSULATOR

Rear coil spring installation

NOTE: The spring insulators should be inspected for cuts or tears. They should be replaced if the vehicle has over 50,000 miles.

10. Snap the upper insulator onto the spring. Position the lower insulator and the spring in the control arm. Install the coil springs so that the upper ends are positioned as shown in the illustration.

11. Installation of the remaining components is in the reverse order of removal. Control arm mounting nuts should not be tightened until the vehicle is unsupported and resting on its wheels at normal trim height.

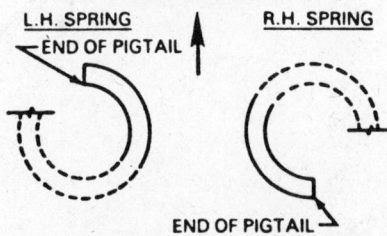

TOP VIEW OF UPPER END OF SPRING

FRONT OF CAR

L.H. SPRING   R.H. SPRING
←END OF PIGTAIL

END OF PIGTAIL→

Rear coil spring positioning

Ball Joint

REMOVAL & INSTALLATION

1. Raise and support the rear of the vehicle and remove the wheels.
2. Disconnect the ELC height sensor link (right control arm) and/or the parking brake cable retaining link (left control arm).

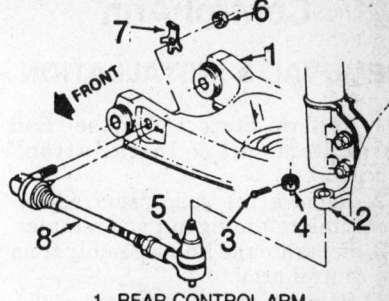

1. REAR CONTROL ARM
2. KNUCKLE
3. COTTER PIN
4. CASTELLATED NUT (50 N·m/37 FT. LB.)
5. OUTER SUSPENSION ADJUSTMENT LINK
6. LINK RETAINING NUT (85 N·m/63 FT. LB.)
7. LINK RETAINER
8. SUSPENSION ADJUSTMENT LINK ASSEMBLY

Rear suspension adjustment link

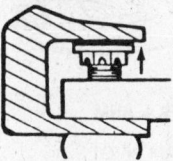

REMOVE CASTELLATED NUT AND REINSTALL WITH FLAT SIDE FACING UPWARD. PLACE J-34505 INTO POSITION AS SHOWN. LOOSEN NUT AND BACK OFF UNTIL...

...THE NUT CONTACTS THE TOOL. CONTINUE BACKING OFF THE NUT UNTIL THE NUT FORCES THE BALL STUD OUT OF THE KNUCKLE.

Separating the ball joint from the steering knuckle

3. Remove the cotter pin and castellated nut from the outer suspension adjustment link.
4. Separate the outer suspension link from the knuckle.
5. Support the control arm with a suitable jack. The lower control arm MUST be supported to prevent the coil spring from forcing the control arm downward.
6. Remove the ball stud cotter pin.
7. Remove the castellated nut and then reinstall it with the flat side facing upward. DO NOT tighten the nut.
8. Install a ball joint separator tool and separate the knuckle from the ball stud by backing off the inverted nut against the tool.
9. Separate the ball joint from the control arm.
10. Installation is in the reverse order of removal. Please note following:
   a. Tighten a NEW castellated nut to 7.5 ft. lbs. (10 Nm). Tighten the nut an additional $\frac{2}{3}$ of a turn.
   b. Align the slot in the nut to the cotter pin hole by tightening only. Do not loosen the nut to align the holes.

Control Arm

REMOVAL & INSTALLATION

1. Perform Steps 1–2 of the "Ball Joint Removal & Installation" procedures.
2. Remove the suspension adjustment link retaining nut and retainer.
3. Separate the link assembly from the control arm.
4. Remove the coil spring as detailed previously.
5. Perform Steps 6–9 of the "Ball Joint Removal & Installation" procedure.
6. Remove the control arm.
7. Installation is in the reverse order of removal.

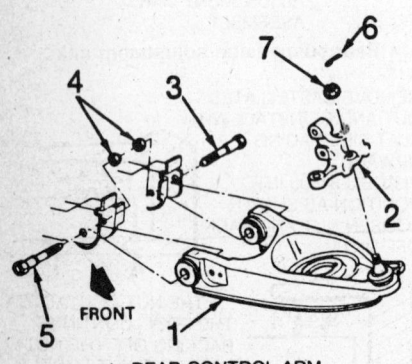

1. REAR CONTROL ARM
2. KNUCKLE
3. CONTROL ARM PIVOT BOLT—REAR
4. CONTROL ARM PIVOT NUTS (115 N·m/85 FT-LB)
5. CONTROL ARM PIVOT BOLT—FRONT
6. COTTER PIN
7. CASTELLATED NUT

Rear control arm

BRAKES

For all brake system removal, installation and adjustment procedures, please refer to "Brakes" in the Unit Repair section.

Master Cylinder

REMOVAL & INSTALLATION

1. Disconnect and plug hydraulic lines, and drain the cylinder.
2. Remove the attaching nuts and remove the master cylinder from the power booster unit.
3. Reverse to install. Bleed the system.

FRONT OF DASH

1. NUT (20 N·m/14 FT. LB.)
2. SEAL
3. POWER BOOSTER
4. CHECK VALVE
5. MASTER CYLINDER
6. VACUUM SWITCH BRACKET (DIESEL)
7. NUT (30 N·m/22 FT. LB.)
8. VACUUM SWITCH (GAS)

Typical master cylinder and power booster mounting

Power Booster

REMOVAL & INSTALLATION

1. From inside the car, detach the brake pushrod from the brake pedal.
2. Disconnect the hydraulic lines from the front of the master cylinder.
3. Remove the nuts from the mounting studs which hold the unit to the dash panel. Remove the unit and clean it prior to installation.
4. Installation is the reverse of removal. Bleed the brake system.

Parking Brake

ADJUSTMENT

1. Depress the parking brake pedal 1½ in. (35mm).
2. Raise the vehicle and support it with jack stands.
3. Tighten the adjusting nut until the left rear wheel can just be turned to the rear with both hands, but is locked when forward rotation is attempted.
4. With the mechanisms totally disengaged, both rear wheels should turn freely in either direction with no brake drag.

— **CAUTION** —
Do not adjust the parking brake cable so tight as to cause brake drag.

5. Lower the vehicle.

STEERING

Steering Wheel

REMOVAL & INSTALLATION

— **CAUTION** —
Disconnect the battery ground cable before removing the steering wheel. When installing a steering wheel, always make sure that the turn signal lever is in the neutral position.

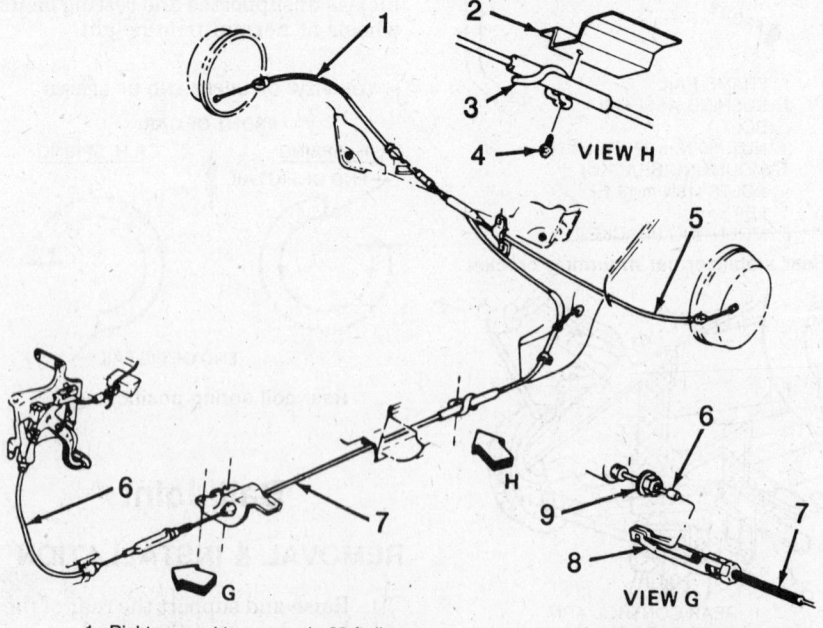

1. Right rear cable
2. Underbody
3. Guide
4. 28 ft. lbs.
5. Left rear cable
6. Front cable assembly
7. Intermediate cable
8. Equalizer assembly
9. Nut

Parking brake cable assembly

1. Remove the trim retaining screws from behind the wheel. On wheels with a center cap, pull off the cap.
2. Lift the trim off and pull the horn wires from the turn signal cancelling cam.
3. Remove the retainer and the steering wheel nut.
4. Mark the wheel-to-shaft relationship, and then remove the wheel with a puller.
5. Install the wheel on the shaft aligning the previously made marks. Tighten the nut.
6. Insert the horn wires into the cancelling cam.
7. Install the center trim and reconnect the battery cable.

Turn Signal Switch

REMOVAL & INSTALLATION

1. Remove the steering wheel as previously outlined. Remove the trim cover.
2. Loosen the cover screws. Pry the cover off with a screwdriver, and lift the cover off the shaft.
3. Position the U-shaped lockplate compressing tool on the end of the steering shaft and compress the lock plate by turning the shaft nut clockwise. Pry the wire snapring out of the shaft groove.
4. Remove the tool and lift the lock plate off the shaft.
5. Slip the cancelling cam, upper bearing preload spring, and thrust washer off the shaft.

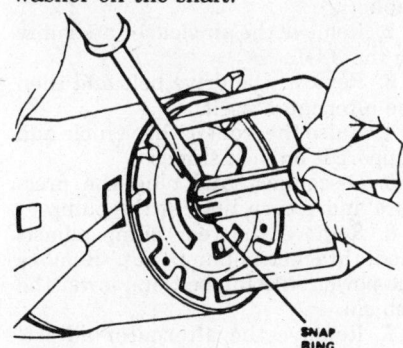

Depress the lockplate and remove the snapring

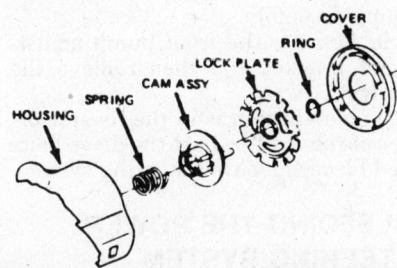

Remove these parts for access to the turn signal switch

6. Remove the turn signal lever. Push the flasher knob in and unscrew it. Remove the button retaining screw and remove the button, spring and knob.
7. Pull the switch connector out the mast jacket and tape the upper part to facilitate switch removal. Attach a long piece of wire to the turn signal switch connector. When installing the turn signal switch, feed this wire through the column first, and then use this wire to pull the switch connector into position. On tilt wheels, place the turn signal and shifter housing in low position and remove the harness cover.
8. Remove the three switch mounting screws. Remove the switch pulling it straight up while guiding the wiring harness cover through the column.
9. Install the replacement switch by working the connector and cover down through the housing and under the bracket. On tilt models, the connector is worked down through the housing, under the bracket, and then the cover is installed on the harness.
10. Install the switch mounting screws and the connector on the mast jacket bracket. Install the column-to-dash trim plate.
11. Install the flasher knob and the turn signal lever.
12. With the turn signal lever in neutral and the flasher knob out, slide the thrust washer, upper bearing preload spring, and cancelling cam onto the shaft.
13. Position the lock plate on the shaft and press it down until a new snapring can be inserted in the shaft groove. Always use a new snapring when assembling.
14. Install the cover and the steering wheel.

Ignition Switch

REMOVAL & INSTALLATION

The switch is located inside the channel section of the brake pedal support and is completely inaccessible without first lowering the steering column. The switch is actuated by a rod and rack assembly. A gear on the end of the lock cylinder engages the toothed upper end of the rod.
1. Lower the steering column; be sure to properly support it.
2. Put the switch in the OFF-UN-LOCKED position. With the cylinder removed, the rod is in LOCK when it is in the next to the uppermost detent. OFF-UNLOCKED is two detents from the top.
3. Remove the two switch screws and remove the switch assembly.

4. Before installing, place the new switch in OFF-UNLOCKED position and make sure the lock cylinder and actuating rod are in OFF-UN-LOCKED (third detent from the top) position.
5. Install the activating rod into the switch and assemble the switch on the column. Tighten the mounting screws. Use only the specified screws since overlength screws could impair the collapsibility of the column.
6. Reinstall the steering column.

Ignition Lock Cylinder

REMOVAL & INSTALLATION

1. Place the lock in the RUN position.
2. Remove the lock plate, turn signal switch and buzzer switch.
3. Remove the screw and lock cylinder.

— CAUTION —
If the screw is dropped on removal, it could fall into the column, requiring complete disassembly to retrieve the screw.

4. Rotate the cylinder clockwise to align cylinder key with the keyway in the housing.
5. Push the lock all the way in.
6. Install the screw. Tighten the screw to 14 inch lbs. for adjustable columns and 25 inch lbs. for standard columns.

Tie-Rod

REMOVAL & INSTALLATION

1. Loosen the jam nut on the steering rack (inner tie-rod).
2. Remove the tie-rod end nut. Separate the tie-rod end from the steering knuckle using a puller.
3. Unscrew the tie-rod end, counting the number of turns.
4. To install, screw the tie-rod end onto the steering rack (inner tie-rod) the same number of turns as counted for removal. This will give approximately correct toe.
5. Install the tie-rod end into the knuckle. Install the nut and tighten to 40 ft. lbs.
6. If the toe must be adjusted, use pliers to expand the boot clamp. Turn the inner tie-rod to adjust. Replace the clamp.
7. Tighten the jam nut to 50 ft. lbs.

Steering Gear

REMOVAL & INSTALLATION

1. Raise and support the front end

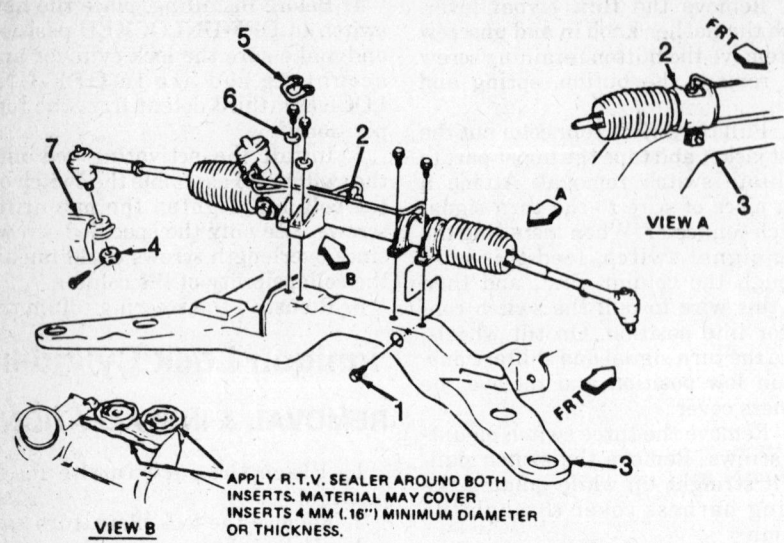

VIEW A

VIEW B

APPLY R.T.V. SEALER AROUND BOTH
INSERTS. MATERIAL MAY COVER
INSERTS 4 MM (.16") MINIMUM DIAMETER
OR THICKNESS.

1. BOLT 68 N·m (50 LBS.
   FT.) AFTER SECOND
   REUSE OF BOLT,
   LOCTITE THREAD
   LOCKING KIT, #1052624
   MUST BE USED
2. STEERING GEAR
3. FRAME
4. 50 N·m (35 LBS. FT.), 70
   N·m (52 LBS. FT.)
   MAXIMUM PERMISSIBLE
   TORQUE TO ALIGN
   COTTER PIN SLOT. (1⁄6

TURN MAXIMUM) DO
NOT BACK OFF FOR
COTTER PIN INSERTION
5. RETAINER
6. WASHER
7. STEERING KNUCKLE

APPLY R.T.V. SEALER
AROUND BOTH INSERTS.
MATERIAL MAY COVER
INSERTS 4 MM (.16")
MINIMUM DIAMETER OR
THICKNESS.

Rack and pinion assembly

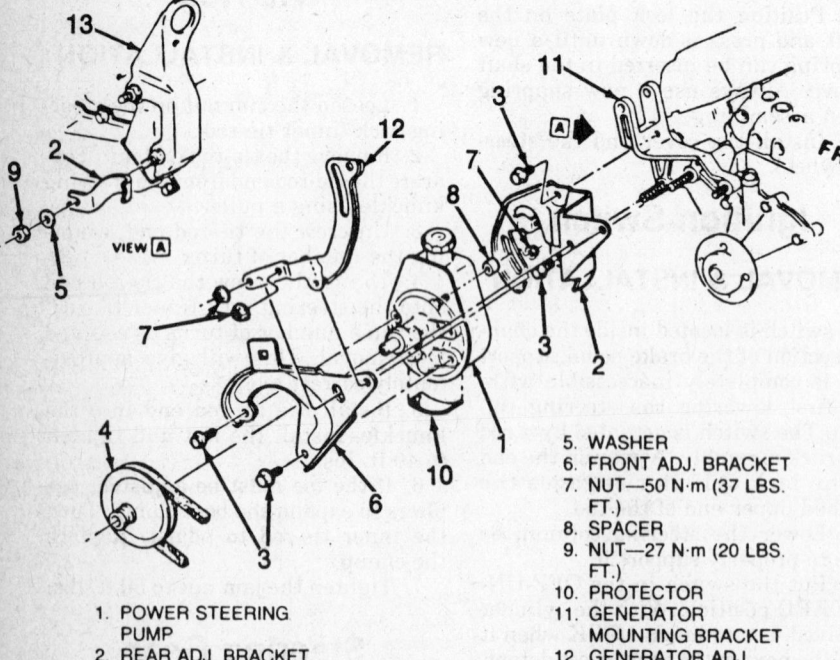

VIEW A

1. POWER STEERING
   PUMP
2. REAR ADJ. BRACKET
3. BOLT—50 N·m (37 LBS.
   FT.)
4. PULLEY

5. WASHER
6. FRONT ADJ. BRACKET
7. NUT—50 N·m (37 LBS.
   FT.)
8. SPACER
9. NUT—27 N·m (20 LBS.
   FT.)
10. PROTECTOR
11. GENERATOR
    MOUNTING BRACKET
12. GENERATOR ADJ.
    BRACKET
13. ENGINE LIFT BRACKET
    & SHIELD

Power steering pump mounting

of the car with jackstands under the frame members. Allow the front suspension to hang freely. Disconnect the power steering hoses from the gear, where equipped.

2. Move the intermediate shaft seal upward and remove the intermediate shaft-to-stub shaft pinch bolt.

3. Remove both front wheels.

4. Remove the cotter pins and nut from both tie-rod ends. Disconnect the tie-rod ends from the steering knuckles.

5. Remove the line retainer.

6. Remove the outlet and pressure hose.

7. Remove the five rack and pinion assembly mounting bolts.

8. Loosen the front engine cradle mounting bolts. Install jack stands and the lower the rear of the cradle about 3 in. (76mm).

CAUTION

Do not lower the rear of the engine cradle too far.
---

9. Remove the rack and pinion assembly.

10. Installation is in the reverse order of removal. Tighten the rack mounting bolts to 50 ft. lbs. (68 Nm). Tighten the tie-rod end nut to 35–52 ft. lbs. (50–70 Nm). Bleed the power steering system and check for leaks.

Power Steering Pump

REMOVAL & INSTALLATION

1. Disconnect the negative battery cable.

2. Remove the air cleaner assembly on the 3.0L.

3. Remove the drive belt and then the alternator itself.

4. Raise the front of the vehicle and support it on jack stands.

5. Disconnect and plug the pressure and return lines at the pump.

6. Remove the rear pump adjustment bracket-to-pump nut. Remove the power steering belt and lower the vehicle.

7. Remove the alternator adjustment bracket and support brace.

8. Remove the rear pump adjustment bracket and then remove the pump assembly.

9. Remove the front pump adjustment bracket and then remove the pulley.

10. Installation is in the reverse order of removal. Adjust the drive belts and bleed the power steering system.

BLEEDING THE POWER STEERING SYSTEM

1. Fill the fluid reservoir.

2. Let the fluid stand undisturbed for two minutes, then crank the engine for about two seconds. Refill reservoir if necessary.

3. Repeat Steps 1 and 2 above until the fluid level remains constant after cranking the engine.

4. Raise the front of the car until the wheels are off the ground, then start the engine. Increase the engine speed to about 1500 rpm.

5. Turn the wheels lightly against the stops to the left and right, checking the fluid level and refilling if necessary.

CHASSIS ELECTRICAL

Headlight Switch

REMOVAL & INSTALLATION

1. Disconnect negative battery cable. Remove the steering column lower cover or the instrument panel trim plate covering the headlamp switch, if a rocker-type headlamp switch is used.

2. Disconnect wiring harness retainer below headlight switch assembly. On some models the switch connector is integral with the instrument panel; simply pull the switch outward to disconnect it.

3. On knob-type switches, depress spring loaded release button on top of headlight switch and remove switch, knob and rod assembly (switch ON).

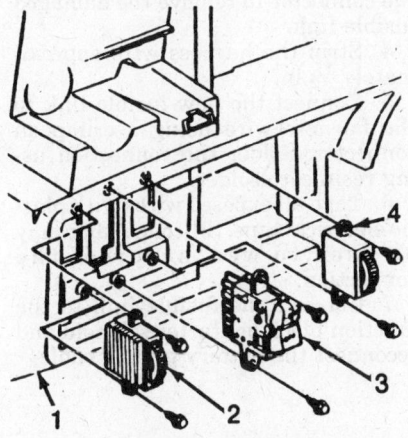

1. Instrument panel
2. Interior light dimmer
3. Headlamp switch
4. Twilight sentinal switch

Light switch assemblies—rocker type

4. Remove screw with ground wire at bottom of switch housing and any other mounting screws.

5. Pull assembly down and rearward, disconnect wiring harness connectors, bulb(s) and remove assembly.

6. Installation is the reverse of removal.

Instrument Cluster

REMOVAL & INSTALLATION

1. Disconnect the negative battery cable.

2. Remove the left sound insulator.

3. Remove the instrument panel insert and applique trim from the instrument panel.

4. Place the shift lever in the Park position and remove the shift indicator clip from the steering column.

5. Remove the nuts securing the steering column to the upper mounting bracket and lower the steering column.

6. Remove the screw securing the upper steering column mounting bracket to the cowl and lower the bracket.

7. Remove the cluster retaining screws, disconnect the speedometer cable, printed circuit connector and remove the cluster.

8. The installation is the reverse of the removal procedure. Be sure the shift indicator is properly aligned.

Speedometer Cable

REMOVAL & INSTALLATION

1. Disconnect the negative battery cable.

2. Remove the screws holding the instrument cluster trim plate, left and right telltale lens, instrument cluster lens and the transaxle shift indicator assembly. Remove the lens and retainers.

3. Remove the temperature indicator and the fuel gauge.

4. Remove the screws holding the speedometer assembly to the housing. Pull the speedometer head out and disconnect the screw holding the vehicle speed sensor.

5. Remove the speedometer cable. Remove the speedometer.

6. Installation is the reverse of the removal procedure.

Windshield Wiper Motor

REMOVAL & INSTALLATION

1. Remove the cowl screen or grille.

2. Loosen the linkage drive link-to-crankarm attaching nuts, and remove the link from the arm.

3. Disconnect the wiring and washer hoses.

4. Remove the three motor attaching screws, guide the crankarm through the hole in the dash, and remove the motor.

5. Reverse the above steps to install.

WIPER BLADE REPLACEMENT

Depending on model and availability, one of three methods is used:

1. A tab on the arm saddle is depressed.

2. A spring type blade clip is depressed.

3. A coil spring retainer is depressed with a screwdriver.

Details can be found in the Maintenance Unit Repair Section.

Heater Blower Motor

REMOVAL & INSTALLATION

1. Disconnect the negative battery cable.

2. Disconnect the electrical connections at the blower motor.

3. Disconnect the cooling hose from the blower motor.

4. Remove the mounting screws and remove the motor.

5. Reverse to install. Use a silicone sealer on the blower motor sealing surfaces.

Heater Core

REMOVAL & INSTALLATION

1. Drain the radiator. Make sure the engine is cool for this procedure.

2. Remove the heater hoses from the core and plug the hoses and the nipples to prevent spillage.

3. Remove the instrument panel.

4. Remove the four defroster nozzle attaching screws at the cowl and the screw on the case and remove the nozzle.

5. Disconnect the vacuum hoses.

6. Disconnect the electrical connector at the programmer.

7. Under the hood, remove the heater case-to-cowl attaching screws.

8. Under the instrument panel, remove the heater case-to-cowl attaching screw.

9. Remove the heater case.

10. Remove the four case-to-core screws and remove the core.

11. Installation is the reverse of removal.

Radio

REMOVAL & INSTALLATION

1. Disconnect the negative battery cable.
2. Remove the screws from the top of the instrument panel center insert.
3. Remove the radio knobs and remove the insert.
4. Remove the rear window defogger switch to gain access to the left side mounting screw if so equipped.
5. Remove the mounting screws.
6. Remove the radio and disconnect the wiring. Reverse to install.

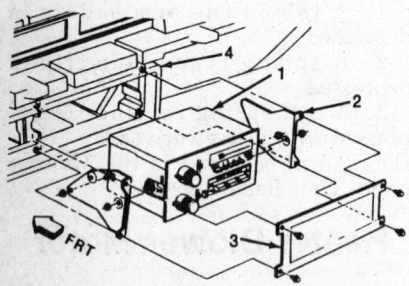

1. Radio
2. Bracket
3. Cover (if no radio)
4. Instrument panel

Typical radio mounting

Fuses

The fuse block is located behind a cover on the instrument panel to the left of the steering column. Fuse holders are labeled as to their service and the correct amperage. Always replace blown fuses with new ones of the correct amperage. Otherwise electrical overloads and possible wiring damage will result.

Electrical Component Location

FUSE PANEL

The fuse panel is located on the left side of the instrument panel assembly. In order to gain access to the fuse panel, it is necessary to first remove the cover plate.

ELECTRONIC CONTROL MODULE

The electronic control module is located on the right side of the vehicle. It is positioned up behind the glovebox. In order to gain access to the assembly you must first remove the trim panel and/or glovebox assembly.

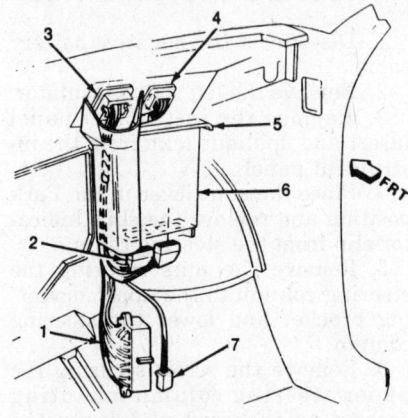

1. ECM dash harness connector
2. ECM connector
3. ECM engine harness connector
4. A/C engine harness
5. Bracket
6. ECM module
7. VSS buffer connector

Typical electronic control module (ECM) mounting

TURN SIGNAL AND HAZARD FLASHER

The turn signal and hazard flasher is located directly under the steering column of the vehicle. It is secured in place with a plastic retainer. In order to gain access to the components, it may first be necessary to remove the under dash padding.

CONVENIENCE CENTER

If equipped, the convenience center is located on the right side of the vehicle. It is positioned under the dash panel. In order to gain access to the convenience center it may be necessary to remove instrument panel sound absorber.

CIRCUIT BREAKER

A circuit breaker is an electrical switch which breaks the circuit during an electrical overload. The circuit breaker will remain open until the short or overload condition in the circuit is corrected. Circuit breakers are located in the fuse box.

FUSIBLE LINKS

Fusible links are sections of wire, with special insulation, designed to melt under electrical overload. Replacements are simply spliced into the wire. There may be as many as five of these in the engine compartment wiring harnesses. These are:
1. Horn relay to fuse panel circuit — one link.
2. Charging circuit, from the starter solenoid to the horn relay — two links.
3. Starter solenoid to ammeter circuit — one link.
4. Horn relay to rear window defroster circuit — one link.
The fusible links are all two wire gauge sizes smaller than the wires they protect. Most models have fusible links at these locations.

REPLACEMENT

1. Disconnect the battery ground cable.
2. Disconnect the fusible link from the junction block or starter solenoid.
3. Cut the harness directly behind the connector to remove the damaged fusible link.
4. Strip the harness wire approximately ½ in.
5. Connect the new fusible link to the harness wire using a crimp on connector. Solder the connection using resin core solder.
6. Tape all exposed wires with plastic electrical tape. Silicone sealer may be used to weatherproof the connection.
7. Connect the fusible link to the junction block or starter solenoid and reconnect the battery ground cable.

GM "J" Body
Cavalier, Cimarron, Firenza, 2000, Skyhawk

YEAR IDENTIFICATION

1982–83 Cavalier

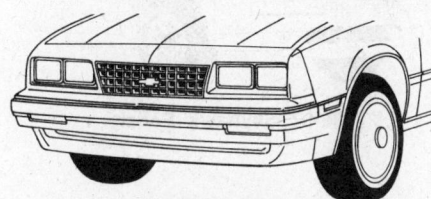

1984-87 Cavalier

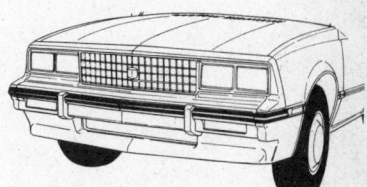

1982–83 Cimarron

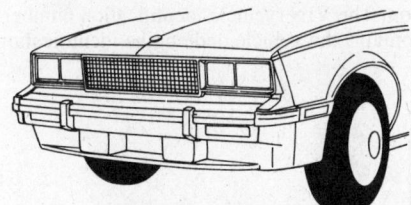

1984–85 Cimarron

1982–83 Firenza

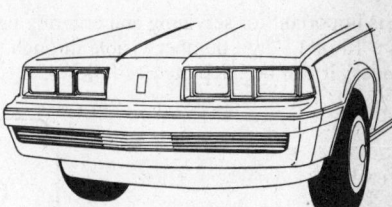

1984 Firenza

1984-87 Firenza GT

1982–83 2000

1984 2000 Sunbird

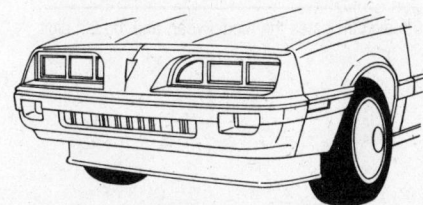

1984–85 2000 Sunbird LE, SE

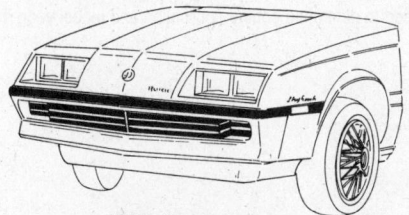

1982–83 Skyhawk

1984–85 Skyhawk

C711

YEAR IDENTIFICATION

1986-87 Skyhawk

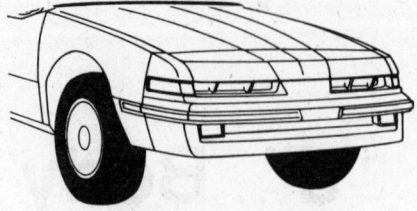

1986-87 Sunbird GT

1986-87 Cimarron

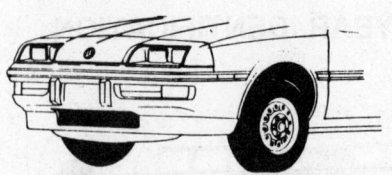

1986-87 Skyhawk Sedan

VEHICLE IDENTIFICATION NUMBER (VIN)

It is important for servicing and ordering parts to be certain of the vehicle and engine identification. The VIN (vehicle identification number) is a 13 or 17 digit number visible through the windshield on the driver's side of the dash and contains the vehicle and engine identification codes. It can be intepreted as follows:

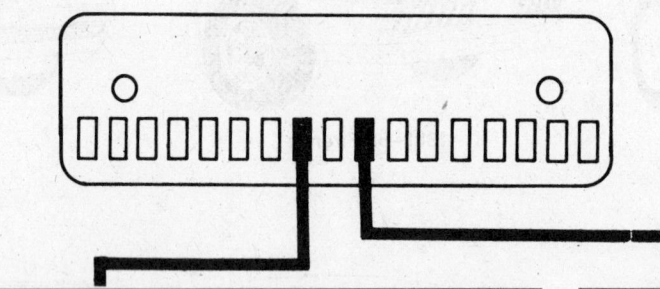

Engine Code						Model Year Code	
Code	Cu. In.	Liters	Cyl.	Carb.	Eng. Mfg.	Code	Year
G	110 (OHV)	1.8	4	2 bbl	Chev.	C	1982
O	110 (OHC)	1.8	4	TBI	Pontiac	D	1983
J	110 (OHC)	1.8	4	MFI (Turbo)	Pontiac	E	1984
B	122	2.0	4	①	Chev.	F	1985
P	122	2.0	4	TBI	Chev.	G	1986
W	173	2.8	V6	MFI	Chev.	H	1987

The seventeen digit Vehicle Identification Number can be used to determine engine application and model year. The 10th digit indicates the model year, and the 8th digit identifies the factory installed engine.
OHV–Overhead valve engine
OHC–Overhead cam engine
TBI–Throttle Body Injection
MFI–Multi-Port Fuel Injection
①–1982: 2 bbl.
   1983 and later: TBI
NOTE: Some 1983–85 Canadian models with the 2.0
Liter engine use a 2 bbl. carburetor

GENERAL ENGINE SPECIFICATIONS

Year	Engine No. Cyl. Displ. Cu. in.	Engine VIN Code	Fuel Delivery System	Engine Mfg.	Horsepower @ rpm	Torque @ rpm (ft. lb.)	Bore × Stroke	Compression Ratio	Oil Pressure 2400 rpm
'82	4-110	G	2-bbl	Chev.	88 @ 5100	100 @ 2800	3.50 × 2.91	9.0:1	45
'83–'87	4-110	O	TBI	Pont.	84 @ 5200	102 @ 2800	3.34 × 3.13	8.8:1	45
	4-110	J	MFI②	Pont.	150 @ 5600	150 @ 2800	3.34 × 3.13	8.0:1	65④
	4-122	P	TBI	Chev.	86 @ 4900	100 @ 3000	3.50 × 3.15	9.3:1	68③
	4-122	B	①	Chev.	90 @ 5100	111 @ 2800	3.50 × 3.15	9.0:1	45
	6-173	W	MFI	Chev.	120 @ 4800	155 @ 3600	3.50 × 2.99	8.9:1	50

① 1982: 2 bbl
1983 and later: TBI

② Turbocharged
③ @ 1200 rpm
④ @ 2500 rpm

TUNE-UP SPECIFICATIONS

When analyzing compression test results, look for uniformity among cylinders rather than specific pressures.

Year	Eng. VIN Code	Engine No. Cyl. Displacement (cu. in.)	Eng. Mfg.	hp	Spark Plugs Orig Type	Spark Plugs Gap (in.)	Ignition Timing (deg) Man. Trans.	Ignition Timing (deg) Auto. Trans.	Intake Valve Opens (deg)■	Fuel Pump Pressure (psi)	Idle Speed (rpm) Man. Trans.	Idle Speed (rpm) Auto. Trans.
'82	G	4-110	Chev.	88	R-42TS	0.045①	12B	12B	30	4.5–6.0	②	②
'83–'86	0	4-110	Pont.	84	R-42XLS6④	0.060	8B	8B	N.A.	9–13	②	②
	B	4-122	Chev.	90	R-42CTS	0.035	—	12B	30	4.5–6.0③	②	②
	P	4-122	Chev	86	R-42CTS	0.035	②	②	N.A.	12	②	②
	J	4-110	Pont.	150	R-42CXLS	0.035	②	②	N.A.	12	②	②
	W	6-173	Chev.	130	R42CTS	0.045	②	②	31	9–13	②	②
'87	All				See Underhood Specifications Stickers							

NOTE: The underhood specifications sticker often reflects tune-up specification changes made in production. Sticker figures must be used if they disagree with those in this chart.

■ All figures Before Top Dead Center
B Before Top Dead Center
Part numbers in this chart are not recommendations

by Chilton for any product by brand name.
① Certain models may use 0.035 in. Gap—see underhood specifications sticker to be sure

② See underhood specifications sticker
③ 1983–84 w/TBI—12 psi
④ 1984–85—R44XLS
N.A.: Not Available

FIRING ORDERS

NOTE: To avoid confusion, always replace spark plug wires one at a time.

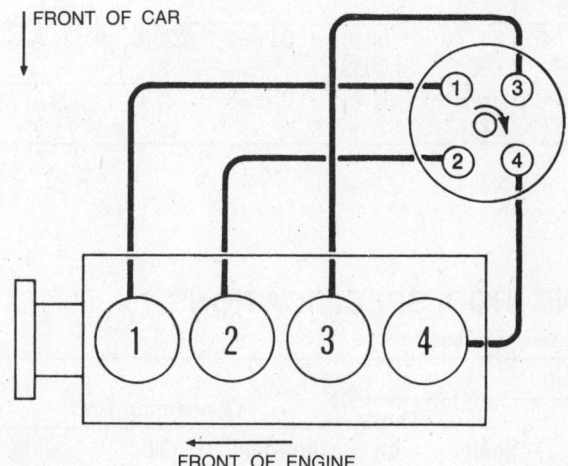

GM (Chevrolet) 110 and 122 overhead valve (OHV)
Engine firing order: 1-3-4-2
Distributor rotation: clockwise

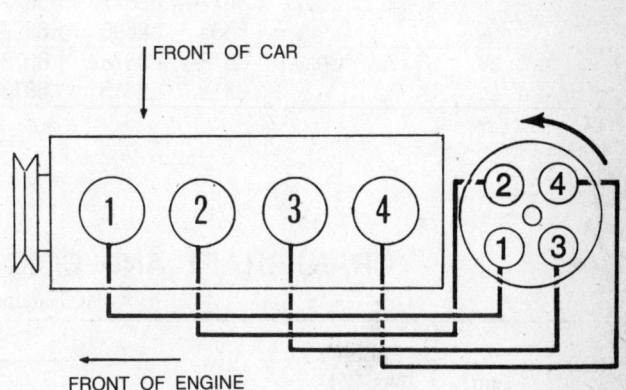

GM (Pontiac) 110 overhead camshaft (OHC)
Engine firing order: 1-3-4-2
Distributor rotation: counterclockwise

CAPACITIES

Year	Eng. VIN Code	Engine Displacement (Cu. In.)	Eng. Mfg.	Crankcase Quarts (Liters) w/filter	wo/filter	Transaxle Pints (L) 4 speed	5 speed	Auto.	Gas Tank Gal (L)	Cooling System Qts (L) w/heater	w/AC
'82	G	110	Chev.	4.0 (3.8)	4.0 (3.8)	5.9 (2.8)	—	10.5 (5.0)	14 (53)	8.0 (7.57)	8.0 (7.57)
'82–'87	O, J	110	Pont.	①	①	—	2.5 (5.3)	10.5 (5.0)	14 (53)	7.8 (7.4)	7.9 (7.5)
	B, P	122	Chev.	4.0 (3.8)	4.0 (3.8)	5.9 (2.8)	—	10.5 (5.0)	14 (53)	8.3 (7.7)	8.3 (7.7)
	W	173	Chev.	4.0 (3.8)	4.0 (3.8)	6 (2.8)	—	8 (3.8)	14 (53)	12.4 (11.7)	12.4 (11.7)

① Add 3 qts, check oil level at dipstick and add as necessary.

VALVE SPECIFICATIONS

Year	Eng. VIN Code	Engine No. Cyl. Displacement (cu. in.)	Eng. Mfg.	Seat Angle (deg)	Face Angle (deg)	Spring Test Pressure (lbs. @ In.)	Spring Installed Height (in.)	Stem-to-Guide Clearance (in.) Intake	Exhaust	Stem Diameter (in.) Intake	Exhaust
'82	G	4-110	Chev.	46	45	183 @ 1.33	1.60	0.0011– 0.0026	0.0014– 0.0031	0.3139– 0.3144	0.3129– 0.3136
'83–'87	O, J	4-110	Pont.	46	46	N.A.	N.A.	0.0006– 0.0016	0.0012– 0.0024	N.A.	N.A.
	B, P	4-122	Chev.	46	45	183 @ 1.33	1.60	0.0011– 0.0026	0.0014– 0.0031	0.3139– 0.3144	0.3129– 0.3136
	W	6-173	Chev.	46	45	195 @ 1.18	1.57	0.0010– 0.0027	0.0010– 0.0027	N.A.	N.A.

N.A.: Not Available

CAMSHAFT SPECIFICATIONS

All measurements in inches

Year	Eng. VIN Code	Engine	Eng. Mfg.	Journal Diameter 1	2	3	4	5	Bearing Clearance	Lobe Lift Intake	Exhaust	Camshaft End Play
'82	G	4-110	Chev.	1.8677– 1.8696	1.8677– 1.8696	1.8677– 1.8696	1.8677– 1.8696	1.8677– 1.8696	0.0010– 0.0039	0.2625	0.2625	N.A.
'82–'87	O, J	4-110	Pont.	1.6714– 1.6720	1.6812– 1.6816	1.6911– 1.6917	1.7009– 1.7015	1.7108– 1.7114	N.A.	0.2409	0.2409	0.016–① 0.064
	B, P	4-122	Chev.	1.8677– 1.8696	1.8677– 1.8696	1.8677– 1.8696	1.8677– 1.8696	1.8677– 1.8696	0.0010– 0.0039	0.2600	0.2600	N.A.
	W	6-173	Chev.	1.8678– 1.8815	1.8678– 1.8815	1.8678– 1.8815	1.8678– 1.8815	1.8678– 1.8815	N.A.	0.2626	0.2732	N.A.

N.A.: Not Available
① 1986–87: 0.04–0.16

CRANKSHAFT AND CONNECTING ROD SPECIFICATIONS

All measurements are given in inches

Year	Eng. VIN Code	Engine No. Cyl Displacement (cu in.)	Eng. Mfg.	Crankshaft Main Brg Journal Dia	Main Brg Oil Clearance	Shaft End-Play	Thrust on No.	Connecting Rod Journal Diameter	Oil Clearance	Side Clearance
'82	G	4-110	Chev.	2.4944– 2.4954②	0.0006– 0.0018③	0.0019– 0.0071	4	1.9983– 1.9993	0.0009– 0.0031	0.0039– 0.0240

CRANKSHAFT AND CONNECTING ROD SPECIFICATIONS

All measurements are given in inches

Year	Eng. VIN Code	Engine No. Cyl Displacement (cu in.)	Eng. Mfg.	Crankshaft			Thrust on No.	Connecting Rod		
				Main Brg Journal Dia	Main Brg Oil Clearance	Shaft End-Play		Journal Diameter	Oil Clearance	Side Clearance
'83–'87	O, J	4-110	Pont.	①	0.0006–0.0016	0.0118–0.0027	3	1.9278–1.9286	0.0007–0.0024	0.0027–0.0095
	B, P	4-122	Chev.	2.4944–2.4954②	0.0006–0.0018③	0.0019–0.0071	4	1.9983–1.9993	0.0009–0.0031	0.0039–0.0240④
	W	6-173	Chev.	2.6473–2.6482	.0016–.0033	.0024–.0083	3	1.9983–1.9994	0.0014–0.0037	0.0063–0.0173

① Bearings are identified by color:
Brown 2.2830–2.2832
Green 2.2827–2.2830
② No. 5: 2.4936–2.4946
③ No. 5: 0.0014–0.0027
④ '84–'85: .004–.015

PISTON AND RING SPECIFICATIONS

All measurements are given in inches.

Year	Eng. VIN Code	Engine No. Cyl. Disp. (cu in.)	Eng. Mfg.	Piston-to-Bore Clearance	Ring Gap			Ring Side Clearance		
					Top Compression	Bottom Compression	Oil Control	Top Compression	Bottom Compression	Oil Control
'82	G	4-110	Chev.	0.0008–0.0018	0.0098–0.0197	0.0098–0.0197	snug	0.0012–0.0027	0.0012–0.0034	0.0078
'83–'87	O, J	4-110	Pont.	0.0008①	0.0010–0.0020	0.0010–0.0020	0.0010–0.0020	0.0020–0.0030	0.0010–0.0024	snug
	B, P	4-122	Chev.	0.0008–0.0018②	0.0098–0.0197	0.0098–0.0197	snug	0.0012–0.0027	0.0012–0.0034	0.0078
	W	6-173	Chev.	0.0007–0.0017	0.0098–0.0197	0.0098–0.0197	0.020–0.055	0.0012–0.0027	0.0016–0.0037	0.0078 max

① Code J: 0.0004–0.0012
② 1984–85: 0.0007–0.0017

TORQUE SPECIFICATIONS

All readings in ft. lbs.

Year	Eng. VIN Code	Engine No. Cyl Displacement (cu. in.)	Liters	Eng. Mfg.	Cylinder Head Bolts	Rod Bearing Bolt	Main Bearing Bolt	Crankshaft Pulley Bolt	Flywheel to Crankshaft Bolts	Manifold	
										Intake	Exhaust
'80–'87	G	4-110	1.8	Chev.	65–75	34–40	63–74	66–84	45	20–25	22–28
	O, J	4-110	1.8	Pont.	①	39	57	115	45	25	15
	B, P	4-122	2.0	Chev.	65–75	34–43	63–77	66–89	②45–63	18–25	20–30
	W	6-173	2.8	Chev.	70	37	68	75	45	23	25

CAUTION: Verify the correct original equipment engine is in the vehicle by referring to the VIN engine code before torquing any bolts.
① Torque bolts to 18 ft. lb., then turn each bolt 60°, in sequence, 3 times for a 180° rotation, then run the engine to normal operating temperature and turn each bolt, in sequence, an additional 30°–50°.
② Auto. trans.: 45–59.

WHEEL ALIGNMENT SPECIFICATIONS

Year	Camber (positive)		Toe	
	Range (degrees)	Preferred (degrees)	Range (degrees)	Preferred (degrees)
'82	1/16 to 1 1/16	9/16	1/4 to 0	1/8 ①
'83	7/32 to 1 7/32	23/32	5/16 to 1/16	1/8 ①
'84–'87	3/16 to 1 3/16	11/16	1/4 to 0	1/8 ①

① Out

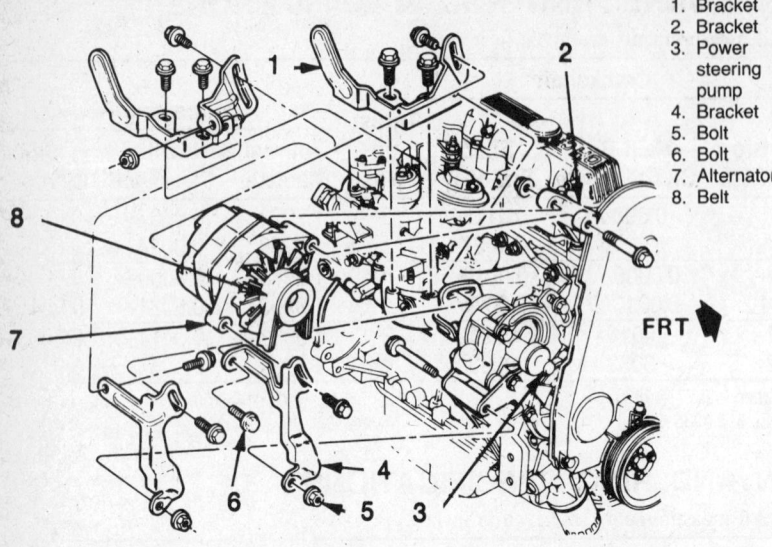

1. Bracket
2. Bracket
3. Power steering pump
4. Bracket
5. Bolt
6. Bolt
7. Alternator
8. Belt

FRT

Alternator installation, OHC engines with power steering

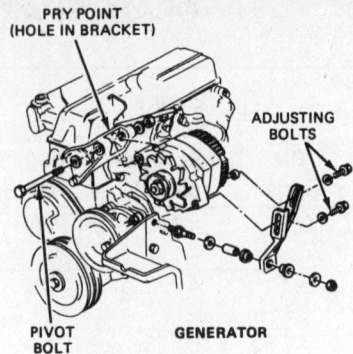

PRY POINT (HOLE IN BRACKET)

ADJUSTING BOLTS

PIVOT BOLT

GENERATOR

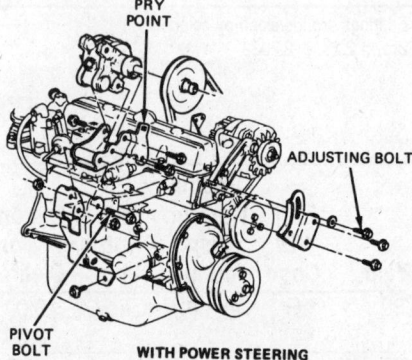

PRY POINT

ADJUSTING BOLT

PIVOT BOLT

WITH POWER STEERING

Drive belt adjustments

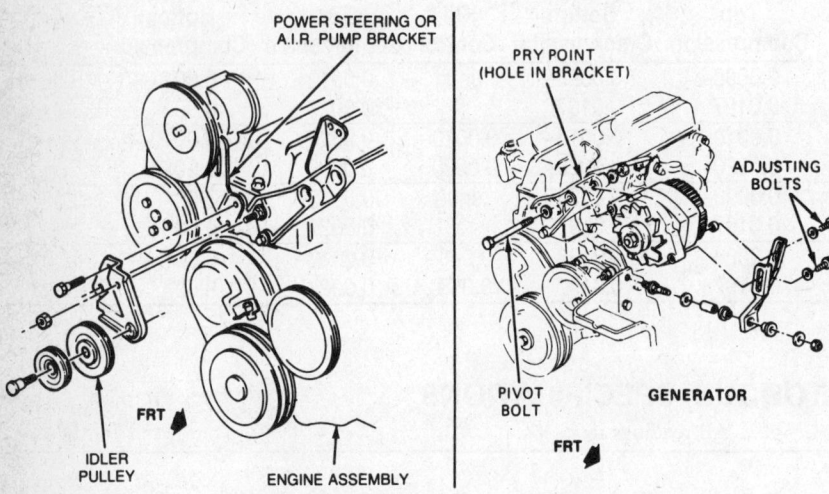

POWER STEERING OR A.I.R. PUMP BRACKET

PRY POINT (HOLE IN BRACKET)

ADJUSTING BOLTS

FRT

IDLER PULLEY

PIVOT BOLT

GENERATOR

ENGINE ASSEMBLY

FRT

Alternator installation, OHV engines

ENGINE ELECTRICAL

Alternator

For further information on the charging system, please refer to "Charging and Starting" in the Unit Repair section.

PRECAUTIONS

1. When installing a battery, make sure that the positive and negative cables are not reversed.

2. When jump-starting the car, be sure that like terminals are connected. This also applies to using a battery charger. Reversed polarity will burn

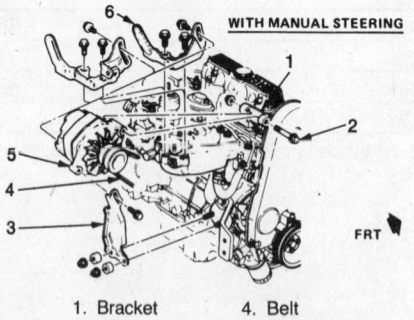

WITH MANUAL STEERING

FRT

1. Bracket
2. Bolt
3. Bracket
4. Belt
5. Alternator
6. Bracket

Alternator installation, OHC engines with manual steering

out the alternator and regulator in a matter of seconds.

3. Never operate the alternator with the battery disconnected or on an otherwise uncontrolled open circuit.

4. Do not short across or ground

any alternator or regulator terminals.

5. Do not try to polarize the alternator.

6. Do not apply full battery voltage to the field (brown) connector.

7. Always disconnect the battery ground cable before disconnecting the alternator lead.

8. Always disconnect the battery (negative cable first) when charging it.

9. Never subject the alternator to excessive heat or dampness. If you are steam-cleaning the engine, cover the alternator.

10. Never use arc-welding equipment on the car with the alternator connected.

REMOVAL & INSTALLATION

1. Disconnect the negative battery cable at the battery.

— **CAUTION** —

Failure to disconnect the negative cable may result in injury from the positive battery lead at the alternator, and may short the alternator and regulator during the removal process.

2. Disconnect and label the two terminal plug and the battery leads from the rear of the alternator.

3. Loosen the mounting bolts. Push the alternator inwards and slip the drive belt off the pulley.

4. Remove the mounting bolts and remove the alternator.

5. To install, place the alternator in its brackets and install the mounting bolts. Do not tighten them yet.

6. Slip the belt back over the pulley. Pull outwards on the unit and adjust the belt tension. Tighten the mounting and adjusting bolts.

7. Install the electrical leads.

8. Install the negative battery cable.

Voltage Regulator

A solid state regulator is mounted within the alternator. All regulator components are enclosed in a solid mold. The regulator is non-adjustable and requires no maintenance.

Starter

For further information on the starting system, please refer to "Charging and Starting" in the Unit Repair section.

REMOVAL & INSTALLATION

OHV Engine

1. Disconnect the negative battery cable at the battery.

2. Label and disconnect the solenoid wires and battery cable.

3. Remove the rear motor support bracket. Remove the A/C compressor support rod (if so equipped).

4. Working under the car, remove the two starter-to-engine bolts, and allow the starter to drop down. Note the location and number of any shims. Remove the starter.

5. Installation is the reverse. Tighten the mounting bolts to 25–35 ft. lbs.

OHC Engines

1. Disconnect the battery ground cable.

2. Remove the air cleaner.

3. Remove the lower starter bolt.

4. Remove the rear starter brace.

5. Remove the wiring from the starter.

6. Remove the upper starter bolt.

7. Raise and support the car on jackstands.

8. Disconnect the speedometer cable.

9. Push the shifter cable up and guide the starter, armature end first, down between the stabilizer bar and the engine.

10. Installation is the reverse of removal.

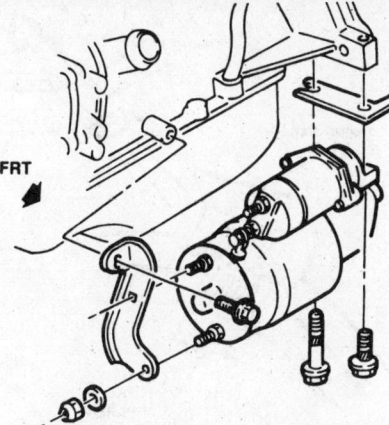

Starter mounting, OHV engines

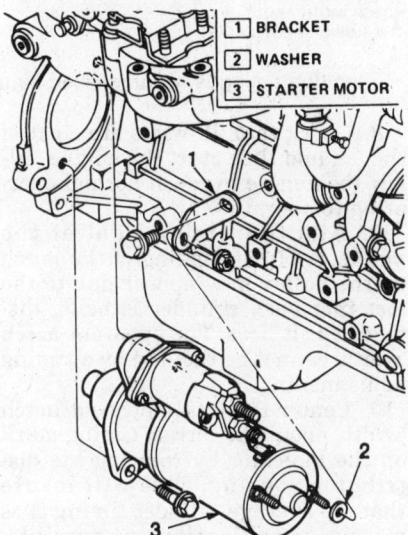

1	BRACKET
2	WASHER
3	STARTER MOTOR

Starter mounting, OHC engines

Distributor

REMOVAL & INSTALLATION

OHV Engines

1. Disconnect the negative battery cable.

2. Tag and disconnect all wires leading from the distributor cap.

3. Remove the air cleaner housing as previously detailed.

4. Remove the distributor cap.

5. Disconnect the AIR pipe-to-exhaust manifold hose at the air management valve.

6. Unscrew the rear engine lift bracket bolt and nut, lift it off the stud and then position the entire assembly out of the way to facilitate better access to the distributor.

7. Mark the position of the distributor, relative to the engine block and then scribe a mark on the distributor body indicating the initial position of the rotor.

8. Remove the hold-down nut and clamp from the base of the distributor. Remove the distributor from the engine. The drive gear on the distributor shaft is helical and the shaft will rotate slightly as the distributor is removed. Note and mark the position of the rotor at this second position. Do not crank the engine while the distributor is removed.

9. To install the distributor, rotate the shaft until the rotor aligns with the second mark you made (when the shaft stopped moving). Lubricate the drive gear with clean engine oil and install the distributor into the engine. As the distributor is installed, the rotor should move to the first mark that you made. This will ensure proper timing. If the marks do not align properly, remove the distributor and try again.

10. Install the clamp and hold-down nut.

NOTE: You may wish to use a magnet attached to an extension bar to position the clamp on the stud.

11. Installation of the remaining components is in the reverse order of removal. Check the ignition timing.

INSTALLATION IF THE ENGINE WAS DISTURBED

If the engine was cranked while the distributor was removed, you will have to place the engine on TDC of the compression stroke to obtain proper ignition timing.

1. Remove the No. 1 spark plug.

2. Place your thumb over the spark plug hole. Crank the engine slowly until compression is felt. It will be easier if you have someone rotate the engine by hand, using a wrench on the crankshaft pulley.

3. Align the timing mark on the crankshaft pulley with the 0° mark on the timing scale attached to the front of the engine. This places the engine at TDC of the compression stroke.

4. Turn the distributor shaft until the rotor points to the No. 1 spark plug tower on the cap.

5. Install the distributor into the engine. Be sure to align the distributor-to-engine block mark made earlier.

6. Perform Steps 10–11 of the preceding removal and installation procedure.

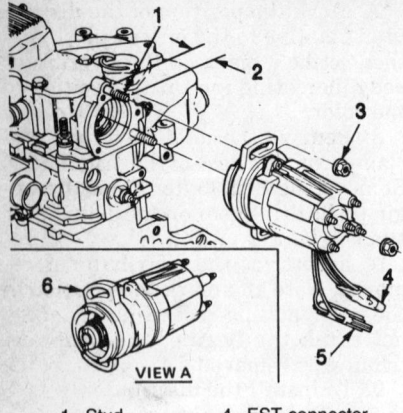

1. Stud
2. 20 ± 1.0
3. Nut
4. EST connector
5. Coil Connector
6. Distributor

Distributor mounting on OHC engines

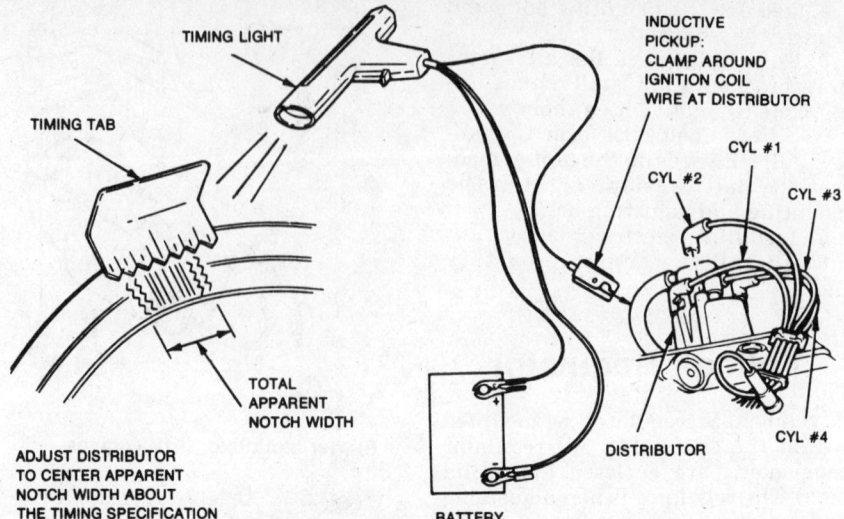

Ignition timing is accomplished by using the averaging method; see the text

OHC Engines

1. Disconnect the battery ground.
2. Mark the spark plug wires and remove the wires and coil.
3. Disconnect the wiring from the distributor.
4. Remove the two distributor hold-down nuts.
5. Remove the distributor.
6. Installation is the reverse of removal. Torque the hold-down nuts to 13 ft. lbs. If the engine was rotated while the distributor was out, see steps 1–5 of the above procedure.

Ignition Timing

ADJUSTMENT

1. Refer to the instructions on the emission control sticker inside the engine compartment. Follow all instructions on the label.
2. Locate the timing marks on the crankshaft pulley and the front of the engine.
3. Clean off the marks so that you can see them. Chalk or white paint will help to make them more visible.
4. Attach a tachometer to the engine as detailed previously.
5. Disconnect the 4-terminal EST connector at the distributor so that the engine will switch to the bypass timing mode.
6. Attach a timing light as per the manufacturer's instructions. Clamp the inductive pick-up around the high tension coil wire (not the No. 1 spark plug wire) at the distributor. Before installing the pick-up on the wire, it will be necessary to peel back the protective plastic cover which encases the wire.
7. Loosen the distributor clamp bolt slightly so that the distributor may be rotated as necessary to adjust timing.

8. Check that all wires are clear of the fan and then start the engine. Allow the engine to reach normal operating temperature.
9. Aim the timing light at the marks. A slight jiggling of the notch on the pulley may appear due to the fact that each cylinder is being displayed as it fires. The apparent notch 'width' cannot be reduced by a timing adjustment.
10. Center the total apparent notch 'width' about the correct timing mark on the indicator by rotating the distributor housing. This will insure that the average cylinder timing is as close to specifications as possible. Once again, the apparent notch 'width' cannot be reduced by timing adjustment.
11. Turn off the engine and tighten the distributor lock bolt. Start the engine and recheck the timing. Sometimes the distributor will move a little during the tightening process. If the ignition timing is within 1° of the correct setting, that is close enough; a tolerance of up to 2° is permitted by the manufacturer.
12. Turn off the engine and disconnect the timing light and the tachometer. Reconnect the 4-terminal EST connector.

FUEL SYSTEM

Fuel Pump

A mechanical fuel pump is used on carbureted engines. It is of the diaphragm-type and because of the design is serviced by replacement only.

No adjustments or repairs are possible. The pump is operated by an eccentric on the camshaft. An electric, in-tank fuel pump is used with fuel injected engines. No adjustments or repairs are possible.

TESTING THE MECHANICAL FUEL PUMP

To determine if the pump is in good condition, tests for both volume and pressure should be performed. The tests are made with the pump installed, and the engine at normal operating temperature and idle speed. Never replace a fuel pump without first performing these simple tests. Be sure that the fuel filter has been changed at the specified interval. If in doubt, install a new filter first.

Pressure Test

1. Disconnect the fuel line at the carburetor and connect a fuel pump pressure gauge. Fill the carburetor float bowl with gasoline.
2. Start the engine and check the pressure with the engine at idle. If the pump has a vapor return hose, squeeze it off so that an accurate reading can be obtained. Pressure should not be below 4.5 psi.
3. If the pressure is incorrect, replace the pump. If it is ok, go on to the volume test.

Volume Test

1. Disconnect the pressure gauge. Run the fuel line into a graduated container.
2. Run the engine at idle until one pint of gasoline has been pumped. One pint should be delivered in 30 seconds or less. There is normally

enough fuel in the carburetor float bowl to perform this test, but refill it if necessary.

3. If the delivery rate is below the minimum, check the lines for restrictions or leaks, then replace the pump.

TESTING THE ELECTRIC FUEL PUMP

Pressure Test

---------- CAUTION ----------

Before performing any tests, do the following to prevent personal injury: Remove the FUEL PUMP fuse from the fuse panel in the passenger compartment. Start the engine and run it until all fuel in the system is used. Crank the engine for an additional 3 seconds to relieve any residual pressure. Turn the ignition to OFF and replace the fuse.

---

1. Remove the air cleaner and plug the thermal vacuum port on the throttle body unit.

2. Remove the steel fuel line from between the throttle body unit and the fuel filter.

3. Install a fuel pressure gauge with at least a 15 psi capacity between the throttle body and the filter.

4. Start the engine and observe the pressure reading. Pressure should be 9–13 psi. If the pressure is not within these limits, one or more of the following could be at fault:

    a. A short in the system

    b. A clogged fuel filter

    c. A shorted or defective oil pressure switch

    d. Defective fuel pump relay

    e. Defective fuel pump

Check each of these components in turn to diagnose the problem before replacing the pump.

5. Follow the Cautions at the start of this procedure to depressurize the system, then remove the pressure gauge and install the fuel line. Torque the nuts to 19–25 ft. lbs.

6. Start the engine and check for leaks.

7. Unplug the thermal vacuum port on the throttle body.

Mechanical Pump

REMOVAL & INSTALLATION

The fuel pump is located at the center rear of the engine.

1. Disconnect the negative battery cable. Raise and support the vehicle safely.

2. Disconnect the inlet hose from the pump. Disconnect the vapor return hose, if equipped.

3. Loosen the fuel line at the carbu-

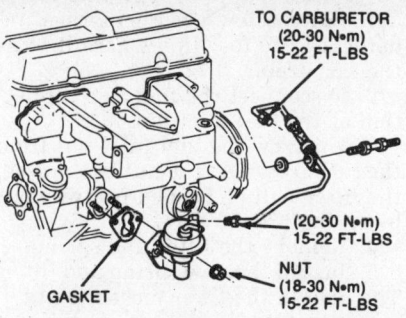

Fuel pump mounting on carbureted engines

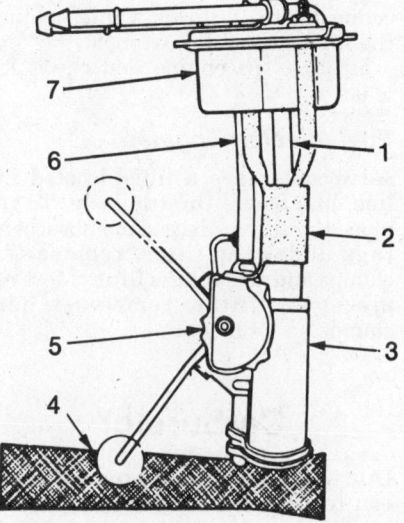

1. Fuel line
2. Rubber coupler and sound insulator
3. Electric fuel pump
4. Filter/strainer
5. Fuel level sender
6. Return tube
7. Splash cup liquid/vapor separator

Fuel injected engine fuel pump

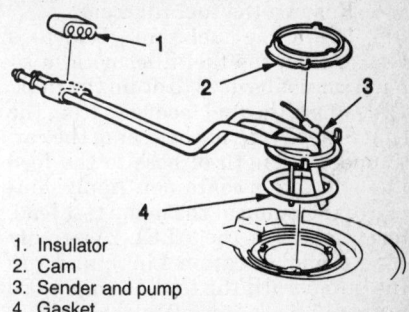

1. Insulator
2. Cam
3. Sender and pump
4. Gasket

Fuel meter removal from the fuel tank on cars with fuel injection

retor, then disconnect the outlet hose from the pump.

4. Remove the two mounting bolts and remove the fuel pump from the engine.

5. To install, place a new gasket on the fuel pump and install the pump to the engine. Tighten the mounting bolts alternately and evenly.

6. Install the pump outlet hose. This is easier if the hose is disconnected from the carburetor. Tighten the fitting while backing up the pump nut with another wrench. Install the hose at the carburetor.

7. Install the inlet and vapor hoses, then lower the vehicle and connect the negative battery cable. Start the engine and check for leaks.

Electric Pump

REMOVAL & INSTALLATION

1. Depressurize the fuel system. See the Pressure Test procedure above.

2. Disconnect the battery ground.

3. Raise and support the car on jackstands.

1. Filler neck   4. Nut   7. Anti-squeak
2. Strap   5. Nut   8. Bolt
3. Clamp   6. Bolt   9. Cap

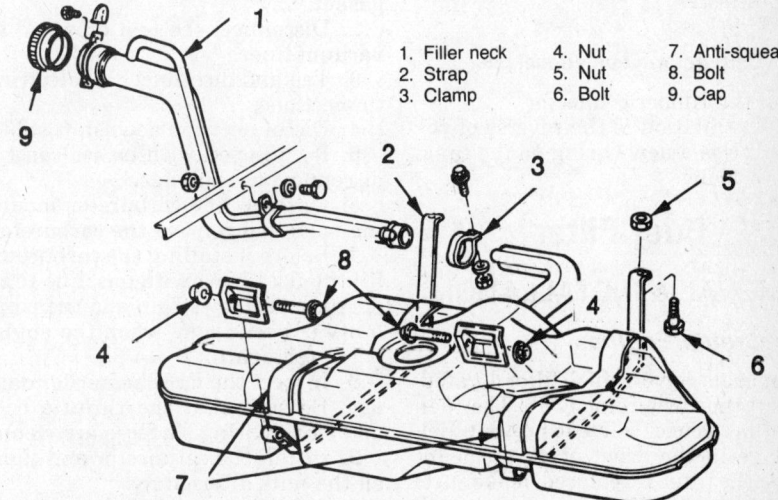

Fuel tank mounting on cars with fuel injection. The tank must be removed to remove the electric fuel pump

4. Remove the fuel filler cap.

5. Drain the fuel tank. Due to a restrictor in the fuel filler neck, a siphon cannot be used to drain the tank. Disconnect the fuel feed hose from the chassis feed pipe at the rear of the car. Connect a length of hose to the feed line and into a container. Apply voltage to the pump at the pump test lead, terminal G on the ALCL (Assembly Line Communication Link), and run the pump until the tank is empty. Do not run the pump after the tank is emptied, as this will damage the pump.

6. Disconnect the wiring from the tank.

7. Disconnect the filler neck hose and the vent hose.

8. Remove the fuel tank strap rear support bolts and lower the tank on a jack, just enough to disconnect the fuel feed line, return and vapor lines from the fuel meter.

9. Remove the tank.

10. Remove the fuel meter/pump assembly by turning the cam lockring counterclockwise. Lift the assembly from the tank and remove the pump from the meter.

11. Pull the pump up onto the attaching hose while pulling outward from the bottom support. Take care that you don't damage the rubber insulator and strainer. After the pump is clear of the bottom support pull it

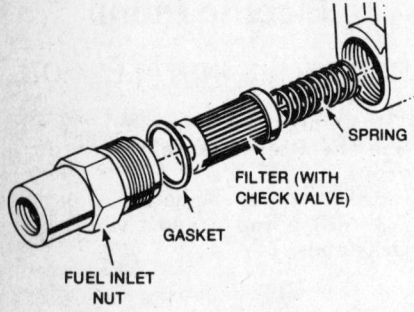

Carburetor-mounted fuel filter

out of the rubber connector.

12. Installation is the reverse of removal. Use a new O-ring on the tank cam lockring.

Fuel Filter

REMOVAL & INSTALLATION

Carbureted Engines

All models have a fuel filter located within the carburetor body. The fuel filter has a check valve to prevent fuel spillage in the event of an accident. When the filter is replaced, make sure the new one is of the same type. All filters are of the paper element type. Replace the filter every 15,000 miles.

1. Place a few absorbent rags underneath the fuel line where it joins the carburetor.

2. Disconnect the fuel line connection at the fuel inlet nut.

3. Unscrew the fuel inlet nut from the carburetor. As the nut is removed, the filter will be pushed partway out by spring pressure.

4. Remove the filter and spring.

5. Install the new spring and filter. The hole in the filter faces the nut.

6. Install a new gasket on the inlet nut and install the nut into the carburetor. Tighten securely.

7. Install the fuel line. Tighten the connector to 18 ft. lbs. while holding the inlet nut with a wrench.

8. Start the engine and check for leaks.

Fuel Injected Engines

All models have a filter located in-line, just before throttle body. To replace the filter, place some absorbent rags under the filter, remove the clamps and replace the filter. Most replacement filters come with new clamps.

Carburetor

Due to the complex nature of modern feedback carburetor systems, comprehensive diagnosis and testing procedures fall outside the confines of this repair manual. For further information on feedback carburetors, please refer to *Chilton's Guide To Fuel Injection And Feedback Carburetors.*

REMOVAL & INSTALLATION

1. Remove the air cleaner and gasket.

2. Disconnect the fuel pipe and all vacuum lines.

3. Tag and disconnect all electrical connections.

4. Disconnect the downshift cable.

5. If equipped with cruise control, disconnect the linkage.

6. Unscrew the carburetor mounting bolts and remove the carburetor.

7. Before installing the carburetor, fill the float bowl with gasoline to reduce the battery strain and the possibility of back-firing when the engine is started again.

8. Inspect the EFE heater for damage. Be sure that the throttle body and EFE mating surfaces are clean.

9. Install the carburetor and tighten the nuts alternately.

10. Installation of the remaining components is in the reverse order of removal.

IDLE SPEED AND MIXTURE ADJUSTMENT— CARBURETED ENGINES

All carbureted J-cars are equipped with an Idle Speed Control (ISC) motor which is in turn controlled by the Electronic Control Module (ECM). All idle speeds are programmed into the ECM's memory and then relayed to the ISC motor as any given situation requires. Curb idle is pre-set at the factory and not routinely adjustable. Although curb idle is not to be adjusted under normal conditions, it can be adjusted, but only upon replacement of the ISC.

The idle mixture screws are concealed under staked-in plugs. Idle mixture is not considered to be a normal tune-up procedure, because of the sensitivity of emission control adjustments. Mixture adjustment requires not only the special tools with which to remove the concealing plugs, but also the addition of an artificial enrichment substance (generally propane) which must be introduced into the carburetor by means of a finely calibrated metering valve. These tools are not generally available and require a certain amount of expertise to use, therefore, mixture adjustments are purposely not covered in this book.

Fuel Injection

Due to the complex nature of modern fuel injection systems, comprehensive diagnosis and testing procedures fall outside the confines of this repair manual. For all fuel injection system diagnosis and testing procedures, please refer to *Chilton's Guide To Fuel Injection And Feedback Carburetors.*

ENGINE COOLING

Radiator

REMOVAL & INSTALLATION

1. Disconnect the negative battery cable.

2. Drain the cooling system.

3. Disconnect the electrical lead at the fan motor.

4. Remove the fan frame-to-radiator support attaching bolts and then remove the fan assembly.

5. Disconnect the upper and lower radiator hoses and the coolant recovery hose from the radiator.

6. Disconnect the transmission oil cooler lines from the radiator and wire them out of the way.

7. Remove the radiator-to-radiator support attaching bolts and clamps. Remove the radiator.

8. Place the radiator in the vehicle so that the bottom is located in the lower mounting pads. Tighten the attaching bolts and clamps.

9. Connect the transmission oil cooler lines and tighten the bolts to 20 ft. lbs.

10. Installation of the remaining components is in the reverse order of removal.

Water Pump

REMOVAL & INSTALLATION

OHV Engines

1. Disconnect the negative battery cable.
2. Drain the cooling system.
3. Remove all accessory drive belts.
4. Remove the alternator.
5. Unscrew the water pump pulley mounting bolts and then pull off the pulley.
6. Remove the mounting bolts and remove the water pump.
7. Place a $\frac{1}{8}$ in. bead of RTV sealant on the water pump sealing surface. While the sealer is still wet, install the pump and tighten the bolts to 13–18 ft. lbs.
8. Installation of the remaining components is in the reverse order of removal.

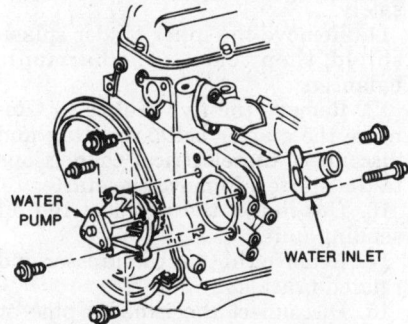

OHV engine water pump installation

OHC Engines

1. Remove the timing belt as described later.
2. Remove the timing belt rear protective covers.
3. Remove the hose from the pump.
4. Unbolt and remove the pump.
5. Installation is in the reverse of removal. Torque the bolts to 19 ft. lbs.

Thermostat

REMOVAL & INSTALLATION

OHV Engines

The thermostat is located inside a housing on the back of the cylinder head. It is not necessary to remove the radiator hose from the thermostat housing when removing the thermostat.

1. Disconnect the negative battery cable.
2. Drain the cooling system and remove the air cleaner.
3. Disconnect the A.I.R. pipe at the upper check valve and the bracket at the water outlet.
4. Disconnect the electrical lead.
5. Remove the two retaining bolts from the thermostat housing and lift up the housing with the hose attached. Lift out the thermostat.
6. Insert the new thermostat, spring end down. Apply a thin bead of silicone sealer to the housing mating surface and install the housing while the sealer is still wet. Tighten the housing retaining bolts to 6 ft. lbs.

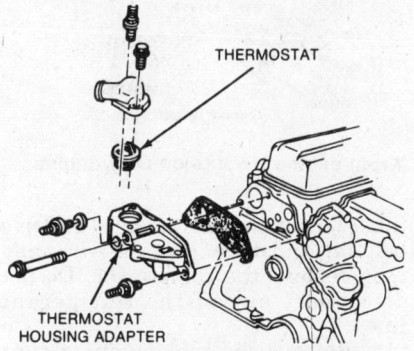

OHV engine thermostat mounting

NOTE: Poor heater output and slow warmup is often caused by a thermostat stuck in the open position; occasionally one sticks shut causing immediate over-heating. Do not attempt to correct a chronic overheating condition by permanently removing the thermostat. Thermostat flow restriction is designed into the system; without it, localized overheating (due to coolant turbulence) may occur, causing expensive troubles.

7. Installation of the remaining components is in the reverse order of removal.

OHC Engines

1. Remove the thermostat housing.
2. Grasp the handle of the thermostat and pull it from the housing.
3. Install the thermostat in the

1. Thermostat housing cap
2. Thermostat
3. Thermostat housing assembly
4. Cylinder head

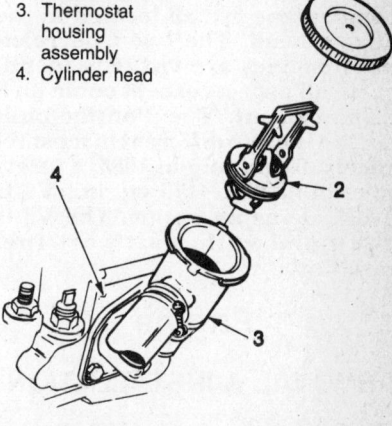

OHC engine thermostat mounting

housing, pushing it down as far as it will go to make sure it's seated.

4. Install the housing on the engine, using a new gasket coated with sealer.

EMISSION CONTROLS

For a description of and service on these devices, see the "Emission Control" Unit Repair section.

Due to the complex nature of modern electronic engine control systems, comprehensive diagnosis and testing procedures fall outside the confines of this repair manual. For complete information on diagnosis, testing and repair procedures concerning all modern engine and emission control systems, please refer to *Chilton's Guide To Electronic Engine Controls*.

ENGINE MECHANICAL

NOTE: J-Cars use three different four cylinder engines. Two are built by Chevrolet, a 1.8L (112 cu.in.) and a 2.0L (122 cu. in.). Both of these Chevrolet-built engines are of the overhead valve configuration (OHV). That means that the camshaft is in the block and the rest of the valve train is on top of the head. The other engine is a

Pontiac-built 1.8L (112 cu. in.) overhead cam engine (OHC). This means that the camshaft and valve components are all located in the engine head. The two Chevrolet-built engines are virtually identical in all aspects except cubic inch displacement. The Pontiac-built engine is quite different in most respects. Beginning in 1985, a Chevrolet-built 2.8L (173 cu. in.) V6 is available as an option. The V6 is equipped with multiport fuel injection.

Engine

REMOVAL & INSTALLATION

OHV 4 Cyl Engines 1982–84

Note: This procedure will require the use of the special powertrain alignment bolt M6XIX65

1. Disconnect the battery cables at the battery, negative cable first.
2. Remove the air cleaner. Drain the cooling system.
3. Remove the power steering pump (if so equipped) and position it out of the way. Leave the lines connected. Remove the windshield washer bottle.
4. If the car is equipped with A/C, remove the relay bracket at the bulkhead connector. Remove the bulkhead connector and then separate the wiring harness connections.
5. If equipped with cruise control, remove the servo bracket and position it out of the way.
6. Tag and disconnect all vacuum hoses and wires.
7. Remove the master cylinder at the vacuum booster.
8. Remove all heater and radiator hoses and position them out of the way.
9. Remove the fan assembly. Remove the horn.
10. Disconnect the carburetor linkage. Raise the front of the car and support it with jackstands.
11. Disconnect the fuel line at the intake manifold.
12. Remove the air conditioning brace (if so equipped).
13. Remove the exhaust shield. Remove the starter.
14. Disconnect the exhaust pipe at the manifold. Remove the wheels.
15. Disconnect the stabilizer bar from the lower control arms. Remove the ball joints from the steering knuckle.
16. Remove the drive axles at the transaxle and then remove the transaxle strut.
17. If equipped with A/C, remove the

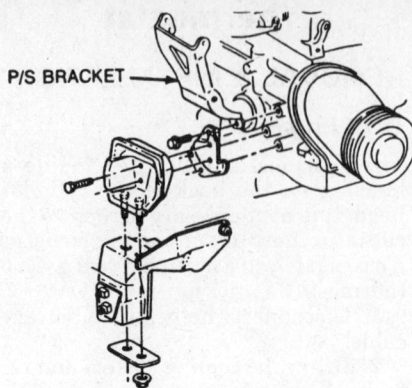

P/S BRACKET

Rear engine mounts on OHV engines

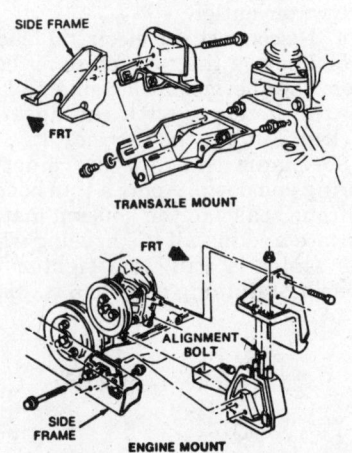

SIDE FRAME

FRT

TRANSAXLE MOUNT

FRT

ALIGNMENT BOLT

SIDE FRAME

ENGINE MOUNT

Front engine mounts on OHV engines

inner fender shield. Remove the drive belt, tag and disconnect the wires and then remove the compressor. Do not disconnect any of the refrigerant lines.
18. Remove the rear engine mount nuts and plate.
19. If equipped with an automatic transaxle, remove the oil filter.
20. Disconnect the speedometer cable and lower the vehicle.
21. If equipped with an automatic transaxle, remove the oil cooler at the transaxle.
22. Remove the front engine mount nuts.
23. Disconnect the clutch cable on the manual transaxle. Disconnect the detent cable on the automatic transaxle.
24. Install an engine lifting device, remove the transaxle mount and bracket. Lift the engine out of the car.
25. Install the engine mount alignment bolt (M6XIX65) to ensure proper power train alignment.
26. Lower the engine into the car, leaving the lifting device attached.
27. Install the transaxle bracket. Install the mount to the side frame and secure with NEW mount bolts.

28. With the weight not yet on the mounts, tighten the transaxle bolts. Tighten the right front mount nuts.
29. Lower the engine fully onto the mounts, remove the lifting device and then raise the front of the car.
30. Installation of the remaining components is in the reverse order of removal. Check the powertrain alignment bolt; if excessive force is required to remove the bolt, loosen the transaxle adjusting bolts and realign the powertrain. Adjust the drive belts and the clutch cable (if equipped with manual transaxle).

2.8L V6 Engine

1. Disconnect the negative battery cable. Drain the cooling system and remove the air cleaner assembly.
2. Remove the air flow sensor. Remove the exhaust crossover heat shield, then remove the crossover pipe.
3. Remove the serpentine belt tensioner and the serpentine belt.
4. Remove the power steering pump mounting bracket. Disconnect the heater pipe at the power steering pump mounting bracket.
5. Disconnect the radiator hoses from the engine.
6. Disconnect the accelerator and throttle valve cable at the throttle valve.
7. Remove the alternator. Tag and disconnect the wiring harness at the engine.
8. Disconnect the fuel hose. Disconnect the coolant bypass and the overflow hoses at the engine.
9. Tag and remove the vacuum hoses to the engine.
10. Raise the vehicle and support it safely.
11. Remove the inner fender splash shield, then remove the harmonic balancer.
12. Remove the flywheel cover. Remove the starter bolts, then tag and disconnect the electrical connections to the starter. Remove the starter.
13. Disconnect the wires at the oil sending unit.
14. Remove the A/C compressor and related brackets.
15. Disconnect the exhaust pipe at the rear of the exhaust manifold.
16. Remove the flex plate-to-torque converter bolts.
17. Remove the transaxle-to-engine bolts. Remove the engine-to-rear mount frame nuts.
18. Disconnect the shift cable bracket at the transaxle. Remove the lower bell housing bolts.
19. Lower the vehicle and disconnect the heater hoses at the engine.
20. Install a suitable engine lifting device and, while supporting the en-

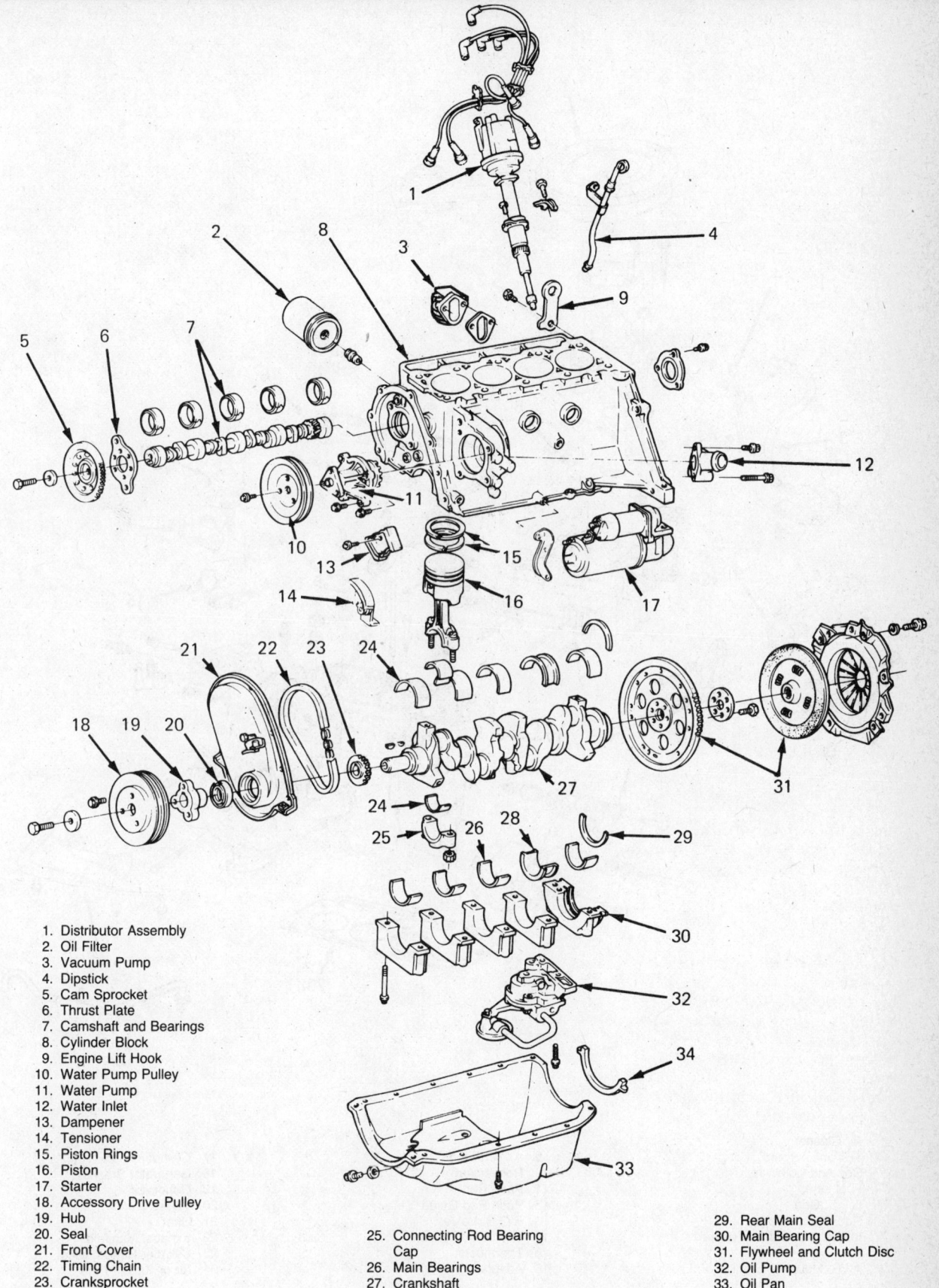

1. Distributor Assembly
2. Oil Filter
3. Vacuum Pump
4. Dipstick
5. Cam Sprocket
6. Thrust Plate
7. Camshaft and Bearings
8. Cylinder Block
9. Engine Lift Hook
10. Water Pump Pulley
11. Water Pump
12. Water Inlet
13. Dampener
14. Tensioner
15. Piston Rings
16. Piston
17. Starter
18. Accessory Drive Pulley
19. Hub
20. Seal
21. Front Cover
22. Timing Chain
23. Cranksprocket
24. Main Bearings

25. Connecting Rod Bearing
    Cap
26. Main Bearings
27. Crankshaft
28. Main Thrust Bearing

29. Rear Main Seal
30. Main Bearing Cap
31. Flywheel and Clutch Disc
32. Oil Pump
33. Oil Pan
34. Seal

Exploded view of the cylinder block—2.0L

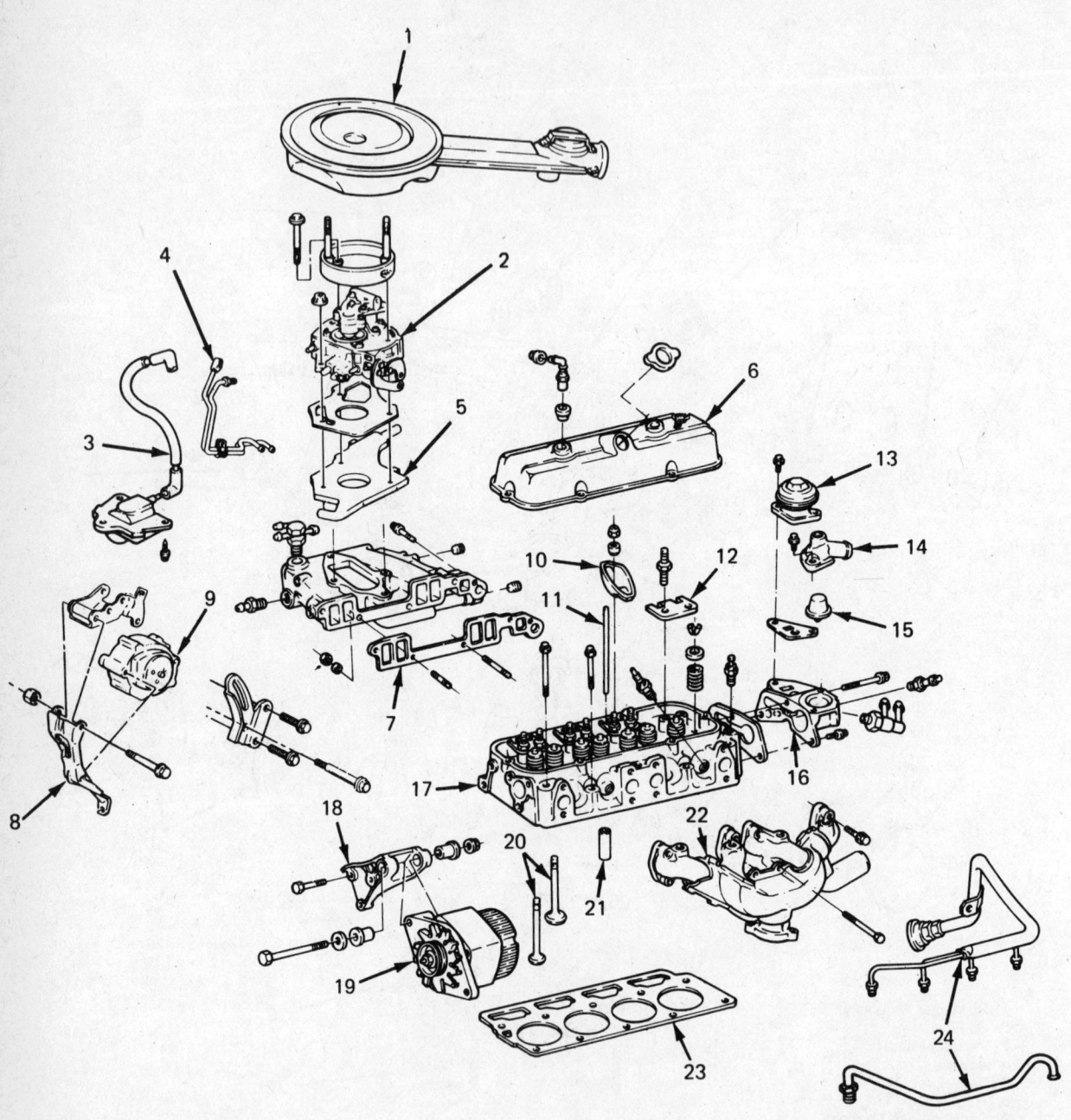

1. Air Cleaner
2. TBI Unit
3. Coil And Coil Wire
4. Fuel Line
5. E.F.E. Grid
6. Rocker Arm Cover
7. Intake Manifold And Gasket
8. A.I.R. Mounting Bracket

9. A.I.R. Pump
10. Rocker Arm
11. Push Rod
12. Push Rod Guide
13. E.G.R. Valve
14. Thermostat Outlet
15. Thermostat
16. Adapter

17. Cylinder Head
18. Generator Bracket
19. Generator
20. Valves
21. Lifter
22. Exhaust Manifold
23. Cylinder Head Gasket
24. Air or Pulsair Pipe

Exploded view of the cylinder head—2.0L

gine and transaxle, remove the upper bell housing bolts.

21. Remove the front mounting bolts.

22. Remove the master cylinder.

23. Remove the engine.

24. Installation is the reverse of removal.

OHC Engines

NOTE: This procedure requires the use of a special tool.

1. Remove battery cables.
2. Drain cooling system.
3. Remove air cleaner.
4. Disconnect engine electrical harness at bulkhead.
5. Disconnect electrical connector at brake cylinder.

6. Remove throttle cable from bracket and EFI assembly.
7. Remove vacuum hoses from EFI assembly.
8. Remove power steering high pressure hose at cut-off switch.

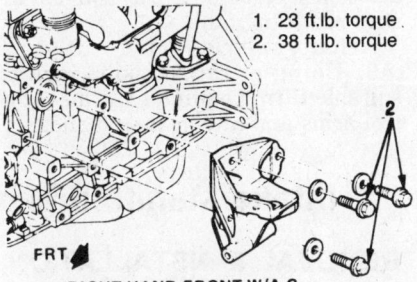

1. 23 ft.lb. torque
2. 38 ft.lb. torque

RIGHT HAND FRONT W/A.C.

Front right engine mount on OHC engines with air conditioning

9. Remove vacuum hoses at map sensor and canister.
10. Disconnect air conditioning relay cluster switches.
11. Remove power steering return hose at pump.
12. Disconnect ECM wire connections, feed harness through bulkhead and lay harness over engine.
13. Remove upper and lower radiator hoses from engine.
14. Remove electrical connections from temperature switch at thermostat housing.
15. Disconnect transmission shift cable at transmission.
16. Raise the car.
17. Remove speedometer cable at transmission and bracket.
18. Disconnect exhaust pipe at exhaust manifold.

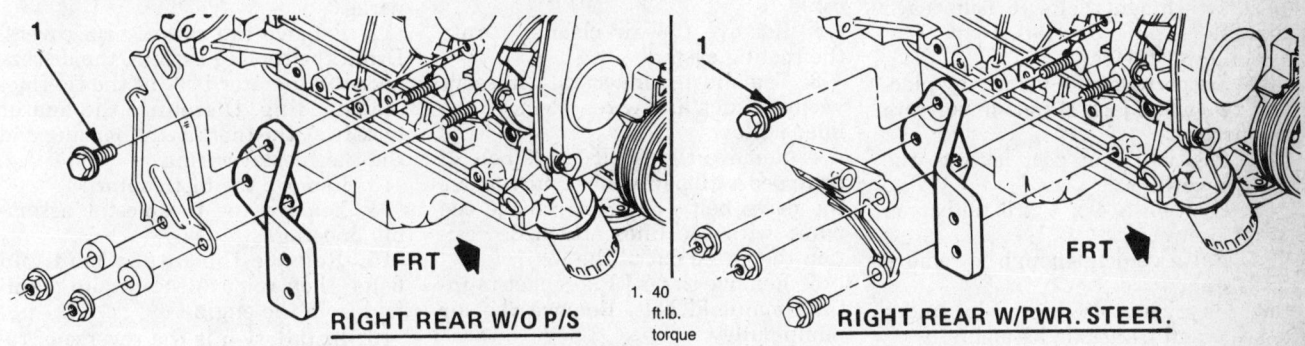

RIGHT REAR W/O P/S

1. 40 ft.lb. torque

RIGHT REAR W/PWR. STEER.

Rear engine mounts on OHC engines

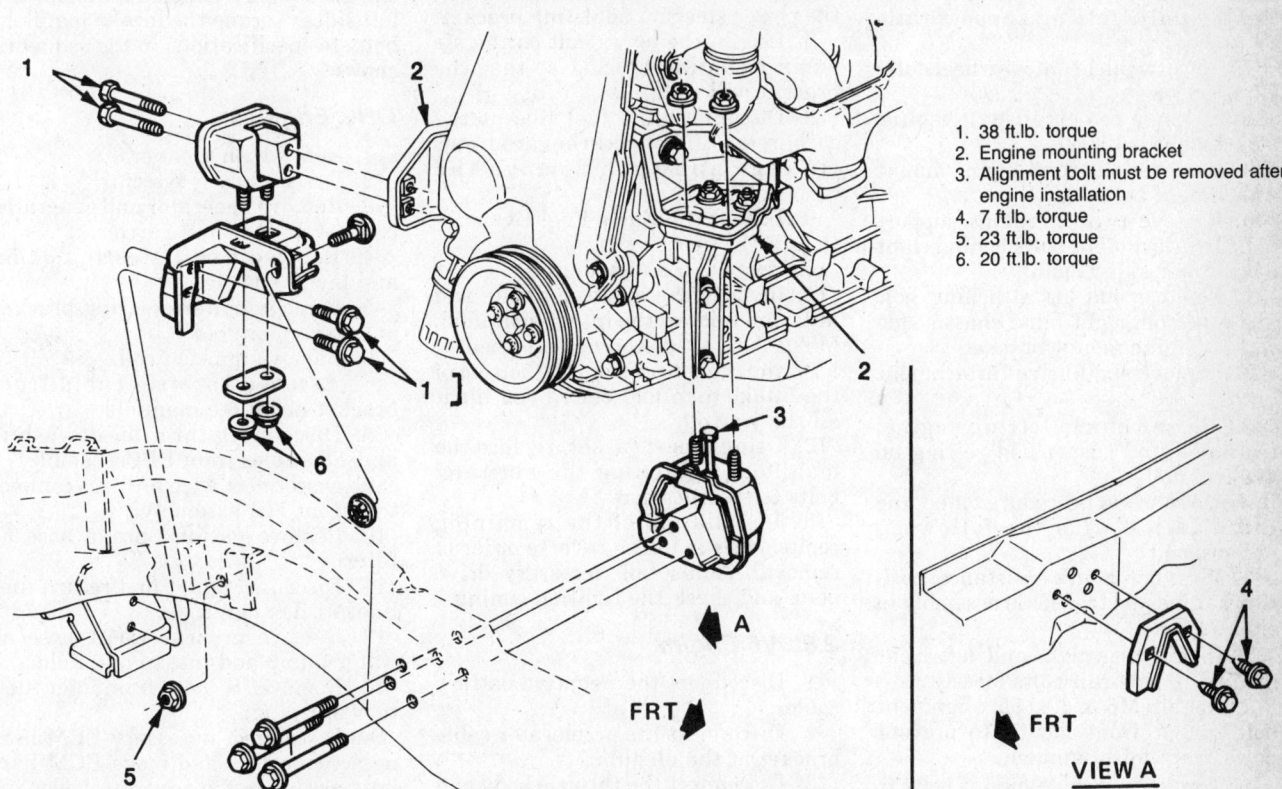

1. 38 ft.lb. torque
2. Engine mounting bracket
3. Alignment bolt must be removed after engine installation
4. 7 ft.lb. torque
5. 23 ft.lb. torque
6. 20 ft.lb. torque

VIEW A

Front engine mounts on OHC engines. The right mount is for cars without air conditioning

19. Remove exhaust pipe from converter.

20. Remove heater hoses from heater core.

21. Remove fuel lines at flex hoses.

22. Remove transmission cooler lines at flex hoses.

23. Remove left and right front wheels.

24. Remove right hand spoiler section and splash shield.

25. Remove right and left brake calipers and support with wire.

26. Remove right and left tie rod ends.

27. Disconnect electrical connections at A/C compressor.

28. Remove A/C compressor and mounting brackets, support A/C compressor with wire in wheel opening.

29. Remove front suspension support attachment bolts (6 bolts each side).

30. Lower the car.

31. Support front of vehicle by placing two short jack stands under core support.

32. Position front post hoist to the rear of cowl.

33. Position a 4 x 4 x 6 timber on front post hoist.

34. Raise vehicle enough to remove jack stands.

35. Position a 4-wheel dolly under engine and transaxle assembly.

36. Position three (3) 4 x 4 x 12 blocks under engine and transaxle assembly only, letting support rails hang free.

37. Lower vehicle onto 4-wheel dolly slightly.

38. Remove rear transaxle mount attachment bolts (2).

39. Remove left front engine mount attachment bolts (3).

40. Remove two (2) engine support to body attachment bolts behind right hand inner axle U-joint.

41. Remove one (1) attaching bolt and nut from right hand chassis side rail to engine mount bracket.

42. Remove six (6) strut attachment nuts.

43. Raise vehicle letting engine, transaxle and suspension resting on 4-wheel dolly.

Reverse removal procedure for engine installation with the following exceptions:

1. With one man's assistance, position engine and transaxle assembly in chassis.

2. Install transaxle and left front mounts to side rail bolts loosely.

3. Install M6 x I x 65 alignment bolt in left front mount to prevent powertrain misalignment.

4. Torque transaxle mount bolts to 42 ft. lbs. and left front mount bolts to 18 ft. lbs.

5. Install right rear mount to body bolts and torque to 38 ft. lbs.

6. Install right rear mount to chassis side rail bolt and nut torque to 38 ft. lbs.

7. Place a floor jack under control arms, jack struts into position and install retaining nuts.

8. Raise vehicle.

9. Using a transmission jack or suitable lifting equipment, raise control arms and attach tie rod ends.

Intake Manifold

REMOVAL & INSTALLATION

OHV 4 Cyl Engines

1. Disconnect the negative battery cable.

2. Remove the air cleaner. Drain the cooling system.

3. Tag and disconnect all necessary vacuum lines and wires. Remove the idler pulley.

4. Remove the A.I.R. drive belt. If equipped with power steering, remove the drive belt and then remove the pump with the lines attached. Position the pump out of the way.

5. Remove the A.I.R. bracket-to-intake manifold bolt. Remove the air pump pulley.

6. If equipped with power steering, remove the A.I.R. thru-bolt and then the power steering adjusting bracket.

7. Loosen the lower bolt on the air pump mounting bracket so that the bracket will rotate.

8. Disconnect the fuel line at the carburetor. Disconnect the carburetor linkage and then remove the carburetor.

9. Lift off the Early Fuel Evaporation (EFE) heater grid.

10. Remove the distributor.

11. Remove the mounting bolts and nuts and remove the intake manifold. Make sure to disconnect the heater hose and condenser from the bottom of the intake manifold before you lift it all the way out.

12. Using a new gasket, replace the manifold, tightening the nuts and bolts to specification.

13. Installation of the remaining components is in the reverse order of removal. Adjust all necessary drive belts and check the ignition timing.

2.8L V6 Engine

1. Disconnect the negative battery cable.

2. Disconnect the accelerator cable bracket at the plenum.

3. Disconnect the throttle body and the EGR pipe from the EGR valve. Remove the plenum assembly.

4. Disconnect the fuel line along the fuel rail.

5. Disconnect the serpentine drive belt. Remove the power steering pump mounting bracket.

6. Remove the heater pipe at the power steering pump bracket.

7. Tag and disconnect the wiring at the alternator and remove the alternator.

8. Disconnect the wires from the cold start injector assembly. Remove the injector assembly from the intake manifold.

9. Disconnect the idle air vacuum hose at the throttle body. Disconnect the wires at the injectors.

10. Remove the fuel rail, breather tube and the fuel runners from the engine.

11. Tag and disconnect the coil wires.

12. Remove the rocker arm covers. Drain the cooling system, the disconnect the radiator hose at the thermostat housing. Disconnec the heater hose from the thermostat housing and the thermostat wiring.

13. Remove the distributor.

14. Remove the thermostat assembly housing.

15. Remove the intake manifold bolts, then remove the intake manifold from the engine.

16. Installation is the reverse of removal. Upon installation, note that the gaskets are marked for right and left sides. Torque the intake manifold bolts to specifications in the sequence shown.

OHC Engines

1. Remove air cleaner.

2. Drain cooling system.

3. Remove generator and generator bracket at camshaft carrier.

4. Remove power steering pump and lay to one side.

5. Remove power steering bracket at intake manifold.

6. Remove ignition coil.

7. Remove throttle cable from bracket at intake manifold.

8. Disconnect throttle, downshift and TV cables from EFI assembly.

9. Disconnect wire harness connectors from TBI assembly.

10. Remove vacuum brake hose at filter.

11. Disconnect inlet and return fuel lines at flex joints.

12. Remove preheat water hose at water pump and intake manifold.

13. Remove "S" hose from inlet tube to water pump.

14. Disconnect necessary ECM harness connectors and move ECM harness assembly for access to lower intake manifold retaining nuts.

15. Remove four (4) lower intake

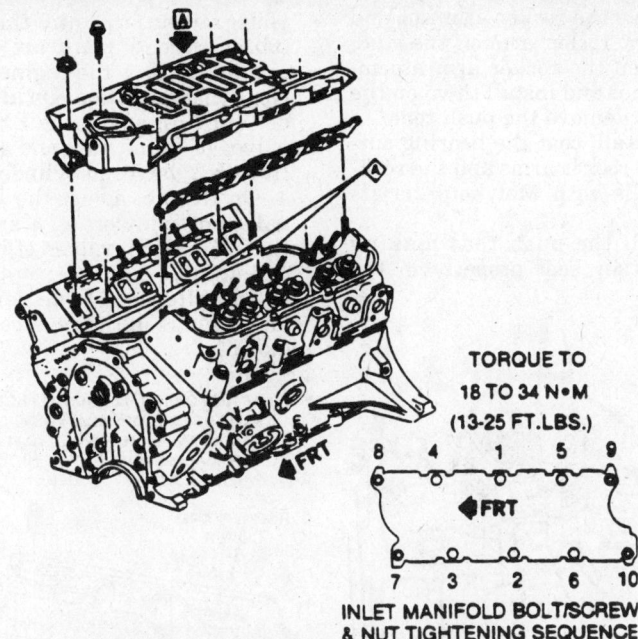

TORQUE TO
18 TO 34 N•M
(13-25 FT.LBS.)

INLET MANIFOLD BOLT/SCREW
& NUT TIGHTENING SEQUENCE

(A) [NOTE] APPLY A SMOOTH CONTINUOUS BEAD
APPROX. 2.0-3.0 WIDE AND 3.0-5.0 THICK
ON BOTH SURFACES. BEAD CONFIGURATION
MUST INSURE COMPLETE SEALING OF WATER
AND OIL. SURFACE MUST BE FREE OF OIL
AND DIRT TO INSURE ADEQUATE SEAL.

2.8L V6 intake manifold installation

manifold retaining nuts and washers.

16. Remove five (5) upper intake manifold retaining nuts and washers and remove intake manifold.

17. Installation is the reverse of removal. Torque the bolts to 16 ft. lbs.

Exhaust Manifold

REMOVAL & INSTALLATION

OHV 4 Cyl Engines

1. Disconnect the negative battery cable.

2. Remove the air cleaner. Remove the exhaust manifold shield. Raise and support the front of the vehicle.

3. Disconnect the exhaust pipe at the manifold and then lower the vehicle.

4. Disconnect the air management-to-check valve hose and remove the bracket. Disconnect the oxygen sensor lead wire.

5. Remove the alternator belt. Remove the alternator adjusting bolts, loosen the pivot bolt and pivot the alternator upward.

6. Remove the alternator brace and the A.I.R. pipes bracket bolt.

7. Unscrew the mounting bolts and remove the exhaust manifold. The manifold should be removed with the A.I.R. plumbing as an assembly. If the

Installing intake manifold gasket on 2.8L V6 engine

manifold is to be replaced, transfer the plumbing to the new one.

8. Clean the mating surfaces on the manifold and the head, position the manifold and tighten the bolts to the proper specifications.

9. Installation of the remaining components is in the reverse order of removal.

2.8L V6 Engine

LEFT SIDE

1. Disconnect the negative battery cable.

2. Remove the air cleaner assembly.

3. Remove the air flow sensor. Remove the engine heat shield.

4. Disconnect the crossover pipe at the manifold.

5. Remove the exhaust manifold bolts.

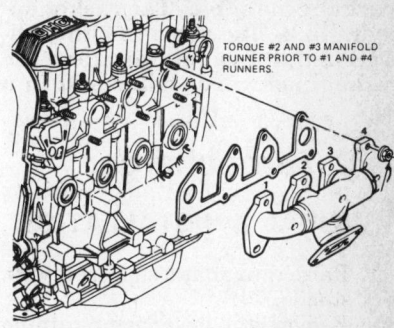

TORQUE #2 AND #3 MANIFOLD RUNNER PRIOR TO #1 AND #4 RUNNERS.

Exhaust manifold torque sequence—1.8 turbocharged engine

6. Remove the exhaust manifold.

7. Installation is the reverse of removal.

RIGHT SIDE

1. Disconnect the negative battery cable.

2. Remove the air cleaner assembly.

3. Remove the air flow sensor. Remove the engine heat shield.

4. Disconnect the crossover pipe at the manifold.

5. Disconnect the accelerator and throttle valve cable at the throttle lever and the plenum. Move aside to gain working clearance.

6. Disconnect the power steering line at the power steering pump.

7. Remove the EGR valve assembly.

8. Raise the vehicle and support it safely.

9. Disconnect the exhaust pipe at the exhaust manifold.

10. Lower the vehicle.

11. Remove the manifold bolts, then remove the exhaust manifold.

12. Installation is the reverse of removal.

OHC Engines

1. Remove air cleaner.

2. Remove spark plug wires and retainers.

3. Remove oil dipstick tube and breather assembly.

4. Disconnect oxygen sensor wire.

5. Disconnect exhaust pipe from manifold flange.

6. Remove exhaust manifold to cylinder head attaching nuts and remove manifold and gasket.

7. Installation is the reverse of removal. Torque the bolts to 16–19 ft. lbs.

NOTE: Before installing a new gasket on the 1.8L MFI Turbo engine (code J), check for the location of the stamped part number on the surface. This gasket should be installed with this number to-

ward the manifold. The gasket appears to be the same in either direction but it is not. Installing the gasket backwards will result in a leak.

Turbocharger

REMOVAL & INSTALLATION

1. Raise the car and support it with jack stands.
2. Remove the lower fan retaining screw.
3. Disconnect the exhaust pipe.
4. Remove the rear A/C support bracket and loosen the remaining bolts.
5. Remove the turbo support bracket bolt to the engine.
6. Disconnect the oil drain hose at the turbo.
7. Lower the vehicle.
8. Disconnect the coolant recovery pipe and move to one side.
9. Disconnect the induction tube.
10. Disconnect the cooling fan.
11. Disconnect the oxygen sensor.
12. Disconnect the oil feed pipe at the union.
13. Disconnect the air intake duct and vacuum hose at the actuator.
14. Remove the exhaust manifold retaining nuts and remove the exhaust manifold and turbocharger.

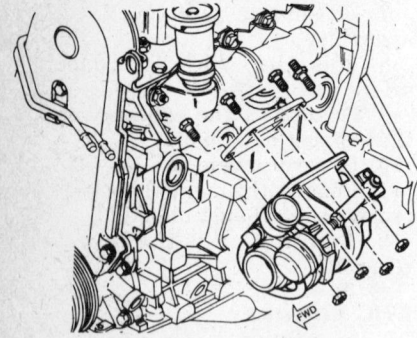

Turbocharger mounting

15. Installation is the reverse of removal. Install a new manifold gasket and tighten retaining bolts to 16 ft. lbs. in sequence shown in the illustration. See note under Exhaust Manifold Removal & Installation for OHC engines.

Rocker Arms and Push Rods

REMOVAL, INSTALLATION AND ADJUSTMENT

OHV Engines

1. Remove the air cleaner. Remove the cylinder head cover.

2. Remove the rocker arm nut and ball. Lift the rocker arm off the stud. Always keep the rocker arm assemblies together and install them on the same stud. Remove the push rods.
3. To install, coat the bearing surfaces of the rocker arms and the rocker arm balls with Molykote® or its equivalent.
4. Install the push rods making sure that they seat properly in the lifter.

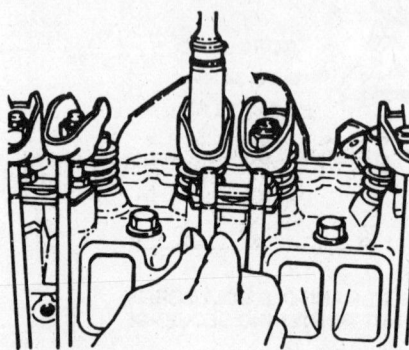

On OHV engines, tighten the rocker arm nut until the pushrod can't be rotated between your fingers

5. Install the rocker arms, balls and nuts. Tighten the rocker arm nuts until all lash is eliminated.
6. Adjust the valves when the lifter is on the base circle of a camshaft lobe:

   a. Crank the engine until the mark on the crankshaft pulley lines up with the '0' mark on the timing tab. Make sure that the engine is in the No. 1 firing position. Place your fingers on the No. 1 rocker arms as the mark on the crank pulley comes near the "0" mark. If the valves are not moving, the engine is in the No. 1 firing position. If the valves move, the engine is in the No. 4 firing position; rotate the engine one complete revolution and it will be in the No. 1 position.

   b. When the engine is in the No. 1 firing position, on all 4 cylinder engines, adjust the EXHAUST valves of cylinders 1 and 3 and the INTAKE valves of cylinders 1 and 2. On V6 engines, adjust the INTAKE valves of cylinders 1, 5 and 6 and the EXHAUST valves of cylinders 1, 2 and 3.

   c. Back the adjusting nut out until lash can be felt at the push rod, then turn the nut until all lash is removed (this can be determined by rotating the push rod while turning the adjusting nut). When all lash has been removed, turn the nut in $1\frac{1}{2}$ additional turns, this will center the lifter plunger.

   d. Crank the engine one com-

plete revolution until the timing tab and the '0' mark are again in alignment. Now the engine is in the No. 4 firing position. On all 4 cylinder engines, adjust the EXHAUST valves of cylinders 2 and 4 and the INTAKE valves of cylinders 3 and 4. On the V6, adjust the INTAKE valves of cylinders 2, 3 and 4 and the EXHAUST valves of cylinders 4, 5 and 6.

7. Installation of the remaining components is in the reverse order of removal.

NOTE: AT TIME OF INSTALLATION, FLANGES MUST BE FREE OF OIL. A ⅛ BEAD OF SEALANT MUST BE APPLIED TO FLANGES AND SEALANT MUST BE WET TO TOUCH WHEN BOLTS ARE TORQUED.

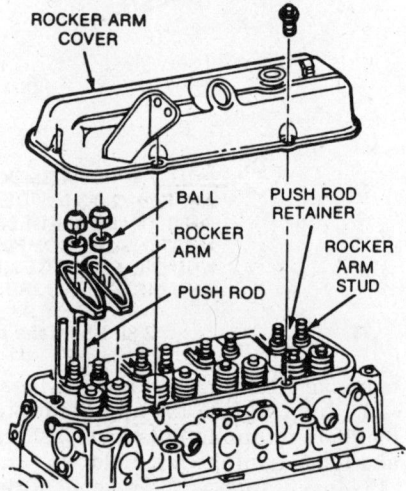

OHV engine rocker arm assembly

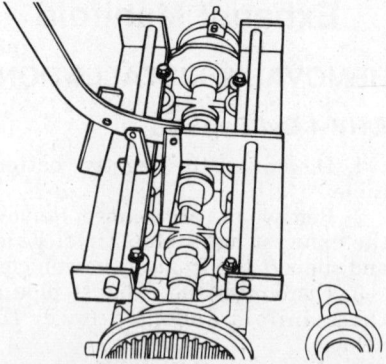

Compressing the valve spring using the valve train compressing tool J-33302, on OHC engines

OHC Engines

NOTE: A special tool is required for this procedure.

1. Remove the camshaft carrier cover.
2. Using a valve train compressing fixture, tool J-33302, depress all the lifters at once.

3. Remove the rocker arms, placing them on the workbench in the same order that they were removed.

4. Remove the hydraulic valve lash compensators keeping them in the order in which they were removed.

5. Installation is in the reverse of removal. Rocker arms and compensators must be replaced in the exact same position as when they were removed.

Cylinder Head

REMOVAL & INSTALLATION

OHV 4 Cyl Engines

NOTE: The engine should be "overnight" cold before removing the cylinder head.

1. Disconnect the negative battery cable.
2. Drain the cooling system into a clean container; the coolant can be reused if it is still good.
3. Remove the air cleaner. Raise and support the front of the vehicle.
4. Remove the exhaust shield. Disconnect the exhaust pipe.
5. Remove the heater hose from the intake manifold an then lower the car.
6. Unscrew the mounting bolts and remove the engine lift bracket (includes air management).
7. Remove the distributor. Disconnect the vacuum manifold at the alternator bracket.
8. Tag and disconnect the remaining vacuum lines at the intake manifold and thermostat.
9. Remove the air management pipe at the exhaust check valve.
10. Disconnect the accelerator linkage at the carburetor or TBI unit and then remove the linkage bracket.
11. Tag and disconnect all necessary wires. Remove the upper radiator hose at the thermostat.
12. Remove the bolt attaching the dipstick tube and hot water bracket.
13. Remove the idler pulley. Remove the A.I.R. and power steering pump drive belts.
14. Remove the A.I.R. bracket-to-intake manifold bolt. If equipped with power steering, remove the air pump pulley, the A.I.R. thru-bolt and the power steering adjusting bracket.
15. Loosen the A.I.R. mounting bracket lower bolt so that the bracket will rotate.
16. Disconnect and plug the fuel line at the carburetor.
17. Remove the alternator. Remove the alternator brace from the head and then remove the upper mounting bracket.

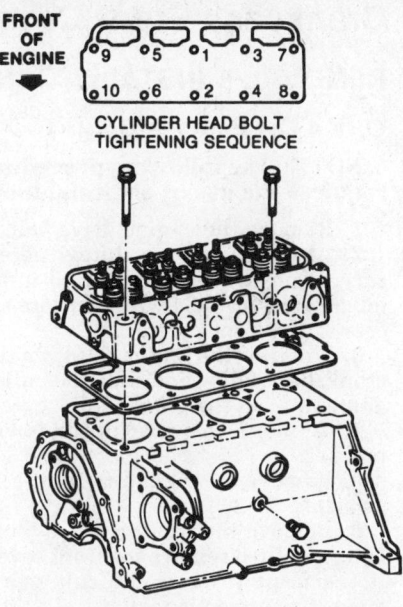

FRONT OF ENGINE

CYLINDER HEAD BOLT TIGHTENING SEQUENCE

OHV engine cylinder head bolt torque sequence

18. Remove the cylinder head cover. Remove the rocker arms and push rods.
19. Remove the cylinder head bolts in the order given in the illustration. Remove the cylinder head with the carburetor or TBI unit, intake and exhaust manifolds still attached. To install, the gasket surfaces on both the head and the block must be clean of any foreign matter and free of any nicks or heavy scratches. Cylinder bolt threads in the block and the bolt must be clean.
20. Place a new cylinder head gasket in position over the dowel pins on the block. Carefully guide the cylinder head into position.
21. Coat the cylinder bolts with sealing compound and install them finger tight.
22. Using a torque wrench, gradually tighten the bolts in the sequence shown in the illustration to the proper specifications.
23. Installation of the remaining components is in the reverse order of removal.

2.8L V6 Engine

1. Disconnect the negative battery cable.
2. Remove the intake manifold.
3. Remove the exhaust manifold.
4. Tag and disconnect the spark plug wires.
5. Remove the pushrods.
6. Remove the cylinder head bolts in the reverse of the tightening sequence, then remove the cylinder head from the engine.
7. Installation is the reverse of the

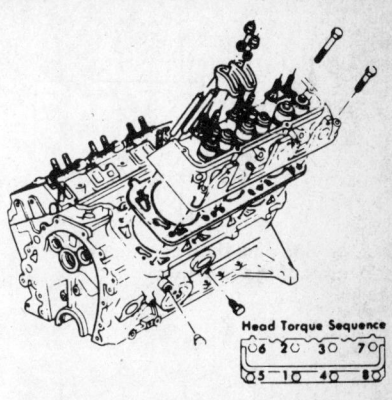

2.8L V6 cylinder head installation

removal procedure. Install the head gasket with the note "This Side Up" showing. Use sealer No. 1052080 or equivalent on the cylinder head bolt threads, and tighten the head bolts to specifications in the sequence shown.

OHC Engine

1. Remove air cleaner.
2. Drain cooling system.
3. Remove generator and pivot bracket at camshaft carrier housing.

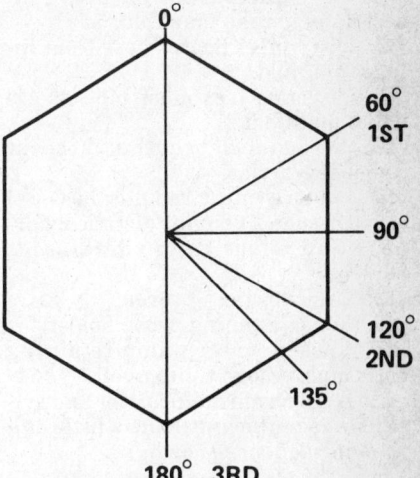

OHC engine cylinder head bolt torque degree sequence

4. Disconnect power steering pump and bracket and lay to one side.
5. Disconnect ignition coil electrical connections and remove coil.
6. Disconnect spark plug wires and distributor cap and remove.
7. Remove throttle cable from bracket at intake manifold.
8. Disconnect throttle cable, downshift cable and T.V. cable from EFI assembly.
9. Disconnect the ECM connectors from the EFI assembly.
10. Remove vacuum brake hose at filter.
11. Disconnect inlet and return fuel lines at flex joints.

C729

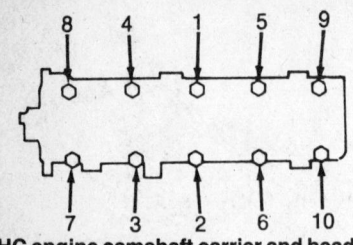

OHC engine camshaft carrier and head bolt tightening sequence

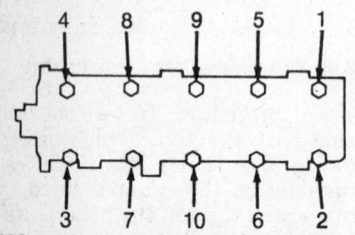

OHC engine camshaft carrier and head bolt loosening sequence

12. Remove water pump bypass hose at intake manifold and water pump.

13. Disconnect ECM harness connectors at intake manifold.

14. Disconnect heater hose from intake manifold.

15. Disconnect exhaust pipe at exhaust manifold.

16. Disconnect breather hose at camshaft carrier.

17. Remove upper radiator hose.

18. Disconnect engine electrical harness and wires from thermostat housing.

19. Remove timing cover.

20. Remove timing probe holder.

21. Loosen water pump retaining bolts and remove timing belt.

22. Loosen camshaft carrier and cylinder head attaching bolts a little at a time in sequence shown.

NOTE: Camshaft carrier and cylinder head bolts should only be removed when engine is cold.

23. Remove camshaft carrier assembly.

24. Remove cylinder head, intake manifold and exhaust manifold as an assembly.

25. Installation is the reverse of removal. Torque head bolts in the sequence shown. Make sure that you follow the note on torquing, at the bottom of the Torque Chart.

OVERHAUL

For all cylinder head overhaul procedures, please refer to "Engine Rebuilding" in the unit repair section.

Crankcase Front Cover

REMOVAL & INSTALLATION

OHV 4 Cyl Engines Only

NOTE: The following procedure requires the use of a special tool.

1. Remove the engine drive belts.

2. Although not absolutely necessary, removal of the right front inner fender splash shield will facilitate access to the front cover.

3. Unscrew the center bolt from the crankshaft pulley and slide the pulley and hub from the crankshaft.

4. Remove the alternator lower bracket.

5. Remove the oil pan-to-front cover bolts.

6. Remove the front cover-to-block bolts and then remove the front cover. If the front cover is difficult to remove, use a plastic mallet.

7. The surfaces of the block and front cover must be clean and free of oil. Apply a $\frac{1}{8}$ in. bead of RTV sealant to the cover. The sealant must be wet to the touch when the bolts are torqued down.

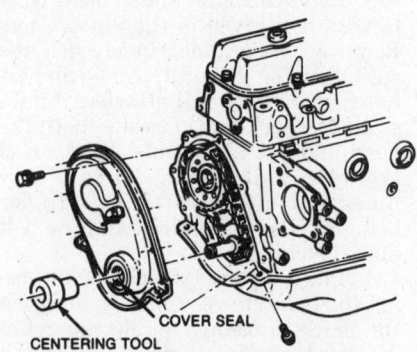

CENTERING TOOL COVER SEAL

Front cover installation on OHV engines; a centering tool will aid in positioning

NOTE: When applying RTV sealant to the front cover, be sure to keep it out of the bolt holes.

8. Position the front cover on the block using a centering tool (J-23042) and tighten the screws.

9. Installation of the remaining components is in the reverse order of removal.

2.8L V6 Engine

1. Disconnect the negative battery cable.

2. Drain the cooling system and remove the coolant recovery tank from the vehicle.

3. Disconnect the manifold and EGR sensor solenoids.

4. Remove the serpentine belt and adjusting pulley.

5. Tag and disconnect the heater hose at the power steering bracket.

6. Tag and disconnect the alternator wiring and remove the alternator.

7. Raise the vehicle and support it safely.

8. Remove the inner fender splash shield. Remove the air conditioner compressor belt.

9. Remove the harmonic balancer if necessary to gain working clearance.

10. Remove the pan to block bolts. Remove the lower cover bolts.

11. Lower the vehicle and disconnect the radiator hoses at the water pump.

12. Remove the heater hose from the thermostat housing.

13. Disconnect the overflow hoses and the canister purge hose.

14. Remove the front cover.

15. Installation is the reverse of removal. Upon installation, apply a 3mm coninuous bead of RTV sealant to the oil pan surface and make sure all mating surfaces are clean of old gasket material.

TIMING COVER OIL SEAL

OHV Engines Only

The oil seal can be replaced with the cover either on or off the engine. If the cover is on the engine, remove the crankshaft pulley and hub first. Pry out the seal using a large screwdriver, being careful not to distort the seal mating surface. Install the new seal so that the open side or helical side is towards the engine. Press it into place with a seal driver made for the purpose. Install the hub if removed.

Timing Chain and Sprockets

REMOVAL & INSTALLATION

OHV 4 Cyl Engines Only

1. Remove the front cover as previously detailed.

2. Place the No. 1 piston at TDC of the compression stroke so that the marks on the camshaft and crankshaft sprockets are in alignment (see illustration).

3. Loosen the timing chain tensioner nut as far as possible without actually removing it.

4. Remove the camshaft sprocket bolts and remove the sprocket and chain together. If the sprocket does not slide from the camshaft easily, a light blow with a soft mallet at the lower edge of the sprocket will dislodge it.

5. Use a gear puller (J-2288-8-20) and remove the crankshaft sprocket.

6. Press the crankshaft sprocket back onto the crankshaft.

7. Install the timing chain over the camshaft sprocket and then around the crankshaft sprocket. Make sure that the marks on the two sprockets are in alignment (see illustration). Lubricate the thrust surface with Molykote® or its equivalent.

8. Align the dowel in the camshaft with the dowel hole in the sprocket and then install the sprocket onto the camshaft. Use the mounting bolts to draw the sprocket onto the camshaft and then tighten them to 27–33 ft. lbs.

9. Lubricate the timing chain with clean engine oil. Tighten the chain tensioner.

10. Installation of the remaining components is in the reverse order of removal.

2.8L V6 Engine

1. Disconnect the negative battery cable.

2. Remove the crankcase cover as described earlier.

3. Position the No. 1 piston at Top Dead Center with the marks on the crankshaft and camshaft sprockets aligned.

4. Remove the camshaft sprocket bolts.

5. Remove the camshaft sprocket and chain from the front of the engine.

NOTE: If the sprocket does not move freely from the camshaft, a light blow using a plastic hammer on the lower edge of the sprocket should dislodge it.

6. Installation is the reverse of removal. Draw the camshaft sprocket onto the camshaft using the mounting bolts. Lubricate the timing chain with engine oil prior to installation.

Timing Belt

REMOVAL & INSTALLATION

OHC Engine Only

NOTE: The following procedure requires the use of a special tool.

1. Remove the timing belt front cover.

2. Rotate the crankshaft so that the timing mark on the crankshaft pulley lines up with the 10° BTDC mark on the indicator scale. The mark on the camshaft sprocket must line up with mark on the camshaft carrier.

3. Remove the crankshaft pulley as previously described.

4. Remove timing probe holder.

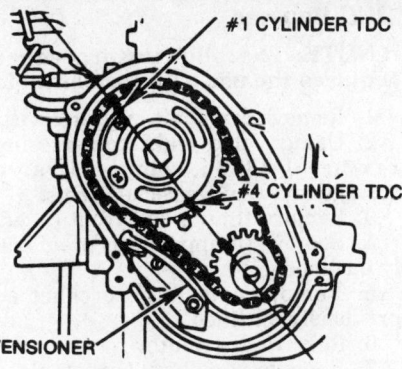

Timing mark alignment on OHV engines

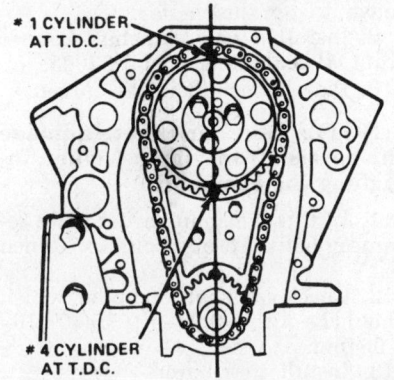

Timing mark alignment on 2.8L V6 engine

5. Loosen the water pump retaining bolts and rotate the water pump to loosen the timing belt.

6. Remove the timing belt.

7. Install timing belt on sprockets.

8. Install the crankshaft pulley.

9. Check if the mark on the camshaft sprocket lines up with mark on the camshaft carrier. The timing mark on the crankshaft pulley should line up at 10° BTDC on the indicator scale.

10. Rotate the water pump clockwise using Tool J-33039 until all slack is removed from the belt. Slightly tighten the water pump retaining bolts.

11. Install Tool J-26486 between the water pump and camshaft sprockets so that the pointer is midway between the sprockets.

NOTE: Whenever a timing belt is replaced on a 1.8L OHC (code 0,J) engine it must be adjusted when the engine is at normal operating temperature (Thermostat Open).

12. If the tension is incorrect, loosen the water pump and rotate it using Tool J-33039 until the proper tension is obtained.

13. Fully torque the water pump retaining bolts to 19 ft. lbs. taking care not to further rotate the water pump.

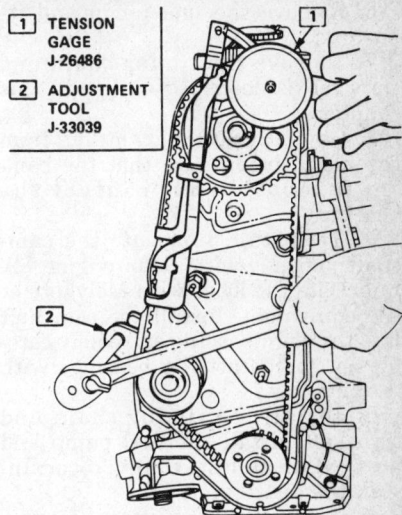

Timing belt tension adjustment on OHC engines

14. Install timing probe holder. Torque nuts to 19 ft. lbs.

15. Install the timing belt front cover and torque the attaching bolts to 5 ft. lbs.

16. Install and adjust the generator and power steering belt. Refill the cooling system, if necessary.

Timing Belt Rear Cover

REMOVAL & INSTALLATION

OHC Engines Only

1. Remove the timing belt from the crankshaft sprocket as previously outlined.

2. Remove the timing belt rear covers attaching bolts and the rear covers.

3. Install the rear covers and torque the attaching bolts to 19 ft. lbs.

4. Install the timing belt and adjust as previously outlined.

Camshaft

REMOVAL & INSTALLATION

OHV 4 Cyl Engines

1. Remove the engine.

2. Remove the intake manifold.

3. Remove the cylinder head cover, pivot the rocker arms to the sides, and remove the pushrods, keeping them in order. Remove the valve lifters, keeping them in order. There are special tools which make lifter removal easier.

4. Remove the front cover.

5. Remove the distributor.

6. Remove the fuel pump and its pushrod.

7. Remove the timing chain and sprocket as described earlier in this chapter.

8. Carefully pull the camshaft from the block, being sure that the camshaft lobes do not contact the bearings.

9. To install, lubricate the camshaft journals with clean engine oil. Lubricate the lobes with Molykote or the equivalent. Install the camshaft into the engine, being extremely careful not to contact the bearings with the cam lobes.

10. Install the timing chain and sprocket. Install the fuel pump and pushrod. Install the timing cover. Install the distributor.

11. Install the valve lifters. If a new camshaft has been installed, new lifters should be used to ensure durability of the cam lobes.

12. Install the pushrods and rocker arms and the intake manifold. Adjust the valve lash after installing the engine. Install the cylinder head cover.

2.8L V6 Engine

1. Disconnect the negative battery cable. Remove the engine assembly from the vehicle.

2. Remove the intake manifold as described earlier.

3. Remove the rocker arm covers. Remove the rocker arm nuts, balls, rocker arms and pushrods.

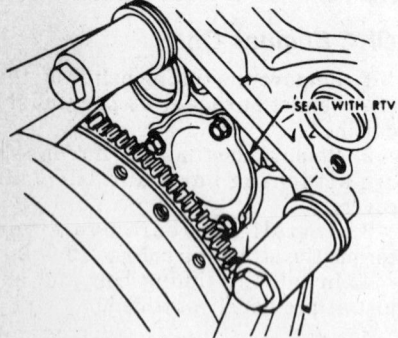

SEAL WITH RTV

Camshaft rear cover on 2.8L V6

4. Remove the upper front cover bolts. Remove the lower cover bolts and the front cover.

5. Remove the camshaft sprocket bolts, camshaft sprocket and timing chain.

6. Remove the camshaft by carefully sliding it out the front of the engine. Measure the camshaft bearing journals using a micrometer and replace the camshaft if the journals exceed .0009 in. (.025mm) out of round.

7. Installation is the reverse of removal. When installing a new camshaft, lubricate the camshaft lobes with GM E.O.S. or equivalent.

OHC Engines

NOTE: The following procedure requires the use of a special tool.

1. Remove camshaft carrier cover.

2. Using valve train compressing Fixture J-33302, compress valve springs and remove rocker arms.

3. Remove timing belt front cover.

4. Remove timing belt as previously outlined.

5. Remove camshaft sprocket as previously outlined.

6. Remove distributor.

7. Remove camshaft thrust plate from rear of camshaft carrier.

8. Slide camshaft rearward and remove it from the carrier.

9. Install a new camshaft carrier front oil seal using Tool J-33085.

10. Place camshaft in the carrier.

NOTE: Take care not to damage the carrier front oil seal when installing the camshaft.

11. Install camshaft thrust plate retaining bolts. Torque bolts to 70 inch lbs.

12. Check camshaft end play, which should be within 0.04–0.16 in. (0.016–0.064mm).

13. Install distributor.

14. Install camshaft sprocket as previously described.

15. Install timing belt as previously described.

16. Install timing belt front cover.

17. Using valve train compressing fixture J-33302, compress valve springs and replace rocker arms.

18. Install camshaft carrier cover as previously described.

Camshaft Carrier

REMOVAL & INSTALLATION

OHC Engines Only

NOTE: Whenever the camshaft carrier bolts are loosened, it is necessary to replace the cylinder head gasket. To do this, see the previous instructions under "Cylinder Head Removal & Installation."

1. Disconnect the positive crankcase ventilation hose from the camshaft carrier.

2. Remove the distributor.

3. Remove the camshaft sprocket as previously outlined.

4. Loosen the camshaft carrier and cylinder head attaching bolts a little at a time in the sequence shown in the "Cylinder Head Removal & Installation" procedure.

NOTE: Camshaft carrier and cylinder head bolts should be loosened only when the engine is cold.

5. Remove the camshaft carrier.

6. Remove the camshaft thrust plate from the rear of the camshaft carrier.

7. Slide the camshaft rearward and remove it from the carrier.

8. Remove the carrier front oil seal.

9. Install a new carrier front oil seal using Tool J-33085.

10. Place the camshaft in the carrier.

NOTE: Take care not to damage the carrier front oil seal when installing the camshaft.

11. Install the camshaft thrust plate and the retaining bolts. Torque the bolts to 70 inch lbs.

12. Check the camshaft end-play which should be within 0.016–0.064 in. (0.04–0.16mm).

13. Clean the sealing surfaces on cylinder head and carrier. Apply a continuous 3mm bead of RTV sealer.

14. Install the camshaft carrier on the cylinder head.

15. Install the camshaft carrier and cylinder head attaching bolts.

16. Torque the bolts a little at a time in the proper sequence, to 18 ft. lbs. Then turn each bolt 60° clockwise in the proper sequence for three times until a 180° rotation is obtained, or equivalent to 1/2 turn. After remainder of installation is completed (with the exception of brackets that attach to carrier), start engine and let it run until thermostat opens. Torque all bolts an additional 30° to 50° in the proper sequence.

17. Install the camshaft sprocket as outlined below.

18. Install the distributor.

19. Connect the positive crankcase ventilation hose to the camshaft carrier.

Camshaft Sprocket

REMOVAL & INSTALLATION

OHC Engines Only

1. Remove the timing belt front cover.

2. Align the mark on camshaft sprocket with mark on camshaft carrier.

3. Remove timing probe holder.

4. Loosen the water pump retaining bolts and remove the timing belt from the camshaft sprocket.

5. Remove the camshaft carrier cover as previously outlined.

6. Hold the camshaft with an open-end wrench. For this purpose a hexagonal is provided in the camshaft. Remove the camshaft sprocket retaining bolt and washer and then the sprocket.

Camshaft sprocket removal on OHC engines

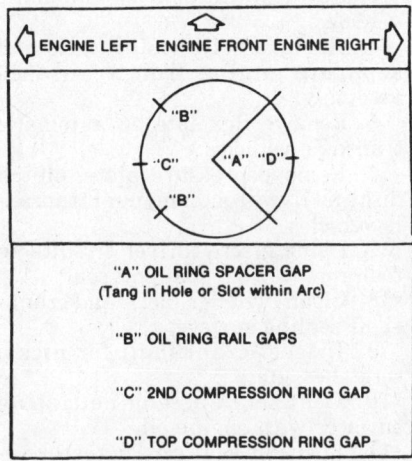

"A" OIL RING SPACER GAP
(Tang in Hole or Slot within Arc)

"B" OIL RING RAIL GAPS

"C" 2ND COMPRESSION RING GAP

"D" TOP COMPRESSION RING GAP

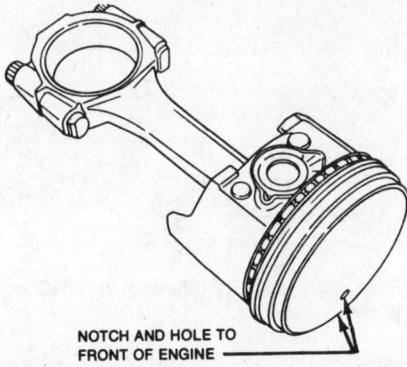

NOTCH AND HOLE TO
FRONT OF ENGINE

Install the piston and rod with the notch and/or hole facing front (engine's front end)

7. Install the camshaft sprocket and align marks on camshaft sprocket and camshaft carrier.

8. Hold the camshaft with a hexagonal open-end wrench. Install the sprocket washer and retaining bolt. Torque to 34 ft. lbs.

9. Install the camshaft carrier cover as previously outlined.

10. Install the timing belt on sprockets and adjust as previously outlined.

11. Install timing probe holder. Torque nuts to 19 ft. lbs.

12. Install timing belt front cover.

Crankshaft Sprocket

REMOVAL & INSTALLATION

OHC Engines Only

1. Remove the timing belt from the crankshaft sprocket as previously described.

2. Remove the crankshaft sprocket to crankshaft attaching bolt and the thrust washer.

3. Remove the sprocket.

4. Position the sprocket over the key on end of crankshaft.

5. Install the thrust washer and the attaching bolt. Torque to 115 ft. lbs.

6. Install the timing belt and adjust as previously described.

Piston and Ring Installation

Pistons are installed with the notch in the top of the piston facing the front end of the engine. See the accompanying illustration for ring positioning. **For engine overhaul procedures, see the Engine Unit Repair Section.**

LUBRICATION SYSTEM

Oil Pan

REMOVAL & INSTALLATION

OHV 4 Cyl Engines

1. Disconnect the negative battery cable.

2. Drain the crankcase. Raise and support the front of the vehicle.

3. Remove the A/C brace if so equipped.

4. Remove the exhaust shield and disconnect the exhaust pipe at the manifold.

5. Remove the starter motor and position it out of the way.

6. Remove the flywheel cover. Remove the oil pan.

NOTE: Prior to oil pan installation, check that the sealing surfaces on the pan, cylinder block and front cover are clean and free of oil. If installing the old pan, be sure that all old RTV has been removed.

7. Apply a ⅛ in. bead of RTV sealant to the oil pan sealing surface. Use

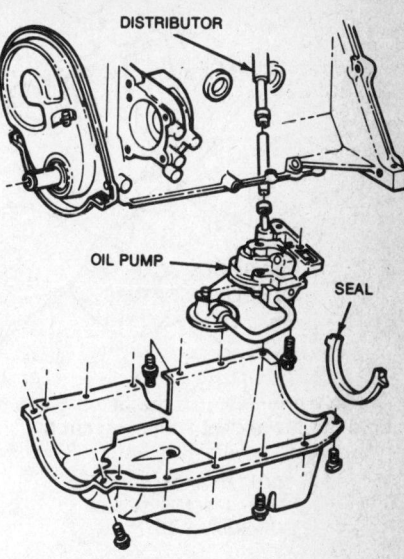

OHV engine oil pan and pump mounting

APPLY RTV SEALER BETWEEN OIL PAN GASKET AND OIL PUMP GASKET

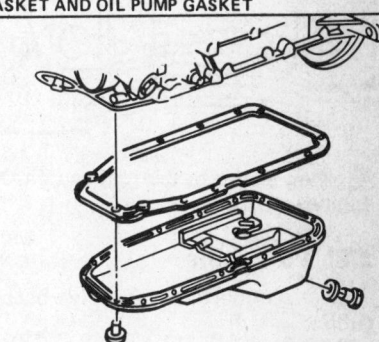

OHC engine oil pan mounting

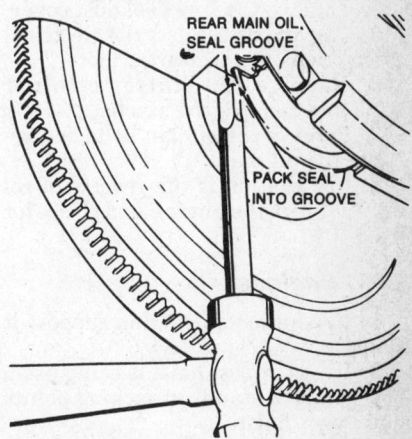

On OHV engines, pack the upper seal into its groove, ¼ inch on each side

a new oil pan rear seal and install the pan in place. Tighten the bolts to 9–13 ft. lbs.

8. Installation of the remaining components is in the reverse order of removal.

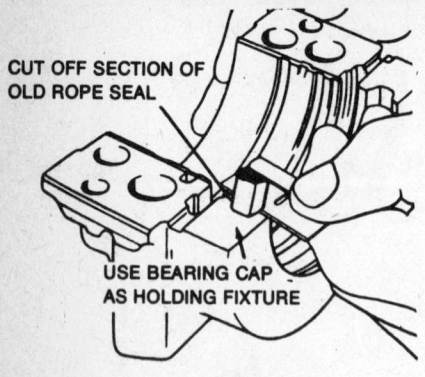

CUT OFF SECTION OF OLD ROPE SEAL

USE BEARING CAP AS HOLDING FIXTURE

On OHV engines, use the bearing cap to hold the lower seal while you cut it

COATED AREA INDICATED WITH #1052357 SEALER OR EQUIVALENT.

SEALER

Applying sealer to the rear cap on OHV engines

2.8L V6 Engine

1. Disconnect the negative battery cable.
2. Raise the vehicle and support it safely.
3. Drain the oil from the crankcase.
4. Remove the flywheel dust cover.
5. Tag and disconnect the electrical connections at the starter motor.
6. Remove the starter retainer bolts and remove the starter.
7. Remove the oil pan bolts and remove the oil pan.
8. Installation is the reverse of removal. Start the engine and check for leaks.

OHC Engines

1. Raise the vehicle and support it safely.
2. If a twin-post hoist is being used, position jack stands at jacking points and lower hoist.
3. Remove right front wheel.
4. Remove right hand splash shield.
5. Remove lower A/C bracket strut rod attachment bolt and swing aside.
6. Remove flywheel dust cover.
7. Remove exhaust pipe to manifold attachment bolts.
8. Drain engine oil.
9. Remove oil pan.
10. Installation is the reverse of removal. Torque the pan bolts to 4 ft. lbs.

Rear Main Oil Seal
REMOVAL & INSTALLATION
OHV 4 Cyl Engines
1980-83 Only

1. Remove the oil pan and pump.
2. Remove the rear main bearing cap.
3. Gently pack the upper seal into the groove approximately $\frac{1}{4}$ inch on each side.
4. Measure the amount the seal was driven in on one side and add $\frac{1}{16}$ in. Cut this length from the old lower cap seal. Be sure to get a sharp cut. Repeat for the other side.
5. Place the piece of cut seal into the groove and pack the seal into the block. Do this for each side.
6. Install a piece of Plastigage or the equivalent on the bearing journal. Install the rear cap and tighten to 75 ft. lbs. Remove the cap and check the gauge for bearing clearance. If out of specification, the ends of the seal may be frayed or not flush, preventing the cap from proper seating. Correct as required.
7. Clean the journal, and apply a thin film of sealer to the mating surfaces of the cap and tighten to 70 ft. lbs. Install the pan and pump.

2.8L V6 Engine

1. Disconnect the negative battery cable.
2. Support the engine and remove the transaxle assembly as described later in this section.
3. Remove the flywheel and verify that the leak is originating from the rear main seal.
4. Remove the seal from the dust lip.

NOTE: Care must be exercised during removal so as not to damage the crankshaft outside diameter area.

5. Clean the cylinder block and crankshaft sealing surface.
6. Inspect the crankshaft for nicks, burrs, scratches, etc.
7. Coat the seal and the engine mating surface with engine oil.
8. Install the new seal using seal installation tool J-34686 or equivalent. Follow the manufacturers instructions supplied with the seal installation tool.
9. For the remainder of the installation, reverse Steps 3 through 1 of the removal procedure.

NOTE: Some 1982 1.8L engines (Code G), experience a rear main

seal oil leak. To correct this condition a new crankshaft part No. 14086053 and a one piece rear main seal kit part No. 14081761 has been released for service. The one piece seal kit contains an installation tool, rear main seal, and an instruction sheet.

OHC Engines

NOTE: The following requires the use of special tools.

1. Remove engine as previously outlined.
2. Remove flywheel dust cover.
3. Remove flexplate to torque converter attachment bolts on automatic vehicles.
4. Remove bellhousing bolts and separate engine from transaxle assembly.
5. Remove flexplate on automatic transaxle vehicles.
6. Remove pressure plate, clutch disc and flywheel on manual transaxle vehicles.
7. Using a screwdriver or suitable tool, remove rear main oil seal.
8. Clean cylinder block and crankshaft sealing surface.
9. Inspect crankshaft for nicks, scratches, etc.
10. Coat seal and engine mating surfaces with engine oil.
11. Position seal on Protector (J-

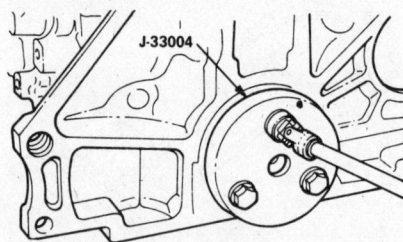

Rear main seal installation on OHC engines

33084-2) and place onto crankshaft flywheel flange.
12. Install Seal Installer (J-33084) on crankshaft flywheel flange, starting the three (3) bolts EVENLY in a rotational sequence until the seal bottoms in the block.
13. Installation is in the reverse order of removal.

Oil Pump

REMOVAL & INSTALLATION

OHV 4 Cyl Engines

1. Remove the engine oil pan.
2. Remove the pump attaching bolts and carefully lower the pump.
3. Install in reverse order. To en-

sure immediate oil pressure on start-up, the oil pump gear cavity should be packed with petroleum jelly. Installation torque is 26–35 ft. lbs.

2.8L V6 Engine

1. Disconnect the negative battery cable.
2. Remove the oil pan as described earlier.
3. Remove the bolt holding the rear main bearing cap.
4. Remove the oil pump and extension shaft.
5. Installation is the reverse of removal.

OHC Engines

1. Remove the crankshaft sprocket.
2. Remove the timing belt rear covers.
3. Disconnect the oil pressure switch wires.
4. Remove the oil pan.
5. Remove the oil filter.
6. Unbolt and remove the oil pick-up tube.
7. Unbolt and remove the oil pump.
8. Installation is the reverse of removal. Use new gaskets in all instances. Torque the oil pump bolts to 5 ft. lbs. Torque the oil pan bolts to 4 ft. lbs., and the oil pickup tube bolts to 5 ft. lbs.

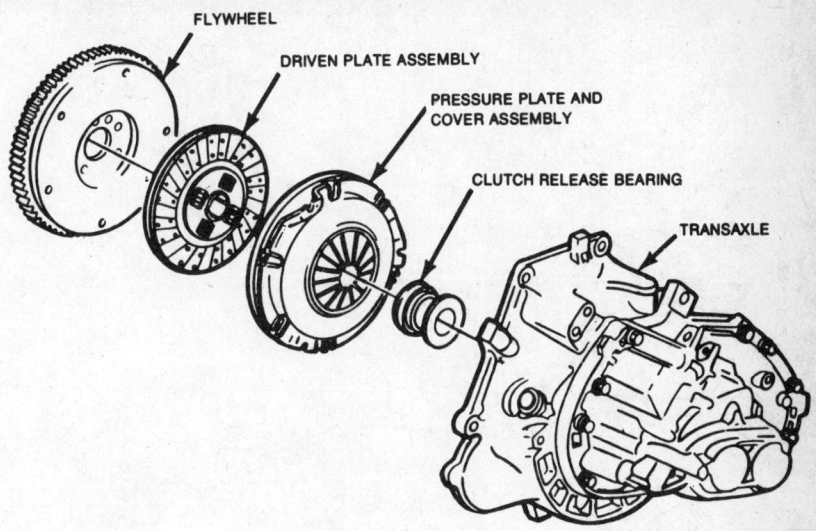

Clutch components

CLUTCH

ADJUSTMENT

The J-cars have a self-adjusting clutch mechanism located on the clutch pedal, eliminating the need for periodic free play adjustments. The self-adjusting mechanism should be inspected periodically as follows:

1. Depress the clutch pedal and look for the pawl on the self-adjusting mechanism to firmly engage the teeth on the ratchet.
2. Release the clutch. The pawl should be lifted off of the teeth by the metal stop on the bracket.

CLUTCH REMOVAL & INSTALLATION

1. Remove the transaxle.
2. Mark the pressure plate assembly and the flywheel so that they can be assembled in the same position. They were balanced as an assembly at the factory.
3. Loosen the attaching bolts one turn at a time until spring tension is relieved.

4. Support the pressure plate and remove the bolts. Remove the pressure plate and clutch disc. Do not disassemble the pressure plate assembly; replace it if defective.
5. Inspect the flywheel, clutch disc, pressure plate, throwout bearing and the clutch fork and pivot shaft assembly for wear. Replace the parts as required. If the flywheel shows any signs of overheating, or if it is badly grooved or scored, it should be refaced or replaced.
6. Clean the pressure plate and flywheel mating surfaces thoroughly. Position the clutch disc and pressure plate into the installed position, and support with a dummy shaft or clutch aligning tool. The clutch plate is assembled with the damper springs offset toward the transaxle. One side of the factory-supplied clutch disc is stamped "Flywheel Side."
7. Install the pressure plate-to-flywheel bolts. Tighten them gradually in a criss-cross pattern.
8. Lubricate the outside groove and the inside recess of the release bearing with high temperature grease. Wipe off any excess. Install the release bearing.
9. Install the transaxle.

Neutral Start Switch

A neutral start switch is located on the clutch pedal assembly; the switch prevents the engine from starting unless the clutch is depressed. If the switch is faulty, it can be unbolted and replaced without removing the pedal assembly from the car. No adjustments for the switch are provided.

Clutch Cable

REPLACEMENT

1. Press the clutch pedal up against the bumper stop so as to release the pawl from the detent. Disconnect the clutch cable from the release lever at the transaxle assembly. Be careful that the cable does not snap back toward the rear of the car as this could damage the detent in the adjusting mechanism.
2. Remove the hush panel from inside the car.
3. Disconnect the clutch cable from the detent end tangs. Lift the locking pawl away from the detent and then pull the cable forward between the detent and the pawl.
4. Remove the windshield washer bottle.
5. From the engine side of the cowl, pull the clutch cable out to disengage it from the clutch pedal mounting bracket. The insulators, dampener and washers may separate from the cable in the process.
6. Disconnect the cable from the transaxle mounting bracket and remove it.
7. Install the cable into both insulators, damper and washer. Lubricate the rear insulator with tire mounting lube or the like to ease installation into the pedal mounting bracket.
8. From inside the car, attach the end of the cable to the detent. Be sure to route the cable underneath the pawl and into the detent cable groove.
9. Press the clutch pedal up against the bumper stop to release the pawl

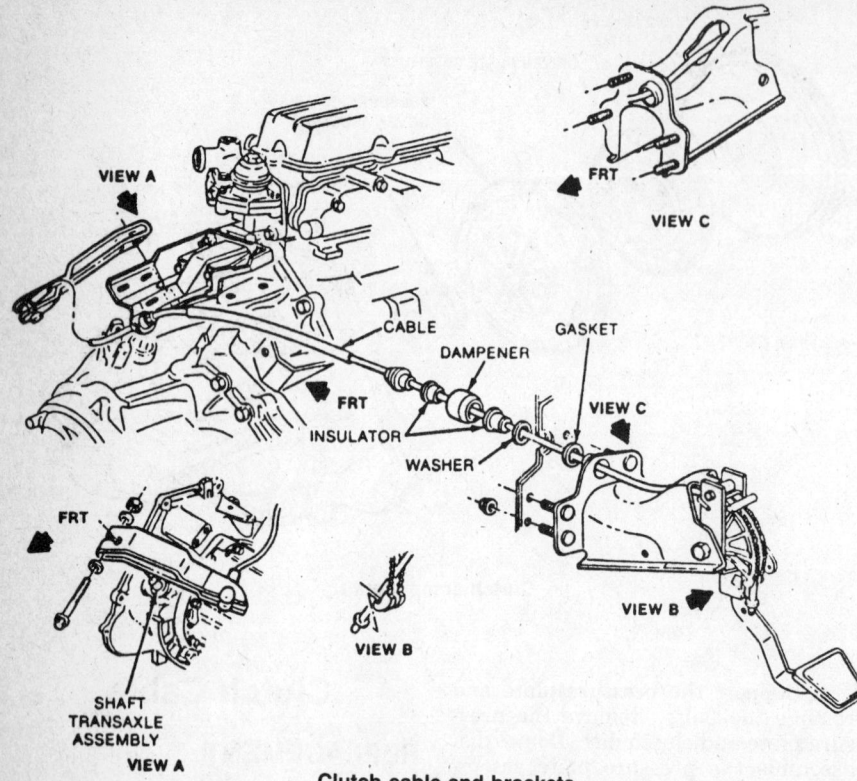

Clutch cable and brackets

from the detent. Install the other end of the cable at the release lever and the transaxle mount bracket.

10. Install the hush panel and the windshield washer bottle.

11. Check the clutch operation.

MANUAL TRANSAXLE

REMOVAL & INSTALLATION

1. Disconnect the negative battery cable.

2. Install an engine holding bar so that one end is supported on the cowl tray over the wiper motor and the other end rests on the radiator support. Use padding and be careful not to damage the paint or body work with the bar. Attach a lifting hook to the engine lift ring and to the bar and raise the engine enough to take the pressure off the motor mounts.

NOTE: If a lifting bar and hook is not available, a chain hoist can be used, however, during the procedure the vehicle must be raised, at which time the chain hoist must be adjusted to keep tension on the engine/transaxle assembly.

3. Remove the heater hose clamp at the transaxle mount bracket. Disconnect the electrical connector and remove the horn assembly.

4. Remove the transaxle mount attaching bolts. Discard the bolts attaching the mount to the side frame; new bolts must be used at installation.

5. Disconnect the clutch cable from the clutch release lever. Remove the transaxle mount bracket attaching bolts and nuts.

6. Disconnect the shift cables and retaining clips at the transaxle. Disconnect the ground cables at the transaxle mounting stud.

7. Remove the four upper transaxle-to-engine mounting bolts.

8. Raise the vehicle and support it on stands. Remove the left front wheel.

9. Remove the left front inner splash shield. Remove the transaxle strut and bracket.

10. Remove the clutch housing cover bolts.

11. Disconnect the speedometer cable at the transaxle.

12. Disconnect the stabilizer bar at the left suspension support and control arm.

13. Disconnect the ball joint from the steering knuckle.

14. Remove the left suspension support attaching bolts and remove the support and control arm as an assembly.

15. Install boot protectors and disengage the drive axles at the transaxle. Remove the left side shaft from the transaxle.

16. Position a jack under the transaxle case, remove the lower two transaxle-to-engine mounting bolts and remove the transaxle by sliding it towards the driver's side, away from the engine. Carefully lower the jack, guiding the right shaft out the transaxle.

17. When installing the transaxle, guide the right drive axle into its bore as the transaxle is being raised. The right drive axle CANNOT be readily installed after the transaxle is connected to the engine. Installation of the remaining components is in the reverse order of removal with the following notes. Tighten the transaxle-to-engine mounting bolts to 55 ft. lbs. Tighten the suspension support-to-body attaching bolts to 75 ft. lbs. and the clutch housing cover bolts to 10 ft. lbs. Using new bolts, install and tighten the transaxle mount-to-side frame to 40 ft. lbs. When installing the bolts attaching the mount-to-transaxle bracket, check the alignment bolt at the engine mount. If excessive effort is required to remove the alignment bolt, realign the powertrain components and tighten the bolts to 40 ft. lbs., and then remove the alignment bolt.

OVERHAUL

For all manual transmission or

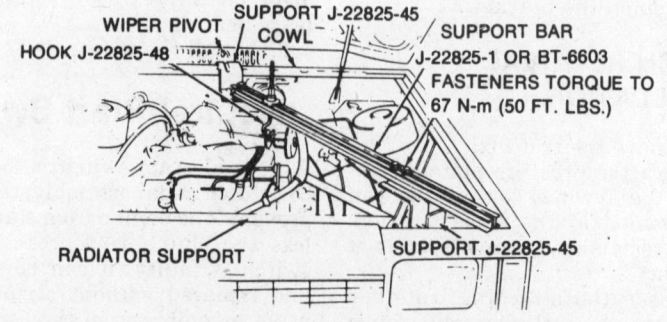

Install an engine holding bar when removing the manual transmission

transaxle overhaul procedures, please refer to "Manual Transmission Overhaul" in the unit repair section.

SHIFT LINKAGE ADJUSTMENT

1. Disconnect the negative battery cable.

2. Place the transaxle in first gear, then loosen the shift cable attaching pins at the transaxle levers on the transaxle case.

3. Remove the shifter boot and retainer.

4. Install a No. 22 ($^5/_{32}$ in.) drill bit into the alignment hole at the side of the shifter assembly. Install a yoke clip between the shifter tower and carrier.

5. Remove the lash from the transaxle by rotating the upright select lever (lever D) while tightening the cable attaching pin nut.

6. Remove the drill bit and yoke at the shifter assembly, install the shifter boot and retainer and connect the negative battery cable.

7. Road test the vehicle to check for good gate feel during shifting. Fine tune the adjustment as necessary.

AUTOMATIC TRANSAXLE

For all adjustments, see the "Automatic Transmission" Unit Repair Section.

REMOVAL & INSTALLATION

1. Disconnect the negative battery cable where it attaches to the transaxle.

2. Insert a $^1/_4$ x 2 in. bolt into the hole in the right front motor mount to prevent any mislocation during the transaxle removal.

3. Remove the air cleaner. Disconnect the T.V. cable at the carburetor.

4. Unscrew the bolt securing the T.V. cable to the transaxle. Pull up on the cable cover at the transaxle until the cable can be seen. Disconnect the cable from the transaxle rod.

5. Remove the wiring harness retaining bolt at the top of the transaxle.

6. Remove the hose from the air management valve and then pull the wiring harness up and out of the way.

7. Install an engine support bar as shown in the illustration. Raise the

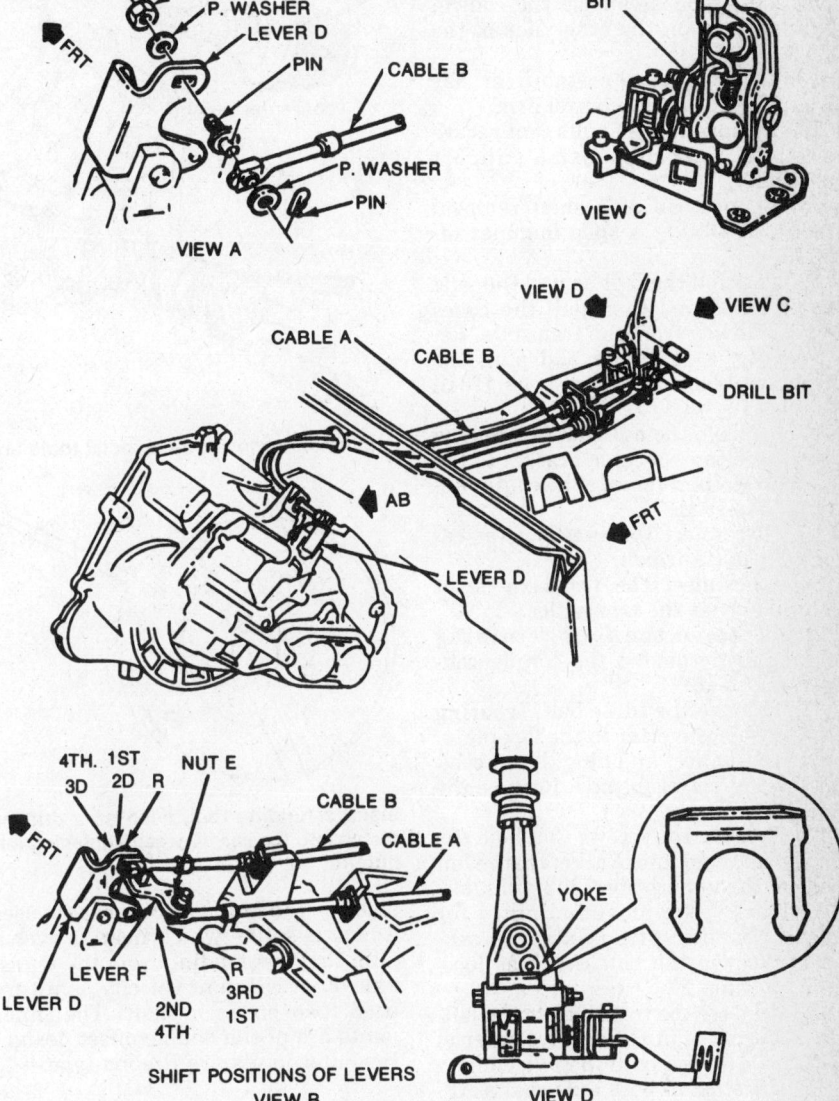

VIEW A

VIEW C

Shift linkage adjustment

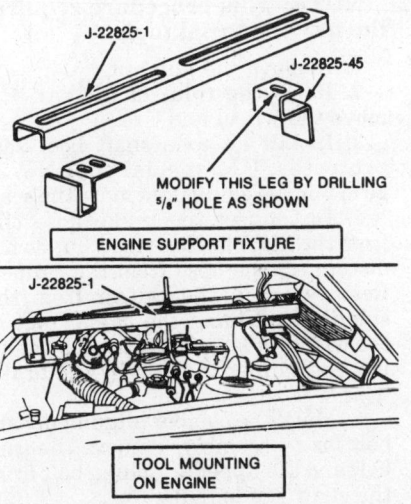

Supporting the engine with a holding bar when removing the automatic transmission

engine just enough to take the pressure off the motor mounts.

—— **CAUTION** ——

The engine support bar must be located in the center of the cowl and the bolts must be tightened before attempting to support the engine.

8. Remove the transaxle mount and bracket assembly. It may be necessary to raise the engine slightly to aid in removal.

9. Disconnect the shift control linkage from the transaxle.

10. Remove the top transaxle-to-engine mounting bolts. Loosen, but do not remove, the transaxle-to-engine bolt nearest to the starter.

11. Unlock the steering column. Raise and support the front of the car. Remove the front wheels.

12. Pull out the cotter pin and loosen

the castellated ball joint nut until the ball joint separates from the control arm. Repeat on the other side of the car.

13. Disconnect the stabilizer bar from the left lower control arm.

14. Remove the six bolts that secure the left front suspension support assembly.

15. Connect an axle shaft removal tool (J-28468) to a slide hammer (J-23907).

16. Position the tool behind the axle shaft cones and then pull the cones out and away from the transaxle. Remove the axle shafts and plug the transaxle bores to reduce fluid leakage.

17. Remove the nut that secures the transaxle control cable bracket to the transaxle, then remove the engine-to-transaxle stud.

18. Disconnect the speedometer cable at the transaxle.

19. Disconnect the transaxle strut (stabilizer) at the transaxle.

20. Remove the four retaining screws and remove the torque converter shield.

21. Remove the three bolts securing the torque converter to the flex plate.

22. Disconnect and plug the oil cooler lines at the transaxle. Remove the starter.

23. Remove the screws that hold the brake and fuel line brackets to the left side of the underbody. This will allow the lines to be moved slightly for clearance during transaxle removal.

24. Remove the bolt that was loosened in Step 10.

25. Remove the transaxle to the left. Installation is in the reverse order of removal. Please note the following:

  a. Reinstall both axle shafts AFTER the transaxle is in position.

  b. When installing the front suspension support assembly you must follow the tightening sequence shown in the illustration.

  c. Check alignment when installation is complete.

DRIVE AXLE

Halfshafts

The J-cars use unequal-length halfshafts, with specific application for automatic and manual transaxle use. All halfshafts except the left-hand inboard joint of the automatic transaxle incorporate a male spline; the shafts interlock with the transaxle gears through the use of barrel-type snap rings. The left-hand inboard

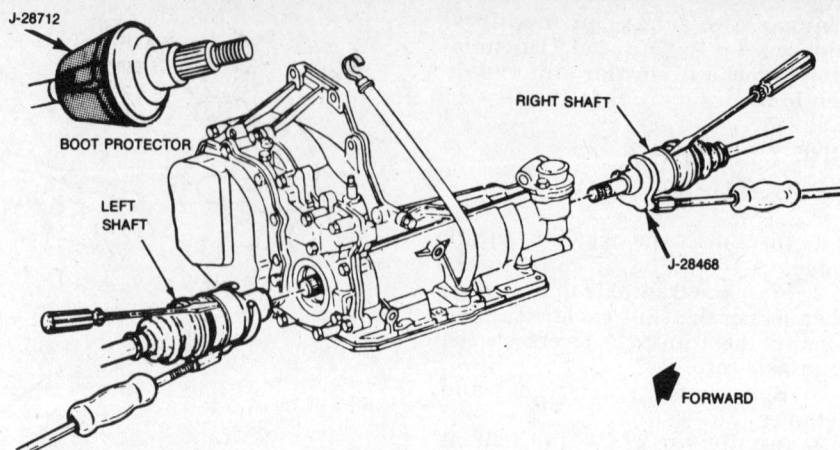

Halfshaft removal; the special tools are attached to slide hammers in this diagram

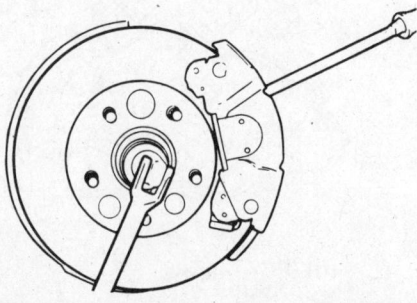

Insert a holding tool, such as a drift or punch into the caliper when tightening the hub nut

shaft on the automatic transaxle uses a female spline which installs over a stub shaft protruding from the transaxle. Four constant velocity joints are used, two on each shaft. The inner joints are of the double offset design; the outer joints are Rzeppa-type.

REMOVAL & INSTALLATION

NOTE: This procedure requires the use of special tools.

1. Remove the hub nut.

2. Raise the front of the car. Remove the wheel and tire.

3. Install an axle shaft boot seal protector, GM special tool No. J-28712 or the equivalent, onto the seal.

4. Disconnect the brake hose clip from the MacPherson strut, but do not disconnect the hose from the caliper. Remove the brake caliper from the spindle, and hang the caliper out of the way by a length of wire. Do not allow the caliper to hang by the brake hose.

5. Mark the camber alignment cam bolt for reassembly. Remove the cam bolt and the upper attaching bolt from the strut and spindle.

6. Pull the steering knuckle assembly from the strut bracket.

7. Using GM special tool J-28468 or the equivalent, remove the axle shaft from the transaxle.

8. Using GM special tool J-28733 or the equivalent spindle remover, remove the axle shaft from the hub and bearing assembly.

9. If a new drive axle is to be installed, a new knuckle seal should be installed first.

10. Loosely install the drive axle into the transaxle and steering knuckle.

11. Loosely attach the steering knuckle to the suspension strut.

12. Install the brake caliper. Tighten the bolts to 30 ft. lbs. (40 Nm).

13. The drive axle is an interference fit in the steering knuckle. Press the axle into place, then install the hub nut. When the shaft begins to turn with the hub, insert a drift through the caliper into one of the cooling slots in the rotor to keep it from turning. Tighten the hub nut to 70 ft. lbs. (100 Nm) to completely seat the shaft.

14. Load the hub assembly by lowering it onto a jackstand. Align the camber cam bolt marks made during removal, install the bolt and tighten to 140 ft. lbs. (190 Nm). Tighten the upper nut to the same value.

15. Install the axle shaft all the way into the transaxle using a screwdriver inserted into the groove provided on the inner retainer. Tap the screwdriver until the shaft seats in the transaxle.

16. Connect the brake hose clip to the strut. Install the tire and wheel, lower the car, and tighten the hub nut to 225 ft. lbs. (305 Nm).

CV-JOINT OVERHAUL

For all CV-Joint overhaul procedures, please refer to "CV-Joint Overhaul" in the unit repair section.

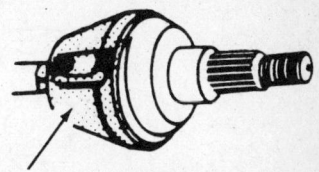

Rear Axle/Axle Shafts
REMOVAL & INSTALLATION

1. Raise the vehicle on a hoist and support assembly with jack stands under the control arms.
2. Remove the stabilizer bar from the axle assembly, if so equipped.
3. Remove the wheel and tire assembly and brake drum.

NOTE: Do not hammer on the brake drum as damage to the bearing could result.

4. Remove shock absorber lower attaching bolts and paddle nuts at axle and disconnect shocks from control arm.
5. Disconnect the parking brake cable from the axle assembly.
6. To assure that the axle assembly is not suspended by the brake lines, disconnect brake line at the brackets from axle assembly.
7. Lower the rear axle and remove the coil springs and insulators.
8. Remove control arm bolts from underbody bracket and lower axle.
9. Remove the hub attaching bolts and remove hub, bearing and backing plate assembly.

NOTE: Be careful not to drop hub and bearing assembly, as damage to the bearing could result. All rear suspension fasteners are important attaching parts in that they could affect the performance of vital parts and systems, and/or could result in major repair expense. They must be replaced with one of the same part number or with a equivalent part if replacement becomes necessary. Do not use a replacement part of lesser quality or substitute design. Torque values must be used as specified during reassembly to assure proper retention of all parts. There is to be no welding as it may result in extensive damage and weakening of the metal.

10. Installation is the reverse of the removal procedure.
11. Torque to the following specifications:

    a. Shock absorber upper attaching nut 13 ft. lbs.

    b. Upper shock mount nuts to shock tower 13 ft. lbs.

    c. Lower attaching bolt to axle 41 ft. lbs.

    d. Stabilizer bar attaching nuts to axle 10 ft.lbs.

    e. Control arm attaching nuts and bolts 13 ft. lbs.

    f. Control arm to body bracket bolt 67 ft. lbs.

    g. Brake line bracket to frame screw 8 ft. lbs.

    h. Bracket to control arm screw 11 ft. lbs.

    i. Hub and bearing assembly to rear axle bolt 39 ft. lbs.

    j. Wheel lug nuts 103 ft. lbs.

12. Bleed the brake system, refill the master cylinder and lower the vehicle.

Front Axle Hub and Bearings

REMOVAL & INSTALLATION

1. Loosen the hub nut.
2. Raise and support the vehicle properly, then remove the wheel and tire.
3. Install boot cover J-33162 on L4 engine with automatic trans.
4. Remove the hub nut.

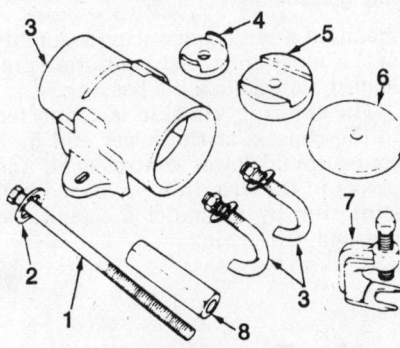

1. Bolt J21474-19
2. Bushing
3. Control Arm bushing remover and installer J29376
4. Remover J29376-6
5. Installer J29376-4
6. Plate J29376-7
7. Wheel stud remover J6627-A
8. Nut J21474-18

Special Tools

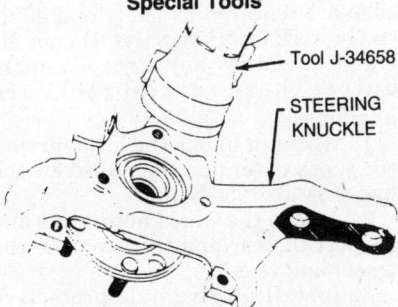

Tool J-34658

STEERING KNUCKLE

Installation of steering knuckle seal

USE J-33162 FOR TRI-POT JOINT

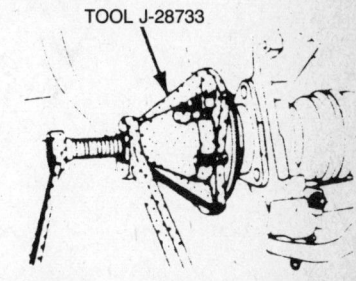

Install drive axle boot protector on the right side, inboard axle boot.

TOOL J-28733

Removal of the hub and bearing assembly

5. Remove the caliper and rotor as shown in the brake section.
6. Remove three hub and bearing mounting bolts.
7. Remove the splash shield.

NOTE: If the bearing assembly is to be re-used, mark attaching bolt and corresponding hole for installation in the same position. Use of a hammer or direct heat should be avoided, as this could result in internal bearing damage.

8. Install hub and bearing assembly remover tool J-28733 and turn bolt to press the hub and bearing assembly off of the drive axle.
9. Disconnect the stabilizer link bolt at the lower control arm.
10. Seperate the ball joint using J-29330.

    a. Place J29330 into position as shown.

    b. Loosen nut and back off until the nut contacts the tool.

    c. Continue backing off the nut until it forces the ball stud out of the knuckle.

11. Remove the drive axle from the knuckle and support it out of the way.
12. Remove the inner knuckle to the inboard side using a brass drift.
13. Clean and inspect the bearing mating surfaces and steering knuckle bore.

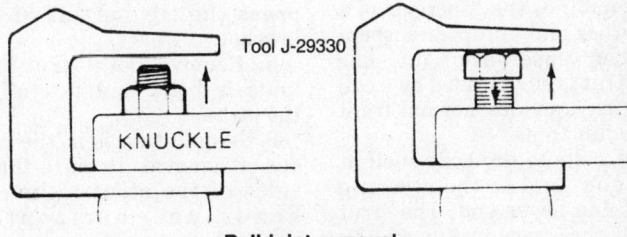

Tool J-29330

KNUCKLE

Ball joint removal

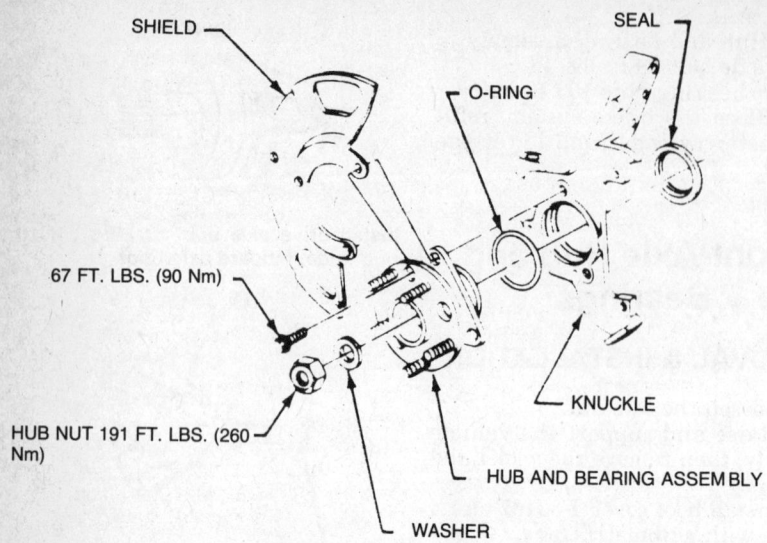

SHIELD

SEAL

O-RING

67 FT. LBS. (90 Nm)

KNUCKLE

HUB NUT 191 FT. LBS. (260 Nm)

HUB AND BEARING ASSEMBLY

WASHER

Typical hub and bearing assembly

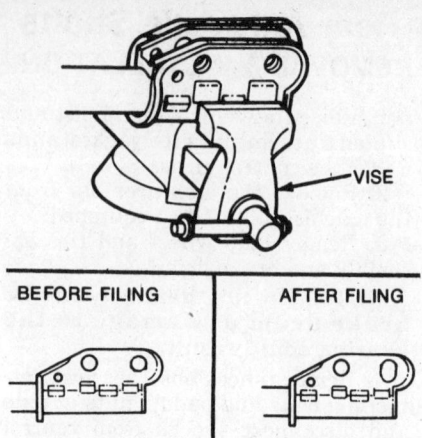

VISE

| BEFORE FILING | AFTER FILING |

Modifying the strut mounting holes

14. Install a new O-ring between the bearing assembly and the knuckle assembly.

15. Install the hub and bearing assembly and torque to 67 FT.lbs.

16. Install a new knuckle seal using J-34658 from the inboard side, generously lubricating the seal lips and cavity between the seal and bearing with GM 9985254 or equvalent wheel bearing grease.

17. Install drive axle to hub and bearing assembly being careful not to damage the seal.

18. Reconnect the lower ball joint and stabilizer link.

19. Install a new hub and bearing nut on drive axle and apply partial torque of 74 ft. lbs.

20. Install the rotor and caliper (see brake section).

21. Install the wheel and tire.

22. Lower vehicle and torque hub nut to 191 ft. lbs.

FRONT SUSPENSION

The J-cars use a MacPherson strut front suspension design. A MacPherson strut combines the functions of a shock absorber and an upper suspension member (upper arm) into one unit. The strut is surrounded by a coil spring, which provides normal front suspension functions.

The strut bolts to the body shell at its upper end, and to the steering knuckle at the lower end. The strut pivots with the steering knuckle by

means of a sealed mounting assembly at the upper end which contains a pre-loaded, non-adjustable bearing.

The steering knuckle is connected to the chassis at the lower end by a conventional lower control arm, and pivots in the arm in a preloaded ball joint stud by means of a castellated nut and cotter pin.

MacPherson Struts

REMOVAL & INSTALLATION

The struts retain the springs under tremendous pressure even when removed from the car. For these reasons, several expensive special tools and substantial specialized knowledge are required to safely and effectively work on these parts. Do not attempt to service any strut assembly unless these special tools are available.

1. Working under the hood, pry off the shock cover and then unscrew the upper strut-to-body nuts.

2. Loosen the wheel nuts, raise and support the car and then remove the wheel and tire.

3. Install a drive axle protective cover (J28712).

4. Use a two-armed puller and press the tie rod out of the strut bracket.

5. Remove both strut-to-steering knuckle bolts and carefully lift out the strut.

6. Installation is in the reverse order of removal. Be sure that the flat sides of the strut-to-knuckle bolt heads are horizontal (see illustration).

OVERHAUL

For all spring and shock absorber removal & installation procedures, and all strut overhaul procedures, please refer to "Strut Overhaul" in the Unit Repair Section.

STRUT MODIFICATION

This modification is made only if a camber adjustment is anticipated.

1. Place the strut in a vise. This step is not absolutely necessary; filing can be accomplished by disconnecting the strut from the steering knuckle.

2. File the holes in the outer flanges so as to enlarge the bottom holes until they match the slots already in the inner flanges.

3. Camber adjustment procedures are detailed later in this chapter.

Coil Springs

REMOVAL & INSTALLATION

———— CAUTION ————

The coil springs are retained under considerable pressure. They can exert enough force to cause serious injury. Exercise extreme caution when disassembling the strut for coil spring removal.

This procedure requires the use of a spring compressor and several other special tools. It cannot be performed without them. If you do not have access to these tools, DO NOT attempt to disassemble the strut.

1. Remove the strut assembly.

2. Clamp the spring compressor (J26584) in a vise. Position the strut assembly in the bottom adapter of the compressor and install the special tool J26584-86 (see illustration). Be sure that the adapter captures the strut and that the locating pins are engaged.

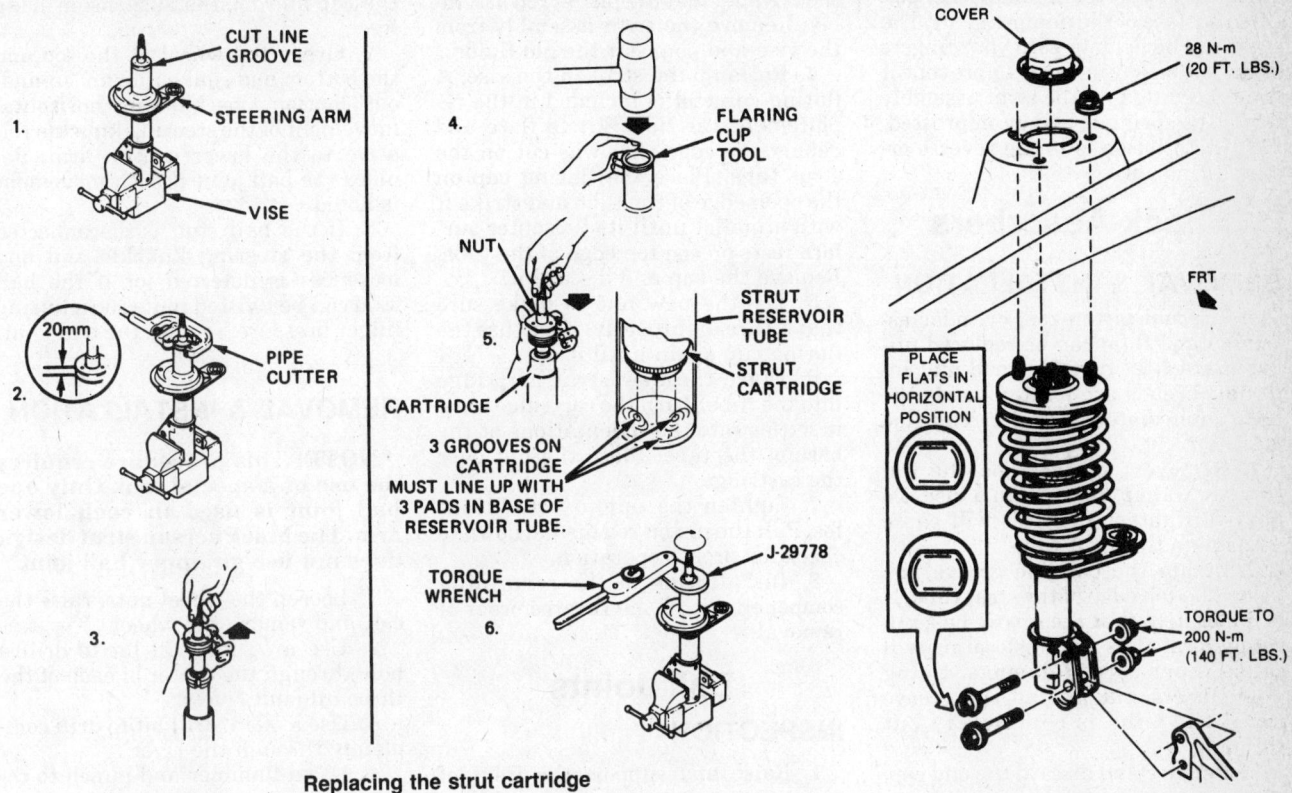

Replacing the strut cartridge

Strut assembly mounting

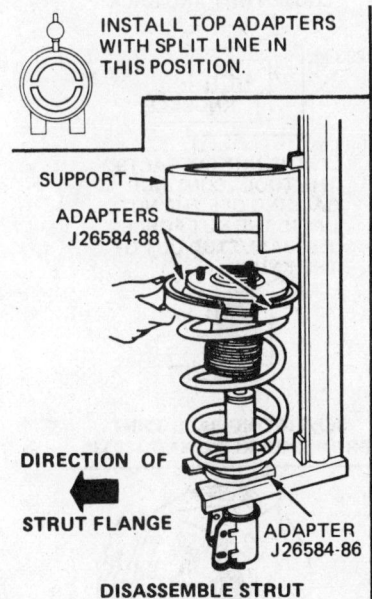

INSTALL TOP ADAPTERS WITH SPLIT LINE IN THIS POSITION.

SUPPORT

ADAPTERS J26584-88

DIRECTION OF

STRUT FLANGE

ADAPTER J26584-86

DISASSEMBLE STRUT

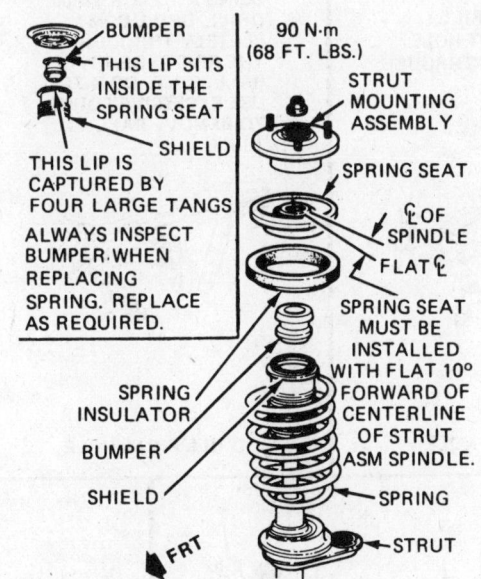

BUMPER

THIS LIP SITS INSIDE THE SPRING SEAT

SHIELD

THIS LIP IS CAPTURED BY FOUR LARGE TANGS

ALWAYS INSPECT BUMPER WHEN REPLACING SPRING. REPLACE AS REQUIRED.

90 N·m (68 FT. LBS.)

STRUT MOUNTING ASSEMBLY

SPRING SEAT

₵ OF SPINDLE

FLAT ₵

SPRING SEAT MUST BE INSTALLED WITH FLAT 10° FORWARD OF CENTERLINE OF STRUT ASM SPINDLE.

SPRING INSULATOR

BUMPER

SHIELD

SPRING

STRUT

FRT

Coil spring removal and installation

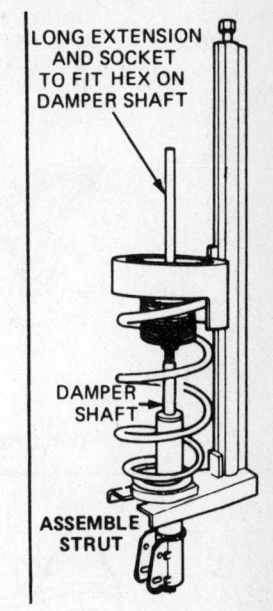

LONG EXTENSION AND SOCKET TO FIT HEX ON DAMPER SHAFT

DAMPER SHAFT

ASSEMBLE STRUT

3. Rotate the strut assembly so that the top mounting assembly lip aligns with the compressor support notch. Insert two top adapters (J26584-88) between the top mounting assembly and the top spring seat. Position the adapters so that the split lines are in the 3 o'clock and 9 o'clock positions.

4. Using a 1 in. socket, turn the screw on top of the compressor clockwise until the top support flange contacts the adapters. Continue turning the screw until the coil spring is compressed approximately ½ in. (4 complete turns). Never bottom the spring or the strut damper rod.

5. Unscrew the nut from the strut

damper shaft and then lift off the top mounting assembly.

6. Turn the compressor adjusting screw counterclockwise until the spring tension has been relieved. Remove the adapters and then remove the coil spring.

7. When installing a new spring, NEVER place a hard tool such as pli-

ers or screwdriver against the polished surface of the damper shaft. The shaft can be held up with your fingers or an extension in order to prevent it from receding into the strut assembly while the spring is being compressed.

8. Installation is in the reverse order of removal.

Shock Absorbers

REMOVAL & INSTALLATION

The internal piston rod, cylinder assembly and fluid can be replaced utilizing a service cartridge and nut. Internal threads are located inside the tube immediately below a cut line groove.

1. Remove the strut and the coil springs. Clamp the strut in a vise. Do not overtighten it as this will cause damage to the strut tube.

2. Locate the cut line groove just below the top edge of the strut tube. It is imperative that the groove be accurately located as any mislocation will cause inner thread damage. Using pipe cutters, cut around the groove until the tube is completely cut through.

3. Remove and discard the end cap,

the cylinder and the piston rod assembly. Remove the strut assembly from the vise and pour out the old fluid.

4. Reclamp the strut in the vise. A flaring cup tool is included in the replacement cartridge kit to flare and deburr the edge that was cut on the strut tube. Place the flaring cup on the open edge of the tube and strike it with a mallet until its flat outer surface rests on the top edge of the tube. Remove the cup and discard it.

5. Try the new nut to make sure that it threads properly. If not, use the flaring cup again until it does.

6. Place the new strut cartridge into the tube. Turn the cartridge until it settles into the indentations at the base of the tube. Place the nut over the cartridge.

7. Tighten the nut to 140–170 ft. lbs. Pull the piston rod up and down to check for proper operation.

8. Installation of the remaining components is in the reverse order of removal.

Ball Joints

INSPECTION

1. Raise and support the front of

the car and let the suspension hang free.

2. Grasp the wheel at the top and the bottom and shake it in an "in-and-out" motion. Check for any horizontal movement of the steering knuckle relative to the lower control arm. Replace the ball joint if such movement is noted.

3. If the ball stud is disconnected from the steering knuckle and any looseness is detected, or if the ball stud can be twisted in its socket using finger pressure, replace the ball joint.

REMOVAL & INSTALLATION

NOTE: This procedure requires the use of a special tool. Only one ball joint is used in each lower arm. The MacPherson strut design does not use an upper ball joint.

1. Loosen the wheel nuts, raise the car, and remove the wheel.

2. Use a ⅛ in. drill bit to drill a hole through the center of each of the three all joint rivets.

3. Use a ½ in. drill bit to drill completely through the rivet.

4. Use a hammer and punch to re-

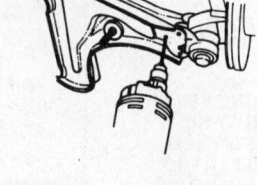

USING 1/8" DRILL, DRILL A PILOT HOLE COMPLETELY THROUGH THE RIVET.

DRILL PILOT HOLE

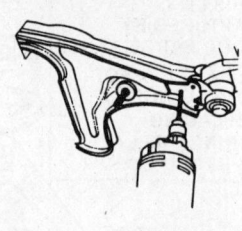

USING A 1/2" OR 13mm DRILL, DRILL COMPLETELY THROUGH THE RIVET. REMOVE BALL JOINT. DO NOT USE EXCESSIVE FORCE TO REMOVE BALL JOINT.

DRILL FINAL HOLE

PLACE J 29330 INTO POSITION AS SHOWN. LOOSEN NUT AND BACK OFF UNTIL . . .

J29330

KNUCKLE

. . . THE NUT CONTACTS THE TOOL. CONTINUE BACKING OFF THE NUT UNTIL THE NUT FORCES THE BALL STUD OUT OF THE KNUCKLE.

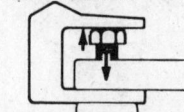

SEPARATING BALL JOINT FROM KNUCKLE USING J29330

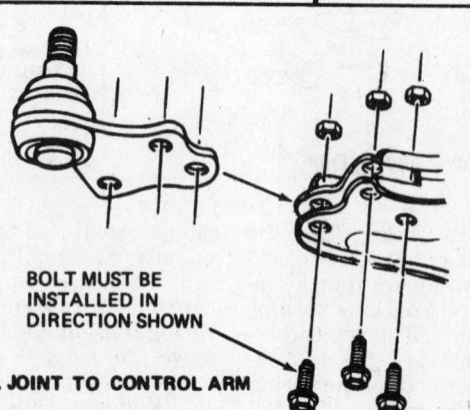

BOLT MUST BE INSTALLED IN DIRECTION SHOWN

INSTALL BALL JOINT TO CONTROL ARM

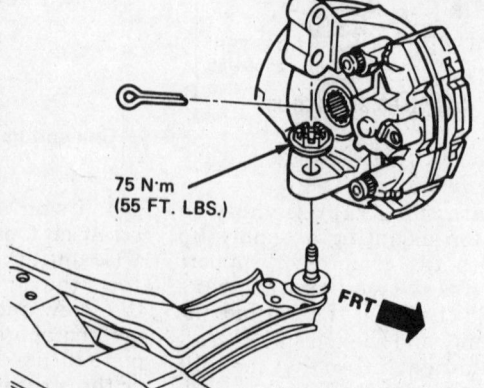

75 N·m (55 FT. LBS.)

FRT

Ball joint removal and installation

move the rivets. Drive them out from the bottom.

5. Use the special tool J29330 or a ball joint removal tool to separate the ball joint from the steering knuckle (see illustration). Don't forget to remove the cotter pin.

6. Disconnect the stabilizer bar from the lower control arm. Remove the ball joint.

7. Install the new ball joint into the control arm with the three bolts supplied as shown. Installation of the remaining components is in the reverse order of removal. Use a new cotter pin when installing the castellated nut on the ball joint. Check the toe setting and adjust as necessary.

Control Arm

REMOVAL & INSTALLATION

1. Raise and support the front of the car. Remove the wheel.
2. Disconnect the stabilizer bar from the control arm and/or support.
3. Separate the ball joint from the steering knuckle as previously detailed.
4. Remove the two control arm-to-support bolts and remove the control arm.
5. If control arm support bar removal is necessary, unscrew the six mounting bolts and remove the support.
6. Installation is in the reverse order of removal. Tighten the control arm support rail bolts in the sequence shown. Check the toe and adjust as necessary.

Wheel Bearings

The front wheel bearings are sealed, non-adjustable units which require no periodic attention. They are bolted to the steering knuckle by means of an integral flange.

REPLACEMENT

NOTE: This procedure requires the use of special tools. You will need a special tool to pull the bearing free of the halfshaft, GM tool No. J-28733 or the equivalent. You should also use a halfshaft boot protector, GM tool No. J-28712 or the equivalent to protect the parts from damage.

1. Remove the wheel cover, loosen the hub nut, and raise and support the car. Remove the front wheel.
2. Install the boot cover, GM part No. J-28712 or the equivalent.
3. Remove and discard the hub nut.

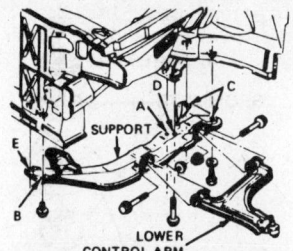
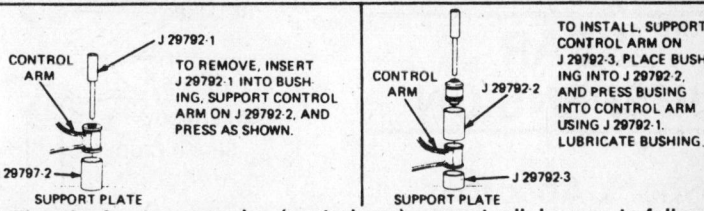

Installing the front suspension (control arm) support rail; be sure to follow the tightening sequence exactly

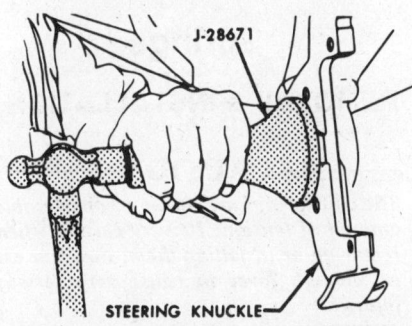

Use a seal driver when installing the new seal into the knuckle

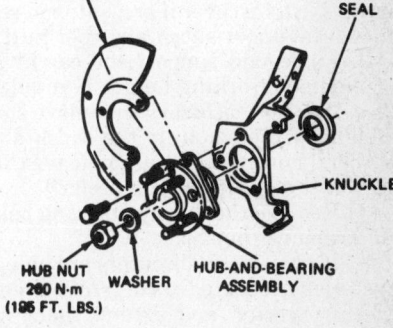

Hub and bearing attachment to the knuckle

Be sure to use a new one on assembly, not the old one.

4. Remove the brake caliper and rotor:
   a. Remove the allen head caliper mounting bolts.
   b. Remove the caliper from the knuckle and suspend from a length of wire. Do not allow the caliper to hang from the brake hose. Pull the rotor from the knuckle.
5. Remove the three hub and bearing attaching bolts. If the old bearing is to be reused, match mark the bolts and holes for installation. The brake rotor splash shield will have to come off, too.
6. Attach a puller, GM part No. J-28733 or the equivalent, and remove the bearing. If corrosion is present, make sure the bearing is loose in the knuckle before using the puller.
7. Clean the mating surfaces of all dirt and corrosion. Check the knuckle bore and knuckle seal for damage. If a new bearing is to be installed, remove the old knuckle seal and install a new one. Grease the lips of the new seal before installation; install with a seal driver made for the purpose, GM tool No. J-28671 or the equivalent.

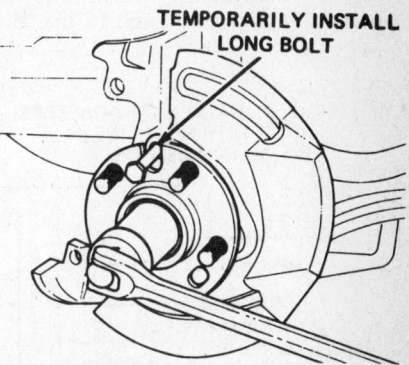

Insert a bolt into the rotor when tightening the hub nut

8. Push the bearing onto the halfshaft. Install a new washer and hub nut.
9. Tighten the new hub nut on the halfshaft until the bearing is seated. If the rotor and hub start to rotate as the hub nut is tightened, insert a long bolt through the cutout in the hub assembly to prevent rotation. Do not apply full torque to the hub nut at this time—just seal the bearing.
10. Install the brake shield and the

bearing retaining bolts. Tighten the bolts evenly to 40 ft. lbs.

11. Install the caliper and rotor. Be sure that the caliper hose isn't twisted. Install the caliper bolts and tighten to 21–35 ft. lbs.

12. Install the wheel. Lower the car. Tighten the hub nut to 185 ft. lbs.

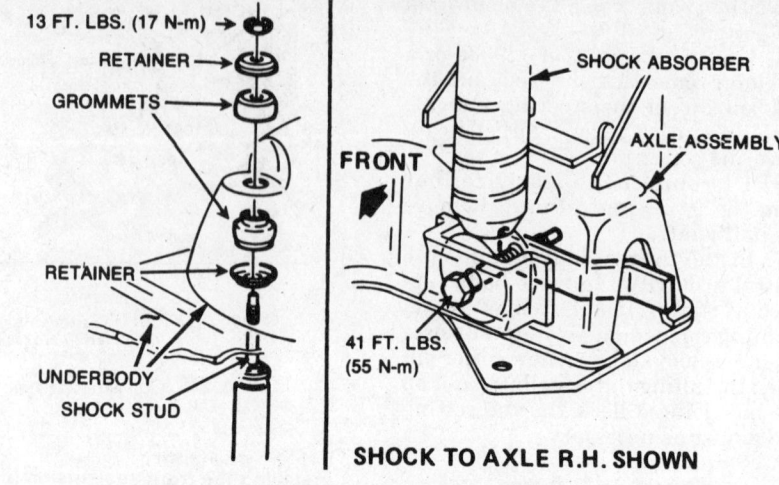

SHOCK TO AXLE R.H. SHOWN

Shock absorber mounting

REAR SUSPENSION

Shock Absorber

REMOVAL & INSTALLATION

1. Open the hatch or trunk lid, remove the trim cover if present, and remove the upper shock absorber nut.

2. Raise and support the car at a convenient working height if you desire. It is not necessary to remove the weight of the car from the shock absorbers, however, so you can leave the car on the ground if you prefer.

3. Remove the lower attaching bolt and remove the shock.

4. If new shock absorbers are being installed, repeatedly compress them while inverted and extend them in their normal upright position. This will purge them of air.

5. Install the shocks in the reverse order of removal. Tighten the lower mount nut and bolt to 55 ft. lbs. the upper to 13 ft. lbs.

Springs

REMOVAL & INSTALLATION

──── CAUTION ────

The coil springs are under a considerable amount of tension. Be very careful when removing or installing them; they can exert enough force to cause very serious injuries.

1. Raise and support the car on a hoist. Do not use a twin-post hoist. The swing arc of the axle may cause it to slip from the hoist when the bolts are removed. If a suitable hoist is not available, raise and support the car on jackstands, and use a jack under the axle.

2. Support the axle with a jack that can be raised and lowered.

3. Remove the brake hose attaching brackets (right and left), allowing the hoses to hang freely. Do not disconnect the hoses.

4. Remove both shock absorber lower attaching bolts from the axle.

5. Lower the axle. Remove the coil spring and insulator.

6. To install, position the spring and insulator on the axle. The leg on the upper coil of the spring must be parallel to the axle, facing the left hand side of the car.

7. Install the shock absorber bolts. Tighten to 41 ft. lbs. Install the brake line brackets. Tighten to 8 ft. lbs.

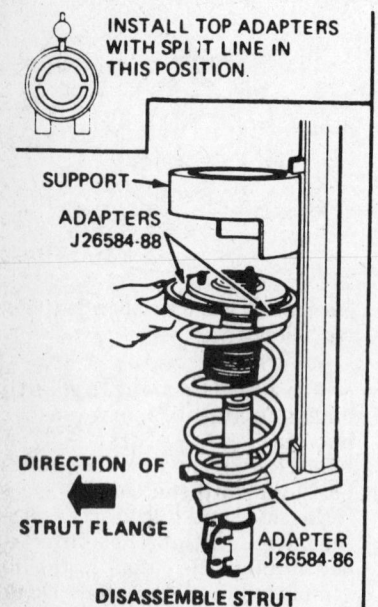

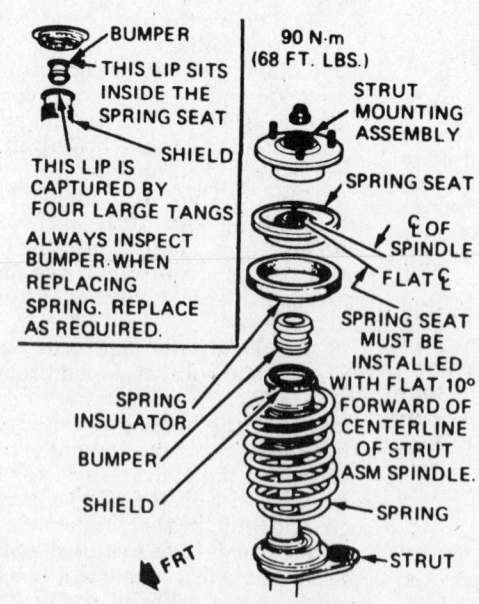

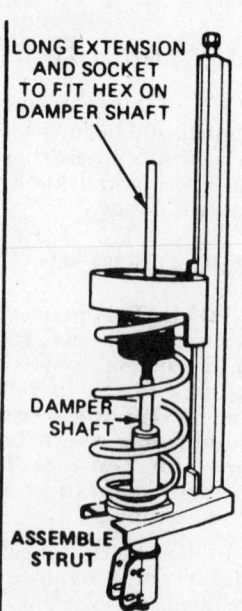

Coil spring removal and installation

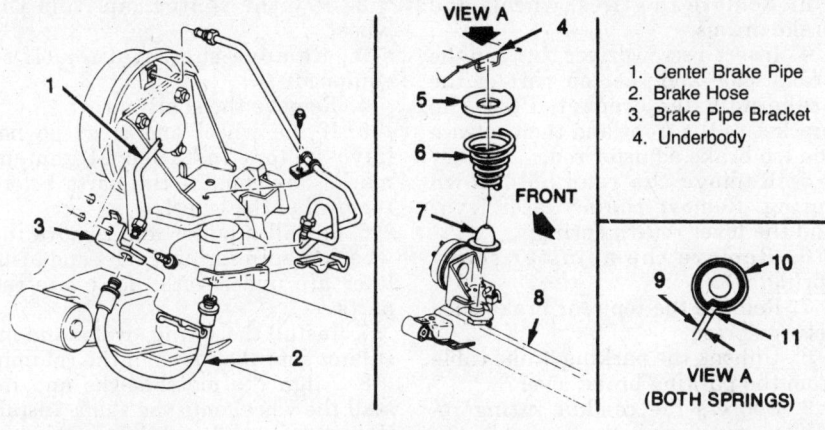

1. Center Brake Pipe
2. Brake Hose
3. Brake Pipe Bracket
4. Underbody

VIEW A
(BOTH SPRINGS)

5. Spring Insulator
6. Spring
7. Compression Bumper
8. Axle Assembly
9. .549 Inch (15mm) Max.
10. Spring
11. Spring Stop Part of Spring Seat

Rear spring mounting

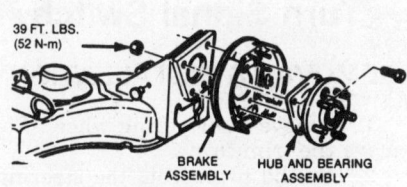

Rear hub and bearing

Rear Hub and Bearing
REMOVAL & INSTALLATION

1. Loosen the wheel lug nuts. Raise and support the car and remove the wheel.
2. Remove the brake drum.

NOTE: Do not hammer on the brake drum to remove; damage to the bearing will result.

3. Remove the four hub and bearing retaining bolts and remove the assembly from the axle. The top rear attaching bolt will not clear the brake shoe when removing the hub and bearing assembly. Partially remove the hub and bearing assembly prior to removing this bolt.
4. Installation is in the reverse. Hub and bearing bolt torque is 39 ft. lbs.

BRAKE SYSTEM

Master Cylinder

REMOVAL & INSTALLATION

1. Unplug the electrical connector from the master cylinder.

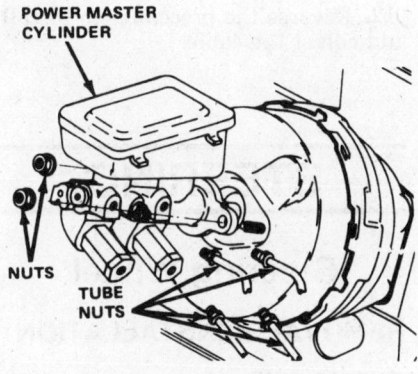

Master cylinder mounting on cars with power brakes

2. Place a number of cloths or a container under the master cylinder to catch the brake fluid. Disconnect the brake tubes from the master cylinder; use a flare nut wrench if one is available. Tape over open ends of the tubes.

NOTE: Brake fluid eats paint. Wipe up any spilled fluid immediately, then flush the area with clear water.

3. Remove the two nuts attaching the master cylinder to the booster or firewall.
4. Remove the master cylinder.
5. To install, attach the master cylinder to the booster with the nuts. Torque to 22–30 ft. lbs.
6. Remove the tape from the lines and connect to the master cylinder. Torque to 10–15 ft. lbs. Connect the electrical lead.
7. Bleed the brakes.

Proportioning Valve

REMOVAL & INSTALLATION

There is a front and a rear proportioning valve located at the lower left side of the master cylinder. To remove the proportioning valves, disconnect the brake lines from the valves, then disconnect the valves from the master cylinder and remove the O-rings. Replace the old O-rings and proportioning valves with new ones and reinstall into the master cylinder. Torque the proportioning valves to 18–30 ft. lbs.

Vacuum Booster

REMOVAL & INSTALLATION

1. Remove the master cylinder from the booster. It is not necessary to disconnect the lines from the master cylinder. Just move the cylinder aside.
2. Disconnect the vacuum booster pushrod from the brake pedal inside the car. It is retained by a bolt. A spring washer is under the bolt head, and a flat washer goes on the other side of the pushrod eye, next to the pedal arm.
3. Remove the four attaching nuts from inside the car. Remove the booster.
4. Install the booster on the firewall. Tighten the mounting nuts to 22–33 ft. lbs.
5. Connect the pushrod to the brake pedal.
6. Install the master cylinder. Mounting torque is 22–33 ft. lbs.

Wheel Cylinder

REMOVAL AND INSTALLATION

1. Raise the rear of the vehicle and support it safely.
2. Remove the rear wheel and brake drum assembly.
3. Disconnect the inlet tube nut and line from the wheel cylinder.
4. Remove the wheel cylinder retainer using two awls or pins 1/8 inch diameter or less.
   a. Insert the awls or pins into the access slots between the wheel cylinder pilot and the retainer locking tabs.
   b. Bend both tabs away simultaneously.
5. Remove the wheel cylinder.
To install:
1. Position the wheel cylinder and hold it in place using a wooden block

placed between the the wheel cylinder and the axle flange.

2. Install a new wheel cylinder retainer over the wheel cylinder abutment using a $1\frac{1}{8}$ inch 12 point socket and extension.

3. Reconnect the inlet tube nut and torque to 12 ft. lbs.

4. Reinstall the brake drum and bleed the brake system.

5. Install the wheels, lower the vehicle and check for leaks.

Parking Brake

ADJUSTMENT

1. Raise and support the car with both rear wheels off the ground.

2. Pull the parking brake lever exactly two ratchet clicks.

NOTE: To prevent damage to the threaded adjusting rod, thoroughly clean and lubricate the threads before turning the adjusting nut.

3. Loosen the equalizer locknut, then tighten the adjusting nut until the left rear wheel can just be turned backward using two hands, but is locked in forward rotation.

4. Tighten the locknut.

5. Release the parking brake. Rotate the rear wheels—there should be no drag.

6. Lower the car.

CABLE REMOVAL & INSTALLATION

Front Cable

1. Place the gear selector in Neutral and apply the parking brake.

2. Remove the center console.

3. Disconnect the parking brake cable from the lever.

4. Remove the cable retaining nut and the bracket securing the front cable to the floor panel.

5. Raise the car and loosen the equalizer nut.

6. Loosen the catalytic converter shield and then remove the parking brake cable from the body.

7. Disconnect the cable from the equalizer and then remove the cable from the guide and the underbody clips.

8. Reverse the procedure and adjust the cable.

Right and Left Rear Cables

1. Raise and support the rear of the car.

2. Back off the equalizer nut until the cable tension is eliminated.

3. Remove the tires, wheels and brake drums.

4. Insert a screwdriver between the brake shoe and the top part of the brake adjuster bracket. Push the bracket to the front and then release the top brake adjuster rod.

5. Remove the rear hold down spring. Remove the actuator lever and the lever return spring.

6. Remove the adjuster screw spring.

7. Remove the top rear brake shoe return spring.

8. Unhook the parking brake cable from the parking brake lever.

9. Depress the conduit fitting retaining tangs and then remove the conduit fitting from the backing plate.

10. Remove the cable end button from the connector.

11. Depress the conduit fitting retaining tangs and remove the conduit fitting from the axle bracket.

12. Reverse the procedure to install and adjust the cable.

STEERING

Steering Wheel

REMOVAL & INSTALLATION

Standard Wheel

1. Disconnect the negative cable at the battery.

2. Pull the pad from the wheel. The horn lead is attached to the pad at one end; the other end of the pad has a wire with a spade connector. The horn lead is disconnected by pushing and turning; the spade connector is simply unplugged.

3. Remove the retainer under the pad (if so equipped).

4. Remove the steering shaft nut.

5. There should be alignment marks already present on the wheel and shaft. If not, matchmark the parts.

6. Remove the wheel with a puller.

7. Install the wheel on the shaft, aligning the matchmarks. Install the shaft nut and tighten to 30 ft. lbs.

8. Install the retainer.

9. Plug in the spade connector, and push and turn the horn lead to connect. Install the pad. Connect the negative battery cable.

Sport Wheel

1. Disconnect the negative cable at the battery.

2. Pry the center cap from the wheel.

3. Remove the retainer (if so equipped).

4. Remove the shaft nut.

5. If the wheel and shaft do not have factory-installed alignment marks, matchmark the parts before removal of the wheel.

6. Install a puller and remove the wheel. A horn spring, eyelet and insulator are underneath; don't lose the parts.

7. Install the spring, eyelet and insulator into the tower in the column.

8. Align the matchmarks and install the wheel onto the shaft. Install the retaining nut and tighten to 30 ft. lbs.

9. Install the retainer. Install the center cap. Connect the negative battery cable.

Turn Signal Switch

REMOVAL & INSTALLATION

1. Remove the steering wheel. Remove the trim cover.

2. Pry the cover from the steering column.

3. Position a U-shaped lockplate compressing tool on the end of the steering shaft and compress the lock plate by turning the shaft nut clockwise. Pry the wire snap-ring out of the shaft groove.

4. Remove the tool and lift the lockplate off the shaft.

5. Slip the cancelling cam, upper bearing preload spring, and thrust washer off the shaft.

6. Remove the turn signal lever. Remove the hazard flasher button retaining screw and remove the button, spring and knob.

7. Pull the switch connector out of the mast jacket and tape the upper part to facilitate switch removal. Attach a long piece of wire to the turn signal switch connector. When installing the turn signal switch, feed this wire through the column first, and then use this wire to pull the switch connector into position. On tilt wheels, place the turn signal and shifter housing in low position and remove the harness cover.

8. Remove the three switch mounting screws. Remove the switch by pulling it straight up while guiding the wiring harness cover through the column.

9. Install the replacement switch by working the connector and cover down through the housing and under the bracket. On tilt models, the connector is worked down through the

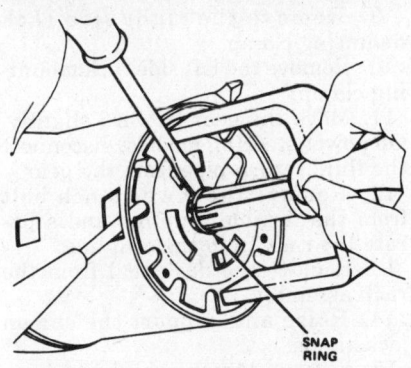

Depress the lockplate and remove the snap ring

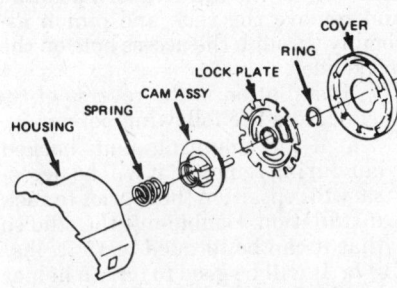

Remove these parts to get at the turn signal switch

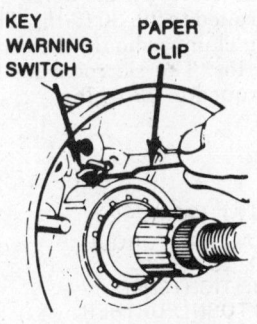

Remove the key warning buzzer switch with a paper clip

housing, under the bracket, and then the cover is installed on the harness.

10. Install the switch mounting screws and the connector on the mast jacket bracket. Install the column-to-dash trim plate.

11. Install the flasher knob and the turn signal lever.

12. With the turn signal lever in neutral and the flasher knob out, slide the thrust washer, upper bearing pre-load spring, and cancelling cam onto the shaft.

13. Position the lock plate on the shaft and press it down until a new snap-ring can be inserted in the shaft groove. Always use a new snap-ring when assembling.

14. Install the cover and the steering wheel.

Ignition Switch

REMOVAL & INSTALLATION

The switch is located inside the channel section of the brake pedal support and is completely inaccessible without first lowering the steering column. The switch is actuated by a rod and rack assembly. A gear on the end of the lock cylinder engages the toothed upper end of the rod.

1. Lower the steering column; be sure to properly support it.

2. Put the switch in the OFF-UN-LOCKED position. With the cylinder removed, the rod is in OFF-UN-LOCKED position when it is in the next to the uppermost detent.

3. Remove the two switch screws and remove the switch assembly.

4. Before installing, place the new switch in OFF-UNLOCKED position and make sure the lock cylinder and actuating rod are in OFF-UN-LOCKED position (second detent from the top).

5. Install the activating rod into the switch and assemble the switch on the column. Tighten the mounting screws. Use only the specified screws, since overlength screws could impair the collapsibility of the column.

6. Reinstall the steering column.

Ignition Lock Cylinder

REMOVAL & INSTALLATION

1. Remove the steering wheel.

2. Turn the lock to the Run position.

3. Remove the lock plate, turn signal switch or combination switch, and the key warning buzzer switch. The warning buzzer switch can be fished out with a bent paper clip.

4. Remove the lock cylinder retaining screw and lock cylinder.

─────── CAUTION ───────

If the screw is dropped on removal, it could fall into the column, requiring complete disassembly to retrieve the screw.

───────────────────────

5. Rotate the cylinder clockwise to align the cylinder key with the keyway in the housing.

6. Push the lock all the way in.

7. Install the screw. Tighten to 15 inch lbs.

8. The rest of installation is in the reverse of removal. Turn the lock to RUN to install the key warning buzzer switch, which is simply pushed down into place.

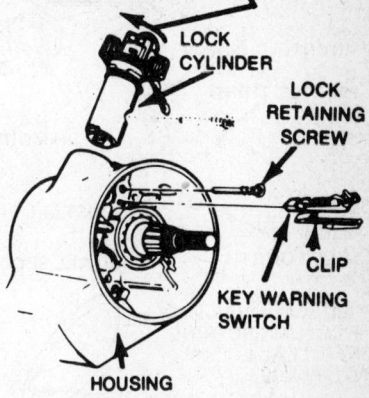

Lock cylinder installation

Power Steering Pump

REMOVAL & INSTALLATION

1. Disconnect the negative battery cable.

2. Disconnect the vent hose at the carburetor.

3. Loosen the adjusting bolt and pivot bolt on the pump, then remove the pump's drive belt.

4. Remove the three pump-to-bracket bolts and remove the adjusting bolt.

5. Remove the high pressure fitting from the pump.

6. Disconnect the reservoir-to-pump hose from the pump.

7. Remove the pump.

8. Installation is in the reverse order of removal. Adjust the belt tension and bleed the system.

Rack and Pinion Unit

REMOVAL & INSTALLATION

1. From the driver's side, remove the sound insulator.

2. From under the instrument panel, pull the seal assembly down from the steering column and remove the upper pinch bolt from the flexible coupling.

3. Remove the air cleaner and the windshield washer jar.

4. On power steering models, disconnect the pressure line from the steering gear and remove the screw securing the pressure line bracket to the cowl. Move the pressure line aside.

5. Raise and support the car on jackstands.

6. Remove both front wheels.

7. Disconnect both tie rods from the struts.

8. Lower the car.

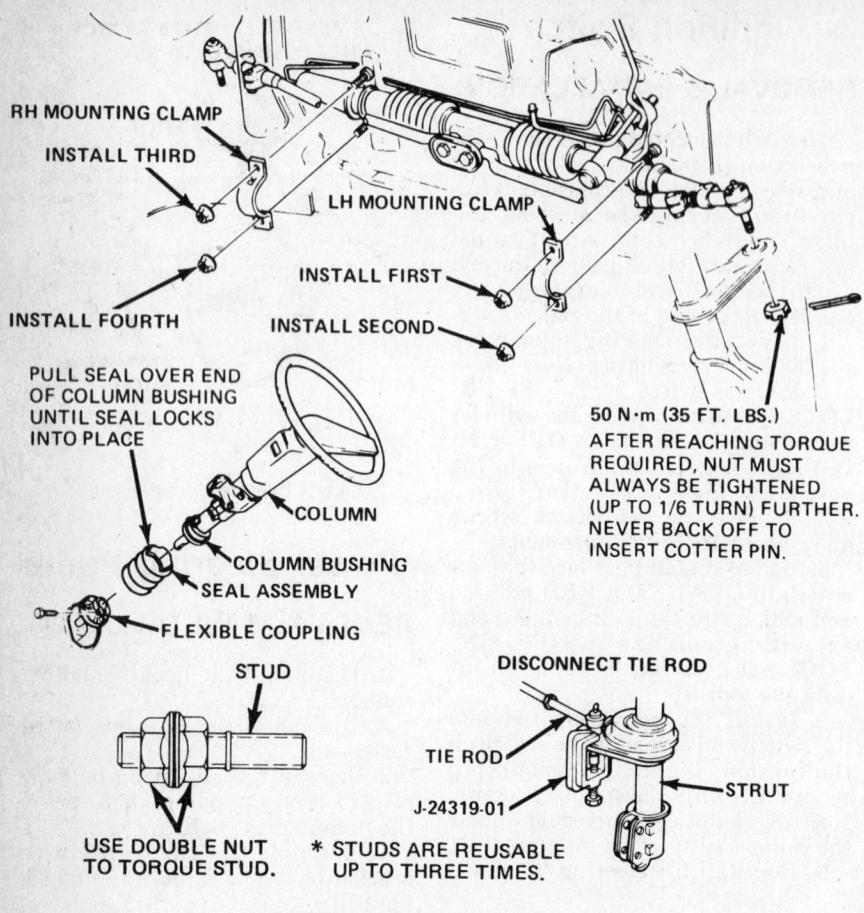

9. Remove the right side rack mounting clamp.

10. Remove the left side rack mounting clamp.

11. Move the gear forward slightly. On power steering models, disconnect the fluid return pipe from the gear.

12. Remove the lower pinch bolt from the flexible coupling and separate the rack from the coupling.

13. Remove the dash seal from the rack assembly.

14. Raise and support the car on jackstands.

15. Remove the splash shield from the inner, left fender.

16. Turn the left knuckle and hub assembly to the full left turn position and remove the rack and pinion assembly through the access hole on the left fender.

17. Installation is the reverse of removal. Note the following points:

   a. If the mounting studs backed out during removal, it will be necessary to reposition them prior to rack installation. Double-nut the stud so that it can be torqued to 15 ft. lbs.

   b. It will be good to have a helper inside the car to guide the flexible coupling onto the stud shaft and onto the steering column.

   c. Both pinch bolts should be torqued to 29–30 ft. lbs. The mounting clamps should be torqued to 28 ft. lbs. The tie rod nuts should be torqued to 35 ft. lbs.

Power rack and pinion unit mounting

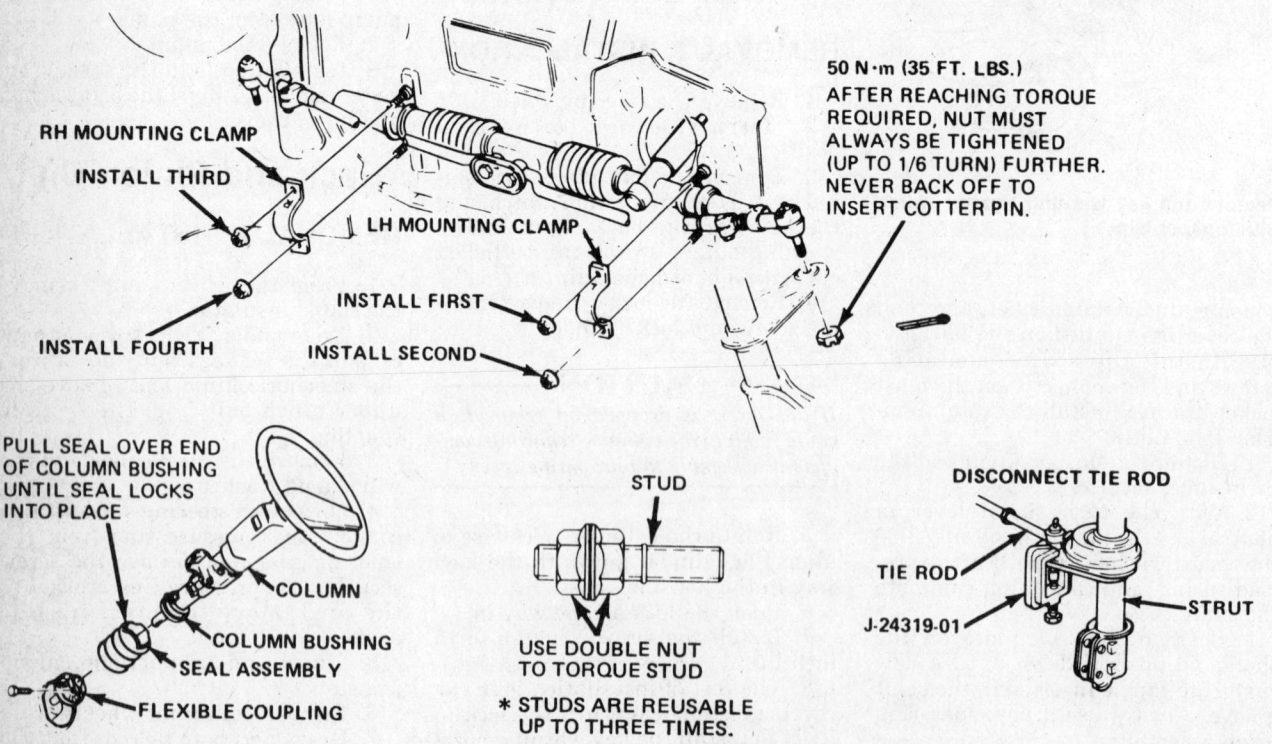

Manual rack and pinion unit mounting

Tie Rod Ends

REMOVAL & INSTALLATION

1. Loosen both pinch bolts at the outer tie rod.

2. Remove the tie rod end from the strut assembly using a suitable removal tool.

3. Unscrew the outer tie rod end from the tie rod adjuster, counting the number of revolutions required before they are disconnected.

4. Install the new tie rod end, screwing it on the same number of revolutions as counted in Step 3. When the tie rod end is installed, the tie rod adjuster must be centered between the tie rod and the tie rod end, with an equal number of threads exposed on both sides of the adjuster nut. Tighten the pinch bolts to 20 ft. lbs.

5. Install the tie rod end to the strut assembly and tighten to 50 ft. lbs. If the cotter pin cannot be installed, tighten the nut up to $\frac{1}{16}$ in. further. Never back off the nut to align the holes for the cotter pin.

6. Have the front end alignment adjusted.

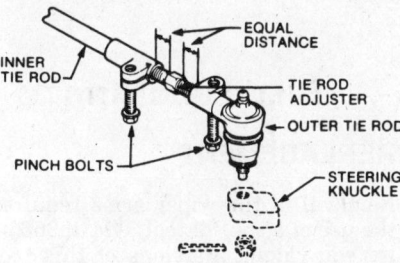

Tie rod end removal and installation

CHASSIS ELECTRICAL

Heater Blower Motor

REMOVAL & INSTALLATION

1. Disconnect the negative battery cable.

2. Disconnect the electrical connections at the blower motor and blower resistor.

3. Remove the plastic water shield from the right side of the cowl.

4. Remove the blower motor retaining screws and then pull the blower motor and cage out.

5. Hold the blower motor cage and

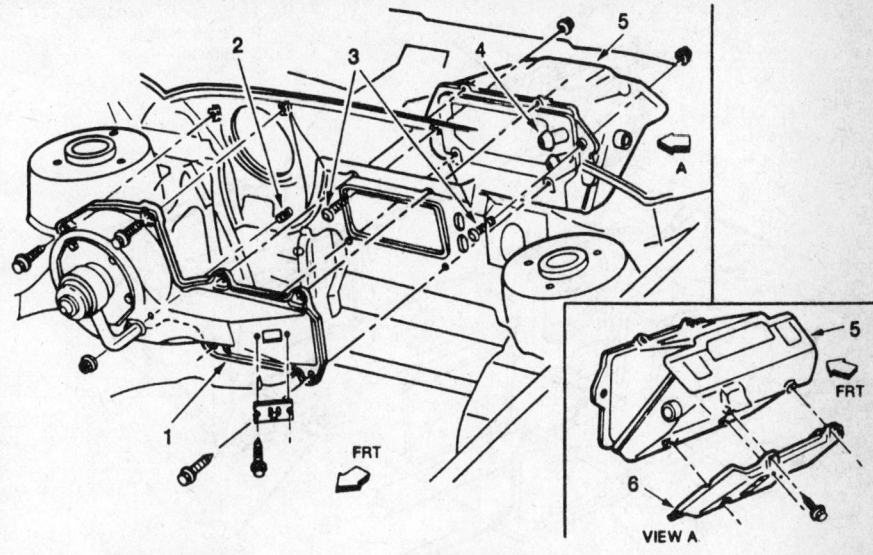

1. Heater case
2. Stud dash panel
3. Locating studs dash panel
4. Heater core
5. Heater module
6. Module cover

Heater assembly on models without air conditioning

remove the cage retaining nut from the blower motor shaft.

6. Remove the blower motor and cage.

7. Installation is in the reverse order of removal.

Heater Core

REMOVAL & INSTALLATION

Cars Without Air Conditioning

1. Disconnect the negative battery cable and drain the cooling system.

2. Remove the heater inlet and outlet hoses from the heater core.

3. Remove the heater outlet deflector.

4. Remove the retaining screws and then remove the heater core cover.

5. Remove the heater core retaining straps and then remove the heater core.

6. Installation is in the reverse order of removal.

Cars With Air Conditioning

1. Disconnect the negative battery cable and drain the cooling system.

2. Raise and support the front of the vehicle.

3. Disconnect the drain tube from the heater case.

4. Remove the heater hoses from the heater core.

5. Lower the car. Remove the right and left hush panels, the steering column trim cover, the heater outlet duct and the glove box.

6. Remove the heater core cover.

Be sure to pull the cover straight to the rear so as not to damage the drain tube.

7. Remove the heater core clamps and then remove the core.

8. Installation is in the reverse order of removal.

Radio

REMOVAL & INSTALLATION

NOTE: Do not operate the radio with the speaker leads disconnected. Operating the radio without an electrical load will damage the output transistors.

1. Disconnect the negative battery cable.

2. Remove the instrument panel trim plate.

3. Check the right side of the radio to determine whether a nut or a stud is used for side retention.

4. If a nut is used, remove the hush panel and then loosen the nut from below on cars without air conditioning. On cars with air conditioning, remove the hush panel, the A/C duct and the A/C control head for access to the nut. Do not remove the nut; loosen it just enough to pull the radio out. If a rubber stud is used, go on to Step 5.

5. Remove the two radio bracket-to-instrument panel attaching screws. Pull the radio forward far enough to disconnect the wiring and antenna and then remove the radio.

6. Installation is in the reverse order of removal.

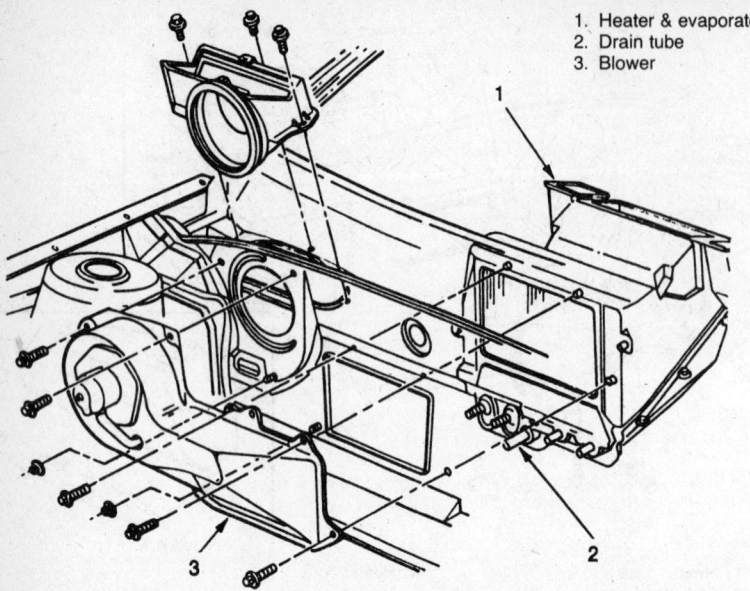

1. Heater & evaporator
2. Drain tube
3. Blower

Heater assembly on models with air conditioning

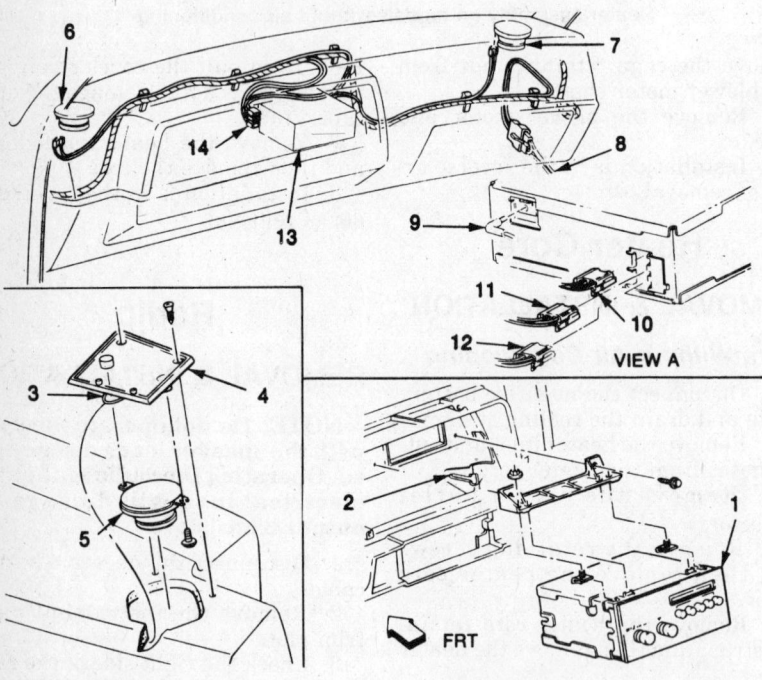

VIEW A

FRT

1. Radio
2. Screw on side of radio fits here
3. Retainer
4. Grille
5. Speaker
6. Front speaker
7. Front speaker
8. Rear speaker wire
9. Antenna
10. Rear speakers
11. Front speakers
12. Instrument panel harness
13. Receiver
14. Instrument panel harness

Radio removal and installation

Windshield Wiper Switch

REMOVAL & INSTALLATION

1. Remove the steering wheel and directional signal switch, It may be necessary to loosen the two column mounting nuts and remove the four bracket-to-mast jacket to allow the connector clip on the ignition switch to be pulled out of the column assembly.

2. Disconnect the washer/wiper switch lower connector.

3. Remove the screws attaching the column housing to the mast jacket. Be sure to note the position of the dimmer switch actuator rod for reassembly in the same position. Remove the column housing-and-switch as an assembly.

NOTE: The tilt and travel columns have a removable plastic cover on the column housing. This provides access to the wiper switch without removing the entire column housing.

4. Turn upside down and use a drift to remove the pivot pin from the washer/ wiper switch. Remove the switch.

5. Place the switch into the position in the housing, then install the pivot pin.

6. Position the housing onto the mast jacket and attach by installing the screws. Install the dimmer switch actuator rod in the same position as noted earlier. Check switch operation.

7. Reconnect lower end of the switch assembly.

8. Install the remaining components in reverse order of removal. Be sure to attach column mounting bracket in original position.

Blade and Arm

REPLACEMENT

Removal of the wiper arms requires the use of a special tool, GM J8966 or its equivalent. Versions of this tool are generally available in auto parts stores.

1. Insert the tool under the wiper arm and lever the arm off the shaft.

2. Disconnect the washer hose from the arm (if so equipped). Remove the arm.

3. Installation is in the reverse order of removal. The proper park position is at the top of the blackout line on the glass. If the wiper arms and blades were in the proper position prior to removal, adjustment should not be required.

ADJUSTMENT

The only adjustment for the wiper arms is to remove an arm from the transmission shaft, rotate the arm the required distance and direction and then install the arm back in position so it is in line with the blackout line on the glass. The wiper motor must be in the Park position.

The correct blade-out wipe position on the driver's side is $1\frac{3}{32}$ in. (28mm)

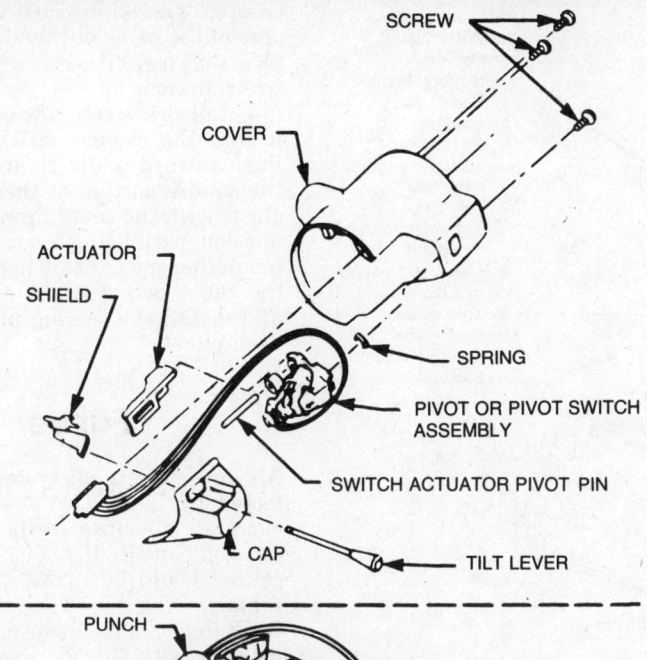

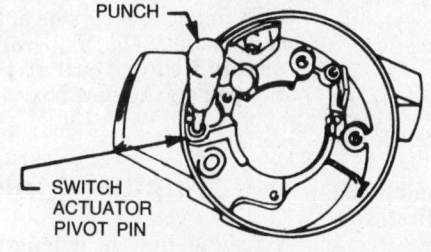

REMOVE AND INSTALL PIVOT AND SWITCH ASSEMBLY

Removal and installation of the washer switch

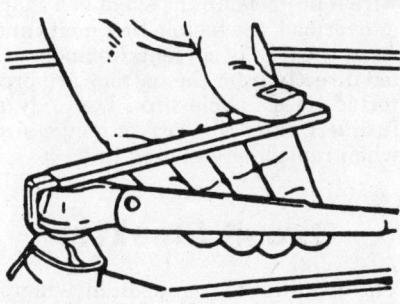

Removing the wiper arm with a special tool

from the tip of the blade to the left windshield pillar moulding. The correct blade-down wipe position on the passenger side of the car is in line with the blackout line at the bottom of the glass.

Linkage

REMOVAL & INSTALLATION

1. Remove the wiper arms.
2. Remove the shroud top vent grille.
3. Loosen (but do not remove) the drive link-to-crank arm attaching nuts.
4. Unscrew the linkage-to-cowl panel retaining screws and remove the linkage.
5. Installation is in the reverse order of removal.

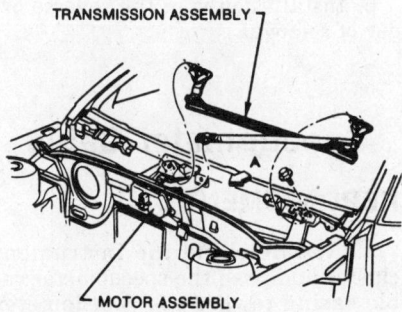

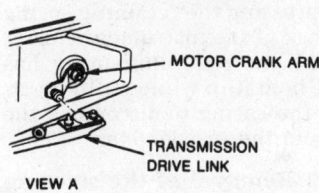

VIEW A

Wiper motor and linkage

Wiper Motor

REMOVAL & INSTALLATION

1. Loosen (but do not remove) the drive link-to-crank arm attaching nuts and detach the drive link from the motor crank arm.
2. Tag and disconnect all electrical leads from the wiper motor.
3. Unscrew the mounting bolts, rotate the motor up and outward and remove it.
4. Guide the crank arm through the opening in the body and then tighten the mounting bolts to 4–6 ft. lbs.
5. Install the drive link to the crank arm with the motor in the park position.
6. Installation of the remaining components is in the reverse order of removal.

Instrument Cluster

REMOVAL & INSTALLATION

1. Disconnect the negative battery cable.
2. Remove the right and left hush panels and the steering column trim cover. Disconnect the vent panels from the bottom of the panel (if so equipped).
3. Remove the glove box. Disconnect the temperature and mode control cables on cars without air conditioning. On cars with air conditioning, remove the lower A/C duct.
4. Remove the three steering column retaining bolts (two at the instrument panel pad and one at the cowl) and lower the steering column.
5. Remove the lower right hand trim plate. Disconnect the cigar lighter and accessory switches.
6. Pull the heater or A/C control head out far enough to disconnect any wiring or vacuum harnesses, then remove the head.
7. Disconnect the front end and engine harnesses from the bulkhead connector in the engine compartment and then remove the bulkhead connector from the cowl (2) screws.
8. Loosen the set screw and remove the hood release handle. Unscrew the retaining nut and pull the hood release cable loose.
9. Unscrew the four upper instrument panel retaining screws (in the defroster duct openings).
10. Unscrew the two lower corner instrument panel retaining nuts. Remove the screw to the instrument panel brace from the left side of the glove box opening.
11. Pull the instrument panel out

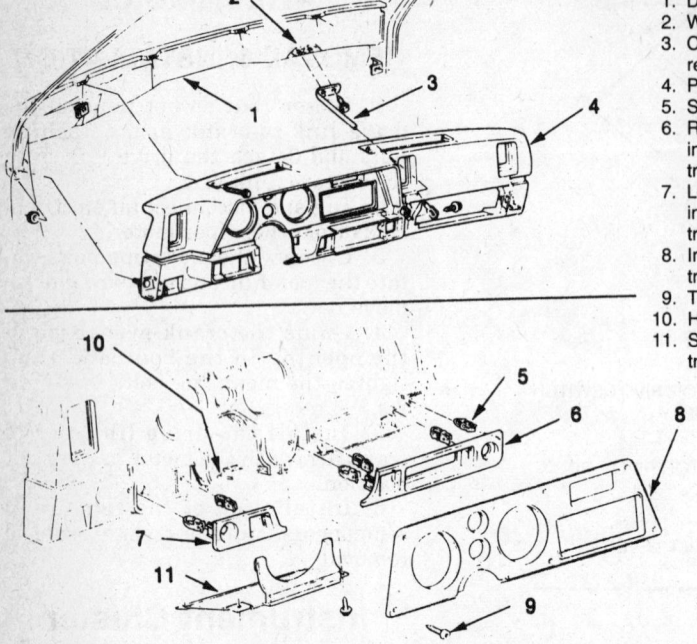

1. Dash panel
2. Weld nuts
3. Center reinforcement
4. Pad
5. Snap-in clips
6. Right lower instrument panel trim plate
7. Left lower instrument panel trim plate
8. Instrument panel trim plate
9. Torx screw
10. Hush panel
11. Steering column trim cover

Instrument panel and trim plate mounting

far enough to disconnect the ignition, the headlight dimmer switch and the turn signal switch. Tag and disconnect all other wiring and vacuum lines.

12. Remove the instrument panel with the wiring harness intact.

13. Installation is in the reverse order of removal.

Center Console

REMOVAL & INSTALLATION

Manual Transmission

1. Place the gear selector in Neutral and apply the parking brake.

2. Lift the ashtray out of the console and then remove the two screws in the opening.

3. Loosen the set screw underneath the shifter knob and remove the knob.

4. Remove the screw under the parking brake handle. Remove the two screws at the rear of the console and lift it off.

5. Installation is in the reverse order of removal.

Automatic Transmission

1. Place the gear selector in Neutral and apply the parking brake.

2. Lift out the ashtray from the front of the console and remove the two screws from the opening.

3. Gently pry the emblem out of the center of the shift knob and remove

the snap ring that secures the knob. Remove the knob.

4. Lift the trim plate assembly out by pulling the front end up first. Disconnect the wiring harness.

5. Remove the three screws under the trim plate and then lift out the rear ashtray and remove the screw under it. Remove the console.

6. Installation is in the reverse order of removal.

Speedometer Cable

REPLACEMENT

1. Reach behind the instrument cluster and push the speedometer cable casing toward the speedometer while depressing the retaining spring on the back of the instrument cluster case. Once the retaining spring has released, hold it in while pulling outward on the casing to disconnect the casing from the speedometer.

NOTE: Removal of the steering column trim plate and/or the speedo cluster may provide better access to the cable.

2. Remove the cable casing sealing plug from the dash panel. Then, pull the casing down from behind the dash and remove the cable.

3. If the cable is broken and cannot be entirely removed from the top, support the car securely, and then unscrew the cable casing connector at

the transmission. Pull the bottom part of the cable out, and then screw the connector back onto the transmission.

4. Lubricate the new cable. Insert it into the casing until it bottoms. Push inward while rotating it until the square portion at the bottom engages with the coupling in the transmission, permitting the cable to move in another inch or so. Then, reconnect the cable casing to the speedometer and install the sealing plug into the dash panel.

Fuses

All major electrical systems are protected by fuses. In the event of an overload, the fuse melts, protecting the component. If a fuse blows, the cause should be investigated before replacing the fuse. The fuse box is located under the left side of the instrument panel. The amperage of each fuse and the circuit it protects is stamped on the fuse box.

Fusible Links

A fusible link is a length (usually about 4 inches) of wire located in the circuit it protects. The wire is usually 4 gauge sizes smaller than the circuit wire it protects. In the event of a short or overload, the fusible link melts and stops the flow of current. Components fed directly from the battery are protected by a fusible link. Use only a fusible link of the correct gauge size when replacing a melted link.

Circuit Breakers

The headlights, windshield wipers, power door locks and power windows are protected by circuit breakers. The CB for the headlights is located in the headlight switch; the one for the wipers is located in the wiper switch; the ones for the power door locks and power windows are located in the fuse box. Breakers reset themselves automatically when the problem is relieved. A convenience center is located on the underside of the instrument panel on later models, providing a central location for various relays, hazard flasher unit and buzzer. All units are plug-in modules. The turn signal flasher is locted directly under the steering column of the vehicle. In order to gain access, it may be necessary to remove the under dash padding panel.

GM "N" Body
Buick Somerset, Skylark, Oldsmobile Calais, Pontiac Grand Am

YEAR IDENTIFICATION

1985–86 Grand Am

1985–86 Calais

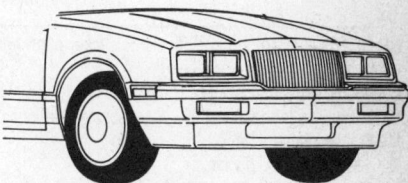

1985–86 Somerset

1986 Calais GT

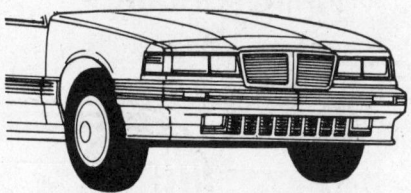

1986-87 Grand Am SE

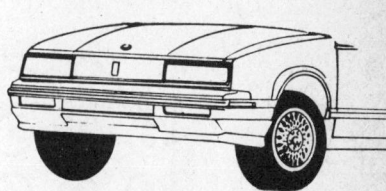

1987 Calais GT

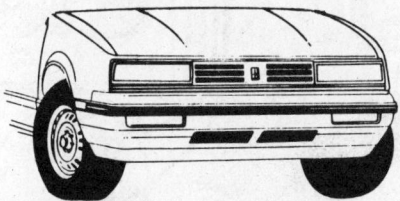

1987 Calais Supreme Cpe

VEHICLE IDENTIFICATION NUMBER (VIN)

It is important for servicing and ordering parts to be certain of the vehicle and engine identification. The VIN (vehicle identification number) is a 17 digit number visible through the windshield on the driver's side of the dash and contains the vehicle and engine identification codes. It can be interpreted as follows:

Engine Code						Model Year Code	
Code	Cu. In.	Liters	Cyl.	Carb	Eng. Mfg.	Code	Year
U	151	2.5	4	TBI	Pont.	F	1985
L	181	3.0	V6	MFI	Buick	G	1986
						H	1987

The seventeen digit Vehicle Identification Number can be used to determine engine application and model year. The 10th digit indicates the model year, and the 8th digit identifies the factory installed engine.
TBI–(Throttle body injection)
MFI–Multi-Port Fuel Injection

GENERAL ENGINE SPECIFICATIONS

Year	Engine No. Cyl. Displ. cu. in.	Engine VIN Code	Fuel Delivery System	Engine Mfg.	Horsepower @ rpm	Torque @ rpm (ft. lbs.)	Bore × Stroke	Compression Ratio	Oil Pressure @ rpm
'85–'86	4-151	U	TBI	Pont.	92 @ 4500	134 @ 2800	4.00 × 3.00	9.0:1	36–41 @ 2000
	6-181	L	MFI	Buick	125 @ 4900	150 @ 2400	3.80 × 2.66	8.45:1	37 @ 2400

TUNE-UP SPECIFICATIONS

(When analyzing compression test results, look for uniformity among cylinders rather than specific pressures.)

Year	Eng. VIN Code	Eng. No. Cyl. Displacement cu. in.	Eng. Mfg.	hp	Spark Plugs Orig. Type	Gap (in.)	Ignition Timing (deg) Man. Trans.	Auto. Trans.	Valves Intake Opens (deg)	Fuel Pump Pressure (psi)	Idle Speed (rpm) Man. Trans.	Auto. Trans.
'84–'86	U	4-151	Pont	54	R43TSX	.060	8B	8B	NA	12	①	①
	L	6-181	Buick	125	R44LTS	.040	15B	15B	NA	34–44	①	①
'87	All				See Underhood Specifications Sticker							

NOTE: The underhood specifications sticker often reflects tune-up specification changes made in production. Sticker figures must be used if they disagree with those in this chart.

B Before Top Dead Center

Part numbers in this chart are not recommendations by Chilton for any product by brand name

NA Not available at time of publication

① See underhood sticker

FIRING ORDERS

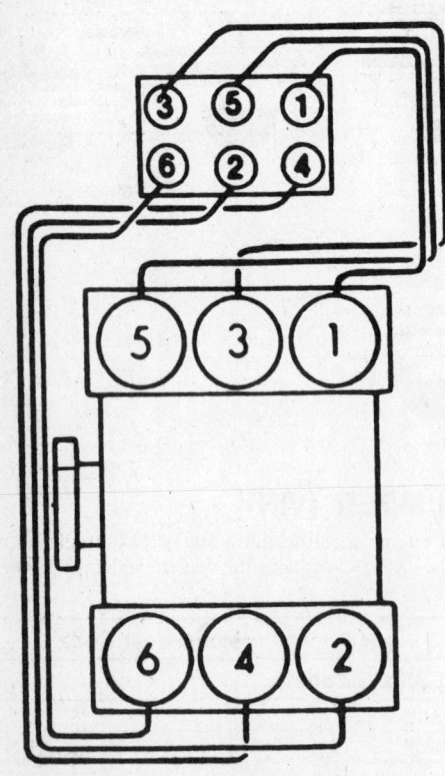

Buick V6 with C3I ignition system, Engine firing order: 1-6-5-4-3-2

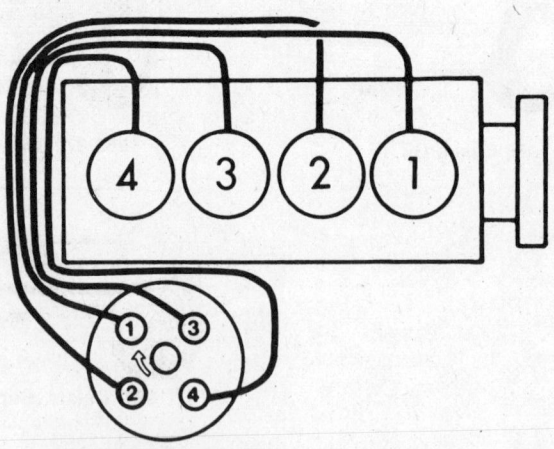

GM (Pontiac) 151-4
Engine firing order: 1-3-4-2
Distributor rotation: clockwise

CAPACITIES

Year	Eng. VIN Code	Engine Displacement (Cu. In.)	Eng. Mfg.	Crankcase Quarts		Transaxle Pints		Gas Tank Gal	Cooling System Qts	
				w/filter	wo/filter	5 speed	Auto.		w/heater	w/AC
'84–'87	U	4-151	Pont.	①	3	5.3	8②	13.6	7.8	7.9
	L	6-181	Buick	①	4	5.3	8②	13.6	7.8	7.9

① Fill to mark on dipstick
② 12 pts. if drained completely

VALVE SPECIFICATIONS

Year	Eng. VIN Code	Engine No. Cyl. Displacement cu. in.	Eng. Mfg.	Seat Angle (deg)	Face Angle (deg)	Spring Test Pressure (lbs. @ In.)	Spring Installed Height (in.)	Stem to Guide Clearance (in.)		Stem Diameter (in.)	
								Intake	Exhaust	Intake	Exhaust
'85–'87	U	4-151	Pont.	46	45	78–86 @ 1.66	1.69	0.0010–0.0027	0.0010–0.0027①	0.342–0.343	0.342–0.343
	L	6-181	Buick	45	45	210–230 @ 1.340	1.727	0.0015–0.0035	0.0015–0.0032	0.3401–0.3412	0.3405–0.3412

① The exhaust valve stem is tapered. The clearance at the top of the guide is shown. The clearance at the bottom of the guide is 0.0020–0.0037

CAMSHAFT SPECIFICATIONS

(All measurements in inches)

Year	Eng. VIN Code	Engine	Eng. Mfg.	Journal Diameter					Bearing Clearance	Lobe Lift		Camshaft End Play
				1	2	3	4	5		Intake	Exhaust	
'85–'87	U	4-151	Pont.	——1.869——					.0007–.0027	.398	.398	.0015–.0050
	L	6-181	Buick	——1.785–1.786——					.0005–.0025	NA	NA	NA

NA Not available at time of publication

CRANKSHAFT AND CONNECTING ROD SPECIFICATIONS

(All measurements are given in inches.)

Year	Eng. VIN Code	Engine No. Cyl. Displacement cu. in.	Eng. Mfg.	Crankshaft				Connecting Rod		
				Main Brg. Journal Dia.	Main Brg. Oil Clearance	Shaft End-Play	Thrust on No.	Journal Diameter	Oil Clearance	Side Clearance
'85–'87	U	4-151	Pont.	2.30	.0005–.0022	.0035–.0085	5	2.00	.0005–.0022	.006–.022
	L	6-181	Buick	2.4995	.0003–.0018	.0030–.0150	2	2.487	.0005–.0026	.003–.015

TORQUE SPECIFICATIONS

(All readings in ft. lbs.)

Year	VIN Code	Engine No. Cyl. Displacement cu. in.	Eng. Mfg.	Cylinder Head Bolts	Rod Bearing Bolts	Main Bearing Bolts	Crankshaft Bolt	Flywheel to Crankshaft Bolts	Manifold	
									Intake	Exhaust
'85–'87	U	4-151	Pont.	92	32	70	200	44	29	44
	L	6-181	Buick	80	40	100	200	60	47	25

PISTON AND RING SPECIFICATIONS

(All measurements are given in inches.)

Year	Eng. VIN Code	Engine No. Cyl. Disp. cu. in.	Eng. Mfg.	Piston-to-Bore Clearance	Ring Gap			Ring Side Clearance		
					Top Compression	Bottom Compression	Oil Control	Top Compression	Bottom Compression	Oil Control
'85–'87	U	4-151	Pont.	.0014–.0022①	.010–.020	.010–.020	.020–.060	.002–.003	.001–.003	.015–.055
	L	6-181	Buick	.0008–.0020②	.010–.020	.010–.020	.015–.055	.010–.020	.010–.020	.015–.055

① Measured 1.8 inches from piston top
② Measured at the top of the skirt

WHEEL ALIGNMENT SPECIFICATIONS

Year	Model	Caster		Camber		Toe-In (in.)
		Range (deg)	Pref. Setting (deg)	Range (deg)	Pref. Setting (deg)	
'85	All	⅔P to 2⅔P	1⅔P	⅕P to 1½P	⅚P	⅛ OUT
'86–'87	All	¾P to 2¾P	1¾P	¼ to 1⅜P	¹³⁄₁₆P	¹⁄₁₆ OUT

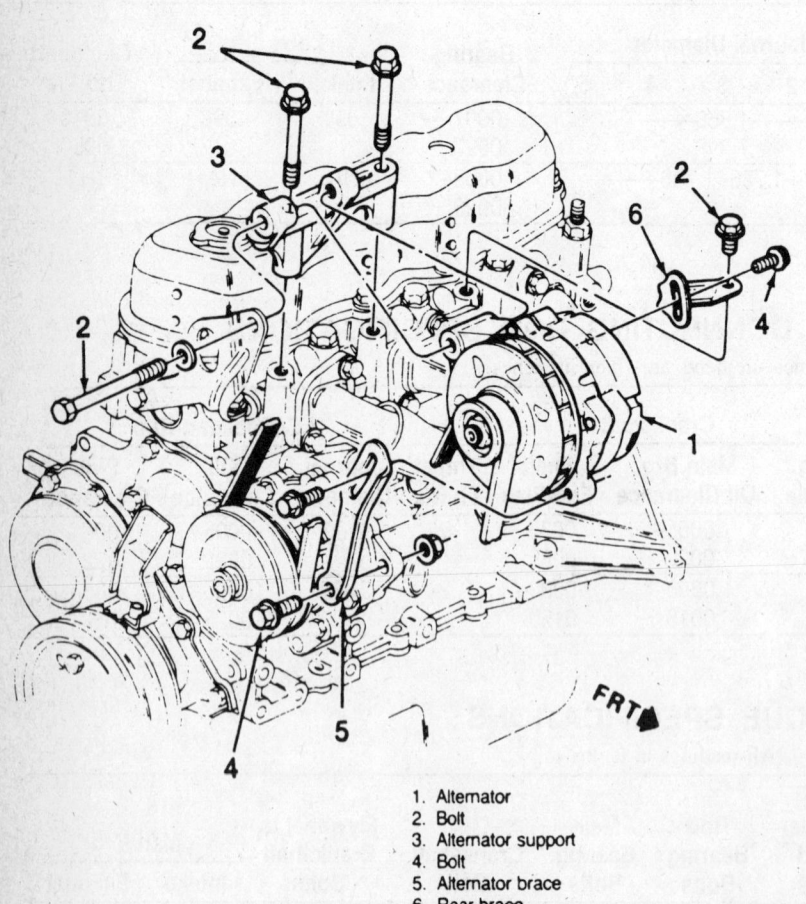

1. Alternator
2. Bolt
3. Alternator support
4. Bolt
5. Alternator brace
6. Rear brace

4 cyl alternator mounting

ENGINE ELECTRICAL

For further information on the charging system, please refer to "Charging and Starting" in the Unit Repair section.

Alternator

REMOVAL & INSTALLATION

1. Disconnect the negative battery cable.
2. Remove the two-terminal plug and the battery lead from the back of the alternator assembly.
3. Loosen the adjusting bolts and remove the alternator belt.
4. Remove the alternator retaining bolts and lift the alternator assembly from the vehicle.
5. Installation is the reverse of removal. Once the alternator is installed, check for proper belt tension. On 4 cylinder engines, adjust the drive belt tension to 70 lbs. on a used belt, or 165 lbs. on a new belt. On V6 engines, adjust the belt tension to 70 lbs. on a used belt or 145 lbs. on a new belt.

NOTE: When adjusting belt tension, apply pressure at the center of the alternator, not against either end frame.

Voltage Regulator

REMOVAL & INSTALLATION

A solid state regulator is mounted within the alternator. The regulator is non-adjustable and requires no maintenance. Should the regulator require service, the alternator must be removed and disassembled.

Starter

For further information on the starting system, please refer to "Charging and Starting" in the Unit Repair section.

REMOVAL & INSTALLATION

1. Disconnect the negative battery cable.
2. Raise the vehicle and support it safely. Disconnect the starter wiring, being careful not to let the positive battery cable end touch ground.
3. Remove the dust cover bolts and pull the dust cover back to gain access to the front starter bolt.
4. Remove the front starter bolt.
5. Remove the rear support bracket.
6. Pull the rear dust cover back to gain access to the rear starter bolt and remove the rear bolt.
7. Push the dust cover back into place and pull the starter assembly back and out. Be sure to note any shims during removal so they may be installed in their original location.
8. Installation is the reverse of removal.

Distributor

NOTE: Two types of ignition systems are used on N-Body vehicles. The HEI system is used on the four cylinder engines and the Computer Controlled Coil Ignition (C3I) system is used on the V6. The C3I system does not use a distributor; instead, a coil pack, ignition module, crankshaft sensor and camshaft sensor make up the ignition system. The following procedure is for the HEI distributor.

REMOVAL & INSTALLATION

1. Disconnect the negative battery cable.
2. Disconnect the ignition switch battery feed wire and the tachometer lead (if equipped) from the distributor cap.
3. Release the coil connectors from the cap. Remove the distributor cap

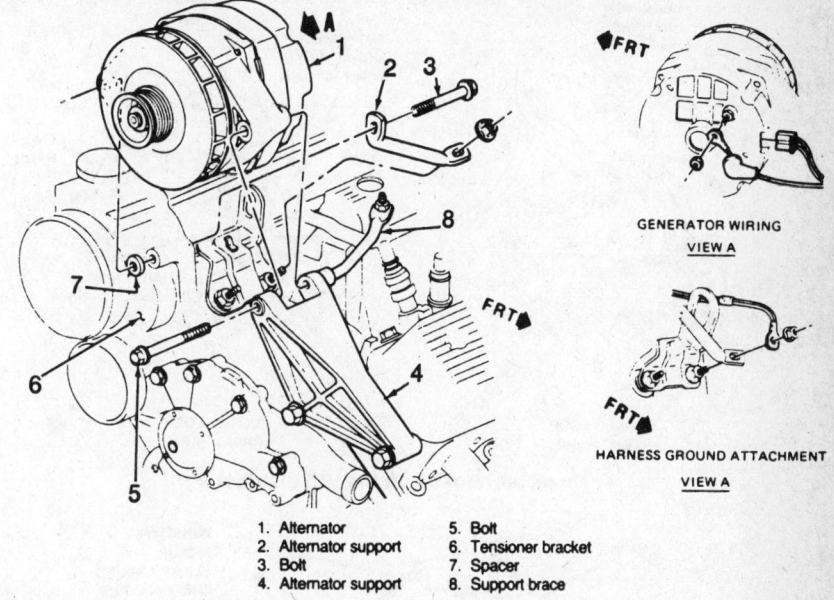

1. Alternator
2. Alternator support
3. Bolt
4. Alternator support
5. Bolt
6. Tensioner bracket
7. Spacer
8. Support brace

GENERATOR WIRING
VIEW A

HARNESS GROUND ATTACHMENT
VIEW A

V6 alternator mounting and wiring attachments

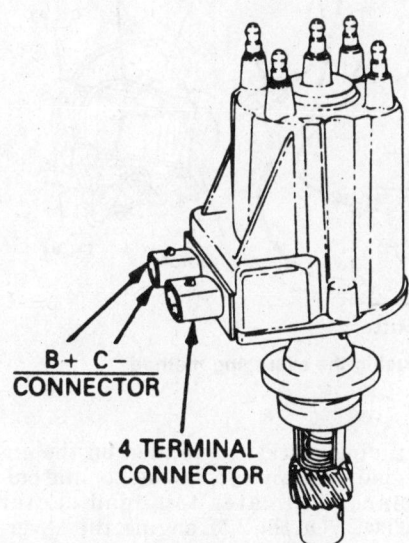

B + C −
CONNECTOR

4 TERMINAL
CONNECTOR

Typical 4 cyl HEI/EST distributor

and position it out of the way. Do not remove the ignition wires from the cap.
4. Disconnect the four-terminal ECM harness from the distributor.
5. Remove the distributor clamp screw and hold down clamp.
6. Mark the rotor position by scribing a mark on the distributor housing before removal. Carefully pull the distributor up until the rotor just stops turning and again mark the rotor position for installation. The drive gear on the shaft is helical and the shaft will rotate slightly as the distributor is removed. Do not crank the engine while the distributor is removed.

7. To install the distributor, rotate the shaft until the rotor aligns with the second mark made. Lubricate the drive gear with clean engine oil and install the distributor into the engine. As the distributor is installed, the rotor should move to the first mark made during removal. This ensures proper timing. If the marks do not align properly, remove the distributor and try again.
8. Install the clamp and hold down nut, then install the remaining components in the reverse order of removal. Start the engine and check the ignition timing.

INSTALLATION IF THE ENGINE WAS DISTURBED

If the engine was accidently cranked while the distributor was removed, use the following procedure for installation. The engine must be set on TDC of the compression stroke to obtain the proper spark timing.

1. Remove the No. 1 spark plug.
2. Place finger over the No. 1 spark plug hole and crank the engine slowly until compression is felt.
3. Align the timing mark on the pulley to "0" on the engine timing indicator.
4. Align the rotor with the marks made earlier and install the distributor as outlined in Step 7, above. If no rotor alignment marks were made during removal, turn the rotor to a point between No. 1 and No. 4 on four cylinder engines.

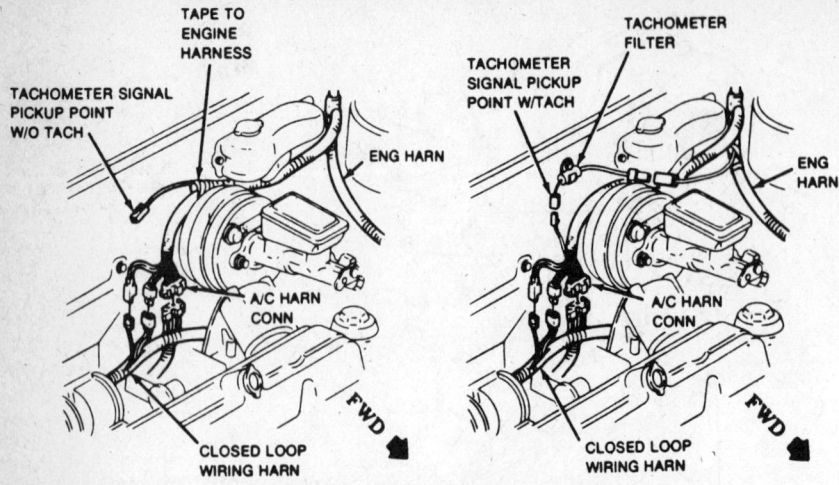

Tachometer hookup on the HEI distributor

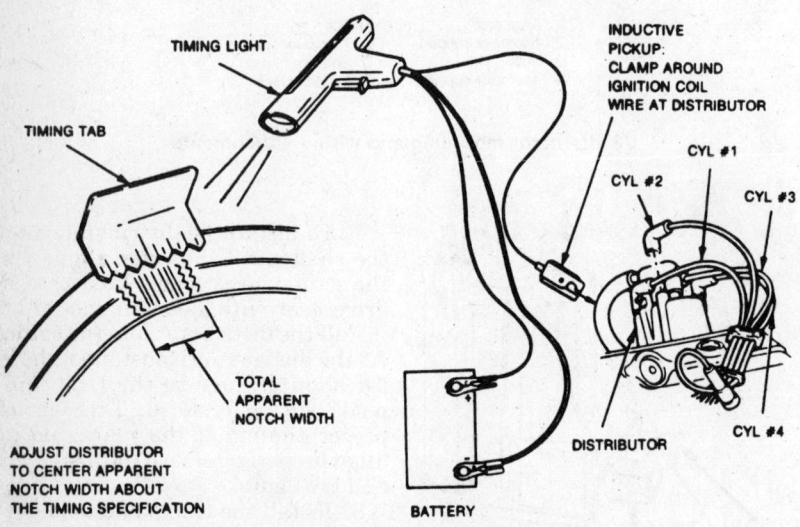

Ignition timing is accomplished using the averaging method

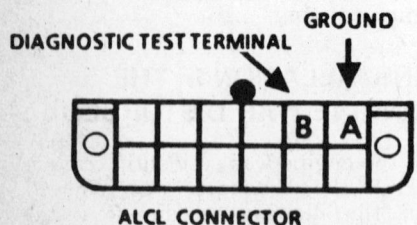

ALCL CONNECTOR

Install a jumper wire to terminals A and B of the ALCL connector to set ignition timing

5. Complete the installation as described in Steps 7 and 8, above. Check and adjust the ignition timing to specifications.

IGNITION TIMING ADJUSTMENT

Timing specifications for each engine are listed in the Tune-Up Specifications Chart and on the underhood emission control label. The ignition timing marks are located on the engine front cover; a saw slot on the balancer indicates top dead center (TDC). On the 2.5L engine, the "Averaging Method" is used to set the ignition timing. This involves the use of BOTH the No. 1 and No. 4 spark plug wires to trigger the timing light. To set the timing on 2.5L engines:

1. See the underhood emission control label and follow all timing instructions given.

2. The engine should be at normal operating temperature, air cleaner installed, air conditioner off, electric cooling fan off and parking brake set firmly. Place the automatic transmission in Drive or the manual in Neutral.

3. Connect an inductive timing light pickup to the No. 1 plug wire and make sure the "Check Engine" light is not on.

4. Ground the ALCL connector under the dash by using a jumper wire to connect terminals A and B. The "Check Engine" light should begin flashing.

5. Using the timing light, check and record the position of the timing mark.

6. Repeat Step 3, but connect the timing light inductive pickup to the No. 4 spark plug wire. Take the total of the two recorded timing marks and divide by two to come up with an average timing. For example: No. 1 timing = 4° and No. 4 timing = 8°; 4 + 8 = 12 ÷ 2 = 6° average timing. If a change is necessary, subtract the average timing from the timing specification to determine the amount of timing change to No. 1 cylinder. For example: if the timing spec is 8° and the average timing is 6°, advance No. 1 cylinder 2° to set the timing.

7. Once the timing is properly set, remove the jumper wire from the ALCL connector. To clear the ECM memory, disconnect the ECM harness from the positive battery pigtail for 10 seconds with the key in the OFF position.

Computer Controlled Coil Ignition (C3I)

COMPONENT REMOVAL

Ignition Coil

NOTE: Two types of C3I ignition coils are used. Type 1 coils have the plug wires arranged three on each side of the coil pack. Type 2 coils group all six plug wires along one side of the coil pack. Type 1 coils are serviced as a complete unit, while Type 2 has individual coils available.

1. Disconnect the negative battery cable.
2. Remove the spark plug wires.
3. Remove the screws holding the coil to the ignition module.
4. Tilt the coil assembly to the rear and remove the coil to module connectors.
5. Remove the coil assembly.
6. Installation is the reverse of removal.

Ignition Module

1. Disconnect the negative battery cable.
2. Disconnect the 14-pin connector at the ignition module.
3. Remove the spark plug wires at the coil assembly.
4. Remove the nuts and washers securing the ignition module assembly to the mounting bracket.
5. Remove the screws securing the ignition module to the coil.

TYPE I   TYPE II

Module/coil assembly

6. Tilt the coil and disconnect the coil to module connectors.

7. Separate the coil and module.

8. Installation is the reverse of removal.

Crankshaft Sensor

NOTE: It is not necessary to remove the sensor bracket.

1. Disconnect the negative battery cable.

2. Disconnect the sensor 3-way connector.

3. Raise the vehicle and support it safely.

4. Rotate the harmonic balancer so the slot in the disc is aligned with the sensor.

5. Loosen the sensor retaining bolt.

6. Slide the sensor outboard and remove through the notch in the sensor housing.

7. Install the new sensor in the housing and rotate the harmonic balancer so that the disc is positioned in the sensor.

8. Adjust the sensor so that there is an equal distance on each side of the disc. There should be approximately .030 in. (.76mm) clearance between the disc and the sensor.

9. Tighten the retaining bolt and recheck the clearance.

10. Install remaining components in the reverse order of removal.

Camshaft Position Sensor

NOTE: If only the camshaft sensor is being replaced, it is not necessary to remove the entire assembly. The sensor is replaceable separately.

1. Disconnect the negative battery cable.

2. Disconnect the ignition module 14-pin connector.

3. Remove the spark plug wires at the coil assembly.

4. Remove the ignition module bracket assembly.

5. Disconnect the sensor 3-way connector.

6. Remove the sensor mounting screws, then remove the sensor.

7. Installation is the reverse of removal.

Camshaft Position Sensor Drive Assembly

1. Follow steps 1–6 of the cam sensor removal procedure. Note the position of the slot in the rotating vane.

2. Remove the bolt securing the drive assembly to the engine.

3. Remove the drive assembly.

4. Install the drive assembly with the slot in the vane. Install mounting bolt.

5. Install the camshaft sensor.

6. Rotate the engine to set the No. 1 cylinder at TDC/compression.

7. Mark the harmonic balancer and rotate the engine to 25 degrees after top dead center.

8. Remove the plug wires from the coil assembly.

9. Using weatherpack removal tool J-28742-A, or equivalent, remove terminal B of the sensor 3-way connector on the module side.

10. Probe terminal B by installing a jumper and reconnecting the wire removed to the jumper wire.

11. Connect a voltmeter between the jumper wire and ground.

12. With the key ON and the engine stopped, rotate the camshaft sensor counterclockwise until the sensor switch just closes. This is indicated by the voltage reading going from a high 5–12 volts to a low 0–2 volts. The low voltage indicates the switch is closed.

13. Tighten the retaining bolt and reinstall the wire into terminal B.

14. Install remaining components.

FUEL SYSTEM

Two types of fuel injection systems are used on N-Body vehicles. The four cylinder utilizes Throttle Body (TBI) Injection and the V6 is equipped with Multi-Port (MFI) Injection. For all information on these fuel injection systems not covered below, please refer to *Chilton's Guide To Fuel Injection And Feedback Carburetors.*

RELIEVING FUEL SYSTEM PRESSURE

On the four cylinder engine with throttle body (TBI) fuel injection, remove the fuse marked "Fuel Pump" from the block in the passenger compartment. Start the engine and allow it to idle until it stalls. Crank the engine an additional three seconds to make sure all fuel pressure is exhausted from the fuel lines, then turn

off the key and reinstall the fuel pump fuse.

On the V6 engine with multiport (MPFI) fuel injection, connect fuel pressure gauge J-34730-1 or equivalent to the fuel pressure valve. Wrap a clean shop towel around the fitting while making connections to catch any fuel spray. Install a bleed hose onto the gauge assembly and place the end into a suitable container. Open the valve on the pressure gauge to bleed the fuel pressure from the system.

FUEL SYSTEM PRESSURE TEST

TBI System

1. Relieve fuel system pressure.

2. Remove the air cleaner and plug the THERMAC vacuum port on the throttle body.

3. Install pressure gauge J-29658 or equivalent on the throttle body side of the fuel filter at the rear of the car near the fuel tank.

4. Start the engine and read the fuel pressure on the gauge. It should be 9–13 psi.

5. Turn the ignition off, relieve the fuel system pressure and remove the fuel pressure gauge. Reconnect all fuel and vacuum lines and install the air cleaner.

MPFI System

Depressurize the fuel system and connect pressure gauge J-34730-1 or equivalent to the fuel pressure test point (Shrader fitting) on the fuel rail. Wrap a clean shop cloth around the fitting to catch any fuel leakage when connecting the gauge. Turn the ignition ON and read the fuel pressure on the gauge. It should be 37–43 psi. Start the engine and again note the fuel pressure one the gauge. With the engine at idle, pressure should be 33–40 psi. This idle pressure will vary somewhat depending on barometric pressure, but in any case it should be lower. Relieve the fuel pressure and disconnect the gauge.

Fuel Pump

REMOVAL & INSTALLATION

1. All fuel injected engines use a pressurized fuel system. The fuel pressure must be relieved before attempting any removal procedures.

2. Disconnect the negative battery cable.

3. Raise the vehicle and support it safely. Drain the fuel tank.

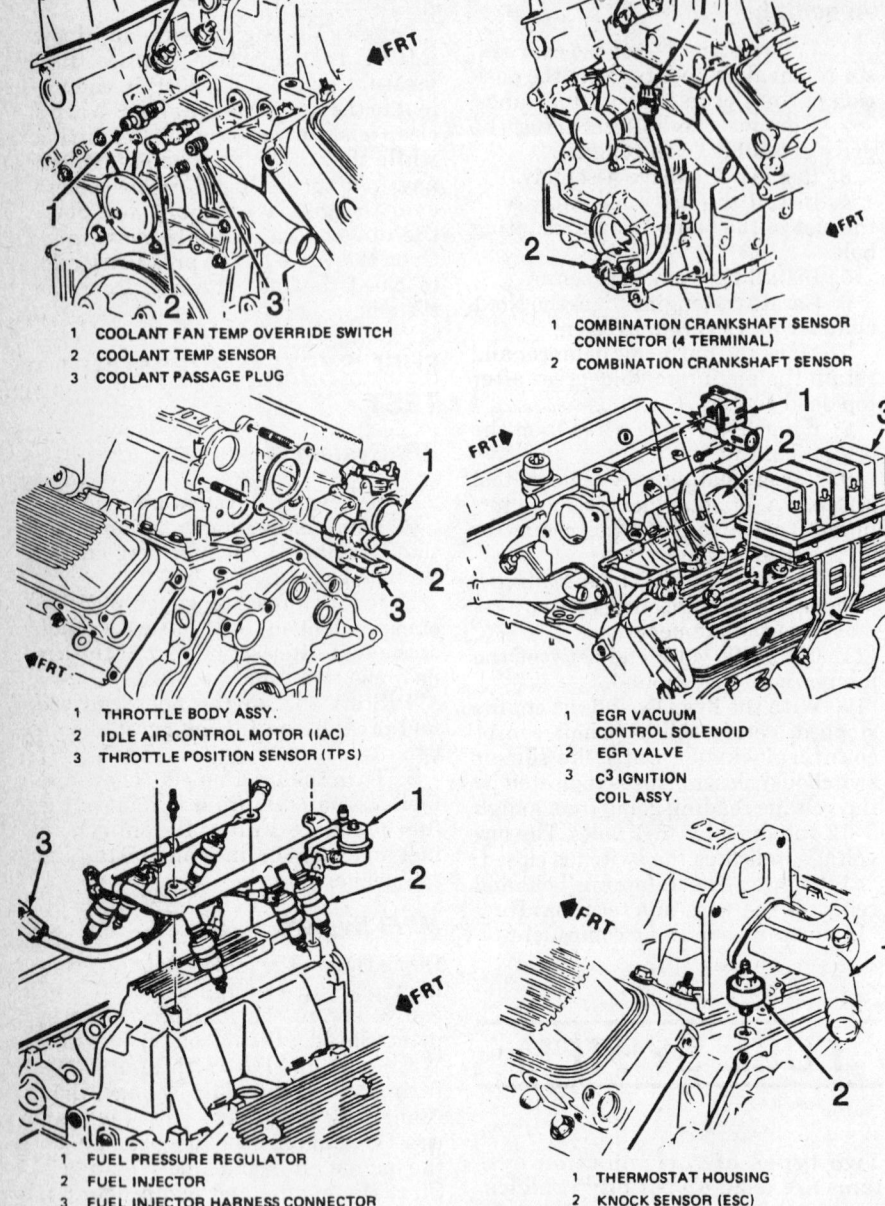

1 COOLANT FAN TEMP OVERRIDE SWITCH
2 COOLANT TEMP SENSOR
3 COOLANT PASSAGE PLUG

1 COMBINATION CRANKSHAFT SENSOR
CONNECTOR (4 TERMINAL)
2 COMBINATION CRANKSHAFT SENSOR

1 THROTTLE BODY ASSY.
2 IDLE AIR CONTROL MOTOR (IAC)
3 THROTTLE POSITION SENSOR (TPS)

1 EGR VACUUM
CONTROL SOLENOID
2 EGR VALVE
3 C³ IGNITION
COIL ASSY.

1 FUEL PRESSURE REGULATOR
2 FUEL INJECTOR
3 FUEL INJECTOR HARNESS CONNECTOR

1 THERMOSTAT HOUSING
2 KNOCK SENSOR (ESC)

3.0L multi—port fuel injection component locations

injected models. It is located on a frame crossmember near the rear of the vehicle. Always use a backup wrench when removing or installing the fuel fittings and depressurize the fuel system by removing the fuel pump fuse and allowing the engine to run until it stalls. Crank the starter for three seconds after the engine stalls to ensure all pressure is relieved from the fuel lines before attempting to remove any fuel lines. Replace the O-rings at the fuel coupling connections and tighten the connections to 22 ft. lbs. (30 Nm). Take precautions to avoid the risk of fire.

COOLING SYSTEM

Radiator

REMOVAL & INSTALLATION

1. Disconnect the negative battery cable.
2. Drain the engine coolant from the radiator.
3. Disconnect the forward strut brace at the radiator and swing it rearward. To prevent shearing of the rubber bushing, loosen the bolt before swinging the brace.
4. Remove the forward lamp harness from the fan frame and disconnect the fan connector.
5. Remove the fan attaching bolts, then remove the fan and frame assembly.
6. Scribe a reference mark on the hood latch assembly, then remove the hood latch from the radiator support.
7. Disconnect the coolant hoses from the radiator and remove the coolant recovery hose from the radiator neck.
8. On models with automatic transmission, disconnect the transmission oil cooler lines from the radiator.
9. Remove the radiator-to-radiator support attaching bolts and clamps, then lift the radiator from the engine compartment. It may be necessary to raise the right hand (driver) side of the radiator first to allow the neck to clear the A/C compressor.
10. Installation is the reverse of removal. Torque the radiator mounting bolts to 7.5 ft. lbs. (10 Nm); transaxle cooling lines to 20 ft. lbs. (27 Nm); and hood latch bolts to 18 ft. lbs. (25 Nm). Refill the cooling system, start the engine and check for leaks.

CAUTION

Do not drain or store fuel in an open container. Serious explosion and fire hazard exists. Empty the contents of the fuel tank into an approved gasoline storage container and take precautions to avoid the risk of fire.

---

4. Remove the fuel tank by supporting it and disconnecting the two retaining straps. Lower the tank enough to disconnect the sending unit wires, hoses and ground strap, if equipped.
5. Lower the fuel tank from the vehicle and remove the sending unit.

The sending unit is retained by a cam lock ring. The fuel pump is attached to the tank sending unit.
6. Installation is the reverse of removal. Replace the O-ring when installing the sending unit and be sure to install the anti-squeak pieces on top of the fuel tank or the assembly will rattle. Tighten the retaining straps.

Fuel Filter

REMOVAL & INSTALLATION

An inline fuel filter is used on all fuel

Water Pump

REMOVAL & INSTALLATION

NOTE: Special pulley removal and installation tools are required to remove and install the water pump pulley on the 2.5L engine.

1. Disconnect the negative battery cable.
2. Drain the cooling system.
3. Remove the drive belts.
4. Remove the fan, pulley and radiator shroud, as required to gain working clearance and access to the water pump bolts.
5. Remove the radiator and heater hose(s) from the water pump.
6. Remove the water pump retaining bolts. On the V6 engine, the long bolt is removed through the access hole that is provided in the body side rail.
7. Remove the water pump from the vehicle.
8. On 2.5L engines, remove the water pump pulley and transfer it to the replacement assembly using the removal tool J2975-A and the installation tool J25033-B.
9. Installation is the reverse of removal. Clean all gasket mating surfaces and place a $\frac{1}{8}$ in. bead of RTV sealant on all sealing surfaces. Water pump mounting bolts must also be coated with RTV sealer to avoid coolant leaks. Torque the water pump mounting bolts to 25 ft. lbs. (34 Nm) on 4 cylinder engines, or 22 ft. lbs. (30 Nm) on V6 engines.

Thermostat

REMOVAL & INSTALLATION

2.5L Engine

——— CAUTION ———

The engine must be cold for this procedure.

1. Remove the thermostat housing cap.
2. Grasp the handle of the thermostat assembly and gently pull upward.
3. Clean the thermostat housing and O-ring.
4. Apply a suitable lubricant to the O-ring, then install the thermostat into the housing, pushing down to ensure that the thermostat is firmly seated.
5. Replace the thermostat housing cap.

3.0L V6 Engine

1. Disconnect the negative battery cable.

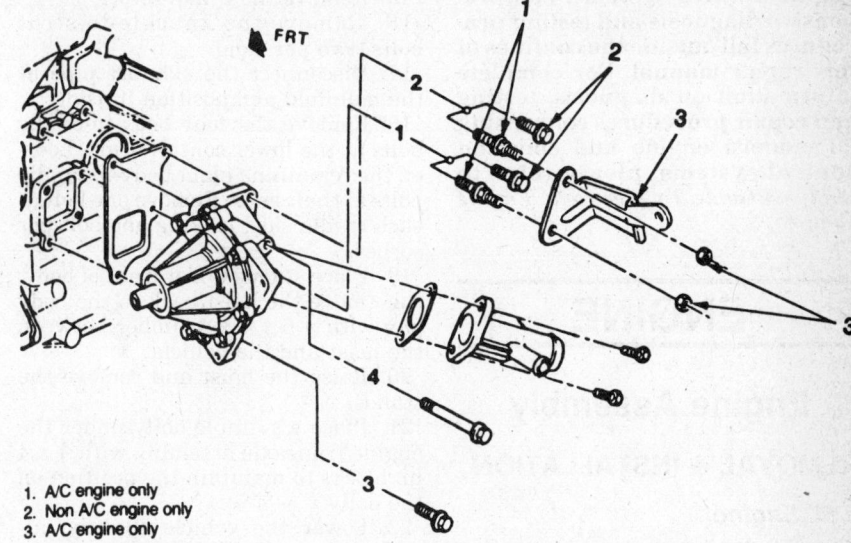

1. A/C engine only
2. Non A/C engine only
3. A/C engine only

Water pump mounting on 2.5L engine

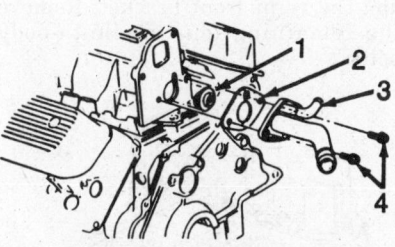

1. Thermostat
2. Gasket
3. Thermostat housing
4. Bolt

V6 thermostat housing

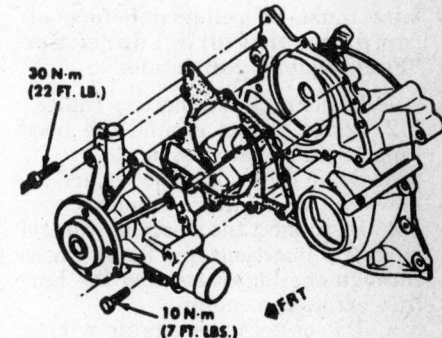

30 N·m (22 FT. LB.)
10 N·m (7 FT. LBS.)
FRT

Water pump mounting on V6

2. Drain the engine coolant below the level of the thermostat housing.
3. Remove the water outlet to thermostat housing attaching bolts.
4. Remove the thermostat housing and lift out the thermostat.
5. Installation is the reverse of removal. Clean all gasket mating surfaces and coat the thermostat housing with a $\frac{1}{8}$ in. bead of RTV sealant. Make sure the thermostat is installed correctly (wax pellet toward the engine) and torque the thermostat housing bolts to 21 ft. lbs. (28 Nm).

EMISSION CONTROLS

For all information on emission control maintenance, please refer to "Emission Controls" in the Unit Repair section. Due to the complex nature of modern electronic

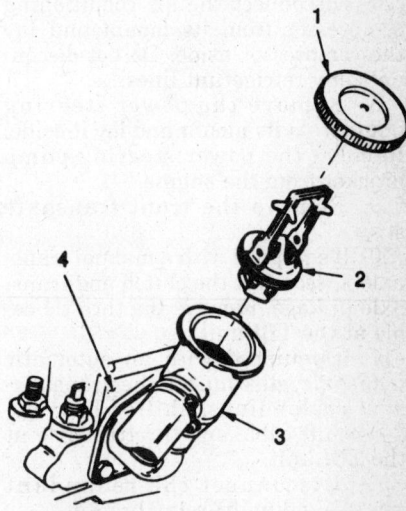

1. Thermostat housing cap
2. Thermostat
3. Thermostat housing assembly
4. Cylinder head

4 cyl thermostat housing

engine control systems, comprehensive diagnosis and testing procedures fall outside the confines of this repair manual. For complete information on diagnosis, testing and repair procedures concerning all modern engine and emission control systems, please refer to *Chilton's Guide To Electronic Engine Controls.*

ENGINE

Engine Assembly

REMOVAL & INSTALLATION

2.5L Engine

NOTE: The following procedure is for removing the engine and transaxle as a unit. The fuel pressure must be relieved before attempting this procedure. See "Fuel System" for details.

1. Disconnect both battery cables.
2. Scribe marks around the hood hinges and remove the hood.
3. Drain the cooling system and remove the air cleaner assembly.
4. Disconnect the electronic control module connections and feed harness through the bulkhead. Lay the harness across the engine.
5. Disconnect the engine wiring harness after tagging all connectors and lay it across the engine.
6. Disconnect the heater hoses, radiator hoses and vacuum lines after tagging them for installation.
7. Disconnect the air conditioning compressor from its mount and lay the compressor aside. Do not disconnect any refrigerant lines.
8. Remove the power steering pump from its mount and lay it aside. Remove the power steering pump bracket from the engine.
9. Remove the front transaxle strut.
10. If equipped with a manual transaxle, disconnect the clutch and transaxle linkage. Remove the throttle cable at the TBI unit.
11. If equipped with an automatic transaxle, disconnect the transmission cooler lines, shifter linkage, downshift cable and throttle cable at the TBI unit.
12. Disconnect the redundant ground and multi-relay bracket.
13. Raise the vehicle and support it safely.
14. Remove the front wheels. Remove the calipers and wire them up out of the way. Do not allow the calipers to hang by the brake hoses.

15. Remove the brake rotors.
16. Remove the knuckle-to-strut bolts (two per side).
17. Disconnect the exhaust pipe at the manifold and position it aside.
18. Remove the four body-to-cradle bolts at the lower control arms. Loosen the remaining eight body-to-cradle bolts at their ends. Remove one bolt at each cradle side, leaving one bolt per corner.
19. Place stands under front of body, then move the hoist back to the body pan with a 6 x 4 x 4 timber between the hoist and the vehicle.
20. Raise the hoist and remove the stands.
21. Place a suitable dolly under the engine/transaxle assembly with 4 x 4 in. blocks to maintain the position on the dolly.
22. Lower the vehicle slightly, allowing the engine/transaxle assembly to rest on the dolly.
23. Remove the engine mount bolts and the right front bracket. Remove the remaining four cradle-to-body bolts.

24. Lower the engine/transaxle assembly from the vehicle, then separate the engine and transaxle with the assembly on the floor.
25. Installation is the reverse of removal procedures.

3.0L V6 Engine

1. Disconnect the battery cables.

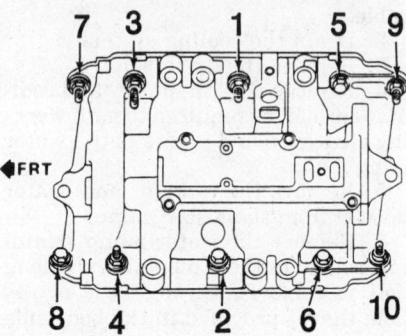

3.0L intake manifold bolt torque sequence

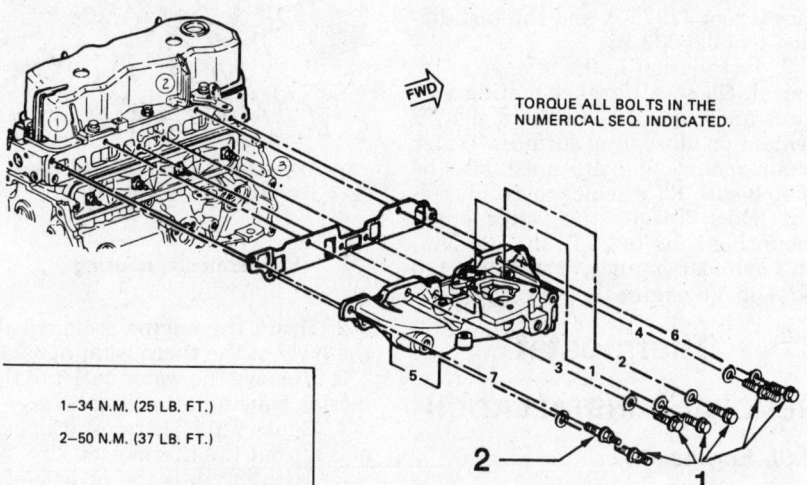

1–34 N.M. (25 LB. FT.)

2–50 N.M. (37 LB. FT.)

TORQUE ALL BOLTS IN THE NUMERICAL SEQ. INDICATED.

2.5L intake manifold bolt torque sequence

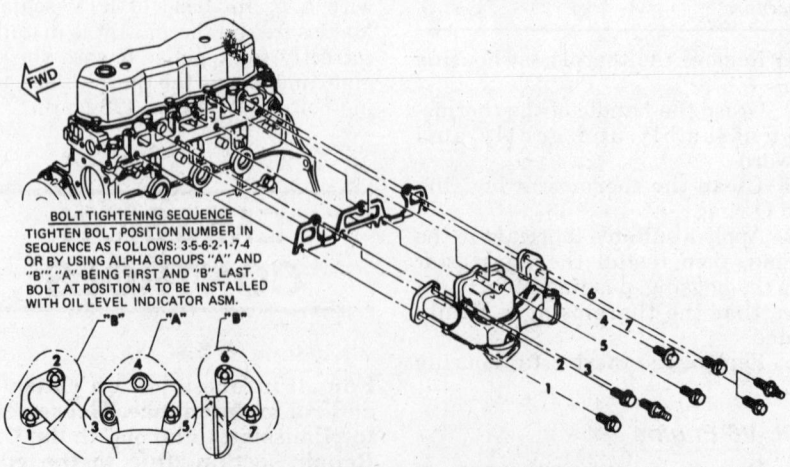

BOLT TIGHTENING SEQUENCE
TIGHTEN BOLT POSITION NUMBER IN SEQUENCE AS FOLLOWS: 3-5-6-2-1-7-4 OR BY USING ALPHA GROUPS "A" AND "B", "A" BEING FIRST AND "B" LAST. BOLT AT POSITION 4 TO BE INSTALLED WITH OIL LEVEL INDICATOR ASM.

2.5 L exhaust manifold bolt torque sequence

2. Matchmark the hood hinges and remove the hood.

3. Raise the vehicle and support it safely. Drain the radiator.

4. Remove the starter and torque converter cover.

5. Remove the torque converter bolts. Matchmark the torque converter to the flywheel for reassembly.

6. Disconnect the A/C compressor wiring. Disconnect the compressor from its engine mount and position it aside. Do not disconnect any refrigerant lines.

7. Disconnect the heater hoses.

8. Remove the lower radiator hose.

9. Remove the front motor mount bolts and remove the right inner fender splash shield.

10. Remove the transaxle-to-engine mount bolt located between the transaxle and the cylinder block.

11. Remove the two right rear motor mount bolts.

12. Disconnect the exhaust pipe from the exhaust manifold flange.

13. Lower the vehicle.

14. Remove the serpentine drive belt, alternator wiring and the alternator. Tag the alternator wires before removal.

15. Remove the power steering pump and the fluid lines.

16. Disconnect the mass air flow sensor and the air intake duct. Disconnect the top radiator hose.

17. Disconnect the electric fan wiring and remove the fan assembly. Remove the radiator.

18. Install an engine lifting device and remove the left upper transaxle mount.

19. Disconnect and remove the master cylinder, if necessary.

20. Disconnect and remove the fuel lines at the fuel rail. See the instructions for depressurizing the fuel system before disconnecting any fuel lines.

21. Disconnect the throttle, T.V. and cruise control cables at the throttle body.

22. Remove the remaining engine to transaxle retaining bolts. Remove the engine from the vehicle using the lifting device.

23. Installation is the reverse of removal procedures.

Intake Manifold

REMOVAL & INSTALLATION

2.5L Engine

1. Disconnect the negative battery cable.

2. Remove the air cleaner and the heat stove pipe.

3. Remove the PCV valve and hose.

4. Drain the cooling system.

5. Tag and remove the vacuum lines. Depressurize the fuel system and remove the fuel lines.

6. Tag and remove the wiring and throttle linkage from the TBI injection system.

7. Disconnect the transaxle downshift linkage and cruise control linkage, if equipped.

8. Disconnect the throttle linkage and bell crank. Position the assembly to the side for clearance.

9. Disconnect the heater hose. Disconnect and remove the upper bracket on the power steering pump.

10. Remove the ignition coil.

11. Remove the intake manifold retaining bolts. Remove the intake manifold.

12. Installation is the reverse of removal. Clean all gasket mating surfaces and torque the intake manifold bolts to specifications in the order illustrated.

3.0L Engine

1. Disconnect the negative battery cable.

2. Disconnect the mass air flow sensor and remove the air intake duct.

3. Remove the serpentine drive belt, alternator and bracket.

4. Remove the C3I ignition module and bracket.

5. Tag and remove all vacuum and wiring connectors as required.

6. Remove the throttle, cruise control and T.V. cables from the throttle body assembly.

7. Drain the coolant and disconnect the heater hoses from the throttle body.

8. Remove the upper radiator hose.

9. Depressurize the fuel system and remove the fuel lines, fuel rain and fuel injectors. Remove the spark plug wires.

10. Remove the intake manifold retaining bolts. Remove the intake manifold.

11. Installation is the reverse of removal. Clean all gasket mating surfaces. Torque all bolts to specifications in the sequence illustrated.

Exhaust Manifold

REMOVAL & INSTALLATION

2.5L Engine

1. Disconnect the negative battery cable.

2. Remove the air cleaner and heat stove tube.

3. Remove the alternator top mounts and position the unit to one side.

4. Disconnect the oxygen sensor connector.

5. Raise the vehicle and support it safely.

6. Disconnect the exhaust pipe from the exhaust manifold retaining flange.

7. Lower the vehicle.

8. Remove the exhaust manifold retaining bolts and lift the exhaust manifold from the engine.

9. Installation is the reverse of removal. Clean all gasket mating surfaces and torque all bolts to specifications.

3.0L Engine

1. Disconnect the negative battery cable.

2. Raise the vehicle and support it safely.

3. Remove the exhaust pipe to manifold flange retaining bolts.

4. Lower the vehicle and disconnect the oxygen sensor electrical connector. Remove the spark plug wires.

5. Remove the two nuts retaining the crossover pipe to the manifold.

6. Remove all necessary components in order to gain access to the exhaust manifold retaining bolts. Tag any wire or cable connections for installation purposes.

7. Remove the manifold retaining bolts and remove the manifold.

8. Installation is the reverse of removal. Torque all exhaust manifold retaining bolts to specifications.

Rocker Arm And Pushrod

REMOVAL & INSTALLATION

2.5L Engine

1. Disconnect the negative battery cable.

2. Remove all necessary components in order to gain access to the valve cover retaining bolts.

3. Remove the valve cover bolts and lift off the valve cover.

4. Remove the rocker arm bolt and ball and lift off the rocker arm. If only the pushrods are being replaced, simply loosen the rocker arm bolt and swing the rocker arm aside. Keep all removed components in order so they may be assembled in their original locations.

5. Install the push rods, making sure they seat properly in the lifter.

6. Install the rocker arms, balls and nuts. Tighten the rocker arm nuts until all lash is eliminated.

7. Adjust the valves when the lifter is on the base circle of the camshaft lobe:

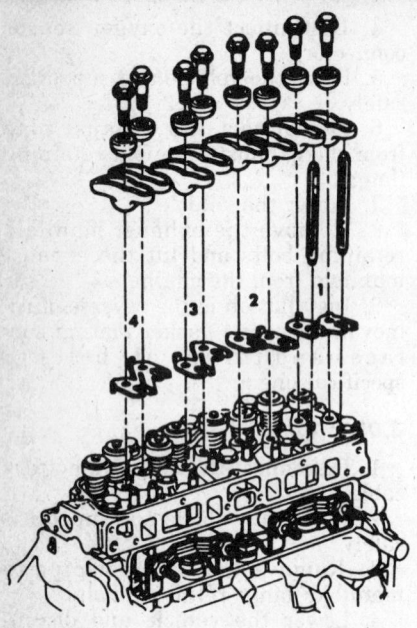

2.5L rocker arm assembly

a. Crank the engine until the mark on the crankshaft pulley lines up with the "O" mark on the timing tab. Make sure that the engine is in the No. 1 firing position (TDC/compression). Place your fingers on the No. 1 rocker arms as the mark on the crank pulley comes near the "O" mark. If the valves are not moving, the engine is in the No. 1 firing position; if the valves move, the engine is in the No. 4 firing position and the engine must be rotated one complete revolution to set up TDC.

b. When the engine is in the No. 1 firing position, adjust the No. 1 and 3 EXHAUST and No. 1 and 2 INTAKE valves.

c. Back the adjusting nut out until lash can be felt at the push rod, then turn the nut until all lash is removed. This can be determined by rotating the push rod while turning the adjusting nut. When all lash has been removed, turn the nut down 1½ additional turns to center the lifter plunger.

d. Crank the engine one complete revolution until the timing tab and the "O" mark are again in alignment. Now the engine is in the No. 4 firing position; adjust the No. 2 and 4 EXHAUST and the No. 3 and 4 INTAKE valves.

8. Installation of the remaining components is in the reverse order of removal.

3.0L Engine

1. Disconnect the negative battery cable.

2. Remove the valve covers. It will be necessary to remove the C3I ignition coil module to gain access to the rear valve cover bolts. It may also be necessary to remove the alternator, alternator brace, engine lift bracket and power steering belt tensioner. Remove any vacuum hoses after tagging them for installation as necessary.

3. Remove the rocker arm pedestal retaining bolts and remove the rocker arm and pedestal assembly. Remove the pushrods.

NOTE: Check the position of the double ended bolts and note for reassembly.

4. Place all removed components on a clean surface in order so they may be installed in their original locations.

5. Installation is the reverse of removal. Torque all rocker arm pedestal bolts to 45 ft. lbs. (60 Nm) and install the remaining components. Make sure the drive belt is tensioned properly.

Cylinder Head

REMOVAL & INSTALLATION

2.5L Engine

NOTE: Relieve fuel system pressure before disconnecting any fuel lines.

1. Disconnect the negative battery cable.

2. Drain the cooling system and remove the dipstick tube.

3. Remove the air cleaner assembly.

4. Disconnect the exhaust pipe to manifold flange.

5. Tag and disconnect the electrical wiring and throttle linkage from the TBI assembly.

6. Remove the heater hose from the intake manifold.

7. Remove the ignition coil. Tag and disconnect all wiring connections from the intake manifold and the cylinder head.

8. Remove the A/C compressor brackets and lay the compressor aside. Do not disconnect any refrigerant lines. Remove the alternator.

9. Remove the power steering pump upper bracket.

10. Remove the radiator hoses.

11. Remove the rocker arm covers, rocker arms and pushrods.

12. Remove the cylinder head retaining bolts, then lift the cylinder head from the engine.

13. Clean all gasket mating surfaces and install a new head gasket.

14. Installation is the reverse of removal. Torque all cylinder head bolts

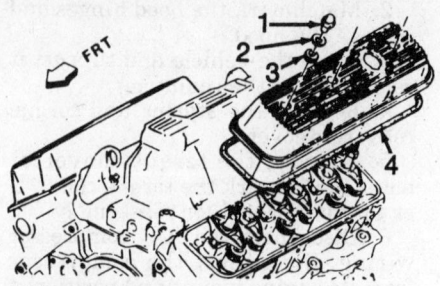

1. Nut 10 N·m (88 lb. in.)
2. Washer
3. Rubber grommet
4. Formed rubber gasket

V6 rocker arm cover and gasket

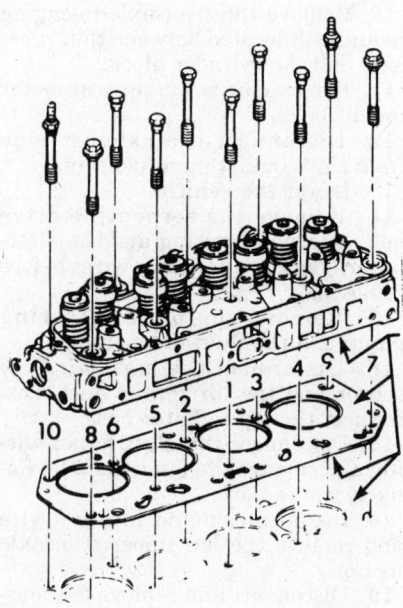

2.5L cylinder head bolt torque sequence

to specifications in the sequence illustrated.

3.0L Engine

1. Disconnect the negative battery cable.

2. Remove the mass air flow sensor and the air intake duct.

3. Remove the C3I ignition module and wiring.

4. Remove the serpentine belt. Remove the alternator and bracket.

5. Tag and remove all necessary vacuum lines and electrical connections.

6. Remove the fuel lines and the fuel rail after first relieving the fuel system pressure. Remove the spark plug wires.

7. Drain the cooling system and remove the heater and radiator hoses from the throttle body and intake

manifold. Remove the radiator and cooling fan.

8. Remove the intake manifold retaining bolts in reverse of the torque sequence and remove the intake manifold from the engine.

9. Remove the valve covers. Remove the rocker arms, pedestals and pushrods. Keep all parts in order so they may be assembled in their original locations.

10. Remove the left exhaust manifold.

11. Remove the power steering pump, dipstick and dipstick tube.

12. Remove the left cylinder head retaining bolts in reverse of the torque sequence. Lift the left cylinder head from the engine.

13. Raise and support the vehicle safely. Remove the right exhaust manifold retaining bolts.

14. Remove the right cylinder head retaining bolts in reverse of the torque sequence. Lift the right cylinder head from the vehicle.

15. Installation is the reverse of the removal procedure. Clean all gasket mating surfaces and tighten the head bolts to specifications in the sequence illustrated. Torque the intake manifold mounting bolts in the sequence illustrated under "Intake Manifold Removal & Installation."

Front Cover and Oil Seal

REMOVAL & INSTALLATION

2.5L Engine

1. Disconnect the negative battery cable.

2. Remove the inner fender splash shield.

3. Remove the fan pulley and the crankshaft hub.

4. Remove the front cover retaining bolts and remove the front cover.

5. Clean all gasket mating surfaces and apply a $\frac{3}{8}$ in. wide by $\frac{3}{16}$ in. thick bead of RTV sealant to the joint of the oil pan and timing gear cover.

6. Apply a $\frac{1}{4}$ in. wide by $\frac{1}{8}$ in. thick bead of RTV sealant to the timing gear cover at the engine block mating surfaces.

7. Install the new front oil seal using tool J-34995 or equivalent.

NOTE: Tool J-34995 is also used as a centering tool which fits over the crankshaft seal and is used to correctly position the front cover during installation.

8. Partially tighten the two timing case cover opposing screws, then tighten the remaining cover screws

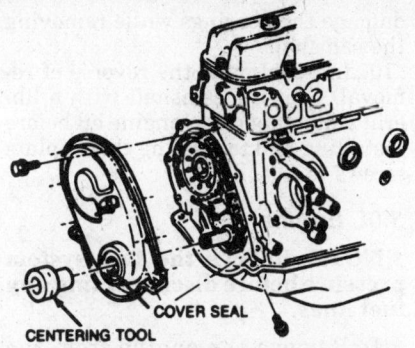

Front cover installation; a centering tool can aid in positioning

and remove the centering tool from the front cover.

9. Install the remaining components in the reverse order of removal.

3.0L Engine

1. Disconnect the negative battery cable.

2. Drain the cooling system.

3. Loosen, but do not remove, the water pump pulley bolts. Remove the serpentine belt and remove the pulley.

4. Remove the water pump retaining bolts and remove the water pump.

5. Raise and support the vehicle safely. Remove the right tire and the right inner fender splash shield.

6. Remove the crankshaft balancer.

7. Drain the engine oil and remove the oil filter.

8. Remove the radiator and heater hoses.

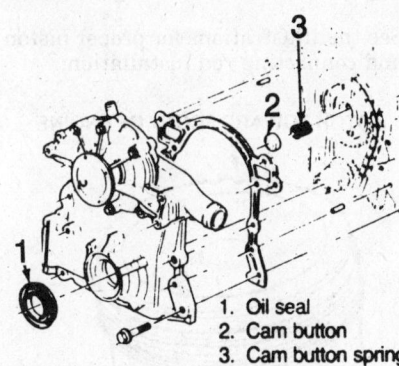

1. Oil seal
2. Cam button
3. Cam button spring

V6 front cover and seal

9. Remove the crankshaft sensor and the engine oil pan.

10. Remove the front cover retaining bolts, then remove the front cover and gasket.

11. Pry out the old oil seal and install the new seal using tool J-35354 or equivalent.

12. Clean all gasket mating surfaces and install the front cover with a new gasket. Coat all front cover bolts with thread sealer prior to installation.

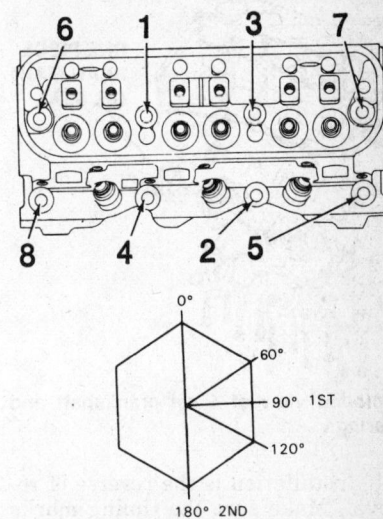

3.0L cylinder head bolt torque sequence and angle—tightening procedure

13. Install the remaining components in the reverse order of removal. Torque the crankshaft balancer bolts to specifications.

Timing Chain or Gears

REMOVAL & INSTALLATION

2.5L Engine

NOTE: If the camshaft gear is to be replaced, the engine must be removed from the vehicle. The crankshaft gear may be replaced with the engine in the vehicle.

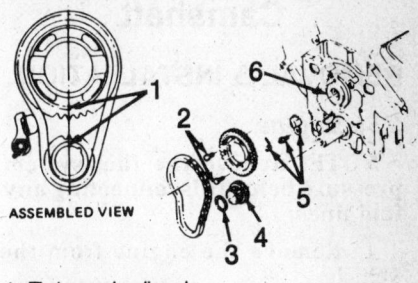

ASSEMBLED VIEW

1. Timing marks aligned
2. 22 ft. lbs. (30 Nm)
3. Seal
4. Crankshaft gear
5. Dampner assembly
6. Camshaft

V6 timing chain and sprocket

1. Disconnect the negative battery cable.

2. If replacing the camshaft gear, remove the engine. Remove the timing cover bolts and remove the cover. Remove the camshaft and press the gear off the cam.

3. If removing the crankshaft gear, remove the timing cover. Remove the retaining bolt and slide the crankshaft gear forward off of the shaft.

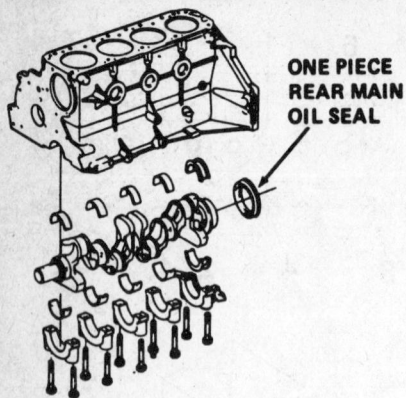

Exploded view of 4 cyl crankshaft and bearings

ONE PIECE REAR MAIN OIL SEAL

4. Installation is the reverse of removal. Make sure the timing marks are aligned properly and torque the crankshaft and/or camshaft retaining bolts to specifications.

3.0L Engine

1. Disconnect the negative battery cable.
2. Remove the timing cover.
3. Align the timing marks on the sprockets.
4. Remove the camshaft sprocket bolts, then remove the camshaft sprocket and chain.
5. Remove the crankshaft sprocket by sliding it forward.
6. Installation is the reverse of removal. Make sure the timing marks are aligned on the sprockets.

Camshaft

REMOVAL & INSTALLATION

2.5L Engine

NOTE: Relieve the fuel system pressure before disconnecting any fuel lines.

1. Remove the engine from the vehicle.
2. Remove the rocker cover, rocker arms and pushrods.
3. Remove the distributor, spark plugs and fuel pump.
4. Remove the pushrod cover and gasket, then remove the lifters.
5. Remove the alternator, alternator lower bracket and the front engine mount bracket assembly.
6. Remove the oil pump driveshaft and gear assembly.
7. Remove the crankshaft hub and timing gear cover.
8. Remove the two camshaft thrust plate screws by working through the holes in the gear.
9. Remove the camshaft and gear assembly by pulling it through the front of the block. Take care not to

damage the bearings while removing the camshaft.
10. Installation is the reverse of removal. Coat the camshaft with a liberal amount of clean engine oil before installing and torque the thrust plate screws to 6 ft. lbs.

3.0L Engine

NOTE: Relieve the fuel system pressure before disconnecting any fuel lines.

1. Remove the engine from the vehicle.
2. Remove the intake manifold.
3. Remove the valve covers, rocker arm assemblies, push rods and lifters. Keep all parts in order for reassembly.
4. Remove the crankshaft balancer.
5. Remove the front cover. Remove the timing chain and sprockets. Make sure the timing marks are aligned properly before removal.
6. Remove the camshaft retainer bolts and slide the camshaft forward out of the engine block. Take care not to damage the bearings while removing the camshaft.
7. Installation is the reverse of removal. Coat the camshaft with a liberal amount of clean engine oil prior to installation.

Piston And Connecting Rod Positioning

See the illustrations for proper piston and connecting rod installation.

NOTCH TOWARD FRONT OF ENGINE

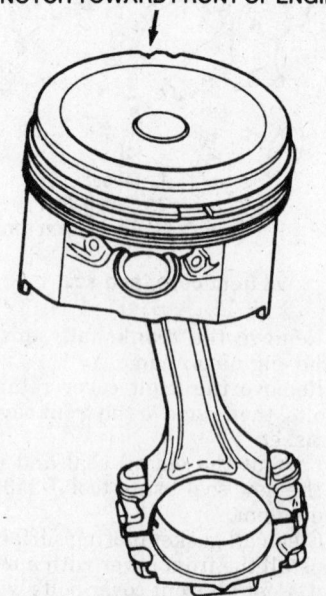

2.5L piston identification

NOTCH TOWARD FRONT OF ENGINE

1

3.0L piston identification

ENGINE LUBRICATION

Oil Pan

REMOVAL & INSTALLATION

2.5L Engine

1. Disconnect the negative battery cable.
2. Raise and support the vehicle safely and drain the engine oil.
3. Remove the cradle to front engine mount bolts.
4. Disconnect the exhaust pipe at the manifold at the rear transaxle mount.
5. Disconnect the starter and remove the flywheel housing inspection cover.
6. Remove the upper alternator bracket.
7. Install engine support J-22825-40 or equivalent and raise the engine.
8. Remove the lower alternator bracket and engine support bracket.
9. Remove the oil pan retaining bolts and remove the oil pan.
10. Clean all gasket mating surfaces.
11. Install the rear oil pan gasket in the rear main bearing cap and apply a small amount of sealer in the depressions where the pan gasket engages into the block.
12. Install the front oil pan gasket on the timing gear cover, pressing the tips into the holes provided on the cover.
13. Install the side gaskets onto the oil pan. Use grease as a retainer, if necessary.
14. Apply a $\frac{1}{8}$ in. by $\frac{1}{4}$ in. long bead of sealer at the split lines of the front and side gaskets.
15. Install the oil pan. The bolts into

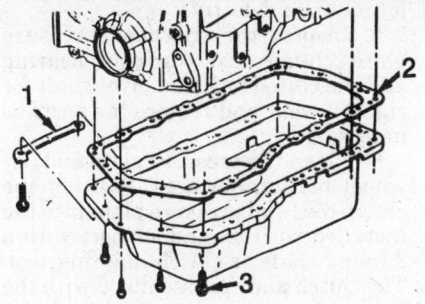

1. Tension spring
2. Rubber gasket
3. Bolt

V6 oil pan installation

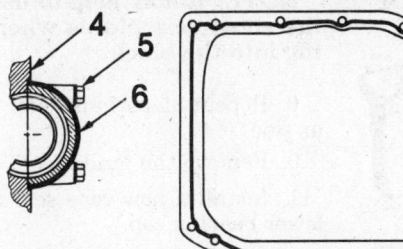

1—OIL PAN

2—APPLY A 3/8" WIDE BY 3/16" THICK BEAD OF RTV SEALER IN AREA INDICATED

3—APPLY A 3/16" WIDE BY 1/8" THICK BEAD OF RTV SEALER IN AREA INDICATED

4—ENGINE BLOCK ASSEMBLY

5—REAR BEARING

6—GROOVE IN MAIN BEARING CAP MUST BE FILLED FLUSH TO 1/8" ABOVE SURFACE WITH RTV

2.5L oil pan installation

the timing gear cover should be installed last; they are installed at an angle and the holes line up after the rest of the pan bolts are snugged up.

16. Install the remaining components in the reverse order of removal. Fill the crankcase with oil, start the engine and check for leaks.

3.0L Engine

1. Disconnect the negative battery cable.
2. Raise and support the vehicle safely.
3. Drain the engine oil and remove the oil filter.
4. Remove the flywheel cover.
5. Remove the oil pan retaining bolts and the oil pan tensioner spring, which is located behind the oil filter adapter.
6. Remove the oil pan from the engine.
7. Clean all gasket mating surfaces and install the oil pan in the reverse order of removal.

Oil Pump

REMOVAL & INSTALLATION

2.5L Engine

1. Disconnect the negative battery cable.
2. Remove the oil pan.
3. Remove the oil pump mounting bolts.
4. Remove the oil pump, pipe and screen as an assembly.
5. Installation is the reverse of removal. Torque the oil pump mounting bolts to 22 ft. lbs. (30 Nm).

3.0L Engine

1. Disconnect the negative battery cable.
2. Remove the front cover from the engine as outlined under "Front Cover and Oil Seal."

3. Remove the oil filter adapter, pressure regulator valve and valve spring.
4. Remove the oil pump cover retaining screws and cover.
5. Remove the oil pump gears.
6. Installation is the reverse of removal.

Rear Main Seal

REMOVAL & INSTALLATION

2.5L Engine

NOTE: The rear main bearing oil seal is a one piece unit and can be replaced without the removal of the oil pan or crankshaft.

1. Remove the transaxle as outlined under "Transaxle Removal & Installation."
2. If equipped with manual transmission, remove the pressure plate and clutch disc.
3. Remove the flywheel retaining bolts and remove the flywheel.
4. Pry out the old seal using a suitable tool.
5. Clean the block and crankshaft-to-seal mating surfaces.
6. Install the new rear main seal into the block using installer tool J-34924 or equivalent. Lubricate the

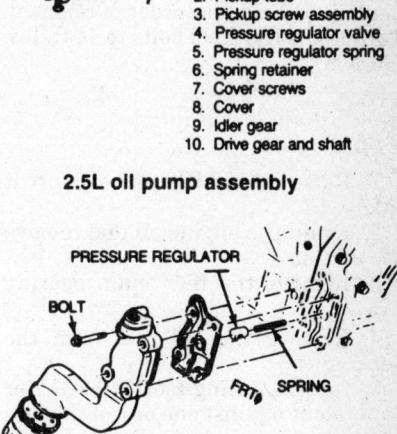

1. Pump body
2. Pickup tube
3. Pickup screw assembly
4. Pressure regulator valve
5. Pressure regulator spring
6. Spring retainer
7. Cover screws
8. Cover
9. Idler gear
10. Drive gear and shaft

2.5L oil pump assembly

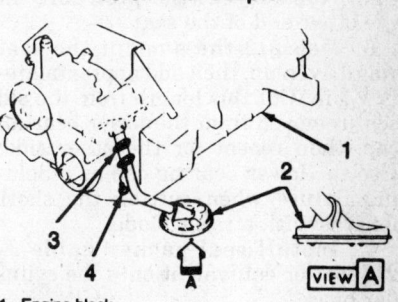

PRESSURE REGULATOR

BOLT

FRT SPRING

V6 oil filter adapter

1. Bolt
2. Oil pump cover
3. Outer gear
4. Inner gear
5. Front cover

V6 oil pump assembly

1. Engine block
2. Oil pump pipe and screen
3. Gasket
4. Bolt

V6 oil pipe and screen assembly

C767

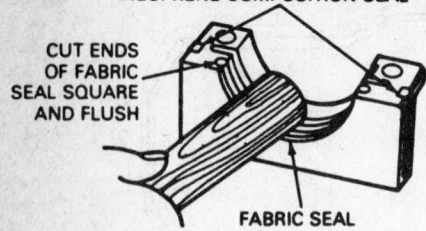

Installing rear main bearing cap oil seal on V6

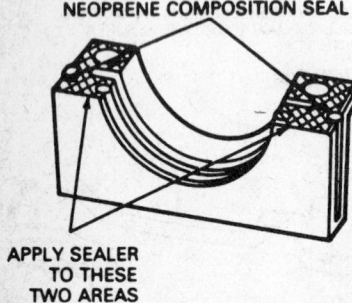

Applying sealer to bearing cap on V6

outside of the seal to aid installation and press the seal in evenly with the tool.

7. Install the remaining components in the reverse order of removal. Torque the flywheel bolts to 44 ft. lbs. (60 Nm).

3.0L Engine

1. Raise the vehicle and support it safely.

2. Drain the engine oil and remove the oil pan.

3. Remove the rear main bearing cap.

4. Remove the old seal from the bearing cap.

5. Insert packing tool J-21526-2 or equivalent against one end of the seal in the cylinder block. Pack the old seal into the groove until it is packed tight, then repeat the procedure on the other end of the seal.

6. Measure the amount the seal was driven up, then add approximately $\frac{1}{16}$ in. Cut this length from the old seal removed from the lower bearing cap, then repeat for the other side. Use the lower bearing cap as a holding fixture when cutting the short lengths with a razor blade.

7. Install seal packer guide J-21526-1 or equivalent onto the cylinder block.

8. Using the packing tool, work the short pieces into the guide tool and pack into the cylinder block until the tool hits the built-in stop.

NOTE: It may help to use oil on the short seal pieces when packing into the block.

9. Repeat Steps 7 and 8 for the other side.

10. Remove the guide tool.

11. Install a new rope seal into the lower bearing cap.

12. Install the lower main bearing cap and torque the bolts to 100 ft. lbs. (135 Nm).

13. Install the remaining components in the reverse order of removal. Fill the crankcase with engine oil, start the engine and check for leaks.

CLUTCH

REMOVAL & INSTALLATION

1. Remove the transaxle.

2. Mark the pressure plate assembly and the flywheel so that they may be assembled in their original position. They are balanced as an assembly at the factory.

3. Loosen the attaching bolts one turn at a time until all spring tension is released.

4. Support the pressure plate and remove the bolts, then remove the pressure plate, clutch disc, throwout bearing and the clutch fork and pivot shaft assembly. Replace any parts

found to be defective.

5. Inspect the flywheel, pressure plate, clutch disc, throwout bearing and the clutch fork and pivot shaft for signs of wear and replace the parts as necessary.

6. Clean the pressure plate and flywheel mating surfaces. Position the clutch disc and pressure plate into the installed position and support with a dummy shaft or clutch aligning tool. The clutch plate is assembled with the damper springs offset toward the transaxle.

7. Install the pressure plate-to-flywheel bolts and tighten gradually in a criss-cross pattern to 15 ft. lbs. (20 Nm).

8. Lubricate the outside grooves and the inside recess of the release bearing with high temperature grease. Wipe off the excess and install the bearing.

9. Install the transaxle.

Clutch Cable

REMOVAL & INSTALLATION

1. Support clutch pedal upward against the bumper stop to release the pawl from the detent. Disconnect the clutch cable from the clutch release lever at the transaxle assembly.

NOTE: Be careful not to let the cable snap rapidly toward the rear of the vehicle. The detent in the adjusting mechanism can be damaged.

2. Remove the hush panel from inside the car.

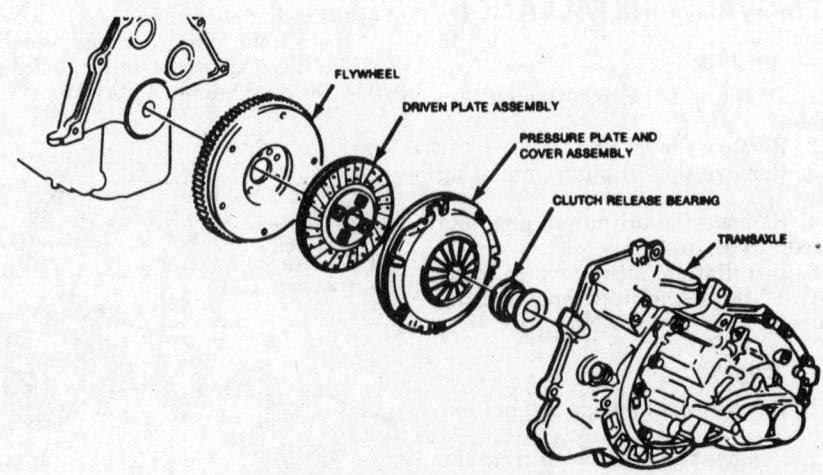

Clutch components

3. Disconnect the clutch cable from the detent end tangs. Lift the locking pawl away from the detent, then slide the cable forward between the detent and locking pawl.

4. Remove the windshield washer bottle.

5. Pull the clutch cable out from the engine side of the cowl, disengaging it from the clutch pedal mounting bracket. The insulators, dampers and washers may separate from the cable when removing. Disconnect the cable from the transaxle mounting bracket and remove the cable.

6. Inspect the clutch cable for fraying, kinks, work ends or excessive friction and replace if necessary.

7. Install the cable into both insulators, damper and washer. Lubricate the rear insulator with rubber lube or soapy water to ease the assembly into the pedal mounting bracket.

8. Working inside the car, route the liner on the cable into the rubber isolator on the pedal bracket and then attach the end of the cable to the detent, being sure to route the cable under the pawl and into the detent cable groove.

9. Install the hush panel.

10. Support the clutch pedal upward against the bumper stop to release the pawl from the detent. Install the other end of the cable to the clutch release lever and transaxle mount bracket.

11. Install the windshield washer bottle and adjust the clutch as outlined below.

ADJUSTMENT

Check the clutch operation and adjust by lifting the clutch pedal up to allow the mechanism to adjust the cable length. Depress clutch pedal slowly several times to set pawl into mesh with the detent teeth. If fork lever travel is not correct, check the adjusting mechanism by depressing the clutch pedal and looking for the pawl to firmly engage with the teeth in the detent.

MANUAL TRANSAXLE

REMOVAL & INSTALLATION

1. Disconnect the negative battery cable.

2. Install and engine holding bar so that one end is supported on the cowl

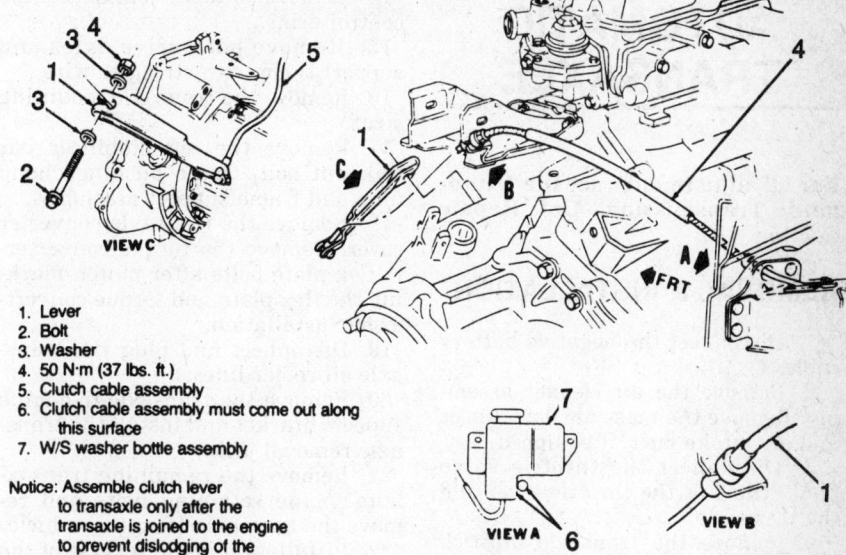

1. Lever
2. Bolt
3. Washer
4. 50 N·m (37 lbs. ft.)
5. Clutch cable assembly
6. Clutch cable assembly must come out along this surface
7. W/S washer bottle assembly

Notice: Assemble clutch lever to transaxle only after the transaxle is joined to the engine to prevent dislodging of the clutch throwout bearing.

Clutch cable routing

tray over the wiper motor and the other end rests on the radiator support. Use padding and be careful not to damage the paint or body work with the bar. Attach a lifting hook to the engine lift ring and to the bar and raise the engine enough to take the pressure off the motor mounts.

NOTE: If a lifting bar is not available a chain hoist can be used. However, during the removal procedure the vehicle must be raised and the chain hoist adjusted to keep tension on the engine/transaxle assembly.

3. Remove the hush panel from inside the vehicle.

4. Remove the transaxle mount attaching bolts. Discard the bolts attaching the mount to the side frame. New bolts must be used upon installation.

5. Disconnect the clutch cable from the clutch release lever. Remove the transaxle mount bracket attaching bolts and nuts.

6. Disconnect the shift cables and retaining clips at the transaxle mounting stud.

7. Remove the four upper transaxle-to-engine mounting bolts. Remove the air management valve attaching bolts in order to gain clearance to remove the right upper transaxle-to-engine bolt.

8. Raise the vehicle and support it safely on jack stands. Remove the left front wheel.

9. Remove the left front inner splash shield. Remove the transaxle strut and bracket.

10. Remove the clutch housing cover bolts.

11. Disconnect the speedometer cable at the transaxle.

12. Disconnect the stabilizer bar at the left suspension support and control arm.

13. Disconnect the ball joint from the steering knuckle.

14. Remove the left suspension support attaching bolts and remove the support and control arm as an assembly.

15. Install boot protectors and disengage the drive axles at the transaxle. Remove the left side shaft from the transaxle.

16. Position a jack under the transaxle case, remove the lower two transaxle-to-engine mounting bolts and remove the transaxle by sliding it toward the driver's side, away from the engine. Carefully lower the jack, guiding the right shaft out of the transaxle.

17. When installing the transaxle, guide the right drive axle into its bore as the transaxle is being raised. The right drive axle cannot be readily installed after the transaxle is connected to the engine. Installation of the remaining components is in the reverse order of removal. Tighten the transaxle-to-engine bolts to 55 ft. lbs., suspension support-to-body attaching bolts to 75 ft. lbs. and the clutch housing cover bolts to 10 ft. lbs. Using new bolts, install and tighten the transaxle mount-to-side frame to 40 ft. lbs. When installing the bolts attaching the mount-to-transaxle bracket, check the alignment bolt at the engine mount.

AUTOMATIC TRANSAXLE

For all adjustments, see the "Automatic Transmission" Unit Repair Section.

REMOVAL & INSTALLATION

1. Disconnect the negative battery cable.
2. Remove the air cleaner assembly. Remove the mass air flow sensor and air intake duct, if equipped.
3. Disconnect the throttle valve (T.V.) cable at the throttle lever and the transaxle.
4. Remove the transaxle dipstick and fill tube.
5. Install engine support tool J-28467 or equivalent. Insert a ¼ x 2 in. bolt in the hole at the right motor mount to maintain driveline alignment.
6. Remove the nut securing the wiring harness to the automatic transaxle assembly. Disconnect the wiring connectors at the speed sensor, TCC connector and back-up lamp switch.

NOTE: When servicing requires that the T-latch type wiring connector be disconnected from the switch, care must be taken to ensure proper reassembly of both the connector and the T-latch. Failure to do so may result in intermittent loss of switch functions.

7. Remove the shift linkage from the transaxle.
8. Remove the top two transaxle-to-engine bolts and the left upper transaxle mount along with the bracket assembly.
9. Remove the rubber hose from the transaxle vent pipe. Remove the remaining upper engine-to-transaxle bolts.
10. Raise and support the vehicle safely, then remove both front tires.
11. Drain the transaxle fluid.
12. Remove the shift linkage and bracket from the transaxle.
13. Install drive axle boot seal protector J-33162 or equivalent on the inner seals.

NOTE: Some vehicles may use a silicone (gray) boot on the inboard axle joint. Use tool J-33162 on these boots. All other boots are made from a thermo-plastic material (black) and do not require use of a boot seal protector.

14. Remove both ball joints from the control arms.
15. Remove both drive axles and support them with string or wire.
16. Remove the transaxle mounting strut.
17. Remove the left stabilizer bar link pin bolt, frame bushing clamp nuts and frame support assembly.
18. Remove the transaxle converter cover. Remove the torque converter-to-flex plate bolts after match marking the flex plate and torque converter for installation.
19. Disconnect and plug the transaxle oil cooler lines.
20. Remove the transaxle-to-engine support bracket and install the transaxle removal jack.
21. Remove the remaining transaxle-to-engine retaining bolts and remove the transaxle from the vehicle.
22. Installation is the reverse of the removal procedure. Be sure to fill the transaxle with the proper grade and type automatic transmission fluid.

DRIVE AXLE

Halfshafts

REMOVAL & INSTALLATION

NOTE: Some vehicles use a silicone (gray) boot on the right hand inboard joint. Use boot protector J-33162 or equivalent on these boots during removal. All other boots are made of thermoplastic material (black) and do not require the use of a boot protector.

1. Raise the vehicle and install jackstands under the engine cradle.
2. Lower the vehicle slightly so that the weight rests on the cradle and not on the lower control arms.
3. Remove the tires and wheels.
4. Install a drift punch through the brake rotor cooling holes to lock the rotor in place. Clean the drive axle

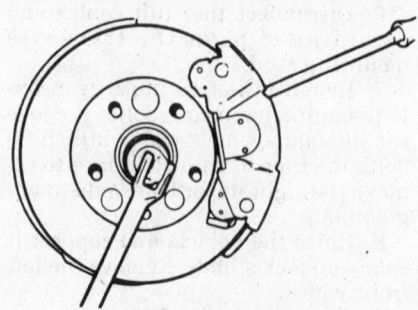

Removing and installing hub nuts

threads of dirt, then lubricate and remove the hub nut and washer using a suitable socket and breaker bar.
5. Remove the caliper mounting bolts and support the caliper with wire or string. Do not let the caliper hang by the brake hose.
6. Remove the brake rotor.
7. Remove the lower ball joint nut.
8. Remove the stabilizer bolt from the lower control arm.
9. Install tool J-28733, or equivalent, and press the drive axle in and away from the hub. The drive axle should only be pressed in until the press fit between the drive axle and hub is loose.

CAUTION
Be careful not to press the drive axle in too far as damage to the joint may occur.

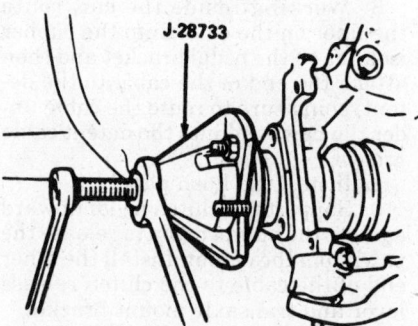

Loosening splines between hub and drive axle

10. Separate and remove the lower ball joint from the steering knuckle.
11. Separate the drive axle from the hub by pulling the hub assembly out away from the drive axle.

NOTE: On vehicles equipped with Tri-Pot joints, care must be taken not to allow the joints to become overextended. When either end or both ends of the shaft are disconnected, overextending the joint could result in separation of internal components. This could cause failure of the joint, so it's important to handle the drive axle in a manner that prevents overextension.

12. Install slide hammer tool and boot protector, if necessary, and remove the drive axle from the differential assembly.
13. Installation is the reverse of removal. Start the splines of the drive axle into the transaxle and push until the axle snaps into place. For all axle shaft overhaul procedures, please refer to the "U-Joint/CV Joint" Unit Repair section.

Constant Velocity Joint

OVERHAUL

Please refer to the "U-Joint/CV Joint" Unit Repair section for all overhaul information.

Outer And Inner CV Boot

REMOVAL & INSTALLATION

Outer Boot

1. Disconnect the negative battery cable.

BOOT PROTECTOR

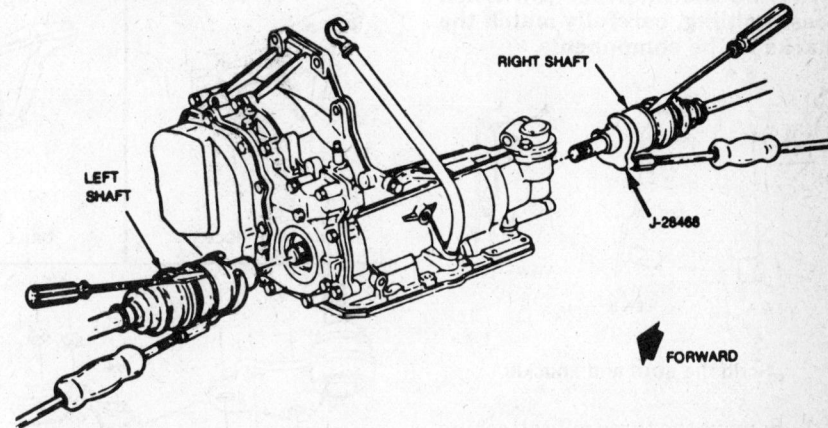

Halfshaft removal; the special tools are attached to a slide hammer in this diagram

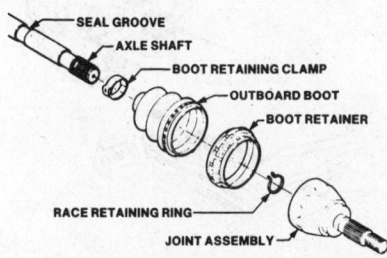

Typical outer CV boot assembly

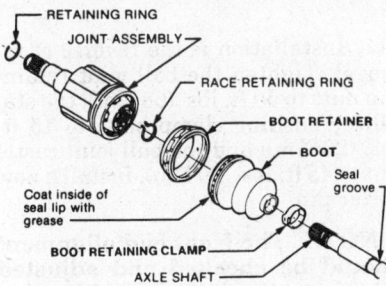

Typical inner CV boot assembly

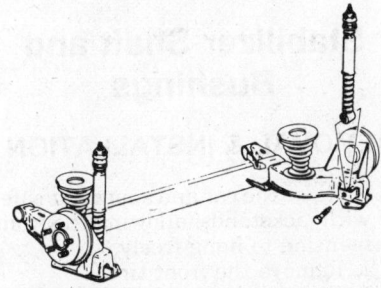

Typical rear axle assembly

2. Raise and support the vehicle on jackstands, then remove the front wheel.

3. Remove the brake caliper and support on a wire, then remove the rotor.

4. Slide the outer CV joint assembly off the axle shaft.

5. Remove the bearing retaining ring, then the boot retainer, the clamp and the outer boot.

6. To install, pack the new boot with grease and reverse the removal procedures.

Inner Boot

1. Disconnect the negative battery cable.

2. Raise and support the vehicle on jackstands, the remove the front wheel.

3. Remove the outer boot assembly.

4. Remove the boot retaining clamps and the spacer ring.

5. Slide the axle and the spider bearing assembly out of the tri-pot housing. Install the spider retainer onto the spider bearing assembly.

6. Remove the spider assembly and the boot from the axle.

7. To install, pack the new boot with grease and reverse the removal procedures.

Front Hub and Bearing

REMOVAL & INSTALLATION

NOTE: Several special tools are required for this procedure.

1. Raise and support the front end on jackstands, allowing the wheels to hang.

2. Remove the front wheels.

3. Install drive axle boot seal protector tool J-28712 on the outer CV joints and J-34754 on the inner Tri-Pot joints.

4. Insert a long punch through the caliper and into the rotor to keep it from turning.

5. Clean the axle threads and lubricate them with a thread lubricant.

6. Remove the hub nut and washer.

7. Remove and support the caliper out of the way.

8. Remove the rotor.

9. Using puller J-28733, loosen the splined fit between the hub and shaft.

10. Remove the three hub attaching bolts, shield, hub and bearing assembly, and O-ring.

NOTE: The hub and bearing are serviced as an assembly only.

11. Remove the bearing seal from the knuckle.

12. Installation is the reverse of removal. Use a new O-ring and bearing seal. Lubricate the new bearing seal and the bearing with wheel bearing grease. Tighten the hub bolts to 40 ft. lbs. (55 Nm); the caliper bolts to 28 ft. lbs. (38 Nm); the hub nut to 185 ft. lbs. (260 Nm).

FRONT SUSPENSION

MacPherson Strut

REMOVAL & INSTALLATION

NOTE: Before removing front suspension components, their positions should be marked so they may assembled correctly. Scribe the knuckle along the lower outboard strut radius (A), the strut flange on the inboard side along the curve of the knuckle (B) and make a chisel mark across the strut/knuckle interface (C). When reassembling, carefully match the marks to the components.

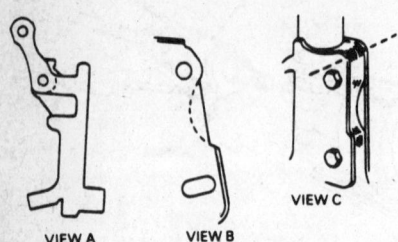

VIEW A   VIEW B   VIEW C

Scrib the strut and knuckle

1. Remove the three nuts attaching the top of the strut assembly to the body.
2. Raise the car and support it safely.
3. Place jackstands under the frame.
4. Lower the car slightly so that the weight rests on the jackstands and not on the control arms.
5. Remove the front tire.

CAUTION

Whenever working near the drive axles, take care to prevent the inner Tri-Pot joints from being overextended. Overextension of the joint could result in separation of internal components which could go undetected and result in failure of the joint.

6. Some vehicles may use a silicone (gray) boot on the in'oard axle joint. Use boot protector J-33162 or equivalent on these boots. All other boots are made from a thermoplastic material (black) and do not require the use of a boot seal protector.
7. Disconnect the brake line bracket from the strut assembly.
8. Remove the strut to steering knuckle bolts.
9. Remove the strut assembly from the vehicle. Care should be taken to avoid chipping or cracking the spring coating when handling the front suspension coil spring assembly.
10. Installation is the reverse of removal.

OVERHAUL

For all MacPherson strut disassembly, shock replacement and overhaul information, please refer to the "Strut Overhaul" Unit Repair section.

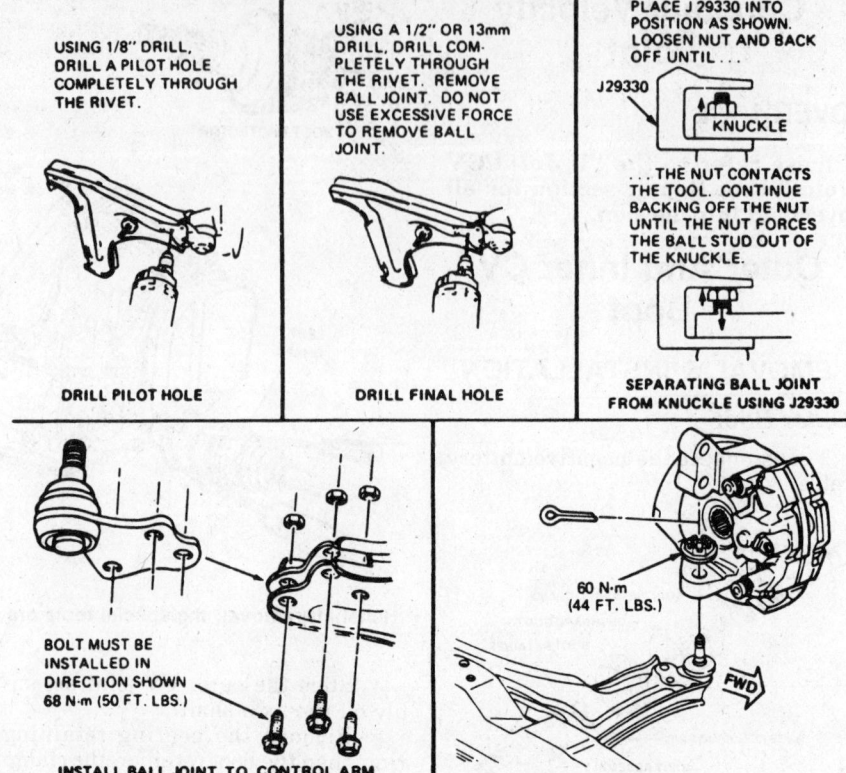

USING 1/8" DRILL, DRILL A PILOT HOLE COMPLETELY THROUGH THE RIVET.

DRILL PILOT HOLE

USING A 1/2" OR 13mm DRILL, DRILL COMPLETELY THROUGH THE RIVET. REMOVE BALL JOINT. DO NOT USE EXCESSIVE FORCE TO REMOVE BALL JOINT.

DRILL FINAL HOLE

PLACE J 29330 INTO POSITION AS SHOWN. LOOSEN NUT AND BACK OFF UNTIL . . .

J29330   KNUCKLE

. . . THE NUT CONTACTS THE TOOL. CONTINUE BACKING OFF THE NUT UNTIL THE NUT FORCES THE BALL STUD OUT OF THE KNUCKLE.

SEPARATING BALL JOINT FROM KNUCKLE USING J29330

BOLT MUST BE INSTALLED IN DIRECTION SHOWN 68 N·m (50 FT. LBS.)

INSTALL BALL JOINT TO CONTROL ARM

60 N·m (44 FT. LBS.)

FWD

Removing ball joint assembly

Ball Joint

REMOVAL & INSTALLATION

1. Raise the car and support it safely.
2. Place jackstands under the frame.
3. Lower the car slightly so the weight rests on the jackstands and not the control arm.
4. Remove the front tire.
5. If a silicone (gray) boot is used on the inboard axle joint, install boot seal protector J-33162 or equivalent. If a thermoplastic (black) boot is used, no protector is necessary.
6. Remove the cotter pin from the ball joint castle nut.
7. Remove the castle nut and disconnect the ball joint from the steering knuckle using ball joint separator J-34505 or equivalent.
8. Drill out the three rivets retaining the ball joint.

CAUTION

Be careful not to damage the drive axle boot when drilling out the ball joint rivets.

9. Loosen the stabilizer shaft bushing assembly nut.
10. Remove the ball joint from the control arm.

11. Installation is the reverse of removal. Tighten the ball joint retaining nuts to 50 ft. lbs. (68 Nm); the stabilizer bushing clamp bolts to 15 ft. lbs. (20 Nm); and the ball joint castle nut to 45 ft. lbs. (60 Nm). Install a new cotter pin.

NOTE: The front end alignment should be checked and adjusted whenever the strut assemblies are removed.

Stabilizer Shaft and Bushings

REMOVAL & INSTALLATION

1. Raise the car and support it safely with jackstands, allowing the front suspension to hang freely.
2. Remove the front tire.
3. Disconnect the stabilizer shaft from the control arms.
4. Disconnect the stabilizer shaft from the support assemblies.
5. Loosen the front bolts and remove the rear and center bolts from the support assemblies to lower them enough to remove the stabilizer shaft.
6. Remove the stabilizer shaft and bushings.

7. Installation is the reverse of removal. Torque the support assembly bolts to 65 ft. lbs. (90 Nm); the stabilizer shaft to support assembly nuts to 18 ft. lbs. (25 Nm); and the stabilizer shaft to control arm nuts to 15 ft. lbs. (20 Nm).

Lower Control Arm

REMOVAL & INSTALLATION

1. Raise the car and support it safely on jackstands. Place the jackstands under the frame so that the suspension hangs freely.
2. Remove the tire.
3. Disconnect the stabilizer shaft from the control arm and/or support assembly.
4. Disconnect the ball joint from the steering knuckle using separator tool J-29330 or equivalent. See "Ball Joint Removal."

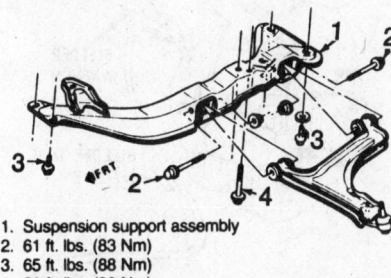

1. Suspension support assembly
2. 61 ft. lbs. (83 Nm)
3. 65 ft. lbs. (88 Nm)
4. 66 ft. lbs. (90 Nm)

Control arm and lower support assembly

5. To remove the support assembly with the control arm attached, remove the bolts mounting the support assembly to the car. To remove the control arm only, remove the control arm to support assembly bolts.
6. Installation is the reverse of removal. Torque the ball joint castle nut to 45 ft. lbs. (60 Nm). If the support assembly was removed, tighten the rear bolts first to 65 ft. lbs. (88 Nm); the center bolts second to 66 ft. lbs. (90 Nm); and the front bolts last to 65 ft. lbs. (88 Nm). Tighten the control arm pivot bolts to 60 ft. lbs. (85 Nm) with the weight of the car on the control arm.

REAR SUSPENSION

Shock Absorber

REMOVAL & INSTALLATION

1. Open the trunk and remove the trim cover (if equipped) over the shock absorber attaching bolts.

2. Remove the upper shock absorber attaching nut.

——— CAUTION ———

Do not remove both shock absorbers at the same time as suspending the rear axle at full length could result in damage to brake lines and hoses.

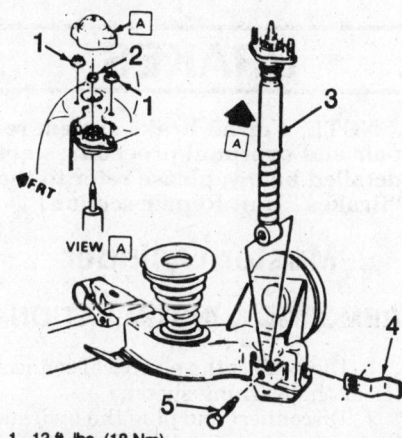

1. 13 ft. lbs. (18 Nm)
2. 28 ft. lbs. (38 Nm)
3. Shock absorber
4. 35 ft. lbs. (48 Nm)
5. Tab nut
A. Arrow should point to left side

Rear shock absorber mounting

3. Raise the vehicle and support it safely.
4. Remove the lower attaching bolt and remove the shock absorber.
5. Installation is the reverse of removal.

Stabilizer Bar

REMOVAL & INSTALLATION

1. Raise the vehicle and support it safely with jackstands.
2. Remove the nuts and bolts at both the axle and control arm attachments and remove the bracket, insulator and stabilizer bar.
3. Install the U-bolts, upper clamp, spacer and insulator in the trailing axle. Position the stabilizer bar in the insulators and loosely install the lower clamp and nuts.
4. Attach the end of the stabilizer bar to the control arms and torque all nuts to 15 ft. lbs. (20 Nm).
5. Tighten the axle attaching nut, then lower the vehicle.

Springs and Insulators

REMOVAL & INSTALLATION

1. Raise the vehicle and support the rear axle with a hydraulic jack.

2. Install jackstands under the frame.
3. Remove the rear tires.
4. Remove the right and left brake line bracket attaching screws from the body and allow the brake line to hang free.
5. Remove both shock absorber lower mounting bolts with the rear axle supported by the hydraulic jack.
6. Carefully lower the rear axle and remove the springs and/or insulators.

——— CAUTION ———

Do not suspend the rear axle by the brake hoses or damage to the hoses could result. Lower the axle just enough to remove the springs and support it during all service procedures.

7. Installation is the reverse of removal. Position the springs and insulators in their seats and raise the axle. The ends of the opper coil on the spring must be positioned in the seat of the body and within the limits. Prior to installing the spring it will be necessary to install the upper insulators to the body with adhesive to keep it in position while raising the axle assembly and springs.

Rear Hub and Bearing Assembly

REMOVAL & INSTALLATION

1. Raise the vehicle and support it safely.
2. Remove the tire.
3. Remove the brake drum. Do not hammer on the brake drum during removal or damage to the assembly could result.

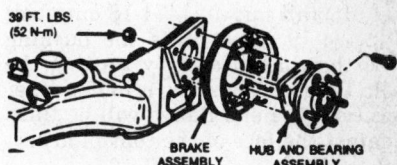

Rear hub and bearing assembly

4. Remove the four hub and bearing assembly to rear axle attaching bolts, then remove the hub and bearing assembly from the axle. The top rear attaching bolt will not clear the brake shoe when removing the hub and bearing assembly. Partially remove the hub prior to removing this bolt.
5. Installation is the reverse of removal.

Control Arm Bushings

REMOVAL & INSTALLATION

NOTE: This procedure requires the use of a number of special tools

1. Raise the vehicle and support it safely.
2. Remove the tire and support the body with jackstands.
3. If replacing the right bushing, disconnect the brake line from the body. If the left bushing is being replaced, disconnect the brake line bracket from the body and the parking brake cable from its hook guide on the body.

NOTE: Replace the bushings one at a time.

4. Remove the nut, bolt and washer from the control arm and bracket attachment and rotate the control arm downward.
5. Install tool J-29376-1 or equivalent on the control arm over the bushing and tighten the attaching nuts until the tool is securely in place.
6. Install J-21474-19 bolt through plate J-29376-7 and install into J-29376-1 receiver.
7. Place J-29376-6 remover into position on the bushing and install nut J-21474-18 onto J-21474-19 bolt. Remove the bushing from the control arm by turning the bolt.
8. To install the bushing, first install tool J-29376-1 on the control arm.
9. Install J-21474-19 bolt through plate J-29376-7 and install into J-29376-1 receiver.
10. Install the bushing on the bolt and position it into the housing. Align bushing installation arrow with the arrow on the receiver for proper indexing of the bushing.
11. Install nut J-21474-18 onto bolt J-21474-19, then press the bushing into the control arm by turning the bolt. When the bushing is in its proper position, the end flange will be flush against the face of the control arm.

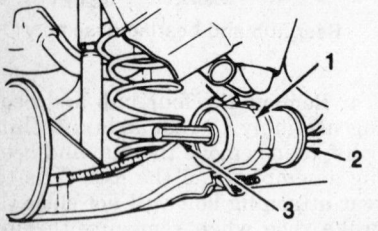

1. Receiver J-29376-1
2. Plate J-29376-7
3. Nut J-21474-18
4. Installer J-29376-4
5. Index marks

Installing control arm bushings

12. Use a screw-type jackstand to position the control arm into the bracket and install the bolt and nut. Do not torque the bolt at this time. It is necessary to torque the bolt with the vehicle at standing height.
13. The remainder of the installation is in the reverse order of removal.

BRAKES

NOTE: For all brake system repair and overhaul procedures not detailed below, please refer to the "Brakes" Unit Repair section.

Master Cylinder

REMOVAL & INSTALLATION

1. Disconnect the electrical connector at the warning switch.
2. Disconnect and plug the hydraulic lines to prevent the entry of dirt into the system.
3. Drain the brake fluid from the master cylinder and discard. Exercise caution when handling the brake fluid as it will damage painted surfaces.
4. Remove the attaching nuts then lift the master cylinder clear of the brake power booster unit.

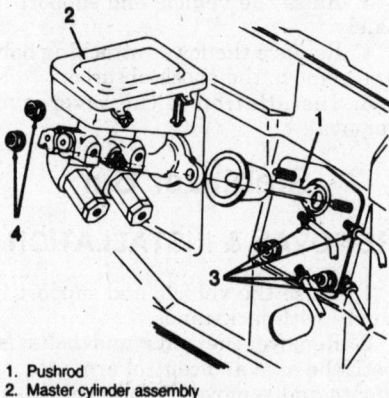

1. Pushrod
2. Master cylinder assembly
3. Tube nut
4. Nut

Master cylinder mounting

5. Installation is the reverse of removal. Bleed the brake system.

BLEEDING THE BRAKE SYSTEM

On diagonally split brake systems, start the manual bleeding procedure with the right rear, then the left front; right rear then right front.

1. Clean the bleeder screw at each wheel.
2. Attach a small rubber hose to the bleed screw and place the end in a clear container of fresh brake fluid.
3. Fill the master cylinder with fresh brake fluid. The master cylinder reservoir should be checked and topped up often during the bleeding procedure.

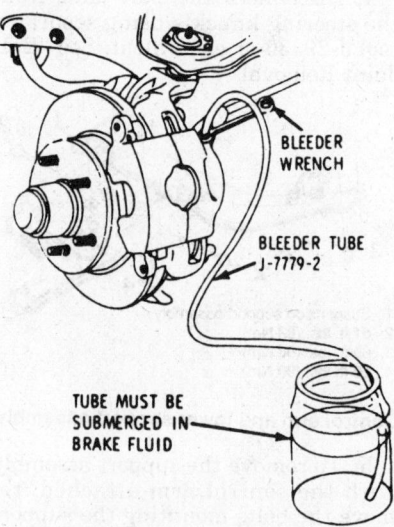

Bleeding the brakes

4. Have an assistant slowly pump up the brake pedal and hold pressure.
5. Open the bleeder screw about one-quarter turn. The pedal should fall to the floor as air and fluid are pushed out. Close the bleed screw while the assistant holds the pedal to the floor, then slowly release the pedal and wait 15 seconds. Repeat the process until no more air bubbles are forced from the system when the brake pedal is applied. It may be necessary to repeat this ten or more times to get all of the air from the system.
6. Repeat this procedure on the remaining wheel cylinders and calipers. Remember to wait 15 seconds between each bleeding and do not pump the pedal rapidly. Rapid pumping of the brake pedal pushes the master cylinder secondary piston down the bore in a manner that makes it difficult to bleed the system.
7. Check the brake pedal for sponginess and the brake warning light for indication of unbalanced pressure.

Repeat the entire bleeding procedure to correct either of these two conditions.

Brake Booster

REMOVAL & INSTALLATION

1. Working inside the car, detach the brake pushrod from the brake pedal.
2. Disconnect the hydraulic lines from the front of the master cylinder and the vacuum line from the engine.
3. Remove the nuts from the mounting studs which hold the unit to the dash panel and remove the booster and master cylinder as an assembly. Continue disassembly to separate the booster and master cylinder on the bench.
4. Installation is the reverse of removal. Clean the mounting surfaces before installing the booster and bleed the brake system.

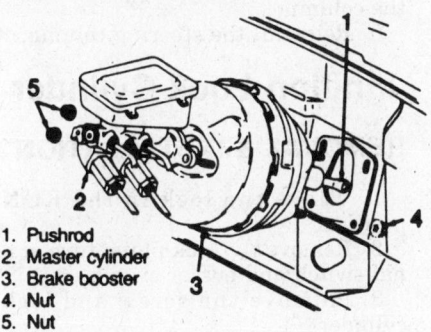

1. Pushrod
2. Master cylinder
3. Brake booster
4. Nut
5. Nut

Brake booster location

Wheel Cylinders

REMOVAL & INSTALLATION

1. Raise the vehicle and support it safely on jackstands.
2. Remove the tire. Mark and remove the brake drum as described above.
3. Remove the brake shoes and springs.
4. Clean the area around the brake line, disconnect and cap the brake line at the wheel cylinder.
5. Insert two awls, $\frac{1}{8}$ in. diameter, into the access slots between the wheel cylinder pilot and the retainer locking tabs. Bend both tabs away simultaneously.
6. Remove the wheel cylinder.
7. To install, position the wheel cylinder in place and hold it there with a wood block between the cylinder and the flange.
8. Install a new retainer on the cylinder using a $1\frac{1}{8}$ in. 12 point socket and extension.

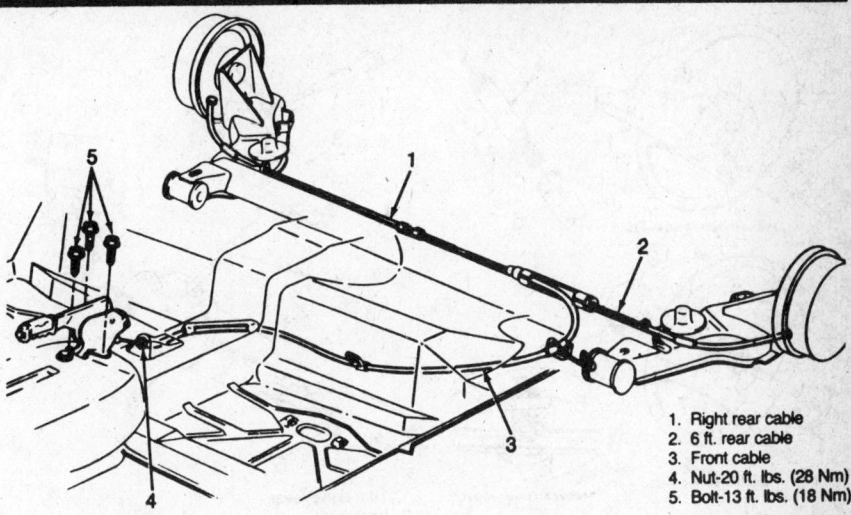

1. Right rear cable
2. 6 ft. rear cable
3. Front cable
4. Nut-20 ft. lbs. (28 Nm)
5. Bolt-13 ft. lbs. (18 Nm)

Parking brake cable routing

9. Install the brake shoes, springs and drum and bleed the brakes as described previously.

Parking Brake

ADJUSTMENT

1. Depress the parking brake pedal exactly three ratchet clicks.
2. Raise the vehicle and support it safely with jackstands.
3. Check that the equalizer nut groove is liberally lubricated with chassis lube. Tighten the adjusting nut until the right rear wheel can just be turned to the rear with both hands, but is locked when forward rotation is attempted.
4. With the mechanism totally disengaged, both rear wheels should turn freely in either direction with no brake drag. Do not adjust the parking brake so tightly as to cause brake drag.

STEERING

Steering Wheel

REMOVAL & INSTALLATION

——— CAUTION ———
Disconnect the negative battery cable before removing the steering wheel. When installing a steering wheel, make sure the turn signal lever is in the neutral position.

1. Remove the trim retaining screws from behind the wheel. On wheel with a center cap, pull off the cap.

2. Lift off the trim and pull the horn wires from the turn signal cancelling cam.
3. Remove the retainer and the steering wheel nut.
4. Mark the wheel-to-shaft relationship and then remove the steering wheel using a suitable puller.
5. To install, place the wheel on the shaft and align the previously make marks. Tighten the mounting nut.
6. Insert the horn wires into the cancelling cam, then install the center trim and reconnect the battery cable.

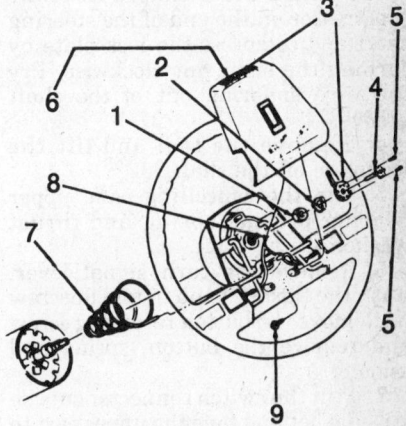

1. Steering wheel nut 41 N·m (30 ft. lbs.)
2. Steering wheel nut retainer
3. Telescoping adjuster lever
4. Steering shaft lock knob bolt
5. Steering shaft lock knob bolt positioning screw (2)
6. Steering wheel pad
7. Horn contact spring
8. Horn lead
9. Fully driven, seated and not stripped

Tilt wheel mounting

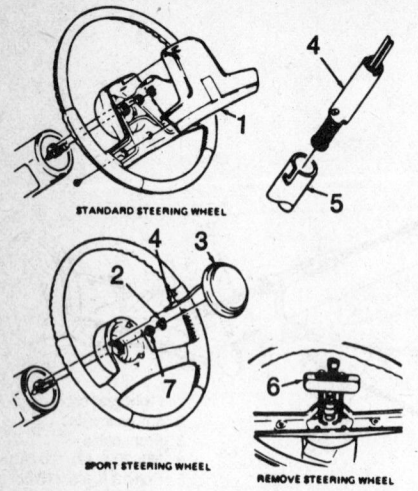

STANDARD STEERING WHEEL

SPORT STEERING WHEEL

REMOVE STEERING WHEEL

1. Pad
2. Retainer
3. Cap
4. Horn lead
5. Cam tower
6. J-1859-03 or BT-61-9
7. Nut 30 ft. lbs.
   (41 Nm)

Steering wheel mounting

Turn Signal Switch

REMOVAL & INSTALLATION

1. Remove the steering wheel as described above.
2. Loosen the cover screws, pry the cover off with a screwdriver, then lift the cover off the shaft.
3. Position the U-shaped lockplate compressor on the end of the steering shaft and compress the lock plate by turning the shaft nut clockwise. Pry the wire snapring out of the shaft groove.
4. Remove the tool and lift the lockplate off the shaft.
5. Slip the cancelling cam, upper bearing preload spring and thrust washer off the shaft.
6. Remove the turn signal lever. Push the flasher knob in and unscrew it. Remove the button retaining screw and remove the button, spring and knob.
7. Pull the switch connector out the mast jacket and tape the upper part to help switch removal. Attach a long piece of wire to the turn signal switch connector. When installing the turn signal switch, feed this wire throught the column first, then use the wire to pull the switch connector into position. On tilt wheels, place the turn signal and shifter housing in low position and remove the harness cover.
8. Remove the three switch mounting screws and remove the switch by

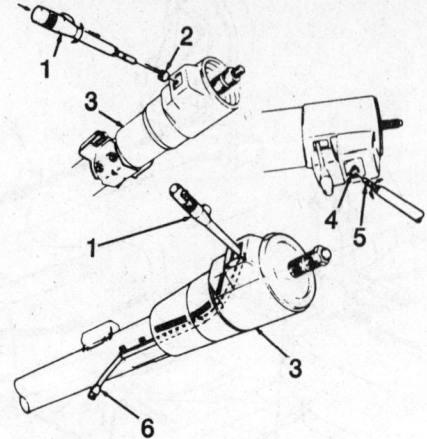

1. Turn signal lever
2. Insulator
3. Housing
4. Switch notch
5. Tang
6. Cruise control wiring

Multi-function switch removal

pulling it straight up while guiding while guiding the wiring harness cover through the column.

9. Install the replacement switch by working the connector and cover down through the housing and under the bracket. On tilt models, the connector is worked down through the housing, under the bracket, and then the cover is installed on the harness.
10. Install the switch mounting screws and the connector on the mast jacket bracket. Install the column-to-dash trim plate.
11. Install the flasher knob and the turn signal lever.
12. With the turn signal lever in neutral and the flasher knob out, slide the thrust washer, upper bearing preload spring and cancelling cam on the shaft.
13. Position the lock plate on the shaft and press it down until a new snapring can be inserted in the shaft groove. Always use a new snapring when assembling.
14. Install the cover and the steering wheel.

Ignition Switch

REMOVAL & INSTALLATION

The switch is located on the upper side of the lower steering column area and is completely inaccessible without first lowering the steering column. The switch is actuated by a rod and rack assembly. A gear on the end of the lock cylinder engages the toothed upper end of the rod.

1. Lower the steering column; be sure to properly support it.

2. Disconnect the wiring from the switch.
3. Remove the two switch screws and remove the switch assembly.
4. Before installing, place the slider on the new switch in one of the following positions, depending on the steering column and accessories:
     a. Standard column with key release—extreme left detent
     b. Standard column with Park Lock—one detent from extreme left
     c. All other standard columns—two detents from extreme left
     d. Adjustable column with key release—extreme right detente
     e. Adjustable column with Park Lock—one detent from extreme right
     f. All other adjustable columns—two detents from extreme right
5. Install the activating rod into the switch and assembly the switch to the column. Tighten the mounting screws. Do not use oversize screws as they could impair the collapsibility of the column.
6. Reinstall the steering column.

Ignition Lock Cylinder

REMOVAL & INSTALLATION

1. Place the lock in the RUN position.
2. Remove the lock plate, turn signal switch and buzzer switch.
3. Remove the screw and lock cylinder.

--- **CAUTION** ---

If the screw is dropped on removal, it could fall into the column, requiring complete disassembly to retrieve the screw.

4. Rotate the cylinder clockwise to align the cylinder key with the keyway in the housing.
5. Push the lock all the way in.
6. Install the screw and tighten to 14 inch lbs. for adjustable columns and 25 inch lbs. for standard columns.

Rack and Pinion Assembly

REMOVAL & INSTALLATION

1. Remove the left sound insulator.
2. Disconnect the upper pinch bolt on the coupling assembly.
3. Disconnect the clamp nuts.
4. Raise the car and support it safely.
5. Remove the clamp nut.
6. Remove both front tires.
7. Disconnect the tie rod ends from the steering knuckles using separator tool J-24319-01 or equivalent.

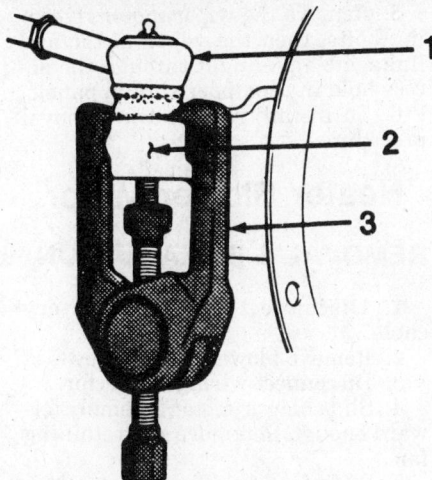

1. Tie rod end
2. Steering knuckle
3. Tie rod end puller J-24319-01

Seperating tie rod end from knuckle

8. Lower the vehicle.
9. Disconnect the fluid line retainer.
10. Disconnect the fluid lines at the steering gear from the pump.
11. Move the steering gear forward and remove the lower pinch bolt on the coupling assembly.
12. Disconnect the coupling from the steering gear.
13. Remove the rack and pinion assembly with the dash seal through the left wheel opening.
14. Installation is the reverse of removal. Refill the power steering pump and bleed the system.

BLEEDING THE POWER STEERING SYSTEM

If the power steering hydraulic system has been serviced, an accurate fluid level reading cannot be obtained unless air is bled from the system.
1. With the wheels turned all the way to the left, add power steering fluid to the COLD mark on the fluid level indicator.
2. Start the engine and check the fluid level at fast idle. Add fluid, if necessary to bring the level up to the COLD mark.
3. Bleed air from the system by turning the wheels from side to side without hitting the stops. Keep the fluid level just above the internal pump casting or at the COLD mark. Fluid with air in it has a light tan or red appearance.
4. Return the wheels to the center position and continue running the engine for two or three minutes.
5. Road test the vehicle to check steering function and recheck the flu-

id level with the system at its normal operating temperature. Fluid should be at the HOT mark.

Power Steering Pump

REMOVAL & INSTALLATION

4 Cyl Engine

1. Remove the drive belt.
2. Disconnect the fluid lines from the pump.
3. Remove the front adjustment bracket-to-rear adjustment bracket bolt.
4. Remove the front adjustment bracket-to-engine bolt and spacer.
5. Remove the pump with the front adjustment bracket attached.
6. Transfer the pulley and front adjustment bracket to the new pump, if necessary.
7. Installation is the reverse of removal. Adjust the belt tension and fill the system with power steering fluid. Bleed the system of air as described above.

V6 Engine

1. Remove the serpentine drive belt.
2. Remove the pump mounting bolts.
3. Pull the pump forward and disconnect the fluid lines.
4. Remove the pump and transfer the pulley as necessary.
5. Installation is the reverse of removal. Install the drive belt, fill the system with power steering fluid and bleed as described above.

CHASSIS ELECTRICAL

Headlight Switch

REMOVAL & INSTALLATION

Calais

1. Disconnect the negative battery cable.
2. Remove the lower steering column collar.
3. Remove the instrument panel cluster trim plate.
4. Remove the headlight switch mounting screws, then pull the switch assembly rearward and unplug both electrical connections.
5. Remove the headlight switch from the vehicle.
6. To install, reverse the removal procedures.

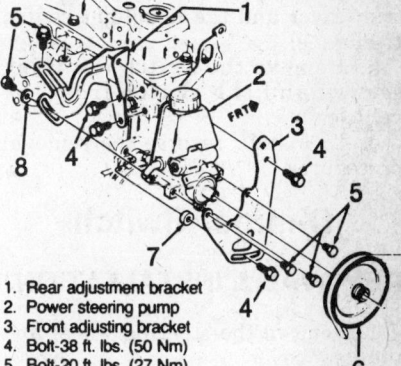

1. Rear adjustment bracket
2. Power steering pump
3. Front adjusting bracket
4. Bolt-38 ft. lbs. (50 Nm)
5. Bolt-20 ft. lbs. (27 Nm)
6. Pulley
7. Spacer
8. Washer

4cyl power steering pump mounting

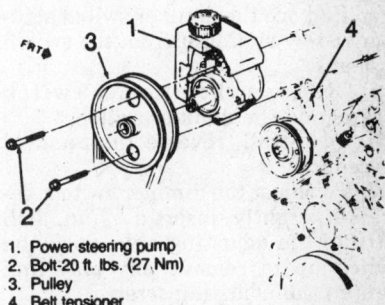

1. Power steering pump
2. Bolt-20 ft. lbs. (27 Nm)
3. Pulley
4. Belt tensioner

V6 power steering pump mounting

Grand Am

1. Disconnect the negative battery cable.
2. Remove the headlight switch trim plate.
3. Remove the head light switch screws, then pull the switch rearward and unplug the electrical connectors.
4. Remove the headlight switch from the vehicle.
5. To install, reverse the removal procedures.

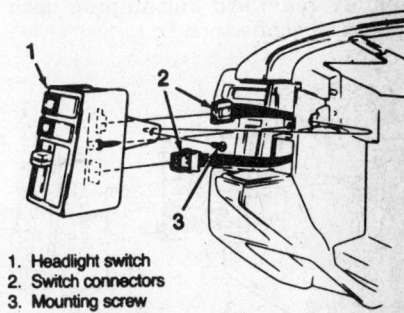

1. Headlight switch
2. Switch connectors
3. Mounting screw

Typical head lamp switch installation

Somerset and Skylark

1. Disconnect the negative battery cable.

2. Remove the instrument panel trim cover and the headlight switch trim panel.

3. Remove the switch retaining screws and the switch from the vehicle.

4. To install, reverse the removal procedures.

Dimmer Switch

REMOVAL & INSTALLATION

1. Remove the steering wheel and the trim cover.

2. Remove the turn signal switch assembly.

3. Remove the ignition switch stud and the screw, then the ignition switch.

4. Remove the dimmer switch actuator rod by sliding it from the switch assembly.

5. Remove the dimmer switch screws and the dimmer switch.

6. To install, reverse the removal procedures.

7. To adjust the dimmer switch, depress it slightly, insert a $3/32$ in. drill bit into the adjusting hole, push the switch up to remove any play and tighten the adjusting screw.

Windshield Wiper Switch

REMOVAL & INSTALLATION

Calais

1. Disconnect the negative battery cable.

2. Remove the lower steering column collar.

3. Remove the instrument panel cluster trim plate.

4. Remove the wiper switch mounting screws, then pull the switch assembly rearward and unplug both electrical connectors.

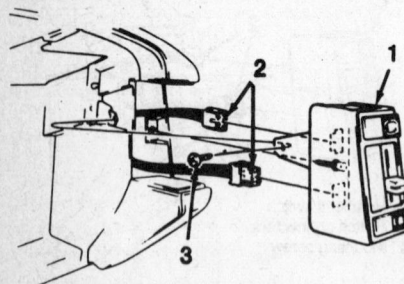

1. Windshield wiper and rear window defogger switch
2. Wiper and rear window defogger switch connector
3. Fully driven, seated and not stripped

Wiper and reart defogger switch—Calais shown

5. Remove the windshield wiper switch from the vehicle.

6. To install, reverse the removal procedures.

Grand Am

1. Disconnect the negative battery cable.

2. Remove the wiper switch trim plate and the screws, then pull the switch assembly rearward and unplug the electrical connectors.

3. Remove the wiper switch from the vehicle.

4. To install, reverse the removal procedures.

Somerset and Skylark

1. Disconnect the negative battery cable.

2. Remove the instrument panel trim cover and the wiper switch trim panel.

3. Remove the wiper switch screws and the switch.

4. To install, reverse the removal procedures.

Windshield Wiper Motor

REMOVAL & INSTALLATION

1. Disconnect the negative battery cable.

2. Remove the wiper arm assemblies.

3. Loosen but do not remove the retaining nuts that secure the transmission drive link to the motor crank arm.

4. Remove the air inlet screw panel, then the transmission drive link from the motor crank arm.

5. Remove the wiper motor retaining bolts, then the wiper motor and linkage by guiding it through the access hole in the upper shroud panel.

6. To install, reverse the removal procedures.

Heater Blower Motor

REMOVAL & INSTALLATION

1. Disconnect negative battery cable.

2. Remove blower motor screws.

3. Disconnect wiring connector.

4. Slide blower motor assembly forward enough to remove nut retaining fan.

5. Slide fan out of housing.

6. Remove blower motor.

7. Remove fan.

8. Installation is the reverse of the removal procedure.

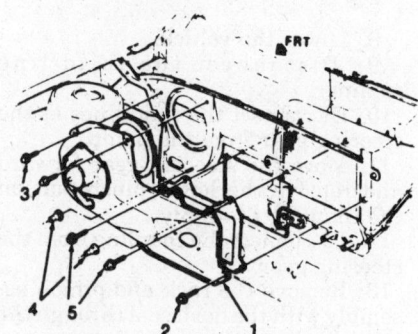

1. Blower assembly
2. Install first
3. Install second
4. 23 inch lbs.
5. 30 inch lbs.

Removing blower motor assembly

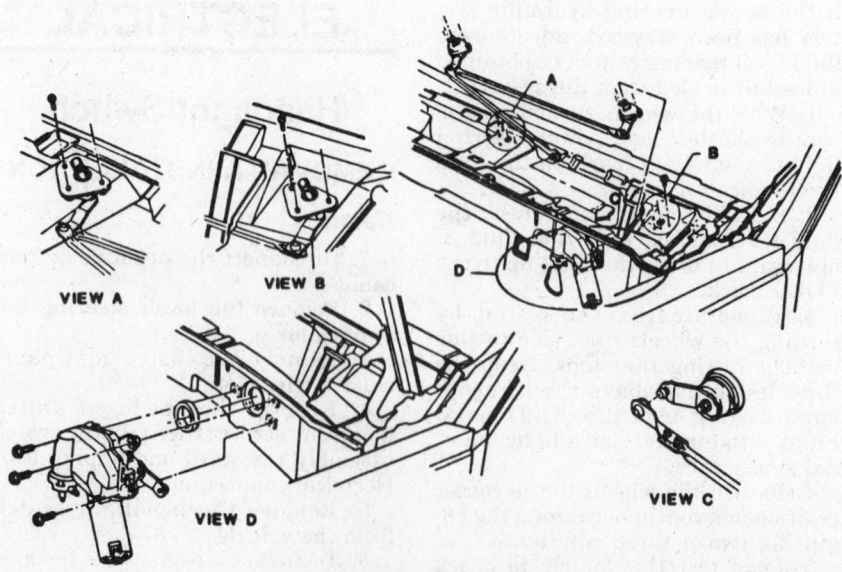

Wiper motor and transmission

Heater Core

REMOVAL & INSTALLATION

1. Disconnect negative battery cable.
2. Drain cooling system.
3. Remove console extensions
4. Remove console ducts.
5. Remove hush panel.
6. Remove plenum.
7. Remove the housing.
8. Raise and support the vehicle safely.
9. Loosen the hoses.
10. Lower the vehicle.
11. Remove the heater core.
12. Installation is the reverse of the removal procedure.

Radio

REMOVAL & INSTALLATION

1. Disconnect the negative battery cable.
2. Remove radio panel trim plate by pulling rearward.
3. Remove two 7mm screws from radio mounting bracket.
4. Pull radio rearward and unplug digital clock connector, radio connector and antenna lead.
5. Installation is the reverse of the removal procedure.

Instrument Cluster

REMOVAL & INSTALLATION

Calais

1. Disconnect the negative battery cable.
2. Remove the steering column collar and the steering column opening filler screws.
3. Remove the cluster trim plate screws and the trim plate.
4. Remove the steering column support bolts, then lower the steering column.
5. Remove the cluster-to-instrument panel pad screws and the cluster by pulling it rearward.
6. To install, reverse the removal procedures.

Grand Am, Somerset and Skylark

1. Disconnect the negative battery cable.
2. Remove the cluster lower trim plate.
3. Lower the steering column as required.
4. Remove the upper cluster trim plate.

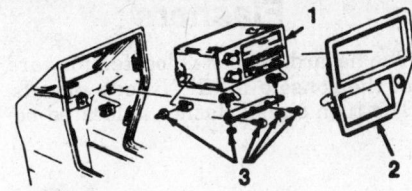

1. Radio
2. Radio trim plate
3. Fully driven, seated and not stripped

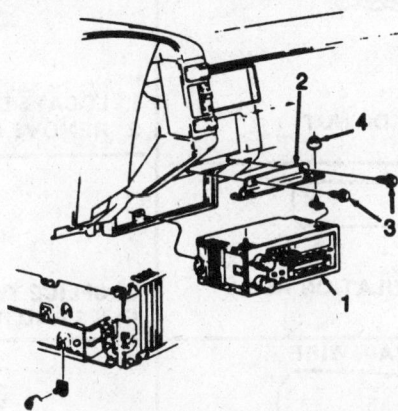

1. Radio
2. Bracket
3. Fully driven, seated and not stripped
4. Tighten to 23 inch lbs.
5. Lower center trim cover
6. Cover used without radio

Typical console radio installation

5. Remove the cluster retaining screws and the cluster, by pulling it rearward.
6. To install, reverse the removal procedures.

Electronic Control Module

The electronic control module is located on the right side of the vehicle, underneath the dash pad. To gain access to this component you may first have to remove the right dash pad trim panel or glovebox assembly.

Fuse Panel

The fuse panel is located on the left side of the vehicle. It is under the instrument panel assembly. In order to gain access to the fuse panel, it may be necessary to first remove the under dash padding.

The amperage of each fuse and the circuit it protects are marked on the fusebox or the fusebox cover. In addition, the amperage of the fuse is marked on the plastic fuse body so that it faces out when installed. Replacing a fuse with one of a higher amperage rating is not recommended and could cause electrical damage. A suspected blown fuse can easily be pulled out and inspected; the clear plastic body gives full view of the element to blade construction for visual inspection.

Fusible Links

Fusible links are used to prevent major wire harness damage in the event of a short circuit or an overload condition in the wiring circuits which are normally not fused, due to carrying high amperage loads or because of their locations within the wiring harness. Each fusible link is of a fixed value for a specific electrical load and should a link fail, the cause of the failure must be determined and repaired prior to installing a new fusible link of the same value.

The N-Body cars have five fusible links. Four are at the starter and one is located behind the battery. They protect the starting and charging circuits, lighting, cooling fan and electronic control module (ECM).

To replace a fusible link, cut off the burned link beyond the original splice. Replace the link with a new one of the same rating. If the splice has two wires, two repair links are required, one for each wire. Connect the new fusible link to the wire, then crimp or solder securely.

— CAUTION —

Use only replacements of the same electrical capacity as the original link, available from a dealer or parts jobber. Replacements of a different electrical value will not provide adequate system protection and could lead to ECM damage.

Circuit Breakers

The headlights are protected by a circuit breaker in the headlamp switch. If the circuit breaker trips, the headlights will either flash on and off or stay off altogether. The circuit breaker resets automatically after the overload is removed. There are also two circuit breakers in the fuse box, one for the power windows and the other for all power accessories.

The windshield wipers are also protected by a circuit breaker. If the motor overheats, the circuit breaker will trip, remaining off until the motor cools or the overload is removed. One common cause of overheating is operation of the wipers in heavy snow.

Flashers

The hazard flasher is located forward of the console in all N-Body models. The turn signal flasher is located behind the instrument panel, on the left hand side of the steering column. In both cases, replacement is accomplished by unplugging the old flasher and plugging in a new one.

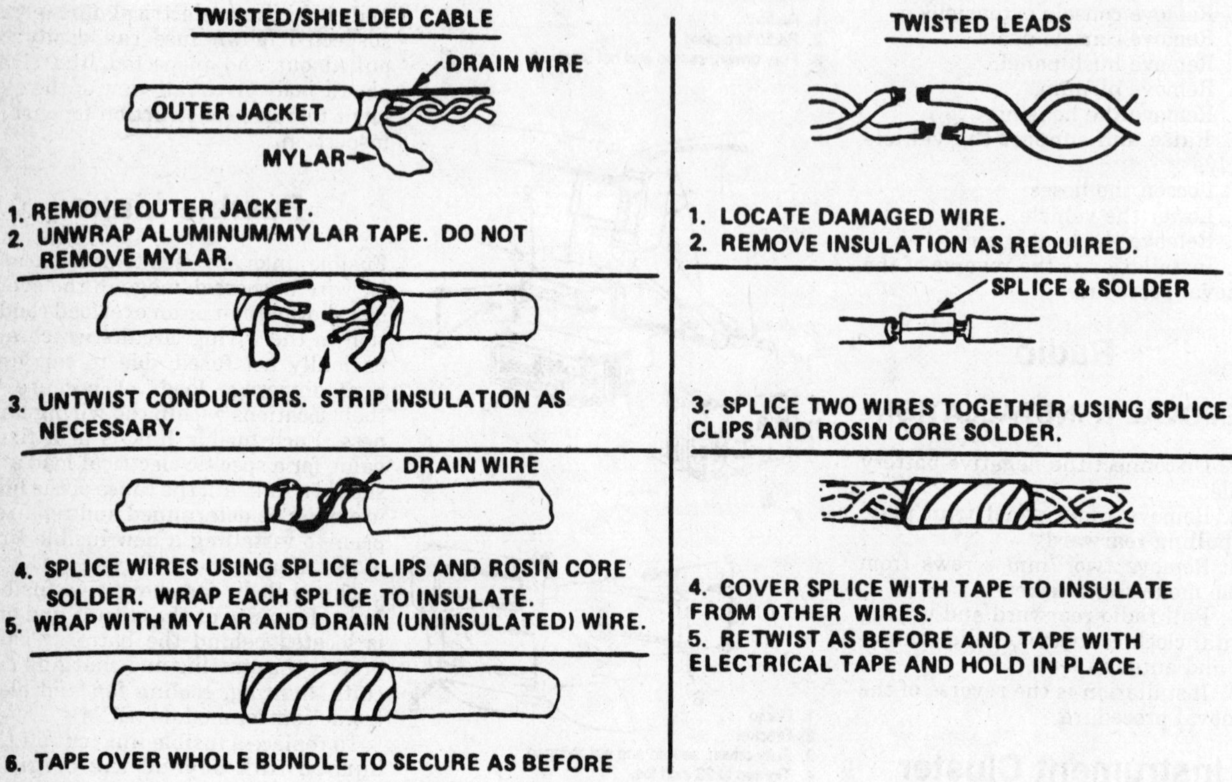
TWISTED/SHIELDED CABLE
DRAIN WIRE
OUTER JACKET
MYLAR
1. REMOVE OUTER JACKET.
2. UNWRAP ALUMINUM/MYLAR TAPE. DO NOT REMOVE MYLAR.
3. UNTWIST CONDUCTORS. STRIP INSULATION AS NECESSARY.
DRAIN WIRE
4. SPLICE WIRES USING SPLICE CLIPS AND ROSIN CORE SOLDER. WRAP EACH SPLICE TO INSULATE.
5. WRAP WITH MYLAR AND DRAIN (UNINSULATED) WIRE.
6. TAPE OVER WHOLE BUNDLE TO SECURE AS BEFORE
TWISTED LEADS
1. LOCATE DAMAGED WIRE.
2. REMOVE INSULATION AS REQUIRED.
SPLICE & SOLDER
3. SPLICE TWO WIRES TOGETHER USING SPLICE CLIPS AND ROSIN CORE SOLDER.
4. COVER SPLICE WITH TAPE TO INSULATE FROM OTHER WIRES.
5. RETWIST AS BEFORE AND TAPE WITH ELECTRICAL TAPE AND HOLD IN PLACE.

Wiring Repair

GM "P" Body
Pontiac Fiero

1984–86 Fiero

1985-87 Fiero GT

VEHICLE IDENTIFICATION NUMBER (VIN)

It is important for servicing and ordering parts to be certain of the vehicle and engine identification. The VIN (vehicle identification number) is a 17 digit number visible through the windshield on the driver's side of the dash and contains the vehicle and engine identification codes. It can be interpreted as follows:

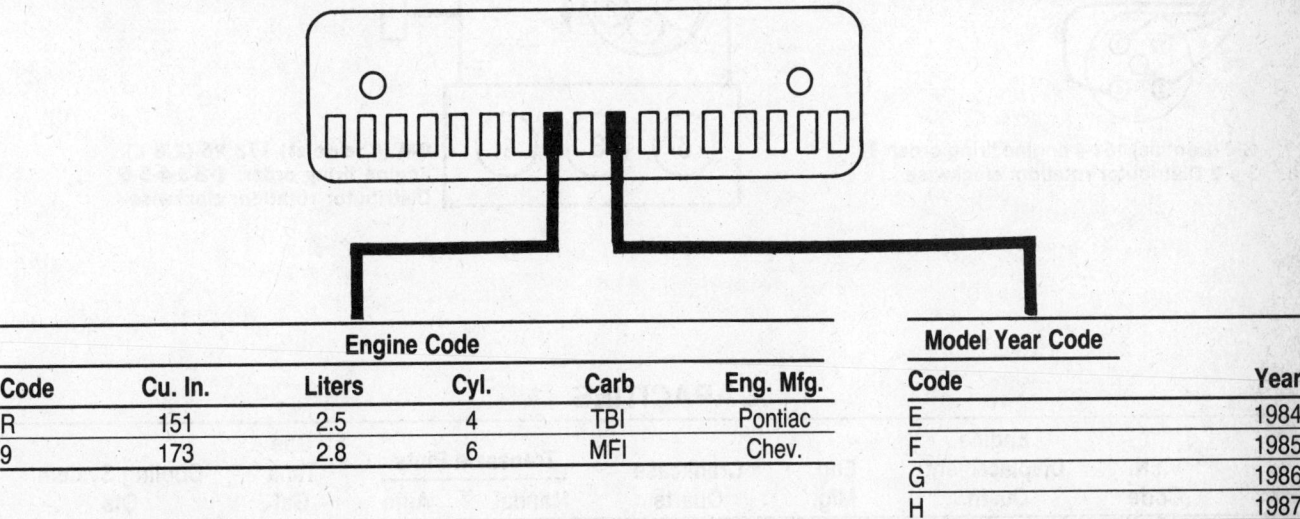

Engine Code						Model Year Code	
Code	Cu. In.	Liters	Cyl.	Carb	Eng. Mfg.	Code	Year
R	151	2.5	4	TBI	Pontiac	E	1984
9	173	2.8	6	MFI	Chev.	F	1985
						G	1986
						H	1987

The seventeen digit Vehicle Identification Number can be used to determine engine application and model year. The 10th digit indicates the model year, and the 8th digit identifies the factory installed engine.
TBI (Throttle body injection)

GENERAL ENGINE SPECIFICATIONS

Year	Eng. V.I.N. Code	Engine Displacement Cu. In.	Eng. Mfg.	Fuel Delivery	Horsepower @ rpm ■	Torque @ rpm (ft lbs) ■	Bore × Stroke (in.)	Compression Ratio	Oil Pressure @ 2000 rpm
'84–'85	R	151	Pont.	TBI	90 @ 4000	132 @ 2800	4.000 × 3.000	9.0:1	36–41
'86–'87	R	151	Pont.	TBI	92 @ 4400	134 @ 2800	4.000 × 3.000	9.0:1	36–41
'85	9	173	Chev.	MFI	130 @ 5400	160 @ 3600	3.500 × 3.000	8.5:1	30–45
'86–'87	9	173	Chev.	MFI	140 @ 5200	170 @ 3600	3.500 × 3.000	8.5:1	30–45

TUNE-UP SPECIFICATIONS

When analyzing compression test results, look for uniformity among cylinders rather than specific pressures.

Year	V.I.N. Code	Eng. No. Cyl. Displ. Cu. in.	Eng. Mfg.	hp	Spark Plugs Orig Type	Spark Plugs Gap (in.)	Ignition Timing (deg)▲ Man. Trans.	Ignition Timing (deg)▲ Auto. Trans.	Intake Valve Opens (deg)■	Fuel Pump Pressure (psi)	Idle Speed (rpm)▲ Man. Trans.	Idle Speed (rpm)▲ Auto. Trans.
'84–'85	R	151	Pont	90	R43CTS	.060	①	①	33	6–8	①	①
'86	R	151	Pont.	92	R43CTS6	.060	①	①	33	6–8	①	①
'85–'86	9	173	Chev	130	R42CTS	.045	①	①	25	6–8	①	①
'87	All				See Underhood Specifications Sticker							

NOTE: The underhood specifications sticker often reflects tune-up specification changes made in production. Sticker figures must be used if they disagree with those in this chart.
① See underhood sticker

FIRING ORDER

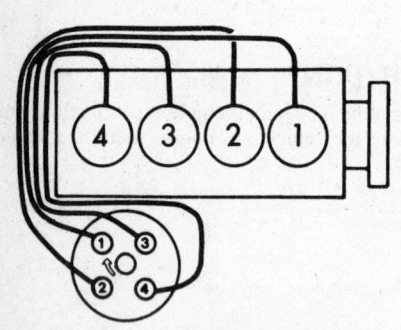

GM (Pontiac) 151-4 engine firing order: 1-3-4-2 Distributor rotation: clockwise

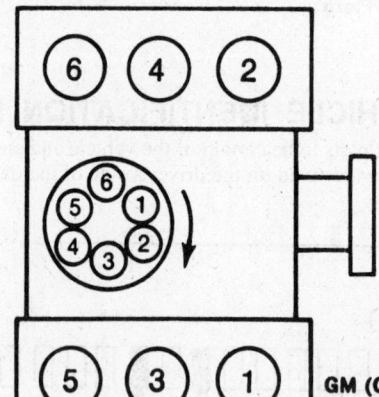

GM (Chevrolet) 173 V6 (2.8 L)
Engine firing order: 1-2-3-4-5-6
Distributor rotation: clockwise

CAPACITIES

Year	V.I.N. Code	Engine Displacement Cu. In.	Eng. Mfg.	Crankcase Quarts	Transaxle Pints Manual	Transaxle Pints Auto	Gas Tank Gal	Cooling System Qts
'84–'87	R	151	Pont.	3①	6.0	8.0	10.5	13.8
'85–'87	9	173	Chev.	4①	5.9	8.0	10.3	13.8

① With or without filter change

VALVE SPECIFICATIONS

Year	V.I.N. Code	Engine No. Cyl. Displacement (cu. in.)	Eng. Mfg.	Seat Angle (deg)	Face Angle (deg)	Spring Test Pressure (lbs. @ in.)	Spring Installed Height (in.)	Stem to Guide Clearance (in.)		Stem Diameter (in.)	
								Intake	Exhaust	Intake	Exhaust
'84–'87	R	4-151	Pont.	45	45	176 @ 1.26	1.69	0.0010–0.0027	0.0010–0.0027	0.3418–0.3425	0.3418–0.3425
'85–'87	9	6-173	Chev.	46	45	195 @ 1.181	1.57	0.0010–0.0027	0.0010–0.0027	0.3410–0.3416	0.3410–0.3416

CRANKSHAFT AND CONNECTING ROD SPECIFICATIONS

All measurements are given in inches.

Year	V.I.N. Code	Engine No. Cyl. Displacement (cu. in.)	Eng. Mfg.	Crankshaft				Connecting Rod		
				Main Brg. Journal Dia.	Main Brg. Oil Clearance	Shaft End-Play	Thrust on No.	Journal Diameter	Oil Clearance	Side Clearance
'84–'87	R	4-151	Pont.	2.2995–2.3005	0.0005–0.0022	0.0035–0.0085	5	1.9995–2.0005	0.0005–0.0026	0.006–0.022
'85–'87	9	6-173	Chev.	2.4937–2.4946	0.0016–0.0031	0.0023–0.0082	3	1.9984–1.9994	0.0014–0.0037	0.006–0.17

CAMSHAFT SPECIFICATIONS

All measurements in inches

Year	VIN Code	Engine	Eng. Mfg.	Journal Diameter					Bearing Clearance	Lobe Lift		Camshaft End Play
				1	2	3	4	5		Intake	Exhaust	
'84–'87	R	4-151	Pont.	—		1.869		—	0.0007–0.0027	0.398	0.398	0.0015–0.0050
'85–'87	9	6-173	Chev.	—		1.869		—	0.0010–0.0040	0.231	0.263	—

PISTON AND RING SPECIFICATIONS

All measurements are given in inches. To convert inches to metric units, refer to the Metric Information section.

Year	V.I.N. Code	Engine Type/ Disp. cu. in.	Eng. Mfg.	Piston-to-Bore Clearance	Ring Gap			Ring Side Clearance		
					Top Compression	Bottom Compression	Oil Control	Top Compression	Bottom Compression	Oil Control
'84	R	4-151	Pont.	0.0025–0.0033	0.010–0.022	0.010–0.027	0.015–0.055	0.002–0.003	0.002–0.003	snug
'85–'87	R	4-151	Pont.	0.0014–0.0022①	0.010–0.020	0.010–0.020	0.020–0.060	0.002–0.003	0.001–0.003	0.015–0.055
'85–'87	9	6-173	Chev.	0.0007–0.0017	0.0098–0.0197	0.0098–0.0197	0.020–0.055	0.0012–0.0028	0.0016–0.0037	.008 max.

① Measured 1.8 inch down from piston top

TORQUE SPECIFICATIONS
All readings in ft. lbs.

Year	V.I.N. Code	Engine No. Cyl. Displacement (cu. in.)	Eng. Mfg.	Cylinder Head Bolts	Rod Bearing Bolts	Main Bearing Bolts	Crankshaft Bolt	Flywheel to Crankshaft Bolts	Manifold	
									Intake	Exhaust
'84–'85	R	4-151	Pont.	92	32	70	200	44	①	44
'86–'87	R	4-151	Pont.	①	32	70	162	44	①	44
'85–'87	9	6-173	Chev.	65–90	34–40	63–74	66–84	45–55	20–25	22–28

① See Text and Illustration for procedure and specifications

WHEEL ALIGNMENT SPECIFICATIONS

Year	Model	Caster		Camber		Toe-In (in.)	Steering Axis (deg) Inclination
		Range (deg)	Pref. Setting (deg)	Range (deg)	Pref. Setting (deg)		
'84–'87	All	3N-7P	5P	5/16N–1 5/16P	½P	1/16 ± 1/32	—

ENGINE ELECTRICAL

Test details can be found in the "Charging and Starting Systems" Unit Repair Section. The voltage regulator is a solid-state, non-adjustable unit integral with the alternator. The alternator must be disassembled to remove the regulator.

Alternator

REMOVAL & INSTALLATION

1. Disconnect the negative battery cable.
2. Remove the air cleaner.
3. Disconnect the upper strut mount.
4. Disconnect the alternator adjusting bolt, upper adjusting bracket and drive belt.
5. Disconnect the wiring from the back of the alternator.
6. Lower the alternator mounting bracket and remove the alternator from the bottom of the vehicle.
7. Installation is the reverse of removal.

INTEGRAL VOLTAGE REGULATOR

An alternator with an integral voltage regulator is standard equipment. There are no adjustments possible with this unit; testing procedures will be found in the "Charging and Starting Systems" Unit Repair Section.

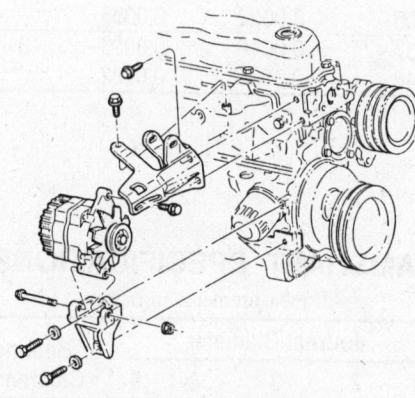

Generator mounting

Starter

Starter motor troubleshooting and repairs are covered in the "Charging and Starting" Unit Repair Section.

REMOVAL & INSTALLATION

1. Disconnect the battery ground cable.
2. Raise and support the vehicle safely.
3. Disconnect all wires at solenoid terminals. Note color coding of wires for installation.
4. Remove starter support mount bolts.

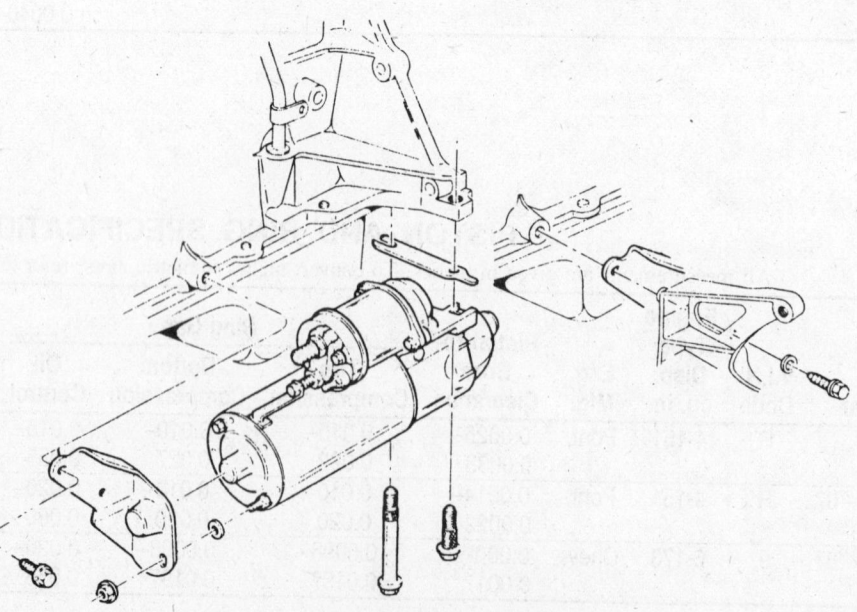

Starter motor mounting

5. Loosen the front bracket bolt or nut and rotate bracket clear. Lower and remove starter. Note the location of any shims so that they may be replaced in the same positions upon installation.

6. Reverse procedure to install.

Distributor

All models with gasoline engines are quipped with HEI distributor and ignition system. This system uses no points and is, therefore, relatively maintenance free.

——— **CAUTION** ———

When using an auxiliary starter switch on HEI systems, the distributor BATT lead must be disconnected. Failure to do this may cause damage to the grounding circuit in the ignition switch.

REMOVAL & INSTALLATION

1. Disconnect the negative battery cable.

2. Tag and disconnect all wires leading from the distributor cap. Do not disconnect the plug wires.

3. Remove the external ignition coil.

4. Remove the distributor cap by turning the four latches counterclockwise. You will need a stubby screwdriver to get at the latches on the four cylinder engine, because there isn't much room between the distributor and the firewall. Remove the distributor cap and set it aside without disconnecting any of the wires.

5. Remove the vacuum hose from the vacuum advance unit. On 1980 models, mark the position of the vacuum advance unit in relation to the engine for correct installation.

6. Remove the hold down and clamp at the base of the V6 distributor. The four cylinder engine has two bolts and a clamp. Remove the outer bolt first, then loosen but do not remove, the inner bolt. Slide the clamp back and remove it.

7. Before removing the distributor, note the position of the rotor. Scribe a mark on the distributor body indicating the initial position of the rotor.

8. Remove the distributor from the engine. The drive gear on the distributor shaft is helical and the shaft will rotate slightly as the distributor is removed. Note and mark the position of the rotor at this second position. Do not crank the engine with the distributor removed.

9. To install the distributor, rotate the distributor shaft until the rotor aligns with the 2nd mark you made (when the shaft stopped moving). Lu-bricate the drive gear with clean engine oil, then install the distributor into the engine.

10. Install the clamp and hold-down bolt. Tighten them until the distributor can just be moved with a little effort.

11. Connect the ignition wire and tachometer wire and install the distributor cap.

12. Set the timing and tighten the hold-down bolt.

INSTALLATION IF THE ENGINE WAS DISTURBED.

If the engine was cranked while distributor was removed, you will have to place the engine on TDC of the compression stroke to obtain proper ignition timing.

1. Remove the No. 1 spark plug.

2. Place your thumb over the spark plug hole. Crank the engine slowly until compression is felt. It will be easier if you have someone rotate the engine by hand, using a wrench on the crankshaft pulley.

3. Align the timing mark on the crankshaft pulley with the 0° mark on the timing scale attached to the front of the engine. This places the engine at TDC of the compression stroke.

4. Turn the distributor shaft until the rotor points between the No. 1 and No. 3 spark plug towers on the cap for the four cylinder engine or between the No. 1 and No. 6 spark plug towers for the V6.

5. Install the distributor into the engine.

6. Perform Steps 9–12 of the preceding removal and installation procedure.

TACHOMETER HOOKUP

The tachometer (TACH) terminal is next to the ignition switch (BAT) connector on the distributor cap.

——— **CAUTION** ———

Never ground the TACH terminal; serious damage will result. If there is any doubt as to the correct tachometer hookup, check with the tachometer manufacturer.

IGNITION TIMING ADJUSTMENT

1. Connect a timing light to the No. 1 spark plug wire according to the light manufacturer's instructions. DO NOT PIERCE THE SPARK PLUG WIRE TO CONNECT THE TIMING LIGHT.

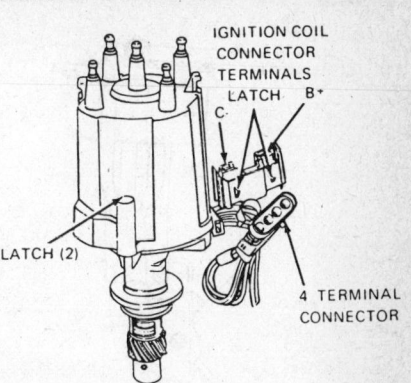

Typical distributor used with a separately mounted coil

2. Follow the instructions on the underhood engine decal.

3. Disconnect the 4-terminal connector at the distributor.

4. Start the engine and run it at idle speed.

5. Aim the timing light at the degree scale just over the harmonic balancer.

6. Adjust the timing by loosening the securing clamp and rotating the distributor until the desired ignition advance is achieved, then tighten the clamp.

7. Loosen the distributor retaining bolt. On some engines it's necessary to slide the clamp back slightly. Do not remove the retaining bolt.

8. Adjust the timing, then replace and tighten the clamp. To advance the timing, rotate the distributor opposite the normal direction of rotor rotation. Retard the timing by rotating the distributor in the normal direction of rotor rotation.

FUEL SYSTEM

The TBI (Throttle Body Injection) system used by Fiero cars with the four cylinder engine uses an electric fuel pump. This pump is located in the gas tank. On 1985 and later models equipped with the V6 engine, a new multi-port fuel injection (MFI) system is available. Fuel is pumped from the tank by a high pressure fuel pump, located inside the fuel tank on this system also.

Fuel Pressure Relief Procedure

——— **CAUTION** ———

Before opening any part of the fuel system, the pressure must be relieved. Follow the procedure below to relieve the pressure:

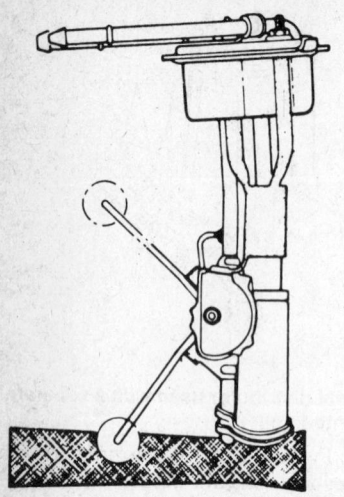

Typical electric fuel pump and sending unit

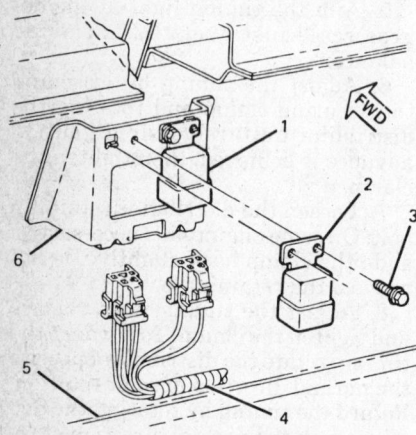

1. Fuel pump relay
2. Relay assy. (ac clutch control)
3. Bolt
4. E.F.I. harness
5. Floor pan
6. Bracket

Fuel pump relay location

1. Remove the fuel pump fuse from the fuse panel.
2. Start the engine and let it run until all fuel in the line is used.
3. Crank the starter an additional three seconds to relieve any residual pressure.
4. With the ignition OFF, replace the fuse.

Electric Fuel Pump

REMOVAL & INSTALLATION

1. Relieve the fuel system pressure (refer to procedure above).
2. Drain the fuel tank.
3. Disconnect wiring from the tank.

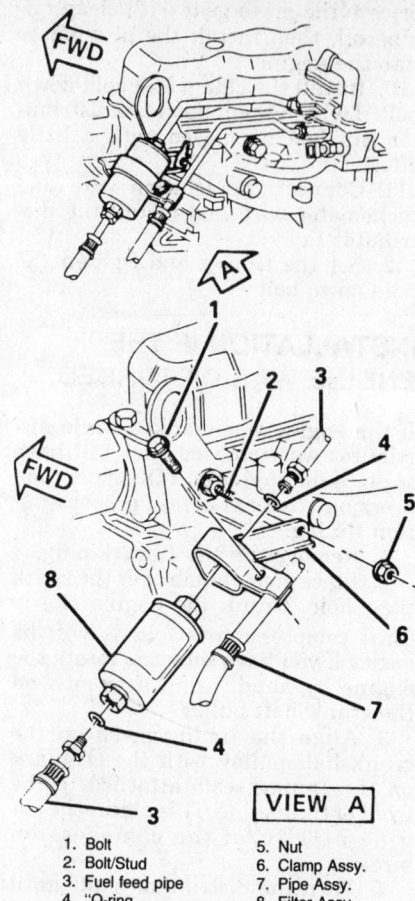

VIEW A

1. Bolt
2. Bolt/Stud
3. Fuel feed pipe
4. "O-ring
5. Nut
6. Clamp Assy.
7. Pipe Assy.
8. Filter Assy.

Fuel filter

4. Remove the ground wire retaining screw from under the body.
5. Disconnect all hoses from the tank.
6. Support the tank on a jack and remove the retaining strap nuts.
7. Lower the tank and remove it.
8. Remove the fuel gauge/pump retaining ring using a spanner wrench such as tool J-24187.
9. Remove the gauge unit and the pump.
10. Installation is the reverse of removal. Always replace the O-ring under the gauge/pump retaining ring.

Fuel Filter

REMOVAL & INSTALLATION

NOTE: Relieve fuel pressure (see caution above).

The filter is an inline unit ahead of the TBI or MFI unit. To remove the filter, make sure the engine is cold, unclamp and remove the fuel hose, then unscrew the filter from the steel fuel line. Installation is the reverse of removal.

Fuel Injection

Idle speed is controlled by the Electronic Control Module (ECM). No adjustments are possible.

ENGINE COOLING

— CAUTION —
Keep hands, tools and clothing away from engine cooling fan to help prevent personal injury. This fan is electric and can come on whether or not the engine is running. The fan can start automatically in response to a heat sensor with the ignition in the ON position.

Radiator

REMOVAL & INSTALLATION

1. Drain the engine coolant.
2. Disconnect the wiring harness from the fan and fan frame.
3. Remove the fan and frame assembly.
4. Disconnect the upper radiator support bracket.
5. Disconnect the coolant hoses at the radiator.
6. Disconnect the transmission/engine oil cooler lines at the radiator.
7. Remove the radiator from the car.
8. Installation is the reverse of removal. After installation run the engine and check for leaks.

Water Pump

REMOVAL & INSTALLATION

1. Disconnect battery negative cable.
2. Remove accessory drive belts. Drain the cooling system through the petcock on the bottom of the radiator.
3. Remove water pump attaching bolts and remove pump. Note that on the V6, there is a nut at the top/left position. The nut screws onto the outer end of a stud that retains the timing cover underneath.
4. If installing a new water pump, transfer pulley from old unit. With sealing surfaces cleaned, place a ⅛ in. (3mm) bead of sealant No. 1052289 or equivalent on the water pump sealing surface. While sealer is still wet, install pump and torque bolts/nut to 6 ft. lbs. on the four cylinder engine. On the V6, torque the two bottom/center

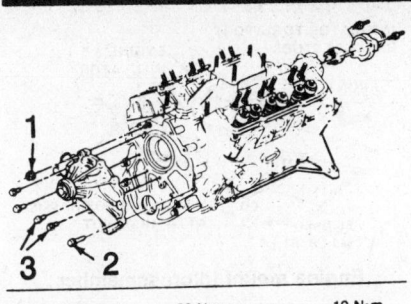

1– 30 N•m (22 FT-LBS) 2– 30 N•m (22 FT-LBS) 3– 10 N•m (7 FT-LBS)

Water pump bolt torques for the V6 engine

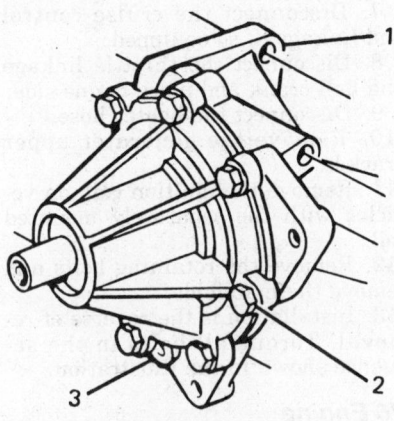

1. Body
2. Bolt
3. Housing

Water pump mounting

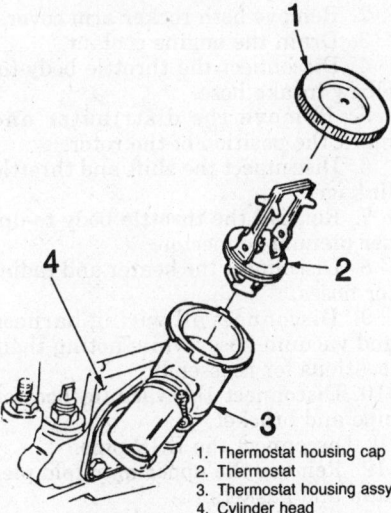

1. Thermostat housing cap
2. Thermostat
3. Thermostat housing assy.
4. Cylinder head

Thermostat and housing

bolts to 7 ft. lbs.; torque the other bolts and the nut to 22 ft. lbs.

5. Install accessory drive belts.
6. Connect battery negative cable.

Thermostat

REMOVAL & INSTALLATION

1. Remove the thermostat cap.

2. Grasp the thermostat handle and gently pull up.

3. Before installing, clean the thermostat housing and O-ring. Apply a suitable lubricant to the O-ring for easier installation.

4. Push the thermostat down into the housing with the handle upward until it is properly seated and install the cap.

EMISSION CONTROLS

For a description and service procedures for the emission control systems, please refer to "Emission Controls" in the Unit Repair section.

ENGINE MECHANICAL

REMOVAL & INSTALLATION

NOTE: The engine assembly is removed from underneath the vehicle.

1. Disconnect the battery cables.
2. Drain the engine coolant.
3. Remove the rear compartment lid and also the side panels on the V6 engine.

NOTE: Do not remove the torsion rod retaining bolts.

4. Remove the air cleaner assembly.
5. Disconnect the throttle and shift cables.
6. Disconnect the heater hose at the intake manifold.
7. Disconnect vacuum hoses from all non-engine components.
8. Disconnect the fuel lines and filter.
9. Disconnect the fuel pump relay and the oxygen sensor.
10. On models equipped with automatic transaxle, disconnect the transaxle cooler lines.
11. Disconnect the slave cylinder from the manual transaxle equipped vehicles.
12. Disconnect the engine to chassis ground strap.
13. Discharge the A/C system if so equipped then disconnect the A/C lines at the compressor and seal the open ends.

— CAUTION —
Do not disconnect any refrigerant lines unless experienced with air conditioning systems. Escaping refrigerant will freeze any surface it contacts, including skin and eyes.

14. Remove the rear console.
15. Remove the ECM harness through the bulkhead.
16. Install an engine support fixture.
17. Remove the engine strut bracket and mark the bolt and bracket for reassembly.
18. Raise the vehicle and support it safely.

1. Place a 4 × 4 at jacking locations
2. 4 wheel support dolly
3. Caliper supported
4. Support control arm on both sides
5. 4 × 4's
6. Wheel chocks

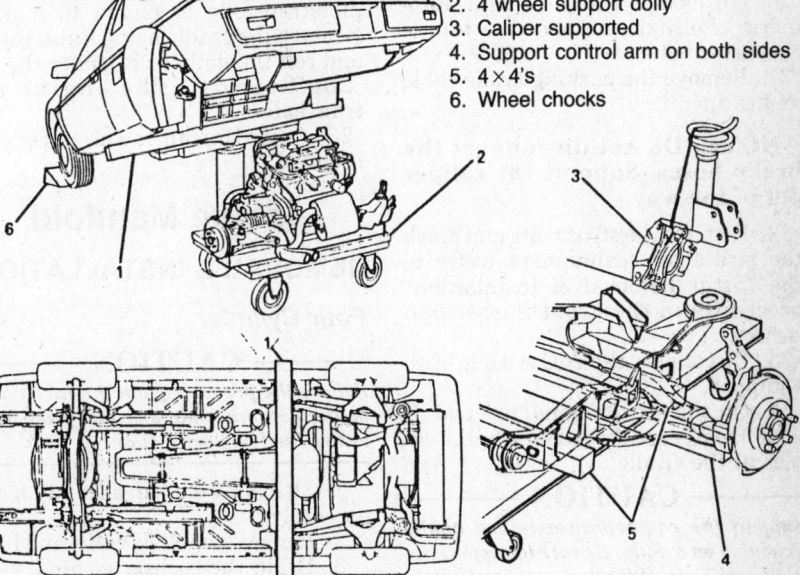

Engine removal and cradle support points

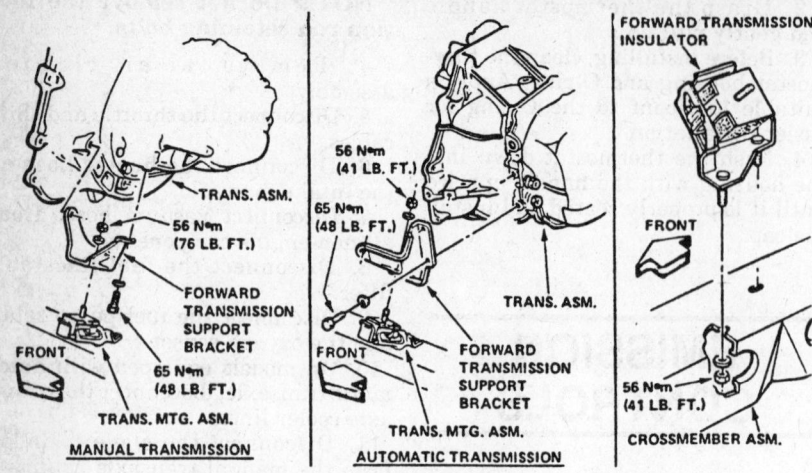

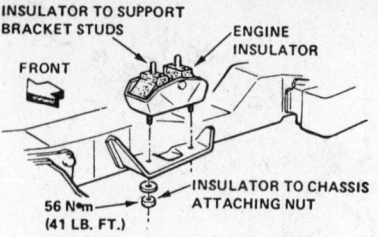

Engine mount to crossmember

Forward transaxle mount and mounting brackets

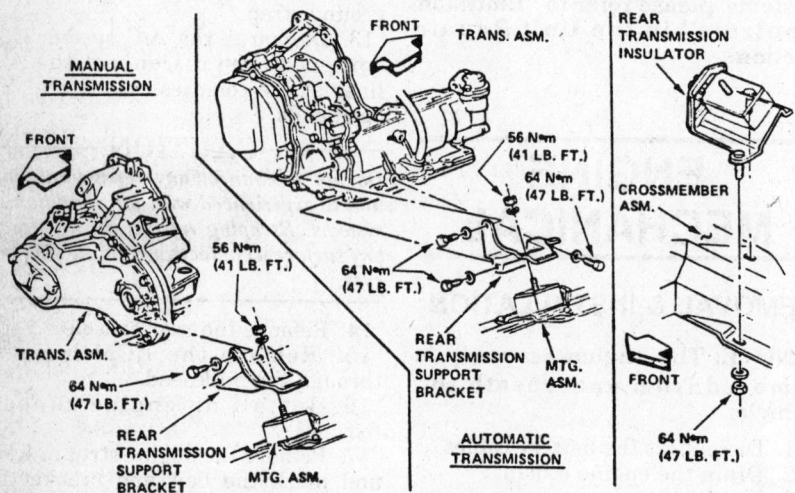

Rear transaxle mount and mounting brackets

19. Remove the rear wheels.
20. On models equipped with automatic transaxle, remove the torque converter bolts.
21. Remove the parking brake cable and calipers.

NOTE: Do not disconnect the brake hoses. Support the caliper out of the way.

22. Remove the strut bolts and mark the struts for realignment (refer to the "Strut Removal & Installation" procedure in the Front Suspension section).
23. Disconnect the A/C wiring, if so equipped.
24. On the 4 cyl engine, use tool J-34065 to release the parking brake cables at the cradle.

——— CAUTION ———
Support the engine/transaxle and cradle assembly on a dolly. Be sure to support the outboard ends of the lower control arms. Disconnect the engine support fixture.

25. Lower the car and attach the engine/transaxle assembly to a dolly. Remove the cradle bolts. Raise the car and roll the dolly from under the car.
26. Separate the engine and transaxle.
27. Installation is the reverse of removal.

Intake Manifold

REMOVAL & INSTALLATION

Four Cylinder

——— CAUTION ———
Relieve the pressure from the fuel system before disconnecting any fuel lines (refer to the Fuel System section).

1. Remove the air cleaner assembly.
2. Remove the PCV valve and hose.
3. Drain the cooling system.
4. Relieve the fuel system pressure and disconnect the fuel lines.

5. Disconnect the vacuum hoses.
6. Disconnect the wiring and the throttle linkage from the throttle body assembly.
7. Disconnect the cruise control and linkage, if so equipped.
8. Disconnect the throttle linkage and bell crank and place to one side.
9. Disconnect the heater hose.
10. Remove the generator upper bracket.
11. Remove the ignition coil, on vehicles with the separately mounted coil.
12. Remove the retaining bolts and remove the manifolds.
13. Installation is the reverse of removal. Torque all bolts in the sequence shown in the illustration.

V6 Engine

NOTE: Refer to the Fuel System section for the procedure on relieving fuel system pressure.

1. Disconnect the negative battery cable.
2. Remove both rocker arm covers.
3. Drain the engine coolant.
4. Disconnect the throttle body-to-elbow intake hose.
5. Remove the distributor and mark the position of the rotor.
6. Disconnect the shift and throttle linkage.
7. Remove the throttle body-to-upper plenum connector.
8. Disconnect the heater and radiator hoses.
9. Disconnect all wiring harness and vacuum hoses while noting their locations for reassembly.
10. Disconnect the vacuum booster pipe and bracket.
11. Disconnect the EGR pipe.
12. Remove the upper manifold plenum and gaskets.
13. Remove the intermediate intake manifold and gasket.
14. Remove the lower intake manifold and gaskets.
15. Clean all gasket surfaces on the intake manifolds and cylinder head.
16. Install the lower intake manifold and gasket and torque in sequence to 19 ft. lbs.
17. Install the intermediate intake manifold and gaskets and torque in sequence to 15 ft. lbs.

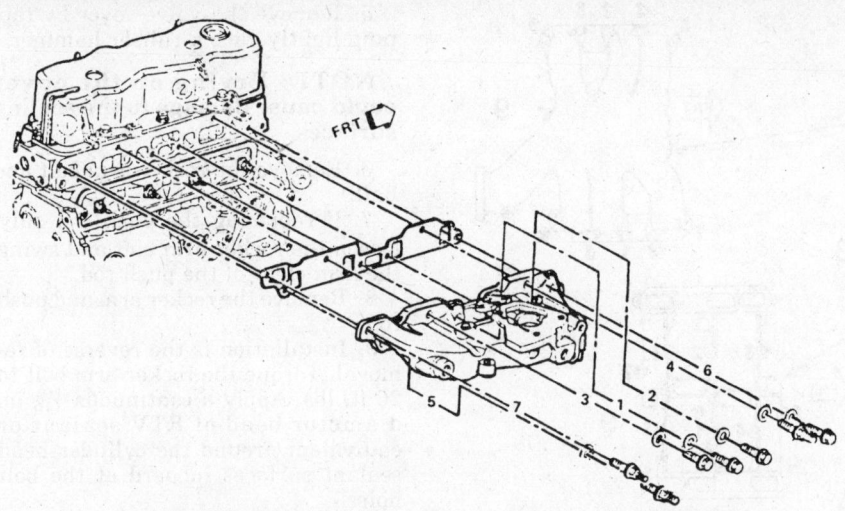

Intake manifold torque sequence—four cyl. engine, 1984–85

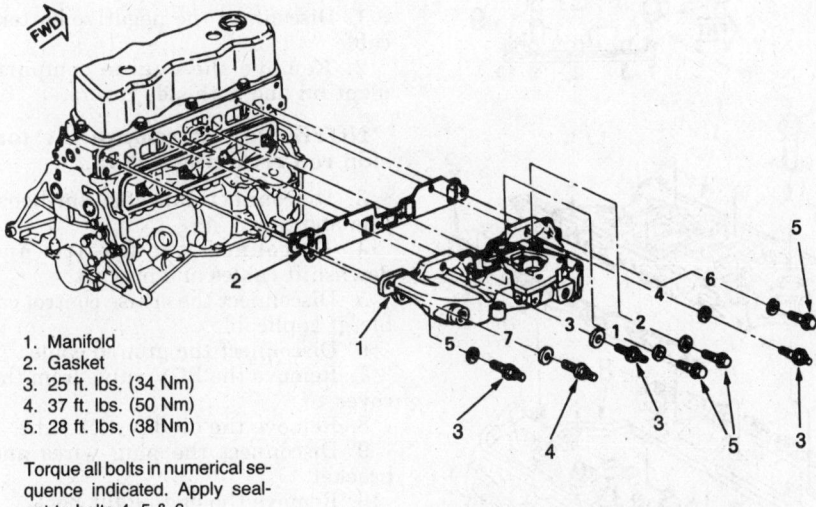

1. Manifold
2. Gasket
3. 25 ft. lbs. (34 Nm)
4. 37 ft. lbs. (50 Nm)
5. 28 ft. lbs. (38 Nm)

Torque all bolts in numerical sequence indicated. Apply sealant to bolts 4, 5 & 6

Intake manifold torque sequence—four cyl. engine, 1986 and later

18. Install the upper manifold plenum and gaskets and torque in sequence.
19. The remainder of the installation is the reverse of removal. Check engine timing, coolant level and for leaks.

Exhaust Manifold

REMOVAL & INSTALLATION

Four Cylinder

1. Remove the air cleaner and the EFI bracket tube.
2. Raise the vehicle and support it with jack stands.
3. Remove the exhaust pipe and lower the vehicle.
4. Remove the retaining bolts and washers and remove the exhaust manifold and gasket.

5. Installation is the reverse of removal. Clean the sealing surfaces and use a new gasket. Torque the retaining bolts to the sequence shown in the illustration. Torque bolts No. 3, 4 and 5 to 37 ft. lbs. Torque bolts No. 1, 2, 6 and 7 to 16. ft. lbs.

Exhaust Manifold And Crossover

REMOVAL & INSTALLATION

V6 Engine
FRONT

1. Disconnect the negative battery cable.
2. Remove the rear compartment lid.

NOTE: Do not remove the torsion rod retaining bolts.

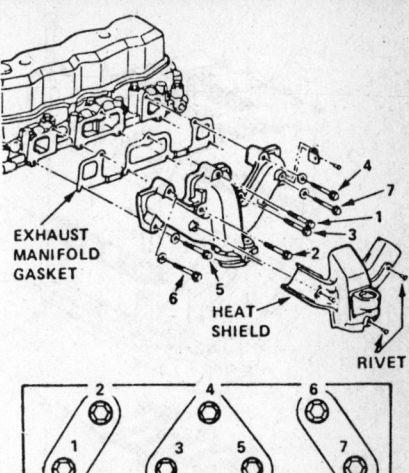

EXHAUST MANIFOLD GASKET
HEAT SHIELD
RIVET

BOLT LOCATIONS

Exhaust manifold torque sequence—four cyl. engine, 1984

3. Remove the brake vacuum hose.
4. Remove the manifold heat shield.
5. Remove the front crossover bolts.
6. Raise the car and remove the front converter heat shield and the lower manifold bolts.
7. Lower the car and remove the upper manifold bolts then remove the manifold.
8. Install the manifold and torque the upper bolts to 18 ft. lbs.
9. Raise the car and torque the lower bolts to 18 ft. lbs.
10. Install the front converter heat shield.
11. Lower the car and torque the crossover bolts to 22 ft. lbs.
12. The remainder of the installation is the reverse of removal. Check for exhaust or vacuum leaks.

REAR

1. Disconnect the manifold-to-crossover bolts.
2. Remove the manifold bolts then remove the manifold.
3. Installation is the reverse of removal. Torque the manifold bolts to 18 ft. lbs. and the manifold-to-crossover bolts to 22 ft. lbs.

Rocker Arm And Push Rod

REMOVAL & INSTALLATION

Four Cylinder

1. Remove the air cleaner.
2. Remove the PCV valve and hose.
3. Remove the valve cover bolts.
4. Disconnect the wires from the spark plugs and clips.

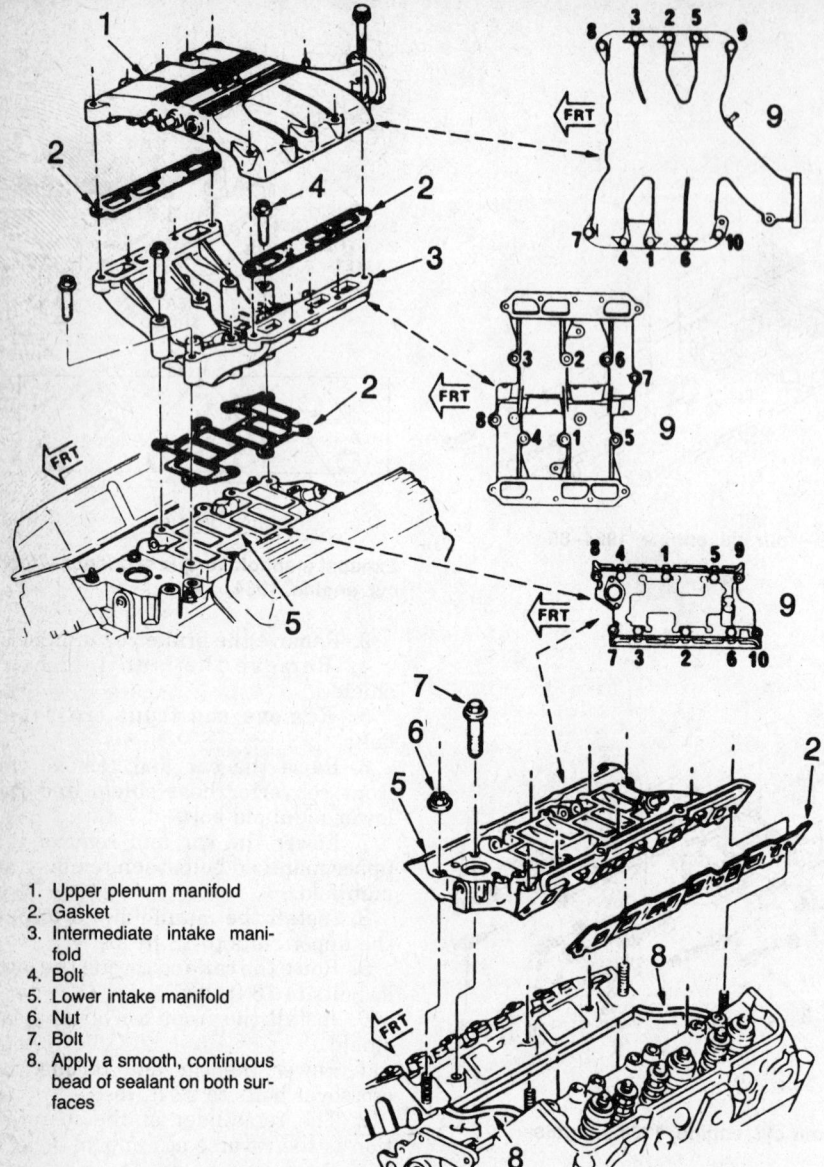

1. Upper plenum manifold
2. Gasket
3. Intermediate intake manifold
4. Bolt
5. Lower intake manifold
6. Nut
7. Bolt
8. Apply a smooth, continuous bead of sealant on both surfaces

Intake manifold installation and torque sequence—V6 engine

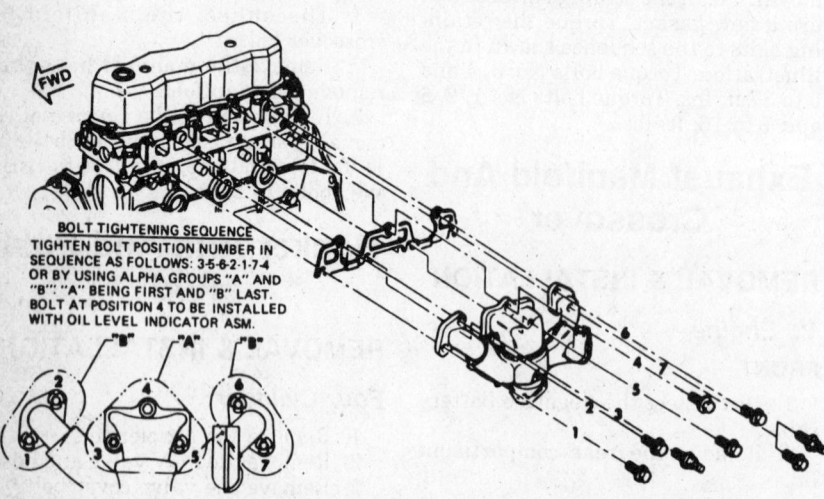

BOLT TIGHTENING SEQUENCE
TIGHTEN BOLT POSITION NUMBER IN SEQUENCE AS FOLLOWS: 3-5-6-2-1-7-4 OR BY USING ALPHA GROUPS "A" AND "B", "A" BEING FIRST AND "B" LAST. BOLT AT POSITION 4 TO BE INSTALLED WITH OIL LEVEL INDICATOR ASM.

Exhaust manifold torque sequence—four cyl. engine, 1985 and later

5. Remove the valve cover by tapping lightly with a rubber hammer.

NOTE: Prying on the cover could cause damage to the sealing surfaces.

6. Remove the rocker arm bolt and ball.
7. If replacing the push rod only, loosen the rocker arm bolt and swing the arm clear of the push rod.
8. Remove the rocker arm and push rod.
9. Installation is the reverse of removal. Torque the rocker arm bolt to 20 ft. lbs. Apply a continuous $\frac{3}{16}$ in. diameter bead of RTV sealant or equivalent around the cylinder head sealant surfaces inboard at the bolt holes.

V6 Engine

1. Disconnect the negative battery cable.
2. Remove the engine compartment lid and both side covers.

NOTE: Do not remove the torsion rod retaining bolts.

3. Disconnect the vacuum boost line and tube.
4. Disconnect the throttle and downshift cables and bracket.
5. Disconnect the cruise control cable, if applicable.
6. Disconnect the ground cable.
7. Remove the PCV valve from the cover.
8. Remove the oil dip stick tube.
9. Disconnect the plug wires and bracket.
10. Remove the engine lift hook.
11. Remove the rocker arm cover bolts and carefully remove the cover by bumping with your hand or a rubber mallet. If prying is necessary do not distort the sealing flange.
12. Remove the rocker arm nuts.

NOTE: Keep all components in order so that they may be installed in the same location. Keep the sealant out of the bolt holes. Keep all components in order so that they may be installed in the same location.

13. Remove the rocker arm pivot balls, arms and pushrods.
14. Before installation, coat the bearing surfaces of the rocker arms and pivot balls with Molykote® or equivalent.
15. Install the push rods, rocker arms and pivot balls. Make sure the push rods are seated in the valve lifters.
16. Adjust the rocker arm nuts until lash is eliminated.
    a. Rotate the engine until the

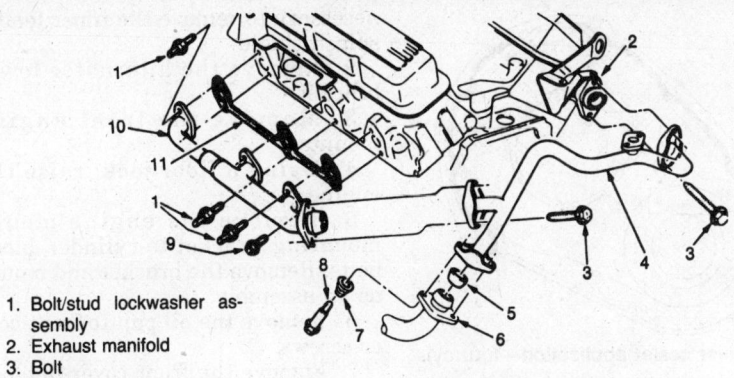

1. Bolt/stud lockwasher assembly
2. Exhaust manifold
3. Bolt
4. Crossover pipe
5. Seal
6. Muffler
7. Spring
8. Bolt

9. Bolt and lockwasher assembly
10. Exhaust manifold
11. Gasket

Exhaust manifold installation—V6 engine

mark on the torsional damper lines up with the "0" mark on the timing tab, with the engine in the No. 1 firing position. This may be determined by placing fingers on the No. 1 rocker arms as the mark on the damper comes near the "0" mark. If the valves are not moving, the engine is in the No. 1 firing position. With the engine in the No. 1 firing position, the following valves may be adjusted: Exhaust—1, 2, 3; Intake—1, 5, 6.

b. Back out the adjusting nut until lash is felt at the pushrod, then turn the adjusting nut until all lash is removed. This can be determined by rotating the pushrod while turning the adjusting nut. When lash has been removed, turn the adjusting nut in 1½ additional turns to center the lifter plunger.

c. Crank the engine one revolution until the timing tab "0" mark and torsional damper mark are again in alignment. This is the No. 4 firing position. With the engine in this position, the following valves may be adjusted: Exhaust—4, 5, 6; Intake—2, 3, 4.

17. Install the rocker arm covers.

a. Clean the surfaces on the cylinder head and rocker arm cover.

b. Place a 3mm diameter (⅛ inch) dot of RTV sealer, at the intake manifold and cylinder head split line.

c. Install the rocker arm cover gasket, using care to line up the holes in the gasket with the bolt holes in the cylinder head.

d. Install the rocker arm cover bolts and torque to 90 inch lbs.

18. The remainder of the installation is the reverse of removal.

Valve Lash

No routine adjustment is necessary. If

it is necessary to adjust the valves on the V6 because of engine wear, follow Steps 16 a-c. in the "Rocker Arm and Pushrod Removal & Installation" procedure immediately above.

Cylinder Head
REMOVAL & INSTALLATION
Four Cylinder

1. Drain the cooling system.
2. Raise the vehicle and support safely with jack stands.
3. Remove the exhaust pipe.
4. Lower the vehicle.
5. Remove the oil level indicator tube.
6. Remove the air cleaner assembly.
7. Disconnect the EFI electrical connections and vacuum hoses. Depressurize the fuel system before disconnecting any fuel lines.

8. Remove the EGR base plate.
9. Remove the heater hose from the intake manifold.
10. Remove the ignition coil lower mounting bolt and wiring connections.
11. Remove all wiring connections from the intake manifold and cylinder head.
12. Remove the engine strut bolt from the upper support.
13. Remove the generator belt.
14. Remove the throttle cables from the intake manifold.
15. Remove the valve cover, rocker arms and push rods. Loosen the cylinder head bolts in reverse of the torque sequence.
16. Remove the cylinder head bolts and remove the cylinder head.
17. Before installing, clean the gasket surfaces of the head and block.
18. Make sure the retaining bolt threads and the cylinder block threads are clean since dirt could affect bolt torque.
19. Install a new gasket over the dowel pins in the cylinder.
20. Install the cylinder head into place over the dowel pins.
21. Coat the cylinder head bolt threads with sealing compound and install finger tight.
22. On 1984–85 models, tighten the cylinder head bolts gradually in the sequence shown in the illustration. Final torque is 92 ft. lbs.
23. On 1986 and later models, torque the head bolts gradually to 18 ft. lbs. in the sequence shown in the illustration. Repeat the sequence, bringing the torque to 22 ft. lbs. on all bolts except number 9. Torque number 9 to 29 ft. lbs. Repeat the sequence. Turn all the bolts, except number 9, 120 de-

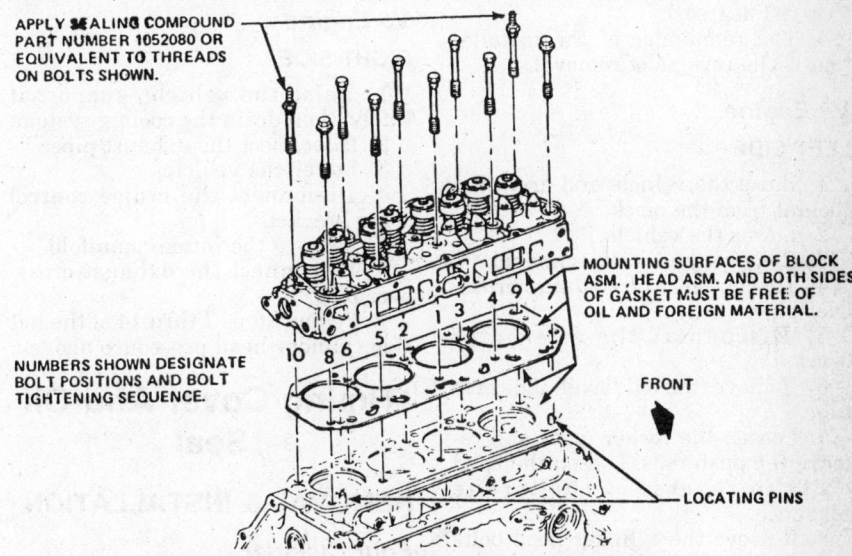

APPLY SEALING COMPOUND PART NUMBER 1052080 OR EQUIVALENT TO THREADS ON BOLTS SHOWN.

NUMBERS SHOWN DESIGNATE BOLT POSITIONS AND BOLT TIGHTENING SEQUENCE.

MOUNTING SURFACES OF BLOCK ASM., HEAD ASM. AND BOTH SIDES OF GASKET MUST BE FREE OF OIL AND FOREIGN MATERIAL.

FRONT

LOCATING PINS

Cylinder head torque sequence—four cyl. engine

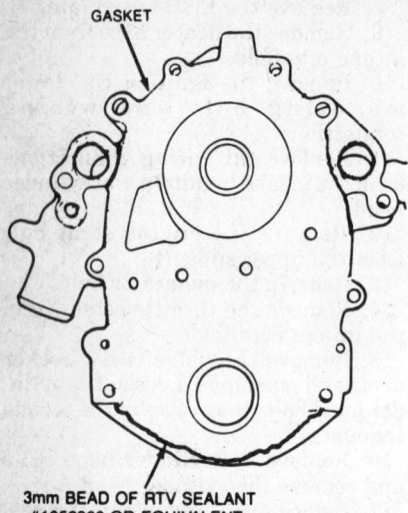

GASKET

3mm BEAD OF RTV SEALANT
#1052366 OR EQUIVALENT

Applying sealer to the front cover on the V6 engine

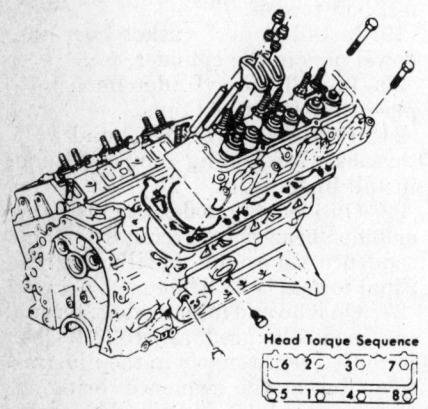

Cylinder head torque sequence—V6 engine

grees (two flats). Turn number 9 $\frac{1}{4}$ turn (90 degrees).

24. The remainder of the installation is the reverse of removal.

V6 Engine

LEFT SIDE

1. Raise the vehicle and drain the coolant from the block.
2. Lower the vehicle.
3. Remove the intake manifold.
4. Disconnect the exhaust crossover pipe.
5. Disconnect the alternator bracket.
6. Remove the oil level indicator tube.
7. Loosen the rocker arms and remove the push rods. Loosen the head bolts in reverse of the torque sequence.
8. Remove the cylinder head bolts then remove the cylinder head.
9. Before installing, clean the gas-

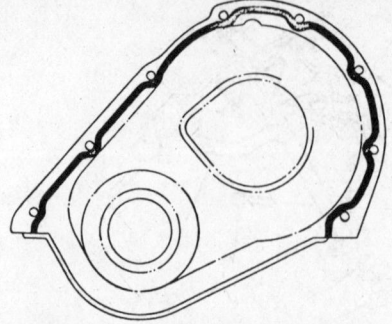

Timing cover sealer application—four cyl. engine

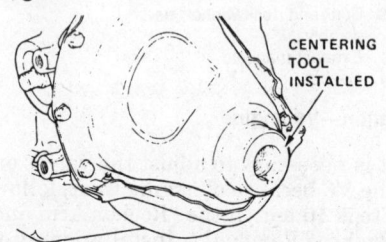

CENTERING TOOL INSTALLED

Front cover centering tool installed—four cyl. engine

ket surfaces on the head, cylinder block and intake manifold.

10. Place the gasket in position over the dowel pins with the note "This Side UP" showing.
11. Place the cylinder head into position.
12. Coat the cylinder head bolt threads with a sealer and install the bolts. Tighten the bolts in sequence to 66 ft. lbs.
13. Install the pushrods and loosely retain with the rocker arms. Make sure the lower ends of the pushrods are in the lifter seats then adjust the valve lash. Refer to the procedure under Rocker Arm, Cover and Push Rod.
14. The remainder of the installation is the reverse of removal.

V6 Engine

RIGHT SIDE

1. Raise the vehicle, support it safely, then drain the cooling system.
2. Disconnect the exhaust pipe.
3. Lower the vehicle.
4. Disconnect the cruise control servo bracket.
5. Remove the intake manifold.
6. Disconnect the exhaust crossover pipe.
7. Follow Steps 7 thru 14 of the left side cylinder head procedure above.

Timing Cover And Oil Seal

REMOVAL & INSTALLATION

Four Cylinder

1. Remove the crankshaft hub. It is

necessary to remove the inner fender splash shield.

2. Remove the alternator lower bracket.
3. Remove the front engine mounts.
4. Using a floor jack, raise the engine.
5. Remove the engine mount mounting bracket-to-cylinder block bolts. Remove the bracket and mount as an assembly.
6. Remove the oil pan-to-front cover screws.
7. Remove the front cover-to-block screws.
8. Pull the cover slightly forward, just enough to allow cutting of the oil pan front seal flush with the block on both sides.
9. Remove the front cover and attached portion of the pan seal.
10. Clean the gasket surfaces throughly.
11. Cut the tabs from the new oil pan front seal.
12. Install the seal on the front cover, pressing the tips into the holes provided.
13. Coat the new gasket with sealer and position it on the front cover.
14. Apply a $\frac{1}{8}$ in. bead of silicone sealer to the joint formed at the oil pan and block.
15. Align the front cover seal with a centering tool and install the front cover. Tighten the screws.
16. Install the hub and torque the hub bolt to 160 ft. lbs.

NOTE: Coat the pulley to hub bolts with a locking sealant.

Front Cover

REMOVAL & INSTALLATION

V6 Engine

1. Disconnect the negative battery cable.
2. Remove the A/C compressor and bracket, without disconnecting the refrigerant lines and position out of the way.
3. Remove the water pump.
4. Raise the vehicle and support it safely.
5. Remove the torsional damper.
6. Remove the oil pan to cover bolts.
7. Lower the vehicle and remove the front cover.
8. Before installing, clean the sealing surfaces on the front cover and cylinder block. Install a new gasket and apply a $\frac{1}{8}$ in. bead of RTV sealer to the oil pan sealing surface of the front cover.
9. Place the front cover on the en-

gine and install the stud bolt and standard bolts.

10. The remainder of the installation is the reverse of removal.

Front Cover Oil Seal

REMOVAL & INSTALLATION

V6 Engine

1. Remove the torsional damper.
2. Pry out the seal using a suitable tool.
3. Before installing, lubricate the seal with clean engine oil.
4. Insert the seal in the front cover with the lip facing the engine.
5. Using tool J-23042 Seal Installer, or equivalent, drive the seal into place.
6. Install the torsional damper and check for leaks.

Camshaft And Timing Gear

REMOVAL

Four Cylinder

1. Remove the engine as previously described.
2. Install the engine on a stand.
3. Remove the rocker arm cover, loosen valve rocker arm bolts and pivot rocker arms clear of push rods.
4. Remove the distributor and fuel pump.
5. Remove push rods cover, push rods and valve lifters.
6. Remove the generator, lower generator bracket and front engine mount bracket assembly.
7. Remove the oil pump drive shaft and gear assembly.
8. Remove the front pulley hub and timing gear cover.
9. Remove the two camshaft thrust plate screws by working through holes in the camshaft gear.
10. Remove the camshaft and gear assembly by pulling it out through the front of the block. Support shaft carefully when removing so as not to damage camshaft bearings.
11. If the gear must be removed from the shaft, use press plate and adaptor tool J-971 on press.
12. Place tools on table of press. Place the camshaft through the opening in the tools. Press shaft out of gear using socket or other suitable tool. Thrust plate must be so positioned that woodruff key in shaft does not damage it when the shaft is pressed out of gear.

Removing the camshaft thrust screws— four cyl. engine

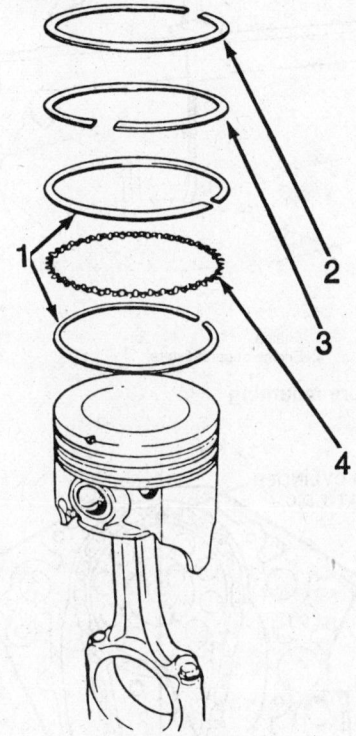

1. Oil rings
2. Top compression ring
3. Second compression ring
4. Expander

Typical piston and rod assembly

INSTALLATION

1. To assemble camshaft gear, thrust plate and gear spacer ring to camshaft, proceed as follows:
   a. Firmly support shaft at back of front journal in an arbor press using press plate adaptors.
   b. Place gear spacer ring and thrust plate over end of shaft, and install woodruff key in shaft keyway.
   c. Install camshaft gear and press it onto the shaft until it bottoms against the gear spacer ring. The end clearance of the thrust plate should be replaced.
2. Throughly coat the camshaft journals with a high quality engine oil supplement.
3. Install the camshaft assembly in

the engine block, being careful not to damage bearings or cam.

4. Turn crankshaft and camshaft so that the teeth will line up. Engine is now in the number 4 cylinder firing position. Install camshaft thrust plate to block screws and tighten 75 inch lbs.
5. Install the timing gear cover and gasket.
6. Line up the keyway in hub with key on crankshaft and slide hub onto shaft. Install the center bolt and torque to 160 ft. lbs.
7. Install as previously described valve lifters, push rods, push rod cover, oil pump shaft and gear assembly and fuel pump.
8. Install the distributor as follows:
   a. Turn the crankshaft 360° to firing position of number one cylinder (number one exhaust and intake valve lifters both on base circle of camshaft and timing mark on harmonic balancer indexed with top dead center mark on timing pad).
   b. Install the distributor in its original position and align shaft so that rotor arm points toward number one cylinder spark plug contact.
9. Pivot the rocker arms over push rods. With lifters on base circle of camshaft, tighten rocker arm bolt to 20 ft. lbs. (27 Nm). Do not overtorque.
10. Install the front mount assembly lower generator bracket and generator.
11. Complete the engine installation as described earlier.

Timing Chain And Sprockets

REMOVAL & INSTALLATION

V6 Engine

1. Remove the crankcase front cover.
2. Place the No. 1 piston at top dead center, with the marks on the camshaft and crankshaft sprockets aligned.
3. Remove the camshaft sprocket and chain.

NOTE: It may be necessary to use a plastic mallet on the lower edge of the sprocket to dislodge it.

4. Remove the camshaft sprocket with tool J-5825.
5. Install the sprocket with tool J-5590.
6. Apply Molykote® or equivalent to the sprocket thrust surface.
7. Hold the sprocket with the chain hanging down and align the marks on the camshaft and crankshaft sprockets.

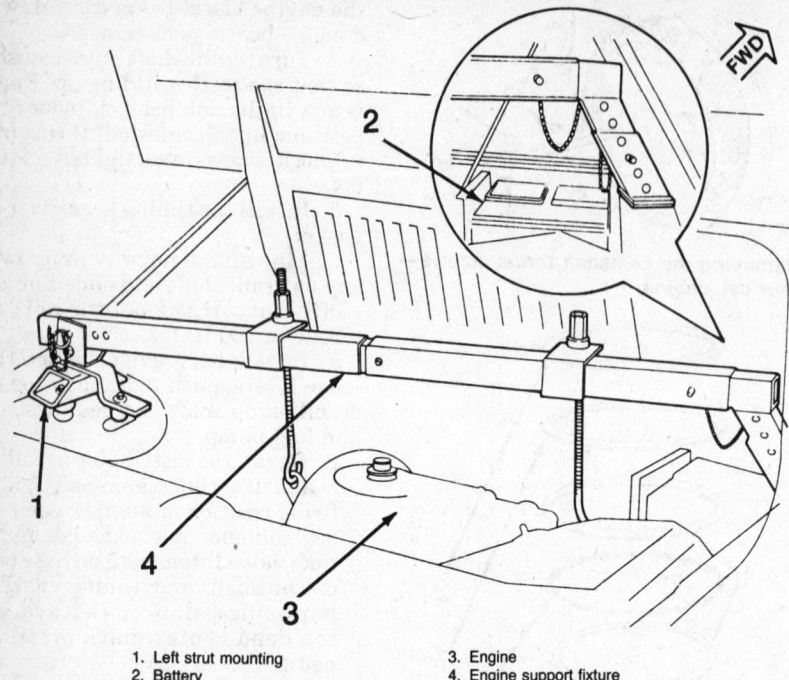

1. Left strut mounting
2. Battery
3. Engine
4. Engine support fixture

Engine holding fixture mounting

V-6 timing chain and sprockets—exploded view

8. Align the dowel in the camshaft with the dowel hole in the camshaft sprocket.

9. Draw the camshaft sprocket onto the camshaft, using the mounting bolts and torque to 15–20 ft. lbs.

10. Lubricate the timing chain with engine oil.

11. Install the crankcase front cover.

Camshaft

REMOVAL & INSTALLATION

V6 Engine

1. Remove the engine (on cradle).
2. Remove the valve lifters.
3. Remove the crankcase front cover.
4. Remove the timing chain and sprocket.
5. Remove the rear cover.
6. Carefully remove the camshaft to avoid damage to the bearings.
7. Before installation, lubricate the camshaft journals with engine oil.

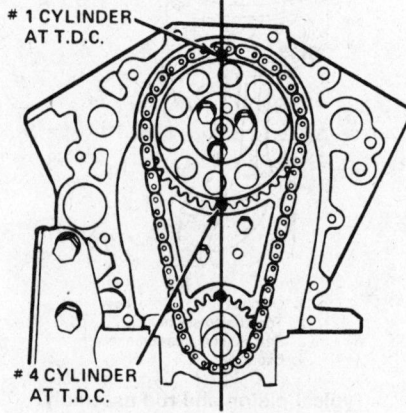

1 CYLINDER AT T.D.C.

4 CYLINDER AT T.D.C.

Align the timing marks for the camshaft and crankshaft sprockets as shown

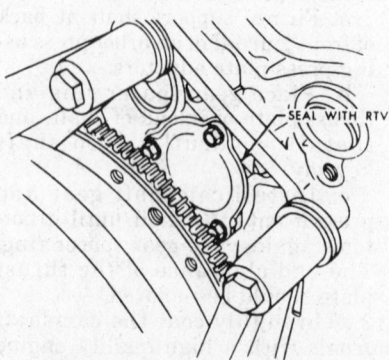

SEAL WITH RTV

The camshaft rear cover on the V6

NOTE: If a new camshaft is to be installed, coat the lobes with an engine oil supplement such as GM E.O.S. or its equivalent.

8. The remainder of installation is the reverse of removal.

Piston and Connecting Rod Position

See the accompanying illustrations to properly install piston and connecting rod assemblies. Align the piston and connecting rod assembly with the piston mark (notch) toward the front of the engine. For further information on piston replacement and overhaul, please refer to the "Engine Rebuilding" Unit Repair section.

ENGINE LUBRICATION

Oil Pan

REMOVAL & INSTALLATION

Four Cylinder

1. Remove the engine cradle. The cradle can be removed from the car without removing the engine or transaxle.

   a. Using engine support fixture tool J-28467 or equivalent, raise the engine enough to take tension off of the engine mounts.

   b. Raise the vehicle and support safely with jack stands.

   c. Remove the exhaust pipe bolts at the manifold.

   d. Remove the rear wheels and tire assemblies.

   e. Remove both lower control arms at the knuckle.

   f. Remove both toe-link rod at the knuckle.

   g. Remove the emergency brake cable at the cradle.

   h. Remove the engine and transmission mounting bolts.

   i. Remove the cradle bolts and remove the cradle assembly.

2. Drain the engine oil.

3. Remove the nuts from the the engine mount to the support bracket.

4. Disconnect the exhaust pipe at the manifold and the rear transaxle mount.

5. Remove the starter and flywheel cover.

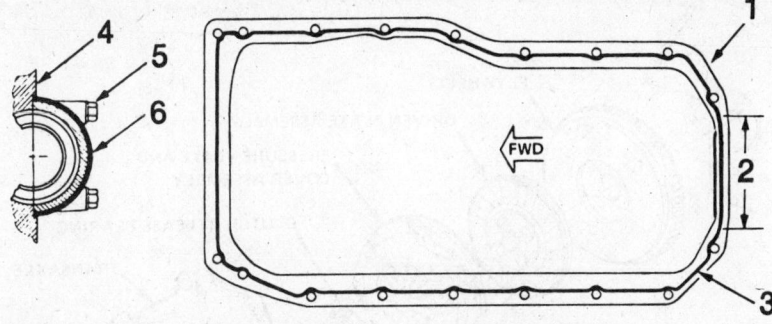

1. Oil pan
2. Apply a ⅜" thick bead of RTV sealer in area indicated
3. Apply a 3/16" wide by ⅛" thick bead of RTV sealer in area indicated
4. Engine block assy.
5. Rear bearing
6. Groove in main bearing cap must be filled flush to ⅛" above surface with RTV

A Oil pan sealer application—four cyl. engine

6. Remove the upper generator bracket.

7. Support the engine with tool J-28467 or equivalent.

8. Remove the lower generator bracket and engine support bracket.

9. Remove the oil pan retaining bolts and remove the oil pan.

10. Installation is the reverse of removal. Apply RTV sealer or equivalent as shown in the illustration. The two bolts in the timing gear cover should be installed last after the pan bolts are tight.

11. When installing the engine cradle torque the following as indicated:

Rear cradle bolts—76 ft. lbs (103 Nm).

Front cradle nut—67 ft. lbs. (90 Nm).

Engine mount assembly—42 ft. lbs. (57 Nm).

Rear mount assembly—18 ft. lbs.(24 Nm).

Front mount assembly—36 ft. lbs. (48 Nm).

Lower control arm at knuckle—33 ft. lbs. (45 Nm).

Lower control arm at cradle—69 ft. lbs. (93 Nm).

V6 Engine

1. Disconnect the negative battery cable.

2. Raise the vehicle and support it safely.

3. Drain the crankcase.

4. Remove the flywheel shield or clutch housing cover.

5. Remove the starter.

6. Remove the oil pan.

7. Before installation, clean all mating surfaces.

8. Place a ⅛ in. bead of RTV sealant on the oil pan sealing flange.

9. Install the oil pan and torque the 1 in. bolts to 6–9 ft. lbs. and the 1.5 inch bolts to 14–22 ft. lbs.

10. The remainder of the installation is the reverse of removal.

Oil Pump

REMOVAL & INSTALLATION

Four Cylinder

1. Remove the oil pan as described earlier.

2. Remove the two flange mounting bolts and the nut from the main bearing cap bolt.

3. Remove the pump and screen as an assembly.

4. Installation is the reverse of removal. Align the pump shaft with the drive shaft tang. Torque the pump retaining bolts to 20 ft. lbs.

V6 Engine

1. Remove the oil pan.

2. Remove the pump and driveshaft extension.

3. To install, engage the driveshaft extension in the cover end of the distributor drive gear.

4. Install the pump-to-rear bearing cap bolt and torque to 26–35 ft. lbs.

5. Install the oil pan and refill with oil.

Rear Main Oil Seal

REMOVAL & INSTALLATION

Four Cylinder

NOTE: This is a one piece seal and can be replaced without removal of the oil pan or crankshaft.

1. Remove the transaxle assembly.

2. Remove the flywheel.

3. If equipped with a manual transaxle, remove the pressure plate and disc.

4. Pry out the rear main seal.

5. Before installing, clean the block and crankshaft-to-seal mating surfaces.

6. Lubricate the outside of the seal

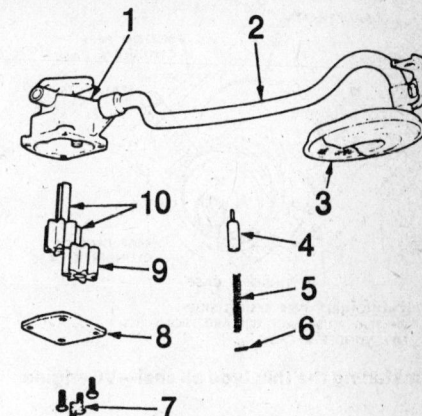

1. Pump body
2. Pickup tube
3. Pickup screen assy.
4. Pressure regulator valve
5. Pressure regulator spring
6. Spring retainer
7. Cover Screws
8. Cover
9. Idler gear
10. Drive gear and shaft

Oil pump - exploded view

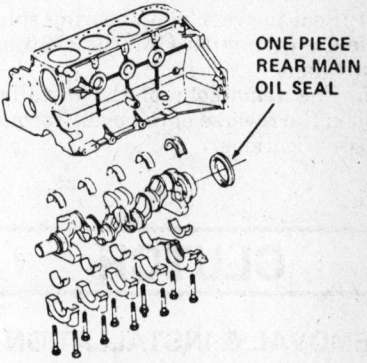

ONE PIECE REAR MAIN OIL SEAL

Crankshaft bearings and rear seal

for ease of installation and press into the block with fingers.

7. Install the flywheel and torque the bolts to 44 ft. lbs.

8. Install the transaxle assembly.

V6 Engine (Thin Seal)

1. Remove the engine and mount it on a suitable stand.

2. Remove the oil pan and oil pump assembly.

3. Remove the front cover, then lock the chain tensioner with a pin.

4. Rotate the crankshaft until the timing marks on the cam and crank sprockets align.

5. Remove the camshaft bolt, cam sprocket and timing chain.

6. Rotate the crankshaft to the horizontal position.

7. Remove the rod bearing nuts, caps and bearings.

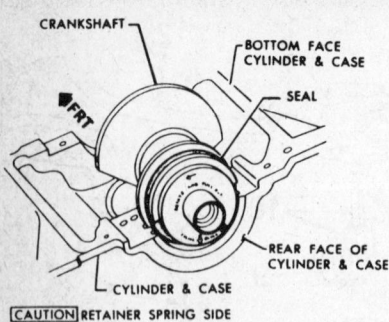

Installing the thin type oil seal—V6 engine

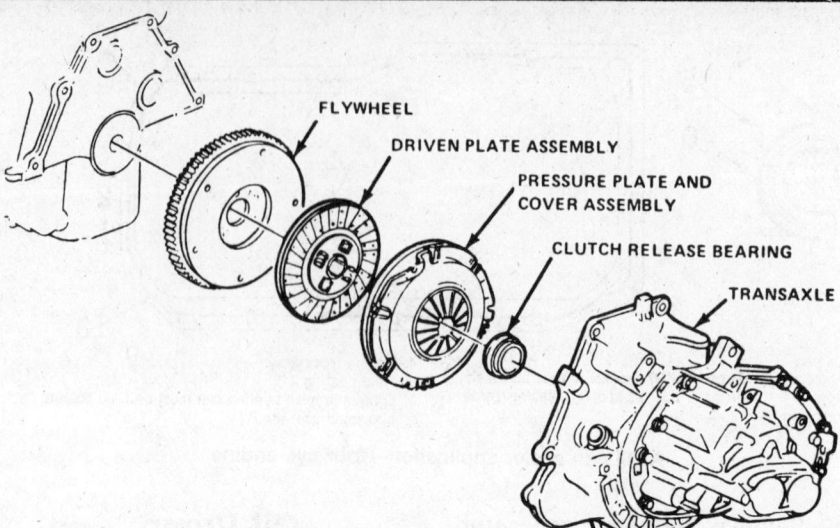

Exploded view of clutch assembly

8. Remove the crankshaft and the old oil seal.

9. Apply a light coat of GM 1052726 or equivalent to the outside of the seal.

10. Install the new seal and tool in the rear area of the crankshaft.

11. Install the crankshaft and tool in the engine.

12. Position the seal tool so that the arrow points toward the cylinder block and remove the tool.

13. Put a light coat of oil on the crankshaft journals.

14. Seal the rear main bearing split line surface with GM 1052726 or equivalent.

15. The remainder of the installation is the reverse of removal. Torque to specifications.

CLUTCH

REMOVAL & INSTALLATION

1. Remove the transaxle.

2. Mark the pressure plate assembly and the flywheel so that they can be assembled in the same position. They were balanced as an assembly at the factory.

3. Loosen the attaching bolts one turn at a time until spring tension is relieved.

4. Support the pressure plate and remove the bolts. Remove the pressure plate and clutch disc. Do not disassemble the pressure plate assembly; replace it if defective.

5. Inspect the flywheel, clutch disc, pressure plate, throwout bearing and the clutch fork and pivot shaft assembly for wear. Replace the parts as required. If the flywheel shows any signs of overheating or if it is badly grooved or scored, it should be replaced.

6. Clean the pressure plate and flywheel mating surfaces throughly. Position the clutch disc and pressure plate into the installed position and support with a dummy shaft or clutch aligning tool. The clutch plate is assembled with the damper springs offset toward the transaxle. One side of the factory-supplied clutch disc is stamped "Flywheel Side".

7. Install the pressure plate-to-flywheel bolts. Tighten them gradually in a crisscross pattern.

8. Lubricate the outside groove and the inside recess of the release bearing with high temperature grease. Wipe off any excess. Install the release bearing.

9. Install the transaxle.

MANUAL AND AUTOMATIC TRANSAXLE

REMOVAL & INSTALLATION

1. Remove the air cleaner assembly.

2. Disconnect the negative battery cable.

3. Disconnect the ground cable at the transaxle.

4. Disconnect the shift and select cable at the transaxle.

5. Remove the upper transaxle to engine bolts.

6. Install an engine support fixture tool J-28467 or equivalent.

7. Hoist the car and support it safely with jack stands.

8. Remove the rear wheels and tires.

9. Remove the axle shafts.

10. Remove the heat shield from the catalytic converter.

11. Disconnect the exhaust pipe at the exhaust manifold.

12. Remove the engine mount to cradle nuts.

13. Support the cradle with an adjustable stand.

14. Remove the rear cradle to body bolts.

15. Remove the forward cradle to body through bolts.

16. Lower the cradle and move out of the way.

17. Remove the starter and inspection cover shields and remove the starter.

18. Remove the flywheel to converter bolts.

19. Disconnect and plug cooler lines, if equipped with Automatic Transaxle.

20. Position a transmission stand under the transaxle.

21. On manual transaxles, remove the lower transaxle to engine bolts and remove the transaxle.

22. On automatic transaxles, remove the transaxle to support mounting bolts on the right side.

23. Install the starter and the inspection cover shields.

24. Hoist the cradle into position.

NOTE: Lower the cradle at the front and raise the car. Work cradle at rear into position on mounts then raise the front into position.

25. The remainder of the installation is the reverse of removal. Torque the retaining nuts to the following specifications:

Starter-to-engine—32 ft. lbs. (43 Nm).

Front cradle-to-body nuts—67 ft. lbs. (90 Nm).

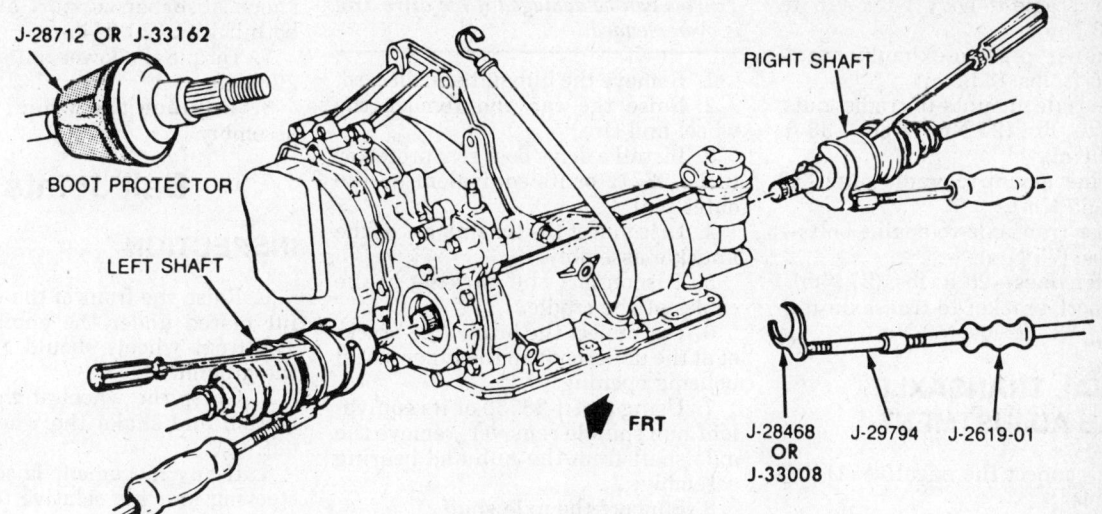

J-28712 OR J-33162

BOOT PROTECTOR

LEFT SHAFT

RIGHT SHAFT

FRT

J-28468
OR
J-33008

J-29794   J-2619-01

Removing the drive axle from the transaxle

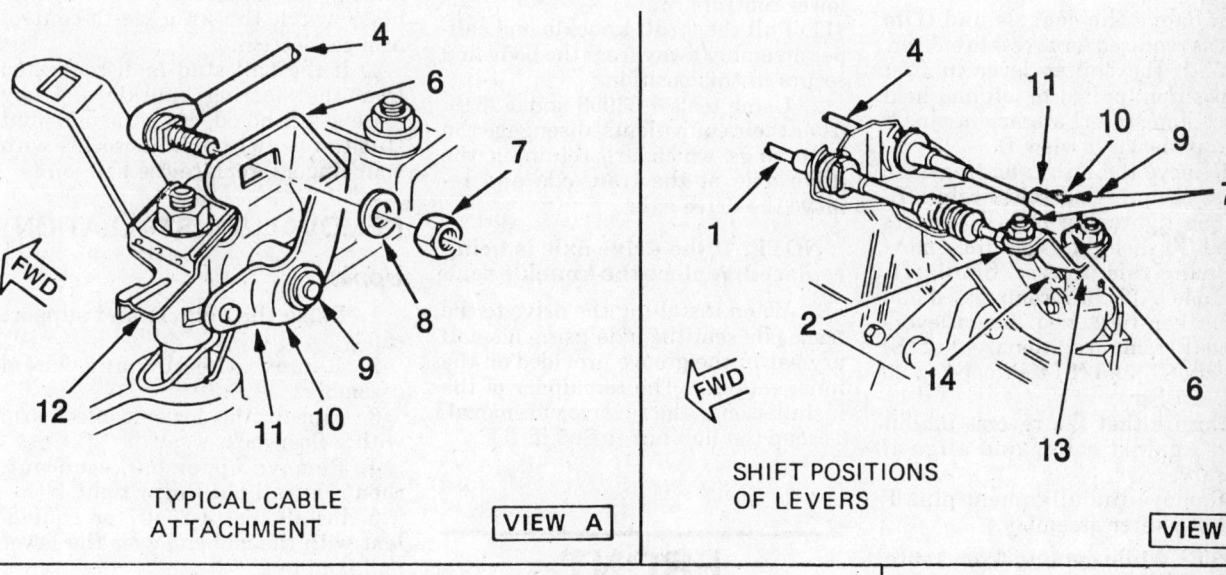

TYPICAL CABLE
ATTACHMENT

VIEW A

SHIFT POSITIONS
OF LEVERS

VIEW B

VIEW C

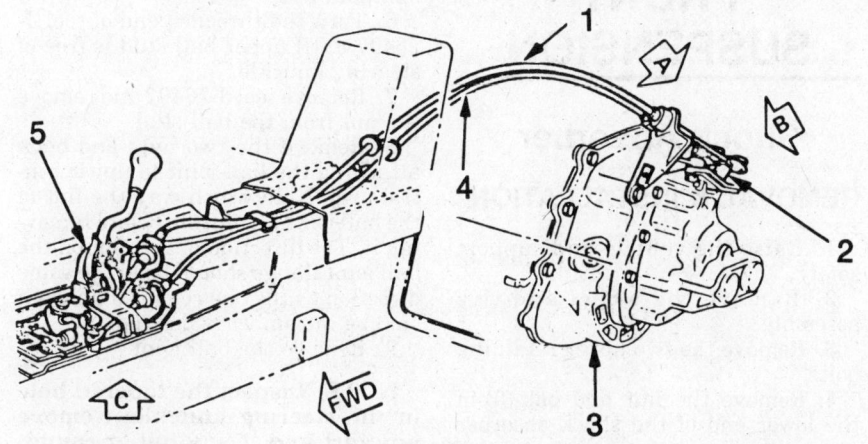

1. Cable A	5. Trans. control assy.	9. R	13. R/3rd/1st
2. Lever F	6. Lever D	10. 1st/2nd	14. 2nd/4th
3. Tansaxle assy.	7. Nut E	11. 4th/3rd	15. Alignment pin F
4. Cable B	8. Washer P	12. Retainer clip J	16. Alignment pin G

Manual transaxle cable adjustment

Rear cradle-to-body bolts—76 ft. lbs. (103 Nm).

Exhaust pipe-to-exhaust manifold—25 ft. lbs. (33 Nm).

Transaxle mounts-to-cradle nuts, rear—8 ft. lbs. (24 Nm); front—36 ft. lbs. (48 Nm).

Engine mount-to-cradle nuts—40 ft. lbs. (55 Nm).

Upper transaxle-to-engine bolts—55 ft. lbs. (75 Nm).

Cooler lines—20 ft. lbs. (27 Nm).

Support bracket-to-transaxle (automatic)—37 ft. lbs. (50 Nm).

MANUAL TRANSAXLE CABLE ADJUSTMENT

1. Disconnect the negative (1) battery cable.
2. Place the transaxle in 1st gear.
3. Loosen the shift cable attaching nuts (E) at the transaxle levers (D) and (F).
4. Remove the console and trim plates as required for access to shifter.
5. With the shifter lever in first gear position (pulled to left and held against stop), insert alignment pins F and G as shown in view D.
6. Remove the lash from transaxle by first compressing select cable (B) and then tightening nut (E). Levers (D) and (F) should be kept from moving during this process. Similarly, shift cable (A) is first compressed and nut (E) then tightened. Again levers (D) and (F) remain stationary. Nut (E) on levers (D) and (F) tightened to 20 ft. lbs. (27 Nm).
7. Ensure that the reverse inhibit cam is against roller and align if necessary.
8. Remove the alignment pins F and G at shifter assembly.

NOTE: While cycling from 1st to 2nd and 2nd to 1st, the select cable should not move. Difficulty in shifting the transaxle to reverse may be corrected by moving select lever (D) inboard toward the 1st-3rd-Reverse position during the shift cable (A) adjustment. For linkage adjustment and fluid and filter change for automatic transaxles, please refer to Automatic Transmission in the Unit Repair section.

DRIVE AXLE

REMOVAL & INSTALLATION

CAUTION
Use care when removing the drive axle.

Tri-pots can be damaged if the drive axle is overextended.

1. Remove the hub nut and discard.
2. Raise the car and remove the wheel and tire.
3. Install a drive boot seal protector tool J-28712 or its equivalent on the outer seal.
4. Disconnect the toe link rod at the knuckle assembly.
5. Disconnect the parking brake cables at the cradle.
6. Disconnect the brake line bracket at the underbody in the inner wheel housing opening.
7. Using tool J-28733 or its equivalent hub spindle remover, remove the axle shaft from the hub and bearing assembly.
8. Support the axle shaft.
9. Remove the clamp bolt from the lower control arm ball stud.
10. Separate the knuckle from the lower control arm.
11. Pull the strut, knuckle and caliper assembly away from the body and secure in this position.
12. Using tools J-33008 and J-2619-01 or their equivalents, disengage the snap rings which are retaining the drive axle at the transaxle and remove the drive axle.

NOTE: If the drive axle is being replaced, replace the knuckle seal.

13. When installing the drive to the transaxle seat the axle using a small pry bar in the groove provided on the inner retainer. The remainder of the installation is the reverse of removal. Torque the hub nut to 225 ft. lbs.

FRONT SUSPENSION

Shock Absorber

REMOVAL & INSTALLATION

1. Raise the vehicle and support safely.
2. Remove the wheel and tire assembly.
3. Remove the two upper retaining bolts.
4. Remove the nut and bolt from the lower end of the shock absorber and remove the shock absorber from the vehicle.
5. To install, place the lower portion of the shock into position and hand tighten the nut and bolt.
6. Extend the shock up into the shock absorber support and torque both bolts to 20 ft. lbs.
7. Torque the lower nut and bolt to 20 ft. lbs.
8. Replace the wheel and tire assembly.

Ball Joints

INSPECTION

1. Raise the front of the car with a lift placed under the engine cradle. The front wheels should be clear of the ground.
2. Grasp the wheel at the top and bottom and shake the wheel in and out.
3. If any movement is seen of the steering knuckle relative to the control arm, the ball joints are defective and must be replaced. Note that movement elsewhere may be due to loose wheel bearings or other troubles; watch the knuckle-to-control arm connection.
4. If the ball stud is disconnected from the steering knuckle and any looseness is noted, often the ball joint stud can be twisted in its socket with your fingers, replace the ball joints.

REMOVAL & INSTALLATION

Upper

1. Raise the vehicle and support safely.
2. Remove the tire and wheel assembly.
3. Support the lower control arm with a floor jack.
4. Remove upper ball stud nut, then reinstall nut finger tight.
5. Install tool J-26407 or equivalent with the cup end over the lower ball stud nut.
6. Turn the threaded end of tool J-26407 until upper ball stud is free of steering knuckle.
7. Remove tool J-26407 and remove the nut from the ball stud.
8. Remove the two nuts and bolts attaching the ball joint to upper control arm. Note which way the flat of the ball joint is pointing before removing it. The direction of this flat on the ball joint flange should be in the same direction as the one removed unless a change in camber is desired.
9. Remove the ball joint.

NOTE: Inspect the tapered hole in the steering knuckle. Remove any dirt and if any out-or-roundness, deformation, or damage is noted, the knuckle must be replaced.

10. Install the bolts and nuts attaching the ball joint to the upper control

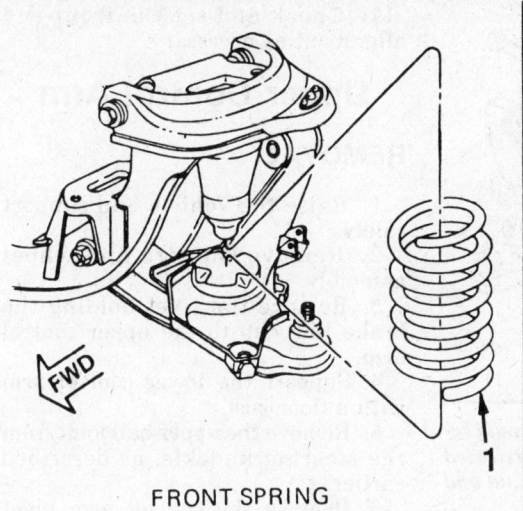

FRONT SPRING

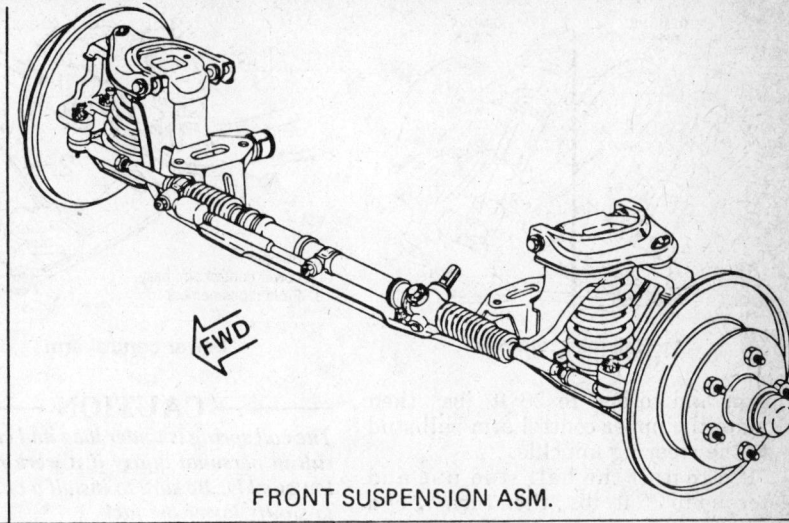

FRONT SUSPENSION ASM.

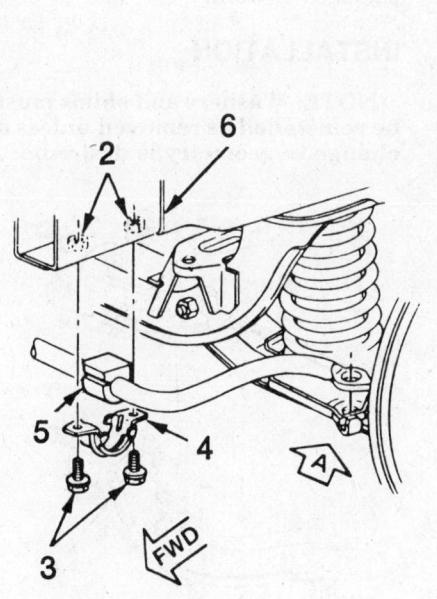

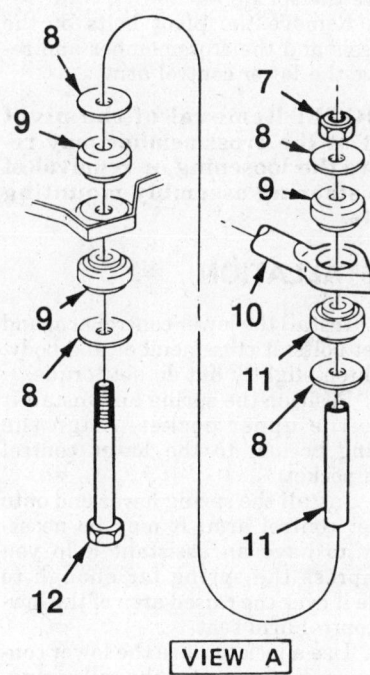

VIEW A

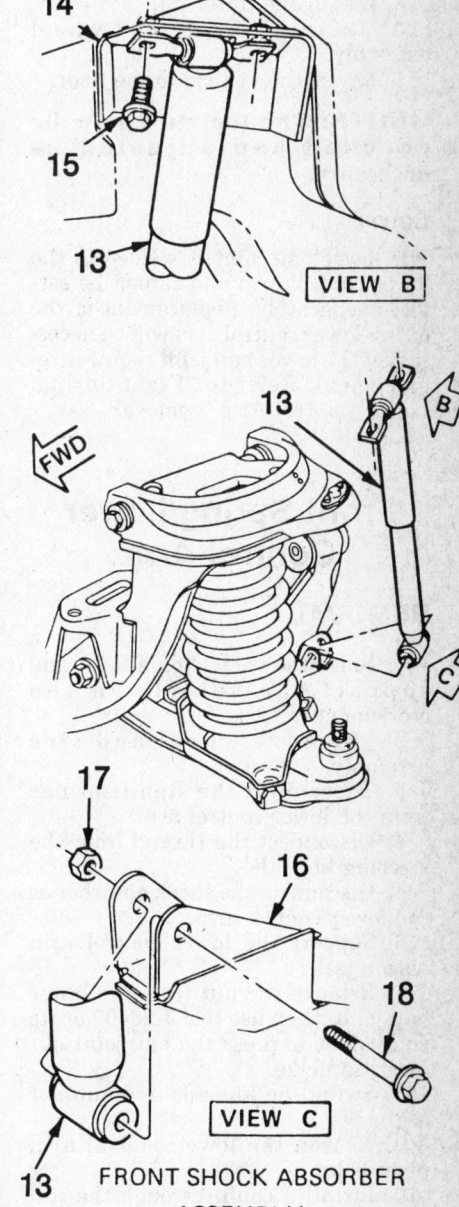

VIEW B

VIEW C

FRONT SHOCK ABSORBER
ASSEMBLY

1. Spring
2. Weld nuts
3. Bolt/screw (15 ft. lbs.)
4. Clamp
5. Bushing
6. Front rail
7. Tighten until nut bottoms on end of bolt thread
8. Washer
9. Grommet
10. Shaft front stabilizer
11. Spacer
12. Torque from this end 13 ft. lbs.
13. Shock absorber
14. Shock absorber support
15. Bolt
16. Control arm
17. Nut
18. Bolt

Front suspension

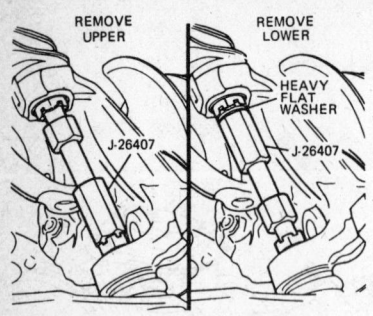

Ball joint removal

arm and torque to 28 ft. lbs., then mate the upper control arm ball stud to the steering knuckle.

11. Install the ball stud nut and torque to 35 ft. lbs., then turn $\frac{1}{6}$ of a turn to align cotter pin.

12. Install the cotter pin.

13. Install the tire and wheel assembly.

14. Lower the vehicle to the floor.

NOTE: The toe must now be checked and adjusted as necessary.

Lower

The lower ball joint is welded to the lower control arm and cannot be serviced separately. Replacement of the entire lower control arm will be necessary if the lower ball joint requires replacement. Refer to "Front Spring/Lower Control Arm" removal.

Front Spring/Lower Control Arm

REMOVAL

1. Raise the vehicle on a hoist and support the vehicle on the crossmember.

2. Remove wheel and tire assembly.

3. Disconnect the stabilizer bar from the lower control arm.

4. Disconnect the tie rod from the steering knuckle.

5. Disconnect the shock absorber at the lower control arm.

6. Support the lower control arm with a jack.

7. Remove the nut from the lower ball joint, then use tool J-26407 or its equivalent to press the ball joint out of the knuckle.

8. Swing the knuckle and hub out of the way.

9. Loosen the lower control arm pivot bolts.

10. Install a chain through the coil spring as a safety precaution.

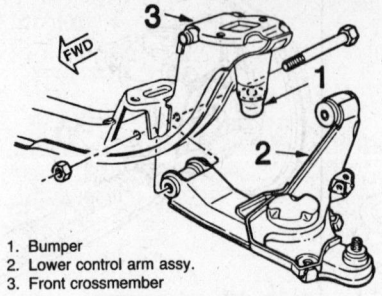

1. Bumper
2. Lower control arm assy.
3. Front crossmember

Lower control arm

——— CAUTION ———

The coil spring is under load and could result in personal injury if it were released too quickly. Be sure to install a chain and to slowly lower the jack.

11. Slowly lower the jack and remove the spring.

12. Remove the pivot bolts at the chassis and the crossmember and remove the lower control arm.

NOTE: Removal of the pivot bolt at the crossmember may require the loosening or removal of the steering assembly mounting bolts.

INSTALLATION

1. Install the lower control arm and pivot bolts at crossmember and body. Tighten slightly but do not torque.

2. Position the spring and install it into the upper pocket. Align the spring bottom to the lower control arm pocket.

3. Install the spring lower end onto lower control arm. It may be necessary to have an assistant help you compress the spring far enough to slide it over the raised area of the lower control arm seat.

4. Use a jack to raise the lower control arm and compress the coil spring.

5. Install the ball joint through the lower control arm and into the steering knuckle. Install nut to ball joint stud and torque to 55 ft. lbs. Install a new cotter pin.

6. Connect the stabilizer bar and torque the bolt to 16 ft. lbs.

7. Connect the tie rod and torque to 29 ft. lbs.

8. Install the shock absorber to the lower control arm and torque the bolt to 35 ft. lbs.

9. If the bolts were removed or loosened at the steering assembly replace with new bolts and torque to 21 ft. lbs.

10. With the suspension system in its normal standing height, torque the lower control arm to body bolt at 62 ft. lbs. and the lower control arm to crossmember nut at 52 ft. lbs.

11. Check and set the front end alignment as necessary.

Upper Control Arm

REMOVAL

1. Raise the vehicle and support safely.

2. Remove the tire and wheel assembly.

3. Remove the rivet holding the brake line clip to the upper control arm.

4. Support the lower control arm with a floor jack.

5. Remove the upper ball joint from the steering knuckle, as described earlier.

6. Remove the control arm pivot bolt and remove the control arm from vehicle.

7. Transfer the ball joint if not damaged or worn.

INSTALLATION

NOTE: Washers and shims must be reinstalled as removed unless a change in geometry is desired.

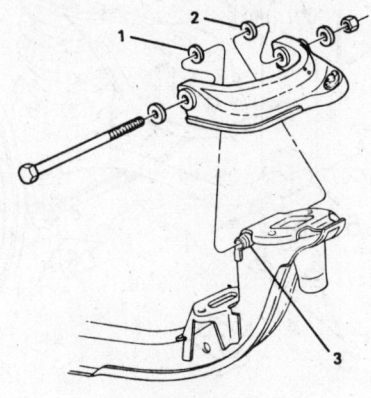

1. Front washer
2. Rear washer
3. Support bracket

Upper control arm

1. Install the upper control arm and pivot bolt to the vehicle. The inner pivot bolt must be installed with the bolt head toward the front.

2. Install the pivot.

3. Position the control arm in a horizontal plane and torque the nut to 66 ft. lbs.

NOTE: The bolt may turn, when torqued to minimum, if nut is not backed up with a wrench. This does not mean the joint is loose.

4. Install the ball joint to the upper control arm and to steering knuckle, as described earlier. Install the nut,

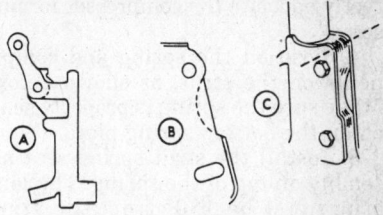

torque to 35 ft. lbs. Install a new cotter pin.

5. Install the wheel and tire.
6. Lower the vehicle.

Steering Knuckle

REMOVAL

1. Raise vehicle on a hoist and support the lower control arm with a jackstand.

——— CAUTION ———

This keeps the coil spring compressed. Use care to support adequately or personal injury could result.

2. Remove tire and wheel assembly.
3. Remove the disc brake caliper. Secure the caliper to the suspension using wire. Do not allow the caliper to hang by the brake hose. Insert a piece of wood between the shoes to hold the piston in the caliper bore. The block of wood should be about the same thickness as the brake disc.
4. Remove the hub and disc.
5. Remove the splash shield.
6. Remove both ball stud nuts (see Ball Joint Removal).
7. Remove the tie rod end from the steering knuckle.
8. Using tool J-26407 or its equivalent, press the upper ball stud from the steering knuckle.
9. Reverse tool J-26407 to the other ball stud and press the lower ball stud from the steering knuckle.
10. Remove the ball stud nuts and remove the steering knuckle.

INSTALLATION

1. Place the steering knuckle in position and insert the upper and lower ball studs into knuckle bosses.
2. Install the ball stud nuts and tighten to specifications. Torque the lower to 55 ft. lbs.; torque the upper to 35 ft. lbs. Install new cotter pins.
3. Install the splash shield to the steering knuckle. Torque to 7 ft. lbs.
4. Install tie rod end to the steering knuckle. Torque to 29 ft. lbs. and install the cotter pin.
5. Repack the wheel bearings. Then install the hub and disc, bearings and nut. Torque to specifications.
6. Install the brake caliper.
7. Install the tire wheel assembly.
8. Remove the jackstand and lower the vehicle to the floor.

FRONT WHEEL BEARING ADJUSTMENT

1. Raise the vehicle and support with jack stands.

2. Remove the wheel.
3. Remove the dust cap from the hub.
4. Remove the cotter pin from spindle and spindle nut.
5. Tighten the spindle nut to 12 ft. lbs. while turning the wheel assembly forward by hand to fully seat the bearings. This will remove any grease or burrs which could cause excessive wheel bearing play later.
6. Back off the nut to the "just loose" position.
7. Hand tighten the spindle nut. Loosen the spindle nut until either hole in the spindle lines up with a slot in the nut (not more than $\frac{1}{2}$ flat).
8. Install new cotter pin. Bend the ends of the cotter pin against nut, cut off extra length to ensure ends will not interfere with the dust cap.
9. Measure the looseness in the hub assembly. There will be from 0.001–0.005 in. end play when properly adjusted.
10. Install the dust cap on the hub.
11. Replace the wheel cover or hub cap.
12. Lower the vehicle to the floor.
13. Perform the same operation for each front wheel.

REAR SUSPENSION

MacPherson Strut

REMOVAL & INSTALLATION

1. Remove the engine compartment cover.
2. Remove the three upper strut nuts and washers.
3. Loosen the wheel lug nuts.
4. Raise the vehicle and support it on jackstands under the frame members. Support the rear control arm with a floor jack.
5. Remove the wheel and tire.
6. Remove the brake line clip.
7. Scribe the strut and knuckle.
   a. Using a sharp tool, scribe the knuckle along the lower outboard strut radius, as in view A.
   b. Scribe the strut flange on the inboard side, along the curve of the knuckle, as in view B.
   c. Make a chisel mark across the strut/knuckle interface, as shown in view C.
8. Remove the two strut mounting nuts and bolts and remove the strut assembly and spacer plate.
9. Installation is the reverse of re-

Scribing strut and knuckle

moval. Align the scribe marks on the strut and knuckle and replace the bolts in the same order in which they were removed. Tighten the strut mounting nuts to 140 ft. lbs. and the upper strut nuts to 18 ft. lbs.

Coil Spring

REMOVAL

NOTE: Special tool J-26584 or its equivalent must be used to disassemble and assemble strut damper. Care must be used not to damage the special coating on the coil springs, or damage could occur to the coils.

1. Clamp tool J-26584 Strut Compressor in vise.
2. Place the strut assembly in bottom adapter of compressor and install tool J-26584-89 (make sure adapter captures the strut and locating pins are engaged).
3. Rotate the strut assembly to align the top mounting assembly lip with strut compressor support notch.
4. Insert tool J-26584-430 top adapter on the top spring seal. Position top adapters so that the long stud is at high location to strut flange.
5. Using a ratchet with 1 in. socket, turn the compressor forcing screw clockwise until the top support flange contacts the tool J-26584-430 top adapter. Continue turning the screw conpressing the strut spring.
6. Place tool J-26584-430 top adapter over the spring seat assembly.
7. Turn the strut compressor forcing the screw counterclockwise until the strut spring tension is relieved. Remove the top adapters, bottom adapter, then remove the strut.

INSTALLATION

1. Clamp the strut compressor body tool J-26584 in vise.
2. Place the strut assembly in bottom adapter compressor and install tool J-26584-89 (make sure adapter captures strut and locating pins are engaged).
3. Rotate the strut assembly until the mounting flange is facing out, di-

rectly opposite the compressor forcing screw.

4. Position the spring and components on the strut, as shown below. Make sure the spring is properly seated on the bottom spring plate.

5. Install the strut spring seat assembly on top of the spring. The long stud must be 180° from the strut mounting flange.

6. Place tool J-26584-403 top adapter over the spring seat assembly.

7. Turn the compressor forcing screw until the compressor top support just contacts the top adapters (do not compress spring at this time).

8. Install tool J-26584-27 Strut Alignment Rod through the top spring seat and thread the rod onto the damper shaft, hand tight.

9. Compress the spring by turning the screw clockwise until enough of the damper shaft is exposed to where the nut can be threaded securely, and thread nut on the damper shaft.

NOTE: Do not compress spring until it bottoms. Be sure that the damper shaft comes through the CENTER of the spring seat opening, or damage could occur.

10. Remove the alignment rod and position the strut mount over the damper shaft and spring seat studs. Install the washer and nut.

11. Turn the forcing screw counterclockwise to back off support and remove the strut assembly from the compressor.

Lower Control Arm

REMOVAL & INSTALLATION

1. Raise the car and support it safely.
2. Remove the ball joint clamping bolt.
3. Separate the knuckle from the ball joint.
4. Remove the lower control arm pivot bolts at the frame and the control arm.
5. Installation is the reverse of removal.

NOTE: The toe-in and camber settings should be checked and adjusted as required.

Lower Ball Joint

REMOVAL & INSTALLATION

1. Raise the car and support it safely, then remove the wheel.
2. Remove the clamp bolt from the lower control arm ball stud.

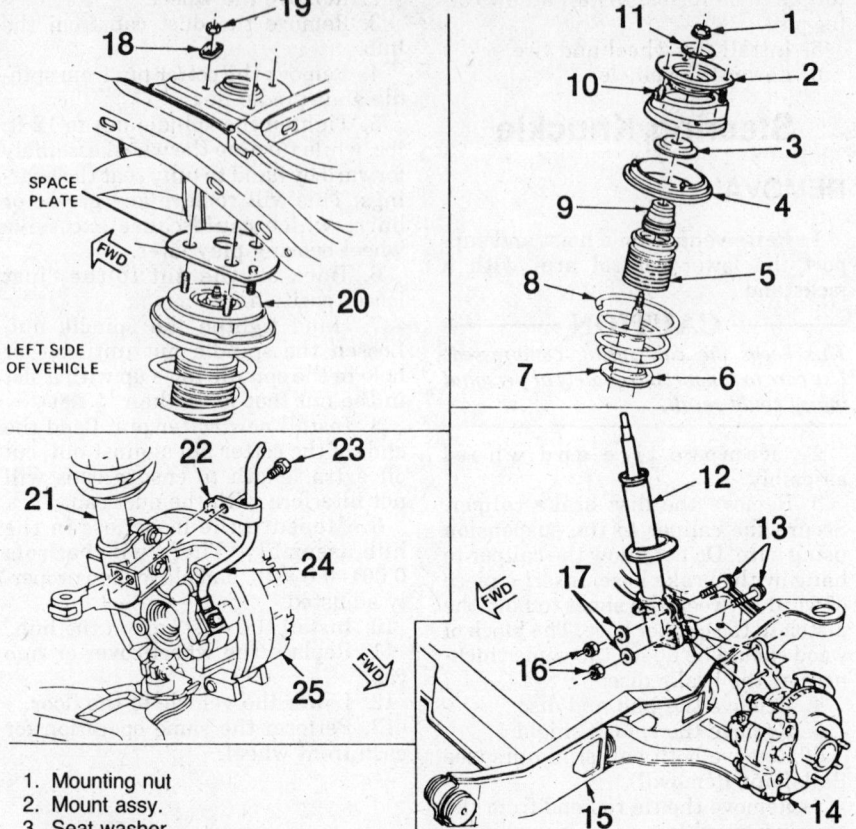

1. Mounting nut
2. Mount assy.
3. Seat washer
4. Upper spring insulator
5. Shield
6. Lower spring insulator
7. Lower spring seat
8. Spring
9. Bumper
10. Seat assy.
11. Upper mount washer
12. Strut assy.
13. Strut mounting bolts
14. Knuckle and hub assy.
15. Cradle assy.
16. Strut mounting nuts
17. Strut lower washers
18. Strut upper washers
19. Strut upper nuts
20. Rear strut mount assy.
21. Strut assy.
22. Brake line clip
23. Brake line clip bolt
24. Rear brake hose
25. Caliper assy.

Exploded view of strut assembly

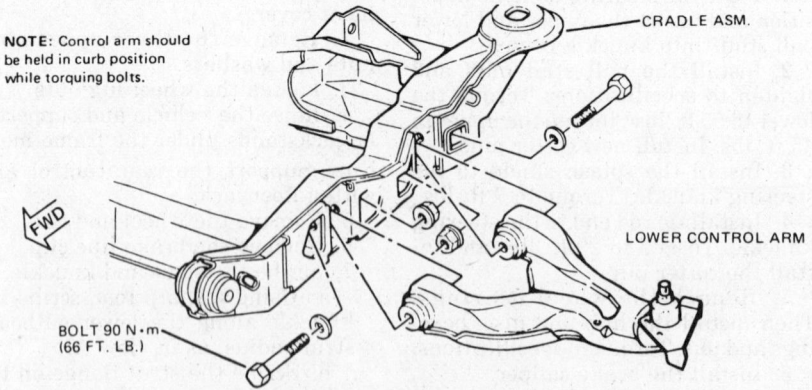

NOTE: Control arm should be held in curb position while torquing bolts.

CRADLE ASM.

LOWER CONTROL ARM

BOLT 90 N·m
(66 FT. LB.)

Lower control arm removal

3. Disconnect the ball joint from the knuckle.

NOTE: It may be necessary to tap the ball stud with a mallet.

4. Using an ⅛ in. drill, drill the rivets approximately ¼ in. deep in the center of the rivet.

5. Use a ½ in. drill bit and drill just deep enough to remove the rivet head.

6. Remove the rivets using a hammer and a punch.

7. The ball joint is replaced using nuts and bolts. Torque to 13 ft. lbs. Check the toe-in setting and adjust as necessary.

BRAKES

To service the brake shoes, drums and wheel cylinders or brake pads and calipers, please refer to "Brakes" in the Unit Repair Section.

Master Cylinder

REMOVAL & INSTALLATION

1. Disconnect the hydraulic lines at the master cylinder.

2. Place a number of cloths or a container under the master cylinder to catch the brake fluid. Disconnect the brake tubes from the master cylinder; use a flare nut wrench if one is available. Tape over the open ends of the tubes.

3. Remove the two nuts attaching the master cylinder to the booster or firewall, then remove the master cylinder.

4. To install, position the master cylinder and install the retaining bolts. Torque to 22–30 ft. lbs.

5. Remove the tape from the lines and connect to the master cylinder. Torque to 10–15 ft. lbs. Connect the electrical lead.

6. Bleed the brakes.

Combination valve

REMOVAL & INSTALLATION

NOTE: The combination valve is not repairable and must be replaced if found defective.

1. Disconnect and plug the hydraulic lines at the combination valve.

2. Disconnect the warning switch wiring harness from the valve switch terminal and remove the combination valve.

3. Installation is the reverse of removal.

4. Bleed the entire brake system.

——— CAUTION ———

Do not move the car until a firm brake pedal is obtained.

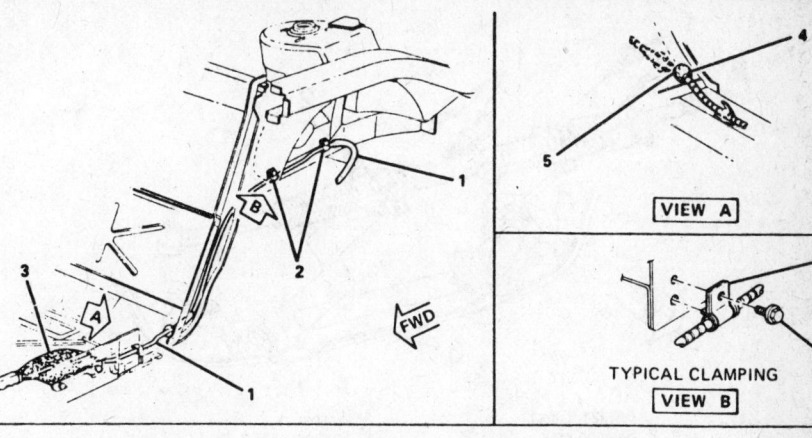

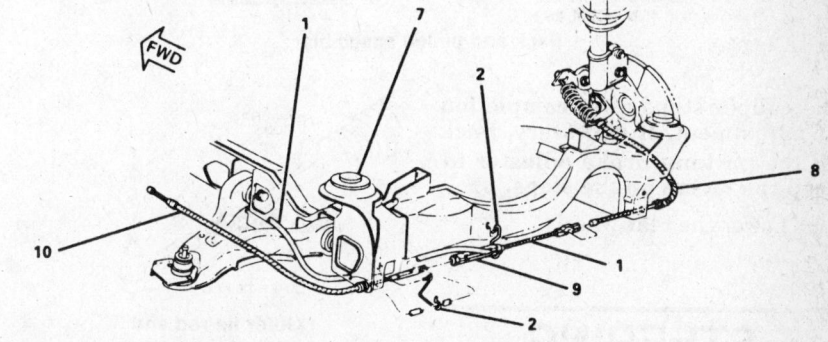

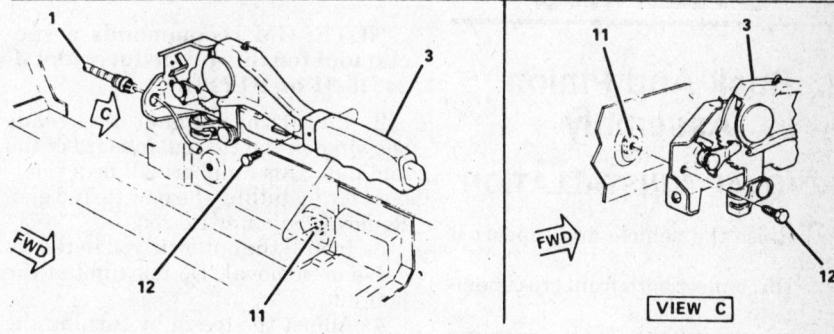

1. Front cable assy.
2. Clip
3. Lever assy.
4. Grommet
5. Hose in floor pan
6. Bolt
7. Frame assy.
8. R.H. cable assy.
9. Equalizer
10. L.H. cable assy.
11. Weld nut
12. Bolt/Screw

Parking brake assembly

Parking Brake Adjustment

Adjustment of parking brake cable is necessary whenever the rear brake cables have been disconnected. Need for the parking brake adjustment is indicated if the hydraulic brake system operates with good reserve but parking brake hand level travel is more than 9 ratchet clicks.

1. Place the parking brake hand lever in the unapplied position.

2. Raise the rear wheels off floor and support safely.

3. Apply lubricant to the groove in the equalizer nut.

4. Hold the brake cable stud from turning and tighten the equalizer nut until cable slack is removed.

5. Make sure the caliper levers are against the stops on the caliper housing after tightening the equalizer nut.

6. If the levers are off the stops, loosen the cable until the levers do return to the stops.

7. Operate the parking brake lever several times to check adjustment. Properly adjusted parking brake shoes and properly adjusted brake cable will result in a parking brake handle movement of 5–8 notches when a force is applied perpendicularly at the handle grip mid-point.

NOTE: The levers must be on

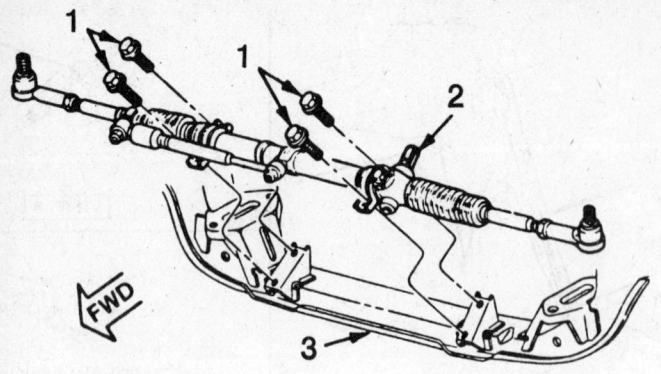

1. Bolt (21 ft. lbs.)
2. Steering assy.
3. Cross member
4. Nut (32 ft. lbs.)
5. Washer
6. Stud assy (36 ft. lbs.)
7. Steering link damper

Rack and pinion assembly

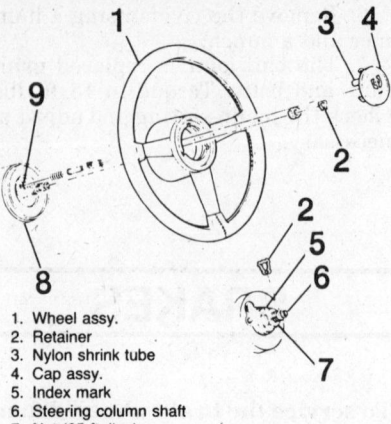

1. Wheel assy.
2. Retainer
3. Nylon shrink tube
4. Cap assy.
5. Index mark
6. Steering column shaft
7. Nut (35 ft. lbs.)
8. Steering column
9. Canceling cam tower centered in slot of lock plate

Steering wheel removal

the caliper stops after completion of adjustment. If necessary, back off the parking brake adjuster to keep the levers on the stops.

8. Lower the rear wheels.

STEERING

Rack And Pinion Assembly

REMOVAL & INSTALLATION

1. Raise the vehicle and support it safely.
2. Disconnect both front crossmember braces.
3. Disconnect the flexible coupling pinch bolt to the shaft.
4. Remove the outer tie rod cotter pins and nuts on both sides.
5. Disconnect the tie rods from the steering knuckle.
6. Remove the four bolts retaining the steering assembly to the crossmember and remove the steering assembly.
7. Installation is the reverse of removal. Tighten the flexible coupling bolt to 46 ft. lbs., the four new steering assembly bolts to 21 ft. lbs., the four crossmember brace bolts to 20 ft. lbs. and the tie rod nut at each knuckle to 29 ft. lbs., then 1/6 turn to align the cotter pin.

Outer Tie-Rod

REMOVAL & INSTALLATION

1. Loosen the jam nut and remove the tie-rod from the steering knuckle.

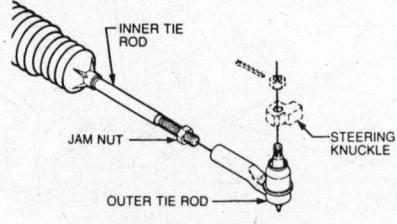

Outer tie rod end

NOTE: GM recommends a special tool for this procedure; tool J-24319-01 or BT7101.

2. Count the number of threads showing on the tie rod, inboard of the jam nut. This number will be a reference for installing the new tie rod end. Remove the outer tie rod.
3. Install the outer tie-rod in the reverse or removal. Do not tighten the jam nut.
4. Adjust the toe-in by turning the inner tie-rod the required number of turns.
5. Make sure the boot is not twisted then torque the jam nut to 50 ft. lbs.

Steering Wheel

REMOVAL & INSTALLATION

1. Pry off the center cap, then remove the retainer clip and nut.
2. Remove the wheel using a steering wheel puller.
3. When installing align the index mark on the steering wheel with index mark on the steering shaft. Torque the retaining nut to 35 ft. lbs.

— **CAUTION** —
The cancelling cam tower must be centered in the slot of the lock plate cover before assembling the wheel.

Turn Signal Switch

REMOVAL & INSTALLATION

1. Remove the steering wheel and the trim cover.
2. Pry the cover from the steering column.
3. Position a U-shaped lockplate compressing tool on the end of the steering shaft and compress the lock plate by turning the shaft nut clockwise. Pry the wire snapring out of the shaft groove.
4. Remove the tool and lift the lockplate off the shaft.
5. Slip the cancelling cam, upper bearing preload spring and thrust washer off the shaft.
6. Remove the turn signal lever, the hazard flasher button retaining screw, the button, the spring and the knob.
7. Pull the switch connector out of the mast jacket and tape the upper part to facilitate switch removal. Attach a long piece of wire to the turn signal switch connector. When installing the turn signal switch, feed this wire through the column first, then use this wire to pull the switch connector into position. On tilt wheels, place the turn signal and shifter housing in low position, then remove the harness cover.
8. Remove the three switch mounting screws. Remove the switch by pulling it straight up while guiding the wiring harness cover through the column.
9. Install the replacement switch by working the connector and cover down through the housing, then under the bracket. On tilt models, the connector is worked down through the housing and under the bracket, then the cover is installed on the harness.
10. Install the switch mounting

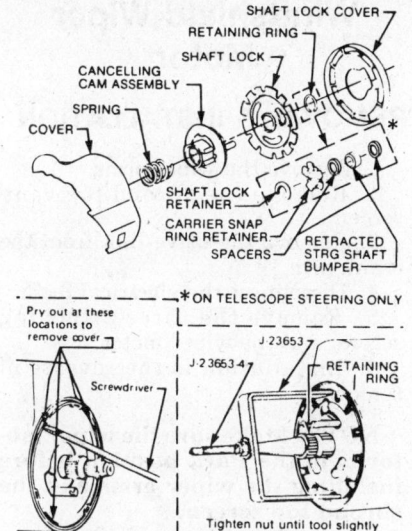

SHAFT LOCK COVER
RETAINING RING
SHAFT LOCK
CANCELLING CAM ASSEMBLY
SPRING
COVER
SHAFT LOCK RETAINER
CARRIER SNAP RING RETAINER
SPACERS
RETRACTED STRG SHAFT BUMPER
*ON TELESCOPE STEERING ONLY

Pry out at these locations to remove cover
Screwdriver
J-23653
J-23653-4
RETAINING RING
Tighten nut until tool slightly depresses shaft lock

These parts must be removed to remove the turn signal switch

screws and the connector on the mast jacket bracket, then the column-to-dash trim plate.

11. Install the flasher knob and the turn signal lever.

12. With the turn signal lever Neutral and the flasher knob Out, slide the thrust washer, the upper bearing preload spring and the cancelling cam onto the shaft.

13. Position the lock plate on the shaft and press it down until a new snapring can be inserted in the shaft groove. Always use a new snapring when assembling.

14. Install the cover and the steering wheel.

Ignition Switch

REMOVAL & INSTALLATION

The switch is located inside the channel section of the brake pedal support and is completely inaccessible without first lowering the steering column. The switch is actuated by a rod and rack assembly. A gear on the end of the lock cylinder engages the toothed upper end of the rod.

1. Lower the steering column; be sure to properly support it.

2. Put the switch in the OFF-UN-LOCKED position when it is in the next to the uppermost detent.

3. Remove the two switch screws and the switch assembly.

4. Before installing, place the new switch in the OFF-UNLOCKED position (2nd detent from the top).

5. Install the activating rod into the switch and assemble the switch on the column. Tighten the mounting screws. Use only the specified screws, since overlength screws could impair the collapsibility of the column.

6. If the dimmer switch was re-moved, install the switch and depress it enough to insert a $^3/_{32}$ inch drill. Force the switch up to remove the lash, then tighten the screw.

7. Reinstall the steering column.

Ignition Lock Cylinder

REMOVAL & INSTALLATION

1. Remove the steering wheel.

2. Turn the lock to the Run position.

3. Remove the lock plate, turn signal switch or combination switch and the key warning buzzer switch. The warning buzzer switch can be fished out with a bent paper clip.

4. Remove the lock cylinder retaining screw and lock cylinder.

— CAUTION —

If the screw is dropped on removal, it could fall into the column, requiring complete disassembly to retrieve the screw.

5. Rotate the cylinder clockwise to align the cylinder key with the keyway in the housing.

6. Push the lock all the way in.

7. Install the screw. Tighten to 15 inch lbs.

8. The rest of installation is the reverse of removal. Turn the lock to RUN to install the key warning buzzer switch, which is simply pushed down into place.

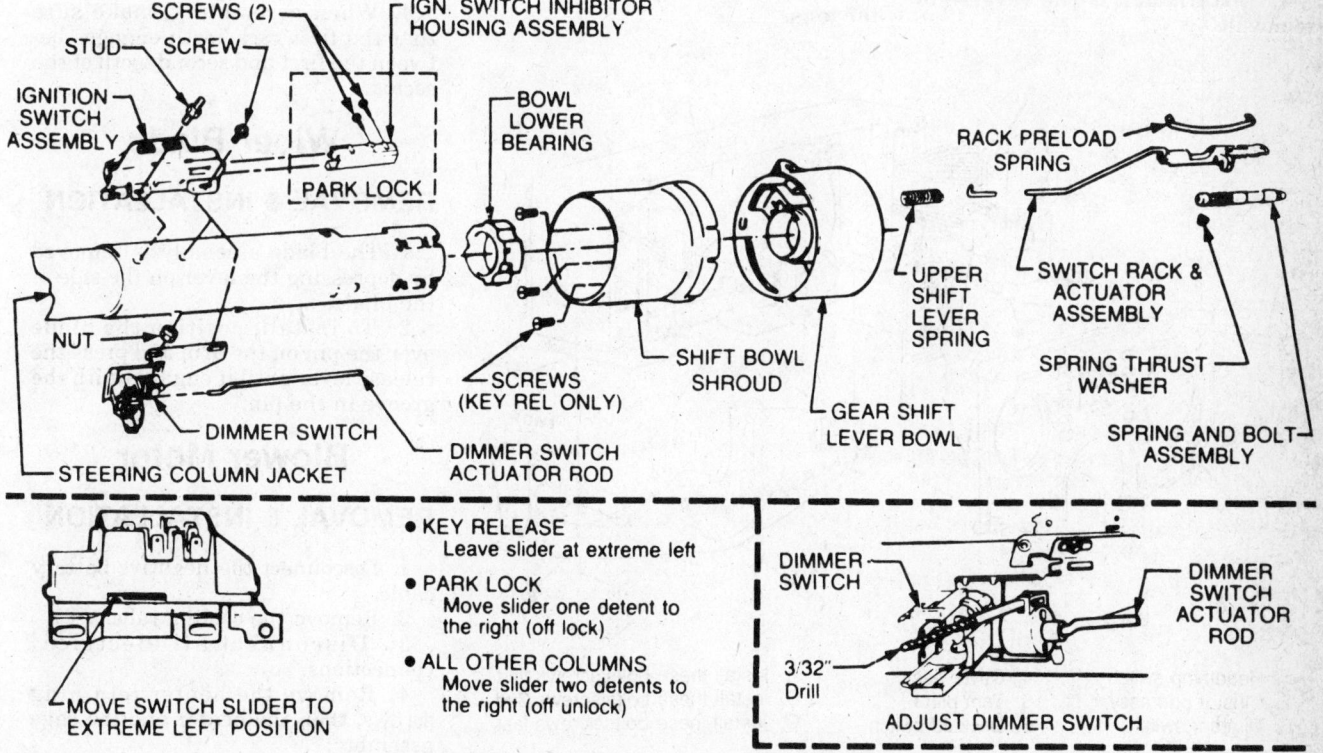

SCREWS (2)
STUD
SCREW
IGN. SWITCH INHIBITOR HOUSING ASSEMBLY
IGNITION SWITCH ASSEMBLY
PARK LOCK
BOWL LOWER BEARING
RACK PRELOAD SPRING
NUT
UPPER SHIFT LEVER SPRING
SWITCH RACK & ACTUATOR ASSEMBLY
SPRING THRUST WASHER
DIMMER SWITCH
SCREWS (KEY REL ONLY)
SHIFT BOWL SHROUD
GEAR SHIFT LEVER BOWL
SPRING AND BOLT ASSEMBLY
STEERING COLUMN JACKET
DIMMER SWITCH ACTUATOR ROD

• KEY RELEASE
Leave slider at extreme left
• PARK LOCK
Move slider one detent to the right (off lock)
• ALL OTHER COLUMNS
Move slider two detents to the right (off unlock)

MOVE SWITCH SLIDER TO EXTREME LEFT POSITION

DIMMER SWITCH
3/32" Drill
DIMMER SWITCH ACTUATOR ROD
ADJUST DIMMER SWITCH

Ignition switch and dimmer switch removal and installation

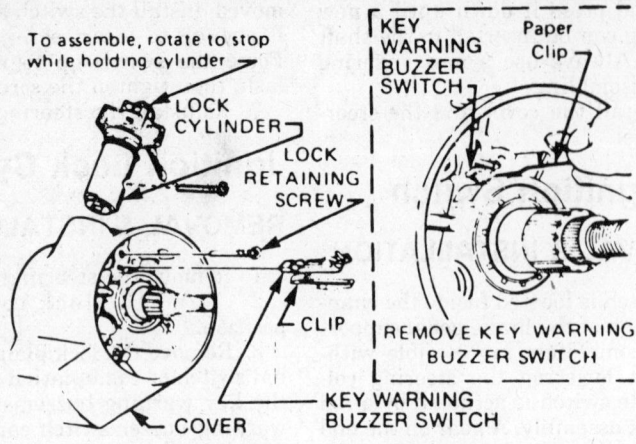

Ignition lock cylinder removal

CHASSIS ELECTRICAL

Headlight Switch

REMOVAL & INSTALLATION

1. Disconnect the negative battery cable.
2. Remove the headlight/dimmer switch trim plate screws.
3. Disconnect the electrical connector and remove the switch assembly.
4. Installation is the reverse of removal.

Instrument Cluster

REMOVAL & INSTALLATION

1. Disconnect the negative battery cable.
2. Remove the rear cluster cover.
3. Remove the front trim plate.
4. Remove the steering column cover.
5. Remove the cluster attaching screws, disconnect the wiring harness and remove the cluster assembly.

NOTE: The speedometer, the tach and the gauges may be serviced by removing the front cluster lens.

Windshield Wiper Motor

REMOVAL & INSTALLATION

1. Remove the wiper arms.
2. Remove the shroud top vent screen.
3. Remove the drive link from the crank arm.
4. Disconnect the electrical leads.
5. Remove the three attaching screws and the wiper motor.
6. Installation is the reverse of removal.

NOTE: Make sure the wiper motor is in the Park position before installing the wiper arms and the shroud top screen.

Windshield Wiper Switch (Pivot Switch)

REMOVAL & INSTALLATION

1. Remove the steering wheel as outlined earlier in the Steering section.
2. Remove the ignition lock cylinder and the ignition switch and dimmer switch as outlined earlier in the Steering section.
3. Remove the related parts shown in the illustration and assemble in the reverse order.
4. When assembling, make sure that the first rack tooth engages between the first and second tooth of the sector.

Wiper Blade

REMOVAL & INSTALLATION

1. The blade assembly is removed by depressing the lever on the side of the blade.
2. To install, position the blade over the pin on the arm and press the release lever until it engages with the groove in the pin.

Blower Motor

REMOVAL & INSTALLATION

1. Disconnect the negative battery cable.
2. Remove the cooling tube.
3. Disconnect all electrical connections.
4. Remove the heater retaining screws, then the heater and the cage assembly.
5. Installation is the reverse of removal.

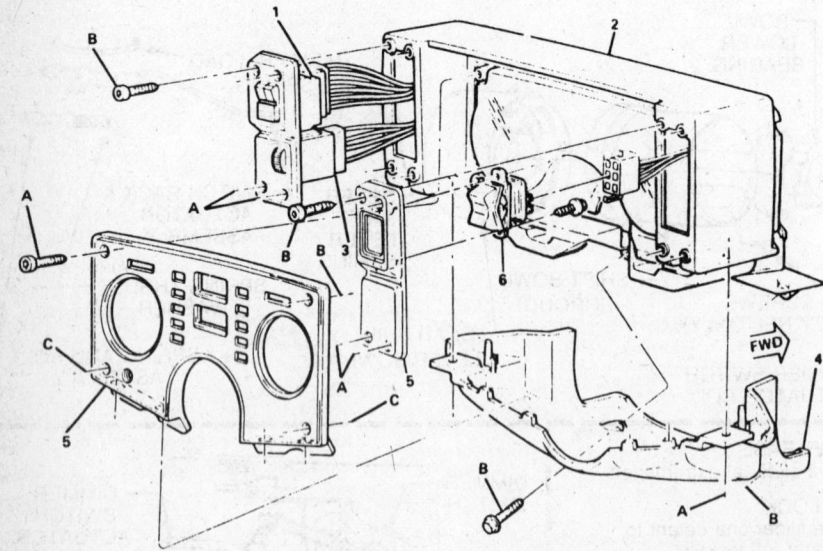

1. Headlamp switch
2. Cluster pad assy.
3. Dimmer switch
4. Cover assy.
5. Trim plate
6. Deck lid switch

A. Install these bolts/screws first
B. Install these bolts/screws 2nd
C. Install these bolts/screws last

Instrument cluster trim plates

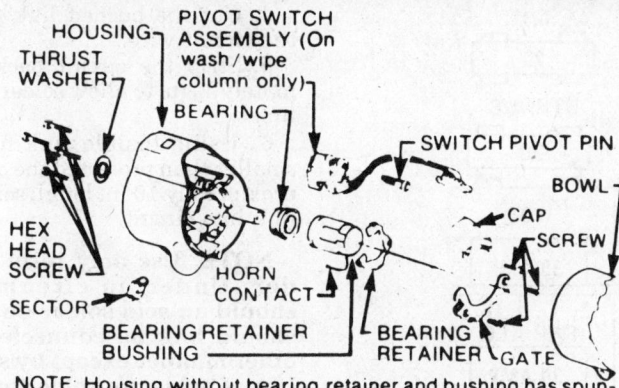

NOTE: Housing without bearing retainer and bushing has spun-in bearing. If repair is necessary, complete housing assembly replacement is necessary.

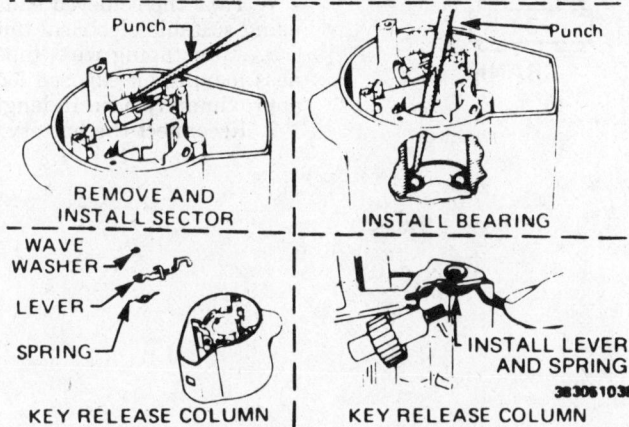

Wiper switch removal—standard columns

Heater Core

REMOVAL & INSTALLATION

With A/C

1. Under the hood, disconnect and plug the heater hoses at the heater.
2. Remove the speaker grille and the speaker.
3. Remove the heater core cover, the retainers and the heater core.
4. Installation is the reverse of removal. Refill the cooling system as required.

Without A/C

1. Disconnect the negative battery cable.
2. Disconnect the following wire connections.
   a. Heater relay.
   b. Heat blower resistor.
   c. Heater blower switch.
   d. Heater ground connection.
   e. Forward courtesy lamp socket.
3. Remove the windshield washer fluid container.
4. Disconnect the heater core inlet and outlet hoses.
5. Remove the heater core grommets.

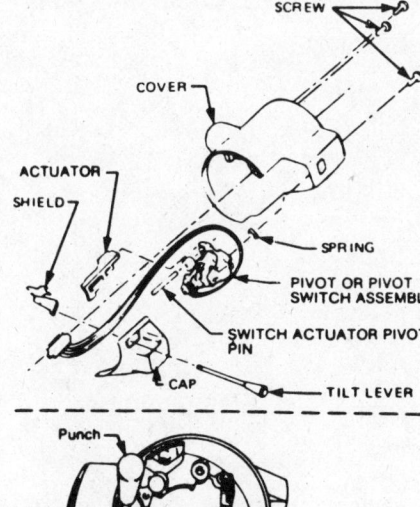

Wiper switch removal—adjustable columns

6. Remove the heater case cover.
7. Remove the heater core retainer and the heater core.
8. Installation is the reverse of removal. Refill the cooling system as required.

Radio

REMOVAL & INSTALLATION

1. Remove the console trim plate assembly.
2. Disconnect the side retaining nuts and the rear retaining bolt.
3. Disconnect the electrical and the antenna connections.
4. Remove the radio out of the front of the console.
5. Installation is the reverse of removal.

NOTE: It is important when doing any radio work to avoid pinching the speaker wires. A short circuit to ground from either wire will cause damage to the output circuit of the radio.

Fuse Block

The fuse block is a swing-down unit located in the underside of the instrument panel left of the steering column. The fuse block uses miniaturized fuses, designed for increased circuit protection and greater reliability. Various convenience connectors, which snap-lock into the fuse block, add to the serviceability of this unit.

Convenience Center

The Convenience Center is a stationary unit. It is located on the right side of the heater or A/C module in the vehicle under the I.P. panel. This location provides easy access to the audio alarm, hazard warnings, the horn relay, the seatbelt key and the headlamp warning alarm. All units are serviced by plug-in replacement.

Fusible Link

Added protection is provided in all battery feed circuits and other selected circuits by a fusible link. This link is a short piece of copper wire approximately 4 in. long inserted in series with the circuit and acts as a fuse. The link is two or more gages smaller in size than the circuit wire it is protecting and will burn out without damage to the circuit in case of current overload.

Fusible Link Replacement

1. Disconnect the battery.
2. Locate the burned out link.
3. Strip away all melted harness insulation.

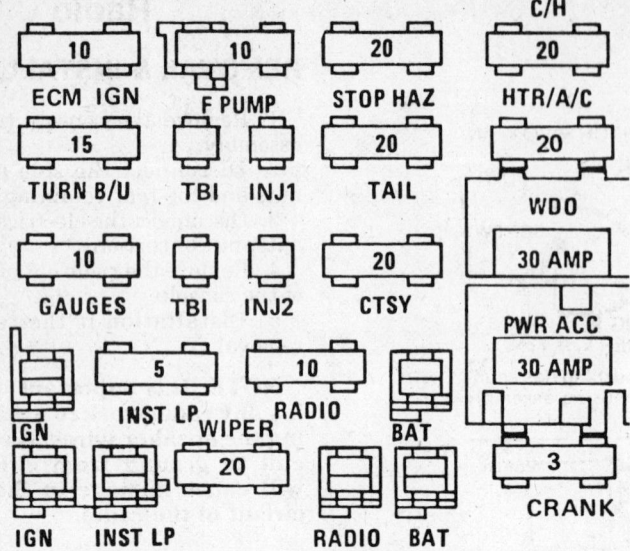

Front face of fuse block

4. Cut the burned link ends from the circuit wire.

5. Strip the circuit back approximately ½ in. to allow soldering of new link.

6. Using fusible link four gages smaller than protected the circuit (approximately 10 in. long), solder a new link into circuit.

NOTE: Use only resin core solder. Under no circumstances should an acid solder be used nor should link be connected in any other manner except by soldering. Use of acid core solder may result in corrosion.

7. Tape the soldered ends securely using suitable electrical tape.

8. After taping wire, tape the harness leaving an exposed loop of wire approximately 5 in. in length.

9. Reconnect the battery.

GM "T" Body
Chevette, Pontiac T1000

YEAR IDENTIFICATION

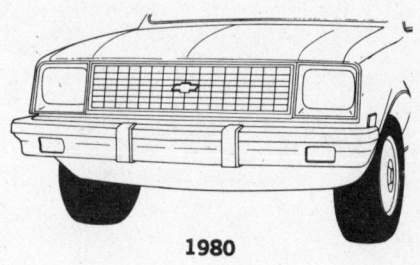

1980

1981-87 Chevette

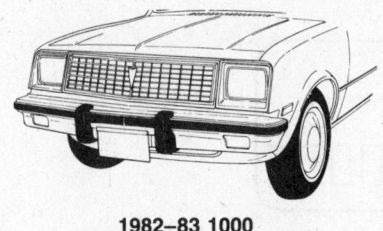

1982–83 1000

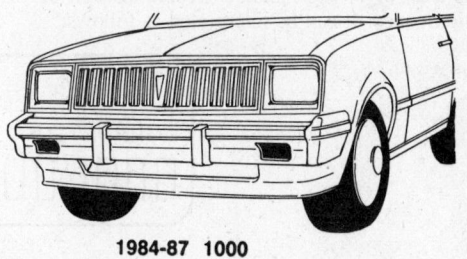

1984-87 1000

VEHICLE IDENTIFICATION NUMBER (VIN)

It is important for servicing and ordering parts to be certain of the vehicle and engine identification. The VIN (vehicle identification number) is a 13 or 17 digit number visible through the windshield on the driver's side of the dash and contains the vehicle and engine identification codes. It can be interpreted as follows:

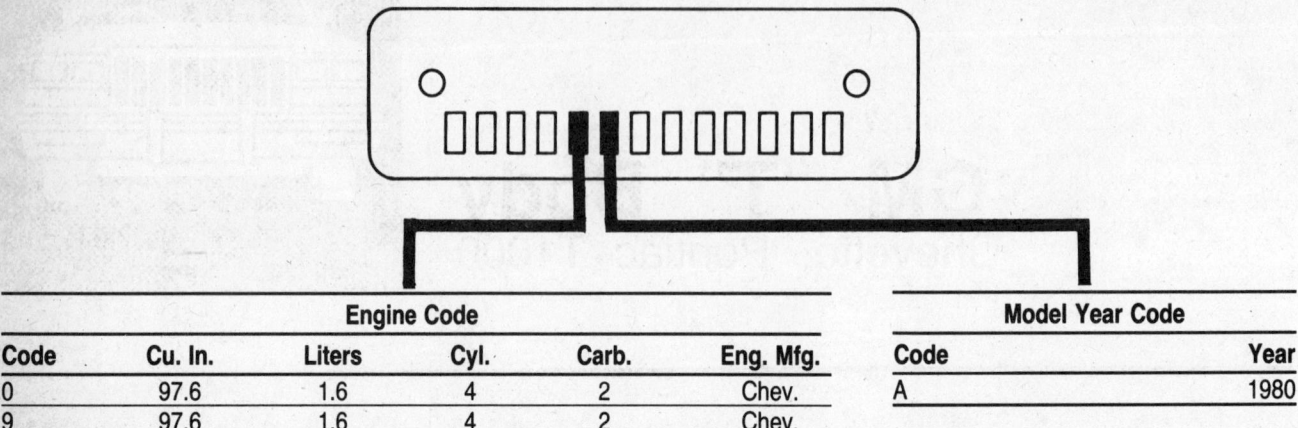

Engine Code						Model Year Code	
Code	Cu. In.	Liters	Cyl.	Carb.	Eng. Mfg.	Code	Year
0	97.6	1.6	4	2	Chev.	A	1980
9	97.6	1.6	4	2	Chev.		

The thirteen digit Vehicle Identification Number can be used to determine engine application and model year. The 6th digit indicates the model year, and the 5th digit identifies the factory installed engine.

VEHICLE IDENTIFICATION NUMBER (VIN)

It is important for servicing and ordering parts to be certain of the vehicle and engine identification. The VIN (vehicle identification number) is a 13 or 17 digit number visible through the windshield on the driver's side of the dash and contains the vehicle and engine identification codes. It can be interpreted as follows:

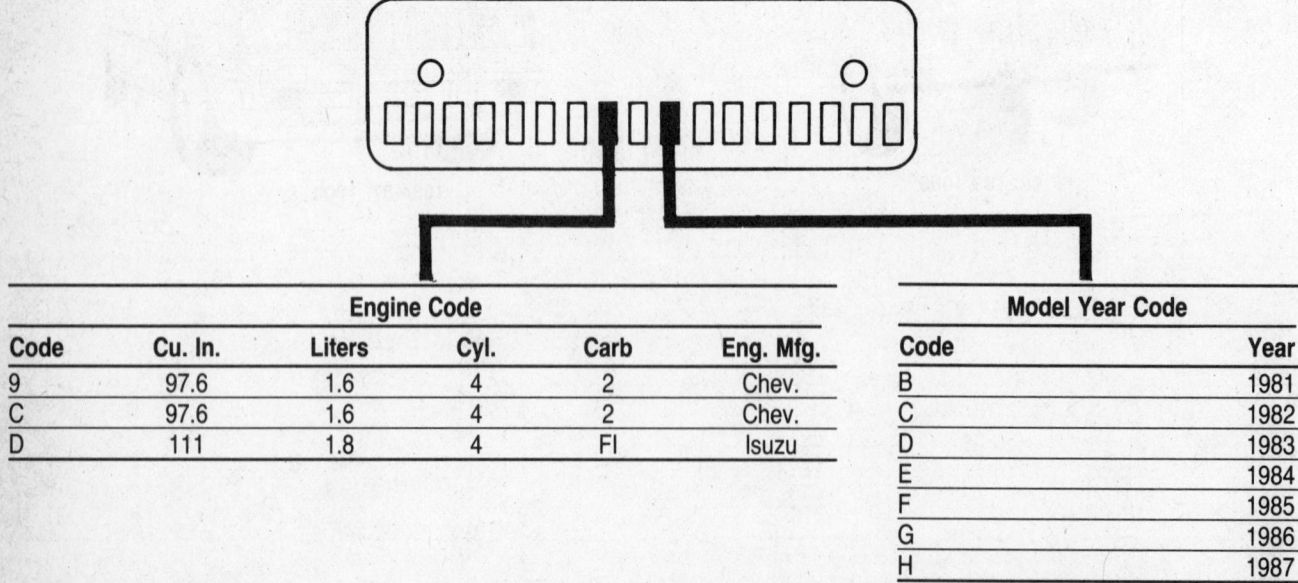

Engine Code						Model Year Code	
Code	Cu. In.	Liters	Cyl.	Carb	Eng. Mfg.	Code	Year
9	97.6	1.6	4	2	Chev.	B	1981
C	97.6	1.6	4	2	Chev.	C	1982
D	111	1.8	4	FI	Isuzu	D	1983
						E	1984
						F	1985
						G	1986
						H	1987

The seventeen digit Vehicle Identification Number can be used to determine engine application and model year. The 10th digit indicates the model year, and the 8th digit identifies the factory installed engine.

GENERAL ENGINE SPECIFICATIONS

Year	Eng. VIN Code	No. Cyl. Displacement liters (cu in.)	Mfg.	Carburetor Type	Horsepower @ rpm	Torque @ rpm (ft lbs)	Bore × Stroke (in.)	Compression Ratio	Oil Pressure @ 2000 rpm
'80	9	4-1.6 (97.6)	Chev.	6510C	70 @ 5200	82 @ 2400	3.228 × 2.980	8.5:1	55
'80	0	4-1.6 (97.6)	Chev.	6510C	74 @ 5200	88 @ 2800	3.228 × 2.980	8.5:1	55
'81–'87	C	4-1.6 (98)	Chev.	6510C	65 @ 5200③	80 @ 2400④	3.228 × 2.980	9.0:1①	55
'81–'87 Diesel	D	4-1.8 (111)	Isuzu	Fuel Injection	51 @ 5000	72 @ 2000	3.310 × 3.230	22.0:1	64②

NOTE: Horsepower and torque are SAE net figures. They are measured at the rear of the transmission with all accessories installed and operating. Since the figures vary when a given engine is installed in different models, some are representative rather than exact.

① '81—8.5:1
② @ 5000
③ '81—70 @ 5200
④ '81—82 @ 2400

FIRING ORDER

NOTE: To avoid confusion, replace spark plugs and wires one at a time.

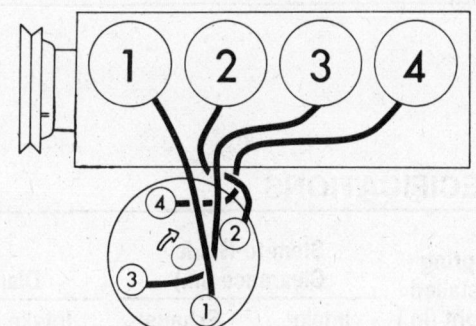

Chevrolet 98 cu. in. (1.6 liter) 4 cyl.
Engine firing order: 1-3-4-2
Distributor rotation: clockwise

TUNE-UP SPECIFICATIONS

When analyzing compression test results, look for uniformity among cylinders rather than specific pressures.

Year	Eng. V.I.N. Code	No. Cyl. Displacement (liters)	Mfg.	Spark Plugs Orig. Type	Spark Plugs Gap (in.)	Distributor Point Dwell (deg)	Distributor Point Gap (in.)	Ignition Timing (deg) Man Trans	Ignition Timing (deg) Auto Trans	Valves Intake Opens (deg)	Fuel Pump Pressure (psi)	Idle Speed (rpm) Man Trans	Idle Speed (rpm) Auto Trans.
'80	9	4-1.6	Chev.	R42TS	.035	Electronic		12B	18B	28B	5–6.5	800	750①
	0	4-1.6 HO	Chev.	R42TS	.035	Electronic		12B	18B	31B	5–6.5	800	750
'81	9	4-1.6	Chev.	R42TS	.035	Electronic		18B	18B	28B	2.5–6.5	800	700
'82–'87	C	4-1.6	Chev.	R42CTS	.035	Electronic		8B	8B	—	5.5–6.5	800	700
'87	All			See Underhood Specifications Sticker									

NOTE: The underhood specifications sticker often reflects tune-up specification changes made in production. Sticker figures must be used if they disagree with those in this chart. Product numbers in this chart are not recommendations by Chilton for any product by brand name.

B Before Top Dead Center
HO High Output
① California: 800 rpm

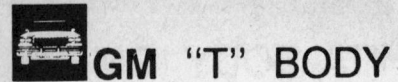

DIESEL TUNE-UP SPECIFICATIONS

Year	No. Cyl.–Displacement (liters)	Static Injection Timing	Fuel Injection Order	Compression (lbs)	Injection Nozzle Opening Pressure (psi)	Intake Valve Opens (deg)	Idle Speed ▲ (rpm) Man.	Idle Speed ▲ (rpm) Auto.
'81–'82	L-4 (1.8)	18 BTDC	1-3-4-2	441①	1707	32	625	725
'83–'84	L-4 (1.8)	11 BTDC	1-3-4-2	441①	1707	32	620	720
'85–'86	L-4 (1.8)	18 BTDC	1-3-4-2	441①	1706–1848	32	625	—
'87	L-4 (1.8)	See Underhood Specifications Sticker						

NOTE: The underhood specifications sticker often reflects changes made in production. Sticker figures must be used if they disagree with those in the above chart.
▲ See underhood sticker for fast idle speed.
BTDC Before Top Dead Center
① At 200 rpm

CAPACITIES

Year	Engine No. Cyl. Displacement (liters)	Engine Crankcase	Transmission Pts. to Refill After Draining Manual 4-Speed	Transmission Pts. to Refill After Draining Manual 5-Speed	Transmission Pts. to Refill After Draining Automatic ●	Drive Axle (pts)	Gasoline Tank (gals)	Cooling System (qts) With Heater	Cooling System (qts) With A/C
'80–'87	4-1.6	4①	3½	4	6	1¾	12.5	9	9¼
'81–'87	4-1.8 Diesel	6①	3½	3¼	6	1¾	12.5	8.5	9.0

● Specifications do not include torque converter
① With filter change

VALVE SPECIFICATIONS

Year	Engine No. Cyl. Displacement (liters)	Seat Angle (deg)	Face Angle (deg)	Spring Test Pressure (lbs @ in.)	Spring Installed Height (in.)	Stem-to-Guide Clearance (in.) Intake	Stem-to-Guide Clearance (in.) Exhaust	Stem Diameter (in.) Intake	Stem Diameter (in.) Exhaust
'80–'87	4-1.6	45	46	173 @ .886	1.25	.0006–.0017	.0014–.0025	.3141	.3133
'81–'87	4-1.8 Diesel	45	45	108 @ 1.24①	1.61	.0015–.0028	.0018–.0030	.3128–.3134	.3126–.3132

① Exhaust 112 @ 1.22; Inner spring test
pressures—intake 58 @ 1.14
exhaust 60 @ 1.12

CRANKSHAFT AND CONNECTING ROD SPECIFICATIONS

All measurements are given in inches

Year	Engine No. Cyl. Displacement (liters)	Crankshaft Main Brg. Journal Dia	Crankshaft Main Brg. Oil Clearance	Crankshaft Shaft End-Play	Crankshaft Thrust on No.	Connecting Rod Journal Diameter	Connecting Rod Oil Clearance	Connecting Rod Side Clearance
'80–'87	4-1.6	2.0078–2.0088	①	.004–.008	4	1.809–1.810	.0014–.0031	.004–.012
'81–'87	4-1.8 Diesel	2.2010–2.2020	.0015–.0027	.0024–.0094	3	1.927–1.928	.0016–.0032	N.A.

N.A.—Not Applicable
① No. 5—.0009–.0026
All others—.0005–.0018

CAMSHAFT SPECIFICATIONS

All measurements in inches. To convert inches to metric units, refer to Metric Information section.

Year	Engine Type/ Disp. L(cu in.)	Journal Diameter					Bearing Clearance	Lobe Lift		Camshaft End Play
		1	2	3	4	5		Intake	Exhaust	
'80–'87	4-1.6(98)	1.7682–1.7697	1.7584–1.7598	1.7485–1.7500	1.7387–1.7402	1.1816 1.1837	.0020–.0044	.2410	.2410	.0067–.0169
'81–'87	4-1.8(111)			NA			.0008–.0035	6.1163	6.1163	.17–.43

NA—Not Available

TORQUE SPECIFICATIONS

All readings in ft. lbs.

Year	Engine No. Cyl. Displacement (liters)	Cylinder Head Bolts	Rod Bearing Bolts	Main Bearing Bolts	Crankshaft Pulley Bolt	Flywheel to Crankshaft Bolts	Manifold	
							Intake	Exhaust
'80–'87	1.6	75	40	40	100	40	18	①
'81–'87	4-1.8 Diesel	②	65	75	N/A	N/A	30	N/A

N/A Not available
① Center bolts—13-18; end bolts—19-25
② First tighten to 21-36 ft. lbs. then retighten to 83-98 (new bolt), 90-105 (reused bolt)

PISTON AND RING SPECIFICATIONS

All measurements are given in inches.

Year	Engine Type/ Disp. L(cu in.)	Piston-to-Bore Clearance	Ring Gap			Ring Side Clearance		
			Top Compression	Bottom Compression	Oil Control	Top Compression	Bottom Compression	Oil Control
'80–'87	4-1.6	.0008–.0016	.009–.019	.008–.018	.015–.055	.0012–.0027	.0012–.0032	.0000–.0050
'81–'82	4-1.8 Diesel	.0006–.0014	.0078–.0157	.0078–.0157	.0078–.0157	.0035–.0049	.0014–.0020	.0012–.0028
'83–'87	4-1.8 Diesel	.0002–.0017	.0078–.0157	.0078–.0157	.0078–.0157	.0035–.0049	.0019–.0033	.0012–.0028

WHEEL ALIGNMENT SPECIFICATIONS

Year	Model	Caster		Camber		Toe-in (in.)
		Range (deg)	Pref. Setting (deg)	Range (deg)	Pref. Setting (deg)	
'80–'81	All	3½P–5½P	4½P	¼P–½P	¼P	1⁄16P
'82–'87	All	4P–6P	5P	¼P–½P	¼P	1⁄16P

N Negative
P Positive

ENGINE ELECTRICAL

A Delcotron 10-SI series alternator is used on gasoline engines. This unit contains a solid state, integrated circuit voltage regulator. The alternator is non-adjustable and requires no periodic maintenance.

The diesel engine is fitted with a Hitachi alternator, which is equipped with an IC regulator and drives a vacuum pump mounted at its rear.

For further testing and overhaul procedures, please refer to "Charging and Starting" in the Unit Repair section.

Alternator

REMOVAL & INSTALLATION

1. Disconnect the negative battery cable.
2. Disconnect the alternator wiring. On the Chevette diesel, remove the fan shroud and fresh air duct, then disconnect the oil and vacuum lines at the vacuum pump.
3. Remove the brace bolt and the drive belt.
4. Support the alternator, remove the mounting bolt, and remove the alternator.

NOTE: On the diesel, the mounting bolts are removed from below the car.

5. Installation is the reverse of removal. Adjust drive belt deflection to $1/2$ in. under moderate thumb pressure.

Starter

Engine cranking is accomplished by a solenoid-actuated starter motor powered by the vehicle battery. The motor on the gasoline engine is a Delco-Remy unit similar to other GM starters. No periodic lubrication of the motor or solenoid is necessary. The diesel engine is equipped with a Hitachi reduction gear starter motor which is solenoid activated.

For further testing and overhaul procedures, please refer to "Charging and Starting" in the Unit Repair section.

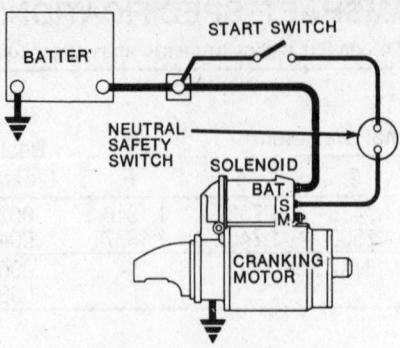

Typical gasoline engine starting system (© Chevrolet Div., G.M. Corp.)

REMOVAL & INSTALLATION

Gasoline Engine

WITHOUT POWER BRAKES

1. Disconnect the battery negative cable.
2. Remove the air cleaner.
3. Disconnect the gas line at the carburetor and move to one side.
4. Disconnect the vacuum hose at the carburetor.
5. Remove the splash shield from the distributor coil and move to one side.
6. Using a 6 inch and 12 inch extension with a universal socket, remove the upper starter bolt.
7. Remove the lower starter bolt.
8. Disconnect the starter wiring.
9. Remove the master cylinder mounting nuts to gain access for removing the starter. Installation is the reverse of removal.

WITH POWER BRAKES

1. Disconnect the battery ground cable.
2. Remove the air cleaner.
3. Disconnect the gas line at the carburetor and move to one side.
4. Remove the splash shield from the distributor coil and move to one side.
5. Using a 6 inch and 12 inch extension with a universal socket, remove the upper starter bolt.
6. Remove the steering column cover screws and remove the cover.
7. Remove the steering column upper nuts and toe pan screw.
8. Raise the vehicle on a hoist and remove the steering shaft from the steering coupling.
9. Lower the vehicle and move the steering column from inside the car to gain access to the starter.
10. Disconnect the starter wiring.
11. Remove the starter lower bolt and remove the starter.
12. Installation is the reverse of removal.

Diesel Engine

1. Disconnect the negative battery cable.
2. Disconnect the starter wiring after labeling. The starter is located at the right rear of the engine.
3. Remove the upper mounting nut and the lower mounting bolt, then remove the starter.

Distributor

All gasoline engine models are equipped with High Energy Ignition (HEI). 1981 and later models are equipped with EST (Electronic Spark Timing) distributors which have no mechanical or vacuum advance mechanisms. The ignition coil is mounted externally, on the left side of the engine, beneath the intake manifold, and is not visible on A/C equipped cars. The coil has a plastic cover.

A conventional ignition system is not needed on the diesel engine, because it uses compression heat rather than a manufactured spark to ignite its air/fuel mixture. An electrically operated glow plug system is used on the diesel engine to preheat the combustion chambers for easy cold start-up. See the Diesel Maintenance Unit Repair Section for additional diesel engine information.

REMOVAL & INSTALLATION

1. Disconnect the negative battery cable.
2. If the vehicle is air conditioned, disconnect the electrical lead at the air conditioning compressor. Remove the compressor mounting thru bolt and two adjusting bolts. Remove two bolts and remove the compressor upper mounting bracket. Raise the vehicle on a hoist. Remove the two bolts securing the compressor lower mounting bracket and pull the bracket outward for clearance. Lower the vehicle.

— CAUTION —
Do not disconnect any A/C refrigerant lines.

3. Remove the air cleaner.
4. Remove the distributor cap and place it aside.
5. Remove the ignition coil cover by prying on the flat on the front edge of the cover.
6. Remove the ignition coil mounting bracket bolts.
7. Disconnect the electrical connector with red and brown wires that goes from the ignition coil to the distributor.
8. Remove the fuel pump, gasket, and push rod, noting the direction in which push rod was installed.

NOTE: It is important that the push rod be installed in exactly the same direction as removed.

9. Scribe a mark on the engine in line with the distributor rotor. Note the approximate position of the distributor housing in relation to the engine.

10. Remove the distributor hold-down bolt and clamp and remove the distributor.

11. If the engine has not been disturbed with the distributor removed, simply reverse the installation procedure to install, aligning the marks made during removal. If the engine has been disturbed, remove the No. 1 spark plug and place your thumb or finger over the spark plug hole. Manually turn the engine in the normal direction of rotation until compression is felt and the timing marks point to Top Dead Center. Align the marks made during removal and install the distributor.

HEI SYSTEM TACHOMETER HOOKUP

Connect a tachometer to the negative terminal on the coil and to a ground; some tachometers must connect to the negative terminal and the battery positive terminal. Check the tachometer manufacturer's instructions to make sure the tachometer is compatible with the HEI system. Never ground the TACH terminal or serious module damage could occur.

Ignition Timing

ADJUSTMENT

Models Without EST Distributor

NOTE: Use an adapter to make timing light connections at the distributor No. 1 terminal. Follow all the instructions on the Vehicle Emissions Control Information label located on the radiator support panel.

1. Bring the engine to normal operating temperature. Stop the engine and connect a tachometer. Disconnect and plug the PCV hose at the vapor canister and the vacuum hose at the distributor vacuum advance unit (on models so equipped). Start the engine and check curb idle speed. Adjust as necessary.

2. Stop the engine, clean the timing marks and mark them with chalk to make them more visible. Connect a timing light.

3. Start the engine and aim the timing light at the timing marks. If the marks align, stop the engine, re-

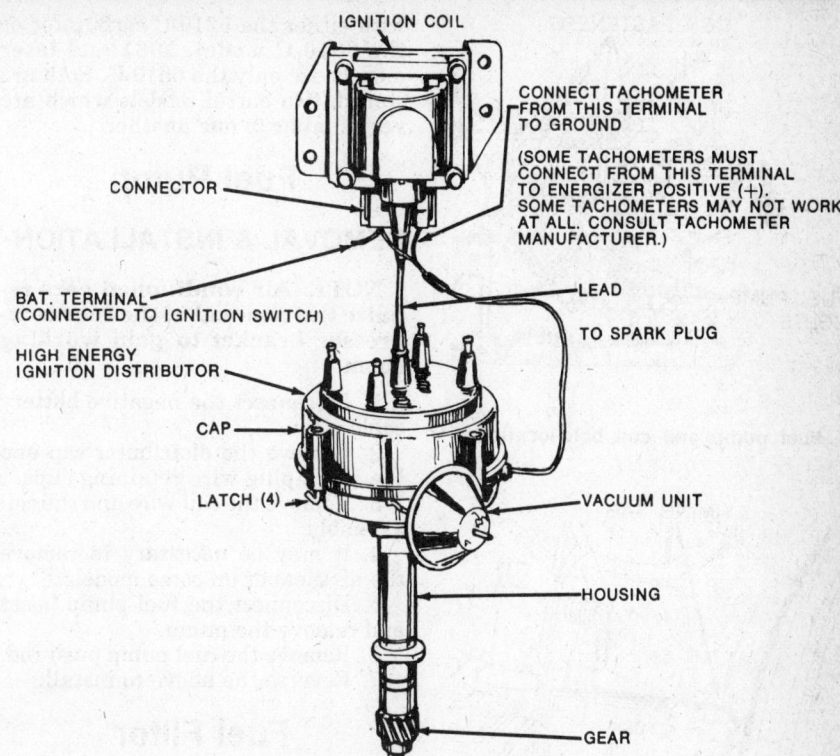

HEI tachometer hookup

connect the PCV and vacuum hoses, and remove the timing light.

4. If adjustment is necessary, loosen the distributor clamp and rotate the distributor to align the marks. Tighten the clamp and recheck the timing.

NOTE: Air conditioned models require removal of the compressor, bracket, and belt to reach the distributor clamp.

———— CAUTION ————
Do not disconnect A/C refrigerant lines. Move compressor and bracket to one side.

5. Reset the curb idle speed if necessary, stop the engine, and remove the tachometer and timing light. Reconnect the PCV and vacuum hoses.

Models with EST Distributor

NOTE: Engines with Electronic Spark Timing (EST) can be identified by the absence of a vacuum and a mechanical spark advance on the distributor. EST allows continuous spark timing adjustments to be made by the ECM (Electronic Control Module).

1. Follow all instructions on the Vehicle Emissions Control information label located on the radiator support panel.

2. Connect the pick-up lead of the timing light to the number one spark plug. Use a jumper lead or adapter between the wire and plug, better yet use a timing light with an inductive type pick-up.

NOTE: Do not pierce the wire or attempt to insert a wire between the boot and the wire. Connect the timing light power leads according to the manufacturer's instructions.

3. Start the engine and make sure it is operating at normal operating temperature.

4. Increase idle and disconnect the four (4) terminal EST connectors at the distributor. This will cause the engine to operate in the bypass timing mode.

NOTE: Steps 3–4 are important, because if the engine is at idle and not warm enough, when the EST terminal is disconnected, the oxygen sensor could cool off, putting the system into an open loop operation resulting in the engine shutting off.

5. With the engine running, aim the timing light at the timing mark.

6. If a change is necessary, loosen the distributor hold-down clamp bolt at the base of the distributor. While observing the mark with the timing light, slightly rotate the distributor until the correct timing is indicated.

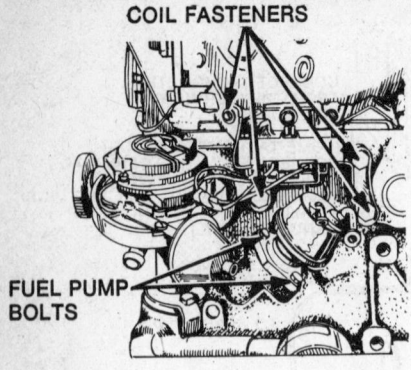

COIL FASTENERS

FUEL PUMP BOLTS

Fuel pump and coil bolt locations

TIMING TAB

BEFORE

CRANKSHAFT PULLEY

Ignition timing marks

Tighten the hold-down bolt, and recheck the timing.

7. Turn off the engine and reconnect all wires.

NOTE: Models with air conditioning require removal of the compressor since it is mounted directly above the distributor. Do not disconnect the refrigerant lines. Move the compressor and bracket to one side.

GASOLINE FUEL SYSTEM

The carburetor identification number is stamped on the float bowl, right next to the fuel inlet nut. When replacing the fuel bowl, be sure to transfer the identification number to the new bowl. 1980 models are equipped

with either the 5210-C carburetor or the 6510-C model. 1981 and later models use only the 6510-C. Both are staged, two barrel models which are very similar to one another.

Fuel Pump

REMOVAL & INSTALLATION

NOTE: Air conditioned cars require the removal of the rear compressor bracket to gain working room.

1. Disconnect the negative battery cable.
2. Remove the distributor cap and the spark plug wire retaining clips.
3. Remove the coil wire and the coil assembly.
4. It may be necessary to remove the air cleaner on some models.
5. Disconnect the fuel pump hoses and remove the pump.
6. Remove the fuel pump push rod.
7. Reverse the above to install.

Fuel Filter

REMOVAL & INSTALLATION

NOTE: Do not perform this operation on a hot engine. Place some rags under the fuel fitting to catch any spilled fuel.

1. Disconnect the small fuel line connection nut, using a flare nut wrench, while holding the large fitting nut with a standard open end wrench. A flared nut wrench is preferred over a standard open end wrench since it will not slip off and round off the corners of the tubing nut.
2. Remove the large filter retaining nut from the carburetor. There is a spring behind the filter. Remove the filter and spring.
3. Install the spring and the filter element.
4. Install the new gasket on the retaining nut and screw it into place. Do not overtighten; the threads are rather soft.
5. Discard the gas-soaked rag safely.

Carburetor

REMOVAL & INSTALLATION

1. Remove air cleaner and gasket.
2. Disconnect the fuel and vacuum lines from the carburetor.
3. Disconnect the accelerator linkage and the electrical connectors.
4. Remove the carburetor attaching nuts and remove the carburetor.

5. Remove the electric Early Fuel Evaporation (EFE) heater (if so equipped) and the insulator gasket.
6. Be sure the throttle body and intake manifold sealing surfaces are clean.
7. Install a new EFE heater (if so equipped) and an insulator gasket on the manifold.
8. Install the carburetor over the manifold studs.
9. Install the vacuum lines and loosely connect the fuel line.
10. Install and tighten the attaching nuts to 12 ft. lbs.
11. Tighten the fuel inlet nut to 25 ft. lbs.
12. Connect the accelerator linkage and the electrical connectors.
13. Check and adjust the idle speed as required.
14. Install the air cleaner and gasket.

For all carburetor adjustments and specifications not detailed below, please refer to "Carburetors" in the Unit Repair section.

IDLE SPEED ADJUSTMENT

Two idle speeds are controlled by a solenoid on models without both automatic transmission and air conditioning. One is normal curb idle speed (solenoid energized). The second is low idle speed (solenoid de-energized) which prevents dieseling when the ignition is turned off. On cars with both automatic transmission and air conditioning the solenoid is energized when air conditioning is on to maintain curb idle speed.

1980 Models with 6510-C Carburetor

1. Adjust the ignition timing as previously outlined.
2. On non-air conditioned models, connect the PCV and the vacuum advance hoses and adjust the idle speed to specification using the idle speed adjusting screw on the carburetor.
3. If the car is equipped with air conditioning:
   a. Disconnect the electrical connection at the A/C compressor.
   b. Disconnect and plug the EGR and PCV hoses.
   c. Turn the air conditioning On and start the engine.
   d. Open the throttle slightly to extend the throttle solenoid on the carburetor. If the speed is incorrect, turn off the engine and turn the solenoid screw to adjust. Start the engine and check the speed. Repeat this until the correct idle speed is obtained.

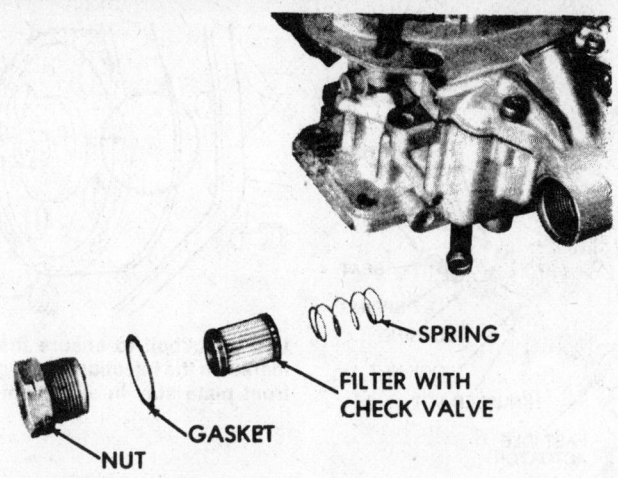

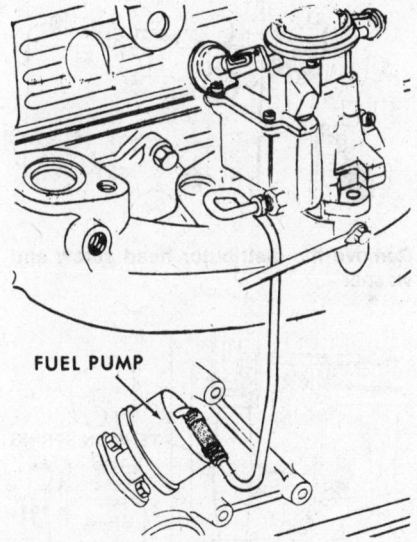

SPRING

FILTER WITH
CHECK VALVE

NUT

GASKET

Fuel filter assembly—2 bbl

Chevette fuel pump location

FUEL PUMP

DIESEL FUEL SYSTEM

The diesel fuel system consists of a high pressure fuel injection pump driven by the camshaft timing belt, four pressure activated fuel injectors installed in the cylinder head and connected by fuel lines to the pump, a fuel filter with built in water separator, drain and hand primer, a fuel tank and connecting fuel feed and return lines. The injection pump is equipped with an electrically operated fuel cut-off solenoid which halts fuel flow (and the engine) whenever the ignition key is turned to the OFF position.

For further information on the diesel fuel system, please refer to "Diesel Maintenance" in the Unit Repair section.

Fuel Filter

REMOVAL & INSTALLATION

1. Disconnect the negative battery cable.
2. Disconnect the water sensor lead at the bottom of the filter, then disconnect the water filter to main body hose.
3. Remove the filter element by turning it counterclockwise using a filter strap wrench. Be careful not to spill any fuel.
4. After draining the filter, unscrew the water sensor from the bottom of the element.
5. Install the sensor in the new filter after apply a thin film of diesel fuel to the sensor O-ring.
6. Clean the filter mounting surface, apply a thin film of diesel fuel to the gasket on the new filter and install the filter. Continue turning the filter an additional $2/3$ turn after it contacts the filter main body.
7. Connect the sensor wire. Disconnect the fuel outlet hose from the injector pump and place in a suitable container, then operate the priming pump handle several times to fill the filter with fuel. Reconnect the hose to the injector pump and start the engine to check for leaks.

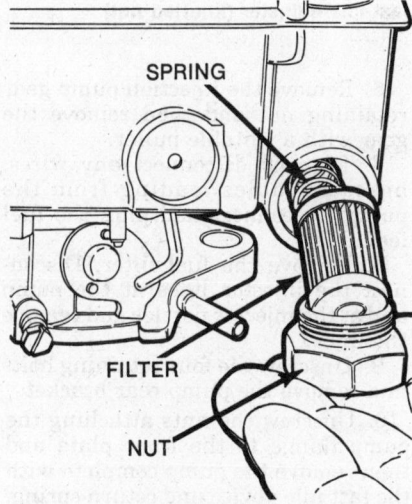

SPRING

FILTER

NUT

Fuel filter assembly—Rochester 1 bbl

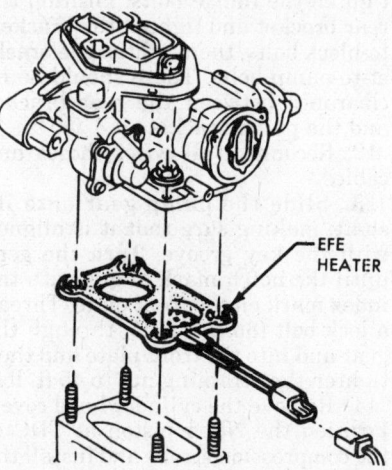

EFE HEATER

EFE heater and insulator gasket location

e. Connect the wiring at the compressor and unplug and connect the hoses.

1981 and Later Models

On these models, the carburetor mixture and idle speed are adjusted by the Computer Command Control (CCC) System. It is possible to adjust the basic idle speed; however, this procedure requires special knowledge and tools.

IDLE MIXTURE ADJUSTMENT

Turning the mixture screw on these carburetors will have no appreciable effect. The factory-recommended idle mixture adjustment procedure on many models through 1980 requires special apparatus to artificially enrich the mixture with propane gas. On

1981 and later models, the idle mixture is adjusted by the CCC system and requires no manual adjustment.

Idle Speed

ADJUSTMENT

1. Set the parking brake and block the wheels.
2. Place the transmission in Neu-

tral. Connect a diesel tachometer as per the manufacturer's instructions.

NOTE: A standard gasoline engine tachometer will not work on a diesel engine.

3. Start the engine and allow it to reach normal operating temperature.

4. Loosen the lock nut on the idle speed adjusting screw and turn the screw to obtain the correct idle speed (see underhood specifications sticker).

5. Tighten the lock nut, turn the engine off and disconnect the tachometer.

Fast Idle Speed

ADJUSTMENT

1. Set the parking brake and block the wheels.

2. Place the transmission in Neutral.

3. Connect a diesel tachometer.

4. Start the engine and allow it to run until it reaches normal operating temperature.

5. Apply vacuum to the fast idle actuator.

6. Loosen the lock nut on the fast idle adjusting screw and adjust the knurled nut to obtain the fast idle speed specified on the emission label. After adjusting, retighten the lock nut.

Injection Pump

REMOVAL & INSTALLATION

NOTE: This procedure will require the use of two special tools: a gear puller (J-22888), and a fixing plate (J-29761). It is a long and complicated procedure and must be performed in conjunction with the following "injection timing" procedure.

1. Disconnect the negative battery cable.

2. Drain the cooling system. Remove the fan shroud, radiator and coolant recovery tank.

3. Disconnect the bypass hose leading from the front cover and then remove the upper half of the front cover.

NOTE: Fan removal may facilitate better access to certain front cover retaining bolts.

4. Loosen the timing belt tension pulley and plate bolts. Slide the tensioner over.

5. Unscrew the two retaining bolts and remove the tension spring from behind the front plate, by the injection pump.

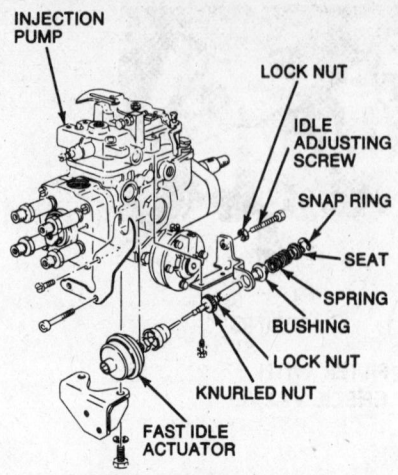

Exploded view of diesel injection pump linkage showing idle adjusting screw and fast idle adjuster (knurled nut)

6. Remove the injection pump gear retaining nut and then remove the gear with a suitable puller.

7. Tag and disconnect any wires, hoses or cables leading from the pump. Disconnect and plug the fuel feed lines.

8. Remove the fuel filter. Disconnect the injector lines at the pump and at the injector nozzles and remove the lines.

9. Unscrew the four retaining bolts and remove the pump rear bracket.

10. Unscrew the nuts attaching the pump flange to the front plate and then remove the pump complete with the fast idle device and return spring. To install:

11. Place the pump in position and tighten the flange bolts. Position the rear bracket and tighten the bracket-to-block bolts, then tighten the bracket-to-pump bolts. There should be no clearance between the rear bracket and the pump bracket.

12. Reconnect all wires, hoses and cables.

13. Slide the pump gear onto its shaft, making sure that it is aligned with the key groove. Turn the gear until the notch mark aligns with the index mark on the front plate. Thread a lock bolt (8mm x 1.25) through the gear and into the front plate and then tighten the retaining nut to 45 ft. lbs.

14. Remove the cylinder head cover. Position the No. 1 piston at TDC of the compression stroke and install the fixing plate into the slot in the rear of the camshaft to prevent it from rotating.

15. Unscrew the cam gear retaining bolt and, using a puller, remove the gear. Reinstall the gear loosely so that it can be turned smoothly by hand.

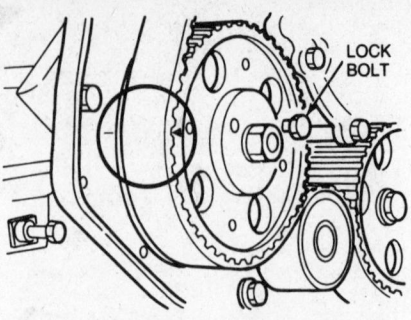

Use a lockbolt to ensure that the index marks on the injection pump gear and the front plate stay in alignment—diesel

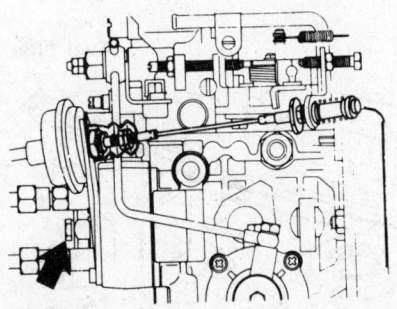

Remove the distributor head screw and washer

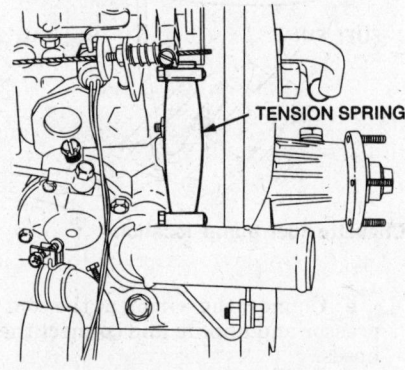

Tension spring, located behind the front plate beside the injection pump on diesel engine

16. Grasp the timing belt on each side near the lower half of the front cover; move it back and forth until the cogs on the belt engage with those on the lower gears. Slide the belt over the pump gear and then over the cam gear (you may need to turn the cam gear slightly to facilitate proper engagement of the cogs).

17. Make sure that any slack in the belt is concentrated around the tension pulley and NOT around or between the two upper gears. Depress the tension pulley with your finger and then install the tension spring.

18. Partially tighten the tension

pulley bolts; first the upper, then the lower. Tighten the cam gear retaining bolt to 45 ft. lbs.

19. Remove the pump gear lock bolt. Remove the fixing plate from the end of the camshaft.

20. Check that the No. 1 piston is still at TDC. Check that the marks on the front plate and the pump gear are still aligned. Check that the fixing plate still fits properly into the rear of the camshaft.

NOTE: If theses three steps do not check out correctly, repeat the entire procedure, DO NOT attempt to compensate by moving the camshaft, pump gear or crankshaft.

21. Loosen the tension pulley and plate bolts. Make sure the belt slack is concentrated around the pulley and then tighten the bolts in the same manner as before. Belt tension should be checked at a point between the cam gear and the pump gear.

22. Installation of the remaining components is in the reverse order of removal.

23. Check the injection timing.

Injection Timing

ADJUSTMENT

1. Check that the No. 1 piston is at TDC of the compression stroke. Make sure that the timing belt is properly tensioned and the timing marks are aligned.

2. Remove the cylinder head cover and check that the fixing plate used in the previous section will still fit smoothly into the slot at the rear of the camshaft.

3. Remove the injection lines as detailed earlier and then remove the distributor head screw and washer.

4. Position a Static Timing Gauge (J-29763) and a dial indicator in the distributor head hole. Set the lift approximately 0.04 in. (1 mm) from the end of the plunger.

5. Turn the crankshaft until the No. 1 piston is 45–60 degrees BTDC and then zero the dial indicator.

NOTE: The damper pulley is notched with eleven lines; four in one position, seven in another. The group of four are to be used for static timing.

6. Turn the crankshaft to the 18° position, loosen the two nuts on the injection pump flange and move the pump until the proper reading is achieved. Swivel the pump up to retard the timing and down to advance the timing. When adjustment is cor-

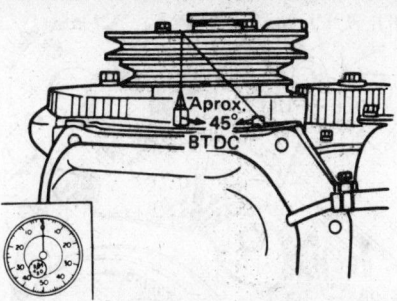

Zeroing the dial indicator

rect, retighten the pump flange nuts.

7. Remove the dial indicator and install the distributor head screw and washer.

8. Install the cylinder head cover, injection lines and fuel filter.

9. Reconnect all necessary wires and hoses. Installation of the remaining components is in the reverse order of removal.

Fuel Injector Nozzle

REMOVAL & INSTALLATION

NOTE: The primary function of an injection nozzle is to distribute fuel in the combustion chamber. Do not, under any circumstances, crank the engine while an injection line or injector is disconnected.

1. Disconnect the negative battery cable.

2. Remove the fresh air duct and disconnect the PCV hose.

3. Disconnect the injection line at the injector nozzle and then loosen it at the injection pump. Carefully move it out of the way.

4. Remove the fuel return line.

5. Unscrew and remove the injector.

6. Installation is in the reverse order of removal.

ENGINE COOLING

A standard pressurized cooling system is used. A permanently lubricated impeller-type water pump forces coolant through engine and cylinder head water jackets and into a cross-flow radiator. Some models use a heavy-duty radiator with a fan shroud. The pressure-type radiator cap pressurizes the cooling system to 15 psi. A 190°F (180°F on Diesel) thermostat in the coolant outlet passage is

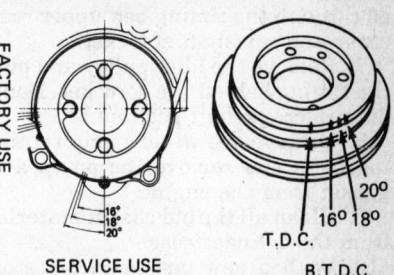

Static timing notches on the damper pulley

used to control coolant flow. A translucent plastic coolant recover reservoir is used to provide for coolant expansion. Coolant level is checked by observing the amount present in the reservoir with the engine at normal operating temperature. Add coolant to the reservoir, not the radiator. A 50/50 mixture of ethylene glycol antifreeze and water yielding freeze protection to −20°F should be used as coolant.

Radiator

REMOVAL & INSTALLATION

1. Disconnect the negative battery cable.

2. Drain the cooling system.

3. Remove the upper radiator support or the upper fan shroud, as necessary.

4. Disconnect the coolant hoses and the automatic transmission cooler lines from the radiator.

5. Remove the radiator.

6. Reverse to install.

Water Pump

REMOVAL & INSTALLATION

Gasoline Engine

1. Disconnect the battery negative cable and remove the alternator, and A/C compressor drive belts.

2. Remove the engine fan, spacer (air conditioned models), and the pulley.

3. Remove the timing belt front cover by removing the two upper bolts, center bolt, and two lower nuts. Remove the timing belt lower cover retaining nut and remove the cover.

4. Drain the coolant from the engine.

5. Remove the lower radiator hose and the heater hose at the water pump.

6. Turn the crankshaft pulley so that the mark on the pulley is aligned with the "O" mark on the timing scale and that a ⅛ in. drill bit can be insert-

ed through the timing belt upper rear cover and camshaft sprocket.

7. Remove the idler pulley and pull the timing belt off the sprocket. Don't disturb crankshaft position.

8. Remove the water pump retaining bolts and remove the pump and gasket from the engine.

9. Clean all the old gasket material from the cylinder case.

10. With a new gasket in place on the water pump, position the water pump in place on the cylinder case and install the water pump retaining bolts.

11. Install the timing belt onto the cam sprocket.

12. Apply sealer to the idler pulley attaching bolt and install the bolt and the idler pulley. Turn the idler pulley counterclockwise on its mounting bolt to remove the slack in the timing belt.

13. Use a tension gauge to adjust timing belt tension. Check belt tension midway between the tensioner and the cam sprocket on the idler pulley side. Correct belt tension is 70 lbs. Torque the idler pulley mounting bolt to 13–18 ft. lbs.

14. Remove the 1/8 in. drill bit from the upper rear timing belt cover and cam sprocket.

15. Install the lower radiator hose and the heater hose to the water pump.

16. Install the timing belt front covers.

17. Install the water pump pulley, spacer (if equipped), and engine fan.

18. Install the engine drive belt(s).

19. Refill the cooling system.

20. Connect the battery negative cable.

21. Start the engine and check for leaks. Run the engine with the heater on until the thermostat opens, then recheck the coolant level.

Diesel Engine

1. Disconnect the negative battery cable and drain the cooling system.

2. Remove the fan shroud, fan assembly and the accessory drive belt.

3. Unscrew the retaining bolts and remove the damper pulley.

4. Remove the upper and lower halves of the front cover and then remove the bypass hose at the pump.

5. Unscrew the pump retaining bolts and remove the pump assembly.

6. Installation is in the reverse order of removal.

Thermostat

REMOVAL & INSTALLATION

1. Drain the radiator and remove the radiator hose at the water outlet.

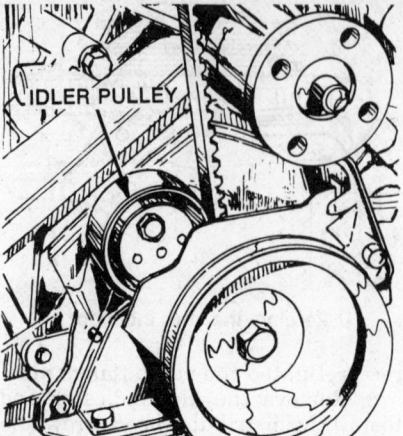

Gasoline engine timing belt idler pulley

2. Remove the thermostat housing bolts and remove the housing, gasket, and thermostat.

3. Install the thermostat. Use a new gasket on the thermostat housing and install the thermostat housing bolts.

4. Install the radiator hose at the water outlet.

5. Fill the cooling system. Run the engine with the heater on until the thermostat opens, then recheck the coolant level.

EMISSION CONTROLS

For information on emission controls, please refer to "Emission Controls" in the Unit Repair Section. Due to the complex nature of modern electronic engine control systems, comprehensive diagnosis and testing procedures fall outside the confines of this repair manual. For complete information on diagnosis, testing and repair procedures concerning all modern engine and emission control systems, please refer to *Chilton's Guide To Electronic Engine Controls*.

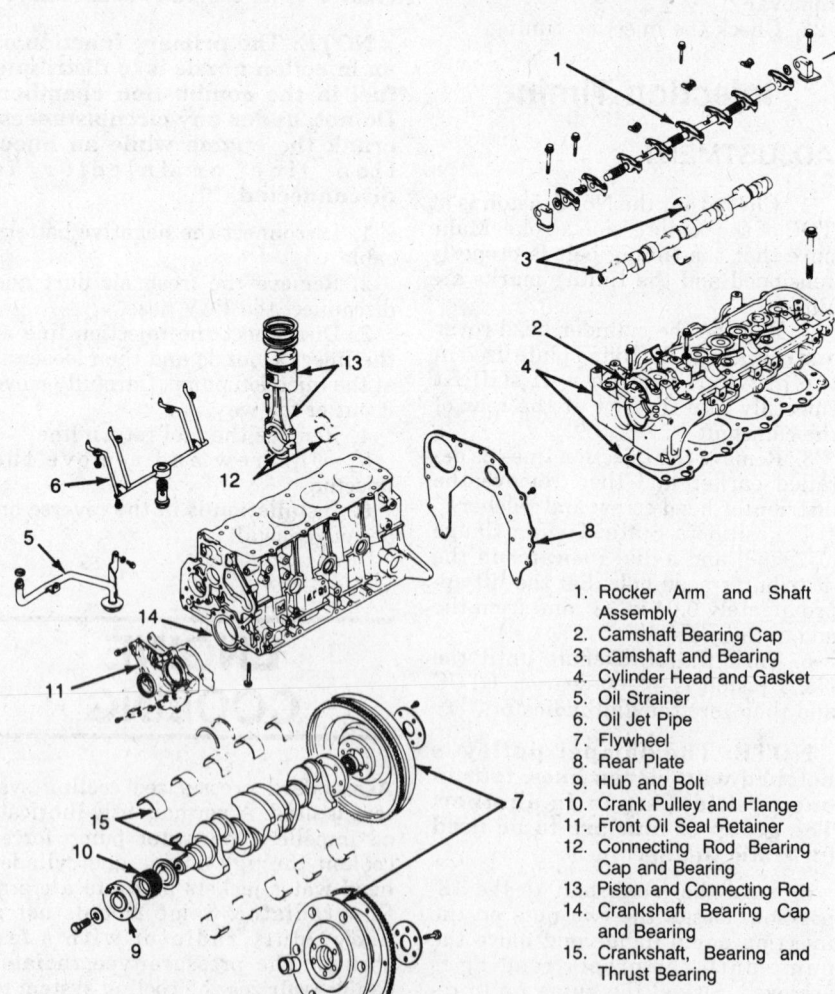

1. Rocker Arm and Shaft Assembly
2. Camshaft Bearing Cap
3. Camshaft and Bearing
4. Cylinder Head and Gasket
5. Oil Strainer
6. Oil Jet Pipe
7. Flywheel
8. Rear Plate
9. Hub and Bolt
10. Crank Pulley and Flange
11. Front Oil Seal Retainer
12. Connecting Rod Bearing Cap and Bearing
13. Piston and Connecting Rod
14. Crankshaft Bearing Cap and Bearing
15. Crankshaft, Bearing and Thrust Bearing

Exploded view of the diesel engine

ENGINE

REMOVAL & INSTALLATION

——— CAUTION ———

Do not discharge the air conditioning compressor or disconnect any of the refrigerant lines unless you have the skill and experience necessary to do so. Personal injury from the freon gas may result.

1. Remove the hood.
2. Disconnect the battery cables.
3. Remove the battery cable clips from the frame rail.
4. Drain the cooling system. Disconnect the radiator hoses from the engine and the heater hoses at the heater.
5. Tag and disconnect any wires leading from the engine.
6. Remove the radiator upper support and remove the radiator and engine fan. On the diesel, you must also remove the oil cooler.
7. Remove the air cleaner assembly.
8. Disconnect the following items:
   a. Fuel line at the rubber hose along the left frame rail. On the diesel, disconnect and plug the fuel lines at the injector pump and position them out of the way.
   b. Automatic transmission throttle valve linkage.
   c. Accelerator cable.
9. On air conditioned cars, remove the compressor from its mount and lay it aside. If equipped with power steering, remove the power steering pump and bracket and lay it aside.
10. Raise the car and support it with jackstands.
11. Remove the engine strut (shock-type) on the diesel.
12. Disconnect the exhaust pipe at the exhaust manifold.
13. Remove the flywheel dust cover on manual transmission cars or the torque converter underpan on automatic transmission cars.
14. On automatic transmission cars, remove the torque converter-to-flywheel bolts.
15. Remove the converter housing or flywheel housing-to-engine retaining bolts and lower the car.
16. Position a floor jack or other suitable support under the transmission.
17. Remove the safety straps from the front engine mounts and remove the mount nuts.
18. Remove the oil filter on the diesel.
19. Install the engine lifting apparatus.

1. Camshaft Cover and Gasket	14. Rocker Arm, Adjuster, Valve Springs, Valve Spring Cap, and Keys	26. Lower Cover
2. Camshaft Sprocket	15. Piston Rings	27. Idler
3. Camshaft Sprocket Guide	16. Piston	28. Crankshaft Sprocket
4. Camshaft Oil Seal	17. Connecting Rod	29. Crankcase Front Cover
5. Camshaft	18. Connecting Rod Bearing and Cap	30. Cylinder Block
6. Exhaust Manifold	19. Piston Pin	31. Engine Mounting Bracket
7. Camshaft Housing	20. Intake Manifold Gasket	32. Crankshaft and Bearings
8. Camshaft Rear Cover Gasket	21. Intake Manifold	33. Flywheel
9. Camshaft Rear Cover	22. Valves	34. Engine Mount
10. Camshaft Housing Cover and Gasket	23. Cylinder Head Gasket	35. Oil Pump Assembly
11. Timing Belt Cover	24. Washer	36. Transmission Mounting and Support
12. Timing Belt	25. Crankshaft Pulley	37. Engine Mounting Plate and Spring
13. Oil Dipstick and Tube		38. Oil Pan and Gasket

Exploded view of the gasoline engine

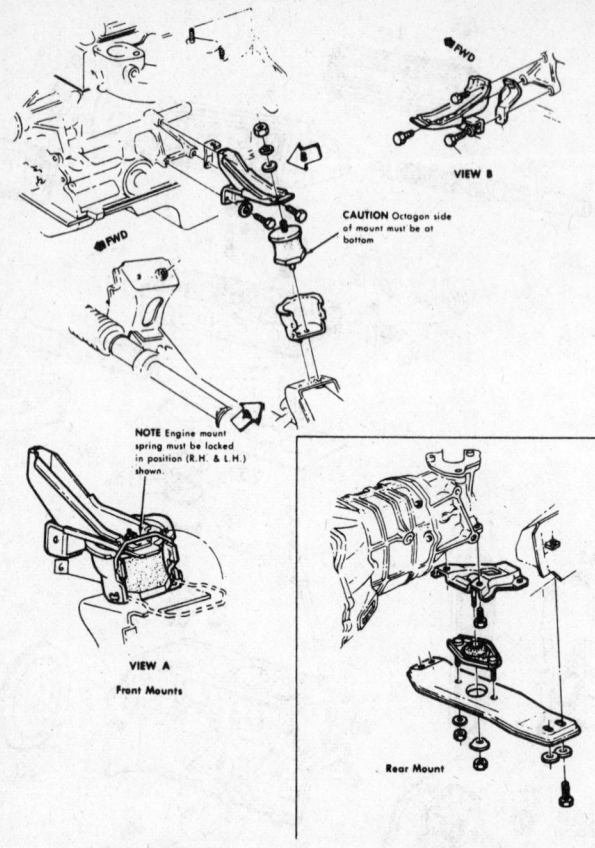

CAUTION Octagon side of mount must be at bottom

VIEW B

@FWD

NOTE Engine mount spring must be locked in position (R.H. & L.H.) shown.

VIEW A
Front Mounts

Rear Mount

Engine mounts—gasoline engine

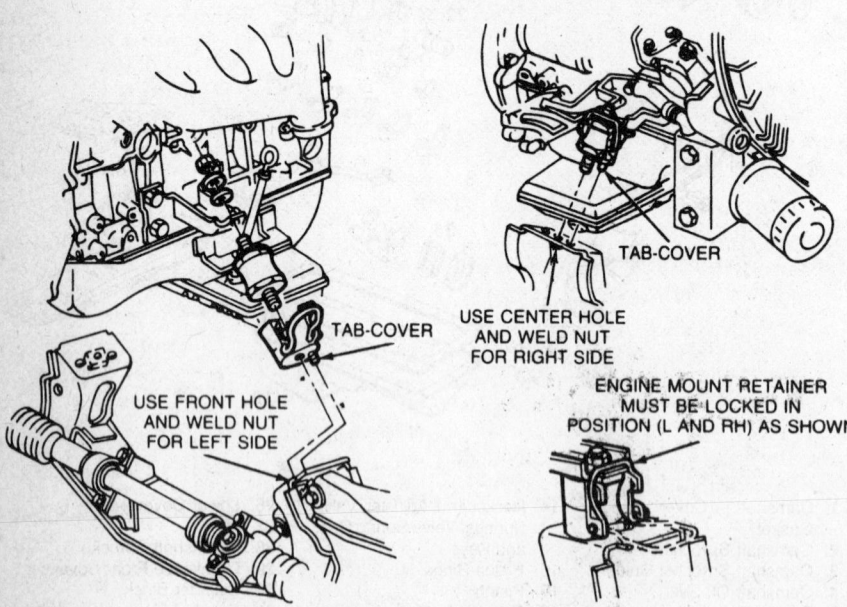

TAB-COVER

TAB-COVER

USE CENTER HOLE AND WELD NUT FOR RIGHT SIDE

USE FRONT HOLE AND WELD NUT FOR LEFT SIDE

ENGINE MOUNT RETAINER MUST BE LOCKED IN POSITION (L AND RH) AS SHOWN

Engine mounts—diesel engine

20. Remove the engine by pulling forward to clear the transmission while lifting slowly. Check to make sure that all necessary disconnections have been made and that proper clearance exists with surrounding components. Remove the lifting apparatus.

To install the engine:
21. Install the engine lifting apparatus and install guide pins in the engine block.
22. Install the engine in the car by aligning the engine with the transmission housing.

23. Install the front engine mount nuts and safety straps.
24. Raise the car and support it with jackstands.
25. Install the engine-to-transmission housing bolts. Tighten to 25 ft. lbs.
26. On automatic transmission cars, install the torque converter to the flywheel. Torque the bolts to 35 ft. lbs.
27. Install the flywheel dust cover or torque converter underpan as applicable.
28. Install the engine strut on the diesel.
29. Install the exhaust pipe to the exhaust manifold and lower the car.
30. Install the air conditioning compressor or the power steering pump if necessary, and adjust drive belt tension.
31. Connect the fuel lines, automatic transmission throttle valve linkage and accelerator cable.
32. Install the air cleaner.
33. Install the engine fan, radiator, and radiator upper support. Install the oil cooler if so equipped.

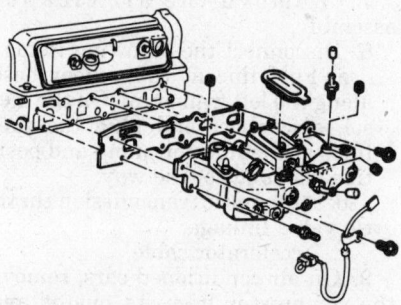

Intake manifold—gasoline engine

34. Connect all wires previously disconnected.
35. Connect the radiator and heater hoses and fill the cooling system.
36. Install the battery cable clips along the frame rail.
37. Install the engine hood.
38. Connect the battery cables, start the engine and check for leaks.

Intake Manifold

REMOVAL & INSTALLATION

Gasoline Engine

1. Disconnect the battery ground.
2. Drain the cooling system.
3. Remove the air cleaner.
4. Disconnect the upper radiator and heater hoses.
5. Remove the EGR valve.
6. Disconnect all electrical wiring, vacuum hoses and the accelerator linkage from the carburetor.
7. Disconnect the fuel line from the carburetor.

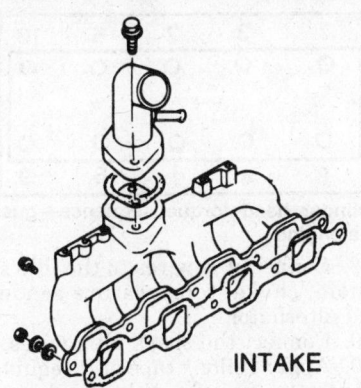

INTAKE

Intake manifold—diesel engine

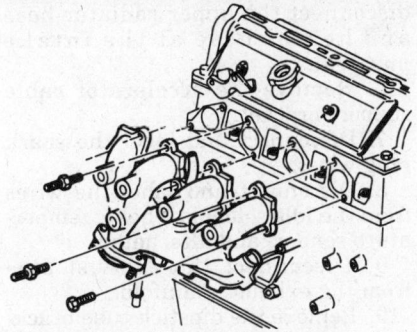

Exhaust manifold—gasoline engine

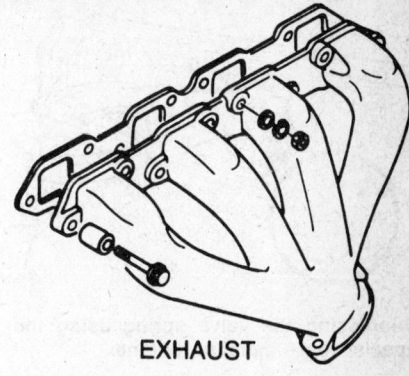

EXHAUST

Exhaust manifold—diesel engine

CYLINDER NO.	1		2		3		4	
VALVES	I	E	I	E	I	E	I	E
STEP. 1	○	○	○			○		
STEP. 2				◎	◎		◎	◎

I: INTAKE VALVE
E: EXHAUST VALVE

Valve adjustment sequence for the diesel engine

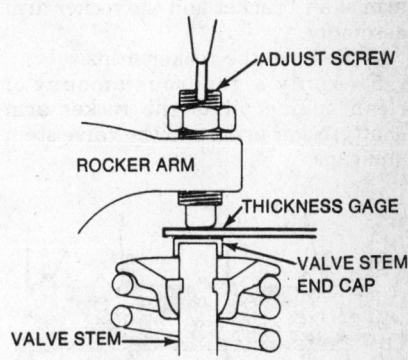

Valve adjustment—diesel engine

8. Remove the coil.
9. Remove the manifold.
10. If installing a new manifold, transfer all good parts. Always use a new gasket. Installation is the reverse of removal. Torque all bracket bolts to 30 ft. lbs. and intake manifold bolts to 15 ft. lbs.

Diesel Engine

1. Disconnect the negative battery cable.
2. Disconnect the fresh air hose and the vent hose. Remove the fuel separator.
3. Tag and disconnect all electrical connectors, the accelerator linkage and the glow plug wires.
4. Disconnect the injector lines at the injection pump and at the injector nozzles. Remove the injector lines and the hold-down clamps.
5. Remove the glow plug line at the cylinder head.
6. If equipped with power steering, remove the drive belt, the idler pulley and the bracket.
7. Remove the upper half of the front cover and the bracket.
8. Unscrew the mounting bolts and remove the intake manifold.
9. Places a new gasket over the mounting studs on the cylinder head and install the manifold. Tighten the bolts to 30 ft. lbs.
10. Installation of the remaining components is in the reverse order of removal.

Exhaust Manifold

REMOVAL & INSTALLATION

1. Disconnect the battery ground.
2. Raise the vehicle and support it on stands.
3. Disconnect the exhaust pipe from the flange.
4. Lower the vehicle.
5. On the diesel, remove the power steering belt, the flex hose and the power steering pump (if so equipped).

6. Remove the carburetor heat tube (gasoline engine only).
7. Remove the pulse air tubing, if so equipped.
8. Remove the manifold.
9. Installation is the reverse of removal. Install the two upper inner bolts first, to properly position the manifold. Tighten the bolts to the specified torque.

Valve Adjustment

GASOLINE ENGINE

Adjustment of the hydraulic valve lash adjusters is not possible. Cleanliness should be exercised when handling the valve lash adjusters. Before installation of lash adjusters, fill them with oil and check the lash adjuster oil hole in the cylinder head to make sure that it is free of foreign matter.

DIESEL ENGINE

NOTE: The rocker arm shaft bracket bolts and nuts should be tightened to 20 ft. lbs. before adjusting the valves.

1. Unscrew the retaining bolts and remove the cylinder head cover.
2. Rotate the crankshaft until the No. 1 or No. 4 piston is at TDC of the compression stroke.
3. Start with the intake valve on the No. 1 cylinder and insert a feeler gauge of the correct thickness (intake–0.01 in.; exhaust–0.014 in.) into the gap between the valve stem cap and the rocker arm. If adjustment is required, loosen the lock nut on top of the rocker arm and turn the adjusting screw clockwise to decrease the gap and counterclockwise to increase it. When the proper clearance is reached, tighten the lock nut and then recheck the gap. Adjust the remaining three valves in this step (see illustration) in the same manner.
4. Rotate the crankshaft one complete revolution and then adjust the

remaining valves accordingly (see illustration).

Rocker Arm

REMOVAL & INSTALLATION

Gasoline Engine

NOTE: A special valve spring compressor is necessary for this procedure. (Tool-J-25477) Also prelubricate new rocker arms with Molykote® or its equivalent.

1. Remove the camshaft cover.
2. Using the special valve spring compressor, compress the valve springs and remove the rocker arms. Keep the rocker arms and guides in order so that they can be installed in their original locations.
3. To install the rocker arms, compress the valve springs and install the rocker arm guides.
4. Position the rocker arms in the

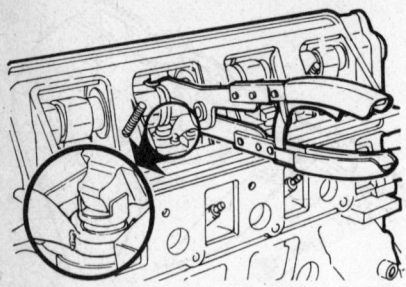

Depressing the valve spring using the special tool—gasoline engine

guides and on the valve lash adjusters.

5. Install the camshaft cover.

Diesel Engine

1. Disconnect the negative battery cable.

2. Remove the cylinder head cover.

3. Remove the rocker arm shaft bracket bolts and nuts in sequence (see illustration). Remove the rocker arm shaft bracket and the rocker arm assembly.

4. Remove the rocker arms.

5. Apply a generous amount of clean engine oil to the rocker arm shaft, rocker arms and the valve stem end caps.

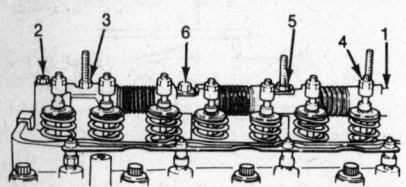

Loosening and tightening sequence for the rocker arm shaft bracket bolts and nuts—Diesel engine

6. Install the rocker arm shaft assembly and then tighten the bolts to 20 ft. lbs. in the same sequence as removal.

7. Adjust the valves as previously detailed and reinstall the cylinder head cover.

Cylinder Head

REMOVAL & INSTALLATION

Gasoline Engine

1. Disconnect the negative battery cable.

2. Remove all accessory drive belts.

3. Remove the engine fan, timing belt cover and the timing belt, as outlined later in this section.

4. Remove the air cleaner and snorkel (silencer) assembly.

5. Drain the cooling system and

disconnect the upper radiator hose and heater hose at the intake manifold.

6. Remove the accelerator cable support bracket.

7. Disconnect and label the spark plug wires.

8. Disconnect and label the wires from the idle solenoid, choke, temperature sender, and alternator.

9. Disconnect the exhaust pipe from the exhaust manifold.

10. Remove the dipstick tube bracket-to-manifold attaching bolt.

11. Disconnect the fuel line at the carburetor.

12. Take off the coil cover. Remove the coil bracket bolts and lay the coil aside.

13. Remove the camshaft cover.

14. Remove the camshaft cover-to-camshaft housing attaching stubs.

15. Remove the rocker arms, rocker arm guides, and valve lash adjusters. Keep the parts in order so that they can be installed in their original locations.

16. Remove the camshaft carrier bolts and remove the camshaft carrier. A sharp wedge may be necessary to separate the camshaft carrier from the cylinder head. Be very cautious not to damage the mating surfaces.

17. Remove the manifold and cylinder head assembly.

18. Install a new cylinder head gasket with the words "This Side Up" facing up over dowel pins in the block. Make sure that the gasket is absolutely clean.

19. Install the manifold and cylinder head assembly.

20. Apply a light, thin continuous bead of sealant to the jointing surfaces of the cylinder head and the camshaft carrier and install the camshaft carrier. Clean any excess sealer from the cylinder head. Apply sealing compound to the camshaft carrier/cylinder head bolts and install the bolts finger-tight. Tighten the bolts a little at a time and in the proper sequence until the final specified torque figure is reached.

21. Install the camshaft cover-to-camshaft housing attaching studs.

22. Install the valve lash adjusters and rocker arm guides. Prelube the rocker arms with engine assembly lubricant and install the rocker arms.

23. Using new gaskets, install the camshaft covers.

24. Install the coil bracket mounting bolt.

25. Connect the fuel line to the carburetor.

26. Install the dipstick tube bracket-to-manifold attaching bolt.

27. Attach the exhaust pipe to the exhaust manifold.

```
FRONT   7   3   2   6   10
        O   O   O   O   O

        O   O   O   O   O
        8   4   1   5   9
```

Cylinder head torque sequence—gasoline engine

28. Connect the wires to the idle solenoid, choke, temperature sender, and alternator.

29. Connect the spark plug wires.

30. Apply Teflon® tape or its equivalent to the threads of the accelerator cable support bracket attaching bolts and install the bracket.

31. Install the air cleaner and snorkel (silencer) assembly.

32. Connect the upper radiator hose and heater hose to the intake manifold.

33. Fill the cooling system.

34. Install the timing belt, timing belt cover, engine fan, drive belts and connect the negative battery cable.

Diesel Engine

1. Disconnect the negative battery cable.

2. Drain the cooling system.

3. Remove the cylinder head cover.

4. Disconnect the bypass hose. Remove the upper half of the front cover.

5. Loosen the tension pulley bolts and then remove the camshaft as detailed later in this section.

6. Tag and disconnect the glow plug resistor wire.

7. Tag and disconnect the glow plug resistor wire.

8. Disconnect the injector lines at the injector pump and at the injector nozzles and then remove the injector lines. Disconnect and plug the fuel leak-off hose.

9. Disconnect the exhaust pipe at the manifold.

10. Remove the oil feed pipe from the rear of the cylinder head.

11. Disconnect the upper radiator hose and position it out of the way.

12. Remove the head bolts in the sequence shown and then remove the cylinder head with the intake and exhaust manifolds installed.

NOTE: The gasket surfaces on both the head and the block must be clean of any foreign matter and free of nicks or heavy scratches. Cylinder bolt threads in the block and on the bolt must also be clean.

13. Place a new gasket over the dowel pins with the word "TOP" facing up.

14. Apply engine oil to the threads and the seating face of the cylinder head bolts, install them and then tighten them in the proper sequence.

15. Install the camshaft and rocker

arm assembly. Loosen the adjusting screws so that the entire rocker arm assembly is held in a free state.

16. Reinstall the timing belt as outlined later in this section.

17. Connect the upper radiator hose and the oil feed pipe.

18. Connect the exhaust pipe to the manifold.

19. Install the fuel leak-off hose. Connect the injector lines.

20. Connect the glow plug resistor wire.

21. Adjust the valve clearance as previously detailed. Install the cylinder head cover.

22. Refill the cooling system.

Timing Belt Cover

REMOVAL & INSTALLATION

Upper Front Cover

1. Disconnect the negative battery cable. Remove the radiator upper mounting panel on models without A/C or fan shroud on models with A/C.

2. Remove engine accessory drive belts on the gasoline engine. Remove the bypass hose on the diesel engine.

3. Remove the engine fan.

4. Remove the cover retaining screws and nuts and remove the cover.

5. To install; align the screw slots on the upper and lower parts of the cover.

6. Install the cover retaining screws and nuts.

7. Install the engine fan.

8. Install the engine accessory drive belts or the bypass hose.

9. Connect the negative battery cable.

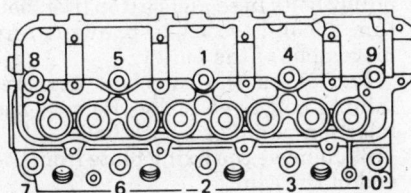

Cylinder head torque sequence—diesel engine

Lower Front Cover

1. Disconnect the negative battery cable.

2. Loosen the alternator and the A/C compressor bolts, if so equipped. Remove the drive belt.

3. Remove the damper pulley-to-crankshaft bolt and washer and remove the pulley.

4. Remove the upper front timing belt cover as outlined previously.

5. Remove the lower cover retaining nut (gasoline) or bolts (diesel). Remove the lower cover.

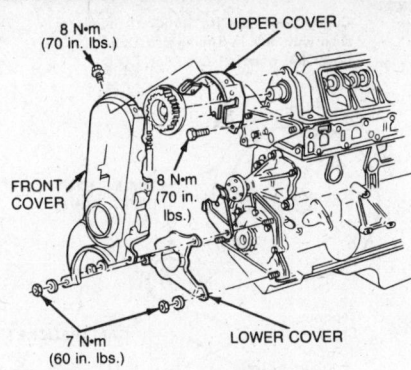

Timing belt front cover fasteners—gasoline engine

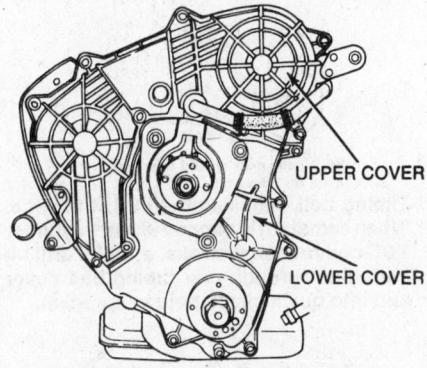

Timing belt covers—diesel engine

6. To install the cover, align the cover with the studs on the engine block.

7. Install the lower front cover retaining nut or bolts.

8. Install the upper front timing belt cover.

9. Install the crankshaft damper pulley. Torque the retaining bolt to the specified torque.

10. Install the drive belt and tighten the alternator and compressor mounting bolts.

11. Connect the negative battery cable.

Upper Rear Cover—Gasoline Engine

1. Crank the engine so that No. 1 cylinder is at TDC of the compression stroke.

2. Disconnect the negative battery cable.

3. Remove the upper and lower front cover, the timing belt, and the camshaft timing sprocket.

4. Remove the three screws retaining the camshaft sprocket cover to the camshaft carrier.

5. Inspect the condition of the cam seal

6. Position and align a new gasket over the end of the camshaft and against the camshaft carrier.

7. Install the three camshaft sprocket cover retaining screws.

8. Install the camshaft sprocket, timing belt, and the upper and lower front covers.

9. Connect the negative battery cable.

Timing Belt & Sprockets

REMOVAL & INSTALLATION

Gasoline Engine

—— CAUTION ——

Do not discharge the air conditioning compressor or disconnect the air conditioning lines. Personal injury could result.

NOTE: Rotate the engine to bring No. 1 cylinder to TDC. The timing mark should be at the 0° mark on the timing scale. With No. 1 cylinder at TDC, a $\frac{1}{8}$ in. drill bit may be inserted through a hole in the timing belt upper rear cover into a hole in the camshaft drive sprocket. These holes are provided to facilitate and verify camshaft timing. Aligning these holes now will make installation of the new belt much easier.

1. Disconnect the negative battery cable.

2. Remove the alternator and air conditioning compressor drive belts.

3. Remove the engine fan and pulley.

4. Remove the engine upper and lower front timing belt covers.

5. Remove the timing belt idler pulley.

6. Remove the timing belt from the camshaft and crankshaft timing sprockets.

7. With the distributor cap off, mark the location of the rotor in the No. 1 spark plug firing position on the distributor housing. On air conditioned cars, remove the compressor and lower its mounting bracket.

8. Remove the camshaft timing sprocket bolt and washer and remove the camshaft sprocket.

9. Remove the crankshaft sprocket.

10. To install; place the crankshaft sprocket on the crankshaft making sure that the locating tabs face outward.

11. Install the crankshaft sprocket.

12. Align the camshaft sprocket dowel with the hole in the end of the camshaft and install the sprocket on the camshaft.

13. Apply thread locking compound to the camshaft sprocket retaining

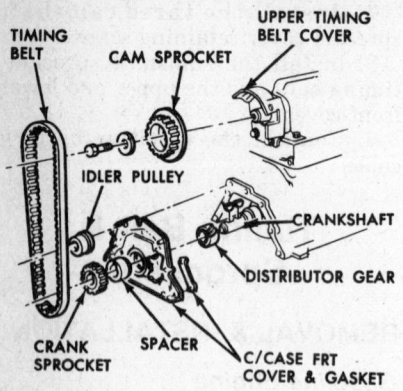

Timing belt and gears—gasoline engine

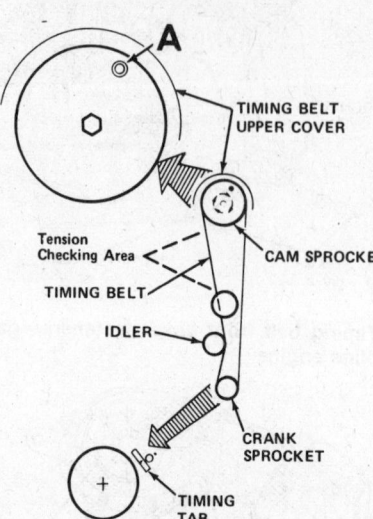

Quick Check Hole (In Sprocket) should align with hole in Timing Belt Upper Cover (A) when #1 Cyl. is at T.D.C.

Pulley timing mark should align with 0° mark on timing tab.

Timing belt installation—1.6 L Chevette. When camshaft is aligned at No. 1 cylinder TDC compression stroke, a 1/8 in. drill bit should fit through rear timing belt cover and into quick check hole in sprocket.

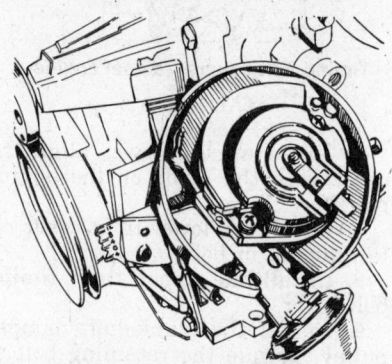

Correct distributor rotor alignment for timing belt installation

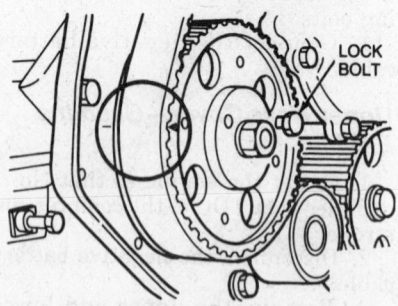

Injection gear setting mark—diesel engine

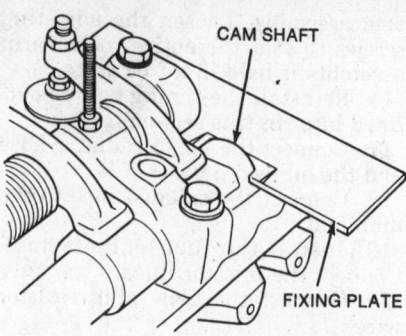

Camshaft fixing plate—diesel engine

bolt and washer and torque to 65–85 ft. lbs.

14. Position the timing belt over the crankshaft sprocket.

15. Install the crankshaft pulley.

16. Align the crankshaft pulley timing mark with the "0" mark on the timing scale and the distributor rotor with the scribed mark on the distributor housing.

17. Align the hole in the camshaft sprocket with the hole in the upper rear timing belt cover. Insert a ⅛ in. drill bit to hold the sprocket in alignment.

18. Install the timing belt on the camshaft and crankshaft sprockets.

19. To adjust timing belt tension see "Timing Belt Adjustment."

20. Install the distributor cap. On air conditioned cars, install the lower compressor bracket and the compressor.

21. Install the upper and lower front timing belt covers.

22. Install the engine fan and pulley.

23. Install the alternator and, if necessary, the air conditioning compressor drive belts.

24. Connect the negative battery cable.

Diesel Engine

NOTE: In order to complete this procedure you will need three special tools. A gear puller (J-22888), a fixing plate (J-29761) and a belt tension gauge (J-26484).

1. Disconnect the negative battery cable.

2. Drain the cooling system.

3. Remove the fan shroud, cooling fan and the pulley.

4. Disconnect the bypass hose and then remove the upper half of the front cover.

5. With the No. 1 piston at TDC of the compression stroke, make sure that the notch mark on the injection pump gear is aligned with the index

mark on the front plate. If so, thread a lock bolt (8mm x 1.25) through the gear and into the front plate.

6. Remove the cylinder head cover and install a fixing plate (J-29761) in the slot at the rear of the cam. This will prevent the cam from rotating during the procedure.

7. Remove the crankshaft damper pulley and check to make sure that the No. 1 piston is still at TDC.

8. Remove the lower half of the front cover and then remove the timing belt holder from the bottom of the front plate.

9. Remove the tension spring behind the front plate, next to the injection pump.

10. Loosen the tension pulley and slide the timing belt off the pulleys.

11. Remove the camshaft gear retaining bolt, install a gear puller and remove the gear.

12. When assembling, reinstall the cam gar loosely so that it can be turned smoothly by hand.

13. Slide the timing belt back over the gears and note the following: the belt should be properly tensioned between the pulleys, the cogs on the belt and the gears should be properly engaged, the crankshaft should not be turned and the belt slack should be concentrated at the two tension pulleys. Push the tension pulley in with your finger and install the tension spring.

14. Partially tighten the tension pulley bolts in sequence (top first, bottom second) so as to prevent any movement of the pulley.

15. Tighten the camshaft gear retaining bolt to 45 ft. lbs. Remove the injection pump gear lock bolt.

16. Remove the fixing plate from the end of the cam.

17. Install the crankshaft damper pulley and then check that the No. 1 piston is still at TDC. Do not try to adjust it by moving the crankshaft.

18. Check that the marks on the injection pump gear and the front plate are still aligned and that the fixing plate still fits properly into the slot on the camshaft.

19. Loosen the tensioner pulley and plate bolts, concentrate the looseness of the timing belt around the tensioner and then tighten the bolts.

20. Belt tension should be 46–63 lbs., checked at a point midway between the upper two pulleys.

21. Remove the damper pulley again

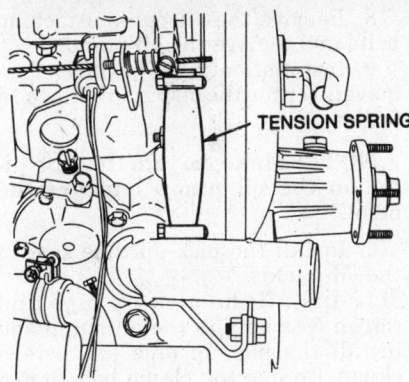

Tension spring—diesel engine

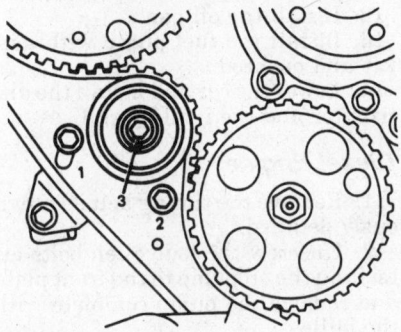

Tighten the tension pulley bolts in sequence—diesel engin

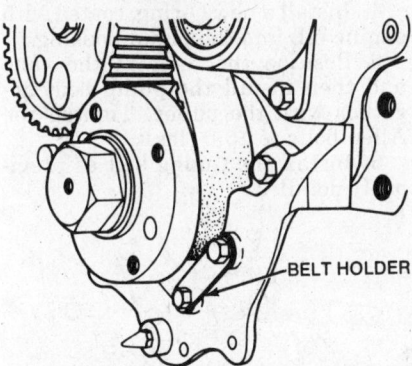

Diesel engine: the timing belt holder must be removed before the timing belt can be taken off

and install the belt holder in position away from the timing belt.

22. Installation of the remaining components is in the reverse order of removal.

TIMING BELT ADJUSTMENT

1. Remove the fan, fan belt, water pump pulley and upper cam belt cover.

2. Rotate the crankshaft clockwise a minimum of one revolution. Stop with No. 1 piston at TDC. DO NOT TURN THE ENGINE BACKWARD!

3. Install a belt tension gauge on the same side as the idler pulley (in-

jection pump pulley on diesel), midway between the cam sprocket and the idler pulley (injection pump pulley on diesel). Be sure that the center finger of the gauge extension fits in a notch between the teeth on the belt. Correct belt tension is 70 lbs. (46–63 lbs. for the diesel).

4. If the tension is incorrect, loosen the idler pulley attaching bolt and using a ¼ in. Allen wrench, rotate the pulley counterclockwise on its attaching bolt until the proper tension is obtained. Torque the bolt to 15 ft. lbs.

5. The remainder of the installation is the reverse of the removal procedure.

Camshaft

REMOVAL & INSTALLATION

Gasoline Engine

NOTE: A special valve spring compressor (tool No. J-25477) is necessary for this procedure. If replacing the camshaft or rocker arms, prelube new parts with engine assembly lubricant.

1. Disconnect the negative battery cable.

2. Remove engine accessory drive belts.

3. Remove the engine fan and pulley.

4. Remove the upper and lower front timing belt covers.

5. Loosen the idler pulley and remove the timing belt from the camshaft sprocket.

6. Remove the camshaft sprocket attaching bolt and washer and remove the camshaft sprocket.

7. Remove the camshaft cover. Using the special valve spring compressor, remove the rocker arms and guides. Keep the rocker arms and guides in order so that they can be installed in their original locations.

8. Remove any components necessary to gain working clearance.

NOTE: The heater assembly will probably have to be removed from the firewall.

9. Remove the camshaft carrier rear cover.

10. Remove the camshaft thrust plate bolts. Slide the camshaft slightly to the rear and remove the thrust plate.

11. Remove the engine mount nuts and wire retainers.

12. Using a floor jack, raise the front of the engine.

13. Remove the camshaft from the camshaft carrier. Heavy pressure will be needed to pull the camshaft and seal forward.

14. Install the camshaft into the camshaft carrier.

15. Lower the engine.

16. Install the engine mount nuts and attach the retaining wires.

17. Slide the camshaft slightly to the rear and install the thrust plate. Slide the camshaft forward and install the carrier rear cover.

18. Position and align a new gasket over the end of the camshaft, against the camshaft carrier.

19. Install any components which were removed to gain working clearance.

20. Install the valve rocker arms and guides in their original locations using the special valve spring compressor. Install the camshaft covers.

21. Align the dowel in the camshaft sprocket with the hole in the end of the camshaft and install the sprocket.

22. Apply thread locking compound to the sprocket retaining bolt threads and install the bolt and washer. Torque the sprocket retaining bolt to 65–85 ft. lbs.

23. Turn the crankshaft clockwise to bring the No. 1 cylinder to top dead center. Make sure that the distributor rotor is in position to fire the No. 1 spark plug. Align the hole in the camshaft sprocket with the hole in the upper rear timing belt cover and install the timing belt on the camshaft sprocket.

24. Adjust timing belt tension as previously outlined.

25. Install the upper and lower front timing belt covers.

26. Install the engine fan and pulley.

27. Install the engine accessory drive belts.

28. Connect the negative battery cable.

Diesel Engine

NOTE: In order to complete this procedure you will need a gear puller (J-22888) and a fixing plate (J-29761).

1. Remove the cylinder head cover.

2. Remove the timing belt as previously detailed. Remove the plug.

3. Install the fixing plate into the slot at the rear of the camshaft.

4. Remove the camshaft gear retaining bolt and then use a puller to remove the cam gear.

5. Remove the rocker arms and shaft as previously detailed.

6. Unscrew the bolts attaching the front head plate and then remove the plate.

7. Unscrew the camshaft bearing cap retaining bolts and remove the bearing caps with the cap side bearings.

8. Lift out the camshaft oil seal and then remove the camshaft.

9. Coat the cam and cylinder head journals with clean engine oil.

10. Position the camshaft back in the cylinder head with a new oil seal.

11. Apply a suitable liquid gasket to the cylinder head face of the No. 1 camshaft bearing cap.

12. Install the remaining bearing caps. Install the rocker arm shaft assembly, leaving the adjusting screws loose.

13. Install the front head plate.

14. Install the timing belt as previously detailed.

15. Adjust the valve clearance to specifications and then install the cylinder head cover.

Piston and Connecting Rod Positioning

Install piston and connecting rod assemblies into their original cylinders. Install the piston and rod assemblies with the notch (arrow-diesel) on the piston crown facing to the front of the engine. The numbers on the connecting rods and bearing caps must be on the same side when installing pistons and connecting rods.

ENGINE LUBRICATION

Oil Pan

REMOVAL & INSTALLATION

Gasoline Engine

1. Disconnect the negative battery cable.

2. Drain the cooling system.

3. Remove the heater housing assembly from the firewall and rest it on top of the engine.

4. Remove the upper radiator support. On cars with A/C, remove the upper half of the fan shroud.

5. Remove the radiator hoses and on cars with automatic transmission, disconnect and plug the cooler lines from the radiator.

6. Remove the radiator.

7. On cars equipped with A/C, remove the condenser from its supporting bracket. Lay the condenser on top of the engine. Do not disconnect any of the refrigerant lines.

8. Remove the motor mount nuts and clips.

9. Raise the car and drain the engine oil.

10. Remove the flywheel splash shield.

11. On all models with the 200 automatic transmission, loosen the catalytic converter-to-exhaust pipe clamp bolts. On other models, disconnect the exhaust pipe at the manifold.

12. Remove the body-to-crossmember braces, if so equipped.

13. Remove the rack and pinion unit from the crossmember and the steering shaft. Pull the unit down and out of the way.

14. With a floor jack and a lifting adapter, raise the front of the engine.

15. Remove the oil pan bolts.

16. Pull the oil pan down and remove the oil pump suction pipe and the screen.

17. Remove the oil pan.

18. Clean all of the old sealer that is loose off the oil pan mating surface. It is not necessary to clean all of the sealer material off. Reverse the above procedure to install. Tighten the oil pan attaching bolts to 5 ft. lbs.

Diesel Engine

1. Remove the engine as detailed earlier in this section.

2. Support the engine in a stand.

3. Unscrew the nuts and bolts attaching the oil pan to the crankcase and then remove the pan.

4. Clean the mating surfaces of the oil pan and the block. Apply a suitable liquid gasket to the front and rear mating surfaces and then install a new gasket.

5. Install the oil pan retaining bolts and tighten them to 5 ft. lbs.

6. Reinstall the engine.

Oil Pump

REMOVAL & INSTALLATION

Gasoline Engine

1. Remove the ignition coil attaching bolts and lay the coil aside.

2. Raise the car and remove the fuel pump, pushrod, and gasket.

3. Lower the car and remove the distributor. On air conditioned cars, remove the compressor mounting bolts and lay it aside. Do not disconnect any refrigerant lines.

4. Raise the car and remove the oil pan as previously outlined.

5. Remove the oil pump pipe and screen assembly clamp and remove the bolts attaching the pipe and screen assembly.

6. Remove the pipe and screen assembly from the oil pump.

7. Remove the pick-up tube seal from the oil pump.

8. Remove the oil pump attaching bolts and remove the oil pump.

9. Installation is the reverse of removal. Torque the oil pump bolts to 15 ft. lbs.

NOTE: Make certain that the pilot on the oil pump engages the case.

10. Install the pick-up tube seal in the oil pump.

11. Install the pick-up pipe and screen assembly in the oil pump and install the pick-up pipe and screen clamp. Torque the clamp bolt to 6–8 ft. lbs. Torque the pick-up tube and screen mounting bolt to 19–25 ft. lbs.

12. Install the oil pan.

13. Install the fuel pump with gasket and pushrod.

14. Lower the car and install the distributor and the ignition coil.

Diesel Engine

1. Remove the timing belt as previously detailed.

2. Unscrew the four allen bolts attaching the oil pump to the front plate and remove the pump complete with the pulley.

3. Coat the vane with clean engine oil and then install it with the taper side toward the cylinder body.

4. Install a new O-ring, coated with engine oil, into the pump housing.

5. Position the rotor in the vane and then install the pump body together with the pulley. Tighten the Allen bolts to 15 ft. lbs.

6. Install the timing belt as previously detailed.

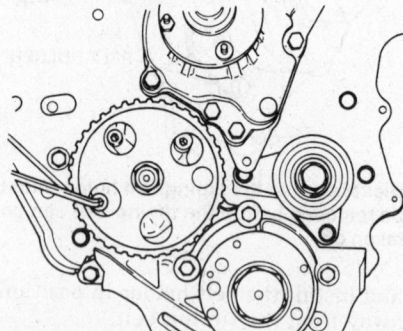

Removing the allen bolts attaching the oil pump to the front pate—diesel engine

Rear Main Oil Seal

REPLACEMENT

Gasoline Engine

1. Remove the engine from the car and place it in a stand.

2. Remove the flywheel or flexplate.

3. Remove the oil pan.

4. Remove the rear main bearing cap.

5. Clean the bearing cap and case.

6. Check the crankshaft seal for excessive wear, etc.

7. Install a new crankshaft seal. Make sure that it is properly seated against the rear main bearing seal bulkhead.

8. Apply RTV sealer or its equivalent to the bearing cap horizontal split line.

9. With the sealer still wet, install the rear main baring cap. Tighten the bearing bolts to 10–12 ft. lbs. Tap the crankshaft toward the rear, then toward the front to be sure everything is properly seated. Tighten the cap bolts to the specified torque.

10. Apply RTV sealer or its equivalent in the vertical grooves of the rear main bearing cap.

11. Remove any excess sealer and install the oil pan. Torque the oil pan bolts to 4–5 ft. lbs.

12. Install the engine.

Diesel Engine

1. Remove the transmission as detailed later in this section. If equipped with a manual transmission, remove the clutch.

2. Loosen the flywheel retaining bolts in a diagonal pattern and then remove the flywheel.

3. Pry off the old oil seal.

4. Coat the lipped portion and the fitting face of the new oil seal with engine oil and install it into the crankshaft bearing. Make sure that the seal is properly seated.

5. Coat the threads of the new mounting bolts with Loctite® and install the flywheel. Tighten the bolts to 40 ft. lbs. in a diagonal sequence. Do not reuse the old bolts, they must be new.

6. Installation of the remaining components is in the reverse order of removal.

CLUTCH

All manual transmission models use a cable-operated diaphragm spring-type clutch. The clutch cable is attached to the clutch pedal at its upper end and is threaded at its lower end where it attaches to the clutch fork. The clutch release fork pivots on a ball stud located opposite the clutch cable attaching point. The pressure plate, clutch disc, and throwout bearing are of conventional design.

Clutch

REMOVAL & INSTALLATION

1. Raise the car on a hoist.

2. Remove the transmission.

3. Remove the throwout bearing from the clutch fork by sliding the fork off the ball stud against spring tension. If the ball stud is to be replaced, remove the locknut and stud from the bellhousing.

4. If the balance marks on the pressure plate and the flywheel are not easily seen, mark them with paint or a centerpunch.

5. Alternately loosen the pressure plate-to-flywheel attaching bolts one turn at a time until spring tension is released.

6. Support the pressure plate and cover assembly, then remove the bolts and the clutch assembly.

--- CAUTION ---

Do not disassemble the clutch cover and pressure plate for repair. If defective, replace the assembly.

7. Check the pressure plate, clutch plate and flywheel for wear. If the flywheel is scored, worn or discolored from overheating, it should be either refaced or replaced. Replace the clutch plate as necessary.

8. Align the balance marks on the clutch disc on the pressure plate with the long end of the splined hub facing

forward and the damper springs inside the pressure plate. Insert a dummy shaft through the cover and clutch disc.

9. Position the assembly against the flywheel and insert the dummy shaft into the pilot bearing in the crankshaft.

10. Align the balance marks and install the pressure plate-to-flywheel bolts finger tight.

--- CAUTION ---

Tighten all bolts evenly and gradually until tight to avoid possible clutch distortion. Torque the bolts to 18 ft. lbs. (14 ft. lbs. on diesel engine) and remove the dummy shaft.

11. Pack the groove on the inside of the throwout bearing with graphite grease. Also coat the fork groove and ball stud depression with the lubricant.

12. Install the throwout bearing and

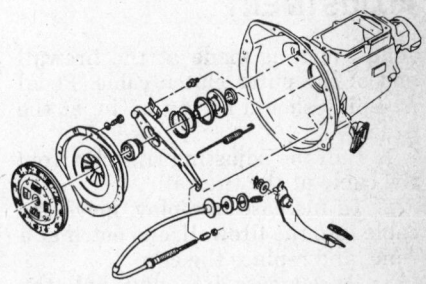

Clutch assembly

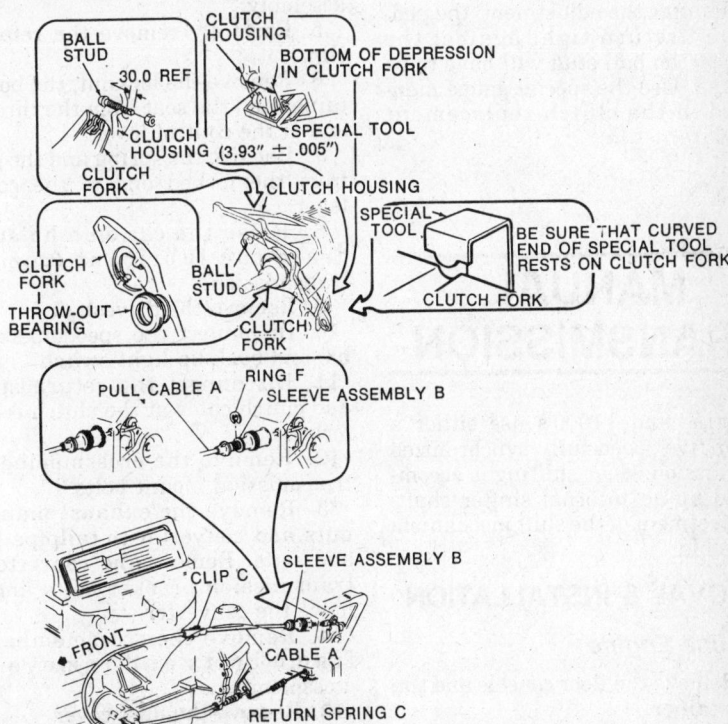

Clutch cable and ball stud adjustment details

release fork assembly in the bellhousing with the fork spring hooked under the ball stud and the fork spring fingers inside the bearing groove.

13. Position the transmission and clutch housing and install the clutch housing attaching bolts and lockwashers. Torque the bolts to 25 ft. lbs.

14. Complete the transmission installation.

NOTE: Check the position of the engine in the front mounts and realign as necessary. A special gauge (J-23644) is necessary to adjust ball stud position if it has been removed.

15. Adjust clutch pedal free-play if necessary.

16. Lower the car and check operation of the clutch and transmission.

CLUTCH PEDAL FREE-PLAY ADJUSTMENT

Adjustment is made at the firewall end of the outer clutch cable. Pedal free-play should be $\frac{1}{2}$ to 1 in. at the pedal.

1. Pull the adjusting ring clip from the cable at the firewall.

2. To increase free-play, move the cable into the firewall, one notch at a time, and replace the clip.

3. To decrease free-play, pull the cable out, one notch at a time, and replace the clip.

4. If, after the adjustment, the pedal won't return tight against the bumper, the ball stud will have to be adjusted. Use the special gauge mentioned in the clutch replacement procedure.

MANUAL TRANSMISSION

Chevettes and T1000's use either a four or five speed fully synchronized transmission. Gear shifting is accomplished by an internal shifter shaft. No adjustment of the shift mechanism is possible.

REMOVAL & INSTALLATION

Gasoline Engine

1. Remove the floor console and the boot retainer.

2. Lift up the boot in order to gain access to the locknut on the shift lever. Loosen the locknut and unscrew

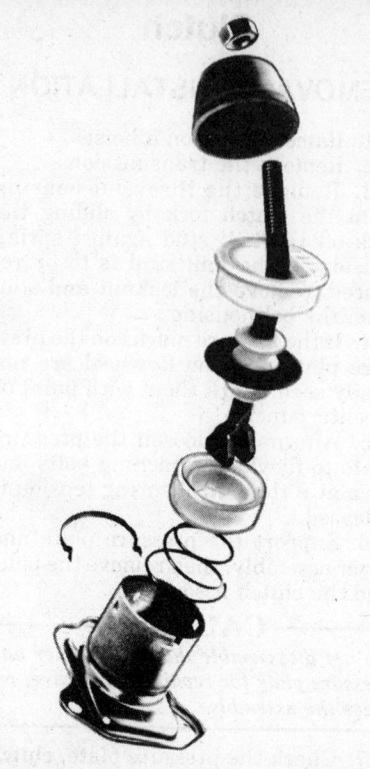

Gasoline engine shift lever components

the upper portion of the shift lever with the knob attached.

3. Remove the foam insulator.

4. Remove the three bolts on the extension and remove the control assembly.

5. Carefully remove the retaining clip.

6. Remove the locknut, the boot retainer and the seat from the threaded end of the control lever.

7. Remove the spring and the guide from the forked end of the control lever.

8. Raise the car on a hoist and drain the lubricant from the transmission.

9. Remove the driveshaft.

10. Disconnect the speedometer cable and back-up light switch.

11. Disconnect the return spring and clutch cable at the clutch release fork.

12. Remove the crossmember-to-transmission mount bolts.

13. Remove the exhaust manifold nuts and converter-to-tailpipe bolts and nuts. Remove the converter-to-transmission bracket bolts and remove the converter.

14. Remove the crossmember-to-frame bolts and remove the crossmember.

15. Remove the dust cover.

16. Remove the clutch housing-to-engine retaining bolts, slide the transmission and clutch housing to

the rear, and remove the transmission.

17. To install, place the transmission in gear, position the transmission and clutch housing, and slide forward. Turn the output shaft to align the input shaft splines with the clutch hub.

18. Install the clutch housing retaining bolts and lockwashers. Torque the bolts to 25 ft. lbs.

19. Install the dust cover.

20. Position the crossmember to the frame and loosely install the retaining bolts. Install the crossmember-to-transmission mounting bolts. Torque the center nuts to 33 ft. lbs.; the end nuts to 21 ft. lbs. Torque the crossmember-to-frame bolts to 40 ft. lbs.

21. Install the exhaust pipe to the manifold and the converter bracket on the transmission.

22. Connect the clutch cable. Adjust clutch pedal free-play.

23. Connect the speedometer cable and back-up light switch.

24. Install the driveshaft.

25. Fill the transmission to the correct level with SAE 80W or SAE 80W–90 GL-5 gear lubricant. Lower the car.

26. Install the shift lever and check operation of the transmission.

Diesel Engine

1. Disconnect the negative battery cable.

2. Unscrew the retaining screws and then remove the shift lever console.

3. Remove the mounting screws and remove the shift lever assembly.

4. Unscrew and remove the upper starter mounting bolts.

5. Raise the front of the car and drain the lubricant from the transmission.

6. Remove the drive shaft as detailed later in this section.

7. Disconnect the speedometer and the back-up light switch wires.

8. Disconnect the return spring and clutch cable at the clutch release fork.

9. Remove the starter lower bolt and support the starter.

10. Unscrew the retaining bolts and disconnect the exhaust pipe from the manifold.

11. Remove the flywheel inspection cover.

12. Unscrew the rear transmission support mounting bolt. Support the transmission underneath the case and then remove the rear support from the frame.

13. Lower the transmission approximately four inches.

14. Remove the transmission hous-

ing-to-engine block bolts. Pull the transmission straight back and away from the engine.

15. Installation of the remaining components is in the reverse order of removal. Please note the following:

a. Be sure to lubricate the drive gear shaft with a light coat of grease before installing the transmission.

b. After installation, fill the transmission to the level of the filler hole with 5W-30SF engine oil.

OVERHAUL

For all overhaul procedures, please refer to "Manual Transmission Overhaul" in the Unit Repair section.

AUTOMATIC TRANSMISSION

All Chevettes and T1000's use either the Turbo Hydra-Matic 180 or 200 transmission.

For all identification and service procedures, please refer to "Automatic Transmissions" in the Unit Repair section.

REMOVAL & INSTALLATION

1. Before raising the car, disconnect the negative battery cable and the T.V./detent cable at the bracket and carburetor or pump.

2. Remove the air cleaner and dipstick.

3. On vehicles with air conditioning, remove the heater core cover screws from the heater assembly. Disconnect the wire connector and with hoses attached, place the heater core cover out of the way.

4. Raise vehicle on hoist and remove propeller shaft.

5. Disconnect speedometer cable, electrical lead to case connector and oil cooler pipes.

6. Disconnect shift control linkage.

7. Support transmission with suitable transmission jack and remove the rear transmission support bolts.

8. Remove the nuts holding the converter bracket to the support.

9. Disconnect exhaust pipe at the rear of the catalytic converter.

10. Disconnect exhaust pipe at manifold and remove the exhaust pipe, catalytic converter and converter bracket as an assembly.

11. Remove the torque converter under pan.

12. Remove converter to flexplate bolts.

13. Lower transmission until jack is barely supporting it and remove transmission to engine mounting bolts.

14. Raise transmission to its normal position, then place a block of wood between the rack-and-pinion housing and the engine oil pan, then support engine with jack and slide transmission rearward from engine and lower it away from vehicle.

NOTE: The use of a converter holding tool J-5384 is necessary when lowering the transmission or keep the rear of the transmission lower than the front so not to lose the converter.

15. Installation is the reverse of removal. Before installing the flex plate to converter bolts, make certain that the weld nuts on the converter are flush with the flex plate and the converter rotates freely by hand in this position. Hand start the three bolts and tighten finger tight, then torque to specifications. This will insure proper converter alignment. Install new oil seal on oil filler tube before installing tube.

DRIVE AXLE

For all U-Joint overhaul procedures, please refer to "U-Joint/CV-Joint Overhaul" in the Unit Repair section.

Driveshaft

REMOVAL & INSTALLATION

1. Raise the car on a hoist. Scribe matchmarks on the driveshaft and the companion flange and disconnect the rear universal joint by removing the trunnion bearing straps.

2. Move the driveshaft to the rear under the axle to remove the slip yoke from the transmission. Watch for leakage from the transmission output shaft housing.

3. Install the driveshaft in the reverse order of removal. Tighten the trunnion strap bolts to 16 ft. lbs.

Axle Shaft, Bearing, and Seal

For further information on the

rear axle, please refer to "Drive Axles" in the Unit Repair section.

REMOVAL & INSTALLATION

1. Raise the car on a hoist. Remove the wheel and tire assembly and the brake drum.

2. Clean the area around the differential carrier cover.

3. Remove the differential carrier cover to drain the rear axle lubricant.

4. Use a metric Allen wrench to unscrew the differential pinion shaft lock-screw and remove the differential pinion shaft. It may be necessary to shorten the Allen wrench to do this.

5. Push the flanged end of the axle shaft toward the center of the car and remove the C-lock from the inner end of the shaft.

6. Remove the axle shaft from the housing making sure not to damage the oil seal.

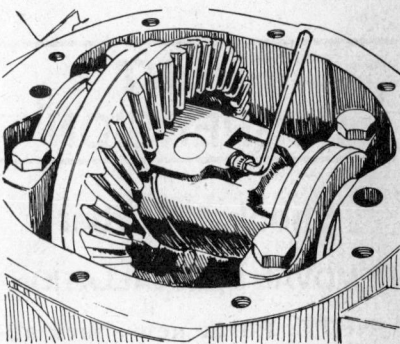

Removing the differential pinion shaft lockscrew

7. If replacing the seal only, remove the oil seal by using the inner end of the axle shaft. Insert the end of the shaft behind the steel case of the oil seal and carefully pry the seal out of the bore.

8. To remove bearings, insert a bearing and seal remover into the bore so that the tool head grasps behind the bearing. Slide the washer against the seal or bearing and turn the nut against the washer. Attach a slide hammer and remove the bearing.

9. Lubricate a new bearing with hypoid lubricant and install it into the housing with a bearing installer tool. Make sure that the tool contacts the end of the axle tube to ensure that the bearing is at the proper depth.

10. Lubricate the cavity between the seal lips with a high melting point wheel bearing grease. Place a new oil seal on the seal installation tool and position the seal in the axle housing bore. Tap the seal into the bore flush with the end of the housing.

11. To install the axle shaft, slide

the axle shaft into place making sure that the splines on the end of the shaft do not damage the oil seal and that they engage the splines of the differential side gear. Install the C-lock on the inner end of the axle shaft and push the shaft outward so that the shaft lock seats in the counterbore of the differential side gear.

12. Position the differential pinion shaft through the case and pinions, aligning the hole in the shaft with the lockscrew hole. Install the lockscrew.

13. Clean the gasket mounting surfaces on the differential carrier and the carrier cover. Install the carrier cover using a new gasket and tighten the cover bolts in a cross-wise pattern to 22 ft. lbs.

14. Fill the rear axle with lubricant to the bottom of the filler hole.

15. Install the brake drum and the wheel and tire assembly.

16. Lower the car.

FRONT SUSPENSION

Shock Absorber

REMOVAL & INSTALLATION

NOTE: Purge new shock absorbers of air by repeatedly extending in the normal position and compressing while inverted.

1. Hold the shock absorber upper stem and remove the nut, upper retainer, and rubber grommet.

2. Raise the car on a hoist.

3. Remove the bolt from the lower end of the shock absorber and remove the shock absorber.

4. With the lower retainer and rubber grommet in position, extend the shock absorber stem and install the stem through the wheelhouse opening.

5. Install and torque the lower bolt to 22 ft. lbs. for 1980 and 35–50 ft. lbs. for 1981 and later.

6. Lower the car.

7. Install the upper rubber grommet, retainer, and nut to the shock absorber stem.

8. Hold the shock absorber upper stem and torque the nut to 7 ft. lbs.

Lower Ball Joint

REMOVAL & INSTALLATION

NOTE: The ball joint studs use a

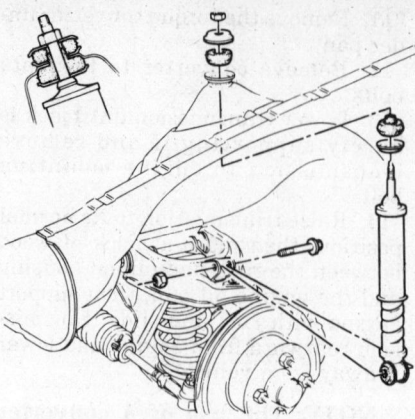

Front shock absorber mounting

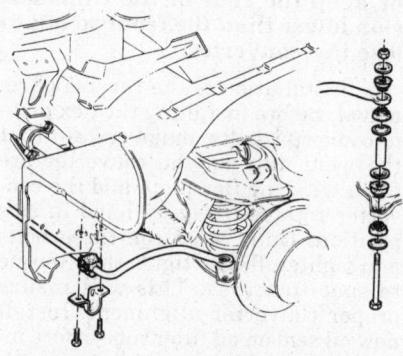

Front suspension stabilizer bar attachment

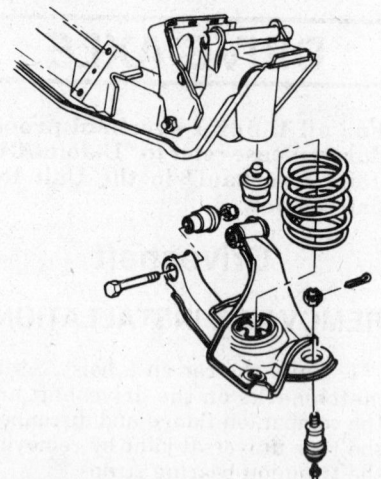

Correct position for front spring installation

special nut which must be discarded whenever loosened and removed. On assembly, use a standard nut to draw the ball joint into position on the knuckle, then remove the standard nut and install a new special nut for final installation.

1. Raise the car on a hoist.

2. Remove the tire and wheel.

3. Support the lower control arm with a hydraulic floor jack.

4. Loosen, but do not remove the lower ball stud nut.

5. Install a ball joint removal tool with the cup end over the upper ball stud nut.

6. Turn the threaded end of the ball joint removal tool until the ball stud is free of the steering knuckle.

7. Remove the ball joint removal tool and remove the nut from the ball stud.

8. Remove the ball joint.

NOTE: Inspect the tapered hole in the steering knuckle. Clean the area. If any out-of-roundness, deformation, or damage is found, the steering knuckle must be replaced.

9. To install the lower ball joint, mate the ball stud through the lower control arm and into the steering knuckle.

10. Install and torque the ball stud nut to 41–54 ft. lbs.

11. Install the tire and wheel.

12. Lower the car.

Lower Control Arm and Coil Spring

REMOVAL & INSTALLATION

NOTE: The ball joint studs use a special nut which must be discarded whenever loosened and removed. On assembly, use a standard nut to draw the ball joint into position on the knuckle, then remove the standard nut and install a new special nut for final installation.

1. Raise the car on a frame contact hoist.

2. Remove the wheel and tire.

3. Disconnect the stabilizer bar from the lower control arm and disconnect the tie-rod from the steering knuckle.

4. Support the lower control arm with a jack.

5. Remove the nut from the lower ball joint, then use a ball joint removal tool to press out the lower ball joint.

6. Swing the knuckle and hub aside and attach them securely with wire.

7. Loosen the lower control arm pivot bolts.

8. As a safety precaution, install a chain through the coil spring.

9. Slowly lower the jack.

10. When the spring is extended as far as possible, use a pry bar to carefully lift the spring over the lower control arm seat. Remove the spring.

11. Remove the pivot bolts and remove the lower control arm.

12. Install the lower control arm and pivot bolts to the underbody brackets. Torque the lower control arm pivot bolts to 49 ft. lbs.

13. Position inspring correctly and install it in the upper pocket. Use tape to hold the insulator onto the spring.

14. Install the lower end of the spring onto the lower control arm. An assistant may be necessary to compress the spring far enough to slide it over the raised area of the lower control arm seat.

15. Use a jack to raise the lower control arm and compress the coil spring.

16. Install the ball joint through the lower control arm and into the steering knuckle. Install the nut on the ball stud and torque to 41–54 ft. lbs.

17. Connect the stabilizer bar to the lower control arm. Connect the tie-rod to the steering knuckle. Install the wheel and tire.

18. Lower the car.

Upper Ball Joint

REMOVAL & INSTALLATION

NOTE: The ball joint studs use a special nut which must be discarded whenever loosened and removed. on assembly, use a standard nut to draw the ball joint into position on the knuckle, then remove the standard nut and install a new special nut for the final installation.

1. Raise the car on a hoist.
2. Remove the tire and wheel.
3. Support the lower control arm with a floor jack.
4. Loosen, but do not remove the upper ball stud nut.
5. Install a ball joint removal tool with the cup end over the lower ball stud nut.
6. Turn the threaded end of the ball joint removal tool until the upper ball stud is free of the steering knuckle.
7. Remove the ball joint removal tool and remove the nut from the ball stud.
8. Remove the two nuts and bolts attaching the ball joint to the upper control arm and remove the ball joint.

NOTE: Inspect the tapered hole in the steering knuckle. Clean the area. If any out-of-roundness, deformation, or damage is found, the steering knuckle must be replaced.

9. To install the upper ball joint, install the nuts and bolts attaching the ball joint to the upper control arm. Torque the nuts to 29 ft. lbs. Then

mate the upper control arm ball stud to the steering knuckle.

10. Install and torque the ball stud nut to 29–36 ft. lbs.
11. Install the tire and wheel.
12. Lower the car.

Upper Control Arm

REMOVAL & INSTALLATION

NOTE: The ball joint studs use a special nut which must be discarded whenever loosened and removed. On assembly, use a standard nut to draw the ball joint into position on the knuckle, then remove the standard nut and install a new special nut for final installation.

1. Raise the car and support it safely.
2. Remove the tire and wheel.
3. Support the lower control arm with a floor jack.
4. Remove the upper ball joint from the steering knuckle as previously described.
5. Remove the upper control arm pivot bolts and remove the upper control arm.
6. To install the upper control arm, install the upper control arm with its pivot bolts.

NOTE: The inner pivot bolt must be installed with the bolt head toward the front.

7. Install the pivot bolt nut.
8. Position the upper control arm in a horizontal plane and torque the nut to 43–50 ft. lbs.
9. Install the ball joint to the upper control arm and to the steering knuckle as previously described. Torque the ball joint-to-upper control arm attaching bolts to 29 ft. lbs. Torque the ball stud nut to 29–36 ft. lbs.
10. Install the tire and wheel.
11. Lower the car.

FRONT WHEEL BEARING ADJUSTMENT

1. Raise the car and support at the front lower control arm.
2. Remove the hub cap or wheel cover from the wheel. Remove the dust cap from the hub.
3. Remove the cotter pin from the spindle and spindle nut.
4. Spin the wheel forward by hand and tighten the spindle nut to 12 ft. lbs. This will fully seat the bearings.
5. Back off the nut to a just loose position.
6. Hand-tighten the spindle nut.

Loosen the spindle nut until either hole in the spindle aligns with a slot in the nut, but not more than $\frac{1}{2}$ flat.

7. Install a new cotter pin, bend the ends of the pin against the nut, and cut off any extra length to avoid interference with the dust cap.
8. Proper bearing adjustment should give 0.001–0.005 in. of endplay.
9. Install the dust cap on the hub and the hub cap or wheel cover on the wheel.
10. Lower the car.
11. Adjust the opposite front wheel bearings.

REAR SUSPENSION

When using a hoist contacting the rear axle, be sure that the stabilizer links and the track rod are not damaged.

Shock Absorber

REMOVAL & INSTALLATION

NOTE: Purge new shock absorbers of air by repeatedly extending in the normal position and compressing while inverted.

1. Raise the car on a hoist.
2. Support the rear axle.
3. Remove the shock absorber upper attaching nut and lower attaching bolt and nut, and remove the shock absorber.
4. Install the retainer and the rubber grommet onto the shock absorber.
5. Place the shock absorber into its installed position and install and tighten the upper retaining nut to 7 ft. lbs.
6. Install the lower shock absorber nut and bolt and torque to 21 ft. lbs.
7. Remove the rear axle supports and lower the car.

Rear Spring

REMOVAL & INSTALLATION

1. Raise the car on a hoist.
2. Support the rear axle with a floor jack.
3. Disconnect both shock absorbers from their lower brackets.
4. Disconnect the rear axle extension center support bracket from the underbody. Use caution when disconnect the extension and safely support it when disconnected.

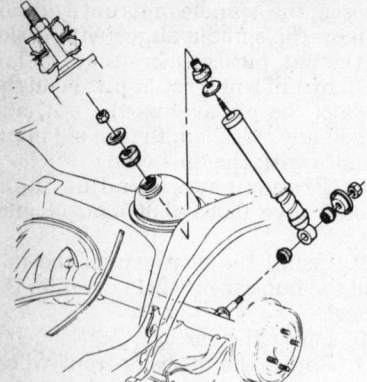

Rear shock absorber mounting

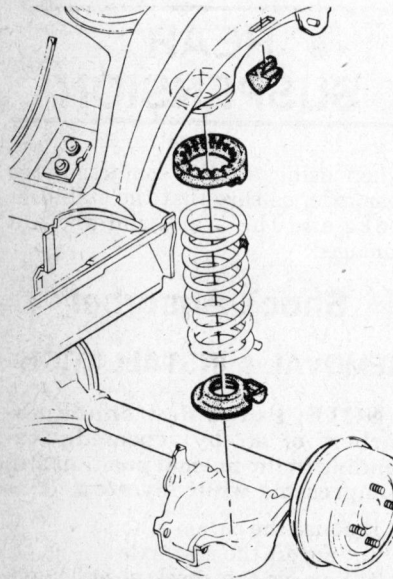

Rear spring installation—position both insulators as shown

5. Lower the rear axle and remove the springs and spring insulators.

CAUTION

Do not stretch the rear brake hoses when lowering the rear axle.

6. To install, place the insulators on top and on the bottom of the springs and position the springs between their upper and lower seats.
7. Raise the rear axle. Connect the rear axle extension center support bracket to the underbody. Torque the bolts to 37 ft. lbs.
8. Connect the shock absorbers to their lower brackets. Torque the nuts to 21 ft. lbs.
9. Remove the jack from the axle.
10. Lower the car.

Rear Wheel Bearings

Please refer to "Axle, Bearing and Seal" in the Drive Axle Section.

BRAKES

To service brake shoes, drums and wheel cylinders or brake pads and calipers, see "Brakes" in the Unit Repair section.

Combination Valve

REMOVAL & INSTALLATION

NOTE: If the combination valve is found defective it must be replaced.

1. Disconnect the negitive battery cable.
2. Clean all the dirt from the switch and connections.
3. Disconnect he electrical lead from the switch connection.
4. Disconnect the hydraulic lines from the connections at the switch. It may be necessary to loosen the line conncections at the master cylinder to loosen lines. Cover the open ends with a clean, lint-free material.
5. Remove the mounting screw and remove the combination valve.
6. Installation is the reverse of removal. Bleed the brake system.

Wheel Cylinder

REMOVAL

1. Clean the area around the inlet tube line and disconnect the tube line.
2. Remove the wheel cylinder retainer by inserting two awls into the access slots between the wheel cylinder pilot and the retainer locking tabs. Bend both tabs away simutaneously.

INSTALLATION

1. Place the wheel cylinder into position and hold in place using a wooden block between the cylinder and axle flange.
2. Use a 1¹⁄₈ inch, 12 point socket and socket extension to aid in installing a new retainer over the wheel cylinder abutment.
3. Connect the inlet tube and torque the nut to 120–180 inch lbs.

Master Cylinder

REMOVAL & INSTALLATION

1. Disconnect the master cylinder pushrod from the brake pedal.
2. Remove the pushrod boot.

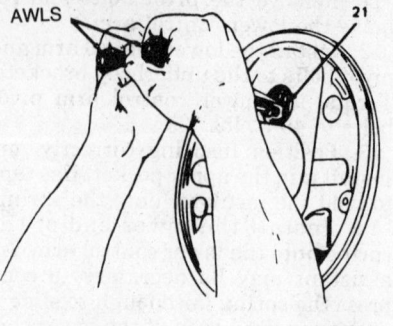

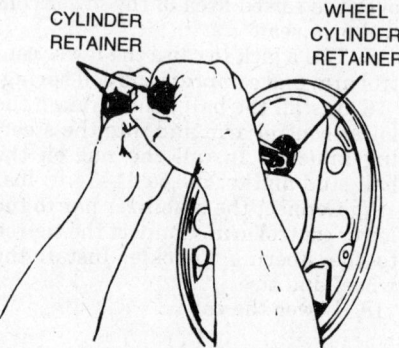

Wheel cylinder removal

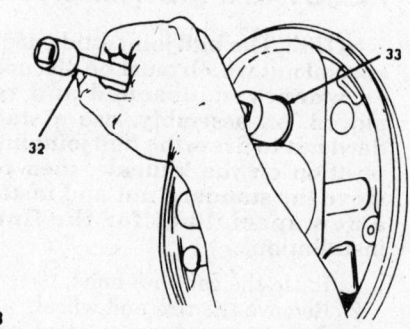

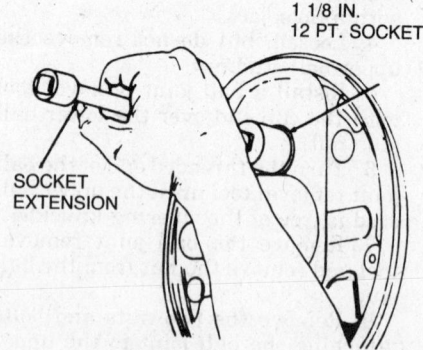

Wheel cylinder installation

3. Remove the air cleaner.
4. Thoroughly clean all dirt from the master cylinder and the brake lines. Disconnect the brake lines from the master cylinder and plug them to prevent the entry of dirt.
5. Remove the master cylinder se-

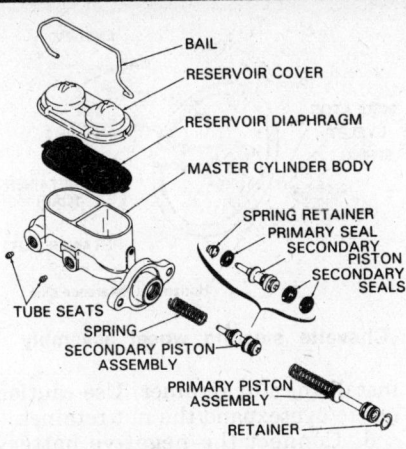

BAIL
RESERVOIR COVER
RESERVOIR DIAPHRAGM
MASTER CYLINDER BODY
SPRING RETAINER
PRIMARY SEAL
SECONDARY PISTON
SECONDARY SEALS
TUBE SEATS
SPRING
SECONDARY PISTON ASSEMBLY
PRIMARY PISTON ASSEMBLY
RETAINER

Dual master cylinder with common reservoir

curing nuts and remove the master cylinder.

6. Install the master cylinder with its spacer. Tighten the securing nuts.

7. Connect the brake lines to their ports.

8. Place the pushrod boot over the end of the pushrod. Secure the pushrod to the brake pedal with the pin and clip.

9. Fill the master cylinder and bleed the entire hydraulic system. After bleeding, fill the master cylinder to within 1/4 in. from the top of the reservoir. Check for leaks.

10. Install the air cleaner.

11. Check brake operation before moving the car.

PARKING BRAKE ADJUSTMENT

1. Raise the car on a hoist.

2. Apply the parking brake three notches from the fully released position.

3. Tighten the parking brake cable equalizer adjusting nut under the car until a light drag is felt when the rear wheels are rotated forward.

4. Fully release the parking brake and rotate the rear wheels. There should be no drag.

5. Lower the car.

Power Brake Booster

REMOVAL & INSTALLATION

1. Remove the air cleaner.

2. Disconnect the vacuum hose from the check valve.

3. Remove the master cylinder brace.

4. Remove the master cylinder-to-power cylinder nut, and pull forward on the master cylinder until it clears the power cylinder mounting studs.

Move the master cylinder aside and support it, being careful of the brake lines.

5. Remove the nuts securing the power cylinder to the firewall.

6. Remove the pushrod-to-pedal retainer and slip the pushrod off the pedal pin. Remove the power cylinder.

7. Installation is the reverse of removal.

STEERING

All models use manual rack and pinion steering which encloses the steering gear and linkage in one unit. Power steering is available as an option in all 1981 and later models with an automatic transmission.

Tie Rod

REMOVAL & INSTALLATION

1. Loosen the jam nut located on the inner tie rod.

2. Remove the outer tie rod cotter pin and nut.

3. Using a tie rod end separating tool, remove the tie rod from the steering knuckle.

4. Remove the outer tie rod from the inner tie rod assembly. Count the number of turns it takes to remove the tie rod end.

5. Install the new outer tie rod end onto the inner tie rod assembly, turning it in the same number of turns as the old tie rod took to remove. Do not tighten the jam nut.

6. Install the outer tie rod into the steering knuckle and torque the nut to 32 ft. lbs. Install a new cotter pin.

7. Set toe-in adjustment to specification by turning the inner tie rod (an alignment rack is necessary for adjustment). Be sure not to twist the boot when making the adjustment. If an alignment rack is not available, tighten the jam nut and have the front end alignment checked as soon as possible.

8. Torque the jam nut to 50 ft. lbs.

Steering Gear

REMOVAL & INSTALLATION

1. Raise the vehicle and support it with jackstands.

2. Remove the bolts and shield.

3. Remove the outer tie-rod cotter pins and nuts on both sides.

4. Using a tie-rod separating tool,

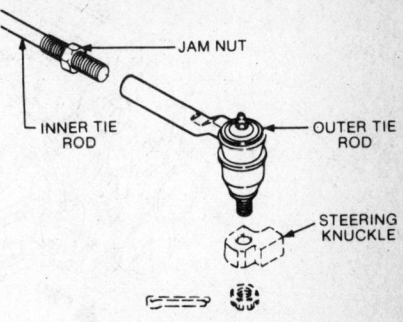

JAM NUT
INNER TIE ROD
OUTER TIE ROD
STEERING KNUCKLE

Tie rod end assembly

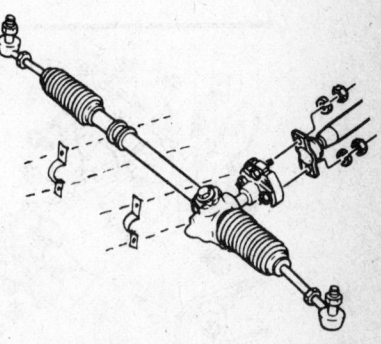

Manual rack and pinion assembly

disconnect the tie-rods from the steering knuckles.

5. On power steering models remove the two hydraulic lines from the steering gear.

6. Remove the flexible coupling pinch bolt to the shaft.

7. Remove the four bolts at the clamps, and remove the assembly from the vehicle.

8. To install, position the assembly to the vehicle with the stub shaft in position with the flexible coupling and install the clamps and four new bolts.

9. Install the flexible coupling pinch bolt to the shaft.

10. Install the tie-rods into the steering knuckles and torque the nuts to 30 ft. lbs. Install a new cotter pin.

11. On power steering models install the two hydraulic hoses and bleed the system.

12. Install the bolts and shield. Remove the jackstands and lower the vehicle.

Power Steering Pump

REMOVAL & INSTALLATION

1. Remove the upper adjusting bolt.

2. Remove the lower brace bolt-to-pump bracket.

3. Remove the left hand crossmember brace to body.

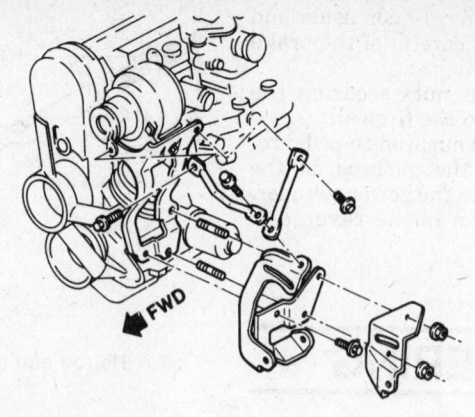

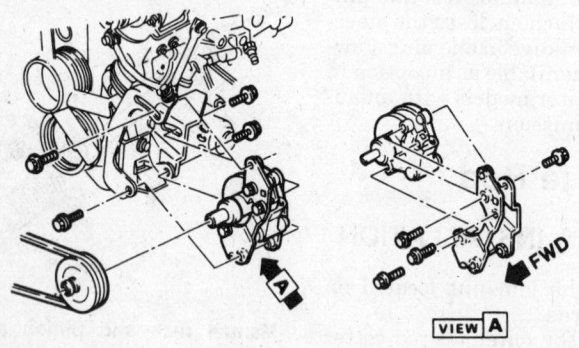

Power steering pump mounting—gasoline engine

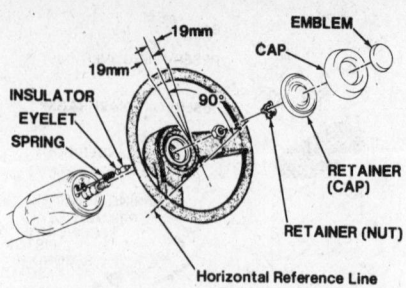

Chevette steering wheel assembly

4. Remove the pressure line and the reservoir line at the pump.

5. Remove the rear pump adjusting bracket.

6. Remove the front pivot bolt at the pump and remove the bolt.

7. Remove the front pump bracket at the bolt-to-engine. Remove the bracket and pump.

8. Installation is the reverse of the removal procedure. In addition, adjust the belt tension, fill the reservoir and bleed the system.

BLEEDING THE POWER STEERING SYSTEM

NOTE: When checking or adjusting the fluid level after service, air must be bled from the system using the following procedure.

1. Install the pump, bracket and all hoses and lines to specifications, EXCEPT the pressure line at the pump outlet.

2. Add fluid to the reservoir until fluid begins leaving the pump at the pressure fittings.

3. Attach the pressure line to the pump.

4. Continue filling the reservoir until the proper level is reached.

5. Road test the car to make sure the steering system functions normally.

6. Recheck the fluid level and top up as necessary with power steering fluid.

Steering Wheel

REMOVAL & INSTALLATION

1. Disconnect the negative battery cable.

2. Pull up on the horn cap to remove it. Remove the horn ring-to-steering wheel attaching screws and remove the ring.

3. Remove the wheel nut retainer and the wheel nut.

——— CAUTION ———
Do not overexpand the retainer.

4. Using a suitable steering wheel puller, thread the puller anchor screws into the threaded holes in the steering wheel. With the center bolt of the puller butting against the steering shaft, turn the center bolt to remove the steering wheel.

5. To install, place the turn signal lever in the neutral position and install the steering wheel. Torque the steering wheel nut to 30 ft. lbs. and install the nut retainer. Use caution not to overexpand the nut retainer.

6. Connect the negative battery cable.

Turn Signal Switch

REMOVAL & INSTALLATION

1. Remove the steering wheel as previously described.

2. Position a small prybar into one of the three cover slots. Pry up and out (at least two slots) to free the cover.

3. Press down on the lockplate, but do not relieve the full load of the spring because the ring will rotate and make removal difficult. Pry the round wire snap-ring out of the shaft groove and discard it. Lift the lockplate off the end of the shaft.

4. Slide the turn signal canceling cam, upper bearing preload spring, and thrust washer off the end of the shaft.

5. Remove the multi-function lever by rotating it clockwise to its stop (off position), then pull the lever straight out to disengage it.

6. Push the hazard warning knob in and unscrew the knob.

7. Remove the two screws, pivot arm, and spacer.

8. Wrap the upper part of the connector with tape to prevent snagging the wires during switch removal.

9. Remove the three switch mounting screws and pull the switch straight up, guiding the wiring harness through the column housing.

——— CAUTION ———
On installation it is extremely important that only the specified screws, bolts, and nuts be used. The use of overlength screws could prevent the steering column from compresssing under impact.

10. Position the switch into the housing.

11. Install the three switch mounting screws. Replace the spacer and pivot arm. Be sure that the spacer protrudes through the hole in the arm. Be sure that the spacer protrudes through the hole in the arm

and that the arm finger encloses the turn signal switch frame.

12. Install the hazard warning knob.

13. Make sure that the turn signal switch is in the neutral position and that the hazard warning knob is out. Slide the thrust washer, upper baring preload spring, and the cancelling cam into the upper end of the shaft.

14. Place the lockplate and a new snapring onto the end of the shaft. Compress the lockplate as far as possible. Slide the new snapring into the shaft groove and remove the lockplate compressor tool.

──── CAUTION ────

On assembly, always use a new snapring.

15. Install the multi-function lever, guiding the wire harness through the column housing. Align the lever pin with the switch slot. Push on the end of the lever until it is seated securely.

16. Install the steering wheel as previously described.

Lock Cylinder

REMOVAL & INSTALLATION

The lock cylinder is located on the right side of the steering column and should be removed only in the Run position. Removal in any other position will damage the key buzzer switch. The lock cylinder cannot be disassembled; if replacement is required, a new cylinder coded to the old key must be installed.

1. Remove the steering wheel and turn signal switch as previously described.

2. Do not remove the buzzer switch or damage to the lock cylinder will result.

3. Place the lock cylinder in the RUN position. Remove the securing screw and remove the cylinder.

4. To install the lock cylinder, hold the cylinder sleeve and rotate knob (key in) clockwise to stop. (This retracts the actuator). Insert the cylinder into the housing bore with the key on the cylinder sleeve aligned with the keyway in the housing. Push the cylinder in until it bottoms and install the retaining screw.

5. Install the turn signal switch and the steering wheel as previously described.

Ignition Switch and Dimmer Switch

REMOVAL & INSTALLATION

The ignition switch is mounted on top of the mast jacket near the front of the

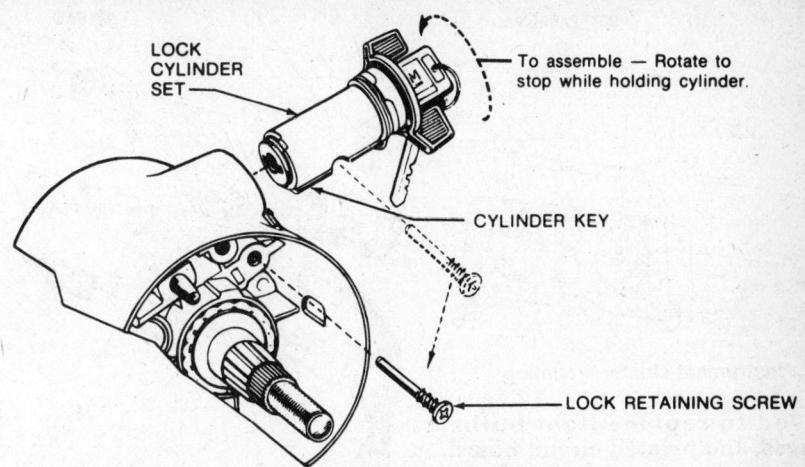

Lock cylinder installation details

instrument panel. The switch is located inside the channel section of the brake pedal support and is completely inaccessible without first lowering the steering column.

1. Disconnect the negative battery cable.

2. Remove the steering wheel as previously described.

3. Move the driver's seat as far back as possible.

4. Remove the floor pan bracket screw.

5. Remove the two column bracket-to-instrument panel nuts and lower the column far enough to disconnect the ignition switch wiring harness.

──── CAUTION ────

Be sure that the steering column is properly supported before proceeding.

6. The switch should be in the Lock position before removal. If the lock cylinder has already been removed, the actuating rod to the switch should be pulled up until there is a definite stop, then moved down one detent to the Lock position.

7. Remove the two mounting screws and remove the ignition and dimmer switch.

8. Refer to the lock cylinder installation procedure previously described in "Lock Cylinder Removal & Installation."

9. Turn the cylinder clockwise to stop and then counterclockwise to stop, then counterclockwise again to stop (OFF-UNLOCK) position.

10. Place the ignition switch in the OFF-UNLOCK position. Move the slider two positions to the right from ACCESSORY to the OFF-UNLOCK position.

11. Fit the actuator rod into the slider hole and install the switch on the column. Be sure to use only the cor-

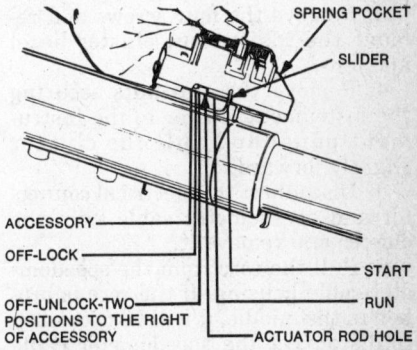

Postioning the ignition switch for installation

rect screws. Be careful not to move the switch out of its detent.

12. Check the dimmer switch adjustment.

13. Connect the ignition switch wiring harness.

14. Loosely install the column bracket-to-instrument panel nuts.

15. Install the floor pan bracket screw and tighten it to 20 ft. lbs.

16. Tighten the column bracket-to-instrument panel nuts to 22 ft. lbs.

17. Install the steering wheel as previously outlined.

18. Connect the battery negative cable.

CHASSIS ELECTRICAL

Instrument Cluster and Speedometer Cable

REMOVAL & INSTALLATION

The instrument cluster must be re-

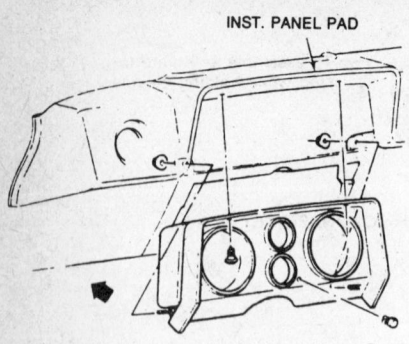

INST. PANEL PAD

Instrument cluster mounting

moved to replace light bulbs, gauges, and printed circuit board.

1. Disconnect the negative battery cable.

2. Remove the clock stem knob (if equipped).

3. Remove the four screws and remove the instrument cluster bezel and lens.

4. Remove the two nuts securing the instrument cluster to the instrument panel and pull the cluster slightly forward.

5. Disconnect the electrical connector and speedometer cable from the cluster and remove it.

6. Pull the core from the speedometer cable housing. If the core is broken in the middle, it will be necessary to disconnect the speedometer cable at the transmission and remove the rest of the core through the bottom of the cable housing.

7. Attach the cable housing to the transmission and insert the new core through the top of the housing.

8. Attach the speedometer cable to the rear of the speedometer.

9. Installation is the reverse of removal procedures.

Headlight Switch

REMOVAL & INSTALLATION

1. Disconnect the negative battery cable.

2. Pull the headlight switch control knob to the On position.

3. Reach up under the instrument panel and depress the switch shaft retainer button while pulling on the switch control shaft knob.

4. Remove the three screws and remove the headlight switch trim plate.

5. Remove the light switch ferrule nut from the front of the instrument panel.

6. Disconnect the multi-contact connector from the bottom of the headlight switch.

7. Installation is the reverse of removal.

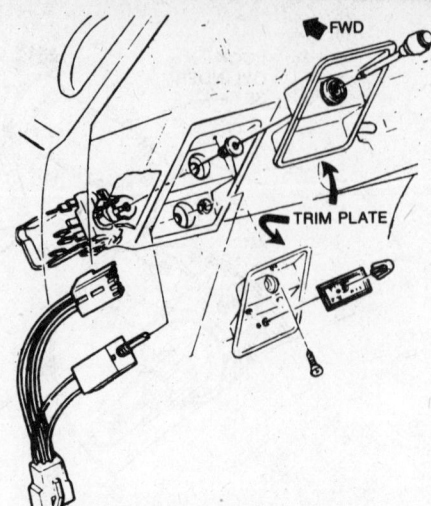

FWD
TRIM PLATE

Headlight switch mounting

Windshield Wiper Motor

REMOVAL & INSTALLATION

1. Working inside the car, reach up under the instrument panel above the steering column and loosen, but do not remove, the transmission drive link-to-motor crank arm attaching nuts.

2. Disconnect the transmission drive link from the wiper rotor crank arm.

3. Raise the hood and disconnect the wiper motor wiring.

4. Remove the three motor attaching bolts.

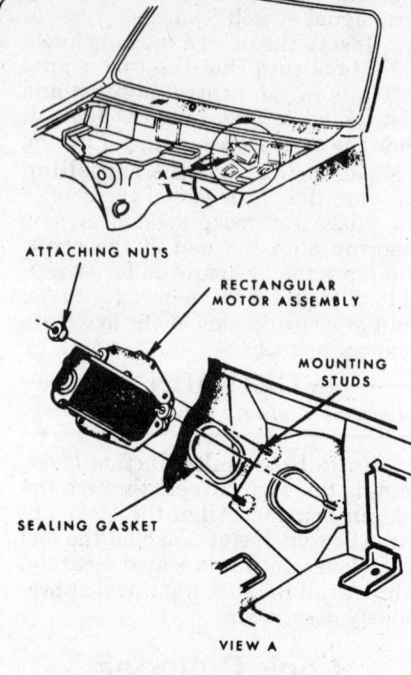

ATTACHING NUTS
RECTANGULAR MOTOR ASSEMBLY
MOUNTING STUDS
SEALING GASKET
VIEW A

Wiper motor mounting

5. Remove the motor while guiding the crank arm through the hole.

6. To install, align the sealing gasket to the base of the motor and reverse the rest of the removal procedure.

NOTE: If the wiper motor-to-firewall sealing gasket is damaged during removal, it should be replaced with a new gasket to prevent possible water leaks.

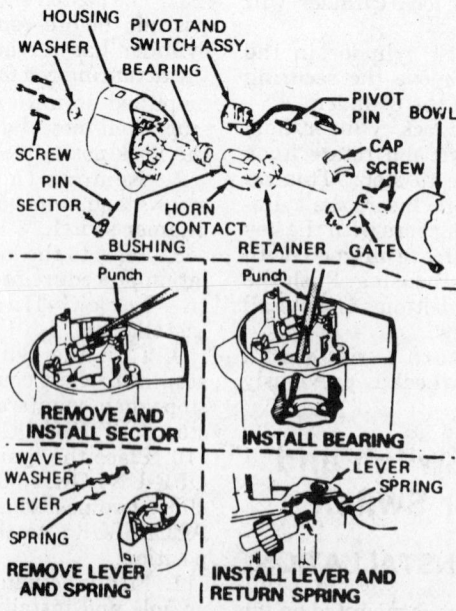

HOUSING
WASHER
PIVOT AND SWITCH ASSY.
BEARING
PIVOT PIN
BOWL
SCREW
CAP SCREW
PIN SECTOR
HORN CONTACT
BUSHING
RETAINER
GATE
Punch
Punch

REMOVE AND INSTALL SECTOR

INSTALL BEARING

WAVE WASHER
LEVER SPRING

LEVER SPRING

REMOVE LEVER AND SPRING

INSTALL LEVER AND RETURN SPRING

Wiper switch removal and installation—standard columns

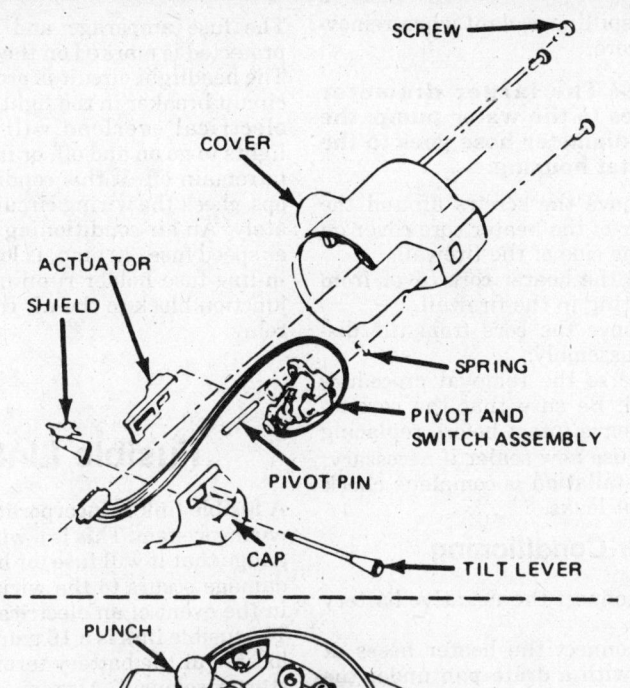

SCREW

COVER

ACTUATOR

SHIELD

SPRING

PIVOT AND
SWITCH ASSEMBLY

PIVOT PIN

CAP

TILT LEVER

PUNCH

PIVOT PIN

REMOVE AND INSTALL PIVOT ASSEMBLY

Wiper switch removal and installation—tilt columns

Wiper Switch

REMOVAL & INSTALLATION

1. Remove the steering wheel and the ignition and dimmer switch as outlined in the "Steering" section.

2. Remove the parts as shown in the illustration.

3. On models equipped with the standard column, assemble the rack so that the first rack tooth engages between the first and second tooth of the sector. The remainder of the installation is the reverse of removal.

Wiper Blade

REPLACEMENT

To remove the blade from the arm, depress the spring type blade clip away from the underside of the arm and slide the arm out of the blade clip. To install the blade, slide the tip end of the arm into the blade clip until the pin on the tip end engages the hole in the clip. The rubber wiper element can be replace separately from the blade; see the "General Maintenance" Unit Repair Section for further details.

Radio

REMOVAL & INSTALLATION

1. Disconnect the negative battery cable.

2. Remove the nut from the mounting stud on the bottom of the radio.

3. Remove all control knobs and/or spacers from the right and left radio control shafts.

4. Remove the four screws from the center trim plate and pull the trim plate and the radio forward slightly.

5. Disconnect the antenna lead from the rear of the radio.

6. Disconnect the speaker and electrical connectors from the radio harness.

7. Disconnect the electrical connectors from the rear window defogger and cigarette lighter.

8. Use a deep well socket to remove the retaining nuts from both control shafts and remove the radio.

9. To install, reverse the removal procedure.

Heater Blower Motor

REMOVAL & INSTALLATION

1. Disconnect the negative battery cable.

2. Disconnect the electrical lead from the blower motor.

3. Scribe a mark to reference the blower motor flange-to-case position.

4. Remove the blower motor-to-case attaching screws and remove the blower motor and wheel as an assembly. Pry the flange gently if the sealer acts as an adhesive.

5. Remove the blower wheel retaining nut and separate the motor and wheel.

6. Reverse Steps 1–5 to install. Be sure to align the scribe marks made during removal.

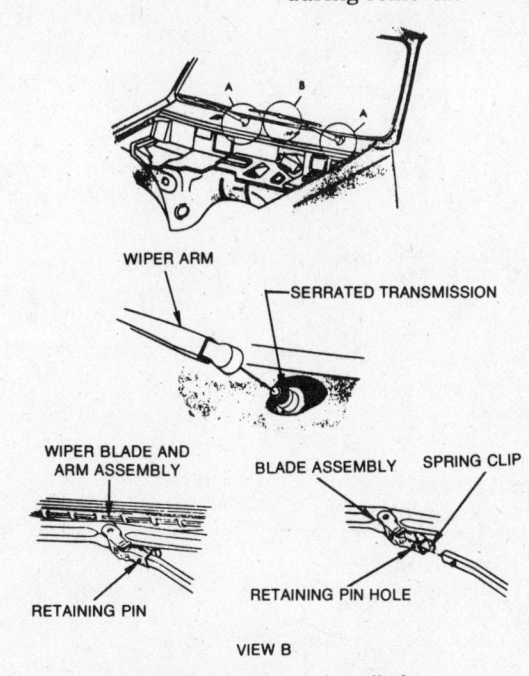

WIPER ARM

SERRATED TRANSMISSION

WIPER BLADE AND
ARM ASSEMBLY

BLADE ASSEMBLY

SPRING CLIP

RETAINING PIN

RETAINING PIN HOLE

VIEW B

Wiper blade and arm installation

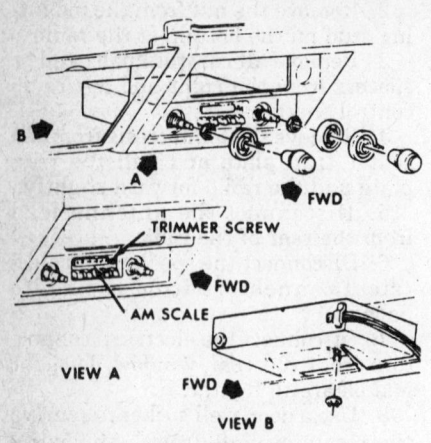

TRIMMER SCREW

AM SCALE

VIEW A

FWD

VIEW B

Radio mounting

NOTE: Assemble the blower wheel to the motor with the open end of the wheel away from the motor. If necessary, replace the sealer at the motor flange.

Heater Core

REMOVAL & INSTALLATION

Without Air Conditioning

1. Disconnect the negative battery cable.
2. Drain the radiator.
3. Disconnect the heater hoses at the heater core tube connections. Use care when removing the hoses as the core tube attachment seams can be easily damaged if too much force is used on them. When the hoses are removed, install plugs in the core tubes to avoid spilling coolant when removing the core.

NOTE: The larger diameter hose goes to the water pump: the smaller diameter hose goes to the thermostat housing.

4. Remove the screws around the perimeter of the heater core cover on the engine side of the firewall.
5. Pull the heater core cover from its mounting in the firewall.
6. Remove the core from the distributor assembly.
7. Reverse the removal procedure to install. Be sure that the core-to-case sealer is intact before replacing the core; use new sealer if necessary. When installation is complete, check for coolant leaks.

With Air Conditioning

1. Disconnect the negative battery cable.
2. Disconnect the heater hoses at the core with a drain pan under the car. Plug the hoses to prevent spillage.
3. Remove the A/C hose bracket.
4. Removes the heater core case cover and remove the core from the case.
5. Installation is the reverse of removal.

Fuses & Circuit Breakers

The fuse panel is located under the left hand side of the instrument panel.

The fuse amperage and the circuit protected is marked on the fuse panel. The headlight circuit is protected by a circuit breaker in the light switch. An electrical overload will cause the lights to go on and off, or in some case to remain off. If this condition develops, check the wiring circuits immediately. An air conditioning high blower speed fuse, 30 amp, is located in an in-line fuse holder running from the junction block to the air conditioning relay.

Fusible Links

A fusible link is incorporated into the wiring system. This is a wire of such a gauge that it will fuse (or melt) before damage occurs to the wiring harness in the event of an electrical overload. The fusible link is a 16 gauge red wire located at the battery terminal of the starter solenoid. Also, on diesel models a 14 gauge brown fusible link is located between the battery and the glow plug relay.

Flashers

The hazard warning flasher is located in the fuse box. The directional signal flasher is located above the brake pedal bracket, to the left of the steering column.

Chevrolet
Spectrum

YEAR IDENTIFICATION

1985-86 Spectrum

1987 Spectrum

VEHICLE IDENTIFICATION NUMBER (VIN)

It is important for servicing and ordering parts to be certain of the vehicle and engine identification. The VIN (vehicle identification number) is a 13 or 17 digit number visible through the windshield on the driver's side of the dash and contains the vehicle and engine indentification codes. It can be interpreted as follows:

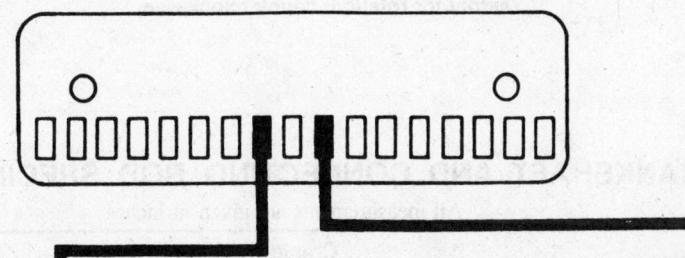

Engine Code							Model Year Code	
Code	Cu. In.	Liters	Cyl.	Carb.	Eng. Mfg.		Code	Year
K	94	1.5	4	2bbl	Isuzu		F	1985
							G	1986
							H	1987

GENERAL ENGINE SPECIFICATIONS

Year	Engine No. Cyl. Displ. (cu. in.)	Engine VIN Code	Fuel Delivery System	Engine Mfg.	Horsepower @ rmp	Torque @ rpm (ft. lb.)	Bore × Stroke	Compression Ratio	Oil Pressure @ 3800 rpm
'85–'87	4–94	K	2bbl	Isuzu	70 @ 5400	87 @ 3400	3.031 × 3.110	9.6 : 1	57–85

TUNE-UP SPECIFICATIONS

When analyzing compression test results, look for uniformity among cylinders rather than specific pressures.

Year	Eng. VIN Code	Engine No. Cyl. Displacement (cu. in.)	Eng. Mfg.	hp	Spark Plugs Orig. Type	Gap (in.)	Ignition Timing (deg BTDC) Man. Trans.	Auto. Trans.	Valves Intake Opens (deg)	Fuel Pump Pressure (psi)	Idle Speed (rpm) Man. Trans.	Auto. Trans.
'85–'86	K	4–94	Isuzu	70	BPR6ES-11	.040	15	10	17B	6–8	700	950
'87	All				See Underhood Specifications Sticker							

NOTE: The underhood specifications sticker often reflects tune-up specification changes made in production. Sticker figures must be used if they disagree with those in this chart.

FIRING ORDER

NOTE: To avoid confusion, replace spark plugs and wires one at a time.

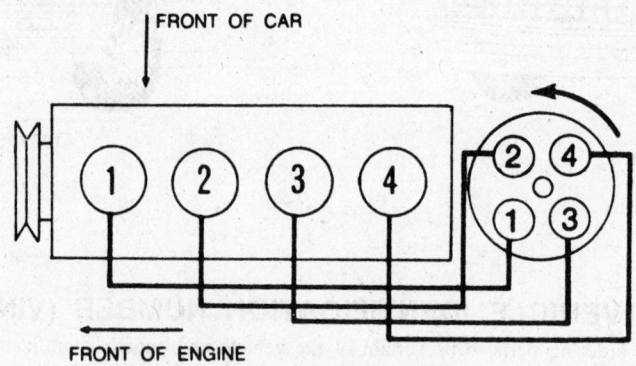

GM (ISUZU) 94–4 (1.5L)
Engine firing order: 1–3–4–2
Distributor rotation: counterclockwise

CRANKSHAFT AND CONNECTING ROD SPECIFICATIONS

All measurements are given in inches

Year	Eng. VIN Code	Engine No. Cyl Displacement (cu. in.)	Eng. Mfg.	Crankshaft Main Brg. Journal Diameter	Main Brg. Oil Clearance	Shaft End-Play	Thrust on No.	Connecting Rod Journal Diameter	Oil Clearance	Side Clearance
'85–'87	K	4–94	Isuzu	1.8865–1.8873	.00079–.00199	.0024–.0095	2	1.5720–1.5726	.00098–.00229	.0079–.0138

CAPACITIES

Year	Eng. VIN Code	Engine Displacement (cu. in.)	Eng. Mfg.	Crankcase Quarts		Transaxle Pints		Gas Tank Gal	Cooling System Quarts	
				w/filter	wo/filter	Manual	Auto.		w/heater	w/AC
'85–'87	K	94	Isuzu	3.4	3.0	5.6	12.6	11	6.7	6.7

VALVE SPECIFICATIONS

Year	Eng. VIN Code	Engine No. Cyl. Displacement (cu. in.)	Eng. Mfg.	Seat Angle (deg)	Face Angle (deg)	Spring Test Pressure (lbs. @ In.)	Spring Installed Height (in.)	Stem-to-Guide Clearance (in.)		Stem Diameter (in.)	
								Intake	Exhaust	Intake	Exhaust
'85–'87	K	4–94	Isuzu	45	45	47 @ 1.57	1.57	.0009–.0022	.00118–.00248	.2740–.2750	.2740–.2744

CAMSHAFT SPECIFICATIONS

All measurements in inches

Year	Eng. VIN Code	Engine (cu. in.)	Eng. Mfg.	Journal Diameter					Bearing Clearance	Lobe Life		Camshaft End Play
				1	2	3	4	5		Intake	Exhaust	
'85–'87	K	94	Isuzu			—1.021—			.00236–.00437		—1.426—	.00394 .00710

TORQUE SPECIFICATIONS

All readings in ft. lbs.

Year	Eng. V.I.N. Code	Engine No. Cyl. Displacement (cu. in.)	Eng. Mfg.	Cylinder Head Bolts	Rod Bearing Bolt	Main Bearing Bolt	Crankshaft Pulley Bolt	Flywheel-to-Crankshaft Bolts	Manifold	
									Intake	Exhaust
'85–'87	K	4–94	Isuzu	②	25	65	108	22①	17	17

① Tighten an additional 45° after torquing
② 1st step: 29 ft. lbs., 2nd step: 58 ft. lbs.

PISTON AND RING SPECIFICATIONS

All measurements are given in inches

Year	Eng. VIN Code	Engine No. Cyl. Disp. (cu. in.)	Eng. Mfg.	Piston-to-Bore Clearance	Ring Gap			Ring Side Clearance		
					Top Compression	Bottom Compression	Oil Control	Top Compression	Bottom Compression	Oil Control
'85–'87	K	4–94	Isuzu	.0012–.0020	.0098–.0137	—	.0039–.0236	.00098–.00256	—	—

WHEEL ALIGNMENT

Year	Model	Caster		Camber		Toe-in (in.)
		Range (deg)	Pref Setting (deg)	Range (deg)	Pref Setting (deg)	
'85–'87	Spectrum	1¾P–2¾P	2¼P	¼N–1½P	⅜P	0 ± .08

ENGINE ELECTRICAL

Alternator

For further information on the charging system, please refer to "Charging and Starting" in the Unit Repair section.

SERVICE PRECAUTIONS

1. When installing a battery, make sure that the positive and negative cables are not reversed.

2. When jump-starting the car, be sure that like terminals are connected. This also applies to using a battery charger. Reversed polarity will burn out the alternator and regulator in a matter of seconds.

3. Never operate the alternator with the battery disconnected or on an otherwise uncontrolled open circuit.

4. Do not short across or ground any alternator or regulator terminals.

5. Do not try to polarize the alternator.

6. Do not apply full battery voltage to the field (brown) connector.

7. Always disconnect the battery ground cable before disconnecting the alternator lead.

8. Always disconnect the battery (negative cable first) when charging it.

9. Never subject the alternator to excessive heat or dampness. If you are steam-cleaning the engine, cover the alternator.

10. Never use arc welding equipment on the car with the alternator connected.

REMOVAL & INSTALLATION

1. Disconnect the negative battery cable at the battery.

——————— CAUTION ———————

Failure to disconnect the negative cable may result in injury from the positive battery lead at the alternator and may short the alternator and regulator during the removal process.

2. Disconnect and label the two terminal plug and the battery leads from the rear of the alternator.

3. Loosen the mounting bolts. Push the alternator inwards and slip the drive belt off the pulley.

4. Remove the mounting bolts and remove the alternator.

5. To install, place the alternator in its brackets and install the mount-

ing bolts. Do not tighten them yet.

6. Slip the belt back over the pulley. Pull outwards on the unit and adjust the belt tension. Tighten the mounting and adjusting bolts.

7. Install the electrical leads and the negative battery cable.

Regulator

A solid state regulator is mounted within the alternator. All regulator components are enclosed in a solid mold. The regulator is non-adjustable and requires no maintenance.

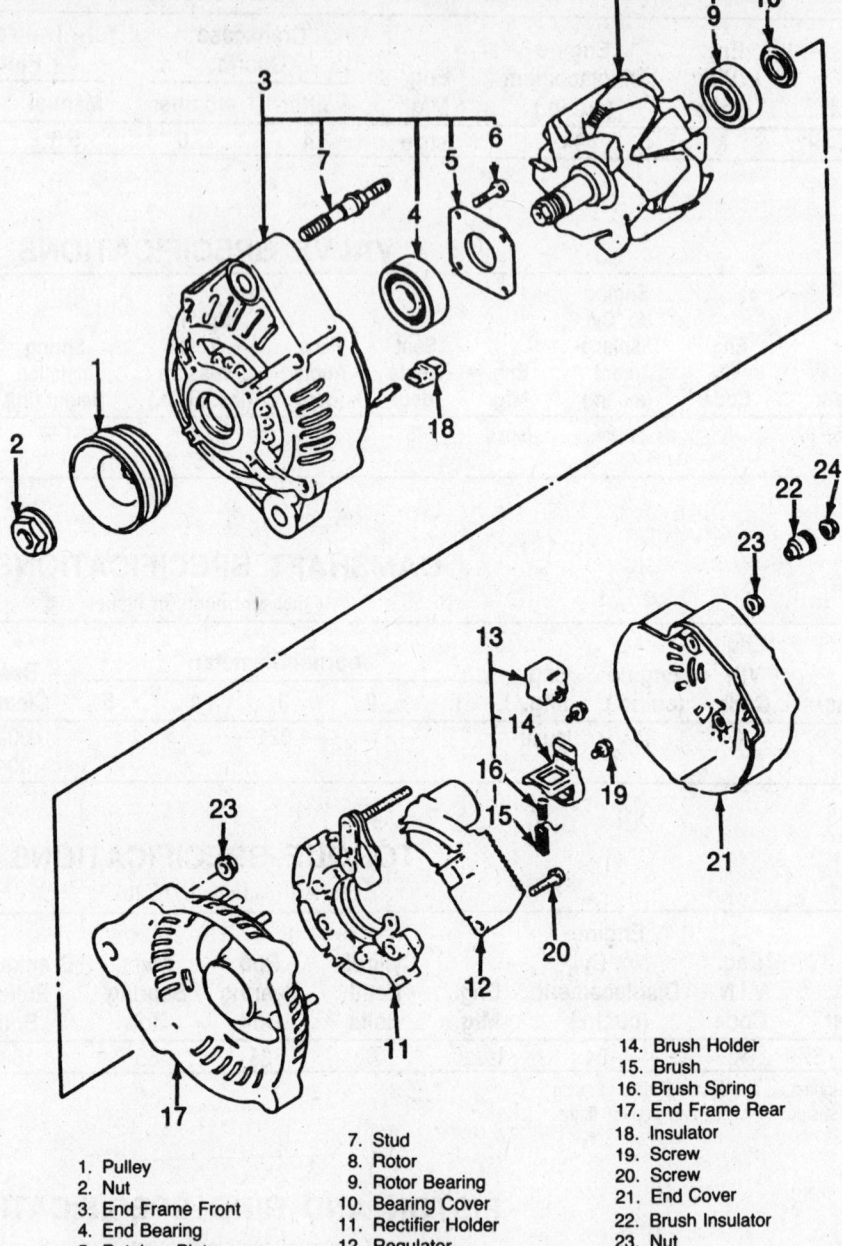

1. Pulley
2. Nut
3. End Frame Front
4. End Bearing
5. Retainer Plate
6. Screw
7. Stud
8. Rotor
9. Rotor Bearing
10. Bearing Cover
11. Rectifier Holder
12. Regulator
13. Brush Holder
14. Brush Holder
15. Brush
16. Brush Spring
17. End Frame Rear
18. Insulator
19. Screw
20. Screw
21. End Cover
22. Brush Insulator
23. Nut
24. Nut

Exploded view of the alternator

Starter

For further information on the starting system, please refer to "Charging and Starting" in the Unit Repair section.

REMOVAL & INSTALLATION

1. Disconnect the negative battery cable at the battery.

2. Disconnect the ignition switch lead wire and the battery cable from the starter motor terminal.

3. Remove the two mounting bolts from the starter and remove the starter.

4. To install, reverse the removal procedure.

Distributor

REMOVAL & INSTALLATION

1. Disconnect the negative battery terminal.

2. Remove the distributor cap.

3. Mark and remove all electrical leads and vacuum lines connected to the distributor assembly.

4. Mark the relationship of the rotor to the distributor housing and the distributor housing to the engine.

5. Remove the hold-down bolt, clamp and distributor.

6. To install, reverse the removal procedure and check the timing.

INSTALLATION IF THE ENGINE WAS DISTURBED

If the engine was cranked while the distributor was removed, you will have to place the engine on TDC of the compression stroke to obtain the proper ignition timing.

1. Remove the No. 1 spark plug.

2. Place your thumb over the spark plug hole. Crank the engine slowly until compression is felt. It will be easier if you have someone rotate the engine by hand, using a wrench on the crankshaft pulley.

3. Align the timing mark on the crankshaft pulley with the "O" degree mark the timing scale attached to the front of the engine. This places the engine at TDC of the compression stroke.

4. Turn the distributor shaft until the rotor points to the No. 1 spark plug tower on the cap.

5. Install the distributor into the engine. Be sure to align the distributor-to-engine block mark made earlier.

6. To complete installation, reverse the removal procedure and check the timing.

IGNITION TIMING ADJUSTMENT

1. Set the parking brake and block the wheels.

2. Place the manual transmission in Neutral or the automatic transmission in Park.

3. Allow the engine to reach normal operating temperature. Make sure that the choke valve is open. Turn off all of the accessories.

4. If equipped with power steering,

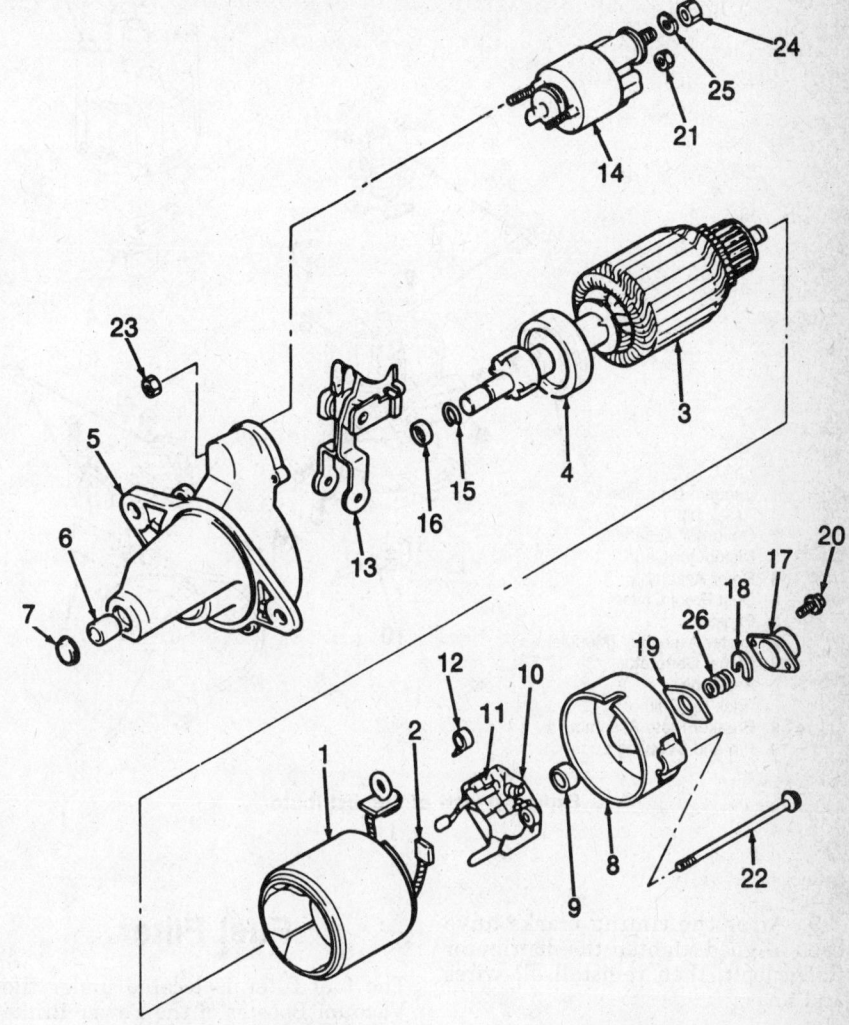

1. Starter yoke assembly	14. Magnetic switch
2. Brush	15. Snap ring
3. Armature	16. Sleeve
4. Overrunning Clutch	17. End frame cover
5. Gear case assembly	18. Lock plate
6. Bearing	19. Seal
7. Gear case cover	20. Screw
8. Frame assembly	21. Nut
9. Bearing	22. Through bolt
10. Brush holder	23. Nut
11. Brush	24. Nut
12. Brush spring	25. Lockwasher
13. Shift lever	26. Brake spring

Exploded view of the starter

place the front wheels in a straight line.

5. Disconnect and plug the distributor vacuum line, the canister purge line, the EGR vacuum line and the ITC valve vacuum line at the intake manifold.

6. Connect a timing light to the No. 1 spark plug wire and a tachometer to the tachometer filter connector on the coil.

NOTE: Check the idle speed. If the speed is not correct, refer to

the "Idle Speed Adjustment" procedure in the section and set the idle speed.

7. Loosen the distributor flange bolt.

8. Using the timing light, align the notch on the crankshaft pulley with the mark on the timing cover by turning the distributor.

NOTE: Adjust the timing to 15 degrees BTDC at 750 rpm (M/T) or 10 degrees BTDC at 1000 rpm (A/T).

C845

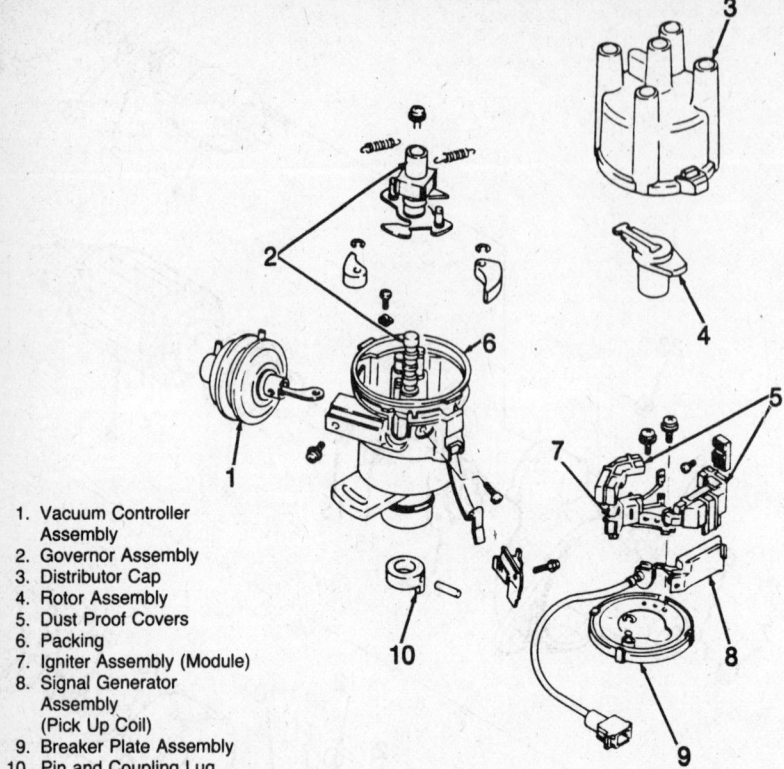

1. Vacuum Controller Assembly
2. Governor Assembly
3. Distributor Cap
4. Rotor Assembly
5. Dust Proof Covers
6. Packing
7. Igniter Assembly (Module)
8. Signal Generator Assembly (Pick Up Coil)
9. Breaker Plate Assembly
10. Pin and Coupling Lug

Exploded view of the distributor

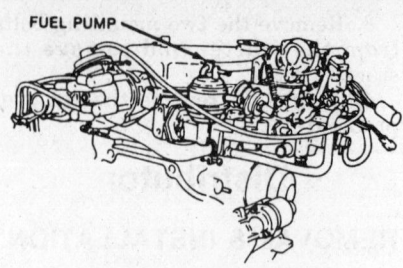

Location of the fuel pump

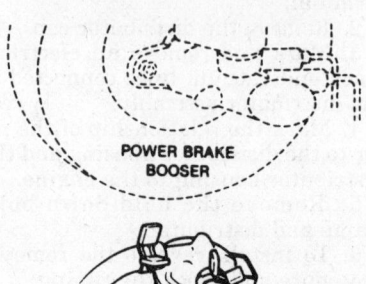

Fuel filter location

9. After the timing marks have been aligned, tighten the distributor flange bolt, then reinstall all wires and lines.

FUEL SYSTEM

Fuel Pump

The mechanical fuel pump is located at the left rear side of the cylinder head intake manifold. It is a diaphragm-type pump which is operated by an eccentric on the camshaft. No adjustment or repairs are possible, it is serviced by replacement only.

REMOVAL & INSTALLATION

1. Disconnect the fuel and air hoses from the fuel pump.
2. Remove the bolts, fuel pump and heat insulator assembly.
3. After removing the fuel pump, cover the mounting face of the cylinder head to prevent oil discharge.
4. To install, reverse the removal procedures and replace the heat insulator assembly.

Fuel Filter

The fuel filter is located under the Vacuum Booster of the Power Brake System and should be replaced every 15,000 miles.

REMOVAL & INSTALLATION

1. Remove the fuel tank cap.
2. Disconnect the hoses from the fuel filter.

NOTE: Cap the fuel hoses to prevent fuel spillage or dirt entry.

3. Remove the fuel filter.
4. Reinstall the fuel tank cap.
5. To install, reverse the removal procedures, securely attach the fuel filter clips and start the engine to check for leakage.

Carburetor

The carburetor is a downdraft type having two stages of operation. The primary side has a small bore, double venturis, a bridge nozzle and a duty solenoid to control the fuel metering. The secondary side has a large bore and a secondary main metering system to control heavy load conditions.

A high altitude emission control device is incorporated to supply clean air to the carburetor at specific altitudes, so over rich air/fuel ratio conditions will not occur.

REMOVAL & INSTALLATION

1. Disconnect the negative battery cable.
2. Drain the coolant.
3. Remove the air cleaner.
4. Disconnect the harness connector and hoses.
5. Remove the accelerator cable from the carburetor.
6. Remove the bolts securing the carburetor to the intake manifold. Remove the carburetor and place a cover over the intake manifold.
7. To install, reverse the removal procedures.

IDLE SPEED ADJUSTMENT

1. Set the parking brake and block the wheels.
2. Place the manual transmission in Neutral or the automatic transmission in Park. Check the float level. Establish a normal operating temperature and make sure that the choke plate is open.
3. Turn off all of the accessories and wait until the cooling fan is not operating.
4. If equipped with power steering, place the wheels in the straight forward position. Remove the air filter.

5. Disconnect and plug the distributor vacuum line, canister purge line, EGR vacuum line and ITC valve vacuum line.

6. Connect a tachometer to the coil tachometer connector and a timing light to the No. 1 spark plug wire. Check the timing and idle speed.

7. If the idle speed needs adjusting, turn the idle speed adjusting screw.

8. If equipped with A/C, adjust the system to Max. Cold and place the blower on "High" position. Set the fast idle speed by turning the Fast Idle Adjusting Screw.

9. When adjustment is completed, turn the engine off, remove the test equipment, install the air filter and vacuum lines.

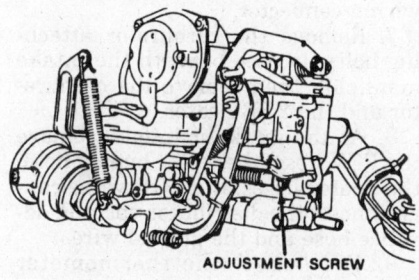

Adjusting the Idle speed screw

IDLE MIXTURE ADJUSTMENT

NOTE: The idle mixture screw is adjusted and sealed at the factory and no adjustment is required.

1. Remove the carburetor from the engine.

2. Using a center punch, make a punch mark on the idle mixture sealing plug. Drill a hole through the plug, insert a threaded screw and pull the plug from the throttle body.

NOTE: If the idle mixture screw is damaged from the drilling process, replace the screw.

3. Lightly seat the idle mixture screw, then back out 3 turns (M/T) or 2 turns (A/T). DO NOT overtighten the idle mixture screw.

4. Reinstall the carburetor and the air cleaner.

5. Adjust the idle speed.

6. Using a dwell meter, connect the positive lead to the duty monitor and the negative lead to ground. Place the meter dial on the 4 or 6 cylinder scale.

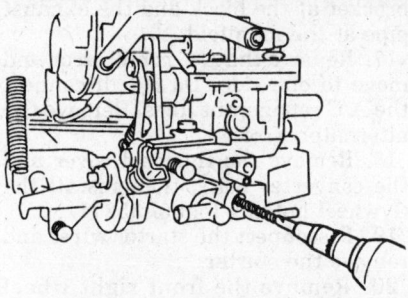

Drilling the idle mixture screw plug

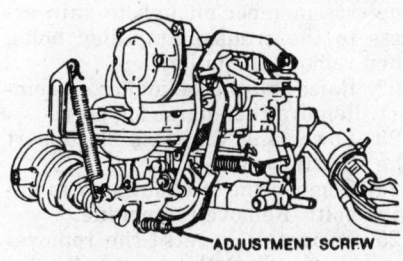

Adjusting the fast idle screw

Turn the idle mixture screw until the dwell meter reads 36 degrees (4 cylinder scale) or 24 degrees (6 cylinder scale).

7. Adjust the throttle adjusting screw to 750 rpm (M/T) or 1000 (A/T), then stop the engine and remove the tachometer.

8. Drive a new idle mixture plug into the throttle body, flush with the throttle body.

IDLE SPEED ADJUSTMENT (WITH A/C)

1. With the engine running, turn the A/C control to MAX COLD and the blower on high.

2. Turn the FIDC adjusting screw, on the tip of the throttle lever, to set the idle speed to 850 rpm (M/T) or 980 rpm (A/T).

3. Check the fast idle speed.

FAST IDLE SPEED ADJUSTMENT

1. Remove the air cleaner.

2. Open the throttle valve slightly, close the choke valve and release the throttle; set on the 1st stage of the fast idle cam.

3. Start the engine. DO NOT touch the accelerator pedal.

4. Connect a tachometer to the engine and check the fast idle speed; 850 rpm (M/T) or 980 rpm (A/T).

NOTE: Adjust the fast idle speed by turning the FIDC screw, located on the tip of the throttle lever.

5. Stop the engine, remove the tachometer and install the air cleaner.

ENGINE COOLING

Radiator

REMOVAL & INSTALLATION

1. Disconnect the negative battery cable.

2. Drain the cooling system.

3. Remove the air intake duct.

4. Remove the fan motor cable from the fan motor socket.

5. Disconnect the thermo switch cable.

6. Remove the fan motor assembly.

7. Remove the radiator hoses at the radiator, the coolant recovery hose at the filler neck and the Auto. Trans. oil cooler lines.

8. Remove the radiator mounting bolts and the radiator.

9. To install, reverse the removal procedures.

Water Pump

REMOVAL & INSTALLATION

1. Drain the cooling system.

2. Loosen the power steering pump adjustment bolts and remove the belt.

3. Remove the timing belt.

4. Remove the tension pulley and spring.

5. Remove the water pump mounting bolts, the water pump and gasket. Clean the mounting surfaces of all gasket material.

6. To install, reverse the removal procedures. Torque the water pump to 17 ft. lbs. and the tension pulley to 30 ft. lbs.

Thermostat

REMOVAL & INSTALLATION

1. Remove the negative battery cable. Drain the cooling system.

2. Remove the top radiator hose from the outlet pipe.

3. Remove the outlet pipe bolts, the outlet pipe, gasket and thermostat from the thermostat housing.

4. To install, reverse the removal procedure. Use 17 ft. lbs. of torque when installing the outlet pipe.

EMISSION CONTROLS

Please refer to "Emission Controls" in the Unit Repair Section for all service and maintenance procedures.

ENGINE MECHANICAL

The Spectrum uses an Isuzu 1.5L (94 cu. in.) overhead cam (OHC) engine. The 4-cylinder, in-line engine utilizes one compression and one oil control ring on each piston. The overhead camshaft, which is driven by the crankshaft through a timing belt, directly drives the rocker arms.

Engine

REMOVAL & INSTALLATION

1. Remove the hood and disconnect the negative battery cable.
2. Drain the cooling system.
3. Remove the air cleaner and the throttle cable at the carburetor.
4. Disconnect the heater hoses at the intake manifold, the coolant hose at the thermostat housing and the thermostat housing at the cylinder head.
5. Remove the distributor from the cylinder head.
6. Disconnect the oxygen sensor electrical connector.
7. Support the engine using a vertical lift and remove the right motor mount.
8. Disconnect the necessary electrical connectors and vacuum hoses.
9. Disconnect the flex hose at the exhaust manifold and the lower radiator hose at the block.
10. Remove the upper A/C compressor bolt and remove the belt.
11. Disconnect the power steering bracket at the block and remove the belt.
12. Disconnect the fuel lines from the fuel pump and the electrical connectors from under the carburetor.
13. Remove the upper starter bolt and raise the vehicle.
14. Drain the oil from the crankcase and remove the oil filter.
15. Disconnect the oil temperature switch connector.
16. Disconnect the exhaust pipe bracket at the block and the exhaust pipe at the manifold.
17. Remove the A/C compressor and move to one side. Do not disconnect the A/C refrigerant lines. Remove the alternator wires.
18. Remove the flywheel cover and the converter bolts, then install the flywheel holding tool (J-35271).
19. Disconnect the starter wires and remove the starter.
20. Remove the front right wheel and inner splash shield.
21. Lower the engine by lowering the crossmember enough to gain access to the crankshaft pulley bolts, then remove the pulley.
22. Raise the engine and crossmember. Remove the engine support.
23. Lower the vehicle and support the transmission.
24. Remove the transmission to engine bolts. Remove the engine.
25. To install, reverse the removal procedure, adjust the drive belts and refill the fluids.

Exhaust Manifold

REMOVAL & INSTALLATION

1. Disconnect the negative battery cable and the oxygen sensor wiring connector.
2. Disconnect the thermostatic air cleaner (TAC) flex hose.
3. Remove the hot air cover and raise the vehicle.
4. Disconnect the exhaust pipe from the exhaust manifold and lower the vehicle.
5. Remove the nuts and bolts securing the exhaust manifold to the cylinder head. Clean the gasket mounting surfaces.
6. To install, use new gaskets and reverse the removal procedures. Torque the exhaust manifold to 17 ft. lbs. and the exhaust pipe to 42 ft. lbs., then start the engine and check for leaks.

Intake Manifold

REMOVAL & INSTALLATION

1. Disconnect the negative battery cable. Drain the engine coolant.
2. Remove the bolt securing the alternator adjusting plate to the engine.
3. Disconnect and label all of the hoses attached to the air cleaner and remove the air cleaner.
4. Disconnect the air inlet temperature switch wiring connector.
5. Disconnect and label the hoses, electrical connectors, and control cable attached to the carburetor.
6. If equipped with A/C, disconnect the FIDC vacuum hose, the pressure tank control valve hose, the distributor/3-way connector hose and the VSV wiring connector.
7. Remove the carburetor attaching bolts (located beneath the intake manifold), then remove the carburetor and the EFE heater.
8. At the intake manifold, remove the PCV hose, the water bypass hose, the heater hoses, the EGR valve/canister hose, the distributor vacuum advance hose and the ground wires.
9. Disconnect the thermometer unit switch wiring connector.
10. Remove the intake manifold attaching nuts/bolts and the intake manifold.
11. Clean the sealing surfaces of the intake manifold and cylinder head.
12. To install, use new gaskets and reverse the removal procedure. Torque the intake manifold to 17 ft. lbs.; then adjust the engine control cable and the alternator belt tension. Refill the engine with coolant and check for leaks.

Rocker Arm Shafts and Rocker Arms

REMOVAL & INSTALLATION

1. Refer to the "Cylinder Head Cov-

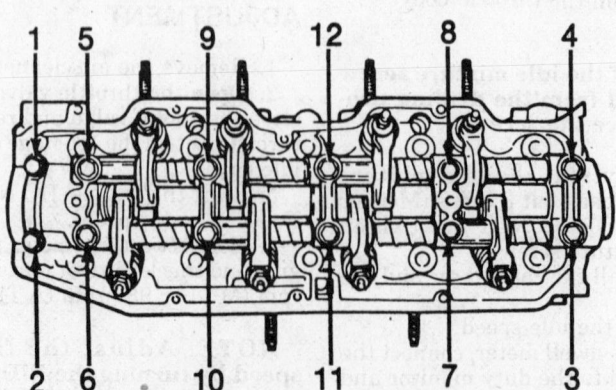

Rocker arm/shaft assembly removal sequence

1. Wing Nut
2. Air Cleaner Assembly
3. Air Duct
4. TCA Flex Hose
5. Carburetor

6. EFE Heater Assembly
7. Packing
8. Head Cover
9. Packing
10. Clip
11. Bolt; Head Cover
12. Bolt; Head Cover
13. Packing
14. Cap; Oil Filler
15. Packing

43. Camshaft
44. Oil Seal; Camshaft
45. Timing Pulley; Camshaft
46. Packing

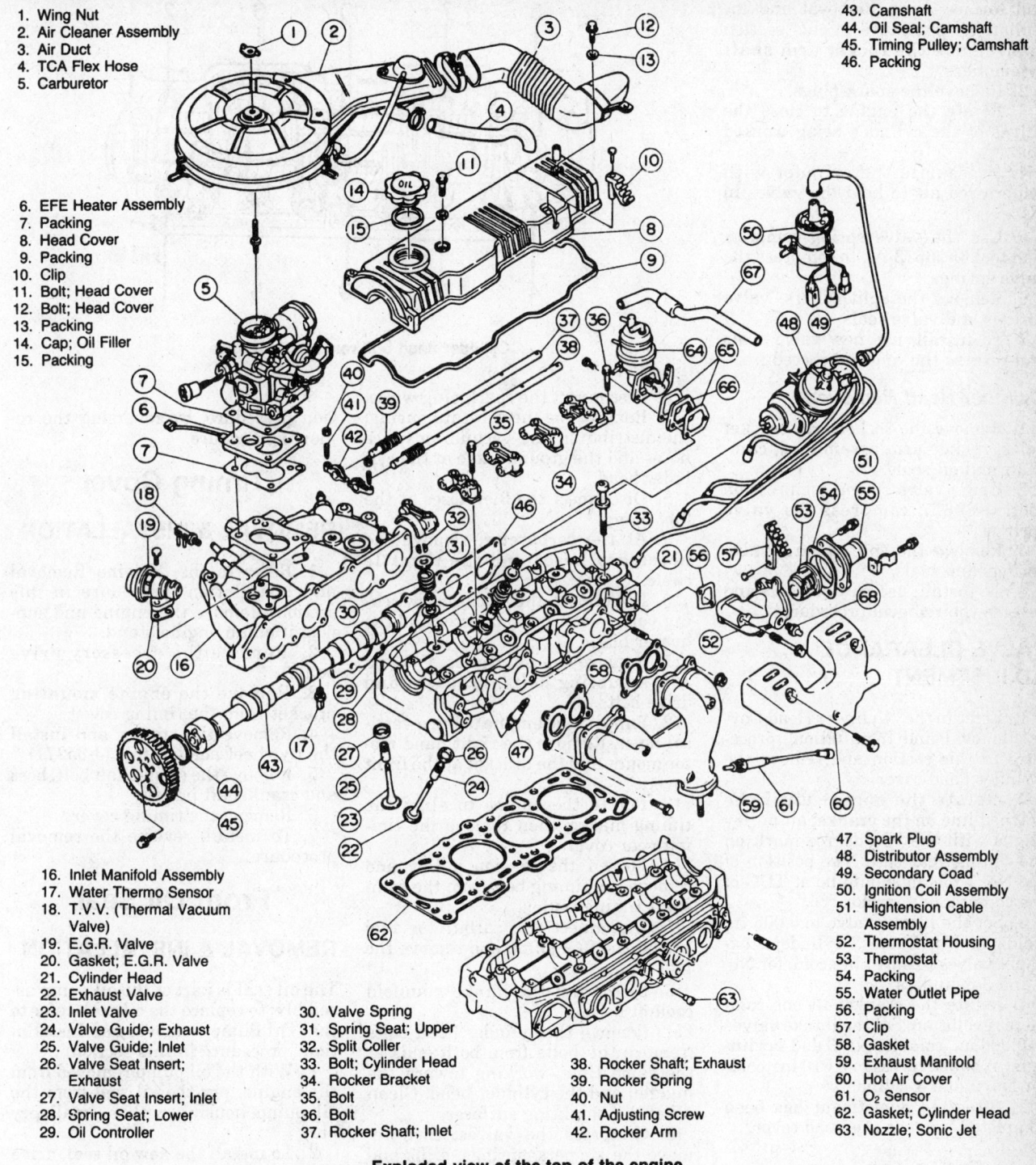

16. Inlet Manifold Assembly
17. Water Thermo Sensor
18. T.V.V. (Thermal Vacuum Valve)
19. E.G.R. Valve
20. Gasket; E.G.R. Valve
21. Cylinder Head
22. Exhaust Valve
23. Inlet Valve
24. Valve Guide; Exhaust
25. Valve Guide; Inlet
26. Valve Seat Insert; Exhaust
27. Valve Seat Insert; Inlet
28. Spring Seat; Lower
29. Oil Controller

30. Valve Spring
31. Spring Seat; Upper
32. Split Coller
33. Bolt; Cylinder Head
34. Rocker Bracket
35. Bolt
36. Bolt
37. Rocker Shaft; Inlet

38. Rocker Shaft; Exhaust
39. Rocker Spring
40. Nut
41. Adjusting Screw
42. Rocker Arm

47. Spark Plug
48. Distributor Assembly
49. Secondary Coad
50. Ignition Coil Assembly
51. Hightension Cable Assembly
52. Thermostat Housing
53. Thermostat
54. Packing
55. Water Outlet Pipe
56. Packing
57. Clip
58. Gasket
59. Exhaust Manifold
60. Hot Air Cover
61. O₂ Sensor
62. Gasket; Cylinder Head
63. Nozzle; Sonic Jet

Exploded view of the top of the engine

er Removal and Installation" procedure in this section and remove the cylinder head cover.

2. Remove the rocker arm bracket bolts in sequence (work from both ends equally, toward the middle).

3. Remove the rocker arm shafts and then the rocker arms from the shafts.

4. To install, reverse the removal procedure.

NOTE: The rocker arm shafts are different from each other, make sure they are installed in the same position that they were removed. Install the rocker arms with the identification marks toward the front of the engine. Apply sealant to the bracket and cylinder head mating surfaces of the front and rear rocker brackets.

5. Mount the rocker assemblies securely to the dowel pins on the cylinder head. Torque the rocker arm bolts to 16 ft. lbs.

Valve Spring and Seal

REMOVAL & INSTALLATION
Cylinder Head On Engine

1. Refer to the "Rocker Arm Shafts

and Rocker Arms Removal and Installation" procedure in this section and remove the rocker arm shaft assemblies.

2. Remove the spark plugs.

3. Rotate the engine to close the valves of the cylinder being worked on.

4. Pressurize the cylinder with compressed air to hold the valves in place.

5. Use the valve spring compression tool (J-26513-A), to compress the valve springs.

6. Remove the split collars, valve springs and valve seals.

7. To install, use new valve seals and reverse the removal procedure.

Cylinder Head Removed

1. Remove the rocker arm bracket bolts, rocker arm assemblies, camshaft and oil seals.

2. Using valve spring compression tool (J-8062), compress the valve springs.

3. Remove the split collars, valve springs and seals.

4. To install, use new oil seals and reverse the removal procedure.

VALVE CLEARANCE ADJUSTMENT

1. Refer to the "Cylinder Head Cover Removal and Installation" procedure in this section and remove the cylinder head cover.

2. Rotate the engine until the notched line on the crankshaft pulley aligns with the "O" degree mark on the timing gear case. The position of the No. 1 piston should be at TDC of the compression stroke.

3. Set the intake valve to 0.006 in. (cold) for No. 1 and 2 cylinders; exhaust valves to 0.010 in. (cold) for No. 1 and 3 cylinders.

4. Rotate the crankshaft one complete revolution. Set the intake valves to 0.006 in. (cold) for No. 3 and 4 cylinders; exhaust valves to 0.010 in. (cold) for No. 2 and 4 cylinders.

5. After the adjustment has been completed, replace the head cover.

Cylinder Head

REMOVAL & INSTALLATION

1. Disconnect the negative battery cable.

2. Drain the cooling system.

3. Remove the air cleaner.

4. Disconnect the flex hose and oxygen sensor at the exhaust manifold.

5. Disconnect the exhaust pipe bracket at the block and the exhaust pipe at the manifold.

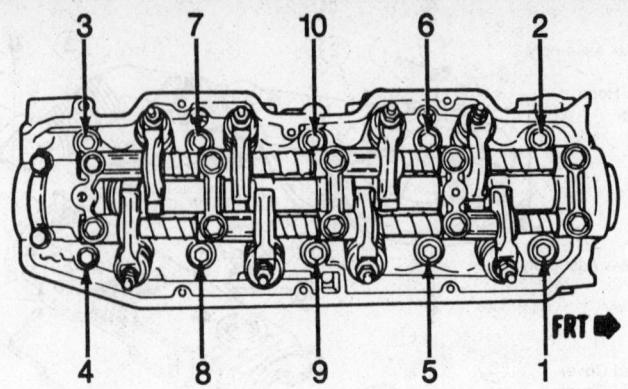

Cylinder head bolt removal sequence

6. Disconnect the spark plug wires.

7. Remove the thermostat housing, the distributor, the vacuum advance hoses and the ground cable at the cylinder head.

8. Disconnect the fuel hoses at the fuel pump.

9. At the carburetor, remove the necessary hoses and the throttle cable.

10. Disconnect the vacuum switching valve electrical connector and the heater hoses.

11. Remove the alternator, P/S and A/C adjusting bolts, brackets and drive belts.

12. Support the engine using a vertical hoist. Remove the right hand motor mount and the bracket at the front cover.

13. Rotate the engine to align the timing marks, then remove the timing gear cover.

14. Loosen the tension pulley and remove the timing belt from the camshaft timing pulley.

15. Disconnect the carburetor fuel line at the fuel pump and remove the fuel pump.

16. Disconnect the intake manifold coolant hoses.

17. Remove the cylinder head bolts (remove the bolts from both ends at the same time, working toward the middle) and the cylinder head. Clean all of the mounting surfaces.

18. Compress the valves; then remove the keepers, springs, seals and valves.

19. To install, use new seals and gaskets, apply oil to the bolt threads and torque the head bolts.

NOTE: When torquing the cylinder head bolts, work from the middle toward both ends at the same time. First, torque the bolts to 29 ft. lbs. and then final torque them to 58 ft. lbs.

20. After torquing, adjust the valve clearance and complete the installa-

tion procedure, by reversing the removal procedure.

Timing Cover

REMOVAL & INSTALLATION

1. Refer to the "Engine Removal and Installation" procedure in this section. Remove the engine and support it on an engine stand.

2. Remove the accessory drive belts.

3. Remove the engine mounting bracket from the timing cover.

4. Remove the starter and install the flywheel holding tool (J-35271).

5. Remove the crankshaft bolt, boss and crankshaft pulley.

6. Remove the timing cover.

7. To install, reverse the removal procedure.

Front Oil Seal

REMOVAL & INSTALLATION

The oil seal is part of the oil pump assembly; to replace the oil seal, refer to the "Oil Pump Removal and Installation" procedure in this section.

1. With the oil pump removed from the engine, pry the oil seal from the oil pump housing with a small pry bar.

2. To install the new oil seal, drive it into the housing using the seal installing tool (J-35269).

Timing Belt

REMOVAL

1. Remove the engine by referring to the "Engine Removal and Installation" procedure in this section. Mount the engine to an engine stand.

2. Remove the accessory drive belts.

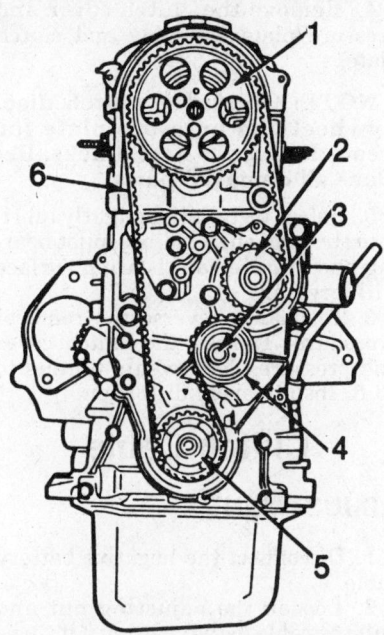

1. Camshaft timing pulley
2. Water pump timing pulley
3. Bolt
4. Tension pulley
5. Crankshaft timing pulley
6. Timing belt

Timing belt assembly

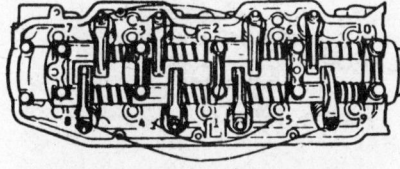

Cylinder head bolt torque sequence

3. Remove the engine mounting bracket from the timing cover.

4. Rotate the crankshaft until the notch on the crankshaft pulley aligns with the "O" degree mark on the timing cover and the No. 4 cylinder is on TDC of the compression stroke.

5. Remove the starter and install the flywheel holding tool (J-35271).

6. Remove the crankshaft bolt, boss and pulley.

7. Remove the timing cover bolts and the timing cover.

8. Loosen the tension pulley bolt.

9. Insert an allen wrench into the tension pulley hexagonal hole and loosen the timing belt by turning the tension pulley clockwise.

10. Remove the timing belt.

NOTE: Inspect the timing belt for signs of cracking, abnormal wear and hardening. Never expose the belt to oil, sunlight or heat. Avoid excessive bending, twisting or stretching.

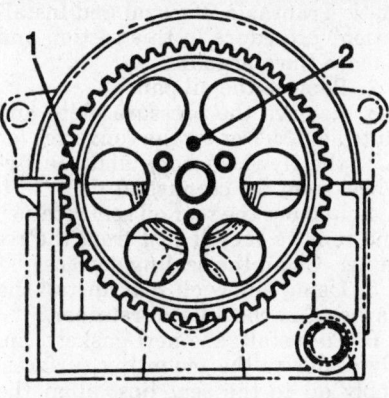

1 ALIGNMENT MARK   2 DOWEL

Alignment of the camshaft pulley

INSTALLATION

1. Position the woodruff key on the crankshaft followed by the crankshaft timing gear. Align the groove on the timing gear with the mark on the oil pump.

2. Align the camshaft timing gear mark with the upper surface of the cylinder head and the dowel pin in its uppermost position.

3. Place the timing belt arrow in the direction of the engine rotation and install the timing belt. Tighten the tension pulley bolt.

4. Turn the crankshaft two complete revolutions and realign the crankshaft timing gear groove with the mark on the oil pump.

5. Loosen the tension pulley bolt and apply tension to the belt with an allen wrench. Torque the pulley bolt to 37 ft. lbs. while holding the pulley stationary.

6. Adjust the valve clearances.

7. To complete the installation, reverse the removal procedure.

Cylinder Head Cover

REMOVAL & INSTALLATION

1. Disconnect the negative battery cable and remove the PCV hoses.

2. Remove the spark plug wires from the mounting clip.

3. Remove the ground wire from the right rear side of the head cover.

4. Support the engine and remove the right side engine mounting rubber, bolts and plate.

5. Remove the mounting bracket on the timing cover.

6. Remove the four bolts holding the timing cover and the two bolts holding the cylinder head cover.

7. Loosen the timing cover and remove the cylinder head cover.

NOTE: If the cylinder head cover sticks, strike the end of the cover with a rubber mallet or pry it from the cylinder head.

8. With the cover removed, clean the sealing surfaces of the cover and the cylinder head.

9. To install, apply sealer to the sealing surfaces and reverse the removal procedures.

10. Start the engine and check for leaks.

Camshaft

REMOVAL & INSTALLATION

1. Disconnect the negative battery cable.

2. Align the crankshaft pulley notch with the "O" degree mark on the timing cover.

3. Remove the cylinder head cover.

4. Remove the timing cover.

5. Loosen the camshaft timing gear bolts (DO NOT rotate the engine).

6. Loosen the timing belt tensioner and remove the timing belt from the camshaft timing gear.

7. Remove the rocker arm shaft/rocker arm assembly.

8. Remove the distributor bolt and the distributor.

9. Remove the camshaft and the camshaft seal.

10. To install, drive a new camshaft seal on the camshaft using the seal installation tool (J-35268), reverse the removal procedure, adjust the valves and the timing belt.

Piston & Rod Positioning

Install the piston and rod assemblies into the same cylinder bore, facing the same direction from which they were removed. Each piston has a front directional mark stamped on the top surface.

ENGINE LUBRICATION

Oil Pan

REMOVAL & INSTALLATION

1. Disconnect the negative battery cable.

2. Raise the vehicle, place it on jack stands and drain the crankcase.

3. Disconnect the exhaust pipe bracket from the block and the exhaust pipe at the manifold.

4. Disconnect the right hand tension rod located under the front bumper.

5. Remove the oil pan bolts and oil pan, then clean the sealing surfaces.

6. To install, use a new gasket, apply sealant to the oil pump housing and the rear retainer housing, reverse the removal procedure.

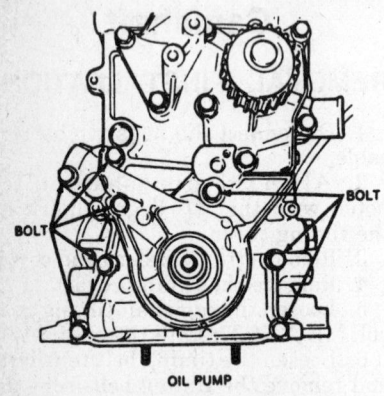

OIL PUMP

Location of the oil pump

Oil Pump

REMOVAL & INSTALLATION

1. Refer to the "Engine Removal and Installation" procedure in this section and remove the engine.

2. Drain the crankcase.

3. Remove the alternator belt and the starter.

4. Install the flywheel holding tool (J-35271) to secure the flywheel.

5. Remove the crankshaft pulley and boss.

6. Remove the timing cover bolts and the timing cover.

7. Loosen the tension pulley and remove the timing belt.

8. Remove the crankshaft timing gear and the tension pulley.

9. Remove the oil pan bolts, oil pan, oil strainer fixing bolt and the oil strainer assembly.

10. Remove the oil pump bolts and the oil pump assembly.

11. Remove the sealing material from the oil pump and engine block sealing surfaces.

12. To install, lubricate the oil pump, use new gaskets, apply sealant to the sealing surfaces and reverse the removal procedure.

Rear Main Seal

REMOVAL & INSTALLATION

1. Refer to the "Manual or Automatic Transaxle Removal and Installation" procedure in this section and remove the transaxle.

2. Remove the oil pan.

3. Remove the pressure plate and clutch for M/T or torque converter for A/T, the flywheel bolts and the flywheel from the crankshaft.

4. Remove the rear oil seal retainer and remove the oil seal from the retainer. Clean the sealing surfaces.

5. Using a new oil seal, install the new seal in the oil seal retainer.

6. To install, use new gaskets, apply sealer to the mounting surfaces, apply oil to the seal lips, align the dowel pins of the retainer with the engine block and reverse the removal procedure.

CLUTCH

REMOVAL & INSTALLATION

1. Refer to the "Manual Transaxle Removal and Installation" procedure in this section and remove the transaxle.

2. Install a pilot shaft tool (J-35282) into the pilot bearing to support the clutch assembly during the removal procedure.

NOTE: Observe the alignment marks on the clutch and the clutch cover and pressure plate assembly. If the markings are not present, be sure to add them.

3. Loosen the clutch cover and pressure plate assembly retaining bolts evenly (one at a time) until the spring pressure is released.

4. Remove the clutch cover and pressure plate assembly and clutch plate.

NOTE: Check the clutch disc, flywheel and pressure plate for wear, damage or heat cracks. Replace all damaged parts.

5. Before installation, lightly lubricate the pilot shaft splines, pilot bearing and pilot release bearing surface with grease.

6. To install, reverse the removal procedure. Tighten the clutch cover and pressure plate evenly (torque to 13 ft. lbs.) to avoid distortion.

Clutch Cable

ADJUSTMENT

1. Disconnect the negative battery cable.

2. Loosen the adjusting nut and pull the cable to the rear until it turns freely.

3. Adjust the cable length by turning the adjusting nut.

4. When the clutch pedal free play travel reaches 0.39–0.79 in. release the cable.

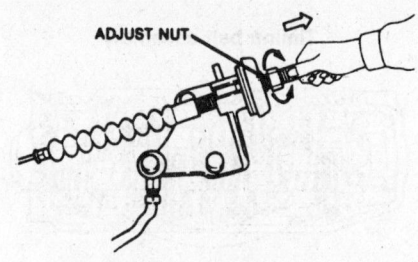

ADJUST NUT

Adjustment of the clutch cable

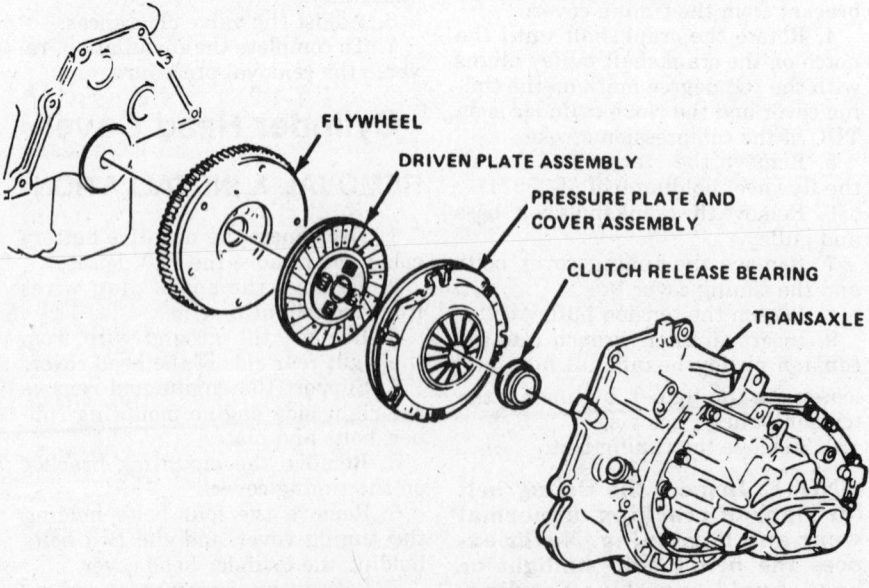

FLYWHEEL

DRIVEN PLATE ASSEMBLY

PRESSURE PLATE AND COVER ASSEMBLY

CLUTCH RELEASE BEARING

TRANSAXLE

Exploded view of the clutch assembly

5. When the adjustment has been completed, tighten the lock nut.

REMOVAL & INSTALLATION

1. Disconnect the negative battery cable.
2. Loosen the clutch cable adjusting nuts. Disconnect the cable from the release arm and cable bracket.
3. At the clutch pedal, remove the cable retaining bolt.
4. Disconnect the cable from the front of the dash.
5. Remove the clutch cable from the vehicle.
6. To install, grease the clutch cable pin and reverse the removal procedure.
7. Adjust the clutch cable.

Clutch Start Switch

REMOVAL & INSTALLATION

The clutch switch is mounted above the clutch pedal. It is connected to the electrical circuit of the starter motor and operates in accordance to the movement of the clutch pedal. To operate the starter, the clutch pedal must be depressed all the way.

1. Disconnect the negative battery cable.
2. Remove the lead wire at the switch.
3. At the clutch pedal stop bracket, remove the switch mounting screw.
4. Remove the switch from the clutch pedal.
5. To install, reverse the removal procedure.

Clutch Release Bearing

REMOVAL & INSTALLATION

1. Refer to the "Manual Transaxle Removal and Installation" procedure in this section and remove the transaxle.
2. Disconnect the return spring from the shaft fork.
3. Remove the clutch release bearing from the pilot shaft bearing retainer.

NOTE: Inspect the bearing for wear, damage or rough rotation. DO NOT place the bearing in solvent or damage may occur to the seals.

4. Lightly lubricate the bearing with grease.
5. To install, reverse the removal procedure.

MANUAL TRANSAXLE

REMOVAL & INSTALLATION

1. Drain the oil from the transaxle.
2. Disconnect the negative battery cable at the battery and the transaxle.
3. Disconnect the wiring connectors, speedometer cable, clutch cable and shift cables from the transaxle.
4. Remove the air cleaner heat tube.
5. Remove the upper transaxle to engine retaining bolts.
6. Raise the vehicle. Remove the left-front wheel assembly and splash shield.
7. Disconnect the left tie rod at the steering knuckle and the left tension rod.
8. Disconnect the drive axles and remove the shafts by pulling them straight out from the transaxle (avoid damaging the oil seals).
9. Remove the dust cover at the clutch housing.
10. Support the transaxle with a floor jack and remove the transaxle-to-engine retaining bolts.
11. While sliding the transaxle away from the engine, carefully lower the jack, guiding the right axle shaft out of the transaxle.

NOTE: The right axle shaft MUST be installed to the transaxle when the transaxle is being installed to the engine.

12. To install, reverse the removal procedure.

Shift Control Assembly

REMOVAL & INSTALLATION

1. Disconnect the negative battery cable.
2. Remove the shifter boot and console.
3. Disconnect the shift cables at the control assembly.
4. Remove the shift control assembly.
5. To install, reverse the removal procedure.

Shift Cables

REMOVAL & INSTALLATION

1. Disconnect the negative battery cable.

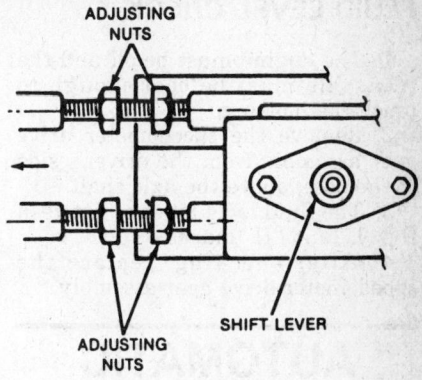

SHIFT AND SELECT CABLE ADJUSTMENT PROCEDURE
1. Place transaxle and shift lever in the neutral position.
2. Turn adjusting nuts until shift lever is in a vertical position.
3. After adjustment, tighten adjusting nuts securely

Adjusting the shift linkage of the transaxle

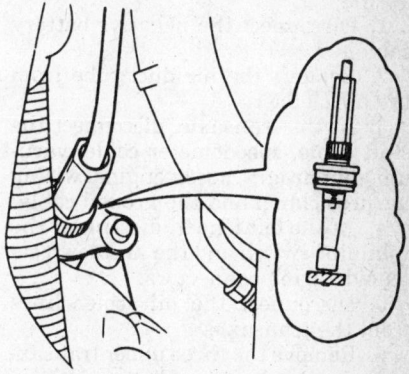

Checking the fluid level of the transaxle

2. Disconnect the retaining clips and the shift cables at the transaxle.
3. Remove the shifter boot and console.
4. Disconnect the shift cables at the control assembly.
5. Pull the carpet back and remove the front left sill plate.
6. Remove the shift cable plate screws.
7. To install, reverse the removal procedure and adjust the shift cables.

Shift Linkage

ADJUSTMENT

1. Loosen the adjusting nuts.
2. Place the transaxle and the shift lever in the Neutral position.
3. Turn the adjusting nuts until the shift lever is in the vertical position.
4. Tighten the adjusting nuts.

FLUID LEVEL CHECK

1. The engine must be off and the transaxle must be cool enough to touch the housing.

2. Remove the speedometer drive gear assembly from the driver's side of the case, above the axle shaft.

3. The fluid level must be between the "L" and "H" marks.

4. After checking, replace the speedometer drive gear assembly.

AUTOMATIC TRANSAXLE

For all automatic transaxle adjustments and service procedures, please refer to the "Automatic Transmission" Unit Repair section.

REMOVAL & INSTALLATION

1. Disconnect the negative battery cable.

2. Remove the air duct tube from the air cleaner.

3. At the transaxle, disconnect the shift cable, speedometer cable, vacuum diaphragm hose, engine wiring harness clamp and the ground cable.

4. At the left fender, disconnect the inhibitor switch and the kickdown solenoid wiring connectors.

5. Disconnect the oil cooler lines from the transaxle.

6. Remove the three upper transaxle-to-engine mounting bolts and raise the vehicle.

7. Remove both front wheels and the left front fender splash shield.

8. Disconnect both tie rod ends at the steering knuckles.

9. Remove both front tension rod brackets and disconnect the rods from the control arms.

10. Disengage the axle shafts from the transaxle.

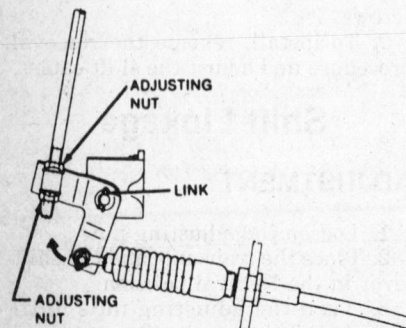

Adjusting the shift cable of the automatic transaxle

11. Remove the flywheel dust cover and the converter-to-flywheel attaching bolts.

12. Remove the transaxle rear mount through bolt.

13. Disconnect the starter wiring and the starter. Support the transaxle.

14. Remove the lower transaxle-to-engine mounting bolts and remove the transaxle.

15. To install, reverse the removal procedure, torque the converter-to-flywheel at 30 ft. lbs., transaxle-to-engine at 56 ft. lbs., adjust the shift linkage and fill the transaxle with Dexron® II automatic transmission fluid.

DRIVE AXLE

Front Axle Shaft

REMOVAL & INSTALLATION

1. Refer to the "Front Hub Removal and Installation" procedure in this section and remove the front hub.

Support the axle shaft with mechanic's wire.

2. Remove the drain plug and drain the oil from the transaxle.

3. Place a large pry bar between the differential case and the inboard constant velocity joint. Pry the axle shaft from the differential case.

4. Remove the front axle assembly.

5. To install, reverse the removal procedure. When installing the axle shaft, press it into the differential case until it locks with with snap ring.

Constant Velocity Joint

For further information on constant velocity joint overhaul, please refer to "U-Joints/CV-Joints" in the Unit Repair section.

Steering Knuckle and Hub Assembly

REMOVAL & INSTALLATION

DO NOT remove the hub from the steering knuckle unless it is absolutely necessary.

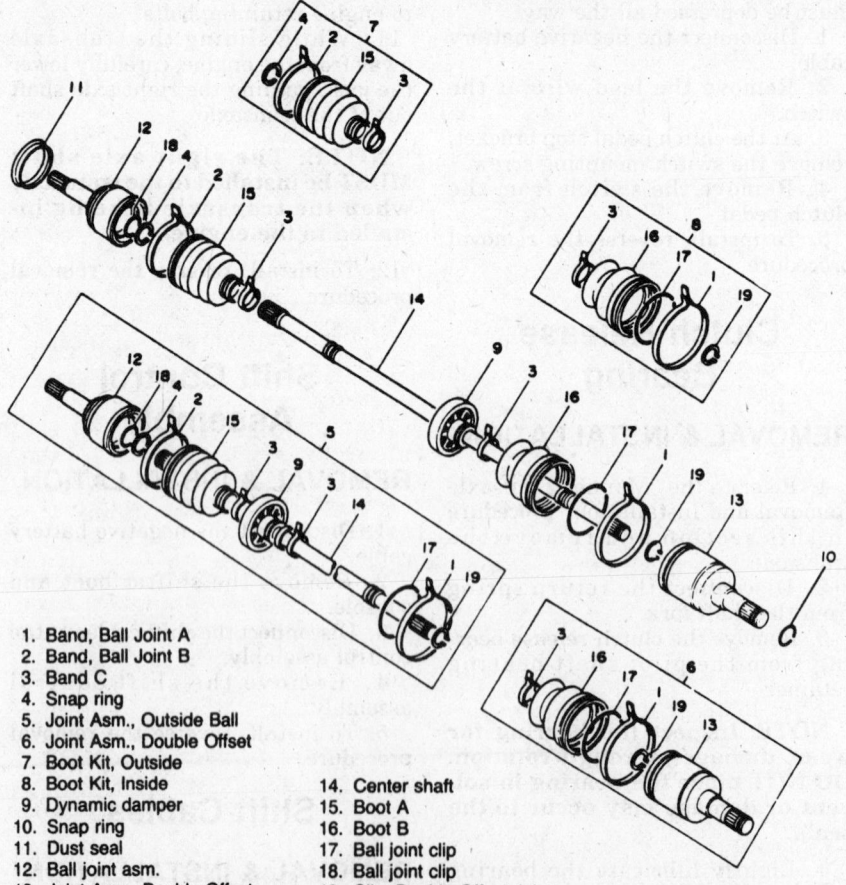

1. Band, Ball Joint A
2. Band, Ball Joint B
3. Band C
4. Snap ring
5. Joint Asm., Outside Ball
6. Joint Asm., Double Offset
7. Boot Kit, Outside
8. Boot Kit, Inside
9. Dynamic damper
10. Snap ring
11. Dust seal
12. Ball joint asm.
13. Joint Asm., Double Offset
14. Center shaft
15. Boot A
16. Boot B
17. Ball joint clip
18. Ball joint clip
19. Clip, Double Offset Joint

Exploded view of the front axle

1. Loosen the wheel nuts. Remove the grease cap, cotter pin, hub nut and thrust washer.

2. Remove the caliper and support it on a wire.

3. Remove the rotor.

4. Remove the tie rod nut. Using a ball joint removal tool, separate the tie rod from the steering knuckle.

5. Remove the two ball joint-to-control arm/tension rod retaining nuts and bolts.

6. Remove the two strut-to-steering knuckle retaining nuts/bolts.

7. Remove the steering knuckle. When removing the axle shaft from the steering knuckle, be careful not to drop it and support it with a wire.

8. To install, reverse the removal procedure and bleed the brake system.

Steering Knuckle, Hub & Bearing

DISASSEMBLY & REASSEMBLY

1. Remove the inner seal and snap ring.

2. Using an arbor press, press the hub from the steering knuckle. If necessary, press the spacer from the hub.

3. Remove the outer seal and snap ring.

4. Using an arbor press, press the bearing from the steering knuckle. Clean and inspect all parts.

5. To assemble, reverse the disassembly procedure.

NOTE: Replace the seals and bearings. Lubricate all parts.

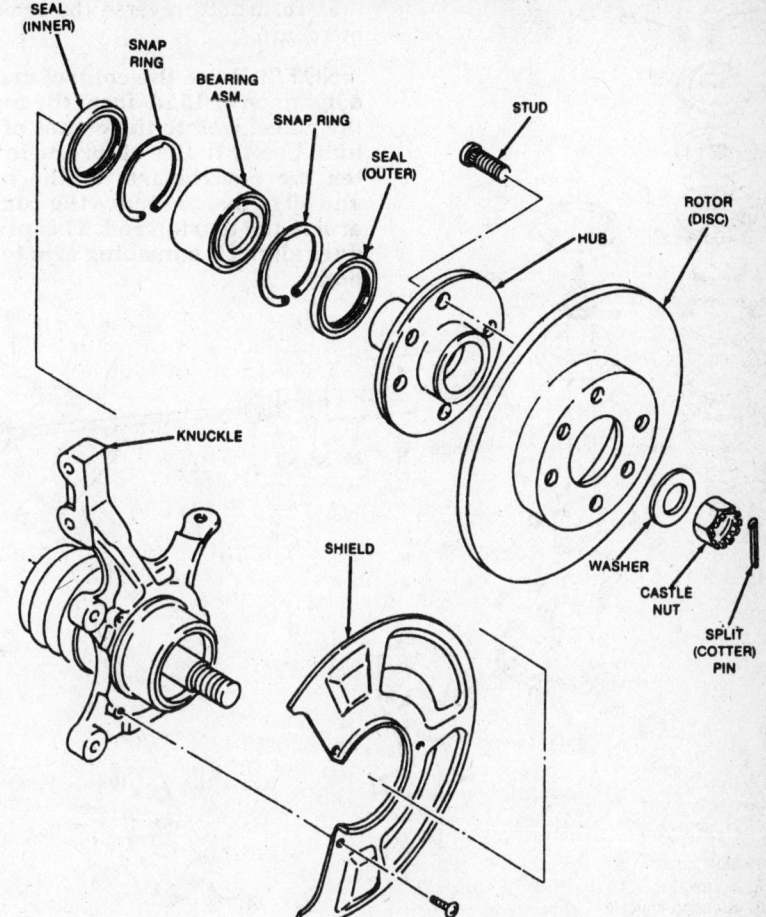

Exploded view of the front hub assembly

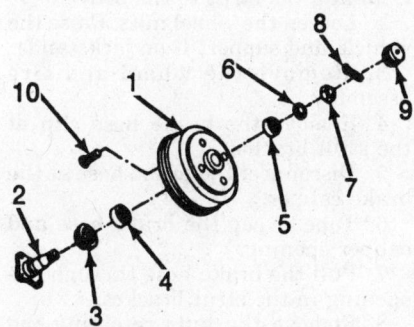

1. Hub & Drum Assembly
2. Knuckle
3. Oil Seal
4. Inner Bearing
5. Outer Bearing
6. Washer
7. Hub Nut 29 Nm (22 ft.lbs.)
8. Cotter Pin
9. Hub Cap
10. Wheel Lug

Exploded view of the rear axle hub assembly

Rear Axle Hub

REMOVAL & INSTALLATION

1. Raise the rear end of the vehicle and support it on jackstands. Remove the rear wheels.

2. Remove the hub cap, cotter pin, hub nut, washer and outer bearing.

3. Remove the hub.

4. To install, reverse the removal procedure.

NOTE: If the cotter pin holes are out of alignment upon reassembly, use a wrench to tighten the nut until the hole in the shaft and a slot of the nut align.

Rear Axle Bearing

REMOVAL & INSTALLATION

1. Refer to the "Rear Axle Hub Removal and Installation" procedure in this section, then remove the hub.

2. Using a slide hammer puller and

attachment, pull the oil seal from the hub. Remove the inner bearing.

3. Using a brass drift and a hammer, drive both bearing races from the hub.

4. Clean, inspect and/or replace all parts.

5. To install, pack the bearings with grease, coat the oil seal lips with grease and reverse the removal procedure.

Rear Axle Shaft

REMOVAL & INSTALLATION

1. Refer to the "Rear Axle Hub Removal and Installation" procedure in this section, then remove the hub.

2. Remove the four bolts retaining the axle shaft to the axle assembly.

3. Separate the brake assembly from the axle shaft and support it with a wire.

4. Remove the axle shaft from the assembly.

5. To install, reverse the removal procedure.

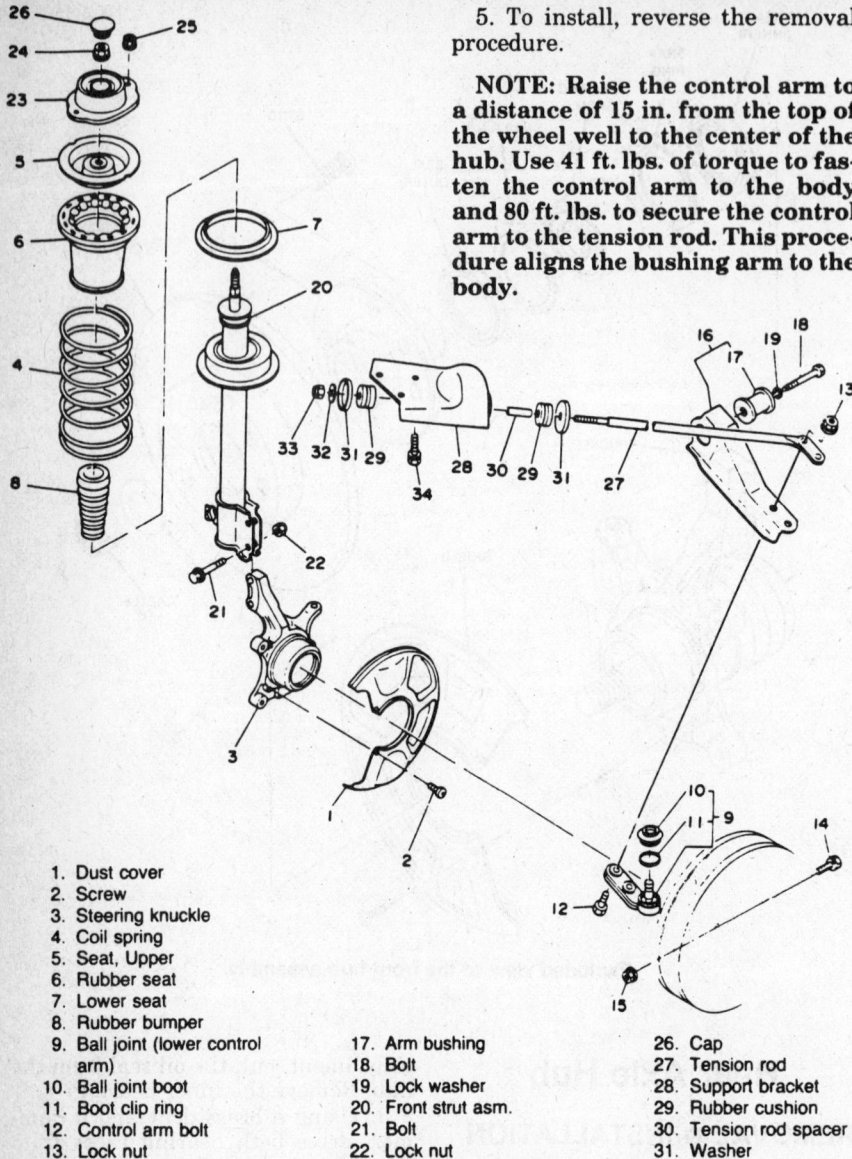

5. To install, reverse the removal procedure.

NOTE: Raise the control arm to a distance of 15 in. from the top of the wheel well to the center of the hub. Use 41 ft. lbs. of torque to fasten the control arm to the body and 80 ft. lbs. to secure the control arm to the tension rod. This procedure aligns the bushing arm to the body.

1. Dust cover
2. Screw
3. Steering knuckle
4. Coil spring
5. Seat, Upper
6. Rubber seat
7. Lower seat
8. Rubber bumper
9. Ball joint (lower control arm)
10. Ball joint boot
11. Boot clip ring
12. Control arm bolt
13. Lock nut
14. Bolt
15. Lock nut
16. Lower arm asm.
17. Arm bushing
18. Bolt
19. Lock washer
20. Front strut asm.
21. Bolt
22. Lock nut
23. Strut upper mount
24. Flange nut (strut shaft)
25. Nut
26. Cap
27. Tension rod
28. Support bracket
29. Rubber cushion
30. Tension rod spacer
31. Washer
32. Washer
33. Lock nut
34. Bolt

Exploded view of the front suspension assembly

FRONT SUSPENSION

Control Arm

REMOVAL & INSTALLATION

1. Raise and support the front of the vehicle.
2. Remove the control arm to tension arm retaining nuts and bolts.
3. Remove the nut/bolt securing the control arm to the body.
4. Remove the control arm and check for cracking or distortion.

Ball Joint

INSPECTION

Before removing the ball joint for replacement, check it and the boot for excessive wear or damage.

REMOVAL & INSTALLATION

1. Loosen the wheel nuts.
2. Raise the vehicle and support it on jackstands.
3. Remove the wheel and tire assembly.
4. Remove the two nuts retaining the ball joint to the tension rod and control arm assembly.

5. Remove the pinch bolt retaining the ball joint to the steering knuckle.
6. Remove the ball joint.
7. To install, reverse the removal procedure.

Tension Rod

REMOVAL & INSTALLATION

1. Raise and support the vehicle on jackstands.
2. If equipped with a stabilizer bar, remove the nuts, bolts and insulators retaining it to the tension rod.
3. Remove the nut and washer retaining the tension rod to the body.
4. Remove the nuts and bolts retaining the tension rod to the control rod.
5. Remove the tension rod.
6. To install, reverse the removal procedure.

Stabilizer Bar

REMOVAL & INSTALLATION

1. Raise and support the vehicle on jack stands.
2. Remove the nuts, bolts and insulators retaining the stabilizer bar to the tension rod.
3. Remove the stabilizer bar.
4. To install, reverse the removal procedure. Align the front side of the insulator edge with the paint mark on the upper rear edge of the tension bar.

MacPherson Strut

REMOVAL & INSTALLATION

1. Open the hood. Remove the nuts retaining the strut to the body.
2. Loosen the wheel nuts. Raise the vehicle and support it on jackstands.
3. Remove the wheel and tire assembly.
4. Remove the brake hose clip at the strut bracket.
5. Disconnect the brake hose at the brake caliper.
6. Tape or cap the brake hose and caliper opening.
7. Pull the brake hose through the opening in the strut bracket.
8. Remove the nuts retaining the strut to the steering knuckle.
9. Remove the strut assembly.
10. To install, reverse the removal procedure and bleed the brake system.

INSPECTION

Check the shock absorber for leaks or defective operation and the coil spring for wear, cracks or distortion.

OVERHAUL

For all overhaul procedures, please refer to "Struts" in the Unit Repair section.

REAR SUSPENSION

Shock Absorber

REMOVAL & INSTALLATION

1. Open the trunk and lift off the trim cover (hatch back models only). Remove the upper shock absorber nut.
2. Remove the lower bolt of the shock absorber.
3. Remove the shock absorber.
4. To install, reverse the removal procedure.

NOTE: When replacing the shock absorber, NEVER reuse the old lower bolt, ALWAYS use a new one.

Axle Assembly

REMOVAL & INSTALLATION

1. Raise the rear end of the vehicle and support it on jack stands.
2. Remove the rear wheels.
3. At the center of the rear axle, remove the brake line, retaining clip and flexible hose.
4. Remove the parking brake tension spring at the rear axle.
5. Disconnect the parking brake cable from the turn buckle and at the cable joint.
6. Support the axle with a jack, then remove the lower shock absorber bolt and disconnect it from the axle.
7. Lower the axle support and remove the coil spring.
8. Remove the bolts retaining the axle to the body and remove the axle assembly.
9. To install, reverse the removal procedure.
10. After installation, bleed the brake system.

NOTE: Raise the axle assembly to a distance of 15.2 in. from the top of the wheel well to the center of the axle hub, then torque the fasteners. ALWAYS replace the lower shock absorber bolt with a new one.

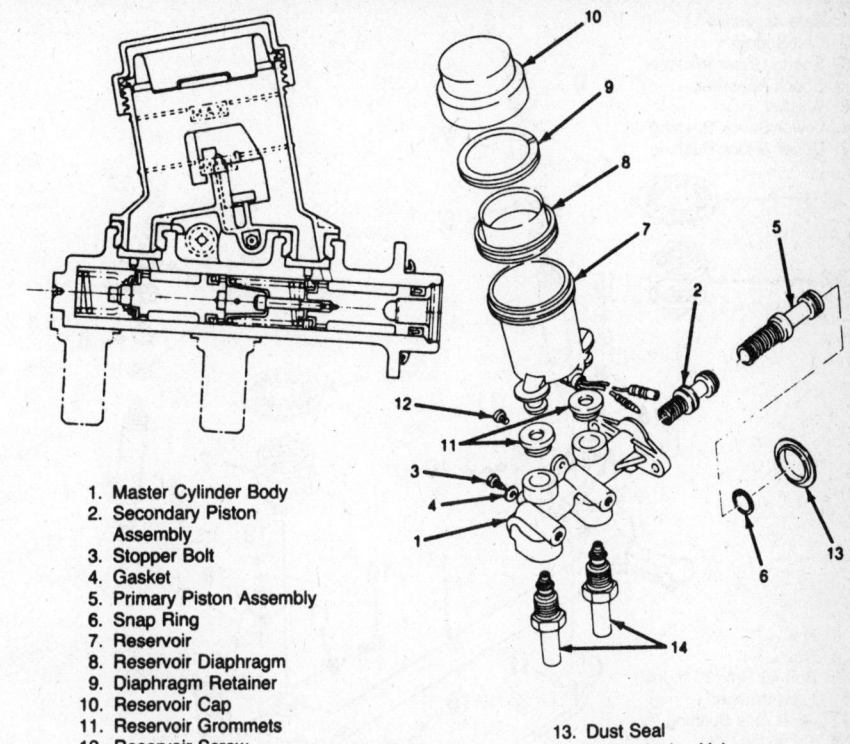

1. Master Cylinder Body
2. Secondary Piston Assembly
3. Stopper Bolt
4. Gasket
5. Primary Piston Assembly
6. Snap Ring
7. Reservoir
8. Reservoir Diaphragm
9. Diaphragm Retainer
10. Reservoir Cap
11. Reservoir Grommets
12. Reservoir Screw
13. Dust Seal
14. Proportioning Valves

Exploded view of the master assembly

Coil Springs

REMOVAL & INSTALLATION

Refer to the "Rear Axle Assembly Removal and Installation" in this section.

Axle Bushings

REMOVAL & INSTALLATION

1. Refer to the "Rear Axle Assembly Removal and Installation" procedures in this section, then remove the axle assembly.
2. With the axle assembly removed, press the bushings from the housings.
3. To install, press new bushings into the housings and reverse the removal procedure.

Stabilizer Bar

REMOVAL & INSTALLATION

1. Remove the bolts retaining the stabilizer bar to the lower ends of the axle assembly.
2. Remove the stabilizer bar.
3. To install, reverse the removal procedure.

BRAKES

To service the brake shoes, drums, brake pads, rotors, calipers and/or wheel cylinders, please refer to the "Brakes" in the Unit Repair Section.

Master Cylinder

REMOVAL & INSTALLATION

1. Remove some brake fluid from the master cylinder with a syringe.
2. Disconnect and cap or tape the openings of the brake tube.
3. Disconnect the brake fluid level warning switch connector.
4. Remove the 2 nuts securing the master cylinder to the power brake booster.
5. Remove the master cylinder from the power brake booster.
6. To install, reverse the removal procedure, add fluid to the reservoir and bleed the brake system.

Power Brake Booster

REMOVAL & INSTALLATION

1. Refer to the "Master Cylinder

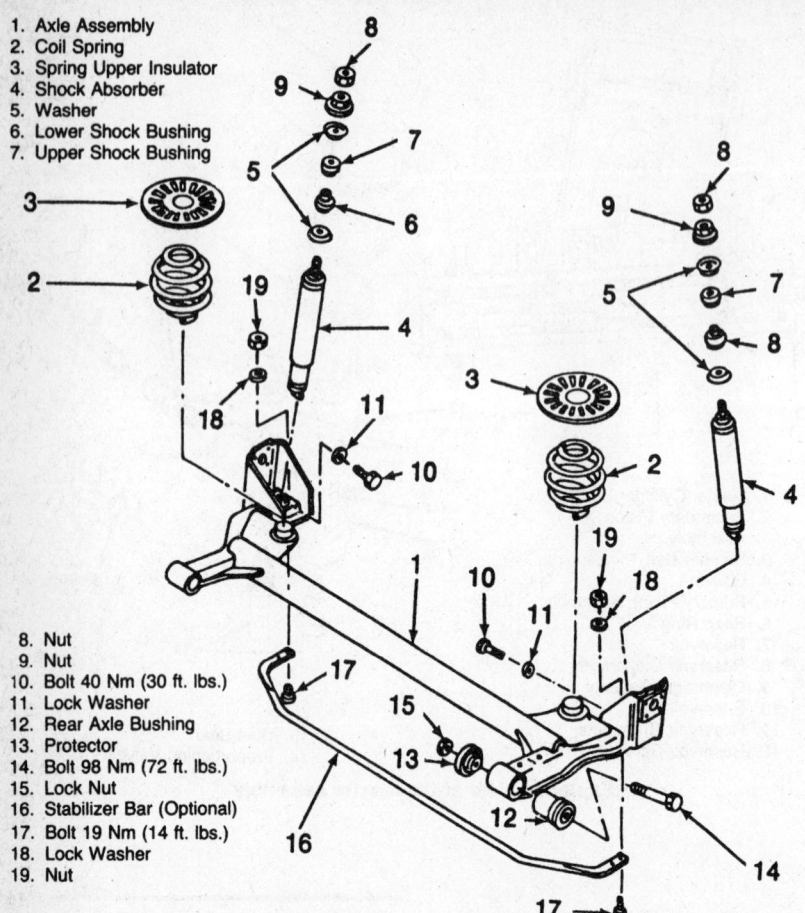

1. Axle Assembly
2. Coil Spring
3. Spring Upper Insulator
4. Shock Absorber
5. Washer
6. Lower Shock Bushing
7. Upper Shock Bushing

8. Nut
9. Nut
10. Bolt 40 Nm (30 ft. lbs.)
11. Lock Washer
12. Rear Axle Bushing
13. Protector
14. Bolt 98 Nm (72 ft. lbs.)
15. Lock Nut
16. Stabilizer Bar (Optional)
17. Bolt 19 Nm (14 ft. lbs.)
18. Lock Washer
19. Nut

Exploded view of the rear axle assembly

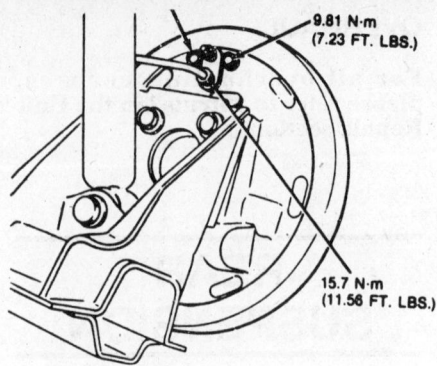

Wheel cylinder attaching bolts

Removal and Installation" procedure in this section and remove the master cylinder.

2. Remove the vacuum hose from the vacuum servo.

3. Remove the clevis pin from the brake pedal.

4. Remove the 4 nuts from the brake assembly under the dash and remove the power booster from the engine compartment.

5. To install, reverse the removal procedure.

Proportioning Valves

REMOVAL & INSTALLATION

1. Clean the area around the reservoir and brake pipe connections.

2. Remove the brake fluid from the master cylinder reservoir with a syringe.

3. Disconnect the brake pipes from the proportioning valves. Cap or tape all openings.

4. While holding the master cylinder, Use a box wrench and remove the proportioning valves from the master cylinder.

NOTE: It may be necessary to remove the master cylinder and place in a vise to sufficiently hold it while removing the proportioning valves.

5. Installation is the reverse of removal. Fill the reservoir and bleed the system.

Wheel Cylinder

REMOVAL & INSTALLATION

1. Remove the bnrake shoe and components toi gain access to the wheel cylinder.

2. Clean the area around the brake pipe and disconnect it from the wheel cylinder. Cap or tape all openings.

3. Remove the two bolts and remove the wheel cylinder.

4. Installation is the reverse of removal. Torque the mounting nuts to 7 ft. lbs.

Parking Brake

ADJUSTMENT

The parking brake adjustment is nor-

mal when the lever moves 7–9 notches at 66 lbs. of force. If it is not within limits, adjust the rear brakes. If this adjustment does not affect the specifications, adjust the parking brake turnbuckle.

STEERING

Tie Rod

REMOVAL & INSTALLATION

1. Raise the vehicle and remove the front wheel.

2. Remove the castle nut from the ball joint. Using a ball joint removal tool, separate the tie rod from the steering knuckle.

3. Disconnect the retaining wire from the inner boot and pull back the boot.

4. Using a chisel, straighten the staked part of the locking washer between the tie rod and the rack.

5. Remove the tie rod from the rack.

6. To install, reverse the removal procedure.

Rack & Pinion

REMOVAL & INSTALLATION

Manual Steering

1. Refer to the "Tie Rod Removal and Installation" procedure in this section. Remove both tie rod ends from the steering knuckles and the left inner tie rod from the rack.

2. Remove the intermediate shaft cover.

3. Loosen the upper pinch bolt and remove the lower pinch bolt at the pinion shaft.

4. Remove the steering gear to body retaining nuts.

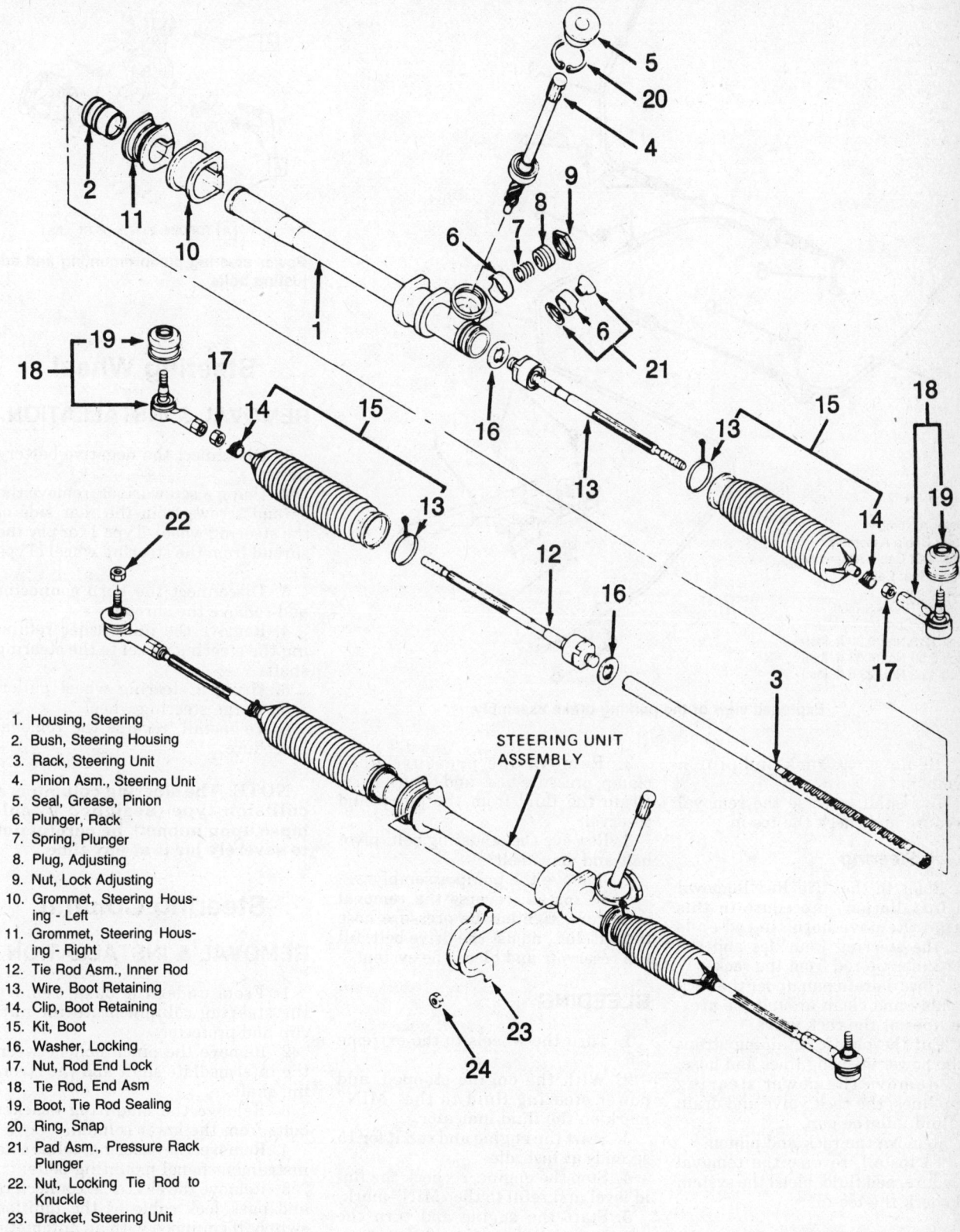

STEERING UNIT
ASSEMBLY

1. Housing, Steering
2. Bush, Steering Housing
3. Rack, Steering Unit
4. Pinion Asm., Steering Unit
5. Seal, Grease, Pinion
6. Plunger, Rack
7. Spring, Plunger
8. Plug, Adjusting
9. Nut, Lock Adjusting
10. Grommet, Steering Housing - Left
11. Grommet, Steering Housing - Right
12. Tie Rod Asm., Inner Rod
13. Wire, Boot Retaining
14. Clip, Boot Retaining
15. Kit, Boot
16. Washer, Locking
17. Nut, Rod End Lock
18. Tie Rod, End Asm
19. Boot, Tie Rod Sealing
20. Ring, Snap
21. Pad Asm., Pressure Rack Plunger
22. Nut, Locking Tie Rod to Knuckle
23. Bracket, Steering Unit
24. Nut, Bracket

Exploded view of the manual rack and pinion assembly

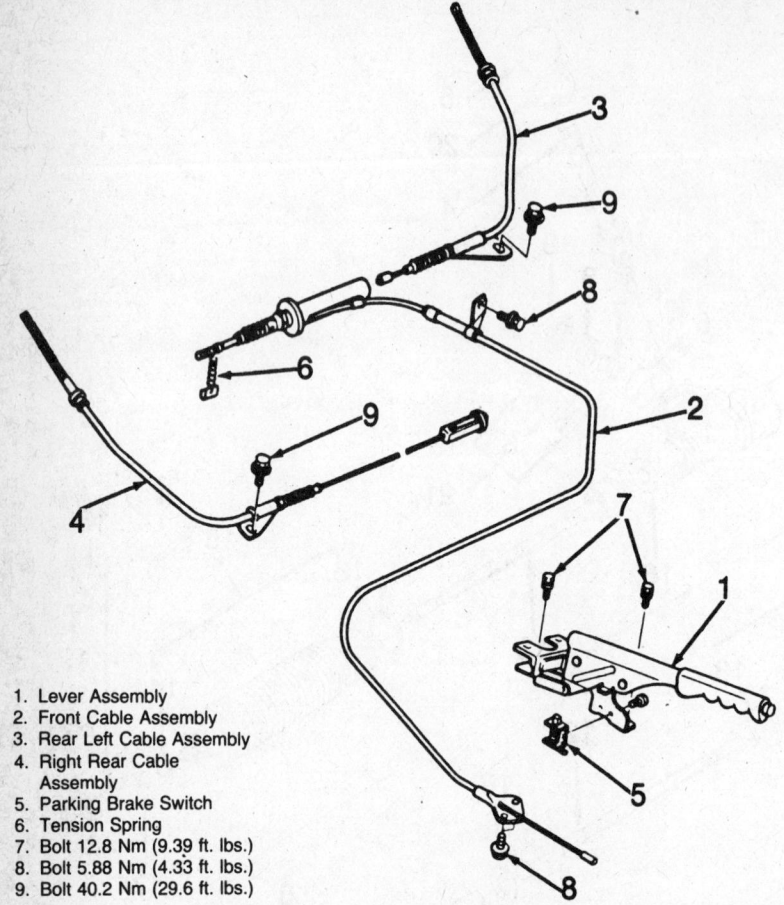

1. Lever Assembly
2. Front Cable Assembly
3. Rear Left Cable Assembly
4. Right Rear Cable Assembly
5. Parking Brake Switch
6. Tension Spring
7. Bolt 12.8 Nm (9.39 ft. lbs.)
8. Bolt 5.88 Nm (4.33 ft. lbs.)
9. Bolt 40.2 Nm (29.6 ft. lbs.)

Exploded view of the parking brake assembly

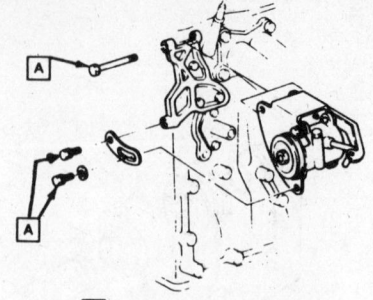

A TORQUE: 20 N·m (15 FT. LBS.)

Power steering pump mounting and adjusting bolts

5. Remove the rack and pinion assembly.
6. To install, reverse the removal procedure and check the toe-in.

Power Steering

1. Refer to the "Tie Rod Removal and Installation" procedure in this section. Remove both tie rod ends from the steering knuckles and the right inner tie rod from the rack.
2. Place a drain pan under the rack assembly and clean around the pressure lines at the rack valve.
3. Cut the plastic retaining straps at the power steering lines and hose.
4. Remove the power steering pump lines, the rack valve and drain the fluid into the pan.
5. Remove the rack and pinion.
6. To install, reverse the removal procedure, add fluid, bleed the system and check the toe-in.

Power Steering Pump

REMOVAL & INSTALLATION

1. Place a drain pan below the pump.

2. Remove the pressure hose clamp, pressure hose and return hose. Drain the fluid from the pump and reservoir.
3. Remove the adjusting bolt, pivot bolt and drive belt.
4. Remove the pump assembly.
5. To install, reverse the removal procedure, tighten the pressure hose to 20 ft. lbs., adjust the drive belt, fill the reservoir and bleed the system.

BLEEDING

1. Turn the wheels to the extreme left.
2. With the engine stopped, add power steering fluid to the "MIN" mark on the fluid indicator.
3. Start the engine and run it for 15 seconds at fast idle.
4. Stop the engine, recheck the fluid level and refill to the "MIN" mark.
5. Start the engine and turn the wheels from side to side (3 times).
6. Stop the engine check the fluid level.

NOTE: If air bubbles are still present in the fluid, the procedure must be repeated.

Steering Wheel

REMOVAL & INSTALLATION

1. Disconnect the negative battery cable.
2. Using a screwdriver, remove the shroud screws from the rear side of the steering wheel (Type 1) or pry the shroud from the steering wheel (Type 2).
3. Disconnect the horn connector and remove the shroud.
4. Remove the nut/washer retaining the steering wheel to the steering shaft.
5. Using a steering wheel puller, remove the steering wheel.
6. To install, reverse the removal procedure.

NOTE: The steering column is a collision type (designed to collapse upon impact), be careful not to severely jar it at any time.

Steering Column

REMOVAL & INSTALLATION

1. From under the dash, remove the steering column protector nut, clip and protector.
2. Remove the pinch bolt between the intermediate shaft and the steering shaft.
3. Remove the mounting bracket bolts from the lower column.
4. Remove the steering column to instrument panel mounting bolts.
5. Remove the electrical connectors and park lock cable at the ignition switch. If equipped with an automatic transaxle, remove the park lock cable bracket.
6. Remove the steering column assembly.
7. To install, reverse the removal procedure.

1. Steering wheel
2. Lower steering wheel cover
3. Horn contact ring
4. Screw
5. Screw
6. Washer
7. Nut
8. Steering shaft column asm.
9. Steering shaft
10. Steering column
11. Steering column bush
12. Plate
13. Bolt
14. Washer
15. Bolt
16. Nut
17. Clip
18. Rubber cushion
19. Washer
20. Steering intermediate shaft
21. Bolt
22. Snap ring
23. Washer
24. Cover Set
25. Screw
26. Screw
27. Protector
28. Clip
29. Seal
30. Nut
31. Boot
32. Boot clip
33. Boot plate
34. Shroud asm.
35. Screw
36. Steering wheel emblem
37. Steering lock asm.
38. Bolt
39. Ignition starter switch
40. Screw
41. Ignition switch
42. Turn signal switch

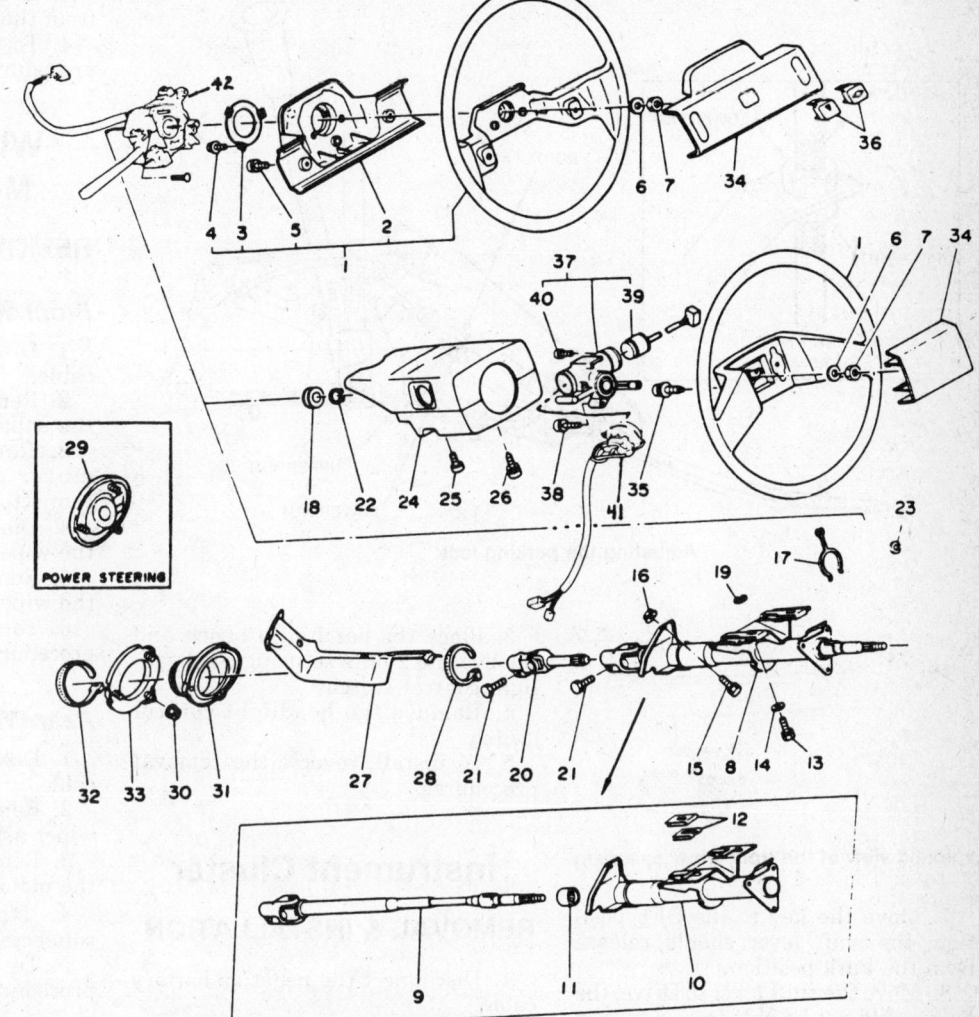

Exploded view of the steering column

Intermediate Shaft

REMOVAL & INSTALLATION

1. Remove the protector nut, clip and protector.
2. Remove the pinch bolts at the pinion shaft and steering shaft.
3. Remove the intermediate shaft.
4. To install, reverse the removal procedure.

Turn Signal/Dimmer Switch

REMOVAL & INSTALLATION

1. Disconnect the negative battery cable.
2. Remove the horn shroud, steering wheel nut/washer and steering wheel assembly.
3. Remove the steering cowl at-

taching screw and steering cowl.
4. Disconnect the combination/starter switch connector.
5. Remove the turn signal/dimmer switch attaching screw and switch.
6. To install, reverse the removal procedure.

Ignition Switch

REMOVAL & INSTALLATION

1. Refer to the "Turn Signal/Dimmer Switch Removal and Installation" procedure in the this section. Remove the turn signal/dimmer switch.
2. Insert the key into the ignition and place the key in the ON position (the lock bar must be pulled all the way in).
3. Remove the snap ring and rubber cushion from the steering shaft.
4. Disconnect the switch wires at the connectors.

5. Remove the 2 screws retaining the ignition/starter switch and remove the switch.
6. To install, reverse the removal procedure.

Parking Lock (Auto Trans)

ADJUSTMENT

1. Remove the shift lever knob screws and the shift lever knob.
2. Remove the 4 console screws and console.
3. Place the shift lever in the Park position and the ignition in the Lock position.
4. Loosen the upper and lower adjusting nuts of the parking lock cable.
5. Adjust the cable so that the shift lever cannot be moved from the Park position.
6. Tighten the upper nut.

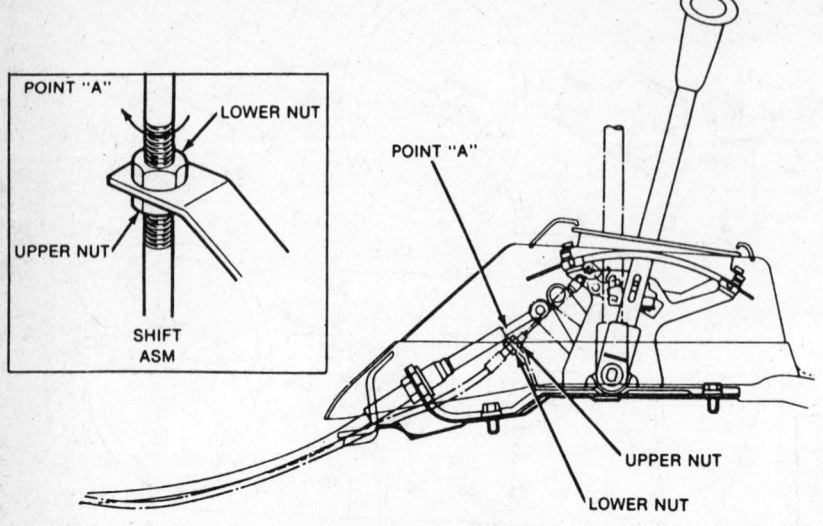

Adjusting the parking lock

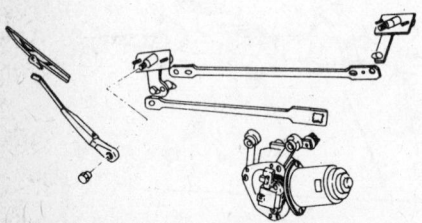

Exploded view of the front wiper assembly

7. Move the key to the OFF position; the shift lever should release from the Park position.

8. Move the shift lever to Drive; the key should not be able to be removed from the ignition assembly.

9. To complete the adjustment, re-install the items that were removed.

CHASSIS ELECTRICAL

Headlight Switch

The headlight control switch is a 3-position, push type switch which is located at the left-side of the instrument panel. The dimmer/passing light switch is a part of and actuated by the turn signal switch.

REMOVAL & INSTALLATION

1. Remove the instrument cluster bezel retaining screw and the bezel.

2. Disconnect the headlight and the windshield wiper control switch electrical connectors.

3. Place the bezel on a bench and remove the 2 nuts securing the headlight control switch.

4. Remove the headlight control switch.

5. To install, reverse the removal procedures.

Instrument Cluster

REMOVAL & INSTALLATION

1. Disconnect the negative battery cable.

2. Remove the instrument cluster bezel retaining screws and bezel.

3. Disconnect the windshield wiper and lighting switch connectors.

4. Remove the instrument cluster retaining screws and pull out the assembly.

5. Remove the trip reset knob and the assembly glass.

6. Remove the buzzer, sockets and bulbs.

7. Remove the speedometer assembly, fuel and temperature gauge.

8. Remove the tachometer, it equipped.

9. To install, reverse the removal procedure.

Speedometer Cable

REMOVAL & INSTALLATION

1. Refer to the "Instrument Cluster Removal and Installation" procedure in this section. Remove the instrument cluster.

2. Disconnect the speedometer from the transaxle.

3. Remove the speedometer cable from the flexible clip and the cowl.

4. To install, reverse the removal procedure.

Windshield Wiper Motor Assembly

REMOVAL & INSTALLATION

Front Window

1. Disconnect the negative battery cable.

2. Remove the lock nuts retaining the wiper arms and the wiper arms.

3. Remove the cowl cover, wiper motor cover and the electrical connector.

4. Disconnect the drive arm from the wiper link.

5. Remove the mounting bolts and the wiper motor.

6. To install, reverse the removal procedure.

Rear Window

1. Disconnect the negative battery cable.

2. Remove the trim pad and the wiper arm assemblies.

3. Remove the mounting bolts and the motor assembly.

4. Disconnect the electrical connector.

5. To install, reverse the removal procedure.

Wiper Blade

REPLACEMENT

Depress the spring type blade clip away from the underside of the arm and slide the arm out of the blade clip. To install the blade, slide the end tip of the arm into the blade clip until the pin on the tip end engages with the hole in the clip.

Wiper Switch

REMOVAL & INSTALLATION

Front

1. Refer to the "Headlight Switch Removal and Installation" procedure in this section and remove the instrument cluster bezel.

2. Remove the wiper switch electrical connector, attaching nuts and bracket.

3. Remove the wiper switch.

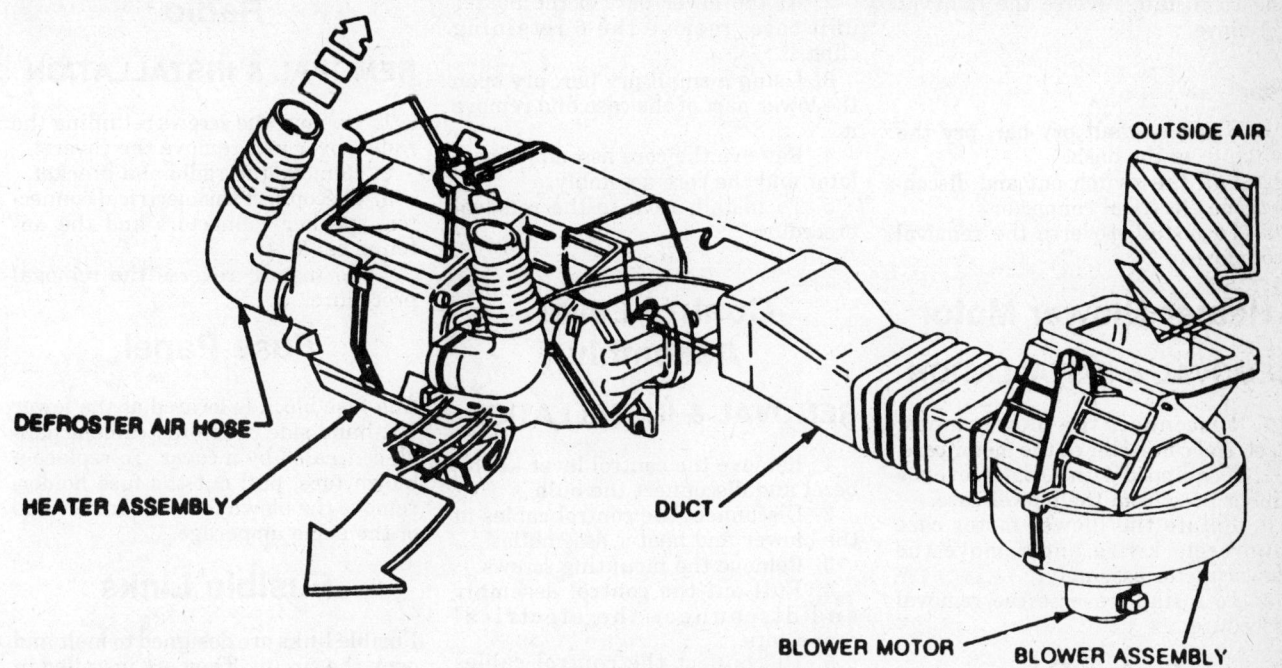

DEFROSTER AIR HOSE

HEATER ASSEMBLY

OUTSIDE AIR

DUCT

BLOWER MOTOR

BLOWER ASSEMBLY

View of the heater assembly

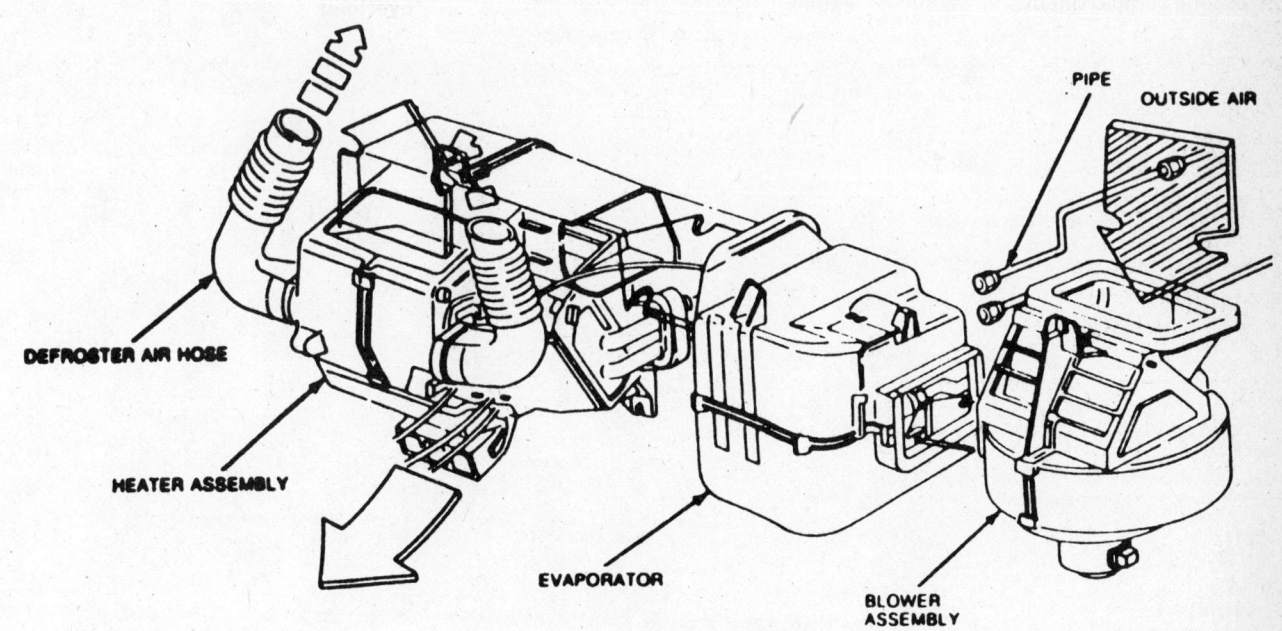

DEFROSTER AIR HOSE

HEATER ASSEMBLY

EVAPORATOR

PIPE

OUTSIDE AIR

BLOWER ASSEMBLY

View of the heater assembly with A/C

4. To install, reverse the removal procedure.

Rear

1. Using a small pry bar, pry the switch from the dash.
2. Pull the switch out and disconnect the electrical connector.
3. To install, reverse the removal procedure.

Heater Blower Motor

REMOVAL & INSTALLATION

1. Disconnect the blower motor electrical connector at the motor case.
2. If equipped with A/C, remove the rubber hose from the blower case.
3. Rotate the blower motor case counterclockwise and remove the blower motor assembly.
4. To install, reverse the removal procedure.

Heater Core

REMOVAL & INSTALLATION

1. Disconnect the heater hoses in the engine compartment.

2. At the lower part of the heater unit case, remove the 6 retaining clips.
3. Using a small pry bar, pry open the lower part of the case and remove it.
4. Remove the core assembly insulator and the core assembly.
5. To install, reverse the removal procedure.

Control Lever Assembly

REMOVAL & INSTALLATION

1. Remove the control lever knobs, bezel and disconnect the bulb.
2. Disconnect the control cables at the blower and heater assemblies.
3. Remove the mounting screws.
4. Pull out the control assembly and disconnect the electrical connectors.
5. Disconnect the control cables from the control assembly.
6. Remove the blower, A/C and heater switches.
7. To install, reverse the removal procedures and adjust the control cables.

Radio

REMOVAL & INSTALLATION

1. Remove the screws retaining the radio cover and remove the cover.
2. Remove the radio and bracket.
3. Disconnect the electrical connector, speaker connectors and the antenna cable.
4. To install, reverse the removal procedure.

Fuse Panel

The fuse block is located at the lower left-hand side of the instrument panel, concealed by a cover. To replace a blown fuse, pull out the fuse holder, remove the blown fuse and install one of the same amperage.

Fusible Links

Fusible links are designed to melt and open the circuit. They are installed in the various circuits to prevent overloads and provide protection for the entire wiring harness. Before replacing a fusible link, be sure to determine and correct the cause of the overload.

Chevrolet
Sprint

YEAR IDENTIFICATION

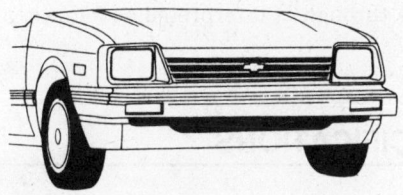

1985–86 Sprint

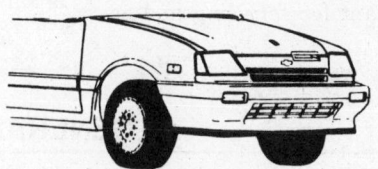

1987 Sprint Turbo

VEHICLE IDENTIFICATION NUMBER (VIN)

It is important for servicing and ordering parts to be certain of the vehicle and engine identification. The VIN (vehicle identification number) is a 17 digit number visible through the windshield on the driver's side of the dash and contains the vehicle and engine identification codes. It can be interpreted as follows:

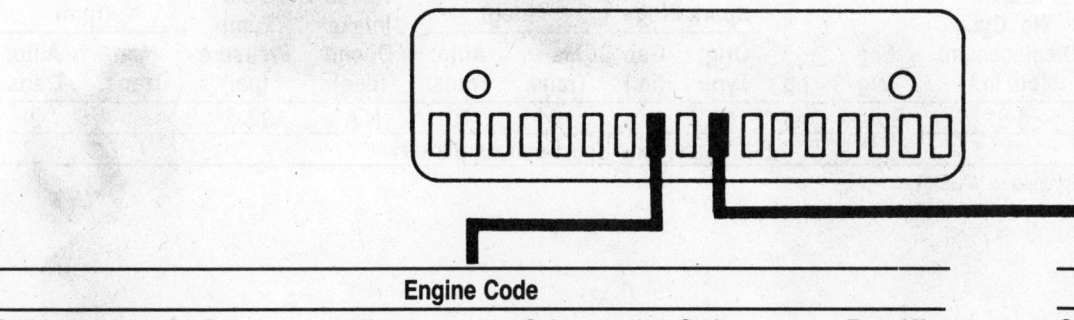

Engine Code						Model Year Code	
Code	Cu. In.	Liters	Cyl.	Carb.	Eng. Mfg.	Code	Year
M	61	1.0	3	2bbl	Suzuki	F	'85
						G	'86
						H	'87

The seventeen digit Vehicle Identification Number can be used to determine engine application and model year. The 10th digit indicates the model year, and the 8th digit identifies the factory installed engine.

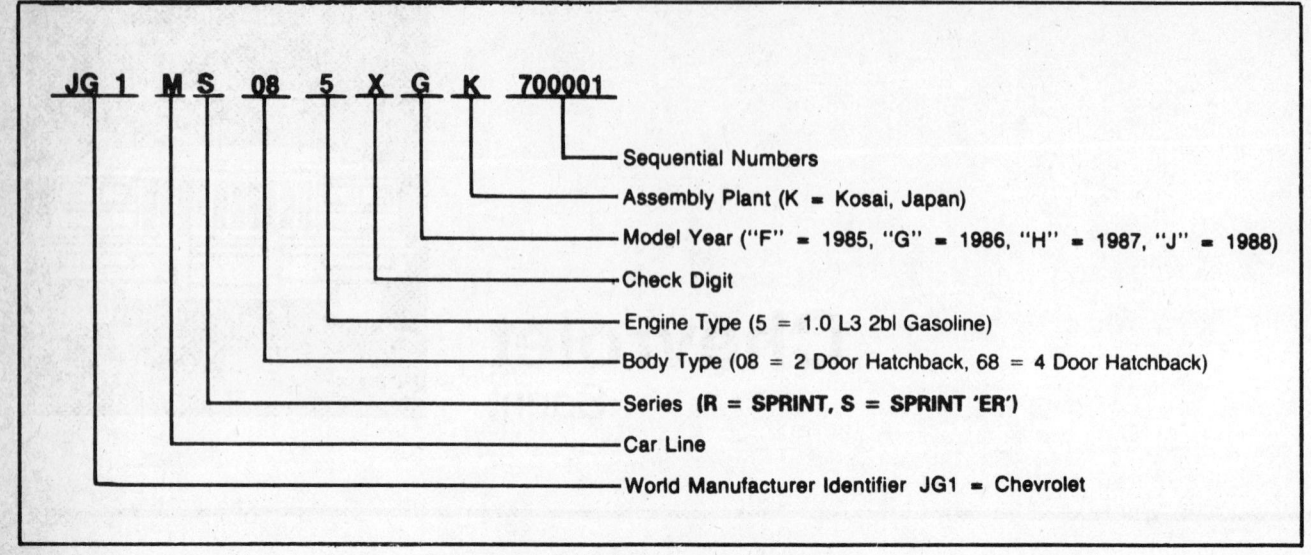

Vehicle Identificaton Number

VEHICLE IDENTIFICATION NUMBER (VIN)

It is important for servicing and or-dering parts to be certain of the vehi-cle and engine identification. The VIN (vehicle identification number) is a 17 digit number visible through the windshield on the driver's side of the dash and contains the vehicle and engine identification codes. It can be interpreted as follows.

GENERAL ENGINE SPECIFICATIONS

Year	Engine No. Cyl. Displ. (cu. in.)	Engine VIN Code	Fuel Delivery System	Engine Mfg.	Horsepower @ rpm	Torque @ rpm (ft. lb.)	Bore × Stroke	Compression Ratio	Oil Pressure 2400 rpm
'85–'87	3-61	M	2bbl	Suzuki	48 @ 5100	57 @ 3200	2.91 × 3.03	9.5:1	48 psi

TUNE-UP SPECIFICATIONS

When analyzing compression test results, look for uniformity among cylinders rather than specific pressures

Year	Eng. VIN Code	Engine No. Cyl. Displacement (cu. in.)	Eng. Mfg.	hp	Spark Plugs Orig. Type	Spark Plugs Gap (in.)	Ignition Timing (deg) Man. Trans.	Ignition Timing (deg) Auto. Trans.	Valves Intake Opens (deg)	Fuel Pump Pressure (psi)	Idle Speed (rpm) Man. Trans.	Idle Speed (rpm) Auto. Trans.
'85–'86	M	3-61	Suzuki	48	①	③	③	③	N.A.	3.5②	③	③
'87					See Underhood Specifications Sticker							

① NGK:BPRGES-11 or Nippondenso WIGEXR-U11
② @ 5,000 rpm
③ Refer to underhood specifications sticker

CAPACITIES

Year	VIN Code	Engine Displacement cu. in.	Eng. Mfg.	Crankcase (qts)	Manual Transaxle (pts)	Gas Tank (gals)	Cooling System (qts)
'85–'87	M	61	Suzuki	3.7	4.8	8.3	4.5

FIRING ORDERS

NOTE: To avoid confusion, replace spark plugs and wires one at a time.

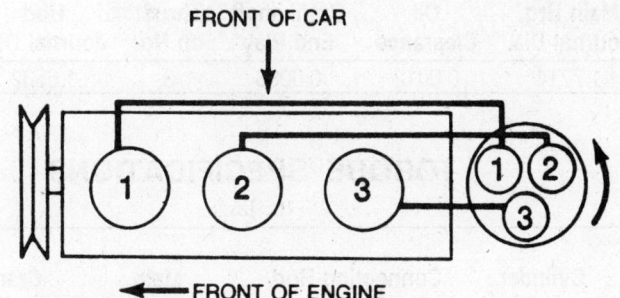

FRONT OF CAR

← FRONT OF ENGINE

GM (SUZUKI) 63–3 (1.0L)
Engine firing order: 1–3–2
Distributor rotation: counterclockwise

CHAMSHAFT SPECIFICATIONS

Year	Engine Displacement (cu. in.)	Journal Diameter				Journal Clearance	Lobe Life	
		1	2	3	4		Intake	Exhaust
'85–'87	61	1.7372–1.7381	1.7451–1.7460	1.7530–1.7539	1.7609–1.7618	0.0029	1.5012	1.5012

PISTON AND RING SPECIFICATIONS

All measurements are given in inches.

Year	VIN Code	Engine No. Cyl. Disp. (cu. in.)	Eng. Mfg.	Piston-to-Bore Clearance	Ring Gap			Ring Side Clearance		
					Top Compression	Bottom Compression	Oil Control	Top Compression	Bottom Compression	Oil Control
'85–'87	M	3-61	Suzuki	0.0008–0.0015	0.0079–0.0129	0.0079–0.0137	0.0079–0.0275	0.0012–0.0027	0.0008–0.0023	—
'86–'87	M	3-61	Suzuki ER Model	0.0008–0.0023	0.0079–0.0157	—	0.0079–0.0275	0.0012–0.0027	—	—

VALVE SPECIFICATIONS

Year	Engine Displacement (cu. in.)	Seat Angle (deg.)	Face Angle (deg.)	Spring Test Pressure (lbs. @ in.)	Spring Installed Height (in.)	Stem-to-Guide Clearance (in.)		Stem Diameter (in.)	
						Intake	Exhaust	Intake	Exhaust
'85–'87	61	45°	45°	60	1.63	0.0014	0.0020	0.2745	0.2740

CRANKSHAFT AND CONNECTING ROD SPECIFICATIONS
(All specifications in inches)

Year	Engine Displacement (cu. in.)	Main Brg. Journal Dia.	Main Brg. Oil Clearance	Crankshaft End Play	Thrust on No.	Connecting Rod Journal Dia.	Rod Bearing Oil Clearance	Rod Bearing Side Clearance
'85–'87	61	1.7714	0.0012	0.0083	3	1.6532	0.0015	0.0058

TORQUE SPECIFICATIONS
(ft. lbs.)

Year	Engine Displacement (cu. in.)	Cylinder Head Bolts	Connecting Rod Bearing Bolts	Main Bearing Bolts	Crankshaft Bolt	Flywheel to Crankshaft Bolts	Camshaft Cap Bolts
'85–'87	61	48	35	38	50	44	—

ENGINE ELECTRICAL

Alternator

For all alternator/regulator troubleshooting, please refer to the "Charging and Starting" Unit Repair section.

REMOVAL & INSTALLATION

1. Disconnect the negative battery cable.
2. Disconnect the wiring connectors from the back of the alternator.
3. Remove the adjusting arm mounting bolt, the lower pivot bolt and the drive belt.
4. Remove the alternator.
6. To install, reverse the removal procedures. Adjust the drive belt to have $^1/_4$ to $^3/_8$ inch play on the longest run of the drive belt.

Voltage Regulator

An integral voltage regulator is part of the alternator and no adjustments are necessary or possible. For all testing procedures, please refer to the "Charging and Starting" Unit Repair section.

Starter Motor

For all starter motor diagnosis and repair procedures, please refer to the "Charging and Starting" Unit Repair section.

REMOVAL & INSTALLATION

1. Disconnect the negative battery cable.

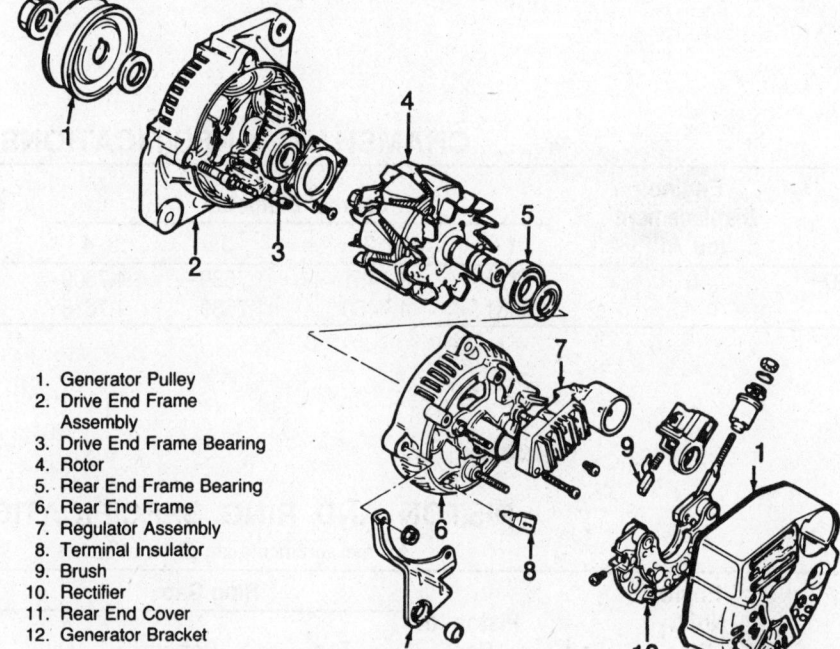

1. Generator Pulley
2. Drive End Frame Assembly
3. Drive End Frame Bearing
4. Rotor
5. Rear End Frame Bearing
6. Rear End Frame
7. Regulator Assembly
8. Terminal Insulator
9. Brush
10. Rectifier
11. Rear End Cover
12. Generator Bracket

Exploded view of the alternator

2. Disconnect the ignition switch wire and the battery cable from the starter.
3. Remove the two engine-to-starter mounting bolts and the starter.
4. To install, reverse the removal procedures.

Distributor

REMOVAL & INSTALLATION

1. Disconnect the negative battery cable.
2. Disconnect the wiring harness at the distributor and the vacuum line at the distributor vacuum unit.

3. Remove the distributor cap.

NOTE: Mark the distributor body in reference to where the rotor is pointing. Mark the distributor hold down bracket and cylinder head for a reinstallation location point.

4. Remove the hold down bolt and the distributor from the cylinder head. DO NOT rotate the engine after the distributor has been removed.
5. To install, aligning all of the reference marks and reverse the removal procedures. Check the timing.

For all testing and service procedures, please refer to the "Elec-

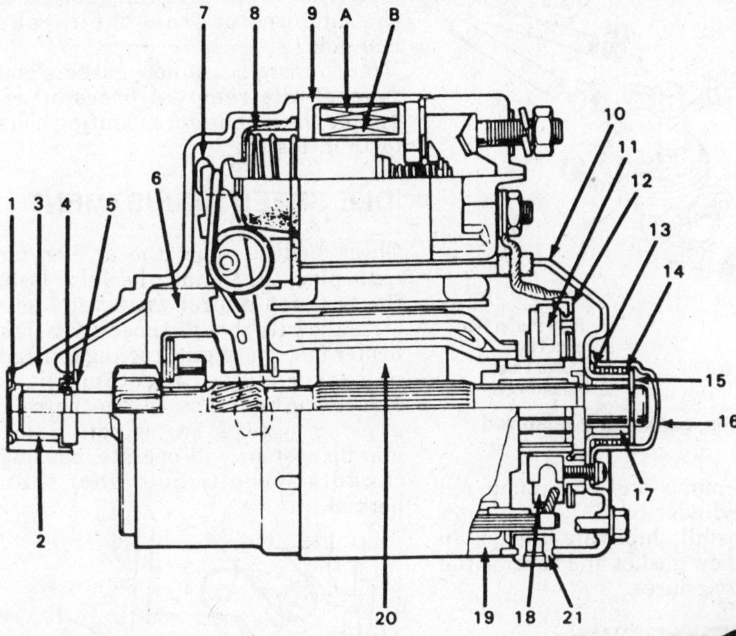

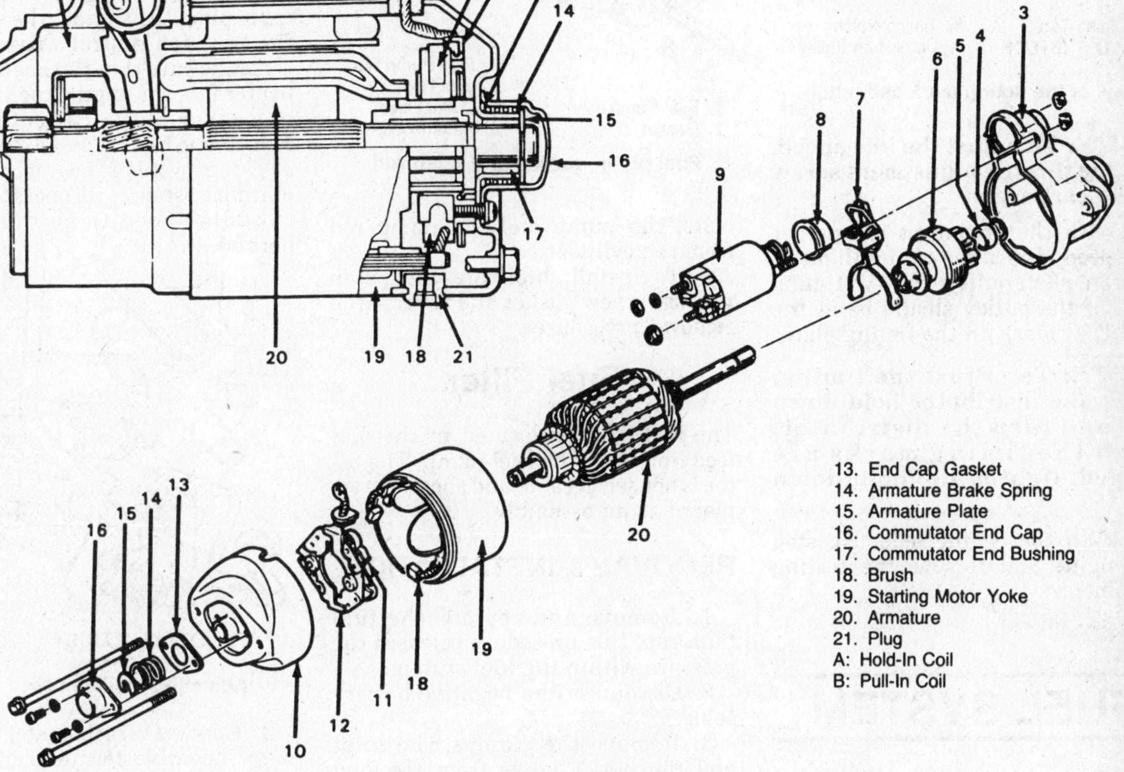

1. Drive Housing Cover
2. Drive Bushing
3. Drive Housing
4. Armature Ring
5. Armature Stop Ring
6. Over-Running Clutch
7. Pinion Drive Lever
8. Switch Cover
9. Magnetic Switch
10. Commutator End Housing
11. Brush Spring
12. Brush Holder

13. End Cap Gasket
14. Armature Brake Spring
15. Armature Plate
16. Commutator End Cap
17. Commutator End Bushing
18. Brush
19. Starting Motor Yoke
20. Armature
21. Plug
A: Hold-In Coil
B: Pull-In Coil

Exploded view of the starter

tronic Ignition" Unit Repair section.

IGNITION TIMING ADJUSTMENT

Make sure that the headlights, the heater fan, the engine cooling fan and any other electrical equipment is turned OFF. If any of the current drawing systems are operating, the idle up system will operate, causing the idle speed to be higher than normal.

1. Connect a tachometer to the primary negative terminal of the ignition coil, a timing light to the No. 1 spark plug and refer to the underhood sticker.

2. Start the engine and bring it to normal operating temperatures.

3. Check and/or adjust the idle speed, it should be 750 rpm for models with manual transaxles and 850 rpm on models with automatic transaxles.

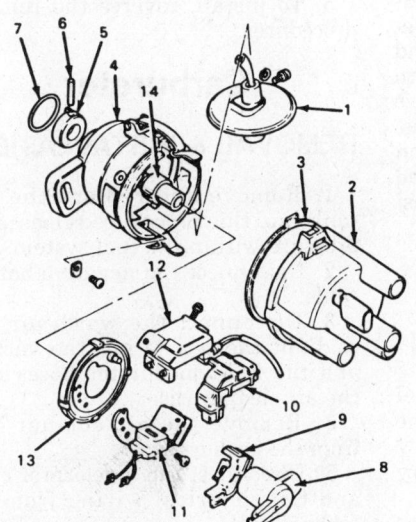

1. Vacuum Advance Unit
2. Distributor Cap
3. Seal
4. Distributor Housing
5. Distributor Coupling
6. Pin
7. Seal
8. Rotor
9. Pick Up Coil Dust Cover
10. Module Dust Cover
11. Pick Up Coil
12. Module
13. Pick Up Coil Base Plate
14. Pole Piece

Exploded view of the distributor

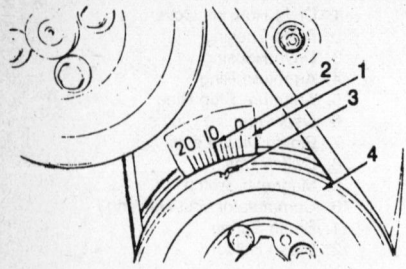

1. Timing Tab
2. (AT) 6° (BTDC)
3. Timing Notch
4. Crankshaft Pulley

View of the timing mark and notch

NOTE: To adjust the idle speed, turn the throttle adjustment screw on the carburetor.

4. With the engine at the proper idle speed, direct the timing light to the crankshaft pulley. The "V" timing mark of the pulley should be at the 10° BTDC mark on the timing plate.

NOTE: To adjust the timing, loosen the distributor hold down bolt and turn the distributor. When the timing marks are aligned, tighten the hold down bolt.

5. With the timing adjusted, stop the engine and remove the testing equipment.

FUEL SYSTEM

An Hitachi electronically controlled, two barrel, downdraft type carburetor is used on all gasoline engines, which incorporates a no pedal depressed automatic choke. A mixture control solenoid controls the primary slow and primary main fuel feed system to maintain the correct air/fuel mixture under all engine operating conditions. A mechanical fuel pump is located on the right side of the cylinder head which supplies fuel at a rate of 3.5 psi @ 5000 rpm.

Fuel Pump

REMOVAL & INSTALLATION

1. Remove and replace the fuel tank cap, this procedure releases the pressure within the fuel system.
2. Disconnect the negative battery cable.
3. Remove the air cleaner from the carburetor.
4. Remove the fuel inlet, outlet and return hoses from the fuel pump.
5. Remove the fuel pump mounting

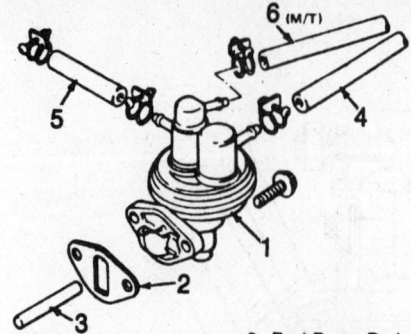

1. Fuel Pump
2. Gasket
3. Fuel Pump Rod
4. Inlet Hose
5. Outlet Hose
6. Return Hose

Fuel pump, gasket and push rod

bolts, the pump and the pump rod from the cylinder head.
6. To install, lubricate the pump rod, use a new gasket and reverse the removal procedures.

Fuel Filter

The fuel filter is located in the fuel feed line, near the fuel pump. The filter is not serviceable and should be replaced as an assembly.

REMOVAL & INSTALLATION

1. Remove and replace the fuel tank cap, this procedure releases the pressure within the fuel system.
2. Disconnect the negative battery cable.
3. Remove the clamps, the inlet and the outlet hoses from the fuel filter.
4. Remove the fuel filter from the bracket.
5. To install, reverse the removal procedures.

Carburetor

REMOVAL & INSTALLATION

1. Remove and replace the fuel tank cap, this procedure releases the pressure within the fuel system.
2. Disconnect the negative battery cable.
3. Disconnect the warm air, the cold air, the second air, the vacuum and the EGR modulator hoses from the air cleaner case.
4. Remove the air cleaner case from the carburetor.
5. Disconnect the accelerator cable and the electrical wiring from the carburetor.
6. Remove the emission control and the fuel hoses from the carburetor.
7. Disconnect the No. 1 and No. 2 choke hoses from the carburetor.

8. Remove the mounting bolts and the carburetor from the intake manifold.
9. To install, use new gaskets and reverse the removal procedures. Torque the carburetor mounting bolts to 18 ft. lbs.

IDLE SPEED ADJUSTMENT

Check and/or adjust the accelerator cable play, the timing, the valve lash, the emission control wiring and hoses. Make sure that the headlights, the heater fan, the engine cooling fan and any other electrical equipment is turned OFF. If any of the current drawing systems are operating, the idle up system will operate, causing the idle speed to be higher than normal.

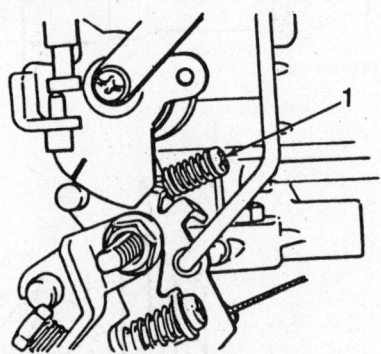

THROTTLE ADJUST SCREW

Adjusting the idle speed screw

1. Connect a tachometer to the primary negative terminal of the ignition coil and refer to the underhood sticker.
2. Place the transaxle in Neutral, set the parking brake and block the wheels.
3. Start the engine and bring it to normal operating temperatures.
4. Check and/or adjust the idle speed, it should be 700–800 rpm with manual transaxles, 800–900 rpm with automatic transaxles.

NOTE: To adjust the idle speed, turn the throttle adjustment screw on the carburetor.

5. With the engine at the proper idle speed, check and/or adjust the idle up speed.
6. Stop the engine and remove the tachometer.

IDLE MIXTURE ADJUSTMENT

The carburetor is adjusted at the factory and no further adjustment should be necessary. However, if the engine performance is poor, the emis-

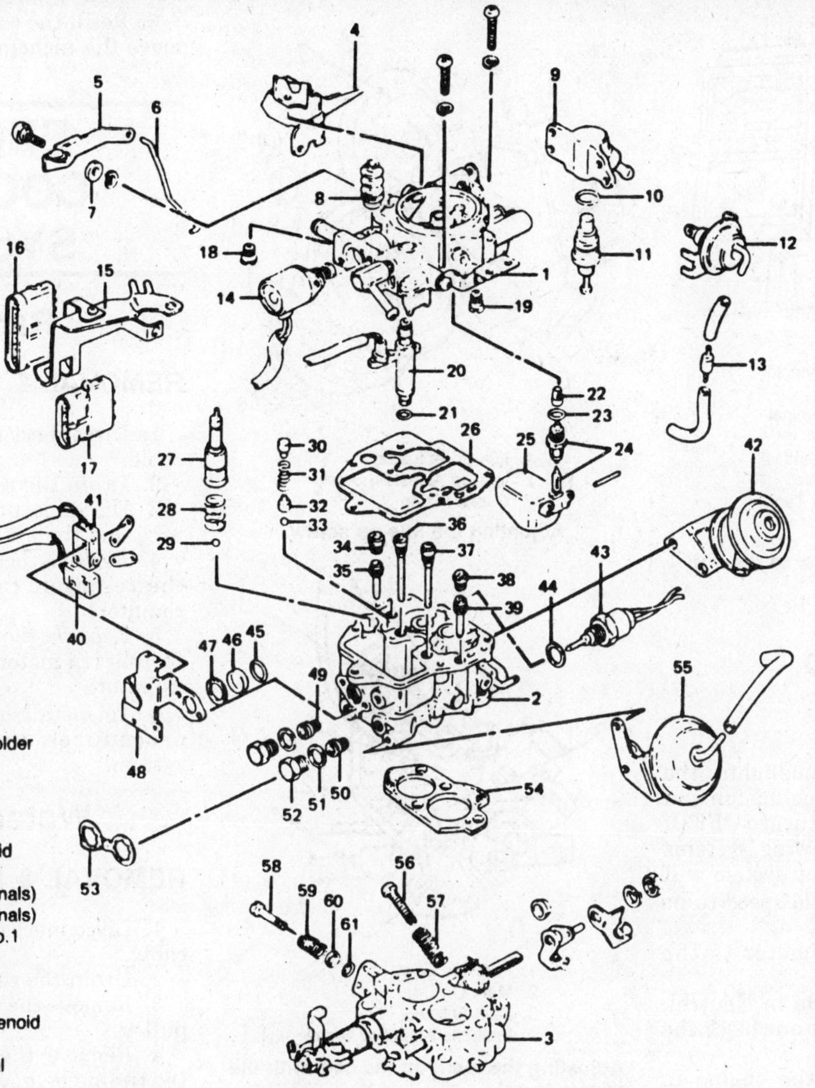

1. Air Horn
2. Float Chamber
3. Throttle Chamber
4. Cable Bracket
5. Pump Lever
6. Pump Rod
7. Washer
8. Choke Lever Guide
9. Thermo Element Holder
10. Seal
11. Thermo Element
12. Choke Piston
13. Delay Valve
14. Switch Vent Solenoid
15. Connector Holder
16. Connector (6 Terminals)
17. Connector (4 Terminals)
18. Primary Slow Air No.1 Bleeder
19. Secondary Slow Air Bleeder
20. Mixture Control Solenoid Valve
21. Solenoid Valve Seal
22. Needle Valve Filter
23. Needle Valve Gasket
24. Needle Valve

25. Float
26. Air Horn Gasket
27. Pump Piston
28. Piston Return Spring
29. Ball
30. Injector Weight
31. Injector Spring
32. Injector Weight
33. Ball
34. Primary Slow Air No. 2 Bleeder
35. Primary Slow Jet
36. Primary Main Air Bleeder
37. Secondary Main Air Bleeder
38. Plug
39. Secondary Slow Jet
40. Idle Micro Switch
41. Wide Open Micro Switch
42. Idle Up Actuator
43. Solenoid Valve (Fuel Cut)
44. Washer
45. Level Gauge Seat
46. Level Gauge
47. Level Gauge Gasket
48. Micro Switch Bracket
49. Primary Main Jet
50. Secondary Main Jet
51. Drain Plug Gasket
52. Drain Plug
53. Lock Plate
54. Insulator
55. Secondary Diaphragm
56. Throttle Adjust Screw
57. Spring
58. Mixture Adjust Screw
59. Spring
60. Washer
61. Seal

Exploded view of the carburetor

sion test fails, or the carburetor has been replaced or overhauled, an idle mixture adjustment is necessary. Before adjusting the idle mixture, check the timing/idle speed, the valve lash and make sure that all electrical accessories are turned OFF.

1. Refer to the "Carburetor Removal & Installation" procedures in this section and remove the carburetor from the intake manifold.

2. Using an $^{11}/_{64}$ inch bit, drill through the idle mixture screw housing, in line with the retaining pin. Use a punch to drive the pin from the housing.

3. Install the carburetor to the intake manifold by reversing the removal procedures.

4. Place the transaxle in Neutral, set the parking brake and block the wheels.

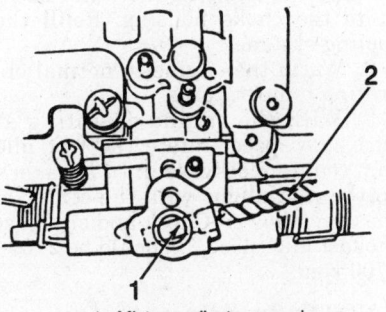

1. Mixture adjust screw pin
2. Drill

Removing the idle mixture pin

5. Start the engine and bring it to normal operating temperatures.

6. Disconnect the Duty Cycle Check connector, located near the water reservoir tank. Connect the positive terminal of a dwell meter to the blue/red wire and the negative terminal to the black/green wire.

7. Set the dwell meter to the 6 cylinder position, make sure that the indicator moves.

8. Check and/or adjust the idle speed.

9. Operate the engine at idle speed and adjust the idle mixture screw, allow the engine to stabilize between adjustments. Adjust the dwell to 21°–27°; recheck the idle speed and adjust, if necessary.

10. After completing the adjustment, stop the engine, disconnect the dwell meter and connect the Duty Cycle Check connector to the coupler.

11. Install a new idle mixture adjust screw pin in the throttle housing, drive it in place.

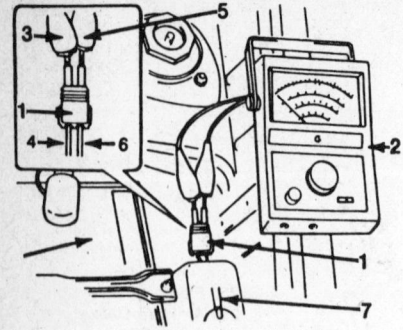

1. Duty Check Connector
2. Dwell Meter
3. Positive (+) Terminal
4. "Blue/Red" Wire
5. Negative (−) Terminal
6. "Black/Green Wire
7. Water Reservoir Tank
8. Battery

Connecting the dwell meter to the duty check connector

Idle Up

ADJUSTMENT

Make sure that the headlights, the heater fan, the engine cooling fan and any other equipment is turned OFF. If any of the current drawing systems are operating, the idle up system will operate, causing the idle speed to be higher than normal.

1. Connect a tachometer to the engine.
2. Place the transaxle in Neutral, apply the parking brake and block the wheels.
3. Start and warm the engine to normal operating temperatures. Check the idle speed.
4. Turn the headlights ON and make sure that the screw on the idle up actuator, moves down.
5. Check and/or adjust the idle up speed, it should be 750–850 rpm manual transaxle, 800–900 rpm automatic.

NOTE: To adjust the idle up speed, turn the screw on the end of the idle up actuator.

6. Turn the headlights OFF and turn ON (separately) other current drawing equipment; check to see if the idle up screw moves down when each piece of equipment is turned ON.
7. Stop the engine and remove the tachometer.

Fast Idle

ADJUSTMENT

This adjustment is to be performed when the ambient temperature is be-

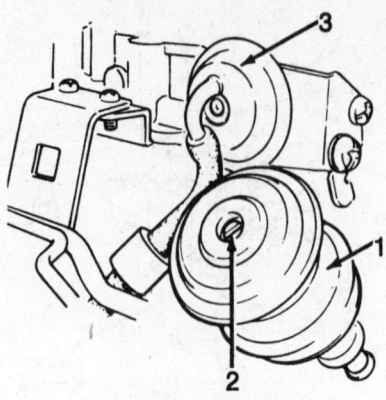

1. Idle-Up Actuator
2. Idle-Up Adjusting Screw
3. Choke Piston

Adjusting the idle-up screw

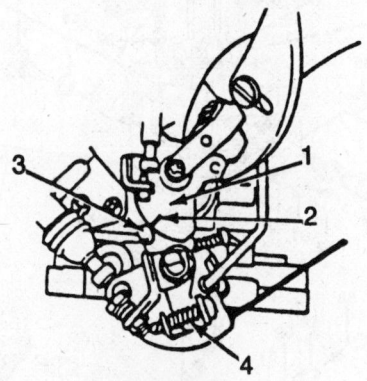

1. Fast Idle Cam
2. Mark On Cam
3. Cam Follower
4. Fast Idle Adjusting Screw

Adjusting the mark on the cam with the cam follower

low 77°F. Adjust the idle speed before performing this procedure.

1. Drain some of the engine coolant, remove and plug the water hoses at to the choke housing. Refill the cooling system.
2. Warm the engine to normal operating temperatures.
3. With the engine operating at curb idle speed, rotate the fast idle cam counterclockwise until the mark on the cam aligns with the center of the cam follower. Check and/or adjust the fast idle speed, it should be 2100–2700 rpm.

NOTE: To adjust the fast idle speed, turn the fast idle adjusting screw.

4. With the adjustment completed, stop and cool off the engine.
5. Drain some of the engine coolant. Remove the plugs from the water hoses and install them to the choke housing.

6. Refill the cooling system and remove the tachometer.

ENGINE COOLING SYSTEM

Radiator

REMOVAL & INSTALLATION

1. Disconnect the negative battery cable.
2. Drain the cooling system.
3. Disconnect the cooling fan motor wire and the air inlet hose.
4. Remove the inlet, the outlet and the reservoir tank hoses from the radiator.
5. Remove the mounting bolts, the cooling fan motor, the shroud and the radiator.
6. To install, reverse the removal procedures and fill the cooling system.

Water Pump

REMOVAL & INSTALLATION

1. Disconnect the negative battery cable.
2. Drain the cooling system.
3. Remove the water pump belt and pulley.
4. Remove the crankshaft pulley, the timing belt outside cover, the timing belt and the tensioner.
5. Remove the mounting bolts and the water pump.
6. Clean the gasket mating surfaces.
7. To install, use a new gasket/sealer and reverse the removal procedures. Torque the water pump bolts to 7.5–9 ft. lbs. Adjust the water pump belt deflection to $1/4 - 3/8$ inch between the water pump and the crankshaft pulleys.

Thermostat

REMOVAL & INSTALLATION

1. Disconnect the negative battery cable.
2. Drain the cooling system to a level below the thermostat.
3. Remove the air cleaner.
4. Disconnect the electrical connector at the thermostat cap.
5. Remove the inlet hose, the cap mounting bolts and the thermostat from the thermostat housing.

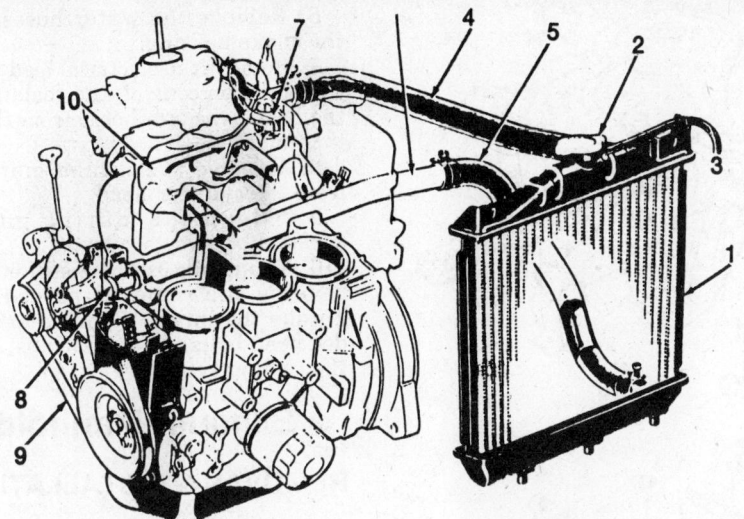

1. Radiator
2. Radiator Cap
3. To Water Reservoir Tank
4. Inlet Hose
5. Outlet Hose
6. Water Intake Pipe
7. Thermostat
8. Water Pump
9. Water Pump Belt
10. Intake Manifold

View of the cooling system

6. Clean the gasket mounting surfaces.

NOTE: Make sure that the thermostat air bleed hole is clear.

7. To install, use a new gasket, place the thermostat spring side down and reverse the removal procedures. Fill the cooling system, run the engine to normal operating temperatures and check for leaks.

EMISSION CONTROLS

Due to the complex nature of modern electronic engine control systems, comprehensive diagnosis and testing procedures fall outside the confines of this repair manual. For complete information on diagnosis, testing and repair procedures concerning all modern engine and emission control systems, please refer to *Chilton's Guide To Electronic Engine Controls.*

ENGINE MECHANICAL

Engine

REMOVAL & INSTALLATION

1. Remove the battery cables.

2. Remove the hood, the battery, the battery tray, the air cleaner and the outside air duct.

3. Drain the cooling system, the engine and the transaxle.

4. Remove the radiator, heater and vacuum hoses from the engine.

5. Disconnect the cooling fan wires.

6. Remove the cooling fan, the shroud and the radiator as an assembly.

7. Remove the fuel hoses from the fuel pump.

8. Remove the brake booster hose from the intake manifold, the accelerator cable from the carburetor and the speed control cable from the transaxle.

9. Remove the clutch cable and the bracket from the transaxle.

10. Remove the necessary wiring from the engine and transaxle.

11. Remove the A/C adjusting bolt and the drive belt splash shield.

12. Raise and support the vehicle on jackstands.

13. Disconnect the exhaust pipe from the exhaust manifold.

14. Remove the A/C pivot bolt, the drive belt and the mounting bracket.

15. Disconnect the gearshift control shaft and extension rod at the transaxle.

16. Disconnect the ball joints.

17. Remove the axle shafts from the transaxle.

18. Remove the engine torque rods and the transaxle mount nut.

19. Lower the vehicle.

20. Remove the engine side mount and the mount nuts.

21. Connect a vertical hoist to the engine, then lift the engine and transaxle assembly from the vehicle.

22. To install, reverse the removal procedures. Refill the engine, the transaxle and the cooling system.

Intake Manifold

REMOVAL & INSTALLATION

1. Disconnect the negative battery cable.

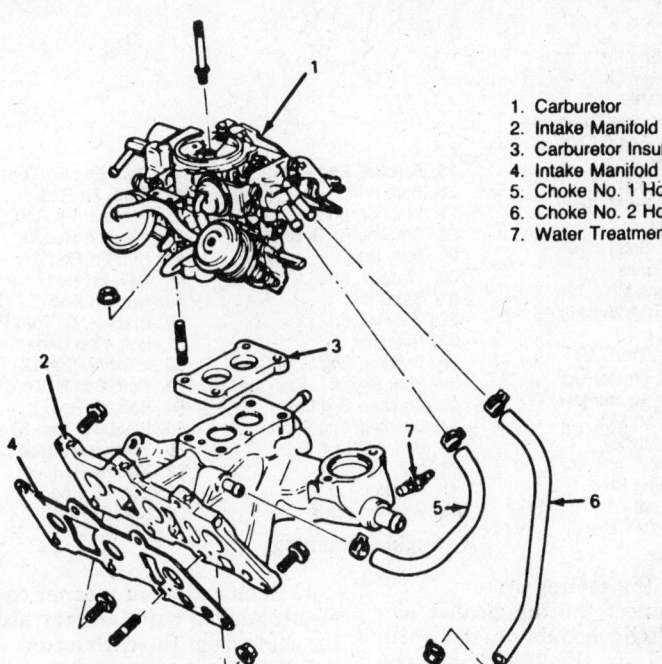

1. Carburetor
2. Intake Manifold
3. Carburetor Insulator
4. Intake Manifold Gasket
5. Choke No. 1 Hose
6. Choke No. 2 Hose
7. Water Treatment Gauge

Exploded view of the intake manifold and carburetor assembly

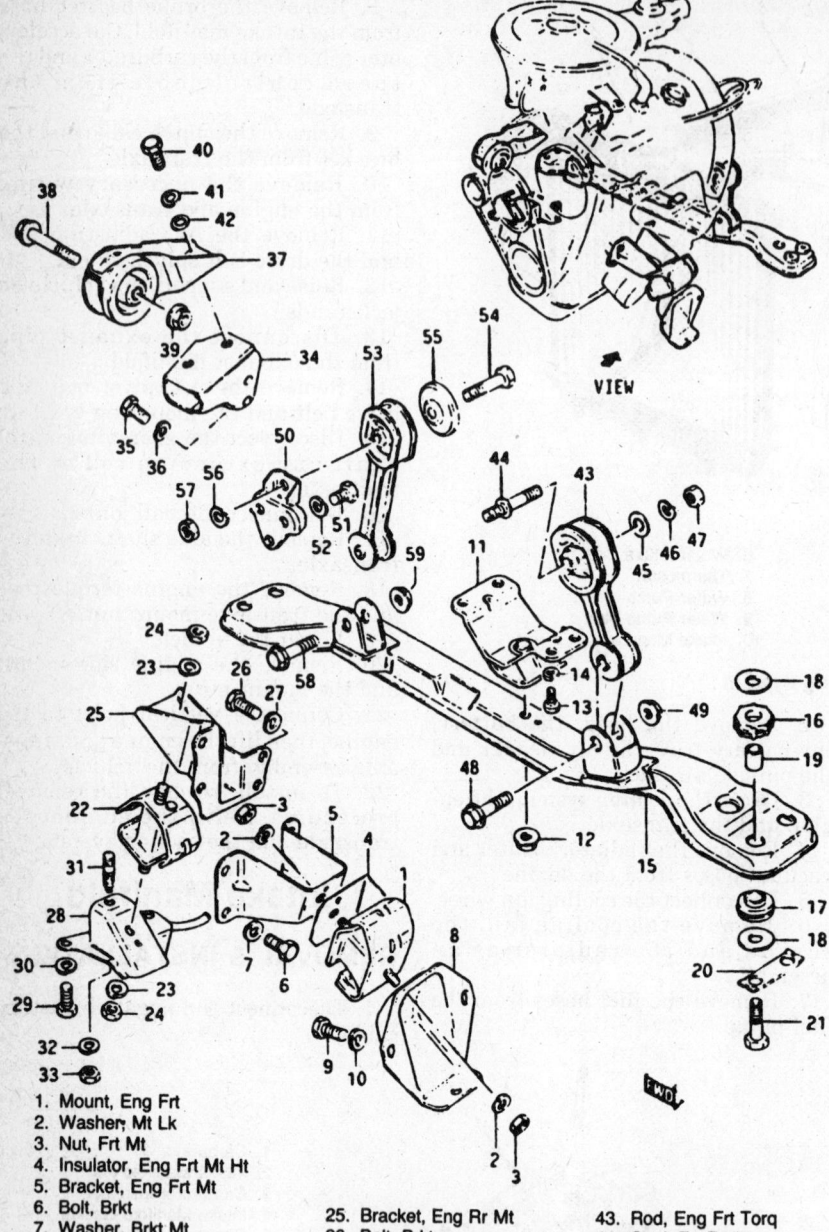

Engine mounting

1. Mount, Eng Frt
2. Washer, Mt Lk
3. Nut, Frt Mt
4. Insulator, Eng Frt Mt Ht
5. Bracket, Eng Frt Mt
6. Bolt, Brkt
7. Washer, Brkt Mt
8. Bracket, Frt Mt Body
9. Bolt, Brkt
10. Washer, Brkt Lk
11. Mount, Trans
12. Nut, Trans Mt
13. Bolt (M8 × 1.25 × 20)
14. Washer
15. Member, Trans Mt
16. Cushion, Mt Mbr Upr
17. Cushion, Mt Mbr Lwr
18. Washer, Mt Mbr
19. Spacer, Mbr
21. Bolt, Mbr
22. Mount, Eng Rr
23. Washer, Mt
24. Nut (M10 × 1.25 × 8)

25. Bracket, Eng Rr Mt
26. Bolt, Brkt
27. Washer, Brkt Lk
28. Bracket, Rr Mt Body
29. Bolt, Brkt
30. Washer, Brkt Lk
31. Stud, Brkt
32. Washer, Brkt Lk
33. Nut, Brkt
34. Bracket, Eng Si Mt
35. Bolt, Si Brkt
36. Washer, Brkt Lk
37. Bushing, Eng Si Mt
38. Bolt, Mt Bush
39. Nut, Mt Bush
41. Washer, Lk
42. Washer, Bush

43. Rod, Eng Frt Torq
44. Stud, Frt Rod
45. Washer, Frt Rod
46. Washer, Rod Lk
47. Nut, Frt Rod
48. Bolt, Frt Rod
49. Nut, Frt Rod
50. Bracket, Rr Torq Rod
51. Bolt, Rod Brkt
52. Washer, Brkt Lk
53. Rod, Eng Rr Torq
54. Bolt, Rr Rod
55. Plate, Rr Torq Stopper
56. Washer, Rr Rod Lk
57. Nut, Rr Rod
58. Bolt, Rr Rod
59. Nut, Rr Rod

2. Drain the cooling system.

3. Disconnect the air cleaner element, the EGR modulator, the warm air, the cool air, the 2nd air and the vacuum hoses from the air cleaner case.

4. Remove the air cleaner case, the electrical lead wires and the accelerator cable from the carburetor.

5. Disconnect the emission control and the fuel hoses from the carburetor.

6. Remove the water hoses from the choke housing.

7. Remove the electrical lead wires, the emission control, the coolant and the brake vacuum hoses from the intake manifold.

8. Remove the intake manifold from the cylinder head.

9. Clean the mating gasket surfaces.

10. To install, use new gaskets and reverse the removal procedures. Torque the intake manifold-to-cylinder head bolts to 14–20 ft. lbs. Refill the cooling system.

Exhaust Manifold

REMOVAL & INSTALLATION

1. Disconnect the negative battery cable.

2. Raise and support the vehicle on jackstands.

3. Remove the exhaust pipe at the exhaust manifold.

4. Remove the lower heat shield bolt and the 2nd air pipe at the exhaust manifold.

5. If equipped, remove the A/C drive belt and the lower adjusting bracket.

6. Lower the vehicle.

7. Remove the spark plug and the oxygen sensor wires.

8. Remove the hot air shroud from the exhaust manifold.

9. Remove the 2nd air valve hoses, the valve and the pipe from the exhaust manifold.

10. Remove the mounting bolts and the exhaust manifold.

11. Clean the gasket mating surfaces.

12. To install, use a new gasket and reverse the removal procedures. Torque the exhaust manifold fasteners to 14–20 ft. lbs. and the exhaust pipe to 30–43 ft. lbs.

Rocker Arms/Shafts

REMOVAL & INSTALLATION

1. Disconnect the negative battery cable.

2. Remove the air cleaner and the cylinder head cover.

3. Remove the distributor cap, then mark the position of the rotor and the distributor housing with the cylinder head. Remove the distributor and the case from the cylinder head.

4. Loosen the rocker arm valve adjusters, turn back the adjusting screws so that the rocker arms move freely.

5. Remove the rocker arm shaft retaining screws and pull out the shafts.

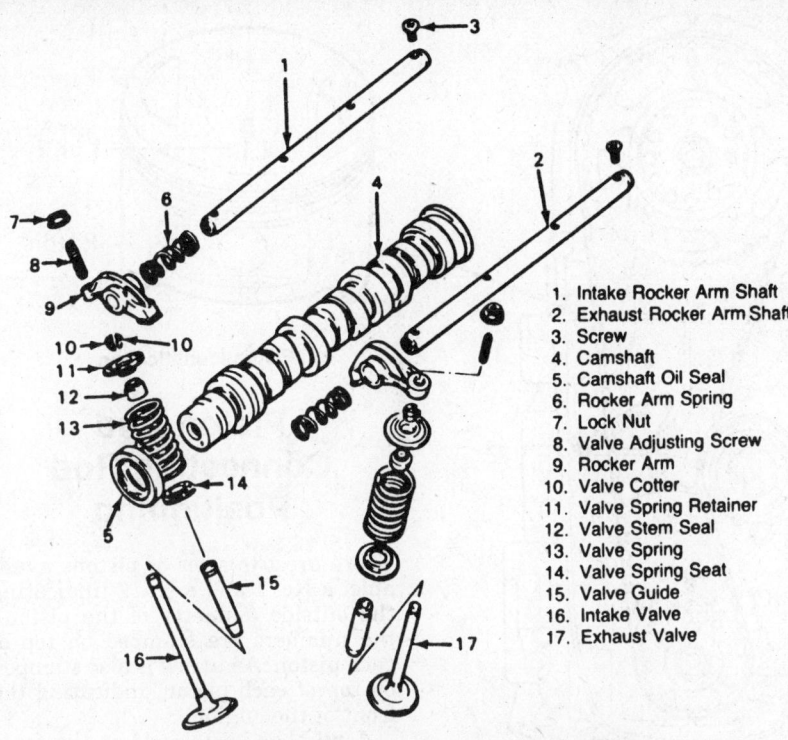

1. Intake Rocker Arm Shaft
2. Exhaust Rocker Arm Shaft
3. Screw
4. Camshaft
5. Camshaft Oil Seal
6. Rocker Arm Spring
7. Lock Nut
8. Valve Adjusting Screw
9. Rocker Arm
10. Valve Cotter
11. Valve Spring Retainer
12. Valve Stem Seal
13. Valve Spring
14. Valve Spring Seat
15. Valve Guide
16. Intake Valve
17. Exhaust Valve

Exploded view of the rocker arm assembly

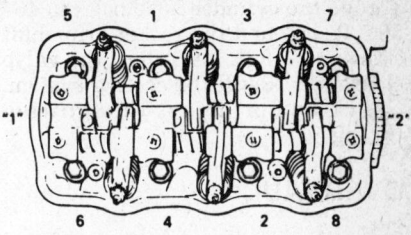

"1" Camshaft Pulley Side
"2" Distributor Side

Cylinder head torque sequence

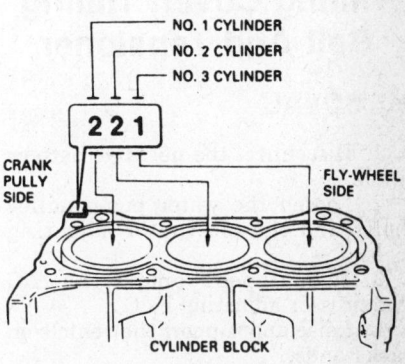

Cylinder identification

Remove the rocker arms and springs from the cylinder head.

NOTE: **Make a note of the differences between the rocker arm shafts. The intake shaft's stepped end is 0.55 inch, which faces the camshaft pulley; the exhaust shaft's stepped end is 0.59 inch, which faces the distributor.**

6. To install, use new gaskets and reverse the removal procedures. Torque the rocker arm shaft screws to 7–9 ft. lbs. Adjust the valve clearances. Check and/or adjust the ignition timing.

Valve Clearance

ADJUSTMENT

1. Remove the air cleaner and the cylinder head cover.
2. Rotate the crankshaft clockwise and align the "V" mark on the crankshaft pulley with the "0" mark on the timing plate.
3. Remove the distributor cap and to make sure that the rotor is facing the fuel pump, if not, rotate the crankshaft 360°.
4. Check and/or adjust the valves of the No. 1 cylinder.

NOTE: **On a COLD engine, adjust the valves to 0.006 in. (intake) and 0.008 in. (exhaust). On a HOT engine, adjust the valves to 0.010 in. (intake) and 0.012 in. (exhaust).**

5. After adjusting the valves of the No. 1 cylinder, rotate the crankshaft pulley 240° (the "V" mark should align with the lower left oil pump mounting bolt, when facing the crankshaft pulley) and adjust the valves of the No. 3 cylinder.
6. After adjusting the valves of the No. 3 cylinder, rotate the crankshaft pulley 240° (the "V" mark should align with the lower right oil pump mounting bolt, when facing the crankshaft pulley) and adjust the valves of the No. 2 cylinder.
7. After the valves have been adjusted, install the removed items by reversing the removal procedures. Torque the valve adjustment locknuts to 11–13 ft. lbs.

Cylinder Head

REMOVAL & INSTALLATION

1. Disconnect the negative battery cable.
2. Drain the cooling system.
3. Remove the air cleaner and the cylinder head cover.
4. Remove the distributor cap, then mark the position of the rotor and the distributor housing with the cylinder head. Remove the distributor and the case from the cylinder head.
5. Remove the accelerator cable from the carburetor. Remove the emission control and the coolant hoses from the carburetor/intake manifold.

6. Remove the electrical lead connectors from the carburetor/intake manifold and the lead wire from the oxygen sensor.
7. Remove the fuel hoses from the fuel pump and the pump from the cylinder head.
8. Remove the brake vacuum hose from the intake manifold.
9. Remove the crankshaft pulley, the outside cover, the timing belt and the tensioner from the front of the engine.
10. Remove the exhaust and the 2nd air pipes from the exhaust manifold.
11. Remove the exhaust/intake manifolds and the engine side mount from the cylinder head.
12. Loosen the rocker arm valve adjusters, turn back the adjusting screws so that the rocker arms move freely. Remove the rocker arm shaft retaining screws and pull out the shafts. Remove the rocker arms and springs from the cylinder head.

NOTE: **Make a note of the differences between the rocker arm shafts. The intake shaft's stepped end is 0.55 inch, which faces the camshaft pulley; the exhaust shaft's stepped end is 0.59 inch, which faces the distributor.**

13. Remove the mounting bolts and the cylinder head from the engine.
14. To install, use new gaskets and reverse the removal procedures.

Torque the cylinder head bolts to 46–50.5 ft. lbs. and the rocker arm shaft screws to 7–9 ft. lbs. Adjust the valve clearances. Refill the cooling system. Check and/or adjust the ignition timing.

OVERHAUL

For all cylinder head overhaul procedures, please refer to "Engine Rebuilding" in the Unit Repair section.

Timing Cover, Timing Belt And Tensioner

REMOVAL

1. Disconnect the negative battery cable.
2. Loosen the water pump pulley bolts and the alternator adjusting bolt.
3. If equipped, remove the A/C compressor adjusting bolt.
4. Raise and support the vehicle on jackstands.
5. Remove the drive belt splash shield, the right fender plug and the drive belts.
6. Remove the crankshaft and the water pump pulleys.
7. Remove the bolts from the bottom of the belt cover.
8. Lower the vehicle.
9. Remove the bolts from the top of the belt cover and the cover.
10. Remove the cylinder head cover and loosen the rocker arm adjusting bolts.
11. Remove the distributor cap.
12. Loosen the tensioner pulley and adjusting stud bolt.
13. Remove the timing belt, the tensioner, the tensioner plate and spring.

INSTALLATION

1. Install the tensioner assembly but DO NOT tighten the bolts.
2. Turn the camshaft pulley clockwise and align the mark on the pulley with the "V" mark on the inside cover.
3. Using a 17 mm wrench, turn the crankshaft clockwise and align the punch mark on the crankshaft pulley with the arrow mark on the oil pump.
4. With the timing marks aligned, install the timing belt so that there is no belt slack on the right side (facing the engine) of the engine, apply belt tension with the tensioner pulley.
5. Turn the crankshaft 1 rotation clockwise to remove the belt slack. Torque the tensioner stud, first, and then the tensioner bolt to 17–21 ft. lbs.
6. To complete the installation, use

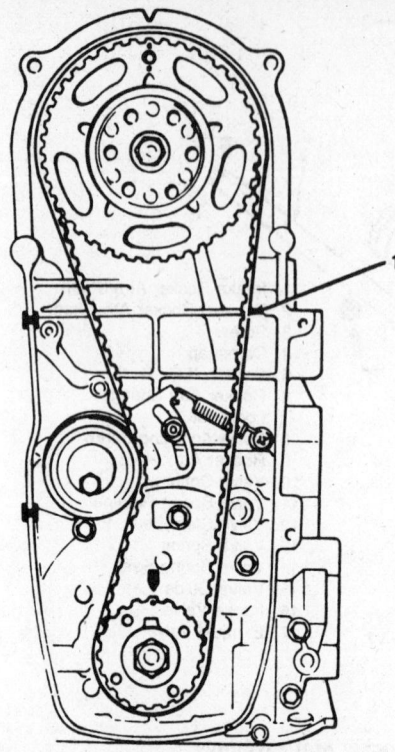

1. DRIVE SIDE OF BELT
View of the timing belt assembly

new gaskets and reverse the removal procedures. Torque the crankshaft pulley to 7–9 ft. lbs. Adjust the valve clearances.

Camshaft

REMOVAL & INSTALLATION

1. Refer to "Timing Cover, Timing Belt and Tensioner Removal & Installation" procedures in this section and remove the timing belt.
2. Remove the air cleaner, the cylinder head cover, the distributor and the distributor case. Remove the rocker arm shafts and the rocker arms.
3. Remove the fuel pump and the push rod from the cylinder head.
4. Using a spanner wrench tool J–34836 to hold the camshaft pulley, remove the camshaft pulley bolt, the pulley, the alignment pin and the inside cover.
5. Carefully slide the camshaft from the rear of the cylinder head.
6. Clean the gasket mounting surfaces. Check for wear and/or damage, replace the parts as necessary.
7. To install, use new gaskets/seals and reverse the removal procedures. Torque the camshaft pulley bolt to 41–46 ft. lbs. Adjust the valve clearances and check the timing.

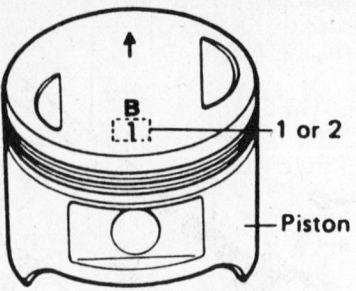

Piston Identification

Piston And Connecting Rod Positioning

There are two sizes of pistons available: a No. 1 and a No. 2 (indicating the outside diameter of the piston), the numbers are stamped on top of each piston. An arrow is also stamped on top of each piston, indicating the front of the engine.

A number is stamped at the front right of the engine block, on the cylinder head gasket surface. The number indicates of the pistons sizes, in order, ranging from the front-to-rear cylinders. Install the correct diameter piston (with the arrow facing the front of the engine) and the connecting rod (with the oil hole facing the intake manifold) into the correct cylinder bore.

For all connecting rod overhaul procedures, please refer to the "Engine Rebuilding" Unit Repair section.

ENGINE LUBRICATION

Oil Pan

REMOVAL & INSTALLATION

1. Remove the negative battery cable.
2. Raise and support the vehicle on jackstands.
3. Drain the engine oil.
4. Remove the flywheel dust cover.
5. Remove the exhaust pipe at the exhaust manifold.
6. Remove the oil pan bolts, the pan and the oil pump strainer.
7. Clean the gasket mating surfaces.
8. To install, use new gaskets and

reverse the removal procedures. Torque the oil pan bolts to 9–12 ft. lbs. Refill the engine oil.

Rear Main Seal

REMOVAL & INSTALLATION

1. Refer to the "Manual Transaxle Removal & Installation" procedures in this section and remove the transaxle.

2. Raise and support the vehicle. Remove the oil pan.

3. Remove the pressure plate, the clutch plate and the flywheel.

4. Remove the mounting bolts and the rear seal housing.

5. Pry the oil seal from the oil seal housing.

6. To install, use new gaskets/seals and reverse the removal procedures. Torque the oil seal housing to 7–9 ft. lbs. and the flywheel to 57–65 ft. lbs.

NOTE: After installing the oil seal housing, trim the gasket flush with the bottom of the case.

Oil Pump And Front Seal

REMOVAL & INSTALLATION

1. Refer to the "Timing Cover, Timing Belt and Tensioner Removal & Installation" procedures in this section and remove the timing belt.

2. Raise and support the vehicle. Remove the oil pan.

3. Use a screwdriver to hold the crankshaft timing belt pulley, remove the crankshaft bolt and pull the timing pulley from the shaft.

4. Remove the alternator mounting bracket and the A/C compressor bracket, if equipped.

5. Remove the alternator adjusting bolt and the upper cover bolt.

6. Remove the oil pump mounting bolts and the oil pump.

7. Pry the crankshaft oil seal from the oil pump.

8. Clean the gasket mounting surfaces. Remove the gear plate from the back of the oil pump and pack the oil pump gears with petroleum jelley.

9. To install, use new gaskets/seals and reverse the removal procedures. Torque the oil pump bolts to 7–9 ft. lbs. and the crankshaft timing pulley bolt to 47–54 ft. lbs. Adjust the valve clearances and check the timing.

NOTE: To install the oil pump to the engine, place the Oil Seal Guide tool J–34853 on the crankshaft and slide the oil pump onto

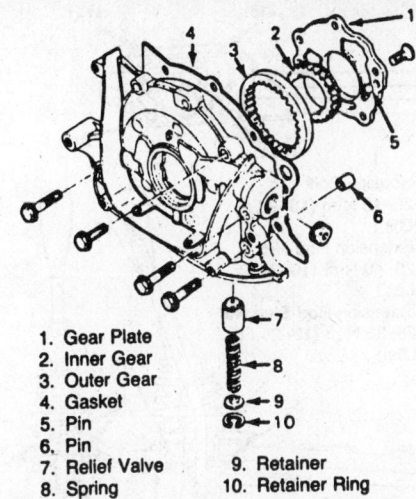

1. Gear Plate
2. Inner Gear
3. Outer Gear
4. Gasket
5. Pin
6. Pin
7. Relief Valve
8. Spring
9. Retainer
10. Retainer Ring

Exploded view of the oil pump assembly

the alignment pins. After installing the oil seal housing, trim the gasket flush with the bottom of the case.

CLUTCH

REMOVAL & INSTALLATION

1. Refer to the "Manual Transaxle Removal & Installation" procedures in this section and remove the transaxle.

2. Install tool J–34860 into the pilot bearing to support the clutch assembly.

NOTE: Look for the "X" mark or white painted number on the clutch cover and the "X" mark on the flywheel. If there are no markings, mark the clutch cover and the flywheel for reassembly purposes.

3. Loosen the clutch cover-to-flywheel bolts, one turn at a time (evenly) until the spring pressure is released.

4. Remove the clutch cover and clutch disc.

5. Inspect the parts for wear, if necessary, replace the parts.

6. To install, reverse the removal procedures. Torque the clutch cover bolts to 14–20 ft. lbs.

FREE PLAY ADJUSTMENT

1. At the transaxle, move the clutch release arm to check the free play, it should be 0.08–0.16 in.

2. If necessary, turn the clutch cable joint nut to adjust the cable length.

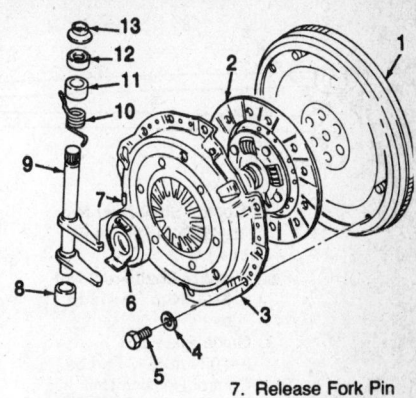

1. Flywheel
2. Disc
3. Clutch Cover
4. Lock Washer
5. Bolt
6. Release Bearing
7. Release Fork Pin
8. No. 2 Bushing
9. Release Shaft
10. Return Spring
11. No. 1 Bushing
12. Shaft Seal
13. Shaft Cover

Exploded view of the clutch assembly

MANUAL TRANSAXLE

REMOVAL & INSTALLATION

1. Disconnect the negative battery cable and the ground strap at the transaxle.

2. Remove the air cleaner and the air pipe.

3. Remove the clutch cable from the clutch release lever.

4. Remove the starter, the speedometer cable, the electrical wires and the wiring harness from the transaxle.

5. Remove the front and rear torque rod bolts at the transaxle.

6. Raise and support the vehicle on jackstands.

7. Drain the transaxle fluid.

8. Remove the exhaust pipe at the exhaust manifold and at the 1st exhaust hanger.

9. Remove the clutch housing lower plate.

10. Disconnect the gear shift control shaft and the extension rod at the transaxle.

11. Remove the left front wheel.

12. Using a pry bar, pry on the inboard joints of the right and left hand axle shafts, to detach the axle shafts from the snap rings of the differential side gears.

13. On the left side, remove the stabilizer bar mounting bolts and the ball joint stud bolt. Push down on the stabilizer bar and remove the ball joint stud from the steering knuckle.

14. Pull the left axle shaft out of the transaxle.

15. Remove the front torque rod.

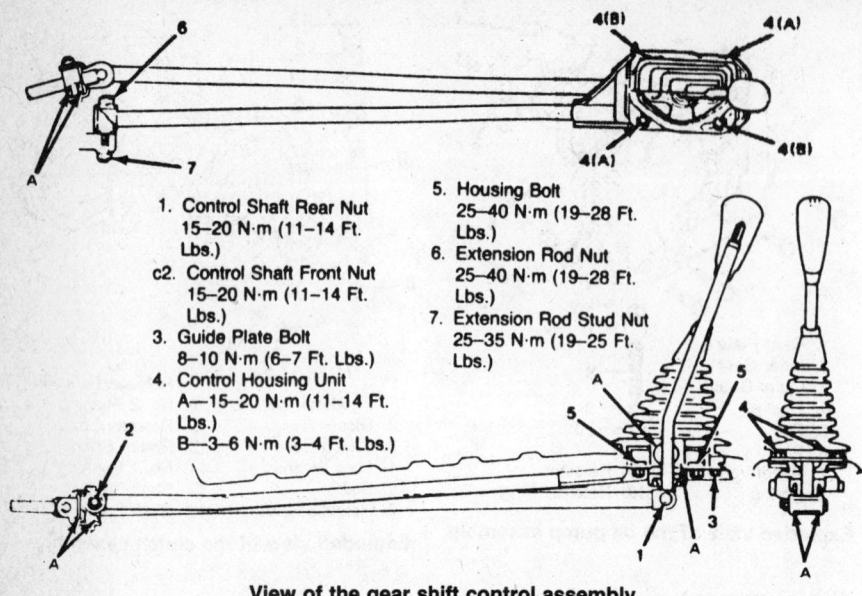

1. Control Shaft Rear Nut
15–20 N·m (11–14 Ft. Lbs.)
c2. Control Shaft Front Nut
15–20 N·m (11–14 Ft. Lbs.)
3. Guide Plate Bolt
8–10 N·m (6–7 Ft. Lbs.)
4. Control Housing Unit
A—15–20 N·m (11–14 Ft. Lbs.)
B—3–6 N·m (3–4 Ft. Lbs.)
5. Housing Bolt
25–40 N·m (19–28 Ft. Lbs.)
6. Extension Rod Nut
25–40 N·m (19–28 Ft. Lbs.)
7. Extension Rod Stud Nut
25–35 N·m (19–25 Ft. Lbs.)

View of the gear shift control assembly

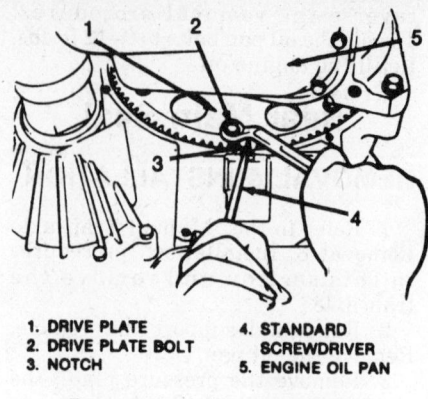

1. DRIVE PLATE
2. DRIVE PLATE BOLT
3. NOTCH
4. STANDARD SCREWDRIVER
5. ENGINE OIL PAN

Removing drive plate bolts

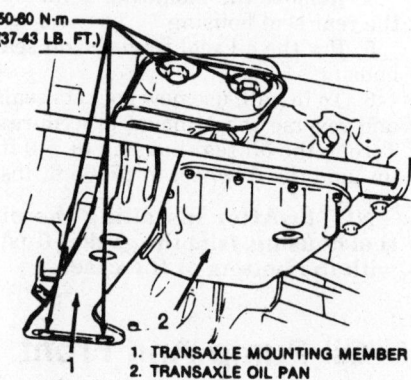

1. TRANSAXLE MOUNTING MEMBER
2. TRANSAXLE OIL PAN

Transaxle mounting member and oil pan

16. Secure and support the transaxle case with a floor jack.

17. Remove the transaxle-to-body mounting bolts and the mounts.

18. Remove the transaxle-to-engine mounting bolts.

19. Disconnect the transaxle from the engine by sliding it to the left side and lower the jack.

NOTE: When removing the transaxle, support the right axle shaft, so it does not become damaged.

20. To install, guide the right axle shaft into the transaxle and reverse the removal procedures. Torque the transaxle-to-engine bolts to 35 ft. lbs.; the transaxle-to-mount bolts to 34 ft. lbs.; the mounting member bolts to 40 ft. lbs.; the stabilizer bar bolts to 30 ft. lbs. and the ball joint stud bolt to 44 ft. lbs. Adjust the clutch cable and refill the transaxle.

OVERHAUL

For all manual transmission or transaxle overhaul procedures, please refer to the "Manual Transmission" Unit Repair section.

SHIFT LINKAGE ADJUSTMENT

1. At the console, loosen the gear shift control housing nuts and the guide plate bolts.

2. Adjust the guide plate, so that the shift lever is centered and at a right angle to the plate.

3. When the guide plate is positioned correctly, torque the guide plate bolts to 7 ft. lbs. and the housing nuts to 4 ft. lbs.

AUTOMATIC TRANSMISSION/ TRANSAXLE

NOTE: Procedures for the adjustment of the linkage and bands and for fluid and filter changes are found in the Unit Repair Section.

REMOVAL & INSTALLATION

1. Disconnect the air suction guide from the air cleaner.

2. Disconnect the negative and the positive battery cables.

3. Remove the battery and the battery bracket tray.

4. Remove the negative cable from the transaxle.

5. Disconnect the solenoid wire coupler and the shift lever switch wire couplers.

6. Remove the wiring harness from the transaxle.

7. Remove the speedometer cable from the transaxle.

8. Disconnect the oil pressure control cable from the accelerator cable, and then the accelerator cable from the transaxle.

9. Remove the select cable from the transaxle.

10. Remove the starter motor.

11. Drain the transaxle fluid.

12. Disconnect the oil outlet and inlet hoses from the oil pipes. After disconnecting, plug the two oil hoses to prevent fluid in the hoses and oil cooler from draining.

13. Raise the vehicle and support it safely.

14. Remove the exhaust No. 1 pipe.

15. Remove the clutch housing lower plate.

16. Remove the six drive plate bolts. To lock the drive plate, engage a screw driver with the drive plate gear through the notch provided at the under side of the transmission case.

17. Remove the left hand front drive axle. See drive axle section for removal procedures.

18. Detach the inboard joint of the right hand drive axle from the differential.

19. Disconnect the transaxle mounting member.

20. Securely support the transaxle with a suitable jack for removal.

21. Remove the transaxle left mounting.

22. Remove the bolts fastening the engine and the transaxle.

23. Disconnect the transaxle from the engine by sliding towards the left side, and then, carefully lower the jack.

NOTE: When removing the transaxle assembly from the engine, move it in parallel with the crankshaft and use care so as not to apply excessive force to the drive plate and torque converter. After removing the transaxle assembly, be sure to keep it so that the oil

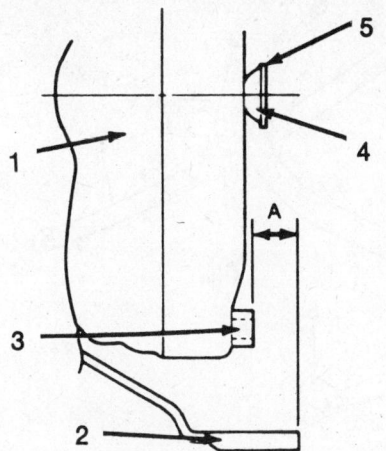

1. TORQUE CONVERTER
2. TRANSAXLE CASE HOUSING
3. FLANGE NUT
4. CUP
5. "APPLY GREASE HERE"
A: MORE THAN 21.4 mm (0.85 IN.)

Torque converter installation

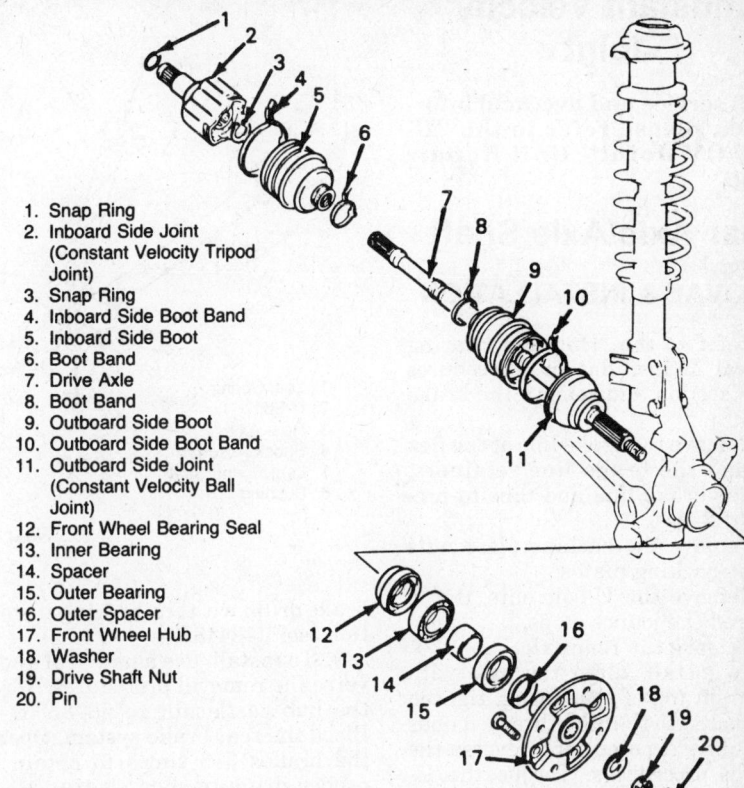

1. Snap Ring
2. Inboard Side Joint (Constant Velocity Tripod Joint)
3. Snap Ring
4. Inboard Side Boot Band
5. Inboard Side Boot
6. Boot Band
7. Drive Axle
8. Boot Band
9. Outboard Side Boot
10. Outboard Side Boot Band
11. Outboard Side Joint (Constant Velocity Ball Joint)
12. Front Wheel Bearing Seal
13. Inner Bearing
14. Spacer
15. Outer Bearing
16. Outer Spacer
17. Front Wheel Hub
18. Washer
19. Drive Shaft Nut
20. Pin

Exploded view of the axle shaft

pan is at the bottom. If the transaxle is tilted, fluid in it may flow out.

24. To install the transaxle, reverse the removal procedure noting the following important steps.

25. Before installing the transaxle assembly apply grease around the cup at the center of the torque converter. Then measure the distance between the torque converter flange nut and the transaxle case housing. The distance should be more than 0.85 in. (21.4 mm). If the distance is less than 0.85 in. (21.4 mm), the torque converter has been installed incorrectly and must be removed and reinstalled correctly.

26. When installing the transaxle, guide the right drive axle into the differential side gear as the transaxle is being raised.

27. After inserting the inboard joints of the right hand and left hand drive axles into the differential side gears, push the inboard joints into the side gears until the snap rings on the drive axles engage the side gears.

28. After connecting the oil pressure control cable to the accelerator cable, check the oil pressure control cable play and adjust if necessary.

29. Install the select cable.

30. Refill the transaxle and check the fluid level.

31. Tighten the following bolts and nuts to specifications.

a. Drive Plate Bolts—14 ft. lbs. (19 Nm).

b. Mounting Member Bolts—40 ft. lbs. (55 Nm).

c. Transaxle Mounting Nuts—33 ft. lbs. (45 Nm).

d. Transaxle Mounting Bolts (8mm)—40 ft. lbs. (55 Nm).

e. Transaxle Mounting Nuts—40 ft. lbs. (55 Nm).

f. Stabilizer Shaft Mounting Bolts—31 ft. lbs. (42 Nm).

g. Ball Stud Bolt—44 ft. lbs. (60 Nm).

h. Wheel Nuts—40 ft. lbs. (55 Nm).

AXLE SHAFTS AND CV-JOINTS

Front Axle Shaft

REMOVAL & INSTALLATION

1. Remove the grease cap, the cotter pin and the axle shaft nut from both front wheels.
2. Loosen the wheel nuts.

3. Raise and support the front of the vehicle on jackstands.
4. Remove the front wheels.
5. Drain the transaxle fluid.
6. Using a pry bar, pry on the inboard joints of the right and left hand axle shafts to detach the axle shafts from the snap rings of the differential side gears.
7. Remove the stabilizer bar mounting bolts and the ball joint stud bolt. Pull down on the stabilizer bar and remove the ball joint stud from the steering knuckle.
8. Pull the axle shafts out of the transaxle's side gear, first, and then from the steering knuckles.

NOTE: To prevent the axle shaft boots from becoming damaged, be careful not to bring them into contact with any parts. If any malfunction is found in the either of the joints, replace the joints as an assembly.

9. To install, snap the axle shaft into the transaxle, first, and then into the steering knuckle.
10. To complete the installation, reverse the removal procedures. Torque the stabilizer bar mounting bolts to 30 ft. lbs.; the ball joint stud bolt to 44 ft. lbs. and the axle shaft nut to 144 ft. lbs.

Constant Velocity Joint

For all service and overhaul information, please refer to the "U-Joint/CV-Joint" Unit Repair section.

Rear Axle/Axle Shaft

REMOVAL & INSTALLATION

1. Refer to the "Hub and Bearing Removal & Installation" procedures in this section and remove the brake drums.

2. Remove the brake line at the flex hose and the brake line retainers. Plug the brake line and tube to prevent fluid loss.

3. Remove the backing plate nuts and the backing plates.

4. Remove the U-bolt nuts, the U bolts and the jounce stops.

5. Remove the rear axle.

6. To install, align the axle with the pin on top of the spring, tighten the U-bolts (so that the threaded ends are equally exposed) and reverse the removal procedures. Torque the U-bolts to 22–32 ft. lbs. and the backing plate nuts to 13–20 ft. lbs.

NOTE: When installing the backing plate, apply watertight sealant to the joint seam of the backing plate and the axle.

Hub And Bearing

REMOVAL & INSTALLATION

1. Raise and support the rear of the vehicle on jackstands.

2. Remove the wheel assembly.

3. Remove the dust cap, the cotter pin, the castle nut and the washer.

4. Loosen the adjusting nuts of the parking brake cable.

5. Remove the plug from the rear of the backing plate. Insert a screwdriver through the hole, making contact with the shoe hold down spring, then push the spring to release the parking brake shoe lever.

6. Using a slide hammer tool J–2619–01 and a brake drum remover tool J–34866, pull the brake drum from the axle shaft.

7. Using a brass drift and a hammer, drive the rear wheel bearings from the brake drum.

NOTE: When installing the wheel bearings, face the sealed sides (numbered sides) outward. Fill the wheel bearing cavity with bearing grease.

8. Drive the new bearings into the

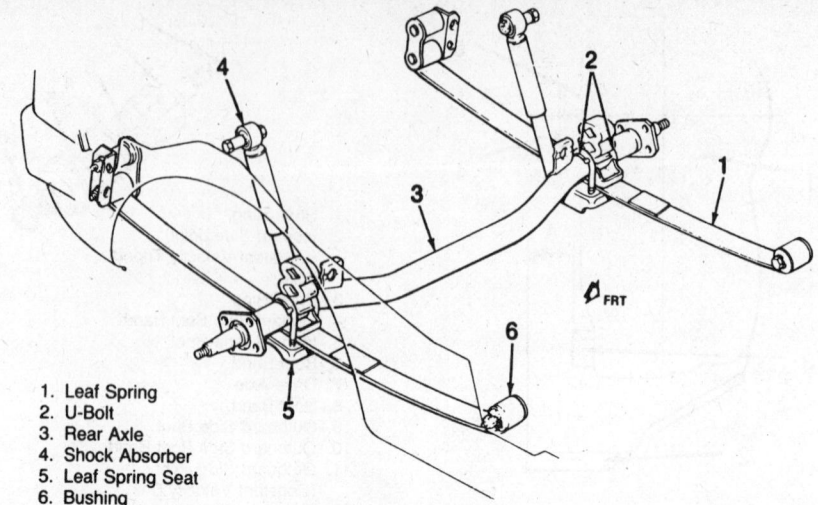

1. Leaf Spring
2. U-Bolt
3. Rear Axle
4. Shock Absorber
5. Leaf Spring Seat
6. Bushing

Rear axle assembly

brake drum with the bearing installation tool J–34482.

9. To install, use a new seal and reverse the removal procedures. Torque the hub castle nut to 58–86 ft. lbs. Bleed the rear brake system. Operate the brakes 3–5 times to obtain the proper drum-to-shoe clearance. Adjust the parking brake cable.

FRONT SUSPENSION

MacPherson Strut

REMOVAL & INSTALLATION

1. Raise and support the front of the vehicle.

2. Remove the wheel assembly.

3. Remove the brake hose securing ring and the hose from the strut.

4. Remove the upper strut support nuts from the engine compartment.

5. Remove the strut-to-steering knuckle bolts and the strut.

6. To install, reverse the removal procedures. Torque the upper mounting nuts to 13–20 ft. lbs. and the strut-to-steering knuckle bolts to 50–65 ft. lbs.

OVERHAUL

For all spring and shock absorber removal and installation procedures, and all strut overhaul procedures, please refer to the "Strut Overhaul" Unit Repair section.

Stabilizer Bar

REMOVAL & INSTALLATION

1. Raise and support the front of the vehicle on jackstands. Remove the front wheel assemblies.

2. Remove the stabilizer bar-to-body mounting bolts.

3. Remove the cotter pin, the castle nut, the washer, the bushing and the stabilizer bar from the lower control arms.

4. To install, reverse the removal procedures. Torque the stabilizer bar-to-control arm to 29–65 ft. lbs. and the stabilizer bar-to-body bolts to 22–39 ft. lbs.

Ball Joint

The ball joint is part of the control arm and can only be serviced as an assembly. To service the ball joint/control arm assembly, please refer to the "Control Arm Removal & Installation" procedures in this section.

Lower Control Arm

REMOVAL & INSTALLATION

1. Raise and support the front of the vehicle on jackstands. Remove the front wheel assembly.

2. Remove the cotter pin, the castle nut, the washer and the bushing from the stabilizer bar.

3. Remove the stabilizer bar-to-body mounting bracket bolts.

4. Remove the ball stud and the control arm bolts.

5. Remove the control arm.

6. To install, reverse the removal

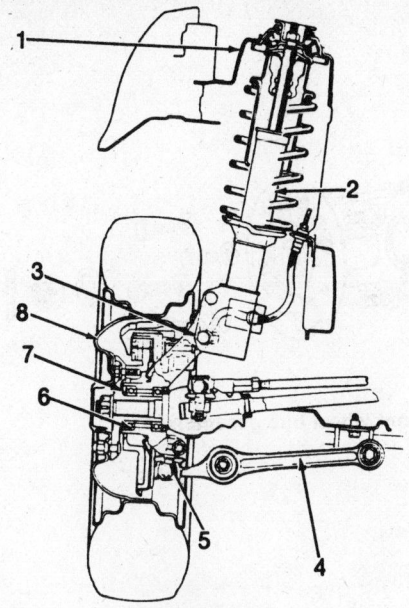

1. Car Body
2. Strut Assembly
3. Steering Knuckle
4. Suspension Control Arm
   (Lower Arm)
5. Ball Stud
6. Wheel Bearing
7. Front Wheel Hub
8. Wheel

Cross-sectional view of the front suspension

procedures. Torque the control arm-to-body bolt to 36–50 ft. lbs.; the ball stud-to-steering knuckle to 36–50 ft. lbs.; the stabilizer bar-to-control arm to 29–65 ft. lbs. and the stabilizer bar-to-body to 22–39 ft. lbs.

Front Hub

REMOVAL & INSTALLATION

1. Raise and support the front of the vehicle on jackstands. Remove the front wheel assembly.
2. Remove the dust cap and the cotter pin from the axle shaft.
3. Loosen the castle nut on the axle shaft.
4. Remove the brake caliper bolts and the brake caliper.

NOTE: When removing the brake caliper, DO NOT remove the brake hose, suspend it on a wire.

5. Remove the castle nut and the washer from the axle shaft.
6. Using a slide hammer tool J–2619–01 and a brake drum remover tool J–34866, pull the hub from the steering knuckle.
7. Remove the spacing ring from the rear of the hub.

1. Knuckle, Strg R.H.
2. Knuckle, Strg L.H.
3. Absorber, W/Strut R.H.
4. Absorber, W/Strut L.H.
5. Bolt, Strut Brkt
6. Washer, Lk
7. Nut
8. Spring, Frt Coil
9. Stopper, Frt Bpr
10. Seat, Frt Spr
11. Seat, Srt Spr Upr
12. Bearing, Frt Strut
13. Seal, Frt Strut Brg Dust
14. Seat, Frt Strut Brg
15. Seat, Strut Mt
16. Support, Frt Strut
17. Washer (3/8 × 11/16)
18. Nut (M8 × 1.25 × 6)
19. Mount, Frt Strut
20. Stopper, Frt Strut
    Rebound
21. Support, Frt Strut Inr
22. Washer
23. Nut
24. Arm, Frt Cont
25. Seal, Ball Stud Dust
26. Clip, Dust Seal
27. Bushing, Cont Arm
28. Bolt, Ball Stud
29. Washer, Ball Stud
30. Nut, Ball Stud Lk
31. Bolt, Cont Arm
32. Washer (3/8 × 11/16)
33. Shaft, Frt Stab
34. Bushing, Stab Shf
35. Washer, Stab Shf
36. Nut (M12 × 1.25 × 10)
37. Pin
38. Mount, Stab Shf
39. Bracket, Stab Shf Mt
40. Bolt
41. Washer, Lk

Front suspension

8. Remove the bolts and separate the hub from the brake disc.
9. To install, place the spacing ring on the hub (install beveled side first) and reverse the removal procedures. Torque the hub-to-brake disc bolts to 29–43 ft. lbs., the brake caliper bolts to 17–26 ft. lbs. and the axle shaft nut to 108–195 ft. lbs.

NOTE: When installing the hub to the steering knuckle, tap it with

a plastic hammer to align the hub, then, using the installation tool J–34856, drive the hub into the steering knuckle.

Steering Knuckle And Wheel Bearings.

REMOVAL & INSTALLATION

1. Refer to the "Front Hub Remov-

al & Installation" procedures in this section and remove the hub from the steering knuckle.

2. Remove the tie rod end cotter pin and nut.

3. Using the ball joint puller tool J-21687-02, remove the ball joint from the steering knuckle.

4. Remove the ball stud bolt from the steering knuckle.

5. Remove the strut-to-steering knuckle bolts.

6. Remove the steering knuckle and support the axle shaft on a wire.

7. Using a brass drift and a hammer, drive the wheel bearings from the steering knuckle.

8. Remove the spacer and clean the steering knuckle cavity.

9. Lubricate the new bearings and the steering knuckle cavity.

10. Using the installation tool J-34856, drive the new bearings (with the internal seals facing outward) into the steering knuckle.

11. Using the seal installation tool J-34881, drive the new seal into the steering knuckle (grease the seal lip).

12. To complete the installation, reverse the removal procedures. Torque the strut-to-steering knuckle bolts to 50–65 ft. lbs.; the ball joints-to-steering knuckle nuts to 22–40 ft. lbs. and the axle shaft castle nut to 108–195 ft. lbs.

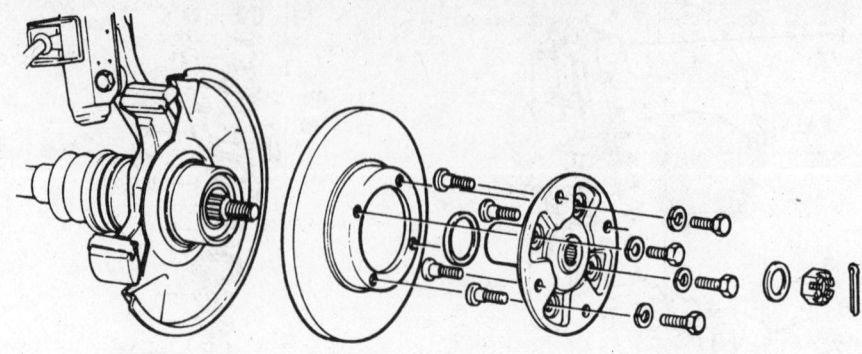

Exploded view of the front wheel hub assembly

REAR SUSPENSION

The rear suspension is a rigid axle type. One tapered leaf spring is longitudinally slung under each side of the axle. The front end of the spring is attached to the body bracket through a rubber bushing and shackle assembly. A shock absorber is positioned inboard of the spring and slanted toward the rear of the vehicle.

Shock Absorber

REMOVAL & INSTALLATION

1. Raise and support the rear of the vehicle on jackstands. Remove the wheel assembly.

2. Remove the lower mounting nut, the lock washer and and the outer washer.

3. Remove the upper mounting bolt, the lock washer and nut.

4. Remove the shock absorber.

5. To install, reverse the removal procedures. Torque the upper mounting bolt to 33–50 ft. lbs. and the lower mounting nut to 8–12 ft. lbs.

1. Plate, Backing R.H.
2. Plate, Backing L.H.
3. Bolt, Bkg Plt
4. Bolt, Whl Cyl
5. Plug, Bkg Plt
6. Shoe, Brk
7. Spring, Shoe Rtn
8. Pin, Shoe Holdn
9. Spring, Shoe Holdn
10. Spring, Shoe Holdn R.H.
11. Spring, SHoe Holdn L.H.
12. Cylinder, Rr Rhl
13. Cup Set, whl Cyl
14. Spring, Whl Cyl
15. Plug, Bleeder
16. Cap, Bleeder Plug
17. Gasket, Whl Cyl
18. Strut, Brk R.H.

19. Strut, Brk L.H.
20. Spring, Quadrant
21. Spring, Anti-Rattle
22. Lever, Pk SHoe R.H.
23. Lever, Pk Shoe L.H.
24. Retainer, Pk Lvr
25. Retainer, Pk Lvr
25. Nut
26. Washer (8.2 × 15.6 × 2.5)
27. Plug, Bkg Plt #2

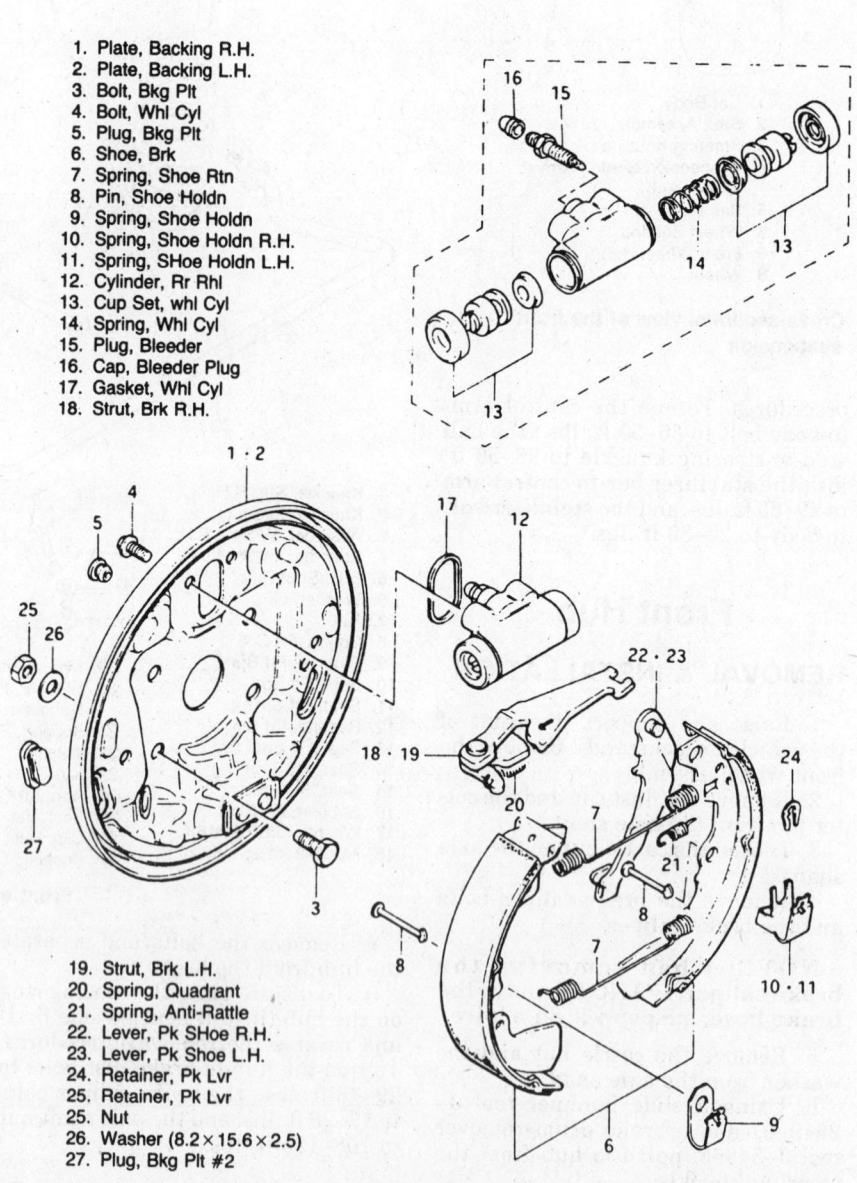

Rear drum brakes

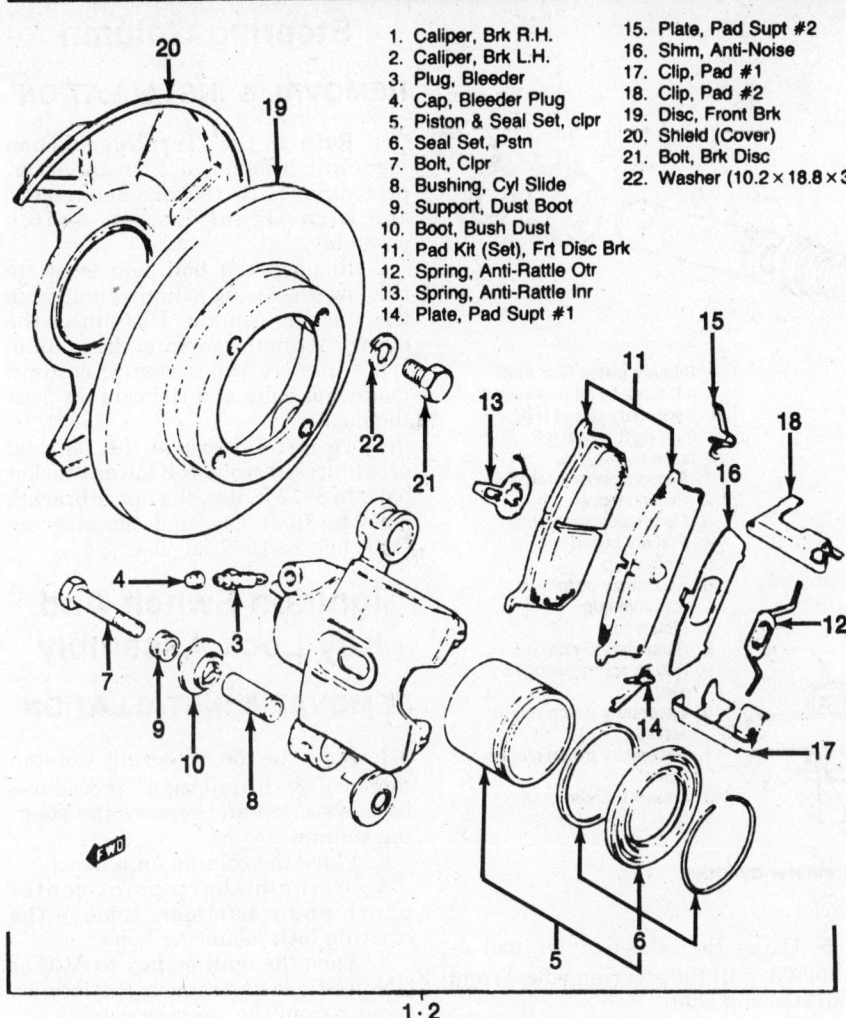

1. Caliper, Brk R.H.
2. Caliper, Brk L.H.
3. Plug, Bleeder
4. Cap, Bleeder Plug
5. Piston & Seal Set, clpr
6. Seal Set, Pstn
7. Bolt, Clpr
8. Bushing, Cyl Slide
9. Support, Dust Boot
10. Boot, Bush Dust
11. Pad Kit (Set), Frt Disc Brk
12. Spring, Anti-Rattle Otr
13. Spring, Anti-Rattle Inr
14. Plate, Pad Supt #1
15. Plate, Pad Supt #2
16. Shim, Anti-Noise
17. Clip, Pad #1
18. Clip, Pad #2
19. Disc, Front Brk
20. Shield (Cover)
21. Bolt, Brk Disc
22. Washer (10.2 × 18.8 × 3)

Front calipers and components

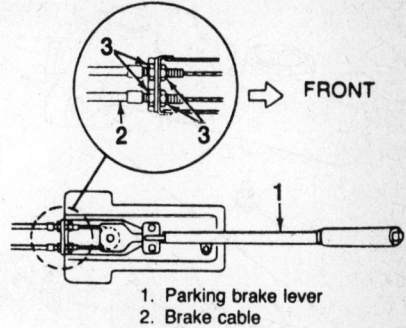

1. Parking brake lever
2. Brake cable
3. Nuts

Adjusting the parking brakes

NOTE: To service the brake shoes, the drums, the wheel cylinders or brake pads and calipers, please refer to the "Brakes" Unit Repair section.

Master Cylinder

REMOVAL & INSTALLATION

1. Clean around the reservoir cap and take some of the fluid out with a syringe.
2. Disconnect and plug the brake tubes from the master cylinder.
3. Remove the mounting nuts and washers.
4. Remove the master cylinder.
5. To install, reverse the removal procedures. Torque the mounting bolts to 8–12 ft. lbs. Bleed the brake system.

Power Brake Booster

REMOVAL & INSTALLATION

1. Refer to the "Master Cylinder Removal & Installation" procedures in this section and remove the master cylinder.
2. Disconnect the push rod clevis pin from the brake pedal arm.
3. Disconnect the vacuum hose from the brake booster.
4. Remove the mounting nuts from under the dash and the booster.
5. To install, reverse the removal procedures. Torque the booster-to-cowl nuts to 14–20 ft. lbs. Bleed the brake system, if necessary.

PARKING BRAKE ADJUSTMENT

1. Remove both door seal plates and the seat belt buckle bolts at the floor.
2. Disconnect the shoulder harness bolts at the floor and the interior, bottom trim panels.

Spring

REMOVAL & INSTALLATION

1. Raise and support the rear of the vehicle on jackstands. Remove the front wheel assembly.
2. Remove the U-bolt nuts.
3. Remove the shackle and leaf spring front nuts.
4. Remove the front spring bolt.
5. Remove the spring from the vehicle.

NOTE: Apply a thin coat of silicone grease to the springs bushings before installation.

6. To install, align the spring pin with the hole in the axle shaft housing and reverse the removal procedures. Torque the front spring bolt to 33–50 ft. lbs.; the rear spring shackle nuts to 22–40 ft. lbs. and the U-bolt nuts to 22–33 ft. lbs.

Rear Wheel Bearings and Rear Axle

REMOVAL & INSTALLATION

Refer to the rear axle and axle shaft shaft removal and installation section for the proper procedures.

BRAKES

Service brakes are vacuum operated hydraulic and self adjusting with the power assist. A single piston, self adjusting, floating caliper disc brake is used on the front. A leading trailing, self adjusting drum brake is used on the rear. Wear indicators are installed on the front brake pads.

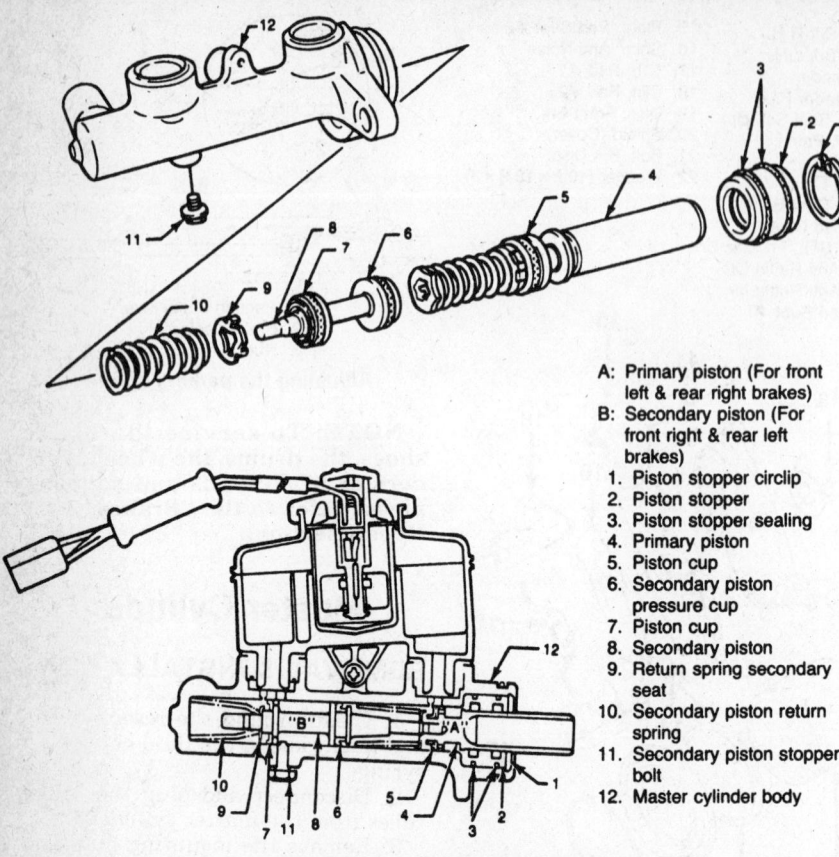

A: Primary piston (For front left & rear right brakes)
B: Secondary piston (For front right & rear left brakes)
1. Piston stopper circlip
2. Piston stopper
3. Piston stopper sealing
4. Primary piston
5. Piston cup
6. Secondary piston pressure cup
7. Piston cup
8. Secondary piston
9. Return spring secondary seat
10. Secondary piston return spring
11. Secondary piston stopper bolt
12. Master cylinder body

Exploded view of the master cylinder

3. Raise the rear seat cushion.

4. Pull up the carpet to gain access to the parking brake lever.

5. Loosen the parking brake cable adjusting nuts.

6. Adjust the parking brake cables, so that they work evenly.

7. Adjust the cable, so that when the parking brake handle is pulled, its travel is between 5–8 notches, with 44 lbs. of force.

8. After adjustment, reverse the removal procedures.

STEERING

Steering Wheel

REMOVAL & INSTALLATION

1. Disconnect the negative battery cable.

2. Loosen the pad screws and remove the the pad.

3. Remove the steering wheel nut.

4. Scribe a matchmark line on the steering wheel and the shaft.

5. Using the wheel puller tool J-1859-03, pull the steering wheel from the steering shaft.

6. To install, reverse the removal procedures. Torque the steering wheel nut to 19–29 ft. lbs.

Turn Signal/Dimmer Switch Assembly

REMOVAL & INSTALLATION

1. Refer to the "Steering Wheel Removal & Installation" procedures in this section and remove the steering wheel.

2. Remove the upper and lower steering column covers.

3. Disconnect the turn signal/dimmer switch assembly electrical connector.

4. Remove the screws and the turn signal/dimmer switch assembly from the steering column.

5. To install, reverse the removal procedures.

NOTE: When installing, be careful that the lead wires do not get caught by the lower cover.

Steering Column

REMOVAL & INSTALLATION

1. Refer to the "Turn Signal/Dimmer Switch Removal & Installation" procedures in this section and remove the turn signal/dimmer switch assembly.

2. Remove the bolt and separate the lower steering column shaft from the steering column. Disconnect the electrical connectors from the column.

3. Remove the steering column mounting bolts and the column from the dash.

4. To install, reverse the removal procedures. Torque the lower bracket bolts to 8–12 ft. lbs., the upper bracket bolts to 10 ft. lbs. and the steering shaft bolt to 15–22 ft. lbs.

Ignition Switch And Key Lock Assembly

REMOVAL & INSTALLATION

1. Refer to the "Steering Column Removal & Installation" procedures in this section and remove the steering column.

2. Place the column on a bench.

3. Using a sharp point center punch and a hammer, remove the steering lock mounting bolts.

4. Turn the ignition key to ACC or ON positions and remove the lock assembly from the steering column.

5. To install, reverse the removal procedures. After installing the lock, turn the key to LOCK position and pull out the key. Turn the steering shaft to make sure the shaft is locked. Install new mounting bolts to the lock housing, tighten until the bolt heads break off. Torque the lower bracket bolts to 8–12 ft. lbs.; the upper bracket bolts to 10 ft. lbs. and the steering shaft bolt to 15–22 ft. lbs.

Steering Gear

REMOVAL & INSTALLATION

1. Refer to the "Tie Rod Removal & Installation" procedures in this section and remove the tie rod ends from the steering knuckles.

2. Under the dash, remove the steering joint cover.

3. Remove the lower steering shaft-to-steering gear clinch bolt and separate the steering shaft from the steering gear.

4. Remove the steering gear mounting bolts, the brackets and the steering gear case from the vehicle.

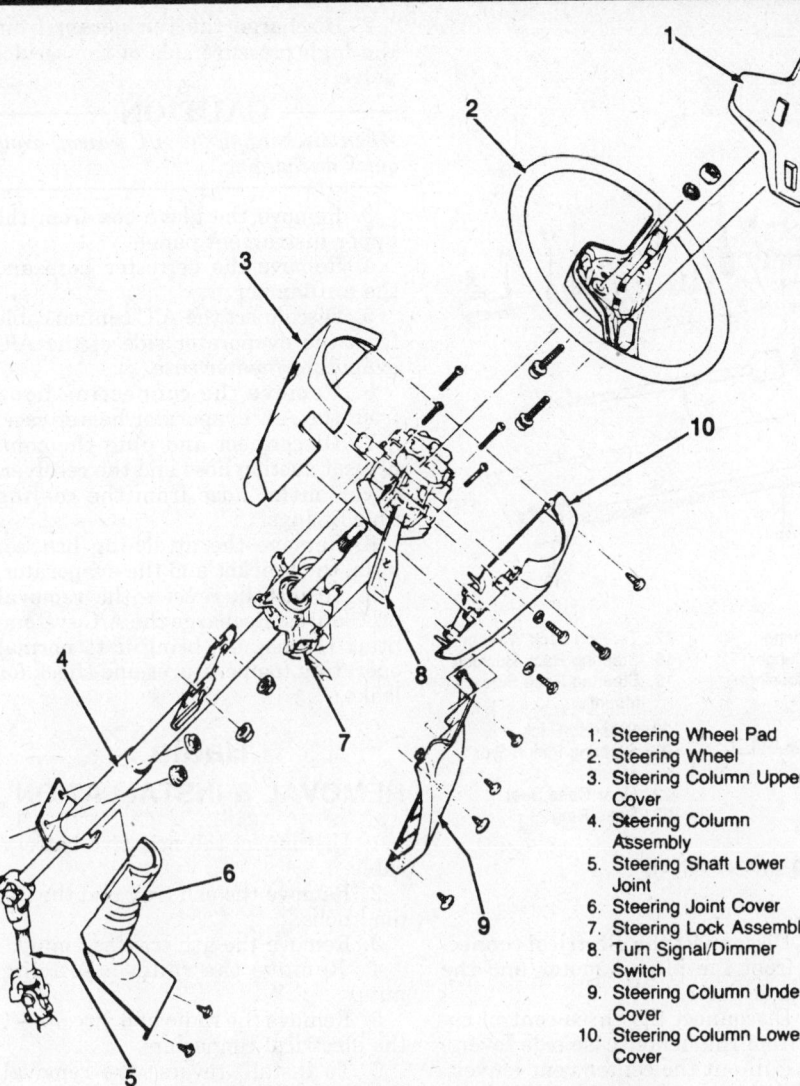

1. Steering Wheel Pad
2. Steering Wheel
3. Steering Column Upper Cover
4. Steering Column Assembly
5. Steering Shaft Lower Joint
6. Steering Joint Cover
7. Steering Lock Assembly
8. Turn Signal/Dimmer Switch
9. Steering Column Under Cover
10. Steering Column Lower Cover

Exploded view of the steering column assembly

5. To install, reverse the removal procedures. Torque the steering gear case bolts to 14–22 ft. lbs.; the steering gear-to-steering shaft bolt to 14–22 ft. lbs. and the tie rod end-to-steering knuckle nut to 22–40 ft. lbs.

Tie Rod

REMOVAL & INSTALLATION

1. Raise and support the front of the vehicle on jackstands. Remove the front wheel assembly.

2. Remove the cotter pin and the castle nut from the tie rod end.

3. Using the ball joint remover tool J-21687-02, remove the tie rod end ball joint from the steering knuckle.

4. Loosen the lock nut on the tie rod end.

5. Unscrew the the tie rod end from the tie rod, count the number of revolutions necessary to remove the tie

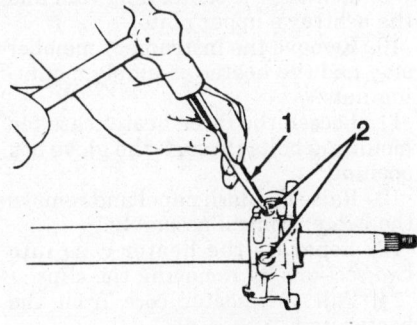

1. Center Punch (With Sharp Point)
2. Steering Lock Mounting Bolts

Removing the ignition switch/key lock assembly

rod end, for installation purposes.

6. At the steering gear, remove the boot clamps and pull the boot back over the tie rod.

7. Using a pair of pliers, bend the

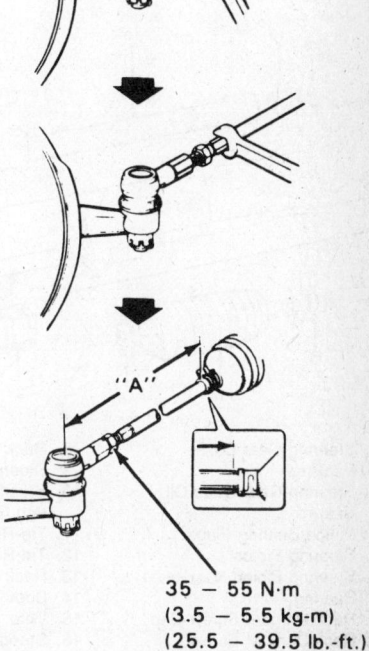

35 – 55 N·m
(3.5 – 5.5 kg-m)
(25.5 – 39.5 lb.-ft.)

Toe adjustment

lock washer back from the tie rod joint.

8. Using two wrenches, hold the steering gear and unscrew the tie rod end.

9. Remove the tie rod and slide the boot from the tie rod.

10. To install, reverse the removal procedures. Torque the tie rod-to-steering gear to 51–72 ft. lbs.; the tie rod end lock nut to 26–40 ft. lbs. and the tie rod end-to-steering knuckle to 22–40 ft. lbs. With the tie rod secured to the steering gear, bend the lock washer over the flat spot on the tie rod ball end.

TOE ADJUSTMENT

Toe is adjusted by changing the tie rod length. Loosen the right and left tie rod end lock nuts first and then turn left and right tie rods by the same amount to align toe to specification. In this adjustment right and left tie rods should be equal in length. Toe adjustment should be (0 ± 0.157 in.).

NOTE: Before turning the tie rods, apply grease between tie rods and rack boots so that the boots won't be twisted. After adjustment, tighten lock nuts to specified torque and make sure that the rack boots are not twisted.

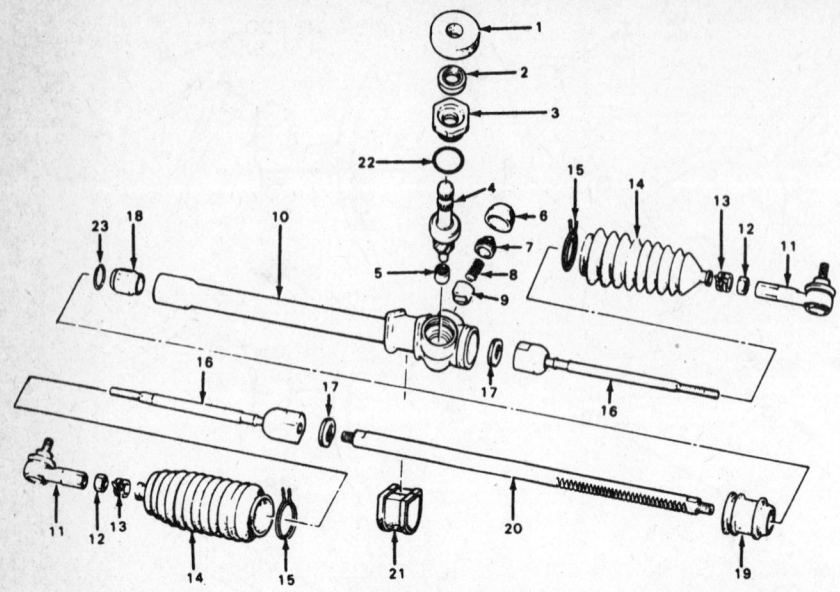

1. Steering Gear Case Packing
2. Steering Gear Gase Oil Seal
3. Pinion Bearing Plug
4. Steering Pinion
5. Steering Pinion Needle Bearing
6. Rack Damper Screw Cap
7. Rack Damper Screw
8. Rack Plunger Spring
9. Steering Rack Plunger
10. Steering Rack Housing and Gear Case
11. Tie-Rod End
12. Tie-Rod End Lock Nut
13. Rack Boot Clip
14. Boot
15. Wire
16. Steering Tie-Rod
17. Tie-Rod Lock Washer
18. Steering Rack Bushing
19. Steering Rack Side Mount
20. Steering Rack
21. Steering Pinion Side Mount
22. Gear Case Seal
23. Snap Ring

Exploded view of the steering gear assembly

CHASSIS ELECTRICAL

Heater Blower Motor

REMOVAL & INSTALLATION

1. Disconnect the negative battery cable.
2. Disconnect the defroster hose on the steering column side.
3. Disconnect the blower motor electrical connector.
4. Remove the 3 mounting screws and the blower motor.
5. To install, reverse the removal procedures.

Heater Core

REMOVAL & INSTALLATION

1. Disconnect the negative battery cable. Drain the cooling system.
2. Disconnect the two water hoses from the radiator at the heater unit.
3. Remove the glove box from the upper instrument panel.
4. Remove the defroster hoses from the heater case.
5. Disconnect the electrical connectors from the blower motor and the heater resistor.
6. Disconnect the three control cables from the heater case side levers.
7. Pull out the center vent louver.
8. Disconnect both side vent ducts from the center duct vent.
9. Remove the center duct vent and the ashtray's upper plate.
10. Remove the instrument member stay and the heater assembly mounting nuts.
11. Loosen the three heater case top mounting bolts through the glove box opening.
12. Raise the dash panel and remove the heater control assembly.
13. Separate the heater case into two sections by removing the clips.
14. Pull the heater core from the heater unit.
15. To install, reverse the removal procedures. Refill the cooling system. Start the engine, bring it to normal operating temperature and check for leaks.

A/C Evaporator Core

REMOVAL & INSTALLATION

1. Disconnect the negative battery cable.
2. Discharge the refrigerant from the high pressure side of the service valve.

—————— **CAUTION** ——————
When discharging the A/C system, avoid quick discharging.

3. Remove the glove box from the upper instrument panel.
4. Remove the defroster hose and the air damper.
5. Disconnect the A/C control cable from the evaporator side of the A/C evaporator/heater case.
6. Remove the connecting band from the A/C evaporator/heater case.
7. Disconnect and plug the compressor suction hose and the receiver/dryer outlet hose from the cooling unit fittings.
8. Remove the attaching bracket nuts, the bracket and the evaporator.
9. To install, reverse the removal procedures. Recharge the A/C system. Start the engine, bring it to normal operating temperatures and check for leaks.

Radio

REMOVAL & INSTALLATION

1. Disconnect the negative battery cable.
2. Remove the ash tray and the radio knobs.
3. Remove the ash tray assembly.
4. Remove the radio mounting nuts.
5. Remove the radio and disconnect the electrical connectors.
6. To install, reverse the removal procedures.

Windshield Wiper Switch

REMOVAL & INSTALLATION

1. Disconnect the negative battery cable.
2. Remove the steering column trim panel.
3. Lower the steering column.
4. Remove the cluster bezel and the bezel.
5. Disconnect the wiper switch connector.
6. Remove the wiper switch.
7. To install, reverse the removal procedures.

Windshield Wiper Motor

REMOVAL & INSTALLATION
Front

1. Disconnect the crank arm from the wiper motor.

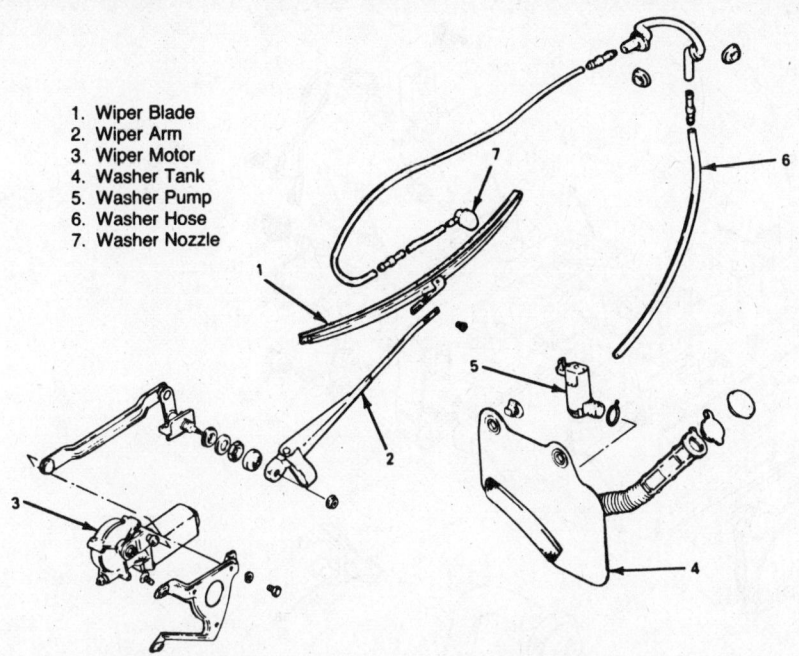

1. Wiper Blade
2. Wiper Arm
3. Wiper Motor
4. Washer Tank
5. Washer Pump
6. Washer Hose
7. Washer Nozzle

Exploded view of the rear wiper system

2. Disconnect the electrical connector from the wiper motor.
3. Remove the wiper motor from the vehicle.
4. To install, reverse the removal procedures.

Rear

1. Remove the electrical connector from the rear wiper motor.
2. Remove the rear motor mounting bracket.
3. Disconnect the motor from the wiper linkage.
4. Remove the motor from the vehicle.
5. To install, reverse the removal procedures.

Wiper Blade

REMOVAL & INSTALLATION

1. Remove the mounting screw from the wiper blade-to-wiper arm.
2. Remove the wiper blade from the wiper arm.

3. To install, reverse the removal procedures.

Instrument Panel

The instrument cluster consists of a speedometer, odometer, trip meter, coolant temperature gauge and a fuel gauge. Warning lamps are used for low oil pressure, alternator, parking brake ON, brake failure warning, high coolant temperature, fasten seat belt, highbeam indicator and oxygen sensor/ECM check.

Headlight Switch

REMOVAL & INSTALLATION

1. Disconnect the negative battery cable.
2. Remove the steering column trim panel.

3. Lower the steering column.
4. Remove the cluster bezel and the bezel.
5. Disconnect the headlight switch connector.
6. Remove the headlight switch.
7. To install, reverse the removal procedures.

Speedometer Cable

REMOVAL & INSTALLATION

1. Disconnect the negative battery cable.
2. Remove the cluster lens and the cluster mounting screws.
3. Disconnect the speedometer cable at the transaxle.
4. Remove the speedometer cable at the instrument cluster.
5. Remove the speedometer cable from the vehicle.
6. To install, reverse the removal procedures.

Instrument Cluster

REMOVAL & INSTALLATION

1. Disconnect the negative battery cable.
2. Remove the steering column trim panel.
3. Lower the steering column.
4. Remove the cluster lens and the cluster mounting screws.
5. Disconnect the speedometer cable at the transaxle and at the instrument cluster.
6. Disconnect and mark the electrical connectors at the instrument cluster.
7. Remove the instrument cluster from the vehicle.
8. To install, reverse the removal procedures.

Fusible Link and Fuses

The main fuse at the battery is a fusible link. The wiring circuits are protected by 14 fuses in the fuse block. The fuse block is located at the lower left of the instrument panel. The cover is built into the instrument panel.

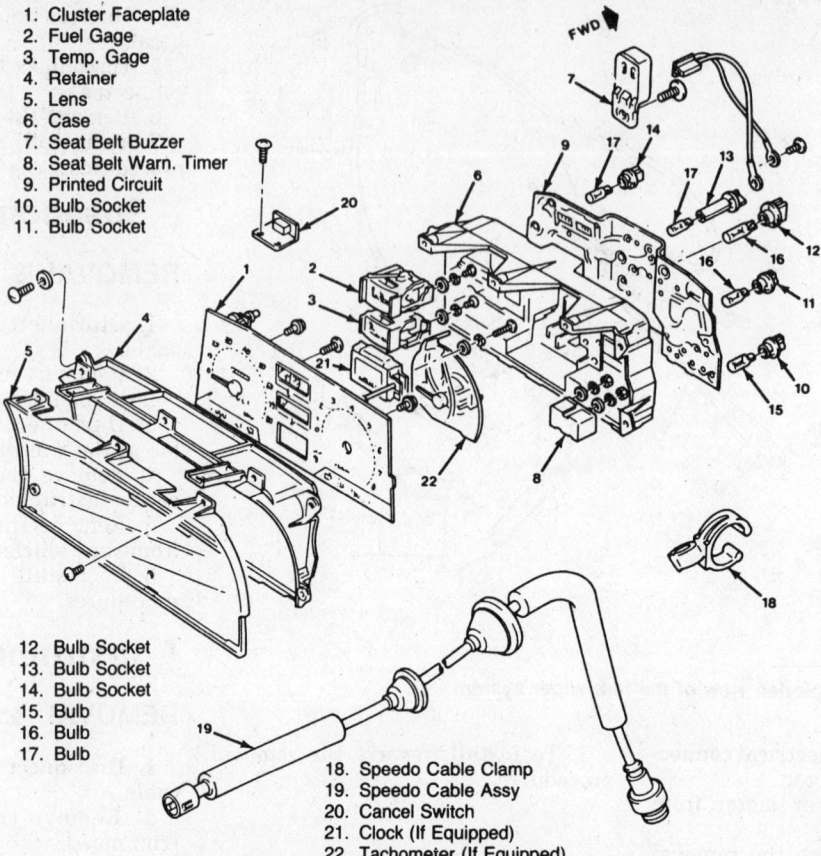

1. Cluster Faceplate
2. Fuel Gage
3. Temp. Gage
4. Retainer
5. Lens
6. Case
7. Seat Belt Buzzer
8. Seat Belt Warn. Timer
9. Printed Circuit
10. Bulb Socket
11. Bulb Socket

12. Bulb Socket
13. Bulb Socket
14. Bulb Socket
15. Bulb
16. Bulb
17. Bulb

18. Speedo Cable Clamp
19. Speedo Cable Assy
20. Cancel Switch
21. Clock (If Equipped)
22. Tachometer (If Equipped)

Exploded view of the instrument cluster

Chevrolet
Nova

YEAR IDENTIFICATION

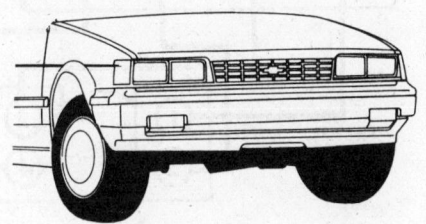

1985-87 Nova

ENGINE IDENTIFICATION

Code	Cu. In.	Liters	Cyl.	Carb.	Eng. Mfg.
4A-C	97	1.5	4	2 bbl	Toyota
4A-LC	97	1.5	4	2 bbl	Toyota

GENERAL ENGINE SPECIFICATIONS

Year	Eng. No. Cyl. Displ. Cu. In.	Engine Code	Fuel Delivery System	Engine Mfg.	Horsepower @ rpm	Torque @ rpm (ft lbs)	Bore × Stroke	Compression Ratio	Oil Pressure @ rpm
'85	4-97	4A-C	2 bbl	Toyota	70 @ 4800	85 @ 2800	3.19 × 3.03	9.0:1	34 @ 2000
'86–'87	4-97	4A-LC	2 bbl	Toyota	74 @ 4800	85 @ 2800	3.19 × 3.03	9.0:1	34 @ 2000

TUNE-UP SPECIFICATIONS

When analyzing compression test results, look for uniformity among cylinders rather than specific pressures

Year	Engine Code	Engine No. Cyl. Displacement (cu. in.)	Eng. Mfg.	hp	Spark Plugs Orig. Type	Spark Plugs Gap (in.)	Ignition Timing (deg) Man. Trans.	Ignition Timing (deg) Auto. Trans.	Fuel Pump Pressure (psi)	Idle Speed (rpm) Man. Trans.	Idle Speed (rpm) Auto. Trans.
'85	4A-C	4-97	Toyota	70	BPR5EY11	.043	5B	5B	2.5–3.5	650	800
'86	4A-LC	4-97	Toyota	74	BPR5E11	.043	5B	5B	2.5–3.5	650	750
'87	4A-LC				See Underhood Specifications Sticker						

NOTE: The underhood specifications sticker often reflects tune-up specification changes made in production. Sticker figures must be used if they disagree with those in this chart.

FIRING ORDER

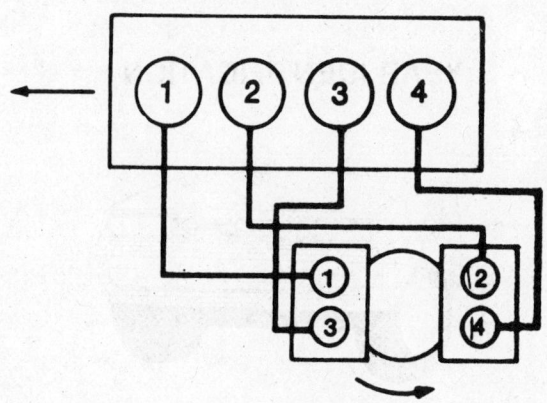

Firing order: 1–3–4–2

CAPACITIES

Year	Eng. Code	Engine Displacement (Cu. in.)	Eng. Mfg.	Crankcase Quarts w/filter	Crankcase Quarts wo/filter	Transaxle Pints Manual	Transaxle Pints Auto.	Gas Tank Gals.	Cooling System Quarts w/heater	Cooling System Quarts w/AC
'85	4A-C	97	Toyota	3.2	3.0	5.4	11.6	13.3	6.2	6.2
'86–'87	4A-LC	97	Toyota	3.5	3.2	5.4	11.6	13.2	6.3	6.3

VALVE SPECIFICATIONS

Year	Engine No. Cyl. Displacement (cu. in.)	Seat Angle (deg)	Face Angle (deg)	Outer Spring Test Pressure (lbs. @ in.)	Spring Installed Height (in.)	Stem-to-Guide Clearance (in.) Intake	Stem-to-Guide Clearance (in.) Exhaust	Stem Diameter (in.) Intake	Stem Diameter (in.) Exhaust
'85	4-97	45	45.5	52 @ 1.52	1.52	0.0010–0.0024	0.0012–0.0026	0.2744–0.2750	0.2742–0.2748
'86–'87	4-97	45	45.5	46.3 @ 1.52	1.52	0.0010–0.0024	0.0012–0.0026	0.2744–0.2750	0.2742–0.2748

CRANKSHAFT AND CONNECTING ROD SPECIFICATIONS

All measurements are given in inches

Year	Eng. Code	Engine No. Cyl. Displacement (cu. in.)	Crankshaft				Connecting Rod		
			Main Brg. Journal Dia.	Main Brg. Oil Clearance	Shaft End-Play	Thrust on No.	Journal Diameter	Oil Clearance	Side Clearance
'85	4A-C	4-97	1.8892–1.8898	0.0005–① 0.0019	0.0008–0.0073	3	1.5742–1.5748	0.0008–0.0020	0.0059–0.0098
'86–'87	4A-LC	4-97	1.8892–1.8898	0.0005–② 0.0015	0.0008–0.0073	3	1.5742–1.5748	0.0008–0.0020	0.12

① Maximum clearance—0.0031
② Maximum clearance is .0039 for 1986 engines, .0031 for 1985 engines

CAMSHAFT SPECIFICATIONS

All measurements are given in inches

Year	Eng. Code	Engine No. Cyl. Displacement (cu. in.)	Journal Diameter					Bearing Clearance	Lobe Lift		Camshaft End Play
			1	2	3	4	5		Intake	Exhaust	
'85	4A-C	4-97			1.1015–1.1022			0.0015–0.0029	1.5528–1.5531	1.5528–1.5531	0.0031–0.0071
'86–'87	4A-LC	4-97			1.1015–1.1022			0.0015–0.0029	1.5409①	1.5409①	0.0031–0.0071

① Minimum lobe height

PISTON AND RING SPECIFICATIONS

All measurements are given in inches.

Year	Eng. Code	Engine No. Cyl. Disp. (cu. in.)	Piston-to-Bore Clearance	Ring Gap			Ring Side Clearance		
				Top Compression	Bottom Compression	Oil Control	Top Compression	Bottom Compression	Oil Control
'85	4A-C	4-97	0.0039–0.0047	0.0079–0.0157	0.0059–0.0138	0.0039–0.0236	0.0016–0.0031	0.0012–0.0028	snug
'86–'87	4A-LC	4-97	0.0035–0.0043	0.0098–0.0185	0.0059–0.0165	0.0018–0.0402	0.0016–0.0031	0.0012–0.0028	snug

TORQUE SPECIFICATIONS

All readings in ft. lbs.

Year	Eng. Code	Engine No. Cyl. Displacement (cu. in.)	Cylinder Head Bolts	Rod Bearing Bolts	Main Bearing Bolts	Crankshaft Bolt	Flywheel-to-Crankshaft Bolts	Manifold	
								Intake	Exhaust
'85	4A-C	4-97	40–47	34–39	40–47	80–94	55–61	15–21	15–21
'86–'87	4A-LC	4-97	43	29	43	80–94	55–61	15–21	15–21

WHEEL ALIGNMENT SPECIFICATIONS

Year		Caster Range (deg)	Caster Pref. Setting (deg)	Camber Range (deg)	Camber Pref. Setting (deg)	Toe (in.)	Steering Axis (deg) Inclination
'85	Front	¼–1¾P	1P	1¼N–¼P	½N	0 ± 0.16	11¾–13¼
	Rear	—	—	1¼N–¼P	½N	0.150 ± 0.16	—
'86–'87	Front	⅙P–1⅔P	⅚P	¾N–¼P	¼P	0 ± 0.078	—
	Rear	—	—	1¼N–¼P	¾N	.075–.233	—

ENGINE ELECTRICAL

Alternator

For all alternator testing and diagnosis, please refer to the "Charging and Starting" Unit Repair section.

REMOVAL & INSTALLATION

1. Disconnect the negative battery cable.

2. Tag and then disconnect each alternator wiring connector. Keep attaching nuts and washer for use on the new unit.

3. Loosen the alternator adjusting lockbolt (located in the slotted bar at the bottom of the unit) and the hinge bolt and nut, located at the top of the unit. Turn the adjusting bolt (located in the end of the slotted bar) to shift the alternator toward the block; remove the belt.

4. Remove the adjusting bolt, then remove the hinge bolt and nut and remove the alternator. Keep plain and lockwashers in their proper locations for reassembly.

5. Install in reverse order. The belt has teeth which run along its length. Make sure all these teeth align with indentations in each of the pulleys; all teeth must ride inside the pulley surface.

Starter

For all starter overhaul procedures, please refer to the "Charging and Starting" Unit Repair section.

REMOVAL & INSTALLATION

1. Disconnect the negative battery cable.

2. If the car has an automatic transmission, remove the fluid filler tube.

3. Raise the vehicle and support it securely.

4. Remove the nut and washer and disconnect the battery cable from solenoid, retaining the parts in a safe place. Disconnect the wire at terminal 50 on the solenoid.

5. Support the starter; then, remove the two mounting bolts from the forward side of the flywheel or torque converter housing and remove the unit.

6. Install in reverse order, torquing the starter mounting bolts to 29 ft. lbs.

Distributor

REMOVAL & INSTALLATION

On this vehicle, the distributor is actually an Integrated Ignition Assembly which includes not only the usual distributor components but also the ignition coil and electronic igniter assembly.

1. Disconnect the negative battery cable. Remove either the valve cover or No. 1 spark plug. Turn the engine over in its normal direction of rotation until you feel air being forcibly expelled from the spark plug hole (if you have removed No. 1 plug); or the rocker arms for No. 1 cylinder indicate both valves are closed as the timing marks approach 0° Top Dead Center (if you've removed the valve cover). Turn the engine so the crankshaft is at the TDC position according to the timing marks, and then replace the valve cover or spark plug.

2. Mark the location of the distributor in relation to the block at the slotted flange where the attaching bolt holds the distributor in position. Now, remove the four attaching screws and remove the distributor cap, with wires attached.

3. Disconnect the electrical connector for the distributor. Note their routing and then disconnect the hoses going to the vacuum advance/retard unit.

4. Remove the hold-down bolt located in the slotted flange, then carefully raise the distributor and remove it.

5. To install the distributor, reverse the removal procedure. Note that you should start by aligning the protrusion at the bottom of the distributor housing with the pin on the side of the distributor drive gear. Oil the drive gear teeth with clean engine oil, align the marks made in Step 2, and slide the distributor down into the block. You may find that the drive gear teeth butt up against those on the camshaft. If this happens, the distributor will resist moving into the block. Press it gently downward while you shift the rotor back and forth slightly to make the gears mesh. Once they have done so, you should be able to push the distributor in until the slotted flange seats against the block. Reverse the remaining procedures to complete the installation. Set the ignition timing.

Installation if Engine Timing has been Disturbed

Perform the portions of Step 1 above to set the engine in TDC No. 1 cylinder firing position. Align the protrusion at the bottom of the distributor housing with the pin on the side of the distributor drive gear. Then, proceed by reversing the remaining removal procedures as necessary. Set the ignition timing.

IGNITION TIMING ADJUSTMENT

1. Install a timing light to the No. 1 (crankshaft pulley end) spark plug wire and to the battery according to the manufacturer's instructions. On inductive timing lights, the induction clip can simply be installed over the plug wire; on most lights, the wire will have to be disconnected by pulling on the boot at the end of the wire and connecting the timing light pickup between the plug and the end of the wire. Connect a tachometer according to the manufacturer's instructions.

2. Note the routing of the two vacu-

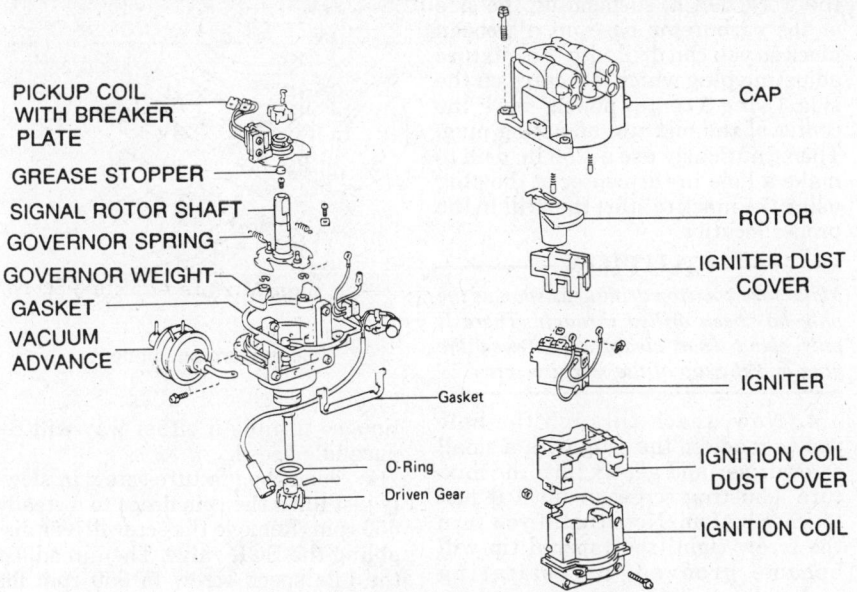

PICKUP COIL
WITH BREAKER
PLATE

GREASE STOPPER

SIGNAL ROTOR SHAFT

GOVERNOR SPRING

GOVERNOR WEIGHT

GASKET

VACUUM
ADVANCE

Gasket

O-Ring

Driven Gear

CAP

ROTOR

IGNITER DUST
COVER

IGNITER

IGNITION COIL
DUST COVER

IGNITION COIL

Exploded view of distributor assembly

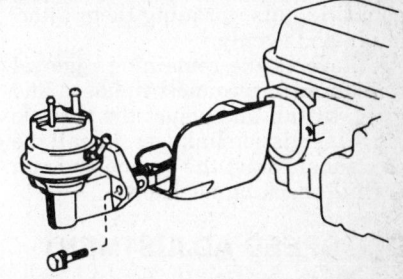

Mechanical fuel pump mounting

um lines going to the distributor vacuum advance/retard unit or mark their locations. Disconnect the two vacuum lines and plug both. Then start the engine and run it until it reaches normal operating temperature. Check the engine rpm; it should be 950 or less with either type of transmission in neutral. If it is too high, adjust it as described later in this section.

3. Aim the timing light at the scale on the front cover near the front pulley. Timing is 5° BTDC. If the timing is not correct, loosen the distributor hold-down bolt until it is just finger tight and turn the distributor slightly to correct it. Once the reading is correct, tighten the hold-down bolt and recheck the timing. Correct it if necessary. Stop the engine, remove the timing light, and unplug and reconnect the hoses.

TACHOMETER HOOKUP

The ignition system is equipped with a service connector plug that shares a common wiring harness with the primary wiring connector leading to the distributor assembly. One lead of the tachometer is connected to that plug. Connect the tachometer leads according to the instruction manual for the instrument. Since not all types of tachometers are compatible with this ignition system, you should consult the instruction book that came with your instrument to make sure it will work with electronic ignition systems, and will not damage the system. Never ground the TACH terminal of

the distrubutor assembly or damage to the ignition system will result.

FUEL SYSTEM

Mechanical Fuel Pump

REMOVAL & INSTALLATION

1. Note the routing of fuel lines and mark if necessary. Then, use a pair of pliers to slide the retaining clips for the fuel lines well off the fuel pump connections. Disconnect the fuel lines, using a twisting motion to break them loose if they are hard to disconnect.

2. Remove the two bolts from the cylinder head, and remove the pump, gasket, and heat shield, noting the position of the shield. Reseal the surface of the block to prevent loss of oil.

3. Scrape the gasket surfaces of the block (and pump, if it's being reused) with a scraper that will not scratch the aluminum surface of the head.

4. Install in reverse order. When reconnecting the fuel lines, make sure the clips are installed to the inside of the bulged sections of the fuel pump connections, and are not right at the ends of the fuel lines. Start the engine and check for leaks.

Fuel Filter

REMOVAL & INSTALLATION

1. Note the routing of inlet and out-

let lines and the direction of flow as marked on the filter. The arrow points toward the carburetor. Then, with a pair of pliers, shift the clips on the inlet and outlet hoses back and well away from the connections on the filter.

2. Disconnect the fuel lines, using a twisting motion to break them loose. Immediately plug the openings to prevent the spillage of fuel and entry of dirt. Pull the filter out of its retaining clip.

3. Install the filter in reverse order. When reconnecting the fuel lines, make sure the clips are installed to the inside of the bulged sections of the fuel filter connections, and are not right at the ends of the fuel lines. Start the engine and check for leaks.

Carburetor

REMOVAL & INSTALLATION

1. Disconnect the emission control hoses from the air cleaner, labeling them or noting their routing as necessary. Remove the wingnut and mounting bolts and remove the air cleaner.

2. Disconnect the accelerator cable at the carburetor. If the car has an automatic transmission, disconnect the transmission cable at the throttle linkage.

3. Disconnect the wiring connector for the carburetor solenoid valves.

4. Label and then disconnect the carburetor emission control hoses. Disconnect the fuel line, draining any fuel into a metal or ceramic container (not a styrofoam cup). Disconnect the evaporative emissions canister hose.

5. Remove the carburetor mounting nuts. Then, remove the cold mixture heater wire clamp and the EGR vacuum control bracket. Remove the carburetor.

6. Remove the insulating gasket from the manifold. Use a clean rag to seal off the intake manifold opening.

7. Clean the sealing surfaces of the carburetor and manifold and install a new insulating gasket. Put the carbu-

retor in position and install the mounting nuts, torquing them alternately and evenly.

8. Reverse the remaining removal procedures to reconnect linkages and hoses. Install and adjust the throttle and transmission linkages. Install the air cleaner. Start the engine and run it while checking for leaks.

IDLE SPEED ADJUSTMENT

1. The idle speed must be adjusted with the engine at normal operating temperature and all emission controls and hoses connected. The choke must be wide open and both manual and automatic transmissions should be in Neutral. Apply the handbrake. Check and, if necessary, set the ignition timing as described above.

2. Connect a tachometer as described above. Adjust the idle speed screw to 650 rpm on manual trans-

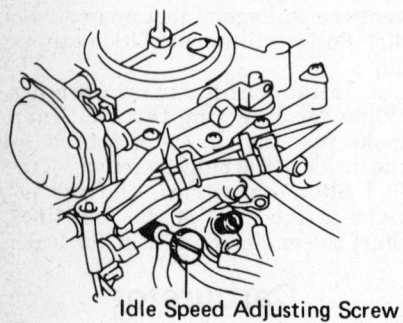

Idle Speed Adjusting Screw

Idle speed adjustment

mission vehicles and 800 rpm on automatic transmission vehicles up to 1985. On 1986 and later models with automatic transmissions, it is 750 rpm.

IDLE MIXTURE ADJUSTMENT

NOTE: Idle mixture does not require adjustment as a matter of routine maintenance. Only if the engine will not idle properly and all vacuum leaks, tune-up and mechanical problems have been eliminated as possible causes of the rough idle or stalling should the mixture adjustment be performed. Performing this procedure requires drills of .256 and .295 in. diameter. You should also have a source of compressed air to remove metal drillings.

1. Remove the carburetor as described above.

2. Plug all the carburetor vacuum ports so drillings will not be able to enter them.

3. Lay the carburetor on its side on

the work bench, suspending the side of the carburetor base on a wooden block so you can drill into the mixture adjusting plug which is located on the side. Using a center punch, mark the center of the mixture adjusting plug. Then, cautiously use a .256 in. drill to make a hole in the center of the plug using the mark to start the drill in the proper location.

— CAUTION —
Make sure you stop drilling as soon as the plug has been drilled through. There is only about 1mm clearance between the plug and the top of the mixture screw.

4. Now, reach through the hole you've made in the plug with a small screwdriver and gently turn the mixture adjusting screw in until it just touches bottom. Note that if you turn the screw tight, the tapered tip will become grooved, necessitating replacement.

5. Drill the plug with a .295 in. drill. The drill will, in effect, grasp the plug and permit you to remove it by pulling upward.

6. Now, use compressed air to remove any metal filings and then remove the mixture screw by screwing it out all the way. Inspect the tip for grooving and the top for damage to the screwdriver groove and replace the screw if necessary.

7. Install the inspected or replaced mixture screw and turn it in slowly and gently until it just touches bottom. Then back it out, counting the turns accurately. Turn it out $3\frac{1}{4}$ turns for U.S. vehicles and $2\frac{1}{2}$ turns for Canadian vehicles.

8. Reinstall the carburetor and all associated parts. Start the engine and operate it until it reaches operating temperature. All accessories must be off, wheels must be in straight ahead position with power steering, the transmission must be in neutral, and the carburetor sight glass should indicate proper float level. Also, the choke should be wide open and ignition timing should be correct.

9. Install a tachometer and slowly adjust the idle mixture screw back and forth to create the highest possible idle speed.

10. Insert a small screwdriver between the EGR valve and the EGR vacuum modulator bracket to disable the EGR valve.

11. Adjust the idle speed screw to obtain 700 rpm. Then repeat the mixture screw adjustment. If idle speed can be increased by optimizing the mixture adjustment, repeat the speed adjustment. Keep going back and forth in this way until the engine idles at 700 rpm and the mixture adjusting screw is at the optimum posi-

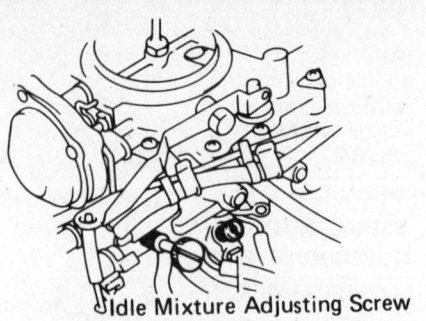

Idle Mixture Adjusting Screw

Idle mixture adjustment

tion, so turning it either way will reduce idle speed.

12. Turn the mixture screw in slowly just until the rpm drops to a steady 650 rpm. Remove the screwdriver disabling the EGR valve. Then, readjust the idle speed screw to 650 rpm for manual transmission cars and 800 rpm for automatics. Turn the engine OFF.

13. Remove the air cleaner and EGR mounting bracket. Using a hammer and drift, tap a new mixture adjusting plug in place with the tapered end inward. Reinstall the air cleaner and EGR vacuum modulator bracket.

COOLING SYSTEM

Radiator

REMOVAL & INSTALLATION

1. Open the drain cock in the lower radiator tank and drain the coolant into a clean container.

2. Disconnect the engine and A/C fan motor electrical connectors. On cars with air conditioning, remove the four bolts going into the top tank and two going into the bottom tank and remove the fan shroud.

3. On cars with automatic transmissions, place a clean pan underneath where you are working and then disconnect the two oil cooler hoses at the radiator and plug the open ends.

4. Disconnect the hose to the coolant reservoir and fasten it in a position that is high enough to keep the coolant in the reservoir from draining.

5. Loosen the clamps, pull them well back from the ends of the hoses, and disconnect the top and bottom hoses at the radiator.

6. Unbolt the two radiator supports at the top, remove the two supports, and remove the radiator.

7. Install the radiator in reverse order. Make sure the radiator fits properly at the bottom so the rubber cushions under the two supports are only slightly compressed.

8. Refill both the automatic transmission and cooling system with approved fluids. Bring the engine to operating temperature, recheck fluid levels, and operate the engine while checking for leaks.

Water Pump

REMOVAL & INSTALLATION

1. Remove the mounting bolts and remove the upper timing cover and gasket. Drain the coolant into a clean container through both the radiator and engine block drain cocks. Once one drain cock is opened and pressure relieved, remove the radiator cap to vent the system and aid draining.

2. Just begin loosening each of the four water pump pulley mounting bolts,. Then loosen the alternator adjusting and mounting bolts, move the alternator toward the engine to remove belt tension, and remove the alternator/water pump drive belt.

3. Remove the water pump pulley mounting bolts and remove the pulley.

4. Remove the water pump inlet mounting bolt from the side of the block, then remove the two nuts attaching the inlet pipe to the rear of the water pump and remove the pipe.

5. Remove the mounting bracket bolt and then remove the dipstick tube. Plug the hole in the block with a clean rag. Remove the right hand under cover. If the car has power steering, remove the power steering adjusting bracket.

6. Remove the three bolts and remove the water pump. Keep engine coolant off the timing belt! Remove the O-ring from the block.

7. To install the pump, first replace the O-ring in the block with a new one. Then, reverse the remaining removal procedures, torquing the three water pump mounting bolts to 11 ft. lbs. When installing the oil dipstick tube, install a new O-ring there also, coating it with oil prior to installation. Refill the cooling system, if necessary adding clean anti-freeze and water in a 50–50 mixture. Operate the engine until it has reached operating temperature and air has bled out, then add coolant until the system is

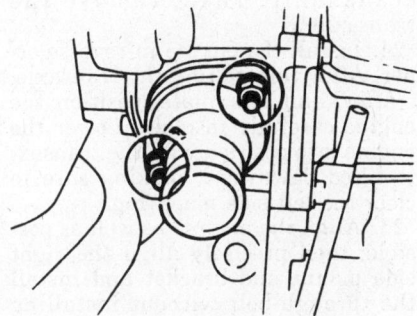

Drain Cock

Location of drains on radiator and engine block

Thermostat housing fasteners

full and replace the cap. Check for leaks.

Thermostat

REMOVAL & INSTALLATION

1. Drain coolant from the radiator drain cock until the level is well below the top tank. After the drain cock is opened and pressure relieved, remove the radiator cap to aid draining.

2. Loosen clamps and disconnect the small and large hoses at the thermostat housing. Then, remove the two bolts and pull the water inlet off the housing base. Remove the thermostat and O-ring.

3. Install a new thermostat with the bellows inward. Make sure the thermostat fits squarely in the indented portion of the housing. Install a new O-ring into the water inlet.

4. Install the water inlet and two bolts, tightening the bolts alternately in small increments.

5. Reconnect hoses and position and tighten the clamps. Refill the cooling system, if necessary adding clean anti-freeze and water in a 50–50 mixture. Operate the engine until it has reached operating temperature and air has bled out, then add coolant until the system is full and replace the cap. Check for leaks.

EMISSION CONTROLS

For all descriptions and service procedures, please refer to the "Emission Control" Unit Repair section. Due to the complex nature of modern electronic engine control systems, comprehensive diagnosis and testing procedures fall outside the confines of this repair manual. For complete information on diagnosis, testing and repair procedures concerning all modern engine and emission control systems, please refer to *Chilton's Guide To Electronic Engine Controls*.

ENGINE

REMOVAL & INSTALLATION

1. Remove the battery cover and battery. Drain coolant into a clean container from both radiator and block drain cocks.

2. Drain the engine crankcase and transaxle fluid.

3. Remove the hood (use an assistant).

4. Remove the air cleaner and associated ducting.

5. Drain and remove the coolant reservoir tank.

6. Remove the radiator and shroud as described earlier in this section.

7. Disconnect the throttle and automatic transaxle linkages at the carburetor.

8. Disconnect the following wires:

    a. Vacuum Control Valve connector

    b. Oxygen sensor connector

    c. Back-up light switch connector on manual transaxle-equipped cars

    d. Neutral safety switch connector on automatic transaxle-equipped cars

    e. Water temperature switch connector

    f. Distributor connector

    g. Starter wiring

    h. Ground cables at the engine and transmission

    i. Oil pressure switch connector

    j. A/C compressor clutch connector (if A/C equipped)

    k. Alternator wiring

    l. Water temperature sending unit connector

    m. Electric (carburetor) bowl vent control valve

n. CMH relay connector (located in the air cleaner)

o. Fuel cut solenoid connector

9. Label and then disconnect all vacuum hoses running between the engine and firewall or fender well mounted accessories.

10. Disconnect both fuel hoses at the pump.

11. Disconnect the heater hose at the water pump inlet housing and the cooling water hose at the rear plate of the cylinder head.

12. Loosen the adjusting and mounting bolts for the power steering pump (if equipped) and then remove the belt. Then, remove these bolts, pull the air pump out of the way and suspend it.

13. Loosen the A/C compressor belt tension adjusting bolt (on cars so-equipped) and remove the belt. Then, remove the four mounting bolts and remove the compressor, suspending it out of the way with hoses still connected.

14. Disconnect the speedometer cable at the transaxle.

15. Working on the clutch slave cylinder:

a. Remove the clip from the hydraulic line bracket

b. Disconnect the mounting bracket at the transaxle

c. Remove the two mounting bolts for the slave cylinder and then move it out of the way, suspending it with the hydraulic line still connected.

16. Disconnect the shift cable by removing the clip and washers and then disconnecting the cable at the outer shift lever or outer selector lever.

17. Raise the vehicle and support it securely.

18. Disconnect the exhaust pipe at the manifold. Then, disconnect the front and rear mounts at the crossmember by first removing the two bolt covers and then removing the two bolts from each mounting. Then, remove the center crossmember (supporting the engine).

19. To disconnect the driveshafts from the transaxle, remove the nuts and bolts. Have someone depress the brake pedal while you loosen the nuts.

20. Lower the vehicle to the ground. Attach a chain hoist to the lift bracket on the engine keeping the wiring harness in front of the chain. Put enough tension on the chain to support the engine.

21. Remove the right hand mount through bolt; then remove the left hand mount through bolt; remove the left hand mount from the transaxle by removing the two bolts. Remove the mounting bracket from the transaxle.

22. Lift the engine/transaxle assembly out of the engine compartment, proceeding slowly in order to clear the right side mount, the power steering housing and neutral safety switch. Make sure wiring, hoses, and cables are clear of the engine.

23. Mount the engine/transaxle assembly on a stand. Then, remove the starter, and engine rear plate. On automatic transmission-equipped cars, remove the six torque converter mounting bolts by turning the crankshaft to gain access to each bolt from underneath. Hold the crankshaft pulley bolt with a wrench to remove these bolts without turning the crankshaft. Then, remove the transaxle.

24. Install the engine in reverse order. After reinstalling the transaxle, starter, and rear plate, position the engine carefully in place. Lower the engine into position with the transaxle tilted downward, making sure to clear the left side mounting.

25. Align the mounts as well as possible; then precisely align the right side mount and bracket and install the through-bolt without installing the nut. Then, install the left hand mounting bracket onto the transaxle. Install the left hand mount onto the bracket, align the left hand bracket with the body bracket and install the mounting bolt and nut hand tight. Disconnect the lifting chain from the engine.

26. Support the vehicle up in the air in a secure manner. Reverse the remaining removal procedures to install. Observe the following torque figures:

a. Driveshaft bolts—27 ft. lbs.

b. Engine mount center crossmember—29 ft. lbs.

c. Exhaust pipe-to-manifold—46 ft. lbs.

d. Power steering pump mounting bolts—29 ft. lbs.

e. Power steering pump adjusting bolt—32 ft. lbs.

Make sure the power steering pump belt runs in all three pulley grooves. Replenish all fluids with approved type at the proper level. Then start the engine and correct any leaks.

Combination Manifold

REMOVAL & INSTALLATION

1. Remove the air cleaner assembly and hoses.

2. Disconnect the accelerator and choke linkages from the carburetor. If necessary, label and then disconnect all vacuum lines going to the carburetor. Disconnect the fuel lines, draining any fuel into a metal container. Disconnect the fuel line from the fuel

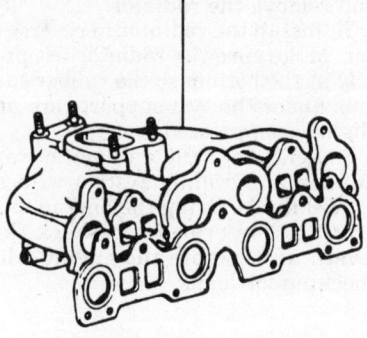

Intake and Exhaust Manifold

Combination manifold and gasket

pump at the pump, as well. Disconnect the electric lines.

3. Disconnect or remove any emission control hardware that is in the way; then, remove the mounting bolts, and remove the carburetor from the manifold.

4. Remove the EFE gasket. Remove the vacuum line and the dashpot bracket. Remove the carburetor heat shield.

5. Raise the vehicle and support it securely. Then, disconnect the exhaust pipe at the manifold, the pipe bracket at the engine, and the hose at the converter pipe. Lower the vehicle.

7. Disconnect the brake vacuum hose. Remove the accelerator and throttle cable brackets.

8. Working from the center outward, remove the manifold retaining nuts in several stages so tension is gradually released.

9. Remove the intake/exhaust manifold from the head as a unit.

10. To install, make sure both gasket surfaces are clean and position a new gasket on the head. Then, position the manifold and install retaining nuts just hand tight. Tighten them alternately is several stages to the specification shown in the Torque Specifications Chart. Reverse the remaining removal procedures. When installing the carburetor, clean both gasket surfaces, use a new gasket and tighten bolts alternately and evenly. When the installation is complete, start the engine and check for leaks.

Rocker Arm and Shaft Assembly

REMOVAL & INSTALLATION

1. Remove the air cleaner and valve cover.

2. Remove the five rocker shaft assembly retaining bolts in several stages--note that they MUST be loos-

ened in the correct sequence. That is: front bolt first; rear bolt second; forward-center bolt third; rearward-center bolt fourth; and center bolt fifth.

3. Remove the rocker arm and shaft assembly.

4. Inspect for wear by attempting to rock the rocker levers on the shaft. If negligible motion is felt, wear is acceptable. If there is noticeable wear, note order of assembly and the fact that there are two types of rockers, and then remove bolts and slide rockers, springs and pedestals off the shaft.

5. Measure the inside diameter of each rocker lever with a dial indicator; measure the shaft diameter at rocker wear areas with a micrometer. Subtract the shaft diameter from the rocker inside diameter; the difference must not exceed 0.0024 in. Replace rockers and, if necessary, the shaft to correct clearance problems.

6. Also inspect the rocker camshaft follower pads. If the surfaces are worn irregularly or are rough replace the parts or grind them smooth.

7. Assemble the pedestals, rockers, springs and bolts in exact reverse order. Oil all wear surfaces thoroughly with clean engine oil. Note that rocker shaft oil holes MUST face downward. Install the assembly on the head and start the bolts, tightening them finger tight. Torque in three stages going in this sequence each pass: center bolt first, center/rearward bolt second, center/forward bolt third, rear bolt fourth, and front bolt fifth. Final torque figure is 18 ft. lbs. Adjust the valves as described below. Install the valve cover using new gaskets and seals and the air cleaner.

VALVE CLEARANCE ADJUSTMENT

1. The valve clearances must be set with the engine hot. If clearances are being set because parts have been disassembled and clearances changed, you should set the valves cold and then reset them with the engine hot. The best procedure is to drive the car a few miles (well past the point where the temperature gauge reaches normal readings) in order to have the engine as warm as possible throughout the procedure. Then, remove the air cleaner and valve cover.

2. Turn the engine over by using a socket wrench on the pulley retaining bolt until the timing marks on the front pulley indicate TDC and both No. 1 (front) cylinder rocker arms are loose. Note that you must adjust the intake valves to one clearance and exhaust valves to another. Half the valves are adjusted with the crank-

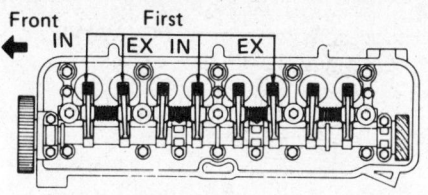

Valve adjustment sequence—step one

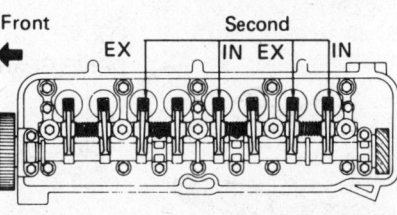

Valve adjustment sequence—step two

shaft in one position; then the other half are adjusted with the crankshaft tuned 360 degrees. It's best to adjust all the intake valves for the first adjustment sequence, then switch feeler gauges and adjust all the exhaust valves. Proceed in the same way for the next sequence. Intake valves are adjusted to 0.008 in. and exhaust to 0.012 in.

4. To adjust each valve, first slide the feeler gauge straight between the rocker and valve tip. The surfaces will just touch, giving a VERY slight pull on the gauge. If the gauge is tight or won't fit, or if the gauge slides with no pull at all, loosen the adjusting nut. Tighten the adjusting stud with a screwdriver to reduce the clearance or loosen it to increase the clearance. Don't tighten the stud past the point where you can feel it start to touch the gauge. Once clearance is correct (gauge has a very slight pull), hold the stud position while you tighten the nut. Then, recheck the clearance and readjust if necessary. On the first sequence, you can adjust the No. 1 and 2 cylinder intakes and No. 1 and 3 cylinder exhaust valves. You can determine whether valves are intakes or exhausts by aligning them with manifold intake or exhaust tubes. The sequence from front to rear is: intake, exhaust, intake, intake, exhaust; exhaust, intake.

5. Once the first adjustment sequence has been completed, proceed with the second. Turn the engine crankshaft just 360 degrees until the TDC marks again align. Then, proceed with the adjustment for the intakes on No. 3 and 4 cylinders and the exhausts on No. 2 and 4.

6. Install the valve cover using new gaskets and seals. Install the air cleaner.

Cylinder Head

REMOVAL & INSTALLATION

1. Disconnect the negative battery connector.

2. Drain the coolant into a clean container, opening both the radiator and cylinder block drain cocks.

3. Remove the air cleaner and all connecting hoses. First label and then disconnect all vacuum hoses.

4. Raise the vehicle and support it securely. Drain the engine oil. Disconnect the exhaust pipe at the manifold. Remove the exhaust pipe bracket at the engine. Remove the hose at the converter pipe.

5. If the car has power steering, loosen the power steering pump pivot bolt. Then, lower the vehicle to the ground.

5. Disconnect the accelerator and throttle cables at the carburetor and at the cable bracket.

6. Disconnect wiring at the cowl, oxygen sensor, and distributor.

7. Disconnect the fuel hoses at the fuel pump.

8. Disconnect the upper radiator hose at the water outlet. Then, remove the water outlet from the head. Remove the heater hose.

9. Remove the power steering pump adjusting bracket, if the car has power steering.

10. Turn the engine over with a socket wrench on the crankshaft so it is at Top Dead Center and both No. 1 cylinder rocker arms are loose. Label the spark plug wires as to their location in the distributor cap. Disconnect the vacuum hoses, electrical connections and spark plug wires at the distributor and remove the distributor.

11. Remove the PCV valve. Remove the wiring harness that passes over the valve cover.

12. Remove the upper timing belt cover bolts. Remove the cylinder head cover and gasket. Remove the alternator belt.

13. Remove the water pump pulley bolts and remove the pulley. Then, remove the upper timing cover and its gasket.

14. Remove both remaining timing covers. Then, matchmark the timing belt and both sprockets for reassembly in the same position; mark an arrow on the belt for rotation in the same direction, also.

15. Loosen the idler pulley bolt. Move it so as to release the timing belt tension, and then snug the idler pulley bolt back up. Then, remove the timing belt. Avoid twisting or bending it.

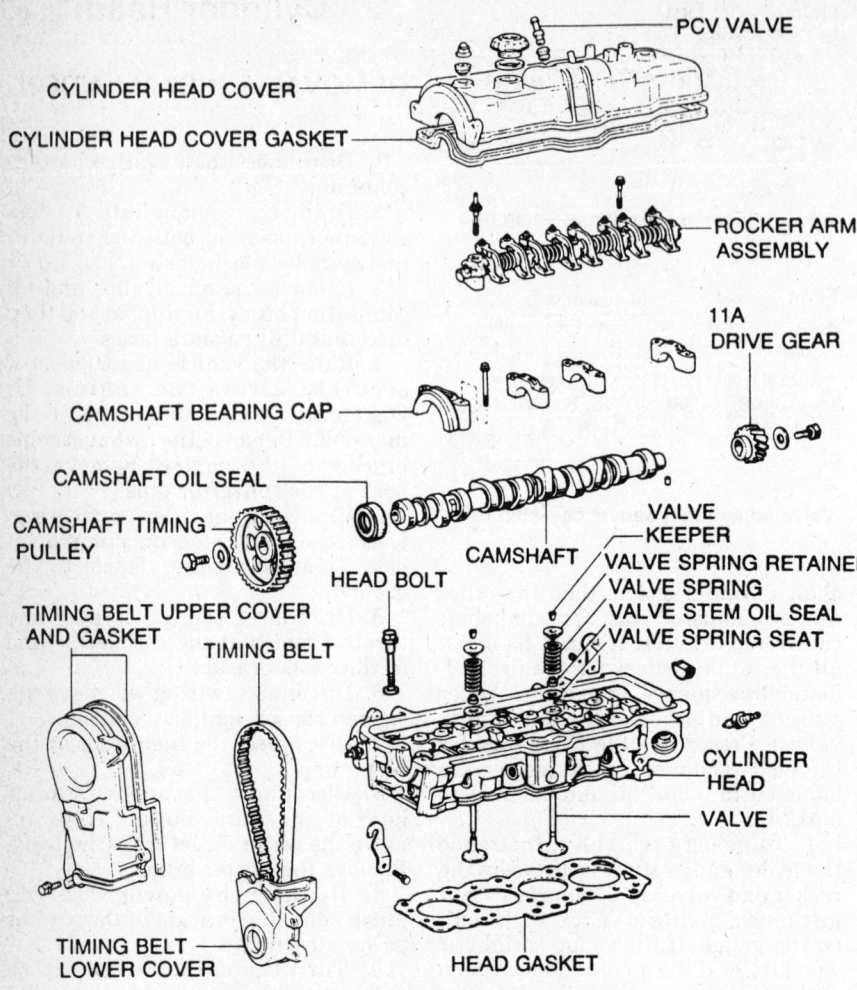

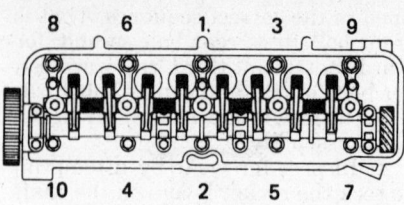

CYLINDER HEAD COVER

CYLINDER HEAD COVER GASKET

PCV VALVE

ROCKER ARM ASSEMBLY

11A DRIVE GEAR

CAMSHAFT BEARING CAP

CAMSHAFT OIL SEAL

CAMSHAFT TIMING PULLEY

HEAD BOLT

CAMSHAFT

VALVE KEEPER

VALVE SPRING RETAINER

VALVE SPRING

VALVE STEM OIL SEAL

VALVE SPRING SEAT

TIMING BELT UPPER COVER AND GASKET

TIMING BELT

CYLINDER HEAD

VALVE

TIMING BELT LOWER COVER

HEAD GASKET

Exploded view of cylinder head assembly

Cylinder head bolt torque sequence

16. Loosen the head bolts in reverse of the torque sequence shown, in three stages, and then remove them. Then, lift the head directly off the block. If it is necessary to pry the head off the block, use a bar between the head and the projection provided on top of the block.

NOTE: Do not pry except at the projection provided. Be careful not to damage the block deck or cylinder head sealing surface.

17. To install the cylinder head, reverse the removal procedure, referring to the "Timing Belt Removal & Installation" procedure. Install and torque the head bolts in the order shown. Observe the following torque figures:
   a. Timing gear idler bolt — 22–32 ft. lbs.
   b. Head bolts — 43 ft. lbs.
Adjust the valves with the engine cold, then replace the valve cover. Run the engine until it is hot and then readjust the valves.

Timing Belt Covers

REMOVAL & INSTALLATION

1. Disconnect the negative battery cable.
2. Loosen the water pump pulley bolts and then remove the alternator/water pump drive belt. If the car has power steering, remove the steering pump drive belt.
3. Remove the bolts and then remove the water pump pulley. Drain the cooling system.
4. Disconnect the upper radiator hose at the water pump outlet. Then, label and disconnect all vacuum hoses that are in the way.
5. If the car has air conditioning, loosen the idler pulley mounting bolt. Loosen the adjusting nut and then remove the A/C drive belt. Then, remove the idler pulley and its adjusting bolt. Remove the alternator.
6. Remove the crankshaft pulley mounting bolt and remove the pulley with a puller.

7. Remove the bolts from the upper timing cover that are accessible from above. Then, raise and support the car securely and remove the lower bolts from this cover and remove the upper cover.
8. Remove the mounting bolts from the lower timing belt cover and then remove that cover and gasket.
9. Lower the vehicle back to its normal position. Finally, remove the center cover bolts, the cover, and gasket.
10. Install in reverse order, torquing the crankshaft pulley bolt to 80–94 ft. lbs. and the center engine mount bolts to 29 ft. lbs.

Front Crankshaft Oil Seal

REMOVAL & INSTALLATION

1. Remove the No. 1 (top) timing belt cover as described earlier. Remove the air conditioning belt if the car has A/C.
2. Turn the crankshaft clockwise until No. 1 cylinder is at Top Center firing position (air is expelled from the spark plug hole as the piston approaches Top Center).
3. Remove the right side under cover. Remove the flywheel cover.
4. Remove the crankshaft pulley with a puller.
5. Remove the No. 2 (lower) timing cover and its gasket. Mark the locations of both timing pulleys and the rotating direction of the belt.
6. Loosen the idler pulley bolt, move the idler so as to release belt tension, and then retighten the bolt to retain the tensioner in the released position.
7. Remove the timing belt guide and then remove the belt from the crankshaft timing pulley. Then, slide the pulley off the crankshaft.
8. Pry out the oil seal from the front using a flat bladed screwdriver.
9. Tap a new seal into the recess using an installer such as GM special tool No. J-35403 or equivalent and a hammer. Make sure the seal sits squarely in the bore; it must not be cocked. Coat the sealing surfaces with Multipurpose grease.
10. Reassemble all parts removed in

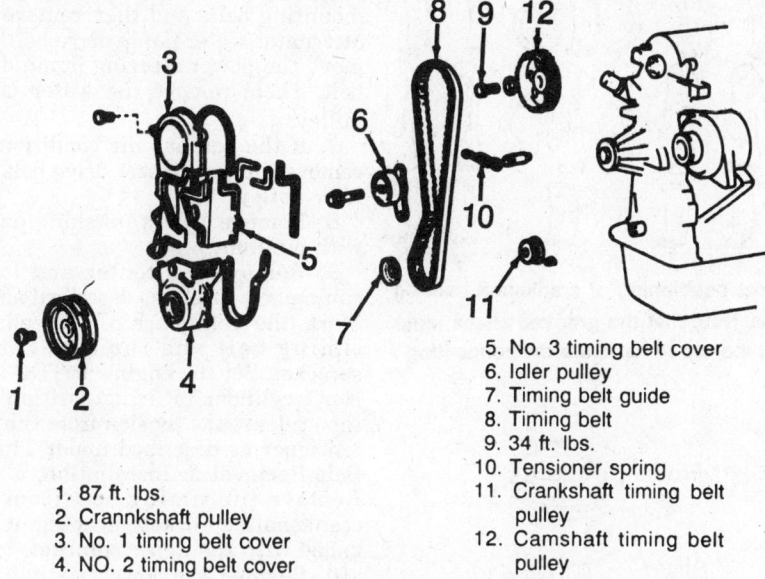

1. 87 ft. lbs.
2. Crannkshaft pulley
3. No. 1 timing belt cover
4. NO. 2 timing belt cover

5. No. 3 timing belt cover
6. Idler pulley
7. Timing belt guide
8. Timing belt
9. 34 ft. lbs.
10. Tensioner spring
11. Crankshaft timing belt pulley
12. Camshaft timing belt pulley

Exploded view of timing belt assembly

reverse order. Make sure to properly adjust the timing belt as described in the next procedure.

Timing Belt and Sprockets

REMOVAL & INSTALLATION

1. Remove the timing belt covers as described above. Set the engine to TDC No. 1 firing position as described in the procedure for removing the covers above.

2. If you may be reusing the timing belt, mark an arrow showing direction of rotation and then matchmark the belt and both pulleys.

3. Loosen the idler pulley mounting bolt and push the idler pulley to the left as far as it will go and hold it; then, retighten the mounting bolt.

4. Remove the timing belt. Then, you may remove the idler pulley mounting bolt and the pulley and return spring if they require service. If the sprockets must be removed, simply pull the crankshaft sprocket and key off. Loosen the camshaft pulley bolt with a socket wrench, using an open end wrench on the camshaft portion with flats to keep the camshaft from turning. Install these sprockets in reverse order, torquing the camshaft pulley retaining bolt to 34 ft. lbs.

5. Be careful not to bend, twist, or turn the belt inside out. Keep grease or water from contacting it as you handle it. Inspect the belt for cracks

or missing teeth, or general wear, and replace it if necessary.

6. Install the idler pulley, return spring and bolt, with the pulley in the retracted position. Install the belt. Align timing marks earlier if the belt is being reused. Point the directional arrow in the right direction. Otherwise, make sure timing marks on both pulleys align with the timing marks on the engine.

7. Loosen the bolt mounting the idler pulley so the spring tensions the belt, then install the crankshaft pulley bolt and turn it with a wrench so the crankshaft turns just two revolutions forward. Recheck the valve timing. If it is incorrect, adjust the position of the belt and then repeat this step.

8. Torque the idler pulley mounting bolt to 27 ft. lbs.

9. Measure the belt deflection with about 4.5 lbs. pressure; it must be .24-.28 in. If necessary, reposition the idler pulley to correct the tension.

10. Install the timing covers and front crankshaft pulley as described above.

Camshaft

REMOVAL & INSTALLATION

1. Follow the procedure for cylinder head removal above up to the point where the valve covers and timing covers have been removed, but leave the timing belt in position (through Step 14).

2. Remove the distributor.

3. Remove the fuel pump.

4. Remove the distributor gear bolt.

5. Unbolt and remove the rocker shaft assembly.

6. Loosen the idler pulley mount bolt and push the pulley as far to the left as it will go. Then, retightn the bolt to hold the pulley there. Now, gently pull the belt off the camshaft timing gear, holding it up so it will not come out of mesh with the crankshaft pulley. Support the belt securely so it will remain in mesh as you work. Also, be careful not to get oil on the belt or drop anything down inside the valve cover.

7. Use a large open-end wrench to hold the camshaft while you remove the sprocket bolt. The flats for the wrench are located between the first and second cam lobes. Remove the sprocket. Then, loosen the distributor drive gear bolt, keeping the camshaft from turning in the same way.

8. Loosen the camshaft bearing cap bolts a little at a time and alternately, using the proper sequence: front cap first, rear cap second, foreward-center cap third, and rearward-center cap last. Once bolt tension is gone, remove the bolts and then the caps, keeping the caps in proper order and in the same direction (you may want to number and arrow them).

9. Remove the camshaft oil seal and the camshaft from the head. Remove the distributor drive gear.

10. To install the camshaft, first install the distributor drive gear and plate washer and bolt.

11. Coat all bearing surfaces with clean engine oil, and then put the camshaft into position. Install bearing caps Nos. 2, 3 and 4 in proper positions and direction.

12. Grease the inside surface of the oil seal and liquid sealer to the outside edge and then slip the seal onto the camshaft. Make sure it is on straight, as a crooked seal will leak.

13. Apply liquid sealer to the bottom surfaces of the No. 1 bearing cap and then put it into position. Install all cap bolts finger tight.

14. Torque the bearing cap bolts alternately and evenly. Repeat the basic sequence, which is the reverse of the removal sequence, until the torque of 9 ft. lbs. is reached. The torquing sequence is: front-center first, rear-center second, front third, and rear last.

15. Recheck the camshaft thrust clearance at the front of the camshaft with a dial indicator. Normal range is 0.0031–0.0071 in. with a limit of 0.0098 in. Then, torque the distributor drive gear bolt to 22 ft. lbs. Install the remaining components in the re-

verse of the "Cylinder Head Removal & Installation" procedure above.

Piston and Connecting Rod Positioning

Note the locations of main bearings. Upper bearings are all grooved for distribution of oil to the connecting rods, while the lower mains are all plain. Thrust is taken by washers on either side of the center (No. 3) bearing cap, with tabs on the lower washers which fit into notches in the lower cap. Note that both the piston crowns and rod sides have marks that must both face forward when the engine is finally assembled. Note also the sequence of ring installation; in fact the upper outside diameter of the No. 2 compression ring is smaller than the lower O.D., while the No. 1 compression ring has an even, barrel face. Note also the positioning of the oil ring expander and side rail.

Oil Pan

REMOVAL & INSTALLATION

1. Disconnect the negative battery cable. Raise the vehicle and support it on axle stands.
2. Drain the oil pan. Remove the stabilizer bar (see the Front Suspension section below). On some models, it may be necessary to remove other steering linkage parts; do this now if it appears other parts will interfere with pan removal.
3. Remove the right side undercover.
4. Disconnect the exhaust pipe at the manifold.
5. Support the engine securely using a floor jack with a wooden block to cushion the impact of the jack on engine parts. Then, remove various engine mount parts as described in the "Timing Belt Covers Removal & Installation" procedure above. Also, remove the engine shock absorber.
6. Raise the engine far enough to provide clearance for removal of the pan. Then, remove the pan bolts and the pan and gasket.
7. Install the oil pan in reverse order. Make sure to scrape both gasket surfaces, and use a new gasket, and an appropriate type of liquid sealer. Torque oil pan bolts to 4 ft. lbs. Refer again to procedures elsewhere to reassemble engine mounts and suspension components.
gine lift points and securely suspend the engine.

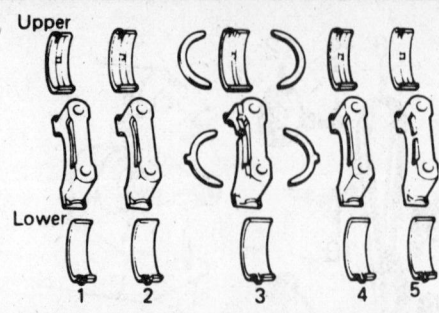

Correct positioning of crankshaft bearing shells. Note that the grooved shells must be at the top to lubricate the connecting rods

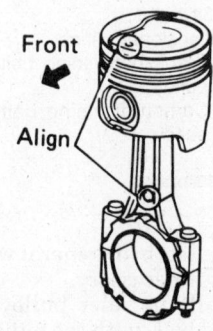

Piston alignment marks

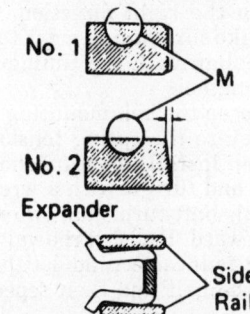

Piston ring installation. Note that the No. 2 compression ring taper faces up and that both compression rings are marked on top. Proper installation of the oil ring and side rail are also shown

Oil Pump

REMOVAL & INSTALLATION

1. Remove the engine under cover. Drain the oil pan. Matchmark and then remove the hood.
2. Unbolt the engine right hand under cover and flywheel cover.
3. Remove the oil pan as described above.
4. Remove the two oil pickup brace bolts (at the block) and the two mounting bolts and remove the oil pickup.
5. Attach a lifting sling to the en-

6. Loosen the water pump pulley mounting bolts and then remove the alternator/water pump drive belt. Remove the power steering pump drive belt. Then, remove the water pump pulley.
7. If the car has air conditioning, remove the compressor drive belt and idler pulley.
8. Remove the crankshaft pulley with a puller.
9. Remove the center and lower timing belt covers as described above. Mark the relationship between the timing belt and the crankshaft sprocket. Set the engine at TDC with No. 1 cylinder in firing position and then release the tension from the belt tensioner as described under Timing Belt Removal & Installation, above. Remove the timing belt from the crankshaft sprocket, keeping it engaged with the upper sprocket.
10. Remove the lower timing belt sprocket. Remove the timing belt idler pulley.
11. Remove the dipstick and dipstick tube.
12. Remove the seven oil pump mounting bolts and remove the oil pump. You can tap lightly on the lower surface of the pump from the rear to loosen it.
13. To install the pump, first install a new gasket and then engage the teeth of the oil pump drive (smaller) gear with the crankshaft gear. There are both small and large spline teeth, so make sure the teeth correspond properly. Then, put the pump in position, and install the seven mounting bolts, torquing to 15 ft. lbs.
14. Perform the remaining steps of installation in reverse of the removal procedure, referring to the "Oil Pan Removal & Installation" and "Timing Belt Removal & Installation" procedures as necessary.

Rear Main Seal

REMOVAL & INSTALLATION

1. Remove the transmission as described below.
2. On manual transmission cars, remove the clutch and then remove the flywheel. On cars with automatic transmissions, remove the torque converter drive plate.
3. Remove the rear end plate. Unbolt and then remove the oil seal retainer.
4. Support the oil seal retainer edges rear side downward on blocks so the entire area below the seal is open. Then, use a screwdriver at various points around the outer edge of the seal to tap the seal downward and out of the retainer.

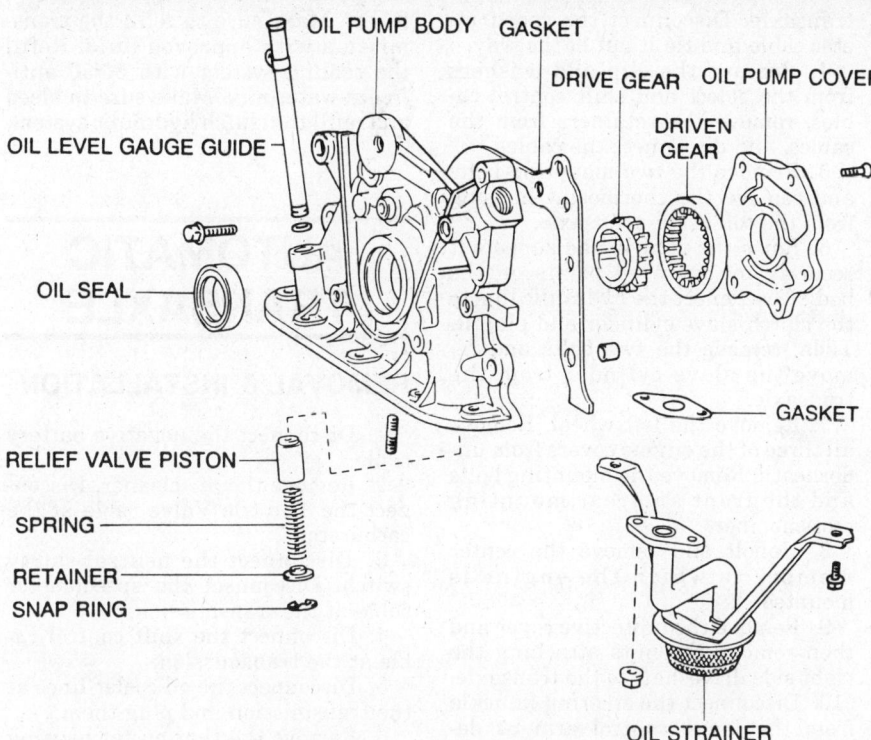

Exploded view of the oil pump assembly

5. Support the seal in a similar way but with the front (or engine) side downward. Coat the outer diameter with Multipurpose grease and then use a seal installer (such as GM Tool J–35388 or equivalent) or large wood block to tap the seal straight into the bore in the retainer.

6. Coat the seal inside diameter with Multipurpose grease and carefully install the seal and retainer straight over the crankshaft rear sealing surface. Use a new gasket under the retainer. Install the remaining parts in reverse order. The flywheel mounting bolts should be torqued in several stages, going back and forth and around the center of the flywheel to get even tightening. Torque them to 58 ft. lbs. Torque the drive plate for automatic transaxle equipped cars in the same way, but use a figure of 61 ft. lbs.

CLUTCH

REMOVAL & INSTALLATION

NOTE: Do not allow grease or oil to contaminate any of the disc, pressure plate, or flywheel friction surfaces.

1. Remove the transmission from the car as described below.

2. Remove the clutch cover and disc from the bellhousing.

3. Unfasten the clips that retain the release fork bearing, then withdraw the release bearing hub complete with the release bearing.

4. Remove the tension spring from the clutch linkage. Remove the release fork and support.

5. Punch matchmarks on the clutch cover and pressure plate so the pressure plate can be returned to its original position when you reinstall it. This is important to maintain the

balance of the flywheel/pressure plate assembly.

6. Loosen the mounting bolts for the pressure plate, turning each only about a half-turn at a time and alternating back and forth across the plate so pressure will be released gradually and evenly all around. If the tension is not released in this way, the tremendous spring pressure behind the plate could be released suddenly and violently! When the spring pressure has been fully released, remove the bolts and remove the pressure plate and clutch disc.

7. To install, insert a clutch alignment tool or an old transmission mainshaft through the clutch disc and then insert the tool or shaft into the pilot bearing. Note that the the disc is installed with the concave side toward the flywheel.

8. Install the pressure plate over the disc with matchmarks aligned and install the bolts, tightening alternately all and evenly to apply even pressure all around. Final torque the bolts to 14 ft. lbs. Remove the centering tool or input shaft.

9. The remaining steps of installation are the reverse of the removal procedure, except that you must lubricate the release bearing hub and release fork contact points with a light coating of Multipurpose grease.

FREE-PLAY ADJUSTMENT

1. Check pedal height as measured from the insulating sheet on the floor to the center of the pedal. It should be 5.65–6.043 in. If height is correct, go to Step 2. If not, remove the instrument lower finish panel and air duct. Then, Loosen the locknut located high and on your side of the pedal lever.

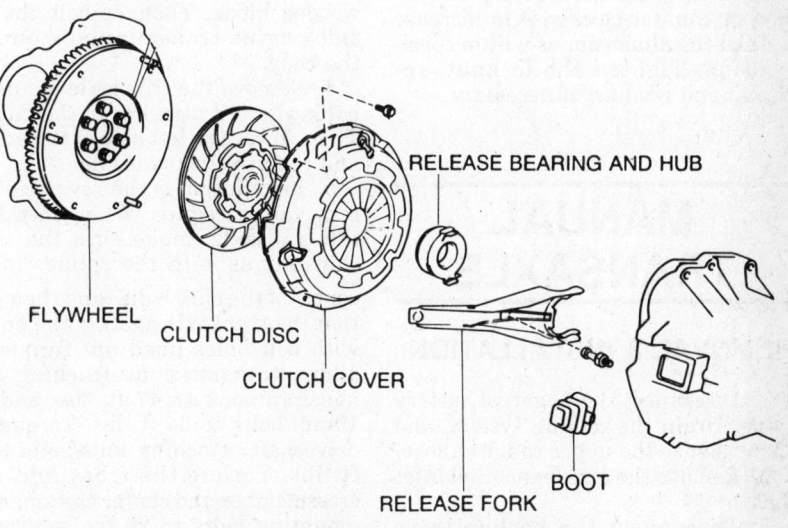

Exploded view of the clutch assembly

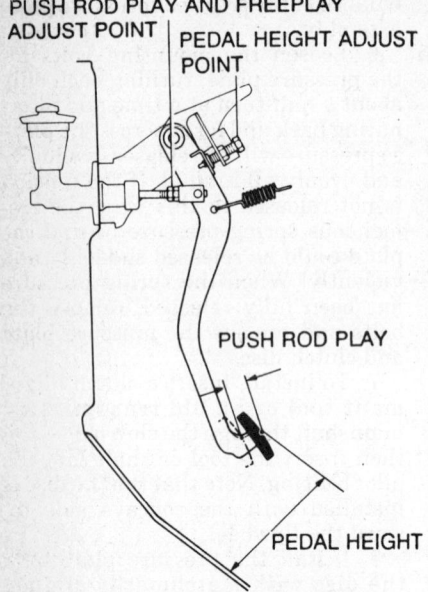

PUSH ROD PLAY AND FREEPLAY
ADJUST POINT | PEDAL HEIGHT ADJUST
POINT

PUSH ROD PLAY

PEDAL HEIGHT

Clutch pedal free play adjustment

Turn the bolt in to decrease or out to increase it until it is within specifications. Tighten the locknut, recheck, and readjust if necessary.

2. Measure the position of the clutch pedal. Now, push in the pedal until you feel increased resistance as the clutch pressure plate springs begin being compressed. Measure the position of the pedal at this point and then subtract the smaller figure from the larger one. This will give the freeplay dimension. The dimension should be .51–.91 in. on cars built before 1986, and .20–.59 in. on '86 and later models. If not, loosen the pushrod locknut, located below the pedal height adjustment locknut and on the cowl side of the pedal lever. Turn the clutch master cylinder pushrod clockwise to decrease the freeplay dimension or counter-clockwise to increase it until the dimension is within specification. Tighten the locknut, recheck, and readjust if necessary.

MANUAL TRANSAXLE

REMOVAL & INSTALLATION

1. Disconnect the negative battery cable. Drain the cooling system, and then remove the upper radiator hose.
2. Remove the air cleaner and inlet duct.
3. Disconnect the backup lamp switch connector at the switch on the transaxle. Disconnect the speedometer cable and tie it out of the way.

4. Remove the clip and washers from the select and shift control cables, remove the retainers from the cables, and disconnect the cables.
5. Remove the two mounting bolts and remove the thermostat housing from the top of the transaxle.
6. Raise the vehicle and support it securely by approved points on the body. Disconnect the hydraulic line to the clutch slave cylinder and plug it. Then, remove the two bolts and remove the slave cylinder from the transaxle.
7. Remove the left wheel. Remove all three of the engine covers from underneath. Remove the mounting bolts and the front and rear mounting crossmember.
8. Unbolt and remove the center member to which the engine is mounted.
9. Remove the protective cover and then remove the nuts attaching the right side driveshaft to the transaxle.
10. Disconnect the steering knuckle from the lower control arm as described later in this section. Remove the left side driveshaft attaching nuts. Pull the steering knuckle outward and remove the left side driveshaft from the transaxle. Repeat both procedures on the opposite side.
11. Disconnect the battery cable and ignition switch wire at the starter. Remove the two bolts and remove the starter.
12. Disconnect the ground cable at the transaxle. Unbolt and remove the plate that covers the lower section of the bell housing.
13. Using a wooden block atop a floor jack, raise the engine slightly with pressure on the bottom of the oil pan. Support the transaxle underneath with another floorjack and wooden block. Then, unbolt the left side engine-transmission mount at the body.
14. Remove the transaxle mounting bolts where it attaches to the engine. Then, lower the left side of the engine and remove the transaxle.
15. Installation is the reverse of the removal procedure. When installing the transaxle, make sure the input shaft aligns with the splines in the center of the clutch disc and then position the transaxle against the engine with bolt holes lined up. Torque the 12mm transmission attaching (bell housing) bolts to 47 ft. lbs. and the 10mm bolts to 34 ft. lbs. Torque the driveshaft attaching nuts/bolts to 27 ft. lbs. Torque the front and rear crossmember and center crossmember mounting bolts to 29 ft. lbs. Torque the left side engine mount bolts to 38 ft. lbs. Make sure to refill the transmission with approved fluid. Refill the cooling system with 50/50 anti-freeze-water mix. Make sure to bleed and refill the clutch hydraulic system.

AUTOMATIC TRANSAXLE

REMOVAL & INSTALLATION

1. Disconnect the negative battery cable.
2. Remove the air cleaner. Disconnect the Throttle Valve cable at the carburetor.
3. Disconnect the neutral safety switch. Disconnect the speedometer cable at the transmission.
4. Disconnect the shift control cable at the transmission.
5. Disconnect the oil cooler lines at the transmission and plug them.
6. Remove the thermostat housing assembly from the top of the transmission.
7. Remove the single upper mount-to-bracket bolt. Remove the two upper bellhousing bolts.
8. Raise and support the vehicle at approved locations on jackstands.
9. Drain the transmission fluid.
10. Remove the left wheel. Remove all three of the engine covers from underneath. Remove the mounting bolts and the front and rear mounting crossmember.
11. Unbolt and remove the center member to which the engine is mounted.
12. Remove the nuts attaching the right side driveshaft to the transaxle. Do the same on the opposite side.
13. Disconnect the steering knuckle from the lower control arm as described later in this section. Remove the left side driveshaft attaching nuts. Pull the steering knuckle outward and remove the left side driveshaft from the transaxle.
14. Disconnect the battery cable and ignition switch wire at the starter. Remove the two bolts and remove the starter.
15. Unbolt and remove the plate that covers the lower section of the bell housing.
16. Remove the six bolts attaching the torque converter to the drive plate. You'll have to rotate the engine to gain access to one bolt at a time.
17. Using a wooden block atop a floor jack, raise the engine slightly with pressure on the bottom of the oil pan. Support the transaxle under-

neath with another floorjack and wooden block. Then, unbolt the left side engine-transmission mount at the body.

18. Remove the transaxle mounting bolts where it attaches to the engine. You'll need a guide pin such as a bolt smaller than the torque converter bolts with the head cut off. Insert the guide pin through one of the accessible torque converter boltholes. Pry on the outer end of the pin to being forcing the transaxle to disengage from the engine. This will ensure that the converter comes off with the transaxle. Then, lower the left side of the engine and remove the transaxle.

19. To install, first apply multipurpose grease to the center of the converter, where it fits inside the crankshaft. Install the guide pin in one of the holes in the converter. Then, align the pin with the corresponding hole in the drive plate. Align the two knock pins in the block with the corresponding holes in the converter housing.

20. Situate the transmission in its installed position. Install one of the converter-to-guide plate bolts. Then install the transmission converter housing -to-engine bolts. Torque the 12mm bolts to 47 ft. lbs. and the 10mm bolts to 34 ft. lbs.

21. Install the left side engine mount and torque the bolts to 38 ft. lbs.

22. Install the torque converter-to-driveplate bolts by rotating the engine and transmission for access.Torque to 20 ft. lbs. Perform the remaining installation steps in reverse of the removal procedure. Torque the driveshaft attaching nuts/bolts to 27 ft. lbs. Torque the front and rear crossmember and center crossmember mounting bolts to 29 ft. lbs. Make sure to refill the transmission with approved fluid.

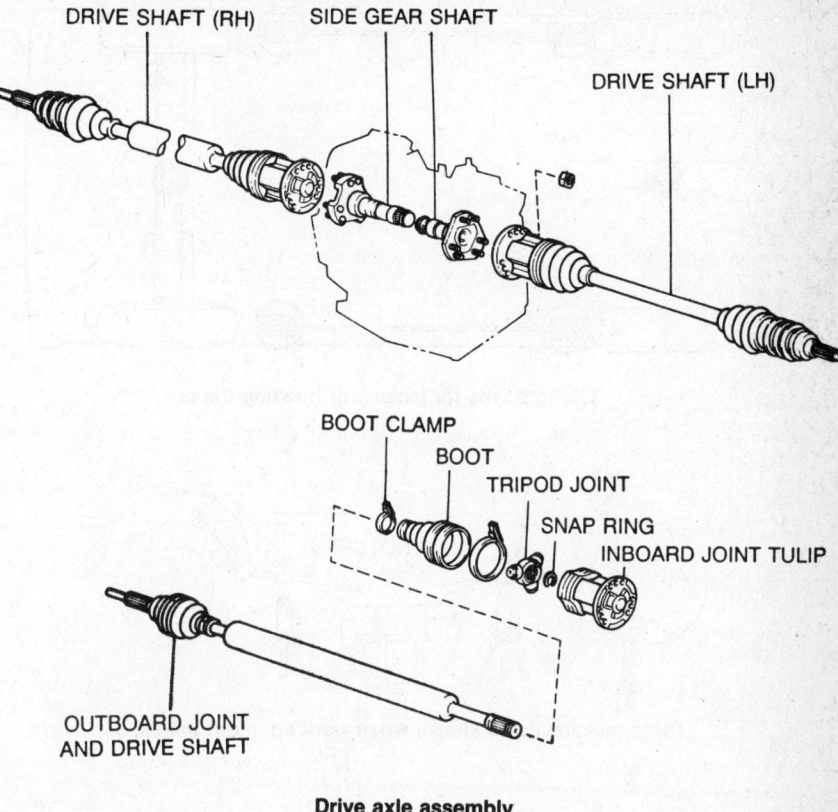

Drive axle assembly

DRIVE AXLE

Axle Shaft

REMOVAL & INSTALLATION

1. Raise and support the vehicle securely and remove the front wheels. Remove the cotter pin, locknut cap, and bearing locknut.

2. Loosen and remove the six nuts fastening the driveshaft to the flange at the transaxle. Have someone depress the brake pedal to keep the shaft from turning as you loosen the nuts.

3. Remove the bolt and two nuts and disconnect the steering knuckle at the lower control arm.

4. Remove the brake caliper as described later in this section and suspend it with wire without disconnecting the hydraulic line. Remove the disc brake rotor.

5. Cover the outboard CV-joint rubber boot with a cloth to prevent damage. Then, using a two-jawed puller, pull the axle hub from the driveshaft. Remove the driveshaft.

6. To install, first insert the outboard joint into the axle hub; then connect the shaft at the inboard side and install the nuts finger tight.

7. Connect the steering knuckle to the lower arm, torquing nuts/bolt to 47 ft. lbs.

8. Install the brake disc to the axle hub.

9. Install the brake caliper to the steering knuckle, torquing the bolts to 65 ft. lbs.

10. Install the bearing locknut; have someone depress the brake pedal while you torque it to 137 ft. lbs.

11. Torque the six nuts fastening the inboard end of the driveshaft at the transaxle (have someone depress the brake pedal while you do this), torquing to 27 ft. lbs.

Constant Velocity Joint

OVERHAUL

For all CV-joint overhaul procedures, please refer to the "U-Joint" Unit Repair section.

REAR AXLE

Axle Hub

REMOVAL & INSTALLATION

1. Raise and support the vehicle in a secure manner and remove the rear wheels. Remove the rear brake drum.

2. Disconnect the brake line at the rear of the backing plate and plug the open end of the line.

3. Remove the four bolts holding the axle hub to the axle carrier, using the holes in the axle flange to gain access. Remove the axle hub and rear brake assembly. Remove the O-ring.

4. Install a new O-ring onto the axle carrier.

5. Perform the remaining steps in

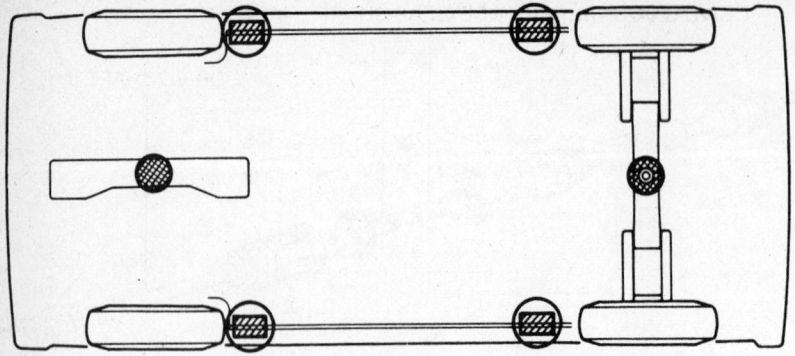

Lifting points for jacking or hoisting the car

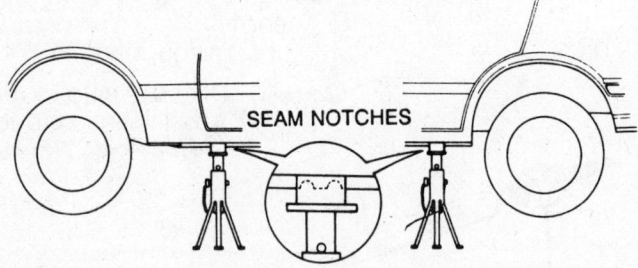

SEAM NOTCHES

Place jackstands as shown when working under the car

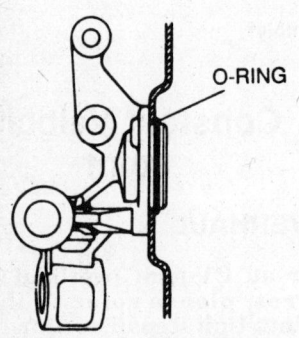

O-RING

Location of axle carrier O-ring

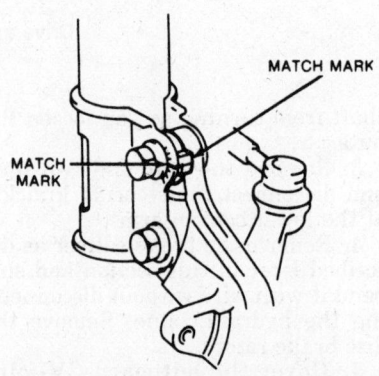

MATCH MARK

MATCH MARK

Mark the camber adjusting cam before removing the knuckle attaching nuts and bolts so camber can be restored without rechecking alignment

reverse order. Torque the axle hub and rear brake assembly mounting bolts to 59 ft. lbs. Refill the brake master cylinder with new, approved fluid and bleed the system.

FRONT SUSPENSION

MacPherson Strut

REMOVAL & INSTALLATION

1. Remove the hubcap and loosen the lugnuts.
2. Unfasten the three nuts which secure the upper shock absorber mounting plate to the top of the wheel arch. Raise the front of the car and support it securely at approved support points.
3. Remove the lugnuts and the front wheel. Detach the front brake line at the clamp on the strut. Disconnect the brake hose to brake pipe connection at the mount on the body. Catch any fluid that drains in a pan. Pull the brake hose back through the opening the the strut bracket. Tape both open ends of the hydraulic system.
4. Remove the brake caliper and wire it up and out of the way. Support the caliper securely so that the brake hose will not be under any strain (you do not need to disconnect the hose from the caliper.).
5. Mark the adjusting cam so the camber adjustment can be restored when the strut is reassembled. Remove the two bolts attaching the lower end of the shock absorber to the steering knuckle lower arm. Remove the strut assembly.
6. Remove the camber adjusting cam from the knuckle. Install a cloth over the drive axle boot to protect it while the strut is removed.
7. Install in reverse order. Note the following points:
    a. When installing the camber adjusting cam back into the knuckle, make sure to restore the alignment adjustment.
    b. Install the strut with the studs at the top passing through the holes in the body. Then, install the strut-to-knuckle attaching bolts and nuts and torque to 105 ft. lbs.
    c. When reconnecting the disconnected brake hose, make sure to route it properly (back through the strut bracket) and to torque the fitting to 11 ft. lbs. Bleed the brake system.
    d. Install the studs on top of the strut through the body. Then, install the nuts and torque them to 13 ft. lbs. Pack the shaft nut area with grease before installing the dust cover.
    e. Lower the vehicle to the ground and torque the wheel lugnuts to 76 ft. lbs.

STRUT OVERHAUL

For all overhaul procedures, please refer to the "Strut Overhaul" Unit Repair section.

FRONT WHEEL BEARING ADJUSTMENT

Front wheel bearings on the Nova are not adjusted as a normal maintenance operation. If the hub is disassembled, it is simply pressed back together until fully assembled using special tools.

Ball Joints

INSPECTION

1. Turn the front wheels so they are straight and chock the rear wheels. Raise the vehicle and place a wooden block of 7–8 in. under it. Then, lower the vehicle onto the block until the spring is compressed to only about

half its compression when the vehicle is resting on it.

2. Attempt to move the lower arm up and down. There should be no noticeable play.

3. If the ball joint is off the vehicle you can check the required rotating torque with and inch pound torque wrench. Flip the ball joint all the way back and forth five times. Then, install the nut and turn the stud with a torque wrench at a rate of about one turn in three seconds. At the fifth turn, measure the required torque; it should be 9–30 inch lbs. If outside these specifications, replace the ball joint.

REMOVAL & INSTALLATION

1. Raise and support the vehicle via approved jacking points. Remove the front wheel.

2. Remove the two nuts and one bolt which attach the ball joint to the lower control arm.

3. Remove the nut attaching the ball joint stud to the steering knuckle. Then, press the stud out of the steering knuckle with an appropriate special tool.

——— CAUTION ———
Use a new self-locking nut in the following step. Failure to do so could create an unsafe front suspension.

4. Install in reverse order. Install a new self-locking nut and torque to 82 ft. lbs. Torque the ball joint-to-lower arm nuts and bolt to 47 ft. lbs.

Lower Control Arm

REMOVAL & INSTALLATION

1. Raise the front of the vehicle and support it with jackstands. Remove the wheel.

2. Remove the two nuts attaching the ball joint to the steering knuckle.

3. Remove the bracket bolts (two each) for the front and rear mounts.

7. Install in reverse order, noting the following points:

   a. First install the lower arm-to-body mounting bolts just finger tight.

   b. Install and torque the ball joint-to-arm nuts and bolt to 47 ft. lbs.

   c. Install the wheel and lower the vehicle. Finally, torque the lower arm bracket bolts to 105 ft. lbs. for the front and 64 ft. lbs. for the rear.

   e. Adjust front end alignment.

REAR SUSPENSION

MacPherson Strut

REMOVAL & INSTALLATION

1. Working inside the car, remove the rear quarter window garnish molding and back window panel.

2. Raise the rear of the vehicle and support it securely at approved points on jackstands. Remove the rear wheel.

3. Disconnect the flexible hose at the strut. Plug the openings. Then, remove the flexible hose and clip from the mounting point on the strut. Finally, reconnect the brake line to the flex hose to prevent an excessive amount of brake fluid from draining from the system.

4. Remove the nuts and bolts mounting the strut onto the axle carrier and then disconnect the strut.

5. Remove the three upper strut mounting nuts and carefully remove the strut assembly.

6. Installation is the reverse of the removal procedure. Note the following points:

   a. Torque the upper strut retaining nuts to 17 ft. lbs.

   b. Torque the strut to axle carrier bolts to 105 ft. lbs.

   c. Torque the nut holding the suspension support to the shock absorber to 36 ft. lbs.

   d. Bleed the brakes as described later in this section.

OVERHAUL

For all strut overhaul procedures, please refer to the "Strut Overhaul" Unit Repair section.

BRAKES

To service brake shoes, drums and wheel cylinders or brake pads and calipers, please refer to the "Brakes" Unit Repair section.

Master Cylinder

REMOVAL & INSTALLATION

——— CAUTION ———
Be careful not to spill brake fluid on the painted surfaces of the vehicle; it will damage the finish.

1. Disconnect the hydraulic lines from the master cylinder and plug the openings. Drain brake fluid from the reservoir with a syringe.

2. Disconnect the hydraulic fluid pressure differential switch wiring connector.

3. If you're planning to disassemble the master cylinder, loosen the master cylinder reservoir mounting (or set) bolts.

4. Remove the nuts and remove the master cylinder from the power brake booster.

5. Install the master cylinder by first cleaning out the groove on the lower installation surface. Confirm that the "UP" mark on the master cyl-

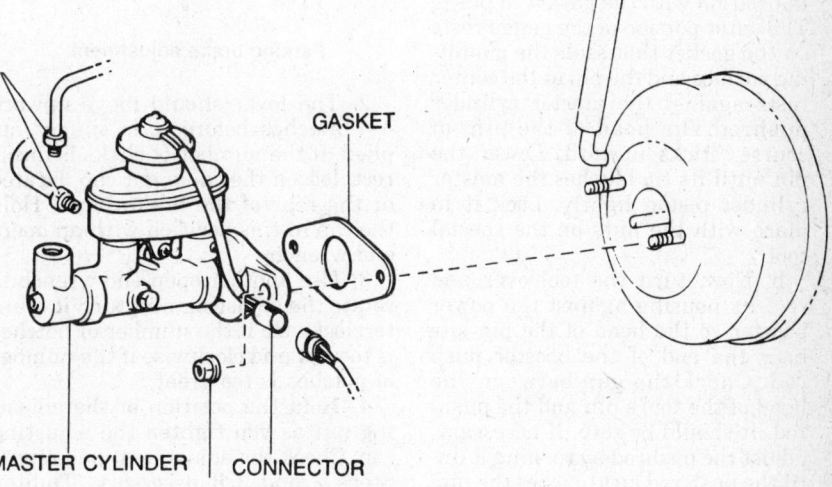

GASKET

MASTER CYLINDER   CONNECTOR

Master cylinder mounting

inder boot is in the correct position (at the top).

6. Adjust the booster pushrod as described below under "Power Brake Booster Removal & Installation."

7. Install the master cylinder to the power brake booster with a new gasket. Torque the attaching nuts to 9 ft. lbs.

8. Connect the brake lines and torque them to 11 ft. lbs. Connect the level warning switch connector.

9. Fill the fluid reservoir and bleed the brake system. Check for fluid leakage and tighten or replace fittings as necessary.

Power Brake Booster

REMOVAL & INSTALLATION

NOTE: To perform this procedure, you will need a booster push rod gauge GM part No. J–34873–A or equivalent to set the booster push rod length.

1. Remove the master cylinder as described above. Disconnect the hydraulic lines at the three-way union and plug all openings. Diusconnect the union from its mount.

3. Pull back the clamp and disconnect the booster vacuum line at the booster.

4. Remove the instrument panel lower finish panel and the air duct. Remove the pedal return spring. Locate the clevis rod where it attaches to the brake pedal underneath the dash. Pull out the clip and then remove the clevis pin.

5. Remove the four attaching nuts and then remove the booster, bracket and gasket.

6. Adjust the booster push rod as follows:

a. Set the booster push rod gauge in position with the gasket in place. The outer portion of the gauge rests on the gasket that seals the mounting surface and the pin at the center rests against the master cylinder pushrod. The head of the pin, of course, sticks upward. Lower the pin until its tip touches the master cylinder piston lightly. Lock it in place with the nuts on the special tool.

b. Now, turn the tool over and rest its housing against the power booster so the head of the pin sits near the end of the booster push rod. Check the gap between the head of the tool's pin and the pushrod. It should be zero. If necessary, adjust the pushrod by turning it until the push rod just touches the pin.

7. Install in reverse order, torquing the mounting nuts to 9 ft. lbs. Make

sure to bleed the brake system thoroughly and check for leaks in the system.

8. Adjust the pedal as follows:

a. Check pedal height from the asphalt sheet on the floor. It should be 5.79–6.18 in. If necessary, adjust it by turning the pedal pushrod. You must remove the instrument lower finish panel and air duct to reach the pushrod. Make sure to then adjust the stop lamp switch until it just touches the pedal stop.

b. Check the pedal freeplay by first removing all vacuum from the system by repeatedly pressing the brake pedal downward with full force. Push in the pedal until resistance is just felt, and then measure the distance from its untouched position to the point where it begins to resist, using an assistant. It must be .12–.24 in. If necessary, adjust the pedal freeplay with the pedal pushrod.

c. Start the engine and confirm that freeplay still exists. Install the air duct and finish panel.

PARKING BRAKE ADJUSTMENT

1. Release the parking brake all the way and release the button. Then, pull the lever upward slowly as you count the clicks. Count two clicks as one notch.

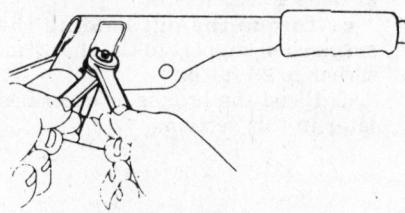

Parking brake adjustment

2. The lever should move upward 4–7 notches before it is snugly applied. If the number of clicks is incorrect, loosen the cable nut cap, located at the rear of the brake lever. Hold the cap in this position with an open-end wrench.

3. Use another open-end wrench to rotate the adjusting nut; turn it counterclockwise if the number of notches is too low and clockwise if the number of notches is too great.

4. Hold the position of the adjusting nut as you tighten the adjusting cap. Check the adjustment and repeat Steps 2 and 3 if necessary. Tighten the adjusting nut securely and ensure that the adjustment is correct.

STEERING

Tie Rod
REMOVAL & INSTALLATION

1. Remove the cotter pin and nut holding the knuckle arm to the tie rod end. Then, use a tool designed for this purpose to press the stud up and out of the steering knuckle.

2. Loosen the tie rod-to-steering rack-end nut. Matchmark the position of the tie rod on the rack-end so you can restore toe-in.

3. Install in reverse order, making sure you align the matchmarks. Torque the tie rod-to-rack-end locknut to 35 ft. lbs. Torque the tie-rod end-to-steering knuckle nut to 36 ft. lbs. and then install a new cotter pin.

Manual Steering Gear

REMOVAL & INSTALLATION

1. Remove the clamp bolts at top and bottom of the universal joint linking the steering box and steering shaft. Then, remove the U-joint.

2. Remove the cotter pin and nut holding the knuckle arm to the tie rod end. Then, use a tool designed for this purpose to press the stud up and out of the steering knuckle.

3. Remove the two bolts and two nuts and remove the steering gear housing and brackets.

4. Install the housing in reverse order. Make sure the clamps are installed squarely over the rubber insulators so they will not be damaged when the nuts and bolts are torqued. Torque the attaching nuts and bolts to 43 ft. lbs. Torque the tie rod end nuts attaching the studs to the knuckle arms to 36 ft. lbs. Torque the U-joint clamp bolts to 26 ft. lbs. If new parts have been installed, have the toe-in and steering wheel center point checked and adjusted.

Power Steering Gear

REMOVAL & INSTALLATION

1. Remove the intermediate steering shaft protector. Loosen the upper shaft pinch bolt and remove the lower one.

2. Open the hood and place a drain pan under the steering gear assembly. Clean the area around the inlet and return lines at the steering gear valve.

3. Loosen the wheel lugnuts, and then raise and securely support the

vehicle. Remove both front wheels and tires.

4. Remove the cotter pins and nuts and press both tie rod ends out of the knuckles.

5. Support the transaxle with a jack. Remove the rear center engine mounting member-to-body mounting bolts.

6. Remove the rear engine mount-to-mount bracket attaching nut and bolt.

7. Disconnect the pressure and return lines at the steering gear. Remove the four steering gear-to-body attaching nuts and bolts, raising and lowering the rear of the transaxle as necessary to gain access to the steering gear-to-body attaching bolts and nuts.

8. Remove the steering gear through the access hole.

9. Installation is the reverse of removal. Add fluid to the pump reservoir and bleed the system.

Power Steering Pump

REMOVAL & INSTALLATION

1. Remove the air cleaner. Loosen the nut at the center of the pump pulley while the belt is still attached.

2. Loosen the pump mounting and adjusting bolts, move the pump so as to reduce belt tension and then remove the belt.

3. Remove the nut, pulley, and key from the pump shaft.

4. Place a drain pan under the pump. Remove the return hose clamp. Disconnect the hoses at the pump reservoir and then tie them up high so fluid will not drain out. In the case of the return line, pull it well back from the fitting, and pull the hose off the pump with a twisting motion.

5. Remove the adjusting bolt. Remove the pivot bolt and disconnect the drive belt. Remove the pump assembly. Remove the pump bracket.

6. Installation is the reverse of removal. If you are replacing the pump, switch the pulley and the mounting nut to the new pump. Be careful not to lose the woodruff key. The nut is tightened after the pump is installed and tension is put on the belt. Torque the pressure hose fitting to 34 ft. lbs. Tension the belt so deflection is .31–.39 in. with moderate thumb pressure (about 20 lbs.) applied in the center of the span. Fill the reservoir with Dexron II automatic transmission fluid and bleed the system as described below. Check for leaks and correct if necessary.

BLEEDING THE POWER STEERING SYSTEM

1. Raise the front of the car and support it securely on jackstands (this will minimize steering effort). Fill the power steering pump reservoir with Dexron® II.

2. With the engine OFF, keep the reservoir full as someone turns the steering wheel from lock to lock several times. Stop with the steering system at one lock.

3. Pull the high tension lead out of the coil. Continue to keep the reservoir full as someone cranks the engine for 30 seconds at a time (with a one minute rest in between) until fluid level remains constant.

4. Turn the steering wheel to the opposite lock and repeat Step 3.

5. Reconnect the high tension lead, start the engine and allow it to idle. Turn the wheel from lock to lock several times. Note the level of the fluid.

6. Lower the car to the ground. Note the fluid level. Repeat Step 5, stopping with the wheel at the centered position.

7. The fluid level should not have risen more than .2 in. If it has, repeat Step 6 until the level does not rise appreciably.

Steering Wheel

REMOVAL & INSTALLATION

1. Remove the screw from the bottom of the steering wheel pad and pull the pad upward and off the wheel.

2. Remove the steering wheel attaching nut from the end of the column. Matchmark the relationship between the end of the column and the wheel.

3. Screw the attaching bolts of a steering wheel puller into the threads on either side of the column and turn the center bolt of the puller to remove the wheel.

4. Install the wheel in reverse order. Once the wheel is started onto the splines (with the matchmarks lined up), the nut can be installed and torqued to force it onto the column. Torque the nut to 25 ft. lbs.

Combination Switch

REMOVAL & INSTALLATION

1. Disconnect the negative battery cable. Remove the instrument lower finish panel, air duct, and column lower cover.

2. Disconnect the ignition and turn signal switch wiring from the connector.

3. Remove the combination switch with the column upper cover.

4. Installation is the reverse of removal.

Ignition Lock and Switch

REMOVAL & INSTALLATION

1. Disconnect the negative battery cable. Unscrew the two retaining bolts and remove the steering column garnish.

2. Remove the upper and lower steering column covers.

3. Turn the key to the "ACC" position.

4. Push in the lock cylinder stop, located near the inner end of the cylinder, with a cotter pin or center punch. Then, pull out the key and lock cylinder. If you have trouble gaining access, it may help to remove the steering wheel and combination switch as described above.

5. Remove the mounting screw and withdraw the ignition switch from the lock housing.

6. To install, first line up the notch on the top of the switch with the projection inside the housing and turn the slot in the switch shaft so it will engage the projection on the steering lock shaft. Then position the switch inside the lock and install the retaining screw.

7. Make sure that both the lock cylinder and the column lock are in "ACC" position. Slide the cylinder into the lock housing until the stop tab engages the hole in the lock.

8. The rest of the installation procedure is the reverse of removal.

CHASSIS ELECTRICAL

Instrument Cluster

REMOVAL & INSTALLATION

1. Disconnect the negative battery cable. Remove the A/C registers and then remove the four screws and remove the meter hood.

2. Remove the six retaining screws and remove the combination meter. Disconnect the speedometer cable and electrical connectors before pulling the meter all the way out.

3. To gain access to the gauges, re-

move the six screws from the rear of the combination meter.

4. Installation is the reverse of removal.

Wiper Motor

REMOVAL & INSTALLATION

1. Disconnect the negative battery cable.

2. Disconnect the wiper motor electrical connector. Pull off the service cover.

3. Remove the wiper motor mounting bolts.

4. Matchmark the relationship between the motor shaft and linkage lever. Remove the nut and pry the lever off the shaft.

—————— CAUTION ——————

Be careful not to bend the linkage. If the motor is actually being replaced, matchmark the shaft of the new motor similarly and install the linkage lever so that it is in the same position.

5. Installation is the reverse of the removal.

Wiper Blade

REMOVAL & INSTALLATION

1. Lift up on the spring release tab on the wiper blade-to-arm connector.

2. Pull the assembly off the arm.

3. Install in reverse order. If you want to replace a wiper insert alone, press the old insert down and away from the blade assembly to free it from the retaining clips on the ends of the blades. Then, slide the insert out of the blade.

4. Slide the new insert into the blade assembly and bend the insert upward slightly to engage the retaining clips.

Heater Blower Motor

REMOVAL & INSTALLATION

1. Remove the three screws attaching the retainer.

2. Remove the glovebox.

3. Remove the duct connecting the blower assembly and the heater assembly.

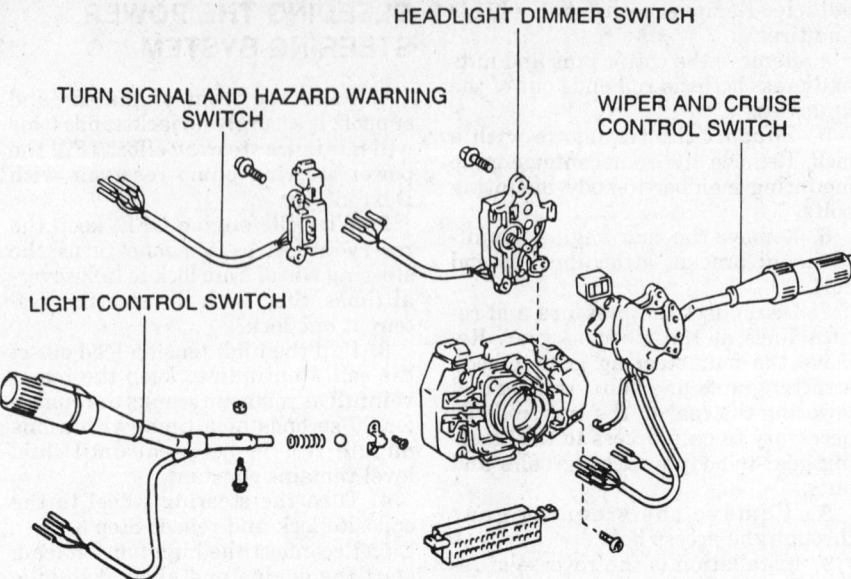

Exploded view of combination switch

4. Disconnect the blower motor wiring connector at the blower case.

5. Disconnect the air source selector control cable at the blower assembly.

6. Remove the two nuts and single bolt that attach the blower assembly to the heater case, and remove it.

7. Installation is the reverse of removal.

Heater Core

REMOVAL & INSTALLATION

1. Disconnect the heater hoses working inside the engine compartment.

2. Remove the six clips that retain the lower heater case, and remove it.

3. Pry the lower portion of the case apart with a screwdriver. Then, remove the core from the case.

4. Installation is the reverse of the removal procedure.

Radio

REMOVAL & INSTALLATION

1. Remove the two screws at the top of the dashboard center trim panel.

2. Remove the center panel out far enough to gain access to the cigarette lighter wiring and disconnect the wiring. Remove the trim panel.

3. Unfasten the screws which secure the radio to the instrument panel braces.

4. Lift the radio out part way and disconnect the leads from it. Remove the radio.

5. Installation is the reverse of removal.

Fuses, Fusible Links and Circuit Breakers

The fusebox is located in a kick panel on the driver's left side. Circuit breakers are located behind that kick panel. To reset breakers, unplug them, insert a needle into the reset hole and press it inward. If this does not restore operation, check continuity between the terminals with an ohmmeter. A large box of relays for the air conditioning compressor clutch, blower motor, headlights, engine main power supply and other electrical functions is located on the left side fender well. Fusible links are located in the main wiring harness leading to the alternator.

Unit Repair Sections

General Maintenance

Introduction

Routine maintenance is probably the most important part of automobile care and the easiest to neglect. A regular program aimed at monitoring essential systems ensures that all components are in good and safe working order, and can prevent small problems from developing into major headaches. Routine maintenance also pays off big dividends in keeping major repair costs at a minimum and extending the life of the car.

The vehicle owner's manual includes a maintenance schedule indicating service intervals in numbers of months or thousands of miles. This schedule should always be followed. We have provided in this section a guide to service intervals based on an averaging of manufacturer's recommendations. In most cases the suggested interval offered here will be close to that given by the manufacturer of your car, but the manufacturer's schedule should always take precedence.

We have divided the maintenance work to be done into three categories: Under Hood, Under Car, and Exterior. The checks in each section require only a few minutes of attention every few weeks and the services to be performed can be easily accomplished in a morning. The most important part of any maintenance program is regularity. The few minutes or occasional morning spent on these seemingly trivial tasks will forestall or eliminate major problems later.

Under Hood Maintenance

AUTOMATIC TRANSMISSION, AUTOMATIC TRANSAXLE

The fluid level in the automatic transmission or transaxle should be checked every three months or 6000 miles. All automatic transmissions have a dipstick for fluid level checks.

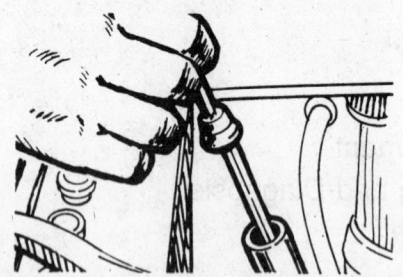

Check the automatic transmission fluid level with the dipstick provided

1. Drive the car until it is at normal operating temperature. The level should not be checked immediately after the car has been driven for a long time at high speed, or in city traffic in hot weather; in those cases, the transmission should be given a half hour to cool down.
2. Stop the car, apply the parking brake, then shift slowly through all gear positions, ending in Park. Leave the engine running.

3. Remove the dipstick, wipe it clean, then reinsert it, pushing it fully home.
4. Pull the dipstick again and, holding it horizontally, read the fluid level.
5. Cautiously feel the end of the dipstick to determine the temperature. Most dipsticks are marked with both cool and hot levels. If the fluid is not up to the correct level, more will have to be added.

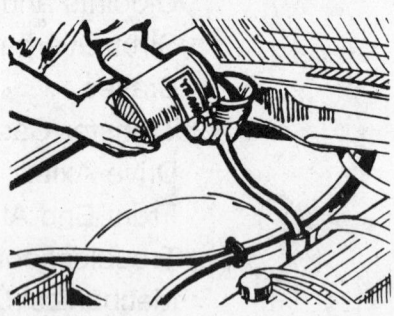

Fill the automatic transmission through the dipstick tube

NOTE: On General Motors Citation, Omega, Phoenix, Skylark, Cavalier, Cimarron, J2000, Celebrity, Cierra and 6000 models, the "Cold" level marks (dimples) are above the "Hot" level area.

6. Fluid is added through the dipstick tube. You will probably need the aid of a spout or a long-necked funnel. Be sure that whatever you pour through is perfectly clean and dry. Fluid recommendations can be found in the owner's manual or the Auto-

matic Transmission Unit Repair Section in this book. Add fluid slowly and in small amounts, checking the level frequently between additions. Do not overfill, which will cause foaming, fluid loss, slippage, and possible transmission damage.

BATTERY

Fluid Level (Except "Maintenance Free" Batteries)

Check the battery electrolyte level at least once a month, or more often in hot weather or during periods of extended car operation. The level can be checked through the case on translucent polypropylene batteries; the cell caps must be removed on other models. The electrolyte level in each cell should be kept filled to the split ring inside, or the line marked on the outside of the case.

If the level is low add only distilled water or colorless, odorless drinking water through the opening until the level is correct. Each cell is completely separate from the others, so each must be checked and filled individually.

If water is added in freezing weather, the car should be driven several miles to allow the water to mix with the electrolyte. Otherwise, the battery could freeze.

Specific Gravity (Except "Maintenance Free" Batteries)

At least once a year, check the specific gravity of the battery. It should be between 1.20 and 1.26 at room temperature. See the "Charging and Starting Systems" Section in this book for details.

Cables and Clamps

Once a year, the battery terminals and the cable clamps should be cleaned. Loosen the clamps and remove the cables, negative cable first. On batteries with posts on top, the use of a puller specially made for the purpose is recommended. These are inexpensive, and available in auto parts stores. Side terminal battery cables are secured with a bolt.

Clean the cable clamps and the battery terminal with a wire brush, until all corrosion, grease, etc. is removed and the metal is shiny. It is especially important to clean the inside of the clamp thoroughly, since a small deposit of foreign material or oxidation there will prevent a sound electrical connection and inhibit either starting or charging. Special tools are available for cleaning these parts, one type for conventional batteries and another type for side terminal batteries.

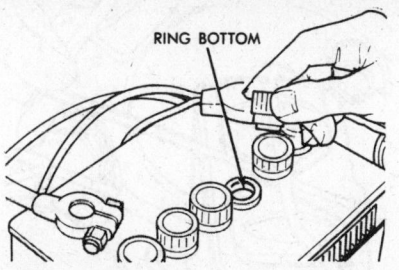

Fill the battery cell to the bottom of the split ring

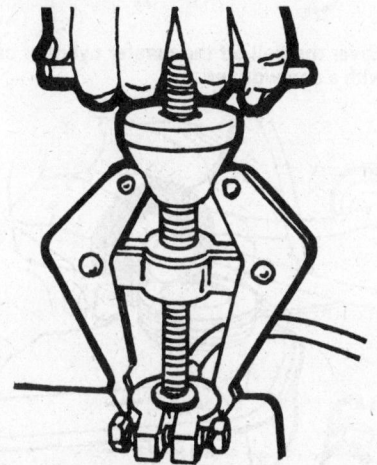

Use a puller to remove the clamp on post-type batteries

Clean the clamp with a wire brush

Before installing the cables, loosen the battery hold-down clamp or strap, remove the battery and check the battery tray. Clear it of any debris, and check it for soundness. Rust should be wire brushed away, and the metal given a coat of anti-rust paint. Replace the battery and tighten the hold-down clamp or strap securely, but be careful not to overtighten, which will crack the battery case.

After the clamps and terminals are clean, reinstall the cables, negative cable last; do not hammer on the clamps to install. Tighten the clamps securely, but do not distort them. Give

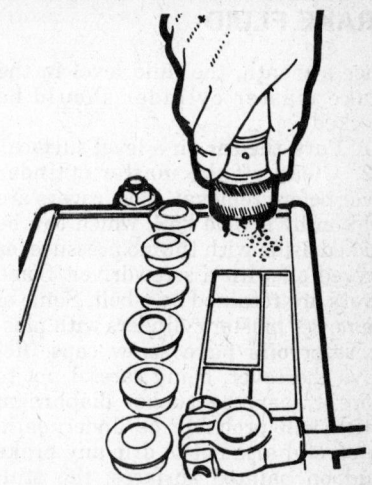

The posts are easily cleaned with a wire brush, or the battery post tool shown

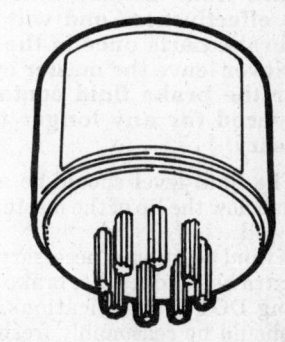

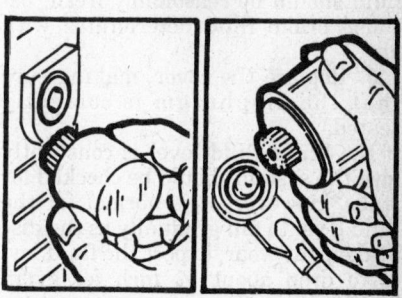

A special tool is required to clean the terminals and clamps on side terminal batteries

the clamps and terminals a thin external coat of grease after installation, to retard corrosion.

Check the cables at the same time that the terminals are cleaned. If the cable insulation is cracked or broken, or if the ends are frayed, the cable should be replaced with a new cable of the same length and gauge.

CAUTION

Keep flame or sparks away from the battery; it gives off explosive hydrogen gas. Battery electrolyte contains sulphuric acid. If you should splash any on your skin or in your eyes, flush the affected area with plenty of clear water; if it lands in your eyes, get medical help immediately.

BRAKE FLUID

Once a month, the fluid level in the brake master cylinder should be checked.

1. Park the car on a level surface.
2. Clean off the master cylinder cover before removal. Most covers are held on by a wire bail, which can be pushed aside with thumb pressure, or levered off with a screwdriver. Some covers are retained by a bolt. Some of the newer master cylinders with plastic reservoirs have screw caps. Remove the cover, being careful not to drop or tear the rubber diaphragm which will probably be underneath. Be careful also not to drip any brake fluid on painted surfaces; the stuff eats paint.

NOTE: Brake fluid absorbs moisture from the air, which reduces effectiveness and will corrode brake parts once in the system. Never leave the master cylinder or the brake fluid container uncovered for any longer than necessary.

3. The fluid level should be about $\frac{1}{4}$ inch below the lip of the master cylinder well.
4. If fluid addition is necessary, use only extra heavy duty disc brake fluid meeting DOT 3 specifications. The fluid should be reasonably fresh, because brake fluid deteriorates with age.
5. Replace the cover, making sure that the diaphragm is correctly seated.

If the brake fluid level is constantly low, the system should be checked for leaks. However, it is normal for the fluid level to fall gradually as the disc brake pads wear; expect the fluid level to drop about $\frac{1}{8}$ inch for every 10,000 miles of wear.

BELT TENSION ADJUSTMENT

Every six months or 12,000 miles, check the water pump, alternator, power steering pump, air pump, and air conditioning compressor drive belts for proper tension. Also look for signs of wear, fraying, separation, glazing and so on, and replace the belts as required.

Belt tension should be checked with a gauge made for the purpose. If a gauge is not available, tension can be checked with moderate thumb pressure applied to the belt at its longest span midway between pulleys. If the belt has a free span less than twelve inches, it should deflect approximately $\frac{1}{8}$–$\frac{1}{4}$ inch. If the span is longer

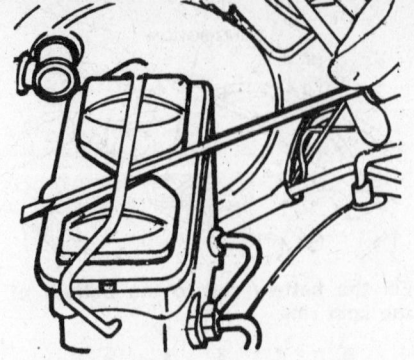

Lever the ball off the master cylinder cap with a screwdriver

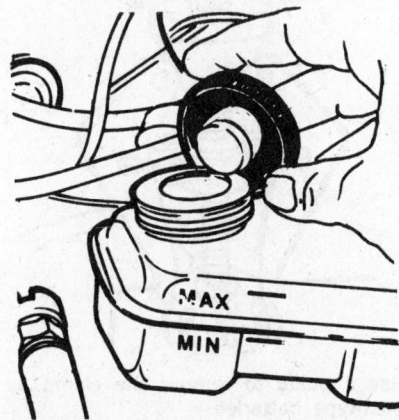

Screw caps are used on some master cylinders

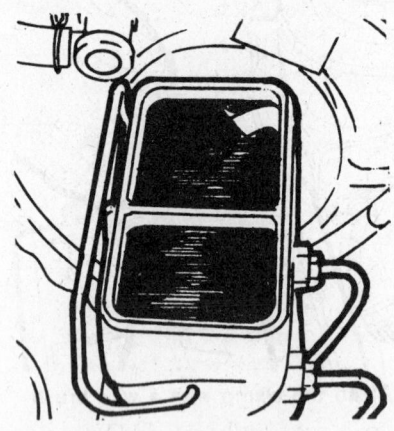

Proper brake fluid level

than twelve inches, deflection can range between $\frac{1}{8}$–$\frac{3}{8}$ inches.

NOTE: On cars except American Motors models which use a one-piece "serpentine" belt to drive all accessories, belt tension is automatically adjusted. On cars which have two "serpentine" belts, or one "serpentine" belt as well as conventional V-belts, and on all American Motors models with the "serpentine" belt, belt

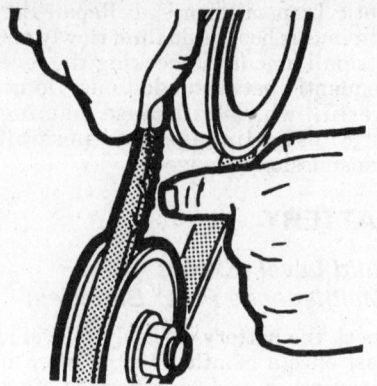

Check the belts for wear

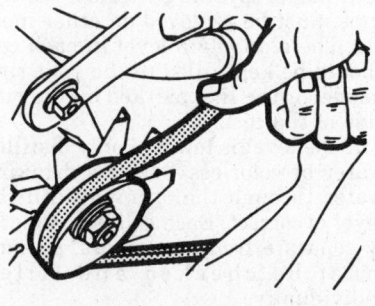

Check the belt tension at the middle of the longest span between pulleys

tensions usually must be checked and adjusted. Belt tension is higher on "serpentine" belts and cannot be tested with thumb pressure. Some Ford models (Thunderbird/XR-7 with AOD transmission) require special tools for adjustment. American Motors "serpentine" belts are adjusted at the alternator.

To adjust or replace belts:

1. Loosen the driven accessory's pivot and mounting bolts. Some air conditioning compressor belts are tensioned by an idler pulley; in this case, loosen the idler pulley and use a $\frac{1}{2}$ in. drive ratchet in the square hole provided to lever the idler pulley up or down.
2. Move the accessory toward or away from the engine until the tension is correct. You can use a wooden hammer handle or broomstick as a lever, but do not use anything metallic.
3. Tighten the bolts and recheck the tension. If new belts have been installed, run the engine for a few minutes, then recheck and readjust as necessary.

NOTE: If the driven component has two drive belts, the belts should be replaced in pairs to maintain proper tension.

It is better to have belts too loose

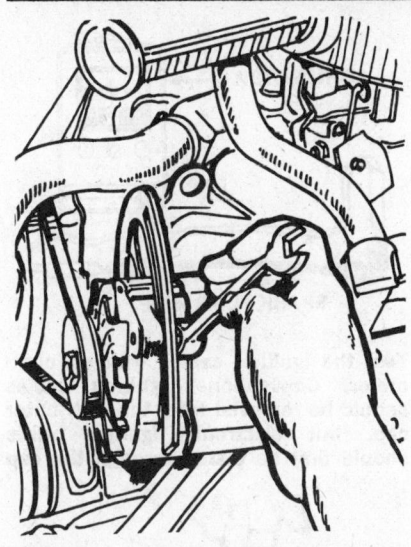

To either adjust or remove a belt, loosen the driven component's adjusting bolt

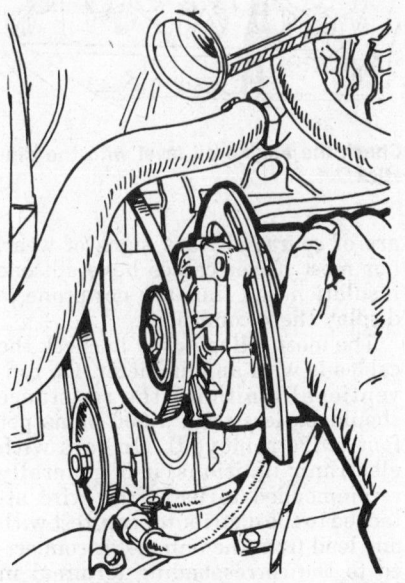

Push the component toward the engine to remove the belt

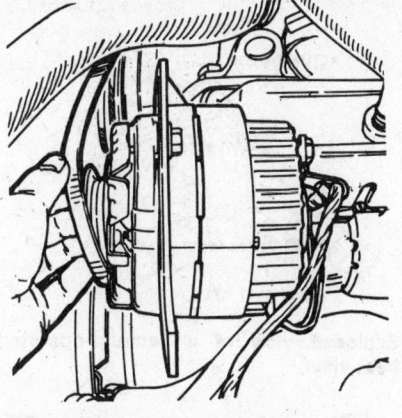

Slip the replacement belt over the pulley

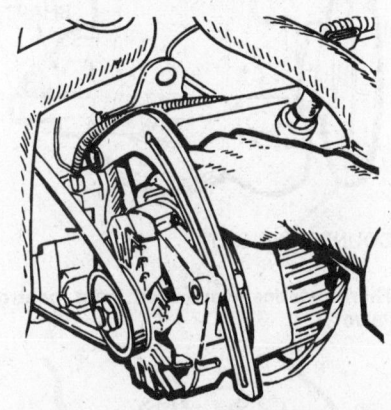

Pull outwards on the component to tension the belt, then tighten the bolts; recheck the belt tension after tightening

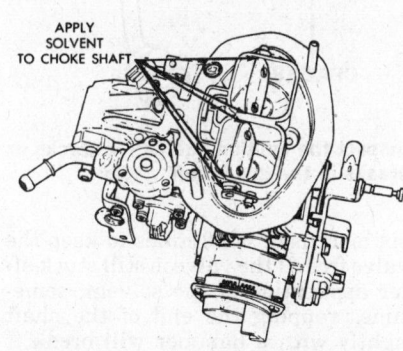

Use a spray solvent on the choke shaft, but do not apply any lubricants

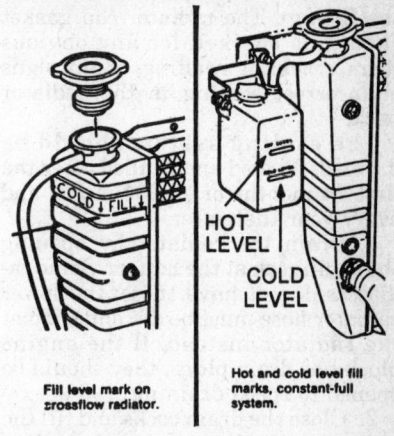

Fill level mark on crossflow radiator.

Hot and cold level fill marks, constant-full system.

Proper coolant level is about one inch below the radiator neck, or between the lines on the recovery tank

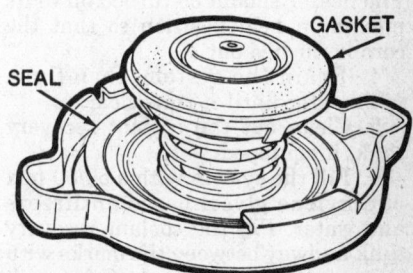

Check the radiator cap gasket and sealing surface

than too tight, because overtight belts will lead to bearing failure, particularly in the water pump and alternator. However, loose belts place an extremely high impact load on the driven component due to the whipping action of the belt.

CARBURETOR AND CHOKE LINKAGE

Every 12 months or 6000 miles, examine the carburetor linkage and choke plate for free movement. The choke plate action can generally be freed, if necessary, with the application of a solvent made for the purpose to the ends of the choke shaft. This solvent will also clean grease and dirt from the throttle linkage.

COOLING SYSTEM

Once a month, the engine coolant level should be checked. On cars without a coolant recovery system, this should only be done when the engine is cold.

Remove the radiator cap, the coolant level should be about one inch below the radiator filler neck.

CAUTION

To avoid injury when working with a hot engine, cover the radiator cap with a thick cloth. Wear a heavy glove to protect your hand. Turn the radiator cap slowly to the first stop, and allow all the pressure to vent (indicated when the hissing noise stops). When the pressure has been released, remove the cap the rest of the way.

On cars with a coolant recovery tank, coolant should be visible within the tank; as long as the coolant is between the markings on the tank, the level is correct.

If coolant is needed, a 50/50 mix of ethylene glycol-based antifreeze and water should always be used, both winter and summer. This is imperative on cars with air conditioning; without the antifreeze, the heater core could freeze when the air conditioning is used. Add coolant to the radiator if the car does not have a coolant recovery system. Add coolant to the recovery tank on cars so equipped.

The radiator hoses and clamps and the radiator cap should be checked at the same time as the coolant level. Hoses which are brittle, cracked, or swollen should be replaced. Clamps should be checked for tightness (screwdriver tight only; do not allow the clamp to cut into the hose or crush the fitting). The radiator cap gasket should be checked for any obvious tears, cracks or swelling, or any signs of incorrect seating in the radiator neck.

The cooling system should be drained, flushed and refilled after the first 24 months or 24,000 miles, and every year thereafter.

1. Drain the radiator by opening the drain cock at the bottom. Some radiators do not have these; the lower radiator hose must be disconnected at the radiator instead. If the engine block has drain plugs, they should be opened to speed draining.

2. Close the drain cocks and fill the system with clear water. A cooling system flushing additive can be used, if desired.

3. Run the engine until it is hot. The heater should be turned on to its maximum heat position so that the core is flushed out.

4. Drain the system, then flush with water until it runs clear.

5. Clean out the coolant recovery tank, if equipped.

6. Fill the system with a 50/50 mix of ethylene glycol-based antifreeze and water. Fill the coolant recovery tank midway between the marks with this mixture also (except G.M. cars, which should be filled to the "Full Cold" mark).

7. Run the engine until it is hot, then let it cool and top up the radiator or coolant recovery tank as necessary with the anti-freeze/water mixture.

HEAT RISER

The heat riser is a thermostatically or vacuum operated valve in the exhaust manifold. (Not all cars have one.) It closes when the engine is warming up, to direct hot exhaust gases to the intake manifold, in order to preheat the incoming fuel/air mixture. If it sticks open, the result will be frequent stalling during warmup, especially in cold and damp weather. If it sticks shut, the result will be a rough idle after the engine is warm.

NOTE: Some 1981 and later GM engines are equipped with an electrically heated ceramic grid mounted below the carburetor which takes the place of a heat riser.

The heat riser should move freely. It can be checked easily when the engine is cold by giving the counterweight on the valve shaft a twirl, or pulling the vacuum rod to open and shut the valve. If the valve is sticking or binding, a quick shot of solvent made for the purpose will free it up. This solvent should be applied every

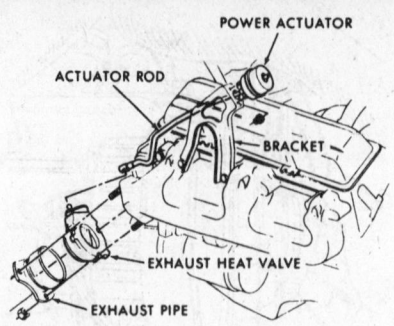

Exploded view of a vacuum-operated heat riser

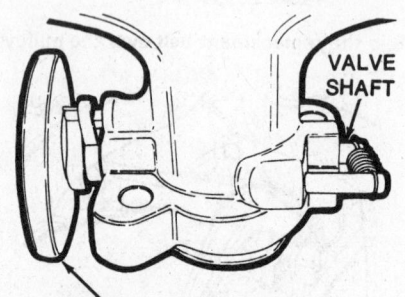

Thermostatically-operated heat control valve

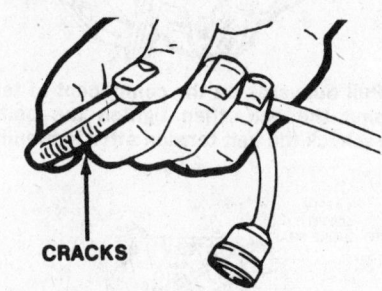

Inspect the ignition cables for cracks or breaks in the insulation

six months or 6000 miles to keep the valve free. If the valve is still stuck after application of the solvent, sometimes rapping the end of the shaft lightly with a hammer will break it loose. Otherwise, the components will have to be removed for further repairs.

IGNITION WIRES

The ignition system receives regular attention in the form of a tune-up, and thus is not covered here. But one of the most commonly overlooked components is the ignition cable, or spark plug wire.

Although they rarely show any visible signs of deterioration, the ignition cables should be checked at every tune-up, and replaced every 50,000 miles. Cracking and embrittlement

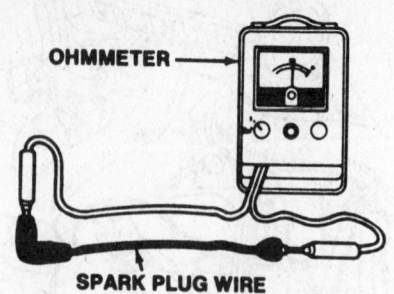

Test the ignition cables with an ohmmeter. Conventional ignition cables should be removed from the distributor cap, but electronic ignition wires should first be tested through the cap

Check the engine oil level with the dipstick

are of course obvious signs of wear, but most newer cables have silicone insulation and thus are not prone to display these conditions.

The most reliable way to check the cables is with an ohmmeter. On conventional ignitions, the resistance should be less than 7,000 ohms per foot (wire removed). On cars with electronic ignitions, it is generally recommended to leave the wire attached to the distributor cap, test with one lead from the ohmmeter connected to the corresponding terminal in the distributor cap, the other lead touched to the disconnected end of the cable at the spark plug. Then, if resistance seems close to the limit, remove the wire from the cap and retest. In general, the spark plug wires on electronic ignitions should be replaced if the total resistance is over 36,000 ohms. (50,000 ohms on Ford and Chrysler products).

Always replace the cables with new ones of the same type. Replace the wires one at a time, working from the longest to the shortest.

OIL LEVEL

The engine oil should be checked on a regular basis, ideally at each fuel stop, or once a week. It is best to check when the engine is at operating temperature, but checking the level immediately after shutting off the engine will give a false reading, because all of the oil will not yet have drained back into the crankcase. The car should be parked on a level surface to obtain an accurate reading.

1. Remove the oil dipstick. Wipe it clean, then replace it, seating it firmly.

2. Remove the dipstick again and hold it horizontally to prevent the oil from running. The level should be between the "Add" and "Full" marks on the dipstick. The dipstick may be marked "Add" and "Safe", or may have lines scribed on it; in any case, the oil level should be above the lower marking.

3. If the oil is below the lower mark, enough oil should be added to the engine to raise the level to the upper mark. The markings are usually spaced so that one-half to one quart of oil will raise the level from the "Add" mark to the "Full" mark. Oil is added through the capped opening in the valve cover. Only oils labeled SE or SF should be used; select a viscosity that will be compatible with the temperatures expected until the next drain interval.

NOTE: The diesel engines used in GM cars require the use of SF/CC or SF/CD type oils only. Do not use oil which is rated for SE or SF use only, or which is rated for CD use. Do not use the oil if the rating CD appears anywhere on the can, either alone or in combination with ratings other than SF, such as SE/CD. The use of CD type oil will void the manufacturer's warranty, and may cause expensive engine damage and leakage.

4. Replace the dipstick, then check the level again after any additions of oil. Be careful not to overfill, which will lead to leakage and seal damage.

POWER STEERING

The power steering fluid level is checked with a dipstick inserted into the pump reservoir. The dipstick may be attached to the reservoir cap, or inserted into a tube on the pump body. The level should be checked at every oil change. On all cars except Ford products, the level can be checked with the fluid either warm or cold; on Fords, the engine must be at operating temperature.

1. On Ford products, with the engine hot and idling, turn the steering wheel back and forth to the full right and full left stops several times, then center the wheels and shut off the engine.

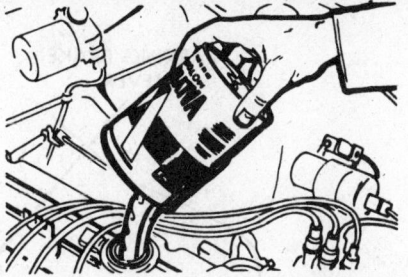

Add oil through the valve cover

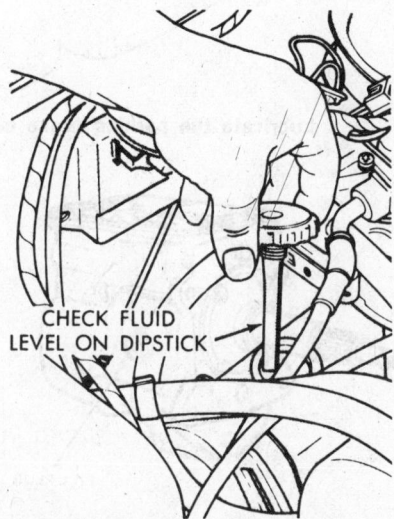

CHECK FLUID LEVEL ON DIPSTICK

The power steering level is checked with the dipstick installed in the reservoir

2. On all cars, with the engine off, pull or unscrew the dipstick and check the level. If the engine is warm, the level should be between the "Hot" and "Cold" marks on the dipstick; on Fords, the level should be between the "Cold Full" and "Hot Full" marks. If the engine is cold, the fluid should be between the "Add" and "Cold" marks; this does not apply to Ford products.

3. If the level is low, add power steering fluid until correct. Be careful not to overfill, which will cause fluid loss and seal damage.

WINDSHIELD WASHER FLUID

Check the fluid level in the windshield washer tank at every oil level check. The fluid can be mixed in a 50% solution with water, if desired, as long as temperatures remain above freezing. Below freezing, the fluid should be used full strength. Never add engine coolant antifreeze to the washer fluid, because it will damage the car's paint.

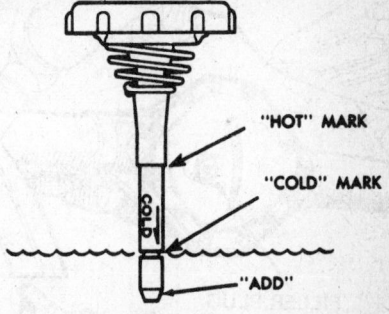

"HOT" MARK
"COLD" MARK
"ADD"

Power steering dipstick markings, typical of all types except Ford

Under Car

AXLE

The fluid level in the drive axle should be checked every 12 months or 12,000 miles. On the front wheel drive Omni, Horizon, Aries, Reliant LeBaron and Dodge 400 with automatic transmission, the drive axle lubricant is separate from the automatic fluid and must be checked separately. The level can be checked through the fill plug in the drive axle housing.

On the American Motors Eagle, SX/4 and Kammback, both drive axles should be checked. Both assemblies have fill plugs for this purpose.

1. With the car parked on a level surface, remove the filler plug. The plug can be found either in the rear cover of the differential, or on the front of the pinion housing.

2. If lubricant dribbles out when the plug is removed, the level is correct. Otherwise, stick in your finger (watch out for sharp threads); the fluid should be even with or just a little below the filler hole.

3. If lubricant is needed, use SAE 80W-90 GL-5 gear oil (SAE 80W GL-5 in very cold climates) to fill standard axles. Limited slip axles require a special lubricant, available in auto parts stores. The Omni, Horizon, Aries, Reliant Dodge 400 and LeBaron drive axles should be filled with DEXRON II ATF fluid.

4. When the level is correct, install the plug and tighten until snug. Do not overtighten.

Drive axles should be drained and refilled according to the manufacturer's maintenance schedule, usually found in the owner's manual. If the unit is used in severe driving conditions (trailer towing, etc.) the lubricant should be changed more often. Some later model drive axles do not require regular draining and refilling. Refer to the owner's manual for information on this subject. The axle may be drained by removing the drain

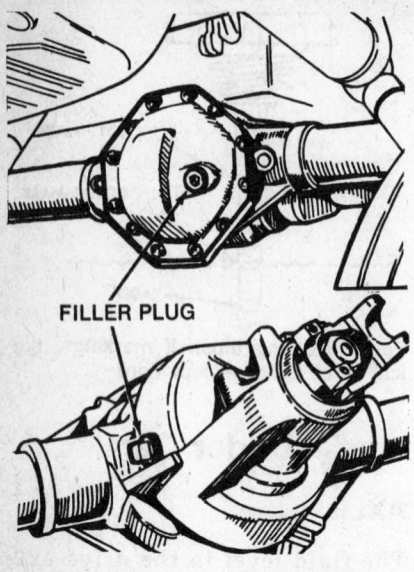

FILLER PLUG

Rear axle filler plug locations

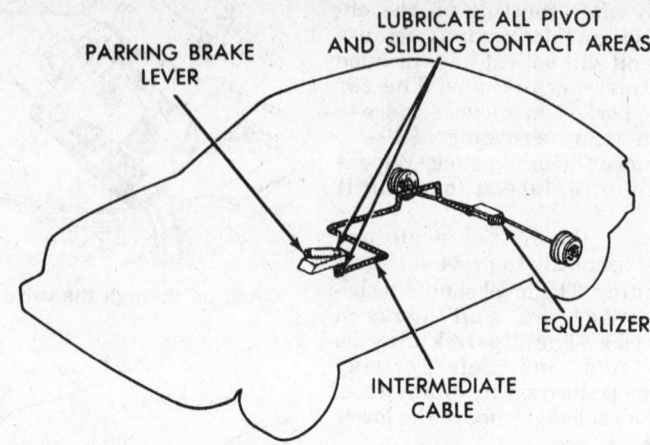

PARKING BRAKE LEVER

LUBRICATE ALL PIVOT AND SLIDING CONTACT AREAS

EQUALIZER

INTERMEDIATE CABLE

Lubricate the parking brake cable with white waterproof grease

SEALANT

Apply a bead of silicone sealer to the rear cover if no gasket is used

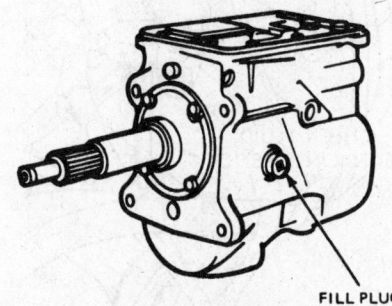

FILL PLUG

MANUAL TRANSMISSION
FILL TO BOTTOM OF FILLER HOLE WITH VEHICLE ON LEVEL GROUND.

Typical manual transmission filler plug location

plug at the bottom of the axle housing, if present. Otherwise the rear cover (if equipped) must be removed or a suction gun used through the filler hole. Always use silicone sealer or a gasket when re-installing the rear cover. Run sealer around the insides of the bolt holes. Tighten the bolts a few turns at a time in a crisscross pattern.

EXHAUST SYSTEM

The exhaust system should be checked twice a year for general soundness. Inspect the pipes for holes, broken welds, leaking seams, or loose connections. Leaks at connections can sometimes be successfully repaired with the use of a commercial exhaust pipe sealer, but holes or breaks warrant replacement of the part. The exhaust pipe hangers and straps should be examined for any breaks or cracks;

replace these as necessary. Some slight cracking of rubber hangers is normal, but deep cracks or cuts are cause for replacement.

——— CAUTION ———
Check the exhaust system only when it is cold. The temperature on an exhaust system using a catalytic converter can reach 1000°F after only a short period of engine operation.

MANUAL TRANSMISSION OR MANUAL TRANSAXLE

The fluid level in the manual transmission (or transaxle on front wheel drive cars) should be checked twice a year, or every 6000 miles.

1. Park the car on a level surface. The transmission should be cool to the touch.

2. Remove the filler plug from the side of the transmission or transaxle. If lubricant trickles out as the plug is removed, the fluid level is correct. If not, stick in your finger (watch out for

sharp threads); the lubricant should be right up to the edge of the filler hole.

3. If lubricant is needed, use SAE 80W-90 GL-5 gear lubricant (SAE 80W GL-5 in extremely cold climates) in manual transmission.

Front wheel drive transaxles use different lubricants. The Omni and Horizon with the A412 transaxle (starter on the radiator side of the engine) require GL-4 hypoid gear lubricant; the same SAE viscosities apply (80W-90 or 80W; 75W in temperatures below -30°F). GL-5 classification lubricants are specifically not recommended. Omnis and Horizons with the A460 transaxle (starter on the firewall side of the engine), and all Aries, Reliant LeBaron and Dodge 400 models use DEXRON II automatic transmission fluid.

The front wheel drive Citation, Omega, Phoenix, Skylark Cavalier, J2000, Cimarron, Celebrity, Cierra and 6000 require DEXRON II automatic transmission fluid. The use of a manual transmission lubricant is specifically not recommended.

The Ford Escort, EXP, and Mercury Lynx and LN-7 use Ford Type F automatic transmission fluid. The use of a manual transmission lubricant is specifically not recommended.

4. When the level is correct, install the filler plug and tighten until snug.

PARKING BRAKE LINKAGE

The parking brake cable assembly should be inspected twice a year for fraying, kinks, and binding. A smooth white waterproof lubricant should be applied at the same time to all pivot points and areas in sliding contact.

SUSPENSION LUBRICATION

Depending on the year of manufacture, there may be as many as twelve grease fittings on the suspension parts, or as few as two. Typical locations for grease nipples are on the ball joints, control arm pivot points, steering linkage, and the tie-rod ends.

Lubricate these fittings with a small hand operated grease gun filled with EP chassis lubricant. Pump grease into the fitting slowly, until it begins to ooze out around the joint, or until the grease begins to expand the rubber boot around the fitting. Be extremely careful not to rupture any seals or boots, as this will lead to lubricant loss and contamination of the parts involved.

Occasionally, the grease nipples may become clogged with dirt or hardened grease. If so, unscrew them with a wrench of the proper size and clean them out with solvent. When reinstalled, they may be covered with plastic caps made for the purpose, or a piece of aluminum foil.

The chassis and suspension parts should be lubricated once a year, or every 7500 miles, whichever comes first.

TRANSFER CASE

If you have a four-wheel drive AMC car, you should check the transfer case lubricant level every 5000 miles.

1. Park the car on a level surface.
2. Check the build date tag on the rear of the transfer case.
3. If the transfer case was built after March 1980, the fill plug will be at location "A" in the illustration. Remove the fill plug. The lubricant should be right up to the edge of the filler hole. Check and correct as necessary.
4. If the transfer case was built before March, 1980, the filler plug may be in any one of the four locations shown in the illustration. Check to see which one you have, then remove the filler plug. Use a length of wire to measure the distance from the bottom edge of the fill hole to the lubricant. The correct distance depends on the location of the hole:
   • "A" 0.56 inch
   • "B" 1.13 inch
   • "C" 1.20 inch
   • "D" 0.56 inch
5. The correct fluid to use is 10W-30 SE or SF motor oil. Capacity is 4.0 pints, regardless of when the transfer case was built. Some early owner's manuals may have listed the capacity as 3.0 pints, but this is incorrect; revised publications call for a capacity of 4.0 pints.

The transfer case should be drained and refilled every 15,000 miles. The drain plug is located at the lower edge of the rear face of the case. Installation torque for the plugs is 18 ft. lbs. The case is made from aluminum, so this figure should not be exceeded.

Exterior

DRAIN HOLES AND UNDERBODY

Most cars have drain holes spaced along the lower edge of the rocker panels and doors. These holes should be cleared of any debris or rust twice a year. A small screwdriver can be used to open plugged drain holes.

Every spring, the underbody should be flushed with clear water to remove deposits of mud, road salt, and debris. It is advisable to loosen any packed-in sediment before flushing to assure a more thorough cleaning.

HINGES AND LOCKS

Once a year, the door, hood, and trunk hinges, and all locks should be lubricated to ensure smooth operation. The hinge points should be lightly oiled. Lock cylinders may be easily lubricated with a shot of silicone spray directed into the keyhole. Silicone lubricant also works well on the door latch mechanisms, and keeps the door, trunk, and window weather seals pliable when applied in a light film.

TIRES

Tires should be checked weekly for proper air pressure. A chart, located either in the glove compartment or on the driver's or passenger's door, gives the recommended inflation pressures. Maximum fuel economy and tire life will result if the pressure is maintained at the highest figure given on the chart.

Pressures should be checked before driving since pressure can increase as much as six pounds per square inch (psi) due to heat buildup. It is a good idea to have your own accurate pressure gauge, because not all gauges on service station air pumps can be trusted. When checking pressures, do not neglect the spare tire. Note that some spare tires require pressures considerably higher than those used in the other tires.

While you are about the task of checking air pressure, inspect the tire treads for cuts, bruises and other damage. Check the air valves to be sure that they are tight. Replace any missing valve caps.

Check the tires for uneven wear that might indicate the need for front end alignment or tire rotation. Tires

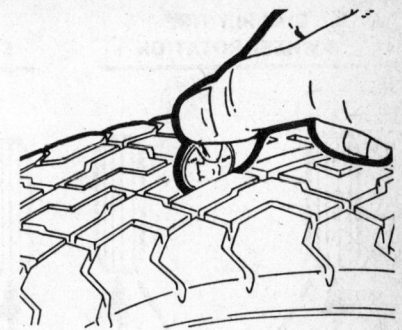

Tire tread depth can be checked with a penny. If the top of Lincoln's head is visible, the tires are due for replacement

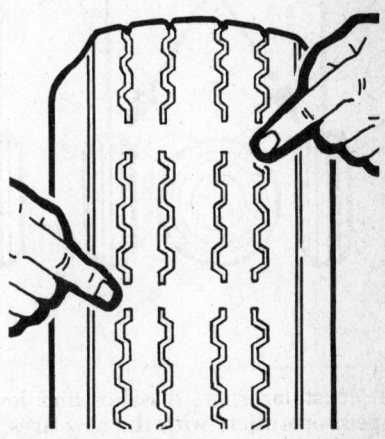

Tread wear indicators will appear as a band across the tire when the tread has worn out.

should be replaced when a tread wear indicator appears as a solid band across the tread.

When buying new tires, give some thought to the following points, especially if you are considering a switch to larger tires or a different profile series:

1. All four tires must be of the same construction type. This rule cannot be violated. Radial, bias, and bias-belted tires must not be mixed.
2. The wheels should be the correct width for the tire. Tire dealers have charts of tire and rim compatibility. A mismatch will cause sloppy handling and rapid tire wear. The tread width should match the rim width (inside bead to inside bead) within an inch. For radial tires, the rim width should be 80% or less of the tire (not tread) width.
3. The height (mounted diameter) of the new tires can change speedometer accuracy, engine speed at a given road speed, fuel mileage, acceleration, and ground clearance. Tire manufacturers furnish full measurement specifications.
4. The spare tire should be usable,

BIAS PLY TIRE 4-WHEEL ROTATION	BIAS PLY TIRE 5-WHEEL ROTATION	RADIAL PLY TIRES 4-WHEEL ROTATION	RADIAL PLY TIRES 5-WHEEL ROTATION

Tire rotation diagrams

at least for short distance and low speed operation, with the new tires.

5. There shouldn't be any body interference when loaded, on bumps, or in turns.

Tire Rotation

Tire rotation is recommended every 6000 miles or so, to obtain maximum tire wear. The pattern you use depends on whether or not your car has a usable spare. Radial tires should not be cross-switched (from one side of the car to the other); they last longer if their direction of rotation is not changed. Snow tires sometimes have directional arrows molded into the side of the carcass; the arrow shows the direction of rotation. They will wear very rapidly if the rotation is reversed. Studded tires will lose their studs if their rotational direction is reversed.

NOTE: Mark the wheel position or direction of rotation on radial tires or studded snow tires before removing them.

Storage

Store the tires at the proper inflation pressure if they are mounted on wheels. Keep them in a cool dry place, laid on their sides. If the tires are stored in the garage or basement, do not let them stand on a concrete floor; set them on strips of wood.

WINDSHIELD WIPERS AND WASHERS

For maximum effectiveness and longest element life, the windshield and wiper blades should be kept clean. Dirt, tree sap, road tar and so on will cause streaking, smearing and blade deterioration if left on the glass. It is advisable to wash the windshield carefully with a commercial glass cleaner at least once a month. Wipe off the rubber blades with the wet rag afterwards.

For access to the blades on wiper systems which park below the hood line, turn the ignition key to "On" and run the wipers to the center of the windshield. Shut the wipers off with the ignition key, not the wiper switch. Do not attempt to move the wipers by hand; damage to the motor and drive mechanism will result.

If the blades are found to be cracked, broken or torn, they should be replaced immediately. Replacement intervals will vary with usage, although ozone deterioration usually limits blade life to about one year. If the wiper pattern is smeared or streaked, or if the blade chatters across the glass, the elements should be replaced. It is easiest and most sensible to replace the elements in pairs.

There are basically three different types of refills, which differ in their

method of replacement. One type has two release buttons, approximately one-third of the way up from the ends of the blade frame. Pushing the buttons down releases a lock and allows the rubber filler to be removed from the frame. The new filler slides back into the frame and locks in place.

The second type of refill has two metal tabs which are unlocked by squeezing them together. The rubber filler can then be withdrawn from the frame jaws. A new refill is installed by inserting the refill into the front frame jaws and sliding it rearward to engage the remaining frame jaws. There are usually four jaws; be certain when installing that the refill is engaged in all of them. At the end of its travel, the tabs will lock into place on the front jaws of the wiper blade frame.

The third type is a refill made from polycarbonate. The refill has a simple locking device at one end which flexes downward out of the groove into which the jaws of the holder fit, allowing easy release. By sliding the new refill through all the jaws and pushing through the slight resistance when it reaches the end of its travel, the refill will lock into position.

Regardless of the type of refill used, make sure that all of the frame jaws are engaged as the refill is pushed into place and locked. The metal blade holder and frame will scratch the glass if allowed to touch it.

TRICO

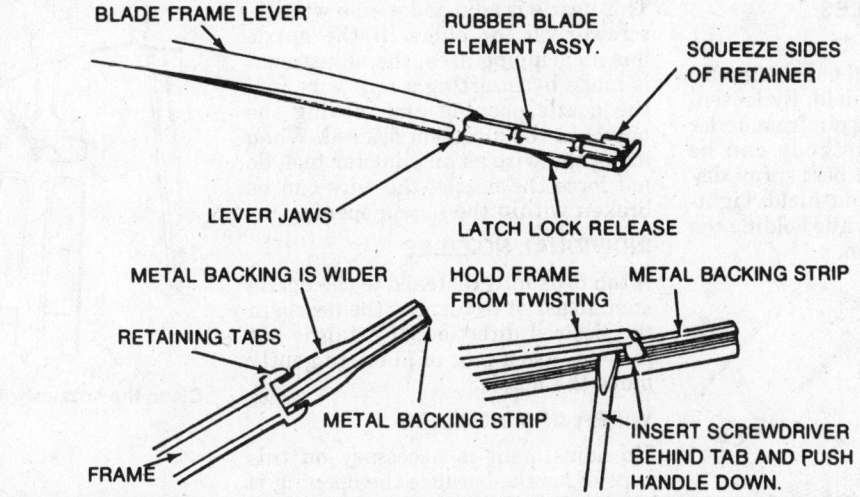

BLADE FRAME LEVER

RUBBER BLADE ELEMENT ASSY.

SQUEEZE SIDES OF RETAINER

LEVER JAWS

LATCH LOCK RELEASE

METAL BACKING IS WIDER

HOLD FRAME FROM TWISTING

METAL BACKING STRIP

RETAINING TABS

METAL BACKING STRIP

FRAME

INSERT SCREWDRIVER BEHIND TAB AND PUSH HANDLE DOWN.

ANCO

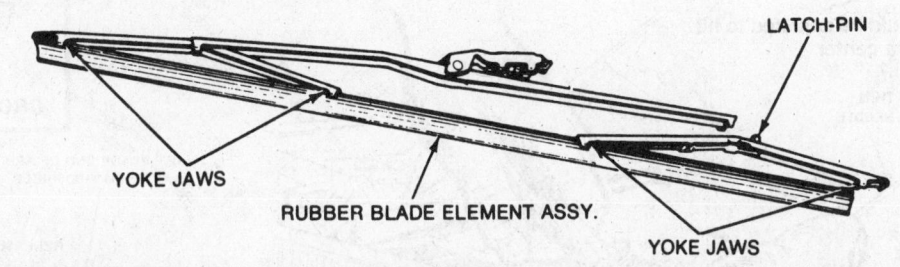

LATCH-PIN

YOKE JAWS

RUBBER BLADE ELEMENT ASSY.

YOKE JAWS

POLYCARBONATE

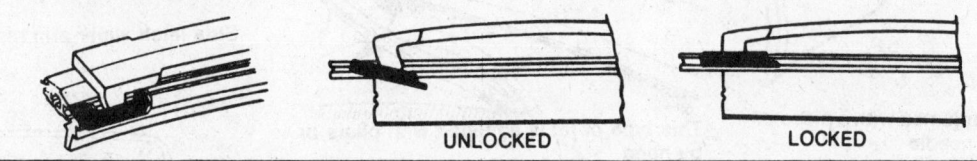

UNLOCKED

LOCKED

TRIDON

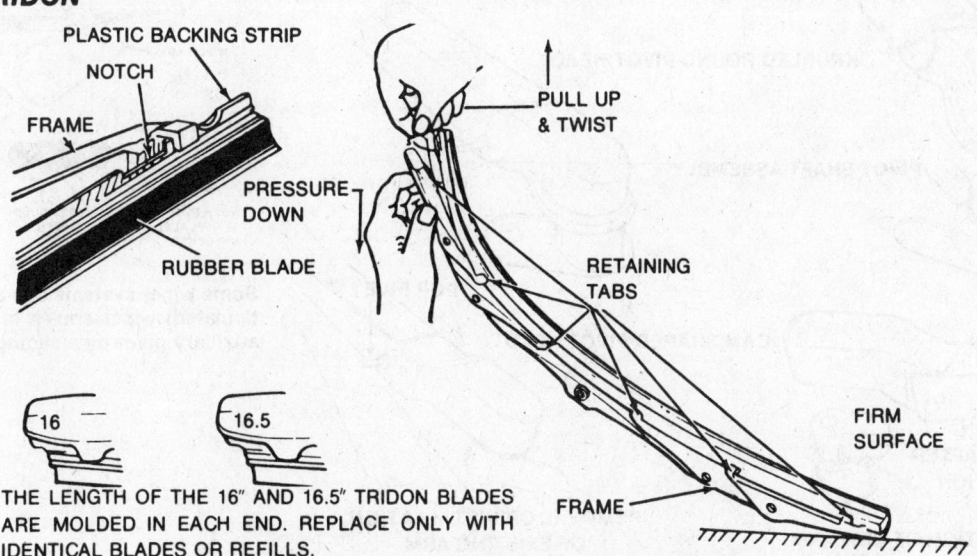

PLASTIC BACKING STRIP

NOTCH

FRAME

PULL UP & TWIST

PRESSURE DOWN

RUBBER BLADE

RETAINING TABS

16   16.5

FIRM SURFACE

FRAME

THE LENGTH OF THE 16" AND 16.5" TRIDON BLADES ARE MOLDED IN EACH END. REPLACE ONLY WITH IDENTICAL BLADES OR REFILLS.

Windshield wiper blade replacement methods

Washer Nozzle Adjustment

CENTERED SINGLE POST—NON-ADJUSTABLE NOZZLES

This type is usually located on the rear center of the hood panel, directly in front of the windshield. By loosening the body retaining nut from under the hood, the nozzle body can be turned to provide the best spray discharge to cover the windshield. Tighten the retaining nut while holding the nozzle body in position.

CENTERED SINGLE POST—ADJUSTABLE NOZZLES

This nozzle is adjusted with a wrench, screwdriver, or pliers. If the nozzle has no gripping area, the adjustment is made by inserting a stiff wire into the nozzle opening and moving the nozzle in the direction desired. When using the wire as an adjuster tool, do not force the nozzle; the wire can be broken within the nozzle opening.

INDIVIDUAL NOZZLES

A tab is usually fastened to the nozzle stem to assist in turning the nozzle in the desired direction. If a tab is not present, use a pair of pliers to gently move the nozzle.

WIPER ARM NOZZLES

No adjustment is necessary on this type of nozzle, because the opening is centered on the wiper arm and moves along with the arm.

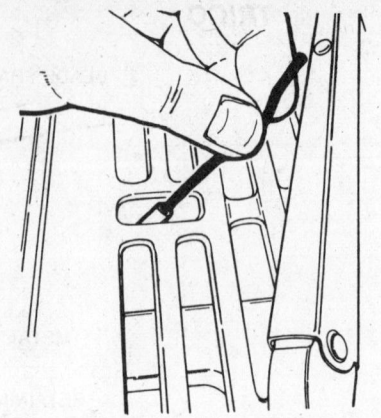

Clean the nozzles with a piece of fine wire

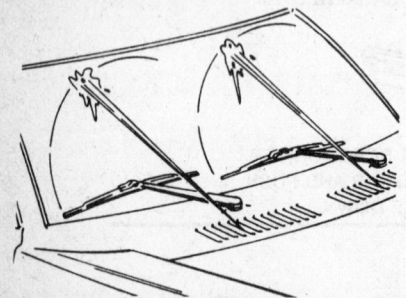

Washer nozzles should be adjusted to hit the windshield above center

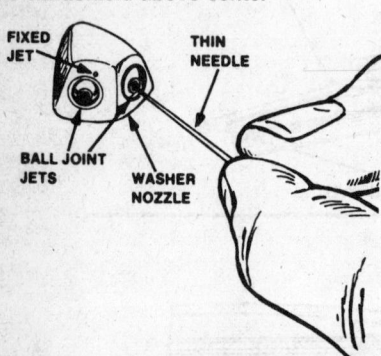

FIXED JET THIN NEEDLE BALL JOINT JETS WASHER NOZZLE

Some jets can be adjusted with a piece of fine wire or a thin needle

This type of jet is adjusted with pliers or by hand

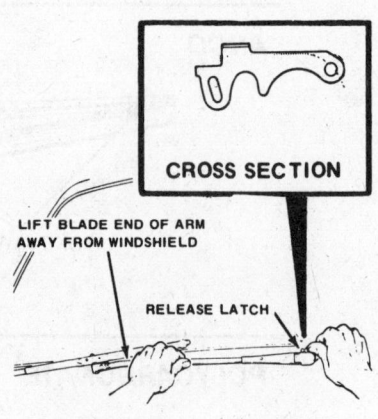

CROSS SECTION

LIFT BLADE END OF ARM AWAY FROM WINDSHIELD

RELEASE LATCH

Side latch wiper arm replacement

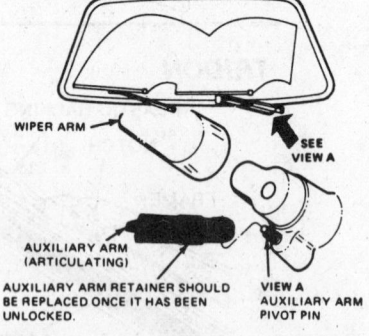

WIPER ARM SEE VIEW A AUXILIARY ARM (ARTICULATING) AUXILIARY ARM RETAINER SHOULD BE REPLACED ONCE IT HAS BEEN UNLOCKED. VIEW A AUXILIARY ARM PIVOT PIN

Some wiper systems use an auxiliary (articulated) wiper arm. It is secured to an auxiliary pivot by a sliding lock.

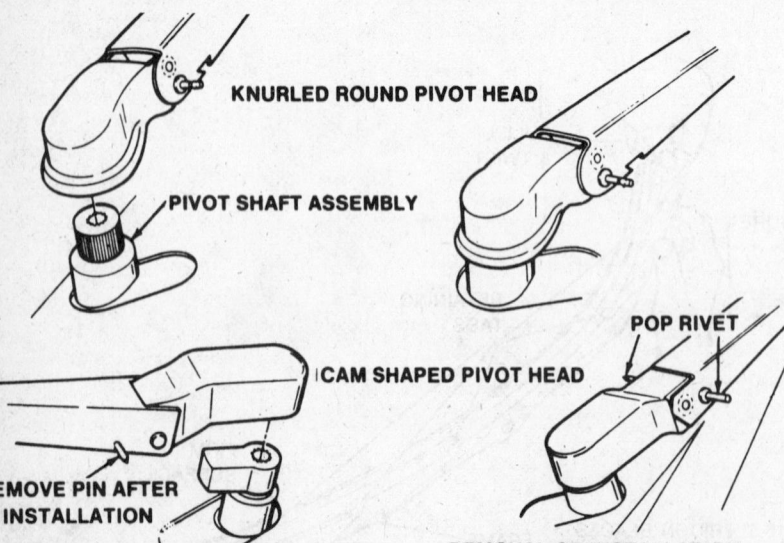

KNURLED ROUND PIVOT HEAD

PIVOT SHAFT ASSEMBLY

CAM SHAPED PIVOT HEAD

POP RIVET

REMOVE PIN AFTER INSTALLATION

INSTALLATION OF NEW ARM

REMOVAL OR INSTALLATION OF EXISTING ARM

Pin and hole type wiper arm replacement

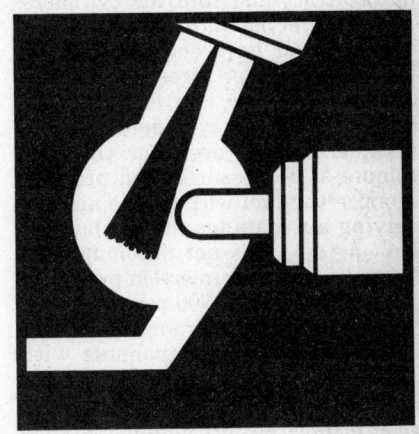

Diesel Maintenance

NOTE: Standard maintenance procedures are given here while component removal, installation and adjustment procedures are given in the appropriate car section.

How The Diesel Engine Works

Four-stroke diesels require four piston strokes for the complete cycle of actions, exactly like a gasoline engine. The difference lies in how the fuel mixture is ignited. A diesel engine does not rely on a conventional spark ignition to ignite the fuel mixture for the power stroke. Instead, a diesel relies on the heat produced by compressing air in the combustion chamber to ignite the fuel and produce a power stroke. This is known as a compression-ignition engine. No fuel enters the cylinder on the intake stroke, only air. At the end of the compression stroke, fuel is sprayed into the precombustion chamber (prechamber). The mixture ignites and spreads out into the main combustion chamber, forcing the piston downward (power stroke). The fuel/air mixture ignites because of the very high combustion chamber temperatures generated by the extraordinarily high compression ratios used in diesel engines. Typically, the compression ratios used in automotive diesels run anywhere from 16:1 to 23:1. A typical spark-ignition engine has a ratio of about 8:1. This is why a spark-ignition engine which continues to run after you have shut off the engine is said to be "dieseling". It is running on combustion chamber heat alone.

Designing an engine to ignite on its own combustion chamber heat poses certain problems. For instance, although a diesel engine has no need for a coil, spark plugs, or a distributor, it does need what are known as "glow plugs". These superficially resemble spark plugs, but are only used to warm the combustion chambers when the engine is cold. Without these plugs, cold starting would be impossible, due to the enormously high compression ratios and the characteristics of the diesel fuel itself.

All diesel engines use fuel injection, because unlike spark-ignited engines, the fuel cannot be drawn through the

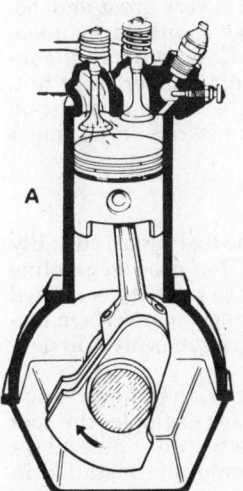
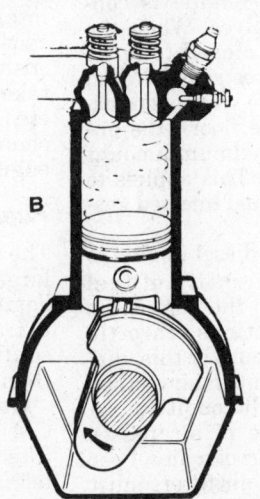
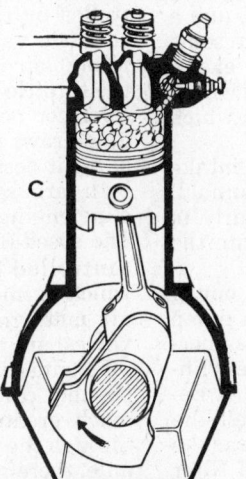
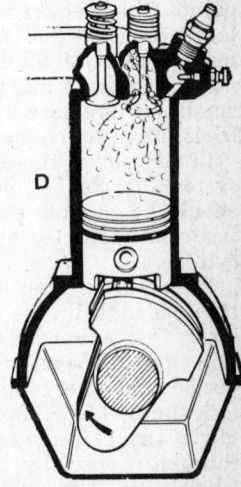

Cycles of a four-stroke cycle diesel engine. (A) Intake stroke: The downward movement of the piston draws air into the cylinder through the open intake valve. (B) Compression: The intake valve closes and the piston moves upward, compressing the air in the cylinder. (C) Ignition: While the piston is at approximately top dead center, fuel is injected into the cylinder. The superheated air (from compression) ignites the fuel, causing an explosion which forces the piston downward. (D) Exhaust: The piston begins moving upward, forcing the exhaust gas out of the cylinder, through the open exhaust valve.

intake tract and into the cylinders. The introduction of fuel into a diesel engine must be precisely timed so that each cylinder "fires" at the proper moment. Also, the fuel injection pressure (at the cylinder) must be great enough to overcome the high compression pressures, and properly atomize the fuel without the aid of a moving air mass (as in a carbureted gas engine). It is not uncommon for diesel engine fuel injection pressures to be set at 1500–1700 psi.

Diesel engines share many of their basic mechanical components with gasoline engines, though the cylinder block, head(s), crankshaft, connecting rods, pistons, etc. are manufactured to be much stronger for use in diesel engines. The additional strength of the components is necessary due to the very high cylinder pressure generated within the diesel engine.

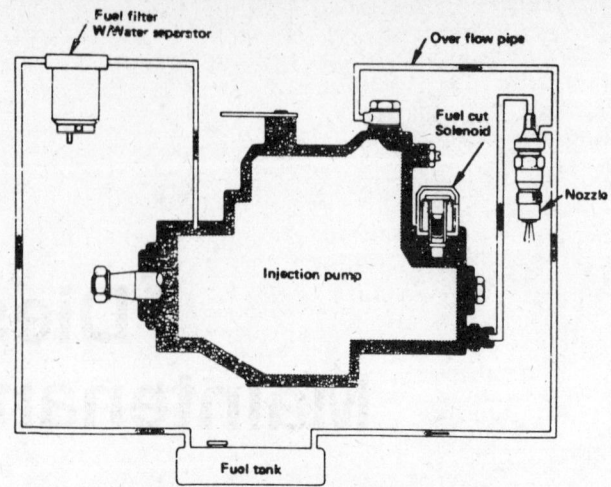

Chevette diesel fuel system schematic

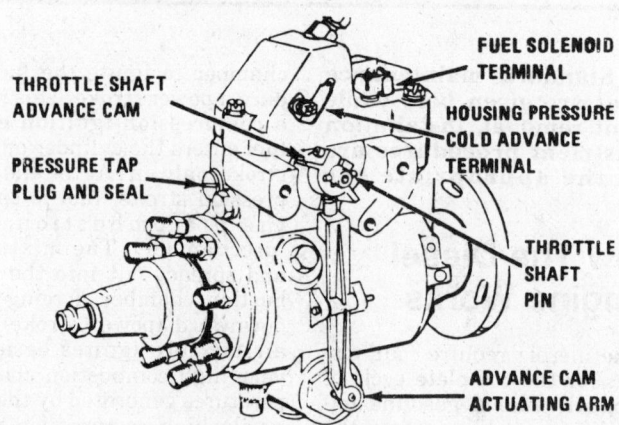

1981 and later GM V8 diesel fuel injection pump with Housing Pressure Cold Advance (HPCA) system

Maintenance Procedures

Maintenance procedures for the diesel engine generally fall into three categories:
1. Fuel system
2. Starting system
3. Engine mechanical systems
Of these, the fuel system is usually the most likely source of engine troubles, and should be high on the list for regular maintenance attention.

FUEL SYSTEM

The typical diesel engine fuel system consists of fuel tank, fuel feed and return lines, mechanical fuel injection pump, fuel injectors and lines, and a large capacity fuel filter. On some models, the GM V8 diesel for example, the engine is also equipped with a small, low pressure fuel pump which feeds the injection pump.

In addition to these, the air intake system (air cleaner, intake manifold) should be checked over regularly to insure unrestricted air flow into the cylinders.

In operation, fuel is drawn out of the fuel tank by the injection pump (or its feed pump) and fed by the injection pump to the injectors in the cylinder head at a very high pressure. Before the fuel is allowed to enter the main injection pump, it passes through a specially built fuel filter which traps solid particles (and water on some models) in the fuel. Fuel that is not used is pumped back to the fuel tank through the fuel return lines. This recirculated fuel helps cool the injection pump.

Air Cleaner

On a gasoline engine, the volume of air taken in by the engine is controlled by throttle valves. When the throttle valves are closed (engine idling), air intake is restricted. When the throttle valves are wide open (accelerator pedal to the floor), the engine draws in the maximum amount of air it possibly can. This applies to both carbureted and fuel injected gasoline engines.

The speed (rpm) of a diesel engine is controlled by the quantity of fuel which is injected into the engine; no air metering restrictions (throttle valves) are used. Because of this, diesel engines ingest as much air as they possibly can under all conditions. A much greater volume of air passes through the air cleaner of a diesel per mile, therefore, diesel air filters must either be larger or the filter replacement intervals more frequent than those of a similarly sized gasoline engine.

One word of caution; never remove the air cleaner on a diesel with the engine running, and never run the engine with the air cleaner removed. The volume of air drawn through the intake manifold is very great and, because the intake manifold is unobstructed, anything drawn into the intake manifold (air cleaner wing nut, etc.) goes straight to the combustion chambers, where it can cause major engine damage.

Fuel Filter

The diesel engine fuel filter is usually larger than the filter used on gasoline engines. The extra capacity is needed to trap the suspended particles in diesel fuel, which is generally "dirtier" than gasoline.

On some engines, the Chevette and GM V6 diesels, for example, the fuel filter looks like a second engine oil filter, and is removed and installed in the same manner as the canister-type oil filter. On GM V8 engines, the fuel filter is located at the rear of the engine and is unbolted from its bracket after its fuel lines are disconnected. See the Chevette car section for diesel fuel filter removal and installation. The fuel filter must be changed according to the manufacturer's suggested interval. See the owner's manual for information.

After installing the fuel filter on GM V8s, start the engine and check for leaks. Run the engine for about two minutes, then stop the engine for the same amount of time to allow any air trapped in the injection system to bleed off. Many diesels also have a small, in-tank filter which is usually maintenance-free.

Water In Fuel

Diesel fuel is a hydrophilic fluid; it naturally attracts water. Since diesel fuel and water do not mix, the water remains floating beneath the fuel at the bottom of the tank. This water must be removed every now and then, or it will be drawn into the fuel circuit and pass through the injection system, causing corrosion and possible component failure (injection pumps can cost up to $1,000). Water in the fuel system will also cause the engine to run poorly, if at all.

Most diesel fuel tanks are equipped with a separator which can isolate from 1 to 3 gallons of water from the fuel. Many GM diesels are also equipped with "Water in Fuel" lights in the dashboard which warn of the presence of H_2O in the fuel tank. These warning systems can be installed by the dealer on GM models not so equipped.

On some diesels, such as the Chevette, there is a water catcher in the bottom of the fuel filter which can easily be bled off. In addition, there are several bolt-on water filters on the market which attach to the fuel line under the hood and separate water from the fuel. Depending on which kind you buy, draining water from the system is simply a matter of opening the petcock at the bottom of the filter and letting the water drain out, or, if money is no object, a separator is available on which water is drained from the filter simply by activating a switch on the dashboard.

Bleeding Water From The Chevette Diesel Fuel Filter

1. Place a 4 pint see-through container at the end of the vinyl hose beneath the drain plug on the filter.
2. Open the drain plug approximately 4 turns.
3. Operate the priming pump handle at the top of the filter by pumping it about 10 times or until all the water is drained out. The water will collect at the bottom of the see-through container and the diesel fuel will float on

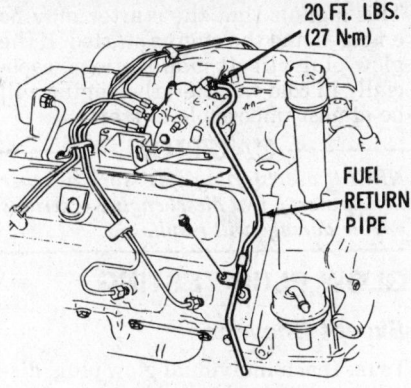

20 FT. LBS. (27 N·m)

FUEL RETURN PIPE

Fuel return pipe

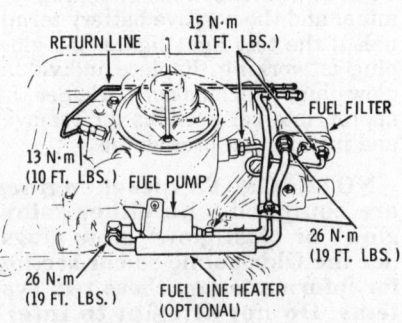

RETURN LINE — 15 N·m (11 FT. LBS.)

FUEL FILTER

13 N·m (10 FT. LBS.) — FUEL PUMP

26 N·m (19 FT. LBS.)

26 N·m (19 FT. LBS.) — FUEL LINE HEATER (OPTIONAL)

Top view of the GM V6 diesel engine. Note the location of the fuel filter.

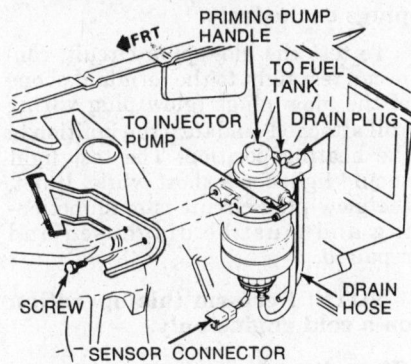

FRT — PRIMING PUMP HANDLE — TO FUEL TANK

TO INJECTOR PUMP — DRAIN PLUG

SCREW — DRAIN HOSE

SENSOR CONNECTOR

Chevette diesel fuel filter assembly, showing drain plug and hose and fuel priming pump

top of it. When the pump is pushing through nothing but diesel fuel system bleeding is complete.

4. Close the drain plug and again operate the pump handle up and down several times to prime the fuel system.
5. Start the engine and check for leaks. Make sure the "Water in Fuel" light in the instrument panel goes off. If it doesn't the water in the fuel tank will have to be drained. See procedure below.

Removing Water From The Fuel Tank

Treat diesel fuel with the same re-

spect you would gasoline, and after the procedure, properly dispose of the fuel.

1. Remove the fuel tank cap.
2. Connect a pump or siphon hose to the $\frac{1}{4}$ in. fuel return hose (smaller of the two fuel hoses) above the rear axle, or under the hood near the fuel pump (on the passenger's side of the engine, near the front).
3. Siphon until all water is removed from the tank. Do not use your mouth to create siphon vacuum, EVER! The best method is to siphon the water into a large capacity see-through container. The water will collect at the bottom of the container. There are several different types of siphon pumps available in auto parts and hardware stores that makes fuel tank draining both easy and safe.
4. When all water has been removed from the tank, be sure to reinstall the fuel return hose and fuel cap.

NOTE: If the entire fuel system (not just the tank) is contaminated by water, the vehicle must be stopped immediately and the fuel system must be purged. This includes draining and removing the fuel tank, blowing low pressure compressed air backwards through the fuel feed and return lines, and bleeding the water out of all injection components. This job should be referred to a qualified technician.

COLD WEATHER FUEL SYSTEM MAINTENANCE

——— **CAUTION** ———
NEVER use "starting aids" such as ether to help start a GM diesel engine—serious engine damage will result.

As will be explained later under "Fuel Recommendations", diesel fuel tends to become "cloudy", or thicker, as the temperature drops. The thicker the diesel fuel becomes, the slower it flows through the fuel system, until finally it stops flowing altogether somewhere near the bottom of the thermometer. One way to fight sluggish fuel flow is to use winterized blends of diesel fuel or straight No. 1 diesel fuel.

Another way is to install an aftermarket fuel system pre-heater. These are generally canisters which connect into the fuel line and use coolant from the engine cooling system to heat the fuel before it reaches the injection pump. The one drawback with this system is the engine must be started before the pre-heater begins to work. Also available are electric fuel warmers. These pre-heat the fuel going into the filter and can be used in conjunction with the coolant-type fuel heater.

For 1981 and later Diesels, GM offers an optional electric diesel fuel heater (V6 and V8 only) and an engine block heater (all GM diesels). The fuel heater is thermostatically controlled to heat the fuel before it enters the fuel filter when fuel temperature is 20°F or lower. The fuel heater works only when the ignition key is in the RUN position. On these models, the fuel tank filter has a bypass valve which allows fuel to flow to the heater when the tank filter is covered with fuel wax. The engine block heater is equipped with an electrical cord wrapped up on the right side of the engine compartment. The cord plugs into regular 110 volt household current. The block heater can be used, according to the type of oil in the crankcase, up to eight hours or overnight to warm up the block. Consult the manufacturer's Diesel Engine Supplement for more information.

1981 and later GM V6 and V8 diesel engine fuel injection pumps are equipped with a Housing Pressure Cold Advance (HPCA) system which advances the injection timing about 3° during cold operation to promote easier cold starts, better idle and less noise when cold. The system should be maintenance free.

Starting System

The diesel starting system includes one (sometimes two) heavy duty battery, the starter, and the glow plug circuit. In addition to the heavy duty battery(ies), the majority of diesel engines also have starters and battery cables designed specifically as heavy duty items for diesel usage only. Because of the high compression of any diesel, the torque required to turn the engine is much greater than a gasoline engine. The starter must be powerful enough to handle the increased load; the battery cables must be thick enough to withstand the heat generated by the starter load.

For battery maintenance, see the regular "Maintenance" section. Jump starting procedures for a dual battery car are given below. Starter maintenance is included in the appropriate car section, or the "Charging and Starting Systems" section.

The glow plug circuit is used on the diesel to initially start the engine. When the ignition switch is turned to the ON position, a light will come on in the instrument panel signalling that the glow plugs are preheating the combustion chambers. After a certain interval (depending on how cold the engine is), the light will go off.

This signals that the starter may be engaged and the engine started. If the glow plug circuit malfunctions, especially in cold weather, the engine will be almost impossible to start.

CAUTION

NEVER use "starting aids" such as ether to help start a GM diesel engine — serious engine damage will result.

GLOW PLUG TESTING

Except Chevette

To test each individual glow plug, disconnect the busbar and/or wire connector from the glow plug and connect a test light between the glow plug terminal and the positive battery terminal. If the test light lights, the glow plug is working. Replace individual glow plugs which do not work. See the appropriate car sections for removal and installation procedures.

NOTE: GM V8 diesel engines are equipped with either "slow glow" or "fast glow" glow plugs. See the Oldsmobile 88 car section for information on these two systems. Do not attempt to interchange any parts of these two glow plug systems. The GM V6 diesel uses the "fast glow" style glow plugs exclusively.

To test the glow plug circuit, connect a test light to the terminal of one of the glow plugs (glow plug wiring still attached) and turn the ignition to the heating position. The test light should light for a short while. If not, the glow plug circuit is malfunctioning and must be diagnosed and repaired.

NOTE: Perform this operation on a cold engine only.

Chevette

To test the glow plugs of the Chevette, check for continuity across the plug terminals and the body with an ohmmeter. If there is continuity between these points, the glow plugs is okay; if not, replace the glow plug.

Jump-Starting a Dual Battery Diesel

Many GM diesels are equipped with two 12 volt batteries. The batteries are connected in parallel circuit (positive terminal to positive terminal, negative terminal to negative terminal). Hooking the batteries up in parallel circuit increases battery cranking power without increasing total battery voltage output (12 volts). On the other hand, hooking two 12 volt batteries up in a series circuit (positive terminal to negative terminal)

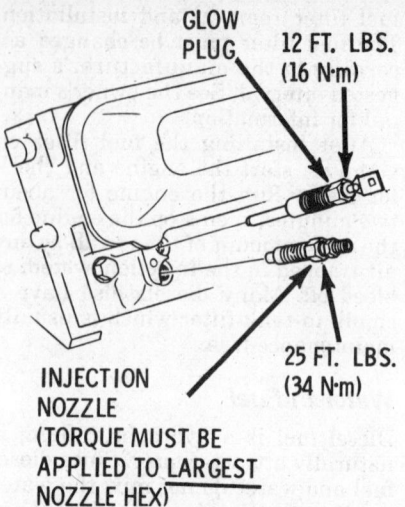

GM V8 diesel engine glow plug and injection nozzle—1980 and later models shown. Nozzles of 1978–79 models are retained by a collar clamp and bolt.

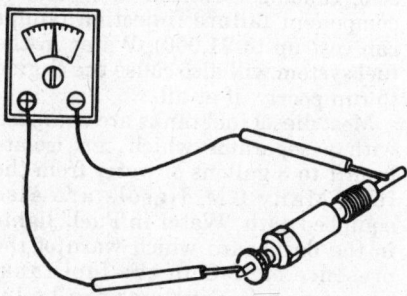

Testing the Chevette diesel glow plug with an ohmmeter

increases total battery output to 24 volts (12 volts + 12 volts).

CAUTION

NEVER hook the batteries up in a series circuit; SEVERE electrical system damage will result.

In the event that a dual battery diesel must be jump started, use the following procedure.

1. Open the hood and locate the batteries. On GM diesels, the manufacturer usually suggests using the battery on the driver's side of the car to make the connection.

2. Position the donor car so that the jumper cables will reach from its battery (must be 12 volt, negative ground) to the appropriate battery in the diesel. Do not allow the cars to touch.

3. Shut off all electrical equipment on both vehicles. Turn off the engine of the donor car, set the parking brakes on both vehicles and block the wheels. Also, make sure both vehicles are in Neutral (manual transmission models) or Park (automatic transmission models).

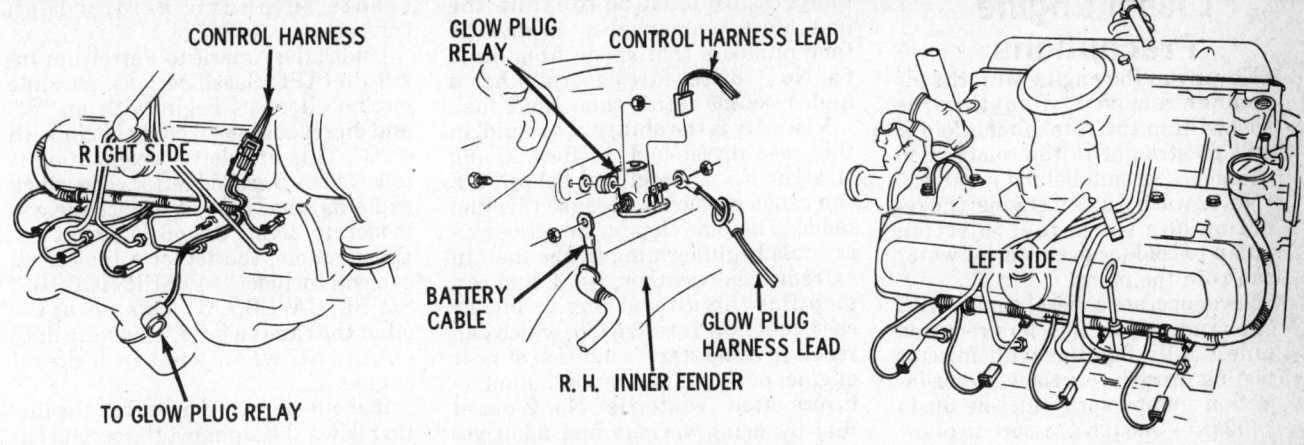

Typical GM V8 diesel glow plug harness arrangement—V6 similar.

4. Using the jumper cables, connect the positive (+) terminal of the donor car battery to the positive terminal of one (not both) of the diesel batteries.

5. Using the second jumper cable, connect the negative (-) terminal of the donor battery to a solid, stationary, metallic point on the diesel (alternator bracket, engine block, etc.). Be very careful to keep the jumper cables away from moving parts (cooling fan, alternator belt, etc.) on both vehicles.

6. Start the engine of the donor car and run it at moderate speed.

7. Start the engine of the diesel.

8. When the diesel starts, disconnect the battery cables in the reverse order of attachment.

Engine Mechanical Systems

Although diesel engines are very low in carbon monoxide (CO) and hydrocarbon (HC) emissions, "particulate" emission output is very high. This is evident from the black smoke emitted by diesels, which is most noticeable during hard acceleration or high engine loads. The particulates are made up of mostly soot (carbon) and sulphur particles. The majority of these particulates are released into the atmosphere. However, some of the particulate matter, because it is produced within the engines cylinders, is left inside the engine and gradually contaminates the engine oil. This contamination makes the oil corrosive, due to the sulphur, and abrasive, due to the carbon. Serious engine damage will result if these contaminants continue to accumulate in the oil. Engine oil and filters of diesel engines must be changed more frequently than those of gasoline engines, due to the increased rate at which the contami-

nants form in the diesel. Consult the "Maintenance" section for oil and filter change procedures. The manufacturer's recommended oil change interval will be given in the owner's manual. An explanation of diesel engine oils is given at the end of this section.

As explained earlier, very high cylinder compression is the key to the operation of the diesel engine. The normal compression of most gasoline engines will rarely exceed 180 psi; whereas with diesel engines, compression pressures of 350–400 psi are commonplace.

—— CAUTION ——

DO NOT attempt to check the compression of a diesel engine with a standard compression gauge—personal injury could result. A special, high pressure compression gauge is needed to safely check the compression of any diesel.

COMPRESSION TEST

GM V6 and V8 Diesel Engines

1. Remove the air cleaner and install air crossover cover (Tool No. J-26996-1).

2. Disconnect the wire from the fuel shutoff solenoid terminal of the injection pump.

3. Disconnect the wires from the glow plugs and remove all glow plugs.

4. Screw compression gauge J-26999 into the glow plug hole in the cylinder being checked.

5. Crank the engine, allowing six "puffs" for each cylinder. The lowest reading cylinder should not be less than 70% of the highest, and no cylinder should be less than 275 pounds.

Chevette Diesel

1. Start the engine and bring it to normal operating temperature.

2. Disconnect or remove the following:

  a. Sensing resistor

  b. Glow plug connector

  c. Glow plugs (4)

  d. Fuel cut-off solenoid connector

  e. Disconnect the in-line fusible link wire of Q.S.S. (Quick Start and Silent idling) system at the connector.

3. Install an adapter (special tool J-29762) into the glow plug hole, then hook a compression gauge (must read to 600 psi) to the adapter.

4. Engage the starter motor to take the reading. Standard compression is 441 psi at 200 rpm or more. Limit is 370 psi at 200 rpm or less.

CONNECTING A TACHOMETER TO A DIESEL ENGINE

As mentioned earlier, the diesel engine does not require an electrical ignition system. Because of this, problems arise when attempts are made to connect a tachometer to the engine for the purpose of idle adjustments, etc. The average gasoline engine tachometer senses the ignition spark pulses and converts them into a readable engine rpm signal. This type of tachometer is useless on the diesel engine, as you may have guessed, because of the diesel's compression ignition system.

There are several magnetic and photoelectric tachometers available from various tool manufacturers (Kent-Moore Corp., Snap-on Tools, etc.) which were designed specifically for use with the diesel engine. These units can run into a little more money than the average do-it-yourselfer may be willing to spend, in which case any adjustments requiring the monitoring of engine rpm should be performed by a competent service technician.

Diesel Engine Precautions

• Never run the engine with the air cleaner removed; if anything is sucked into the intake manifold it will go straight to the combustion chambers, or jam behind a valve.

• Never wash a diesel engine; the reaction of a warm fuel injection pump to cold (or even warm) water can ruin the pump.

• Never operate a diesel engine with one or more fuel injectors removed unless fully familiar with injector testing procedures; some diesel injection pumps spray fuel at up to 1400 psi—enough pressure to allow the fuel to penetrate your skin.

• Do not skip engine oil and filter changes.

• Strictly follow the manufacturer's oil and fuel recommendations as given in the owner's manual.

• Do not use home heating oil as fuel for any diesel unless it's a dire emergency.

• Do not use "starting aids" such as ether in the automotive diesel engine, as these can cause severe internal engine damage.

• Do not run a diesel engine with the "Water in Fuel" warning light on in the dashboard.

• If removing water from the fuel tank, use the same caution as when working around gasoline engine fuel components.

• Do not allow diesel fuel to come in contact with rubber hoses or components on the engine, as it can damage them.

Fuel and Oil Recommendations

FUEL

Fuel makers produce two grades of diesel fuel, No. 1 and No. 2, for use in automotive diesel engines. Generally speaking, No. 2 fuel is recommended over No. 1 for driving in temperatures above 20°F. In fact, in many areas, No. 2 diesel is the only fuel available. By comparison, No. 2 diesel fuel is less volatile than No. 1 fuel, and gives better fuel economy. No. 2 fuel is also a better injection pump lubricant. Two important characteristics of diesel fuel are its cetane number and its viscosity.

The cetane number of a diesel fuel refers to the case with which a diesel fuel ignites. High cetane numbers mean that the fuel will ignite with relative ease or that it ignites well at low temperatures. Naturally, the lower the cetane number, the higher the temperature must be to ignite the fuel. Most commercial fuels have cetane numbers that range from 35 to 65. No. 1 diesel fuel generally has a higher cetane rating than No. 2 fuel.

Viscosity is the ability of a liquid, in this case diesel fuel, to flow. Using straight No. 2 diesel fuel below 20°F can cause problems, because this fuel tends to become cloudy, meaning wax crystals begin forming in the fuel. In extreme cold weather, No. 2 fuel can stop flowing altogether. In either case, fuel flow is restricted, which can result in a "no start" condition or poor engine performance. Fuel manufacturers often "winterize" No. 2 diesel fuel by using various fuel additives and blends (No. 1 diesel fuel, kerosene, etc.) to lower its winter-time viscosity. Generally speaking, though, No. 1 diesel fuel is more satisfactory in extremely cold weather.

NOTE: No. 1 and No. 2 diesel fuels will mix and burn with no ill effects, although the engine manufacturer will undoubtedly recommend one or the other. Consult the owner's manual for information.

Depending on local climate, most fuel manufacturers make winterized No. 2 fuel available seasonally. Many automobile manufacturers (Oldsmobile, for example) publish pamphlets giving the locations of diesel fuel stations nationwide. Contact the local dealer for information.

Do not substitute home heating oil for automotive diesel fuel. While in some cases, home heating oil refinement levels equal those of diesel fuel, many times they are far below diesel engine requirements. The result of using "dirty" home heating oil will be a clogged fuel system, in which case the entire system may have to be dismantled and cleaned.

One more word on diesel fuels. Don't thin diesel fuel with gasoline in cold weather. The lighter gasoline, which is more explosive, will cause rough running at the very least, and may cause extensive engine damage if enough is used. In addition, the combination of diesel fuel and gasoline produces an extremely explosive mixture that is even more volatile than gasoline, making the simple act of adding gasoline to a diesel fuel tank very dangerous.

OIL

Diesel engines require different engine oil from those used in gasoline engines. Besides doing the things gasoline engine oil does, diesel oil must also deal with increased engine heat and the diesel blow-by-gases, which create sulphuric acid, a high corrosive.

Under the American Petroleum Institute (API) classifications, gasoline engine oil codes begin with an "S", and diesel engine oil codes begin with a "C". This first letter designation is followed by a second letter code which explains what type of service (heavy, moderate, light) the oil is meant for. For example, the top of a typical oil can will include: "API SERVICES SC, SD, SE, CA, CB, CC". This means the oil in the can is a good, moderate duty engine oil when used in a diesel engine.

It should be noted here that the further down the alphabet the second letter of the API classification is, the greater the oil's protective qualities are (CD is the severest duty diesel engine oil, CA is the lightest duty oil, etc.). The same is true for gasoline engine oil classifications (SF is the severest duty gasoline engine oil, SA is the lightest duty oil, etc.).

Many diesel manufacturers recommend an oil with both gasoline and diesel engine API classifications. Consult the owner's manual for specifications.

The top of the oil can will also contain an SAE (Society of Automotive Engineers) designation, which gives the oil's viscosity. A typical designation will be: SAE 10W–30, which means that the oil is a "winter" viscosity oil, meaning it will flow and give protection at low temperatures.

On the diesel engine, oil viscosity is critical, because the diesel is much harder to start (due to its higher compression) than a gasoline engine. Obviously, if you fill the crankcase with a very heavy oil during winter (SAE 20W–50, for example), the starter is going to require a lot of current from the battery to turn the engine. And, since batteries don't function well in cold weather in the first place, you may find yourself stranded some morning. Consult the owner's manual for recommended oil specifications for the climate you live in.

Aftermarket Fuel System Accessories

Due to reasons described previously, most diesel engine problems can be attributed to either fuel contamination or cold weather fuel performance characteristics. Diesel-engined vehicle manufacturers have designed and installed various systems to combat these problems, but ultimately, their best efforts are limited by cost.

Inconvenience is a major concern to diesel owners. Excepting the Chevette, for example, if water accumulates (in substantial quantities) in the diesel fuel system, the fuel and water must be siphoned from the fuel tank and purged from the remainder of the fuel system. It goes without saying that this operation is a messy, time-consuming process. Even if the vehicle is equipped with a water/fuel separator having a drain valve, the owner must manually open the valve from either under the hood or beneath the vehicle. Although the fuel filter installed by the manufacturer offers adequate performance when maintained properly, the addition of another, separate diesel fuel filter is a wise improvement.

If you live in an extremely cold climate, you've probably experienced cold starting problems due to fuel "waxing", plugged filters "gelled" fuel, etc. If your vehicle is not factory-equipped with the optional fuel line or cylinder block heaters, these heaters can be purchased from the aftermarket (retail auto parts manufacturers). The installation of either of these items can improve cold-starting dramatically.

WATER/FUEL SEPARATORS

Centrifugal Action

Sometimes referred to as a "cyclonic" water/fuel separator, this device uses baffles which spin the fuel as it comes through the separator intake. Since water is heavier than diesel fuel, the water will spin away from the fuel, sink to the bottom of the separator, and collect in the sediment bowl.

This type of separator is most efficient in dealing with large water droplets. If the water is an emulsion with the fuel, that is, if the water is equally dispersed through the fuel in very small droplets, some of the water will remain with the fuel to travel through the fuel system.

Coalescing Action

In this type of separator, the fuel must pass through a coalescent filtering media before proceeding through the fuel system. The idea behind the coalescent media is to trap even the smallest droplets of water on the media. As the small droplets combine into larger, heavier droplets, gravity acts on the droplets to pull them downward, off of the media and into the sediment bowl.

FUEL FILTER/SEPARATOR COMBINATION UNITS

Most separators of either the centrifugal or coalescent types are available with disposeable fuel filtering elements which are built into the separator unit. If your car already has a large, disposeable filter, it would probably be more cost-effective to stay with a separator only, and to change the factory-equipped filter at the recommended intervals. Should your vehicle have a fairly small filter, and/or an inconveniently located water drain (or none at all), choose the filter/separator combination. The filter/separator offers both increased fuel filtering ability and efficient water separation.

CONVENIENCE ADD-ONS

Available with many separators and filter/separators are items such as dash-mounted water-in-fuel indicator lamps, audible water-in-fuel alarms, and dash-controlled water ejection systems. A properly chosen system would warn you of water in the fuel, and allow you to eject the water by simply flipping a dash-mounted switch.

Installing a Separator

Clear installation instructions and the necessary installation parts will be provided with the separator kit. Follow those instructions exactly. A general list of suggestions follows:

1. Fuel additives should not be used unless approved by the separator manufacturer.

2. Do not install a separator less than 4 in. away from any exhaust system component.

3. If plastic fittings are supplied with the kit, do not replace them with metal fittings. Also, use extreme caution when tightening the fittings, especially those made of plastic.

4. Use a fuel-proof sealer on all fitting threads, only if the threads are not factory-coated with sealer.

5. Use only fuel-proof hoses for the installation.

6. Do not eliminate the original equipment fuel filter, even if a filter/separator is installed.

7. For new car warranty purposes, a filter/separator should be located BEFORE the original equipment filter. The fuel must pass through the original filter last, before entering the fuel injection pump.

8. If any type of fuel line heater is installed, it is best to position the heater between the fuel tank and the separator intake.

9. To ease the job of the separator,

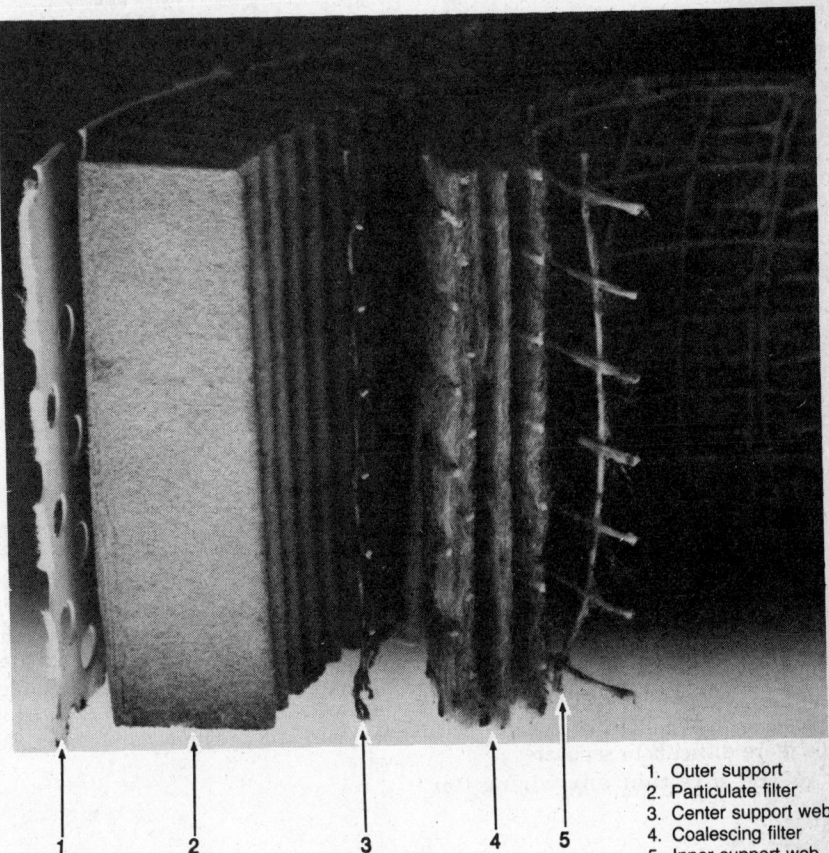

1. Outer support
2. Particulate filter
3. Center support web
4. Coalescing filter
5. Inner support web

Cross section of a coalescing filter/fuel filter combination (© CR Industries)

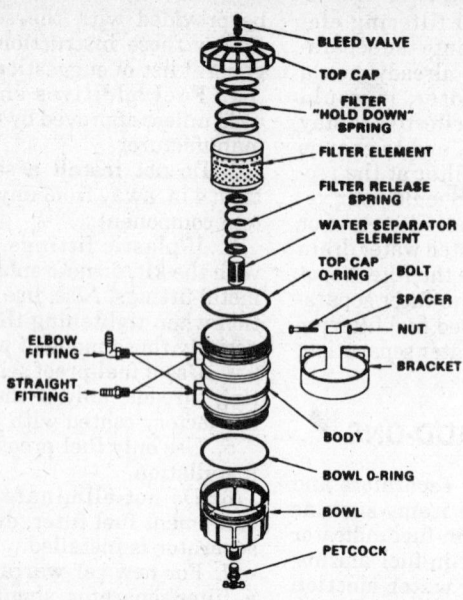

BLEED VALVE
TOP CAP
FILTER "HOLD DOWN" SPRING
FILTER ELEMENT
FILTER RELEASE SPRING
WATER SEPARATOR ELEMENT
TOP CAP O-RING — BOLT
SPACER
NUT
ELBOW FITTING
STRAIGHT FITTING — BRACKET
BODY
BOWL O-RING
BOWL
PETCOCK

Exploded view of a fuel filter/water separator combination unit (© CR Industries)

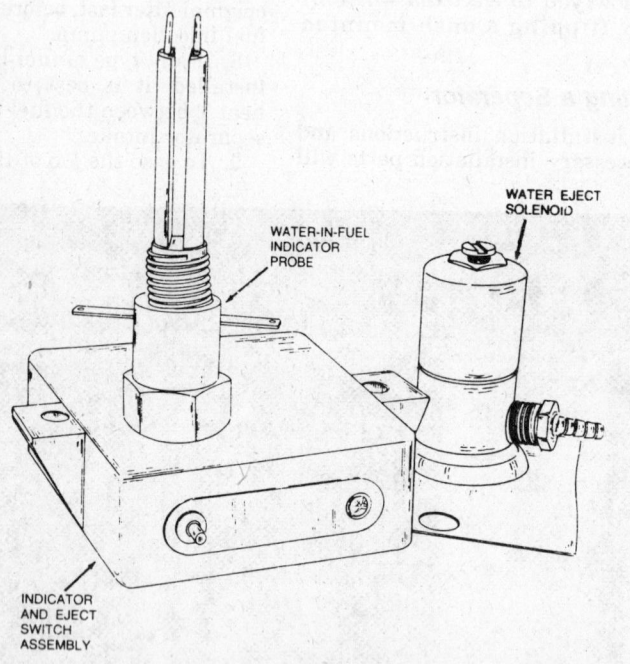

WATER-IN-FUEL INDICATOR PROBE

WATER EJECT SOLENOID

INDICATOR AND EJECT SWITCH ASSEMBLY

Typical convenience accessories for a water-in-fuel detection system (© CR Industries)

the separator should be installed between the fuel transfer pump and the tank (unless the separator manufacturer specifies otherwise). Fuel and water which have been churned through the fuel transfer pump will be more difficult to separate.

10. Be sure that any wiring (for warning lamps, water ejection, etc.) is routed and connected properly. If the wiring must pass through a drilled hole, be sure to use a rubber grommet between the drilled component(s) and the wire to prevent damage to the wire.

FUEL LINE HEATERS

Two popular types of fuel line heaters are available for diesel passenger cars. Both types raise the temperature of the fuel to prevent "waxing" and "gelling" of the fuel in the lines during cold weather operation. One type uses engine coolant as a heating source. In order for this type to heat the fuel, the engine must first be started and allowed to run until the coolant temperature increases. Though this type of heater will usually increase fuel mileage, it offers no aid in starting ability.

The other type of heater uses a 12V DC electric heating element. This type is recommended, due to its ability to warm the fuel BEFORE the engine is started. This type of heater will also usually increase the overall fuel mileage.

Installation

Follow the manufacturer's instructions exactly. Also, see suggestions 5, 8, and 9 under Separator Installation.

CYLINDER BLOCK HEATERS

A cylinder block heater electrically (usually 110V house current) heats the engine coolant, which in turn warms the cylinder block, heads, and engine oil. In this case, the warmth is not used to alter the characteristics of the fuel. Block heaters offer two main advantages when starting a diesel in cold weather.

1. The reduced viscosity (thinning) of the engine oil from the warmth allows the engine to be "turned over" easier (and faster) by the starter. Less strain is imposed on the starting system.

2. Because the diesel relies on the heat of compression to ignite the fuel, the increase in the base combustion chamber temperature results in a higher temperature during compression. This allows the fuel to ignite easier than if just the glow plugs were used.

Installation

Most cylinder block heaters replace one of the existing freeze (or expansion) plugs of the cylinder block. Follow the manufacturers installation instructions exactly. Also, refer to the manufacturers recommendations for usage.

Tools and Equipment

In addition to the normal assortment of screwdrivers and pliers, automotive service work requires an investment in wrenches, sockets and the handles needed to drive them, and various measuring tools such as torque wrenches and feeler gauges.

The best approach to gathering the required equipment is to proceed slowly, buying high-quality tools as they are needed. An initial investment should be made in a set of quality wrenches, ranging in size from $\frac{1}{4}$ inch to one inch, if your car has standard bolts, or from 5mm to 19mm if your car has metric fasteners. High quality forged wrenches are available in three styles; open end, box end, and combination open/box end. The combination tools are generally the most desirable as a starter set; the wrenches shown in the illustration are of the combination type.

NOTE: Many later model American cars use both metric and standard nuts and bolts.

The other set of tools inevitably required is a ratchet handle and socket set. This set should have the same size range as your wrench set. The ratchet, extension, and flex drives fro the sockets are available in many sizes; it is advisable to choose a $\frac{3}{8}$ inch drive set initially. One break in the inch/metric sizing war is that metric-sized sockets sold in the U.S. have inch-sized drive ($\frac{1}{4}$, $\frac{3}{8}$, $\frac{1}{2}$, etc.). Sockets are available in six and twelve point versions; six point types are generally cheaper and are a good choice for a first set.

The choice of a drive handle for the sockets should be made with some care. If this is your first set, take the plunge and invest in a flexhead ratchet; it will get into many places otherwise accessible only through a long chain of universal joints, extensions and adapters. An alternative is a flex handle; such a tool is shown in the illustration, below the ratchet handle. In addition to the range of sockets mentioned, a rubber-lined spark plug socket should be purchased. Spark plugs have either a $\frac{13}{16}$ or a $\frac{5}{8}$ inch hex; get the correct socket for the plugs in your car.

The most important thing to consider when purchasing hand tools is quality. Don't be misled by the low cost of "bargain" tools. Forged wrenches, tempered screwdriver blades, and fine tooth ratchets are a much better investment than their less expensive counterparts. The skinned knuckles and frustration inflicted by poor quality tools make any job an unhappy core. Another consideration is that quality tools sold by reputable firms come with an on-the-spot replacement guarantee; if the tool breaks, you get a new one, no questions asked.

The tools needed for basic maintenance jobs, in addition to those just mentioned, include:

1. Jackstands, for support;
2. Oil filter wrench;
3. Oil filler spout or funnel;
4. Grease gun;
5. Battery hydrometer;
6. Battery post and clamp cleaner;
7. Container for draining oil;
8. Many rags for the inevitable spills.

In addition to these items there are several others which are not absolutely necessary, but handy to have around. These include a transmission funnel and filler tube, a drop (trouble) light on a long cord, an adjustable wrench (crescent wrench), and slip joint pliers.

A more extensive list of tools, suitable for tune-up work, can be drawn up easily. While the tools involved are slightly more sophisticated, they need not be outrageously expensive. For example, there are several inexpensive tach/dwell meters on the market that are every bit as good for the average mechanic as a $100.00 professional model. The key to these purchases is to make them with an eye towards adaptability and wide range. Using the tach/dwell meter example again, if the model you buy runs up to at least 1,500 rpm on the tachometer scale, the dwell meter works on 4, 6, or 8 cylinder engines, and the tachometer unit is adaptable to both conventional and electronic ignitions, it will serve for a long time on a variety of automobiles. A basic list of tune-up tools could include:

1. A tach/dwell meter;
2. Spark plug gauge and gapping tool;
3. Feeler blades;
4. Timing light.

In this list, the choice of a timing light should be made carefully. A light which works on the DC current supplied by the car battery is the best choice; it should have a xenon tube for brightness. If your car has electronic ignition, the light should have an inductive pick-up (the timing light illustrated has one of these), and since nearly all cars will have electronic ignition in the future, this feature is a reasonable one to look for.

In addition to these basic tools, there are several other tools and gauges you may find useful. These include:

1. A compression gauge. The screw-in type is slower to use, but eliminates the possibility of a faulty reading due to escaping pressure.
2. A manifold vacuum gauge.
3. A test light.
4. An induction meter. This is used

U21

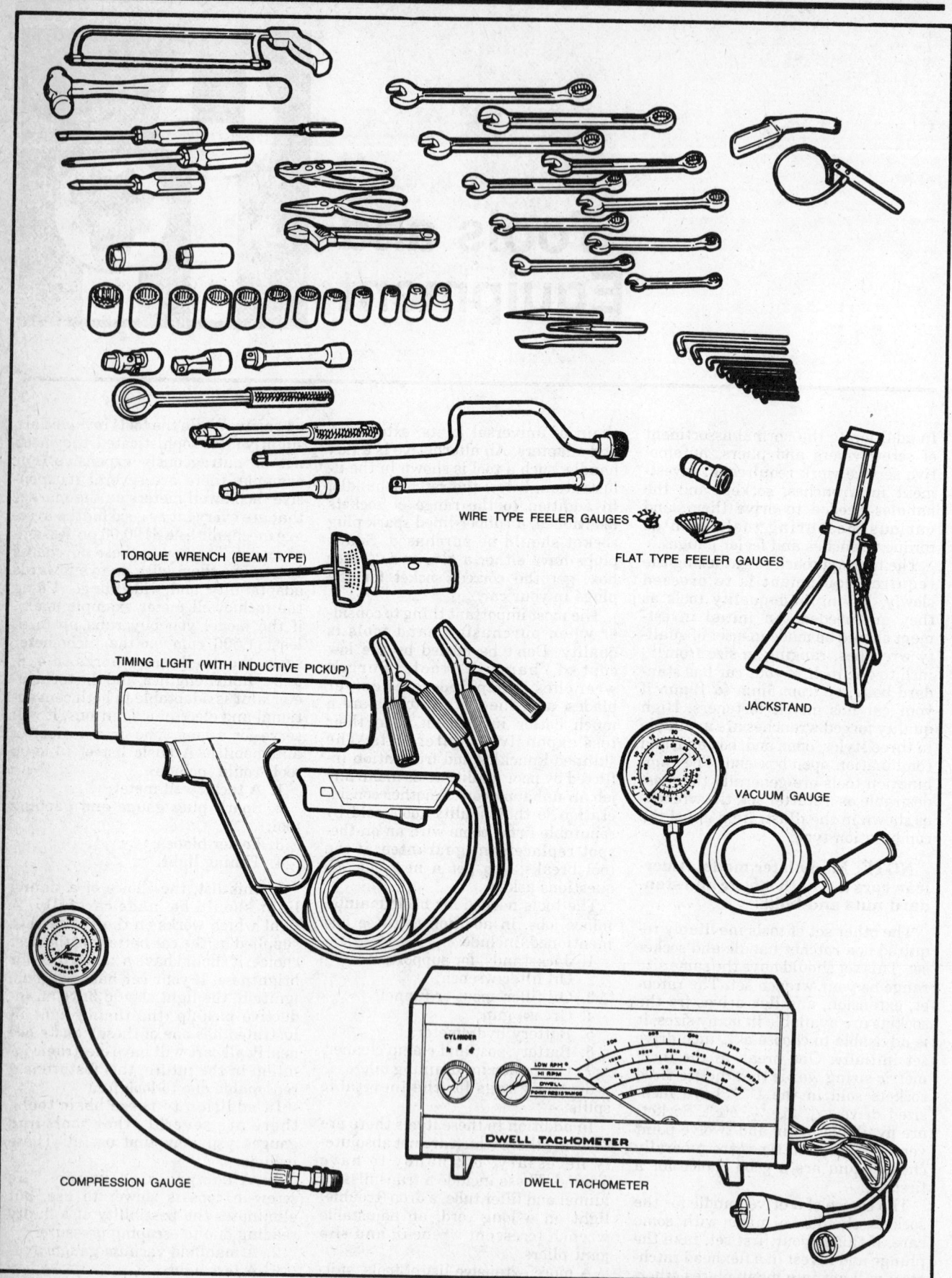

WIRE TYPE FEELER GAUGES

FLAT TYPE FEELER GAUGES

TORQUE WRENCH (BEAM TYPE)

JACKSTAND

TIMING LIGHT (WITH INDUCTIVE PICKUP)

VACUUM GAUGE

COMPRESSION GAUGE

DWELL TACHOMETER

A basic tool collection will handle almost any automotive repair work

to determine whether or not there is current flowing in a wire, and thus is extremely helpful in electrical troubleshooting.

Finally, you will probably find a torque wrench necessary for all but in the most basic of work. The beam type models are perfectly adequate, although the newer click (break-away) type are more precise. Whichever type you choose, plan on having it recalibrated every once in a while.

Special Tools

Several procedures in this manual refer to special tools needed to make repairs or adjustments. These tools can be purchased from the following companies:

AMC, GM
Special Tool Division
Kent-Moore Corp.
29784 Little Mack
Roseville, MI 48066

Ford
Owatonna Tool Co.
Owatonna, MN 55060

Chrysler
Miller Special Tools
A Division of Utica Tool Co.
32615 Park Lane
Garden City, MI 48135

SPECIAL TEST EQUIPMENT

A variety of diagnostic tools are available to help troubleshoot and repair computerized engine and emission control systems. The most sophisticated of these devices are the console-type engine analyzers that usually occupy a garage service bay, but there are several types of aftermarket electronic testers available that will allow quick circuit tests of the engine control system by plugging directly into a special test connector located in the engine compartment or under the dashboard. Several tool and equipment manufacturers offer simple, hand-held testers that measure various circuit voltage levels on command to check all system components for proper operation. Although these testers usually cost about $300–500, consider that the average computer-controlled carburetor can cost twice as much and the money saved by not replacing perfectly good sensors in an attempt to correct a problem could justify the purchase price of a special diagnostic tester.

These testers can allow quick and easy test measurements while the en-

Aftermarket hand-held testers can make diagnosing computer-controlled systems easier

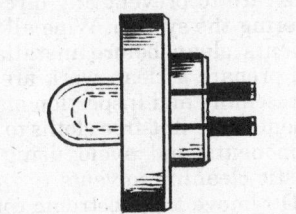

Throttle body fuel injector tester

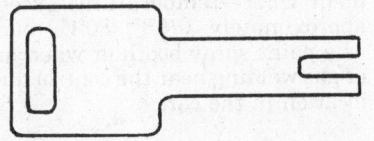

Special key for activating GM on-board diagnosis system. Insert the prongs into the diagnostic test terminals located under the dash

gine is operating or while the car is being driven. In addition, the on-board computer memory can be read to access any stored trouble codes; in effect allowing the computer to tell you where it hurts and aid trouble diagnosis by pinpointing exactly which circuit or component is malfunctioning. In the same manner, repairs can be tested to make sure the problem has been corrected. The biggest advantage these special testers have is their relatively easy hookups that minimize or eliminate the chances of making the wrong connections and getting false voltage readings or damaging the on-board computer.

NOTE: It should be remembered that these testers check voltage levels in circuits; they don't detect mechanical problems or failed components if the circuit voltage falls within the preprogrammed limits stored in the tester PROM unit. Also, most of the

hand-held testes are designed to work only on one or two systems made by a specific manufacturer.

A variety of aftermarket testers are available to help diagnose different computerized engine control systems. Owatonna Tool Company (OTC), for example, markets a device called the OTC Monitor 2000 which plugs directly into the assembly line diagnostic link (ALDL). When the correct manufacturer cartridge is plugged into the unit, the OTC tester makes diagnosis a simple matter of pressing the correct buttons. An adapter is supplied with the tester to allow connection to all types of ALDL links, regardless of the number of pin terminals used.

Servicing Your Car Safely

It is virtually impossible to anticipate all of the hazards involved with automotive maintenance and service, but care and common sense will prevent most accidents. The rules of safety for mechanics range from "don't smoke around gasoline," to "use the proper tool for the job." The trick to avoiding injuries is to develop safe work habits and take every possible precaution.

Any computer-based electronic engine control system is extremely sensitive to electrical voltages and cannot tolerate careless or haphazard testing or service procedures. An inexperienced individual can literally do major damage looking for a minor problem by using the wrong kind of test equipment or connecting test leads or connectors with the ignition switch ON. When selecting test equipment, make sure the manufacturers instructions state that the tester is compatible with whatever type of electronic control system is being serviced. Read all instructions carefully and double check all test points before installing probes or making any connections.

Aftermarket electronic testers are available from a variety of sources, as well as from the manufacturer, but care should be taken that the test equipment being used is designed to diagnose a particular system accurately without damaging the control unit (ECU) or components being tested.

DO'S

• DO keep a fire extinguisher and first aid kit within easy reach.
• DO wear safety glasses or goggles when cutting, drilling, grinding or prying, even if you have 20-20 vi-

sion. If you wear glasses for the sake of vision, they should be made of hardened glass that can serve also as safety glasses, or wear safety goggles over your regular glasses.

• DO shield your eyes whenever you work around the battery. Batteries contain sulphuric acid. In case of contact with the eyes or skin, flush the area with water or a mixture of water and baking soda and get medical attention immediately.

• DO remove the battery cables before charging the battery. Never use a high-output charger on an installed battery or attempt to use any type of "hot shot" (24 volt) starting aid.

• DO use safety stands for any undercar service. Jacks are for raising vehicles; safety stands are for making sure the vehicle stays raised until you want it to come down. Whenever the car is raised, block the wheels remaining on the ground and set the parking brake.

• DO use adequate ventilation when working with any chemicals or hazardous materials. Follow the manufacturer's directions for usage. Brake fluid, anti-freeze, solvents, paints, etc. are all deadly poisons if taken internally. Seal the containers tightly after use and store them safely, out of the reach of children.

• DO use caution when working on clutches or brakes. The asbestos used in the friction material will cause lung cancer if inhaled. Wipe the component with a damp rag to remove dust, and dispose of the rag after use.

• DO disconnect the negative battery cable when working on the electrical system. The secondary ignition system can contain up to 40,000 volts.

• DO properly maintain your tools. Loose hammerheads, mushroomed punches and chisels, frayed or poorly grounded electrical cords, excessively worn screwdrivers, spread open-end wrenches, cracked sockets, slipping ratchets, or faulty droplight sockets can cause accidents.

• DO use the proper size and type of tool for the job being done.

• DO when possible, pull on a wrench handle rather than push on it, and adjust your stance to prevent a fall.

• DO be sure that adjustable wrenches are tightly closed on the nut or bolt and pulled so that the face is on the side of the fixed jaw.

• DO select a wrench or socket that fits the nut or bolt. The wrench or socket should sit straight, not cocked.

• DO strike squarely with a hammer; avoid glancing blows.

• DO set the parking brake and block the drive wheels if the work requires the engine running.

• DO depressurize the fuel system before attempting to disconnect any fuel lines. Although only fuel injection vehicles use a pressurized fuel system, it's a good idea to exercise caution whenever disconnecting any fuel line or hose during service procedures. Take precautions to avoid a fire hazard.

• DO use clean rags and tools when working on an open fuel system and take care to prevent any dirt from entering the system. Wipe all components clean before installation and prepare a clean work area for disassembly and inspection of components. Use lint-free cloths to wipe components and avoid using any caustic cleaning solvents.

• DO remove the electronic control unit (on-board computer) if the vehicle is to be placed in an environment where temperatures exceed approximately 176°F (80°C), such as a paint spray booth or when arc or gas welding near the control unit location in the car.

DON'TS

• DON'T run an engine in a garage or anywhere else without proper ventilation—EVER! Carbon monoxide is poisonous; it takes a long time to leave the human body and you can build up a deadly supply of it in your system by simply breathing in a little every day. You may not realize you are slowly poisoning yourself. Always use power vents, windows, fans or open the garage doors.

• DON'T work around moving parts while wearing a necktie or other loose clothing. Short sleeves are much safer than long, loose sleeves; hard-toed shoes with neoprene soles protect your toes and give a better grip on slippery surfaces. Jewelry such as watches, rings, fancy belt buckles, beads or body adornment

of any kind is not safe working around a car. Long hair should be hidden under a hat or cap.

• DON'T use pockets for toolboxes. A fall or bump can drive a screwdriver deep into your body. Even a wiping cloth hanging from the back pocket can wrap around a spinning shaft or fan.

• DON'T smoke when working around gasoline, cleaning solvent or other flammable material.

• DON'T use gasoline to wash your hands; there are excellent soaps available. Gasoline may contain lead, and lead can enter the body through a cut, accumulating in the body until you are very ill. Gasoline also removes all the natural oils from the skin so that bone dry hands will suck up oil and grease.

• DON'T service the air conditioning system unless you are equipped with the necessary tools and training. The refrigerant, R-12, is extremely cold when compressed, and when released into the air will instantly freeze any surface it contacts, including your eyes. Although the refrigerant is normally non-toxic, R-12 becomes a deadly poisonous gas in the presence of an open flame. One good whiff of the vapors from burning refrigerant can be fatal.

• DON'T install or remove battery cables with the key ON or the engine running. Jumper cables should be connected with the key OFF to avoid power surges that can damage electronic control units. Engines equipped with computer controlled systems should avoid both giving and getting jump starts due to the possibility of serious damage to components from arcing in the engine compartment when connections are made with the ignition ON.

• DON'T remove or attach wiring harness connectors with the ignition switch ON, especially to the electronic control unit.

• DON'T drop any components during service procedures and never apply 12 volts directly to any component (like a fuel injector) unless instructed specifically to do so. Some component electrical windings are designed to safely handle only 4 or 5 volts and can be destroyed in seconds if 12 volts are applied directly to the connector.

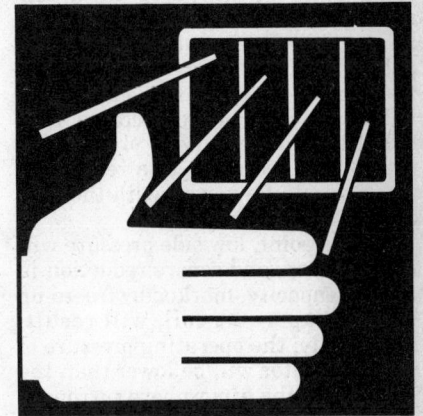

Air Conditioning

AIR CONDITIONING SYSTEMS

Automotive air conditioning systems are basic in design and operation, but many different components are used by the vehicle manufacturers to operate and control the systems to their specifications.

Basic System

The basic air conditioning system utilizes the compressor, condenser, evaporator, receiver-drier, expansion valve and a thermostatic or ambient type switch to control evaporator freeze-up. The controls are manually operated and the unit is basic in design. This system is usually installed as an add-on or after-market unit. A sight glass may be used in the system.

GENERAL SERVICING PROCEDURES

The most important aspect of air conditioning service is the maintenance of a pure and adequate charge of refrigerant in the system. A refrigeration system cannot function properly if a significant percentage of the charge is lost. Leaks are common be-

cause the severe vibration encountered in an automobile can easily cause a sufficient cracking or loosening of the air conditioning fittings; as a result, the extreme operating pressures of the system force refrigerant out.

The problem can be understood by considering what happens to the system as it is operated with a continuous leak. Because the expansion valve regulates the flow of refrigerant to the

evaporator, the level of refrigerant there is fairly constant. The receiver-drier stores any excess of refrigerant, and so a loss will first appear there as a reduction in the level of liquid. As this level nears the bottom of the vessel, some refrigerant vapor bubbles will begin to appear in the stream of liquid supplied to the expansion valve. This vapor decreases the capacity of the expansion valve very little as the valve opens to compensate for

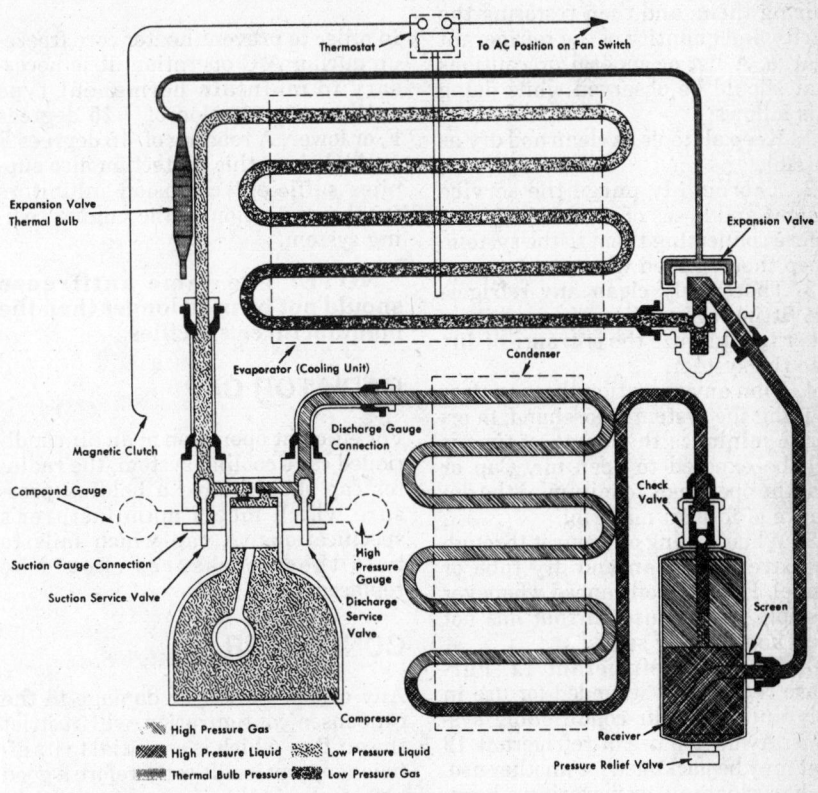

Basic air conditioning system

its presence. As the quantity of liquid in the condenser decreases, the operating pressure will drop there and throughout the high side of the system. As the R-12 continues to be expelled, the pressure available to force the liquid through the expansion valve will continue to decrease, and, eventually, the valve's orifice will prove to be too much of a restriction for adequate flow even with the needle fully withdrawn.

At this point, low side pressure will start to drop, and severe reduction in cooling capacity, marked by freeze-up of the evaporator coil, will result. Eventually, the operating pressure of the evaporator will be lower than the pressure of the atmosphere surrounding it, and air will be drawn into the system wherever there are leaks in the low side.

Because all atmospheric air contains at least some moisture, water will enter the system and mix with the R-12 and the oil. Trace amounts of moisture will cause sludging of the oil, and corrosion of the system. Saturation and clogging of the filter-drier, and freezing of the expansion valve orifice will eventually result. As air fills the system to a greater and greater extent, it will interfere more and more with the normal flows of refrigerant and heat.

From this description, it should be obvious that much of the repairman's time will be spent detecting leaks, repairing them, and then restoring the purity and quantity of the refrigerant charge. A list of general precautions that should be observed while doing this follows:

1. Keep all tools as clean and dry as possible.

2. Thoroughly purge the service gauges and hoses of air and moisture before connecting them to the system. Keep them capped when not in use.

3. Thoroughly clean any refrigerant fitting before disconnecting it, in order to minimize the entrance of dirt into the system.

4. Plan any operation that requires opening the system beforehand, in order to minimize the length of time it will be exposed to open air. Cap or seal the open ends to minimize the entrance of foreign material.

5. When adding oil, pour it through an extremely clean and dry tube or funnel. Keep the oil capped whenever possible. Do not use oil that has not been kept tightly sealed.

6. Use only refrigerant 12. Purchase refrigerant intended for use in only automatic air conditioning systems. Avoid the use of refrigerant 12 that may be packaged for another use, such as cleaning, or powering a horn, as it is impure.

7. Completely evacuate any system that has been opened to replace a component, or that has leaked sufficiently to draw in moisture and air. This requires evacuating air and moisture with a good vacuum pump for at least one hour.

If a system has been open for a considerable length of time it may be advisable to evacuate the system for up to 12 hours (overnight).

8. Use a wrench on both halves of a fitting that is to be disconnected, so as to avoid placing torque on any of the refrigerant lines.

9. When overhauling a compressor, pour some of the oil into a clean glass and inspect it. If there is evidence of dirt or metal particles, or both, flush all refrigerant components with clean refrigerant before evacuating and recharging the system. In addition, if metal particles are present, the compressor should be replaced.

10. Schrader valves may leak only when under full operating pressure. Therefore, if leakage is suspected but cannot be located, operate the system with a full charge of refrigerant and look for leaks from all Schrader valves. Replace any faulty valves.

Additional Preventive Maintenance Checks

ANTIFREEZE

In order to prevent heater core freeze-up during A/C operation, it is necessary to maintain permanent type antifreeze protection of +15 degrees F, or lower. A reading of -15 degrees F is ideal since this protection also supplies sufficient corrosion inhibitors for the protection of the engine cooling system.

NOTE: The same antifreeze should not be used longer than the manufacturer specifies.

RADIATOR CAP

For efficient operation of an air conditioned car's cooling system, the radiator cap should have a holding pressure which meets manufacturer's specifications. A cap which fails to hold these pressures should be replaced.

CONDENSER

Any obstruction of or damage to the condenser configuration will restrict the air flow which is essential to its efficient operation. It is therefore a good rule to keep this unit clean and in proper physical shape.

NOTE: Bug screens are regarded as obstructions.

CONDENSATION DRAIN TUBE

This single molded drain tube expels the condensation, which accumulates on the bottom of the evaporator housing, into the engine compartment. If this tube is obstructed, the air conditioning performance can be restricted and condensation buildup can spill over onto the vehicle's floor.

Safety Precautions

Because of the importance of the necessary safety precautions that must be exercised when working with air conditioning systems and R-12 refrigerant, a recap of the safety precautions are outlined.

1. Avoid contact with a charged refrigeration system, even when working on another part of the air conditioning system or vehicle. If a heavy tool comes into contact with a section of copper tubing or a heat exchanger, it can easily cause the relatively soft material to rupture.

2. When it is necessary to apply force to a fitting which contains refrigerant, as when checking that all system couplings are securely tightened, use a wrench on both parts of the fitting involved, if possible. This will avoid putting torque on refrigerant tubing. (It is advisable, when possible, to use tube or line wrenches when tightening these flare nut fittings.)

3. Do not attempt to discharge the system by merely loosening a fitting, or removing the service valve caps and cracking these valves. Precise control is possible only when using the service gauges. Place a rag under the open end of the center charging hose while discharging the system to catch any drops of liquid that might escape. Wear protective gloves when connecting or disconnecting service gauge hoses.

4. Discharge the system only in a well ventilated area, as high concentrations of the gas can exclude oxygen and act as an anaesthetic. When leak testing or soldering, this is particularly important, as toxic gas is formed when R-12 contacts any flame.

5. Never start a system without first verifying that both service valves are back-seated, if equipped, and that all fittings throughout the system are snugly connected.

6. Avoid applying heat to any refrigerant line or storage vessel. Charging may be aided by using wa-

ter heated to less than 125° to warm the refrigerant container. Never allow a refrigerant storage container to sit out in the sun, or near any other source of heat, such as a radiator.

7. Always wear goggles when working on a system to protect the eyes. If refrigerant contacts the eyes, it is advisable in all cases to see a physician as soon as possible.

8. Frostbite from liquid refrigerant should be treated by first gradually warming the area with cool water, and then gently applying petroleum jelly. A physician should be consulted.

9. Always keep refrigerant drum fittings capped when not in use. Avoid sudden shock to the drum, which might occur from dropping it, or from banging a heavy tool against it. Never carry a drum in the passenger compartment of a car.

10. Always completely discharge the system before painting the vehicle (if the paint is to be baked on), or before welding anywhere near refrigerant lines.

AIR CONDITIONING TOOLS AND GAUGES

Test Gauges

Most of the service work performed in air conditioning requires the use of a set of two gauges, one for the high (head) pressure side of the system, the other for the low (suction) side.

The low side gauge records both pressure and vacuum. Vacuum readings are calibrated from 0 to 30 inches and the pressure graduations read from 0 to no less than 60 psi.

The high side gauge measures pressure from 0 to at least 600 psi. Both gauges are threaded into a manifold that contains two hand shut-off valves. Proper manipulation of these valves and the use of the attached test hoses allow the user to perform the following services:

1. Test high and low side pressures.
2. Remove air, moisture, and contaminated refrigerant.
3. Purge the system (of refrigerant).
4. Charge the system (with refrigerant).

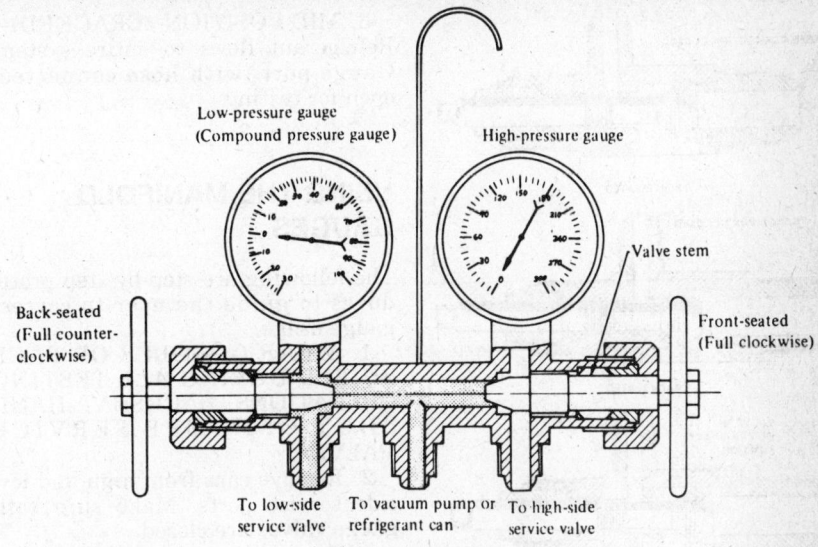

Typical manifold gauge set

NOTE: Chrysler Corp. requires the use of a third gauge on those units that have an evaporator pressure regulator (EPR) valve mounted on the suction side of the compressor.

The manifold valves are designed so they have no direct effect on gauge readings, but serve only to provide for, or cut off, flow of refrigerant through the manifold. During all testing and hook-up operations, the valves are kept in a closed position to avoid disturbing the refrigeration system. The valves are opened only to purge the system of refrigerant or to charge it.

When purging the system, the center hose is uncapped at the lower end, and both valves are cracked open slightly. This allows refrigerant pressure to force the entire contents of the system out through the center hose. During charging, the valve on the high side of the manifold is closed, and the valve on the low side is cracked open. Under these conditions, the low pressure in the evaporator will draw refrigerant from the relatively warm refrigerant storage container into the system.

Service Valves

For the user to diagnose an air conditioning system he or she must gain "entrance" to the system in order to observe the pressures. There are two types of terminals for this purpose, the hand shut off type and the familiar Schrader valve.

The Schrader valve is similar to a tire valve stem and the process of connecting the test hoses is the same as threading a hand pump outlet hose to a bicycle tire. As the test hose is threaded to the service port the valve

core is depressed, allowing the refrigerant to enter the test hose outlet. Removal of the test hose automatically closes the system.

Extreme caution must be observed when removing test hoses from the Schrader valves as some refrigerant will normally escape, usually under high pressure. (Observe safety precautions.)

Some systems have hand shut-off valves (the stem can be rotated with a special ratcheting box wrench) that can be positioned in the following three ways:

1. FRONT SEATED – Rotated to full clockwise position.

   a. Refrigerant will not flow to

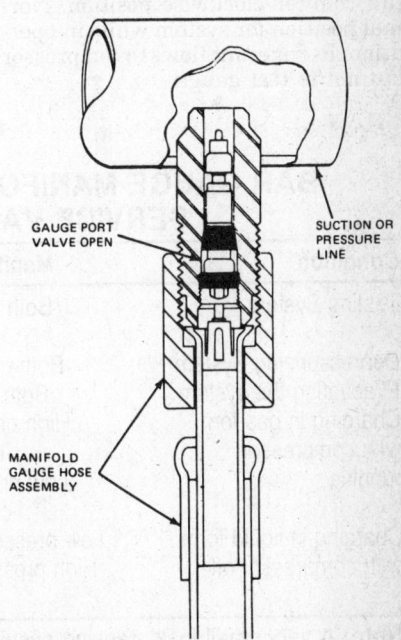

Manifold gauge hose connected to a Schraeder type service port

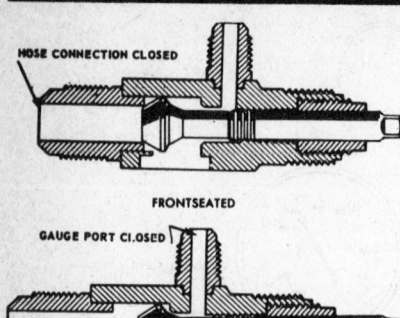

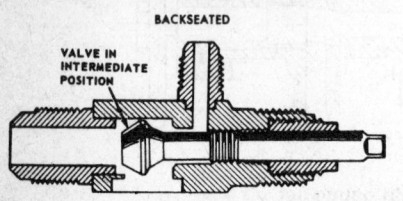

Manual service valve positions

compressor, but will reach test gauge port. COMPRESSOR WILL BE DAMAGED IF SYSTEM IS TURNED ON IN THIS POSITION.

b. The compressor is now isolated and ready for service. However, care must be exercised when removing service valves from the compressor as a residue of refrigerant may still be present within the compressor. Therefore, remove service valves slowly observing all safety precautions.

2. BACK SEATED — Rotated to full counter clockwise position. Normal position for system while in operation. Refrigerant flows to compressor but not to test gauge.

3. MID-POSITION (CRACKED) — Refrigerant flows to entire system. Gauge port (with hose connected) open for testing.

USING THE MANIFOLD GAUGES

The following are step-by-step procedures to guide the user to correct gauge usage.

1. WEAR GOGGLES OR FACE SHIELD DURING ALL TESTING OPERATIONS. BACKSEAT HAND SHUT-OFF TYPE SERVICE VALVES.

2. Remove caps from high and low side service ports. Make sure both gauge valves are closed.

3. Connect low side test hose to service valve that leads to the evaporator (located between the evaporator outlet and the compressor).

4. Attach high side test hose to service valve that leads to the condenser.

5. Mid-position hand shutoff type service valves.

6. Start engine and allow for warm-up. All testing and charging of the system should be done after engine and system have reached normal operation temperatures (except when using certain charging stations).

7. Adjust air conditioner controls to maximum cold.

8. Observe gauge readings.

When the gauges are not being used it is a good idea to:

a. Keep both hand valves in the closed position.

b. Attach both ends of the high and low service hoses to the manifold, if extra outlets are present on the manifold, or plug them if not.

Also, keep the center charging hose attached to an empty refrigerant can. This extra precaution will reduce the possibility of moisture entering the gauges. If air and moisture have gotten into the gauges, purge the hoses by supplying refrigerant under pressure to the center hose with both gauge valves open and all openings unplugged.

DISCHARGING, EVACUATING AND CHARGING

Discharging the System

——— CAUTION ———
Perform operation in a well-ventilated area.

When it is necessary to remove (purge) the refrigerant pressurized in the system, follow this procedure:

1. Operate air conditioner for at least 10 minutes.

2. Attach gauges, shut off engine and air conditioner.

3. Place a container or rag at the outlet of the center charging hose on the gauge. The refrigerant will be discharged there and this precaution will avoid its uncontrolled exposure.

4. Open low side hand valve on gauge slightly.

5. Open high side hand valve slightly.

NOTE: Too rapid a purging process will be identified by the appearance of an oily foam. If this occurs, close the hand valves a little more until this condition stops.

6. Close both hand valves on the gauge set when the pressures read 0 and all the refrigerant has left the system.

Evacuating the System

Before charging any system it is necessary to purge the refrigerant and draw out the trapped moisture with a suitable vacuum pump. Failure to do so will result in ineffective charging and possible damage to the system.

Use this hook-up for the proper evacuation procedure:

1. Connect both service gauge hoses to the high and low service outlets.

BAR GAUGE MANIFOLD AND COMPRESSOR SERVICE VALVE SETTINGS

Condition	Manifold Valves	Compressor Valves
Testing System	Both fully closed	Both cracked off backseat
Depressurizing System	Both cracked open	Both at mid position
Evacuating the system	Both wide open	Both at mid position
Charging in gas form with compressor running	High pressure valve closed Low pressure valve cracked	High pressure valve cracked off backseat Low pressure valve at mid position
Charging in liquid form with compressor off	Low pressure valve closed High pressure valve wide open	Both valves mid positioned

Note: A very small leak, causing system discharge about every two weeks, can be caused by a leaky Schrader type service valve. Check these valves with extra care when testing for a small leak.

2. Open high and low side hand valves on gauge manifold.

3. Open both service valves a slight amount (from back seated position), allow refrigerant to discharge from system.

4. Install center charging hose of gauge set to vacuum pump.

5. Operate vacuum pump for at least one hour. (If the system has been subjected to open conditions for a prolonged period of time it may be necessary to "pump the system down" overnight. Refer to "System Sweep" procedure.)

NOTE: If low pressure gauge does not show at least 28" hg. within 5 minutes, check the system for a leak or loose gauge connectors.

6. Close hand valves on gauge manifold.

7. Shut off pump.

8. Observe low pressure gauge to determine if vacuum is holding. A vacuum drop may indicate a leak.

System Sweep

An efficient vacuum pump can remove all the air contained in a contaminated air conditioning system very quickly, because of its vapor state. Moisture, however, is far more difficult to remove because the vacuum must force the liquid to evaporate before it will be able to remove it from the system. If a system has become severely contaminated, as, for example, it might become after all the charge was lost in conjunction with vehicle

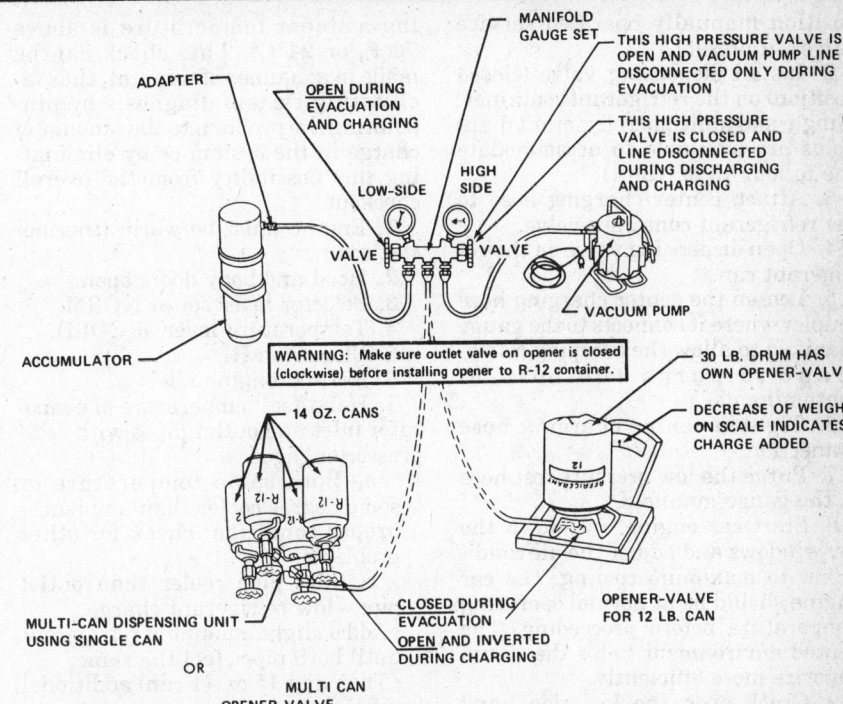

WARNING: Make sure outlet valve on opener is closed (clockwise) before installing opener to R-12 container.

Typical gauge connections for discharge, evacuation and charging the system

accident damage, moisture removal is extremely time consuming. A vacuum pump could remove all of the moisture only if it were operated for 12 hours or more.

Under these conditions, sweeping the system with refrigerant will speed the process of moisture removal considerably. To sweep, follow the following procedure:

1. Connect vacuum pump to

gauges, operate it until vacuum ceases to increase, then continue operation for ten more minutes.

2. Charge system with 50% of its rated refrigerant capacity.

3. Operate system at fast idle for ten minutes.

4. Discharge the system.

5. Repeat twice the process of charging to 50% capacity, running the system for ten minutes, and discharging it, for a total of three sweeps.

6. Replace drier.

7. Pump system down as in Step 1.

8. Charge system.

Charging the System

— **CAUTION** —

Never attempt to charge the system by opening the high pressure gauge control while the compressor is operating. The compressor accumulating pressure can burst the refrigerant container, causing sever personal injuries.

BASIC SYSTEM

In this procedure the refrigerant enters the suction side of the system as a vapor while the compressor is running. Before proceeding, the system should be in a partial vacuum after adequate evacuation. Both hand valves on the gauge manifold should be closed.

1. Attach both test hoses to their respective service valve ports. Mid-

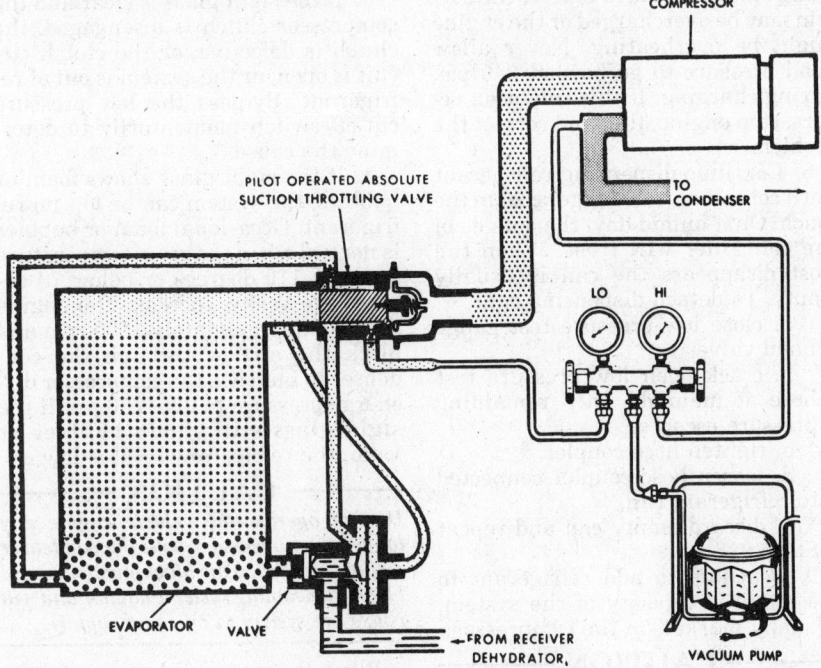

Schematic for evacuating the system

position manually operated service valves, if present.

2. Install dispensing valve (closed position) on the refrigerant container. (Single and multiple refrigerant manifolds are available to accommodate one to four 15 oz. cans.)

3. Attach center charging hose to the refrigerant container valve.

4. Open dispensing valve on the refrigerant can.

5. Loosen the center charging hose coupler where it connects to the gauge manifold to allow the escaping refrigerant to purge the hose of contaminants.

6. Tighten center charging hose connection.

7. Purge the low pressure test hose at the gauge manifold.

8. Start car engine, roll down the car windows and adjust the air conditioner to maximum cooling. The car engine should be at normal operating temperature before proceeding. The heated environment helps the liquid vaporize more efficiently.

9. Crack open the low side hand valve on the manifold. Manipulate the valve so that the refrigerant that enters the system does not cause the low side pressure to exceed 40 psi. Too sudden a surge may permit the entrance of unwanted liquid to the compressor. Since liquids cannot be compressed, the compressor will suffer damage if compelled to attempt it. If the suction side of the system remains in a vacuum the system is blocked. Locate and correct the condition before proceeding any further.

NOTE: Placing the refrigerant can in a container of warm water (no hotter than 125°F) will speed the charging process. Slight agitation of the can is helpful too, but be careful not to turn the can upside down.

Some manufacturers allow for a partial charging of the A/C system in the form of a liquid (can inverted and compressor off) by opening the high side gauge valve only, and putting the high side compressor service valve in the middle position (if so equipped). The remainder of the refrigerant is then added in the form of a gas in the normal manner, through the suction side only.

SYSTEMS WITHOUT SIGHT GLASS, EXCEPT CCOT SYSTEM

The following procedure can be used to quickly determine whether or not an air conditioning system has the proper charge of refrigerant (providing ambient temperature is above 70°F, or 21°C). This check can be made in a manner of minutes, thus facilitating system diagnosis by pinpointing the problem to the amount of charge in the system or by eliminating this possibility from the overall checkout.

1. Engine must be warm (thermostat open).

2. Hood and body doors open.

3. Selector lever set at NORM.

4. Temperature lever at COLD.

5. Blower on HI.

6. Normal engine idle.

7. Hand-feel temperature of evaporator inlet and outlet pipes with compressor engaged.

 a. Both same temperature or some degree cooler than ambient—proper condition: check for other problems.

 b. Inlet pipe cooler than outlet pipe—low refrigerant charge.
• Add a slight amount of refrigerant until both pipes feel the same.
• Then add 15 oz. (1 can) additional refrigerant.

 c. Inlet pipe has frost accumulation—outlet pipe warmer: proceed as in Step b above.

If during the charging process the head pressure exceeds 200 psi, place an electric fan in front of the car and direct the turbulent air to the condenser. If no fan is available, repeatedly pour cool water over the top of the condenser. These cooling actions may be necessary on an extremely warm day to help dissipate the heat emitted by the engine during idle.

If this fails and pressure on the discharge side continues to rise, the system may be overcharged or the engine might be overheating. Never allow head pressure to go beyond 240 psi. during charging. If this condition occurs, stop engine, find and correct the problem.

8. Continue dispensing refrigerant until container is no longer cool to the touch. On a humid day, the outside of the container will frost. When the frost disappears the can is usually empty. To detach dispensing can:

 a. close low pressure test gauge hand valve.

 b. crack open low pressure test hose at manifold until remaining pressure escapes.

 c. tighten hose coupler.

 d. loosen hose coupler connected to refrigerant can.

 e. discard empty can and repeat Steps 2–8.

9. Continue to add refrigerant to the required capacity of the system. (Usually marked on the compressor).

———— CAUTION ————
DO NOT OVERCHARGE. This condition

is usually indicated by an abnormally high side pressure reading and a noisy compressor resulting in ineffective cooling and damage to the system.

SYSTEMS WITH A SIGHT GLASS

The air conditioning systems that use a sight glass as a means to check the refrigerant level, should be carefully checked to avoid under or over charging. The gauge set should be attached to the system for verification of pressures.

To check the system with the sight glass, clean the glass and start the vehicle engine. Operate the air conditioning controls on maximum for approximately five minutes to stabilize the system. The room temperature should be above 70 degrees. Check the sight glass for one of the following conditions:

1. If the sight glass is clear, the compressor clutch is engaged, the compressor discharge line is warm and the compressor inlet line is cool, the system has a full charge of refrigerant.

2. If the sight glass is clear, the compressor clutch is engaged and there is no significant temperature difference between the compressor inlet and discharge lines, the system is empty or nearly empty. By having the gauge set attached to the system a measurement can be taken. If the gauge reads less than 25 psi, the low pressure cutoff protection switch has failed.

3. If the sight glass is clear and the compressor clutch is disengaged, the clutch is defective, or the clutch circuit is open, or the system is out of refrigerant. By-pass the low pressure cut-off switch momentarily to determine the cause.

4. If the sight glass shows foam or bubbles, the system can be low on refrigerant. Occasional foam or bubbles is normal when the room temperature is above 110 degrees or below 70 degrees. To verify, increase the engine speed to approximately 1500 rpm and block the airflow through the condenser to increase the compressor discharge pressure to 225–250 psi. If the sight glass still shows bubbles or foam, the refrigerant level is low.

———— CAUTION ————
Do not operate the vehicle engine any longer than necessary with the condenser airflow blocked. This blocking action also blocks the cooling system radiator and will cause the system to overheat rapidly.

When the system is low on refrigerant, a leak is present or the system was not properly charged. Use a leak detector and locate the problem area and repair. If no leakage is found, charge the system to its capacity. (Refer to the refrigerant capacity chart at the end of this section).

CAUTION

It is not advisable to add refrigerant to a system utilizing the suction throttling valve and a sight glass, because the amount of refrigerant required to remove the foam or bubbles will result in an overcharge and potentially damaged system components.

CCOT SYSTEM

When charging the CCOT system, attach only the low pressure line to the low pressure gauge port, located on the accumulator. Do not attach the high pressure line to any service port or allow it to remain attached to the vacuum pump after evacuation. Be sure both the high and the low pressure control valves are closed on the gauge set. To complete the charging of the system, follow the outline supplied.

1. Start the engine and allow to run at idle, with the cooling system at normal operating temperature.
2. Attach the center gauge hose to a single or multi-can dispenser.
3. With the multi-can dispenser inverted, allow one pound or the contents of one or two 14 oz. cans to enter the system through the low pressure side by opening the gauge low pressure control valve.
4. Close the low pressure gauge control valve and turn the A/C system on to engage the compressor. Place the blower motor in its high mode.
5. Open the low pressure gauge control valve and draw the remaining charge into the system. Refer to the capacity chart at the end of this section for the individual vehicle or system capacity.
6. Close the low pressure gauge control valve and the refrigerant source valve, on the multi-can dispenser. Remove the low pressure hose from the accumulator quickly to avoid

	Amount of refrigerant / Check item	Almost no refrigerant	Insufficient	Suitable	Too much refrigerant
Temperature of high pressure and low pressure lines.		Almost no difference between high pressure and low pressure side temperature.	High pressure side is warm and low pressure side is fairly cold.	High pressure side is hot and low pressure side is cold.	High pressure side is abnormally hot.
State in sight glass.		Bubbles flow continuously. **Bubbles will disappear and something like mist will flow when refrigerant is nearly gone.**	The bubbles are seen at intervals of 1 - 2 seconds.	Almost transparent. Bubbles may appear when engine speed is raised and lowered. **No clear difference exists between these two conditions.**	No bubbles can be seen.
Pressure of system.		High pressure side is abnormally low.	Both pressure on high and low pressure sides are slightly low.	Both pressures on high and low pressure sides are normal.	Both pressures on high and low pressure sides are abnormally high.
Repair.		**Stop compressor immediately** and conduct an overall check.	Check for gas leakage, repair as required, replenish and charge system.		Discharge refrigerant from service valve of low pressure side.

Using a sight glass to determine the relative refrigerant charge

loss of refrigerant through the Schrader valve.

7. Install the protective cap on the gauge port and check the system for leakage.
8. Test the system for proper operation.

Leak Testing the System

There are several methods of detecting leaks in an air conditioning system; among them, the two most popular are (1) halide leak-detection or the "open flame method," and (2) electronic leak-detection.

The halide leak detection is a torch like device which produces a yellowgreen color when refrigerant is introduced into the flame at the burner. A purple or violet color indicates the presence of large amounts of refrigerant at the burner.

An electronic leak detector is a small portable electronic device with an extended probe. With the unit activated the probe is passed along those components of the system which contain refrigerant. If a leak is detected, the unit will sound an alarm signal or activate a display signal depending on the manufacturer's design. It is advisable to follow the manufacturer's instructions as the design and function of the detection may vary significantly.

CAUTION

Caution should be taken to operate either type of detector in well ventilated areas, so as to reduce the chance of personal injury, which may result from coming in contact with poisonous gases produced when R-12 is exposed to flame or electric spark.

REFRIGERANT CAPACITIES CHART

	1980		1981		1982–87	
Auto Manufacturer	Models	Recharge Capacities (lbs.)①	Models	Recharge Capacities (lbs.)①	Models	Recharge Capacities (lbs.)①
AMERICAN MOTORS CORP.	Concord	2	Concord	2	All	2.00
	Pacer	2⅛	Spirit	2		
	Spirit	2	Eagle	2		
	Eagle	2				

AIR CONDITIONING

REFRIGERANT CAPACITIES CHART

Auto Manufacturer	1980		1981		1982–87	
	Models	Recharge Capacities (lbs.) ①	Models	Recharge Capacities (lbs.) ①	Models	Recharge Capacities (lbs.) ①
BUICK MOTOR DIVISION	Skyhawk	2½	Skyhawk	2½	Electra, LeSabre	3.50
	Skylark	2¾	Skylark	2¾	Regal, Century, Skylark	2.75
	Electra, LeSabre	3¾	Electra, LeSabre	3¾		
	Century, Regal, Riviera	3½	Century, Regal, Riviera	3½	Riviera, Somerset Regal	3.25
					Skyhawk	2.50
CADILLAC MOTOR CAR DIVISION	Seville, Eldorado	3½	Seville, Eldorado	3½	Cimarron	1.87
	All others	3¾	All others	3¾	Seville	2.75
					All others	3.50
CHEVROLET MOTOR DIVISION	Monza, Nova	3½	Monza, Nova	3½	Corvette, Camaro	3.00
	Chevette	2¼	Chevette	2¼	Caprice, Impala	3.50
	Camaro	3¼	Camaro	3¼	Monte Carlo	3.25
	Corvette	3	Corvette	3	Malibu, Celebrity, Citation, Cavalier	2.75
	Citation	2¾	Citation	2¾		
	Malibu, Impala, Caprice	3¾	Malibu, Impala, Caprice	3¾	Chevette	2.25
CHRYSLER CORPORATION	All, except below	2⅝	All, except below	2⅝	Diplomat, Gran Fury, New Yorker, Mirada, Cordoba, Imperial	2.62
	Omni, Horizon	2⅛	Omni, Horizon, Aries, Reliant	2⅛		
					Aries, Reliant, LeBaron, E-Class, 400, 600	2.37 ③
					Omni, Horizon, 024, TC3, Charger, Turismo	2.12
FORD DIVISION	Pinto	2¼	Escort, EXP	2½	Thunderbird, Fairmont, Granada, Mustang, Escort, EXP	2.56
	All others ②	3½	All others ②	3½		
					LTD, Crown Victoria	3.25
LINCOLN-MERCURY DIVISION	Bobcat	2¼	Lynx, LN7	2½	Lincoln Continental, XR-7, Zephyr, Cougar, Capri, Lynx, LN7	2.56
	All others ②	3½	All others ②	3½		
					Lincoln, Mark VI	3.00
					Marquis, Grand Marquis	3.25
OLDSMOBILE DIVISION	Cutlass, 88, 98	3¾	Cutlass, 88, 98	3¾	88, 98	3.50
	Toronado	3½	Toronado	3½	Cutlass, Cutlass Ciera, Omega	2.75
	Starfire	2½	Starfire	2½		
	Omega	2¾	Omega	2¾	Toronado	3.25
					Firenza	2.50
PONTIAC MOTOR DIVISION	Catalina, Bonneville	3½	Catalina, Bonneville Lemans, Gran Am	3½	Firebird	3.00
	Firebird	3¼	Firebird	3¼	Bonneville, Parisienne, Grand Am	3.30
	Sunbird	2½	Sunbird	2½		
	Phoenix	2¾	Phoenix	2¾	Grand Prix	3.30
					6000, Phoenix	2.75
					J-2000, Fiero, Sunbird	2.50
					T-1000	2.25

① All refrigerant charges listed are approximate and represent a minimal reserve with moderate head pressures. Check label on or near compressor for correct charge.

② Ford, Mercury and Lincoln vehicles using Frigidaire 6 cylinder compressor–4¼ lbs.

③ 1985 and later Aries, Reliant, LeBaron, E class, 400 & 600 recharge capacities are 2.12 lbs.

Gauges and Indicators

There are various systems used to indicate values of heat, pressure, vacuum, current flow, and fuel supply. The following are the more popular systems used.

Bourdon Tube

This gauge consists of a flattened tube that is bent to form a curve. The curve tends to straighten under internal pressure caused by engine oil pressure. The curved tube is geared or linked to an indicator needle which may be read on a calibrated scale.

Bourdon tube oil pressure gauges are used on some Corvettes and the optional instrument panels on some Chevrolet sport models. This type of gauge may be easily distinguished from the electrical type by the small copper or nylon tube running from the gauge to the engine.

Bi-Metallic or Thermal

This gauge is activated by the difference in the expansion factors of a bi-metal bar. A sending unit, consisting of a variable resistance conductor, influences current flow to a voltage limiter, or directly to a heating element coiled around a bi-metal bar in the gauge. A bi-metallic gauge pointer will move slowly to its gauging position.

Magnetic

In this system, the indicator needle is moved by changing the balance between the magnetic pull of two coils built in the gauge. When the ignition switch is in the "off" position, the pointer may rest any place on the gauge dial. Balance is controlled by the action of a sending unit or a tank unit containing a rheostat, the value of which varies with temperature, pressure or movement of a float arm. A magnetic gauge will snap to its position when turned on.

Vacuum Gauges

The gauge operates by monitoring engine

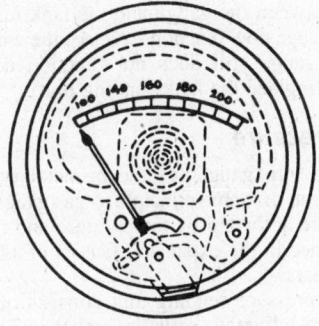

Bourdon tube gauge

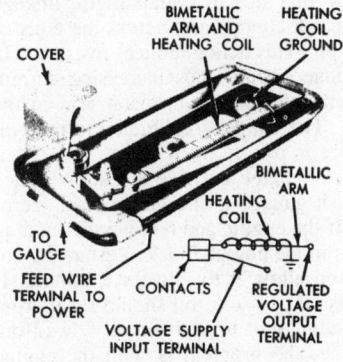

Constant voltage regulator

vacuum. High engine vacuum draws the needle to the high side of the gauge against internal spring tension. As engine vacuum decreases, the spring tension overcomes the vacuum pull and the needle moves to the low side of the gauge.

Warning Lights

This system is quite popular and may be used to indicate heat, low pressure or as a battery discharge indicator. General Motors uses a two-light temperature indicator version of this unit in some models.

Electronic Instruments

These systems use vacuum fluorescent displays to replace the conventional indicator needle type gauge. Typical instrumentation of this type is used for the chronometer, odometer, speedometer, temperature, and fuel gauges. These systems vary, and both digital and graphic displays are used. They are often connected to a microcomputer and used to "call up" additional information. Display information may be shown in English or Metric.

SECTION 1
BOURDON TUBE

Oil Pressure

The gauge is the pressure expansion type and is activated by oil pressure developed by the oil pump, acting directly on the mechanism of the gauge. The gauge is connected by a small tube to the main oil passage in the engine oiling system. This design registers the full pressure of the oil pump.

TESTING

A gauge pointer that flutters is usually an indicator that oil has entered the gauge tube. The tube should contain trapped air to cushion the pulsations of the oil pump and relief valve. Oil can work up into the gauge lines as a result of a gauge or tube leak or improper installation. To correct this condition, renew the unit or correct the leak; then, with the gauge line disconnected at both ends, blow the line clear. Connect line at gauge first and then at the engine.

If the gauge reads too low or reads no pressure, test for a possible obstruction by disconnecting the line at the gauge. Hold the end of the line over an empty container, then start the engine. After a few bubbles, oil should flow steadily.

If oil does not flow satisfactorily, first make sure that the oil level is correct and that the oil pump is functioning. Should the engine oil system be operating correctly, the problem is either with the gauge or the line. Check the line for kinks, leaks, or blockage which would prevent oil from reaching the gauge. If the line is unobstructed, remove the gauge unit from the instrument panel. Check to make sure that the hole leading to the Bourdon tube is clear and be sure that the lever linkage and pointer gears operate freely. If none of these points is at fault, the Bourdon tube itself is defective and the gauge must be replaced.

SECTION 2
BI-METAL

Fuel

Bi-metal or thermal type gauges operate on the principle of constant applied voltage and are sensitive only to changes originating at the sending unit.

The fuel gauge system consists of a sending unit, located in the fuel tank, and a registering unit mounted in the instrument cluster. The sending unit is a rheostat that varies its resistance depending on the amount of fuel in the tank.

TESTING THE DASH GAUGE

— **CAUTION** —
Gauge systems using constant voltage regulators should not be grounded while testing. An excess of 5 volts is likely to burn out the unit.

To safely test this type of voltage regulated system:

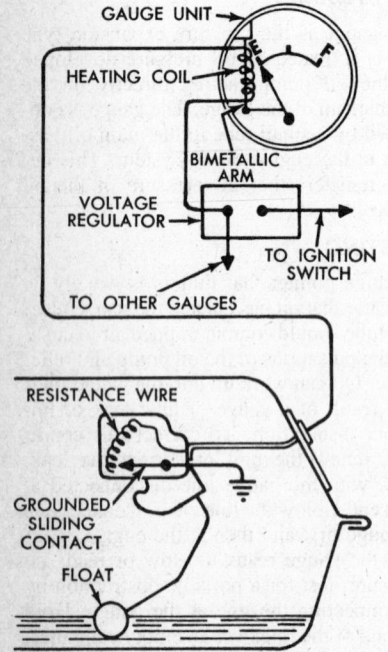

Bi-metallic fuel gauge system

1. Have the ignition switch in the "off" position.
2. Connect the terminals of four, series-connected. D-type flashlight batteries (total of six volts) to the terminals of the gauge to be tested. Three volts should cause the gauge to read approximately half-scale.

If the gauge reads half-full and was not working properly before, the sending unit in the tank is probably defective.

If the gauge is inaccurate or does not register, replace it.

If both the fuel gauge and temperature gauge are in error, in the same manner, the constant voltage regulator is probably at fault.

While working under the dash, be careful not to ground any of the gauges. A full flow of current through the regulator to ground is likely to burn out the regulator.

TESTING THE SENDING UNIT

If the dash gauge test shows that unit to be satisfactory, the sending unit or gauge system wiring is faulty. Substitute a jumper wire between the gauge and the tank unit. If the gauge now functions, replace the wire. If the gauge still does not function correctly, replace the tank sending unit.

Oil Pressure

Oil pressure gauges of the bi-metal type operate on the same principle as gas gauges. They are activated by temperature and the difference in the expansion factors of a bi-metal bar.

The pressure sending unit consists of a pressure-activated variable resistor. This sealed unit is usually screwed into the engine oil pressure circuit. As pressure is applied to one side of a diaphragm, linkage advances a contact arm across the coils of a resistor. This action reduces resistance in the gauge circuit, thus increasing current flow and heat to the bi-metal arm in the gauge. The gauge is calibrated to read oil pressure in psi.

Run the engine and have an assistant watch the dash gauge. If the gauge reads zero, turn off the engine and remove the sending unit from the engine block. Restart the engine and allow it to idle for a minute. If there is oil pressure, oil should surge from the sending unit hole. If no oil flows from the hole, the problem is with the engine lubricating system. If oil flows, the fault lies with the sending unit, the wiring, or the dash gauge.

Bi-metallic oil gauge circuit

Check the gauge by grounding the connecting wire for an instant with the ignition switch turned on. A good gauge will go to the top of its scale.

— **CAUTION** —
Grounding the connecting wire for any longer than a moment will damage the dash units.

If the gauge did not move when grounded, check the wiring to the dash unit for continuity. If the wiring is not faulty and the gauge doesn't register when grounded, replace the gauge. If the gauge functions when grounded, replace the sending unit.

Temperature

The temperature gauge consists of a sending unit, mounted in the cylinder head or block, and a remote resistor unit (temperature gauge) mounted on the instrument panel. The principle of operation is essentially the same as the bi-metallic fuel gauge, the exception being that the resistance of the sending unit is influenced by engine temperature instead of tank fuel level, as with the fuel gauge.

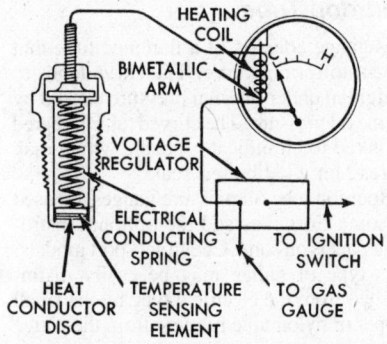

Bi-metallic temperature gauge circuit

The temperature sending unit is constructed with a coil spring and sensing disc. Current passing through this coil encounters increased resistance, proportional to an increase in temperature. The gauge registers this resistance change and is calibrated to indicate the temperature.

TESTING THE DASH GAUGE

Connect four D-cells (total of 6 volts) in series with the dash gauge, with the ignition switched off. A good gauge will register ½ on the scale. Replace the gauge if it does not move.

TESTING THE SENDING UNIT

Bring the engine to normal operating temperature (check with a thermometer). If the gauge doesn't register, disconnect the connecting wire from the engine sending unit and ground the connecting wire for an instant and have an assistant observe the gauge.

— **CAUTION** —
Grounding the wire for any longer than a moment will damage the dash units.

If the gauge shows no reading, replace the connecting wire. If the gauge registers when grounded, replace the sending unit.

SECTION 3
MAGNETIC

Fuel

The magnetic fuel gauge consists of two units, the dash unit and the sending unit in the fuel tank. One terminal of the dash unit is connected to the ignition switch so that the system is active only when the ignition is on. With the ignition off, the pointer may come to rest at any position on the dial.

The gauge pointer is moved by varying the magnetic pull of two coils in the unit. The magnetic pull is controlled by the action of the tank unit which contains a variable rheostat, the value of which varies with movement of a float and arm.

When the ignition switch is on and the tank unit arm is in the full position, the current flow to ground is through the resistor, battery coil and the ground coil. Because the ground coil has more windings than the battery coil, it builds up a stronger magnetic field and the pointer is pulled to the full position.

When the tank unit arm is in the empty position, the current flow is through the resistor, the battery coil and the wire to ground at the tank unit. The pointer is thus pulled to the empty position. The resistor in series with the battery coil balances resistance between the two coils in the dash unit.

TESTING THE DASH GAUGE

Disconnect the wire from the tank unit. Using a tank unit of known accuracy, clip a test wire from the body of the test unit to ground. Clip another test wire from the connector of the test unit to the tank unit wire. With the ignition on, moving the float arm through its entire range should cause the gauge to respond proportionally. If the dash gauge does not correspond to the movement of the test unit and the wiring to the gauge is OK, the dash unit is bad.

TESTING THE TANK UNIT

If tests indicate that the trouble lies in the tank unit, remove the unit and check for mechanical failure. The unit may have either a ruptured or binding float.

An electrical check for circuit continuity may be made throughout the unit's range.

Temperature

The temperature gauge system consists of a magnetic dash unit and a resistance-type sending unit screwed into the water jacket of the cylinder head or the engine block.

The dash unit has two magnetic poles. One of the windings is connected to the ignition switch and ground. This electromagnet exerts a steady pull to hold the gauge pointer to the left or ''cold'' position when the ignition is on.

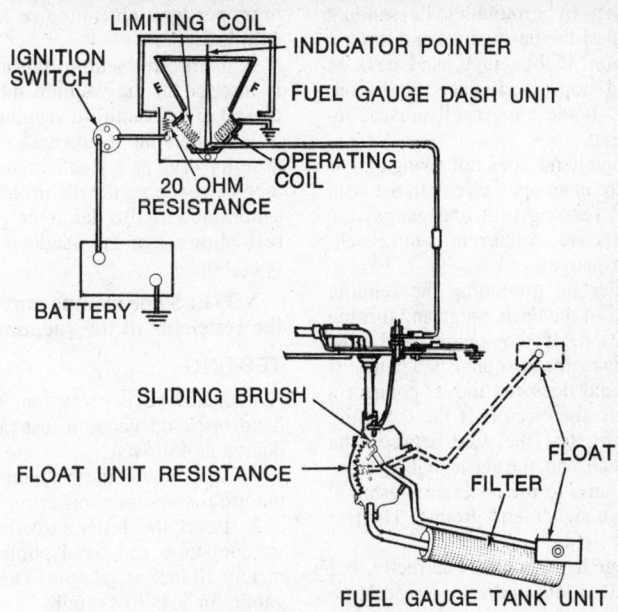

Magnetic fuel gauge circuit

The other winding in the dash unit connects to a ground through the engine sending unit. This electromagnet exerts a steady pull on the gauge pointer toward the right, or ''hot'' side of the gauge. The strength of this pull is dependent upon the current allowed to pass through the engine unit (sending unit) resistor.

The sending unit, located in the engine cooling system, contains a flat disc (thermistor) that changes resistance as its temperature varies.

NOTE: This sending unit, while similar in appearance, is different and is not interchangeable with the unit used in systems using bi-metal or thermal dash gauges. The resistance of the thermistor disc is maximum when the temperature is cold and minimum when hot. The decrease in resistance allows more current to flow through the electromagnet connected to the engine unit. The resulting increase in magnetic pull causes the gauge pointer to move to the right, or ''hot'' side.

TESTS

1. Disconnect the wire at the sending unit and turn on the ignition switch. The gauge hand should stay against the cold side stop pin.

2. Ground the wire disconnected from the sending unit. With the ignition switch still on, the gauge hand should swing across the dial to the hot stop pin.

CORRECTIVE MEASURES

If the gauge hand does not stay to the left, either the wire is grounded between the dash unit and the engine unit or the dash unit is defective.

Test further by disconnecting the sending unit wire at the gauge. Turn on the ignition. If the gauge hand stays on the left-hand stop pin, replace the disconnected wire. If the gauge still moves, replace the gauge.

If the gauge hand does not swing across the dial, there is an open circuit in the wire between the sending unit and gauge, the gauge is defective, or current is not reaching the dash gauge.

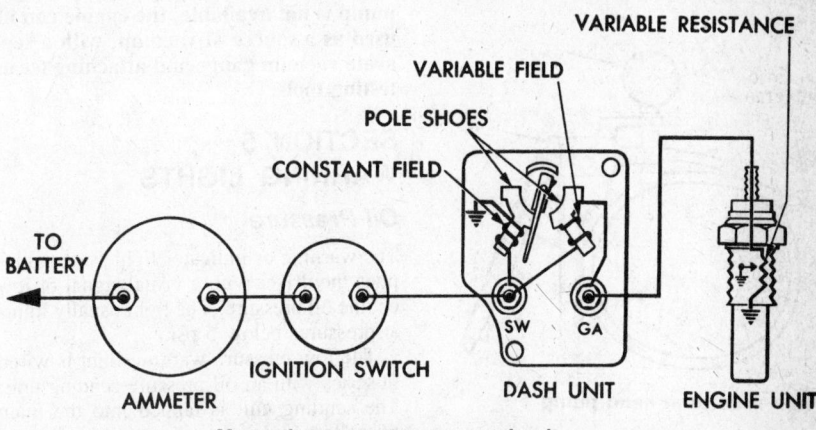

Magnetic temperature gauge circuit

Test further by grounding the sending unit terminal of the dash gauge and turning on the ignition. If the gauge stays on the left-hand stop pin, replace the disconnected wire. If the gauge still moves, replace the gauge.

If the gauge hand does not swing across the dial, there is an open circuit in the wire between the sending unit and gauge, the gauge is defective, or current is not reaching the dash gauge.

Test further by grounding the sending unit terminal of the dash gauge and turning on the ignition. If the gauge hand now moves, replace the disconnected wire. If the gauge hand does not move, connect a test lamp into the circuit. If the test lamp does not light, test the wire between the ignition switch and the dash unit by connecting the lamp to the accessory terminal at the ignition switch and ground. The test lamp should light.

If the gauge hand operates correctly, but the gauge does not indicate temperature correctly, either the sending unit is defective or the dash gauge is out of calibration. Replace sending unit with one of known accuracy. If gauge reading is still incorrect, replace the gauge.

If the gauge hand is at maximum at all times, and tests 1 and 2 indicate that the wiring and the dash unit are good, the sending unit must be replaced.

If the gauge hand will not move, the dash unit is bad, or incorrectly installed. Correct the installation or replace the gauge.

SECTION 4
VACUUM AND ECONOMY GAUGES

The fuel economy gauge indicates engine manifold vacuum, as a function of throttle position and engine load. The face of the gauge dial is divided into three segments; Poor (low vacuum), Good (normal vacuum for cruise), and Decelerate (high vacuum). Although the gauge is not intended as a close tolerance vacuum indicator, it may be assumed that a gauge reading continuously below the Good band (normal cruise or idle) may mean poor engine performance due to improper ignition timing or manifold vacuum leakage.

A manifold vacuum pulsation restrictor is inserted in the vacuum tube at the end closest to the manifold vacuum connection. This enables the inside area of the vacuum hose to serve as a small vacuum reservoir, thereby reducing the manifold vacuum pulsations and to also damp the gauge reading restriction against sudden accelerator operation.

NOTE: Some manufacturers do not use the restrictor in the vacuum line.

TESTING

A standard vacuum system test, using a hand operated vacuum test pump, is conducted as follows:

1. Disconnect the vacuum tube at the manifold vacuum connection.

2. Insert the tester into the end of the vacuum tube and hand pump to approximately 20 inches vacuum. Observe the test gauge for loss in vacuum.

   a. If the tester vacuum gauge indicates a loss in vacuum, remove the vacuum tube connector from the threaded vacuum connector on the back of the gauge. Apply a short length of teflon tape around the threads of the vacuum connection and reinstall the vacuum tube connector on the threaded vacuum connection. Recheck the gauge with the hand vacuum pump. If the tester gauge still indicates a loss in vacuum, replace the gauge assembly.

   b. If the vacuum reading remains steady, the vacuum tube and gauge are OK. Check the end of the tube to be sure the pulsation restrictor is installed; then reconnect the manifold vacuum connection.

Connect the vacuum tube of the test pump directly to the economy gauge tube connector. Pump the tester to approximately 20 inches vacuum and observe the tester gauge for a loss in vacuum.

If the tester gauge indicates a loss in vacuum, replace the economy gauge assembly.

If the tester gauge reading remains steady, the hose to the engine manifold vacuum port must be repaired or replaced.

NOTE: If a hand operated vacuum pump is not available, the engine can be used as a source of vacuum, with a separate vacuum gauge and attaching tee as testing tools.

SECTION 5
WARNING LIGHTS
Oil Pressure

The warning or indicator light system supplies the driver with a visual signal of low engine oil pressure. The light usually lights at pressures below 5 psi.

The low pressure warning light is wired in series with an oil pressure sending unit. The sending unit is tapped into the main oil gallery and is sensitive to oil pressure.

The unit contains a diaphragm, spring linkage and electrical contacts. When the ignition switch is on, the warning light circuit is energized and the circuit is completed through the closed contacts in the sending unit. When the engine starts, oil pressure will compress the diaphragm, opening the contact points and breaking the circuit.

TESTS

The light should light when the engine is not running and the ignition switch is turned on. If the light does not go on, first substitute a new bulb. If there is still no light, check the wire from the light to the switch. If the wire is not at fault, disconnect the wire at the sending unit and ground it. Replace the sending unit if the light now lights.

Temperature

This system employs a heat sending unit with either one or two sets of contacts. Some systems use a green light to indicate subnormal, and a red light to warn of abnormal heat. The more common system, however, uses a simple make-and-break heat-sensitive sending unit screwed into the engine cooling system, and wired in series with the hot indicator light in the instrument panel.

The two-light system uses a bi-metal element mounted between two signal circuits. Normal operating temperature (somewhere between 120°F. and 250°F.) will cause the bi-metal bar to assume a position of no contact between the low and the high temperature circuit. When the ignition switch is turned on, with a cold engine, the cold (green) circuit is complete. If the engine becomes hot enough to move the bi-metal bar so that it touches the contacts of the hot circuit, the hot (red) light comes on. This hot signal indicates that temperatures are in the area of 250°F. in the sealed cooling system.

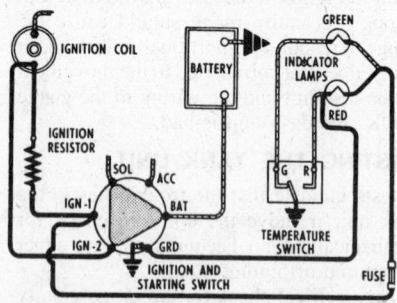

Cold and hot temperature indicator circuit

TESTS

Use the same testing procedure given for oil pressure.

Charge Indicator

A light is used to indicate general charging system operation. When output is below battery potential, a red light is shown. When output is above battery potential, other factors (wiring, voltage regulator, etc.) being

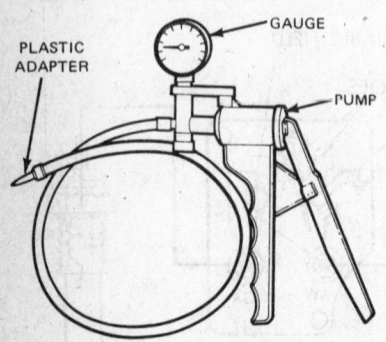

Vacuum hand pump

PLASTIC ADAPTER

GAUGE

PUMP

normal, the light is out.

The charge indicator bulb is connected to the charging circuit, obtaining its ground through the voltage regulator. When the output rises above battery potential, the current flow causes the light to go out.

When an alternator is used, it is necessary to supply a small amount of excitation current to the alternator field, due to the small amount of residual magnetism. Current can be supplied from the battery, through the indicator light, and to the regulator terminal on the alternator. This current has a value of about 12 volts at .25 amperes and will cause the indicator light to come on. Most systems have a resistor in parallel with the bulb to provide excitation if the bulb burns out and to prevent the light from glowing dimly during normal operation.

When the alternator starts to supply current, an output voltage is developed at the regulator terminal. When this voltage exceeds the battery voltage, current will pass from the alternator to the battery and to the system. This current is flowing in the reverse direction of the voltage supplied by the battery. The current flow coming from the alternator exceeds the battery current by a regulated 1 or 2 volts. This is not enough to light the indicator light, therefore, the light will go out when the alternator is supplying sufficient current.

If the alternator output current should drop below battery voltage, current will begin to flow in the opposite direction. If it exceeds 2 or 3 volts, the light will glow indicating that the alternator is not operating properly.

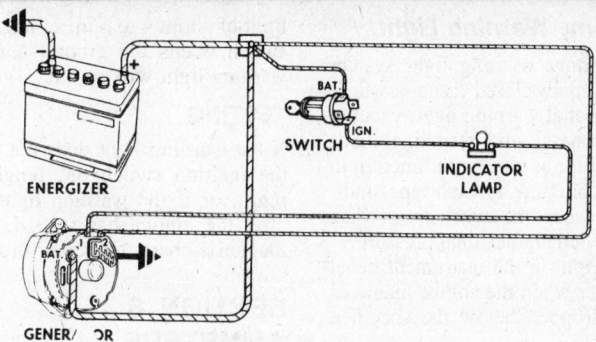

Charging indicator light circuit

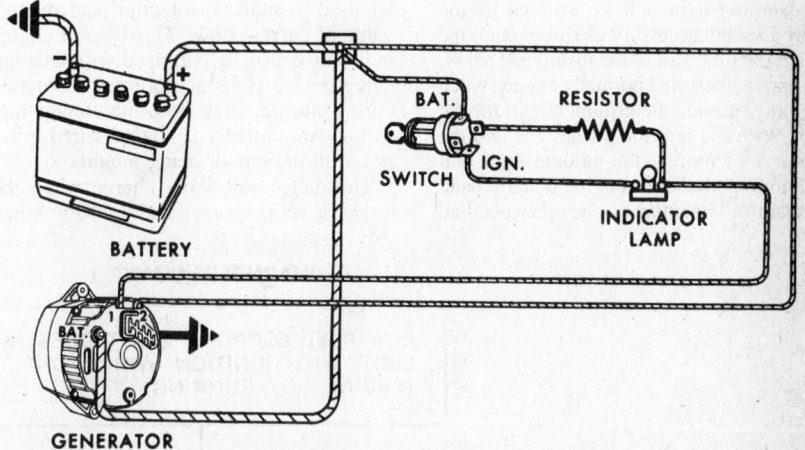

Simplified view of a typical charging system schematic showing the resistor wire in parallel with the charge indicator lamp

Coolant Level Indicator

Some GM models have a warning light which comes on if the coolant level in the radiator drops below a predetermined level. The coolant level indicator consists of three units; a sending unit which is threaded into the side tank of the radiator, a module which is mounted behind the instrument cluster, and a warning light. As a bulb test, the light is wired so that it comes on when the key is turned to the "START" position.

LIGHT DOESN'T COME ON

Perform the following checks if the warning light won't come on when the key is turned to the "START" position:

1. With the ignition switch in the "ON" position, unfasten the lead from the coolant level sending unit. If the light comes on replace the sending unit.

2. If the light didn't come on in step 1, check the light in the indicator and replace it, if necessary.

3. If the bulb is OK, check the wiring between the sending unit and module, and then between the module and light. If the wiring is not "open," replace the module.

LIGHT WON'T GO OUT

Perform the following checks if the light won't go out when the coolant is at the specified level:

1. Detach the lead from the coolant level

sending unit and ground the lead connector with a jumper wire. Turn the ignition switch to the "ON" position.

2. If the light doesn't come on, replace the sending unit. If the light remains on disconnect the jumper wire and proceed with the next step.

3. Check for a short in the sending unit-to-module wiring. If there is no short, replace the module.

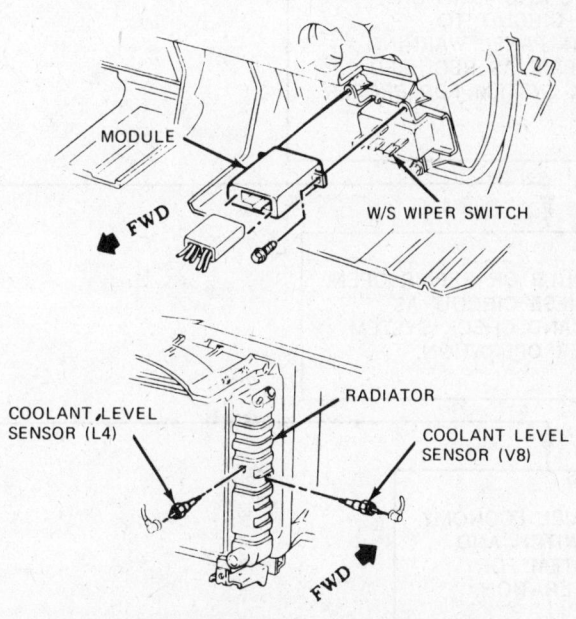

Coolant level sending unit and module

Fuel Economy Warning Light

The fuel economy warning light system consists of a normally closed vacuum switch, an instrument panel warning light, vacuum hose, wire harness, and attaching hardware. Its operation is similar in function to that of the oil pressure (switch type) indicating system, except the switch opens when vacuum is applied, rather than pressure.

A warning light in the instrument panel warns the driver when the engine manifold vacuum has dropped below the specified limit.

Electrical Circuit

The warning light bulb is powered by the ignition switch accessory circuit through the printed circuit board of the instrument panel. The wire harness and normally closed switch assembly provide the ground circuit for the bulb. With the ignition switch ON and the engine not running, the colored light will be illuminated. As the engine is started and the manifold vacuum reaches the specified limit of about 4 to 6 in. of hg., the vacuum switch opens the ground circuit and the warning light will go out.

TESTING

If the warning light does not operate with the ignition switch on, (engine not running), or if the warning light remains on after the engine has started, refer to the diagnosis charts for repair procedures.

SECTION 6
AMMETERS

The automotive ammeter is a gauge or meter used to indicate direction and relative value of current flow. This type of charge indicator is usually equipped with a dampening device to reduce pointer fluctuation during current surge from the voltage regulator. An ammeter is always wired in series with the circuit being monitored.

The meter will show charge when the battery is being charged and discharge when the battery is being discharged. It merely gives an indication of the state of charge of the battery, since it shows a relatively high charging rate when the battery is low, and a low charging rate when the battery is near full charge. An ammeter does not give a complete report of battery condition, whereas a voltmeter does. Just after cranking the engine, the meter will swing toward the charge side for a short time, if lights and accessories are turned off. As the energy spent in cranking is restored to the battery, the pointer will gradually move back toward center but should stay on the charge side. If the battery charge is low, however, the indicator will show a high charging rate for an indeterminate length of time.

The ammeter does not show the charging rate of the alternator.

At speeds above 30–35 mph, with all lights and accessories on, the indicator should show a reading somewhere on the charge side, depending on the state of the battery. Above this speed, the indicator should never show a discharge reading; if it does, the

DIAGNOSIS CHART 1

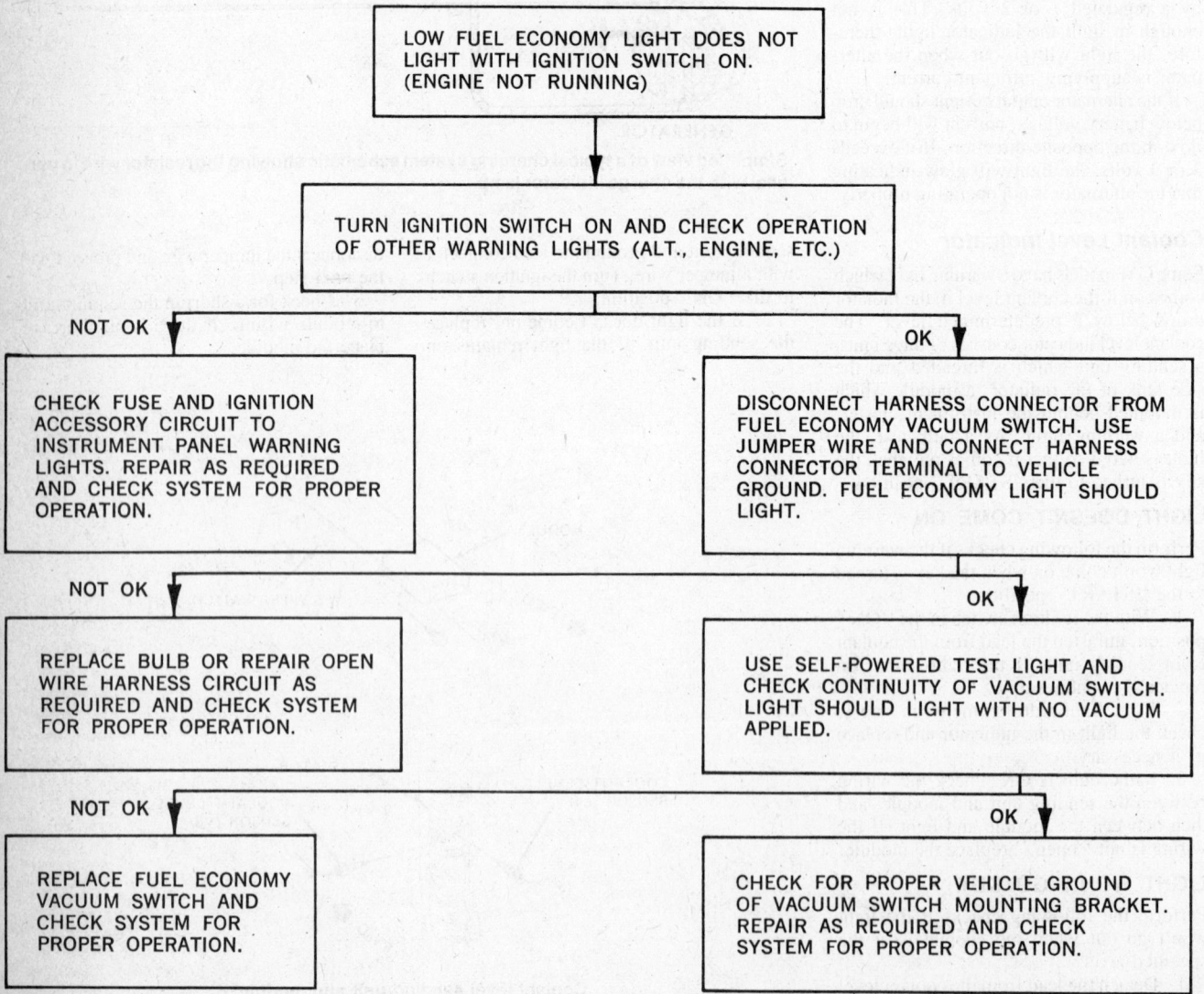

DIAGNOSIS CHART 2

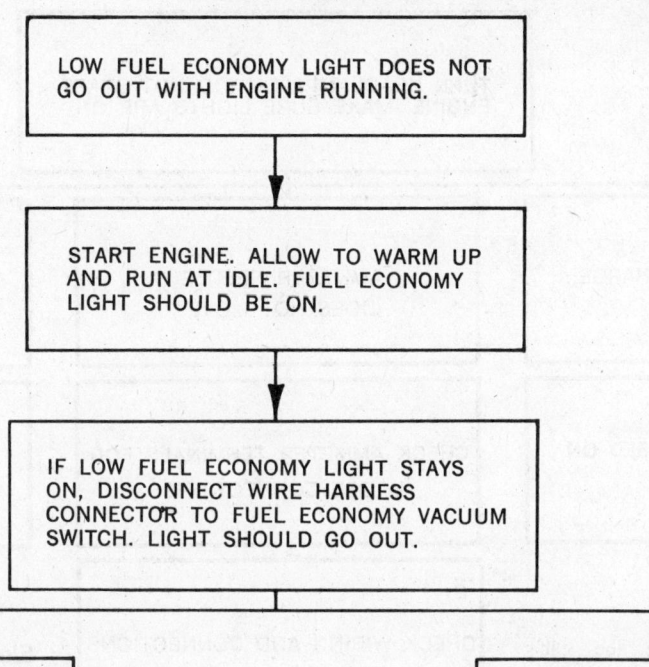

LOW FUEL ECONOMY LIGHT DOES NOT GO OUT WITH ENGINE RUNNING.

↓

START ENGINE. ALLOW TO WARM UP AND RUN AT IDLE. FUEL ECONOMY LIGHT SHOULD BE ON.

↓

IF LOW FUEL ECONOMY LIGHT STAYS ON, DISCONNECT WIRE HARNESS CONNECTOR TO FUEL ECONOMY VACUUM SWITCH. LIGHT SHOULD GO OUT.

NOT OK

CHECK WIRE HARNESS CIRCUIT FOR GROUND OR SHORT. REPAIR AS REQUIRED AND CHECK SYSTEM FOR PROPER OPERATION.

OK

CHECK VACUUM HOSE TO VACUUM SWITCH FOR LEAK. PINCHED OR PLUGGED CONDITION.

NOT OK

REPAIR OR REPLACE VACUUM HOSE AS REQUIRED AND CHECK SYSTEM FOR PROPER OPERATION.

OK

USE HAND VACUUM PUMP AND CHECK VACUUM SWITCH FOR PROPER OPERATION. REPLACE SWITCH, IF REQUIRED, AND CHECK SYSTEM FOR PROPER OPERATION.

alternator and regulator should be tested. See "Charging and Starting Systems" for troubleshooting.

SECTION 7
VOLTMETERS

A voltmeter is used on some cars, instead of an ammeter. The voltmeter indicates regulated voltage, which shows the charging system's ability to keep the battery charged. A voltmeter is always wired in parallel with the circuit being monitored. Voltmeter readings that are continuously high or low, may indicate a defective regulator, broken or slipping alternator drivebelt, a faulty alternator, or a defective battery. For testing and service of these items, see "Charging and Starting Systems."

If a faulty voltmeter is suspected, check the voltage regulator output with a test voltmeter of known accuracy (See "Charging and Starting Systems"). If the voltage indicated on the test instrument is within specifications, and disagrees with the car's

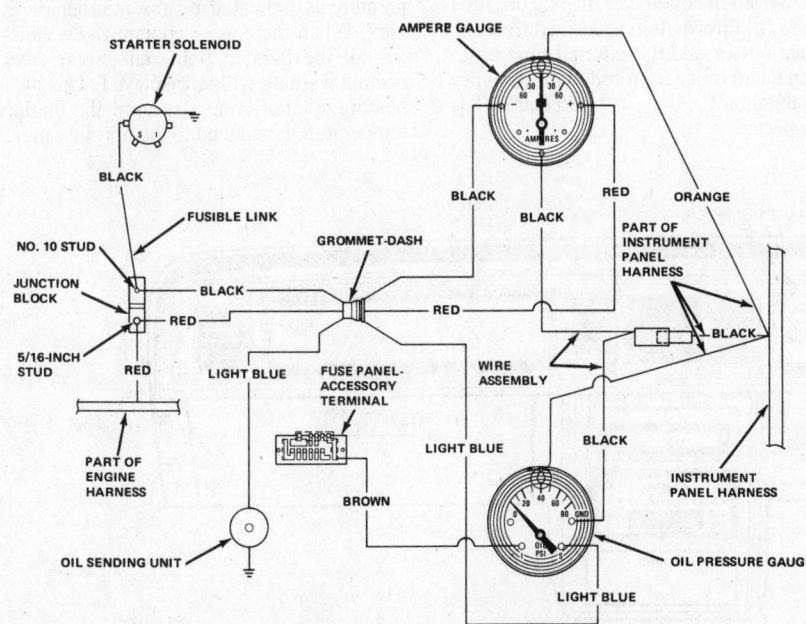

Typical ammeter and oil pressure gauge wiring

AMMETER DIAGNOSIS

```
┌─────────────────────────────────┐
│  TURN HEADLAMPS ON (DO NOT START │
│  ENGINE. MAKE SURE LIGHTS ARE ON)│
└─────────────────────────────────┘
```

AMMETER SHOWS CHARGE	AMMETER NEEDLE DOES NOT MOVE	AMMETER SHOWS DISCHARGE
CONNECTIONS REVERSED ON AMMETER	CHECK AMMETER TERMINALS FOR LOOSE CONNECTIONS	AMMETER OK
	CHECK WIRING AND CONNECTIONS	
	FAULTY AMMETER	

voltmeter reading, replace the car's voltmeter.

SECTION 8
ELECTRONIC
INSTRUMENTS

Electronic Chronometer

Digital chronometers may be capable of displaying the time of day, month, the day of the month, and on some models an elapsed time function is built-in.

Electronic Digital Speedometer

Digital speedometers are as easy to read as a digital clock, with vehicle speed shown clearly in large numbers. On some models a mode selection may be made to display either miles or kilometers per hour.

Electronic Odometer

Vehicle and trip distance, or kilometers and

average speed, are available at the touch of a button from the odometer display on some models. All information is derived from the distance sensor and the internal time base. Accumulated mileage stored in the memory is maintained even if the battery is disconnected.

Electonic Fuel Gauge or Display

The electronic fuel gauge is a bar graph with a row of bar segments. The fuel level is indicated by the number of lit bars. The bars individually turn off as the fuel is consumed.

The electronic fuel display on some models offers multiple functions. In addition to fuel remaining in the tank it may also measure fuel comsumption for range, present fuel efficiency, and trip fuel efficiency. Also a mode selection may be made to display either gallons or liters.

Electronic Temperature Gauge

The electronic temperature gauge is a bar graph with a row of bar segments. The temperature is indicated by the number of lit bars. When the engine coolant rises, causing all the bars to light, an engine over heating warning will be displayed. This over heating symbol continues until the engine temperature is reduced to the normal range.

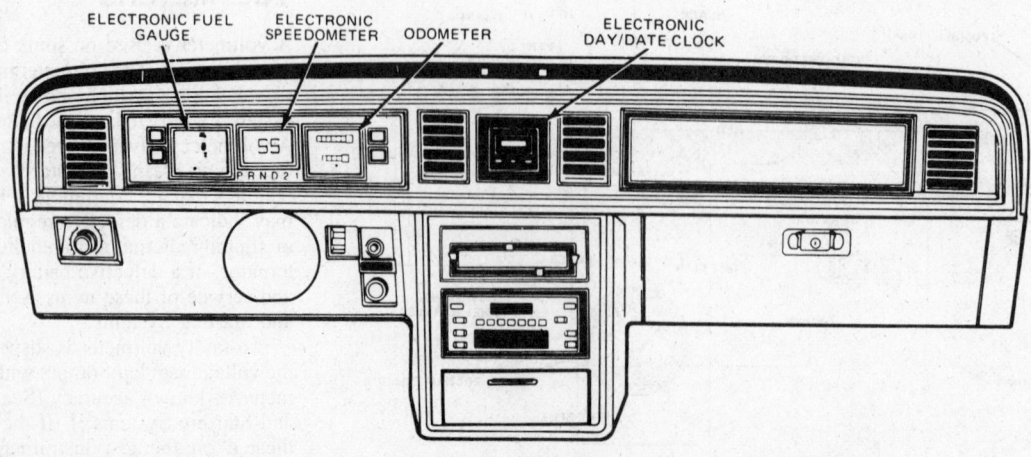

Typical electronic instrument cluster

Charging and Starting

BATTERIES

The modern battery is a 12-volt lead-acid unit having a particular ampere hours capacity; it is dependent upon the required work load (radio, air conditioning, electric windows, tailgate and etc.). The different sizes and shapes are specified by the vehicle manufacturer and are matched to the vehicle's electrical needs.

The primary purpose of the battery is to supply a source of energy for cranking the engine; it also provides the necessary power for the ignition system. The battery can (for a limited time) supply adequate current to satisfy the electrical demands during periods when requirements exceed the alternator output.

REPLACING THE BATTERY

Before deciding on a particular battery, consider some of the essentials that may put the replacement battery in a different category from the original unit, supplied with the vehicle. When the original battery wears out, resistance in the wiring circuits is probably increased and the starter may be less efficient, along with the ignition system. There is also the likelihood that electrical accessories have been added.

All of the above reasons are justifications for choosing a greater capacity battery than the one supplied by the manufacturer. It should be noted, however, that vehicles equipped with one or more on-board computers must use a replacement battery that is identical to original equipment specifications. On these cars, the battery terminals must never be disconnected with the ignition in the ON position, or computer damage could result.

Preparation

If a dry "charged" battery is selected as a replacement, place the new battery on a bench or a work table. Never activate the battery when installed in the vehicle. Remove the vent caps from all the cells, then fill each cell carefully, using sulfuric acid and distilled water (electrolyte) at a strength of 1.250–1.265 specific gravity to about $^3/_8$ in. above the top of the separators or to the indicated level mark.

> **CAUTION**
>
> *Because the electrolyte is extremely corrosive to metals and many other materials, do not pour it into sinks or drains. If the battery acid is spilled on the battery (during filling or charging), the bench or the clothing, immediately flush it off with generous amounts of water and baking soda or ammonia.*

Place a battery type thermometer in one of the center cells, then check the specific gravity of the electrolyte with a battery hydrometer. The battery temperature must be above 80°F and the specific gravity must be above 1.250 prior to installing the battery.

When charging a 12 volt battery, set the charging rate at 35 amperes until the electrolyte has reached 80°F and the electrolyte gravity is 1.250 or higher. Lower charging rates also may be used to obtain 80°F and 1.250 specific gravity. When charging, do not allow the electrolyte temperature to exceed 125°F. Normally, 10–15 minutes charging will be sufficient; however, in colder climates a little longer time may be necessary. When the battery is removed from the charger, top it off with electrolyte (if necessary) and replace the vent plugs.

Many replacement batteries are of the "maintenance free" design, which are prefilled and may require charging before installation. Refer to the card supplied with the battery or the label attached to the top for preinstallation charging instructions. Under no circumstances should a high-output battery charger be used on any vehicle equipped with an on-board computer. When installing, make sure that both ends of the battery cables are clean and securely tightened, observing correct polarity.

> **CAUTION**
>
> *Be careful not to install the battery with the cables reversed; reversed polarity can destroy the alternator and/or the regulator in a very short time.*

TROUBLESHOOTING THE BATTERY

Problems with the battery can have one or more of the following causes:

1. Battery too small for the job (accessories, etc.).
2. Tired battery (worn out).
3. Corroded battery connections.
4. Alternator not charging.
5. Alternator charging rate too low.
6. Regulator defective.
7. Regulator out of adjustment.
8. Regulator has poor ground.
9. Alternator inoperative.

10. Loose alternator drive belt.

11. Constant drain of current due to short circuit.

Corrections

1. The battery demand may exceed the capacity; additional accessories, too frequent use of the starter or low operational speeds, require a greater source of electrical supply. Install a larger capacity battery.

2. Age or abuse is usually the cause of a tired battery. Since no amount of charging will offer more than temporary relief, install a new battery of proper capacity if the plates are sulfated.

3. Corroded battery posts or side terminal cable retaining bolts and connections, result from the chemical reaction between dissimilar metals and the battery electrolyte. Excessive corrosion at a battery post (or connection) is usually an indication of the failure of a seal between the post and the battery cover.

——— **CAUTION** ———

Always disconnect the negative battery terminal first and make sure the ignition key is switched OFF.

NOTE: Remove the cable and seal the post-to-battery cover with rubber cement or silicone sealer. Clean the post and the cable clamp, install the cable on the post, tighten the clamp and apply a thin coat of non-metalic grease to retard corrosion. Felt washers impregnated with an anti-corrosive substance are available; these are slipped over the post prior to cable installation. If the battery is equipped with a side terminal, wire brush the bolt, the battery connector plate and the cable connector, then install the connector bolt and apply a thin coat of non-metalic grease to the cable end.

4. A non-charging alternator may be caused by a defective alternator or other system component. Check the entire charging system and correct the fault.

5. A low charging rate may be caused by a loose drive belt, a loose or poor battery post connections, a high resistance in the charging circuit, a poor or improperly adjusted regulator (if adjustable).

6. A defective regulator may have burned points (if mechanical), an open circuit in the control system, be out of adjustment (if adjustable) or have a poor ground.

7. The inoperative alternator may have damaged diodes, poor internal connections, an open, grounded or shorted field circuit and/or stator windings.

8. A loose drive belt will cause low or partial charging; correct by adjusting the drive belt.

9. A constant battery current drain may be caused by frayed wiring insulation; forming a short circuit. There is also the possibility of a light (in the trunk, glove box, under the hood and etc.) or other electrical accessories remaining on after the ignition is turned off. To correct the situation:

a. Using a sensitive ammeter, determine if there is a current drain by opening the circuit at either battery post connection, then hook the ammeter in series and check for a current drain.

b. If the meter registers a drain, isolate the leak by reconnecting the battery, then (one by one) check each circuit at the fuse block. This is a tedious but unavoidable procedure which consists of removing each fuse and testing that circuit with the probes of the ammeter (in series). The circuit which activates the meter is the guilty one; identify the trouble spot by the process of elimination. Correct the trouble by repairing the short, replacing the switch or the electrical component(s).

If the fuse block test does not indicate the trouble, check the circuits which are protected with circuit breakers (headlamps, parking lamps, seat window controls and etc.).

SPECIFIC GRAVITY (HYDROMETER) TEST

Before attempting any electrical

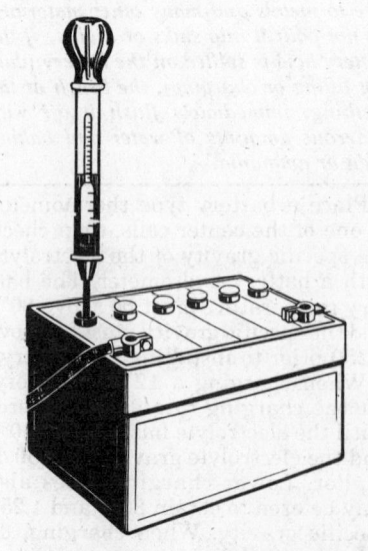

Testing battery specific gravity

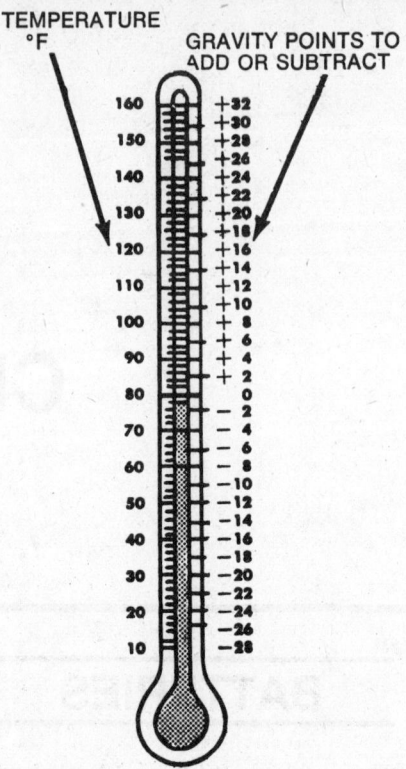

Hydrometer temperature correction chart

checks, it is important to check the condition of the battery. While not technically exact, a practical measurement of the chemical condition of the battery is indicated by measuring the specific gravity of the acid (electrolyte) contained in each cell. The electrolyte in a fully charged battery is usually 1.260–1.280 times heavier than pure water at 80°F. Since variations in the specific gravity readings of a fully charged battery may differ, it is most important that all battery cells produce an equal reading.

As the battery discharges, a chemical change takes place within each cell. The sulfate factor of the electrolyte combines chemically with the battery plates, reducing the weight of the electrolyte. A reading of the electrolyte specific gravity will be less than that of a fully charged one.

The hydrometer is the instrument used to determine the specific gravity of liquids; it is available from many sources, including local auto replacement parts stores. The following chart gives an indication of the specific gravity values, related to the battery charge condition. If, after charging, the specific gravity between any two cells varies more than 50 points (0.050), the battery should be replaced.

Specific Gravity Reading	Charged Condition
1.260–1.280	Fully charged
1.230–1.250	three-quarter charged
1.200–1.220	One-half charged
1.170–1.190	One-quarter charged
1.140–1.160	Just about flat
1.110–1.130	All the way down

TESTING BATTERY POLARITY

The battery polarity is very important, for permanent alternator diode damage will result from reversing the polarity. To determine the battery polarity, place the voltmeter selector on the high scale and connect the voltmeter leads to the battery posts.

NOTE: If the gauge needle moves in the correct direction, the positive lead of the meter is on the positive (+) post of the battery; if it moves in the wrong direction, polarity is reversed.

TESTING THE "MAINTENANCE FREE" BATTERY

Vehicles equipped with a sealed top or "maintenance free" battery do not require the usual maintenance. Since the battery has a greater amount of electrolyte and a reduced need for water, the top of the battery has no filler caps and is sealed. A small vent is provided at one edge of the battery top. There are two types of sealed batteries; one with a charge indicator eye and the other without. Both types may be tested in the following manner:

1. Check the condition of the battery case; if it is damaged so that loss of electrolyte is possible, the battery must be replaced.

2. If the battery has a charge indicator eye, check the following:

a. If the eye is dark, the battery has enough electrolyte. If the eye is lighted, the electrolyte level is too low and the battery must be replaced.

b. If a green dot appears in the middle of the eye, the battery is sufficiently charged; proceed to Step 4. If the green dot is not visible, charge the battery according to Step 3.

3. If the green dot is not visible in the eye or if it has no eye, charge it at the following rates:

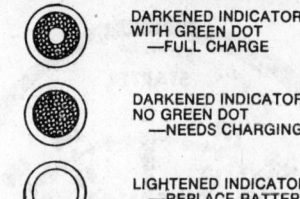

Delco sealed battery indicator conditions

Amps	Time
75	40 min
50	1 hr
25	2 hr
10	5 hr

CAUTION

If the green dot appears or electrolyte squirts out of the vent, stop the charging and proceed to Step 4.

4. Either disconnect the high tension coil wire or the engine harness (electronic ignition) and crank the starter motor for 15 seconds, to remove the surface charge.

5. Connect a voltmeter and a 230 amp load across the battery terminals.

6. Take a voltmeter reading after the load has been connected for 15 seconds, then disconnect the load.

7. Consult the following chart. If the battery voltage is that specified (or more) for the given ambient temperature, the battery is good. If the voltage falls below that specified, then the battery is bad and must be replaced.

Ambient Temperature (°F)	Minimum Voltage
70 (or above)	9.6
60	9.5
50	9.4
40	9.3
30	9.1
20	8.9
10	8.7
0	8.5

Test Equipment

OHMMETER

An ohmmeter is used to measure electrical resistance in a unit or circuit. The ohmmeter has a self-contained power supply. In use, it is connected across (or in parallel with) the terminals of the unit being tested.

AMMETER

An ammeter is used to measure cur-

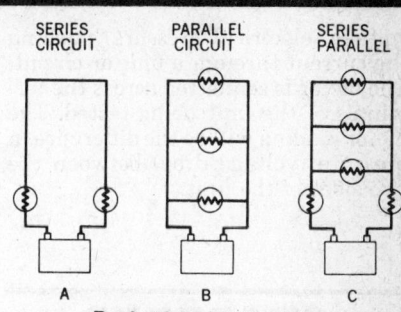

Basic electrical circuits

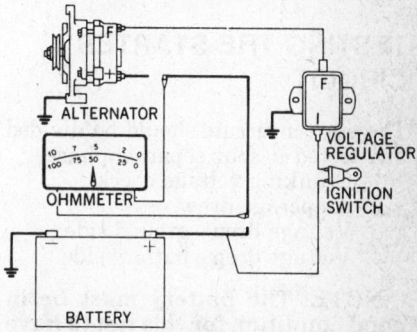

Ohmmeter connected to test wire resistance—typical (ohmmeter has self-contained power supply)

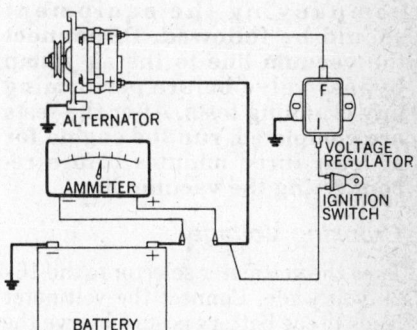

Ammeter connected in series circuits

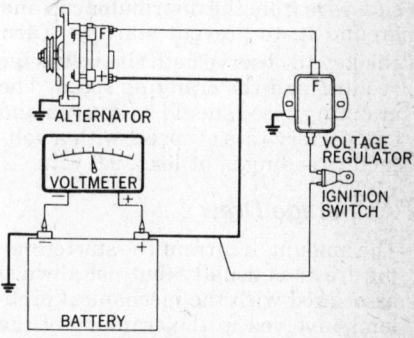

Voltmeter connected in parallel circuits

rent (amount of electricity) flowing through a unit or circuit. Ammeters are always connected in the line (in series) with the unit or circuit being tested.

VOLTMETER

A voltmeter is used to measure the

voltage (electrical pressure) pushing the current through a unit or circuit. The meter is connected across the terminals of the unit being tested. The meter reading will be the difference in pressure (voltage drop) between the two sides of the unit.

STARTERS

TESTING THE STARTER CIRCUIT

The starter circuit should be divided and tested in four separate phases:
1. Cranking voltage check.
2. Amperage draw.
3. Voltage drop—ground side.
4. Voltage drop—battery side.

NOTE: The battery must be in good condition for this test to have significance. To accurately check the battery condition, use equipment designed to measure its capacity under load. Instructions accompanying the equipment should be followed. Disconnect the vacuum line to the air pump bypass valve before performing any cranking tests. After the tests are completed, run the engine for at least three minutes before reconnecting the vacuum line.

Cranking Voltage

Turn the voltmeter selector to the 16–20 volt scale. Connect the voltmeter leads to the battery posts (observe the polarity and reverse the meter leads, if necessary). Remove the high tension wire from the distributor cap and ground it, to prevent starting. Turn the key to observe both the voltmeter reading and the cranking speed. The cranking speed should be an even and satisfactory rate of speed, with a voltmeter reading of at least 9.6 volts.

Amperage Draw

The amount of current the starter motor draws is usually (but not always) associated with the mechanical problems involved in the cranking of the engine. (Mechanical trouble in the engine, frozen or worn starter parts, misaligned starter or starter components and etc.). Because the starter motor amperage draw is directly influenced by anything restricting the free turning of the engine or the starter, it is important that the engine and all of the components be at operating temperatures. When measuring the starter current draw, remove the high

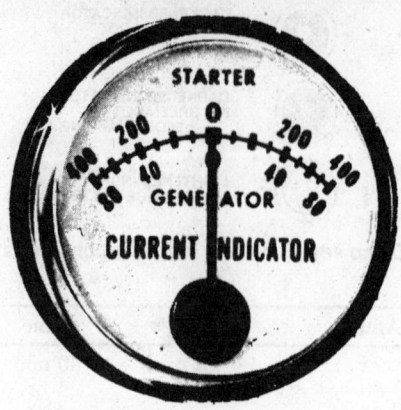

Current indicator

tension wire from the center of the distributor cap and ground it.

NOTE: If equipped with an electronic ignition, disconnect the control box from the distributor (harness).

A very simple and inexpensive starter current indicator is available at auto parts stores. This indicator is an induction-type gauge and shows, without disconnecting any wires, starter current draw.

Place the yoke of the meter directly over the insulated starter supply cable (cable must be straight for a minimum of 2 in.). Close the starter switch for about 20 seconds, watch the meter dial and record the average reading. If the indicator swings in the wrong direction, reverse the position of the meter. The cranking amperage draw can vary from 150–400 amperes, depending on the engine size, the compression and the starter design.

More accurate but complex equipment is available from name brand manufacturers. This equipment consists of a combination voltmeter, ammeter and carbon pile rheostat. When using this equipment, follow the equipment manufacturer's procedures and recommendations. High amperage and lazy performance indicates an excessively tight engine, friction in the starter or starter drive or a grounded starter field or armature. Normal amperage and lazy performance indicates a high resistance or possibly poor connections in the starter circuit. Low amperage and lazy or no performance suggest a poor battery condition or bad cables and/or connections.

Voltage Drop—Grounded Side

Place a voltmeter on the 3.0 volt scale and connect the negative lead to the grounded battery post and the positive test lead to the starter motor

housing. Close the starter switch and note the voltmeter reading. If the reading is the same as the battery reading, the ground circuit is open somewhere between the battery and the starter; in many cases the reading will be very small.

NOTE: The reading shown will indicate voltage drop (loss) between battery ground post and starter housing; it should not exceed 0.2 volt. If the voltage drop is above the specified amount, the next step is to isolate and correct the cause. It can be a bad cable or a connection in the battery-to-starter ground circuit. A check of this type should progress along the various points of possible trouble, between the battery ground post and the starter motor housing, until the trouble spot has been located.

Voltage Drop—Battery Side

Poor starter cranking may result from bad connections or faulty components of the battery or the hot phase of the circuit. Without disconnecting any wires, connect one voltmeter lead to the hot post of the battery and the other lead to the starter field terminal; the meter should be set to the 16–20 volt scale.

NOTE: Before closing the starter switch, the voltmeter reading will be that of the battery.

Connect a remote control starter switch between the battery and the starter relay solenoid terminal. Set the voltmeter to 3.0 volt scale, then crank the engine with the remote control and watch the voltmeter; it should not register more than 0.5 volt. If more than this, check each part of the circuit for the voltage drop to isolate the trouble (high resistance).

Without disturbing the voltmeter-to-battery hook-up, move the free voltmeter lead to the battery terminal of the relay (solenoid) and crank the engine. The voltmeter should show less than 0.1 volt. If this reading is correct, move the same voltmeter lead to the starting motor terminal of the relay (solenoid). While the engine is being cranked, the voltmeter should show less than 0.3 volt; if it does, the trouble lies in the relay. If the reading is correct, the trouble is in the cable or connections between the relay and the starter motor.

NOTE: Due to the design of the Chrysler reduction gear starter, testing is limited to measuring voltage drop in starter cable connection.

STARTER MOTOR DIAGNOSIS

Starter Won't Crank The Engine

1. Dead battery.
2. Open starter circuit, such as:
   a. Broken or loose battery cables.
   b. Inoperative starter motor solenoid.
   c. Broken or loose wire from ignition switch to solenoid.
   d. Poor solenoid or starter ground.
   e. Bad ignition switch.
3. Defective starter internal circuit, such as:
   a. Dirty or burnt commutator.
   b. Stuck, worn or broken brushes.
   c. Open or shorted armature.
   d. Open or grounded fields.
4. Starter motor mechanical faults, such as:
   a. Jammed armature end bearings.
   b. Bad bearings, allowing armature to rub fields.
   c. Bent shaft.
   d. Broken starter housing.
   e. Bad starter drive mechanism.
   f. Bad starter drive or flywheel-driven gear.
5. Engine hard or impossible to crank, such as:
   a. Hydrostatic lock (water in combustion chamber).
   b. Crankshaft seizing in bearings.
   c. Piston or ring seizing.
   d. Bent or broken connecting rod.
   e. Seizing of connecting rod bearings.
   f. Flywheel jammed or broken.

Starter Spins Free—Won't Engage

1. Sticking or broken drive mechanism.
2. Damaged ring gear.

Solenoid And Neutral Safety Switch

IDENTIFICATION

Solenoids Without Relays

This type of starter solenoid is always mounted on the starter. It makes electrical contact for the starter, it pulls the starter and the drive clutch into mesh with the flywheel. The Chrysler reduction gear starter has this solenoid embodied in the starter housing, however an internal relay is integral to the brush plate. The ignition bypass terminal is usually marked "R" or "IGN", if it is used.

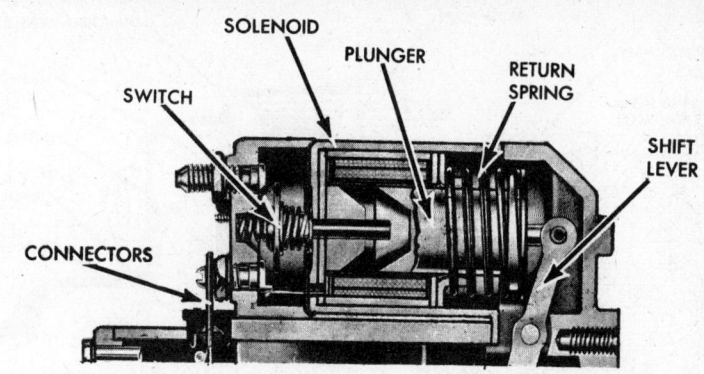

Starter solenoid mounted on starter motor

Solenoids With Separate Relays

The solenoid is always mounted on the starter. In addition to making contact for the starter, it also pulls the starter drive clutch gear into mesh with flywheel. A single control terminal is used on the solenoid. The relay is usually found mounted to the inner fender panel or on the firewall.

Solenoids With Built-In Relays

These units are mounted on the starter and are connected, through linkage, to the starter clutch. The relay portion is built into an integral with the solenoid assembly.

Neutral Safety Switches

The purpose of the neutral safety switch is to prevent the starter from cranking the engine except when the transmission is in the Neutral or the Park positions. On some vehicles, the neutral safety switch is located on the transmission; it serves to ground the solenoid or magnetic switch, whichever is used. On other vehicles, the neutral safety switch is located either at the bottom of the steering column (where it contacts the shift mechanism), on the steering column, underneath the dash or on the shift linkage (console).

Some manual transmission models have a clutch linkage safety switch to prevent the starter operation unless the clutch pedal is depressed. On most cars, the neutral safety switch and the back-up light switch are combined into a single switch mechanism.

TROUBLESHOOTING NEUTRAL SAFETY SWITCHES—QUICK TEST

If the starter fails to function and the neutral safety switch is to be checked, a jumper can be placed across its terminals. If the starter then functions the safety switch is defective. In the case of the neutral safety switches having one wire, the wire must be grounded for testing purposes. If the starter works with the wire grounded, the switch is defective.

NEUTRAL SAFETY SWITCH— BACK-UP LIGHT SWITCH

When the neutral safety switch is built in combination with the back-up light switch, the easiest way to tell which terminals are for the back-up lights is to place a jumper and cross every pair of wires. The pair of wires which light the back-up lamps should be ignored when testing the neutral safety switch. Once the back-up light wires have been located, jump the other pair of wires to test the neutral safety switch. If the starter functions only when the jumper is placed across these two wires, the neutral safety switch is defective or requires adjustment.

Reduction-Gear Starter Motor

CHRYSLER CORPORATION

NOTE: Refer to the following separate sections for the Nippondenso and the Mitsubishi reduction gear type starter repair procedures.

Disassembly

1. Support the assembly in a vise equipped with soft jaws; do not clamp. Care must be used not to distort or damage the die cast aluminum.
2. Remove the thru-bolts and the end housing.
3. Carefully pull the armature up and out of the gear housing, then the starter frame and the field assembly. Remove the steel and the fiber thrust washer.

NOTE: On V8 engines the starting motors have the wire of the shunt field coil soldered to the brush terminal. The 6 cyl engines have the four coils in series and do

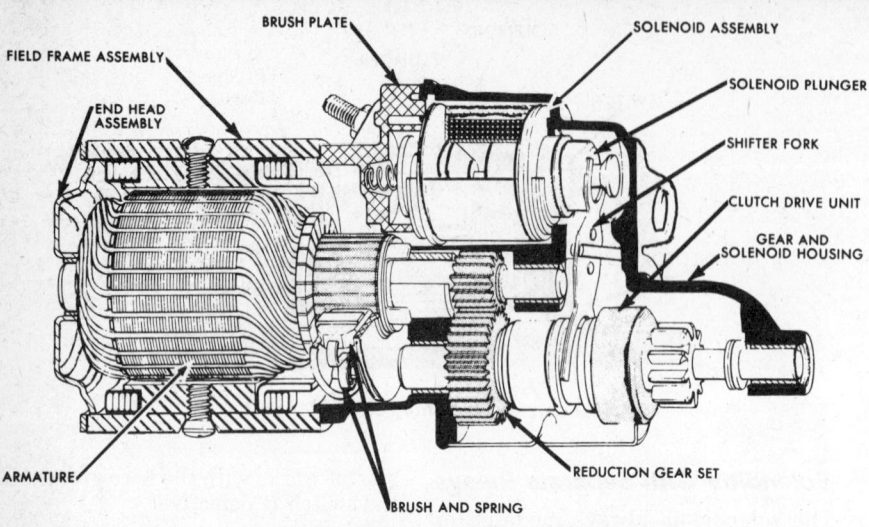

Rear wheel drive reduction gear starter—Chrysler Corp.

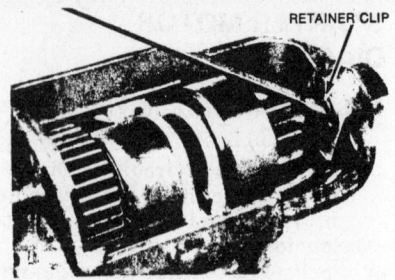

Removing the retainer ring—rear wheel drive reduction gear motor—Chrysler Corp.

not have a wire soldered to the brush terminal. One pair of brushes is connected to this terminal, while the other pair is attached to the series field coils by means of a terminal screw. Carefully pull the frame and the field assembly up enough to expose the terminal screw and the solder connection of the shunt field at the

brush terminal. Place two wooden blocks between the starter frame and gear housing to facilitate removal of the terminal screw and unsoldering of the shunt field wire at the brush terminal.

4. Support the brush terminal with a finger behind the terminal and remove the screw.

5. On the V8 engine starters, unsolder the shunt field coil lead from the brush terminal and the housing.

6. The brush holder plate with the

terminal, the contact and the brushes is serviced as an assembly.

7. Clean the old sealant from around the plate and the housing, then remove the brush holder attaching screw.

8. On the shunt type, unsolder the solenoid winding from the brush terminal, then remove the $^{11}/_{32}$ in. nut, the washer and the insulator from solenoid terminal.

9. Remove the brush holder plate with the brushes as an assembly.

10. Remove gear housing ground screw, then the solenoid assembly from the well. Remove the nut, the washer and the seal from starter (battery) terminal, then the terminal from plate.

11. Remove the solenoid contact and the plunger from solenoid, then the coil sleeve. Remove the solenoid return spring, the coil retaining washer, the retainer and the dust cover from the gear housing.

12. Release the snapring which locates the driven gear pinion shaft, then the front retaining ring. Push the pinion shaft rearward, then remove the snapring, the thrust washers, the clutch and the pinion, then the two shift fork nylon actuators.

13. Remove the driven gear and the friction washer. Pull the shifting fork forward and remove the moving core.

14. Remove the fork retainer pin and the shifting fork assembly. The gear housing with bushings is serviced as an assembly.

Replacement Of Brushes

1. The brushes that are worn more than $1/2$ the length of new brushes or are oil-soaked, should be replaced.

2. When resoldering the shunt field and the solenoid lead, make a strong low-resistance connection using a high-temperature solder and a resin flux.

—— CAUTION ——

Do not use acid or acid-core solder. Do not break the shunt field wire units when removing and installing the brushes.

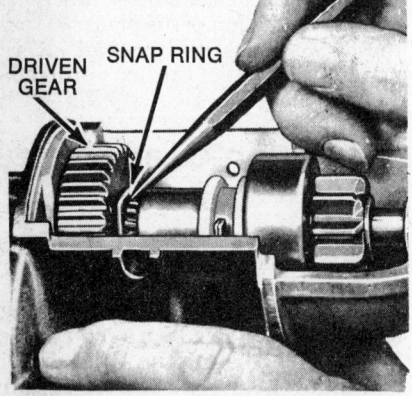

Removing the drive gear snap-ring—rear wheel drive reduction gear motor Chrysler Corp.

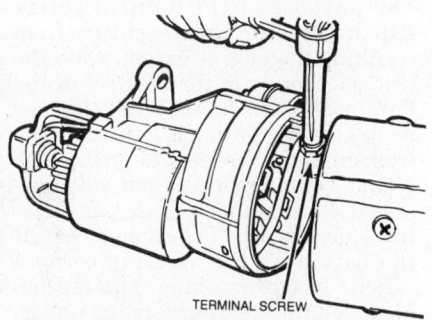

Removing terminal screw—rear wheel drive reduction gear starter

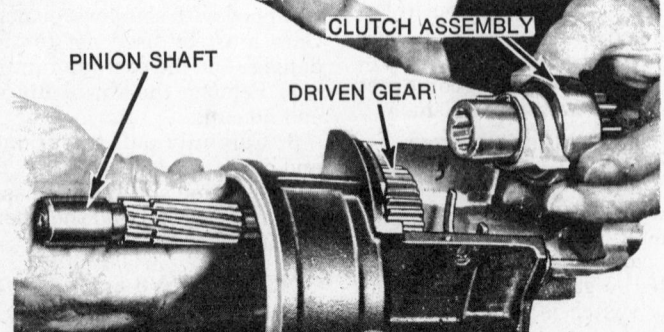

Removing the clutch assembly—rear wheel drive reduction gear motor—Chrysler Corp.

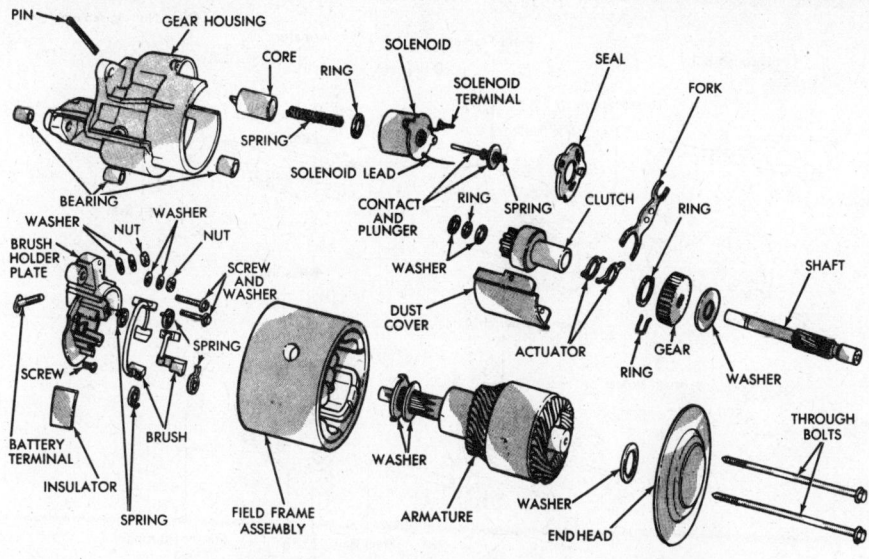

Rear wheel drive reduction gear motor—Chrysler Corp.

Starter Clutch and Pinion Gear Inspection

1. Do not immerse the starter clutch unit in a cleaning solvent. The outside of the clutch and the pinion must be cleaned with a cloth so as not to wash the lubricant from the inside of the clutch.

2. Rotate the pinion; the pinion gear should rotate smoothly and in one direction only. If the starter clutch unit does not function properly or if the pinion is worn, chipped or burred, replace the starter clutch unit.

Commutator Inspection

1. Inspect the commutator and the brush contact surface when the starter is assembled, for flat spots, out-of-roundness or excessive wear.

2. Reface the commutator (if necessary), by removing only a sufficient amount of metal to provide a smooth, even surface.

3. Using light pressure, clean the grooves of the face of the commutator with a pointed tool; neither remove any metal nor widen the grooves.

Assembly

1. The shifter fork consists of two spring steel plates held together by two rivets. Before assembling the starter, check the plates for side movement. After lubricating the plates with a small amount of SAE 10 engine oil, they should have about $\frac{1}{16}$ in. side movement to insure proper pinion gear engagement.

2. Position the shift fork in the drive housing and install the shifting fork retainer pin.

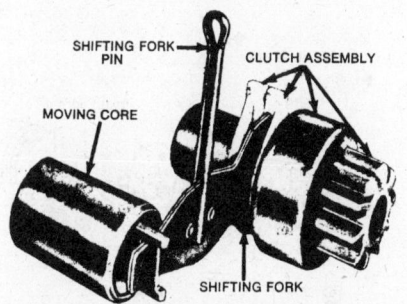

Shift fork and clutch arrangement—rear wheel drive reduction gear motor—Chrysler Corp.

NOTE: One tip of the pin should be straight and the other bent at a 15° angle away from the housing. The fork and retainer pin should operate freely after bending the tip of the pin.

3. Install the solenoid by moving the core and engaging the shifting fork.

4. Place the pinion shaft into the drive housing, then install the friction washer and the drive gear.

5. Install the clutch/pinion assembly, the thrust washer and the retaining washer. Engage the shifting fork with the clutch actuators.

CAUTION
The friction washer must be positioned on the shoulder of the pinion shaft splines before the driven gear is positioned.

6. Install the driven gear snapring and the pinion shaft retaining ring.

7. The starter solenoid return spring can now be inserted in the movable core.

8. Install the solenoid contact plunger assembly into the solenoid and reform the double wires, so they can be curved around the contactor; this will allow the terminal stud to enter the brush holder properly.

CAUTION
The contactor must not touch the double wires after assembly is completed.

9. Assemble the battery terminal stud into the brush holder, then position the seal on the brush holder plate.

10. Run the solenoid lead wire through the brush holder hole and attach the solenoid stud, the insulating washer, the flat washer and the nut. Wrap the solenoid lead wire tightly around the brush terminal post and solder it.

11. Fix the brush holder to the solenoid attaching screws, then gently lower the solenoid coil and the brush plate into the gear housing. Position the brush plate assembly into the starter gear housing, then install the nuts and tighten.

12. Solder the shunt coil lead wire to the starter brush terminal and install the brush terminal screw.

13. Position the field frame onto the gear housing and start the armature into the housing, carefully engaging the splines on the shaft with the reduction gear by rotating the armature.

14. Install the fiber thrust washer and the steel washer onto the armature shaft. Replace the starter end housing and the starter through-bolts, then tighten securely.

NIPPONDENSO/MITSUBISHI REDUCTION STARTER

Disassembly

1. Disconnect the wire terminal from the field coil stud and move the rubber shield from the wire end. Remove the two end frame through bolts.

2. Remove the two end frame cap screws, the upper left solenoid screw and the wire retainer.

3. Remove the end shield, the two brush plate field frame brushes and the brush plate, then slide the armature out of the field frame and remove the field frame.

4. Remove the two gear housing screws, the gear housing from the solenoid, the clutch rollers, the retainer, the pinion and the clutch.

5. Remove the solenoid steel ball, the spring, the solenoid cover screws, the solenoid cover and the solenoid plunger.

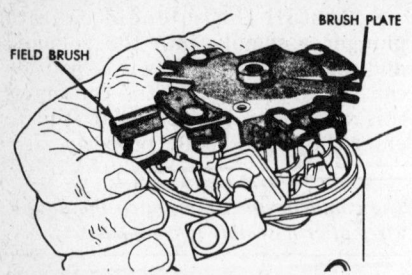

Field brush removal from Bosch starter—Nippondenso similar

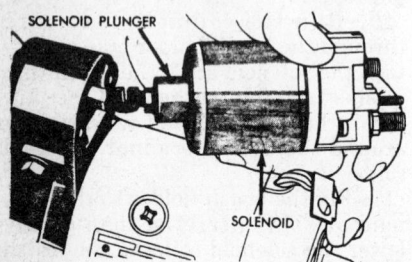

Nippondenso solenoid removal—Bosch similar

Inspection and Service

1. DO NOT immerse the parts in cleaning solvent; immersing the field frame, the coil assembly and/or armature will damage the insulation. Wipe these parts with a cloth only.

2. DO NOT immerse the drive unit in cleaning solvent.

NOTE: The drive clutch is pre-lubricated at the factory and solvent will wash the lubrication from the clutch.

3. The drive unit may be cleaned with a brush moistened with cleaning solvent and wiped dry with a cloth.

4. The brushes that are worn more than $\frac{1}{2}$ the length of a new brush or are oil soaked, should be replaced; new brushes are $\frac{11}{16}$ in. long.

5. The field brushes are serviced as part of the field and the frame assembly. The ground brushes and the springs come as part of the brush plate assembly.

Assembly

To install the parts, reverse the removal procedures.

Nippondenso Gear Reduction Starter

1985–87 GM NOVA

Disassembly

1. Remove the nut and disconnect the motor wire from the magnetic switch terminal.

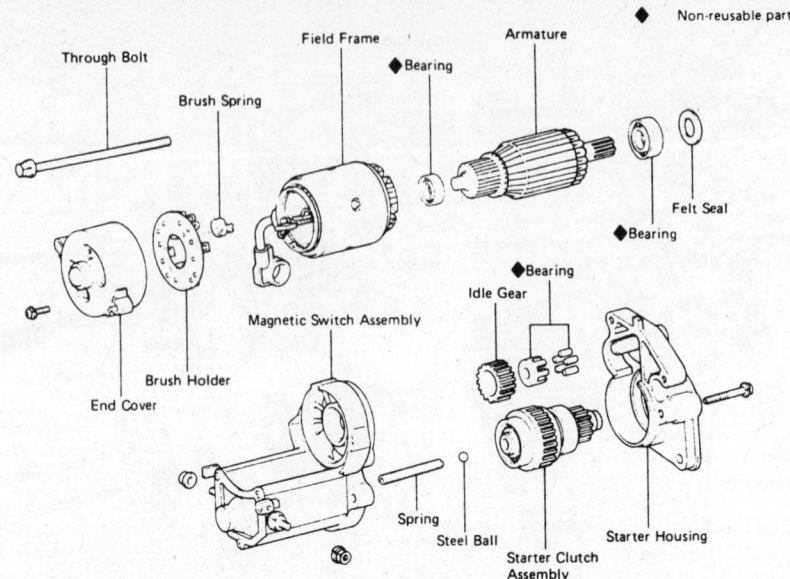

Exploded view of the 1.0 kW type starter—G.M. reduction type

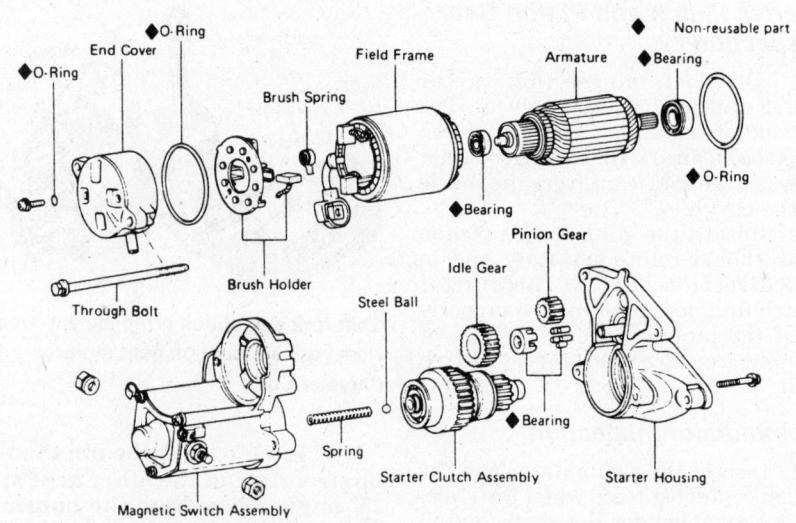

Exploded view of the 1.4 kW type starter—G.M. reduction type

2. Remove the two through bolts and pull the field frame (with the armature) from the magnetic switch assembly.

3. On the 1.0 kW type, remove the felt seal from the armature shaft; on the 1.4 kW type, remove the field frame-to-magnetic switch assembly O-ring.

4. Remove the two starter housing-to-magnetic switch assembly bolts and separate the housing from the assembly.

NOTE: On the 1.0 kW type, remove the idler gear and the clutch assembly; on the 1.4 kW type, remove the pinion gear, the idler gear and the clutch assembly.

5. Using a magnetic finger, remove the spring and the steel ball from the hole in the clutch assembly shaft.

6. Remove the field frame end cover.

NOTE: On the 1.4 kW type, remove the large O-ring.

7. Using a small pry bar, separate the brush springs, then remove the brushes from the brush holder and pull the brush holder from the field frame.

8. Remove the armature from the field frame.

Armature Coil Inspection

1. Using an ohmmeter, make sure that there is no continuity between the commutator and the armature coil core.

NOTE: If there is continuity, replace the armature.

2. Using an ohmmeter, check for continuity between the commutator segments.

NOTE: If there is no continuity between any of the segments, replace the armature.

Commutator Inspection

If the commutator is dirty, burnt or the runout exceeds 0.0020 in., use a lathe to clean the surface; do not machine the diameter to less than 1.14 in. diameter. Make sure that the undercut depth between the segments is 0.020–0.031 in., clean and free of foreign material. If NOT, use a scraping tool (a broken hacksaw blade works fine) to scrape out the insulating material.

Field Coil Inspection

1. Using an ohmmeter, make sure that there is continuity between the lead wire and the brush lead of the field coil.

NOTE: If there is no continuity, replace the field frame.

2. Using an ohmmeter, make sure that there is no continuity between the field coil and the field frame.

NOTE: If there is continuity, replace the field frame.

Brushes Inspection

If the brush length is less than 0.335 in. (1.0 kW) or 0.394 in. (1.4 kW), replace the brush and dress with emery cloth.

Clutch and Gears Inspection

Check the gear teeth for wear or damage, if damaged, replace them. Turn the clutch assembly pinion clockwise and make sure that it rotates freely; try to turn the pinion counterclockwise and make sure that it locks. If the pinion does not respond correctly, replace it.

Bearings Inspection

While applying inward force on the bearings, turn each by hand; if resistance or sticking is noticed, replace the bearings. To replace the bearings, use the tool No. 09286-46011 (or equivalent) to pull the bearing(s) from the armature shaft. Using the tool No. 09285-76010 (or equivalent) and an arbor press, press the new bearing(s) onto the armature shaft.

Magnetic Switch Inspection

Using an ohmmeter, check for conti-

nuity between the grounded terminal and the insulated terminal, then between the grounded terminal and the housing. If there is no continuity in either case, replace the magnetic switch assembly.

Assembly

NOTE: Before installing the gears and the bearings, lubricate them with high temperature grease.

1. Install the armature into the field frame.

2. Using a steel wire, pull the brush springs back and install the brushes into the brush holder.

NOTE: Make sure that the positive brush wires are not grounded. For the 1.4 kW type, install the large O-ring onto the field frame.

3. Install the end cover onto the field frame.

4. Install the steel ball and the spring into the clutch assembly shaft hole.

5. On the 1.0 kW type, install the clutch assembly, the idler gear and the bearing into the starter housing. On the 1.4 kW type, install the clutch assembly, the idler gear, the bearing and the pinion gear into the starter housing.

6. Install the starter housing to the magnetic switch assembly, then secure with the two screws.

NOTE: On the 1.0 kW type, place the felt washer onto the armature shaft; on the 1.4 kW type, place the O-ring onto the field frame.

7. Assemble the field frame to the magnetic switch assembly (by matching the protrusions of both housings) and install the two through bolts.

8. Reconnect the coil lead wire to the magnetic switch assembly's terminal.

1985–87 GM SPRINT (A/T)

Disassembly

1. Remove the nut and disconnect the motor wire from the magnetic switch terminal.

2. Remove the two through bolts and pull the field frame (with the armature) from the magnetic switch assembly.

3. Remove the two starter housing-to-magnetic switch assembly bolts and separate the housing from the assembly.

4. Remove the pinion gear, the pinion retainer/bearings and the clutch assembly.

5. Using a magnetic finger, remove

the spring and the steel ball from the hole in the clutch assembly shaft.

6. Remove the field frame end cover.

7. Using a small pry bar, separate the brush springs, then remove the brushes from the brush holder and pull the brush holder from the field frame.

8. Remove the armature from the field frame.

Armature Coil Inspection

1. Using an ohmmeter, make sure that there is no continuity between the commutator and the armature coil core. If there is continuity, replace the armature.

2. Using an ohmmeter, check for continuity between the commutator segments. If there is no continuity between any of the segments, replace the armature.

Commutator Inspection

If the commutator is dirty, burnt or the runout exceeds 0.002 in., use a lathe to clean the surface; do not machine the diameter to less than 1.14 in. diameter. Make sure that the undercut depth between the segments is 0.018–0.030 in., clean and free of foreign material. If NOT, use a scraping tool (a broken hacksaw blade works fine) to scrape out the insulating material.

Field Coil Inspection

1. Using an ohmmeter, make sure that there is continuity between the lead wire and the brush lead of the field coil. If there is no continuity, replace the field frame.

2. Using an ohmmeter, make sure that there is no continuity between the field coil and the field frame. If there is continuity, replace the field frame.

Brushes Inspection

If the brush length is less than 0.394 in., replace the brush and dress with emery cloth.

Clutch and Gears Inspection

Check the gear teeth for wear or damage, if damaged, replace them. Turn the clutch assembly pinion clockwise and make sure that it rotates freely; try to turn the pinion counterclockwise and make sure that it locks. If the pinion does not respond correctly, replace it.

Bearings Inspection

While applying inward force on the bearings, turn each by hand; if resistance or sticking is noticed, replace

1. Drive housing
2. Screw
3. Lock washer
4. Starter clutch
5. Clutch drive ball
6. Clutch drive spring
7. Starter pinion

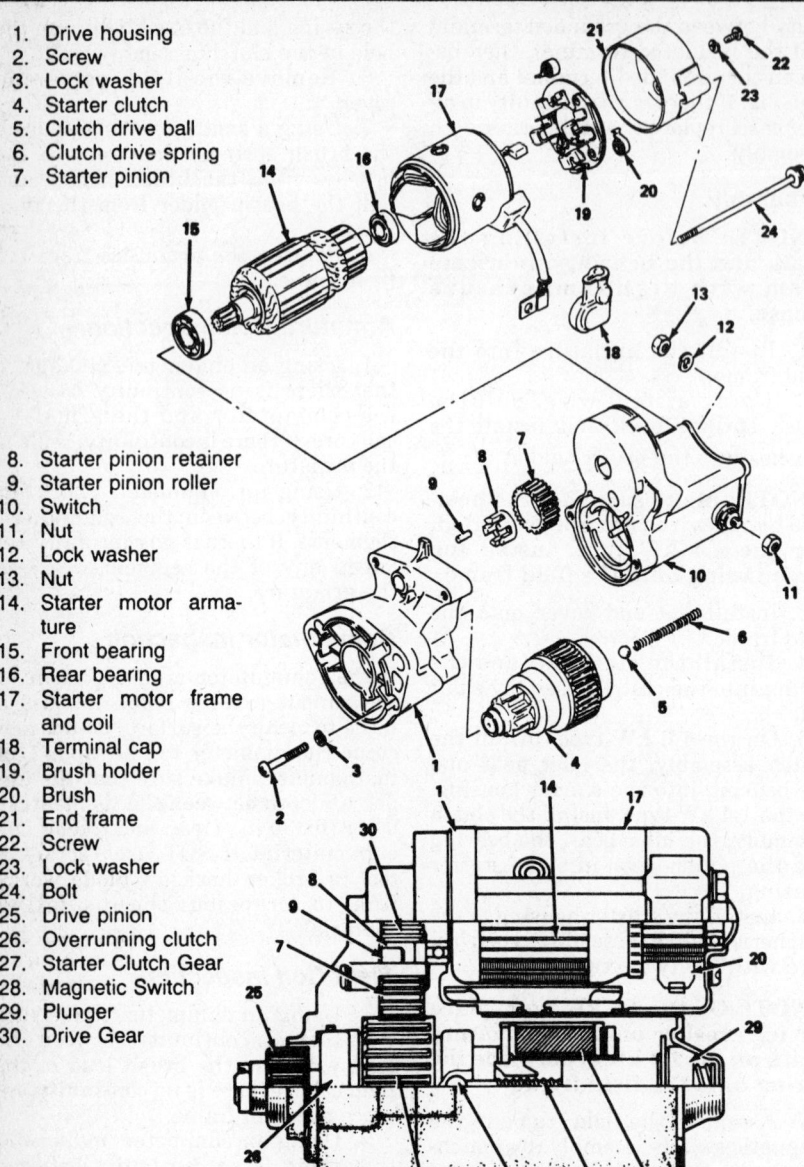

8. Starter pinion retainer
9. Starter pinion roller
10. Switch
11. Nut
12. Lock washer
13. Nut
14. Starter motor armature
15. Front bearing
16. Rear bearing
17. Starter motor frame and coil
18. Terminal cap
19. Brush holder
20. Brush
21. End frame
22. Screw
23. Lock washer
24. Bolt
25. Drive pinion
26. Overrunning clutch
27. Starter Clutch Gear
28. Magnetic Switch
29. Plunger
30. Drive Gear

Exploded view of the reduction gear starter—1985–87 Sprint (A/T)

the bearings. To replace the bearings, use the tool No. 09286-46011 (or equivalent) to pull the bearing(s) from the armature shaft. Using the tool No. 09285-76010 (or equivalent) and an arbor press, press the new bearing(s) onto the armature shaft.

Magnetic Switch Inspection

Using an ohmmeter, check for continuity between the grounded terminal and the insulated terminal, then between the grounded terminal and the housing. If there is no continuity in either case, replace the magnetic switch assembly.

Assembly

NOTE: Before installing the

gears and the bearings, lubricate them with a high temperature grease.

1. Install the armature into the field frame.
2. Using a steel wire, pull the brush springs back and install the brushes into the brush holder.

NOTE: Make sure that the positive brush wires are not grounded.

3. Install the end cover onto the field frame.
4. Install the steel ball and the spring into the clutch assembly shaft hole.
5. Install the clutch assembly, the pinion gear and the pinion retainer/bearing into the starter housing.

6. Install the starter housing to the magnetic switch assembly, then secure with the two screws.
7. Assemble the field frame to the magnetic switch assembly (by matching the protrusions of both housings) and install the two through bolts.
8. Reconnect the coil lead wire to the magnetic switch assembly's terminal.

Hitachi Gear Reduction Starter

GM CHEVETTE DIESEL

Disassembly

1. Disconnect the wire lead at the solenoid. Remove the solenoid-to-starter bolts and the solenoid from the shift lever.
2. Remove the torsion spring from the solenoid, then the starter through bolts and the rear cover.
3. Remove the four brushes from the brush holder, then the frame, the armature and the brush holder as a unit, from the gear case.
4. Carefully remove the brushes and the commutator, do not allow them to contact the adjacent parts.
5. Remove the brush holder and pull the armature assembly from the frame.
6. Remove the bearing retainer and the pinion from the gear case.
7. Remove the retaining clip, then disassemble the pinion assembly.

Inspection And Repair

Inspect the component parts, replace any that are damaged or worn.

Assembly

To assemble, apply lubricant to the pinion assembly and reverse the removal procedures. After the armature has been installed, raise the end of the brush springs and install the brushes. Install the brush holder by aligning it with the frame.

Direct Drive Starters

NIPPONDENSO, BOSCH & MITSUBISHI—CHRYSLER CORPORATION

Disassembly

1. Disconnect the field coil wire from the solenoid terminal.
2. Remove the solenoid mounting screws (the solenoid Bosch auto. trans. models) and work the solenoid (plunger Bosch auto. trans. models) off the shift fork.

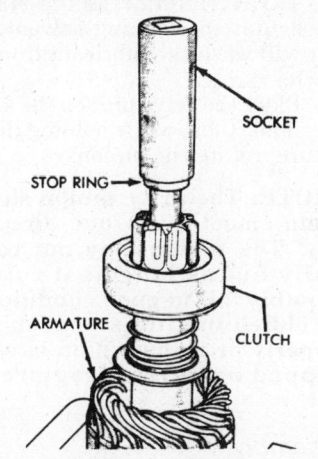

Pressing the stop ring off the snap ring using a socket—Chrysler Corp.

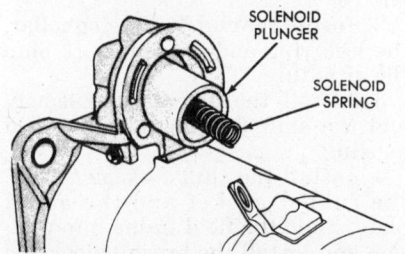

Removing the solenoid plunger and the spring on the Bosch starter—used with Chrysler automatic transmission.

1. Starter yoke assembly
2. Brush
3. Armature
4. Overrunning clutch
5. Gear case assembly
6. Bearing
7. Gear case cover
8. Frame assembly
9. Bearing
10. Brush holder
11. Brush
12. Brush spring
13. Shift lever
14. Magnetic switch
15. Snap ring
16. Snap ring retainer
17. End frame cover
18. Lock plate
19. Seal
20. Screw
21. Nut
22. Through bolt
23. Nut
24. Nut
25. Lockwasher
26. Brake spring

Exploded view of the reduction gear starter—1985–87 Spectrum

3. On the Nippondenso units, remove the bearing cover, the armature shaft lock, the washer, the spring and the seal.

4. On the Bosch units, remove the two end shield bearing cap screws, the cap and the washers.

5. Remove the two commutator end frame cover thru-bolts, the cover, the two brushes and the brush plate.

6. Slide the field frame off over the armature. Remove the shift lever pivot bolt, the rubber gasket and the metal plate.

7. For the Bosch (auto. trans.) and all Nippondenso starters, remove the armature assembly and the shift lever from the drive end housing. For the Bosch (man. trans.) starter, press the stop collar off the snap ring, then remove the snap ring, the clutch assembly, the clutch assembly and the drive end housing from the armature.

8. For all, except the Bosch (man. trans.) starters, press the stop collar off the snap ring, then remove the snap ring, the stop collar and the clutch.

Inspection and Service

1. The brushes that are worn more than ½ the length of new brushes or are oil-soaked, should be replaced; the new brushes are ¹¹⁄₁₆ in. long.

2. DO NOT immerse the starter clutch unit in cleaning solvent; solvent will wash the lubricant from the clutch.

3. Place the drive unit on the armature shaft then, while holding the armature, rotate the pinion.

NOTE: The drive pinion should rotate smoothly in one direction only. The pinion may not rotate easily but as long as it rotates smoothly it is in good condition. If the clutch unit does not function properly or if the pinion is worn, chipped or burred, replace the unit.

Assembly

1. Lubricate the armature shaft and the splines with SAE 10 or 30 W oil.

2. On all starters, except the Bosch (man. trans.), install the clutch, the stop collar, the lock ring and the shift fork on the armature. On the Bosch (man. trans.) starter, fit the drive end

housing onto the armature, then install the clutch, the stop collar and the snap ring onto the armature.

3. On all starters, except the Bosch (man. trans.), install the armature assembly and the shift fork in the drive end housing.

4. Install the shift fork pivot bolt, the rubber gasket and the metal plate. Slide the field frame into position and install the brush holder and the brushes.

5. Position the commutator end frame cover and the thru-bolts.

6. On the Nippondenso units, install the seal, the spring, the washer, the armature shaft lock and the bearing cover.

7. On the Bosch units, install the shim and the armature shaft lock. Check the end play (0.002–0.012 in.), then install the bearing cover.

8. Assemble the solenoid (or plunger on Bosch auto. trans. models) to the shift fork, then install the solenoid with its mounting bolts. Connect the field wire to the solenoid.

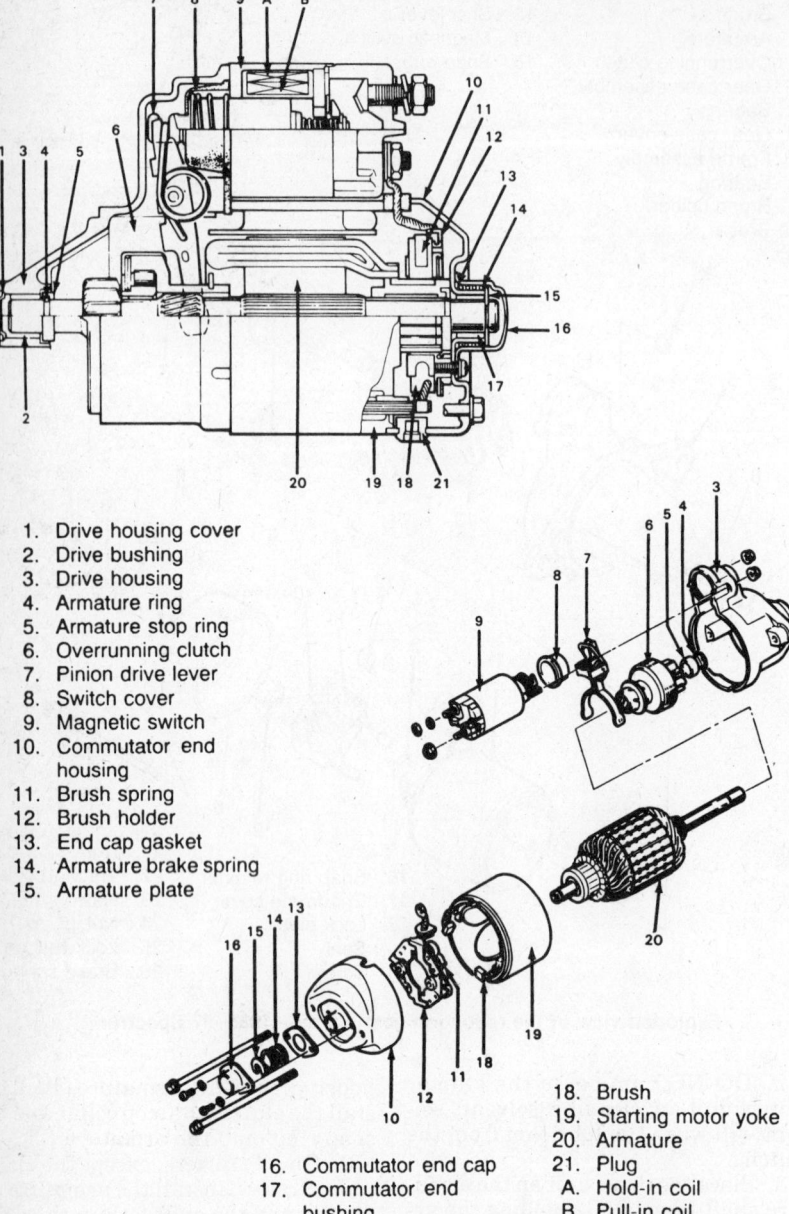

1. Drive housing cover
2. Drive bushing
3. Drive housing
4. Armature ring
5. Armature stop ring
6. Overrunning clutch
7. Pinion drive lever
8. Switch cover
9. Magnetic switch
10. Commutator end housing
11. Brush spring
12. Brush holder
13. End cap gasket
14. Armature brake spring
15. Armature plate

16. Commutator end cap
17. Commutator end bushing

18. Brush
19. Starting motor yoke
20. Armature
21. Plug
A. Hold-in coil
B. Pull-in coil

Exploded view of the reduction gear starter—1985–87 Sprint (M/T)

NIPPONDENSO – 1985–87 GM SPECTRUM/SPRINT MODELS

Disassembly

1. Disconnect the field coil wire from the solenoid terminal.
2. Remove the solenoid mounting screws and work the solenoid off the shift fork.
3. Remove the bearing cover, the armature shaft lock, the washer, the spring and the seal.
4. Remove the two commutator end frame cover thru-bolts, the cover, the brushes and the brush plate.

5. Slide the field frame off over the armature. Remove the shift lever pivot bolt, the rubber gasket and the metal plate.
6. Remove the armature assembly and the shift lever from the drive end housing.
7. Press the stop collar off the snap ring, then remove the snap ring, the stop collar and the clutch assembly.

Inspection and Service

1. The brushes that are worn more than $\frac{1}{2}$ the length of new brushes or are oil-soaked, should be replaced; the new brushes are 0.63 in. long.

2. DO NOT immerse the starter clutch unit in cleaning solvent; solvent will wash the lubricant from the clutch.
3. Place the drive unit on the armature shaft, then, while holding the armature, rotate the pinion.

NOTE: The drive pinion should rotate smoothly in one direction only. The pinion may not rotate easily but as long as it rotates smoothly it is in good condition. If the clutch unit does not function properly or if the pinion is worn, chipped or burred, replace the unit.

Assembly

1. Lubricate the armature shaft and the splines with SAE 10 or 30 W oil.
2. Install the clutch, the stop collar, the lock ring and the shift fork onto the armature.
3. Install the armature assembly and the shift fork in the drive end housing.
4. Install the shift fork pivot bolt, the rubber gasket and the metal plate. Slide the field frame into position and install the brush holder and the brushes.
5. Position the commutator end frame cover and the thru-bolts.
6. Install the seal, the spring, the washer, the armature shaft lock and the bearing cover.
7. Assemble the solenoid onto the shift fork, then install the solenoid with its mounting bolts. Connect the field wire to the solenoid.

Autolite/Motorcraft Positive Engagement Starter Motor

FORD MOTOR CO. AND AMERICAN MOTORS

The starting motor is a series-parallel wound, four pole, four brush unit, equipped with an overrunning clutch drive pinion; the clutch drive pinion engages with the flywheel ring gear by an actuating lever, operated by a movable pole piece. The pole piece is hinged to the starter frame and can drop into position through an opening in the frame. The three conventional field coils are located at the three pole piece positions. The 4th field coil is designed to serve as an engaging coil and a hold-in coil, for the operation of the drive pinion.

When the ignition switch is placed into the START position, the starter relay is energized and current flows from the battery to the starter motor terminal. This prime surge of current first flows through the starter engaging coil, creating a very strong magnetic field. The magnetism draws the movable pole piece down toward the starter frame, which causes the attached lever to move the starter pinion, engaging it with the flywheel ring gear. When the movable pole shoe is fully seated, the field coil grounding contacts are opened; the starter is then in normal operation. During the engine cranking operation, a holding coil is used to hold the movable pole shoe in the fully seated position.

Vehicles equipped with automatic transmissions usually have a starter neutral circuit control; this is to prevent operating of the starter if the selector lever is not in the Neutral or the Park positions. Manual transmission models may have a clutch switch which only allows the engine to start when the clutch is fully depressed.

Disassembly

1. Remove the cover screw, the cover thru-bolts, the starter drive end housing and the starter drive plunger lever return spring.

2. Remove the starter gear plunger lever pivot pin, the lever and the armature. Remove the stop ring retainer and the stop ring from the armature shaft (discard the ring), then the starter drive gear assembly.

3. Remove the brush end plate, the insulator assembly and the brushes from the plastic holder, then lift out the brush holder. For reassembly, note the position of the brush holder with respect to the end terminal.

4. Remove the two ground brush-to-frame screws.

5. Bend up the sleeve's edges which are inserted in the frame's rectangular hole, then remove the sleeve and the retainer. Detach the field coil ground wire from the copper tab.

6. Remove the three coil retaining screws. Cut the field coil connection at the switch post lead, then remove the pole shoes and the coils from the frame.

7. Cut the positive brush leads from the field coils (as close to the field connection point as possible).

8. Check the armature and the armature windings for broken or burned insulation, open circuits or grounds.

9. Check the commutator for runout; if it is rough, has flat spots or is more than 0.005 in. out of round, reface the commutator.

10. Inspect the armature shaft and the two bearings for scoring and ex-

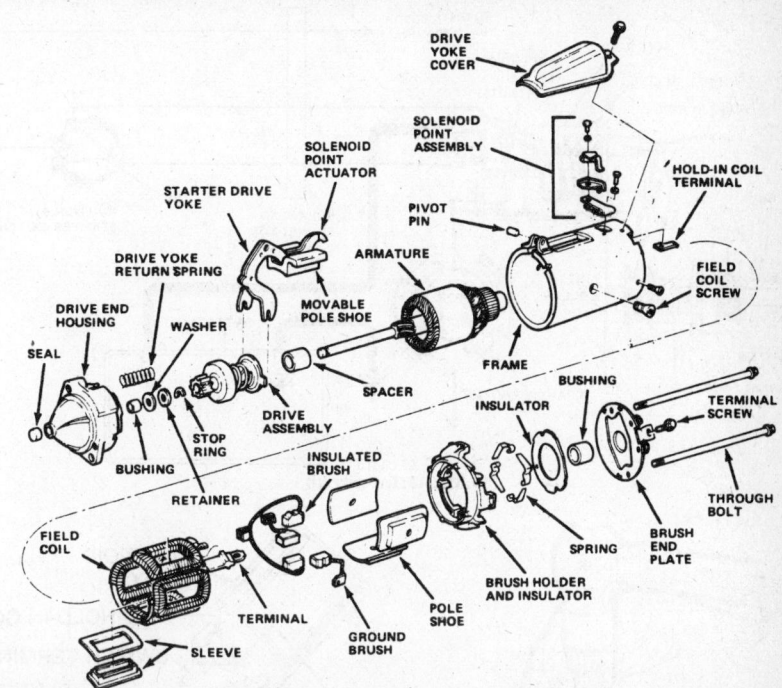

Exploded view of the Ford positive engagement starter motor—AMC similar

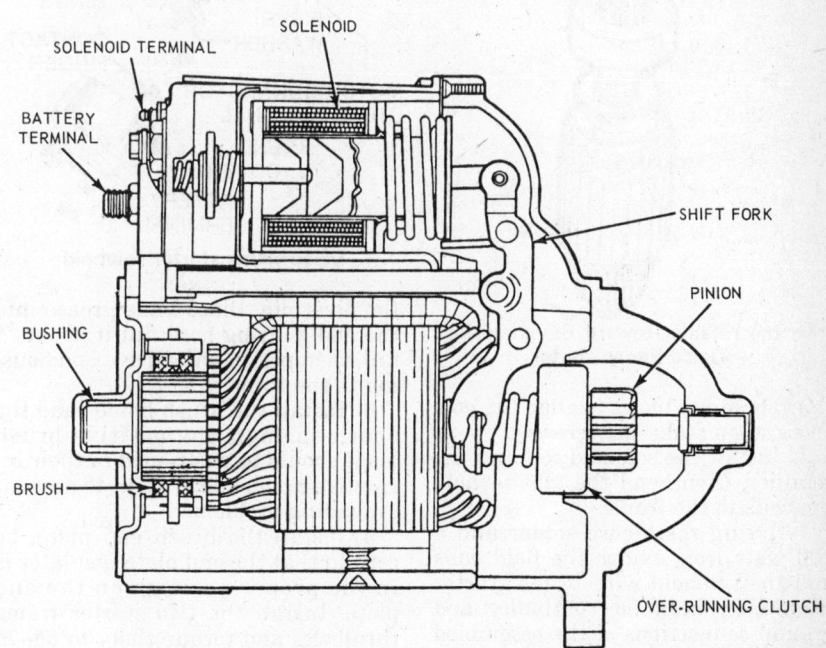

Cut away view of the Ford solenoid actuated starter motor

cessive wear, then replace (if necessary).

11. Inspect the starter drive; if the gear teeth are pitted, broken or excessively worn, replace the starter drive.

NOTE: The factory brush length is ½ in.; the wear limit is ¼ in.

Assembly

1. Install the starter terminal, the

insulator, the washers and the nut in the frame.

NOTE: Be sure to position the screw slot perpendicular to the frame end surface.

2. Position the coils and the pole pieces, with the coil leads in the terminal screw slot, then install the screws. When tightening the pole screws, strike the frame with several

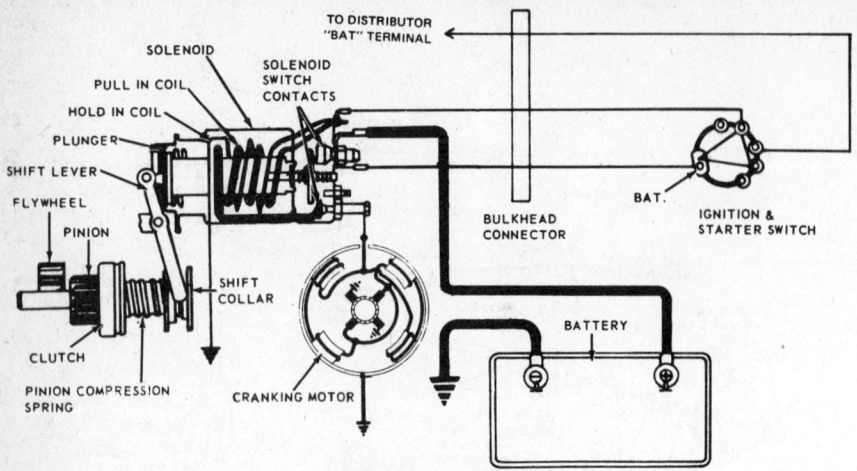

G.M. starter circuit

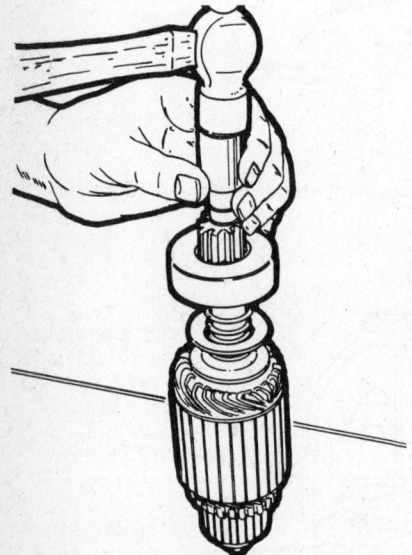

Drive the retainer toward the snap-ring —Delco-Remy starter

SOLENOID BODY
TO HOLD-IN COIL
SWITCH TERMINAL
TO PULL-IN COIL
PLUNGER
CONTACT RINGS
FIBER WASHER
CONTACT FINGER
MOTOR CONNECTOR STRAP TERMINAL
END COVER
BATTERY TERMINAL

Delco-Remy starter solenoid

sharp hammer blows to align the pole shoes, then stake the screws.

3. Install the solenoid coil and the retainer, then bend the tabs to hold the coils to the frame.

4. Using rosin-core solder and a 300 watt iron, solder the field coils and the solenoid wire to the starter terminal. Check for continuity and ground connections of the assembled coils.

5. Position the solenoid coil ground terminal over the nearest ground screw hole and the ground brushes-to-starter frame, then install the screws.

6. Apply a thin coating of Lubriplate® on the armature shaft splines. Install the starter motor drive gear assembly-to-armature shaft, followed by a new stop ring and retainer. Install the armature in the starter frame.

7. Position the starter drive gear plunger lever to the frame and the starter drive assembly, then install

the pivot pin. Place some grease into the end housing bore; fill it about ¼ full, then position the drive end housing to the frame.

8. Install the brush holder and the brush springs; the positive brush leads should be positioned in their respective brush holder slots, to prevent grounding problems.

9. Install the brush end plate; be certain that the end plate insulator is in the proper position on the end plate. Install the two starter frame thru-bolts and torque them to 55–75 inch lbs.

10. Install the starter drive plunger lever cover and tighten the retaining screw.

Delco Remy Starter Motor

GENERAL MOTORS CORP. AND AMERICAN MOTORS

Disassembly

1. Detach the field coil connectors from the motor solenoid terminal.

NOTE: If equipped, remove solenoid mounting screws.

2. Remove the thru-bolts, the commutator end frame, the field frame and the armature assembly from drive housing.

NOTE: On diesel equipped vehicles, the starter has an end frame insulator and the armature will remain in the drive end frame. To disassemble, remove the shift lever pivot bolt and the center bearing screws.

3. Remove the overrunning clutch from the armature shaft as follows:

   a. Slide the two piece thrust collar off the end of the armature shaft.

   b. Slide a standard ½ in. pipe coupling or other spacer onto the shaft, so that the coupling end butts against the retainer edge.

   c. Using a hammer, tap the coupling end, driving the retainer towards the armature end of the snapring.

   d. Using snap ring pliers, remove the snapring from its groove in the shaft, then slide the retainer and the clutch from the shaft.

4. Disassemble the field frame brush assembly by releasing the V-spring and removing the support pin. The brush holders, the brushes and the springs can now be pulled out as a unit and the leads disconnected.

NOTE: On the integral frame units, remove the brush holder from the brush support and the brush screw.

5. If equipped, separate the solenoid from the lever housing.

Cleaning and Inspection

1. Clean the parts with a rag; DO NOT immerse the parts in a solvent.

—— CAUTION ——

Immersion in a solvent will dissolve the grease that is packed in the clutch mechanism; it will damage the armature and the field coil insulation.

2. Test the overrunning clutch action; the pinion should turn freely in the overrunning direction but must not slip in the cranking direction. Check that the pinion teeth have not been chipped, cracked or excessively worn; replace the unit (if necessary).

3. Inspect the armature commutator; if the commutator is rough or out of round, it should be machined and undercut.

NOTE: Undercut the insulation between the commutator bars by 1/32 in. The undercut must be the

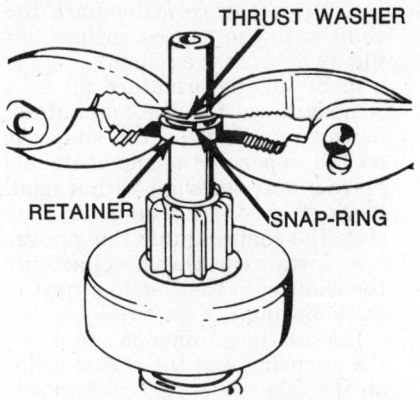

Squeeze the snap-ring into its groove

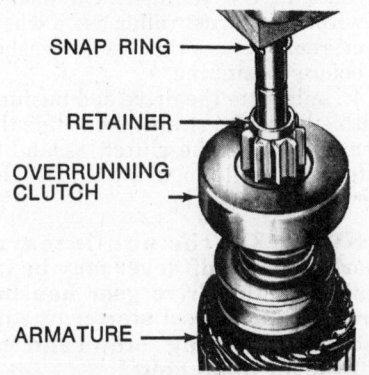

Forcing the snap ring over the armature shaft—Delco-Remy motor

full width of the insulation and flat at the bottom; a triangular groove will not be satisfactory. Most late model starter motors use a molded armature commutator design; no attempt to undercut the insulation should be made or serious damage may result to the commutator.

Assembly

1. Install the brushes into the holders, then install solenoid (if equipped).

2. Assemble the insulated and the grounded holder together. Using the V-spring, position and assemble the unit on the support pin. Push the holders and the spring to bottom of the support, then rotate the spring to engage the slot in the support. Attach the ground wire to the grounded brush and the field lead wire to the insulated brush, then repeat this procedure for other brush sets.

3. Assemble the overrunning clutch to the armature shaft as follows:

   a. Lubricate the drive end of the shaft with silicone lubricant.

   b. Slide the clutch assembly onto the shaft with the pinion outward. On the diesel starter, install the center bearing and the fiber washer first.

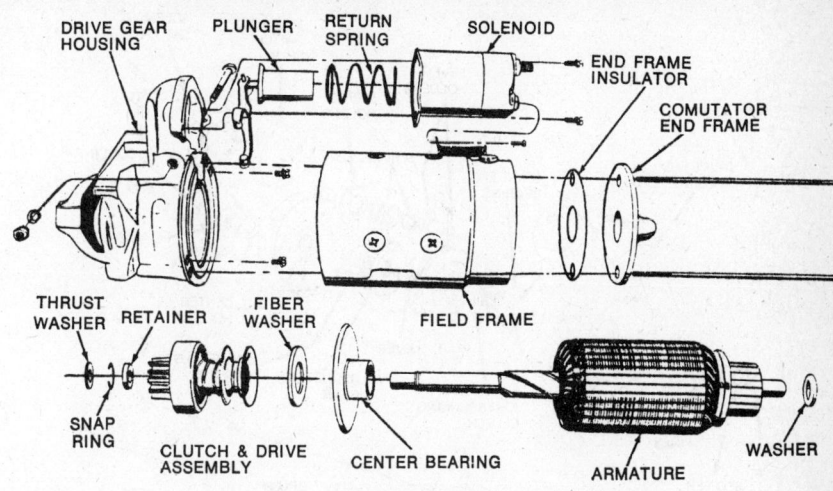

Exploded view of the GM 27 MT starter

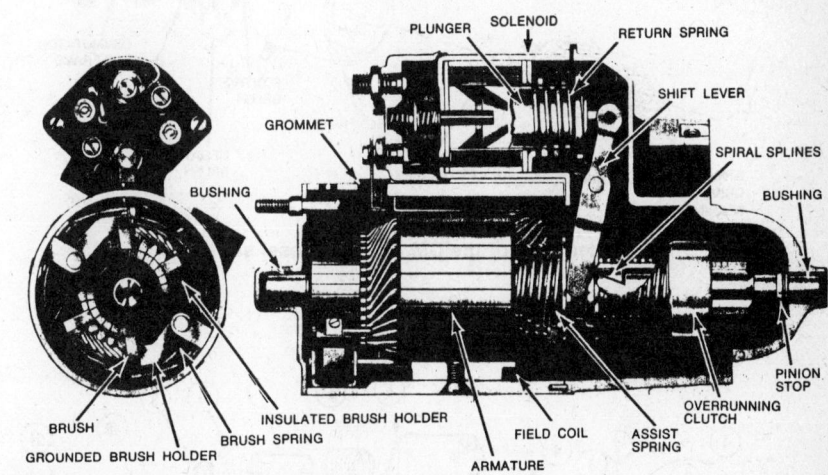

Typical Delco-Remy starter motor, using an assist spring—light duty Chevrolet illustrated

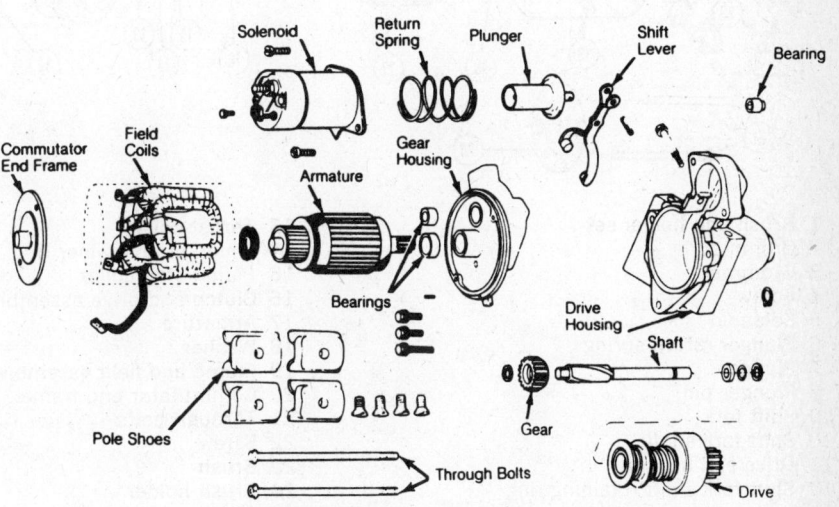

Exploded view of the GM 15MT/GR starter

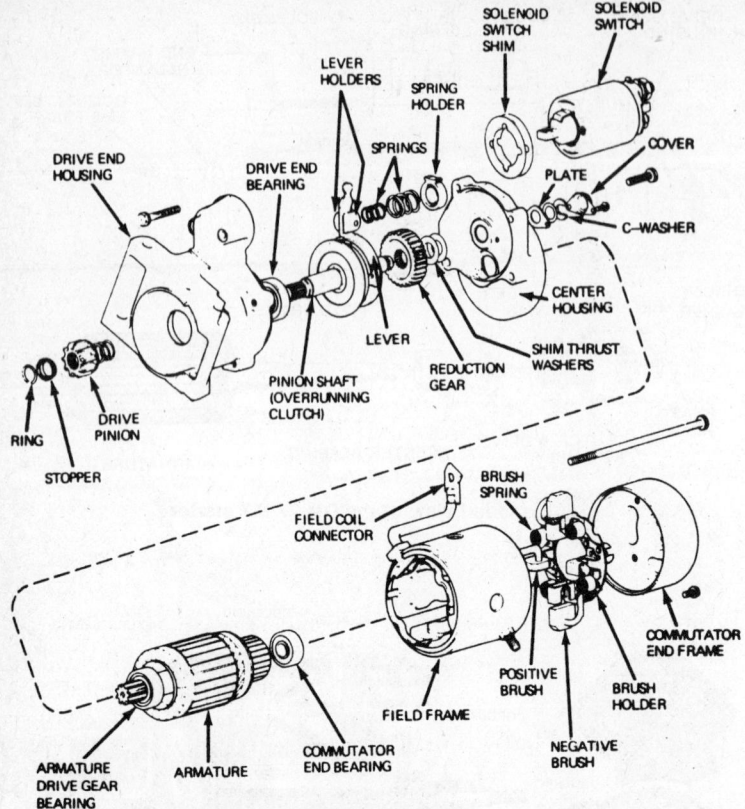

Exploded view of the GM ALU/GR diesel starter

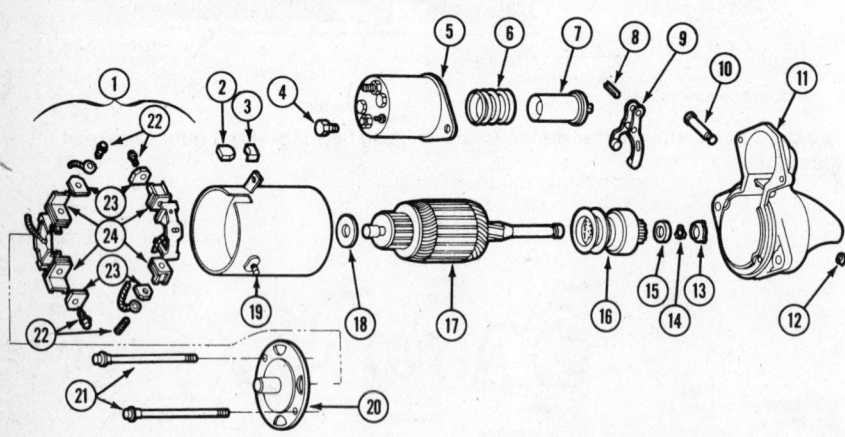

1 Brush and holder set	13 Thrust collar
2 Grommet	14 Pinion stop retainer ring
3 Grommet	15 Pinion stop collar
4 Screw	16 Clutch and drive assembly
5 Solenoid	17 Armature
6 Plunger return spring	18 Washer
7 Plunger	19 Frame and field assembly
8 Plunger pin	20 Commutator end frame
9 Shift fork	21 Through bolts
10 Shift fork shaft	22 Screw
11 Drive end housing	23 Brush
12 Shift fork shaft retaining ring	24 Brush holder

Exploded view of the Delco-Remy 5 MT starter, typical

c. Slide the retainer onto the shaft with the cupped surface facing away from the pinion.

d. Stand the armature up on a wood surface with the commutator downward. Position the snapring on the upper end of the shaft and drive it onto the shaft with a small block of wood and a hammer, then slide the snapring into the groove.

e. Install the thrust collar onto the shaft with the shoulder next to the snapring.

f. With the retainer on one side of the snapring and the thrust collar on the other side, squeeze two sets together (with pliers) until the ring seats in the retainer. On models without a thrust collar use a washer; remember to remove the washer before continuing.

4. Lubricate the drive end bushing with silicone lubricant, then slide the armature and the clutch assembly into place, while engaging the shift lever with the clutch.

NOTE: On the non-integral starters, the shift lever may be installed in the drive gear housing first. On the diesel starter, install the center bearing screws and the shift lever pivot bolt.

5. Position the field frame over the armature and apply sealer (silicone) between the frame and the solenoid case. Position the frame against the drive housing, making sure the brushes are not damaged in the process.

6. Lubricate the commutator end bushing with silicone lubricant, place a washer on the armature shaft and slide the commutator end frame onto the shaft. Install the thru-bolts and tighten.

NOTE: On the diesel starter, install the insulator and the end frame.

7. Reconnect the field coil connections to the solenoid motor terminal. Install the solenoid mounting screws (if equipped).

8. Check the pinion clearance; it should be 0.010–0.140 in. with the pinion in the cranking position, on all models.

ALTERNATORS

PRELIMINARY CHARGING SYSTEM INSPECTION

NOTE: Before performing any tests on the charging system, these precautions should be taken to en-

sure the accuracy of the tests in this section.

1. Check the condition of the alternator belt and tighten (if necessary).

2. Clean the battery cable connections at the battery; make sure that the connections between the battery wires and the battery clamps are good. Reconnect the negative terminal only and proceed to the next step.

3. With the key turned OFF, disconnect the positive battery terminal clamp, then insert a test light between the positive battery terminal and the disconnected positive battery terminal clamp. If the test light glows, there is a short in the electrical system; the short must be repaired before proceeding. If the light fails to glow, reconnect the clamp and proceed to the next step.

NOTE: The some alternators draw a slight current even when the key is turned OFF. To properly check these systems for a short, the regulator must be disconnected. Also, if equipped with an electric clock, disconnect the lead wire from the clock.

4. Check the charging system wiring for brakes or shorts.

5. Make sure that the battery is fully charged and in good condition.

Chrysler Alternator/ Electronic Voltage Regulator

1980–83 MODELS

Charging Circuit Resistance Test

The charging circuit resistance test will show the amount of voltage drop between the alternator output "Bat" terminal wire and the battery.

1. Disconnect the negative battery cable and the battery lead at the alternator "Bat" terminal.

2. Connect a 0–100 amps scale, DC ammeter in series between the alternator "Bat" terminal and the disconnected "Bat" lead wire. Connect the ammeter's positive lead to the "Bat" terminal and the negative lead to the disconnect "Bat" lead.

3. Connect the positive voltmeter lead to the connected "Bat" lead wire and the negative voltmeter lead to the battery positive post.

4. Remove the voltage regulator (green) from the alternator, then connect a jumper wire from the alternator field terminal to a ground.

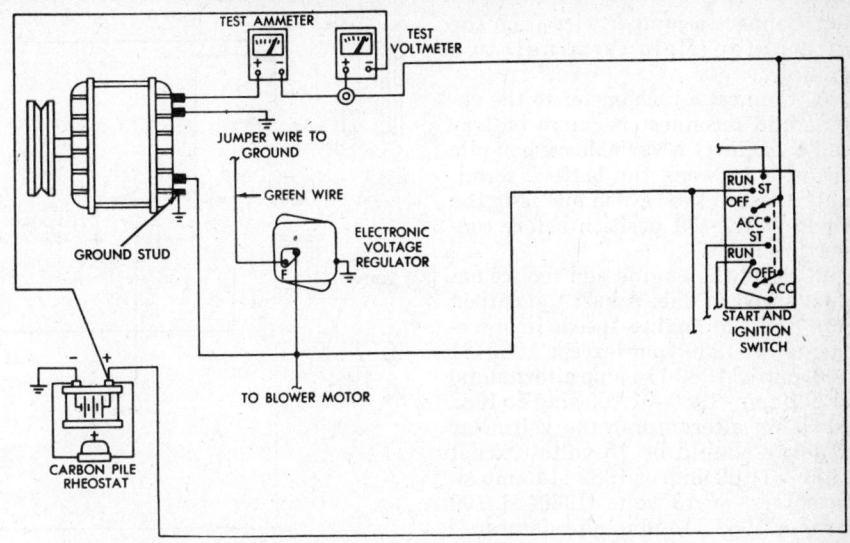

Charging circuit resistance test connections—1980–83 Chrysler

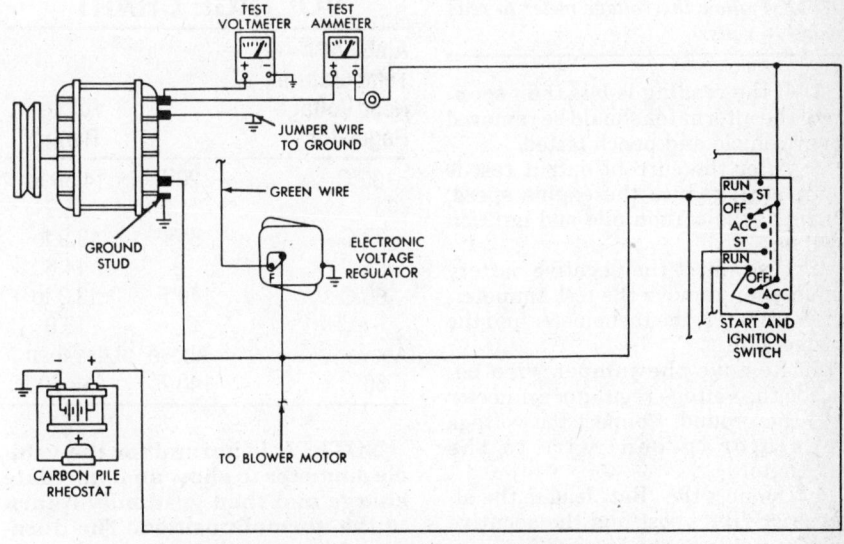

Current output test connections—1980–83 Chrysler

5. Connect a tachometer to the engine and reconnect the negative battery cable.

6. Connect a variable carbon pile rheostat to the battery terminals; be sure the carbon pile is in the Open or the Off position before connecting the leads.

7. Start the engine and reduce the engine speed to idle. Adjust the engine speed and the carbon pile to maintain 20 amps flowing in the circuit. Observe the voltmeter reading; it should not exceed 0.7 volts.

NOTE: If a higher voltage drop is indicated, inspect, clean and tighten the connections in the charging circuit. A voltage drop test may be performed at each connection to locate the connection with excessive resistance. If the charging circuit resistance tests satisfactorily, reduce the en-

gine speed, then turn Off the carbon pile and the ignition switch.

Current Output Test

The current output test determines whether or not the alternator is capable of delivering its rated current output.

1. Disconnect the negative battery cable and the "Bat" lead wire at the alternator output terminal.

2. Connect an ammeter (range 0–100a) in series between the alternator "Bat" terminal and the disconnected "Bat" lead wire; the positive lead to disconnected "Bat" wire.

3. Connect the positive voltmeter lead (range 0–15 volts) to the "Bat" terminal of the alternator and the negative voltmeter lead to a good ground.

4. Disconnect the green wire (from the voltage regulator) at the alterna-

tor. Connect a jumper wire from the alternator (field terminal) to a ground.

5. Connect a tachometer to the engine and reconnect negative battery cable. Connect a variable carbon pile rheostat between the battery terminals; be sure the carbon pile is in the Open or the Off position before connecting leads.

6. Start the engine and reduce engine speed to idle. Adjust the carbon pile and the engine speed (in increments) to 1250 rpm (except 1980–81 100 amp or 1982 114 amp alternators) or 900 rpm (1980–81 100 amp or 1982 114 amp alternators); the voltmeter reading should be 15 volts (except 1980–81 100 amp or 1982 114 amp alternators) or 13 volts (1980–81 100 amp or 1982 114 amp alternators).

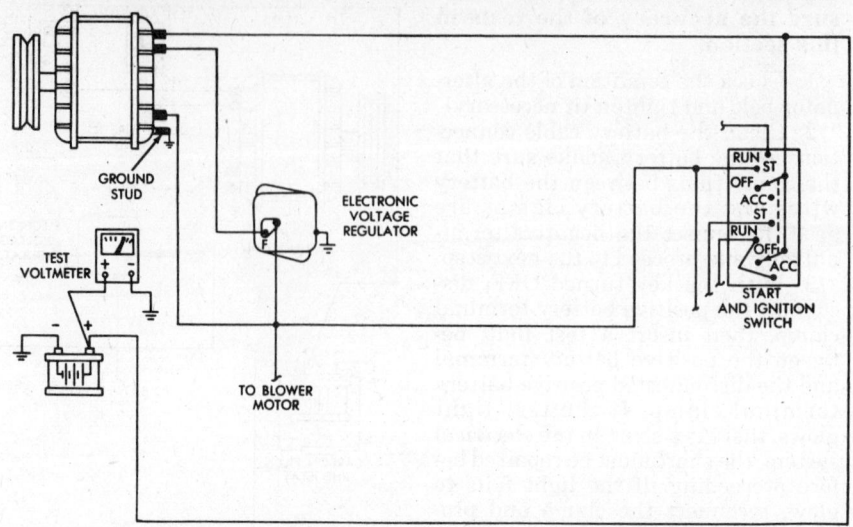

Voltage regulator test connections—1980–83 Chrysler

——— CAUTION ———

DO NOT allow the voltage meter to read above 16 volts.

7. If the reading is less than specified, the alternator should be removed from vehicle and bench tested.

8. After the current output test is completed, reduce the engine speed, turn Off the carbon pile and ignition switch.

9. Disconnect the negative battery cable, then remove the test ammeter, the voltmeter, the tachometer and the carbon pile.

10. Remove the jumper wire between the voltage regulator connector and the ground. Connect the voltage regulator (green) wire to the alternator.

11. Connect the "Bat" lead to the alternator "Bat" post and the negative battery cable to the battery.

Voltage Regulator Test

1. Clean the battery terminals and check the specific gravity; it should be above 1.220 to allow a properly regulated voltage check.

NOTE: If the specific gravity is below 1.220, charge or use another battery but DO NOT leave the uncharged battery in the circuit.

2. Connect the positive voltmeter lead to the positive battery post and the negative voltmeter lead to a good ground.

3. Connect a tachometer to engine, then start and operate engine at 1250 rpm with all lights and accessories turned Off. Check the voltmeter, the regulator is working properly if the voltage readings are in accordance with the voltage chart.

VOLTAGE CHART

Ambient Temperature Near Voltage Regulator		Voltage Range
−30°C	−20°F	14.39 to 15.9
27°C	80°F	13.9 to 14.6
60°C	140°F	13.3 to 13.9
Above 60°C	Above 140°F	Less than 13.60

NOTE: It is normal for the vehicle ammeter to show an immediate charge and then gradually return to the normal position. The duration the ammeter hand remains to the right will be dependent on the length of the cranking time.

4. If the voltage is below the limits or is fluctuating, proceed as follows:

a. Check for a good voltage regulator ground. The voltage regulator ground is obtained through the regulator case, to the mounting screws and to the body. This ground circuit should be checked for opens.

b. Turn OFF the ignition switch and disconnect the voltage regulator connector; be sure the connector terminals have not spread open to cause an open or an intermittent connection.

c. DO NOT start the engine or distort the terminals with the voltmeter probe: turn ON the ignition switch and check for battery voltage at the wiring harness terminal. Both the blue and the green leads should read with battery voltage. Turn OFF the ignition switch.

d. If the previous steps 4a through 4c tested satisfactory, replace the regulator and repeat the test.

5. If the voltage is above the limits shown on the chart, proceed as follows:

a. Turn OFF the ignition switch and disconnect the voltage regulator connector; be sure the terminals on the connector have not spread open.

b. DO NOT start the engine or distort the terminals with the voltmeter probe: turn ON the ignition switch, check for battery voltage at the wiring harness terminal. Both the blue and the green leads should read with battery voltage. Turn OFF the ignition switch.

c. If the previous steps 5a and 5b tested satisfactory, replace the regulator and repeat the test.

Chrysler/Bosch Alternator And Electronic Voltage Regulator

1984 AND LATER MODELS

Charging Circuit Resistance Test

The charging circuit resistance test will show the amount of "Voltage Drop" between the alternator output "Bat" terminal wire and the battery.

1. Disconnect the negative battery cable and the battery lead at the alternator "Bat" terminal.

2. Connect a 0–100 amps scale, DC

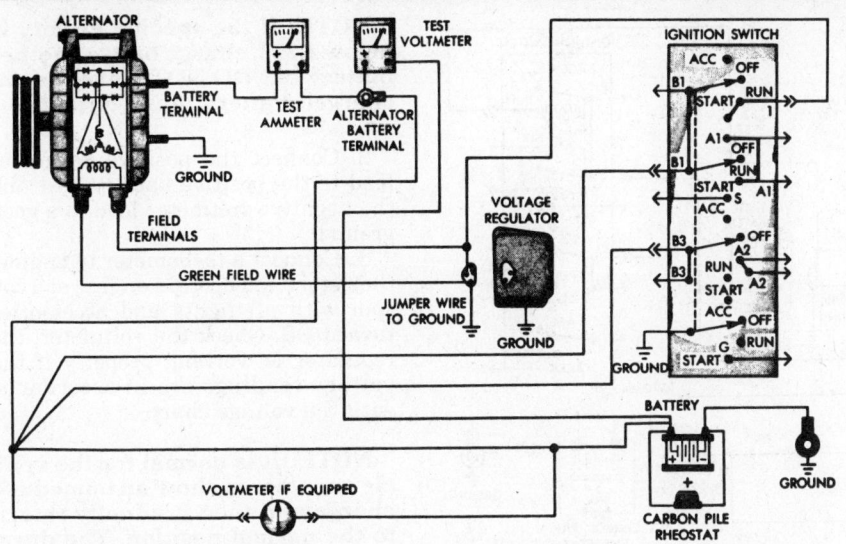

Charging circuit resistance test connections—Chrysler 60/78 amp alternator

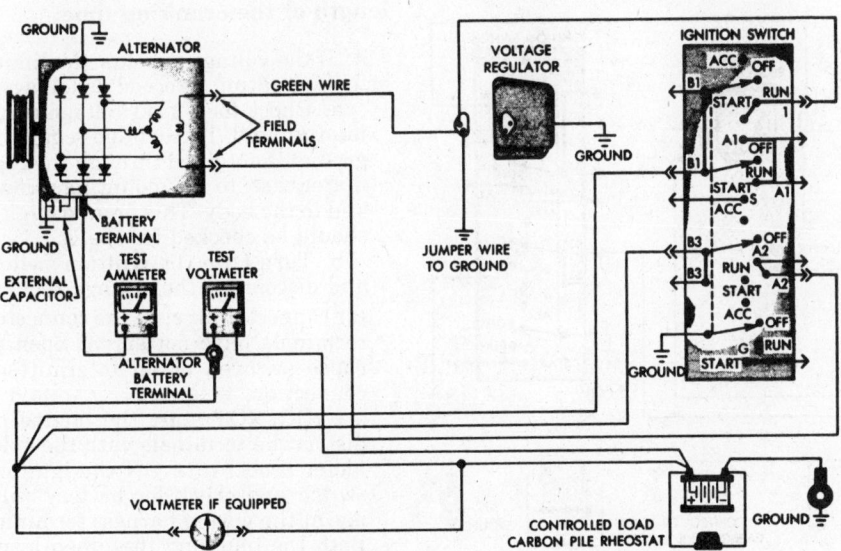

Charging circuit resistance test connections—Bosch alternator

Current Output Test

The current output test determines whether or not the alternator is capable of delivering its rated current output.

1. Disconnect the negative battery cable and the "Bat" lead wire at the alternator output terminal.

2. Connect an ammeter (range 0–100a) in series between the alternator "Bat" terminal and the disconnected "Bat" lead wire; the positive lead to disconnected "Bat" wire.

3. Connect the positive voltmeter lead (range 0–18 volts) to the "Bat" terminal of the alternator and the negative voltmeter lead to a good ground.

4. Disconnect the electrical connector from the voltage regulator. Connect a jumper wire from the electrical connector (green wire) to a ground.

— CAUTION —

DO NOT connect the electrical connector blue (J2) wire to a ground.

5. Connect a tachometer to the engine and reconnect negative battery cable. Connect a variable carbon pile rheostat between the battery terminals; be sure the carbon pile is in the Open or the Off position before connecting leads.

6. Start the engine and reduce engine speed to idle. Adjust the carbon pile and the engine speed (in increments) to 1250 rpm; the voltmeter reading should be 15 volts.

— CAUTION —

DO NOT allow the voltage meter to read above 16 volts.

7. If the reading is less than specified, the alternator should be removed from vehicle and bench tested.

8. After the current output test is completed, reduce the engine speed, turn Off the carbon pile and ignition switch.

9. Disconnect the negative battery cable, then remove the test ammeter, the voltmeter, the tachometer and the carbon pile.

10. Remove the jumper wire between the voltage regulator connector and the ground. Connect the electrical connector to the voltage regulator.

11. Connect the "Bat" lead to the alternator "Bat" post and the negative battery cable to the battery.

Voltage Regulator Test

1. Clean the battery terminals and check the specific gravity; it should be above 1.220 to allow a properly regulated voltage check.

ammeter in series between the alternator "Bat" terminal and the disconnected "Bat" lead wire. Connect the ammeter's positive lead to the "Bat" terminal and the negative lead to the disconnect "Bat" lead.

3. Connect the positive voltmeter lead to the connected "Bat" lead wire and the negative voltmeter lead to the battery positive post.

4. Remove the wiring connector from the voltage regulator, then connect a jumper wire from the electrical connector (green wire) to a ground.

— CAUTION —

DO NOT connect the blue (J2) lead of the wiring connector to a ground.

5. Connect a tachometer to the engine and reconnect the negative battery cable.

6. Connect a variable carbon pile rheostat to the battery terminals; be sure the carbon pile is in the Open or the Off position before connecting the leads.

7. Start the engine and reduce the engine speed to idle. Adjust the engine speed and the carbon pile to maintain 20 amps flowing in the circuit. Observe the voltmeter reading; it should not exceed 0.5 volts.

NOTE: If a higher voltage drop is indicated, inspect, clean and tighten the connections in the charging circuit. A voltage drop test may be performed at each connection to locate the connection with excessive resistance. If the charging circuit resistance tests satisfactorily, reduce the engine speed, then turn Off the carbon pile and the ignition switch.

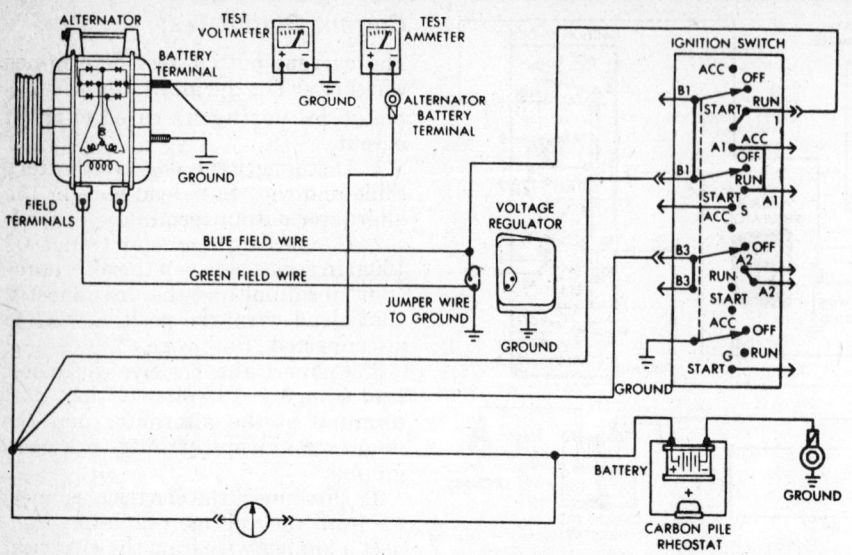

Current output test connections—Chrysler Corp.

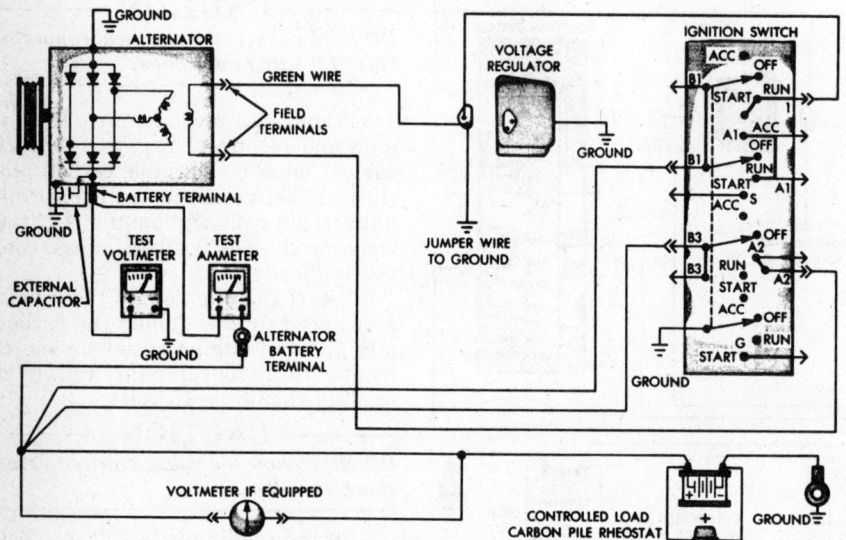

Current output test connections—Bosch alternator

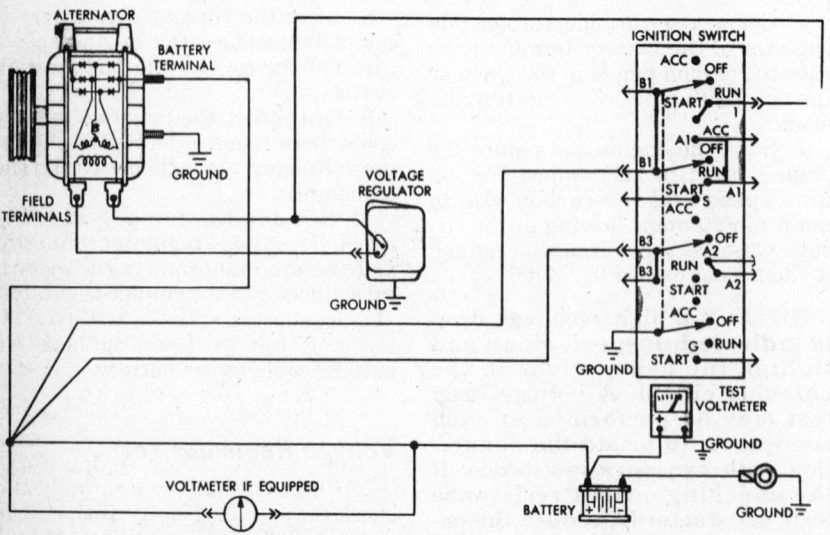

Voltage regulator test connections—Chrysler 60/78 amp alternator

NOTE: If the specific gravity is below 1.220, charge or use another battery but DO NOT leave the uncharged battery in the circuit.

2. Connect the positive voltmeter lead to the positive battery post and the negative voltmeter lead to a good ground.

3. Connect a tachometer to engine, then start and operate engine at 1250 rpm with all lights and accessories turned Off. Check the voltmeter, the regulator is working properly if the voltage readings are in accordance with the voltage chart.

NOTE: It is normal for the vehicle ammeter to show an immediate charge and then gradually return to the normal position. The duration the ammeter hand remains to the right will be dependent on the length of the cranking time.

4. If the voltage is below the limits or is fluctuating, proceed as follows:

a. Check for a good voltage regulator ground. The voltage regulator ground is obtained through the regulator case, to the mounting screws and to the body. This ground circuit should be checked for opens.

b. Turn OFF the ignition switch and disconnect the voltage regulator connector; be sure the connector terminals have not spread open to cause an open or an intermittent connection.

c. DO NOT start the engine or distort the terminals with the voltmeter probe: turn ON the ignition switch and check for battery voltage at the wiring harness terminal. Both the blue and the green leads should read with battery voltage. Turn OFF the ignition switch.

d. If the previous steps 4a through 4c tested satisfactory, replace the regulator and repeat the test.

5. If the voltage is above the limits shown on the chart, proceed as follows:

a. Turn OFF the ignition switch and disconnect the voltage regulator connector; be sure the terminals on the connector have not spread open.

b. DO NOT start the engine or distort the terminals with the voltmeter probe: turn ON the ignition switch, check for battery voltage at the wiring harness terminal. Both the blue and the green leads should read with battery voltage. Turn OFF the ignition switch.

c. If the previous steps 5a and 5b tested satisfactory, replace the regulator and repeat the test.

VOLTAGE CHART

AMBIENT TEMPERATURE NEAR VOLTAGE REGULATOR		VOLTAGE RANGE
−30°C	−20°F	14.9 to 15.8
27°C	80°F	13.9 to 14.4
60°C	140°F	13.0 to 13.7
Above 60°C	Above 140°F	Less than 13.60

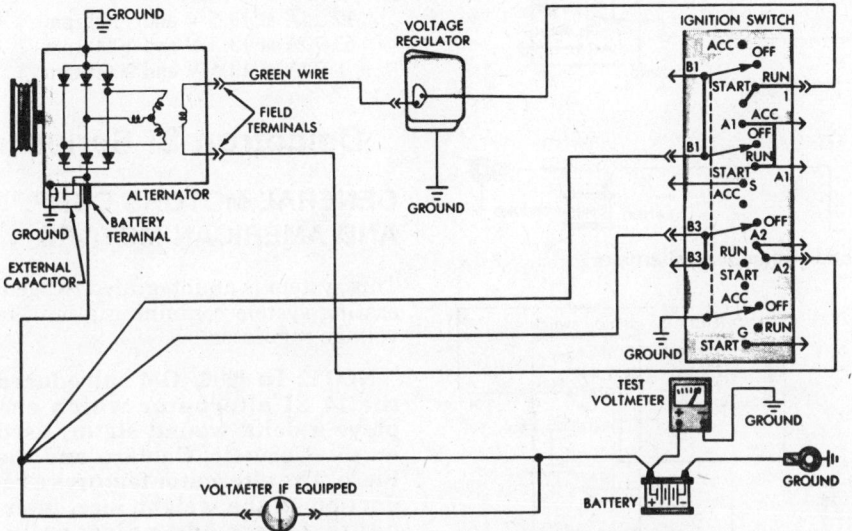

Voltage regulator test connections—Bosch alternator

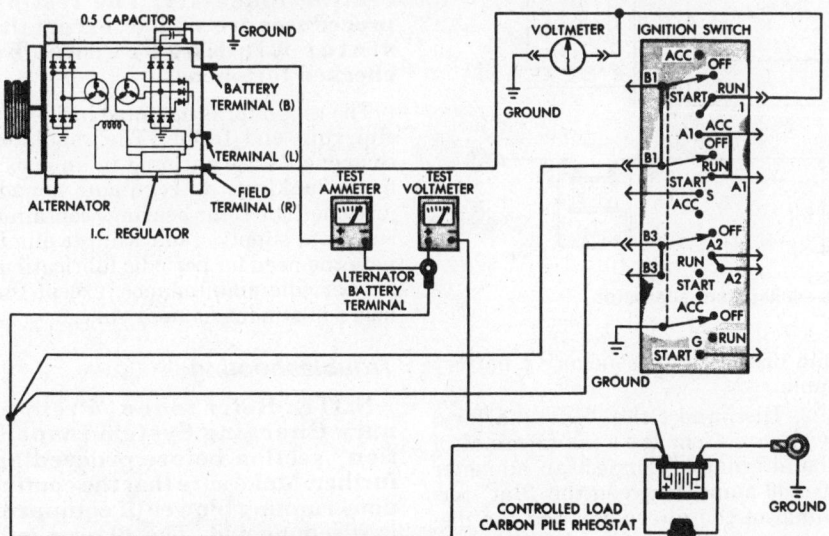

Charging circuit resistance test connections—Mitsubishi alternator

Mitsubishi Alternator/ Electronic Voltage Regulator

CHRYSLER MODELS

Troubleshooting

NOTE: Refer to the "Preliminary Charging System Inspec- tion" section before proceeding further. Make sure that the continuous running blower (if equipped) is disconnected. The blower will run with the key On or the blower control Off, unless disconnected.

Charging Circuit Resistance Test

The charging circuit resistance test will show the amount of voltage drop between the alternator output "Bat" terminal wire and the battery.

1. Turn the ignition switch Off and disconnect the negative battery terminal.

2. At the alternator output terminal, disconnect the "Bat" wire.

3. Using an ammeter (0–100 amps), connect the positive lead to the alternator "Bat" terminal and the negative lead to the "Bat" wire.

4. Using a voltmeter (0–18 volts), connect the positive lead to the disconnected alternator "Bat" terminal wire and the negative lead to the battery positive cable.

5. Connect a variable carbon pile rheostat between the battery terminals; be sure the control switch is in the Open or Off position. Start the engine and allow it to run at idle.

6. Adjust the carbon pile and the engine speed to maintain a 20 amp flow in the circuit. The voltmeter reading should not exceed 0.5 volts.

NOTE: If a higher voltage drop is indicated, clean and tighten the charging circuit connections. A voltage drop test may be performed at each connection.

Voltage Regulator Test

1. With the ignition switch OFF, disconnect the positive battery cable and connect an ammeter between the cable and the battery's positive terminal.

2. Connect a voltmeter between terminal "L" of the alternator and a ground. The voltmeter reading should be zero; if voltage is present, suspect a defective alternator.

3. Turn the ignition switch ON, but DO NOT start the engine. The voltmeter reading should be considerably lower than the battery voltage; if it is close to the battery voltage, suspect a defective alternator.

4. Using a jumper wire, connect it to the ammeter terminals (short circuit the ammeter), then start the engine; make sure that when the engine is started, no starting current is applied to the ammeter.

5. Remove the jumper wire (short circuit) from across the ammeter terminals and increase the engine speed immediately to 2000–3000 rpm, then record the ammeter reading.

6. If the ammeter reading is 5.0 amps (1983), 10.0 amps (1984–87) or less, take the voltmeter reading without changing the engine speed (2000–3000 rpm). The reading will be the charging voltage (14.1–14.7 volts @ 68°F).

NOTE: Since the voltage regulator is a temperature compensation

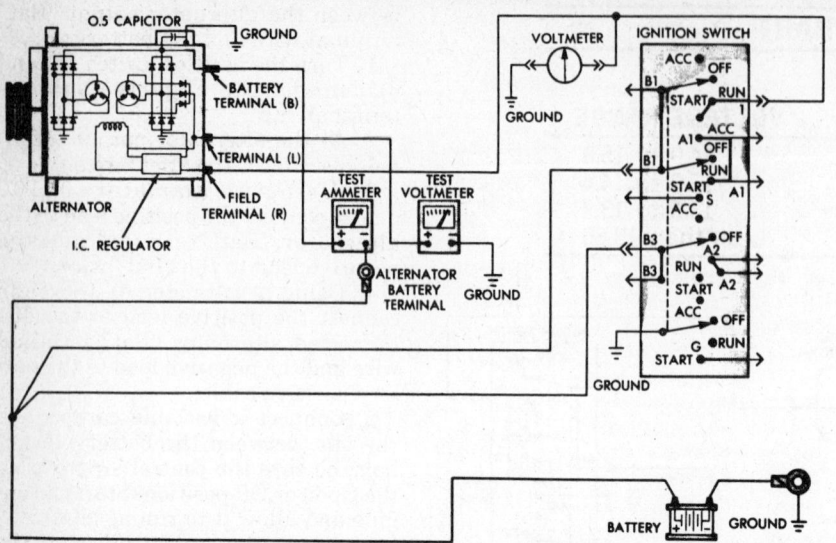

Voltage regulator test connections—Mitsubishi alternator

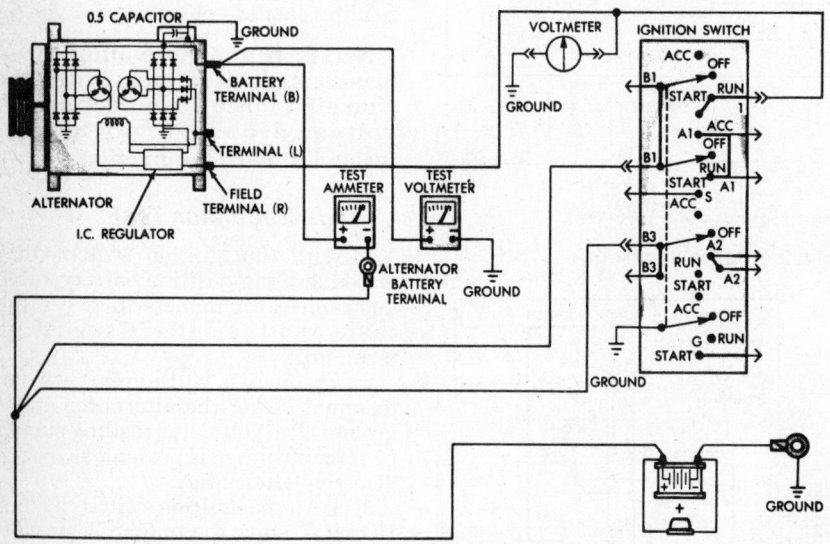

Current output test connections—Mitsubishi alternator

the engine at the rpms listed in the chart below. The current output should be close to the specifications.

— CAUTION —
DO NOT allow the voltage meter to read above 16 volts.

---

7. If the readings are below specifications, internal trouble is indicated. Remove the alternator for further testing.

17-25A at 13.5 V and 500 rpm
63-70A at 13.5 V and 1000 rpm
74A at 13.5 V and 2000 rpm

Delcotron SI Series

GENERAL MOTORS CORP. AND AMERICAN MOTORS

This system is an integrated AC generating system containing a built-in voltage regulator.

NOTE: In 1986, GM introduced the 17 SI alternator which employs a delta wound stator, used on the Corvette, Camaro and the Fiero. The alternator features a reduction in the weight, size, internal fasteners and noise; an increase in amperage output and operating longevity. The testing procedures are similar, except the stator windings cannot be checked for opens.

The regulator is mounted inside the slip ring end frame. The regulator components are enclosed in an epoxy mold, making the regulator nonadjustable. The rotor bearings contain a sufficient supply of lubricant to eliminate the need for periodic lubrication. No periodic maintenance, except the belt adjustment, is necessary.

Troubleshooting

NOTE: Refer to the "Preliminary Charging System Inspection" section before proceeding further. Make sure that the continuous running blower (if equipped) is disconnected. The blower will run with the key ON when the blower control is Off, unless disconnected.

Fusible Links

All GM vehicles are equipped with fusible links. The links are made of a piece of wire several gauges smaller than the supply wire that they are connected to. Their function is similar to that of a fuse, protecting the wiring in the event of an overload or a short circuit. They will usually melt before

---

type, the charging voltage varies with the temperature. The temperature around the rear bracket must be measured and the charging voltage adjusted to the temperature. The temperature compensation gradient is (−)0.7–0.13 volts @ 50°F.

7. If the ammeter reading is more than 5.0 amps (1983) or 10.0 amps (1984–87), charge the battery until the reading falls to less than 5.0 amps (1983) or 10.0 amps (1984–87) or replace the battery with a fully charged one.

Current Output Test

The current output test determines whether or not the alternator is able to delivery its rated output.

1. Turn OFF the ignition switch

and disconnect the negative battery cable.

2. Disconnect the "Bat" (battery) lead wire from the rear terminal of the alternator. Connect an ammeter (0–100 amps) between the "Bat" terminal of the alternator and the disconnect lead wire.

3. Connect the positive lead of a voltmeter (0–18 volts) to the "Bat" terminal of the alternator and the negative lead to a good ground.

4. Connect a tachometer to the engine and reconnect the negative battery cable.

5. Connect a variable carbon pile rheostat between the battery terminals; be sure the control switch is in the Open or Off position. Start the engine and allow it to run at idle.

6. Adjust the carbon pile and run

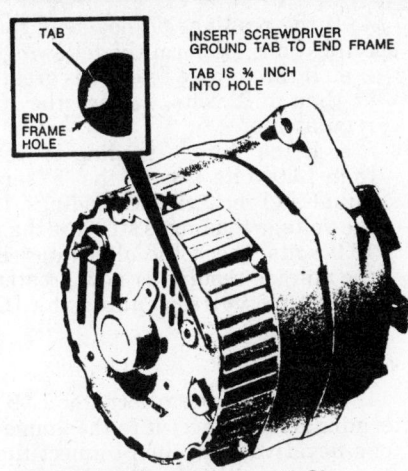

End view of the Delcotron 10-SI alternator—G.M.

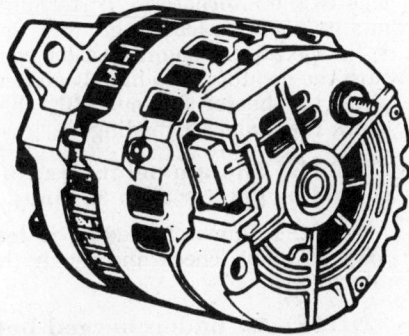

View of the 17SI alternator used on the GM Corvette, Camaro and Fiero

the wiring is damaged in the circuit. These links must be inspected before continuing with troubleshooting procedures.

Charging System Operation

1. With the engine running and the accessories turned Off, place a current indicator over the positive battery cable.

2. If a charge of about 5.0 amps is recorded, the charging system is working; if a draw of about 5.0 amps is recorded, the system is not working. The needle moves toward the battery when a charge condition is indicated and away from the battery when a draw condition is indicated. If a draw is indicted, proceed with further testing. If an excessive charge (10–15 amps) is indicated, check for an overcharge, caused by a faulty regulator.

Indicator Light Circuit Check

Check the indicator light for normal operation:

If the alternator light is operating properly, proceed to the next section. If one of the following condition exists, proceed as directed:

Ignition Switch Condition	Light Condition	Engine Condition
Off	Off	Stopped
On	On	Stopped
On	Off	Running

1. The ignition switch Off, the light stays On: disconnect the leads from the No. 1 and 2 terminals. If the light remains On, there is a short between these two leads. If the lamp turns Off, replace the rectifier bridge.

2. The switch is On, the light is On and the engine running; the causes for this condition are covered in Charging System Test.

3. The ignition switch On, the light Off, the engine not running: this condition may be caused by the defects listed in Step a., by reversal of the No. 1 and 2 leads at the alternator or by an open circuit. If the circuit is open, proceed as follows:

a. Connect a voltmeter from the No. 2 alternator terminal to ground. If a reading is obtained, proceed to the next step. If a zero reading is obtained, repair the circuit between the No. 2 terminal and the battery. If the light turns On, no further testing is necessary.

b. With the ignition switch ON and with the No. 1 and 2 terminals disconnected at the alternator, momentarily ground the No. 1 terminal lead.

CAUTION
DO NOT ground the No. 2 Lead.

c. If the light still doesn't turn On, check for a blown fuse or fusible link, a burned out bulb, a defective bulb socket or an open No. 1 lead circuit between the generator and the ignition switch.

d. If the lamp turns On, remove the ground at the No. 1 terminal, then with the No. 1 and 2 terminals connected to the alternator, insert a screwdriver into the test hole at the back of the alternator, to ground the winding.

e. If the light does not come On, check the connection between the wiring harness and the No. 1 terminal of the alternator. If the connection is OK, disassemble at the alternator and check the brushes, the slip rings and the field winding.

f. If a light now turns On and a reading was obtained in Step 1, check the resistor in line (circuit board), if the resistor checks out all right, replace the regulator.

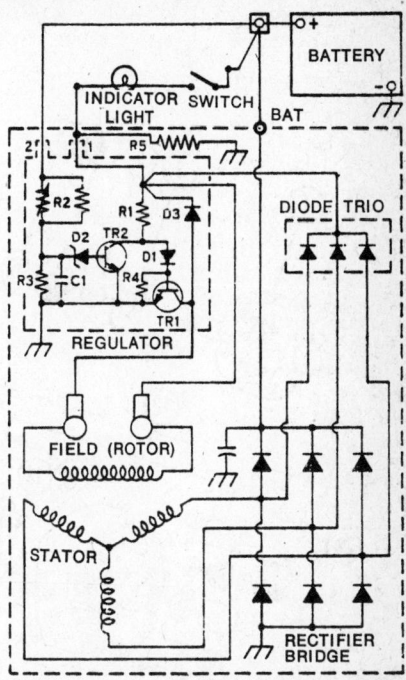

Typical 10-SI charging system circuitry

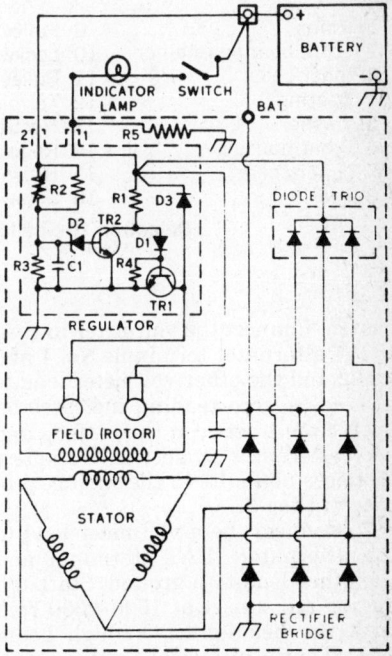

Schematic of the 17SI alternator. Note the delta stator windings

Charging System Test

After the battery conditions, the drive belt tension, the wiring terminals and the connections have been checked, charge the battery fully and perform the following test:

1. Connect a voltmeter between the alternator "BAT" terminal and a ground, then turn On the ignition

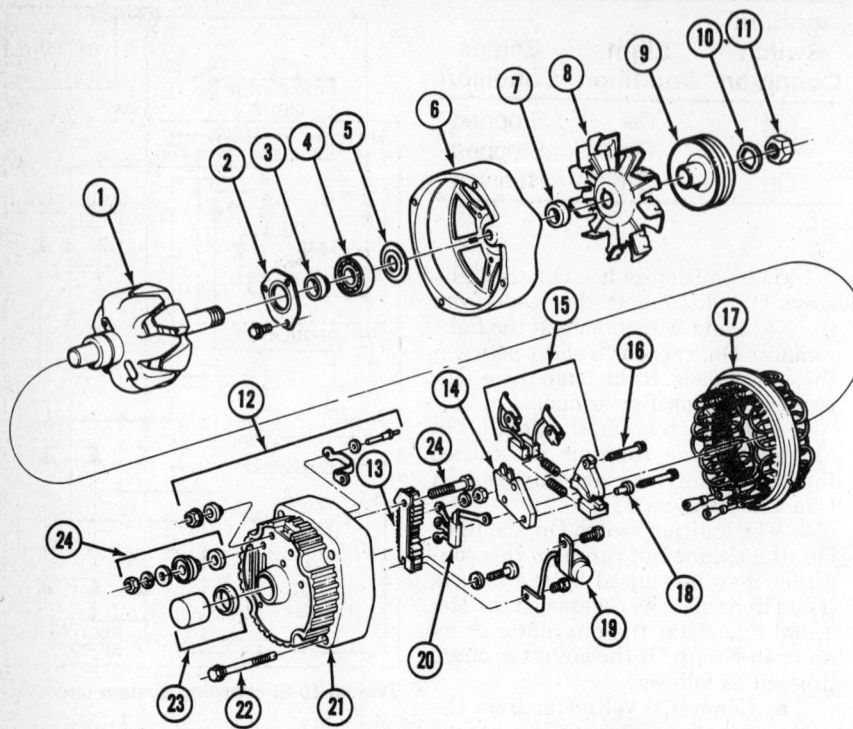

1 Rotor
2 Front bearing retainer
3 Inner collar
4 Bearing
5 Washer
6 Front housing
7 Outer collar
8 Fan
9 Pulley
10 Lockwasher
11 Pulley nut
12 Terminal assembly
13 Rectifier bridge
14 Regulator
15 Brush assembly
16 Screw
17 Stator
18 Insulating washer
19 Capacitor
20 Diode trio
21 Rear housing
22 Through bolt
23 Bearing and seal assembly
24 Terminal assembly

Delcotron 10-SI alternator—exploded view

switch. Connect the voltmeter in turn to the alternator terminals No. 1 and No. 2 and the other voltmeter lead to ground. A zero reading indicates an open circuit between the battery and each alternator connection. If the test discloses no faults in the wiring, proceed to Step 3.

2. Connect the a voltmeter lead to the alternator "BAT" terminal and the other lead to a ground. Start the engine and run it at 1500–2000 rpm with the headlights On high beam and all of the electrical accessories on High. If the voltmeter reads 12.8 volts, ground the field winding by inserting a screwdriver into the test hole in the end frame.

— **CAUTION** —

Do not force the tab more than ¾ in. into end frame.

a. If the voltage increases to 13 volts or more, the regulator unit is defective.

b. If the voltage does not increase significantly, the alternator is defective.

Nippondenso Alternator/Electronic Voltage Regulator

1985–87 GM NOVA, 1985–87 SPECTRUM/SPRINT

Charging Circuit Check
WITHOUT LOAD

1. Disconnect the alternator's "B" terminal and connect it to the ammeter's negative lead, then connect the ammeter's positive lead to the alternator's "B" terminal.

2. Connect the voltmeter's positive lead to the alternator's "B" terminal and its negative lead to ground.

3. Operate the engine from idle to 2000 rpm and check the meter readings.

NOTE: The amperage reading should be less than 10 amps and the voltage reading should be 13.9–15.1 volts.

4. If the readings do not meet the specifications, perform the following:

a. If the voltage reading is greater than 15.1 volts, replace the IC regulator.

b. If the voltage reading is less than 13.9 volts, ground the "F" terminal and recheck the readings. If the voltage reading is still less than 13.9 volts, check the alternator; if the voltage reading is now greater than 15.1 volts, replace the IC regulator.

WITH LOAD

1. Disconnect the alternator's "B" terminal and connect it to the ammeter's negative lead, then connect the ammeter's positive lead to the alternator's "B" terminal.

2. Connect the voltmeter's positive lead to the alternator's "B" terminal and its negative lead to ground.

3. Operate the engine at 2000 rpm, turn the headlights to high beam and place the heater fan on "HI", then check the ammeter reading.

NOTE: The ammeter reading should be greater than 30 amps.

4. If the ammeter reading is less than 30 amps, check and repair the alternator.

NOTE: An undercharged battery will sometimes cause a reduced ammeter reading.

Rotor Inspection

1. Using an ohmmeter, place the probes on the slip rings and check for continuity. If there is no continuity, replace the rotor.

2. Using an ohmmeter, place one probe on the rotor and the other on the slip ring. If there is continuity, replace the rotor.

3. Check the slip rings for damage or wear; they should be smooth and even. Using a micrometer, check the diameter of the slip rings; if they are less than 0.551 in., replace the rotor.

Stator Inspection

1. Using an ohmmeter, inspect all of the leads for continuity; if there is no continuity, replace the drive end frame assembly.

2. Using an ohmmeter, place one probe on the leads and the other on the frame; if there is continuity, replace the drive end frame assembly.

Brushes and Brush Holder Inspection

1. If either brush is less that 0.177 in., replace the brush.

2. To replace a brush in the holder, perform the following steps:

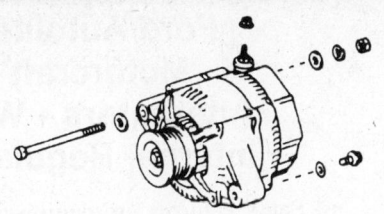

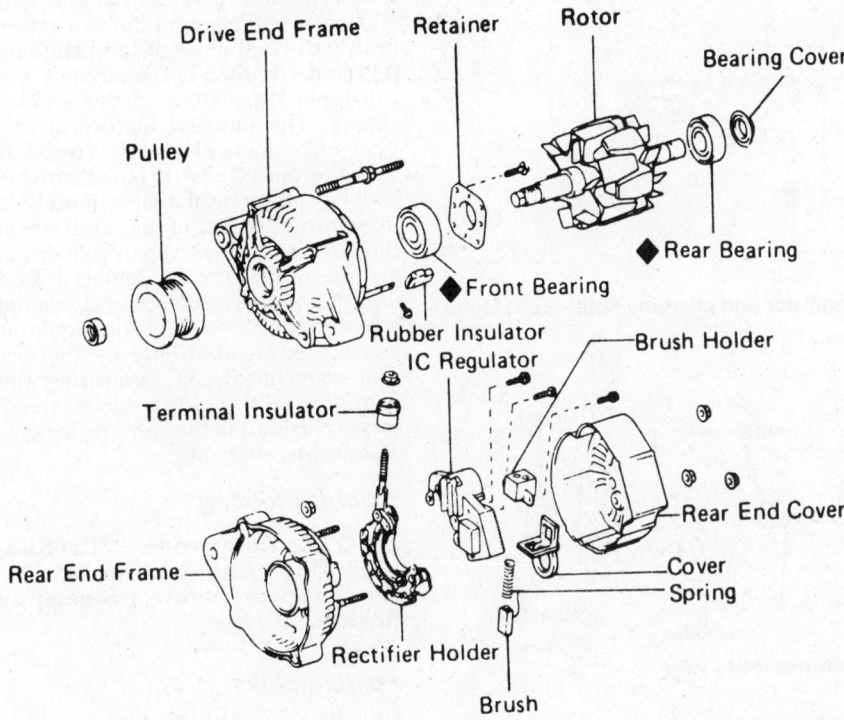

Exploded view of the Nova alternator—Sprint and Spectrum are similar

◆ Non-reusable part

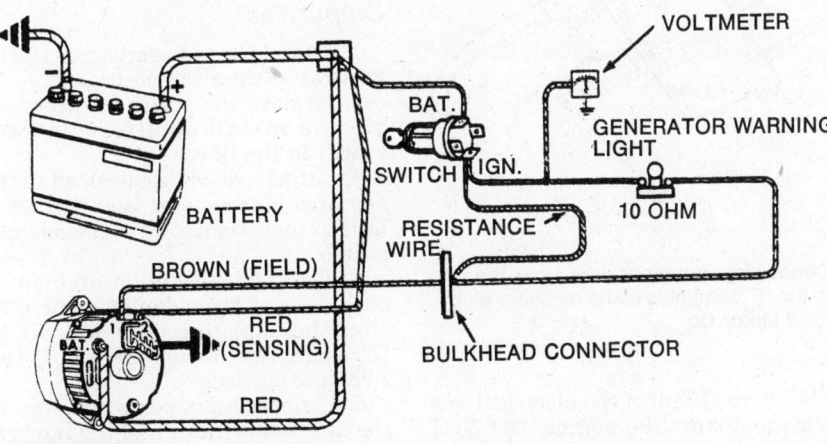

Typical 10–SI alternator charging circuit

a. Unsolder the wire, then remove the brush and the spring.

b. Place the new brush wire through the spring, then insert it into the brush holder.

c. Expose 0.413 in. of the wire

through the holder and solder into place.

d. Make sure that the brush moves freely in the holder, then cut off the excess length.

Bearing Inspection

Make sure that the bearings are not rough of worn.

1. To remove the front bearing, remove the four screws and the bearing retainer, then replace the bearing.

2. To remove the rear bearing, perform the following:

 a. Using the tool No. 09820-00021 (or equivalent), pull the bearing with the bearing cover from the rotor shaft.

 b. Using the tool No. 09612-10092 (or equivalent), and an arbor press, press the new bearing (with the bearing cover) onto the rotor shaft.

Autolite/Motorcraft Alternator/External Regulator

FORD MOTOR CO. AND AMERICAN MOTORS

Troubleshooting

NOTE: Refer to the "Preliminary Charging System Inspection" section before proceeding further.

Fusible Links

Check the fusible link located between the starter relay and the alternator. Replace the link if it is burned or open.

Charging System Operation

NOTE: If the current indicator is to give an accurate reading, the battery cables must be of the same gauge and length as the original equipment.

1. With the engine running and the accessories turned Off, place a current indicator over the positive battery cable.

2. If a charge of about 4.0 amps or a draw of about 5.0 amps is recorded, the charging system is not working.

 a. The needle moves toward the battery when a charge condition is indicted.

 b. If a draw is indicated, continue to the next testing procedure.

 c. If an overcharge of 10–15 amps is indicted, check for a faulty regulator or a bad ground at the regulator or the alternator.

Testing The Ignition Switch To Regulator Circuit

1. Disconnect the regulator wiring harness from the regulator.

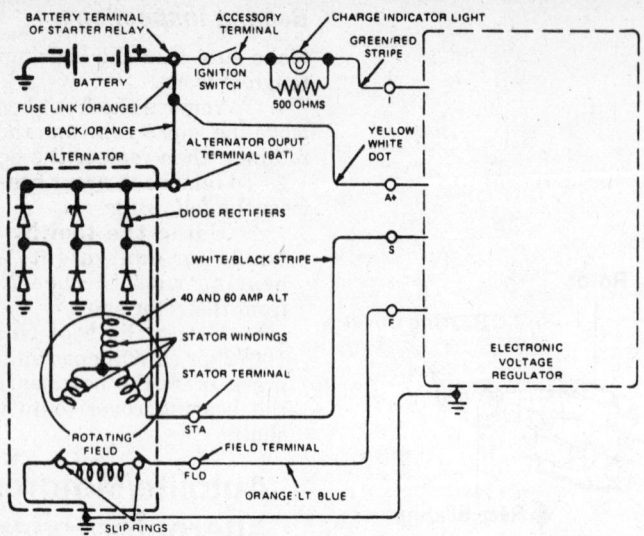

Charging system schematic with electronic regulator and charging light—Ford Motor Co.

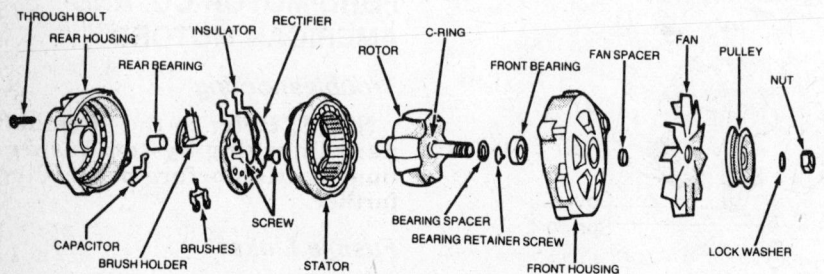

Ford side terminal alternator—exploded view

Ford/Autolite/ Motorcraft Alternators—With Internal Regulator

Some vehicles are equipped with an Autolite alternator having an integral regulator mounted to the rear end housing. The regulator is a hybrid unit featuring to use of solid state integrated circuits. These circuits will consist of transistors, diodes and resistors. The unusual feature of this type of a micro-electronic circuit is that the entire circuit is constructed on a silicone crystal approximately $\frac{1}{8}$ in. square. Because of the small size of the circuit, it is not repairable or adjustable and must be replaced as a unit, if found to be defective. It should be noted that the size of the regulator housing is dictated only by the fact that some means of connecting the regulator to the alternator is necessary. Overhaul is the same as for other Autolite alternators.

Troubleshooting

NOTE: Refer to the "Preliminary Charging System Inspection" section before proceeding further.

Fusible Links

Check the fusible link located between the starter relay and the alternator. Replace the link if it is burned or open.

Output Test

1. Place the transmission in the Neutral or the Park position.
2. Remove the positive battery cable and install a battery adapter switch in the line.
3. Attach one voltmeter lead to the negative battery post and the other lead to the circuit side of the adapter switch.
4. Connect an ammeter lead to each side of the adapter switch, so that the charging current will go through the ammeter when the switch is opened.
5. Connect a jumper wire between the alternator frame and the integral regulator field terminal (cover plug removed).
6. Close the adapter switch and start the engine, then open the adapter switch.
7. Running engine at 2,000 rpm, observe the voltmeter and the ammeter. At 15 volts indicated, the ammeter should read 50–57 amps; if no-charge condition still exists, the regu-

2. Turn ON the key. Using a test light or a voltmeter, check for voltage between the No. 1 wire and a ground. Check for voltage between the No. A wire and a ground. If voltage is present at this part of the system, the circuit is OK. If there is no voltage at the No. 1 wire, check for a burned-out charge indicator bulb, a burned-out resistor, a break or a short in the wiring. If there is no voltage present at the No. A wire, check for a bad connection at the starter relay, a break or a short in the wire.

Isolation Test

This test determines whether the regulator or the alternator is faulty, after the circuit is found to be in good working order.

1. Disconnect the regulator wiring harness from the regulator.
2. Connect a jumper wire from the No. A wire to the No. F wire in the wiring harness plug.
3. Connect a voltmeter to the battery; the positive voltmeter lead goes to the positive terminal and the negative lead to the negative terminal. Record the reading on the voltmeter.

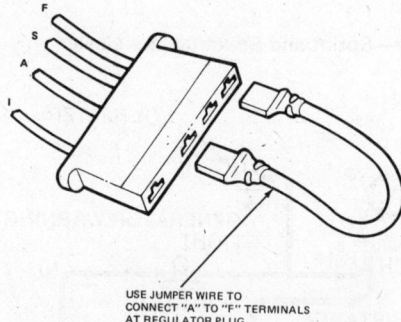

Connecting a jumper wire from the "A" to the "F" terminals of the regulator plug— Ford Motor Co.

4. Turn Off all of the electrical systems and start the engine; DO NOT race the engine.
5. Gradually increase the engine speed to 1500–2000 rpm. The voltmeter reading should increase above the previously recorded battery voltage reading by at least 1–2 volts. If there is no increase, the alternator is not working correctly. If there is an increase, the voltage regulator needs to be replaced.

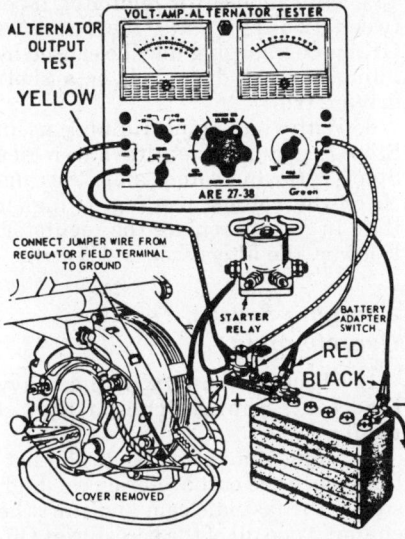

Output test connection—Ford integral regulator

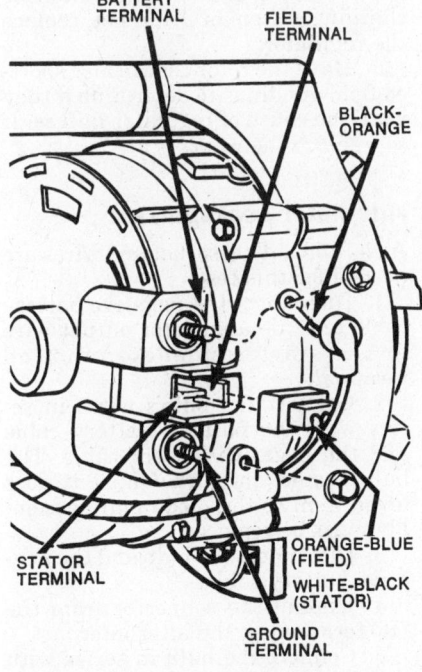

Wiring connections—Ford side terminal alternator

lator is probably faulty and must be replaced. An output of 42–48 amps usually indicates an open diode rectifier, while an output 10–15 amps below minimum specifications usually indicates a shorted diode. An alternator with a shorted diode usually will whine at idle speed.

Field Test (Voltmeter)

1. Turn the ignition switch to the OFF position.
2. Remove the wire from the regulator supply terminal.

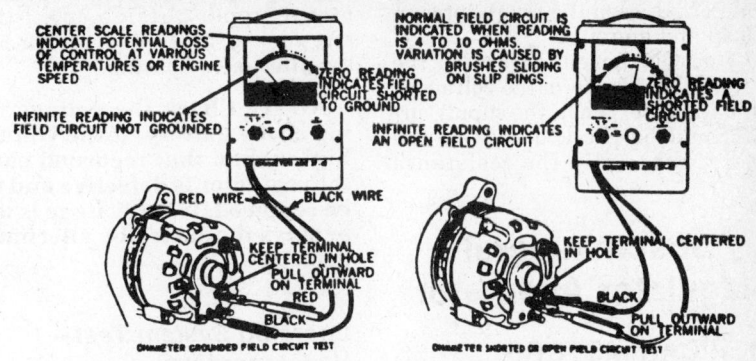

Field circuit connection with ohmmeter—Ford integral regulator

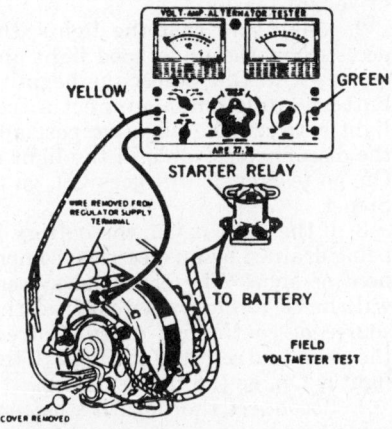

Voltmeter connection for field test—Ford integral regulator

3. Remove the cover plug from the regulator field terminal and connect one voltmeter lead to this terminal. A $\frac{1}{4}$ ohm resistor should be in the circuit.
4. Connect the other voltmeter lead to a good ground.
5. The voltmeter should read 12 volts; if no voltage is present, the field circuit is open or grounded.
6. If the voltmeter reads more than 1 volt, but still less than the battery voltage, a partial ground may exist in the alternator field circuit and the circuit should be checked with an ohmmeter.

Field Test (Ohmmeter)

1. Disconnect the negative battery cable and remove the alternator from the vehicle.
2. Remove the regulator from the alternator (covered later).
3. Make the ohmmeter tests as illustrated. If any of the tests indicates a field circuit problem, disassemble the alternator to further isolate the trouble, as follows:

   a. Contact each ohmmeter probe to a slip ring. The resistance should be 4–5 ohms.

NOTE: A higher reading indicates a damaged slip ring solder connection or a broken wire. A lower reading indicates a shorted wire or the slip ring assembly.

   b. Connect one ohmmeter probe to a slip ring and the other probe to the rotor shaft. Any reading other than infinite ohms indicates a short to ground. If neither of these tests (A and B) isolates the trouble, the brushes or the brush assembly are the probable cause.

Voltage Limiter Test

1. Check the battery specific gravity. If it is not at least 1.230, charge the battery or install a charged battery for the test.
2. Make sure the lights and the accessories are turned Off, including such items as dome lights.
3. Make the test connections as illustrated.
4. Place the transmission in the Neutral or the Park positions, close battery adapter switch and start the engine.
5. Open the battery adapter switch and operate engine at 2000 rpm for 5 minutes; the voltmeter should read 13.3–15.3 volts.
6. If the voltage does not rise above 12 volts, perform a regulator supply voltage test to determine whether or not the regulator is getting voltage from the battery.

NOTE: Before replacing the regulator, check the wiring of the entire charging system for shorts, opens or high resistance connections.

Regulator Supply Voltage Test

The regulator is turned On by the application of battery voltage through a 10 ohm resistor wire. If the supply circuit is defective, the regulator will not function and the alternator will not produce current.

1. Connect a 12 volt test light or a

voltmeter between the regulator supply lead and a ground.

2. Turn ON the ignition switch; the test light should glow or the voltmeter should indicate. If not, the supply circuit should be checked back to the battery, especially the resistance wire.

Bosch Integral Regulator Alternator

AMERICAN MOTORS

The Bosch charging system is a conventional 12 volt, negative ground unit, consisting of the alternator, the regulator and the battery.

The alternator rotor is supported by ball bearings which are permanently lubricated and require no periodic service. The stator windings are wrapped on a laminated core which forms part of the alternator frame. The six diodes are used to convert the AC voltage to DC, supplied to the output terminal. The alternator field current is supplied through a diode trio, which is also connected to the stator windings. A capacitor mounted on the end housing is used to protect the diode plate assembly from high voltages; it also suppresses radio noise and requires no periodic maintenance.

The voltage regulator is a solid state unit mounted on the end plate; it retains the brushes in an integral holder. The unit is attached to the end frame with two screws and can be replaced without disturbing the alternator. The regulator is non-adjustable and must be replaced as a unit (if defective).

Troubleshooting

NOTE: Refer to the "Preliminary Charging System Inspection" section before proceeding further.

Indicator Lamp Test

The indicator lamp will only glow when a no-charge condition exists at the alternator (it also lights during starting as a bulb check). To diagnose:
1. Check the alternator belt tension.
2. Start the engine, then measure and record the battery voltage at the battery.
3. Raise the engine speed to fast idle.
4. There is a grounding sleeve on the voltage regulator; it is a metal tab

on the upper outside edge. Use a screwdriver to ground the sleeve to the alternator housing.

NOTE: Check the voltage reading at the battery. If the voltage is higher than that recorded earlier, the regulator is defective and must be replaced. If the voltage is lower or stays the same, the alternator is defective.

Charging System Test— Undercharging

1. Check and adjust the alternator drive belt tension.
2. Make sure that the lights, the accessories, the underhood light and etc. are Off. Disconnect the negative battery cable, then connect a test light between the negative post and the disconnected cable. If the light is On, go to Step 3; if it goes Off, go to Step 4.
3. If the light is On, the battery is being drained by an electrical component or short. The electrical system will have to be traced to find the source of continuous drain. Correct the cause and retest as in Step 2. If the light is On, go the the next step.
4. Reconnect the negative battery cable. Connect a jumper wire between the negative coil terminal and a ground. Crank the engine and obtain a stabilized voltage reading. DO NOT crank the engine more than 15 seconds at a time, to avoid starter damage. If the reading is above 9 volts, go to Step 6; if below 9 volts, go to the next step.
5. Check the battery voltage while cranking the engine. If within 0.5 volts of the alternator reading, test the battery using the full load procedure. If the battery is good, go to Step 6. If not, replace the battery, then go to Step 6. If the battery voltage is not within 0.5 volts of alternator voltage, check and correct the battery-to-alternator circuit resistance.
6. Disconnect the jumper wire at the coil. Connect a voltmeter to the battery and record the reading. Place the carburetor on the high step of the fast idle cam, start the engine, turn On all of the accessories (headlights on high beam, A/C on High, radio and blower On) and check the voltage reading. If lower, go to Step 8; if higher, go to the next step.
7. Turn Off the accessories, allow the engine warm up (heat in the upper radiator hose) and allow the voltmeter reading to stabilize. If under 12.5 volts, go to the next step. If over

15.5 volts, replace the regulator. If between 12.5–15.5 volts, the system is OK; undercharging may be caused by idling, a loose drive belt or a short driving trip.
8. With the engine running as in Step 7, ground the alternator (see Step 4 of the Indicator Lamp Test) and check the voltage reading. If higher than in Step 6, replace the regulator. If lower, the alternator is defective.

Charging System Test— Overcharging

1. Check the battery with a heavy load test. Replace the battery if required.
2. Connect a voltmeter to the battery, place the carburetor on the high step of the fast idle cam and start the engine. Turn all of the accessories Off. With the engine warm (heat in the upper radiator hose) and voltmeter reading stabilized, check the reading. If the reading is 12.5–15.0 volts, the charging system is OK; if not, replace the regulator.
3. Have the rotor checked for shorted field windings to determine if they were the cause of regulator failure. If so, replace the rotor.

Alternator Leakage Test

A No. 158 bulb, a socket and wires are needed for this test.
1. Disconnect the negative battery cable and the alternator output wire at the starter solenoid junction terminal.
2. Connect the bulb's wires in series with the positive battery cable and the negative battery cable. The bulb should not turn On; if it does (even dimly), the diode plate assembly must be replaced.
3. Disconnect the bulb and the negative battery cable.
4. Unplug the connector from the "R" terminal at the alternator.
5. Connect the bulb in series with the "R" terminal and the positive battery cable, then with the negative battery cable. The bulb should not turn On; if it does (even dimly), the diode plate or the regulator may be defective. Test and replace as necessary.

Regulator Replacement

1. Remove the regulator/brush holder retaining screws and washers.
2. Tip the assembly and lift it from the rear housing.
3. To install, reverse the removal procedures.

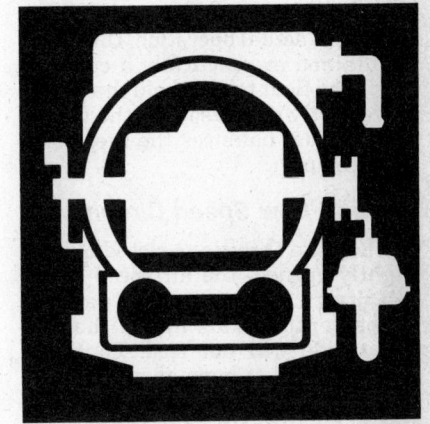

Carburetors

FUNCTIONS

Gasoline is the source of fuel for power in the automobile engine and the carburetor is the mechanism which automatically mixes liquid fuel with air in the correct proportions to provide the desired power output from the engine. The carburetor performs this function by metering, atomizing, and mixing fuel with air flowing through the engine. A carburetor also regulates the volume of air-to-fuel mixture which enters the engine. It is the carburetor's regulation of the mixture flow which gives the operator control of the engine speed.

Metering

The automotive internal combustion engine operates efficiently within a relatively small range of air-to-fuel ratios. It is the function of the carburetor to meter the fuel in exact proportions to the air flowing into the engine, so that the optimum ratio of air-to-fuel is maintained under all operating conditions. Regulations governing exhaust gas emissions have made the proper metering of fuel by the carburetor an increasingly important factor. Too rich a mixture will result in poor economy and increased emissions, while too lean a mixture will result in loss of power and generally poor performance. Carburetors are matched to engines so that metering can be accomplished by using carefully calibrated metering jets which allow fuel to enter the engine at a rate proportional to the engine's ability to draw air.

Atomization

The liquid fuel must be broken up into small particles so that it will more readily mix with air and vaporize. The more contact the fuel has with the air, the better the vaporization. Atomization can be accomplished in two ways; air may be drawn into a stream of fuel which will cause a turbulence and break the solid stream of fuel into smaller particles; or a nozzle can be positioned at the point of highest air velocity in the carburetor and the fuel will be torn into a fine spray as it enters the air stream.

Distribution

The carburetor is the primary device involved in the distribution of fuel to the engine. The more efficiently fuel and air are combined in the carburetor, the smoother the flow of vaporized mixture through the intake manifold to each combustion chamber. Hence, the importance of the carburetor in fuel distribution.

PRINCIPLES

Vacuum

All carburetors operate on the basic principle of pressure difference. Any pressure less than atmospheric pressure is considered vacuum or a low pressure area. In the engine, as the piston moves down on the intake stroke with the intake valve open, a partial vacuum is created in the intake manifold. The farther the piston travels downward, the greater the vacuum created in the manifold. As vacuum increases in the manifold, a difference in pressure occurs between the carburetor and cylinder. The carburetor is positioned in such a way that the high pressure above it, and the vacuum or low pressure above it, and the vacuum or low pressure beneath it, causes air to be drawn through it. Fuel and air always move from high to low pressure areas.

Venturi Principle

To obtain greater pressure drop at the tip of the fuel nozzle so that fuel will flow, the principle of increasing the air velocity to create a low pressure area is used. The device used to increase the velocity of the air flowing through the carburetor is called a venturi. A venturi is a specially designed restriction placed in the air flow. In order for the air to pass through the restriction, it must accelerate causing a pressure drop or vacuum as it passes.

CARBURETOR CIRCUITS

Float Circuit

The float circuit includes the float, float bowl, and a needle valve and seat. This circuit controls the amount of gas allowed to flow into the carburetor. As the fuel level rises, it causes the float to rise which pushes the needle valve into its seat. As soon as the valve and seat make contact, the flow of gas is cut off from the fuel inlet.

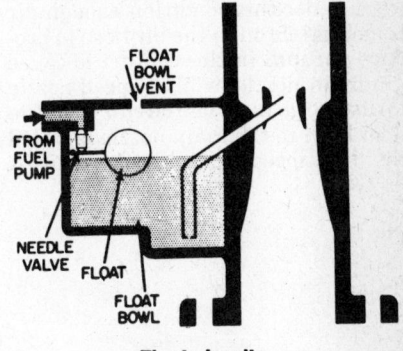

Float circuit

When the level of fuel drops, the float sinks and releases the needle valve from its seat which allows the gas to flow in. In actual operation, the fuel is maintained at practically a constant level. The float tend to hold the needle valve partly closed so that the incoming fuel just balances the fuel being withdrawn.

Idle and Low Speed Circuit

When the throttle is closed or only slightly opened, the air speed is low and practically no vacuum develops in the venturi. This means that the fuel nozzle will not feed. Thus, the carburetor must have another circuit to supply fuel during operation with a closed or slightly opened throttle. This circuit is called the idle and low speed circuit. It consists of passages in which air and gas can flow beneath the throttle plate. With the throttle plate closed, there is high vacuum from the intake manifold. Atmospheric pressure pushes the air/fuel mixture through the passages of the idle and low speed circuit and past the tapered point of the idle adjustment screw, which regulates engine idle mixture volume.

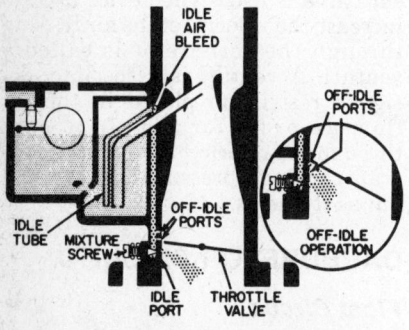

Idle and low speed circuit

High Speed Partial Load Circuit

When the throttle plate is opened sufficiently, there is little difference in vacuum between the upper and lower part of the air horn. Thus, little air/fuel mixture will discharge from the low speed and idle circuit. However, under this condition enough air is moving through the air horn to produce vacuum in the venturi to cause the main nozzle or high speed nozzle to discharge fuel. The circuit from the float bowl to the main nozzle is called the high speed partial load circuit. A

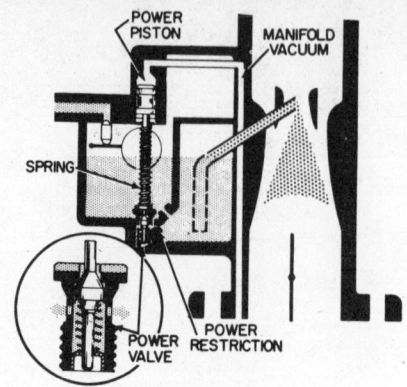

Power circuit

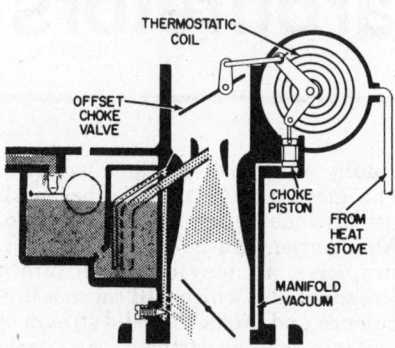

Choke system

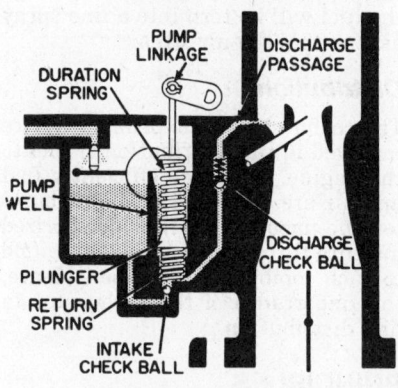

Accelerator pump circuit

nearly constant air/fuel ratio is maintained by this circuit from part to full-throttle.

High Speed Full Power Circuit

For high-speed, full-power, wide open throttle operation, the air/fuel mixture must be enriched; this is done either mechanically or by intake manifold vacuum.

Full Power Circuit (Mechanical)

This circuit includes a metering rod jet and a metering rod. The rod has two steps of different diameters and is attached to the throttle linkage. When the throttle is wide open, the metering rod is lifted bringing the smaller diameter of the rod into the jet. When the throttle is partly closed, the larger diameter of the metering rod is in the jet. This restricts fuel flow to the main nozzle but adequate amounts of fuel do flow for part-throttle operation.

Full Power Circuit (Vacuum)

This circuit is operated by intake manifold vacuum. It includes a vacuum diaphragm or piston linked to a valve. When the throttle is opened so that intake manifold vacuum is reduced, the spring raises the diaphragm or piston. This allows more fuel to flow in, either by lifting a metering rod or by opening a power valve.

Accelerator Pump Circuit

For acceleration, the carburetor must deliver additional fuel. A sudden inrush of air is caused by rapid acceleration or applying full throttle. When the throttle is opened, the pump lever pushes the plunger down and this forces fuel to flow through the accelerator pump circuit and out the pump jet. This fuel enters the air passage through the carburetor to supply additional fuel demands.

Choke

When starting an engine, it is necessary to increase the amount of fuel delivered to the intake manifold. This increase is controlled by the choke. The choke consists of a valve in the top of the air horn controlled mechanically by an automatic device. When the choke valve is closed, only a small amount of air can get past it.

When the engine is cranked, a fairly high vacuum develops in the air horn. This vacuum causes the main nozzle to discharge a heavy stream of fuel. The quantity delivered is sufficient to produce the correct air/fuel mixture needed for starting the engine. The choke is released either manually or by heat from the engine.

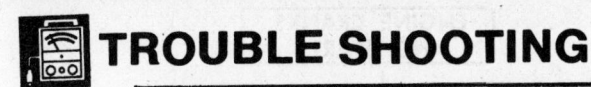 **TROUBLE SHOOTING**

ENGINE HESITATES ON ACCELERATION

NOTE: Carburetor problems cannot be isolated effectively unless all other engine systems are functioning correctly and the engine is properly tuned.

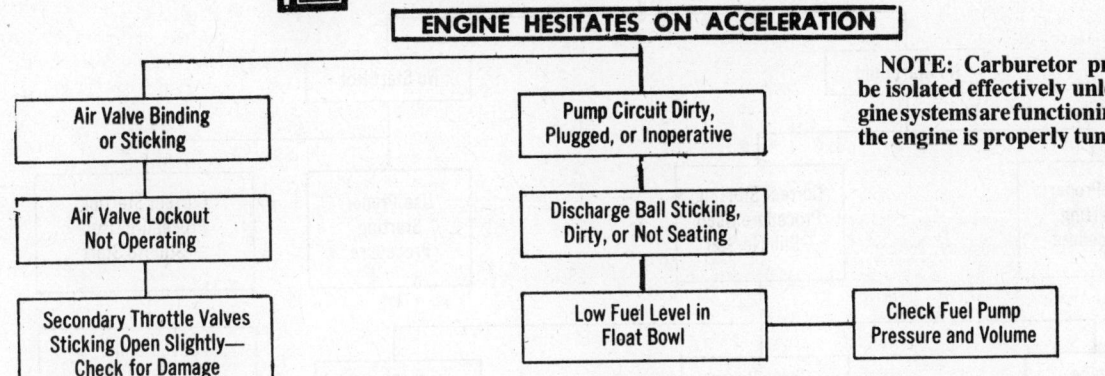

- Air Valve Binding or Sticking
- Air Valve Lockout Not Operating
- Secondary Throttle Valves Sticking Open Slightly— Check for Damage

- Pump Circuit Dirty, Plugged, or Inoperative
- Discharge Ball Sticking, Dirty, or Not Seating
- Low Fuel Level in Float Bowl — Check Fuel Pump Pressure and Volume

ENGINE FEELS SLUGGISH OR FLAT ON ACCELERATION

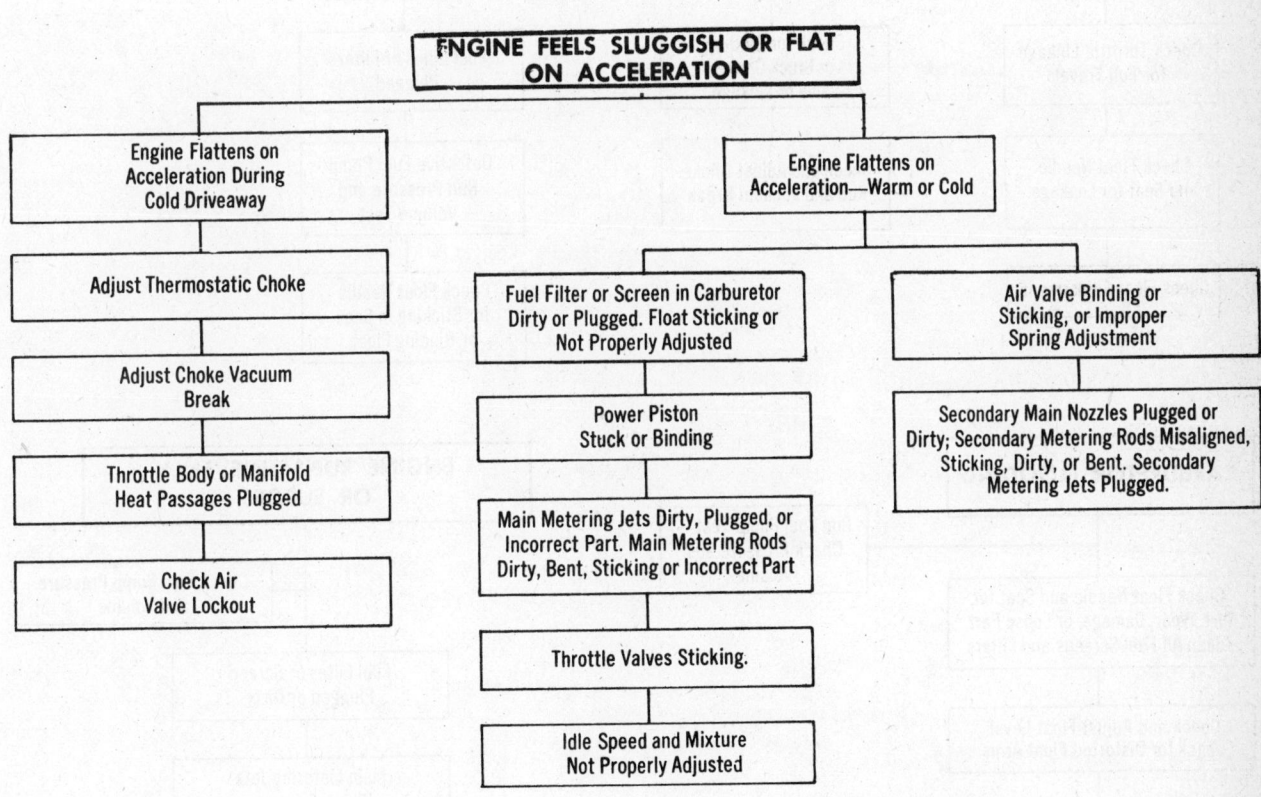

- Engine Flattens on Acceleration During Cold Driveaway
- Adjust Thermostatic Choke
- Adjust Choke Vacuum Break
- Throttle Body or Manifold Heat Passages Plugged
- Check Air Valve Lockout

- Fuel Filter or Screen in Carburetor Dirty or Plugged. Float Sticking or Not Properly Adjusted
- Power Piston Stuck or Binding
- Main Metering Jets Dirty, Plugged, or Incorrect Part. Main Metering Rods Dirty, Bent, Sticking or Incorrect Part
- Throttle Valves Sticking.
- Idle Speed and Mixture Not Properly Adjusted

- Engine Flattens on Acceleration—Warm or Cold
- Air Valve Binding or Sticking, or Improper Spring Adjustment
- Secondary Main Nozzles Plugged or Dirty; Secondary Metering Rods Misaligned, Sticking, Dirty, or Bent. Secondary Metering Jets Plugged.

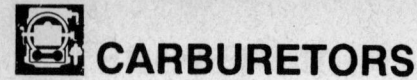

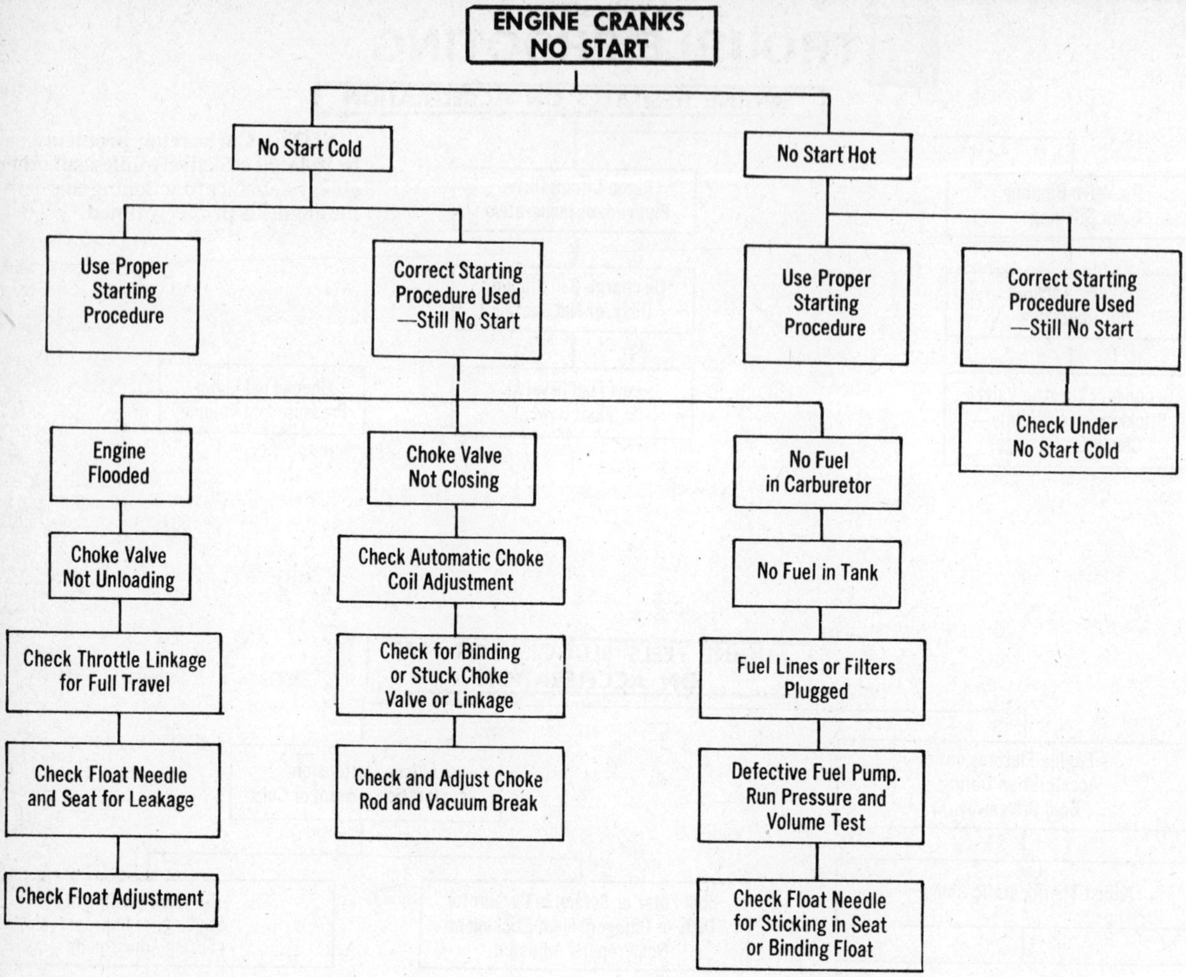

ENGINE CRANKS NO START

No Start Cold

- Use Proper Starting Procedure
- Correct Starting Procedure Used —Still No Start
  - Engine Flooded
    - Choke Valve Not Unloading
    - Check Throttle Linkage for Full Travel
    - Check Float Needle and Seat for Leakage
    - Check Float Adjustment
  - Choke Valve Not Closing
    - Check Automatic Choke Coil Adjustment
    - Check for Binding or Stuck Choke Valve or Linkage
    - Check and Adjust Choke Rod and Vacuum Break

No Start Hot

- Use Proper Starting Procedure
- Correct Starting Procedure Used —Still No Start
  - Check Under No Start Cold
- No Fuel in Carburetor
  - No Fuel in Tank
  - Fuel Lines or Filters Plugged
  - Defective Fuel Pump. Run Pressure and Volume Test
  - Check Float Needle for Sticking in Seat or Binding Float

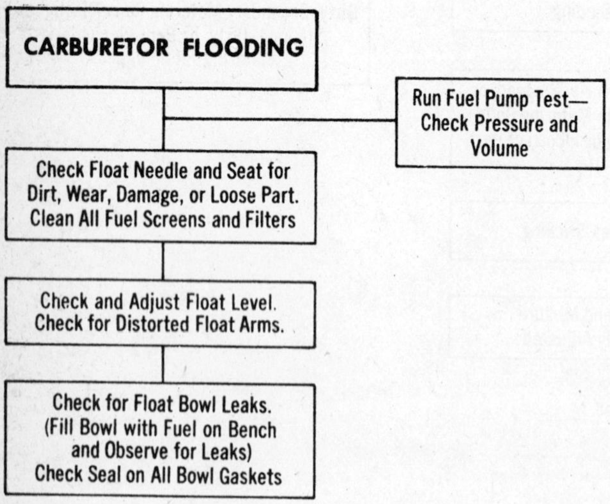

CARBURETOR FLOODING

- Run Fuel Pump Test— Check Pressure and Volume
- Check Float Needle and Seat for Dirt, Wear, Damage, or Loose Part. Clean All Fuel Screens and Filters
- Check and Adjust Float Level. Check for Distorted Float Arms.
- Check for Float Bowl Leaks. (Fill Bowl with Fuel on Bench and Observe for Leaks) Check Seal on All Bowl Gaskets

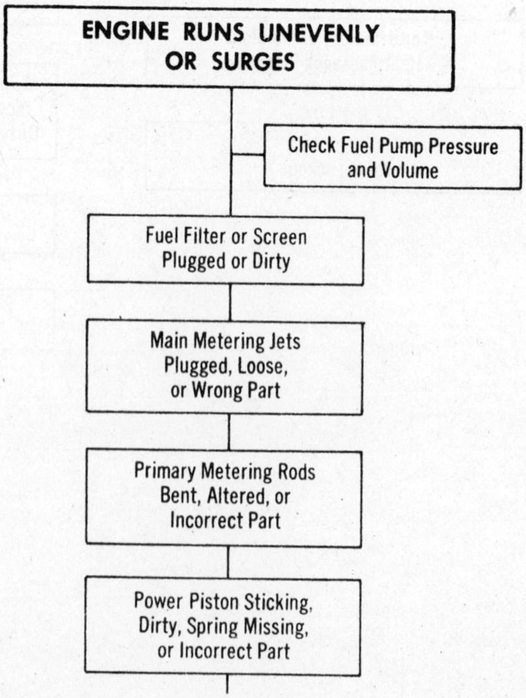

ENGINE RUNS UNEVENLY OR SURGES

- Check Fuel Pump Pressure and Volume
- Fuel Filter or Screen Plugged or Dirty
- Main Metering Jets Plugged, Loose, or Wrong Part
- Primary Metering Rods Bent, Altered, or Incorrect Part
- Power Piston Sticking, Dirty, Spring Missing, or Incorrect Part

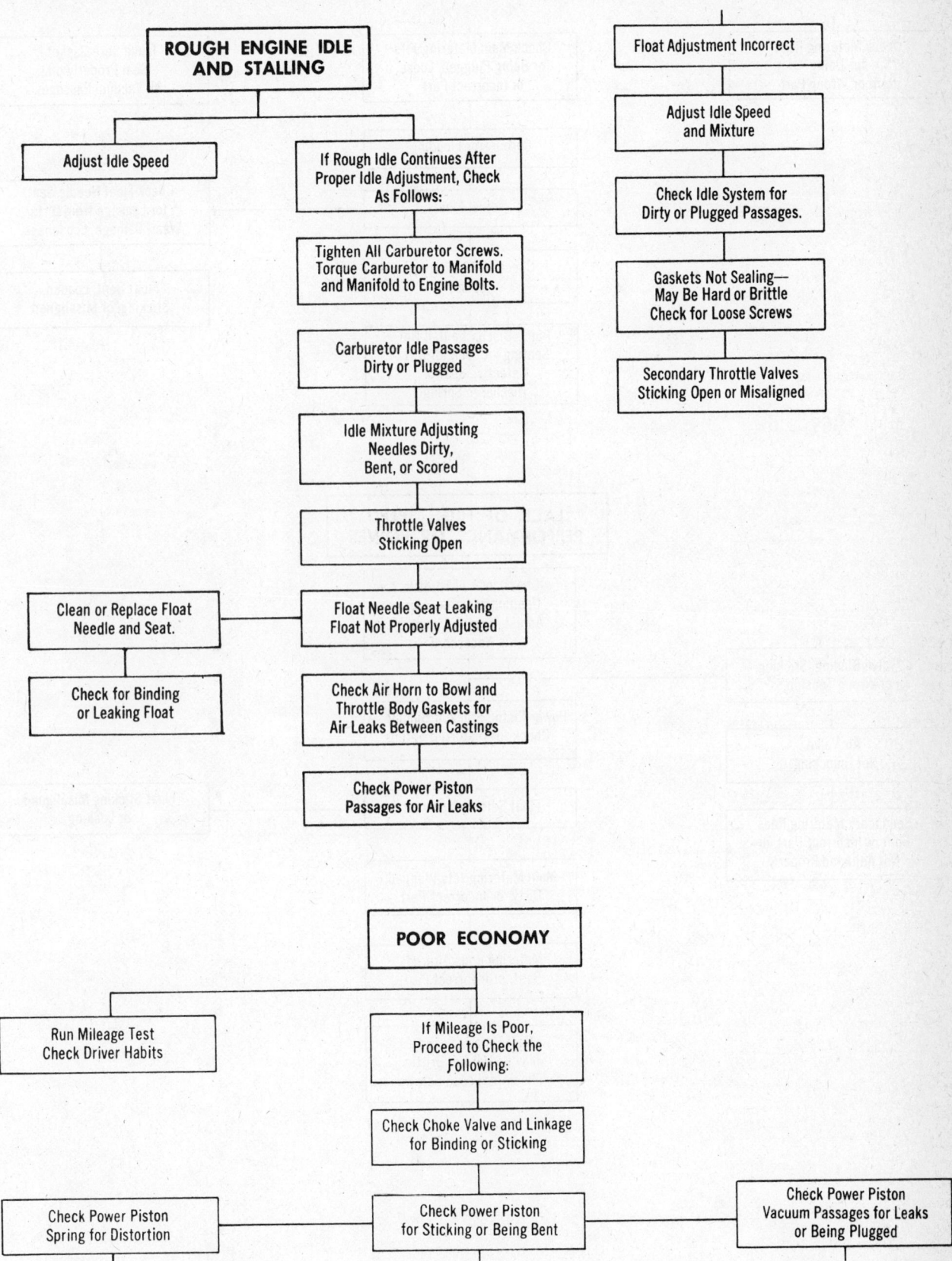

ROUGH ENGINE IDLE AND STALLING

Adjust Idle Speed

If Rough Idle Continues After Proper Idle Adjustment, Check As Follows:

Tighten All Carburetor Screws. Torque Carburetor to Manifold and Manifold to Engine Bolts.

Carburetor Idle Passages Dirty or Plugged

Idle Mixture Adjusting Needles Dirty, Bent, or Scored

Throttle Valves Sticking Open

Clean or Replace Float Needle and Seat.

Float Needle Seat Leaking Float Not Properly Adjusted

Check for Binding or Leaking Float

Check Air Horn to Bowl and Throttle Body Gaskets for Air Leaks Between Castings

Check Power Piston Passages for Air Leaks

Float Adjustment Incorrect

Adjust Idle Speed and Mixture

Check Idle System for Dirty or Plugged Passages.

Gaskets Not Sealing— May Be Hard or Brittle Check for Loose Screws

Secondary Throttle Valves Sticking Open or Misaligned

POOR ECONOMY

Run Mileage Test Check Driver Habits

If Mileage Is Poor, Proceed to Check the Following:

Check Choke Valve and Linkage for Binding or Sticking

Check Power Piston Spring for Distortion

Check Power Piston for Sticking or Being Bent

Check Power Piston Vacuum Passages for Leaks or Being Plugged

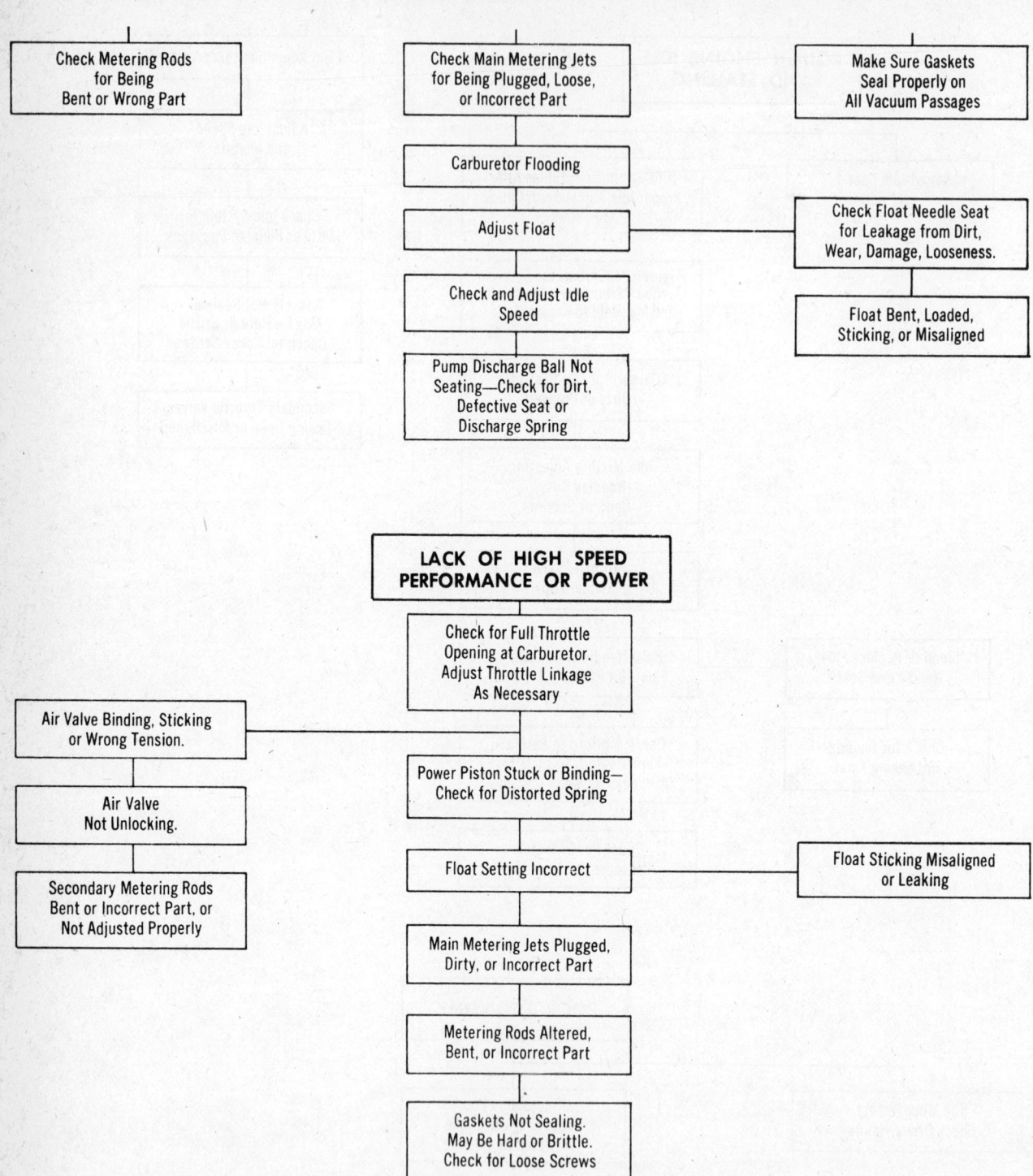

Check Metering Rods
for Being
Bent or Wrong Part

Check Main Metering Jets
for Being Plugged, Loose,
or Incorrect Part

Make Sure Gaskets
Seal Properly on
All Vacuum Passages

Carburetor Flooding

Adjust Float

Check Float Needle Seat
for Leakage from Dirt,
Wear, Damage, Looseness.

Check and Adjust Idle
Speed

Float Bent, Loaded,
Sticking, or Misaligned

Pump Discharge Ball Not
Seating—Check for Dirt,
Defective Seat or
Discharge Spring

**LACK OF HIGH SPEED
PERFORMANCE OR POWER**

Check for Full Throttle
Opening at Carburetor.
Adjust Throttle Linkage
As Necessary

Air Valve Binding, Sticking
or Wrong Tension.

Power Piston Stuck or Binding—
Check for Distorted Spring

Air Valve
Not Unlocking.

Float Setting Incorrect

Float Sticking Misaligned
or Leaking

Secondary Metering Rods
Bent or Incorrect Part, or
Not Adjusted Properly

Main Metering Jets Plugged,
Dirty, or Incorrect Part

Metering Rods Altered,
Bent, or Incorrect Part

Gaskets Not Sealing.
May Be Hard or Brittle.
Check for Loose Screws

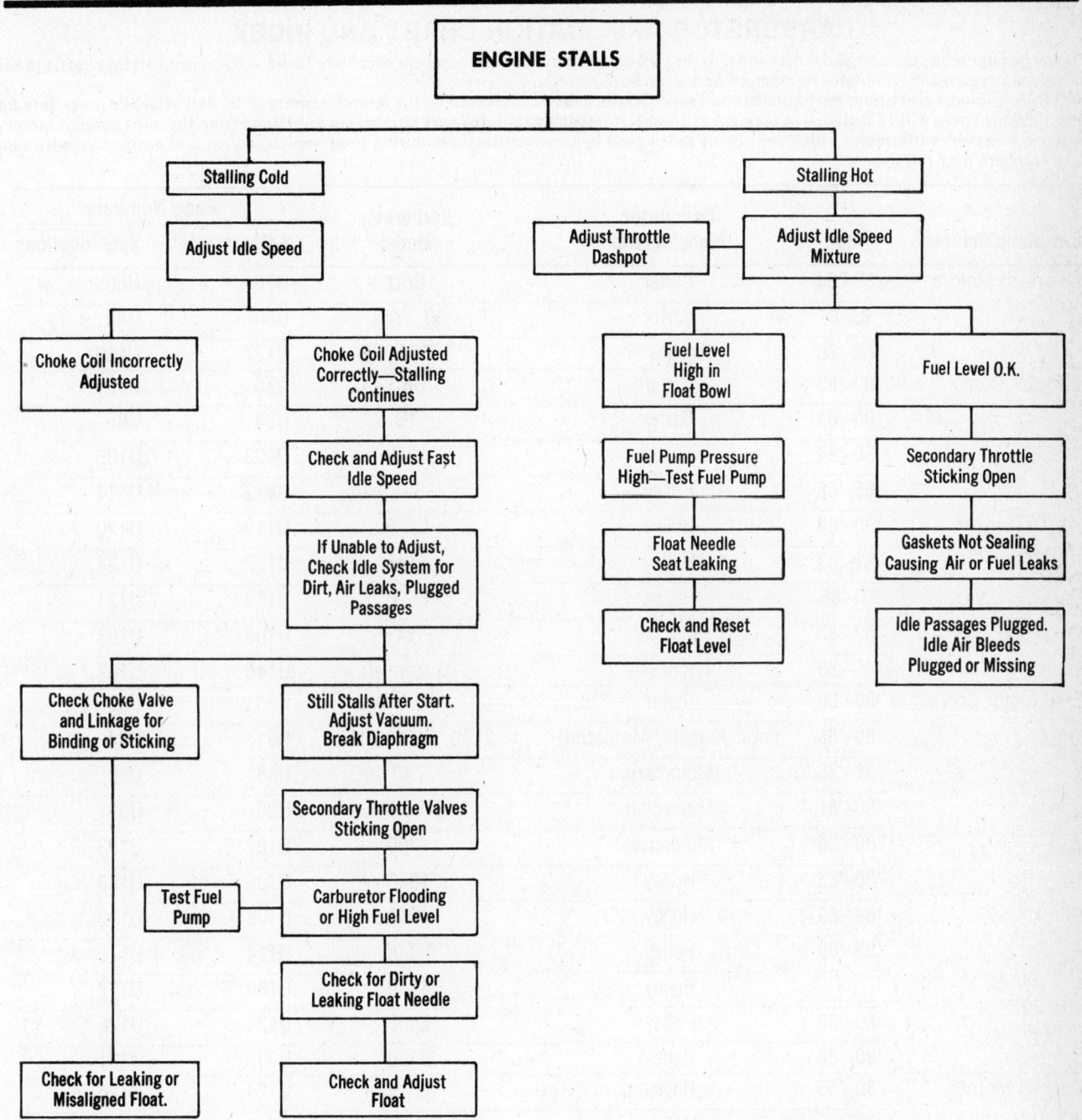

ENGINE STALLS

Stalling Cold
- Adjust Idle Speed
  - Choke Coil Incorrectly Adjusted
    - Check Choke Valve and Linkage for Binding or Sticking
      - Check for Leaking or Misaligned Float.
  - Choke Coil Adjusted Correctly—Stalling Continues
    - Check and Adjust Fast Idle Speed
    - If Unable to Adjust, Check Idle System for Dirt, Air Leaks, Plugged Passages
      - Still Stalls After Start. Adjust Vacuum. Break Diaphragm
      - Secondary Throttle Valves Sticking Open
      - Carburetor Flooding or High Fuel Level — Test Fuel Pump
      - Check for Dirty or Leaking Float Needle
      - Check and Adjust Float

Stalling Hot
- Adjust Idle Speed Mixture — Adjust Throttle Dashpot
  - Fuel Level High in Float Bowl
    - Fuel Pump Pressure High—Test Fuel Pump
    - Float Needle Seat Leaking
    - Check and Reset Float Level
  - Fuel Level O.K.
    - Secondary Throttle Sticking Open
    - Gaskets Not Sealing Causing Air or Fuel Leaks
    - Idle Passages Plugged. Idle Air Bleeds Plugged or Missing

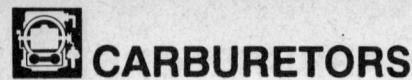
CARBURETORS

CARBURETOR APPLICATION CHART AND INDEX

The carburetor manufacturer and model and also the carburetor identification numbers which are listed in the specifications chart, appear either on a tag on the carburetor or stamped on the carburetor body.

NOTE: New model carburetor part numbers and specifications are not released by the manufacturers until well after the press date for this manual. These will be included in the next edition. New model carburetor part numbers are obtained from the most current factory sources, however, carburetors which are new or redesigned by the manufacturer during the production year and designated with new part numbers may not appear.

Car Manufacturer	Year	Carburetor Manufacturer	Carburetor Model	Page Numbers	
				Adjustments	Specifications
American Motors	'80–'84	Carter	BBD	U76	U80
	'83–86	Carter	YF, YFA	U80	U83
	'80–'86	Rochester	2SE, E2SE	U127	U136
Chrysler Corp.	'80–'83	Carter	BBD	U76	U79
	'80–'84	Carter	TQ	U84	U86
	'80–'84	Holley	1945	U103	U105
	'85–'86	Holley	2280, 6280	U112	U114
	'80–'86	Holley	5220	U119	U120
	'81–'83	Holley	6145	U122	U124
	'81–86	Holley	6520	U119	U121
	'82–'85	Mikuni	NA	U160	U163
	'85–'86	Rochester	Quadrajet	U146	U151
Ford Motor Co.	'80–'86	Carter	YF, YFA	U80	U83
	'80–'86	Ford, Autolite, Motorcraft	2100, 2150, 2150A	U91	U94
	'81–'86	Motorcarft	740	U86	U90
	'80–'81	Motorcraft	2700 VV	U96	U99
	'80–'86	Motorcraft	7200	U102	U103
	'80–'83	Holley	1946	U106	U108
	'84–'85	Holley	1949	U108	U112
	'83–'85	Holley	4180C	U114	U116
	'84	Holley	6149	U108	U112
	'80–'82	Holley	6500	U124	U126
	'80–'82	Holley	5200	U99	U101
General Motors	'80–'86	Holley	6510C	U124	U126
	'80–'86	Holley	5210C	U117	U119
	'80–'86	Rochester	2SE, E2SE	U127	U137
	'80–'86	Rochester	2MC, M2MC, M2ME, E2ME, E2MC	U141	U145
	'80–'86	Rochester	Quadrajet	U146	U151
	'85–'86	Spectrum	2bbl	U165	U165
	'85–'86	Sprint	2bbl	U167	U167
	'85–'86	Nova	2bbl	U169	U169

NA – Not Available

CARTER CARBURETORS
Model BBD

The BBD carburetor is a two barrel unit. It is equipped with a dashpot on some applications.

VACUUM STEP-UP PISTON ADJUSTMENT

1. Remove the dust cover.
2. Be sure not to disturb the adjusting screw on top of the piston. If it is disturbed, reset the gap at the top of the piston to 0.035–0.040 in.
3. Back off the curb idle adjustment until the throttle valves are completely closed. Count the number of turns so that the screw can later be returned to the original position. Then turn the idle screw in one full turn on AMC products only.
4. Fully depress the step-up piston while holding moderate pressure on the rod lifter tab and loosen and tighten the rod lifter lockscrew.
5. Release the piston and rod lifter; return the curb idle screw to its original position.
6. Replace the dust cover, unless the accelerator pump is to be adjusted.

ACCELERATOR PUMP STROKE ADJUSTMENT

1. Back off the idle adjusting screw.

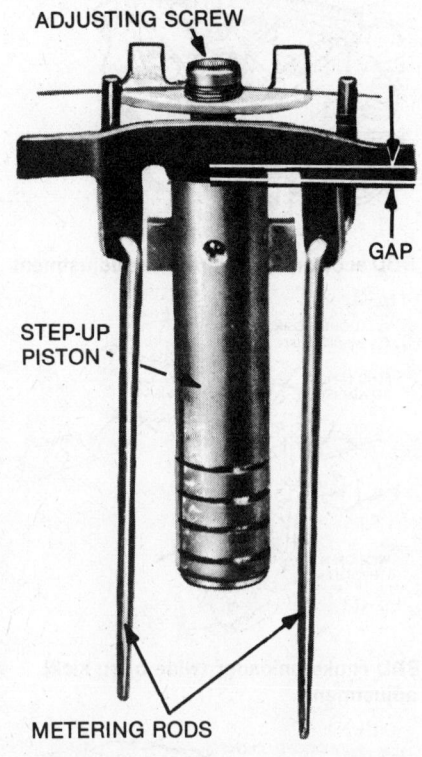

BBD vacuum step-up piston and metering rod assembly

Open the choke valve so that the fast idle cam allows the throttle valves to close. Be sure that the accelerator pump "S" link is in the outer hole of the pump arm if there are two holes.
2. Turn the idle adjusting screw in two complete turns after it contacts the stop.
3. Remove the dust cover. With the throttle valves closed tightly, measure the distance between the top of the air horn and the top of the pump plunger shaft. If the dimension is not as specified, loosen the pump arm adjusting lockscrew (near the plunger shaft) and rotate the sleeve to obtain the correct dimension.

FAST IDLE CAM POSITION ADJUSTMENT

1. With the fast idle speed adjusting screw contacting the second highest speed step on the fast idle cam, move the choke valve toward the closed position with light pressure on the choke shaft lever. On AMC, loosen the choke cover and turn ¼ turn rich.
2. Insert the specified drill (refer to Specifications), between the top of the choke valve and the wall of the air horn. An adjustment will be neces-

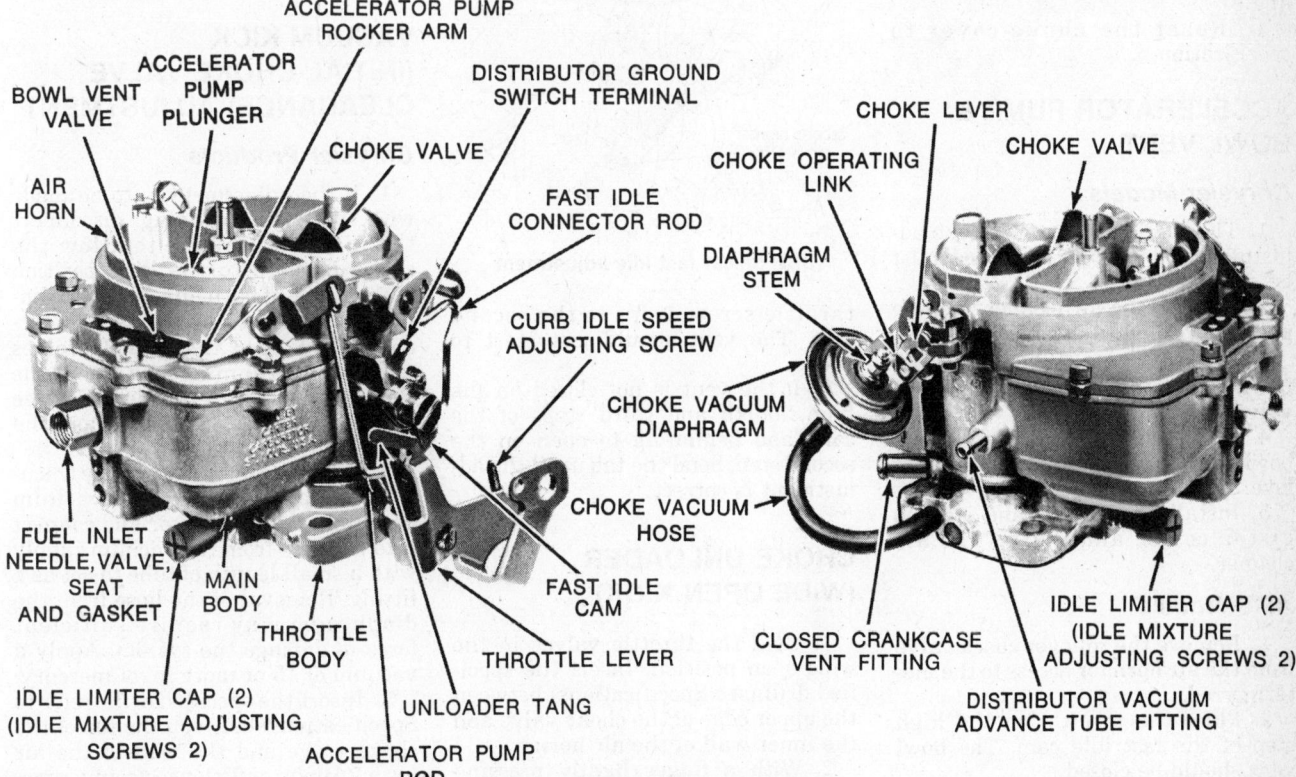

BBD carburetor assembly—typical

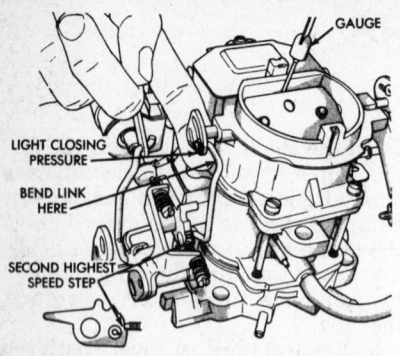

BBD fast idle cam position adjustment

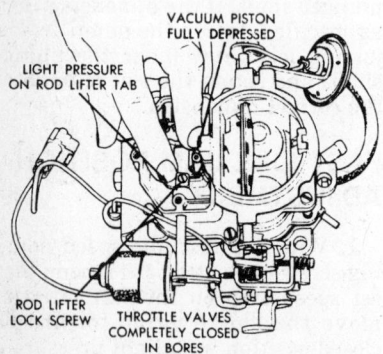

BBD vacuum step-up piston adjustment

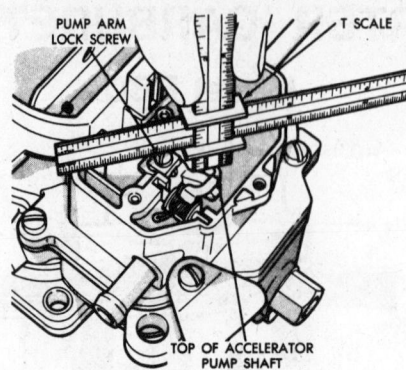

BBD accelerator pump stroke adjustment

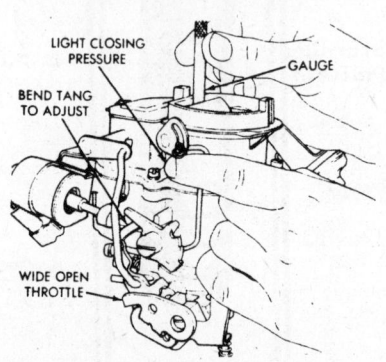

BBD choke unloader (wide open kick) adjustment

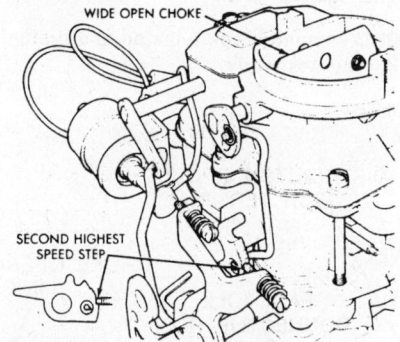

BBD on-car fast idle adjustment

sary if a slight drag is not obtained as the drill is being removed.

3. If an adjustment is required, bend the fast idle connector rod at the angle.

4. Reset the choke cover to specification.

ACCELERATOR PUMP & BOWL VENT

Chrysler Models

1. The accelerator pump stroke adjustment and the curb idle speed must be adjusted first.

2. Remove the air cleaner, step-up piston cover, and the gasket.

3. Insert the specified gauge (0.080 in.) between the top of the bowl vent valve and the seat.

4. If adjustment is needed, bend the bowl vent lever tab. Support the vent lever before bending the tab.

5. Install the gasket and step-up piston cover, and install the air cleaner.

AMC Models

1. Remove the rollover check valve from the air horn for access to the metering rod.

2. Place the throttle on the high step of the fast idle cam. The bowl vent should be closed.

3. Move the fast idle cam until the throttle screw drops to the second step. The vent should just start to open.

4. If the vent is not closed on the high, fourth and third steps of the cam, and beginning to open on the second step, bend the tab until the adjustment is correct.

CHOKE UNLOADER (WIDE OPEN KICK)

1. Hold the throttle valves in the wide open position. Insert the specified drill (see Specifications) between the upper edge of the choke valve and the inner wall of the air horn.

2. With a finger lightly pressing against the control lever, a slight drag should be felt as the drill is being withdrawn. If an adjustment is necessary, bend the unloader tang on the throttle lever until the correct opening has been obtained.

FAST IDLE SPEED (ON VEHICLE)

1. On Chrysler products, disconnect and plug the connections for the heated air control. EGR, and OSAC valve or distributor. On 1980–81 Chrysler products with ESA (Electronic Spark Advance), ground the idle switch. Do not disconnect the vacuum hose to the vacuum transducer. Disconnect the EGR and TCS solenoid on AMC cars through 1980. On 1981–84 AMC disconnect the EGR. With the engine off and the transmission in Park or Neutral position, open the throttle slightly.

2. Close the choke valve until the fast idle screw can be positioned on the second highest speed step of the fast idle cam.

3. Start the engine and let the idle stabilize. Turn the fast idle speed screw in or out to obtain the specified speed.

4. Stopping the engine between adjustments is not necessary. However, reposition the fast idle speed screw on the cam after each speed adjustment to provide the correct throttle closing torque.

VACUUM KICK (INITIAL CHOKE VALVE CLEARANCE) ADJUSTMENT

Chrysler Products

1. If the adjustment is to be made with the engine running, disconnect the fast idle linkage to allow the choke to close to the kick position with engine at curb idle. If an auxiliary vacuum source is to be used, as recommended, open the throttle valves (engine not running) and move the choke to the closed position. Release the throttle first, then release the choke.

2. When using an auxiliary vacuum source, disconnect the vacuum hose from the carburetor and connect it to the hose from the vacuum supply with a small length of tube to act as a fitting. Removal of the hose from the diaphragm may require sufficient force to damage the system. Apply a vacuum of 15 or more in. of mercury.

3. Insert the specified drill (refer to Specifications) between the top of the choke valve and the wall of the air horn. Apply sufficient closing pressure on the lever to which the choke

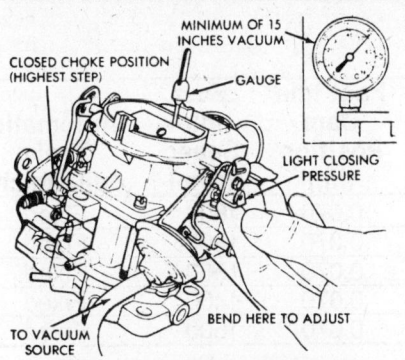

BBD vacuum kick adjustment

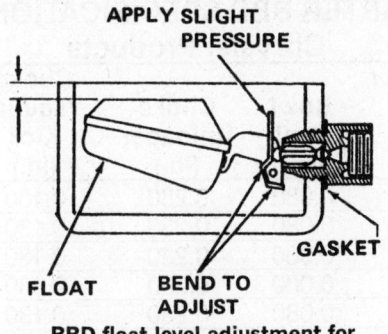

BBD float level adjustment for AMC products

rod attaches to provide a minimum choke valve opening without distortion of the diaphragm link. Note that the cylindrical stem of the diaphragm will extend as the internal spring is compressed. This spring must be fully compressed for proper measurement of the vacuum kick adjustment.

4. An adjustment will be necessary if a slight drag is not obtained as the drill is being removed. Shorten or lengthen the diaphragm link to obtain the correct choke opening. Length changes should be made carefully by bending (opening or closing) the U-bend provided in the diaphragm link.

—— **CAUTION** ——

Do not apply twisting or bending force to the diaphragm.

5. Reinstall the vacuum hose on the correct carburetor fitting. Return the fast idle linkage to its original condition if it was disturbed, as suggested in Step 1.

6. Make the following check: with no vacuum applied to the diaphragm, the choke valve should move freely between the open and closed positions. If its movement is not free, examine the linkage for misalignment or interference caused by the bending operation. Repeat the adjustment if necessary to provide proper link operations.

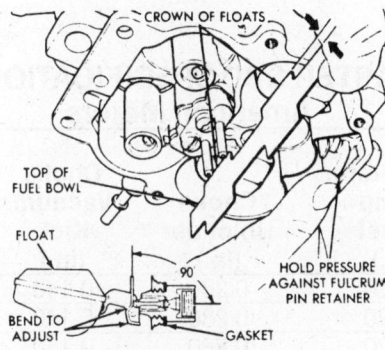

BBD float level adjustment

AMC Models

This adjustment is called Initial Choke Valve Clearance Adjustment on AMC products.

1. Loosen the choke cover, turn ¼ turn rich, and tighten one cover screw.

2. Apply a vacuum of at least 19 in. Hg to pull the diaphragm in against the stop.

3. Open the throttle valve slightly to place the fast idle screw on the high step of the cam.

4. Measure the clearance between the choke plate upper edge and the air horn wall.

5. Adjust the clearance between the choke plate upper edge and the air horn wall.

5. Adjust the clearance by bending the diaphragm connector link at the angle. Reset the choke or replace the cover.

FLOAT LEVEL

Chrysler Models

1. Invert the carburetor so that the weight of the floats is the only force on the needle and seat.

2. Use a T-scale to check the float level. Measure from the surface of the fuel bowl to the crown of each float at center.

3. To adjust, hold the floats on the bottom of the bowl and bend the float lip to give the specified dimension.

AMC Models

1. Remove the air horn.

2. Hold the float lip gently against the needle to raise the float.

3. Place a straightedge across the float bowl to measure the float level at the top of the float.

4. To adjust, bend the float lip, being careful not to exert pressure on the synthetic needle tip.

DASHPOT ADJUSTMENT

Chrysler Models

The dashpot is used on manual transmission models only.

1. Make sure that the curb idle speed is correctly adjusted.

2. Start the engine. Position the throttle lever so that the actuating tab is just contacting the dashpot plunger stem. Let the engine speed stabilize for 30 seconds.

3. The speed should be 2500 rpm.

4. Adjust the setting by loosening the locknut and moving the dashpot.

CARTER BBD SPECIFICATIONS
Chrysler Products

Year	Model ②	Float Level (in.)	Accelerator Pump Travel (in.)	Bowl Vent (in.)	Choke Unloader (in.)	Choke Vacuum Kick (in.)	Fast Idle Cam Position (in.)	Fast Idle Speed (rpm)	Automatic Choke Adjustment
'80	8233S	¼	0.500 ①	0.080	0.280	0.130	0.070	1500	Fixed
	8235S	¼	0.500 ①	0.080	0.280	0.130	0.070	1700	Fixed
	8237S	¼	0.500 ①	0.080	0.280	0.110	0.070	1500	Fixed
	8239S	¼	0.500 ①	0.080	0.280	0.110	0.070	1500	Fixed
	8286S	¼	0.500 ①	0.080	0.280	0.100	0.070	1400	Fixed
'81–'82	8290S	¼	0.500 ①	0.080	0.280	0.100	0.070	1600	Fixed
	8291S	¼	0.500 ①	0.080	0.280	0.130	0.070	1400	Fixed
	8292S	¼	0.500 ①	0.080	0.280	0.130	0.070	1600 ③	Fixed

CARTER BBD SPECIFICATIONS
Chrysler Products

Year	Model ②	Float Level (in.)	Accelerator Pump Travel (in.)	Bowl Vent (in.)	Choke Unloader (in.)	Choke Vacuum Kick (in.)	Fast Idle Cam Position (in.)	Fast Idle Speed (rpm)	Automatic Choke Adjustment
'83	8290S	1/4	0.470 ①	0.080	0.280	0.100	0.070	1600	Fixed
	8291S	1/4	0.470 ①	0.080	0.280	0.130	0.070	1400	Fixed
	8369S	1/4	0.500 ①	0.080	0.280	0.130	0.070	1500	Fixed
'84	8385S	1/4	0.470 ①	0.080	0.280	0.130	0.070	1400	Fixed
	8369S	1/4	0.500	0.080	0.280	0.130	0.070	1500	Fixed

① At idle
② Models numbers located on tag or casting
③ 1982: 1500 rpm

CARTER BBD SPECIFICATIONS
American Motors

Year	Model ①	Float Level (in.)	Accelerator Pump Travel (in.)	Choke Unloader (in.)	Choke Vacuum Kick (in.)	Fast Idle Cam Position (in.)	Fast Idle Speed (rpm)	Automatic Choke Adjustment
'80	8216	1/4	0.520	0.280	0.140	0.090	1850	2 Rich
	8246	1/4	0.520	0.280	0.140	0.095	1850	2 Rich
	8247	1/4	0.520	0.280	0.150	0.095	1700	1 Rich
	8248	1/4	0.520	0.280	0.150	0.095	1700	1 Rich
	8253	1/4	0.470	0.280	0.128	0.095	1850	2 Rich
	8256	1/4	0.470	0.280	0.128	0.093	1850	2 Rich
	8278	1/4	0.542	0.280	0.140	0.093	1850	Index
'81	8310	1/4	0.525	0.280	0.140	0.095	1850	Index
	8302	1/4	0.500	0.280	0.128	0.095	1850	1 Rich
	8303	1/4	0.500	0.280	0.128	0.090	1700	1 Rich
	8306	1/4	0.500	0.280	0.128	0.090	1700	1 Rich
	8307	1/4	0.500	0.280	0.128	0.095	1850	1 Rich
	8308	1/4	0.500	0.280	0.128	0.095	1850	2 Rich
	8309	1/4	0.520	0.280	0.128	0.093	1700	2 Rich
'82	8338	1/4	0.520	0.280	0.140	0.095	1850	1 Rich
	8339	1/4	0.520	0.280	0.140	0.095	1850	1 Rich
'83	8360	1/4	0.520	0.280	0.140	0.095	1850	Fixed
	8364	1/4	0.520	0.280	0.140	0.095	1700	Fixed
	8367	1/4	0.520	0.280	0.140	0.095	1700	Fixed
	8362	1/4	0.520	0.280	0.140	0.095	1850	Fixed
'84-'85	8383	1/4	0.520	0.280	0.140	0.095	1850	1/2-1 1/2 Rich
	8384	1/4	0.520	0.280	0.140	0.095	1700	1/2-1 1/2 Rich

① Model numbers located on the tag or casting

Model YF, YFA

The YF carburetor is a single barrel downdraft carburetor with a diaphragm type accelerator pump and diaphragm operated metering rods.

FLOAT ADJUSTMENT

1. Invert the air horn assembly and check the clearance from the top of the float to the surface of the air horn with a T-scale. The air horn should be held at eye level when gauging and the float arm should be resting on the needle pin.

2. Do not exert pressure on the needle valve when measuring or adjusting the float. Bend the float arm as necessary to adjust the float level.

——— CAUTION ———

Do not bend the tab at the end of the float arm as it prevents the float from striking the bottom of the fuel bowl when empty and keep the needle in place.

METERING ROD ADJUSTMENT

1. Remove the air horn. Back out

the idle speed adjusting screw until the throttle plate is seated fully in its bore.

2. Press down on the upper end of the diaphragm shaft until the diaphragm bottoms in the vacuum chamber.

3. The metering rod should contact the bottom of the metering rod well. The lifter link at the outer end nearest the springs and at the supporting link should be bottomed.

4. On models not equipped with an adjusting screw, adjust by bending the lip of the metering rod is attached.

5. On models with an adjusting screw, turn the screw until the metering rod just bottoms in the body casting. For final adjustment, turn the screw one additional turn clockwise.

FAST IDLE CAM ADJUSTMENT

1. Put the fast idle screw on the second highest step of the fast idle cam against the shoulder of the high step.

2. Adjust by bending the choke plate connecting rod to obtain the specified clearance between the lower edge of the choke plate and the air horn wall.

CHOKE UNLOADER ADJUSTMENT

With the throttle valve held wide open and the choke valve held in the closed position, bend the unloader tang on the throttle lever to obtain the specified clearance between the lower edge of the choke valve and the air horn wall.

AUTOMATIC CHOKE ADJUSTMENT

Loosen the choke cover retaining screws, then turn the choke cover so that the index mark on the cover lines up with the specified mark on the choke housing.

CHOKE PLATE PULLDOWN ADJUSTMENT

1983–84 Piston Type Choke

NOTE: This adjustment requires that the thermostatic spring housing and gasket (choke cap) are removed. Refer to the "Choke Cap" removal procedure below.

1. Remove the air cleaner assembly, then the choke cap.

2. Bend a 0.026 in. diameter wire gauge at a 90 degree angle approxi-

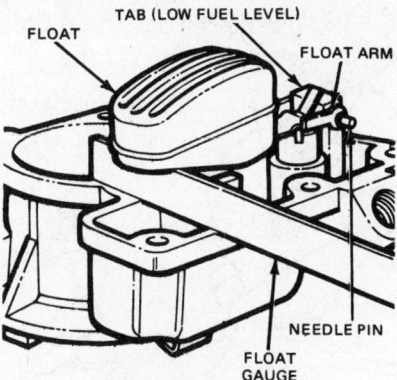

YFA float level adjustment

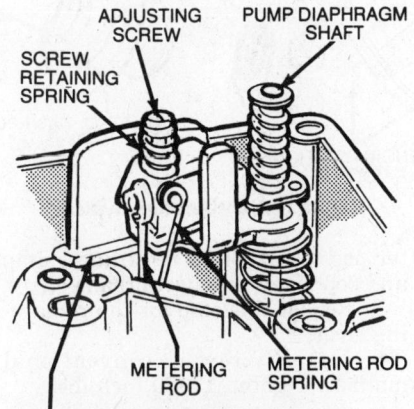

ROD ACTION CAUSED BY SCREW ACTING AS PIVOT POINT FOR LEVER
YFA metering rod adjustment

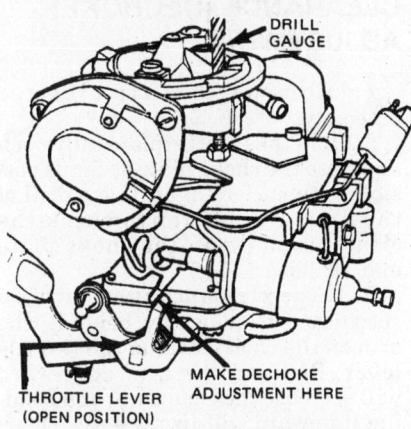

YFA choke unloader adjustment

mately ⅛ in. from one end. Insert the bent end of the gauge between the choke piston slot and the right hand slot in the choke housing. Rotate the choke piston lever counterclockwise until the gauge is shut in the piston slot.

3. Apply light pressure on the choke piston lever to hold the gauge in place, then measure the clearance between the lower edge of the choke plate and the carburetor bore using a drill with the diameter equal to the specified pulldown clearance.

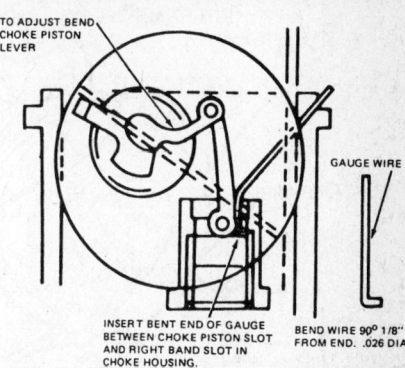

Choke plate pulldown—piston type choke—YFA

4. Bend the choke piston lever to obtain the proper clearance.

5. Install the choke cap.

Diaphragm Type Choke

1. Activate the pulldown motor by applying an external vacuum source.

2. Close the choke plate as far as possible without forcing it.

3. Using a drill of the specified size, measure the clearance between the lower edge of the choke plate and the air horn wall.

4. If adjustment is necessary bend the choke diaphragm link as required.

CHOKE CAP REMOVAL

1983–84

NOTE: The automatic choke has two rivets and a screw, retaining the choke cap in place. There is a locking and indexing plate to prevent misadjustment.

1. Remove the air cleaner assembly from the carburetor.

2. Check choke cap retaining ring rivets to determine if mandrel is well below the rivet head. If mandrel appears to be at or within the rivet head thickness, drive it down or out with a ⅙ inch diameter punch.

3. Use a ⅛ inch diameter or No. 32 drill (.128 inch diameter) for drilling the rivet heads. Drill into the rivet head until the rivet head comes loose from the rivet body.

4. After the rivet head is removed, drive the remaining portion of the rivet out of the hole with a ⅛ inch diameter punch.

NOTE: This procedure must be followed to retain the hole size.

5. Repeat Steps 1–4 for the remaining rivet.

6. Remove the screw in the conventional manner.

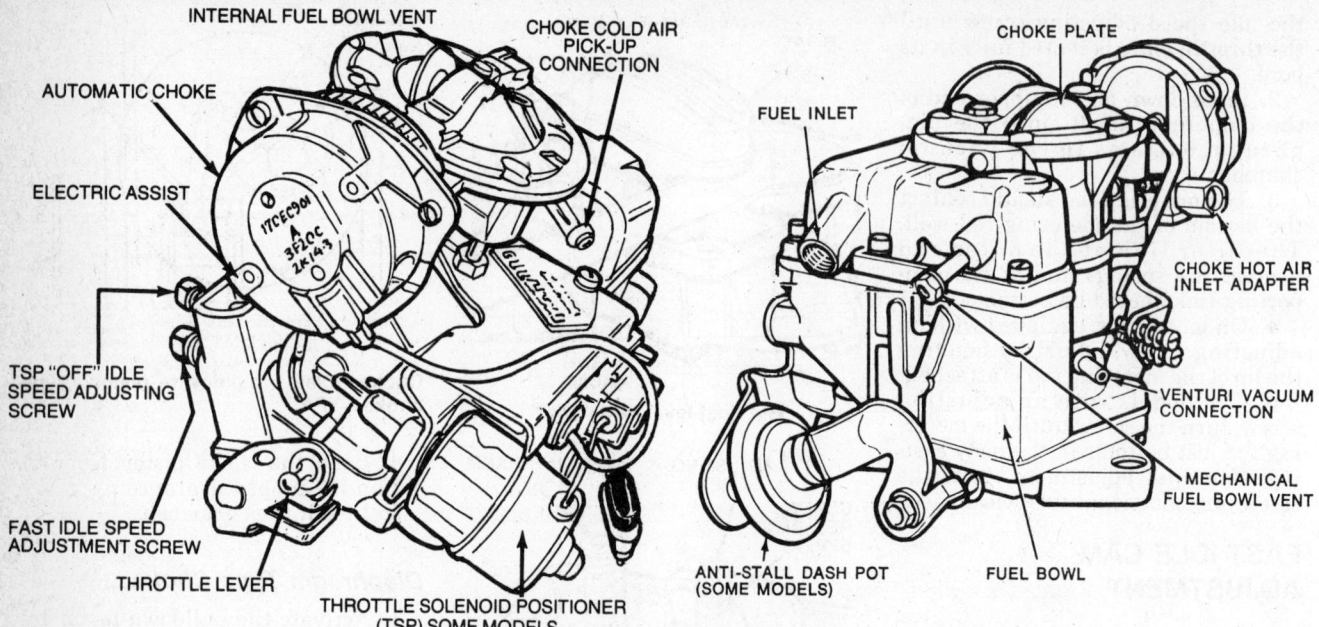

Carter YFA carburetor—typical

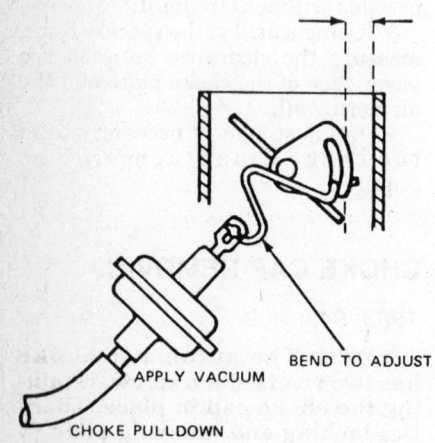

Choke plate pulldown—diaphragm type choke—YFA

CHOKE CAP INSTALLATION

1. Install choke cap gasket.
2. Install the locking and indexing plate.
3. Install the notched gasket.
4. Install choke cap, making certain that bimetal loop is positioned around choke lever tang.
5. While holding cap in place, actuate choke plate to make certain bimetal loop is properly engaged with lever tang. Set retaining clamp over choke cap and orient clamp to match holes in casting (holes are not equally spaced). Make sure retaining clamp is not upside down.
6. Place rivet in rivet gun and trigger lightly to retain rivet (¹⁄₈ inch diameter x ½ inch long x ¼ inch diameter head).
7. Press rivet fully into casting af-

ter passing through retaining clamp and pop rivet (mandrel breaks off).
8. Repeat this step for the remaining rivet.
9. Install screw in conventional manner. Tighten 17–20 inch lbs.

CHOKER PLATE CLEARANCE (DECHOKE) ADJUSTMENT

1. Remove the air cleaner assembly.
2. Hold the throttle plate fully open and close the choke plate as far as possible without forcing it. Use a drill of the proper diameter to check the clearance between the choke plate and air horn.
3. If the clearance is not within specification, adjust by bending the arm on the choke lever of the throttle lever. Bending the arm downward will decrease the clearance, and bending it upward will increase the clearance. Always recheck the clearance after making any adjustment.

MECHANICAL FUEL BOWL VENT ADJUSTMENT

1. Start the engine and wait until it has reached normal operating temperature before proceeding.
2. Check engine idle rpm and set to specifications.
3. Check DC motor operation by opening throttle off idle. The DC motor should extend. Release the throttle, and the DC motor should retract

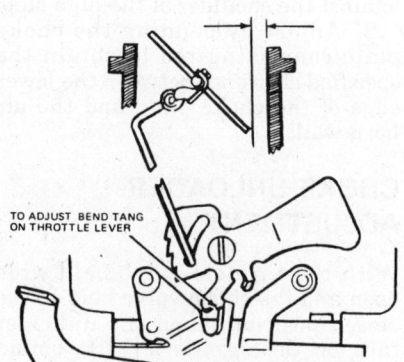

Dechoke adjustment—YFA

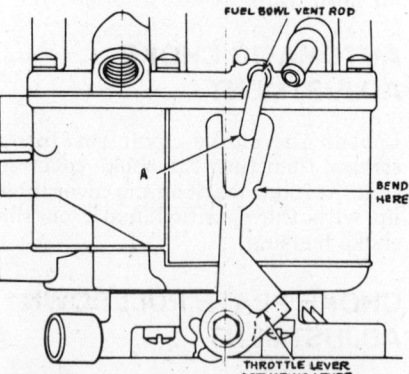

Mechanical fuel bowl vent adjustment—YFA

when in contact with the throttle lever.
4. Disconnect the idle speed motor in the idle position.
5. Turn engine OFF.
6. Open the throttle lever so that the throttle lever actuating lever does not touch the fuel bowl vent rod.
7. Close the throttle lever to the

idle set position and measure the travel of the fuel bowl vent rod at point A. The distance measured represents the travel of the vent rod from where there is no contact with the ac-

tuating lever to where the actuating lever moves the vent rod to the idle set position. The travel of the vent rod at point A should be 0.100–0.150 in. (2.54–3.81mm).

8. If adjustment is required, bend the throttle actuating lever at notch shown.

9. Reconnect the idle speed control motor.

CARTER YF, YFA SPECIFICATIONS
American Motors

Year	Model ①	Float Level (in.)	Fast Idle Cam (in.)	Unloader (in.)	Choke
'83–'85	7700	0.600	0.175	0.370	Fixed
	7701	0.600	0.175	0.370	Fixed
	7702	0.600	0.175	0.370	Fixed
	7703	0.600	0.175	0.370	Fixed

① Model numbers located on the tag or casting

CARTER YF, YFA, YFA-FB SPECIFICATIONS
Ford Motor Co.

Year	Model ①	Float Level (in.)	Fast Idle Cam (in.)	Choke Plate Pulldown (in.)	Unloader (in.)	Dechoke (in.)	Choke
'80	DEDE-GA,HA EODE-JA,NA, LA,MA	25/32	0.140	—	0.250	—	2 Rich
'83	E3ZE-LA	0.650	0.140	0.260	—	0.220	—
	E3ZE-MA	0.650	0.140	0.260	—	0.220	—
	E3ZE-TB	0.650	0.140	0.240	—	0.220	—
	E3ZE-UA	0.650	0.140	0.240	—	0.220	—
	E3ZE-VA	0.650	0.140	0.260	—	0.220	—
	E3ZE-YA	0.650	0.140	0.260	—	0.220	—
	E3ZE-NB	0.650	0.160	0.260	—	0.220	—
	E3ZE-PB	0.650	0.160	0.260	—	0.220	—
	E3ZE-ASA	0.650	0.160	0.260	—	0.220	—
	E3ZE-APA	0.650	0.140	0.240	—	0.220	—
	E3ZE-ARA	0.650	0.140	0.240	—	0.220	—
	E3ZE-ADA	0.650	0.140	0.260	—	0.220	—
	E3ZE-AEA	0.650	0.140	0.260	—	0.220	—
	E3ZE-ACA	0.650	0.140	0.260	—	0.220	—
	E3ZE-ATA	0.650	0.160	0.260	—	0.220	—
	E3ZE-ABA	0.650	0.140	0.260	—	0.220	—
	E3ZE-UB	0.650	0.140	0.240	—	0.220	—
	E3ZE-TC	0.650	0.140	0.240	—	0.220	—
'84	E4ZE-HC,DB	0.650	0.140	0.260	—	0.270	—
	E4ZE-MA,NA	0.650	0.140	0.240	—	0.270	—
	E4ZE-PA,RA	0.650	0.140	0.260	—	0.270	—
'85	E5ZE-CA	0.650	0.140	0.260	—	0.270	—
	E5ZE-AA	0.650	0.140	0.260	—	0.270	—
'86	E5ZE-AA	0.650	0.140	0.260	—	0.270	—
	E5ZE-AB	0.650	0.140	0.260	—	0.270	—
	E5ZE-CA	0.650	0.140	0.260	—	0.270	—
	E5ZE-CB	0.650	0.140	0.260	—	0.270	—
	E6ZE-EA	0.650	0.140	0.260	—	0.270	—
	E6ZE-DA	0.650	0.140	0.260	—	0.270	—

① Model number located on the tag or casting

Model TQ

The TQ (Thermo-Quad) has a fuel bowl made of phenolic resin. This acts as a heat insulator. Fuel is kept 20 degrees cooler than in metal carburetors. It also has a suspended design metering system which aids in cooling. All the calibration points are in the upper aluminum casting or air horn and are in effect suspended in the cavities in the main body.

FLOAT ADJUSTMENT

1. With the bowl cover inverted, the gasket installed, and the floats resting on the seated needle, the dimension of each float from the bottom side of the float to the cover gasket should be as shown in the specifications chart.
2. To adjust, bend the float lever. Do not allow the float lever lip to be pressed against the needle during adjustment.

SECONDARY THROTTLE LINKAGE

1. Block the choke valve in the wide open position and invert the carburetor.
2. Slowly open the primary throttle valves until the secondary valves start to open. Measure between the lower edge of the primary valve and its bore. Open the throttle to the wide open position. The primary and secondary levers should contact the stops at the same time.
3. If it is necessary to adjust, bend the secondary throttle operating rod at the lower angle until the correct dimension is obtained.

SECONDARY AIR VALVE OPENING

1. With the air valve in the closed position, the opening along the air valve at its long side must be at its maximum and parallel with the air horn gasket surface.
2. With the air valve wide open, the opening of the air valve at the short side and the air horn must match the dimensions in the Specifications Charts. The corner of the air valve is notched for adjustment. Bend the corner with a pair of pliers to give proper opening.

ACCELERATOR PUMP STROKE ADJUSTMENT

1. Make sure the throttle connector

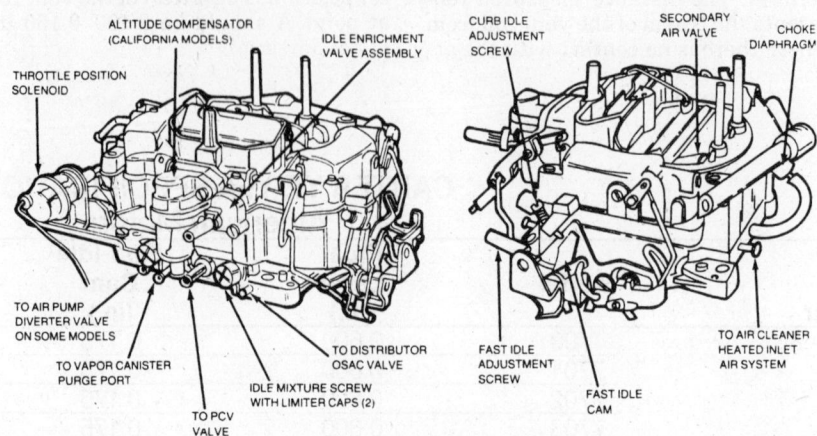

TQ carburetor assembly—typical

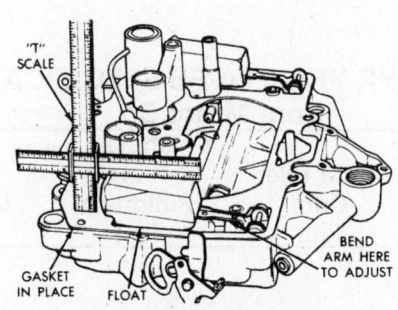

TQ float adjustment

TQ secondary throttle adjustment

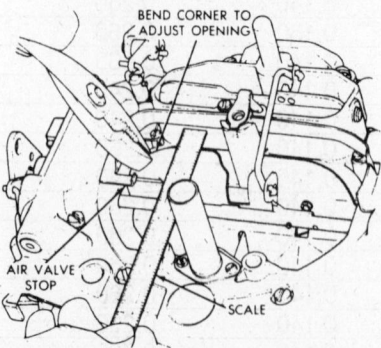

TQ secondary air valve adjustment

First Stage

1. Make sure the throttle connector rod is in the correct pump arm slot.

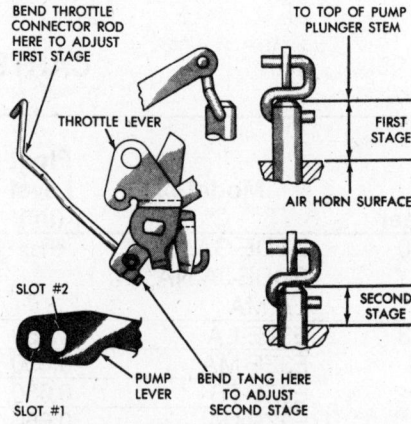

TQ accelerator pump stroke adjustment

rod is in the correct hole of the pump arm.
2. Measure the height of the accelerator pump plunger at curb idle. The ignition switch must be on if there is an idle stop solenoid.
3. Adjust plunger height by bending the throttle connector rod.

NOTE: Carburetors with staged pump systems require a second height measurement at the throttle position related to a secondary throttle lockout.

2. Use a scale to measure the height of the accelerator pump plunger stem at curb idle.
3. Adjust the pump plunger height by bending the throttle connector rod.

Second Stage

1. Open the choke then open the throttle until the secondary lockout is just applied. The plunger downward travel stops at that point.
2. Use a scale to measure the accelerator pump plunger height.
3. Adjust by bending the tang.

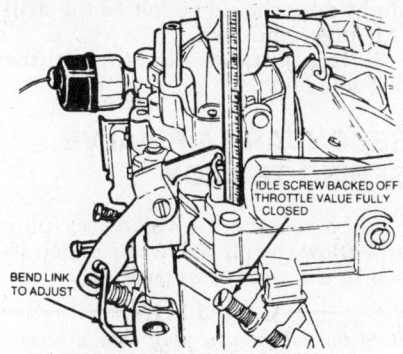

TQ accelerator pump adjustment

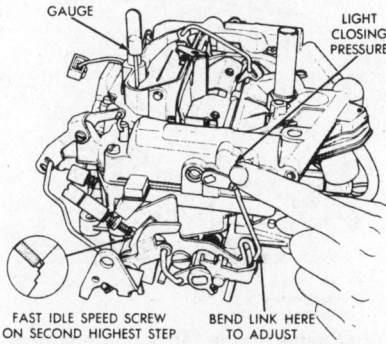

TQ vacuum kick adjustment

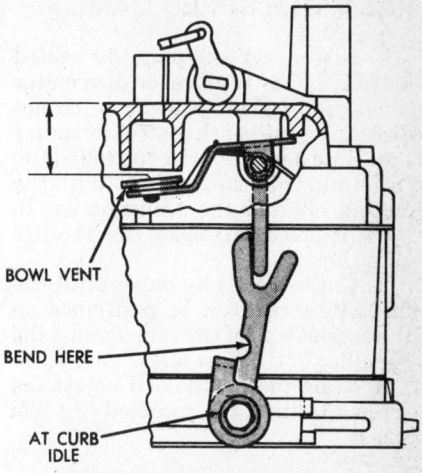

TQ bowl vent adjustment

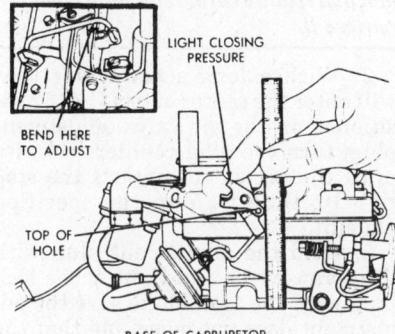

TQ choke control lever

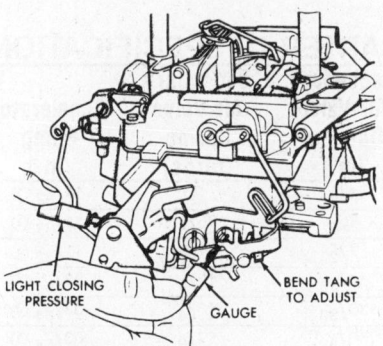

TQ fast idle cam linkage adjustment

CHOKE CONTROL LEVER

1. Disconnect the diaphragm rod.
2. Close the choke by pushing on the choke lever with the throttle partly open.
3. Measure the vertical distance from the top of the rod hole in the control lever down to the carburetor base. The dimension should be as shown in the Specifications Chart.
4. To adjust, bend the link which connects the two choke shafts. If an adjustment is needed, the vacuum kick, fast idle cam, and choke unloader must be readjusted.

CHOKE VACUUM KICK ADJUSTMENT

NOTE: The test can be made on or off the vehicle.

1. If the adjustment is to be made with the engine running, back off the fast idle speed screw until the choke can be closed to the kick position with the engine at curb idle. Note the number of screw turns required so that the fast idle can be returned to the original adjustment.
2. If an auxiliary vacuum source is to be used, open the throttle valve (engine not running) and move the choke to the closed position. Release the throttle first, then release the choke. When using an auxiliary vacuum

TQ secondary throttle lockout adjustment

source, disconnect the vacuum hose from the carburetor and connect it to the hose from the vacuum supply with a small length of tube to act as a fitting. Removal of the hose from the diaphragm may require sufficient force to bend the bracket. Apply a vacuum of 15 or more inches of mercury.
3. Insert the specified drill between the long side, lower edge, of the choke valve and the air horn wall.
4. Apply sufficient pressure on the choke control lever to provide a minimum choke valve opening. The spring connecting the control lever to the adjustment lever must be fully extended for proper adjustment.
5. Bend the tang to change contact with the end of the diaphragm rod. Do not adjust the diaphragm rod. A slight

drag should be felt as the drill is being removed.

FAST IDLE CAM LINKAGE

1. With the fast screw on the second fastest step of the cam against the shoulder of the first step, there should be 0.100 in. between the air horn wall and edge of the choke valve.
2. To adjust, bend the fast idle connector rod at the lower angle.

SECONDARY THROTTLE LOCKOUT

1. Move the choke control lever to the open choke position.
2. Measure the clearance between the lockout lever and the stop.
3. Bend the tang on the fast idle control lever to provide the proper clearance. Clearance should be 0.060–0.090 in.

BOWL VENT VALVE ADJUSTMENT

1. Remove the air cleaner. Disconnect the hose to the solenoid bowl vent diaphragm.
2. Connect an auxiliary vacuum source. With 15 in. Hg applied, the valve should move down. This can be observed down through the air horn vent tube.
3. Turn the ignition switch on and disconnect the auxiliary vacuum source. The valve should remain down. With the ignition off, the valve should move back up.
4. If the valve does not move down when vacuum is applied, the diaphragm is leaking and must be replaced. If the valve does not stay down with the ignition on and the vacuum removed, the solenoid or the wiring is defective.

FAST IDLE SPEED CAM

1. Disconnect and plug the heated air, EGR, OSAC valve, or distributor connections. With lean burn, do not disconnect the spark control computer hose. Use a jumper wire to ground the carburetor idle stop switch. With the engine off and the transmission in Park or Neutral, open the throttle slightly.

2. Close the choke valve until the fast idle screw can be positioned on the second step of the cam against the shoulder of the first step.

3. Start the engine and adjust the screw to obtain the specified fast idle speed.

CHOKE UNLOADER ADJUSTMENT

1. Hold the throttle valves in the wide open position and insert the specified drill between the bottom of the choke valve and inner wall of the air horn.

2. With a finger pressing lightly against the choke control lever, a

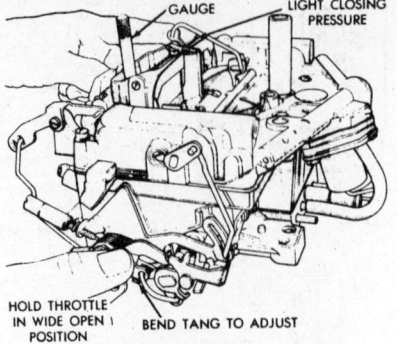

TQ choke unloader adjustment

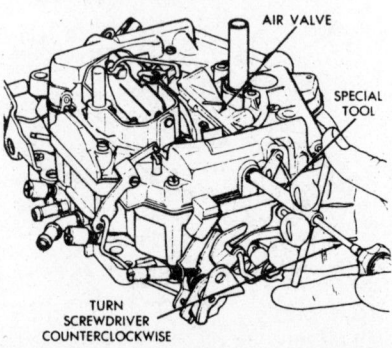

TQ air valve spring tension adjustment

slight drag should be felt as the drill is being withdrawn.

3. To adjust, bend the tang on the fast idle lever.

SECONDARY AIR VALVE SPRING TENSION

1. Loosen the air valve lock plug and allow the air valve to position itself in the wide open position.

――――――― **CAUTION** ―――――――
Hold the adjustment plug with a screwdriver when loosening the lock plug. If you don't, the spring may snap out of position and require carburetor disassembly to retrieve it.

2. With a long screwdriver that will enter the center of tool C-4152 positioned on the air valve adjustment plug, turn the plug counterclockwise until the air valve contacts the stop lightly, then tighten the specified amount.

3. Hold the adjustment plug with the screwdriver and tighten the lock plug with the tool. Make sure the adjustment does not move and that the air valve moves freely.

CARTER TQ SPECIFICATIONS

Year	Model ①	Float Setting (in.)	Secondary Throttle Linkage (in.)	Secondary Air Valve Opening (in.)	Secondary Air Valve Spring (turns)	Accelerator Pump (in.)	Choke Control Lever (in.)	Choke Unloader (in.)	Vacuum Kick (in.)	Fast Idle Speed (rpm)
'80	9235S	$29/32$	②	$1/2$	3	$11/32$ ③	$3^3/8$	0.310	0.100	1600
	9243S	$29/32$	②	$1/2$	$2^5/8$	$11/32$ ④	$3^3/8$	0.310	0.100	1600
	9244S	$29/32$	②	$1/2$	$2^1/2$	$11/32$ ④	$3^3/8$	0.310	0.100	1600
'81	9372S	$29/32$	②	$13/32$	$1^3/4$	$33/64$ ④	$3^3/8$	0.312	0.130	1400
	9373S	$29/32$	②	$13/32$	$1^3/4$	$33/64$ ④	$3^3/8$	0.312	0.130	1400
	9364S	$29/32$	②	$13/32$	$1^7/8$	$33/64$ ③	$3^3/8$	0.312	0.100	1500
'82	9372S	$29/32$	②	$13/32$	$1^3/4$	$33/64$ ④	$3^3/8$	0.310	0.130	1400
'83	9374S	$29/32$	②	$13/32$	$1^3/4$	⑤	$3^3/8$	0.310	0.130	1400
	9385S	$29/32$	②	$13/32$	$1^3/4$	$33/64$ ④	$3^3/8$	0.310	0.130	1400
'84	93895	$29/32$	②	$13/32$	$1^3/4$	⑤	—	0.310	0.130	1400

NOTE: All choke settings are fixed.
① Model numbers located on the tag or on the casting
② Adjust link so primary and secondary stops both contact at same time
③ Slot #1
④ Slot #2
⑤ First stage—$33/64$
   Second stage—$25/64$

FORD, AUTOLITE, MOTORCRAFT CARBURETORS

Model 740

The model 740 has five basic systems: choke system, idle system, main metering system, acceleration system and power enrichment system. The choke system is used for cold starting and features a bi-metallic spring and an electric heater for fast cold starts and improved warm-up. The idle system is a separate and adjustable system for the correct air/fuel mixture for both idle and low speed performance.

The main metering system provides the correct air/fuel mixture for normal cruising speeds. A main metering system is provided for both primary

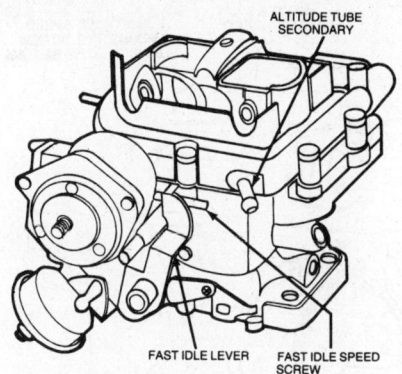

Model 740 carburetor—¾ front view

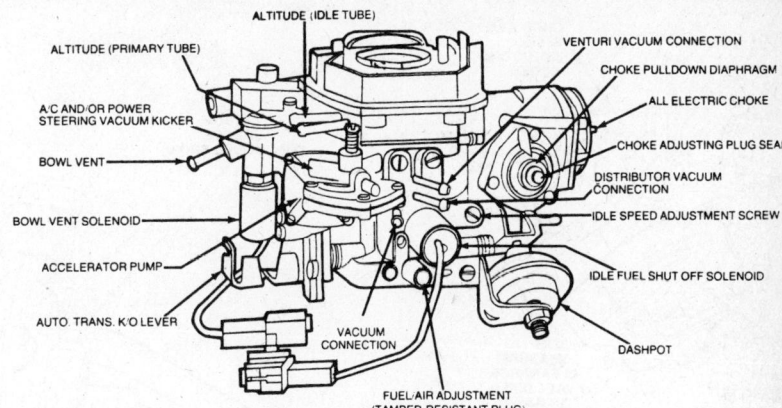

Model 740 carburetor—full rear

and secondary stage operation. The accelerating system is mechanically operated from the primary throttle linkage and provides fuel to the primary stage during acceleration. Fuel is provided by a diaphragm-type pump. The power enrichment system consists of a vacuum operated power valve and an airflow-regulated pull-over system in the secondary. This system is used along with the main metering system to provide satisfactory performance during moderate to heavy acceleration. Distributor and EGR vacuum ports are located in the primary venturi area of the carburetor.

FAST IDLE CAM

1. Set the fast idle screw on the kickdown step of the cam against the shoulder of the top step.
2. Manually close the primary choke plate, and measure the distance between the downstream side of the choke plate and the air horn wall.
3. Adjust the right fork of the choke bimetal shaft, which engages the fast idle cam, by bending the fork up and down to obtain the specified clearance.

FAST IDLE

1. Place the transmission in Neutral or Park.
2. Bring the engine to normal operating temperature.
3. Disconnect and plug the vacuum hose at the EGR and purge valves.
4. Identify the vacuum source to the air by-pass section of the air supply control valve. If a vacuum hose is connected to the carburetor, disconnect the hose and plug the hose at the air supply control valve.
5. Place the fast idle adjustment on the second step of the fast idle cam. Run the engine until the cooling fan comes on.
6. While the cooling fan is on,

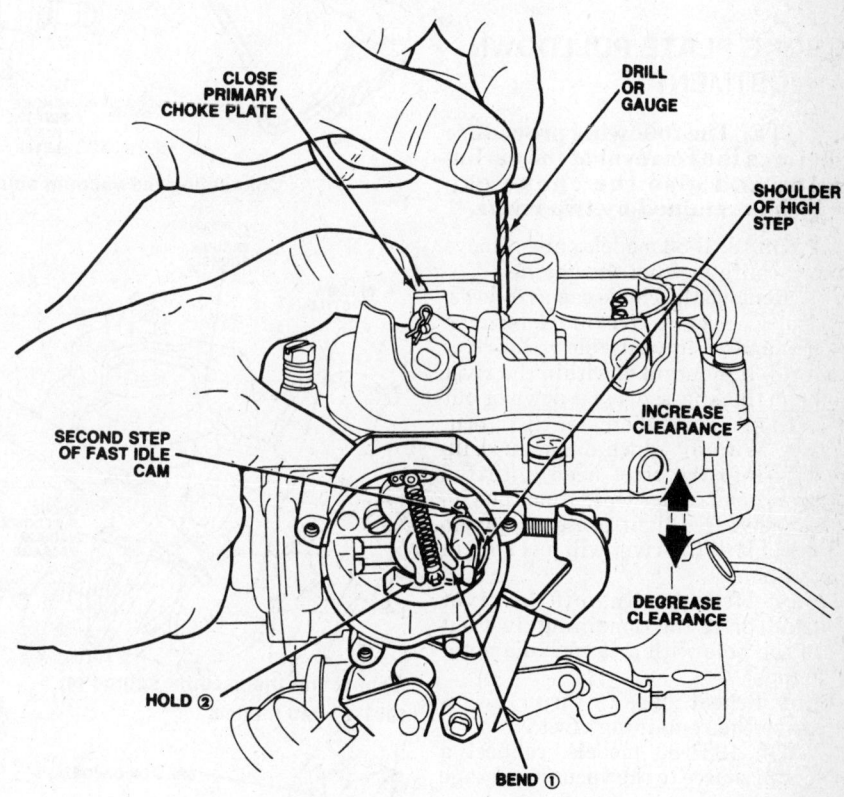

Fast idle cam adjustment—model 740

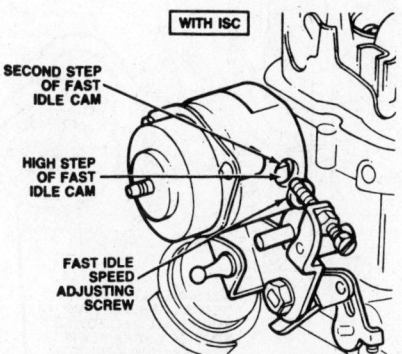

Fast idle speed adjusting screw and fast idle cam—model 740

check the fast idle rpm. If adjustment is necessary, loosen the locknut and adjust to specification on underhood decal.

7. Remove all plugs and reconnect hoses to their original position.

DASHPOT

With the throttle set at the curb idle position, fully depress the dashpot stem and measure the distance between the stem and the throttle lever. Adjust by loosening the locknut and turning the dashpot.

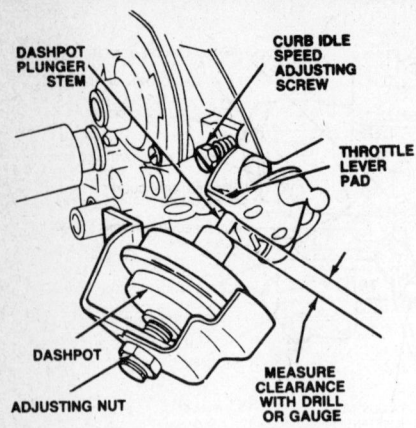

Dashpot assembly — model 740

CHOKE PLATE PULLDOWN ADJUSTMENT

NOTE: The following procedure requires the removal of the carburetor and also the choke cap which is retained by two rivets.

1. On 1981–83 models only, remove the carburetor from the engine.

2. Remove the choke cap as follows:

   a. Check the rivets to determine if mandrel is well below the rivet head. If mandrel is within the rivet head thickness, drive it down or out with a $\frac{1}{16}$ inch diameter tip punch.

   b. With a $\frac{1}{8}$ inch diameter drill, drill into the rivet head until the rivet head comes loose from the rivet body. Use light pressure on the drill bit or the rivet will just spin in the hole.

   c. After drilling off the rivet head, drive the remaining rivet out of the hole with a $\frac{1}{8}$ inch diameter punch.

   d. Repeat steps (a thru c) to remove the remaining rivet.

3. On 1981-83 models, connect a vacuum source to the vacuum passage adjacent to the primary throttle bore. On 1984 and later models connect a vacuum source to the vacuum tube on the choke pulldown cover.

4. Set the fast idle adjusting screw on the high step of the fast idle cam by temporarily opening the throttle lever and rotating the choke bimetal shaft lever counterclockwise until the choke plates are in the fully closed position.

5. While applying the external vacuum, lightly force the choke thermostat actuating lever counterclockwise.

6. Using the drill diameter specified in the carburetor specifications table at the end of this section, measure the clearance between the downstream side of the choke plate and the air horn wall.

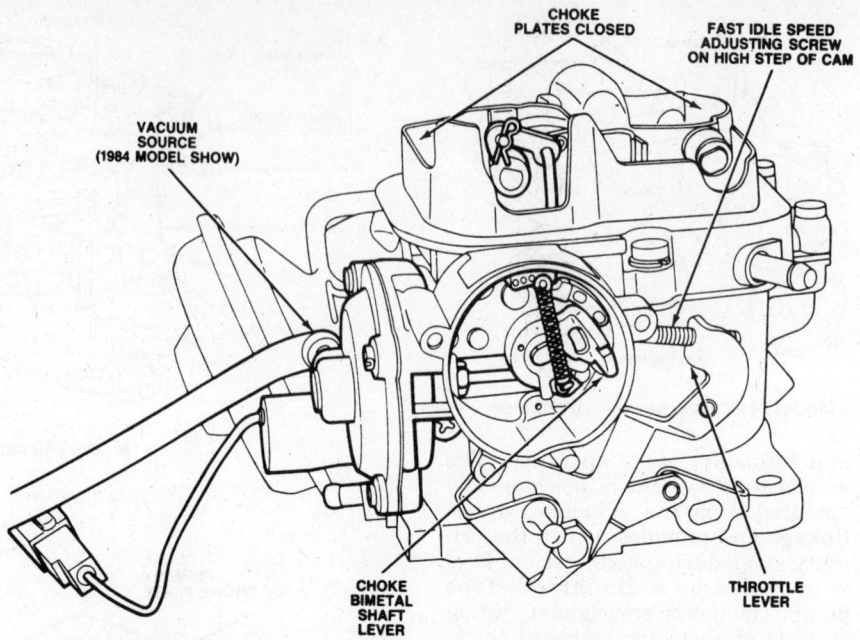

Connecting the vacuum source on 1984 and later 740 models

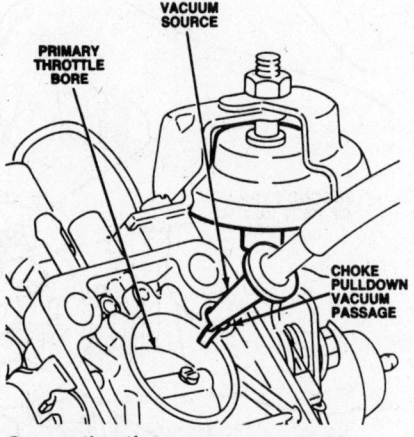

Connecting the vacuum source on 1981–83 740 models

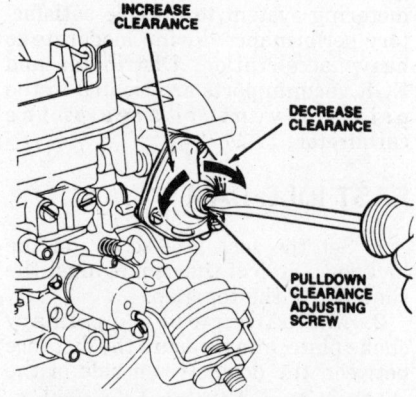

Adjusting choke plate pulldown clearance on 1981–83 740 models

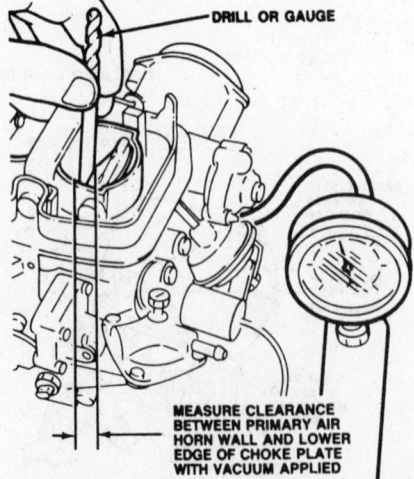

Measuring choke plate pulldown clearance on 1981–83 740 models

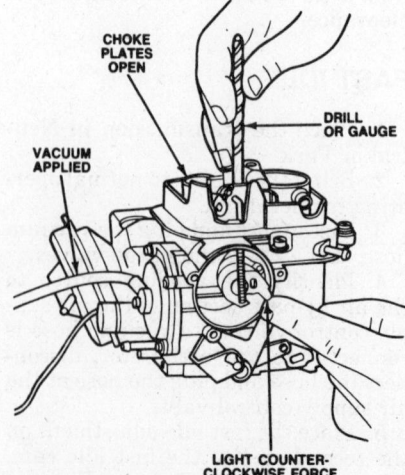

Measuring choke plate pulldown clearance on 1984 and later 740 models

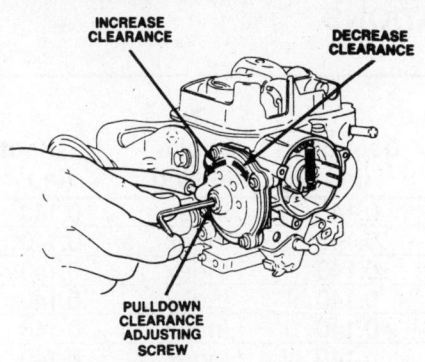

Adjusting choke plate pulldown clearance on 1984 and later 740 models

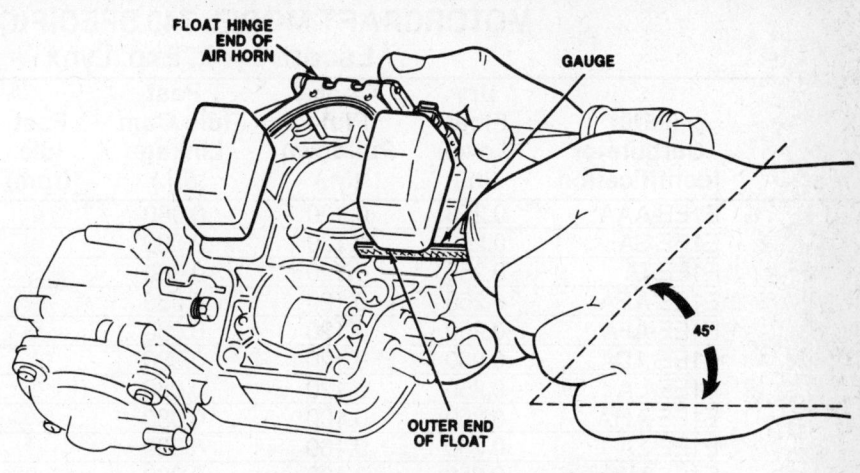

Measuring float clearance—model 740

7. If an adjustment is necessary, turn the vacuum diaphragm adjusting screw in or out as required.

NOTE: On 1984 and later models the choke pulldown adjustment screw is sealed with a limiting plug. Refer to the procedure which follows for removal.

Choke Pulldown Limiting Plug Removal

1984 and Later Models Only

1. Remove the choke pulldown diaphragm cover.
2. Using pliers, grasp the back of the adjustment screw and turn out of the cover.
3. Drive the plugs out of the cover, using a punch and a hammer.

—————— CAUTION ——————
Wear eye protection when driving out plugs.

DRY FLOAT ADJUSTMENT

1. Place the air horn assembly upside down and at a 45 degree angle with the air horn gasket in place. The float tang should rest lightly on the inlet needle.
2. Measure the clearance with a suitable gauge at the extreme end or toe of the float.
3. Remove float and adjust to specification by bending the float level adjusting tang up or down.

NOTE: Care must be taken not to scratch or damage the float tang while adjusting.

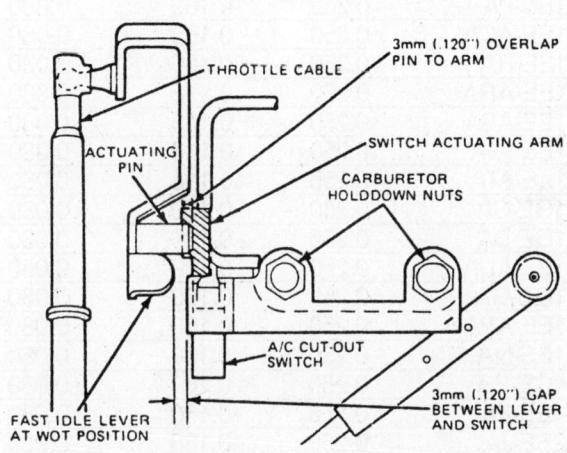

(WOT) A/C cut-off switch adjustment—Model 740

FLOAT DROP ADJUSTMENT

1. Suspend air horn assembly in normal position with air horn gasket in position.
2. The distance from the air horn gasket to the bottom of the float should be 1.69 ± 0.31 in. (43 ± 8mm).
3. Remove float and adjust to specification by bending the float drop tang.

WIDE OPEN THROTTLE (WOT) A/C CUT-OUT SWITCH

A visual inspection is required to ensure adequate pin and actuating arm overlap with the carburetor linkage in the WOT position.

Adjustments to the switch position are made by bending its support bracket outboard. A 0.120 in. (3mm)

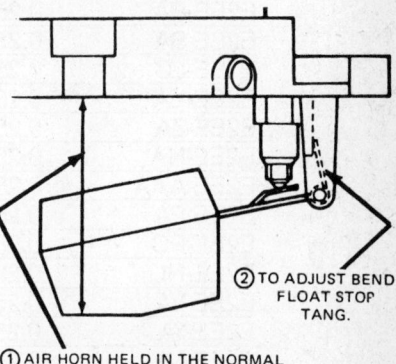

② TO ADJUST BEND FLOAT STOP TANG.

① AIR HORN HELD IN THE NORMAL POSITION. FLOAT HANGING IN FULL DOWN POSITION. MEASURE FROM AIR HORN COVER GASKET TO FLOAT, AS SHOWN. (43mm ± 8mm SETTING)

Float drop adjustment—Model 740

minimum overlap is desired. Precaution is required to ensure adequate clearance between the tip of the carburetor fast idle lever and switch housing.

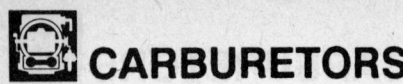

CARBURETORS

MOTORCRAFT MODEL 740 SPECIFICATIONS
Escort, Lynx, Exp, Lynx

Year	(9510)* Carburetor Identification	Dry Float Level (in.)	Choke Plate Pulldown (in.)	Fast Idle Cam Linkage (in.)	Fast Idle (rpm)	Dechoke (in.)	Choke Setting	Dashpot (in.)
'81	E1EE-AAA	0.250	0.120	0.080	①	0.140	Index	0.140
	E1EE-SA	0.250	0.120	0.080	①	0.140	Index	0.140
	E1EE-TA	0.250	0.120	0.080	①	0.140	Index	0.140
	E1EE-AEA	0.250	0.120	0.080	①	0.140	Index	0.140
	E1EE-AFA	0.250	0.120	0.080	①	0.140	Index	0.140
	E1EE-ADA	0.250	0.120	0.080	①	0.140	Index	0.140
	E1EE-LA	0.250	0.120	0.080	①	0.140	Index	0.140
	E1EE-AHA	0.250	0.100	0.080	①	0.140	Index	0.160
	E1EE-ZA	0.250	0.160	0.080	①	0.140	1 Lean	0.160
	E1EE-MA	0.250	0.160	0.080	①	0.140	1 Lean	0.160
	E1EE-NA	0.250	0.160	0.080	①	0.140	1 Lean	0.160
	E1EE-PA	0.250	0.160	0.080	①	0.140	1 Lean	0.160
	E1EE-ACA	0.250	0.160	0.080	①	0.140	1 Lean	0.160
	E1EE-RA	0.250	0.160	0.080	①	0.140	1 Lean	0.160
	E1EE-ARA	0.250	0.118	0.080	①	0.140	Index	0.140
	E1EE-ASA	0.250	0.118	0.080	①	0.140	Index	0.140
	E1EE-AVA	0.250	0.118	0.080	①	0.140	Index	0.140
	E1EE-ATA	0.250	0.118	0.080	①	0.140	Index	0.140
'82	E1GE-CA	0.250	0.120	0.080	2400	0.140	Index	0.140
	E1GE-DA	0.250	0.120	0.080	2400	0.140	Index	0.140
	E1EE-ALA	0.250	0.160	0.080	2400	0.140	1 Lean	0.160
	E1GE-GA	0.250	0.160	0.080	2400	0.140	1 Lean	0.160
	E1EE-APA	0.250	0.160	0.080	2400	0.140	1 Lean	0.160
	E1EE-NA	0.250	0.160	0.080	2400	0.140	1 Lean	0.160
	E1GE-EA	0.250	0.160	0.080	2400	0.140	1 Lean	0.160
	E1EE-ZA	0.250	0.160	0.080	2400	0.140	1 Lean	0.160
	E2EE-JA	0.250	0.138	0.080	2400	0.140	Index	0.060
	E2EE-GA	0.250	0.138	0.080	2200	0.140	Index	0.060
	E2EE-GC	0.250	0.138	0.080	2200	0.140	Index	0.140
	E2EE-EA	0.250	0.138	0.080	2400	0.140	Index	0.160
	E2EE-SA	0.250	0.138	0.080	2400	0.140	Index	0.060
	E2EE-LC	0.250	0.177	0.080	①	0.140	Index	0.160
	E2EE-LA	0.250	0.138	0.080	2400	0.140	Index	0.160
	E2EE-ZA	0.250	0.138	0.080	2400	0.140	2 Rich	0.160
	E2EE-NA	0.250	0.138	0.080	2400	0.140	Index	0.160
	E2EE-AAA	0.250	0.138	0.080	2400	0.140	2 Rich	0.160
	E2EE-PA	0.250	0.138	0.080	2400	0.140	Index	0.160
	E2EE-PC	0.250	0.177	0.080	2200	0.140	Index	0.160
	E2EE-NC	0.250	0.177	0.080	2200	0.140	Setting	0.160
	E2EE-VA	0.250	0.138	0.080	2400	0.140	1 Lean	0.160
	E2EE-YA	0.250	0.138	0.080	2400	0.140	1 Lean	0.160
	E2EE-MC	0.250	0.177	0.080	2200	0.140	Index	0.160
	E2EE-MA	0.250	0.138	0.080	2400	0.140	Index	0.160
'83	E3EE-CA	0.300	0.320	0.080	①	0.140	NA	0.140
	E3EE-EA	0.300	0.320	0.080	①	0.140	NA	0.140
	E3EE-DA	0.300	0.340	0.080	①	0.140	NA	0.140
	E3EE-AA	0.300	0.140	0.080	①	0.140	NA	0.140
	E3EE-JA	0.300	0.140	0.080	①	0.140	NA	0.140
	E3EE-BA	0.300	0.140	0.080	①	0.140	NA	0.140
	E3EE-KA	0.300	0.140	0.080	①	0.140	NA	0.140
	E3EE-GB	0.300	0.312	0.080	①	0.140	NA	0.095
	E3EE-NA	0.300	0.140	0.080	①	0.140	NA	—
	E3EE-PA	0.300	0.140	0.080	①	0.140	NA	0.140
	E3GE-DA	0.300	0.170	0.080	①	0.140	NA	0.140

MOTORCRAFT MODEL 740 SPECIFICATIONS
Escort, Lynx, Exp, Lynx

Year	(9510)* Carburetor Identification	Dry Float Level (in.)	Choke Plate Pulldown (in.)	Fast Idle Cam Linkage (in.)	Fast Idle (rpm)	Dechoke (in.)	Choke Setting	Dashpot (in.)
'83	E3GE-HA	0.300	0.170	0.080	①	0.140	NA	0.140
	E3GE-FA	0.300	0.170	0.080	①	0.140	NA	0.140
	E3GE-JA	0.300	0.170	0.080	①	0.140	NA	0.160
	E3GE-PA	0.300	0.260	0.080	①	0.140	NA	0.140
	E3GE-SA	0.300	0.260	0.080	①	0.140	NA	0.140
	E3GE-RA	0.300	0.260	0.080	①	0.140	NA	0.140
	E3GE-MA	0.300	0.280	0.093	①	0.140	NA	0.160
	E3GE-UA	0.300	0.280	0.093	①	0.140	NA	0.160
	E3GE-NA	0.300	0.140	0.093	①	0.140	NA	0.160
	E3GE-KB	0.300	0.300	0.080	①	0.140	NA	0.160
	E3GE-KD	0.300	0.300	0.080	①	0.140	NA	0.160
	E3GE-LA	0.300	0.300	0.080	①	0.140	NA	0.160
	E3GE-LC	0.300	0.300	0.080	①	0.140	NA	0.160
	E3GE-DC	0.300	0.170	0.080	①	0.140	NA	—
	E3GE-FC	0.300	0.170	0.080	①	0.140	NA	—
	E3GE-JC	0.300	0.170	0.080	①	0.140	NA	—
'84–'85	E4EE-YA	0.300	0.320	0.110	①	0.140	NA	0.095
	E4EE-ACA	0.300	0.320	0.080	①	0.140	NA	0.080
	E4EE-ADA	0.300	0.260	0.080	①	0.140	NA	0.140
	E4EE-ABA	0.300	0.320	0.110	①	0.140	NA	0.095
	E4EE-AAA	0.300	0.320	0.095	①	0.140	NA	0.095
	E4EE-AFA	0.300	0.218	0.080	①	0.140	NA	—
	E4GE-LA	0.300	0.300	0.080	①	0.140	NA	0.160
	E4GE-KA	0.300	0.300	0.080	①	0.140	NA	0.160
	E4GE-SA	0.300	0.280	0.100	①	0.140	NA	0.160
	E4GE-MA	0.300	0.300	0.180	①	0.140	NA	0.160
	E4GE-UA	0.300	0.325	0.080	①	0.140	NA	—
	E4GE-TA	0.300	0.300	0.125	①	0.140	NA	—
	E4GE-RA	0.300	0.325	0.130	①	0.140	NA	—
	E4GE-ACA	0.300	0.140	0.080	①	0.140	NA	0.080
	E4GE-ZA	0.300	0.250	0.108	①	0.140	NA	—
'86	E5GE-AAA	0.300	0.300	0.110	①	0.140	NA	0.060
	E5GE-ADA	0.300	0.300	0.100	①	0.140	NA	0.020
	E5GE-ACA	0.300	0.300	0.100	①	0.140	NA	0.020
	E5GE-AEC	0.300	0.280	0.080	①	0.140	NA	0.080
	E5GE-AFC	0.300	0.280	0.080	①	0.140	NA	0.060

*Basic carburetor number for Ford Carburetors
NA—Not available
① See underhood decal.

Models 2100, 2150

The Model 2100 and 2150 two barrel carburetor are basically the same in construction. Adjustments are performed in the same manner for both carburetors.

FLOAT LEVEL (DRY)

The dry float level measurement is a preliminary check and must be followed by a wet float level measurement with the carburetor mounted on the engine.

1. With the air horn removed gently raise the float to seat the inlet needle by applying light finger pressure at the float tab. Lower the float by reducing finger pressure until a light step is felt. Measure the distance between the main body gasket surface (gasket removed) and the top of the float. This measurement should be taken near the center of the float at a point $\frac{1}{8}$ in. from the free-end of the float.

NOTE: 1983 and later carburetors are equipped with a spring loaded fuel inlet needle and the ball must not be depressed when the fuel level check is being made.

2. If necessary, bend the float tab to obtain the correct level.

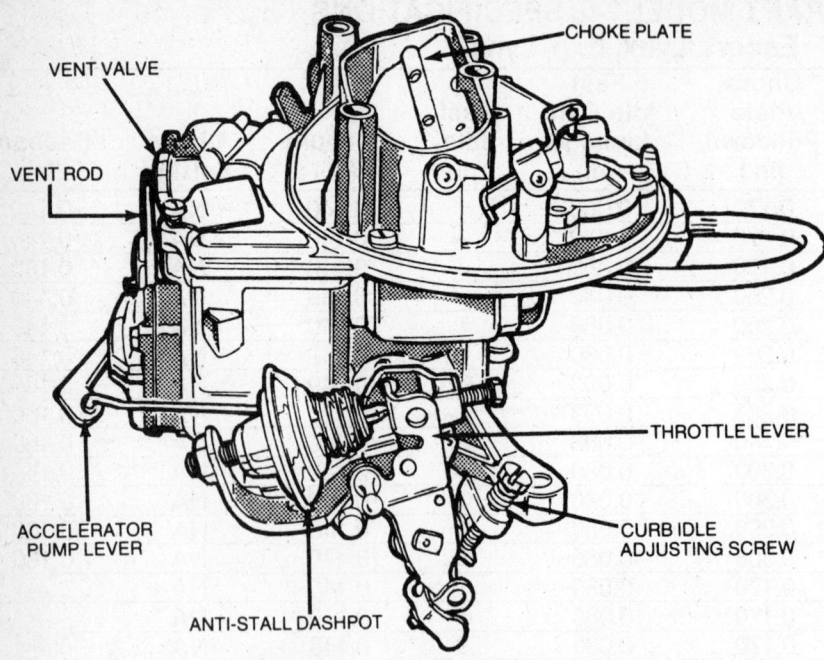

VENT VALVE

VENT ROD

ACCELERATOR
PUMP LEVER

ANTI-STALL DASHPOT

CHOKE PLATE

THROTTLE LEVER

CURB IDLE
ADJUSTING SCREW

Model 2100 two barrel carburetor

FLOAT LEVEL (WET)

1. Remove the screws that hold the air horn to the main body and break the seal between the air horn and main body. Leave the air horn and gasket loosely in place on top of the main body.

2. Start the engine and allow it to idle for at least three minutes.

3. After the engine has idled long enough to stabilize the fuel level, remove the air horn assembly.

4. With the engine idling, use a T-scale to measure the distance from the top of the fuel bowl machined surface to the surface of the fuel. The scale must be held at least $\frac{1}{4}$ in. away from any vertical surface to ensure proper measurement.

5. If any adjustment is required, stop the engine to avoid a fire from fuel spraying on the engine.

6. Bend the float tab upward to raise the level and downward to lower the level.

CAUTION
Be sure to hold the fuel inlet needle off its seat when bending the float tab so as not to damage the Viton® tip.

7. Each time the float level is changed, the air horn must be temporarily positioned and the engine started to stabilize the fuel level before again checking it.

CHOKE PLATE PULLDOWN

Model 2100

1. Loosen the screws on the choke

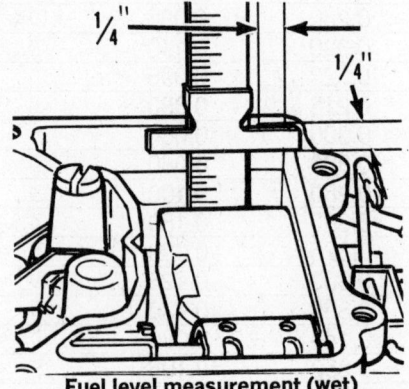

$\frac{1}{4}$" $\frac{1}{4}$"

Fuel level measurement (wet)

cover and rotate the cover $\frac{1}{4}$ turn counterclockwise (rich), then tighten the screws.

2. Operate the throttle to allow full closing of the choke plate.

3. Press down on the choke modulator arm until the choke modulator diaphragm is bottomed and then measure the distance from the lower edge of the choke plate to the inside air horn wall.

4. Adjustment is achieve by turning the diaphragm stop screw on the underside of the air horn.

5. Turn the screw clockwise to decrease clearance and counterclockwise to increase clearance.

NOTE: Do not reset the choke cover until the fast idle cam adjustment is made.

Model 2150

1. Remove the air cleaner assembly.

2. Set the throttle on the top step of the fast idle cam.

3. Noting the position of the choke housing cap, loosen the retaining screws and rotate the cap 90 degrees in the rich (closing) direction.

4. Activate the pull-down motor by manually forcing the pull-down control diaphragm link in the direction of applied vacuum or by applying vacuum to the external vacuum tube.

5. Using a drill gauge of the specified diameter, measure the clearance between the choke plate and the center of the air horn wall nearest the fuel bowl.

6. To adjust, reset the diaphragm stop on the end of the choke pull-down diaphragm.

NOTE: Loctite® was applied to the adjusting screw during manufacture and this will have to be loosened before the adjustment can be made. Heat the area around the screw with an electric soldering gun until the Loctite® softens enough to permit the screw to turn freely.

7. After adjusting, check and adjust the fast idle cam. Check and reset fast idle speed, if necessary. Install the air cleaner.

FAST IDLE CAM

1. The choke setting should still be 90° rich, as in Step 1 of the "Pulldown" procedure. Press and release the throttle to set the fast idle cam.

2. Activate the choke pulldown mechanism as in step 4 of the pulldown procedure.

3. Press and release the throttle to set the fast idle cam. It should drop to the kickdown step, and the fast idle speed screw should be opposite the V notch in the cam.

4. To adjust, turn the hex head screw on the plastic fast idle cam lever. After adjustment, allow the choke plate to close and check that it closes tightly. Reset the choke cover and connect the vacuum hose if removed.

CHOKE UNLOADER (DECHOKE)

1. With the throttle held completely open, move the choke plate to the closed position.

2. Measure the distance between the lower edge of the choke plate and the air horn wall.

3. Adjust by bending the tang on the fast idle speed lever which is located on the throttle shaft.

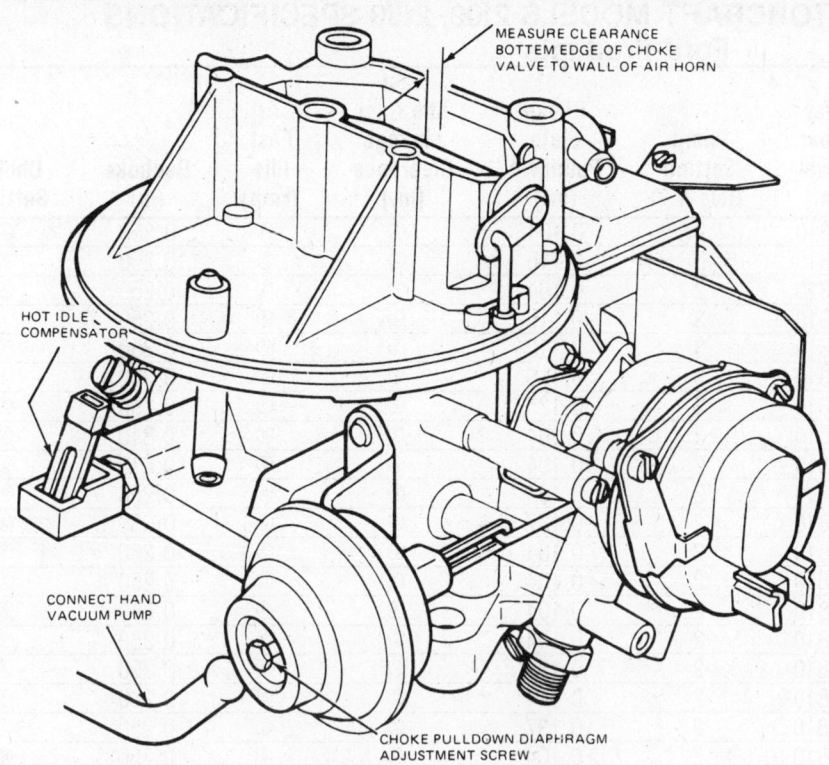

MEASURE CLEARANCE
BOTTOM EDGE OF CHOKE
VALVE TO WALL OF AIR HORN

HOT IDLE
COMPENSATOR

CONNECT HAND
VACUUM PUMP

CHOKE PULLDOWN DIAPHRAGM
ADJUSTMENT SCREW

Adjusting choke plate pulldown

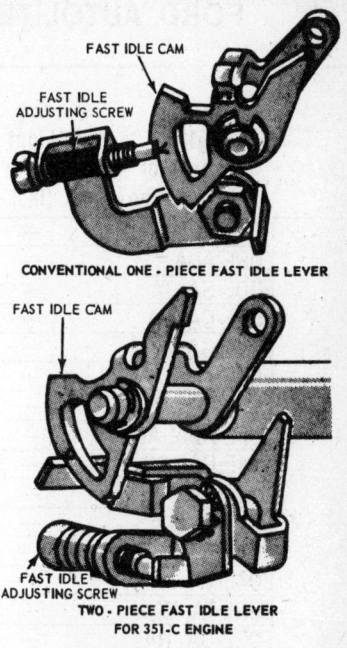

FAST IDLE CAM

FAST IDLE
ADJUSTING SCREW

CONVENTIONAL ONE - PIECE FAST IDLE LEVER

FAST IDLE CAM

FAST IDLE
ADJUSTING SCREW

**TWO - PIECE FAST IDLE LEVER
FOR 351-C ENGINE**

Fast idle adjustment

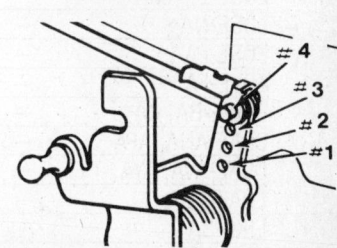

#4
#3
#2
#1

Accelerator pump stroke adjustment

NOTE: Final unloader adjustment must be performed on the car and the throttle should be opened by using the accelerator pedal of the car. This is to be sure that full throttle operation is achieved.

ACCELERATOR PUMP

The accelerator pump operating rod must be positioned in the proper holes of the accelerator pump lever and the throttle over-travel lever to assure correct pump travel. If adjusting is required, additional holes are provided in the throttle over-travel lever.

DASHPOT ADJUSTMENT

With the throttle set at the curb idle position, fully depress the dashpot stem and measure the distance between the stem and the throttle lever. Adjust by loosening the locknut and turning the dashpot.

FAST IDLE

Adjust the fast idle with the engine at

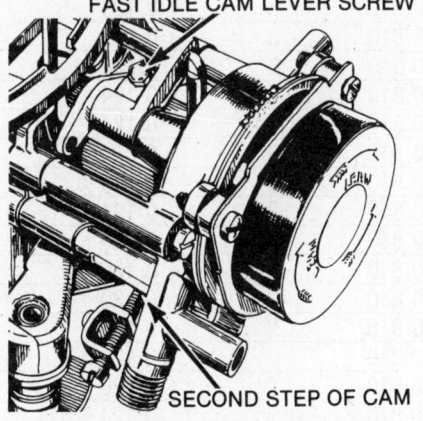

FAST IDLE CAM LEVER SCREW

SECOND STEP OF CAM

2100, 2150 fast idle cam linkage adjustment

normal operating temperature. On AMC cars, plug the spark port on the carburetor, and remove the EGR vacuum line at the valve and plug it. On Ford cars, if the engine is equipped with a spark delay valve, remove it and reroute the partial throttle vacuum signal line directly to the advance side of the distributor. If the distributor is a dual diaphragm type, leave

the manifold vacuum line connected to the retard side of the distributor, and remove and plug the line to the advance side.

If an EGR/PVS valve or cold weather modulator is located in the vacuum hose routing, disconnect and plug the hose at the EGR valve. If the engine does not have a cold weather modulator or an EGR/PVC valve, leave the EGR hose attaching. Trace the thermactor (air pump) dump valve vacuum hose from the dump valve to the carburetor; disconnect the dump valve vacuum hose nearest the carburetor, plug the original vacuum source and connect the dump valve directly to manifold vacuum. The fast idle screw should be resting against the second step of the fast idle cam on all models except Fords with the 302 engine, which have the screw set on the high step of the cam. Adjust the fast idle speed by turning the fast idle screw.

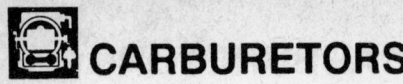

CARBURETORS

FORD, AUTOLITE, MOTORCRAFT MODELS 2100, 2150 SPECIFICATIONS
Ford Products

Year	(9510)* Carburetor Identification	Dry Float Level (in.)	Wet Float Level (in.)	Pump Setting Hole # ①	Choke Plate Pulldown (in.)	Fast Idle Cam Linkage Clearance (in.)	Fast Idle (rmp)	Dechoke (in)	Choke Setting
'80	EO4E-PA, RA	—	0.810	2	0.104	②	③	0.250	③
	EOBE-AUA	—	0.810	3	0.116	②	③	0.250	③
	EODE-SA, TA	—	0.810	2	0.104	②	③	0.250	③
	EOKE-CA, DA	—	0.810	3	0.116	②	③	0.250	③
	EOKE-GA, HA	—	0.810	3	0.116	②	③	0.250	③
	EOKE-JA, KA	—	0.810	3	0.116	②	③	0.250	③
	D84E-TA, UA	—	0.810	2	0.125	②	③	0.250	③
	EO4E-ADA, AEA	—	0.810	2	0.104	②	③	0.810	③
	EO4E-CA	—	0.810	2	0.104	②	③	0.810	③
	EO4E-EA, FA	—	0.810	2	0.104	②	③	0.250	③
	EO4E-JA, KA	—	0.810	2	0.137	②	③	0.250	③
	EO4E-SA, TA	—	0.810	2	0.104	②	③	0.250	③
	EO4E-VA, YA	—	0.810	2	0.104	②	③	0.250	③
	EODE-TA, VA	—	0.810	2	0.104	②	③	0.250	③
	EOSE-GA, HA	—	0.810	2	0.104	②	③	0.250	③
	EOSE-LA, MA	—	0.810	2	0.104	②	③	0.250	③
	EOSE-NA	—	0.810	2	0.104	②	③	0.250	③
	EOSE-PA	—	0.810	2	0.137	②	③	0.250	③
	EOVE-FA	—	0.810	2	0.104	②	③	0.250	⑨
	EOWE-BA, CA	—	0.810	2	0.137	②	③	0.250	③
	D9AE-ANA, APA	—	0.810	3	0.129	②	③	0.250	③
	D9AE-AVA, AYA	—	0.810	3	0.129	②	③	0.250	③
	EOAE-AGA	—	0.810	3	0.159	②	③	0.250	③
'81	EIKE-CA	7/16	0.810	3	0.124	②	③	0.250	④
	EIKE-EA	7/16	0.810	3	0.124	②	③	0.250	④
	EIKE-DA	7/16	0.810	3	0.124	②	③	0.250	④
	EIKE-FA	7/16	0.810	3	0.124	②	③	0.250	④
	EIKE-GA	7/16	0.810	3	0.124	②	③	0.250	④
	EIKE-SA	7/16	0.810	3	0.124	②	③	0.250	④
	EIKE-RA	7/16	0.810	3	0.120	②	③	0.250	④
	EIKE-HA	7/16	0.810	3	0.120	②	③	0.250	④
	EIWE-FA	7/16	0.810	2	0.120	②	③	0.250	④
	EIWE-EA	7/16	0.810	2	0.120	②	③	0.250	④
	EIWE-CA	7/16	0.810	2	0.120	②	③	0.250	④
	EIWE-DA	7/16	0.810	2	0.120	②	③	0.250	④
	EIAE-AKA	7/16	0.810	3	0.124	②	③	0.250	④
	EIAE-AJA	7/16	0.810	3	0.124	②	③	0.250	④
	EIAE-YA	7/16	0.810	3	0.124	②	③	0.250	④
	EIAE-ZA	7/16	0.810	3	0.124	②	③	0.250	④
	EIAE-ADA	7/16	0.810	3	0.124	②	③	0.250	④
	EIAE-AEA	7/16	0.810	3	0.124	②	③	0.250	④
	EIAE-TA	—	0.810	3	0.104	②	③	0.250	④
	EIAE-UA	—	0.810	2	0.104	②	③	0.250	④
	EIDE-LA	7/16	0.810	2	0.120	②	③	0.250	④
	EIDE-KA	7/16	0.810	2	0.120	②	③	0.250	④
	EIDE-JA	7/16	0.810	2	0.120	②	③	0.250	④
	EIDE-HA	7/16	0.810	2	0.120	②	③	0.250	④
'82	E2BE-UA	7/16	0.810	2	0.110	②	2200	0.250	④
	E2BE-AAA	7/16	0.810	2	0.110	②	2200	0.250	④
	E2BE-VA	7/16	0.810	2	0.113	②	2200	0.250	④
	E2BE-ABA	7/16	0.810	2	0.113	②	2200	0.250	④
	E2BE-AGA	7/16	0.810	2	0.113 ⑤	②	2200	0.250	④
	E2BE-AHA	7/16	0.810	2	0.113	②	2200	0.250	④

FORD, AUTOLITE, MOTORCRAFT MODELS 2100, 2150 SPECIFICATIONS
Ford Products

Year	(9510)* Carburetor Identification	Dry Float Level (in.)	Wet Float Level (in.)	Pump Setting Hole # ①	Choke Plate Pulldown (in.)	Fast Idle Cam Linkage Clearance (in.)	Fast Idle (rmp)	Dechoke (in)	Choke Setting
'82	E2VE-CA	7/16	0.810	2	0.113	②	2200	0.250	④
	E24E-CA	7/16	0.810	2	0.110	②	1200	0.250	④
	E24E-DA	7/16	0.810	2	0.110	②	1200	0.250	④
	E24E-AA	7/16	0.810	2	0.110	②	2100	0.250	④
	E24E-BA	7/16	0.810	2	0.110	②	2100	0.250	④
	E24E-EA	7/16	0.810	2	0.110	②	③	0.250	④
	E24E-FA	7/16	0.810	2	0.110	②	③	0.250	④
	E2KE-AA	7/16	0.810	2	0.140	②	1500	0.250	④
	E2KE-BA	7/16	0.810	2	0.140	②	1500	0.250	④
	E2WE-EA	7/16	0.810	2	0.137	②	1500	0.250	④
	E2WE-FA	7/16	0.810	2	0.137	②	1500	0.250	④
	E2DE-JA	7/16	0.810	2	0.137	②	1600	0.250	④
	E2DE-KA	7/16	0.810	2	0.137	②	1600	0.250	④
	E2DE-LA	7/16	0.810	2	0.137	②	1700	0.250	④
	E2DE-MA	7/16	0.810	2	0.137	②	1700	0.250	④
	E25E-DA	7/16	0.810	2	0.144	②	1500	0.250	④
	E2AE-SA	7/16	0.810	2	0.172	②	1550	0.250	④
	E25E-CA	7/16	0.810	2	0.137	②	1700	0.250	④
	E2ZE-BAA	13/32	0.780	2	0.172 ⑤	②	1400	0.250	④
	E2ZE-BBA	13/32	0.780	2	0.172 ⑤	②	1400	0.250	④
	E3CE-LA	7/16	0.810	3	0.103	②	2200	0.250	④
	E3CE-MA	7/16	0.810	3	0.103	②	2200	0.250	④
	E3CE-JA	7/16	0.810	3	0.103	②	2200	0.250	④
	E3CE-KA	7/16	0.810	3	0.103	②	2200	0.250	④
	E3CE-NA	7/16	0.810	3	0.120	②	2100	0.250	④
	E3CE-PA	7/16	0.810	3	0.120	②	2100	0.250	④
'83	E3CE-AA	7/16	0.810	3	0.103	②	2200	0.250	④
	E3CE-BA	7/16	0.810	3	0.103	②	2200	0.250	④
	E3CE-GA	7/16	0.810	3	0.103	②	2200	0.250	④
	E3CE-HA	7/16	0.810	3	0.103	②	2200	0.250	④
	E3CE-EA	7/16	0.810	3	0.113	②	2100	0.250	④
	E3CE-FA	7/16	0.810	3	0.113	②	2100	0.250	④
	E3SE-ATA	7/16	0.810	3	0.113	②	2200	0.250	④
	E3SE-AUA	7/16	0.810	3	0.113	②	2200	0.250	④
	E3SE-ALA	7/16	0.810	3	0.107	②	2200	0.250	④
	E3SE-AMA	7/16	0.810	3	0.107	②	2200	0.250	④
	E3SE-BDA	7/16	0.810	3	0.107	②	2200	0.250	④
	E3SE-BEA	7/16	0.810	3	0.107	②	2200	0.250	④
	E3SE-ANA	7/16	0.810	3	0.101	②	2200	0.250	④
	E3SE-APA	7/16	0.810	3	0.101	②	2200	0.250	④
	E3SE-AJA								
	E3SE-BFA	7/16	0.810	3	0.107	②	2200	0.250	④
	E3SE-BGA	7/16	0.810	3	0.107	②	2200	0.250	④
	E3SE-EA	7/16	0.810	3	0.113	②	2200	0.250	④
	E3SE-FA	7/16	0.810	3	0.113	②	2200	0.250	④
	E3SE-LA	7/16	0.810	3	0.107	②	2200	0.250	④
	E3SE-MA	7/16	0.810	3	0.107	②	2200	0.250	④
	E3SE-JA	7/16	0.810	3	0.101	②	2200	0.250	④
	E3SE-KA	7/16	0.810	3	0.101	②	2200	0.250	④
	E3SE-NA	7/16	0.810	3	0.107	②	2200	0.250	④
	E3SE-PA	7/16	0.810	3	0.107	②	2200	0.250	④
	E3SE-GA	7/16	0.810	3	0.120	②	2100	0.250	④
	E3SE-HA	7/16	0.810	3	0.120	②	2100	0.250	④

FORD, AUTOLITE, MOTORCRAFT MODELS 2100, 2150 SPECIFICATIONS
Ford Products

Year	(9510)* Carburetor Identification	Dry Float Level (in.)	Wet Float Level (in.)	Pump Setting Hole # ①	Choke Plate Pulldown (in.)	Fast Idle Cam Linkage Clearance (in.)	Fast Idle (rmp)	Dechoke (in)	Choke Setting
'83	E3AE-TA	7/16	0.810	3	0.103	②	2200	0.250	④
	E3AE-ADA	7/16	0.810	3	0.103	②	2200	0.250	④
	E3AE-UA	7/16	0.810	3	0.103	②	2200	0.250	④
	E3AE-AEA	7/16	0.810	3	0.103	②	2200	0.250	④
	E3AE-TA	7/16	0.810	3	0.103	②	2200	0.250	④
	E3AE-UA	7/16	0.810	3	0.103	②	2200	0.250	④
	E3AE-RA	7/16	0.810	3	0.103	②	2200	0.250	④
	E3AE-SA	7/16	0.810	3	0.103	②	2200	0.250	④
	E3AE-EA	7/16	0.810	2	—	②	1550	0.250	④
'84	E3EA-EA	7/16	0.810	2	—	②	1550	0.250	④
	E4CE-AA	7/16	0.810	3	0.103	②	2200	0.250	2NR
	E4CE-BA	7/16	0.810	3	0.103	②	2200	0.250	2NR
	E4SE-CA	7/16	0.810	3	0.103	②	2200	0.250	④
	E4SE-DA	7/16	0.810	3	0.103	②	2200	0.250	④
'85	E4SE-CA	3/32	0.810	3	0.103	—	③	0.250	4NR
	E4SE-DA	3/32	0.810	3	0.103	—	③	0.250	4NR
	E5SE-CA E3AE-EA(Alt)	7/16	0.810	2	—	④	③	0.250	④

*Basic carburetor number for Ford products
① With link in inboard hole of pump lever
② Opposite "V" notch; see text
③ See underhood decal
④ V-notch
⑤ ± .010"

Model 2700 VV

Since the design of the 2700 VV (variable venturi) carburetor differs considerably from the other carburetors in the Ford lineup, an explanation in the theory and operation is presented here.

In exterior appearance, the variable venturi carburetor is similar to conventional carburetors and, like a conventional carburetor, it uses a normal float and fuel bowl system. However, the similarity ends there. In place of a normal choke plate and fixed area venturis, the 2700VV carburetor has a pair of small oblong castings in the top of the upper carburetor body where you would normally expect to see the choke plate. These castings slide back and forth across the top of the carburetor in response to fuel-air demands. Their movement is controlled by a spring-loaded diaphragm valve regulated by a vacuum signal taken below the venturis in the throttle bores. As the throttle is opened, the strength of the vacuum signal increases, opening the venturis and allowing more air to enter the carburetor.

Fuel is admitted into the venturi area by means of tapered metering rods that fit into the main jets. These rods are attached to the venturis, and, and the venturis open or close in response to air demand, the fuel needed to maintain the proper mixture increases or decreases as the metering rods slide in the jets. In comparison to a conventional carburetor with fixed venturis and a variable air supply, this system provides much more precise control of the fuel-air supply during all modes of operation. Because of the variable venturi principle, there are fewer fuel metering systems and fuel passages. The only auxiliary fuel metering systems required are an idle trim, accelerator pump (similar to a conventional carburetor), starting enrichment, and cold running enrichment.

NOTE: Adjustment, assembly and disassembly of this carburetor require special tools for some of the operations. These tools are available (see the Tools and Equipment Section). Do not attempt any operations on this carburetor without first checking to see if you need the special tools for that particular operation. The adjustment and repair procedures given here mention when and if you will need the special tools.

Some 1980–81 models equipped with the 2700 VV carburetor experienced engine stalling, stumbling and poor performance. According to a Ford Motor Company service bulletin No. 81–9–10 issued in May of 1981, this condition may be caused by fluids trapped in the venturi valve diaphragm cavity which eventually deteriorate the diaphragm, resulting in a leak. A quick check to verify if the above condition exists is to visually observe the venturi valve action while running the engine. Throttle movement from idle to just above idle should show a corresponding movement of the venturi valves. If no venturi valve action is observed and the valves are not sticking, there may be a leak in the diaphragm (stalls may be encountered while performing this check). If the above condition is suspected, it is advised that the car be serviced at a Ford or Mercury dealer using the procedure stated in the service bulletin.

FLOAT LEVEL ADJUSTMENT

1. Remove and invert the upper part of the carburetor, with the gasket in place.

2. Measure the vertical distance between the carburetor body, outside the gasket, and the bottom of the float.

3. To adjust, bend the float operating lever that contacts the needle valve. Make sure that the float remains parallel to the gasket surface.

FLOAT DROP ADJUSTMENT

1. Remove and hold upright the upper part of the carburetor.

2. Measure the vertical distance between the carburetor body, outside the gasket, and the bottom of the float.

3. Adjust by bending the stop tab on the float lever that contacts the hinge pin.

FAST IDLE SPEED ADJUSTMENT

1. With the engine warmed up and idling, place the fast idle lever on the step of the fast idle cam specified on the engine compartment sticker or in the specifications chart. Disconnect and plug the EGR vacuum line.

2. Make sure the high speed cam positioner lever is disengaged.

3. Turn the fast idle speed screw to adjust to the specified speed.

FAST IDLE CAM ADJUSTMENT

You will need a special tool for this job; Ford calls it a stator cap (No. T77L-9848-A). It fits over the choke thermostatic lever when the choke cap is removed.

1. Remove the choke coil cap. On 1980 California model and all 1981 and later models, the choke cap is riveted in plate. The top rivets will have to be drilled out; the bottom rivet will have to be driven out from the rear. New rivets must be used upon installation.

2. Place the fast idle lever in the corner of the specified step of the fast idle cam (the highest step is first)

with the high speed cam positioner retracted.

3. If the adjustment is being made

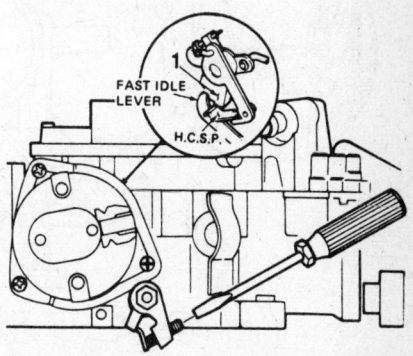

2700 VV fast idle speed adjustment

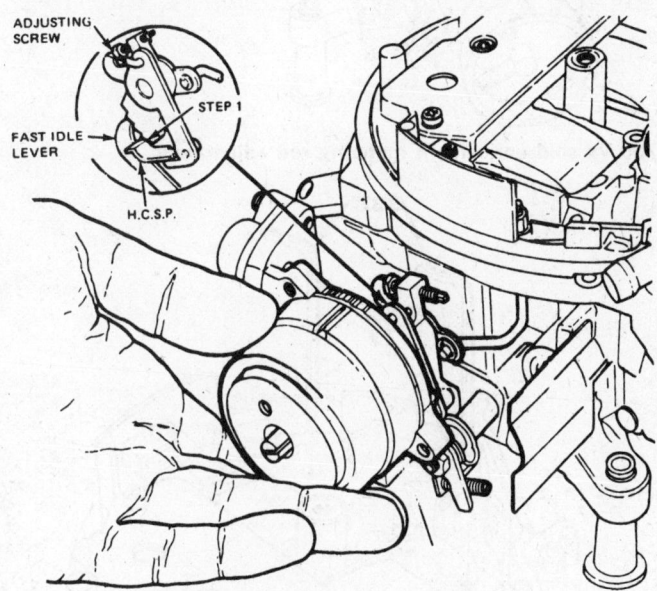

2700 VV fast idle cam adjustment

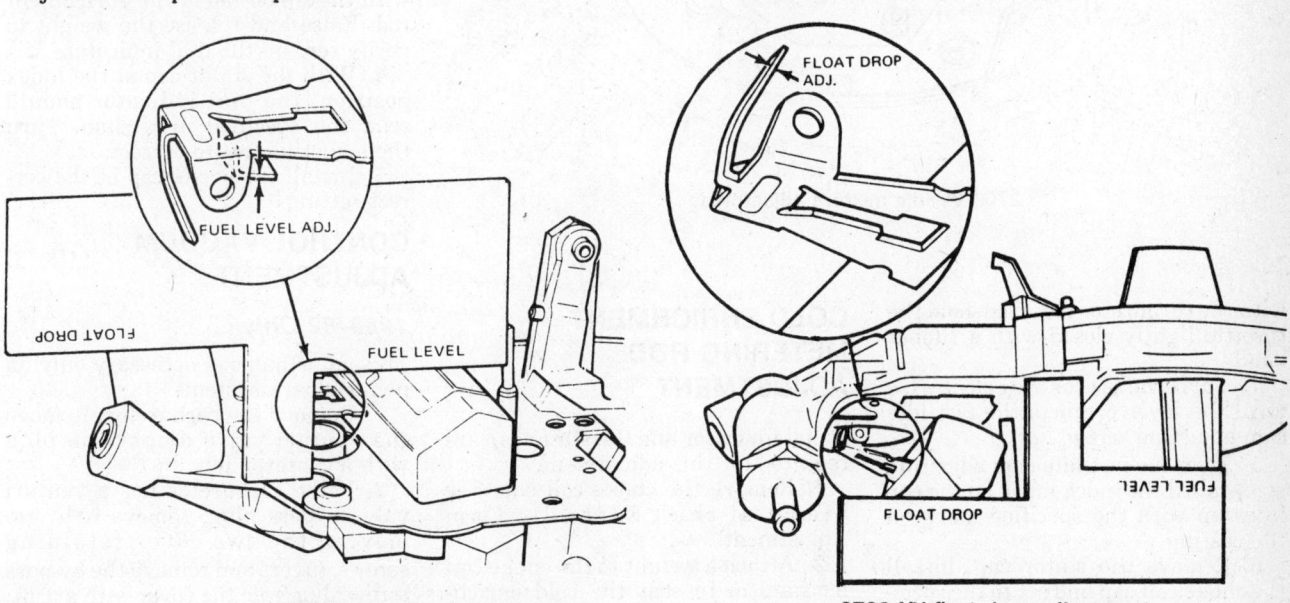

2700 VV float level adjustment

2700 VV float drop adjustment

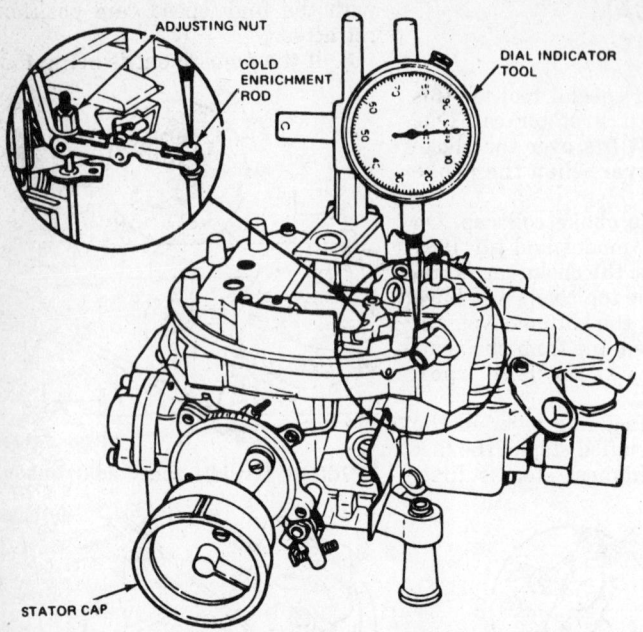

2700 VV cold enrichment metering rod adjustment

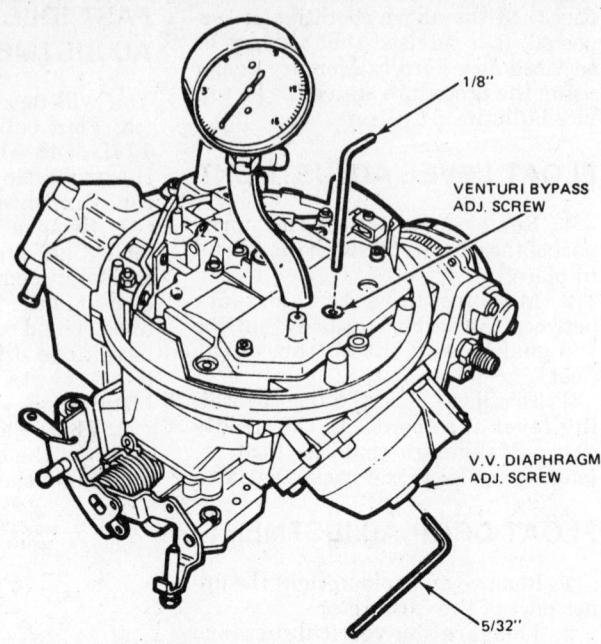

2700 VV control vacuum adjustment

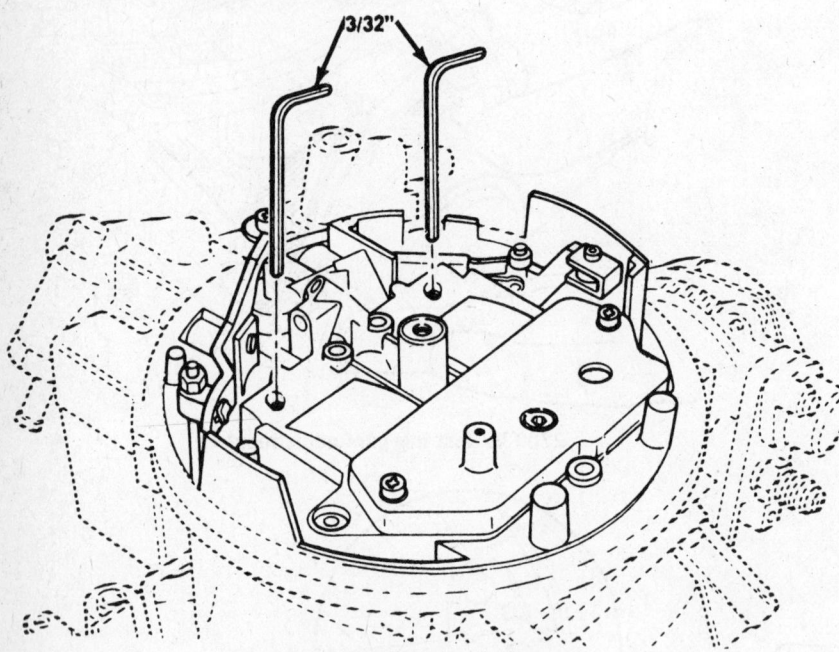

2700 VV idle mixture adjustment

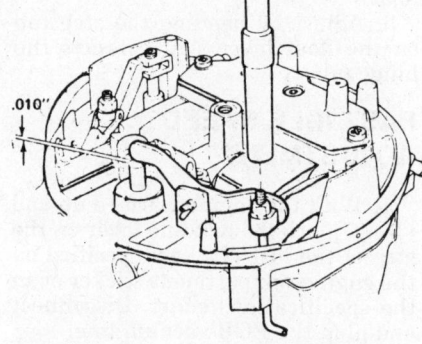

2700 VV internal vent adjustment

3. Install and zero a dial indicator with the tip on top of the enrichment rod. Raise and release the weight to verify zero on the dial indicator.

4. With the stator cap at the index position, the dial indicator should read the specified dimension. Turn the adjusting nut to correct.

5. Install the choke cap at the correct setting.

CONTROL VACUUM ADJUSTMENT

1980–82 Only

This adjustment is necessary only on non-feedback systems.

1. Remove the carburetor. Remove the venturi valve diaphragm plug with a centerpunch.

2. If the carburetor has a venturi valve bypass plug, remove it by removing the two cover retaining screws; invert and remove the by-pass screw plug from the cover with a drift. Install the cover.

with the carburetor removed, hold the throttle lightly closed with a rubber band.

4. Turn the stator cap clockwise until the lever contacts the fast idle cam adjusting screw.

5. Turn the fast idle cam adjusting screw until the index mark on the cap lines up with the specified mark on the casting.

6. Remove the stator cap. Install the choke coil cap and set to the specified housing mark.

COLD ENRICHMENT METERING ROD ADJUSTMENT

A dial indicator and the stator cap are required for this adjustment.

1. Remove the choke coil cap. See Step 1 of the "Fast Idle Cam Adjustment".

2. Attach a weight to the choke coil mechanism to seat the cold enrichment rod.

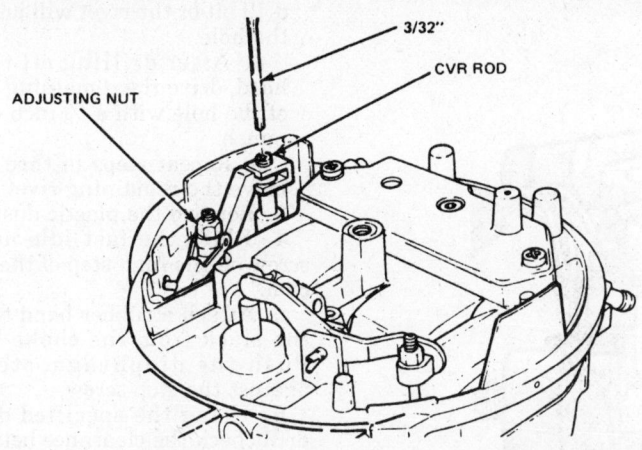

ADJUSTING NUT

3/32"

CVR ROD

2700 VV control vacuum regulator adjustment

valve to the wide open position and insert an Allen wrench into the stop screw hole. Turn clockwise to increase the gap. Remove the wrench and check the gap again.

6. Replace the wide open stop screw and turn it clockwise until it contacts the valve.

7. Push the venturi valve wide open and check the gap. Turn the stop screw to bring the gap to specifications.

8. Reassemble the carburetor with a new expansion plug.

CONTROL VACUUM REGULATOR (CVR) ADJUSTMENT

The cold enrichment metering rod adjustment must be checked and set before making this adjustment.

1. After adjusting the cold enrichment metering rod, leave the dial indicator in place but remove the stator cap. Do not re-zero the dial indicator.

2. Press down on the CVR rod until it bottoms on its seat. Measure this amount of travel with the dial indicator.

3. If the adjustment is incorrect, hold the $\frac{3}{8}$ in. CVR adjusting nut with a box wrench to prevent it from turning. Use a $\frac{3}{32}$ in. Allen wrench to turn the CVR rod; turning counterclockwise will increase the travel, and vice versa.

3. Install the carburetor. Start the engine and allow it to reach normal operating temperature. Connect a vacuum gauge to the venturi valve cover. Set the idle speed to 500 rpm with the transmission in Drive.

4. Push and hold the venturi valve closed. Adjust the bypass screw to obtain a reading of 8 in. H_2O on the vacuum gauge. Make sure the idle speed remains constant. Open and close the throttle and check the idle speed.

5. With the engine idling, adjust the venturi valve diaphragm screw to obtain a reading of 6 in. H_2O. Set the curb idle to specification. Install new venturi valve bypass and diaphragm plugs.

VENTURI VALVE LIMITER ADJUSTMENT

1. Remove the carburetor. Take off the venturi valve cover and the two rollers.

2. Use a center punch to loosen the expansion plug at the rear of the carburetor main body on the throttle side. Remove it.

3. Use an Allen wrench to remove the venturi valve wide open stop screw.

4. Hold the throttle wide open.

5. Apply a light closing pressure on the venturi valve and check the gap between the valve and the air horn wall. To adjust, move the venturi

MOTORCRAFT MODEL 2700 VV SPECIFICATIONS
Ford Products

Year	Model	Float Level (in.)	Float Drop (in.)	Fast Idle Cam Setting (notches)	Cold Enrichment Metering Rod (in.)	Control Vacuum (in. H_2O)	Venturi Valve Limiter (in.)	Choke Cap Setting (notches)	Control Vacuum Regulator Setting (in.)
'80	All	1³/₆₄	1¹⁵/₃₂	1 Rich/4th step	.125	①	②	③	.075
'81	EIAE-AAA	1.010–1.070	1.430–1.490	1 Rich/4th step	④	①	②	Index	—
	D9AE-AZA	1.015–1.065	1.435–1.485	1 Rich/4th step	.125	①	②	Index	—

① See text
② Opening gap: 0.99–1.01
   Closing gap: 0.94–0.98
③ See underhood decal
④ 0°F—0.490 @ starting position
   75°F—0.475 @ starting position

Model 5200

The 5200 carburetor is a two-stage, two-venturi carburetor in which the secondary venturi is the larger. The secondary system is mechanically operated.

FAST IDLE CAM

1980–82

1. Place the fast idle screw on the second step of the fast idle cam

against the shoulder of the top step.

2. Apply light pressure (downward) on the choke lever tang, and, using the proper size drill, measure the clearance between the lower edge of

LIGHT CLOSING PRESSURE ON CHOKE LEVER

GAUGE

FAST IDLE SCREW ON SECOND HIGHEST STEP OF CAM

Fast idle cam adjustment—Holley 1945

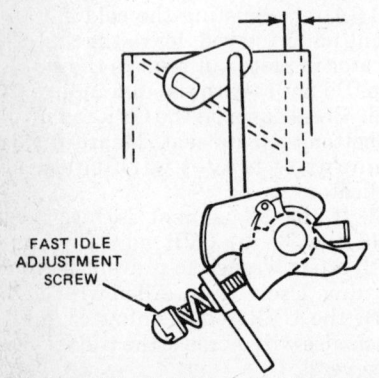

FAST IDLE ADJUSTMENT SCREW

1980 and later fast idle cam adjustment— measure the clearance between the lower edge of the choke plate and the air horn wall

the choke plate and the air horn wall.

3. Bend the choke lever tang down to increase clearance and up to decrease the clearance.

NOTE: On 1982 and later models Ford recommends that if an adjustment is necessary, the choke lever should be replaced since the lever tang is hardened. This may also be necessary for 1980–81 models.

CHOKE PLATE PULLDOWN

1980

1. Remove the choke thermostatic spring cover.
2. Pull the water cover and the thermostatic spring cover assembly or the electric choke assist assembly out of the way.
3. Set the fast idle cam on the second step.
4. Push the diaphragm stem against its stop and insert the speci-

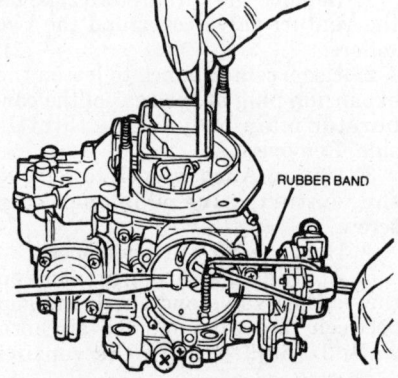

RUBBER BAND

Choke plate pulldown adjustment

fied gauge between the lower edge of the choke valve and the air horn wall.

5. Apply sufficient pressure to the upper edge of the choke valve to take up any slack in the choke linkage.

6. Turn the adjusting screw in or out to adjust the choke plate-to-air horn clearance.

1981–82

NOTE: The following procedure requires the removal of the carburetor and also the choke cap which is retained by two rivets.

1. Remove the carburetor from the engine.

2. Remove the choke cap as follows:

   a. Check the rivets to determine if mandrel is well below the rivet head. If mandrel is within the rivet head thickness, drive it down or out with a $\frac{1}{16}$ inch diameter tip punch.

   b. With a $\frac{1}{8}$ inch diameter drill, drill into the rivet head until the rivet head comes loose from the rivet body. Use light pressure on the

drill bit or the rivet will just spin in the hole.

   c. After drilling off the rivet head, drive the remaining rivet out of the hole with a $\frac{1}{8}$ inch diameter punch.

   d. Repeat steps (a thru c) to remove the remaining rivet.

3. Remove the plastic dust cover.

4. Place the fast idle adjusting screw on the high step of the fast idle cam.

5. Attach a rubber band to remove the slack from the choke linkage. Push the diaphragm stem back against the stop screw.

6. Using the specified diameter drill check the clearance between the lower edge of the choke plate and the air horn wall.

7. If adjustment is necessary, obtain a replacement kit containing a new choke pulldown diaphragm cover, adjusting screw and cup plug.

8. After installing the adjusting screw in the cover, adjust the pulldown by turning the screw clockwise to decrease and counterclockwise to increase the setting.

9. After making the adjustment, install a new plug in the choke pulldown adjustment access opening.

10. Remove the rubber band and reinstall the choke cap using rivets ($\frac{1}{8}$ inch diameter x $\frac{1}{2}$ inch long with a $\frac{1}{4}$ inch diameter head).

DECHOKE (UNLOADER) ADJUSTMENT

Dechoke clearance adjustment is controlled by the fast idle cam adjustment. The figures in the specification chart refer to choke plate clearance between the plate and the air horn wall. Clearance can be measured as follows:

1. Hold the throttle wide open. Remove any slack from the choke linkage by applying pressure to the upper edge of the choke valve.

2. Measure the distance between the lower edge of the choke plate and the air horn wall.

3. Adjust by bending the tab on the fast idle lever where it touches the cam.

FAST IDLE SPEED

Set the fast idle speed with the fast idle screw positioned on the second step of the fast idle cam and with the engine at operating temperature. Remove the EGR line at the valve and plug it. If the car is equipped with a spark delay valve, remove the valve and route the distributor advance

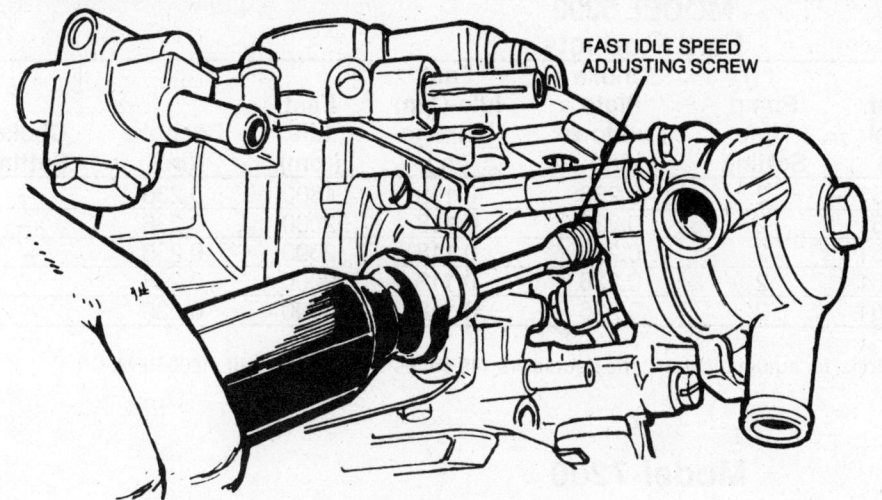

Fast idle adjustment

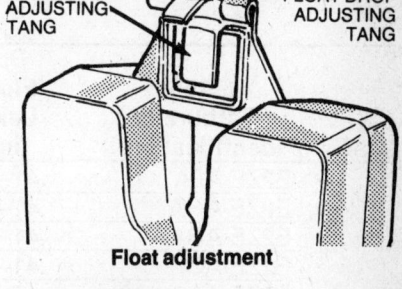

Float adjustment

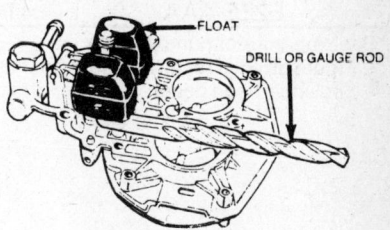

Checking the float level

vacuum signal directly to the distributor advance diaphragm. On all manual transmission models, remove and plug the vacuum line to the distributor. If the distributor also has a retard diaphragm, leave the hose connected to it alone. If the engine has a deceleration valve, remove this hose at the carburetor and plug it. Finally, if the car has air conditioning it must be off before adjusting the fast idle.

FLOAT LEVEL ADJUSTMENT

With the bowl cover held upside down and the float tang resting lightly on the spring loaded fuel inlet needle, measure the clearance between the edge of the float and the bowl cover. To adjust the level, bend the float tang up or down as required. Adjust both floats equally.

SECONDARY THROTTLE STOP SCREW

1. Turn the secondary throttle stop screw counterclockwise until the secondary throttle plate seats in its bore.
2. Turn the screw clockwise until it touches the tab on the secondary throttle lever.
3. Add ¼ turn clockwise for four cylinder engines.

MODEL 5200
Ford Products

Year	(9510)* Carburetor Identification ①	Dry Float Level (in.)	Pump Hole Setting	Choke Plate Pulldown (in.)	Fast Idle Cam Linkage (in.)	Fast Idle (rpm)	Dechoke (in.)	Choke Setting
'80	D9EE-APA, ANA	0.460	2	0.236	0.118	②	0.236	1 Rich
	EOEE-GA, RA	0.460	2	0.196	0.078	②	0.196	②
	EOEE-JA, TA	0.460	2	0.196	0.078	②	0.196	②
	EOEE-JC, TC	0.460	—	0.196	0.078	②	0.196	②
	EOEE-JD, TD	0.460	2	0.177	0.078	②	0.196	②
	EOEE-AEA, AFA	0.460	2	0.196	0.078	②	0.196	②
	EOZE-ACB	0.460	—	0.275	0.157	②	0.236	②
	EOZE-AZA	0.460	2	0.275	0.157	②	0.393	②
	EOZE-AAA	0.460	3	0.275	0.157	②	0.236	②
	EOZE-ACA	0.460	2	0.275	0.157	②	0.236	②
	EOZE-ATA	0.460	2	0.275	0.118	②	0.236	②
'81	EIZE-YA	.41-.51	2	0.200	.080	②	0.200	②
	EOEE-RB	.41-.51	2	0.200	.080	②	0.200	②
	EIZE-VA	.41-.51	2	0.200	.080	②	0.200	②
	D9EE-ANA	.41-.51	2	0.240	0.720	②	0.200	②
	D9EE-APA	.41-.51	2	0.240	0.120	②	0.200	②
'82	E1ZE-ADB	.41-.51	3	0.275	0.240	1600	0.393	—
	E1ZE-ACA	.41-.51	2	0.200	.080	1800	0.196	—
	E1BE-RA, GA	.41-.51	2	0.200	.080	1800	0.196	—
	E1ZE-YA	.41-.51	2	0.200	.080	2000	0.196	—
	E1ZE-VA	.41-.51	2	0.200	.080	2000	0.196	—

MODEL 5200
Ford Products

Year	(9510)* Carburetor Identification ①	Dry Float Level (in.)	Pump Hole Setting	Choke Plate Pulldown (in.)	Fast Idle Cam Linkage (in.)	Fast Idle (rpm)	Dechoke (in.)	Choke Setting
'82	E2ZE-AFA	.41–.51	2	0.236	0.118	1800	0.236	—
	E2ZE-AHA	.41–.51	2	0.236	0.118	2000	0.236	—
	E2ZE-ABA	.41–.51	2	0.236	0.118	2000	0.236	—
	E2ZE-AGA	.41–.51	2	0.236	0.118	2000	0.236	—
	E2ZE-AAA	.41–.51	2	0.236	0.118	2000	0.236	—

*Basic carburetor number
① Figure given is for all manual transmissions; for automatic trans. the figures are: (49 states) 2000 RPM; (Calif.) 1800 RPM.
② See underhood decal

Model 7200

The Motorcraft model 7200 variable venturi (VV) carburetor shares most of its design features with the model 2700 VV. The major difference between the two is that the 7200 is designed to work with Ford's EEC (electronic engine control) feedback system. The feedback system precisely controls the air/fuel ratio by varying signals to the feedback control monitor located on the carburetor, which opens or closes the metering valve in response. This expands or reduces the amount of control vacuum above the fuel bowl, leaning or richening the mixture accordingly.

For further information on feedback carburetors, please refer to *Chilton's Guide To Fuel Injection And Feedback Carburetors.*

FLOAT LEVEL, FLOAT DROP, FAST IDLE SPEED ADJUSTMENTS

These adjustments are performed in the same manner as for the 2700 VV. See that section for procedures.

FAST IDLE CAM ADJUSTMENT

This procedure is the same as for the 2700 VV. Use the procedure in that section. The 7200 VV used on California models has a choke cover held on with rivets. The carburetor must be removed to remove the rivets. With the carburetor removed, the top two rivets can be drilled out with a ⅛ in. drill bit. Drill only through the rivet head. The bottom rivet is located in a blind hole and must be removed by lightly tapping the backside of the retainer ring with a punch. The cover must be installed with replacement rivets, Ford part No. 388575, or the equivalent.

COLD ENRICHMENT METERING ROD ADJUSTMENT

This adjustment is made in the same manner as for the 2700 VV. See the paragraph under "Fast Idle Cam Adjustment," concerning the riveted choke cover used on California models.

INTERNAL VENT, VENTURI VALVE LIMITER ADJUSTMENTS

These adjustments are the same as for the 2700 VV. See that section for details.

CONTROL VACUUM REGULATOR (CVR) ADJUSTMENT

Use the procedure in the 2700 VV section. Note that the control vacuum is not adjustable on any 7200 carburetor; only the regulator is adjustable.

HIGH SPEED CAM POSITIONER, IDLE MIXTURE ADJUSTMENTS

Procedures are the same as for the 2700 VV. See that section for details. Like the 2700 VV, the 7200 idle trim is preset at the factory and non-adjustable.

MOTORCRAFT MODEL 7200 VV SPECIFICATIONS

Year	Model	Float Level (in.)	Float Drop (in.)	Fast Idle Cam Setting (notches)	Cold Enrichment Metering Rod (in.)	Control Vacuum (in. H₂O)	Venturi Valve Limiter (in.)	Choke Cap Setting (notches)
'80	All	1³⁄₆₄	1¹⁵⁄₃₂	1 Rich/3rd step	.125	②	③	④
'81	D9AE-AZA	1.015–1.065	1.435–1.485	1 Rich/3rd step	.125	②	⑤	Index
	EIAE-LA	1.015–1.065	1.435–1.485	0.360/2nd step	⑦	②	⑥	1 Rich
	EIAE-SA	1.015–1.065	1.435–1.485	0.360/2nd step	⑦	②	⑥	1 Rich
	EIAE-KA	1.010–1.070	1.430–1.490	0.360/2nd step	⑩	②	⑭	Index
	EIDE-AA	1.010–1.070	1.430–1.490	0.360/2nd step	⑩	②	⑭	Index
	EIVE-AA	1.015–1.065	1.435–1.485	0.360/2nd step	⑦	②	③	Index

MOTORCRAFT MODEL 7200 VV SPECIFICATIONS

Year	Model	Float Level (in.)	Float Drop (in.)	Fast Idle Cam Setting (notches)	Cold Enrichment Metering Rod (in.)	Control Vacuum (in. H_2O)	Venturi Valve Limiter (in.)	Choke Cap Setting (notches)
'82	E2AE-LB	1.010–1.070	1.430–1.490	0.360/2nd step	⑧	②	⑨	Index
	E2DE-NA	1.010–1.070	1.430–1.490	0.360/2nd step	⑧	②	⑨	Index
	E2AE-LC	1.010–1.070	1.430–1.490	0.360/2nd step	⑧	②	⑨	Index
	E25E-FA	1.010–1.070	1.430–1.490	0.360/2nd step	⑧	②	⑨	Index
	E25E-GB	1.010–1.070	1.430–1.490	0.360/2nd step	⑧	②	⑨	Index
	E2SE-GA	1.010–1.070	1.430–1.490	0.360/2nd step	⑧	②	⑨	Index
	E2AE-RA	1.010–1.070	1.430–1.490	0.360/2nd step	⑩	②	⑨	Index
	E1AE-ACA	1.010–1.070	1.430–1.490	0.360/2nd step	⑩	②	⑨	Index
	E2SE-DB	1.010–1.070	1.430–1.490	0.360/2nd step	⑪	②	⑨	Index
	E2SE-DA	1.010–1.070	1.430–1.490	0.360/2nd step	⑪	②	⑨	Index
	E1AE-SA	1.010–1.070	1.430–1.490	0.360/2nd step	⑫	②	⑬	1 Rich
	E2AE-MA	1.010–1.070	1.430–1.490	0.360/2nd step	⑫	②	⑬	1 Rich
	E2AE-MB	1.010–1.070	1.430–1.490	0.360/2nd step	⑫	②	⑬	1 Rich
	E2AE-TA	1.010–1.070	1.430–1.490	0.360/2nd step	⑫	②	⑬	Index
	E2AE-TB	1.010–1.070	1.430–1.490	0.360/2nd step	⑫	②	⑬	Index
	E25E-AC	1.010–1.070	1.430–1.490	0.360/2nd step	⑪	②	⑨	Index
	E1AE-AGA	1.010–1.070	1.430–1.490	0.360/2nd step	⑫	②	⑨	Index
	E2AE-NA	1.010–1.070	1.430–1.490	0.360/2nd step	⑫	②	⑨	Index
'83	E2AE-NA	1.010–1.070	1.430–1.490	0.360/2nd step	⑫	②	⑨	Index
	E2AE-AJA	1.010–1.070	1.430–1.490	0.360/2nd step	⑫	②	⑨	Index
	E2AE-APA	1.010–1.070	1.430–1.490	0.360/2nd step	⑫	②	⑨	Index
'84–'86	E2AE-AJA	1.010–1.070	1.430–1.490	0.360/2nd step	⑫	②	⑨	Index
	E2AE-APA	1.010–1.070	1.430–1.490	0.360/2nd step	⑫	②	⑨	Index

① Not used
② See text
③ Opening gap: 0.99–1.01
   Closing gap: 0.39–0.41
④ See underhood decal
⑤ Maximum opening: .99/1.01
   Wide open on throttle: .94/.98

⑥ Maximum opening: .99/1.01
   Wide open on throttle: .74/.76
⑦ 0°F—0.490 @ starting position
   75°F—0.475 @ starting position
⑧ 0°F—0.525 @ starting position
   75°F—0.445 @ starting position
⑨ Maximum opening: .99/1.01
   Wide open on throttle: .39/.41

⑩ 0°F—0.490 @ starting position
   75°F—0.445 @ starting position
⑪ 0°F—0.525 @ starting position
   75°F—0.475 @ starting position
⑫ 0°F—0.490 @ starting position
   75°F—0.460 @ starting position
⑬ Maximum opening: .99/1.01
   Wide open on throttle: .74/.76
⑭ Maximum opening: .99/1.01
   Wide open on throttle: .48/.52

HOLLEY CARBURETORS

Model 1945

The model 1945 carburetor is a concentric downdraft single barrel carburetor with an internal float bowl which completely surrounds the venturi. The unit uses dual nitrophyl floats which permit operation at extreme angles. It is used on Chrysler Corporation six cylinder engines.

FLOAT ADJUSTMENT

1. Remove the float bowl cover and invert the bowl. Hold the retaining spring in place.
2. Place a straight-edge across the surface of the bowl. The gasket should be in place. The straight-edge should just clear the toes of the floats by the specified measurement.
3. If the adjustment is necessary, bend the float tang to obtain the correct adjustment.

FAST IDLE ADJUSTMENT

1. Remove the air cleaner and disconnect the vacuum lines to the heated air control and the OSAC (Orifice Spark Advance Control) valve. If there is no OSAC valve, disconnect the hose to the distributor and the EGR hose. Cap all carburetor vacuum fittings.
2. With the engine off, transmission in Neutral and the parking brake set, open the throttle and close the choke.
3. Close the throttle. This will place the fast idle speed screw on the highest step.
4. Move the fast idle cam until the screw drops to the second highest speed step.
5. Start the engine and stabilize the engine speed. Rotate the fast idle speed screw to obtain the specified setting. See Specifications Chart.

FAST IDLE CAM ADJUSTMENT

1. Place the fast idle speed adjust-

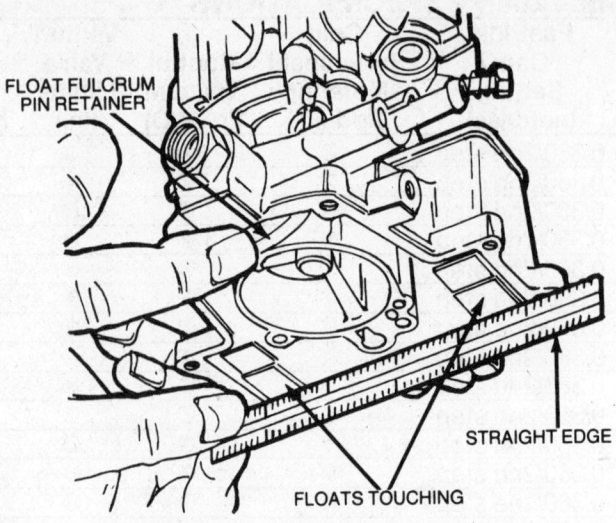

FLOAT FULCRUM PIN RETAINER

STRAIGHT EDGE

FLOATS TOUCHING

Checking the float adjustment—Holley 1945

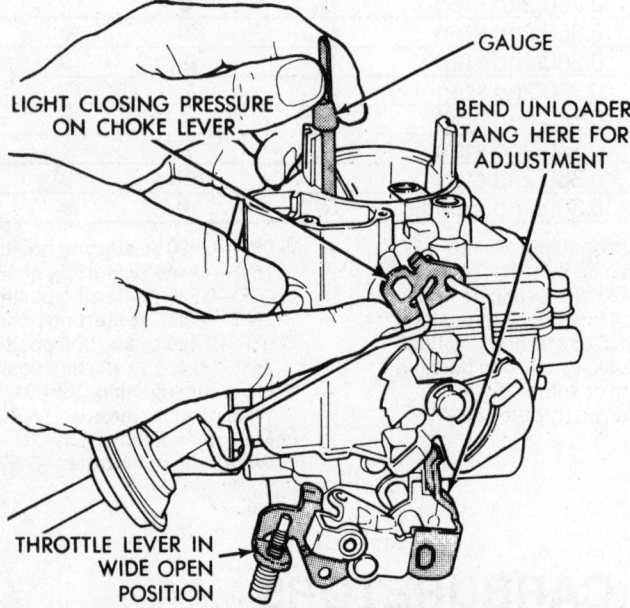

GAUGE

LIGHT CLOSING PRESSURE ON CHOKE LEVER

BEND UNLOADER TANG HERE FOR ADJUSTMENT

THROTTLE LEVER IN WIDE OPEN POSITION

Choke unloader adjustment—Holley 1945.

ing screw on the second highest step of the fast idle cam.

2. Place light pressure on the choke shaft lever to move the choke valve towards the close position.

3. Insert the specified gauge between the top of the choke and the air horn wall at the throttle lever side.

4. To adjust bend the fast idle connector rod at angle until the correct valve opening is obtained.

CHOKE UNLOADER ADJUSTMENT

1. Hold the throttle valves wide-open and insert the specified gauge between the upper edge of the choke valve and the inner wall of the air horn.

2. Place slight pressure against the control lever and attempt to remove the gauge. There should be a slight drag as the gauge is being withdrawn. If adjustment is necessary, bend the unloader tang on the throttle lever until the correct opening has been obtained.

CHOKE VACUUM KICK ADJUSTMENT

1. With the engine not running, open the throttle and move the choke to the closed position. Release the throttle first and then the choke.

2. If an auxiliary vacuum source is used, disconnect the vacuum hose

from the carburetor and connect it to the hose from the vacuum supply with an extra length of tube. Apply a vacuum of 15 or more in. Hg.

3. Insert the correct gauge (see Specifications chart) between the choke valve upper edge and the wall of the air horn. Close and hold the choke rod lever with light pressure. The cylindrical stem of the diaphragm will extend as the internal spring is compressed. This spring must be fully compressed for proper measurement of the vacuum kick.

4. If adjustment is necessary, shorten or lengthen the diaphragm link to obtain the correct opening on models through 1981. On 1982 and later models insert a $5/64$ inch Allen wrench into the vacuum diaphragm and turn to adjust.

─────── CAUTION ───────
Do not twist or bend the diaphragm.

5. Install the vacuum hose on the correct carburetor fitting and connect the fast idle linkage.

6. Check the operation in the following manner. With vacuum applied to the diaphragm, the choke valve should move freely between the open and closed positions. If there is binding, examine the linkage for misalignment or interference caused by bending.

ACCELERATOR PUMP ADJUSTMENT

1. With the throttle in the curb idle position, measure the distance between the pump link pivot and the link connection to the throttle lever. 1980 models have two holes in the throttle lever. Make sure the link is in the correct hole or slot.

2. If the measurement is incorrect, the link may be bent at the "U" to adjust.

NOTE: If the pump link is adjusted, the "Bowl Vent Adjustment" must be checked and, if necessary, reset.

BOWL VENT ADJUSTMENT

1. With the throttle set at curb idle speed, measure the distance from the cover support surface down to the flat on the bowl vent leer.

2. If adjustment is necessary, turn the bowl vent lever adjusting screw with a screwdriver.

3. Install the bowl vent spring and cover plate.

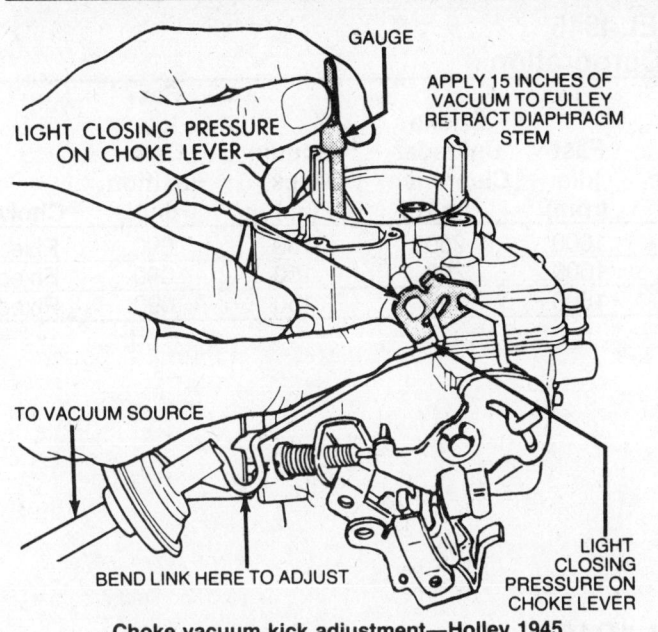

Choke vacuum kick adjustment—Holley 1945

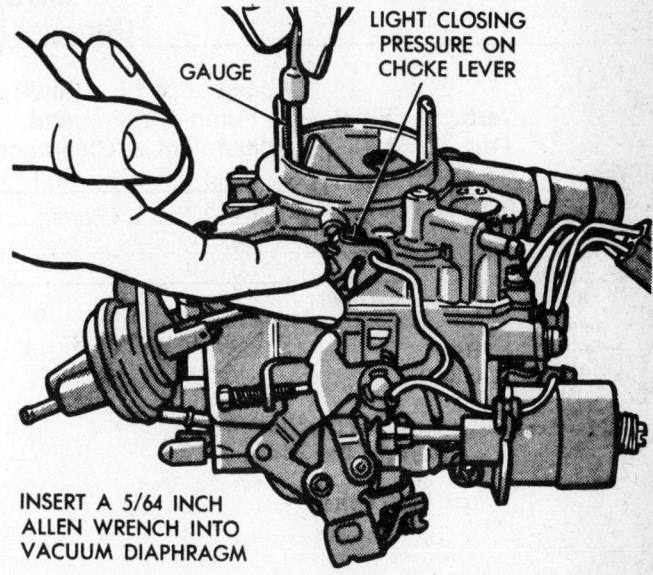

Choke vacuum kick adjustment, 1982—Holley 1945

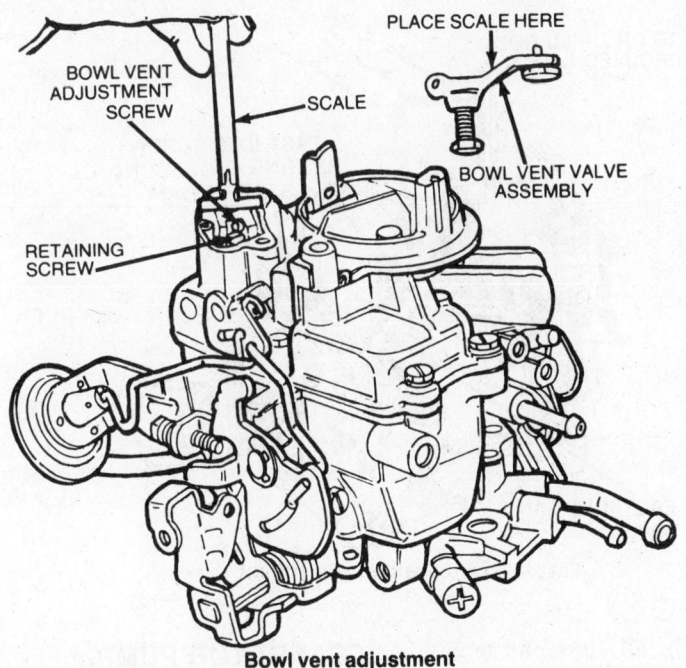

Bowl vent adjustment

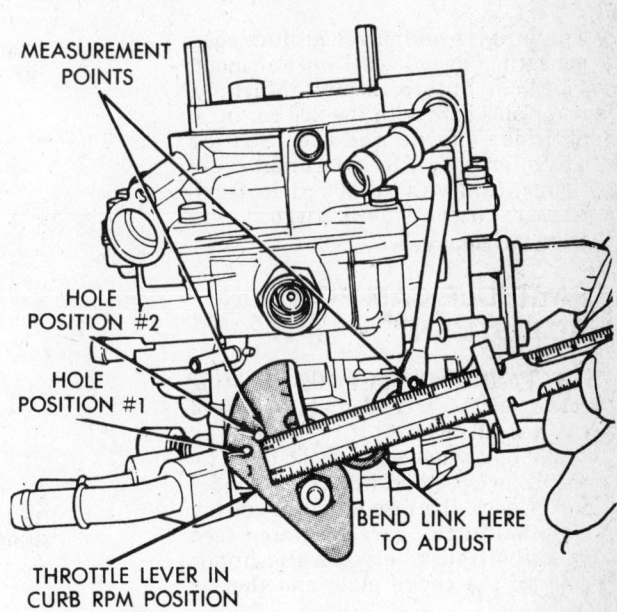

Accelerator pump adjustment

MODEL 1945
Chrysler Corporation

Year	Carb. Part No. ②	Float Level (in.)	Accelerator Pump Adjustment (in.)	Bowl Vent Clearance (in.)	Fast Idle (rpm)	Choke Unloader Clearance (in.)	Vacuum Kick (in.)	Fast Idle Cam Position (in.)	Choke
'80	R-8718-A	①	1.70 ③	1/16	1400	.250	.150	.090	Fixed
	R-8831-A	①	1.615 ④	1/16	1600	.250	.140	.090	Fixed
	R-8832-A	①	1.70 ③	1/16	1400	.250	.110	.090	Fixed
	R-8833-A	①	1.615 ④	1/16	1600	.250	.110	.090	Fixed

MODEL 1945
Chrysler Corporation

Year	Carb. Part No. ②	Float Level (in.)	Accelerator Pump Adjustment (in.)	Bowl Vent Clearance (in.)	Fast Idle (rpm)	Choke Unloader Clearance (in.)	Vacuum Kick (in.)	Fast Idle Cam Position (in.)	Choke
'81	R-9253-A	⑤	1.615 ④	—	1600	.250	.150	.090	Fixed
'82	R-9627A	⑤	1.615 ④	—	1600	.250	.150	.090	Fixed
	R-9628A	⑤	1.615 ④	—	1800	.250	.150	.090	Fixed

① Flush with the top of the bowl cover gasket, plus or minus 1/32
② Located on a tag attached to the carburetor.
③ Position #1
④ Position #2
⑤ Flush with the top of the main body casting to 0.050″ above

Model 1946

This unit is a one barrel, altitude compensating model used on Fairmont, Fairmont Futura, Zephyr, Mustang, and Capri cars with the 200 cu. in., 6 cylinder engine and the 1981–82 Thunderbird, XR-7, Granada and Cougar cars with the 200 cu. in. 6 cylinder engine and automatic transmission.

FAST IDLE CAM POSITION ADJUSTMENT

1. Position the fast idle adjusting screw on the second highest step of the fast idle cam.
2. Lightly move the choke plate toward the closed position.
3. Check the fast idle cam setting by placing the correct gauge (see specifications) between the upper edge of the choke plate and the air horn wall.
4. If the setting is not as specified, bend the fast idle cam link.

FAST IDLE ADJUSTMENT

1. Remove the spark delay valve, if so equipped, and route the distributor vacuum hose directly to the advance side of the distributor.
2. Trace the EGR signal vacuum hose from the EGR valve to the carburetor. If an EGR/PVS valve or cold weather modulator is located in the hose, disconnect the EGR hose at the EGR valve and plug the hose. If not equipped with EGR/PVS or a cold weather modulator, do not detach the hose except on 1980 models; disconnect and plug the EGR hose on all

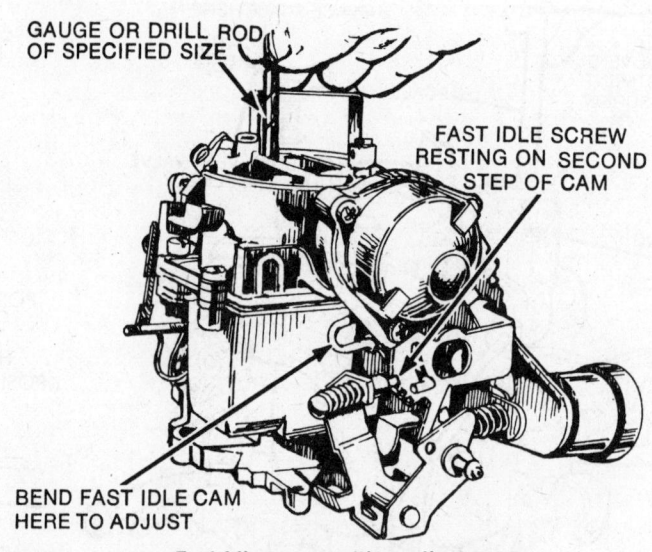

GAUGE OR DRILL ROD OF SPECIFIED SIZE

FAST IDLE SCREW RESTING ON SECOND STEP OF CAM

BEND FAST IDLE CAM HERE TO ADJUST

Fast idle cam position adjustment

1980 models. On all 1981 and later models disconnect and plug the vacuum hoses at the EGR and purge valves.
3. Run the engine to normal operating temperature. With the choke plate fully open and the transmission in Park, place the fast idle screw on the next to the highest step of the fast idle cam. Allow the engine speed to stabilize and adjust the speed to the fast idle speed specification found on the underhood sticker.
4. Run the engine at 2500 rpm for about 15 seconds and recheck the fast idle speed.
5. When the speed is properly adjusted, turn off the engine and re-route the vacuum lines.

ACCELERATOR PUMP STROKE

The accelerator pump stroke is preset at the factory and should not be adjusted to improve driveability.

DECHOKE ADJUSTMENT

1. With the engine OFF, hold the throttle in the wide open position.
2. Insert the specified gauge between the upper edge of the choke plate and the wall of the air horn.
3. With a slight pressure against the choke shaft a slight drag should be felt when the gauge is withdrawn.
4. To adjust, bend the unloader tab

5. Activate the pulldown diaphragm by applying vacuum to the external tube.

6. Make sure that the pulldown diaphragm is fully retracted.

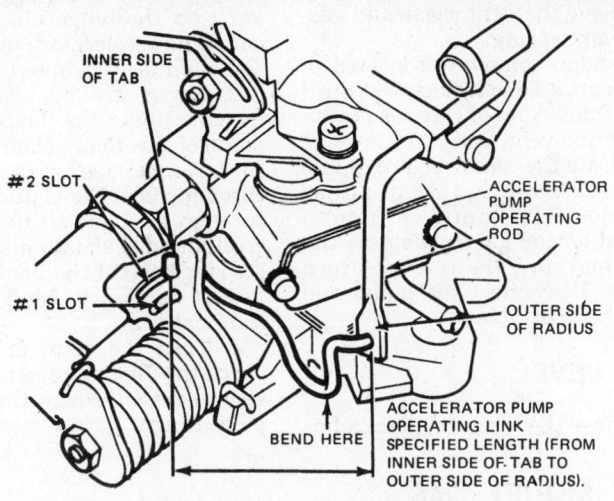

Accelerator pump adjustment

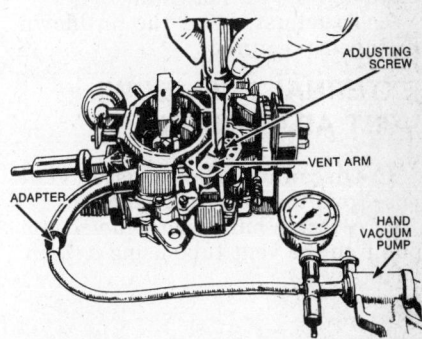

External fuel bowl vent adjustment

on the throttle lever until the correct opening is obtained.

CHOKE PULLDOWN 1980

NOTE: On 1981 and later models this adjustment is preset at the factory and protected by a tamper resistant plug.

1. Set the fast idle screw on the highest step of the fast idle cam.
2. Cool the choke housing until the plate is fully closed.
3. Mark the choke setting for later resetting.

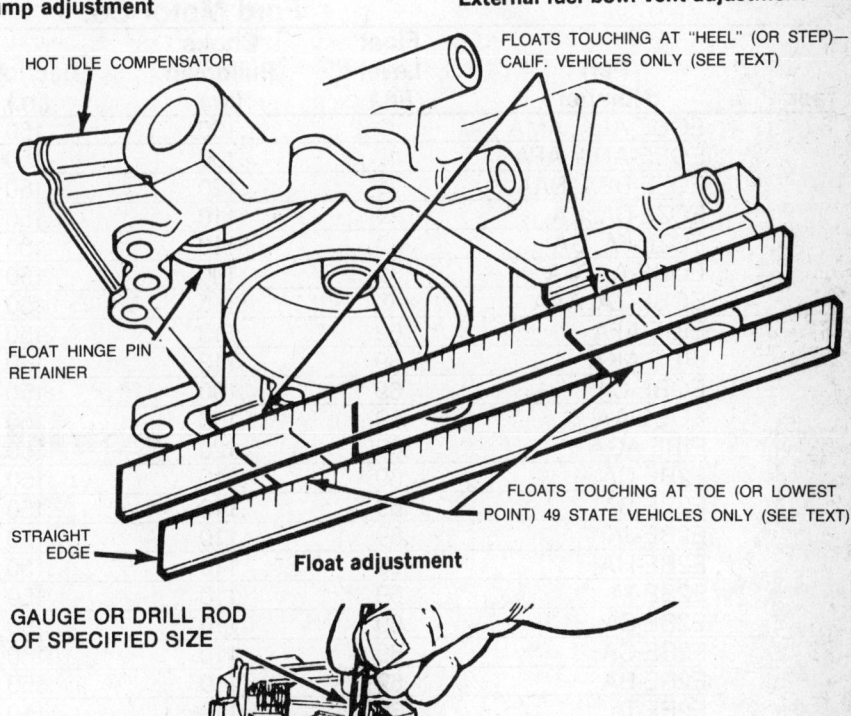

Float adjustment

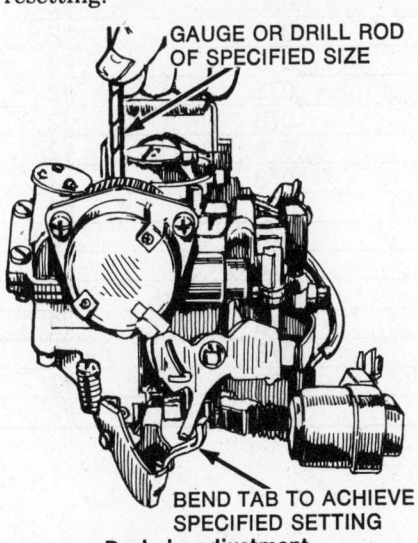

Dechoke adjustment

4. On 1980 California models, removes the choke thermostat housing, retaining ring and screws. Temporarily remove the index spacer. Reinstall the housing, retainer, and screws. Then, on all models, loosen the choke housing screws and rotate the choke cap 90° in the rich (closed) direction. Tighten the screws.

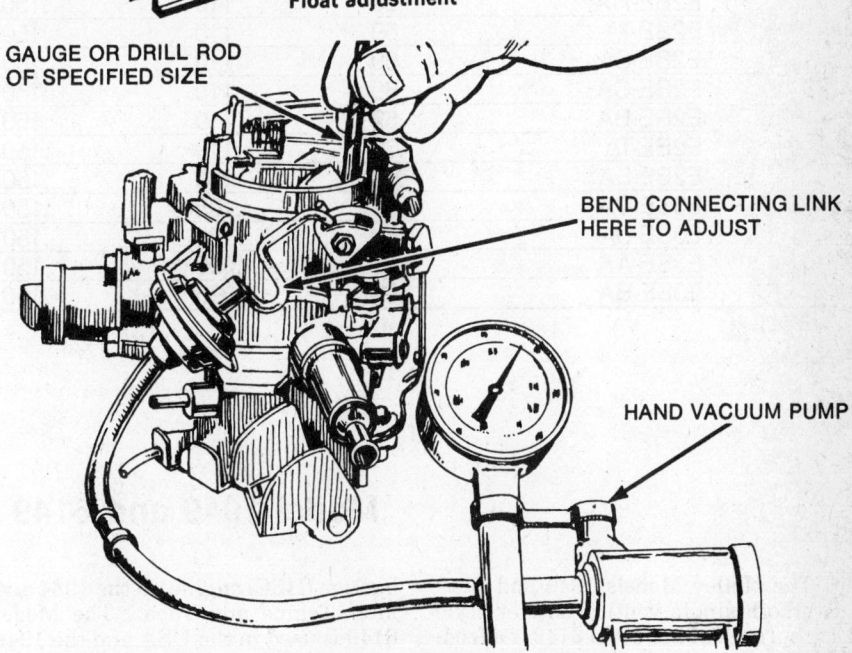

Choke pulldown adjustment

7. If the motor does not fully retract with vacuum, test it for leakage. Replace it if it leaks.

8. Insert the specified gauge between the upper edge of the choke plate and the air horn wall.

9. To adjust, bend the pulldown linkage as required.

EXTERNAL FUEL BOWL VENT ADJUSTMENT

1. Disconnect the canister vent hose from the fuel bowl vent.

2. Attach a hand operated vacuum pump to the vent tube using a ³⁄₈ in. adapter.

3. Remove the vent cover and gasket and vent spring.

4. The adjusting screw is located on the nylon arm. Turn it clockwise until no more than ¹⁄₈ in. of threads is visible above the vent arm.

5. Operate the hand vacuum pump and turn the screw ¹⁄₈ turn at a time counterclockwise, until vacuum is registered on the gauge. Release the vacuum and turn the screw ¹⁄₂ turn clockwise. Disconnect the pump and replace the vent cover.

FLOAT LEVEL

1. Remove the air horn, place a finger over the hinge pin retainer and catch the accelerator pump ball when the main body is inverted.

2. Lay a straight edge across the housing under the floats. The lowest point of the floats should just touch the straight edge for 49 states models through 1981. For California Models, through 1981 and all 1982 models the straight edge should just contact the step (or heel) of the float.

3. If necessary, bend the tang on the float arm.

4. Turn the main body back and check the float alignment. No binding should exist through the float movement range.

MODEL 1946
Ford Motor Co.

Year	Part Number	Float Level (in.)	Choke Pulldown (in.)	Dechoke (in.)	Fast Idle Cam (in.)	Accelerator Pump Stroke Slot
'80	EOBE-ALA, AMA	①	.100	.150	.070	#2
	EOEE-ANA, APA	①	.100	.150	.070	#2
	EOZE-BBA, BAA	①	.120	.150	.086	#2
	EOZE-DA, EA	①	.110	.150	.070	#2
	EOZE-FA, GA	①	.110	.150	.070	#2
	EOBE-AA, CA	①	.100	.150	.070	#2
	EOBE-ZA, AAA	①	.115	.150	.090	#1
'81	EIBE-AFA	.69	.113	.150	.082	#2
	EIBE-AKA	.69	.113	.150	.082	#2
	EOBE-CA	.69	.100	.150	.070	#2
	EOBE-AA	.69	.100	.150	.070	#2
'82	EIBE-AGA	.69	.120	.150	.086	#2
	E2BE-CA	.69	.110	.150	.078	#2
	E2BE-BA	.69	.110	.150	.078	#2
	E2BE-JA	.69	.110	.150	.078	#2
	E2BE-HA	.69	.110	.150	.078	#2
	E2BE-TA	.69	.110	.150	.078	#2
	E2BE-SA	.69	.110	.150	.078	#2
'83	E2BE-CA	.69	.110	.150	.078	#2
	E2BE-BA	.69	.110	.150	.078	#2
	E2BE-TA	.69	.110	.150	.078	#2
	E2BE-SA	.69	.110	.150	.078	#2
	E3SE-CA	.69	.105	.150	.078	#2
	E3SE-DA	.69	.105	.150	.078	#2
	E3SE-AA	.69	.095	.150	.078	#2
	E3SE-BA	.69	.095	.150	.078	#2

① See text

Model 1949 and 6149

The Holley Models 1949 and 6149 are both single venturi booster style carburetors. The Model 6149 is a feedback carburetor. Both carburetors are used on the 2.3 liter High Swirl Combustion (HSC) engine, in the 1984 and later Tempo and Topaz. The Model 6149 is used in the USA and the 1949 in Canada. Both models are used with either manual or automatic transaxles. The Model 6149 carburetor uses twelve basic systems. The Model 1949 carburetor uses thirteen systems. Ten systems are common to both carburetors.

DRY FLOAT LEVEL ADJUSTMENT

1. Remove the carburetor air horn.
2. With the air horn assembly removed, place a finger over float hinge pin retainer, and invert the main body. Catch the accelerator pump check ball and weight.
3. Using a straight edge, check the position of the floats. The correct dry float setting is that both pontoons at the extreme outboard edge by flush with the surface of the main body casting (without gasket). If adjustment is required, bend the float tabs to raise or lower the float level.
4. Once adjustment is correct, turn main body right side up, and check the float alignment. The float should move freely throughout its range without contacting the fuel bowl walls. If the float pontoons are misaligned, straighten by bending the float arms. Recheck the float level adjustment.
5. During assembly, insert the check ball first and then the weight.

AUXILIARY MAIN JET/ PULLOVER VALVE ADJUSTMENT

The length of the auxiliary main jet/ pullover valve adjustment screw which protrudes through the back side (side opposite the adjustment screw head) of the throttle pick-up lever must be 0.345 ± 0.010 in. (8.76mm). To adjust, turn screw in or out as required.

MECHANICAL FUEL BOWL VENT ADJUSTMENT (LEVER CLEARANCE)

Off Vehicle Adjustment

1. Secure the choke plate in the wide-open position.

2. Set the throttle at the TSP Off position.
3. Turn the TSP Off idle adjustment screw counterclockwise until the throttle plate is closed in the throttle bore.
4. Fuel bowl vent clearance: Dimension A should be within 0.120 ± 0.010 in. (3.05mm).
5. If out of specification, bend the bowl vent actuator lever at the adjustment point to obtain the required clearance.

— CAUTION —
Do not bend fuel bowl vent arm and/or adjacent portion of the actuator lever.

NOTE: TSP Off rpm must be set after carburetor installation.

On Vehicle Adjustment

NOTE: This adjustment must be performed after curb idle speed has been set to specification.

1. Secure the choke plate in the wide open position.
2. Turn ignition key to the On position to activate the TSP (engine not running). Open throttle so that the TSP plunger extends.
3. Verify that the throttle is in the idle set position (contacting the TSP plunger). Measure the clearance of the fuel bowl vent arm to the bowl vent actuating lever.
4. Fuel bowl vent clearance: Dimension A should be within 0.020–0.040 in.

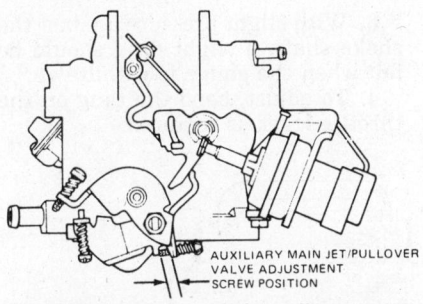

Accelerator pump stroke adjustment— Models 1949 and 6149

NOTE: There is a difference in the on vehicle and off vehicle specification.

5. If out of specification, bend the bowl vent actuator lever at the adjustment point to obtain the required clearance.

— CAUTION —
Do not bend fuel bowl vent arm and/or adjacent portion of the actuating lever.

ACCELERATOR PUMP STROKE ADJUSTMENT

1. Check the length of the accelerator pump operating link from its inside edge at the accelerator pump operating rod to its inside edge at the throttle lever hole. The dimension should be 2.15 ± .010 in. (54.61 ± .25 mm).
2. Adjust to proper length by bending loop in operating link.

CHOKE PLATE PULLDOWN ADJUSTMENT

NOTE: This adjustment is preset at the factory and protected by a tamper resistant plug.

FAST IDLE CAM INDEX ADJUSTMENT

1. With the engine cool, position the fast idle screw on the high step of the fast idle cam.
2. Activate the pulldown motor by applying an external vacuum source of 15–20 inches Hg.
3. Apply light pressure to the upper edge of the choke plate in the closing direction to remove clearance between the pulldown motor clevis and the modulator stem.
4. Open the throttle slightly and allow the fast idle cam to drop.
5. Close the throttle and measure

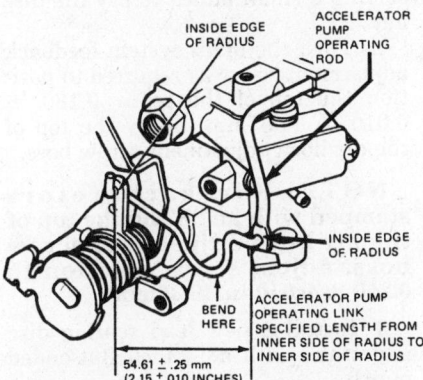

Auxiliary main jet/Pullover valve (timing adjustment)—Models 1949 and 6149

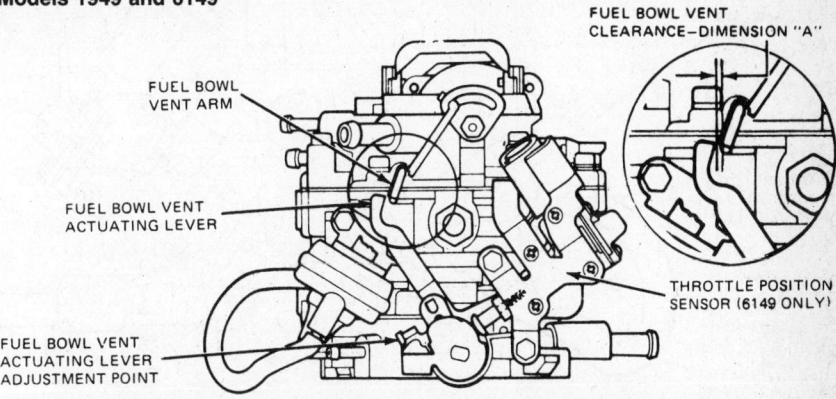

Mechanical fuel bowl vent adjustment—Models 1949 and 6149

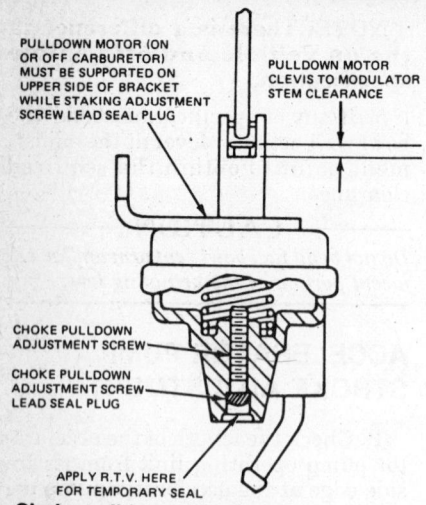

PULLDOWN MOTOR (ON OR OFF CARBURETOR) MUST BE SUPPORTED ON UPPER SIDE OF BRACKET WHILE STAKING ADJUSTMENT SCREW LEAD SEAL PLUG

PULLDOWN MOTOR CLEVIS TO MODULATOR STEM CLEARANCE

CHOKE PULLDOWN ADJUSTMENT SCREW

CHOKE PULLDOWN ADJUSTMENT SCREW LEAD SEAL PLUG

APPLY R.T.V. HERE FOR TEMPORARY SEAL

Choke pulldown adjustment—Models 1949 and 6149

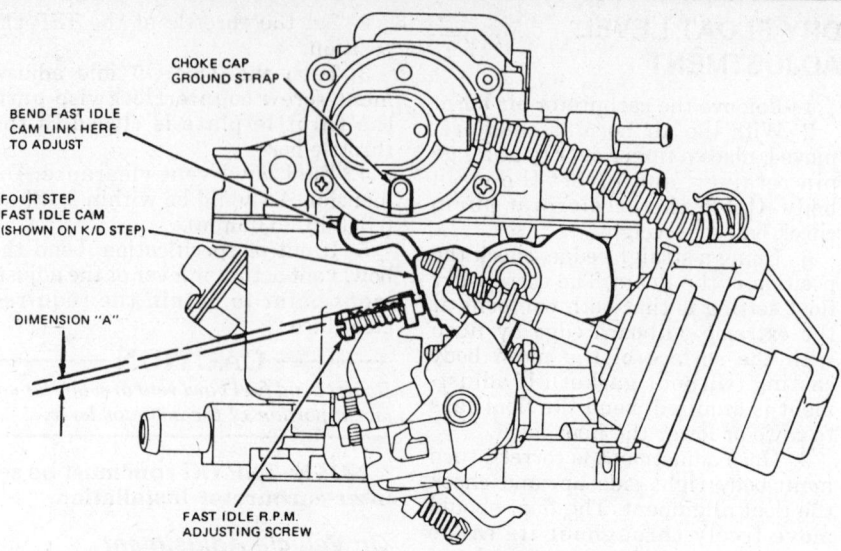

CHOKE CAP GROUND STRAP

BEND FAST IDLE CAM LINK HERE TO ADJUST

FOUR STEP FAST IDLE CAM (SHOWN ON K/D STEP)

DIMENSION "A"

FAST IDLE R.P.M. ADJUSTING SCREW

Fast idle cam index adjustment—Models 1949 and 6149

the clearance between the top edge of the fast idle rpm adjusting screw and the shoulder of the fast idle cam high step (dimension A is the fast idle cam index shown in the illustration). Refer to the specifications table.

6. Remove the light closing pressure from the upper edge of the choke plate.

7. Open the throttle to the wide open position and return slowly.

8. The fast idle adjustment screw must contact the lower end of the fast idle cam kickdown step by at least half of its diameter four carburetors with four step cams or must contact the third step by at least half of its diameter without contacting the second or fourth steps for carburetors with five step cams.

9. If Steps 5 and 8 are okay, the fast idle cam index is within specification. If adjustment is necessary bend the

fast idle cam link at the loop to obtain the correct specification at Dimension A (see the specifications table).

DECHOKE ADJUSTMENT

1. With the engine off and cool, hold the throttle in the wide open position.

2. Use a drill of the specified size and measure the clearance between the upper edge of the choke plate and the air horn wall.

3. With slight pressure against the choke shaft, a slight drag should be felt when the gauge is withdrawn.

4. To adjust, bend the tang on the throttle lever as required.

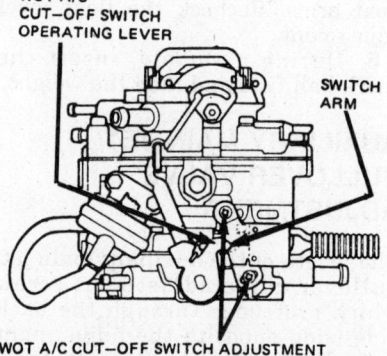

WOT A/C CUT-OFF SWITCH OPERATING LEVER

SWITCH ARM

WOT A/C CUT-OFF SWITCH ADJUSTMENT POINT—LOOSEN SCREWS AND ROTATE BRACKET

WOT A/C cut-off switch adjustment screws—Model 1949

FEEDBACK SYSTEM DIAPHRAGM ADJUSTMENT

Model 6149

1. Remove the main system feedback diaphragm adjustment screw lead sealing disc from the air horn screw boss by drilling a $\frac{3}{32}$ inch diameter hole through the disc, then inserting a small punch to pry the disc out.

2. Turn the main system feedback adjustment screw as required to position the top of the screw 0.180 ± 0.010 in. (4.57mm) below the top of the air horn adjustment screw boss.

NOTE: For carburetors stamped with an "S" on the top of the air horn adjustment screw boss, adjust screw position to 0.250 ± 0.010 in. (6.35mm).

3. Install a new lead sealing disc and stake with a $\frac{1}{4}$ inch flat-ended punch.

4. Apply an external vacuum source (hand vacuum pump, 10 in. Hg

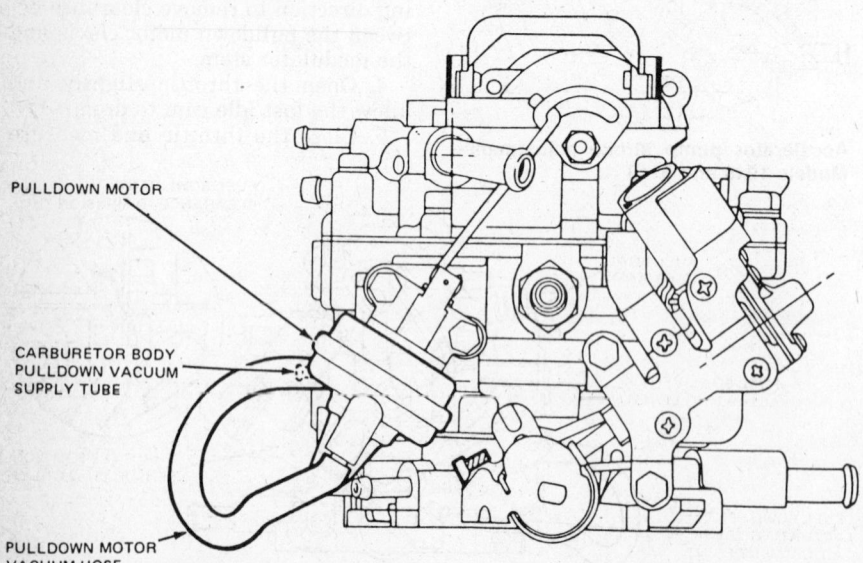

PULLDOWN MOTOR

CARBURETOR BODY PULLDOWN VACUUM SUPPLY TUBE

PULLDOWN MOTOR VACUUM HOSE

Choke plate pulldown motor—Models 1949 and 6149

maximum) and check for leaks, diaphragm should hold vacuum.

WOT A/C CUT-OFF SWITCH ADJUSTMENT

Model 1949

The WOT A/C cut-off switch is a normally closed switch (allowing current to flow at any throttle position other than wide-open throttle).

1. Disconnect the wiring harness at the switch connector.

2. Connect a 12 volt DC power supply and test lamp. With the throttle at curb idle, TSP off idle or fast idle position, the test light must be ON. If the test lamp does not light, replace the switch assembly.

3. Rotate the throttle to the wide-open position. The test lamp must go OFF, indicating an open circuit.

4. If the lamp remains ON, insert a 0.165 in. drill or gauge between the throttle lever WOT stop and the WOT stop boss on the carburetor main body casting. Hold the throttle open as far as possible against the gauge. Loosen

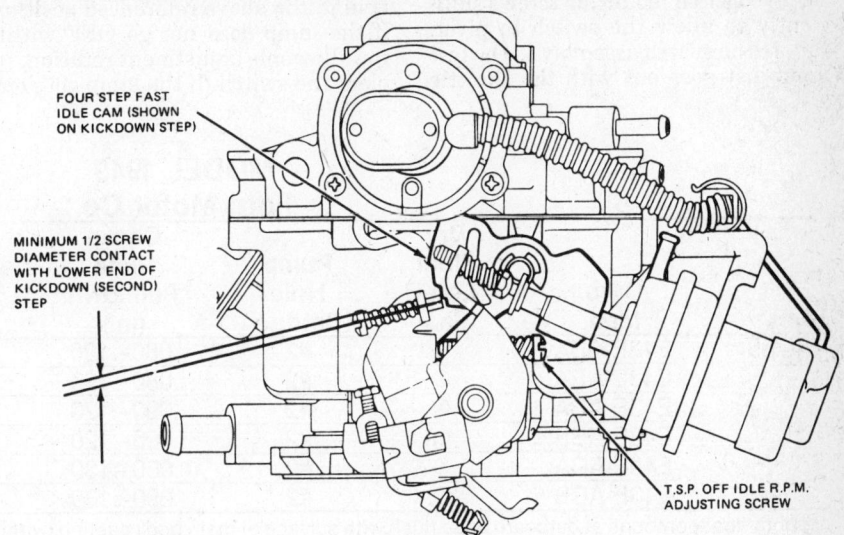

Fast idle cam index (four step idle cams)—Models 1949 and 6149

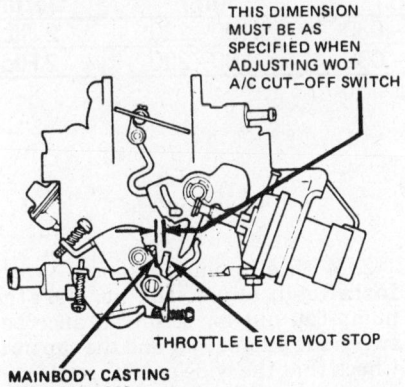

WOT A/C cut-off switch adjustment (clearance)—Model 1949

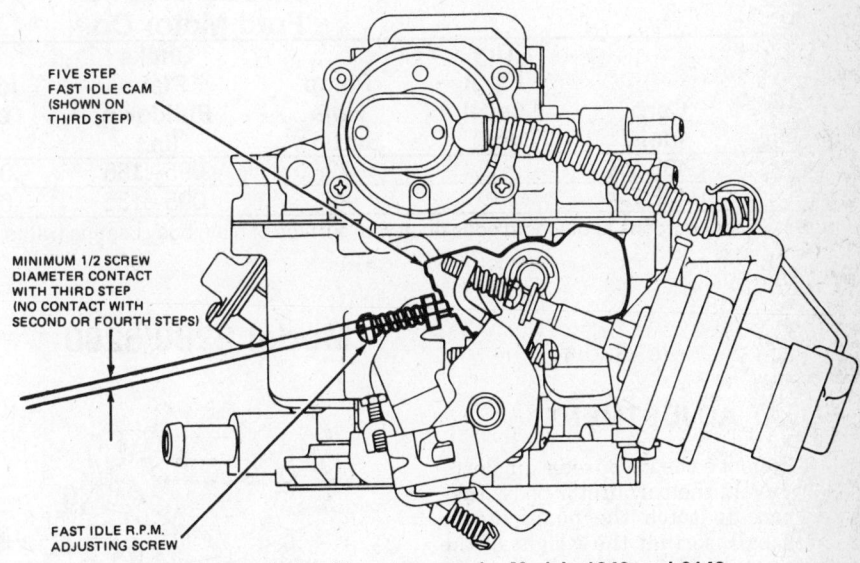

Fast idle cam index (five step cams)—Models 1949 and 6149

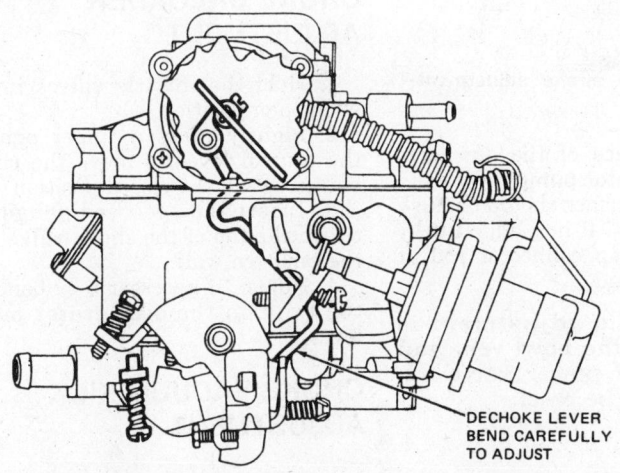

Dechoke adjustment—Models 1949 and 6149

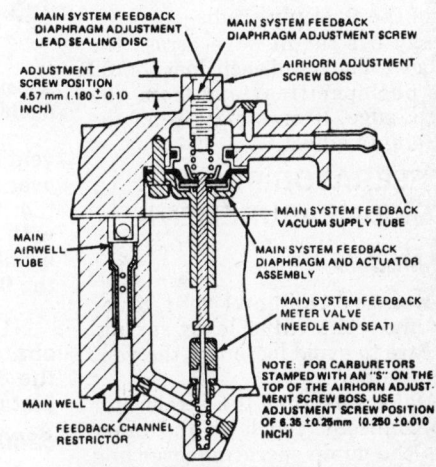

Diaphragm adjustment—Model 6149

the two switch mounting screws sufficiently to allow the switch to pivot. Rotate the switch assembly so the test lamp just goes out with the throttle held in the above referenced position. If the lamp does not go OFF within the allowable adjustment rotation, replace the switch. If the lamp goes out, tighten the two switch bracket-to-carburetor screws to 45 inch lbs. (5 Nm) and remove drill or gauge and repeat Step 3.

MODEL 1949
Ford Motor Co.

Year	Carb. Iden.	Dry Float Level (in.)	Pump Hole Setting	Choke Plate Pulldown (in.)	Fast Idle Cam Linkage (in.)	Dechoke (in.)	Choke Setting
'84–'85	E43E-ADA	①	#2	.080–.120	.020–.030	.180–.220	2 Rich
	E43E-AEA	①	#2	.080–.120	.020–.030	.180–.220	2 Rich
	E43E-ABA	①	#2	.090–.120	.020–.030	.180–.220	1 Rich
	E43E-ABB	①	#2	.090–.120	.020–.030	.180–.220	1 Rich
	E43E-ACA	①	#2	.090–.130	.020–.030	.180–.220	1 Rich
	E43E-ACB	①	#2	.090–.130	.020–.030	.180–.220	1 Rich

① Both float pontoons at outboard edge flush with surface of main body casting (without gasket).

MODEL 6149-FB
Ford Motor Co.

Year	Carb. Iden.	Dry Float Level (in.)	Pump Hole Setting	Choke Plate Pulldown (in.)	Fast Idle Cam Linkage (in.)	Dechoke (in.)	Choke Setting
'84	E43E-VA	①	#2	.095–.135	.020–.030	.180–.220	2 Rich
	E43E-ZA	①	#2	.095–.135	.020–.030	.180–.220	2 Rich

① Both float pontoons at outboard edge flush with surface of main body casting (without gasket).

Model 2280/6280

FLOAT ADJUSTMENT

1. Remove the carburetor air horn.
2. Invert the carburetor body, taking care to catch the pump intake check ball, so that the weight of the floats only is forcing the needle against the seat. Hold a finger against the hinge pin retainer to fully seat the float in the float pin cradle.
3. Lay a straight edge across the float bowl. The toe of each float should be as per specifications from the straight-edge. If necessary, bend the float tang to adjust.

ACCELERATOR PUMP STROKE MEASUREMENT

2280 Models

1. Remove the bowl vent cover plate and vent valve lever spring. Take care to avoid loosening the vent valve retainer.
2. Make sure that the accelerator pump connector rod is in the inner hole of the pump operating lever and the throttle is at curb idle.
3. Place a straight edge on the bowl

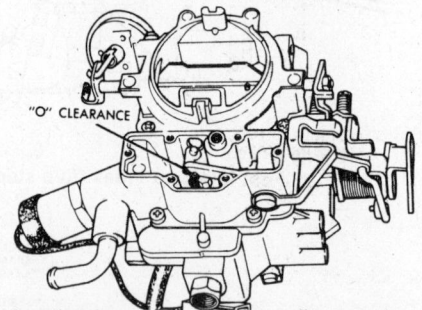

Accelerator pump stroke adjustment—model 6280

vent cover surface of the air horn, over the accelerator pump lever.
4. The lever surface should be flush with the air horn. If not, adjust it by bending the pump connector rod at the 90 degree bend.

NOTE: If this adjustment is changed, both the bowl vent and the mechanical power valve adjustments must be reset.

6280 Models

1. Remove the bowl vent cover plate and gasket.

2. With all pump links and levers installaed, adjust the accelerator pump cap nut for zero clearance between the pump lever and the cap nut. Check that the wide open throttle can be reached without binding.
3. Install the gasket and the bowl vent cover plate.

CHOKE UNLOADER ADJUSTMENT

1. Hold the throttle valves in the wide open position.
2. Lightly press a finger against the control lever to move the choke valve toward the closed position.
3. Insert the specified gauge between the top of the choke valve and the air horn wall.
4. Adjust, if necessary, by bending the tang on the accelerator pump lever.

CHOKE VACUUM KICK ADJUSTMENT

1. Open the throttle, close the choke, then close the throttle to trap

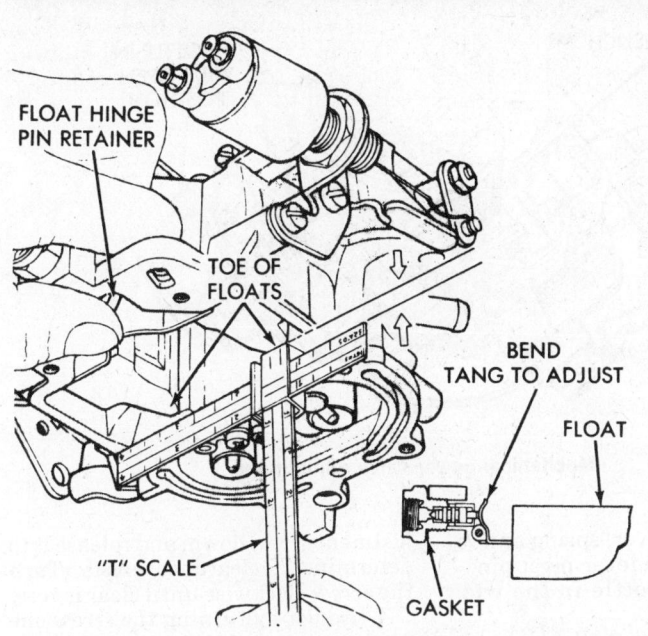

Float adjustment

Accelerator pump stroke adjustment—model 2280

the fast idle cam at the closed choke position.

2. Disconnect the vacuum hose from the carburetor and connect it to an auxiliary vacuum source with a length of hose. Apply at least 15 in. Hg.

3. Completely compress the choke lever spring in the diaphragm stem without distorting the linkage.

4. Insert the specified gauge between the top of the choke valve and the air horn wall.

5. Adjust by bending the diaphragm link. Check for free movement. Replace the vacuum hose.

FAST IDLE CAM POSITION ADJUSTMENT

1. Position the adjusting screw on the second highest step of the fast idle cam.

2. Move the choke towards the closed position with light finger pressure.

3. Insert the specified gauge between the choke valve and the air horn wall.

4. Adjust by opening or closing the U-bend in the fast idle connector link.

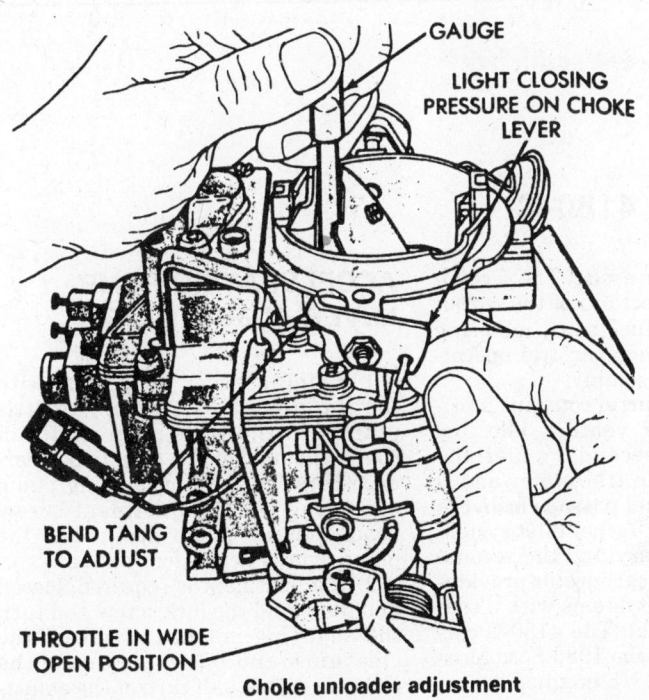

Choke unloader adjustment

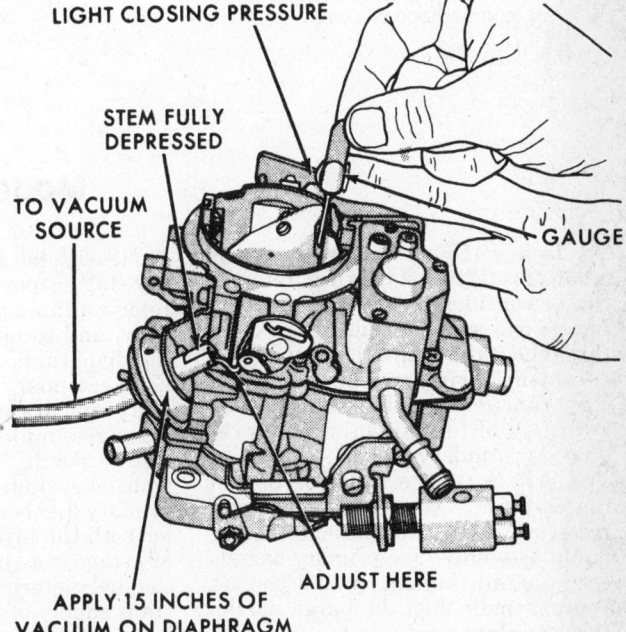

Choke vacuum kick adjustment

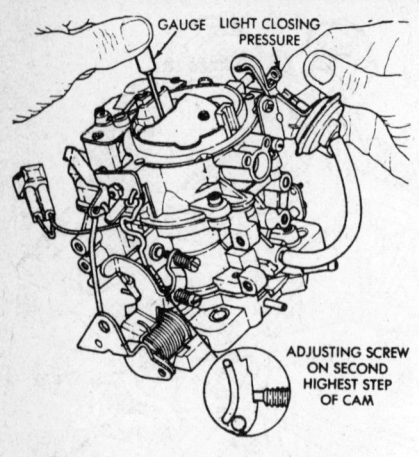

Fast idle cam position adjustment

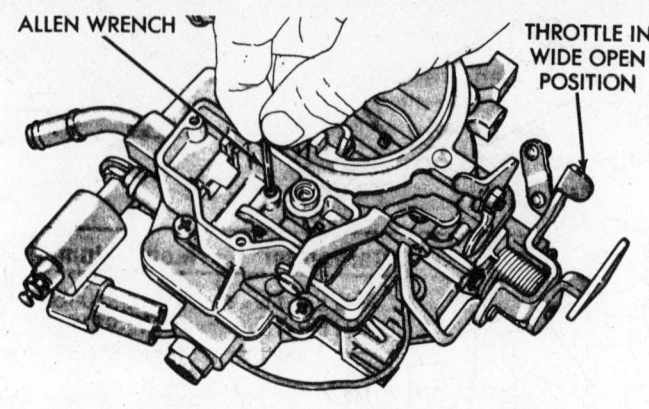

Mechanical power valve adjustment

MECHANICAL POWER VALVE ADJUSTMENT

2280 Models Only

1. Remove the bowl vent cover plate, vent valve lever, spring and retainer. Remove the lever pivot pin.

2. Hold the throttle in the wide open position.

3. Using a $\frac{5}{64}$ in. Allen wrench, press the mechanical power valve adjustment screw down, and release it to determine if clearance exists. Turn the screw clockwise until clear is zero.

4. Adjust by turning the screw one turn counterclockwise.

5. Install all parts.

MODEL 2280/6280
Chrysler Corporation

Year	Carb. Part No.	Float Level (in.)	Accelerator Pump Adjustment (in.)	Fast Idle (rpm)	Choke Unloader Clearance (in.)	Vacuum Kick (in.)	Fast Idle Cam Position (in.)	Choke
'85	R-40121-A	9/32	①	1700	.280	.130	.060	Fixed
	R-40157-A	9/32	①	1600	.200	.140	.052	Fixed
'86	R-40276A	9/32	①	②	.280	.130	.060	Fixed
	R-40245A	9/32	①	②	.200	.140	.052	Fixed

① Flush with top of bowl vent casting
② Refer to underhood sticker

Model 4180-C

The Holley 4180-C 4 bbl carburetor is a downdraft, two-stage carburetor, It can be considered as two dual carburetors; one supply a fuel/air mixture throughout the entire range of engine operating (primary stage), and the other functioning only when a greater quantity of fuel/air mixture is required (secondary stage).

The primary stage (front section of the carburetor contains a fuel bowl, metering block, and an accelerating pump assembly. The primary barrels each contain a primary and booster venturi, main fuel discharge nozzle, throttle plate, and idle fuel passage. The Model 4180-C uses an electric choke with hot air assist. The secondary stage, (rear section) of the carburetor contains a fuel bowl, metering body, and secondary throttle operating diaphragm assembly.

Each secondary barrel contains a primary and booster venturi, idle fuel passages, main secondary fuel discharge nozzle, throttle plate, and a transfer system fuel passage from the primary fuel bowl. A fuel inlet system for both the primary and the secondary stages for the carburetor provides the fuel metering systems with a constant supply of fuel. The 4180-C carburetor is used on the 1983 Ford Mustang with the 302 V8 engine.

ACCELERATING PUMP LEVER ADJUSTMENT

1. Using a feeler gauge and with the throttle plates (primary throttle plates) in the wide open position, there should be the specified clearance between the accelerating pump operating lever adjustment screw head and the pump arm when the pump arm is depressed manually.

2. If adjustment is required, loosen and then hold the lock screw and turn the adjusting nut in to increase the clearance and out to decrease the clearance. One half turn of the adjust-

ing nut is equal to approximately 0.015 in. (0.381mm). When the proper adjustment has been obtained hold the adjustment in position with a wrench and tighten the nut.

DRY FUEL LEVEL FLOAT ADJUSTMENT

The dry float adjustment is a preliminary fuel level adjustment only. The final adjustment ("Wet Fuel Level Adjustment") must be performed after the carburetor is installed on the engine. With the fuel bowls and float assemblies removed, adjust the floats so that the floats are parallel to the fuel bowls, with the top of the fuel bowls inverted.

WET FUEL LEVEL ADJUSTMENT

The fuel pump pressure and volume must be to specifications prior to performing the following adjustments.

1. Operate the engine to normalize engine temperatures and place the vehicle on a flat surface, as near level as possible. Remove the air cleaner, if it was not previously removed.
2. Run engine at 1000 rpm for about 30 seconds to stabilize fuel level.
3. Stop engine and remove sight plug on side of primary carburetor bowl.
4. Check fuel level. It should be at bottom of sight plug hole. If fuel spills out when sight plug is removed, lower fuel level. If fuel level is below sight plug hole, raise fuel level.

---- CAUTION ----

Do not loosen lock screw or nut or attempt to adjust fuel level with sight plug removed or engine running because fuel may spray out creating a fire hazard.

---

5. Adjust the front level as necessary by loosening the lock screw, and turning the adjusting nut clockwise to lower fuel level or counterclockwise to raise fuel level ($\frac{1}{16}$ turn adjusting nut will change fuel level approximately $\frac{1}{32}$ inch). Tighten lock screw and install sight plug, using old gasket. Start engine and run at 1000 rpm for about 30 seconds to stabilize fuel level.
6. Stop engine, remove sight plug and check fuel level. Repeat Step 5 until fuel level is at bottom of sight plug hole. When fuel level is at bottom of sight plug hole, install sight plug using new adjusting plug gasket.
7. Repeat Steps 3–6 for secondary fuel bowl.

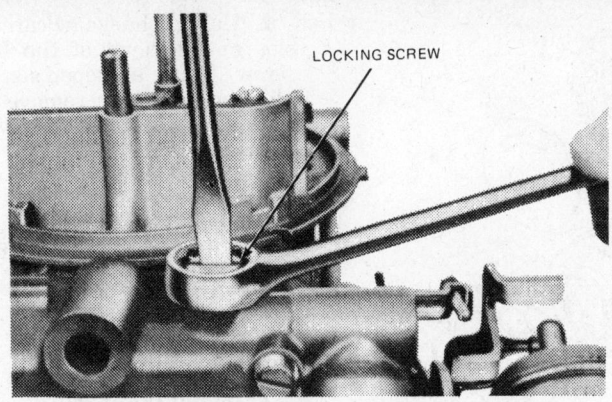

Fuel level adjustment—wet—Holley 4180C

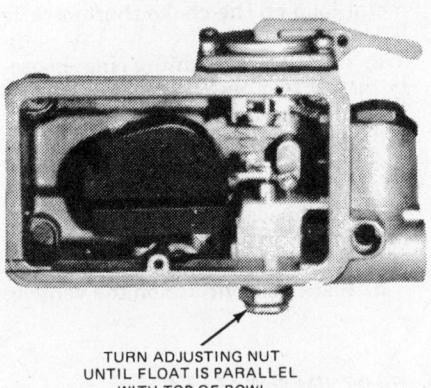

TURN ADJUSTING NUT
UNTIL FLOAT IS PARALLEL
WITH TOP OF BOWL
(HOLDING BOWL UPSIDE DOWN)

Dry float adjustment—Holley 4180C

NOTE: **The secondary throttle must be used to stabilize the fuel level in the secondary fuel bowl.**

SECONDARY THROTTLE PLATE ADJUSTMENT

1. With carburetor off the engine, hold the secondary throttle plates closed.
2. Turn the secondary throttle shaft lever adjusting screw (stop screw) out (counterclockwise) until the secondary throttle plates seat in the throttle bores.
3. Turn the screw in clockwise until the screw JUST contacts the secondary lever, then turn screw in (clockwise) $\frac{1}{4}$ turn.

CHOKE PULLDOWN ADJUSTMENT

1. Remove the choke thermostat housing, gasket and retainer. See the choke cap removal and installation proccedure below.
2. Insert a piece of wire into the choke piston bore to move the piston down against the stop screw. Main-

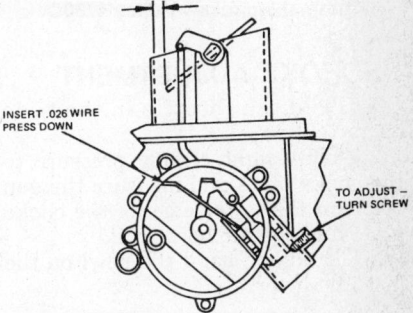

Choke pulldown adjustment—Holley 4180C

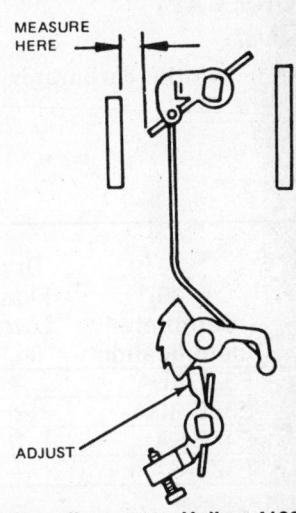

Dechoke adjustment—Holley 4180C

tain light closing pressure on the choke plate and measure the gap between the lower edge of the choke plate and the air horn wall.
3. To adjust, remove the putty covering the adjustment screw and turn the screw clockwise to decrease or counter-clockwise to increase the gap setting. Take care to close the choke plate during screw adjustment. Screw may be turned into side of piston, resulting in damage to piston.
4. Reinstall the choke thermostatic housing, gasket, and retainer.

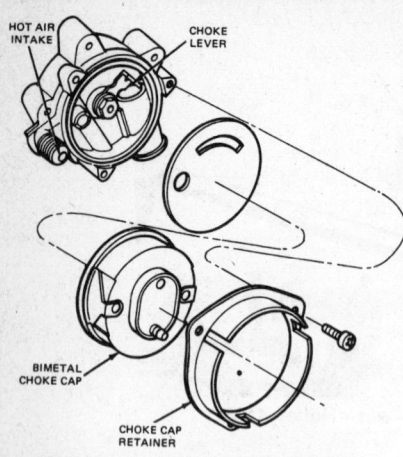

Automatic choke—Holley 4180C

DECHOKE ADJUSTMENT

1. Hold the throttle in the wide open position.
2. Apply light closing pressure to the choke plate and measure the gap between the lower edge of the choke plate and the air horn wall.
3. To adjust, bend the pawl on the fast idle lever.

CHOKE THERMOSTATIC SPRING HOUSING (CHOKE CAP)

Removal

1. Remove the carburetor from vehicle.

2. Using a hacksaw carefully cut a slot in the head of the breakaway screw. Using a proper sized straight blade screw driver, remove the breakaway screw in the conventional manner.
3. Repeat Step 2 for the remaining breakaway screw.
4. Remove the remaining standard screw. Remove the retaining ring, choke cap and gasket.

Installation

1. Install the choke cap gasket. Install the choke cap by engaging the bimetal loop on the choke thermostatic lever.
2. Install the retaining ring. Loosely install two new breakaway screws and one standard screw.
3. Align the choke cap to the proper index mark.
4. Tighten the breakaway screws until the heads break off. Tighten the remaining screw to 16–18 inch lbs. (1.8–2.0 N m).
5. Install carburetor on the vehicle.

FAST IDLE CAM SET

1. Rotate the choke cap 45 degrees counterclockwise (rich) to close the choke plate. Tighten the attaching screw at the time.

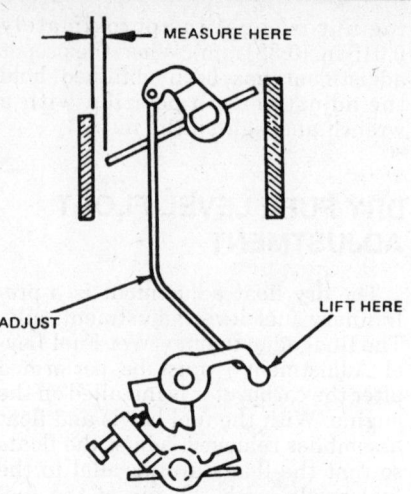

Fast idle cam set—Holley 4180C

2. Open and close the throttle to place the fast idle screw on the top step of the cam.
3. Place a pulldown gauge between the lower edge of the choke plate and the air horn wall, then open and close the throttle to allow the fast idle cam to drop.
4. Press upward on the fast idle cam. There should be little or no movement indicating that the fast idle screw is on the kickdown (2nd) step of the cam, against the first step.

MODEL 4180C
Ford Motor Company

Year	(9510)* Carburetor Identification	Dry Float Level (in.)	Wet Float Level (in.)	Pump Setting Hole	Choke Plate Pulldown (in.)	Fast Idle Cam Linkage Clearance (in.)	Fast Idle (rpm)	Dechoke (in)	Choke Setting
'83	E3ZE-AUA	②	①	#1	.195–.215	NA	③	.300	3 Rich
	E3ZE-BGA	②	①	#1	.195–.215	NA	③	.300	3 Rich
'84	E4ZE-SA	②	①	#1	.195–.215	NA	③	.300	1 Lean
'85	E5ZE-GA	—	①	#1	.168–.188	—	③	.300	2 Lean

NA—not available
① Bottom of sight plug
② See text
③ See Underhood sticker

Model 5210-C

The Holley 5210-C is a progressive two barrel carburetor with an automatic choke system which is activated by a water heated thermostatic coil. An electrically heated choke is used on most later models. It also has an exhaust gas recirculation system with the valve located in the intake manifold. It is used on 1980 Chevettes (USA). 1980–86 Chevettes (Canada).

FLOAT LEVEL

1. With the carburetor air horn inverted, and the float tang resting lightly on the inlet needle, insert the specified gauge between the air horn and the float.
2. Bend the float tang if an adjustment is needed.

FAST IDLE CAM ADJUSTMENT

1. Place the fast idle screw on the second step of the fast idle cam and against the shoulder of the high step.
2. Place the specified drill or gauge on the down side of the choke plate.
3. To adjust, bend the choke lever tang.

CHOKE PLATE PULLDOWN (VACUUM BREAK) ADJUSTMENT

1980–86 Models

1. Attach a hand vacuum pump to the vacuum break diaphragm; apply vacuum and seat the diaphragm.
2. Push the fast idle cam lever down to close the choke plate.
3. Take any slack out of the linkage in the open choke position.
4. Insert the specified gauge between the lower edge of the choke plate and the air horn wall.
5. If the clearance is incorrect, turn the vacuum break adjusting screw, located in the break housing, to adjust.

CHOKE UNLOADER ADJUSTMENT

1. Position the throttle lever at the wide open position.
2. Insert a gauge of the size specified in the chart between the lower edge of the choke valve and the air horn wall.
3. Bend the unloader tang for adjustment.

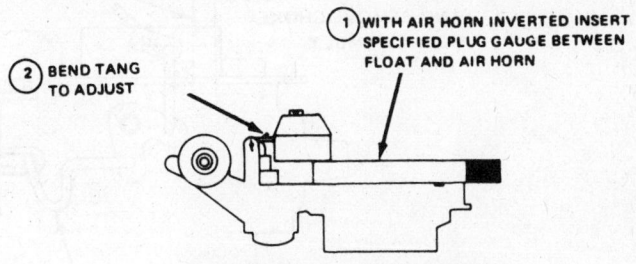

5210-C Float level adjustment

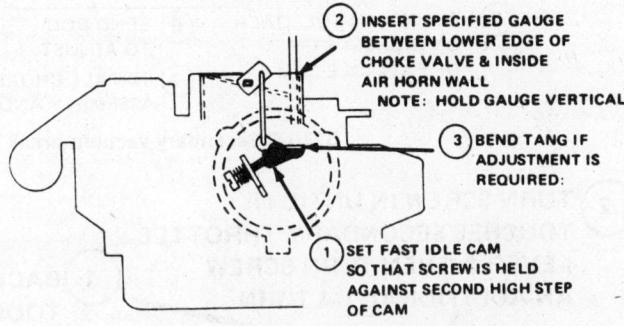

5210-C Fast idle cam adjustment

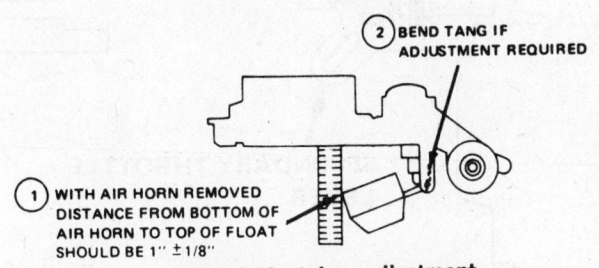

5210-C Float drop adjustment

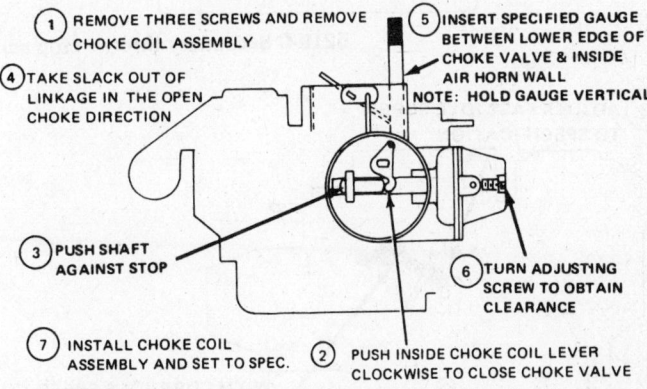

5210-C Vacuum break (choke plate pulldown) adjustment

FAST IDLE SPEED ADJUSTMENT

1. The engine must be at normal operating temperature with the air cleaner off.
2. With the engine running, position the fast idle screw on the high step of the cam for GM cars, or on the second step against the shoulder of the high step for AMC cars. Plug the EGR Port on the carburetor.
3. Adjust the speed by turning the fast idle screw.

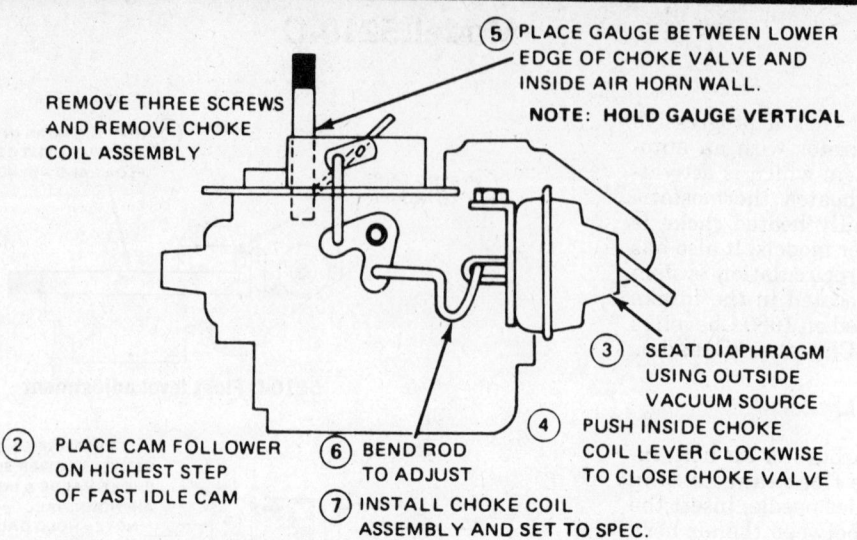

⑤ PLACE GAUGE BETWEEN LOWER EDGE OF CHOKE VALVE AND INSIDE AIR HORN WALL.

NOTE: HOLD GAUGE VERTICAL

REMOVE THREE SCREWS AND REMOVE CHOKE COIL ASSEMBLY

③ SEAT DIAPHRAGM USING OUTSIDE VACUUM SOURCE

② PLACE CAM FOLLOWER ON HIGHEST STEP OF FAST IDLE CAM

⑥ BEND ROD TO ADJUST

④ PUSH INSIDE CHOKE COIL LEVER CLOCKWISE TO CLOSE CHOKE VALVE

⑦ INSTALL CHOKE COIL ASSEMBLY AND SET TO SPEC.

5210-C Secondary vacuum break adjustment

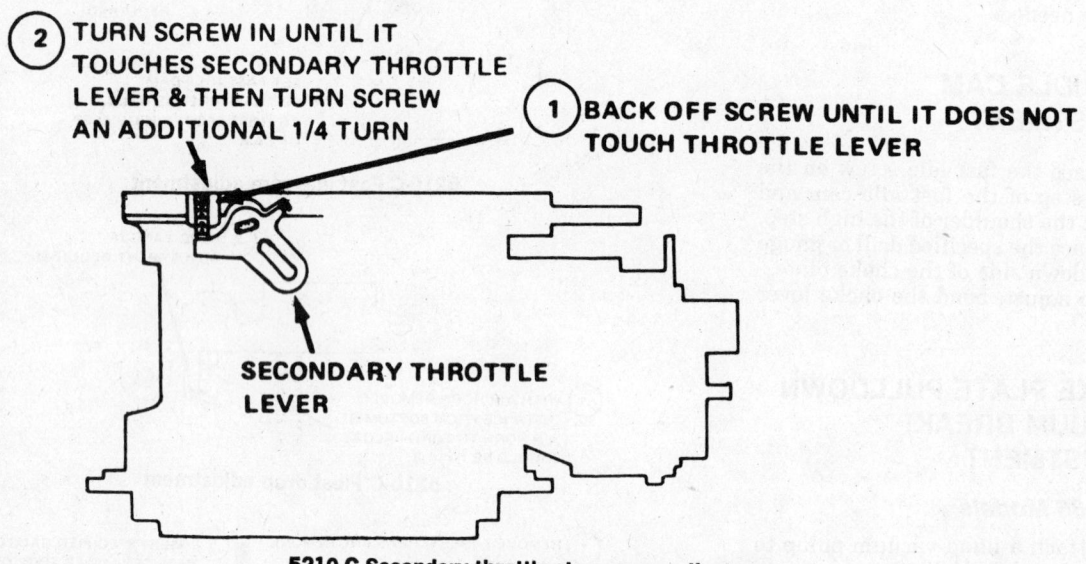

② TURN SCREW IN UNTIL IT TOUCHES SECONDARY THROTTLE LEVER & THEN TURN SCREW AN ADDITIONAL 1/4 TURN

① BACK OFF SCREW UNTIL IT DOES NOT TOUCH THROTTLE LEVER

SECONDARY THROTTLE LEVER

5210-C Secondary throttle stop screw adjustment

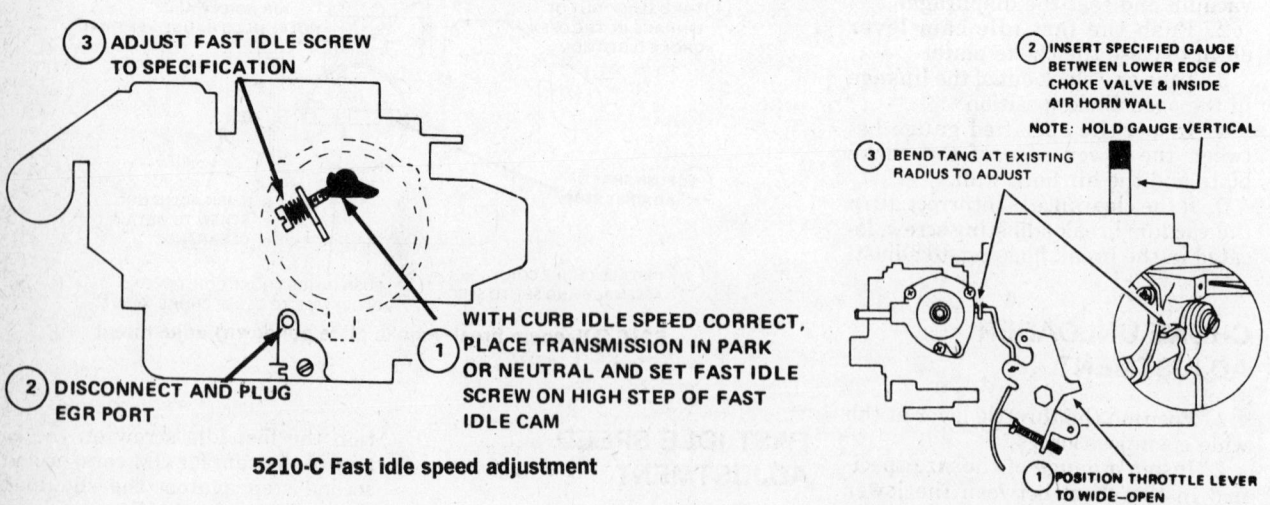

③ ADJUST FAST IDLE SCREW TO SPECIFICATION

② DISCONNECT AND PLUG EGR PORT

① WITH CURB IDLE SPEED CORRECT, PLACE TRANSMISSION IN PARK OR NEUTRAL AND SET FAST IDLE SCREW ON HIGH STEP OF FAST IDLE CAM

5210-C Fast idle speed adjustment

② INSERT SPECIFIED GAUGE BETWEEN LOWER EDGE OF CHOKE VALVE & INSIDE AIR HORN WALL

NOTE: HOLD GAUGE VERTICAL

③ BEND TANG AT EXISTING RADIUS TO ADJUST

① POSITION THROTTLE LEVER TO WIDE-OPEN

5210-C Choke unloader adjustment

MODEL 5210-C
Chevrolet Chevette

Year	Carb. Part No. ① ②	Float Level (Dry) (in.)	Fast Idle Cam (in.)	Secondary Vacuum Break (in.)	Fast Idle Setting (rpm)	Choke Unloader (in.)	Choke Setting
'80	All	0.50	0.110	0.120	2500	0.350	Fixed
'81	14032301	0.50	0.110	0.120	2500	0.350	Fixed
	14032302	0.50	0.110	0.120	2500	0.275	Fixed
'82	14043392	0.50	0.110	0.120	2500	0.275	Fixed
	14043393	0.50	0.110	0.120	2500	0.350	Fixed
'83 (Canada)	All	0.50	0.090	③	④	0.275	Fixed
'84–'85 (Canada)	14076317	0.50	0.110	⑤	④	0.350	Fixed
	14076318	0.50	0.110	⑤	④	0.300	Fixed
	14076319	0.50	0.120	⑥	④	0.350	Fixed
'86 (Canada)	14076393	0.50	0.100	⑤	④	0.325	Fixed
	14076394	0.50	0.090	⑤	④	0.275	Fixed

① Located on tag attached to the carburetor, or on the casting or choke plate
② GM identification numbers are used in place of the Holley numbers
③ Hot: 0.280 Cold: 0.100
④ See underhood sticker
⑤ Hot: 0.250 Cold: 0.100
⑥ Hot: 0.290 Cold: 0.110

Model 5220, 6520

Both these models are staged dual venturi carburetors. The model 6520 has the electronic feedback system. On the 6520 always check the condition of hoses and related wiring before making carburetor adjustments.

For further information on feedback carburetors, please refer to *Chilton's Guide To Fuel Injection And Feedback Carburetors.*

FLOAT SETTING AND FLOAT DROP ADJUSTMENT

1. Remove and invert the air horn.

2. Insert a 0.480 inch gauge between the air horn and float.

3. If necessary, bend the tang on the float arm to adjust.

4. Turn the air horn right side up and allow the float to hang freely. Measure the float drop from the bottom of the air horn to the bottom of the float. It should be exactly $1\frac{7}{8}$ in. Correct by bending the float tang.

VACUUM KICK ADJUSTMENT

1. Open the throttle, close the choke, then close the throttle to trap the fast idle system at the closed choke position.

2. Disconnect the vacuum hose to the carburetor and connect it to an auxiliary vacuum source.

3. Apply at least 15 in. Hg vacuum to the unit.

4. Apply sufficient force to close the choke valve without distorting the linkage.

5. Insert a gauge (see Specification Chart) between the top of the choke plate and the air horn wall.

6. Adjust by rotating the Allen screw in the center diaphragm housing.

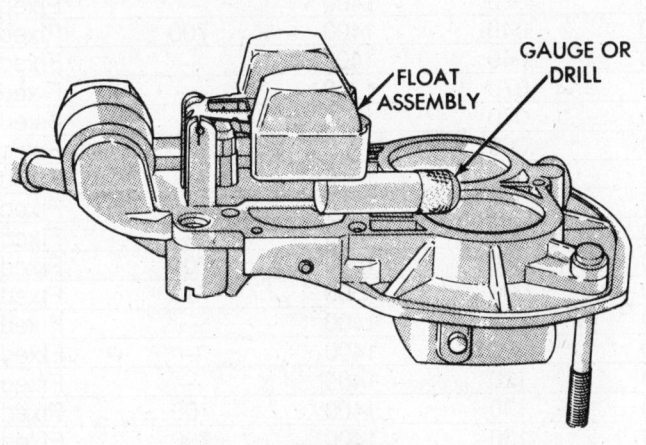

Float setting adjustment

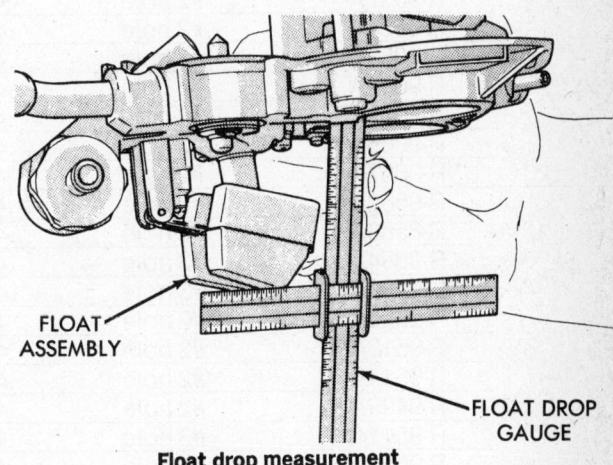

Float drop measurement

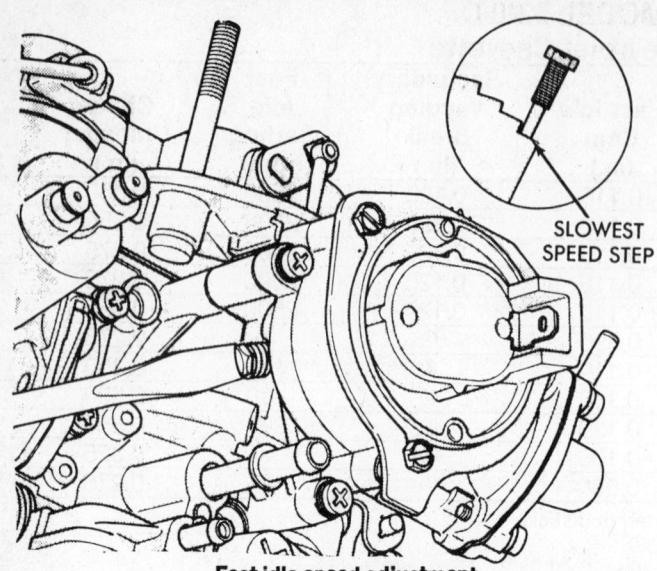

Fast idle speed adjustment

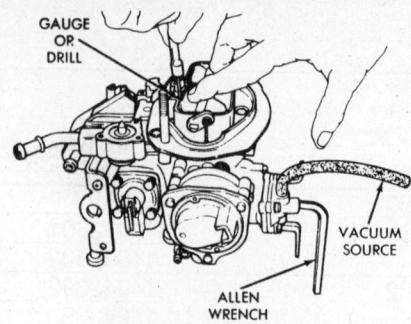

Vacuum kick adjustment

7. Replace the vacuum hose.

FAST IDLE SPEED ADJUSTMENT

1. Remove the air cleaner, disconnect and plug the EGR line, but do not disconnect the spark control computer vacuum line. Turn the air conditioning off.

2. Disconnect the radiator fan electrical connector and use a jumper wire to complete the circuit at the fan. Do not short to ground, as this will damage the system.

3. With the parking brake set and the transmission in Neutral (engine still off), open the throttle and place the fast idle screw on the slowest step of the cam.

4. Start the engine and check the idle speed. If it continues to rise slowly, the idle stop switch is not grounded properly.

5. Adjust the fast idle with the screw, moving the screw off the cam each time to adjust. Allow the screw to fall back against the cam and the speed to stabilize between each adjustment.

MODEL 5220
Chrysler Corporation

Year	Carb. Part No.	Accelerator Pump	Dry Float Level (in.)	Vacuum Kick (in.)	Fast Idle RPM (w/fan)	Throttle Stop Speed RPM	Choke
'80	R8838A, 8839A, 9110A, 9111A, 9325A, 9327A	#2 hole	.480	.040	1700	700	Fixed
	R8726A, 8727A, 8837A, 9108A, 9321A, 9323A	#2 hole	.480	.070	1400	700	Fixed
	R9109A	#2 hole	.480	.100	1400	700	Fixed
'81	R-9056A	#2 hole	.480	.070	1400	700	Fixed
	R-9057A	#2 hole	.480	.070	1400	—	Fixed
	R-9058A	#2 hole	.480	.040	1400	700	Fixed
	R-9059A	#2 hole	.480	.040	1400	—	Fixed
	R-9064A	#2 hole	.480	.070	1300	—	Fixed
	R-9065A	#2 hole	.480	.070	1300	—	Fixed
	R-9066A	#2 hole	.480	.060	1300	700	Fixed
	R-9067A	#2 hole	.480	.060	1300	—	Fixed
'82	R-9582A	#3 hole	.480	.060	1200	700	Fixed
	R-8583A	#3 hole	.480	.060	1200	—	Fixed
	R-9584A	#3 hole	.480	.060	1500	700	Fixed
	R-9585A	#3 hole	.480	.060	1500	700	Fixed
	R-9820A	#2 hole	.480	.080	1400	—	Fixed
	R-9513A	#2 hole	.480	.120	1400	—	Fixed
	R-9514A	#2 hole	.480	.120	1400	—	Fixed
	R-9499A	#2 hole	.480	.130	1400	700	Fixed
	R-9511A	#3 hole	.480	.130	1400	700	Fixed
	R-9512A	#2 hole	.480	.130	1400	—	Fixed

MODEL 5220
Chrysler Corporation

Year	Carb. Part No.	Accelerator Pump	Dry Float Level (in.)	Vacuum Kick (in.)	Fast Idle RPM (w/fan)	Throttle Stop Speed RPM	Choke
'83	R-40020A	#3 hole	.480	.055	1500	—	Fixed
	R-40022A	#3 hole	.480	.055	1500	—	Fixed
	R-40023A	#2 hole	.480	.070	1400	700	Fixed
	R-40024A	#2 hole	.480	.070	1400	700	Fixed
	R-40025A	#2 hole	.480	.070	1400	700	Fixed
	R-40026A	#7 hole	.480	.070	1400	700	Fixed
'84	R-400601A	#2 hole	.480	.055	1200	—	Fixed
	R-400851A	#2 hole	.480	.040	1500	—	Fixed
	R-40170A	#3 hole	.480	.060	1650	—	Fixed
	R-40171A	#3 hole	.480	.060	1700	—	Fixed
	R-400671A	#3 hole	.480	.070	1500	—	Fixed
	R-400681A	#3 hole	.480	.070	1700	—	Fixed
	R-400581A	#2 hole	.480	.070	1400	—	Fixed
	R-401071A	#2 hole	.480	.070	1600	—	Fixed
'85	R-40060-A	#2 hole	.480	.050	①	—	Fixed
	R-40116-A	#3 hole	.480	.095	①	—	Fixed
	R-40117-A	#3 hole	.480	.095	①	—	Fixed
'86	R-40060-2A	—	.480	.055	①	—	Fixed
	R-40116-A	—	.480	.095	①	—	Fixed
	R-40117-A	—	.480	.095	①	—	Fixed

① See underhood sticker

MODEL 6520
Chrysler Corporation

Year	Carb. Part No. ①	Accelerator Pump	Dry Float Level (in.)	Float Drop (in.)	Vacuum Kick (in.)	Fast Idle RPM
'81	R-9052A	#2 hole	.480	1.875	.070	1400 ②
	R-9053A	#2 hole	.480	1.875	.070	1400 ②
	R-9054A	#2 hole	.480	1.875	.040	1400 ②
	R-9055A	#2 hole	.480	1.875	.040	1400 ②
	R-9060A	#2 hole	.480	1.875	.030	1100 ②
	R-9061A	#2 hole	.480	1.875	.030	1100 ②
	R-9602A	#2 hole	.480	1.875	.035	1500 ②
	R-9603A	#2 hole	.480	1.875	.035	1500 ②
	R-9125A	#2 hole	.480	1.875	.030	1200 ②
	R-9126A	#2 hole	.480	1.875	.030	1200 ②
	R-9604A	#2 hole	.480	1.875	.035	1600 ②
	R-9605A	#2 hole	.480	1.875	.035	1600 ②
'82	R-9822A	#2 hole	.480	1.875	.080	1400
	R-9823A	#2 hole	.480	1.875	.080	1400
	R-9824A	#2 hole	.480	1.875	.065	1400
	R-9503A	#3 hole	.480	1.875	.085	1300
	R-9504A	#3 hole	.480	1.875	.085	1300
	R-9505A	#3 hole	.480	1.875	.100	1600
	R-9506A	#3 hole	.480	1.875	.100	1600
	R-9750A	#3 hole	.480	1.875	.085	1300
	R-9751A	#3 hole	.480	1.875	.085	1300
	R-9509A	#3 hole	.480	1.875	.085	1600
	R-9510A	#3 hole	.480	1.875	.085	1600

MODEL 6520
Chrysler Corporation

Year	Carb. Part No. ①	Accelerator Pump	Dry Float Level (in.)	Float Drop (in.)	Vacuum Kick (in.)	Fast Idle RPM
'82	R-9752A	#3 hole	.480	1.875	.100	1600
	R-9753A	#3 hole	.480	1.875	.100	1600
	R-9507A	#3 hole	.480	1.875	.085	1300
	R-9508A	#3 hole	.480	1.875	.085	1300
'83	R-40003A	#3 hole	.480	1.875	.070	1400
	R-40004A	#3 hole	.480	1.875	.080	1500
	R-40005A	#3 hole	.480	1.875	.080	1350
	R-40006A	#3 hole	.480	1.875	.080	1275
	R-40007A	#3 hole	.480	1.875	.070	1400
	R-40008A	#3 hole	.480	1.875	.070	1600
	R-40010A	#3 hole	.480	1.875	.080	1500
	R-40012A	#3 hole	.480	1.875	.070	1600
	R-40014A	#3 hole	.480	1.875	.080	1275
	R-40080A	#2 hole	.480	1.875	.045	1400
	R-40081A	#3 hole	.480	1.875	.045	1400
'84	R-400641A	#3 hole	.480	1.875	.080	1500
	R-400651A	#3 hole	.480	1.875	.080	1600
	R-400811A	#2 hole	.480	1.875	.080	1500
	R-400821A	#2 hole	.480	1.875	.080	1600
	R-40071A	#3 hole	.480	1.875	.080	1500
	R-40122A	#2 hole	.480	1.875	.080	1500
'85	R-40058A	#2 hole	.480	1.875	.070	1400
	R-40134A	#3 hole	.480	1.875	.075	1700
	R-40135A	#3 hole	.480	1.875	.075	1850
	R-40138A	#3 hole	.480	1.875	.075	1700
	R-40139A	#3 hole	.480	1.875	.075	1850
'86	R-40058-1A	—	.480	1.875	.070	③
	R-40134-A	—	.480	1.875	.075	③
	R-40135-1A	—	.480	1.875	.075	③
	R-40138-1A	—	.480	1.875	.075	③
	R-40139-1A	—	.480	1.875	.075	③

① Located on tag attached to the carburetor
② With radiator fan running
③ Refer to underhood sticker

Model 6145

FLOAT ADJUSTMENT

1. With the gasket in place invert the bowl and place a straight edge across the gasket surface. The portion of the floats, farthest from the fuel inlet, should just touch the straight edge.
2. If adjustment is necessary, bend the float tang.

CHOKE VACUUM KICK ADJUSTMENT

1. Open the throttle, close choke then close throttle so that the fast idle cam is at the closed position.
2. Disconnect the vacuum hose from the carburetor and connect to a hose of an auxiliary vacuum source with an extra length of tube. Apply a vacuum of 15 or more inches of mercury.
3. Apply light pressure on the choke lever to close the choke and measure the distance between the choke valve and the air horn wall on the throttle lever side with the specified gauge.
4. On 1981 models bend the diaphragm link at the U-bend to adjust on 1982–83 models insert a $5/64$ in. Allen wrench into the choke diaphragm and turn to adjust the choke vacuum kick.
5. Reconnect the vacuum hose after adjustment.

FAST IDLE CAM ADJUSTMENT

1. Position the fast idle speed adjusting screw on the second highest step of the fast idle cam.
2. Using light pressure on the choke shaft lever, move the choke towards the closed position.
3. Insert the specified gauge between the top of the choke valve and the air horn wall at the throttle lever side.
4. If an adjustment is necessary, bend the fast idle connecting rod at the angle until the correct valve opening is obtained.

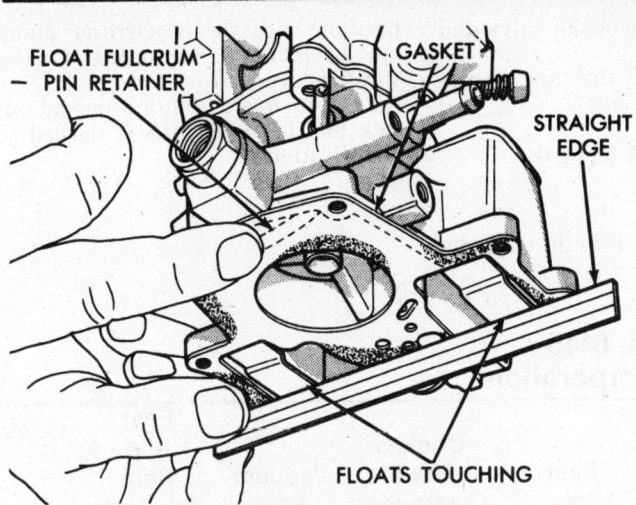

Float level adjustment—Holley 6145

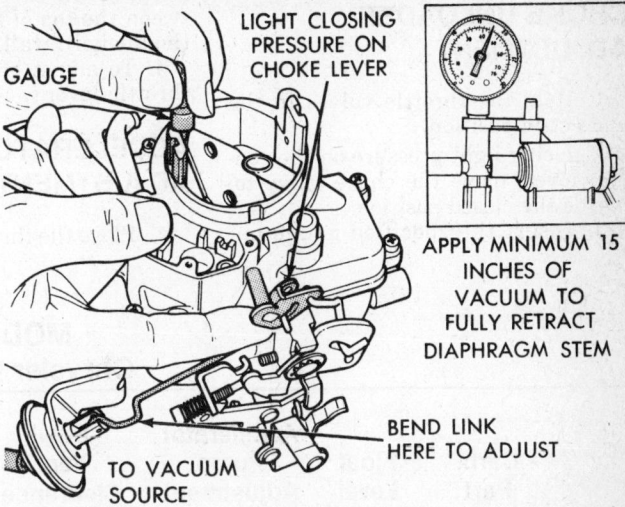

APPLY MINIMUM 15 INCHES OF VACUUM TO FULLY RETRACT DIAPHRAGM STEM

Choke vacuum kick adjustment, 1981—Holley 6145

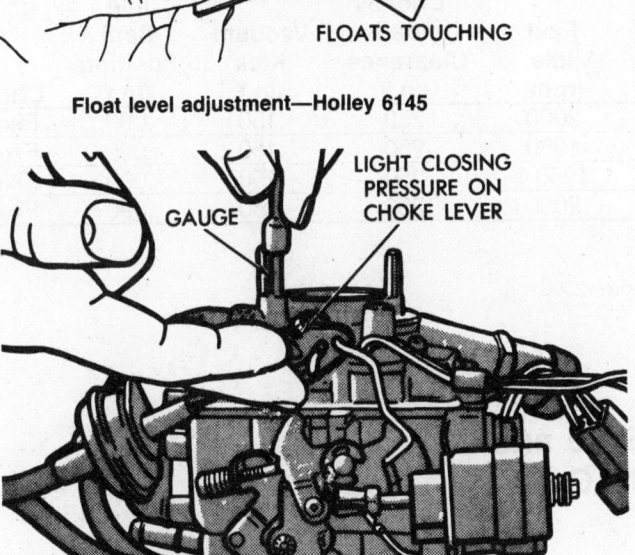

INSERT A 5/64 INCH ALLEN WRENCH INTO VACUUM DIAPHRAGM TO ADJUST VACUUM KICK

Choke vacuum kick adjustment, 1982–83—Holley 6145

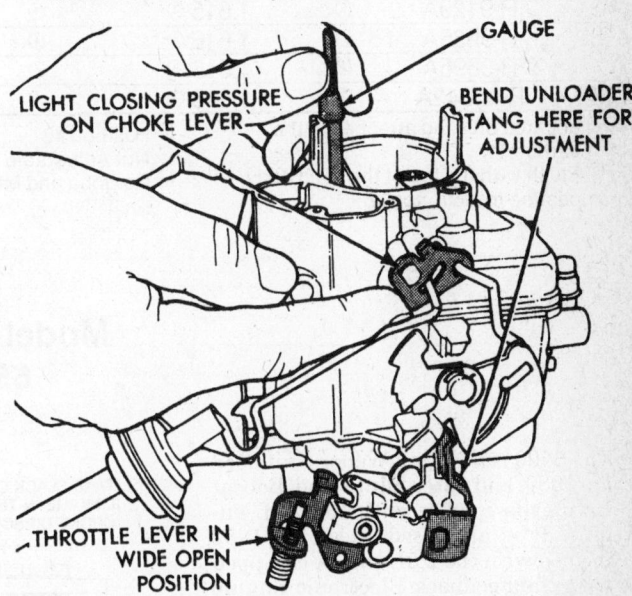

Choke unloader adjustment, typical—Holley 6145

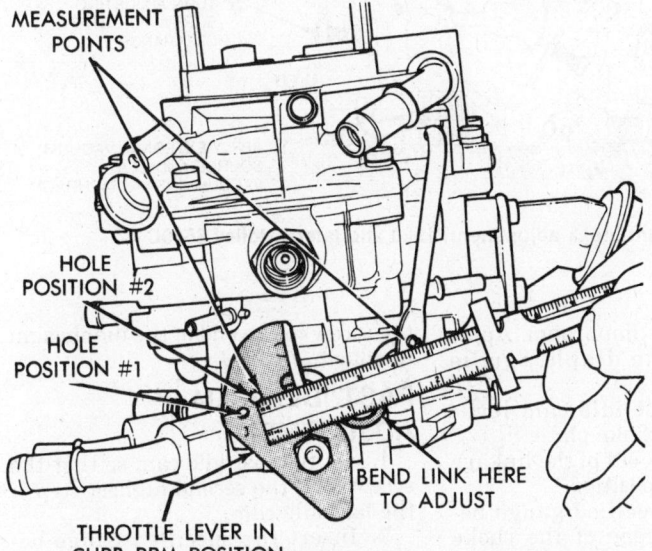

Accelerator pump adjustment, typical—Holley 6145

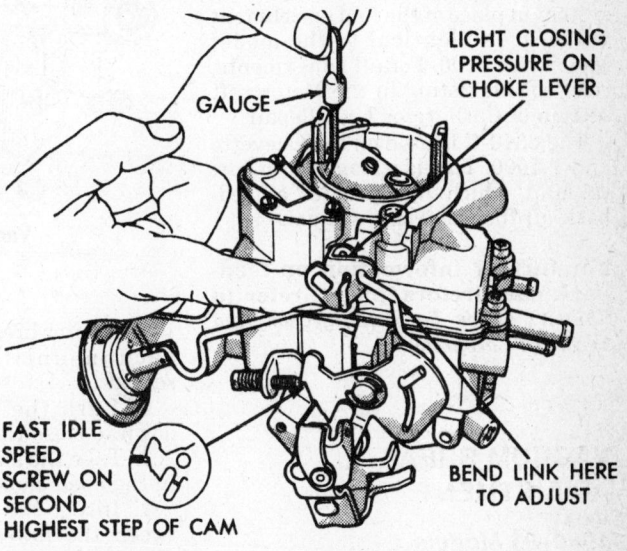

Fast idle cam adjustment, typical—Holley 6145

CHOKE UNLOADER ADJUSTMENT

1. Hold the throttle valves in the wide open position.
2. Using light pressure on the control lever, move the choke valve towards the closed position.
3. Insert the specified gauge between the top of the choke valve and the air horn wall.
4. To adjust bend the tang on the throttle lever.

ACCELERATOR PUMP ADJUSTMENT

1. Place the throttle in the curb idle position with the accelerator pump operating link in the proper slot in the throttle lever.
2. Measure the pump operating link and bend the link if needed to specifications.

MODEL 6145
Chrysler Corporation

Year	Carb. Part No. ①	Float Level (in.)	Accelerator Pump Adjustment (in.)	Bowl Vent Clearance (in.)	Fast Idle (rpm)	Choke Unloader Clearance (in.)	Vacuum Kick (in.)	Fast Idle Cam Position (in.)	Choke
'81	R-9129A	②	1.615 ③	④	2000	.250	.150	.090	Fixed
'82	R-9936A	②	1.616 ③	④	1950	.250	.150	.090	Fixed
	R-9695A	②	1.615 ③	④	1950 ⑤	.250	.150	.090	Fixed
'83	R-40042A	②	1.615 ③	④	2000	.250	.150	.090	Fixed

① Located on a tag attached to the carburetor
② Flush with the top of the main body casting to .050″ above
③ Position #2
④ Not Adjustable
⑤ Cordoba and Mirada—2000 rpm

Model 6500 and 6510-C

The 6500 is a Holley-Weber Unit used on 1980 and later Pinto and Bobcat California models with the 2.3L engine. It is also used on all 1981–82 models with the 2.3L engine equipped with the Feedback Electronic Engine Control System. With the exception of an externally variable fuel metering system in place of the fuel enrichment valve, it is identical to the model Motorcraft 5200. For all adjustments, refer to this listing in the Motorcraft section of Carburetor Unit Repair.

The 6510-C is used on the Chevette and T-1000. This is a staged, two barrel unit which incorporates a feedback air/fuel metering system.

For further information on feedback carburetors, please refer to *Chilton's Guide To Fuel Injection And Feedback Carburetors.*

Vacuum break adjustment, 1980 and later—Holley 6510C

VACUUM BREAK ADJUSTMENT

1980–83 Models

1. Attach a hand vacuum pump to the vacuum break diaphragm. Apply vacuum until the diaphragm is seated.
2. Push the fast idle cam lever down to close the choke plate.
3. Take the slack out of the linkage in the open choke position.
4. Insert the specified gauge between the lower edge of the choke plate and the air horn wall.
5. If the clearance is incorrect, turn the screw in the end of the diaphragm to adjust.

FAST IDLE CAM ADJUSTMENT

1. Set the fast idle cam so that the screw is on the second highest step of the fast idle cam.
2. Insert the specified gauge between the lower edge of the choke valve and the air horn wall.

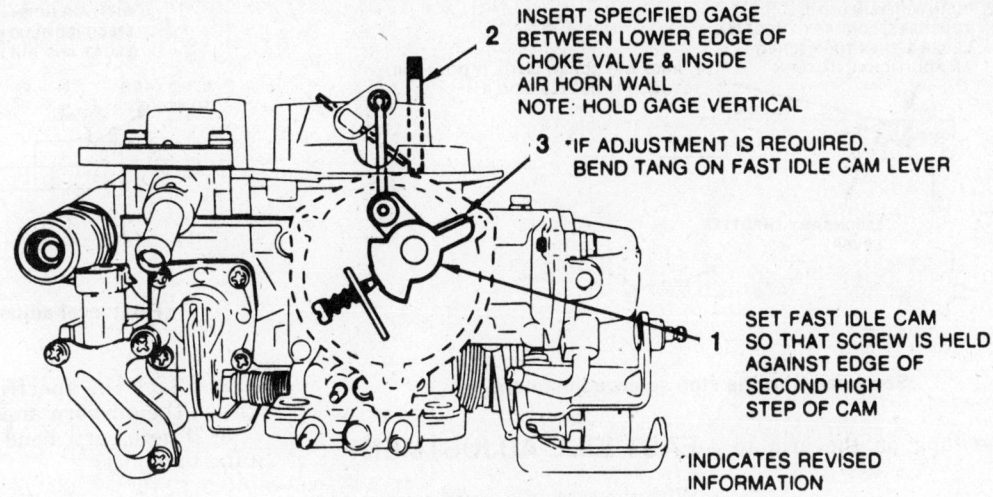

2 INSERT SPECIFIED GAGE
BETWEEN LOWER EDGE OF
CHOKE VALVE & INSIDE
AIR HORN WALL
NOTE: HOLD GAGE VERTICAL

3 *IF ADJUSTMENT IS REQUIRED.
BEND TANG ON FAST IDLE CAM LEVER

1 SET FAST IDLE CAM
SO THAT SCREW IS HELD
AGAINST EDGE OF
SECOND HIGH
STEP OF CAM

*INDICATES REVISED
INFORMATION

Fast idle cam adjustment—Holley 6510C

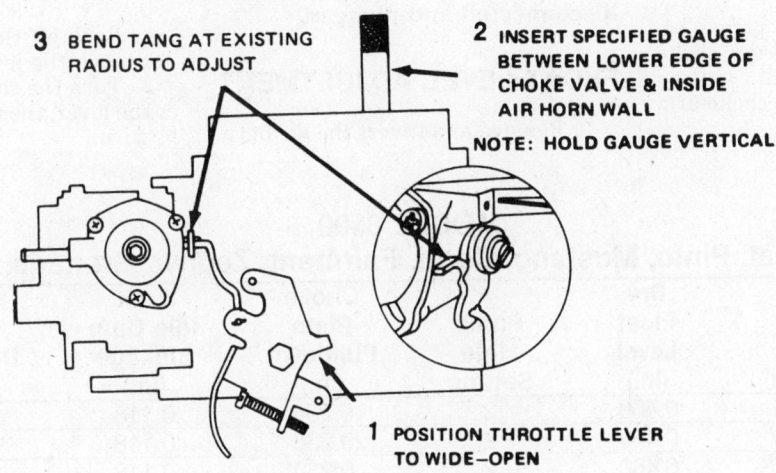

3 BEND TANG AT EXISTING
RADIUS TO ADJUST

2 INSERT SPECIFIED GAUGE
BETWEEN LOWER EDGE OF
CHOKE VALVE & INSIDE
AIR HORN WALL

NOTE: HOLD GAUGE VERTICAL

1 POSITION THROTTLE LEVER
TO WIDE–OPEN

Choke unloader adjustment

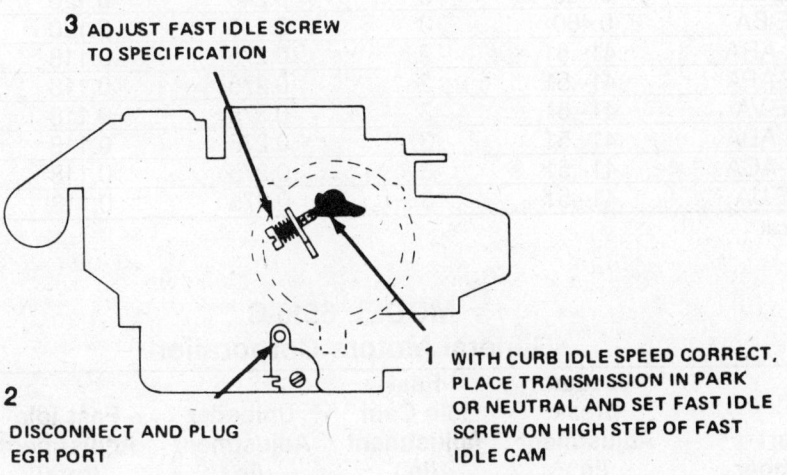

3 ADJUST FAST IDLE SCREW
TO SPECIFICATION

2 DISCONNECT AND PLUG
EGR PORT

1 WITH CURB IDLE SPEED CORRECT,
PLACE TRANSMISSION IN PARK
OR NEUTRAL AND SET FAST IDLE
SCREW ON HIGH STEP OF FAST
IDLE CAM

Fast idle speed adjustment

2 TURN SCREW IN UNTIL IT TOUCHES SECONDARY THROTTLE LEVER & THEN TURN SCREW AN ADDITIONAL 1/4 TURN

1 BACK OFF SCREW UNTIL IT DOES NOT TOUCH THROTTLE LEVER

SECONDARY THROTTLE LEVER

Secondary throttle stop screw adjustment

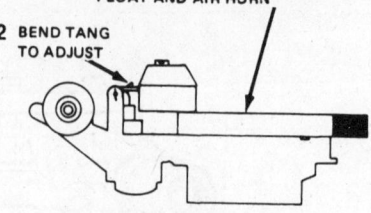

1 WITH AIR HORN INVERTED INSERT SPECIFIED PLUG GAUGE BETWEEN FLOAT AND AIR HORN

2 BEND TANG TO ADJUST

Float level adjustment

3. Bend the tang on the arm to adjust.

UNLOADER ADJUSTMENT

1. Place the throttle in the wide open position.
2. Insert a 0.350 inch gauge between the lower edge of the choke valve and the air horn wall.
3. Bend the tang on the choke arm to adjust.

FAST IDLE ADJUSTMENT

With the curb idle speed correct, place the fast idle screw on the highest cam step and adjust to the specified rpm.

NOTE: The EGR line must be disconnected and plugged.

FLOAT LEVEL ADJUSTMENT

1. Remove and invert the air horn.

2. Place the specified gauge between the air horn and the float.
3. If necessary, bend the float arm tang to adjust.

SECONDARY THROTTLE STOP SCREW ADJUSTMENT

1. Back off the screw until it does not touch the lever.
2. Turn the screw in until it touches the lever, then turn it an additional $\frac{1}{4}$ turn.

MODEL 6500
Ford Bobcat, Pinto, Mustang, Capri, Fairmont, Zephyr, Granada, Cougar

Year	Carburetor Identification	Dry Float Level (in.)	Pump Hole Setting	Choke Plate Pulldown (in.)	Fast Idle Cam Linkage (in.)	Dechoke (in.)	Choke Setting
'80	EOEE-NA, VA	0.460	2	0.236	0.118	0.393	①
	EOEE-NC, NV	0.460	2	0.236	0.118	0.157	①
	EOEE-ND, VD	0.460	2	0.236	0.118	0.393	①
	EOZE-AFA, SA	0.460	2	0.236	0.118	0.393	①
	EOZE-AFC, SC	0.460	—	0.236	0.118	0.393	①
'81	EIZE-RA	0.460	3	0.240	0.120	0.400	—
	EIZE-SA	0.460	3	0.240	0.120	0.400	—
	EIDE-DA	0.460	3	0.240	0.120	0.400	—
	EIDE-EA	0.460	3	0.240	0.120	0.400	—
'82	E2ZE-ARA	.41–.51	2	0.275	0.118	0.393	—
	E2ZE-APA	.41–.51	2	0.275	0.118	0.393	—
	E2ZE-VA	.41–.51	3	0.275	0.118	0.393	—
	E2ZE-ADA	.41–.51	3	0.275	0.118	0.393	—
	E2ZE-ACA	.41–.51	3	0.275	0.118	0.393	—
	E2ZE-UA	.41–.51	3	0.275	0.118	0.393	—

① See underhood decal

MODEL 6510-C
General Motors Corporation

Year	Part Number	Vacuum Break Adjustment (in.)	Fast Idle Cam Adjustment (in.)	Unloader Adjustment (in.)	Fast Idle Adjustment (rpm)	Float Level Adjustment (in.)	Choke Setting
'80	All w/manual	.275	.130	.350	2600	.500	Fixed
	All w/automatic	.300	.130	.350	2500	.500	Fixed

MODEL 6510-C
General Motors Corporation

Year	Part Number	Vacuum Break Adjustment (in.)	Fast Idle Cam Adjustment (in.)	Unloader Adjustment (in.)	Fast Idle Adjustment (rpm)	Float Level Adjustment (in.)	Choke Setting
'81	14004768	.300	.130	.350	①	.500	Fixed
	14004769	.300	.130	.350	①	.500	Fixed
	14004770	.300	.130	.350	①	.500	Fixed
	14004771	.300	.130	.350	①	.500	Fixed
	14004777	.300	.130	.350	①	.500	Fixed
'82	14032364	.270	.080	.350	①	.500	Fixed
	14032365	.270	.080	.350	①	.500	Fixed
	14032366	.270	.080	.350	①	.500	Fixed
	14032367	.270	.080	.350	①	.500	Fixed
	14032368	.270	.080	.350	①	.500	Fixed
	14032369	.270	.080	.350	①	.500	Fixed
	14032370	.270	.080	.350	①	.500	Fixed
	14032371	.270	.080	.350	①	.500	Fixed
	14033392	.270	.080	.350	①	.500	Fixed
	14033393	.270	.080	.350	①	.500	Fixed
	14047072	.270	.080	.350	①	.500	Fixed
'83	14048827	.270	.080	.350	①	.500	Fixed
	14048828	.300	.080	.350	①	.500	Fixed
	14048829	.270	.080	.350	①	.500	Fixed
'84-'86	14068690	.270	.080	.350	①	.500	Fixed
	14068691	.270	.080	.350	①	.500	Fixed
	14068692	.300	.080	.350	①	.500	Fixed
	14076363	.300	.080	.350	①	.500	Fixed

① See underhood decal

ROCHESTER CARBURETORS

Angle Degree Tool

An angle degree tool is recommended by Rochester Products Division, for use to confirm adjustments to the choke valve and related linkages on their late model two and four barrel carburetors, in place of the plug type gauges. Decimal and degree conversion charts are provided for use by technicians who have access to an angle gauge and not plug gauges. It must be remembered that the relationship between the decimal and the angle readings are not exact, due to manufacturers tolerances.

To use the angle gauge, rotate the degree scale until zero (0) is opposite the pointer. With the choke valve completely closed, place the gauge magnet squarely on top of the choke valve and rotate the bubble until it is centered. Make the necessary adjustments to have the choke valve at the specified degree angle opening as read from the degree angle tool.

NOTE: The carburetor may be off the engine for adjustments. Be sure the carburetor is held firmly during the use of the angle gauge.

Model Identification

General Motors Rochester carburetors are identified by their model number. The first number indicates the number of barrels, while one of the last letters indicates the type of choke used. These are V for the manifold mounted choke coil, C for the choke coil mounted on the carburetor, and E for electric choke, also mounted on the carburetor. Model numbers ending in A indicate an altitude-compensating carburetor.

Models 2SE and E2SE

The Rochester 2SE and E2SE Varajet II carburetors are two barrel, two stage downdraft units. Most carburetor components are aluminum, although a zinc choke housing is used on four cylinder engines installed in 1980 models. The E2SE is used both in conventional installations and in the Computer Controlled Catalytic Converter System. In that installation the E2SE is equipped with an electrically operated mixture control solenoid, controlled by the Electronic Control Module. The 2SE and E2SE are also used on the AMC four cylinder in 1980–83.

For further information on feedback carburetors, please refer to *Chilton's Guide To Fuel Injection And Feedback Carburetors.*

FLOAT ADJUSTMENT

1. Remove the air horn from the throttle body.
2. Use your fingers to hold the retainer in place, and to push the float down into light contact with the needle.
3. Measure the distance from the toe of the float (furtherest from the hinge) to the top of the carburetor (gasket removed).

ANGLE DEGREE TO DECIMAL CONVERSION
Model 4MV Carburetor

Angle Degrees	Decimal Equiv. Top of Valve	Angle Degrees	Decimal Equiv. Top of Valve
5	.019	33	.158
6	.022	34	.164
7	.026	35	.171
8	.030	36	.178
9	.034	37	.184
10	.038	38	.190
11	.042	39	.197
12	.047	40	.204
13	.051	41	.211
14	.056	42	.217
15	.060	43	.225
16	.065	44	.231
17	.070	45	.239
18	.075	46	.246
19	.080	47	.253
20	.085	48	.260
21	.090	49	.268
22	.095	50	.275
23	.101	51	.283
24	.106	52	.291
25	.112	53	.299
26	.117	54	.306
27	.123	55	.314
28	.128	56	.322
29	.134	57	.329
30	.140	58	.337
31	.146	59	.345
32	.152	60	.353

PLUGGING AIR BLEED HOLES

PUMP CUP OR VALVE STEM SEAL

TAPE HOLE IN TUBE

TAPE END OF COVER

Vacuum break information—E2SE

4. To adjust, remove the float and gently bend the arm to specification. After adjustment, check the float alignment in the chamber.

NOTE: Some models have a float stabilizer spring. If used, remove the spring with float. Use care when removing.

PUMP ADJUSTMENT

1. With the throttle closed and the fast idle screw off the steps of the fast idle cam, measure the distance from the air horn casting to the top of the pump stem.

2. To adjust, remove the retaining screw and washer and remove the pump lever. Bend the end of the lever to correct the stem height. Do not twist the lever or bend it sideways.

3. Install the lever, washer and screw and check the adjustment. When correct, open and close the

throttle a few times to check the linkage movement and alignment.

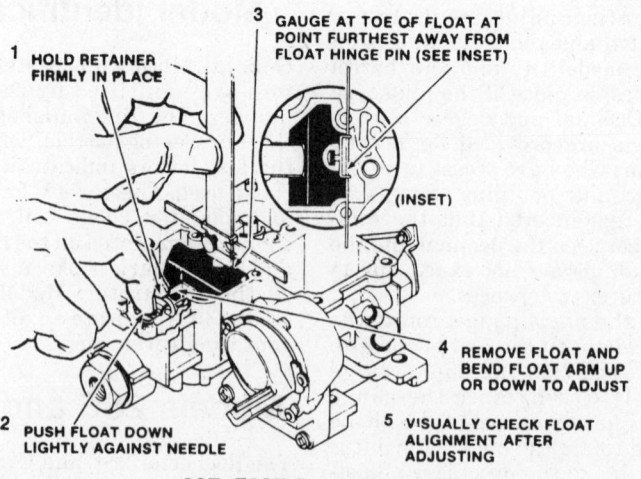

1 HOLD RETAINER FIRMLY IN PLACE

2 PUSH FLOAT DOWN LIGHTLY AGAINST NEEDLE

3 GAUGE AT TOE OF FLOAT AT POINT FURTHEST AWAY FROM FLOAT HINGE PIN (SEE INSET)

(INSET)

4 REMOVE FLOAT AND BEND FLOAT ARM UP OR DOWN TO ADJUST

5 VISUALLY CHECK FLOAT ALIGNMENT AFTER ADJUSTING

2SE, E2SE float adjustment

NOTE: No pump adjustment is required on 1981 and later models.

ANGLE DEGREE TO DECIMAL CONVERSION
Model M2MC, M2ME and M4MC Carburetor

Angle Degrees	Decimal Equiv. Top of Valve	Angle Degrees	Decimal Equiv. Top of Valve
5	.023	33	.203
6	.028	34	.211
7	.033	35	.220
8	.038	36	.227
9	.043	37	.234
10	.049	38	.243
11	.054	39	.251
12	.060	40	.260
13	.066	41	.269
14	.071	42	.277
15	.077	43	.287
16	.083	44	.295
17	.090	45	.304
18	.096	46	.314
19	.103	47	.322
20	.110	48	.332
21	.117	49	.341
22	.123	50	.350
23	.129	51	.360
24	.136	52	.370
25	.142	53	.379
26	.149	54	.388
27	.157	55	.400
28	.164	56	.408
29	.171	57	.418
30	.179	58	.428
31	.187	59	.439
32	.195	60	.449

NOTE: ON MODELS USING A CLIP TO RETAIN PUMP ROD IN PUMP LEVER, NO PUMP ADJUSTMENT IS REQUIRED. ON MODELS USING THE "CLIPLESS" PUMP ROD, THE PUMP ADJUSTMENT SHOULD NOT BE CHANGED FROM ORIGINAL FACTORY SETTING UNLESS GAUGING SHOWS OUT OF SPECIFICATION. THE PUMP LEVER IS MADE FROM HEAVY DUTY, HARDENED STEEL MAKING BENDING DIFFICULT. DO NOT REMOVE PUMP LEVER FOR BENDING UNLESS ABSOLUTELY NECESSARY.

1 THROTTLE VALVES COMPLETELY CLOSED. MAKE SURE FAST IDLE SCREW IS OFF STEPS OF FAST IDLE CAM.

3 IF NECESSARY TO ADJUST, REMOVE PUMP LEVER RETAINING SCREW AND WASHER AND REMOVE PUMP LEVER BY ROTATING LEVER TO REMOVE FROM PUMP ROD. PLACE LEVER IN A VISE, PROTECTING LEVER FROM DAMAGE, AND BEND END OF LEVER (NEAREST NECKED DOWN SECTION).

NOTE: DO NOT BEND LEVER IN A SIDEWAYS OR TWISTING MOTION.

2 GAUGE FROM AIR HORN CASTING SURFACE TO TOP OF PUMP STEM. DIMENSION SHOULD BE AS SPECIFIED.

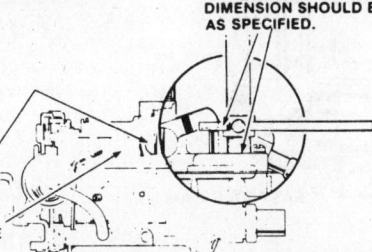

5 OPEN AND CLOSE THROTTLE VALVES CHECKING LINKAGE FOR FREEDOM OF MOVEMENT AND OBSERVING PUMP LEVER ALIGNMENT.

4 REINSTALL PUMP LEVER, WASHER AND RETAINING SCREW. RECHECK PUMP ADJUSTMENT ① AND ②. TIGHTEN RETAINING SCREW SECURELY AFTER THE PUMP ADJUSTMENT IS CORRECT.

2SE, E2SE pump adjustment

FAST IDLE ADJUSTMENT

1. Set the ignition timing and curb idle speed, and disconnect and plug hoses as directed on the emission control decal.
2. Place the fast idle screw on the highest step of the cam.
3. Start the engine and adjust the engine speed to specification with the fast idle screw.

NOTE: On models using a clip to retain pump rod in pump lever, no pump adjustment is required. On models using the "CLIPLESS" pump rod, the pump rod adjustment should not be changed from the origional factory setting unless gauging shows out of specification. The pump lever is made from heavy duty, hardened steel making bending difficult. Do not remove pump lever for bendsing unless absolutely necessary.

CHOKE COIL LEVER ADJUSTMENT

1. Remove the three retaining screws and remove the choke cover and coil. On models with a riveted choke cover, drill out the three rivets and remove the cover and choke coil.

NOTE: A choke stat cover retainer kit is required for reassembly.

2. Place the fast idle screw on the high step of the cam.
3. Close the choke by pushing in on the intermediate choke lever. On front wheel drive models, the intermediate choke lever is behind the choke vacuum diaphragm.
4. Insert a drill or gauge of the specified size into the hole in the choke housing. The choke lever in the housing should be up against the side of the gauge.
5. If the lever does not just touch the gauge, bend the intermediate choke rod to adjust.

FAST IDLE CAM (CHOKE ROD) ADJUSTMENT

1980–82 Models

NOTE: A special angle gauge should be used.

1. Adjust the choke coil lever and fast idle first.
2. Rotate the degree scale until it is zeroed.
3. Close the choke and install the

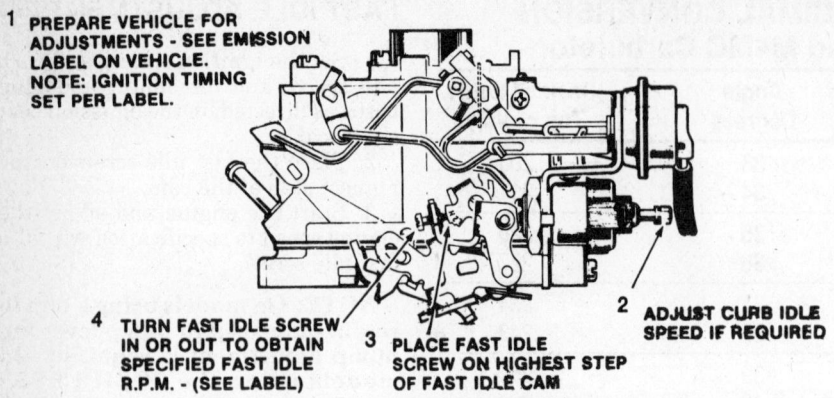

1 PREPARE VEHICLE FOR ADJUSTMENTS - SEE EMISSION LABEL ON VEHICLE. NOTE: IGNITION TIMING SET PER LABEL.

4 TURN FAST IDLE SCREW IN OR OUT TO OBTAIN SPECIFIED FAST IDLE R.P.M. - (SEE LABEL)

3 PLACE FAST IDLE SCREW ON HIGHEST STEP OF FAST IDLE CAM

2 ADJUST CURB IDLE SPEED IF REQUIRED

2SE, E2SE fast idle adjustment

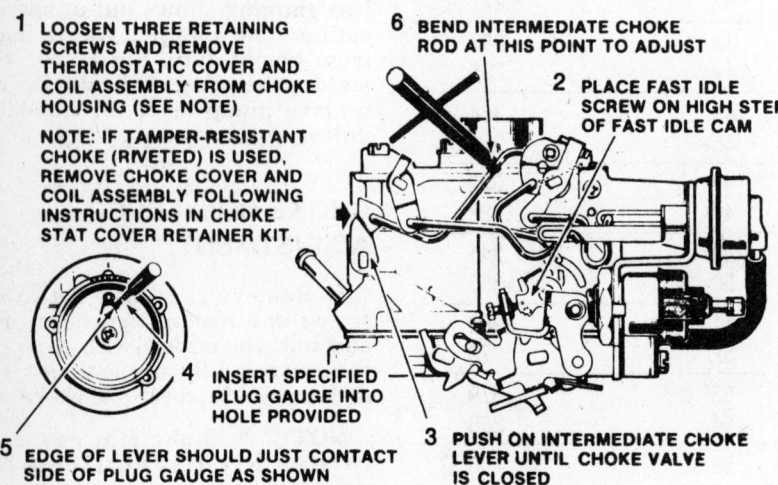

1 LOOSEN THREE RETAINING SCREWS AND REMOVE THERMOSTATIC COVER AND COIL ASSEMBLY FROM CHOKE HOUSING (SEE NOTE)

NOTE: IF TAMPER-RESISTANT CHOKE (RIVETED) IS USED, REMOVE CHOKE COVER AND COIL ASSEMBLY FOLLOWING INSTRUCTIONS IN CHOKE STAT COVER RETAINER KIT.

6 BEND INTERMEDIATE CHOKE ROD AT THIS POINT TO ADJUST

2 PLACE FAST IDLE SCREW ON HIGH STEP OF FAST IDLE CAM

3 PUSH ON INTERMEDIATE CHOKE LEVER UNTIL CHOKE VALVE IS CLOSED

4 INSERT SPECIFIED PLUG GAUGE INTO HOLE PROVIDED

5 EDGE OF LEVER SHOULD JUST CONTACT SIDE OF PLUG GAUGE AS SHOWN

2SE, E2SE choke coil lever adjustment

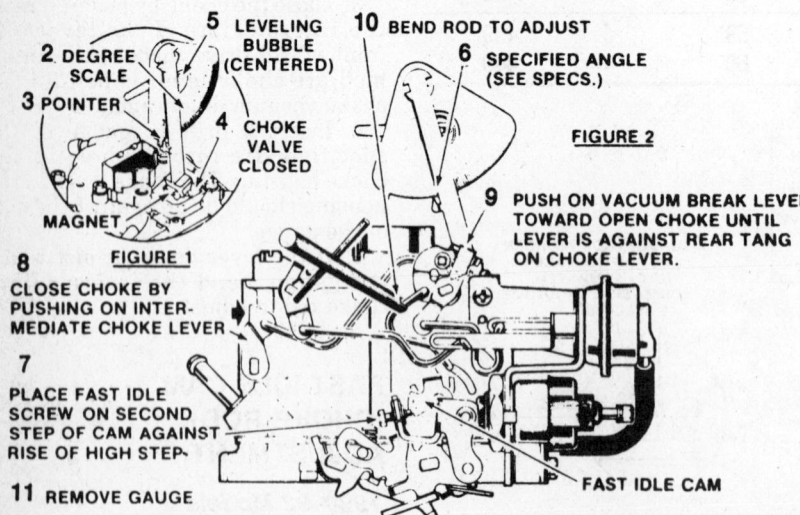

5 LEVELING BUBBLE (CENTERED)

2 DEGREE SCALE

3 POINTER

4 CHOKE VALVE CLOSED

MAGNET

FIGURE 1

8 CLOSE CHOKE BY PUSHING ON INTERMEDIATE CHOKE LEVER

7 PLACE FAST IDLE SCREW ON SECOND STEP OF CAM AGAINST RISE OF HIGH STEP

11 REMOVE GAUGE

10 BEND ROD TO ADJUST

6 SPECIFIED ANGLE (SEE SPECS.)

FIGURE 2

9 PUSH ON VACUUM BREAK LEVER TOWARD OPEN CHOKE UNTIL LEVER IS AGAINST REAR TANG ON CHOKE LEVER.

FAST IDLE CAM

2SE, E2SE fast idle cam adjustment—models through 1982

degree scale onto the choke plate. Center the leveling bubble.

4. Rotate the scale so that the specified degree is opposite the scale pointer.

5. Place the fast idle screw on the second step of the cam (against the high step). Close the choke by pushing in the intermediate lever.

6. Push on the vacuum break lever

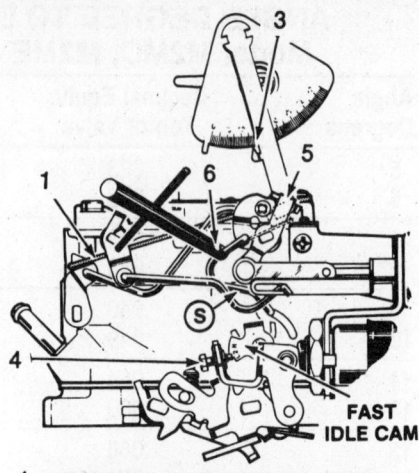

FAST IDLE CAM

1 ATTACH RUBBER BAND TO INTERMEDIATE CHOKE LEVER.

2 OPEN THROTTLE TO ALLOW CHOKE VALVE TO CLOSE.

3 SET UP ANGLE GAGE AND SET ANGLE TO SPECIFICATIONS.

4 PLACE FAST IDLE SCREW ON SECOND STEP OF CAM AGAINST RISE OF HIGH STEP.

5 PUSH ON CHOKE SHAFT LEVER TO OPEN CHOKE VALVE AND TO MAKE CONTACT WITH BLACK CLOSING TANG.

6 SUPPORT AT "S" AND ADJUST BY BENDING FAST IDLE CAM ROD UNTIL BUBBLE IS CENTERED.

E2SE fast idle cam (choke rod) adjustment—1983 and later

in the direction of opening choke until the lever is against the rear tang on the choke lever.

7. Bend the fast idle cam rod at the U to adjust angle to specifications.

1983–84 Models

Refer to the illustration for adjustment procedure on these models.

AIR VALVE ROD ADJUSTMENT

1980 Models

1. Seat the vacuum diaphragm with an outside vacuum source. Tape over the purge bleed hole if present.

2. Close the air valve.

3. Insert the specified gauge between the rod and the end of the slot in the plunger on fours, or between the rod and the end of the slot in the air valve on V6s.

4. Bend the rod to adjust the clearance.

1981–82 Models

1. Align the zero degree mark with the pointer on an angle gauge.

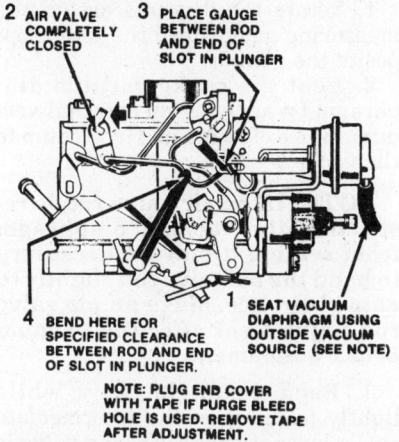

2SE and E2SE air valve rod adjustment—1980 G.M. models, 1980–82 American Motors

2. Close the air valve and place a magnet on top of it.

3. Rotate the bubble until it is centered.

4. Rotate the degree scale until the specified degree mark is aligned with the pointer.

5. Seat the vacuum diaphragm using an external vacuum source.

6. On four cylinder models plug the end cover. Unplug after adjustment.

7. Apply light pressure to the air valve shaft in the direction to open the air valve until all the slack is removed between the air link and plunger slot. 8.Bend the air valve link until the bubble is centered.

1983–84 Models

Refer to the illustration for the adjustment procedure on these models.

PRIMARY SIDE VACUUM BREAK ADJUSTMENT

1980 GM Models and 1980–83 AMC Models

1. Follow Steps 1–4 of the "Fast Idle Cam Adjustment" procedure.

2. Seat the choke vacuum diaphragm with an outside vacuum source.

3. Push in on the intermediate choke lever to close the choke valve, and hold closed during adjustment.

4. Adjust by bending the vacuum break rod until the bubble is centered.

1981–82 GM Models

NOTE: Prior to adjustment, remove the vacuum break from the carburetor. Place the bracket in a vise and using the proper safety precautions, grind off the adjustment screw cap then reinstall the vacuum break.

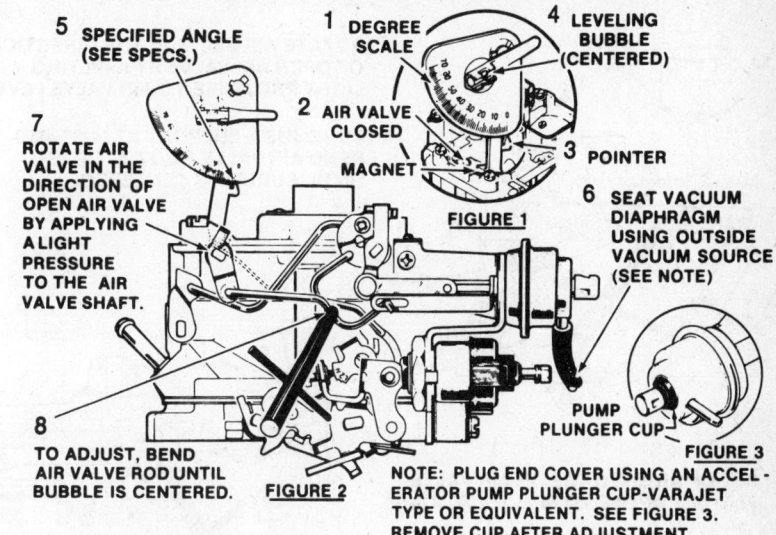

E2SE air valve adjustment—1981–82 4 cyl. except G.M. "J" series

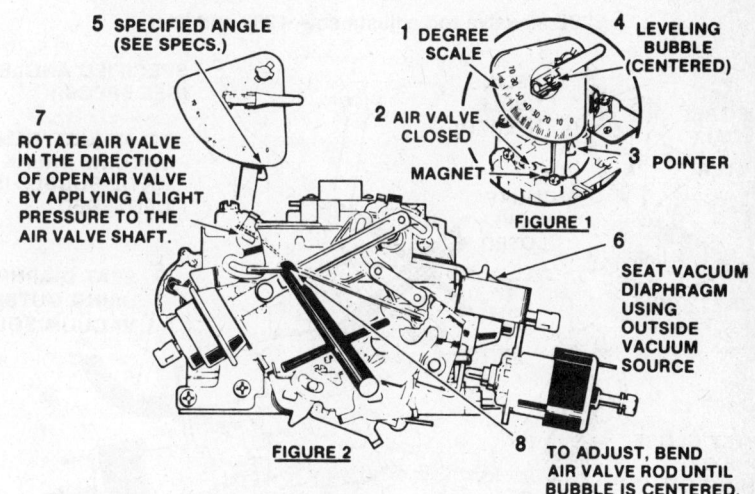

E2SE air valve adjustment—1981–82 V6 engine

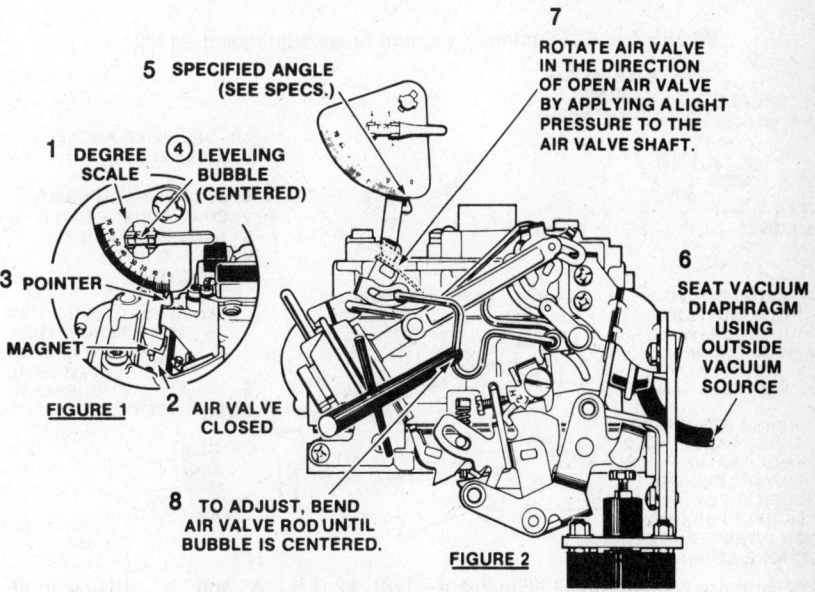

E2SE air valve adjustment—1982 G.M. J series

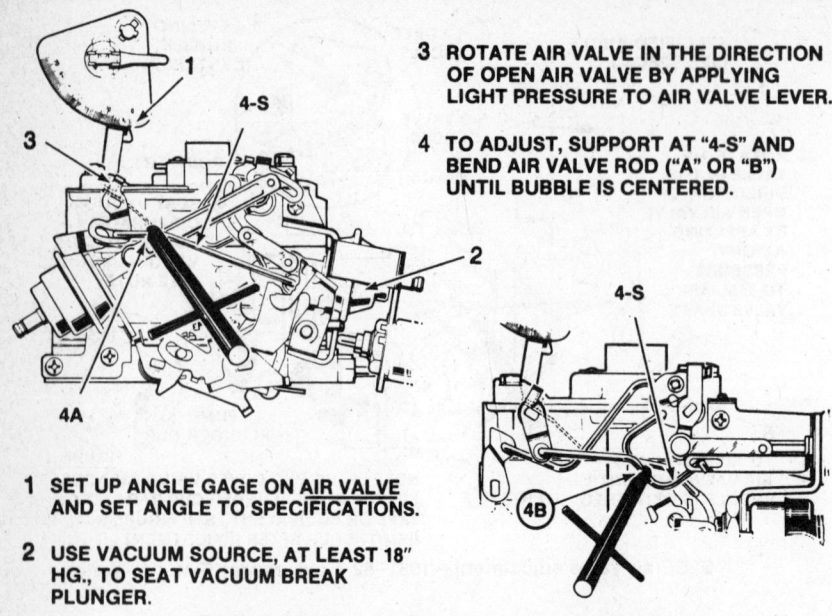

3 ROTATE AIR VALVE IN THE DIRECTION OF OPEN AIR VALVE BY APPLYING LIGHT PRESSURE TO AIR VALVE LEVER.

4 TO ADJUST, SUPPORT AT "4-S" AND BEND AIR VALVE ROD ("A" OR "B") UNTIL BUBBLE IS CENTERED.

1 SET UP ANGLE GAGE ON AIR VALVE AND SET ANGLE TO SPECIFICATIONS.

2 USE VACUUM SOURCE, AT LEAST 18" HG., TO SEAT VACUUM BREAK PLUNGER.

E2SE air valve rod adjustment—1983 and later

1 DEGREE SCALE
2 POINTER
3 CHOKE VALVE CLOSED
4 LEVELING BUBBLE (CENTERED)
5 SPECIFIED ANGLE (SEE SPECS.)
6 SEAT DIAPHRAGM USING OUTSIDE VACUUM SOURCE
7 LIGHTLY CLOSE CHOKE BY PUSHING ON INTERMEDIATE CHOKE LEVER
8 TO ADJUST, BEND VACUUM BREAK ROD UNTIL BUBBLE IS CENTERED
MAGNET

V6 2SE and E2SE primary vacuum break adjustment—1980

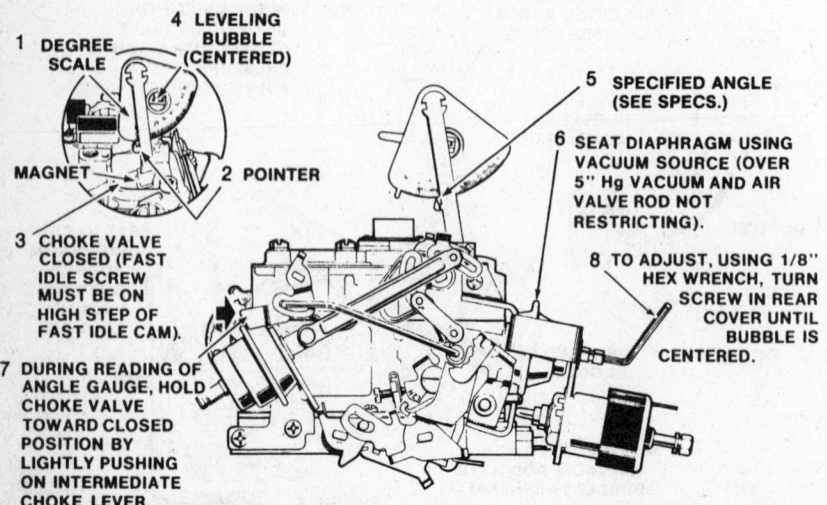

1 DEGREE SCALE
2 POINTER
3 CHOKE VALVE CLOSED (FAST IDLE SCREW MUST BE ON HIGH STEP OF FAST IDLE CAM).
4 LEVELING BUBBLE (CENTERED)
5 SPECIFIED ANGLE (SEE SPECS.)
6 SEAT DIAPHRAGM USING VACUUM SOURCE (OVER 5" Hg VACUUM AND AIR VALVE ROD NOT RESTRICTING).
7 DURING READING OF ANGLE GAUGE, HOLD CHOKE VALVE TOWARD CLOSED POSITION BY LIGHTLY PUSHING ON INTERMEDIATE CHOKE LEVER.
8 TO ADJUST, USING 1/8" HEX WRENCH, TURN SCREW IN REAR COVER UNTIL BUBBLE IS CENTERED.
MAGNET

E2SE primary vacuum break adjustment—1981–82 G.M. "A" and "X" series with V6 engine

1. Rotate the degree scale on the measuring gauge until the zero is opposite the pointer.

2. Seat the choke vacuum diaphragm by applying an external vacuum source of over 5 in. Hg vacuum to the vacuum brake.

NOTE: If the air valve rod is restricting the vacuum diaphragm from seating it may be necessary to bend the air valve rod slightly to gain clearance. Make an air valve rod adjustment after the vacuum break adjustment.

3. Read the angle gauge while lightly pushing on the intermediate choke lever so that the choke valve is toward the close position.

4. Use a $\frac{1}{8}$ in. hex wrench and turn the screw in the rear cover until the bubble is centered. Apply a silicone sealant over the screw head to seal the setting.

1983–84 GM Models

Refer to the illustration for the adjustment procedure on these models.

ELECTRIC CHOKE SETTING

This procedure is only for those carburetors with choke covers retained by screws. Riveted choke covers are preset and nonadjustable.

1. Loosen the three retaining screws.

2. Place the fast idle screw on the high step of the cam.

3. Rotate the choke cover to align the cover mark with the specified housing mark.

NOTE: The specification "index" which appears in the specification table refers to the mark between "1 notch lean" and "1 notch rich".

SECONDARY VACUUM BREAK ADJUSTMENT

1980 Models

This procedure is for V6 installations in front wheel drive models only.

1. Follow Steps 1–4 of the "Fast Idle Cam Adjustment" procedure.

2. Seat the choke vacuum diaphragm with an outside vacuum source.

3. Push in on the intermediate choke lever to close the choke valve, and hold closed during adjustment. Make sure the plunger spring is compressed and seated, if present.

4. Bend the vacuum break rod at the U next to the diaphragm until the bubble is centered.

NOTE: Prior to adjustment, remove the vacuum break from the carburetor. Place the bracket in the vise and using the proper safety precautions, grind off the adjustment screw cap then reinstall the vacuum break.

1981–82 GM Models

NOTE: Plug the end cover using an accelerator pump plunger cup or equivalent. Remove the cup after the adjustment (A and X series only).

1. Rotate the degree scale on the measuring gauge until the zero is opposite the pointer.

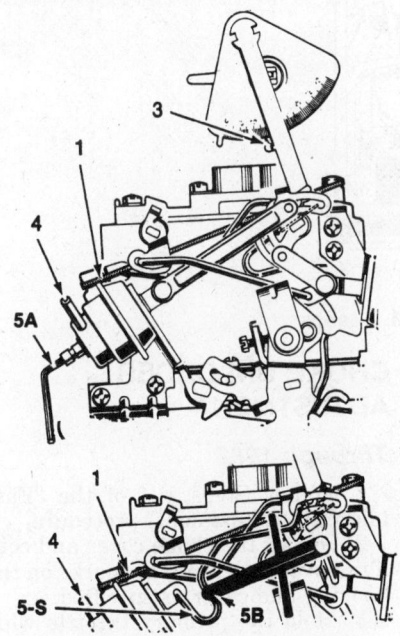

1 ATTACH RUBBER BAND TO INTERMEDIATE CHOKE LEVER.

2 OPEN THROTTLE TO ALLOW CHOKE VALVE TO CLOSE.

3 SET UP ANGLE GAGE AND SET ANGLE TO SPECIFICATION.

4 RETRACT VACUUM BREAK PLUNGER USING VACUUM SOURCE, AT LEAST 18" HG. PLUG AIR BLEED HOLES WHERE APPLICABLE.
WHERE APPLICABLE, PLUNGER STEM MUST BE EXTENDED FULLY TO COMPRESS PLUNGER BUCKING SPRING.

5 TO CENTER BUBBLE, EITHER:

A. ADJUST WITH 1/8" (3.175 mm) HEX WRENCH (VACUUM STILL APPLIED)
-OR-

B. SUPPORT AT "5-S", BEND WIRE-FORM VACUUM BREAK ROD (VACUUM STILL APPLIED)

E2SE secondary vacuum break adjustment—1983 and later

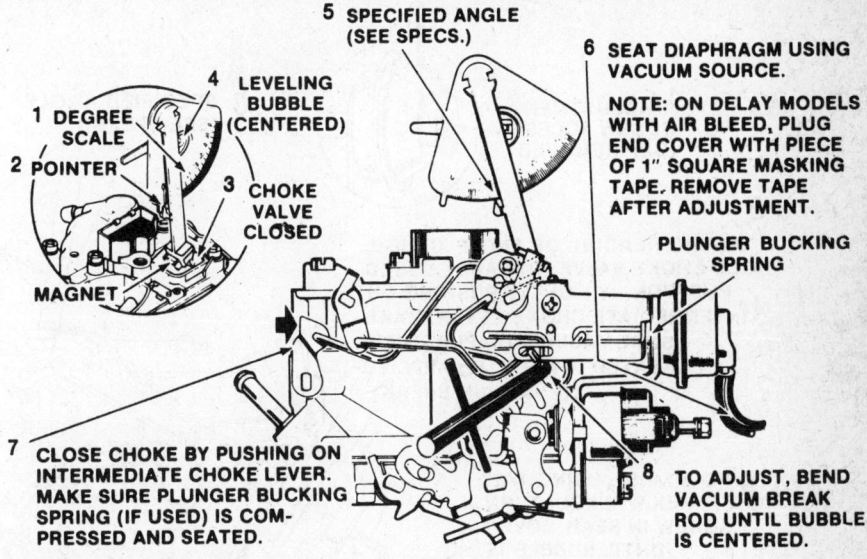

5 SPECIFIED ANGLE (SEE SPECS.)

4 LEVELING BUBBLE (CENTERED)

1 DEGREE SCALE

2 POINTER

3 CHOKE VALVE CLOSED

MAGNET

6 SEAT DIAPHRAGM USING VACUUM SOURCE.

NOTE: ON DELAY MODELS WITH AIR BLEED, PLUG END COVER WITH PIECE OF 1" SQUARE MASKING TAPE. REMOVE TAPE AFTER ADJUSTMENT.

PLUNGER BUCKING SPRING

7 CLOSE CHOKE BY PUSHING ON INTERMEDIATE CHOKE LEVER. MAKE SURE PLUNGER BUCKING SPRING (IF USED) IS COMPRESSED AND SEATED.

8 TO ADJUST, BEND VACUUM BREAK ROD UNTIL BUBBLE IS CENTERED.

2SE, E2SE primary vacuum break adjustment—1980 G.M. and 1980–83 American Motors with 4 cyl. engines

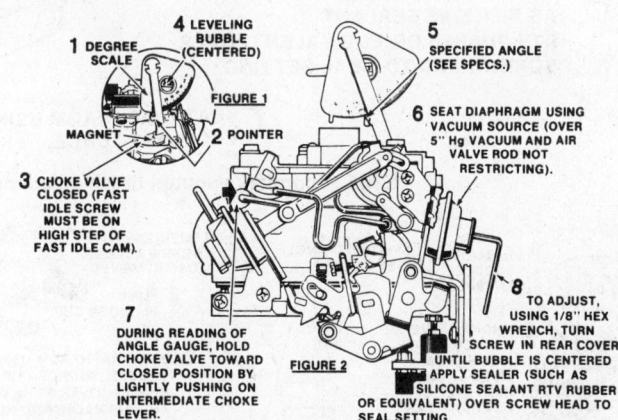

4 LEVELING BUBBLE (CENTERED)

1 DEGREE SCALE

FIGURE 1

MAGNET

2 POINTER

3 CHOKE VALVE CLOSED (FAST IDLE SCREW MUST BE ON HIGH STEP OF FAST IDLE CAM).

7 DURING READING OF ANGLE GAUGE, HOLD CHOKE VALVE TOWARD CLOSED POSITION BY LIGHTLY PUSHING ON INTERMEDIATE CHOKE LEVER.

FIGURE 2

5 SPECIFIED ANGLE (SEE SPECS.)

6 SEAT DIAPHRAGM USING VACUUM SOURCE (OVER 5" Hg VACUUM AND AIR VALVE ROD NOT RESTRICTING).

8 TO ADJUST, USING 1/8" HEX WRENCH, TURN SCREW IN REAR COVER UNTIL BUBBLE IS CENTERED APPLY SEALER (SUCH AS SILICONE SEALANT RTV RUBBER OR EQUIVALENT) OVER SCREW HEAD TO SEAL SETTING.

E2SE primary vacuum break adjustment—4 cyl.—1982 G.M. J series

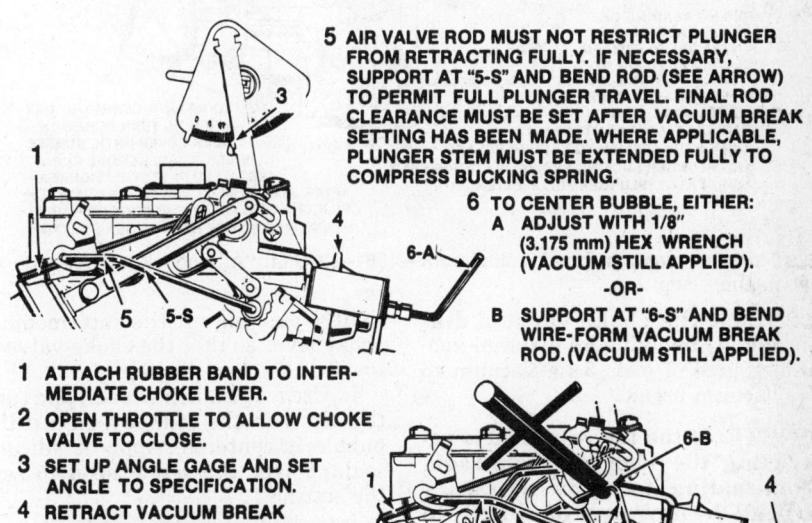

5 AIR VALVE ROD MUST NOT RESTRICT PLUNGER FROM RETRACTING FULLY. IF NECESSARY, SUPPORT AT "5-S" AND BEND ROD (SEE ARROW) TO PERMIT FULL PLUNGER TRAVEL. FINAL ROD CLEARANCE MUST BE SET AFTER VACUUM BREAK SETTING HAS BEEN MADE. WHERE APPLICABLE, PLUNGER STEM MUST BE EXTENDED FULLY TO COMPRESS BUCKING SPRING.

6 TO CENTER BUBBLE, EITHER:
A ADJUST WITH 1/8" (3.175 mm) HEX WRENCH (VACUUM STILL APPLIED)
-OR-

B SUPPORT AT "6-S" AND BEND WIRE-FORM VACUUM BREAK ROD. (VACUUM STILL APPLIED).

1 ATTACH RUBBER BAND TO INTERMEDIATE CHOKE LEVER.

2 OPEN THROTTLE TO ALLOW CHOKE VALVE TO CLOSE.

3 SET UP ANGLE GAGE AND SET ANGLE TO SPECIFICATION.

4 RETRACT VACUUM BREAK PLUNGER USING VACUUM SOURCE, AT LEAST 18" HG. PLUG AIR BLEED HOLES WHERE APPLICABLE.

E2SE primary vacuum break adjustment—1983 and later

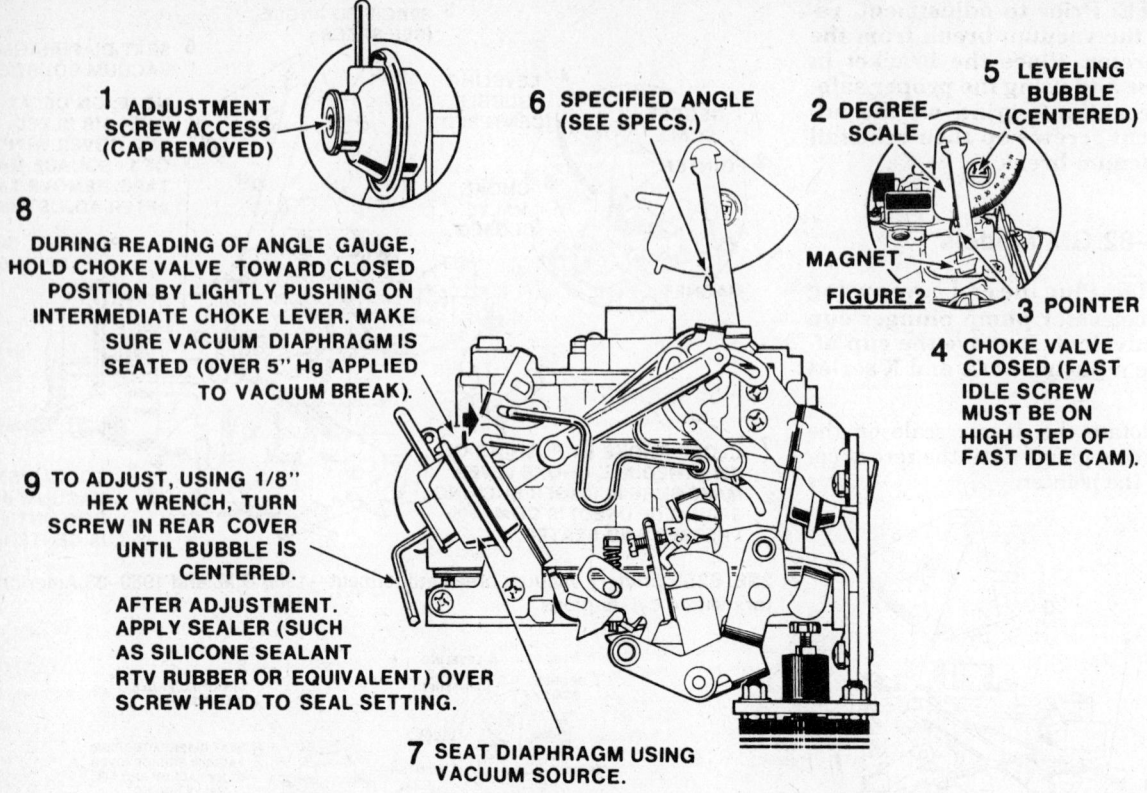

1 ADJUSTMENT SCREW ACCESS (CAP REMOVED)

6 SPECIFIED ANGLE (SEE SPECS.)

5 LEVELING BUBBLE (CENTERED)

2 DEGREE SCALE

MAGNET

FIGURE 2

3 POINTER

4 CHOKE VALVE CLOSED (FAST IDLE SCREW MUST BE ON HIGH STEP OF FAST IDLE CAM).

8 DURING READING OF ANGLE GAUGE, HOLD CHOKE VALVE TOWARD CLOSED POSITION BY LIGHTLY PUSHING ON INTERMEDIATE CHOKE LEVER. MAKE SURE VACUUM DIAPHRAGM IS SEATED (OVER 5" Hg APPLIED TO VACUUM BREAK).

9 TO ADJUST, USING 1/8" HEX WRENCH, TURN SCREW IN REAR COVER UNTIL BUBBLE IS CENTERED.

AFTER ADJUSTMENT. APPLY SEALER (SUCH AS SILICONE SEALANT RTV RUBBER OR EQUIVALENT) OVER SCREW HEAD TO SEAL SETTING.

7 SEAT DIAPHRAGM USING VACUUM SOURCE.

E2SE secondary vacuum break adjustment—1982 G.M. J series

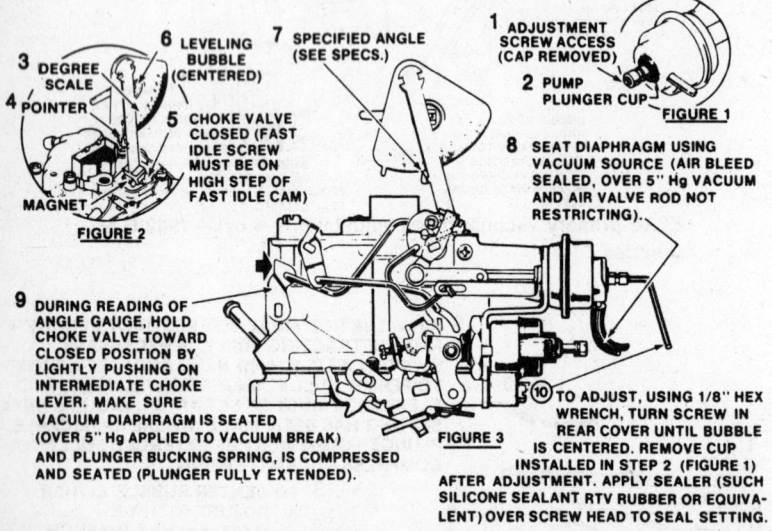

3 DEGREE SCALE

6 LEVELING BUBBLE (CENTERED)

7 SPECIFIED ANGLE (SEE SPECS.)

1 ADJUSTMENT SCREW ACCESS (CAP REMOVED)

2 PUMP PLUNGER CUP

FIGURE 1

4 POINTER

5 CHOKE VALVE CLOSED (FAST IDLE SCREW MUST BE ON HIGH STEP OF FAST IDLE CAM).

MAGNET

FIGURE 2

8 SEAT DIAPHRAGM USING VACUUM SOURCE (AIR BLEED SEALED, OVER 5" Hg VACUUM AND AIR VALVE ROD NOT RESTRICTING).

9 DURING READING OF ANGLE GAUGE, HOLD CHOKE VALVE TOWARD CLOSED POSITION BY LIGHTLY PUSHING ON INTERMEDIATE CHOKE LEVER. MAKE SURE VACUUM DIAPHRAGM IS SEATED (OVER 5" Hg APPLIED TO VACUUM BREAK) AND PLUNGER BUCKING SPRING, IS COMPRESSED AND SEATED (PLUNGER FULLY EXTENDED).

FIGURE 3

10 TO ADJUST, USING 1/8" HEX WRENCH, TURN SCREW IN REAR COVER UNTIL BUBBLE IS CENTERED. REMOVE CUP INSTALLED IN STEP 2 (FIGURE 1) AFTER ADJUSTMENT. APPLY SEALER (SUCH SILICONE SEALANT RTV RUBBER OR EQUIVALENT) OVER SCREW HEAD TO SEAL SETTING.

E2SE primary vacuum break adjustment—1981–82 G.M. "A" and "X" series with 4 cyl engine

2. Seat the choke vacuum diaphragm by applying an external vacuum source of over 5 in. vacuum to the vacuum break.

NOTE: If the air valve rod is restricting the vacuum diaphragm from seating it may be necessary to bend the air valve rod slightly to gain clearance. Make an air valve rod adjustment after the vacuum break adjustment.

3. Read the angle gauge while lightly pushing on the intermediate choke lever so that the choke valve is toward the close position.

4. Use a 1/8 in. hex wrench and turn the screw in the rear cover until the bubble is centered. Apply a silicone sealant over the screw head to seal the setting.

1983–84 GM Models

Refer to the illustration for the adjustment procedure on these models.

CHOKE UNLOADER ADJUSTMENT

Through 1982

1. Follow Steps 1–4 of the "Fast Idle Cam Adjustment" procedure.

2. Install the choke cover and coil, if removed, aligning the marks on the housing and cover as specified.

3. Hold the primary throttle wide open.

4. If the engine is warm, close the choke valve by pushing in on the intermediate choke lever.

5. Bend the unloader tang until the bubble is centered.

1983–84 Models

Refer to the illustration for the adjustment procedure on these models.

SECONDARY LOCKOUT ADJUSTMENT

1. Pull the choke wide open by pushing out on the intermediate choke lever.

2. Open the throttle until the end of the secondary actuating lever is opposite the toe of the lockout lever.

3. Gauge clearance between the lockout lever and secondary lever should be as specified.◊

4. To adjust, bend the lockout lever where it contacts the fast idle cam.

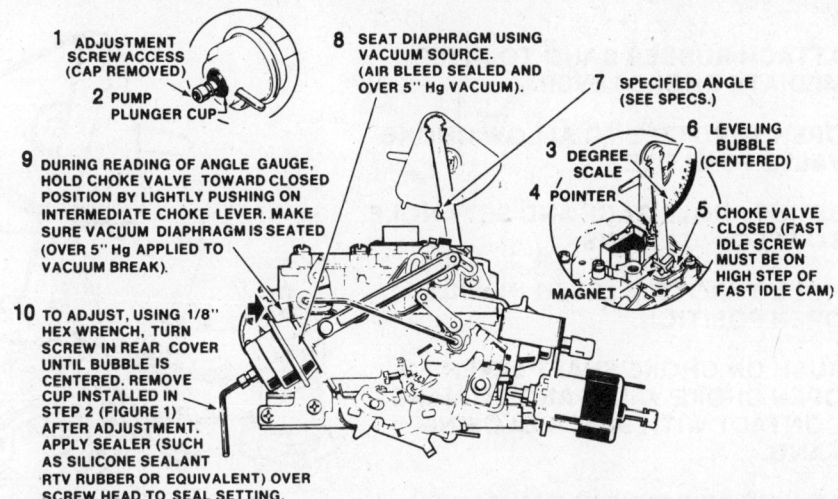

1 ADJUSTMENT SCREW ACCESS (CAP REMOVED)

2 PUMP PLUNGER CUP

8 SEAT DIAPHRAGM USING VACUUM SOURCE. (AIR BLEED SEALED AND OVER 5" Hg VACUUM).

7 SPECIFIED ANGLE (SEE SPECS.)

6 LEVELING BUBBLE (CENTERED)

3 DEGREE SCALE

4 POINTER

5 CHOKE VALVE CLOSED (FAST IDLE SCREW MUST BE ON HIGH STEP OF FAST IDLE CAM)

MAGNET

9 DURING READING OF ANGLE GAUGE, HOLD CHOKE VALVE TOWARD CLOSED POSITION BY LIGHTLY PUSHING ON INTERMEDIATE CHOKE LEVER. MAKE SURE VACUUM DIAPHRAGM IS SEATED (OVER 5" Hg APPLIED TO VACUUM BREAK).

10 TO ADJUST, USING 1/8" HEX WRENCH, TURN SCREW IN REAR COVER UNTIL BUBBLE IS CENTERED. REMOVE CUP INSTALLED IN STEP 2 (FIGURE 1) AFTER ADJUSTMENT. APPLY SEALER (SUCH AS SILICONE SEALANT RTV RUBBER OR EQUIVALENT) OVER SCREW HEAD TO SEAL SETTING.

E2SE secondary vacuum break adjustment—1981 and later G.M. A and X series

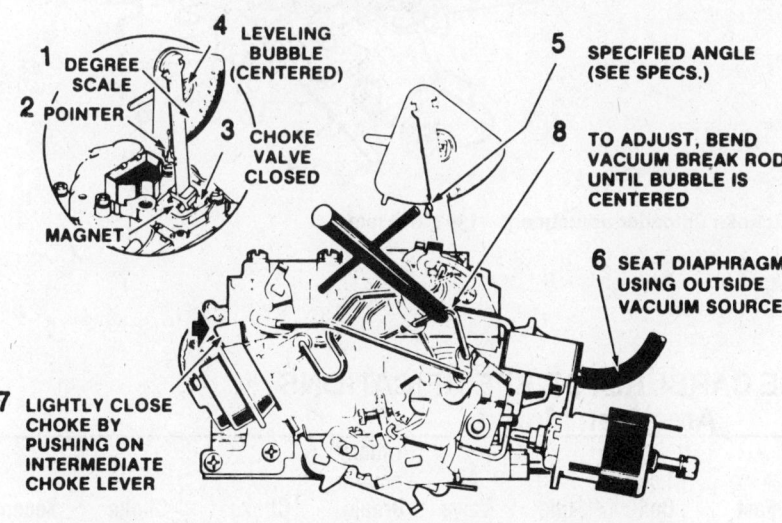

1 DEGREE SCALE

2 POINTER

4 LEVELING BUBBLE (CENTERED)

3 CHOKE VALVE CLOSED

5 SPECIFIED ANGLE (SEE SPECS.)

8 TO ADJUST, BEND VACUUM BREAK ROD UNTIL BUBBLE IS CENTERED

6 SEAT DIAPHRAGM USING OUTSIDE VACUUM SOURCE

MAGNET

7 LIGHTLY CLOSE CHOKE BY PUSHING ON INTERMEDIATE CHOKE LEVER

E2SE secondary vacuum break adjustment—1980 models

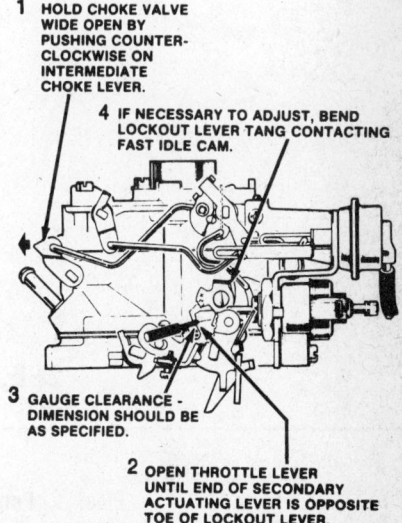

1 HOLD CHOKE VALVE WIDE OPEN BY PUSHING COUNTER-CLOCKWISE ON INTERMEDIATE CHOKE LEVER.

4 IF NECESSARY TO ADJUST, BEND LOCKOUT LEVER TANG CONTACTING FAST IDLE CAM.

3 GAUGE CLEARANCE - DIMENSION SHOULD BE AS SPECIFIED.

2 OPEN THROTTLE LEVER UNTIL END OF SECONDARY ACTUATING LEVER IS OPPOSITE TOE OF LOCKOUT LEVER.

2SE and E2SE secondary lockout adjustment—typical

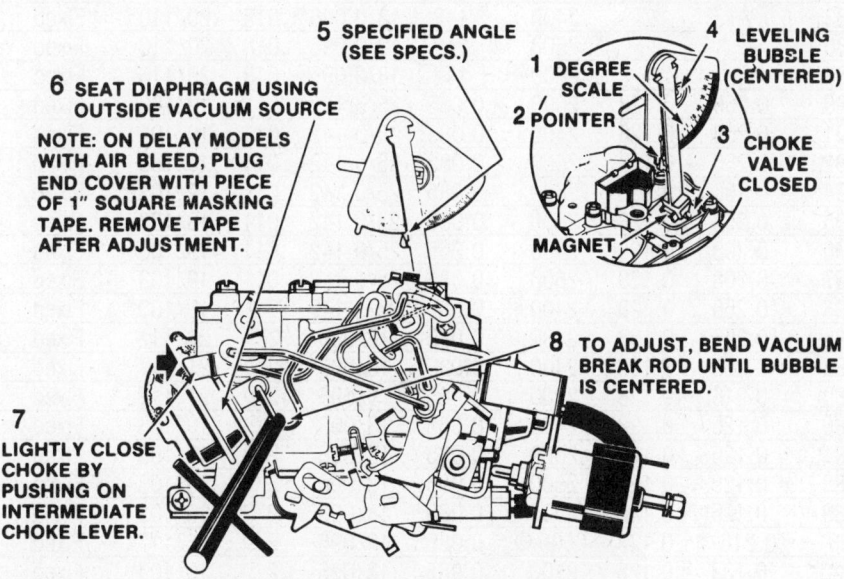

5 SPECIFIED ANGLE (SEE SPECS.)

6 SEAT DIAPHRAGM USING OUTSIDE VACUUM SOURCE

NOTE: ON DELAY MODELS WITH AIR BLEED, PLUG END COVER WITH PIECE OF 1" SQUARE MASKING TAPE. REMOVE TAPE AFTER ADJUSTMENT.

1 DEGREE SCALE

2 POINTER

4 LEVELING BUBBLE (CENTERED)

3 CHOKE VALVE CLOSED

MAGNET

8 TO ADJUST, BEND VACUUM BREAK ROD UNTIL BUBBLE IS CENTERED.

7 LIGHTLY CLOSE CHOKE BY PUSHING ON INTERMEDIATE CHOKE LEVER.

E2SE choke unloader adjuster—typical

1 **ATTACH RUBBER BAND TO INTER-MEDIATE CHOKE LEVER.**

2 **OPEN THROTTLE TO ALLOW CHOKE VALVE TO CLOSE.**

3 **SET UP ANGLE GAGE AND SET ANGLE TO SPECIFICATIONS.**

4 **HOLD THROTTLE LEVER IN WIDE OPEN POSITION.**

5 **PUSH ON CHOKE SHAFT LEVER TO OPEN CHOKE VALVE AND TO MAKE CONTACT WITH BLACK CLOSING TANG.**

6 **ADJUST BY BENDING TANG UNTIL BUBBLE IS CENTERED.**

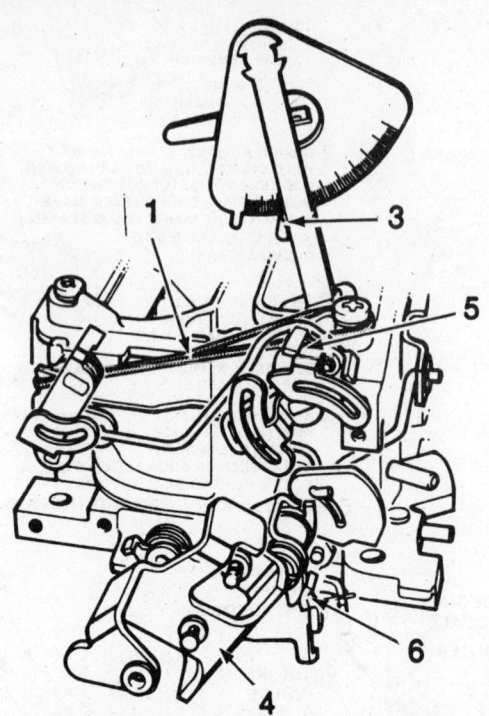

E2SE choke unloader adjustment—1983 and later

2SE, E2SE CARBURETOR SPECIFICATIONS
American Motors

Year	Carburetor Identification	Float Level (in.)	Pump Rod (in.)	Fast Idle (rpm)	Choke Coil Lever (in.)	Fast Idle Cam (deg./in.)	Air Valve Rod (in.)	Primary Vacuum Break (deg./in.)	Choke Setting (notches)	Choke Unloader (deg./in.)	Secondary Lockout (in.)
'80	17080681	3/16	17/32	2400	.142	18/0.096	.018	20/.110	Fixed	32/.195	N.A.
	17080683	3/16	1/2	2400	.142	18/0.096	.018	20/.110	Fixed	32/.195	N.A.
	17080686	3/16	1/2	2600	.142	18/0.096	.018	20/.110	Fixed	32/.195	N.A.
	17080688	3/16	1/2	2600	.142	18/0.096	.018	20/.110	Fixed	32/.195	N.A.
'81	17081790	0.256	0.128	2600	0.085	25/0.142	.011	19/.103	Fixed	32/.195	0.065
	17081791	0.256	0.128	2400	0.085	25/0.142	.011	19/.103	Fixed	32/.195	0.065
	17081792	0.256	0.128	2400	0.085	25/0.142	.011	19/.103	Fixed	32/.195	0.065
	17081794	0.256	0.128	2600	0.085	25/0.142	.011	19/.103	Fixed	32/.195	0.065
	17081795	0.256	0.128	2600	0.085	25/0.142	.011	19/.103	Fixed	32/.195	0.065
	17081796	0.208	0.128	2400	0.065	25/0.142	.011	19/.103	Fixed	32/.195	0.065
	17081797	0.208	0.128	2600	0.085	25/0.142	.011	19/.103	Fixed	32/.195	0.085
	17081793	0.256	0.128	2400	0.085	25/0.142	.011	19/.103	Fixed	32/.195	0.065
'82	17082385	0.256	0.128	2400	0.085	18/.096	2①	21/.117	Fixed	34/.211	0.065
	17082383	0.256	0.128	2400	0.085	18/.096	2①	21/.117	Fixed	34/.211	0.065
	17082380	0.216	0.128	2400	0.085	18/.096	2①	21/.117	Fixed	34/.211	0.065
	17082386	0.125	0.128	2400	0.065	18/.096	2①	19/.103	Fixed	34/.211	0.065
	17082387	0.125	0.128	2600	0.085	18/.096	2①	19/.103	Fixed	34/.211	0.065
	17082388	0.125	0.128	2500	0.085	18/.096	2①	19/.103	Fixed	34/.211	0.065
	17082389	0.125	0.128	2500	0.085	18/.096	2①	19/.103	Fixed	34/.211	0.065
'83–'84	1982380	0.216 ②	0.128	2500 ③	0.085	18/.096	2①	21/.117	Fixed	34/.211	0.065
	1983384	0.138	0.128	2700	0.085	18/.096	2①	19/.103	Fixed	34/.211	0.065
	1983385	0.138	0.128	2700	0.085	18/.096	②①	19/.103	Fixed	34/.211	0.065

2SE, E2SE CARBURETOR SPECIFICATIONS
American Motors

Year	Carburetor Identification	Float Level (in.)	Pump Rod (in.)	Fast Idle (rpm)	Choke Coil Lever (in.)	Fast Idle Cam (deg./in.)	Air Valve Rod (in.)	Primary Vacuum Break (deg./in.)	Choke Setting (notches)	Choke Unloader (deg./in.)	Secondary Lockout (in.)
'85–'86	17085006	4/32	0.128	④	0.085	22/.123	1 ①	21/.117	Fixed	40/.260	0.025
	17085380	5/32	0.128	④	0.085	22/.123	1 ①	26/.149	Fixed	40/.260	0.025
	17085381	5/32	0.128	④	0.085	22/.123	1 ①	26/.149	Fixed	40/.260	0.025
	17085382	5/32	0.128	④	0.085	22/.123	1 ①	26/.149	Fixed	40/.260	0.025
	17085383	5/32	0.128	④	0.085	22/.123	1 ①	26/.149	Fixed	40/.260	0.025
	17085385	5/32	0.128	④	0.085	22/.123	1 ①	26/.149	Fixed	40/.260	0.025
	17085388	4/32	0.128	④	0.085	22/.123	1 ①	21/.117	Fixed	30/.179	0.025
	17086081	4/32	0.128	④	0.085	22/.123	1 ①	25/.142	Fixed	30/.179	0.025

N.A.: Not Available
① Degrees—see procedure
② Auto. trans.—.138
③ Auto. trans.—2700
④ See underhood decal

2SE, E2SE CARBURETOR SPECIFICATIONS
General Motors—U.S.A.

Year	Carburetor Identification	Float Lever (in.)	Pump Rod (in.)	Fast Idle (rpm)	Choke Coil Lever (in.)	Fast Idle Cam (deg./in.)	Air Valve Rod (in.)	Primary Vacuum Break (deg./in.)	Choke Setting (notches)	Secondary Vacuum Break (deg./in.)	Choke Unloader (deg./in.)	Secondary Lockout (in.)
'80	17059614	3/16	1/2	2600	.085	18/.096	.025	17/.090	Fixed	—	36/.227	.120
	17059615	3/16	5/32	2600	.085	18/.096	.025	19/.103	Fixed	—	36/.227	.120
	17059616	3/16	1/2	2600	.085	18/.096	.025	17/.090	Fixed	—	36/.227	.120
	17059617	3/16	5/32	2600	.085	18/.096	.025	19/.103	Fixed	—	36/.227	.120
	17059618	3/16	1/2	2600	.085	18/.096	.025	17/.090	Fixed	—	36/.227	.120
	17059619	3/16	5/32	2600	.085	18/.096	.025	19/.103	Fixed	—	36/.227	.120
	17059620	3/16	1/2	2600	.085	18/.096	.025	17/.090	Fixed	—	36/.227	.120
	17059621	3/16	5/32	2600	.085	18/.096	.025	19/.103	Fixed	—	36/.227	.120
	17059650	3/16	3/32	2600	.085	27/.157	.025	30/.179	Fixed	38/.243	30/.179	.120
	17059651	3/16	3/32	1900	.085	27/.157	.025	22/.123	Fixed	23/.120	30/.179	.120
	17059652	3/16	3/32	2000	.085	27/.157	.025	30/.179	Fixed	38/.243	30/.179	.120
	17059653	3/16	3/32	1900	.085	27/.157	.025	22/.123	Fixed	23/.120	30/.179	.120
	17059714	11/16	5/32	2600	.085	18/.096	.025	23/.129	Fixed	—	32/.195	.120
	17059715	11/16	3/32	2200	.085	18/.096	.025	25/.142	Fixed	—	32/.195	.120
	17059716	11/16	5/32	2600	.085	18/.096	.025	23/.129	Fixed	—	32/.195	.120
	17059717	11/16	3/32	2200	.085	18/.096	.025	25/.142	Fixed	—	32/.195	.120
	17059760	1/8	5/64	2000	.085	17.5/.093	.025	20/.110	Fixed	33/.203	35/.220	.120
	17059762	1/8	5/64	2000	.085	17.5/.093	.025	20/.110	Fixed	33/.203	35/.220	.120
	17059763	1/8	5/64	2000	.085	17.5/.093	.025	20/.110	Fixed	33/.203	35/.220	.120
	17059774	5/32	1/2	①	.085	18/0.096	.018	19/.103	Fixed	—	32/.195	.012
	17059775	5/32	17/32	①	.085	18/0.096	.018	21/.117	Fixed	—	32/.195	.012
	17059776	5/32	1/2	①	.085	18/0.096	.018	19/.103	Fixed	—	32/.195	.012
	17059777	5/32	17/32	①	.085	18/0.096	.018	21/.117	Fixed	—	32/.195	.012
	17080674	3/16	1/2	①	.085	18/0.096	.018	19/.103	Fixed	—	32/.195	.012
	17080675	3/16	1/2	①	.085	18/0.096	.018	21/.117	Fixed	—	32/.195	.012
	17080676	3/16	1/2	①	.085	18/0.096	.018	19/.103	Fixed	—	32/.195	.012
	17080677	3/16	1/2	①	.085	18/0.096	.018	21/.117	Fixed	—	32/.195	.012
'81	17081650	1/4	Fixed	2600	.085	17/.090	1 ②	25/.142	Fixed	34/.211	35/.220	.012
	17081651	1/4	Fixed	2400	.085	17/.090	1 ②	29/.171	Fixed	35/.220	35/.220	.012
	17081652	1/4	Fixed	2600	.085	17/.090	1 ②	25/.142	Fixed	34/.211	35/.220	.012
	17081653	1/4	Fixed	2600	.085	17/.090	1 ②	29/.171	Fixed	35/.220	35/.220	.012
	17081670	5/32	Fixed	2600	.085	18/.096	1 ②	19/.103	Fixed	—	32/.195	.012
	17081671	5/32	Fixed	2600	.085	33.5/.207	1 ②	21/.117	Fixed	—	32/.195	.012

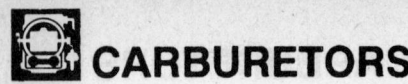

CARBURETORS

2SE, E2SE CARBURETOR SPECIFICATIONS
General Motors—U.S.A.

Year	Carburetor Identification	Float Lever (in.)	Pump Rod (in.)	Fast Idle (rpm)	Choke Coil Lever (in.)	Fast Idle Cam (deg./in.)	Air Valve Rod (in.)	Primary Vacuum Break (deg./in.)	Choke Setting (notches)	Secondary Vacuum Break (deg./in.)	Choke Unloader (deg./in.)	Secondary Lockout (in.)
'81	17081672	5/32	Fixed	2600	.085	18/.096	1 ②	19/.103	Fixed	—	32/.195	.012
	17081673	5/32	Fixed	2600	.085	33.4/.207	1 ②	21/.117	Fixed	—	32/.195	.012
	17081740	1/4	Fixed	2400	.085	17/.090	1 ②	25/.142	Fixed	35/.220	35/.220	.012
	17081742	1/4	Fixed	2400	.085	17/.090	1 ②	25/.142	Fixed	35/.220	35/.220	.012
'82	17081600	5/16	Fixed	①	③	24/.136	1 ②	20/.110	Fixed	27/.157	35/.220	③
	17081601	5/16	Fixed	①	③	24/1.36	1 ②	20/.110	Fixed	27/.157	35/.220	③
	17081607	5/16	Fixed	①	③	24/.136	1 ②	20/.110	Fixed	27/.157	35/.220	③
	17081700	5/16	Fixed	①	③	24/.136	1 ②	20/.110	Fixed	27/.157	35/.220	③
	17081701	5/16	Fixed	①	③	24/.136	1 ②	20/.110	Fixed	27/.157	35/.220	③
	17082196	5/16	Fixed	①	.085	18/.096	1 ②	21/.117	Fixed	19/.103	27/.157	③
	17082316	1/4	Fixed	2600	.085	17/.090	1 ②	30/.179	Fixed	34/.211	45/.304	③
	17082317	1/4	Fixed	2600	.085	17/.090	1 ②	30/.179	Fixed	35/.220	45/.304	③
	17082320	1/4	Fixed	2800	.085	25/.142	1 ②	30/.179	Fixed	35/.220	45/.304	③
	17082321	1/4	Fixed	2600	.085	25/.142	1 ②	30/.179	Fixed	35/.220	45/.304	③
	17082390	13/32	Fixed	2500	.085	17/.090	1 ②	26/.149	Fixed	34/.211	35/.220	.011–.040
	17082391	13/32	Fixed	2600	.085	25/.142	1 ②	29/.171	Fixed	35/.220	35/.220	.011–.040
	17082490	13/32	Fixed	2500	.085	17/.090	1 ②	26/.149	Fixed	34/.211	35/.220	.011–.040
	17082491	13/32	Fixed	2600	.085	25/.142	1 ②	29/.171	Fixed	35/.220	35/.220	.011–.040
	17082640	1/4	Fixed	2600	.085	17/.090	1 ②	30/.179	Fixed	34/.211	45/.304	③
	17082641	1/4	Fixed	2400	.085	17/.090	1 ②	30/.179	Fixed	35/.220	45/.304	③
	17082642	1/4	Fixed	2800	.085	25/.142	1 ②	30/.179	Fixed	35/.220	45/.304	③
'83	17083356	13/32	Fixed	①	.085	22/.123	1 ②	25/.142	Fixed	35/.220	30/.179	.025
	17083357	13/32	Fixed	①	.085	22/.123	1 ②	25/.142	Fixed	35/.220	30/.179	.025
	17083358	13/32	Fixed	①	.085	22/.123	1 ②	25/.142	Fixed	35/.220	30/.179	.025
	17083359	13/32	Fixed	①	.085	22/.123	1 ②	25/.142	Fixed	35/.220	30/.179	.025
	17083368	13/32	Fixed	①	.085	22/.123	1 ②	25/.142	Fixed	35/.220	30/.179	.025
	17083369	13/32	Fixed	①	.085	22/.123	1 ②	25/.142	Fixed	35/.220	30/.179	.025
	17083370	13/32	Fixed	①	.085	22/.123	1 ②	25/.142	Fixed	35/.220	30/.179	.025
	17083391	13/32	Fixed	①	.085	28/.164	1 ②	30/.179	Fixed	35/.220	38/.243	.025
	17083392	13/32	Fixed	①	.085	28/.164	1 ②	30/.179	Fixed	35/.220	38/.243	.025
	17083393	13/32	Fixed	①	.085	28/.164	1 ②	30/.179	Fixed	35/.220	38/.243	.025
	17083394	13/32	Fixed	①	.085	28/.164	1 ②	30/.179	Fixed	35/.220	38/.243	.025
	17083395	13/32	Fixed	①	.085	28/.164	1 ②	30/.179	Fixed	35/.220	38/.243	.025
	17083396	13/32	Fixed	①	.085	28/.164	1 ②	30/.179	Fixed	35/.220	38/.243	.025
	17083397	13/32	Fixed	①	.085	28/.164	1 ②	30/.179	Fixed	35/.220	38/.243	.025
	17083450	1/4	Fixed	①	.085	28/.164	1 ②	27/.157	Fixed	35/.220	45/.304	.025
	17083451	1/4	Fixed	①	.085	28/.164	1 ②	27/.157	Choke Setting	35/.220	45/.304	Secondary Lockout
	17083452	1/4	Fixed	①	.085	28/.164	1 ②	27/.157	Setting	35/.220	45/.304	.025
	17083453	1/4	Fixed	①	.085	28/.164	1 ②	27/.157	(notches)	35/.220	45/.304	.025
	17083454	1/4	Fixed	①	.085	28/.164	1 ②	27/.157	Fixed	35/.220	45/.304	.025
	17083455	1/4	Fixed	①	.085	28/.164	1 ②	27/.157	Fixed	35/.220	45/.304	.025
	17083456	1/4	Fixed	①	.085	28/.164	1 ②	27/.157	Fixed	35/.220	45/.304	.025
	17083630	1/4	Fixed	①	.085	28/.164	1 ②	27/.157	Fixed	35/.220	45/.304	.025
	17083631	1/4	Fixed	①	.085	28/.164	1 ②	27/.157	Fixed	35/.220	45/.304	.025
	17083632	1/4	Fixed	①	.085	28/.164	1 ②	27/.157	Fixed	35/.220	45/.304	.025
	17083633	1/4	Fixed	①	.085	28/.164	1 ②	27/.157	Fixed	35/.220	45/.304	.025
	17083634	1/4	Fixed	①	.085	28/.164	1 ②	27/.157	Fixed	35/.220	45/.304	.025
	17083635	1/4	Fixed	①	.085	28/.164	1 ②	27/.157	Fixed	35/.220	45/.304	.025
	17083636	1/4	Fixed	①	.085	28/.164	1 ②	27/.157	Fixed	35/.220	45/.304	.025
'84	17072683	9/32	Fixed	①	.085	28/.164	1 ②	25/.142	Fixed	35/.220	45/.304	.025
	17074812	9/32	Fixed	①	.085	28/.164	1 ②	25/.142	Fixed	35/.220	45/.304	.025
	17084356	9/32	Fixed	①	.085	22/.123	1 ②	25/.142	Fixed	30/.179	30/.179	.025
	17084357	9/32	Fixed	①	.085	22/.123	1 ②	25/.142	Fixed	30/.179	30/.179	.025
	17084358	9/32	Fixed	①	.085	22/.123	1 ②	25/.142	Fixed	30/.179	30/.179	.025

2SE, E2SE CARBURETOR SPECIFICATIONS
General Motors—U.S.A.

Year	Carburetor Identification	Float Lever (in.)	Pump Rod (in.)	Fast Idle (rpm)	Choke Coil Lever (in.)	Fast Idle Cam (deg./in.)	Air Valve Rod (in.)	Primary Vacuum Break (deg./in.)	Choke Setting (notches)	Secondary Vacuum Break (deg./in.)	Choke Unloader (deg./in.)	Secondary Lockout (in.)
'84	17084359	9/32	Fixed	①	.085	22/.123	1 ②	25/.142	Fixed	30/.179	30/.179	.025
	17084368	1/8	Fixed	①	.085	22/.123	1 ②	25/.142	Fixed	30/.179	30/.179	.025
	17084370	1/8	Fixed	①	.085	22/.123	1 ②	25/.142	Fixed	30/.179	30/.179	.025
	17084430	11/32	Fixed	①	.085	15/.077	1 ②	26/.149	Fixed	30/.179	30/.179	.025
	17084431	11/32	Fixed	①	.085	15/.077	1 ②	26/.149	Fixed	38/.243	42/.277	.025
	17084434	11/32	Fixed	①	.085	15/.077	1 ②	26/.149	Fixed	38/.243	42/.277	.025
	17084435	11/32	Fixed	①	.085	15/.077	1 ②	26/.149	Fixed	38/.243	42/.277	.025
	17084452	5/32	Fixed	①	.085	28/.164	1 ②	25/.142	Fixed	38/.243	42/.377	.025
	17084453	5/32	Fixed	①	.085	28/.164	1 ②	25/.142	Fixed	35/.220	45/.304	.025
	17084455	5/32	Fixed	①	.085	28/.164	1 ②	25/.142	Fixed	35/.220	45/.304	.025
	17084456	5/32	Fixed	①	.085	28/.164	1 ②	25/.142	Fixed	35/.220	45/.304	.025
	17084458	5/32	Fixed	①	.085	28/.164	1 ②	25/.142	Fixed	35/.220	45/.304	.025
	17084532	5/32	Fixed	①	.085	28/.164	1 ②	25/.142	Fixed	35/.220	45/.304	.025
	17084534	5/32	Fixed	①	.085	28/.164	1 ②	25/.142	Fixed	35/.220	45/.304	.025
	17084535	5/32	Fixed	①	.085	28/.164	1 ②	25/.142	Fixed	35/.220	45/.304	.025
	17084537	5/32	Fixed	①	.085	28/.164	1 ②	25/.142	Fixed	35/.220	45/.304	.025
	17084538	5/32	Fixed	①	.085	28/.164	1 ②	25/.142	Fixed	35/.220	45/.304	.025
	17084540	5/32	Fixed	①	.085	28/.164	1 ②	25/.142	Fixed	35/.220	45/.304	.025
	17084542	1/8	Fixed	①	.085	28/.164	1 ②	25/.142	Fixed	35/.220	45/.304	.025
	17084632	9/32	Fixed	①	.085	28/.164	1 ②	25/.142	Fixed	35/.220	45/.304	.025
	17084633	9/32	Fixed	①	.085	28/.164	1 ②	25/.142	Fixed	35/.220	45/.304	.025
	17084635	9/32	Fixed	①	.085	28/.164	1 ②	25/.142	Fixed	35/.220	45/.304	.025
	17084636	9/32	Fixed	①	.085	28/.164	1 ②	25/.142	Fixed	35/.220	45/.304	.025
'85	17084534	5/32	Fixed	①	.085	28/.164	1 ②	25/.142	Fixed	35/.220	45/.304	—
	17084535	5/32	Fixed	①	.085	28/.164	1 ②	25/.142	Fixed	35/.220	45/.304	—
	17084540	5/32	Fixed	①	.085	28/.164	1 ②	25/.142	Fixed	35/.220	45/.304	—
	17084542	4/32	Fixed	①	.085	28/.164	1 ②	25/.142	Fixed	35/.220	45/.304	—
	17085356	9/32	Fixed	①	.085	22/.123	1 ②	25/.142	Fixed	30/.179	30/.179	—
	17085357	9/32	Fixed	①	.085	22/.123	1 ②	25/.142	Fixed	30/.179	30/.179	—
	17085358	9/32	Fixed	①	.085	22/.123	1 ②	25/.142	Fixed	30/.179	30/.179	—
	17085359	9/32	Fixed	①	.085	22/.123	1 ②	25/.142	Fixed	30/.179	30/.179	—
	17085368	4/32	Fixed	①	.085	22/.123	1 ②	25/.142	Fixed	30/.179	30/.179	—
	17085369	9/32	Fixed	①	.085	22/.123	1 ②	25/.142	Fixed	30/.179	30/.179	—
	17085370	4/32	Fixed	①	.085	22/.123	1 ②	25/.142	Fixed	30/.179	30/.179	—
	17085371	9/32	Fixed	①	.085	22/.123	1 ②	25/.142	Fixed	30/.179	30/.179	—
	17085452	5/32	Fixed	①	.085	28/.164	1 ②	25/.142	Fixed	35/.220	45/.304	—
	17085453	5/32	Fixed	①	.085	28/.164	1 ②	25/.142	Fixed	35/.220	45/.304	—
	17085458	5/32	Fixed	①	.085	28/.164	1 ②	25/.142	Fixed	35/.220	45/.304	—
'86	17084534	5/32	Fixed	①	.085	28/.164	1 ②	25/.142	Fixed	35/.220	45/.304	—
	17084535	5/32	Fixed	①	.085	28/.164	1 ②	25/.142	Fixed	35/.220	45/.304	—
	17084540	5/32	Fixed	①	.085	28/.164	1 ②	25/.142	Fixed	35/.220	45/.304	—
	17084542	5/32	Fixed	①	.085	28/.164	1 ②	25/.142	Fixed	35/.220	45/.304	—

① See underhood decal
② Measurement in degrees
③ Not available

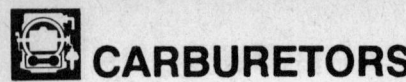

CARBURETORS

2SE, E2SE CARBURETOR SPECIFICATIONS
General Motors—Canada

Year	Carburetor Identification	Float Lever (in.)	Pump Rod (in.)	Fast Idle (rpm)	Choke Coil Lever (in.)	Fast Idle Cam (deg./in.)	Air Valve Rod (in.)	Primary Vacuum Break (deg./in.)	Choke Setting (notches)	Secondary Vacuum Break (deg./in.)	Choke Unloader (deg./in.)	Secondary Lockout (in.)
'81	17059660	1/4	17/32	①	.085	24/.136	1	30/.179	Fixed	32/.195	30/.179	②
	17059662	1/4	17/32	①	.085	24/.136	1	30/.179	Fixed	37/.195	30/.179	②
	17059651	1/4	17/32	①	.085	24/.136	1	30/.179	Fixed	32/.195	30/.179	②
	17059666	1/4	17/32	①	.085	24/.136	1	26/.149	Fixed	32/.195	30/.179	②
	17059667	1/4	17/32	①	.085	24/.136	1	26/.149	Fixed	32/.195	30/.179	②
	17059622	5/32	17/32	①	.085	18/.096	1	17/.090	Fixed	—	36/.227	②
	17059623	5/32	17/32	①	.085	18/.096	1	19/.103	Fixed	—	36/.227	②
	17059624	5/32	17/32	①	.085	18/.096	1	17/.090	Fixed	—	36/.227	②
'82	17082440	1/4	19/32	①	.085	24/.136	1	30/.179	Fixed	32/.195	45/.304	②
	17082441	1/4	19/32	①	.085	24/.136	1	30/.179	Fixed	32/.195	45/.304	②
	17082443	1/4	19/32	①	.085	24/.136	1	30/.179	Fixed	32/.195	45/.304	②
	17082460	1/4	19/32	①	.085	18/.096	1	21/.117	Fixed	—	36/.227	②
	17082461	1/4	19/32	①	.085	18/.096	1	21/.117	Fixed	—	36/.227	②
	17082462	1/4	19/32	①	.085	18/.096	1	21/.117	Fixed	—	36/.227	②
	17082464	1/8	19/32	①	.085	18/.096	1	21/.117	Fixed	—	36/.227	②
	17082465	1/8	19/32	①	.085	18/.096	1	21/.117	Fixed	—	36/.227	②
	17082466	1/8	19/32	①	.085	18/.096	1	21/.117	Fixed	—	36/.227	②
	17082620	7/16	19/32	①	.085	24/.136	1	30/.179	Fixed	32/.195	45/.304	②
	17082621	7/16	19/32	①	.085	24/.136	1	30/.179	Fixed	32/.195	45/.304	②
	17082622	7/16	19/32	①	.085	24/.136	1	30/.179	Fixed	32/.195	45/.304	②
	17082623	7/16	19/32	①	.085	24/.136	1	30/.179	Fixed	32/.195	45/.304	②
'83	17083311	5/16	Fixed	①	.085	24/.136	1	18/.096	Fixed	20/.110	35/.220	.025
	17083314	5/16	Fixed	①	.085	24/.136	1	16/.083	Fixed	20/.110	35/.220	.025
	17083401	5/16	Fixed	①	.085	24/.136	1	18/.096	Fixed	20/.110	35/.220	.025
	17083440	1/4	19/32	①	.085	24/.136	1	28/.164	Fixed	32/.195	40/.260	.025
	17083441	1/4	19/32	①	.085	24/.136	1	28/.164	Fixed	32/.195	40/.260	.025
	17083442	1/4	19/32	①	.085	24/.136	1	28/.164	Fixed	32/.195	40/.260	.025
	17083443	1/4	19/32	①	.085	24/.136	1	28/.164	Fixed	32/.195	40/.260	.025
	17083444	1/4	19/32	①	.085	24/.136	1	28/.164	Fixed	32/.195	40/.260	.025
	17083445	1/4	19/32	①	.085	24/.136	1	28/.164	Fixed	32/.195	40/.260	.025
	17083460	1/4	19/32	①	.085	18/.096	1	19/.103	Fixed	—	36/.227	.025
	17083461	1/4	19/32	①	.085	18/.096	1	18/.096	Fixed	—	36/.227	.025
	17083462	1/4	19/32	①	.085	18/.096	1	19/.103	Fixed	—	36/.227	.025
	17083464	1/8	19/32	①	.085	18/.096	1	19/.103	Fixed	—	36/.227	.025
	17083465	1/8	19/32	①	.085	18/.096	1	20/.110	Fixed	—	36/.227	.025
	17083466	1/8	19/32	①	.085	18/.096	1	19/.103	Fixed	—	36/.227	.025
	17083620	7/16	19/32	①	.085	24/.136	1	28/.164	Fixed	32/.195	40/.260	.025
	17083621	7/16	19/32	①	.085	24/.136	1	28/.164	Fixed	32/.195	40/.260	.025
	17083622	7/16	19/32	①	.085	24/.136	1	28/.164	Fixed	34/.195	40/.260	.025
	17083623	7/16	19/32	①	.085	24/.136	1	28/.164	Fixed	32/.195	40/.260	.025
'84	17084312	5/16	Fixed	①	.085	24/.136	1	18/.096	Fixed	20/.110	35/.220	.025
	17084314	5/16	Fixed	①	.085	29/.171	1	16/.083	Fixed	20/.110	30/.179	.025
	17084480	1/4	Fixed	①	.085	24/.136	1	28/.164	Fixed	32/.195	45/.304	.025
	17084481	1/4	Fixed	①	.085	24/.136	1	28/.164	Fixed	32/.195	45/.304	.025
	17084482	1/4	Fixed	①	.085	24/.136	1	28/.164	Fixed	32/.195	45/.304	.025
	17084483	1/4	Fixed	①	.085	24/.136	1	28/.164	Fixed	32/.195	45/.304	.025
	17084484	1/4	Fixed	①	.085	24/.136	1	28/.164	Fixed	32/.195	45/.304	.025
	17084485	1/4	Fixed	①	.085	24/.136	1	28/.164	Fixed	32/.195	45/.304	.025
	17084486	1/4	Fixed	①	.085	24/.136	1	28/.164	Fixed	32/.195	45/.304	.025
	17084487	1/4	Fixed	①	.085	24/.136	1	28/.164	Fixed	32/.195	45/.304	.025
	17084620	7/16	Fixed	①	.085	24/.136	1	26/.149	Fixed	32/.195	45/.304	.025
	17084621	7/16	Fixed	①	.085	24/.136	1	26/.149	Fixed	32/.195	45/.304	.025
	17084622	7/16	Fixed	①	.085	24/.136	1	26/.149	Fixed	32/.195	45/.304	.025
	17084623	7/16	Fixed	①	.085	24/.136	1	26/.149	Fixed	32/.195	45/.304	.025

2SE, E2SE CARBURETOR SPECIFICATIONS
General Motors—Canada

Year	Carburetor Identification	Float Lever (in.)	Pump Rod (in.)	Fast Idle (rpm)	Choke Coil Lever (in.)	Fast Idle Cam (deg./in.)	Air Valve Rod (in.)	Primary Vacuum Break (deg./in.)	Choke Setting (notches)	Secondary Vacuum Break (deg./in.)	Choke Unloader (deg./in.)	Secondary Lockout (in.)
'85	17084312	5/16	Fixed	①	.085	—	1	18/.096	Fixed	20/.110	35/.220	—
	17084314	5/16	Fixed	①	.085	—	1	16/.083	Fixed	20/.110	30/.179	—
	17085484	12/32	Fixed	①	.085	—	1	28/.164	Fixed	32/.195	45/.304	—
	17085485	12/32	Fixed	①	.085	—	1	28/.164	Fixed	32/.195	45/.304	—
	17085482	12/32	Fixed	①	.085	—	1	28/.164	Fixed	32/.195	45/.304	—
	17085483	12/32	Fixed	①	.085	—	1	28/.164	Fixed	32/.195	45/.304	—
	17085484	12/32	Fixed	①	.085	—	1	28/.164	Fixed	32/.195	45/.304	—
	17085485	12/32	Fixed	①	.085	—	1	28/.164	Fixed	32/.195	45/.304	—
	17085486	12/32	Fixed	①	.085	—	1	28/.164	Fixed	32/.195	45/.304	—
	17085487	12/32	Fixed	①	.085	—	1	28/.164	Fixed	32/.195	45/.304	—
'86	17086484	12/32	Fixed	①	.085	—	1	28/.164	Fixed	32/.195	45/.304	—
	17086485	12/32	Fixed	①	.085	—	1	28/.164	Fixed	32/.195	45/.304	—
	17086486	4/32	Fixed	①	.085	—	1	28/.164	Fixed	32/.195	45/.304	—
	17086487	4/32	Fixed	①	.085	—	1	28/.164	Fixed	32/.195	45/.304	—

① See underhood decal
② Not available

Models 2MC, M2MC, M2ME and E2ME

The Rochester model 2MC carburetor is a two-barrel single stage carburetor which incorporates the design features of the primary side of the Rochester Quadrajet four-barrel carburetor. It is used on small displacement V8s. The M2MC version with front and rear vacuum brake diaphragms, was introduced on the 301 V8.

The Dualjet E2ME Model 210 is a variation of the M2ME, modified for use with the Electronic Fuel Control System (also called the Computer Controlled Catalytic Converter, or C-4, System). An electrically operated mixture control solenoid is mounted in the float bowl. Mixture is thus controlled by the Electronic Control Module, in response to signals from the oxygen sensor mounted in the exhaust system upstream of the catalytic converter.

For further information on feedback carburetors, please refer to *Chilton's Guide To Fuel Injection And Feedback Carburetors*.

FLOAT LEVEL ADJUSTMENT

See the illustration for float level adjustment for all carburetors. The E2ME procedure is the same except for adjustment (step 4 in the figure). For the E2ME only, if the float level is too high, hold the retainer firmly in place and push down on the center of the float to adjust.

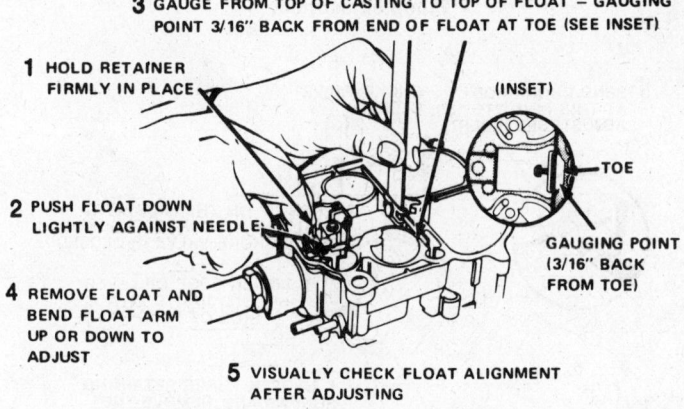

3 GAUGE FROM TOP OF CASTING TO TOP OF FLOAT – GAUGING POINT 3/16" BACK FROM END OF FLOAT AT TOE (SEE INSET)

1 HOLD RETAINER FIRMLY IN PLACE

(INSET)

TOE

2 PUSH FLOAT DOWN LIGHTLY AGAINST NEEDLE

GAUGING POINT (3/16" BACK FROM TOE)

4 REMOVE FLOAT AND BEND FLOAT ARM UP OR DOWN TO ADJUST

5 VISUALLY CHECK FLOAT ALIGNMENT AFTER ADJUSTING

2MC, M2MC, M2ME, E2ME float level adjustment—typical

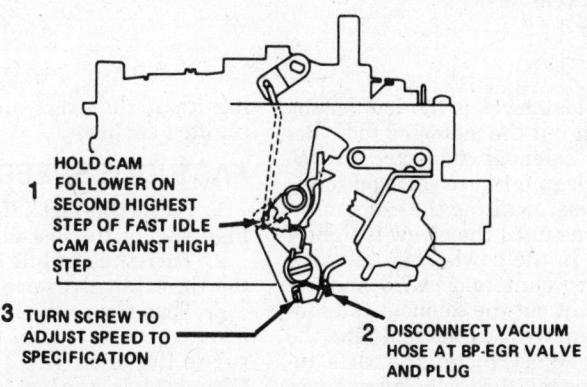

1 HOLD CAM FOLLOWER ON SECOND HIGHEST STEP OF FAST IDLE CAM AGAINST HIGH STEP

3 TURN SCREW TO ADJUST SPEED TO SPECIFICATION

2 DISCONNECT VACUUM HOSE AT BP-EGR VALVE AND PLUG

M2MC and E2ME fast idle speed adjustment—typical

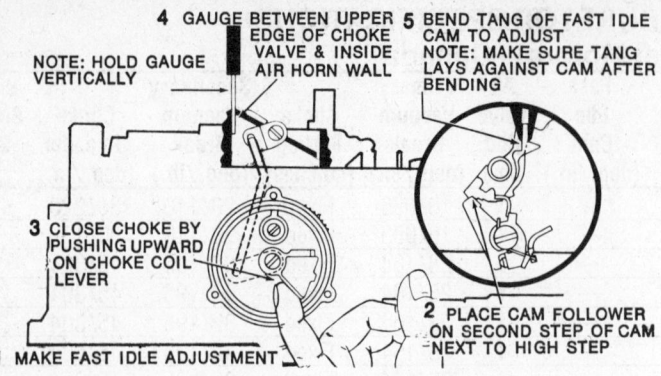

2MC, M2MC, M2ME, E2ME fast idle cam adjustment—typical

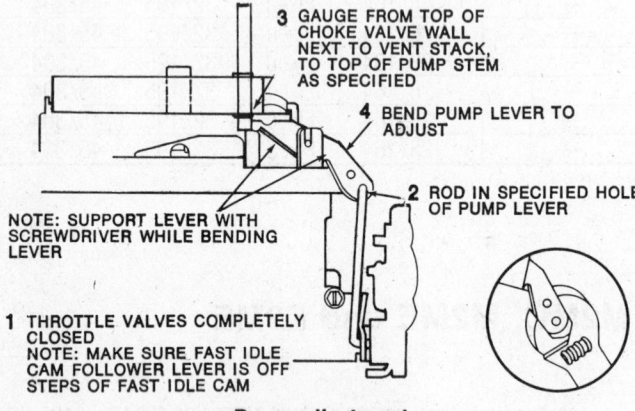

Pump adjustment

5 BEND CHOKE ROD AT THIS POINT TO ADJUST (SEE INSERT)

CHOKE VALVE CLOSED

2 PUSH UP ON THERMOSTATIC COIL TANG (COUNTERCLOCKWISE) UNTIL CHOKE VALVE IS CLOSED

4 LOWER EDGE OF LEVER SHOULD JUST CONTACT SIDE OF PLUG GAUGE

1 LOOSEN THREE RETAINING SCREWS AND REMOVE THE THERMOSTATIC COVER AND COIL ASSEMBLY FROM CHOKE HOUSING

3 INSERT SPECIFIED PLUG GAUGE

2MC, M2MC, M2ME, E2MC choke coil lever adjustment—typical

FAST IDLE CAM (CHOKE ROD) ADJUSTMENT

1. Adjust the fast idle speed.
2. Place the cam follower lever on the second step of the fast idle cam, holding it firmly against the rise of the high step.
3. Close the choke valve by pushing upward on the choke coil lever inside the choke housing, or by pushing up on the vacuum break lever tang.
4. Gauge between the upper edge of the choke valve and the inside of the air horn wall.
5. Bend the tang on the fast idle cam to adjust.

PUMP ADJUSTMENT

This adjustment is not required on E2ME carburetors used in conjunction with the computer controlled systems.

1. With the fast idle cam follower off the steps of the fast idle cam, back out the idle speed screw until the throttle valves are completely closed.
2. Place the pump rod in the proper hole of the lever.
3. Measure from the top of the choke valve wall, next to the vent stack, to the top of the pump stem.
4. Bend the pump lever to adjust.

CHOKE COIL LEVER ADJUSTMENT

1. Remove the choke cover and thermostatic coil from the choke housing. On models with a fixed choke cover, drill out the rivets and remove the cover. A stat cover kit will be required for assembly.
2. Push up on the coil tang (counterclockwise) until the choke valve is closed. The top of the choke rod should be at the bottom of the slot in the choke valve lever. Place the fast idle cam follower on the high step of the cam.
3. Insert a 0.120 in. plug gauge in the hole in the choke housing.
4. The lower edge of the choke coil lever should just contact the side of the plug gauge.
5. Bend the choke rod to adjust.

2MC LEAN/RICH VACUUM BRAKE ADJUSTMENT

1. Place the cam follower on the highest step of the fast idle cam.
2. Seat the vacuum break diaphragm by using an outside vacuum source. Tape over the bleed hole, if any, under the rubber cover on the diaphragm.
3. Remove the choke cover and

If the float level is too low on the E2ME, lift out the metering rods. Remove the solenoid connector screws. Turn the lean mixture solenoid screw in clockwise, counting the exact number of turns until the screw is lightly bottomed in the bowl. Then turn the screw out counterclockwise and remove it. Lift out the solenoid and connector. Remove the float and bend the arm up to adjust. Install the parts, installing the mixture solenoid screw in until it is lightly bottomed, then turn-

ing it out the exact number of turns counted earlier.

FAST IDLE SPEED

1. Place the fast idle lever on the high step of the fast idle cam.
2. Turn the fast idle screw out until the throttle valves are closed.
3. Turn the screw in to contact the lever, then turn it in the number of turns listed in the specifications. Check this preliminary setting against the sticker figure.

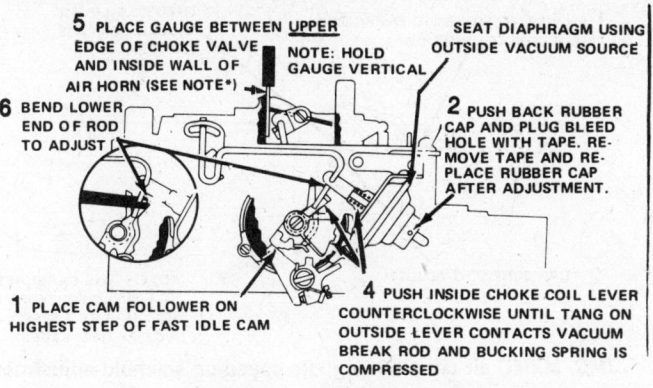

5 PLACE GAUGE BETWEEN UPPER EDGE OF CHOKE VALVE AND INSIDE WALL OF AIR HORN (SEE NOTE*)

6 BEND LOWER END OF ROD TO ADJUST

SEAT DIAPHRAGM USING OUTSIDE VACUUM SOURCE

NOTE: HOLD GAUGE VERTICAL

2 PUSH BACK RUBBER CAP AND PLUG BLEED HOLE WITH TAPE. REMOVE TAPE AND REPLACE RUBBER CAP AFTER ADJUSTMENT.

1 PLACE CAM FOLLOWER ON HIGHEST STEP OF FAST IDLE CAM

4 PUSH INSIDE CHOKE COIL LEVER COUNTERCLOCKWISE UNTIL TANG ON OUTSIDE LEVER CONTACTS VACUUM BREAK ROD AND BUCKING SPRING IS COMPRESSED

2MC rich vacuum break setting

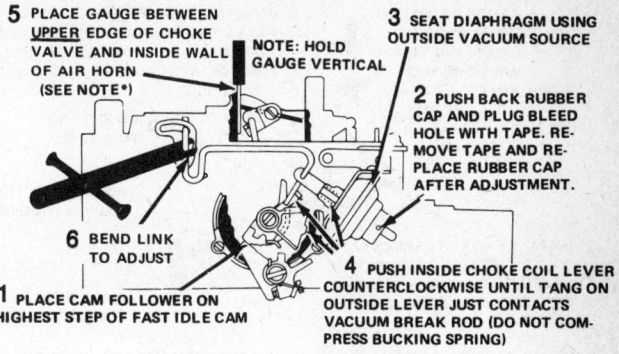

5 PLACE GAUGE BETWEEN UPPER EDGE OF CHOKE VALVE AND INSIDE WALL OF AIR HORN (SEE NOTE*)

NOTE: HOLD GAUGE VERTICAL

3 SEAT DIAPHRAGM USING OUTSIDE VACUUM SOURCE

2 PUSH BACK RUBBER CAP AND PLUG BLEED HOLE WITH TAPE. REMOVE TAPE AND REPLACE RUBBER CAP AFTER ADJUSTMENT.

6 BEND LINK TO ADJUST

1 PLACE CAM FOLLOWER ON HIGHEST STEP OF FAST IDLE CAM

4 PUSH INSIDE CHOKE COIL LEVER COUNTERCLOCKWISE UNTIL TANG ON OUTSIDE LEVER JUST CONTACTS VACUUM BREAK ROD (DO NOT COMPRESS BUCKING SPRING)

2MC lean vacuum break setting

thermostatic coil and push up on the coil lever inside the choke housing until the tang on the vacuum break lever contacts the tang on the vacuum break plunger stem. Do not compress the bucking spring for lean adjustment. Compress the bucking spring for rich adjustment.

4. With the choke rod in the bottom of the slot in the choke lever, gauge between the upper edge of the choke valve and the inside wall of the air horn.

5. Bend the link rod at the vacuum break plunger stem to adjust the rich setting. Bend the link rod at the opposite end from the diaphragm to adjust the lean setting.

FRONT/REAR VACUUM BRAKE ADJUSTMENT
M2MC, M2ME and E2ME (1980)

1. Sat the front diaphragm, using an outside vacuum source. If there is an air bleed hole on the diaphragm, tape it over.

2. Remove the choke cover and coil. Rotate the inside coil lever counterclockwise. On models with a fixed choke cover (riveted), push up on the vacuum break lever tang and hold it in position with a rubber band.

3. Check that the specified gap is present between the top of the choke valve and the air horn wall.

4. Turn the front vacuum break adjusting screw to adjust.

5. To adjust the rear vacuum break diaphragm, perform Steps 1–3 on the rear diaphragm, but make sure that the plunger bucking spring is compressed and seated in Step 2. Adjust by bending the link at the bend nearest the diaphragm.

1981–84 Models

On these models a choke valve measuring gauge J-26701 or equivalent is used to measure angle (degrees instead of inches). See illustration for procedure.

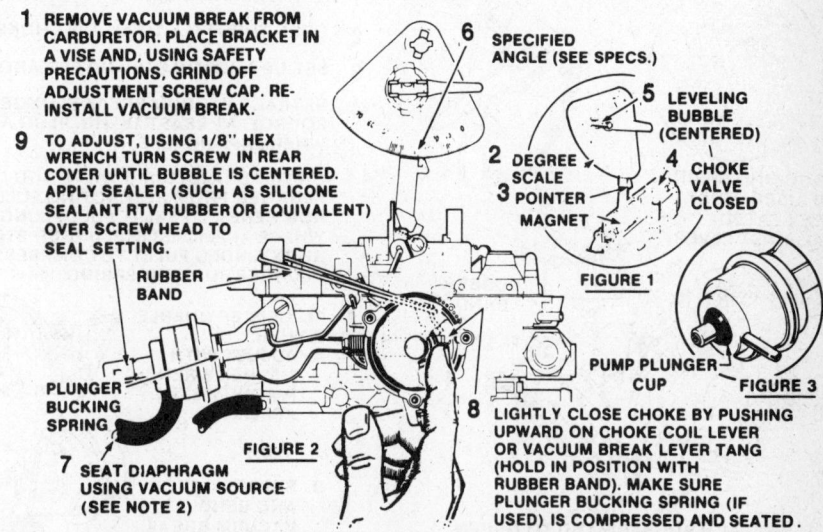

1 REMOVE VACUUM BREAK FROM CARBURETOR. PLACE BRACKET IN A VISE AND, USING SAFETY PRECAUTIONS, GRIND OFF ADJUSTMENT SCREW CAP. REINSTALL VACUUM BREAK.

9 TO ADJUST, USING 1/8" HEX WRENCH TURN SCREW IN REAR COVER UNTIL BUBBLE IS CENTERED. APPLY SEALER (SUCH AS SILICONE SEALANT RTV RUBBER OR EQUIVALENT) OVER SCREW HEAD TO SEAL SETTING.

RUBBER BAND

PLUNGER BUCKING SPRING

7 SEAT DIAPHRAGM USING VACUUM SOURCE (SEE NOTE 2)

FIGURE 2

6 SPECIFIED ANGLE (SEE SPECS.)

5 LEVELING BUBBLE (CENTERED)

2 DEGREE SCALE
3 POINTER
MAGNET

4 CHOKE VALVE CLOSED

FIGURE 1

PUMP PLUNGER CUP   FIGURE 3

8 LIGHTLY CLOSE CHOKE BY PUSHING UPWARD ON CHOKE COIL LEVER OR VACUUM BREAK LEVER TANG (HOLD IN POSITION WITH RUBBER BAND). MAKE SURE PLUNGER BUCKING SPRING (IF USED) IS COMPRESSED AND SEATED.

NOTE 2: ON DELAY MODELS, PLUG END COVER USING AN ACCELERATOR PUMP PLUNGER CUP - 2G TYPE (FIGURE 3) OR EQUIVALENT. SEAT VACUUM DIAPHRAGM MAKING SURE VACUUM IS ABOVE 5" Hg WHEN READING GAUGE (STEP 9). REMOVE CUP AFTER ADJUSTMENT.

NOTE 1: MAKE CHOKE COIL LEVER ADJUSTMENT AND FAST IDLE ADJUSTMENT. DO NOT REMOVE RIVETS AND CHOKE COVER TO PERFORM THIS ADJUSTMENT. USE RUBBER BAND ON VACUUM BREAK LEVER TANG TO HOLD CHOKE VALVE CLOSED (STEP 8).

E2ME rear vacuum break adjustment—1981–82

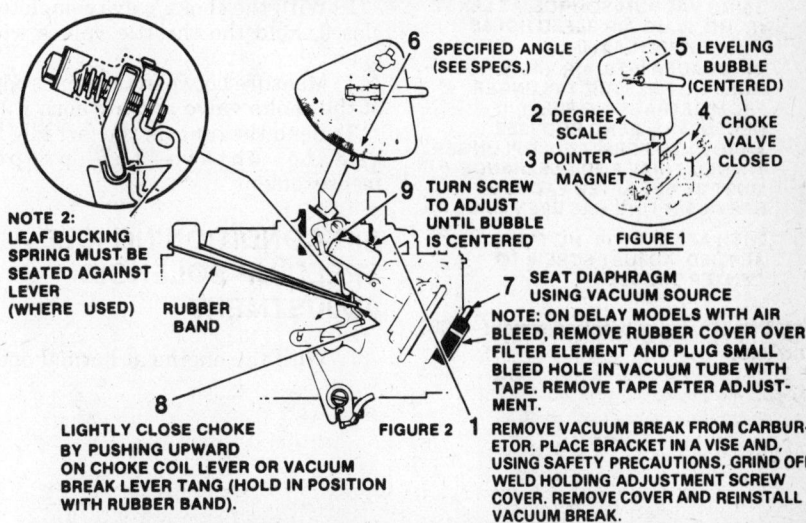

NOTE 2: LEAF BUCKING SPRING MUST BE SEATED AGAINST LEVER (WHERE USED)

RUBBER BAND

8 LIGHTLY CLOSE CHOKE BY PUSHING UPWARD ON CHOKE COIL LEVER OR VACUUM BREAK LEVER TANG (HOLD IN POSITION WITH RUBBER BAND).

FIGURE 2

6 SPECIFIED ANGLE (SEE SPECS.)

5 LEVELING BUBBLE (CENTERED)

2 DEGREE SCALE
3 POINTER MAGNET

4 CHOKE VALVE CLOSED

9 TURN SCREW TO ADJUST UNTIL BUBBLE IS CENTERED

FIGURE 1

7 SEAT DIAPHRAGM USING VACUUM SOURCE

NOTE: ON DELAY MODELS WITH AIR BLEED, REMOVE RUBBER COVER OVER FILTER ELEMENT AND PLUG SMALL BLEED HOLE IN VACUUM TUBE WITH TAPE. REMOVE TAPE AFTER ADJUSTMENT.

1 REMOVE VACUUM BREAK FROM CARBURETOR. PLACE BRACKET IN A VISE AND, USING SAFETY PRECAUTIONS, GRIND OFF WELD HOLDING ADJUSTMENT SCREW COVER. REMOVE COVER AND REINSTALL VACUUM BREAK.

NOTE 1: MAKE CHOKE COIL LEVER AND FAST IDLE ADJUSTMENT (BENCH OR ON-THE-CAR SETTING). DO NOT REMOVE RIVETS AND CHOKE COVER TO PERFORM THIS ADJUSTMENT. USE RUBBER BAND ON VACUUM BREAK LEVER TANG TO HOLD CHOKE VALVE CLOSED (STEP 8).

E2ME front vacuum break adjustment—1981–82

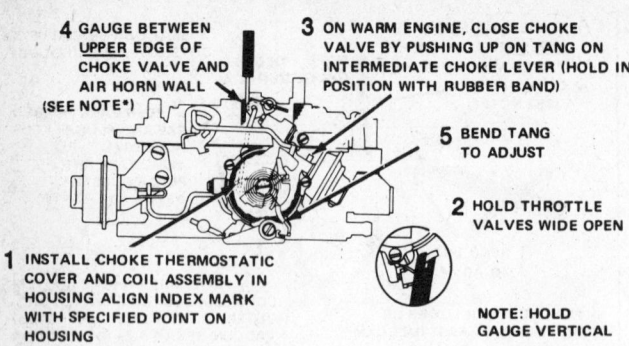

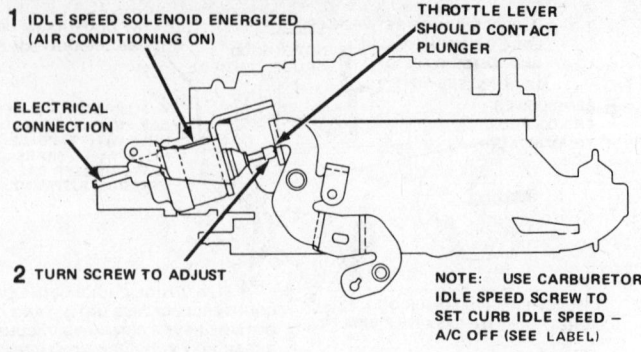

4 GAUGE BETWEEN <u>UPPER</u> EDGE OF CHOKE VALVE AND AIR HORN WALL (SEE NOTE*)

3 ON WARM ENGINE, CLOSE CHOKE VALVE BY PUSHING UP ON TANG ON INTERMEDIATE CHOKE LEVER (HOLD IN POSITION WITH RUBBER BAND)

5 BEND TANG TO ADJUST

2 HOLD THROTTLE VALVES WIDE OPEN

1 INSTALL CHOKE THERMOSTATIC COVER AND COIL ASSEMBLY IN HOUSING ALIGN INDEX MARK WITH SPECIFIED POINT ON HOUSING

NOTE: HOLD GAUGE VERTICAL

1 IDLE SPEED SOLENOID ENERGIZED (AIR CONDITIONING ON)

THROTTLE LEVER SHOULD CONTACT PLUNGER

ELECTRICAL CONNECTION

2 TURN SCREW TO ADJUST

NOTE: USE CARBURETOR. IDLE SPEED SCREW TO SET CURB IDLE SPEED — A/C OFF (SEE LABEL)

2MC, M2MC, M2ME, E2ME unloader adjustment—typical

2MC, M2MC air conditioning idle speed-up solenoid adjustment

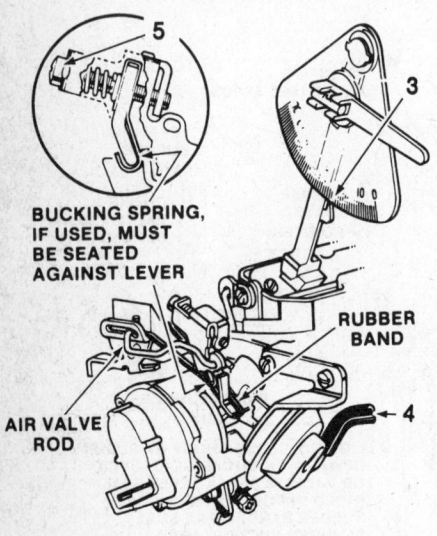

BUCKING SPRING, IF USED, MUST BE SEATED AGAINST LEVER

RUBBER BAND

AIR VALVE ROD

1 ATTACH RUBBER BAND TO GREEN TANG OF INTERMEDIATE CHOKE SHAFT

2 OPEN THROTTLE TO ALLOW CHOKE VALVE TO CLOSE

3 SET UP ANGLE GAGE AND SET TO SPECIFICATION

4 RETRACT VACUUM BREAK PLUNGER USING VACUUM SOURCE, AT LEAST 18" HG. PLUG AIR BLEED HOLES WHERE APPLICABLE ON QUADRAJETS, AIR VALVE ROD MUST NOT RESTRICT PLUNGER FROM RETRACTING FULLY. IF NECESSARY, BEND ROD (SEE ARROW) TO PERMIT FULL PLUNGER TRAVEL. FINAL ROD CLEARANCE MUST BE SET AFTER VACUUM BREAK SETTING HAS BEEN MADE.

5 WITH AT LEAST 18" HG STILL APPLIED, ADJUST SCREW TO CENTER BUBBLE

E2ME front vacuum break adjustment—1983 and later

1 ATTACH RUBBER BAND TO GREEN TANG OF INTERMEDIATE CHOKE SHAFT.

2 OPEN THROTTLE TO ALLOW CHOKE VALVE TO CLOSE.

3 SET UP ANGLE GAGE AND SET ANGLE TO SPECIFICATION.

4 RETRACT VACUUM BREAK PLUNGER, USING VACUUM SOURCE, AT LEAST 18" HG. PLUG AIR BLEED HOLES WHERE APPLICABLE.

4A ON QUADRAJETS, AIR VALVE ROD MUST NOT RESTRICT PLUNGER FROM RETRACTING FULLY. IF NECESSARY. BEND ROD HERE TO PERMIT FULL PLUNGER TRAVEL. WHERE APPLICABLE, PLUNGER STEM MUST BE EXTENDED FULLY TO COMPRESS PLUNGER BUCKING SPRING.

5 TO CENTER BUBBLE, EITHER:
A. ADJUST WITH 1/8" HEX WRENCH (VACUUM STILL APPLIED)

-OR-

B. SUPPORT AT "S" AND BEND VACUUM BREAK ROD (VACUUM STILL APPLIED)

E2ME rear vacuum break adjustment— 1983 and later

UNLOADER ADJUSTMENT

1. With the choke valve completely closed, hold the throttle valves wide open.

2. Measure between the upper edge of the choke valve and air horn wall.

3. Bend the tang on the fast idle lever to obtain the proper measurement.

AIR CONDITIONING IDLE SPEED-UP SOLENOID ADJUSTMENT

1. With the engine at normal oper-ating temperature and the air condi-tioning turned on but the compressor clutch lead disconnected, the solenoid should be electrically energized (plunger stem extended). Open the throttle slightly to allow the solenoid plunger to fully extend.

2. Adjust the plunger screw to ob-tain the specified idle speed.

3. Turn off the air conditioner. The solenoid plunger should move away from the tang on the throttle lever.

4. Adjust the curb idle speed with the idle speed screw, if necessary.

NOTE: Do not adjust if carbure-tor is computer controlled.

2MC, M2MC, M2ME, E2ME CARBURETOR SPECIFICATIONS
General Motors—U.S.A.

Year	Carburetor Identification ①	Flat Level (in.)	Choke Rod (in.)	Choke Unloader (in.)	Vacuum Break Lean or Front (deg./in.)	Vacuum Break Rich or Rear (deg./in.)	Pump Rod (in.)	Choke Coil Lever (in.)	Automatic Choke (notches)
'80	17080108, 110 17080130, 131	3/8	.243	.243	.142	—	5/16 ②	.120	Fixed
	17080132, 133, 147, 148, 149	5/16	.243	.243	.142	—	5/16 ②	.120	Fixed
	17080138, 140	3/8	.243	.243	.142	—	5/16 ②	.120	Fixed
	17080150, 152, 153	3/8	.071	.220	.243	.157	11/32 ③	.120	Fixed
	17080160	5/16	.110	.243	.168	.207	1/4 ②	.120	Fixed
	17080190, 192	9/32	.074	.243	.123	.110	1/4 ②	.120	Fixed
	17080191	11/32	.139	.243	.096	.096	1/4 ②	.120	Fixed
	17080195, 197	9/32	.139	.243	.103	.071	1/4 ②	.120	Fixed
	17080490, 492	5/16	.139	.243	.117	.203	1/4 ②	.120	Fixed
	17080491	5/16	.139	.243	.117	.220	1/4 ②	.120	Fixed
	17080493, 495	5/16	.139	.243	.117	.179	3/8	.120	Fixed
	17080494	5/16	.139	.243	.117	.179	1/4 ②	.120	Fixed
	17080496, 498	5/16	.139	.243	.117	.203	Fixed	.120	Fixed
'81	17080185, 187	9/32	.139	.243	19/.103	14/.071	1/4 ②	.120	Fixed
	17080191	11/32	.139	.243	18/.096	18/.096	1/4 ②	.120	Fixed
	17081130, 131, 132, 133	11/32	.110	.243	25/.142	—	Fixed	.120	Fixed
	17081138, 140	11/32	.110	.260	25/.142	—	Fixed	.120	Fixed
	17081150, 152	13/32	.071	.220	24/.136	36/.227	Fixed	.120	Fixed
	17081160	11/32	.074	.220	24/.136	37/.234	④	.120	Fixed
	17081196	5/16	.139	.243	28/.164	24/.136	④	.120	Fixed
	17081190, 193	5/16	.139	.243	21/.117	31/.187	Fixed	.120	Fixed
	17081191, 194	5/16	.139	.243	28/.164	24/.136	④	.120	Fixed
	17081198	3/8	.139	.243	28/.164	24/.136	④	.120	Fixed
	17081192, 197	5/16	.139	.243	21/.117	30/.179	④	.120	Fixed
	17081199	3/8	.096	.243	18/.096	24/.136	Fixed	.120	Fixed
'82	17082130, 132, 138, 140	3/8	.110	.164	27/.157	—	④	④	Fixed
	17082150	13/32	.071	.220	24/.136	38/.243 ⑤	④	④	Fixed
	17082182, 184	5/16	.096	.195	28/.164	24/.136	④	④	Fixed
	17082192, 194	5/16	.096	.195	28/.164	24/.136	④	④	Fixed
	17082196	5/16	.096	.157	21/.117	19/.103	④	④	Fixed
	17082497	5/16	.113	.195	28/.164	24/.136	④	.120	Fixed
'83	17082130, 132	3/8	.110	.243	27/.157	—	④	.120	Fixed
	17083190, 192	5/16	.096	.195	28/.164	24/.136	④	.120	Fixed
	17083193	5/16	.090	.157	23/.129	28/.164	④	.120	Fixed
	17083194	5/16	.090	.220	27/.157	25/.142	④	.120	Fixed
'84	17082130	3/8	.110	.243	27/.157	None	④	.120	Fixed
	17082132	3/8	.110	.243	27/.157	None	④	.120	Fixed
	17084191	5/16	.096	.195	28/.164	24/.136	④	.120	Fixed
	17084193	5/16	.090	.220	27/.157	25/.142	④	.120	Fixed
	17084194	5/16	.090	.220	27/.157	25/.142	④	.120	Fixed
	17084195	5/16	.090	.220	27/.157	25/.142	④	.120	Fixed
'85	17085190	10/32	.096	.195	28/.164	24/.136	④	.120	Fixed
	17085192	11/32	.090	.220	27/.157	25/.142	④	.120	Fixed
	17085194	11/32	.090	.220	27/.157	25/.142	④	.120	Fixed
'86	17086190	10/32	.096	.195	28/.164	24/.136	④	.120	Fixed

① The carburetor identification number is stamped on the float bowl, next to the fuel inlet nut.
② Inner hole
③ Outer hole
④ Not Adjustable
⑤ High altitude—0.206

2MC, M2MC, M2ME, E2ME CARBURETOR SPECIFICATIONS
General Motors—Canada

Year	Carburetor Identification ①	Float Level (in.)	Choke Rod (in.)	Choke Unloader (in.)	Vacuum Break Lean or Front (deg./in.)	Vacuum Break Rich or Rear (deg./in.)	Pump Rod (in.)	Choke Coil Lever (in.)	Automatic Choke (notches)
'81	17080191	11/32	.139	.243	18/.096	18/.096	1/4 ②	.120	Fixed
	17081492	9/32	.139	.243	17/.090	19/.103	1/4 ②	.120	Fixed
	17081493	9/32	.139	.243	17/.090	19/.103	1/4 ②	.120	Fixed
	17081170	13/32	.110	.243	25/.142	—	1/4 ②	.120	Fixed
	17081171	13/32	.110	.243	25/.142	—	1/4 ②	.120	Fixed
	17081174	9/32	.110	.243	25/.142	—	1/4 ②	.120	Fixed
	17081175	9/32	.110	.243	25/.142	—	1/4 ②	.120	Fixed
'82	17082174	9/32	.110	.243	25/.142	—	5/16 ②	.120	Fixed
	17082175	9/32	.110	.243	25/.142	—	5/16 ②	.120	Fixed
	17082492	9/32	.139	.243	17/.090	19/.103	1/4 ②	.120	Fixed
	17082172	9/32	.110	.243	25/.142	—	5/16 ②	.120	Fixed
	17082173	9/32	.110	.243	25/.142	—	5/16 ②	.120	Fixed
'83–'84	17083172	9/32	.139	.243	17/.090	19/.103	1/4 ②	.120	Fixed
'85	17085170	9/32	.139	.243	17/.090	19/.103	9/32 ②	.120	Fixed
'86	17086170	9/32	.139	.243	17/.090	19/.103	9/32 ②	.120	Fixed

① The carburetor identification number is stamped on the float bowl, next to the fuel inlet nut.
② Inner hole

Quadrajet

The Rochester Quadrajet carburetor is a two stage, four-barrel downdraft carburetor. It has been built in many variations designated as 4MC, 4MV, M4MC, M4MCA, M4ME, M4MEA, E4MC, and E4ME. See the beginning of the Rochester section for an explanation of these designations.

The primary side of the carburetor is equipped with two primary bores and a triple venturi with plain tube nozzles. During off idle and part throttle operation, the fuel is metered through tapered metering rods operating in specially designed jets positioned by a manifold vacuum responsive piston.

The secondary side of the carburetor contains two secondary bores. An air valve is used on the secondary side for metering control and supplements the primary bore. The secondary air valve operates tapered metering rods which regulate the fuel in constant proportion to the air being supplied.

FAST IDLE SPEED

1. Position the fast idle lever on the high step of the fast idle cam.
2. Be sure that the choke is wide open and the engine warm. Plug the EGR vacuum hose. Disconnect the vacuum hose to the front vacuum break unit, if there are two.
3. Make a preliminary adjustment by turning the fast idle screw out un-

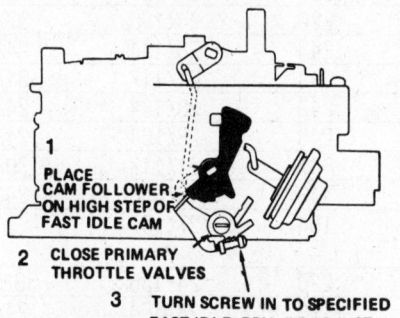

1 PLACE CAM FOLLOWER ON HIGH STEP OF FAST IDLE CAM
2 CLOSE PRIMARY THROTTLE VALVES
3 TURN SCREW IN TO SPECIFIED FAST IDLE RPM TO ADJUST

Quadrajet fast idle adjustment

til the throttle valves are closed, then screwing it in the specified number of turns after it contacts the lever (see the carburetor specifications).

4. Use the fast idle screw to adjust the fast idle to the speed, and under the conditions, specified on the engine compartment sticker or in the specifications chart.

CHOKE ROD (FAST IDLE CAM)

1. Adjust the fast idle and place the cam follower on the second step of the fast idle cam against the shoulder of the high step.
2. Close the choke valve by exerting counter-clockwise pressure on the external choke lever. Remove the coil assembly from the choke housing and

push upon the choke coil lever. On models with a fixed (riveted) choke cover, push up on the vacuum brake lever tang and hold in position with a rubber band.

3. Insert a gauge of the proper size between the upper edge of the choke valve and the inside air horn wall.
4. To adjust, bend the tang on the fast idle cam. Be sure that the tang rests against the cam after bending.

PRIMARY (FRONT) VACUUM BREAK ADJUSTMENT

1980–81 Models

1. Seat the front vacuum diaphragm using an outside vacuum source. If there is a diaphragm unit bleed hole, tape it over.
2. Push up on the inside choke coil lever until the tang on the vacuum brake lever contacts the tang on the vacuum break plunger. On models with a fixed choke coil cover, push up on the vacuum brake lever tang.
3. Place the proper size gauge between the upper edge of the choke valve and the inside of the air horn wall.
4. To adjust, turn the adjustment screw on the vacuum break plunger lever.
5. Install the vacuum hose to the vacuum brake unit.

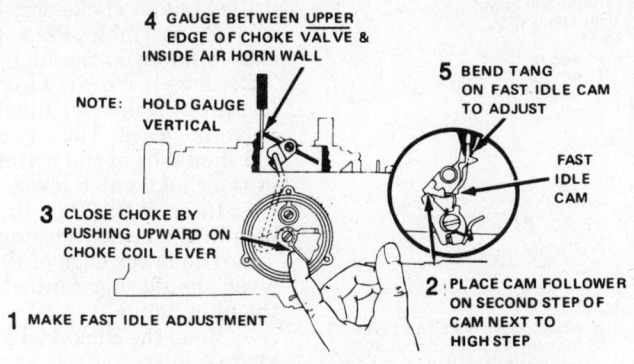

Quadrajet choke rod (fast idle cam) adjustment—typical

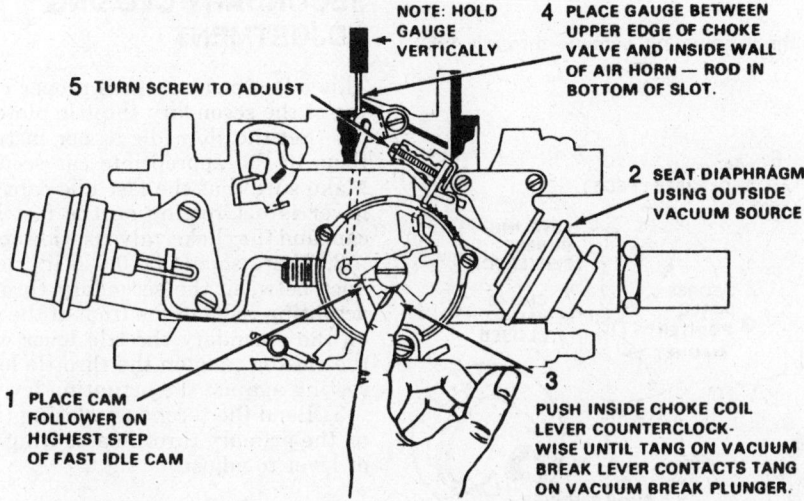

Quadrajet front vacuum break adjustment—typical through 1981

1 ATTACH RUBBER BAND TO GREEN TANG OF INTERMEDIATE CHOKE SHAFT

2 OPEN THROTTLE TO ALLOW CHOKE VALVE TO CLOSE

3 SET UP ANGLE GAGE AND SET TO SPECIFICATION

4 RETRACT VACUUM BREAK PLUNGER USING VACUUM SOURCE, AT LEAST 18" HG. PLUG AIR BLEED HOLES WHERE APPLICABLE

ON QUADRAJETS, AIR VALVE ROD MUST NOT RESTRICT PLUNGER FROM RETRACTING FULLY. IF NECESSARY, BEND ROD (SEE ARROW) TO PERMIT FULL PLUNGER TRAVEL. FINAL ROD CLEARANCE MUST BE SET AFTER VACUUM BREAK SETTING HAS BEEN MADE.

5 WITH AT LEAST 18" HG STILL APPLIED, ADJUST SCREW TO CENTER BUBBLE

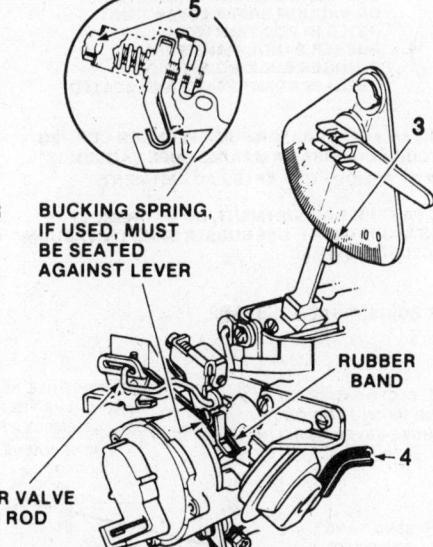

Quadrajet front vacuum break adjustment—1982 and later

1982–84 Models

On these models a choke valve measuring gauge J-26701 or equivalent is used to measure angle (degrees instead of inches). See illustration for procedure.

SECONDARY (REAR) VACUUM BRAKE ADJUSTMENT
1980 Models

1. Tape over the bleed hole in the rear vacuum break diaphragm and seat the diaphragm using an outside vacuum source. Make sure the diaphragm plunger bucking spring, if any, is compressed. On delay models (1980), plug the end cover with a pump plunger cup or equivalent and remove after adjustment.

2. Close the choke by pushing up on the choke coil lever inside the choke housing. On models with a fixed choke coil cover, push up on the vacuum break lever tang and use a rubber band to hold in place.

3. With the choke rod in the bottom of the slot in the choke lever, measure between the upper edge of the choke valve and the air horn wall with a wire-type gauge.

4. To adjust, bend the vacuum brake rod at the first bend near the diaphragm except on 1980 models with a screw at the rear of the diaphragm; on those models, turn the screw to adjust.

5. Remove the tape covering the bleed hole of the diaphragm and connect the vacuum hose.

1981–84 Models

On these models a choke valve measuring gauge J-26701 or equivalent is used to measure angle (degrees instead of inches). See illustration for procedure.

CHOKE UNLOADER

1. Push up on the vacuum break lever to close the choke valve, and fully open the throttle valves.

2. Measure the distance from the upper edge of the choke valve to the air horn wall.

3. To adjust, bend the tang on the fast idle lever.

4MV CHOKE COIL ROD

1. Close the choke valve by rotating the choke coil lever counterclockwise.

2. Disconnect the thermostatic coil rod from the upper lever.

3. Push down on the rod until it contacts the bracket of the coil.

4. The rod must fit in the notch of the upper lever.

5. If it does not, it must be bent on the curved portion just below the upper lever.

MC, ME CHOKE COIL LEVER ADJUSTMENT

1. Remove the choke cover and thermostatic coil from the choke housing. On models with a fixed (rivet) choke cover, the rivets must be

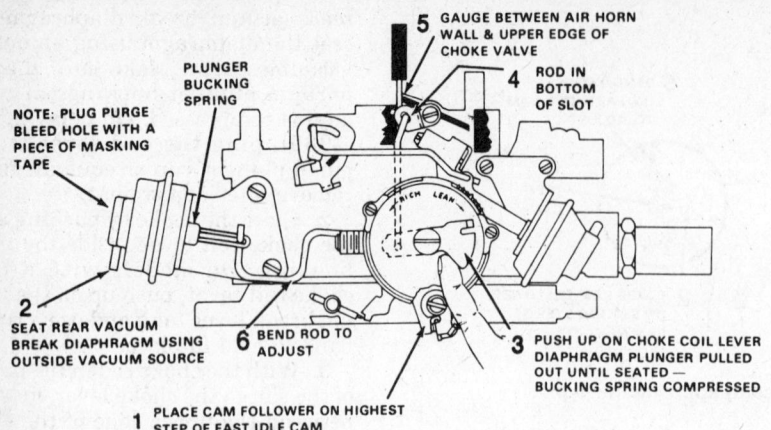

Quadrajet rear vacuum break adjustment (without adjusting screw)—through 1980

1. REMOVE VACUUM BREAK FROM CARBURETOR. PLACE BRACKET IN A VISE AND, USING SAFETY PRECAUTIONS, GRIND OFF ADJUSTMENT SCREW CAP. REINSTALL VACUUM BREAK.

9. TO ADJUST, USING 1/8" HEX WRENCH TURN SCREW IN REAR COVER UNTIL BUBBLE IS CENTERED. APPLY SEALER (SUCH AS SILICONE SEALANT RTV RUBBER OR EQUIVALENT) OVER SCREW HEAD TO SEAL SETTING.

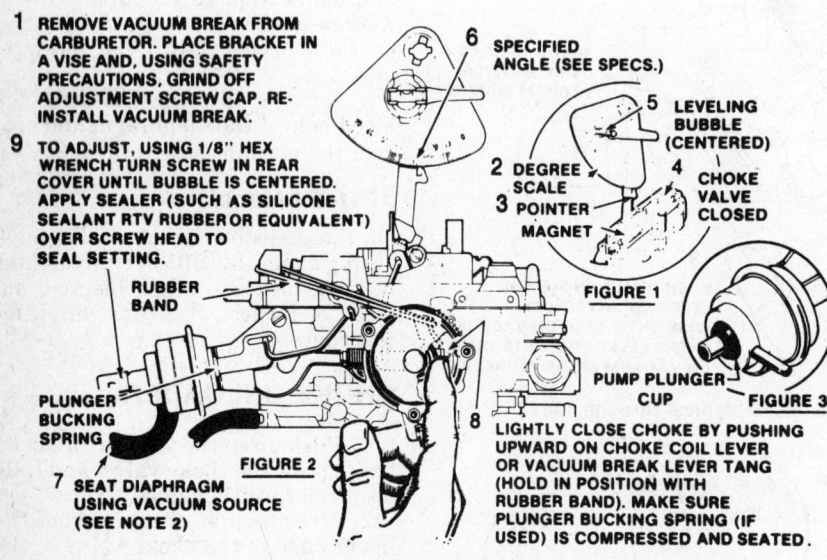

6. SPECIFIED ANGLE (SEE SPECS.)
5. LEVELING BUBBLE (CENTERED)
2. DEGREE SCALE
3. POINTER MAGNET
4. CHOKE VALVE CLOSED
FIGURE 1

RUBBER BAND
PLUNGER BUCKING SPRING
7. SEAT DIAPHRAGM USING VACUUM SOURCE (SEE NOTE 2)
FIGURE 2

PUMP PLUNGER CUP   FIGURE 3

8. LIGHTLY CLOSE CHOKE BY PUSHING UPWARD ON CHOKE COIL LEVER OR VACUUM BREAK LEVER TANG (HOLD IN POSITION WITH RUBBER BAND). MAKE SURE PLUNGER BUCKING SPRING (IF USED) IS COMPRESSED AND SEATED.

NOTE 2: ON DELAY MODELS, PLUG END COVER USING AN ACCELERATOR PUMP PLUNGER CUP - 2G TYPE (FIGURE 3) OR EQUIVALENT. SEAT VACUUM DIAPHRAGM MAKING SURE VACUUM IS ABOVE 5" Hg WHEN READING GAUGE (STEP 9). REMOVE CUP AFTER ADJUSTMENT.

NOTE 1: MAKE CHOKE COIL LEVER ADJUSTMENT AND FAST IDLE ADJUSTMENT. DO NOT REMOVE RIVETS AND CHOKE COVER TO PERFORM THIS ADJUSTMENT. USE RUBBER BAND ON VACUUM BREAK LEVER TANG TO HOLD CHOKE VALVE CLOSED (STEP 8).

Quadrajet rear vacuum break adjustment—1981–82

drilled out. A choke stat kit is necessary for assembly. Place the fast idle cam follower on the high step.

2. Push up on the coil tang (counter-clockwise) until the choke valve is closed. The top of the choke rod should be at the bottom of the slot in the choke valve lever.

3. Insert a 0.120 in. drill bit in the hole in the choke housing.

4. The lower edge of the choke coil lever should just contact the side of the plug gauge.

5. Bend the choke rod at the top angle to adjust.

SECONDARY CLOSING ADJUSTMENT

This adjustment assures proper closing of the secondary throttle plates.

1. Set the slow idle as per instructions in the appropriate car section. Make sure that the fast idle cam follower is not resting on the fast idle cam and the choke valve is wide open.

2. There should be 0.020 in. clearance between the secondary throttle actuating rod and the front of the slot on the secondary throttle lever with the closing tang on the throttle lever resting against the actuating lever.

3. Bend the secondary closing tang on the primary throttle actuating rod or lever to adjust.

SECONDARY OPENING ADJUSTMENT

1. Open the primary throttle valves until the actuating link contacts the upper tang on the secondary lever.

2. With two point linkage, the bottom of the link should be in the center of the secondary lever slot.

3. With three point linkage, there should be 0.070 in. clearance between the link and the middle tang.

4. Bend the upper tang on the secondary lever to adjust as necessary.

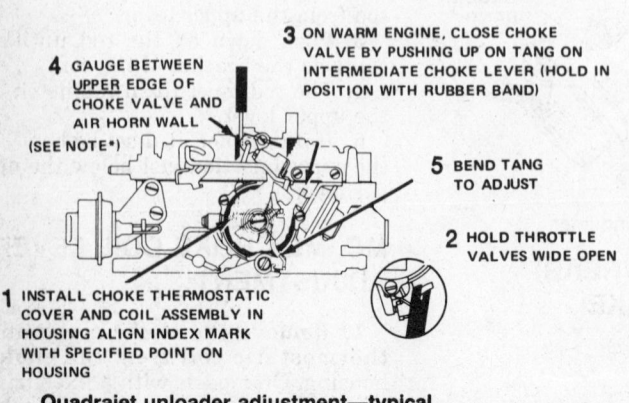

Quadrajet unloader adjustment—typical

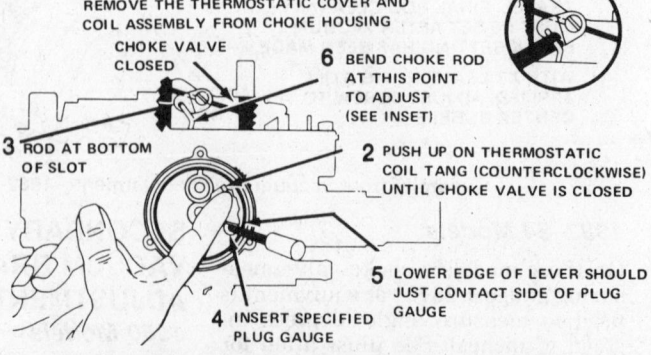

Quadrajet choke coil lever adjustment—typical

1. ATTACH RUBBER BAND TO GREEN TANG OF INTERMEDIATE CHOKE SHAFT.

2. OPEN THROTTLE TO ALLOW CHOKE VALVE TO CLOSE.

3. SET UP ANGLE GAGE AND SET ANGLE TO SPECIFICATION.

   RETRACT VACUUM BREAK PLUNGER, USING VACUUM SOURCE, AT LEAST 18" HG. PLUG AIR BLEED HOLES WHERE APPLICABLE.

4A. ON QUADRAJETS, AIR VALVE ROD MUST NOT RESTRICT PLUNGER FROM RETRACTING FULLY. IF NECESSARY. BEND ROD HERE TO PERMIT FULL PLUNGER TRAVEL. WHERE APPLICABLE, PLUNGER STEM MUST BE EXTENDED FULLY TO COMPRESS PLUNGER BUCKING SPRING.

5. TO CENTER BUBBLE, EITHER:
   A. ADJUST WITH 1/8" HEX WRENCH (VACUUM STILL APPLIED)
   -OR-
   B. SUPPORT AT "S" AND BEND VACUUM BREAK ROD (VACUUM STILL APPLIED)

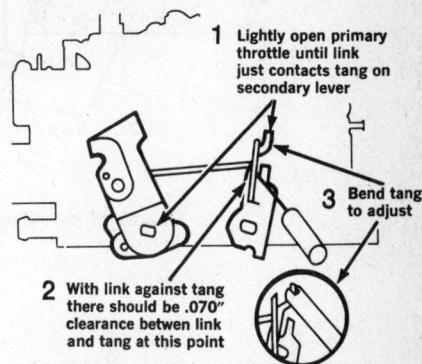

Quadrajet rear vacuum break adjustment—typical 1983 and later

and connector. Remove the float and bend the arm up to adjust. Install the parts, turning the mixture solenoid screw in until it is lightly bottomed, then unscrewing it the exact number of turns counted earlier.

ACCELERATOR PUMP

The accelerator pump is not adjust-

1. Lightly open primary throttle until link just contacts tang on secondary lever

2. With link against tang there should be .070" clearance betwen link and tang at this point

3. Bend tang to adjust

Secondary opening adjustment—three point linkage

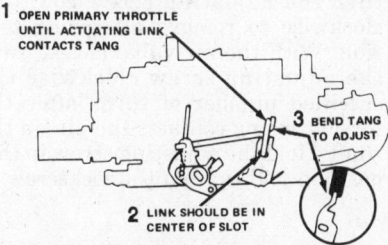

1. OPEN PRIMARY THROTTLE UNTIL ACTUATING LINK CONTACTS TANG

2. LINK SHOULD BE IN CENTER OF SLOT

3. BEND TANG TO ADJUST

Quadrajet secondary opening adjustment, two point linkage

FLOAT LEVEL

With the air horn assembly removed, measure the distance from the air horn gasket surface (gasket removed) to the top of the float at the toe ($3/16$ in. back from the toe).

NOTE: Make sure the retaining pin is firmly held in place and that the tang of the float is lightly held against the needle and seat assembly.

Remove the float and bend the float arm to adjust except on carburetors used with the computer controlled systems (E4MC and E4ME). For those carburetors, if the float level is too high, hold the retainer firmly in place and push down on the center of the float to adjust. If the float level is too low on models with the computer controlled system, lift out the metering rods. Remove the solenoid connector screw. Turn the lean mixture solenoid screw in clockwise, counting and recording the exact number of turns until the screw is lightly bottomed in the bowl. Then turn the screw out clockwise and remove. Lift out the solenoid

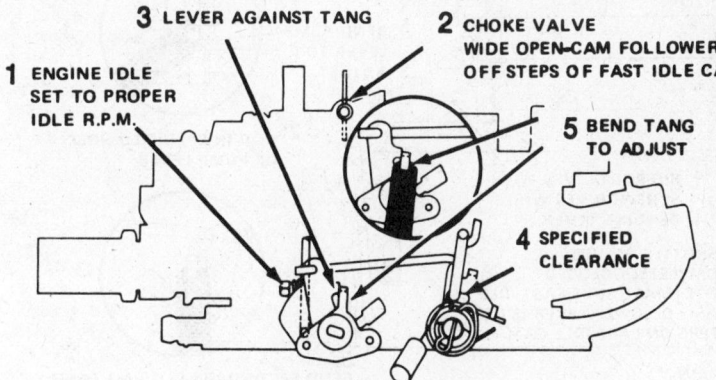

1. ENGINE IDLE SET TO PROPER IDLE R.P.M.

2. CHOKE VALVE WIDE OPEN-CAM FOLLOWER OFF STEPS OF FAST IDLE CAM

3. LEVER AGAINST TANG

4. SPECIFIED CLEARANCE

5. BEND TANG TO ADJUST

Quadrajet Secondary Closing Adjustment

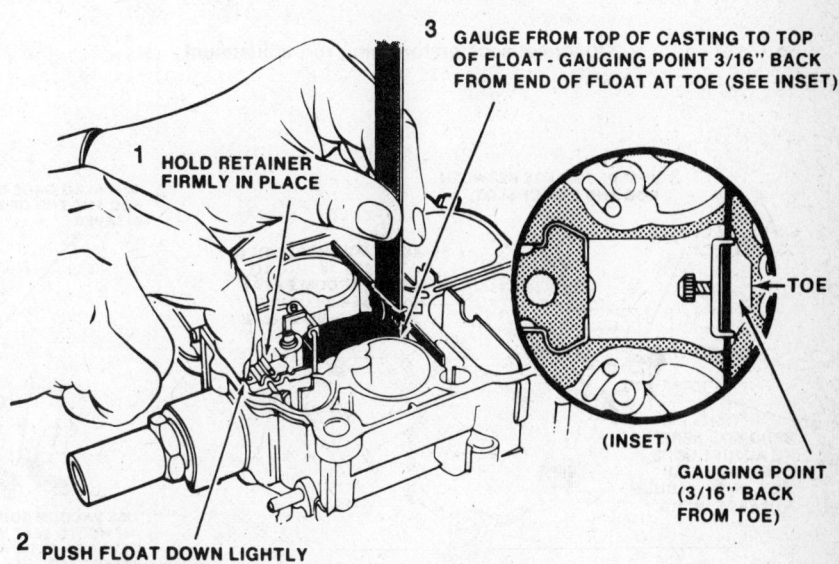

3. GAUGE FROM TOP OF CASTING TO TOP OF FLOAT - GAUGING POINT 3/16" BACK FROM END OF FLOAT AT TOE (SEE INSET)

1. HOLD RETAINER FIRMLY IN PLACE

2. PUSH FLOAT DOWN LIGHTLY AGAINST NEEDLE

TOE

(INSET)

GAUGING POINT (3/16" BACK FROM TOE)

Quadrajet float level adjustment—typical

WITH LOCK SCREW LOOSENED AND WITH
AIR VALVE CLOSED, TURN ADJUSTING
SCREW HALF TURN AFTER SPRING
CONTACTS PIN.
TIGHTEN LOCK SCREW

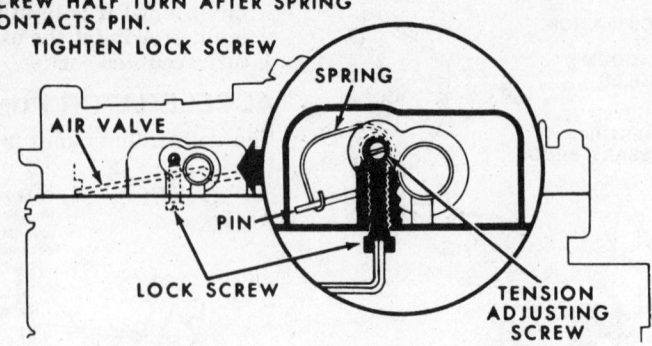

Quadrajet air valve spring setting—typical

able on computer controlled carburetors (E4MC and E4ME).

1. Close the primary throttle valves by backing out the slow idle screw and making sure that the fast idle cam follower is off the steps of the fast idle cam.

2. Bend the secondary throttle closing tang away from the primary throttle lever, if necessary, to insure that the primary throttle valves are fully closed.

3. With the pump in the appropriate hole in the pump lever, measure from the top of the choke valve wall to the top of the pump stem.

4. To adjust, bend the pump lever.

5. After adjusting, readjust the secondary throttle tang and the slow idle screw.

AIR VALVE SPRING ADJUSTMENT

To adjust the air valve spring windup, loosen the Allen head lockscrew and turn the adjusting screw counterclockwise to remove all spring tension. With the air valve closed, turn the adjusting screw clockwise the specified number of turns after the torsion spring contacts the pin on the shaft. Hold the adjusting screw in this position and tighten the lockscrew.

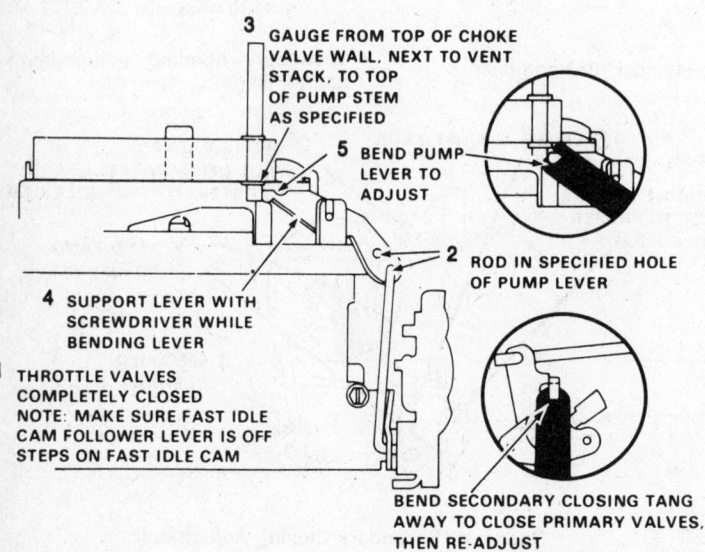

Quadrajet accelerator pump rod adjustment

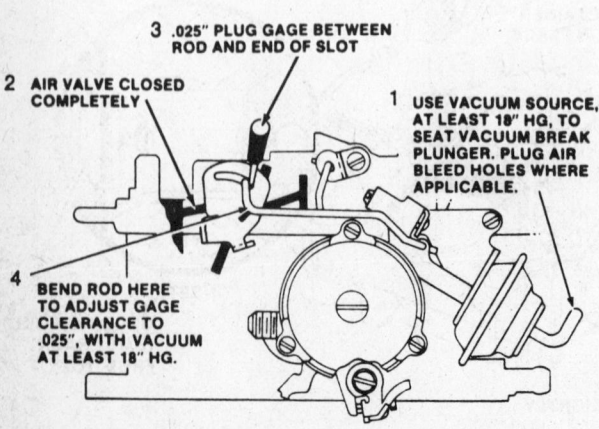

Air valve rod adjustment, Front—E4ME, E4MC

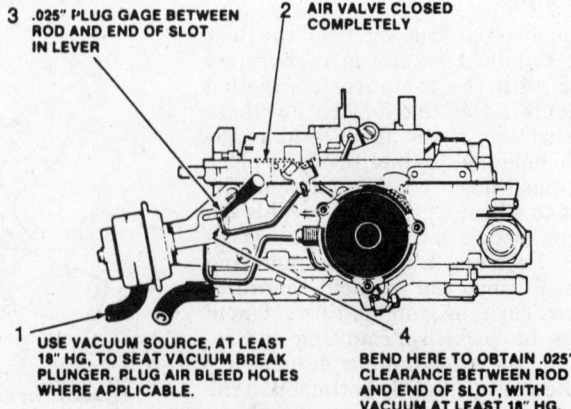

Air valve rod adjustment, Rear—E4ME, E4MC

QUADRAJET CARBURETOR SPECIFICATIONS
Chrysler Products

Year	Carburetor Identification ①	Float Level (in.)	Air Valve Spring (turn)	Pump Rod (in.)	Primary Vacuum Break (in./ deg.)	Secondary Vacuum Break (in./deg.)	Secondary Opening (in.)	Choke Rod (in.)	Choke Unloader (in.)	Fast Idle Speed ④ (rpm)
'85	17085407	14/32	7/8	—	.193/25°	—	—	—	.250	1450
'86	17085433	14/32	7/8	—	.140/25	—	—	.120	.179	①

① Refer to the underhood sticker

QUADRAJET CARBURETOR SPECIFICATIONS
Cadillac

Year	Carburetor Identification ①	Float Level (in.)	Air Valve Spring (turn)	Pump Rod (in.)	Primary Vacuum Break (in./ deg.)	Secondary Vacuum Break (in./deg.)	Secondary Opening (in.)	Choke Rod (in.)	Choke Unloader (in.)	Fast Idle Speed (rpm)
'80	17080230	7/16	1/2	9/32 ②	0.149	0.136	③	0.083	0.220	1450
	17080530	17/32	1/2	Fixed	0.142	0.400	③	0.083	0.260	1350
'81	17081248	3/8	5/8	Fixed	0.164	0.136	③	0.139	0.243	④
	17081289	13/32	5/8	Fixed	0.164	0.136	③	0.139	0.243	④
'82	17082246	3/8	5/8	Fixed	0.149/26	0.149/26	③	0.139	0.195	④
	17082247	13/32	5/8	Fixed	0.164/28	0.136/24	③	0.139	0.243	④
'83	17082266	3/8	5/8	Fixed	0.149/26	0.149/26	③	0.071	0.195	④
	17082267	3/8	5/8	Fixed	0.149/26	0.149/26	③	0.071	0.195	④

① The carburetor identification number is stamped on the float bowl, near the secondary throttle lever.
② Inner hole
③ No measurement necessary on two point linkage; see text.
④ See underhood decal.

QUADRAJET CARBURETOR SPECIFICATIONS
Buick

Year	Carburetor Identification ①	Float Level (in.)	Air Valve Spring (turn)	Pump Rod (in.)	Primary Vacuum Break (in./ deg.)	Secondary Vacuum Break (in./deg.)	Secondary Opening (in.)	Choke Rod (in.)	Choke Unloader (in/degrees)	Fast Idle Speed ④ (rpm)
'80	17080240	3/16	9/16	9/32 ③	0.083	0.083	②	0.074	0.179	⑥
	17080241	7/16	3/4	9/32 ③	0.129	0.114	②	0.096	0.243	⑥
	17080242	13/32	9/16	9/32 ③	0.077	0.096	②	0.074	0.220	⑥
	17080243	3/16	9/16	9/32 ③	0.083	0.083	②	0.074	0.179	⑥
	17080244	5/16	5/8	9/32 ③	0.096	0.071	②	0.139	0.243	⑥
	17080249	7/16	3/4	9/32 ③	0.129	0.114	②	0.096	0.243	⑥
	17080253	13/32	1/2	9/32 ③	0.149	0.211	②	0.090	0.220	⑥
	17080259	13/32	1/2	9/32 ③	0.149	0.211	②	0.090	0.220	⑥
	17080270	15/32	5/8	3/8 ⑦	0.149	0.211	②	0.074	0.220	⑥
	17080271	15/32	5/8	3/8 ⑦	0.142	0.211	②	0.110	0.203	⑥
	17080272	15/32	5/8	3/8 ⑦	0.129	0.175	②	0.074	0.203	⑥
	17080502	1/2	7/8	Fixed	0.136	0.179	②	0.110	0.243	⑥
	17080504	1/2	7/8	Fixed	0.136	0.179	②	0.110	0.243	⑥
	17080540	3/8	9/16	Fixed	0.103	0.129	②	0.074	0.243	⑥
	17080542	3/8	9/16	Fixed	0.103	0.066	②	0.074	0.243	⑥
	17080543	3/8	9/16	Fixed	0.103	0.129	②	0.074	0.243	⑥
	17080553	15/32	1/2	Fixed	0.142	0.220	②	0.090	0.220	⑥
	17080554	15/32	1/2	Fixed	0.142	0.211	②	0.090	0.220	⑥
'81	17081202 204	11/32	7/8	Fixed	0.157 ⑧	—	②	0.110	0.243	⑩

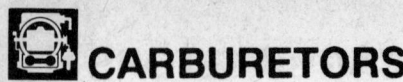

QUADRAJET CARBURETOR SPECIFICATIONS
Buick

Year	Carburetor Identification ①	Float Level (in.)	Air Valve Spring (turn)	Pump Rod (in.)	Primary Vacuum Break (in./deg.)	Secondary Vacuum Break (in./deg.)	Secondary Opening (in.)	Choke Rod (in.)	Choke Unloader (in/degrees)	Fast Idle Speed ④ (rpm)
'81	17081203 207	11/32	7/8	Fixed	0.157 ⑧	—	②	0.110	0.243	⑩
	17081216 218	11/32	7/8	Fixed	0.157 ⑧	—	②	0.110	0.243	⑩
	17081242	3/8	9/16	Fixed	0.090 ⑧	0.077 ⑨	②	0.139	0.243	⑩
	17081243	5/16	9/16	Fixed	0.103 ⑧	0.090 ⑨	②	0.139	0.243	⑩
	17081245	3/8	5/8	Fixed	0.164 ⑧	0.136 ⑨	②	0.139	0.243	⑩
	17081247	3/8	5/8	Fixed	0.164 ⑧	0.136 ⑨	②	0.139	0.243	⑩
	17081248 249	3/8	5/8	Fixed	0.164 ⑧	0.136 ⑨	②	0.139	0.243	⑩
	17081253 254	15/32	1/2	Fixed	0.142 ⑧	0.227 ⑨	②	0.071	0.220	⑩
	17081270	7/16	5/8	Fixed	0.136 ⑧	0.211 ⑨	②	0.074	0.220	⑩
	17081272	5/8	5/8	Fixed	0.136 ⑧	0.260 ⑨	②	0.074	0.220	⑩
	17081274	5/8	5/8	Fixed	0.136 ⑧	0.220 ⑨	②	0.083	0.220	⑩
	17081289	5/8	5/8	Fixed	0.164 ⑧	0.136 ⑨	②	0.139	0.243	⑩
'82	17082202	11/32	7/8	Fixed	0.110/20	—	②	0.110	0.243	⑤
	17082204	11/32	3/8	Fixed	0.110/20	—	②	0.110	0.243	⑤
	17082244	7/16	9/16	Fixed	0.117/21	0.083/16	②	0.139	0.195	⑤
	17082245	3/8	5/8	Fixed	0.149/26	0.149/26	②	0.139	0.195	⑤
	17082246	3/8	5/8	Fixed	0.149/26	0.149/26	②	0.139	0.195	⑤
	17082247	13/32	5/8	Fixed	0.164/28	0.136/24	②	0.139	0.243	⑤
	17082248	13/32	5/8	Fixed	0.164/28	0.136/24	②	0.139	0.243	⑤
	17082251	15/32	1/2	Fixed	0.142/25	0.304/45	②	0.071	0.220	⑤
	17082253	15/32	1/2	Fixed	0.142/25	0.227/36	②	0.071	0.220	⑤
	17082264	7/16	9/16	Fixed	0.117/20	0.083/16	②	0.139	0.195	⑤
	17082265	3/8	5/8	Fixed	0.149/26	0.149/26	②	0.139	0.195	⑤
	17082266	3/8	5/8	Fixed	0.149/26	0.149/26	②	0.139	0.195	⑤
	17082267	3/8	5/8	Fixed	0.164/28	0.136/24	②	0.139	0.243	⑤
	17082268	13/32	5/8	Fixed	0.164/28	0.136/24	②	0.139	0.243	⑤
'83	17082265	3/8	5/8	Fixed	0.149/26	0.149/26	②	0.139	0.195	⑪
	17082266	3/8	5/8	Fixed	0.149/26	0.149/26	②	0.139	0.195	⑪
	17082267	3/8	5/8	Fixed	0.149/26	0.149/26	②	0.096	0.195	⑪
	17082268	3/8	5/8	Fixed	0.149/26	0.149/26	②	0.096	0.195	⑪
	17083242	9/32	9/16	Fixed	0.110/20	—	②	0.139	0.243	⑪
	17083244	1/4	9/16	Fixed	0.117/21	0.083/16	②	0.139	0.195	⑪
	17083248	3/8	5/8	Fixed	0.149/26	0.149/26	②	0.139	0.195	⑪
	17083250	7/16	1/2	Fixed	0.157/27	0.271/42	②	0.071	0.220	⑪
	17083253	7/16	1/2	Fixed	0.157/27	0.269/41	②	0.071	0.220	⑪
	17083553	7/16	1/2	Fixed	0.157/27	0.269/41	②	0.071	0.220	⑪
'84	17084201	11/32	7/8	Fixed	0.157/27	—	②	0.110	0.243	⑪
	17084205	11/32	7/8	Fixed	0.157/27	—	②	0.243	0.243	⑪
	17084208	11/32	7/8	Fixed	0.157/27	—	②	0.110	0.243	⑪
	17084209	11/32	7/8	Fixed	0.157/27	—	②	0.243	0.243	⑪
	17084210	11/32	7/8	Fixed	0.157/27	—	②	0.110	0.243	⑪
	17084240	5/16	1	Fixed	0.136/24	—	②	—	0.195	⑪
	17084244	5/16	1	Fixed	0.136/24	—	②	—	0.195	⑪
	17084246	5/16	1	Fixed	0.123/22	0.136/24	②	—	0.195	⑪
	17084248	5/16	1	Fixed	0.136/24	—	②	—	0.195	⑪
	17084252	7/16	1/2	Fixed	0.157/27	0.269/41	②	—	0.220	⑪
	17084254	7/16	1/2	Fixed	0.157/27	0.269/41	②	—	0.220	⑪

QUADRAJET CARBURETOR SPECIFICATIONS
Buick

Year	Carburetor Identification ①	Float Level (in.)	Air Valve Spring (turn)	Pump Rod (in.)	Primary Vacuum Break (in./deg.)	Secondary Vacuum Break (in./deg.)	Secondary Opening (in.)	Choke Rod (in.)	Choke Unloader (in/degrees)	Fast Idle Speed ④ (rpm)
'85	17085202	11/32	7/8	Fixed	0.157/27	—	②	—	38°	⑪
	17085203	11/32	7/8	Fixed	0.157/27	—	②	—	38°	⑪
	17085204	11/32	7/8	Fixed	0.157/27	—	②	—	38°	⑪
	17085208	11/32	7/8	Fixed	0.157/27	—	②	—	38°	⑪
	17085218	11/32	7/8	Fixed	0.157/27	—	②	—	38°	⑪
	17085282	11/32	1/2	Fixed	0.142/25	0.273/43	②	—	35°	⑪
	17085502	14/32	7/8	Fixed	0.149/26	0.227/36	②	—	39°	⑪
	17085503	14/32	7/8	Fixed	0.149/26	0.227/36	②	—	39°	⑪
	17085506	14/32	1	Fixed	0.157/27	0.227/36	②	—	36°	⑪
	17085508	14/32	1	Fixed	0.157/27	0.227/36	②	—	36°	⑪
	17085524	14/32	1	Fixed	0.142/25	0.227/36	②	—	36°	⑪
	17085526	14/32	1	Fixed	0.142/25	0.227/36	②	—	36°	⑪
	17085554	14/32	1/2	Fixed	0.157/27	0.269/41	②	—	35°	⑪

① The carburetor identification number is stamped on the float bowl, near the secondary throttle lever.
② No measurement necessary on two point linkage; see text
③ Inner hole
④ On high step of cam, automatic in Park
⑤ 3 turns after contacting lever for preliminary setting
⑥ 2 turns after contacting lever for preliminary setting
⑦ Outer hole
⑧ Front
⑨ Rear
⑩ 4½ turns after contacting lever for preliminary setting
⑪ See underhood decal

QUADRAJET CARBURETOR SPECIFICATIONS
Chevrolet

Year	Carburetor Identification ①	Float Level (in.)	Air Valve Spring (turn)	Pump Rod (in.)	Primary Vacuum (deg./in.)	Secondary Vacuum (deg./in.)	Secondary Opening (in.)	Choke Rod (in.)	Choke Unloader (in.)	Fast Idle Speed ④ (rpm)
'80	17080202	7/16	7/8	1/4 ⑧	0.157	—	⑤	0.110	0.243	⑩
	17080204	7/16	7/8	1/4 ⑧	0.157	—	⑤	0.110	0.243	⑩
	17080207	7/16	7/8	1/4 ⑧	0.157	—	⑤	0.110	0.243	⑩
	17080228	7/16	7/8	9/32 ⑧	0.179	—	⑤	0.110	0.243	⑩
	17080243	3/16	9/16	9/32 ⑧	0.016	0.083	⑤	0.074	0.179	⑩
	17080274	15/32	5/8	5/16 ⑨	0.110	0.164	⑤	0.083	0.203	⑩
	17080282	7/16	7/8	11/32 ⑨	0.142	—	⑤	0.110	0.243	⑩
	17080284	7/16	7/8	11/32 ⑨	0.142	—	⑤	0.110	0.243	⑩
	17080502	1/2	7/8	Fixed	0.136	0.179	⑤	0.110	0.243	⑩
	17080504	1/2	7/8	Fixed	0.136	0.179	⑤	0.110	0.243	⑩
	17080542	3/8	9/16	Fixed	0.103	0.066	⑤	0.074	0.243	⑩
	17080543	3/8	9/16	Fixed	0.103	0.129	⑤	0.074	0.243	⑩
'81	17081202	11/32	7/8	Fixed	0.149	—	⑤	0.110	0.243	⑪
	17081203	11/32	7/8	Fixed	0.149	—	⑤	0.110	0.243	⑪
	17081204	11/32	7/8	Fixed	0.149	—	⑤	0.110	0.243	⑪
	17081207	11/32	7/8	Fixed	0.149	—	⑤	0.110	0.243	⑪
	17081216	11/32	7/8	Fixed	0.149	—	⑤	0.110	0.243	⑪
	17081217	11/32	7/8	Fixed	0.149	—	⑤	0.110	0.243	⑪
	17081218	11/32	7/8	Fixed	0.149	—	⑤	0.110	0.243	⑪
	17081242	5/16	9/16	Fixed	0.090	0.077	⑤	0.139	0.243	⑪
	17081243	1/4	9/16	Fixed	0.103	0.090	⑤	0.139	0.243	⑪
'82	17082202	11/32	7/8	Fixed	0.157	—	⑤	0.110	0.243	⑫
	17082204	11/32	7/8	Fixed	0.157	—	⑤	0.110	0.243	⑫
	17082203	11/32	7/8	Fixed	0.157	—	⑤	0.243	0.243	⑫
	17082207	11/32	7/8	Fixed	0.157	—	⑤	0.243	0.243	⑫

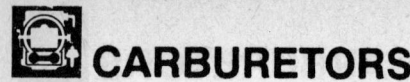

CARBURETORS

QUADRAJET CARBURETOR SPECIFICATIONS
Chevrolet

Year	Carburetor Identification ①	Float Level (in.)	Air Valve Spring (turn)	Pump Rod (in.)	Primary Vacuum (deg./in.)	Secondary Vacuum (deg./in.)	Secondary Opening (in.)	Choke Rod (in.)	Choke Unloader (in.)	Fast Idle Speed ④ (rpm)
'83	17083202	¹¹/₃₂	⁷/₈	Fixed	—	27/.157	⑤	0.110	0.243	⑬
	17083203	¹¹/₃₂	⁷/₈	Fixed	—	27/.157	⑤	0.243	0.243	⑬
	17083204	¹¹/₃₂	⁷/₈	Fixed	—	27/.157	⑤	0.110	0.243	⑬
	17083207	¹¹/₃₂	⁷/₈	Fixed	—	27/.157	⑤	0.243	0.243	⑬
	17083216	¹¹/₃₂	⁷/₈	Fixed	—	27/.157	⑤	0.110	0.243	⑬
	17083218	¹¹/₃₂	⁷/₈	Fixed	—	27/.157	⑤	0.110	0.243	⑬
	17083236	¹¹/₃₂	⁷/₈	Fixed	—	27/.157	⑤	0.110	0.243	⑬
	17083506	⁷/₁₆	⁷/₈	Fixed	27/.157	36/.227	⑤	0.110	0.227	⑬
	17083508	⁷/₁₆	⁷/₈	Fixed	27/.157	36/.227	⑤	0.110	0.227	⑬
	17083524	⁷/₁₆	⁷/₈	Fixed	25/.142	36/.227	⑤	0.110	0.227	⑬
	17083526	⁷/₁₆	⁷/₈	Fixed	25/.142	36/.227	⑤	0.110	0.227	⑬
'84	17084201	¹¹/₃₂	⁷/₈	Fixed	.157/27	—	⑤	0.110	0.243	⑬
	17084205	¹¹/₃₂	⁷/₈	Fixed	.157/27	—	⑤	0.243	0.243	⑬
	17084208	¹¹/₃₂	⁷/₈	Fixed	.157/27	—	⑤	0.110	0.243	⑬
	17084209	¹¹/₃₂	⁷/₈	Fixed	.157/27	—	⑤	0.243	0.243	⑬
	17084210	¹¹/₃₂	⁷/₈	Fixed	.157/27	—	⑤	0.110	0.243	⑬
	17084507	⁷/₁₆	1	Fixed	.157/27	.227/36	⑤	0.110	0.227	⑬
	17084509	⁷/₁₆	1	Fixed	.157/27	.227/36	⑤	0.110	0.227	⑬
	17084525	⁷/₁₆	1	Fixed	.142/25	.227/36	⑤	0.110	0.227	⑬
	17084527	⁷/₁₆	1	Fixed	.142/25	.227/36	⑤	0.110	0.227	⑬
'85	17085202	¹¹/₃₂	⁷/₈	Fixed	0.157/27	—	②	0.110	0.243/38°	⑪
	17085203	¹¹/₃₂	⁷/₈	Fixed	0.157/27	—	②	⑭	0.243/38°	⑪
	17085204	¹¹/₃₂	⁷/₈	Fixed	0.157/27	—	②	0.110	0.243/38°	⑪
	17085207	¹¹/₃₂	⁷/₈	Fixed	0.157/27	—	②	0.243	0.243/38°	⑪
	17085218	¹¹/₃₂	⁷/₈	Fixed	0.157/27	—	②	0.110	0.243/38°	⑪
	17085282	¹¹/₃₂	¹/₂	Fixed	0.142/25	0.273/43	②	0.110	0.220/35°	⑪
	17085502	¹⁴/₃₂	⁷/₈	Fixed	0.149/26	0.227/36	②	0.110	0.251/39°	⑪
	17085503	¹⁴/₃₂	⁷/₈	Fixed	0.149/26	0.227/36	②	0.110	0.251/39°	⑪
	17085506	¹⁴/₃₂	1	Fixed	0.157/27	0.227/36	②	0.110	0.227/36°	⑪
	17085508	¹⁴/₃₂	1	Fixed	0.157/27	0.227/36	②	0.110	0.227/36°	⑪
	17085524	¹⁴/₃₂	1	Fixed	0.142/25	0.227/36	②	0.110	0.227/36°	⑪
	17085526	¹⁴/₃₂	1	Fixed	0.142/25	0.227/36	②	0.110	0.227/36°	⑪
	17085554	¹⁴/₃₂	¹/₂	Fixed	0.157/27	0.269/41	②	0.071	0.220/35°	⑪
'86	17086003	¹¹/₃₂	⁷/₈	Fixed	0.157/27	—	②	—	0.243/38°	⑪
	17086004	¹¹/₃₂	⁷/₈	Fixed	0.157/27	—	②	—	0.243/38°	⑪
	17086005	¹¹/₃₂	⁷/₈	Fixed	0.157/27	—	②	—	0.243/38°	⑪
	17086006	¹¹/₃₂	⁷/₈	Fixed	0.157/27	—	②	—	0.243/38°	⑪

① The carburetor identification number is stamped on the float bowl, near the secondary throttle lever.
② Without vacuum advance.
③ With automatic transmission; vacuum advance connected and EGR disconnected and the throttle positioned on the high step of cam.
④ With manual transmission; without vacuum advance and the throttle positioned on the high step of cam.
⑤ No measurement necessary on two point linkage; see text.
⑥ 3 turns after contacting lever for preliminary setting.
⑦ 2 turns after contacting lever for preliminary setting.
⑧ Inner hole
⑨ Outer hole
⑩ 4 turns after contacting lever for preliminary setting.
⑪ 4½ turns after contacting lever for preliminary setting
⑫ 3⅛ turns after contacting lever for preliminary setting
⑬ See underhood sticker
⑭ 3 step cam: 0.110, 2 step cam: 0.243

QUADRAJET CARBURETOR SPECIFICATIONS
Oldsmobile

Year	Carburetor Identification ①	Float Level (in.)	Air Valve Spring (turn)	Pump Rod (in.)	Primary Vacuum Break (in./deg.)	Secondary Vacuum Break (in./deg.)	Secondary Opening (in.)	Choke Rod (in.)	Choke Unloader (in.)	Fast Idle Speed ④ (rpm)
'80	17080202	7/16	7/8	1/4 ⑦	0.157	—	④	0.110	0.243	⑤
	17080204	7/16	7/8	1/4 ⑦	0.157	—	④	0.110	0.243	⑤
	17080250	13/32	1/2	9/32 ⑦	0.149	0.211	④	0.090	0.220	⑤
	17080251	13/32	1/2	9/32 ⑦	0.149	0.211	④	0.090	0.220	⑤
	17080252	13/32	1/2	9/32 ⑦	0.149	0.211	④	0.090	0.220	⑤
	17080253	13/32	1/2	9/32 ⑦	0.149	0.211	④	0.090	0.220	⑤
	17080259	13/32	1/2	9/32 ⑦	0.149	0.211	④	0.090	0.220	⑤
	17080260	13/32	1/2	9/32 ⑦	0.149	0.211	④	0.090	0.220	⑤
	17080504	1/2	7/8	⑧	0.136	0.179	④	0.110	0.243	⑤
	17080553	15/32	1/2	⑧	0.142	0.220	④	0.090	0.220	⑤
	17080554	15/32	1/2	⑧	0.142	0.211	④	0.090	0.220	⑤
'81	17081250	13/32	1/2	9/32 ⑦	0.149 ⑨	0.211 ⑩	④	0.090	0.220	⑤
	17081253	15/32	1/2	⑧	0.142 ⑨	0.227 ⑩	④	0.071	0.220	⑤
	17081254	15/32	1/2	⑧	0.142 ⑨	0.227 ⑩	④	0.071	0.220	⑤
	17081248	3/8	—	⑧	0.164 ⑨	0.136 ⑩	④	0.139	0.243	⑤
	17081289	13/32	—	⑧	0.164 ⑨	0.136 ⑩	④	0.139	0.243	⑤
'82	17082202	11/32	7/8	Fixed	0.110/20	—	④	0.110	0.243	⑤
	17082204	11/32	3/8	Fixed	0.110/20	—	④	0.110	0.243	⑤
	17082244	7/16	9/16	Fixed	0.117/21	0.083/16	④	0.139	0.195	⑤
	17082245	3/8	5/8	Fixed	0.149/26	0.149/26	④	0.139	0.195	⑤
	17082246	3/8	5/8	Fixed	0.149/26	0.149/26	④	0.139	0.195	⑤
	17082247	13/32	5/8	Fixed	0.164/28	0.136/24	④	0.139	0.243	⑤
	17082248	13/32	5/8	Fixed	0.164/28	0.136/24	④	0.139	0.243	⑤
	17082251	15/32	1/2	Fixed	0.142/25	0.304/45	④	0.071	0.220	⑤
	17082253	15/32	1/2	Fixed	0.142/25	0.227/36	④	0.071	0.220	⑤
	17082264	7/16	9/16	Fixed	0.117/20	0.083/16	④	0.139	0.195	⑤
	17082265	3/8	5/8	Fixed	0.149/26	0.149/26	④	0.139	0.195	⑤
	17082266	3/8	5/8	Fixed	0.149/26	0.149/26	④	0.139	0.195	⑤
	17082267	3/8	5/8	Fixed	0.164/28	0.136/24	④	0.139	0.243	⑤
	17082268	13/32	5/8	Fixed	0.164/28	0.136/24	④	0.139	0.243	⑤
'83	17082265	3/8	5/8	Fixed	0.149/26	0.149/26	②	0.139	0.195	⑪
	17082266	3/8	5/8	Fixed	0.149/26	0.149/26	②	0.139	0.195	⑪
	17082267	3/8	5/8	Fixed	0.149/26	0.149/26	②	0.096	0.195	⑪
	17082268	3/8	5/8	Fixed	0.149/26	0.149/26	②	0.096	0.195	⑪
	17083242	9/32	9/16	Fixed	0.110/20	—	②	0.139	0.243	⑪
	17083244	1/4	9/16	Fixed	0.117/21	0.083/16	②	0.139	0.195	⑪
	17083248	3/8	5/8	Fixed	0.149/26	0.149/26	②	0.139	0.195	⑪
	17083250	7/16	1/2	Fixed	0.157/27	0.271/42	②	0.071	0.220	⑪
	17083253	7/16	1/2	Fixed	0.157/27	0.269/41	②	0.071	0.220	⑪
	17083553	7/16	1/2	Fixed	0.157/27	0.269/41	②	0.071	0.220	⑪
'84	17084201	11/32	7/8	Fixed	0.157/27	—	②	0.110	0.243	⑪
	17084205	11/32	7/8	Fixed	0.157/27	—	②	0.243	0.243	⑪
	17084208	11/32	7/8	Fixed	0.157/27	—	②	0.110	0.243	⑪
	17084209	11/32	7/8	Fixed	0.157/27	—	②	0.243	0.243	⑪
	17084210	11/32	7/8	Fixed	0.157/27	—	②	0.110	0.243	⑪
	17084240	5/16	1	Fixed	0.136/24	—	②	—	0.195	⑪
	17084244	5/16	1	Fixed	0.136/24	—	②	—	0.195	⑪
	17084246	5/16	1	Fixed	0.123/22	0.136/24	②	—	0.195	⑪
	17084248	5/16	1	Fixed	0.136/24	—	②	—	0.195	⑪
	17084252	7/16	1/2	Fixed	0.157/27	0.269/41	②	—	0.220	⑪
	17084254	7/16	1/2	Fixed	0.157/27	0.269/41	②	—	0.220	⑪

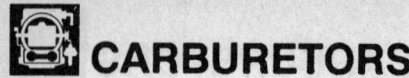

CARBURETORS

QUADRAJET CARBURETOR SPECIFICATIONS
Oldsmobile

Year	Carburetor Identification ①	Float Level (in.)	Air Valve Spring (turn)	Pump Rod (in.)	Primary Vacuum Break (in./deg.)	Secondary Vacuum Break (in./deg.)	Secondary Opening (in.)	Choke Rod (in.)	Choke Unloader (in.)	Fast Idle Speed ④ (rpm)
'85	17084282	11/32	1/2	9/32	0.142/25	0.278/43	④	0.110	0.220	⑤
	17085554	14/32	1/2	9/32	0.157/27	0.269/41	④	0.110	0.220	⑤
'86	17086008	11/32	1/2	Fixed	0.142/25	0.287/43	④	0.171	0.220	⑤
	17086009	14/32	1/2	Fixed	0.142/25	0.287/43	④	0.171	0.220	⑤

① The carburetor identification number is stamped on the float bowl, next to the secondary throttle lever.
② 1800 rpm on Omega and 400 cu. in. engines with the cam follower on the highest step of the fast idle cam; 900 rpm on all others with the fast idle cam follower on the lowest step of the fast idle cam.
④ No measurement necessary on two point linkage; see text.
⑤ 3 turns after contacting lever for preliminary setting.
⑥ 2 turns after contacting lever for preliminary setting.
⑦ Inner hole
⑧ Not Adjustable
⑨ Front
⑩ Rear
⑪ See underhood sticker

QUADRAJET CARBURETOR SPECIFICATIONS
Pontiac

Year	Carburetor Identification ①	Float Level (in.)	Air Valve Spring (turn)	Pump Rod (in.)	Primary Vacuum Break (in./deg.)	Secondary Vacuum Break (in./deg.)	Secondary Opening (in.)	Choke Rod (in.)	Choke Unloader (in./deg.)	Fast Idle Speed ② (rpm)
'80	17080249	7/16	3/4	9/32 ⑥	0.129	0.114	④	0.096	0.243	③
	17080270	15/32	5/8	3/8 ⑦	0.149	0.211	④	0.074	0.220	③
	17080272	15/32	5/8	3/8 ⑦	0.129	0.175	④	0.074	0.203	③
	17080274	15/32	5/8	5/16 ⑥	0.110	0.164	④	0.083	0.203	③
	17080502	1/2	7/8	⑧	0.136	0.179	④	0.110	0.243	③
	17080504	1/2	7/8	⑧	0.136	0.179	④	0.110	0.243	③
	17080553	15/32	1/2	⑧	0.142	0.220	④	0.090	0.220	③
'81	17081202, 204	11/32	7/8	⑧	0.157 ⑩	—	④	0.110	0.243	⑨
	17081203, 207	11/32	7/8	⑧	0.157 ⑩	—	④	0.110	0.243	⑨
	17081216, 217, 218	11/32	7/8	⑧	0.157 ⑩	—	④	0.110	0.243	⑨
	17081242	3/8	9/16	⑧	0.090 ⑩	0.077	④	0.139	0.243	⑨
	17081243	5/16	9/16	⑧	0.103 ⑩	0.090	④	0.139	0.243	⑨
	17081245	3/8	5/8	⑧	0.164 ⑩	0.136	④	0.139	0.243	⑨
	17081247	3/8	5/8	⑧	0.164 ⑩	0.136	④	0.139	0.243	⑨
	17081248, 249	3/8	5/8	⑧	0.164 ⑩	0.136	④	0.139	0.243	⑨
	17081253, 254	15/32	1/2	⑧	0.142 ⑩	0.227	④	0.071	0.220	⑨
	17081270	7/16	5/8	⑧	0.136 ⑩	0.211	④	0.074	0.220	⑨
	17081272	7/16	5/8	⑧	0.136 ⑩	0.260	④	0.074	0.220	⑨
	17081274	7/16	5/8	⑧	0.136 ⑩	0.220	④	0.083	0.220	⑨
	17081289	13/36	5/8	⑧	0.164 ⑩	0.136	④	0.139	0.243	⑨
'82	17082202	11/32	7/8	Fixed	0.110/20 ⑭	—	④	0.110	0.243	⑫ ⑮
	17082204	11/32	3/8 ⑬	Fixed	0.110/20 ⑭	—	④	0.110	0.243	⑫ ⑮
	17082203	11/32	7/8	Fixed	0.157/27	—	④	0.243	0.243	⑮
	17082207	11/32	7/8	Fixed	0.157/27	—	④	0.243	0.243	⑮
	17082244	7/16	9/16	Fixed	0.117/21	0.083/16	④	0.139	0.195	⑫
	17082245	3/8	5/8	Fixed	0.149/26	0.149/26	④	0.139	0.195	⑫
	17082246	3/8	5/8	Fixed	0.149/26	0.149/26	④	0.139	0.195	⑫
	17082247	13/32	5/8	Fixed	0.164/28	0.136/24	④	0.139	0.243	⑫
	17082248	13/32	5/8	Fixed	0.164/28	0.136/24	④	0.139	0.243	⑫
	17082251	15/32	1/2	Fixed	0.142/25	0.304/45	④	0.071	0.220	⑫
	17082253	15/32	1/2	Fixed	0.142/25	0.227/36	④	0.071	0.220	⑫

QUADRAJET CARBURETOR SPECIFICATIONS
Pontiac

Year	Carburetor Identification ①	Float Level (in.)	Air Valve Spring (turn)	Pump Rod (in.)	Primary Vacuum Break (in./deg.)	Secondary Vacuum Break (in./deg.)	Secondary Opening (in.)	Choke Rod (in.)	Choke Unloader (in./deg.)	Fast Idle Speed ② (rpm)
'82	17082264	7/16	9/16	Fixed	0.117/20	0.083/16	④	0.139	0.195	⑫
	17082265	3/8	5/8	Fixed	0.149/26	0.149/26	④	0.139	0.195	⑫
	17082266	3/8	5/8	Fixed	0.149/26	0.149/26	④	0.139	0.195	⑫
	17082267	3/8	5/8	Fixed	0.164/28	0.136/24	④	0.139	0.243	⑫
	17082268	13/32	5/8	Fixed	0.164/28	0.136/24	④	0.139	0.243	⑫
'83	17082265	3/8	5/8	Fixed	0.149/26	0.149/26	②	0.139	0.195	⑯
	17082266	3/8	5/8	Fixed	0.149/26	0.149/26	②	0.139	0.195	⑯
	17082267	3/8	5/8	Fixed	0.149/26	0.149/26	②	0.096	0.195	⑯
	17082268	3/8	5/8	Fixed	0.149/26	0.149/26	②	0.096	0.195	⑯
	17083242	9/32	9/16	Fixed	0.110/20	—	②	0.139	0.243	⑯
	17083244	1/4	9/16	Fixed	0.117/21	0.083/16	②	0.139	0.195	⑯
	17083248	3/8	5/8	Fixed	0.149/26	0.149/26	②	0.139	0.195	⑯
	17083250	7/16	1/2	Fixed	0.157/27	0.271/42	②	0.071	0.220	⑯
	17083253	7/16	1/2	Fixed	0.157/27	0.269/41	②	0.071	0.220	⑯
	17083553	7/16	1/2	Fixed	0.157/27	0.269/41	②	0.071	0.220	⑯
'84	17084201	11/32	7/8	Fixed	0.157/27	—	②	0.110	0.243	⑯
	17084205	11/32	7/8	Fixed	0.157/27	—	②	0.243	0.243	⑯
	17084208	11/32	7/8	Fixed	0.157/27	—	②	0.110	0.243	⑯
	17084209	11/32	7/8	Fixed	0.157/27	—	②	0.243	0.243	⑯
	17084210	11/32	7/8	Fixed	0.157/27	—	②	0.110	0.243	⑯
	17084240	5/16	1	Fixed	0.136/24	—	②	—	0.195	⑯
	17084244	5/16	1	Fixed	0.136/24	—	②	—	0.195	⑯
	17084246	5/16	1	Fixed	0.123/22	0.136/24	②	—	0.195	⑯
	17084248	5/16	1	Fixed	0.136/24	—	②	—	0.195	⑯
	17084252	7/16	1/2	Fixed	0.157/27	0.269/41	②	—	0.220	⑯
	17084254	7/16	1/2	Fixed	0.157/27	0.269/41	②	—	0.220	⑯
'85	17085202	11/32	7/8	Fixed	0.157/27	—	②	0.110	0.243/38	⑯
	17085203	11/32	7/8	Fixed	0.157/27	—	②	⑰	0.243/38	⑯
	17085204	11/32	7/8	Fixed	0.157/27	—	②	0.110	0.243/38	⑯
	17085207	11/32	7/8	Fixed	0.157/27	—	②	0.243	0.243/38	⑯
	17085218	11/32	7/8	Fixed	0.157/27	—	②	0.110	0.243/38	⑯
	17085282	11/32	1/2	Fixed	0.142/25	0.273/43	②	0.110	0.220/35	⑯
	17085502	14/32	7/8	Fixed	0.149/26	0.227/36	②	0.110	0.251/39	⑯
	17085503	14/32	7/8	Fixed	0.149/26	0.227/36	②	0.110	0.251/39	⑯
	17085506	14/32	1	Fixed	0.157/27	0.227/36	②	0.110	0.227/36	⑯
	17085508	14/32	1	Fixed	0.157/27	0.227/36	②	0.110	0.227/36	⑯
	17085524	14/32	1	Fixed	0.142/25	0.227/36	②	0.110	0.227/36	⑯
	17085526	14/32	1	Fixed	0.142/25	0.227/36	②	0.110	0.227/36	⑯
	17085554	14/32	1/2	Fixed	0.157/27	0.269/41	②	0.071	0.220/35	⑯
'86	17086003	11/32	7/8	Fixed	0.157/27	—	②	0.110	0.243/38	⑯
	17086004	11/32	7/8	Fixed	0.157/27	—	②	0.110	0.243/38	⑯
	17086005	11/32	7/8	Fixed	0.157/27	—	②	0.243	0.243/38	⑯
	17086006	11/32	7/8	Fixed	0.157/27	—	②	0.110	0.243/38	⑯
	17086007	11/32	1/2	Fixed	0.142/25	0.287/43	②	0.071	0.220/35	⑯
	17086008	11/32	1/2	Fixed	0.142/25	0.287/43	②	0.071	0.220/35	⑯
	17086040	11/32	7/8	Fixed	0.157/27	—	②	0.110	0.243/38	⑯

① The carburetor identification number is stamped on the float bowl, near the secondary throttle lever.
② On highest step.
③ 1½ turns after contacting lever for preliminary setting
④ No measurement necessary on two point linkage; see text.
⑤ 2 turns after contacting lever for preliminary setting.
⑥ Inner hole
⑦ Outer hole
⑧ Not Adjustable
⑨ 4½ turns after contacting lever for preliminary setting
⑩ Front
⑪ Rear
⑫ 3 turns after contacting lever for preliminary setting
⑬ Firebird—7/8
⑭ Firebird—0.157 in./27°
⑮ Firebird—3⅛ turns after contacting lever for preliminary setting
⑯ See underhood sticker
⑰ 3 step cam: 0.110, 2 step cam: 0.243

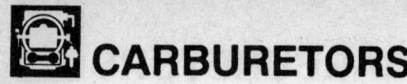

CARBURETORS

QUADRAJET CARBURETOR SPECIFICATIONS
All Canadian Models

Year	Carburetor Identi-fication [1]	Float Level (in.)	Air Valve Spring (turn)	Pump Rod (in.)	Primary Vacuum Break (deg./in.)	Secondary Vacuum Break (deg./in.)	Secondary Opening (in.)	Choke Rod (in.)	Choke Unloader (deg./in.)	Fast Idle Speed [2] (rpm)
'81	17080201	15/32	7/8	9/32 [2]	—	23/0.129	[4]	0.314	0.277	[5]
	17080205	15/32	7/8	9/32 [2]	—	23/0.129	[4]	0.314	0.277	[5]
	17080206	15/32	7/8	9/32 [2]	—	23/0.129	[4]	0.314	0.277	[5]
	17080290	15/32	7/8	9/32 [2]	—	26/0.149	[4]	0.314	0.277	[5]
	17080291	15/32	7/8	9/32 [2]	—	26/0.149	[4]	0.314	0.277	[5]
	17080292	15/32	7/8	9/32 [2]	—	26/0.149	[4]	0.314	0.277	[5]
	17080213	3/8	1	9/32 [2]	23/0.129	30/0.179	[4]	0.234	0.260	[5]
	17080215	3/8	1	9/32 [2]	23/0.129	30/0.179	[4]	0.234	0.260	[5]
	17080298	3/8	1	9/32 [2]	23/0.129	30/0.179	[4]	0.234	0.260	[5]
	17080507	3/8	1	9/32 [2]	23/0.129	30/0.179	[4]	0.234	0.260	[5]
	17080513	3/8	1	9/32 [2]	23/0.129	30/0.179	[4]	0.234	0.260	[5]
	17081250	13/32	1/2	9/32 [2]	26/0.149	34/0.211	[4]	0.090	0.220	[5]
	17080260	13/32	1/2	9/32 [2]	26/0.149	34/0.211	[4]	0.090	0.220	[5]
	17081276	15/32	5/8	5/16 [2]	20/0.110	28/0.164	[4]	0.083	0.203	[5]
	17081286	13/32	1/2	9/32 [2]	18/0.096	34/0.211	[4]	0.077	0.220	[5]
	17081287	13/32	1/2	9/32 [2]	18/0.096	34/0.211	[4]	0.077	0.220	[5]
	17081282	3/8	5/8	9/32 [2]	20/0.110	—	[4]	0.110	0.243	[5]
	17081283	3/8	7/8	9/32 [2]	20/0.110	—	[4]	0.110	0.243	[5]
	17081284	1/2	7/8	9/32 [2]	20/0.110	—	[4]	0.110	0.243	[5]
	17081285	1/2	7/8	9/32 [2]	20/0.110	—	[4]	0.110	0.243	[5]
	17080243	3/16	9/16	9/32 [2]	14.5/0.075	16/0.083	[4]	0.075	0.179	[5]
	17081295	13/32	9/16	9/32 [2]	14.5/0.075	13/0.066	[4]	0.075	0.220	[5]
	17081294	5/16	5/8	9/32 [2]	24.5/0.139	14/0.071	[4]	0.139	0.243	[5]
	17081290	13/32	7/8	9/32 [2]	46/0.314	24/0.136	[4]	0.314	0.277	[5]
	17081291	13/32	7/8	9/32 [2]	46/0.314	24/0.136	[4]	0.314	0.277	[5]
	17081292	13/32	7/8	9/32 [2]	46/0.314	24/0.136	[4]	0.314	0.277	[5]
	17081506	13/32	7/8	9/32 [2]	46/0.314	36/0.227	[4]	0.314	0.277	[5]
	17081508	13/32	7/8	9/32 [2]	46/0.314	36/0.227	[4]	0.314	0.277	[5]
	17080202	7/16	7/8	1/4 [2]	20/0.110	—	[4]	0.110	0.243	[5]
	17080204	7/16	7/8	1/4 [2]	20/0.110	—	[4]	0.110	0.243	[5]
	17080207	7/16	7/8	1/4 [2]	20/0.110	—	[4]	0.110	0.243	[5]
'82	17082280	3/8	7/8	9/32 [2]	25/0.142	—	[4]	0.110	0.243	[5]
	17082281	3/8	7/8	9/32 [2]	25/0.142	—	[4]	0.110	0.243	[5]
	17082282	3/8	7/8	9/32 [2]	25/0.142	—	[4]	0.110	0.243	[5]
	17082283	3/8	7/8	9/32 [2]	25/0.142	—	[4]	0.110	0.243	[5]
	17082286	13/32	1/2	9/32 [2]	22/0.123	34/0.211	[4]	0.077	0.243	[5]
	17082287	13/32	1/2	9/32 [2]	22/0.123	34/0.211	[4]	0.077	0.243	[5]
	17082288	3/8	7/8	9/32 [2]	25/0.142	—	[4]	0.110	0.243	[5]
	17082289	3/8	7/8	9/32 [2]	25/0.142	—	[4]	0.110	0.243	[5]
	17082296	1/2	7/8	9/32 [2]	25/0.142	—	[4]	0.110	0.243	[5]
	17082297	1/2	7/8	9/32 [2]	25/0.142	—	[4]	0.110	0.243	[5]
'83	17080213	3/8	1	9/32	23/.129	30/.179	[4]	0.234	0.260	[5]
	17082213	9/32	1	9/32	23/.129	30/.179	[4]	0.234	0.260	[5]
	17082282	3/8	7/8	9/32	25/.142	—	[4]	0.110	0.243	[5]
	17082283	3/8	7/8	9/32	25/.142	—	[4]	0.110	0.243	[5]
	17082286	13/32	1/2	9/32	23/.129	34/.211	[4]	0.107	0.220	[5]
	17082287	13/32	1/2	9/32	23/.129	34/.211	[4]	0.107	0.220	[5]
	17082296	1/2	7/8	9/32	25/.142	—	[4]	0.110	0.243	[5]
	17082297	1/2	7/8	9/32	25/.142	—	[4]	0.110	0.243	[5]
	17083280	3/8	7/8	9/32	25/.142	—	[4]	0.110	0.243	[5]
	17083281	3/8	7/8	9/32	25/.142	—	[4]	0.110	0.243	[5]
	17083282	3/8	7/8	9/32	25/.142	—	[4]	0.110	0.243	[5]
	17083283	3/8	7/8	9/32	25/.142	—	[4]	0.110	0.243	[5]
	17083290	13/32	7/8	9/32	—	24/.136	[4]	0.314	0.251	[5]

QUADRAJET CARBURETOR SPECIFICATIONS
All Canadian Models

Year	Carburetor Identi- fication ①	Float Level (in.)	Air Valve Spring (turn)	Pump Rod (in.)	Primary Vacuum Break (deg./in.)	Secondary Vacuum Break (deg./in.)	Secondary Opening (in.)	Choke Rod (in.)	Choke Unloader (deg./in.)	Fast Idle Speed ② (rpm)
'83	17083292	13/32	7/8	9/32	—	24/.136	④	0.314	0.251	⑤
	17083298	3/8	1	9/32	23/.129	30/.179	④	0.234	0.260	⑤
'84	17084280	3/8	7/8	9/32 ②	23/.129	—	④	0.110	0.243	⑤
	17084281	3/8	7/8	9/32 ②	23/.129	—	④	0.110	0.243	⑤
	17084282	3/8	7/8	9/32 ②	23/.129	—	④	0.110	0.243	⑤
	17084283	3/8	7/8	9/32 ②	23/.129	—	④	0.110	0.243	⑤
	17084284	3/8	7/8	9/32 ②	23/.129	—	④	0.110	0.243	⑤
	17084285	3/8	7/8	9/32 ②	23/.129	—	④	0.110	0.243	⑤
	17084286	13/32	1/2	9/32 ②	23/.129	34/.211	④	0.107	0.220	⑤
	17084287	13/32	1/2	9/32 ②	23/.129	34/.211	④	0.107	0.220	⑤
	17084288	3/8	7/8	9/32 ②	23/.129	—	④	0.110	0.243	⑤
	17084289	3/8	7/8	9/32 ②	23/.129	—	④	0.110	0.243	⑤
	17084296	1/2	7/8	9/32 ②	23/.129	—	④	0.110	0.243	⑤
	17084297	1/2	7/8	9/32 ②	23/.129	—	④	0.110	0.243	⑤
'85	17080213	3/8	1	9/32 ②	23/.129	30/0.179	④	0.234	40/0.260	⑤
	17080298	3/8	1	9/32 ②	23/.129	30/0.179	④	0.234	40/0.260	⑤
	17082213	3/8	1	9/32 ②	23/.129	30/0.179	④	0.234	40/0.260	⑤
	17083298	3/8	1	9/32 ②	23/.129	30/0.179	④	0.234	40/0.260	⑤
	17085247	13/32	7/8	9/32 ②	20/0.110	—	④	0.096	30/0.179	⑤
	17085246	13/32	7/8	9/32 ②	20/0.110	—	④	0.096	30/0.179	⑤
	17085249	13/32	7/8	9/32 ②	20/0.110	—	④	0.096	30/0.179	⑤
	17085248	13/32	7/8	9/32 ②	20/0.110	—	④	0.096	30/0.179	⑤
	17085580	3/8	7/8	9/32 ②	21/0.117	—	④	0.077	30/0.179	⑤
	17085582	3/8	7/8	9/32 ②	21/0.117	—	④	0.077	30/0.179	⑤
	17085581	3/8	7/8	9/32 ②	21/0.117	—	④	0.077	30/0.179	⑤
	17085583	3/8	7/8	9/32 ②	21/0.117	—	④	0.077	30/0.179	⑤
	17085584	3/8	7/8	9/32 ②	21/0.117	—	④	0.077	30/0.179	⑤
	17085586	3/8	7/8	9/32 ②	21/0.117	—	④	0.077	30/0.179	⑤
	17085592	13/32	1/2	9/32 ②	21/0.117	34/.211	④	0.077	35/0.220	⑤
	17085594	13/32	1/2	9/32 ②	21/0.117	34/.211	④	0.077	28/0.164	⑤
	17085588	3/8	7/8	9/32 ②	21/0.117	—	④	0.077	30/0.179	⑤
	17085590	3/8	7/8	9/32 ②	21/0.117	—	④	0.077	30/0.179	⑤
	17085596	1/2	7/8	9/32 ②	23/0.129	—	④	0.077	38/0.243	⑤
	17085598	1/2	7/8	9/32 ②	23/0.129	—	④	0.077	38/0.243	⑤
'86	17086246	13/32	7/8	9/32 ②	20/0.110	—	④	0.096	30/0.179	⑤
	17086247	13/32	7/8	9/32 ②	20/0.110	—	④	0.096	30/0.179	⑤
	17086248	13/32	7/8	9/32 ②	20/0.110	—	④	0.096	30/0.179	⑤
	17086249	13/32	7/8	9/32 ②	20/0.110	—	④	0.096	30/0.179	⑤
	17086580	12/32	7/8	9/32 ②	21/0.117	—	④	0.077	30/0.179	⑤
	17086581	12/32	7/8	9/32 ②	21/0.117	—	④	0.077	30/0.179	⑤
	17086582	12/32	7/8	9/32 ②	21/0.117	—	④	0.077	30/0.179	⑤
	17086583	12/32	7/8	9/32 ②	21/0.117	—	④	0.077	30/0.179	⑤
	17086584	12/32	7/8	9/32 ②	21/0.117	—	④	0.077	30/0.179	⑤
	17086586	12/32	7/8	9/32 ②	21/0.117	—	④	0.077	30/0.179	⑤
	17086588	12/32	7/8	9/32 ②	21/0.117	—	④	0.077	30/0.179	⑤
	17086590	12/32	7/8	9/32 ②	21/0.117	—	④	0.077	30/0.179	⑤
	17086596	16/32	7/8	9/32 ②	21/0.117	—	④	0.077	30/0.179	⑤
	17086598	16/32	7/8	9/32 ②	21/0.117	—	④	0.077	30/0.179	⑤

① The carburetor identification number is stamped on the float bowl, near the secondary throttle lever.
② Inner hole
③ Outer hole
④ No measurement necessary on two point linkage; see text.
⑤ See underhood decal

MIKUNI CARBURETORS

2.6L Feedback Carburetor

All Federal and California 2.6L Mitsubishi engines are equipped with a two barrel downdraft carburetor designed for electronic fuel control and closed loop operation. With the closed loopsystem of mixture control, this carburetor includes the special feature to provide optimum air-fuel control during all ranges of engine operation. Fuel metering is accomplished through the use of three solenoid valves which reduce or add fuel to the engine.

There are eight basic systems in the feedback carburetor: fuel inlet, primary metering, secondary metering, accelerator pump, choke, jet mixture, enrichment, and fuel cut-off. The first five systems are basically the same between standard and feedback carburetors. The remaining three are unique to feedback carburetors. The enrichment system consist of an enrichment solenoid and a metering jet.

This system is used to provide additional fuel to the main metering system. Activation of the enrichment valve is controlled by the length of time that current is supplied by the solenoid valve. The jet mixture system supplies fuel to the engine through the jet mixture jet and passages. This system is calibrated by the jet mixture solenoid valve, which responds to a signal from the ECU.

The closed loop system provides the capability to perform closed loop fuel control in response to various sensor signals. The throttle position sensor (TPS) provides angle information to the ECU. The (TPS) is mounted on the carburetor. The idle position switch is installed on the carburetor and is "ON" when the throttle plate is at the closed or idle position. It provides information to the ECU and is used to adjust idle speed.

Standard Carburetor

All front wheel drive models (USA and Canadian), with the 2.6 liter (156 cubic inch) Mitsubishi engine are equipped with a conventional downdraft two barrel compound type carburetor. The automatic choke is a thermowax type which is controlled by engine coolant temperature. This carburetor also features a diaphragm type accelerator pump, bowl vent, fuel cut-off solenoid, air switching valve (ASV), sub EGR valve, coasting air valve (CAV), jet air control valve (JACV) and a high altitude compensation (HAC) system (California only). The air switching valve system is activated by ported carburetor vacuum and supplies additional air to the low speed passage by cutting off fuel flow to the bypass holes and pilot outlet.

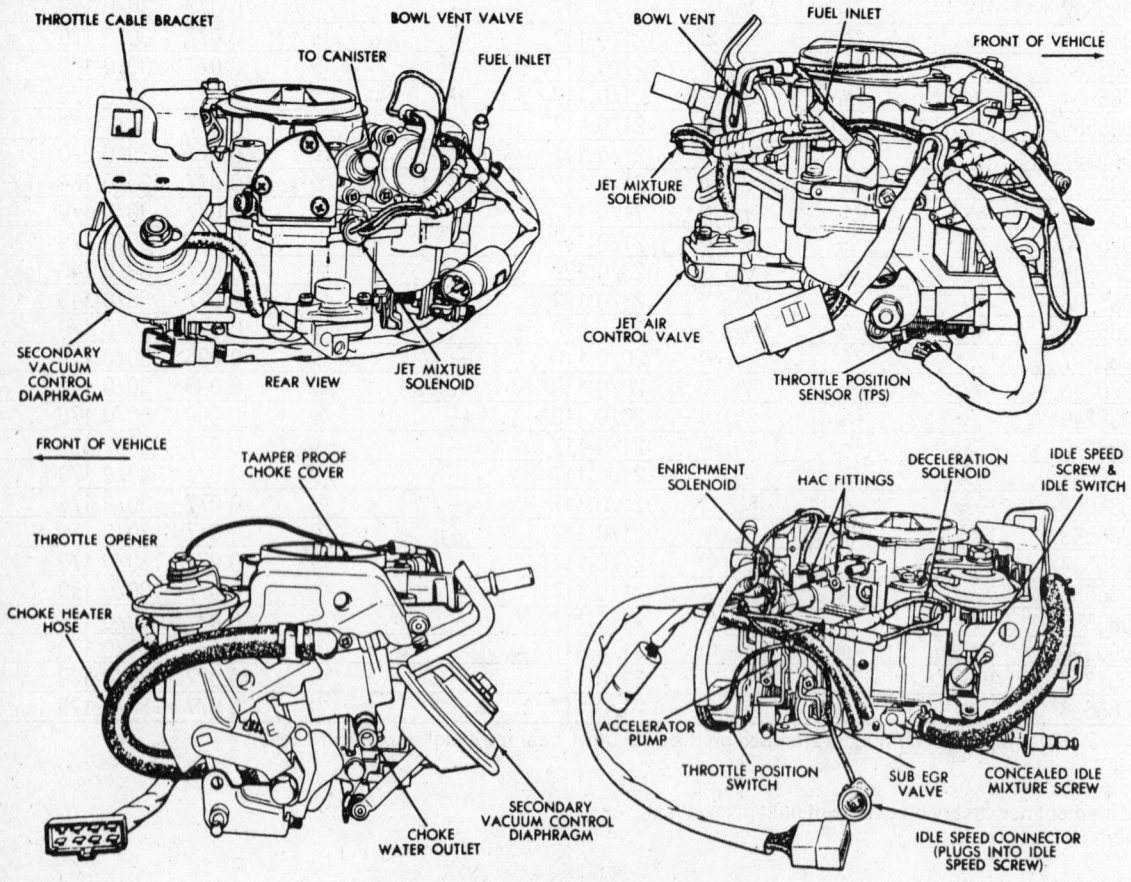

Mikuni feedback carburetor assembly

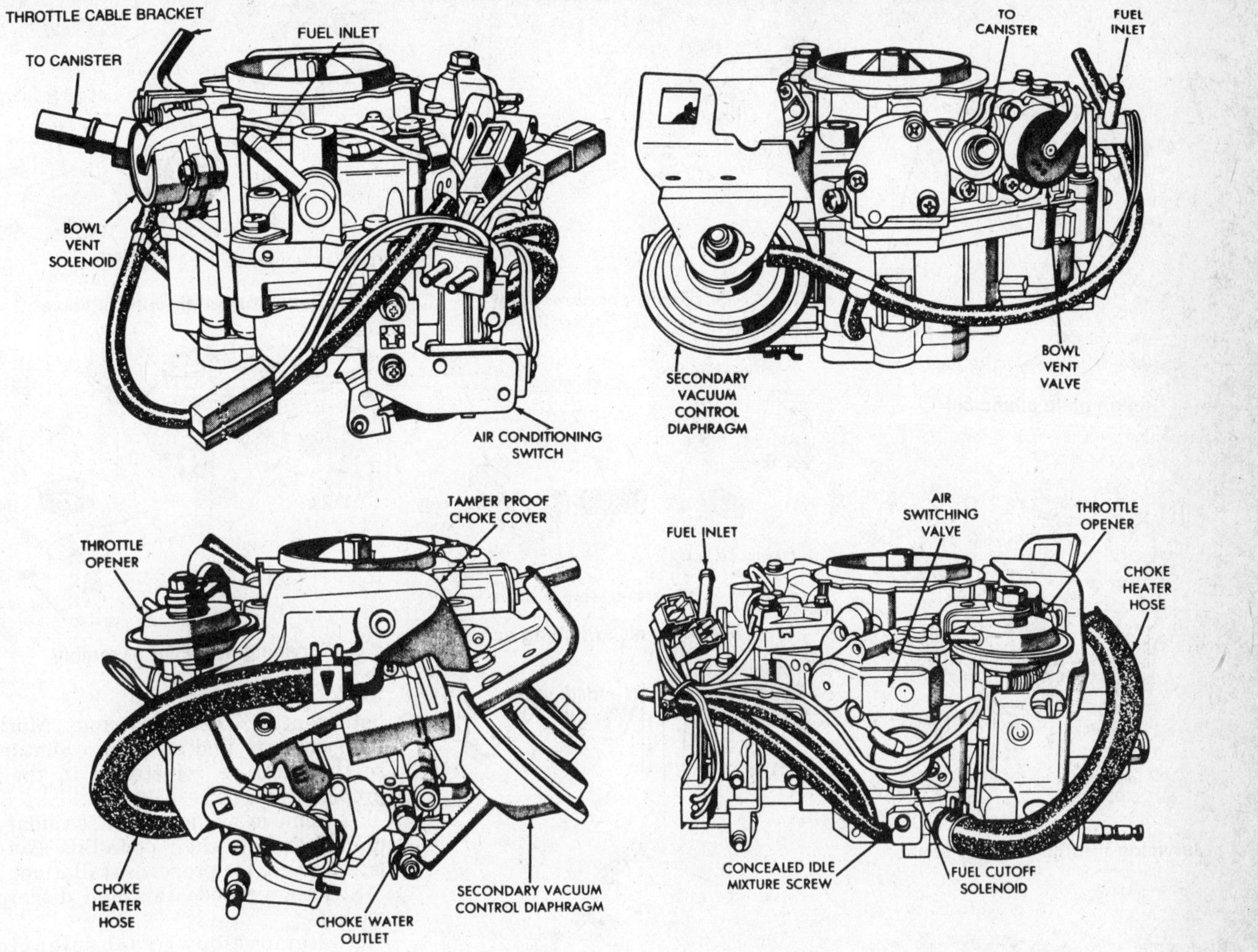

THROTTLE CABLE BRACKET

TO CANISTER

FUEL INLET

BOWL VENT SOLENOID

AIR CONDITIONING SWITCH

TO CANISTER

FUEL INLET

SECONDARY VACUUM CONTROL DIAPHRAGM

BOWL VENT VALVE

THROTTLE OPENER

TAMPER PROOF CHOKE COVER

CHOKE HEATER HOSE

CHOKE WATER OUTLET

SECONDARY VACUUM CONTROL DIAPHRAGM

FUEL INLET

AIR SWITCHING VALVE

THROTTLE OPENER

CHOKE HEATER HOSE

CONCEALED IDLE MIXTURE SCREW

FUEL CUTOFF SOLENOID

Mikuni non—feedback carburetor assembly

DISASSEMBLY

1. Compress clamps and remove water hose from choke assembly.
2. Drill out staked portions of staked covered screws.
3. Using a small hammer and a pointed punch, gently tap the edge of the remaining screw counterclockwise until the screw is removed. Remove the choke cover.
4. Note the relationship between the punched mark and the scribed line on the choke pinion plate. During reassembly the line and punch mark must be aligned to this position.
5. Remove "E" clip from throttle opener link. Remove throttle opener screws and set opener aside.
6. Remove ground wire from fuel cut-off solenoid (if equipped), then remove the mounting screw and set solenoid aside.
7. Remove throttle return spring and damper spring.
8. Remove "E" clips and choke unloader link from carburetor.

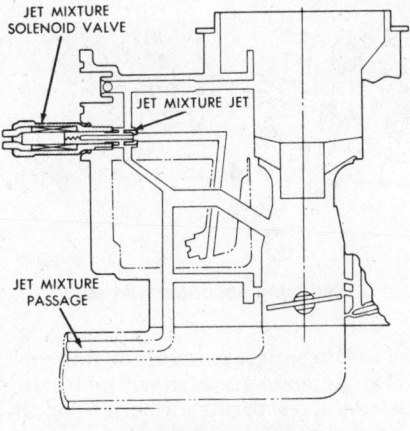

JET MIXTURE SOLENOID VALVE

JET MIXTURE JET

JET MIXTURE PASSAGE

Jet mixture system

9. Disconnect vacuum hose and link from vacuum chamber, remove mounting screws and set chamber aside.
10. On feedback models, remove screws securing throttle position sensor and set sensor aside. Remove vacuum connector hoses from carburetor on all models.

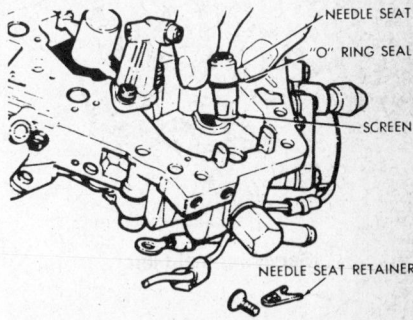

NEEDLE SEAT

"O" RING SEAL

SCREEN

NEEDLE SEAT RETAINER

Servicing needle seat assembly

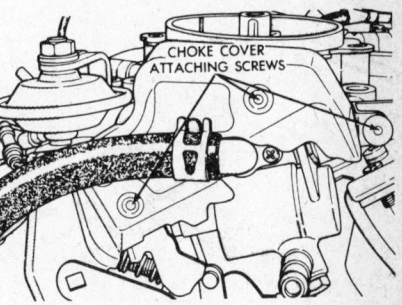

CHOKE COVER ATTACHING SCREWS

Feedback model choke cover screws

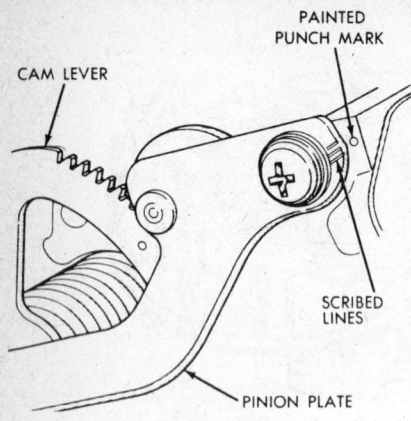

Pinion plate alignment

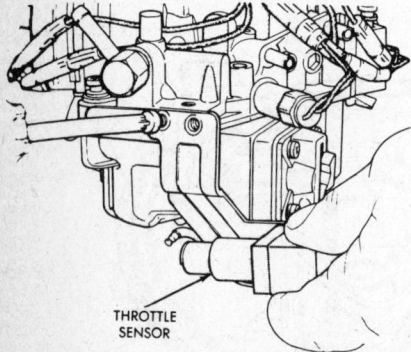

Servicing throttle position sensor

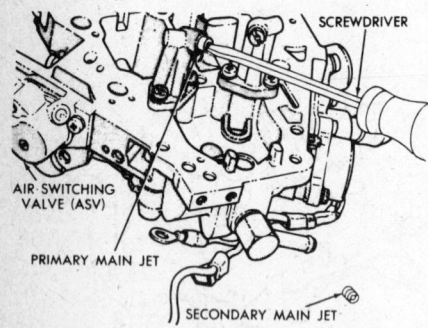

Servicing main jets

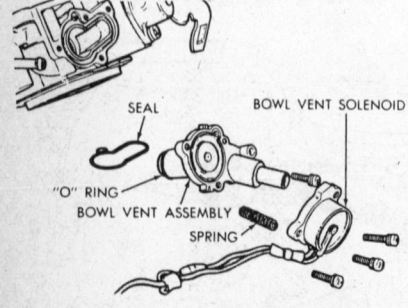

Servicing bowl vent

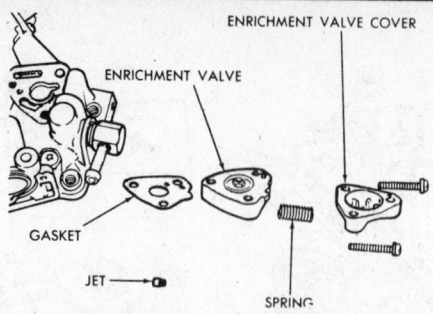

Servicing enrichment jet

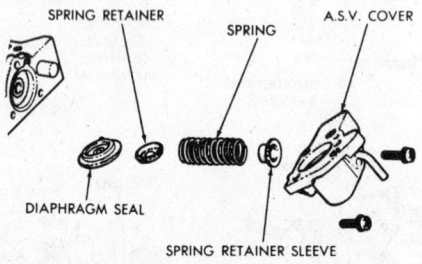

Servicing air switching valve

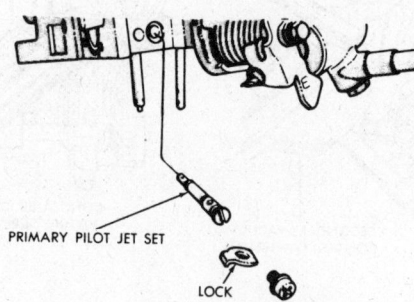

Servicing primary jet set

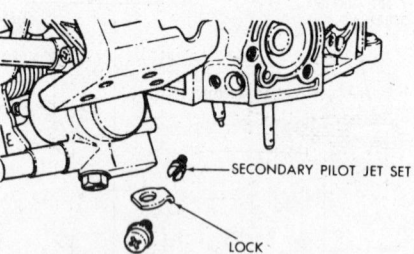

Servicing secondary jet set

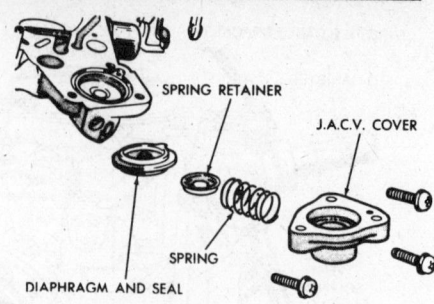

Servicing jet air control valve

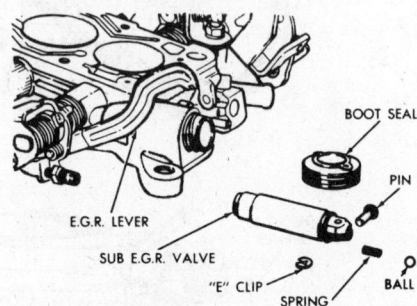

Servicing sub EGR assembly

11. Disconnect accelerator linkage.
12. Remove six airhorn mounting screws and seperate air horn from carburetor body.
13. Slide out float pivot pin and remove float assembly.
14. Unscrew retainer and remove needle seat and screen assembly. Do not lose shim from under needle seat assembly.
15. On feedback models, disconnect solenoid wiring and unscrew solenoid from air horn.
16. Remove venturi retainers and both venturi. Discard O-rings. Mark both primary and secondary venturi so they can be reinstalled in their proper positions.
17. Remove primary and secondary main jets from their pedestals. Note jet numbers for proper installation.
18. Remove pedestals and discard gaskets.
19. Remove bowl vent solenoid mounting screws, seperate bowl vent from air horn, and discard O-ring and seal.
20. Remove enrichment valve screws and seperate valve from air horn. Remove jet.
21. On non feedback models, remove air switching valve screws and seperate valve from air horn.
22. Remove screw, lock and primary jet set.
23. Remove screw, lock, and secondary jet set.
24. Remove primary and secondary air bleed jets from top of air horn. Note sizes for proper reinstallation.
25. Invert air horn carefully and drop out pump weight, check ball and hex nut.
26. Remove accelerator pump screws and seperate pump from air horn.
27. Remove jet air control valve screws and seperate valve from throttle body.
28. Remove "E" clip from sub EGR lever. Carefully slide pin from lever and sub EGR valve. The lever will be under spring tension caused by a steel ball and spring in the sub EGR valve and the lever. Be careful not to lose

the ball and spring when removing lever. Remove valve from throttle boby. Reverse procedure to Assemble the carburetor.

───── CAUTION ─────

Priming a carburetor by pouring gasoline into the air horn is dangerous and should be avoided. Cranking the engine, and then depressing the accelerator several times is the recommended way to prime the carburetor.

FLOAT LEVEL ADJUSTMENT

1. Invert the air horn assembly without a gasket.
2. With a gauge, measure the distance from the bottom of the float to the surface of the air horn. The distance should be 0.0787–0.779 in. (17.8–20.8mm).
3. If the reading is not within this range the shim under the needle seat must be changed. Shim kits are available which have three shims: 0.0118 in. (0.3mm), 0.0157 in. (0.4mm), 0.0196 in. (0.5mm). Adding or removing a shim will change the float level by three times the thickness of the shim.

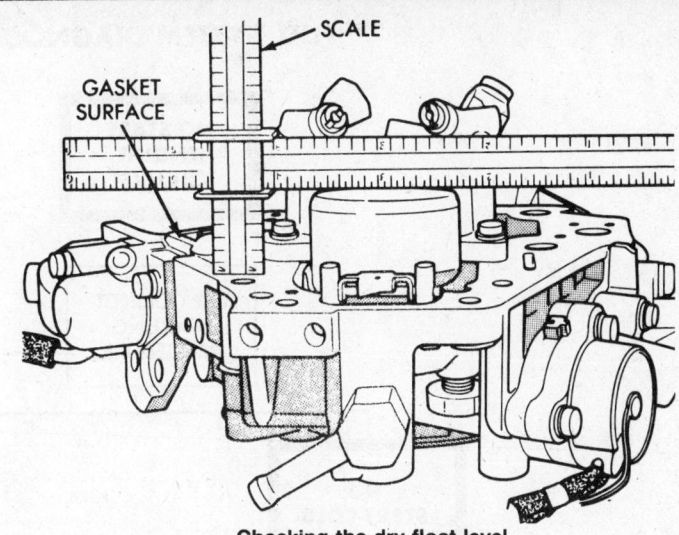

Checking the dry float level

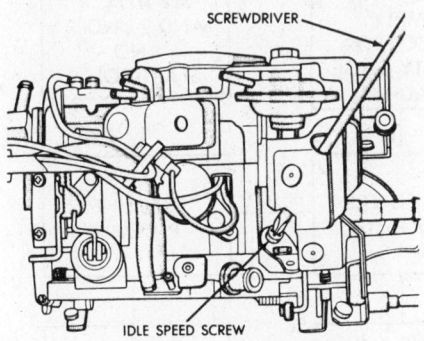

Idle speed adjustment

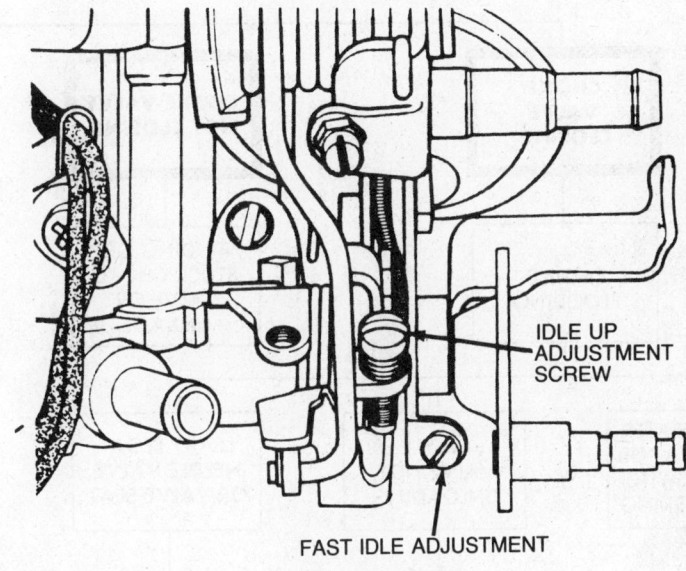

Idle up adjustment

CAM LEVER ALIGNMENT

Refer to illustration for adjustment.

IDLE SPEED ADJUSTMENT

Before adjusting idle speed, check ignition timing and adjust if necessary.

1. Set the parking brake and place the car in neutral. Turn off all lights and accessories. Disconnect the radiator fan. Connect tachometer to engine.
2. Start and run engine until it reaches normal operating temperature.
3. Open throttle and raise engine rpm to 2500 for ten seconds, then return engine to idle.
4. Wait two minutes and rpm indicated on the tachometer. If rpm is dif-

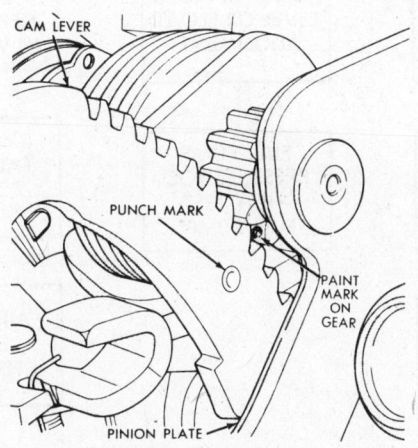

Cam lever alignment

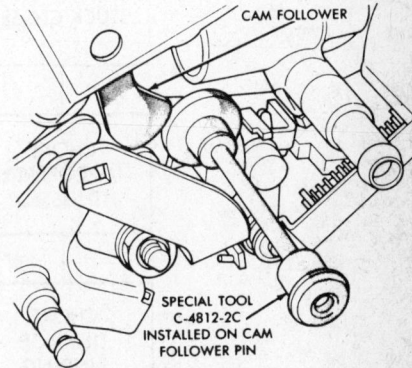

Tool C—4812—2C installed on feedback models

ferent from that specified on the underhood sticker, turn adjusting screw until correct rpm is obtained. On feedback models, the idle switch connector must be removed before making this adjustment. On air conditioned models, set temperation control lever to coldest position and turn A/C on. With air compressor running,

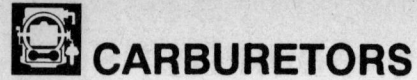

FUEL SYSTEM DIAGNOSIS

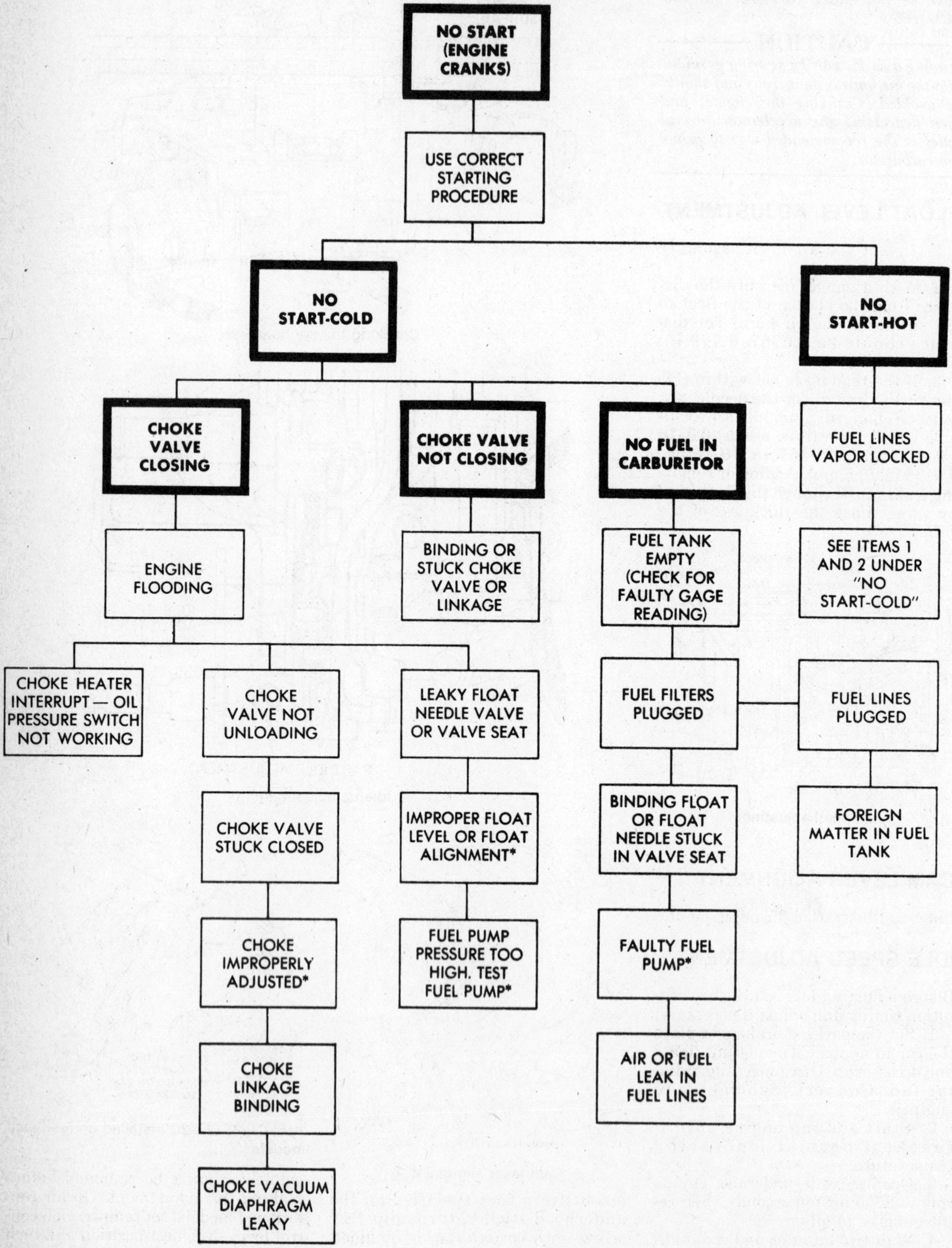

```
                        ┌─────────────────┐
                        │    NO START     │
                        │    (ENGINE      │
                        │    CRANKS)      │
                        └─────────────────┘
                                 │
                        ┌─────────────────┐
                        │  USE CORRECT    │
                        │   STARTING      │
                        │   PROCEDURE     │
                        └─────────────────┘
```

NO START-COLD

NO START-HOT

CHOKE VALVE CLOSING

CHOKE VALVE NOT CLOSING

NO FUEL IN CARBURETOR

FUEL LINES VAPOR LOCKED

ENGINE FLOODING

BINDING OR STUCK CHOKE VALVE OR LINKAGE

FUEL TANK EMPTY (CHECK FOR FAULTY GAGE READING)

SEE ITEMS 1 AND 2 UNDER "NO START-COLD"

CHOKE HEATER INTERRUPT — OIL PRESSURE SWITCH NOT WORKING

CHOKE VALVE NOT UNLOADING

LEAKY FLOAT NEEDLE VALVE OR VALVE SEAT

FUEL FILTERS PLUGGED

FUEL LINES PLUGGED

CHOKE VALVE STUCK CLOSED

IMPROPER FLOAT LEVEL OR FLOAT ALIGNMENT*

BINDING FLOAT OR FLOAT NEEDLE STUCK IN VALVE SEAT

FOREIGN MATTER IN FUEL TANK

CHOKE IMPROPERLY ADJUSTED*

FUEL PUMP PRESSURE TOO HIGH. TEST FUEL PUMP*

FAULTY FUEL PUMP*

CHOKE LINKAGE BINDING

AIR OR FUEL LEAK IN FUEL LINES

CHOKE VACUUM DIAPHRAGM LEAKY

set engine speed to 900 rpm with idle up screw.

5. Turn off engine, reconnect fan, disconnect tachometer, and reconnect idle switch connector.

FAST IDLE SPEED ADJUSTMENT

1. Set the parking brake and place the car in Neutral. Turn off all lights and accessories. Disconnect the radiator fan. Connect tachometer to engine.

2. Start and run engine until it reaches normal operating temperature.

3. Disconnect and plug vacuum advance hose at distributor. Disconnect radiator fan.

4. Open throttle slightly and install tool C-4812-2C on choke cam follower pin.

5. Release throttle lever and adjust fast idle speed to specification on the underhood sticker.

6. Remove tool, turn off engine, reconnect fan, unplug and reconnect vacuum advance hose, and remove tachometer.

Spectrum Two Barrel Carburetor

PRIMARY THROTTLE VALVE OPENING (FULL OPENING)

1. Inspect the angle of the primary throttle valve when the throttl valve has been fully opened. The valve angle should be 90° from the horozontal plane.

2. If adjustment is needed, bend the throttle adjust arm.

SECONDARY THROTTLE VALVE OPENING

1. Open the throttle lever, fully open the secondary throttle valve and inspect the angle. The valve angle should be 87° from the horozontal plane.

2. If needed bend the secondary shaft lever to adjust.

CHOKE VALVE ADJUSTMENT (THIRD STAGE)

1. Check the choke valve in the third stage of the fast idle cam.

a. Set the choke valve to full open.

b. Slowly open the throttle lever while lightly pushing the choke valve in the closing direction with your fingers and set the choke valve to the third stage of the fast idle cam.

c. The choke valve clearance should be 0.093 in.

d. If adjustment is needed, remove the rivet of the automatic choke and adjust by bending the choke lever in the housing. Reinstall the choke lever by riveting.

PRIMARY THROTTLE VALVE OPENING (SECOND STAGE)

1. Check the clearance of the primary throttle valve in the second stage of the fast idle cam.

a. Set the choke valve to full open.

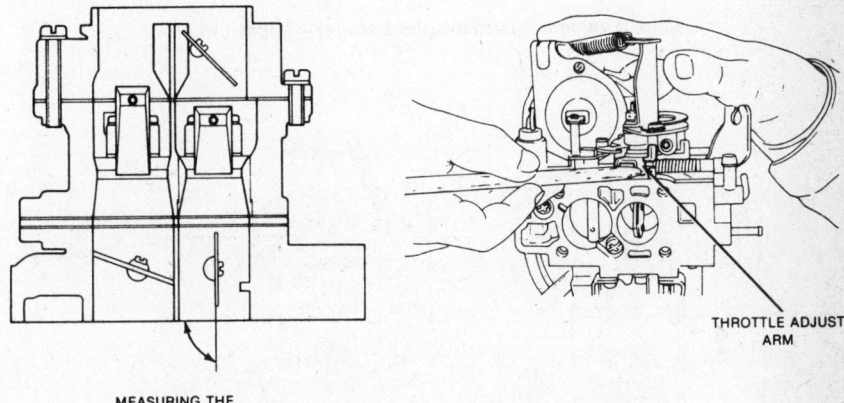

MEASURING THE ANGLE

THROTTLE ADJUST ARM

ADJUSTING

Primary throttle valve angle (full open)—Spectrum

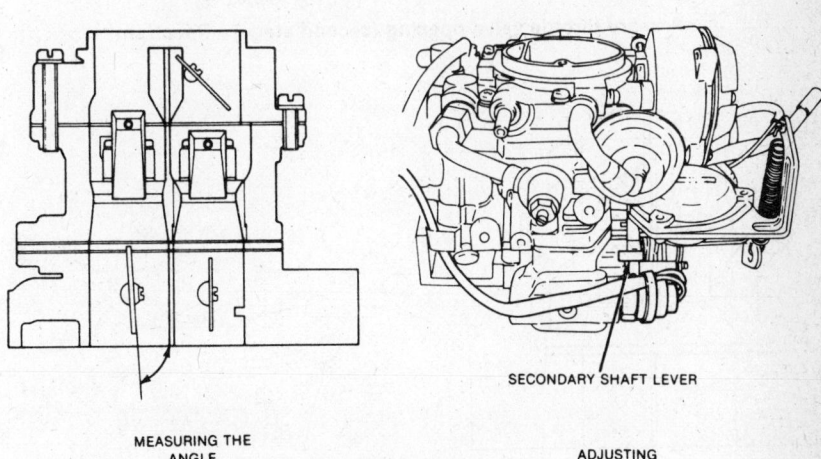

MEASURING THE ANGLE

SECONDARY SHAFT LEVER

ADJUSTING

Secondary throttle valve opening—Spectrum

b. Open the throttle valve slowly while pushing the choke valve lightly in the closing direction with your fingers and set the choke valve to the second stage of the fast idle cam.

c. The primary throttle valve clearance should be:A/T-0.692 In.,M/T-0.543 in.

d. Adjustment is made with the fast idle screw.

UNLOADER ADJUSTMENT

1. Check the clearance of the choke valve when the primary throttle valve has been fully opened. The clearance should be 0.071 in.

2. If adjustment is needed remove the rivit of the automatic choke and adjust by bending the choke lever in

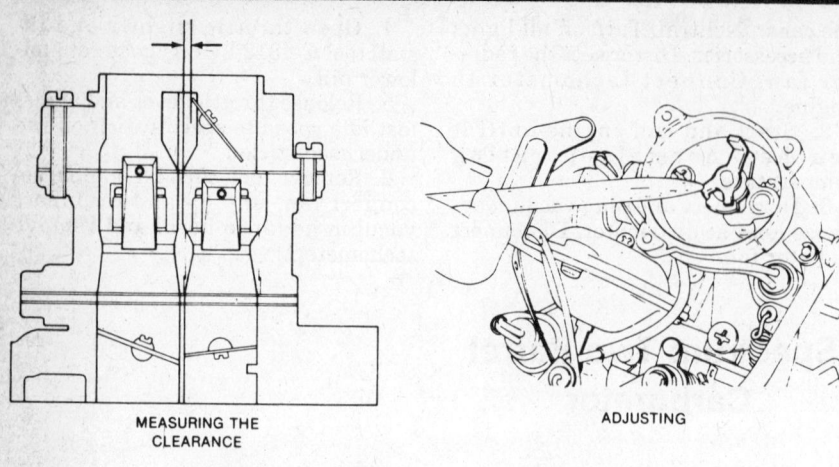

MEASURING THE CLEARANCE   ADJUSTING

Choke valve adjustment (third stage) – Spectrum

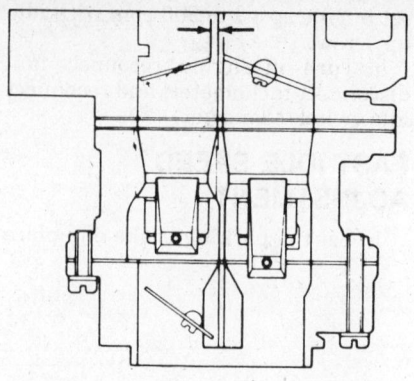

Throttle position sensor adjustment – Spectrum

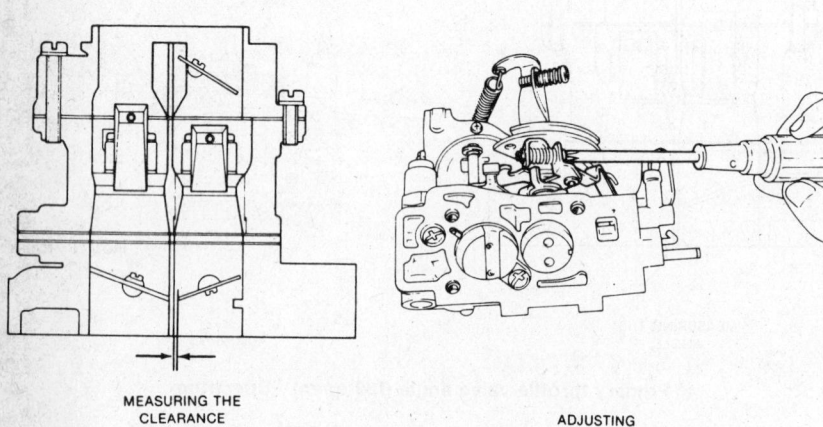

MEASURING THE CLEARANCE   ADJUSTING

Primary throttle valve opening (second stage) – Spectrum

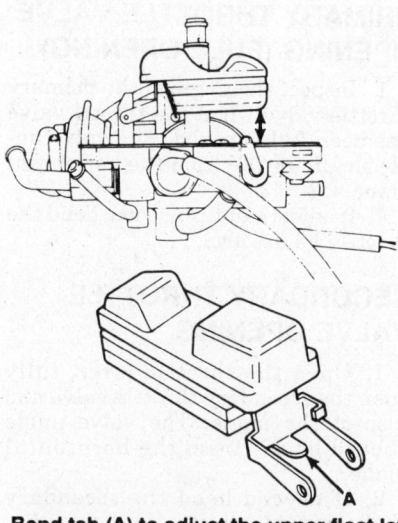

Bend tab (A) to adjust the upper float level – Spectrum

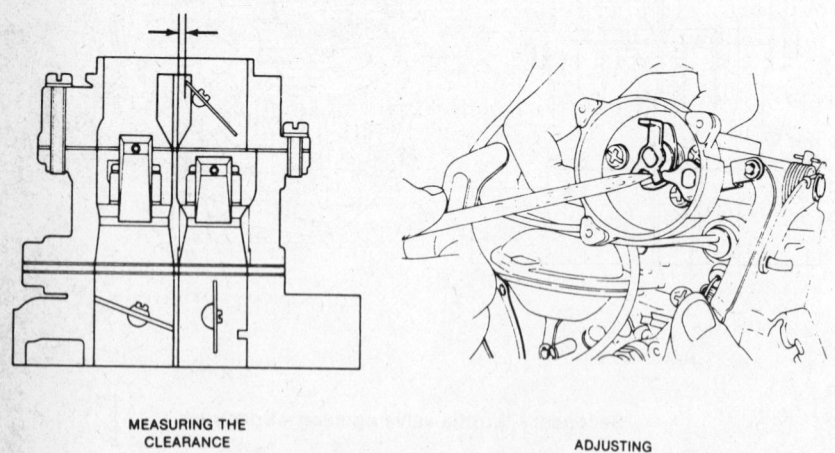

MEASURING THE CLEARANCE   ADJUSTING

Unloader adjustment – Spectrum

the housing. Reinstall the choke lever by riveting.

CHOKE BREAKER ADJUSTMENT

1. Apply a vacuum of about 400mm Hg to the choke breaker diaphragm unit.

2. Lightly push the choke valve to the closing side. The clearance should be 1985: 0.053 in., 1986: 0.057 in.

3. Adjust by bending the choke lever.

THROTTLE POSITION SENSOR (TPS) TEST AND ADJUSTMENT

NOTE: After the connection of the ohmmeter is made to the TPS, this test should be performed in as short of time as possible.

1. Check that the TPS bracket screws are tight.

2. Check that there is no play in the TPS arm and primary throttle valve arm.

3. Connect an ohmmeter to the green and black leads of the TPS.

4. Open the throttle lever about one-third (no continuity in this case) and then gradually close the lever and check that there is continuity when the primary slot valve reaches the the prescribed clearance of .015(A/T), .011(M/T).

5. Adjust by loosening the TPS screws. After adjustment check the clearance as in step 4.

SECONDARY TOUCH ANGLE

1. Measure the primary throttle

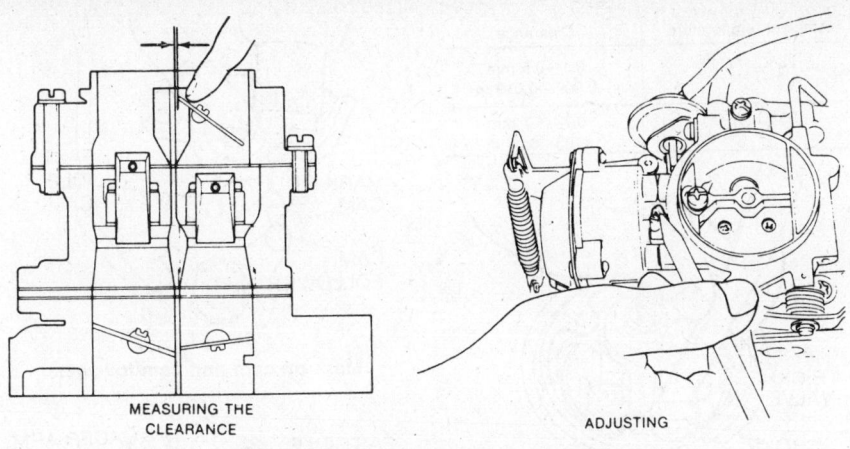

MEASURING THE CLEARANCE

ADJUSTING

Choke breaker adjustment – Spectrum

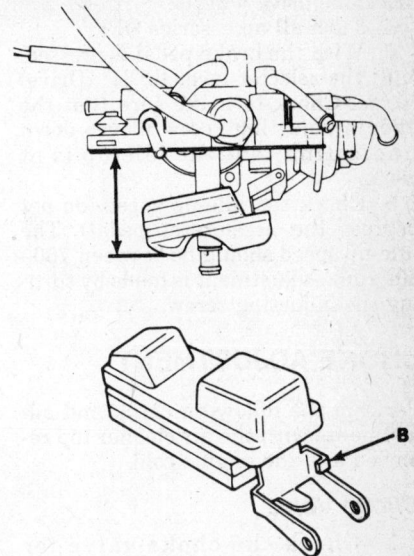

Bend tab (B) to adjust the lower float level – Spectrum

valve opening at the same time the secondary throttle valve starts to open.

2. The clearance should be 0.023 in. Adjust by bending the throttle adjusting arm.

FLOAT LEVEL ADJUSTMENT

1. Measure the clearance between

the float top and gasket when the float is in the raised position. The clearance should be 0.059 in.

2. Bend tab (A) to adjust.

NOTE: Care should be taken not to damage the needle valve when adjusting the float level.

3. Measure the clearance between the float bottom and gasket at the

lowered position of the float. the clearance should be 1.7 in. Adjust by bending (B) shown in the illustration.

Sprint Two Barrel Carburetor (MRO8)

FLOAT LEVEL ADJUSTMENT

The fuel level in the float chamber should be within the round mark at the center of the level gauge. If it is not check and adjust the float level as follows:

1. Remove and invert the air horn.
2. Measure the distance between the float and the gasketed surface of the choke chamber. The measured distance is the float level and it should be 0.21–0.24 in. The measurement should be made without the gasket on the air horn.
3. Adjustment is made by bending the tounge up and down.

IDLE-UP ADJUSTMENT

The idle-up actuator operates even when the cooling fan is running. Therefore the idle-up adjustment must be performed when the cooling fan is not running.

M/T Models

1. Warm up the engine to normal operating temperature.

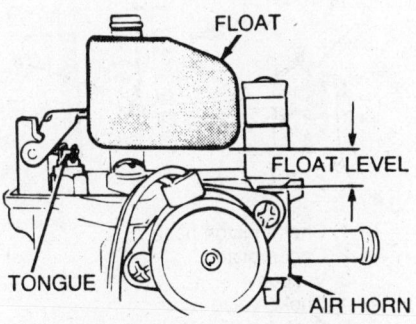

Float level adjustment – Sprint

2. After warming up, run the engine at idle speed.
3. Check to make sure that the idle-up adjusting screw moves down (indicating that the idle-up is at work) when the lights are turned ON.
4. With the lights turned ON, check the engine rpm (idle-up speed). Be sure that the heater fan, rear defogger (if equipped), engine cooling fan, and air conditioner (if equipped) are all turned OFF. The idle-up speed should be 750–850 rpm. Adjust by turning the adjusting screw.
5. After making the idle-up adjust-

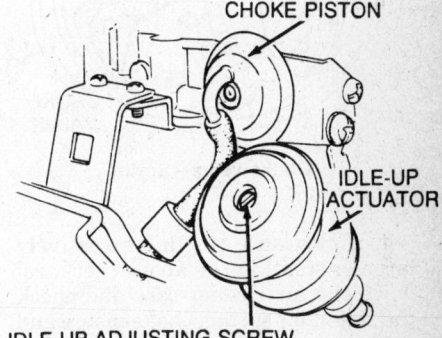

Idle-up adjusting screw – Sprint

ment, make sure the idle-up adjusting screw moves as in Step 3 when only the heater fan is operated and then only the rear defogger or engine cooling fan is operated (lights should be off).

A/T Models

1. Warm up the engine to normal operating temperature.
2. After warming up, run the engine at idle speed.
3. Apply the parking brake and

block the drive wheels.

4. Turn all accessories OFF.

5. With the brake pedal depressed, shift the selector lever to "D" (Drive) range. Check to make sure that the idle-up adjusting screw moves down (indicating that the idle-up is at work).

6. Check the idle-up speed (do not depress the accelerator pedal). The Idle-up speed should be between 700–800 rpm. Adjustment is made by turning the adjusting screw.

CHOKE ADJUSTMENT

Perform the following check and adjustments with the air cleaner top removed and the engine cold.

Choke Valve

1. Check the choke valve for smooth movement by pushing it with a finger.

2. Make sure that the choke valve is closed almost completely when ambient temperature is below 77°F and the engine is cold.

3. Check to see that the choke valve to carburetor bore clearance is within specifications when the ambient temperature is above 77°F and the engine is cool.

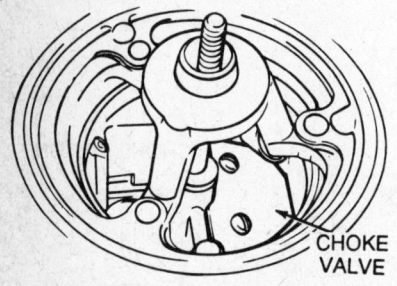

Choke valve – Sprint

4. If clearance is found excessively large or small in the above check, remove the air cleaner case and check the strangler spring, choke piston and each link in the choke system for smooth operation. Lubricate the choke valve shaft and each link with a spray lubricant if necessary. Do not remove the rivetted choke lever guide.

5. If after lubrication the clearance is still out of specification, remove the carburetor from the intake manifold and remove the idle-up actuator from the carburetor. Turn the fast idle cam counterclockwise and insert an available pin into the holes on the cam and bracket to lock the cam. In this state, bend the choke lever up or down with pliers. Bending up causes the choke valve to close, and vice versa.

Ambient temperature	Clearance
25°C (77°F)	0.1—0.5 mm 0.004—0.019 in
35°C (95°F)	0.7—1.7 mm 0.03—0.06 in

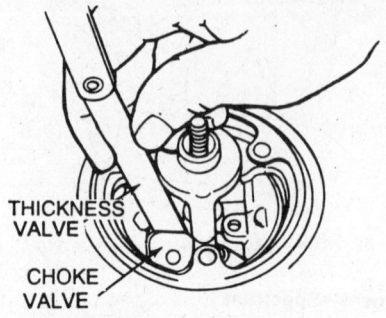

Choke valve to bore clearance – Sprint

Choke Piston

1. Disconnect the choke piston hose at the throttle chamber.

2. While lightly pushing down on the choke valve to the closing position with your finger, apply vacuum to the choke piston hose, and check to make sure that the choke valve to the carburetor bore clearance is 0.09–0.10 in.

3. With vacum applied as in step 2, move the choke piston rod with a small tool and check to see that the choke valve to carburetor bore clearance is within 0.16–0.18 1n.

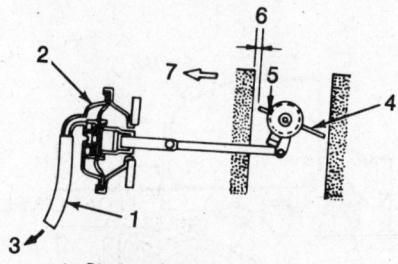

1. Choke piston hose
2. Choke piston
3. Vacuum
4. Choke valve
5. Push here lightly
6. Choke valve to bore clearance
7. Forward

Checking choke piston – Sprint

FAST IDLE CAM ADJUSTMENT

NOTE: Ambient temperature must be between 72–82°F before performing this check.

1. Drain the cooling system when the engine is cold, and remove the carburetor from the intake manifold.

2. Leave the carburetor in a place

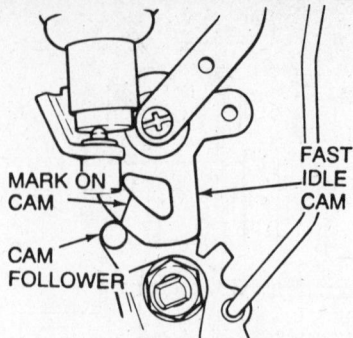

Mark on cam and cam follower

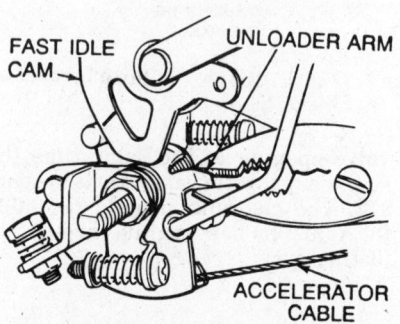

Unloader level arm – Sprint

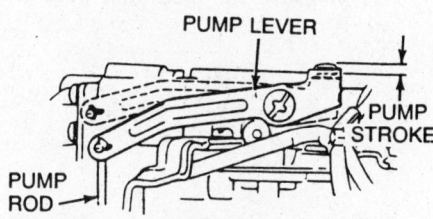

Pump stroke – Sprint

where the ambient temperature is between 72–82°F for an hour.

3. After an hour, make sure that the mark on the cam and the center of the cam follower are in alignment.

UNLOADER ADJUSTMENT

NOTE: Perform this check and adjustment when the engine is cool.

1. Remove the air cleaner cover.

2. Make sure that the choke valve is closed.

3. Fully open the throttle valve and check the choke valve to carburetor bore clearance is within 0.10–0.12 in.

4. If the clearance is out of specification adjust by bending the unloader arm.

PUMP STROKE ADJUSTMENT

1. Warm up the engine to normal operating temperature.

2. Stop the engine and remove the air cleaner.

3. Depress the accelerator pedal all the way from idle position to wide open throttle and take the measurement of the pump stroke. The pump stroke should be 0.16–0.18 in. If out of specification check the pump lever and pump rod for smooth movement.

Nova Two Barrel Carburetor

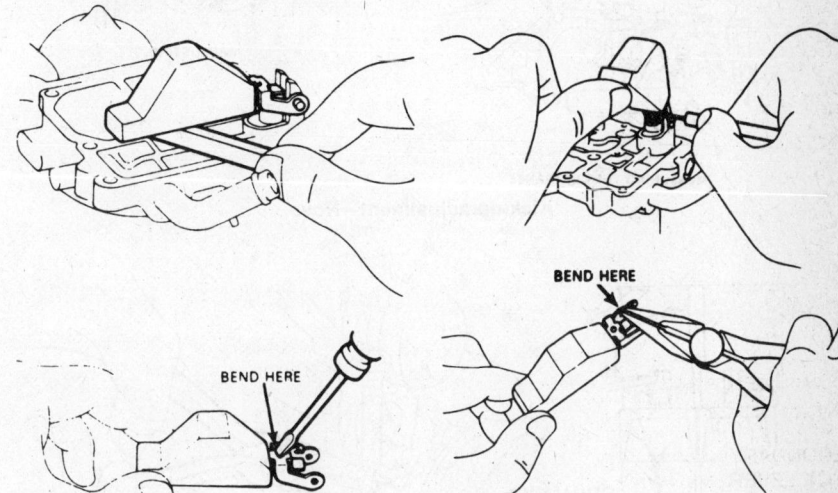

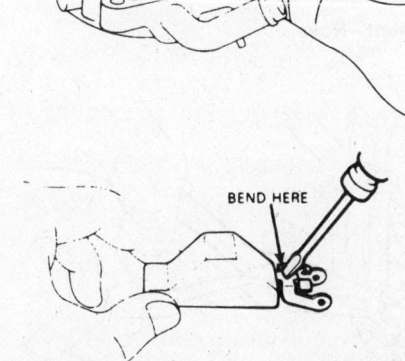

Checking the float level in the upper position – Nova

Checking the float level in the lower position – Nova

FLOAT ADJUSTMENT

1. Allow the float the hang down by its own weight. Check the clearance between the float tip and air horn. The float level should be 0.075 in.

NOTE: This measurement should be made without a gasket on the air horn.

2. Adjust by bending a portion of the float lip.

3. Lift up the float and check the clearance between the needle valve plunger and the float lip. The float level in the lowered position should be 0.0657–0.0783 in.

4. Adjust by bending a portion of the float lip.

THROTTLE VALVE OPENING

1. Check the full opening angle of the primary throttle valve, with a T scale. The standard angle should be 90° from the horizontal plane.

2. Adjust by bending the 1st throttle lever stopper.

3. Check the full opening clearance between the secondary throttle valve and the body. The standard clearance should be 0.500 in.

4. Adjust by bending the secondary throttle lever stopper.

KICK-UP ADJUSTMENT

1. With the primary throttle valve fully opened, check the clearance between the secondary throttle valve and the body. The clearance should be 0.006 in.

2. Adjust by bending the secondary throttle lever.

SECONDARY TOUCH ADJUSTMENT

1. Check the primary throttle valve opening clearance at the same time the 1st kick lever just touches the 2nd kick lever. The clearance should be 1985: 0.170 in., 1986: 0.230 in.

2. Adjust by bending the 1st kick lever.

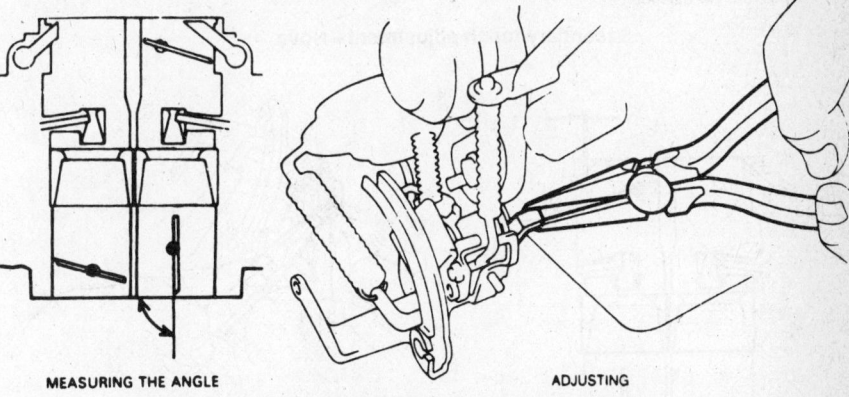

Primary throttle valve adjustment – Nova

UNLOADER ADJUSTMENT

1. With the primary throttle valve fully opened, check that the choke valve clearance is 0.120 in.

2. Adjust by bending the fast idle lever.

CHOKE BREAKER ADJUSTMENT

1. Set the idle cam. While holding the throttle slightly open, push the choke valve closed, and hold it closed as you release the throttle valve.

2. Apply vacuum to the choke breaker 1st diaphragm.

3. Check the choke valve clearance. It should be 0.095 in.

4. Adjust by bending the relief lever.

5. Apply vacuum to choke diaphragms 1st and 2nd.

6. Check the choke valve clearance. It should be 0.245 in.

7. Adjust by turning the diaphragm adjusting screw.

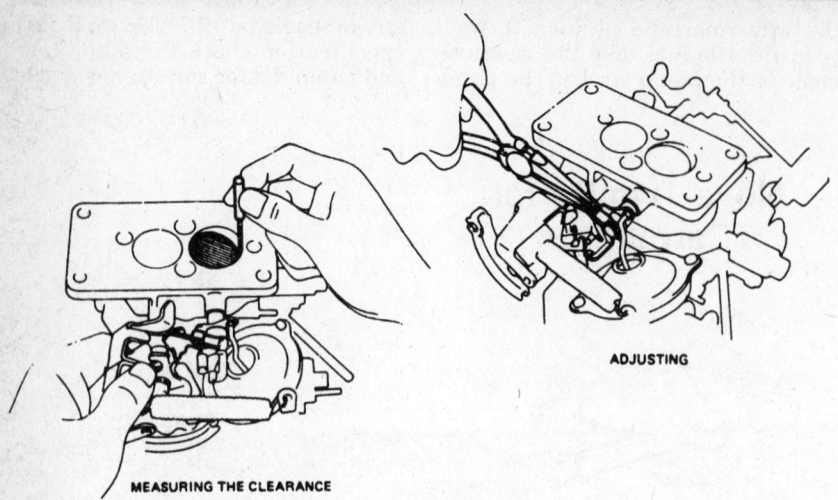

ADJUSTING

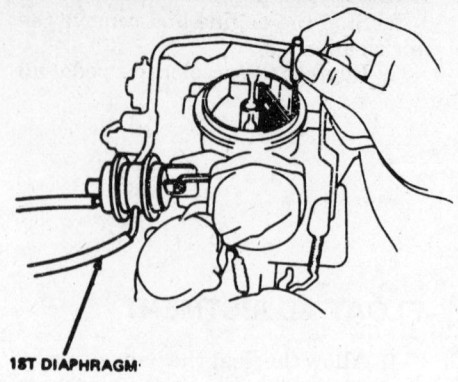

1ST DIAPHRAGM

MEASURING THE CLEARANCE

Kick-up adjustment — Nova

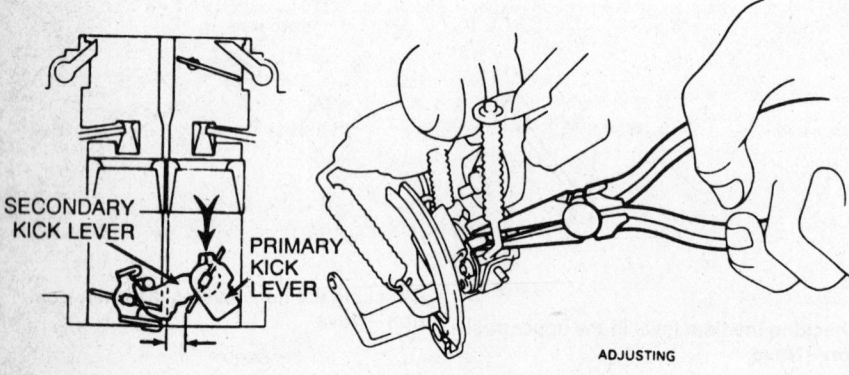

SECONDARY KICK LEVER

PRIMARY KICK LEVER

ADJUSTING

MEASURING THE CLEARANCE

Secondary touch adjustment — Nova

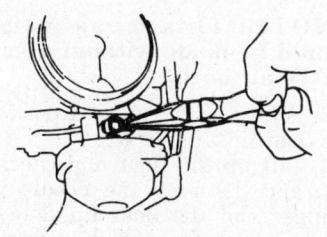

ADJUSTING

Choke breaker 1st diaphragm adjustment — Nova

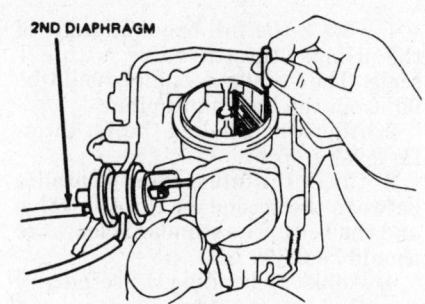

2ND DIAPHRAGM

MEASURING THE CLEARANCE

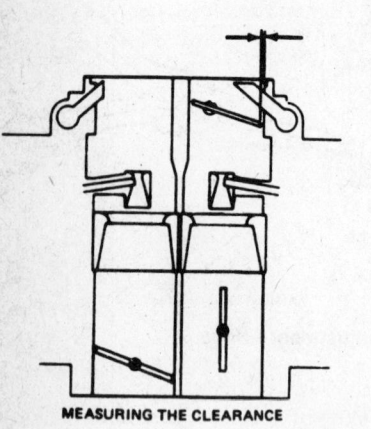

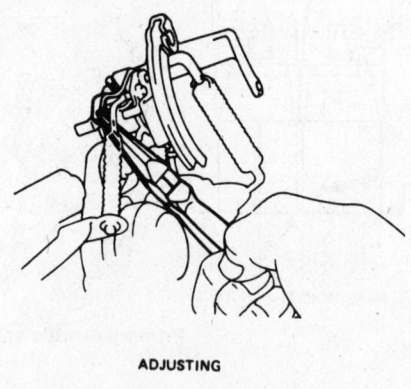

ADJUSTING

MEASURING THE CLEARANCE

Unloader adjustment — Nova

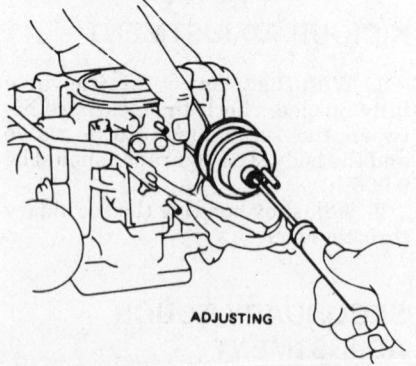

ADJUSTING

PUMP STROKE ADJUSTMENT

1. With the choke fully opened, measure the length of the stroke. 1985: 0.157 in., 1986: 0.079 in.

2. Adjust the pump stroke by bending the connecting link.

Choke breaker 1st and 2nd diaphragm adjustment — Nova

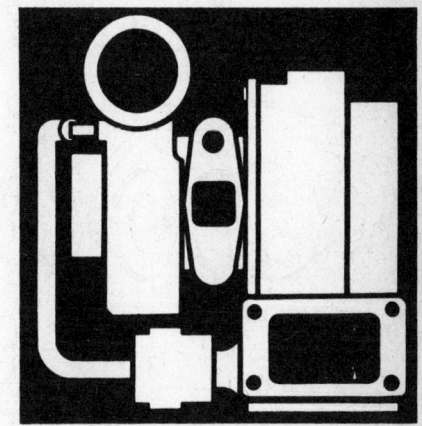

Turbocharging

Theory

The internal combustion engine can be thought of as an air pump. The action of the pistons moving down or up in their cylinders when the intake or exhaust valves are open alternately draws air and fuel into the engine or expels burnt gases into the atmosphere. The amount of air and fuel pulled into the engine (known as an engine's volumetric efficiency) is governed by the drawing efficiency of the piston as it descends in its cylinder, and by the scavenging effect of the exiting exhaust gases, which act to pull additional air/fuel mixture in through the open intake valves during valve overlap periods. The more air and fuel each cylinder pulls in, the more power the engine will produce.

Theoretically, a normally aspirated engine should be able to draw in an amount of air and fuel equal to its displacement (e.g. a 350 cu in. engine should draw in 350 cu in. of air and fuel). In practice, however, only about 80% of the displacement capacity is drawn through because of flow restrictions, the slight pressure drop through the carburetor, and the inability of the exhaust stroke to drive out all of the burnt gases.

There are several ways to increase an engine's drawing power (volumetric efficiency). These include increasing valve overlap, increasing engine bore and/or stroke, supercharging the engine, or (the most popular approach) turbocharging.

In effect, the turbocharger is an air pump which crams more air/fuel mixture into the cylinders than they could possibly draw in by themselves.

In doing so, the turbocharger increases the engine's volumetric efficiency past its normal 80%, which proportionately increases engine horsepower and torque output.

Perhaps the most advantageous aspect of the turbocharger is that it does not require usable engine horsepower to operate. By comparison, say a car is climbing a steep hill and the driver decides to turn on the air conditioner. The moment the air conditioner is turned on, a power drain on the engine can usually be felt. That's because some of the power that was being used to drive the car up the hill is now being used to turn the air conditioner compressor. A turbocharger, on the other hand, does not drain power from the engine to operate because it uses the free energy of the exhaust gases as they are blown out of the en-

The COMPRESSOR is a centrifugal, radial outflow type. It comprises a cast compressor wheel, backplate assembly, and specially-designed housing that encloses the wheel and directs the air/fuel mixture through the compressor.

The CENTER HOUSING supports the compressor and turbine wheel shaft in bearings which contain oil holes for directing lubrication to the bearing bores and shaft journals.

The OUTLET ELBOW ASSEMBLY contains the WASTEGATE ASSEMBLY, or bypass valve, which allows a portion of the exhaust gas to bypass the turbine wheel so boost pressure can be controlled.

The ACTUATOR is a spring-loaded diaphragm device that senses the outlet pressure of the compressor.

The TURBINE is a centripetal, radial inflow type. It comprises a cast turbine wheel, wheel shroud, and specially-designed housing that encloses the wheel and directs the exhaust gas through the turbine.

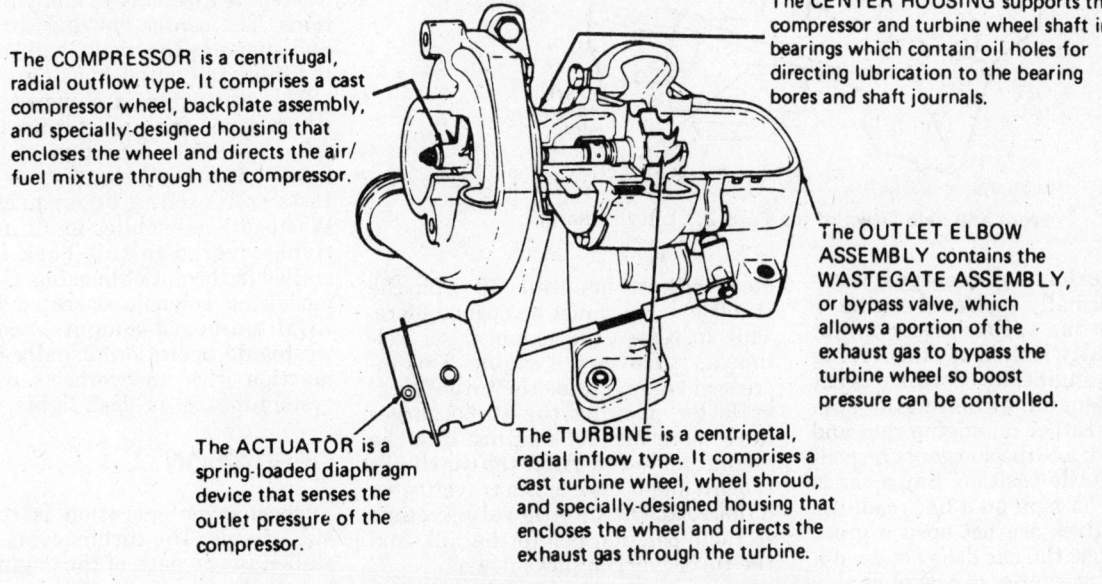

Turbocharger components, typical of all models

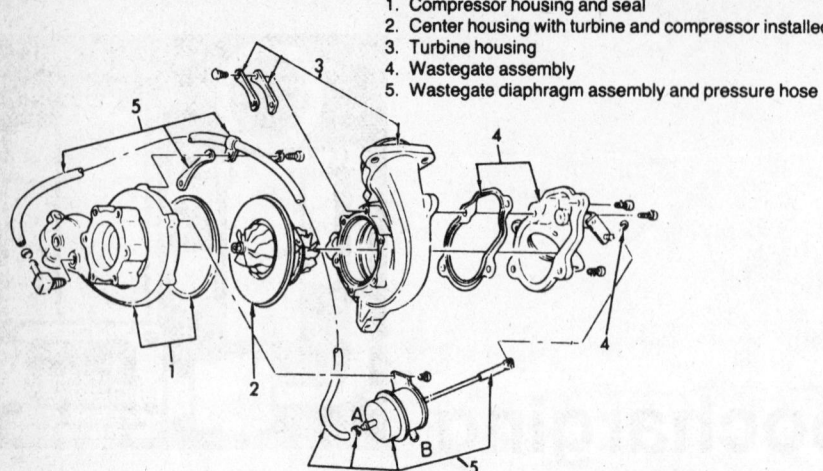

1. Compressor housing and seal
2. Center housing with turbine and compressor installed
3. Turbine housing
4. Wastegate assembly
5. Wastegate diaphragm assembly and pressure hose

Typical GM 3.8L (231 cu In.) engine turbocharger. "A" is pressure side of wastegate diaphragm, "B" is vacuum side

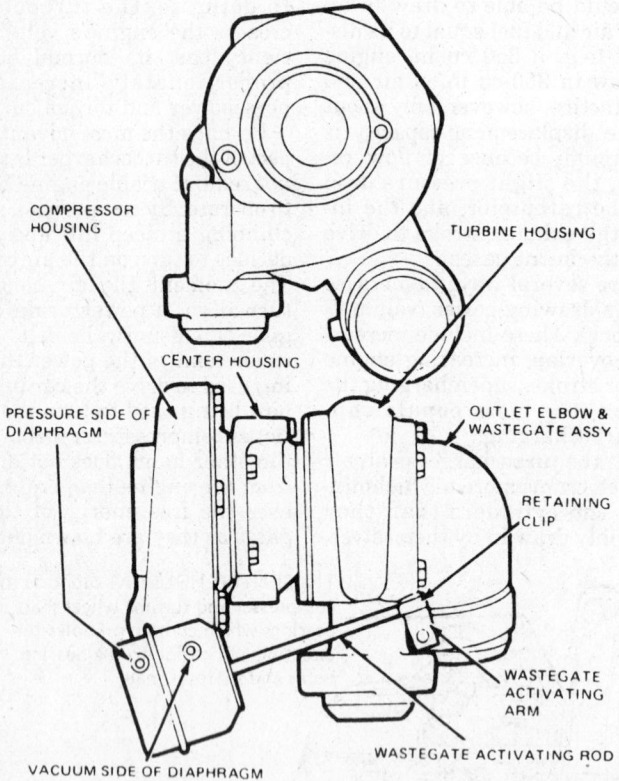

COMPRESSOR HOUSING

TURBINE HOUSING

CENTER HOUSING

PRESSURE SIDE OF DIAPHRAGM

OUTLET ELBOW & WASTEGATE ASSY

RETAINING CLIP

WASTEGATE ACTIVATING ARM

WASTEGATE ACTIVATING ROD

VACUUM SIDE OF DIAPHRAGM

Front and side views of the Ford 2.3 L turbocharger

gine. This exhaust gas energy is wasted on a normally aspirated engine.

Because the turbocharger is not mechanically linked to the driving part of the engine, its operation is not directly dependent on engine rpm alone, but rather on engine rpm and engine load: a turbocharger is responsive to throttle position. Say a car is driving at 55 mph on a flat road: the throttle valves are not open a great deal, because the car does not need a great deal of energy to travel at this speed. Soon the car starts to climb a

steep hill: to maintain 55 mph the throttle valves must be opened more. This increases the exhaust gas volume as it leaves the engine. This increased volume spins the turbocharger faster, making the turbocharger force more air/fuel mixture into the engine, and so on. After the car climbs the hill and is once again traveling on a flat road, the throttle valves return to their position before the hill, and the turbocharger slows down.

An adequate supply of clean engine oil is essential for cooling and lubrica-

tion and to maintain the turbocharger bearing assembly. The turbocharger wheels routinely operate at 130,000–140,000 rpm during boost and any interruption in the oil supply to the bearing assembly can result in major turbocharger damage. Contamination of the engine oil can also cause serious damage. Any time a basic engine bearing (main, connecting rod or camshaft) is replaced due to damage, the oil and oil filter must be changed and the turbocharger flushed with clean engine oil to remove any contamination. In addition, any time the turbocharger is removed for service or as part of another procedure, the oil and oil filter should be changed. When first starting the engine after removing the turbocharger, fill the turbocharger oil passage with clean engine oil and crank the engine a few times to allow oil pressure to build up. It's also a good idea to allow the engine to idle for one minute before shutting it off, especially when running at freeway speeds for long periods of time, to prevent the possibility of turbocharger bearing damage due to sudden oil starvation.

COMPONENTS

The turbocharger unit consists of two vaned wheels (compressor and turbine) connected by a common axle (shaft), and a housing which can be sub-divided into three sections: inlet (or compressor), center, and outlet (or turbine). The inlet housing surrounds the compressor wheel, and connects to the air intake and the intake manifold. The outlet housing surrounds the turbine wheel, and connects to the exhaust system; it also houses the wastegate assembly in many installations. The center housing surrounds and supports the shaft, and connects the inlet and outlet housings.

The wastegate is a bypass valve, which opens at a predetermined pressure. It shunts a portion of the exhaust gas around the turbine wheel, thus controlling boost pressure. Wastegate assemblies in all installations covered in this book are installed in the outlet housing. On some models, a solenoid operated by the ECM (on-board computer) controls wastegate operation, usually in conjunction with an overboost warning system (buzzer or dash light).

OPERATION

Turbocharger operation is remarkably simple. The turbine wheel is installed in the path of the engine's exhaust gas, and the compressor wheel is installed in the intake path. Exhaust gas is directed through the turbine housing, causing the turbine wheel to spin. This spinning motion is transferred by the connecting shaft to the compressor wheel. As the compressor wheel spins, it packs the intake charge into a dense mass, which is fed into the engine. Combustion converts the charge into exhaust. The exhaust charge is directed through the turbine housing, where it spins the turbine wheel, and then out through the turbine housing discharge into the exhaust system.

Thus, turbocharger operation is self-perpetuating. However, unchecked turbocharger operation will increase compressor pressure (called boost pressure) beyond the design limits of the engine, and will seriously damage internal engine components. Boost pressure is controlled by the wastegate. When boost pressure rises to a predetermined value, the wastegate opens, bypassing exhaust flow around the turbine.

Greater volumetric efficiency is a benefit of the turbocharging process, but increased cylinder pressure is a drawback, because it raises the engine's octane requirement. The two are inseparable, so a method must be devised to compensate for the increased octane requirement to avoid detonation (spark knock). Water injection, alcohol injection, low boost pressures, charge intercoolers, ignition spark retardation, and alcohol fuels have all been used to control detonation, with varying degrees of success.

Ford controls detonation by limiting boost and by spark retardation. Wastegate operation begins at five psi, and enough exhaust gas is routed around the turbine to limit boost to a maximum of six psi. The electronic ignition system has been modified in the turbocharged engine to include two spark retardation points. When boost pressure reaches approximately one-half to one psi, a switch in the intake manifold sends a signal to the ignition module, which retards ignition timing six degrees. A second manifold switch sends its signal when boost reaches four psi, resulting in an additional six degrees of retard.

The General Motors system of detonation control is slightly different. Boost is limited to a maximum of approximately six psi. In addition, a detonation sensor is installed in the engine block (V6) or intake manifold (V8). Vibrations caused by detonation are transmitted to the sensor, which sends a signal to the Electronic Spark Control (ESC) module. The module processes this signal, and sends a command signal to the HEI distribu-

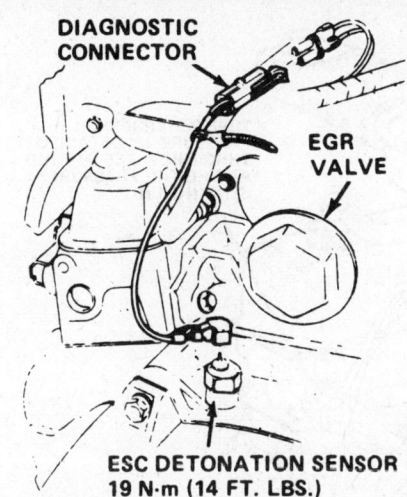

Buick 231 V6 (3.8 L) detonation sensor installation

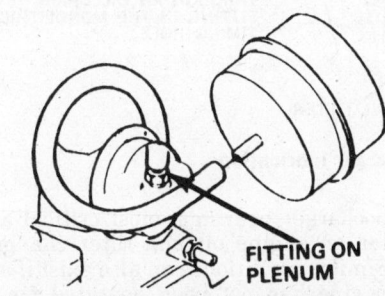

New type GM wastegate diaphragm uses plenum vacuum only—1981 and later Buick Regal unit shown

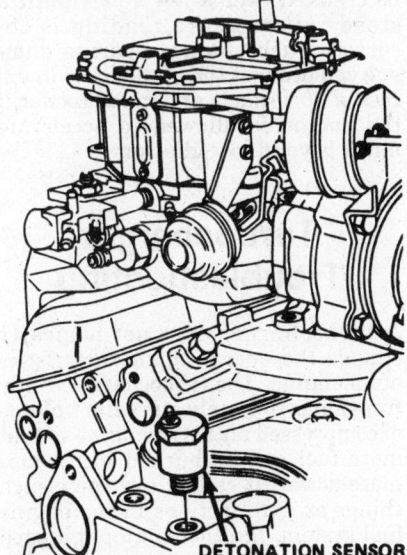

Pontiac turbocharged V8 detonation sensor location

tor to retard timing. Timing retard ranges up to 22° on V6s, or 15° on V8s.

LUBRICATION

The turbocharger shaft spins in bearings lubricated by engine oil. Turbine

speeds routinely reach 120,000–140,000 rpm, making an adequate and well-filtered oil supply critical for proper operation. Any interruption or contamination of the oil supply will result in engine damage as well. Ford cautions that accelerating the engine to top rpm immediately after starting can result in engine and turbocharger damage (due to the lack of oil pressure). Immediately shutting down the engine after it has been operated at high rpm for an extended period can also result in turbocharger damage, since oil pressure will be shut off, but the turbine will continue to spin for a few moments. Shutting the throttle abruptly when the engine is at high speed can also cause extensive damage, but for a different reason: sudden closed throttle operation causes the mixture to become very lean, resulting in detonation, high engine temperature, and consequent damage.

General Motors recommends the following procedure before starting the engine when changing the oil and filter, or performing any operation which results in oil drainage or loss:

1. Disconnect the ignition switch connector (pink wire) from the HEI distributor module.
2. Crank the engine several times until the oil light goes out. Do not crank the engine for more than thirty seconds at a time to avoid starter damage.
3. Reconnect the pink wire. Start the engine.

Turbocharger Maintenance

Proper maintenance is important, particularly regarding air and oil filtration, to maximize the service life and performance of the turbocharger. Experience has shown that the main cause of turbocharger failure is due to oil lag, restriction or lack of oil flow and dirt in the oil. The second principle cause of failure is foreign objects entering the compressor and/or turbine wheels.

AIR INTAKE SYSTEM

Dust or sand entering the turbocharger compressor housing from a leaky air inlet system can seriously erode the compressor wheel blades and will result in deterioration of turbocharger and engine performance. The wearing away of the blades, if uneven, can induce shaft motion which will pound out the turbocharger shaft bearings. Ingestion of sand or dust will also cause excessive wear on engine parts, such as pistons, rings, valves, etc.

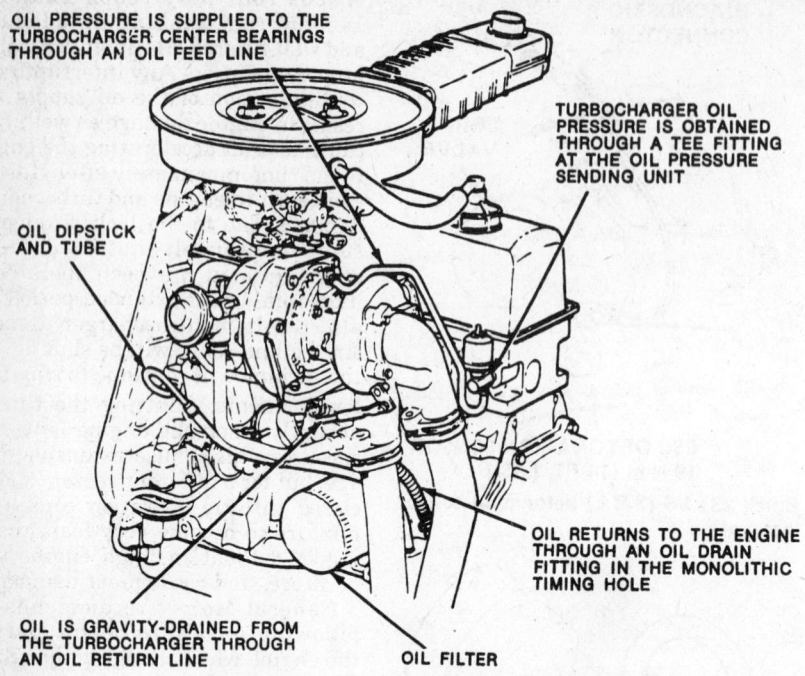

OIL PRESSURE IS SUPPLIED TO THE TURBOCHARGER CENTER BEARINGS THROUGH AN OIL FEED LINE

TURBOCHARGER OIL PRESSURE IS OBTAINED THROUGH A TEE FITTING AT THE OIL PRESSURE SENDING UNIT

OIL DIPSTICK AND TUBE

OIL RETURNS TO THE ENGINE THROUGH AN OIL DRAIN FITTING IN THE MONOLITHIC TIMING HOLE

OIL IS GRAVITY-DRAINED FROM THE TURBOCHARGER THROUGH AN OIL RETURN LINE

OIL FILTER

Ford 2.3 L turbocharger lubrication

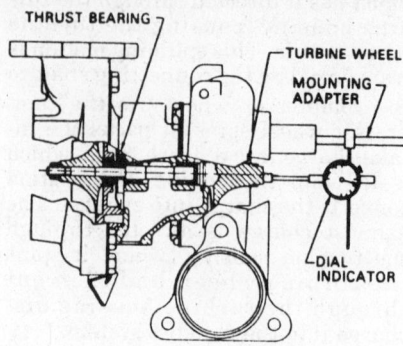

THRUST BEARING — TURBINE WHEEL — MOUNTING ADAPTER — DIAL INDICATOR

Thrust bearing clearance measurement

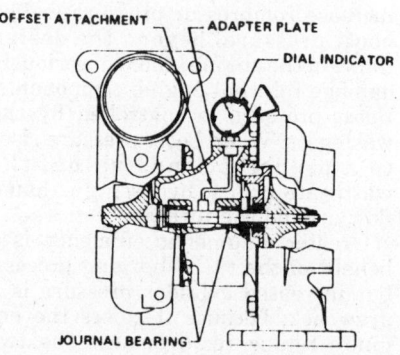

OFFSET ATTACHMENT — ADAPTER PLATE — DIAL INDICATOR — JOURNAL BEARING

Journal bearing clearance measurement

Plugged or restricted air cleaner systems (due to neglected air filter changes) will reduce air pressure and volume at the compressor air inlet and cause the turbocharger to lose performance. The restricted air cleaner and the resultant air pressure drop between cleaner and turbocharger can, during engine idle periods, cause oil pullover at the compressor end of the turbocharger and result in an oil leak at the seal.

LUBRICATION SYSTEM

Dirt or foreign material, when introduced into the turbocharger bearing system by the lube oil, causes wear on the center housing bearing bore surfaces. Contaminents act as abrasives and will eventually cause the shaft hub and either or both wheels to rub on the housings, causing the rotating assembly to turn slower. Engine power loss, excessive smoke, excessive noise and appearance of oil at either or both ends of the turbocharger could be noted. Contaminated and dirty oil problems can be eliminated by regular oil and filter changes.

A turbocharger should never be operated under engine load conditions with less than 30 psi oil pressure. The turbocharger is much more sensitive to a limited oil supply than an engine, due to the high rotational speed of the shaft and relatively small area of the bearing surfaces. Oil pressure and flow lag during engine starting can have a detrimental effect on the tur-

bocharger bearings, most critical after an engine oil and filter change. Similar conditions can also exist if an engine has not been operated for a long period of time, since engine lube systems tend to bleed down. Before allowing the engine to start, it should be cranked over a few times until a steady oil pressure reading is observed. Turbocharger bearing damage can occur if the oil delay is in excess of 30 seconds and much sooner if the engine is allowed to accelerate much beyond low idle rpm.

Turbocharger Troubleshooting

A turbocharger does not basically change the operating characteristics of an engine. The turbocharger's only function is to supply a greater volume of compressed air to the engine so that more fuel can be burned to produce more power. It cannot overcome such things as malfunctions in the engine fuel system, ignition timing, plugged air cleaner elements, etc. If a turbocharged engine system has malfunctioned and the turbocharger has been inspected and determined to be functioning normally, proceed with troubleshooting as though the engine were naturally aspirated (non-turbocharged). Simply replacing a good turbocharger with another will not correct engine deficiencies. Always inspect and asses turbocharger condi-

tion before removing it from the engine as follows:

1. Remove the inlet and exhaust ducts from the turbocharger.

2. Inspect both turbocharger wheels for blade damage caused by foreign material entering the turbocharger. The wheels can be visually checked by simply looking through the compressor housing inlet opening while holding the the throttle blade open. A light is necessary when examining the turbine wheel blade tips since they are positioned inside the turbine housing. Look between the turbine wheel blades from the exhaust outlet end of the turbine housing.

3. Inspect the outer blade tip edges on both wheels adjacent to their respective housing bores and check for wheel rub.

4. Rotate the shaft wheel assembly by hand and feel for drag or binding conditions. Push the shaft to one side, rotate it and feel for rub. It should turn smoothly.

5. Lift both ends of the shaft up and down at the same time and feel for excessive journal bearing clearance. If clearance is normal, very little shaft movement will be detected. Actual shaft end play can be measured with a dial indicator without removing the turbocharger from the engine.

6. If the shaft assembly rotates

freely and no wheel damage, binding or rub has been noted, it can be assumed that the turbocharger is not in need of service.

─────── CAUTION ───────

Operation of the turbocharger without all normally installed inlet ducts and filters connected can result in personal injury and equipment damage from foreign objects entering the turbocharger.

TESTING WASTEGATE OPERATION

As noted before, the wastegate is a safety valve for the engine. If the wastegate sticks shut, boost pressure will build until the air/fuel mixture charge becomes too powerful for the mechanical components (pistons, bearings, etc.) and causes engine damage.

If the wastegate sticks open, little or no boost will be received from the turbocharger, which translates into mediocre engine performance. The simplest wastegate test is to remove the pressure hose at the wastegate diaphragm unit, connect a pressure pump (such as the type used for cooling system testing) and apply pressure. At the specified opening pressure (7 psi for Ford, 8.5–9.5 for GM), the link between the wastegate and its diaphragm unit will just move (about .015 in). The movement is not great, but it should be easy to see.

If the wastegate does not move, try to operate the linkage by hand. It should move under moderate hand pressure. If it moves, the problem is probably in the diaphragm unit (broken diaphragm). To test the diaphragm, remove the vacuum hose from the diaphragm, hook up a manual vacuum pump and apply 25 in. Hg of vacuum to the diaphragm unit. If the vacuum drops below 18 in. Hg within one minute, replace the diaphragm unit.

NOTE: Some 1981 and later GM turbos have a new type of diaphragm which opens the wastegate during idle and part throttle, when there's no boost, to reduce engine backpressure and improve fuel economy. To test this type of unit, apply about 20 in. Hg of vacuum to the diaphragm unit: the wastegate link should move slightly. This unit operates solely with plenum vacuum and can be identified by the absence of a boost pressure signal line on the diaphragm unit.

TESTING OPERATION OF GM DETONATION SENSOR

Connect a tachometer and timing light to the engine, run the engine at 1800–2500 rpm and tap on the intake manifold next to the detonation sensor.

NOTE: Be careful to keep all wires, clothing and tools away from moving engine parts.

Rap continuously, quickly and moderately hard. This should trigger the detonation sensor. When it triggers, engine speed should drop at least 200 rpm and timing should retard at least 4°, probably more.

TURBOCHARGER TROUBLESHOOTING

Problem	Cause	How To Check	Solution
No boost	Gasket leak, hole in exhaust system	Temporarily block tailpipe with engine running. Any exhaust leaks in the system will be heard.	Repair leaks (usually at gasket surfaces)
	Dirty air filter	Remove air filter and check	Replace or clean filter
	Blocked air intake	Visually inspect for blockage	Clear intake
	Worn valves or rings	Compression test engine	Repair
	Throttle valves not opening completely	Manually operate throttle linkage, check valve movement	Adjust linkage, repair carburetor
	Exhaust blockage	Check catalytic converter for melted and blocked catalyst, check muffler and exhaust pipes for debris	Replace catalytic converter, repair exhaust system
	Wastegate stuck open	Test wastegate operation	Repair or replace wastegate assembly
Fuel odor under boost	Leak at compressor or intake manifold	Look for fuel stains at fittings	Tighten fittings or replace gaskets
Ignition miss at high speed, under load	Spark plug gap too large	Remove spark plugs, measure gap	Reduce gap
	Faulty coil	Test Coil	Replace
Ignition miss (often)	Excessive resistance in ignition cables	Check cable resistance (see Tune-Up Unit Repair section)	Replace cables as necessary
Oil leaks into turbine	Blocked oil return hose	Remove hose and check for blockage or crimps	Repair or replace hose

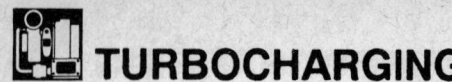

TURBOCHARGER TROUBLESHOOTING

Problem	Cause	How To Check	Solution
Detonation	Fuel octane rating too low	Check octane rating of fuel used against that recommended by manufacturer (consult owner's manual)	Switch to higher octane unleaded fuel
	Faulty sensor	Check G.M. as instructed here; have Ford system checked by qualified technician	Replace as necessary
	Faulty ignition retard unit	Refer to qualified technician	Repair or replace as necessary
	Engine overheating	Check coolant level, debris clogged radiator, no coolant circulation, blocked thermostat	Repair or replace as necessary
Poor idle	Air leak between compressor and carburetor	Listen at joints for hissing sound while the engine idles	Repair

Emission Controls

EMISSION CONTROLS

Emission control devices are designed to eliminate the chemical compounds that escape from the engine crankcase, from the exhaust and from evaporation of fuel out of the tank and carburetor. With the growing use of onboard computers, it has become possible for car manufacturers to meet strict Federal emission standards by using electronic engine controls to monitor operating conditions and adjust engine calibrations for the best possible performance and economy with minimum emissions.

Engine calibration has a big effect on emissions out the tailpipe. The calibration consists of spark timing, fuel mixture, choke setting, idle speed and spark plug gap. Calibrations are not a service problem as long as the engine is adjusted to the factory specifications, which are found on a sticker in the engine compartment. Engines must be adjusted to these factory specifications or emissions will be high. Additionally, emission control systems have become such an integral part of the overall engine design that best engine performance is dependent on best emission control system performance. This is especially true for computer controlled systems.

NOTE: Any attempt to disconnect or bypass any OEM emission device is a violation of federal law.

The latest emission control systems use electronic instead of vacuum devices and are much more sensitive to malfunctions in any component. Following is a description of each group of controls and how they work to reduce emissions. Due to the complex nature of modern electronic engine control systems, comprehensive diagnosis and testing procedures fall outside the confines of this repair manual. For complete information on diagnosis, testing and repair procedures concerning all modern engine and emission control systems, please refer to *Chilton's Guide To Electronic Engine Controls.*

Emission Service Indicators

RESET PROCEDURES

Indicator lights or flags will periodically appear on or near instrument cluster to alert the driver that various emission control components need to be serviced or replaced. Most of the reminder lights are triggered at preset mileages programmed into either mechanical or electronic counter or odometer switches. The mechanical counter switches are normally operated by the speedometer cable, while the electronic counter switches are pulsed by a speed sensor usually located in the speedometer assembly. After servicing the indicated emission system (EGR valve, Oxygen sensor, etc.). the service indicator device will have to be reset to eliminate the light or flag. Follow the appropriate procedure outlined below for each year and model listed.

American Motors

The reset switch is located under the hood on the left side of the firewall, between the upper and lower speedometer cables. There is a rest screw on the unit that must be rotated one-quarter turn to the detent position.

Chrysler Corporation

1980 MODELS

Chrysler models use either electronic or mechanical service counters. The electronic switch is located under the instrument panel somewhere near the lower left instrument cluster. It is usually covered with a green plastic case.

To reset the mechanical switch, remove it from the mounting bracket and then remove the plastic case. Insert a small screwdriver or rod into the hole in the switch body to close the contacts and turn off the indicator light. To reset the mechanical switch, first locate the unit between the upper and lower speedometer cables. Turn the screw on the upper side of the switch to reset.

General Motors

1980 CADILLAC

To reset the switch, remove the lower steering column cover and locate the reset cable at the lower left side of the speedometer cluster. Pull the cable lightly to reset the switch, then replace the column cover. Do not pull hard on the cable or damage to the cable and switch will occur.

1980 AND LATER MODELS

An emission indicator flag will appear in the odometer window when service is necessary on some 1980 and later GM models. The flags are marked SENSOR, EMISSION, or

FLAG WINDOW IN SPEEDOMETER FACE

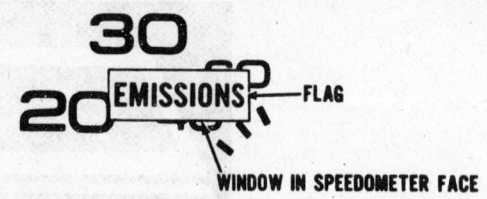

RESETTING FLAG WITH DOWNWARD MOVEMENT

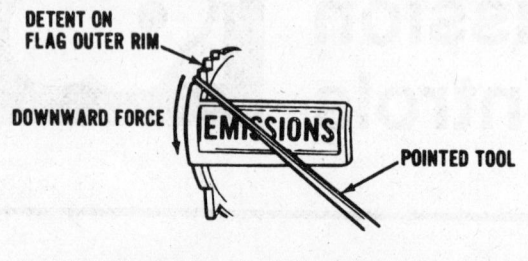

FLAG IN RESET POSITION

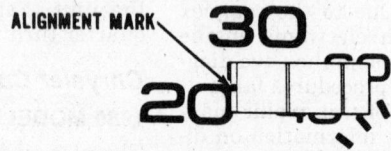

Resetting Emissions flag on 1980 GM models—typical

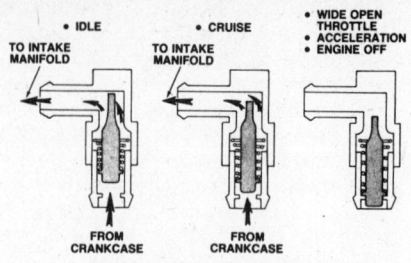

PCV valve operation

CATALYST, depending on the device that is schedule for regular maintenance. To reset the flag, first remove the instrument panel trim plate and instrument cluster cover lens. There are reset notches on the driver's side of the indicator flag.

Insert a long, pointed probe diagonally into the detents on the upper left side and rotate the flag downward until an alignment mark becomes visible in the left side of the odometer window. Once the flag has been lowered, replace the cluster lens and trim plate.

CRANKCASE CONTROLS

PCV System

Ventilation of a crankcase is necessary because of the compression blowby past and piston rings. This blow-by is mostly unburned gasoline. If allowed to stay in the crankcase, it dilutes the oil and increases engine wear. The PCV system uses engine vacuum to draw out the crankcase fumes. The crankcase or the rocker arm cover is connected by a hose to engine vacuum at the intake manifold or carburetor. When the engine is running, the crankcase fumes are drawn into the engine and burned in the combustion chamber. Fresh air enters the crankcase through the oil filler cap on the open system. When the oil filler cap is connected to the air cleaner, it is known as a closed system.

At wide open throttle, there is little vacuum in the engine, so the PCV system doesn't pull any fumes out of the crankcase. Because the hose connection from the crankcase to the intake manifold acts like a vacuum leak, there has to to be some kind of control to limit the air flow. The PCV valve is the control. It can be an actual valve, with an internal plunger, or a simple orifice without any moving parts. In the plunger types, a spring moves the plunger against engine vacuum, allowing less flow at high vacuum and more flow at low vacuum. In the event of a backfire, the plunger moves to close the PCV valve and prevent a possible crankcase explosion.

Fresh air enters the air cleaner and goes through a hose to the crankcase or rocker cover. The fumes exit the crankcase and enter the intake manifold, either through a hose or some other type of connection, usually with a PCV valve controlling the flow. Most systems use some kind of PCV filter, usually mounted at the end of the hose in the air cleaner. The filter keeps dust from entering the crankcase and also prevents oil fumes from ruining the air cleaner element.

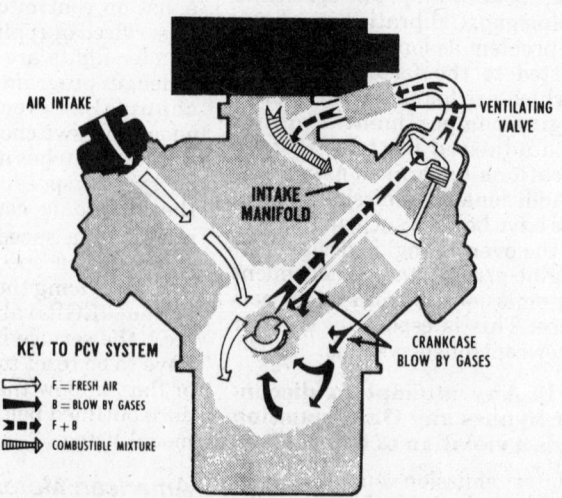

KEY TO PCV SYSTEM

F = FRESH AIR
B = BLOW BY GASES
F + B
COMBUSTIBLE MIXTURE

Typical open crankcase ventilation system

GM diesel V8 engines are equipped with one of two different crankcase ventilation systems. The first system uses a crankcase vacuum regulator valve to meter the flow of crankcase gases back into the engine. The regulator limits crankcase vacuum as the gases are drawn from the valve covers through the regulator, and into the air crossover. This sytem is used on 1981 and later non-California models. Other models use a crankcase flow control valve to meter the blow-by gases back into the engine. On these models, a ventilation hose runs from each valve cover and connects the flow control valve, which is screwed into the back of the air crossover.

TESTING PCV SYSTEMS

NOTE: Do not attempt to test the crankcase controls on GM V8 diesels. Instead, clean the valve cover filter assemblies and vent pipes and check rubber fittings every 15,000 miles, and replace or clean the breather cap assembly and ventilation regulator valve (if equipped) every 30,000 miles.

Checking crankcase vacuum is the most effective way to test any PCV sytem. If there is a vacuum in the crankcase, then the major part of the system has to be working. Inspect the system to find out where the fresh air enters the engine. This is usually through a hose attached to the air cleaner, but is may be through the oil filler cap on some models. If the fresh air entry is separate from the oil filler cap, simply remove the cap.

On all models, use a piece of paper or a PCV tester to measure the crankcase vacuum at the oil filler cap, with the cap removed and the engine idling in Park or Neutral. It may take a few seconds for the vacuum to build up enough to suck the piece of paper against the oil filler hole. If the vacuum does not build up, check to be sure you have plugged the fresh air entry. An alternate method on some cars is to use the piece of paper or PCV tester on the end of the fresh air entry hose. When you do it that way, the oil filler cap must be the solid type and you must leave it in place.

If there is no crankcase vacuum, pull the PCV valve from the crankcase and hold your finger over the end of it. You should feel full manifold vacuum with the engine idling. If not, the valve is plugged or there is an obstruction in a hose or passageway. On some designs the valve may be screwed into its mounting, with a hose leading to the rocker cover or crankcase. If the valve has good suc-

tion, but there is no crankcase vacuum, check the hose to be sure it is open. PCV valves that are restricted or plugged must be replaced, unless they are the type that will come apart for cleaning. Lack of crankcase vacuum can also be caused by vacuum leaks at rocker cover, oil pan, or other engine gaskets. Usually, tightening the bolts will stop the leak.

In some extreme cases, usually on high mileage engines, the PCV system is in good shape, but the blow-by past the rings is so much that the system can't handle it, and the engine will blow smoke out the oil filler hole. Switching to a PCV valve with a higher flow may temporarily correct the problem. But the only good solution is to overhaul the engine. If the motor oil is contaminated with gasoline, the PCV system will pick up the unburned vapors, add them to the intake mixture and cause the engine to run excessively rich. After checking crankcase vacuum, always check the condition of the fresh air filter and hose, to be sure they are clean and not clogged.

NOTE: The PCV system operation is not computer controlled, but if inoperative it will directly affect the operation of any computerized emission system. If poor performance is a problem, check the PCV system first.

running, a hose to the intake manifold or carburetor base allows engine vacuum to pull fresh air through the canister, drawing the vapors into the engine where they are burned. Fresh air enters the canister through a filter, which keeps the charcoal clean.

When the engine is running, air must enter the tank to replace the fuel that is used up and prevent a vacuum. On all makes of canister storage models, air enters the tank through the filter in the canister, but air can also enter the tank through the pressure-vacuum tank cap.

All evaporation control systems use some sort of vapor separator at the fuel tank to prevent liquid fuel from traveling along the vent line to the canister. The early models had very elaborate separators mounted separately from the tank, but now they are simpler and usually attached to the top of the tank. The only periodic servicing required on evaporation controls is replacement of the canister filter on those models on which it is replaceable.

NOTE: If the vent lines become blocked, it is possible for some evaporation control systems to pull liquid fuel from the tank into the charcoal canister. If any charcoal canister is found to be fuel-soaked it should be replaced and all hoses checked for obstructions.

FUEL EVAPORATION CONTROLS

Charcoal Canister Vapor System

Evaporation controls are made up of hoses which allow the tank and carburetor vapors to go to a canister filled with charcoal. When the engine is

EXHAUST CONTROLS

Thermostatic Air Cleaner (TAC)

Fresh air supplied to the air cleaner comes either from the normal snorkle, or from a tube connected to an exhaust manifold stove. A door in the snorkle regulates the source of incoming air so that a warm engine always

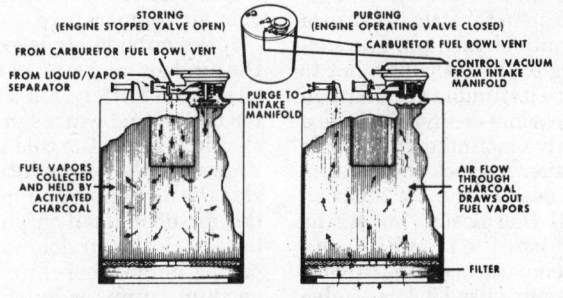

Vapor storage canister operation—typical

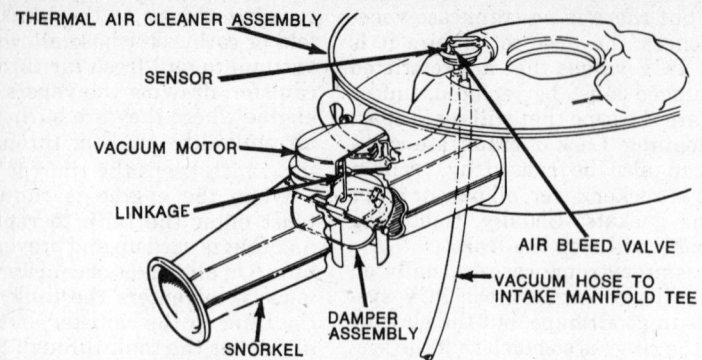

THERMAL AIR CLEANER ASSEMBLY
SENSOR
VACUUM MOTOR
LINKAGE
AIR BLEED VALVE
VACUUM HOSE TO INTAKE MANIFOLD TEE
DAMPER ASSEMBLY
SNORKEL

Vacuum controlled thermostatic air cleaner.

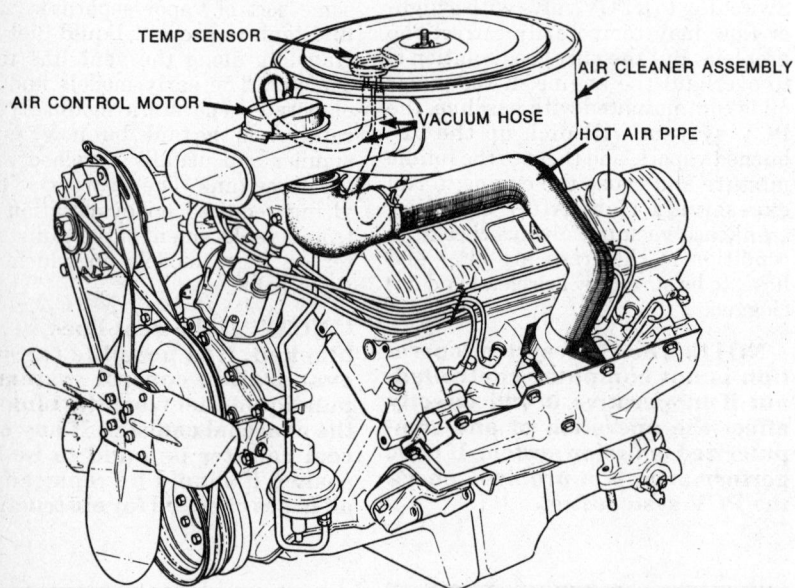

TEMP SENSOR
AIR CONTROL MOTOR
AIR CLEANER ASSEMBLY
VACUUM HOSE
HOT AIR PIPE

A typical heated air cleaner system, with the hot air pipe connected to the left exhaust manifold

takes in warm air, approximately 100°F. The snorkel door may be controlled in any number of ways, but most are vacuum operated. The vacuum operated designs use a thermostatic bimetal switch inside the air cleaner that bleeds off vacuum as the engine warms up and regulates the position of the air door.

Vacuum operated air doors are all designed so that the air cleaner takes in cold air when there is no vacuum. This means that an air door in the hot air position will switch to the cold position at wide open throttle because of the loss of manifold vacuum. The sudden switching of the door from hot to cold may cause a stumble or misfire in the engine, so some designs include a modulator valve mounted on the side of the air cleaner to block the vacuum and hold the door in the hot air position. A small thermostat inside the modulator opens it when the underhood temperatures reach normal. Other designs used a delay valve that allows the air door to move to the

cold position slowly, to prevent stumble.

TESTING TAC OPERATION

To test the vacuum type of heated air cleaner, inspect the air door with the engine off. It should be in the cold air position. Start the engine. If the engine is cold, the air door should move to the hot air position. As the engine warms up, the air door should move to a mid position, depending on the outside air temperature.

If the outside air is extremely cold, the air door may stay in the hot air position indefinitely. On a warm day, after the engine warms up, the air door should move to the cold air position. If it doesn't, the temperature sensor inside the air cleaner might be faulty, or the air door itself might be hanging up. Check the air door (a small mirror can be helpful here) by using a hand vacuum pump, or by running a hose from manifold vacuum to the vacuum

motor. Connect and disconnect the hose to see if the air door moves freely. If the air door is free, check out the hoses for leaks or blockage. If the hoses are okay, the trouble must be in the temperature sensor, and it should be replaced.

Both General Motors and Ford use a modulator in the air cleaner vacuum line on some engines. The modulator mounts on the side of the air cleaner and has two hose connections, one to the air cleaner temperature sensor, and the other to the vacuum motor. Below 50–80°F. the modulator is a one-way check valve, which allows vacuum to move the air door to the hot air position, but traps the vacuum so the door will not jump back to the cold air position during acceleration. This prevent a stumble.

After the module warms up the check valve unseats so that the vacuum can pass freely in either direction, and the air door then operates normally. The connections from the modulator are important. The connection in the center (usually the larger diameter) goes to the vacuum motor, and the connection on the edge goes to the vacuum source, which is the temperature sensor.

To test the modulator on a cold engine, apply enough vacuum to the edge port to move the air door to the hot position. Then remove the hose from the port, and the air door should stay in the hot position. Make the same test when the engine is warmed up, and the air door should move to the cold position when you pull off the hose.

Exhaust Gas Recirculation (EGR)

Gasoline Engines

NOx (oxides of nitrogen) is a tailpipe emission caused by the oxidation of nitrogen in the combustion chamber. When the peak combustion temperatures go over 2500°F, NOx is formed in excessive amounts. To keep the combustion temperatures down, exhaust gas is recirculated by allowing intake manifold vacuum to draw exhaust gas into the intake manifold. The lower combustion temperatures also help control spark knock (ping).

An EGR valve is used to control the flow of exhaust gas into the intake manifold. All EGR valves look similar and are operated by vacuum. When the vacuum is off, the valve is closed. Several different types of controls are used to turn the vacuum to the EGR valve on and off. Most of them have to do with engine temperature, as de-

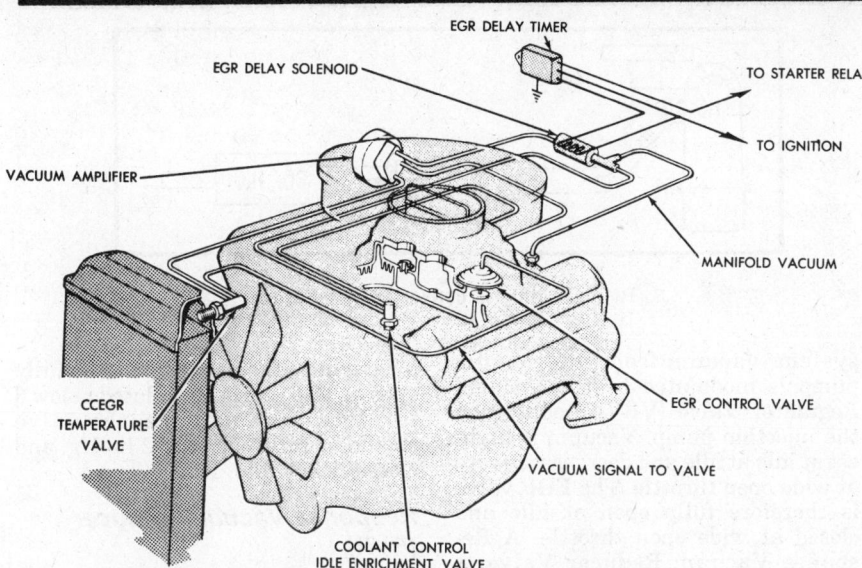

EGR DELAY TIMER

EGR DELAY SOLENOID

TO STARTER RELAY

TO IGNITION

VACUUM AMPLIFIER

MANIFOLD VACUUM

CCEGR TEMPERATURE VALVE

EGR CONTROL VALVE

VACUUM SIGNAL TO VALVE

COOLANT CONTROL IDLE ENRICHMENT VALVE

Venturi vacuum exhaust gas recirculation

scribed later. On computerized control systems, EGR operation is regulated by the electronic control unit.

NOTE: All EGR systems are designed to cut off exhaust recirculation when the engine is cold, at idle, or under hard acceleration. If the EGR valve is stuck open, the engine won't idle.

TESTING EGR SYSTEM

—————— **CAUTION** ——————
The EGR valve gets hot during normal operation. Take normal precautions to avoid accidental burns.

Testing of EGR systems should verify that when the engine is at normal operating temperature the EGR valve is closed at idle, open above idle, and that the exhaust gas is actually recirculating. If the EGR valve sticks open at idle, the engine will run very rough, or may not even start. If this happens the valve should be removed and cleaned or replaced. To check for valve operating above idle, check with a mirror to see if the diaphragm or stem moves when the engine is at fast idle in Park or Neutral. If the diaphragm does not move when the throttle is opened, there is either a problem with vacuum, or the valve is stuck closed. With a vacuum gauge hooked up to the EGR port, you should see vacuum on the gauge when the throttle is opened. EGR valves should not leak when tested with a hand vacuum pump. If they do they must be replaced.

NOTE: The EGR valve should open when about 3–5 in. Hg. is ap-

plied with a hand vacuum pump. **Back pressure operated EGR valves cannot be vacuum tested.**

To find out if the exhaust gas is actually recirculating, use a hand vacuum pump to open the EGR valve with the engine idling. If the engine runs rough or dies, the exhaust gas is recir-

culating. If the engine does not run rough, make a second test of 2500 rpm. Opening the EGR valve at that rpm should cause a change in engine speed. If it does, the exhaust gas is recirculating. To make the 2500 rpm test, remove and plug the hose from the EGR port. Attach the suction hose to the EGR valve before running the engine at 2500 rpm. Simply pulling off the EGR hose at 2500 rpm is not a valid test, because the extra air entering the engine through the hose could cause a speed change by itself.

If the exhaust is not recirculating, it means that a passageway or the valve itself is clogged up. The only way to fix it is to clean out the clogging as best you can, replace the clogged part, or replace the EGR valve. Many EGR valves have a back pressure sensor built into the valve. This sensor is a pressure operated bleed that disables the EGR valve and keeps it closed when there is no exhaust pressure. This type of valve cannot be tested with a hand vacuum pump with the engine off because the bleed is open. The only practical way to test these new valves is by substitution of a known good valve. If a valve is not available, the suspect valve can be removed, and the mounting holes temporarily taped shut. If this cor-

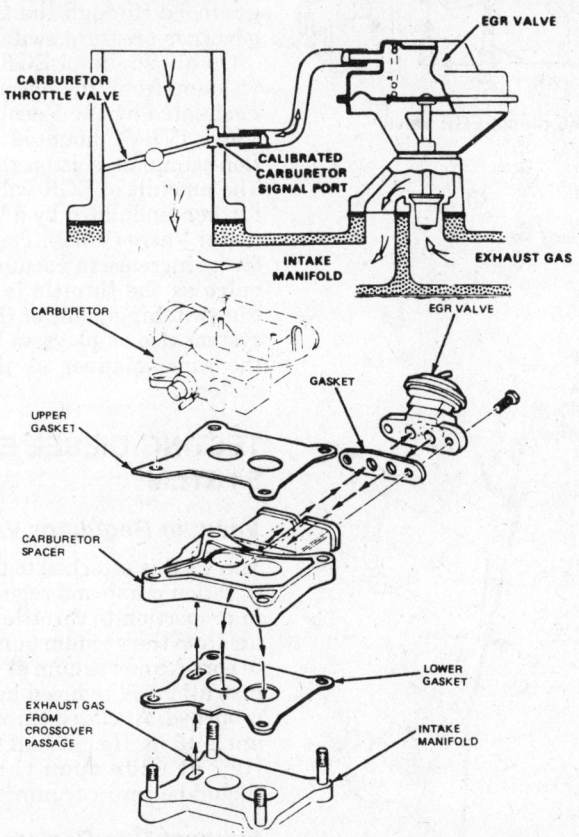

EGR VALVE

CARBURETOR THROTTLE VALVE

CALIBRATED CARBURETOR SIGNAL PORT

INTAKE MANIFOLD

EXHAUST GAS

CARBURETOR

EGR VALVE

GASKET

UPPER GASKET

CARBURETOR SPACER

LOWER GASKET

EXHAUST GAS FROM CROSSOVER PASSAGE

INTAKE MANIFOLD

Most cars use an EGR system with a valve and a ported vacuum signal, as shown here. Some cars use the venturi vacuum with a separate amplifier to operate the valve.

rects the problem, then a new valve should be installed.

Diesel EGR Systems

GM V6 and V8 Engines

GM has equipped its V8 and V6 diesel engines with EGR systems. The diesel EGR systems work in the same basic manner as gasoline engine EGR systems: exhaust gases are introduced into the combustion chambers to reduce combustion temperatures, and thus lower the formation of nitrogen oxides (NOx). There are two systems used on the V8 diesels. One is used on the B (large body) type station wagons, and one system is used on all other cars.

On the B-body station wagon EGR

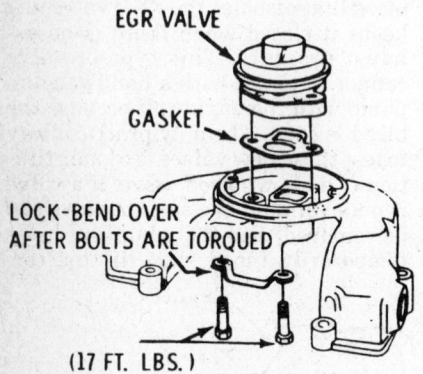

GM V6 diesel EGR valve

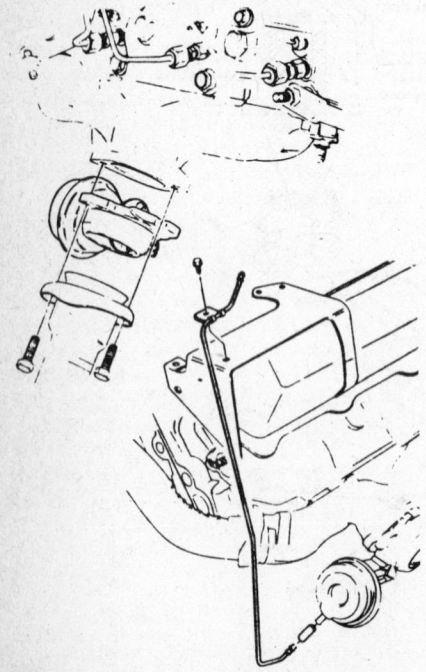

GM V6 diesel Exhaust Pressure Regulator Valve (EPR)

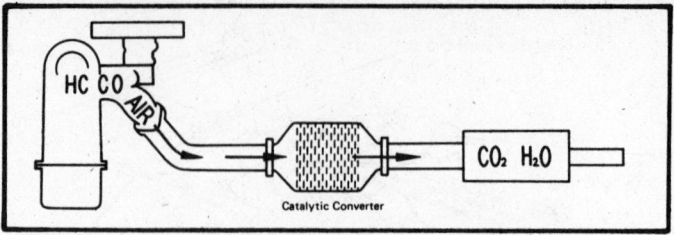

Typical catalytic converter installation

system, vacuum from the vacuum pump is modulated by the Vacuum Regulator Valve (VRV) mounted on the injection pump. Vacuum is highest at idle at idle and decreases to zero at wide open throttle. The EGR valve is therefore fully open at idle and closed at wide open throttle. A Response Vacuum Reducer Valve is used between the VRV and the EGR valve to allow the EGR valve to change position quickly as throttle position is changed.

On all other V8 diesel engines, the EGR system is the same as used on the B-body wagon, except a solenoid is added to the system that shuts off vacuum to the EGR valve when the Torque Converter Clutch is engaged. This solenoid is fed 12V from the TCC switch portion of the VRV and is grounded through the transmission's governor pressure switch.

On all V6 diesel EGR systems, the vacuum from the vacuum pump is modulated by the Vacuum Regulator Valve (VRV) mounted on the injection pump, as it is on the V8 diesels. The amount of EGR valve opening is further modulated by a Vacuum Modulator Valve (VMV). The VMV allows for an increase in vacuum to the EGR valve as the throttle is closed, up to the switching point of the VMV. The system also employs an VRV valve in the same manner as the V8 diesel system.

TESTING DIESEL EGR SYSTEM

Vacuum Regulator Valve (VRV)

The VRV is attached to the side of the injection pump and regulates vacuum in proportion to throttle angle. Vacuum from the vacuum pump is supplied to port A and vacuum at port B (see illustration) is reduced as the throttle is opened. At closed throttle the vacuum is 15 in. Hg.; at half throttle, 6 in. Hg.; at wide open throttle there should be zero vacuum.

Exhaust Gas Recirculation (EGR) VALVE

Apply vacuum to the vacuum port. On

V8 engines, the valve should be fully open at 10.5 in. Hg. and closed below 6 in. Hg. On V6 engines, the valve should be fully open at 12 in. Hg. and closed below 6 in. Hg.

Response Vacuum Reducer (RVR)

Connect a vacuum gauge to the port marked "To EGR valve or TCC solenoid". Connect a hand operated vacuum pump to the VRV port. Draw 15 in. of vacuum on the pump and the reading on the vacuum gauge should be .75 in. Hg. lower than the vacuum pump reading on all except High Altitude V8 engines. On High Altitude V8 engines ONLY, the reading should be 2.5 in. Hg. lower.

Exhaust Pressure Regulator Valve (V6 Diesels)

Apply vacuum to the vacuum port of the valve. The valve should be fully closed at 12 in. Hg. and open below 6 in. Hg.

Vacuum Modulator Valve (VMV)

To test the VMV, block the drive wheels, and apply the parking brake. With the shift lever in Park, start the engine and run at a slow idle. Connect a vacuum gauge to the hose that connects to the port marked "MAN". There should be at least 14 in. Hg. of vacuum. If not, check the vacuum pump, VRV, RVR, solenoid, and all connecting hoses. Reconnect the hose to the "MAN" port. Connect a vacuum gauge to the "DIST" port on the VMV. The vacuum reading should be 12 in. Hg. except on High Altitude cars, which should be 9 in. Hg.

Catalytic Converters

Two main types of converters are used on today's vehicles. The first is an oxidation type converter containing two precious (noble) metals, platinum and palladium to effectively catalyze the oxidation of the hydrocarbons (HC) and carbon monoxide (CO). The second type converter used is considered a three-way catalyst, containing plat-

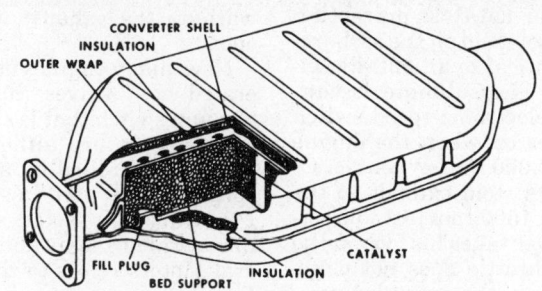

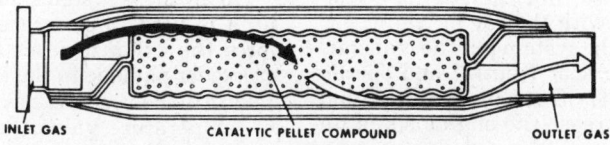

Pellet type catalytic converter

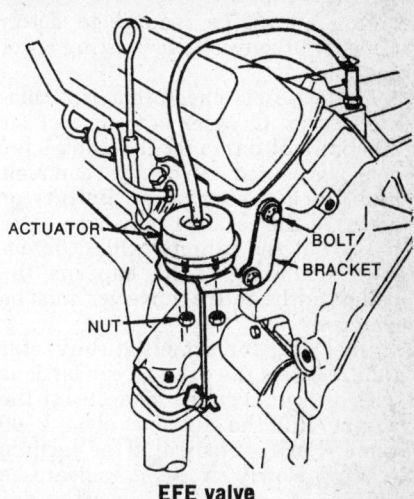

EFE valve

NOTE: Some computer controlled systems use an air management valve to increase converter efficiency by routing air to the exhaust system under certain conditions.

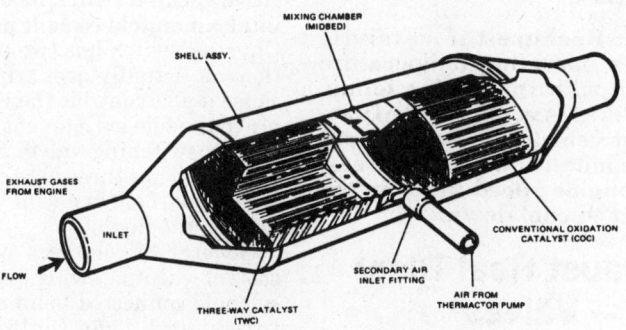

Sectional view of three-way catalytic converter

inum and rhoduim in the front part of the converter to reduce the oxides of nitrogen (NOx), while platinum and palladium are used in the rear section to oxidize the hydrocarbons (HC) and carbon monoxide (CO), as was done in the two-way converters.

Oxidizing Catalytic Converters

These converters do not operate unless there is sufficient oxygen in the exhaust stream. It is extremely important that the proper amount of oxygen is supplied at all times. This is accomplished by a secondary air source, provided by either an air pump system or a pulse air type system. The catalytic converter system is protected by several devices that block out the secondary air supply when the engine is laboring under any abnormal hot or cold operating situation, preventing converter overheating and burnout. Converter temperatures are normally between 900 and 1500°F, with peak temperatures around 1800°F, so the converterss must be hot to properly perform their functions. Should the converter be

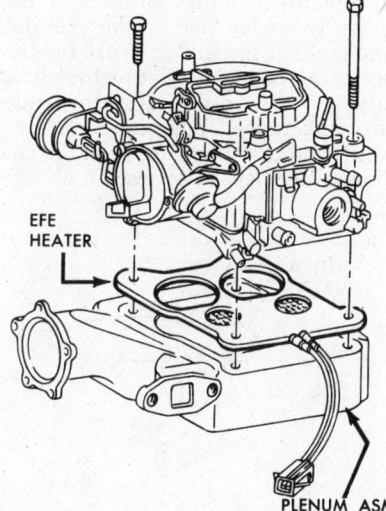

1981 G.M. electric early fuel evaporative heater

supplied too rich a mixture of hydrocarbons (HC), such as would result from a misfiring spark plug or stuck choke valve, along with an oversupply of fresh air, the converter temperature would increase sharply, causing a burnout of the catalyst material.

Three-Way Catalytic Converters

The three-way catalytic converters use a combination of catalysts which produce two different chemical reactions, oxidation and reduction. By adding fresh air to the unburned hydrocarbons (HC) and carbon monoxide (CO) within the converter, the oxidizing or combustion process takes place.

Just the reverse process is required to lower the oxides of nitrogen (NOx) emissions. The oxides of nitrogen (NOx) already contains excessive oxygen and the process of separating the excess oxygen from the nitrogen is called a reducing reaction. This reducing or reduction process is done in the front section of the converter while the oxidizing process is accomplished in the rear section. A fresh air connector is located on the center of the converter shell to add fresh air from the air system as required.

To enable the three-way converter to operate properly, the engine air/fuel ratio must be held within a tight range, called a "Stoichiometric" range. This is accomplished with the use of the closed loop, feedback fuel management systems incorporating the latest electronic controls.

TESTING

There is no way to test a catalytic converter in the field to see if it is actually working. Tailpipe readings may be used to set carburetor idle mixtures, when the car maker requires it, but taking a tailpipe reading to determine if the converter is working is not possible.

The one field check that is recommended in all cases is to inspect for mechanical damage. If a converter gets overheated, the catalyst can melt and block the exhaust. Pellets or pieces of the catalyst may even come flying out the tailpipe while the engine is running. If this happens, the pellets or the entire converter must be changed.

Checking for a melted converter that restricts the exhaust can be done with a vacuum gauge connected to the engine. Run the engine at about 2500 rpm in Park or Neutral. If the vacuum reading slowly drops, it indicates a buildup of pressure in the exhaust.

The use of leaded fuel will slowly destroy the efficiency of the catalyst. If used long enough, leaded fuel can even cause catalyst plugging to the point where the engine will not run. If you know that a car has been run on several tanks or leaded fuel, then you can be sure that the catalyst is ruined. The only thing you can do is change the catalyst or install a new converter.

NOTE: Do not change the catalyst if the car has been run on only one tank or less of leaded fuel. Switching back to unleaded will allow the catalyst to recover and be almost as efficient as it was.

CONVERTER OVERHEAT PROTECTION

Some cars have overheat protection systems for the converter. Ford Motor Co. sometimes uses a heat sensitive switch mounted in the floorpan above the converter. The switch turns a vacuum to the air pump bypass valve. When the vacuum is shut off the bypass valve dumps the pump air into the atmosphere, diverting the air away from the exhaust system. Without the air in the exhaust, the converter's catalytic heat reaction slows and the system cools down.

Chrysler Corporation cars use an overheat protection system that holds the throttle open to prevent high speed closed throttle deceleration. Any engine decelerating on closed throttle is usually running rich, because the high vacuum pulls so much fuel out of the carburetor bowl through the idle circuit. This rich mixture can cause the catalytic reaction to speed up, increasing the heat generated to dangerous levels. Holding the throttle open slightly while decelerating allows more air into the engine and eliminates the problem.

The Chrysler catalyst protection system uses a solenoid on the carburetor that is identical to an anti-dieseling solenoid. The solenoid is controlled by an electronic speed switch and only comes on when the engine speed is above 2000 rpm. When the solenoid is on, its stem extends to the equivalent of a 1500 rpm fast idle setting. If the driver takes his foot off the throttle, the throttle does not close, but rests against the extended solenoid stem. The solenoid goes off below 2000 rpm so that the engine doesn't run away with the car in traffic.

To test the system put the transmission in Park or Neutral and operate the throttle from under the hood. Slowly increase the engine speed until it is above 2000 rpm. The solenoid stem should extend. As the speed drops below 2000 rpm, the stem should retract

NOTE: Because the catalytic converter operating temperature increases with the engine idling, DO NOT allow any catalyst-equipped vehicle to idle for more than five minutes without increasing the engine speed to allow the converter to cool down.

Exhaust Heat Riser Valves

Exhaust heat riser valves have been used for many years to force part of the engine exhaust through a passageway under the intake manifold and preheat the fuel mixture to allow better atomization of the fuel droplets. The heat valve was spring loaded into the closed position, but heat would make the spring relax so that during high speed operation or after warmup the exhaust would push it open.

Now, many engines use vacuum-operated heat valves, controlled by a vacuum switch that is sensitive to engine temperature (although the above system, controlled by a thermostatic spring, is still used in some engines). Ford calls their system simply a vacuum operated exhaust heat valve. General Motors refers to theirs as Early Fuel Evaporation, and Chrysler calls theirs a Power Heat Control Valve.

On all these systems, manifold vacuum is used to close the valve, and force the exhaust gases through the crossover passage in the intake manifold. All the systems have some kind of temperature valve that shuts the vacuum off when the engine warms up. Both Chrysler and Ford products use a simple coolant temperature-sensitive vacuum switch mounted on the intake manifold coolant passage. The Chrysler switch has two hose connections. It actually does triple duty because it also controls the vacuum supply to the idle enrichment system and the air switching valve. Ford's vacuum switch has three hose connections, but one of them is a vent with a filter to keep the dirt out.

General Motors cars use either a coolant vacuum switch, or a vacuum solenoid connected to an oil temperature switch. The coolant vacuum switch has two hose connections and a vent when it controls the heat valve only. When it is tied into other emission control systems, it can have as many as five hose connections, and a vent. Many General Motors cars also have a check valve in the hose so that vacuum will be trapped in the heat valve actuator when the engine is accelerated. This keeps the heat valve

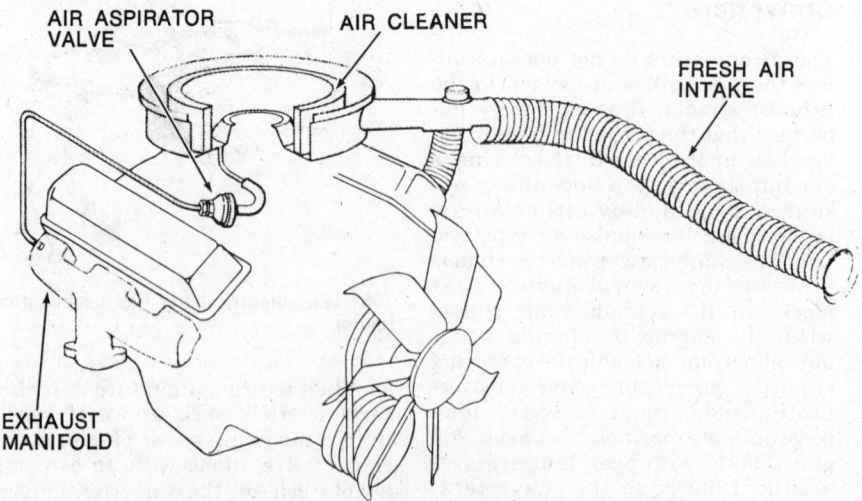

Chrysler Air Aspirator system

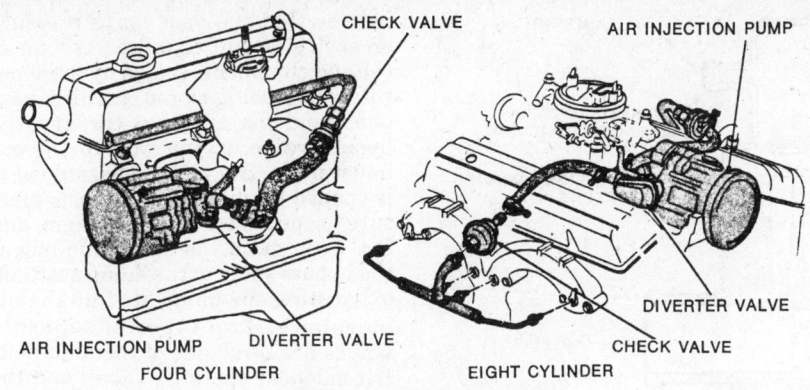

Chevrolet air pump system

CHECK VALVE
AIR INJECTION PUMP
AIR INJECTION PUMP
DIVERTER VALVE
FOUR CYLINDER
DIVERTER VALVE
CHECK VALVE
EIGHT CYLINDER

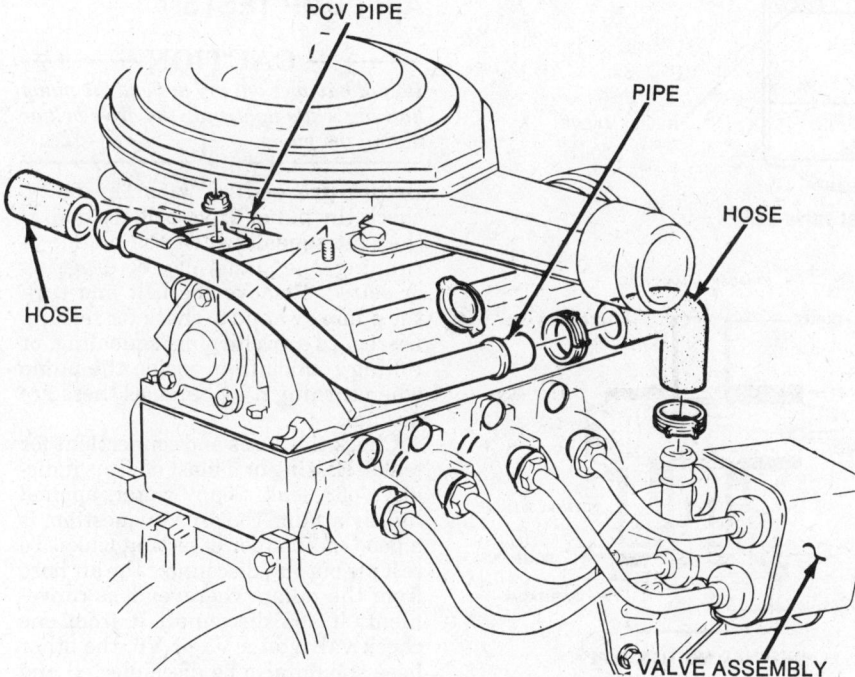

PCV PIPE
PIPE
HOSE
HOSE
HOSE
VALVE ASSEMBLY

Chevette Pulse Air pipe and hose

TESTING EFE HEATER

To check the resistance of the heater, turn the ignition OFF, disconnect the heater electrical connector, using a ohmmeter, measure the resistance across the two terminals of the heater connector. If resistance is under 2 ohms, the heater is good. If not, replace the heater.

Air Injection Systems

On these systems, a belt-driven air pump supplies air to small tubes positioned in the exhaust port near each exhaust valve. The air mixes with any unburned hydrocarbons in the exhaust and the hydrocarbons burn up in the exhaust system. On late model engines, air may not be pumped to every exhaust port, and some engines have only a single air injection fitting on the exhaust pipe near its connection to the exhaust manifold. Air injection systems are frequently used on engines with catalytic converters so that the converter gets enough air to keep the reaction going.

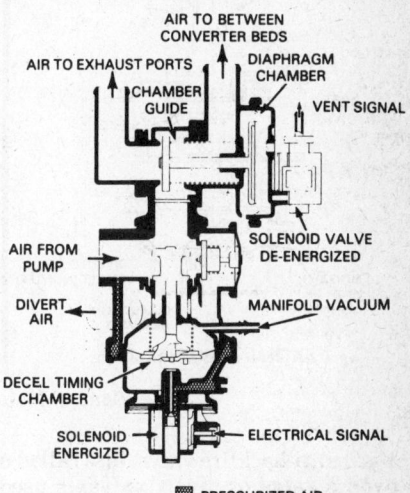

AIR TO BETWEEN
CONVERTER BEDS
AIR TO EXHAUST PORTS
DIAPHRAGM CHAMBER
CHAMBER GUIDE
VENT SIGNAL
AIR FROM PUMP
SOLENOID VALVE DE-ENERGIZED
DIVERT AIR
MANIFOLD VACUUM
DECEL TIMING CHAMBER
SOLENOID ENERGIZED
ELECTRICAL SIGNAL

▓ PRESSURIZED AIR

AIR SWITCHING VALVE CROSS SECTION
(AIR TO CONVERTER OPERATION)

Typical GM air switching valve (ASV) assembly

in the closed position and prevents a rattle.

TESTING

Testing the vacuum operated heat riser valve is a matter of making sure it closes and opens freely. You can move it by hand to see if it works, on a warm engine. On a cold engine, the valve should be closed, and disconnecting the hose should allow it to open (engine idling). On a cold engine, there should be vacuum at the vacuum actuator, and on a warm engine the vacuum should be shut off.

GM Early Fuel Evaporation (EFE) System

The electrically operated EFE system

used on some 1981 and later GM engines performs the same function as the vacuum operated heat riser on other engines, which is to preheat the engine induction system during cold driveway. Rapid heating is desirable because it provides quick fuel evaporation and more uniform fuel distribution to aid cold driveability.

The electrically heated EFE system has a ceramic heater grid located underneath the primary bore(s) of the carburetor which is part of the carburetor insulator. When the ignition is turned on and engine coolant temperature is low, voltage is applied to the EFE relay, which in turn transfers the voltage to the EFE heater in the ceramic grid. When temperature increases, a thermal valve switch de-energizes the relay and the heater is turned off.

Plumbing on air injection systems varies considerably. At first, all the plumbing was external, with individual tubes inserted into each exhaust port either through the cylinder head or the exhaust manifold. Now many engines have internal passageways to duct the air to the exhaust port. A check valve is used between the pump and the exhaust port nozzle to keep hot exhaust gases from traveling up the plumbing and destroying the pump. Some V8 and V6 engines use two check valves.

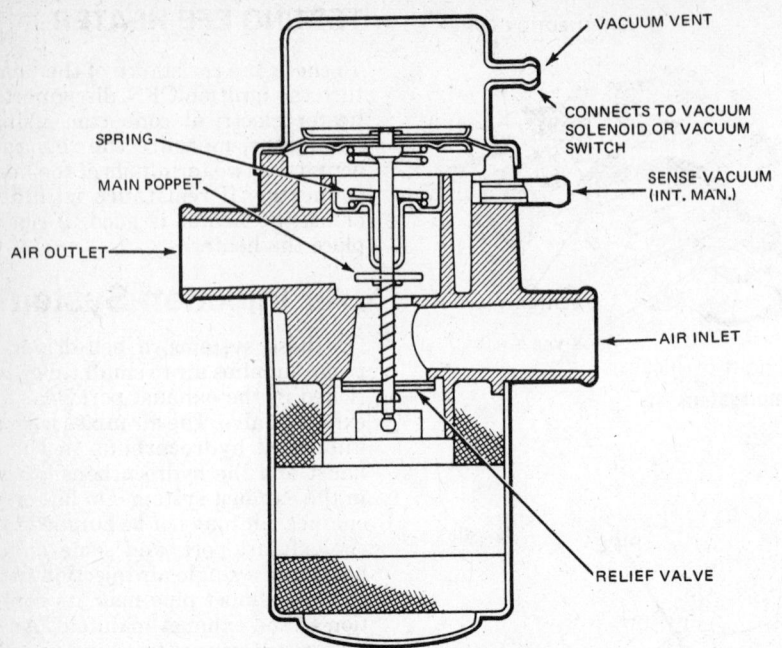

Vacuum differential valve-VDV

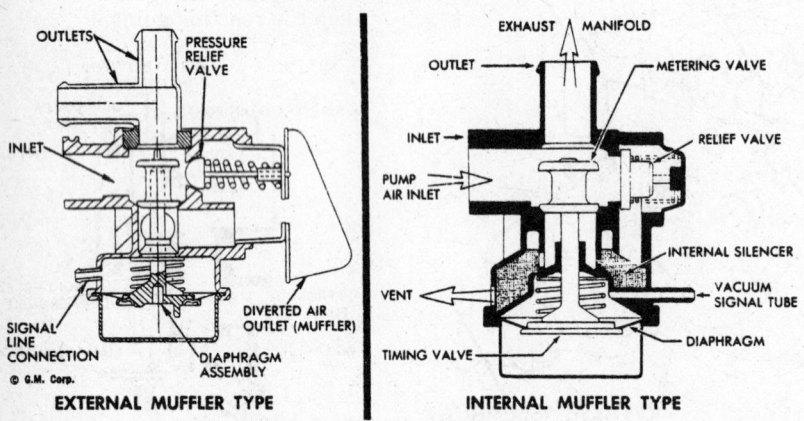

General Motors Diverter Valves

EXTERNAL MUFFLER TYPE   INTERNAL MUFFLER TYPE

hole so that the vacuum or pressure on each side will equalize. As long as the end chamber is sealed by the separate valve being closed, nothing happens, and the air flows through the bypass valve on the way to the exhaust ports. But if the separate valve is opened it admits atmospheric pressure to one side of the diaphragm, and the vacuum on the other side moves the bypass valve to the dump position, exhausting the pump air into the atmosphere. Two types of separate valves are used, one of them an electric solenoid operated valve, and the other a vacuum-operated valve.

AIR PUMP TESTS

─── CAUTION ───

Do not hammer on, pry or bend the pump housing while tightening the drive belt or testing the pump.

Before proceeding with the tests, check the pump drive belt tension. If the belt squeals when the engine is running, the pump may be dragging or seized. Remove the belt and turn the pump by hand to check for seizure. Disregard any chirping, squealing, or rolling sounds from inside the pump when turning it by hand, as these are normal.

Check the hoses and connections for leaks. Hissing or a blast of air is indicative of a leak. Soapy water, applied lightly around the area in question, is a good method for detecting leaks. To test air output, disconnect the air hose from the pump wherever it is convenient. If you disconnect it from one check valve on a V8 or V6, the other hose should also be disconnected and plugged for the test. Run the engine at idle and feel the blast of air from the hose with your hand. Increase the engine speed to 1500 rpm and feel the blast of air again. If the blast increases and is steady, the pump is okay.

Pump Noise Diagnosis

The air pump is normally noisy. As engine speed increases, the noise of the pump will rise in pitch. The rolling sound the pump bearings make is normal. However, if this sound becomes objectionable at certain speeds, the pump is defective and will have to be replaced. A continual hissing sound from the air pump pressure relief valve at idle indicates a defective valve. Replace the relief valve.

If the pump rear bearing fails, a continual knocking sound will be hard. Since the rear bearing is not separately replaceable, the pump will have to be replaced as an assembly.

An anti-backfire valve, also called a bypass valve or divert valve, is used between the pump and the check valve. Usually, the diverter valve is mounted on the pump or near it. A small sensing hose connects the diverter valve to intake manifold vacuum. When the vacuum rises during deceleration, the diverter valve opens and vents the pump air into the atmosphere. This prevents an over-rich fuel mixture in the exhaust system from exploding or backfiring out the tailpipe. Some systems have a delay valve, similar to a spark delay valve, in the sensing hose. This delays for a few seconds the drop in vacuum when the throttle closes, so that the air is not dumped every time the driver takes his foot off the throttle in traffic.

Temperature controls are also used in the sensing hose hookup. Usually, the temperature valve shuts the vacuum off when the engine is cold, so that the pump air doesn't go to the engine exhaust ports until the engine warms up. Some cars have a temperature sensor mounted under the car above the catalytic converter. If the converter overheats, the sensor turns off a solenoid which shuts off the air to the diverter valve. The diverter valve then goes to the dump position, shutting off the air to the exhaust to keep the converter from melting or burning up.

Ford Motor Company 4-cylinder, V6, and some inline 6 engines use a unique air bypass valve, with two small sensing hoes connected to it. Each of the hoses connects to one side of a diaphragm in the valve. The hose on the body of the valve connects to manifold vacuum, and the hose closer to the end connects to a separate on-off valve. The diaphragm has a small

DIVERTER (ANTI-BACKFIRE) VALVE TEST

Detach the hose, which runs from the bypass valve to the check valve. Connect a tachometer to the engine. With the engine running at normal idle speed, check to see that air is flowing from the bypass valve hose connection. Increase the engine speed to 1500-2000 rpm and allow the throttle to snap shut. The flow of air from the bypass valve at the check valve hose connection should stop momentarily and air should then flow from the exhaust port on the valve body or the silencer assembly.

Let the throttle snap shut several times. If the flow of air is not diverted into the atmosphere from the valve exhaust port or if it fails to stop flowing from the hose connection, check the vacuum lines and connections. If these are tight, either the bypass valve or one of the accessory valves in the small sensing hose is defective and must be replaced. A leaking diaphragm will cause the air to flow out both the hose connection and the exhaust port at the same time. If this happens, replace the valve.

NOTE: Late model systems should stop flowing at idle, as described earlier. If not, the bypass valve or accessory valve is defective.

CHECK VALVE TEST

Remove the hose from the check valve. With the engine running at 1500 rpm in Park or Neutral, hold the back of your hand near the check valve to test for exhaust gas leakage. If the valve leaks, it must be replaced.

NOTE: Vibration and flutter of the valve at idle is a normal condition caused by exhaust pulsations. It does not mean that the valve is defective.

VACUUM DIFFERENTIAL VALVE TEST

Disconnect the small sensing hose at the bypass valve and connect a vacuum gauge to the hose. With the engine idling in Park or Neutral, the gauge should read full manifold vacuum. Run the engine at a steady 2500 rpm in Park or Neutral, and release the throttle. As the engine decelerates, the vacuum gauge should drop close to zero, then return to full manifold vacuum as the engine speed drops to idle. If not, the VDV is defective and must be replaced.

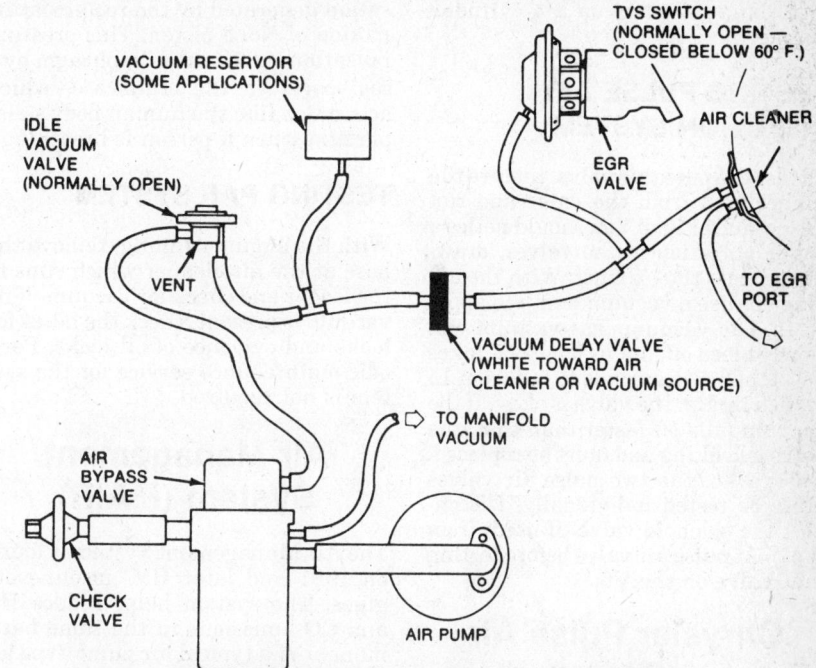

Air pump system using a timed air by-pass valve vacuum vent

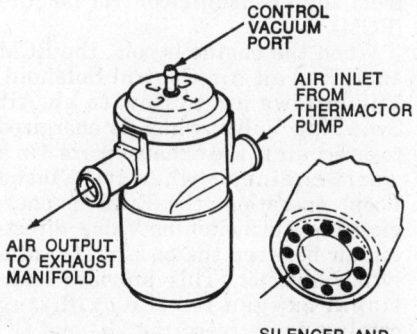

Catalyst cars use a different air bypass valve, with small hose connecting to the end.

NOTE: The small hose nozzle should be connected to manifold vacuum.

Chrysler Air Aspirator System (AAS)

Chrysler Corporation cars which use this system have done away with the air pump. The complete air aspirator system consists of a hose from the clean side of the air cleaner, the aspirator valve mounted on top of the engine, and a tube connecting the valve with the exhaust manifold. The suction in the exhaust draws in air through the air cleaner and this extra air helps the catalytic converter burn up the pollutants. The aspirator valve is similar to the check valve used with all air pump systems. It keeps the exhaust from flowing back into the air cleaner, but allows clean air to go into the exhaust.

TESTING AAS VALVE

Testing the air aspirator valve is done by disconnecting the hose from the air cleaner and checking for slight suction at idle with a piece of paper over the end of the valve. Speeding the engine up slightly will show if the valve is leaking. Exhaust should not come out of the valve. Vibration of the valve diaphragm is normal due to exhaust impulses.

GM Pulse Air Injection System

This system is similar to Chrysler's Air Aspirator. A hose from the clean side of the air cleaner connects to the pulse air valve. Tubes connect the pulse air valve to each cylinder's exhaust port. Suction in the exhaust draws fresh air from the air cleaner into the exhaust, and the air helps the catalytic converter burn up the pollutants. The pulse air valve consists of four or six check valves built into a housing. It allows each exhaust port to suck in fresh air independently of the other ports. The check valves only open when there is suction in the exhaust. If there is any back pressure, the check valves close to prevent exhaust flow back into the air cleaner. On some applications the pulse air valve is connected to only three of the four exhaust ports on a 4-cylinder engine.

TESTING PULSE AIR INJECTION SYSTEM

To test the pulse air valve, remove the rubber hose from the valve and run the engine at idle. You should notice a slight pulsation of the valves, drawing air into the exhaust. With the engine off, use a vacuum pump to apply 15 in. Hg. vacuum. The vacuum will slowly bleed off, but as long as it takes more than two seconds to fall from 15 in. to 5 in. Hg. the valve is okay. If the vacuum falls off faster than that, the valve is leaking and must be replaced. On the V6, the two pulse air valves must be tested individually. Disconnect the solenoid valve (if used) from the front pulse air valve before testing that valve on the V6.

Chrysler Pulse Air Feeder (PAF) System

The PAF system supplies secondary air into the exhaust system between the front and rear catalytic converters, which promotes oxidation of exhaust emissions in the rear catalytic converter. The system consists of a pulse air feeder, which contains two reed valve assemblies, a hose which links the pulse air feeder to the air cleaner, and a tube which runs from the feeder to the exhaust system. At the bottom of the feeder there are two tubes, one which runs into the oil sump and one which connects to No. 3 cylinder crankcase above the oil level. The main reed valve is actuated by a diaphragm in the feeder which, in turn, is activated by the pressure pulsation generated by the reciprocating motion of No. 3 piston. This pressure pulsation is fed to the diaphragm by a seal cover in the crankcase, which acts much like the human body's diaphragm when a person is breathing.

TESTING PAF SYSTEM

With the engine running, remove the hose at the air cleaner which runs to the feeder and check for vacuum. If no vacuum is present, check the hoses for leaks and evidence of oil leaks. Periodic maintenance service for the system is not required.

Air Management System (MAIR)

The Air Management System is found on 1981 and later GM gasoline engines. The system helps reduce HC and CO emissions in the same basic manner of a typical air pump-type air injection system, except that the MAIR system is controlled by signals from the electronic control module (ECM).

When the engine is cold, the ECM energizes an Air Control Solenoid. This allows air to flow to an Air Switching Valve, which is energized to direct air to the exhaust ports. On a warm engine or when in "Closed Loop" operation, the ECM de-energizes the Air Switching Valve, directing air between the beds of the catalytic converter. This provides additional oxygen for the oxidizing catalyst to decrease the HC and CO levels. If the Air Control Valve detects a rapid increase in manifold vacuum (deceleration, etc.), certain operating modes (wide open throttle, etc.), or the ECM self-diagnostic system detects any problem in the MAIR system as a whole, air is diverted (divert mode) to the air cleaner or directly into the atmosphere.

The air flow and control hoses transmit pressurized air to the catalytic converter or to the exhaust ports through internal (intake manifold) passages or external piping. The check valves prevent backflow of exhaust gas into the air distribution system. The valve prevents backflow when the air pump "bypasses" at high speed and loads, or in case the air pump malfunctions.

NOTE: Due to the complex nature of modern electronic engine control systems, comprehensive diagnosis and testing procedures fall outside the confines of this repair manual. For complete information on diagnosis, testing and repair procedures concerning all modern engine and emission control systems, please refer to *Chilton's Guide To Electronic Engine Controls*.

Ford Pulse Air (Thermactor II) System

Some Ford engines are equipped with an air injection system which does not use an air pump. Instead, natural pulses present in the exhaust system are used to pull the air into the system through the pulse air valves. The pulse valve is connected to the exhaust manifold by a tube and to the air cleaner or silencer with a hose. Make sure air can flow freely through the air cleaner or silencer to the check valve.

Engine Rebuilding

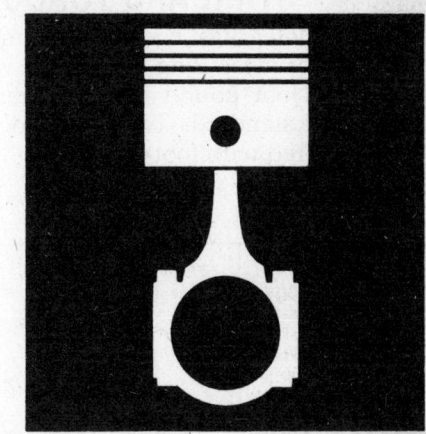

This section describes, in detail, the procedures involved in rebuilding a typical engine. The procedures are basically identical to those used in rebuilding engines of nearly all design and configurations.

The section is divided into two parts. The first, Cylinder Head Reconditioning, assumes that the cylinder head is removed from the engine, all manifolds are removed, and the cylinder head is on a workbench. The camshaft should be removed from overhead cam cylinder heads. The second section, Cylinder Block Reconditioning, covers the block, pistons, connecting rods and crankshaft. It is assumed that the engine is mounted on a work stand, and the cylinder head and all accessories are removed.

Procedures are identified as follows:

Unmarked—Basic procedures that must be performed in order to successfully complete the rebuilding process.

Starred (*)—Procedures that should be performed to ensure maximum performance and engine life.

Double starred (**)—Procedures that may be performed to increase engine performance and reliability.

In many cases, a choice of methods is also provided. Methods are identified in the same manner as procedures. The choice of method for a procedure is at the discretion of the user.

The tools required for the basic rebuilding procedure should, with minor exceptions, be those included in a mechanic's tool kit. An accurate torque wrench, and a dial indicator (reading in thousandths) mounted on a universal base should be available. Special tools, where required, all are readily available from the major tool suppliers. The services of a competent automotive machine shop must also be readily available.

When assembling the engine, any parts that will be in frictional contact must be prelubricated, to provide protection on initial start-up. Any product specifically formulated for this purpose may be used. NOTE: *Do not use engine oil*. Where semi-permanent (locked but removable) installation of bolts or nuts is desired, threads should be cleaned and coated with Loctite® or a similar product (non-hardening).

Aluminum has become increasingly popular for use in engines, due to its low weight and excellent heat transfer characteristics. The following precautions must be observed when handling aluminum engine parts:

—Never hot-tank aluminum parts.

—Remove all aluminum parts (identification tags, etc.) from engine parts before hot-tanking (otherwise they will be removed during the process).

—Always coat threads lightly with engine oil or anti-seize compounds before installation, to prevent seizure.

—Never over-torque bolts or spark plugs in aluminum threads. Should stripping occur, threads can be restored using any of a number of thread repair kits available (see next section).

Magnaflux and Zyglo are inspection techniques used to locate material flaws, such as stress cracks. Magnafluxing coats the part with fine magnetic particles, and subjects the part to a magnetic field. Cracks cause breaks in the magnetic field, which are outlined by the particles. Since Magnaflux is a magnetic process, it is applicable only to ferrous materials. The Zyglo process coats the material with a fluorescent dye penetrant, and then subjects it to blacklight inspection, under which cracks glow brightly. Parts made of any material may be tested using Zyglo. While Magnaflux and Zyglo are excellent for general inspection, and locating hidden defects, specific checks of suspected cracks may be made at lower cost and more readily using spot check dye. The dye is sprayed onto the suspected area, wiped off, and the area is then sprayed with a developer. Cracks then will show up brightly. Spot check dyes will only indicate surface cracks; therefore, structural cracks below the surface may escape detection. When questionable, the part should be tested using Magnaflux or Zyglo.

REPAIRING DAMAGED THREADS

Several methods of repairing damaged threads are available. Heli-Coil® (shown here), Keenserts® and Microdot® are among the most widely used. All involve basically the same principle—drilling out stripped threads, tapping the hole and installing a prewound insert— making welding, plugging and oversize fasteners unnecessary.

Two types of thread repair inserts are usually supplied—a standard type for most Inch Coarse, Inch Fine, Metric Coarse and Metric Fine thread sizes and a spark plug type to fit most spark plug port sizes. Consult the individual manufacturer's catalog to determine exact applications. Typical thread repair kits will contain a selection of prewound threaded inserts, a tap (corresponding to the outside diameter threads of the insert) and an installation tool. Most manufacturers also supply blister-packed thread repair inserts separately and a master kit with a variety of taps and inserts plus installation tools.

Before effecting a repair to a threaded hole, remove any snapped, broken or damaged bolts or studs. Penetrating oil can be used to free frozen threads; the offending item can be removed with locking pliers or with a screw or stud extractor. After the hole is clear, the thread can be repaired as follows.

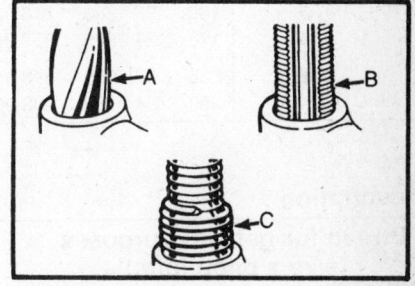

A. Drill out the damaged threads with the specified drill. Drill completely through the hole or to the bottom of a blind hole.

B. With the tap supplied tap the hole to receive the threaded insert. Keep the tap well oiled and back it out frequently to avoid clogging the threads.

C. Screw the threaded insert onto the installation tool until the tang engages the slot. Screw the insert into the tapped hole until it is ¼–½ turn below the top surface. After installation, break the tang off with a hammer and punch.

STANDARD TORQUE SPECIFICATIONS AND CAPSCREW MARKINGS

Newton-Meter has been designated as the world standard for measuring torque and will gradually replace the foot-pound and kilogram-meter torque measuring standard. Torquing tools are still being manufactured with foot-pounds and kilogram-meter scales, along with the new Newton-Meter standard. To assist the repairman, foot-pounds, kilogram-meter and Newton-Meter are listed in the following charts, and should be followed as applicable.

U.S. BOLTS

SAE Grade Number	1 or 2			5			6 or 7			8		
Capscrew Head Markings Manufacturer's marks may vary. Three-line markings on heads below indicate SAE Grade 5. 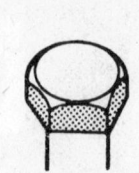												
Usage	Used Frequently			Used Frequently			Used at Times			Used at Times		
Quality of Material	Indeterminate			Minimum Commercial			Medium Commercial			Best Commercial		
Capacity Body Size	Torque			Torque			Torque			Torque		
(inches)–(thread)	Ft-Lb	kgm	Nm	Ft-Lb	kgm	Nm	Ft-Lb	kgm	Nm	Ft-Lb	kgm	Nm
1/4–20	5	0.6915	6.7791	8	1.1064	10.8465	10	1.3630	13.5582	12	1.6596	16.2698
–28	6	0.8298	8.1349	10	1.3830	13.5582				14	1.9362	18.9815
5/16–18	11	1.5213	14.9140	17	2.3511	23.0489	19	2.6277	25.7605	24	3.3192	32.5396
–24	13	1.7979	17.6256	19	2.6277	25.7605				27	3.7341	36.6071
3/8–16	18	2.4894	24.4047	31	4.2873	42.0304	34	4.7022	46.0978	44	6.0852	59.6560
–24	20	2.7660	27.1164	35	4.8405	47.4536				49	6.7767	66.4351
7/16–14	28	3.8132	37.9629	49	6.7767	66.4351	55	7.6065	74.5700	70	9.6810	94.9073
–20	30	4.1490	40.6745	55	7.6065	74.5700				78	10.7874	105.7538
1/2–13	39	5.3937	52.8769	75	10.3725	101.6863	85	11.7555	115.2445	105	14.5215	142.3609
–20	41	5.6703	55.5885	85	11.7555	115.2445				120	16.5860	162.6960
9/16–12	51	7.0533	69.1467	110	15.2130	149.1380	120	16.5960	162.6960	155	21.4365	210.1490
–18	55	7.6065	74.5700	120	16.5960	162.6960				170	23.5110	230.4860
5/8–11	83	11.4789	112.5329	150	20.7450	203.3700	167	23.0961	226.4186	210	29.0430	284.7180
–18	95	13.1385	128.8027	170	23.5110	230.4860				240	33.1920	325.3920
3/4–10	105	14.5215	142.3609	270	37.3410	366.0660	280	38.7240	379.6240	375	51.8625	508.4250
–16	115	15.9045	155.9170	295	40.7985	399.9610				420	58.0860	568.4360
7/8–9	160	22.1280	216.9280	395	54.6285	535.5410	440	60.8520	596.5520	605	83.6715	820.2590
–14	175	24.2025	237.2650	435	60.1605	589.7730				675	93.3525	915.1650
1–8	236	32.5005	318.6130	590	81.5970	799.9220	660	91.2780	894.8280	910	125.8530	1233.7780
–14	250	34.5750	338.9500	660	91.2780	849.8280				990	136.9170	1342.2420

METRIC BOLTS

Description					
	Torque ft-lbs. (Nm)				
Thread for general purposes (size x pitch (mm))	Head Mark 4			Head Mark 7	
6 x 1.0	2.2 to 2.9	(3.0 to 3.9)		3.6 to 5.8	(4.9 to 7.8)
8 x 1.25	5.8 to 8.7	(7.9 to 12)		9.4 to 14	(13 to 19)
10 x 1.25	12 to 17	(16 to 23)		20 to 29	(27 to 39)
12 x 1.25	21 to 32	(29 to 43)		35 to 53	(47 to 72)
14 x 1.5	35 to 52	(48 to 70)		57 to 85	(77 to 110)
16 x 1.5	51 to 77	(67 to 100)		90 to 120	(130 to 160)
18 x 1.5	74 to 110	(100 to 150)		130 to 170	(180 to 230)
20 x 1.5	110 to 140	(150 to 190)		190 to 240	(160 to 320)
22 x 1.5	150 to 190	(200 to 260)		250 to 320	(340 to 430)
24 x 1.5	190 to 240	(260 to 320)		310 to 410	(420 to 550)

CAUTION: Bolts threaded into aluminum require much less torque

NOTE: This engine rebuilding section is a guide to accepted rebuilding procedures. Typical examples of standard rebuilding procedures are illustrated.

CYLINDER HEAD RECONDITIONING

Procedure	Method
Identify the valves:	Invert the cylinder head, and number the valve faces front to rear, using a permanent felt-tip marker.
Remove the rocker arms (OHV engines only):	Remove the rocker arms with shaft(s) or balls and nuts. Wire the sets of rockers, balls and nuts together, and identify according to the corresponding valve.
Remove the camshaft (OHC engines only):	See the engine service procedures earlier in this book for details concerning specific engines.
Remove the valves and springs:	Using an appropriate valve spring compressor (depending on the configuration of the cylinder head), compress the valve springs. Lift out the keepers with needlenose pliers, release the compressor, and remove the valve, spring, and spring retainer.
Remove glow plugs and fuel injectors (Diesel engines only):	Label and remove all fuel injectors and glow plugs from the head. Glow plugs unscrew. See the appropriate car section for injector removal. Inspect glow plugs for bulges, cracks or signs of melting. Clean injector tips with a steel brush, then inspect for evidence of melting.
**Remove pre-combustion chamber inserts (Diesel engines only):	**Remove the pre-combustion chambers using a hammer and a thin, blunt brass drift, inserted through the injector hole (or glow plug hole, whichever is more convenient). If chamber is to be reused, carefully remove all carbon from it. NOTE: *Remove chamber only if being replaced, if a glow plug tip has broken off and must be removed, or if chamber is obviously damaged or loose.*

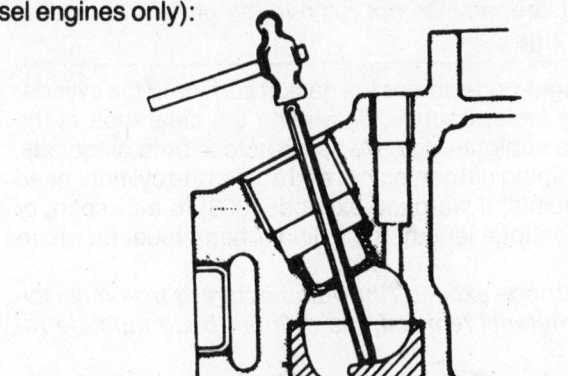

Removing pre-combustion chamber with a drift (© G.M. Corp.)

Check the valve stem-to-guide clearance:	Clean the valve stem with lacquer thinner or a similar solvent to remove all gum and varnish. Clean the valve guides using solvent and an expanding wire-type valve guide cleaner. Mount a dial indicator so that the stem is at 90° to the valve stem, as close to the valve guide as possible. Move the valve off its seat, and measure the valve guide-to-stem clearance by rocking the stem back and forth to actuate the dial indicator. Measure the valve stems using a micrometer, and compare to specifications, to determine whether stem or guide wear is responsible for excessive clearance.

DIAL INDICATOR

VALVE STEM

Checking the valve stem-to-guide clearance

CYLINDER HEAD RECONDITIONING

Procedure	Method
De-carbon the cylinder head and valves:	Chip carbon away from the valve heads, combustion chambers, and ports, using a chisel made of hardwood. Remove the remaining deposits with a stiff wire brush. NOTE: *Ensure that the deposits are actually removed, rather than burnished.*

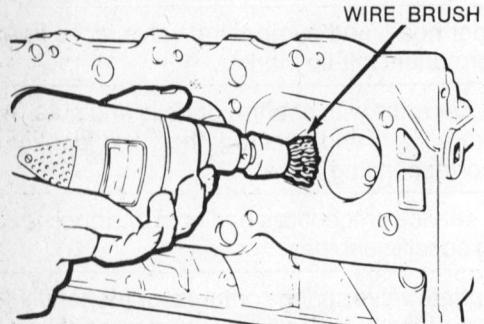

WIRE BRUSH

Removing carbon from the cylinder head

Procedure	Method
Hot-tank the cylinder head (cast iron heads only): CAUTION: *Do not hot-tank aluminum parts.*	Have the cylinder head hot-tanked to remove grease, corrosion, and scale from the water passages. NOTE: *In the case of overhead cam cylinder heads, consult the operator to determine whether the camshaft bearings will be damaged by the caustic solution.*
Degrease the remaining cylinder head parts:	Using solvent (i.e., Gunk), clean the rockers, rocker shaft(s) (where applicable), rocker balls and nuts, springs, spring retainers, and keepers. Do not remove the protective coating from the springs.
Check the cylinder head for warpage:	Place a straight-edge across the gasket surface of the cylinder head. Using feeler gauges, determine the clearance at the center of the straight-edge. Measure across both diagonals, along the longitudinal centerline, and across the cylinder head at several points. If warpage exceeds .003' in a 6' span, or .006' over the total length, the cylinder head must be resurfaced. NOTE: *If warpage exceeds the manufacturer's maximum tolerance for material removal, the cylinder head must be replaced.* When milling the cylinder heads of V-type engines, the intake manifold mounting position is altered, and must be corrected by milling the manifold flange a proportionate amount.

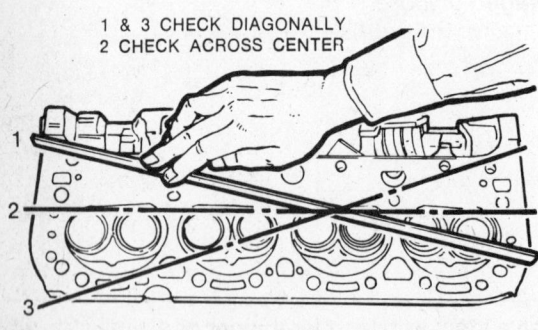

1 & 3 CHECK DIAGONALLY
2 CHECK ACROSS CENTER

Checking cylinder head for warpage

Procedure	Method
**Porting and gasket matching:	**Coat the manifold flanges of the cylinder head with Prussian blue dye. Glue intake and exhaust gaskets to the cylinder head in their installed position using rubber cement and scribe the outline of the ports on the manifold flanges. Remove the gaskets. Using a small cutter in a hand-held power tool gradually taper the walls of the port out to the scribed outline of the gasket. Further enlargement of the ports should include the removal of sharp edges and radiusing of sharp corners. Do not alter the valve guides. NOTE: *The most efficient port configuration is determined only by extensive testing. Therefore, it is best to consult someone experienced with the head in question to determine the optimum alterations.*

CYLINDER HEAD RECONDITIONING

Procedure	Method

*Knurling the valve guides:

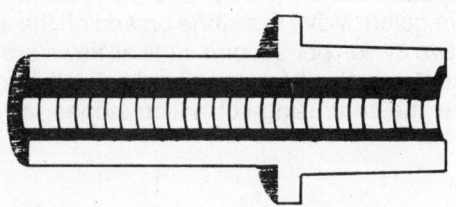

Cut-away view of a knurled valve guide

*Valve guides which are not excessively worn or distorted may, in some cases, be knurled rather than replaced. Knurling is a process in which metal is displaced and raised, thereby reducing clearance. Knurling also provides excellent oil control. The possibility of knurling rather than replacing valve guides should be discussed with a machinist.

Replacing the valve guides:
NOTE: *Valve guides should only be replaced if damaged or if an oversize valve stem is not available.*

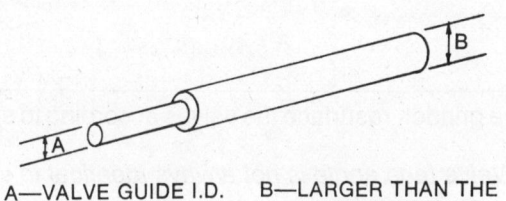

A—VALVE GUIDE I.D.   B—LARGER THAN THE VALVE GUIDE O.D.

Valve guide removal tool

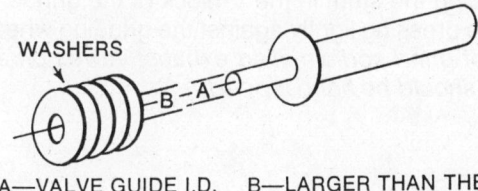

WASHERS

A—VALVE GUIDE I.D.   B—LARGER THAN THE VALVE GUIDE O.D.

Valve guide installation tool (with washers used for installation)

Depending on the type of cylinder head, valve guides may be pressed, hammered, or shrunk in. In cases where the guides are shrunk into the head, replacement should be left to an equipped machine shop. In other cases, the guides are replaced as follows: Press or tap the valve guides out of the head using a stepped drift (see illustration). Determine the height above the boss that the guide must extend, and obtain a stack of washers, their I.D. similar to the guide's O.D., of that height. Place the stack of washers on the guide, and insert the guide into the boss.
NOTE: *Valve guides are often tapered or beveled for installation.*
Using the stepped installation tool (see illustration), press or tap the guides into position. Ream the guides according to the size of the valve stem.

Replacing valve seat inserts:

Replacement of valve seat inserts which are worn beyond resurfacing or broken, if feasible, must be done by a machine shop.

Resurfacing the valve seats using reamers:

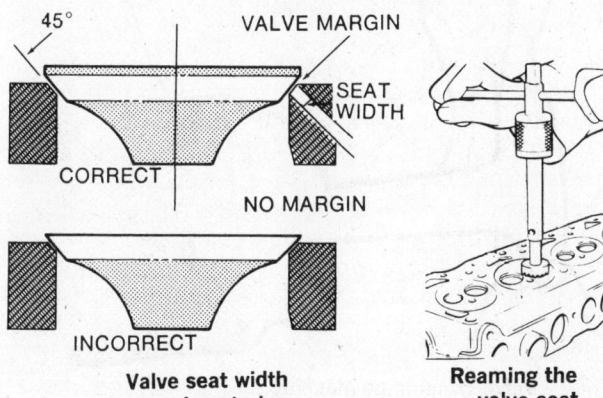

45°   VALVE MARGIN

SEAT WIDTH

CORRECT

NO MARGIN

INCORRECT

Valve seat width and centering

Reaming the valve seat

Select a reamer of the correct seat angle, slightly larger than the diameter of the valve seat, and assemble it with a pilot of the correct size. Install the pilot into the valve guide, and using steady pressure, turn the reamer clockwise.
CAUTION: *Do not turn the reamer counterclockwise.*
Remove only as much material as necessary to clean the seat. Check the concentricity of the seat (see below). If the dye method is not used, coat the valve face with Prussian blue dye, install and rotate it on the valve seat. Using the dye marked area as a centering guide, center and narrow the valve seat to specifications with correction cutters.
NOTE: *When no specifications are available, minimum seat width for exhaust valves should be 5/64", intake valves 1/16".*
After making correction cuts, check the position of the valve seat on the valve face using Prussian blue dye.
NOTE: *Do not cut induction hardened seats; they must be ground.*

CYLINDER HEAD RECONDITIONING

Procedure	Method

*Resurfacing the valve seats using a grinder:

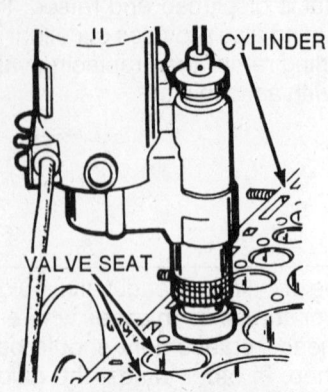

Grinding a valve seat

*Select a pilot of the correct size, and a coarse stone of the correct seat angle. Lubricate the pilot if necessary, and install the tool in the valve guide. Move the stone on and off the seat at approximately two cycles per second, until all flaws are removed from the seat. Install a fine stone, and finish the seat. Center and narrow the seat using correction stones, as described above.

Resurfacing (grinding) the valve face:

Using a valve grinder, resurface the valves according to specifications.
CAUTION: *Valve face angle is not always identical to valve seat angle.*
A minimum margin of 1/32" should remain after grinding the valve. The valve stem top should also be squared and resurfaced, by placing the stem in the V-block of the grinder, and turning it while pressing lightly against the grinding wheel.
NOTE: *Do not grind sodium filled exhaust valves on a machine. These should be hand lapped.*

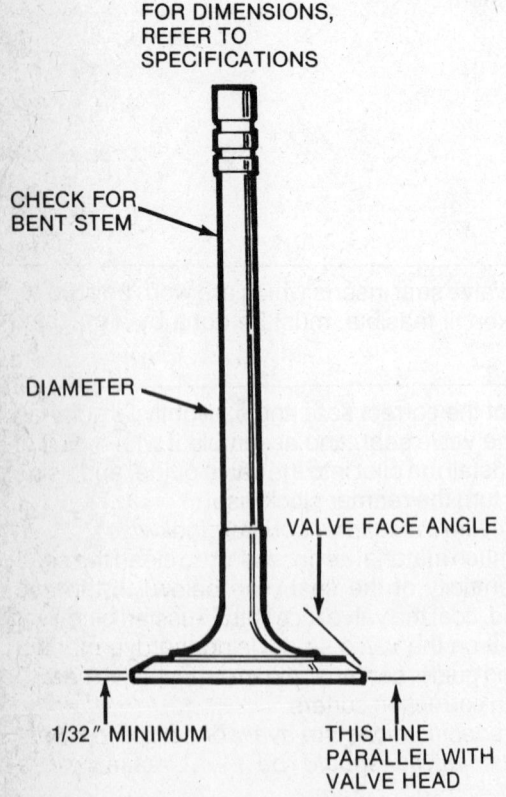

FOR DIMENSIONS, REFER TO SPECIFICATIONS

CHECK FOR BENT STEM

DIAMETER

VALVE FACE ANGLE

1/32" MINIMUM

THIS LINE PARALLEL WITH VALVE HEAD

Critical valve dimensions

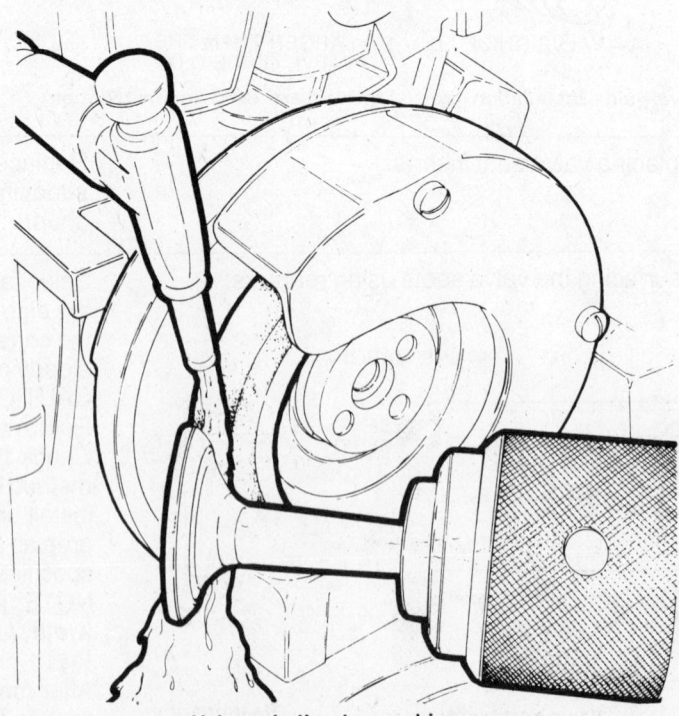

Valve grinding by machine

CYLINDER HEAD RECONDITIONING

Procedure	Method

Checking the valve seat concentricity:

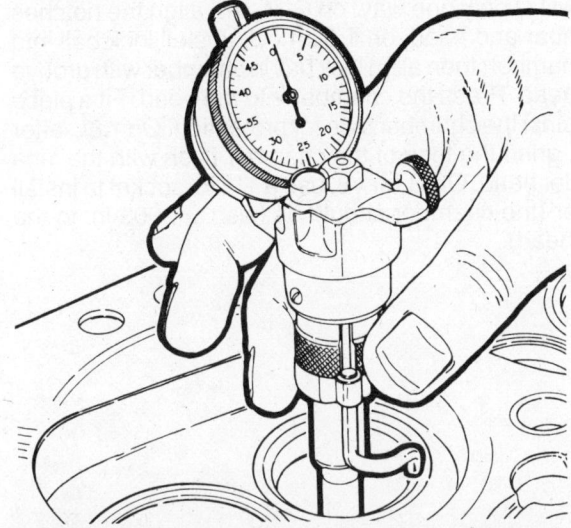

Checking valve seat concentricity using a dial gauge

Coat the valve face with Prussian blue dye, install the valve, and rotate it on the valve seat. If the entire seat becomes coated, and the valve is known to be concentric, the seat is concentric.

*Install the dial gauge pilot into the guide, and rest the arm on the valve seat. Zero the gauge, and rotate the arm around the seat. Run-out should not exceed .002".

---

*Lapping the valves:
NOTE: *Valve lapping is done to ensure efficient sealing of resurfaced valves and seats.*

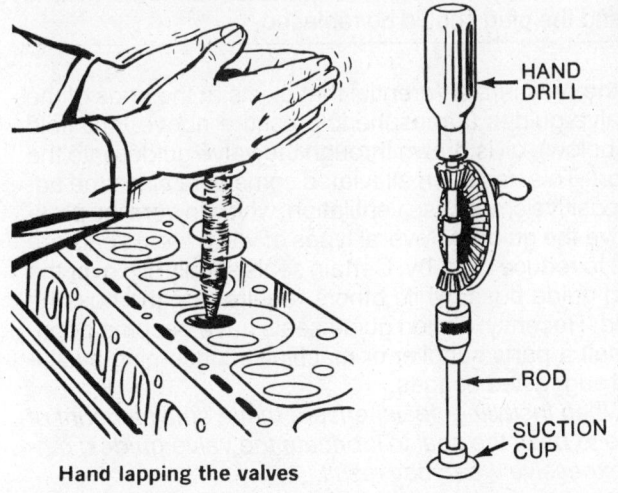

Hand lapping the valves

HAND DRILL

ROD

SUCTION CUP

Home made mechanical valve lapping tool

*Invert the cylinder head, lightly lubricate the valve stems, and install the valves in the head as numbered. Coat valve seats with fine grinding compound, and attach the lapping tool suction cup to a valve head.
NOTE: *Moisten the suction cup.*
Rotate the tool between the palms, changing position and lifting the tool often to prevent grooving. Lap the valve until a smooth, polished seat is evident. Remove the valve and tool, and rinse away all traces of grinding compound.
**Fasten a suction cup to a piece of drill rod, and mount the rod in a hand drill. Proceed as above, using the hand drill as a lapping tool.
CAUTION: *Due to the higher speeds involved when using the hand drill, care must be exercised to avoid grooving the seat.* Lift the tool and change direction of rotation often.

Check the valve springs:

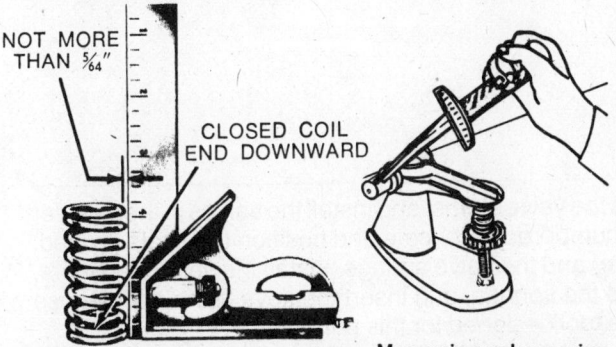

NOT MORE THAN $\frac{5}{64}$"

CLOSED COIL END DOWNWARD

Checking valve spring free length and squareness

Measuring valve spring test pressure

Place the spring on a flat surface next to a square. Measure the height of the spring, and rotate it against the edge of the square to measure distortion. If spring height varies (by comparison) by more than $\frac{1}{16}$" or if distortion exceeds $\frac{1}{16}$", replace the spring.
**In addition to evaluating the spring as above, test the spring pressure at the installed and compressed (installed height minus valve lift) height using a valve spring tester. Springs used on small displacement engines (up to 3 liters) should be ∓ 1 lb. of all other springs in either position. A tolerance of ∓ 5 lbs. is permissible on larger engines.

CYLINDER HEAD RECONDITIONING

Procedure	Method

Install pre-combustion chambers (Diesel engines only)

Pre-combustion chambers are press-fit into the head. The chambers will fit only one way: on G.M. V8, align the notches in the chamber and head; on 1.8L 4 cyl., install lock ball into groove in chamber, then align lock ball in chamber with groove in cylinder head. Press the chamber into the head. Fit a piece of metal against the chamber face for protection. On 1.8L, after installation, grind the face of the chamber flush with the face of the cylinder head. On G.M. V8, use a 1¼ in. socket to install the chamber (the chamber should be flush ± .003 in. to the face of the head).

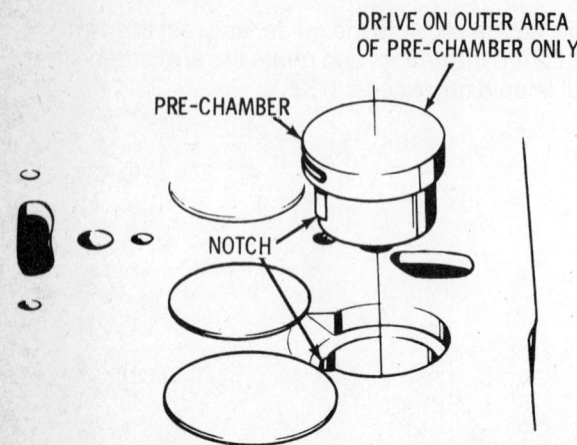

DRIVE ON OUTER AREA OF PRE-CHAMBER ONLY

PRE-CHAMBER

NOTCH

Align the notches to install the pre-combustion chamber (© G.M. Corp.)

Install fuel injectors and glow plugs (Diesel engines)

Before installing glow plugs, check for continuity across plug terminals and body. If no continuity exists, the heater wire is broken and the plug should be replaced.

---

*Install valve stem seals:

*Due to the pressure differential that exists at the ends of the intake valve guides (atmospheric pressure above, manifold vacuum below), oil is drawn through the valve guides into the intake port. This has been alleviated somewhat since the addition of positive crankcase ventilation, which lowers the pressure above the guides. Several types of valve stem seals are available to reduce blow-by. Certain seals simply slip over the stem and guide boss, while others require that the boss be machined. Recently, Teflon guide seals have become popular. Consult a parts supplier or machinist concerning availability and suggested usages.

NOTE: *When installing seals, ensure that a small amount of oil is able to pass the seal to lubricate the valve guides; otherwise, excessive wear may result.*

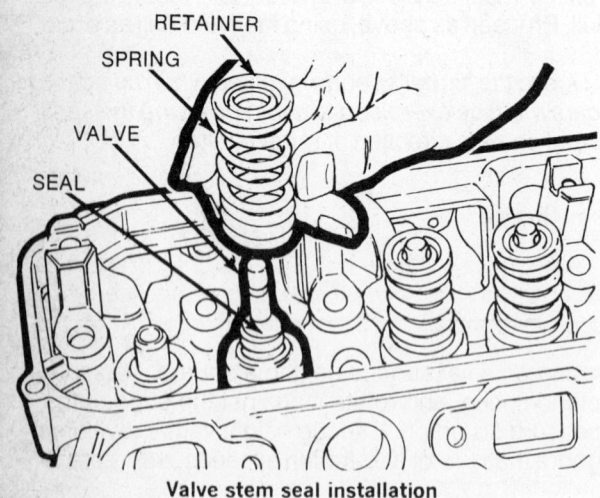

RETAINER

SPRING

VALVE

SEAL

Valve stem seal installation

Install the valves:

Lubricate the valve stems, and install the valves in the cylinder head as numbered. Lubricate and position the seals (if used, see above) and the valve springs. Install the spring retainers, compress the springs, and insert the keys using needlenose pliers or a tool designed for this purpose.

NOTE: *Retain the keys with wheel bearing grease during installation.*

CYLINDER HEAD RECONDITIONING

Procedure	Method

Check valve spring installed height:

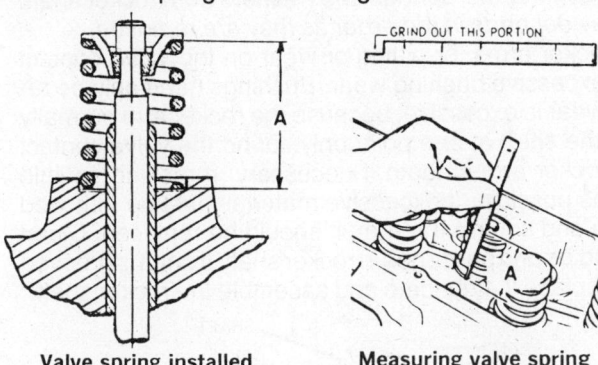

Valve spring installed
height dimension

Measuring valve spring
installed height

Measure the distance between the spring pad and the lower edge of the spring retainer, and compare to specifications. If the installed height is incorrect, add shim washers between the spring pad and the spring.
CAUTION: *Use only washers designed for this purpose.*

Install the camshaft (OHC engines only) and check end play:

See the engine service procedures earlier in this book for details concerning specific engines.

Inspect the rocker arms, balls, studs, and nuts (OHV engines only):

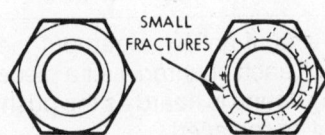

Stress cracks in the rocker nuts

Visually inspect the rocker arms, balls, studs, and nuts for cracks, galling, burning, scoring or wear. If all parts are intact, liberally lubricate the rocker arms and balls, and install them on the cylinder head. If wear is noted on a rocker arm at the point of valve contact, grind it smooth and square, removing as little material as possible. Replace the rocker arm if excessively worn. If a rocker stud shows signs of wear, it must be replaced (see below). If a rocker nut shows stress cracks, replace it. If an exhaust ball is galled or burned, substitute the intake ball from the same cylinder (if it is intact), and install a new intake ball.
NOTE: *Avoid using new rocker balls on exhaust valves.*

Replacing rocker studs (OHV engines only):

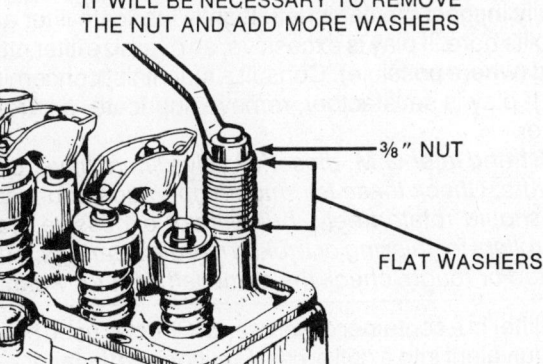

AS STUB BEGINS TO PULL UP, IT WILL BE NECESSARY TO REMOVE THE NUT AND ADD MORE WASHERS

⅜″ NUT

FLAT WASHERS

Extracting a pressed-in rocker stud

In order to remove a threaded stud, lock two nuts on the stud, and unscrew the stud using the lower nut. Coat the lower threads of the new stud with Loctite®, and install.
Two alternative methods are available for replacing pressed in studs. Remove the damaged stud using a stack of washers and a nut (see illustration). In the first, the boss is reamed .005–.006″ oversize, and an oversize stud pressed in. Control the stud extension over the boss using washers, in the same manner as valve guides. Before installing the stud, coat it with white lead and grease. To retain the stud more positively drill a hole through the stud and boss, and install a roll pin. In the second method, the boss is tapped, and a threaded stud installed. Retain the stud using Loctite® Stud and Bearing Mount.

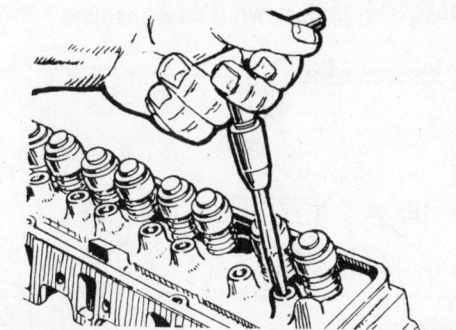

Reaming the stud bore for oversize rocker studs

CYLINDER HEAD RECONDITIONING

Procedure	Method

Inspect the rocker shaft(s) and rocker arms (OHV engines only):

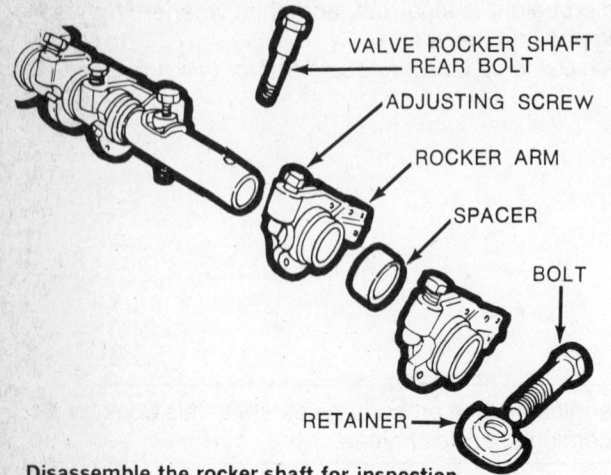

Disassemble the rocker shaft for inspection

Remove rocker arms, springs and washers from rocker shaft. NOTE: *Lay out parts in the order as they are removed.* Inspect rocker arms for pitting or wear on the valve contact point, or excessive bushing wear. Bushings need only be replaced if wear is excessive, because the rocker arm normally contacts the shaft at one point only. Grind the valve contact point of rocker arm smooth if necessary, removing as little material as possible. If excessive material must be removed to smooth and square the arm, it should be replaced. Clean out all oil holes and passages in rocker shaft. If shaft is grooved or worn, replace it. Lubricate and assemble the rocker shaft.

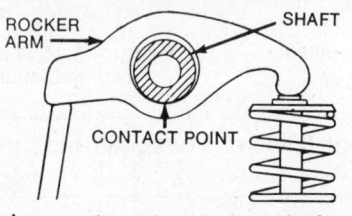

Rocker arm-to-rocker shaft contact area

Inspect the camshaft bushings and the camshaft (OHC engines):

See next section.

Inspect the pushrods (OHV engines only):

Remove the pushrods, and, if hollow, clean out the oil passages using fine wire. Roll each pushrod over a piece of clean glass. If a distinct clicking sound is heard as the pushrod rolls, the rod is bent, and must be replaced.

*The length of all pushrods must be equal. Measure the length of the pushrods, compare to specifications, and replace as necessary.

Inspect the valve lifters (OHV engines only):

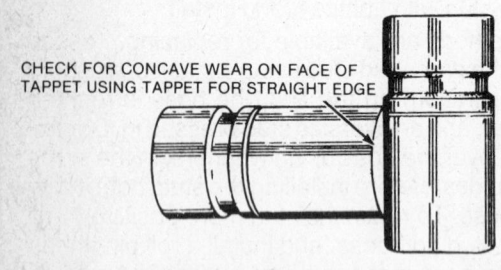

CHECK FOR CONCAVE WEAR ON FACE OF TAPPET USING TAPPET FOR STRAIGHT EDGE

Checking the lifter face

Remove lifters from their bores, and remove gum and varnish, using solvent. Clean walls of lifter bores. Check lifters for concave wear as illustrated. If face is worn concave, replace lifter, and carefully inspect the camshaft. Lightly lubricate lifter and insert it into its bore. If play is excessive, an oversize lifter must be installed (where possible). Consult a machinist concerning feasibility. If play is satisfactory, remove, lubricate, and reinstall the lifter.
NOTE: *1981 and later G.M. diesel V8 valve lifters have roller cam followers. Check these for smooth operation and wear. The roller should rotate freely, but without excessive play. Check the rollers for missing or broken needle bearings. If the roller is pitted or rough, check the camshaft lobe for wear.*

---

*Testing hydraulic lifter leak down (OHV gasoline engines only):

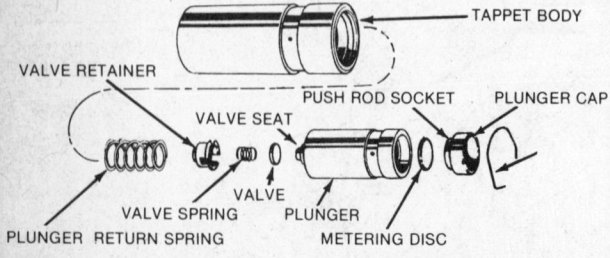

Typical exploded view of hydraulic valve lifter

Submerge lifter in a container of kerosene. Chuck a used pushrod or its equivalent into a drill press. Position container of kerosene so pushrod acts on the lifter plunger. Pump lifter with the drill press, until resistance increases. Pump several more times to bleed any air out of lifter. Apply very firm, constant pressure to the lifter, and observe rate at which fluid bleeds out of lifter. If the fluid bleeds very quickly (less than 15 seconds), lifter is defective. If the time exceeds 60 seconds, lifter is sticking. In either case, recondition or replace lifter. If lifter is operating properly (leak down time 15–60 seconds), lubricate and install it.

CYLINDER HEAD RECONDITIONING

Procedure	Method
Bleed the hydraulic lifters (diesel engines only):	After the cylinder heads are installed on G.M. V8 diesels, the valve lifters must be bled down before the crankshaft is turned. Failure to bleed down the lifters will cause damage to the valve train. See diesel engine rocker arm replacement procedure in Oldsmobile 88, 98, etc. car section for procedures. NOTE: *When installing new lifters, prime by working the lifter plunger while submerged in clean kerosene or diesel fuel.*

CYLINDER BLOCK RECONDITIONING

Procedure	Method
Checking the main bearing clearance: 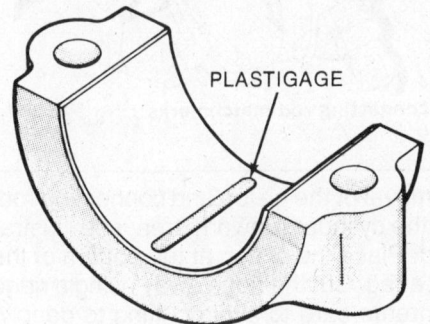 Plastigage® installed on the lower bearing shell 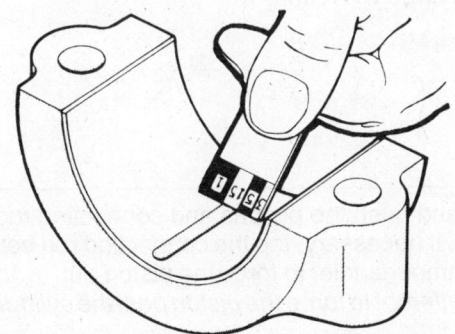 Measuring Plastigage® to determine bearing clearance	Invert engine, and remove cap from the bearing to be checked. Using a clean, dry rag, thoroughly clean all oil from crankshaft journal and bearing insert. NOTE: *Plastigage is soluble in oil; therefore, oil on the journal or bearing could result in erroneous readings.* Place a piece of Plastigage along the full length of journal, reinstall cap, and torque to specifications. Remove bearing cap, and determine bearing clearance by comparing width of Plastigage to the scale on Plastigage envelope. Journal taper is determined by comparing width of the Plastigage strip near its ends. Rotate crankshaft 90° and retest, to determine journal eccentricity. NOTE: *Do not rotate crankshaft with Plastigage installed.* If bearing insert and journal appear intact, and are within tolerances, no further main bearing service is required. If bearing or journal appear defective, cause of failure should be determined before replacement. *Remove crankshaft from block (see below). Measure the main bearing journals at each end twice (90° apart) using a micrometer, to determine diameter, journal taper and eccentricity. If journals are within tolerances, reinstall bearing caps at their specified torque. Using a telescope gauge and micrometer, measure bearing I.D. parallel to piston axis and at 30° on each side of piston axis. Subtract journal O.D. from bearing I.D. to determine oil clearance. If crankshaft journals appear defective, or do no meet tolerances, there is no need to measure bearings; for the crankshaft will require grinding and/or undersize bearings will be required. If bearing appears defective, cause for failure should be determined prior to replacement.
Checking the connecting rod bearing clearance:	Connecting rod bearing clearance is checked in the same manner as main bearing clearance, using Plastigage. Before removing the crankshaft, connecting rod side clearance also should be measured and recorded. *Checking connecting rod bearing clearance, using a micrometer, is identical to checking main bearing clearance. If no other service is required, the piston and rod assemblies need not be removed.

CYLINDER BLOCK RECONDITIONING

Procedure	Method

Removing the crankshaft:

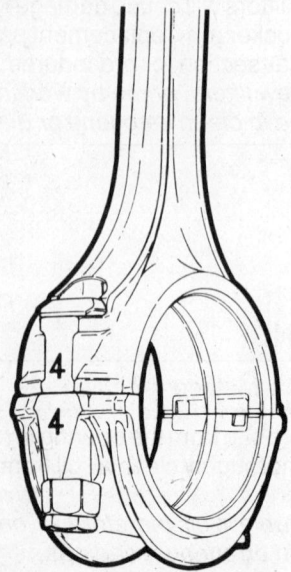

Connecting rod matched to cylinder with a number stamp

Using a punch, mark the corresponding main bearing caps and saddles according to position (i.e., one punch on the front main cap and saddle, two on the second, three on the third, etc.). Using number stamps, identify the corresponding connecting rods and caps, according to cylinder (if no numbers are present). Remove the main and connecting rod caps, and place sleeves of plastic tubing over the connecting rod bolts, to protect the journals as the crankshaft is removed. Lift the crankshaft out of the block.

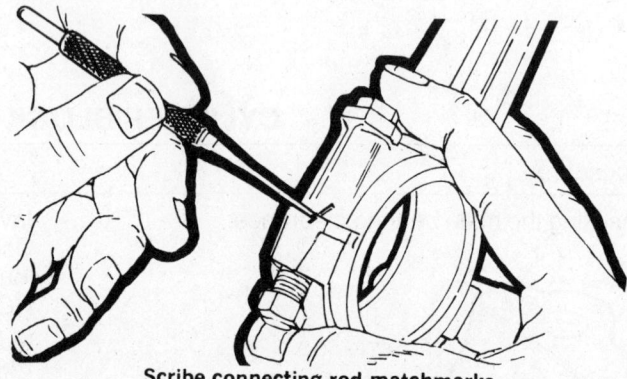

Scribe connecting rod matchmarks

Remove the ridge from the top of the cylinder:

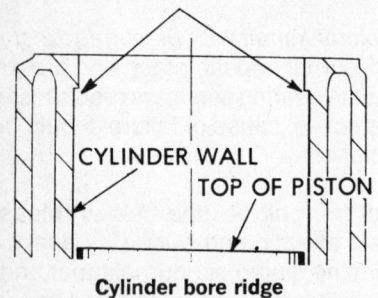

RIDGE CAUSED BY CYLINDER WEAR

CYLINDER WALL
TOP OF PISTON

Cylinder bore ridge

In order to facilitate removal of the piston and connecting rod, the ridge at the top of the cylinder (unworn area; see illustration) must be removed. Place the piston at the bottom of the bore, and cover it with a rag. Cut the ridge away using a ridge reamer, exercising extreme care to avoid cutting to deeply. Remove the rag, and remove cuttings that remain on the piston.
CAUTION: *If the ridge is not removed, and new rings are installed, damage to rings will result.*

Removing the piston and connecting rod:

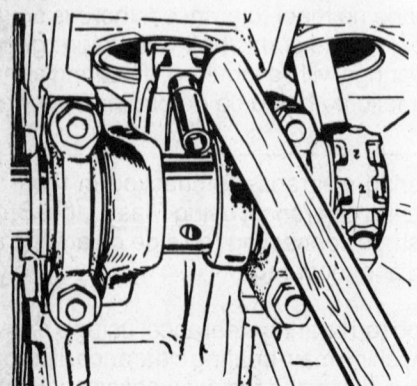

Removing the piston

Invert the engine, and push the pistons and connecting rods out of the cylinders. If necessary, tap the connecting rod boss with a wooden hammer handle, to force the piston out.
CAUTION: *Do not attempt to force the piston past the cylinder ridge* (see above).

CYLINDER BLOCK RECONDITIONING

Procedure	Method
Service the crankshaft:	Ensure that all oil holes and passages in the crankshaft are open and free of sludge. If necessary, have the crankshaft ground to the largest possible undersize. **Have the crankshaft Magnafluxed, to locate stress cracks. Consult a machinist concerning additional service procedures, such as surface hardening (e.g., nitriding, Tuftriding) to improve wear characteristics, cross drilling and chamfering the oil holes to improve lubrication, and balancing.
Removing freeze plugs:	Drill a small hole in the middle of the freeze plugs. Thread a large sheet metal screw into the hole and remove the plug with a slide hammer.
Remove the oil gallery plugs:	Threaded plugs should be removed using an appropriate (usually square) wrench. To remove soft, pressed in plugs, drill a hole in the plug, and thread in a sheet metal screw. Pull the plug out by the screw using pliers.
Hot-tank the block: NOTE: *Do not hot-tank aluminum parts.*	Have the block hot-tanked to remove grease, corrosion, and scale from the water jackets. NOTE: *Consult the operator to determine whether the camshaft bearings will be damaged during the hot-tank process.*

Check the block for cracks:	Visually inspect the block for cracks or chips. The most common locations are as follows:
	    Adjacent to freeze plugs.
	    Between the cylinders and water jackets.
	    Adjacent to the main bearing saddles.
	    At the extreme bottom of the cylinders.
	Check only suspected cracks using spot check dye (see introduction). If a crack is located, consult a machinist concerning possible repairs.
	
	**Magnaflux the block to locate hidden cracks. If cracks are located, consult a machinist about feasibility of repair.

Install the oil gallery plugs and freeze plugs:	Coat freeze plugs with sealer and tap into position using a piece of pipe, slightly smaller than the plug, as a driver. To ensure retention, stake the edges of the plugs. Coat threaded oil gallery plugs with sealer and install. Drive replacement soft plugs into block using a large drift as a driver. *Rather than reinstalling lead plugs, drill and tap the holes, and install threaded plugs.
*Check the deck height:	*The deck height is the distance from the crankshaft centerline to the block deck. To measure, invert the engine, and install the crankshaft, retaining it with the center main cap. Measure the distance from the crankshaft journal to the block deck, parallel to the cylinder centerline. Measure the diameter of the end (front and rear) main journals, parallel to the centerline of the cylinders, divide the diameter in half, and subtract it from the previous measurement. The results of the front and rear measurements should be identical. If the difference exceeds .005″, the deck height should be corrected. NOTE: *Block deck height and warpage should be corrected at the same time.*

CYLINDER BLOCK RECONDITIONING

Procedure	Method

Check the block deck for warpage:

Using a straightedge and feeler gauges, check the block deck for warpage in the same manner that the cylinder head is checked (see Cylinder Head Reconditioning). If warpage exceeds specifications, have the deck resurfaced.

NOTE: *In certain cases a specification for total material removal (Cylinder head and block deck) is provided. This specification must not be exceeded.*

Check the bore diameter and surface:

Measuring the cylinder bore with a dial gauge

Visually inspect the cylinder bores for roughness, scoring, or scuffing. If evident, the cylinder bore must be bored or honed oversize to eliminate imperfections, and the smallest possible oversize piston used. The new pistons should be given to the machinist with the block, so that the cylinders can be bored or honed exactly to the piston size (plus clearance). If no flaws are evident, measure the bore diameter using a telescope gauge and micrometer, or dial guage, parallel and perpendicular to the engine centerline, at the top (below the ridge) and bottom of the bore. Subtract the bottom measurements from the top to determine taper, and the parallel to the centerline measurements from the perpendicular measurements to determine eccentricity. If the measurements are not within specifications, the cylinder must be bored or honed, and an oversize piston installed. If the measurements are within specifications the cylinder may be used as is, with only finish honing (see below).

NOTE: *Prior to boring, check the block deck warpage, height and bearing alignment.*

CAUTION: *The 4 cyl. 140 G.M. engine cylinder walls are impregnated with silicone. Boring or honing can be done only by a shop with the proper equipment.*

TELESCOPE GAUGE 90°
FROM PISTON PIN

Measuring cylinder bore with a
telescope gauge

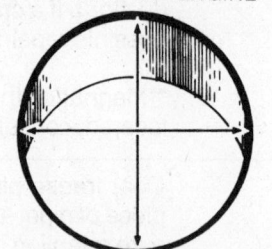

← CENTERLINE OF ENGINE →

A—AT RIGHT ANGLE TO
CENTERLINE OF ENGINE
B—PARALLEL TO
CENTERLINE OF ENGINE
Cylinder bore measuring points

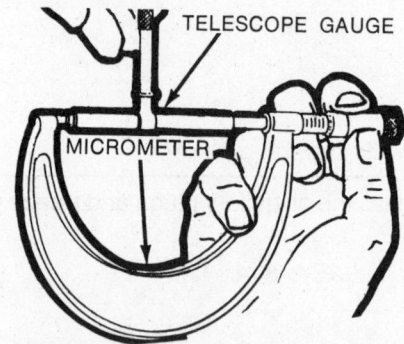

TELESCOPE GAUGE

MICROMETER

Determining cylinder bore by measuring
telescope gauge with a micrometer

Check the cylinder block bearing alignment:

Remove the upper bearing inserts. Place a straightedge in the bearing saddles along the centerline of the crankshaft. If clearance exists between the straightedge and the center saddle, the block must be alignbored.

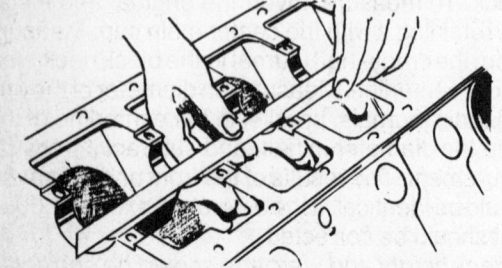

Checking main bearing saddle alignment

CYLINDER BLOCK RECONDITIONING

Procedure	Method

Clean and inspect the pistons and connecting rods:

Using a ring expander, remove the rings from the piston. Remove the retaining rings (if so equipped) and remove piston pin.

NOTE: *If the piston pin must be pressed out, determine the proper method and use the proper tools; otherwise the piston will distort.*

Clean the ring grooves using an appropriate tool, exercising care to avoid cutting too deeply. Thoroughly clean all carbon and varnish from the piston with solvent.

CAUTION: *Do not use a wire brush or caustic solvent on pistons.*

Inspect the pistons for scuffing, scoring, cracks, pitting, or excessive ring groove wear. If wear is evident, the piston must be replaced. Check the connecting rod length by measuring the rod from the inside of the large end to the inside of the small end using calipers (see illustration). All connecting rods should be equal length. Replace any rod that differs from the others in the engine.

*Have the connecting rod alignment checked in an alignment fixture by a machinist. Replace any twisted or bent rods.

*Magnaflux the connecting rods to locate stress cracks. If cracks are found, replace the connecting rod.

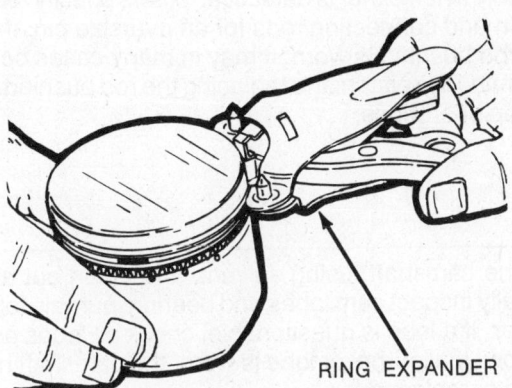

RING EXPANDER

Removing the piston rings

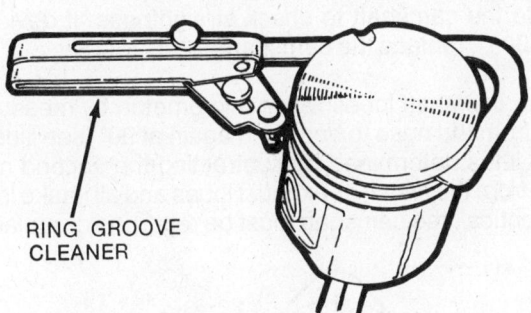

RING GROOVE CLEANER

Cleaning the piston ring grooves

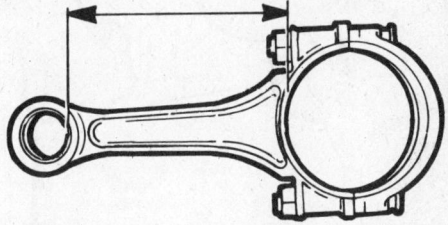

Check the connecting rod length (arrow)

Fit the pistons to the cylinders:

Using a telescope gauge and micrometer, or a dial gauge, measure the cylinder bore diameter perpendicular to the piston pin, 2½° below the deck. Measure the piston perpendicular to its pin on the skirt. The difference between the two measurements is the piston clearance. If the clearance is within specifications or slightly below (after boring or honing), finish honing is all that is required. If the clearance is excessive, try to obtain a slightly larger piston to bring clearance within specifications. Where this is not possible, obtain the first oversize piston, and hone (or if necessary, bore) the cylinder to size.

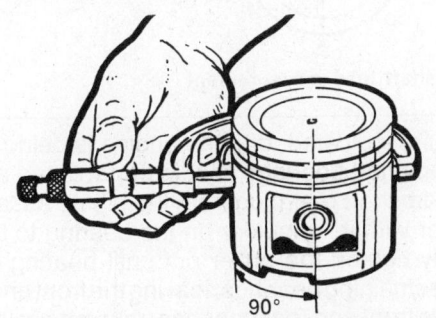

90°

Measuring the piston prior to fitting

Assemble the pistons and connecting rods:

Inspect piston pin, connecting rod small end bushing, and piston bore for galling, scoring, or excessive wear. If evident, replace defective part(s). Measure the I.D. of the piston boss and connecting rod small end, and the O.D. of the piston pin. If within specifications, assemble piston pin and rod.

CAUTION: *If piston pin must be pressed in, determine the proper method and use the proper tools; otherwise the piston will distort.*

CYLINDER BLOCK RECONDITIONING

Procedure	Method

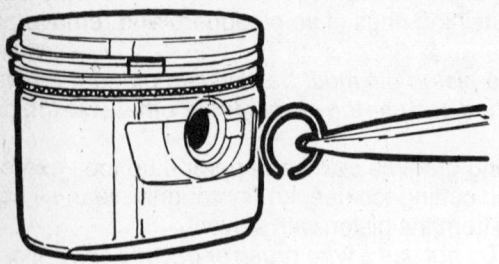

Installing piston pin lock rings

Install the lock rings; ensure that they seat properly. If the parts are not within specifications, determine the service method for the type of engine. In some cases, piston and pin are serviced as an assembly when either is defective. Others specify reaming the piston and connecting rods for an oversize pin. If the connecting rod bushing is worn, it may in many cases be replaced. Reaming the piston and replacing the rod bushing are machine shop operations.

Clean and inspect the camshaft:

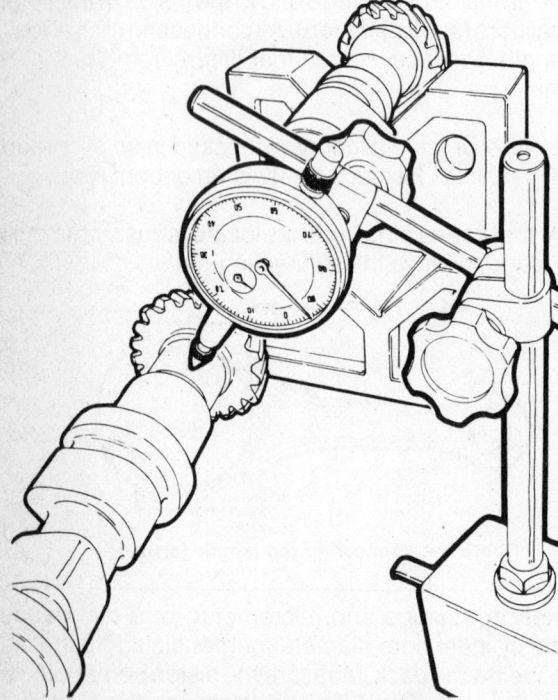

Checking the camshaft for straightness

Degrease the camshaft, using solvent, and clean out all oil holes. Visually inspect cam lobes and bearing journals for excessive wear. If a lobe is questionable, check all lobes as indicated below. If a journal or lobe is worn, the camshaft must be reground or replaced.

NOTE: *If a journal is worn, there is a good chance that the bushings are worn.*

If lobes and journals appear intact, place the front and rear journals in V-blocks, and rest a dial indicator on the center journal. Rotate the camshaft to check straightness. If deviation exceeds .001°, replace the camshaft.

*Check the camshaft lobes with a micrometer, by measuring the lobes from the nose to base and again at 90° (see illustration). The lift is determined by subtracting the second measurement from the first. If all exhaust lobes and all intake lobes are not identical, the camshaft must be reground or replaced.

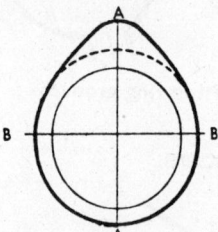

Camshaft lobe measurement

Replace the camshaft bearings (OHV engines only):

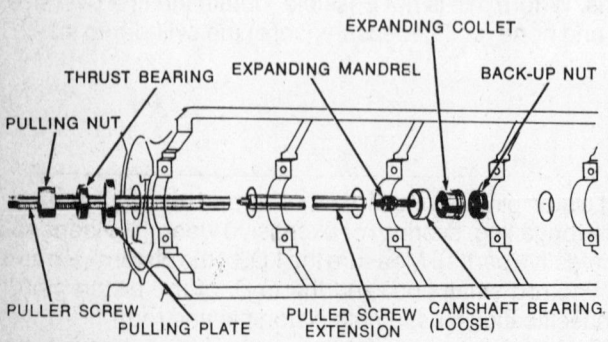

Camshaft removal and installation tool (typical)

If excessive wear is indicated, or if the engine is being completely rebuilt, camshaft bearings should be replaced as follows: Drive the camshaft rear plug from the block. Assemble the removal puller with its shoulder on the bearing to be removed. Gradually tighten the puller nut until bearing is removed. Remove remaining bearings, leaving the front and rear for last. To remove front and rear bearings, reverse position of the tool, so as to pull the bearings in toward the center of the block. Leave the tool in this position, pilot the new front and rear bearings on the installer, and pull them into position: Return the tool to its original position and pull remaining bearings into postion.

NOTE: *Ensure that oil holes align when installing bearings.*

Replace camshaft rear plug, and stake it into position to aid retention.

CYLINDER BLOCK RECONDITIONING

Procedure	Method

Finish hone the cylinders:

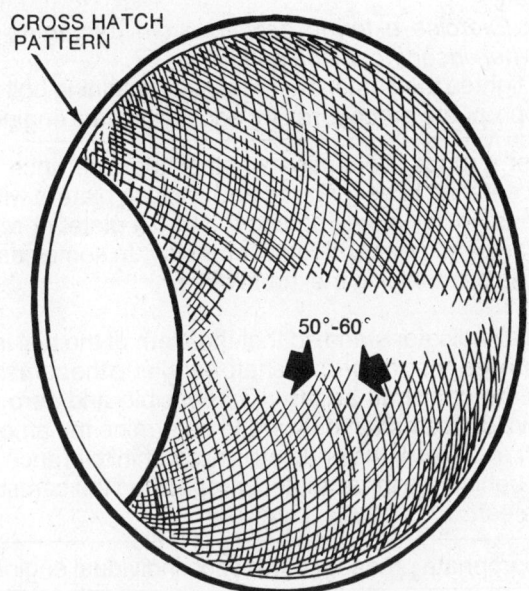

CROSS HATCH PATTERN

50°-60°

Chuck a flexible drive hone into a power drill, and insert it into the cylinder. Start the hone, and move it up and down the cylinder at a rate which will produce approximately a 60° cross-hatch pattern (see illustration).

NOTE: *Do not extend the hone below the cylinder bore.*

After developing the pattern, remove the hone and recheck piston fit. Wash the cylinders with a detergent and water solution to remove abrasive dust, dry, and wipe several times with a rag soaked in engine oil.

Check piston ring end-gap:

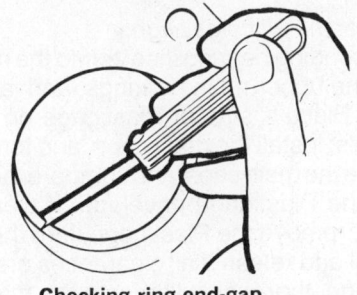

Checking ring end-gap

Compress the piston rings to be used in a cylinder, one at a time, into that cylinder, and press them approximately 1″ below the deck with an inverted piston. Using feeler gauges, measure the ring end-gap, and compare to specifications. Pull the ring out of the cylinder and file the ends with a fine file to obtain proper clearance.

CAUTION: *If inadequate ring end-gap is utilized, ring breakage will result.*

Install the piston rings:

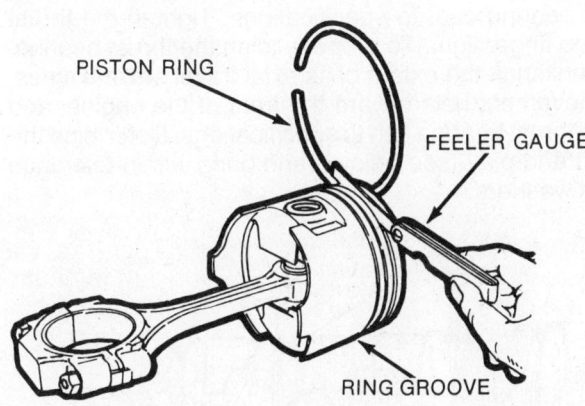

PISTON RING

FEELER GAUGE

RING GROOVE

Checking ring side clearance

Inspect the ring grooves in the piston for excessive wear or taper. If necessary, recut the groove(s) for use with an overwidth ring or a standard ring and spacer. If the groove is worn uniformly, overwidth rings, or standard rings and spacers may be installed without recutting. Roll the outside of the ring around the groove to check for burrs or deposits. If any are found, remove with a fine file. Hold the ring in the groove, and measure side clearance. If necessary, correct as indicated above.

NOTE: *Always install any additional spacers above the piston ring.*

The ring groove must be deep enough to allow the ring to seat below the lands (see illustration). In many cases, a "go-no-go" depth gauge will be provided with the piston rings. Shallow grooves may be corrected by recutting, while deep grooves require some type of filler or expander behind the piston. Consult the piston ring supplier concerning the suggested method. Install the rings on the piston, lowest ring first, using a ring expander.

NOTE: *Position the ring markings as specified by the manufacturer (see car section).*

CYLINDER BLOCK RECONDITIONING

Procedure	Method
Install the camshaft (OHV engines only):	Liberally lubricate the camshaft lobes and journals, and install the camshaft. CAUTION: *Exercise extreme care to avoid damaging the bearings when inserting the camshaft.* Install and tighten the camshaft thrust plate retaining bolts. See the appropriate procedures for each individual engine.
Check camshaft end-play (OHV engines only): 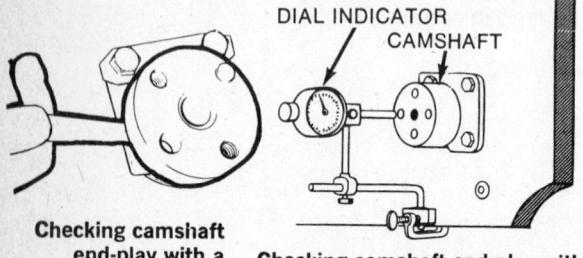 Checking camshaft end-play with a feeler gauge Checking camshaft end-play with a dial indicator	Using feeler gauges, determine whether the clearance between the camshaft boss (or gear) and backing plate is within specifications. Install shims behind the thrust plate, or reposition the camshaft gear and retest end-play. In some cases, adjustment is by replacing the thrust plate. *Mount a dial indicator stand so that the stem of the dial indicator rests on the nose of the camshaft, parallel to the camshaft axis. Push the camshaft as far in as possible and zero the gauge. Move the camshaft outward to determine the amount of camshaft endplay. If the endplay is not within tolerance, install shims behind the thrust plate, or reposition the camshaft gear and retest.
Install the rear main seal (where applicable):	See the appropriate procedures for each individual engine.
Install the crankshaft: 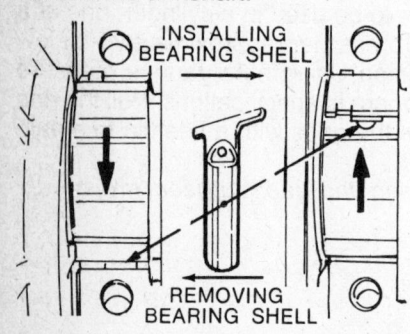 Removal and installation of upper bearing insert using a roll-out pin 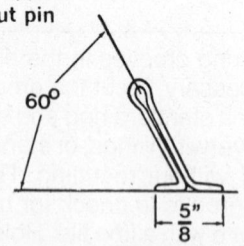 Home-made bearing roll-out pin	Thoroughly clean the main bearing saddles and caps. Place the upper halves of the bearing inserts on the saddles and press into position. NOTE: *Ensure that the oil holes align.* Press the corresponding bearing inserts into the main bearing caps. Lubricate the upper main bearings, and lay the crankshaft in position. Place a strip of Plastigage on each of the crankshaft journals, install the main caps, and torque to specifications. Remove the main caps, and compare the Plastigage to the scale on the Plastigage envelope. If clearances are within tolerances, remove the Plastigage, turn the crankshaft 90°, wipe off all oil and retest. If all clearances are correct, remove all Plastigage, thoroughly lubricate the main caps and bearing journals, and install the main caps. If clearances are not within tolerance, the upper bearing inserts may be removed, without removing the crankshaft, using a bearing roll out pin (see illustration). Roll in a bearing that will provide proper clearance, and retest. Torque all main caps, excluding the thrust bearing cap, to specifications. Tighten the thrust bearing cap finger tight. To properly align the thrust bearing, pry the crankshaft the extent of its axial travel several times, the last movement held toward the front of the engine, and torque the thrust bearing cap to specifications. Determine the crankshaft end-play (see below), and bring within tolerance with thrust washers.

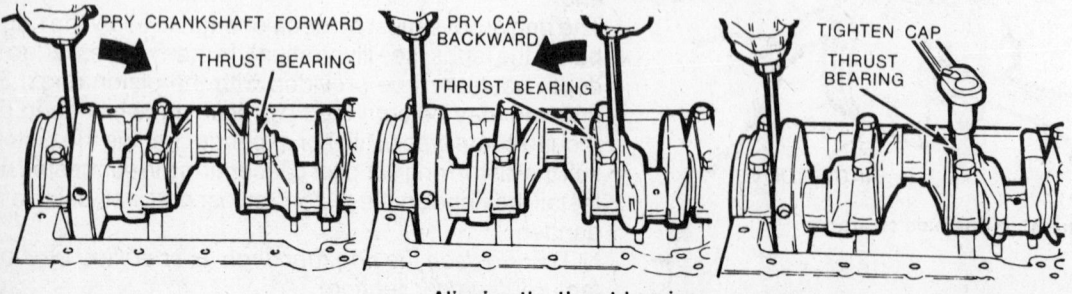

Aligning the thrust bearing

CYLINDER BLOCK RECONDITIONING

Procedure   **Method**

Measure crankshaft end-play:

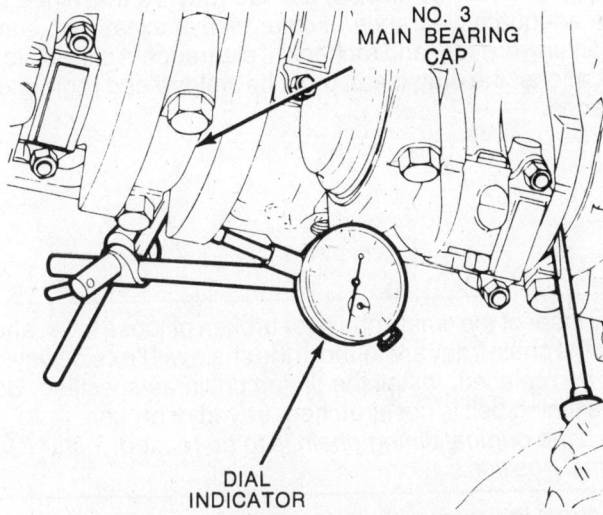

Checking crankshaft end-play with a dial indicator

Mount a dial indicator stand on the front of the block, with the dial indicator stem resting on the nose of the crankshaft, parallel to the crankshaft axis. Pry the crankshaft the extent of its travel rearward, and zero the indicator. Pry the crankshaft forward and record crankshaft end-play.

NOTE: *Crankshaft end-play also may be measured at the thrust bearing, using feeler gauges* (see illustration).

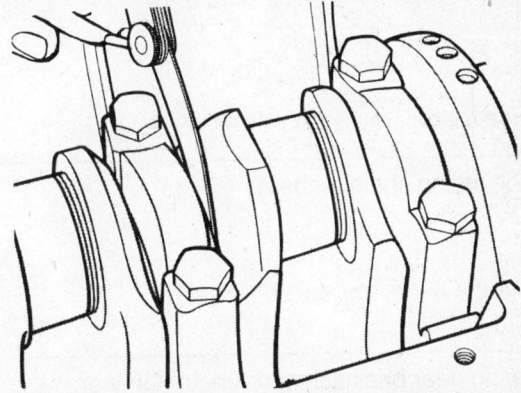

Checking crankshaft end-play with a feeler gauge

Install the pistons:

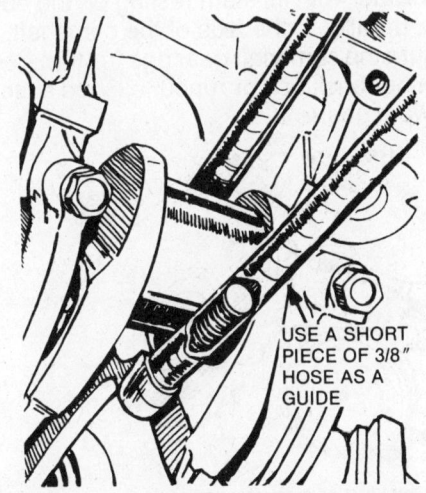

Tubing used to protect crankshaft journals and cylinder walls during piston installation

Press the upper connecting rod bearing halves into the connecting rods, and the lower halves into the connecting rod caps. Position the piston ring gaps according to specifications (see car section), and lubricate the pistons. Install a ring compressor on a piston, and press two long (8″) pieces of plastic tubing over the rod bolts. Using the tubes as a guide, press the pistons into the bores and onto the crankshaft with a wooden hammer handle. After seating the rod on the crankshaft journal, remove the tubes and install the cap finger tight. Install the remaining pistons in the same manner. Invert the engine and check the bearing clearance at two points (90° apart) on each journal with Plastigage.

NOTE: *Do not turn the crankshaft with Plastigage installed.*

If clearance is within tolerances, remove *all* Plastigage, thoroughly lubricate the journals, and torque the rod caps to specifications. If clearance is not within specifications, install different thickness bearing inserts and recheck.

CAUTION: *Never shim or file the connecting rods or caps.*

Always install plastic tube sleeves over the rod bolts when the caps are not installed, to protect the crankshaft journals.

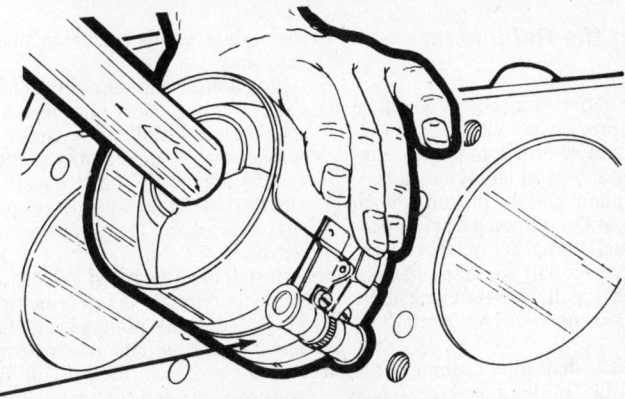

RING COMPRESSOR — Installing a piston

CYLINDER BLOCK RECONDITIONING

Procedure	Method
Check connecting rod side clearance: 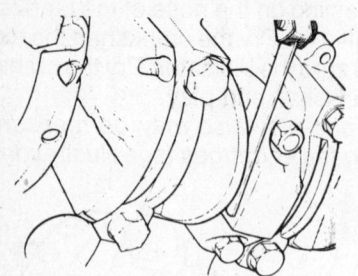 Checking connecting rod side clearance	Determine the clearance between the sides of the connecting rods and the crankshaft, using feeler gauges. If clearance is below the minimum tolerance, the rod may be machined to provide adequate clearance. If clearance is excessive, substitute an unworn rod, and recheck. If clearance is still outside specifications, the crankshaft must be welded and reground, or replaced.
Inspect the timing chain (or belt):	Visually inspect the timing chain for broken or loose links, and replace the chain if any are found. If the chain will flex sideways, it must be replaced. Install the timing chain as specified. Be sure the timing belt is not stretched, frayed or broken. NOTE: *If the original timing chain is to be reused, install it in its original position.*
Check timing gear backlash and runout (OHV engines): 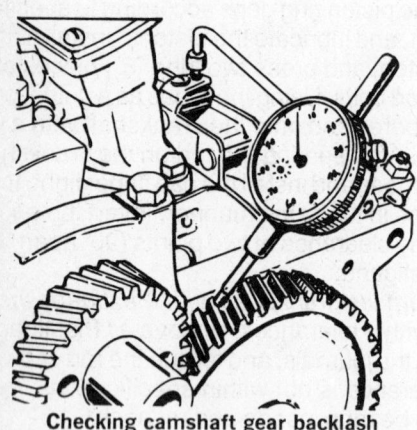 Checking camshaft gear backlash	Mount a dial indicator with its stem resting on a tooth of the camshaft gear (as illustrated). Rotate the gear until all slack is removed, and zero the indicator. Rotate the gear in the opposite direction until slack is removed, and record gear backlash. Mount the indicator with its stem resting on the edge of the camshaft gear, parallel to the axis of the camshaft. Zero the indicator, and turn the camshaft gear one full turn, recording the runout. If either backlash or runout exceed specifications, replace the worn gear(s). 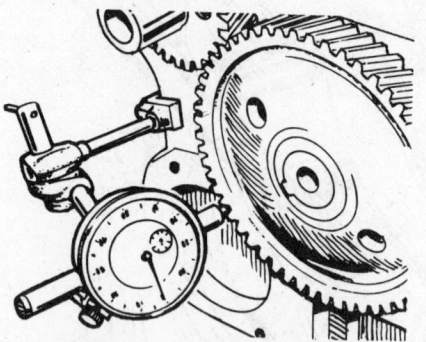 Checking camshaft gear runout

Completing the Rebuilding Process

Following the above procedures, complete the rebuilding process as follows:

Fill the oil pump with oil, to prevent cavitating (sucking air) on initial engine start up. Install the oil pump and the pickup tube on the engine. Coat the oil pan gasket as necessary, and install the gasket and the oil pan. Mount the flywheel and the crankshaft vibration damper or pulley on the crankshaft. NOTE: *Always use new bolts when installing the flywheel.*

Inspect the clutch shaft pilot bushing in the crankshaft. If the bushing is excessively worn, remove it with an expanding puller and a slide hammer, and tap a new bushing into place.

Position the engine, cylinder head side up. Lubricate the lifters, and install them into their bores. Install the cylinder head, and torque it as specified. Insert the pushrods (where applicable), and install the rocker shaft(s) (if so equipped) or position the rocker arms on the pushrods. Adjust the valves.

Install the intake and exhaust manifolds, the carburetor(s), the distributor and spark plugs. Adjust the point gap and the static ignition timing. Mount all accessories and install the engine in the car. Fill the radiator with coolant, and the crankcase with high quality engine oil.

Break-in Procedure

Start the engine, and allow it to run at low speed for a few minutes, while checking for leaks. Stop the engine, check the oil level, and fill as necessary. Restart the engine, and fill the cooling system to capacity. Check the point dwell angle and adjust the ignition timing and the valves. Run the engine at low to medium speed (800–2500 rpm) for approximately ½ hour, and retorque the cylinder head bolts. Road test the car, and check again for leaks.

Follow the manufacturer's recommended engine break-in procedure and maintenance schedule for new engines.

Manual Transmissions

HOW TO IDENTIFY A TRANSMISSION

Refer to the Basic Manual Transmission Application chart to determine if more than one type of transmission could have been used in your vehicle. If only one type is applicable, refer to the page index of the Manual Transmission Identification and Page Index chart, which will tell you where the overhaul procedures for your transmission are located. Several special tools may be required for disassembly or assembly of various transmission components; whenever necessary, these special tools are identified by their OEM part number. By using the OEM part number, an equivalent special tool can be cross-referenced to various aftermarked tool suppliers.

Should more than one type of transmission be listed under the same basic description (e.g. two different 4 speeds), then refer to the Manual Transmission Identification and Page Index chart. This chart contains identifying characteristics of each transmission which will enable you to make a positive identification. In many cases, it will be necessary to view the transmission from beneath the vehicle to correctly identify a particular model.

CAUTION
Care must be exercised during the disassembly and assembly of the manual transmission due to the usage of metric nuts and bolts. Proper wrenches and sockets should be used to avoid damage to the transmission and fasteners. DO NOT attempt to interchange metric threaded fasteners with U.S. Fine or Standard fasteners as damage can result.

BASIC MANUAL TRANSMISSION APPLICATIONS

Manufacturer	Model	Year	Transmission Type 3 Speed	4 Speed	5 Speed
American Motors	All	1980–83	–	6	–
	Eagle	1982–87	–	21	22
Chrysler Corporation	Rear Wheel Drive	1980	1	3	–
	Front Wheel Drive	1980–83	–	18	–
		1981–86	–	4	–
		1983–87	–	–	19
Ford Motor Company	Rear Wheel Drive	1980–87	–	5	–
		1984–87	–	–	7
		1980–82	–	6	–
		1980–83	–	8	–
		1981–83	–	–	9
	Front Wheel Drive	1981–87	–	2	2

BASIC MANUAL TRANSMISSION APPLICATIONS

Manufacturer	Model	Year	Transmission Type		
			3 Speed	4 Speed	5 Speed
General Motors	Rear Wheel Drive except Chevette (64–87), Corvette	1980–81	10	–	–
		1980–84	–	11,24	–
		1982–86	–	–	22
		1980–83	–	14	–
	Corvette	1984–87	–	15	–
	Chevette	1980–81	–	12	–
		1982–87 ①	–	12	22
		1982–87 ②	–	–	23
	Front Wheel Drive except Nova, Sprint	1980–87	–	20	17
	Nova	1985–87	–	–	16
	Sprint	1985–87	–	–	13

① Gasoline engines only
② Diesel engines only

MANUAL TRANSMISSION IDENTIFICATION

Chilton Type	Transmission Designation				Identifying Characteristics
	AMC	Chrysler	Ford	GM	
1	–	A-230	–	–	A,D,F
2	–	–	MTX	–	E,G,H
3	–	Overdrive–4	–	–	A,D,G or G_1
4	–	A–460	–	–	E,G,L
5	–	–	ET	–	B,C,G,I
6	SR4	–	RAD	–	B,C,G,J
7	–	–	T50D	–	B,C,H
8	–	–	RUG	–	B,C,G_1
9	–	–	RAP	–	B,C,H_1
10	–	–	–	Saginaw	A,D,F
11	–	–	–	76mm	A_2,D,G,K
12	–	–	–	70mm	C,G,I
13	–	–	–	MTX	E,H
14	–	–	–	Warner T–10	A_1,D,G,P
15	–	–	–	83mm	G_1
16	–	–	–	C51	E,H
17	–	–	–	76mm	E,H,L
18	–	A–12	–	–	E,G,R
19	–	A–465/A–525	–	–	E,G,S,H
20	–	–	–	76mm	E,G,L
21	Warner T–4	–	–	–	B,C,G,N

MANUAL TRANSMISSION IDENTIFICATION

Chilton Type	Transmission Designation				Identifying Characteristics
	AMC	Chrysler	Ford	GM	
22	Warner T–5	–	–	77mm	B,C,H$_1$,N
23	–	–	–	69.5mm	C,H$_1$,O
24	–	–	–	Muncie	A$_2$,D,G,L,Q

A Side cover
A$_1$ 6 bolt side cover
A$_2$ 7 bolt side cover
B Top cover
B$_1$ 6 bolt top cover
B$_2$ 9 bolt top cover
C Internal shift linkage
D External linkage shift
E Transaxle
F Three-speed
G Four-speed
G$_1$ Four-speed overdrive
H Five-speed

H$_1$ Five-speed overdrive
I The belklhousing is bolted to the transmission case from INSIDE the bellhousing—bolts not exposed while installed.
J Transmission case-to-bellhousing bolts are accesible with the transmission installed. Also, reverse is positioned at the upper left portion of the gearshift pattern.
K Cast iron case
L Aluminun case
M Overdrive unit only

N Reverse is positioned at the right rear portion of the gearshift pattern[
O No cover hon transmission case: bellhousing intergral with transmission case.
P Side cover has curved bottom
Q Side cover has straight bottom
R Starter motor located on radiator side of the engine compartment, near the radiator fan motor.
S Starter motor located near the firewall side of the engine compartment.
NA Not available

CHILTON TYPE 1

Transmission Case

DISASSEMBLY

Shift Housing and Mechanism

1. Shift to the 2nd gear.
2. Unbolt and remove the side cover with the shift mechanism.
3. If the shaft O-ring seals need replacement:
  a. Pull the shaft forks out of the shafts.
  b. Remove the nuts and the operating levers from the shafts.
  c. Deburr and remove the shafts, then remove the O-ring retainers and the O-rings.

Drive Pinion Retainer and Extension Housing

1. At the front of the transmission case, remove the pinion bearing retainer bolts, the retainer and the gasket, then pry off the retainer oil seal.
2. Using a brass drift, tap drive pinion forward as far as possible. Rotate the cut away part of the 2nd gear to face to the countershaft gear. Shift the 2nd–3rd synchronizer sleeve forward.
3. Remove the speedometer pinion adapter retainer, then work the adapter and the pinion out of the extension housing.
4. Remove the extension housing bolts. Using a plastic hammer, break the housing loose and carefully remove it.

Idler Gear and Mainshaft

1. Insert a dummy shaft into the case to push the Reverse idler shaft and key out of the case.
2. Remove the dummy shaft and the idler gear together to prevent losing the roller bearings.
3. Remove both tanged idler gear thrust washers.
4. Remove the mainshaft assembly through the rear of the case.

Countershaft Gear and Drive Pinion

1. Using a mallet and a dummy shaft, tap the countershaft rearward enough to remove the key, then drive the countershaft out of the case.

NOTE: When removing the countershaft, maintain contact between the countershaft and the dummy shaft so that the washers will not drop out.

2. Lower the countershaft gear to the bottom of the case.
3. Remove the snap ring from the pinion bearing outer race (outside front of the case).
4. Using a plastic hammer, drive the pinion shaft into the case, then remove the assembly through the rear of the case.
5. If the bearing is to be replaced, remove the snap ring and press off the bearing.
6. Remove the countershaft gear and the dummy shaft out through the rear of the case.

Mainshaft

1. Remove the snap ring from the front end of the mainshaft and the 2nd gear stop ring. Remove the synchronizer and the 2nd gear from the mainshaft.
2. Spread the snap ring in the mainshaft bearing retainer and slide the retainer back off the bearing race.
3. Remove the snap ring at the rear of the mainshaft. Support the front side of the Reverse gear and press the bearing off the mainshaft.

NOTE: When removing the bearing from the mainshaft, be careful not to let the parts drop when the bearing clears the shaft.

4. Remove the mainshaft bearing and the Reverse gear from the shaft.
5. Remove the snap ring from the rear of the shaft, then slide the 1st–Reverse synchronizer assembly off the splines (remove rearward). Remove the stop ring and the 1st gear through the rear.

ASSEMBLY

Countershaft Gear

1. Slide the dummy shaft into the countershaft gear.
2. Slide one roller thrust washer over the dummy shaft and into the gear, followed by the 22 greased rollers.
3. Repeat Step 2, adding one roller thrust washer on the end.
4. Repeat Steps 2 and 3 at other end of the countershaft gear. There are a total of 88 rollers and 6 thrust washers.
5. Place the greased front thrust washer on the dummy shaft against the gear with the tangs forward.

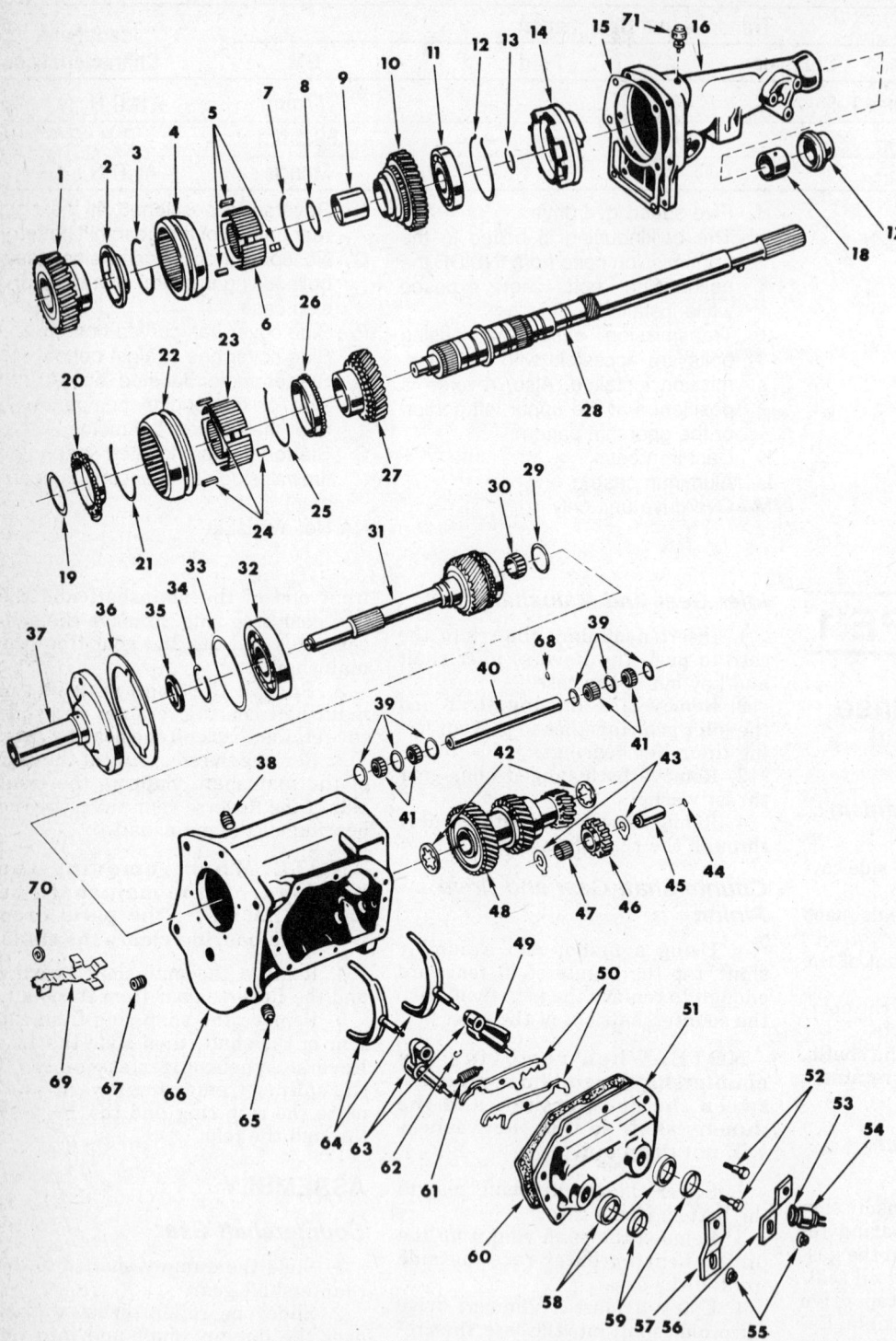

1. 1st speed gear
2. Ring
3. Spring
4. Sleeve
5. Struts
6. Clutch gear
7. Spring
8. Snap ring
9. Reverse gear bushing
10. Reverse gear
11. Output shaft bearing
12. Snap ring
13. Snap ring
14. Bearing retainer
15. Extension housing gasket
16. Extension seal
17. Seal
18. Extension bushing
19. Snap ring
20. Synchronizer ring
21. Spring
22. Sleeve
23. Clutch gear
24. Struts
25. Spring
26. Ring
27. 2nd speed gear
28. Output shaft
29. Snap ring
30. Roller
31. Main drive gear (pinion)
32. Drive pinion bearing
33. Snap ring
34. Snap ring
35. Drive pinion seal
36. Retainer gasket
37. Bearing retainer
38. Plug
39. Countershaft bearing washer
40. Countershaft
41. Roller
42. Thrust washer
43. Rev. idler gear shaft
44. Key
45. Rev. idler gear shaft
46. Reverse idler gear
47. Rev. roller
48. Countershaft gear
49. Low and rev. lever
50. Interlock lever
51. Housing
52. Bolt
53. Gasket
54. Back-up light switch
55. Locking nut
56. Lever
57. Lever
58. Seal
59. Retainer
60. Housing gasket
61. Interlock spring
62. Snap ring
63. Lever w/shaft
64. Gearshift fork
65. Plug
66. Case
67. Expansion plug
68. Countershaft key
69. Clip
70. Magnet
71. Extension vent

Exploded view of the Chrysler A-230 (Chilton Type 1) transmission

6. Grease the rear thrust washer and install it in the case, with the tangs rearward. Place the countershaft gear assembly in the bottom of the transmission case until the drive pinion is installed.

Pinion Gear

1. Press the new bearing onto the pinion shaft with the snap ring groove forward, then install the new snap ring.

2. Install the 15 rollers and the retaining ring in the drive pinion gear.
3. Install the drive pinion and the bearing assembly into the case.
4. To install the countershaft gear

assembly, position it and the thrust washers so that the countershaft can be tapped into position, then insert the key into the countershaft.

NOTE: Be careful to keep the countershaft against the dummy shaft to keep the parts from falling between them.

5. Tap the drive pinion forward for clearance.

Mainshaft

1. Place a stop ring flat on the bench, followed by a clutch gear and a sleeve. Drop the struts into their slots and snap in a strut spring placing the tang inside one strut. Turn the assembly over and install 2nd strut spring (with the tang in a different strut).
2. Slide the 1st gear and the stop ring over the rear of the mainshaft and against the thrust flange between the 1st and 2nd gears on the shaft.
3. Slide the 1st–Reverse synchronizer assembly over the rear of the mainshaft, indexing the hub slots to the 1st gear stop ring lugs.
4. Install the 1st–Reverse synchronizer clutch gear snap ring on the mainshaft.
5. Slide the Reverse gear and the mainshaft bearing into place. Press the bearing onto the shaft, by supporting the inner race of the bearing. Be sure the snap ring groove on the outer race is forward.
6. Install the bearing retaining snap ring on the mainshaft. Spread the snap ring in the retainer groove and slide it over the bearing, then seat the ring in the groove.

NOTE: The snap ring is selected for minimum end play. There are several thicknesses available.

7. Place the 2nd gear over the front of the mainshaft with the thrust surface against the flange.
8. Install the stop ring and the 2nd–3rd synchronizer assembly against the 2nd gear, followed by the 2nd–3rd synchronizer clutch gear snap ring onto the shaft.
9. Move the 2nd–3rd synchronizer sleeve forward as far as possible. Install the front stop ring inside the sleeve with lugs indexed to the struts. Coat the stop ring with grease to hold it in position.
10. To provide clearance, rotate the cut out on the 2nd gear toward the countershaft gear.
11. Insert the mainshaft assembly into the case, then tilt the assembly to clear the cluster gears and insert the pilot rollers into the drive pinion gear.

NOTE: If the assembly is installed correctly, the bearing re-

tainer will bottom in the case without force; if not, check for a misplaced strut, a pinion roller or a stop ring.

Reverse Idler Gear

1. Install the dummy shaft into the idler gear followed by the 22 greased rollers.
2. Position the Reverse idler thrust washers in case with grease.
3. Position the idler gear and dummy shaft assembly in the case, then install the idler shaft and key.

Extension Housing

1. Remove the extension housing yoke seal and drive the bushing out of the housing.
2. Align the oil hole in the bushing with the oil slot in the housing, then drive the bushing and the new seal into the housing.
3. Install the extension housing with the gasket to hold the mainshaft and the bearing retainer in place.

Drive Pinion Bearing Retainer

1. Install the outer snap ring drive pinion bearing, then tap the assembly back until the snap ring contacts the case.
2. Install a new seal in the retainer bore.
3. Position the main drive pinion bearing retainer with gasket onto the front of the case. Coat the threads with sealing compound, then install and torque the bolts to 30 ft. lbs.

Gearshift Mechanism and Housing

1. If removed, place the 2 interlock levers on the pivot pin with the spring hangers offset toward each other, so that the spring is installed in a straight line. Place the E-clip on the pivot pin.
2. Grease and install new the O-ring seals on both shift shafts. Grease the housing bores and push each shaft into its bore.
3. Install the spring on the interlock lever hangers.
4. Rotate each shift shaft fork bore into the vertical position and install the shift forks through the bores and under both interlock levers.
5. Position the 2nd–3rd synchronizer sleeve to the rear, in the 2nd gear position. Position the 1st–Reverse synchronizer sleeve to the middle of the travel, in the Neutral position. Place the shift forks in the Neutral positions.
6. Install the gasket and the gearshift mechanism, then torque the bolts to 15 ft. lbs.

NOTE: The bolt with the extra long shoulder must be installed at the center rear of the case.

7. Install the speedometer drive pinion gear and the adapter.

NOTE: The range number on the adapter, which represents the number of teeth on the gear, should be in the 6 o'clock position.

CHILTON TYPE 2

Transaxle Case

DISASSEMBLY

4 Speed

1. Insert a drift into the input shift shaft hole and shift the transaxle into Neutral by either pushing or pulling the shaft into the center detent position.
2. Place the transaxle on a bench with the clutch housing face down and drain the transaxle fluid.
3. Remove the Reverse idler shaft retaining bolt.
4. Remove the detent plunger retaining screw. Using a magnet, remove the detent spring and the plunger.

NOTE: Label these parts for they appear similar to the input shift shaft plunger and spring contained in the clutch case.

5. Using a 19mm socket, remove the shift fork interlock sleeve retaining pin.
6. Using a 10mm socket, remove the clutch housing-to-transaxle case.
7. Tap the transaxle case with a plastic tipped hammer to break the seal between the case halves.

— **CAUTION** —
While separating the case halves, be careful that the tapered roller bearing cups or shims do not drop out. DO NOT insert pry bars between the case halves.

8. Remove the case magnet. Remove the Reverse idler shaft and gear by lifting the shaft straight upward.
9. Using a 4mm Allen wrench, remove the set screw from the shift lever assembly.
10. Using a pair of pliers, rotate the shift lever shaft 90° to disengage the Reverse inhibitor plunger from the detent notch in the shift lever shaft. Slide the shaft toward the differential (away from the expansion plug in the clutch housing) and remove the shift lever assembly.

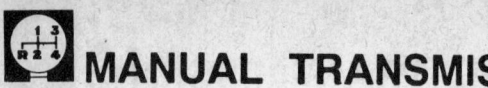

NOTE: If equipped with a 4.05:1 final drive ratio, it may be necessary to tilt the differential assembly slightly, to remove the shift lever assembly.

11. Remove the main shaft assembly, the input cluster shaft assembly and the main shift control shaft assembly as one unit.

12. Lift the differential and final drive gear assembly from the clutch housing case.

5 Speed

NOTE: The following procedures are used in conjunction with the 4 speed procedures.

1. Remove the C-clip retaining ring from the 5th gear shift relay lever.

2. Remove the 5th gear shift relay lever.

3. Using a punch, drive the roll spring pin from the shift lever shaft.

4. Gently pry the shift shaft lever to remove it from the case.

5. To disengage the inhibitor, insert a punch into the shift shaft lever hole and rotate the shaft 90°.

6. To prevent the ball and inhibitor spring from shooting out, hold a rag over the hole in the lever.

7. Remove the shift shaft, then the ball and spring from the inhibitor.

8. Remove the shift lever and the 3rd–4th bias spring as an assembly.

9. Remove the 5th gear shaft and gear fork assembly from the case.

ASSEMBLY

1. Place the differential and the final drive gear assembly into the clutch housing case.

2. Position the main shift control shaft assembly, so that the shift forks engage their respective slots in the synchronizer sleeves, on the main shaft assembly.

3. Mesh the main shaft assembly with the input cluster shaft assembly. Hold the input cluster shaft assembly, the main shaft and the main shift control shaft in their respective working positions, lower them into their bores in the clutch housing case as a unit.

4. Position the shift lever assembly in its working position, with one shift lever pin located in the socket of the input shift shaft selector plate arm assembly and the other in the socket of the main shift control shaft block. Slide the shift lever shaft through the shift lever and into its clutch housing bore. Rotate the shift lever shaft so the Reverse inhibitor notch faces the Reverse inhibitor plunger.

5. Position the shift lever shaft so the set screw hole on the shaft aligns with the hole in the shift lever.

6. Make sure the selector pin is in the Neutral gate of the control selector plate, then position the fork selector arm finger so that it is partially engaged with the 1st–2nd fork and the 3rd–4th fork.

7. Engage the Reverse idler gear groove with the pin at the end of the Reverse relay lever, then slide the shaft through the gear and into its bore. Align the retaining screw hole in the case.

NOTE: This will allow proper alignment between the Reverse idler shaft retaining screw hole in the transaxle case when the case is placed over the assembly.

8. Install the magnet in the clutch housing case pocket.

9. Apply a $\frac{1}{16}$ in. wide bead of sealer to the clean surface of the clutch housing. Carefully lower the transaxle case over the clutch housing case, then move it gently until the shift control shaft, the main shaft and the input cluster shaft align with their respective bores in the transaxle case. Gently slide the transaxle case over the dowels and flush onto the clutch housing case. Make sure that the case does not bind on the magnet.

10. Install the 14 transaxle case-to-clutch housing bolts and torque to 13–17 ft. lbs. using a 10mm socket.

11. Using a drift, align the bore in the Reverse idler shaft with the retaining screw hole in the transaxle case.

12. Install the Reverse idler shaft retaining bolt and torque to 16–20 ft. lbs.

13. Apply Teflon pipe sealant to the threads of the interlock sleeve retaining pin. Using a drift, align the interlock sleeve slot with the hole in the transaxle case, then install the retaining pin and torque to 12–15 ft. lbs.

14. Apply Teflon pipe sealant to the threads of the detent plunger retaining screw. Install the detent plunger and spring, then torque the retaining screw to 9–12 ft. lbs.

15. Place the transaxle in an upright position. Insert a drift through the hole in the input shift shaft, then shift the transaxle in and out of the gears to check the assembly.

Mainshaft

DISASSEMBLY

1. Using a puller and an arbor press, remove the tapered roller bearing from the pinion end of the main shaft, then label the bearing for proper installation.

NOTE: The bearing does not have to be removed to disassemble the main shaft, only to replace it (if damaged).

2. Remove and label the bearing on the 4th gear end of the shaft.

3. Remove the 4th gear and synchronizer blocking ring.

4. Remove the 3rd–4th synchronizer retaining ring.

5. Slide the 3rd–4th gear synchronizer assembly, the blocking ring and the 3rd speed gear from the shaft.

6. Remove the 2nd–3rd thrust washer retaining ring and the two piece 2nd–3rd gear thrust washer.

7. Remove the 2nd speed gear and the blocking ring.

8. Remove the 1st–2nd synchronizer retaining ring.

9. Slide the 1st–2nd synchronizer assembly, the blocking ring and the 1st speed gear off the shaft.

NOTE: If equipped with a 5 speed, proceed with the following steps:

1. Remove the bearing from the gear end of the shaft.

2. Remove the 5th gear, the blocking ring and the synchronizer assembly.

3. Press the bearing from the pinion end of the shaft.

NOTE: This bearing must be pressed on and off.

Assembly

1. Clean, inspect and lightly oil the parts with the appropriate transaxle fluid.

NOTE: Before assembling the synchronizers note the following points:

a. All index marks must be aligned.

b. Place the tab of the synchronizer spring into the groove on one of the inserts and snap the spring into place. Place the tab of the other spring into the same insert on the other side of the synchronizer assembly, then rotate the spring in the opposite direction and snap it into place.

c. The sleeve and the hub have an extremely close fit and must be held square to prevent jamming. Do not force the sleeve onto the hub.

2. Slide the blocking ring and the 1st speed gear onto the main shaft. Slide the 1st–2nd synchronizer assembly into place, making sure that the shift fork groove on the reversing slide gear faces the 1st speed gear. Install the synchronizer retaining ring.

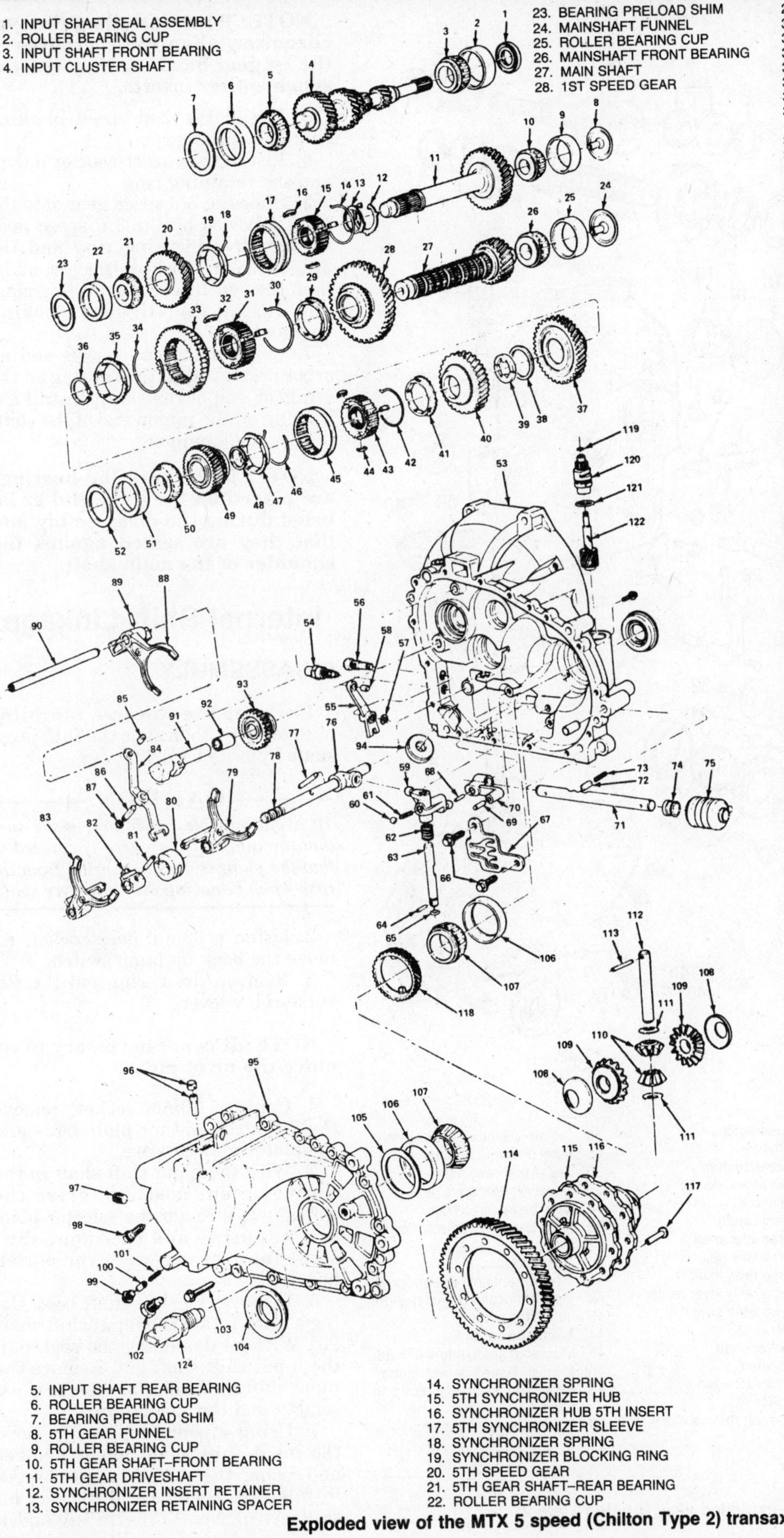

1. INPUT SHAFT SEAL ASSEMBLY
2. ROLLER BEARING CUP
3. INPUT SHAFT FRONT BEARING
4. INPUT CLUSTER SHAFT

23. BEARING PRELOAD SHIM
24. MAINSHAFT FUNNEL
25. ROLLER BEARING CUP
26. MAINSHAFT FRONT BEARING
27. MAIN SHAFT
28. 1ST SPEED GEAR

29. SYNCHRONIZER BLOCKING RING
30. SYNCHRONIZER SPRING
31. 1ST/2ND SYNCHRONIZER HUB
32. SYNCHRONIZER HUB 1ST/2ND INSERT
33. REVERSE SLIDING GEAR
34. SYNCHRONIZER SPRING
35. SYNCHRONIZER BLOCKING RING
36. 1ST/2ND SYNCHRONIZER RETAINING RING
37. 2ND SPEED GEAR
38. 2ND/3RD THRUST WASHER RETAINING RING
39. 2ND/3RD GEAR THRUST WASHER
40. 3RD SPEED GEAR
41. SYNCHRONIZER BLOCKING RING
42. SYNCHRONIZER SPRING
43. 3RD/4TH SYNCHRONIZER HUB
44. SYNCHRONIZER HUB 3RD/4TH INSERT
45. 3RD/4TH SYNCHRONIZER SLEEVE
46. SYNCHRONIZER SPRING
47. SYNCHRONIZER BLOCKING RING
48. 3RD/4TH SYNCHRONIZER RING
49. 4TH SPEED GEAR
50. MAINSHAFT REAR BEARING
51. ROLLER BEARING CUP
52. BEARING PRELOAD SHIM
53. CLUTCH HOUSING CASE
54. SWITCH ASSEMBLY BACK-UP LAMPS
55. REVERSE RELAY LEVER
56. REVERSE RELAY LEVER PIVOT PIN
57. EXTERNAL RETAINING RING
58. REVERSE RELAY LEVER PIN
59. SHIFT LEVER
60. 10.319mm BALL
61. 5TH/REVERSE INHIBITOR SPRING
62. 3RD/4TH SHIFT BIAS SPRING
63. SHIFT LEVER SHAFT
64. SHIFT LEVER PIN
65. SHIFT LEVER SHAFT SEAL
66. SHIFT GATE ATTACHING BOLTS
67. SHIFT GATE PLATE
68. SELECTOR ARM PIN
69. SHIFT GATE SELECTOR PIN
70. SHIFT GATE SELECTOR ARM
71. INPUT SHIFT SHAFT
72. SHIFT SHAFT DETENT PLUNGER
73. SHIFT SHAFT DETENT SPRING
74. ASSEMBLY–SHIFT SHAFT SEAL
75. SHIFT SHAFT BOOT
76. FORK CONTROL SHAFT BLOCK
77. REVERSE RELAY LEVER ACTUATING PIN
78. MAIN SHIFT FORK CONTROL SHAFT
79. 1ST/2ND FORK
80. FORK INTERLOCK SLEEVE
81. SPRING PIN
82. FORK SELECTOR ARM
83. 3RD/4TH FORK
84. 5TH SHIFT RELAY LEVER
85. REVERSE SHIFT RELAY LEVER PIN
86. 5TH RELAY LEVER PIVOT PIN
87. EXTERNAL RETAINING RING
88. 5TH FORK
89. 5TH FORK RETAINING PIN
90. 5TH FORK CONTROL SHAFT
91. REVERSE IDLER GEAR SHAFT
92. REVERSE IDLER GEAR BUSHING
93. REVERSE IDLER GEAR
94. CASE MAGNET
95. TRANSAXLE CASE
96. VENT ASSEMBLY
97. FILL PLUG
98. REVERSE SHAFT RETAINING BOLT
99. DETENT PLUNGER RETAINING SCREW
100. SHIFT SHAFT DETENT PLUNGER
101. SHIFT SHAFT DETENT SPRING
102. FORK INTERLOCK SLEEVE RETAINING PIN
103. TRANSAXLE CASE BOLT
104. SEAL ASSEMBLY (LH) DIFFERENTIAL
105. SHIM DIFFERENTIAL BEARING PRELOAD
106. DIFFERENTIAL BEARING CUP
107. DIFFERENTIAL BEARING ASSEMBLY
108. SIDE GEAR THRUST WASHER
109. SIDE GEAR
110. PINION GEAR
111. PINION GEAR THRUST WASHER
112. PINION GEAR SHAFT
113. PINION GEAR SHAFT RETAINING PIN
114. FINAL DRIVE GEAR
115. DIFFERENTIAL (LH) CASE
116. DIFFERENTIAL (RH) CASE
117. CASE AND DRIVE GEAR ATTACHING RIVET
118. SPEEDO DRIVE GEAR
119. 5.16mm × 1.6 O-RING SEAL
120. SPEEDO GEAR RETAINER
121. SPEEDO RETAINER-TO-CASE SEAL
122. SPEEDO DRIVEN GEAR
123. CASE-TO-CLUTCH HOUSING DOWEL
124. TRANSAXLE NEUTRAL SENSING SWITCH

5. INPUT SHAFT REAR BEARING
6. ROLLER BEARING CUP
7. BEARING PRELOAD SHIM
8. 5TH GEAR FUNNEL
9. ROLLER BEARING CUP
10. 5TH GEAR SHAFT–FRONT BEARING
11. 5TH GEAR DRIVESHAFT
12. SYNCHRONIZER INSERT RETAINER
13. SYNCHRONIZER RETAINING SPACER

14. SYNCHRONIZER SPRING
15. 5TH SYNCHRONIZER HUB
16. SYNCHRONIZER HUB 5TH INSERT
17. 5TH SYNCHRONIZER SLEEVE
18. SYNCHRONIZER SPRING
19. SYNCHRONIZER BLOCKING RING
20. 5TH SPEED GEAR
21. 5TH GEAR SHAFT–REAR BEARING
22. ROLLER BEARING CUP

Exploded view of the MTX 5 speed (Chilton Type 2) transaxle

NOTE: When installing the synchronizer, align the 3 grooves in the 1st gear blocking ring with the synchronizer inserts.

3. Install the 2nd speed blocking ring and speed gear.

4. Install the thrust washer halves and the retaining ring.

5. Slide the 3rd speed gear onto the shaft followed by the 3rd speed gear synchronizer blocking ring and the 3rd–4th gear synchronizer assembly, then the synchronizer retaining ring.

6. Install the 4th gear blocking ring and gear.

7. Using a $1\frac{1}{16}$ in. socket and an arbor press, install the bearing on the 4th gear end of the shaft. Install the bearing on the pinion end of the shaft in the same manner.

NOTE: Make sure the bearings are placed on the same end as labeled during the disassembly and that they are seated against the shoulder of the main shaft.

Internal Shift Linkage

DISASSEMBLY

1. Cover the Reverse inhibitor plunger bore, then slide the shift lever shaft from its bore.

—————— **CAUTION** ——————

To avoid possible eye injury, make sure that the inhibitor bore area is covered so that the plunger does not spring from the case when removing the shift lever shaft.

2. Using a 30mm deep socket, remove the back up lamp switch.

3. Remove the C-clip and the Reverse relay lever.

NOTE: It is not necessary to remove the pivot pin.

4. Using a 10mm socket, remove the 2 control selector plate bolts and the plate from the case.

5. With the input shift shaft in the center detent position, drive the spring pin through the selector plate arm assembly and the input shift shaft into the recess of the clutch housing.

6. Remove the shift shaft boot. Using a drift, rotate the input shift shaft 90°. Without damaging the seal, pull the input shift shaft out. Remove the input shift shaft selector plate arm assembly and the spring pin.

7. Using a pencil magnet, remove the input shift shaft detent plunger and spring, then label them for proper installation. Using a seal remover/installer tool No. T77F-7288-A and a

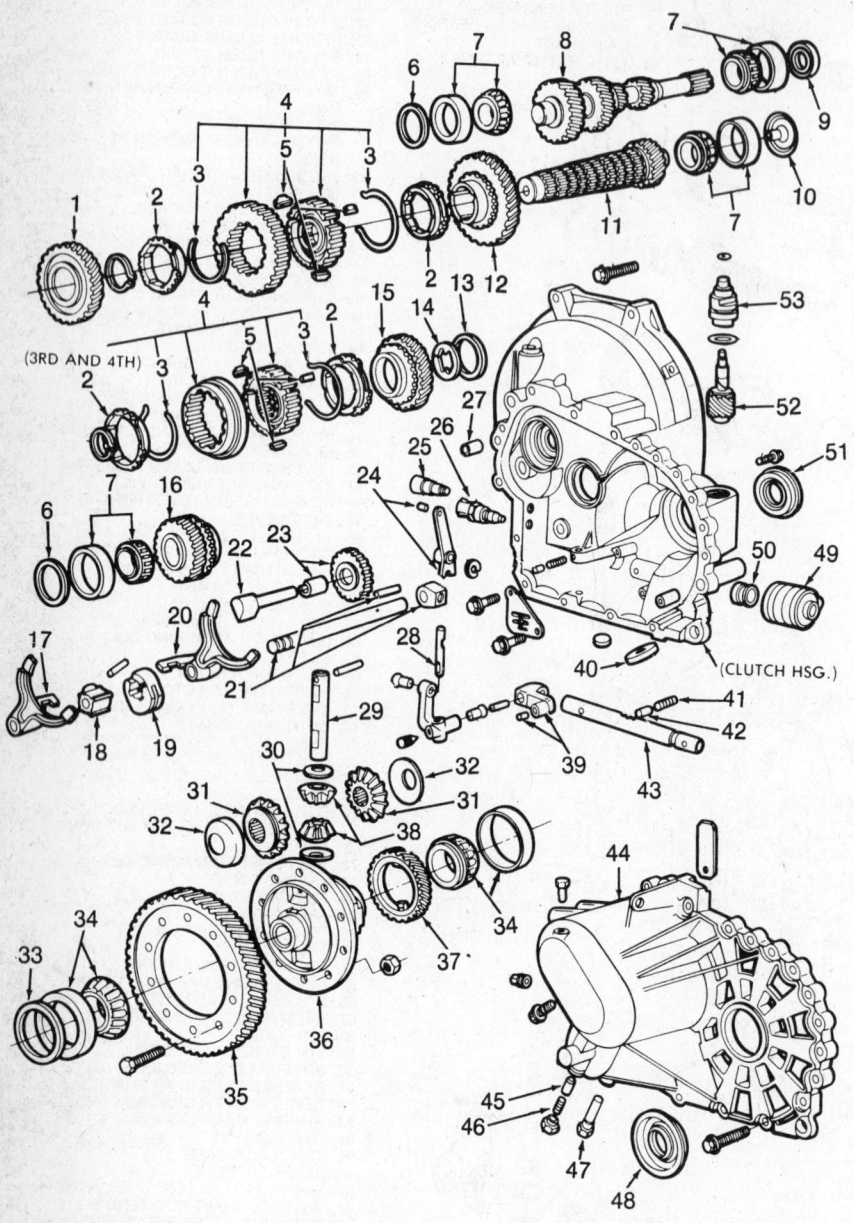

1. 2nd speed gear
2. Synchronizer blocking ring
3. Synchronizer spring
4. 1st and 2nd synchronizer assy.
5. Synchronizer hub 1st/2nd insert
6. Input shaft seal
7. Input shaft bearing and cup-front and rear
8. Input cluster shaft
9. Input shaft seal
10. Mainshaft funnel
11. Main shaft
12. 1st speed gear
13. 2nd/3rd gear thrust washer retaining ring
14. 2nd/3rd gear thrust washer
15. 3rd speed gear
16. 4th speed gear
17. 3rd/4th fork
18. Fork selector arm
19. Fork interlock sleeve
20. 1st/2nd fork
21. Main shift shaft
22. Reverse idler shaft
23. Reverse idler gear
24. Reverse relay lever
25. Reverse relay lever pivot pin
26. Back-up lamp switch
27. Dowel
28. Shift lever shaft
29. Pinion shaft
30. Pinion thrust washer
31. Side gear kit
32. Side gear thrust washer
33. Shim
34. Differential bearing assembly
35. Final drive output gear
36. Transaxle case assy.
37. Speedometer drive gear
38. Differential pinion gear
39. Input shift shaft selector plate arm
40. Case magnet
41. Input detent shift shaft spring
42. Input shift shaft detent plunger
43. Input shift shaft
44. Transaxle case
45. Main shift shaft detent plunger
46. Main shift shaft detent spring
47. Fork interlock sleeve retaining spring
48. Differential seal assembly

MTX 4 speed transaxle—exploded view (Chilton Type 2)

Slide Hammer tool No. T50T-100-A, remove the transaxle input shift shaft oil seal assembly.

ASSEMBLY

1. Grease the lip of the new oil seal and install it, using the tools described in the disassembly procedure.
2. Using a small drift, force the detent spring and plunger down into the bore while sliding the input shift shaft into its bore and over the plunger.

NOTE: Be careful not to damage the shift shaft oil seal.

3. Install the selector plate arm in its working position and slide the shaft through the selector plate arm. Align the hole in the selector plate arm with the hole in the shaft and install the roll pin, then install the boot.

NOTE: When properly installed the pin on the selector arm will be facing up. make sure the notches in the shift shaft face the detent plunger.

4. Install the selector plate and torque the attaching bolts to 6–8 ft. lbs.

NOTE: The pin in the selector arm must ride in the cut out of the gate in the selector plate.

5. If removed, apply Teflon® type sealant to the threads of the Reverse relay lever pivot pin. Install the Reverse relay lever and the retaining clip.

NOTE: Make sure the pin at the end of the lever faces outward.

6. Apply Teflon® type sealant to the threads of the back up lamp switch and install the switch.
7. Using a small drift, depress the Reverse inhibitor plunger and slide the shift lever shaft (with the oil relief flat first) through the case pedestal. Install the shaft so that the main shaft assembly or the differential will not interfere with the shift lever shaft.

Main Shift Control Shaft

DISASSEMBLY

1. Rotate the 3rd–4th shift fork on the shaft until the notch in the fork is located over the interlock sleeve. Rotate the 1st–2nd shift fork on the shaft until the notch in the fork is located over the selector arm finger.

With the forks in position, slide the 3rd–4th fork and interlock sleeve off the shaft.

2. Using a 5mm punch, remove the selector arm retaining pin.
3. Remove the selector arm and the 1st–2nd shift fork from the shaft.
4. If equipped with a 5 speed, remove the roll pin (using a drift punch), then slide the fork from the shaft.

ASSEMBLY

1. Clean and lightly oil the parts with the appropriate transaxle fluid.
2. Install the 1st–2nd shift fork and the selector arm on the shaft.
3. Align the hole in the selector arm with the hole in the shaft. Make sure the selector arm finger is aligned with the oil relief flats on the detent end of the shift shaft, then install the retaining pin.
4. Position the slot in the 1st–2nd fork over the fork selector arm finger. Position the slot in the 3rd–4th fork over the interlock sleeve. Slide the 3rd–4th fork and interlock sleeve onto the main shift control shaft, then align the interlock sleeve spines on the fork selector arm and slide into position.
5. If equipped with a 5 speed, hold the shaft with the hole on the left, then install the fork (the protruding spline must point toward the long end of the shaft) and the roll pin.

Input Cluster Shaft Seal

REMOVAL & INSTALLATION

1. Working from the outside of the case, remove the input shaft seal using a seal remover tool No. T77F-7050-A and a hammer. Position the remover tool against the seal by placing it in the slot cut in the case.
2. To install, lightly oil the input shaft seal and tap into place with a 1¼ in. socket and a hammer.

Input Cluster Shaft Bearings

REMOVAL

1. Using a Bearing Puller/Installer No. D79L-4621-A and an arbor press, remove the bearing cone and the roller assemblies.
2. Label the bearings for correct installation.

INSTALLATION

1. Thoroughly clean and lightly oil the bearings.
2. Using the tools used during removal, press the bearings on the proper end.

Bearing Cups

The input cluster shaft and the main shaft are supported at each end by tapered roller bearings. The cups, which support the bearings, are located in the transaxle case and the clutch housing case; they can be removed and installed by hand. The shims, used to preload the tapered roller bearings, are located behind the bearing cups in the transaxle case. It is important to keep the preload shim with its matching cup during disassembly. Also, label the bearing cups if they are removed from the case. Prior to installation, lightly grease the bearing cups.

CHILTON TYPE 3

Gearshift Housing And Shift Mechanism

DISASSEMBLY

1. If available, mount the transmission in a work stand.
2. Remove the Reverse operating lever from the shift lever shaft.
3. Position the operating levers in their Neutral positions. Remove the gearshift housing retaining bolts and the housing from the transmission.

NOTE: If either the 1st–2nd or 3rd–4th shift forks stick to their respective synchronizer sleeves, carefully work the forks away from the sleeves, then remove the assembly. Note that Steps 4, 5 and 6 need to be performed ONLY if oil leakage is noticed around the gearshift lever shafts.

4. Remove the operating lever-to-shift lever nuts and washers (if equipped). Disengage and remove the levers from the flats on the shafts.

NOTE: Before performing Step 5, the lever shafts must be free of burrs; the housing bore damage and oil leakage will result if burred shafts are pulled through the housing bores.

5. Remove the shift levers from the housing.

6. Remove the retainers and the O-rings from the housing.

7. Remove the ring from the interlock pivot pin, then the interlock levers and the spring from the housing.

8. Remove the Reverse detent spring and the ball from the bore of the transmission case.

ASSEMBLY

1. Install the interlock levers on the pivot pin and retain them with the E-clip. Using pliers, install the interlock spring onto the interlock levers.

2. Grease the shift lever bores of the gearshift housing, then install the shift levers into their respective bores. Install the shift lever shaft O-rings and the retainers.

3. Install the operating levers onto their respective shift lever shafts and torque the retaining nuts to 18 ft. lbs.

4. Rotate each operating lever until the fork bores of the shift levers are positioned straight up. Install the 3rd–4th shift fork into its bore and under both interlock levers.

5. Position both of the synchronizer sleeves in their Neutral positions (centered in their travel). Position the 1st–2nd shift fork into the groove of the 1st–2nd synchronizer sleeve.

6. Slide the Reverse idler gear to its Neutral position, then lay the transmission on its right side. Coat a new extension housing gasket with grease and place it on the transmission case.

7. Install the Reverse lever detent ball and spring (in order) into the bore in the side of the case.

8. Lower the gearshift housing onto the transmission case, guiding the 3rd–4th shift fork into the 3rd–4th synchronizer groove and the 1st–2nd shift fork shaft into the 1st–2nd shift lever bore.

NOTE: Hold the Reverse interlock link against the 1st–2nd shift lever to provide clearance.

9. Using a screwdriver, raise the interlock lever against its spring tension; this will allow the 1st–2nd shift fork shaft to slip under the levers. Make sure that the Reverse detent spring is positioned in the cover bore. The shift housing should now seat against the transmission case.

10. Install the housing bolts and tighten them finger tight. Shift the transmission through all gears to make sure that it is operating properly, without any binding. Torque the housing bolts (evenly) to 15 ft. lbs.

NOTE: Make sure that the housing bolts are installed in their original locations: 8 of the bolts are shoulder bolts which are used for

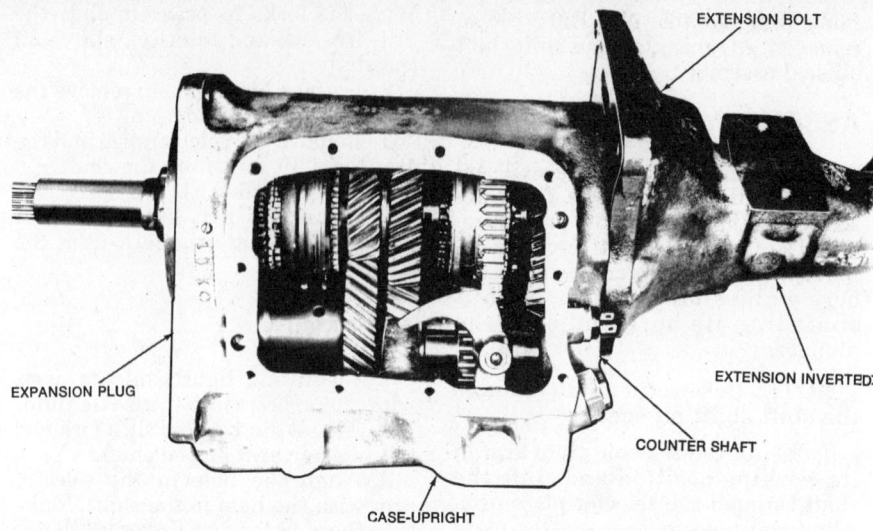

Rotating the extension housing on an Overdrive-4 (Chilton Type 3)

locating purposes; one bolt has a longer shoulder. The longer bolt is installed in the center hole along the rearward housing surface. The two standard style bolts are placed above and below the long shouldered bolt.

11. Grease the Reverse shaft, then attach the operating lever with the nut. Torque the nut to 18 ft. lbs.

12. Install the speedometer drive pinion gear and the adapter, making sure that the stamped range number is at the 6 o'clock position.

13. Install the transmission as outlined in the appropriate car section.

Extension Housing, Mainshaft And Drive Pinion

DISASSEMBLY

NOTE: This procedure is to be performed ONLY after the gearshift housing and the shift mechanism have been removed.

1. Remove the speedometer pinion adapter bolt and retainer, then carefully remove the adapter and the pinion from the extension housing.

2. Remove the extension housing-to-transmission case bolts.

3. Rotate the extension housing around the mainshaft in order to expose the rear of the countershaft.

NOTE: One of the extension attaching bolts may be installed to hold the extension in this inverted position.

4. Drill or center punch a hole in the front countershaft expansion plug.

5. Place a punch in the expansion plug hole and push the countershaft rearward until the Woodruff key is exposed, then remove the key.

6. Push the countershaft forward against the expansion plug, then tap the shaft with a brass drift and a hammer (in the same direction) to drive the expansion plug from the transmission case.

7. Using a countershaft arbor (Chrysler tool No. C-3938), push the countershaft out of the rear of the case.

— **CAUTION** —

Be careful not to allow the countershaft washers to fall out of position.

8. Lower the cluster gear to the bottom of the transmission case.

9. Move the extension housing back to its original position.

10. Remove the drive pinion bearing retainer bolts, the retainer and the gasket, then remove the oil seal from the retainer.

— **CAUTION** —

Be careful not to damage the seal bore of the retainer; oil leakage could occur.

11. Using a brass drift and a hammer, tap the pinion and bearing assembly forward, then remove the assembly from the front of the transmission case.

12. Slide the 3rd–4th synchronizer sleeve slightly forward, then slide the Reverse idler gear to the center of its shaft. Using a soft faced hammer, tap the extension housing rearward. Remove the housing and the mainshaft from the transmission case as an assembly.

13. Remove the snap ring which retains the 3rd–4th synchronizer to the

mainshaft, then remove the synchronizer assembly.

14. Slide the 4th speed gear and the stop ring off of the mainshaft.

NOTE: Do not separate the synchronizer parts unless replacement of a particular part is required due to damage.

15. In the extension housing, compress the mainshaft bearing snap ring; while holding the snap ring compressed, pull the mainshaft assembly and the bearing from the extension housing.

16. Remove the mainshaft bearing-to-shaft snap ring. Using an arbor press, press the bearing from the mainshaft.

17. Remove the following parts from the mainshaft:

    a. The bearing.
    b. The bearing retainer ring.
    c. The 1st speed gear.
    d. The 1st speed stop ring.

18. Remove the 1st–2nd synchronizer assembly-to-mainshaft snap ring and separate the assembly from the mainshaft. Refer to the note which follows Step 14.

ASSEMBLY

1. If disassembled, partially assemble the synchronizers as follows:

    a. Place a stop ring on the workbench, followed by the clutch gear and the sleeve.

    b. Drop the struts into their slots. Snap in a strut spring by placing the tang inside one of the struts.

    c. Turn the assembly over on the stop ring, then install the 2nd strut spring, with the tang inside a different strut than the first spring.

2. Install the 2nd speed gear on the mainshaft with the synchronizer cone facing the rear. The gear must be pushed against the shoulder on the mainshaft.

3. Install the 1st–2nd synchronizer assembly (including the stop ring, with the lugs indexed in the hub slots) on the mainshaft, then down against the 2nd gear cone. Secure the assembly with a new snap ring. Slide the next stop ring over the shaft and index the lugs into the hub slots.

4. Install the mainshaft bearing retaining ring followed by the mainshaft rear bearing. Using an arbor press, press the bearing into place, then secure it with a new (selective) snap ring.

5. Install the partially assembled mainshaft until the bearing retaining ring engages the slot in the extension housing.

6. Compress the bearing retaining ring until the bearing can be bottomed against the shoulder in the extension housing. Release the tension on the snap ring and make sure that it fully seats in the snap ring groove.

7. Slide the 4th speed gear onto the mainshaft (with the synchronizer cone facing the front), followed by the 4th speed gear stop ring.

8. Install the 3rd–4th synchronizer clutch gear assembly onto the mainshaft, with the shift fork slot positioned toward the rear. Be sure to index the rear stop ring with the clutch gear struts.

9. Install the retaining snap ring, then position the front stop ring (greased) on the clutch gear, again indexing the ring lugs with the struts.

10. Using a new extension housing gasket, coat it with grease and position it on the extension housing.

11. Slide the Reverse idler gear to the center of the Reverse idler shaft, then move the 3rd–4th synchronizer sleeve forward as far as possible without dislodging the struts.

12. Carefully insert the mainshaft assembly into the transmission case, tilting it as required to clear the gears.

13. Position the 3rd–4th synchronizer sleeve in the Neutral position.

14. Rotate the extension housing in order to expose the rear of the countershaft, then install a bolt to secure the extension housing in this position.

15. Install the drive pinion and bearing assembly into the front of the transmission case; seat the bearing in the case bore. If necessary, tap the assembly into place (lightly) using a soft faced hammer.

NOTE: The bearing's outer snap ring should bottom against the case face with a minimal amount of effort. If the snap ring will not seat, check for a mispositioned strut, pinion roller or stop ring.

16. While holding the countershaft gear to prevent damage, turn the transmission assembly upside down. Lower the countershaft gear assembly into place, making sure that the thrust washers stay in position. The thrust washer tangs must be aligned with the corresponding case slots.

17. Install the countershaft about half way into the bore (in a forward motion), then install the Woodruff key. Push the countershaft forward until the end of the shaft is flush with the rear case face. Remove the arbor tool.

18. Rotate the extension housing to its proper position and install the extension housing bolts. Torque the bolts to 30 ft. lbs.

19. Rotate the transmission to an upright position. Install the pinion bearing retainer. Coat the retainer bolts with sealing compound, then install the bolts and torque them to 30 ft. lbs.

20. Install a new countershaft bore expansion plug.

Drive Pinion And Countershaft Gear

DISASSEMBLY

NOTE: This procedure is to be performed only after the extension housing and the mainshaft have been removed from the transmission case.

1. Remove the inner snap ring from the pinion bearing, then press the bearing off of the pinion.

2. Remove the snap ring and the bearing rollers (16) from the drive pinion cavity.

3. Remove the countershaft gear from the bottom of the transmission case.

4. Remove the arbor, the needle bearings (76), the thrust washers and the spacers from the center of the countershaft gear.

5. Using Chrysler tool No. C-3638, attach a $\frac{7}{16}$ in. socket ($\frac{1}{4}$ or $\frac{3}{8}$ in. drive) onto the smaller hex of the tool. Position the tool in the transmission case with the blunt end of the tool is against the case and the socket end against the Reverse idler gear shaft.

6. Expand the tool so that it pushes the Reverse idler gear shaft out of the transmission case. It may be necessary to remove the tool and use a socket with an extension to push the shaft completely out of the case. Remove the Reverse idler gear.

7. If oil leakage is noted around the Reverse idler shift fork bore, perform the following:

    a. Remove any burrs from the lever shaft to prevent bore damage as the shift lever is withdrawn.

    b. Push the lever toward the inside of the transmission case and remove the fork.

    c. Remove the retainer and the O-ring from the bore of the transmission case.

8. Clean, inspect and lubricate the parts prior to assembly. Replace the parts which are irreparably damaged.

ASSEMBLY

1. Coat the inside ends of the countershaft gear bore with grease. Install the roller bearing spacer and the arbor tool (previously used) into the countershaft gear bore.

2. Install the 38 roller bearings and

the two spacer rings into the countershaft gear bore (19 rollers and one spacer per side).

3. Coat the countershaft thrust washers with grease and install them over the arbor.

NOTE: When installing the countershaft washers, make sure that the tanged sides of the washers face the case bosses.

4. Place the countershaft assembly at the bottom of the transmission case, making sure that the thrust washers stay in position.

5. With the outer snap ring groove of the drive pinion bearing facing forward, press the bearing onto the pinion shaft until the bearing is fully seated against the drive gear shoulder.

6. Install a new snap ring onto the bearing, making sure that the snap ring is fully seated.

NOTE: The pinion bearing snap ring is selective; various thicknesses are available to alter the end play.

7. Install the 16 bearing rollers (coated with grease) into the cavity of the drive pinion gear. Install the bearing retaining snap ring.

8. Carefully drive a new oil seal into the pinion bearing retainer.

Reverse Gear, Lever And Fork

ASSEMBLY

1. If removed, install the Reverse lever, the O-ring (lubricated) and the retainer.

2. Drive the Reverse idler gear shaft in far enough for the Reverse idler gear to be positioned on the shaft. Install the Reverse idler gear (position the fork slot toward the rear), engaging the slot with the Reverse shift fork.

3. Drive the Reverse gear shaft into the case until the Woodruff key may be installed. Install the Woodruff key and drive the shaft inward until it is flush with the end of the case.

4. If removed, install the back up lamp switch (with a new gasket) and torque to 15 ft. lbs.

Extension Housing Bushing

REPLACEMENT

NOTE: To perform this procedure, the extension housing must be removed from the transmission.

1. Remove the extension housing yoke seal, then drive the bushing out of the housing using an appropriate driver (Chrysler tool No. C-3974).

2. Obtain a new bushing and position the bushing on the driver (used during removal). Align the oil hole of the bushing with the oil slot of the extension housing, then drive the bushing into place.

3. Carefully tap a new extension housing oil seal into place.

CHILTON TYPE 4

Transaxle Case

DISASSEMBLY

The Chrysler designed and built A-460 (4 speed) fully synchronized manual transaxle combines gear reduction, ratio selection and differential functions in one unit housed in a die cast aluminum case.

1. With the transaxle removed from the vehicle, remove the differential cover bolts and the stud nuts, then remove the cover.

2. Remove the differential bearing retainer bolts.

3. Using tool No. L-4435, rotate the differential bearing retainer to remove it.

4. Remove the extension housing bolts, then the differential assembly and extension housing.

5. Remove the selector shaft housing bolts, then the selector shaft housing.

6. Remove the stud nuts and the bolts from the rear end cover, then using a small pry bar, pry off the rear end cover.

7. Using snap ring pliers, remove the large snap ring from the intermediate shaft rear ball bearing.

8. Remove the bearing retainer plate by tapping it with a plastic hammer.

9. Remove the 3rd-4th shift fork rail.

10. Remove the Reverse idler gear shaft and gear.

1. Bearing retainer plate	9. Differential bearing retainer	17. Input shaft seal
2. Rear cover	10. Bearing retainer seal	18. Input shaft spacer
3. Magnet	11. Extension	19. Reverse idler shaft
4. Oil pan	12. Extension O-ring	20. Reverse idler gear
5. Transaxle case	13. Retainer	21. Reverse idler spacer
6. Cup (bearing)	14. Bearing cup	22. Reverse gearshift lever
7. Spacer	15. Extension seal	
8. Oil feed baffle	16. Bearing retainer	

Chrysler A-460 transaxle case (Chilton Type 4)

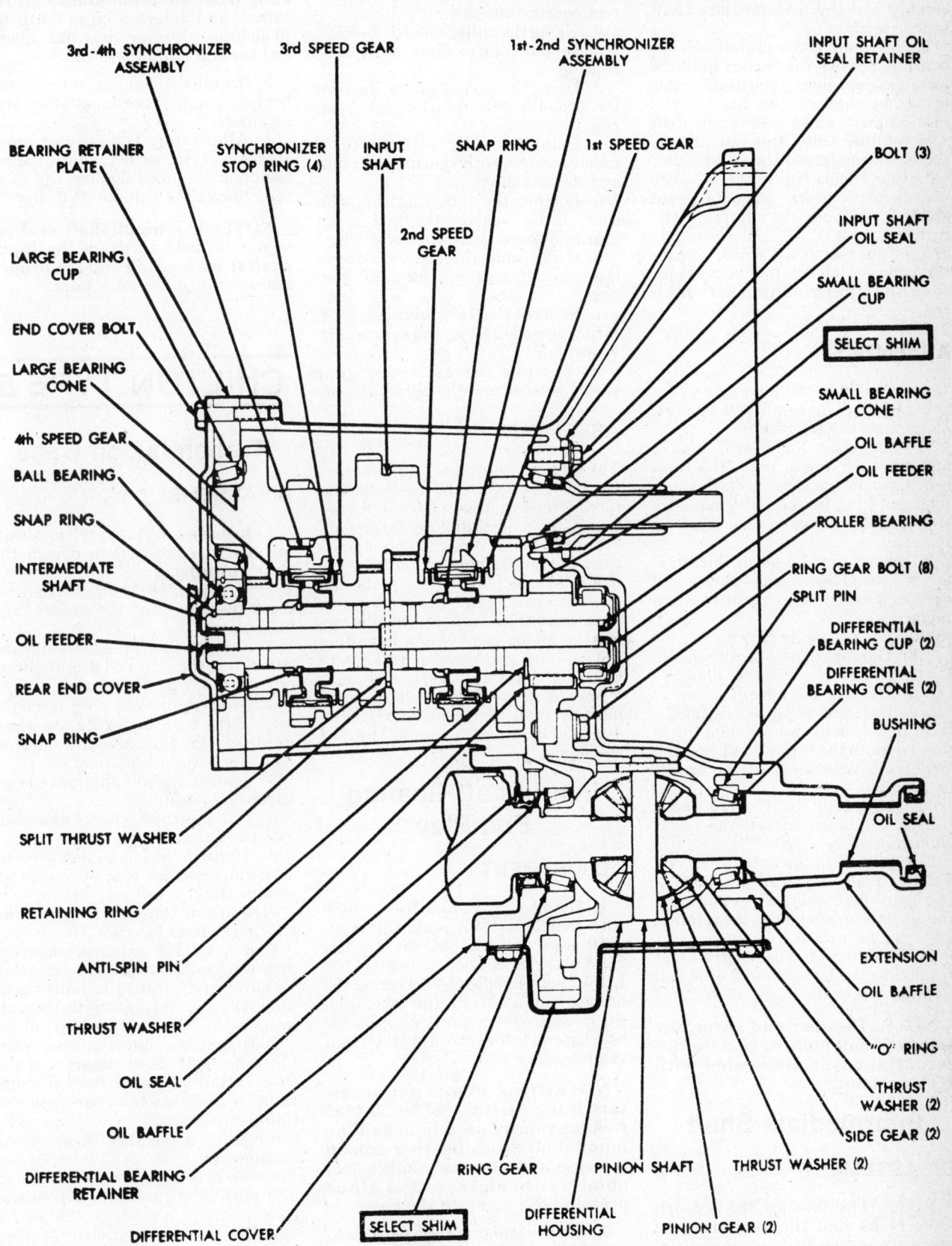

3rd-4th SYNCHRONIZER ASSEMBLY

3rd SPEED GEAR

1st-2nd SYNCHRONIZER ASSEMBLY

INPUT SHAFT OIL SEAL RETAINER

BEARING RETAINER PLATE

SYNCHRONIZER STOP RING (4)

INPUT SHAFT

SNAP RING

1st SPEED GEAR

BOLT (3)

2nd SPEED GEAR

INPUT SHAFT OIL SEAL

LARGE BEARING CUP

SMALL BEARING CUP

END COVER BOLT

SELECT SHIM

LARGE BEARING CONE

SMALL BEARING CONE

4th SPEED GEAR

OIL BAFFLE

BALL BEARING

OIL FEEDER

SNAP RING

ROLLER BEARING

INTERMEDIATE SHAFT

RING GEAR BOLT (8)

SPLIT PIN

OIL FEEDER

DIFFERENTIAL BEARING CUP (2)

REAR END COVER

DIFFERENTIAL BEARING CONE (2)

SNAP RING

BUSHING

SPLIT THRUST WASHER

OIL SEAL

RETAINING RING

ANTI-SPIN PIN

EXTENSION

THRUST WASHER

OIL BAFFLE

OIL SEAL

"O" RING

OIL BAFFLE

THRUST WASHER (2)

DIFFERENTIAL BEARING RETAINER

SIDE GEAR (2)

RING GEAR

PINION SHAFT

THRUST WASHER (2)

DIFFERENTIAL COVER

SELECT SHIM

DIFFERENTIAL HOUSING

PINION GEAR (2)

Cross-sectional view of the A-460 transaxle (Chilton Type 4)

11. Remove the input shaft gear assembly and the intermediate shaft gear assembly.

12. To remove the clutch release bearing, remove the E-clips from the clutch release shaft, then disassemble the clutch shaft components.

13. Remove the three input shaft seal retainer bolts, the seal, the retainer assembly and the select shim.

14. Using tools No. C-4171, C-4656 and an arbor press, press the input shaft front bearing cup from the transaxle case.

15. Using tool No. C-4660, remove the two bearing retainer strap bolts, then the intermediate shaft front bearing.

ASSEMBLY

The assembly of the transaxle is the reverse of disassembly; however, please note the following:

1. Using tools No. C-4657, C4171 and an arbor press, press the front bearing onto the intermediate shaft; the input shaft front bearing cup is installed with the same tools used for removal.

2. Determine the shim thickness for the correct bearing end play only if any of the following parts are replaced:

   a. The transaxle case.
   b. The input shaft seal retainer.
   c. The bearing retainer plate.
   d. The rear end cover.
   e. The input shaft or bearings.

3. To determine proper shim thickness, refer to the Input Shaft Bearing End Play Adjustment at the end of this section.

4. Using tool No. C-4674 and a plastic hammer, install the input shaft oil seal.

5. Using a $\frac{1}{16}$ inch bead of RTV sealant, place it around the edge of the input shaft seal retainer and making sure the drain hole of the retainer is facing downward.

6. The differential bearing retainer is installed with the same special tool used for removal.

NOTE: The rear end cover, the selector shaft housing and the differential cover are sealed with RTV sealant.

Intermediate Shaft

DISASSEMBLY

NOTE: The 1st–2nd, the 3rd–4th shift forks and the synchronizer stop rings are interchangeable. However, if parts are to be reused, reassemble them in their original position.

1. Remove the intermediate shaft rear bearing snap ring.

2. Using the puller tool No. C-4693, remove the intermediate shaft rear bearing.

3. Using snap ring pliers, remove the 3rd–4th synchronizer hub snap ring.

4. Using the puller tool No. L-4534, remove the 3rd–4th synchronizer hub and the 3rd speed gear.

5. Remove the retaining ring, the split thrust washer, the 2nd speed gear and the synchronizer stop ring.

6. Using snap ring pliers, remove the 1st–2nd synchronizer hub snap ring.

7. Remove the 1st speed gear, the stop ring and the 1st–2nd synchronizer assembly.

8. Remove the 1st speed gear thrust washer and the anti-spin pin.

ASSEMBLY

The assembly of the intermediate shaft is the reverse of the disassembly; however, please note the following: When assembling the intermediate shaft, make sure the speed gears turn freely and have a minimum of 0.003 in. end play. When installing the 1st speed gear thrust washer make sure the chamfered edge is facing the pinion gear. When installing the 1st–2nd synchronizer make sure the relief faces the 2nd speed gear. Use an arbor press to install the intermediate shaft rear bearing, the 3rd–4th synchronizer hub and the 3rd speed gear.

Input Shaft Bearing End Play

ADJUSTMENT

1. Using special tool No. L-4656 with handle C-4171, press the input shaft front bearing cup slightly forward in the case. Then, using tool No. L-4655 with handle C-4171, press the bearing cup back into the case, from the front, to the properly position the bearing cup before checking the input shaft endplay.

NOTE: This step is not necessary if the special tool No. L-4655 was previously used to install the input shaft front bearing cup in the case and no input shaft select shim has been installed since pressing the cup into the case.

2. Select a gauging shim which will give 0.001–0.020 in. (0.025–0.254mm) end play.

NOTE: Measure the original

shim from the input shaft seal retainer and select a shim 0.010 in. (0.254mm) thinner than the original for the gauging shim.

3. Install the gauging shim on the bearing cup and the input shaft seal retainer.

4. Alternately tighten the input shaft seal retainer bolts until the retainer is bottomed against the case, then torque the bolts to 21 ft. lbs.

NOTE: The input shaft seal retainer is used to draw the input shaft front bearing cup the proper distance into the case bore.

CHILTON TYPE 5

Transmission Case

DISASSEMBLY

1. Remove the clutch release bearing and the lever, then detach the clutch housing.

2. Drain the lubricant, then remove the cover and the gasket from the case.

3. Remove the threaded plug, the spring and the shift rail detent plunger from the front of the case.

4. Drive the access plug from the rear of the case. Drive the interlock retaining pin from the case and remove the interlock plate.

5. Remove the roll pin from the selector lever arm.

6. Tap the front end of the shift rail, to displace the plug at the rear of the extension housing. Remove the shift rail from the rear of the extension housing.

7. Remove the selector arm and shift forks from the case.

8. Remove the extension housing attaching bolts. Loosen the extension housing and rotate the housing to align the countershaft with the cutaway in the extension housing flange.

9. Drive the countershaft rearward until the shaft clears the front of the case. Install a dummy shaft through the case and into the gear until the countershaft gear can be lowered to the bottom of the case. Remove the countershaft.

10. Remove the extension housing and mainshaft assembly from the case.

11. Remove the input shaft bearing retainer bolts, the bearing retainer and the input shaft from the case.

12. Remove the Reverse idler gear and shaft from the rear of the case.

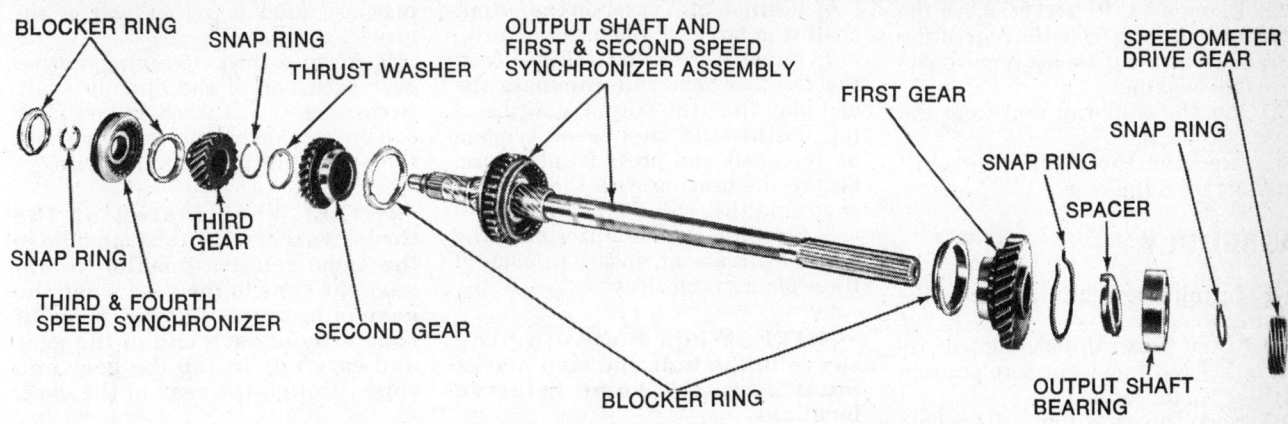

Output shaft and related components (Chilton Type 5) transmission

13. Remove the bearing retainers, the bearings and the dummy shaft from the countershaft gear.

14. Remove the retainer and the pilot bearing from the input shaft gear.

15. Do not remove the ball bearing from the input shaft unless replacement is necessary. To remove it, take off the snap ring and press the bearing off the shaft.

16. Pry the input shaft seal out of the bearing retainer.

17. Remove the 4th speed gear blocking ring from the front of the output shaft.

18. Remove the snap ring from the forward end of the output shaft.

19. Support the 3rd speed gear (on press plates), the output shaft and the extension housing in a press. Push the output shaft out of the 3rd–4th synchronizer and the 3rd speed gear, while supporting the extension housing and the output shaft from beneath. Remove the snap ring, the washer, the 2nd speed gear and the blocking ring from the output shaft.

20. Disassemble the synchronizer assembly by pulling the sleeve from the hub, then remove the inserts and spring.

21. Remove the output shaft bearing-to-extension housing snap ring.

22. Using a plastic hammer, tap the output shaft assembly from the extension housing.

23. Measure or scribe the speedometer gear location on the output shaft and press the gear off.

24. Position press plates behind 1st speed gear and place the assembly in a press. The 1st–2nd synchronizer are serviced as an assembly.

NOTE: NO attempt should be made to separate the hub from the shaft. The only serviceable parts are the springs and inserts. If the hub or sleeve is worn, the shaft and synchronizer must be replaced as an assembly.

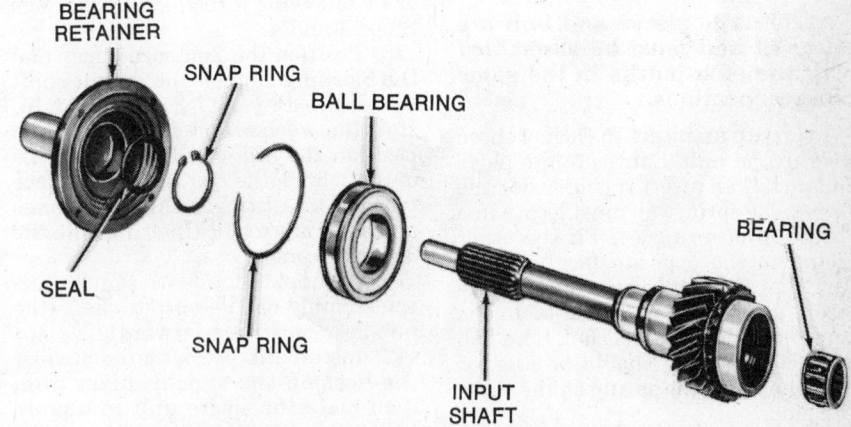

Input shaft and related components (Chilton Type 5) transmission

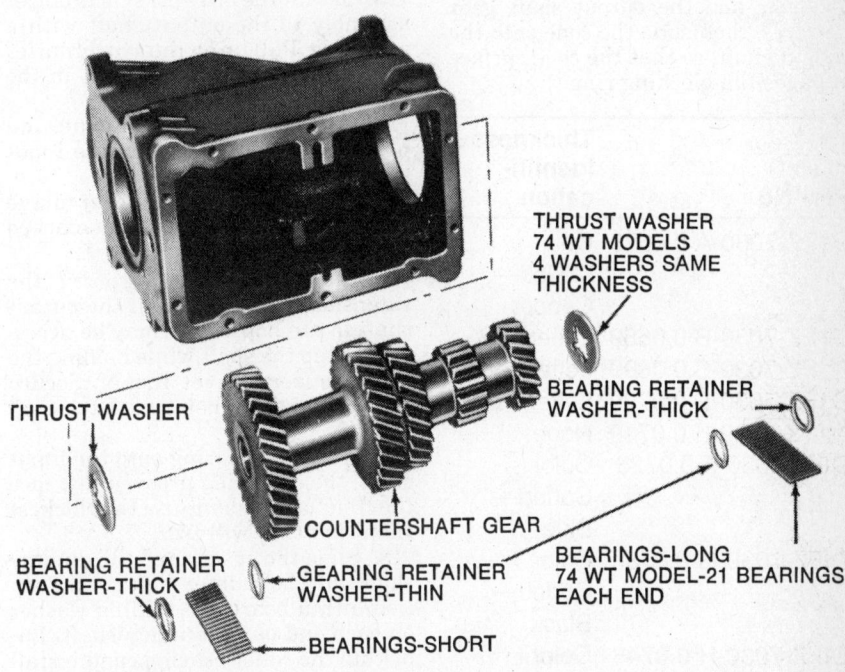

Countershaft and related components (Chilton Type 5) transmission

25. Using a $^9/_{16}$ in. socket, drive the shift rail bushing from the rear of the extension housing. Do not remove serviceable bushings.

26. Pry the shift rail seal from the rear of the case.

27. Remove the remaining shift linkage from the case.

ASSEMBLY

1. Install a new shift rail seal in the rear of the case.

2. If the shift rail bushing was removed, drive a new one into position with a $^9/_{16}$ in. socket.

3. Slide the synchronizer hub onto the shaft, making sure that the shift fork groove is facing the front of the shaft.

NOTE: The sleeve and hub are select fit and must be assembled with the etch marks in the same relative locations.

4. Locate an insert in each of three slots in the hub. Lubricate the parts and install an insert spring inside the sleeve; the spring tab must locate in a U-section of an insert. Fit the other spring to the opposite face, making sure that the tab locates in the same insert. Both springs should be in the same rotational direction. The tab end of one spring should be aligned with the tab of the spring on the opposite side.

5. Assemble a blocking ring on the 1st gear side of the 1st–2nd synchronizer. Lubricate the cone surface of 1st gear and the output shaft gear journals, then slide the cone onto the output shaft, so that the cone surface engages the blocking ring.

Part No.	Thickness	Identification
D1FZ-7030-A	0.0679	Color Coded—Copper
D1FZ-7030-B	0.0689	Letter—W
D1FZ-7030-C	0.0699	Letter—V
D1FZ-7030-D	0.0709	Letter—U
D1FZ-7030-E	0.0719	None
DIFZ-7030-F	0.0728	Color Coded—Blue
DIFZ-7030-G	0.0738	Color Coded—Black
DIFZ-7030-H	0.0748	Color Coded—Brown

6. Position the spacer on the output shaft (the larger diameter rearward).

7. Install a snap ring (selected from the chart) which will eliminate the end play from the output shaft bearing. Position the output shaft bearing on the shaft and press it into place. Secure the bearing with the thickest snap ring that will fit the groove.

8. Slide the synchronizer onto the hub and locate an insert in each of three slots in the sleeve.

NOTE: When installing the sleeve to the hub, the etch marks must be in the same relative locations.

9. Lightly oil the parts and complete the assembly of the synchronizer by following directions in previous Steps 3 and 4.

10. Position the 2nd speed gear and the blocking ring on the output shaft (the dog teeth must face rearward. Install the washer and snap ring, then position the 3rd speed gear onto the output shaft (the dog teeth must face forward). Lubricate the gear cones and assemble a blocking ring onto the 3rd gear cone.

11. Position the 3rd–4th synchronizer assembly on the output shaft (the hub boss must face forward).

12. Install the press plates against the boss on the synchronizer hub, then place the entire unit in a press (the extension end up) and press the synchronizer assembly onto the output shaft as far as possible.

13. Retain the 3rd–4th synchronizer assembly to the output shaft with a snap ring. Pull up on the synchronizer so that the snap ring is tight in the groove.

14. Lubricate the gear cone and place the blocking ring on the input shaft gear cone.

15. Press the speedometer drive gear onto the shaft to the marked location.

16. Lubricate the bearing bore of the extension housing. Install the output shaft in the housing; it may be necessary to tap the shaft while holding the synchronizer sleeves firmly. Secure the shaft to the housing with the snap ring.

17. Press the bearing onto the input shaft; the snap ring groove must face the front of the shaft (use the thickest snap ring that will fit).

18. Slide the spacer and the dummy shaft into the countershaft gear. Position a thin bearing retaining washer on each end of the dummy shaft. Lubricate the roller bearings and install the long bearings in the small end of the gear and the short bearings in the long end of the gear (21 needle bearings are used at either end of the gear).

19. Place a thick retaining washer over each end of the dummy shaft. Grease the thrust washers and place one on each end of the dummy shaft, then lower the gear into the case.

NOTE: When installing the thrust washers, the tabs must be in the same relative position to engage the slots in the case when the gear is lowered. Loop a piece of rope around each end of the gear and carefully install the gear and rope through the rear of the case.

20. Lubricate the Reverse idler gear shaft. Position the selector lever relay on the pivot pin and secure with a spring clip. Hold the gear in the lever (with the long hub facing the rear of the case) and slide the Reverse idler shaft into place. Seat the shaft in the case with a brass hammer.

21. Install a new seal in the input shaft bearing retainer and the input shaft in the case with a new bearing retainer O-ring. Tap on the outer race of the bearing to seat the outer snap ring.

——— CAUTION ———
Use a soft hammer and do not tap on the input shaft itself.

22. Carefully slide the 3rd–4th synchronizer sleeve into the 4th speed position.

23. Place a new gasket on the extension housing.

24. Lubricate and install the input shaft pilot bearing on the shaft. Slide the extension housing and output shaft into place, being careful not to disturb the 3rd–4th synchronizer.

25. Align the cutaway in the extension housing flange with the countershaft bore in the rear of the case.

26. Move the countershaft gear into place and install the countershaft, making sure that the thrust washers remain in place. The flat on the countershaft should be parallel to the top of the case. Tap the shaft with a brass hammer until the front of the shaft is flush with the case.

27. Rotate the extension housing to align the bolt holes and loosely install the bolts. Apply sealer to the attaching bolts and torque to 33–36 ft. lbs.

NOTE: When installing the extension housing-to-case, make sure that the rail slides freely in its bore. Binding is remedied by slightly rotating the extension housing to free the rail, then push the housing into the case.

28. Place the shift forks in the syn-

chronizer sleeves and install the interlock lever and new retaining pin. Lubricate the shift rail oil seal and slide the shift rail through the extension housing, the case and the 1st–2nd speed shift fork. Position the selector arm on the rail and slide the rail through the 3rd–4th shift fork. Slide the shift rail through the front of the case until the center detent bore is aligned with the detent plunger bore, then install a new retaining pin in the selector arm.

29. Install the detent plunger, the spring and the plug with sealer. Install a new access plug in the rear of the case. Position a new oil seal with a tension spring and lip facing in the direction of the case and drive the seal in until it bottoms.

30. Position a new O-ring in the groove in the case. Position the input shaft bearing retainer with the groove in the retainer aligned with the oil passage in the case and install the retaining bolts finger tight.

31. Install the flywheel housing, then torque the housing and the front bearing retainer bolts. Coat the retainer with grease.

32. Install the clutch release arm and the bearing.

33. Install a new extension housing plug, using sealer.

34. Install a new cover gasket and cover, with the vent facing the rear. Apply sealer to the left front cover attaching bolt and torque to 8–10 ft. lbs.

CHILTON TYPE 6

Transmission Case

DISASSEMBLY

1. Remove the lower extension housing bolt and drain the lubricant.

2. Drive the access plug from the rear of the extension housing. Remove the offset lever assembly nut and washer, then the lever assembly.

3. Remove the remaining extension housing bolts and washers, the extension from the case and discard the old gasket.

4. Remove the cover-to-case bolts, the cover, the shifter fork and the shift rod assembly, then discard the old cover gasket.

5. Remove the front bearing retainer-to-case bolts and washers, then the front bearing retainer and gasket.

6. Remove the Reverse lever assembly-to-pivot bolt spring clip, then the pivot and the Reverse lever assembly.

7. Remove the input bearing-to-input shaft snap ring, the input bearing outer snap ring and pull the bearing out.

8. Remove the speedometer drive gear-to-output shaft snap ring, then slide the gear off and remove the lock ball from the shaft.

9. Remove the output shaft bearing-to-shaft snap ring. Use the outer snap ring to pull the output shaft bearing from the shaft and the case, then remove the snap ring from the bearing.

10. Remove the input shaft through the front bearing hole in the case. Carefully lift the output shaft and gear train from the top of the case. Slide the Reverse idler gear shaft through the rear of the case and remove Reverse gear.

11. Insert a dummy shaft through the front of the case to drive the countershaft out through the rear of the case. Lift the countershaft gear, the thrust washers and the dummy shaft through the top of the case. Remove the cluster gear thrust washers.

12. Clean and inspect the parts. If the back up light switch is damaged, remove it.

ASSEMBLY

1. Position the Reverse idler gear and shaft in place.

2. Coat the surfaces of the countershaft thrust washers with a thin film of grease and position in the case (the plastic washer goes in front, the bronze one at the rear). Position the cluster gear assembly in the bottom of the case.

3. Place the transmission in the vertical position. Align the countershaft gear bore and the thrust washers with the case bore, then install the countershaft from the rear of the case. Return the transmission to the horizontal position.

4. Position the output shaft assembly into the case through the cover opening. With the snap ring groove facing rearward, place the rear bearing on the output shaft. Place the transmission in the vertical position and install the bearing. Position the 1st gear thrust washer on the roll pin, holding it tightly during bearing installation. Install the rear bearing snap rings.

5. Install the input shaft and the blocking ring through the front of the case. Make sure that the blocking ring notches engage the synchronizer insert.

6. Using a new gasket, install the front bearing retainer. Apply gasket sealer to the bolt threads and torque to 11–15 ft. lbs.

7. Install the Reverse idler gear lever assembly, taking care to insert the fork in the Reverse idler gear groove.

8. Apply gasket sealer to the Reverse lever pivot bolt threads and install the bolt. Align the lever on the pivot bolt and torque the bolt to 15–25 ft. lbs. Install the Reverse lever retaining spring clip to the Reverse gear pivot bolt. Tilt the transmission forward and pour a light coating of gear lube over the gear train.

9. Using a new cover gasket, install the cover assembly. Install the bolts and the wiring clips, then torque.

NOTE: The two shouldered locating bolts must be installed first. Position the shift rail in the 1st or 3rd gear.

10. Insert the speedometer drive gear lock ball into its hole. While holding the ball, slide the speedometer drive gear into place and secure it with a new snap ring.

11. Using a new gasket, install the extension housing to the case. Using gasket sealer on the bolts, torque them to 18–27 ft. lbs. Take care not to damage the extension yoke seal.

12. Install the offset lever assembly onto the shift shaft, then secure the assembly with a nut and flat washer. Use sealer on the shift shaft threads and torque to 8–12 ft. lbs.

13. Install the gearshift lever and check its operation in each gear position.

14. Using a soft mallet, install the access plug into the rear of the extension housing.

Cover Assembly

DISASSEMBLY

1. Remove the detent screw, the spring and the plunger.

2. Pull the shifter shaft rod rearward while rotating it counterclockwise.

3. Remove the selector/interlock-to-shifter shaft spring pin.

4. Remove the shifter shaft from the cover (DO NOT damage the seal).

5. Remove the manual selector and interlock plate.

6. Remove the 1st–2nd speed shifter fork and the 3rd–4th speed shifter fork.

7. Clean and inspect the parts. Replace the shifter shaft seal and the welch plug, if damaged.

ASSEMBLY

1. Assemble the two plastic inserts

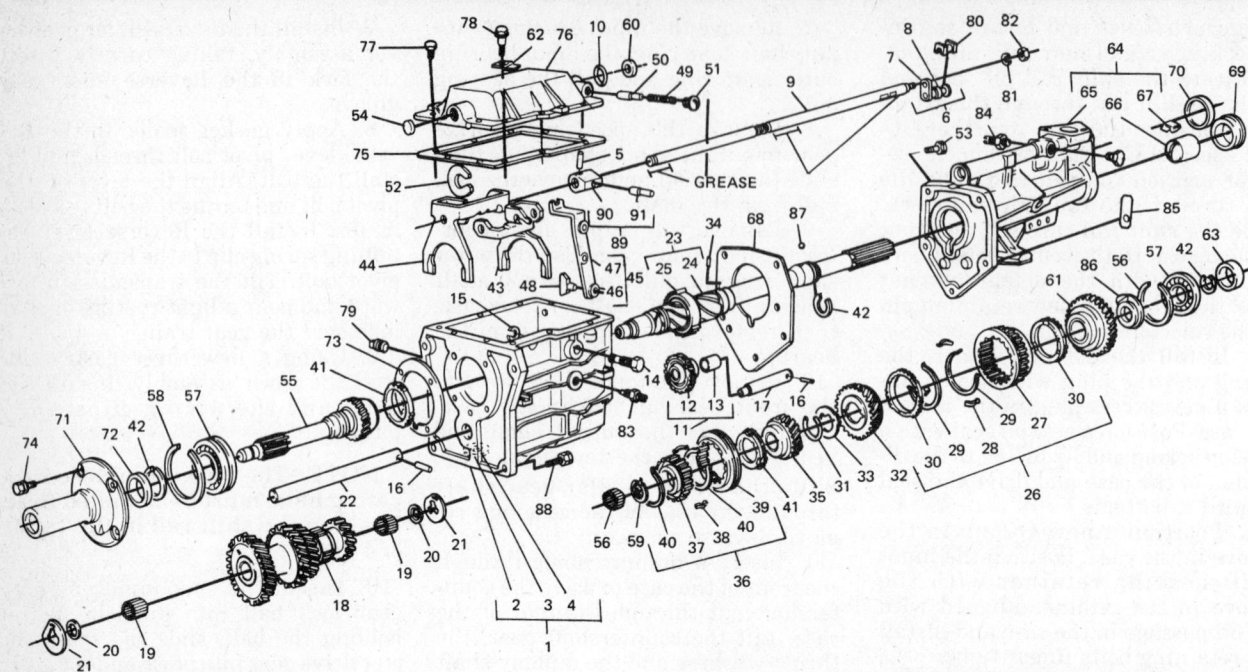

1 Case assembly—transmission
2 Case—transmission
3 Magnet—transmission case chip
4 Nut spring 9/64
5 Pin—3/16 diameter x 13/16 rolled spring
6 Lever assembly—transmission gearshift shaft offset
7 Lever transmission gearshift shaft offset
8 Pin—transmission gearshift shaft offset lever
9 Shaft—transmission shifter
10 Seal—O-ring
11 Gear & bush assembly—transmission reverse idler sliding
12 Gear—transmission reverse idler sliding
13 Bushing—transmission reverse idler gear
14 Pin—transmission reverse gear selector fork pivot
15 Ring—7/16 retaining
16 Pin—1/4 x 1 spring
17 Shaft—transmission reverse idler gear
18 Gear—transmission countershaft
19 Roller—transmission countershaft bearing
20 Washer—208/.918 flat
21 Washer—transmission countershaft gear thrust
22 Countershaft—transmission
23 Shaft assembly—transmission output
24 Shaft—transmission output
25 Hub—transmission synchronizer 1st & 2nd gear cluster
26 Shaft and gear assembly—transmission output
27 Gear—transmission reverse sliding
28 Insert—transmission synchronizer hub
29 Spring—transmission synchronizer retaining
30 Ring—transmission synchronizer blocking
31 Ring—transmission 2nd speed gear retaining snap
32 Gear—transmission 2nd speed
33 Washer—transmission 2nd speed gear thrust
34 Pin—1/8 x 1/4 rolled spring
35 Gear—transmission 3rd speed
36 Synchronizer assembly—3rd & 4th speed
37 Hub—transmission synchronizer
38 Insert—transmission synchronizer hub
39 Sleeve—transmission 3rd & 4th gear clutch hub

40 Spring—transmission synchronizer retaining
41 Ring—transmission synchronizer blocking
42 Ring—transmission m/d gear bearing shaft snap
43 Fork—transmission 1st & 2nd gear shift
44 Fork—transmission 3rd & 4th gear shift
45 Lever assembly—transmission reverse gear shaft relay

46 Retaining—transmission reverse gear shaft relay lever
47 Lever—transmission reverse gear shaft relay
48 Fork—transmission reverse gear shift
49 Spring—transmission shifter interlock
50 Plunger—transmission meshlock
51 Screw—m12 x 10 round head flat
52 Plate—transmission gear selector interlock
53 Screw & washer assembly—m10 x 30 hex head
54 Plug—3/4 diameter welch type
55 Shaft—transmission input
56 Roller—transmission mainshaft bearing
57 Bearing assembly—transmission m/d gear ball
58 Ring—m/d gear bearing retaining snap
59 Ring—1.00 retaining
60 Seal—transmission shift shaft
61 Gear—transmission 1st speed
62 Clip—spark control switch wire retaining
63 Gear—speedometer drive
64 Extension assembly—transmission
65 Extension—transmission
66 Bushing—transmission extension
67 Stop—transmission gear shift lever reverse
68 Gasket—transmission extension
69 Seal assembly—transmission extension oil
70 Plug—transmission extension
71 Retainer—transmission input shaft gear bearing
72 Seal assembly—transmission input shaft oil
73 Gasket—transmission input shaft bearing retainer
74 Bolt—M8 x 20 hex head-lock
75 Gasket—transmission case cover
76 Cover—transmission case
77 Screw—m6 x 20 hex head
78 Bolt—m6 x 32 hex washer HD shoulder
79 Plug—1/2·14 pipe (filler)
80 Bushing—transmission gear shift damper
81 Washer—spring lock
82 Nut—hexagon
83 Switch assembly—back-up lamp
84 Switch assembly—transmission seat belt warning sensor
85 Tag—transmission service identification
86 Washer—transmission 1st gear thrust
87 Ball—.25 diameter
88 Screw & lockwasher assembly—m12 x 40
89 Arm assembly—transmission control selector
90 Arm—transmission control selector
91 Pin—transmission gear shift

Exploded view of the Ford RAD and Warner SR4 (Chilton Type 6) transmission

to each shift fork; the two projections on the inside of the inserts fit into the blind holes in the ends of the shift forks. Insert the selector arm plates into the shift forks.

2. Install the 3rd–4th shifter fork and the 1st–2nd shifter fork into the cover, then lubricate the shifter shaft bore with grease.

3. Install the manual selector arm through the interlock plate and position the two pieces into the cover, with the wide leg of the interlock plate facing the inside of the transmission case.

4. Align the shifter shaft in the cover, then insert the shaft through the shifter forks and manual selector. Coat the shifter shaft with a light coating of grease. Make sure the detent grooves face the plunger side of the cover.

5. Align the pin holes in the manual selector arm and shifter shaft and install the spring pin flush with the surface of the selector arm.

6. Install the detent plunger, the spring and the plug. Torque the plug to 8–12 ft. lbs.

7. Check the operation of the shift forks in each gear position.

Output Shaft

DISASSEMBLY

1. Scribe alignment marks on the synchronizer and the blocking rings. Remove the front output shaft snap ring. Slide the 3rd–4th synchronizer assembly, the blocking rings and the 3rd gear off the shaft.

2. Remove the snap ring and the 2nd speed gear thrust washer from the shaft. Slide 2nd speed gear and the blocking ring off the shaft, taking care not to lose the sliding gear from the 1st–2nd synchronizer assembly.

NOTE: The 1st–2nd synchronizer hub cannot be removed from the output shaft.

3. Remove the 1st gear thrust washer (oil slinger) from the rear of the output shaft. Remove the spring pin retaining 1st speed gear to the shaft.

4. Slide 1st speed gear off the output shaft and remove the blocking ring. Take care not to lose the sliding gear from the 1st–2nd synchronizer assembly.

5. Clean and inspect the parts.

ASSEMBLY

1. Place a blocking ring on the 1st speed gear cone, then slide the gear and ring assembly onto the output shaft.

NOTE: Make sure that the inserts in the synchronizer engage in the blocking ring notches.

2. Install the 1st speed gear-to-output shaft spring pin.

3. Install a blocking ring on the 2nd speed gear cone, then slide the gear and ring assembly onto the output shaft.

NOTE: Make sure that the inserts in the synchronizer engage in the blocking ring notches.

4. Install the 2nd speed gear thrust washer and new snap ring on the shaft.

5. Install a blocking ring on the 3rd speed gear cone, then slide the gear and ring assembly onto the output shaft. Install the 3rd–4th synchronizer.

NOTE: Make sure that the inserts in the synchronizer engage in the blocking ring notches.

6. Install a new 3rd–4th synchronizer snap ring.

7. Place the 1st gear thrust washer (oil slinger) on the shaft and on the 1st gear spring pin.

——— CAUTION ———
The oil grooves must be positioned against the gear.

Countershaft Gear Bearing

REPLACEMENT

1. Remove the dummy shaft, the bearing retainer washers and needle bearings from the countershaft gear. Clean and inspect the parts.

2. Coat the bore at each end of the countershaft gear with grease to retain the needle bearings.

3. While holding the dummy shaft in the gear, install the needle bearings and the retainer washers in each end of the gear.

Input Shaft Bearing

REPLACEMENT

1. Remove the roller bearings from the input shaft.

2. Remove the input shaft bearing snap ring and press the input shaft out of the bearing. Clean and inspect the parts.

3. Press the input shaft bearing onto the input shaft, making sure that the snap ring groove faces the front of the shaft and install a new snap ring.

4. Lightly coat the bore of the input shaft with grease.

NOTE: If a thick film of grease, such as wheel bearing grease, is applied to the shaft, the lubrication holes may become clogged, thereby preventing transmission oil from reaching the bearings, possibly resulting in premature bearing failure.

5. Install the roller bearings in the bore.

Synchronizer

REPLACEMENT

1. Scribe alignment marks on the hub and sleeve of the synchronizer.

2. Push the synchronizer sleeve from each synchronizer hub.

NOTE: The 1st–2nd synchronizer hub cannot be removed from the output shaft.

3. Separate the inserts and the springs from the hubs, taking care not to mix the 1st–2nd synchronizer parts with the 3rd–4th synchronizer. Clean and inspect the parts.

4. Position the sleeve on the hub, making sure that the alignment marks are aligned.

5. Position the three inserts on the hub. Install the insert springs, taking care not to seat the bent tab in one of the inserts. The springs must face in opposite directions.

CHILTON TYPE 7

Transmission Case

DISASSEMBLY

1. Using the Bench Mounted Holding Fixture tool No. T57L-5000-B, secure the transmission to the bench.

2. Remove the drain plug from the lower right-side of the main case and drain any excess oil from the transmission.

3. Place the shift lever in the Neutral position, then remove the turret cover-to-transmission bolts.

4. Using a medium pry bar, pry the turret cover from the extension housing.

5. Using a $^3/_{16}$ inch (5mm) pin punch and a hammer, remove the offset lever-to-shifter shaft roll pin, then the damper sleeve.

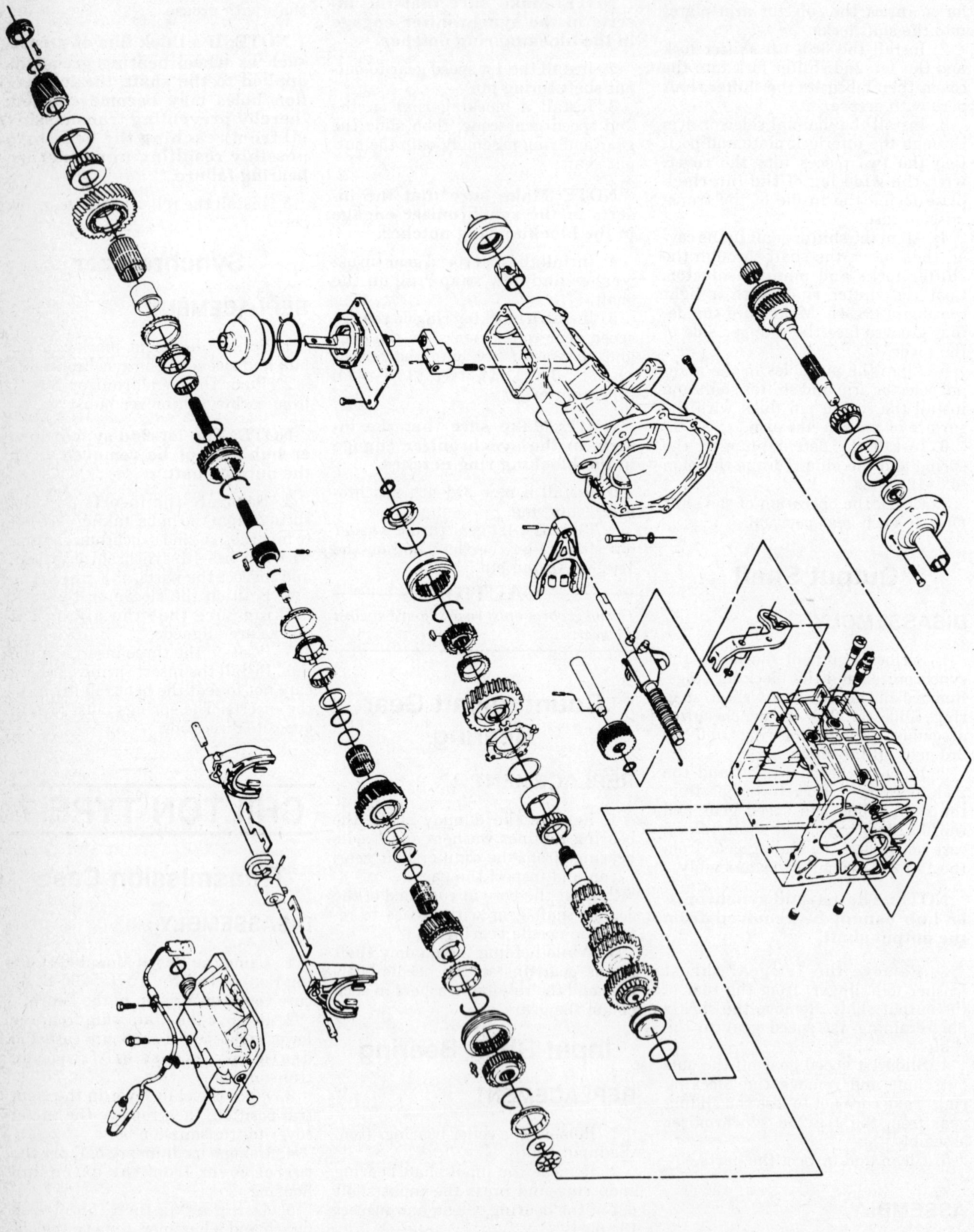

Exploded view of the Ford T5OD transmission (Chilton Type 7)

6. Remove the extension housing-to-main case bolts. Using a medium pry bar, pry the extension housing (break the seal) from the main case. Remove the extension housing/offset lever assembly by sliding it rearward.

7. Remove the offset lever, the roll pin, the detent spring/ball from the extension housing detent plate.

8. Remove the shift cover-to-main case bolts. Using a medium pry bar, pry the shift cover from the main case, then lift it slightly and slide it towards the filler plug side of the transmission. When shift forks clear groove in the 5th/Reverse shift lever, continue lifting the cover.

9. Using a pair of needle-nose pliers, remove the 5th/Reverse shift lever-to-lever pivot pin C-clip.

10. Using the T50 Torx® Driver tool, remove the 5th/Reverse shift lever pivot pin but do not remove the 5th/Reverse shift lever. Remove the back-up lamp switch.

11. Using a pair of snap ring pliers, remove the 5th gear synchronizer snap ring/spacer from the rear of the countershaft.

12. Remove the 5th gear, the synchronizer, the shift fork and the shift rail by gripping the components as an assembly and pulling them rearward from the main case.

NOTE: To disengage the 5th/Reverse shift rail, work it until it is free of the shift rail.

13. To remove the speedometer gear, press downward on the speedometer gear retaining clip and slide the gear from the output shaft, then remove the retaining clip.

14. Remove the front bearing retainer-to-main case bolts. Using a medium pry bar, pry the bearing retainer housing from the main case.

15. To remove the input shaft, rotate it until the flat on the clutch teeth aligns with the countershaft, then pull it from the main case; be careful not to drop the roller bearings, the thrust bearing or the race from the rear of the input shaft.

16. Remove the 4th gear blocking ring from the 3rd/4th synchronizer.

17. Pull the output shaft rearward, until the 1st gear stops against the case, then remove the output shaft bearing race.

NOTE: If the race sticks, work the shaft back and forth until it is free.

18. Tilt the output shaft so that the gear/synchronizer assembly end may be lifted up and out of the main case.

19. From the main case, remove the 5th/Reverse shift fork, the Reverse shift fork and the inhibitor spring.

20. Using a 3/16 inch (5mm) pin punch and a hammer, drive the roll pin from the Reverse idler shaft.

21. Through the back of the main case, slide out the Reverse idler shaft, then remove the Reverse idler gear and the overtravel rubber stop.

22. Using a hammer and a punch or chisel, flatten the countershaft retainer tabs (all four corners). Remove the countershaft retainer-to-main case bolts, the retainer, the shims and the bearing race.

NOTE: If the race sticks, work the shaft back and forth until it is free.

23. Using the Puller tool No. T81P-1104-C1 and the Puller Clamp tool No. D84L-1123-A, press the bearing from the rear of the countershaft.

24. Move the countershaft rearward, tilt the assembly upward and remove it from the case.

25. Clean all of the parts in solvent and inspect for damage or wear; replace the parts as necessary. Remove the front bearing from the countershaft.

ASSEMBLY

1. Using an arbor press and the Bearing Installation tool No. T57L-4621-B, press a new bearing onto the front of the countershaft, then position the countershaft in the main case.

2. Using an arbor press and the Bearing Installation tool No. T83P-7025-AH, press the rear bearing onto the countershaft.

NOTE: When pressing the rear bearing onto the countershaft, place two pieces of 1/4 inch bar stock inside the main case, between the countergear front and the main case to support it.

——— CAUTION ———

During installation, if the countershaft is not properly supported, permanent distortion/damage may result to the main case.

3. Install the rear bearing race onto the countershaft. Install the countershaft bearing retainer and torque the retainer-to-main case bolts to 10–15 ft. lbs. (15–20 Nm).

NOTE: Initially, when installing the countershaft bearing retainer, do not use any shims.

4. Using a Dial Indicator and the Bracketry tool No. D78P-4201-F, measure the countergear end play; it should be 0.001–0.005 in. (0.0254–0.127mm). If the end play is excessive, remove the countershaft bearing retainer and install shims.

5. After reinstalling the countershaft bearing retainer, bend the retaining tabs over the mounting bolts.

6. Install the Reverse idler gear in the main case with the shift lever groove facing the rear of the case, then the Reverse idler shaft and the rubber overtravel stop.

7. Using a 3/16 in. pin punch, drive the Reverse idler shaft roll pin into the idler shaft to secure the shaft.

8. Position the Reverse shifting fork and the 5th/Reverse shifting lever into the main case.

9. Install the output shaft assembly into the main case.

10. Using Polyethylene Grease No. DOAZ-19584-A, coat the input shaft roller bearings (place the bearings into the input shaft), the thrust bearing and the bearing race.

11. Install the 4th gear blocking ring; align the blocking ring notches with the inserts of the 3rd/4th synchronizer.

12. To install the input shaft, align the flat on the synchronizer teeth the with the countershaft, then install the input shaft.

13. Install the input shaft bearing race into the input shaft bearing retainer; do not install the shims. Install the bearing retainer (inner notch facing upwards) onto the main case; do not use sealant. Torque the bearing retainer-to-main case bolts to 11–20 ft. lbs. (15–27 Nm).

14. Install the output shaft rear bearing race; if necessary, tap the bearing into place using a plastic tipped hammer.

15. Install the 5th gear onto the countershaft. Install the shifting rail/5th gear shifting fork assembly into the main case.

NOTE: When installing the shifting rail/5th gear shifting fork assembly, align the shift rail fork and slide the rail through the fork, stop after the rod passes through the fork.

16. Place the shift lever return spring in the main case and slide the shifting rail through it; the long end of the spring MUST face the rear of the main case.

17. Install the blocking ring and the 5th gear synchronizer into the 5th

gear shifting fork, then slide the fork rail assembly into position.

18. Using a pair of snap ring pliers, install the 5th gear synchronizer retainer and snap ring.

19. Using a pair of needlenose pliers, connect the lever return spring to the front of the main case.

20. Apply Teflon Pipe Sealant No. D8AZ-19554-A to the 5th/Reverse shift lever pivot pin and the back-up light switch. Position the Reverse shift fork pin and the 5th gear shift rail pin so that they are engaged with the shift lever, then install the shift lever pivot pin. Using the T50 Torx® Driver tool, torque the pivot pin-to-shift lever to 23–32 ft. lbs. and the back-up light switch-to-transmission to 12–18 ft. lbs.

21. Install the speedometer gear onto the output shaft; make sure that the retainer clip engages a hole in the output shaft.

22. Using Silicone Rubber Sealant No. D6AZ-19562-A, apply a $^1/_8$ in. bead to the shift cover assembly. Position the synchronizers and the shifting cover into the Neutral positions, then install the cover assembly (shifting forks engaging the synchronizers). Torque the shift cover-to-main case bolts to 6–11 ft. lbs. (8–15 Nm).

23. Using Silicone Rubber Sealant No. D6AZ-19562-A, apply a $^1/_8$ in. bead to the extension housing mating surface and the lubrication funnel in the extension housing.

24. Coat the offset lever's detent spring, the detent and the detent ball (place in the Neutral position) with petroleum jelly; position and install these parts into the extension housing (be sure to position the offset lever with the spring over the detent ball.

25. Install the extension housing and shift lever to the main case; be sure the lubrication funnel engages into the 5th gear synchronizer.

26. To install the offset lever, push it downward to compress the detent spring and to push the lever and the housing into position. Install the extension housing-to-main case bolts and torque to 20–45 ft. lbs. (27–61 Nm).

27. Using a $^3/_{16}$ in. (5mm) pin punch, drive the roll pin into the offset lever-to-shifter shaft hole. Install the damper sleeve into the offset lever.

28. To measure the output shaft end play, perform the following procedures:

a. Position the transmission so that the extension is facing upwards.

b. Using a Dial Indicator, secure it to the extension housing and position it so that it rides on the end of the output shaft.

c. Rotate the input and the output shafts, then zero the dial indicator.

d. Using a wooden block, push upwards on the input shaft and note the dial indicator reading.

NOTE: A shim must be installed that is the thickness of the dial indicator reading, which will provide a zero end play.

––––––– **CAUTION** –––––––

DO NOT overload the bearings with too thick of a shim; a ± 0.002 in. (± 0.050mm) is acceptable.

29. Place the transmission on a level surface and remove the input bearing retainer and the bearing race from the retainer; install the shim under the bearing race.

30. Using Silicone Rubber Sealant No. D6AZ-19562-A, apply a $^1/_8$ in. bead to the bearing retainer, install the retainer and check the end play.

NOTE: When applying sealant to the bearing retainer, sealant must not cover the notch on the inner edge of the retainer. Be sure to position the retainer with the inner notch facing upwards.

31. Using Silicone Rubber Sealant No. D6AZ-19562-A, apply a $^1/_8$ in. bead to the turret cover. Place the cover onto the extension housing and torque the bolts to 11–15 ft. lbs. (15–20 Nm).

32. Install and torque the drain plug-to-main case to 15–30 ft. lbs. (20–41 Nm).

CHILTON TYPE 8

DISASSEMBLY

1. Remove the lower extension housing bolt to drain the lubricant.

2. Remove the cover bolts and the cover, then discard the gasket.

3. Using a magnet, remove the screw, the detent spring and the plug from the case.

4. Drive the roll pin from the shifter shaft.

5. Remove the backup lamp switch, the snap ring and the dust cover from the rear of the extension housing.

6. Remove the shifter shaft from the turret assembly.

7. Remove the extension housing bolts and the housing, then discard the gasket.

8. Remove the speedometer gear snap ring, the slide the gear from the shaft and the drive ball.

9. Remove the output shaft bearing snap ring and the bearing.

10. Using a dummy shaft, push the countershaft out through the rear of the case, then lower the countershaft gear to the bottom of the case.

11. Remove the input shaft bearing retainer bolts, then slide the retainer and the gasket from the input shaft; discard the gasket.

12. Remove the input shaft bearing snap ring and the bearing.

13. Remove the input shaft and the blocking ring (including roller bearings) from the case.

14. Remove the overdrive shift pawl, the gear selector and the interlock plate. Remove the 1st–2nd gearshift selector arm plate and the roll pin from the 3rd–Overdrive shift fork.

15. Drive the 3rd–Overdrive shift rail and expansion plug from the rear of the case, then remove the mainshaft.

16. Remove the 1st–2nd gear shift fork and the 3rd–Overdrive shift fork.

17. Remove the countershaft gear and the thrust washers from the case.

18. Remove the snap ring from the front of the output shaft. Slide the 3rd gear, the O.D. synchronizer, the blocking ring and the gear from the shaft.

19. Remove the next snap ring and washer, then remove 2nd speed gear. Remove next snap ring and the 1st–2nd synchronizer. Slide the 1st gear and the blocking ring from the rear of the shaft.

20. Remove the roll pin from the Reverse fork, then slide the Reverse shifter rail through the rear of the case. Remove the Reverse gearshift fork and spacer.

21. Drive the Reverse gear shaft out through the rear of the case.

22. Remove the Reverse idler gear, the thrust washers and the roller bearings.

23. Remove the retaining clip, the Reverse gearshift relay lever and the Reverse gear selector fork pivot pin. Remove the O.D. shift control link assembly. Remove the shift shaft seal from the rear of the case and the expansion plug from the front of the case.

ASSEMBLY

To assemble, reverse the disassembly procedures. Torque the extension housing bolts in a criss-cross pattern to 42–50 ft. lbs. The bearing rollers, the extension housing bushing, the shifter shaft and the gear shift damper bushing are to be lubricated with grease before assembly (Ford ESW-MIC109-A). The gear shift shaft sleeve and the turret cover assembly should be coated with sealer prior to installation. The intermediate and high rail welch plug must be seated firmly; it must not protrude above the front face of the case, nor seat below 0.6 in. below the front face.

With the 1st gear thrust washer clamped tightly against the output shaft shoulder, the 1st gear end play is 0.005–0.024 in. The 2nd gear end play is 0.003–0.021 in., O.D. end play is 0.009–0.023 in. and the countershaft gear end play, checked after installation between the thrust washers, is 0.004–0.018 in.

NOTE: When the gearshift selector arm plate is seated in the 1st–2nd shift fork plate slot, the shifter shaft must pass freely through the bore without binding.

CHILTON TYPE 9

Transmission Case

DISASSEMBLY

1. Using a 10mm wrench, remove the ten cover bolts and lift off the cover.
2. Drain the lubricant.
3. Using a pencil size magnet, remove the shift rail detent plug, the spring and the plunger from the upper left side of the case.
4. Working through the shift turret opening in the extension housing, remove the access plug from the rear of the housing.
5. After shifting into Reverse gear, remove the roll pin from the gear shift shaft offset lever, then slide the offset lever and bushing off the shaft.
6. Remove the 5th speed interlock pilot bolt from the front top of the extension housing.
7. Remove the 6 extension housing bolts, then slide the housing and the gasket off the output shaft.
8. Remove the snap ring, the speedometer drive gear and the drive ball from the output shaft.
9. Remove the 5th gear synchronizer snap ring from the output shaft, then slide the retaining spacer off the output shaft.
10. Shift the transmission into 1st gear. Using a hammer and a punch, drive out the roll pin located inside the case, which secures the 1st, 2nd, 3rd, 4th and Reverse selector pin, then remove the selector pin.
11. Slide the shifter shaft, the 5th speed shift fork and the synchronizer

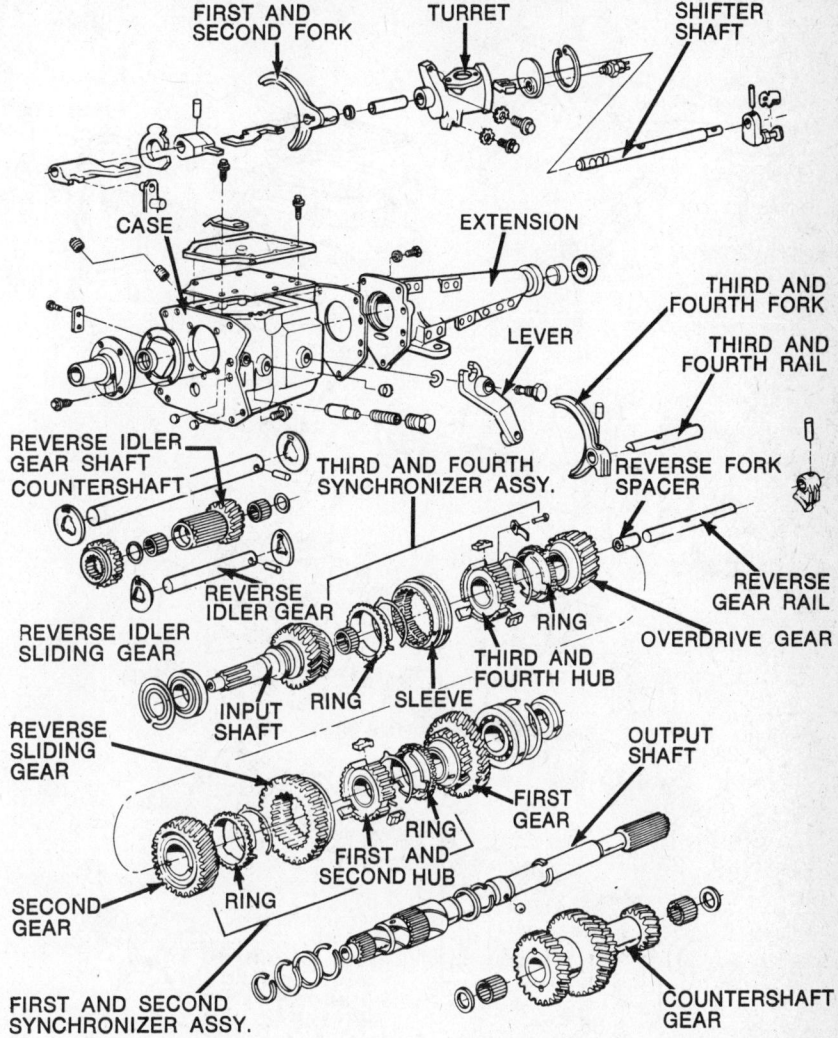

Exploded view of the Ford single rail 4-speed overdrive (Chilton Type 8)

from the output shaft as an assembly.
12. Remove the interlock sleeve bolt from the right side of the case.
13. Remove the interlock sleeve, the 3rd–4th speed shift fork and the 1st–2nd speed shift fork from the case.
14. Working inside the case, remove the C-clip from the Reverse gear selector fork pivot pin. Remove the pivot pin, then lift the Reverse gear selector fork relay lever, the spring and the Reverse gearshift fork from the case.
15. Slide the 5th speed maindrive gear off the output shaft.
16. Remove the snap ring located at the rear of the 5th speed cluster gear.
17. Using a puller, remove the 5th speed cluster gear.
18. Remove the output shaft rear bearing snap ring and the bearing cup.
19. Remove the 4 input shaft bearing retainer bolts, the bearing retain-

er, the seal, the shim and the O-ring from the case.
20. Without loosing the roller bearings, the thrust washers and the thrust bearing, rotate the input shaft so that the teeth recess face the countershaft gear, then lift the input gear from the case.
21. Remove the output shaft assembly through the top of the transmission case.
22. Remove the snap ring from the rear of the case and the countershaft gear rear bearing cup from the case.
23. Remove the three bolts, the bearing retainer, the gasket, the shim and the front bearing cup from the case.
24. Lift the countershaft gear through the top of the case.
25. Remove the Reverse idler gear and shaft by removing the roll pin that secures the shaft to the case.

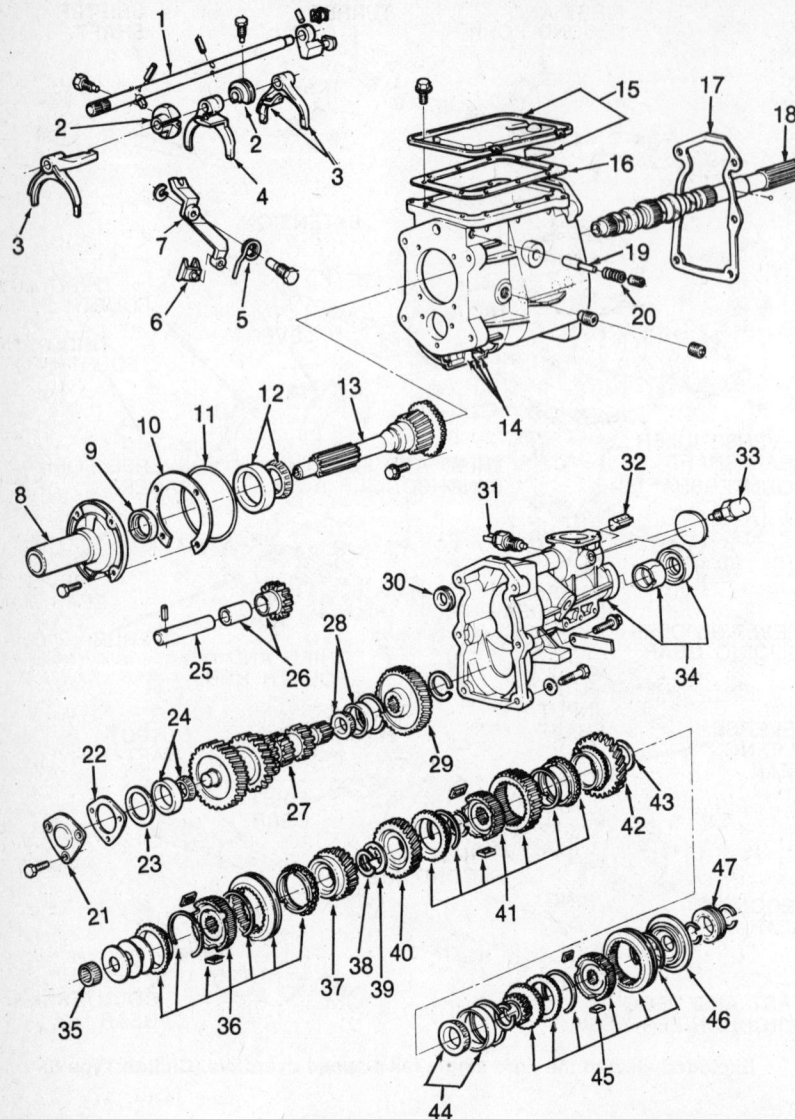

1. Shifter shaft
2. Gear selector interlock
3. 3rd/4th shifter fork
4. 1st/2nd shifter fork
5. Shift lever return spring
6. Reverse shifter fork
7. Reverse shift lever
8. Bearing retainer
9. Input shaft oil seal
10. Input shaft front bearing shim
11. Bearing retainer seal
12. Input shaft bearing assembly
13. Input shaft
14. Case assy.
15. Cover assy.
16. Cover gasket
17. Extension gasket

18. Output shaft
19. Meshlock plunger
20. Interlock shifter spring
21. Countershaft gear front retainer
22. Retainer gasket
23. Front bearing shim
24. Front bearing assy.
25. Reverse idler gear shaft
26. Reverse idler gear and bushing
27. Countershaft cluster gear
28. Rear bearing assy.
29. 5th speed cluster gear
30. Shifter shaft seal
31. Back-up switch
32. Lever reverse stop

33. 5th speed inhibitor plunger
34. Extension housing assy.
35. Mainshaft roller bearing
36. 3rd/4th synchronizer assy.
37. 3rd speed gear
38. Snap ring
39. Thrust washer
40. 2nd speed gear
41. 1st/2nd synchronizer assy.
42. 1st speed gear
43. Thrust washer
44. Output shaft bearing assy.
45. 5th speed synchronizer assy.
46. Retaining spacer
47. Speedometer drive gear

Ford RAP five speed overdrive (Chilton Type 9) transmission

ASSEMBLY

NOTE: Coat the bolts and plugs used throughout the case with a thread sealant to prevent leakage.

1. Hold the Reverse idler gear into position with the long end of the hub facing to the rear of the case. Slide the idler gear shaft through the case and gear and align the roll pin holes, then secure the shaft with the roll pin.

2. Lower the countershaft and bearings into place and install the rear bearing cup, then secure with the snap ring.

3. Position the front bearing cup, the shim, a new gasket and the bearing retainer to the front of the case. Install the bearing retainer cap bolts and torque to 7–10 ft. lbs. (while rotating the gear). If the gear rotating effort increases while torquing the bearing retainer, replace the shim with a thinner one.

4. The correct end play is 0.001–0.005 inch.

NOTE: Decrease the shim thickness to increase the end play and increase the shim thickness to reduce end play.

5. Lower the main shaft into the case through the case cover opening.

6. Apply a coat of polyethylene grease to the thrust washers and the thrust bearing. Place the thrust washer on the 3rd–4th speed synchronization thrust surface. Place the thrust bearing and the remaining thrust washer on the 3rd–4th speed synchronizer.

7. Without disturbing the roller bearings, carefully install the input shaft assembly in the case with the blank portion of the teeth facing the countershaft gear.

8. Coat a new input shaft O-ring with polyethylene grease and position it in the bearing retainer groove.

9. Install the output shaft bearing cup and the snap ring into the rear of the case.

10. Position the shim and bearing retainer into the case. Install the bearing retainer bolts and torque to 8–10 ft. lbs. (while rotating the input shaft).

NOTE: If the input shaft turning effort increases when torquing the bearing retainer bolts, replace the shim with a thicker one.

11. Install a dial indicator on the case. Pry the output shaft toward the dial indicator and zero the indicator. Pry the output shaft in the opposite direction. The end play should be between 0.001–0.005 inch. Increase shim thickness to decrease end play or decrease shim thickness to increase end play. Remove the dial indicator.

12. Install the spring and the Reverse fork on the relay lever. Position the relay lever assembly in the case, then install the pivot pin in the case and lever assembly. Secure the lever with a C-clip.

13. Install the 5th speed cluster gear and secure with a snap ring.

14. Slide the 5th speed main drive gear onto the output shaft. Coat the

blocking ring with polyethylene grease and position it on the main drive gear.

15. Position the 1st–2nd and the 3rd–4th shift forks onto the main shaft assembly.

16. Place the interlock gear selector sleeve between the 2 shifter forks and install the interlock pilot bolt in the right side of the case.

17. With the synchronizer thrust surface facing the rear of the output shaft, install the shifter shaft, the 5th speed shift fork and the 5th speed synchronizer as an assembly.

18. Working through the cover opening in the case, install the gearshift selector pin in the shifter shaft and secure with a roll pin.

19. Slide the 5th speed synchronizer retaining plate onto the output shaft and secure with a snap ring.

20. Secure the speedometer drive gear ball to the output shaft with polyethylene grease then slide the speedometer drive gear onto the shaft over the ball and secure with a snap ring.

21. Using a new gasket, position the extension housing on the case. Install the two pilot bolts, one in the upper left side of the housing and the other in the lower right corner. Install the four remaining bolts and torque to 40–60 ft. lbs.

22. Install the 5th gear pilot bolt in the top of the extension housing.

23. Shift the into Reverse gear. Install the offset lever on the rear of the shifter shaft and secure with a roll pin.

24. Install the detent plunger, the spring and the plug in the upper right side of the case, then torque the plug to 12–14 ft. lbs.

25. Install the access plug in the rear of the extension housing.

26. Using a new gasket place the cover on the case and torque the bolts to 8–10 ft. lbs.

Output Shaft

DISASSEMBLY

1. Slide the 3rd–4th speed synchronizer off the front end of the output shaft.

2. Slide the 3rd speed gear off the front of the output shaft.

3. Remove the snap ring and the 2nd speed gear thrust washer from the output shaft. Slide the 2nd speed gear and the synchronizer blocking ring off the output shaft.

4. Remove the 1st–2nd speed synchronizer snap ring from the output shaft, then press the synchronizer off the output shaft.

5. Remove the snap ring from the rear of the output shaft. Place the output shaft in a press, then remove the 1st speed gear, the thrust washer and the output shaft rear bearing.

ASSEMBLY

1. Position the 1st gear thrust washer and the bearing on the rear of the output shaft. Apply pressure on the bearing inner race until the bearing is bottomed on the spacer and shaft.

2. Select a snap ring that will not allow any clearance between the bearing race and the ring groove. Then press the 1st–2nd gear synchronizer and the Reverse sliding gear into place, then secure with the snap ring.

3. Slide the 2nd gear and the thrust washer into position, then secure with the snap ring.

4. Slide the 3rd gear and the 3rd–4th synchronizer into place. Make sure the thrust surface of the synchronizer hub is facing forward.

Input Shaft

DISASSEMBLY

1. If not previously removed, remove the roller bearings from the input shaft.

2. Place the input gear in a press and press the input gear from the bearing.

ASSEMBLY

1. With the taper toward the front of the gear, apply pressure on the inner race and press the bearing onto the input gear until it is bottomed.

2. Apply a heavy coat of polyethylene on the inner bearing surface of the gear. Insert the 15 roller bearings into the gear.

Countershaft Gear

DISASSEMBLY

1. Place the countershaft gear in a press and remove the rear bearing.

2. Place the countershaft in a vise protected with wood blocks and pry the front bearing from the countershaft.

ASSEMBLY

1. With the taper facing outward, exert pressure on the inner race of the front bearing and press the bearing until it is bottomed on the gear.

2. Install the rear bearing in the same manner.

Input Shaft Gear Bearing Retainer

DISASSEMBLY

1. Place the bearing retainer in a vise.

2. Using a slide impact type puller, remove the seal from the bearing retainer.

ASSEMBLY

Install the seal in the retainer with the lip facing forward. Make sure the seal is bottomed in the retainer.

Extension Housing

DISASSEMBLY

1. Carefully remove the seal from the extension housing.

2. Using a suitable driver, remove the bushing.

ASSEMBLY

Install the bushing and the seal using a suitable driver.

CHILTON TYPE 10

Transmission Case

DISASSEMBLY

1. Remove the side cover assembly and the shift forks.

2. Remove the clutch gear bearing retainer.

3. Remove the clutch gear bearing-to-gear stem snap ring. Pull the clutch gear outward until a screwdriver can be inserted between the bearing and the case. Remove the clutch gear bearing.

4. Remove the speedometer driven gear and the extension bolts.

5. Remove the Reverse idler shaft snap ring.

6. Remove the mainshaft and the extension assembly through the rear of the case.

7. Remove the clutch gear and the 3rd speed blocking ring from inside the case. Remove the 14 roller bearings from the clutch gear.

1 Synchronizer retainer ring
2 Synchronizer blocking ring
3 Synchronizer assembly
4 Second speed gear
5 Main shaft
6 Synchronizer assembly
7 Gear assembly
8 Thrust washer
9 Retainer clip

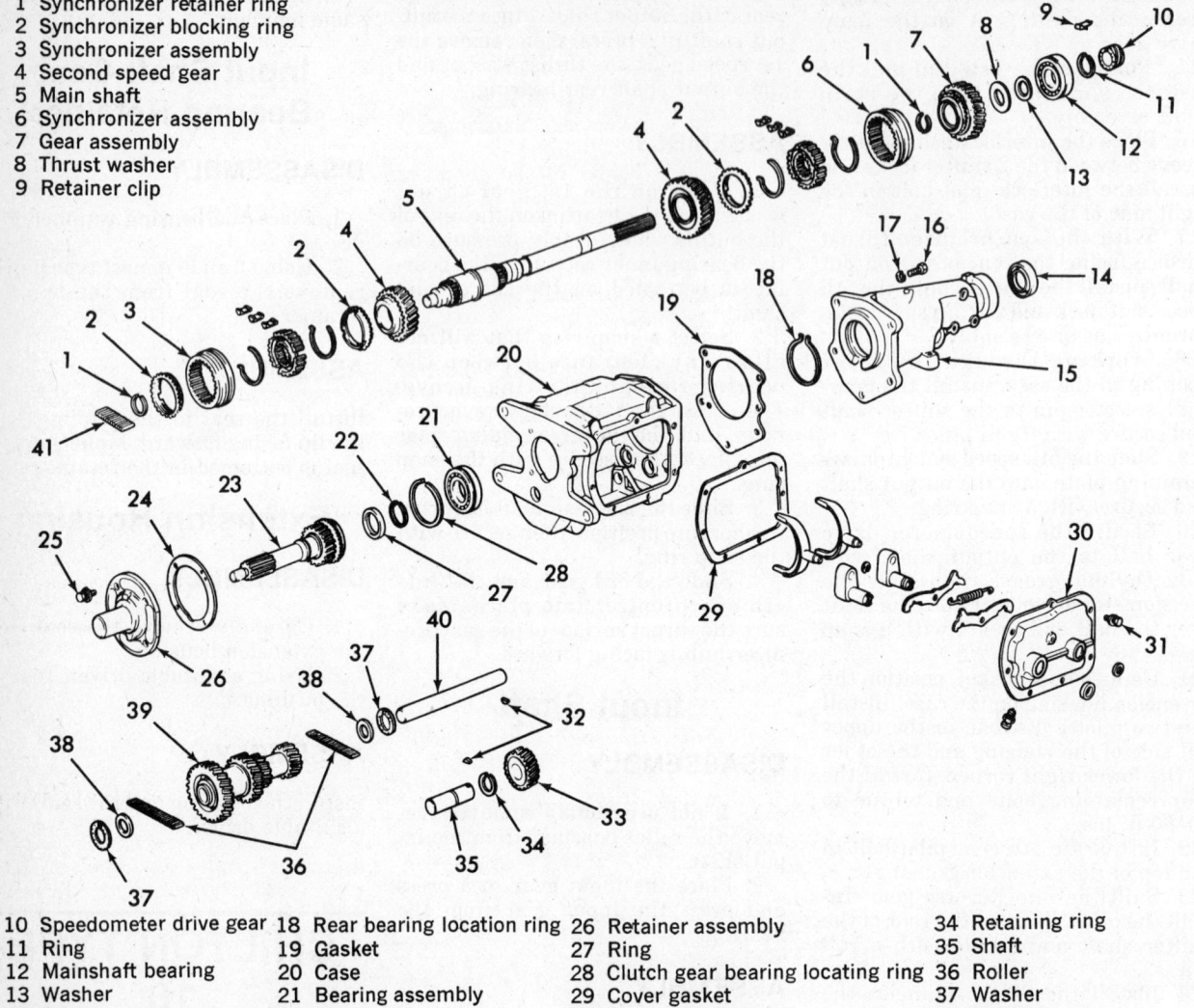

10 Speedometer drive gear	18 Rear bearing location ring	26 Retainer assembly	34 Retaining ring
11 Ring	19 Gasket	27 Ring	35 Shaft
12 Mainshaft bearing	20 Case	28 Clutch gear bearing locating ring	36 Roller
13 Washer	21 Bearing assembly	29 Cover gasket	37 Washer
14 Seal	22 Ring	30 Cover assembly	38 Washer
15 Extension housing	23 Clutch gear	31 Bolt and lockwasher	39 Gear assembly
16 Bolt	24 Gasket	32 Woodruff keys	40 Counter gear shaft
17 Washer	25 Bolt and lockwasher	33 Gear assembly	41 Mainshaft bearing roller

Exploded view of the Saginaw 3-speed (Chilton Type 10) transmission

8. Expand the snap ring which retains the mainshaft rear bearing and remove the extension.

9. Using a dummy shaft, drive the countershaft and the key out through the rear of the case. Remove the gear, the two tanged thrust washers and the dummy shaft. Remove the bearing washer and the 27 roller bearings from each end of the countergear.

10. Using a long drift, drive the Reverse idler shaft and key through the rear of the case.

11. Remove the Reverse idler gear and the tanged steel thrust washer.

ASSEMBLY

1. Using a dummy shaft, grease and load a row of 27 roller bearings

and a thrust washer at each end of countergear.

2. Place the countergear assembly into the case from the rear. Place a tanged thrust washer (tang away from the gear) at each end. Install the countershaft and the key, making sure that the tangs align with the notches in the case.

3. Install the Reverse idler gear thrust washer, the gear and the shaft with a key from the rear of the case.

NOTE: Be sure the thrust washer is between the gear and the rear of the case with the tang toward the notch in the case.

4. Expand the snap ring in the extension housing. Assemble the extension over the rear of the mainshaft

and onto the rear bearing. Seat the snap ring in the rear bearing groove.

5. Install the 14 mainshaft pilot bearings into the clutch gear cavity. Assemble the 3rd speed blocking ring onto the clutch gear clutching surface with the teeth toward the gear.

6. Place the clutch gear, the pilot bearings and the 3rd speed blocking ring assembly over the front of the mainshaft assembly; be sure the blocking rings align with the keys in the 2nd–3rd synchronizer assembly.

7. Stick the extension gasket to the case with grease. Install the clutch gear, the mainshaft and the extension together; be sure the clutch gear engages the teeth of the countergear anti lash plate. Torque the extension bolts to 45 ft. lbs.

8. Place the bearing over the stem of the clutch gear and into the front case bore. Install the front bearing to the clutch gear snap ring.

9. Install the clutch gear bearing retainer and the gasket. The retainer oil return hole must be at the bottom. Torque the retainer bolts to 10 ft. lbs.

10. Install the Reverse idler gear shaft E-ring.

11. Shift the synchronizer sleeves to the Neutral positions. Install the cover, the gasket and the forks; aligning the forks with the synchronizer sleeve grooves. Torque the side cover bolts to 10 ft. lbs.

12. Install the speedometer driven gear.

Mainshaft

DISASSEMBLY

1. Remove the 2nd–3rd speed sliding clutch hub snap ring from the mainshaft. Remove the clutch assembly, the 2nd speed blocking ring and the 2nd gear from front of the mainshaft.

2. Depress the speedometer drive gear retaining clip and remove the gear. Some units have a metal speedometer driver gear which must be pulled off.

3. Remove the rear bearing snap ring.

4. Support the Reverse gear and press on the rear of the mainshaft. Remove the Reverse gear, the thrust washer, the spring washer, the rear bearing and the snap ring.

NOTE: When pressing off the rear bearing, be careful not to cock the bearing on the shaft.

5. Remove the 1st and Reverse sliding clutch hub snap ring. Remove the clutch assembly, 1st speed blocking ring and the 1st gear; sometimes the synchronizer hub and gear must be pressed off.

ASSEMBLY

1. Turn the front of the mainshaft up.

2. Install the 2nd gear with the clutching teeth up; the rear face of the gear butts against the flange on the mainshaft.

3. Install a blocking ring with the clutching teeth down. The three blocking rings are the same.

4. Install the 2nd–3rd speed synchronizer assembly with the fork slot down; press it onto the mainshaft splines.

NOTE: Both synchronizer as-semblies are the same. Be sure that the blocking ring notches align with the synchronizer assembly keys.

5. Install the synchronizer snap ring; both synchronizer snap rings are the same.

6. Turn the rear of the shaft up, then install the 1st gear with the clutching teeth up; the front face of the gear butts against the flange on the mainshaft.

7. Install a blocking ring with the clutching teeth down.

8. Install the 1st–Reverse synchronizer assembly with the fork slot down, then press it onto the mainshaft splines; be sure the blocking ring notches align with the synchronizer assembly keys.

9. Install the snap ring.

10. Install the Reverse gear with the clutching teeth down.

11. Install the steel Reverse gear thrust washer and the spring washer.

12. Press the rear ball bearing onto the shaft with the snap ring slot down.

13. Install the snap ring.

14. Install the speedometer drive gear and the retaining clip; press on the metal speedometer drive gear.

Clutch Keys and Springs

REPLACEMENT

The keys and the springs may be replaced if worn or broken, but the hubs and sleeves are matched pairs, they must be kept together.

1. Mark the hub and sleeve for reassembly.

2. Push the hub from the sleeve, then remove the keys and the springs.

3. Place the three keys and the two springs (one on each side of hub) in position, so the three keys are engaged by both springs; the tanged ends of the springs should not be installed into the same key.

4. Slide the sleeve onto the hub by aligning the marks.

NOTE: A groove around the outside of the synchronizer hub marks the end that must be opposite the fork slot in the sleeve when assembled.

Extension Oil Seal and Bushing

REPLACEMENT

1. Remove the seal.

2. Using the bushing removal and installation tool, drive the bushing into the extension housing.

3. Drive the new bushing in from the rear. Lubricate the inside of the bushing and the seal. Install a new oil seal with the extension seal installation tool or other suitable tool.

Clutch Bearing Retainer Oil Seal

REPLACEMENT

1. Pry the old seal out.

2. Install the new seal using the seal installer. Seat the seal in the bore.

CHILTON TYPE 11

Transmission Case

DISASSEMBLY

1. Drain the lubricant. Remove the side cover and the shift forks.

2. Remove the clutch gear bearing retainer. Remove the bearing-to-gear stem snap ring and pull out on the clutch gear until a small pry bar can be inserted between the bearing, the large snap ring and case to pry the bearing off.

NOTE: The clutch gear bearing is a slip fit on the gear and in the case. Removal of the bearing will provide clearance for the clutch gear and the mainshaft removal.

3. Remove the extension housing bolts, then remove the clutch gear, the mainshaft and the extension as an assembly.

4. Spread the snap ring which holds the mainshaft rear bearing and remove the extension case.

5. Using a dummy shaft, drive the countershaft and its woodruff key out through the rear of the case. Remove the countergear assembly and the bearings.

6. Using a long drift, drive the Reverse idler shaft and the woodruff key through the rear of the case.

7. Expand and remove the 3rd–4th speed sliding clutch hub snap ring from the mainshaft. Remove the clutch assembly, the 3rd gear blocking ring and the 3rd speed gear from the front of the mainshaft.

8. Press in the speedometer gear

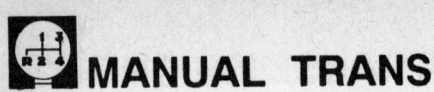

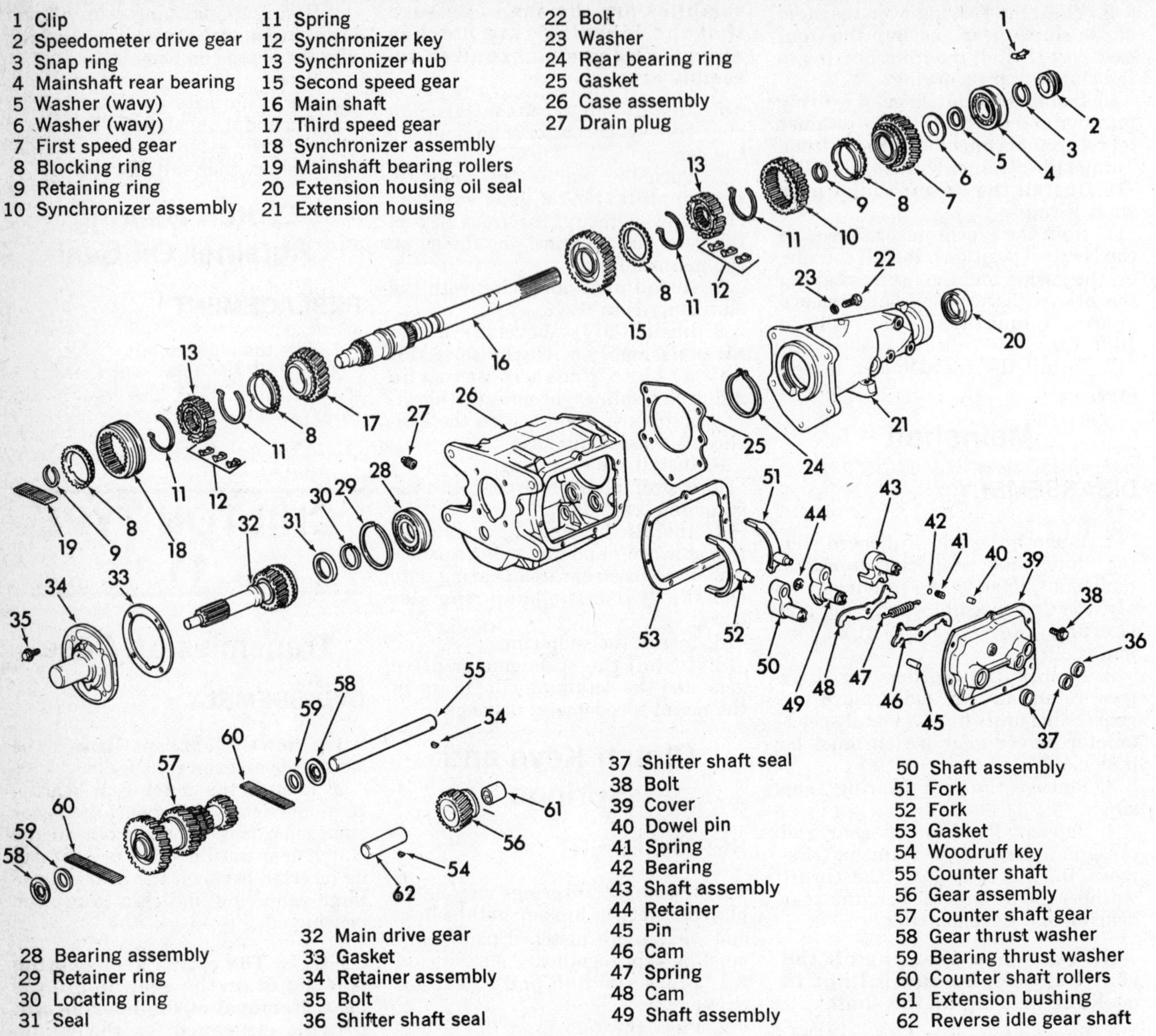

1 Clip
2 Speedometer drive gear
3 Snap ring
4 Mainshaft rear bearing
5 Washer (wavy)
6 Washer (wavy)
7 First speed gear
8 Blocking ring
9 Retaining ring
10 Synchronizer assembly
11 Spring
12 Synchronizer key
13 Synchronizer hub
15 Second speed gear
16 Main shaft
17 Third speed gear
18 Synchronizer assembly
19 Mainshaft bearing rollers
20 Extension housing oil seal
21 Extension housing
22 Bolt
23 Washer
24 Rear bearing ring
25 Gasket
26 Case assembly
27 Drain plug

28 Bearing assembly
29 Retainer ring
30 Locating ring
31 Seal
32 Main drive gear
33 Gasket
34 Retainer assembly
35 Bolt
36 Shifter shaft seal

37 Shifter shaft seal
38 Bolt
39 Cover
40 Dowel pin
41 Spring
42 Bearing
43 Shaft assembly
44 Retainer
45 Pin
46 Cam
47 Spring
48 Cam
49 Shaft assembly

50 Shaft assembly
51 Fork
52 Fork
53 Gasket
54 Woodruff key
55 Counter shaft
56 Gear assembly
57 Counter shaft gear
58 Gear thrust washer
59 Bearing thrust washer
60 Counter shaft rollers
61 Extension bushing
62 Reverse idle gear shaft

Exploded view of the Saginaw 4-speed (Chilton Type 11) transmission

retaining clip and slide the gear off the mainshaft. Remove the rear bearing snap ring from the mainshaft.

9. Using an arbor press, support the 1st gear on press plates, then press the 1st gear, the thrust washer, the spring washer, the rear bearing and snap ring from the rear of the mainshaft.

—— **CAUTION** ——

Be sure to center the gear, the washers, the bearings and the snap ring when pressing the rear bearing.

10. Expand and remove the 1st–2nd sliding clutch hub snap ring from the mainshaft, then remove the clutch assembly, the 2nd speed blocking ring and the 2nd speed gear from the rear of the mainshaft.

NOTE: After thoroughly cleaning the parts and the transmission case, inspect and replace the damaged or worn parts. When checking the bearings, do not spin them at high speeds. Clean and rotate the bearings by hand to detect the roughness or unevenness. Spinning can damage the balls and the races.

ASSEMBLY

1. Grease both inside ends of the countergear. Install a dummy shaft into the countergear, then load a row of roller bearings (27) and thrust washers at each end of the countergear.

2. Position the countergear assembly into the case through the rear opening. Place a tanged thrust washer at each end of the countergear.

3. Install the countergear shaft and woodruff key from the rear of the case.

NOTE: Make sure that the shaft engages both thrust washers and that the tangs align with their notches in the case.

4. Install the Reverse idler gear, the shaft and the woodruff key. Install the extension-to-rear bearing snap ring. Assemble the extension housing over the rear of the mainshaft and onto the rear bearing.

5. Install the 14 mainshaft pilot bearings into the clutch opening and the 4th speed blocking ring onto the

clutching surface of the clutch gear (with the clutching teeth facing the gear).

6. Assemble the clutch gear, the pilot bearings and the 4th speed blocking ring unit over the front of the mainshaft. Do not assemble the bearing to the gear at this point.

— CAUTION —

Be sure that the blocking ring notches align with the 3rd–4th synchronizer assembly keys.

7. Install the extension-to-case gasket and secure it with grease. Install the clutch gear, the mainshaft and the extension housing as an assembly. Install the extension-to-case bolts (apply sealer to the bottom bolt) and torque to 45 ft. lbs.

8. Install the outer snap ring on the front bearing and place the bearing over the stem of the clutch gear and into the case bore.

9. Install the snap ring to the clutch gear stem. Install the clutch gear bearing retainer and the gasket, with the retainer oil return hole at the bottom.

10. Place the synchronizer sleeves into the Neutral positions and install the cover, the gasket and the fork assemblies to the case; be sure the forks align with the synchronizer sleeve grooves. Torque the cover bolts to 22 ft. lbs.

Mainshaft

ASSEMBLY

Install the following parts with the front of the mainshaft facing up:

1. Install the 3rd speed gear with the clutching teeth up; the rear face of the gear will abut with the mainshaft flange.

2. Install a blocking ring (with the clutching teeth down) over the 3rd speed gear synchronizing surface.

NOTE: The four blocking rings are the same.

3. Press the 3rd–4th synchronizer assembly (with the fork slot down) onto the mainshaft splines until it bottoms.

— CAUTION —

The blocking ring notches must align with the synchronizer assembly keys.

4. Install the synchronizer hub-to-mainshaft snap ring; both synchronizer snap rings are the same.

Install the following parts with the rear of the mainshaft facing up:

5. Install the 2nd speed gear with the clutching teeth up; the front face

of the gear will abut with the flange on the mainshaft.

6. Install a blocking ring (with the clutching teeth down) over the 2nd speed gear synchronizing surface.

7. Press the 1st–2nd synchronizer assembly (with the fork slot down) onto the mainshaft.

— CAUTION —

The blocking ring notches must align with the synchronizer assembly keys.

8. Install the synchronizer hub-to-mainshaft snap ring.

9. Install a blocking ring with the notches down so they align with the 1st–2nd synchronizer assembly keys.

10. Install the 1st gear with the clutching teeth down. Install the 1st gear thrust washer and the spring washer.

11. Press the rear ball bearing (with the slot down) onto the mainshaft. Install the snap ring. Install the speedometer gear and clip.

CHILTON TYPE 12

Transmission Case

DISASSEMBLY

1. Place the transmission so that it is resting on the bell housing.

2. Drive the spring pin from the shifter shaft arm assembly and the shifter shaft, then remove the shifter shaft arm assembly.

3. Remove the five extension housing-to-case bolts and the extension housing.

4. Press down on the speedometer gear retainer, then remove the gear and the retainer from the mainshaft.

5. Remove the snap rings from the shifter shaft, then the Reverse shifter shaft cover, the shifter shaft detent cap, the spring, the ball and the interlock lock pin.

6. Pull the Reverse lever shaft outward to disengage the Reverse idler, then remove the idler shaft with the gear attached.

7. Remove the Reverse gear snap ring, the Reverse countershaft gear and the gears.

8. Turn the case on its side and remove the clutch gear bearing retainer bolts, the retainer and the gasket.

9. Remove the clutch gear ball bearing-to-bell housing snap ring, then the bell housing-to-case bolts.

10. Turn the case so that it rests on

the bell housing, then expand the mainshaft bearing snap ring and remove the case by lifting it off the mainshaft.

NOTE: Make sure that the mainshaft assembly, the countergear and shifter shaft assembly stay with the bell housing.

11. Lift the entire mainshaft assembly complete with shifter forks and countergear from the bell housing.

ASSEMBLY

1. Using a press, install the shielded ball bearing to the clutch gear shaft with the snap ring groove up.

2. Install the snap ring on the clutch gear shaft. Place the pilot bearings into the clutch gear cavity, using heavy grease to hold them in place.

3. Assemble the clutch gear to the mainshaft and the detent lever to the shift shaft with the roll pin.

4. Position the 1st–2nd gear shifter so that it engages the detent lever.

5. Assemble the 3rd–4th gear shifter fork to the detent bushing and slide the assembly on the shift shaft to place it below the 1st–2nd shifter fork arm.

6. Install the shifter assembly to the synchronizer sleeve grooves on the mainshaft.

5. With the front of the bell housing resting on wooden blocks, place a thrust washer over the hole for the countergear shaft. The thrust washer must be placed in the holes in the bellhousing.

6. Mesh the countershaft gears to the mainshaft gears and install this assembly into the bellhousing.

7. Turn the bellhousing on its side, then install the snap ring to the ball bearing on the clutch gear and the bearing retainer to the bell housing. Use sealant on the four retaining bolts.

8. Turn the bell housing (so that it is resting on the blocks) and install the Reverse lever to the case using grease to hold it in place. When installing the Reverse lever, the screwdriver slot should be parallel to the front of the case.

9. Install the Reverse lever snap ring and the roller bearing-to-countergear opening with the snap ring groove inside of the case.

10. Using rubber cement, install the gasket on the bell housing. Before installing the case, make sure the synchronizers are in the Neutral position, the detent bushing slot is facing outward and the Reverse lever is flush with the inside wall of the case.

11. Expand the snap ring in the

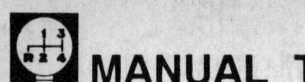

1	Bolt	6	Wire assembly	11	Cap
2	Bearing retainer	7	Switch assembly (TCS)	12	Ball
3	Seal assembly	8	Gasket assembly	13	Gasket
4	Gasket	9	Case assembly	14	Cap
5	Clutch housing	10	Spring	15	Retainer

16 Back-up light switch
17 Plug
18 Cap
19 Bolt
20 Retaining ring
21 Locating ring
22 Bearing assembly
23 Bearing assembly
24 Bolt
25 Main drive gear
26 Bearing rollers
27 Shift fork
28 Pin
29 Bushing
30 Detent lever
31 Shift fork
32 Shift shaft
33 Pin
34 Lock ring
35 Extension assembly
36 Gasket
37 Arm assembly
38 Pin
39 Bushing
40 Seal
41 Reverse shaft and lever
42 Lock ring
43 Clip
44 Retaining ring
45 Synchronizer assembly
46 Mainshaft
47 Second speed gear
48 Synchronizer assembly
49 Synchronizer blocking ring
50 Synchronizer spring
51 Synchronizer key
52 Third speed gear
53 First speed gear
54 Locating ring
55 Mainshaft rear bearing
56 Reverse gear
57 Retaining ring
58 Speedometer drive gear
59 Retainer ring
60 Thrust washer
61 Countershaft gear
62 Locating ring
63 Bearing race
64 Bearing assembly
65 Countershaft reverse gear
66 Reverse idler shaft
67 Retainer ring
68 Thrust washer
69 Reverse idler gear

Exploded view of the GM 70 mm (Chilton Type 12) transmission

mainshaft case opening and let it slide over the bearing.

12. Install the interlock lock pin with locking compound to hold the shifter shaft in place and the idler shaft so it engages with the Reverse lever inside the shaft.

13. Install the cover over the screwdriver arm to hold the Reverse lever in place.

14. Install the detent ball, the spring and the cap in the case, then the Reverse gear (with the chamfer on the gear teeth facing up). Push the Reverse gear onto the splines and secure with a snap ring.

15. Install the smaller Reverse gear on the countergear shaft (with the shoulder resting against the countergear bearing) and secure with a snap ring.

16. Install the snap ring, the thrust washer and the Reverse idler gear (with the gear teeth chamfer facing down) to the idler shaft, then secure with the thrust washer and the snap ring.

17. Install the shifter shaft snap rings and engage the speedometer gear retainer in the hole in the mainshaft (with the retainer loop toward the front), then slide the speed-

ometer gear over the mainshaft and into position.

NOTE: Before installation, heat the gear to 175°F; use an oven or heat lamp, not a torch.

18. Place the extension housing and the gasket on the case, then loosely install the two pilot bolts (one in the top right hand corner and the other in the bottom left hand corner) and the other three bolts. The pilot bolts must be installed in the right holes to prevent splitting the case.

19. Assemble the shifter shaft arm over the shifter shaft, align with the

drilled hole near the end of the shaft, then drive the spring pin into the shifter shaft arm and shaft.

20. Turn the case on its side and loosely install the two pilot bolts through the bell housing and then the four retaining bolts.

Mainshaft

DISASSEMBLY

1. Separate the shift shaft assembly and countergear from the mainshaft.
2. Remove the clutch gear and the blocking ring from the mainshaft; make sure you don't lose any of the clutch gear roller bearings.
3. Remove the 3rd–4th gear synchronizer hub snap ring and the hub, using an arbor press (if necessary).
4. Remove the blocking ring and the 3rd speed gear. Using an arbor press and press plates, remove the ball bearing from the rear of the mainshaft. Remove the remaining parts from the mainshaft keeping them in order for later reassembly.

ASSEMBLY

1. With the rear of the mainshaft turned up, install the 2nd speed gear with the clutching teeth facing upward; the rear face of the gear will butt against the flange of the mainshaft.
2. Install a blocking ring (with the clutching teeth down) over the 2nd speed gear.
3. Install the 1st–2nd synchronizer assembly (with the fork slot down), then press it onto the splines on the mainshaft until it bottoms.

NOTE: Make sure the notches of the blocking ring align with the keys of the synchronizer assembly.

4. Install the synchronizer hub-to-mainshaft snap ring, then install a blocking ring (with the notches facing down) so they align with the keys of the 1st–2nd gear synchronizer assembly.
5. Install the 1st speed gear (with the clutching teeth down), then the rear ball bearing (with the snap ring groove down) and press it into place on the mainshaft.
6. Turn the mainshaft up and install the 3rd speed gear (with the clutching teeth facing up); the front face of the gear will butt against the flange on the mainshaft.
7. Install a blocking ring (with the clutching teeth facing down) over the

synchronizer surface of the 3rd speed gear.

8. Install the 3rd–4th gear synchronizer assembly (with the fork slot facing down); make sure the notches of the blocking ring align with the keys of the synchronizer assembly.
9. Install the synchronizer hub-to-mainshaft snap ring and a blocking ring (with the notches facing down) so that they align with the keys of the 3rd–4th gear synchronizer assembly.

Synchronizer Keys And Springs

REPLACEMENT

1. The synchronizer hubs and the sliding sleeves are an assembly which should be kept together as an assembly; the keys and the springs can be replaced.
2. Mark the position of the hub and the sleeve for reassembly.
3. Push the hub from the sliding sleeve; the keys will fall out and the springs can be easily removed.
4. Place the new springs in position (with one on each side of the hub) so that the three keys are engaged by both springs.

5. Place and hold the keys in position, then slide the sleeve into the hub aligning the marks made during disassembly.

Extension Oil Seal

REPLACEMENT

1. Pry the oil seal and drive the bushing from rear of the extension housing.
2. Coat the inside diameter of the seal and bushing with transmission fluid and install them.

Drive Gear Bearing Oil Seal

REPLACEMENT

Pry out the old seal and install a new one making sure that it bottoms properly in its bore.

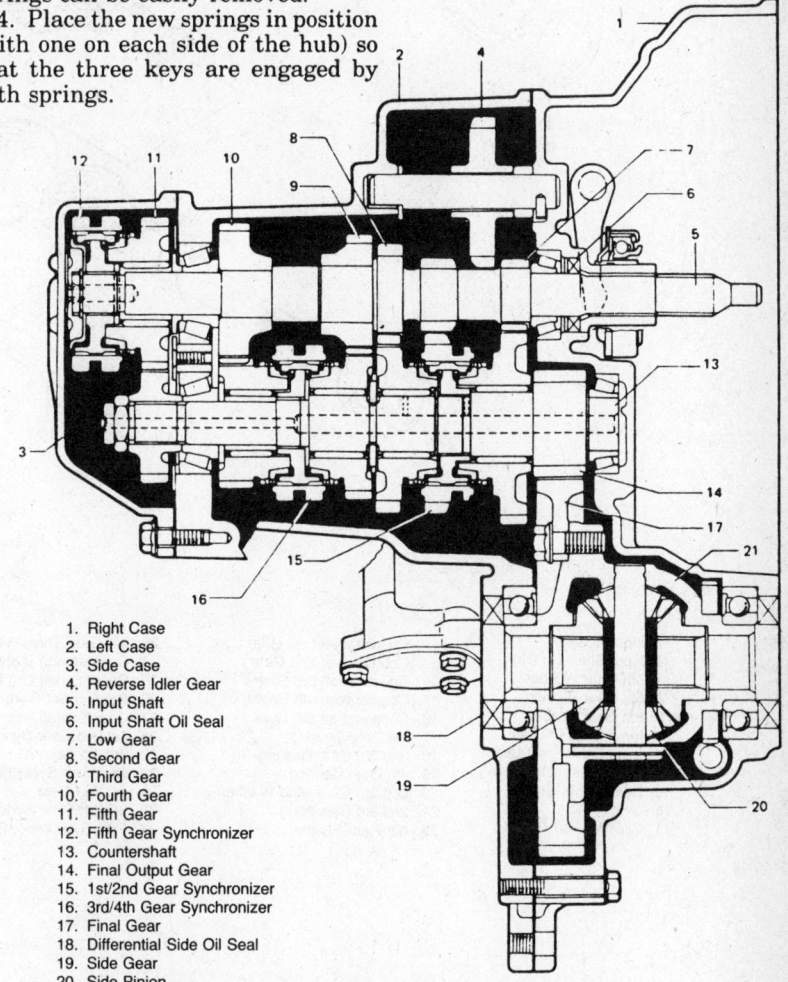

1. Right Case
2. Left Case
3. Side Case
4. Reverse Idler Gear
5. Input Shaft
6. Input Shaft Oil Seal
7. Low Gear
8. Second Gear
9. Third Gear
10. Fourth Gear
11. Fifth Gear
12. Fifth Gear Synchronizer
13. Countershaft
14. Final Output Gear
15. 1st/2nd Gear Synchronizer
16. 3rd/4th Gear Synchronizer
17. Final Gear
18. Differential Side Oil Seal
19. Side Gear
20. Side Pinion
21. Differential Carrier

Cut away view of the 5-speed transaxle (Chilton Type 13)

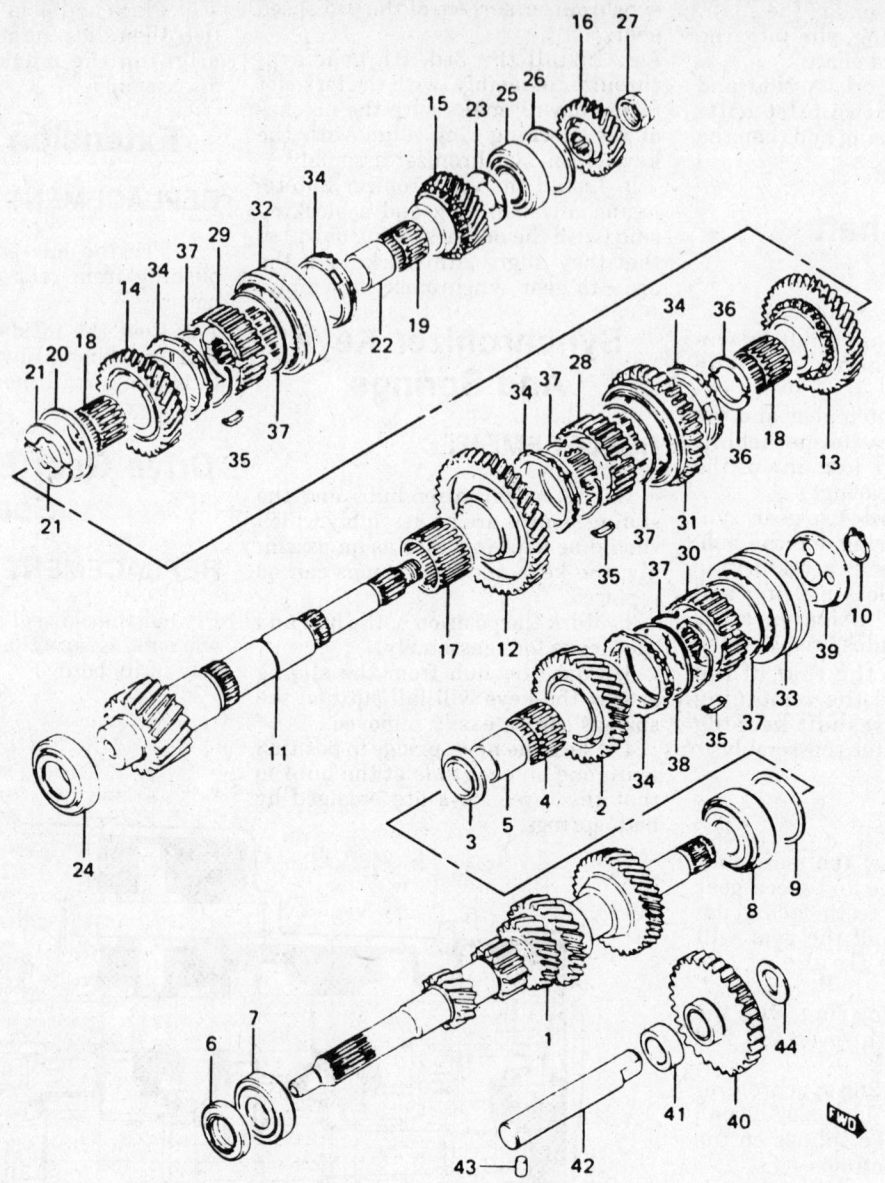

Exploded view of the 5-speed gear assembly (Chilton Type 13)

1. Input Shaft
2. Input Shaft 5th Gear
3. 5th Gear Washer
4. 5th Gear Bearing
5. 5th Gear Spacer
6. Input Shaft Oil Seal
7. Input Shaft Right Bearing
8. Input Shaft Left Bearing
9. Bearing Set Shim
10. Snap Ring
11. Countershaft

12. Countershaft Low Gear
13. Countershaft 2nd Gear
14. Countershaft 3rd Gear
15. Countershaft 4th Gear
16. Countershaft 5th Gear
17. 1st Gear Bearing
18. 2nd/3rd Gear Bearing
19. 4th Gear Bearing
20. 2nd/3rd Gear Ring Washer
21. 2nd/3rd Gear Ring
22. 4th Gear Spacer

23. 4th Gear Thrust Washer
24. Countershaft Right Bearing
25. Countershaft Left Bearing
26. Bearing Set Shim
27. Countershaft Nut
28. 1st/2nd Gear Sync. Hub
29. 3rd/4th Gear
30. Over Top Sync. Hub
31. 1st/2nd Gear
32. 3rd/4th Gear Sync. Sleeve
33. Fifth Gear Sync. Sleeve

34. Synchronizer Ring
35. Synchronizer Shifting Key
36. 1st/2nd Gear Sync. Hub Ring
37. Synchronizer Spring
38. Sync. Ring 5th Spring
39. Fifth Gear Sync. Hub Plate
40. Reverse Idle Gear
41. Idle Gear Spacer
42. Reverse Gear Shaft
34. Pin
44. Washer

CHILTON TYPE 13

Transaxle Case

DISASSEMBLY

1. Remove the backup light switch, then the 7 side case bolts and the case from the transaxle.

2. Remove the snap ring and the hub plate from the input shaft, then the shift fork screw and the 5th gear shift fork guide ball. Place the 5th gear in the Neutral position and remove the roll pin from the 5th gear shift fork.

3. Push the gear shift shaft In, to place the transaxle in gear, then slide the 5th gear synchronizer to engage the 5th gear; the transaxle should now be locked in two gears. Remove the 5th gear lock nut staked mark, then the nut.

4. Remove the 5th gear shift fork, the sleeve, the hub, the synchronizer ring, the spring and the keys as an assembly.

5. Remove the 5th drive gear, the bearing, the spacer and the washer from the input shaft, then the 5th driven gear from the countershaft.

6. Remove the bearing/shim plate-to-left case bolts and the plate.

NOTE: Mark or tag the input shaft and the countershaft shims for reassembly reference.

7. Remove the three left case cap bolts and the cap, then the gear shift yoke-to-gear shift/selector shaft roll pin.

NOTE: Using a drift punch, install it into the gear shift/selector shaft roll pin hole, then raise the shaft to remove the yoke and the roll pin.

8. Remove the Reverse spring and the detent ball bolt, the spring and the ball from the case. Remove the gear shift/selector shaft locating bolt, the gear shift guide-to-case bolts, the gear shift/selector shaft assembly and the low speed selector spring from the case.

9. Remove the three bolts holding the detent balls and springs of the shift fork shafts. Remove the 13 case bolts, then insert a small pry bar between the case halves and separate the cases.

10. Remove the left case; the the input shaft, the countershaft, the differ-

ential and other parts should remain on the right case.

11. Raise the 5th and the Reverse shift shafts. Remove the Reverse idler shift lever-to-case bolts and the lever, then the 5th and the Reverse gear shift shafts (together) as an assembly.

12. Remove the input shaft, the countershaft, the 1st–2nd and the 3rd–4th shift shafts (together) as an assembly.

NOTE: To remove the components in Step 12, it may be necessary to raise the differential assembly.

13. Remove the differential assembly from the right case. Using tool No. J-29369-2 and a slide hammer, pull the input shaft and the countershaft bearing cups from the right case.

14. Using tool No. J-23907, remove the input shaft seal from the right case. Remove the differential side oil seals from both cases.

15. Position the gear shift arm above the square recess in the right case (by moving the shift shaft). Using a drift punch, remove the gear shift arm roll pin.

16. Remove the shifter shaft retaining bolt, the detent ball and the spring. Slide the shifter shaft out, then remove the arm, the roll pin and the shaft from the case.

17. Remove the shifter shaft boot from the oil seal flange. Using a pair of pliers, remove the shifter shaft seal from the case.

18. Remove the input shaft and the countershaft bearing cups from the left case.

ASSEMBLY

NOTE: Before installing the oil seals, apply a coat of grease to the seal lips.

1. Using tools No. J-34855 and J-7079-2, install the input shaft seal to the right case. Using tools No. J-34849 and J-7079-2, install the input shaft and the countershaft bearing cups to the right case.

2. Using tools No. J-29130 and J-8092, install the differential oil seals to both cases.

3. Apply locking cement to the bolt threads and install the Reverse gear shift lever, then torque the bolts to 14–20 ft. lbs.

4. Using tool No. J-34857, drive the new gear shift shaft seal into the case, then install the gear shift lever and the boot.

5. Install the detent ball, the spring, the gasket and the bolt to the shift shaft, then the gear shift arm and the roll pin to the gear shift shaft.

6. Install the differential assembly, then the 5th and the Reverse gear shift shafts into the right case. Install the input shaft, the countershaft, the 1st–2nd and the 3rd–4th gear shift shafts into the right case.

7. Install the idler gear, the shaft, the pin, the spacer, the washer and the magnet into the right case.

8. Apply Loctite® 518 to the left case mounting surface, then install the left case to the right case and torque the bolts to 14–20 ft. lbs.

NOTE: After installing the cases, check the input shaft and the countershaft for smooth turning (turn by hand).

9. Install the three shift fork shaft detent balls, the springs, the gaskets and the bolts, then torque the bolts to 8–11 ft. lbs.

10. Install the gear shift yoke onto the gear shift arm. Install the gear shift/selector shaft assembly, by guiding the shaft into the gear shift yoke hole. Align the yoke hole with the shaft hole and install a new roll pin.

11. Install the gear shift locating bolt and washer, then torque the bolt to 30–43 ft. lbs. Install the gear shift guide case and torque the bolts to 6–7 ft. lbs.

12. Install the Reverse check ball, the spring, the washer and the bolt. Install the case cap with a new O-ring to the left case.

13. Using tool No. J-34858, measure and determine the bearing shim size for the input shaft and the countershaft as follows:

   a. Install the left bearing cups for the input shaft and the countershaft. Using finger pressure, push the countershaft's left bearing cup against the bearing rollers. Rotate the countershaft 3–4 times to seat the bearings. Install the nut (from tool No. J-34858) onto the shaft, followed by the countershaft nut, then torque the nut to 44–58 ft. lbs.

   b. Install the Dial Gauge tool No. J-29763 on the Shim Selector tool No. J-34858, then place the tools on a flat surface and zero the gauge. Place the tools on the countershaft and the left case, then press down on the selector tool and read the dial indicator.

NOTE: The reading on the dial indicator will indicate the size of the shim to be used. The shim stock has 12 selective thicknesses, the size is stamped on one side; select the correct size and install it on the back side of the left hand countershaft bearing cup.

c. For the input shaft, repeat the above procedures; select the correct size and install it on the back side of the left hand input shaft bearing cup.

14. Install the case plate by fitting the plate's protrusion into the groove of the gear shift guide shaft. Apply locking cement to the screw threads and torque to 5 ft. lbs.

15. Install tool No. J-34852 onto the input shaft and connect a spring balance to the wire of the tool. Place the transaxle in the Neutral and the 4th gear positions, then pull the spring balance to check the preload.

NOTE: If the preload is not to specifications, change the shims of the input shaft and the countershaft.

16. With the transaxle in the 4th gear, install the 5th gear onto the input shaft and the countershaft, then engage the synchronizer so the transaxle will be locked in two gears. Install and stake the countershaft nut after torquing it to 44–57 ft. lbs.

17. Disengage the 5th gear synchronizer and shift the transaxle in the Neutral position.

18. Install the shift guide ball and the shift fork screw to the shifting fork; after torquing the screw to 6–8 ft. lbs., stake the screw.

19. Using a drift punch and a hammer, drive a new roll pin into the 5th and the Reverse gear shift shaft. Install the synchronizer hub plate and a new snap ring.

20. Apply Loctite® 518 to the side case mounting surface and the install the side case. Fit the side case oil receiver cup into the input shaft hole, then install the bolts and torque to 14–20 ft. lbs.

21. Install the backup light switch.

Input Shaft

DISASSEMBLY

NOTE: The 3rd and the 4th gears should not be removed from the input shaft. If the gears are damaged, replace the input shaft.

Using tool No. J-34843 and an arbor press, press the right bearing (small) and the left bearing (large) from the input shaft.

ASSEMBLY

Using tool No. J-34844 and an arbor press, press the right and the left bearings onto the input shaft.

CHILTON TYPE 14

Transmisson Case

DISASSEMBLY

1. Drain the lubricant and shift into the 2nd gear position. Remove the side cover and the shift controls.

2. Remove the four front bearing retainer bolts, the retainer and the gasket.

3. If equipped with an output companion flange, remove it.

4. At the Reverse shifter lever boss, drive the lock pin up, then pull the shift shaft out about $\frac{1}{8}$ in. to disengage the shifter fork from the Reverse gear.

5. Remove the five extension housing bolts and tap the extension (with soft hammer) rearward. When the idler gear shaft is out (as far as it will go), move the extension housing to the left so that the Reverse fork clears the Reverse gear, then remove extension housing and the gasket.

6. Remove the speedometer gear outer snap ring, then tap or slide the speedometer from the mainshaft.

7. Remove the 2nd snap ring.

8. Remove the Reverse gear from the mainshaft and the rear part of the Reverse idler gear from the case.

9. Remove the front bearing snap ring and the spacer washer.

10. Pull the front bearing from the case.

11. Remove the rear retainer lock bolt.

12. Shift the 1st–2nd and the 3rd–4th clutch sliding sleeves forward for clearance.

13. Remove the mainshaft and the rear bearing retainer assembly from the case.

14. Remove the front Reverse idler gear and the thrust washer from the case (the gear teeth must face forward).

15. Using a dummy shaft, drive the countergear shaft out through the rear of the case. Remove the countergear and the tanged thrust washers.

ASSEMBLY

1. Place the case on its side and install the countergear tanged washers with the tangs in the thrust face notches, secure them with grease.

2. Install the countergear (with a dummy shaft) in the case. Drive the countergear shaft through the rear of

the case, forcing the dummy shaft out through the front. Install the shaft key and tap the shaft in until it is flush with the rear face of the case.

3. Install the front Reverse idler gear with the teeth forward, using grease to secure the thrust washer in place.

4. Use heavy grease to secure the 16 roller bearings and the washer in the main drive gear, mate the main drive gear with the mainshaft. Hold them together by moving the 3rd–4th clutch sliding sleeve forward.

5. Place a new gasket on the rear of the case. Install the mainshaft and drive gear assembly into the case.

6. Align the rear bearing retainer with the case, then install the locating pin and the locking bolt.

7. Install the bearing snap ring on the front main bearing. Tap the bearing into the case, then install the spacer washer and the thickest snap ring that can be fitted.

8. Install the front bearing retainer and the gasket, using sealer on the bolts.

9. Install the rear Reverse idler gear, engaging the splines with the portion of the gear in the case.

10. Slide the Reverse gear onto the shaft. Install the speedometer gear and the two thickest snap rings that can be fitted.

11. Install the idler shaft into the extension housing until the hole in the shaft aligns with the lockpin hole, then drive the lockpin and a sealant coated plug into place.

12. Place the Reverse shifter shaft and the detent into the extension housing, using grease to hold the Reverse shift fork in position. Install the shaft O-ring, after the shaft is in place.

13. Put the tanged thrust washer on the Reverse idler shaft; the tang must be in the notch of the extension housing thrust face.

14. Place the 1st–2nd and the 3rd–4th clutch sliding sleeves in the Neutral positions. Pull the Reverse shift shaft part way out and push the Reverse shift fork in as far as possible. Start the extension housing onto the mainshaft and push in on the shifter shaft to engage the shift fork with the Reverse gear collar. When the fork engages, turn the shifter shaft to let the Reverse gear go to the rear and the extension housing to fit in place.

15. Install the Reverse shift shaft lockpin.

16. Install the extension housing bolts.

NOTE: Use sealant on the upper left side bolt.

17. Position the 1st–2nd clutch slid-

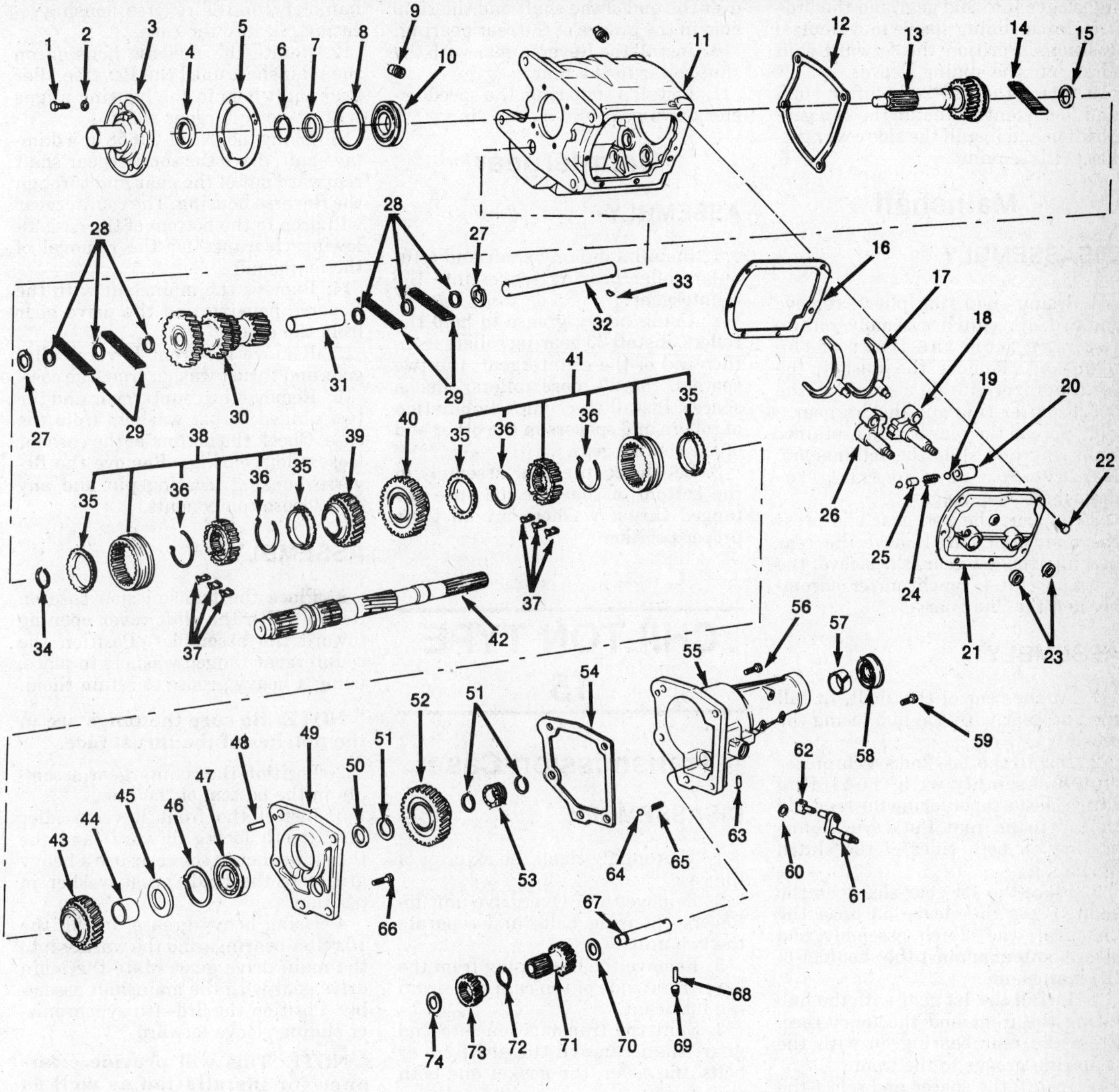

Exploded view of the Borg Warner 4-speed (Chilton Type 14) transmission

1. Bolt	20. Interlock sleeve		57. Extension bushing
2. Lock washer	21. Transmission cover	39. 3rd speed gear	58. Extension seal
3. Bearing retainer	22. Bolt	40. 2nd speed gear	59. Bolt
4. Seal	23. Shifter shaft seal	41. 1st and 2nd synchronizer	60. Shifter shaft oil seal
5. Gasket	24. Rev. lever poppet spring	42. Transmission main shaft	61. Reverse shifter shaft
6. Snap ring	25. Interlock pin	43. 1st speed gear	62. Reverse fork
7. Ring	26. 3rd and 4th shaft assy.	44. 1st speed sleeve	63. Lock pin
8. Snap ring	27. Washer	45. 1st speed thrust washer	64. Lever poppet spring
9. Drain plug	28. Countershaft bearing washer	46. Snap ring	65. Ball
10. Bearing	29. Roller	47. Main shaft rear bearing	66. Bolt
11. Transmission case	30. Countershaft gear	48. Pin	67. Shaft
12. Front gasket	31. Spacer	49. Retainer assy.	68. Pin
13. Clutch gear	32. Counter shaft	50. Ring	69. Plug
14. Roller	33. Key (1/8″ × 5/8″)	51. Retainer ring	70. Rev. idler gear washer
15. Spacer	34. Clutch hub retainer ring	52. Reverse gear	71. Reverse idler gear
16. Side gasket	35. Blocking ring	53. Speedometer drive gear	72. Front ring
17. 1st, 2nd, 3rd and 4th fork	36. Synchronizer spring	54. Gasket	73. Rev. idler gear (front)
18. 1st and 2nd shaft assy.	37. Shifting key	55. Extension assy.	74. Washer
19. Balls	38. 3rd and 4th synchronizer	56. Bolt	

ing sleeve into 2nd gear and the 3rd–4th clutch sliding sleeve into Neutral positions. Position the forward shift forks into the sliding sleeves.

18. Place the 1st–2nd shifter shaft and the detent plate into the 2nd gear position and install the side cover gasket, with sealant.

Mainshaft

DISASSEMBLY

1. Using snap ring pliers, remove the 3rd–4th clutch assembly retaining ring from the front of the mainshaft. Remove the washer, the syncrhonizer/clutch assembly, the synchronizer ring and the 3rd gear.

2. Spread the rear bearing retainer snap ring and slide the retainer off. Remove the rear bearing-to-mainshaft snap ring.

3. Support the 2nd gear and press the mainshaft out; remove the rear bearing, the 1st gear, the sleeve, the 1st–2nd clutch/synchronizer assembly and the 2nd gear.

ASSEMBLY

1. At the rear of the shaft, install the 2nd gear with the hub facing the rear.

2. Install the 1st–2nd synchronizer clutch assembly with the sliding clutch sleeve taper facing the rear and the hub to the front. Put a synchronizer ring on both sides of the clutch assemblies.

3. Place the 1st gear sleeve on the shaft. Press the sleeve on until the 2nd gear, the clutch assembly and sleeve bottom against the shoulder of the mainshaft.

4. Install the 1st gear with the hub facing the front and the inner race. Press the rear bearing on with the snap ring groove to the front.

5. Install the spacer and select the thickest snap ring that can be fitted into the mainshaft behind the rear bearing.

6. Install the 3rd gear with the hub to the front and the 3rd gear synchronizing ring with the notches to the front.

7. Install the 3rd–4th gear clutch assembly with the taper facing the front.

NOTE: Make sure that the keys in the hub match the notches in the 3rd gear synchronizing ring.

8. Install the thickest snap ring that will fit in the mainshaft groove in front of the 3rd–4th clutch assembly.

9. Place the rear bearing retainer

over the end of the shaft and the snap ring in the groove of the rear bearing.

10. Install the Reverse gear with the shift collar to the rear.

11. Install a snap ring, the speedometer drive gear and a snap ring.

Countergear

ASSEMBLY

1. Install a dummy shaft and a tubular roller bearing spacer into the countergear.

2. Using heavy grease to hold the rollers, install 20 bearing rollers in either end of the countergear, the two spacers, the 20 more rollers, then a spacer. Install the same combination of rollers and spacers in the other end of the countergear.

3. Set the countergear assembly in the bottom of the case, be sure the tanged thrust washers are in their proper position.

CHILTON TYPE 15

Transmission Case

DISASSEMBLY

1. Thoroughly clean the exterior of the case.

2. Remove the 7 Overdrive unit-to-Reverse housing bolts and separate the two units.

3. Remove the drain plug from the lower right side of the case and drain the lubricant.

4. Shift the transmission into 2nd gear, then remove the shift cover bolts, the cover, the gasket and both shift forks.

5. Remove the backup switch from the Reverse housing.

6. Rotate the Reverse shifter shaft, then remove the shift fork and gear from the mainshaft.

7. Remove the lock pin from the Reverse shift lever boss and pull the shaft from the housing.

8. Remove the drive gear bearing retainer bolts, the retainer and the gasket from the front of the case.

9. Remove the front bearing snap ring, the selective fit snap ring and the spacer washer.

10. Using tool No. J-6654-01 and J-8433-1, pull the drive gear bearing from the case.

11. Remove the six Reverse housing-to-case bolts. Using a small drift and a

hammer, tap the reverse housing locating pin into the case.

12. Rotate the Reverse housing on the mainshaft until the Reverse idler gear shaft hole in the housing aligns with the countergear shaft.

13. Using tool No. J-24658 or a dummy shaft, drive the countergear shaft rearward out of the gear and through the Reverse housing. The countergear will drop to the bottom of the case allowing clearance for the removal of the mainshaft.

14. Remove the mainshaft with the Reverse housing and the drive gear from the case.

15. Remove the front Reverse idler gear and thrust washer from the case.

16. Remove the countergear and the two tanged thrust washers from the case. Check the bottom of the case for loose pilot bearings. Remove the Reverse housing locating pin and any other loose components.

ASSEMBLY

1. Place the transmission case on its side with the shift cover opening toward the assembler. Position the countergear tanged washers in place, using a heavy grease to retain them.

NOTE: Be sure the tangs are in the notches of the thrust face.

2. Position the countergear assembly in the bottom of the case.

3. Install the front Reverse idler gear (teeth facing forward) and the thrust washer in the case; use a heavy grease to the hold thrust washer in position.

4. Using heavy grease, install the 16 roller bearings and the washer into the main drive gear. Mate the main drive gear with the mainshaft assembly. Position the 3rd–4th synchronizer sliding sleeve forward.

NOTE: This will provide clearance for installation as well as hold the assembly together.

5. Position a new Reverse housing to the case gasket on the rear of the case.

6. Install the mainshaft and drive gear assembly into the case.

7. Place the bearing snap ring on the front main bearing. Using tool No. J-5590 or a hollow tube, position the front main bearing into the case opening and tap into place. Install the spacer washer and the snap ring to secure the main drive bearing.

8. Raise the countergear in the case, aligning the holes in the case with the center of the gear. With the thrust washers in place, slide the countershaft through the rear of the case. Install the woodruff key and tap

the shaft into the case, until it is flush with the rear face of the case.

9. Align the Reverse housing and the gasket to the case. Install the locating pin for the Reverse housing and tap the pin in until it is flush with the housing.

10. Install the 6 Reverse housing-to-case bolts.

11. Install the Reverse shift shaft and the O-ring into the housing, then the retaining pin.

12. Install the Reverse gear and the shift fork. Slide the gear and the fork forward on the mainshaft until the shift fork and the shifter shaft can be indexed into position.

13. Position the drive gear bearing retainer and the gasket to the front of the case. Apply sealant to the bolts.

14. Install the rear Reverse idler gear. Align the splines on the rear gear with the front gear and slide together.

15. Assemble the overdrive unit to the Reverse housing. Guide the idler shaft on the O/D unit into the idler gears and align the splines on the mainshaft with the splines in the input sun gear. Slide the units together and install the retaining bolts.

16. Slide the 1–2 synchronizer forward into the 2nd gear. Install the shift forks into the grooves of the synchronizers. Place the side cover with a gasket on the transmission. Guide the shift forks into the cover and install the bolts.

17. Check the operation of the transmission by manually shifting the transmission into all gears.

Mainshaft

DISASSEMBLY

1. Using snap ring pliers, remove the 3–4 synchronizer assembly retaining ring from front of the mainshaft, then slide the washer, the synchronizer assembly and the synchronizer ring 3rd speed gear from the mainshaft.

2. Spread the rear bearing retainer snap ring and slide the retainer from the mainshaft.

3. Remove the rear bearing-to-mainshaft snap ring.

4. Support the mainshaft under the 2nd gear and press the mainshaft from the rear bearing, the 1st gear and sleeve, the 1–2 synchronizer assembly and the 2nd gear.

ASSEMBLY

1. At the rear of mainshaft, assemble the 2nd speed gear with the hub of the gear facing the rear of the shaft.

2. Install the 1st–2nd synchronizer assembly (sliding the synchronizer sleeve taper toward the rear, the hub to the front) on the mainshaft together with a synchronizer ring on both sides of the synchronizer assemblies.

3. Position the 1st gear sleeve on the shaft and press the sleeve onto the mainshaft until the 2nd gear, the synchronizer assembly and the sleeve bottom against the shoulder of the mainshaft.

4. Install the 1st speed gear (with the hub facing the front) and support the inner race, then press the rear bearing onto the mainshaft with the snap ring groove toward the front of the case.

5. Install the spacer and the new snap ring (thickest one that will fit) in the mainshaft behind the rear bearing.

6. Install the 3rd speed gear (the hub to the front of the case) and the 3rd speed gear synchronizing ring (with the notches to the front of the case).

7. Install the 3rd–4th speed gear synchronizer assembly (hub and sliding sleeve) with the taper facing the front.

NOTE: Make sure that the keys in hub correspond to the notches in the 3rd speed gear synchronizing ring.

8. Install the new snap ring (the thickest that will fit) in the groove of the mainshaft in front of the 3rd–4th speed synchronizer assembly.

9. Install the rear bearing retainer (Reverse housing) over the end of the mainshaft. Spread the snap ring to drop around the rear bearing, then release the snap ring when it aligns with the groove in the rear bearing.

1. Overdrive Valve Body GASKET
2. Overdrive Valve Body Spacer PLATE
3. Overdrive Valve BODY ASSEMBLY
4. Steel BALL
5. Overdrive Oil Screen TUBE

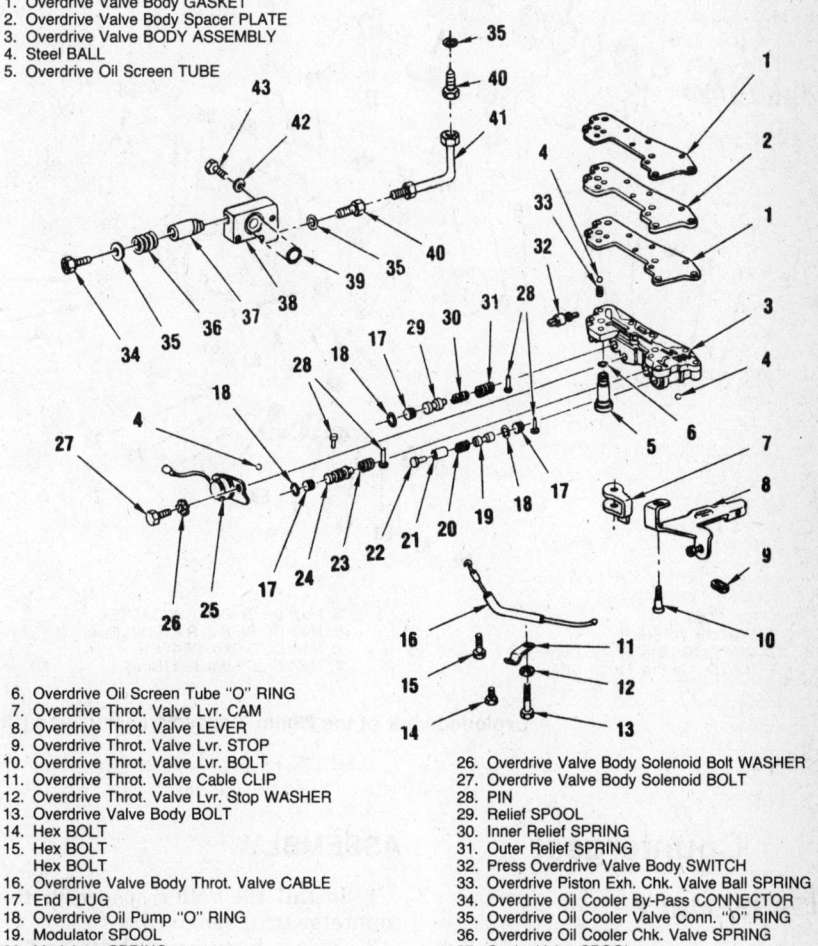

6. Overdrive Oil Screen Tube "O" RING
7. Overdrive Throt. Valve Lvr. CAM
8. Overdrive Throt. Valve LEVER
9. Overdrive Throt. Valve Lvr. STOP
10. Overdrive Throt. Valve Lvr. BOLT
11. Overdrive Throt. Valve Cable CLIP
12. Overdrive Throt. Valve Lvr. Stop WASHER
13. Overdrive Valve Body BOLT
14. Hex BOLT
15. Hex BOLT
    Hex BOLT
16. Overdrive Valve Body Throt. Valve CABLE
17. End PLUG
18. Overdrive Oil Pump "O" RING
19. Modulator SPOOL
20. Modulator SPRING
21. Piston BUSHING
22. Modulator PISTON
23. Shift SPRING
24. Shift SPOOL
25. Overdrive Valve Body SOLENOID

26. Overdrive Valve Body Solenoid Bolt WASHER
27. Overdrive Valve Body Solenoid BOLT
28. PIN
29. Relief SPOOL
30. Inner Relief SPRING
31. Outer Relief SPRING
32. Press Overdrive Valve Body SWITCH
33. Overdrive Piston Exh. Chk. Valve Ball SPRING
34. Overdrive Oil Cooler By-Pass CONNECTOR
35. Overdrive Oil Cooler Valve Conn. "O" RING
36. Overdrive Oil Cooler Chk. Valve SPRING
37. Cooler Valve SPOOL
38. Overdrive Oil Cooler VALVE
39. Overdrive Oil Cooler Valve "O" RING
40. Oil Cooler FITTING
41. Overdrive Oil Cooler TUBE
42. Overdrive Oil Cooler Valve Bolt Lk. WASHER
43. Overdrive Oil Cooler Valve BOLT

Exploded view of the valve body (Chilton Type 15)

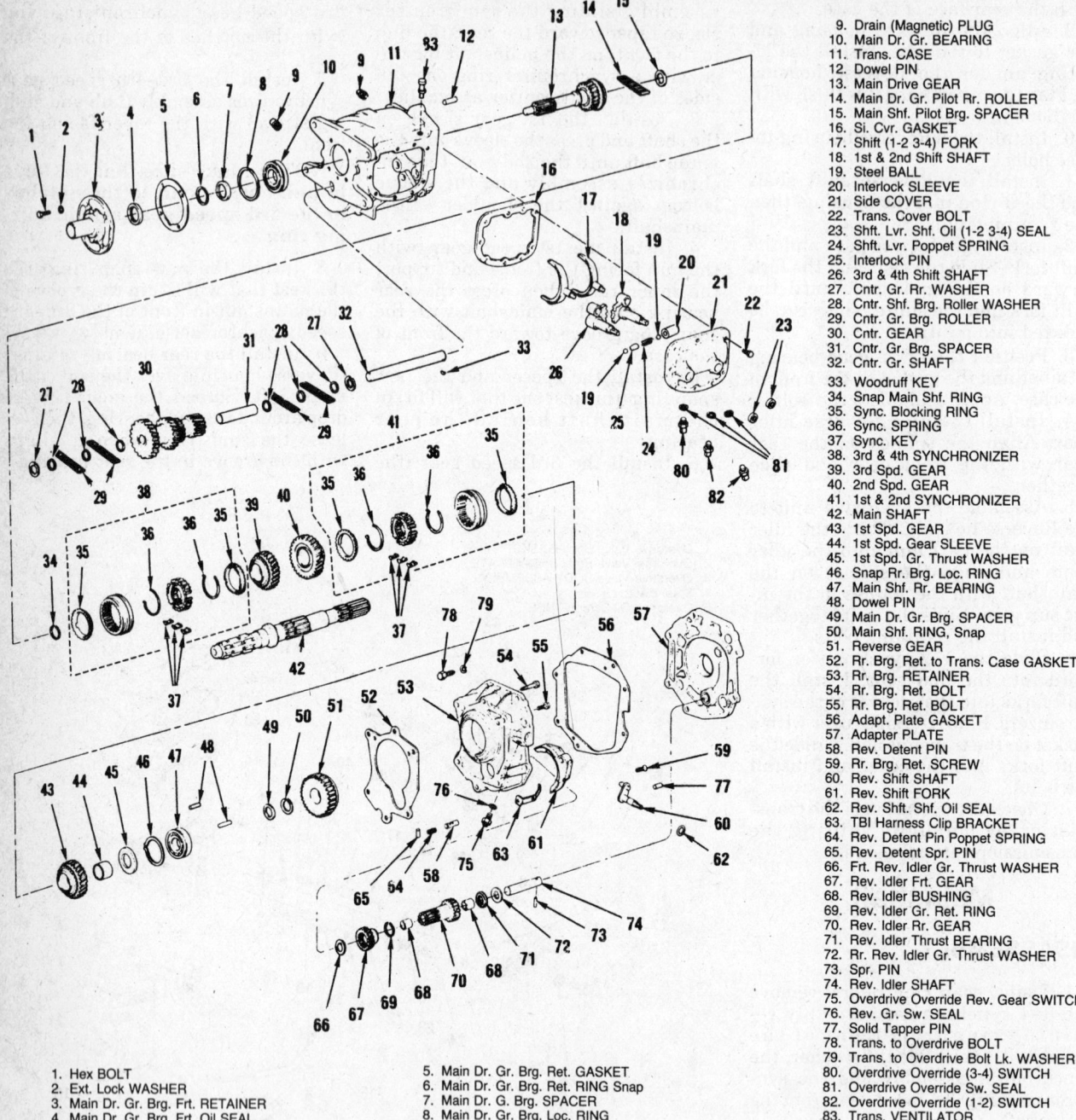

9. Drain (Magnetic) PLUG
10. Main Dr. Gr. BEARING
11. Trans. CASE
12. Dowel PIN
13. Main Drive GEAR
14. Main Dr. Gr. Pilot Rr. ROLLER
15. Main Shf. Pilot Brg. SPACER
16. Si. Cvr. GASKET
17. Shift (1-2 3-4) FORK
18. 1st & 2nd Shift SHAFT
19. Steel BALL
20. Interlock SLEEVE
21. Side COVER
22. Trans. Cover BOLT
23. Shft. Lvr. Shf. Oil (1-2 3-4) SEAL
24. Shft. Lvr. Poppet SPRING
25. Interlock PIN
26. 3rd & 4th Shift SHAFT
27. Cntr. Gr. Rr. WASHER
28. Cntr. Shf. Brg. Roller WASHER
29. Cntr. Gr. Brg. ROLLER
30. Cntr. GEAR
31. Cntr. Gr. Brg. SPACER
32. Cntr. Gr. SHAFT
33. Woodruff KEY
34. Snap Main Shf. RING
35. Sync. Blocking RING
36. Sync. SPRING
37. Sync. KEY
38. 3rd & 4th SYNCHRONIZER
39. 3rd Spd. GEAR
40. 2nd Spd. GEAR
41. 1st & 2nd SYNCHRONIZER
42. Main SHAFT
43. 1st Spd. GEAR
44. 1st Spd. Gear SLEEVE
45. 1st Spd. Gr. Thrust WASHER
46. Snap Rr. Brg. Loc. RING
47. Main Shf. Rr. BEARING
48. Dowel PIN
49. Main Dr. Gr. Brg. SPACER
50. Main Shf. RING, Snap
51. Reverse GEAR
52. Rr. Brg. Ret. to Trans. Case GASKET
53. Rr. Brg. RETAINER
54. Rr. Brg. Ret. BOLT
55. Rr. Brg. Ret. BOLT
56. Adapt. Plate GASKET
57. Adapter PLATE
58. Rev. Detent PIN
59. Rr. Brg. Ret. SCREW
60. Rev. Shift SHAFT
61. Rev. Shift FORK
62. Rev. Shft. Shf. Oil SEAL
63. TBI Harness Clip BRACKET
64. Rev. Detent Pin Poppet SPRING
65. Rev. Detent Spr. PIN
66. Frt. Rev. Idler Gr. Thrust WASHER
67. Rev. Idler Frt. GEAR
68. Rev. Idler BUSHING
69. Rev. Idler Gr. Ret. RING
70. Rev. Idler Rr. GEAR
71. Rev. Idler Thrust BEARING
72. Rr. Rev. Idler Gr. Thrust WASHER
73. Spr. PIN
74. Rev. Idler SHAFT
75. Overdrive Override Rev. Gear SWITCH
76. Rev. Gr. Sw. SEAL
77. Solid Tapper PIN
78. Trans. to Overdrive BOLT
79. Trans. to Overdrive Bolt Lk. WASHER
80. Overdrive Override (3-4) SWITCH
81. Overdrive Override Sw. SEAL
82. Overdrive Override (1-2) SWITCH
83. Trans. VENTILATOR

1. Hex BOLT
2. Ext. Lock WASHER
3. Main Dr. Gr. Brg. Frt. RETAINER
4. Main Dr. Gr. Brg. Frt. Oil SEAL

5. Main Dr. Gr. Brg. Ret. GASKET
6. Main Dr. Gr. Brg. Ret. RING Snap
7. Main Dr. G. Brg. SPACER
8. Main Dr. Gr. Brg. Loc. RING

Exploded view of the 83mm 4 speed transmission (Chilton Type 15)

Countergear

DISASSEMBLY

1. Remove the dummy shaft or the tool No. J-24658 from the countergear.

2. Tip the countergear on end and let the six spacers, the 112 rollers and the roller sleeve slide out of the gear.

ASSEMBLY

1. Install the roller spacer in the countergear (if removed).

2. Insert a dummy shaft or the loading tool No. J-24658 into the countergear.

3. Using heavy grease to retain the rollers, install the spacer, the 28 rollers, a spacer, the 28 rollers and a spacer in either end of the countergear.

Repeat the procedure in the other end of the countergear.

CHECKING COUNTERGEAR END PLAY

1. Rest the transmission case on its side with the side cover opening toward the assembler. Put the countergear tanged thrust washers in

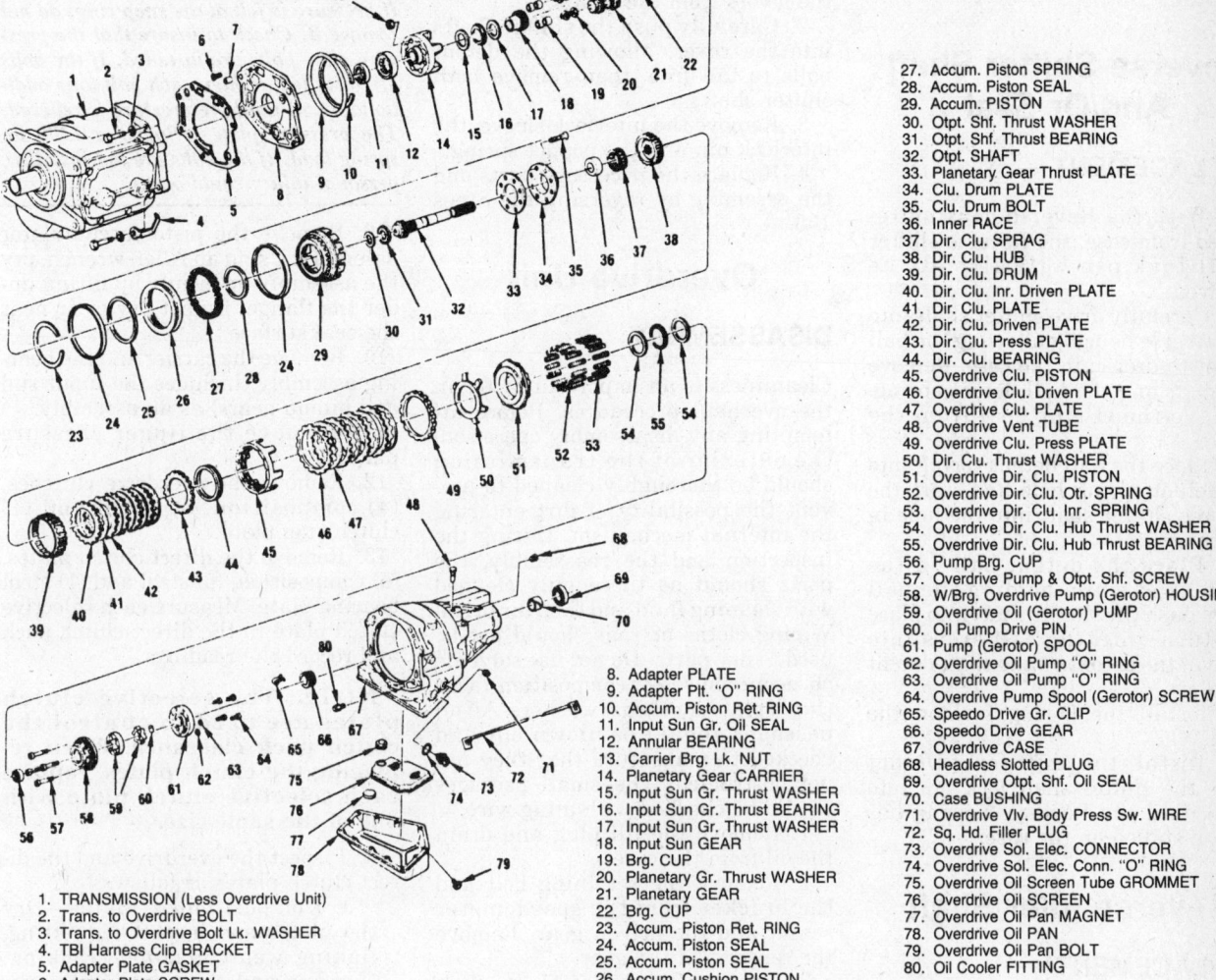

27. Accum. Piston SPRING
28. Accum. Piston SEAL
29. Accum. PISTON
30. Otpt. Shf. Thrust WASHER
31. Otpt. Shf. Thrust BEARING
32. Otpt. SHAFT
33. Planetary Gear Thrust PLATE
34. Clu. Drum PLATE
35. Clu. Drum BOLT
36. Inner RACE
37. Dir. Clu. SPRAG
38. Dir. Clu. HUB
39. Dir. Clu. DRUM
40. Dir. Clu. Inr. Driven PLATE
41. Dir. Clu. PLATE
42. Dir. Clu. Driven PLATE
43. Dir. Clu. Press PLATE
44. Dir. Clu. BEARING
45. Overdrive Clu. PISTON
46. Overdrive Clu. Driven PLATE
47. Overdrive Clu. PLATE
48. Overdrive Vent TUBE
49. Overdrive Clu. Press PLATE
50. Dir. Clu. Thrust WASHER
51. Overdrive Dir. Clu. PISTON
52. Overdrive Dir. Clu. Otr. SPRING
53. Overdrive Dir. Clu. Inr. SPRING
54. Overdrive Dir. Clu. Hub Thrust WASHER
55. Overdrive Dir. Clu. Hub Thrust BEARING
56. Pump Brg. CUP
57. Overdrive Pump & Otpt. Shf. SCREW
58. W/Brg. Overdrive Pump (Gerotor) HOUSING
59. Overdrive Oil (Gerotor) PUMP
60. Oil Pump Drive PIN
61. Pump (Gerotor) SPOOL
62. Overdrive Oil Pump "O" RING
63. Overdrive Oil Pump "O" RING
64. Overdrive Pump Spool (Gerotor) SCREW
65. Speedo Drive Gr. CLIP
66. Speedo Drive GEAR
67. Overdrive CASE
68. Headless Slotted PLUG
69. Overdrive Otpt. Shf. Oil SEAL
70. Case BUSHING
71. Overdrive Vlv. Body Press Sw. WIRE
72. Sq. Hd. Filler PLUG
73. Overdrive Sol. Elec. CONNECTOR
74. Overdrive Sol. Elec. Conn. "O" RING
75. Overdrive Oil Screen Tube GROMMET
76. Overdrive Oil SCREEN
77. Overdrive Oil Pan MAGNET
78. Overdrive Oil PAN
79. Overdrive Oil Pan BOLT
80. Oil Cooler FITTING

8. Adapter PLATE
9. Adapter Plt. "O" RING
10. Accum. Piston Ret. RING
11. Input Sun Gr. Oil SEAL
12. Annular BEARING
13. Carrier Brg. Lk. NUT
14. Planetary Gear CARRIER
15. Input Sun Gr. Thrust WASHER
16. Input Sun Gr. Thrust BEARING
17. Input Sun Gr. Thrust WASHER
18. Input Sun GEAR
19. Brg. CUP
20. Planetary Gr. Thrust WASHER
21. Planetary GEAR
22. Brg. CUP
23. Accum. Piston Ret. RING
24. Accum. Piston SEAL
25. Accum. Piston SEAL
26. Accum. Cushion PISTON

1. TRANSMISSION (Less Overdrive Unit)
2. Trans. to Overdrive BOLT
3. Trans. to Overdrive Bolt Lk. WASHER
4. TBI Harness Clip BRACKET
5. Adapter Plate GASKET
6. Adapter Plate SCREW
7. Dowel PIN

Exploded view of the 83mm 4 speed transmission overdrive unit (Chilton Type 15)

place, retaining them with heavy grease.

NOTE: Make sure the tangs are resting in the notches of the case.

2. Set the countergear in place in the bottom of the case.

3. Position the case to rest on its front face.

4. Lubricate and insert the countershaft (pushing the loading tool No. J-24658 out through the front of the case) until woodruff key slot is in its installed position (do not install key).

5. Attach a dial indicator and check end play of the countergear. If the end play is greater than 0.025 in., a new thrust washer must be installed.

Synchronizer Keys And Springs

REPLACEMENT

The synchronizer hubs and the sliding sleeves are a selected assembly and should be kept together as originally assembled, but the keys and the two springs may be replaced if worn or broken.

1. If the relation of the hub and the sleeve are not already marked, mark for assembly purposes.

2. Push the hub from the sliding sleeve, the keys will fall free and the springs may be easily removed.

3. Place the springs in position (one on each side of hub) so the keys are engaged by both springs.

NOTE: Place the keys in position and while holding them in place, slide the sleeve onto the hub, aligning the marks made before disassembly.

Drive Gear Bearing Retainer Oil Seal

REPLACEMENT

1. Pry out the old seal.

2. Using a new seal, drive it into the retainer using tool No. J-23096 until it bottoms in bore. Lubricate the

I.D. of the seal with transmission lubricant.

Reverse Shifter Shaft And/Or Seal

REPLACEMENT

1. With the Reverse housing removed from case, the Reverse shifter shaft lock pin will already be removed.

2. Carefully drive shifter shaft into the Reverse housing allowing the ball detent to drop into the case. Remove the shaft and the ball detent spring. Remove the O-ring seal from the shaft.

3. Place the ball detent spring into the detent spring hole and start the Reverse shifter shaft into the hole in the boss.

4. Place the detent ball on the spring and while holding the ball down, push the shifter shaft into place and turn until the ball drops into place in the detent on the shaft detent plate.

5. Install the O-ring seal on the shaft.

6. Install the shift fork. Do not drive the shifter shaft lock pin into place until the Reverse housing has been installed on the case.

Reverse Idler Shaft

REPLACEMENT

1. Place a small punch into the front cover hole of the overdrive unit and drive the pin into the shaft until the shaft can be pulled from front cover.

2. Insert a new idler shaft into cover until the hole in the shaft aligns with the hole in the boss.

3. Insert the roll pin into the boss opening and drive the pin into the cover until the shaft is securely locked in place.

Side Cover

ASSEMBLY

Although the service of the side cover is covered here, the transmission does not have to be removed to perform these operations. To remove the side cover in the vehicle, simply drain the transmission, disconnect the electrical leads at the side cover switches, disconnect the 1st–2nd and the 3rd–4th linkage, then remove the attaching bolts.

1. Remove the outer shifter lever

nuts and the lockwasher, then pull the levers from the shafts.

2. Carefully push the shifter shafts into the cover, allowing the detent balls to fall free, then remove both shifter shafts.

3. Remove the interlock sleeve, the interlock pin and the poppet spring.

4. Replace the necessary parts and the assembly by reversing the Steps 1–3.

Overdrive Unit

DISASSEMBLY

Cleanliness is an important factor in the overhaul procedures. Before attempting any disassembly operation, the exterior of the transmission should be thoroughly cleaned to prevent the possibility of dirt entering the internal mechanism. During the inspection and the reassembly, the parts should be thoroughly cleaned with cleaning fluid and then air dried. Wiping cloths or rags should not be used to dry parts. Do not use solvents on neoprene seals, composition faced clutch plates or thrust washers. All oil passages should be blown out and checked to make sure that they are not obstructed. The small passages should be checked with a tag wire.

1. Remove the fill plug and drain the oil from the case.

2. Remove the retaining bolt and the bracket from the speedometer sensor and the driven gear. Remove the sensor and the gear.

3. Remove the three $1/8$ in. pipe plugs from the rear of the unit.

4. Using tool No. J-34681, install the pressure plate bolts until they are flush with the case. Turn the bolts two additional turns, by rotating each bolt one turn at a time.

NOTE: This sequence must be followed in order to prevent the pressure plate from cocking and causing damage to the unit.

5. Remove the 4 adapter plate-to-case Allen head bolts.

6. Using a plastic hammer and a small pry bar, remove the adapter plate. Tap the adapter plate to separate it from the case.

NOTE: Do not pry between the case and the adapter plate, damage to the sealing surfaces could occur.

7. Bolt the overdrive unit to tool No. J-34162. Mount the holding fixture to the base plate tool No. J-3389-20.

8. Remove the large snap ring from the O/D unit forward of the accumulator piston.

9. Remove the piston/accumulator assembly. Using an Allen wrench, pry the assembly up evenly by lifting under the flange. Do not pry at or near the seal surface.

10. Remove the carrier and the bearing assembly (includes the input sun and pinion gears) as an assembly.

11. Remove the finger pressure plate.

12. Remove the overdrive clutches, (4) composition, (4) steel and (1) clutch stop plate.

13. Remove the direct clutch plates, (5) composition, (5) steel and (1) steel bearing plate. Measure each selective clutch plate in the direct clutch pack and record the readings.

NOTE: The selective clutch plates are used to control the clutch pack clearance. When replacing the clutch plates, replace each selective clutch plate with one of the same size.

14. Inspect the overdrive and the direct clutch plates as follows:

a. Compositioned Plates – Dry the plates and inspect for pitting, flaking wear, glazing, cracking, charring and chips or metal particles imbedded in lining. If a plate shows any of the above conditions, replacement is required.

b. Steel Plates – Wipe the plates dry and check for discoloration. If the surface is smooth and even color, reuse the plate. If severe heat spot discoloration or surface scuffing is indicated, the plate must be replaced.

15. Remove the thrust washer and the bearing from the output sun gear; the thrust washer may stick to the input sun gear hub.

16. Remove pump housing Allen head bolts by rotating the hub to gain access to the bolts.

17. Remove the output shaft assembly (the output sun gear, the sprag clutch, the clutch hub, the gerotor pump and the speedometer drive gear.

18. Remove the pressure plate and the springs by positioning tool No. J-21420-2 on the pressure plate (with the bolt from tool No. J-23327) through the center of the plate. Next position tool No. J-23327 on the rear of the case and install the retaining

nut. Remove the three retaining bolts and the tool No. J-34681, from the rear of the case. Loosen the retaining nut on the tool No. J-23327 bolt to relieve the spring pressure.

19. Remove the cooler valve assembly by loosening the nuts on the tube and then the valve-to-case bolts.

20. Remove the 12 oil pan bolts and then pry the pan from the case.

21. Remove the oil filter and the tube from the valve body.

22. Disconnect the T.V. cable from the lever, then remove the cable retaining bolt and the cable assembly.

23. Remove the T.V. lever bolt and the lever from the valve body.

24. Remove the remaining valve body bolts and the valve body with the spacer plate.

NOTE: There are two check balls, one on each side of the spacer plate. One ball is located in the case and the other is spring loaded in the valve body.

ASSEMBLY

1. Install the pressure plate springs into the transmission case pockets.

2. Place the pressure plate on top of the springs and seat the springs into the pockets of pressure plate.

3. Position the plate tool No. J-21420-2 on top of the pressure plate with the bolt from tool No. J-23327 through the center of the plate. Next, position tool No. J-23327 on the rear of the case and install the retaining nut. Tighten the nut until the pressure plate is drawn approximately $\frac{1}{8}$ in. below the step for the overdrive clutch plates. Install the three pressure plate bolts (J-34681). Remove the tools No. J-21420-2 and J-23327 from the case.

4. Install the output shaft assembly into the case. Install the pump bolts and torque.

NOTE: Be sure the O-rings are positioned properly on pump cover before installing the output shaft assembly.

5. Install the thrust bearing on the output sun gear.

6. Install the tanged direct clutch thrust washer with the tabs facing pressure plate.

7. Install the direct clutch thrust bearing and the thrust washer.

NOTE: The thrust washer will have a tooth missing from its outer edge. The side of the thrust washer with the circular grind pattern must face the thrust bearing. The side with the grind pattern can be identified by the notch ground into the tooth.

8. Install one composition clutch disc and then a selective clutch plate.

NOTE: The selective clutch plates come in five sizes (0.080–0.120 in.) and is used to control the clutch pack clearance; a 0.050–0.070 in. clearance must be maintained in the direct clutch pack. Excessive or insufficient amount of clutch travel will cause failure to the clutch plates and discs.

9. Alternate the remaining clutch discs and plates until all five plates and discs are installed.

10. Install the lower half of the carrier assembly onto the direct clutch pack; index the carrier until all of the clutch plates are engaged.

11. Install the steel overdrive stop clutch plate and then alternate with a disc and a plate until the four plates and disc are installed.

12. Install the finger pressure plate.

13. Install the carrier thrust plate with tabs facing the sprag clutch.

14. Install the two pinion gears with the index mark on the gears facing inward or towards each other. Install the other two pinion gears with the index mark 90° from the first two gears.

15. Install the output sun thrust washer into the rear of the input sun gear; use petrolatum to retain the thrust washer to the input sun gear.

16. Install the input sun gear.

NOTE: If the input sun gear spreads the pinion gears when installing, the pinion gears are not indexed properly.

17. Install the selective thrust washer with the washer oil grooves facing the input sun gear.

18. Install the thrust bearing on the input sun gear.

19. Install the carrier thrust washer to the cover; use petrolatum to retain the thrust washer to the cover.

20. Install the four pinion gear thrust washers onto the carrier cover; use petrolatum to retain washers to the cover.

21. Install the carrier cover, the four new nuts and torque.

NOTE: If the pinion gears are not indexed properly, the four cover bolt holes will not align with the lower half of the carrier bolts.

22. Measure the end play for the overdrive unit as follows:

a. Place the straight edge tool No. J-34673 across the face of the overdrive unit. Using the Depth Micrometer tool No. J-34672, mea-
sure the distance from the bearing to the top of the bar. Next, using tool No. J-34673 and a 0–1 micrometer, measure the thickness of the bar and subtract this from the reading of the depth micrometer tool No. J-34672 and record this reading.

b. Place the straight edge tool No. J-34673 across the rear of the adapter plate. Using the Depth Micrometer tool No. J-34672, measure the distance from the top of the bar to the adapter plate mounting surface and record the reading.

c. Next, measure the distance from the top of the bar to the bearing seat in the adapter plate and record the reading.

d. Subtract the reading of Step (c) from Step (b) and record the difference.

e. Next, subtract the difference of step (d) from step (a). The difference will be the end play, it should be –0.003–0.003 in.

NOTE: If the results of your measurements are not within the specifications, it will be necessary to remove the carrier cover and change the input sun selective thrust washer. The selective thrust washers are available in eight sizes, in 0.005 in. increments, ranging from 0.123–0.158 in.

23. Install the accumulator and the piston assembly; coat the lips of the seals with clean Dexron® II automatic transmission fluid before installing.

24. Install the large snap ring at the front of the overdrive unit.

25. Place the a new seal on the tool No. J-34523 and install it at the front side of the adapter plate.

26. Place seal protector tool No. J-34621 on the input sun gear and install the adapter plate. Apply a light coating of RTV Sealant No. 1052366 around the heads of the adapter plate bolts. Install the adapter plate bolts and torque.

27. Remove the seal protector.

28. Remove the first $\frac{1}{8}$ in. pipe plug from the left side of the overdrive unit. Install the air line fitting tool No. J-34742 into the plug hole and torque.

29. Measure the clutch pack clearance as follows:

a. Loosen the pressure plate bolts (J-34681) evenly until the spring pressure is released.

b. Assemble the Dial Indicator tool No. J-8001 to the rear of the overdrive unit.

c. Apply a minimum of 100 psi to the air line fitting tool No. J-34742 and read the dial indicator, it should be between 0.050–0.070 in.

NOTE: If the reading does not fall within the specification, it will be necessary to disassemble the overdrive unit to change the direct clutch selective clutch plates. The selective clutch plates are available in five sizes, they are in 0.010 in. increments ranging from 0.080–0.120 in. If the clutch pack clearance is within specification, remove the clutch pack retaining bolts (J-34681).

30. Coat the three $\frac{1}{8}$ in. pipe plugs with anti-sieze compound, then install and torque the plugs.
31. Remove the air line adapter tool No. J-34742. Coat the plug with anti seize compound, then install and torque the plug.
32. Install the speedometer gear and the sensor.
33. Using tool No. J-21426, install a new output seal; coat the lip of the seal with Dexron® II transmission fluid.
34. Install the valve body as follows:
   a. Install the check ball into the case.
   b. Position the gaskets, one on each side of the separator plate and position the separator plate on the valve body.
   c. Position the valve to the case, then install and torque the bolts.
   d. Install the T.V. cable, the retaining clip and the bolt, then torque bolt.
   e. Install the T.V. lever and torque bolt, then connect T.V. cable to the lever.
   f. Install the Throttle Setting Gage tool No. J-34671-1 into the T.V. cable bore on the side of the case. Set the T.V. cable hook onto the high step of the gauge. Place the valve body cam stop as close to the lever as possible, then install and torque the bolt.
   g. Set the T.V. cable hook onto the lower step of the gauge. Place the tool No. J-34671-2 between the piston and the solenoid bracket, then adjust the T.V. lever screw/bolt until the bolt makes contact with the cam stop.
35. Install the pickup tube and the oil filter on the valve body.
36. Apply a bead of RTV Sealant No. 1052366 or equivalent to the oil pan flange and assemble it wet. Install the magnet in the oil pan. The bead of RTV should be applied around the inside of the bolt holes. Install the pan bolts and torque.
37. Assemble the Overdrive unit to the Reverse housing. Guide the idler shaft on the adapter plate into the idler gears and align the mainshaft splines with the input sun gear

splines. Slide the units together, then install the bolts and torque the bolts.

Valve Body

DISASSEMBLY

NOTE: In the following procedures, use tool No. J-34529 to relieve the valve pressures.

1. Relieve the shift valve pressure, then remove the pin, the spring and the valve.
2. Relieve the pressure relief valve pressure, then remove the pin, the spring and the valve.
3. Relieve the accumulator valve pressure, then remove the pin, the spring, the valve, the plug, the sleeve and the plunger.
4. Disconnect the solenoid electrical lead at the pressure switch, then remove the solenoid bolts, the solenoid and the check ball.
5. Disconnect the other electrical lead at the pressure switch, then remove the switch from the valve body.
6. To assemble, reverse the removal procedures. Coat the components with clean Dexron® II automatic transmission fluid before assembling.

Output Shaft

DISASSEMBLY

1. Remove the speedometer gear retaining clip and the gear.
2. Remove the pump cover-to-pump housing Allen head bolts and the cover.
3. Mark the pump gears with a grease pencil; the gears must be installed in same direction as removed.
4. Position the output shaft with the splines down, then rotate the pump housing until gears slide out.
5. Remove the drive pin from the output shaft.
6. Remove the pump housing from the output shaft and the thrust washer from the pump housing.
7. Remove the thrust bearing and the washer from the clutch hub and the clutch hub from the output shaft.

NOTE: Record the direction of the hub on the shaft. The oil grooves face the sprag clutch or forward on shaft.

8. Remove the sprag clutch from the output shaft.

NOTE: Record the direction of the sprag clutch; the lip on the

sprag clutch cage faces the oil grooves on the clutch hub.

ASSEMBLY

NOTE: Before installation, coat the parts with clean Dexron® II automatic transmission fluid.

1. Install the sprag clutch on the output shaft; the lip on the sprag clutch cage faces rearward or towards the oil grooves on the clutch hub.
2. Install the clutch hub on the output shaft; the oil grooves on the hub face the sprag clutch or forward on the shaft.
3. Install the thrust washer and the thrust bearing on the clutch hub.
4. Install the thrust washer on the pump housing; use petrolatum to retain the thrust washer to the housing.
5. Install the pump housing and the pin to the output shaft.
6. Install the pump gears in the housing; the gears must be installed in same direction as removed.
7. Place the pump cover on the housing, then install the cover-to-pump housing bolts and torque.
8. Install the speedometer gear and the retaining clip on the output shaft.
9. Install new O-rings on the pump; use petrolatum to retain the O-rings to the cover.

Carrier Assembly

DISASSEMBLY

1. Remove the four carrier cover nuts and the cover.
2. Remove the thrust washer, the thrust bearing, the selective washer and the input sun gear.
3. Remove the four pinion gears.
4. Remove the steel thrust plate from the carrier.
5. Clean and inspect the parts, then replace any parts that are cracked, chipped or show excessive wear. The carrier assembly must be reassembled in the transmission case.

Piston/Accumulator Assembly

DISASSEMBLY

1. Remove the accumulator-to-piston snap ring, the accumulator and the 24 springs from the piston.
2. Remove the two accumulator and the two piston O-rings.
3. To assemble, reverse the removal procedures. Coat the O-rings with clean Dexron® II automatic transmission fluid before installing.

CHILTON TYPE 16

Transaxle case

DISASSEMBLY

1. Remove the speedometer drive gear and the front bearing retainer, then the case cover.
2. Remove the selecting bellcrank assembly bolts, the set bolt, the shift and the selector lever assembly.
3. Move the shift levers into the Lock position, then remove the output shaft lock nut and return the transaxle to the Neutral position.
4. Remove the No. 3 shifting fork bolt and the lock washer. Using two small pry bars and a hammer, tap out the input shaft snap ring.
5. Using tool No. 09602-10010, remove the No. 3 hub sleeve assembly and the shifting fork.
6. Using tool No. 09950-20015, remove the 5th driven gear, then the synchronizer ring, the needle roller bearing and the spacer.
7. Remove the rear bearing retainer and the Reverse idler shaft lock bolt. Using snap ring pliers, remove the two snap rings.
8. Using two small pry bars and a hammer, tap out the No. 2 shifting fork snap ring.
9. Using tool No. 09313-30021, remove the three plugs and the lock ball assembly. Using a magnet, remove the four seats, the springs and the balls.
10. Remove the 16 case-to-housing bolts and separate the case from the housing, using a plastic hammer.
11. Remove the two Reverse shift arm bracket bolts and the bracket, then pull out the Reverse idler gear and the shaft.
12. Using two small pry bars and a hammer, tap out the 3 snap rings. Remove the shifting fork set bolts, then the No. 2 shifting fork shaft and the head. Using a magnet, remove the two balls.
13. Remove the No. 3 shifting fork shaft and the reverse shifting fork. Remove the No. 1 shifting shaft, then the No. 1 and the No. 2 shifting forks.
14. Remove the input and the output shafts (together), the differential assembly, the magnet and the oil receiver.
15. Using two small pry bars and a hammer, tap the snap ring from the input shaft.
16. Using an arbor press, press the radial ball bearing and the 4th speed gear from the input shaft, then remove the needle roller bearing and the synchronizer ring.
17. Using snap ring pliers, remove the snap ring. Using an arbor press, press the No. 2 hub sleeve, the 3rd gear, the synchronizer ring and the needle roller bearings as an assembly.
18. Using an arbor press, remove the radial ball bearing and the 4th driven gear from the output shaft, then the spacer.
19. Shift the No. 1 hub sleeve into the 1st gear position. Using an arbor press, press the 3rd driven and the 2nd gears from the output shaft, then remove the needle roller bearing, the spacer and the synchronizer ring.
20. Using two small pry bars and a hammer, tap the snap ring from the output shaft.
21. Using an arbor press, press the No. 1 hub sleeve, the 1st gear and the synchronizer ring from the output shaft, then remove the needle roller bearing, the thrust washer and the locking ball.

ASSEMBLY

NOTE: Before installing the roller bearings and other moving parts, coat with a multipurpose grease.

1. Install the clutch hub and the keys onto the No. 2 hub sleeve, then the springs under the keys.

--- **CAUTION** ---
The springs must be positioned so that the end gaps are not aligned with each other.

2. Install the needle roller bearings onto the input shaft and the synchronizer ring onto the No. 2 hub sleeve assembly (align the ring slots with the keys).
3. Using a arbor press, press the 3rd gear and the No. 2 hub sleeve assembly onto the input shaft, then secure with a snap ring (select one which will allow the minimum axial play).

NOTE: Using a feeler gauge, check the 3rd gear thrust clearance; it should be 0.004–0.014 in.

4. Install the synchronizer ring onto the No. 2 hub sleeve assembly (align the ring slots with the keys), then the 4th gear and the needle roller bearing onto the input shaft.
5. Using tool No. 09608-20011 and an arbor press, press the radial ball bearing onto the input shaft, then secure with a snap ring (select one which will allow the minimum axial play).

NOTE: Using a feeler gauge, check the 4th gear thrust clearance; it should be 0.004–0.022 in. If the output shaft was replaced, drive the slotted spring 0.236 in. into the new shaft.

6. Install the clutch hub and the keys onto the No. 1 hub sleeve, then the springs under the keys.

--- **CAUTION** ---
The springs must be positioned so that the end gaps are not aligned with each other.

7. On the output shaft, install the locking ball and fit the thrust washer securely over it. Using an arbor press, press the 1st gear, the synchronizer ring, the hub and the sleeve assembly onto the output shaft, then install the snap ring (select one which will allow the minimum axial play).

NOTE: Using a feeler gauge, check the 1st gear thrust clearance; it should be 0.004–0.016 in.

8. Install the spacer, the synchronizer ring, the needle roller bearing and the 2nd gear. Using an arbor press, press the 3rd driven gear onto the output shaft.

NOTE: Using a feeler gauge, check the 2nd gear thrust clearance; it should be 0.004–0.018 in.

9. Install the spacer. Using an arbor press and the tool No. 09608-12010, press the 4th driven gear and the radial ball bearing onto the output shaft.
10. Install the magnet and the oil receiver (using two bolts).
11. Install the thinnest side bearing shim into the case. Using tool No. 09608-20011, drive the outer side bearing race into the case.
12. Install the differential and the case, then torque the case bolts to 22 ft. lbs. Using tool No. 09564-32011 and a torque meter, measure the bearing preload; it should be 6.9–13.9 inch lbs. (new) or 4.3–8.7 inch lbs. (used).

NOTE: When adjusting the shim thickness, the preload will change 2.6–3.5 inch lbs. with each new shim thickness.

13. Remove the case and install the input and the output shafts at the same time.
   a. Install the No. 1 and No. 2 shift forks into the No. 1 and the No. 2 hub sleeve grooves.
   b. Install the No. 1 fork shaft into the No. 1 shift fork hole.
   c. Install the 2 interlock balls into the Reverse shift fork hole.
   d. Install the No. 3 and the Reverse fork shaft.

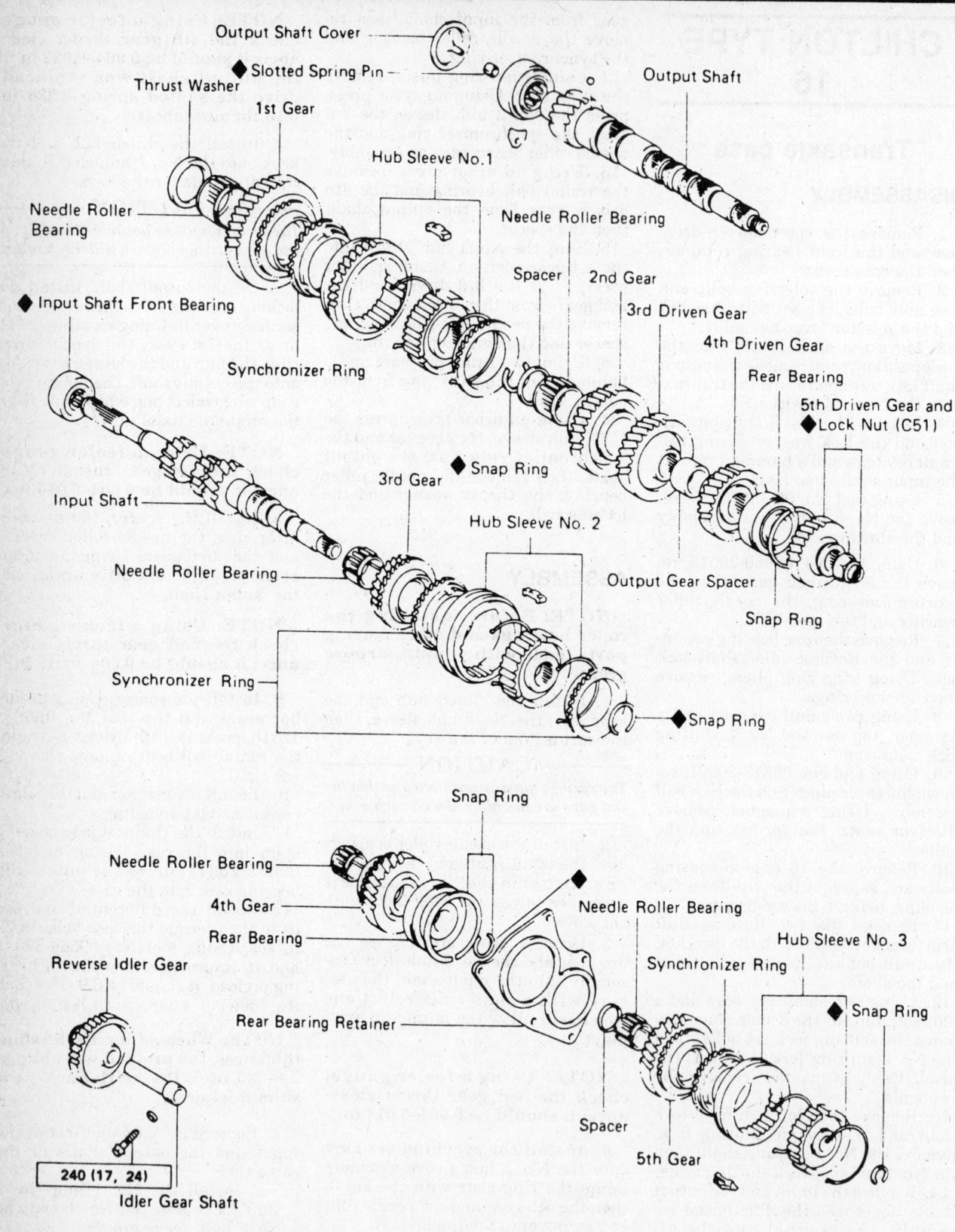

Output Shaft Cover

◆ Slotted Spring Pin

Thrust Washer

1st Gear

Hub Sleeve No. 1

Output Shaft

Needle Roller Bearing

Spacer

2nd Gear

Needle Roller Bearing

3rd Driven Gear

◆ Input Shaft Front Bearing

4th Driven Gear

Rear Bearing

5th Driven Gear and
◆ Lock Nut (C51)

Synchronizer Ring

3rd Gear

◆ Snap Ring

Input Shaft

Hub Sleeve No. 2

Needle Roller Bearing

Output Gear Spacer

Snap Ring

Synchronizer Ring

◆ Snap Ring

Snap Ring

Needle Roller Bearing

4th Gear

Needle Roller Bearing

Rear Bearing

Hub Sleeve No. 3

Reverse Idler Gear

Synchronizer Ring

◆ Snap Ring

Rear Bearing Retainer

Spacer

5th Gear

240 (17, 24)

Idler Gear Shaft

kg-cm (ft-lb, N·m) : Tightening torque

◆ : Non-reusable part

Exploded view of the C51 5-speed gear assembly (Chilton Type 16)

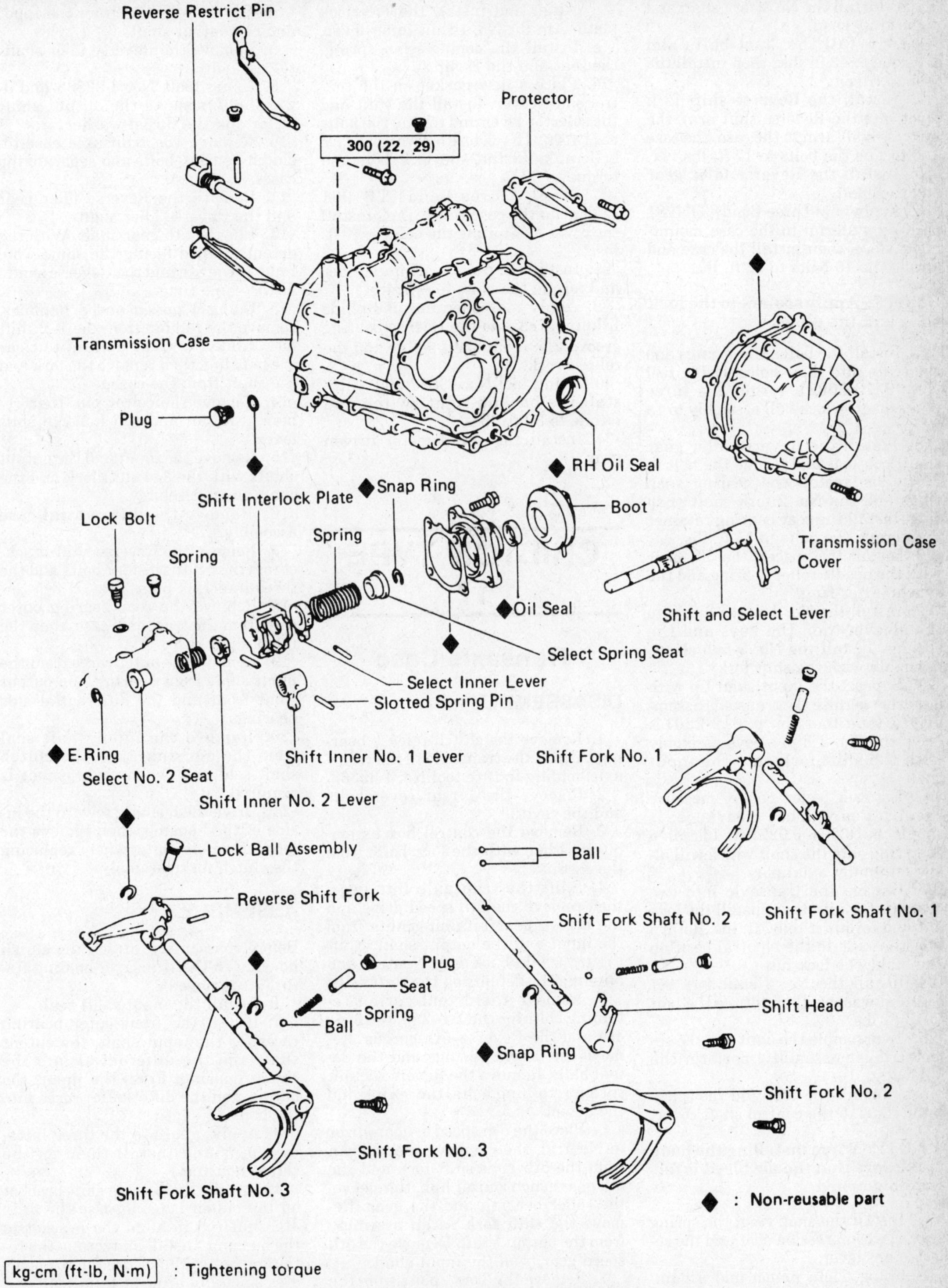

Reverse Restrict Pin

Protector

300 (22, 29)

Transmission Case

Plug

RH Oil Seal

Boot

Transmission Case Cover

Shift Interlock Plate

Snap Ring

Spring

Spring

Oil Seal

Shift and Select Lever

Lock Bolt

Select Spring Seat

Spring

Select Inner Lever

Slotted Spring Pin

Shift Fork No. 1

E-Ring

Shift Inner No. 1 Lever

Select No. 2 Seat

Shift Inner No. 2 Lever

Shift Fork Shaft No. 1

Lock Ball Assembly

Ball

Reverse Shift Fork

Shift Fork Shaft No. 2

Plug

Seat

Shift Head

Spring

Ball

Snap Ring

Shift Fork No. 2

Shift Fork No. 3

Shift Fork Shaft No. 3

♦ : Non-reusable part

kg-cm (ft-lb, N·m) : Tightening torque

Exploded view of the C51 5-speed transaxle case (Chilton Type 16)

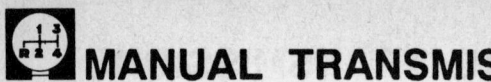

e. Install the No. 2 fork shaft and the shift head.

f. Install the 3 set bolts and torque to 13 ft. lbs., then install the snap rings.

14. Install the Reverse shift fork pivot into the Reverse shift arm, the Reverse shift arm to the case, the case and torque the bolts to 17 ft. lbs.

15. Install the Reverse idler gear and the shaft.

16. Apply new Three Bond® TB1281 packing material to the case mounting surface, then install the case and torque the 16 bolts to 22 ft. lbs.

NOTE: Apply sealant to the lock balls and the plugs.

17. Install the balls, the springs and the seats into their holes. Using tool No. 09313-30021, torque the three plugs and the lock ball assembly to 18 ft. lbs.

18. Install the Reverse idler gear shaft lock bolt and torque the bolt to 29 ft. lbs. Install the bearing snap rings and the No. 2 fork shaft snap ring. Install the rear bearing retainer and torque the bolts to 14 ft. lbs.

19. Install the spacer, the 5th gear with the needle roller bearing and the synchronizer ring.

20. Install the No. 3 clutch hub into the sleeve, then the keys and the springs. Install the No. 3 hub sleeve assembly with the shift fork.

21. Support the input shaft tip with a spacer (to raise the transaxle assembly). Using tool No. 09612-22011, drive the No. 3 hub sleeve assembly with the shift fork onto the input shaft.

22. Using a feeler gauge, measure the 5th gear thrust clearance; is should be 0.004–0.022 in. Install a snap ring onto the shaft which will allow minimum axial play.

23. Engage the transaxle into two gears (to lock it), then install the lock nut and torque it to 87 ft. lbs. Return the transaxle to the Neutral position and stake the lock nut.

24. Install the No. 3 shift fork set bolt and washer, then torque the bolt to 13 ft. lbs.

25. To assemble the shift and the selector lever assembly, perform the following:

a. Install the boot and the selector shaft to the control shaft cover.

NOTE: When installing the boot, make sure that the air bleed is facing downward.

b. Install the snap ring, the spring seat, the compression seat and the select inner lever.

c. Using a pin punch and a hammer, drive in the slotted spring.

d. Align and install the interlock plate with the No. 1 shift inner lever.

e. Install the compression spring, the seat and the E-clip.

26. Place a new gasket on the control shaft cover, install the shift and the selector lever and torque the bolts to 14 ft. lbs., then install the bellcrank. Install the lock bolt and torque to 22 ft. lbs.

27. Apply Three Bond® TB1281 sealant to the case cover, then install the cover and torque the bolts to 13 ft. lbs.

28. Install the front bearing retainer and torque the bolts to 8 ft. lbs.

29. Apply molybdenum disulphide lithium grease to the release bearing groove, the input shaft spline and the release fork.

30. Using tool No. 09817-16011, install the backup light switch and torque to 30 ft. lbs.

31. Install the speedometer driven gear.

CHILTON TYPE 17

Transaxle Case

DISASSEMBLY

1. Remove the clutch release bearing. Attach the transaxle to the transaxle holding fixture tool No. J-33366.

2. Remove the 7 rear cover bolts and the cover.

3. Remove the control box assembly together with the four bolts from the case.

4. Shift the transaxle into gear, then remove the 5th speed drive and the driven gear retaining nuts from the input and the output shaft. Shift the transaxle back into Neutral and aligning the detents on the shift rails.

5. Remove the detent spring retaining bolts for the 1st–2nd, the 3rd–4th and the Reverse–5th speeds. Remove the detent springs and the detent balls. Remove the Reverse detent spring retaining bolts, the spring and the detent.

6. Place the 5th speed synchronizer in Neutral, then remove the roll pin from the 5th gear shift fork and the 5th gear synchronizer hub, the sleeve, the roller bearing and the gear. Remove the shift fork as an assembly from the output shaft. Remove the 5th speed gear from the input shaft.

7. Remove the Torx® bolts from the bearing retainer, then the bearing re-

tainer and the shims from the input and the output shafts.

8. Remove the Reverse idler shaft-to-case bolt.

9. Using tools No. J-22888 and J-22888-30, remove the output shaft collar and the thrust washer.

10. Remove the transaxle case-to-clutch housing bolts and separate the cases.

11. Remove the Reverse idler gear and the Reverse idler shaft.

12. Lift the 5th gear shaft. With the detent aligned facing the same way, remove the 5th and the Reverse shafts at the same time.

13. Using a punch and a hammer, remove the roll pin from the 1–2 shift fork. Slide the shaft upward to clear the housing, then remove the fork and the shaft from the case.

14. Remove the cotter pin, then remove the pin and the Reverse shift lever.

15. Remove the input and the output shafts with the 3–4 shift fork and the shaft as an assembly.

16. Remove the differential case assembly.

17. Remove the Reverse shift bracket together with the four bolts and the three interlock pins.

18. Remove the rear bearing outer race from the transaxle case, then the input shaft race.

19. Remove the outer races from the input shaft front bearing, the output shaft front and the differential side bearings.

20. Remove the input shaft seal from the housing, then the clutch shaft seal only when replacement is required.

21. Drive the bushing toward the inside of the housing, then remove the fork assembly only when replacing the clutch fork assembly.

ASSEMBLY

Before reassembly, attach the clutch housing to the transaxle holding fixture (if removed).

1. Install the input shaft seal.

2. Install the front outer bearing races for the input shaft, the output shaft and the differential into the clutch housing. Press the input, the output and the differential races into the housing.

3. Apply grease to the three interlock pins and install them on the clutch housing.

4. Install the Reverse shift bracket on the clutch housing. Use the 3rd–4th shift rod to align the bracket to the housing. Install and torque the retaining bolts. Make sure the rod operates smoothly after installation.

5. Install the differential assembly

1. CLUTCH AND DIFF. HOUSING
2. CLUTCH SHAFT BUSHING
3. INPUT SHAFT OIL SEAL
4. DRIVE SHAFT OIL SEAL
5. STRAIGHT KNOCK PIN
6. TRANSAXLE CASE
7. DRAIN PLUG
8. GASKET
9. MAGNET
10. BEARING RETAINER
11. REAR COVER
12. GASKET
13. INPUT SHAFT
14. INPUT SHAFT FRONT BEARING
15. 3RD GEAR ASSEMBLY
16. 3RD/4TH SYNCHRONIZER ASM.
17. SYNCHRONIZER SLEEVE
18. CLUTCH HUB
19. INSERT
20. INSERT SPRING
21. 3RD/4TH BLOCKER RING
22. 4TH GEAR ASSEMBLY
23. 3RD NEEDLE BEARING
24. 4TH NEEDLE BEARING
25. 4TH COLLAR
26. 4TH GEAR THRUST WASHER
27. INPUT SHAFT REAR BEARING
28. 5TH GEAR
29. INPUT SHAFT END NUT
30. OUTPUT SHAFT
31. OUTPUT SHAFT FRONT BEARING
32. 1ST GEAR ASSEMBLY
33. 1ST/2ND SYNCHRONIZER ASSEMBLY
34. REVERSE GEAR
35. CLUTCH HUB
36. INSERT
37. INSERT SPRING
38. 1ST/2ND BLOCKER RING
39. 2ND GEAR ASSEMBLY
40. 1ST NEEDLE BEARING
41. 2ND NEEDLE BEARING
42. 2ND COLLAR
43. 3RD/4TH OUTPUT GEAR
44. KEY
45. OUTPUT SHAFT REAR BEARING
46. INPUT SHAFT BEARING SHIM
47. OUTPUT SHAFT BEARING SHIM
48. 5TH GEAR THRUST WASHER
49. 5TH NEEDLE BEARING
50. 5TH COLLAR
51. 5TH GEAR ASSEMBLY
52. 5TH SYNCHRONIZER ASSEMBLY
53. SYNCHRONIZER SLEEVE
54. CLUTCH HUB
55. INSERT
56. INSERT SPRING
57. 5TH BLOCKER RING
58. INSERT STOPPER PLATE
59. OUTPUT SHAFT END NUT
60. REVERSE IDLER GEAR ASM.
61. REVERSE IDLER SHAFT

62. STRAIGHT PIN
63. REVERSE IDLER SHAFT BOLT
64. GASKET
65. CLUTCH FORK SHAFT ASM.
66. CLUTCH RELEASE BEARING

67. RELEASE BEARING SPRING
68. CLUTCH SHAFT BUSHING
69. CLUTCH SHAFT SEAL
70. CLUTCH PRESSURE PLATE ASM.
71. CLUTCH DISK ASSEMBLY

Exploded view of the 5-speed (76mm) transaxle (Chilton Type 17)

first, then the input and the output shaft with the 3rd–4th shift fork and the shaft together as an assembly into the clutch housing.

NOTE: Make sure the interlock pin is in the 3rd–4th shifter shaft before installing.

6. The 3rd–4th shift shaft is installed into the raised collar of the Reverse shift lever bracket.

7. Install the 1–2 shift fork onto the synchronizer sleeve and insert the shifter shaft into the Reverse shift le-ver bracket. Align the hole in the fork with the shaft and install the roll pin.

8. Install the Reverse lever on the shift bracket.

9. Install the Reverse and the 5th gear shifter shaft; engage the Reverse shaft with the Reverse shift lever at the same time.

NOTE: Make sure the interlock pin is in the 5th gear shifter shaft before installing.

10. Install the Reverse idler shaft with the gear into the clutch housing.

NOTE: Make sure the Reverse lever is engaged in the gear collar.

11. Using tool No. J-33373, measure and determine the shim size.

a. Position the outer bearing races on the input, the output and the differential bearings. Position the shim selection gauges on the bearing races. The three gauges are identified: Input, Output and Differential.

b. Place the 7 spacers (provided with the tool No. J-33373) evenly around the clutch housing perimeter.

c. Install the bearing and the shim retainer on the transaxle case. Torque the bolts to 11–16 ft. lbs.

d. Carefully position the transaxle case over the gauges and on the spacers. Install the bolts (provided in the tool kit) and tighten the bolts alternately until the case is seated on the spacers, then torque the bolts to 10 ft. lbs.

e. Rotate each gauge to seat the bearings. Rotate the differential case through three revolutions in each direction.

f. With the three gauges compressed, measure the gap between the outer sleeve and the base pad using the available shim sizes. Use the largest shim that can be placed into the gap and drawn through without binding; this will be the correct shim for the bearing being measured.

g. When each of the three shims selected, remove the transaxle case, the spacers and the three gauges.

12. Position the shim selected for the input, the output and the differential into the bearing race bores in the transaxle case.

13. Using tool No. J-24256-A, J-8092 and an arbor press, install the rear input shaft bearing race; press the bearing until it is seated in its bore.

14. Using tool No. J-33370, J-8092 and an arbor press, install the rear output shaft bearing; press the bearing until it is seated in its bore.

15. Using tool No. J-8611-01, J-8092 and an arbor press, install the rear differential case bearing race; press the bearing until it is seated in its bore.

16. Apply a $\frac{1}{8}$ in. bead of Loctite® 514 to the mating surfaces of the clutch housing and the transaxle case.

17. Be sure the magnet is installed in the transaxle case.

18. Install the case on the clutch housing and the Reverse idle shaft bolt into the case, then torque the bolt to 22–33 ft. lbs.

19. Install the 14 case bolts and torque them to 22–33 ft. lbs. (in a diagonal sequence).

20. Install the drive axle seals.

21. Install the thrust washer and the collar to the output shaft.

22. Install the 5th gear to the input shaft. Install the needle bearing, the 5th gear, the blocking ring, the hub/sleeve assembly (with the shift fork in its groove) and the backing plate on the output shaft. Align the shift fork on the shifter shaft and install the roll pin.

23. Install the Reverse detent balls and the springs, then the 1st–2nd, the 3rd–4th and the 5th speed gears. Install the bolts and torque to 15–21 ft. lbs.

24. Apply Loctite® 262 to the input and the output shaft threads. Install new retaining nuts and torque to 87–101 ft. lbs.; stake the nuts after reaching the final torque.

25. Install the gasket and the control box assembly on the transaxle case, then torque the bolts to 11–16 ft. lbs.

NOTE: Make sure the transaxle shifts properly before installing the rear cover.

26. Install the gasket and the rear cover with the 7 bolts, then torque the bolts to 11–16 ft. lbs.

27. Install the clutch fork assembly (if removed). Using tool No. J-28412, install the bushing into the upper hole. Install the oil seal. Before installing the bushing, apply grease to both the interior and the exterior.

28. Install the clutch release bearing.

Input Shaft

DISASSEMBLY

1. Using tool No. J-22912-01 and an arbor press, remove the front bearing.

2. Pull out the rear bearing 4th gear, the 3rd–4th synchronizer assembly and 3rd gear as an assembly.

NOTE: This procedure requires a arbor press and tool No. J-22912-01.

3. Remove the outer parts from, the input shaft.

Output Shaft

DISASSEMBLY

1. Using tool No. J-22227-A and an arbor press, remove the front bearing.

2. Using tool No. J-22912-01 and an arbor press, remove the rear bearing and the 3rd–4th gear as an assembly.

3. Remove the key, the 2nd gear, the needle bearing and the blocking ring.

4. Using an arbor press, remove the collar, the Reverse gear assembly and the 1st gear as an assembly.

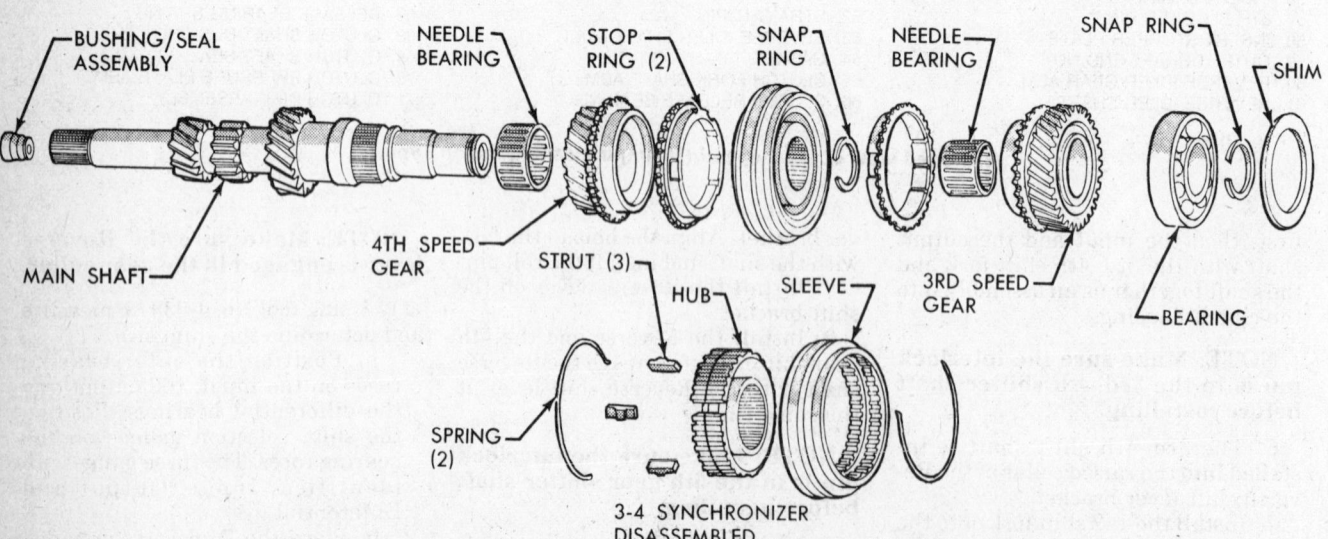

Exploded view of the Chrysler A-412 mainshaft (Chilton Type 18) transaxle

CHILTON TYPE 18

Transaxle Case

DISASSEMBLY

NOTE: Final mainshaft adjustment requires a measurement made with a special tool. Check Step 16 of the assembly procedure before disassembly.

1. Remove the clutch pushrod, being careful not to bend it.
2. Unscrew the selector shaft plug from the case. Remove the detent spring assembly and rubber boot, then tap out the selector shaft and pry out the oil seal.
3. Using a small pry bar, pry out the two mainshaft bearing retaining nut rubber plugs.
4. Remove the four bolts and the clutch release bearing end cover. Hold the clutch release lever upwards while removing the cover to avoid loading or damage to the case threads. Take out the release bearing and plastic sleeve.
5. Using two small pry bars, push the circlip off the clutch torque shaft. Pull the torque shaft out of the case, then remove the pedal return spring and release lever. Pry out the torque shaft oil seal.
6. Remove the three mainshaft bearing retainer nuts; two were under the rubber covers removed earlier and the 3rd is inside the clutch release housing. The three studs and clips will drop into the case. Remove the Reverse idler set screw (bolt) and the backup light switch.
7. Remove the ten case bolts and the four stud nuts, then the transmission case.

NOTE: The factory uses a special tool to do this; it pushes against the end of the mainshaft. Make sure to tag the shims for reuse.

8. Remove the two bolts, the Reverse shift fork and the supports.
9. Remove the snap ring from the end of the pinion shaft.
10. Pull off the bearing and the 4th gear from the end of the mainshaft. Remove the 4th gear needle bearing needle bearing.
11. Using a small pry bar, pry off the shift rail E-clips, then remove the shift forks assembly.
12. Remove the mainshaft assembly; it can be disassembled by removing the snap rings and the components. The clutch pushrod seal and bushing assembly can be driven out of the shaft with a ⅜ in. diameter brass rod. Replace it by driving it with a plastic hammer.
13. Remove the pinion shaft snap ring and the 3rd gear, then the 2nd gear and its needle bearing.
14. Pry or pull out the Reverse idler gear shaft.
15. Using a puller, remove the 1st gear and the 1st–2nd synchronizer assembly from the pinion shaft.

NOTE: The inner sleeve for the 2nd and the 1st gear are removed together.

16. Remove the 1st gear needle bearing and scribe a mark across the 1st–2nd synchronizer for reassembly.
17. Remove the pinion shaft bolts, the retainer, the thrust washer (the flat side goes up) and the pinion shaft.

ASSEMBLY

1. Place a 0.65mm shim in the bearing housing and press the small bearing cup into the clutch housing, then move the pinion up and down, measuring the end play with a dial indicator.

NOTE: Do not rotate the shaft while moving it up and down.

2. The correct preload is determined by adding 0.20mm to the reading obtained from the dial indicator in Step 1, along with the shim thickness, 0.65mm. For example: if the measurement is 0.30mm, the correct shim to

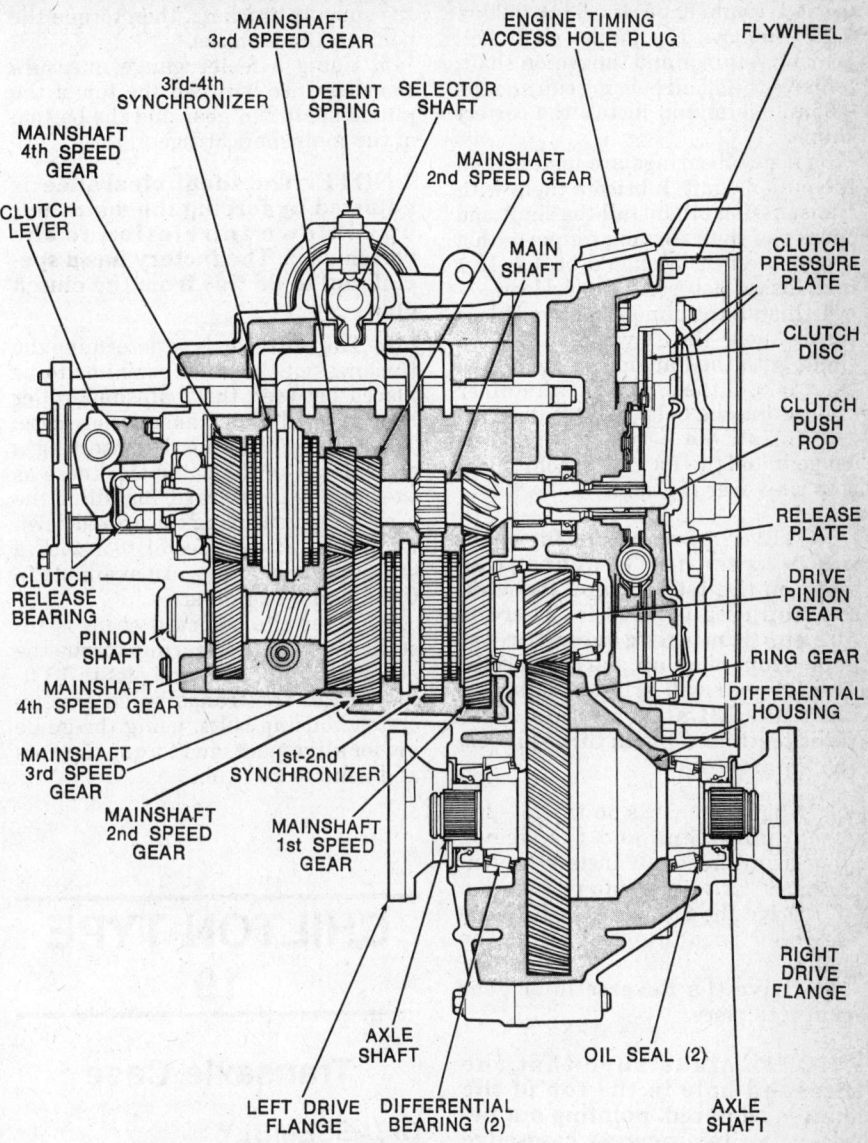

Cutaway view of the Chrysler A-412 (Chilton Type 18) transaxle

use is 1.15mm (0.65 + 0.03 + 0.20 = 1.15). Remove the pinion shaft ball bearing retainer and the pinion shaft. Remove the small bearing cup and the 0.65mm shim and install the correct shim.

3. If new bearings are installed on the pinion shaft, lubricate them with transmission oil, install the shaft and check the shaft turning torque with a torque wrench; it should be 4.4–13.1 inch lbs., if not, reset the preload.

4. Install the pinion shaft and place the 1st gear thrust washer over the shaft, with the flat side up facing the gear. Install the pinion shaft retainer and torque the bolts to 29 ft. lbs.

5. Install the needle bearing, the 1st gear and the 1st gear synchronizer stop ring over the shaft.

NOTE: The wear limit for spacing between the synchronzier teeth on the 1st gear and those on the stop ring is 0.019 in. There is one tooth missing from the 1st gear stop ring on early models. The 1st gear will grind, if this ring isn't used. Later models have three teeth missing in three places, 120° apart.

6. Align the marks on the 1st–2nd synchronizer hub and the sleeve, made on disassembly. Install the synchronizer, driving it into place.

7. Drive the 2nd gear needle bearing inner race into place over the shaft.

8. Drive the Reverse idler gear shaft into place.

NOTE: Make sure that the threaded hole in the top of the shaft is centered, pointing out between the two nearest case edge bolt holes.

9. Place the 2nd gear needle bearing over the pinion shaft, then the 2nd gear stop ring, the 2nd gear and the 3rd gear onto the shaft; make sure that the 3rd gear has the thrust face down.

10. Install the 3rd gear sanp ring. Measure the end play between 3rd gear and the snap ring with a feeler gauge; it should be 0–0.004 in. The snap rings are available in thicknesses from 0.098–0.118 in. for adjustment. Replace the snap ring with the one selected.

11. Install the mainshaft assembly.

12. Install the shift fork assemblies and the E-clips.

13. Install the 4th gear needle bearing over the mainshaft and place the 4th gear synchronizer stop ring in place. Install the 4th gear and the snap ring.

14. Install the Reverse shift fork and

the support brackets, then torque the bolts to 105 inch lbs.

15. Using a feeler gauge, measure the clearance between the top of the pinion shaft 2nd gear and the bottom of the mainshaft 3rd gear.

NOTE: The ideal clearance is adjusted by forcing the mainshaft up or down in relation to the clutch case. The factory has a special tool to do this from the clutch end.

16. The next step is to determine the thickness of the shim or shims to be placed between the mainshaft roller bearing and the transmission case. The factory does this by inserting a special tool of the same thickness as the bearing in the case, installing the case, then measuring the up and down movement of the special tool with a dial indicator. Shims are available in 0.012 and 0.024 in. sizes.

17. After the selected shim is installed behind the bearing, torque the bearing retainer clamp bolts to 13 ft. lbs. Install the transmission case-to-clutch housing bolts, using the guide pin for alignment, and torque the nuts and bolts to 20 ft. lbs.

CHILTON TYPE 19

Transaxle Case

DISASSEMBLY

The Chrysler designed and built A–465 (1983–84) and the A–525 (1984 and later) fully synchronized 5 speed manual transaxles combine gear reduction, ratio selection and differential functions in one unit housed in a die cast aluminum case. The A–525 has a close ratio gearset with different 2nd, 3rd and 4th gear ratios than the A–465, to provide better performance through the gears, while the 1st and the 5th gear ratios are the same as the A–465 to maintain the same launch and top gear characteristics.

1. With the transaxle removed from the vehicle, remove the differential cover bolts and the stud nuts, then remove the cover.

2. Remove the differential bearing retainer bolts.

3. Using tool No. L-4435, rotate the differential bearing retainer to remove it.

4. Remove the extension housing

bolts, then the differential assembly and extension housing.

5. Remove the selector shaft housing bolts, then the selector shaft housing.

6. Remove the stud nuts and the bolts from the rear end cover, then using a small pry bar, pry off the rear end cover.

7. Using snap ring pliers, remove the large snap ring from the intermediate shaft rear ball bearing.

8. Remove the bearing retainer plate by tapping it with a plastic hammer.

9. Remove the 3rd–4th shift fork rail.

10. Remove the Reverse idler gear shaft and gear.

11. Remove the input shaft gear assembly and the intermediate shaft gear assembly.

12. To remove the clutch release bearing, remove the E-clips from the clutch release shaft, then disassemble the clutch shaft components.

13. Remove the three input shaft seal retainer bolts, the seal, the retainer assembly and the select shim.

14. Using tools No. C-4171, C-4656 and an arbor press, press the input shaft front bearing cup from the transaxle case.

15. Using tool No. C-4660, remove the two bearing retainer strap bolts, then the intermediate shaft front bearing.

16. Remove the 5th speed shifter pin, the 5th speed detent ball and the spring.

17. Remove the 5th speed synchronizer strut retainer plate snap ring and the 5th speed synchronizer strut retainer plate.

18. Remove the 5th speed synchronizer assembly and shift fork with shift rail.

19. Remove the intermediate shaft 5th speed gear, the input shaft 5th speed gear snap ring and the 5th speed gear.

20. Remove the bearing support plate bolts and pry off the bearing support plate.

ASSEMBLY

The assembly of the transaxle is the Reverse of disassembly; however, please note the following:

1. Using tools No. C-4657, C4171 and an arbor press, press the front bearing onto the intermediate shaft; the input shaft front bearing cup is installed with the same tools used for removal.

2. Determine the shim thickness for the correct bearing end play only if any of the following parts are replaced:

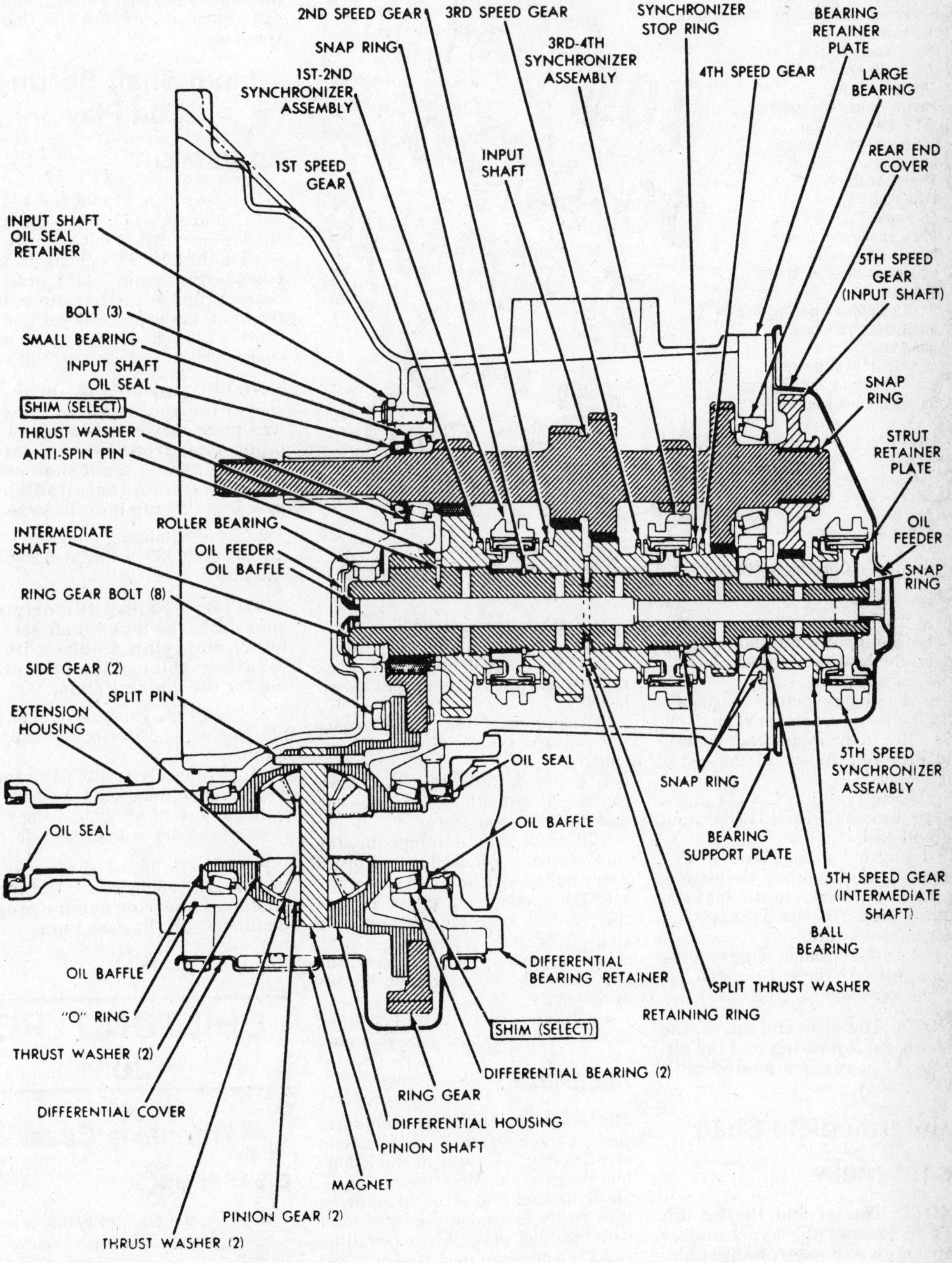

2ND SPEED GEAR

3RD SPEED GEAR

SNAP RING

1ST-2ND SYNCHRONIZER ASSEMBLY

3RD-4TH SYNCHRONIZER ASSEMBLY

SYNCHRONIZER STOP RING

BEARING RETAINER PLATE

1ST SPEED GEAR

INPUT SHAFT

4TH SPEED GEAR

LARGE BEARING

REAR END COVER

INPUT SHAFT OIL SEAL RETAINER

BOLT (3)

SMALL BEARING

INPUT SHAFT OIL SEAL

SHIM (SELECT)

THRUST WASHER

ANTI-SPIN PIN

5TH SPEED GEAR (INPUT SHAFT)

SNAP RING

STRUT RETAINER PLATE

OIL FEEDER

SNAP RING

INTERMEDIATE SHAFT

ROLLER BEARING

OIL FEEDER

OIL BAFFLE

RING GEAR BOLT (8)

SIDE GEAR (2)

SPLIT PIN

EXTENSION HOUSING

OIL SEAL

OIL SEAL

OIL BAFFLE

DIFFERENTIAL BEARING RETAINER

5TH SPEED SYNCHRONIZER ASSEMBLY

SNAP RING

BEARING SUPPORT PLATE

5TH SPEED GEAR (INTERMEDIATE SHAFT)

BALL BEARING

SPLIT THRUST WASHER

RETAINING RING

OIL BAFFLE

"O" RING

THRUST WASHER (2)

DIFFERENTIAL COVER

SHIM (SELECT)

DIFFERENTIAL BEARING (2)

RING GEAR

DIFFERENTIAL HOUSING

PINION SHAFT

MAGNET

PINION GEAR (2)

THRUST WASHER (2)

Cross-sectional view of the A-465 and the A-525 transaxle (Chilton Type 19)

1. 5th gear
2. 5th speed stop ring
3. 5th speed synchronizer
4. 5th speed synchronizer retainer plate
5. 3rd gear
6. 4th speed stop ring
7. 3rd & 4th synchronizer
8. 4th speed gear
9. Interm. shaft rear bearing
10. 1st & 2nd stop ring
11. 1st & 2nd synchronizer
12. 2nd gear
13. Thrust washer
14. Retaining ring
15. Oil feeder
16. Inter. shaft roller bearing
17. Intermediate shaft
18. Low gear thrust washer
19. 1st gear
20. Input shaft front bearing cup
21. Input shaft front bearing
22. Input shaft
23. Input shaft rear bearing
24. Input shaft rear bearing cup
25. 5th gear input shaft

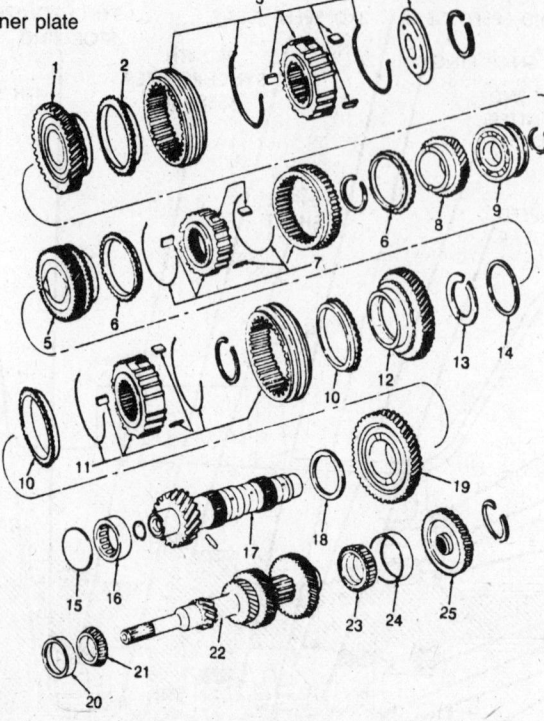

Chrysler A-465 and A-525 gear train (Chilton Type 19)

a. The transaxle case.
b. The input shaft seal retainer.
c. The bearing retainer plate.
d. The rear end cover.
e. The input shaft or bearings.

3. To determine proper shim thickness, refer to the Input Shaft Bearing End Play Adjustment at the end of this section.

4. Using tool No. C-4674 and a plastic hammer, install the input shaft oil seal.

5. Using a $\frac{1}{16}$ inch bead of RTV sealant, place it around the edge of the input shaft seal retainer and making sure the drain hole of the retainer is facing downward.

6. The differential bearing retainer is installed with the same special tool used for removal.

NOTE: The rear end cover, the selector shaft housing and the differential cover are sealed with RTV sealant.

Intermediate Shaft

DISASSEMBLY

NOTE: The 1st–2nd, the 3rd–4th shift forks and the synchronizer stop rings are interchangeable. However, if parts are to be reused reassemble in the original position.

1. Remove the intermediate shaft rear bearing snap ring.

2. Using the puller tool No. C-4693, remove the intermediate shaft rear bearing.

3. Using snap ring pliers, remove the 3rd–4th synchronizer hub snap ring.

4. Using the puller tool No. L-4534, remove the 3rd–4th synchronizer hub and the 3rd speed gear.

5. Remove the retaining ring, the split thrust washer, the 2nd speed gear and the synchronizer stop ring.

6. Using snap ring pliers, remove the 1st–2nd synchronizer hub snap ring.

7. Remove the 1st speed gear, the stop ring and the 1st–2nd synchronizer assembly.

8. Remove the 1st speed gear thrust washer and the anti-spin pin.

ASSEMBLY

The assembly of the intermediate shaft is the Reverse of the disassembly; however, please note the following: When assembling the intermediate shaft, make sure the speed gears turn freely and have a minimum of 0.003 in. end play. When installing the 1st speed gear thrust washer make sure the chamfered edge is facing the pinion gear. When installing the 1st–2nd synchronizer make sure

the relief faces the 2nd speed gear. Use an arbor press to install the intermediate shaft rear bearing, the 3rd–4th synchronizer hub and the 3rd speed gear.

Input Shaft Bearing End Play

ADJUSTMENT

1. Using special tool No. L-4656 with handle C-4171, press the input shaft front bearing cup slightly forward in the case. Then, using tool No. L-4655 with handle C-4171, press the bearing cup back into the case, from the front, to the properly position the bearing cup before checking the input shaft endplay.

NOTE: This step is not necessary if the special tool No. L-4655 was previously used to install the input shaft front bearing cup in the case and no input shaft select shim has been installed since pressing the cup into the case.

2. Select a gauging shim which will give 0.001–0.020 in. (0.025–0.254mm) end play.

NOTE: Measure the original shim from the input shaft seal retainer and select a shim 0.010 in. (0.254mm) thinner than the original for the gauging shim.

3. Install the gauging shim on the bearing cup and the input shaft seal retainer.

4. Alternately tighten the input shaft seal retainer bolts until the retainer is bottomed against the case, then torque the bolts to 21 ft. lbs.

NOTE: The input shaft seal retainer is used to draw the input shaft front bearing cup the proper distance into the case bore.

CHILTON TYPE 20

Transaxle Case

DISASSEMBLY

1. Place the transaxle onto a work stand, with the shaft assemblies facing up.

2. Remove the 15 clutch housing-to-case bolts. The cover is assembled with RTV sealer, if removal is diffi-

cult, rap the cover with a soft hammer.

3. Lift out the ring gear/differential assembly and move them aside; this procedure does not cover differential overhaul.

4. Shift into the Neutral position. Bend back the lock tab, then remove the bolt, the shifter shaft and the shift fork shaft from the synchronizer forks.

5. Disengage, then remove the Reverse shift fork from the guide pin and interlock bracket. Unscrew the lock bolt and remove the Reverse idler gear shaft, the gear and the spacer assembly.

6. Remove the detent shift lever and the interlock assembly, leaving the shift forks engaged with the synchronizers.

7. Lift the input and the output shafts from the case as an assembly.

NOTE: When removing the shafts, mark the location and the position of the shift forks, then remove them from the shafts.

ASSEMBLY

1. Place the input and the output shafts together on the workbench. Install the shift forks onto the shafts, then carefully lower them into the case as an assembly.

2. Place the interlock bracket onto a dummy shaft (make sure the bracket engages the shift fork fingers), then place the detent shift lever into the interlock.

3. Install the shifter shaft through the interlock bracket and the detent shift lever (do not push through any farther). Install the Reverse shift fork onto the dummy shaft and engage the fork with the interlock bracket.

4. Install the Reverse idler gear, the shaft and the install the spacer.

NOTE: When installing the Reverse idler shaft, make sure the long end of the shaft points upward; the large chamfered ends of the idler gear teeth should also be facing up. The flat on the Reverse idler shaft should be facing the input shaft.

5. Push the shifter shaft through the Reverse shift fork until it fits into the inhibitor spring spacer, then remove the dummy shaft. Shift into the Neutral position, then install the shifter shaft bolt and the lock through the detent shift lever. Bend the lock tab over the bolt head.

6. Install the fork shaft through the synchronizer forks and into the case bore.

7. Carefully install the ring gear

and differential case assembly.

8. Install the magnet into the case. Apply a thin bead of RTV silicone sealer to the clutch cover and install the cover. Tap the cover gently with a soft hammer to seat it. Install the 15 attaching bolts and torque (in two sequence steps) to 16 ft. lbs.

9. Torque the idler shaft retaining bolt to 7 ft. lbs., then shift through the gears to check operation.

Input Shaft

DISASSEMBLY

1. Install the support plates under the 4th gear, then press the gear and the left hand bearing from the shaft.

2. Remove the brass blocking ring and the 3rd–4th synchronizer snap ring.

3. Install the support plates behind the 3rd gear, then press the 3rd gear and the synchronizer from the shaft. Press the right hand bearing from the shaft.

ASSEMBLY

1. Using a long piece of pipe or GM tool No. J-28406, press the right hand bearing onto the shaft.

2. Place the 3rd gear onto the shaft; it should have its synchronizer portion facing up towards the 3rd–4th synchronizer. Install the brass blocking ring onto the gear, then press the 3rd–4th gear synchronizer into place.

3. Using a piece of pipe, which will contact the synchronizer hub near the shaft, install the snap ring with the beveled edges away from the synchronizer.

——— **CAUTION** ———
When installing the snap ring, do not press on the outside of the hub.

4. Install the brass blocking ring. Press the 4th gear onto the shaft with its synchronizer portion facing the synchronizer, then press the left hand bearing into place.

Output Shaft

DISASSEMBLY

1. Install the support plates behind the 4th gear. Use a rod or a pilot which will fit the through the left hand bearing to press off the bearing and the 4th gear.

2. Remove the 3rd gear snap ring, then slide the 1st–2nd synchronizer into 1st position. Support the 2nd with the plates, then press the 2nd and the 3rd gear off the output shaft.

3. Remove the brass blocking ring and the 1st–2nd synchronizer snap ring.

4. Use the press plates to support the 1st gear, then press the gear and the synchronizer from the shaft. Press the right hand bearing from the shaft.

5. Pry out the synchronizer springs, being careful not to distort them. Scribe a mark across the hub and the sleeve, then separate the hub, the sleeve and the three keys, mark their locations.

6. Replace the parts as necessary. Assemble the hub and the sleeve according to the scribed marks. The extruded lip on the hub faces away from the shift fork groove in the sleeve.

7. Install one retaining spring, then carefully pull it away from the key positions (one at a time) and install the keys. The spring must be caught on the keys. Install the other spring on the other side in the same way, but be sure the open segment is in a different position (staggered) relative to the opening in the first spring installed.

ASSEMBLY

1. Press the right hand bearing into place. Install the 1st gear and its brass blocking ring onto the shaft, then (using a long pipe) press the 1st–2nd synchronizer into position.

NOTE: When pressing the hub onto the shaft; do not press on the outer edges of the hub or the sleeve.

2. Install the snap ring and the brass blocking ring over the synchronizer.

3. Place the 2nd gear onto the shaft. Press the 3rd gear into place (with its hub away from the 2nd gear); press on the gear close to the shaft–do not press on its outer edges. Install the 3rd gear snap ring.

4. Press the 4th gear into place, with its hub facing the 3rd gear. Press the left hand bearing into place.

Case

OVERHAUL

1. Remove the Reverse inhibitor fitting from the outside of the case, then the spring, the pilot and the spacer from the inside.

2. Using a bearing puller, remove the input and the output shaft bearing cups, then slip out the oil slingers.

3. Check the interlock bracket, the Reverse shift fork guide pins and the case magnet for wear or damage. Clean the sealant from the case.

1. Case assembly
2. Axle shaft seal assembly
3. Case locating pin
4. Chip collecting magnet
5. Vent assembly
6. Synchronizer key
7. Oil shield
8. Bearing assembly
9. 4th speed input gear
10. Blocking ring
11. Synchronizer spring
12. Synchronizer assembly
13. 3rd speed input gear
14. Oil shield sleeve
15. Input cluster gear
16. Input gear bearing
17. Input gear seal
18. Input gear retainer assembly
19. Retainer seal
20. Throwout bearing assembly
21. Reverse idler shaft
22. Reverse idler shaft
23. Reverse idler shaft gear
24. Reverse inhibitor spring seat
25. Reverse inhibitor spring
26. Reverse inhibitor spring pin
27. Reverse shift lever
28. Detent lever assembly
29. Detent spring
30. Shift shaft
31. Shift shaft seal assembly
32. Shift interlock
33. 3rd-4th shift fork
34. 1st-2nd shift fork
35. Shift fork shaft
36. Oil guide
37. Clutch fork shaft seal assembly
38. Clutch fork shaft bearing
39. Clutch fork shaft assembly
40. Clutch and differential housing assembly
41. Speedometer driven gear sleeve
42. Speedometer driven gear sleeve seal
43. Speedometer driven gear
44. Case bearing oil shield
45. Case bearing assembly
46. 4th speed gear
47. 3rd speed output gear
48. 2nd speed output gear

49. Synchronizer blocking ring
50. Synchronizer spring
51. Synchronizer key
52. 1st-2nd synchronizer assembly
53. 1st speed output gear
54. Oil shield sleeve
55. Output gear
56. Output bearing assembly

57. Output gear bearing shim
58. Output gear bearing oil shield
59. Output gear bearing oil shield retainer
60. Differential assembly
61. Differential ring gear
62. Differential bearing assembly
63. Differential case

64. Differential pinion shaft
65. Speedometer drive gear
66. Differential bearing assembly
67. Housing bearing shim
68. Side gear thrust washer
69. Differential side gear
70. Pinion thrust washer
71. Differential pinion gear

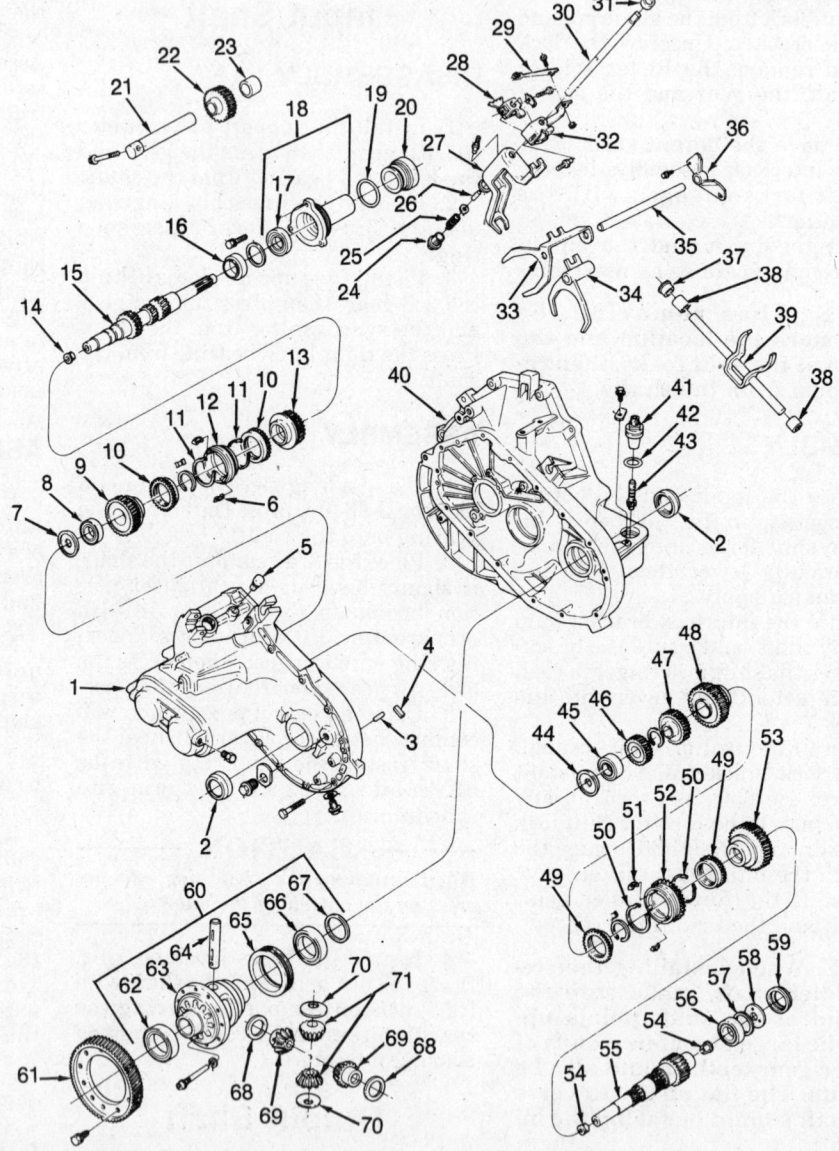

Exploded view of the GM125-4 4-speed (Chilton Type 20) transaxle

NOTE: The preload shims must be selected before final assembly.

4. Install the three left hand bearing cups, then the input shaft, the output shaft and the differential assemblies into position in the case. Install the three right hand bearing cups onto their bearings.

5. Place the GM gauge tools No. J-26935-2 on the input bearing, J-26935-4 on the output bearing and J-26935-3 on the differential bearing.

NOTE: When installing the gauge tools, make sure they fit smoothly and completely over the bearings.

6. Install the metal oil shield retainer over the tool No. J-26935-4 on the output shaft.

7. Install the spacers supplied with the spacer kit around the perimeter of the transaxle case. Carefully install the clutch cover over the gauges and spacers. Install the long bolts provided, then torque evenly and in rotation to 10 ft. lbs.

8. Rotate each gauge to seat the

bearings. Rotate the differential case through three revolutions in each direction.

NOTE: The gap between the outer sleeve and the base pad is the correct thickness for the preload shim at each location. The largest shim which can be placed in the gap and drawn through without binding is the correct one for the assembly.

9. Remove the clutch cover, the spacers and the gauges. Place the selected shims in their respective bores in the clutch cover and add the metal shield, then install the bearing cups.

CHILTON TYPE 21

Transmission Case

DISASSEMBLY

1. Drain the transmission lubricant.

NOTE: The 2WD models do not have a drain plug; the fluid must be siphoned from these transmissions.

2. Using a pin punch and a hammer, remove the offset lever-to-shift rail roll pin.
3. Remove the extension (2WD) or the adapter (4WD) housing-to-case bolts. Remove the housing and the offset lever as an assembly.

——— CAUTION ———
Do not attempt to remove the offset lever while the housing is still in place.

4. Remove the detent ball and the spring from the offset lever.
5. Remove the roll pin from the extension/adapter housing or the offset lever.
6. Remove the countershaft rear thrust bearing and race.
7. Remove the transmission cover-to-case bolts, then lift the cover and the shift fork assembly from the transmission.

NOTE: Two of the transmission cover bolts are alignment type dowel bolts. Mark their location so that they may be reinstalled in their original locations.

8. Remove the Reverse lever-to-reverse lever pivot bolt C-clip.
9. Remove the Reverse lever pivot

bolt, then the Reverse lever and the fork as an assembly.
10. Mark the position of the front bearing cap-to-case, then remove the bearing cap bolts and the cap.
11. Remove the front bearing race and the shims from the bearing cap. Using a small pry bar, pry the oil seal out of the bearing cap.
12. Rotate the main drive gear shaft until the flat portion of the gear faces the countershaft, then remove the main drive gear shaft assembly.
13. Remove the thrust bearing and the 15 roller bearings from the clutch shaft.
14. Remove the output shaft bearing race.

NOTE: If the race is stubborn, tap the front of the output shaft with a plastic mallet to remove the race.

15. Tilt the output shaft assembly upward and remove the assembly from the case.
16. Using a brass drift and an arbor press, remove the countershaft rear bearing.

NOTE: Mark the position of the bearing so that it may be reinstalled properly.

17. Move the countershaft rearward, then tilt it upward and remove it from the case.
18. Remove the countershaft rear bearing spacer.
19. Remove the Reverse idler shaft roll pin, then the Reverse idler shaft and gear.

NOTE: Mark the position of the gear so that it may be reinstalled properly.

20. Using an arbor press, remove the countershaft front bearing.
21. Using the Kent-Moore tools No. J-29721 and J-22912, remove the bearing from the main drive gear shaft.
22. Remove the extension/adapter housing seal using the appropriate tools.
23. Remove the back up light switch from the transmission case.

ASSEMBLY

NOTE: If a replacement fastener must be used, be sure that it matches the original exactly. Many metric fasteners are used in this transmission.

1. Apply a coat of Loctite® 601 (or equivalent) to the outer cage of the front countershaft bearing, then press the bearing into its bore until it is flush with the case.

2. Apply a coat of petroleum jelly to the tabbed countershaft thrust washer, then install the washer so that its tab engages the corresponding depression in the case.
3. Tip the case on end and install the countershaft into the front bearing bore.
4. Install the countershaft rear bearing spacer and coat the rear countershaft bearing with petroleum jelly. Using the Kent-Moore installer tool No. J-29895 and the sleeve protector tool No. J-33032, install the rear countershaft bearing.

NOTE: When properly installed, the rear bearing will extend 0.125 in. beyond the case surface.

5. Position the Reverse idler gear into the case (the shift lever groove must face rearward) and install the Reverse idler shaft from the rear of the case.
6. Install the shaft retaining pin.
7. Install the output shaft assembly into the transmission case.
8. Using the Kent-Moore tool No. J-2995 and an arbor press, install the main drive gear bearing (if removed) onto the main drive gear shaft.
9. Coat the 15 main drive gear roller bearings with petroleum jelly and install them into the recess of the main drive gear.
10. Install the thrust bearing and the race into the recess of the main drive gear.
11. Install the 4th gear blocking ring onto the output shaft.
12. Install the rear output shaft bearing race.
13. Install the main drive gear assembly into the case, engaging the 3rd–4th synchronizer blocking ring.
14. Evenly and carefully tap a new front bearing cap seal into place.
15. Install a new oil seal into the adapter housing (4WD) in the same manner as in Step 14.
16. Install the front bearing race into the front bearing cap; do not install the front bearing cap shims.
17. Temporarily install the front bearing cap without sealant.
18. Install the Reverse lever, the pivot pin (coat the threads with nonhardening sealant) and the retaining C-clip.

NOTE: Be sure the Reverse lever fork is engaged with the Reverse idler gear.

19. Coat the countershaft rear bearing race and the thrust bearing with petroleum jelly, then install these parts into the extension/adapter housing.
20. Temporarily install the extension/adapter housing without sealant,

then tighten but do not final torque the bolts.

21. Turn the case on end and mount a dial indicator onto the extension/adapter housing, so that the indicator needle contacts the end of the output shaft.

22. Rotate the main drive gear and the output shafts, then zero the dial indicator.

23. To remove the end play, pull up on the output shaft, then read the indicator and record the reading.

NOTE: To eliminate the total end play, the bearings must be preloaded from 0.001–0.005 in.

24. Select a shim pack which measures 0.001–0.005 in. thicker than the end play reading obtained during Step 23.

25. Move the transmission to a horizontal position, then remove the front bearing cap and the race. Install the shim pack and the bearing race.

26. Apply an ⅛ in. bead of RTV sealant on the case-to-front bearing cap. Align the case and the cap matchmarks, then install the bearing cap and torque the bolts to 15 ft. lbs.

27. Recheck the end play; no end play should exist.

28. Remove the extension/adapter housing.

29. Move the shift forks and the synchronizer sleeves to their Neutral positions.

30. Apply an ⅛ in. bead of RTV sealer to the cover-to-case mating surface. While aligning the shift forks with the synchronizer sleeves, carefully

lower the cover assembly into place on the transmission.

31. Center the cover in order to engage the Reverse relay lever and install the alignment type (dowel) cover attaching bolts. Install the remaining cover bolts and torque to 9 ft. lbs.

NOTE: The offset lever-to-shift rail roll pin hole must be positioned vertically; if not, repeat Steps 29–31.

32. Apply a ⅛ in. bead of RTV sealer to the extension/adapter housing-to-case mating surface and install the housing over the output shaft.

NOTE: The shift rail must be positioned so that it just enters the shift cover opening.

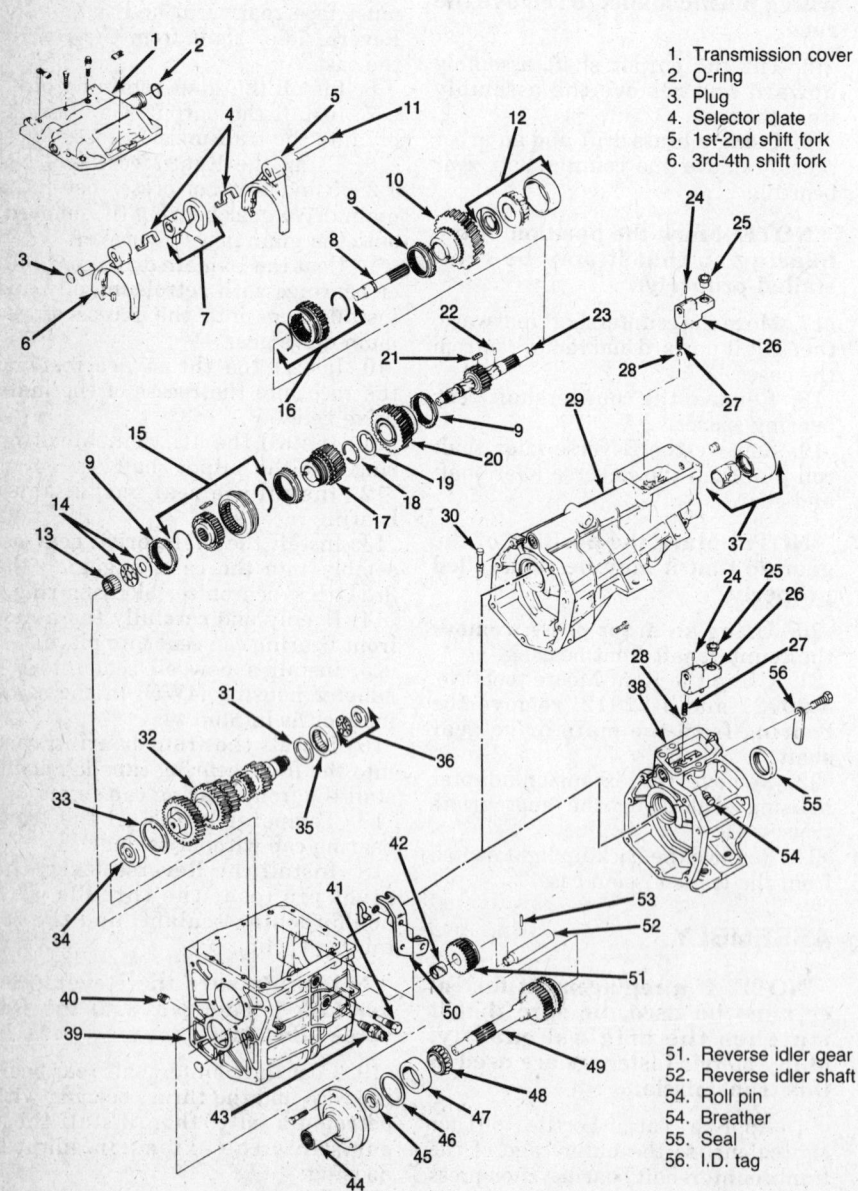

1. Transmission cover
2. O-ring
3. Plug
4. Selector plate
5. 1st-2nd shift fork
6. 3rd-4th shift fork
7. Selector arm interlock plate and pin
8. Mainshaft
9. Blocking ring
10. 1st speed gear
11. Shift rail
12. Thrust washer, rear bearing and cup
13. Clutch shaft needle bearings
14. Needle thrust bearing and race
15. 3rd-4th synchronizer assembly
16. Reverse sliding gear and insert springs
17. 3rd speed gear
18. Snap-ring
19. Thrust washer
20. 2nd speed gear
21. Synchronizer insert
22. Roll pin
23. Mainshaft
24. Roll pin
25. Damper sleeve
26. Offset lever
27. Detent spring
28. Detent ball
29. Extension housing (2WD)
30. Breather
31. Rear countershaft spacer
32. Countershaft gear unit
33. Front countershaft thrust washer
34. Front countershaft bearing
35. Rear countershaft bearing
36. Needle thrust bearing and race
37. Bushing and oil seal
38. Adapter housing (4WD)
39. Transmission case
40. Plug
41. Reverse lever pivot bolt
42. Reverse idler bushing
43. Back-up lamp switch
44. Front bearing cap
45. Oil seal
46. Shim
47. Front bearing cup
48. Front bearing (main)
49. Main drive gear and shaft
50. Reverse lever
51. Reverse idler gear
52. Reverse idler shaft
54. Roll pin
54. Breather
55. Seal
56. I.D. tag

Exploded view of the AMC T4 (Chilton Type 21) transmission

33. Install the detent spring into the offset lever and place the steel ball into the Neutral guide plate detent. Apply pressure to the steel ball with the detent spring and the offset lever, then slide the offset lever on the shift rail and seat the extension/adapter housing against the case. Torque the housing bolts to 25 ft. lbs.

34. Install the roll pin into the offset lever and shift rail.

35. Install the damper sleeve in the offset lever. Coat the back up lamp switch threads with RTV sealer and install the switch into the case. Torque the switch to 15 ft. lbs.

Output Shaft Geartrain

DISASSEMBLY

1. Remove the thrust bearing washer from the front of the output shaft.

2. Scribe matchmarks on the hub and the sleeve of the 3rd–4th synchronizer so that these parts may be reassembled properly.

3. Remove the 3rd–4th synchronizer blocking ring, the sleeve and the hub as an assembly.

4. Remove the insert springs and the inserts from the 3rd–4th synchronizer, then separate the sleeve from the hub.

5. Remove the 3rd speed gear from the shaft.

6. Remove the 2nd speed gear-to-output shaft snap ring, the tabbed thrust washer and the 2nd speed gear.

7. Using the Kent-Moore puller set J-29721 and adapters 293-39, remove the output shaft bearing.

8. Remove the 1st gear thrust washer, the roll pin, the 1st speed gear and the blocking ring.

NOTE: Diagonal cutters may be used to carefully remove the roll pin.

9. Scribe matchmarks on the 1st–2nd synchronizer sleeve and the output shaft.

10. Remove the insert spring and the inserts from the 1st–Reverse sliding gear, then the gear from the output hub.

------- **CAUTION** -------

DO NOT attempt to remove the 1st–2nd–Reverse hub from the output shaft. The shaft and the hub are machined and assembled as a matched set during manufacture.

ASSEMBLY

1. Coat the output shaft and the gear bores with transmission lubricant.

2. Using the matchmarks made during disassembly, align and install the 1st–2nd synchronizer sleeve on the output shaft hub.

3. Install the three inserts and two springs into the 1st–Reverse synchronizer sleeve.

NOTE: The tanged end of each spring should be positioned on the same insert but that the open face of each spring should be opposite each other.

4. Install the blocking ring and the 2nd speed gear onto the output shaft.

5. Install the tabbed thrust washer and the 2nd gear snap ring on the output shaft; be sure that the washer tab is properly seated in the output shaft notch.

6. Install the blocking ring and the 1st speed gear onto the output shaft, then install the 1st gear roll pin.

7. Using an arbor press and the Kent-Moore tool No. J-2995, install the rear bearing onto the output shaft.

8. Install the remaining output shaft components in the following order:

    a. The 1st gear thrust washer.

    b. The 3rd speed gear.

    c. The 3rd–4th synchronizer hub inserts and the sleeve (the hub offset must face forward).

    d. The thrust bearing washer (on the forward end of the output shaft).

Cover and Shift Forks

DISASSEMBLY

1. Place the selector arm plates and the shift rail in the Neutral position (centered).

2. Rotate the shift rail counterclockwise until the selector arm disengages from the selector arm plates; the selector arm roll pin should now be accessible.

3. Pull the shift rail rearward until the selector contacts the 1st–2nd shift fork.

4. Using a $3/16$ in. pin punch, remove the selector arm roll pin and the shift rail.

5. Remove the shift forks, the selector arm, the roll pin and the interlock plate.

6. Using a small pry bar, remove the shift rail oil seal and the O-ring.

7. Remove the nylon inserts and the selector arm plates from the shift forks.

NOTE: Mark the position of the parts so that they may be reinstalled properly.

ASSEMBLY

1. Attach the nylon inserts to the selector arm plates and through the shift forks.

2. If previously removed, coat the edges of the shift rail plug with sealant and install the plug.

3. Coat the shift rail and the rail bores with petroleum jelly, then slide the shift rail into the cover until the end of the rail is flush with the inside edge of the cover.

4. Position the 1st–2nd shift fork into the cover, with the offset of the shift fork facing the rear of the cover then push the shift rail through the fork.

NOTE: The 1st–2nd fork is the larger of the two forks.

5. Position the selector arm and the C-shaped interlock plate into the cover, then push the shift rail through the arm.

NOTE: The widest portion of the interlock plate must face away from the cover and the selector arm roll pin must face downward, towards the rear of the cover.

6. Position the 3rd–4th shift fork into the cover, with the fork offset facing the rear of the cover.

NOTE: The 3rd–4th shift fork selector arm plate must be positioned under the 1st–2nd shift fork selector arm plate.

7. Push the shift rail through the 3rd–4th shift fork and into the front cover rail bore.

8. Rotate the shift rail until the forward selector arm plate faces away from parallel to the cover.

9. Align the roll pin holes of the selector arm and the shift rail, then install the roll pin.

NOTE: The roll pin must be installed flush with the surface of the selector arm to prevent selector arm plate-to-pin interference.

10. Install the O-ring into the groove of the shift rail oil seal, then install the oil seal assembly as follows:

    a. Using the Kent-Moore tool No. J-26628-2, install an oil seal protector over the threaded end of the shift rail.

    b. Lubricate the oil seal lip with the petroleum jelly, then slide it over the protector and onto the shift rail.

    c. Seat the oil seal into the cover.

CLEANING AND INSPECTION

The parts (except the nylon or plastic)

should be thoroughly cleaned in cleaning solvent; the nylon or plastic parts (which are to be reused) should just be wiped clean with a cloth. The assembled roller bearings should be dried with compressed air.

CAUTION

Do not spin the bearings with the compressed air, as they could shatter and cause personal injury. Individual roller bearings, washers and thrust bearings, should be air dried after cleaning, though they may be wiped with a clean cloth.

Inspect the parts for excessive wear and/or damage such as scoring, cracks, nicks and rough edges; replace the defective parts. To check the condition of the assembled bearings, first clean and dry them, then coat the bearings with light engine oil. Slowly spin the bearings by hand and check for any signs of roughness; if the bearing does not feel perfectly smooth, it should be replaced.

NOTE: AMC recommends that if any gear of the mainshaft must be replaced, the countershaft gear should also be replaced, to avoid noisy operation and maintain proper gear mesh.

While the transmission is out of the vehicle, it is good practice to check the condition of the clutch assembly and the throwout bearing. Also, replace the transmission gaskets and the seals during assembly of the transmission.

CHILTON TYPE 22

Transmission Case

DISASSEMBLY

NOTE: Many special tools and an arbor press are required to properly disassemble and assemble this transmission. Read the entire procedure carefully before starting the job.

1. Remove the transmission from the vehicle as outlined in the appropriate car section, then drain the lubricant from the transmission. On AMC Spirit and Concord models, the fluid must be siphoned from the transmission, as these models are not equipped with drain plugs.
2. Using a pin punch and a hammer, carefully the offset lever-to-shift rail roll pin.

3. Remove the extension or adapter (AMC 4WD) housing-to-case bolts. Remove the housing and offset lever as an assembly.

CAUTION

DO NOT attempt to remove the housing with the offset lever in place.

4. Remove the detent ball and the spring from the offset lever, then the roll pin from the extension housing or offset lever.
5. Remove the plastic funnel, the thrust bearing race and thrust bearing from the rear of the countershaft.

NOTE: The funnel and race may be found inside the extension.

6. Remove the cover-to-case bolts and lift the cover assembly off of the case.

NOTE: Two of the cover bolts are alignment type dowel bolts. Mark the location of these bolts so that they may be reinstalled in their original locations.

7. Place a wooden block under the 5th gear shift fork and drive the roll pin from the fork. The wood must be used to prevent damage to the shift rail.
8. Remove the following items from the rear of the countershaft.
   a. The 5th gear synchronizer snap ring.
   b. The shift fork.
   c. The 5th gear synchronizer sleeve.
   d. The blocking ring.
   e. The 5th speed drive gear.
9. Remove the 5th gear synchronizer springs and the inserts from the sleeve and the hub. Mark the sleeve and the hub so that they may be properly reassembled.
10. Remove the snap ring from the 5th speed driven gear. Using the Kent-Moore puller tool No. J-25215, remove the driven gear.
11. Mark the front bearing cap and the case, so that the cap may be reinstalled in its proper position, then remove the front bearing cap.
12. Remove the front bearing race and the end play shim(s) from the bearing cap. Using a small pry bar, carefully pry the oil seal from the cap.
13. Rotate the clutch shaft until the flat surface on the main drive gear faces the countershaft, then remove the clutch shaft and the main gear unit from the case.

NOTE: The clutch shaft bearing is pressed on; an arbor press must be used to replace the bearing if the bearing is rough.

14. Remove the mainshaft rear bearing race and tilt the shaft upward, then remove the output shaft assembly from the case.
15. On AMC models, unhook the overcenter spring from the rear of the case, then use a piece of welding rod (bent into a hook) to grab and pull the spring. On the Chevette, unhook the overcenter spring from the front of the case.
16. Remove the Reverse lever C-clip (all models) and the pivot bolt (Chevette).
17. Rotate the 5th–Reverse shift rail clockwise, to disengage it from the Reverse lever, then remove the rail from the rear of the case.
18. On AMC models, remove the Reverse lever pivot pin and the lever from the Reverse idler gear. On all models, remove the Reverse lever and the fork assembly from the case.
19. On the Chevette models, drive the roll pin from the forward end of the Reverse idler shaft, then remove the Reverse idler shaft, the rubber O-ring and the gear from the case.
20. Remove the gear countershaft snap ring and the spacer.
21. Insert a brass drift through the main drive gear opening in the front of the case, so that it contacts the countershaft gear assembly. Using an arbor press (positioned at the other end of the drift), carefully press the countershaft gear rearward (just enough) to remove the countershaft rear bearing.

NOTE: During the assembly, note that the bearing identification numbers should face outward.

22. Move the countershaft assembly rearward, tilt it upward, then remove the assembly from the case. Mark the position of the front countershaft thrust washer (so that it may be reinstalled properly), then remove the washer from the case.
23. Remove the countershaft rear bearing spacer.
24. Drive the roll pin from the front of the Reverse idler shaft, then remove the shaft and the gear from the case.

NOTE: Mark the position of the gear, so that it may be reinstalled properly.

25. Using an arbor press, remove the countershaft front bearing from the case.
26. Remove the clutch shaft front bearing.
27. Using a flat drift and a hammer, carefully tap out the rear extension and the adapter housing seal.

ASSEMBLY

NOTE: If a replacement fastener is used, be sure that it matches the original EXACTLY. Many metric fasteners are used in this transmission.

1. Apply a coat of Loctite® 601 to the outer cage of the front countershaft bearing, then press the bearing into its bore until it is flush with the case.

2. Apply a coat of petroleum jelly to the tabbed countershaft thrust washer, then install the washer so that its tab engages the corresponding depression in the case.

3. Tip the case on end and install the countershaft into the front bearing bore.

4. Install the countershaft rear bearing spacer and coat the rear countershaft bearing with petroleum jelly. Using the Kent-Moore installer tool No. J-29895 and the sleeve protector tool No. J-33032, install the rear countershaft bearing.

NOTE: When properly installed, the rear bearing will extend 0.125 in. beyond the case surface.

5. Position the Reverse idler gear into the case (with the shift lever groove facing rearward) and install the Reverse idler shaft from the rear of the case.

6. Install the shaft retaining pin.

7. Install the mainshaft assembly and the rear mainshaft bearing race into the case.

8. Using an arbor press, install the clutch shaft/main gear bearing (if removed).

9. Coat the main drive gear roller bearings with petroleum jelly and install them into the rear of the clutch shaft/main gear unit.

10. Install the thrust bearing and the race into the rear of the clutch shaft/main gear unit.

11. Install the 4th gear blocking ring onto the mainshaft.

12. Install the clutch shaft/main gear unit into the case, engaging the 3rd–4th synchronizer blocking ring.

13. Evenly and carefully tap a new front bearing cap seal into place.

14. Install the front bearing race into the front bearing cap; do not install the front bearing cap shims.

15. Temporarily install the front bearing cap, without sealant.

16. Install the following:
    a. The 5th–Reverse lever.
    b. The pivot bolt.
    c. The C-clip retainer.

NOTE: Coat the pivot bolt threads with nonhardening sealant (RTV is preferred). Also, be sure to engage the Reverse lever fork in the Reverse idler gear.

17. On AMC models, install the 5th speed driven gear onto the rear of the mainshaft assembly, then the install the snap ring.

18. Install the countershaft rear bearing spacer and the snap ring.

19. Install the 5th speed gear onto the mainshaft.

20. Install the 5th–Reverse rail through the rear case opening and into the 5th–Reverse lever, then rotate the rail to engage it with the lever.

21. On AMC models only:
    a. Install the overcenter spring.
    b. Assemble the 5th gear synchronizer sleeve, the insert springs and the insert retainer; use the matchmarks made during disassembly to align.
    c. Install the plastic inserts on the 5th speed shift fork.

22. Position the 5th speed synchronizer assembly on the 5th speed shift fork, then slide the assembly onto the countershaft and the 5th–Reverse rail.

NOTE: The 5th–Reverse rail roll pin hole must be aligned with the hole of the 5th speed shift fork.

23. Support the 5th speed shift fork rail and the fork with a block of wood, then drive the roll pin into place.

24. Install the:
    a. Thrust race against the 5th speed synchronizer hub, then retain with the snap ring.
    b. Needle type thrust bearing against the thrust race on the countershaft (coat the bearing and the race with petroleum jelly).
    c. Lipped thrust race over the needle type thrust bearing.
    d. Plastic funnel into the hole in the end of the countershaft gear.

25. Temporarily install the extension/adapter housing and the bolts.

26. Turn the case on end. Mount a dial indicator on the extension/adapter housing, so that the indicator needle contacts the end of the mainshaft, then zero the indicator needle.

27. Pull upward on the mainshaft to remove the end play, then read the indicator and record the reading.

28. Select a shim pack which measures 0.001–0.005 in. thicker than the end play reading obtained during Step 28.

29. Position the case horizontally. Remove the front bearing cap and the bearing race, then install the shim pack. Reinstall the bearing race.

30. Apply a $\frac{1}{8}$ in. bead of RTV sealer on the front bearing cap-to-case mating surface. Align the case and the cap matchmarks. Install the bearing cap and torque the bolts to 15 ft. lbs.

31. Recheck the end play; no play should be evident.

32. Remove the extension and adapter housing, then carefully drive a new housing seal into place.

33. Move the shift forks of the cover and the synchronizer sleeves to their Neutral positions.

34. Apply a $\frac{1}{8}$ in. bead of RTV sealer to the cover-to-case mating surface. While aligning the shift forks with the synchronizer sleeves, carefully lower the cover assembly into case.

35. Center the cover and to engage the Reverse relay lever, then install the alignment type (dowel) cover attaching bolts. Install the remaining cover bolts and torque to 10 ft. lbs.

NOTE: The offset lever-to-shift rail roll pin hole must be positioned vertically; if not, repeat Steps 34, 35 and 36.

36. Apply a $\frac{1}{8}$ in. bead of RTV sealer to the extension/adapter housing-to-case mating surface and install the housing over the mainshaft.

NOTE: The shift rail must be positioned so that it just enters the shift cover opening.

37. Install the detent spring into the offset lever and place the steel ball into the Neutral guide plate detent. Apply pressure on the steel ball with the detent spring and the offset lever, then slide the offset lever on the shift rail and seat the extension/adapter against the case.

38. Install the extension/adapter housing bolts and torque to 25 ft. lbs.

39. Install the roll pin into the offset lever and the shift rail.

40. Install the damper sleeve in the offset lever. Coat the back up lamp switch threads with RTV sealant, then install the switch into the case and torque to 15 ft. lbs.

Mainshaft

DISASSEMBLY

AMC Models

1. Remove the thrust bearing and the washer from the front of the mainshaft.

2. Scribe matchmarks on the 3rd–4th synchronizer hub and the sleeve to indicate their relationship for proper reassembly.

3. Remove the 3rd–4th synchronizer blocking ring, the sleeve and hub from the mainshaft as an assembly. Mark the positions of these items, so that they may be properly reassembled.

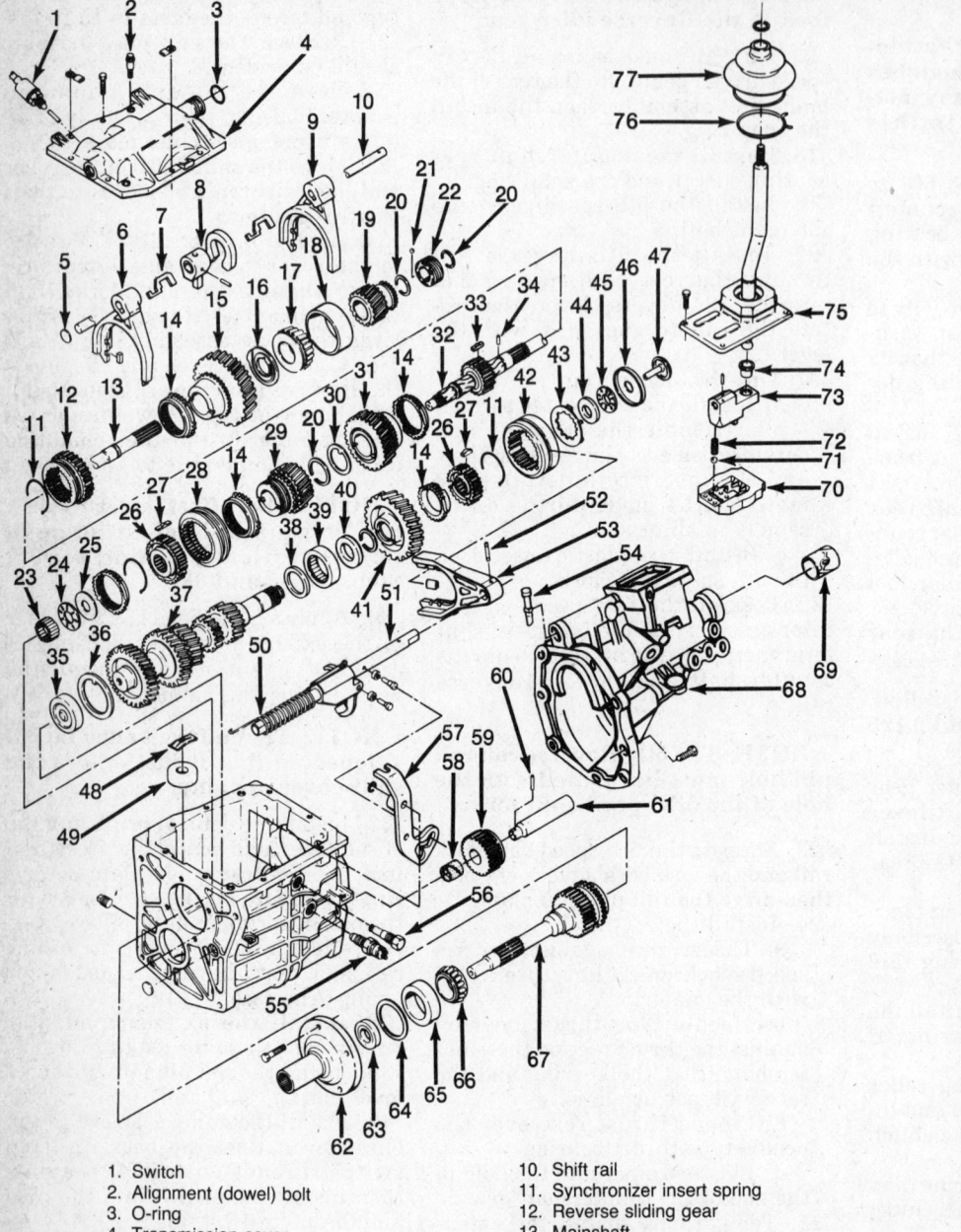

20. Snap-ring
21. Ball
22. Speedometer drive gear
23. Clutch shaft needle bearing
24. Needle thrust bearing
25. Thrust bearing race
26. 3rd-4th synchronizer hub
27. Synchronizer insert
28. 3rd-4th synchronizer sleeve
29. 3rd speed gear
30. Thrust washer
31. 2nd speed gear
32. Mainshaft
33. Insert
34. Roll pin
35. Front countershaft thrust bearing
36. Thrust washer
37. Countergear unit
38. Bearing spacer
39. Rear countershaft bearing
40. Spacer
41. 5th speed gear
42. 5th synchronizer sleeve
43. Thrust washer
44. Bearing race
45. Needle thrust bearing
46. Thrust race
47. Plastic funnel
48. Retainer
49. Magnet
50. Overcenter spring
51. 5th-reverse shift rail
52. Roll pin
53. 5th-reverse shift fork
54. Breather
55. Back-up lamp switch
56. Reverse lever pivot bolt
57. Reverse lever
58. Reverse idler gear bushing
59. Reverse idler gear
60. Roll pin
61. Reverse idler shaft
62. Front bearing cap
63. Oil seal
64. Shim
65. Front bearing cup
66. Front bearing
67. Main drive gear and shaft unit
68. Extension housing
69. Bushing
70. Detent plate
71. Detent ball
72. Spring
73. Offset lever
74. Bushing
75. Lever assembly
76. Retainer
77. Boot

1. Switch
2. Alignment (dowel) bolt
3. O-ring
4. Transmission cover
5. Plug
6. 3rd-4th shift fork
7. Selector plate
8. Selector arm and interlock plate
9. 1st-2nd shift fork
10. Shift rail
11. Synchronizer insert spring
12. Reverse sliding gear
13. Mainshaft
14. Blocking ring
15. 1st speed gear
16. Thrust washer
17. Rear bearing
18. Bearing cup
19. 5th speed driven gear

Exploded view of the Chilton Type 22 transmission used in the Chevette

4. Remove the 3rd–4th synchronizer insert springs, then remove the synchronizer inserts and the sleeve from the hub.

5. Remove the 3rd speed gear from the mainshaft.

6. Remove the 2nd gear snap ring, the tabbed thrust washer and the 2nd speed gear from the mainshaft.

7. Remove the rear mainshaft bearing.

8. Remove the 1st gear thrust washer, the roll pin (using diagonal cutters), the 1st speed gear and the blocking ring.

9. Scribe matchmarks on the 1st–2nd synchronizer sleeve and the mainshaft shaft to indicate their relationship for proper reassembly.

10. Remove the insert spring and the inserts from the 1st–Reverse sliding gear, then remove the gear from the mainshaft hub.

CAUTION

Do not attempt to remove the 1st–2nd–Reverse hub from the mainshaft as these parts are machined as a matched set from the factory.

Chevette

1. Follow Steps 1–3 of the previous AMC procedure.

2. Remove the snap ring, the

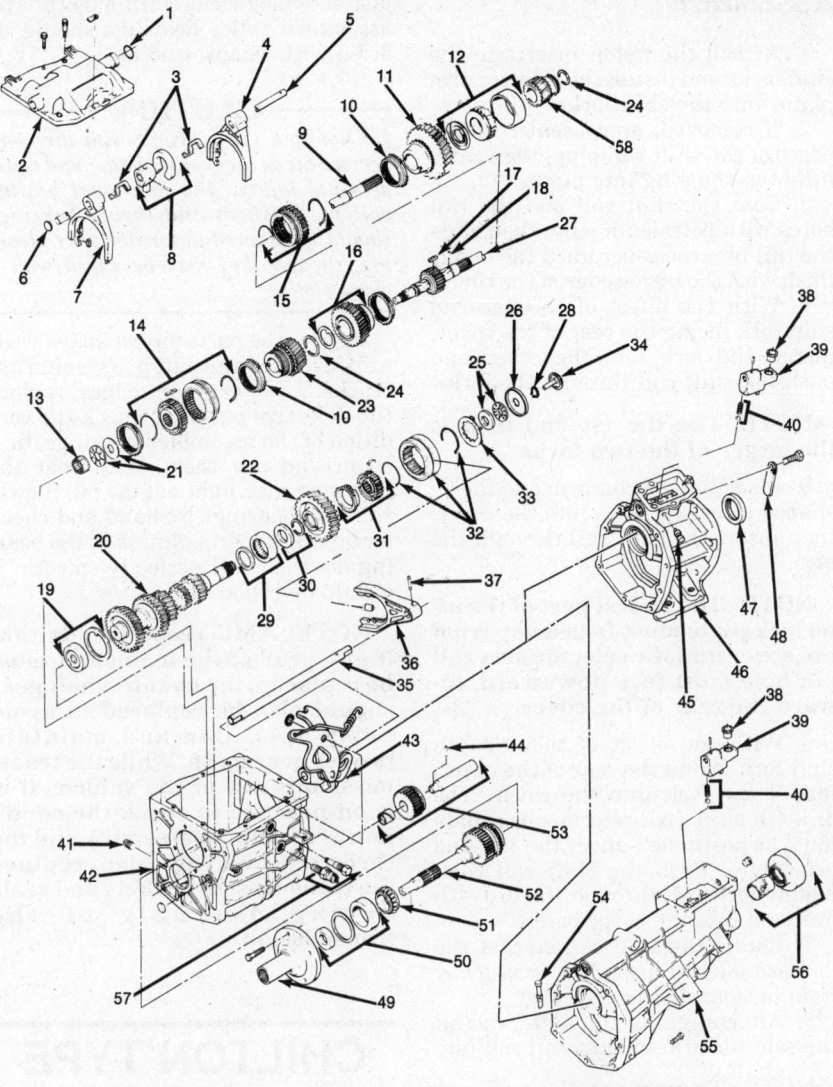

13. Clutch shaft needle bearing
14. 3rd-4th synchronizer assembly
15. Reverse sliding gear and insert springs
16. 2nd speed gear and thrust washer
17. Synchronizer insert
18. 1st gear roll pin
19. Front countershaft bearing and thrust washer
20. Countershaft gear unit
21. Needle thrust bearing and race
22. 5th speed gear
23. 3rd speed gear
24. Snap-ring
25. Needle thrust bearing and race
26. Thrust race
27. Mainshaft
28. Snap-ring
29. Rear countershaft bearing and spacer
30. Snap-ring and spacer
31. 5th gear synchronizer blocking ring, hub and insert
32. 5th gear synchronizer insert springs and sleeve
33. Insert retainer
34. Plastic funnel
35. Reverse seal
36. 5th speed shift fork
37. Roll pin
38. Damper sleeve
39. Offset lever
40. Detent spring and ball
41. Plug
42. Transmission case
43. 5th-reverse shift lever
44. Roll pin
45. Breather
46. Adapter housing (4WD)
47. Seal
48. I.D. tag
49. Front bearing cap
50. Oil seal, shim and cup
51. Front bearing
52. Main drive gear and shaft
53. Reverse idler bushing, gear and shaft
54. Breather
55. Extension housing (2WD)
56. Bushing and oil seal
57. Back-up lamp switch and 5th-reverse lever pivot bolt
58. Fifth speed driven gear

1. O-ring
2. Transmission cover
3. Selector plate
4. 1st-2nd shift fork
5. Shift rail
6. Plug
7. 3rd-4th shift fork
8. Selector arm, interlock plate and pin
9. Mainshaft
10. Blocking ring
11. 1st speed gear
12. Thrust washer, rear bearing and clip

Exploded view of the AMC T5 (Chilton Type 22) transmission

tabbed thrust washer and the 2nd gear from the mainshaft.

3. Using an arbor press and the puller tool, remove the 5th speed gear from the mainshaft.

4. Slide the rear mainshaft bearing off of the mainshaft.

5. Remove the 1st gear thrust washer, the roll pin, the 1st speed gear and the blocking ring.

6. Scribe matchmarks on the 1st–2nd synchronizer hub and the sleeve for reassembly purposes.

7. Remove the insert spring and the inserts from the 1st–Reverse slid-

ing gear, then the gear from the mainshaft.

— CAUTION —

Do not attempt to remove the 1st–2nd–Reverse hub from the mainshaft as these parts are machined as a matched set from the factory.

ASSEMBLY

AMC Models

1. Lubricate the mainshaft and the gear bores with a liberal coating of transmission lubricant.

2. Align and install the 1st–2nd synchronizer sleeve on the mainshaft, using the matchmarks made during disassembly.

3. If removed, install the synchronizer inserts and the springs into the 1st–2nd synchronizer sleeve.

NOTE: The tanged end of each spring should be positioned on the same insert but that the open face of each spring should be opposite the other.

4. Install the blocking ring and the 2nd speed gear onto the mainshaft.

5. Install the tabbed thrust washer and the 2nd speed gear snap ring onto the mainshaft.

NOTE: Be sure that the washer tab is fully seated into the mainshaft notch.

6. Install the blocking ring and the 1st speed gear onto the mainshaft.

7. Carefully drive the 1st gear roll pin into place.

8. Using an arbor press and the special tools, press the rear bearing onto the mainshaft.

9. Install the 1st gear thrust washer.

10. Install the 3rd speed gear, the 3rd–4th synchronizer inserts and the sleeve onto the mainshaft; the offset of the hub must face forward.

11. Install the thrust bearing washer onto the forward end of the mainshaft.

Chevette

1. Follow Steps 1–6 of the previous AMC procedure.

2. Carefully drive the 1st gear roll pin into place, then install the 1st gear thrust washer.

3. Slide the mainshaft rear bearing onto the mainshaft.

4. Using an arbor press and the special tools, press the 5th speed gear onto the mainshaft.

5. Install the 3rd speed gear, the 3rd–4th synchronizer assembly and the thrust bearing onto the mainshaft; the synchronizer hub offset must face forward.

Cover

DISASSEMBLY

1. Place the selector arm plates and the shift rail in the Neutral position (centered).

2. Rotate the shift rail counterclockwise until the selector arm disengages from the selector arm plates. The selector arm roll pin will now be accessible.

3. On AMC models, pull the selector arm rearward until the arm contacts the 1st–2nd shift fork.

4. Using a ³⁄₁₆ in. pin punch and a hammer, carefully drive out the selector arm roll pin, then remove the shift rail.

5. Remove the shift forks, the selector arm plates, the selector arm, the roll pin and the interlock plate.

6. Remove the nylon inserts and the selector arm plates from the shift forks. Mark the positions of the inserts and the plates so they may be properly reinstalled.

ASSEMBLY

1. Attach the nylon inserts to the shift forks and install the selector arm plates into the shift forks.

2. If removed, apply sealer to the edges of the shift rail plug, then carefully tap the plug into place.

3. Coat the shift rail and the rail bores with petroleum jelly, then slide the rail into the cover until the end is flush with the inside edge of the cover.

4. With the offset of the 1st–2nd shift fork facing the rear of the cover, install the fork into the cover and push the shift rail through the fork.

NOTE: The the 1st–2nd fork is the larger of the two forks.

5. Place the selector arm and the C-shaped interlock plate into the cover, then insert the shift rail through the arm.

NOTE: The widest part of the interlock plate must face away from the cover and the selector arm roll pin hole must face downward, toward the rear of the cover.

6. With the offset of the 3rd–4th shift fork facing the rear of the cover, install the fork into the cover. The 3rd–4th shift fork selector arm plate must be positioned under the 1st–2nd arm plate. Push the shift rail completely forward, through the 3rd–4th fork and into the cover bore.

7. Rotate the shift rail so that the forward selector arm plate faces away from but parallel to the cover.

8. After aligning the holes, install the selector arm-to-shift rail roll pin.

NOTE: To prevent the roll pin from contacting the selector arm plates when shifting, the roll pin must be installed flush with the surface of the selector arm.

9. Install the O-ring into the groove of the shift rail oil seal.

10. Installation of the shift rail oil seal should be performed as follows:
   a. Install the Kent-Moore tool No. J-26628-2 over the threaded end of the shift rail.
   b. Lubricate the lip of the oil seal with petroleum jelly.
   c. Slide the seal over the protector tool and onto the shift rail.
   d. Using the Kent-Moore seal installer tool No. J-26628-1, seat the seal in the cover.

CLEANING AND INSPECTION

The parts (except the nylon or plastic) should be thoroughly cleaned in cleaning solvent. The nylon or plastic parts, which are to be reused should just be wiped clean with a cloth. The assembled roller bearings should be dried with compressed air.

----- **CAUTION** -----

Do not spin the bearings with the compressed air as they could shatter and cause personal injury. The individual bearing rollers, washers and thrust bearings should be allowed to air dry after cleaning, though they may be wiped with a clean cloth.

Inspect the parts for excessive wear and/or damage such as scoring, cracks, nicks or rough edges; replace the defective parts. To check the condition of the assembled bearings, first clean and dry them, then coat the bearings with light engine oil. Slowly spin the bearings by hand and check for any signs of roughness. If the bearing does not feel perfectly smooth, it should be replaced.

NOTE: AMC recommends that if any gear of the mainshaft must be replaced, the countershaft gear should also be replaced to avoid noisy operation and maintain proper gear mesh. While the transmission is out of the vehicle, it is good practice to check the condition of the clutch assembly and the throwout bearing. Also, replace the transmission gaskets and seals during assembly of the transmission.

CHILTON TYPE 23

Transmission Case

DISASSEMBLY

NOTE: The use of an arbor press is required to properly disassemble and assemble this transmission. Read the entire procedure carefully before starting the job.

1. Remove the drain plug and allow the lubricant to drain from the case.

2. Remove the throwout bearing and the fork from the transmission as outlined in the appropriate car section.

3. Remove the drive gear bearing retainer. If damaged, remove the ball stud.

4. Remove the Belleville spring from the front of the drive gear bearing.

5. Remove the bolt, the retainer and the speedometer driven gear from the side of the case.

6. Remove the shift lever quadrant from the extension housing.

7. Remove the back up light switch.

8. Remove the extension housing bolts and the housing.

9. Remove the snap rings, the speedometer drive gear, the spacer and the bearing from the mainshaft.

10. Remove the snap ring, then the thrust washer and the lock ball from the driveshaft.

11. Remove the large snap ring from the main drive gear bearing.

12. Remove the following components from the case as an assembly:
   a. The center support
   b. The mainshaft
   c. The countergear
   d. The drive gear

13. Using a drift punch and a hammer, carefully drive the roll pins from the 1st/2nd, the 3rd/4th and the 5th/Reverse shift forks. Support the shaft ends with a bar or a block of wood to prevent damage to these components.

14. Remove the detent spring plate mounting bolts, the detent spring plate, the 3 springs and the balls from the center support.

15. Remove the shifter shafts from the center support, the shift forks from the shafts and the interlock pins from the center support.

16. Move the 1st/2nd synchronizer sleeve to the 1st gear position and the 3rd/4th synchronizer sleeve into the 3rd gear position.

17. Using the Kent-Moore holdling fixture tool No. J-29768, install it on the end of the drive gear shaft and countergear. Remove the countergear retaining nut and the washer.

18. Using a puller, remove the ball bearing and the 5th speed gear from the countershaft.

19. Remove the 5th gear, the blocking ring and the needle bearing from the mainshaft.

20. Remove the self locking nut from the Reverse idler gear shaft.

21. Remove the thrust washers and the Reverse idler gear from the Reverse idler gear shaft.

22. Bend the locking retainer of the mainshaft nut away from the nut, then remove the mainshaft nut.

23. Remove the 5th–Reverse synchronizer locking retainer, the 4th–Reverse synchronizer assembly, the Reverse gear, the needle bearing, the collar and the thrust washer from the mainshaft.

24. Remove the Reverse gear from the countergear and the holding fixture (installed during Step 17).

25. Move the synchronizer sleeves back to their Neutral positions.

26. Expand the countergear bearing snap ring (using snap ring pliers) and gently tap on the front of the center support. Expand the mainshaft bearing snap ring and move the mainshaft inward then remove the countergear and the mainshaft.

27. Remove the drive gear, the needle bearing and the blocking ring from the end of the mainshaft.

ASSEMBLY

1. If removed, install the countergear and the mainshaft snap rings into the center support. Also, install the Reverse idler shaft into the center support.

2. Install the drive gear onto the front of the mainshaft and engage it with the countergear.

3. Install the holding fixture in the same manner as the disassembly, in Step 17.

4. With the mainshaft and the countergears meshed, slide the center support onto the mainshaft. Expand the mainshaft snap ring and continue to push the center support on until the mainshaft bearing groove aligns with the snap ring. Release the mainshaft snap ring to lock the mainshaft bearing into place. Repeat the same procedure to seat the countergear snap ring.

5. Move the synchronizer sleeves to engage both the 1st and the 3rd gear tangs in order to lock the gears.

6. Install the Reverse gear onto the countergear.

7. Install the thrust washer on the mainshaft (with the oil groove facing the rear), then install the collar, the needle bearing and the Reverse gear onto the mainshaft.

8. Install the 5th/Reverse synchronizer (with the face of the higher clutch hub boss facing the Reverse gears).

9. Install the locking retainer and the mainshaft nut onto the mainshaft, then torque the nut to 94 ft. lbs. and then bend the locking retainer tabs to lock the nut.

10. Install the thrust washers and the Reverse idler gear on the Reverse idler gear shaft. Thread a new self locking nut onto the Reverse idler shaft and torque the nut to 80 ft. lbs.

NOTE: The flange of the plate side thrust washer must be fitted to the center support.

11. Install the synchronizer blocking ring and the 5th speed gear (with the needle bearings) onto the mainshaft.

12. Install the 5th speed gear (of the countergear), the ball bearing, the washer and a new self locking nut onto the rear of the countergear, then torque the nut to 80 ft. lbs.

13. Remove the holding fixture and move the synchronizer sleeves back to their Neutral positions.

14. Grease the interlock pins and install them into the center support.

15. Place the shift forks into position on the synchronizer sleeves. Install the shifter shafts through their respective forks from the rear of the center support, except the 5th/Reverse shaft, which is installed from the front of the support.

16. Install the three detent balls, the springs, the detent plate gasket and the detent plate, then torque the detent plate bolts to 14 ft. lbs.

17. Using a drift punch and a hammer, carefully install the retaining pins into the shift forks. Remember to support the shafts with a bar or a block of wood to prevent damage.

18. If removed, lubricate the countergear needle bearing and install it into the front of the case. The bearing should be driven into place while a socket is positioned on the outer bearing race.

19. Install a new center support-to-case gasket on the case and the center support/mainshaft/countergear/drive gear assembly into the case.

20. Install the large snap ring onto the shaft of the drive gear bearing.

21. Install the lock ball, the thrust washer and the snap ring onto the mainshaft.

22. Using a feeler gauge, check the clearance between the 5th speed gear (of the mainshaft) and its thrust washer. The clearance should be 0.010–0.016 in. If necessary, adjust the clearance by purchasing a thrust washer of the correct thickness which will replace the existing washer. The thrust washers are available in thickness ranging from 0.307–0.327 in. in 0.003 in. increments.

NOTE: Use care when removing and installing the snap ring; it must be replaced if it becomes distorted.

23. Install these parts on the mainshaft behind the 5th gear snap ring, in this order:
   a. The ball bearing.
   b. The snap ring.
   c. The speedometer gear clip.
   d. The speedometer drive gear.

24. Using a new gasket, attach the extension housing to the center support and torque the bolts to 27 ft. lbs.

25. Using a new gasket, install the shift lever quadrant onto the extension housing and torque the bolts to 14 ft. lbs.

26. Install the speedometer driven gear and torque the bolt to 14 ft. lbs.

27. Install the back up light switch into the extension housing.

28. Install the belleville washer in the front of the drive gear bearing.

NOTE: The dished side of the washer should face the bearing.

29. Using a new gasket, install the drive gear bearing retainer. Before installing the three lower bearing retainer bolts, coat the threads with Permatex® No. 2 sealer (or its equivalent). Torque the bearing retainer bolts to 14 ft. lbs.

30. Install the throw out bearing and the fork.

31. Install the transmission according to the procedure in the Chevette car section.

Mainshaft

DISASSEMBLY

1. Using an arbor press and the Kent-Moore tool No. J-22912-01, remove the mainshaft rear bearing.

2. Remove the thrust washer, the 1st speed gear, the needle bearings and the spacer.

3. Remove the 1st/2nd synchronizer assembly, the 2nd speed gear and the needle bearings.

4. Remove the snap ring from in front of the 3rd/4th synchronizer and slide the synchronizer off the mainshaft.

Drive Gear Bearing

REMOVAL & INSTALLATION

1. Remove the snap ring from the drive gear shaft.

2. Using an arbor press and the Kent-Moore tool No. J-22912-01, press the drive gear shaft through the bearing.

3. To install, reverse the removal procedures.

Countergear Bearing

REMOVAL & INSTALLATION

The bearing is removed in the same manner as the drive gear bearing, using the same special tool and an arbor press. Note that the groove on the bearing is installed towards the rear of the transmission.

Extension Housing Or Drive Gear Retainer Seals

REMOVAL & INSTALLATION

Using a small pry bar, pry the seals from the housing. Install the new seals by carefully tapping them into place.

Mainshaft

ASSEMBLY

1. Install the 3rd speed gear (with the needle bearings) onto the front of the mainshaft.

NOTE: The coned side of the gear is installed toward the front of the mainshaft.

2. Install the 3rd/4th synchronizer assembly onto the mainshaft, with the large chamfered end facing the front of the case; retain the synchronizer with the snap ring.

3. Install the 2nd speed gear (with the needle bearings) onto the rear of the mainshaft.

NOTE: The coned side of the gear is installed facing the rear of the mainshaft.

4. Install the 1st/2nd synchronizer assembly onto the mainshaft, with the large chamfered end facing the rear of the case.

5. Install the 1st speed gear (with the spacer and the needle bearings), with the coned end facing the front of the case.

6. Install the 1st gear thrust washer, with the slots of the washer facing the gear.

7. Press the rear bearing onto the mainshaft according to the previous procedure under "Removal & Installation".

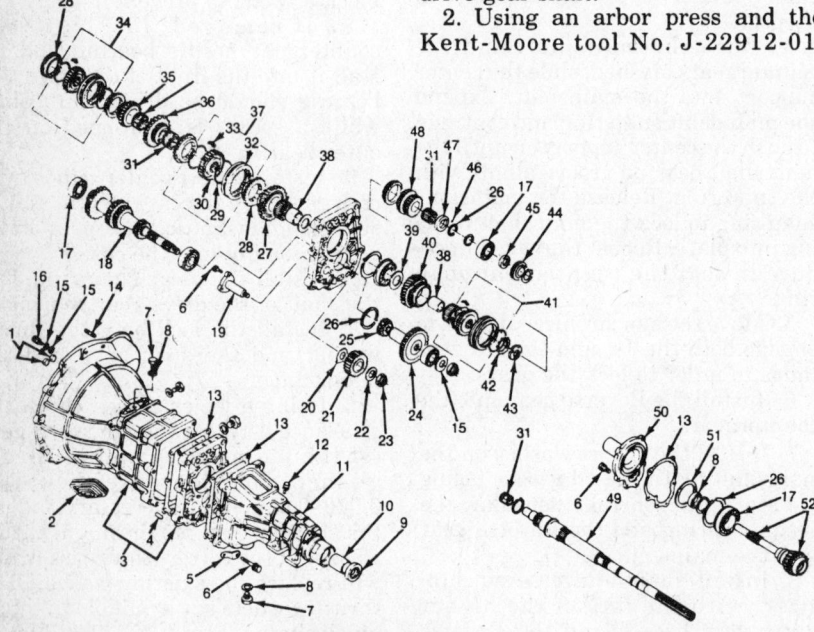

1. Shift rod plug
2. Dust cover
3. Case with center support
4. Pin
5. Return spring bracket
6. Bolt
7. Plug
8. Seal
9. Seal
10. Bushing
11. Extension
12. Ventilator
13. Gasket
14. Starter locating stud
15. Washer
16. Shift fork support
17. Ball bearing assembly
18. Countergear
19. Reverse idler shaft
20. Reverse idler front thrust washer
21. Reverse idler gear

22. Reverse idler rear thrust washer
23. Self-locking nut
24. 5th speed gear
25. Reverse gear
26. Snap-ring
27. 1st speed gear assembly
28. Blocking ring
29. Synchronizer insert spring
30. Synchronizer hub
31. Needle bearing
32. Synchronizer sleeve
33. Synchronizer insert (key)
34. 3rd-4th synchronizer assembly
35. 3rd gear assembly
36. 2nd gear assembly
37. 1st-2nd synchronizer assembly
38. Collar

39. Thrust washer
40. Reverse gear (mainshaft)
41. 5th-reverse synchronizer assembly
42. Mainshaft lockwasher
43. Mainshaft nut
44. Speedometer drive gear
45. Spacer
46. Ball
47. Selective thrust washer
48. 5th speed gear (mainshaft)
49. Mainshaft
50. Drive gear bearing retainer
51. Belleville spring
52. Main drive gear and shaft assembly

Transmission case and geartrain components of the Chilton Type 23 transmission

CHILTON TYPE 24

Transmission Case

DISASSEMBLY

1. Remove the side covers and the shift controls after draining case.
2. Remove the bearing retainer lock strips, the bolts, the retainer and the gasket.
3. Lock up case by shifting into two gears and remove main drive gear retaining nut.

NOTE: This nut may have left hand threads.

4. Return the gears to Neutral position, then remove the lock pin from the Reverse shifter lever boss and pull the shaft out about $\frac{1}{8}$ in.; this will disengage the shift fork from the Reverse gear.
5. Remove the extension housing bolts and tap the housing (with a soft hammer) rearward. When the idler shaft is out (as far as it will go), move extension housing left so the Reverse fork clears the gear, then remove housing and the gasket.
6. Remove the Reverse idler gear, the flat washer, the shaft and the roll spring pin.
7. Remove the speedometer and the Reverse gears.

NOTE: Slide the 3rd/4th synchronizer clutch sleeve to the 4th speed gear position (forward) before removing the mainshaft assembly from the case.

8. Remove the rear bearing retainer and the mainshaft assembly from the case by tapping the bearing retainer with a soft hammer.
9. Unload the bearing rollers from the main drive gear and remove the 4th gear synchronizer blocking ring.
10. Lift the front half of the Reverse idler gear with the tanged thrust washer from the case.
11. Press the main drive gear down from the bearing.
12. Tap the front bearing and the snap ring from the case.
13. From the front of the case, press out the countershaft, then remove the countershaft gear and both tanged washers.
14. Remove the 112 rollers, the six spacers and the roller spacer from the countergear.
15. Remove the front mainshaft snap ring, then slide the 3rd/4th speed clutch, the 3rd speed gear and the synchronizer ring from the front of the mainshaft.
16. Spread the rear bearing retainer snap ring and press the mainshaft out of the retainer.
17. Remove the mainshaft snap ring. Support the 2nd speed gear and press on the rear of the mainshaft to remove the rear bearing, the 1st speed gear, the sleeve, the 1st speed synchronizing ring, the 1st/2nd speed synchronizer clutch, the 2nd speed synchronizer ring and the 2nd speed gear.

NOTE: Thoroughly clean the case and the parts; make a thorough inspection and replace the required parts. When checking the bearings, do not spin at high speeds but rather rotate by hand to detect roughness and unevenness. Spinning can damage the balls and the races.

ASSEMBLY

1. Rest the case on its side with the cover opening toward you. Install the countergear tanged thrust washers with heavy grease; make sure the tangs are in the proper notches.
2. Set the countergear in place; use care not to disturb the tanged washers.
3. Position the case so that it rests on the front face.
4. Lubricate and insert the countershaft in the rear. Turn the countershaft so the flat on end of the shaft is horizontal and facing the bottom of the case.

NOTE: The flat on the shaft must be horizontal and toward the bottom to mate with the rear bearing retainer, when installed.

5. Align the countergear with the shaft in the rear and the hole in the front of the case (pushing the dummy shaft out through the front of the case) until the flat on the shaft is flush with the rear of the case; be sure the thrust washers remain in place.
6. Check the end play of the countergear with a dial indicator. If the end play is more than 0.025 in., install a new thrust washer.
7. Install the cage and the 17 roller bearings into main drive gear; use heavy grease to hold bearings.
8. Install the main drive gear with the bearings through the side opening of the case and into position in the front bore.
9. Place the gasket in position on the rear bearing retainer.
10. Install the 4th speed synchronizing ring onto the main drive gear (notches toward the rear).
11. Position the tanged thrust washer for the Reverse idler on the machined face. Position the front of the Reverse idler gear next to the thrust washer (with the hub facing the rear of the case).

CAUTION
Before attempting to the the install mainshaft to the case, slide the 3rd/4th synchronizer clutch sleeve forward into the 4th speed detent position.

---

12. Lower the mainshaft assembly into the case.

NOTE: Be sure the notches on the 4th speed synchronizer ring correspond to the keys in the clutch assembly.

13. With the guide pin in the rear bearing retainer aligned with the hole in the rear of the case, tap the rear bearing retainer into position with a soft hammer.
14. From the rear of the case, insert the Reverse idler gear, engaging the splines with a portion of the front gear in the case.
15. Place the gasket into position on the rear face of the bearing retainer.
16. Install the remaining flat washer on the Reverse idler shaft.
17. Install the Reverse idler shaft, the roll pin and the thrust washer into the gears and the front boss of the case; make sure to pick up the front tanged thrust washer.
18. Pull the Reverse shifter shaft to the left side of the extension housing and rotate the shaft to bring the Reverse shift fork forward in the housing (reverse the detent position). Start the extension housing onto the case, while slowly pushing in on the shifter shaft to engage the shift fork with the Reverse gear shift collar. Then, pilot the Reverse idler shaft into the extension housing, permitting the extension to slide into the case.
19. Install the extension and the retainer-to-case bolts.
20. Push or pull the Reverse shifter shaft to align the grooves in the shaft with the holes in the boss and drive in the lockpin, then install the shift lever.
21. Press the bearing onto the main drive gear (with the snap ring groove in the front) and into the case until the main drive gear retaining nut threads are exposed.
22. Lock the case by shifting into two gears, then install the main drive gear nut onto the gear shaft and draw it up tight; be sure the bearing is completely seated against the shoulder. Torque the retaining nut to 40 ft. lbs. and lock it in place by staking the main drive gear shaft hold with a

punch; do not damage the shaft threads.

23. Install the main drive gear bearing retainer, the gasket bolts and the boltlock retainers; use sealant on the bolts. Torque the bolts to 20 ft. lbs.

24. Shift the mainshaft 3rd/4th sliding clutch sleeve into the Neutral position and 1st/2nd sliding clutch into the 2nd gear (forward) detent position. Shift the side cover 3rd/4th shift lever into the Neutral detent and the 1st/2nd shift lever into the 2nd gear detent position.

25. Install the side cover, with a new gasket and carefully position it into place. A dowel pin provides the proper alignment position. Install the bolts and torque evenly to 20 ft. lbs.

Mainshaft

ASSEMBLY

1. From the rear of the shaft, assemble the 2nd speed gear (hub of the gear toward rear of the shaft).

2. Install the 1st/2nd synchronizer clutch assembly onto the mainshaft (with the sleeve taper toward the rear

and the hub to the front), together with a synchronizing ring on each side of the clutch assembly, so that the keyways align with the clutch keys.

3. Press the 1st speed sleeve onto the mainshaft; a $1\frac{3}{4}$ in. or a $1\frac{5}{8}$ in. ID pipe, cut to a convenient length makes a suitable tool.

4. Install the 1st speed gear (with the hub facing forward) and press it onto the rear bearing with the snap ring grooves facing forward. Be sure the bearing is firmly seated.

5. Choose a selective fit snap ring (0.087, 0.090, 0.093 or 0.096 in.) and install it into the groove on the mainshaft behind the rear bearing.

NOTE: The maximum clearance of the snap ring and the rear face should be between 0.000–0.005 in. Always use new snap ring.

6. Install the 3rd speed gear (with the hub to the front of the case) and the 3rd speed gear synchronizing ring (with the notches to the front).

7. Install the 3rd/4th speed gear clutch assembly with both the sleeve taper and the hub facing forward.

8. Install the snap ring onto the mainshaft in front of the 3rd/4th speed clutch, with the snap ring ends seated behind the spline teeth.

9. Install the rear bearing retainer. Spread the snap ring in the plate, to allow the ring to drop around the rear bearing and press on the mainshaft end until the snap ring engages the groove in the rear bearing.

10. Install the Reverse gear (with the shift collar to the rear).

11. Install the speedometer drive gear.

Countergear

ASSEMBLY

1. Install the roller spacer into the countergear.

2. With heavy grease to assist, install a spacer in either end of the countergear, the 28 roller bearings, a spacer and 28 more rollers, then another spacer. In the other end of the countergear, do the same.

3. Insert the dummy shaft into the countergear.

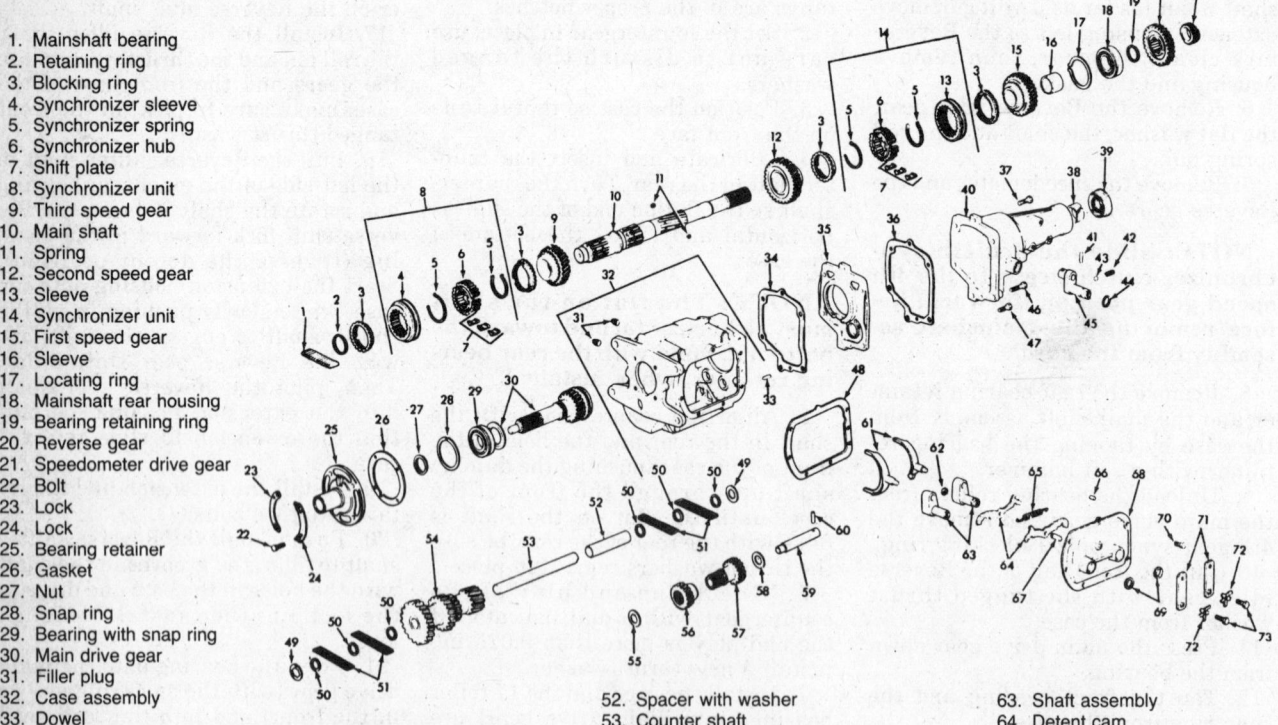

1. Mainshaft bearing
2. Retaining ring
3. Blocking ring
4. Synchronizer sleeve
5. Synchronizer spring
6. Synchronizer hub
7. Shift plate
8. Synchronizer unit
9. Third speed gear
10. Main shaft
11. Spring
12. Second speed gear
13. Sleeve
14. Synchronizer unit
15. First speed gear
16. Sleeve
17. Locating ring
18. Mainshaft rear housing
19. Bearing retaining ring
20. Reverse gear
21. Speedometer drive gear
22. Bolt
23. Lock
24. Lock
25. Bearing retainer
26. Gasket
27. Nut
28. Snap ring
29. Bearing with snap ring
30. Main drive gear
31. Filler plug
32. Case assembly
33. Dowel
34. Gasket
35. Bearing retainer
36. Gasket
37. Bolt
38. Extension bushing
39. Seal assembly
40. Extension assembly
41. Pin
42. Spring
43. Ball
44. Seal
45. Reverse shift fork
46. Bolt
47. Bolt
48. Gasket
49. Washer
50. Washer
51. Roller
52. Spacer with washer
53. Counter shaft
54. Counter shaft gear
55. Washer
56. Reverse idler front gear
57. Reverse idler gear assembly
58. Washer
59. Reverse idler gear shaft
60. Pin
61. Fork
62. Ring
63. Shaft assembly
64. Detent cam
65. Spring
66. Dowel pin
67. Locating pin
68. Cover
69. Shifter shaft lever
70. Bolt
71. Shifter shaft lever
72. Lock
73. Bolt

Exploded view of a Muncie 4 speed (Chilton Type 24) transmission

Automatic Transmissions

TRANSMISSION IDENTIFICATION BY PAN GASKET

AMC
TORQUE COMMAND 727

CHRYSLER
TORQUEFLITE 727

AMC
TORQUE COMMAND 904, 998

CHRYSLER
TORQUEFLITE 904

CHRYSLER
TORQUEFLITE
TRANSAXLE

FORD FMX

FORD C3

FORD C4, C5

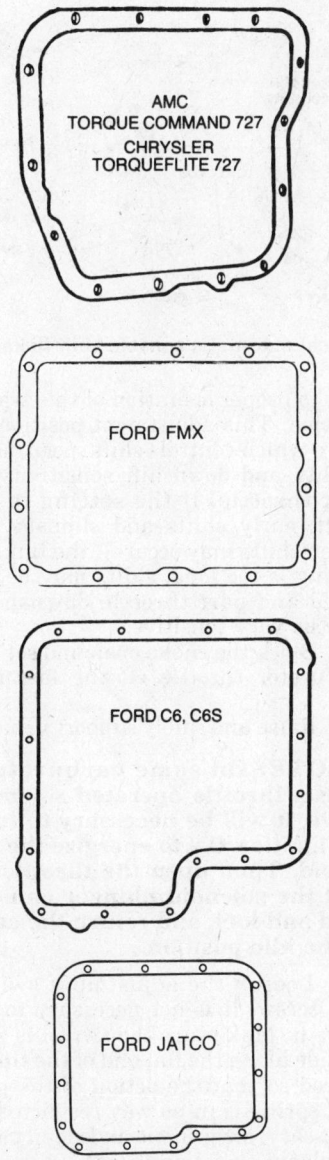

FORD C6, C6S

FORD
AUTOMATIC OVERDRIVE
TRANSMISSION (AOT)

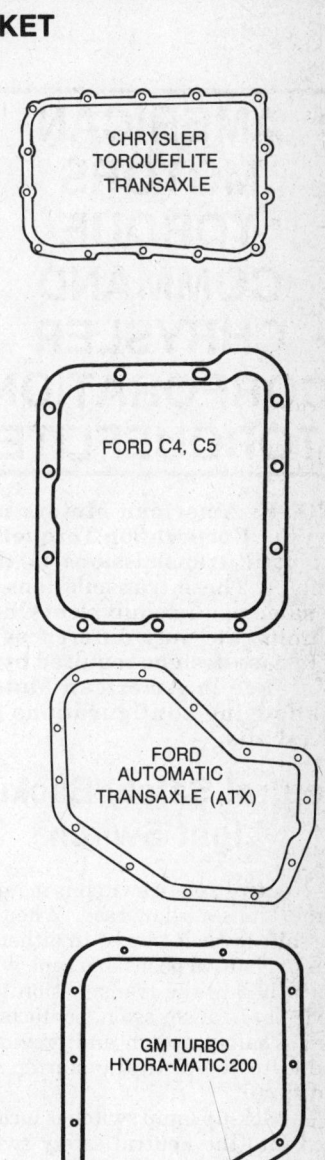

FORD
AUTOMATIC
TRANSAXLE (ATX)

FORD JATCO

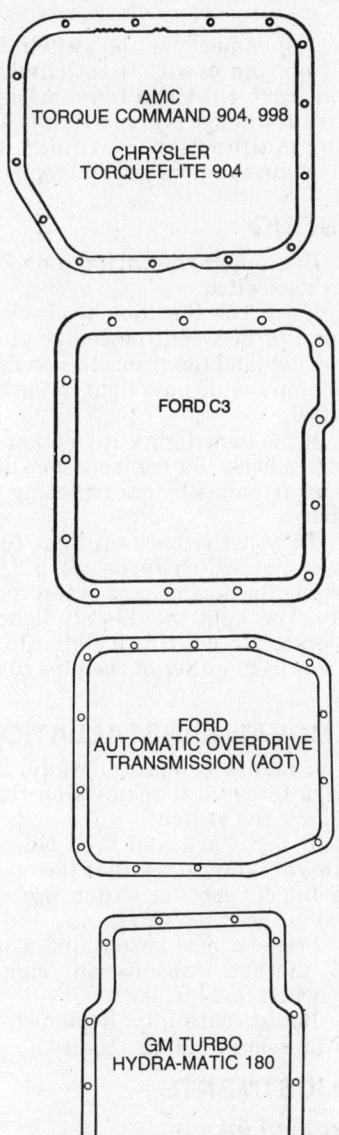

GM TURBO
HYDRA-MATIC 180

GM TURBO
HYDRA-MATIC 200

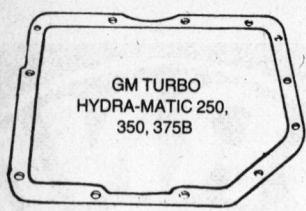

GM TURBO HYDRA-MATIC 250, 350, 375B

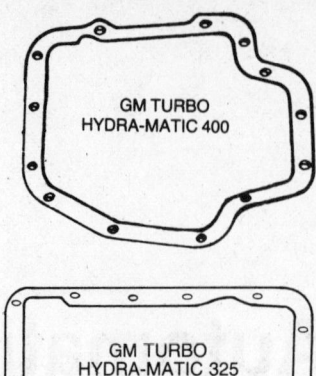

GM TURBO HYDRA-MATIC 400

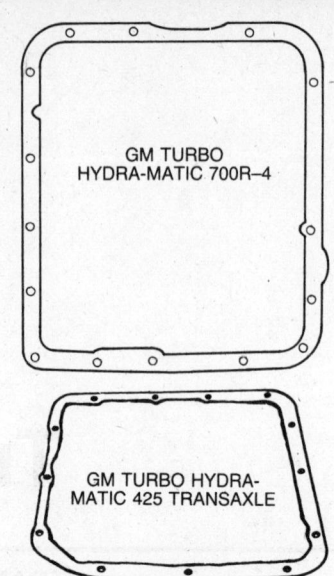

GM TURBO HYDRA-MATIC 700R–4

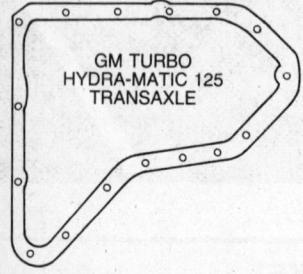

GM TURBO HYDRA-MATIC 125 TRANSAXLE

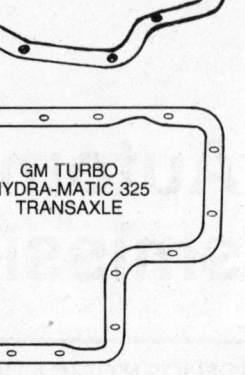

GM TURBO HYDRA-MATIC 325 TRANSAXLE

GM TURBO HYDRA-MATIC 425 TRANSAXLE

AMERICAN MOTORS TORQUE COMMAND CHRYSLER CORPORATION TORQUEFLITE

NOTE: American Motors uses Chrysler Corporation TorqueFlite automatic transmissions in their vehicles. These transmissions are the same as the equivalent Chrysler units, the only differences being in case designs required by the difference in American Motors' bell housing configurations and driveshafts.

Neutral Safety/Backup Light Switch

The Neutral safety switch is mounted in the transmission case. When the gearshift lever is placed in either the Park of Neutral position a cam, which is attached to the transmission lever inside the transmission, contacts the neutral safety switch and provides a ground to complete the starter solenoid circuit.

The back-up lamp switch is incorporated into the neutral safety switch. The center terminal is for the neutral safety switch and the two outer terminals are for the back-up lamps. There

is no adjustment for the switch. If a malfunction occurs, first check to make sure that the transmission gearshift linkage is properly adjusted. If the malfunction continues, the switch must be removed and replaced.

TESTING

1. Disconnect the wiring connector from the switch.
2. Use a 12V test lamp to check for continuity between the center pin of the switch and the transmission case. The lamp should only light in Park or Neutral.
3. If the lamp lights up in other positions, check the transmission linkage adjustments before replacing the switch.
4. To test the back-up light function of the switch repeat Step 2, by bridging the outside pins to test continuity. The light should only light in Reverse. No continuity should be present from either of the pins to the case.

REMOVAL & INSTALLATION

1. Place a container under the switch to catch transmission fluid. Unscrew the switch.
2. Select Park and then Neutral while checking to see that the operation fingers for the switch are centered in the case opening.
3. Screw a new switch and a new seal into the transmission. Tighten the switch to 24 ft. lbs.
4. Retest continuity. Replenish the transmission fluid, as required.

ADJUSTMENTS

Throttle Linkage

Throttle linkage adjustment is impor-

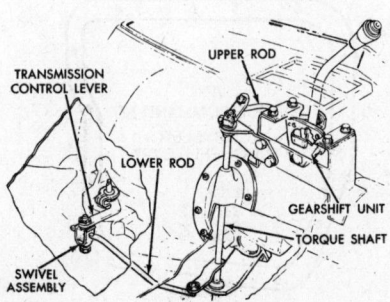

Typical automatic console shift linkage

tant to proper operation of this transmission. This adjustment positions a valve which controls shift speed, shift quality and downshift sensitivity at part throttle. If the setting is too short, early shifts and slippage between shifts may occur. If the linkage setting is too long, shifts may be delayed and part throttle downshifts may be very sensitive.

1. Block the choke open and set the carburetor throttle off the fast idle cam.
2. Raise and safely support vehicle.

NOTE: On some carburetors with a throttle operated solenoid valve, it will be necessary to turn the ignition ON to energize the solenoid. Then open the throttle so that the solenoid plunger can extend and lock and return the carb to the idle position.

3. Loosen the adjustment swivel lock screw. It is not necessary to remove it. Make sure the swivel is free to slide along the flat end of the throttle rod so that the action of the preload spring is in no way restricted. If necessary, disassemble and clean parts in solvent so action is free.

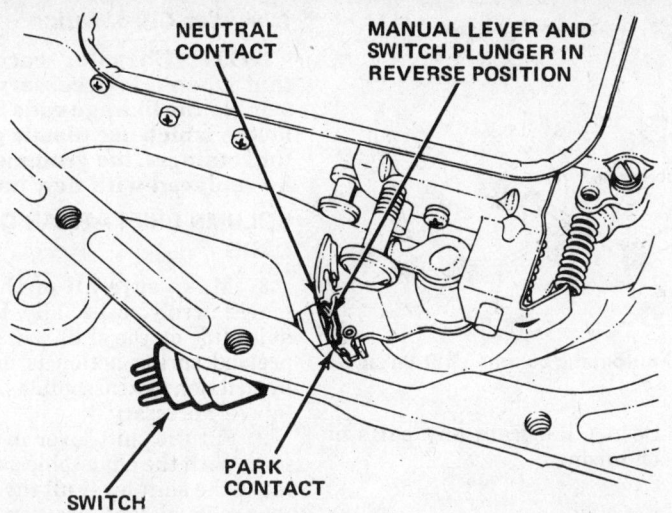

NEUTRAL CONTACT

MANUAL LEVER AND SWITCH PLUNGER IN REVERSE POSITION

SWITCH

PARK CONTACT

Torque-Command and TorqueFlite neutral start and backup light switch; pan removed, looking up

NOTE: Make sure to note proper assembly sequence as you take the mechanism apart, as improper assembly could prove dangerous.

4. Hold the transmission lever forward against its internal stop, then tighten the lock screw. If possible, use an inch pound torque wrench and torque to 100 inch lbs.

5. Restore freedom to the choke linkage if it has been blocked. Test the freedom of operation of the linkage by moving the throttle rod rearward and then slowly releasing it. Confirm that it will return fully forward.

Throttle Cable

1980–85 4 CYLINDER ENGINE

1. Remove the air cleaner assembly.

2. Remove the spark plug wire separator from the throttle cable bracket and move the separator and wires aside.

3. Raise and safely support the vehicle.

4. Remove the strut rod bushing heat shield to gain access to the transmission throttle control lever.

5. Hold the throttle lever rearward against its stop. Use the spare spring to hold the lever. Hook one end of spring to the lever and hook the opposite end of the spring to a convenient attachment point.

6. Lower the vehicle.

7. Block the choke open and set the carburetor linkage completely off the fast idle cam.

8. On four cylinder vehicles without air conditioning; turn the ignition key to the On position to energize the throttle stop solenoid.

9. Unlock the throttle control cable

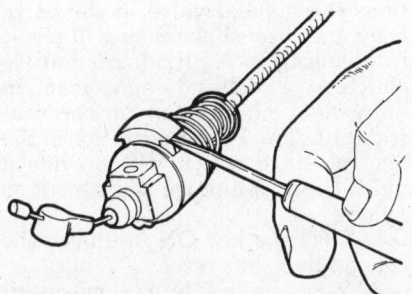

Throttle cable used with the AMC 151 engine. The snap-lock is shown in the raised (unlocked) position.

by releasing the T-shaped cable adjuster clamp. Release the clamp by lifting upward with a small screwdriver.

10. Grasp the cable outer sheath and move the cable and sheath forward to remove any cable load on the throttle bellcrank.

NOTE: The bell crank is part of the carburetor throttle linkage.

11. Adjust the cable by moving the cable and sheath rearward until there is zero lash between the plastic cable end and the bellcrank ball.

12. When zero lash between the cable end and bellcrank is achieved, lock the cable by pressing the T-shaped cable adjuster clamp downward until the clamp snaps into place.

13. Turn the ignition OFF. Install the spark plug wires and separator, connect the throttle stop solenoid on air conditioned vehicles and install the air cleaner.

14. Remove the holding spring from the transmission throttle control lever. Install the strut rod bushing heat shield and lower the vehicle.

15. Road test the vehicle and check the transmission operation. Readjust the throttle cable if necessary.

NOTE: Some V8 cars use a slightly different arrangement, in that the adjusting link is pushed instead of pulled to remove the slack. However, the end result should be the same, and no slack

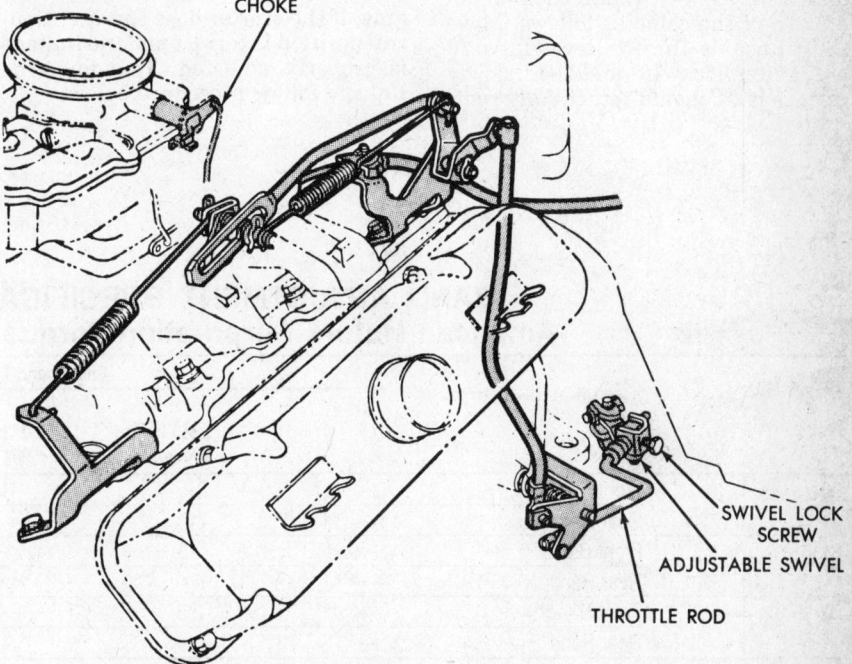

CHOKE

SWIVEL LOCK SCREW

ADJUSTABLE SWIVEL

THROTTLE ROD

V8 throttle rod adjustment—6 cyl. similar

or lash should be permitted. Many Chrysler Corp. vehicles use a lower bellcrank with a short throttle rod and adjustable swivel to hook up to the transmission throttle lever. In these cases, make sure the swivel is free to slide along the throttle rod so that the small preload spring action is not impaired. If necessary, clean and lightly lubricate. Again, the throttle lever must be held firmly forward against its internal stop. In this case, the linkage slack, or backlash was automatically removed by the small preload spring.

1986–87 4 CYLINDER ENGINE

1. Either run the engine until it is hot and verify that the fast idle cam has no effect on idle speed or disconnect the choke mechanism so the throttle will be at the normal hot idle position.

2. Loosen the locking screw that fastens the cable mounting bracket in place. Position the bracket so that both of its alignment tabs touch the case surface of the transaxle. Hold it in this position and then torque the mounting screw to 104 inch lbs.

3. Release the cross lock on the cable housing by pulling it upward. Make sure the cable is free to slide all the way toward the engine against its stop. Then, move the transaxle throttle lever all the way clockwise so it is against its internal stop, and then press the cross-lock downward into the locked position.

4. Reconnect the choke, if it was disconnected. Test the freedom of operation of the cable as follows: Move the transaxle throttle lever forward or counterclockwise and then slowly release it. It should return fully rear-

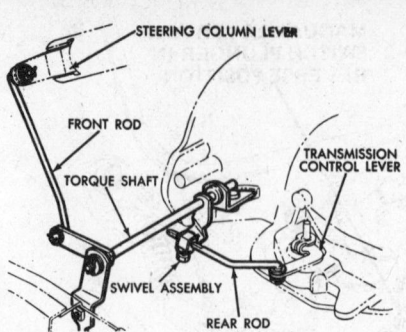

Typical automatic column shift linkage

ward. Do not lubricate any parts of the mechanism.

Gear Shift Linkage

The gear shift linkage adjustment is important because the linkage positions the manual valve in the valve body. Incorrect adjustment will result in creeping in Neutral, premature clutch wear, delayed engagement in any gear or a no-start in Park or Neutral condition. Proper operation of the neutral start switch will provide a quick check of linkage adjustment as follows:

1. Turn the key ON to unlock the column and shift lever.

2. Move the shift lever slowly until it clicks into the Park detent. Try to start the engine. If the starter does operate, Park position is correct.

3. Stop the engine. Repeat the above steps, only this time move the lever to Neutral. Try to start the engine. If the starter does operate, Neutral position is correct and the linkage is properly adjusted. To adjust the linkage follow procedure below.

Chrysler Corporation

NOTE: Chrysler recommends that when it is necessary to disassemble the linkage rods from their levers which use plastic grommets for retainers, the grommets should be replaced with new ones.

COLUMN SHIFT—REAR DRIVE CARS

1. Make sure all of the linkage moves freely, especially the adjustable slide on the shift rod, so that the preload spring action is not reduced by friction. Disassemble, clean and lube if necessary.

2. Put the shift lever in Park.

3. With the adjustable swivel loose, move the shift lever all the way to the rear-most detent position, which is Park.

4. Tighten the swivel lock bolt. Torque it to 90 inch lbs.

5. Verify that the vehicle will only start in Park or Neutral.

CONSOLE SHIFT—REAR DRIVE CARS

1. Adjustment is similar to above, but no preload spring is used. Make sure that with the shift handle in Park, the transmission lever is in the rear-most detent position, which is Park.

2. Tighten the swivel lock bolt with no load applied in either direction on the linkage.

3. Verify that the vehicle will only start in Park or Neutral.

FRONT WHEEL DRIVE CARS

1. Put the gearshift into the "P" (PARK) position.

2. Loosen the clamp bolt on the gearshift cable bracket. Then, pull

BAND ADJUSTMENT SPECIFICATIONS
American Motors Corporation Torque Command

		2L 904	2.5L 904	232 CID 904	258 CID Std-904	304 CID 998
'80	Front Band①	—	2	—	2	2
	Rear Band①	—	7②	—	7②	4
'81	Front Band①	—	2.5	—	2.5	2.5
	Rear Band①	—	7②	—	7	4
'82–87	Front Band①	—	2.5	—	2.5	2.5
	Rear Band①	—	7②	—	7	4

Engine and Transmission Models

NOTE: Numbers represent back-off turns from specified torque. Torque lock-nut to 35 ft. lbs when adjustment is completed.
① Backed off from 72 inch pounds
② Backed off from 41 inch pounds

BAND ADJUSTMENT SPECIFICATIONS
Chrysler Corporation TorqueFlite

		Engine and Transmission Models						
		A904		A904-LA				
		225	225	318	360-2 BBL 360-4 BBL	318	360-2 BBL 360-4 BBL	A727 360-HP
'80	Front Band①	2	2③	2.5③	2.5③	2.5	2.5	2.5
	Rear Band①	7②	4③	4③	4③	2	2	2
'81–'87	Front Band①	2.5③	—	2.5③	—	2.5	—	—
	Rear Band①	7②③	—	4③	—	2	—	—

NOTE: Numbers respresent back-off turns from specified torque. Torque lock-nut to 35 ft. lbs. when adjustment is completed.
① Backed off from 72 inch pounds torque
② Backed off from 41 inch pounds torque
③ With wide ratio gears.

the shift lever all the way to the front detent position. Tighten the lock bolt to 90 inch lbs.

3. Check the adjustment by making sure that: the detent positions for neutral and drive correspond with the stops for the lever gate; and that the starter will operate only with the shift lever in park and neutral positions.

American Motors

1. From under car, loosen the nuts on the trunnions (swivels).
2. Disengage the trunnion and shift rod at the bellcrank.
3. Place the shift lever in Park and lock the steering column.
4. Move the transmission lever to the rear-most detent position, which is Park.
5. Eliminate the backlash by pulling downward on the shiftrod and pressing upward on the outer bellcrank.
6. Adjust the trunnion on the shift rod to be a free fit into bellcrank arm, then tighten the jamnuts, making sure that the shift rod does not turn while tightening nuts.
7. Verify that the vehicle will only start in Park or Neutral.

Band Adjustments

REAR WHEEL DRIVE

Kickdown (Front) Band

The kickdown band adjusting screw is located on the left side of the transmission case above the throttle and shift linkage levers. On 4WD AMC models, it may be necessary to remove the front axle driveshaft to gain proper access to the adjusting screw and locknut.

1. Raise and safely support the front of the vehicle. Loosen the locknut and back off about five turns. Be sure the adjusting screw is free in the case.
2. Use a torque wrench and $5/16$ inch square socket, to tighten the adjusting screw to exactly 72 inch lbs.
3. Back off the adjusting screw exactly to specification. Hold the adjusting screw so that it does not turn and tighten the locknut to 35 ft. lbs.

Rear Band

The rear band adjustment is an inside adjustment so the pan must be removed.

1. Raise and safely support the vehicle.
2. Remove the oil pan and drain the fluid.
3. Look carefully at the fluid, filter and pan bottom for a heavy accumulation of friction material or metal particles. A little accumulation can be considered normal, but a heavy concentration indicates damaged or worn parts.
4. Adjust the band by loosening the locknut, then tightening adjusting screw to the specified torque, using a small torque wrench and a $1/4$ inch hex head socket.
5. Back off the adjusting screw to the specified amount of turns. Refer to the specification chart.
6. Install the locknut, tighten to 35 ft. lbs. making sure adjusting screw does not turn.

NOTE: Install a new transmission filter. Torque the three screws to 35 inch pounds.

7. Using a new gasket on the pan, install and torque bolts evenly to 150 inch pounds.
8. Lower the vehicle and fill the

transmission with the specified amount of Dexron® II type fluid.

FRONT WHEEL DRIVE

Kickdown (Front) Band
A-404

The kickdown band (front band) has its adjusting screw located on the top front (left side) of the transaxle case. Adjustment is as follows:

1. Loosen the locknut and back off about five turns.
2. Tighten the adjusting screw to 72 inch pounds.
3. Back off adjusting screw 3 turns for 1980 and later models, from the 72 inch pound torque, then hold this position and tighten the lock nut to 35 ft. lbs.

A-413, A-415 and A-470

The kickdown band has its adjusting screw located on the front left top of the transaxle case in the same location as the A-404 transaxle. The adjustment is as follows:

1. Loosen the locknut and back off approximately five turns.
2. Tighten the band adjusting screw to 72 inch lbs.
3. Back off the adjusting screw the correct number of turns as shown on chart.
4. Hold this position on the adjusting screw and tighten the lock nut to 35 ft. lbs.

Low-Reverse (Rear) Band
A-404

The low-reverse band (rear band) is not adjustable in this unit. The band lining itself needs to be inspected to determine the need for replacement. The grooves must be no less than

BAND ADJUSTMENT SPECIFICATIONS
Chrysler Corporation Transaxle

		Engine and Transmission Models				
		A-404			A-413, A-470	
		1.7L	2.2L	2.6L	2.2L	2.6L
'80	Front Band	3②	2.75②	—	—	—
	Rear Band	①	3.5③	—	—	—
'81–'83	Front Band	3②	3.0②	3.0②	2.0②	2.0②
	Rear Band	①	①	①	3.5③	3.5③
'84–'87	Front Band	—	—	—	2.5②④	2.5②④
	Rear Band	—	—	—	3.5③④	3.5③④

① Not adjustable
② Backed off from 72 inch pounds torque
③ Backed off from 41 inch pounds torque
④ 1.6L engine w/A415: 3.0
　Rear Band: Non-adjustable

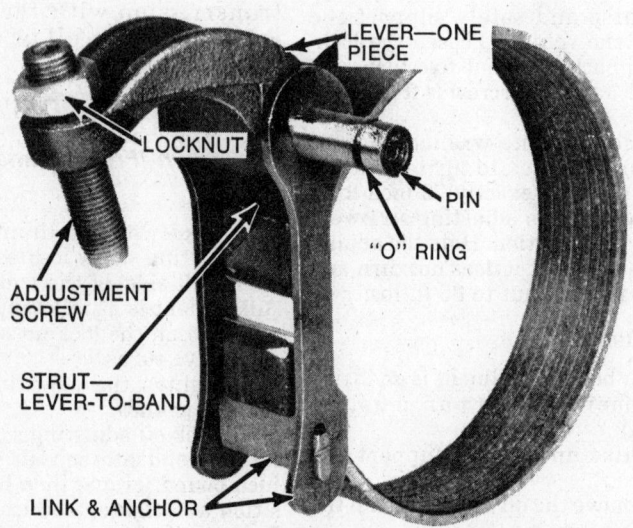

A-904 and AMC 998 low-reverse band components—typical

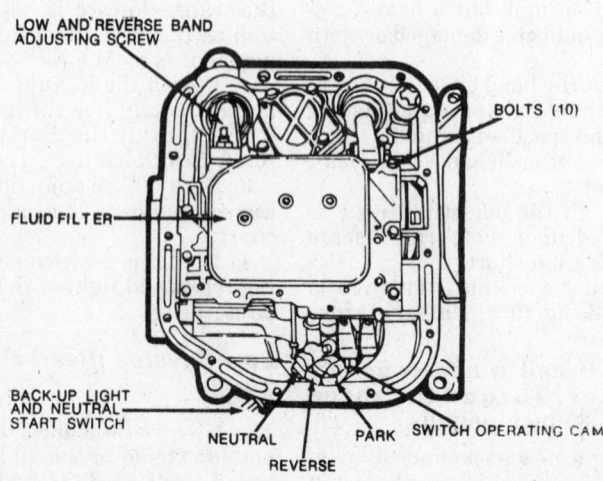

Low and reverse band adjusting screw location

0.008 in. (0.2mm) deep at any point to still be usable. With a 100 pound force applied to band around drum, the end gap must not be less than 0.020 in. (0.5mm).

A-413, A-415 and A-470

The low-reverse band (rear) is not adjustable on the A-415 model. On the A-413 and A-470 the band is adjustable with the bottom oil pan off. Before attempting the band adjustment, the low-reverse band should be checked for proper end gap as follows:

1. Remove the lower oil pan and pressurize the low-reverse servo with 30 psi air pressure.

2. Measure the gap between the band ends. If the gap is less than 0.080 in., the band is worn excessively and should be replaced. To adjust the low-reverse band, proceed as follows.

3. Loosen and back off the locknut approximately 5 turns.

4. Tighten the adjusting screw to 41 inch lbs. true torque.

3. Back off the adjusting screw the correct number of turns as shown on chart, hold adjusting screw position and tighten the locknut to 10 ft. lbs.

4. Reinstall the oil pan and fill the unit with correct type fluid.

Oil Pan

DRAINING, REMOVAL & INSTALLATION

No fluid or filter changes are required for the life of the car if it is used in normal service. Severe service (trailer towing, commercial use, police or taxi use) requires a fluid and filter change every 15,000 miles for Chrysler cars, or every 25,000 miles for AMC cars (refer to your owners manual). Band adjustments should be performed at the same intervals for cars used in severe service.

1. Drive the car until the transmission fluid is at normal operating temperature. Raise and safely support the vehicle.

2. Unbolt the pan. Be ready with a large container to catch the fluid.

NOTE: If the fluid smells burnt or is discolored, serious transmission troubles, probably due to overheating, should be suspected.

3. When the fluid is drained, remove the pan.

4. Unscrew and discard the filter.

5. Install a new filter. The proper torque is 28 inch lbs. for AMC, or 35 inch lbs. for Chrysler.

6. Clean out the pan, being extremely careful not to leave any lint from rags inside.

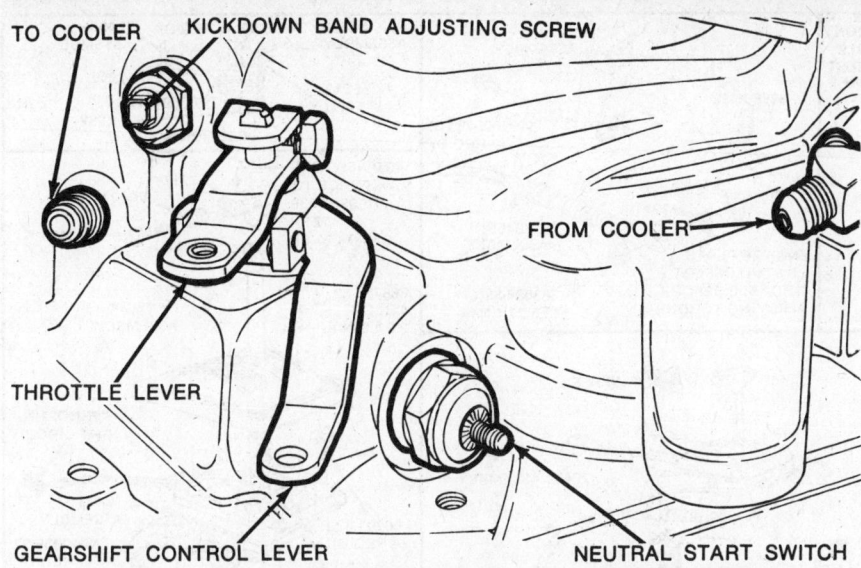

Torque-Command and TorqueFlite external adjustments and controls

7. Replace the pan, using a new gasket. On most front wheel drive models, RTV silicone sealer is used instead of a gasket. Peel off the old sealer and apply a ⅛ in. bead of silicone sealer to the pan flange. Run the bead around the inside of the bolt holes. Tighten the pan bolts to 10–12 ft. lbs. in a criss-cross pattern.

8. It is a good idea to measure the amount of fluid drained from the transmission, because some fluid will remain inside. Initially pour three quarts of Dexron® II automatic transmission fluid through the dipstick tube.

9. Start the engine in Neutral and allow it to idle for two minutes. Do not race the engine. Set the parking brake and shift through each position slowly, then move the lever to Park. Add fluid as necessary until correct level is reached.

FORD MOTOR COMPANY

Automatic Overdrive (AOD)

ADJUSTMENTS

Throttle Valve (TV) Control Linkage System

The throttle valve (TV) control linkage system consists of the linkage lever on the carburetor, the transmission control rod assembly and the external TV control lever on the transmission.

The TV control linkage is set to its proper length during initial assembly using the sliding trunnion block at the transmission end of the TV control rod assembly. Under normal circumstances, it should not be necessary to alter this adjustment. Any required adjustment of the TV control linkage can normally be accomplished using the adjustment screw on the linkage lever at the carburetor. Major linkage adjustment (sliding trunnion on rod) may only be required after maintenance involving the removal and/or replacement of the carburetor. TV control rod assembly or the transmission. Minor linkage adjustment (adjustment screw on linkage lever) may be required after idle speed adjustments greater than 50 rpm and to correct complaints of poor transmission shift quality.

When the linkage is properly adjusted, the TV control lever on the transmission will be at its internal idle stop position (lever up as far as it will travel) when the carburetor is at its hot idle stop with the engine off. There will be a light contact force between the throttle lever and end of the linkage lever adjustment screw. Due to flexibility in the linkage system, the linkage lever adjustment screw would have to be backed out approximately three turns before a gap between the screw and throttle lever could be detected.

At wide open throttle, the TV control lever on the transmission may or may not be at its wide open stop. The wide open throttle position must not be used as the reference point in adjusting the linkage.

Linkage Adjustment at Carburetor

The TV control linkage may be adjusted at the carburetor using the following procedure.

1. De-cam the fast idle cam on the carburetor so that the throttle lever is at its idle stop. Place the shift lever in Neutral (not PARK) and set the parking brake. Turn the engine OFF. On models other than 1985 and later cars with the 3.8 liter engine, skip to Step 3.

2. On 1985 and later models, the 3.8 liter engine has a DC motor idle speed control. The plunger on this control must be retracted before adjusting the TV linkage; you cannot perform the adjustment by merely turning the ignition off! Locate the Self-Test Connector and Self-Test Input Connector in the engine compartment, near the right side fender apron. Connect a jumper wire between the Self-Test Input Connector and signal return ground on the Self Test Connector. Turn the ignition switch to the RUN position without starting the engine, and leave it on for a full 10 seconds, to give the plunger time to retract fully. Then, turn off the ignition switch and remove the jumper wire.

3. Back out the linkage lever adjusting screw all the way (screw end is flush with lever face).

4. Turn in the adjusting screw until a thin shim (0. 005 in. max) or piece of writing paper fits snug between the end of the screw and the throttle lever. To eliminate the effect of friction, push the linkage lever forward (tending to close gap). Release it before checking clearance between the end of the screw and the throttle lever. Do not apply any load on the levers with tools or hands while checking the gap.

5. Turn in the adjusting screw an additional three turns. (Three turns are preferred. One turn minimum is permissible if screw travel is limited).

6. If it is not possible to turn in adjusting screw at least one additional turn or if there was insufficient screw adjusting capacity to obtain an initial gap in Step 2, refer to "Linkage Adjustment at Transmission."

Linkage Adjustment at Transmission

The linkage lever adjustment screw has a limited adjustment capability. If it is not possible to adjust the TV linkage using this screw, the length of the TV control rod assembly must be readjusted using the following procedure. This procedure must also be followed whenever a new TV control rod assembly is installed.

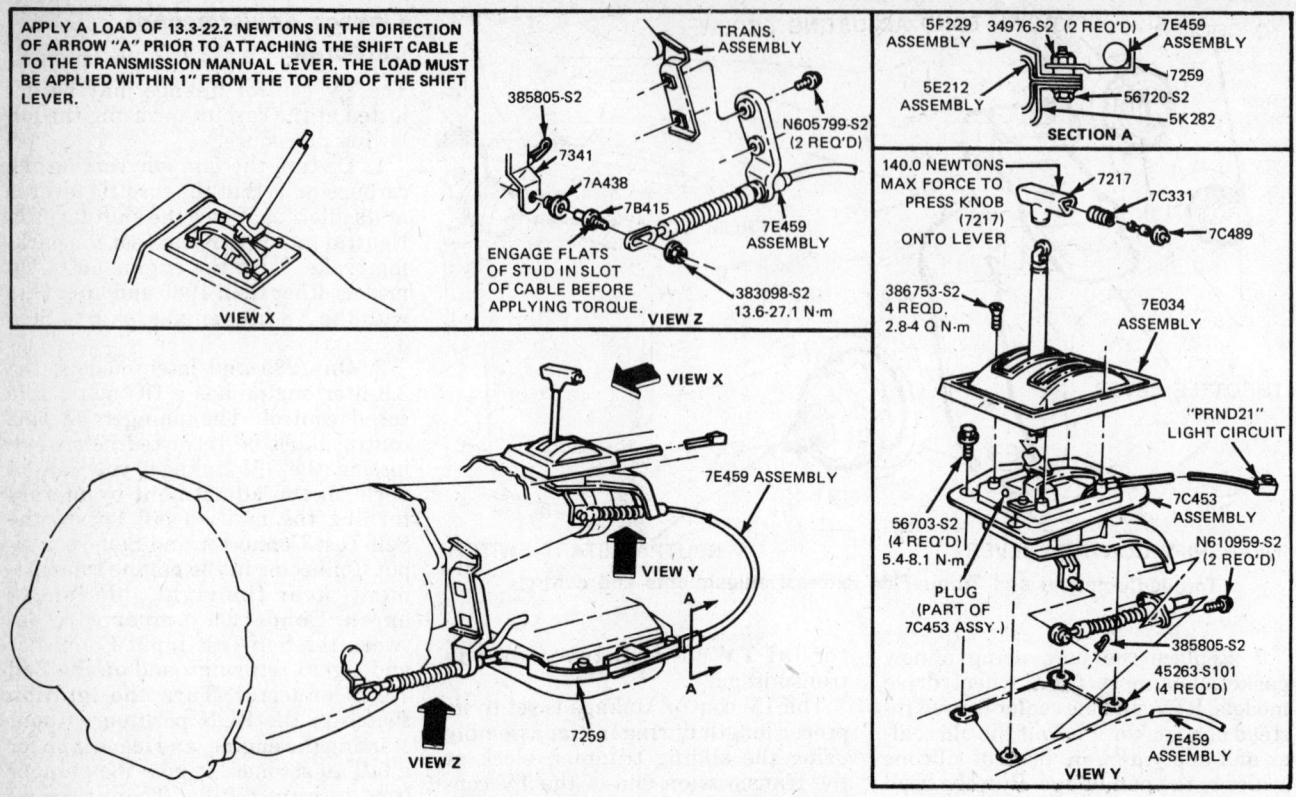

APPLY A LOAD OF 13.3-22.2 NEWTONS IN THE DIRECTION OF ARROW "A" PRIOR TO ATTACHING THE SHIFT CABLE TO THE TRANSMISSION MANUAL LEVER. THE LOAD MUST BE APPLIED WITHIN 1" FROM THE TOP END OF THE SHIFT LEVER.

VIEW X

TRANS. ASSEMBLY

385805-S2

7341

7A438

7B415

ENGAGE FLATS OF STUD IN SLOT OF CABLE BEFORE APPLYING TORQUE.

VIEW Z

N605799-S2 (2 REQ'D)

7E459 ASSEMBLY

383098-S2 13.6-27.1 N·m

5F229 ASSEMBLY

34976-S2 (2 REQ'D)

7E459 ASSEMBLY

5E212 ASSEMBLY

7259

56720-S2

5K282

SECTION A

140.0 NEWTONS MAX FORCE TO PRESS KNOB (7217) ONTO LEVER

7217

7C331

7C489

386753-S2 4 REQD 2.8-4 Q N·m

7E034 ASSEMBLY

"PRND21" LIGHT CIRCUIT

56703-S2 (4 REQ'D) 5.4-8.1 N·m

7C453 ASSEMBLY

N610959-S2 (2 REQ'D)

PLUG (PART OF 7C453 ASSY.)

385805-S2

45263-S101 (4 REQ'D)

7E459 ASSEMBLY

VIEW Y

VIEW X

7E459 ASSEMBLY

VIEW Y

VIEW Z

7259

A A

Typical cable-type floor shift mechanism

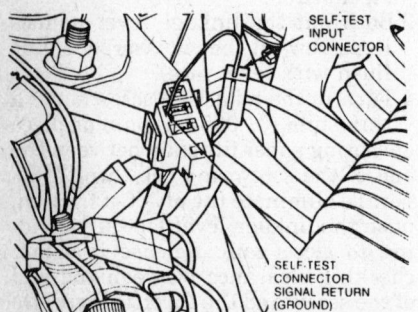

SELF-TEST INPUT CONNECTOR

SELF-TEST CONNECTOR SIGNAL RETURN (GROUND)

To retract the Idle Speed Control Solenoid plunger, jumper, the Self-Test Input Connector and the signal return ground of the Self-Test Connector

1. Set the engine curb idle speed to specification.

2. With the engine OFF, de-cam the fast idle cam on the carburetor so that the throttle lever is against the idle stop. Place the shift lever in Neutral and set the parking brake.

3. Set the linkage lever adjustment screw at approximately mid-range.

4. If a new TV control rod assembly is being installed, connect the rod to the linkage lever at the carburetor.

5. Loosen the bolt on the sliding trunnion block on the TV control rod assembly. Remove any corrosion from the control rod and free-up the trunnion block so that it slides freely on the control rod.

6. Push up on the lower end of the control rod to insure that the linkage lever at carburetor is firmly against the throttle lever. Release force on the rod. The rod must stay up.

7. Push the TV control lever on the transmission up against its internal stop with a firm force (approximately 5 pounds) and tighten the bolt on the trunnion block. Do not relax the force on the lever until the bolt is tightened.

Linkage Adjustment Using TV Control Pressure

The following procedure may be used to check and/or adjust the TV control linkage using TV control pressure.

1. Place the shift selector lever in Neutral and disconnect the idle kicker solenoid. Set the parking brake.

2. Attach a 0–100 psi pressure gauge to the TV Port on the transmission. It's best to use a gauge with eight feet of hose so you can read it while the engine is running. You may have to rig a piping elbow to keep the hose away from the exhaust system.

3. Operate the engine until normal operating temperature is reached and the throttle lever is off fast idle. Leave the engine running.

4. Verify that the throttle lever is at its idle stop. Follow the specified procedure below, depending on the model year of the car:
1980–81
   a. Observing the pressure gauge,

pull back slowly on the Throttle Valve rod at the carburetor. Watch for the pressure to increase suddenly approximately 15–30 psi with only a slight motion of the TV rod. This is the "breakpoint". It, ideally, should occur when the gap is $3/16$–$1/4$ in. The maximum specification is $1/16$–$5/16$ in. An effective means of measuring the gap is to insert standard size drills into the gap between the linkage lever adjustment screw and the throttle lever. If the breakpoint occurs at less than $1/16$ in., the TV linkages is set too long; if it occurs at greater than a $5/16$ in. gap, the linkage is set too short. Adjust the linkage as described in Step 6.
1982–83
   a. Follow the procedure for 1980–81 cars, but note that the ideal adjustment range is $5/32$–$7/32$.
1984–87
   a. Fabricate a block .397 plus or minus .007 in. thick, or use a Letter X drill or 10mm or $25/32$ in. drill. Insert the gauge block or drill shank between the throttle lever and adjusting screw on the transmission linkage lever. Then, note the transmission pressure. It must be 30–40 psi. Proceed to the next step to adjust.

6. Correct a long setting or low pressure by backing out the linkage lever adjustment screw. Turn in the

adjusting screw for a short rod or high pressure condition. If insufficient adjusting capacity is available, the TV control rod length must be reset using the procedure described in "Linkage Adjustment at Transmission". On 1984 and later models only: Remove the gauge block or drill, and recheck the pressure. It must be less then 5 psi. If necessary, back out the adjusting screw until pressure just drops below 5 psi. Reinstall the gauge block and verify that the pressure is still 30–40 psi.

7. If the limits specified cannot be obtained, diagnosis of the transmission control pressure system is required.

Shift Linkage

1. Loosen the linkage adjustment screw located in the engine compartment of Ford/Mercury full size and Lincoln/Mark VI, and under the car on Cougar/Thunderbird models.
2. Move the shift selector firmly against the overdrive gate stop. Hang a weight on the lever to hold it in place.
3. Locate the overdrive detent in the transmission, third from the front.
4. Tighten the adjustment screw. Recheck the adjustment before driving the vehicle.

Neutral Safety Switch

The neutral safety switch in the transmission is merely an on-off switch actuated by the manual linkage detent mechanism within transmission. The switch is replaceable if defective, but is non-adjustable.

Oil Pan

DRAINING, REMOVAL & INSTALLATION

It is not necessary, under normal operating conditions, to periodically change the transmission fluid. Only under severe conditions (police or taxi) or in the case of a major overhaul is it recommended that the fluid be changed.

When the reason for a fluid change is internal transmission damage, it is necessary to also flush the transmission cooler and cooler lines to remove any traces of abrasive matter from the system.

—— CAUTION ——
When flushing torque converters, use only professional equipment designed for this purpose. Traces of solvent in the converter could cause severe transmission damage in the future.

1. Raise and safely support the vehicle on jackstands.
2. Place a drain pan under transmission.
3. Loosen the transmission pan attaching bolts to drain fluid above pan level.
4. When the fluid has drained to the level of the pan mounting flange, remove the attaching bolts beginning at the rear of the pan. Gradually drop the pan and drain slowly.
5. After draining the fluid, remove and thoroughly clean the transmission pan. Discard the filter, valve body to filter gasket, and transmission pan gasket.
6. Install new filter and filter to valve body gasket. Do not attempt to clean and reuse the old filter.
7. Using a new gasket, install the pan on the transmission.
8. Add three quarts of the specified fluid Ford CJ or Dexron® II, through the fill tube.
9. Check and adjust the fluid level as necessary.

FMX Transmission

ADJUSTMENTS

Vacuum Diaphragm

The vacuum diaphragms used in production are nonadjustable. Adjustable units are available for installation in the transmission, allowing changes in the control pressures.

An adjusting screw is located in the vacuum nipple of the diaphragm and is accessible after removing the vacuum supply line from the diaphragm. Using a small screwdrive and turning the screw clockwise will increase the control pressure, while turning the screw counter-clockwise will decrease the control pressure. One complete turn of the adjusting screw will change the control pressure approximately 2–3 psi.

NOTE: The diaphragm should not be adjusted to provide pressures below the specified ranges to change shift engagement feel, as soft or slipping shift points could result and damage to the transmission could occur.

Throttle Linkage

1. Position the carburetor in the wide open throttle position (WOT).
2. Hold the kickdown rod downward with a $4\frac{1}{2}$ pound weight against the through detent stop.
3. Adjust the kickdown adjusting screw to obtain 0.010–0.080 in. clearance between the screw and the throttle arm.
4. Return the system to idle.

FRONT BAND ADJUSTMENT

1. Raise the vehicle and support on jackstands. Drain the transmission fluid from the transmission unit.
2. Remove the transmission pan, then remove the fluid screen and clip from the transmission. Remove all gasket material from the transmission pan and pan mounting surface of the transmission case.

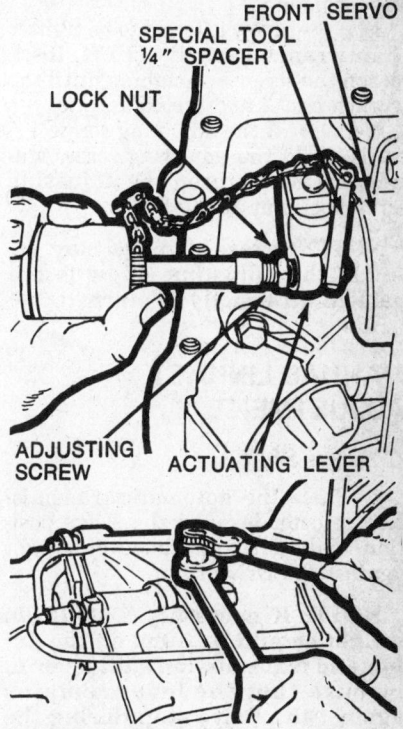

FMX front band (top) and rear band (bottom) adjustments

3. Loosen the front servo adjusting rod, and insert the $\frac{1}{4}$ in. spacer between the adjusting screw and the servo piston stem.
4. Tighten the adjusting screw to 10 inch pounds. Remove the spacer and tighten the adjusting screw an additional $\frac{3}{4}$ turn. Hold the adjusting screw stationary and tighten the locknut. Torque the locknut 20–25 ft. lbs.
5. Install the transmission fluid screen and clip. Install a new transmission pan gasket on the transmission pan. Mate the transmission pan to the transmission case and install the pan bolts.
6. Refill the transmission with the proper type of automatic transmission fluid.
7. Lower the vehicle. Start the engine and check for leaks. Correct as necesary.

REAR BAND ADJUSTMENT

1. Raise the vehicle and support on jackstands. Remove all dirt and for-

eign material away from the adjusting screw threads. Oil the threads.

2. Loosen the reverse band adjusting screw locknut. Using the proper tools, Ford Motor Co. tool numbers T700-7B200-B and T70P-7B200-A or equivalent, tighten the adjusting screw until the tool handle clicks. The tool is a preset torque wrench which clicks and breaks torque when the torque on the adjusting screw reaches 10 foot pounds.

3. If the screw is found to be tighter than wrench capacity (10 ft. lbs.), loosen the screw and tighten until the wrench clicks and breaks torque.

4. Back off the adjusting screw 1½ turns. Hold the adjusting screw stationary and tighten the adjusting screw lock nut 35–40 ft. lbs.

NOTE: Severe damage may result if the adjusting screw is not backed off exactly 1½ turns.

MANUAL LINKAGE ADJUSTMENT

Column Shift

1. Place the automatic transmission selector lever in the Drive position. Make sure that the lever is tight against the Drive stop.

NOTE: If necessary a 10–15 lb. weight should be hung on the automatic transmission shift lever to be sure that the lever remains against the Drive stop during the adjusting procedure.

2. Loosen the shift rod adjusting nut. On vehicles equipped with a shift cable, remove the nut, and remove the cable from the automatic transmission manual lever stud.

3. Shift the automatic transmission manual lever into the Drive position.

4. On vehicles equipped with a shift cable, place the cable end on the automatic transmission manual lever stud, using care to align the flats on the stud with the flats on the shift cable. Start the adjustment nut.

5. Make sure that the automatic transmission lever has not moved from the Drive stop. Torque the adjusting nut to 15 ft. lbs.

6. Check the transmission operation for all selector lever detent positions.

Floor Shift

1. Position the automatic transmission selector lever in the Drive position against the rearward Drive stop.
2. Raise the vehicle and loosen the manual lever shift rod retaining nut.

Move the automatic transmission manual lever to the Drive position.

3. With the automatic transmission selector lever and the manual lever in the Drive positions, torque the attaching nut to 15 ft. lbs.

4. Check the operation of the transmission in each selector lever position. Lower the vehicle.

NEUTRAL SAFETY SWITCH ADJUSTMENT

1980 Models

1. With the automatic transmission manual lever properly adjusted, loosen the two neutral safety switch bolts.

2. With the automatic transmission manual lever in neutral, rotate the switch and insert the gauge pin (No. 43 drill bit) into the gauge pin holes of the neutral safety switch.

NOTE: The gauge pin has to be inserted to a full ³¹⁄₆₄ in. into the three holes of the neutral safety switch.

3. Torque the neutral safety switch bolts to 55–75 inch lbs. Remove the gauge pin from the neutral safety switch.

4. Check the operation of the neutral safety switch. The engine should start with the automatic transmission selector lever in Park or Neutral only.

Vacuum Diaphragm

REMOVAL & INSTALLATION

1. Raise the vehicle and support it safely on jackstands.
2. Disconnect the vacuum diaphragm hoses. Remove the vacuum unit with tool No. S8696-A or equivalent.
3. Remove the vacuum unit control rod from the transmission case.
4. Place the vacuum control rod in the transmission case.
5. Thread the vacuum diaphragm unit into the transmission case using the special tool. Using a torque wrench, torque the unit to 15–23 ft. lbs.
6. Reinstall the vacuum hoses to their proper location. Lower the vehicle. Road test as required and correct as necessary.

Oil Pan

DRAINING, REMOVAL & INSTALLATION

1. Raise the vehicle and support it

safely on jackstands. Place a drain pan under the transmission.

2. Loosen the automatic transmission oil pan bolts and drain the automatic transmission fluid from the unit.

3. When the transmission fluid has drained to the level of the oil pan flange, remove the rest of the transmission oil pan bolts working from the rear and both sides of the pan to allow the transmission pan to drop and the fluid to drain slowly.

4. After the fluid has stopped draining, remove the oil pan. Clean all the old gasket material from the pan and the transmission case. Thoroughly clean the inside of the transmission pan.

5. Install a new transmission pan gasket on the transmission pan and place it against the transmission case.

6. Install all the transmission oil pan bolts. Torque the bolts to 12–17 ft. lbs.

7. Install three quarts of automatic transmission fluid Type "F" (converter not drained). Be sure to use the proper type transmission fluid. Failure to do this could result in serious internal transmission damage. When refilling a dry transmission and torque converter install five quarts of the proper type automatic transmission fluid.

8. Start the engine and operate it at idle speed for approximately two minutes and then raise the engine speed to approximately 1200 rpm until the engine/transmission assembly reaches normal operating temperature.

NOTE: Do not overspeed the engine during warm-up.

6. Check the fluid level after moving the gear selector through all shift ranges. Correct the fluid level as necessary.

C3 Transmission

ADJUSTMENTS

Vacuum Diaphragm

The vacuum diaphragms used in production are nonadjustable. Adjustable type units are available for installation in the transmission, allowing changes in the control pressures.

An adjusting screw is located in the vacuum nipple of the diaphragm and is accessible after removing the vacuum supply line from the diaphragm. Using a small screwdriver and turning the screw clockwise will increase the control pressure, while turning the screw counter-clockwise will decrease the control pressure. One complete turn of the adjusting screw will change the control pressure approximately 2–3 psi.

—— CAUTION ——

The diaphragm should not be adjusted to provide pressures below the specified ranges to change shift engagement feel, as soft or slipping shift points could result and damage to the transmission could occur.

Throttle Linkage

Throttle pressure control linkage is not used on the C3 automatic transmission. A vacuum-operated diaphragm assembly is used to control and modulate the throttle pressure in proportion to the road speed, throttle opening and internal oil pressures.

A downshift control rod is used and is connected to the carburetor control linkage. Both linkages must be properly adjusted to actuate the downshift system.

Accelerator and Downshift Linkage

1. With the engine OFF, fully depress the accelerator pedal and hold in place. Inspect the carburetor for wide open throttle plates and full accelerator linkage travel. Adjust as necessary.
2. With the accelerator fully depressed and adjusted, push the downshift control rod downward to its fully depressed position (downshift valve in the transmission fully depressed).
3. A clearance of .010–.080 in. between the tip of the kickdown adjusting screw and the throttle lever should exist. Adjust the screw as necessary.
4. Release the accelerator linkage. The linkage and the downshift control rod must return to their closed position by return spring tension.

Band Adjustment

NOTE: The intermediate band is the only adjustment needed during normal operation. The reverse band is adjusted internally during assembly or at times of overhaul.

1. Locate the adjusting screw on the left side of the transmission case, in front of the manual control lever.
2. Remove the downshift control rod from the downshift control lever to gain access to the adjusting screw and locknut.
3. Clean the dirt and foreign material from the locknut area. Loosen, remove and discard the locknut from the adjusting screw.
4. Install a new locknut on the adjusting screw.

5. Tighten the adjusting screw to 10 ft. lbs. of torque.

NOTE: A special wrench with a present "click," "break" or "overrun," can be used to tighten the adjusting screw.

6. Back the adjusting screw off exactly $1\frac{1}{2}$ turns; 2 turns for 1984 and later models.
7. Hold the adjustment and tighten the locknut to 35–45 ft. lbs. torque.
8. Install the downshift control rod to the downshift control lever.

Manual Linkage
COLUMN SHIFT

1. Place the selector lever in the "D" position against the "D" stop of the shift gate.

NOTE: The selector lever should be held by hand or by a weight, against the "D" stop of the shift gate during any adjustments.

2. Loosen the shift rod adjusting nut.
3. Position the transmission manual lever in the "D" detent, which is third from the front.

NOTE: The control rod may have to be disconnected at the adjusting nut and bolt to properly engage the transmission detent.

4. Recheck the selector lever so that it is against the "D" stop of the shift gate. Tighten the shift rod adjusting nut securely.

NOTE: Engage the flats of the stud in the control rod slot before tightening the nut, if equipped.

5. Operate the gear selector through the detents and check for proper alignment of the lever indicator to the shift detents. Readjust as necessary.

CONSOLE OR FLOOR SHIFT

1. Place the selector lever in the "D" position and against the rearward "D" stop of the shift gate. Hold in position while any adjustments are made.
2. Loosen the manual shift linkage adjusting nut and move the selector lever to the "D" position on the transmission (third position from front).
3. With the selector lever and the manual lever in the "D" positions respectively, tighten the shift linkage adjusting nut to 10–15 ft. lbs.

Neutral Safety Switch

NOTE: The C3 automatic transmission has a screwtype neutral

start switch, located above the manual control lever on the transmission case. The internal shift detent operates the switch contacts to allow the engine to start in either Park or Neutral positions and to operate the back-up lights when the transmission is in the Reverse position.

REMOVAL & INSTALLATION

1. Raise the vehicle, support safely. Disconnect the wire connector from the switch.
2. Use a thin wall socket and unscrew the switch from the transmission case.

—— CAUTION ——

Use only a thin wall socket and not a wrench to avoid crushing the switch during removal or installation.

3. Install a new O-ring on the switch and install the unit into the transmission case. Torque the switch 12–15 ft. lbs.
4. Install the wire connector to the switch.
5. Check the operation of the switch in each detent position. The engine should start in Neutral and Park positions only, and the back-up lights should be on when the transmission is in reverse.

Oil Pan

DRAINING, REMOVAL & INSTALLATION

1. Raise the vehicle and support safely.
2. Position a drain pan beneath the transmission pan and starting at the rear, loosen, but do not remove the pan bolts.
3. Loosen the pan from the transmission case allow the fluid to drain gradually.
4. Remove all pan bolts except two at the front of the pan and allow the fluid to continue draining.
5. Remove the pan; clean the old gasket from the pan and transmission case.
6. Install a new gasket on the pan and install it to the transmission case.
7. Install all pan bolts and torque to 12–17 ft. lbs.
8. Install three quarts of transmission fluid Type "F" prior to 1981;"CJ" or Dexron® II, 1981 and later into the filler tube (converter not drained). When refilling a dry transmission and converter, install five quarts of fluid into the transmission.

9. Start the engine and operate the engine at idle speed for approximately two minutes. Then raise the engine speed to approximately 1200 rpm until the engine/transmission assembly reaches normal operating temperature.

—— **CAUTION** ——

Do not overspeed the engine during warmup.

10. Check the fluid level after moving the gear selector through all ranges. Correct the fluid level as necessary.

Vacuum Diaphragm

REMOVAL & INSTALLATION

1. Raise the vehicle and support safely. Disconnect the vacuum hoses(s) from the diaphragm unit.
2. Remove the retaining bracket and bolt holding the diaphragm unit to the transmission case.

—— **CAUTION** ——

Do not pry or bend the retainer bracket.

3. Pull the vacuum diaphragm, the actuating pin and the throttle valve from the transmission case. Remove the O-ring from the assembly.
4. Install a new O-ring on the diaphragm unit.
5. Install the throttle valve, the actuating pin and the vacuum diaphragm tubes towards the transmission case and install the assembly into the case.
6. Install the retaining bracket and bolt and torque to 15–23 inch lbs.

C4 Transmission

ADJUSTMENTS

Vacuum Diaphragm

The vacuum diaphragm used in production is non-adjustable. Adjustable type units are available for installation in the transmission, allowing changes in the control pressures.

An adjusting screw is located in the vacuum nipple of the diaphragm and is accessible after removing the vacuum supply line from the diaphragm. Using a small screwdriver and turning the screw clockwise will increase the control pressure, while turning the screw counterclockwise will decrease the control pressure. One complete turn of the adjusting screw will change the control pressure approximately 2–3 psi. After adjustments are made reinstall the vacuum supply line.

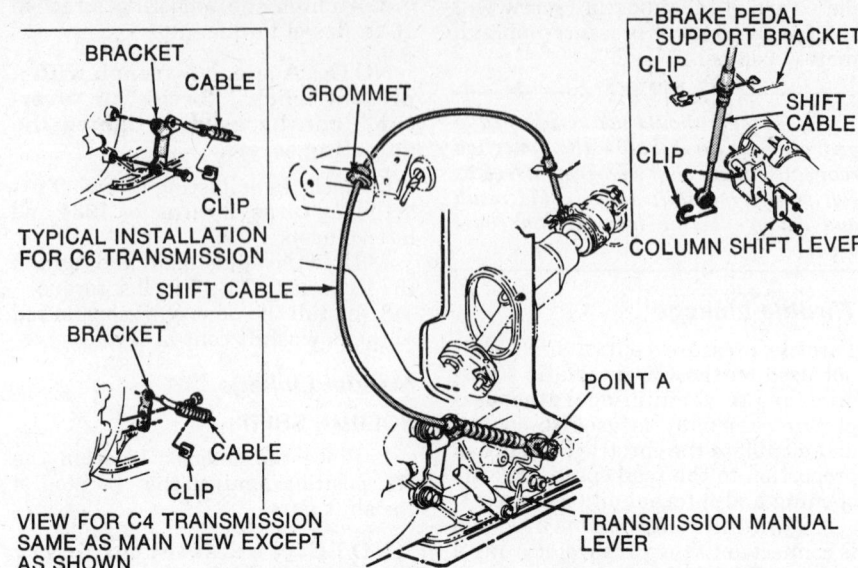

Typical cable-type column shift mechanism

NOTE: The diaphragm should not be adjusted to provide pressures below the specified ranges to change shift engagement feel, as soft or slipping shift points could result and damage to the transmission could occur.

Throttle Linkage

1. Hold the carburetor in the wide open throttle position (WOT).
2. Hold the kickdown rod downward with the specified weight against the through detent stop.
3. Adjust the kickdown screw to obtain 0.010–0.080 in. clearance between the screw and the throttle arm.
4. Return the system to idle position.

Band Adjustment

INTERMEDIATE BAND

1. Clean all dirt from the band adjusting screw area. Remove and discard the locknut.
2. Install a new locknut on the adjusting screw.
3. Torque the adjusting screw to 10 ft. lbs.
4. Back off the adjusting screw exactly 1¾ turns.
5. Hold the adjusting screw from turning and torque the locknut to 35–45 ft. lbs.

LOW AND REVERSE BAND

1. Clean all dirt from the band adjusting screw area. Remove and discard the locknut.
2. Install a new locknut on the adjusting screw.
3. Torque the adjusting screw to 10 ft. lbs.

4. Back off the adjusting screw exactly 3 full turns.
5. Hold the adjusting screw from turning and torque the locknut to 35–45 ft. lbs.

NOTE: In order to tighten the locknut on Pinto applications, special Ford tool T70P-7B200-B or equivalent may be needed.

Manual Linkage

COLUMN SHIFT

1. Place the automatic transmission selector lever in the Drive position. Make sure that the lever is tight against the drive stop.
2. Loosen the shift rod adjusting nut.

NOTE: On vehicles equipped with a shift cable, remove the nut, and remove the cable from the automatic transmission manual lever stud.

3. Shift the automatic transmission manual lever into the Drive position (at the transmission).
4. On vehicles equipped with a shift cable, place the cable end on the automatic transmission manual lever stud, using care to align the shift cable. Start the adjustment nut.
5. Make sure that the automatic transmission lever has not moved from the drive stop. Torque the adjusting nut to 15 ft. lbs.
6. Check the transmission operation for all selector lever detent positions.

FLOOR SHIFT

1. Position the automatic transmis-

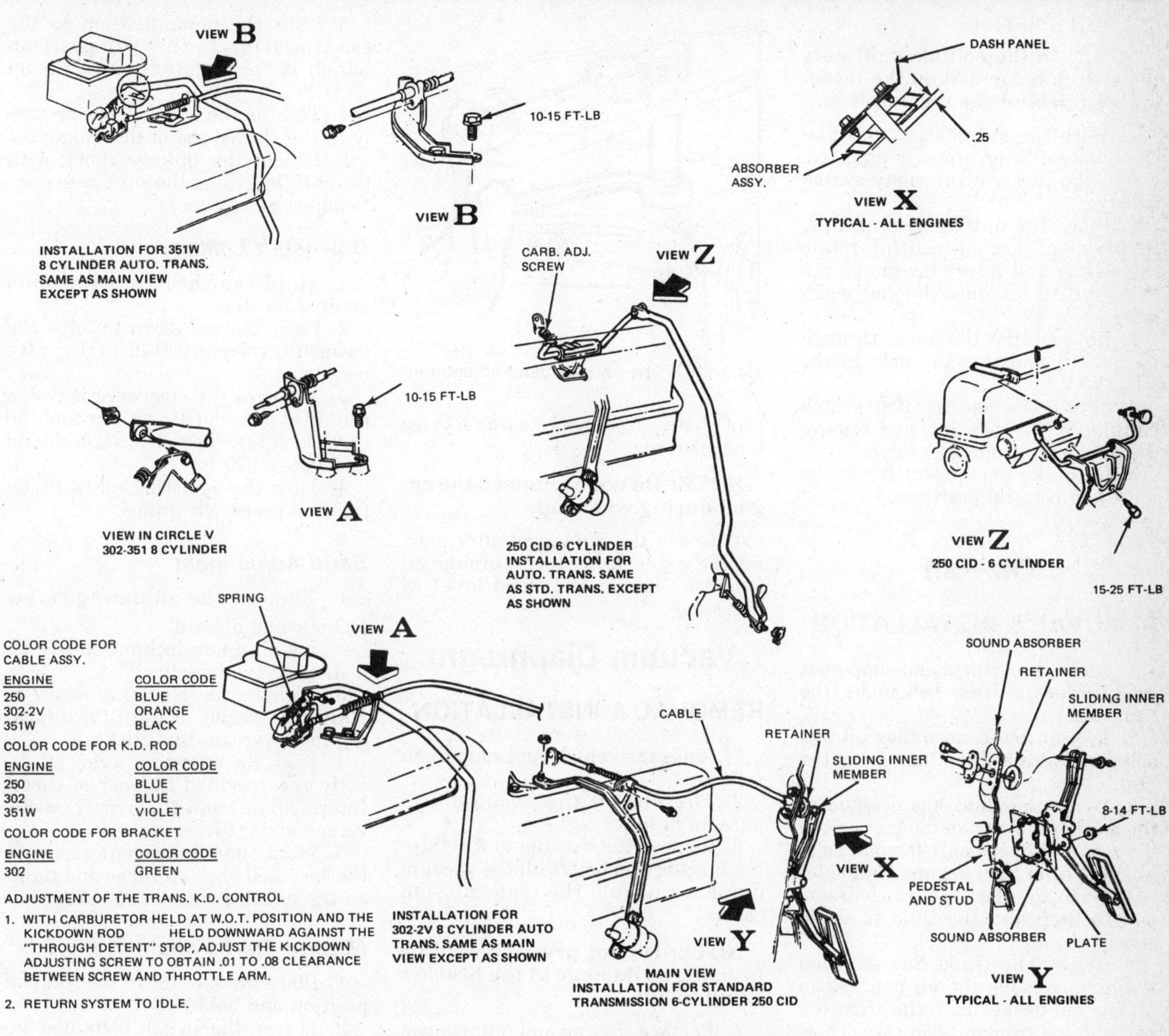

VIEW **B**

INSTALLATION FOR 351W
8 CYLINDER AUTO. TRANS.
SAME AS MAIN VIEW
EXCEPT AS SHOWN

10-15 FT-LB

VIEW **B**

DASH PANEL

.25

ABSORBER
ASSY.

VIEW **X**
TYPICAL - ALL ENGINES

CARB. ADJ.
SCREW

VIEW **Z**

10-15 FT-LB

VIEW **A**

VIEW IN CIRCLE V
302-351 8 CYLINDER

250 CID 6 CYLINDER
INSTALLATION FOR
AUTO. TRANS. SAME
AS STD. TRANS. EXCEPT
AS SHOWN

VIEW **Z**
250 CID - 6 CYLINDER

15-25 FT-LB

SPRING

VIEW **A**

COLOR CODE FOR
CABLE ASSY.

ENGINE	COLOR CODE
250	BLUE
302-2V	ORANGE
351W	BLACK

COLOR CODE FOR K.D. ROD

ENGINE	COLOR CODE
250	BLUE
302	BLUE
351W	VIOLET

COLOR CODE FOR BRACKET

ENGINE	COLOR CODE
302	GREEN

ADJUSTMENT OF THE TRANS. K.D. CONTROL

1. WITH CARBURETOR HELD AT W.O.T. POSITION AND THE
KICKDOWN ROD   HELD DOWNWARD AGAINST THE
"THROUGH DETENT" STOP, ADJUST THE KICKDOWN
ADJUSTING SCREW TO OBTAIN .01 TO .08 CLEARANCE
BETWEEN SCREW AND THROTTLE ARM.

2. RETURN SYSTEM TO IDLE.

INSTALLATION FOR
302-2V 8 CYLINDER AUTO
TRANS. SAME AS MAIN
VIEW EXCEPT AS SHOWN

CABLE

RETAINER

SLIDING INNER
MEMBER

VIEW **X**

VIEW **Y**

MAIN VIEW
INSTALLATION FOR STANDARD
TRANSMISSION 6-CYLINDER 250 CID

SOUND ABSORBER

RETAINER

SLIDING INNER
MEMBER

8-14 FT-LB

PEDESTAL
AND STUD

SOUND ABSORBER   PLATE

VIEW **Y**
TYPICAL - ALL ENGINES

Typical throttle and downshift linkage—except ATX

sion lever in the Drive position
against the drive stop.

2. Raise the vehicle and loosen the
manual lever shift rod retaining nut.
Move the automatic transmission
manual lever to the Drive Position.

3. With the automatic transmission selector lever and the manual lever in the Drive positions, torque the
attaching nut to 15 ft. lbs.

4. Check the operation of the transmission in each selector lever
position.

Neutral Safety Switch
FLOOR SHIFT

1. With the manual linkage properly adjusted and the engine turned
off, place the selector lever in the
Neutral position.

2. Remove the selector lever han-

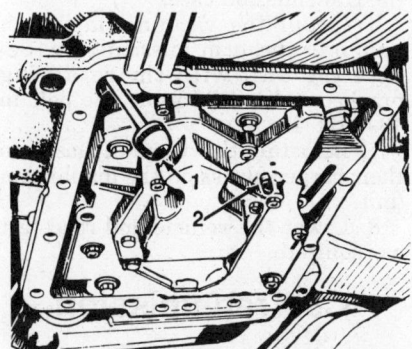

1. If found, this plug may be discarded (see text)
2. Throttle pressure limit valve and spring

**View of C4 with transmission oil pan re-
moved**

dle attaching screw and remove the
handle.

3. Remove the dial housing attaching screws and remove the housing.

4. Take out the two pointer backup
shield attaching screws and remove
the shield.

5. Loose the two screws securing
the neutral start switch to the selector
lever housing.

6. Place the selector lever in the
Park position; hold it against the forward stop.

7. Move the switch rearward to the
end of its travel.

8. Hold the switch in the rearward
position and tighten the two attaching screws.

9. With the selector lever in the
Park position, check the operation of
the switch. The engine should start in
the Park position. If the engine does
not start replace the switch.

10. Install the pointer back-up
shield, dial housing, and selector lever handle.

AUTOMATIC TRANSMISSIONS

COLUMN SHIFT

NOTE: In the column shift cars the switch is located on the transmission behind the manual lever.

1. With the automatic transmission manual lever properly adjusted, loosen the two neutral safety switch bolts.

2. Place the automatic transmission manual lever in neutral, rotate the switch and insert the gauge pin (No. 43 drill bit) into the gauge pin holes.

3. Be sure that the pin is through the switch and into the hole in the other wall.

4. Torque the neutral safety switch bolts to 55–75 inch lbs. and remove the gauge pin from the switch.

5. Check for engine starting in the Park and Neutral positions.

Oil Pan

REMOVAL & INSTALLATION

1. Raise the vehicle and support it safely. Place a drain pan under the transmission.

2. Loosen the transmission oil pan bolts and drain the fluid from the unit.

3. When the fluid has drained to the level of the oil pan flange, remove the rest of the transmission pan bolts working from the rear and both sides of the pan to allow the transmission pan to drop and the fluid to drain slowly.

4. After the fluid has stopped draining, remove the oil pan. Clean all the old gasket material from the pan and the transmission case. Thoroughly clean the inside of the transmission pan and screen.

5. Install a new pan gasket on the transmission pan and place the pan against the case.

6. Install all of the transmission pan bolts. Torque the bolts to 12–17 ft. lbs.

7. If the converter has not been drained, put three quarts of automatic transmission fluid in and run the transmission. Type "F" prior to 1980, "CJ" or Dexron® II, 1980 and later.

8. Check the fluid at normal operating temperature and add fluid if necessary.

9. When filling a dry transmission and torque converter install five quarts of the proper type fluid into the transmission.

10. Start the engine and operate at idle speed for approximately two minutes and then raise the engine speed to approximately 1200 rpm until the engine and transmission assembly

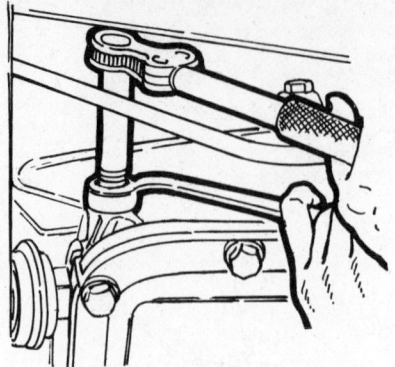

C4 and C5 low-reverse band adjustment

reaches normal operating temperature.

NOTE: Do not overspeed the engine during warm-up.

11. Check the fluid level after moving the gear selector lever through all ranges. Correct the fluid level as necessary.

Vacuum Diaphragm

REMOVAL & INSTALLATION

1. Raise the vehicle and support on jackstands.

2. Disconnect the vacuum diaphragm hoses.

3. Remove the vacuum unit retaining bracket and bolt. Pull the vacuum diaphragm from the transmission case.

NOTE: Do not pry or bend this bracket, as damage to the bracket could result.

4. Remove the vacuum unit control rod from the transmission case.

5. Place the vacuum control rod in the transmission case.

6. Install the vacuum diaphragm unit into the transmission case. Secure the unit with the retaining bracket and bolt. Torque the bolt to 13–16 ft. lbs.

7. Reinstall the vacuum hoses to their proper places on the diaphragm unit.

8. Lower the vehicle and road test as required.

C5 Transmission

ADJUSTMENTS

Manual Linkage

1. Loosen the nut or screw at the slotted rod in the linkage.

2. Put the selector in the "D" position, firmly against the shift gate stop.

3. Shift the manual lever on the transmission to the "D" position which is three detents away from "PARK".

4. Tighten the nut or screw securely in the slotted rod of the linkage.

5. Repeat the linkage check with the shift lever and the shift gate stop. Readjust as required.

Downshift Linkage

1. Hold the throttle wide open against its stop.

2. Push the rod down to force the downshift valve to bottom in the valve body.

3. Measure the clearance between the tip of the adjusting screw and the throttle lever. The clearance should be 0.050–0.070 in.

4. Turn the adjusting screw to obtain the proper clearance.

Band Adjustment

1. Remove the adjusting screw locknut and discard.

2. Install a new locknut on the adjusting screw, loosely.

3. Tighten the adjusting screw to 10 ft. lbs. torque, or until the adjusting tool overruns and clicks.

4. Back the adjusting screw off exactly the specified number of turns: Intermediate band $4\frac{1}{4}$ turns: Low-Reverse band 3 turns.

5. Hold the adjustment screw at the specified back-off turns and tighten the new locknuts to 35–45 ft. lbs.

Neutral Start Switch

1. Place the selector in the Neutral position and hold.

2. Loosen the switch bolts and insert a $\frac{3}{32}$ inch gauge pin or drill through the hole in the switch.

3. Wiggle the switch until the drill seats in the case.

4. Tighten the switch bolts to 55–75 inch lbs. torque and remove the drill.

Vacuum Diaphragm

Adjustment of the vacuum diaphragm is controlled by the installation of longer or shorter throttle valve rods to obtain the proper line pressure. Five selective rods are used.

NOTE: The following procedure will determine if a change in the length of the rod is required.

1. Attach a tachometer to the engine.

2. Attach a hand vacuum pump to the transmission vacuum diaphragm unit.

3. Attach a hydraulic pressure

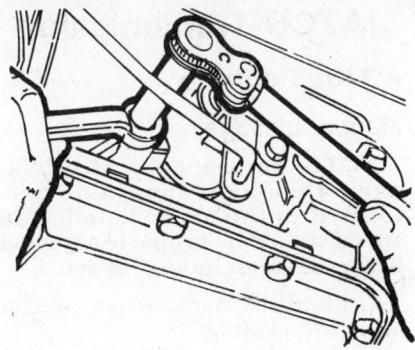

C4, C5 and C6 intermediate band adjustment

Length	Color Code
1.5925–1.5875	Green
1.6075–1.6025	Blue
1.6225–1.6175	Orange
1.6375–1.6325	Black
1.6585–1.6535	Pink and White

gauge to the control pressure outlet on the transmission.

4. Firmly apply the parking brake. On vehicles equipped with a vacuum brake release, apply the service brakes. Otherwise the parking brake will release when the selector is moved to Drive.

5. Start the engine, allow it to reach normal operating temperature.

6. Set the engine idle speed to the specified rpm.

7. Adjust engine speed to 1000 rpm and apply 10 inches of vacuum to vacuum diaphragm unit. Read and record the control pressure in all selector positions.

8. Compare the pressure readings from Step 7 to the specified pressure in the line Pressure chart and Proceed as follows: Pressure within specification no change required. Pressure below specification use the next longest rod. Pressure above specification use the next shortest rod. If the length of the rod is not known, it should be measured with a micrometer.

Oil Pan

DRAINING, REMOVAL & INSTALLATION

1. Raise the vehicle and support safely.

2. Loosen the oil pan retaining bolts, removing only enough bolts to tilt the oil pan and drain the fluid.

3. Carefully remove the remaining bolts and the oil pan from the transmission. Pour out the remaining fluid from the oil pan.

LINE PRESSURE CHART

Transmission Model	Range	10" Vacuum
PEN-C,G,J,K	D	#90–101
PEM-AL,AM	2,1	123–136
	R	151–168
PEP-E,F,G,H,P,N	D	#87–97
	2,1	119–132
	R	145–162
PEP-B,D	D	#86–99
	2,1	120–132
	R	143–165

NOTE: Refer to the ID tag for the transmission model.

NOTE: Discard the nylon shipping plug found in the bottom of the oil pan.

4. Thoroughly clean and remove all gasket material from the oil pan and the pan mounting surface of the transmission case.

5. Install new gasket and mount the pan to the transmission case.

6. Install the pan retaining bolts and torque to 12–16 ft. lbs.

7. Lower the vehicle and fill transmission with fluid. Type "H" only. Start the engine and recheck the fluid level. Correct as required.

C6 Transmission

ADJUSTMENTS

Throttle Linkage

1. Position the carburetor in the wide open throttle position (WOT).

2. Hold the kickdown rod downward with a 4½ lb. weight against the through detent stop.

3. Adjust the kickdown adjusting screw to obtain 0.010–0.080 in. clearance between the screw and the throttle arm.

4. Return the system to idle.

Band Adjustment

NOTE: The only adjustment that is needed on the C6 automatic transmission is the intermediate band adjustment.

1. Raise the vehicle and support on jackstands.

2. Clean all the grease and dirt away from the band adjusting screw area. Remove and discard the locknut.

3. Install a new locknut and torque

the adjusting screw to 10 ft. lbs. Back off the adjusting screw 1½ turns.

4. Hold the adjusting screw, so that it does not turn, and torque the locknut to 40 ft. lbs.

5. Lower the vehicle. Road test and correct as necessary.

Manual Linkage

COLUMN SHIFT

1. Place the transmission selector level in the (D) Drive position. Make sure that the lever is tight against the (D) Drive stop.

NOTE: If necessary a 10–15 lb. weight should be hung on the transmission shift lever to be sure that the lever remains against the Drive stop during the adjusting procedure.

2. Loosen the shift rod adjusting nut. On vehicles equipped with a shift cable, remove the nut, and remove the cable from the transmission manual lever stud.

3. Shift the transmission manual lever into the (D) Drive position.

4. On vehicles equipped with a shift cable, place the cable end on the transmission manual lever stud, using care to align the flats on the stud with the flats on the shift cable. Start the adjustment nut.

5. Make sure that the transmission lever has not moved from the (D) Drive stop. Torque the adjusting nut to 15 ft. lbs.

6. Check the transmission operation for all selector lever detent positions.

FLOOR SHIFT

1. Position the transmission selector lever in the (D) Drive position against the rearward (D) Drive stop.

2. Raise the vehicle and loosen the manual lever shift rod retaining nut. Move the transmission manual lever to the (D) Drive position.

3. With the transmission selector lever and the manual lever in the (D) Drive position, torque the attaching nut to 15 ft. lbs.

4. Check the operation of the transmission in each selector lever position. Lower the vehicle.

Neutral Safety Switch

ALL MODELS

1. With the transmission manual lever properly adjusted, loosen the two neutral safety switch bolts.

2. With the transmission manual lever in neutral, rotate the switch and insert the gauge pin (No. 43 drill bit) into the gauge pin holes of the neutral safety switch.

NOTE: The gauge pin has to be inserted to a full $^{31}/_{64}$ inch into the three holes of the neutral safety switch.

3. Torque the neutral safety switch bolts to 55–75 ft. lbs. Remove the gauge pin from the neutral safety switch.

4. Check the operation of the neutral safety switch.

NOTE: The engine should start with the transmission selector lever in (P) Park or (N) Neutral only.

Vacuum Diaphragm (Modulator)

ADJUSTABLE VACUUM DIAPHRAGM (MODULATOR)

1. Remove the vacuum hose from the vacuum nipple and insert a small screwdriver into the nipple end and engage the adjusting screw.

2. Adjust the pressure by turning the adjusting screw in to increase the pressure or by turning the adjusting screw out to decrease the pressure.

NOTE: One complete turn of the adjusting screw should change the pressure 2–3 psi, either higher or lower, depending upon the direction of rotation.

NON-ADJUSTABLE VACUUM DIAPHRAGM (MODULATOR)

1. Remove the modulator assembly and measure the selective pushrod.

NOTE: The pushrods can be identified by length, color code or part numbers.

2. Install a longer push rod to increase the pressure or a shorter push rod to decrease the pressure.

3. Each selective push rod increment should change the pressure 5–6 psi.

Oil Pan

DRAINING, REMOVAL & INSTALLATION

1. Raise the vehicle and support on jackstands. Place a drain pan under the transmission.

2. Loosen the automatic transmission oil pan bolts and drain the automatic transmission fluid from the unit.

3. When the transmission fluid has drained to the level of the oil pan flange, remove the rest of the transmission oil pan bolts working from the rear and both sides of the pan to allow the transmission pan to drop and the fluid to drain slowly.

4. When the fluid has drained to

the level of the pan flange, remove the rest of the pan bolts. Work from the rear and both sides of the pan to allow it to drop and drain slowly.

5. Install a new transmission pan gasket on the transmission pan and place it against the transmission case.

6. Install all the transmission oil pan bolts. Torque the bolts to 12–17 ft. lbs.

7. Install three quarts of automatic transmission fluid (converter not drained). Be sure to use the proper grade and type transmission fluid; failure
to do this could result in serious internal transmission damage. Use "CJ" or Dexron® II fluid.

8. When refilling a dry transmission and torque converter install five quarts of the proper grade and type automatic transmission fluid into the transmission.

9. Start the engine and operate the engine at idle speed for approximately two minutes and then raise the engine speed to approximately 1200 rpm until the engine/transmission assembly reaches normal operating temperature.

NOTE: Do not overspeed the engine during warm-up.

10. Check the fluid level after moving the gear selector through all ranges. Correct the fluid level as necessary.

Vacuum Diaphragm

REMOVAL & INSTALLATION

1. Raise the vehicle and support on jackstands.

2. Disconnect the vacuum diaphragm hoses.

3. Remove the vacuum unit retaining bracket and bolt. Pull the vacuum diaphragm from the transmission case.

NOTE: Do not pry or bend this bracket, as damage to the bracket could result.

4. Remove the vacuum unit control rod from the transmission case.

5. Place the vacuum control rod in the transmission case.

6. Install the vacuum diaphragm unit into the transmission case. Secure the unit with the retaining bracket and bolt. Torque the bolt 12–16 ft. lbs.

7. Reinstall the vacuum hoses to their proper places on the diaphragm unit.

8. Lower the vehicle and road test as required.

JATCO Transmission

ADJUSTMENT

Manual Linkage

NOTE: The manual control linkage adjustment should be performed in the order listed. Idle speed should be properly adjusted before manual linkage is set.

1. Place the transmission selector in the "N" position.

2. Raise the vehicle and disconnect the clevis from the lower end of the selector lever operating arm.

3. Move the transmission manual lever to the neutral position, third detent position from the back of the transmission.

4. Loosen the two clevis retaining nuts and adjust the clevis so that it freely enters the selector lever operating arm hole. Tighten the clevis retaining nuts to secure the adjustment.

5. Connect the clevis to the lever and secure it with the spring washer, flat washer and retaining clip.

6. Lower the vehicle and check the operation of the transmission in each selector lever position.

Band Servo

1. Raise and safely support the front of the vehicle on jackstands.

2. Remove the servo cover attaching bolts and take off the servo cover.

3. Loosen the band adjusting screw lock nut and tighten the adjusting screw to 9–11 ft. lbs. torque.

4. Back off the adjusting screw two turns. Hold the adjusting screw stationary and tighten the adjusting screw locknut to 22–29 ft. lbs. torque.

5. Lower the vehicle and check the transmission fluid level.

Neutral Safety Switch

1. With the manual linkage properly adjusted, place the transmission manual lever in the "N" position, third detent from the back of the transmission.

2. Remove the transmission manual lever retaining nut and lever.

3. Loosen the two neutral safety switch attaching bolts. Remove the screw from the alignment pin hole at the bottom of the switch.

4. Rotate the switch and insert a 0.079 in. diameter alignment pin through the alignment pin hole into the hole in the internal switch.

5. Tighten the two attaching bolts and remove the alignment pin.

6. Reinstall the alignment pin hole screw in the switch body.

7. Position the transmission manual lever on the manual lever shaft

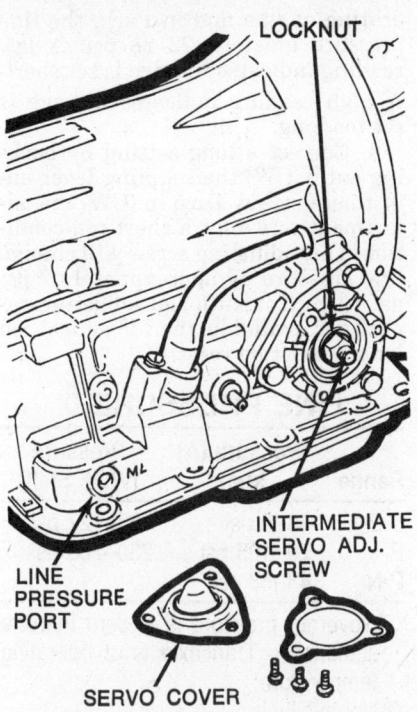

JATCO intermediate band adjustment

and secure it with the flat washer and nut.

8. Check the operation of the switch. The engine should start only with selector lever in Neutral or Park.

Vacuum Diaphragm

The vacuum diaphragms used in production are nonadjustable. If the adjustable type is on the transmission and is in need of adjustment follow the normal adjusting procedures.

An adjusting screw is located in the vacuum nipple of the diaphragm and is accessible after removing the vacuum supply line from the diaphragm. Using a small screwdriver and turning the screw clockwise will increase the control pressure, while turning the screw counterclockwise will decrease the control pressure. One complete turn of the adjusting screw changes the control pressure approximately 2–3 psi.

——— CAUTION ———
The diaphragm should not be adjusted to provide pressures below the specified ranges to change shift engagement fall, as soft or slipping shift points could result in damage to the transmission.

Downshift (Kickdown Switch)

1. Move the ignition switch to the ON position.
2. Loosen the kickdown switch to engage when the accelerator pedal is between $7/8 - 15/16$ of full travel. The

downshift solenoid will click when the switch engages.
3. Tighten the attaching nut and check for proper switch operation.

Oil Pan

DRAINING, REMOVAL & INSTALLATION

1. Raise the vehicle and support safely on jackstands.
2. Position a drain pan beneath the transmission pan and starting at the rear, loosen, but do not remove the pan bolts.
3. Loosen the pan from the transmission case and allow the fluid to drain gradually.
4. Remove all pan bolts except two at the front of the pan and allow the fluid to continue draining.
5. Remove the pan. Clean the remains of the old gasket from the pan and transmission case.
6. Install a new gasket on the pan and install on the transmission.
7. Install all pan bolts and torque to 4–5 ft. lbs.
8. Install three quarts of transmission fluid. "CJ" or Dexron® II, into the filler tube (converter not drained).
9. Start the engine and operate the engine at idle speed for approximately two minutes. Then raise the engine speed to approximately 1200 rpm until the engine and transmission reach operating temperature.

——— CAUTION ———
Do not overspeed the engine during warmup.

10. Check the fluid level after moving the gear selector through all ranges. Correct level as necessary.

Vacuum Diaphragm

REMOVAL & INSTALLATION

1. Raise the vehicle and support safely. Disconnect the vacuum hoses(s) from the diaphragm unit.
2. Turn the threaded diaphragm unit to remove it from the transmission case.
3. Pull the actuating pin and the throttle valve from the transmission case.
4. Remove the O-ring from the assembly.
5. Install a new O-ring on the diaphragm unit.
6. Install the throttle valve, the actuating pin and the vacuum diaphragm tubes toward the transmission case and install the assembly into the case.

7. Install and tighten the vacuum diaphragm, connect vacuum hose(s).

ZF Transmission

ADJUSTMENTS

Manual Lever Linkage

1. Put the selector lever in the Overdrive position, tight against the stop and retain it with a weight. With a floorshift model, block the lever rearward.
2. Disconnect the linkage at the manual lever on the transmission or bellcrank.
3. Shift the lever fully counterclockwise; then shift back three detents (to the Overdrive position). The ZF transmission has an extra detent which is not used, do not be confused.
4. Connect the linkage. Check that the gear selector is still in the Overdrive position and tighten the nut securely.

Kickdown Cable

1. Set the injector pump top lever at the full throttle position.
2. Loosen the front adjusting nut. Tighten the rear adjusting nut on the threaded barrel until a gap of 1.54–1.57 inch exists between the edge of the crimped bead on the cable closest to the barrel and the end of the threaded barrel.
3. Tighten the forward adjusting nut to lock the cable assembly to the bracket and recheck the adjustment gap.

Pan and Filter

DRAINING, REMOVAL & INSTALLATION

1. Raise and support the vehicle safely on jackstands.
2. Remove the drain plug and catch the fluid in a suitable container.
3. Remove the bolt that attaches the filler stub tube to the converter housing and disconnect the tube from the oil pan.
4. Remove the bolts and clamps that secure the pan to the transmission.
5. Use a Torx® bit No. 27 to remove the three bolts that attach the oil screen to the valve body.
6. Install a new O-ring on the screen and install.
7. Position a new gasket on the cleaned oil pan and install with edge clamps in position. Connect the filler tube. Add three quats of Dexron®II fluid, run the engine and fill the transmission to the correct level.

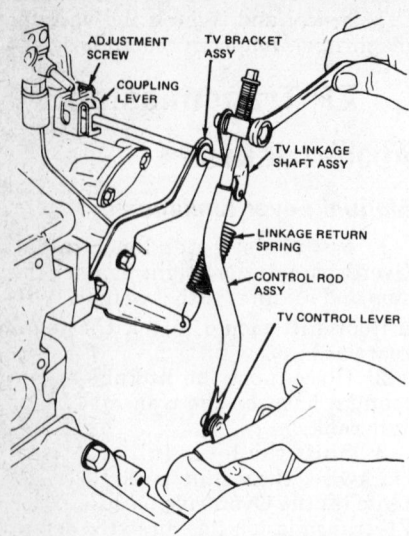

Throttle linkage adjustment—all models with an ATX

Automatic Transaxle (ATX)

ADJUSTMENTS

Manual Linkage

NOTE: This is a critical adjustment. Be sure the "D" detent in the transaxle corresponds exactly with the stop in the console.

1. Raise and support the vehicle safely on jackstands. Position the selector lever in the drive position, against the rearward stop. Hold in position while the adjustment is being done.
2. Loosen the manual lever to control cable retaining nut and move the transmission manual lever to the Drive position, second detent from the most rearward position.
3. Having both the transaxle detent and the shift stop in the console correspond, tighten the retaining nut 10–15 ft.lbs.
4. Lower the vehicle and verify the adjustment. Be sure the park mechanism and the neutral start switch function properly.

Throttle Linkage
MANUAL

NOTE: The TV control linkage must be adjusted at the TV control rod assembly sliding trunnion block using the following procedure.

1. Set the engine curb idle speed to specification.
2. After the curb idle check, turn the engine off and insure that the carburetor throttle lever is against the hot engine curb idle stop. The choke must be off.

NOTE: The linkage cannot be properly set if the choke is allowed to cool and the throttle lever allowed to be on the choke fast idle cam.

3. Set the coupling lever adjustment screw at its approximate mid-range. Insure that the TV linkage shaft assembly is fully seated upward into the coupling lever.
4. Loosen the bolt on the sliding trunnion block on the TV control rod assembly one turn minimum.

CAUTION

The following steps involve working in proximity to the EGR system. Allow the EGR system to cool before proceeding.

5. Free-up the trunnion block so that it slides freely on the control rod.
6. Rotate the transaxle TV control lever up using one finger and a light force, approximately 5 pounds, to insure that the TV control lever is against its internal idle stop. Without relaxing the force on the TV control lever, tighten the bolt on the trunnion block to specification (7–11 ft. lbs.).
7. Verify that the carburetor throttle lever is still against the hot engine curb idle stop.

USING LINE PRESSURE

NOTE: The following procedure may be used to check and/or adjust the TV control linkage using a line pressure gauge.

1. Place the shift selector lever in the Park position.
2. Apply the emergency brake.
3. Attach a 0–300 psi pressure gauge to the line press port on the transaxle with sufficient flexible hose to make gauge accessible while operating engine.
4. Operate the engine until normal operating temperature is reached and the throttle lever is against the hot engine curb idle stop (with A/C off, if so equipped).
5. Verify that the coupling lever adjusting screw is in contact with the TV linkage shaft assembly. If not, then the linkage must first be readjusted.
6. Verify that the carburetor throttle lever is against its hot engine curb idle stop. With the engine operating at idle and in Park, line pressure must be 43–59 psi. If line pressure is greater than 59 psi, the TV control linkage is set too long.
7. Place a 4mm drill (a $\frac{5}{32}$ in. drill or 0.157 gauge pin) between the coupling lever adjustment screw and the TV linkage shaft. With the engine op-

erating at idle and in Park, the line pressure must be 72–88 psi. A low reading indicates linkage is set short. A high reading indicates linkage is set too long.

8. Correct a long setting by backing out (CCW) the coupling lever adjustment screw. Turn in (CW) the adjustment screw for a short rod condition. This adjusting screw will change line pressure by approximately 2 psi per turn. If insufficient adjusting capacity is available, the TV control rod length must be reset.

LINE PRESSURE①

Range	Pressure (At Idle)	Pressure (WOT Stall)
D-2-1	43–5 psi	105–127 psi
R	70–105 psi	230–285 psi
P-N	43–58 psi	②

① Governor pressure is at zero (vehicle stationary). Transaxle is at operating temperature.
② Not available.

Neutral Safety Switch

1. Loosen the two switch retaining bolts and place the manual lever in the Neutral position.
2. Insert a $\frac{3}{32}$ inch drill bit through the hole in the neutral start switch.
3. Move the neutral switch until the drill seats in the case.
4. Torque the neutral start switch retaining bolts to 7–9 ft. lbs.
5. Remove the drill from the switch.

Band Adjustment

NOTE: The band adjustment is done during a transaxle overhaul with the use of special tools. Selective sized servo pistons are used to correctly position the band for its application.

Pan & Filter

DRAINING, REMOVAL & INSTALLATION

1. Raise the vehicle and support on jackstands.
2. Place a drain pan under the transaxle.
3. Loosen the pan attaching bolts and drain the fluid from transaxle.
4. When the fluid has drained to the level of the pan flange, remove the rest of the pan bolts. Work from the rear and both sides of the to allow it to drop and drain slowly.
5. When all of the fluid has drained from the transaxle, remove and thor-

oughly clean the pan. Discard the gasket.

6. Remove the three retaining bolts and remove the filter. Discard the seal.

7. Install a new oil filter and seal. Tighten the bolts 7–9 ft. lbs.

8. Place a new gasket on the oil pan and install the oil pan on the transaxle case. Tighten the retaining bolts to 15–19 ft. lbs.

9. Fill the transaxle to the correct level with Dexron® II automatic transmission fluid.

Neutral Start Switch

REMOVAL & INSTALLATION

1. Disconnect the negative battery cable.

2. Remove the two managed air valve supply rear hoses and all vacuum lines from the managed air valve.

3. Remove the managed air valve supply hose band to intermediate shift control bracket attaching screw.

4. Remove the air cleaner.

5. Disconnect the neutral start switch connector.

6. Remove the two neutral start switch attaching bolts.

7. Remove the neutral start switch.

8. Install the neutral start switch on the manual shaft.

9. Loosely install the two neutral start switch attaching bolts and washers.

10. Using a No. 43 drill (0.089 in.), set the neutral start switch.

11. Tighten the attaching bolts to 7–9 ft. lbs.

12. Connect the neutral start switch connector.

13. Install the managed air valve supply hose band to intermediate shift control bracket attaching screw.

14. Connect the two managed air valve supply rear hoses and all vacuum hoses to the managed air valve.

15. Install the air cleaner.

16. Connect the battery.

17. Start the engine in both Park and Neutral.

GENERAL MOTORS CORPORATION

NOTE: To identify the transmission used in your vehicle, refer to the transmission oil pan outlines at the beginning of this section.

LINKAGE ADJUSTMENT PROCEDURE REFERENCE

COLUMN SHIFT	
Rear Wheel Drive Models	
All	Procedure 1
Front Wheel Drive Models	
Except Eldorado, Riviera, Seville, Toronado	Procedure 2
Eldorado, Riviera, Seville, Toronado	Procedure 3
w/shift linkage①	—
w/shift cable②	—
FLOOR SHIFT	
Rear Wheel Drive Models	
Except Corvette w/TH400, and Chevette/T1000	Procedure 4
Corvette w/TH400	Procedure 6
Chevette/T1000	Procedure 5
Front Wheel Drive Models	
All w/TH125, 125C, and 440-T4	Procedure 2

① Procedure 1 may be used, but note that the adjusting clamp location is different than that of the rear wheel drive cars.
② Not adjustable.

SHIFT LINKAGE/CABLE ADJUSTMENT

NOTE: Refer to the accompanying Linkage Adjustment Reference Chart to determine the procedure which must be used for your vehicle.

Procedure 1

1. Loosen the screw on the shift linkage clamp.

2. Set the lever on the transmission into Neutral by moving it counterclockwise to the L1 detent, then clockwise three detent positions to Neutral.

3. Place the transmission selector lever (in the car) in Neutral as determined by the stop in the steering column. Don't use the indicator pointer for reference.

4. Tighten the shift linkage screw.

5. Check that the key cannot be removed and the steering wheel is not locked with the key in Run and the transmission in Reverse. Check that the key can be removed and that the steering wheel and transmission linkage is locked when the key is in Lock and the transmission is in Park. Be sure that the car will start only in Park and Neutral. If it starts in any gear, the neutral start switch must be adjusted. Start the engine and check for proper shifting into all ranges.

Procedure 2

1. Place the shift selector in Neutral.

2. From under the hood, loosen the nut which attaches the shift cable end to the transmission lever pin.

3. Place the transmission lever in Neutral by turning it clockwise from the Park position. The Neutral position is the second detent FROM the Park (fully counterclockwise) position.

4. With both the shift selector and transmission levers in Neutral, assemble the shift cable to the transmission lever pin and tighten to 15 ft. lbs. (use 20 ft. lbs. on 1986 and later models).

NOTE: Pull the transmission lever out of Park and hold it stationary while tightening the nut or an incorrect adjustment will result.

5. Check for proper operation.

Procedure 3

1. Pull the relay rod fully upward (Park), then push the rod down to the third detent (Neutral) position. The rod should be centered in this position.

2. Loosen the adjustment screw of the relay lever rod clamp.

3. Position the shift selector lever in Neutral.

4. Tighten the clamp adjusting screw to 20 ft. lbs. Refer to the Caution which follows Step 4 of the previous procedure.

5. Check for proper operation.

Procedure 4

NOTE: All models in this group use a cable operated shift selector control. Some models also use an

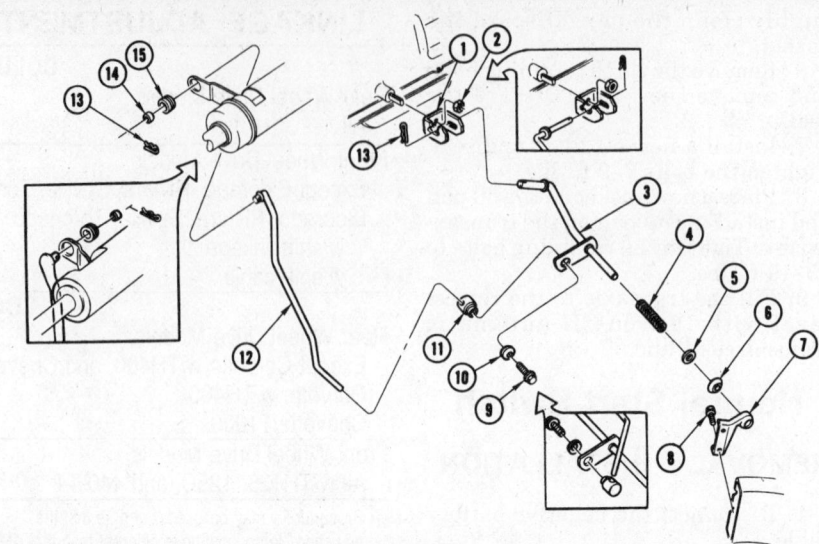

1. Lever
2. Nut (⅜"–16)
3. Equalizer shaft assembly
4. Anti-rattle spring
5. Flat washer
6. Pivot bushing
7. Shaft-to-frame bracket
8. Screw (5/16"–12)
9. Screw (5/16"–18 × ½")
10. Washer
11. Swivel clamp
12. Gearshift control rod
13. Retainer
14. Control rod sleeve
15. Bushing

Typical rear wheel drive column shift linkage. Note that the location of the swivel clamp (11) may differ slightly between models.

adjustable vertical linkage rod which is used for the steering column interlock. For models with an interlock rod, start at Step 1. Models that do not use an interlock, start at Step 7.

1. Models With an Interlock Rod: Place the shift selector lever in the park position.

2. Loosen the cable pin nut at the transmission lever. The pin must be able to slide freely within the slot of the lever.

3. Loosen the swivel clamp (bolt) of the column interlock rod. The rod must be able to slide freely within the clamp.

4. Push the transmission lever fully into the Park detent, then tighten the cable pin nut.

5. Lightly pull the interlock rod downward against the stop, then tighten the swivel clamp bolt. Refer to the Caution following Step 4 of "Procedure 2".

6. Check for proper operation.

7. Models Without an Interlock Rod; Place the shift selector in the Neutral position.

8. Loosen the cable pin nut at the transmission lever. The pin must be able to slide freely within the slot of the lever.

9. Verify that the transmission lever is in the Neutral position by turning the lever fully clockwise, then bring it back 2 detents counterclockwise.

10. Tighten the cable pin nut and check the operation of the shifter mechanism.

Procedure 5

NOTE: The shifter assembly

1. Transmission control cable
2. Cable yoke
3. Cable clip
4. Screw
5. Cable pin
6. Washer
7. Bolt (M8 × 1.25 × 20)
8. Cable mounting bracket
9. Nut (M10 × 1.5)
10. Washer
11. Selector lever

12. Selector lever pin
13. Shift lever
14. Pin

Typical front wheel drive column shift cable and related components—except Eldorado, Riviera, Seville, and Toronado. Note that the cable attachment at the transaxle is the same for floor shift models.

and linkage is basically the same on all models, though the point of adjustment is different between the THM180 and THM200 transmissions.

1. Place the shift selector lever in Neutral.

2. Disconnect the threaded end of the shift selector rod from either the

transmission lever (THM180's) or the shifter lever (THM200's).

3. Rotate the transmission lever fully clockwise (Park), then bring it back (counterclockwise) 2 detents, to Neutral.

4. The shift selector rod link (THM180's) or the clevis (THM200's) should align with its respective mat-

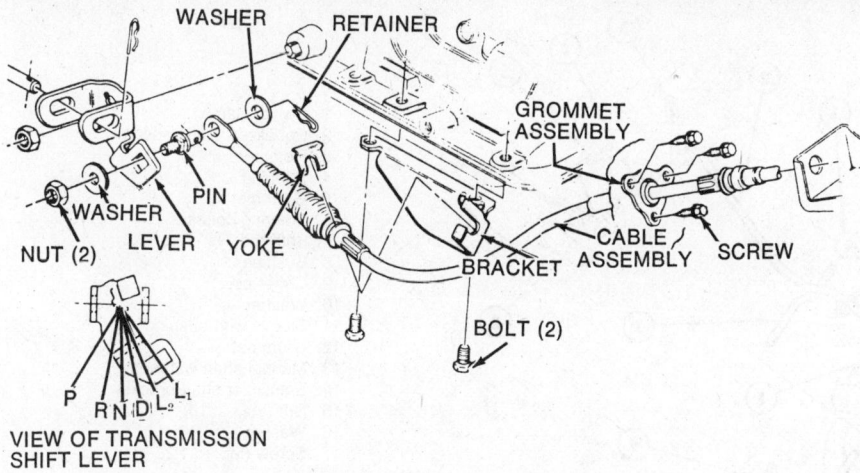

VIEW OF TRANSMISSION SHIFT LEVER

Typical floorshift cable control

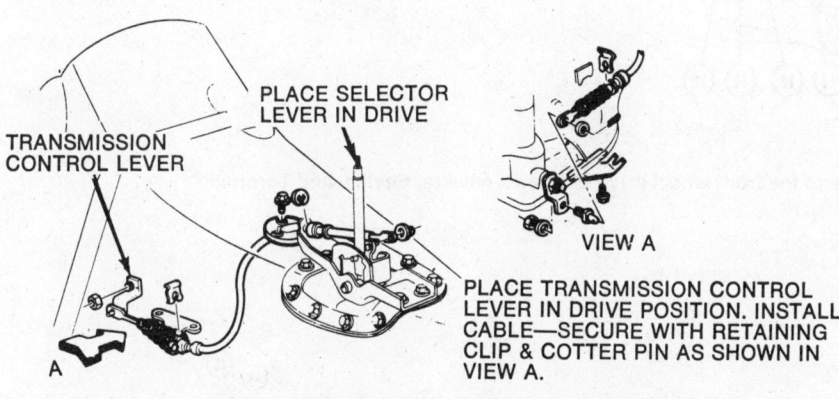

PLACE SELECTOR LEVER IN DRIVE

TRANSMISSION CONTROL LEVER

VIEW A

PLACE TRANSMISSION CONTROL LEVER IN DRIVE POSITION. INSTALL CABLE—SECURE WITH RETAINING CLIP & COTTER PIN AS SHOWN IN VIEW A.

Typical Corvette shift cable adjustment

ing part. If necessary, turn the link (THM180's) or clevis (THM200's) to lengthen or shorten the selector rod as required.

5. Connect the rod after the proper alignment is obtained.

6. Check for proper operation.

Procedure 6

1. Loosen the nut on the transmission lever so that the pin can move in the slot. Remove the console cover.

2. Move the transmission lever counterclockwise to the L1 position and then clockwise five detents to Park.

3. Place the shift lever in Park and insert a 0.40 in. spacer in front of the pawl.

4. Tighten the nut on the transmission lever to 20 ft. lbs.

5. Turn the ignition switch to Lock with the shift lever in Park.

6. Remove the cotter pin and washer from the backdrive cable at the column lever. Disconnect the cable.

7. Working the dash, remove the two nuts at the steering column-to-dash column-to-dash bracket.

8. Turn the lock tube lever counter-

clockwise (when viewed from the front of the column) to remove any free-play from the column.

9. Move the bracket until the cable eye passes freely over the retaining pin on the bracket.

10. While holding the bracket in place, have an assistant tighten the bracket retaining nuts.

11. Install the cotter pin and washer to retain the cable to the lever retaining pin.

THROTTLE VALVE (TV) or DETENT CABLE ADJUSTMENT

NOTE: On all transmissions except the THM250, THM350 and TH400/425 models, a TV cable is used to control hydraulic pressures, shift points, shift feel, and downshifting. The THM250 and THM350 use a detent cable. TH400/425 models use an electrical detent (downshift) switch. Adjustment of this switch is covered separately, after cable adjustment.

Though the detent cable is virtually identical in appearance to the TV cable, the detent cable controls only the downshift functions of the transmission. The difference in terminology (TV vs. detent) is due to the function of the cable, not the design of the cable itself.

Before attempting adjustment, identify the style of the cable which is used in your vehicle (Type One or Type Two). The Type One Cable uses a snap-lock assembly which is integral with the cable. The snap-lock is located next to the cable mounting bracket at the engine. In its normal position, the snap-lock is pushed fully downward (locked) so that it is flush with the snap-lock assembly. Adjustment is made with the snap lock in the raised (unlocked) position, as outlined in the Type One adjustment procedure.

The Type Two cable uses a different type of locking mechanism. Though it is located in the same position as the Type One snap-lock, the Type Two lock tab is set when released (upward) and adjusted when pushed downward (unlocked). On 1981 and earlier models, the Type Two cable can be easily identified by the presence of a spring, visible beneath the lock tab. The spring is not exposed on 1982 and later cable assemblies. Refer to the Type Two adjusting procedure to adjust the cable.

Type One

1. Remove the engine air cleaner assembly.

2. Push up on the bottom of the snap-lock at the cable bracket. Make sure that the cable is free to slide through the snap-lock. On diesels, disconnect the throttle rod from the throttle lever.

3. On all but 125 automatic transaxle with the L-4 engine, move the carburetor lever to the wide open throttle position and hold it there. On the 125 automatic transaxle with the L-4 engine, rotate the carburetor idler lever to "full travel stop" (carburetor open) and hold it in this position. On diesels, rotate the throttle lever to the "full travel stop" and hold it in this position.

4. Push the snap-lock fully downward and let the lever return to the closed position.

5. If the adjustment does not correct late shifting or no part throttle downshift, a transmission fluid pressure test should be made by a qualified mechanic.

6. On diesels, reconnect the throttle rod. Reinstall the air cleaner assembly.

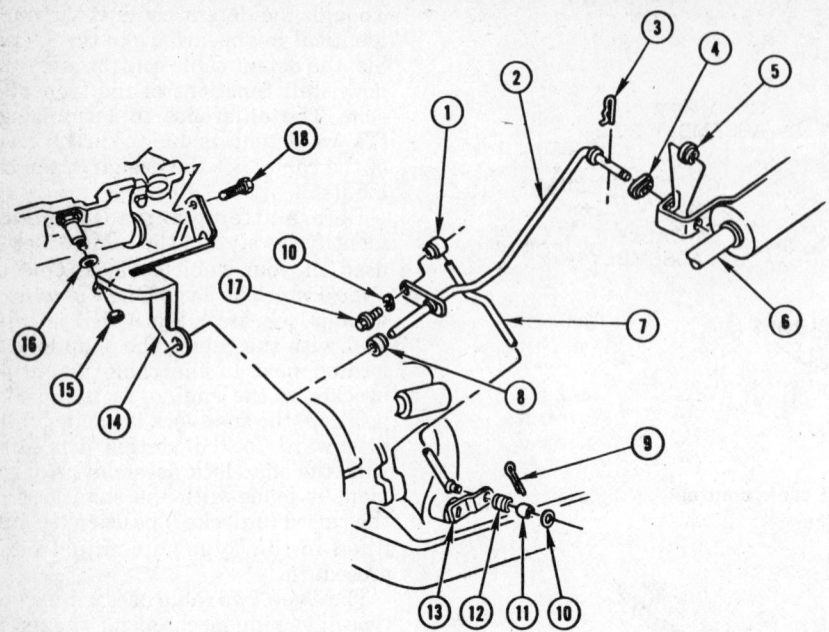

1. Swivel clamp
2. Equalizer shaft assembly
3. Retainer
4. Bushing
5. Grommet
6. Steering column
7. Shift rod
8. Bushing
9. Cotter pin
10. Washer
11. Sleeve with bushing
12. Grommet
13. Manual shaft with seal
14. Equalizer shaft bracket
15. Nut (5/16″ × 18)
16. Washer
17. Screw (M8 × 1.25 × 13.5)
18. Bolt (3/8″–16 × 3/4″)

Column shift linkage of the front wheel drive Eldorado, Riviera, Seville, and Toronado

1. Retainer
2. Washer
3. Lever
4. Park sleeve
5. Detent cable assembly
6. Cable seal
7. Cable retainer
8. Screw (M6.3 × 16 × 9.8)
9. Indicator lamp housing
10. Screw (M4.2 × 1.41 × 10)
11. Nut
12. Pointer
13. Shifter assembly
14. Washer
15. Bushing
16. Nut
17. Handle assembly
18. Nut
19. Back-up lamp switch
20. Shaft housing assembly
21. Shifter lever
22. Rod adjusting link
23. Shift selector rod assembly
24. Nut (M10 × 1.5)
25. Shift selector rod clevis pin
26. Shift selector rod clevis

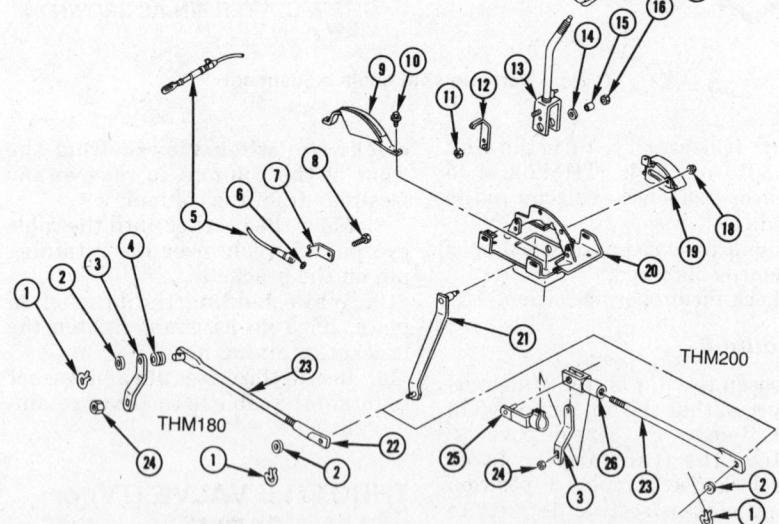

Chevette and T1000 floor shift linkage. Note the difference between the THM180 and THM200 linkage

Type Two

1. Depress and hold the metal lock tab on the TV cable. On diesels up to and including 1983, disconnect the throttle rod from the throttle lever. On later diesels, adjust the pump rod as described later in this section.

2. Move the slider back through the fitting in the direction away from the throttle body until the slider stops against fitting.

3. Release the metal lock tab.

4. Open the throttle lever to "full throttle stop" position. This will automatically adjust the slider on the cable to the correct setting.

5. On diesels, reconnect the throttle rod.

DETENT (DOWNSHIFT) SWITCH ADJUSTMENT

Turbo Hydra-Matic® 400/425

NOTE: All General Motors divi-sions, except Cadillac, use the same detent switch on TH400 and TH425 equipped models. The switch is mounted on the accelerator pedal bracket and is for all intents and purposes, selfadjusting. If a new switch is installed, a preliminary adjustment should be performed according to the accompanying illustration.

Cadillac has used two different types of switches; the majority use a

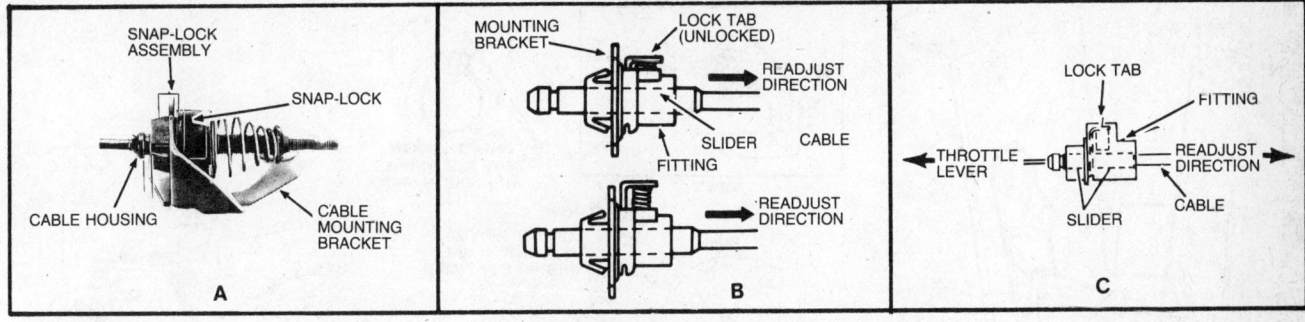

TV/detent cables used with GM automatic transmissions: A. type one; B. type two ('81 and earlier); C. type two ('82 and later).

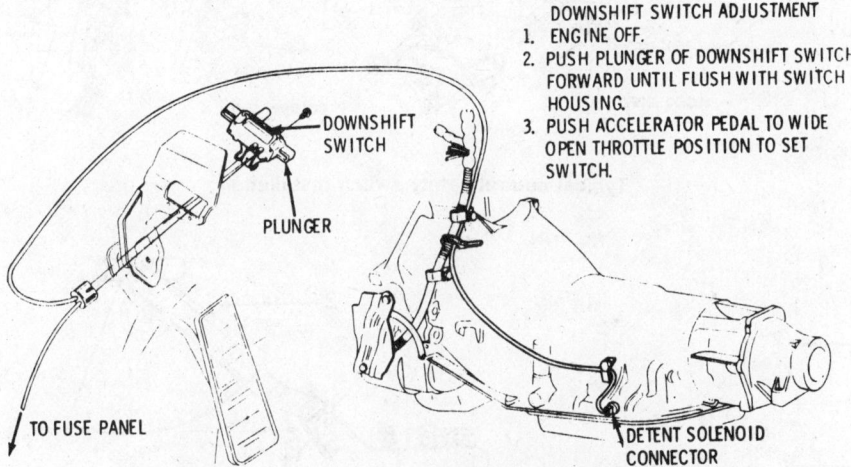

DOWNSHIFT SWITCH ADJUSTMENT
1. ENGINE OFF.
2. PUSH PLUNGER OF DOWNSHIFT SWITCH FORWARD UNTIL FLUSH WITH SWITCH HOUSING.
3. PUSH ACCELERATOR PEDAL TO WIDE OPEN THROTTLE POSITION TO SET SWITCH.

Detent (downshift) switch adjustment—except Cadillac

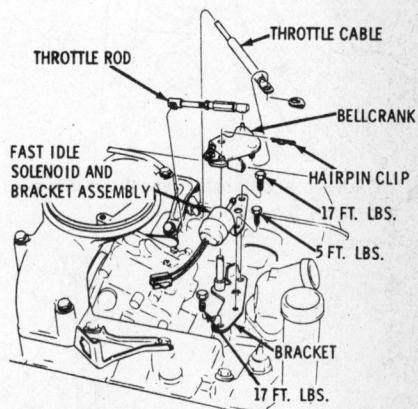

GM diesel throttle linkage—typical, except Chevette

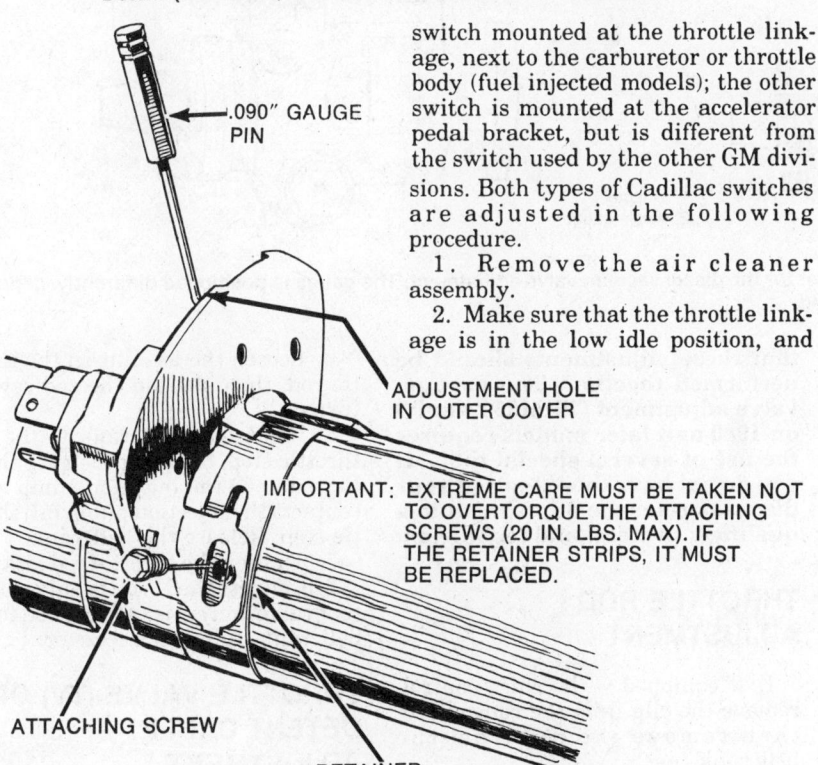

switch mounted at the throttle linkage, next to the carburetor or throttle body (fuel injected models); the other switch is mounted at the accelerator pedal bracket, but is different from the switch used by the other GM divisions. Both types of Cadillac switches are adjusted in the following procedure.

1. Remove the air cleaner assembly.

2. Make sure that the throttle linkage is in the low idle position, and

IMPORTANT: EXTREME CARE MUST BE TAKEN NOT TO OVERTORQUE THE ATTACHING SCREWS (20 IN. LBS. MAX). IF THE RETAINER STRIPS, IT MUST BE REPLACED.

Neutral safety switch adjustment—column shift

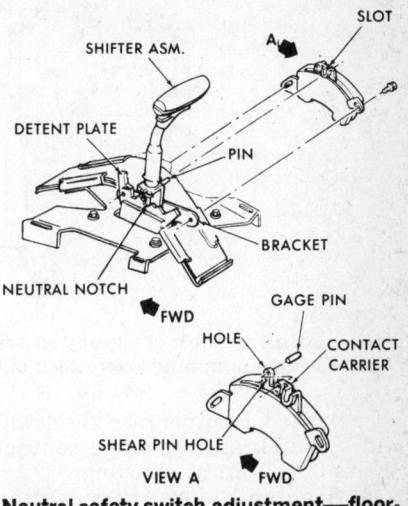

Neutral safety switch adjustment—floor-shift

that the idle speed is set to specifications.

3. Loosen the detent switch mounting screws and insert a .094 in. wire gauge into the hole located below the lower terminal of the switch.

4. For switches mounted at the engine, adjust the position of the switch so that the switch lever just touches the throttle linkage arm. If the switch is mounted at the accelerator pedal, adjust the switch position so that the switch lever just touches the accelera-

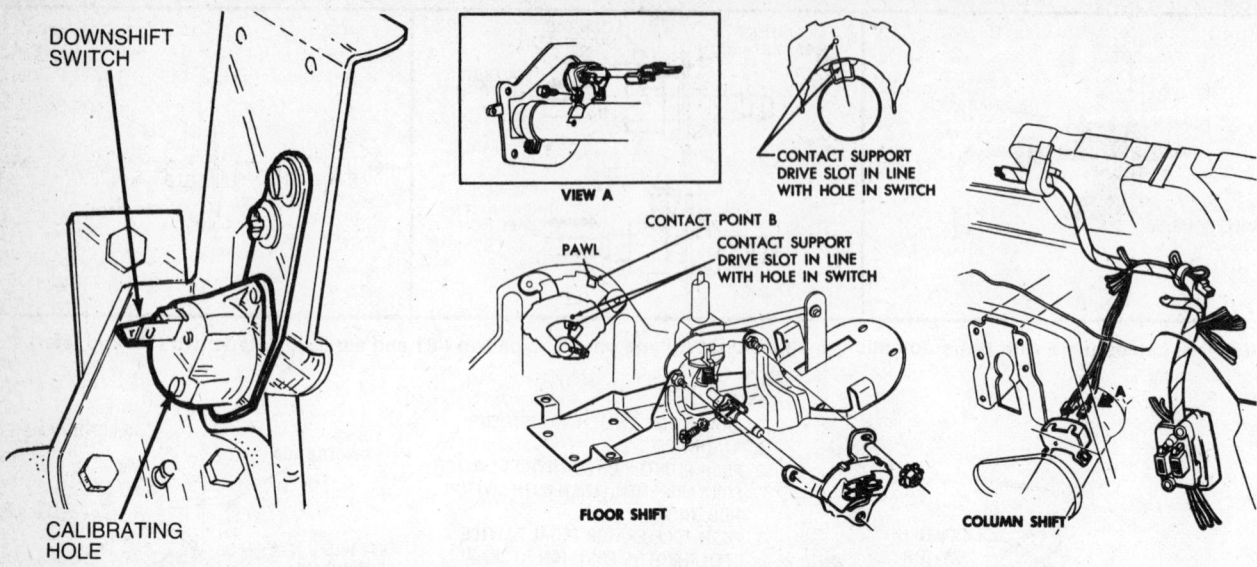

Cadillac accelerator pedal-mounted detent (downshift) switch

Typical neutral safety switch installation

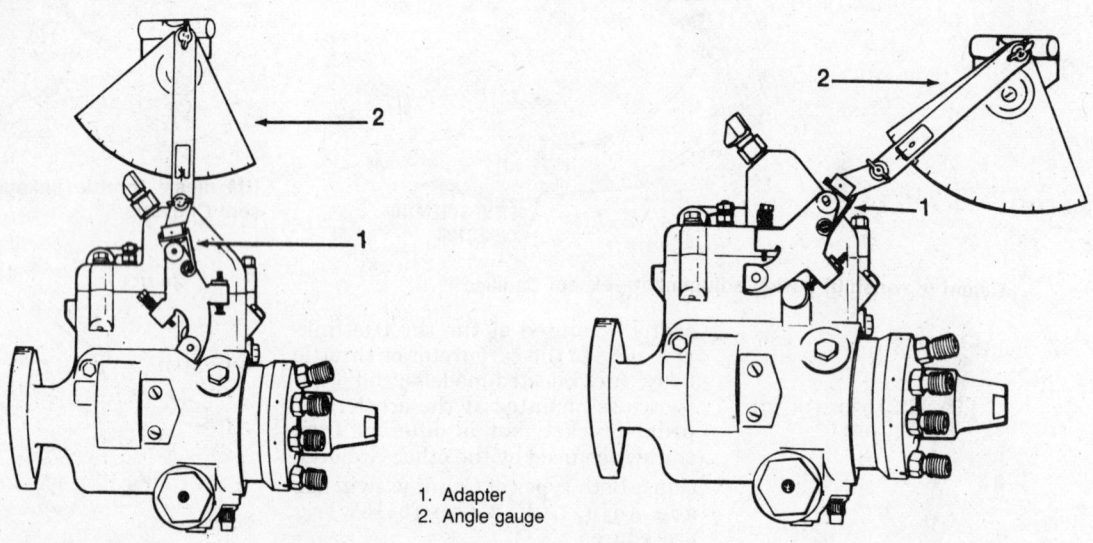

1. Adapter
2. Angle gauge

Installation of an angle gauge and an adapter for the diesel vacuum valve adjustment. The gauge is positioned differently, depending upon the type of throttle lever which is used.

tor relay rod. In either case, the downshift switch should make contact above 60° of throttle opening.

5. Tighten the switch mounting screws and remove the gauge.

6. Reinstall the air cleaner assembly.

DIESEL ENGINE TRANSMISSION LINKAGE ADJUSTMENTS

NOTE: These adjustments are for all GM cars with the V6 and V8 engines. Only the TV cable is adjustable on the diesel Chevette. Before making any linkage adjustments, check the injection timing, and adjust if necessary. Also note

that these adjustments should be performed together. The vacuum valve adjustment (THM350's only) on 1980 and later models requires the use of several special tools. If you do not have these tools at your disposal, refer the adjustment to a qualified, professional technician.

THROTTLE ROD ADJUSTMENT

1. If equipped with cruise control, remove the clip from the control rod, then remove the rod from the bellcrank.

2. Remove the throttle valve cable (THM125, 200, 325) or detent cable (THM350) from the bellcrank.

3. Loosen the locknut on the throttle rod, then shorten the rod several turns.

4. Rotate the bellcrank to the full throttle stop, then lengthen the throttle rod until the injection pump lever contacts the injection pump full throttle stop. Release the bellcrank.

5. Tighten the throttle rod locknut.

6. Connect the throttle valve or detent cable and cruise control rod to the bellcrank. Adjust if necessary.

THROTTLE VALVE (TV) OR DETENT CABLE ADJUSTMENT

NOTE: Refer to the previous cable adjustment procedures for gas engines. Adjust according to the type of cable which is used.

TRANSMISSION VACUUM VALVE ADJUSTMENT

1. Remove the air cleaner assembly.

2. Remove the air intake crossover from the intake manifold. Cover the intake manifold passages to prevent foreign material from entering the engine.

3. Disconnect the throttle rod from the injection pump throttle lever.

4. Loosen the transmission vacuum valve-to-injection pump bolts.

5. Mark and disconnect the vacuum lines from the vacuum valve.

6. Attach a carburetor angle gauge adapter (Kent-Moore tool J-26701-15 or its equivalent) to the injection pump throttle lever. Attach an angle gauge (J-26701 or its equivalent) to the gauge adapter.

NOTE: To service the V6 diesel, it may be necessary to file the gauge adapter in order for it to fit the thicker throttle lever of the V6 injection pump.

7. Turn the throttle lever to the wide open throttle position. Set the angle gauge to zero degrees.

8. Center the bubble in the gauge level.

9. Set the angle gauge to one of the following settings, according to the year and type of engine.

10. Attach a vacuum gauge to Port 2 and a vacuum source (e.g., hand-held vacuum pump) to Port 1 of the vacuum valve (as illustrated).

11. Apply 18–22 in. of vacuum to the valve. Slowly rotate the valve until the vacuum reading drops to one of the following values.

12. Tighten the vacuum valve retaining bolts.

Year	Engine	Setting
1980	V8	49–50°
1981	V8—Calif.	49–50°
1981	V8—non-Calif.	58°
1982 and later	V8	58°

Year	In. Hg.
1980	7
1981 Calif.	7–8
1981 non-Calif.	8½—9
1982 and later	10½

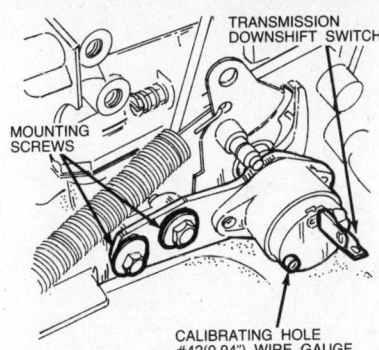

Cadillac carburetor-mounted detent (downshift) switch

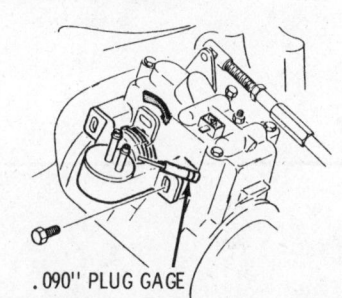

Transmission vacuum valve adjustment— diesel models

13. Reconnect the original vacuum lines in the vacuum valve.

14. Remove the angle gauge and adapter.

15. Connect the throttle rod to the throttle lever.

16. Install the air intake crossover, using new gaskets.

17. Install the air cleaner assembly.

BAND ADJUSTMENTS

NOTE: Only the THM250 has an externally adjustable band. Band adjustments are not externally possible on other transmissions.

Intermediate Band–Turbo Hydramatic® 250

NOTE: The intermediate band must be adjusted with every required fluid change or whenever there is slippage.

1. Position the shift lever in Neutral.

2. Loosen the locknut on the right side of the transmission. Tighten the adjusting screw to 30 inch lbs.

3. Back the screw out three turns and then tighten the locknut to 15 ft. lbs.

Pan & Filter

DRAINING, REMOVAL & INSTALLATION

NOTE: The fluid should be changed with the engine and transmission at normal operating temperature. If the car is raised, the transmission should be level. Be careful when draining, because the fluid will be hot.

1. Raise and safely support the vehicle.

2. On some models, it may be necessary to remove the transmission supporting crossmember in order to gain access to all of the pan bolts. Support the transmission with a jack before removing the crossmember.

3. Place a large pan underneath the transmission to catch the fluid. Loosen all the pan screws, then pull down one corner to drain most of the fluid. Be careful; the fluid will be hot. Do not pry between the pan and the transmission with a screwdriver or the like to remove the pan, as this will damage the mating surfaces. The pan can be tapped with a rubber mallet to loosen its grip.

4. Remove the pan bolts and empty out the pan. The pan can be cleaned with solvent but it must be air dried thoroughly before replacement. Be very careful not to leave any lint or threads from rags in the pan.

NOTE: It is normal to find a SMALL amount of metal shavings in the pan. An excessive amount of metal shavings indicates transmission damage which must be investigated.

5. Remove the filter or strainer retaining bolts (two on the Turbo Hydra-Matic 180, 200, 250 and 350). A reuseable strainer is used on the Turbo Hydra-Matic 180, 200 and 250. The strainer may be cleaned in solvent and thoroughly air dried. Filters are to be replaced. On the 400, 425 and 700R-4 Turbo Hydra-Matic, remove the filter retaining bolt(s), filter, and intake Pipe O-ring (or gasket). Trans-axle models 125, 325 and 440-T4 have strainers and O-rings.

NOTE: Various models use RTV sealant instead of a pan gasket. Clean all surfaces and apply a continuous bead of sealant along the pan mounting flange.

6. Install the new filter or cleaned strainer with a new gasket (or O-ring). Tighten the screws to 12 ft. lbs. On the 400 and 425, install a new intake pipe O-ring and a new filter, tightening the retaining bolts to 10 ft. lbs. Install a new strainer and O-ring on the 125, locating the strainer against the dipstick stop.

7. Install the pan with a new gasket. Tighten the bolts evenly in a crisscross pattern to 12 ft. lbs.

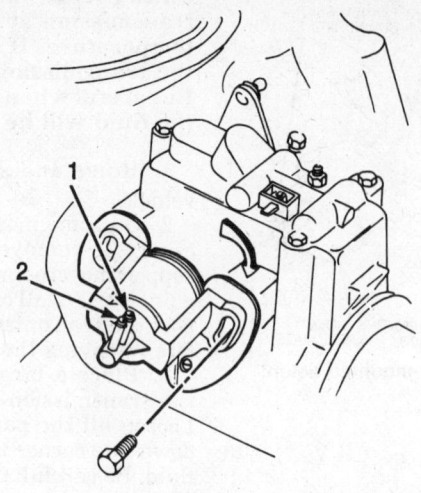

1980–81

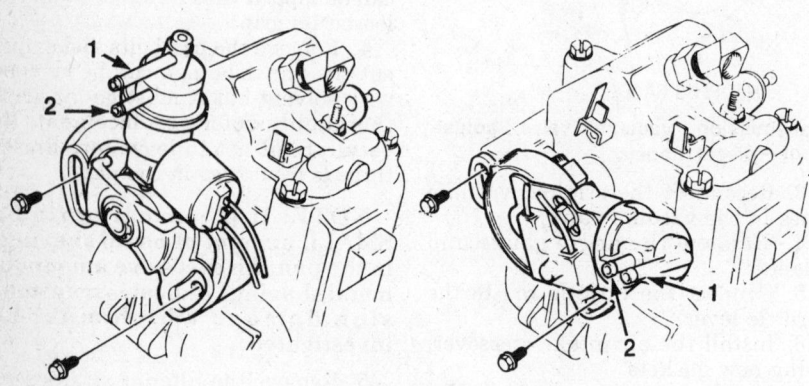

1982 AND LATER

1. Attach vacuum source here
2. Attach vacuum gauge here

Transmission vacuum valve adjustment— diesel models

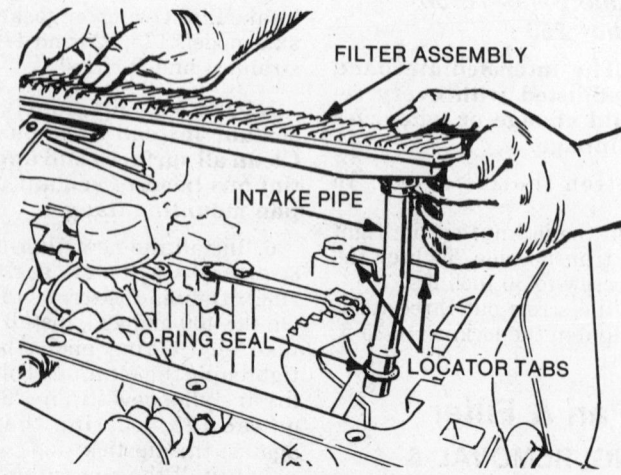

Removing the filter, intake pipe and O-ring on
Turbo Hydra-Matic 400 transmission

8. Replace the crossmember if removed.

9. Lower the car. Add Dextron® II fluid through the dipstick tube.

10. Start the engine and let it idle. Do not race the engine. Shift into each lever position, holding the brakes. Check the fluid level with the engine idling in Park. The level should be between the two dimples on the dipstick, about $\frac{1}{4}$ in. below the ADD mark. Add fluid as necessary.

11. Check the fluid level after the car has been driven enough to thoroughly warm up the transmission. The level should be at the FULL mark on the dipstick. If the transmission is overfilled, the excess must be drained off. Overfilling causes aerated fluid, resulting in transmission slippage and probable damage.

Vacuum Modulator

REMOVAL & INSTALLATION

1. On models equipped, remove the vacuum hose at the modulator. Inspect the hose for signs of transmission fluid that would indicate a leaking modulator.

2. Remove the modulator holddown bracket. Pull the modulator straight back from the case. Install a new O-ring and install the modulator in the reverse order.

TROUBLESHOOTING THE TORQUE CONVERTER CLUTCH (TCC)

NOTE: Most GM cars after the 1984 model year use the Electronic Control Module to activate the TCC. The ECM responds to a great number of different signals to accomplish this. This means that, in many cases, testing for a malfunctioning system is extremely complex; in many others, it requires the use of special test equipment to activate the ECM in various ways for the test. If your car's transmission is not covered in these procedures, it is because the test procedure is beyond the scope of this book.

GM Turbo Hydramatic 1980–84

NOTE: Before diagnosing the TCC system as being at fault in the case of rough shifting or other malfunctions, make sure that the engine is in at least a reasonable state of tune. Also, the following points should be checked.

1. Check the transmission fluid level and correct as necessary.

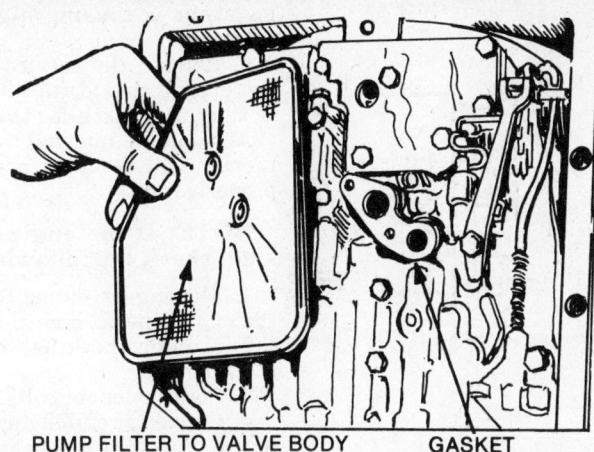

PUMP FILTER TO VALVE BODY   GASKET

Removing the filter and gasket on a Turbo Hydra-Matic 200, 350, or 375B

2. Check the manual linkage adjustment and correct as necessary.

3. Road test the vehicle to verify the complaint. Make sure that the vehicle is at normal operating temperature. The road test should consist of the following steps:

a. Put the gear selector lever in "Overdrive", and accelerate the vehicle. Observe the 1–2, 2–3, and 3–4 upshifts; note that they vary with throttle position, increased throttle giving later shifts.

b. Check for converter clutch engagement and observe the point at which it occurs. It should occur between 35–50 mph, varying with the car model.

c. Observe 4–3 part throttle downshift by depressing the accelerator to ¾ open position at a speed of 45–70 mph. The transaxle should downshift to third gear immediately.

NOTE: If it has been determined that there is a problem with the TCC system, the next step is to determine if the problem is internal or external. The following procedure can be used.

1. Disconnect the electrical connector at the transmission case.

2. Raise and safely support vehicle.

3. Start engine and adjust the speed to 2000 rpm, gear selector in Neutral.

4. Test for 12 volts at the connector using a volt-ohmmeter or a test light. If 12 volts are present, the problem is internal. If no voltage (or low voltage according to the meter) is present at the connector, the problem is external.

NOTE: If the problem is internal (12 volts at the connector) the following steps can be taken.

1. With the wire to the transmission case disconnected, take a 12 volt test light and connect it to the female connector and ground to the male transmission connector.

2. Start the engine and adjust the speed to 2000 rpm, gear selector in Park.

3. If the test light comes on, the governor switch or the internal wiring is shorted to ground. The oil pan will have to be removed, the wiring checked and/or the governor switch replaced.

4. If the test light now comes on, the internal hydraulic/mechanical controls will have to be checked. Refer to the Hydraulic/Mechanical Controls section.

5. If the test light still does not light, there is a problem with the solenoid or governor switch.

NOTE: To test for solenoid or governor switch electrical malfunction, the following steps can be used.

1. Drain the transmission fluid and remove the oil pan.

2. Using an external 12 volt source, (self-powered test light or small lantern battery, etc.) connect a positive lead to the case connector. Remove the lead wire from the governor pressure switch and connect it to the ground lead of the external 12 volt source.

─── **CAUTION** ───

Do not reverse the leads or the solenoid diode will be destroyed by the reverse voltage. Do not use an automobile battery for this test. A self-powered test light is best for these tests.

───────────────

3. If the solenoid clicks, it can be considered serviceable; replace the governor switch.

4. If the solenoid does not click, check the wiring. If the wiring appears to be good, replace the solenoid and recheck.

1986–87 Turbohydramatic 125C

NOTE: Before diagnosing the TCC system as being at fault in the case of rough shifting or other malfunctions, make sure that the engine is in at least a reasonable state of tune. Also, the following points should be checked.

1. Check the transmission fluid level and correct as necessary.

2. Check the manual linkage adjustment and correct as necessary.

3. Road test the vehicle to verify the complaint. Make sure that the vehicle is at normal operating temperature. The road test should consist of the following steps:

a. Put the gear selector lever in "Overdrive", and accelerate the vehicle. Observe the 1–2, 2–3, and 3–4 upshifts; note that they vary with throttle position, increased throttle giving later shifts.

b. Check for converter clutch engagement and observe the point at which it occurs. It should occur between 35–50 mph, varying with the car model.

c. Observe 4–3 part throttle downshift by depressing the accelerator to 3/4 open position at a speed of 45–70 mph. The transaxle should downshift to third gear immediately.

4. Disconnect the ALCL connector and connect a test lamp between terminal "F" and a good ground. Raise and support the vehicle in a secure manner so the drive wheels can rotate freely. Start the engine and allow it to idle. Put the transmission in Drive.

5. Check to see if the test light is lit. If it is lit, the transmission third gear apply switch is defective. If it is not lit, have someone depress the accelerator and increase wheel speed to 25 mph. This should close the third gear apply switch and light the test lamp. If it does not, proceed with the remaining tests in this step. If it does, go on to Step 6.

a. Check for a blown fuse in the appropriate circuit, and replace the fuse if necessary. If this does not fix the problem, disconnect the connector at the transmission and connect a test light between harness connectors "A" and "D". Turn the engine off, and then turn the ignition back on. If the light is on, proceed with the rest of this sub-step. If the light is off, go to "b". Check for a ground in circuit 422. If there is no ground, replace the ECM.

b. Connect a test light from terminal "A" to ground. If the light is

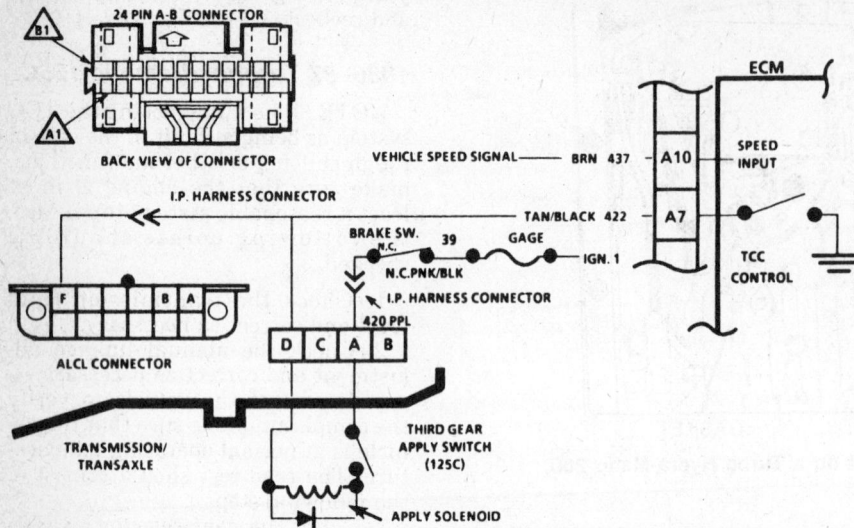

Connecting the test lamp to test the TCC system on the 1986–87 Turbohydromatic 125C. Connect the lamp to the F terminal of the ALCL connector and to a good ground. Other connections to be made are described in the text

off, repair an open circuit in the TCC brake switch or circuit or adjust the switch. If the light is on, ground the TCC test point and again connect the test lamp between harness connector terminals "A" and "D". If the light is now off, repair an open circuit from the wire from the transmission to the ALCL test point, terminal F. If it is on, repair a defective transmission TCC connector, TCC solenoid, or third gear apply switch.

6. Depress the brake pedal. The test light should go out. If not replace a faulty brake switch or adjust it so it works properly. If it does go out, proceed with the next step.

7. Shut the engine off and then turn the ignition switch back on. Connect the test lamp between a 12 volt source and the ALCL terminal "F". Ground the diagnostic terminal and then check to see whether or not the light is on. If the light is on, proceed to Step 8. If it is off, proceed with the rest of this step. Check for an open circuit between the ALCL 422 wire connector ("F") and the terminal on the ECM connector. If this circuit is complete, the ECM is faulty. Before replacing it, have a repair shop check the resistance of each ECM controlled relay and solenoid coil.

8. If the light was on in the step above, check:
    a. The coolant level in the cooling system.
    b. The opening temperature of the thermostat. Make sure the engine runs at above 160°F.
    c. If all these conditions are met, you should have a repair shop check out the Vehicle Speed Sensor. If that proves out, they should then

check for the correct PROM in the ECM.

INTERNAL HYDRAULIC/ MECHANICAL CONTROLS CHECK

NOTE: Since part of these checks involves the governor system, obtain, if possible, a test governor of the same type used in the transmission being serviced. Cut two pieces of $5/32$ in. O.D. rubber vacuum hose to $3/8$ in. long. Put one piece of the cut-off hose under each weight of the governor. Remove the engine vacuum switch electrical connector. The vacuum switch should be mounted on an inner fender wall. Using a jumper wire, connect both terminals of the connector together. As a check that the proper connection has been made, turn the ignition to ON, raise the vehicle and check for 12 volts at the transmission case female connector. There should be 12 volts on a voltmeter. Then, proceed as follows.

1. Remove the transmission governor and replace with the test governor.
2. With the vehicle's wheels off the ground, apply the parking brake. The rear wheels must not be able to turn.
3. Start the engine, selector in Park, and allow to idle.
4. Step on the brake pedal. This will interrupt the flow of current to the transmission.
5. Place the selector in Drive. The transmission should automatically shift into 3rd gear because the test

governor is causing high governor pressure.

6. Release the service brakes and the engine should stall immediately. If the engine stalls, the converter clutch and the internal hydraulic and mechanical controls are operating properly.

NOTE: If the engine does not stall, check the following.

1. Missing or damaged O-ring at the end of the turbine shaft.
2. Missing check ball or O-ring at the solenoid.
3. Loose solenoid bolts.
4. Converter clutch apply passages in the pump blocked or restricted.
5. Defective converter.

NOTE: After the test has been completed remove the test governor and replace it with the original vernor. If the original governor is used for the test, remove the rubber bushings that were installed and be sure the weight springs are in the correct position before reinstalling.

EXTERNAL CONTROLS CHECK

1. Turn the ignition switch to the ON position.
2. Check the vacuum switch connector for 12 volts.
3. If no voltage is present at the vacuum switch, check the fuse block for a blown fuse, the brake switch and the wiring to the vacuum switch.
4. If there is a reading of 12 volts at the switch, reconnect the electrical connector to the vacuum switch. Using a hand vacuum pump with a gauge, apply 2.5–7 in. of vacuum to the vacuum switch.
5. With the ignition switch in the ON position, check for 12 volts at the female end of the transmission connector.
6. If no or low (as read on a voltmeter) voltage is present, look for a break in the wire between the vacuum switch and the transmission. Further vacuum switch checks are given below.

VACUUM SWITCH CHECK

1. Disconnect the vacuum hose and the electrical connector from the vacuum switch.
2. Attach one lead of a test light to either one of the terminals of the vacuum switch. Ground the other vacuum switch terminal.
3. Apply 12 volts to the other test light lead.
4. Attach the hand vacuum pump and gauge to the vacuum switch port.

5. Turn the ignition switch to the ON position.

6. The test light should be off. Apply vacuum with the hand pump until the gauge reads 2.45–7 in. of vacuum. The light should come on.

7. Bleed off some vacuum slowly. The light should remain on until the vacuum drops to 1.5–2.5 in. of vacuum.

8. If the vacuum switch does not turn the test light on and off at the specified vacuum readings, the switch is bad and should be replaced.

NOTE: The high vacuum limit, which is the point where the test light goes out, and the low vacuum limit, which is the point at which the test light comes back on, must have at least 4 in. of vacuum difference. If the above checks of the vacuum switch verify the proper operation of the switch, then the trouble is elsewhere, possibly at the thermal vacuum valve.

THERMAL VACUUM VALVE CHECK

1. Disconnect the vacuum hose at the vacuum switch and install a vacuum gauge to the hose.

2. Start the engine and check the vacuum reading, gear selector in Park. With the engine cold (coolant temperature below 130°F.), vacuum at idle should be zero. Adjust engine speed to 200 rpm. The vacuum should still be zero.

3. With the engine warm, after about five minutes running at fast idle, the coolant temperature should be above 130°F. The vacuum at idle should still be zero while the vacuum at 2000 rpm should be 10 in. of vacuum minimum.

SOLENOID DIODE CHECK

——— CAUTION ———

Do not use an automotive battery for troubleshooting solenoids. Solenoids must not be bench tested by touching the leads of an automotive battery. The internal diode will be destroyed when the leads are reversed, as they must be to check the diode.

NOTE: Remember that a diode allows electricity to flow freely in one direction, and prevents or at least restricts the flow of current in the opposite direction. To check the solenoid diode, an ohmmeter should be used. Use only a meter reading type of ohmmeter since electronic digital-type will often give a false indication. Use the X1 scale on the ohmmeter, and use the following procedure.

1. Verify that the ohmmeter is set to the X1 scale. Zero the meter.

2. Attach the positive solenoid lead (red) to the positive meter lead, and the negative solenoid lead (black) to the negative lead. The meter should read 20–40 ohms, depending on the solenoid temperature. If this reading is obtained, neither the coil or the diode is shorted and they should be considered useable. If the meter reads 0 ohms, then there is a short. The solenoid must be replaced. An open reading again indicates a bad coil in the solenoid, and it must be replaced.

3. Reverse the solenoid lead attachment. If the meter now reads lower (usually reads 2–15 ohms), the solenoid is good. If the reading is the same as before, the diode is bad and the solenoid will have to be replaced.

NOTE: On the Turbo-Hydra-Matic 250/350 transmissions, the solenoid is mounted on the valve body and the removal and replacement includes only R&R of the oil pan and the solenoid. However, on the Turbo Hydra-Matic 200 transmissions, the solenoid is mounted on the inside face of the oil pump. Therefore solenoid replacement for this model transmission will require the removal of the entire transmission, the torque converter and the oil pump. Be careful when diagnosing the solenoid so as not to burn out a good one. Transmissions using the TCC system are also used in vehicles equipped with diesel engines. Due to the different vacuum characteristics of a diesel engine, a slightly different control system is used. Incorporated into the diesel system are a Low and a High Vacuum Switch. These are usually mounted on the engine, just above the right hand valve cover. These switches can also be checked with a 12 volt test light and a hand vacuum pump.

LOW AND HIGH VACUUM SWITCH CHECK (DIESEL)

1. Disconnect the vacuum hose and the electrical connector from the switch. The test is run on both switches, but in this procedure, start with the Low Vacuum Switch. The Low Vacuum Switch should be at the rearmost of the two, with the High Vacuum Switch to the front.

2. Attach one lead of a test light to either one of the terminals of the Low Vacuum Switch and ground the remaining terminal of the vacuum switch.

3. Attach the remaining lead of the

test light to the hot (+ 12 volt) side of the vacuum switch connector. Attach the hand vacuum pump and gauge to the vacuum port of the vacuum switch.

4. Turn the ignition switch to the ON position. If using a self-powered test light, the ignition does not have to be turned on.

5. With the vacuum pump, apply 5.5 in. of vacuum. The Low Vacuum Switch should keep the test light off, and it should remain off until the gauge climbs to a reading of 5.5. Bleed off some vacuum slowly. The Low Vacuum Switch should keep the test light on until the vacuum drops to approximately 4 inches of vacuum. If the Low Vacuum Switch does not turn on the test light at 5.5 on the gauge, and off at 4 inches of vacuum, the Low Vacuum Switch is malfunctioning.

6. Using the same electrical and vacuum hook-up, move to the other vacuum switch, which should be the front one, the High Vacuum Switch. The High Vacuum Switch should light the test lamp and keep the lamp on as vacuum begins to be applied with the hand pump. The light should stay on until the gauge reads approximately 12.5 in. of vacuum, and then the lamp should go out. Bleed the vacuum slowly. The lamp should come back on at 12.5 in. of vacuum. If the High Vacuum Switch does not turn on at 12.5 in. of vacuum and off as the vacuum goes higher, the High Vacuum Switch is malfunctioning.

HIGH VACUUM SWITCH ADJUSTMENT (DIESEL)

NOTE: The High Vacuum Switch on the diesel engine must be adjusted any time the throttle rod, transmission vacuum valve or high idle speed adjustments are altered. The following steps can be used for adjustment.

1. Disconnect the electrical connector from the High Vacuum Switch. This should be the front switch of the two. It has an adjustment port opposite the electrical connector.

2. Using a self-powered test light, connect one lead to either one of the terminals on the switch and connect the probe of the test light to the other switch terminal.

3. Start the engine and allow to run at high idle speed. To do this, actuate the fast idle solenoid. A pink and green wire connector goes to the coolant switch located on the left rear of the engine on the intake manifold, and this connector should be pulled off the coolant switch to produce the fast idle.

1. Transmission control cable
2. Bracket
3. Cable pin
4. Washer
5. Nut

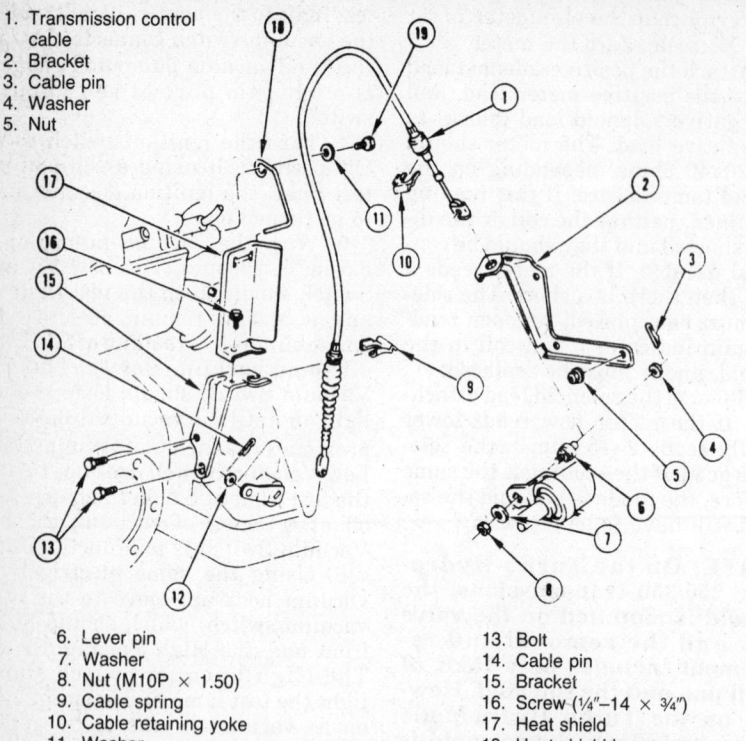

6. Lever pin
7. Washer
8. Nut (M10P × 1.50)
9. Cable spring
10. Cable retaining yoke
11. Washer
12. Washer

13. Bolt
14. Cable pin
15. Bracket
16. Screw (1/4"–14 × 3/4")
17. Heat shield
18. Heat shield brace
19. Bolt (3/8"–16 × 3/4")

Column shift cable and related components—front wheel drive Eldorado, Riviera, Seville, and Toronado

4. Remove the small dust cap from the back of the High Vacuum Switch.

5. The High Vacuum Switch must be closed before making the adjustment. This means that the test light should be on. If the test light is off, the contacts are open. Take a $\frac{5}{64}$ in. Allen wrench and turn the adjustment screw clockwise until the contacts close and the test light comes on.

6. Adjust the vacuum switch by turning the adjustment screw slowly counterclockwise until the switch contacts just open and the test light goes off. Turn the screw slowly so that the screw is not turned past this position.

7. Put the dust cap back on, reconnect the High Vacuum Switch electrical connector, and reconnect the coolant switch connector.

BRAKE SWITCH CHECK

1. Remove the electrical connector from the rear of the brake switch. These rear terminals are for the cruise control and converter clutch release. Turn the ignition switch to the ON position.

2. Ground one of the terminals of the brake release switch.

3. Connect one lead of the test light to the remaining brake release switch terminal. Attach the other lead of the test light to the brake connector wire. The test lamp should light.

4. Apply the brakes. The test light should go out. If the test light is off before applying the brakes or if it comes on during brake application, the switch is bad and should be replaced.

U-Joints and CV-Joints

UNIVERSAL JOINTS

U-Joint is mechanic's jargon for universal joint. U-Joints should not be confused with U-bolts, which are U-shaped bolts used to connect U-joints to the differential pinion flange.

Universal joints provide flexibility between the driveshaft and axle housing to accommodate changes in the angle between them. Changes of length are accommodated by the sliding splined yoke between the driveshaft and transmission. The engine and transmission are mounted rigidly on the car frame. The angles between the transmission, driveshaft and axle change constantly as the car responds to various road conditions.

To give flexibility and still transmit power as smoothly as possible, several types of universal joints are used. The most common type of universal joint is the cross and yoke type. Yokes are used on the ends of the driveshaft with the yoke arms opposite each other. Another yoke is used opposite the driveshaft and when placed together, both yokes engage a center member, or cross, with four arms spaced 90° apart. The U-joint cross is alternately referred to as a spider, and the arms are called trunnions. A bearing cup (or cap) is used on each arm of the cross to accommodate movement as the driveshaft rotates. The bearings used are needle bearings.

A conventional universal joint will cause the driveshaft to speed up and slow down through each revolution and cause a corresponding change in the velocity of drive shaft. This change in speed causes natural vibrations to occur through the driveline, necessitating a third type of universal joint: The constant velocity joint. A rolling ball moves in a curved groove, located between two yoke-and-cross universal joints, connected to each other by a coupling yoke. The result is a uniform motion as the driveshaft rotates, avoiding the fluctuations in driveshaft speed. This type of joint is found in cars with sharp driveline angles, or where the extra measure of isolation is desirable.

Cross And Yoke U-Joint

OVERHAUL

There are two types of cross and yoke U-joints. One type retains the cross within the yoke with C-shaped snap rings. This type is found on all American Motors, Chrysler, and Ford Cars. GM cars generally use the second type of joint, which is held together by injection molded plastic retainer rings. The second type cannot be reassembled with the same parts, once disassembled. However, repair kits are available.

Snapring Type

1. Remove the driveshaft. For the correct procedure, see the car section for the model you are working on.
2. If the front yoke is to be disassembled, matchmark the driveshaft and sliding splined yoke (transmission yoke) so that driveline balance is preserved upon reassembly. Remove the snap rings which retain the bearing caps.

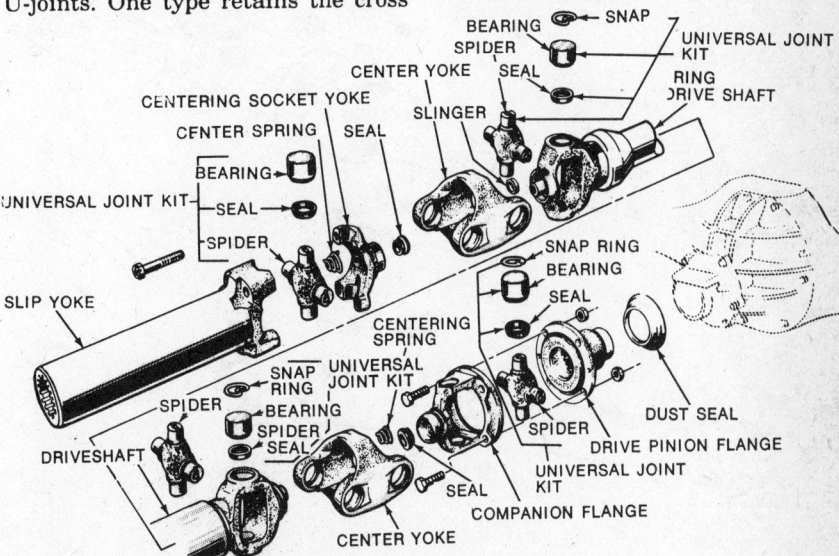

Typical driveshaft with cardan type U–joints

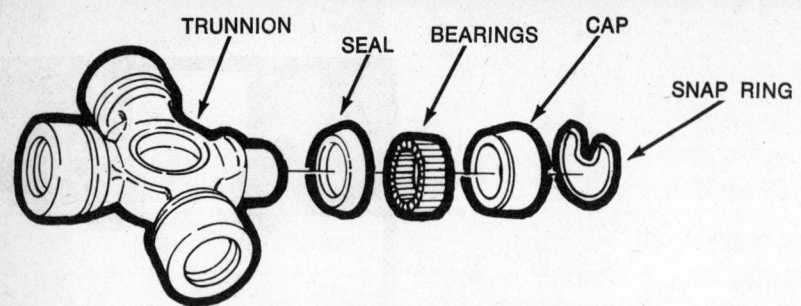

TRUNNION SEAL BEARINGS CAP SNAP RING

Snap ring type universal joint

3. Select two sockets, one small enough to pass through the yoke holes for the bearing caps, the other large enough to receive the bearing cap.

4. Using a vise or a press, position the small and large sockets on either side of the U-joint. Press in on the smaller socket so that it presses the opposite bearing cap out of the yoke and into the larger socket. If the cap does not come all the way out, grasp it with a pair of pliers and work it out.

5. Reverse the position of the sockets so that the smaller socket presses on the cross. Press the other bearing cap out of the yoke.

6. Repeat the procedure on the other bearings.

7. To install, grease the bearing caps and needles throughly if they are not pregreased. Start a new bearing cap into one side of the yoke. Position the cross in the yoke.

8. Select two sockets small enough to pass through the yoke holes. Put the sockets against the cross and the cap, and press the bearing cap ¼ inch below the surface of the yoke. If there is a sudden increase in the force needed to press the cap into place, or if the cross starts to bind, the bearings are cocked, They must be removed and restarted in the yoke. Failure to do so will greatly reduce the life of the bearing.

9. Install a new snap ring.

10. Start a new bearing into the opposite side. Place a socket on it and press in until the opposite bearing contacts the snap ring.

11. Install a new snap ring. It may be necessary to grind the facing surface of the snap ring slightly to permit easier installation.

12. Install the other bearings in the same manner.

13. Check the joint for free movement. If binding exists, smack the yoke ears with a brass or plastic faced hammer to seat the bearing needles. Do not strike the bearings, and support the shaft firmly. Do not install the driveshaft until free movement exists at all joints.

Plastic Retainer Type

Remove and install the bearing caps and trunnion (cross) as described for the snap-ring type universal joints. On an original universal joint, however, the bearing caps will be secured in the yokes with injected plastic. The plastic will shear when the bearing caps are pressed. Service snap-rings are installed in the groove on the inside (of yoke) of the installed caps.

NOTE: The plastic which retains the bearing will be sheared when the bearing cup is pressed out. Be sure to remove the remains of the plastic retainer from the ears of the yoke. It is easier to remove the remains if a small pin or punch is first driven through the injection holes in the yoke. Failure to remove all of the plastic remains may prevent the bearing cups from being pressed into place and the bearing retainers from being properly seated.

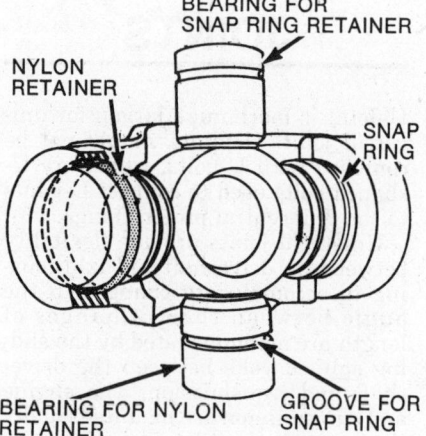

BEARING FOR SNAP RING RETAINER

NYLON RETAINER

SNAP RING

BEARING FOR NYLON RETAINER

GROOVE FOR SNAP RING

U-joint locking methods

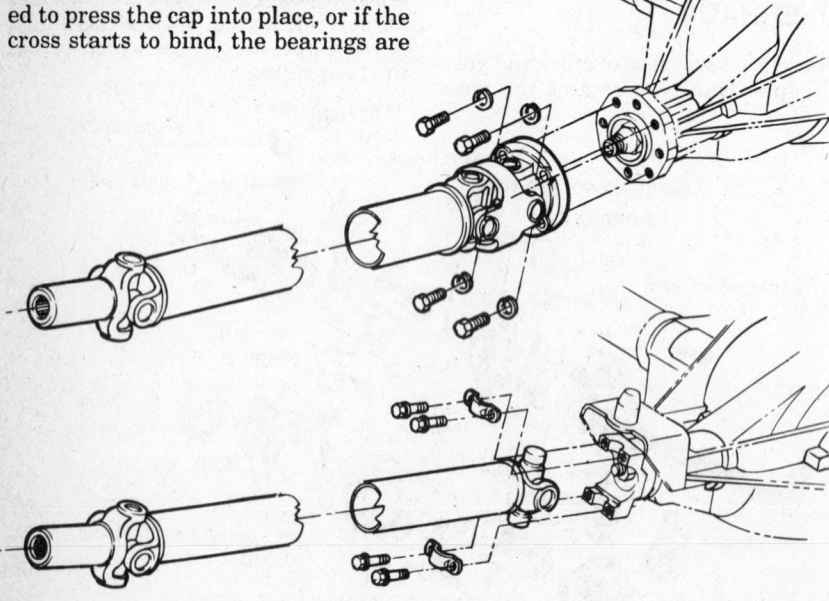

The driveshaft may be retained to the differential pinion by a flange (top) or by U-bolts or straps (bottom)

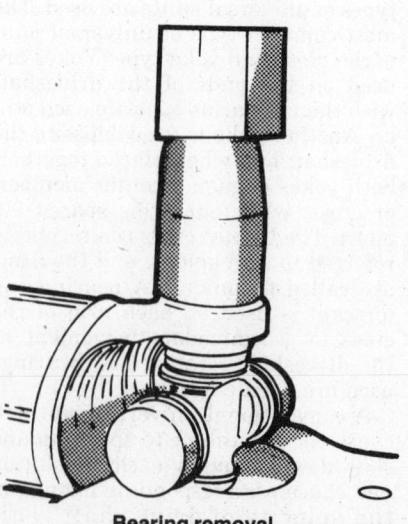

Bearing removal

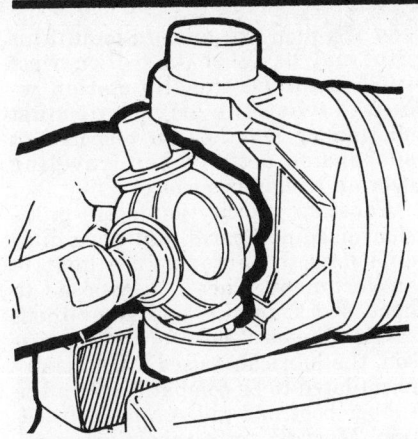

Press a bearing cap into the yoke, then install the cross

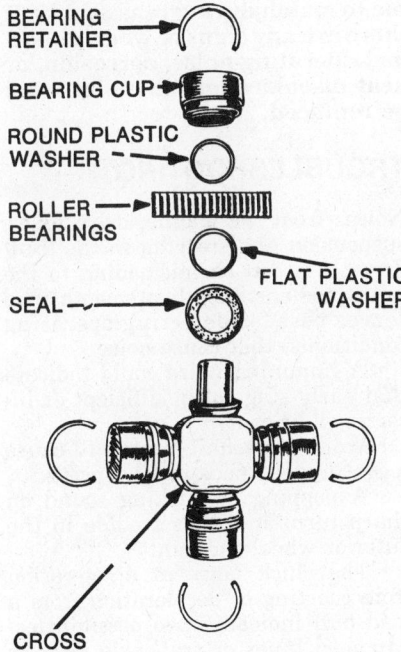

BEARING RETAINER

BEARING CUP

ROUND PLASTIC WASHER

ROLLER BEARINGS

SEAL

FLAT PLASTIC WASHER

CROSS

Plastic retainer U-joint repair kit components

Cardan Type U-Joint

OVERHAUL

Ford and Chrysler products with Cardan type U-joints use snap rings to retain the bearing cups in the yokes. Most GM cars have plastic retainers. Be sure to obtain the correct rebuilding kit.

1. Use a punch to mark the coupling yoke and the adjoining yokes before disassembly, to ensure proper reassembly and driveline balance.

2. It is easiest to remove the bearings from the coupling yoke first. Follow the order indicated in the illustration.

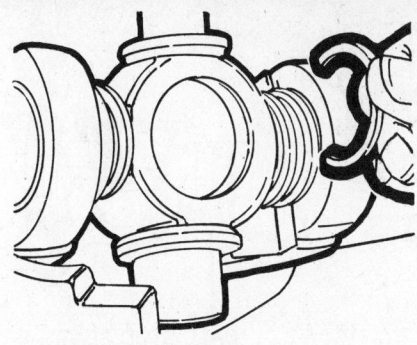

Service snap rings are installed inside the yoke

3. Support the driveshaft horizontally on a press stand, or on the workbench if a vise is being used.

4. If snap rings are used to retain the bearing cups, remove them. Place the rear ear of the coupling yoke over a socket large enough to receive the cup. Place a smaller socket, or a cross press made for the purpose, over the opposite cup. Press the bearing cup out of the coupling yoke ear. If the cup is not completely removed, insert a spacer and complete the operation, or grasp the cup with a pair of slip joint pliers and work it out. If the cups are retained by plastic, this will shear the retainers. Remove any bits of plastic.

5. Rotate the driveshaft and repeat the operation on the opposite cup.

6. Disengage the trunnions of the spider, still attached to the flanged yoke, from the coupling yoke, and pull the flanged yoke and spider from the center ball on the ball support tube yoke.

NOTE: The joint between the shaft and coupling yoke can be

serviced without disassembly of the joint between the coupling yoke and flanged yoke.

7. Pry the seal from the ball cavity, remove the washers, spring and three seats. Examine the ball stud seat and the ball stud for scores or wear. Worn parts can be replaced with a kit. Clean the ball seat cavity and fill it with grease. Install the spring, washer, ball seats, and spacer (washer) over the ball.

8. To assemble, insert one bearing cup part way into one ear of the ball support tube yoke and turn this cup to the bottom.

9. Insert the spider (cross) into the tube so that the trunnion (arm) seats freely in the cup.

10. Install the opposite cup part way, making sure that both cups are straight.

11. Press the cups into position, making sure that both cups squarely engage the spider. Back off if there is a sudden increase in resistance, indicating that a cup is cocked or a needle bearing is out of place.

12. As soon as one bearing retainer groove clears the yoke, stop and install the retainer (plastic retainer models). On models with snap rings, press the cups into place, then install the snap rings over the cups.

13. If difficulty is encountered installing the plastic retainers or the snap rings, smack the yoke sharply with a hammer to spring the ears slightly.

14. Install one bearing cup part way into the ear of the coupling yoke, Make sure that the alignment marks are matched, then engaged the coupling yoke over the spider and press

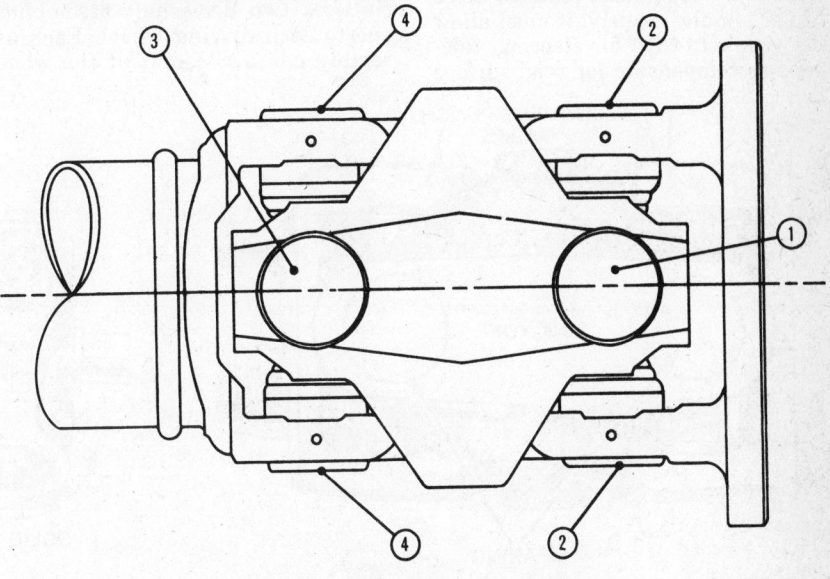

Cardan joint disassembly sequence

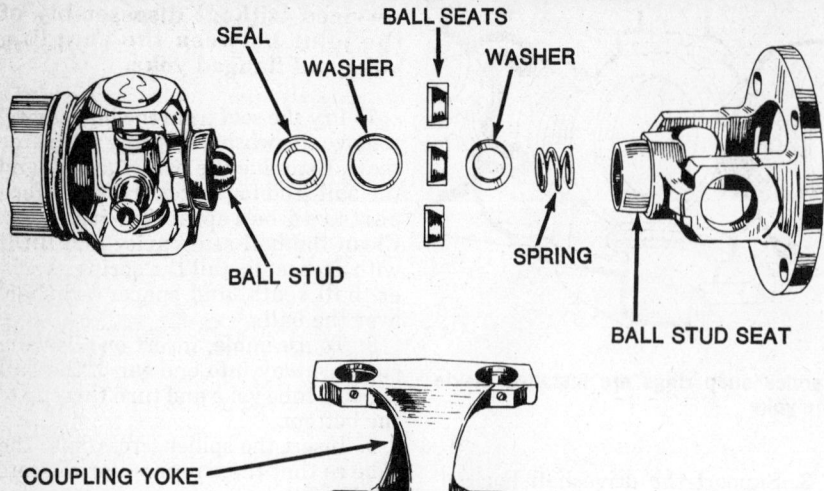

Cardan type joint

in the cups, installing the retainers or snap rings as before.

15. Install the cups and spider into the flanged yoke as with the previous yoke.

NOTE: The flange yoke should snap over center to the right or left and up or down by the pressure of the ball seat spring.

CONSTANT VELOCITY JOINTS

Front wheel drive vehicles present several unique problems to engineers because the driveshaft must do three things, simultaneously. It must allow the wheels to turn for steering, telescope to compensate for road surface vibrations, and it must transmit torque continuously without vibration.

To compensate for these three factors a two-joint driveshaft allows the front wheels to perform these functions. This driveshaft mates disc type straight groove ball joint design with the bell type Rzeppa CV universal joint.

The Rzeppa joint on the outboard end of each driveshaft provides steering ability by allowing drive wheels to steer up to 43° while transmitting all available torque to the wheels. The inboard joint allows telescoping (up to $1\frac{1}{2}$ in.) through the rolling actions of balls in straight grooves and operates at angles up to 20°. The combined action of these two ball type U-joints eliminates vibration.

The typical front wheel drive vehicle uses two driveshaft assemblies-one to each driving wheel. Each assembly has a CV-joint at the wheel end is called the inboard joint. This joint may be either the ball or tripot type. It allows the slip motion required when the driveshaft must shorten or lengthen in response to suspension action when traveling over an irregular surface.

Constant velocity joints are precision machined parts that have difficult jobs to perform in a hostile enviornment. They are exposed to heat, shock, torque, and many thousands of miles of service. For this reason, the lubricants used are specially formulated to be compatible with the rubber boot and give proper lubrication. Most CV-joint repair kits have this special lubricant included.

NOTE: Wear patterns in a used ball or tripot CV-joint are impossible to match during reassembly. If there are any signs of wear, abnormal operating noise, corrosion, or heat discoloration, the joint must be replaced.

TROUBLESHOOTING

Noises from the engine, drive axles, suspension and steering in the front drive cars can be misleading to the untrained ear. Ideally a smooth road serves best for detecting operating condition(s) that cause noise.

• A humming noise could indicate that early stage of insufficient or incorrect lubricant.

• Worn driveshaft joints will cause a continuous knock at low speeds.

• A popping or clicking sound on sharp turns indicates trouble in the outer or wheel end joint.

• The cluck noise at acceleration from coasting or deceleration from a load pull indicated two possibilities-damaged inner or transaxle joint or differential problem(s).

• An inner joint will create a vibra-

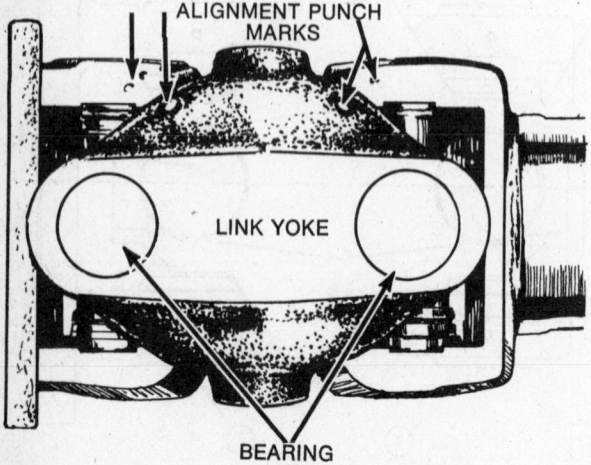

Match marks for double cardan joint

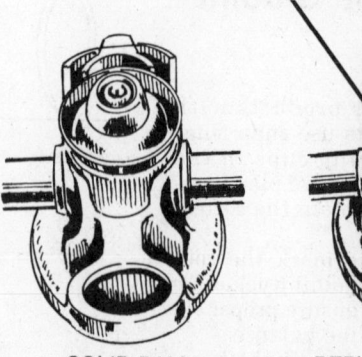
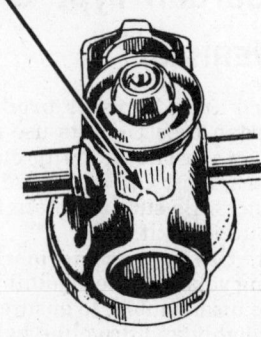

Solid and replaceable U Joint balls

BOOT PROTECTOR
RIGHT SHAFT
LEFT SHAFT
SLIDE HAMMER
WITH ADAPTER

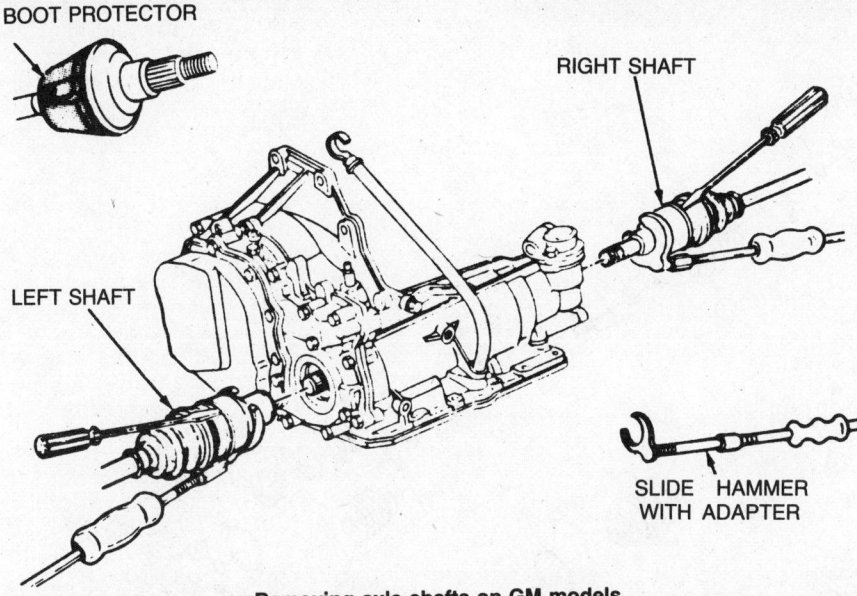

Removing axle shafts on GM models

tion during acceleration due to plunging action hanging up and releasing repeatedly. Probable cause would be foreign particles or lack of lubrication, or improper assembly.

• Remember that tires, suspension, engine, and exhaust system are all up front to add their noises.

• Make a check with front wheels elevated off ground. Spin the wheels by hand to determine if wheel bearing could be noisy or if out of round tires are causing vibration. Many wheel bearings are prelubed and sealed at the factory.

CAUTION

Personal injury can occur from spinning wheels by engine power. Spinning a wheel at excess speed may cause damage to CV-joints that could be operating at angles too steep when wheels are allowed to hang. Overspeeding might also cause damage to tires and the differential.

SHAFT REMOVAL

1. Remove the hub nut and discard it.
2. Drain the lubricant from the transaxle. Remove the differential cover (Chrysler only).
3. The speedometer pinion gear assembly must be removed before the right driveshaft can be removed (automatic transaxles only).
4. Rotate the driveshaft to view the circlip.
5. Compress the circlip tangs with needle nose pliers as you pry into the side gear. This compresses the circlip in position for shaft removal later. Keep an awl between the differential pinion shaft and the end face of the shaft to prevent circlip reentry to the groove.

NOTE: This applies to Chrysler cars only.

6. Remove the ball joint clamp bolt. Drop the lower arm too allow clearance. This will permit the front wheel to swing free.
7. Pull the outer splined shaft from the wheel hub away. Do not pull on the shaft. Grasp the joint housing.
8. Remove the inner joint by pulling outward on the inner joint housing. Do not pull the shaft.

NOTE: Do not allow the assembly to hang at either end. This can jam the CV-joint and cause vibration during operation. If necessary, support the shaft at either end by rope or wire.

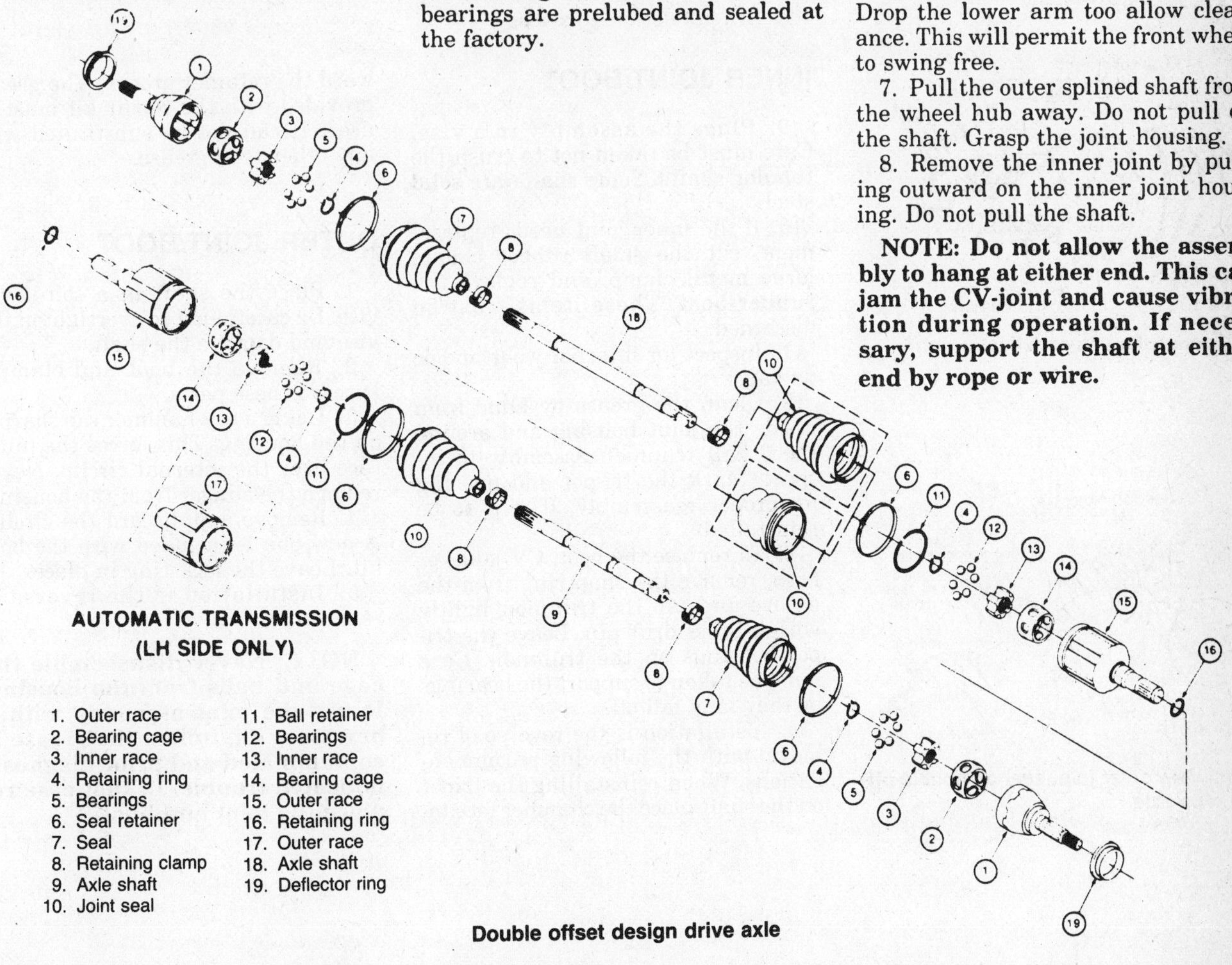

AUTOMATIC TRANSMISSION
(LH SIDE ONLY)
1. Outer race
2. Bearing cage
3. Inner race
4. Retaining ring
5. Bearings
6. Seal retainer
7. Seal
8. Retaining clamp
9. Axle shaft
10. Joint seal
11. Ball retainer
12. Bearings
13. Inner race
14. Bearing cage
15. Outer race
16. Retaining ring
17. Outer race
18. Axle shaft
19. Deflector ring

Double offset design drive axle

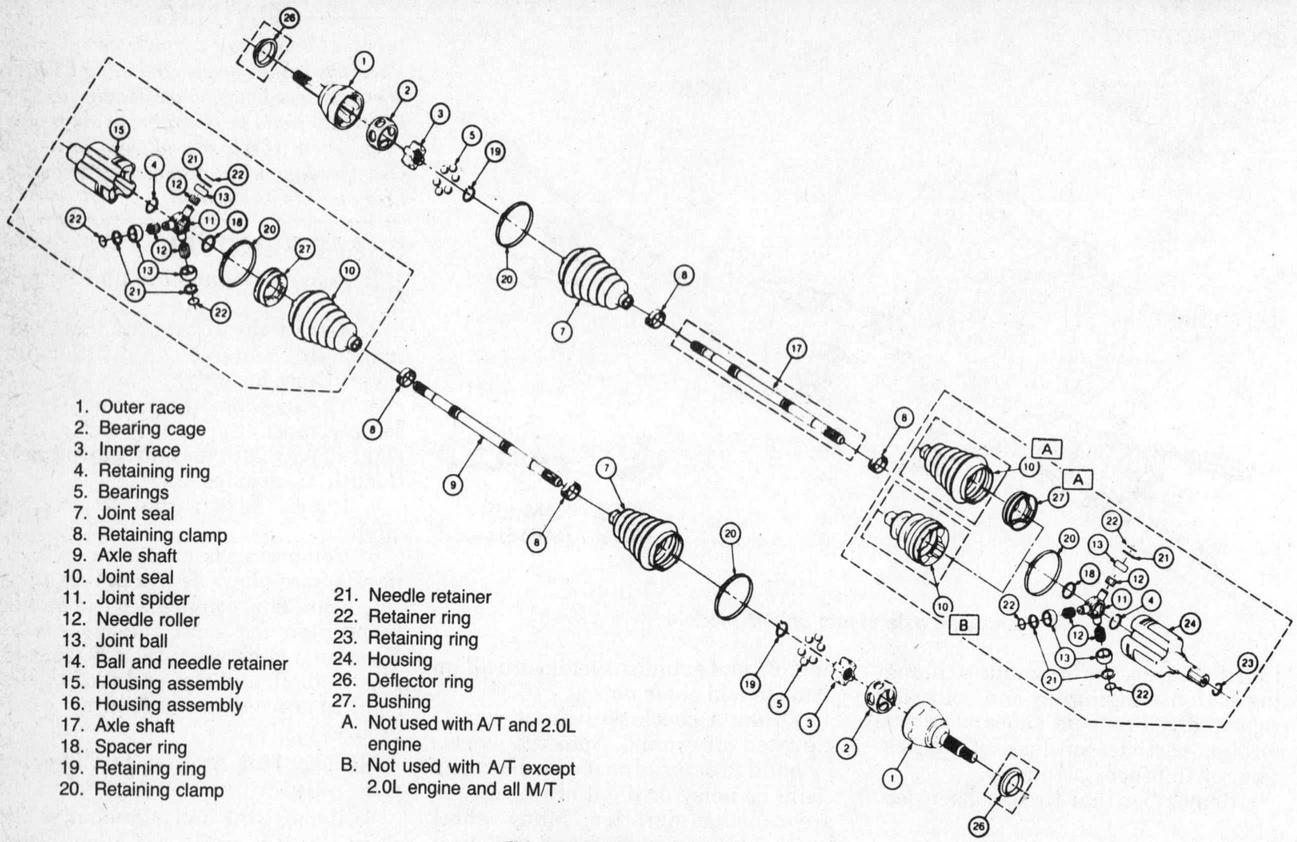

1. Outer race
2. Bearing cage
3. Inner race
4. Retaining ring
5. Bearings
7. Joint seal
8. Retaining clamp
9. Axle shaft
10. Joint seal
11. Joint spider
12. Needle roller
13. Joint ball
14. Ball and needle retainer
15. Housing assembly
16. Housing assembly
17. Axle shaft
18. Spacer ring
19. Retaining ring
20. Retaining clamp

21. Needle retainer
22. Retainer ring
23. Retaining ring
24. Housing
26. Deflector ring
27. Bushing
A. Not used with A/T and 2.0L engine
B. Not used with A/T except 2.0L engine and all M/T

Tri-pot design drive axle

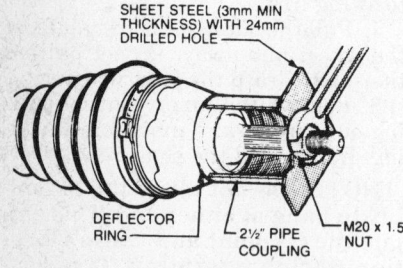

SHEET STEEL (3mm MIN THICKNESS) WITH 24mm DRILLED HOLE

DEFLECTOR RING — 2½" PIPE COUPLING — M20 x 1.5 NUT

Installing steel deflector ring

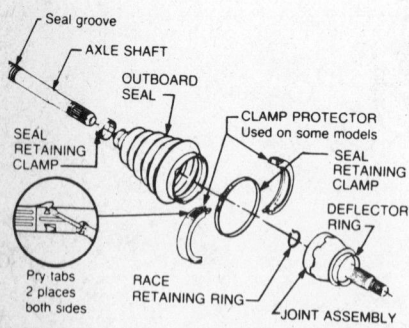

Seal groove
AXLE SHAFT
OUTBOARD SEAL
CLAMP PROTECTOR Used on some models
SEAL RETAINING CLAMP
SEAL RETAINING CLAMP
DEFLECTOR RING
Pry tabs 2 places both sides
RACE RETAINING RING
JOINT ASSEMBLY

Removing outer joint seal on double off-set type axle

INNER JOINT/BOOT

9. Place the assembly in a vise. Care must be taken not to crush the tubular shafts. Some shafts are solid steel.

10. If the inner joint needs replacement, cut the small rubber clamp, large metal clamp, and remove the rubber boot. These items must be discarded.

11. Inspect for internal wear and/or damage.

12. Clean the grease by hand from inside the joint housing and around the 3 ball trunnion assembly to inspect. Mark the tri-pot and housing for proper reassembly, If it is to be reinstalled.

13. To replace the boot, CV-joint, or both, remove the snap ring from the groove and tap the trunnion lightly with a brass drift pin. Leave the tripot bearings on the trunnion. Care must be taken to support the bearings as they may fall off.

14. Installation is the reverse of removal with the following recommendations. When reinstalling the tripot on the shaft place the chamber face toward the retainer groove. The grease provided with the repair kit must be used. It can not be substituted with any other type grease.

OUTER JOINT/BOOT

1. Place the shaft in a soft-jawed vise. Be careful not to overtighten the vise and damage the shaft.

2. Remove the boot and clamps. Discard these parts.

3. Using a soft hammer rap sharply on the housing. This forces the inner race over the internal circlip. Never remove the slinger from the housing.

4. Remove and discard the circlip. A new one is included with the boot kit. Leave the lock ring in place.

5. Installation is the reverse of removal.

NOTE: Never disassemble the cage and balls from the housing. Reuse the joint assembly with a new boot kit, unless the grease is contaminated and prior diagnosis indicated trouble. In that case replace the joint and boot.

Strut Overhaul

STRUT SERVICE AND REPAIR

MacPherson struts are appearing on the front (and rear) wheels of more and more cars. The strut design takes up less room in the engine compartment, compared to a conventional upper and lower arm with shock absorber arrangement. The trend toward smaller, lighter and more efficient packaging mandates the use of a strut suspension to permit more room for engine accessories and front wheel drive components.

Strut Suspension Design

In a conventional front suspension, the wheel is attached to a spindle, which is in turn, connected to upper and lower control arms through upper and lower ball joints. A coil spring between the control arms (sometimes on top of the upper arm) supports the weight of the vehicle and a shock absorber controls rebound and dampens oscillations.

In a strut type suspension, the strut performs a shock dampening function, like a shock absorber, but unlike a conventional shock absorber, the strut is a structural part of the vehicle's suspension.

The strut assembly usually contains a spring seat to retain the coil spring that supports the vehicle's weight. The shock absorber is built into the body of the strut housing. The strut is normally attached at the bottom to the lower control arm and at the top to the car body. The upper mount usually features a bearing that permits the

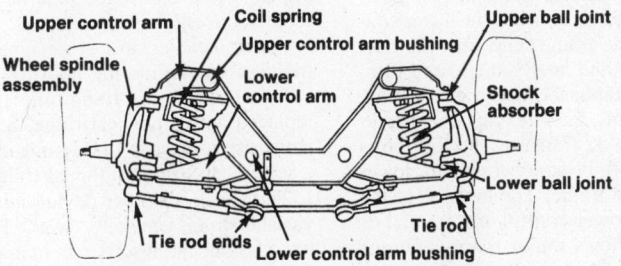

Conventional upper and lower arm suspension

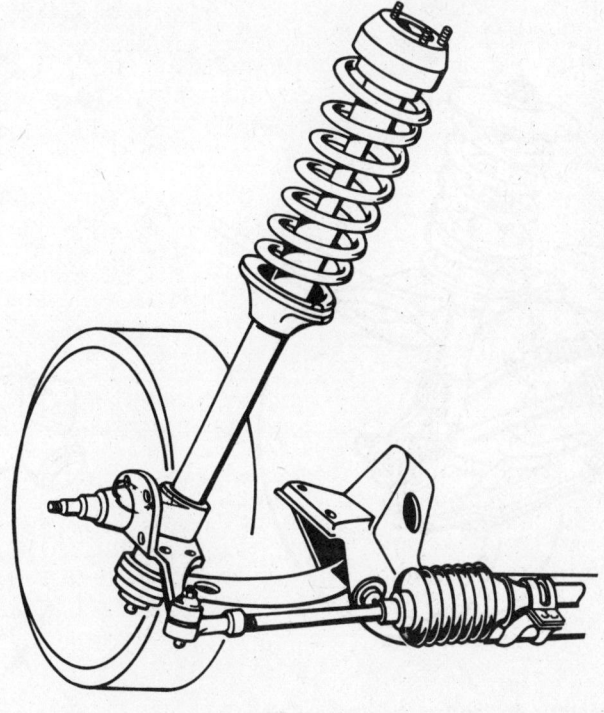

Strut with concentric coil spring (rear wheel drive)

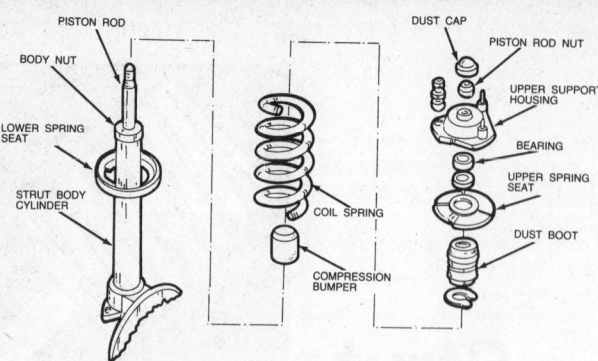

Exploded view of a typical strut

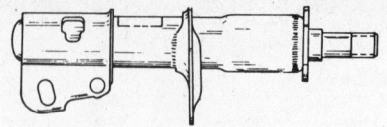

A sealed strut has no body nut and is serviceable by replacement

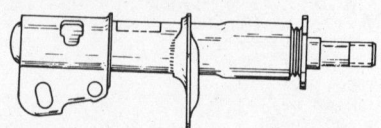

Serviceable struts have a removeable body nut to allow replacement of the strut cartridge

coil spring to rotate as the wheels turn for smoother steering. The entire design eliminates the need for the upper control arm, upper ball joint and many of the conventional suspension bushings. The lower ball joint is no longer a load carrying unit, because it is isolated from the weight of the vehicle.

Domestic struts have taken 2 forms—a concentric coil spring around the strut itself and a spring located between the lower control arm and the frame. GM and Chrysler (except for '82 and later Camaro and Firebird) use the traditional concentric coil spring around the strut. Ford (except the Escort and Lynx) and '82 Camaros and Fire-birds use the spring off the strut between the lower control arm and frame. The location of the spring on the lower control arm instead of on the strut, allows minor road vibrations to be absorbed through the chassis rather than be fed back to the driver through the steering system.

Serviceability

Struts fall into 2 broad categories—serviceable and sealed units. A sealed strut is designed so that the top closure of the strut assembly is permanently sealed. There is no access to the shock absorber cartridge inside the strut housing and no means of replacing the cartridge. It is necessary to replace the entire strut unit.

A serviceable strut is designed so that the cartridge inside the housing, that provides the shock absorbing function, can be replaced with a new cartridge. Serviceable struts use a threaded body nut in place of a sealed cap to retain the cartridge.

The shock absorber device inside a serviceable strut is generally "wet". This means that the shock absorber contains oil that contacts and lubricates the inner wall of the strut body. The oil is sealed inside the strut by the body nut, O-ring and piston rod seal.

Servicing a "wet" strut with the equivalent components involves a thorough cleaning of the inside of the strut body, absolute cleanliness and great care in reassembly.

Cartridge inserts were developed to simplify servicing "wet" struts. The insert is a factory sealed replacement for the strut shock absorber. The replacement cartridge is simply substituted for the original shock absorber cartridge and retained with the body nut, avoiding the near laboratory-like conditions required to service a "wet" strut with "wet" service components.

Most OEM domestic struts are serviced by replacement of the entire unit. There is no strut cartridge to replace. Exceptions to this general rule are the struts used on GM front wheel drive J-cars and A-cars, which feature an internally threaded housing, accessible by removing the OEM cap from the housing. Once the old cartridge is removed, a new cartridge can be threaded

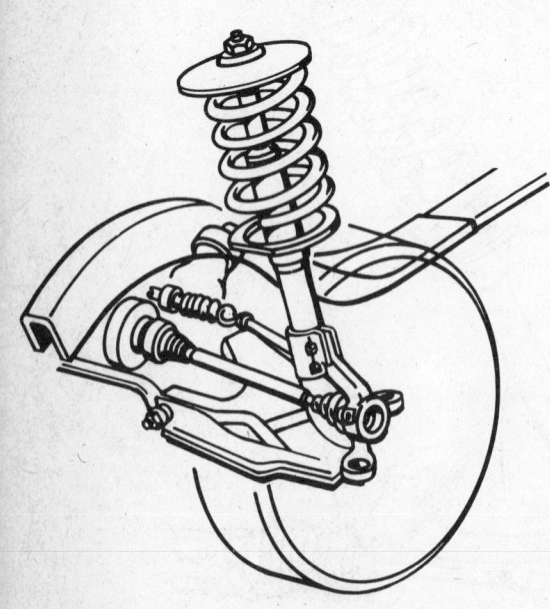

Strut with concentric coil spring (front wheel drive)

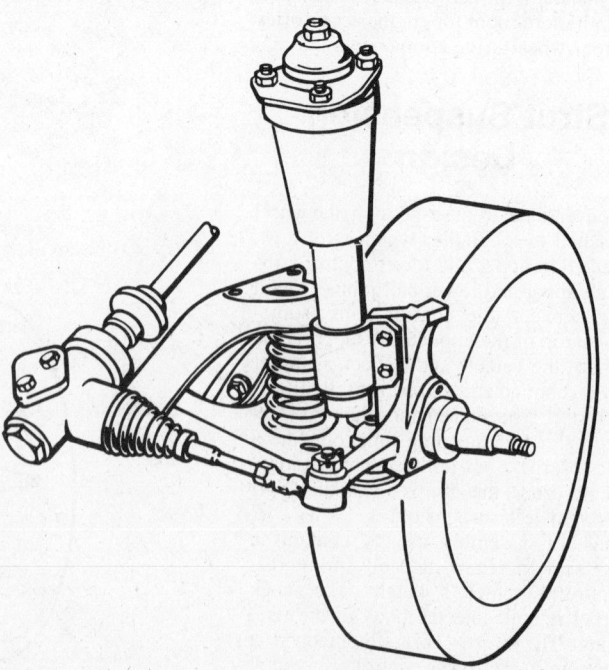

Modified MacPherson strut design with coil spring on the lower arm

into the housing.

Sealed, OEM units can also be serviced by replacement with an aftermarket unit, that will permit future servicing by cartridge replacement.

WHEEL ALIGNMENT

It is not always necessary to re-align the wheels after struts are serviced. If care is taken matchmarking affected components and in reassembling, alignment may be unaffected. However, if wheels were not in proper alignment prior to service, or if the entire strut assembly was replaced, a wheel alignment check should be made. Generally, only camber is adjustable, and then only within a narrow range.

Do not attempt to bend components to correct wheel alignment.

Since the majority of OEM struts are serviced by replacement, most manufacturers recommend wheel alignment following strut replacement.

Tools

Without the right tools, a strut job will take longer than necessary and can be dangerous.

A normal selection of hand tools such as open end and box wrenches, sockets, pliers, screwdrivers and hammers are necessary to work on struts. Extensions and universal joints will help reach tight spots. Be sure to have both metric and inch-sized wrenches on hand. Two big time-savers are "crowsfeet" and ratcheting box wrenches in assorted sizes. Torx fasteners are also showing up more and more in chassis fasteners.

In addition to the normal handtools, some sort of spanner is necessary to remove the body nut on serviceable struts. Sometimes a pipe wrench can be used successfully.

Strut and cartridge replacement requires a spring compressor.

Makeshift tools for compressing coil springs—threaded rod, chains, wire or other methods—should never be used. The coil spring is under tremendous compression and can fly off causing personal injury and damage to equipment. Use only a good quality spring compressor such as described below.

Economy, or manual, spring compressors are the least expensive but more time consuming to use. Angle hooks grasp the spring coils and must be compressed with a wrench. For those who service struts infrequently, this is probably the wisest investment for purchase.

Other manual spring compressors (jaws type) are faster to operate, have a more positive gripping action and can be used on or off the car. These types are probably not cost effective for the do-it-yourselfer, but can be rented from auto supply stores for single-time use.

For volume work, compressors that are pneumatically or hydraulically operated are

MAINTAINING WHEEL ALIGNMENT

The location and method of adjusting wheel alignment determines the components that must be match-marked to maintain wheel alignment. There are 4 basic methods of adjusting wheel alignment. Almost all cars use one of these or a slight variation.

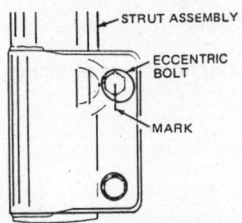

Mark the eccentric (camber adjusting bolt) relative to the clevis mounting bracket.

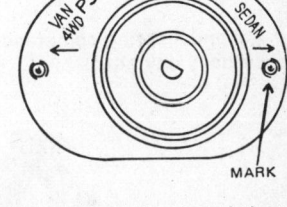

Mark the mounting stud that faces the front of the vehicle. This type of bracket is reversible for varying applications.

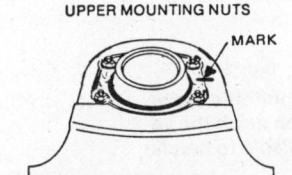

Mark the upper support housing relative to the inner fender before removing the strut from the upper mount.

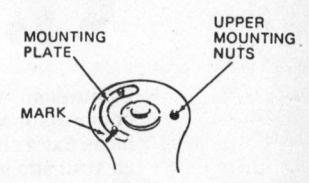

Mark the location of the mounting plate relative to the location on the inner fender.

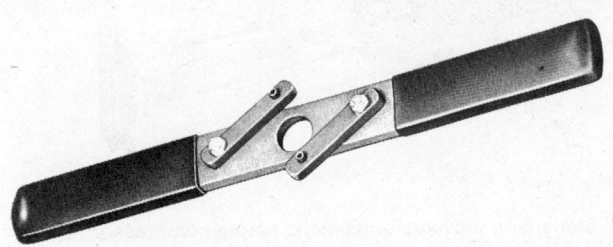

A simple spanner wrench designed for use with body nuts equipped with recessed lugs. A pipe wrench is a frequent substitute

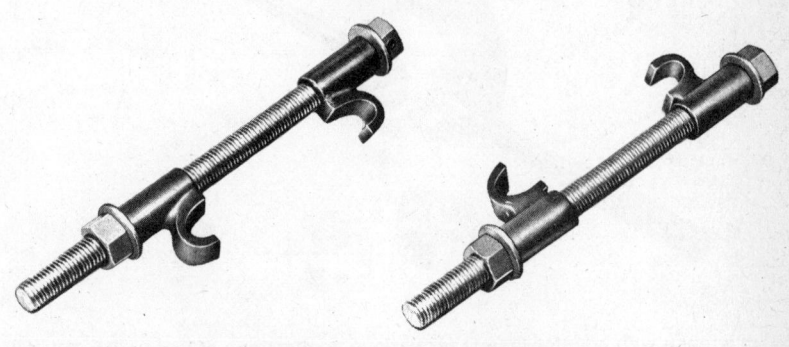

An economical manual spring compressor

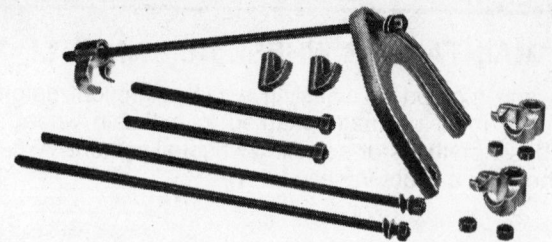

A manual spring compressor with plates or hooks for servicing virtually any strut

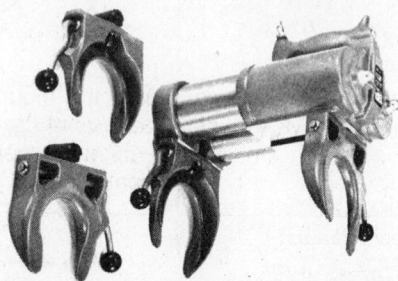

Lightweight, air operated, portable spring compressor can be used on or off the vehicle. Extra shoes are available to handle all strut applications

Stationary, universal pneumatic spring compressor

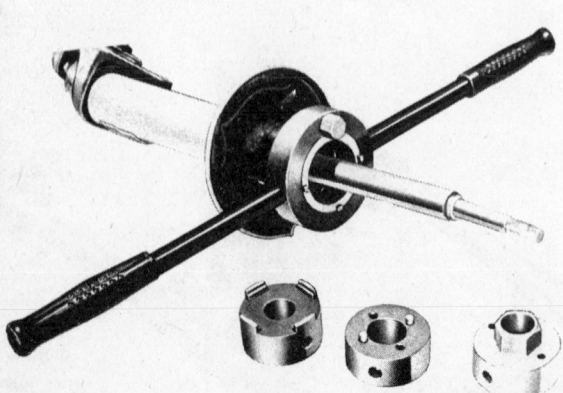

Spanner wrench with adaptor inserts for various applications of body nuts. This type of spanner can be used with a torque wrench for retorqueing the body nut

best. Air operated compressors are suitable for all types of struts (through use of adaptors), are lightweight and can be used on or off the vehicle. Bench mounted hydraulically operated units are probably the safest, but are also the most expensive and require that the strut be removed from the vehicle, which means separating brake lines and other connections which can be time consuming.

There are also universal kits that fit all struts in either the manual or air operated types.

Regardless of what type of spring compressor you're using, GM front wheel drive A-, J-, and X-cars as well as Chrysler Corp. Omni, Horizon and K-cars, require the use of a special spring compressor with self-leveling plates to grasp the spring seats as the spring is compressed. Likewise, the portable, pneumatic units have extra wide shoe sets suitable for these cars. The shoes are also epoxy coated to avoid scratching the coated springs on these models.

GM front wheel drive A-, J- and X-cars also make use of a camber assist tool, that makes camber adjustment a one man job.

A tube cutter is necessary on GM J-cars to cut the welded top from the strut housing for cartridge replacement.

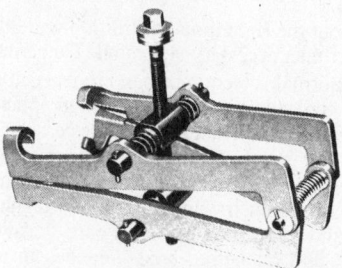

"Jaws" type spring compressor

Spring compressor for GM and Chrysler product applications

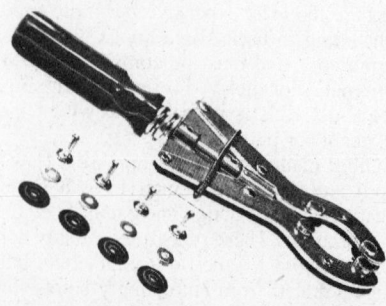

A tube cutter allows opening of the GM J-car struts for cartridge replacement

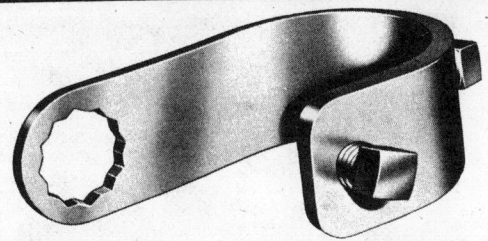

A camber assist tool makes GM cars a one-man job

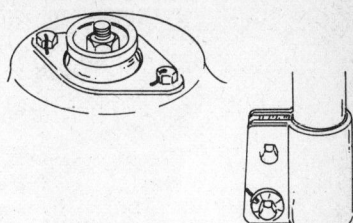

Mark the position of the attachments that control wheel alignment. See Maintaining Wheel Alignment earlier in this section

Repair Tips

1. Make sure you have all the tools you'll need. NEVER IMPROVISE A SPRING COMPRESSOR.

2. Normally both front struts should be repaired or replaced at the same time.

3. The easiest way to work on most struts is to remove the entire unit from the vehicle, unless you have access to an air operated spring compressor. Some struts, however, can, and should, be repaired while installed on the vehicle.

4. Always read the instructions packaged with any replacement parts. In particular, note whether the body nut is supplied new or re-used.

5. Mark the position(s) of any bearing plate nuts or cam bolts to assure proper alignment after installation.

6. Be sure to protect the rubber boot on the drive axle of front wheel drive cars.

7. If necessary to remove the brake caliper, do not let the caliper hang by the brake hose. Suspend the caliper from a wire hook or rope.

8. Be careful in clamping a strut in a vise. Special fixtures are available to hold struts in a vise, but are not necessary if care is used to be sure the housing is not crushed or dented. A block of soft wood on either side of the housing will prevent most damage.

9. Use a spring compressor to relieve tension from the spring. Be sure to clean and lubricate the screw threads, particularly on hand operated (manual) spring compressors.

Some springs have a special coating that should not be scuffed.

10. If you are replacing the strut cartridge, clean the inside of the strut housing and the body nut threads before replacing the oil and installing a new cartridge.

11. Be sure to use OEM quality fasteners any time a fastener is replaced.

STRUT OVERHAUL (OFF-CAR)

Following is a typical overhaul procedure of a serviceable MacPherson strut, after having removed the strut from the vehicle. The vehicle should be firmly supported. If it is necessary, to separate the brake line from the strut for strut removal, the brakes will have to be bled after reinstallation. See the manufacturer's car section for specific MacPherson strut removal and installation procedures.

Photos Courtesy Gabriel Div., Maremont Corp.

Step 1. Examine the strut assembly for damage, dented strut body, spring seat, broken or missing strut mounting parts. Any of these will require replacement of the complete assembly. Also inspect other suspension components for wear or damage

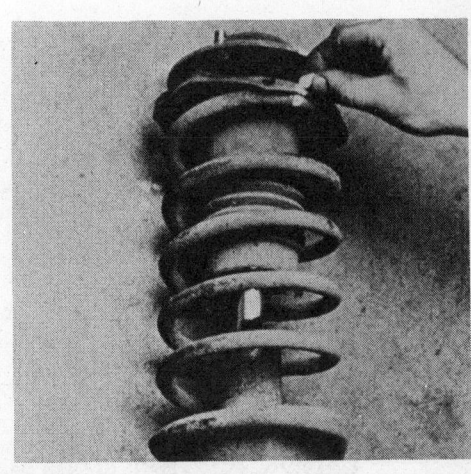

Step 2. Matchmark the upper end of the coil spring and bearing plate to avoid confusion during reassembly

Step 3. To make servicing easier, clamp the strut in a strut vise. The strut vise is designed to clamp the strut tight without damage to strut cylinder. It is very handy for strut work and can be used in your shop vise or mounted to any bench

Step 4. Before using the manual spring compressor, lubricate both sides of the thrust washers and the threads with a light coat of grease

Step 5. Install the compressor hooks on opposite sides of the coil spring with the hooks attached to the upper-most and lower-most spring coils. To avoid possible slippage, use tape or small hose clamps on either side of the compressor hooks

Step 6. Alternately tighten the bolts a few turns at a time until all tension is removed from the spring seat

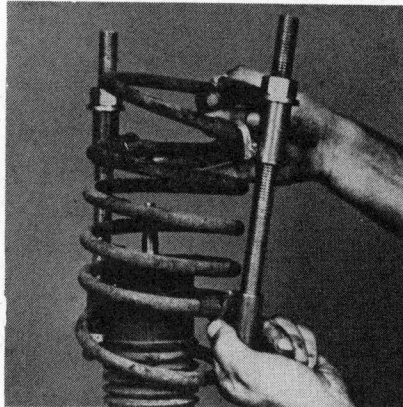

Step 7. Remove the piston rod nut and disassemble the upper mounting parts, keeping them in order for reassembly. Remove the coil spring. There is no need to remove the compressor from the coil spring

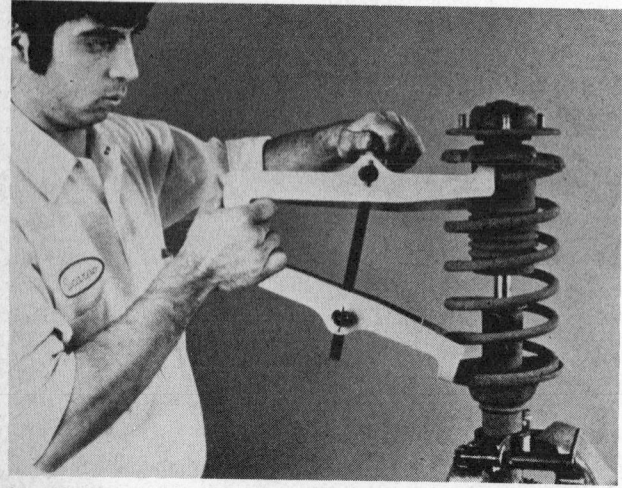

Step 8. An alternative to the manual compressor is the "jaws" type. Turn the load screw to open or close the compressor until the maximum number of spring coils can be engaged

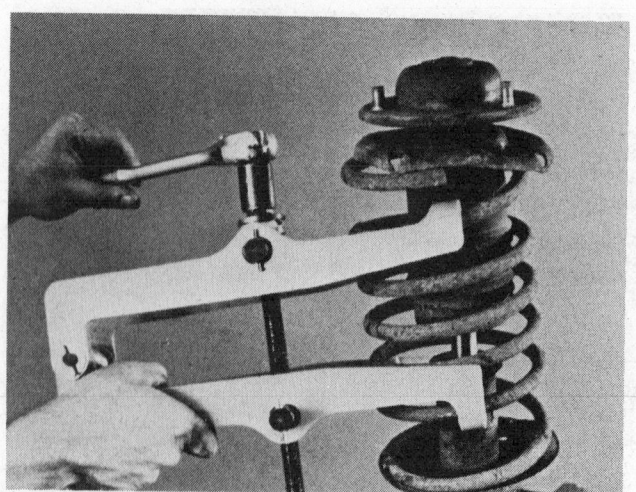

Step 9. Tighten the load screw until the coil spring is loose from the spring seats. There is no need to compress the spring any further

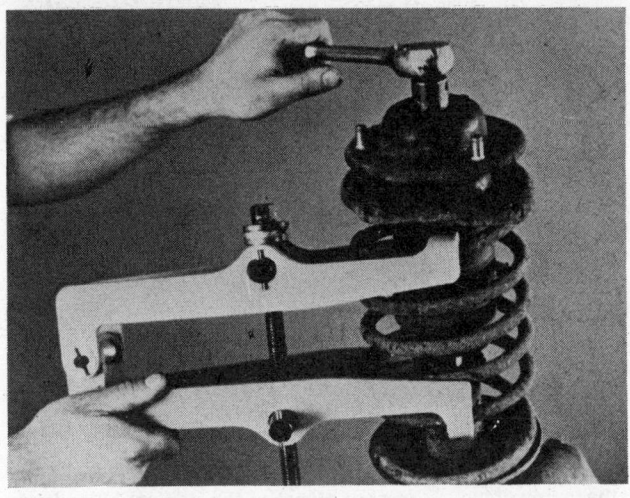

Step 10. Remove the piston rod nut and disassemble the upper mounting parts

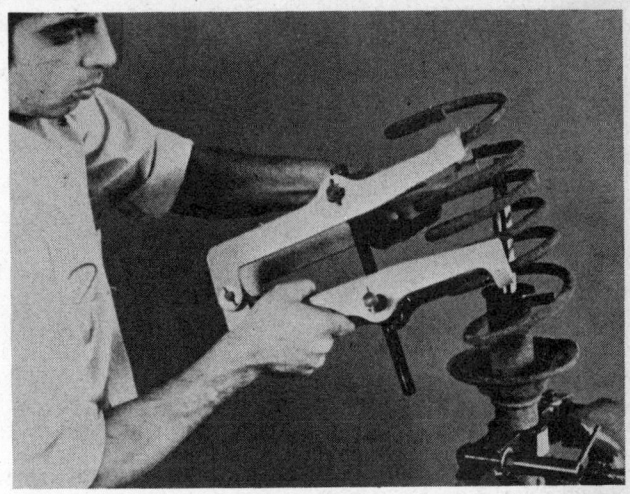

Step 11. Like the manual compressor, there is no need to remove the compressor from the coil spring. Remove the coil spring and compressor

Step 12. Keep the upper mounting parts in order of their removal. They'll be re-assembled in reverse order

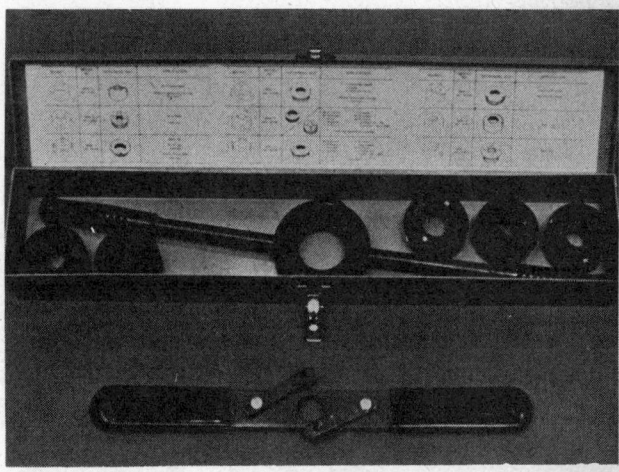

Step 13. A spanner wrench is necessary to remove body nuts, although a pipe wrench will do the job

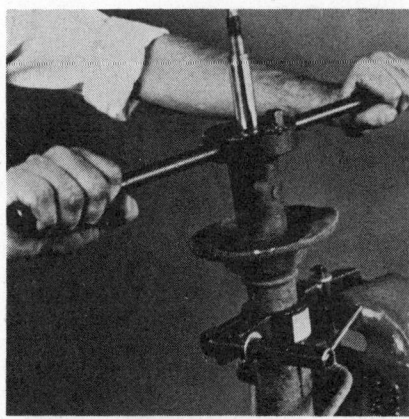

Step 14. Use the spanner wrench or pipe wrench to loosen the body nut

Step 15. Remove the body nut and discard if a new body nut came with the replacement cartridge. If not, save the body nut

Step 16. Use a scribe or suitable tool to remove the O-ring from the top of the housing

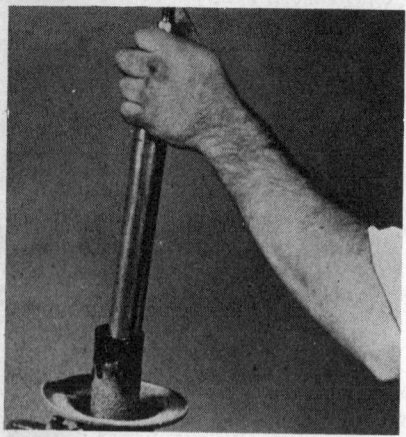

Step 17. Grasp the piston rod and pull cartridge out of the housing. Remove it slowly to avoid splashing oil. Be sure all pieces come out of the housing

Step 18. Pour all of the strut fluid into a suitable container, clean the inside of the strut cylinder, and inspect the cylinder for dents and to insure that all loose parts have been removed from inside of strut body

Step 19. Refill the cylinder with one ounce (a shot glass) of the original oil or fresh oil. The oil helps dissipate internal cartridge heat during operation and results in a cooler running, longer lasting unit. Do not put too much oil in—otherwise the oil may leak at the body nut after it expands when heated

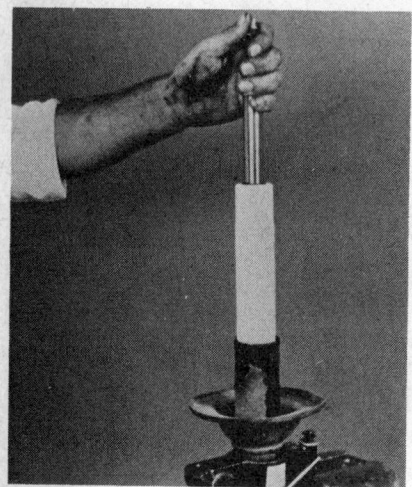

Step 20. Insert the new replacement cartridge into the strut body

Step 21. Push the piston rod *all* the way down, to avoid damage to the piston rod if the spaner wrench slips, and start the body nut by hand. Be sure it is not crossthreaded

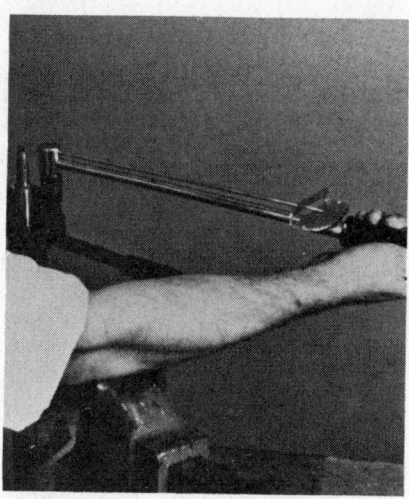

Step 22. Tighten the body nut securely

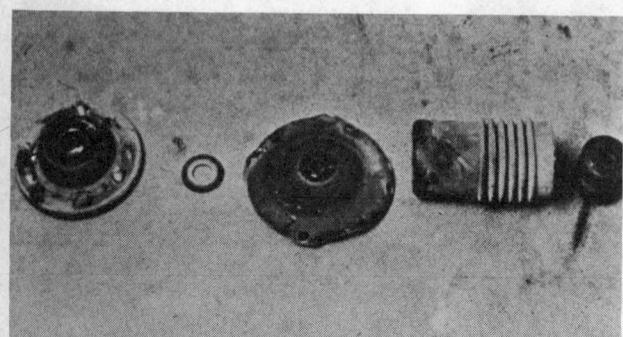

Step 23. Inspect the loose parts prior to re–assembly. Note the chalk mark location for proper seating of the upper spring seat

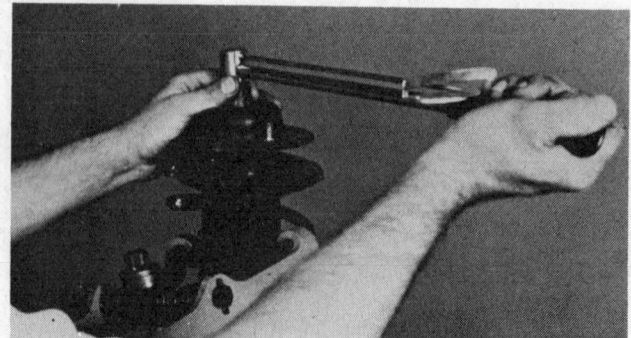

Step 24. Re–assemble the coil spring and upper mounting parts in reverse order. Tighten the piston rod nut and remove the spring compressor. Install the dust cap. Install the strut in the vehicle. See the car section for details

STRUT OVERHAUL

Most domestic car OEM MacPherson struts are sealed units and not repairable. The exceptions are GM front wheel drive A- and J-cars, which use replaceable cartridges. All other cars must use aftermarket struts to be serviceable at a future date. The following procedures cover disassembly of the strut, installation of a serviceable strut, reassembly and cartridge replacement on GM front wheel drive A- and J-models. Consult the applicable manufacturer's car section for removal and installation procedures.

Photos Courtesy Gabriel Div., Maremont Corp.

Step 1. Most domestic cars are serviced initially by replacing the entire strut rather than by using a replacement cartridge. This is necessary because the original equipment struts are sealed shut and cannot be serviced with a replacement cartridge. After-market struts are designed with serviceable threaded body nuts which means they can be serviced in the future by installing a replacement cartridge, using normal cartridge service methods, rather than by replacing the entire strut

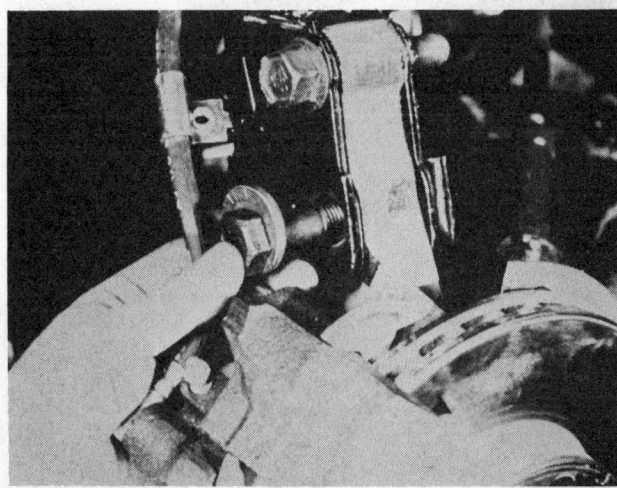

Step 2. An X-car is shown, but the lower mount on the Citation is typical of many vehicles. They all have two bolt clevis mounts and the position of the strut determines the camber adjustment. This means that if you are replacing a sealed strut, front end alignment is necessary because the original alignment is eliminated when you change the strut. If the car has a serviceable strut, you can retain the alignment by marking the position of the mounting bolt relative to the strut. GM has made a running change on the lower mount of their X-Car. The earlier type had an eccentric bolt for camber adjustment. Camber on the latest type is adjusted by pushing or pulling on the wheel with the bolts loosened slightly, but the eccentric can be installed on later cars

Step 3. A special type spring compressor is required for the GM cars and Chrysler K and L cars. A compressor should be used that does not damage the protective coating on the coil spring. Virtually any compressor can be used on other car lines/models

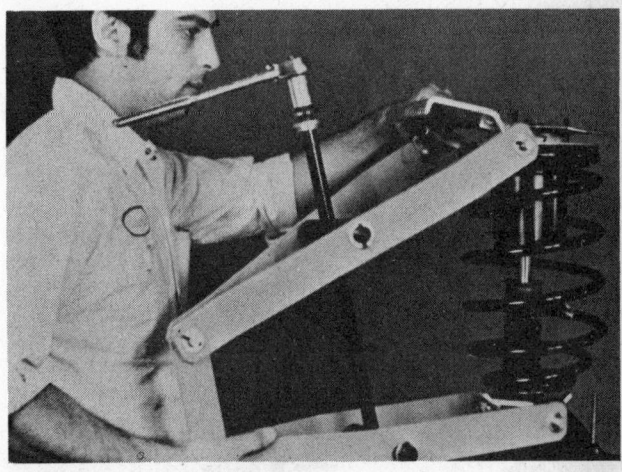

Step 4. Secure the strut in the strut vise; turn the load screw counter-clockwise until the lower plate can be fitted under the lower spring seat and the upper plate can be fitted between the upper spring seat and support housing

STRUT OVERHAUL

Step 5. Make sure that the crescent shaped bars on the upper compression plate are located inside the upper spring seat

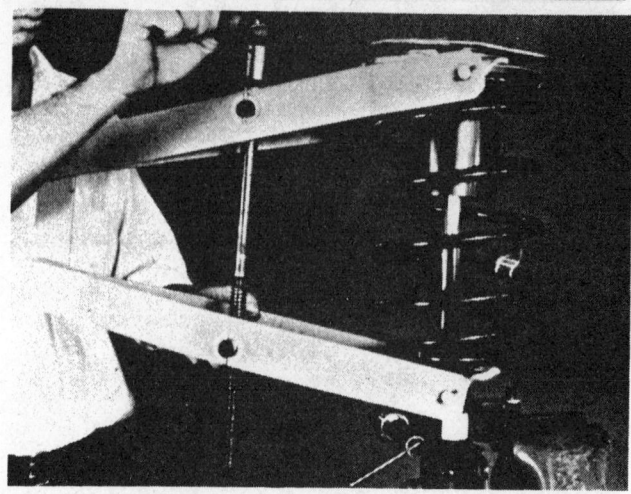

Step 6. Turn the load screw clockwise enough to tighten the compression plates on the spring seats. Stop and make sure that the coil spring will not arch, and that the pivot points are aligned with the center-line of the coil spring

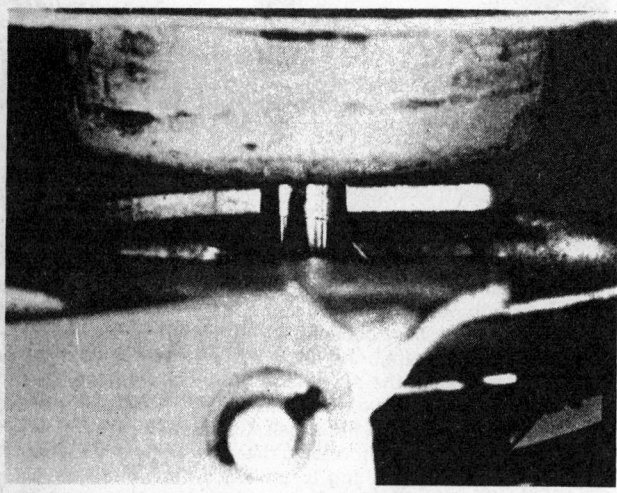

Step 7. Continue to tighten the load screw until the upper support housing can be pulled up to expose about ½ inch of piston rod. This assures that the spring load has been removed from upper spring seat

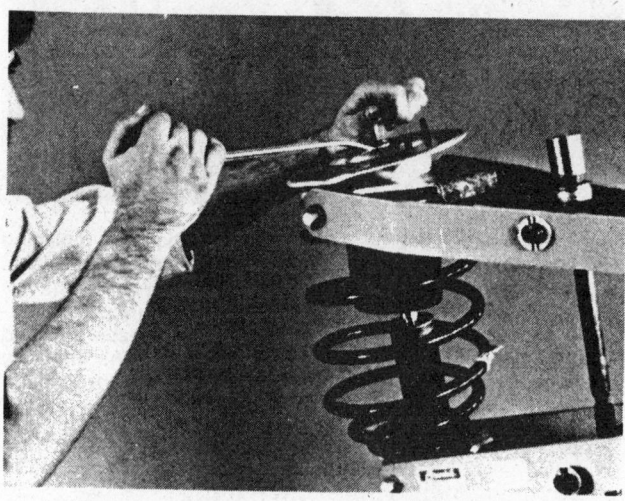

Step 8. Remove the piston rod nut with the aid of a wrench to keep the piston rod from turning and remove upper support housing

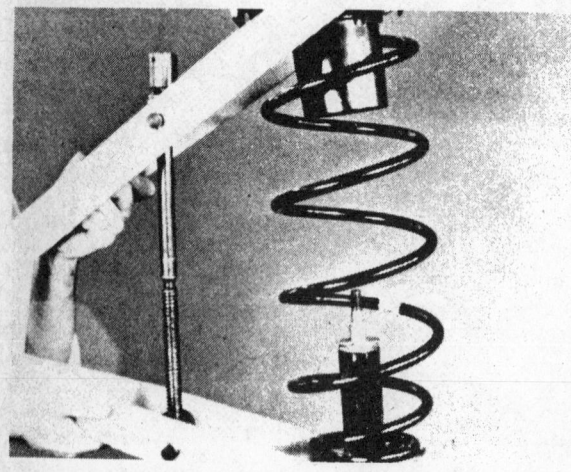

Step 9. Turn the load screw counter-clockwise until the spring tension is completely relieved. Remove the compressor, coil spring and upper support housing from the strut.

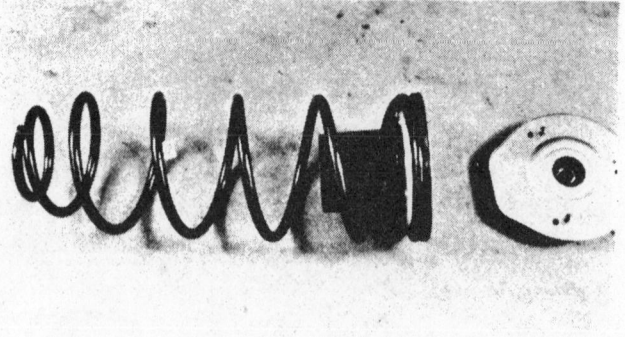

Step 10 Assemble the upper mounting parts in order of their removal. They'll be re-assembled in reverse order

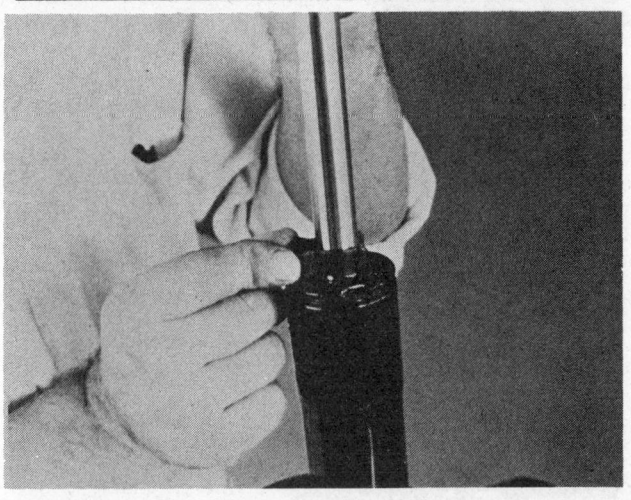

Step 11. Place the new strut in the vise and extend piston rod fully and install clip (spring type clothes pin will do) as shown. This keeps the piston rod extended while assembling the spring and upper mounting parts.

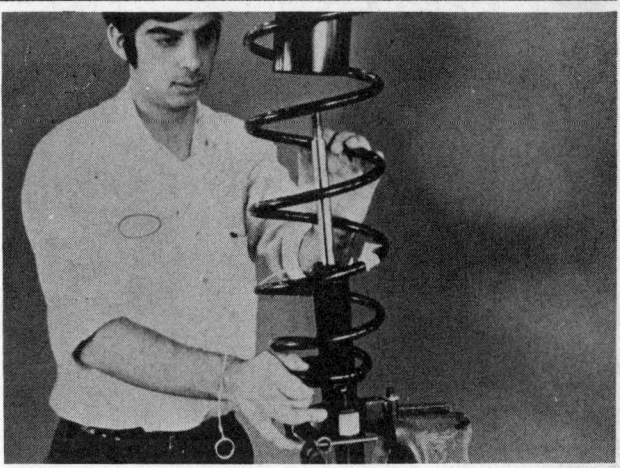

Step 12. Install the coil spring and upper spring seat on the new strut

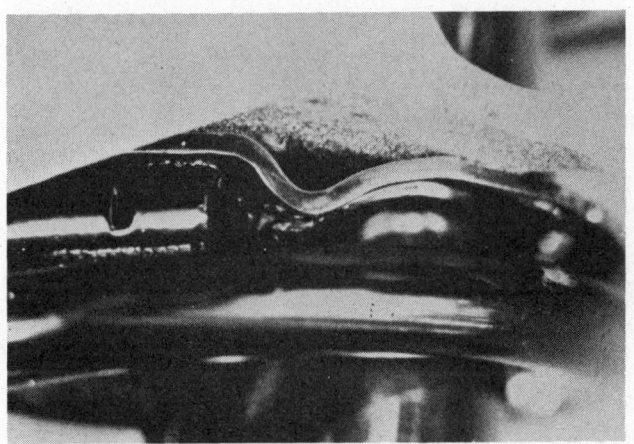

Step 13. Make sure that the spring helix is aligned with the lower spring seat

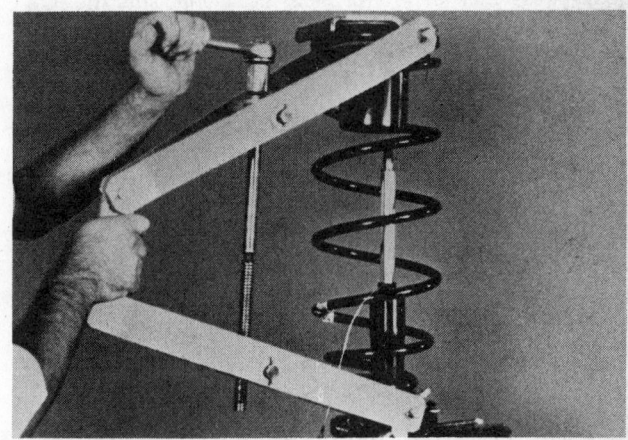

Step 14. Locate upper and lower compression plates on spring seat

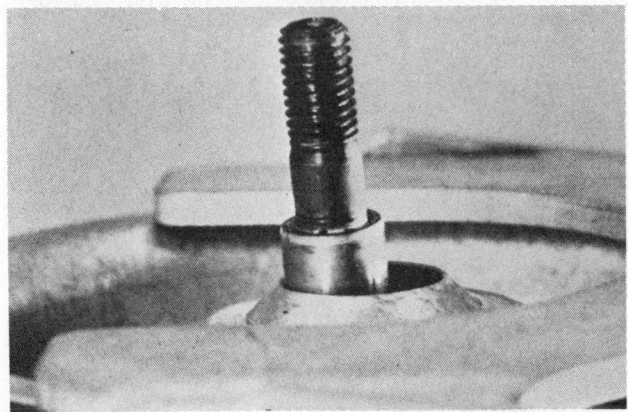

Step 15. Make sure that the crescent shaped bars on the upper compression plate are located on the upper spring seat as shown. Turn the load screw clockwise enough to tighten the compression plates on the upper and lower spring seats. Stop. Again, to assure that the coil spring will not arch, make sure that the pivot points are aligned with the centerline of the coil spring. Then continue turning the load screw clockwise until about 1½ inches of piston rod is showing above the upper spring seat

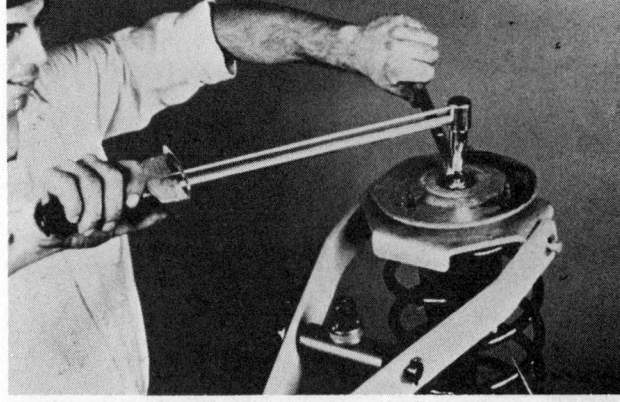

Step 16. Install upper support housing on piston rod. Tighten the piston rod nut and remove the compressor from the strut and the strut from the vise. Install the strut. See the car section for details

STRUT OVERHAUL

GM J- AND A-CARS ONLY

Step 1. Place the strut assembly in a vise, and compress the coil spring. Remove the piston rod, upper support housing, spring seat and coil spring. If the universal pneumatic spring compressor is used, an adaptor provided with the compressor should be fastened to the strut under the steering arm. The ears of the adaptor should be aligned with steering arm. The adaptor provides a square seating surface for the strut while it is being compressed

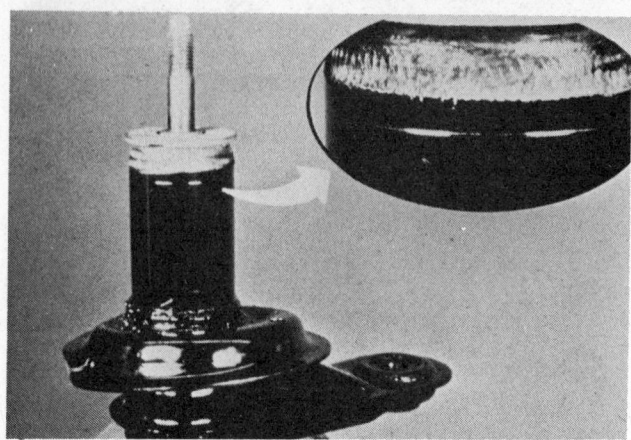

Step 2. J and A–car struts have a welded upper closure, but the strut is designed so the damping mechanism can be replaced with a cartridge insert. Just below the spin weld there is a cut-line scribed in the strut body

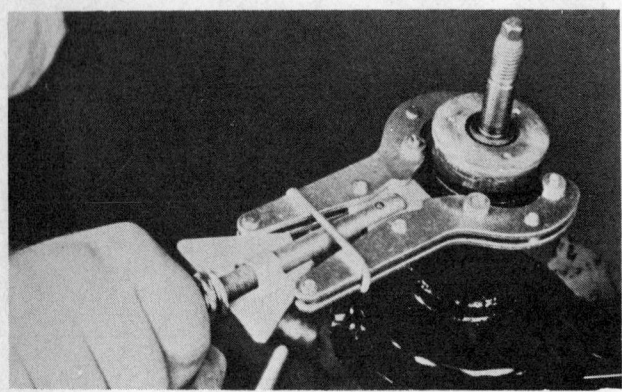

Step 3. Using a pipe cutter, cut open the strut body at the scribed line. (Note: *It is important that the cut be made on the cut-line*)

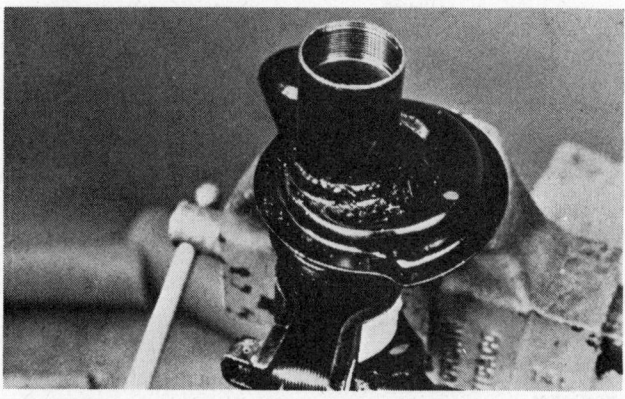

Step 4. Remove the cartridge and oil from the strut. Note the threads on the inside of the strut. Deburr the top of the strut body if necessary

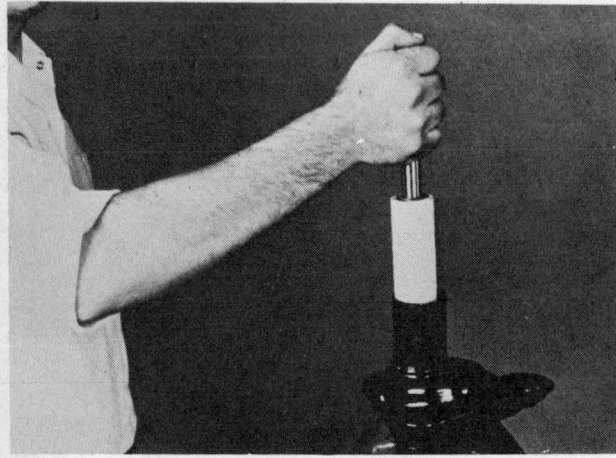

Step 5. Pour about one ounce of oil into the strut body and insert the replacement cartridge. Push the piston rod down and start the body nut by hand. Tighten the nut securely

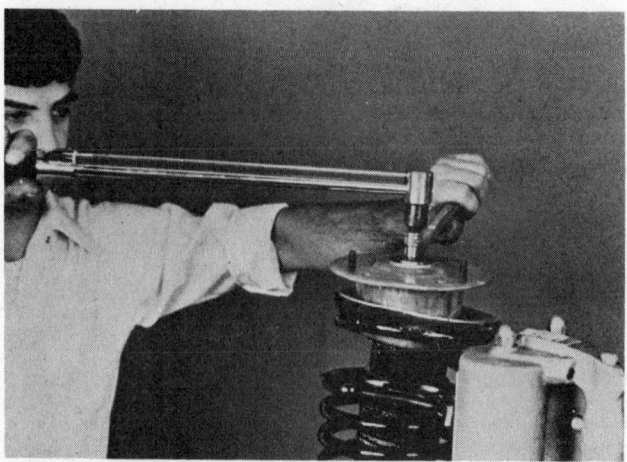

Step 6. Assemble spring, upper spring seat, and upper support housing on the strut, and tighten the new piston rod nut. The renewed strut is now ready to install on the vehicle. Release the spring tension

— MACPHERSON STRUT PROBLEM DIAGNOSIS —

Problems with MacPherson struts generally fall into 3 main categories: suspension, tire wear and steering. In general, the symptoms encountered are not significantly different from those encountered on conventional suspensions.

Suspension

Sag

Vehicle "sag" is a visible tilt of the car from one side to the other or one end to the other while parked on a level surface.

Weak or damaged strut springs could cause this condition and should be repaired immediately.

Sag will also cause steering and tire wear problems to be more pronounced and vehicle instability on rough roads. Front wheel alignment will not solve the problem.

Weak strut springs increase vehicle sag. See "Tire Cupping".

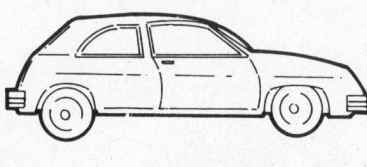

Cartridge Leaks

Strut cartridge leaks (not seepage) indicate the need for cartridge or strut replacement. Be sure the leakage is coming from the strut, and not from elsewhere on the vehicle.

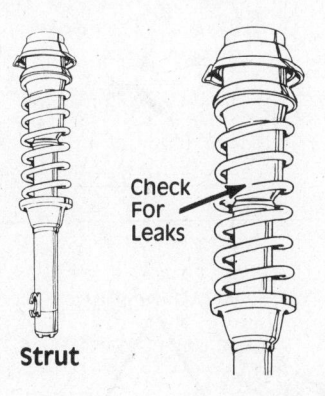

Check For Leaks

Strut

Abnormal Tire Wear

Wear on One Side

One sided tire wear indicates incorrect camber. Check the causes in the accompanying illustration and be sure the wheel alignment is correct.

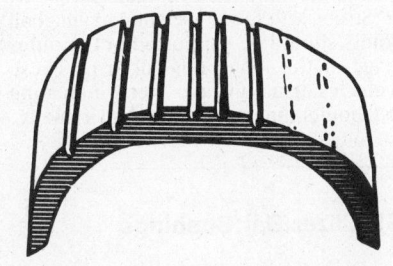

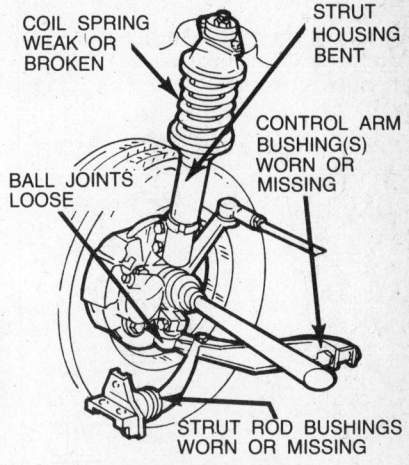

COIL SPRING WEAK OR BROKEN

STRUT HOUSING BENT

CONTROL ARM BUSHING(S) WORN OR MISSING

BALL JOINTS LOOSE

STRUT ROD BUSHINGS WORN OR MISSING

Tire "Cupping"

Cupped tires indicate any or all of the following problems.

1. A weak strut cartridge can be verified by bouncing each corner of the car vigorously and letting go. The car should not bounce more than once, if the shock absorber cartridges are good.

2. Weak strut springs allow sag to increase with only a slight amount of downward pressure. A visual inspection will reveal any broken springs or shiny spots.

3. Check for loose or worn wheel bearings with the weight of the car off of the wheel.

4. Check the wheel balance.

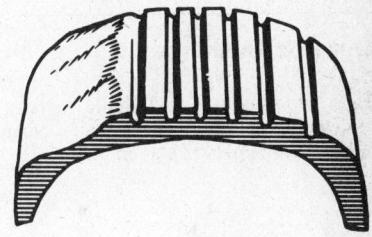

Tread Edge Wear

Wear along tread edges (feathering) indicates a suspension or steering system problem.

1. Strut rod bushings are worn or missing.

2. Tie rod end wear can be determined by grabbing the tie rod end firmly and forcing it up, down or sideways to check for lost motion.

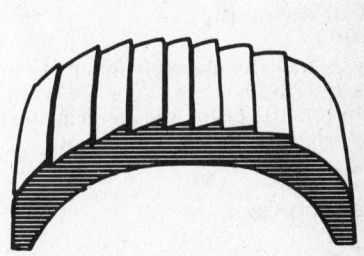

— MACPHERSON STRUT PROBLEM DIAGNOSIS —

Problems with MacPherson struts generally fall into 3 main categories: suspension, tire wear and steering. In general, the symptoms encountered are not significantly different from those encountered on conventional suspensions.

Tires

Both front tires should match and both rear tires should match. Be sure air pressure is correct.

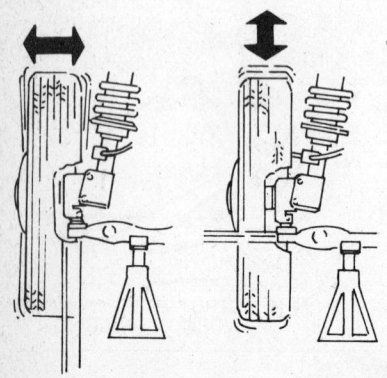

Strut Rod Bushings

Grasp the strut rod and shake it. Any noticeable play indicates excessive wear and need for parts replacement.

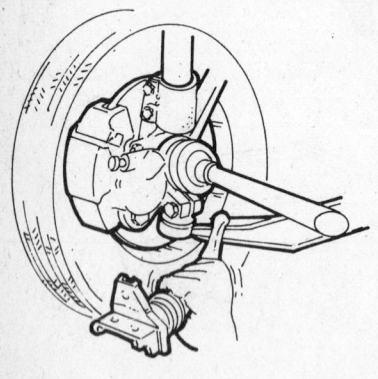

Steering

Ball Joints

Support the car under the frame or crossmember so that the jack does not interfere with the control arm. Rock the tire in and out and up and down. Excessive movement means that both ball joints should be replaced.

Struts with lower weight-carrying ball joints should be supported at the outer edge of the lower control arm. These vehicles usually have wear indicating ball joints that can be checked visually.

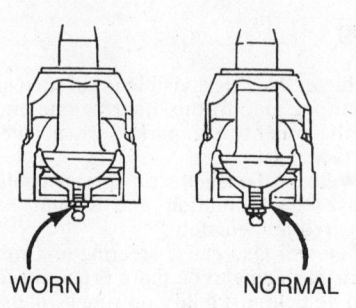

WORN   NORMAL

Stabilizer Bar Bushings

Check for worn bushings or lost motion with the vehicle level and the weight evenly distributed on all wheels.

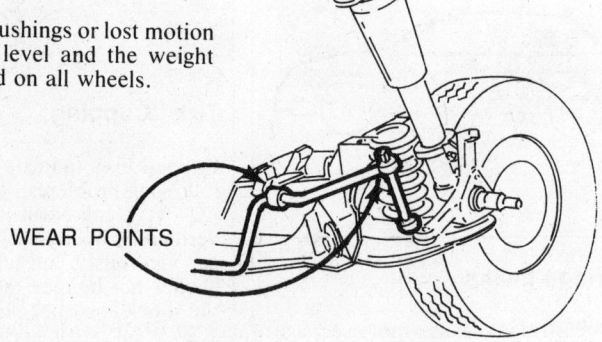

WEAR POINTS

Control Arm Bushings

Support the car under the frame or body and remove the weight from the wheel and control arm. Check for free-play in the bushings at the pivot point, using a pry bar.

NOTE: Some control arm bushings are serviceable only by replacing the entire arm.

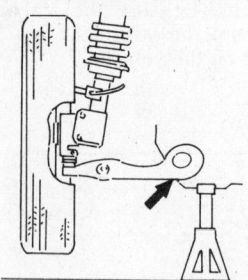

Strut Assembly

Check the strut assembly for cracks or dents in the housing. Look for worn, bent or loose piston rods or dents that will inhibit piston rod movement.

Steering Gear

Check for worn steering gear or loose or worn mounting bolts and bushings.

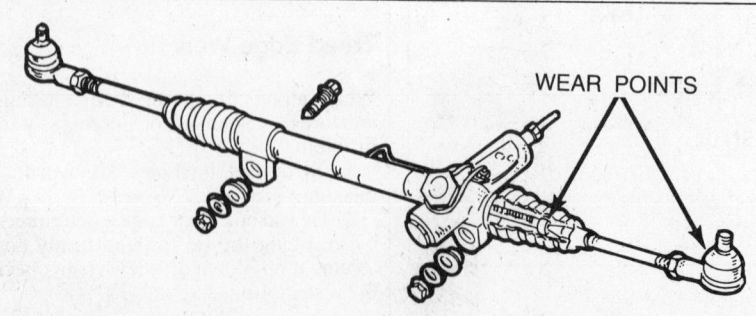

WEAR POINTS

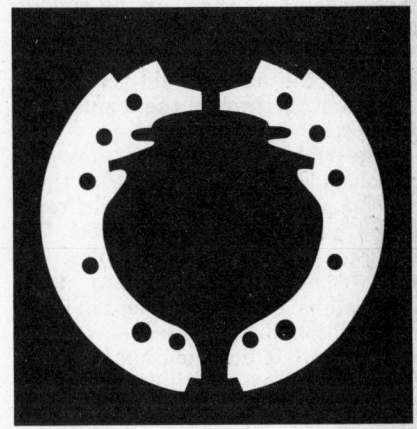

Brakes

BRAKES

Hydraulic Brake Component Service

BASIC OPERATING PRINCIPLES

The hydraulic brake system transports the power required to force the frictional surfaces of the braking system together from the pedal to the individual brake units at each wheel. A hydraulic system is used for two reasons. First, fluid under pressure can be carried to all parts of an automobile by small hoses (some of which are flexible) without taking up a significant amount of room or posing routing problems. Second, a great mechanical advantage can be given to the brake pedal end of the system and the foot pressure required to actuate the brakes can be reduced by making the surface area of the master cylinder pistons smaller than that of any of the pistons in the wheel cylinders or calipers.

The master cylinder consists of a double reservoir and piston assembly as well as other springs, fittings, etc. Double (dual) master cylinders are designed to separate two wheels from the others. The standard approach has been have separate circuits for the front and rear wheels. Newer models may have a diagonally split system; i.e. one front wheel and the opposite side rear wheel are in a separate circuit from the other front and rear wheel.

Steel lines carry the brake fluid to a point on the vehicles frame near each wheel. A flexible hose usually carries the fluid to the disc caliper or wheel cylinder. The flexible line allows for suspension and steering movement.

The rear wheel cylinders contain two pistons each, one at either end, which push outward in opposite directions. The brake calipers usually contain one piston, however in some cases they contain four.

All pistons employ some type of seal, usually made of rubber, to minimize fluid leakage. A rubber dust boot seals the outer end of the cylinder against dust and dirt. The boot fits around the outer end of the piston on disc brake calipers and around the brake actuating rod on the wheel cylinders.

The hydraulic system operates as follows: When at rest, the entire system, from the piston(s) in the master cylinder to those in the wheel cylinders or calipers, is full of brake fluid. Upon application of the brake pedal, fluid trapped in front of the master cylinder piston(s) is forced through the lines to the wheel cylinders and calipers. Here, it forces the pistons outward, in the case of drum brakes and inward toward the disc, in the case of disc brakes. The motion of the pistons is opposed by return springs mounted outside the cylinders in drum brakes and by internal springs or spring seals, in disc brakes.

Upon release of the brake pedal, a spring located inside the master cylinder immediately returns the master cylinder pistons to the normal position. The pistons contain check valves and the master cylinder has compensating ports drilled in it. These are uncovered as the pistons reach their normal position. The piston check valves allow fluid to flow toward the

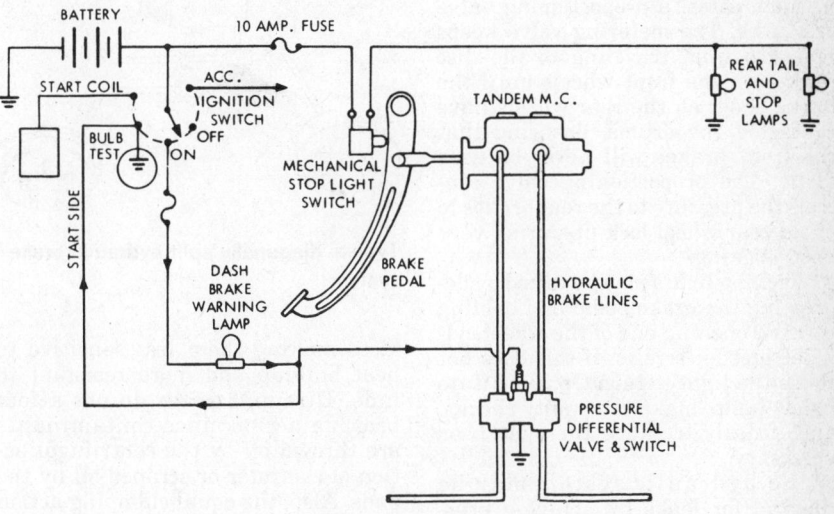

Typical dual brake system

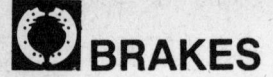
wheel cylinders or calipers as the pistons withdraw. Then, as the rubber boot/seal or return springs force the brake pads or shoes into the released position, the excess fluid returns to the reservoir through the compensating ports.

The dual master cylinder has two pistons, located one behind the other. The primary piston is actuated directly by mechanical linkage from the brake pedal. The secondary piston is actuated by fluid trapped between the two pistons. If a leak develops in front of the secondary piston, it moves forward until it bottoms against the front of the master cylinder. The fluid trapped between the pistons will operate one side of the split system. If the other side of the system develops a leak, the primary piston will move forward until direct contact with the secondary piston takes place and it will force the secondary piston to actuate the other side of the split system. In either case the brake pedal drops closer to the floor board and less braking power is available.

The brake system uses a switch to warn the driver when only half of the brake system is operational. This switch is usually located in a valve body which is mounted on the firewall or the frame below the master cylinder. A hydraulic piston receives pressure from both circuits, each circuit's pressure being applied to one end of the piston. When the pressures are in balance, the piston remains stationary. When one circuit has a leak, however, the greater pressure in that circuit during brake application will push the piston to one side, closing the switch and activating the brake warning light.

In disc brake systems, this valve body contains a metering valve and, in some cases, a proportioning valve or valves. The metering valve keeps pressure from traveling to the disc brakes on the front wheels until the brake shoes on the rear wheels have contacted the drums, ensuring that the front brakes will never be used alone. The proportioning valve controls the pressure to the rear brakes to avoid rear wheel lock-up during very hard braking.

Warning lights may be tested by depressing the brake pedal and holding it while opening one of the wheel cylinder bleeder screws. If this does not cause the light to turn On, substitute a new lamp, make continuity checks, and, finally, replace the switch as necessary.

The hydraulic system may be checked for leaks by applying pressure to the pedal gradually and steadily. If the pedal sinks very slowly to the floor, the system has a leak. This is not to be confused with a springy or spongy feel due to the compression of air within the lines. If the system leaks, there will be a gradual change in the position of the pedal when a constant pressure is applied.

Check for leaks along all lines and at wheel cylinders or calipers. If no external leaks are apparent, the problem is inside the master cylinder.

DISC BRAKES

Disc brake systems utilize a disc (rotor) with brake pads positioned on either side of it. Braking effect is achieved in a manner similar to the way you would squeeze a spinning phonograph record between your fingers. The disc (rotor) is a casting which may be equipped with cooling fins between the two braking surfaces. The fins (if equipped) enable air to circulate between the braking sur-

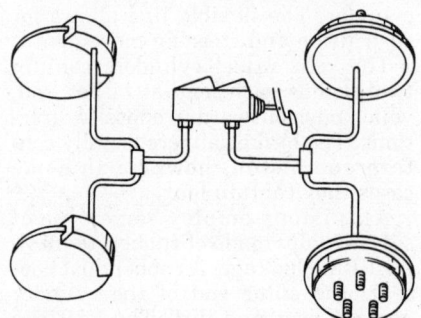

Typical front/rear split hydraulic brake system

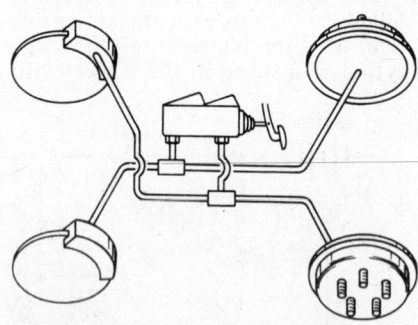

Typical diagonally split hydraulic brake system

faces making them less sensitive to heat buildup and more resistant to fade. Dirt and water do not affect braking action since contaminants are thrown off by the centrifugal action of the rotor or scraped off by the pads. Also, the equal clamping action of the two brake pads tends to ensure uniform, straightline stops. Disc brakes are inherently self-adjusting.

DRUM BRAKES (REAR)

Drum brakes employ two brake shoes mounted on a stationary backing plate. These shoes are positioned inside a circular drum which rotates with the wheel assembly. The shoes are held in place by springs, this allows them to slide toward the drums (when they are applied) while keeping the linings and drums in alignment. The shoes are actuated by a wheel cylinder which is mounted at the top of the backing plate. When the brakes are applied, hydraulic pressure forces the wheel cylinder's actuating links outward. Since these links bear directly against the top of the brake shoes, the tops of the shoes are then forced against the inner side of the drum. This action forces the bottoms of the two shoes to contact the brake drum by rotating the entire assembly slightly (known as servo action). When pressure within the wheel cylinder is relaxed, return springs pull the shoes back away from the drum.

Rear drum brakes are (in most cases) designed to self-adjust themselves during application. Motion causes both shoes to rotate very slightly with the drum, rocking an adjusting lever, thereby causing rotation of the adjusting screw or lever.

POWER BRAKE SYSTEM

Power brakes operate just as standard brake systems except in the actuation of the master cylinder pistons. A vacuum diaphragm is located on the front of the master cylinder and assists the driver in applying the brakes, reducing both the effort and travel he must put into moving the brake pedal.

The vacuum diaphragm housing is connected to the intake manifold by a vacuum hose. A check valve is placed at the point where the hose enters the diaphragm housing, so that during periods of low manifold vacuum brake assist vacuum will not be lost.

Depressing the brake pedal closes off the vacuum source and allows atmospheric pressure to enter on one side of the diaphragm. This causes the master cylinder pistons to move and apply the brakes. When the brake pedal is released, vacuum is applied to both sides of the diaphragm, and the return springs return the diaphragm and the master cylinder pistons to the released position. If the vacuum fails, the brake pedal rod will butt against the end of the master cylinder actuating rod and direct mechanical application will occur as the pedal is depressed.

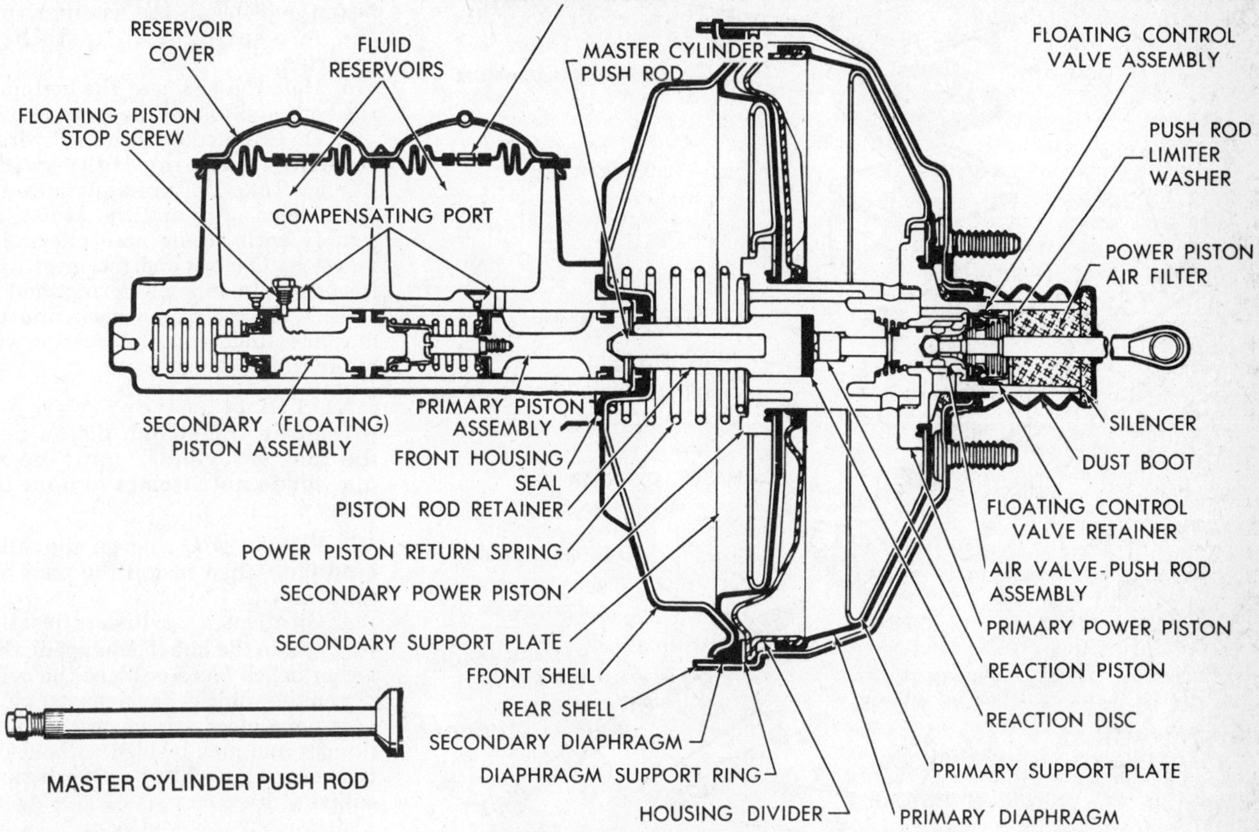

RESERVOIR DIAPHRAGM

RESERVOIR COVER

FLUID RESERVOIRS

MASTER CYLINDER PUSH ROD

FLOATING CONTROL VALVE ASSEMBLY

FLOATING PISTON STOP SCREW

PUSH ROD LIMITER WASHER

COMPENSATING PORT

POWER PISTON AIR FILTER

SECONDARY (FLOATING) PISTON ASSEMBLY

SILENCER

PRIMARY PISTON ASSEMBLY

DUST BOOT

FRONT HOUSING SEAL

FLOATING CONTROL VALVE RETAINER

PISTON ROD RETAINER

POWER PISTON RETURN SPRING

AIR VALVE-PUSH ROD ASSEMBLY

SECONDARY POWER PISTON

PRIMARY POWER PISTON

SECONDARY SUPPORT PLATE

REACTION PISTON

FRONT SHELL

REACTION DISC

REAR SHELL

SECONDARY DIAPHRAGM

PRIMARY SUPPORT PLATE

DIAPHRAGM SUPPORT RING

PRIMARY DIAPHRAGM

HOUSING DIVIDER

MASTER CYLINDER PUSH ROD

Typical dual master cylinder

HYDRAULIC CYLINDERS AND VALVES

Master Cylinders

— **CAUTION** —

The master cylinder unit is a highly calibrated unit specifically designed for the vehicle it is on. Although cylinders may look alike there are many differences in calibration. If replacement is necessary, make sure the replacement unit is the one specified for the vehicle.

NOTE: Some 1981 and later GM vehicles are equipped with "Quick Take-Up" master cylinders which provide a large volume of fluid to the brakes at low pressure when the brake pedal is initially applied. This large volume of fluid is needed because self retracting piston seals are used on the caliper pistons. The piston seals pull the pistons into the calipers after the brakes are released, thereby preventing the brake pads from causing a drag on the rotors.

The "Quick Take-Up" master cylinder has a hydraulically operated brake warning light switch incorporated in the master cylinder body. The piston is accessible by removing the large plug at the front of the master cylinder body. Only remove the plug when overhauling the cylinder, as brake fluid will escape.

Overhaul procedures on these master cylinders are basically the same as those on conventional master cylinders.

SERVICING MASTER CYLINDERS

NOTE: Plastic reservoirs need to be removed only for the following reasons: Reservoir is damaged or the rubber grommet(s) between the reservoir and bore is leaking. Removal of stop pin from Chrysler style plastic reservoir master cylinder to allow removal of pistons. Pin is located underneath front reservoir nipple. Service "Quick Take-up" valve on GM quick take-up master cylinders. The reservoir should be removed by first clamping the cylinder flange in a vice. Next remove the reservoir for the Chrysler style. Grasp the reservoir base on one end and pull away from the body. GM reservoirs must be removed by prying between the reservoir and casting with a pry bar. Grommets can be

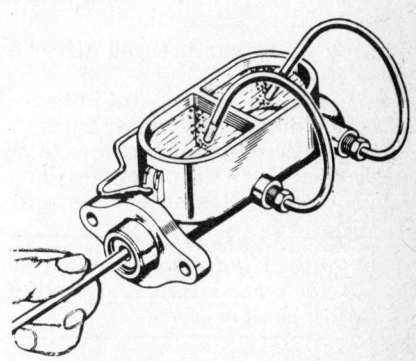

Pre-bleeding master cylinder

reused if they are in good condition. Whether or not the reservoir is removed, it and the cover or caps should be thoroughly cleaned.

1. Remove the cylinder from the vehicle and drain the brake fluid.

2. Mount the cylinder in a vise so that the outlets are up and remove the rubber boot seal from the hub.

3. Remove the stop pin or screw from the bottom of the front reservoir, if present.

4. Remove the snapring from the front of the bore and the primary piston assembly.

5. Remove the secondary piston as-

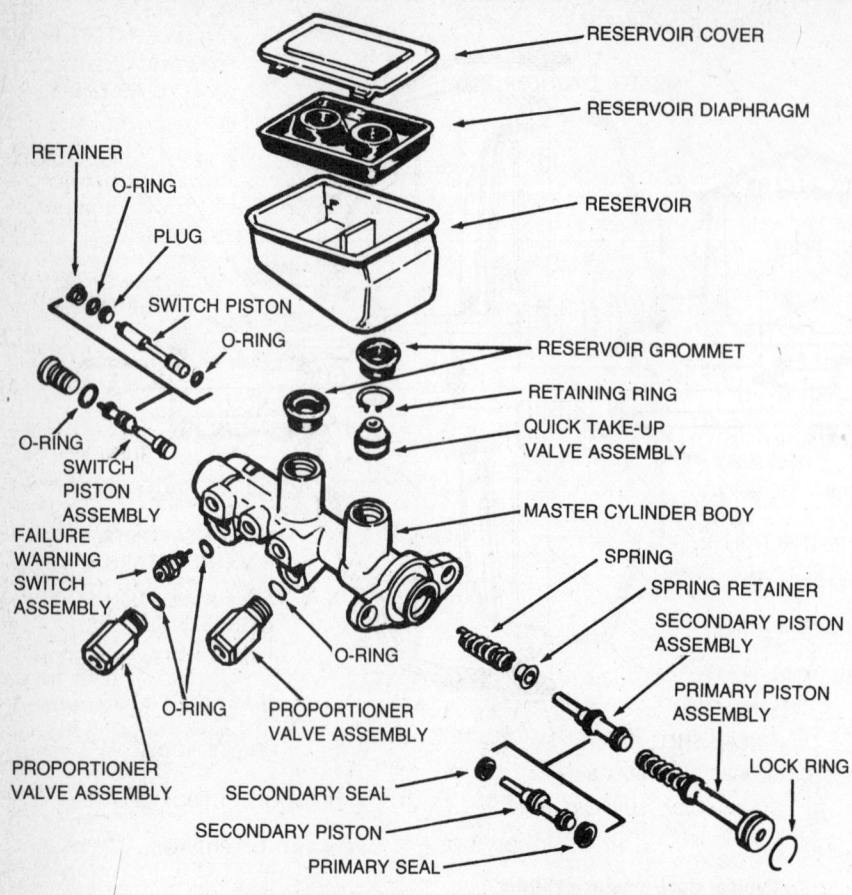

RESERVOIR COVER

RESERVOIR DIAPHRAGM

RESERVOIR

RETAINER
O-RING
PLUG
SWITCH PISTON
O-RING

RESERVOIR GROMMET
RETAINING RING
QUICK TAKE-UP
VALVE ASSEMBLY

O-RING
SWITCH
PISTON
ASSEMBLY
FAILURE
WARNING
SWITCH
ASSEMBLY

MASTER CYLINDER BODY

SPRING
SPRING RETAINER
SECONDARY PISTON
ASSEMBLY
PRIMARY PISTON
ASSEMBLY

LOCK RING

O-RING
PROPORTIONER
VALVE ASSEMBLY

O-RING

PROPORTIONER
VALVE ASSEMBLY
SECONDARY SEAL
SECONDARY PISTON
PRIMARY SEAL

GM "Quick Take Up" master cylinder

sembly using compressed air or a piece of wire.

6. Clean the metal parts in brake fluid and discard the rubber parts.

7. Inspect the bore for damage or wear, then check the pistons for damage and proper clearance in the bore.

——— CAUTION ———

Aluminum cylinder bores cannot be honed. The cylinder must be replaced if the bore is pitted or scored.

8. If the bore is only slightly scored or pitted it may be honed. (See CAUTION). Always use hones that are in good condition and completely clean the cylinder with brake fluid when the honing is completed. If any sign of wear or corrosion is apparent on "Quick Take-Up" master cylinder bores, the master cylinder must be replaced; it cannot be honed. If any evidence of contamination exists in the master cylinder the entire hydraulic system should be flushed and refilled with clean brake fluid. Blow out the passages with compressed air.

NOTE: Most rebuilding kits provide a primary and secondary piston assembly. If the kit you are using only provides seals, see Steps 9-13.

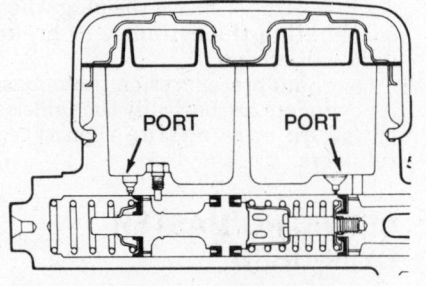

PORT   PORT

Feed and return ports

9. Install new secondary seals in the two grooves in the flat end of the front piston. The lips of the seals will be facing away from each other.

10. Install a new primary seal and the seal protector on opposite end of the front piston with the lips of the seal facing outward.

11. Coat the seals with brake fluid. Install the spring on the front piston with the spring retainer in the primary seal.

12. Insert the piston assembly, spring end first, into the bore and use a wooden rod to seat it.

13. Coat the rear piston seals with brake fluid and install them into the piston grooves with the lips facing the spring end.

14. Assemble the spring onto the piston and install the assembly into the bore spring first. Install the snapring.

15. Hold the piston at the bottom of the bore and install the stop screw.

16. On GM models with the hydraulic brake warning light switch ("Quick Take-Up" units), remove the Allen head plug and the switch assembly with needle nose pliers. Remove the O-rings and retainers from the piston. Install new O-rings and retainers, fit the piston back into the master cylinder after lubricating with brake fluid.

NOTE: If any corrosion is present in the switch piston bore the master cylinder must be replaced: do not attempt to hone the bore.

17. Fit a new O-ring on the Allen head plug, then install the plug and tighten.

18. On all master cylinders, install a new seal in the hub (if equipped), then either bench bleed or bleed the cylinder on the vehicle. Some master cylinders have bleed screws on the outlet flanges and may be bled without disturbing the wheel cylinders or calipers.

MASTER CYLINDER PUSH ROD ADJUSTMENT

Models Equipped with Adjustable Push Rod

After assembly of the master cylinder to the power section, the piston cup in the hydraulic cylinder should just clear the compensating port hole when the brake pedal is full released. If the push rod is too long, it will hold the piston over the port. A push rod that is too short, will give too much loose travel (excessive pedal play). Apply the brakes and release the pedal all the way observing the brake fluid flow back into the master cylinder. A full flow indicates the piston is coming back far enough to release the fluid. A slow return of the fluid indicates the piston is not coming back far enough to clear the ports. The push rod adjustment is too tight and should be shortened.

Disc Brake Calipers

NOTE: Caliper disc brakes can be divided into three types: the four-piston, fixed-caliper type; the single-piston, floating-caliper type and the single-piston sliding-caliper type. Refer to the Brake Specifications Chart for applications.

In the four piston type (two in each side of the caliper), the braking effect is achieved by hydraulically pushing both shoes against the disc sides.

With the single piston floating-caliper type the inboard shoe is pushed hydraulically into contact with the disc, while the reaction force thus generated is used to pull the outboard shoe into frictional contact (made possible by letting the caliper move slightly along the axle centerline).

In the sliding caliper (single piston) type, the caliper assembly slides along the machined surfaces of the anchor plate. A steel key located between the machined surfaces of the caliper and the machines surfaces of the anchor plate is held in place with either a retaining screw or two cotter pins. The caliper is held in place against the anchor plate with one or two support springs.

SERVICING THE CALIPER ASSEMBLY

NOTE: The following is a general caliper service procedure. Before proceeding, check under the individual disc brake section for your vehicle (Delco Moraine, Bendix, etc.) for any special servicing procedures.

1. Raise and support the front of the vehicle on jackstands, then remove the front wheels.

2. Working on one side at a time only, disconnect the hydraulic inlet line from the caliper and plug the end. Remove the caliper mounting bolts or pins and the shims (if used), then slide the caliper off the disc.

3. Remove the disc pads from the caliper or mounting adapter. If the old ones are to be reused, make them so that they can be reinstalled in their original positions.

4. Open the caliper bleed screw and drain the fluid. Clean the outside of the caliper and mount it in a vise with padded jaws.

——— CAUTION ———

When cleaning any brake components, use only brake fluid or denatured (Isopropyl) alcohol. Never use a mineral-based solvent, such as gasoline or paint thinner, since it will swell and quickly deteriorate the rubber parts.

5. Remove the bridge bolts (fixed type), separate the caliper halves and remove the two O-ring seals from the transfer holes.

6. Pry the lip on (each) piston dust boot from its groove, then remove the piston assemblies and spring(s) from the bore(s). If necessary, air pressure

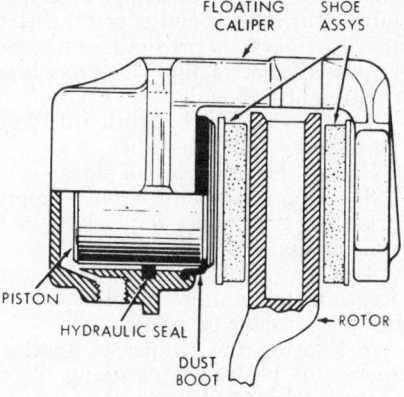

Floating caliper disc brake

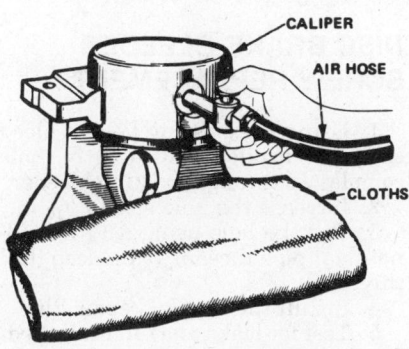

Removing piston pneumatically

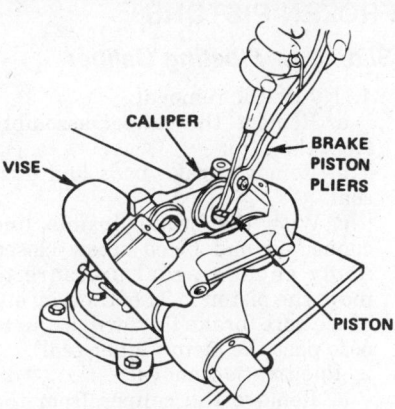

Removing pistons

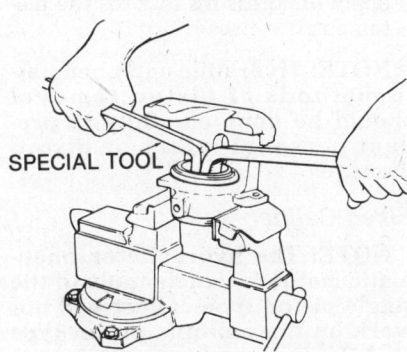

Removing hollow end piston

may be used to force the pistons(s) out of the bore(s), using care to prevent the piston from popping out of control.

7. Remove the boot(s) and seal(s) from the piston(s), then clean the piston(s) in brake fluid. Blow out the caliper passages with an air hose.

8. Inspect the cylinder bore(s) for scoring, pitting or corrosion. Corrosion is a pitted or rough condition not to be confused with staining. Light rough spots may be removed by rotating crocus cloth, using finger pressure, in the bores. DO NOT polish with an in and out motion or use any other abrasive.

9. If the piston(s) are pitted, scored or worn, they must be replaced. A corroded or deeply scored caliper should also be replaced.

10. Check the clearance of the piston(s) in the bores using a feeler gauge. Clearance should be 0.002–0.006 in. If there is excessive clearance the caliper must be replaced.

11. Replace all rubber parts and lubricate with brake fluid. Install the seals (or square cut rings) and boots in the grooves in each piston. The seal should be installed in the groove closest to the closed end of the piston with the seal lips facing the closed end. The lip on the boot should be facing the seal.

12. Lubricate the piston and bore with brake fluid. Position the piston return spring (if equipped), large coil first, in the piston bore.

13. Install the piston in the bore, taking great care to avoid damaging the seal lip as it passes the edge of the cylinder bore.

14. Compress the lip on the dust boot into the groove in the caliper. Be sure the boot is full seated in the groove, as poor sealing will allow contaminants to ruin the bore.

15. On fixed calipers: Position the O-rings in the cavities around the caliper transfer holes, and fit the caliper halves together. Install the bridge bolts (lubricated with brake fluid) and be sure to torque to specification.

16. Install the disc pads in the caliper or adapter and remount the caliper on the hub (see "Disc Pad Replacement"). Connect the brake line to the caliper and bleed the brakes (see "Brake Bleeding"). Replace the wheels. Recheck the brake fluid level, check the brake pedal travel and road test the vehicle.

OVERHAUL TIPS

Field reports indicate that two factors determine whether to replace or rebuild calipers: Can the piston or pistons be removed? Will the bleed screw break off when removal is attempted?

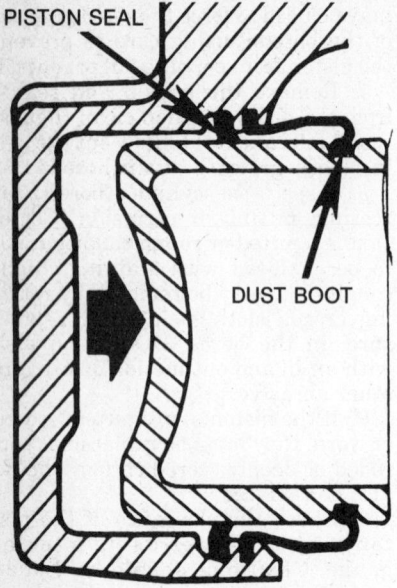

Brake applied

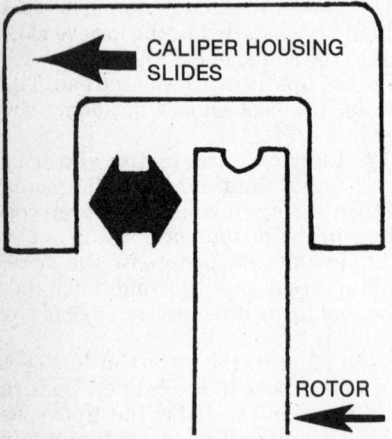

Floating (or sliding) caliper type

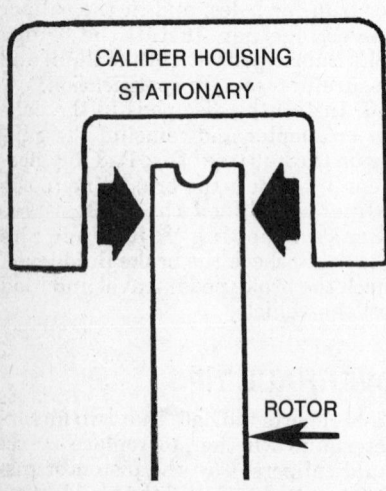

Fixed caliper type

(Rebuilders will not accept a caliper with a broken bleed screw.) Since there is no way to predict how a bleed screw will react, follow this procedure to attempt removal.

1. Insert a drill shank into the bleed screw hole (snug fit).
2. Tap the screw on all sides.
3. Using a six-point wrench, apply pressure gently while working the drill up and down slightly.
4. If the drill starts to bind, the screw is beginning to collapse and cannot be removed intact.
5. Heating the caliper is another successful, but time consuming, bleed screw removal technique. Remove the caliper from the vehicle. Heat the caliper. Shrink the bleed screw by applying dry ice and attempt to remove.

DISC BRAKE BLEEDER SCREW REPLACEMENT

1. Using the existing hole in bleed screw for a pilot, drill a $\frac{1}{4}$ in. hole completely through existing bleeder.
2. Increase the hole size to $\frac{7}{16}$ in.
3. Tap the hole using a $\frac{1}{4}$ in. (18-national pipe thread) $\frac{1}{2}$ in. deep (full thread).
4. Install the bleeder repair kit.
5. Test for leaks and full brake pedal pressure.

FROZEN PISTONS

Sliding or Floating Caliper

1. Hydraulic removal:
   a. Remove the caliper assembly from the rotor.
   b. Remove brake pads and dust seal.
   c. With the brake flexible line connected and bleed screw closed apply enough pedal pressure to move the piston most of the way out of the bore (brake fluid will begin to ooze past the piston inner seal).
2. Pneumatic removal:
   a. Remove the caliper from the vehicle.
   b. With the bleed screw closed, apply air pressure to force the piston out.

NOTE: Hydraulic and pneumatic methods of piston removal should be done carefully to prevent personal injury or piston damage.

Fixed Caliper

NOTE: The hydraulic or pneumatic methods which apply to the single piston type caliper will not work on the multiple piston type brake caliper.

1. Remove the caliper from the ve-

hicle with the two halves separated.
2. Mount in a vise and use a piston puller (many types available) to remove the pistons.

CALIPER CLEANING

NOTE: Castings may be cleaned with any type cleaning fluid after all the rubber seals have been removed.

It is important that all traces of cleaning fluid be completely removed from the caliper casting. Rubber components are compatible with alcohol and/or brake fluid. Use a lint free wiping cloth to clean the caliper and parts. Black stains on the pistons or walls, caused by the seals, will not do harm; however, extreme cleanliness is essential. Blow out the passages with compressed air. A fine grade of crocus cloth may be used to correct minor imperfections in the cylinder bore. Slide crocus cloth with finger pressure in a circular rather than a lengthwise motion. DO NOT use any form of abrasive on a plated piston. Discard a piston which is pitted or has signs of plating wear.

REBUILDING CALIPERS

NOTE: If a fine stone honing of a caliper bore is necessary it should be done with skill and caution. Some vehicles can develop 800 psi hydraulic pressure on severe application so the honing must never exceed 0.003 in. Also the dust seal groove must be free of rust or nicks so that a perfect mating surface is possible on the piston and casting.

Installing Stroking Type Seals and Boots

Stretch the boot and seal over the piston and seat them. The seal lip on the Bendix and Delco styles, faces toward hydraulic pressure; boot lips face toward the brake shoe. Locate the return spring (if used) in the cylinder and carefully start the piston into the cylinder to avoid nicking the seal. Alignment tools are available for inserting the lip cup seals. Fully depress the piston into the bore in order to fasten the boot lip to the caliper housing. On the Delco types, use a wooden drift or a special seating tool to seat the boot ring in the caliper counterbore. It must be flush or below the caliper machined surface.

Installing Fixed Position (Rectangular Ring) Seals and Boots

Insert a rectangular ring seal into

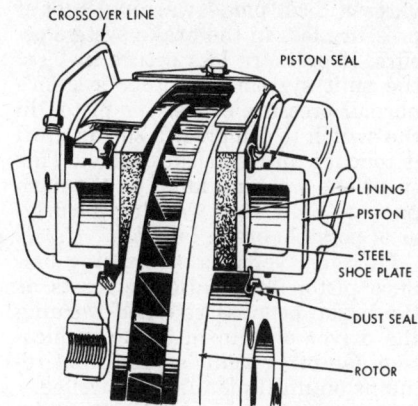

Fixed caliper disc brake

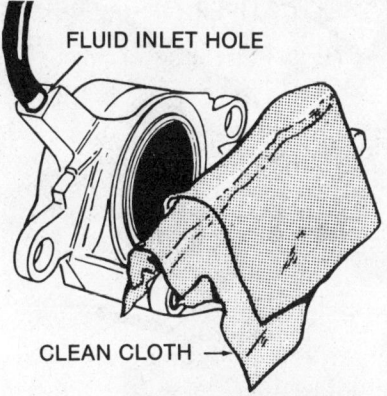

Removing piston hydraulically

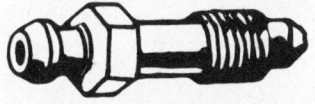

Bleed screw

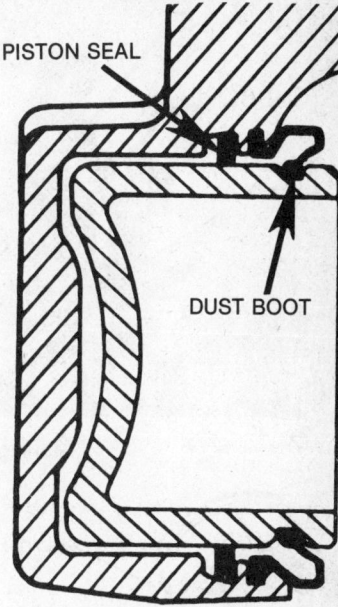

Brake released

Replacing disc brake bleeder screw

Installing Fixed-Caliper Bridge Bolts

If the caliper contains internal fluid crossover passages, be sure to install new O-ring seals at the joints. Mate the caliper halves and install high tensile strength bridge bolts. Never replace the bridge bolts with ordinary standard hardware bolts.

Wheel Cylinders

Wheel cylinders contain a pair of opposed pistons fitted with rubber cups, compression spring and sometimes expander washers to keep the cups tight against the pistons.

SERVICING

1. Raise and support the vehicle on jackstands. Remove the wheel and drum assemblies from the side to be serviced.
2. Remove the brake shoes, then clean the backing plate and the wheel cylinder. Rebuilding can be done on the vehicle, depending on the design of the brake backing plate. If the backing plate is recessed to the point that it is impossible to get a hone into the cylinder, the cylinder has to be removed.
3. To remove the cylinder; disconnect the brake line from the rear of the cylinder, remove the mounting bolts or retainers and the cylinders.

NOTE: On some models, the wheel cylinder is contained by a retaining ring. In order to remove

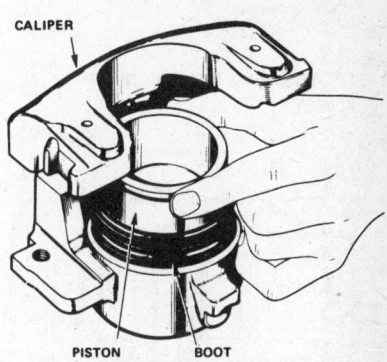

Installing piston

the rear wheel cylinders, remove the wheel cylinder retainer. Insert two pin punches or equivalent tools into the access slots and bend both tabs at the same time thereby releasing the cylinder. Use a new retainer when reinstalling the wheel cylinder. The new retainer can be driven on using a $1\frac{1}{8}$ in. socket with an extension bar.

4. Remove the rubber boots (dust covers) from the ends of the cylinder. Remove the pistons, the piston cups (expanders, if equipped) and the spring from the inside of the cylinder. Remove the bleeder screw and make sure it is not clogged.
5. Discard all of the parts that the rebuilding kit will replace.
6. Examine the inside of the cylinder. If it is severely rusted, pitted or scratched install a new or rebuilt cylinder.
7. If the condition of the cylinder indicates that it can be rebuilt, hone the bore. Light honing will provide a

bore and at any location, push the ring into the seal groove. From this area, with a finger, gently work around the bore until the ring is seated in this channel. Be sure the ring does not twist or roll in the groove. When the boot lip is retained inside the cylinder bore, insert the boot in the same manner. Then work the inside of the boot over the pressure end of the piston, stretching the boot with a small plastic tool and pressing the piston through the seal, straight in, until it bottoms. The inside of the boot should slide on the piston and come to rest in the boot groove. If the boot lip is retained outside of the cylinder bore, first stretch boot over the piston and seat it in its groove, then press the piston through the seal. Fully depress the piston to 50–100 lbs. in order to fasten the boot lip in place. On the Delco-Moraine types, use a wooden drift or a special seating tool to seat the metal boot ring in the caliper counterbore below the face of the caliper.

SPECIAL HONE — CALIPER
PISTON BORE

Honing cylinder bore

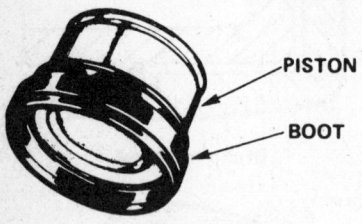

PISTON
BOOT

Assembling boot on piston

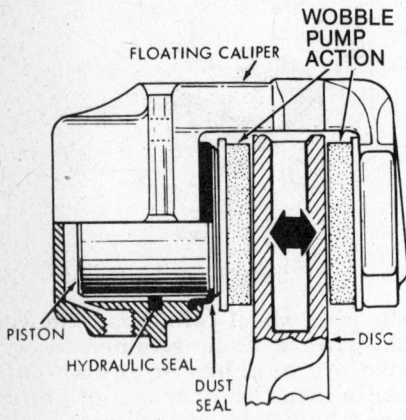

WOBBLE PUMP ACTION
FLOATING CALIPER
PISTON
HYDRAULIC SEAL
DUST SEAL
DISC

Wobble pump action

new surface on the inside of the cylinder which promotes better cup sealing.

8. Wash out the cylinder with brake fluid after honing. Reassemble the cylinder using the new parts provided in the kit. When assembling the cylinder dip all parts in brake fluid.

9. Install the cylinder on the vehicle. Reinstall the brakes, drum/wheel and bleed the brake system.

Hydraulic Control Valves

PRESSURE DIFFERENTIAL VALVE

The pressure differential valve acti-

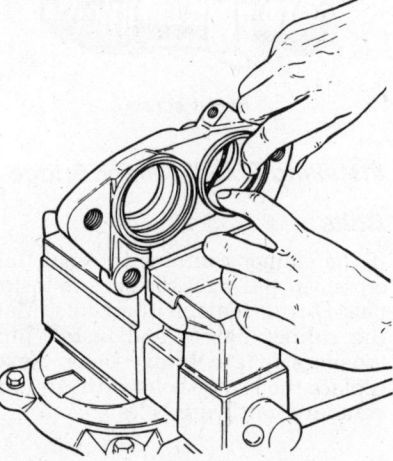

Installing fixed position rectangular ring seal (seal lip toward pressure side)

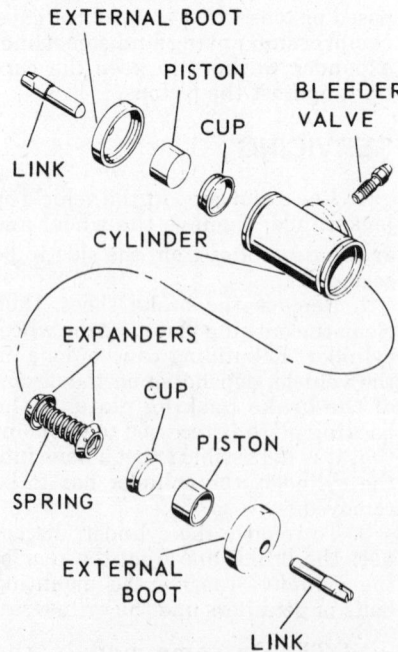

EXTERNAL BOOT
PISTON
CUP
BLEEDER VALVE
LINK
CYLINDER
EXPANDERS
CUP
PISTON
SPRING
EXTERNAL BOOT
LINK

Wheel cylinder components

vates a dash panel warning light if pressure loss in the brake system occurs. If pressure loss occurs in $\frac{1}{2}$ of the split system the other system's normal pressure causes the piston in the switch to compress a spring until it touches an electrical contact. This turns the warning lamp on the dash panel to light, thus warning the driver of possible brake failure.

On some vehicles, the spring balance piston automatically resets as the brake pedal is released warning the driver only upon brake application. On other vehicles, the light remains on until manually cancelled.

Valves may be located separately, as part of a combination valve, or incorporated into the master cylinder.

Resetting Valves

On some vehicles, the valve piston(s) remain off center after failure until necessary repairs are made. The valve will automatically reset itself (after repairs) when pressure is equal on both sides of the system.

If the light does not go out, bleed the brake system that is opposite the failed system. If front brakes failed, bleed the rear brakes, this should force the light control piston toward center.

If this fails, remove the terminal switch. If brake fluid is present in the electrical area, the seals are gone, replace the complete valve assembly.

METERING VALVE

The metering valve's function is to improve braking balance between the front disc and rear drum brakes, especially during light brake application.

The metering valve prevents the application of the front disc brakes until the rear brakes overcome the return spring pressure. Thus, when the front disc pads contact the rotor, the rear shoes will contact the brake drum at the same time.

Inspect the metering valve each time the brakes are serviced. A slight amount of moisture inside the boot does not indicate a defective valve, however, fluid leakage indicates a damaged or worn valve. If fluid leakage is present the valve must be replaced.

The metering valve can be checked very simply. With the vehicle stopped, gently apply the brakes. At about an inch of travel a very small change in pedal effort (like a small bump) will be felt if the valve is operating properly. Metering valves are not serviceable and must be replaced (if defective).

PROPORTIONING VALVE

The proportioning (pressure control) valve is used, on some vehicles, to reduce the hydraulic pressure to the rear wheels to prevent skidding during heavy brake application and to provide better brake balance. It is usually mounted in line to the rear wheels.

When the brakes are serviced the valve should be inspected for leakage. Premature rear brake application during lighting braking can mean a bad proportioning valve. Repair is by replacement of the valve. Make sure the valve port marked "R" is connected toward the rear wheels.

On GM "Quick Take-Up" master cylinders, the proportioning valve(s) is (are) screwed into the master cylinder. Since these vehicles have a diagonally split brake system, two valves are required. One rear brake line screws into each valve. The early type valves (GM front wheel drive) were steel and silver colored, an occasional "clunking" noise was encountered on some early models, but does not affect brake efficiency. Replacement valves are now made of aluminum. Never mix an aluminum valve with a steel valve, always use two aluminum valves.

COMBINATION VALVE

The combination valve may perform two or three functions. They are: metering, proportioning and brake failure warning.

Variations of the two-way combination valve are: proportioning and brake failure warning or metering and brake failure warning.

A three-way combination valve directs the brake fluid to the appropriate wheel, performs necessary valving and contains a brake failure warning.

The combination valve is usually mounted under the hood close to the master cylinder, where the brake lines can easily be connected and routed to the front or rear wheels.

The combination valve is non-serviceable and must be replaced if malfunctioning.

Brake Bleeding

The hydraulic brake system must be free of air to operate properly. Air can enter the system when hydraulic parts are disconnected for servicing or replacement, or when the fluid level in the master cylinder reservoirs is very low. Air in the system will give the brake pedal a spongy feeling upon application.

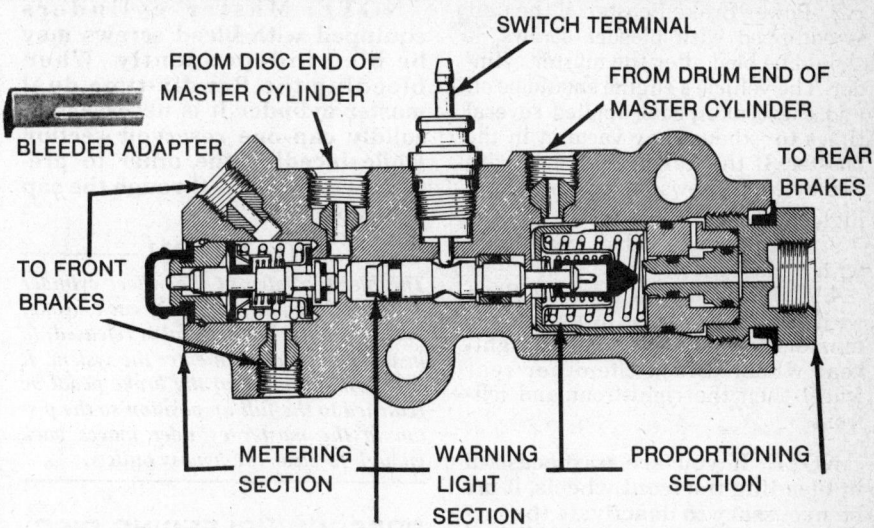

Push valve in when pressure bleeding-not necessary when using pedal bleed method

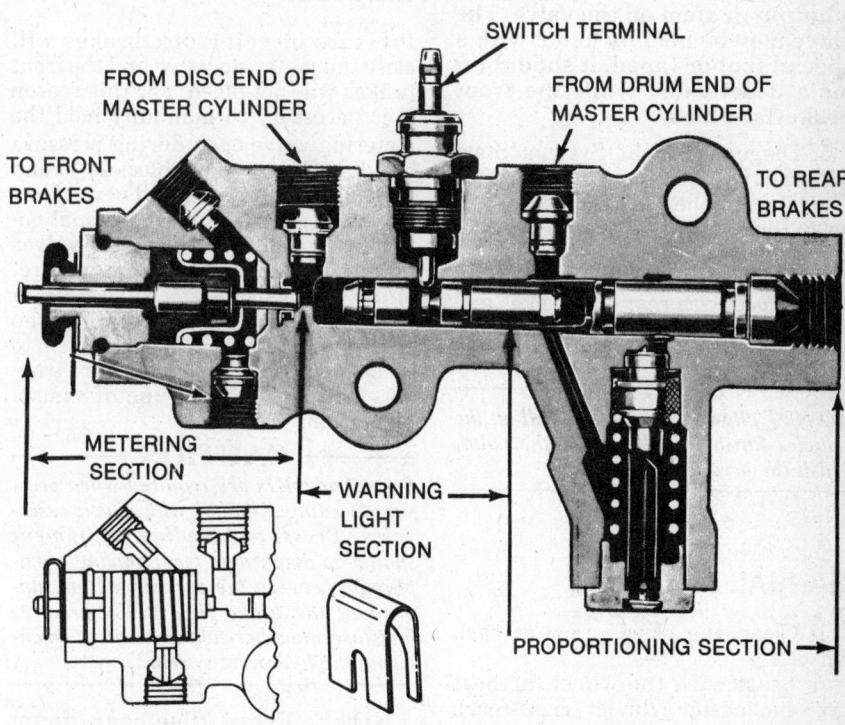

Hold valve out .060 in pressure bleed only-not necessary when using pedal bleed method

The quickest and easiest of the two ways for system bleeding is the pressure method but special equipment is needed to externally pressurize the hydraulic system. The other, more commonly used method of brake bleeding is done manually.

BLEEDING SEQUENCE

Bleeding may be required at only one or two wheels or at the master cylinder, depending upon what point the system was opened to air. If after bleeding the cylinder caliper that was rebuilt or replaced and the pedal still has a spongy feeling upon application, it will be necessary to bleed the entire system. Bleed the system in the following order:

1. Master cylinder: If the cylinder is not equipped with bleeder screws, open the brake line(s) to the wheels slightly while pressure is applied to the brake pedal. Be sure to tighten the line before the brake pedal is released. The procedure for bench bleeding the master cylinder is in the following section.

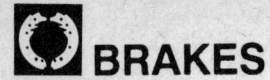

2. Power Brake Booster: If the unit is equipped with bleeder screws, it should be bled after the master cylinder. The vehicle's engine should be off and the brake pedal applied several times to exhaust any vacuum in the booster. If the unit is equipped with two bleeder screws, always bleed the higher one first.

3. Combination Valve: If equipped with a bleeder screw.

4. Front/Back Split Systems: Start with the wheel farthest away from the master cylinder, usually the right-rear wheel. Bleed the other rear wheel then the right-front and left-front.

NOTE: If you are unsuccessful in bleeding the front wheels, it me be necessary to deactivate the metering valve. This is accomplished by either pushing in, or pulling out a button or stem on the valve. The valve may be held by hand, with a special tool or taped, it should remain deactivated while the front brakes are bled.

5. Diagonally Split System: Start with the right-rear then the left-front. The left-rear then the right-front (refer to the following "GM Quick Take-Up Master Cylinder" section).

6. Rear Disc Brakes: If the vehicle is equipped with rear disc brakes and the calipers have two bleeder screws, bleed the inner first then the outer.

—————— CAUTION ——————
DO NOT allow brake fluid to spill on the vehicles finish, it will remove the paint. Flush the area with water.

MANUAL BLEEDING

1. Clean the bleed screw at each wheel.

2. Start with the wheel farthest from the master cylinder (right-rear).

3. Attach a small rubber hose to the bleed screw and place the end in a clear container of brake fluid.

4. Fill the master cylinder with brake fluid. (Check often during bleeding). Have an assistant slowly pump up the brake pedal and hold pressure.

5. Open the bleed screw about one-quarter turn, press the brake pedal to the floor, close the bleed screw and slowly release the pedal. Continue until no more air bubbles are forced from the cylinder on application of the brake pedal.

6. Repeat procedure on remaining wheel cylinders and calipers, still working from cylinder/caliper farthest from the master cylinder.

NOTE: Master cylinders equipped with bleed screws may be bled independently. When bleeding the Bendix-type dual master cylinder it is necessary to solidly cap one reservoir section while bleeding the other to prevent pressure loss through the cap vent hole.

—————— CAUTION ——————
The bleeder valve at the wheel cylinder must be closed at the end of each stroke, and before the brake pedal is released, to insure that no air can enter the system. It is also important that the brake pedal be returned to the full up position so the piston in the master cylinder moves back enough to clear the bypass outlets.

PRESSURE BLEEDING DISC BRAKES.

Pressure bleeding disc brakes will close the metering valve and the front brakes will not bleed. For this reason it is necessary to manually hold the metering valve open during pressure bleeding. Never use a block or clamp to hold the valve open and never force the valve stem beyond its normal position. Two different types of valves are used. The most common type requires the valve stem to be held in while bleeding the brakes, while the second type requires the valve stem to be held out (0.060 in. minimum travel). Determine the type of visual inspection.

—————— CAUTION ——————
Special adapters are required when pressure bleeding cylinders with plastic reservoirs. Pressure bleeding equipment should be diaphragm type; placing a diaphragm between the pressurized air supply and the brake fluid. This prevents moisture and other contaminants from entering the hydraulic system.

NOTE: Front disc/rear drum equipped vehicles use a metering valve which closes off pressure to the front brakes under certain conditions. These systems contain manual release actuators, which must be engaged to pressure bleed the front brakes.

1. Connect the tank hydraulic hose and adapter to the master cylinder.

2. Close the hydraulic valve on the bleeder equipment.

3. Apply air pressure to the bleeder equipment following the equipment manufacturer's recommendations for correct air pressure.

4. Open the valve to bleed air out of the pressure hose to the master cylinder. Never bleed this system using the secondary piston stopscrew on the bottom of many master cylinders.

5. Open the hydraulic valve and bleed each wheel cylinder or caliper. Bleed the rear brake system first when bleeding both front and rear systems.

FLUSHING HYDRAULIC BRAKE SYSTEMS

Hydraulic brake systems must be totally flushed if the fluid becomes contaminated with water, dirt or other corrosive chemicals. To flush, simply bleed the entire system until all of the fluid has been replaced with the correct type of new fluid.

BENCH BLEEDING MASTER CYLINDER

Bench bleeding the master cylinder before installing it on the vehicle reduces the possibility of getting air into the lines.

1. Connect two short pieces of brake line to the outlet fittings, bend them until the free end is below the fluid level in the master cylinder reservoirs.

2. Fill the reservoirs with fresh brake fluid. Pump the piston until no more air bubbles appear in the reservoir(s).

3. Disconnect the two short lines, refill the master cylinder and securely install the cylinder cap(s).

4. Install the master cylinder on the vehicle. Attach the lines but do not completely tighten them. Force any air that might have been trapped in the connection by slowly depressing the brake pedal. Tighten the lines before releasing the brake pedal.

GM QUICK TAKE-UP SYSTEM BLEEDING

Bleed the master cylinder as follows: disconnect the left-front brake line from the master cylinder. Fill the cylinder with fluid until it flows from the opened port. Connect the line and tighten the fitting. Apply the brake pedal slowly one time and keep it applied. Loosen the same brake line fitting to allow any air to escape. Retighten the fitting and release the brake pedal slowly. Wait 15 seconds and repeat the procedure until all of the air is expelled. Bleed the right-front connection in the same manner. Bleed the cylinders and calipers after you are sure all the air is out of the master cylinder.

———— CAUTION ————
Rapid pumping will move the secondary piston down the bore and make it difficult to bleed the system. Always apply slow pedal pressure.

Power Brakes

Vacuum Operated Booster

Power brakes operate just as standard brake systems except in the actuation of the master cylinder pistons. A vacuum diaphragm is located on the front of the master cylinder and assist the drive in applying the brakes, reducing both the effort and travel he must put into moving the brake pedal.

The vacuum diaphragm housing is connected to the intake manifold by a vacuum hose. A check valve is placed at the point where the hose enters the diaphragm housing, so that during periods of low manifold vacuum brake assist vacuum will not be lost.

Depressing the brake pedal closes off the vacuum source and allows atmospheric pressure to enter on one side of the diaphragm. This causes the master cylinder pistons to move and apply the brakes. When the brake pedal is released, vacuum is applied to both sides of the diaphragm, the return springs return the diaphragm and master cylinder pistons to the released position. If the vacuum fails, the brake pedal rod will butt against the end of the master cylinder actuating rod and direct mechanical application will occur as the pedal is depressed.

The hydraulic and mechanical problems that apply to conventional brake systems also apply to power brakes should be checked if the tests and chart below do not reveal the problem. Tests for a system vacuum leak as described below:

1. Operate the engine at idle with the transmission in Neutral without touching the brake pedal for at least one minute.
2. Turn Off the engine and wait one minute.
3. Test for the presence of assist vacuum by depressing the brake pedal and releasing it several times. Light application will produce less and less pedal travel, if vacuum was present. If there is no vacuum, air is leaking into the system somewhere.
4. Test the system operation as follows:
   a. Pump the brake pedal (with engine off) until the supply vacuum is entirely gone.
   b. Put light, steady pressure on the pedal. Start the engine and operate it at idle with the transmission in Neutral.

c. If the system is operating, the brake pedal should fall toward the floor when constant pressure is maintained on the pedal.

NOTE: Power brake systems may be tested for hydraulic leaks just as ordinary systems are tested, except that the engine should be idling with the transmission in Neutral throughout the test.

POWER BRAKE BOOSTER TROUBLESHOOTING

NOTE: The following items are in addition to those listed in the "General Troubleshooting" section. Check those items first.

Hard Pedal

1. Faulty vacuum check valve.
2. Vacuum hose kinked, collapsed, plugged leaky or improperly connected.
3. Internal leak in unit.
4. Damaged vacuum cylinder.
5. Damaged valve plunger.
6. Broken or faulty springs.
7. Broken plunger stem.

Grabbing Brakes

1. Damaged vacuum cylinder.
2. Faulty vacuum check valve.
3. Vacuum hose leaky or improperly connected.
4. Broken plunger stem.

Pedal Goes to Floor

Generally, when this problem occurs, it is not caused by the power brake booster. In rare cases, a broken plunger stem may be at fault.

Overhaul

Most power brake boosters are serviced by replacement only. In many cases, repair parts are not available. A good many special tools are required for rebuilding these units. For these reasons, it would be most practical to replace a failed booster with a new or remanufactured unit.

Hydro-Boost, Hydro-Boost II

Hydro-Boost differs from conventional power brake systems, in that it operates from the power steering pump fluid pressure rather than intake manifold vacuum.

The Hydro-Boost unit contains a spool valve with an open center which controls the strength of pump pressure when braking occurs. A lever assembly controls the valve's position. A boost piston provides the force nec-

essary to operate the conventional master cylinder on the front of the booster.

A reserve of at least two assisted brake applications is supplied by an accumulator which is spring loaded on earlier and pneumatic on later models. The accumulator is an integral part of the Hydro-Boost II unit. The brakes can be applied manually if the reserve system is depleted.

All system checks, tests and troubleshooting procedure are the same for the two systems.

HYDRO-BOOST SYSTEM CHECKS

1. A defective Hydro-Boost cannot cause any of the following conditions: Noisy brakes, fading pedal or pulling brakes. If any of these occur, check elsewhere in the brake system.
2. Check the fluid level in the master cylinder. It should be within $\frac{1}{4}$ in. of the top; if not, add only DOT-3 or DOT-4 brake fluid until the correct level is reached.
3. Check the fluid level in the power steering pump. The engines should be at normal running temperature and stopped. The level should register on the pump dipstick. Add power steering fluid to bring the reservoir level up to the correct level. Low fluid level will result in both poor steering and stopping ability.

———— CAUTION ————
The brake hydraulic system uses brake fluid only, while the power steering and Hydro-Boost systems use power steering fluid only. Don't mix the two.

4. Check the power steering pump belt tension and inspect all of the power steering/Hydro-Boost hoses for kinks or leaks.
5. Check and adjust the engine idle speed, as necessary.
6. Check the power steering pump fluid for bubbles. If air bubbles are present in the fluid, bleed the system. Fill the power steering pump reservoir to specifications with the engine at normal operating temperature. With the engine running, rotate the steering wheel through its normal travel 3–4 times, without holding the wheel against the stops. Check the fluid level again.
7. If the problem still exists, go on to the Hydro-Boost test sections and troubleshooting chart.

HYDRO-BOOST TESTS

Functional Test

1. Check the brake system for

leaks or low fluid level. Correct as necessary.

2. Place the transmission in Neutral and stop the engine. Apply the brakes 4–5 times to empty the accumulator.

3. Keep the pedal depressed with moderate (25–40 lbs.) pressure and start the engine.

4. The brake pedal should fall slightly and then push back up against your foot. If no movement is felt, the Hydro-Boost system is not working.

Accumulator Leak Test

1. Run the engine at normal idle. Turn the steering wheel against one of the stops; hold it there for no longer than 5 seconds. Center the steering wheel and stop the engine.

2. Keep applying the brakes until a "hard" pedal is obtained. There should be a minimum of 2 power (1 on Hydro-Boost II) assisted brake applications when pedal pressure of 20–25 lbs. is applied.

3. Start the engine and allow it to idle. Rotate the steering wheel against the stop. Listen for a light "hissing" sound; this is the accumulator being charged. Center the steering wheel and stop the engine.

4. Wait one hour and apply the brakes without starting the engine. As in Step 2, there should be at least 2 (1 on Hydro-Boost II) stops with power assist. If not, the accumulator is defective and must be replaced.

Hydro-Boost System Bleeding

NOTE: The system should be bled whenever the booster is removed and installed.

1. Fill the power steering pump until the fluid level is at the base of the pump reservoir neck. Disconnect the battery lead from the distributor.

NOTE: On diesel engines remove the electrical lead to the fuel solenoid terminal on the injection pump before cranking the engine.

2. Raise the front of the vehicle, turn the wheels all the way to the left and crank the engine for a few seconds.

3. Check the steering pump fluid level. If necessary, add fluid to the "Add" mark on the dipstick.

4. Lower the vehicle, connect the battery lead and start the engine. Check the fluid level and add fluid to the "Add" mark if necessary. With the engine running, turn the wheels from side-to-side to bleed air from the system. Make sure that the fluid level stays above the internal pump casting.

5. The Hydro-Boost system should now be fully bled. If the fluid is foaming after bleeding, stop the engine, let the system set for one hour, then repeat the 2nd part of Step 4.

6. The preceding procedures should be effective in removing excess air from the system, however, sometimes air may still remain trapped. When this happens the booster may make a "gulping" noise when the brake is applied. Lightly pumping the brake pedal with the engine running should cause this noise to disappear. After the noise stops, check the pump fluid level and add as necessary.

HYDRO-BOOST TROUBLESHOOTING

High Pedal and Steering Effort (Idle)

1. Loose/broken power steering pump belt
2. Low power steering fluid level
3. Leaking hoses or fittings
4. Low idle speed
5. Hose restriction
6. Defective power steering pump

High Pedal Effort (Idle)

1. Binding pedal/linkage
2. Fluid contamination
3. Defective Hydro-Boost unit

Poor Pedal Return

1. Binding pedal linkage
2. Restricted booster return line
3. Internal return system restriction

Pedal Chatter/Pulsation

1. Power steering/pump drivebelt slipping
2. Low power steering fluid level
3. Defective power steering pump
4. Defective Hydro-Boost unit

Brakes Oversensitive

1. Binding pedal/linkage
2. Defective Hydro-Boost unit

Noise

1. Low power steering fluid level
2. Air in the power steering fluid
3. Loose power steering pump drivebelt
4. Hose restrictions

OVERHAUL

Ford Motor Company services the Hydro-Boost unit with a replacement new or rebuilt unit only. No provisions are made for overhaul of the unit. GM Hydro-Boost units may be overhauled by qualified mechanics.

— CAUTION —

DO NOT attempt to interchange the parts between the Hydro-Boost units of different makes of vehicles, because of pressure differentials and differences of the tolerances of the internal parts. Pressure could exceed the normal accumulator release pressure of 1400 psi and injury or damage could result.

Disc Brake Rotors

RUNOUT

Manufacturers differ widely on permissible runout but too much can sometimes be felt as a pulsation at the brake pedal. A wobble pump effect is created when a rotor is not perfectly smooth and the pad hits the high spots forcing fluid back into the master cylinder. This alternating pressure causes a pulsating feeling which can be felt at the pedal when the brakes are applied.

To check the actual runout of the rotor, perform the following procedures:

1. Tighten the wheel spindle nut to a snug bearing adjustment, end-play removed.

2. Fasten a dial indicator on the suspension at a convenient place so that the indicator stylus contacts the rotor face approximately one inch from its outer edge.

3. Set the dial at zero. Check the total indicator reading while turning the rotor one full revolution. If the rotor is warped beyond the runout specification, it is likely that it can be successfully remachined.

"Lateral Runout": A wobbly movement of the rotor from side-to-side as it rotates. Excessive lateral runout causes the rotor faces to knock back the disc pads and can result in chatter, excessive pedal travel, pumping or fighting pedal and vibration during the braking action.

"Parallelism" (lack of): Refers to the amount of variation in the thickness of the rotor. Excessive variation can cause pedal vibration or fight, front end vibrations and possible "grab" during the braking action; a condition comparable to an "out-of-round brake drum." Check parallelism with a micrometer. "Mike" the thickness at eight or more equally spaced points, equally distant from the outer edge of the rotor, preferably at mid-points of the braking surface. Parallelism then is the amount of variation between maximum and minimum measurements.

"Surface or Micro-inch finish, flatness, smoothness": Different from parallelism, these terms refer to the

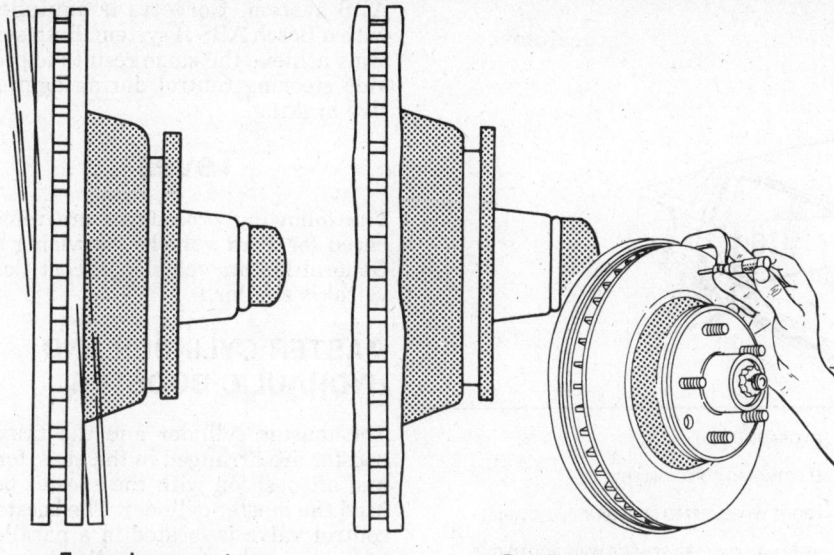

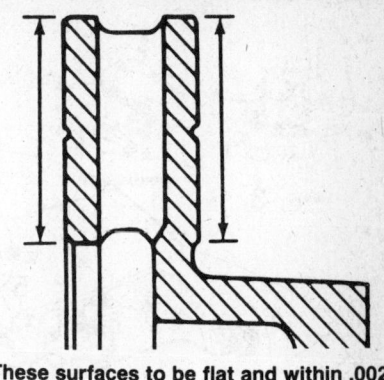

These surfaces to be flat and within .002 in.

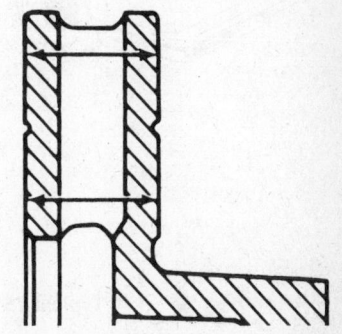

Taper variation not to exceed .003 in.

Excessive runout

Parallelism

degree of perfection of the flat surface on each side of the rotor; that is, the minute hills, valleys and swirls inherent in machining the surface. In a visual inspection, the remachined surface should have a find ground polish with, at most, only a faint trace of nondirectional swirls.

SERVICING THE DISC ROTOR

Disc Replacement

1. Raise and support the vehicle on jackstands, then remove the wheel/tire assembly.

2. Remove the caliper. Secure the caliper out of the way suspended by wire, DO NOT allow the caliper to hang by the brake hose.

3. Remove the wheel bearing nut from the spindle and the outer wheel bearing from the hub.

4. Remove the hub and disc assembly from the spindle.

5. To install, reverse the removal procedures.

NOTE: The disc is removable from the hub on the Eldorado, Toronado and Corvette (rear only). To separate the rear disc and hub on a Corvette the three hub-to-disc attaching rivets must be drilled out. This can be done with the hub and rotor mounted on the vehicle. It is not necessary to install new rivets when the disc is installed.

Anti-Lock Braking System (ABS)

OPERATION

The Anti-Lock Braking System (ABS)

is essentially a brake system enhancement. The purpose of ABS is to increase the driver's control over a vehicle during braking, especially steering control. When a vehicle equipped with a conventional brake system must brake suddenly, one or more wheels may lock up offering little or no steering control to avoid hazards. The ABS is designed to prevent braked wheels from locking. The advantages of the system are considerable. For instance, during a high-speed stop while entering a curve, the ABS is designed to allow the driver to steer through the curve while decelerating. Additionally, the ABS is designed to enhance the braking action of each front wheel independently and the two rear wheels independent of the front wheels. This allows controlled braking even if one or more

wheels encounters a slippery surface. In this situation, the ABS will automatically sense the initial loss of adhesion in any one wheel and reduce or prevent further hydraulic pressure on that wheel's brake caliper or if the rear wheels-both calipers until adhesion is regained.

COMPONENTS

The ABS is essentially the familiar split circuit hydraulic four wheel disc

Ideal rotor surface condition

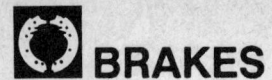

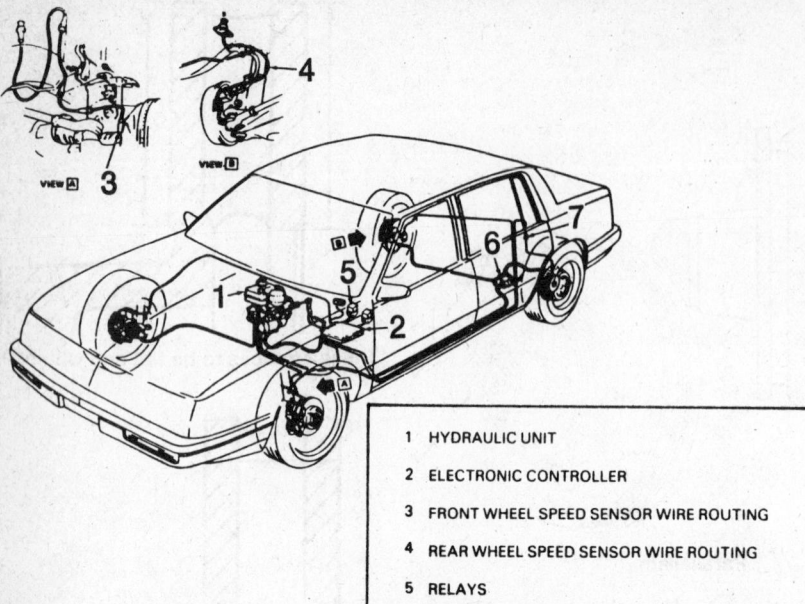

1	HYDRAULIC UNIT
2	ELECTRONIC CONTROLLER
3	FRONT WHEEL SPEED SENSOR WIRE ROUTING
4	REAR WHEEL SPEED SENSOR WIRE ROUTING
5	RELAYS
6	PROPORTIONER VALVE
7	SENSOR CONNECTION TO HARNESS (TRUNK AREA)

Typical TEVES ABS system.

BRAKING SYSTEM

HYDRAULIC BRAKING SYSTEM	3	HYDRAULIC MODULATOR
ABS ELECTRO—HYDRAULICS	3a	SOLENOID VALVE
ABS ELECTRONICS	3b	HYDRAULIC ACCUMULATOR
SENSORS WHEEL, LAT ACCEL.	3c	RETURN PUMP
1 WHEEL-SPEED SENSORS	4	BRAKE MASTER CYLINDER
2 WHEEL BRAKE CYLINDERS	5	ELECTRONIC CONTROLLER
	6	ABS WARNING LIGHT

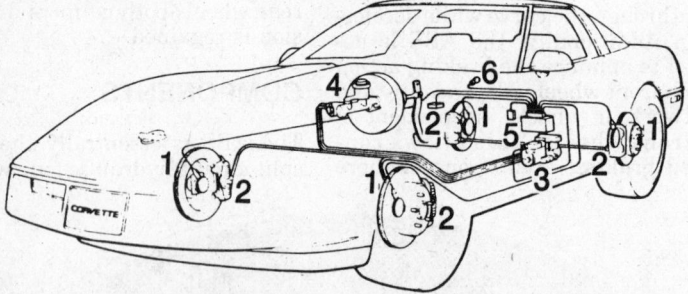

Typical BOSCH II ABS system.

brake system to which a sophisticated electronic and mechanical override system has been carefully mated. Wheel speed sensors, an electronic control unit and a hydraulic unit that incorporates solenoid operated brake line valves are the major components of the system. The sensors monitor the rotation speed of the wheels and provide data about wheel acceleration and deceleration over very small intervals of time. The signals from the sensors are transmitted to the control unit. The control unit monitors the signals and compares them to a contained program. If one of the sensors suddenly shows a deceleration rate that exceeds the threshold values of the programmed system (indicating that a wheel is about to lock and skid) the computer activates the hydraulic control unit to maintain the optimum brake pressure in that wheel or both rear wheels to prevent lock-up. If, for any reason, the ABS should malfunction the brakes will operate as a normal system without ABS and a warning light will turn On indicating service is required.

SYSTEMS

Ford and General Motors, with the exception of the Corvette use the Teves ABS system. Corvette is equipped with a Bosch ABS II system. Both systems achieve the same results to provide steering control during aggressive braking.

Teves

The following procedures are referenced for Ford vehicles, servicing of General Motors vehicles (except Corvette) is similar.

MASTER CYLINDER AND HYDRAULIC BOOSTER

The master cylinder and the brake booster are arranged in the basic fore and aft position with the booster behind the master cylinder. The booster control valve is located in a parallel bore above the master cylinder centerline and is operated by a lever connecting the brake pedal pushrod.

ELECTRIC PUMP AND ACCUMULATOR

The electric pump is a high pressure pump design that runs at frequent intervals for a short period to charge the hydraulic accumulator that supplies the brake system. The accumulator is a gas filled pressure chamber that is part of the pump and motor assembly. The electric motor, pump and accumulator assembly is shock mounted to the master cylinder booster assembly.

VALVE BODY ASSEMBLY

The valve body assembly incorporates three pairs of solenoid valves. One pair for each front wheel and a third pair for both the rear wheels combined. These solenoid valves are inlet-outlet valves with inlet valve normally open and the outlet valve normally closed. The valve body itself is bolted to the inboard side of the master cylinder-boost assembly.

FLUID LEVEL WARNING SWITCHES

These two integral fluid level switches are incorporated in the brake fluid reservoir cap assembly with two electrical connectors, one for each end of the cap, for wire harness connections.

WHEEL SENSORS

The sensors are four variable reluctance electronic sensor assemblies, each with a 104 tooth ring in the antilock system. The sensors are connect-

ed to the electronic controller through a wiring harness. The front sensors are bolted to brackets which are bolted to the front spindles. The front toothed sensor rings are pressed onto the inside of the front rotors. The rear sensors are bolted to brackets, that in turn are bolted to the rear disc brake axle adapters. The toothed rear sensor rings are pressed to the axle shafts, inboard of the axle shaft flange.

ELECTRONIC CONTROLLER

The controller is a self contained non-serviceable unit, which consists of two microprocessors and the necessary circuitry for their operation. The function of the controller is to monitor the system operations during normal driving and during anti-lock braking. Any malfunction of the anti-lock brake system will cause the controller to shut off and bypass the anti-lock brake system. When the anti-lock brake system is bypassed, the normal power assisted braking will still remain.

CHECK ANTI-LOCK BRAKE WARNING LIGHT

The four wheel anti-lock system is self-monitoring. when the ignition switch is in the run position, the electronic controller will perform a preliminary self check on the anti-lock electrical system, this is indicated by a 3–4 second energizing of the amber. Check Anti-Lock Brakes lamp in the over head console. this light will turn Off after the 3–4 second interval, unless there is a malfunction in the anti-lock brake system. If there is a malfunction the check anti-lock brake and/or the brake lamp will stay lit and the diagnostic tests will then pin point the exact component needing service.

BLEEDING THE BRAKE SYSTEM

The front brakes can be bled in the conventional manner, with or without the accumulator being changed. When bleeding the rear brakes the accumulator must be fully charged or the system has to be pressure bled as previously outlined in this section.

Bleeding The Brake System With a Charged Accumulator

NOTE: Be careful when opening the rear caliper bleeder screws, due to the high pressure in the system from a fully charged accumulator at the bleeder screws.

1. With the accumulator fully charged, have someone hold the brake pedal in the applied position, place the ignition switch in the Run position and open the rear brake caliper bleed screws for 10 seconds at a time.
2. Repeat this procedure until the air is cleared from the brake fluid and close the brake caliper bleed screws.
3. Do this to all of the brake calipers and after the bleed screws are closed. Pump the brake pedal a couple of times to complete the bleeding procedure.
4. Adjust the brake fluid level in the reservoir to the maximum level with a fully charged accumulator.

MASTER CYLINDER

Removal & Installation

NOTE: The hydraulic pressure must be discharged from the brake system before removing the master cylinder. To discharge the system, turn the ignition key to the Off position and pump the brake pedal at least 20 times until an increase in pedal force is clearly felt.

1. Disconnect the negative battery cable and the electrical connectors from the master cylinder reservoir cap, main valve, solenoid valve body, pressure warning switch, the hydraulic pump motor and ground connector from the master cylinder.
2. Disconnect the brake lines from the solenoid valve body and plug the line openings in the valve body to prevent fluid loss.

NOTE: DO NOT allow the brake fluid to leak or spill onto any of the electrical connectors.

3. From inside the vehicle, disconnect the hydraulic booster pushrod from the brake pedal in the following order: Disconnect the stop light switch wires at the connector on the brake pedal. Remove the hairpin clip from the stop light switch on the brake pedal and move the switch off of the pedal pin far enough for the switch outer hole to clear the pin. Using a twisting motion, remove the switch, but be careful not to damage the switch during it's removal. Remove the four retaining nuts at the dash panel and from inside the engine compartment. Remove the booster from the dash panel.
4. To install, reverse the removal procedures, then bleed the brake system as previously outlined in this section.

HYDRAULIC ACCUMULATOR

Removal & Installation

1. Discharge the pressure in the brake system and disconnect the electrical connection from the hydraulic pump motor.
2. Using an 8mm hex wrench or equivalent, unscrew the accumulator; be sure that no dirt falls into the open port.
3. Using the same hex wrench or equivalent, remove the accumulator adapter block bolt and the block, if necessary.
4. To install, reverse the removal procedures and be sure to observe the following:
   a. Install new O-rings on the accumulator and adapter block.
   b. Torque the adapter block bolt to 25–34 ft. lbs. and the accumulator to 30–34 ft. lbs.
   c. After installation, place the ignition switch in the On position and check that the check Anti-Lock Brake light turns Off after a maximum of one minute.
   d. Top off the master cylinder reservoir to the Max. mark with brake fluid.

HYDRAULIC PUMP MOTOR

Removal & Installation

1. Discharge the pressure in the brake system and disconnect the negative battery cable.
2. Disconnect the electrical connections from the hydraulic pump motor and pressure warning switch.
3. Remove the accumulator as previously outlined.
4. Remove the suction line between the reservoir and the pump at the reservoir by twisting the hose and pulling on it lightly.

NOTE: To prevent fluid loss, a large vacuum nipple can be slipped over the reservoir opening as the hose is removed.

5. Remove the retaining bolt on the pump high pressure line to the hydraulic booster housing at the housing.

NOTE: Make sure to save the two O-rings that are on both sides of the retaining bolt.

6. Remove the Allen head bolt that holds the pump and motor assembly to the extension housing, located directly under the accumulator. Save the thick spacer between the extension housing and the shock mount.
7. Slide the pump assembly inboard to remove the assembly from

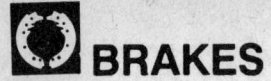

the retainer pin located on the inboard side of the extension housing.

8. To install, reverse the removal procedures. Bleed the brake system and check that the Check Anti-Lock Brake light turns Off after a maximum of one minute.

ELECTRONIC CONTROLLER

Removal & Installation

1. Disconnect the negative battery cable. Disconnect the 35 pin connector from the electronic controller, which is located in the trunk of the vehicle in front of the forward trim panel.
2. Remove the three retaining screws holding the electronic controller to the seat back brace and remove the controller.
3. To install, reverse the removal procedures.

PRESSURE SWITCH

Removal & Installation

1. Discharge the pressure from the brake system and disconnect the negative battery cable.
2. Disconnect the solenoid valve body seven pin connector and remove the pressure switch using special tool No. T85P-20215-B or equivalent, a ½–⅜ inch adapter and a ⅜ inch drive ratchet.
3. To install, reverse the removal procedures and replace the O-ring on the switch. Torque the switch to 15–25 ft. lbs.

FRONT WHEEL SENSOR

Removal & Installation

1. Disconnect the sensor electrical connector for the right or left-front sensor from inside the engine compartment.
2. Raise and support vehicle safely and disengage the wire grommet at the right or left-hand shock tower.
3. Pull the sensor cable connector through the hole and be careful not to damage the connector.
4. Remove the sensor wire from the bracket on the shock tower and the side rail.
5. Loosen the set screw holding the sensor to the sensor bracket post. Remove the sensor through the hole in the disc brake splash shield.
6. Remove the sensor bracket or the sensor bracket post, if it has been damaged by removing the caliper, hub and rotor assembly (as previously outlines). Remove the two brake

splash shield attaching bolts which hold the sensor bracket.

7. Install the sensor bracket with the sensor bracket post, if it has been removed. Torque the post retaining bolt to 40–60 inch lbs. and the splash shield attaching bolts to 10–15 ft. lbs. Install the hub and rotor assembly as previously outlined.

NOTE: If the sensor is going to be reused, the pole face must be cleaned of all dirt and grease, the pole face has to be scraped with a dull knife or equivalent, so that the sensor slides freely on the post. Also glue a new front spacer paper on the pole face.

8. Install a new or old sensor through the hole in the brake shield onto the sensor bracket post. Make sure the paper spacer does not come off during installation.
9. Push the sensor toward the toothed sensor ring until the new paper sensor contacts the ring. Hold the sensor against the sensor ring and torque the set screw to 21–26 inch lbs.
10. Insert the cable into the bracket on the shock strut, rail bracket and then through the inner fender apron to engine compartment and the seat the grommet.
11. Lower the vehicle and reconnect the sensor electrical connection.
12. Check the function of the sensor by road testing the vehicle and making sure the Check Anti-Lock Brake light turns Off.

REAR WHEEL SENSOR

Removal & Installation

1. Disconnect the wheel sensor electrical connector located behind the forward luggage compartment trim panel in the trunk.
2. Lift the carpet in the trunk and push the sensor wire grommet through the hole in the floor of the trunk.
3. Raise and support the vehicle safely and remove the appropriate wheel.
4. Remove the wheel sensor wiring from the axle shaft housing. The wiring harness has three different types of retainers, they are as follows:
   a. One inboard retainer clip is located on the top of the differential housing. Just bend the clip far enough to remove the wiring harness.
   b. The second retainer clip is a C-clip and is located in the center of

the axle shaft housing. Pull rearward on the clip to disengage the clip from the axle housing.
   c. The third clip is at the connection between the rear housing wheel brake tube and flexible hose. Remove the hold down bolt and open the clip to remove the wiring harness.

NOTE: Be careful not to bend the C-clip open beyond the amount needed to remove the clip from the axle housing, because it could break.

5. Remove the rear wheel caliper and rotor assemblies as previously outlined in this section.
6. Remove the wheel sensor retaining bolt, slip the grommet out of the rear brake splash shield and pull the sensor wire outward through the hole.
7. If the sensor bracket is damaged, remove the bracket attaching screws and replace the bracket.
8. Install the sensor bracket, if it was removed and torque the screws to 11–15 ft. lbs.
9. If the sensor is going to be used again, it must be cleaned just as the front wheel sensors were cleaned.
10. Insert the sensor into the large hole in the sensor bracket and install the retaining bolt into the sensor bracket and torque the retaining bolt to 40–60 inch lbs.
11. Push the sensor toward the toothed ring until the new paper sensor touches the sensor ring, hold the sensor against the toothed ring and torque the set screw to 21–26 inch lbs.
12. Install the caliper and rotor assemblies as previously outlined.
13. Push the wire and connector through the splash shield hole and engage the grommet into the shield eyelet. Install the sensor wire in the retainers along the axle housing.
14. Push the connector through the hole in the trunk and seat the grommet in the trunk floor pan.
15. Reconnect the cable electrical connector and install the carpet in the trunk. Check the function of the sensor by road testing the vehicle and checking to see if the Check Anti-Lock Brake light turns Off.

NOTE: If the toothed ring sensor is found to be malfunctioning on either the front wheel or the rear wheel, the rotor assembly has to be removed and the toothed ring sensor has to be pressed out and the new one pressed in.

DISC BRAKE APPLICATION CHART & SPECIFICATIONS

Manufacturer	Make/Model	Year	Text Ref- erence Type	Caliper Style	Manu- facturer	Anchor Bolt (ft lbs)	Bridge, Pin or Key Bolts (ft lbs.)	Wheel Lugs (ft lbs)	Minimum Thickness			Rotor Parallel Varia- tion	Max. Run- out
									Normal Standard	Machine To	Discard At		
American Motors	Eagle	'82–'87	1	Sliding	Bendix	100	30	75	.880	.815	.810	.0005	.003
	Concord, Spirit	'82–'83	1	Sliding	Bendix	85	30	75	.880	.815	.810	.0005	.003
	All	'80–'81	1	Sliding	Bendix	80	15	75	.880	.815	.810	.0005	.003
Chrysler Corporation Front Wheel Drive	Aries, Reliant, LeBaron, Dodge 400 ('83), 600 without H.D. brakes	'83–'87	12 or 14	Floating	ATE	70–100	18–22	80	.935	.912	.882	.0005	.004
	H.D. brakes	'83–'87	12 or 14	Floating	K/H	70–100	25–35	80	.935	.912	.882	.0005	.004
	E Class, New Yorker, Town & Country, Daytona, Laser	'83–'87	12 or 14	Floating	ATE or K/H	70–100	ATE: 18–22 K/H: 25–35	80	.935	.912	.882	.0005	.004
	Omni, Horizon, Charger, Turismo	'83–'87	11	Floating	K/H	70–100	25–40	80	.500	.461	.431	.0005	.004
	Aries, Reliant, LeBaron, Dodge 400	'81–'82	12	Floating	ATE	70–100	18–22	85	.935	.912	.882	.0005	.004
	Omni, Horizon	'80–'82	11	Floating	K/H	70–100	25–40	85	.500	.461	.431	.0005	.004
Rear Wheel Drive	Cordoba ('80–'83), Diplomat, Gran Fury, Mirada ('80–'83), New Yorker, Imperial ('80–'83)	'80–'87	6	Sliding	Chrysler	95–125	15–20	85	1.010	.955	.940	.0005	.004
Ford Motor Co. Front Wheel Drive	Escort, Lynx, LN7, EXP, Tempo, Topaz, Sable ('86–'87), Taurus ('86–'87)	'81–'87	10	Sliding	Ford	—	18–25	80–105	.945	.896	.882	.0005	.003
Rear Wheel Drive	Lincoln Continental, Mark VII Front	'82–'87	13	Sliding	Ford	—	40–60	80–105	1.030	—	.972	.0005	.003
	Rear	'82–'87	7	Sliding	K/H	85–115	15–20	80–105	.945	—	.895	.0004	.004
	Lincoln Town Car, Crown Victoria, Grand Marquis	'83–'87	13	Sliding	Ford	—	40–60	80–105	1.030	—	.972	.0005	.003
	Ford, Mercury, Lincoln	'80–'82	13	Sliding	Ford	—	40–60	80–105	1.030	—	.972	.0005	.003
	All models, except noted	'80–'87	13	Sliding	Ford	—	30–40	80–105	.870	—	.810	.0005	.003
	Granada, Monarch Versailles	'80	1	Sliding	K/H	105U–65L	12–16	80–105	.870	—	.810	.0005	.003
	Pinto, Bobcat	'80	1	Sliding	K/H	105U–65L	12–16	80–105	.870	—	.810	.0005	.003
	Rear Disc Brakes	'80–'82	7	Sliding	K/H	85–115	15–20	80–105	.945	—	.895	.0004	.003
General Motors Buick	Electra Limited, Park Ave (Front Wheel Drive)	'85–'87	2	Floating	Delco	—	35	70	1.043	.972	.957	.0005	.004
	Electra, Estate Wagon (Rear Wheel Drive)	'80–'84	2	Floating	Delco	—	35	80①	1.037	.980	.965	.0005	.004
	Riviera Front	'80–'87	2	Floating	Delco	—	35	100	1.037	.980	.965	.0005	.004
	Rear	'80–'87	9	Floating	Delco	35	30	100	—	.980	.965	.0005	.004
	Century w/H.D.	'83–'87	2	Floating	Delco	—	28	100	1.043	.972	.957	.0005	.004
	exc. H.D.	'82–'87	2	Floating	Delco	—	28	100	.885	.830	.815	.0005	.004
	Skyhawk w/vented disc	'82–'87	2	Floating	Delco	—	28	100	.885	.830	.815	.0005	.004
	w/solid disc	'82	2	Floating	Delco	—	28	100	—	.444	.429	.0005	.004
		'80	3	Floating	Delco	—	—	80	.885	.830	.815	.0005	.005
	Regal, LeSabre	'82–'87	2	Floating	Delco	—	35	70–80	1.043	.980	.965	.0005	.004
	Skylark	'80–'87	2	Floating	Delco	—	21–35	102	.885	.830	.815	.0005	.003
	Century, Regal, LeSabre	'80–'81	2	Floating	Delco	—	35	80	—	.960	.965	.0005	.004

BRAKES

DISC BRAKE APPLICATION CHART & SPECIFICATIONS

Manufacturer	Make/Model	Year	Text Reference Type	Caliper Style	Manu-facturer	Anchor Bolt (ft lbs)	Bridge, Pin or Key Bolts (ft lbs.)	Wheel Lugs (ft lbs)	Minimum Thickness Normal Standard	Machine To	Discard At	Rotor Parallel Varia-tion	Max. Run-out
Cadillac	Cimarron	'82–'87	19	Floating	Delco	—	28	100	.885	.830	.815	.0005	.004
	Fleetwood, DeVille (Front Wheel Drive)	'85–'87	2	Floating	Delco	—	35	70	1.043	.972	.957	.005	.004
	Fleetwood, DeVille, (Rear Wheel Drive) Front	'80–'84	2	Floating	Delco	—	30	100	1.037⑥	.980	.965	.0005	.004
	Rear	'80–'84	9	Floating	Delco	35	30	100	1.250	.910	.905	.0005	.003
	CC, Limousine	'80–'87	2	Floating	Delco	—	.30	100	1.250	1.230	1.215	.0005	.004
	Eldorado, Seville Front	'80–'87	2	Floating	Delco	—	28	100	1.035	.965	.957	.0005	.004
	Eldorado, Seville Rear	'80–'87	9	Floating	Delco	35	30	100	1.035	.965	.957	.0005	.004
Chevrolet	Full Size	'80–'87	2	Floating	Delco	—	35	80②	1.030	.980	.965	.0005	.004
	Malibu, Monte Carlo,	'80–'87	2	Floating	Delco	—	35	80③	1.030	.960	.965	.0005	.004
	Camaro Front	'80–'87	2	Floating	Delco	—	21–35	80	1.030	.980	.965	.0005	.004
	Rear	'80–'87	9	Floating	Delco	—	30–45	80	1.030	.980	.965	.0005	.004
	Corvette Front	'80–'82	4	Fixed	Delco	70	130	70④	1.285	1.230	1.215	.0005	.004
	Rear	'80–'82	4	Fixed	Delco	70	60	70④	1.285	1.230	1.215	.0005	.004
	Front	'84–'87	16	Floating	Girlock	70	24	100	.780	.739	.724	.0005	.006
	Rear	'84–'87	16	Floating	Girlock	44	24	100	.780	.739	.724	.0005	.006
	Celebrity, Cavalier	'82–'87	19	Floating	Delco	—	28	100	Vented .885 / Solid .490	.830 / .444	.815 / .429	.0005	.004
	Citation	'80–'87	19	Floating	Delco	—	28	102	.885	.830	.815	.0005	.003
	Nova	'85–'87	5	Floating	—	65	18	76	.531	.507	.472	.0005	.006
	Spectrum	'85–'87	17	Floating	—	40	36	65	.433	.393	.378	.0005	.006
	Sprint	'85–'87	18	Floating	—	—	26	50	.394	.330	.315	.0005	.003
	Monza	'80	3	Floating	Delco	—	—	80③	.885	.830	.815	.0005	.005
	Chevette	'80–'82	8	Floating	Delco	70	28	70	.440	.390	.374	.0005	.005
		'83–'87	19	Floating	Delco	—	21–25	70	—	.390	.374	.0005	.005
Oldsmobile	98 Regency, Brougham (Front Wheel Drive)	'85–'87	2	Floating	Delco	—	35	70	1.0443	.972	.957	.0005	.004
	Full Size	'80–'87	2	Floating	Delco	—	35	80①	1.040	.960	.965	.0005	.005
	Toronado Front	'80–'87	2	Floating	Delco	—	35	100	1.040	.980	.965	.0005	.004
	Rear	'80–'84	9	Floating	Delco	32	30	100	1.040	.980	.965	.0005	.004
	Cutlass Supreme, Cutlass	'80–'87	2	Floating	Delco	—	35	80	1.040	.980	.965	.0005	.005
	Ciera, Firenza	'82–'87	2	Floating	Delco	—	28	100	Vented 1.043 / Solid .490	.972 / .444	.957 / .429	.0005	.004
	Omega	'80–'84	2	Floating	Delco	—	28	103	.885	.830	.815	.0005	.003
	Starfire	'80	3	Floating	Delco	—	—	80	.880	.830	.815	.0005	.003
Pontiac	Full Size	'80–'87	2	Floating	Delco	—	35	80①	1.040	.980	.965	.0005	.004
	Grand Prix, Grand Am, LeMans	'80–'87	2	Floating	Delco	—	35	80	1.030	.980	.965	.0005	.004
	Firebird Front	'80–'87	2	Floating	Delco	—	21–35	80	1.030	.980	.965	.0005	.004
	Rear	'80–'87	9	Floating	Delco	—	30–45	80	1.030	.980	.965	.0005	.004
	6000, J2000	'82–'87	2	Floating	Delco	—	28	100	Vented 1.043 / Solid .490	.972 / .444	.957 / .429	.0005	.004

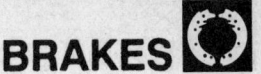
DISC BRAKE APPLICATION CHART & SPECIFICATIONS

Manufacturer	Make/Model	Year	Text Reference Type	Caliper Style	Manufacturer	Anchor Bolt (ft lbs)	Bridge, Pin or Key Bolts (ft lbs.)	Wheel Lugs (ft lbs)	Minimum Thickness			Rotor Parallel Variation	Max. Runout
									Normal Standard	Machine To	Discard At		
	Phoenix, Ventura	'80–'85	2	Floating	Delco	—	28	103	.885	.830	.815	.0005	.003
	Sunbird	'80	3	Floating	Delco	—	—	80	.885	.830	.815	.0005	.005
	T1000	'81–'82	8	Floating	Delco	70	28	70	.440	.390	.374	.0005	.005
		'83–'87	2	Floating	Delco	—	21–25	70	.440	.390	.374	.0005	.005
	Fiero Front	'84–'87	2	Floating	Delco	—	35	81	—	.444	.390	.0005	.004
	Rear	'84–'87	15	Floating	Delco	—	35	81	—	.444	.390	.0005	.004

① 100 with 1/2 in studs
② 100 on s/w
③ 90 with aluminum wheels
④ 80 with aluminum wheels

DISC BRAKE SERVICE

INSPECTION

Disc pads (lining and shoe assemblies) should be replaced in axle sets (both wheels) when the lining on any pad is worn to $\frac{1}{16}$ in. at any point. If the lining is allowed to wear past $\frac{1}{16}$ in. minimum thickness, severe damage to the disc may result. However, State Inspection specifications take precedence over these general recommendations. Note that disc pads in floating caliper type brakes may wear at an angle and measurement should be made at the narrow end of the taper. Tapered linings should be replaced if the taper exceeds $\frac{1}{8}$ in. from end-to-end (the difference between the thickest and thinnest points).

——— **CAUTION** ———

To prevent costly paint damage, remove some brake fluid (don't reuse) from the reservoir and install the reservoir cover before replacing the disc pads. When replacing the pads, the piston is depressed and fluid is forced back through the lines to squirt out of the fluid reservoir. When the caliper is unbolted from the hub, DO NOT let it dangle by the brake hose; it can be rested on a suspension member or wired onto the frame. All disc brake systems are self-adjusting and have no provision for manual adjustment.

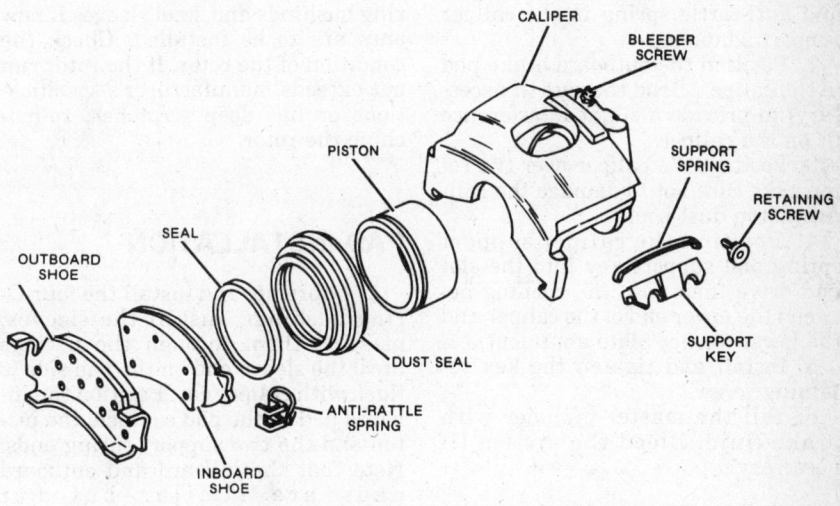

Type one Bendix floating caliper disc brakes (single Piston)

Type One: Kelsey-Hayes or Bendix Sliding Caliper Disc Brakes (Single Piston)

PAD REMOVAL

1. Remove half of the brake fluid from the master cylinder.
2. Remove the retaining screw holding the caliper support key.
3. Using a hammer and a drift punch, drive the caliper retaining key and support spring from the anchor plate.
4. Lift the caliper from of the rotor.

5. Support caliper so it doesn't hang by the brake hose.
6. Using a large C-clamp, push the piston back into its bore, being careful not to scratch the piston or bore and being careful not to cut or tear the dust boot.
7. Remove the inboard pad and anti-rattle spring from the caliper support adapter.
8. Remove the outboard pad from the caliper. Check the condition of the rotor. If the rotor run out exceeds the manufacturer's specifications or has deep scratches, have the rotor resurfaced.
9. Clean all sliding surfaces on the adapter and caliper.

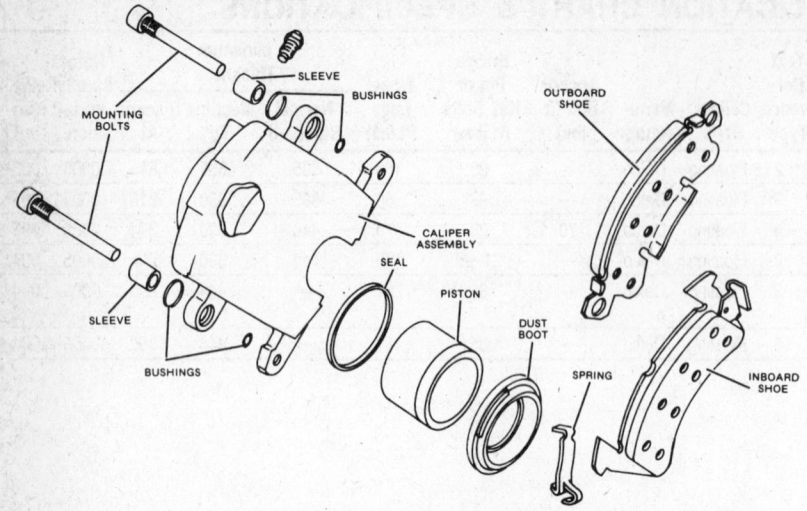

Type two Delco floating disc brakes (single piston)

PAD INSTALLATION

1. Position the inboard brake pad and anti-rattle spring in the caliper support adapter.
2. Position the outboard brake pad in the caliper. Bend the ears (if necessary) to provide a slight interference fit on the caliper.
3. Position the caliper over the rotor; take care not to damage the caliper piston dust boot.
4. Position the caliper support spring and support key into the slot and drive them into the opening between the lower end of the caliper and the lower anchor plate abutment.
5. Install and tighten the key retaining screw.
6. Fill the master cylinder with brake fluid. Bleed the system (if necessary).

Type Two: Delco Floating Caliper (Single Piston)

PAD REMOVAL

1. Remove $1/2$ of the brake fluid from the master cylinder.
2. Position a large C-clamp over the caliper with the screw end against the outboard brake pad. Tighten the clamp until the caliper is pushed out enough to bottom the piston.
3. Remove the C-clamp. Remove the two caliper guide pin mounts and lift the caliper from the rotor.
4. Support the caliper so there is no strain on the brake hose.
5. Press the outboard pad inward, then lift it from the caliper.

6. Press the inboard pad outward, then lift it from the caliper.
7. Remove and discard the four O-ring bushings and steel sleeves if new ones are to be installed. Check the condition of the rotor. If the rotor run out exceeds manufacturer's specifications or has deep scratches, re-machine the rotor.

PAD INSTALLATION

1. Lubricate and install the four O-ring bushings, install the sleeves, pressing them through the O-rings until the sleeve end on the pad side is flush with caliper ear. Position the inboard pad so the pad contacts the piston and the two support spring ends. Note that the inboard and outboard pads are similar but not interchangeable.
2. Press down on the ears at the top of the inboard pad until the pad lies flat and the spring ends are just inside the lower edge of the pad.
3. Position the outboard pad with the ears toward the positioning pin holes and the tab on the inner edge of the pad resting in the notch in the edge of the caliper. Bend the ears if necessary to provide slight interference fit on the caliper.
4. Press the outboard pad tightly into position and use a pair of pliers to clinch the ears of the outboard pad over the outboard caliper half.
5. Position the caliper over the rotor.
6. Install the caliper mounting bolts and tighten to specification.
7. Fill the master cylinder with brake fluid.

Type Three: Delco Floating Caliper (Sleeve Type)

PAD REMOVAL

1. Remove $1/2$ of the brake fluid from the master cylinder.
2. Remove two stamped nuts from the mounting pins and pins.
3. Lift the caliper from the rotor.
4. Support the caliper so there is no strain on the brake hose.
5. Using a large C-clamp, push the piston back into its bore, being careful not to cut or tear the dust boot.
6. Slide the pads past the mounting sleeve openings, then remove the pads, sleeves and bushing assemblies.
7. Check the condition of the rotor. If rotor run out exceeds manufacturer's specifications or has deep scratches, re-machine the rotor.

PAD INSTALLATION

1. Install the sleeves with shouldered ends of bushings to outside of vehicle.
2. Install the pads on the caliper with the ears over the sleeves.
3. Position the caliper on the rotor.
4. Install the mounting pins.
5. Install the stamped nuts on the mounting pins using a small socket to press them on as far as possible.
6. Fill the master cylinder with fluid.

Type Four: Delco Fixed Caliper (Four Pistons)

PAD REMOVAL & INSTALLATION

1. Remove $1/2$ of the brake fluid from the master cylinder.
2. Remove the brake pad retaining pins.
3. Push the pistons back into their bores (being careful to push both pistons at once so as not to force one out of its bore) and lift out one pad by tipping it down at the rear and up at the front.
4. Hold the rear piston in and slide the rear end of the new pad into place, being careful not to force out the front piston.
5. Check the condition of the rotor. If rotor run out exceeds 0.003 in. or has deep scratches, re-machine the rotor.
6. Push the front piston back into

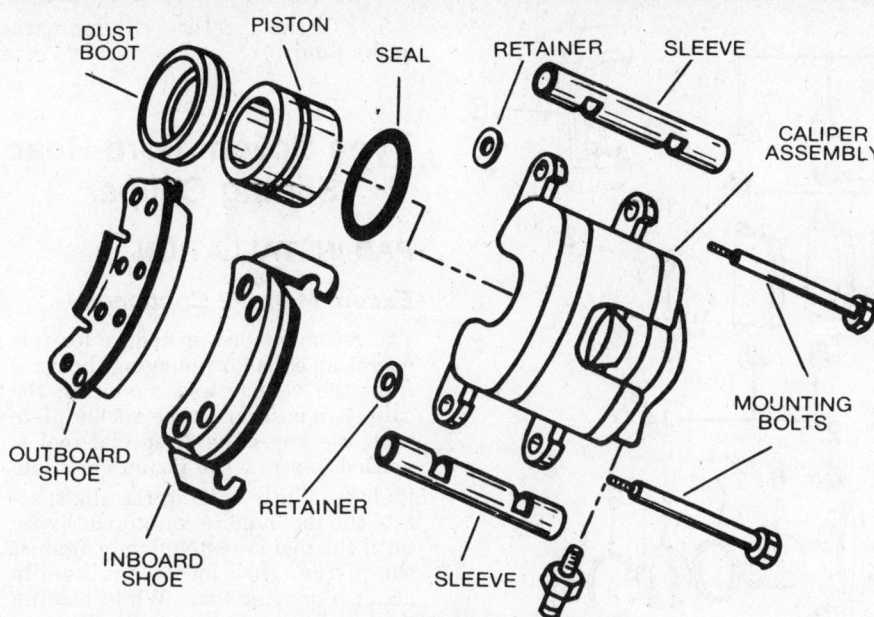

Type three-Disc floating caliper (single piston)

two wheel lug nuts to retain the rotor.

3. Remove the caliper-to-mounting bracket pins.

4. Lift the caliper from the mounting bracket. Using a wire, support the caliper so there is no strain on the brake hose.

5. Using a small pry bar or a C-clamp, force the piston back into the bore, being careful not to scratch the piston and/or bore. Be careful not to cut or tear the dust boots.

6. From the caliper, remove the brake pads, the wear indicator plates, the anti-squeal shims and the four support plates.

7. Check the rotor thickness and runout.

NOTE: If the rotor runout exceeds the manufacturer's specifications or has deep scratches, machine or replace it.

PAD INSTALLATION

1. Clean and lubricate (using silicone grease) the caliper guide pins and guide surfaces.

2. Install new support plates to the mounting bracket and new pad wear indicator plates to each pad.

its bore and slide the front of the new pad into position.

7. Change the other pad in the same manner.

8. Reinstall the retaining pin through the caliper holes and through the holes in the pads.

9. Fill the master cylinder with fresh brake fluid.

Type Five: Chevrolet Nova

PAD REMOVAL

1. Remove $\frac{2}{3}$ of the brake fluid from the master cylinder.

2. Raise and support the front of the vehicle on jackstands. Mark the relationship of the wheel to the axle hub. Remove the front wheel; install

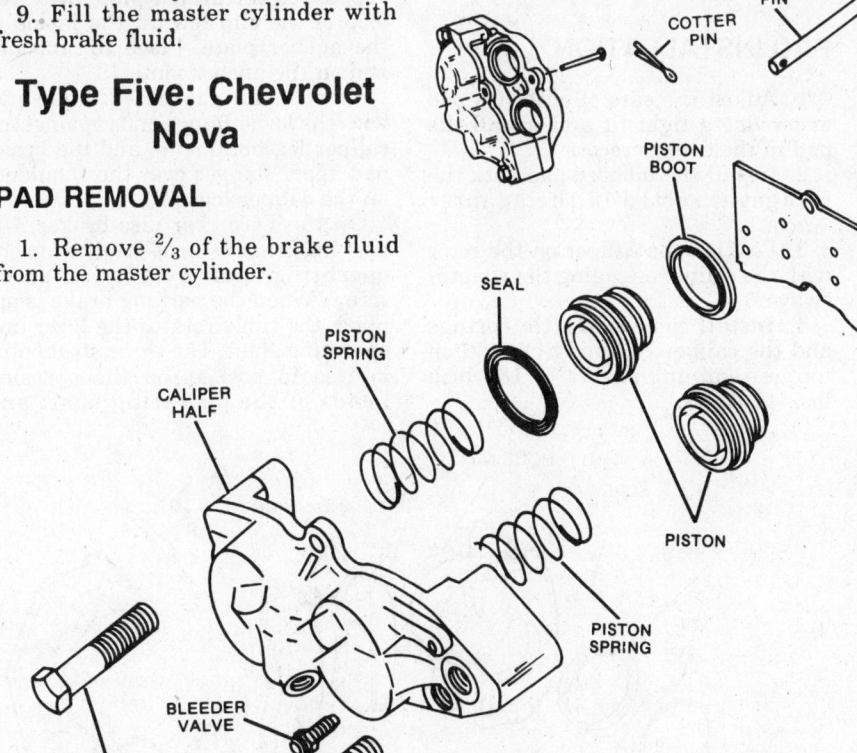

NOTE: When installing the pad wear indicators, be sure the arrow on the indicator is facing in the direction of rotation.

3. Install a new anti-squeal shim to the backside of each pad.

4. Position the new pads on the mounting bracket.

NOTE: The inboard and outboard pads are identical and interchangeable.

5. Position the caliper on the mounting bracket.

6. Align the guide pin holes of the adapter and the caliper. Torque the mounting bracket-to-steering knuckle bolts to 65 ft. lbs. and guide pins to 18 ft. lbs.

Type four Delco fixed caliper

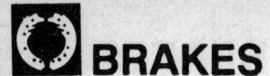
1. Mounting bolt
2. Dust Boot
3. Collar
4. Slide bushing
5. Caliper housing
6. Piston
7. Piston seal
8. Boot
9. Set ring
10. Cap
11. Bleeder screw
12. Anti-squeal shim
13. Pad

14. Pad support plate
15. Pad wear indicator plate
16. Mounting bracket

Exploded view of the caliper assembly – GM Nova

7. Align and install the wheel to the axle hub, then lower the vehicle.

8. Refill the master cylinder to the proper level. If necessary, bleed the brake system.

Type Six: Kelsey-Hayes/Chrysler Sliding Caliper

PAD REMOVAL

1. Remove ½ of the brake fluid from the master cylinder.

2. Remove caliper retaining clips and anti-rattle springs.

3. Lift the caliper from the rotor.

4. Support the caliper so there is no strain on the brake hose.

5. Use a large C-clamp to force the piston back into its bore, being careful not to scratch the piston or bore and not to cut or tear the dust boot.

6. Pry the outboard pad from caliper.

7. Remove inboard pad from the adapter.

8. Check the condition of the rotor. If rotor run out exceeds the manufacturer's specifications or has deep scratches, re-machine the rotor.

PAD INSTALLATION

1. Adjust the ears of outboard pad to provide a tight fit and install the pad in the caliper recess.

2. Install the inboard pad with the flanges inserted in the adapter "ways."

3. Position the caliper on the rotor with the caliper engaging the adapter "ways."

4. Install the anti-rattle springs and the caliper retaining clips, then torque retaining screws to 180 inch lbs. (15 ft. lbs.).

5. Fill the master cylinder with brake fluid.

Type Seven: Ford Rear Sliding Caliper

PAD INSTALLATION

Except Mark VII Continental

The recommended procedure for this operation calls for removing the rotor from the vehicle and mounting the caliper in position in the anchor plate with the key only. A special tool is needed to screw the piston back into its bore. While holding the shaft, rotate the tool handle counterclockwise until the tool is seated firmly against the piston. Now loosen the handle about a quarter turn. While holding the handle, rotate the tool shaft clockwise until the piston is fully bottomed in its bore.

Once the piston is bottomed, remove the caliper from the mounting plate and the tool from the caliper, then reinstall the rotor. Now the new pads can be installed. Make sure that the brake pad anti-rattle clip is in place in the lower inner brake pad support on the anchor plate, with the loop of the clip toward the inside of the anchor plate. Place the inboard pad on the anchor plate.

Now install the outer brake pad with the lower flange ends against the caliper leg abutments and the brake pad upper flanges over the shoulders on the caliper legs.

On the Ford rear disc brakes, the parking brake lever is attached to the operating shaft by a nylon-patch screw. When the parking brake is applied, the cable rotates the lever and operating shaft. The three steel balls, located in pockets on the opposing heads of the operating shaft and

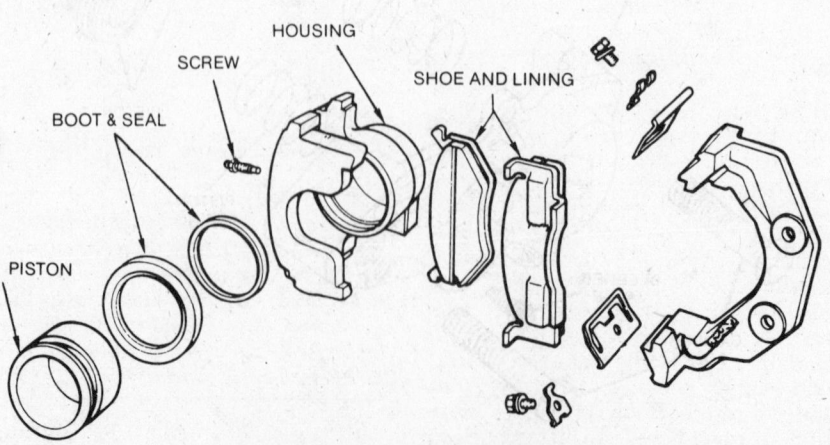

Type six Kelsey Hayes sliding caliper

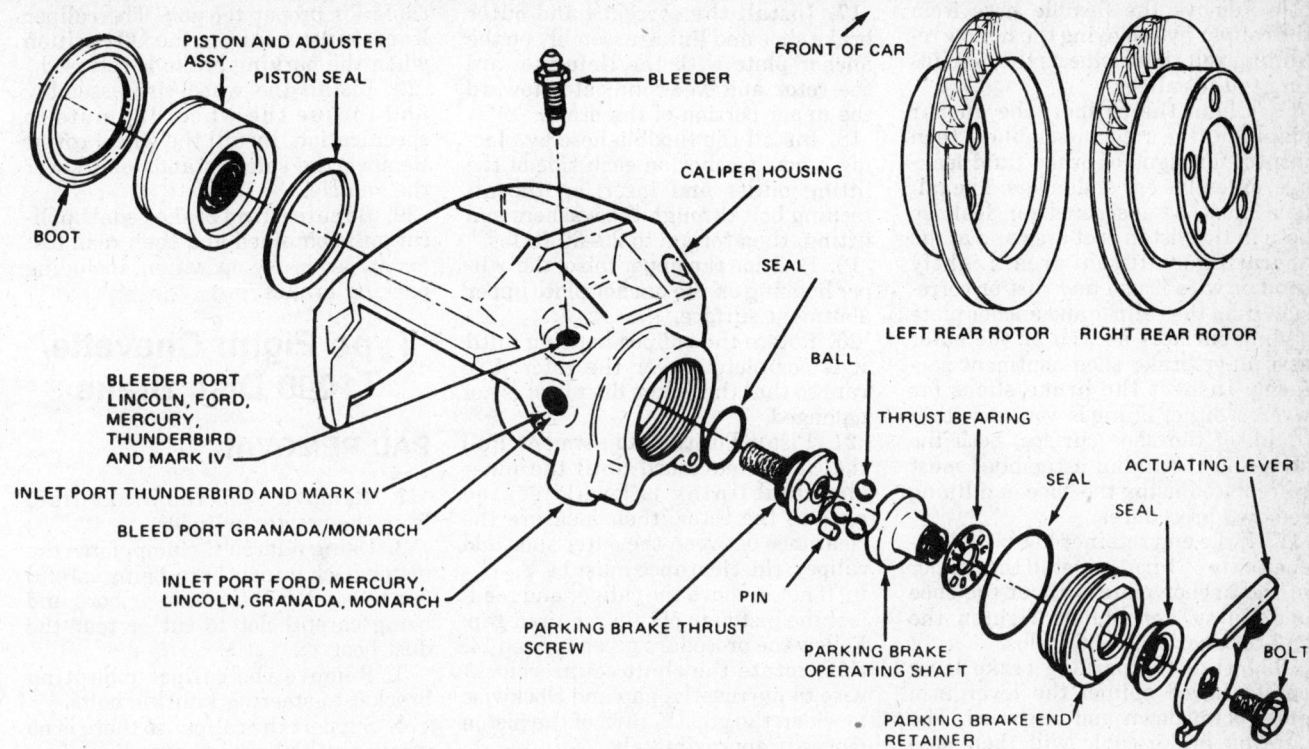

Type seven—Ford optional rear disc brakes

thrust screw, roll between ramps formed in these ball pockets. The balls force the thrust screw away from the operating shaft, driving he piston and pad against the rotor, creating the parking brake force.

Mark VII Continental

1. Raise and support the vehicle on jackstands, then lock both front wheels.

2. Remove the wheel assemblies.

3. Disconnect the parking brake cable from the lever and bracket. Use care to avoid kinking or cutting the cable or return spring.

4. Remove the caliper locating pins.

5. Lift the caliper assembly away from the anchor plate by pushing the caliper upward toward the anchor plate, then rotate the lower end out of the anchor plate.

6. If insufficient clearance between the caliper, the shoe and the lining assemblies prevents removal of the caliper, it is necessary to loosen the caliper end retaining ½ turn, maximum, to allow the piston to be forced back into its bore. To loosen the end retainer, remove the parking brake lever, then mark or scribe the end retainer and caliper housing to be sure that the end retainer is not loosened more than ½ turn. Force the piston back in its bore and remove the caliper.

1. Caliper housing
2. Mounting bolt
3. Bracket
4. Spring
5. Parking brake lever
6. Nut
7. Boot and seal
8. Sleeve
9. Bushing
10. Bleeder valve
11. Fitting
12. Spring
13. Brake pads
14. Actuator assembly
15. Assembly screw
16. Spring

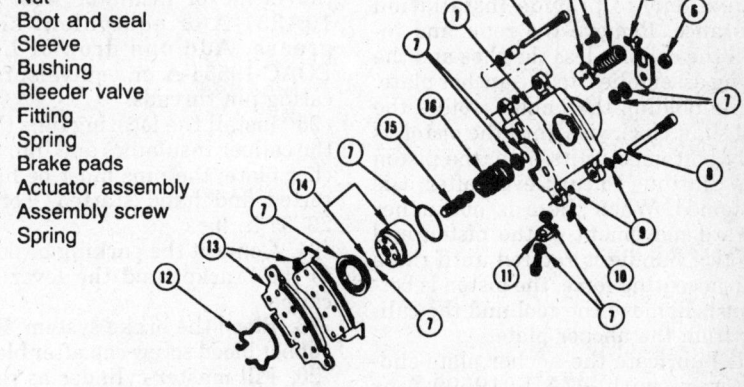

Type seven—Ford optional rear disc brakes

CAUTION

If the retainer must be loosened more than ½ turn, the seal between the thrust screw and the housing may be broken and the brake fluid may leak into the parking brake mechanism chamber. In this case, the end retainer must be removed, then the internal parts cleaned and lubricated.

7. Remove the outer shoe and lining assembly from the anchor plate.

8. Remove the two rotor retainer nuts and the rotor from the axle shaft.

9. Remove the inner brake shoe and the lining assembly from the anchor plate. Remove anti-rattle clip from the anchor plate.

NOTE: If no further service than pad replacement is required, the brake hose removal is not necessary. DO NOT support the caliper by the brake hose.

10. Remove the flexible hose from the caliper by removing the hollow retaining bolt that connects the hose fitting to the caliper.

11. Clean the caliper, the anchor plate and the rotor assemblies, then inspect for signs of brake fluid leakage, excessive ear or damage. The caliper must be inspected for leakage both in the piston boot area and at the operating shaft seal area. Lightly sand or wire brush any rust or corrosion from the caliper and anchor plate sliding surfaces as well as the outer and inner brake shoe abutment surfaces. Inspect the brake shoes for wear. If either lining is worn to within $\frac{1}{8}$ in. of the shoe surface, both the shoe and the lining assemblies must be replaced using the shoe and lining removal procedures.

12. If the end retainer has been loosened only $\frac{1}{2}$ turn, reinstall the caliper in the anchor plate without the shoe and lining assemblies. Tighten the end retainer to 75–96 ft. lbs.

13. Install the parking brake lever on its keyed spline; the lever arm must point down and rearward. The parking brake cable will then pass freely under the axle. Tighten the retainer screw to 16–22 ft. lbs. The parking brake lever must rotate freely after tightening the retainer screw. Remove the caliper from the anchor plate.

14. If new shoe and lining assemblies are to be installed, the piston must be screwed back into the caliper bore, using Tool No. T5P-2588-B or equivalent, to provide installation clearance. Remove the rotor and install the caliper, less the shoe and the lining assemblies, in the anchor plate. While holding the handle, rotate the tool shaft clockwise until the piston is fully bottomed in its bore; the piston will continue to turn even after it is bottomed. When there is no further inward movement of the piston and the tool handle is rotated until there is firm seating force, the piston is bottomed. Remove the tool and the caliper from the anchor plate.

15. Lubricate the anchor plate sliding ways with D7AE-019590-A or equivalent grease. Use only specified grease because a lower temperature type of lubricant may melt and contaminate the brake pads. Use care to prevent any lubricant from getting on the braking surface. Install the anti-rattle clip on the lower rail of the anchor plate.

16. Install the inner brake shoe and the lining assembly on the anchor plate with the lining toward the rotor. Be sure shoes are installed in their original positions as marked for identification before removal. Install the rotor and the two retainer nuts.

17. Install the correct hand outer brake shoe and lining assembly on the anchor plate with the lining toward the rotor and wear indicator toward the upper portion of the brake.

18. Install the flexible hose by placing a new washer on each side of the fitting outlet and inserting the attaching bolt through the washers and fitting, then torque to 20–30 ft. lbs.

19. Position the upper tab of the caliper housing on the anchor plate upper abutment surface.

20. Rotate the caliper housing until it is completely over the rotor. Use care so that the piston dust boot is not damaged.

21. Piston Position Adjustment: Pull the caliper outboard until the inner shoe and lining is firmly seated against the rotor, then measure the clearance between the outer shoe and caliper; the clearance must be $\frac{1}{32}$–$\frac{3}{32}$ in. If not, remove the caliper and readjust the piston to obtain required gap. Follow the procedure given in Step 14, then rotate the shaft counterclockwise to narrow the gap and clockwise to widen the gap ($\frac{1}{4}$ turn of the piston moves it approximately $\frac{1}{16}$ in.).

─────── **CAUTION** ───────

A clearance greater than $\frac{3}{32}$ in. may allow the adjuster to be pulled out of the piston when the service brake is applied. This will cause the parking brake mechanism to fail to adjust. It is then necessary to replace the piston/adjuster assembly.

────────────────────────

22. Lubricate the locating pins and the inside of insulator with D7AZ-19A331-A or equivalent silicone grease. Add one drop of Locite® EOAC-19554-A or equivalent, to locating pin threads.

23. Install the locating pins through the caliper insulators and into the anchor plate; the pins must be hand inserted and hand started. Torque to 29–37 ft. lbs.

24. Connect the parking brake cable to the bracket and the lever on the caliper.

25. Bleed the brake system. Replace rubber bleed screw cap after bleeding.

26. Fill master cylinder as required to within $\frac{1}{8}$ inch of the top of the reservoir.

27. Caliper Adjustment: With the engine running, pump the service brake lightly (approximately 14 lbs. pedal effort) about 40 times. Allow at least one second between pedal applications. As an alternative, with the engine Off, pump the service brake lightly (approximately 87 lbs. pedal effort) about 30 times. Now check the parking brake for excessive travel or very light effort. In either case, repeat the pumping the service brake or (if necessary) check the parking brake

cable for proper tension. The caliper levers must return to the Off position when the parking brake is released.

28. Install the wheel/tire assembly and torque the wheel lug nuts to specification. Install the wheel cover. Remove the safety stands and lower the vehicle.

29. Be sure a firm brake pedal application is obtained and then road test for proper brake operation, including parking brakes.

Type Eight: Chevette/ T1000 Disc Brake

PAD REMOVAL

1. Remove $\frac{1}{2}$ of the brake fluid from the master cylinder.

2. Using a large C-clamp, force the piston back into its bore, being careful not to scratch the piston or bore and being careful not to cut or tear the dust boot.

3. Remove the caliper mounting bracket-to-steering knuckle bolts.

4. Support the caliper so there is no strain on the brake hose.

NOTE: DO NOT remove the socket head retainer bolt.

5. Remove the old shoe and lining assemblies. If the retaining spring does not come out with the inboard shoe, remove the spring from the piston.

6. Check the condition of the rotor. If rotor run out exceeds the manufacturer's specifications or has deep scratches, re-machine the rotor.

PAD INSTALLATION

1. Before installing the inboard shoe, make sure that the shoe retaining spring is properly installed. Push the tab on the single-leg end of the spring down into the shoe hole, then snap the other two legs over the edge of the shoe notch.

2. Position the caliper over the rotor, align the bracket mounting holes and install the mounting bolts.

3. Clinch the outboard shoe to the caliper. After clinching, radial and end play of the outboard shoe should be 0–0.005 inch (0–0.127mm).

Type Nine: GM Rear Disc Brake

PAD REMOVAL

NOTE: Calipers must be removed to replace linings.

1. Remove $\frac{2}{3}$ of the fluid in the front master cylinder.

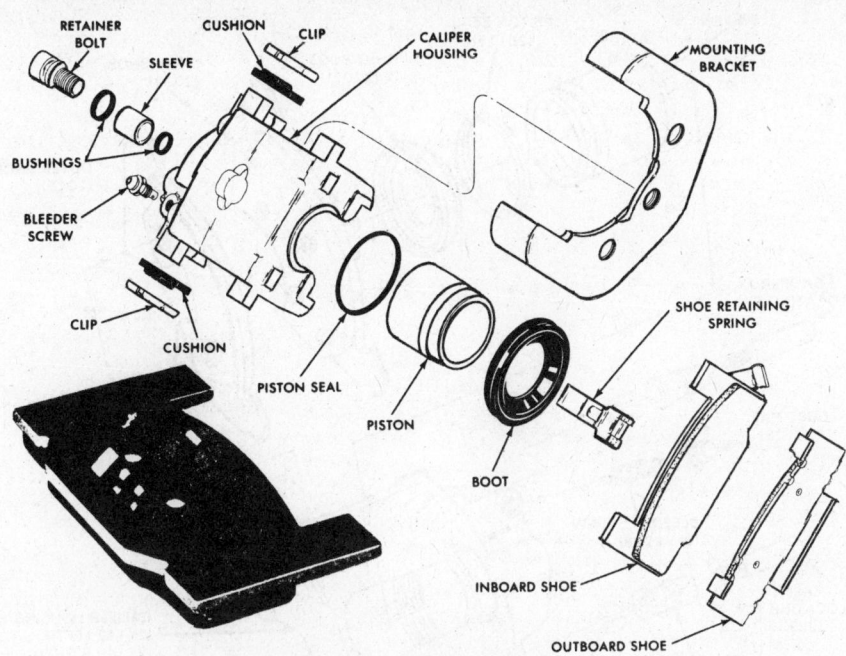
2. Remove the wheel/tire assembly and reinstall one wheel mounting nut, flat side toward rotor, to prevent rotor from falling when caliper is removed.

3. Loosen the tension on the parking brake cable at the equalizer and remove the cable from the parking brake lever at the caliper.

4. Remove the return spring, the lock nut, the lever, the lever seal and the anti-friction washer. The lever must be held in place while removing nut.

5. Using a C-clamp with the solid end of the lever stop and the screw end of the back of the outboard lining assembly, tighten the clamp until the piston bottoms in the caliper.

NOTE: DO NOT position C-clamp on the actuator screw.

6. Before removing the clamp, lube the caliper housing surface (under the lever seal) with silicone grease.

7. Install a new anti-friction washer, a new lever seal and the lever. Be certain to install lever on hex with the arm pointing downward.

8. Rotate lever toward the front of the vehicle; while holding it in this position install the nut and torque it to 25 ft. lbs. Rotate the lever back to the stop.

9. Install the lever return spring and remove C-clamp. The springs are color coded: Red for the right-side of the caliper and black for the left-side.

10. Remove the brake line from the caliper and plug the openings to retain the fluid.

NOTE: If the brake line nut is seized, the brass bolt and block on the caliper can be removed with the brake line attached by removing the bolt and block the copper washers after removing the caliper mounting bolts.

11. Remove the caliper mounting bolts, the caliper and the brake shoes.

12. Remove the two caliper mounting sleeves and the four bushings, then install new parts using a silicone lubricant (the sleeves are installed in the inner bushings).

13. Position the new inboard shoe assembly on the piston. The "D" shaped tab must fit in the indentation provided in the piston.

14. Install the new outboard shoe assembly.

15. To reinstall the caliper, replace any corroded caliper mounting bolts with new parts. Wire brushing or sanding will damage the bolt plating.

16. If the brass bolt and the block were removed with a brass pipe, unplug the fittings, then install the bolt and the block using two new copper

Type eight—Chevette/T1000

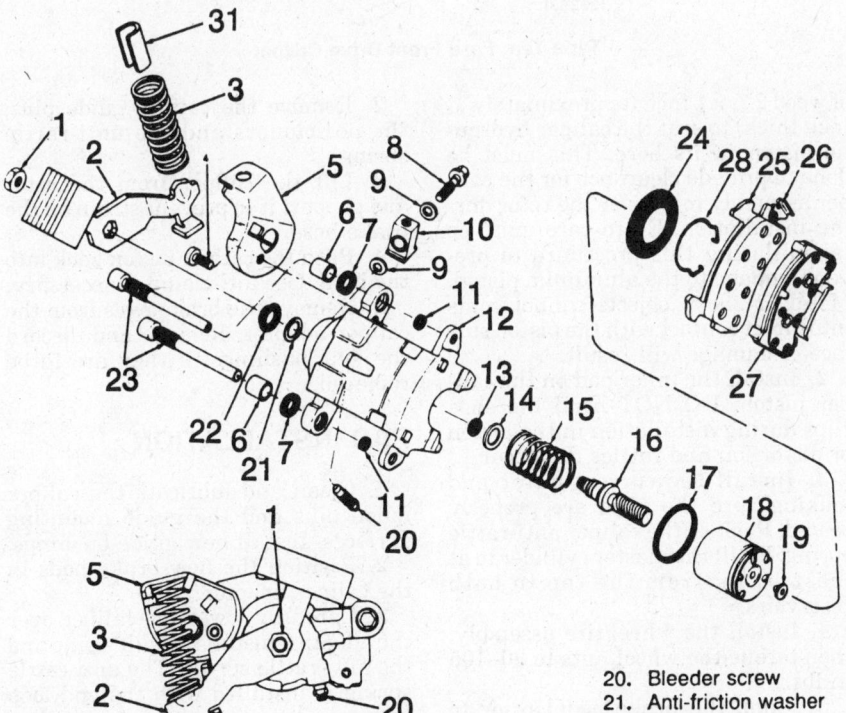

1. Nut
2. Lever
3. Return spring
4. Bolt
5. Bracket
6. Sleeve
7. Bushing
8. Bolt
9. Washer
10. Fitting
11. Bushing
12. Caliper housing
13. Shaft seal
14. Thrust washer
15. Balance spring
16. Actuator screw
17. Piston seal
18. Piston assembly
19. Two-way check valve
20. Bleeder screw
21. Anti-friction washer
22. Lever seal
23. Mounting bolt
24. Boot
25. Inboard shoe and lining
26. Wear sensor
27. Outboard shoe and lining
28. Shoe dampening spring
31. Damper

Exploded view of the rear disc brake assembly—GM

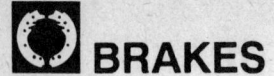

gaskets. Torque to 30 ft. lbs. Be sure that the sleeves and pins are well lubricated with silicone (the mounting bolt should go under the inboard shoe ears).

17. Install the brake line tube nut into the caliper and pump the brake pedal to seat the lining against the rotor.

18. Clinch the upper ear of the outboard shoe by positioning a pair of pliers with one jaw on top of the upper ear and the other jaw in the notch or bottom of shoe, opposite the upper ear. After clinching, there should be no radial clearance between the shoe ears and the caliper housing. Repeat the clinching procedure (if necessary).

19. Connect and adjust the parking brake cables and bleed the rear brake system.

20. Install the wheel and the assembly. Torque the steel mounting nuts to 130 ft. lbs.

Type Ten: Ford Front Drive Sliding Caliper

PAD REMOVAL

1. Remove the master cylinder cap and check the fluid level in the reservoirs. Remove the brake fluid until each reservoir is ½ full. Discard the removed fluid.

2. Remove the wheel/tire assembly from the rotor mounting face. Use care to avoid damage or interference with the caliper splash shield or the bleeder screw fitting.

3. Remove the brake caliper anti-rattle spring by applying upward pressure to the center portion of the spring until the spring tabs are free of the caliper holes.

4. Back out the caliper locating pins. DO NOT remove the pins completely unless the new bushings are to be installed. Reinstalling the pins after complete removal can be difficult.

5. Lift the caliper assembly from the integral knuckle, the anchor plate and the rotor. Remove the outer shoe and the lining assembly from the caliper assembly.

6. Remove the inner shoe and the lining assembly, then inspect both rotor braking surfaces. Minor scoring or build-up of the lining material does not require machining or replacement of the rotor.

7. Suspend the caliper inside the fender housing. Use care not to damage the caliper or stretch the brake hose.

PAD INSTALLATION

1. Use a 4-inch C-clamp and a block

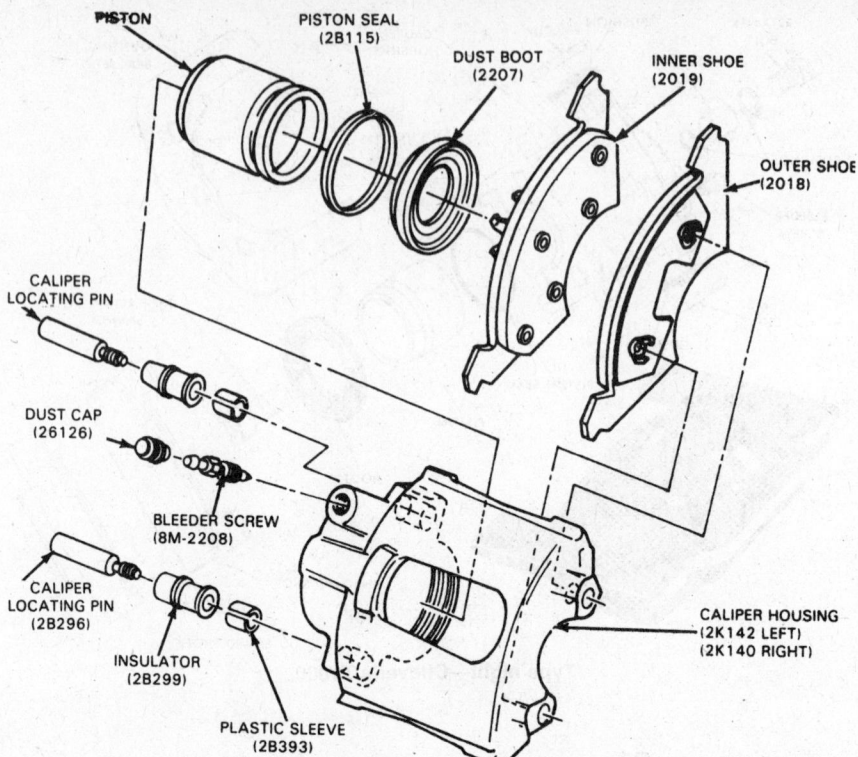

Type Ten; Ford Front Drive Caliper

of wood 2¾ x 1 inch (approximately ¾ inch thick) to seat the caliper hydraulic piston in its bore. This must be done to provide clearance for the caliper assembly to fit over the rotor during installation. Extra care must be taken during this procedure to prevent damage to the aluminum piston. Metal or sharp objects cannot come into direct contact with the piston surface or damage will result.

2. Install the inner pad on the caliper piston. DO NOT bend the shoe clips during installation in the piston or distortion and rattles can occur.

3. Install the correct outer pad making sure the clips are properly seated. Replace the caliper anti-rattle spring. Refill the master cylinder to at least ¼ in. from the top in both reservoirs.

4. Install the wheel/tire assembly, then torque the wheel nuts to 80–105 ft. lbs.

5. Pump the brake pedal prior to moving the vehicle to position the brake pads.

6. Road test the vehicle.

Type Eleven: Kelsey-Hayes Floating Caliper

PAD REMOVAL

1. Remove ½ of the brake fluid from each master cylinder reservoir.

2. Remove the caliper guide pins, the positioners and the anti-rattle spring.

3. Lift the caliper from the rotor and support it to prevent strain on the brake hose.

4. Push the caliper piston back into the bore. Use a C-clamp if necessary.

5. Remove the brake pads from the caliper adaptor. Remove and discard the four bushings, if they are to be replaced.

PAD INSTALLATION

1. Clean and lubricate the caliper guide pins and the guide mounting surfaces. Install new guide bushings.

2. Position the new brake pads in the caliper adaptor.

3. Carefully lower the caliper over the adaptor. Install the guide pins and the anti-rattle spring. The anti-rattle spring is installed with the end loop inboard on the caliper lug.

4. Fill the master cylinder with new fluid. Bleed the brakes if necessary.

Type Twelve: ATE Floating Caliper

PAD REMOVAL

1. Remove the guide pin(s) and the anti-rattle clips or springs.

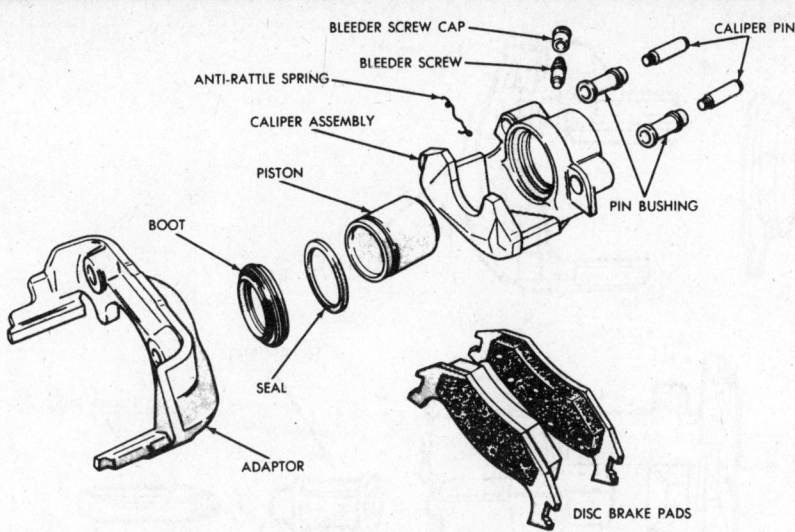

Type Eleven; Kelsey-Hayes Floating Caliper

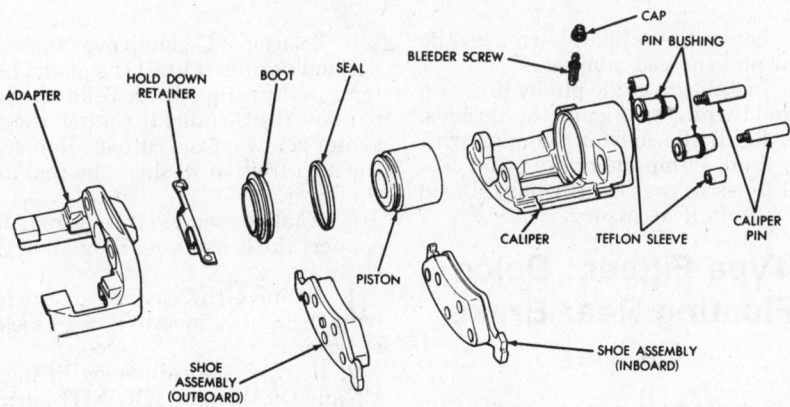

Type Twelve; ATE Floating Caliper

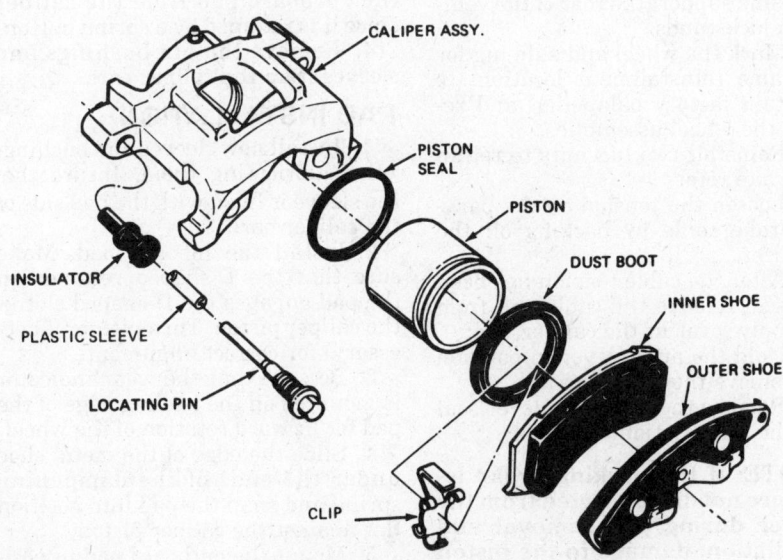

Type Thirteen—Ford sliding caliper

2. Remove the caliper from the rotor by slowly sliding it up and away. Support the caliper so there is no strain on the brake hose. Late model calipers may be pivoted on the anchor bolt.

3. Remove the pads from the adaptor or caliper. In some cases the rotor must be removed to replace the inboard pad.

4. Push the caliper piston back into its bore.

PAD INSTALLATION

1. Install the pads and hardware into the adapter or caliper.

2. Position the caliper over the rotor, then install the guide pin(s), the anti-rattle springs or clips. Fill the master cylinder with new brake fluid. Bleed the brake system if necessary.

Type Thirteen: Ford Sliding Caliper

PAD REMOVAL

1. Remove $\frac{1}{2}$ of the brake fluid from the master cylinder reservoirs.

2. Remove the caliper guide pins.

3. Lift the caliper assembly from the rotor. Support the caliper so there is no strain on the brake hose.

4. Remove the outboard pad from the caliper. Remove the inboard pad from the piston.

NOTE: Step 6 can now be accomplished by using a C-clamp against the inboard pad.

5. Remove the insulators and inserts from the guide pin holes, if they are to be replaced.

6. Push the caliper piston back into its bore.

PAD INSTALLATION

1. Install the new guide bushings and insulators, if they are to be replaced.

2. Install the inboard pad into the piston. Install the outboard pad making sure the buttons are seated into the caliper body. The wear indicator faces toward the front of the vehicle.

3. Lower the caliper assembly onto the anchor plate and slide the guide pins through the holes in the caliper. When the guide pins reach the rubber insulators, they will require more pressure. After the pins bottom thread them into the hole.

—————— **CAUTION** ——————
Take care not to cross thread the guide pins.

U351

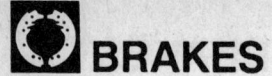

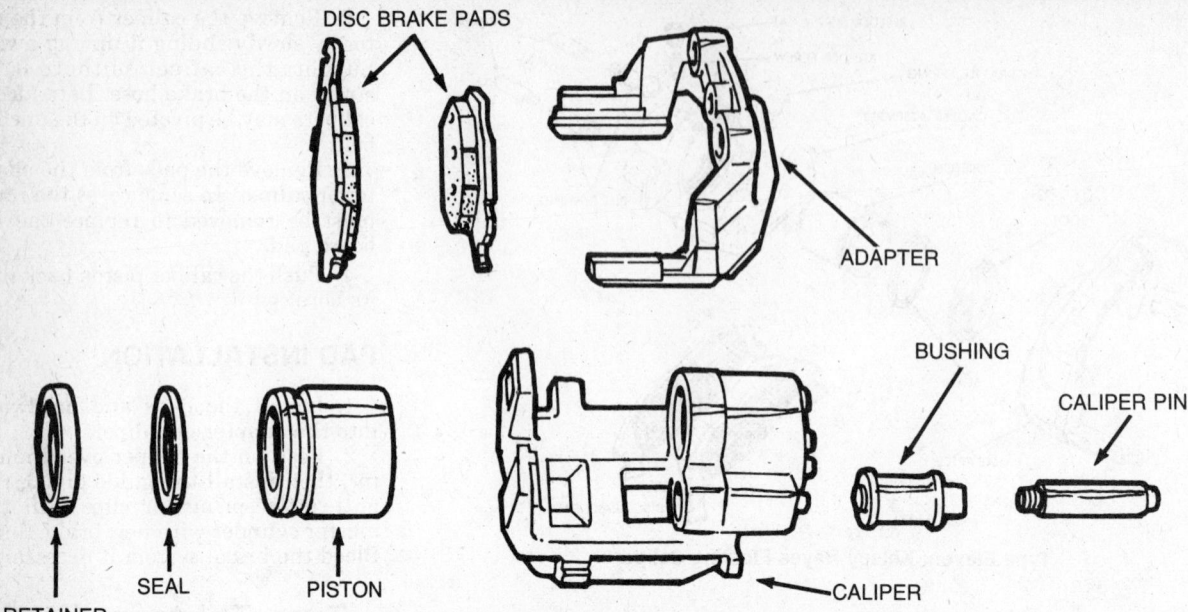

DISC BRAKE PADS

ADAPTER

BUSHING

CALIPER PIN

RETAINER   SEAL   PISTON

CALIPER

Type fourteen—Kelsey–Hayes floating caliper

4. Refill the master cylinder with new brake fluid. Bleed the brake system if necessary.

Type Fourteen: Kelsey-Hayes Floating Caliper

PAD REMOVAL

1. Remove $\frac{1}{2}$ of the brake fluid from the master cylinder reservoir.
2. Remove the caliper guide pin, then slide the caliper up and way from the rotor. Secure the caliper out of the way with a wire. Avoid strain on the brake hose.
3. Remove the outboard pad from the caliper adapter.

NOTE: There are three retaining springs used. One spring is located at the top of the outboard caliper adapter, one at the bottom of the outboard pad and the last on the top of the inboard pad. Pay attention to the shape and location of these springs.

4. Remove the disc brake rotor by sliding it from the hub. Remove the inboard pad.
5. Push the caliper piston back into the caliper bore.

PAD INSTALLATION

1. Slide the inboard pad into position with the spring installed on the adapter.
2. Install the disc brake rotor. Slide the outboard pad and springs into position.

3. Lower the caliper down over the rotor and the pad adapter.
4. Install the guide pin by pressing in and turning to engage the threads.
5. Refill the master cylinder with new fluid. Pump the brake pedal several times to position the pads. Bleed the system if necessary.

Type Fifteen: Delco Floating Rear Brake Caliper

PAD REMOVAL

1. Remove $\frac{2}{3}$ of the brake fluid from the master cylinder.
2. Loosen the rear wheel lugs. Raise and support the rear of the vehicle on jackstands.
3. Mark the wheel and axle lug for the same reinstallment location (to maintain factory balancing) and remove the wheel assemblies.
4. Reinstall two lug nuts to retain the brake rotor.
5. Loosen the tension on the parking brake cable by backing off the equalizer.
6. After the cable tension has been released, remove the cable end from the apply lever at the caliper.
7. Hold the apply lever in position and remove the retaining nut.
8. Remove the lever, the lever seal and the anti-friction washer.

NOTE: If the parking brake levers are not disconnected from the caliper during pad removal and installation, damage to the piston assembly will occur when it is moved back into the caliper bore.

9. Position a C-clamp over the caliper and tighten it until the piston bottoms in the caliper bore. Take care not to allow the C-clamp to contact the actuator screw on the caliper. Reinstall the anti-friction washer, the seal and the lever.
10. If caliper service is required, disconnect the brake line. Plug all of the openings.
11. Remove the caliper mounting bolts using a $\frac{3}{8}$ in. Allen head socket or wrench.
12. Remove the caliper by lifting it Up and Off the rotor. DO NOT permit the caliper to be suspended by the brake hose.
13. Remove the pads from the caliper. A suitable tool is required to pry the outboard pad from the caliper since it is retained by a spring button.
14. Remove the pin bushings and sleeves from the caliper ears.

PAD INSTALLATION

1. Install new sleeves and bushings after lubricating them. Insure that the sleeve is flush with the pad side of the caliper ear.
2. Install the inboard pad. Make sure that the D-shaped retainer on the pad engages the D-shaped slot in the caliper piston. Turn piston (if necessary) for correct alignment.
3. Be sure that the wear indicator is mounted on the leading edge of the pad for forward rotation of the wheel.
4. Slide the edge of the metal shoe under the ends of the dampening spring and snap the pad into position flat against the caliper piston.
5. Mount the outboard pad in position. Be sure it snaps into the caliper recess.

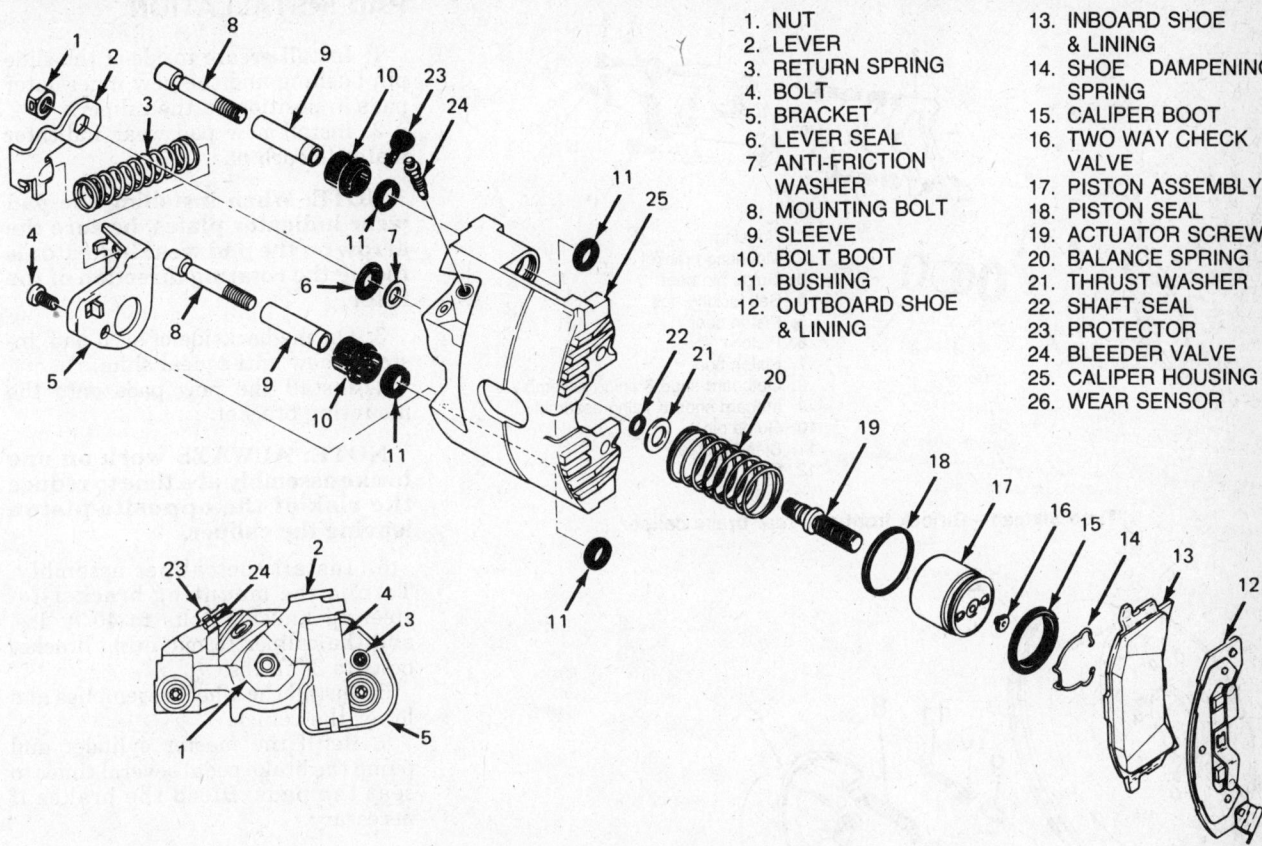

1. NUT
2. LEVER
3. RETURN SPRING
4. BOLT
5. BRACKET
6. LEVER SEAL
7. ANTI-FRICTION WASHER
8. MOUNTING BOLT
9. SLEEVE
10. BOLT BOOT
11. BUSHING
12. OUTBOARD SHOE & LINING
13. INBOARD SHOE & LINING
14. SHOE DAMPENING SPRING
15. CALIPER BOOT
16. TWO WAY CHECK VALVE
17. PISTON ASSEMBLY
18. PISTON SEAL
19. ACTUATOR SCREW
20. BALANCE SPRING
21. THRUST WASHER
22. SHAFT SEAL
23. PROTECTOR
24. BLEEDER VALVE
25. CALIPER HOUSING
26. WEAR SENSOR

Type Fifteen—Delco floating rear brake caliper

6. Install the caliper over the disc rotor in the reverse order of removal. Apply the brakes several times to seat the linings, after filling the master cylinder. Bleed the brakes if necessary.

Type Sixteen: Girlock Floating Front and Rear Calipers

PAD REMOVAL

1. Remove ⅔ of the brake fluid from the master cylinder.
2. Loosen the rear lugs, then raise and support the rear of the vehicle on jackstands. Remove the wheel assemblies. Install two lug nuts to hold the brake disc rotor in position.
3. Position a C-clamp over the caliper, one end on the outboard pad, the other on the inlet fitting bolt head.
4. Tighten the clamp to push the caliper piston until it bottoms in the bore.
5. Remove and discard the upper caliper self-locking bolt. Rotate the caliper on the lower bolt to expose the brake pads.

7. Remove the inner and outer pads from the caliper.
8. Clean the pad mounting frame on the caliper. Inspect the caliper for signs of fluid leakage. Remove and service the caliper if necessary.

PAD INSTALLATION

1. Install the new inner and outer pad into position on the caliper.
2. Rotate the caliper back into position over the disc brake rotor.
3. Install a new self-locking bolt and tighten to 22–25 ft. lbs.
4. Install the wheel assemblies and lower the vehicle.
5. Fill the master cylinder and pump the brake pedal several times to seat the pads. Bleed the brakes if necessary.

PARKING BRAKE SHOES REPLACEMENT

1. Refer to the "Pad Removal & Installation" procedures in this section and remove the rear brake calipers and pads.
2. Remove the caliper support-to-axle hub bolts and the caliper support. Remove the rotor from the axle hub.

3. Spread the brake shoes and remove the star wheel adjuster (inspect the threads) and the star wheel adjuster spring.
4. Remove the hold down springs and pins from the shoes.
5. Using a pair of pliers, remove the shoe return springs.
6. Remove the primary/secondary shoes and the lining assemblies.
7. Clean and inspect the wear bracket, the shims, the springs and the rubber boots, replace the parts as necessary.
8. Using GM 5450032 or equivalent grease, lubricate the wear shims and the wear bracket.
9. To install, use new brake shoes and reverse the removal procedures. Adjust the parking brake shoes.

Type Seventeen: Chevrolet Spectrum Front Disc Brake

PAD REMOVAL

1. Remove ⅔ of the brake fluid from the master cylinder.
2. Loosen the wheel lugs, then

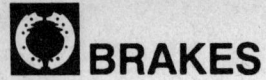
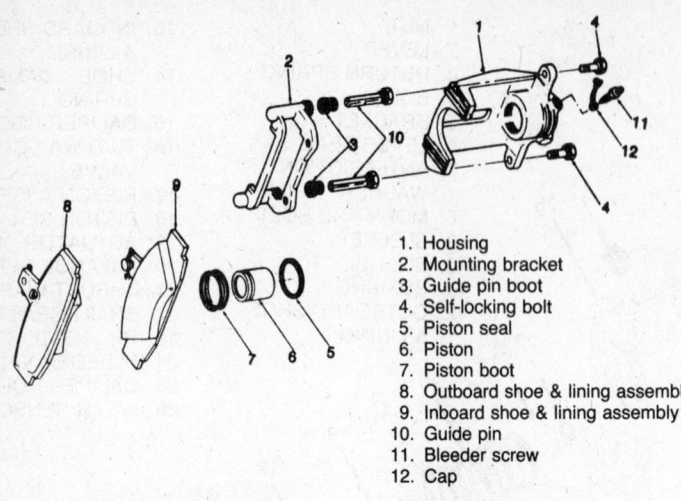

1. Housing
2. Mounting bracket
3. Guide pin boot
4. Self-locking bolt
5. Piston seal
6. Piston
7. Piston boot
8. Outboard shoe & lining assembly
9. Inboard shoe & lining assembly
10. Guide pin
11. Bleeder screw
12. Cap

Type Sixteen—Girlock front and rear brake caliper

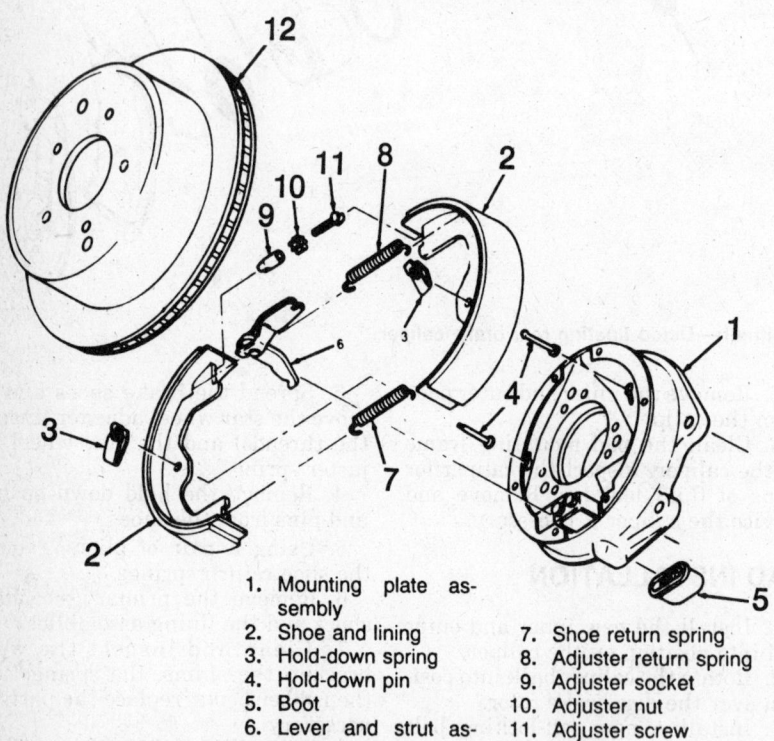

1. Mounting plate assembly
2. Shoe and lining
3. Hold-down spring
4. Hold-down pin
5. Boot
6. Lever and strut assembly
7. Shoe return spring
8. Adjuster return spring
9. Adjuster socket
10. Adjuster nut
11. Adjuster screw
12. Rotor

Exploded view of the parking brake assembly—GM Corvette

raise and support the front of the vehicle on jackstands.

3. Remove the front wheel assemblies. Install two lug nuts to hold the brake rotor in position.

4. Position a C-clamp over the caliper, one end on the outboard pad, the other on the inlet fitting bolt head.

5. Tighten the clamp to push the caliper piston to the bottom of the bore.

6. Remove the caliper-to-mounting bracket bolts and lift the caliper from the steering knuckle.

NOTE: When removing the caliper, DO NOT disconnect the brake hose from the caliper.

7. Using a wire, support the caliper from the vehicle.

8. From the caliper, remove the inner/outer pads, the pad wear indicators, the anti-squeal shims and the retainers.

9. Clean the pad mounting frame on the caliper. Inspect the caliper for signs of fluid leakage, then the rotor thickness and runout.

PAD INSTALLATION

1. Install grease inside of the slide pin bushing and the new inner/outer pads in position on the caliper.

2. Install new pad wear indicator plates to each pad.

NOTE: When installing the pad wear indicator plates, be sure the arrow on the pad wear indicator is facing the rotating direction of the rotor.

3. On the backside of each pad, install a new anti-squeal shim.

4. Install the new pads onto the mounting bracket.

NOTE: ALWAYS work on one brake assembly at a time to reduce the risk of the opposite piston leaving the caliper.

5. Install the caliper assembly. Torque the mounting bracket-to-steering knuckle bolts to 40 ft. lbs. and the caliper to mounting bracket bolts to 36 ft. lbs.

6. Install the wheel assemblies and lower the vehicle.

7. Refill the master cylinder and pump the brake pedal several times to seat the pads. Bleed the brakes if necessary.

Type Eighteen: Chevrolet Sprint Front Disc Brake

PAD REMOVAL

1. Remove ⅔ of the brake fluid from the master cylinder.

2. Loosen the wheel lugs, then raise and support the front of the vehicle on jackstands.

3. Remove the front wheel assemblies. Install two lug nuts to hold the brake rotor in position.

4. Position a C-clamp over the caliper, one end on the outboard pad, the other on the inlet fitting bolt head.

5. Tighten the clamp to push the caliper piston to the bottom of the bore.

6. Remove the caliper-to-steering knuckle bolts and lift the caliper from the steering knuckle.

NOTE: When removing the caliper, DO NOT disconnect the brake hose from the caliper.

7. Using a wire, support the caliper from the vehicle.

8. From the caliper, remove the inner/outer pads and the anti-squeal shims.

9. Clean the pad mounting frame

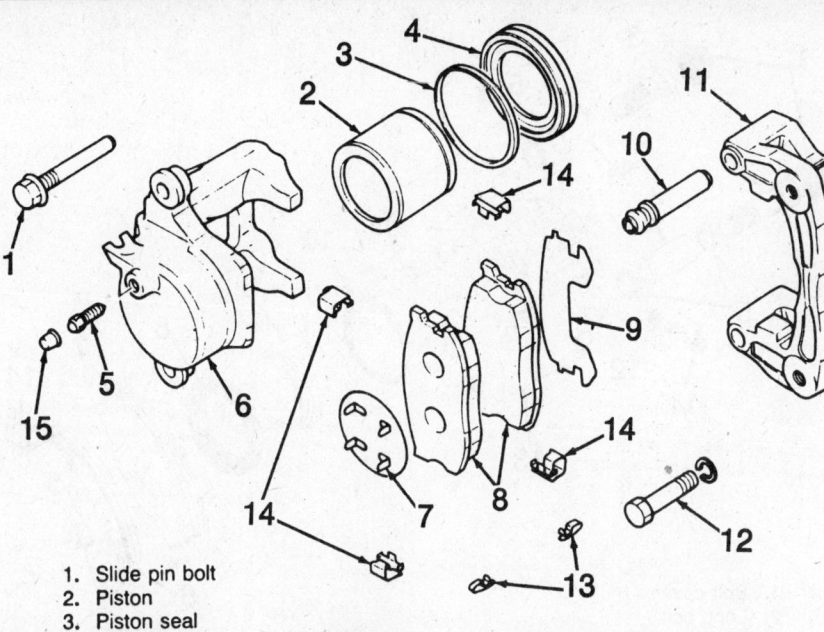

1. Slide pin bolt
2. Piston
3. Piston seal
4. Piston boot
5. Bleeder screw
6. Caliper body
7. Inner shim
8. Pads
9. Outer shim
10. Slide pin boot
11. Bracket
12. Bolt
13. Wear indicator
14. Retainer
15. Cap

Exploded view of the caliper assembly—GM Spectrum

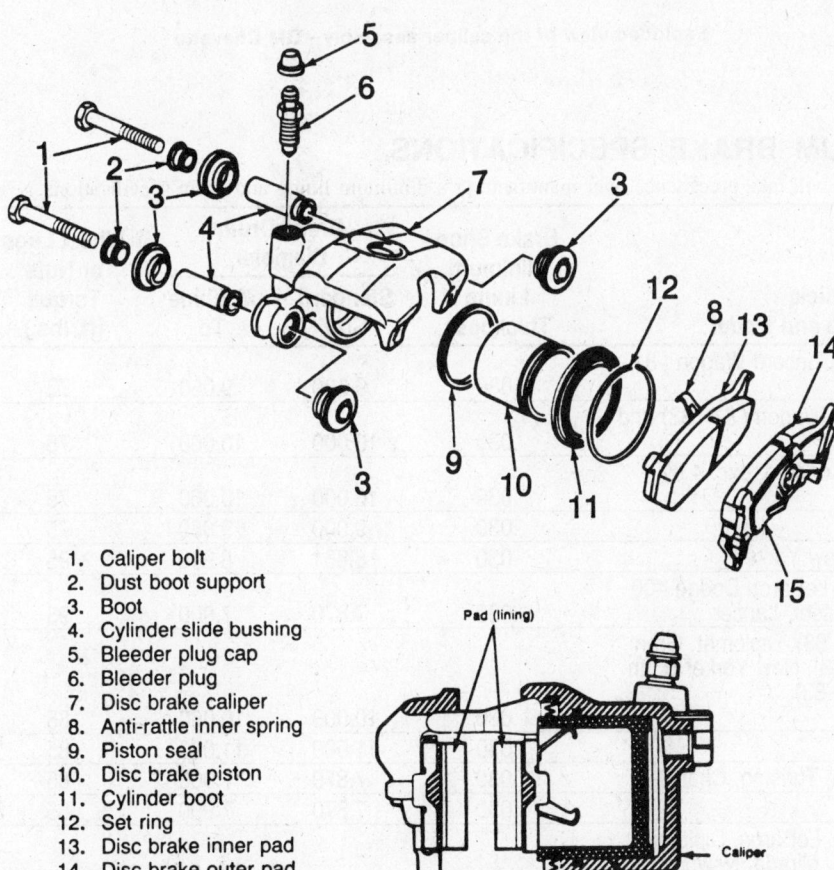

1. Caliper bolt
2. Dust boot support
3. Boot
4. Cylinder slide bushing
5. Bleeder plug cap
6. Bleeder plug
7. Disc brake caliper
8. Anti-rattle inner spring
9. Piston seal
10. Disc brake piston
11. Cylinder boot
12. Set ring
13. Disc brake inner pad
14. Disc brake outer pad
15. Anti-rattle outer spring

Exploded view of the caliper assembly—GM Sprint

on the caliper. Inspect the caliper for signs of fluid leakage, the rotor thickness and runnout. Remove and service the caliper if necessary.

PAD INSTALLATION

1. Install new inner/outer pads into position on the caliper.
2. On the backside of each pad, install a new anti-squeal shim.
3. Install the new outside pad onto the mounting bracket.
4. Install the new inside pad onto the caliper.

NOTE: ALWAYS work on one brake assembly at a time to reduce the risk of the opposite piston leaving the caliper.

5. Install the caliper assembly onto the mounting bracket and the springs onto the caliper. Torque the caliper-to-mounting bracket bolts to 17–26 ft. lbs.
6. Install the wheel assemblies and lower the vehicle.
7. Refill the master cylinder and pump the brake pedal several times to seat the pads. Bleed the brakes if necessary.

Type Nineteen: Chevrolet Chevette Front Disc Brake

PAD REMOVAL

1. Remove $\frac{2}{3}$ of the brake fluid from the master cylinder.
2. Loosen the wheel lugs, then raise and support the front of the vehicle on jackstands.
3. Remove the front wheel assemblies. Install two lug nuts to hold the brake rotor in position.
4. Remove the caliper-to-mounting bracket covers and the bolts.
5. Position a C-clamp over the caliper, one end on the outboard pad, the other on the inlet fitting bolt head.
6. Tighten the clamp to push the caliper piston to the bottom of the bore.
7. Lift the caliper from the steering knuckle.

NOTE: When removing the caliper, DO NOT disconnect the brake hose from the caliper.

8. Using a wire, support the caliper from the vehicle.
9. From the caliper, remove the inner/outer pads and the anti-squeal shims.
10. Clean the pad mounting frame on the caliper. Inspect the caliper for signs of fluid leakage, the rotor thick-

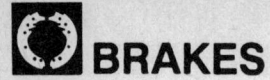

ness and runnout. Remove and service the caliper (if necessary).

PAD INSTALLATION

1. Using silicone grease, fill the
caliper housing bushing cavities and
install the bushing sleeves.

2. Install new inner/outer pads in
position on the caliper.

**NOTE: ALWAYS work on one
brake assembly at a time to reduce
the risk of the opposite piston
leaving the caliper.**

3. Install the caliper assembly onto
the mounting bracket. Torque the caliper-to-mounting bracket bolts to 30–
45 ft. lbs.

4. Install the wheel assemblies and
lower the vehicle.

5. Refill the master cylinder and
pump the brake pedal several times to
seat the pads.

6. Using an 8 oz. ball-peen hammer
and a 16 oz. brass hammer, position
the ball-peen hammer on the outboard pad tabs, then strike the ball-
peen hammer with the brass hammer
to bend the pad tabs at 45° to the caliper; this clinches the outboard pad to
the caliper. Bleed the brake system, if
necessary.

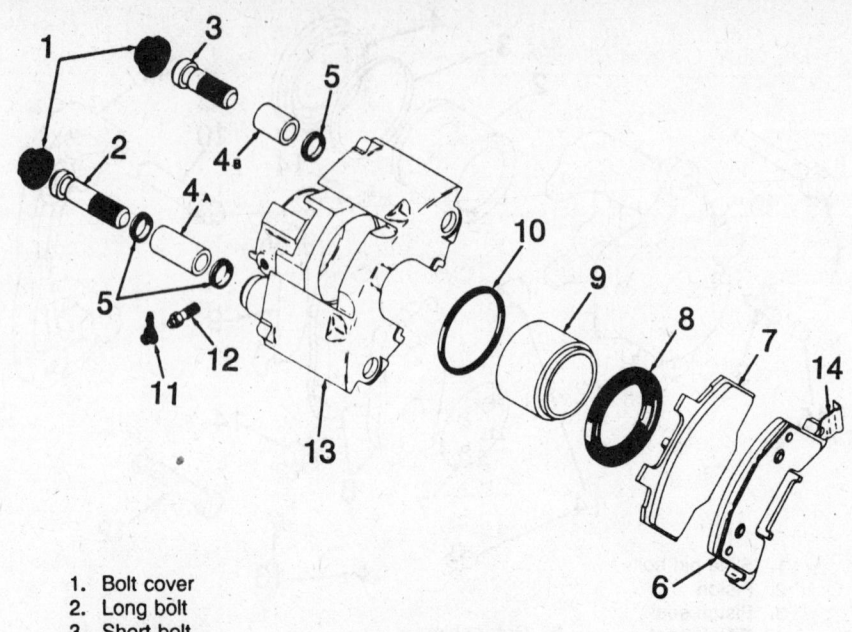

1. Bolt cover
2. Long bolt
3. Short bolt
4A. Long sleeve
4B. Short sleeve
5. Bushing
6. Outboard shoe and lining
7. Inboard shoe and lin-ing
8. Boot
9. Piston
10. Piston seal
11. Port protector
12. Bleeder valve
13. Caliper housing
14. Wear sensor

Exploded view of the caliper assembly—GM Chevette

DRUM BRAKE SPECIFICATIONS

(Note: State and local inspection regulation will take precedence over manufacturer's minimum lining and drum specifications.)

Manufacturer	Vehicle Year, Make and Model	Brake Shoe Minimum Lining Thickness	Brake Drum Diameter		Wheel Lugs or Nuts Torque (ft. lbs.)
			Standard Size	Machine To	
American Motors	'81–'85 All exc. 6 cyl. Concord Wagon ('81–'83) and Eagle	.030	9.000	9.060	75
	'81–'87 6 cyl. Concord Wagon ('81–'83) and Eagle	.030	10.000	10.060	75
	'80 Spirit exc. 4 cyl., Concord, exc. 4 cyl. Pacer, Eagle, AMX	.030	10.000	10.060	75
	'80 Spirit w/4 cyl.	.030	9.000	9.060	75
Chrysler Corp.	'84–'87 Dodge 600, New Yorker	.030	8.861	8.920	95
	'82–'87 Aries, Reliant, LeBaron Dodge 400 (82–'83) Daytona, Laser, Lancer	.030	7.870	7.900	95
	'82–'87 Cordoba ('80–'83), Diplomat, Gran Fury, Mirada ('82–'83), New Yorker (Fifth Ave.), Imperial ('80–'83) w/10" rear brake	.030	10.000	10.060	85
	w/11" rear brake	.030	11.000	11.060	85
	'80–'87 Omni, Horizon, Turismo, Charger	.030	7.870	7.900	85
	'81 Aries, Reliant	.030	7.870	7.900	85
	'80–'81 Aspen, Volare, LeBaron, Diplomat, Cordoba, Gran Fury, Mirada, Newport, New Yorker, Imperial w/10" rear brakes	.030	10.000	10.060	85
	w/11" rear brakes	.030	11.000	11.060	85

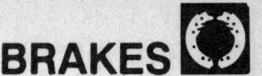
DRUM BRAKE SPECIFICATIONS

(Note: State and local inspection regulation will take precedence over manufacturer's minimum lining and drum specifications.)

Manufacturer	Vehicle Year, Make and Model	Brake Shoe Minimum Lining Thickness	Brake Drum Diameter Standard Size	Brake Drum Diameter Machine To	Wheel Lugs or Nuts Torque (ft. lbs.)
	'86–'87 Shadow, Sundance	①	7.835	7,935	95
Ford Motor Co.	'82–'86 Lincoln Continental				
	front	—	—	—	80–105
	rear	—	—	—	80–105
	'81–'86 Thunderbird, Cougar				
	w/9" rear brake	.030	9.000	9.060	80–105
	w/10" rear brake	.030	10.000	10.060	80–105
	'81–'83 Granada ('81–'82)				
	w/9" rear brake	.030	9.000	9.060	80–105
	w/10" rear brake	.030	10.000	10.060	80–105
	'81–'87 Front wheel drive,				
	w/7" rear brake	.030	7.000	7.060	80–105
	w/8" rear brake	.030	8.000	8.060	80–105
	'80–'87 Mustang, Capri, Fairmont ('80–'83), Zephyr ('80–'83)				
	w/9" rear brake	.030	9.000	9.060	80–105
	w/10" rear brake	.030	10.000	10.060	80–105
	'80–'87 Lincoln Town Car, Mark VI, Mark VII, LTD, Marquis				
	w/10" rear brakes	.030	10.000	10.060	80–105
	w/11" rear brakes	.030	11.030	11.090	80–105
	'80 Thunderbird, Cougar	.030	9.000	9.060	80–105
	'80 Granada, Monarch, Versailles				
	w/o rear disc brakes	.030	10.000	10.060	80–105
	w/rear disc brakes	—	—	—	80–105
	'86–'87 Sable, Taurus (Front Wheel Drive)				
	Sedan, rear	①	8.850	8.909	80–105
	Wagon, rear	①	9.840	9.899	80–105
General Motors Corp.—Buick	'82–'87 Century, Skyhawk	①	7.880	7.899	100
	'82–'87 Regal, LeSabre	①	9.500	9.560	80③
	'80–'85 Riviera				
	w/o rear disc brakes	①	9.500	9.560	100
	w/rear disc brakes	—	—	—	100
	'80–'84 Electra, Estate Wagon	①	11.000	11.060	100
	'80–'87 Skylark	①	7.880	7.899	103
	'80–'81 Century, Regal, LeSabre	①	9.500	9.560	80③
	'80 Skyhawk	①	9.500	9.560	80
General Motors Corp.—Cadillac	'82–'87 Cimarron	.030	7.880	7.899	100
	'82–'87 Fleetwood	.030	11.00	11.060	100
	'82–'87 Eldorado, Seville				
	front	—	—	—	100
	rear	—	—	—	100
	'80–'87 Fleetwood Limo, Commercial Chassis	①	12.00	12.060	100
	'80–'81 Fleetwood, Brougham, DeVille, Seville, (RWD)				
	w/o rear disc brakes	①	11.000	11.060	100
	'80–'81 Eldorado, Seville (FWD)				
	front disc	—	—	—	100
	rear disc	—	—	—	100
General Motors Corp.—Chevrolet	'82–'87 Celebrity, Cavalier	①	7.880	7.899	100
	'82–'87 Camaro				
	w/rear drum brakes	①	9.500	9.560	80†
	w/rear disc brakes	—	—	—	80†

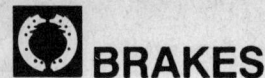

DRUM BRAKE SPECIFICATIONS

(Note: State and local inspection regulation will take precedence over manufacturer's minimum lining and drum specifications.)

Manufacturer	Vehicle Year, Make and Model	Brake Shoe Minimum Lining Thickness	Brake Drum Diameter		Wheel Lugs or Nuts Torque (ft. lbs.)
			Standard Size	Machine To	
	'86–'87 Sprint	.110	7.090	29–50	
	'86–'87 Spectrum	.039	7.090	65	
	'85–'87 Nova	.039	7.913	76	
	'82–'87 Malibu, '82–'83 Monte Carlo	①	9.500	9.560	80†
	'80–'85 Ciitation	①	7.880	7.899	103
	'80–'87 Impala, '80–'85 Caprice w/9 1/2" rear brakes	①	9.500	9.560	80
	w/11" rear brakes	①	11.000	11.060	100
	'80–'87 Chevette	①	7.874	7.899	70
	'80–'87 Corvette front	—	—	—	70④
	rear	—	—	—	70④
	'80–'81 Malibu, Camaro, Monte Carlo	①	9.500	9.560	80④
	'80 Monza	①	9.500	9.560	80④
General Motors Corp.—Oldsmobile	'82–'87 Ciera, Firenza	①	7.880	7.899	100
	'82–'87 Cutlass Supreme, 88	①	9.500	9.560	100⑤
	'80–'84 Omega	①	7.880	7.899	103
	'80–'87 Toronado w/o rear disc	①	9.500	9.560	100
	w/rear disc	—	—	—	100
	'80–'87 Custom Cruiser, 88 (w/403), 98 w/9.5" rear brake	①	9.500	9.560	100
	w/11" rear brake	①	11.000	11.060	100
	'80 Starfire	①	9.500	9.560	80
General Motors Corp.—Pontiac	'82–'87 A6000, J2000	①	7.880	7.899	100
	'82–'87 Firebird w/rear drum brakes	①	9.500	9.560	80④
	w/rear disc brakes	—	—	—	80④
	'81–'87 T 1000, 1000	①	7.874	7.899	70
	'80–'84 Phoenix (F.W.D.)	—	7.880	7.899	103
	'80–'84 Bonneville, Catalina ('80–'81) LeMans, Grand Prix, Grand Am w/9.5" rear brakes	①	9.500	9.560	
	w/11" rear brakes	①	11.000	11.060	80②
	'80–'81 Firebird, Ventura, Phoenix, RWD	①	9.500	9.560	80
	w/rear disc	—	—	—	80
	'80 Sunbird	①	9.500	9.560	80

① .030" over rivet head, if bonded lining use .062"
② w/1/2" stud 100 ft/lbs.
③ w/Aluminum whls. LeSabre 90 ft/lbs., Regal 100 ft/lbs
④ Aluminum whls; Corvette 80, Camaro 105, others 90.
⑤ 88 w/7/16" stud; 80 ft/lbs.

DRUM BRAKES

Brake Drums

BRAKE DRUM TYPES

The FULL-CAST drum has a cast iron web (back) of $^{3}/_{16}$–$^{1}/_{4}$ in. thickness (passenger vehicle sizes) whereas the COMPOSITE drum has a steel web approximately $^{1}/_{8}$ in. thick. These two types of drums, with few exceptions are not interchangeable.

BRAKE DRUM DEPTH

Rest a straight edge across the drum diameter on the open side. The actual drum depth then is the measurement at a right angle from the straight edge to that part of the web which mates against the hub mounting flange.

ALUMINUM DRUMS

When replaced by other types, aluminum drums must be replaced in pairs.

METALLIC BRAKES

Drums designed for use with standard brake linings should not be used with metallic brakes.

REMOVING TIGHT DRUMS

Difficulty removing a brake drum can be caused by shoes which are expanded beyond the drum's inner diameter or shoes which have cut into and ridged the drum. In either case back off the adjuster to obtain sufficient clearance for removal.

BRAKE DRUM INSPECTION

The condition of the brake drum surface is just as important as the surface to the brake lining. All drum surfaces should be clean, smooth, free from hard spots, heat checks, score marks and foreign matter imbedded in the drum surface. They should not be out of round, bell-mouthed or barrel shaped. It is recommended that all

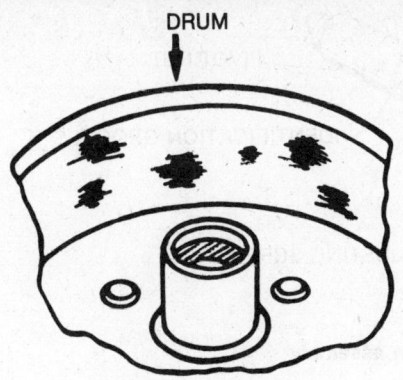

Hard or Chill Spots

LOOK HERE FOR TURNED DRUM TOOL MARK RIDGE

0.60"

Oversize drum

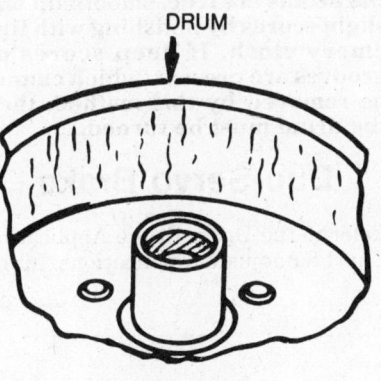

Heat checks

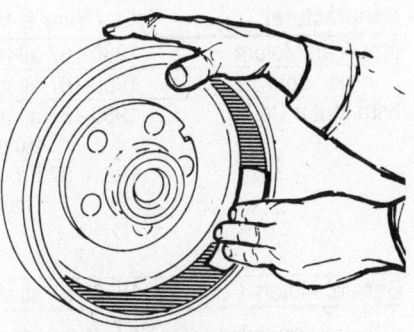

Sanding brake drums

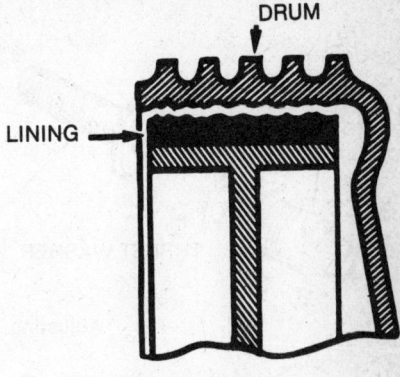

Scored drum surface

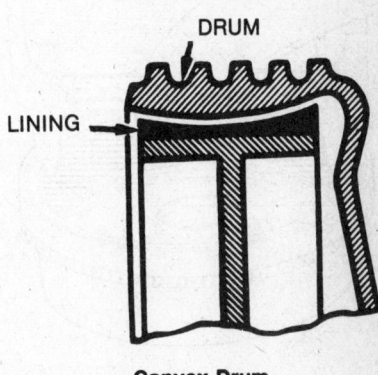

Convex Drum

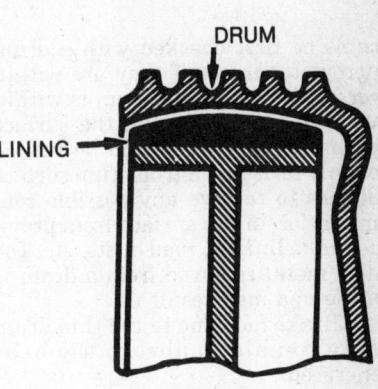

Concave Drum

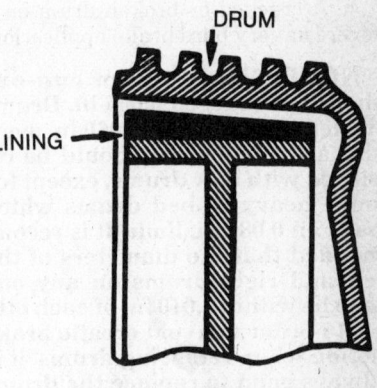

Bellmouth Drum

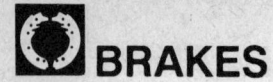

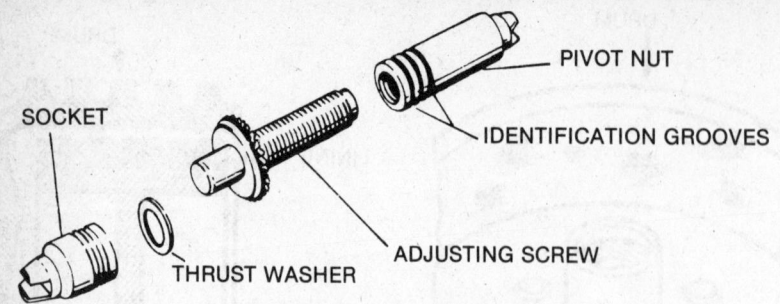

SOCKET

PIVOT NUT

IDENTIFICATION GROOVES

THRUST WASHER

ADJUSTING SCREW

Adjusting screw assembly

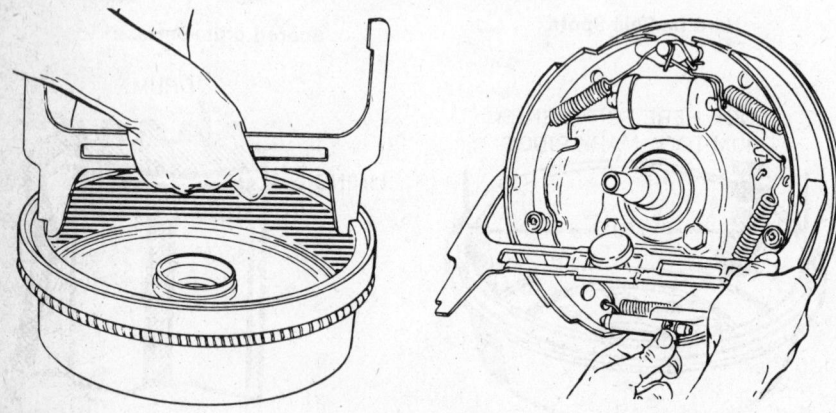

Brake drum guage

Duo-Servo design, the force which the wheel cylinder applies to the shoes is supplemented by the tendency of the shoes to wrap or twist into the drum during braking. Thus two braking forces are applied at each drum every time the brakes are activated.

ADJUSTMENT

The duo-servo brake, with star and screw type self-adjusters, is used on most late model American vehicles. The same basic brake unit has been used on all vehicles. General Motors vehicles use a rod-operated lever to turn the star-wheel, while all others use a cable-operated lever. This is the only difference, other than size, among units used on different models. The drum brakes, used on today's vehicles, are usually self-adjusting. They require manual adjustment only when the shoes have been replaced or when the star and screw adjuster has been disturbed.

NOTE: The drum brakes on most vehicles can be initially adjusted by removing the brake drum, measuring its internal diameter, then adjusting the shoes to that measurement and installing the drum. Use a vernier gauge to make the measurements. This method can be used on all models and may be preferable to punching out the covering over the access hole in the backing plate or brake drum edge.

1. Remove the access slot plug from the backing plate or front of drum. On some vehicles no access slot in the backing plate or in the front of the drums is provided. Some have been filled in and must be punched out to

drums be first checked with a drum micrometer to see if they are within oversize limits. If the drum is within safe limits, even though the surface appears smooth, it should be turned not only to assure a true drum surface but also to remove any possible contamination in the surface from previous brake linings, road dusts, etc. Too much metal removed from a drum is unsafe and may result in:

1. Brake fade due to the thin drum being unable to absorb the heat generated.

2. Poor and erratic brake due to distortion of drums.

3. Noise due to possible vibration caused by thin drums.

4. A cracked or broken drum on a severe or very hard brake application.

NOTE: Brake drum run-out should not exceed 0.005 in. Drums turned to more than 0.060 in. oversize are unsafe and should be replaced with new drums, except for some heavy ribbed drums which have an 0.080 in. limit. It is recommended that the diameters of the left and right drums on any one axle be within 0.010 in. of each other. In order to avoid erratic brake action when replacing drums, it is always good to replace the drums on both wheels at the same time. If

the drums are true, smooth up any slight scores by polishing with fine emery cloth. If deep scores or grooves are present, which cannot be removed by this method, then the drum must be turned.

Duo-Servo Brake

Refer to the Drum Brake Application Chart for adjuster applications. In the

DRUM BRAKE APPLICATION CHART

Manufacturer	Year & Model	Brake Type	Self-Adjuster Type
American Motors	1980–'87 all models	Duo-Servo	Star & Screw
Chrysler Corp.	1980–'87 all models ①	Duo-Servo	Star & Screw
Ford Motor Co.	1980–'87 all models except below	Duo-Servo	Star & Screw
	1981–'87 Front Wheel Drive (Ford)		Star & Screw (8 in. brake) Strut & Pin (7 in. brake)
		Non-Servo-	
General Motors Corp.	1980–'87 all models	Duo-Servo	Star & Screw

① The rear-drum brakes on Chrysler front wheel drive cars, through 1982 are not automatically adjusted.

gain access to the adjuster. Complete the adjustment and cover the hole with a plug to prevent entrance of dirt and water.

2. Using a brake adjusting spoon or screwdriver, pry downward on the end of the tool (starwheel teeth moving up) to tighten the brakes or upward on the end of the tool (starwheel teeth moving down) to loosen the brakes.

NOTE: It will be necessary to use a small rod or suitable tool to hold the adjusting lever away from the starwheel. Be careful not to bend the adjusting lever.

3. When the brakes are tight almost to the point of being locked, back Off the starwheel until the wheel is able to rotate freely. The starwheel on each set of brakes (front or rear) must be backed off the same number of turns to prevent brake pull from side-to-side.

4. After adjustment, check brake pedal travel and then make several stops, while backing the vehicle up, to equalize both wheel systems.

TESTING ADJUSTER

1. Raise and support the vehicle on jackstands. Have a helper handy to apply the brakes.
2. On models with access plugs in the backing plate, loosen the brakes by holding the adjuster lever away from the starwheel and backing off the starwheel approximately 30 notches. On models without access plugs in the backing plate, remove wheel and drum, loosen the adjuster, then reinstall the drum and wheel.
3. Spin the wheel and brake drum in reverse and apply the brakes. The movement of the secondary shoe should pull the adjuster lever up and when the brakes are released the lever should snap down and turn the starwheel.

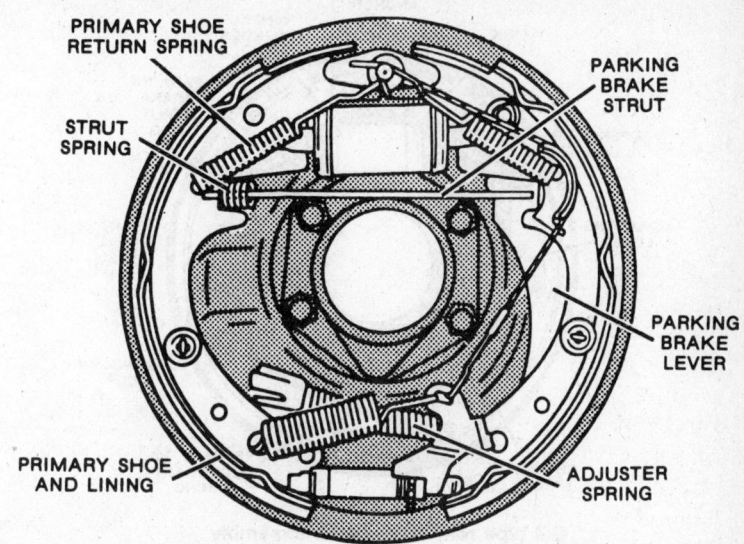

AMC, Ford type rear drum brake assembly

4. If the automatic adjuster doesn't work, the drum must be removed and the adjuster components inspected carefully for breakage, wear or improper installation.

BRAKE SHOE REMOVAL

NOTE: If you are not thoroughly familiar with the procedures involved in brake replacement, disassemble and assemble one side at a time, leaving the other wheel intact, as a reference.

1. Remove the brake drum.
2. Place the hollow end of a brake spring service tool on the brake shoe anchor pin and twist it to disengage one of the brake shoe return springs. Repeat this operation to remove the other return spring.

―――――― CAUTION ――――――
Be careful that the springs do not slip off the tool during removal, as the spring could break loose and cause personal injury.
―――――――――――――――――――――――

3. Reach behind the brake backing plate and place a finger on the end of one of the brake hold-down mounting pins. Using a pair of pliers or special brake pin retainer tool, grasp the washer on the top of the hold-down spring that corresponds to the pin that you are holding. Push down on the pliers and turn them 90° to align the slot in the washer with the head on the spring mounting pin. Remove the spring and washer, then repeat this operation on the hold-down spring of the other brake shoe.

4. Step 4 varies according to the manufacturer: On Ford and American Motors vehicles, place the tip of a screwdriver on the top of the brake adjusting screw and move the screwdriver upward to lift up on the brake adjusting lever. When there is enough slack in the automatic adjuster cable, disconnect the loop on the top of the cable from the anchor. Back off the adjusting screw while holding the adjustment lever away from the screw. Grasp the top of each brake shoe and move them outward to disengage

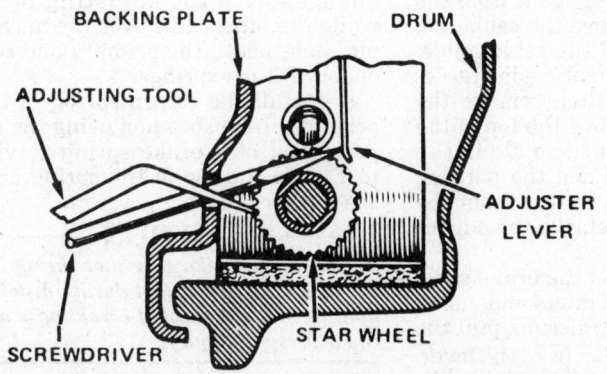

All cars except Chrysler and GM "H" body (Chevette)

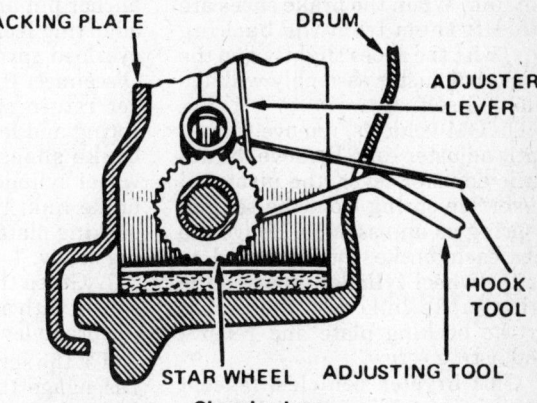

Chrysler type

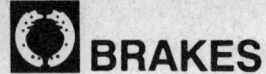

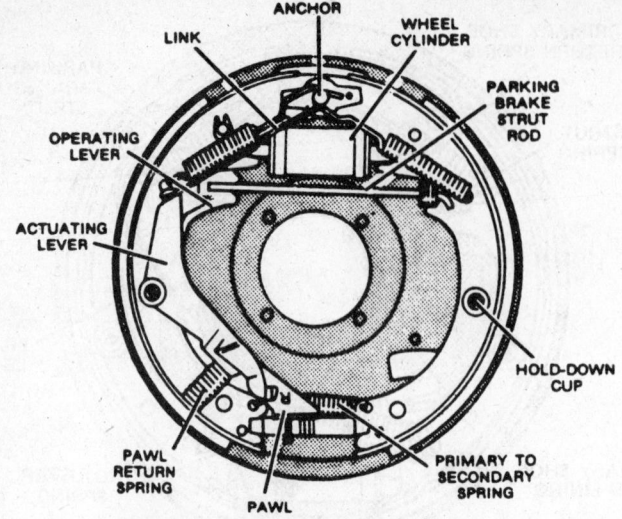

GM type rear drum brake assembly

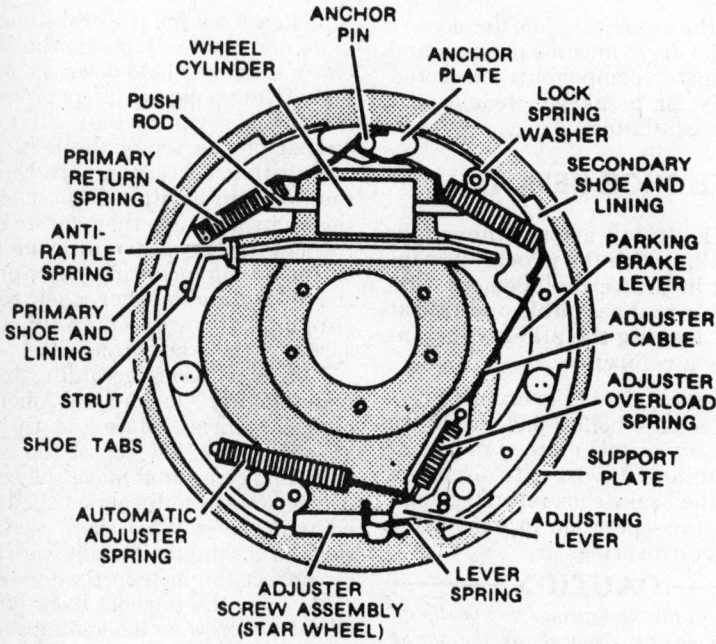

Chrysler type rear drum brake assembly

BRAKE SHOE INSTALLATION

1. The brake cable must be connected to the secondary brake shoe before the shoe is installed on the backing plate. To do this, transfer the parking brake lever from the old secondary shoe to the new one. This is accomplished by spreading the bottom of the horseshoe clip and disengaging the lever. Position the lever on the new secondary shoe, then install the spring washer and the horseshoe clip. Close the bottom of the clip after installing it. Grasp the metal tip of the parking brake cable with a pair of pliers. Position a pair of side cutters on the end of the cable coil spring and using the pliers as a fulcrum, pull the coil spring back with the side cutters. Position the cable in the parking brake lever.

2. Apply a light coating of high temperature grease to the brake shoe contact points on the backing plate. Position the primary brake shoe on the front of the backing plate, then install the hold-down spring and washer over the mounting pin. Install the secondary shoe on the rear of the backing plate.

3. If working on rear brakes, install the parking brake link between the primary brake shoe and the secondary brake shoe.

4. Step 4 varies according to the manufacturer.

5. On Ford and American Motors vehicles, install the automatic adjuster cable loop end on the anchor pin. Make sure that the crimped side of the loop faces the backing plate.

6. On GM vehicles, assemble the automatic adjuster lever, the pivot and the override spring, then install to the secondary springs as an assembly.

7. On Chrysler, (except some front wheel drive models) install the automatic adjuster lever and the return spring. Install the adjuster overload spring and cable. One end of the cable engages with the adjusting lever while the other slips over the anchor pin underneath the primary and secondary return springs.

8. Install the return spring in the primary brake shoe and using the tapered end of a brake spring service tool, slide the top of the spring onto the anchor pin.

CAUTION

Be careful to make sure that the spring does not slip off the tool during installation, as the spring could break loose and cause personal injury.

9. Install the automatic adjuster cable guide in the secondary brake

from the wheel cylinder and parking brake link. When the brake shoes are clear, lift them from the backing plate. Twist the shoes slightly and the automatic adjuster assembly will disassemble itself.

5. On GM vehicles, remove the automatic adjuster link. Remove the automatic adjuster lever, the pivot and the override spring from the secondary spring as an assembly. Move the top of each brake shoe outward to clear the wheel cylinder pins and the parking brake link. Lift the brakes from the backing plate and remove the adjusting screw.

6. On Chrysler vehicles, (except some front wheel drive models), slide

the automatic adjuster cable from the anchor pin and disengage it from the adjusting leer. Remove the cable, the overload spring and the cable guide. Disconnect the automatic adjuster lever return spring, then remove the spring and lever. Move the top of the brake shoes outward to clear the wheel cylinder pins and the parking brake link. Lift the brakes from the backing plate and remove the adjusting screw.

7. Grasp the end of the brake cable spring with a pair of pliers and, using the brake lever as a fulcrum, pull the end of the spring away from the lever. Disengage the cable from the brake lever.

shoe, making sure that the flared hole in the cable guide is inside the hole in the brake shoe. Fit the cable into the groove in the top of the cable guide.

10. Install the secondary shoe return return spring through the hole in the cable guide and the brake shoe. Using the brake spring tool, slide the top of the spring onto the anchor pin.

11. Clean the threads on the adjusting screw and apply a light coating of high-temperature grease to the threads. Screw the adjuster closed, then open it ½ turn.

12. Install the adjusting screw between the brake shoes with the star wheel nearest to the secondary shoe. Make sure that the star wheel is in a position that is accessible from the adjusting slot in the backing plate.

13. Install the short, hooked end of the automatic adjuster spring in the proper hole in the primary brake shoe.

14. Connect the hooked end of the automatic adjuster cable and the free end of the automatic adjuster spring in the slot in the top of the automatic adjuster lever.

15. Pull the automatic adjuster lever (the lever will pull the cable and spring with it) downward and to the left, and engage the pivot hook of the lever in the hole in the secondary brake shoe.

16. Check the entire brake assembly to make sure everything is installed properly. Make sure that the shoes engage the wheel cylinder properly and are flush on the anchor pin. Make sure that the automatic adjuster cable is flush on the anchor pin and in the slot on the block of cable guide. Make sure that the adjusting lever rests on the adjusting screw star wheel. Pull upward on the adjusting cable until the adjusting lever is free of the star wheel, then release the cable. The adjusting lever should snap back into place on the adjusting screw star wheel and turn the wheel one tooth.

17. Expand the brake adjusting screw until the brake drum will just fit over the brake shoes. Install the wheel and drum and adjust the brakes.

Non-Servo Brakes

On the non-servo brake system, each brake shoe is separately anchored and their action is not compounded. The leading shoe does the majority of the work, stopping forward motion. The trailing shoe works in the same manner for rearward motion. Non-servo brakes may or may not be equipped with self-adjusters.

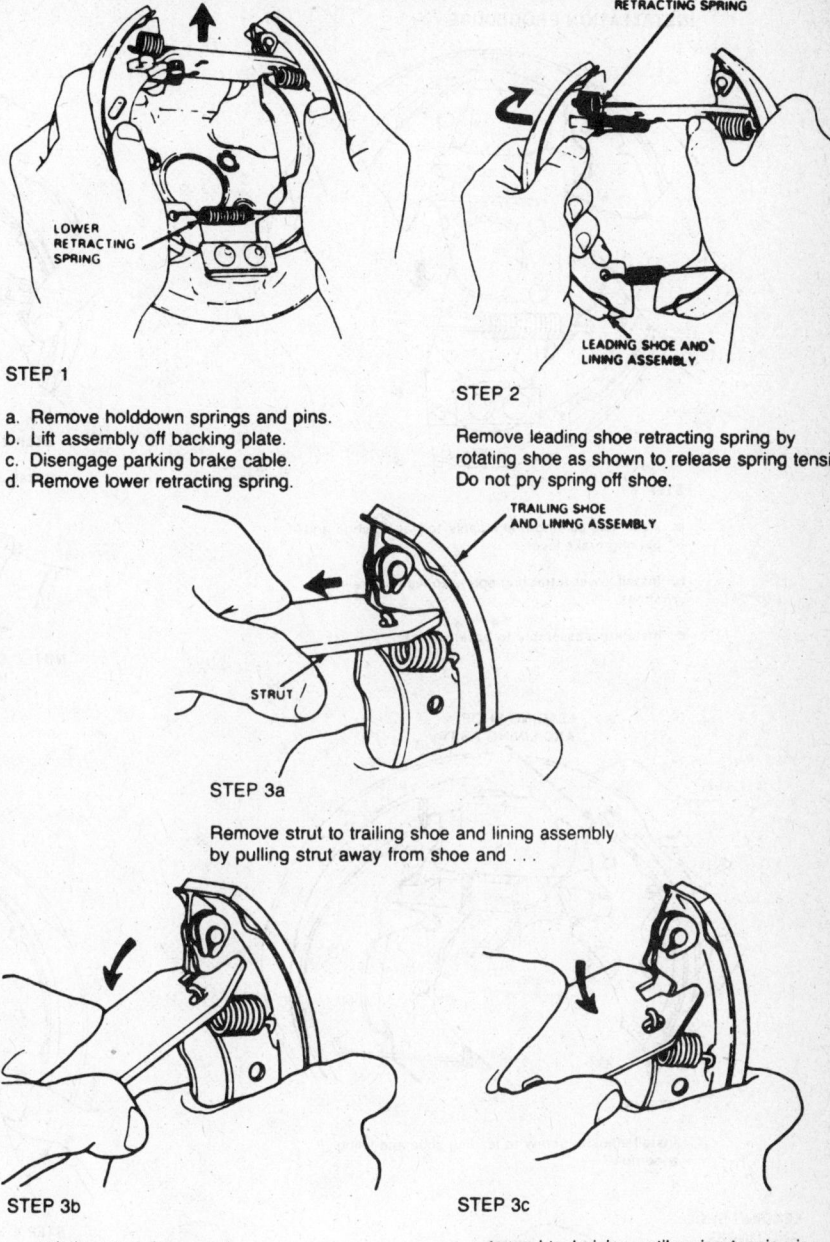

STEP 1

a. Remove holddown springs and pins.
b. Lift assembly off backing plate.
c. Disengage parking brake cable.
d. Remove lower retracting spring.

STEP 2

Remove leading shoe retracting spring by rotating shoe as shown to release spring tension. Do not pry spring off shoe.

STEP 3a

Remove strut to trailing shoe and lining assembly by pulling strut away from shoe and . . .

STEP 3b

. . . twisting strut downward

STEP 3c

. . . toward technician until spring tension is released. Remove spring from slots.

Non-servo; 7 in. Ford rear brakes

FORD NON-SERVO

The star and screw adjuster is used on models with 8 in. diameter brake drums while the strut and pin adjuster is used on 7 in. diameter drums. Normal shoe adjustments are automatic, however, when the shoes have been replaced or the adjuster has been disturbed, the shoes should be initially adjusted by hand.

1. Raise and support the rear of the vehicle on jackstands, then remove the wheels and drums. Drums are removed by releasing the parking brake, removing the dust cap, the cotter pin, the adjusting nut and the wheel bearing, then pulling off the drum.

2. On 7 inch drums with strut and pin adjuster, pivot the adjuster quadrant until it meshes with the knurled pin and is in the 3rd or 4th notch of the outboard end of the quadrant. Install the brake drum and wheel and adjust the wheel bearings by tightening the adjusting nut to 17–25 ft. lbs. while rotating the drum, then back off the adjusting nut about 100° and install the nut retainer and cotter pin.

3. The 8 inch drums are adjusted in the same manner as the star and

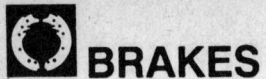
INSTALLATION PROCEDURE

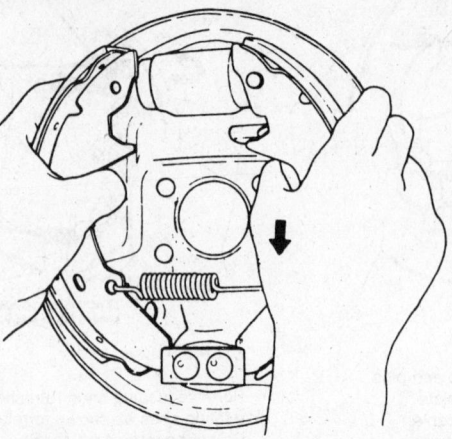

STEP 1

a. Assemble parking brake cable to trailing shoe and parking brake lever.

b. Install lower retracting spring to leading-trailing shoes.

c. Install this assembly to backing plate.

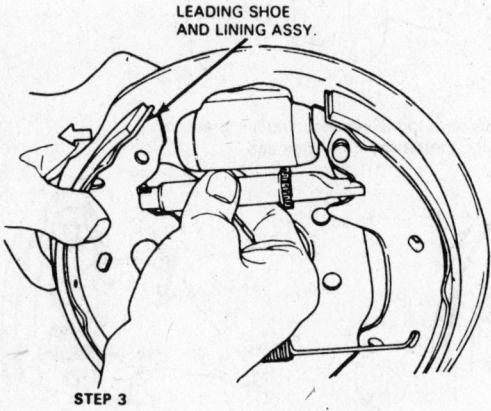

LEADING SHOE AND LINING ASSY.

STEP 3

Install adjuster screw to leading shoe and lining assembly.

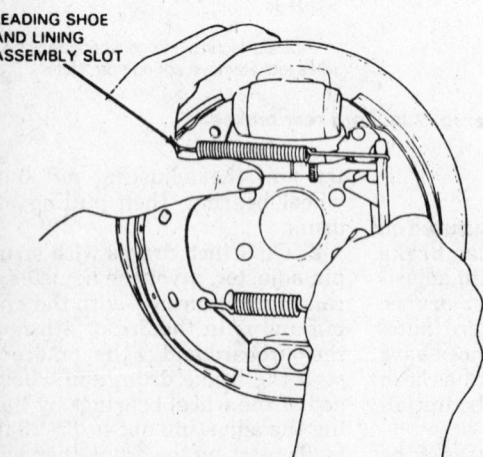

LEADING SHOE AND LINING ASSEMBLY SLOT

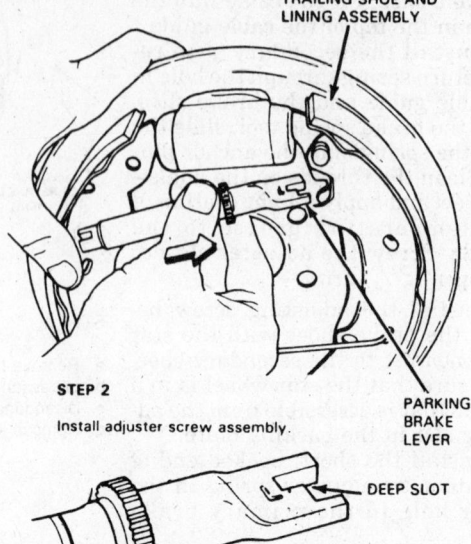

TRAILING SHOE AND LINING ASSEMBLY

STEP 2

Install adjuster screw assembly.

PARKING BRAKE LEVER

DEEP SLOT

NOTE: Socket Blade marked R and L. Install letter in upright position to insure proper slot engagement to parking brake lever.

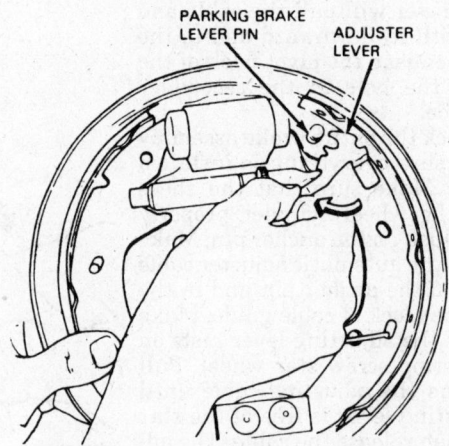

PARKING BRAKE LEVER PIN — ADJUSTER LEVER

STEP 4

Install the adjuster lever in groove of parking brake lever pin.

STEP 5

a. Install shoe holddown springs and pins.

b. Install upper retracting spring to leading shoe slot — stretch spring to install to trailing shoe. If adjuster lever does not contact star wheel after spring installation check adjuster socket installation.

Rear drum brakes

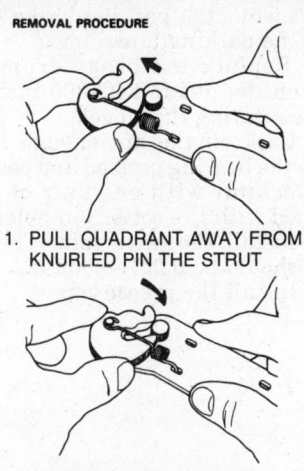

REMOVAL PROCEDURE

1. PULL QUADRANT AWAY FROM KNURLED PIN THE STRUT

2. ROTATE QUADRANT UNTIL TEETH ARE NO LONGER MESHED WITH PIN.

3. REMOVE THE SPRING AND SLIDE QUADRANT OUT OF STRUT—BE CAREFUL NOT TO OVERSTRESS SPRING.INSTALL ADJUSTER QUADRANT PIN INTO SLOT IN STRUT. TURN ASSEMBLY OVER AND INSTALL SPRING

INSTALLATION PROCEDURE

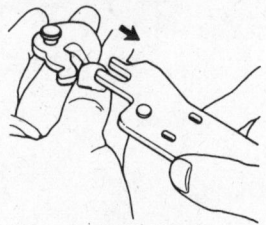

Quadrant removal and installation; 7 in. Ford non-servo brakes

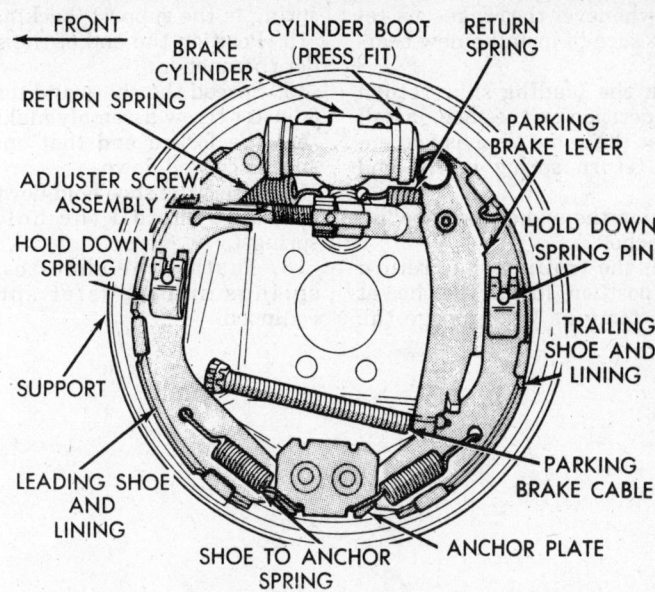

Chrysler non-servo rear brakes (non-self-adjusting)

screw adjuster drums described under "Duo-Servo" brakes, above. See that section for procedure.

4. Complete adjustment by applying the brakes several times.

BRAKE SHOE REMOVAL & INSTALLATION

1. Remove the wheel and hub. Adjusters can be backed off through the back of the brake backing plate with a screwdriver if the drum will not come off.

2. Remove the hold-down springs and pins. Lift the assembly off the brake backing plate and disengage the parking brake cable.

3. On 7 inch drums, remove the lower retracting spring. On 8 inch drums, remove all retracting springs and the adjuster lever.

4. The following removal procedures are for 7 inch drums only. Remove the leading shoe retracting spring by rotating shoe to release spring tension. DO NOT pry the spring off the shoe. Remove the strut to trailing shoe assembly by pulling the strut away from the shoe and twisting the shoe downward until spring tension is released. Remove the spring from the slots.

5. To install, reverse the removal procedures. See adjustment procedure, above, for special information on initial adjustment techniques. Wheel bearings on 8 inch drums are adjusted in the same manner as 7 inch drums. See Step 2 of Adjustment Procedure.

CHRYSLER NON-SERVO BRAKES

The rear brakes on the 1980–82 models are manually adjusted; the 1983 and later models are equipped with self-adjusters.

1. Remove the access slot plug from the upper part of the backing plate.

2. Using a thin brake adjusting spoon pry downward (left-side) or upward (right-side) on the end of the tool (starwheel teeth moving up) to tighten the brakes. The opposite applies to loosen the brakes.

3. When the brakes are tight almost to the point of being locked, back off on the starwheel 10 clicks. The starwheel on each side must be backed off the same number of turns to provide for even braking.

REMOVAL & INSTALLATION

1. Remove the brake drum.

2. Disconnect the parking brake cable from the secondary (trailing) shoe.

3. Remove the shoe-to-anchor retracting spring(s) and the upper spring (if equipped).

4. Remove the shoe holddown springs; compress them slightly and slide them off of the hold-down pins or push in and twist them from the mount pin.

5. Remove the adjuster screw assembly by spreading the shoes apart. Disconnect the adjuster spring from the trailing shoes on self-adjuster models. The adjuster nut must be fully backed off.

6. Raise the parking brake lever. Pull the secondary (trailing) shoe away from the backing plate so pull-back spring tension is released.

7. Remove the secondary (trailing) shoe and disengage the spring end from the backing plate.

8. Raise the primary (leading) shoe to release spring tension. Remove the shoe and disengage the spring end from the backing plate.

9. Inspect the brakes (see procedures under "Brake Drum Inspection).

10. Lubricate the six shoe contact areas on the brake backing plate and the web end of the brake shoe which contacts the anchor plate. Use a multi-purpose lubricant or a high temperature brake grease made for the purpose.

11. Chrysler recommends that the rear wheel bearings be cleaned and

repacked whenever the brakes are renewed. Be sure to install a new bearing seal.

12. With the leading shoe return spring in position on the shoe, install the shoe at the same time as you engage the return spring in the end support.

13. Position the end of the shoe under the anchor.

14. With the trailing shoe return spring in position, install the shoe at the same time as you engage the

spring in the support (backing plate).

15. Position the end of the shoe under the anchor.

16. Spread the shoes and install the adjuster screw assembly making sure that the forked end that enters the shoe is curved down.

17. Insert the shoe hold-down spring pins and install the hold-down springs.

18. Install the shoe-to-anchor springs and adjuster spring (if equipped).

19. Install the parking brake cable onto the parking brake lever.

20. Replace the brake drum and tighten the nut to 240–300 inch lbs. while rotating the wheel.

21. Back off the nut enough to release the bearing preload and position the locknut with one pair of slots aligned with the cotter pin hole.

22. Install the cotter pin. The end play should be 0.001–0.003 in.

23. Install the grease cap.

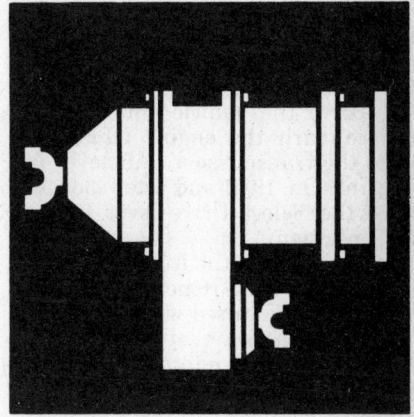

Transfer Cases

AMC Eagle

Two different models of transfer cases have been used in 4WD American Motors passenger cars. All 1980-81 vehicles use the model 119 transfer case; later vehicles use the model 129 unit with the Select Drive 4WD system. Pars usage and operation are basically the same for both units, though the 129 is equipped for use in the 2WD mode. Minor differences between the two transfer cases will be noted as necessary throughout the section.

NOTE: Refer to the AMC car section for services not listed here.

LUBRICATION

The lubricant capacities and types are as follows:

    1980 model 119 — 3 pints of 10W-30 engine oil

    1981 model 119 — 4 pints of 10W-30 engine oil

    1982 and later model 129 — 7 pints of Dextron® II

NOTE: Lubricant capacities for 1982 and later transfer cases were increased from 6 to 7 pints as per an AMC service bulletin dated 5-84.

— CAUTION —
DO NOT use any type of lubricant additive in AMC transfer cases, as their use could cause transfer case damage

TROUBLESHOOTING

Vehicle Wanders or Pulls From a Straight-Ahead Position

1. Check the tire pressures (tires cold). The pressures must be within specification and must not vary more than 1 psi between sides.

2. Check that the tires are all of the same size and type. Replace the tire(s) if necessary until all are matched properly.

3. Check that the lubricant in the transfer case is of the correct type. Drain and refill the transfer case with the correct lubricant, if necessary.

Severe Low Speed Shudder

1. Low level of viscous silicone fluid in the viscous coupling. First, perform the in-car Torque Bias Test as outlined later. Follow the instructions with that procedure to remedy the problem. Check the transfer case fluid; if it is contaminated with viscous fluid, the transfer case must be completely disassembled for inspection. While the unit is disassembled, check for a cracked viscous coupling and/or damaged seal. Replace damaged parts as required.

2. Check that the front and rear axles are of the same gear ratio. Mismatched differential ratios can cause failure of the viscous coupling. Replace one of the gear sets in order to match the other.

Noisy Operation

NOTE: Do not compare the noise of a 4WD vehicle to that of a 2WD vehicle, as 4WD is inherently noisier due to the use of additional drive line components.

1. Check that the lubricant in the transfer case is of the correct type and quantity. Drain and refill, or add fluid as required. Check for fluid leakage.

2. Check the tire pressures, sizes, and styles as mentioned previously. Adjust the tire pressures or replace the tires as required.

In-Vehicle Services

LUBRICANT CHANGE

1. Raise the vehicle and support it safely with jackstands.

2. Place a drain an beneath the transfer case.

3. Remove both the drain and fill plugs and allow the lubricant to drain completely.

4. Install the drain plug and tighten it to 18 ft. lbs.

5. Fill the transfer case with the correct type of lubricant (10W-30 engine oil — 119 unit: Dextron® II automatic transmission fluid — 129 unit). The lubricant level should be up to the lower edge of the fill plug hole. Install the fill plug.

6. Drive the vehicle 8-10 miles to circulate the fluid throughout the transfer case.

7. Remove the fill plug and top off transfer case to the bottom edge of the fill plughole.

8. Install the fill plug and tighten it to 18 ft. lbs. (1980–81), 25 ft. lbs. (1982 and later).

9. Remove the drain pan from beneath the vehicle.

10. Remove the jackstands and lower the vehicle.

TORQUE BIAS TEST

This test may be performed to determine the condition of the viscous coupling, which is the "heart" of the AMC transfer case. Note that if a malfunction in the coupling is observed, the coupling cannot be repaired in any way—it must be replaced if defective. The following procedure is an in-vehicle test. If the transfer case is to be disassembled, test the coupling as outlined within the overhaul procedure (bench test).

1. Drive the vehicle onto a level surface, turn the engine OFF, and place the transmission shift lever in Neutral. On 1982 and later models, place the Select Drive lever in the 4WD position.

2. Raise one of the front wheels off of the floor, then remove the wheel cover from the raised wheel.

3. Attach a socket (of the same size as the lug nuts) to a torque wrench to any one of the lug nuts of the raised wheel.

4. Rotate the wheel with the torque wrench, and note the amount of torque required to turn the wheel.

5. A reading of 45 ft. lbs. minimum should be obtained. If a reading of less than 45 ft. lbs. was obtained, the transfer case must be disassembled, and the bench test of the coupling should be performed. If the reading was 45 ft. lbs. or more, the coupling is operating properly.

6. Remove the torque wrench and socket, then install the wheel cover and lower the vehicle.

PARTS REPLACEMENT

The following parts of the transfer case may be serviced while the unit is installed in the vehicle:

    a. Front and rear yokes
    b. Yoke seals
    c. Rear bearing and retainer
    d. Speedometer gear

The combined procedure which follows covers the replacement of these items.

1. Raise the vehicle and support it safely with jackstands.

2. Remove the transfer case skid plate.

3. Matchmark the propeller shafts and their respective yokes so that these parts may be properly aligned during assembly.

4. Disconnect the propeller shafts from the yokes, then tie the shafts out of the way.

5. Remove the speedometer cable and adapter from the rear retainer.

NOTE: Discard the adapter seal. Use a NEW seal during assembly.

6. Support the engine and transmission using a small hydraulic jack with a block of wood placed between the jack and the transmission.

7. Remove the rear crossmember attaching nuts. Carefully lower the transfer case, just enough to gain access to the rear bearing retainer bolts.

8. Matchmark the retainer and the case so that the retainer may be properly reinstalled. Remove the rear yoke nut and seal washer.

9. Remove the rear yoke.

NOTE: If necessary, the front yoke may be removed in the same manner as the rear yoke.

10. Remove the rear retainer bolts. Tap the retainer lightly to loosen and remove it.

11. Remove the differential shim(s) and the speedometer gear from the rear output shaft.

12. If so equipped, remove the output bearing snap-ring. Remove the output bearing from the retainer.

13. Remove the yoke seal from the retainer. Clean the retainer thoroughly and remove the old sealant from the retainer and case mating surfaces.

Assemble the components in the following manner.

1. Install the output bearing into the retainer, making sure that the shielded side of the bearing faces the interior of the transfer case.

2. If so equipped, install the bearing retaining snap-ring.

3. Install a new yoke seal (or both seals, if removed), using Kent-Moore tool J-29162 or its equivalent.

4. Install the speedometer gear and differential shim(s).

5. Coat the mating surface of the retainer and the retainer bolt threads with sealer (Loctite 515 is recommended) and install the retainer. Be sure to align the matchmarks made during disassembly to locate the retainer.

6. Install and tighten the retainer bolts to 23 ft. lbs.

7. Install the yoke(s), new sealing washer(s), and new yoke nut(s). Torque the yoke nut(s) to 120 ft. lbs.

——— **CAUTION** ———
NEVER reuse the old yoke nuts.

8. Align the front propeller shaft and yoke matchmarks (which were made during disassembly) and attach the propeller shaft to the yoke. Torque the clamp strap bolts to 15 ft. lbs.

9. Install the transfer case drain plug and add the proper amount and type of lubricant to the transfer case.

10. Install the fill plug. Both of the plugs should be torqued to 18 ft. lbs.

11. Raise the transmission, transfer case, and rear crossmember enough to install the crossmember attaching nuts.

12. Install the crossmember attaching nuts and torque them to 30 ft. lbs.

13. Remove the hydraulic jack and wood block.

14. Align the matchmarks, then attach the rear propeller shaft to the yoke. Torque the clamp bolts to 15 ft. lbs.

15. Install a new O-ring seal on the speedometer adapter. Do not reuse the old seal. Install the adapter and cable into the rear retainer.

16. Install the transfer case skid plate.

17. Lower the vehicle.

TRANSFER CASE DISASSEMBLY

1. Remove the transfer case as outlined in the appropriate car section.

2. Remove the drain plug and drain the lubricant from the transfer case.

3. On 1982 and later models only, remove the nut and bolt which attaches the shift motor bracket to the transfer case. Remove the motor and bracket as an assembly.

4. Remove the nuts which attach the yokes. Discard the sealing washers. Remove the yokes.

5. Mount the transfer case on wood blocks which have V-notches cut into them to clear the front transfer case mounting studs.

6. Mark the relationship between the rear retainer and the case. Remove the retainer attaching bolts.

7. Pry the retainer off of the case using two screwdrivers placed into the two slots provided in the retainer for this purpose.

8. Remove the differential shim(s) and the speedometer gear from the rear output shaft.

9. Remove the front case-to-rear case bolts, then pry the cases apart with two screwdrivers.

——— **CAUTION** ———
Screwdriver slots are provided at each end of the rear case for this purpose.
DO NOT attempt to wedge the two halves apart.

10. Remove the thrust bearing and races from the front output shaft. Note their relationships so that these parts may be reinstalled properly.

11. Remove the oil pump from the rear output shaft, noting its position for reassembly.

12. Remove the rear output shaft from the viscous coupling.

13. Remove the pilot bearing rollers from the shaft coupling. Set the rollers aside in a group.

14. Remove the mainshaft O-ring from the end of the shaft.

15. Remove the viscous coupling from the mainshaft and side gear.

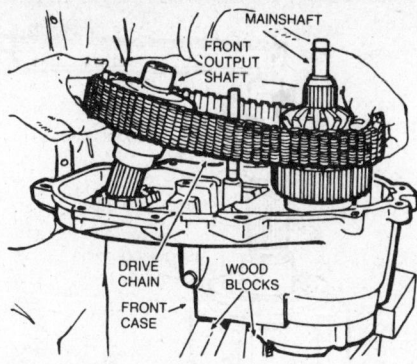

Removing the front output shaft, driven sprocket, and drive chain

16. Lift the front output shaft, sprocket, and chain upward, then tilt the front output shaft toward the mainshaft. Slide the chain off of the mainshaft drive sprocket and remove the assembly.

17. Remove the front thrust bearing assembly. The bearing will be positioned on either the front output shaft or the case.

18. Remove the drive chain from the front output shaft and sprocket.

19. Remove the driven sprocket-to-front output shaft snap-ring. Mark the sprocket and shaft so they may be reassembled properly, then remove the sprocket from the shaft.

20. Remove the mainshaft, side gear, clutch gear, drive sprocket and spline gear as an assembly. Set the assembly aside until disassembly of the front case has been completed.

21. Remove the range fork, rail and clutch sleeve as an assembly. Mark the sleeve and fork so that they be reassembled properly, then remove the sleeve from the fork.

22. On model 119 units, slide the rail out of the fork guide, On model 129 units, remove the pin to separate the fork and the rail, if necessary.

23. Inspect the rail, bracket, and fork for excessive wear, scoring, distortion, etc. Replace any part which is damaged.

24. Slide the rail through the range fork, and on 129 units, install the retaining pin. Set the assembly aside until transfer case assembly.

25. Remove the mainshaft thrust washer from the input gear, then remove the input gear, thrust bearing and race.

26. Remove the detent ball, spring and bolt.

27. Remove the retaining nut and washers from the range sector shaft. Tap the sector shaft with a plastic mallet to remove it from the case.

28. Remove the O-ring seal and seal retainer from the sector shaft bore in the case. The following Steps (29–35)

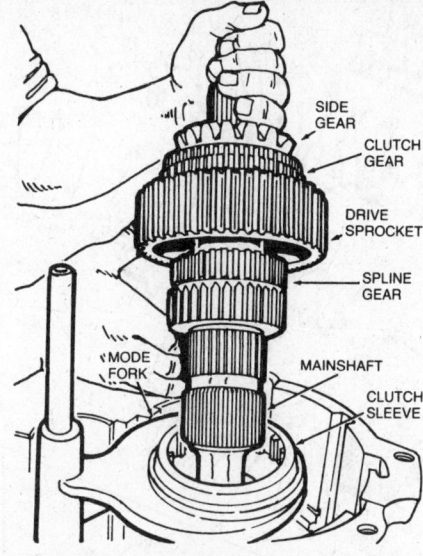

Removing the mainshaft and related components as an assembly

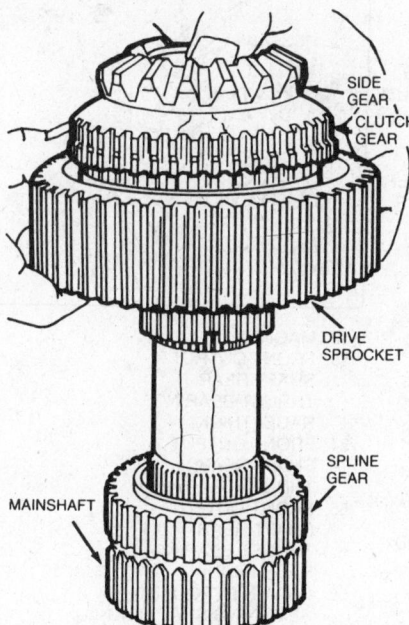

Removing the drive sprocket, clutch gear, side gear, and sprocket carrier

are used to disassemble the mainshaft and gears.

29. Pull the drive sprocket, clutch gear and side gear upward and off of the mainshaft.

30. Remove the needle bearings (82) and two bearing spacers from the mainshaft. Note the position of the spacers so that they may be reinstalled properly.

31. Remove the spline gear and thrust washer from the mainshaft.

32. Remove the side gear, clutch gear and thrust washer from the sprocket carrier and sprocket.

33. Remove the clutch gear and thrust washer from the side gear.

34. Remove the sprocket carrier snap-ring, then remove the drive sprocket from the carrier. Mark the sprocket and the carrier so that they may be reassembled in their proper relationship.

35. Remove the bearing spacers (3) and the sprocket carrier needle bearings (120) from the carrier.

— **CAUTION** —

DO NOT intermix the mainshaft needle bearings (Step 30) with the sprocket carrier needle bearings, as they are of different sizes.

36. Remove the rear output bearing and rear yoke seal from the rear retainer. Note that one side of the bearing is shielded—the bearing must be reinstalled in the same directions.

37. Remove the input gear and yoke seals from the front case.

CLEANING

All parts must be thoroughly washed with clean solvent. Be sure that all of the old lubricant and foreign matter has been removed from all transfer case parts. Verify that the oil feed ports and channels of both case halves are clear by flowing compressed air through them. Inspect all components according to the accompanying Transfer Case Inspection chart.

BEARING, BUSHING, AND SEAL REPLACEMENT

— **CAUTION** —

All bearings must be correctly positioned in the transfer case to avoid oil feed hole blockage. Always be sure that the feed holes are not blocked after any bearing has been replaced.

Front Output Shaft/Front Bearing

This bearing may be removed and installed with a bearing driver. Be sure that the driver contacts the bearings squarely, then check that the oil feed hole is not blocked after the bearing is in place.

Front Output Shaft/Rear Bearing

Replacement is performed in the same manner as the "Front Output Shaft - Front Bearing." Be sure that the bearing is seated flush with the edge of the bore in the case to allow room for the thrust bearing.

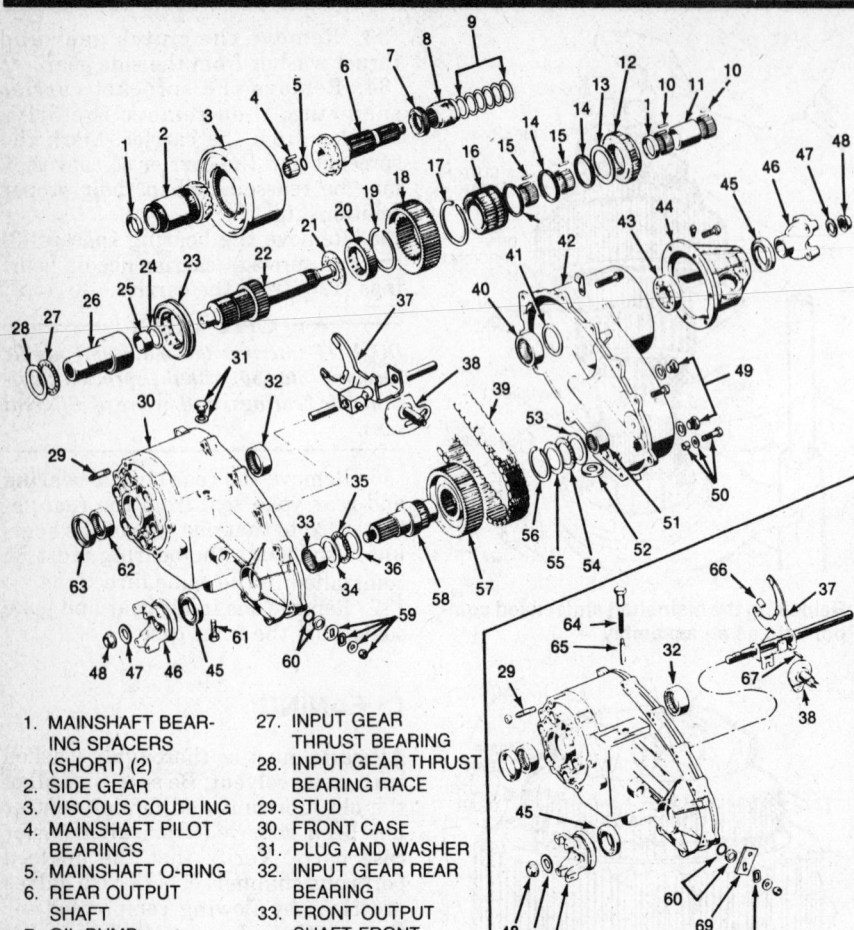

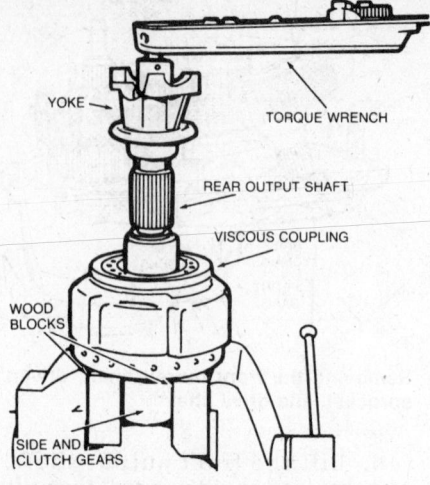

Bench-testing the torque bias of the viscous coupling

1. MAINSHAFT BEARING SPACERS (SHORT) (2)
2. SIDE GEAR
3. VISCOUS COUPLING
4. MAINSHAFT PILOT BEARINGS
5. MAINSHAFT O-RING
6. REAR OUTPUT SHAFT
7. OIL PUMP
8. SPEEDOMETER DRIVE GEAR
9. DIFFERENTIAL SHIMS
10. MAINSHAFT NEEDLE BEARINGS (82)
11. MAINSHAFT NEEDLE BEARING SPACER (LONG) (1)
12. CLUTCH GEAR
13. CLUTCH GEAR THRUST WASHER
14. SPROCKET CARRIER NEEDLE BEARING SPACER (3)
15. SPROCKET CARRIER NEEDLE BEARINGS (120)
16. SPROCKET CARRIER
17. SPROCKET CARRIER SNAP RING
18. DRIVE SPROCKET
19. SPROCKET CARRIER SNAP RING
20. SPLINE GEAR
21. MAINSHAFT THRUST WASHER
22. MAINSHAFT
23. CLUTCH SLEEVE
24. MAINSHAFT THRUST WASHER
25. MAINSHAFT BUSHING
26. INPUT GEAR
27. INPUT GEAR THRUST BEARING
28. INPUT GEAR THRUST BEARING RACE
29. STUD
30. FRONT CASE
31. PLUG AND WASHER
32. INPUT GEAR REAR BEARING
33. FRONT OUTPUT SHAFT FRONT BEARING
34. FRONT OUTPUT SHAFT FRONT THRUST BEARING RACE (THICK)
35. FRONT OUTPUT SHAFT FRONT THRUST BEARING
36. FRONT OUTPUT SHAFT FRONT THRUST BEARING RACE (THIN)
37. RANGE FORK AND RAIL
38. RANGE SECTOR
39. DRIVE CHAIN
40. REAR OUTPUT SHAFT BEARING
41. REAR OUTPUT SHAFT BEARING SEAL
42. REAR CASE
43. REAR OUTPUT BEARING
44. REAR RETAINER
45. YOKE SEAL
46. YOKE
47. SEAL WASHER
48. YOKE NUT
49. FILL AND DRAIN PLUGS
50. ALIGNMENT DOWEL, WASHER AND BOLT
51. FRONT OUTPUT SHAFT REAR BEARING
52. MAGNET
53. FRONT OUTPUT SHAFT REAR THRUST BEARING RACE (THICK)
54. FRONT OUTPUT SHAFT REAR THRUST BEARING
55. FRONT OUTPUT SHAFT REAR THRUST BEARING RACE (THIN)
56. DRIVEN SPROCKET RETAINING SNAP RING
57. DRIVEN SPROCKET
57. DRIVEN SPROCKET
58. FRONT OUTPUT SHAFT
59. RANGE SECTOR SHAFT RETAINING LOCKNUT AND WASHERS
60. RANGE SECTOR SHAFT SEAL AND RETAINER
61. POSITIVE LOCK DETENT BOLT
62. INPUT GEAR FRONT BEARING
63. INPUT GEAR SEAL
64. SPRING
65. PLUNGER
66. SHIFT FORK PAD
67. PIN
68. RANGE LEVER COLLAR
69. RANGE LEVER

Exploded view of the AMC 4WD transfer case. Note the minor differences in parts usage for the model 129 unit (shown in inset).

Input Gear Front and Rear Bearings

1. Drive both bearings out at the same time, using an appropriate bearing driver.

2. Drive the new bearings into place one at a time (rear bearing first).

3. After installation, check that the oil feed holes are not blocked, and that the bearings are flush with the case bore surface.

4. Carefully drive a new oil seal into place.

Rear Output Shaft Bearing Race

1. Pull the race from the transfer case bore using a slide hammer and an appropriate adapter.

2. Using a small screwdriver, carefully pry out the rear output lip seal.

3. Install a new output lip seal.

4. Carefully drive a new bearing race into place, using a bearing driver.

5. Remove the tool, then check to make sure that the oil feed hole is clear.

Mainshaft Pilot Bushing

1. Position the input gear on an opened vise (bushing facing downward). The vise must be opened enough for the bushing to be clear of the vise jaws when pulled downward.

2. Using a slide hammer-type bushing puller, remove the bushing.

3. Drive the new bushing into place, making sure that the oil feed hole is properly aligned.

Rear Retainer Bearing and Seal

1. Remove the bearing using a brass drift and a hammer. The seal is removed in the same manner.

TRANSFER CASE INSPECTION

Inspect the:	For the following conditions:
Gear teeth	1,2
Gear splines	1,2,3,4
Snap-rings and thrust washers	1,2,5
Case halves and rear retainer	5,6,7,8,9
Viscous coupling	10,11,12
Needle, roller, ball, and thrust bearings	1,2,6,13,14
Bearing bores	1,2,5,6

1. Excessive wear
2. Damage
3. Burrs
4. Nicks (minor scratches may be smoothed with an oil-stone)
5. Distortion
6. Cracks
7. Porosity
8. Damaged mating surfaces
9. Stripped bolt threads
10. Perform bench-test of torque-bias
11. Damaged pinions and/or carrier
12. Fluid leakage
13. Note that the front output shaft thrust bearing surfaces are heat-treated—a brown or blue discoloration should be considered NORMAL.
14. Rotational roughness

Removing the mainshaft pilot bushing

2. Drive the new bearing into place, making sure that the shielded side of the bearing faces the interior of the transfer case.

3. Carefully drive a new seal into the retainer.

BENCH-TESTING THE VISCOUS COUPLING

This torque bias test should be performed while the transfer case is disassembled for any reason, as the viscous coupling is the key to the operation of the AMC 4WD system. If a viscous coupling problem was previously diagnosed by the in-vehicle torque bias test, the coupling should again be tested according to the following procedure, for fault verification purposes.

1. Install the clutch gear onto the side gear.
2. Install the clutch gear/side gear assembly into the viscous coupling.
3. Mount the coupling and gear assembly in a vise, with wood blocks between the side gear and the vise jaws. Clamp the side gear firmly.
4. Make sure that the clutch gear is firmly engaged in the coupling, then install the rear output shaft in the viscous coupling.
5. Install the yoke on the rear output shaft and attach with the retaining nut.
6. Attach a socket (of the same size as the yoke nut) to a torque wrench. With the socket engaged to the yoke nut, rotate the rear output shaft and note the torque reading obtained with the torque wrench. The minimum acceptable rotational torque reading is 25 ft. lbs. If the reading is at or above 25 ft. lbs., the coupling is okay. If a lower torque reading is obtained, the coupling is defective and must be replaced.

7. Remove the yoke nut and yoke from the rear output shaft, then remove the coupling assembly from the vise.

TRANSFER CASE ASSEMBLY

NOTE: All parts should be lubricated prior to assembly, with either the specified lubricant (10W-30 engine oil for the model 119 unit; Dexron® II automatic transmission fluid for the 129), or petroleum jelly if stated within the procedure. DO NOT use any type of heavy grease (e.g. chassis lubricant) during assembly of the transfer case as lubricant of this nature can block the oil passages.

1. Install new yoke oil seals.
2. Install a new O-ring and retainer into the range sector shaft bore of the case.
3. Install the range sector. On model 119 units, install the washers and locknut on the sector shaft. On model 129 units, install the O-ring seal, retainer, range lever, washer, and locknut on the sector shaft.
4. Torque the sector shaft locknut to 17 ft. lbs.
5. Install the thrust bearing and race on the input gear. Install the gear into the front of the case.
6. Install the mainshaft thrust washer into the input gear.
7. Assemble the range fork, rail and clutch sleeve, then install the assembly into the case. Make sure that the rail is fully seated in the case bore.

NOTE: The rail bore of the front case must be perfectly dry. A small amount of oil in the bore will prevent proper sealing of the rail.

8. Install a thrust washer and a new O-ring on the mainshaft.
9. Coat the mainshaft needle bearing surface with petroleum jelly, then install the needle bearings and spacers in the following order.
   a. Install the short bearing spacer on the shaft.
   b. Install the first 41 needle bearings.
   c. Install the long bearing spacer.
   d. Install the remaining 41 needle bearings.
   e. Install the remaining short spacer.

NOTE: When installing the spacers, be careful not to disturb the needle bearings. If necessary, use additional petroleum jelly to hold the bearings in place.

10. Install the splined gear on the mainshaft, being careful not to disturb the bearings.

11. Install the sprocket carrier in the drive sprocket, being sure to align the carrier-to-sprocket reference marks made during disassembly.

NOTE: The tapered carrier teeth must be positioned on the same side as the deep recess of the drive sprocket.

12. Install the sprocket carrier snaprings.

13. Install the sprocket carrier needle bearings and spacers in the following manner.

    a. Coat both the sprocket carrier recess and the needle bearings with petroleum jelly.

    b. Install the center spacer.

    c. Install 60 needle bearings into each end of the sprocket carrier.

    d. Install the remaining two spacers, one at each side of the carrier. If necessary, use additional petroleum jelly to hold the needle bearings in place.

14. Install the sprocket carrier and drive sprocket assembly onto the mainshaft, being careful not to disturb the bearings. Note that the recessed side of the drive sprocket must face upward.

15. Position the clutch gear thrust washer on the thrust surface of the sprocket carrier.

16. Install the clutch gear on the gear side, with the tapered edge of the clutch gear facing the side gear teeth.

17. Install the side gear and clutch gear assembly onto the mainshaft, being careful not to disturb the bearings. The side gear must be fully seated in the sprocket carrier.

18. Install the mainshaft and gear assembly into the case, being sure that the mainshaft is fully seated in the input gear.

19. Install the driven sprocket on the front output shaft, being sure to align the sprocket-to-shaft reference marks which were made during disassembly. Install the sprocket snapring.

20. Install the thick front thrust bearing race into the case, followed by the bearing, then the thin race.

21. Install the drive chain on the driven sprocket.

22. Raise and tilt the driven sprocket and chain in order to attach the opposite end of the chain to the drive sprocket.

23. Align the front output shaft with the bore of the case, then install the shaft. Make sure that the front shaft thrust bearing assembly is fully seated in the case.

24. Install the thin race of the front output shaft rear thrust bearing, fol-

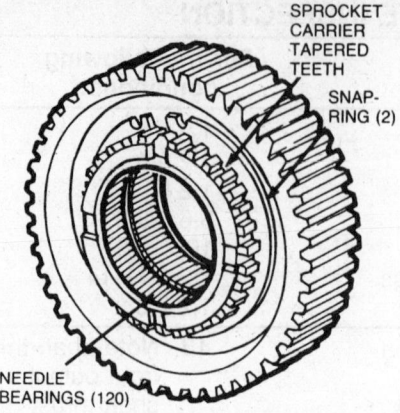

SPROCKET CARRIER TAPERED TEETH

SNAP-RING (2)

NEEDLE BEARINGS (120)

Sprocket carrier and drive sprocket assembly

lowed by the bearing, then the thick rear thrust bearing race.

25. Install the viscous coupling on the side gear and clutch gear. The coupling must be fully seated on the clutch gear, the clutch gear must be flush with the coupling, and the gear teeth should not be visible.

26. Coat the pilot bearing surface of the mainshaft and all of the pilot roller bearings with petroleum jelly. Install the pilot roller bearings on the shaft, using additional petroleum jelly to hold the bearings in place, if necessary.

27. Install the rear output shaft on the mainshaft, then into the viscous coupling. Be careful not to disturb the bearings during shaft installation. The shaft must be fully seated in the coupling; if necessary, tap the shaft with the plastic mallet to seat it.

28. Install the oil pump on the rear output shaft.

29. Install a new rear output shaft bearing oil seal.

30. Apply a bead of sealer (Loctite® 515 is recommended) to the mating surface of the rear case. If removed, reinstall the case magnet.

31. Install the front case to the rear case, being sure to align the dowels at the front case with the bolt holes of the rear case. Seat the rear case onto the front case.

NOTE: If the rear case will not seat completely into the front case, check for the following conditions:

    a. Oil present in the range fork rail bore.

    b. Rear thrust bearing assembly of the front output shaft is not aligned with the rear case.

    c. Mainshaft not completely seated.

    d. Rear case not aligned with the oil pump.

32. Install the rear case-to-front case bolts. Be sure to use flat washers on the bolts at the case ends where the alignment dowles are located. Torque the bolts to 23 ft. lbs.

33. Install the speedometer drive gear and the differential shims on the rear output shaft.

34. Align and temporarily install the rear retainer. Tighten, but do not "final-torque" the bolts.

35. Install the front and rear output shaft yokes. Install the original yoke nuts, then finger-tighten them only.

36. Mount a dial indicator on the rear retainer so that the indicator stylus contacts the rear yoke nut. The stylus must be in line with the rear output shaft.

37. Rotate the front output shaft 10–12 revolutions. Zero the dial indicator and rotate the front shaft one more full revolution. Note the dial indicator reading, which should be 0.002–0.010 in. If the end-play is okay, proceed to the next step. If the end-play is not within specifications, remove the rear retainer and add or subtract differential shims as required. Reinstall the rear retainer, yoke, and nut, the recheck the end play. Repeat if necessary until the end play is correct.

38. Remove the front and rear yokes, then discard the original yoke nuts.

39. Remove the rear retainer. Apply sealer (Loctite® 515 is recommended) to the retainer mating surface and all of the retainer bolt threads. Install the retainer and the bolts. Torque the retainer bolts to 23 ft. lbs.

40. Install the front and rear yokes, using new sealing washers and yoke nuts. Tighten the yoke nuts to 120 ft. lbs.

41. Install the detent ball, spring, and bolt. Apply sealer to the bolt threads and tighten the bolt to 23 ft. lbs.

42. Install the drain plug and washer.

43. Fill the transfer case with the proper types and amount of lubricant (see "Lubrication" at the beginning of this section).

44. Install the fill plug and washer, then tighten both the drain and fill plugs to 18 ft. lbs. (1980–81) 25 ft. lbs. (1982 and later).

45. If removed, install the plug and washer in the front case. Tighten to 18 ft. lbs.

46. On 1982 and later models, install the shift motor and bracket.

47. Reinstall the transfer case according to the procedure in the appropriate car section.

Drive Axles

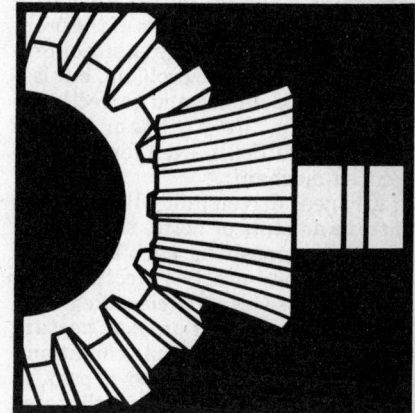

Front Wheel Drive

Front wheel drive cars do not have conventional rear axles or drive shafts. Instead, power is transmitted from the engine to a transaxle, or a combination of transmission and drive axle, in one unit. Both the transmission and drive axle accomplish the same function as their counterparts in a front-engine/rear-drive axle design. The difference is in the location of the components.

In place of a conventional driveshaft, a front-wheel-drive design uses two driveshafts, sometimes called halfshafts, which couple the drive axle portion of the transaxle to the wheels. Universal joints or constant velocity joints are used just as they would in a rear-wheel drive design. See the "U-Joint/CV-Joint" unit repair section.

Rear Wheel Drive

The rear axle must transmit power through 90°. To accomplish this, straight cut bevel gears or spiral bevel gears were used. This type of gear is satisfactory for differential side gears, but since the centerline of the gears must intersect, they rapidly became unsuited for ring and pinion gears. The lowering of the driveshaft brought about a variation of the bevel gear, which is called the hypoid gear. This type of gear does not require a metting of the gear centerlines and can therefore be underslung, relative to the centerline of the ring gear.

GEAR RATIOS

The drive axle of a vehicle is said to have a certain axle ratio. This number (usually a whole number and a decimal fraction) is actually a comparison of the number of gear teeth on the ring gear and the pinion gear. For example, a 4.11 rear means that theoretically, there are 4.11 teeth on the ring gear and one tooth on the pinion. Actually, on a 4.11 rear, there are 36 teeth on the ring gear and nine teeth on the pinion gear. By dividing the number of teeth on the ring gear, the numerical axle ration (4.11) is obtained. This also provides a good method of ascertaining exactly which axle ratio one is dealing with.

DIFFERENTIAL OPERATION

The differential is an arrangement of gears that permits the rear wheels to turn at different speeds when cornering and divides the torque between the axle shafts. The differential gears are mounted on a pinion shaft and the gears are free to rotate on this shaft. The pinion shaft is fitted in a bore in the differential case and is at right angles to the axle shafts.

Power flow through the differential is as follows. The drive pinion, which is turned by the driveshaft, turns the ring gear. The ring gear, which is bolted to the differential case, rotates the case. The differential pinion forces the pinion gears against the side gears. In cases where both wheels have equal traction, the pinion gears do not rotate on the pinion shaft, because the input force of the pinion gear is divided equally between the two side gears. Consequently the pinion gears revolve with the pinion shaft itself. The side gears, which are splined to the axle shafts, and meshed with the pinion gears, rotate the axle shafts.

When it becomes necessary to turn a corner, the differential becomes effective and allows the axle shafts to rotate at different speeds. As the inner wheel slows down, the side gear splined to the inner wheel axle shaft also slows down. The pinion gears act as balancing leavers by maintaining equal tooth loads to both gears while allowing unequal speeds of rotation at the axle shafts. If the vehicle speed remains constant, and the inner wheel slows down to 90 percent of vehicle speeds, the outer wheel will speed up to 110 percent.

LIMITED SLIP DIFFERENTIAL OPERATION

Limited-slip differentials provide driving force to the wheel with the best traction before the other wheel begins to spin. This accomplished through clutch plates or cones. The clutch plates or cones are located between the side gears and inner wall of the differential case. When they are squeezed together through spring tension and outward force from the side gears, there reactions occur. Resistance on the side gears causes more torque to be exerted on the clutch packs or clutch cones. Rapid one-wheel spin cannot occur, because the side gear is forced to turn at the same speed as the case. Most important, with the side gear and the differential case turning at the same speed, the other wheel is forced to rotate in the same direction and at the same speed as the differential case. Thus driving force is applied to the wheel with the better traction.

Differential Diagnosis

The most essential part of rear axle service is proper diagnosis of the problem. Bent or broken axle shafts or broken gears pose little problem, but isolating an axle noise and correctly interpreting the problem can be extremely difficult, even for an experienced mechanic.

Any gear driven unit will produce a certain amount of noise, therefore, a specific diagnosis for each individual unit is the best practice. Acceptable or normal noise can be classified as a slight noise heard only at certain speeds or under unusual conditions. This noise tends to reach a peak at 40–60 mph, depending on the road condition, load, gear ratio and tire size. Frequently, other noises are mistakenly diagnosed as coming from the rear axle. Vehicle noised from tires, transmission, driveshaft, U-joints and front and rear wheel bearings will often be mistaken as emanating from the rear axle. Raising the tire pressure to eliminate tire noise (although this will not silence mud or snow treads), listening for noise at varying speeds and road conditions and listening for noise at drive and coast conditions will aid in diagnosing alleged rear axle noises.

EXTERNAL NOISE ELIMINATION

It is advisable to make a through road test to determine whether the noise originates in the rear axle or whether it originates from the tires, engine transmission, wheel bearings or road surface. Noise originating from other places cannot be corrected by overhauling the rear axle.

Road Noise

Brick roads or rough surfaced concrete, may cause a noise which can be mistaken as coming from the rear axle. Driving on a different type of road (smooth asphalt or dirt) will determine whether the road is the cause of the noise. Road noise is usually the same on drive or coast conditions.

Tire Noise

Tire noise can be mistaken as rear axle noises, even though the tires on the front are at fault. Snow tread and mud tread tires or tires worn unevenly will frequently cause vibrations which seem to originate elsewhere; temporarily, and for test purposes only, inflate the tires to 40–50 lbs. This will significantly alter the noise produced by the tires, but will not alter noise from the rear axles. Noises

from the rear axle will normally cease at speeds below 30 mph on coast, while tire noise will continue at lower tone as car speed is decreased. The rear axle noise will usually change from drive conditions to coast conditions, while tire noise will not. Do not forget to lower the tire pressure to normal after the test is complete.

Engine and Transmission Noise

Engine and transmission noises also seem to originate in the rear axle. Road test the vehicle and determine at which speeds the noise is most pronounced. Stop the car in a quiet place to avoid interfering noises. With the transmission in neutral, run the engine slowly through the engine speeds corresponding to the car speed at which the noise was most noticeable. If a similar noise was produced with the car standing still, the noise is not in the rear axle, but somewhere in the engine or transmission.

Front Wheel Bearing Noise

Front wheel bearing noises, sometimes confused with rear axle noises, will not change when comparing drive and coast conditions. While holding the car speed steady, lightly apply the footbrake. This will often cause wheel bearing noise to lessen, as some of the weight is taken off the bearing. Front wheel bearings are easily checked by jacking up the wheels and spinning the wheels. Shaking the wheels will also determine if the wheel bearings are excessively loose.

Rear Axle Noises

If a logical test of the vehicle shows that the noise is not caused by external items, it can be assumed that the noise originates from the rear axle. The rear axle should be tested on a smooth level road to avoid road noise. It is not advisable to test the axle by jacking up the rear wheels and running the car.

True rear axle noises generally fall into two classes-gear noise and bearing noises, and can be caused by a faulty driveshaft, faulty wheel bearings, worn differential or pinion shaft bearings, U-joint misalignment, worn differential side gears and pinions, or mismatched, improperly adjusted, or scored ring and pinion gears.

Rear Wheel Bearing Noise

A rough rear wheel bearing causes a vibration or growl which will continue with the car coasting or in neutral. A brinelled rear wheel bearing will also cause a knock or click approximately every two revolutions of the

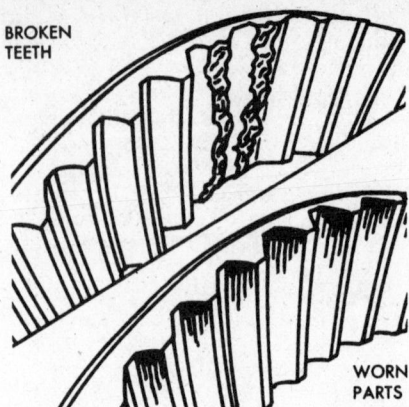

BROKEN TEETH

WORN PARTS

Two types of damage which cause gear noise

rear wheel, due to the fact that the bearing rollers do not travel at the same speed as the rear wheel and axle. Jack up the rear wheels and spin the wheel slowly, listening for signs of a rough or brinelled wheel bearing.

Differential Side Gear and Pinion Noise

Differential side gears and pinions seldom cause noise since their movement is relatively slight on straight ahead driving. Noise produced by these gears will be more noticeable on turns.

Pinion Bearing Noise

Pinion bearing failures can be distinguished by their speed of rotation, which is higher than side bearings or axle bearings. Rough or brinelled pinion bearings cause a continuous low pitch whirring or scraping noise beginning at low speeds.

Side Bearing Noise

Side bearings produce a constant rough noise, which is slower than the pinion bearing noise. Side bearing noise may also fluctuate in the above rear wheel bearing test.

Gear Noise

Two basic types of gear noise exist. First is the type produced by bent or broken gear teeth which have been forcibly damaged. The noise from this type of damage is audible over the entire speed range. Scoring or damage to the hypoid gear teeth generally results from insufficient lubricant, improper lubricant, improper breakin, insufficient gear backlash, improper ring and pinion gear alignment or loss of torque on the drive pinion nut. If not corrected, the scoring will lead to eventual erosion or fracture of the gear teeth. Hypoid gear tooth fracture can also be caused by extended overloading of the gear set (fatigue fracture) or by shock overloading (sudden failure). Differential and side gears rarely give trouble, but common causes of differential failure are shock loading, extended overloading and differential pinion seizure at the cross-shaft resulting from excessive wheel spin and consequent lubricant breakdown.

The second type of gear noise pertains to the mesh pattern between the ring and pinion gears. This type of abnormal gear noise can be recognized as a cycling pitch or whine audible in either drive, float or coast conditions. Gear noises can be recognized as they tend to peak out in a narrow speed range and remain constant in pitch, whereas bearing noises tend to vary in pitch with vehicle speeds. Noises produced by the ring and pinion gears will generally follow the pattern below

A. Drive Noise: Produced under vehicle acceleration.

B. Coast Noise: Produced while the car coasts with a closed throttle.

C. Float Noise: Occurs while maintaining constant car speed (just enough to keep constant) on a level road.

D. Drive, Coast and Float Noise: These noises will vary in tone with speed and be very rough or irregular if the differential or pinion shaft bearings are worn.

Bearing Diagnosis

This section will help in the diagnosis of bearing failure and the causes. Bearing diagnosis can be very helpful in determining the cause of rear axle failure.

When disassembling a rear axle, the general condition of all bearings should be noted and classified where possible. Proper recognition of the cause will help in correcting the problem and avoiding a repetition of the failure. Some of the common causes of bearing failure are:

a. Abuse during assembly or disassembly.

b. Improper assembly methods.

c. Improper or inadequate lubrication.

d. Bearing contact with dirt or water.

e. Wear caused by dirt or metal chips.

f. Corrosion or rust.

g. Seizing due to overloading.

h. Overheating.

i. Wear of the bearing seats.

j. Brinelling from impact or shock loading.

k. Manufacturing defects.

l. Pitting due to fatigue.

To avoid damage to the bearing from improper handling, it is best to treat a used bearing the same as a new bearing. Always work in a clean area with clean tools. Remove all outside dirt from the housing before exposing a bearing and clean all bearing seats before installing a bearing.

— CAUTION —

Never spin a bearing, either by hand or with compressed air, as this will lead to almost certain bearing failure.

Limited-Slip Differential Diagnosis

LUBRICATION

The use of proper lubricant is very important in limited-slip type drive axles. The forces applied when cornering tend to apply the clutch pack or clutch cones. The use of the wrong lubricant can cause the clutch surfaces to grab and chatter while turning. Always follow the manufacturer's recommendations regarding drive axle lubrication. When chatter is encountered, the differential lubricant should be drained and refilled with the specified lubricant.

TESTING

The clutch operation on all limited-slip type axles can be tested as follows. Refer to the manufacturer in question.

American Motors "Twin-Grip"

1. With the engine off and the transmission in neutral, jack up one rear wheel.

2. Block the other wheel to prevent it from moving.

3. With a socket and torque wrench on the axle shaft nut, turn the raised wheel forward.

4. The torque required to move the wheel should be 70–100 ft. lbs. for $8\frac{7}{8}$ in. axles or 80–120 ft. lbs. for $7\frac{9}{16}$ in. axles.

5. A breakaway torque which is less than the specified figure, indicates a need for repair or replacement.

Cadillac Controlled Differential

This unit should not be serviced. If a malfunction exists that cannot be cured by changing the fluid, remove the unit and install a new one.

Chrysler Corp Sure-Grip

1. Place the vehicle on a hoist with the engine off and the automatic transmission in Park (manual transmission in low gear).

2. Attempt to rotate the wheel the Sure-Grip differential can be assumed to be performing satisfactorily.

4. If it is relatively easy to continuously turn either rear wheel, the unit should be removed and replaced.

— CAUTION —

The Sure-Grip differential is serviced as a unit only. Under no circumstances should the unit be disassemnbled and reinstalled.

Ford Motor Company Equa-Lok

1. Jack up one rear wheel and remove the wheel cover.

2. Block the other wheel front and rear to prevent the car from moving.

3. Using a 200 ft. lbs. capacity torque wrench on one of the wheel lug nuts, measure the torque required to continuously rotate the wheel. The breakaway torque reading can be disregarded. The minimum torque to continuously rotate the wheel should be as follows.

a. All axles except integral carrier type: 75 ft. lbs.

b. Integral carrier type axles: 50 ft. lbs.

4. If the minimum torque is not as specified, the differential should be checked for improper assembly.

Ford Motor Company Traction-Lok

Follow the procedure for the Ford Motor Company Equa-Lok rear. The minimum torque to continuously rotate the wheel (disregarding the breakaway torque) should be at least 30 ft. lbs.

General Motors Corp. (Except Cadillac) Positraction

1. Place the transmission in neutral.

2. Raise one rear wheel off the floor and block the other rear wheel (front and rear) to prevent the car from moving.

3. Install a torque wrench and extension on the lug nut and note the torque required to continuously rotate one rear wheel. Disregard the breakaway torque figure, as this may be a great deal higher.

4. The minimum torque to continuously rotate the rear wheel should be at least 35 ft. lbs. If it is not, the rear axle is in need of service.

General Diagnosis

Improper operation of a limited-slip type rear axle is generally indicated by clutch slippage or grabbing, which will sometimes produce a whirring or chatter sound. Occasionally, this condition is induced by improper lubrication. Check the unit for the wrong type of lubricant or lubricant which has broken down or become contaminated. Replace the lubricant with the type specified by the manufacturer.

During normal operation, i.e., straight-ahead driving, both wheels are rotating at equal speeds, and the driving force is distributed equally between both wheels. When cornering, the inside wheel delivers extra driving force, causing slippage in both clutch packs. Therefore, if the wheel rotation of both rear wheels is not equal, the unit will constantly be functioning as if the car were cornering. This will cause constant slippage and lead to eventual failure of the unit. It is important that there be no excessive differences in wheel and tire size, wear pattern, or tire pressures between both rear wheels. Swerving on acceleration is an indication of one or more of the above conditions. Before attempting an overhaul or replacement operation, check both rear wheels for identical tire sizes, tire pressure, tire tread depth, and wear pattern.

DRIVE AXLE DISASSEMBLY ANALYSIS

Testing the Gear Tooth Contact Pattern

Once it has been established that the differential is indeed in need of service, the worst procedure is to simply plunge ahead and remove the differential and disassemble the parts. Prior to disassembly, a tooth contact pattern test should be made. However, it is worthwhile to first know the nomenclature associated with hypoid gear teeth.

The thick end of the tooth is called the heel and the thin end of the tooth is called the toe. The base half of the tooth is called the flank and the other end of the tooth is known as the face. The imaginary line at the halfway point between the face and flank is known as the pitch line. The space between the meshed pinion and ring gear tooth is known as backlash.

A gear tooth contact pattern can be made with the carrier in or out of the housing depending on the type of carrier. On integral carrier models, the lubricant must be drained and the rear cover removed. The ring gear will now be exposed and the test can be made with the carrier still in the housing. On removable carrier models, drain the lubricant and remove the carrier from the housing. The test can be made on the bench.

Unlike simple spur gears, hypoid

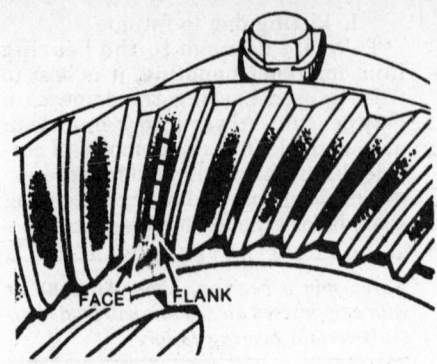

Gear tooth face and flank showing oval gear tooth contact pattern

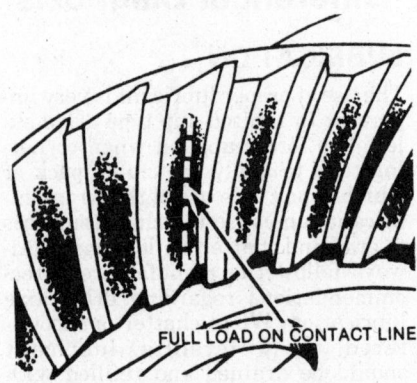

Gear tooth contact pattern showing load centered on gear tooth

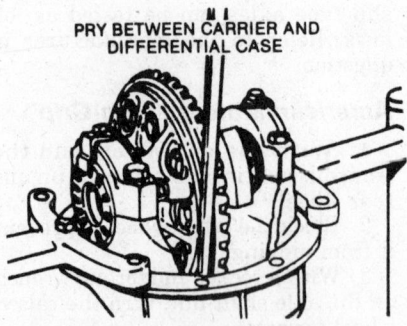

Applying a load to the differential case

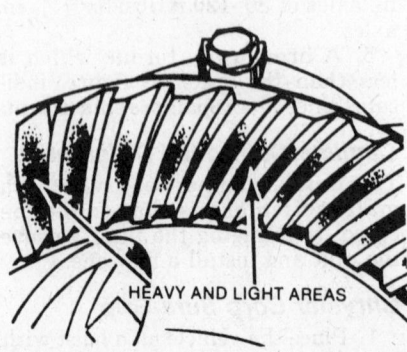

Excessive run-out will cause an uneven pattern

gear teeth leave a complex pattern on the ring gear. When hypoid gears turn, the line contact between pinion and ring gear teeth has the same wiping motion as with spur gear teeth. Because of the complicated movement of hypoid gear teeth, the contact area takes an oval shape as opposed to the rectangular shape lefty by spur gear teeth. Actually, the tooth contact test shows where each gear tooth has been wiped by the movement of the contact line, so that you can tell whether the gears are set correctly. With a properly adjusted ring and pinion (with properly adjusted ring and pinion depth and backlash) the tooth contact will be close to center. In this case, the load is borne by the strongest part of the tooth. If the gear setting is off, the contact line may reach any part of the edge of a tooth, and the metal will be overloaded at that point. When overload occurs, rapid deterioration of the gears will follow.

PREPARING THE TEST

Coat the drive gear teeth with a metallic base artist oil color such as zinc white or titanium white. The tooth coating material must be smooth and firm enough to spread without running. A consistency somewhat like toothpaste works well. If it is necessary to thicken the material, add a small amount of cup grease.

NOTE: Prussian blue dye does not work well, since the blue tends to smear the pattern.

Throughly clean the ring gear and pinion before applying the testing material. Any gear lube left on the teeth will make the pattern quite unreadable. Coat the drive and coast sides of all the ring gear teeth, but leave the pinion gear teeth clean. Do not apply the coating too thickly as the pattern will be smeared.

Because the axle gears are normally easy to rotate, turning resistance must be applied to produce pressure between the pinion and ring gear teeth to make a legible pattern. On a removable carrier type axle, insert a large screwdriver between the carrier housing and the differential case rim. Apply the load squarely against the case rim while prying out against the upper or lower section of the carrier housing. On integral carrier models, apply the parking brake to a point where it requires approximately 50 ft. lbs. to turn the pinion with a torque wrench. Since the shape and position of the contact pattern will vary, depending on the load, try to use the same load for each test or the results can be misleading. This is especially true when testing after an overhaul.

Once the gears have a load applied, obtain a tooth contact pattern by rotating the ring gear and pinion one complete turn in each direction. This will produce a constant pattern on the coast and drive side of each tooth. Do not rotate the ring gear more than one revolution in each direction as this will tend to obscure the pattern.

NOTE: If the pattern does not look right on the first try, try again.

INTERPRETING GEAR TOOTH CONTACT PATTERNS

The tooth contact pattern should be the same on every tooth. If the pattern shows heavy and light areas on different teeth, check the ring gear and differential case for excessive run-out.

NOTE: Run-out can be cured in many cases by removing the ring gear from the case, rotating it 90° or 180°, and remounting it.

Since you can only apply test load pressure to the gears, the contact pattern will be less distinct toward the tooth ends. But, when the ring gear and pinion are under operating loads in the vehicle, the tooth contact area spreads out, especially towards the heel end of the tooth. For this reason, do not try to "get by" with a tooth contact pattern that is centered, but favors the heel end of the teeth. This will only lead to overloading at the heel ends of the gear teeth. On the other hand, a contact pattern which is reasonably centered, but favors the toe end of the teeth, is acceptable.

Assuming that the tooth contact pattern is even on all teeth, the main problems is to get the most distinct part of of the pattern centered on both sides of each tooth. In some cases, the pattern will be centered on the drive side and off the center on the coast side, or vice versa. The off center pattern can be moved to a more acceptable position by slightly altering the backlash. This procedure will not seriously affect the other pattern. More often, however, the pattern will be off center on both sides of the teeth. The basic cause of this condition is an improperly adjusted pinion.

ADJUSTING PINION DEPTH

It is necessary to understand that an incorrect pinion depth setting moves the contact pattern away from the center on both sides of the tooth in opposite directions. This means that

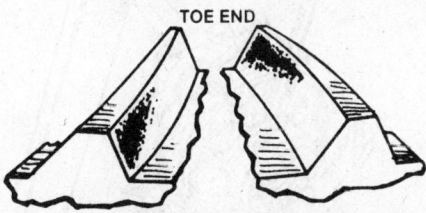

THICKER SPACER NEEDED

TOE END

HEEL END-DRIVE SIDE (CONVEX)   HEEL END-COAST SIDE (CONCAVE)

Tooth contact patterns high on the tooth side

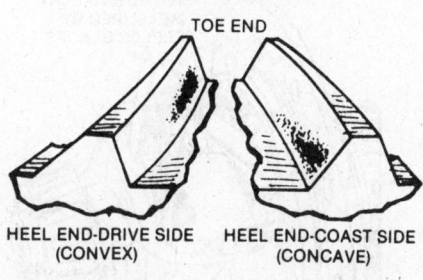

THINNER SPACER NEEDED

TOE END

HEEL END-DRIVE SIDE (CONVEX)   HEEL END-COAST SIDE (CONCAVE)

Gear contact pattern low on tooth side

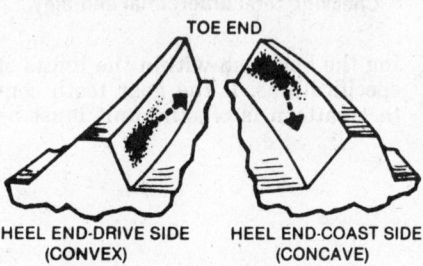

PATTERN MOVES TOWARD CENTER AND DOWN

TOE END

HEEL END-DRIVE SIDE (CONVEX)   HEEL END-COAST SIDE (CONCAVE)

A thicker spacer moves the pattern in and down

PATTERN MOVES INWARD AND UP

TOE END

HEEL END-DRIVE SIDE (CONVEX)   HEEL END-COAST SIDE (CONCAVE)

A thinner spacer will move the pattern up and inward

when you install a thicker or thinner washer under the pinion head you bring the pattern into the center of the tooth from opposite ends.

When the contact pattern is high on the heel end of the drive side and low on the toe end of the coast side, a thicker washer is needed to bring the

pinion in, toward the center of the drive gear. Increasing the thickness of the spacer washer will bring the pattern in, toward the center of the drive gear teeth, and also will move the pattern down from the tooth face. However, this movement is less than the in-or-out movement.

When tooth contact is low on the toe end of the drive side and high on the heel end of the coast side, the pinion must be moved out, by installing a thinner washer under the pinion head. This will move the pattern inward toward the center, and will also result in slight movement of the pattern up from the tooth flank.

A factory service facility will use special tools and gauge blocks to determine the thickness of the spacer under the pinion head. In the absence of such specialized equipment, the following procedure may be used. Bear in mind that with the "hit-or-miss" method, each time you are wrong with the pinion depth, the unit must be disassembled, the spacer thickness changed, and the unit must be completely set up again.

Gather a handful of spacers to cover any thickness and several collapsible

One example of pinion markings

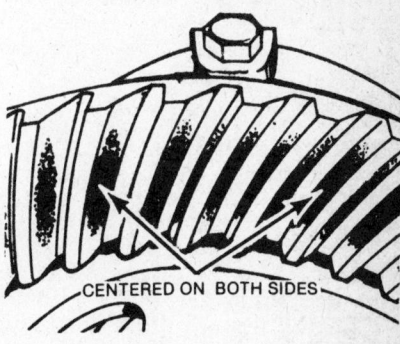

CENTERED ON BOTH SIDES

Gear tooth contact pattern showing load centered on gear tooth

pinion spacers (if the unit uses them). Assembly the unit. If the original gear set is being reused, and the tooth contact pattern is reasonably correct, install a new spacer of the same thickness as the old one. This will provide a reasonable starting point. If the gear contact test indicates a need for movement of the pinion, use a new spacer 0.001–0.002 in. thicker or thinner, depending on the direction the pinion must go. If a new gear set is being used, the thickness of the spacer will have to be determined in the following manner. Compare the markings on the old and new pinion. It will usually be marked with a number preceded by a plus (+) or minus (–) sign. This number indicates the production deviation from the nominal pinion, which are known as "zero pinions." In service, zero pinions are rare. Assume that the old pinion is marked with a plus two (+2). Assume that the new pinion is marked with a +3. By comparing the pinion markings, find the numerical difference between the two pinions, in this case +1. With a micrometer, measure the thickness of the original spacer. We will assume that the old spacer is 0.030 in. thick. If the numerical difference is a negative number (say, –1) then the spacer should be increased by 0.001 in., to 0.031 in. total. This will only provide a reasonable beginning point.

It is rare that this method works out the first time. Assemble the pinion, differential, and ring gear with the spacer of calculated thickness. The side bearing preload, backlash, pinion nut torque, and pinion rotating torque must all be set correctly. Obtain a gear tooth pattern on the ring gear teeth and analyze the results. Small deviations from the acceptable pattern can usually be made by vary-

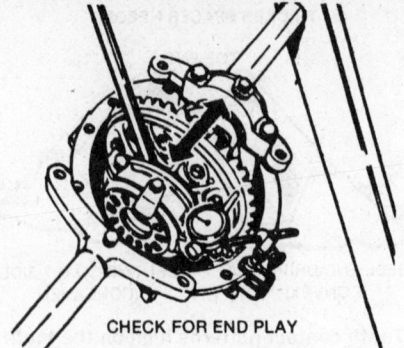

CHECK FOR END PLAY

Checking differential bearing end-play

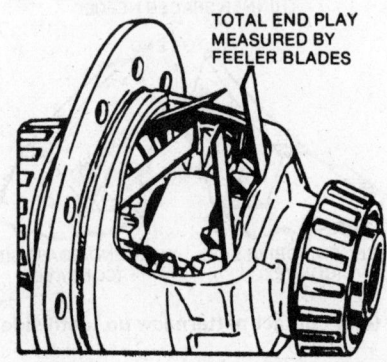

TOTAL END PLAY MEASURED BY FEELER BLADES

Checking total differential end-play

ing the backlash within the limits of specifications. If the gear tooth contact pattern is off, the unit must be

disassembled and another spacer installed. This spacer must be of suitable thickness to compensate for the contact pattern test.

NOTE: Without special tools, there is absolutely no way of determining exactly how much to increase or decrease the thickness of the pinion shim; it must be estimated.

After estimating the thickness of the new shim, assemble the unit again, setting all preloads and backlash. Check the contact pattern again and act accordingly. If the unit uses a collapsible spacer, be sure a new one is installed each time it is disassembled. Crushed spacers can not be used again. It is well to note that the unit may have to be assembled and disassembled several times before an acceptable contact pattern is obtained.

ADJUSTING BACKLASH

The tooth contact pattern can be altered slightly, by varying the backlash sdjustment within the limits of the specifications. The backlash adjustment can be used to alter a pattern which is slightly off center on either side of the tooth, but should not be used as a substitute for pinion depth adjustment. This adjustment must always be made after the pinion depth has been adjusted.

Front End Alignment

SERVICE PROCEDURE INDEX

(Numbers refer to types in text)

Manufacturer/Car	Year	Camber, and Toe in
AMERICAN MOTORS		
American Motors (All Except Pacer)	'80-'86	3
Pacer	'80	9
CHRYSLER CORPORATION		
All Front Wheel Drive	'80-'86	4
All Rear Wheel Drive	'80-'86	5
FORD MOTOR COMPANY		
Ford Fairmont. Mustang: Mercury Capri. Zephyr. '80-'85 Cougar. '82-'85 Lincoln Continental. '80-'85 Thunderbird. '81-'83 Granada: '82-'85 LTD: '82-'85 Marquis. Mark VII	'80-'86	8
Ford Pinto. Mustang II: Mercury Bobcat	'80	8
Ford Granada: Mercury Monarch. Lincoln Versailles	'80	3
Ford (Full Size). Mercury (Full Size)	'80-'86	7
Lincoln Mk IV, Mk V. Mk VI	'80-'83	7
Lincoln Continental (Through 1980). Town Car	'80-'86	7
All Front Wheel Drive	'81-'86	8

Manufacturer/Car	Year	Camber, and Toe in
GENERAL MOTORS		
Chevrolet Chevette. Pontiac 1000 (GM T-Body)	'80-'86	6
Chevrolet Vega. Monza. Pontiac Astre. Sunbird. Oldsmobile Starfire. Buick Skyhawk (GM H-Body)	'80-'86	9
Chevrolet Camaro. Pontiac Firebird	'82-'86	13
Pontiac Fiero	'84-'86	6
All rear wheel drive (Except GM T Body. GM H Body, Pontiac Fiero, 1982-86 Camaro and Firebird	'80-'86	1
Chevrolet Citation. Celebrity. Buick Skylark, Century (FWD), Oldsmobile Omega. Ciera, Pontiac Phoenix, 6000	'80-'86	11
Chevrolet Cavalier, Pontiac 2000, Cadillac Cimarron, Oldsmobile Firenza	'82-'86	12
Oldsmobile Toronado, Buick Riviera. Cadillac Eldorado. Seville (Front Wheel Drive)	'80-'86	10
Buick Electra. Cadillac Fleetwood and Deville, Olsmobile 98 (Front Wheel Drive)	'85-'86	2
Buick Somerset, Oldsmobile Calais, Pontiac Grand Am	'85-'86	12

FRONT END ALIGNMENT

Front wheel alignment is the position of the front wheels relative to each other and to the vehicle. This preset relationship provides safe, accurate steering, directional stability, and minimum tire wear. Many factors are involved in wheel alignment, and adjustments are provided to return those that might change due to normal wear to their original value. The factors which determine wheel alignment are dependent on one another; therefore, when one of the factors is adjusted, the others must be adjusted to compensate.

Descriptions of these factors and their affects on the car are provided below. Adjustment specifications for each model year are given at the beginning of each car section. It should be noted that an alignment rack is necessary to correctly adjust any front end, and worn ball joints or other front end parts should be replaced before any alignment procedure is attempted.

Camber

Camber angle is the number of degrees that the centerline of the wheel is inclined from the vertical when viewed from the front. A small degree of positive camber reduces loading of the outer wheel bearing, and allows for easier steering.

Caster

Caster angle is the number of degrees that a line drawn through the steering knuckle pivots is inclined from the vertical, toward the front or rear of the car. A small degree of positive caster improves directional stability and increases resistance to cross winds or road surface deviations.

Steering Axis Inclination

Steering axis inclination is the number of degrees that a line drawn through the steering knuckle pivots is inclined to the vertical, when viewed from the front of the car. This, in combination with caster, is responsible for directional stability and self-centering of the steering. As the steering knuckle swings from lock to lock, the spindle generates an arc (see illustration), the high point being the straight ahead position of the wheel. Due to this arc, as the wheel turns, the front of the car is raised. The weight of the car acts against this lift, and attempts to return the spindle to the high point of the arc, resulting in

self-centering when the steering wheel is released, and straight line stability.

Included Angle

Included angle is the sum of the camber angle and the steering axis inclination. This angle is determined by the design of the steering knuckle forging and must remain constant. Therefore, if a different camber angle is necessary to make the included angle on both sides identical, a bent spindle or steering knuckle is indicated. If so, the damaged suspension member must be replaced to permit accurate front wheel alignment. Since steering knuckle damage is most commonly due to impact on the lower portion of the wheel (i.e., hitting curb), the side with the greater included angle (camber angle same on each side) will often be found to have a bent spindle.

Toe-In

Toe-in is the difference of the distance between the centers of the front and rear of the front wheels, measured at spindle height. It is most commonly measured in inches, but is occasionally referred to as an angle between the wheels. Toe-in means the front of the tires are closer together than the rear; toe-out is the opposite condition. Toe-in compensates for the tendency of the wheels to deflect out while in motion. Due to this tendency, the wheels of a car with properly adjusted toe-in are traveling straight forward when the car itself is moving straight forward, resulting in directional stability and minimum tire wear. Front wheel drive and four wheel drive cars are often set with toe-out, to compensate for the drive axles' tendency to pull the front wheels together.

Steering wheel spoke misalignment is often is often an indication of incorrect front end alignment. Care should be exercised when aligning the front end to maintain steering wheel spoke position. When adjusting the tie rod ends, adjust each an equal amount (in the opposite direction) to increase or decrease toe. If, following toe adjustment, further adjustments are necessary to center the steering wheel spokes, adjust the tie rod ends an equal amount in the same direction.

Steering Radius

When a car is negotiating a turn, the outer wheel follows the path of a circle of a larger radius than the inner wheel. For this reason, the inner wheel must be steered to a somewhat larger angle than the outer wheel. This value (known as the Ackerman

effect) is designed into the steering linkage; therefore, if alignment is adjusted properly, and the steering radius (or toe-out on turns) appears to be incorrect, the steering arms or the linkage is bent.

Tracking

Tracking is the relationship between the paths traveled by the front and rear wheels when the vehicle is traveling in a straight line. When a car is tracking correctly, the path of the rear wheels will duplicate, or evenly straddle the path of the front wheels. Observing the car from the rear as it is driven away in a straight line will often make incorrect tracking evident.

If incorrect tracking is indicated, check as follows: Drop a plumb line from each lower ball joint, and from a point at each end of the rear axle, and mark the points on the ground with chalk. Measure these points from front to rear and diagonally. If the diagonal measurements are different (a tolerance of $+ \frac{1}{4}$ in. is acceptable), but the longitudinal measurements are the same, the frame is swayed (diamond shaped). If the diagonal and longitudinal measurements are both different, the rear is misaligned. If both diagonal and longitudinal measurements are different, but the car does not appear to be tracking incorrectly, a kneeback condition is indicated. Kneeback implies that one side of the front suspension is bent or pushed back. It is possible to align the front end to specifications, and, if kneeback exists, have very poor handling characteristics.

Ride Height Adjustment

This adjustment is required before adjusting front end alignment on cars with torsion bar front suspension.

NOTE: The car must be on a level floor with the gas tank full and the tire properly inflated. There should be no unusual loads in the car. On all models, check the measurements against those given with the Front End Alignment Specifications in the car sections.

Chrysler Corporation

Rock the car at the centers of the front and rear bumpers five times and allow it to settle.

For 1980–81 Gran Fury, St. Regis and Newport/New Yorker, measure from the bottom of the front frame rail, between the radiator yoke and the forward edge of the front suspen-

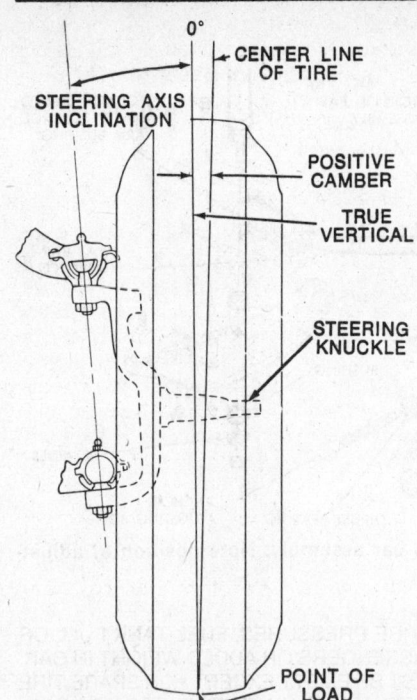

Camber and steering axis inclination angles

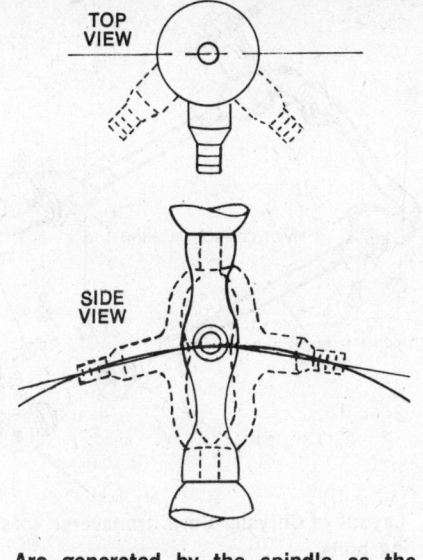

Arc generated by the spindle as the steering knuckle turns

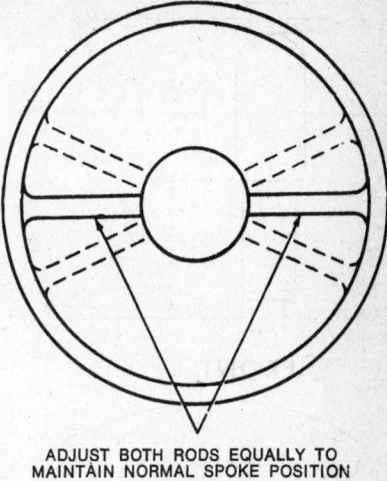

ADJUST BOTH RODS EQUALLY TO MAINTAIN NORMAL SPOKE POSITION

Steering wheel spoke alignment

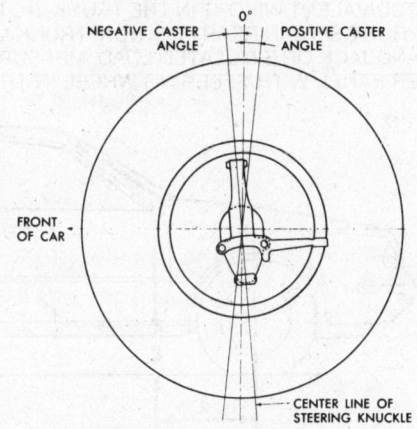

Caster angle

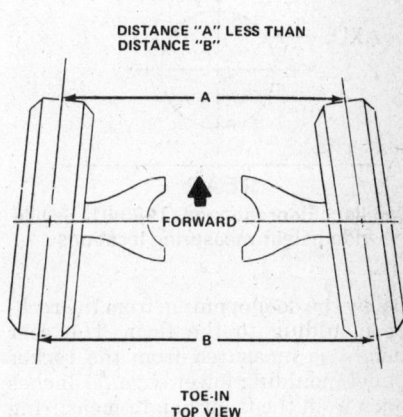

Toe-in

sion crossmember, to ground. For all other 1980–83 torsion bar front suspension models, measure from the head of the front suspension crossmember front isolator bolt to ground. Adjust the height by turning the torsion bar adjusting bolt clockwise to raise or counterclockwise to lower. The height should not vary more than $\frac{1}{4}$ in. from side to side.

NOTE: A change of front tire size can change front end height on these cars.

Cadillac Eldorado, 1980 and Later Seville, Oldsmobile Toronado, 1980 and later Buick Riviera

Front ride height is controlled by the settings of the torsion bar adjusting bolts. The height is adjusted by turning the adjusting bolt clockwise to raise, and counterclockwise to lower. 1980 and later models have electronic level control. 1980–83 are the only model Sevilles equipped with torsion bar front suspension.

Cadillac Eldorado, Seville

Front ride height is measured from the lower edge of the shock absorber dust cover to the centerline of the lower shock mounting bolt. Rear ride height is measured from the frame to the control arm flange.

Oldsmobile Toronado

The front height is measured directly

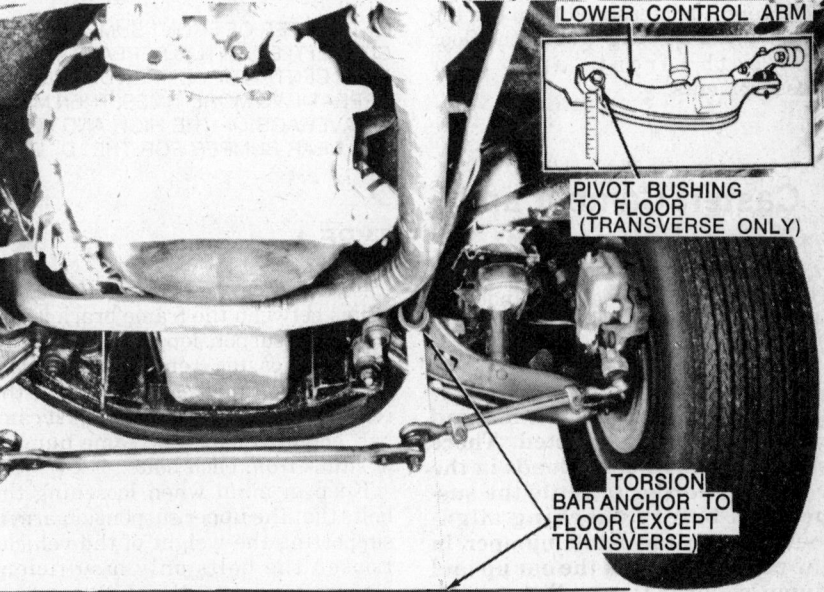

Height measuring location for Chrysler Corp. cars with transverse torsion bars

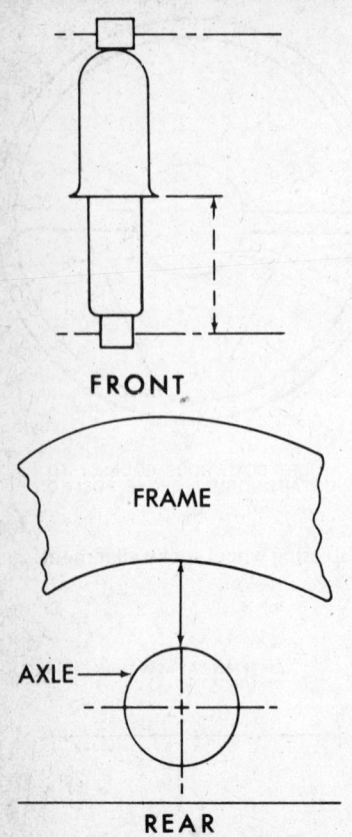

FRONT

FRAME

AXLE →

REAR

Cadillac Eldorado and 1980-81 Seville ride height measuring locations

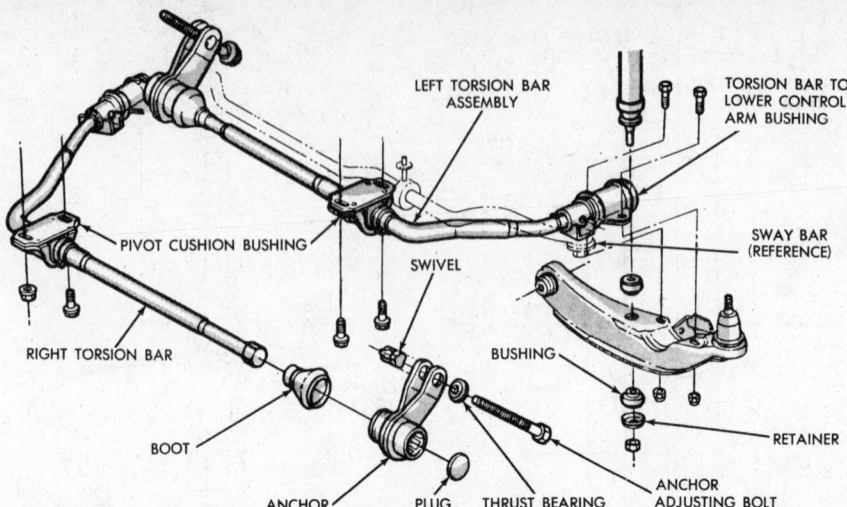

Layout of Chrysler Corp. transverse torsion bar assembly. Note position of adjusting bolts

TRIM HEIGHTS CHECKED WITH CORRECT TIRE PRESSURES, FUEL TANK FULL OR EQUIVALENT WEIGHT IN THE TRUNK. NO PASSENGERS OR ADDED WEIGHT IN CAR. FRONT SEAT IN REAR POSITION. TRUNK MUST BE EMPTY EXCEPT FOR SPARE TIRE AND JACK OR SIMULATED LOAD. MEASURE FROM KNOWN LEVEL FLOOR TO ROCKER PANEL WITH STEERING WHEEL IN THE CENTERED POSITION.

below the door opening, from the rocker moulding to the floor. The rear height is measured from the rocker panel moulding lower edge, 71 inches back from the front height measuring point, to the floor.

Riviera

Ride height is measured from the top of the wheel opening arch to the floor for both front and rear measurements.

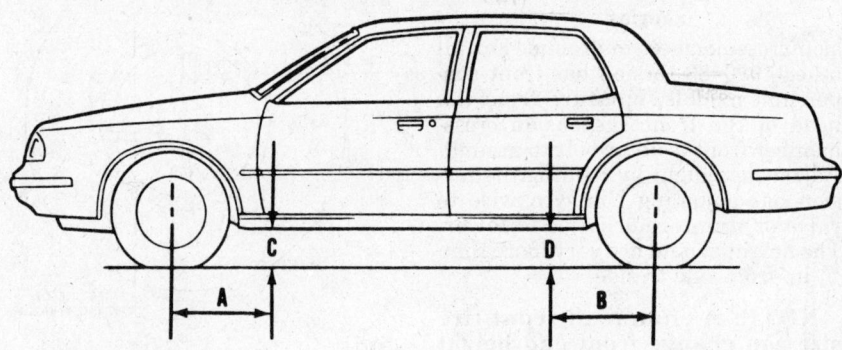

C & D DIMENSION

LIFT CENTER OF FRONT BUMPER UP APPROXIMATELY 1½ in. (38mm) AND LET VEHICLE SETTLE GENTLY. REPEAT TWO MORE TIMES, THEN MEASURE "C" DIMENSIOM. PUSH CENTER OF BUMPER DOWN 1½ in. (38mm) AND LET VEHICLE SETTLE GENTLY. REPEAT TWO MORE TIMES, THEN MEASURE "C" DIMENSION. THE "C" DIMENSION IS AN AVERAGE OF THE HIGH AND LOW MEASUREMENTS. REPEAT PROCEDURE ON THE REAR BUMPER FOR THE "D" DIMENSION.

Caster, Camber and Toe Adjustment

Use the Service Procedure Index at the start of this section to relate these type numbers to makes and models.

NOTE: The car must be on a level floor with the gas tank full and the tire properly inflated. There should be no unusual loads in the vehicle. In order to settle the suspension before checking alignment, grasp the front bumper in the center and rock the car up and down several times. Repeat the procedure on the rear bumper.

TYPE 1

Caster and camber are controlled by shims between the frame bracket and the upper suspension arm pivot shaft. To adjust caster, remove shims from the front bolt and replace them at the rear bolt, or vice versa. To adjust camber, add or remove the same number of shims from each bolt.

Keep in mind when loosening the bolts that the upper suspension arm is supporting the weight of the vehicle. Loosen the bolts only a sufficient amount to remove the shims.

Adjust toe-in by loosening the clamps on the sleeves at the outer ends of the tie rod, and turning the sleeves an equal amount in the opposite direction, to maintain steering wheel spoke alignment while adjusting toe-in.

TYPE 2

To adjust camber on these models loosen both strut-to-knuckle nuts. Attach camber adjusting tool No. J-29862 and set camber to specifications. Retighten both strut-to-knuckle nuts to 144 ft. lbs. To adjust caster loosen, but do not remove, two of the

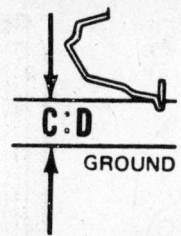

C:D

GROUND

INCORRECT

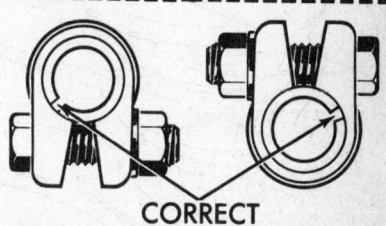

CORRECT

Typical tie rod clamp to sleeve position

OLDSMOBILE RIDE HEIGHT ADJUSTMENT

Maximum variation side-to-side and front-to-rear is 3/4 in. (19mm)

Model	View A	View B	View C	View D
Cutlass 4 dr with P195/75R14, P205/70R14 P205/75R14	24 7/32 (623 mm)	20 15/16 (532 mm)	10 3/16 (259 mm)	10 3/16 (259 mm)
Cutlass Salon	24 7/32 (623 mm)	20 15/16 (532 mm)	10 3/16 (259 mm)	10 3/16 (259 mm)
Cutlass 2 dr with P195/75R14, P205/70R14	24 7/32 (623 mm)	20 15/16 (532 mm)	10 (255 mm)	10 (255 mm)
Cutlass 442	24 7/32 (623 mm)	20 15/16 (532 mm)	10 (255 mm)	10 (255 mm)
Delta 88	29 21/64 (745 mm)	18 11/16 (475 mm)	10 3/8 (246 mm)	10 9/16 (269 mm)
Custom Cruiser	29 21/64 (745 mm)	18 11/16 (475 mm)	10 1/2 (267 mm)	10 5/8 (271 mm)
Toronado	22 27/32 (580 mm)	22 11/64 (563 mm)	9 5/8 (245 mm)	9 5/8 (245 mm)

NOTE: All measurements in inches unless noted.

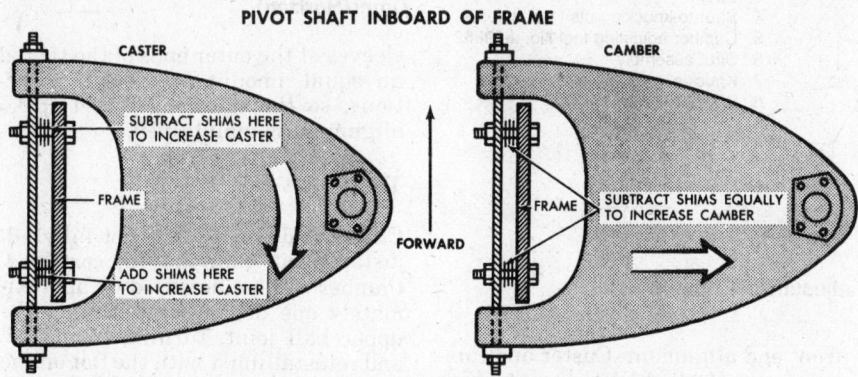

PIVOT SHAFT INBOARD OF FRAME

CASTER — CAMBER

SUBTRACT SHIMS HERE TO INCREASE CASTER

FRAME — FRAME

FORWARD

SUBTRACT SHIMS EQUALLY TO INCREASE CAMBER

ADD SHIMS HERE TO INCREASE CASTER

Typical caster and camber adjustment, Type 1 (reverse the procedure for shims on the opposite side of the frame)

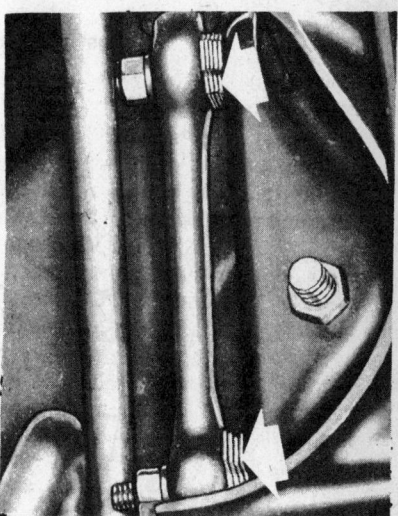

Typical Type 1 caster and camber adjusting shim location

top three strut attaching nuts covering the slotted mounting holes. Remove the nut over the remaining oval strut mounting hole, then move the washer away from the oval strut mounting hole. Lift the front of the car by the body to separate the strut from the inner wheel house. Drill $^{11}/_{32}$ in. holes at the front and rear of the oval strut mounting hole and file the excess metal. Lower the front of the car and reinstall the washer and nut.

Caster adjustment is made by moving the top of the strut forward or rearward to specifications. Tighten the strut attaching nuts to 18 ft. lbs.

Front toe is adjusted by loosening the locknut on both inner tie rods then turning the inner tie rod to specifications. Tighten the locknut on the inner tie rods to 50 ft. lbs.

TYPE 3

Caster is adjusted by lengthening or shortening the struts at the frame crossmember. To adjust, turn both nuts an equal number of turns in the same direction. Caster adjustments should be with $^1/_4°$ of the opposing side of the car.

To adjust camber, loosen the lower control arm pivot bolt and rotate the eccentrics.

Adjust toe-in by loosening the clamp bolts, and turning the adjuster sleeves at the outer ends of the tie rod. Turn each sleeve an equal amount in the opposite direction, in order to maintain steering wheel spoke alignment.

TYPE 4

Ride height should be checked before front end alignment. Ride height is not adjustable on front wheel drive models.

Caster and camber are controlled by eccentric (cam) bolts; only camber is adjustable on the front wheel drive models. The cam bolts are located at the ends of the upper control arm shafts on all models except with front wheel drive. There is only one eccentric cam for each side of the front wheel drive models, on the top bolt connecting the strut to the steering knuckle. To adjust the caster, loosen the eccentric (cam) bolt nuts and turn either of the eccentric bolts. Camber is adjusted by turning both eccentrics

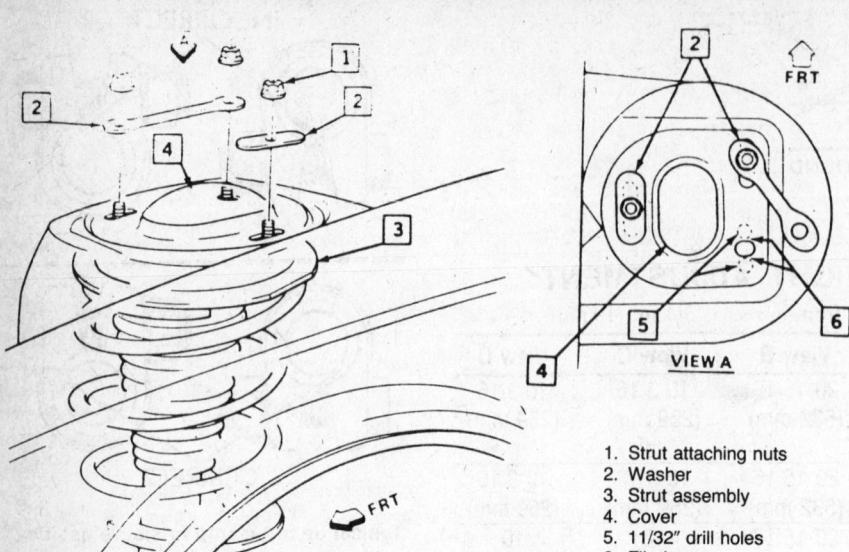

1. Strut attaching nuts
2. Washer
3. Strut assembly
4. Cover
5. 11/32" drill holes
6. File here

Caster adjustment—Type 2

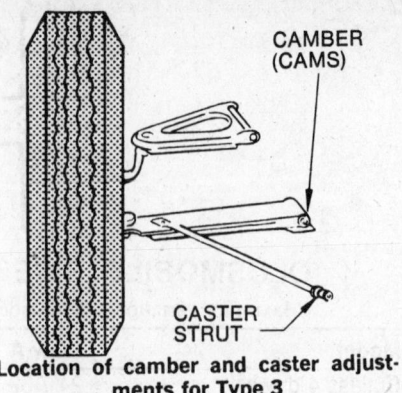

Location of camber and caster adjustments for Type 3

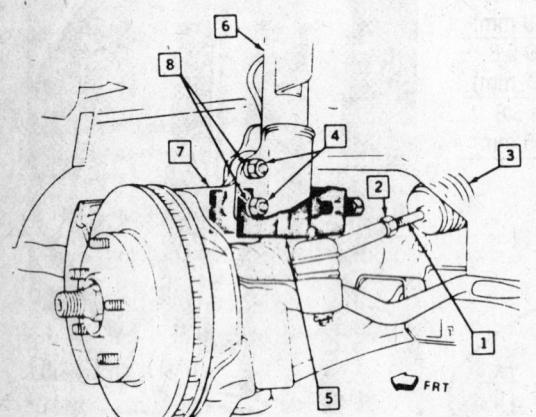

1. Inner tie-rod
2. Inner tie-rod lock nut (50 ft. lbs.)
3. Boot
4. Strut-to-knuckle nuts (144 ft. lbs.)
5. Camber adjusting tool No. J-29862
6. Strut assembly
7. Knuckle
8. Washer

Camber and toe adjustment—Type 2

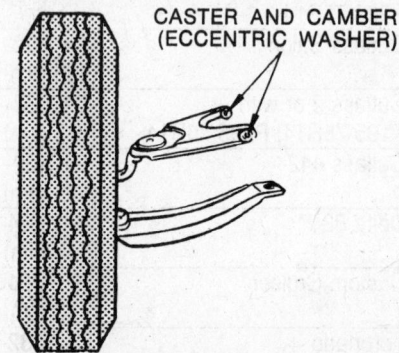

Location of caster and camber adjustments for Type 4 (except Imperial and Omni/Horizon)

an equal amount, except on the front wheel drive models. For those models, loosen the cam and through bolts, and rotate the upper cam bolt to move the knuckle and wheel in or out to specification. Recheck caster after setting camber. Torque the eccentric (cam) bolts to 90 ft. lbs. on the front wheel drive models, and 65–70 ft. lbs. on all others.

To adjust toe-in (toe-out on the front wheel drive models), loosen the tie rod clamp bolts (jam nuts on the front wheel drive models) and turn the adjuster sleeves (tie rods, on the front wheel drive models) at the outer ends of the tie rod an equal amount in opposite directions so that steering wheel spoke alignment is maintained.

TYPE 5

Ride height should be checked before front end alignment. Caster and camber are controlled by the positioning of the upper control arm pivot bar adjusting bolts. To adjust caster, loosen one of the pivot bar adjusting bolts or nuts and slide one end of the bar either inboard or outboard in its elongated mounting hole in the crossmember. Camber is adjusted loosening both the pivot bar adjusting bolts or nuts and sliding both ends of the bar an equal amount.

NOTE: Chrysler recommends the use of a special pry bar no. C–4576, 1980 and later for the adjusting operation on the upper control arm pivot bar.

Recheck caster after setting camber. Torque the pivot bar adjusting bolts or nuts to 160 ft. lbs.

To adjust toe-in, loosen the tie rod clamp bolts and turn the adjuster

sleeves at the outer ends of the tie rod an equal amount in opposite directions, so that steering wheel spoke alignment is maintained.

TYPE 6

Caster and camber are not fully adjustable, but they may be corrected. Camber can be increased by approximately one degree by removing the upper ball joint, turning it around, and reinstalling it with the flat on the upper flange on the inboard side of the control arm. Caster can be changed one degree by changing the position of the washers between the legs of the upper control arm. Placing the thinner washer in front will increase caster, while placing it at the back will reduce caster.

Toe-in is adjusted by loosening the nuts at the steering knuckle end of each tie rod and the rubber cover at the other end, then turning the rod.

TYPE 7

Install Ford tool T79P-3000-A (1980 and later) or its equivalent on the frame rail, position the hooks around the upper control arm pivot shaft, and tighten the adjusting nuts of the tool slightly. Loosen the pivot shaft retaining bolts to permit adjustment.

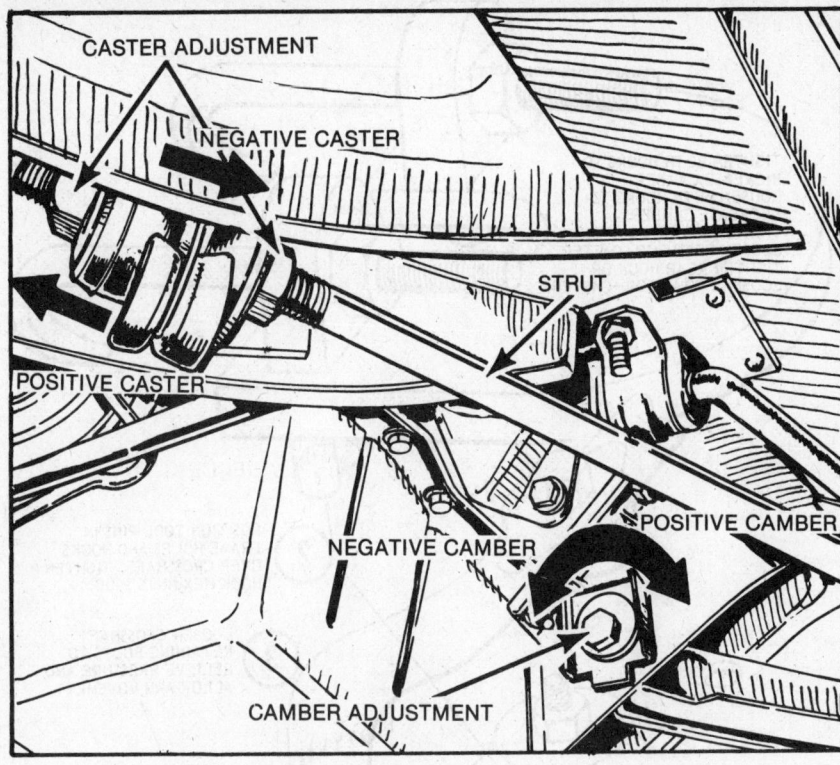

Location of caster and camber adjustments for Type 3

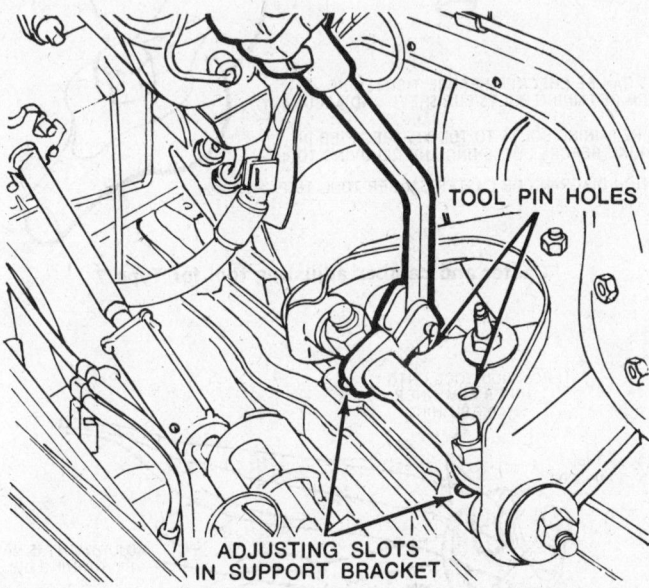

Type 5 upper control arm showing location of pivot bar adjusting nuts and bolts with adjusting pry bar in place.

To adjust caster, loosen or tighten either the front or rear adjusting nut. After adjusting caster, adjust camber by loosening or tightening both nuts an equal amount. Tighten the shaft retaining bolts to specifications, remove the tool, and recheck the adjustments.

Adjust toe-in by loosening the clamp bolts, and turning the adjuster sleeves at the outer ends of the tie rod. Turn the sleeves an equal amount in the opposite direction, to maintain steering wheel alignment.

TYPE 8

NOTE: Camber and caster on

the Fairmont, Zephyr, Capri, 1980 and later Mustang, Cougar and Thunderbird, 1981–82 Granada, and Monarch 1982–84 Lincoln Continental and all FWD models is permanently set at the factory. Only toe-in can be adjusted.

Position one Ford tool T74P-3000 or its equivalent at each end of the upper control arm, pivot shaft with the leg of the tools through the holes in the sheet metal (see illustration). Turn the adjusting bolts until they are solidly contacting sheet metal, and loosen the pivot shaft retaining bolts.

Caster is adjusted by turning the front and rear adjusting bolts in the opposite direction. Camber is adjusted by turning both bolts an equal amount in the same direction. Following the adjustment, tighten the pivot shaft retaining bolts, remove the adjusting tools, and recheck caster and camber.

Prior to adjusting toe-in, align the straight ahead marks at the base of the steering wheel and the head of the steering column. Loosen both the clamp at the outer end of the rack bellows and the tie rod jam nuts. Turn the inner tie rod shafts to adjust toe-in. Turn the shafts an equal amount in the opposite direction, to maintain steering wheel spoke alignment. Following the adjustment, hold the inner shafts with pliers, and tighten the jam nuts to 35–50 ft. lbs.

Camber and caster are adjusted using eccentrics on the lower control arm pivot bolts. Camber is adjusted first, by loosening the front pivot nut and rotating the eccentric. Tighten the front, and loosen the rear pivot nuts. Adjust caster by rotating the rear eccentric, and tighten the rear pivot nut while holding the bolt in position. Recheck camber and caster.

To adjust toe-in, loosen the clamps on the adjusting sleeves at the outer ends of the tie rod, and turn each sleeve an equal amount in the opposite direction, to maintain steering wheel spoke alignment while adjusting toe-in.

TYPE 10

Ride height should be checked and corrected before front end alignment. Caster and camber are adjusted by eccentric cam bolts on the frame end of the upper control arms. Loosen the cam bolt nuts to permit adjustment.

Adjust camber by turning the front cam bolt to make half the necessary correction. Turn the rear cam bolt in the same direction for the other half of the correction.

Adjust caster by turning the front

U385

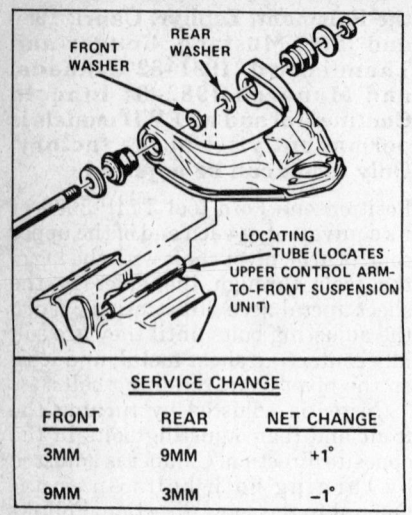

SERVICE CHANGE		
FRONT	REAR	NET CHANGE
3MM	9MM	+1°
9MM	3MM	−1°

Type 6 caster adjustment

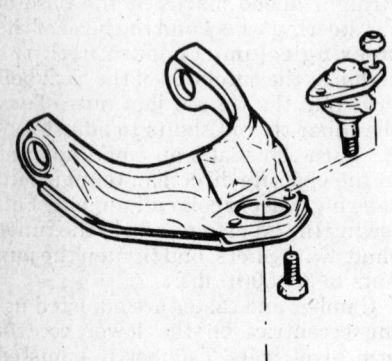

NOTE: TO INCREASE CAMBER, DISCONNECT UPPER BALL JOINT, ROTATE 180° TO POSITION "FLAT" OF FLANGE INBOARD, THEN RECONNECT BALLJOINT.

Type 6 camber adjustment

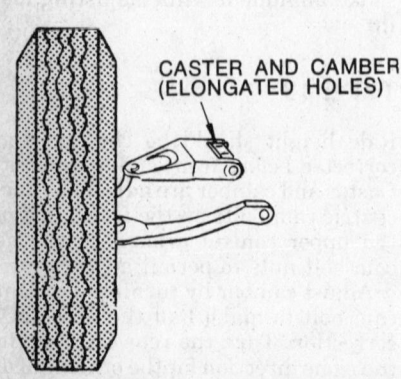

Location of caster and camber adjustments for Type 7

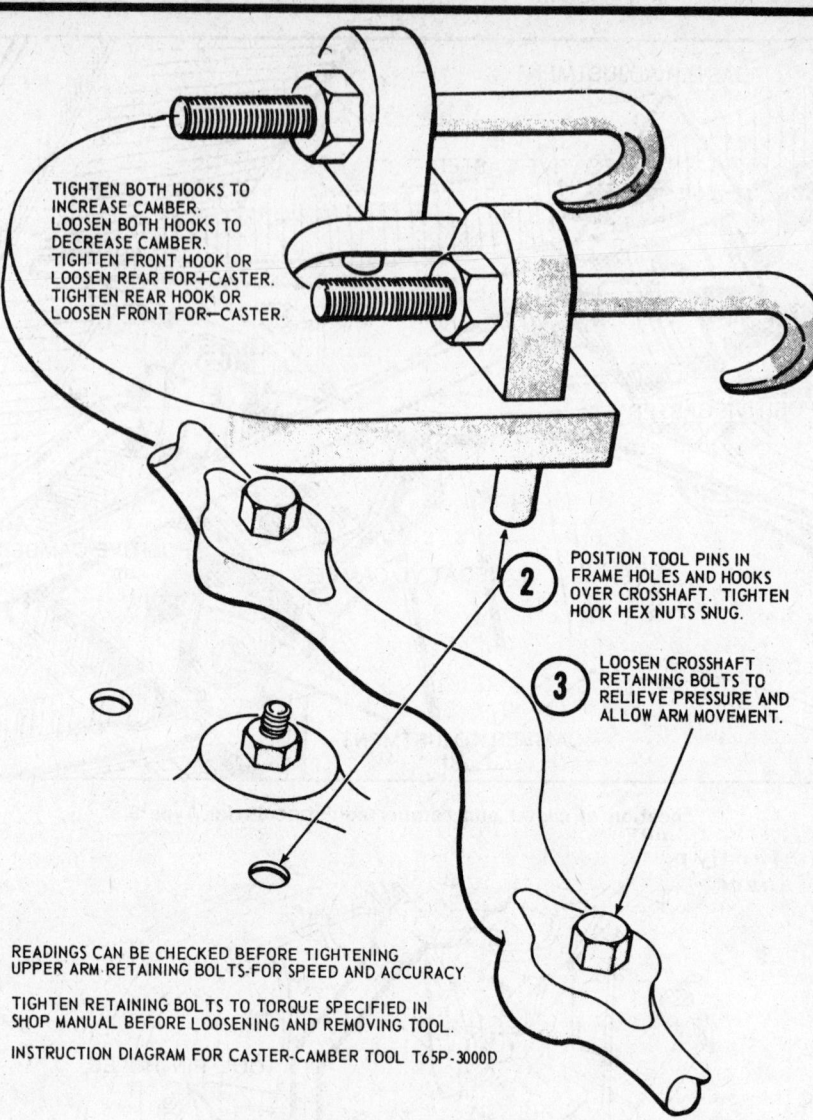

TIGHTEN BOTH HOOKS TO INCREASE CAMBER. LOOSEN BOTH HOOKS TO DECREASE CAMBER. TIGHTEN FRONT HOOK OR LOOSEN REAR FOR +CASTER. TIGHTEN REAR HOOK OR LOOSEN FRONT FOR −CASTER.

② POSITION TOOL PINS IN FRAME HOLES AND HOOKS OVER CROSSHAFT. TIGHTEN HOOK HEX NUTS SNUG.

③ LOOSEN CROSSHAFT RETAINING BOLTS TO RELIEVE PRESSURE AND ALLOW ARM MOVEMENT.

READINGS CAN BE CHECKED BEFORE TIGHTENING UPPER ARM RETAINING BOLTS-FOR SPEED AND ACCURACY

TIGHTEN RETAINING BOLTS TO TORQUE SPECIFIED IN SHOP MANUAL BEFORE LOOSENING AND REMOVING TOOL.

INSTRUCTION DIAGRAM FOR CASTER-CAMBER TOOL T65P-3000D

Caster and camber adjusting tool for Type 7

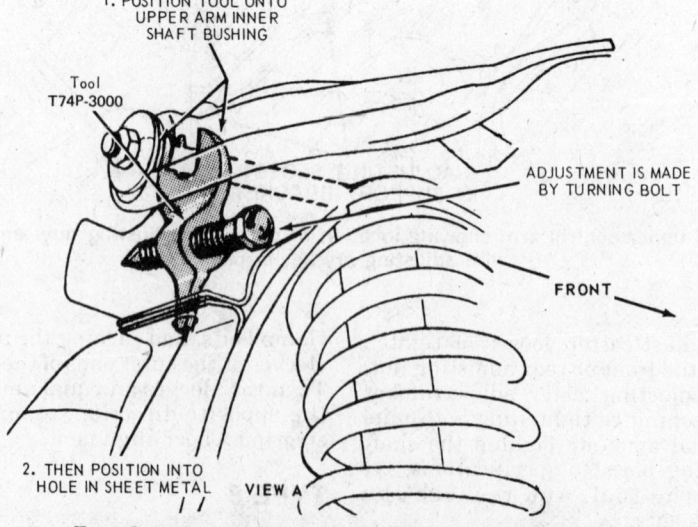

Type 8 caster and camber adjusting tool installation

cam bolt to make a quarter of the necessary correction. Turn the rear cam bolt to bring the camber back to the correct setting. The caster should now be correct. Tighten the cam bolts to 80 ft. lbs. for 1980 and later Toronado. Hold the bolts when tightening the nuts to prevent the settings from changing.

Adjust toe-out by centering the steering wheel, loosening the tie rod sleeve clamps, and turning the adjusting sleeves. Turn the sleeves an equal amount in opposite directions to maintain steering wheel alignment. Position the sleeve clamps up to avoid linkage interference. Tighten the clamps to 15 ft. lbs.

TYPE 11

Camber is adjusted by loosening the cam and through bolts on the Mac-Pherson strut-to-knuckle bolts and rotating the cam bolt to move the upper knuckle and wheel in or out. Tighten the bolts to 140 ft. lbs. after adjustment and check that the cam is seated between the inner and outer guide surfaces.

NOTE: It may be necessary to apply only partial torque because of inaccessibility to the bolts. Torque just enough to hold the correct camber position, then remove the wheel and tire and apply final torque.

Caster is not adjustable.

Toe is adjusted with the steering linkage tie rods. Loosen the jamnuts at the steering knuckle end of the tie rods, and remove the boot clamps. Rotate the tie rods to align the toe. Tighten the jamnuts to 40 ft. lbs., and replace the boot clamps.

TYPE 12

NOTE: Before the camber can be adjusted the strut must be modified by filling the holes in the outer flanges to enlarge the bottom holes until they match the slots in the inner flanges. This can be accomplished by disconnecting the strut from the knuckle.

Camber is adjusted by reaching around both sides of the tire and loosening both strut-to-knuckle bolts just enough to allow movement between the strut and the knuckle. Grasp the top of the tire firmly, and move the tire in or out until the correct camber reading is obtained. Carefully reach around the tire and tighten both bolts enough to hold the correct camber, then remove the wheel and tire and

Type 9 camber (left) and caster (right) adjustments

Typical caster and camber cam locations for Type 10

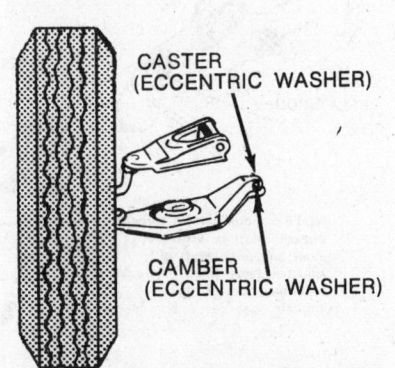

Location of caster and camber adjustments for Type 9

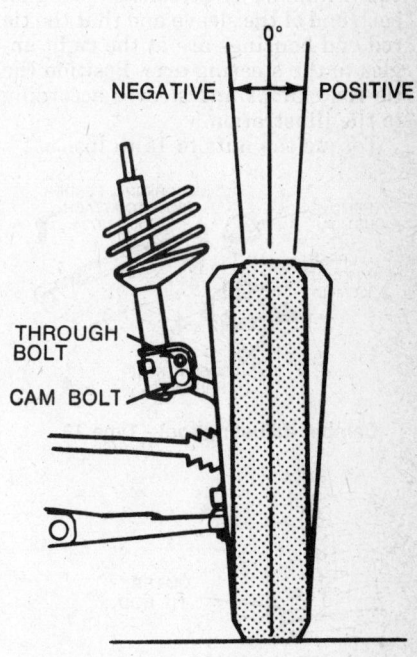

Location of camber adjusting bolts on Type 11

torque both bolts to 140 ft. lbs. Reinstall the tire and wheel.

Caster is not adjustable.

Toe is controlled by the tie rod position. To adjust toe setting, loosen the clamp bolts at the outer end of the tie rod, rotate the adjuster to toe specifications, then tighten the clamp bolts to 15 ft. lbs.

TYPE 13

Caster and camber is adjusted by moving the position of the upper strut mount assembly forward/rearward for caster or inboard/outboard for camber.

The position of the mount can be changed by removing the dust cap and loosening the three upper mount attaching nuts. The weight of the vehicle will normally cause the strut assembly to move to fill the inboard position. A special tool No. J-29724 is available which is attached to a fender bolt and hooked to the upper mount assembly. By tightening the turnbuckle on the tool the proper camber can be obtained. If a caster adjustment is necessary the mount can be tapped forward or rearward with a rubber mallet. Tighten the three nuts to 20 ft. lbs. after adjustment.

Toe-in can be increased or decreased by turning the tie rod adjusting sleeves after loosening the clamp bolts. When adjusting has been completed, check to see that there are an equal number of threads showing on each end of the sleeve and that the tie rod end housings are at the right angles to the steering arm. Position the tie rod clamps and sleeves according to the illustration.

Torque the nuts to 15 ft. lbs.

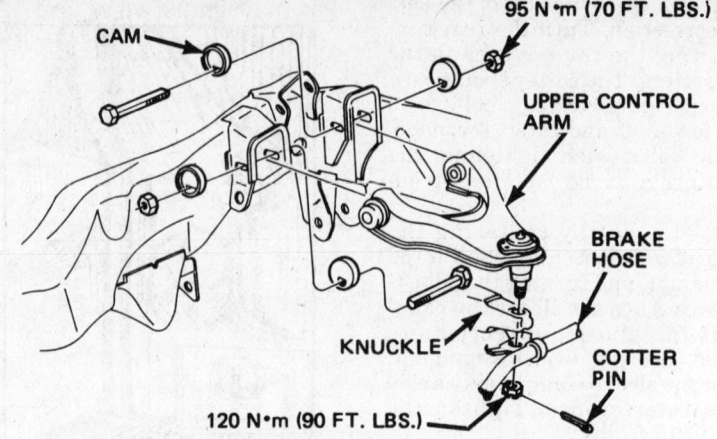

Typical caster and camber cam locations for Type 10

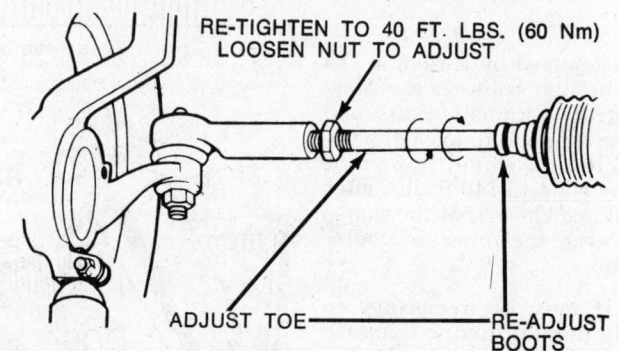

Toe adjustment, Type 11

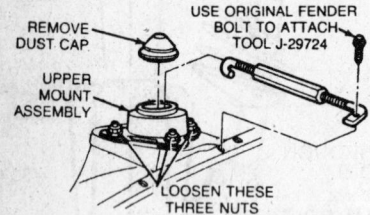

Camber adjusting tool - Type 13

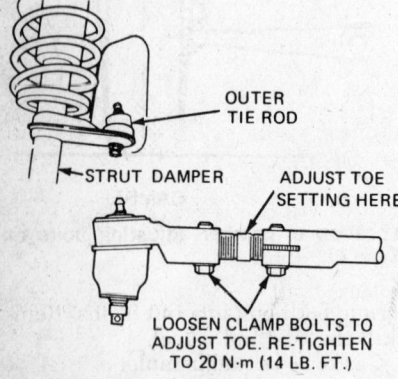

Toe adjustment, Type 12

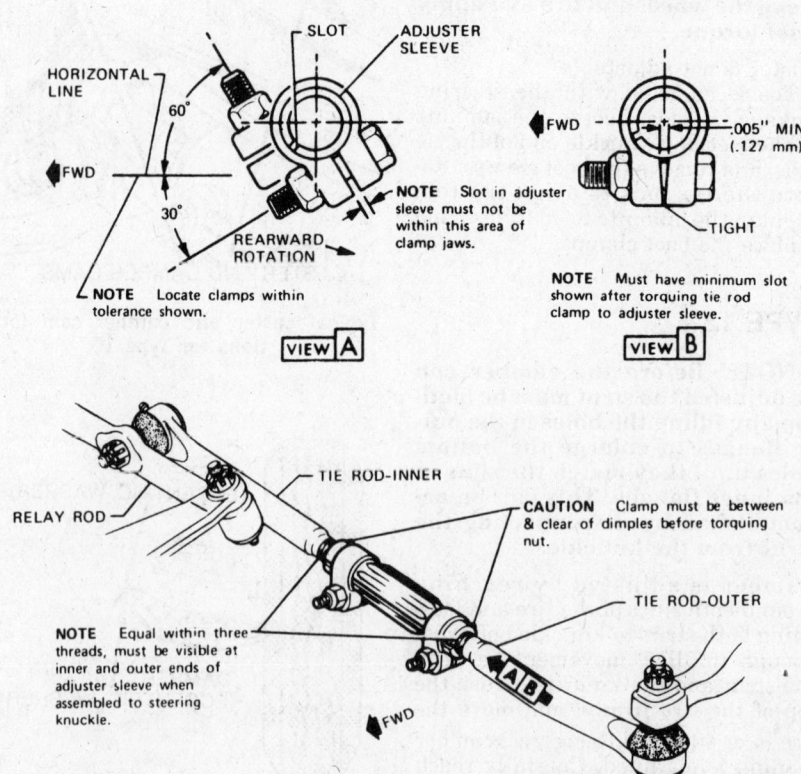

Tie rod clamp and sleeve positioning - Type 13

Troubleshooting and Diagnosis

ENGINE

Gasoline Engine Troubleshooting

See applicable Car or Unit Repair section for specific service procedures

INDEX TO PROBLEMS

Problem Symptom	Begin at Specific Diagnosis, Number
Engine Won't Start	
Starter doesn't turn	1.1, 2.1
Starter turns, engine doesn't	2.1
Starter turns engine very slowly	1.1, 2.4
Starter turns engine normally	3.1, 4.1
Starter turns engine very quickly	6.1
Engine fires intermittently	4.1
Engine fires consistently	5.1, 6.1
Engine Runs Poorly	
Hard starting	3.1, 4.1, 5.1, 8.1
Rough idle	4.1, 5.1, 8.1
Stalling	3.1, 4.1, 5.1, 8.1
Engine dies at high speeds	4.1, 5.1
Hesitation (on acceleration from standing stop)	5.1, 8.1
Poor pickup	4.1, 5.1, 8.1
Lack of power	3.1, 4.1, 5.1, 8.1
Backfire through the carburetor	4.1, 8.1, 9.1
Backfire through the exhaust	4.1, 8.1, 9.1
Blue exhaust gases	6.1, 7.1
Black exhaust gases	5.1
Running on (after the ignition is shut off)	3.1, 8.1
Susceptible to moisture	4.1
Engine misfires under load	4.1, 7.1, 8.4, 9.1
Engine misfires at speed	4.1, 8.4
Engine misfires at idle	3.1, 4.1, 5.1, 7.1, 8.4

TROUBLESHOOTING AND DIAGNOSIS

SAMPLE SECTION

Test and Procedure	Results and Indications	Proceed to
4.1 Check for spark: Hold each spark plug wire approximately $\frac{1}{4}''$ from ground with gloves or a heavy, dry rag. Crank the engine and observe the spark.	If no spark is evident	**4.2**
	If spark is good in some cases	**4.3**
	If spark is good in all cases	**4.6**

SPECIFIC DIAGNOSIS

This section is arranged so that following each test, instructions are given to proceed to another, until a problem is diagnosed.

SECTION 1—BATTERY

Test and Procedure	Results and Indications	Proceed to
1.1 Inspect the battery visually for case condition (corrosion, cracks) and water level.	If case is cracked, replace battery.	**1.4**
	If the case is intact, remove corrosion with a solution of baking soda and water. **(CAUTION: Do not get the solution into the battery).** Fill with water.	**1.2**

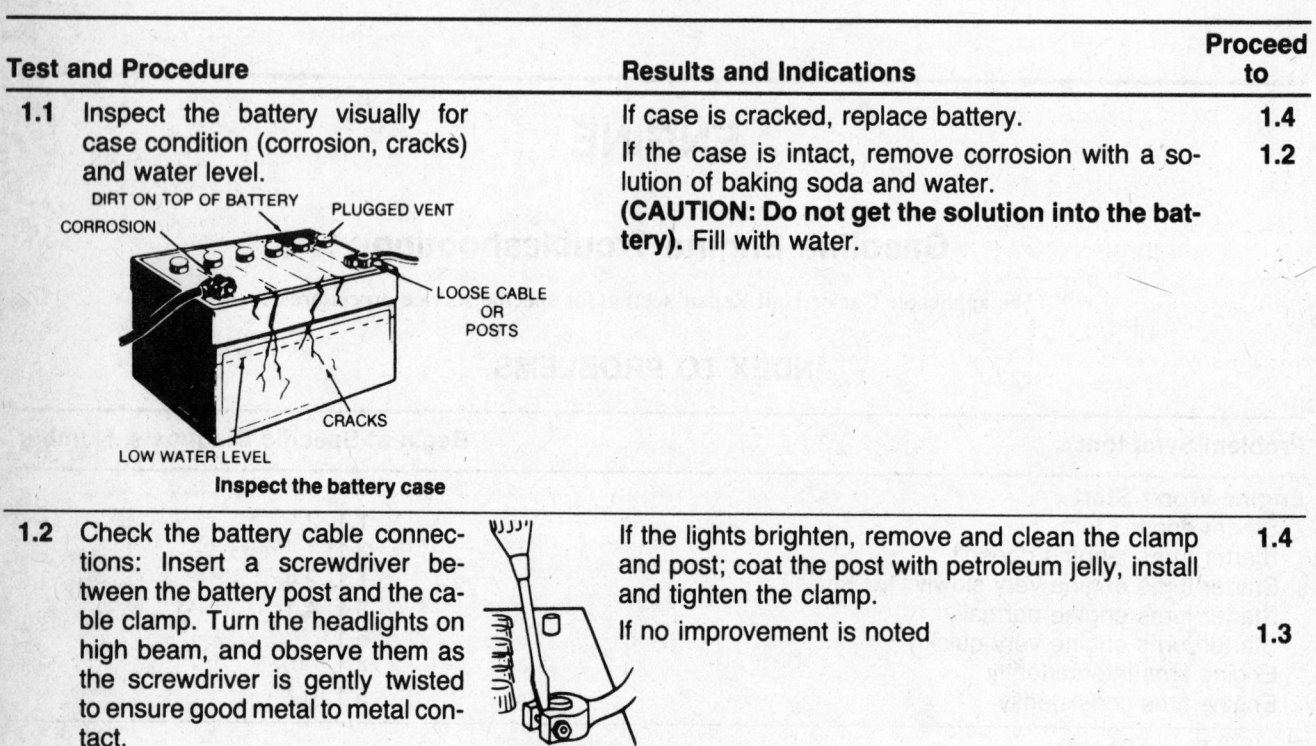

DIRT ON TOP OF BATTERY
PLUGGED VENT
CORROSION
LOOSE CABLE OR POSTS
CRACKS
LOW WATER LEVEL

Inspect the battery case

1.2 Check the battery cable connections: Insert a screwdriver between the battery post and the cable clamp. Turn the headlights on high beam, and observe them as the screwdriver is gently twisted to ensure good metal to metal contact.	If the lights brighten, remove and clean the clamp and post; coat the post with petroleum jelly, install and tighten the clamp.	**1.4**
	If no improvement is noted	**1.3**

TESTING BATTERY CABLE CONNECTIONS USING A SCREWDRIVER

1.3 Test the state of charge of the battery using an individual cell tester or hydrometer.	If indicated, charge the battery. **NOTE: If no obvious reason exists for the low state of charge (i.e., battery age, prolonged storage), proceed to:**	**1.4**

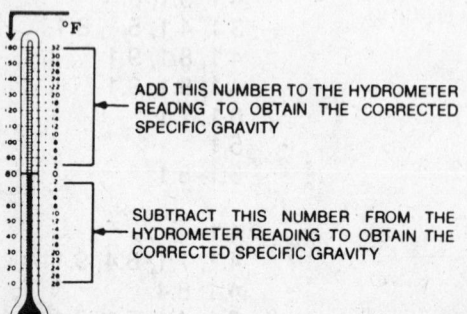

°F

ADD THIS NUMBER TO THE HYDROMETER READING TO OBTAIN THE CORRECTED SPECIFIC GRAVITY

SUBTRACT THIS NUMBER FROM THE HYDROMETER READING TO OBTAIN THE CORRECTED SPECIFIC GRAVITY

Specific Gravity (@ 80° F.)

Minimum	Battery Charge
1.260	100% Charged
1.230	75% Charged
1.200	50% Charged
1.170	25% Charged
1.140	Very Little Power Left
1.110	Completely Discharged

The effects of temperature on battery specific gravity (left) and amount of battery charge in relation to specific gravity (right)

Test and Procedure	Results and Indications	Proceed To
1.4 Visually inspect battery cables for cracking, bad connection to ground, or bad connection to starter.	If necessary, tighten connections or replace the cables.	2.1

SECTION 2—STARTING SYSTEM

Test and Procedure	Results and Indications	Proceed to
Note: Tests in Group 2 are performed with coil high tension lead disconnected to prevent accidental starting.		
2.1 Test the starter motor and solenoid: Connect a jumper from the battery post of the solenoid (or relay) to the starter post of the solenoid (or relay).	If starter turns the engine normally	2.2
	If the starter buzzes, or turns the engine very slowly	2.4
	If no response, replace the solenoid (or relay).	3.1
	If the starter turns, but the engine doesn't, ensure that the flywheel ring gear is intact. If the gear is undamaged, replace the starter drive.	3.1
2.2 Determine whether ignition override switches are functioning properly (clutch start switch, neutral safety switch), by connecting a jumper across the switch(es), and turning the ignition switch to "start".	If starter operates, adjust or replace switch.	3.1
	If the starter doesn't operate	2.3
2.3 Check the ignition switch "start" position: Connect a 12V test lamp or voltmeter between the starter post of the solenoid (or relay) and ground. Turn the ignition switch to the "start" position, and jiggle the key.	If the lamp doesn't light or the meter needle doesn't move when the switch is turned, check the ignition switch for loose connections, cracked insulation, or broken wires. Repair or replace as necessary.	3.1
	If the lamp flickers or needle moves when the key is jiggled, replace the ignition switch.	3.3

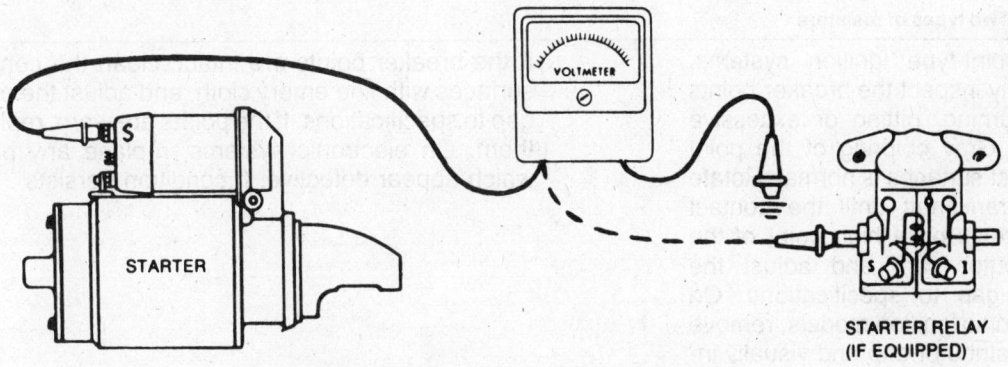

Checking the ignition switch "start" position

2.4 Remove and bench test the starter, according to specifications in the car section.	If the starter does not meet specifications, repair or replace as needed	3.1
	If the starter is operating properly	2.5

Test and Procedure	Results and Indications	Proceed To
2.5 Determine whether the engine can turn freely: Remove the spark plugs, and check for water in the cylinders. Check for water on the dipstick, or oil in the radiator. Attempt to turn the engine using an 18" flex drive and socket on the crankshaft pulley nut or bolt.	If the engine will turn freely only with the spark plugs out, and hydrostatic lock (water in the cylinders) is ruled out, check valve timing.	**9.2**
	If engine will not turn freely, and it is known that the clutch and transmission are free, the engine must be disassembled for further evaluation.	**See Car Section**

SECTION 3—PRIMARY ELECTRICAL SYSTEM

Test and Procedure	Results and Indications	Proceed to
3.1 Check the ignition switch "on" position: Connect a jumper wire between the distributor side of the coil and ground, and a 12V test lamp between the switch side of the coil and ground. Remove the high tension lead from the coil. Turn the ignition switch on and jiggle the key.	If the lamp lights	**3.2**
	If the lamp flickers when the key is jiggled, replace the ignition switch.	**3.3**
	If the lamp doesn't light, check for loose or open connections. If none are found, remove the ignition switch and check for continuity. If the switch is faulty, replace it.	**3.3**

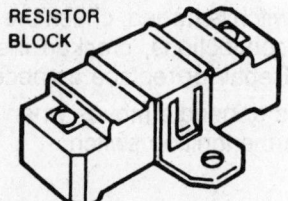

Checking the ignition switch "on" position

3.2 Check the ballast resistor or resistance wire for an open circuit, using an ohmmeter.	Replace the resistor or resistance wire if the resistance is zero. **NOTE: Some ignition systems have no ballast resistor.**	**3.3**

Two types of resistors

3.3 On point-type ignition systems, visually inspect the breaker points for burning, pitting or excessive wear. Gray coloring of the point contact surfaces is normal. Rotate the crankshaft until the contact heel rests on a high point of the distributor cam and adjust the point gap to specifications. On electronic ignition models, remove the distributor cap and visually inspect the armature. Ensure that the armature pin is in place, and that the armature is on tight and rotates when the engine is cranked. Make sure there are no cracks, chips or rounded edges on the armature.	If the breaker points are intact, clean the contact surfaces with fine emery cloth, and adjust the point gap to specifications. If the points are worn, replace them. On electronic systems, replace any parts which appear defective. If condition persists	**3.4**

Test and Procedure	Results and Indications	Proceed To
3.4 On point-type ignition systems, connect a dwell-meter between the distributor primary lead and ground. Crank the engine and observe the point dwell angle. On electronic ignition systems, conduct a stator (magnetic pickup assembly) test. See Electronic Ignition Unit Repair Section.	On point-type systems, adjust the dwell angle if necessary. **NOTE: Increasing the point gap decreases the dwell angle and vice-versa.** If the dwell meter shows little or no reading On electronic ignition systems, if the stator is bad, replace the stator. If the stator is good, proceed to the other tests in The Electronic Ignition Unit Repair Section.	3.6 3.5

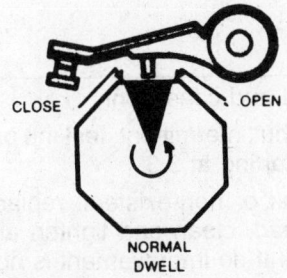 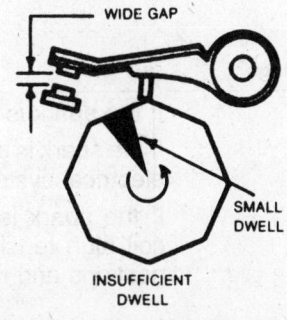 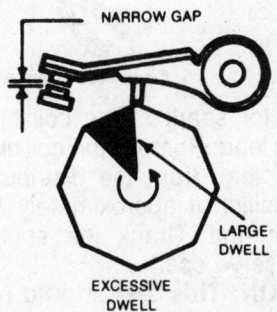

Dwell is a function of point gap

3.5 On the point-type ignition systems, check the condenser for short: connect an ohmeter across the condenser body and the pigtail lead.	If any reading other than infinite is noted, replace the condenser	3.6

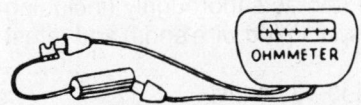

Checking the condenser for short

3.6 Test the coil primary resistance: On point-type ignition systems, connect an ohmmeter across the coil primary terminals, and read the resistance on the low scale. Note whether an external ballast resistor or resistance wire is used. On electronic ignition systems, test the coil primary resistance.	Point-type ignition coils utilizing ballast resistors or resistance wires should have approximately 1.0 ohms resistance. Coils with internal resistors should have approximately 4.0 ohms resistance. If values far from the above are noted, replace the coil.	4.1

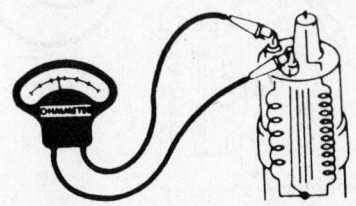

Checking the coil primary resistance

SECTION 4—SECONDARY ELECTRICAL SYSTEM

Test and Procedure	Results and Indications	Proceed to
4.1 Check for spark: Hold each spark plug wire approximately ¼″ from ground with gloves or heavy, dry rag. Crank the engine, and observe the spark.	If no spark is evident	4.2
	If spark is good in some cylinders	4.3
	If spark is good in all cylinders	4.6

Check for spark at the plugs

4.2 Check for spark at the coil high tension lead: Remove the coil high tension lead from the distributor and position it approximately ¼″ from ground. Crank the engine and observe spark. **CAUTION: This test should not be performed on engines equipped with electronic ignition.**	If the spark is good and consistent	4.3
	If the spark is good but intermittent, test the primary electrical system starting at 3.3.	3.3
	If the spark is weak or non-existent, replace the coil high tension lead, clean and tighten all connections and retest. If no improvement is noted	4.4

4.3 Visually inspect the distributor cap and rotor for burned or corroded contacts, cracks, carbon tracks, or moisture. Also check the fit of the rotor on the distributor shaft (where applicable).	If moisture is present, dry thoroughly, and retest per 4.1.	4.1
	If burned or excessively corroded contacts, cracks, or carbon tracks are noted, replace the defective part(s) and retest per 4.1.	4.1
	If the rotor and cap appear intact, or are only slightly corroded, clean the contacts thoroughly (including the cap towers and spark plug wire ends) and retest per 4.1.	
	If the spark is good in all cases	4.6
	If the spark is poor in all cases	4.5

CORRODED OR LOOSE WIRE

HIGH RESISTANCE CARBON

EXCESSIVE WEAR OF BUTTON

ROTOR TIP BURNED AWAY

Inspect the distributor cap and rotor

4.4 Check the coil secondary resistance: On point-type systems connect an ohmmeter across the distributor side of the coil and the coil tower. Read the resistance on the high scale of the ohmmeter. On electronic ignition systems, see The Electronic Ignition Unit Repair Section for specific tests.	The resistance of a satisfactory coil should be between 4,000 and 10,000 ohms. If resistance is considerably higher (i.e., 40,000 ohms) replace the coil and retest per 4.1. **NOTE: This does not apply to high performance coils.**

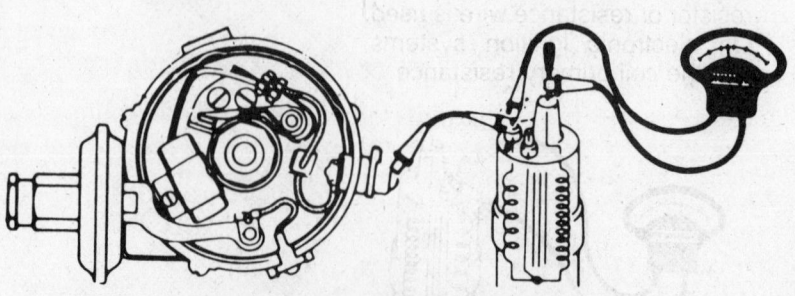

Testing the coil secondary resistance

Test and Procedure	Results and Indications	Proceed To
4.5 Visually inspect the spark plug wires for cracking or brittleness. Ensure that no two wires are positioned so as to cause induction firing (adjacent and parallel). Remove each wire, one by one, and check resistance with an ohmmeter.	Replace any cracked or brittle wires. If any of the wires are defective, replace the entire set. Replace any wires with excessive resistance (over 8000 Ω per foot for suppression wire), and separate any wires that might cause induction firing.	**4.6**

Misfiring can be the result of spark plug leads to adjacent, consecutively firing cylinders running parallel and too close together

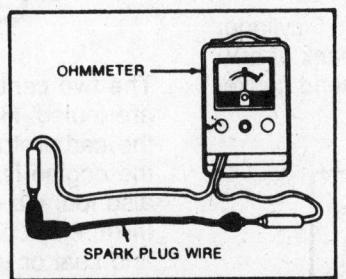

On point-type ignition systems, check the spark plug wires as shown. On electronic ignitions, do not remove the wire from the distributor cap terminal; instead, test through the cap

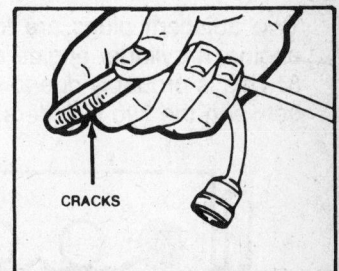

Spark plugs wires can be checked visually by bending them in a loop over your finger. This will reveal any cracks, burned or broken insulation. Any wire with cracked insulation should be replaced

4.6 Remove the spark plugs, noting the cylinders from which they were removed, and evaluate according to the chart in this section.	See chart.	**See Chart**

4.7 Reinstall the spark plugs.
NOTE: Modern electronic ignition systems generate extremely high voltages and high heats. The spark plug boots can soften and actually fuse to the ceramic insulator of the spark plugs after long exposures to high temperature and voltage. If this happens, the boot (and possibly the wire) must be replaced.

To help alleviate this condition, many manufacturers are recommending new silicone compounds to slow the deterioration. The compounds are generally nonconductive, protective lubricants that will not dry out, harden, or melt away. They form a weather-tight seal between rubber or plastic and metal and are found in several typical locations: Inside the insulating boots of spark plug wires, inside primary ignition circuit cable connectors, on distributor and rotor cap electrodes, and under the GM HEI control module.

4.8

Application Point	Silicone Compound
GENERAL MOTORS: Under HEI module	Supplied with new module, or use GE-642 or DC-340
FORD MOTOR COMPANY: Inside spark plug boots, on end of cable when installing new boot, and on rotor and cap electrodes	Ford part number D7AZ-19A331-A or use GE-627 or DC-111
CHRYSLER CORPORATION: ¼" deep within spark control computer connector cavity coating rotor electrode	Use Mopar part number 2932524 or NLGI Grade 2 EP (not a silicone) supplied with new rotor, or use GE-628 or DC-111
AMERICAN MOTORS (Prestolite system): Distributor primary connector—coat male terminal, fill female ¼ full	AMC part number 8127445 or GE-623

GE: General Electric
DC: Dow Corning

Test and Procedure	Results and Indications	Proceed To

4.8 Examine the location of all the plugs.

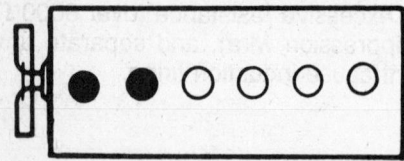

Two adjacent plugs are fouled in a 6-cylinder engine, 4-cylinder engine or either bank of a V-8. This is probably due to a blown head gasket between the two cylinders.

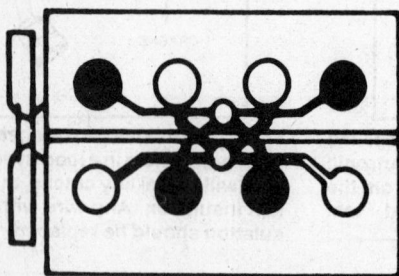

An unbalanced carburetor is indicated. Following the fuel flow on this particular design shows that the cylinders fed by the right-hand barrel are fouled from overly rich mixture, while the cylinders fed by the left-hand barrel are normal.

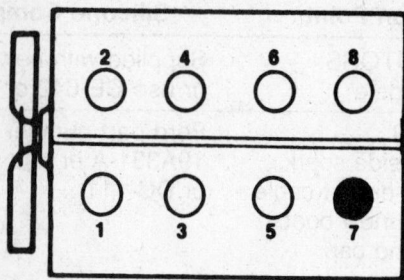

Finding one plug overheated may indicate an intake manifold leak near the affected cylinder. If the overheated plug is the second of two adjacent, consecutively firing plugs, it could be the result of ignition cross-firing. Separating the leads to these two plugs will eliminate cross-fire.

The following diagrams illustrate some of the conditions that the location of plugs will reveal.

4.9

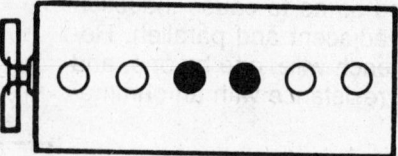

The two center plugs in a 6-cylinder engine are fouled. Raw fuel may be "boiled" out of the carburetor into the intake manifold after the engine is shut-off. Stop-start driving can also foul the center plugs, due to overly rich mixture. Proper float level, a new float needle and seat or use of an insulating spacer may help this problem.

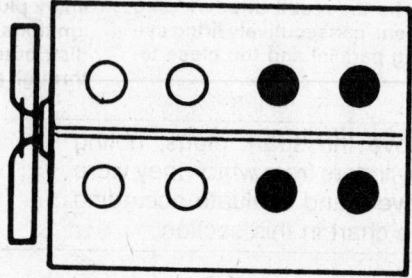

If the four rear plugs are overheated, a cooling system problem is suggested. A thorough cleaning of the cooling system may restore coolant circulation and cure the problem.

Occasionally, the two rear plugs in large, lightly used V-8's will become oil fouled. High oil consumption and smoky exhaust may also be noticed. It is probably due to plugged oil drain holes in the rear of the cylinder head, causing oil to be sucked in around the valve stems. This usually occurs in the rear cylinders first, because the engine slants that way.

Test and Procedure	Results and Indications	Proceed To
4.9 Determine the static ignition timing. Using the crankshaft pulley timing marks as a guide, locate top dead center on the compression stroke of the number one cylinder.	The rotor should be pointing toward the No. 1 tower in the distributor cap, and, on electronic ignitions, the armature spoke for that cylinder should be lined up with the stator.	4.10
4.10 Check coil polarity: Connect a voltmeter negative lead to the coil high tension lead, and the positive lead to ground. **NOTE: Reverse the hook-up for positive ground systems.** Crank the engine momentarily.	If the voltmeter reads up-scale, the polarity is correct.	5.1
	If the voltmeter reads down-scale, reverse the coil polarity (switch the primary leads).	5.1

Checking coil polarity

SECTION 5—FUEL SYSTEM

Test and Procedure	Results and Indications	Proceed to
5.1 Determine that the air filter is functioning efficiently: Hold paper elements up to a strong light, and attempt to see light through the filter.	Clean permanent air filters in solvent (or manufacturer's recommendation), and allow to dry. Replace paper elements through which light cannot be seen.	5.2
5.2 Determine whether a flooding condition exists: Flooding is identified by a strong gasoline odor, and excessive gasoline present in the throttle bore(s) of the carburetor.	If flooding is not evident	5.3
	If flooding is evident, permit the gasoline to dry for a few moments and restart.	
	If flooding doesn't recur	5.7
	If flooding is persistent	5.5

If the engine floods repeatedly, check the choke butterfly flap

5.3 Check that fuel is reaching the carburetor: Detach the fuel line at the carburetor inlet. Hold the end of the line in a cup (not styrofoam), and crank the engine.	If fuel flows smoothly	5.7
	If fuel doesn't flow	5.4
	If fuel flows erratically.	
	NOTE: Make sure that there is fuel in the tank	

Check the fuel pump by disconnecting the output line (fuel pump-to-carburetor) at the carburetor and operating the starter briefly

Test and Procedure	Results and Indications	Proceed To
5.4 Test the fuel pump: Disconnect all fuel lines from the fuel pump. Hold a finger over the input fitting, crank the engine (with electric pump, turn the ignition or pump on); and feel for suction.	If suction is evident, blow out the fuel line to the tank with low pressure compressed air until bubbling is heard from the fuel filler neck. Also blow out the carburetor fuel line (both ends disconnected).	5.7
	If no suction is evident, replace or repair the fuel pump. **NOTE: Repeated oil fouling of the spark plugs, or a no-start condition, could be the result of a ruptured vacuum booster pump diaphragm, through which oil or gasoline is being drawn into the intake manifold (where applicable).**	5.7
5.5 Occasionally, small specks of dirt will clog the small jets and orifices in the carburetor. With the engine cold, hold a flat piece of wood or similar material over the carburetor, where possible, and crank the engine.	If the engine starts, but runs roughly the engine is probably not run enough.	
	If the engine won't start.	5.9
5.6 Check the needle and seat: Tap the carburetor in the area of the needle and seat.	If flooding stops, a gasoline additive (e.g., Gumout) will often cure the problem.	5.7
	If flooding continues, check the fuel pump for excessive pressure at the carburetor (according to specifications). If the pressure is normal, the needle and seat must be removed and checked, and/or the float level adjusted.	5.7
5.7 Test the accelerator pump by looking into the throttle bores while operating the throttle.	If the accelerator pump appears to be operating normally	5.8
	If the accelerator pump is not operating, the pump must be reconditioned. Where possible, service the pump with the carburetor(s) installed on the engine. If necessary, remove the carburetor. Prior to removal	5.8

Check for gas at the carburetor by looking down the carburetor throat while someone moves the accelerator

Test and Procedure	Results and Indications	Proceed To
5.8 Determine whether the carburetor main fuel system is functioning: Spray a commercial starting fluid into the carburetor while attempting to start the engine.	If the engine starts, runs for a few seconds, and dies	5.9
	If the engine doesn't start	6.1
5.9 Uncommon fuel system malfunctions: See below:	If the problem is solved	6.1
	If the problem remains, remove and recondition the carburetor.	

Condition	Indication	Test	Prevailing Weather Conditions	Remedy
Vapor lock	Engine will not re-start shortly after running.	Cool the components of the fuel system until the engine starts. Vapor lock can be cured faster by draping a wet cloth over a mechanical fuel pump.	Hot to very hot	Ensure that the exhaust manifold heat control valve is operating. Check with the vehicle manufacturer for the recommended solution to vapor lock on the model in question.
Carburetor icing	Engine will not idle, stalls at low speeds.	Visually inspect the throttle plate area of the throttle bores for frost.	High humidity, 32–40° F.	Ensure that the exhaust manifold heat control valve is operating, and that the intake manifold heat riser is not blocked.
Water in the fuel	Engine sputters and stalls; may not start.	Pump a small amount of fuel into a glass jar. Allow to stand, and inspect for droplets of a layer of water.	High humidity, extreme temperature changes.	For droplets, use one or two cans of commercial gas line anti-freeze. For a layer of water, the tank must be drained, and the fuel lines blown out with compressed air.

SECTION 6—ENGINE COMPRESSION

Test and Procedure	Results and Indications	Proceed to
6.1 Test engine compression: Remove all spark plugs. Block the throttle wide open. Insert a compression gauge into a spark plug port, crank the engine to obtain the maximum reading, and record.	If compression is within limits on all cylinders	7.1
	If gauge reading is extremely low on all cylinders	6.2
	If gauge reading is low on one or two cylinders: (If gauge readings are identical and low on two or more adjacent cylinders, the head gasket must be replaced.)	6.2

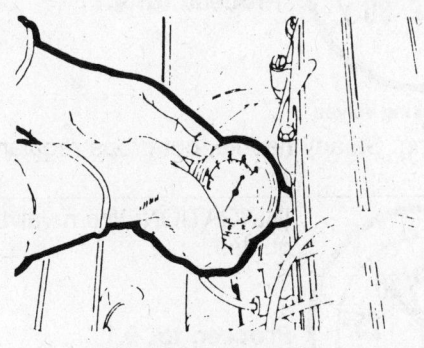

Checking compression

6.2 Test engine compression (wet): Squirt approximately 30 cc. of engine oil into each cylinder, and retest per 6.1.	If the readings improve, worn or cracked rings or broken pistons are indicated:	**See Car Section**
	If the readings do not improve, burned or excessively carboned valves or a jumped timing chain are indicated.	**7.1**
	NOTE: A jumped timing chain is often indicated by difficult cranking.	

SECTION 7—ENGINE VACUUM

Test and Procedure	Results and Indications	Proceed to
7.1 Attach a vacuum gauge to the intake manifold beyond the throttle plate. Start the engine, and observe the action of the needle over the range of engine speeds.	See below.	**See below**

INDICATION: Normal engine in good condition

Proceed to: 8.1

Normal engine

Gauge reading: Steady, from 17–22 in./Hg.

INDICATION: Sticking valves or ignition miss

Proceed to: 9.1, 8.3

Sticking valves

Gauge reading: Intermittent fluctuation at idle

INDICATION: Late ignition or valve timing, low compression, stuck throttle valve, leaking carburetor or manifold gasket

Proceed to: 6.1

Incorrect valve timing

Gauge reading: Low (10–15 in./Hg) but steady

INDICATION: Improper carburetor adjustment or minor intake leak.

Proceed to: 7.2

Carburetor requires adjustment

Gauge reading: Drifting needle

INDICATION: Ignition miss, blown cylinder head gasket, leaking valve or weak valve spring

Proceed to: 8.3, 6.1

Blown head gasket

Gauge reading: Needle fluctuates as engine speed increases

INDICATION: Burnt valve or faulty valve clearance: Needle will fall when defective valve operates

Proceed to: 9.1

Burnt or leaking valves

Gauge reading: Steady needle, but drops regularly

INDICATION: Choked muffler, excessive back pressure in system

Proceed to: 10.1

Clogged exhaust system

Gauge reading: Gradual drop in reading at idle

INDICATION: Worn valve guides

Proceed to: 9.1

Worn valve guides

Gauge reading: Needle vibrates excessively at idle, but steadies as engine speed increases

White pointer = steady gauge hand

Black pointer = fluctuating gauge hand

Test and Procedure	Results and Indications	Proceed To
7.2 Attach a vacuum gauge per 7.1, and test for an intake manifold leak. Squirt a small amount of oil around the intake manifold gaskets, carburetor gaskets, plugs and fittings. Observe the action of the vacuum gauge.	If the reading improves, replace the indicated gasket, or seal the indicated fitting or plug:	8.1
	If the reading remains low:	7.3
7.3 Test all vacuum hoses and accessories for leaks as described in 7.2. Also check the carburetor body (dashpots, automatic choke mechanism, throttle shafts) for leaks in the same manner.	If the reading improves, service or replace the offending part(s):	8.1
	If the reading remains low:	6.1

SECTION 8—SECONDARY ELECTRICAL SYSTEM

Test and Procedure	Results and Indications	Proceed to
8.1 Remove the distributor cap and check to make sure that the rotor turns when the engine is cranked. Visually inspect the distributor components.	Clean, tighten or replace any components which appear defective.	8.2
8.2 Connect a timing light (per manufacturer's recommendation) and check the dynamic ignition timing. Disconnect and plug the vacuum hose(s) to the distributor if specified, start the engine, and observe the timing marks at the specified engine speed.	If the timing is not correct, adjust to specifications by rotating the distributor in the engine: (Advance timing by rotating distributor opposite normal direction of rotor rotation, retard timing by rotating distributor in same direction as rotor rotation.)	8.3
8.3 Check the operation of the distributor advance mechanism(s): To test the mechanical advance, disconnect the vacuum lines from the distributor advance unit and observe the timing marks with a timing light as the engine speed is increased from idle. If the mark moves smoothly, without hesitation, it may be assumed that the mechanical advance is functioning properly. To test vacuum advance and/or retard systems, alternately crimp and release the vacuum line, and observe the timing mark for movement. If movement is noted, the system is operating.	If the systems are functioning	8.4
	If the systems are not functioning, remove the distributor, and test on a distributor tester.	8.4
8.4 Locate an ignition miss: With the engine running, remove each spark plug wire, one at a time, until one is found that doesn't cause the engine to roughen and slow down. CAUTION: Do not pull on the wire to remove the boot from the plug. Be sure your hand is insulated from the wire.	When the missing cylinder is identified	4.1

SECTION 9—VALVE TRAIN

Test and Procedure	Results and Indications	Proceed to
9.1 Evaluate the valve train: Remove the valve cover, and ensure that the valves are adjusted to specifications. A mechanic's stethoscope may be used to aid in the diagnosis of the valve train. By pushing the probe on or near push rods or rockers, valve noise often can be isolated. A timing light also may be used to diagnose valve problems. Connect the light according to manufacturer's recommendations, and start the engine. Vary the firing moment of the light by increasing the engine speed (and therefore the ignition advance), and moving the trigger from cylinder to cylinder. Observe the movement of each valve.	Sticking valves or erratic valve train motion can be observed with the timing light. The cylinder head must be disassembled for repairs.	**See Car Section**
9.2 Check the valve timing: Locate top dead center of the No. 1 piston, and install a degree wheel or tape on the crankshaft pulley or damper with zero corresponding to an index mark on the engine. Rotate the crankshaft in its direction of rotation, and observe the opening of the No. 1 cylinder intake valve. The opening should correspond with the correct mark on the degree wheel according to specifications.	If the timing is not correct, the timing cover must be removed for further investigation.	**See Car Section**

SECTION 10—EXHAUST SYSTEM

Test and Procedure	Results and Indications	Proceed to
10.1 Determine whether the exhaust manifold heat control valve is operating: Operate the valve by hand to determine whether it is free to move. If the valve is free, run the engine to operating temperature and observe the action of the valve, to ensure that it is opening.	If the valve sticks, spray it with a suitable solvent, open and close the valve to free it, and retest. If the valve functions properly	**10.2**
	If the valve does not free, or does not operate, replace the valve.	**10.2**
10.2 Ensure that there are no exhaust restrictions: Visually inspect the exhaust system for kinks, dents, or crushing. Also note that gases are flowing freely from the tailpipe at all engine speeds, indicating no restriction in the muffler or resonator.	Replace any damaged portion of the system.	**11.1**

SECTION 11—COOLING SYSTEM

Test and Procedure	Results and Indications	Proceed to
11.1 Visually inspect the fan belt for glazing, cracks, and fraying, and replace if necessary. Tighten the belt so that the longest span has approximately ½″ play at its midpoint under thumb pressure (see Maintenance Section).	Replace or tighten the fan belt as necessary.	**11.2**

Checking belt tension

Test and Procedure	Results and Indications	Proceed to
11.2 Check the fluid level of the cooling system.	If full or slightly low, fill as necessary.	**11.5**
	If extremely low	**11.3**
11.3 Visually inspect the external portions of the cooling system (radiator, radiator hoses, thermostat elbow, water pump seals, heater hoses, etc.) for leaks. If none are found, pressurize the cooling system to 14–15 psi.	If cooling system holds the pressure	**11.5**
	If cooling system loses pressure rapidly, reinspect external parts of the system for leaks under pressure. If none are found, check dipstick for coolant in crankcase. If no coolant is present, but pressure loss continues	**11.4**
	If coolant is evident in crankcase, remove cylinder head(s), and check gasket(s). If gaskets are intact, block and cylinder head(s) should be checked for cracks or holes.	
	If the gasket(s) is blown, replace, and purge the crankcase of coolant.	**12.6**
	NOTE: Occasionally, due to atmospheric and driving conditions, condensation of water can occur in the crankcase. This causes the oil to appear milky white. To remedy, run the engine until hot, and change the oil and oil filter.	
11.4 Check for combustion leaks into the cooling system: Pressurize the cooling system as above. Start the engine, and observe the pressure gauge. If the needle fluctuates, remove each spark plug wire, one at a time, noting which cylinder(s) reduce or eliminate the fluctuation.	Cylinders which reduce or eliminate the fluctuation, when the spark plug wire is removed, are leaking into the cooling system. Replace the head gasket on the affected cylinder bank(s).	**See Car Section**

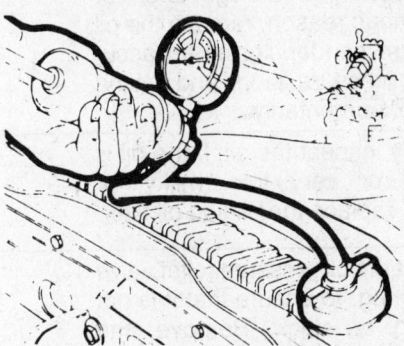

Pressurizing the cooling system

Test and Procedure	Results and Indications	Proceed To
11.5 Check the radiator pressure cap: Attach a radiator pressure tester to the radiator cap (wet the seal prior to installation). Quickly pump up the pressure, noting the point at which the cap releases.	If the cap releases within ±1 psi of the specified rating, it is operating properly.	11.6
	If the cap releases at more than ±1 psi of the specified rating, it should be replaced.	11.6

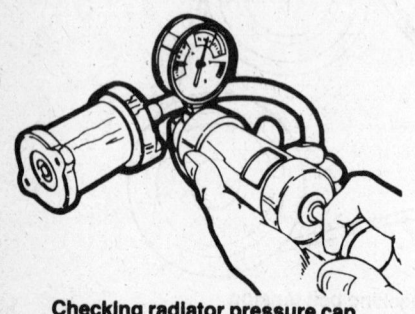

Checking radiator pressure cap

Test and Procedure	Results and Indications	Proceed To
11.6 Test the thermostat: Start the engine cold, remove the radiator cap, and insert a thermometer into the radiator. Allow the engine to idle. After a short while, there will be a sudden, rapid increase in coolant temperature. The temperature at which this sharp rise stops is the thermostat opening temperature.	If the thermostat opens at or about the specified temperature	11.7
	If the temperature doesn't increase (If the temperature increases slowly and gradually, replace the thermostat.)	11.7
11.7 Check the water pump: Remove the thermostat elbow and the thermostat, disconnect the coil high tension lead (to prevent starting), and crank the engine momentarily.	If coolant flows, replace the thermostat and retest per 11 6.	11.6
	If coolant doesn't flow, reverse flush the cooling system to alleviate any blockage that might exist. If system is not blocked, and coolant will not flow, replace the water pump.	See Car Section

SECTION 12—LUBRICATION

Test and Procedure	Results and Indications	Proceed to
12.1 Check the oil pressure gauge or warning light: If the gauge shows low pressure, or the light is on for no obvious reason, remove the oil pressure sender. Install an accurate oil pressure gauge and run the engine momentarily.	If oil pressure builds normally, run engine for a few moments to determine that it is functioning normally, and replace the sender.	—
	If the pressure remains low	12.2
	If the pressure surges	12.3
	If the oil pressure is zero	12.3
12.2 Visually inspect the oil: If the oil is watery or very thin, milky, or foamy, replace the oil and oil filter.	If the oil is normal	12.3
	If after replacing oil the pressure remains low	12.3
	If after replacing oil the pressure becomes normal	—
12.3 Inspect the oil pressure relief valve and spring, to ensure that it is not sticking or stuck. Remove and thoroughly clean the valve, spring, and the valve body.	If the oil pressure improves	—
	If no improvement is noted	12.4

Test and Procedure	Results and Indications	Proceed To
12.4 Check to ensure that the oil pump is not cavitating (sucking air instead of oil): See that the crankcase is neither over nor underfull, and that the pickup in the sump is in the proper position and free from sludge.	Fill or drain the crankcase to the proper capacity, and clean the pickup screen in solvent if necessary. If no improvement is noted	**12.5**
12.5 Inspect the oil pump drive and the oil pump:	If the pump drive or the oil pump appear to be defective, service as necessary and retest per 12.1. If the pump drive and pump appear to be operating normally, the engine should be disassembled to determine where blockage exists.	**12.1**
12.6 Purge the engine of ethylene glycol coolant: Competely drain the crankcase and the oil filter. Obtain a commercial butyl cellosolve base solvent, designated for this purpose, and follow the instructions precisely. Following this, install a new oil filter and refill the crankcase with the proper weight oil. The next oil and filter change should follow shortly thereafter (1000 miles).		

Diesel Engine Troubleshooting

NOTE: The following troubleshooting procedures cover problems usually associated with diesel engines. Those problems common to both gasoline and diesel engines are covered in the gasoline engine troubleshooting procedures.

INDEX TO PROBLEMS

Problem/Symptom	Begin at Specific Diagnosis, Number
Fuel System	Section 1
Engine Starting Difficulty:	
Feed pump does not feed fuel	1.1
Injection pump does not feed fuel	1.2
Incorrect injection timing	1.3
Defective injection nozzles	1.4
Engine Operating Instability:	
Engine shuts off immediately after starting	1.5
Uneven idling	1.6
Engine will not reach maximum rated speed	1.7
Engine exceeds maximum rated speed	1.8
Loss of power	1.9
Engine Knock:	
Associated with exhaust gas problems	1.10
Not associated with exhaust gas problems	1.11
Engine Mechanical	Section 2
Engine Starting Difficulty	2.1
Unusual Noises	2.2
Engine Operating Instability	2.3
Loss of Power	2.4
Exhaust gas Problem	2.5
Engine Shut-Off	2.6
Loss of Oil Pressure	2.7
Oil Leakage	2.8
Compression Pressure Leakage	2.9

SECTION 1—Fuel System

Test and Procedure		Results and Indication	Proceed To
1.1a	Check for pressure at the outlet of the feed pump	If pressure exists, there is a clog in the supply line. Clean or replace it. If there is little or no pressure at the outlet, the filter is clogged. Clean or replace the filter. If the filter is clear, the feed pump piston is inoperative. Relace it.	1.1b
1.1b	Check the feed pump valves	If the inlet and outlet valves do not operate, the check valve or spring is broken. Replace it.	1.2a
1.2a	Check for fuel leakage at the overflow or return line	A clogged filter can result in high pressure causing leakage. Replace the filter.	1.2b
1.2b	Check for fuel in the filter leaking at the overflow valve	If leakage is found, the overflow valve is damaged. Replace it.	1.2c
1.2c	Check for leakage at the injection pump overflow valve	If leakage is found, it is caused by: damaged overflow valve, sticking plunger, or sticking delivery valve. Replace the defective part(s).	1.2d
1.2d	Check the injection pump plunger feed pressures.	If pressure at the plungers is low, replace the plunger(s).	1.2e

Test and Procedure		Results and Indication	Proceed To
1.2e	Check to make sure the injection pump is operating	An inoperative pump is caused by: a damaged or missing shaft key, or a damaged drive gear train.	1.3a
1.3a	Check that the pump timing marks are correctly aligned in the gear train	Incorrect timing marks alignment must be corrected.	1.3b
1.3b	Check that the injection pump is properly mounted	Remove and install the pump correctly	1.4a
1.4a	Install an injection nozzle on a tester and make sure that fuel is continuously ejected	A broken or intermittent stream is caused by a damaged spring or a sticking nozzle needle	1.4b
1.4b	With the nozzle on the tester as in 1.4a, check that shutoff is clean with no dribble or afterdrip	Dribble is caused by a defective nozzle valve seat. Replace the nozzle.	1.4c
1.4c	Using a tester, check injection pressure	Low pressure is a result of a weak spring. Replace the spring or adjust the initial injection pressure.	1.5a
1.5a	See 1.2a	Proceed as in 1.2a	1.5b
1.5b	Check for water in the fuel	Drain and clean the tank	1.5c
1.5c	Check for air in the fuel lines	Air can be introduced through a damaged fuel inlet line, a loose inlet line connector or a damaged gasket	1.5d
1.5d	Check for insufficient fuel feed	Insufficient fuel feed is caused by: a damaged feed pump, a clogged tank vent, or a clogged filter. Replace or repair as necessary.	1.6a
1.6a	Check the control rack action for smooth operation	Uneven control rack operation is caused by: a sticking plunger, improper meshing of the rack and pinion, poor seating of the plunger spring, insufficient clearance between the plunger and lower spring seat, or an overly tight delivery valve holder. Replace or adjust as necessary.	1.6b
1.6b	Check that the injection pump discharge is uniform	If the output is uneven, adjust as necessary	1.6c
1.6c	Check that the injection pump discharge volume is adequate	An inadequate discharge volume is caused by a worn plunger or a broken spring	1.6d
1.6d	Check for even low speed engine performance	If the engine performs unevenly or erratically at low speed only, a worn feed pump piston or defective feed pump valve is the cause.	1.6e
1.6e	Check for smooth engine operation throughout the operating range	This problem is usually caused by mechanical governor defects such as: a defective low speed spring, defective damper spring, or excessive friction among moving parts. Replace the defective parts.	1.6f
1.6f	Check the injectors on a tester	Improper nozzle operation should be corrected accordingly	1.7a
1.7a	Check the operating governor	A broken or weak spring in the governor will prevent full speed operation.	1.7b
1.7b	Check the injectors on a tester for a drop in injector output	A drop in output is caused by a sticking needle or a dirty nozzle. Replace or clean as necessary.	1.8a

Test and Procedure		Results and Indications	Proceed To
1.8a	Check the injection pump for proper rack and pinion action	A catching or dirty rack and pinion will cause overspeeding.	1.8b
1.8b	Check the governor adjustment	An improperly adjusted governor will cause overspeeding. Adjust.	1.9a
1.9a	Check the injection pump output	Low output can be caused by: Incorrect adjustment—Adjust Loose delivery valve—Tighten Broken delivery valve seal—Replace Poor valve seat contact—Replace Broken/weak delivery valve spring—Replace	1.9b
1.9b	Check for unusual noise at the injection pump	A noisy pump is an indication of a broken plunger spring	1.9c
1.9c	Check plunger operation	A sticking injection pump plunger will cause power loss. Replace.	1.9d
1.9d	Check the injection timer	A lag in injection timing is caused by large clearances in the timer due to wear. Replace.	1.9e
1.9e	Check for air or water in the fuel	Bleed the air or drain the fuel and clean the tank and lines	1.9f
1.9f	Check the injection timing	Readjust timing if necessary	1.10a
1.10a	Check the initial injection timing	Adjust if necessary	1.10b
1.10b	Check the injection pressure	High pressure will cause knock. Adjust as necessary	1.10c
1.10c	Check the injector nozzle	A clogged nozzle causes knock. Clean or replace the nozzle.	1.11a
1.11a	Check the injection pump output and timing	Excessive output, coupled with incorrect timing causes knock. Adjust as necessary	1.11b
1.11b	Check the delivery valve seat	Replace a defective seat	1.11c
1.11c	Check the pump plungers	Replace badly worn plungers	1.11d
1.11d	Check injector opening pressure on a tester	Adjust as necessary	1.11e
1.11e	Check the injector	Replace a broken nozzle spring or sticking needle.	

SECTION 2—ENGINE MECHANICAL

Test and Procedure		Results and Indications	Proceed To
2.1a	Check for piston seizing	Seized pistons are caused by low oil pressure, oil breakdown, or overheating. Replace the pistons and liners.	2.1b
2.1b	Check for a damaged flywheel ring gear	A damaged ring gear will cause poor meshing with the starter. Replace the ring gear.	2.1c
2.1c	Make a compression check	Low compression can be caused by: sticking rings, worn rings, worn liners. Replace the rings or liners.	2.2a
2.2a	A knocking noise at idle or during acceleration can be caused by a variety of wear problems.	Use a stethoscope or similar listening device to try to pinpoint the source of the noise. Among other reasons for knocking are: piston pins, rod bearings, loose rod caps, crankshaft journals and/or bearings, crankshaft thrust washer. Replace any worn parts.	2.2b
2.2b	An infrequently encountered noise is a continuous growl during acceleration	This problem is usually caused by problems in the engine timing gears. Poor contact, excessive backlash or loose gears are usually at fault.	2.2c
2.2c	Intermittent noises are the hardest to find. They are usually caused by broken moving parts.	Check the gear train for a chipped or cracked gear; the oil pan for broken parts or foreign objects or the cylinder head for a broken valve or valve spring.	2.3
2.3	Check for oil in the combustion chambers	Oil entering the combustion chambers will cause the engine to overspeed if the amount of oil is too great, or run unevenly. Check for broken or sticking rings, bad head gasket(s) or worn valve guides.	2.4
2.4	Check the compression	Low compression is the main cause of power loss. The main causes for low compression are: worn rings or liners, cracked valves, warped head or block, and bad head gasket.	2.5
2.5	A large amount of black exhaust is caused by low compression	See 2.4 above	2.6
2.6	If the engine stops suddenly during operation, the cause is usually sudden damage	Check the pistons, main bearings or rod bearings for lack of lubrication. A seized camshaft is also a result of low or no lubrication. Check the timing gears for damage.	2.7
2.7	Check for excessive clearance between the bearings and journals on both the mains and rod bearings. Check the oil pressure.	Replace as necessary. Replace the pump as necessary.	2.8
2.8	Aside from the usual leaking gasket problems, check the condition of the combustion chamber O-rings.	Replace as necessary	2.9
2.9	Compression leakage is usually caused by a seal defect between the head and the block	Check the head gasket; check for loose head bolts; check for head or block warpage. Replace or repair as necessary.	

TROUBLESHOOTING AND DIAGNOSIS

Engine Overheating Troubleshooting

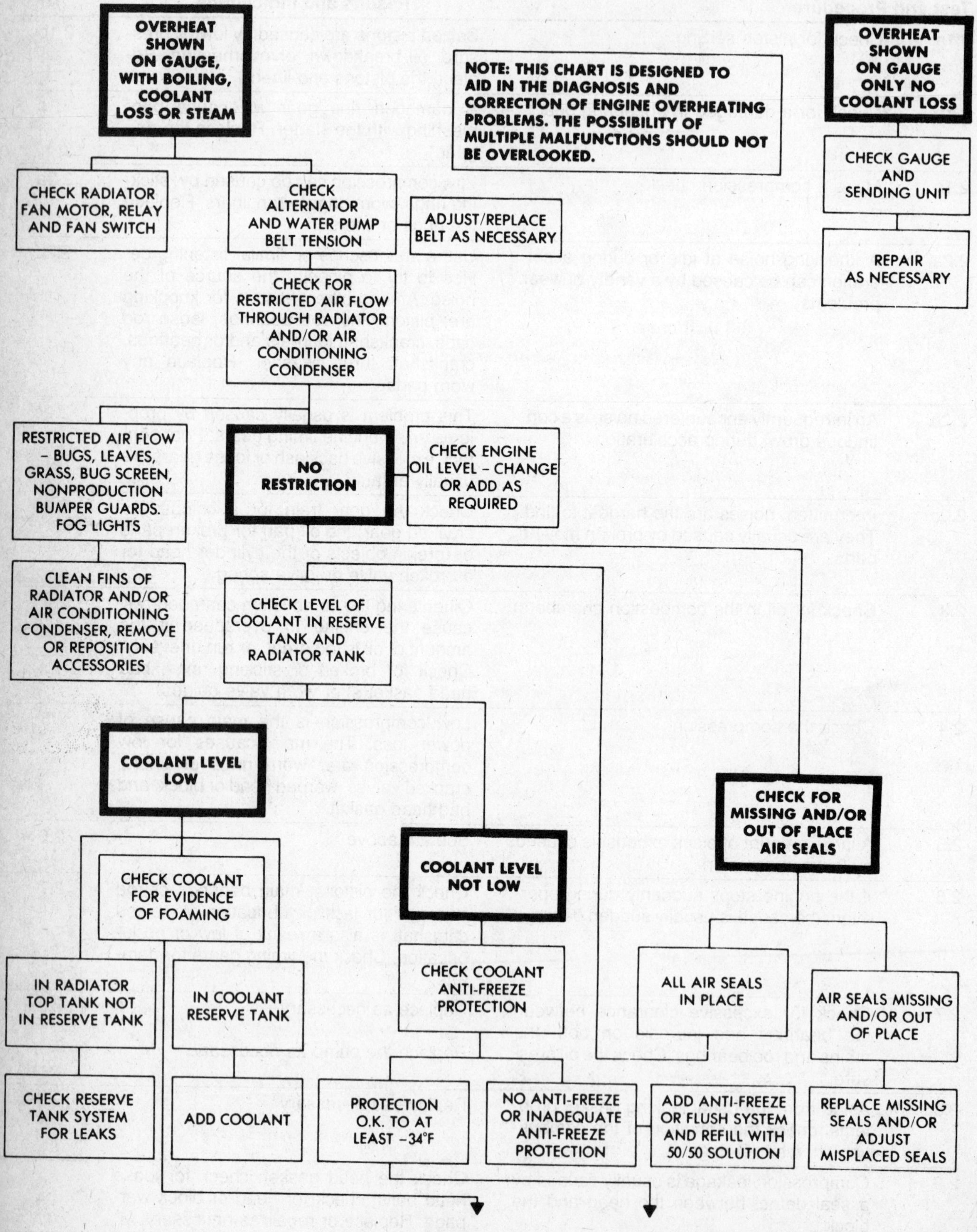

OVERHEAT SHOWN ON GAUGE, WITH BOILING, COOLANT LOSS OR STEAM

NOTE: THIS CHART IS DESIGNED TO AID IN THE DIAGNOSIS AND CORRECTION OF ENGINE OVERHEATING PROBLEMS. THE POSSIBILITY OF MULTIPLE MALFUNCTIONS SHOULD NOT BE OVERLOOKED.

OVERHEAT SHOWN ON GAUGE ONLY NO COOLANT LOSS

CHECK RADIATOR FAN MOTOR, RELAY AND FAN SWITCH

CHECK ALTERNATOR AND WATER PUMP BELT TENSION

ADJUST/REPLACE BELT AS NECESSARY

CHECK GAUGE AND SENDING UNIT

REPAIR AS NECESSARY

CHECK FOR RESTRICTED AIR FLOW THROUGH RADIATOR AND/OR AIR CONDITIONING CONDENSER

RESTRICTED AIR FLOW – BUGS, LEAVES, GRASS, BUG SCREEN, NONPRODUCTION BUMPER GUARDS. FOG LIGHTS

NO RESTRICTION

CHECK ENGINE OIL LEVEL – CHANGE OR ADD AS REQUIRED

CLEAN FINS OF RADIATOR AND/OR AIR CONDITIONING CONDENSER, REMOVE OR REPOSITION ACCESSORIES

CHECK LEVEL OF COOLANT IN RESERVE TANK AND RADIATOR TANK

COOLANT LEVEL LOW

CHECK FOR MISSING AND/OR OUT OF PLACE AIR SEALS

CHECK COOLANT FOR EVIDENCE OF FOAMING

COOLANT LEVEL NOT LOW

IN RADIATOR TOP TANK NOT IN RESERVE TANK

IN COOLANT RESERVE TANK

CHECK COOLANT ANTI-FREEZE PROTECTION

ALL AIR SEALS IN PLACE

AIR SEALS MISSING AND/OR OUT OF PLACE

CHECK RESERVE TANK SYSTEM FOR LEAKS

ADD COOLANT

PROTECTION O.K. TO AT LEAST –34°F

NO ANTI-FREEZE OR INADEQUATE ANTI-FREEZE PROTECTION

ADD ANTI-FREEZE OR FLUSH SYSTEM AND REFILL WITH 50/50 SOLUTION

REPLACE MISSING SEALS AND/OR ADJUST MISPLACED SEALS

Engine Overheating Troubleshooting

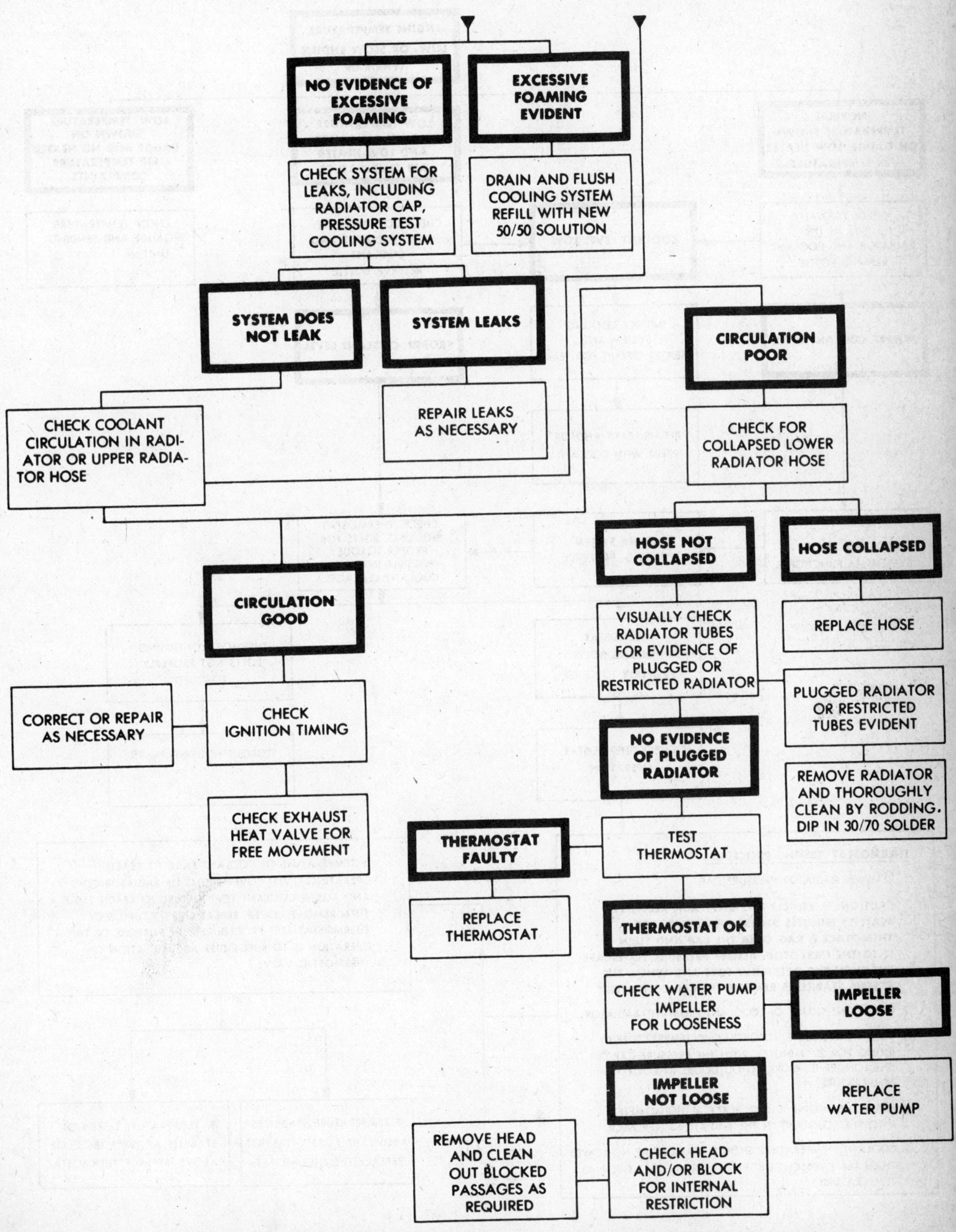

NO EVIDENCE OF EXCESSIVE FOAMING

CHECK SYSTEM FOR LEAKS, INCLUDING RADIATOR CAP, PRESSURE TEST COOLING SYSTEM

EXCESSIVE FOAMING EVIDENT

DRAIN AND FLUSH COOLING SYSTEM REFILL WITH NEW 50/50 SOLUTION

SYSTEM DOES NOT LEAK

SYSTEM LEAKS

REPAIR LEAKS AS NECESSARY

CIRCULATION POOR

CHECK FOR COLLAPSED LOWER RADIATOR HOSE

CHECK COOLANT CIRCULATION IN RADIATOR OR UPPER RADIATOR HOSE

HOSE NOT COLLAPSED

VISUALLY CHECK RADIATOR TUBES FOR EVIDENCE OF PLUGGED OR RESTRICTED RADIATOR

HOSE COLLAPSED

REPLACE HOSE

PLUGGED RADIATOR OR RESTRICTED TUBES EVIDENT

CIRCULATION GOOD

CORRECT OR REPAIR AS NECESSARY

CHECK IGNITION TIMING

NO EVIDENCE OF PLUGGED RADIATOR

REMOVE RADIATOR AND THOROUGHLY CLEAN BY RODDING. DIP IN 30/70 SOLDER

CHECK EXHAUST HEAT VALVE FOR FREE MOVEMENT

THERMOSTAT FAULTY

TEST THERMOSTAT

REPLACE THERMOSTAT

THERMOSTAT OK

CHECK WATER PUMP IMPELLER FOR LOOSENESS

IMPELLER LOOSE

REPLACE WATER PUMP

IMPELLER NOT LOOSE

REMOVE HEAD AND CLEAN OUT BLOCKED PASSAGES AS REQUIRED

CHECK HEAD AND/OR BLOCK FOR INTERNAL RESTRICTION

Low Engine Temperature Troubleshooting

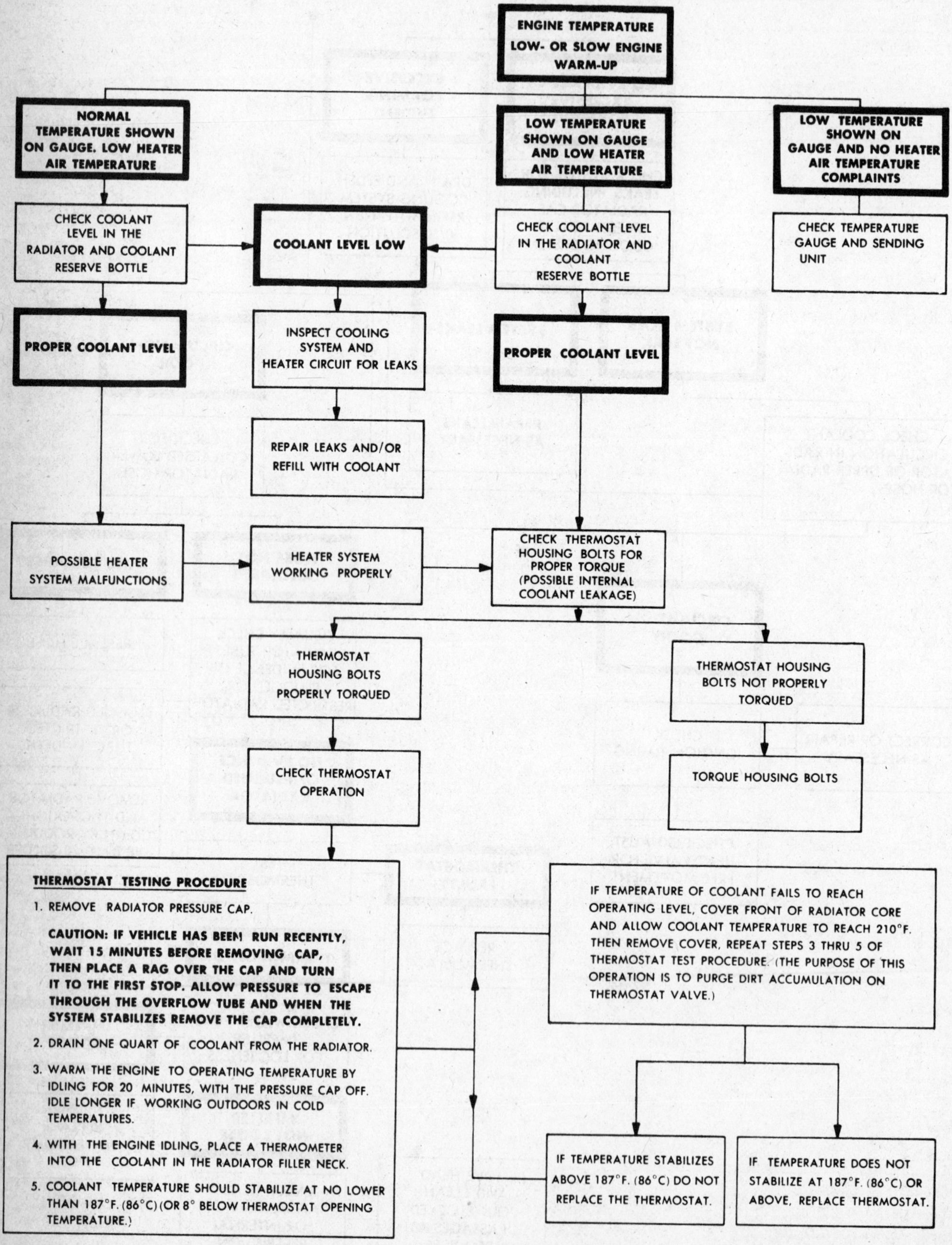

ENGINE TEMPERATURE LOW- OR SLOW ENGINE WARM-UP

NORMAL TEMPERATURE SHOWN ON GAUGE. LOW HEATER AIR TEMPERATURE

LOW TEMPERATURE SHOWN ON GAUGE AND LOW HEATER AIR TEMPERATURE

LOW TEMPERATURE SHOWN ON GAUGE AND NO HEATER AIR TEMPERATURE COMPLAINTS

CHECK COOLANT LEVEL IN THE RADIATOR AND COOLANT RESERVE BOTTLE

COOLANT LEVEL LOW

CHECK COOLANT LEVEL IN THE RADIATOR AND COOLANT RESERVE BOTTLE

CHECK TEMPERATURE GAUGE AND SENDING UNIT

PROPER COOLANT LEVEL

INSPECT COOLING SYSTEM AND HEATER CIRCUIT FOR LEAKS

PROPER COOLANT LEVEL

REPAIR LEAKS AND/OR REFILL WITH COOLANT

POSSIBLE HEATER SYSTEM MALFUNCTIONS

HEATER SYSTEM WORKING PROPERLY

CHECK THERMOSTAT HOUSING BOLTS FOR PROPER TORQUE (POSSIBLE INTERNAL COOLANT LEAKAGE)

THERMOSTAT HOUSING BOLTS PROPERLY TORQUED

THERMOSTAT HOUSING BOLTS NOT PROPERLY TORQUED

CHECK THERMOSTAT OPERATION

TORQUE HOUSING BOLTS

THERMOSTAT TESTING PROCEDURE

1. REMOVE RADIATOR PRESSURE CAP.

   CAUTION: IF VEHICLE HAS BEEN RUN RECENTLY, WAIT 15 MINUTES BEFORE REMOVING CAP, THEN PLACE A RAG OVER THE CAP AND TURN IT TO THE FIRST STOP. ALLOW PRESSURE TO ESCAPE THROUGH THE OVERFLOW TUBE AND WHEN THE SYSTEM STABILIZES REMOVE THE CAP COMPLETELY.

2. DRAIN ONE QUART OF COOLANT FROM THE RADIATOR.

3. WARM THE ENGINE TO OPERATING TEMPERATURE BY IDLING FOR 20 MINUTES, WITH THE PRESSURE CAP OFF. IDLE LONGER IF WORKING OUTDOORS IN COLD TEMPERATURES.

4. WITH THE ENGINE IDLING, PLACE A THERMOMETER INTO THE COOLANT IN THE RADIATOR FILLER NECK.

5. COOLANT TEMPERATURE SHOULD STABILIZE AT NO LOWER THAN 187°F. (86°C) (OR 8° BELOW THERMOSTAT OPENING TEMPERATURE.)

IF TEMPERATURE OF COOLANT FAILS TO REACH OPERATING LEVEL, COVER FRONT OF RADIATOR CORE AND ALLOW COOLANT TEMPERATURE TO REACH 210°F. THEN REMOVE COVER, REPEAT STEPS 3 THRU 5 OF THERMOSTAT TEST PROCEDURE. (THE PURPOSE OF THIS OPERATION IS TO PURGE DIRT ACCUMULATION ON THERMOSTAT VALVE.)

IF TEMPERATURE STABILIZES ABOVE 187°F. (86°C) DO NOT REPLACE THE THERMOSTAT.

IF TEMPERATURE DOES NOT STABILIZE AT 187°F. (86°C) OR ABOVE, REPLACE THERMOSTAT.

DRIVELINE
Clutch System Troubleshooting

Condition	Possible Cause	Corrective Action
Clutch chatter	1. Grease on driven plate (disc) facing. 2. Binding clutch linkage. 3. Loose, damaged facings on driven plate (disc). 4. Engine mounts loose. 5. Incorrect height adjustment of pressure plate release levers. 6. Clutch housing or housing to transmission adapter misalignment. 7. Loose driven plate hub.	1. Replace plate. 2. Check for worn, bent, broken parts. Replace as required. Lube linkage. 3. Replace driven plate. 4. Tighten mounts. Replace if damaged. 5. Adjust release lever height. 6. Check bore and face run out. Correct as required. 7. Replace driven plate.
Clutch grabbing	1. Oil, grease on driven plate (disc) facing. 2. Broken pressure plate. 3. Warped or binding driven plate. Driven plate binding on clutch shaft.	1. Replace driven plate. 2. Replace pressure plate. 3. Replace warped driven plate. Replace clutch shaft if defective, scored, worn.
Clutch slips	1. Lack of lubrication in clutch linkage (linkage binds, causes incomplete engagement. 2. Incorrect pedal, or linkage adjustment. 3. Broken pressure plate springs. 4. Weak pressure plate springs. 5. Grease on driven plate facings (disc).	1. Lubricate linkage. 2. Adjust as required. 3. Replace pressure plate. 4. Replace pressure plate. 5. Replace driven plate.
Incomplete clutch release	1. Incorrect pedal or linkage adjustment or linkage binding. 2. Incorrect height adjustment on pressure plate release levers. 3. Loose, broken facings on driven plate (disc). 4. Bent, dished, warped driven plate caused by overheating.	1. Adjust as required. Lubricate linkage. 2. Adjust release lever height. 3. Replace driven plate. 4. Replace driven plate.
Grinding, whirring grating noise when pedal is depressed	1. Worn or defective throwout bearing. 2. Starter drive teeth contacting flywheel ring gear teeth.	1. Replace throwout bearing. 2. Look for milled or polished teeth on ring gear. Align clutch housing, replace starter drive or drive spring as required.
Squeal, howl, trumpeting noise when pedal is being released (occurs during first inch to inch and one-half of pedal travel)	1. Pilot bushing worn or lack of lubricant.	1. Replace worn bushing. If bushing appears OK, polish bushing with emery, soak lube wick in oil, lube bushing with oil, apply film of chassis grease to clutch shaft pilot hub, reassemble. NOTE: Bushing wear may be due to misalignment of clutch housing or housing to transmission adapter.
Vibration or clutch pedal pulsation with clutch disengaged (pedal fully depressed)	1. Worn or defective engine transmission mounts. 2. Flywheel run out, or damaged or defective clutch components.	1. Inspect and replace as required. 2. Replace components as required. (Flywheel run out at face not to exceed 0.005").

Manual Transmission Troubleshooting

Condition	Probable Cause
Jumping out of high gear	1. Misalignment of transmission case or clutch housing. 2. Worn pilot bearing in crankshaft. 3. Bent transmission shaft. 4. Worn high speed sliding gear. 5. Worn teeth in clutch shaft. 6. Insufficient spring tension on shifter rail plunger. 7. Bent or loose shifter fork. 8. End-play in clutch shaft. 9. Gears not engaging completely. 10. Loose or worn bearings on clutch shaft or mainshaft.
Sticking in high gear	1. Clutch not releasing fully. 2. Burred or battered teeth on clutch shaft. 3. Burred or battered transmission mainshaft. 4. Frozen synchronizing clutch. 5. Stuck shifter rail plunger. 6. Gearshift lever twisting and binding shifter rail. 7. Battered teeth on high speed sliding gear or on sleeve. 8. Lack of lubrication. 9. Improper lubrication. 10. Corroded transmission parts. 11. Defective mainshaft pilot bearing.
Jumping out of second gear	1. Insufficient spring tension on shifter rail plunger. 2. Bent or loose shifter fork. 3. Gears not engaging completely. 4. End-play in transmission mainshaft. 5. Loose transmission gear bearing. 6. Defective mainshaft pilot bearing. 7. Bent transmission shaft. 8. Worn teeth on second speed sliding gear or sleeve. 9. Loose or worn bearings on transmission mainshaft. 10. End-play in countershaft.
Sticking in second gear	1. Clutch not releasing fully. 2. Burred or battered teeth on sliding sleeve. 3. Burred or battered transmission mainshaft. 4. Frozen synchronizing clutch. 5. Stuck shifter rail plunger. 6. Gearshift lever twisting and binding shifter rail. 7. Lack of lubrication. 8. Second speed transmission gear bearings locked will give same effect as gears stuck in second. 9. Improper lubrication. 10. Corroded transmission parts.
Jumping out of low gear	1. Gears not engaging completely. 2. Bent or loose shifter fork. 3. End-play in transmission mainshaft. 4. End-play in countershaft. 5. Loose or worn bearings on transmission mainshaft. 6. Loose or worn bearings in countershaft. 7. Defective mainshaft pilot bearing.
Sticking in low gear	1. Clutch not releasing fully. 2. Burred or battered transmission mainshaft. 3. Stuck shifter rail plunger. 4. Gearshift lever twisting and binding shifter rail. 5. Lack of lubrication. 6. Improper lubrication. 7. Corroded transmission parts.

Condition	Probable Cause
Jumping out of reverse gear	1. Insufficient spring tension on shifter rail plunger. 2. Bent or loose shifter fork. 3. Badly worn gear teeth. 4. Gears not engaging completely. 5. End-play in transmission mainshaft. 6. Idler gear bushings loose or worn. 7. Loose or worn bearings on transmission mainshaft. 8. Defective mainshaft pilot bearing.
Sticking in reverse gear	1. Clutch not releasing fully. 2. Burred or battered transmission mainshaft. 3. Stuck shifter rail plunger. 4. Gearshift lever twisting and binding shifter rail. 5. Lack of lubrication. 6. Improper lubrication. 7. Corroded transmission parts.
Failure of gears to synchronize	1. Binding pilot bearing on mainshaft, will synchronize in high gear only. 2. Clutch not releasing fully. 3. Detent spring weak or broken. 4. Weak or broken springs under balls in sliding gear sleeve. 5. Binding bearing on clutch shaft. 6. Binding countershaft. 7. Binding pilot bearing in crankshaft 8. Badly worn gear teeth. 9. Scored or worn cones. 10. Improper lubrication. 11. Constant mesh gear not turning freely on transmission mainshaft. Will synchronize in that gear only.
Gears spinning when shifting into gear from neutral	1. Clutch not releasing fully. 2. In some cases an extremely light lubricant in transmission will cause gears to continue to spin for a short time after clutch is released. 3. Binding pilot bearing in crankshaft.
Noisy in all gears	1. Insufficient lubricant. 2. Worn countergear bearings. 3. Worn or damaged main drive gear or countergear. 4. Damaged main drive gear or mainshaft bearings. 5. Worn or damaged countergear anti-lash plate.
Noisy in high gear	1. Damaged main drive gear bearing. 2. Damaged mainshaft bearing. 3. Damaged high speed gear synchronizer.
Noisy in neutral	1. Damaged main drive gear bearing. 2. Damaged or loose mainshaft pilot bearing. 3. Worn or damaged countergear anti-lash plate. 4. Worn countergear bearings.
Noisy in all reduction gears	1. Insufficient lubricant. 2. Worn or damaged drive gear or countergear.
Noisy in second only	1. Damaged or worn second gear constant mesh gears. 2. Worn or damaged countergear rear bearings. 3. Damaged or worn second gear synchronizer.
Noisy in second only	1. Damaged or worn second gear constant mesh gears. 2. Worn or damaged countergear rear bearings. 3. Damaged or worn second gear synchronizer.
Noisy in third only (four speed)	1. Damaged or worn third gear constant mesh gears. 2. Worn or damaged countergear bearings.

Condition	Probable Cause
Noisy in reverse only	1. Worn or damaged reverse idler gear or idler bushing. 2. Worn or damaged mainshaft reverse gear. 3. Worn or damaged reverse countergear. 4. Damaged shift mechanism.
Excessive backlash in all reduction gears	1. Worn countergear bearings. 2. Excessive end–play in countergear.

Automatic Transmission Troubleshooting

Keeping alert to changes in the operating characteristics of the transmission (changing shift points, noises, etc.) can prevent small problems from becoming large ones. If the problem cannot be traced to loose bolts, fluid level, misadjusted linkage, clogged filters or similar problems, you should probably seek professional service.

TRANSMISSION FLUID INDICATIONS

The appearance and odor of the transmission fluid can give valuable clues to the overall condition of the transmission. Always note the appearance of the fluid when you check the fluid level or change the fluid. Rub a small amount of fluid between your fingers to feel for grit and smell the fluid on the dipstick.

If The Fluid Appears	It Indicates
Clear and red colored	Normal operation
Discolored (extremely dark red or brownish) or smells burned	Band or clutch pack failure, usually caused by an overheated transmission. Hauling very heavy loads with insufficient power or failure to change the fluid often results in overheating. Do not confuse this appearance with newer fluids that have a darker red color and a strong odor (though not a burned odor).
Foamy or aerated (light in color and full of bubbles)	The level is too high (gear train is churning oil) An internal air leak (air is mixing with the fluid). Have the transmission checked professionally.
Solid residue in the fluid	Defective bands, clutch pack or bearings. Bits of band material or metal abrasives are clinging to the dipstick. Have the transmission checked professionally.
Varnish coating on the dipstick	The transmission fluid is overheating

Problem	Possible Cause	Correction
Slow initial engagement	1. Improper fluid level. 2. Damaged or improperly adjusted linkage. 3. Contaminated fluid. 4. Faulty clutch and band application, or oil control pressure system.	1. Add fluid as required. 2. Repair or adjust linkage. 3. Perform fluid level check. 4. Perform control pressure test.
Rough initial engagement in either forward or reverse	1. Improper fluid level. 2. High engine idle. 3. Looseness in the driveshaft, U-joints or engine mounts. 4. Incorrect linkage adjustment. 5. Faulty clutch or band application, or oil control pressure system. 6. Sticking or dirty valve body.	1. Perform fluid level check. 2. Adjust idle to specifications. 3. Repair as required. 4. Repair or adjust linkage. 5. Perform control pressure test. 6. Clean, repair or replace valve body.

Problem	Possible Cause	Correction
No drive, slips or chatters in first gear in D. All other gears normal.	1. Faulty one-way clutch.	1. Repair or replace one-way clutch.
No drive, slips or chatters in second gear.	1. Improper fluid level. 2. Damaged or improperly adjusted linkage. 3. Intermediate band out of adjustment. 4. Faulty band or clutch application, or oil pressure control system. 5. Faulty servo and/or internal leaks. 6. Dirty or sticking valve body. 7. Polished, glazed intermediate band or drum.	1. Perform fluid level check. 2. Repair or adjust linkage. 3. Adjust intermediate band. 4. Perform control pressure test. 5. Perform air pressure test. 6. Clean, repair or replace valve body. 7. Replace or repair as required.
No drive in any gear.	1. Improper fluid level. 2. Damaged or improperly adjusted linkage. 3. Faulty clutch or band application, or oil control pressure system. 4. Internal leakage. 5. Valve body loose. 6. Faulty clutches. 7. Sticking or dirty valve body.	1. Perform fluid level check. 2. Repair or adjust linkage. 3. Perform control pressure test. 4. Check and repair as required. 5. Tighten to specification. 6. Perform air pressure test. 7. Clean, repair or replace valve body.
No drive forward—reverse OK.	1. Improper fluid level 2. Damaged or improperly adjusted linkage. 3. Faulty clutch or band application, or oil pressure control system. 4. Faulty forward clutch or governor. 5. Valve body loose 6. Dirty or sticking valve body.	1. Perform fluid level check. 2. Repair or adjust linkage. 3. Perform control pressure test. 4. Perform air pressure test. 5. Tighten to specification. 6. Clean, repair or replace valve body.
No drive, slips or chatters in reverse—forward OK.	1. Improper fluid level 2. Damaged or improperly adjusted linkage. 3. Looseness in the drivehsaft, U-joints or engine mounts. 4. Bands or clutches out of adjustment. 5. Faulty oil pressure control system. 6. Faulty reverse clutch or servo. 7. Valve body loose. 8. Dirty or sticking valve body.	1. Perform fluid level check. 2. Repair or adjust linkage. 3. Repair as required. 4. Adjust as necessary. 5. Perform control pressure test. 6. Perform air pressure test. 7. Tighten to specifications. 8. Clean, repair or replace valve body.
Starts in high—in D drag or lockup at 1–2 shift point or in 2 or 1.	1. Improper fluid level. 2. Damaged or improperly adjusted linkage. 3. Faulty governor. 4. Faulty clutches and/or internal leaks. 5. Valve body loose. 6. Dirty, sticking valve body. 7. Poor mating of valve body to case mounting surfaces.	1. Perform fluid level check. 2. Repair or adjust linkage. 3. Repair or replace governor, clean screen. 4. Perform air pressure test. 5. Tighten to specifications. 6. Clean, repair or replace valve body. 7. Replace valve body or case.

Problem	Possible Cause	Correction
Starts up in 2nd or 3rd but no lockup at 1-2 shift points.	1. Improper fluid level. 2. Damaged or improperly adjusted linkage. 3. Improper band and/or clutch application, or oil pressure control system. 4. Faulty governor. 5. Valve body loose. 6. Dirty or sticking valve body. 7. Cross leaks between valve body and case mating surface.	1. Perform fluid level check. 2. Repair or adjust linkage. 3. Perform control pressure test. 4. Perform governor check. Replace or repair governor, clean screen. 5. Tighten to specification. 6. Clean, repair or replace valve body. 7. Replace valve body and/or case as required.
Shift points incorrect.	1. Improper fluid level. 2. Improper vacuum hose routing or leaks. 3. Improper operation of EGR system. 4. Linkage out of adjustment. 5. Improper speedometer gear installed. 6. Improper clutch or band application, or oil pressure control system. 7. Faulty governor. 8. Dirty or sticking valve body.	1. Perform fluid level check. 2. Correct hose routing. 3. Repair or replace as required. 4. Repair or adjust linkage. 5. Replace gear. 6. Perform shift test and control pressure test. 7. Repair or replace governor—clean screen. 8. Clean, repair or replace valve body.
No upshift at any speed in D.	1. Improper fluid level. 2. Vacuum leak to diaphragm unit. 3. Linkage out of adjustment. 4. Improper band or clutch application, or oil pressure control system. 5. Faulty governor. 6. Dirty or sticking valve bdy.	1. Perform fluid level check. 2. Repair vacuum line or hose. 3. Repair or adjust linkage. 4. Perform control pressure test. 5. Repair or replace governor, clean screen. 6. Clean, repair or replace valve body.
Shifts 1-3 in D.	1. Improper fluid level. 2. Intermediate band out of adjustment. 3. Faulty front servo and/or internal leaks. 4. Polished, glazed band or drum. 5. Improper band or clutch application, or oil pressure control system. 6. Dirty or sticking valve body.	1. Perform fluid level check. 2. Adjust band. 3. Perform air pressure test. Repair front servo and/or internal leaks. 4. Repair or replace band or drum. 5. Perform control pressure test. 6. Clean, repair or replace valve body.
Engine over-speeds on 2-3 shift.	1. Improper fluid level. 2. Linkage out of adjustment. 3. Improper band or clutch application, or oil pressure control system. 4. Faulty high clutch and/or intermediate servo. 5. Dirty or sticking valve body.	1. Perform fluid level check. 2. Repair or adjust linkage. 3. Perform control pressure test. 4. Perform air pressure test. Repair as required. 5. Clean repair or replace valve body.
Mushy 1-2 shift.	1. Improper fluid level 2. Incorrect engine idle and/or performance. 3. Improper linkage adjustment. 4. Intermediate band out of adjustment.	1. Perform fluid level check. 2. Tune, adjust engine idle as required. 3. Repair or adjust linkage. 4. Adjust intermediate band. 5. Perform control pressure test.

Problem	Possible Cause	Correction
Mushy 1-2 shift.	5. Improper band or clutch application, or oil pressure control system. 6. Faulty high clutch and/or intermediate servo release. 7. Polished, glazed band or drum. 8. Dirty or sticking valve body.	6. Perform air pressure test. Repair as required. 7. Repair or replace as required. 8. Clean, repair or replace valve body.
Rough 1-2 shift.	1. Improper fluid level. 2. Incorrect engine idle or performance. 3. Intermediate band out of adjustment. 4. Improper band or clutch application, or oil pressure control system. 5. Faulty intermediate servo. 6. Dirty or sticking valve body.	1. Perform fluid level check. 2. Tune, and adjust engine idle. 3. Adjust intermediate band. 4. Perform control pressure test. 5. Air pressure check intermediate servo. 6. Clean, repair or replace valve body.
Rough 2-3 shift	1. Improper fluid level. 2. Incorrect engine idle or performance. 3. Improper band or clutch application, or oil control pressure system. 4. Faulty intermediate servo apply and release and high clutch piston check ball. 5. Dirty or sticking valve body.	1. Perform fluid level check. 2. Tune and adjust engine idle. 3. Perform control pressure test. 4. Air pressure test the intermediate servo apply and release and the high clutch piston check ball. Repair as required. 5. Clean, repair or replace valve body.
Rough 3-1 shift at closed throttle in D.	1. Improper fluid level. 2. Incorrect engine idle or performance. 3. Improper linkage adjustment. 4. Improper clutch or band application or oil pressure control system. 5. Faulty governor operation. 6. Dirty or sticking valve body.	1. Perform fluid level check. 2. Tune, and adjust engine idle. 3. Repair or adjust linkage. 4. Perform control pressure test. 5. Perform governor test. Repair as required. 6. Clean, repair or replace valve body.
No forced downshifts.	1. Improper fluid level. 2. Linkage out of adjustment. 3. Improper clutch or band application, or oil pressure control system. 4. Faulty internal kickdown linkage. 5. Dirty or sticking valve body.	1. Perform fluid level check. 2. Repair or adjust linkage. 3. Perform control pressure test. 4. Repair internal kickdown linkage. 5. Clean, repair or replace valve body.
No 3-1 shift in D.	1. Improper fluid level. 2. Incorrect engine idle, or performance. 3. Faulty governor. 4. Dirty or sticking valve body.	1. Perform fluid level check. 2. Tune, and adjust engine idle. 3. Perform govenor check. Repair as required. 4. Clean, repair or replace valve body.
Runaway engine on 3-2 downshift.	1. Improper fluid level. 2. Linkage out of adjustment. 3. Intermediate band out of adjustment. 4. Improper band or clutch application, or oil pressure control system.	1. Perform fluid level check. 2. Repair or adjust linkage. 3. Adjust intermediate band. 4. Perform control pressure test. 5. Air pressure test check the intermediate servo. Repair servo and/or seals.

Problem	Possible Cause	Correction
Runaway engine on 3-2 downshift.	5. Faulty intermediate servo. 6. Polished, glazed band or drum. 7. Dirty or sticking valve body.	6. Repair or replace as required. 7. Clean, repair or replace valve body.
No engine braking in manual first gear.	1. Improper fluid level. 2. Linkage out of adjustment. 3. Bands or clutches out of adjustment. 4. Faulty oil pressure control system. 5. Faulty reverse servo. 6. Polished, glazed band or drum.	1. Perform fluid level check. 2. Repair or adjust linkage. 3. Adjust as necessary. 4. Perform control pressure test. 5. Perform air pressure test of reverse servo. Repair reverse clutch or rear servo as required. 6. Repair or replace as required.
No engine braking in manual second gear.	1. Improper fluid level. 2. Linkage out of adjustment. 3. Intermediate band out of adjustment. 4. Improper band or clutch application, or oil pressure control system. 5. Intermediate servo leaking. 6. Polished or glazed band or drum.	1. Perform fluid level check. 2. Repair or adjust linkage. 3. Adjust intermediate band. 4. Perform control pressure test. 5. Perform air pressure test of intermediate servo for leakage. Repair as required. 6. Repair or replace as required.
Transmission noisy—valve resonance.	1. Improper fluid level. 2. Linkage out of adjustment. 3. Improper band or clutch application, or oil pressure control system. 4. Cooler lines grounding. 5. Dirty sticking valve body. 6. Internal leakage or pump cavitation.	1. Perform fluid level check. 2. Repair or adjust linkage. 3. Perform control pressure test. 4. Free up cooler lines. 5. Clean, repair or replace valve body. 6. Repair as required.
Transmission overheats.	1. Improper fluid level. 2. Incorrect engine idle, or performance. 3. Improper clutch or band application, or oil pressure control system. 4. Restriction in cooler or lines. 5. Seized one-way clutch. 6. Dirty or sticking valve body.	1. Perform fluid level check. 2. Tune, or adjust engine idle. 3. Perform control pressure test. 4. Repair restriction. 5. Replace one-way clutch. 6. Clean, repair or replace valve body.
Transmission fluid leaks.	1. Improper fluid level. 2. Leakage at gasket, seals, etc. 3. Vacuum diaphragm unit leaking.	1. Perform fluid level check. 2. Remove all traces of lube on exposed surfaces of transmission. Check the vent for free breathing. Operate transmission at normal temperatures and inspect for leakage. Repair as required. 3. Replace diaphragm.

Mechanics' Data

SI METRIC TABLES

The following tables are given in SI (International System) metric units. SI units replace both customary (English) and the older gavimetric units. The use of SI units as a new world-wide standard was set by the International Committee of Weights and Measures in 1960. SI has since been adopted by most countries as their national standard.

These tables are general conversion tables which will allow you to convert customary units, which appear in the text, into SI units.

The following are a list of SI units and the customary units, used in this book, which they replace:

To measure:	Use SI units:	Which replace (customary units):
mass	kilograms (kg)	pounds (lbs)
temperature	Celsius (°C)	Fahrenheit (°F)
length	millimeters (mm)	inches (in.)
force	newtons (N)	pounds force (lbs)
capacities	liters (l)	pints/quarts/gallons (pts/qts/gals)
torque	newton-meters (N·m)	foot pounds (ft lbs)
pressure	kilopascals (kPa)	pounds per square inch (psi)
volume	cubic centimeters (cm³)	cubic inches (cu in.)
power	kilowatts (kW)	horsepower (hp)

If you have had any prior experience with the metric system, you may have noticed units in this chart which are not familiar to you. This is because, in some cases, SI units differ from the older gravimetric units which they replace. For example, newtons (N) replace kilograms (kg) as a force unit, kilopascals (kPa) replace atmospheres or bars as a unit of pressure, and, although the units are the same, the name Celsius replaces centigrade for temperature measurement.

If you are not using the SI tables, have a look at them anyway; you will be seeing a lot more of them in the future.

ENGLISH TO METRIC CONVERSION: MASS (WEIGHT)

Current mass measurement is expressed in pounds and ounces (lbs. & ozs.). The metric unit of mass (or weight) is the kilogram (kg). Even although this table does not show conversion of masses (weights) larger than 15 lbs, it is easy to calculate larger units by following the data immediately below.

To convert ounces (oz.) to grams (g): multiply th number of ozs. by 28
To convert grams (g) to ounces (oz.): multiply the number of grams by .035

To convert pounds (lbs.) to kilograms (kg): multiply the number of lbs. by .45
To convert kilograms (kg) to pounds (lbs.): multiply the number of kilograms by 2.2

lbs	kg	lbs	kg	oz	kg	oz	kg
0.1	0.04	0.9	0.41	0.1	0.003	0.9	0.024
0.2	0.09	1	0.4	0.2	0.005	1	0.03
0.3	0.14	2	0.9	0.3	0.008	2	0.06
0.4	0.18	3	1.4	0.4	0.011	3	0.08
0.5	0.23	4	1.8	0.5	0.014	4	0.11
0.6	0.27	5	2.3	0.6	0.017	5	0.14
0.7	0.32	10	4.5	0.7	0.020	10	0.28
0.8	0.36	15	6.8	0.8	0.023	15	0.42

ENGLISH TO METRIC CONVERSION: TEMPERATURE

To convert Fahrenheit (°F) to Celsius (°C): take number of °F and subtract 32; multiply result by 5; divide result by 9
To convert Celsius (°C) to Fahrenheit (°F): take number of °C and multiply by 9; divide result by 5; add 32 to total

Fahrenheit (F)		Celsius (C)		Fahrenheit (F)		Celsius (C)		Fahrenheit (F)		Celsius (C)	
°F	°C	°C	°F	°F	°C	°C	°F	°F	°C	°C	°F
−40	−40	−38	−36.4	80	26.7	18	64.4	215	101.7	80	176
−35	−37.2	−36	−32.8	85	29.4	20	68	220	104.4	85	185
−30	−34.4	−34	−29.2	90	32.2	22	71.6	225	107.2	90	194
−25	−31.7	−32	−25.6	95	35.0	24	75.2	230	110.0	95	202
−20	−28.9	−30	−22	100	37.8	26	78.8	235	112.8	100	212
−15	−26.1	−28	−18.4	105	40.6	28	82.4	240	115.6	105	221
−10	−23.3	−26	−14.8	110	43.3	30	86	245	118.3	110	230
−5	−20.6	−24	−11.2	115	46.1	32	89.6	250	121.1	115	239
0	−17.8	−22	−7.6	120	48.9	34	93.2	255	123.9	120	248
1	−17.2	−20	−4	125	51.7	36	96.8	260	126.6	125	257
2	−16.7	−18	−0.4	130	54.4	38	100.4	265	129.4	130	266
3	−16.1	−16	3.2	135	57.2	40	104	270	132.2	135	275
4	−15.6	−14	6.8	140	60.0	42	107.6	275	135.0	140	284
5	−15.0	−12	10.4	145	62.8	44	112.2	280	137.8	145	293
10	−12.2	−10	14	150	65.6	46	114.8	285	140.6	150	302
15	−9.4	−8	17.6	155	68.3	48	118.4	290	143.3	155	311
20	−6.7	−6	21.2	160	71.1	50	122	295	146.1	160	320
25	−3.9	−4	24.8	165	73.9	52	125.6	300	148.9	165	329
30	−1.1	−2	28.4	170	76.7	54	129.2	305	151.7	170	338
35	1.7	0	32	175	79.4	56	132.8	310	154.4	175	347
40	4.4	2	35.6	180	82.2	58	136.4	315	157.2	180	356
45	7.2	4	39.2	185	85.0	60	140	320	160.0	185	365
50	10.0	6	42.8	190	87.8	62	143.6	325	162.8	190	374
55	12.8	8	46.4	195	90.6	64	147.2	330	165.6	195	383
60	15.6	10	50	200	93.3	66	150.8	335	168.3	200	392
65	18.3	12	53.6	205	96.1	68	154.4	340	171.1	205	401
70	21.1	14	57.2	210	98.9	70	158	345	173.9	210	410
75	23.9	16	60.8	212	100.0	75	167	350	176.7	215	414

ENGLISH TO METRIC CONVERSION: LENGTH

To convert inches (ins.) to millimeters (mm): multiply number of inches by 25.4

To convert millimeters (mm) to inches (ins.): multiply number of millimeters by .04

Inches	Decimals	Milli-meters	Inches to millimeters inches	mm	Inches	Decimals	Milli-meters	Inches to millimeters inches	mm
1/64	0.051625	0.3969	0.0001	0.00254	33/64	0.515625	13.0969	0.6	15.24
1/32	0.03125	0.7937	0.0002	0.00508	17/32	0.53125	13.4937	0.7	17.78
3/64	0.046875	1.1906	0.0003	0.00762	35/64	0.546875	13.8906	0.8	20.32
1/16	0.0625	1.5875	0.0004	0.01016	9/16	0.5625	14.2875	0.9	22.86
5/64	0.078125	1.9844	0.0005	0.01270	37/64	0.578125	14.6844	1	25.4
3/32	0.09375	2.3812	0.0006	0.01524	19/32	0.59375	15.0812	2	50.8
7/64	0.109375	2.7781	0.0007	0.01778	39/64	0.609375	15.4781	3	76.2
1/8	0.125	3.1750	0.0008	0.02032	5/8	0.625	15.8750	4	101.6
9/64	0.140625	3.5719	0.0009	0.02286	41/64	0.640625	16.2719	5	127.0
5/32	0.15625	3.9687	0.001	0.0254	21/32	0.65625	16.6687	6	152.4
11/64	0.171875	4.3656	0.002	0.0508	43/64	0.671875	17.0656	7	177.8
3/16	0.1875	4.7625	0.003	0.0762	11/16	0.6875	17.4625	8	203.2
13/64	0.203125	5.1594	0.004	0.1016	45/64	0.703125	17.8594	9	228.6
7/32	0.21875	5.5562	0.005	0.1270	23/32	0.71875	18.2562	10	254.0
15/64	0.234375	5.9531	0.006	0.1524	47/64	0.734375	18.6531	11	279.4
1/4	0.25	6.3500	0.007	0.1778	3/4	0.75	19.0500	12	304.8
17/64	0.265625	6.7469	0.008	0.2032	49/64	0.765625	19.4469	13	330.2
9/32	0.28125	7.1437	0.009	0.2286	25/32	0.78125	19.8437	14	355.6
19/64	0.296875	7.5406	0.01	0.254	51/64	0.796875	20.2406	15	381.0
5/16	0.3125	7.9375	0.02	0.508	13/16	0.8125	20.6375	16	406.4
21/64	0.328125	8.3344	0.03	0.762	53/64	0.828125	21.0344	17	431.8
11/32	0.34375	8.7312	0.04	1.016	27/32	0.84375	21.4312	18	457.2
23/64	0.359375	9.1281	0.05	1.270	55/64	0.859375	21.8281	19	482.6
3/8	0.375	9.5250	0.06	1.524	7/8	0.875	22.2250	20	508.0
25/64	0.390625	9.9219	0.07	1.778	57/64	0.890625	22.6219	21	533.4
13/32	0.40625	10.3187	0.08	2.032	29/32	0.90625	23.0187	22	558.8
27/64	0.421875	10.7156	0.09	2.286	59/64	0.921875	23.4156	23	584.2
7/16	0.4375	11.1125	0.1	2.54	15/16	0.9375	23.8125	24	609.6
29/64	0.453125	11.5094	0.2	5.08	61/64	0.953125	24.2094	25	635.0
15/32	0.46875	11.9062	0.3	7.62	31/32	0.96875	24.6062	26	660.4
31/64	0.484375	12.3031	0.4	10.16	63/64	0.984375	25.0031	27	690.6
1/2	0.5	12.7000	0.5	12.70					

ENGLISH TO METRIC CONVERSION: TORQUE

To convert foot-pounds (ft. lbs.) to Newton-meters: multiply the number of ft. lbs. by 1.3

To convert inch-pounds (in. lbs.) to Newton-meters: multiply the number of in. lbs. by .11

in lbs	N-m	in lbs	N-m	in lbs	N-m	in lbs	N-m	in lbs	N-m
0.1	0.01	1	0.11	10	1.13	19	2.15	28	3.16
0.2	0.02	2	0.23	11	1.24	20	2.26	29	3.28
0.3	0.03	3	0.34	12	1.36	21	2.37	30	3.39
0.4	0.04	4	0.45	13	1.47	22	2.49	31	3.50
0.5	0.06	5	0.56	14	1.58	23	2.60	32	3.62
0.6	0.07	6	0.68	15	1.70	24	2.71	33	3.73
0.7	0.08	7	0.78	16	1.81	25	2.82	34	3.84
0.8	0.09	8	0.90	17	1.92	26	2.94	35	3.95
0.9	0.10	9	1.02	18	2.03	27	3.05	36	4.0/

MECHANIC'S DATA

ENGLISH TO METRIC CONVERSION: TORQUE

Torque is now expressed as either foot-pounds (ft./lbs.) or inch-pounds (in./lbs.). The metric measurement unit for torque is the Newton-meter (Nm). This unit—the Nm—will be used for all SI metric torque references, both the present ft./lbs. and in./lbs.

ft lbs	N-m	ft lbs	N-m	ft lbs	N-m	ft lbs	N-m
0.1	0.1	33	44.7	74	100.3	115	155.9
0.2	0.3	34	46.1	75	101.7	116	157.3
0.3	0.4	35	47.4	76	103.0	117	158.6
0.4	0.5	36	48.8	77	104.4	118	160.0
0.5	0.7	37	50.7	78	105.8	119	161.3
0.6	0.8	38	51.5	79	107.1	120	162.7
0.7	1.0	39	52.9	80	108.5	121	164.0
0.8	1.1	40	54.2	81	109.8	122	165.4
0.9	1.2	41	55.6	82	111.2	123	166.8
1	1.3	42	56.9	83	112.5	124	168.1
2	2.7	43	58.3	84	113.9	125	169.5
3	4.1	44	59.7	85	115.2	126	170.8
4	5.4	45	61.0	86	116.6	127	172.2
5	6.8	46	62.4	87	118.0	128	173.5
6	8.1	47	63.7	88	119.3	129	174.9
7	9.5	48	65.1	89	120.7	130	176.2
8	10.8	49	66.4	90	122.0	131	177.6
9	12.2	50	67.8	91	123.4	132	179.0
10	13.6	51	69.2	92	124.7	133	180.3
11	14.9	52	70.5	93	126.1	134	181.7
12	16.3	53	71.9	94	127.4	135	183.0
13	17.6	54	73.2	95	128.8	136	184.4
14	18.9	55	74.6	96	130.2	137	185.7
15	20.3	56	75.9	97	131.5	138	187.1
16	21.7	57	77.3	98	132.9	139	188.5
17	23.0	58	78.6	99	134.2	140	189.8
18	24.4	59	80.0	100	135.6	141	191.2
19	25.8	60	81.4	101	136.9	142	192.5
20	27.1	61	82.7	102	138.3	143	193.9
21	28.5	62	84.1	103	139.6	144	195.2
22	29.8	63	85.4	104	141.0	145	196.6
23	31.2	64	86.8	105	142.4	146	198.0
24	32.5	65	88.1	106	143.7	147	199.3
25	33.9	66	89.5	107	145.1	148	200.7
26	35.2	67	90.8	108	146.4	149	202.0
27	36.6	68	92.2	109	147.8	150	203.4
28	38.0	69	93.6	110	149.1	151	204.7
29	39.3	70	94.9	111	150.5	152	206.1
30	40.7	71	96.3	112	151.8	153	207.4
31	42.0	72	97.6	113	153.2	154	208.8
32	43.4	73	99.0	114	154.6	155	210.2

ENGLISH TO METRIC CONVERSION: FORCE

Force is presently measured in pounds (lbs.). This type of measurement is used to measure spring pressure, specifically how many pounds it takes to compress a spring. Our present force unit (the pound) will be replaced in SI metric measurements by the Newton (N). This term will eventually see use in specifications for electric motor brush spring pressures, valve spring pressures, etc.

To convert pounds (lbs.) to Newton (N): multiply the number of lbs. by 4.45

lbs	N	lbs	N	lbs	N	oz	N
0.01	0.04	21	93.4	59	262.4	1	0.3
0.02	0.09	22	97.9	60	266.9	2	0.6
0.03	0.13	23	102.3	61	271.3	3	0.8
0.04	0.18	24	106.8	62	275.8	4	1.1
0.05	0.22	25	111.2	63	280.2	5	1.4
0.06	0.27	26	115.6	64	284.6	6	1.7
0.07	0.31	27	120.1	65	289.1	7	2.0
0.08	0.36	28	124.6	66	293.6	8	2.2
0.09	0.40	29	129.0	67	298.0	9	2.5
0.1	0.4	30	133.4	68	302.5	10	2.8
0.2	0.9	31	137.9	69	306.9	11	3.1
0.3	1.3	32	142.3	70	311.4	12	3.3
0.4	1.8	33	146.8	71	315.8	13	3.6
0.5	2.2	34	151.2	72	320.3	14	3.9
0.6	2.7	35	155.7	73	324.7	15	4.2
0.7	3.1	36	160.1	74	329.2	16	4.4
0.8	3.6	37	164.6	75	333.6	17	4.7
0.9	4.0	38	169.0	76	338.1	18	5.0
1	4.4	39	173.5	77	342.5	19	5.3
2	8.9	40	177.9	78	347.0	20	5.6
3	13.4	41	182.4	79	351.4	21	5.8
4	17.8	42	186.8	80	355.9	22	6.1
5	22.2	43	191.3	81	360.3	23	6.4
6	26.7	44	195.7	82	364.8	24	6.7
7	31.1	45	200.2	83	369.2	25	7.0
8	35.6	46	204.6	84	373.6	26	7.2
9	40.0	47	209.1	85	378.1	27	7.5
10	44.5	48	213.5	86	382.6	28	7.8
11	48.9	49	218.0	87	387.0	29	8.1
12	53.4	50	224.4	88	391.4	30	8.3
13	57.8	51	226.9	89	395.9	31	8.6
14	62.3	52	231.3	90	400.3	32	8.9
15	66.7	53	235.8	91	404.8	33	9.2
16	71.2	54	240.2	92	409.2	34	9.4
17	75.6	55	244.6	93	413.7	35	9.7
18	80.1	56	249.1	94	418.1	36	10.0
19	84.5	57	253.6	95	422.6	37	10.3
20	89.0	58	258.0	96	427.0	38	10.6

ENGLISH TO METRIC CONVERSION: LIQUID CAPACITY

Liquid or fluid capacity is presently expressed as pints, quarts or gallons, or a combination of all of these. In the metric system the liter (l) will become the basic unit. Fractions of a liter would be expressed as deciliters, centiliters, or most frequently (and commonly) as milliliters.

To convert pints (pts.) to liters (l): multiply the number of pints by .47
To convert liters (l) to pints (pts.): multiply the number of liters by 2.1
To convert quarts (qts.) to liters (l): multiply the number of quarts by .95

To convert liters (l) to quarts (qts.): multiply the number of liters by 1.06
To convert gallons (gals.) to liters (l): multiply the number of gallons by 3.8
To convert liters (l) to gallons (gals.): multiply the number of liters by .26

gals	liters	qts	liters	pts	liters
0.1	0.38	0.1	0.10	0.1	0.05
0.2	0.76	0.2	0.19	0.2	0.10
0.3	1.1	0.3	0.28	0.3	0.14
0.4	1.5	0.4	0.38	0.4	0.19
0.5	1.9	0.5	0.47	0.5	0.24
0.6	2.3	0.6	0.57	0.6	0.28
0.7	2.6	0.7	0.66	0.7	0.33
0.8	3.0	0.8	0.76	0.8	0.38
0.9	3.4	0.9	0.85	0.9	0.43
1	3.8	1	1.0	1	0.5
2	7.6	2	1.9	2	1.0
3	11.4	3	2.8	3	1.4
4	15.1	4	3.8	4	1.9
5	18.9	5	4.7	5	2.4
6	22.7	6	5.7	6	2.8
7	26.5	7	6.6	7	3.3
8	30.3	8	7.6	8	3.8
9	34.1	9	8.5	9	4.3
10	37.8	10	9.5	10	4.7
11	41.6	11	10.4	11	5.2
12	45.4	12	11.4	12	5.7
13	49.2	13	12.3	13	6.2
14	53.0	14	13.2	14	6.6
15	56.8	15	14.2	15	7.1
16	60.6	16	15.1	16	7.6
17	64.3	17	16.1	17	8.0
18	68.1	18	17.0	18	8.5
19	71.9	19	18.0	19	9.0
20	75.7	20	18.9	20	9.5
21	79.5	21	19.9	21	9.9
22	83.2	22	20.8	22	10.4
23	87.0	23	21.8	23	10.9
24	90.8	24	22.7	24	11.4
25	94.6	25	23.6	25	11.8
26	98.4	26	24.6	26	12.3
27	102.2	27	25.5	27	12.8
28	106.0	28	26.5	28	13.2
29	110.0	29	27.4	29	13.7
30	113.5	30	28.4	30	14.2

ENGLISH TO METRIC CONVERSION: PRESSURE

The basic unit of pressure measurement used today is expressed as pounds per square inch (psi). The metric unit for psi will be the kilopascal (kPa). This will apply to either fluid pressure or air pressure, and will be frequently seen in tire pressure readings, oil pressure specifications, fuel pump pressure, etc.

To convert pounds per square inch (psi) to kilopascals (kPa): multiply the number of psi by 6.89

Psi	kPa	Psi	kPa	Psi	kPa	Psi	kPa
0.1	0.7	37	255.1	82	565.4	127	875.6
0.2	1.4	38	262.0	83	572.3	128	882.5
0.3	2.1	39	268.9	84	579.2	129	889.4
0.4	2.8	40	275.8	85	586.0	130	896.3
0.5	3.4	41	282.7	86	592.9	131	903.2
0.6	4.1	42	289.6	87	599.8	132	910.1
0.7	4.8	43	296.5	88	606.7	133	917.0
0.8	5.5	44	303.4	89	613.6	134	923.9
0.9	6.2	45	310.3	90	620.5	135	930.8
1	6.9	46	317.2	91	627.4	136	937.7
2	13.8	47	324.0	92	634.3	137	944.6
3	20.7	48	331.0	93	641.2	138	951.5
4	27.6	49	337.8	94	648.1	139	958.4
5	34.5	50	344.7	95	655.0	140	965.2
6	41.4	51	351.6	96	661.9	141	972.2
7	48.3	52	358.5	97	668.8	142	979.0
8	55.2	53	365.4	98	675.7	143	985.9
9	62.1	54	372.3	99	682.6	144	992.8
10	69.0	55	379.2	100	689.5	145	999.7
11	75.8	56	386.1	101	696.4	146	1006.6
12	82.7	57	393.0	102	703.3	147	1013.5
13	89.6	58	399.9	103	710.2	148	1020.4
14	96.5	59	406.8	104	717.0	149	1027.3
15	103.4	60	413.7	105	723.9	150	1034.2
16	110.3	61	420.6	106	730.8	151	1041.1
17	117.2	62	427.5	107	737.7	152	1048.0
18	124.1	63	434.4	108	744.6	153	1054.9
19	131.0	64	441.3	109	751.5	154	1061.8
20	137.9	65	448.2	110	758.4	155	1068.7
21	144.8	66	455.0	111	765.3	156	1075.6
22	151.7	67	461.9	112	772.2	157	1082.5
23	158.6	68	468.8	113	779.1	158	1089.4
24	165.5	69	475.7	114	786.0	159	1096.3
25	172.4	70	482.6	115	792.9	160	1103.2
26	179.3	71	489.5	116	799.8	161	1110.0
27	186.2	72	496.4	117	806.7	162	1116.9
28	193.0	73	503.3	118	813.6	163	1123.8
29	200.0	74	510.2	119	820.5	164	1130.7
30	206.8	75	517.1	120	827.4	165	1137.6
31	213.7	76	524.0	121	834.3	166	1144.5
32	220.6	77	530.9	122	841.2	167	1151.4
33	227.5	78	537.8	123	848.0	168	1158.3
34	234.4	79	544.7	124	854.9	169	1165.2
35	241.3	80	551.6	125	861.8	170	1172.1
36	248.2	81	558.5	126	868.7	171	1179.0

ENGLISH TO METRIC CONVERSION: PRESSURE

The basic unit of pressure measurement used today is expressed as pounds per square inch (psi). The metric unit for psi will be the kilopascal (kPa). This will apply to either fluid pressure or air pressure, and will be frequently seen in tire pressure readings, oil pressure specifications, fuel pump pressure, etc.

To convert pounds per square inch (psi) to kilopascals (kPa): multiply the number of psi by 6.89

Psi	kPa	Psi	kPa	Psi	kPa	Psi	kPa
172	1185.9	216	1489.3	260	1792.6	304	2096.0
173	1192.8	217	1496.2	261	1799.5	305	2102.9
174	1199.7	218	1503.1	262	1806.4	306	2109.8
175	1206.6	219	1510.0	263	1813.3	307	2116.7
176	1213.5	220	1516.8	264	1820.2	308	2123.6
177	1220.4	221	1523.7	265	1827.1	309	2130.5
178	1227.3	222	1530.6	266	1834.0	310	2137.4
179	1234.2	223	1537.5	267	1840.9	311	2144.3
180	1241.0	224	1544.4	268	1847.8	312	2151.2
181	1247.9	225	1551.3	269	1854.7	313	2158.1
182	1254.8	226	1558.2	270	1861.6	314	2164.9
183	1261.7	227	1565.1	271	1868.5	315	2171.8
184	1268.6	228	1572.0	272	1875.4	316	2178.7
185	1275.5	229	1578.9	273	1882.3	317	2185.6
186	1282.4	230	1585.8	274	1889.2	318	2192.5
187	1289.3	231	1592.7	275	1896.1	319	2199.4
188	1296.2	232	1599.6	276	1903.0	320	2206.3
189	1303.1	233	1606.5	277	1909.8	321	2213.2
190	1310.0	234	1613.4	278	1916.7	322	2220.1
191	1316.9	235	1620.3	279	1923.6	323	2227.0
192	1323.8	236	1627.2	280	1930.5	324	2233.9
193	1330.7	237	1634.1	281	1937.4	325	2240.8
194	1337.6	238	1641.0	282	1944.3	326	2247.7
195	1344.5	239	1647.8	283	1951.2	327	2254.6
196	1351.4	240	1654.7	284	1958.1	328	2261.5
197	1358.3	241	1661.6	285	1965.0	329	2268.4
198	1365.2	242	1668.5	286	1971.9	330	2275.3
199	1372.0	243	1675.4	287	1978.8	331	2282.2
200	1378.9	244	1682.3	288	1985.7	332	2289.1
201	1385.8	245	1689.2	289	1992.6	333	2295.9
202	1392.7	246	1696.1	290	1999.5	334	2302.8
203	1399.6	247	1703.0	291	2006.4	335	2309.7
204	1406.5	248	1709.9	292	2013.3	336	2316.6
205	1413.4	249	1716.8	293	2020.2	337	2323.5
206	1420.3	250	1723.7	294	2027.1	338	2330.4
207	1427.2	251	1730.6	295	2034.0	339	2337.3
208	1434.1	252	1737.5	296	2040.8	240	2344.2
209	1441.0	253	1744.4	297	2047.7	341	2351.1
210	1447.9	254	1751.3	298	2054.6	342	2358.0
211	1454.8	255	1758.2	299	2061.5	343	2364.9
212	1461.7	256	1765.1	300	2068.4	344	2371.8
213	1468.7	257	1772.0	301	2075.3	345	2378.7
214	1475.5	258	1778.8	302	2082.2	346	2385.6
215	1482.4	259	1785.7	303	2089.1	347	2392.5